McGraw-Hill Encyclopedia of CHEMISTRY

McGraw-Hill Encyclopedia of CHEMISTRY

Sybil P. Parker
Editor in Chief

McGraw Hill Book Company

New York St. Louis San Francisco

Auckland Bogotá Guatemala Hamburg Johannesburg
Lisbon London Madrid Mexico Montreal
New Delhi Panama Paris San Juan São Paulo
Singapore Sydney Tokyo Toronto

1234567890 KPKP 89876543

ISBN 0-07-045484-1

Library of Congress Cataloging in Publication Data

McGraw-Hill encyclopedia of chemistry.

"All of the material in this volume has been published previously in the McGraw-Hill encyclopedia of science & technology, fifth edition"—T.p. verso.
Bibliography: p.
Includes index.
1. Chemistry—Dictionaries. I. Parker, Sybil P.
II. McGraw-Hill Book Company. III. McGraw-Hill
encyclopedia of science & technology. 5th ed.
QD5.M36 1983 540′.3′21 82-21665
ISBN 0-07-045484-1

Editorial Staff

Sybil P. Parker, Editor in Chief

Jonathan Weil, Editor
Betty Richman, Editor

Edward J. Fox, Art director

Ann D. Bonardi, Art production supervisor
Cynthia M. Kelly, Art/traffic

Joe Faulk, Editing manager

Ruth L. Williams, Editing supervisor
Dorick Byard, Editing supervisor

Patricia W. Albers, Senior editing assistant
Judith Alberts, Editing assistant
Barbara Begg, Editing assistant

Project Consultants

Dr. Joseph J. Katz
Senior Scientist
Argonne National Laboratory

Prof. Kenneth D. Kopple
Department of Chemistry
Illinois Institute of Technology

Field Consultants

Prof. D. Allan Bromley
Henry Ford II Professor and Director of
A. W. Wright Nuclear Structure Laboratory
Yale University

Prof. Todd M. Doscher
Department of Petroleum Engineering
University of Southern California

Prof. Roland H. Good, Jr.
Department of Physics
Pennsylvania State University

Prof. Richard C. Jarnagin
Department of Chemistry
University of North Carolina

Dr. Royce W. Murray
Kenan Professor of Chemistry
University of North Carolina

Preface

The science of chemistry is concerned with the composition and structure of matter—that is, its physical and chemical properties and its reactions. In this regard, chemistry is closely related to physics, especially in the study of the atom, since all matter is composed of elements whose properties are attributable to the nature of the atomic structure. This volume, therefore, not only covers the fundamental principles of chemistry but also includes relevant topics in physics that are essential for the understanding of modern chemistry.

The *Encyclopedia of Chemistry* is a comprehensive reference that will provide the reader with the most up-to-date information available on each of the major divisions of theoretical chemistry—inorganic, organic, physical, and analytical. In addition, articles explore the nature of matter from the perspective of atomic physics, quantum theory, statistical mechanics, and thermodynamics. Space limitations have precluded coverage of related fields of study such as biochemistry, geochemistry, agricultural chemistry, and chemical technology. The history of chemistry is discussed only if it applies to the development of subjects in individual articles.

This Encyclopedia is unique in both its scope and depth of coverage. Each article begins with a definition and presents a complete yet concise explanation of the subject in language as simple as the topic permits without omitting important technical information. The articles range from subjects as simple as nomenclature and the elements to sophisticated theoretical concepts such as molecular orbital theory. The Encyclopedia as a whole surveys the science of chemistry in nearly 800 articles, including Stereochemistry, Activation analysis, Heterogeneous catalysis, Polymer, Vacuum fusion, Acid-base indicator, Valence, Electron spin, and Organometallic compound.

The articles were selected from the *McGraw-Hill Encyclopedia of Science and Technology* (5th ed., 1982). The text is supplemented by hundreds of structural formulas, chemical reactions, graphs, tables, drawings, and photographs which clarify the presentation. All information is readily accessible through the detailed index and by the use of cross-references throughout. Bibliographies list publications that provide further information.

We wish to thank the many Contributors whose work has been carried over to this specialized volume; and especially the Field Consultants, including Messrs. Bromley, Doscher, Good, Jarnagin, and Murray, whose expertise was invaluable in organizing the 1982 parent project. Two other Field Consultants, Messrs. Katz and Kopple, provided fresh impetus in developing the present volume, as unstinting Project Consultants.

This Encyclopedia is indispensable for chemists and other scientists, students, librarians, science writers, and anyone else needing basic information about chemistry in a readily accessible format.

Sybil P. Parker
EDITOR IN CHIEF

McGraw-Hill Encyclopedia of CHEMISTRY

Absorption-Zirconium

Absorption

Either the taking up of matter in bulk by other matter, as in the dissolving of a gas by a liquid; or the taking up of energy from radiation by the medium through which the radiation is passing. In the first case, an absorption coefficient is defined as the amount of gas dissolved at standard conditions by 1 cm^3 of the solvent. Absorption in this sense is a volume effect: The absorbed substance permeates the whole of the absorber. In absorption of the second type, attenuation is produced which in many cases follows Lambert's law and adds to the effects of scattering if the latter is present.

Absorption of electromagnetic radiation can occur in several ways. For example, microwaves in a wave guide lose energy to the walls of the guide: For nonperfect conductors, the wave penetrates the guide surface and energy in the wave is transferred to the atoms of the guide. Light is absorbed by atoms of the medium through which it passes, and in some cases, this absorption is quite distinctive: Selected frequencies from a heterochromatic source are strongly absorbed, as in the absorption spectrum of the Sun. Electromagnetic radiation can be absorbed by the photoelectric effect, where the light quantum is absorbed and an electron of the absorbing atom is ejected, and also by Compton scattering. Electron-positron pairs may be created by the absorption of a photon of sufficiently high energy. Photons can be absorbed by photoproduction of nuclear and subnuclear particles, analogous to the photoelectric effect.

Sound waves are absorbed at suitable frequencies by particles suspended in the air (wavelength of the order of the particle size), where the sound energy is transformed into vibrational energy of the absorbing particles.

Absorption of energy from a beam of particles can occur by the ionization process, where an electron in the medium through which the beam passes is removed by the beam particles. The finite range of protons and α-particles in matter is a result of this process. In the case of low-energy electrons, scattering is as important as ionization, so that range is a less well-defined concept. Particles themselves may be absorbed from a beam. For example, in a nuclear reaction an incident particle X is absorbed into nucleus Y, and the result may be that another particle Z, or a photon, or particle X with changed energy comes out. Low-energy positrons are quickly absorbed by annihilating with electrons in matter to yield two γ-rays.

[MC ALLISTER H. HULL, JR.]

Acetals

Stable ethers usually prepared by acid-catalyzed alcoholation of the carbonyl group of aldehydes. The Y-shaped general formula for these compounds is shown below. (R represents an organic

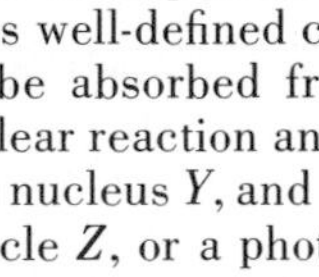

radical.) Cyclic acetals (*meta*-dioxolanes, *meta*-dioxanes, and so forth) are formed from diols. Alcoholation of the carbonyl function of ketones yields analogous substances, ketals. Acetals may also be obtained through the Prins reaction of olefins with aldehydes, usually formaldehyde. The acetals are stable toward alkalies, but they hydrolyze in the presence of acids. They are sometimes used to protect the aldehyde group during reactions and are useful substitutes for the simple ethers as solvents in synthetic chemistry, for example, in the Grignard reactions. Their thermal decomposition yields vinyl ethers. Few acetals have any large-scale industrial applications. Their hydrolytic stability to alkali commends the use of some in soaps as fragrances, and their solvency characteristics, in which they are similar to both esters and ethers, make them convenient solvents and plasticizers. Poly(formaldehyde) and certain copolymers of trioxane with epoxides or dioxolanes are called acetal resins because their main valence chains are sequences of acetal linkages. *See* ALDEHYDE; GLYCOL; KETONE.

[FRANK WAGNER]

Bibliography: D. Barton and W. D. Ollis (eds.), *Comprehensive Organic Chemistry*: *The Synthesis and Reactions of Organic Compounds*, vol. 1, 1979.

Acetate

One of two types of compounds derived from acetic acid, $HC_2H_3O_2$. One type is obtained by the reaction of acetic acid and bases to give salts containing the negative acetate ion, $C_2H_3O_2^-$. The second type of compound is an ester which is derived from acetic acid and an alcohol, for example, ethyl acetate. Acetate is the official name (Federal Trade Commission) for the textile fiber produced from partially hydrolyzed cellulose acetate and once called acetate rayon.

All the metal acetates are water-soluble except silver acetate. The acetate ion is colorless; it may be identified by heating a sample with sulfuric acid to give the odor of acetic acid. The fruitlike odor of the ester formed when ethyl alcohol and a trace of sulfuric acid are heated with the sample is also characteristic. *See* ACETIC ACID; CARBOXYLIC ACID; ESTER.

[E. EUGENE WEAVER]

Acetic acid

A colorless, pungent liquid, CH_3COOH, melting at 16.7° C and boiling at 118.0°C. Acetic acid is the sour principle in vinegar. Concentrated acid is called glacial acetic acid because of its readiness to crystallize at cool temperatures. Acetic acid in vinegar arises through an aerobic fermentation of dilute ethanol solutions, such as wine, cider, and beer, with any of several varieties of *Acetobacter*. Though there are several feasible ways to concentrate acetic acid from fermentation broths, none is currently considered economical.

Production. Acetic acid was formerly manufactured from pyroligneous acid obtained in destructive distillation of wood. These processes are of historical interest because many modern chemical engineering operations developed through the study of acetic acid production. Today acetic acid is manufactured by three main routes: butane liquid-phase catalytic oxidation in acetic acid solvent, palladium–copper salt–catalyzed oxidation of ethylene in aqueous solution, and methanol carbonylation in the presence of rhodium catalyst. Large quantities of acetic acid are recovered in the manufacture of cellulose acetate and polyvinyl alcohol. Some acetic acid is produced in the oxidation of higher olefins, aromatic hydrocarbons, ketones, and alcohols. *See* OXIDATION PROCESS.

Chemical properties. Pure acetic acid is completely miscible with water, ethanol, diethyl ether, and carbon tetrachloride, but is not soluble in carbon disulfide. Freezing of acetic acid is accompanied by a remarkable volume contraction: the molar volume of liquid acetic acid at the freezing point is 57.02 cm³/mole, but at the same temperature the crystalline solid is 47.44 cm³/mole. It is a strongly proton-donating solvent with a relatively small dipole moment and a low dielectric constant. In a water solution, acetic acid is a typical weakly ionized acid ($K_a = 1.8 \times 10^{-5}$). *See* CARBOXYLIC ACID.

The vapor density of acetic acid indicates a molecular weight considerably higher than would be expected for a compound with a formula weight of 60.05. The acid probably exists largely as the dimer in the vapor and liquid states.

Acetic acid neutralizes many oxides and hydroxides, and decomposes carbonates to furnish acetate salts, which are used in textile dyeing and finishing, as pigments, and as pesticides; examples are verdigris, white lead, and paris green. *See* ARSENIC; LEAD.

Over two-thirds of the acetic acid manufactured is used in production of either vinyl acetate or cellulose acetate. Acetic anhydride, the key intermediate in making cellulose acetate, is prepared commercially by pyrolysis of acetic acid in the presence of trialkyl phosphate catalyst. Considerable amounts of acetic acid are consumed in the manufacture of terephthalic acid by liquid-phase oxidation of xylene, and in the preparation of esters for lacquer solvents, paints and varnishes, pharmaceuticals, and herbicides. *See* ESTER; SOLVENT. [FRANK WAGNER]

Bibliography: T. A. Geissman, *Principles of Organic Chemistry*, 4th ed., 1977; C. R. Noller, *Chemistry of Organic Compounds*, 3d ed., 1965; J. D. Roberts and M. C. Caserio, *Basic Principles of Organic Chemistry*, 2d ed., 1977.

Acetone

A chemical compound, CH_3COCH_3. A colorless liquid with an ethereal odor, it is the first member of the homologous series of aliphatic ketones. Its physical properties include boiling point 56.2°C, melting point −94.8°C, and specific gravity 0.791. Acetone is an extremely important, low-cost raw material that is used for production of other chemicals.

Acetone is used as a solvent for cellulose ethers, cellulose acetate, cellulose nitrate, and other cellulose esters. Cellulose acetate is spun from acetone solution. Lacquers, based on cellulose esters, are used in solution in mixed solvents including acetone. Acetylene is safely stored in cylinders under pressure by dissolving it in acetone, which is absorbed on inert material such as asbestos. It has a low toxicity.

Production. The principal method of acetone production uses propylene, obtained from the cracking of petroleum. Addition of sulfuric acid to propylene yields isopropyl hydrogen sulfate, which upon hydrolysis yields isopropyl alcohol. Oxidation or dehydrogenation over metal catalysts, such as copper, converts the alcohol to acetone, as shown in reaction (1).

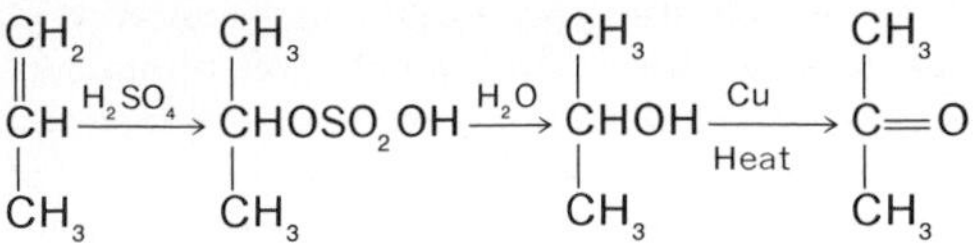
(1)

Acetone is also produced by passage of acetic acid vapor over metallic oxide catalysts at 400–450°C, by partial oxidation of the lower alkane hydrocarbons, and by the decomposition of cumene hydroperoxide, as shown in reaction (2).

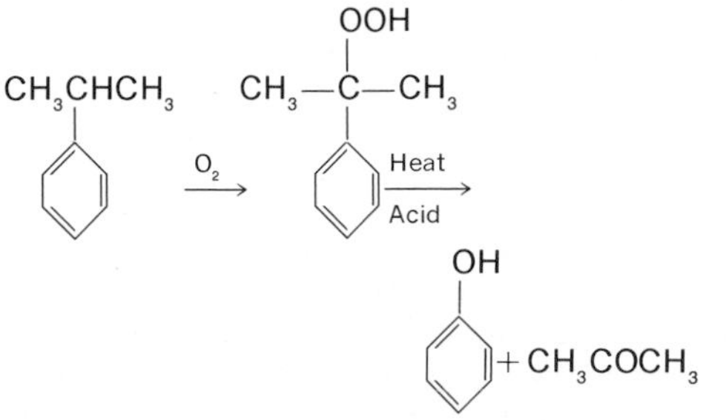
(2)

Phenol is the other product of this last process, which has been in operation in the United States since 1954.

Chemical uses. Pyrolysis of acetone vapor at 700°C produces ketene, which reacts with acetic acid to produce acetic anhydride, as shown in reaction (3).

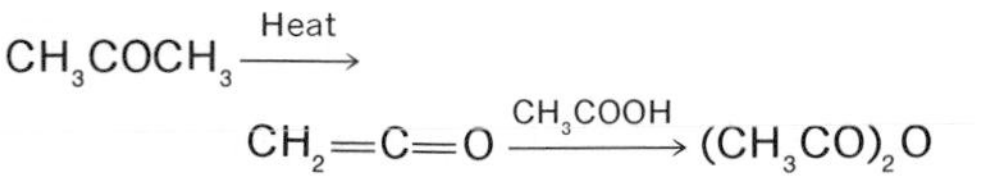
(3)

Aldol-type condensation of acetone with $Ba(OH)_2$ yields (*a*) diacetone alcohol, (*b*) mesityl oxide, and (*c*) phorone, shown in reaction (4). Catalytic hydrogenation of mesityl oxide gives methyl isobutyl ketone. All these products are of commercial use as solvents or as chemical intermediates. *See*

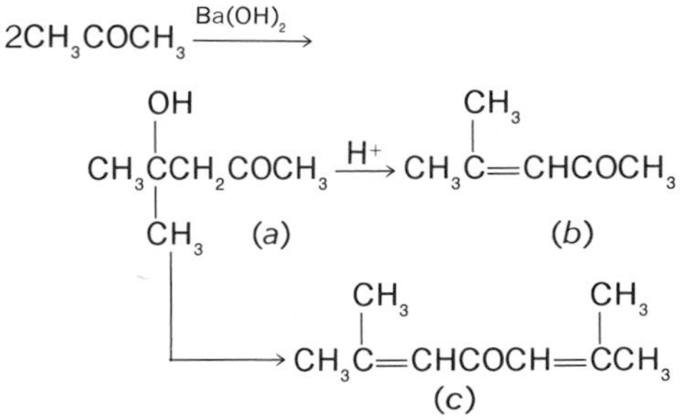
(4)

[DAVID A. SHIRLEY]

Acetylation

The introduction of an acetyl group

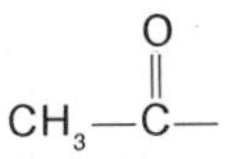

into an organic compound containing the alcoholic or phenolic hydroxyl (—OH) or the amino (—NH_2) and substituted amino groups to yield esters or substituted amides, respectively. The reaction can be used as a quantitative determination of the hydroxyl or amino group, and in qualitative organic analyses as a determination of the class of the amine. Acetylation can be carried out with acetic anhydride, acetyl chloride, or glacial acetic acid (in order of decreasing importance) and with or without an inert solvent, such as benzene or toluene. Often glacial acetic acid itself is used as a solvent and sulfuric acid is used as a catalyst. Ease of reaction is in the order of acetyl chloride > acetic anhydride > acetic acid. The first two often react at room temperature, whereas direct esterification with acetic acid usually requires more drastic conditions. Acetylation of acetylene is accomplished catalytically in the vapor phase or liquid phase to produce vinyl acetate. In general, acetylation reactions are exothermic.

The most important acetates formed by acetylation are cellulose acetate, vinyl acetate, and ethyl acetate. Cellulose acetate is prepared for use as a fiber (acetate rayon), safety movie film, and a tough plastic. Vinyl acetate is used in polymerization or copolymerization reactions to produce vinyl resins and cyclic acetals (safety glass). Ethyl acetate is one of many esters which are often prepared by a continuous process for use as a solvent.

Other important acetylation reactions include (1) production of pharmaceuticals such as aspirin, phenacetin, and derivatives; (2) protection of active groups during organic syntheses, particularly the amino and substituted amino group; (3) determination of the hydroxy acid content of fats and oils (acetyl number); (4) preparation of solid derivatives from liquids for qualitative identification; (5) differentiation of tertiary from primary and secondary amines; (6) preparation of α,β-unsaturated acids and coumarin derivatives (Perkin reaction); (7) preparation of acid anhydrides with acetic anhydride and an organic acid by an exchange reaction; and (8) production of other chemical intermediates. *See* ACID ANHYDRIDE; ACID HALIDE; AMINE. [ELBERT H. HADLEY]

Bibliography: S. H. Pine et al., *Organic Chemistry*, 4th ed., 1980.

Acetylene

A colorless, aromatic gas that burns with a highly luminous flame, sublimes at −83.4°C/760 mmHg, and is shock-sensitive at low temperatures and partial pressures above 20–30 psia (138–207 kPa absolute). Acetylene is safely stored in cylinders under pressure, using acetone as a solvent. Acetylene burns with oxygen to produce the highest achievable flame temperature, over 3300°C, of any carbonaceous fuel. Oxyacetylene flames are of major importance in the welding and cutting of metals.

Reactions. Acetylene has the formula HC≡CH. The triple bond undergoes addition reactions to produce ethylene and ethane derivatives. Unsymmetrical reagents such as hydrogen chloride add to produce vinyl chloride and 1,1-dichloroethane; hydrogen cyanide produces acrylonitrile; acetic acid reacts to form vinyl acetate; and alcohols yield vinyl ethers. Water adds to acetylene in the presence of mercuric salts and strong acids to produce acetaldehyde, and chlorine adds to form 1,2-dichloroethene and 1,1,2,2-tetrachloroethane. The hydrogen atoms of acetylene undergo replacement reactions, demonstrating their moderate acidity. Aqueous solutions of copper and silver salts produce explosive metal acetylides, for example, AgC≡CAg. Reactive metals, for example sodium, on contact with acetylene form stable mono- and disodium salts with the generation of hydrogen. Formaldehyde reacts to produce 1,4-butynediol, which can be hydrogenated to the commercially important 1,4-butanediol. Acetylene reacts with carbon monoxide and alcohols in the presence of nickel carbonyl to form acrylate esters.

Production. Acetylene was first produced commercially by contacting water and calcium carbide, a product of roasting calcium oxide with coke. In 1978 only about 5% of the world's production of 900,000 metric tons of acetylene was produced from calcium carbide. Modern processes generate acetylene by high-temperature hydrocarbon cracking reactions. Energy for the processes comes from externally generated steam, from internal partial oxidation of the feedstock, or by electric arc furnaces still in use in Europe. Raw materials range from methane through naphtha, and experimental processes have used crude oil and coal. About 15% of the acetylene produced in the United States is a by-product of ethylene manufacture from hydrocarbon gas cracking processes. The more favorable economics of ethylene production from hydrocarbons has caused acetylene to be replaced as a major raw material for many important commodity chemicals. As a consequence, since the mid-1960s world production of acetylene has been on the decline. By the late 1970s in the United States only 4% of the vinyl chloride and 10% of the vinyl acetate were manufactured from acetylene, ethylene being the preferred raw material.

The major growing use for acetylene is in the manufacture of 1,4-butanediol. This product is used as a chain extender for polyurethanes and in polybutylene terephthalate resins. Vinyl chloride, acrylic acid and esters, and vinyl acetate still constitute major uses for acetylene. *See* ALKYNE; THERMOCHEMISTRY.

[ROBERT K. BARNES]

Bibliography: S. A. Miller, *Acetylene*, 2 vols., 1965, 1966; C. A. Hampel and G. G. Hawley, *The Encyclopedia of Chemistry*, 3d ed., 1973.

Acetylide

A derivative of acetylene formed by the replacement of one or both of the hydrogens by a metal. General formulas are H—C≡C—M and M—C≡C—M, the latter being known as carbides (although carbides such as SiC, B_6C, and WC are not acetylides). Acetylides are prepared by the action of acetylene on active metals or metal compounds, or by the action of metals or metal compounds on carbon at high temperatures (electric furnace). Alkali and alkaline-earth acetylides are relatively stable, but most heavy metal derivatives are thermodynamically unstable and may explode when dry. They are readily decomposed by dilute acid. Acetylides are salts of a very weak acid and therefore hydrolyze readily. The most important acetylide is calcium carbide, which is used to prepare acetylene. Cuprous acetylide is used in ethynylation and high-pressure reactions. *See* ACETYLENE; CARBIDE.

[ELBERT H. HADLEY]

Acid and base

Two interrelated classes of chemical compounds, the precise definitions of which have varied considerably with the development of chemistry. These changing definitions have led to frequent controversies, some of which are still unresolved. Acids initially were defined only by their common properties. They were substances which had a sour taste, which dissolved many metals, and which reacted with alkalies (or bases) to form salts. For a time, following the work of A. L. Lavoisier, it was believed that a common constituent of all acids was the element oxygen, but gradually it became clear that, if there were an essential element, it was hydrogen, not oxygen. In fact, the definition of an acid, formulated by J. von Liebig in 1840, as "a hydrogen-containing substance which will generate hydrogen gas on reaction with metals" proved to be satisfactory for about 50 years.

Bases initially were defined as those substances which reacted with acids to form salts (they were the "base" of the salt). The alkalies, soda and potash, were the best-known bases, but it soon became clear that there were other bases, notably ammonia and the amines.

Acids and bases are among the most important chemicals of commerce. The inorganic acids are often known as mineral acids, and among the most important are sulfuric, H_2SO_4; phosphoric, H_3PO_4; nitric, HNO_3; and hydrochloric, HCl (sometimes called muriatic). Among the many important organic acids are acetic, CH_3COOH, and oxalic, $H_2C_2O_4$, acids, and phenol, C_6H_5OH. The important inorganic bases are ammonia, NH_3; sodium hydroxide or soda. NaOH; potassium hydroxide, KOH; calcium hydroxide or lime, $Ca(OH)_2$; and sodium carbonate, Na_2CO_3. There are also many organic bases, mostly derivatives of ammonia. Examples are pyridine, C_5H_5N, and ethylamine, $C_2H_5NH_2$.

Arrhenius-Ostwald theory. When the concept of ionization of chemical compounds in water solution became established, some considerably different definitions of acids and bases became popular. Acids were defined as substances which ionized in aqueous solution to give hydrogen ions, H^+, and bases were substances which reacted to give hydroxide ions, OH^-. These definitions are sometimes known as the Arrhenius-Ostwald theory of acids and bases and were proposed separately by S. Arrhenius and W. Ostwald. Their use makes it possible to discuss acid and base equilibria and also the strengths of individual acids and bases. The ionization of an acid in water can be written as Eq. (1). Qualitatively, an acid is strong if

$$HA = H^+ + A^- \tag{1}$$

this reaction goes extensively toward the ionic products and weak if the ionization is only slight. A quantitative treatment of this ionization or dissociation can be given by utilizing the equilibrium expression for the acid, as shown in Eq. (2), where

$$\frac{[H^+][A^-]}{[HA]} = K_{HA} \tag{2}$$

the brackets mean concentration in moles per liter and the constant K_{HA} is called the dissociation constant of the acid. This dissociation constant is a large number for a strong acid and a small number for a weak acid. For example, at 25°C and with water as the solvent, K_{HA} has the value 1.8×10^{-5} for a typical weak acid, acetic acid (the acid of vinegar), and this value varies only slightly in dilute solutions as a function of concentration. Dissociation constants vary somewhat with temperature. They also change considerably with changes in the solvent, even to the extent that an acid which is fully ionized in water may, in some other less basic solvent, become decidedly weak. Almost all the available data on dissociation constants are for solutions in water, partly because of its ubiquitous character, and partly because it is both a good ionizing medium and a good solvent.

Acetic acid has only one one ionizable hydrogen and is called monobasic. Some other acids have two or even three ionizable hydrogens and are called polybasic. An example is phosphoric acid, which ionizes in three steps, shown in Eqs. (3), (4), and (5), each with its own dissociation constant.

Ionization reaction	K_{HA}, 25°C	
$H_3PO_4 = H^+ + H_2PO_4^-$	1.5×10^{-3}	(3)
$H_2PO_4^- = H^+ + HPO_4^{--}$	6.2×10^{-8}	(4)
$HPO_4^{--} = H^+ + PO_4^{---}$	4×10^{-13}	(5)

A similar discussion can be given for the ionization of bases in water. However, the concentrations of the species H^+ and OH^- in a water solution are not independently variable. This is because water itself is both a weak acid and a weak base, ionizing very slightly according to Eq. (6). For pure

$$H_2O = H^+ + OH^- \tag{6}$$

water, the concentrations of H^+ and OH^- are equal. At ordinary temperatures, roughly $2 \times 10^{-7}\%$ of the water is present as ions. As a result of this ionization, the ionic concentrations are related through Eq. (7). At 25°C and with concen-

$$\frac{[H^+][OH^-]}{[H_2O]} = K \tag{7}$$

trations in moles per liter, the product $[H^+][OH^-]$ is equal to 1×10^{-14}.

A major consequence of this interdependence is that measurement of the concentration of either H^+ or OH^- in a water solution permits immediate calculation of the other. This fact led S. P. L. Sϕrenson in 1909 to propose use of a logarithmic pH scale for the concentration of hydrogen ions in water. Although there are some difficulties in giving an exact definition of pH, it is very nearly correct for dilute solutions in water to write Eq. (8). It

$$pH = -\log [H^+] \tag{8}$$

then turns out that pH values of 0–14 cover the range from strongly acidic to strongly basic solutions. The pH of pure water at ordinary temperature is 7.

For many situations, it is desirable to maintain the H^+ and OH^- concentration of a water solution at low and constant values. A useful device for this is a mixture of a weak acid and its anion (or of a weak base and its cation). Such a mixture is called a buffer. A typical example is a mixture of sodium acetate and acetic acid. From the treatment just given, it is evident that for this case Eq. (9)

$$[H^+] = \frac{[CH_3COOH]}{[CH_3COO^-]} \times 1.8 \times 10^{-5} \tag{9}$$

can be formed. In the equation $[CH_3COOH]$ and $[CH_3COO^-]$ represent the concentrations of acetic acid and acetate ion, respectively. Thus, if the concentrations of acetic acid and acetate ion are both 0.1 mole per liter, the H^+ concentration will be 1.8×10^{-5} mole per liter and OH^- will be 5.5×10^{-10} mole per liter. The pH of this solution will be about 4.7. Constant acidity is a most important aspect of blood and other life fluids; these invariably contain weak acids and bases to give the necessary buffering action.

Brönsted theory. The Arrhenius, or water, theory of acid and bases has many attractive features, but it has also presented some difficulties. A major difficulty was that solvents other than water can be used for acids and bases and thus need consideration. For many of the solvents of interest, the necessary extensions of the water theory are both obvious and plausible. For example, with liquid ammonia as the solvent, one can define NH_4^+ as the acid ion and NH_2^- as the base ion, and the former can be thought of as a hydrogen ion combined with a molecule of the solvent. However, for a hydrogenless (aprotic) solvent, such as liquid sulfur dioxide, the extensions are less obvious. Consideration of such systems has led to some solvent-oriented theories of acids and bases to which the names of E. C. Franklin and A. F. D. Germann often are attached. The essence of these theories is to define acids and bases in terms of what they do to the solvent. Thus, one definition of an acid is that it gives rise to "a cation which is characteristic of the solvent," for example, SO^{++} from sulfur dioxide. These theories have been useful in emphasizing the need to consider nonaqueous systems. However, they have not been widely adopted, at least partly because a powerful and wide-ranging protonic theory of acids and bases was introduced by J. N. Brönsted in 1923 and was rapidly accepted by many other scientists. Somewhat similar ideas were advanced almost simultaneously by T. M. Lowry, and the new theory is occasionally called the Brönsted-Lowry theory.

This theory gives a unique role to the hydrogen ion, and there appeared to be justification for this. One justification, of course, was the historically important role already given to hydrogen in defining acids. A rather different justification involved the unique structure of hydrogen ion. It is the only common ion which consists solely of a nucleus, the proton. As a consequence, it is only about 10^{-14} cm in diameter. All other ordinary ions have peripheral electron clouds and, as a result, are roughly 10^6 times larger than the proton. The small size of the latter makes it reasonable to postulate that protons are never found in a free state but rather always exist in combination with some base. The Brönsted theory emphasizes this by proposing that all acid-base reactions consist simply of the transfer of a proton from one base to another.

The Brönsted definitions of acids and bases are: An acid is a species which can act as a source of protons; a base is a species which can accept protons. Compared to the water theory, this represents only a slight change in the definition of an acid but a considerable extension of the term base. In addition to hydroxide ion, the bases now include a wide variety of uncharged species, such as ammonia and the amines, as well as numerous

Conjugate acid-base pairs

	Acids		Bases	
Strong acids ↑	H_2SO_4	———	HSO_4^-	Weak bases ↑
	HCl	———	Cl^-	
	H_3O^+	———	H_2O	
	HSO_4^-	———	SO_4^{--}	
	$HF_{(aq)}$	———	F^-	
	CH_3COOH	———	CH_3COO^-	
	NH_4^+	———	NH_3	
	HCO_3^-	———	CO_3^{--}	
↓ Weak acids	H_2O	———	OH^-	↓ Strong bases
	C_2H_5OH	———	$C_2H_5O^-$	

charged species, such as the anions of weak acids. In fact, every acid can generate a base by loss of a proton. Acids and bases which are related in this way are known as conjugate acid-base pairs, and the table lists several examples. By these definitions, such previously distinct chemical processes as ionization, hydrolysis, and neutralization become examples of the single class of proton transfer or protolytic reactions. The general reaction is expressed as Eq. (10). This equation can be

$$\text{acid}_1 + \text{base}_2 = \text{base}_1 + \text{acid}_2 \quad (10)$$

considered to be a combination of two conjugate acid-base pairs, and the pairs below can be used to construct a variety of typical acid-base reactions. For example, the ionization of acetic acid in water becomes Eq. (11). Water functions here as a base

$$\underset{a_1}{CH_3COOH} + \underset{b_2}{H_2O} = \underset{b_1}{CH_3COO^-} + \underset{a_2}{H_3O^+} \quad (11)$$

to form the species H_3O^+, the oxonium ion (sometimes called the hydronium ion). However, water can also function as an acid to form the base OH^-, and this dual or amphoteric character of water is one reason why so many acid-base reactions occur in it.

As the table shows, strengths of acids and bases are not independent. A very strong Brönsted acid implies a very weak conjugate base and vice versa. A qualitative ordering of acid strength or base strength, as above, permits a rough prediction of the extent to which an acid-base reaction will go. The rule is that a strong acid and a strong base will react extensively with each other, whereas a weak acid and a weak base will react together only very slightly. More accurate calculations of acid-base equilibria can be made by using the ordinary formulation of the law of mass action. A point of some importance is that, for ionization in water, the equations reduce to the earlier Arrhenius-Ostwald type. Thus, for the ionization of acetic acid in water Eq. (11) leads to Eq. (12). Remembering that

$$\frac{[H_3O^+][CH_3COO^-]}{[CH_3COOH][H_2O]} = K \quad (12)$$

the concentration of water will be almost constant since it is the solvent, this can be written as Eq. (13), where K_{HAc} is just the conventional dissociation constant for acetic acid in water.

$$\frac{[H_3O^+][CH_3COO^-]}{[CH_3COOH]} = K[H_2O] \equiv K_{HAc} \quad (13)$$

One result of the Brönsted definitions is that for a given solvent, such as water, there is only a single scale of acid-base strength. Put another way, the relative strength of a set of acids will be the same for any base. Hence, the ordinary tabulation of ionization constants of acids in water permits quantitative calculation for a very large number of acid-base equilibria.

The Brönsted concepts can be applied without difficulty to other solvents which are amphoteric in the same sense as water, and data are available for many nonaqueous solvents, such as methyl alcohol, formic acid, and liquid ammonia. An important practical point is that relative acid (or base) strength turns out to be very nearly the same in these other solvents as it is in water. Brönsted acid-base reactions can also be studied in aprotic solvents (materials such as hexane or carbon tetrachloride which have virtually no tendency to gain or lose protons), but in this case, both the acid and the base must be added to the solvent.

A fact which merits consideration in any theory of acids and bases is that the speeds of large numbers of chemical reactions are greatly accelerated by acids and bases. This phenomenon is called acid-base catalysis, and a major reason for its wide prevalence is that most proton transfers are themselves exceedingly fast. Hence, reversible acid-base equilibria can usually be established very rapidly, and the resulting conjugate acids (or bases) then frequently offer favorable paths for the overall chemical reaction. The mechanisms of many of these catalyzed reactions are known. Some of them are specifically catalyzed by solvated protons (hydrogen ions); others, by hydroxide ions. Still others are catalyzed by acids or bases in the most general sense of the Brönsted definitions. The existence of this general acid and base catalysis constituted an important item in the wide acceptance of the Brönsted definitions.

Lewis theory. Studies of catalysis have, however, played a large role in the acceptance of a set of quite different definitions of acids and bases, those due to G. N. Lewis. These definitions were originally proposed at about the same time as those of Brönsted, but it was not until Lewis restated them in 1938 that they began to gain wide consideration. The Lewis definitions are: An acid is a substance which can accept an electron pair from a base; a base is a substance which can donate an electron pair. (These definitions are very similar to the terms popularized around 1927 by N. V. Sidgwick and others: electron donors, which are essentially Lewis bases, and electron acceptors, which are Lewis acids.) Bases under the Lewis definition are very similar to those defined by Brönsted, but the Lewis definition for acids is very much broader. For example, virtually every cation is an acid, as are such species as $AlCl_3$, BF_3, and SO_3. An acid-base reaction now typically becomes a combination of an acid with a base, rather than a proton transfer. Even so, many of the types of reactions which are characteristic of proton acids also will occur between Lewis acids and bases, for example, neutralization and color change of indicators as well as acid-base catalysis. Furthermore, these new definitions have been useful in suggesting new interrelations and in predicting new reactions, particularly for solid systems and for systems in nonaqueous solvents.

For several reasons, these definitions have not

been universally accepted. One reason is that the terms electron donor and electron acceptor had been widely accepted and appear to serve similar predicting and classifying purposes. A more important reason is unwillingness to surrender certain advantages in precision and definiteness inherent in the narrower Brönsted definitions. It is a drawback of the Lewis definitions that the relative strengths of Lewis acids vary widely with choice of base and vice versa. For example, with the Brönsted definitions, hydroxide ion is always a stronger base than ammonia; with the Lewis definitions, hydroxide ion is a much weaker base than ammonia when reacting with silver ion but is stronger than ammonia when reacting with hydrogen ion. Another feature of the Lewis definitions is that some substances which have long been obvious examples of acids, for example, HCl and H_2SO_4, do not naturally fit the Lewis definition since they cannot plausibly accept electron pairs. Certain other substances, for example, carbon dioxide, are included by calling them secondary acids. These substances, too, tend to have electronic structures in which the ability to accept electron pairs is not obvious, but the more important distinction between them and primary acids is that their rates of neutralization by bases are measurably slow. However, in spite of these difficulties, the use of the Lewis definitions is increasing. Since there does not appear to be any simultaneous tendency to abandon the Brönsted definitions, chemistry seems to be entering a period when the term acid needs a qualifying adjective for clarity, for example, Lewis acid or proton acid.

Hard and soft acids and bases. As pointed out above, one of the drawbacks of such a broad definition as the Lewis one is that it is difficult to systematize the behavior of acids and bases toward each other. Attempts have been made to classify Lewis acids and bases into categories with respect to their mutual behavior. R. G. Pearson in 1963 proposed a simple and lucid classification scheme, based in part on earlier methods, that appears to be promising in its application to a wide variety of Lewis acid-base behavior.

Lewis bases (electron donors) are classified as soft if they have high polarizability, low electronegativity, are easily oxidized, or possess low-lying empty orbitals. They are classified as hard if they have the opposite tendencies. Some bases, spanning the range of hard to soft and listed in order of increasing softness, are H_2O, OH^-, OCH_3^-, F^-, NH_3, C_5H_5N, NO_2^-, NH_2OH, N_2H_4, C_6H_5SH, Br^-, I^-, SCN^-, SO_3^{-2}, $S_2O_3^{-2}$, and $(C_6H_5)_3P$. Acids are divided more or less distinctly into two categories, hard and soft, with a few intermediate cases. Hard acids are of low polarizability, small size, and high positive oxidation state, and do not have easily excitable outer electrons. Soft acids have several of the properties of high polarizability, low or zero positive charge, large size, and easily excited outer electrons, particularly *d* electrons in metals. Some hard acids are H^+, Li^+, Na^+, K^+, Be^{+2}, Mg^{+2}, Ca^{+2}, Sr^{+2}, Mn^{+2}, Al^{+3}, Sc^{+3}, Cr^{+3}, Co^{+3}, Fe^{+3}, As^{+3}, Ce^{+3}, Si^{+4}, Ti^{+4}, Zr^{+4}, Pu^{+4}, $BeMe_2$ (Me is the methyl group), BF_3, BCl_3, $B(OR)_3$, $Al(CH_3)_3$, AlH_3, SO_3, and CO_2. Examples of soft acids are Cu^+, Ag^+, Au^+, Tl^+, Hg^+, Cs^+, Pd^{+2}, Cd^{+2}, Pt^{+2}, Hg^{+2}, CH_3Hg^+, Tl^{+3}, BH_3, $CO(CN)_5^{-2}$, I_2, Br_2, ICN, chloranil, quinones, tetracyanoethylene, O, Cl, Br, I, N, metal atoms, and bulk metals. Intermediate acids are Fe^{+2}, Co^{+2}, Ni^{+2}, Cu^{+2}, Pb^{+2}, Sn^{+2}, $B(CH_3)_3$, SO_2, NO^+, and R_3C^+.

The rule for correlating acid-base behavior is then stated as: Hard acids prefer to associate with hard bases and soft acids with soft bases. Application, for example, to the problem of OH^- and NH_3, mentioned earlier (and recognizing that OH^- is hard compared with NH_3), leads to Eq. (14), which is unfavorable compared with Eq. (15). However, the reaction shown in Eq. (16) is unfavorable compared with the reaction shown in Eq. (17), which is in agreement with experiment.

$$OH^- \text{(hard)} + Ag^+ \text{(soft)} \rightarrow AgOH \qquad (14)$$

$$NH_3 \text{(soft)} + Ag^+ \text{(soft)} \rightarrow Ag(NH_3)^+ \qquad (15)$$

$$NH_3 \text{(soft)} + H^+ \text{(hard)} \rightarrow NH_4^+ \qquad (16)$$

$$OH^- \text{(hard)} + H^+ \text{(hard)} \rightarrow H_2O \qquad (17)$$

The rule is successful in correlating general acid-base behavior of a very wide variety of chemical systems, including metal-ligand interaction, charge-transfer complex formation, hydrogen bond formation, complex ion formation, carbonium ion formation, covalent bond formation, and ionic bond formation.

It is to be emphasized that the hard-and-soft acid-and-base concept is a means of classification and correlation and is not a theoretical explanation for acid-base behavior. The reasons why hard acids prefer hard bases and soft prefer soft are complex and varied. The already well-developed concepts of ionic and covalent bonds appear to be helpful, however. Hard acids and hard bases with small sizes and high charge would be held together with stable ionic bonds. Conversely, the conditions for soft acids and soft bases would be favorable for good covalent bonding. Existing theories of π-bonding also fit into the scheme.

Usanovich theory. Another comprehensive theory of acids and bases was proposed by M. Usanovich in 1939 and is sometimes known as the positive-negative theory. Acids are defined as substances which form salts with bases, give up cations, and add themselves to anions and to free electrons. Bases are similarly defined as substances which give up anions or electrons and add themselves to cations. Two examples of acid-base reactions under this scheme are Eqs. (18) and (19).

$$Na_2O + SO_3 = 2Na^+ + SO_4^{--} \qquad (18)$$

$$2Na + Cl_2 = 2Na^+ + 2Cl^- \qquad (19)$$

In the first, SO_3 is an acid because it takes up an anion, O^{--}, to form SO_4^{--}. In the second example, Cl_2 is an acid because it takes up electrons to form Cl^-. Using conventional terminology, this second reaction is an obvious example of oxidation-reduction. The fact that oxidation-reduction can also be included in the Usanovich scheme is an illustration of the extensiveness of these definitions. So far, this theory has had little acceptance, quite possibly because the definitions are too broad to be very useful.

Generation of chemical species. The acidic or basic character of a solvent can be used to stabilize interesting chemical species, which would otherwise be difficult to obtain. For example, carbonium ions have been thought to be important intermediates in many organic reactions, but because of their fleeting existence as intermediates, their properties have been difficult to study. Most carbonium ions are very strong bases and would, for example, react, as shown by reaction (20). Accordingly, the equilibrium would lie far to the right.

$$R_3C^+ + H_2O \rightarrow R_3COH + H^+ \qquad (20)$$

However, use of a very strongly acidic solvent reverses the reaction, and measurable amounts of carbonium ion are then found. Concentrated sulfuric acid has found use in this connection. The very high acidity of SbF_5 by itself, as a Lewis acid, and in mixtures with other Lewis acids, such as SO_2, or protonic acids, such as HF and FSO_3H, makes possible the study of many otherwise unstable carbonium ions. *See* SUPERACIDS.

Acidity functions. A very different approach to the definition of acids, or perhaps better, to the definition of acidity, is to base the definition on a particular method of measurement. (As one example, it is probably true that the most nearly exact definition of pH is in terms of the electromotive force of a particular kind of galvanic cell.) It is possible to define various acidity functions in this way, and several have been proposed. One of the earliest and also one of the most successful is the H_0 acidity function of L. P. Hammett. This defines an acidity in terms of the observed indicator ratio for a particular class of indicators, those which are uncharged in the basic form B. Suppose there is available a set of such indicators, and suppose further that the values of the dissociation constants of the acid forms BH^+ are known. Then the h_0 acidity of a solution is defined as Eq. (21), where K_{BH^+} is the dissociation constant for the particular indicator employed, and where $[BH^+]/[B]$ is the experimentally observed ratio of concentrations of the conjugate acid and conjugate base forms of the indicator. To have a logarithmic scale (analogous to pH), the further definition is expressed in Eq. (22). The virtues of this scale are that measurements are relatively simple and can be made for concentrated solutions and for solutions in nonaqueous or mixed solvents, situations where the pH scale offers difficulties. A further point is that in dilute aqueous solutions this new acidity becomes identical to pH. Although it has been found that this measure of acidity is fairly consistent within a class of indicators used, different classes can give somewhat different measures of acidity. Hence, caution must be used in interpretation of acidity measured by this technique.

$$h_0 = K_{BH^+} \frac{[BH^+]}{[B]} \qquad (21)$$

$$H_0 = -\log h_0 \qquad (22)$$

For a discussion of measurement of acidity *see* HYDROGEN ION. *See also* BASE; BUFFERS; IONIC EQUILIBRIUM; OXIDATION-REDUCTION; SOLUTION; SOLVENT.

[FRANKLIN A. LONG; RICHARD H. BOYD]

Bibliography: J. F. Coetzee and C. D. Ritchie, *Solute-Solvent Interactions*, 1969; J. J. Lagowski, *The Chemistry of Non-Aqueous Solvents*, vol. 1–3, 1966–1970, vol. 4, 1976; G. Olah, *Chem. Eng. News*, 45:77, Mar. 27, 1967; R. G. Pearson, Acids and bases, *Science*, 151:172, 1966.

Acid anhydride

One of an important class of reactive organic compounds derived from acids via formal intermolecular dehydration; thus, acetic acid,

$$CH_3-\overset{\overset{\displaystyle O}{\|}}{C}-OH$$

on loss of water forms acetic anhydride,

$$CH_3-\overset{\overset{\displaystyle O}{\|}}{C}-O-\overset{\overset{\displaystyle O}{\|}}{C}-CH_3$$

Anhydrides of straight-chain acids containing from 2 to 12 carbon atoms are liquids with boiling points higher than those of the parent acids. They are relatively insoluble in cold water and are soluble in alcohol, ether, and other common organic solvents. The lower members are pungent, corrosive, and weakly lacrimatory. Anhydrides from acids with more than 12 carbon atoms and cyclic anhydrides from dicarboxylic acids are crystalline solids.

Preparation. Because the direct intermolecular removal of water from organic acids is not practicable, anhydrides must be prepared by means of indirect processes. A general method involves interaction of an acid salt with an acid chloride, reaction (1).

$$CH_3-\overset{\overset{\displaystyle O}{\|}}{C}-O^-\,Na^+ + CH_3-\overset{\overset{\displaystyle O}{\|}}{C}-Cl \rightarrow NaCl + (CH_3CO)_2O \qquad (1)$$

Acetic anhydride, the most important aliphatic anhydride, is manufactured by air oxidation of acetaldehyde, using as catalysts the acetates of copper and cobalt, shown in reaction (2); peracetic acid apparently is an intermediate. The anhydride is separated from the by-product water by vacuum distillation.

$$CH_3\overset{\overset{\displaystyle O}{\|}}{C}-H + O_2 \xrightarrow{\text{Catalyst}} CH_3\overset{\overset{\displaystyle O}{\|}}{C}OOH + CH_3CHO \rightarrow (CH_3CO)_2O + H_2O \qquad (2)$$

Another important process utilizes the thermal decomposition of ethylidene acetate (made from acetylene and acetic acid), reaction (3).

$$2CH_3COOH + HC{\equiv}CH \xrightarrow[\text{catalyst}]{\text{Mercury salt}} CH_3-\underset{\diagdown OOCCH_3}{\overset{\diagup OOCCH_3}{C}}-H \xrightarrow[ZnCl_2]{\text{Heat}} CH_3CHO + (CH_3CO)_2O \qquad (3)$$

Acetic anhydride has been made by the reaction of acetic acid with ketene, reaction (4).

$$CH_3COOH + CH_2{=}C{=}O \rightarrow (CH_3CO)_2O \qquad (4)$$

Mixed anhydrides composed of two different radicals are unstable, and disproportionate to give the two simple anhydrides. Direct use is made of this in the preparation of high-molecular-weight anhydrides, as seen in reaction (5). The two simple

$$2CH_2{=}C{=}O + 2CH_3(CH_2)_{14}COOH \rightarrow$$
Ketene Palmitic acid

$$2CH_3{-}\overset{\displaystyle O}{\overset{\|}{C}}{-}O{-}\overset{\displaystyle O}{\overset{\|}{C}}{-}(CH_2)_{14}CH_3 \rightarrow$$
Mixed anhydride

$$(CH_3CO)_2O + [CH_3(CH_2)_{14}CO]_2O \quad (5)$$
Acetic anhydride Palmitic anhydride

anhydrides are easily separable by distillation in a vacuum.

Cyclic anhydrides are obtained by warming succinic or glutaric acids, either alone, with acetic anhydride, or with acetyl chloride. Under these conditions, adipic acid first forms linear, polymeric anhydride mixtures, from which the monomer is obtained by slow, high-vacuum distillation. Cyclic anhydrides are also formed by simple heat treatment of cis-unsaturated dicarboxylic acids, for example, maleic and glutaconic acids; and of aromatic 1,2-dicarboxylic acids, for example, phthalic acid. Commercially, however, both phthalic (I) and maleic (II) anhydrides are primary products of manufacture, being formed by vapor-phase, catalytic (vanadium pentoxide), air oxidation of naphthalene and benzene, respectively; at the reaction temperature, the anhydrides form directly.

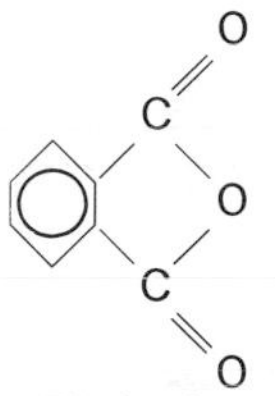
Phthalic anhydride (I)

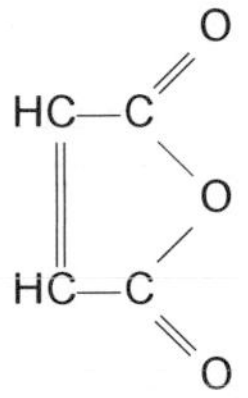
Maleic anhydride (II)

Uses. Large quantities of anhydrides are used in the preparation of esters. Ethyl acetate and butyl acetate (from butyl alcohol and acetic anhydride) are excellent solvents for cellulose nitrate lacquers. Acetates of high-molecular-weight alcohols are used as plasticizers for plastics and resins. Cellulose and acetic anhydride give cellulose acetate, used in acetate rayon and photographic film. The reaction of anhydrides with sodium peroxide forms peroxides (acetyl peroxide is violently explosive), used as catalysts for polymerization reactions and for addition of alkyl halides to alkenes. In Friedel-Crafts reactions, anhydrides react with aromatic compounds, forming ketones such as acetophenone.

Maleic anhydride reacts with many dienes to give hydroaromatics of various complexities (Diels-Alder reaction). Much maleic anhydride is used commercially in the manufacture of alkyd resins from polyhydric alcohols. Soil conditioners are produced by basic hydrolysis of the copolymer of maleic anhydride with vinyl acetate.

Phthalic anhydride and alcohols form esters (phthalates) used as plasticizers for plastics and resins. Condensed with phenols and sulfuric acid, phthalic anhydride yields phthaleins, such as phenolphthalein; with *m*-dihydroxybenzenes under the same conditions, xanthene dyes form, for example, fluorescein. Much phthalic anhydride is used in manufacturing glyptal resins (from the anhydride and glycerol) and in manufacturing anthraquinone. Heating phthalic anhydride with ammonia gives phthalimide, used in Gabriel's synthesis of primary amines, amino acids, and anthranilic acid (*o*-aminobenzoic acid). With alkaline hydrogen peroxide, phthalic anhydride yields monoperoxyphthalic acid, used along with benzoyl peroxide as polymerization catalysts, and as bleaching agents for oils, fats, and other edibles.

Anhydrides react with water to form the parent acid, with alcohols to give esters, and with ammonia to yield amides; and with primary or secondary animes, they furnish *N*-substituted and *N,N*-disubstituted amides, respectively. *See* ACID HALIDE; ACYLATION; CARBOXYLIC ACID; DIELS-ALDER REACTION; ESTER; FRIEDEL-CRAFTS REACTION.

[PAUL E. FANTA]

Bibliography: N. L. Allinger et al., *Organic Chemistry*, 2d ed., 1976.

Acid-base indicator

A substance that reveals, through characteristic color changes, the degree of acidity or basicity of solutions. Indicators are weak organic acids or bases which exist in more than one structural form (tautomers) of which at least one form is colored. Intense color is desirable so that very little indicator is needed; the indicator itself will thus not affect the acidity of the solution.

The equilibrium reaction of an indicator may be regarded typically by giving it the formula HIn. It dissociates into H^+ and In^- ions and is in equilibrium with a tautomer InH which is either a nonelectrolyte or at most ionizes very slightly. In the overall equilibrium shown as Eq. (1) the simplify-

$$\underset{(I)}{InH} \rightleftharpoons \underset{(II)}{HIn \rightleftharpoons H^+ + In^-} \quad (1)$$

ing assumption that the indicator exists only in forms (I) and (II) leads to no difficulty. The addition of acid will completely convert the indicator to form (I), which is therefore called the acidic form of the indicator although it is functioning as a base. A hydroxide base converts the indicator to form (II) with the formation of water; this is called the alkaline form. For the equilibrium between (I) and (II) the equilibrium constant is given by Eq. (2). In

$$K_{In} = \frac{[H^+][In^-]}{[InH]} \quad (2)$$

a manner similar to the pH designation of acidity, that is, $pH = -\log[H^+]$, the K_{In} is converted to pK_{In} with the result shown in Eq. (3). It is seen that the

$$pK_{In} = pH - \log\frac{[In^-]}{[InH]} \quad (3)$$

pK of an indicator has a numerical value approximately equal to that of a specific pH level.

Use of indicators. Acid-base indicators are commonly employed to mark the end point of an acid-base titration or to measure the existing pH of a solution. For titration the indicator must be so chosen that its pK is approximately equal to the pH of the system at its equivalence point. For pH

Common acid-base indicators

Common name	pH range	Color change (acid to base)	pK	Chemical name	Structure	Solution
Methyl violet	0–2, 5–6	Yellow to blue violet to violet		Pentamethylbenzyl-pararosaniline hydrochloride	Base	0.25% in water
Metacresol purple	1.2–2.8, 7.3–9.0	Red to yellow to purple	1.5	*m*-Cresolsulfonphthalein	Acid	0.73% in *N*/50 NaOH, dilute to 0.04%
Thymol blue	1.2–2.8, 8.0–9.6	Red to yellow to blue	1.7	Thymolsulfonphthalein	Acid	0.93% in *N*/50 NaOH, dilute to 0.04%
Tropeoline 00 (Orange IV)	1.4–3.0	Red to yellow		Sodium *p*-diphenyl-aminoazobenzene-sulfonate	Base	0.1% in water
Bromphenol blue	3.0–4.6	Yellow to blue	4.1	Tetrabromophenol-sulfonphthalein	Acid	1.39% in *N*/50 NaOH, dilute to 0.04%
Methyl orange	2.8–4.0	Orange to yellow	3.4	Sodium *p*-dimethyl-aminoazobenzene-sulfonate	Base	0.1% in water
Bromcresol green	3.8–5.4	Yellow to blue	4.9	Tetrabromo-*m*-cresol-sulfonphthalein	Acid	0.1% in 20% alcohol
Methyl red	4.2–6.3	Red to yellow	5.0	Dimethylaminoazo-benzene-*o*-carboxylic acid	Base	0.57% in *N*/50 NaOH, dilute to 0.04%
Chlorphenol red	5.0–6.8	Yellow to red	6.2	Dichlorophenolsulfon-phthalein	Acid	0.85% in *N*/50 NaOH, dilute to 0.04%
Bromcresol purple	5.2–6.8	Yellow to purple	6.4	Dibromo-*o*-cresolsulfon-phthalein	Acid	1.08% in *N*/50 NaOH, dilute to 0.04%
Bromthymol blue	6.0–7.6	Yellow to blue	7.3	Dibromothymolsulfon-phthalein	Acid	1.25% in *N*/50 NaOH, dilute to 0.04%
Phenol red	6.8–8.4	Yellow to red	8.0	Phenolsulfonphthalein	Acid	0.71% in *N*/50 NaOH, dilute to 0.04%
Cresol red	2.0–3.0, 7.2–8.8	Orange to amber to red	8.3	*o*-Cresolsulfonphthalein	Acid	0.76% in *N*/50 NaOH, dilute to 0.04%
Orthocresol-phthalein	8.2–9.8	Colorless to red		—	Acid	0.04% in alcohol
Phenolphthalein	8.4–10.0	Colorless to pink	9.7	—	Acid	1% in 50% alcohol
Thymolphthalein	10.0–11.0	Colorless to red	9.9	—	Acid	0.1% in alcohol
Alizarin yellow GG	10.0–12.0	Yellow to lilac		Sodium *p*-nitrobenzene-azosalicylate	Acid	0.1% in warm water
Malachite green	11.4–13.0	Green to colorless		*p,p'*-Benzylidenebis-*N,N*-dimethylaniline	Base	0.1% in water

measurement, the indicator is added to the solution and also to several buffers. The pH of the solution is equal to the pH of that buffer which gives a color match. Care must be used to compare colors only within the indicator range. A color comparator may also be used, employing standard color filters instead of buffer solutions.

Indicator range. This is the pH interval of color change of the indicator. In this range there is competition between indicator and added base for the available protons; the color change, for example, yellow to red, is gradual rather than instantaneous. Observers may, therefore, differ in selecting the precise point of change. If one assumes arbitrarily that the indicator is in one color form when at least 90% of it is in that form, there will be uncertain color in the range of 90–10% InH (that is, 10–90% In^-). When these arbitrary limits are substituted into the pK equation, the interval between definite colors is shown by Eqs. (4). Thus the

$$\text{pH} = \text{pK} + \log\frac{10}{90} \quad \text{to} \quad \text{pH} = \text{pK} + \log\frac{90}{10} \tag{4}$$

pH uncertainty of the indicator is from pK + 1 to pK − 1 (approximately), and pK ± 1 is called the range of the indicator. The experimentally observed ranges may differ from this prediction somewhat because of variations in color intensity.

Examples. The table lists many of the common indicators, their chemical names, pK values, ranges of pH, and directions for making solutions. Many of the weak-acid indicators are first dissolved in *N*/50 NaOH to the concentration shown, then further diluted with water. The weak-base indicators show some temperature variation of pH range, following approximately the temperature change of the ionization constant of water. Weak-acid indicators are more stable toward temperature change. The colors are formed by the usual chromophoric groups, for example, quinoid and azo-. One of the most common indicators is phenolphthalein, obtained by condensing phthalic anhydride with phenol. The acid form is colorless. It is converted by OH^- ion to the red quinoid form, as shown by Eq. (5).

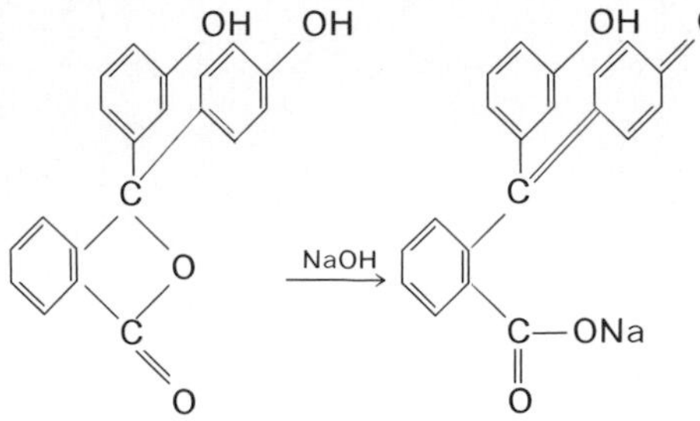

(5)

Other indicators such as methyl orange and methyl red are sodium salts of azobenzene derivatives containing sulfonic and carboxylic groups respectively. *See* ACID AND BASE; HYDROGEN ION; TITRATION. [ALLEN L. HANSON]

Acid halide

One of a large group of organic substances possessing the halocarbonyl group,

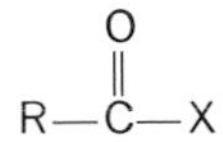

in which X stands for fluorine, chlorine, bromine, or iodine. The terms acyl and aroyl halides refer to aliphatic or aromatic derivatives, respectively.

The great inherent reactivity of acid halides precludes their free existence in nature; all are made by synthetic processes. In general, acid halides have low melting and boiling points and show little tendency toward molecular association. With the exception of the formyl halides (which do not exist), the lower members are pungent, corrosive, lacrimatory liquids that fume in moist air. The higher members are low-melting solids.

Preparation. Acid chlorides are prepared by replacement of carboxylic hydroxyl of organic acids by treatment with phosphorus trichloride, phosphorus pentachloride, or thionyl chloride [reactions (1)–(3)].

$$3RCOOH + PCl_3 \rightarrow 3RCOCl + H_3PO_3 \quad (1)$$
$$RCOOH + PCl_5 \rightarrow RCOCl + POCl_3 + HCl \quad (2)$$
$$RCOOH + SOCl_2 \rightarrow RCOCl + SO_2 + HCl \quad (3)$$

Although acid bromides may be prepared by the above methods (especially by use of PBr_3), acid iodides are best prepared from the acid chloride treatment with either CaI_2 or HI, and acid fluorides from the acid chloride by interaction with HF or antimony fluoride.

Reactions and uses. The reactivity of acid halides centers upon the halocarbonyl group, resulting in substitution of the halogen by appropriate structures. Thus, as shown by reactions (4), with

$$RCOCl + \begin{cases} HOH \rightarrow HCl + RCOOH \\ C_2H_5OH \rightarrow HCl + RCOOC_2H_5 \\ 2NH_3 \rightarrow NH_4Cl + RCONH_2 \\ 2HNR_2 \rightarrow NH_2R_2Cl + RCONR_2 \end{cases} \quad (4)$$

substances containing active hydrogen atoms (for example, water, primary and secondary alcohols, ammonia, and primary and secondary amines), hydrogen chloride is formed together with acids, esters, amides, and *N*-substituted amides, respectively.

The industrially prepared acetyl and benzoyl chlorides are much used in reactions of the above type, particularly in the acetylating or benzoylating of amines and amino acids, and alcohols, especially polyalcohols such as glycerol, the sugars, and cellulose, to form amides, esters, and polyesters, respectively. In reactions (4) the by-product hydrogen chloride must be neutralized. With aliphatic acid halides, this must be done by using an excess of the amine (as shown); aromatic acid chlorides, being insoluble in water, can be used in the presence of aqueous sodium hydrozide (Schotten-Baumann reaction). An alternative technique uses the tertiary amine, pyridine, both as solvent and as neutralizing agent.

Interaction of acid chlorides with sodium salts of organic acids furnishes a general method for the preparation of acid anhydrides, as shown by reaction (5). Analogously, aromatic acyl peroxides,

$$RCOCl + RCOONa \rightarrow NaCl + (RCO)_2O \quad (5)$$

used as bleaching agents for flour, fats, and oils, and as polymerization catalysts, may be prepared from sodium peroxide and the acid halide [reaction (6)]. *See* PEROXIDE.

$$2C_6H_5COCl + Na_2O_2 \rightarrow 2NaCl + \underset{\text{Benzoyl peroxide}}{C_6H_5COOOOCC_6H_5} \quad (6)$$

Acid halides are used to prepare ketones, either by reaction with alkyl or aryl cadmium reagents [reaction (7)] or by the aluminum chloride catalyzed Friedel-Crafts reaction [reaction (8)].

$$RCOCl + RCH_2CdCl \rightarrow RCOCH_2R + CdCl_2 \quad (7)$$
$$RCOCl + C_6H_6 \xrightarrow{AlCl_3} RCOC_6H_5 + HCl \quad (8)$$

Reduction of acid halides is easily effected. In the Rosenmund method, used mainly for aromatic acid halides, hydrogen and a poisoned palladium catalyst are employed [reaction (9)]. The product

$$RCOCl + H_2 \xrightarrow[\text{poisoned}]{Pd} HCl + RCHO \quad (9)$$

here is the aldehyde. For reduction to the alcohol stage, the vigorous reagent lithium aluminum hydride reduces both aliphatic and aromatic acid halides [reaction (10)].

$$4RCOCl + LiAlH_4 \rightarrow LiAlCl_4 + (RCH_2O)_4LiAl \xrightarrow{4HCl} 4RCH_2OH + LiAlCl_4 \quad (10)$$

Direct substitution of halogen (chlorine fastest, bromine more slowly) into acid aliphatic halides is relatively easy, compared with substitution in the parent acid, and takes place on the carbon next to the carbonyl group. The product is an α-halo acid halide, RCHXCOX. These compounds interact with carboxylic acids, via an equilibrium, to form an α-halo acid and an acid halide [reaction (11)].

$$RCHXCOX + RCH_2COOH \rightleftharpoons RCHXCOOH + RCH_2COX \quad (11)$$

Thus, if a small amount of acyl halide is added to a large amount of carboxylic acid, the latter can be chlorinated or brominated to completion by treatment with either chlorine or bromine (Hell-Volhard-Zelinski reaction). *See* ACETYLATION;

ACYLATION; CARBOXYLIC ACID; FRIEDEL-CRAFTS REACTION. [PAUL E. FANTA]

Bibliography: N. L. Allinger et al., *Organic Chemistry*, 2d ed., 1976.

Acidolysis

A chemical reaction involving the decomposition of a molecule with the addition of the elements of the acid. It is sometimes called acyl exchange. The reaction is comparable to hydrolysis or alcoholysis, in which water or an alcohol, respectively, is used in place of the acid. Acidolysis is most comonly an exchange reaction that is catalyzed by concentrated sulfuric acid, zinc chloride, or boron trifluoride. In many cases such reactions are reversible and thus will not go far toward completion unless one of the materials produced is removed as fast as it is formed by distillation or by reaction with a third component. Acidolysis is most commonly applied to the reaction of an organic acid and acetic anhydride to yield an acid anhydride and acetic acid. The acidolysis of esters and acyl halides is a convenient method for the preparation of other esters and acyl halides. It is also a useful reaction to convert neutral esters of dibasic acids to acid esters. The usual method for the preparation of acyl halides from an acid and phosphorus trichloride or thionyl chloride can also be considered to be acidolysis if these reagents are considered the acyl halides of phosphorous acid and sulfurous acid, respectively. *See* ALCOHOLYSIS; HYDROLYSIS.

[ELBERT H. HADLEY]

Acridine

One of a group of organic heterocyclic compounds (also called 9-azaanthracenes) containing benzene rings fused to the 2,3 and 5,6 positions of pyridine. Acridine (I) is a typical member of the group. Several important dyes and medicinals are acridine derivatives. Acridine itself was first noted as a minor contaminant of coal tar anthracene. *See* DYE; HETEROCYCLIC COMPOUNDS.

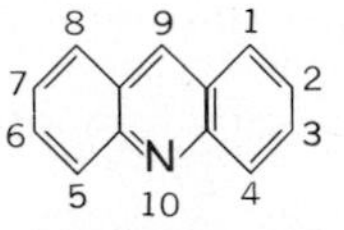

(I)

Acridine is a faintly yellow solid, mp 110–110.5°C, that shows marked fluorescence. The compound, as a tertiary amine, is a weak base (pK_a5.60 at 20°) forming simple and quaternary salts as well as *N*-oxides. Acridine shows aromatic character. It is stable to heat, alkali, and acid, and it undergoes substitution reactions, such as bromination, nitration, and sulfonation.

Quinacrine (IV), also called mepacrine, Atebrin, or 2-methoxy-6-chloro-9-(4′-diethylamino-1′-methylbutyl)-aminoacridine, is a valuable antimalarial drug. A commercial synthesis cyclizes the *N*-arylanthranilic acid (II) to the 9-chloroacridine (III), which with the appropriate amine gives the 9-amino compound, quinacrine. Although quinacrine has been widely used as a clinical suppressive, it is not the ideal antimalarial drug. It is not a preventative nor a true curative agent; with repeated use, it gradually dyes the skin yellow, and it occasionally causes secondary irritations. Other acridines have attracted attention as medicinals, for example, acriflavine (3,6-diamino-10-methylacridinium chloride) against sleeping sickness and as an antibacterial agent, and Rivanol (3,9-diamino-7-ethoxyacridine) against amebic dysentery.

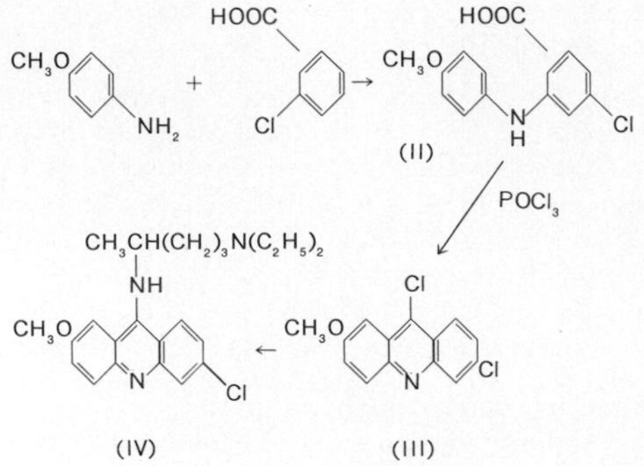

[WALTER J. GENSLER]

Bibliography: R. M. Acheson, *Acridines*, 1956; A. Albert, *The Acridines*, 2d ed., 1966; M. H. Palmer, *The Structure and Reactions of Heterocyclic Compounds*, 1967.

Acrylonitrile

An explosive, poisonous, flammable liquid, boiling at 77.3°C, partly soluble in water. The formula $CH_2=CH-C\equiv N$ indicates it may be regarded as vinyl cyanide, and its systematic name is 2-propenonitrile. Acrylonitrile is prepared by ammoxidation of propylene, according to the reaction below, over various sorts of catalysts, chiefly

$$2\,CH_2{=}CHCH_3 + 2NH_3 + 3O_2 \rightarrow 2\,CH_2{=}CHC{\equiv}N + 6H_2O$$

metallic oxides. Older processes, such as catalytic addition of hydrogen cyanide to acetylene, and the catalytic reaction of propylene with nitric oxide, are little used today. Over 1,600,000,000 lb (7.2×10^8 kg) of acrylonitrile are produced annually in the United States. Approximately the same quantity is produced in the remainder of the world.

Most of the acrylonitrile produced is consumed in the manufacture of acrylic and modacrylic fibers. Substantial quantities are used in acrylonitrile-butadiene-styrene (ABS) resins, in nitrile elastomers, and in the synthesis of adiponitrile by electrodimerization. The adiponitrile is subsequently hydrogenated to hexamethylenediamine, a constituent of nylon. Smaller amounts of acrylonitrile are used in cyanoethylation reactions, in the synthesis of drugs, dyestuffs, and pesticides, and as co-monomers with vinyl acetate, vinylpyridine, and similar monomers.

Acrylonitrile undergoes spontaneous polymerization, often with explosive force. It polymerizes violently in the presence of suitable alkaline substances. *See* CYANOETHYLATION; NITRILE; POLYMERIZATION.

[FRANK WAGNER]

Bibliography: C. Matasa and E. Tonca, *Basic Nitrogen Compounds: Chemistry, Technology and Applications*, 3d ed., 1973; M. Sittig, *Acrylonitrile*, 1965.

Actinide elements

The series of elements beginning with actinium (atomic number 89) and including thorium, protactinium, uranium, and the transuranium elements through the element lawrencium (atomic number 103). These elements, chemically similar, have a strong chemical resemblance to the lanthanide, or rare-earth, elements of atomic numbers 57 to 71. Their atomic numbers, names, and chemical symbols are: 89, actinium (Ac), the prototype element, sometimes not included as an actual member of the actinide series; 90, thorium (Th); 91, protactinium (Pa); 92, uranium (U); 93, neptunium (Np); 94, plutonium (Pu); 95, americium (Am); 96, curium (Cm); 97, berkelium (Bk); 98, californium (Cf); 99, einsteinium (Es); 100, fermium (Fm); 101, mendelevium (Md); 102, nobelium (No); 103, lawrencium (Lr).

Studies of the chemical and physical properties of actinide and lanthanide elements and their compounds have indicated that their electronic structure must be similar; an inner electron shell of fourteen $5f$ electrons in the case of the actinides, and fourteen $4f$ electrons in the case of the lanthanides, is filled in progressing across the series. Except for thorium and uranium, the actinide elements are not present in nature in appreciable quantities. The transuranium elements were discovered and investigated as a result of their synthesis in nuclear reactions. All are radioactive, and except for thorium and uranium, weighable amounts must be handled with special precautions.

The uranium isotopes ^{233}U and ^{235}U and the plutonium isotope ^{239}Pu undergo nuclear fission with slow neutrons with the liberation of large amounts of energy. Thorium can be converted to ^{233}U, and the isotope ^{238}U to ^{239}Pu by neutron irradiation; hence thorium and natural uranium can be used indirectly as nuclear fuels in breeder reactors.

Ion-exchange chromatography has been an important experimental technique in the study of the chemistry of the actinide elements. This method together with the analogy between corresponding actinides and lanthanides was the key to the discovery of the transcurium elements.

The actinide elements are very similar chemically. Most have the following in common: trivalent cations which form complex ions and organic chelates; soluble sulfates, nitrates, halides, perchlorates, and sulfides; and acid-insoluble fluorides and oxalates.

The characteristic oxidation state in aqueous solution is the III state, with higher oxidation states prominent in the early and the lower II state appearing in the latter part of the series (see table).

Oxidation states of actinide elements in aqueous solution

Oxidation state	Elements*
II	Cf, Es, Fm, Md, No
III	Ac, Pa, U, Np, Pu, Am, Cm, Bk, Cf, Es, Fm, Md, No, Lr
IV	Th, Pa, U, Np, Pu, Bk
V	Pa
V (as MO_2^{+} ions)	U, Np, Pu, Am
VI (as MO_2^{2+} ions)	U, Np, Pu, Am
VII (as MO_5^{3-} ions)	Np, Pu

*The most stable states are underscored.

Many solid compounds including hydrides, oxides, and halides have been prepared by dry chemical methods. Binary compounds with carbon, nitrogen, silicon, and sulfur are of interest because of their stability at high temperatures. *See* ACTINIUM; LAWRENCIUM; PERIODIC TABLE; PROTACTINIUM; THORIUM; TRANSURANIUM ELEMENTS; URANIUM.

[GLENN T. SEABORG]

Actinium

A chemical element, Ac, atomic number 89, and atomic weight 227.0. Actinium was discovered by A. Debierne in 1899 as a fraction in uranium residues. However, it was difficult to isolate in an appreciable quantity. Milligram quantities of the

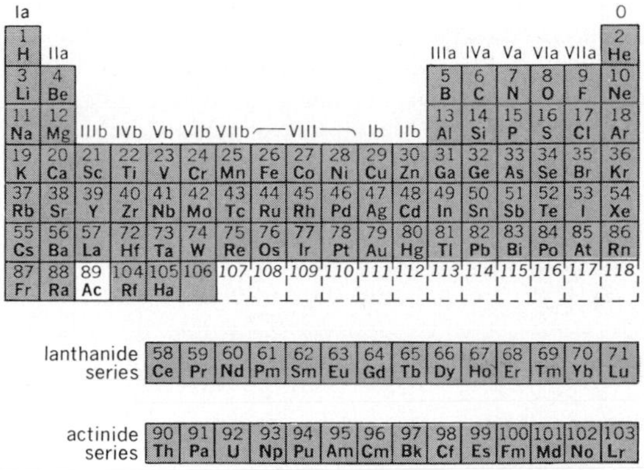

element are now available by irradiation of radium in a nuclear reactor [reaction (1)]. Actinium-

$$Ra^{226}(n,\gamma)Ra^{227};\ Ra^{227} \xrightarrow[\text{short}]{\beta^-} Ac^{227} \qquad (1)$$

227 is a β-emitting element whose half-life is 22 years. Six other radioisotopes of actinium with half-lives ranging from 10 days to less than 1 minute have been identified. X-ray diffraction and microchemical methods have permitted the preparation and identification of several compounds of actinium.

The relationship of actinium to the element lanthanum, the prototype rare earth, is striking. In every case, the actinium compound can be prepared by the method used to form the corresponding lanthanum compound with which it is isomorphous in the solid, anhydrous state. Examples of reactions by means of which some actinium compounds have been formed are given in reactions (2)–(5). In many cases, the compounds were pre-

$$Ac_2O_3 + 6HF \rightarrow 2AcF_3 + 3H_2O \qquad (2)$$
$$AcF_3 + H_2O \rightarrow AcOF + 2HF \qquad (3)$$
$$Ac_2O_3 + 3CCl_4 \rightarrow 2AcCl_3 + 3COCl_2 \qquad (4)$$
$$Ac_2O_3 + 2AlBr_3 \rightarrow 2AcBr_3 + Al_2O_3 \qquad (5)$$

pared in fine glass or quartz capillaries, were separated from the reaction products by sublimation, and were subjected to x-ray diffraction analysis in the capillary. This method has proved extremely advantageous; the disintegration products of actinium are highly radioactive, and containment within a sealed capillary minimizes the hazard. Furthermore, the process of sublimation disposes a thin film of material on the walls of the capillary in such a way that it produces a maximum effect in diffraction of x-rays. As little as 10 μg of material has been identified in this manner.

The actinium 3^+ ion is the largest and most basic of the 3^+ ions, and resists hydrolysis to an unusual extent. *See* ACTINIDE ELEMENTS; RADIOACTIVITY.

[SHERMAN FRIED]

Bibliography: S. Ahrland et al., *Chemistry of the Actinides*, 1975; E. K. Hyde et al., *The Nuclear Properties of the Heavy Elements*, 3 vols., rev. ed. reprint, 1971; J. J. Katz and G. T. Seaborg, *The Chemistry of the Actinide Elements*, 1957.

Activation analysis

A technique in which a neutron, charged particle, or gamma photon is captured by a stable nuclide to produce a different, radioactive nuclide which is then measured. The technique is specific, highly sensitive, and applicable to almost every element in the periodic table. Because of these advantages, activation analysis has been applied to chemical identification problems in many fields of interest.

Neutron method. Neutron activation analysis (NAA) is the most widely used form of activation analysis. In neutron activation analysis the sample to be analyzed is placed in a nuclear reactor where it is exposed to a flux of thermal neutrons. Some of these neutrons are captured by isotopes of elements in the sample; this results in the formation of a nuclide with the same atomic number, but with one more mass unit of weight. A prompt gamma ray is immediately emitted by the new nuclide, hence the term (n,γ) reaction, which can be expressed as reaction (1), where z refers to the

$$^{z}_{m}A(n,\gamma)\rightarrow {}^{z}_{m+1}A + \gamma \qquad (1)$$

atomic number and m the atomic weight. Usually the product nuclide $({}^{z}_{m+1}A)$ is radioactive, and by measuring its decay products one can identify and quantify the amount of target element in the sample. The basic activation equation is given by Eq. (2), and enables one to calculate the number of

$$A = Nf\sigma\left(1 - e\frac{-0.693t}{t_{1/2}}\right) \qquad (2)$$

where A = activity of product nuclide (disintegrations per second)
N = atoms of target element
f = flux of neutrons (neutrons per cm^2-s)
σ = cross section of the target nuclide (cm^2)
$t_{1/2}$ = half-life of induced radioactive nuclide
t = time of irradiation

atoms of an unknown target element by measuring the radioactivity of the product. These radioactive products usually decay by emission of a beta particle (negative electron) followed by a gamma ray (uncharged). Cross sections and half-lives are well known, and neutron fluxes can be measured by irradiation and measurement of known materials. When such techniques are used, the method is known as absolute activation analysis. A simple (and older) method is to simultaneously irradiate and compare with the unknown sample a standard containing known amounts of the elements in question. This is called the comparator technique.

Measurement techniques. Measurement of the induced radioactivities is the key to activation analysis. This is usually obtained from the gamma-ray spectra of the induced radionuclides. Gamma rays from radioactive isotopes have unique, discrete energies, and a device that converts such rays into electronic signals that can be amplified and displayed as a function of energy is a gamma-ray spectrometer. It consists of a detector [germanium doped with lithium, GeLi, or sodium iodide doped with thallium, NaI(Tl)] and associated electronics. Gamma rays interact in the detector to form photoelectrons, and these are then amplified and sorted according to energy. Peaks in the resulting gamma-ray spectra are called gamma-ray photo-peaks. By taking advantage of the different half-lives and different gamma-ray energies of the induced radionuclides, positive identification of many elements can be made. A computer, or provision for recording the spectrometric data in computer-compatible form, is almost a necessity. Calibration of counting conditions with a particular detector enables the activation analyst to relate the area under each gamma-ray photo-peak to an absolute disintegration rate; this supplies the A in Eq. (2). Such techniques are multielement, instrumental, and absolute. Where the sought element captures a neutron to produce a non-gamma-emitting nuclide, the analyst must make chemical separations and then beta-count the sample, or must go to another technique.

Charged-particle method. Activation analysis can also be performed with charged particles (protons or He^{3+} ions, for example), but because fluxes of such particles are usually lower than reactor neutron fluxes and cross sections are much smaller, charged-particle methods are usually reserved for special samples. Charged particles penetrate only a short distance into samples, which is another disadvantage. A variant called proton-induced x-ray emission (PIXE) has been highly successful in analyzing air particulates on filters. Here the samples are all similar, and are low in total mass, and many of the elements of interest such as S, Ca, Fe, Zn, and Pb are not easy to determine by neutron activation analysis. The protons excite prompt x-rays characteristic of the element, and these are measured. Prompt gamma rays have been used for measurement in some neutron activation analysis studies, but that method has had only limited success. Photon activation, using photons produced by electron bombardment of high-z targets such as tungsten, is another special variant of rather minor interest. Neutron sources other than reactors are sometimes used: ^{252}Cf is an element that emits neutrons as it decays, and can be used for neutron activation analysis; and small accelerators called 14-MeV neutron generators produce a low flux of high-energy neutrons which are used primarily for determination of oxygen and nitrogen.

Applications. Activation analysis has been applied to a variety of samples. It is particularly useful for small (1 mg or less) samples, and one irradiation can provide information on 30 or more elements. Samples such as small amounts of pollutants, fly ash, very pure experimental alloys, and biological tissue have been successfully studied by neutron activation analysis. Of particular interest has been its use in forensic studies; paint, glass, tape, and other specimens of physical evidence have been assayed for legal purposes. In addition, the method has been used for authentication of art objects and paintings where only a small sample is available. Activation analysis services are available in numerous university, private, commercial, and United States government laboratories, and while it is not an inexpensive method, for many special samples it is the best and cheapest method of acquiring necessary data. *See* TRACE ANALYSIS.

[W. S. LYON]

Bibliography: J. R. DeVoe (ed.), *1968 International Conference on Modern Trends in Activation Analysis*, vols. 1 and 2, NBS Spec. Pub. 312, 1969; P. Kruger, *Principles of Activation Analysis*, 1971; W. S. Lyon (ed.), *Guide to Activation Analysis*, 1964; W. S. Lyon, T. Braun, and E. Bujdoso (eds.), *Nuclear and Atomic Activation Analysis*, Akademiai Kiado, Budapest, 1976; W. S. Lyon and H. H. Ross, Nucleonics, *Anal. Chem. (Annu. Rev.)*, 50(5): 80, 1978; 1972 International Conference on Modern Trends in Activation Analysis, *J. Radioanal. Chem.*, vol. 16, no. 1 and 2, 1973; 1976 International Conference on Modern Trends in Activation Analysis, *J. Radioanal. Chem.*, vols. 37, 38, and 39, 1977.

Acylation

The process whereby the acyl group is incorporated into a molecule by substitution. If the acetyl group is incorporated, the process is acetylation; if the benzoyl group is substituted, it is benzoylation.

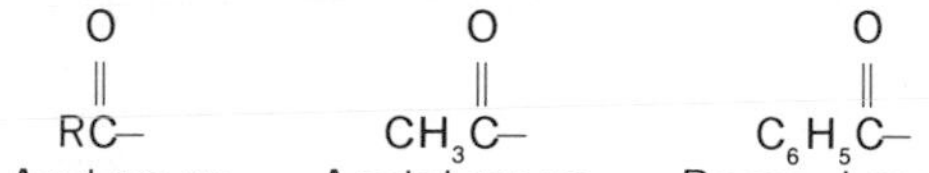

Acyl group   Acetyl group   Benzoyl group

For the process to occur, the acylating agent must attack a molecule containing one or more active or easily replaceable hydrogen atoms. This limits the reaction, for all practical purposes, to simple or polyhydroxy alcohols, phenols, thiols, ammonia, primary and secondary amines, amino acids, esters of malonic acid. β-keto esters, and β-diketones. The common reagents are acid halides, preferably the chloride and occasionally the bromide, and acid anhydrides. For acetylation the highly reactive substance ketene may be used. A related reaction is the Friedel-Crafts acylation, in which the acyl group replaces the hydrogen atom of an aromatic nucleus. *See* FRIEDEL-CRAFTS REACTION.

Alcohols. Primary and secondary alcohols furnish esters on acylation with an acid chloride, using the reagent alone or in pyridine solution. Reactions (1) and (2) apply.

$$\underset{\text{Ethanol}}{C_2H_5OH} + \underset{\text{Acetyl chloride}}{CH_3COCl} \rightarrow HCl + \underset{\text{Ethyl acetate}}{CH_3COOC_2H_5} \qquad (1)$$

$$\underset{\text{Isopropyl alcohol}}{(CH_3)_2CHOH} + \underset{\text{Acetyl chloride}}{CH_3COCl} \rightarrow HCl + \underset{\text{Isopropyl acetate}}{CH_3COOCH(CH_3)_2} \qquad (2)$$

Tertiary alcohols react abnormally, their hydroxyl group exchanging with the halogen of the acid halide. For this reason tertiary alcohols are often acetylated with ketene, as shown by reaction (3).

$$\underset{\textit{tert}\text{-Butyl alcohol}}{(CH_3)_3COH} + CH_2{=}C{=}O \rightarrow \underset{\textit{tert}\text{-Butyl acetate}}{CH_3COOC(CH_3)_3} \qquad (3)$$

Acid anhydrides, in the presence of catalytic amounts of sulfuric acid, boron trifluoride, or sodium acetate, also react with primary and secondary alcohols to form esters, shown by reaction (4).

$$C_2H_5OH + (CH_3CO)_2O \rightarrow CH_3COOH + CH_3COOC_2H_5 \qquad (4)$$

Because acetic acid is formed in the reaction, excess alcohol is used to increase the yield of ester. Under the above conditions tertiary alcohols do not usually react; they may thus be distinguished by this reaction from primary or secondary isomers. *See* ESTER; KETENE.

Other groups. Acetylation of phenols and thiols can be carried out as described above by using the acid anhydride. In the laboratory preparation of aromatic esters of phenols and thiols, however, aromatic acid halides are used. The usual procedure utilizes a mixture of the phenol, the acid halide, and aqueous sodium hydroxide (Schotten-Baumann reaction). This procedure is much used in organic qualitative analysis as a means of identifying acids, alcohols, phenols, and amines.

Acylation and aroylation (with an aromatic radical) are valuable in work with both primary and secondary amines. Reaction (5) is an example of

$$(CH_3)_2NH + C_6H_5COCl + Na^+OH^- \rightarrow \underset{N,N\text{-Dimethyl benzamide}}{C_6H_5CON(CH_3)_2} + Na^+Cl^- + H_2O \qquad (5)$$

aroylation of a secondary amine. The reaction is frequently used to protect the primary amino group of an amino acid while reactions involving the carboxyl group are carried out.

Mechanism of acylation. Acylation reactions always involve a compound containing an active hydrogen atom attached to an electron-rich atom (oxygen, nitrogen, or sulfur); such a compound is a base in the Lewis sense; that is, it is an electron donor. The electron acceptor (Lewis acid) is the acylating agent, the acid anhydride or acid halide, or ketene. The initial step is thus an acid-base reaction, proceeding via nucleophilic attack by the base (II) on the carbonyl dipole of the acid anhydride or acid halide (I), as in reaction (6). In (6) note

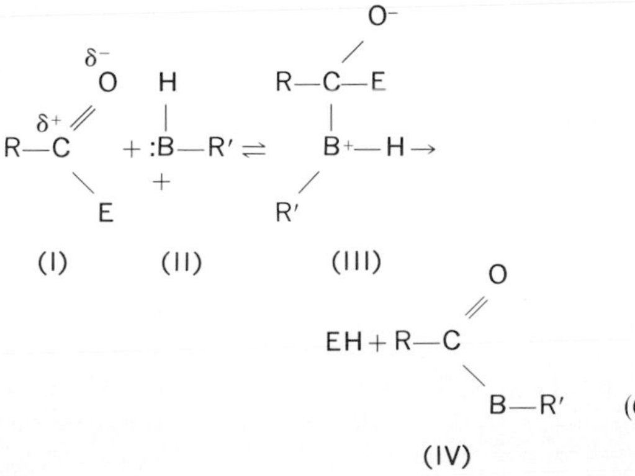

(6)

that R can be aliphatic or aromatic; R′ can be H or aliphatic or aromatic; E can be halogen (Cl, Br, or I), or it can be

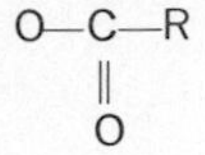

as in anhydride; and B can be O, N, or S. The reaction proceeds to the right, destroying the equilibrium, to give compound (IV) (which can be an acid, ester, thioester, or amide). This occurs when compound EH is thermodynamically stable and is permitted to escape, or is otherwise removed from the reaction as with hydroxide ion (Schotten-Baumann reaction), or when excess amine is its own neutralizing agent.

Applications of acetylation. The most important acylation reaction is acetylation, which is much used industrially. Acetylation of acetylene is accomplished catalytically by the addition of acetic acid to form vinyl acetate, which is an important monomer in the plastics industry [reaction (7)].

$$CH_3COOH + HC{\equiv}CH \rightarrow \underset{\text{Vinyl acetate}}{CH_3\overset{\overset{\displaystyle O}{\|}}{C}OCH{=}CH_2} \qquad (7)$$

The reaction of cellulose with acetic anhydride in the presence of sulfuric acid gives cellulose triacetate, used in the manufacture of the fiber known as acetate rayon.

The important analgesic aspirin, or acetylsalicylic acid, is prepared by the reaction of salicylic acid with acetic anhydride [reaction (8)].

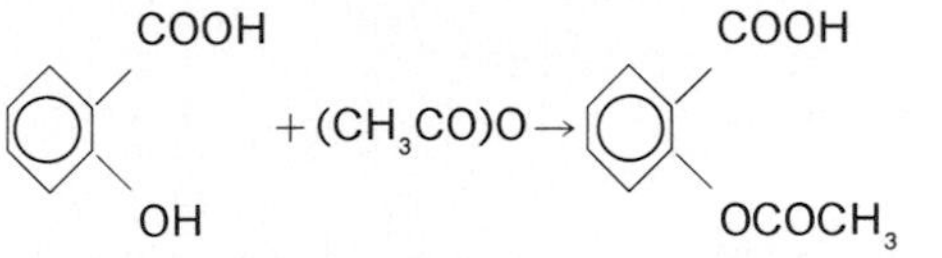

(8)

See ACID ANHYDRIDE; ACID HALIDE; AMINE.

[PAUL E. FANTA]

Bibliography: N. L. Allinger et al., *Organic Chemistry*, 2d ed., 1976.

Adamantane

The $C_{10}H_{16}$ alicyclic hydrocarbon whose structure (Fig. 1*a*) has the same arrangement of carbon atoms as does the basic unit of the diamond lattice (Fig. 1*b*). Originally conceived as a hypothetical molecule because of this analogy with the diamond lattice, adamantane was first obtained in 1933 in minute quantities by laborious separation from Czechoslovakian petroleum. The spherical, rigid, cagelike structure intrigued chemists for many years before synthetic routes to adamantane were developed, which enabled detailed investigations of the chemistry of the compound to be made. Although the first successful syntheses were developed as early as 1941, adamantane remained largely a laboratory curiosity until 1957, since these routes were relatively laborious and sufficient quantities of adamantane could not easily be obtained. At that time, however, the discovery was made that the readily available tetrahydrodicyclopentadiene (Fig. 1*c*) rearranges to adamantane in the presence of the Lewis acids $AlCl_3$ or $AlBr_3$. The impetus of this discovery led to investigations of the reactions of adamantane, which in turn have made important contributions to present knowledge concerning the relationships of molecular structure and chemical reactivity. In addition, many derivatives of adamantane exhibit useful biological properties.

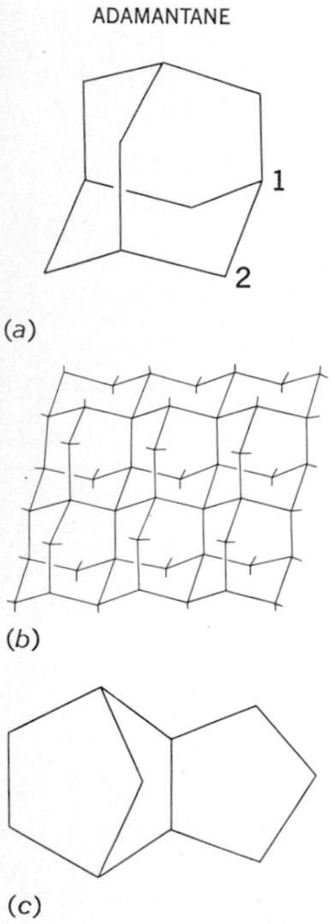

Fig. 1. Structures. (*a*) Adamantane, showing bridgehead carbon labeled 1 and alternate position for substitution at carbon 2. (*b*) Diamond lattice. (*c*) Tetrahydrodicyclopentadiene.

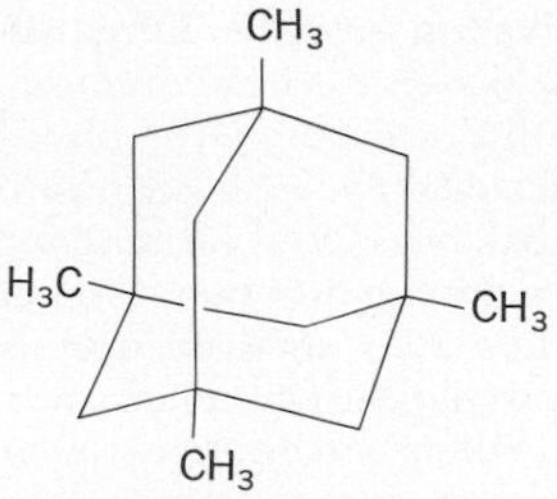

Fig. 2. Structure of 1,3,5,7-tetramethyladamantane.

Preparation. The aluminum halide–catalyzed rearrangement of polycyclic hydrocarbons has been found to be an extremely general and useful approach to the preparation of a number of adamantane-related systems. For example, 1,3,5,7-tetramethyladamantane (Fig. 2) may be readily prepared by the rearrangement of perhydroanthracene (Fig. 3). The ubiquity of the rearrangement route to adamantanes is best exemplified, perhaps, by the discovery that many simple compounds, even mineral oil, rearrange to a mixture of alkyl-substituted adamantanes under more drastic conditions.

Physical properties. Adamantane is composed of three interlocking, chair-form cyclohexane rings. The molecular structure is highly symmetrical and compact. All bond angles are found by x-ray crystallography to be nearly 109.5°, and the bonds around all adjacent carbons are ideally staggered. These structural features afford a high degree of stability to the adamantane molecule. In fact, adamantane is the most stable tricyclic $C_{10}H_{16}$ hydrocarbon possible. This high thermodynamic stability accounts for the success and versatility of the rearrangement route to adamantane systems. The highly symmetrical structure of adamantane also accounts for its extremely high melting point (269°C), the highest known for aliphatic hydrocarbons, and for the seemingly contradictory ease with which it sublimes. *See* CONFORMATIONAL ANALYSIS.

Chemical properties. The unique geometry of the adamantane molecule also gives rise to some striking chemical properties. Although the generation of bridgehead cations in many polycyclic hydrocarbons is extremely difficult, the 1-adamantyl cation has been found to form with relative ease. Hence, cationic processes readily permit the introduction of functional groups at the bridgehead (1-) position of adamantane (Fig. 1*a*). For example,

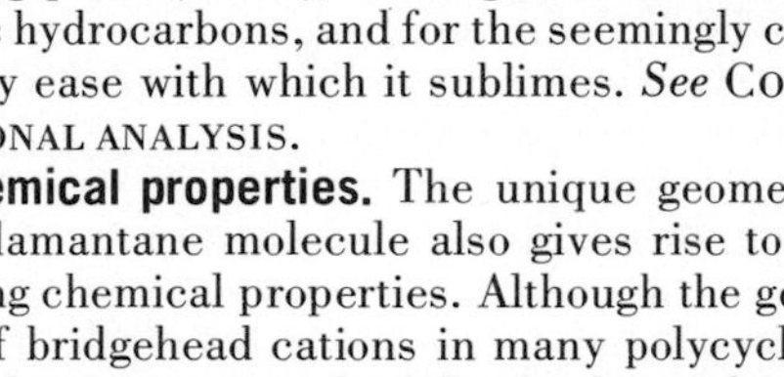

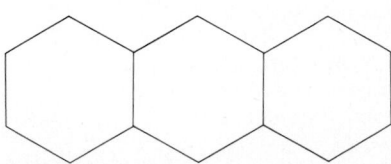

Fig. 3. Perhydroanthracene structure.

bromination in the absence of the usual free radical catalysts gives excellent yields of 1-bromoadamantane.

Medicinal uses. Possibly the most interesting and significant uses of adamantane compounds are found in the field of pharmacology. Several adamantane derivatives have been found to exhibit antiviral activity, whereas the incorporation of an adamantane moiety into many already useful drugs has been found to significantly enhance their efficacy. *See* ALICYCLIC HYDROCARBON; ALIPHATIC HYDROCARBON; ELECTROPHILIC AND NUCLEOPHILIC REAGENTS.

[PAUL E. FANTA]

Bibliography: R. C. Fort, Jr., and P. von R. Schleyer, *Chem. Rev.*, 64:277, 1964; R. C. Fort, Jr., and P. von R. Schleyer, in H. Hart and G. J. Karabatsos (eds.), *Advances in Alicyclic Chemistry*, 1966; E. C. Herrmann, Jr., in C. K. Cain (ed.), *Annual Reports in Medicinal Chemistry 1966*, 1967.

Adsorption

The property of an interface between two immiscible phases (solid, liquid, or vapor) to attract and concentrate components of either phase or both phases as an adsorbed interfacial film. Vapors may adsorb onto solids or liquids, and solutes may adsorb at liquid-vapor, liquid-liquid, liquid-solid, and solid-solid interfaces. Adsorption is a basic thermodynamic property of interfaces, resulting from a discontinuity in intermolecular or interatomic forces. It is also important in nearly all industrial processes and products.

Some definitions that describe adsorption are as follows: The adsorbent is the solid or liquid which adsorbs. The adsorbate is the solid, liquid, or gas which is adsorbed as molecules, atoms, or ions. Physical adsorption or physisorption is reversible adsorption by weak interactions only; no covalent bonds occur between the adsorbent and adsorbate; heats of physical adsorption are usually less than 15–20 kcal/mole (63–84 kJ/mole). Chemical adsorption or chemisorption is adsorption involving stronger interaction between adsorbate and adsorbent usually accompanied by rearrangement of atoms within or between adsorbates; reaction occurs between the surface of the adsorbent and the adsorbate; heats of chemisorption are usually in excess of 20–30 kcal/mole (84–126 kJ/mole).

Applications. Heterogeneous catalysis, in which gas or liquid reactants are specifically adsorbed to a dissimilar phase and chemically altered during their brief retention time, is basic to many industrial processes in the petrochemical, polymer, and chemical industries.

Purification by adsorption is perhaps the oldest known application; examples are wine and beer clarification, color removal in sugar processing, industrial wastewater treatment, and toxic gas adsorption in gas masks.

Adsorption is the basic phenomenon of chromatographic separations, which separate and concentrate components of mixtures according to strength of adsorption onto adsorbents in chromatographic columns.

Adsorption of surface-active substances is the key process in the use of soaps, detergents, emulsifiers, wetting agents, dyes, lubricants, and surface treatments. Other industries which are dependent on adsorption processes include agriculture, mining, petroleum recovery, papermaking, printing, and photography. *See* CATALYSIS; CHEMICAL SEPARATION TECHNIQUE; CHROMATOGRAPHY.

Physical adsorption of vapor. Nearly all vapors tend to adsorb onto inorganic solids at temperatures not too much above their boiling point. The intermolecular attractive forces which cause vapors to adsorb (or condense) are generally dominated by the London dispersion forces, an attraction caused by the perturbation of electron orbits by adjacent atoms. These forces, generally proportional to the square of the polarizability per unit volume, are much stronger for most inorganic solids than for water or organic materials, and that is why inorganic solids are the stronger adsorbents.

Another attractive force important in vapor adsorption is the interaction of electron-donor (basic) sites of vapor molecules with electron-acceptor (acidic) sites of adsorbents, or vice versa. These short-range attractions are much stronger than dipole interactions. Silica, an acidic adsorbent, adsorbs basic vapors (water, ammonia, and so forth) much more strongly than acidic vapors (chloroform, CO_2, NO_2, and so forth) regardless of the dipole moments.

The adsorption of water is dominated by hydrogen bonding, an intermolecular acid-base interaction which permits multilayers to absorb onto acidic or basic adsorbents but prevents absorption

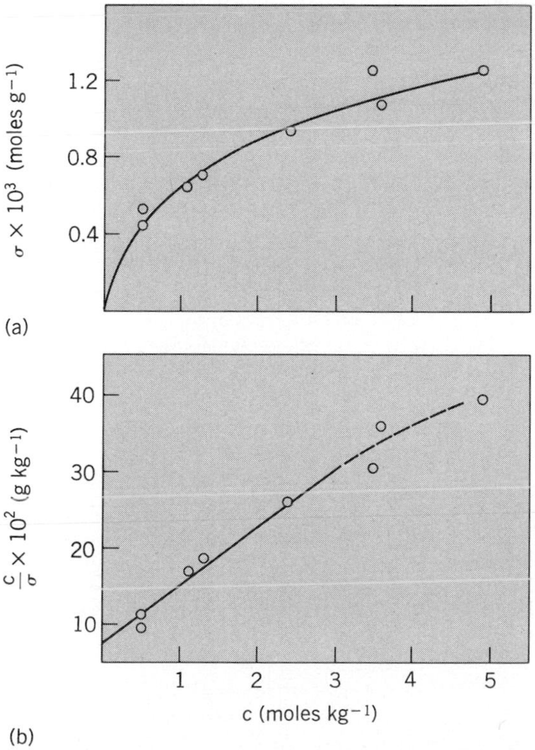

Fig. 1. Adsorption isotherm for benzene on carbon from heptane solution. *(a)* The data are plotted as σ versus c. *(b)* The data are plotted according to Eq. (1). *(Data from M. van der Waarden, J. Colloid Sci., 6:443, 1951)*

onto neutral surfaces such as graphite or polyethylene, except for the acidic or basic sites provided by impurities on these neutral surfaces.

The amount of vapor (or solute) adsorbed per unit mass of adsorbent, σ, is determined as a function of vapor pressure p (or solute concentration c) at a given temperature: this is the adsorption isotherm. If only a monolayer adsorbs, the results tend to follow the Langmuir adsorption isotherm [Eq. (1)], as shown in Fig. 1. In the equation

$$\frac{c}{\sigma}=\frac{1}{K\sigma_m}+\frac{c}{\sigma_m} \qquad (1)$$

σ_m is the amount of adsorbed vapor per unit mass of adsorbent necessary for a complete monolayer, and K is the equilibrium constant for adsorption. However, if multilayers adsorb, as is usually observed in the physical adsorption of small molecules, the amount adsorbed is best plotted by the Brunauer-Emmet-Teller (BET) equation [Eq. (2)],

$$\frac{p}{\sigma(p_0-p)}=\frac{1}{\sigma_m c}+\frac{(c-1)}{\sigma_m c}\left(\frac{p}{p_0}\right) \qquad (2)$$

as shown in Fig. 2. In the equation, c is a constant characteristic of the strength of adsorption, and p_0 is the saturation pressure, a parameter specifying the pressure for many layers of adsorbate to be present on the surface. Several other adsorption isotherm equations have been developed for physical adsorption, and each has a special usefulness.

The free energy of adsorption of uniform multilayers on an adsorbent is designated by π_e and is called the spreading pressure of an adsorbed film in equilibrium with the vapor pressure p_0 of its liquid. For adsorption of a vapor on a solid, Eq. (3) applies, where γ_S, γ_{SL}, γ_{SV}, and γ_L are sur-

$$\pi_e=\gamma_S-\gamma_{SV}=\gamma_S-(\gamma_{SL}+\gamma_L) \qquad (3)$$

face tensions of the solid, solid-liquid, solid-vapor, and liquid surfaces. The spreading pressure is related to the free energy of interaction per unit area at the absorbent-liquid interface, $-W_{SL}$, given by Eq. (4). If the adsorption is in uniform multilayers,

$$W_{SL}=\gamma_S+\gamma_L-\gamma_{SL}=2\gamma_L+\pi_e \qquad (4)$$

π_e can be determined from measurements of moles of adsorbed vapor per unit area (Γ) versus vapor pressure p, using the Gibbs-Bangham equation, Eq. (5), where R is the gas constant and T is the absolute temperature.

$$\pi_e=RT\int_0^{p_0}\Gamma\, d(\ln p) \qquad (5)$$

The free energy of interaction between two phases at an interface depends on the sum of the interaction energies contributed by the different forces operating at the interface, such as dispersion forces (d) [Eq. (6a)], and acid-base interactions (ab) [Eq. (6b)], and for the dispersion forces

$$W_{SL}=W_{SL}{}^{d}+W_{SL}{}^{ab}+\cdots \qquad (6a)$$

$$\gamma_L=\gamma_L{}^{d}+\gamma_L{}^{ab}+\cdots \qquad (6b)$$

the relationship has been established as given in Eq. (7). Thus for the absorption of saturated hydro-

$$W_{SL}{}^{d}=2\sqrt{\gamma_S{}^{d}\gamma_L{}^{d}} \qquad (7)$$

carbons in which $\gamma_L=\gamma_L{}^{d}$, the spreading pressure, π_e, is given by Eq. (8). Therefore, the London at-

$$\pi_e=2\sqrt{\gamma_S{}^{d}\gamma_L{}^{d}}-2\gamma_L \qquad (8)$$

tractive forces of an adsorbent can be fully characterized by its value of $\gamma_S{}^{d}$. Representative values are 98 mJ/m^2 for graphite, 78 mJ/m^2 for silica, and 33.1 mJ/m^2 for polyethylene.

Adsorbents may also be quantitatively characterized for their acidic or basic interactions; the number of acid or basic sites per unit area and their strength can be used to predict such interactions quantitatively.

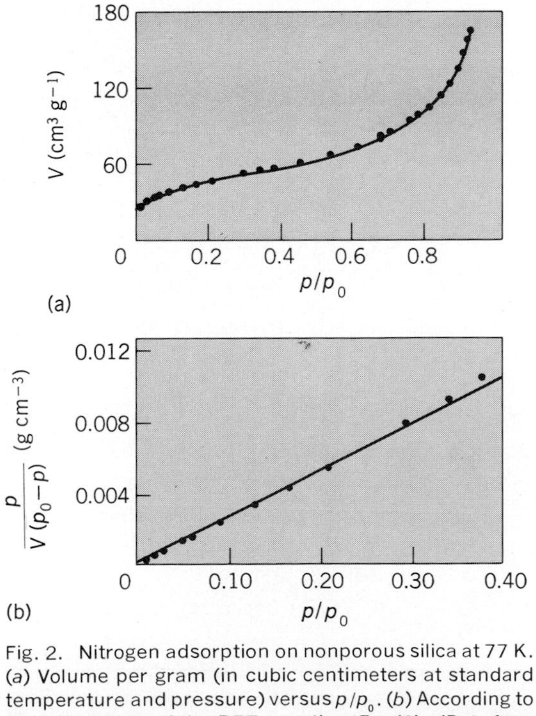

Fig. 2. Nitrogen adsorption on nonporous silica at 77 K. (*a*) Volume per gram (in cubic centimeters at standard temperature and pressure) versus p/p_0. (*b*) According to the linear form of the BET equation [Eq. (2)]. (*Data from D. H. Everett et al., J. Appl. Chem. Biotechnol., 24:199, 1974*).

Chemisorption of vapors. The strong interactions of chemisorption lead to surface compounds with various degrees of covalent bond character. The adsorbed layers are only one molecule thick because covalent bonds exist only between adjacent atoms. Chemisorption occurs on metals and semiconductors and on oxides and sulfides, but is most often observed on transition metals such as silver, nickel, cobalt, platinum, rhodium, and tungsten. Chemisorption is a necessary step in catalysis by these materials.

Because chemisorbed films are tightly bound, high-vacuum techniques such as field-ion microscopy or low-energy electron diffraction (LEED) can be used to determine the two-dimensional architecture of chemisorbed films on some surfaces, and the chemical bonding can be partially elucidated by ultraviolet and x-ray photoelectron spectroscopies (UPS and XPS). Details of the atomic composition of surfaces can be studied by Auger electron spectroscopy (AES), XPS, secondary ion mass spectrometry (SIMS), or ion scattering spectroscopy (ISS). *See* SECONDARY ION MASS SPECTROMETRY (SIMS); SPECTROSCOPY.

Adsorption of solutes from aqueous solutions. Soaps and detergents adsorb at the surface of their aqueous solutions and at all interfaces in contact with them. The sorption is driven mainly by rejection of the dissolved soaps and detergents from solution because of changes in water structure induced by these solutes. Adsorption of anionic surfactants such as sodium dodecyl sulfate or sodium dodecylbenzene sulfonate gives surfaces a negative charge, and adsorption of cationic surfactants such as *n*-hexadecyl trimethyl ammonium chloride gives surfaces a positive charge. Such adsorbed films cause similarly charged particles to repel one another, thereby stabilizing emulsions and dispersions and promoting detergency.

Inorganic solutes also adsorb from aqueous solutions, but the mechanism may not be the same as for surfactants. Anions such as borates, silicates, phosphates, and polyphosphates adsorb strongly onto fabrics, clays, and metals, conferring negative potentials, an important aspect of detergency.

Water-soluble polymers (such as polyethylene oxides, polyvinyl alcohol, hydrolyzed polyacrylamide, and the naturally occurring polysacharrides and other water-soluble gums) adsorb from aqueous solution onto inorganic surfaces, fabrics, or paper. These adsorbed films can be thick enough to prevent particle-particle adhesion and provide steric stabilization to dispersions. Higher-molecular-weight linear water-soluble polymers can adsorb simultaneously on several particles and thereby flocculate dispersions.

Adsorption from nonaqueous solutions. Oil-soluble surfactants have large hydrocarbon groups and at one end a basic or acidic group which can anchor the molecule to acidic or basic sites on the adsorbent. Strong and rapid adsorption is observed when a strong acid-base interaction occurs at the interface. However, very little or no adsorption occurs without some acid-base interaction.

Oil-soluble polymers (such as polyalkylmethacrylates, or block copolymers with polybutadiene or polyisobutylene blocks) adsorb onto inorganic surfaces, mainly by acid-base interaction. The olefin groups of rubber are weak bases, but their large number per polymer molecule causes strong adsorption onto acidic surfaces such as acidic carbon blacks and clays; this strong adsorption is the cause of the reinforcement of rubber by such inorganic particles. Such acid-base interactions in polymer adsorption also are the governing factor in the adhesion to metals of paints and adhesives, including those which are polymerized in places on surfaces.

Kinetics. In many cases of adsorption, the rates depend mainly on the rate of arrival of molecules at surfaces. However, not every molecule sticks upon collision, and sticking coefficients vary from unity to less than 10^{-9}, depending on temperature, coverage, and the particular interface.

In the physical adsorption of gases, sticking coefficients are near unity on bare surfaces, but much smaller sticking coefficients are often observed in chemisorption.

In adsorption from solution, sticking coefficients are unity at liquid surfaces; but at solid surfaces, the sticking coefficients decrease rapidly with increasing surface coverage, often to 10^{-7} or 10^{-8} for the last 20% of a monolayer. Thus, several hours are usually required for adsorption of a complete monolayer of surfactant on solid surfaces, depending on the nature of the system and especially on the temperature.

Kinetics of adsorption have received little attention so far, and they are not very well understood; however, they are very important in many industrial processes. *See* ABSORPTION; CHEMICAL DYNAMICS; INTERMOLECULAR FORCES.

[FREDERICK M. FOWKES]

Bibliography: A. W. Adamson, *The Physical Chemistry of Surfaces*, 3d ed., 1976; S. R. Morrison, *The Chemical Physics of Surfaces*, 1977; M. J. Rosen, *Surfactants and Interfacial Phenomena*, 1978; G. A. Somorjai, *Principles of Surface Chemistry*, 1972.

Alcoholysis

A chemical reaction involving the decomposition of a molecule with the addition of the elements of an alcohol. The reaction is comparable to hydrolysis or acidolysis, in which water or an acid, respectively, is used in place of the alcohol. This is most commonly an exchange reaction. The term is frequently applied to the reaction between an alcohol and an ester to produce a different alcohol and a different ester (transesterification). In such cases, higher-molecular-weight alcohols readily replace those of lower molecular weight; tertiary alcohols are more readily replaced than primary alcohols. In many cases, such reactions are reversible and thus will not go far toward completion unless a large excess of the reacting alcohol is present or one of the products is removed as fast as it is formed. Acids, bases, and sodium alkoxides act as positive catalysts. This type of reaction is useful for the preparation of simple esters of fatty acids from natural glycerides.

The term alcoholysis also applies to reactions involving alcohols and (1) nitriles to form ortho esters at lower temperatures and simple esters at higher temperatures; (2) acid anhydrides to yield esters (particularly useful with anhydrous dibasic acids to form monoesters); (3) alkyl halides to yield other alcohols; (4) acyl halides to yield esters; (5) β-keto esters to yield two esters or an ester and a ketone (a reverse Claisen condensation); (6) lactones to produce hydroxy or alkoxy acids; (7) amides to yield esters; (8) carbohydrates to yield glycosides; and (9) ethylene oxide to form monoalkyl ethers (cellosolves). *See* ACIDOLYSIS; HYDROLYSIS; TRANSESTERIFICATION.

[ELBERT H. HADLEY]

Aldehyde

One of a class of organic chemical compounds represented by the general formula RCHO. Formaldehyde, the simplest aldehyde, has the formula HCHO, where R is hydrogen. For all other aldehydes, R is a hydrocarbon radical which may be substituted with other groups such as halogen or hydroxyl (see table). Aldehydes are closely related to ketones, since both types of compounds contain the carbonyl (>C=O) group; in the case of ketones, the carbonyl group is attached to two hydrocarbon radicals. Because of their high chemical reactivity, aldehydes are important intermediates for the manufacture of resins, plasticizers, solvents, dyes, and pharmaceuticals.

Aldehydes and their formulas

Compound	Formula
Formaldehyde	$HCHO$
Acetaldehyde	CH_3CHO
Acetaldol	$CH_3CHOHCH_2CHO$
Propionaldehyde	C_2H_5CHO
n-Butyraldhyde	$CH_3(CH_2)_2CHO$
Isobutyraldehyde	$(CH_3)_2CHCHO$
Acrolein	$CH_2{=}CHCHO$
Crotonaldehyde	$CH_3CH{=}CHCHO$
Chloral	CCl_3CHO
Chloral hydrate	$CCl_3CH(OH)_2$
Benzaldehyde	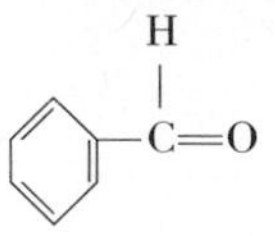
Cinnamaldehyde	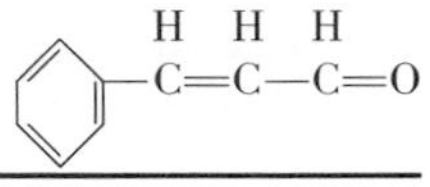

Physical properties. At room temperature formaldehyde is a colorless gas, which is handled commercially in the form of a 37% aqueous solution or as its polymer, paraformaldehyde $(CH_2O)_x$. The other low-molecular-weight aldehydes are colorless liquids having characteristic, somewhat acrid odors. The unsaturated aldehydes acrolein and crotonaldehyde are powerful lacrimators.

Chemical properties. The important reactions of aldehydes include oxidation, reduction, aldol condensation, Cannizzaro reaction, and reactions with compounds containing nitrogen.

Oxidation. Aldehydes are oxidized to carboxylic acids by a wide variety of oxidizing agents, such as air, oxygen, hydrogen peroxide, permanganate, peracids, silver oxide, and nitric acid, as in reaction (1).

$$RC(=O)H + (O) \rightarrow RC(=O){-}OH \qquad (1)$$

The catalytic oxidation of acetaldehyde is used for the production of the industrially important compounds acetic acid, acetic anhydride, and peracetic acid, as in reaction (2), where the product depends on the conditions of the reaction.

$$CH_3CHO \xrightarrow[\text{catalyst}]{O_2\,(\text{air})} \underset{\text{Acetic acid}}{CH_3COOH},\ \underset{\text{Acetic anhydride}}{(CH_3CO)_2O},\ \underset{\text{Peracetic acid}}{CH_3C(=O)OOH} \qquad (2)$$

Reduction. Aldehydes are reduced to primary alcohols cheaply by reaction with hydrogen in the presence of a platinum or nickel catalyst, or in the laboratory by treatment with aluminum isopropoxide or with complex metal hydrides such as lithium aluminum hydride ($LiAlH_4$) or sodium borohydride ($NaBH_4$), as in reaction (3).

$$RCHO + (H) \longrightarrow RCH_2OH \qquad (3)$$

Aldehydes react with hydrogen in the presence of ammonia and a catalyst to form amines. This process [reaction (4)], known as reductive amination, is used commercially to produce many important amines.

$$\underset{\text{Aldehyde}}{RCH{=}O} + NH_3 + H_2 \xrightarrow{\text{Pt or Ni}} \underset{\text{Amine}}{RCH_2NH_2} + H_2O \qquad (4)$$

Aldol condensation. Two molecules of an aldehyde react in the presence of alkaline, acidic, or amine catalysts to form a hydroxyaldehyde. This reaction proceeds only when the carbon adjacent to the carbonyl group has at least one hydrogen atom [reaction (5)].

$$\underset{\text{Aldehyde}}{RCH_2CHO} + RCH_2CHO \xrightarrow{\text{catalyst}} \underset{\text{Aldol (a hydroxy aldehyde)}}{RCH_2CH(OH)CH(R)CHO} \qquad (5)$$

Aldols are readily dehydrated to form unsaturated aldehydes, and the latter can be hydrogenated to form saturated, primary alcohols. This reaction sequence is the basis of the commercial preparation of crotonaldehyde and 1-butanol (*n*-butyl alcohol) from acetaldehyde [reaction (6)].

$$2CH_3CHO \xrightarrow[\text{catalyst}]{\text{alkaline}} \underset{\text{Acetaldol}}{CH_3CHOHCH_2CHO} \xrightarrow{\text{heat}}$$

$$\underset{\text{Crotonaldehyde}}{CH_3CH{=}CH{-}CHO} \xrightarrow{H_2} \underset{\text{l-Butanol}}{CH_3CH_2CH_2OH} \qquad (6)$$

Cannizzaro reaction. On treatment with strong alkali, aldehydes which lack a hydrogen attached to the carbon adjacent to the carbonyl group undergo a mutual oxidation and reduction, yielding one molecule of an alcohol and one molecule of the carboxylic acid salt. The reaction of benzaldehyde with aqueous potassium hydroxide is typical [reaction (7)].

$$2C_6H_5CHO + KOH \xrightarrow{H_2O} C_6H_5CH_2OH + C_6H_5COOK \qquad (7)$$

Acetaldehyde and formaldehyde in the presence of calcium hydroxide react by an aldol condensation followed by a "crossed" Cannizzaro reaction to give the industrially useful polyhydroxy compound pentaerythritol [reaction (8)].

$$CH_3CHO + \underset{\text{Formaldehyde}}{4CH_2O} + Ca(OH)_2 + H_2O \rightarrow$$

$$\underset{\text{Pentaerythritol}}{HOCH_2{-}C(CH_2OH)_2{-}CH_2OH} + \underset{\text{Calcium formate}}{(HCOO)_2Ca} \qquad (8)$$

Reaction with ammonia and amines. The addition of ammonia to the carbonyl group of an aldehyde is reversible, and the resulting aminoalcohol usually cannot be isolated [reaction (9)].

$$RCHO + NH_3 \rightleftarrows RCH(OH)NH_2 \qquad (9)$$

However, aromatic aldehydes such as benzal-

Adiabatic process

A thermodynamic process in which the system undergoing the change exchanges no heat with its surroundings. Reversible adiabatic processes are also isentropic; that is, they take place with no change in entropy. *See* ENTROPY; ISENTROPIC PROCESS.

By the first law of thermodynamics the change of the internal energy U in any process is equal to the sum of the heat Q gained and the work W done on the system. For an adiabatic process the change in internal energy when the system goes from state 1 to state 2 is equal to the external work performed *on* the system (which brings about the change). In the equation below, if U_2 is less than U_1, then W is negative and $-W$ is the work done *by* the system.

$$U_2 - U_1 = W$$

The events inside an engine cylinder are nearly adiabatic because the wide fluctuations in temperature take place rapidly compared to the speed with which the cylinder surfaces can conduct heat. Similarly, fluid flow through a nozzle may be so rapid that negligible exchange of heat between fluid and nozzle takes place. The compressions and rarefactions of a sound wave are rapid enough to be considered adiabatic. *See* THERMODYNAMIC PROCESSES. [PHILIP E. BLOOMFIELD]

Admittance

The reciprocal of the impedance of an electric circuit. Admittance is expressed in the unit mho, coined from the inverse spelling of ohm, the unit of impedance. Admittance is used primarily in computations of parallel alternating-current circuits.

By using admittance Y, current I can be expressed as $I = EY$, where E is the voltage across the impedance Z. In terms of complex quantities where R is the total circuit resistance, X the total circuit reactance, G the conductance, and B the susceptance. However, these are not simple conductance $(1/R)$ and susceptance $(1/X)$. As seen from the equation, both G and B are combinations

$$Y = \frac{1}{Z} = \frac{1}{R \pm jX} = \frac{R}{R^2 + X^2} \mp j\frac{X}{R^2 + X^2} = G \pm jB$$

of resistance and reactance. *See* CONDUCTANCE; SUSCEPTANCE. [BURTIS L. ROBERTSON]

Aerodynamics

The branch of aeromechanics dealing with the properties and characteristics of, and the forces exerted by, air and other gases in motion. The field of aerodynamics includes the science of a gas itself in motion and the science of bodies immersed in a gas between which there exists a relative motion. *See* GAS DYNAMICS.

Aerodynamics is a broad field with numerous specializations and applications, some of which extend into apparently unrelated fields of science and engineering. Perhaps the most frequently practiced function of the aerodynamicist is the analysis of the forces and moments exerted on a solid body in motion through the air.

Of fundamental significance in the term aerodynamics is the prefix, aero, which refers to the air of the Earth's atmosphere. Until man can achieve flight within the atmospheres of other planets, the limits of aerodynamic flight and the majority of practical considerations will be confined to the limits of the Earth's atmosphere as defined in aerodynamic terms.

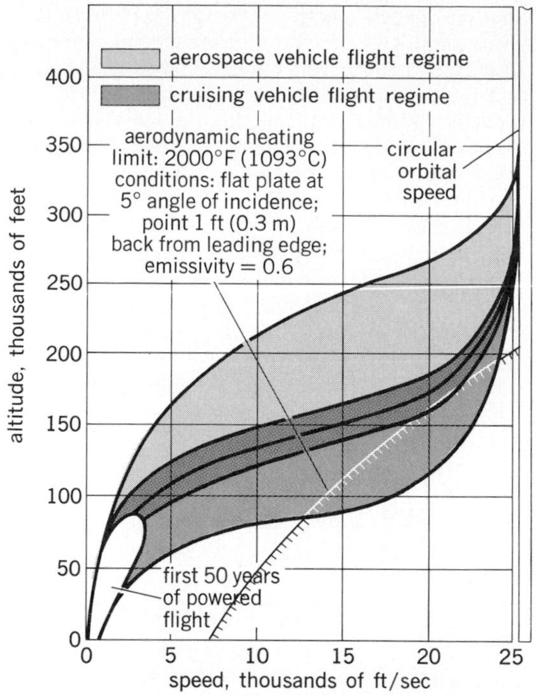

Fig. 1. Practical limits of aerodynamic flight within the atmosphere. 1 ft = 0.3048 m. (*Aero/Space Eng.*)

Figure 1 presents the practical limits of aerodynamic flight within the Earth's atmosphere based on a quantitative analysis of the governing factors. To a large extent the significant aerodynamic reactions of missiles in passing through the atmosphere also occur below the upper boundary shown in Fig. 1.

Two overlapping flight regimes are shown. The upper regime, defined as the aerospace-vehicle flight regime, indicates the operating region of reentry vehicles and space vehicles which use aerodynamic lift during their descent through the atmosphere. Its upper and lower boundaries are defined by a wing loading of 20 lb/ft² (1.0 kPa) and 40 lb/ft² (1.9 kPa), respectively. The lower flight regime in Fig. 1 is defined as the cruising-vehicle flight regime (sometimes referred to as the corridor of continuous flight). The upper and lower boundaries here are 10 lb/ft² (0.5 kPa) and 200 lb/ft² (10 kPa), respectively. The portion of this flight regime penetrated in the first 50 years of powered flight is also indicated.

The aerodynamic heating limit cuts off access to a large portion of the cruising flight regime between 12,000 and 24,000 ft/sec (3.7 and 7.3 km/sec). Actually, this temperature represents the approximate upper extreme of human engineering ability to penetrate what is sometimes called the thermal thicket; it cannot be accurately described as a barrier.

Strictly ballistic reentry vehicles could not reach the surface of the Earth wholly within the boundaries of the flight regimes just defined. The ballistic path of reentry in Fig. 1 would pass through alkaline cupric tartrate complex) is especially valuable for the characterization of reducing sugars. A positive test is indicated by the formation of a red precipitate of cuprous oxide. In the infrared absorption spectrum, the aldehyde group has a characteristic strong band at 1660–1740 cm^{-1} due to the carbonyl stretching vibration accompanied by two weak bands at 2700–2900 cm^{-1} due to the aldehydic carbon-hydrogen stretching. In the proton magnetic resonance spectrum the aldehydic protons have a unique absorption region at 0.0–0.6 τ.

Synthesis. Because of the importance of aldehydes as chemical intermediates, many industrial and laboratory syntheses have been developed. The more important of these methods are illustrated by the following examples.

Catalytic dehydrogenation of primary alcohols. Formaldehyde is produced on a large scale industrially by the catalytic dehydrogenation of methanol. Acetaldehyde is produced similarly by the dehydrogenation of ethanol over a copper catalyst at 250–300°C with formation of hydrogen as a by-product.

Oxidation of primary alcohols. Acetaldehyde is produced on a large scale industrially by passing a mixture of ethanol, oxygen, and steam over silver gauze at 480°C to give an overall yield of 85–90% of the aldehyde [reaction (23)].

$$CH_3CH_2OH + \tfrac{1}{2}O_2 \xrightarrow{Ag} CH_3CHO + H_2O \qquad (23)$$

On a laboratory scale, oxidizing agents such as chromic acid of manganese dioxide have been used to produce aldehydes from primary alcohols, as in reaction (24). The reaction must be carefully controlled to avoid oxidation of the aldehyde to the carboxylic acid.

$$RCH_2OH \xrightarrow{(O)} RCHO \qquad (24)$$

Oxidation of olefins. Ethylene is oxidized directly to acetaldehyde by the Wacker process, employing a palladium–cupric chloride catalyst [reaction (25)].

$$H_2C{=}CH_2 + \tfrac{1}{2}O_2 \xrightarrow{Pd,\ CuCl_2} CH_3CHO \qquad (25)$$

Acrolein, prepared industrially by the oxidation of propylene over a copper oxide catalyst at 350°C, as in reaction (26), is used in the manufacture of glycerol and acrylic acid. *See* GLYCEROL.

$$\underset{\text{Propylene}}{CH_2{=}CHCH_3} + O_2 \xrightarrow[300-400°C]{catalyst} \underset{\text{Acrolein}}{CH_2{=}CHCH{=}O} + H_2O \qquad (26)$$

In the laboratory, olefins are treated with ozone to form an ozonide which decomposes with water to form aldehydes and hydrogen peroxide [reaction (27)].

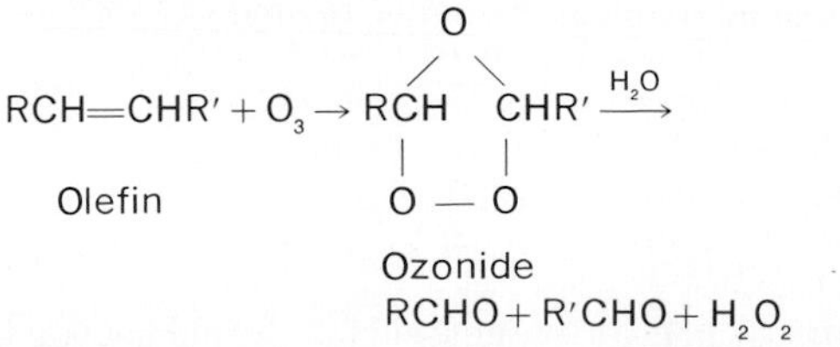

(27)

Hydroformylation of olefins. Olefins may be hydroformylated by reaction with carbon monoxide and hydrogen in the presence of a catalyst, usually cobalt carbonyl, at 180°C and 6000 psi (1 psi = 6895 Pa) as shown in reaction (28). This reaction is also known as the oxo process.

$$\underset{\text{Ethylene}}{CH_2{=}CH_2} + CO + H_2 \xrightarrow{Co_2(CO)_8} \underset{\text{Propionaldehyde}}{CH_3CH_2CHO} \qquad (28)$$

Chlorination of a methyl group attached to a benzene ring. The chlorination of toluene in the presence of strong light (which serves as a catalyst) proceeds stepwise to form benzyl chloride and benzal chloride at 135–175°C, by reactions (29) and (30), respectively.

$$\underset{\text{Toluene}}{C_6H_5CH_3} + Cl_2 \rightarrow \underset{\text{Benzyl chloride}}{C_6H_5CH_2Cl} + HCl \qquad (29)$$

$$C_6H_5CH_2Cl + Cl_2 \rightarrow \underset{\text{Benzal chloride}}{C_6H_5CHCl_2} + HCl \qquad (30)$$

Hydrolysis of benzal chloride gives benzaldehyde, as in reaction (31). This is one industrial process for the production of benzaldehyde.

$$C_6H_5CHCl_2 + H_2O \rightarrow \underset{\text{Benzaldehyde}}{C_6H_5CHO} + 2HCl \qquad (31)$$

Oxidation of a methyl group attached to a benzene ring. Certain substituted benzaldehydes, such as *p*-nitrobenzaldehyde, are readily prepared by oxidation of the corresponding toluene derivatives with chromic acid. Acetic anhydride is used to esterify the aldehyde groups and prevent further oxidation. Subsequent hydrolysis of the diacetate produces the aldehyde [reaction (32)].

$$\underset{p\text{-Nitrotoluene}}{O_2NC_6H_4CH_3} + O_2 + \underset{\text{Acetic anhydride}}{2(CH_3CO)_2O} \rightarrow \underset{\text{Diacetate}}{O_2NC_6H_4CH(OOCCH_3)_2} \xrightarrow{H_2O} \underset{p\text{-Nitrobenzaldehyde}}{O_2NC_6H_4CHO} + \underset{\text{Acetic acid}}{2HOOCCH_3} \qquad (32)$$

One commercial synthesis of benzaldehyde involves the passing of a mixture of toluene and air over a heated metallic oxide catalyst at temperatures above 500°C [reaction (33)].

$$\underset{\text{Toluene}}{C_6H_5CH_3} + O_2 \xrightarrow{catalyst} \underset{\text{Benzaldehyde}}{C_6H_5CHO} + H_2O \qquad (33)$$

Reduction of acid chlorides. Acid chlorides react with hydrogen in the presence of a specially prepared palladium catalyst to form aldehydes and hydrogen chloride, as in reaction (34). This reaction, known as the Rosenmund synthesis, is a useful laboratory method, since the acid chlorides can be prepared from the corresponding carboxylic acids.

$$\underset{\text{Acid chloride}}{RCOCl} + H_2 \xrightarrow{Pd} \underset{\text{Aldehyde}}{RCHO} + HCl \qquad (34)$$

Reaction of Grignard reagents. Aldehydes are formed by the reaction of alkylmagnesium halides

(Grignard reagents) with ethyl orthoformate or diethyl formamide [reactions (35) and (36)].

$$RMgX + HC(OCH_2CH_3)_3 \rightarrow RCH(OCH_2CH_3)_2 \xrightarrow{H_3O^-} RCHO \quad (35)$$

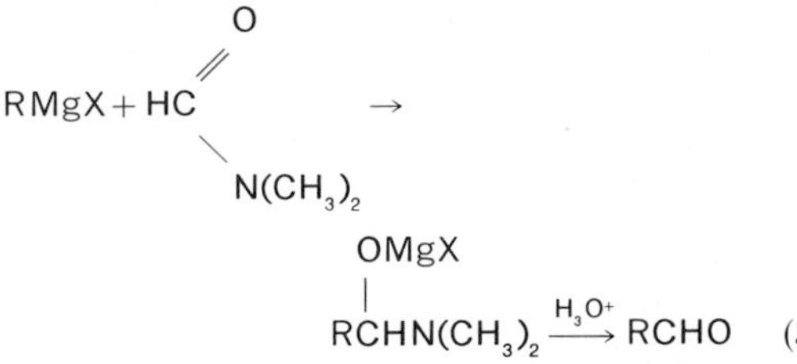

(36)

Reimer-Tiemann synthesis. When phenols are treated with chloroform and strong alkali, *o*-hydroxybenzaldehydes (salicylaldehydes) are formed [reaction (37)].

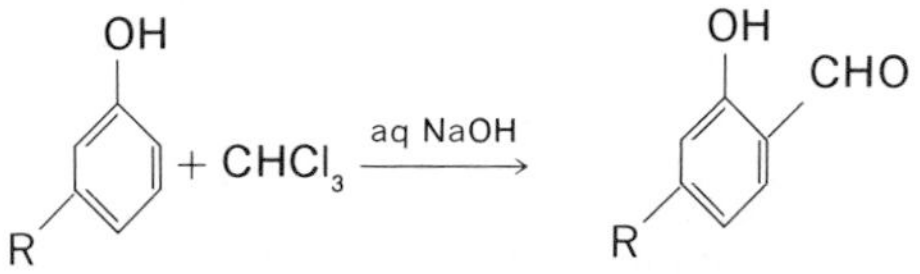

(37)

Phenols

o-Hydroxybenzaldehydes (salicylaldehydes)

Furfural production. Cornstalks, corncobs, grain hulls, and similar farm wastes produce furfural upon heating with dilute sulfuric acid. This reaction (38), which is employed on a large scale

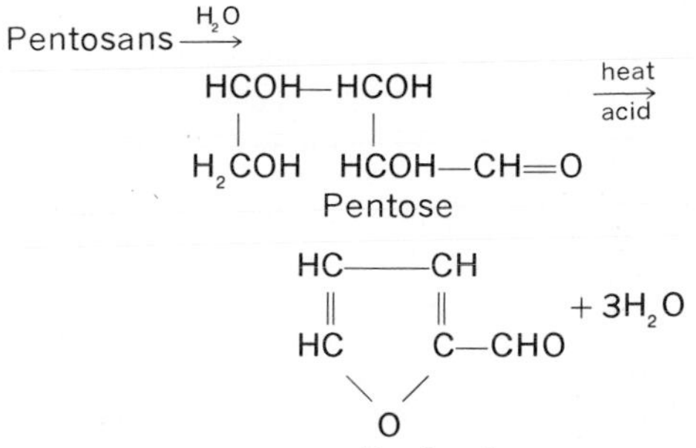

Pentose (38)

Furfural

industrially, proceeds from pentosans found in the agricultural residues.

Chloral production. Chloral is an important intermediate for the production of DDT. It is formed by chlorination of anhydrous ethanol [reaction (39)].

$$\underset{\text{Ethanol}}{2CH_3CH_2OH} + \underset{\text{Chlorine}}{8Cl_2} \rightarrow \underset{\text{Chloral}}{2Cl_3CCHO} + \underset{\text{Hydrogen chloride}}{10HCl} \quad (39)$$

See CONDENSATION REACTION; FORMALDEHYDE; FURFURAL.

[PAUL E. FANTA]

Bibliography: N. L. Allinger et al., *Organic Chemistry*, 2d ed., 1976; S. Patai (ed.), *Chemistry of the Carbonyl Group*, 1966.

Alicyclic hydrocarbon

Organic compounds containing only carbon and hydrogen atoms joined to form one or more rings. The chemical properties of alicyclic hydrocarbons resemble those of the aliphatic hydrocarbons. Synonyms are cycloaliphatic compounds, cycloalkanes, or cycloparaffins. Unsaturated alicyclic hydrocarbons are known as cycloolefins, cycloalkenes, or cycloalkynes, depending on the type of unsaturation.

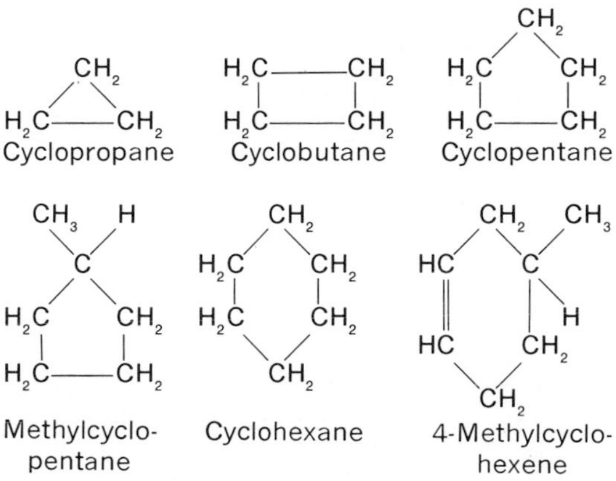

Uses. A number of alicyclic hydrocarbons have industrially important uses. Cyclopropane is an anesthetic and may be prepared as shown in reaction (1).

$$ClCH_2CH_2CH_2Br + Zn \rightarrow$$

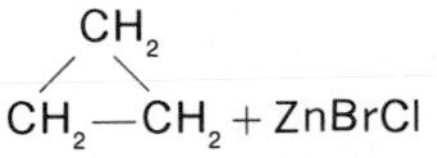

(1)

Cyclopentadiene is used in the preparation of the persistent insecticide chlordane. Cyclohexane is the key intermediate in the preparation of nylon-6 from cyclohexanone and caprolactam. Adipic acid, one of the constituents of nylon-66, can be obtained by the oxidation of cyclohexane. The naturally occurring terpenes, limonene and α-pinene, are used in flavors and perfumes and as starting materials in the preparation of synthetic camphor, which is a cyclic ketone, and racemic menthol, which is a cyclic alcohol. Structural formulas of typical compounds are shown in Fig. 1.

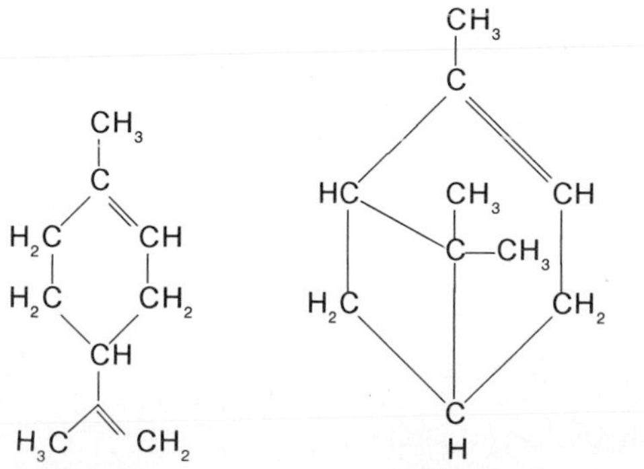

Fig. 1. Structural formulas of alicyclic hydrocarbons.

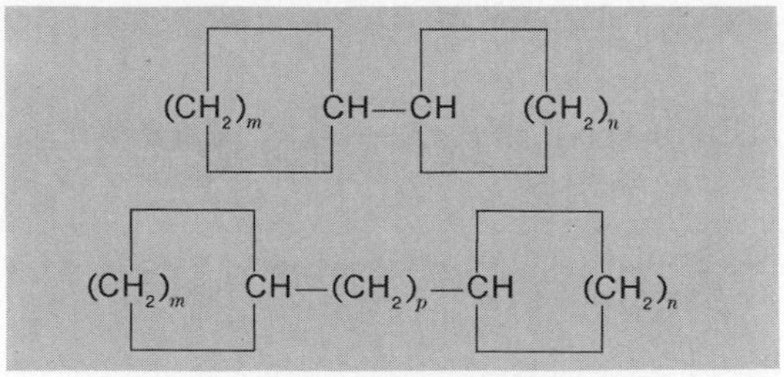

Fig. 2. Structures of discrete ring compounds.

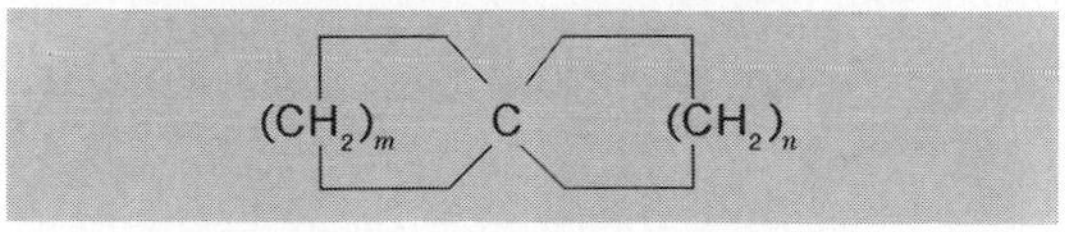

Fig. 3. Structural formula of the spiranes.

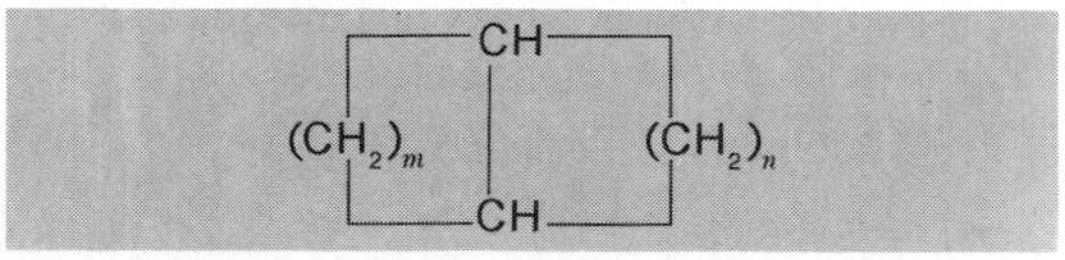

Fig. 4. Structure of a generalized fused ring compound.

Occurrence and nomenclature. Alicyclic hydrocarbons occur widely in nature. Cyclopentanes and cyclohexanes are found in petroleum and are called naphthenes. An important group of alicyclic hydrocarbons are the cyclic terpenes, such as limonene, derived from citrus fruits, and α-pinene, the principal constituent of turpentine, which is the volatile portion of the gum that exudes from incisions in trunks of living pine trees.

Polynuclear alicyclic hydrocarbons are those which contain more than one ring. These can be divided into four subgroups. (1) compounds containing discrete rings joined directly or by a carbon chain, as in Fig. 2; (2) compounds with one carbon atom common to two rings (spiranes), as in Fig. 3; (3) fused or condensed rings where two carbon atoms are common to two rings, as in Fig. 4; and (4) bridged rings where two nonadjacent carbon atoms in a ring are joined by a bridge of one or more carbon atoms, as in Fig. 5.

ALICYCLIC HYDROCARBON

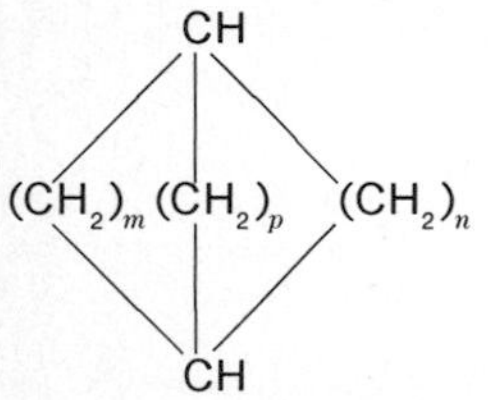

Fig. 5. Generalized structure of an alicyclic hydrocarbon of bridged ring type.

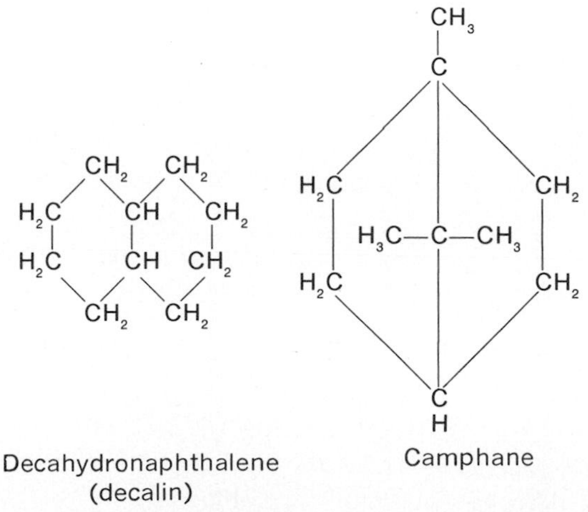

Fig. 6. Structural formulas of decalin and camphane.

The alicyclic hydrocarbons are systematically named by denoting the number of carbon atoms in the nucleus, as in the straight-chain series, and adding the prefix "cyclo." The fused and bridged two-ring alicyclic hydrocarbons utilize the prefix "bicyclo," and inserted in brackets, in decreasing order, the number of ring members joined to either side of the common carbon atom. Thus decahydronaphthalene (decalin) is bicyclo[4.4.0]decane and camphane is 1,7,7-trimethylbicyclo[2.2.1]heptane. Figure 6 depicts the structures of these compounds as skeleton formulas, in which each intersection represents a carbon atom; the hydrogen atoms attached to the ring carbon atoms are not explicitly shown.

Spiranes are named by the use of the prefix "spiro" followed, in decreasing order, by the number of atoms connected to the central atom. This is illustrated by Fig. 7.

Molecular structure. Up to the early 1880s substituted cyclohexanes were the only alicyclic hydrocarbons known. They were formed by the addition of hydrogen to the corresponding aromatic hydrocarbons and for this reason have been called hydroaromatic compounds. It also had been stated that smaller or larger ring compounds were incapable of existence. Between 1880 and 1885, however, representatives of the cyclopropane, cyclobutane, and cyclopentane ring systems were prepared. It was soon apparent that the different ring systems exhibited wide variations in stability. Cyclopropane showed many of the characteristic reactions of unsaturated compounds, and the cyclobutane ring proved quite reactive when compared to cyclopentane and cyclohexane, whose derivatives showed great resistance to ring opening.

In his strain theory, J. F. A. von Baeyer was the first to account for these differences in reactivity. Based on the tetrahedral model of the carbon valences proposed by J. A. Le Bel and J. H. van't Hoff, the normal angle subtended between any two of the four valences of the carbon atom would be 109°28′. If all of the carbon atoms in a ring are in the same plane, any deviation from the standard valence angle would set up a condition of strain. The angular deviations may be readily computed and are given in Table 1. *See* BOND ANGLE AND DISTANCE.

Baeyer's theory proved misleading by implying that large rings would be incapable of existence. Stable rings containing 30 or more members are now known and are indeed strainless, since it has been shown that in rings containing more than four members the carbon atoms are not in a plane but assume a strain-free puckered configuration. The heats of combustion given in Table 1 indicate the

Table 1. Angular strain in planar alicyclic hydrocarbons

Number of C atoms in ring	Angle, degrees	Degree of strain	Heat of combustion/CH_2, kcal
3	60	+24°44′	168.5
4	90	+9°44′	165.5
5	108	+0°44′	159.0
6	120	−5°16′	158.0
7	129	−9°51′	158.0
15	155	−23°16′	157.0

freedom of strain in the larger rings, whereas the three- and four-membered rings do show appreciable strain. Examples of strainless rings are the two forms of cyclohexane known as the chair and boat (Fig. 8). Each point of intersection represents one of the six carbon atoms. These two forms are interconvertible, the chair form being the more stable by 5.6 kcal/mole. The low energy of interconversion precludes isolation of either form in the pure state at room temperature.

Physical properties. The alicyclic hydrocarbons boil 10–20°C higher than the corresponding aliphatic hydrocarbons. A comparison of the relationship between the chain length and physical properties of alkanes and cycloalkanes suggests that in rings containing up to 14 carbon atoms, the molecule is roughly spherical in shape, while in the higher rings the sides become parallel and the properties approach those of the straight-chain hydrocarbons. The physical properties of some alicyclic hydrocarbons are given in Table 2.

Chemical properties. Alicyclic hydrocarbons in general exhibit a stability toward heat and chemical attack comparable to that of the corresponding open-chain compounds. Only in cyclopropanes, and in cyclobutanes to a lesser extent, is there any marked ring instability. Cyclopropane undergoes catalytic hydrogenation readily to *n*-propane, as seen by reaction (2).

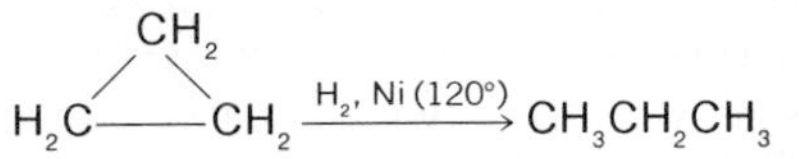

(2)

The ring can also be opened by many reagents, such as bromine, hydrogen bromide, or sulfuric acid, that would normally add to double bonds.

The multiple bond in cycloolefins is generally more reactive than its straight-chain counterpart and, when cleaved, provides a convenient source of difunctional organic derivatives, as shown by reactions (3) and (4).

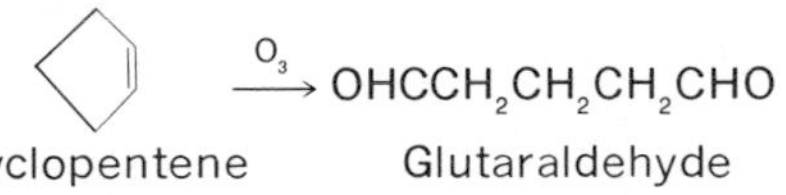

(3)

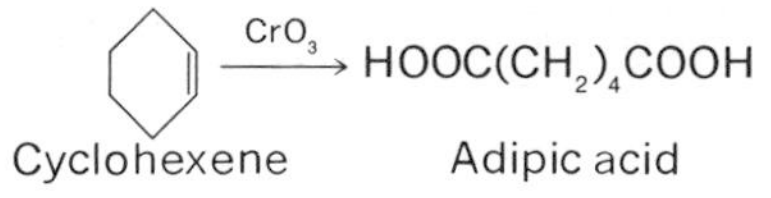

(4)

Certain polynuclear alicyclic compounds provide the best examples of the phenomenon of valence tautomerism. The most remarkable such molecule is tricyclo[3.3.2.0[46]]deca-2,7,9-triene, popularly known as bullvalene. At 80°C, the rapid rearrangement of bullvalene, reaction (5), results

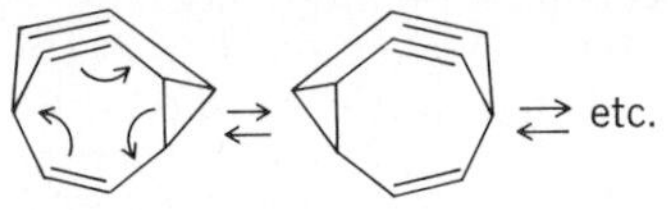

(5)

in the equivalence of each of the C-H groups, as shown by the presence of a single sharp peak in the proton magnetic resonance spectrum.

Stereochemistry. The nature of alicyclic rings allows polysubstituted alicyclic hydrocarbons to exist as geometrical isomers. Thus, 1,2-dimethylcyclohexane exists as cis and trans isomers, as shown in Fig. 9. Analogously, systems of two or more fused rings may also be found in cis and trans forms, as illustrated by the isomeric decalins, which differ only in the manner of ring fusion (Fig. 10).

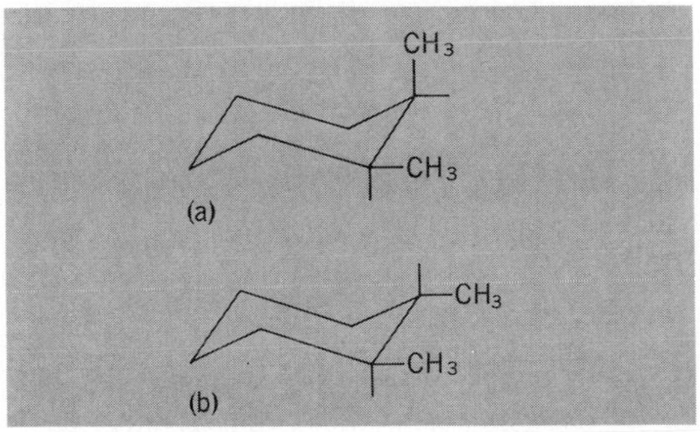

Fig. 9. Stereoisomers of 1,2-dimethylcyclohexane: (*a*) cis; (*b*) trans.

ALICYCLIC HYDROCARBON

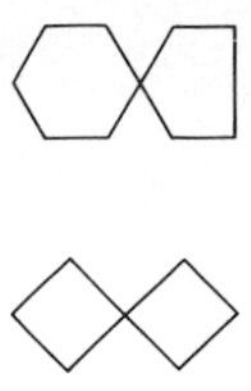

(a)

(b)

Fig. 7. Structures of spiranes: (*a*) spiro [5.4] decane; (*b*) spiro [3.3] heptane.

ALICYCLIC HYDROCARBON

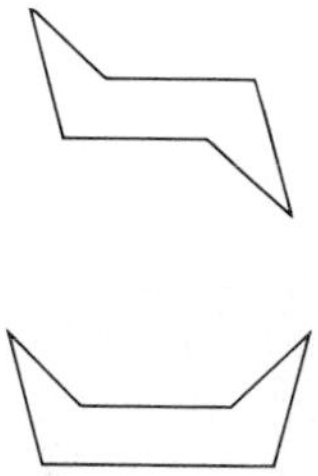

(a)

(b)

Fig. 8. Forms of cyclohexane: (*a*) chair; (*b*) boat.

Table 2. Physical properties of some alicyclic hydrocarbons

Compound	Formula	Melting point, °C	Boiling point, °C	Density
Cyclopropane	C_3H_6	−127	−34.5	0.688
Cyclobutane	C_4H_8	−50	13.1	0.733
Cyclopentane	C_5H_{10}	−93.3	49.5	0.745
Cyclohexane	C_6H_{12}	6.5	80.3	0.779
Cycloheptane	C_7H_{14}	−12	118–120	0.828
Cyclooctane	C_8H_{16}	4.3	148.9	0.836
Cyclodecane	$C_{10}H_{20}$	9.7	69/12 mm	0.853
Cyclotetradecane	$C_{14}H_{28}$	54	131/11 mm	0.863
Cycloeicosane	$C_{20}H_{40}$	46	–	0.850
Tetrahydronaphthalene	$C_{10}H_{12}$	−30	207	0.971
Cyclohexene	C_6H_{10}	−103.7	83	0.810
Cyclopentadiene	C_5H_6	−85	40.2	0.807
Cyclooctyne	C_8H_{12}	–	72–76/100 mm	0.844

Preparative methods. Alicyclic hydrocarbons are generally synthesized by three general methods: (1) ring expansions, (2) ring closures, and (3) ring contractions, each followed by removal of any existing functional group. Some examples are given in reactions (6) to (8). The yields in these reac-

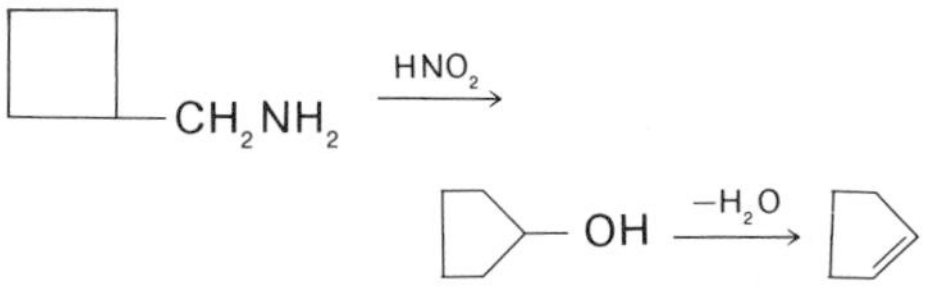

(6)

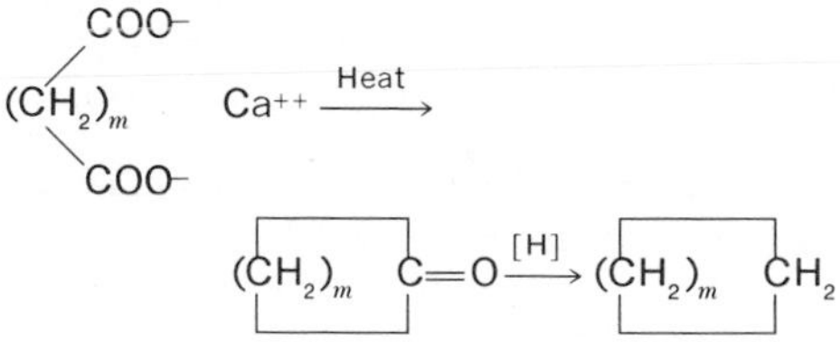

(7)

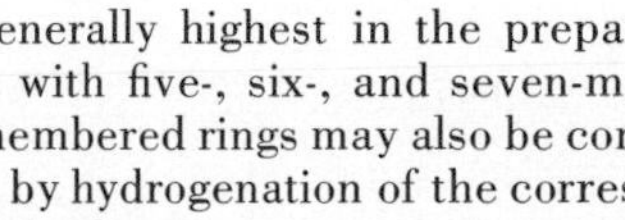

(8)

tions are generally highest in the preparation of compounds with five-, six-, and seven-membered rings. Six-membered rings may also be conveniently prepared by hydrogenation of the corresponding aromatic derivatives.

ALICYCLIC HYDROCARBON

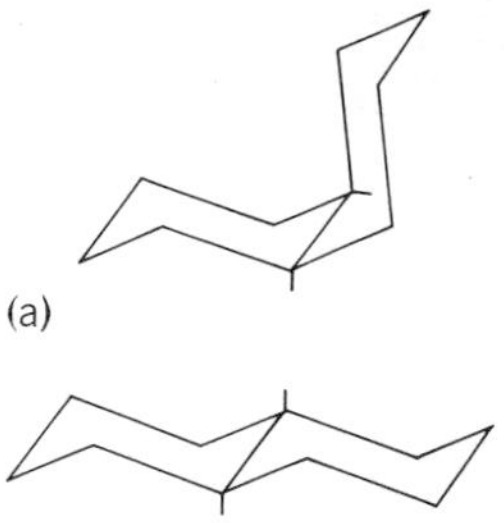

(a)

(b)

Fig. 10. Stereoisomers of decalin: (*a*) cis; (*b*) trans.

Unsaturated alicyclic hydrocarbons can often be synthesized by standard methods for introducing the double or triple bond in open-chain aliphatic compounds. Because of geometrical considerations, certain structural features cannot be accommodated in alicyclic rings below a certain size. Thus cyclopentadiene is the smallest alicyclic diene that has been isolated in pure form; *trans*-cyclooctene is the smallest *trans*-cycloalkene; cyclooctyne is the smallest cycloalkyne; and 1,2-cyclononadiene is the smallest 1,2-diene. 1,-5-

Cyclooctatetraene can be made by the tetramerization of acetylene over nickel cyanide catalyst in tetrahydrofuran at 60–70°C and 15–20 atm (1.5–2.0 MPa) pressure, as in reaction (9). It is not

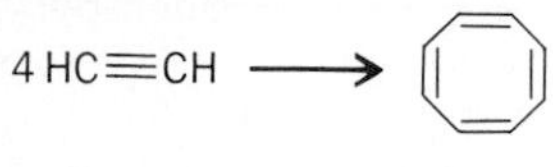

(9)

an aromatic hydrocarbon, and behaves like a typical alkene.

1,5-Cyclooctadiene is made by the 1,4-dimerization of butadiene over a complex nickel catalyst, as in reaction (10).

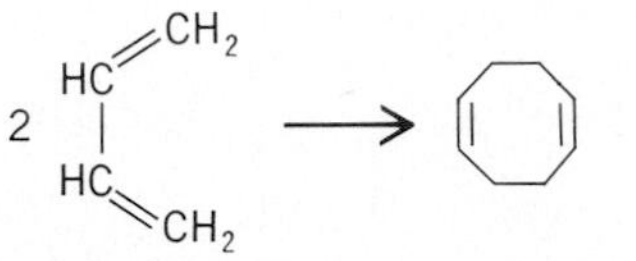

(10)

By using a Ziegler-type complex metal catalyst, butadiene may be trimerized to either *trans,trans,trans*-1,5,9-cyclododecatrine (I) or the *trans,trans,cis*-stereoisomer (II).

(I) (II)

See ADAMANTANE; ALIPHATIC HYDROCARBON; ALKENE; AROMATIC HYDROCARBON; CONFORMATIONAL ANALYSIS; HYDROCARBON; TROPOLONE.

[PAUL E. FANTA]

Bibliography: S. Coffey (ed.), *Rodd's Chemistry of Carbon Compounds* 2d ed., vol. 2, pts. A, B, and C, 1967, 1968, 1969; H. Hart (ed.), *Advances in Alicyclic Chemistry*, vols. 1, 2, and 3, 1966, 1968, 1971; D. Lloyd (ed.), *Topics in Carbocyclic Chemistry*, 1969.

Aliphatic hydrocarbon

One of a group of hydrocarbons in which the carbon atoms are joined in open chains. They may be classified in accordance with their composition and their chemical behavior. Two major classes are the saturated and the unsaturated, the latter including several homologous series depending on the number of double and triple bonds present.

The systematic names adopted by the International Union of Pure and Applied Chemistry for the aliphatic hydrocarbons are formed by adding a suffix (-ane, -ene, -yne, -adiene, and so forth) indicating the type of compound to a prefix indicating the number of carbon atoms present. The first four prefixes are meth-, eth-, prop-, and but-. The succeeding prefixes are derived from the Greek or Latin word for the number: pent-, hex-, hept-, oct-, and so forth.

The saturated compounds, which are known as paraffin hydrocarbons, or more systematically, as alkanes, fit the empirical formula C_nH_{2n+2}. The members having four or more carbon atoms exist in straight-chain (or normal) and branched-chain isomers. Not counting optical isomers, there are 2 butanes with different carbon skeletons, 3 pentanes, 5 hexanes, 9 heptanes, 18 octanes, and 35 nonanes, all of which are known. Calculations have shown that it is theoretically possible to have more than 300,000 eicosanes ($C_{20}H_{42}$) and more than 4,000,000,000 triacontanes ($C_{30}H_{62}$); only a very minute fraction of the number of possible isomers of such hydrocarbons has been prepared.

Compounds containing one double bond and having the formula C_nH_{2n} are olefins (or alkenes). The first two members of the series, ethylene and propene, both exist in one form only. The next higher homolog, C_4H_8, has two straight-chain isomers (1-butene and 2-butene) and one branched-chain isomer (isobutylene or methylpropene, sometimes incorrectly called isobutene). There are two geometrical isomers of 2-butene, namely, *cis*-2-butene and *trans*-2-butene. As the length of the chain increases in both the straight-chain and the branched-chain olefins, the number of isomers formed by a change in the position of the double bond increases rapidly, as shown in the figure. *See* MOLECULAR ISOMERISM.

There are two homologous series of aliphatic hydrocarbons having the formula C_nH_{2n-2}: (1) acetylenes (or alkynes), which contain one triple bond, and (2) diolefins (or alkadienes), which contain two double bonds. The latter may be classified according to the relative positions of the double bonds: (*a*) allenic double bonds (both double bonds on a single carbon atom) as in allene, $CH_2{=}C{=}CH_2$; (*b*) conjugated double bonds (double bonds separated by a single bond) as in 1,3-butadiene, $CH_2{=}CH{-}CH{=}CH_2$; (*c*) nonconjugated, isolated, double bonds (double bonds separated by at least two single bonds) as in 1,4-pentadiene, $CH_2{=}CH{-}CH_2{-}CH{=}CH_2$.

Aliphatic hydrocarbons having three double bonds per molecule are termed triolefins or alkatrienes; those containing a larger number of double

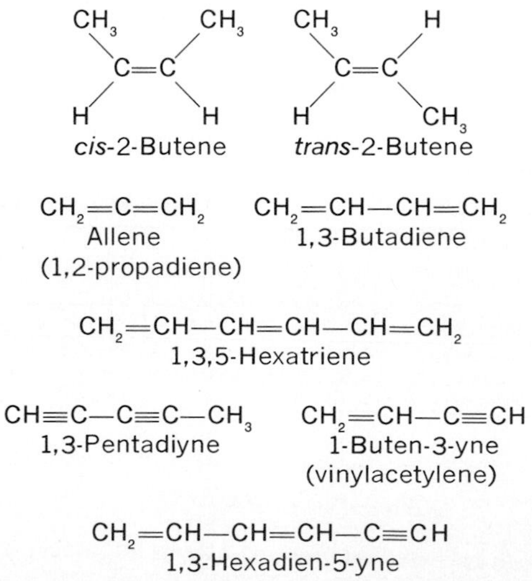

Formulas of some alkenes, alkadienes, and alkynes.

Aliphatic hydrocarbons

Number of carbon atoms	Alkane	Alkene	Alkyne
1	CH_4 Methane	–	–
2	$CH_3—CH_3$ Ethane	$CH_2═CH_2$ Ethylene	$CH≡CH$ Acetylene
3	$CH_3—CH_2—CH_3$ Propane	$CH_2═CH—CH_3$ Propene	$CH≡C—CH_3$ Propyne
4	$CH_3—CH_2—CH_2—CH_3$ Butane	$CH_2═CH—CH_2—CH_3$ 1-Butene	$CH≡CH_2—CH_2—CH_3$ 1-Butyne
	–	$CH_3—CH═CH—CH_3$ 2-Butene	$CH_3—C≡C—CH_3$ 2-Butyne
	$CH_3—CH(CH_3)—CH_3$ Isobutane (methylpropane)	$CH_2═C(CH_3)—CH_3$ Isobutylene (methylpropene)	–

bonds are named in analogous fashion. Similarly, the open-chain hydrocarbons containing two triple bonds have the generic names alkadiynes, while those groups of hydrocarbons containing both double and triple bonds are identified by names such as alkenynes, alkadienynes, or alkenediynes.

The systematic names and the structural formulas of aliphatic hydrocarbons with 1 to 4 carbon atoms are presented in the table. *See* ALKANE; HYDROCARBON.

[LOUIS SCHMERLING]

Alkali

Broadly, any compound having highly basic properties, that is, a compound that readily ionizes in aqueous solution to yield hydroxyl ions so that the pH is high (above 7) or so that litmus paper changes from red to blue. Specifically, an alkali is a hydroxide of one of the alkali metals, although an ammonium salt may also be referred to as an alkali. Alkalies are caustic in that they produce a relatively high concentration of OH ions in solution. Caustics, or escharotics, destroy tissue by abstracting fluid or by corrosive deoxidation. External burns from an alkali are treated by immediate, copious washing with water. *See* ALKALI METALS.

Historically, an alkali was any caustic hydroxide that would neutralize an acid or form a soluble soap with a fatty acid. Thus, an alkali may be an inorganic alcohol in combination with a metal, having none of the characteristics of an alcohol. Alkalies are used commercially in soap manufacture, soluble oils, cutting compounds, cleaning solutions, and aluminum etching.

Alkalies are soluble in water, tarnish in air, and in concentrated form are corrosive to the touch. In commerce, alkali usually refers to sodium carbonate in the form of soda ash, either from natural deposits or manufactured from salt and either sulfuric acid or ammonium bicarbonate. Soda ash is also produced electrolytically from salt, steam, and carbon dioxide to treat water whose hardness arises from the presence of compounds of alkaline-earth metals. Other alkali products include lye, potash, caustic soda, potassium hydroxide, water glass, and bicarbonate of soda. Alkali soils, usually in arid regions where water has not leached out the caustic compounds, have limited agricultural use. *See* ACID AND BASE; ALKALINE-EARTH METALS; HYDROXIDE; SODIUM; WATER SOFTENING.

[FRANK H. ROCKETT]

Alkali metals

The elements of group Ia in the periodic table. Of the alkali metals, lithium differs most from the rest of the group, and tends to resemble the alkaline-earth metals (group IIa of the periodic table) in many ways. In this respect lithium behaves as do

Isotopes of the alkali metals

Element	Normal at. wt	Mass no.	Radio-active	Half-life*
Lithium, Li	6.939	5	Yes	10^{-21}s
		6	No	Stable (7.5)
		7	No	Stable (92.5)
		8	Yes	0.83s
		9	Yes	0.17s
Sodium, Na	22.9898	20	Yes	0.23s
		21	Yes	23.0s
		22	Yes	2.6y
		23	No	Stable (100)
		24	Yes	15.0h
		25	Yes	60s
Potassium, K	39.102	37	Yes	1.2s
		38	Yes	7.7m
		39	No	Stable (93.1)
		40	Yes	1.2×10^9y
		41	No	Stable (6.9)
		42	Yes	12.4h
		43	Yes	22h
		44	Yes	27m
Rubidium, Rb	85.47	81	Yes	4.7h
		82	Yes	6.3h
		83	Yes	80d
		84	Yes	23m
		85	No	Stable (72.2)
		86	Yes	19d
		87	Yes	6.2×10^{10}y (27.8)
		88	Yes	18m
		89	Yes	15m
		90	Yes	2.7m
		91	Yes	14m
Cesium, Cs	132.905	127–132	Yes	Short
		133	No	Stable (100)
		134	Yes	3×10^6y
		135	Yes	2.3y
		136	Yes	13d
		137	Yes	37y
		138–145	Yes	Short
Francium, Fr		223	Yes	21m

*Figures in parentheses indicate the percentage occurrence in nature.

many other elements that are the first members of groups in the periodic table; these tend to resemble the elements in the group to the right rather than those in the same group. Francium, the heaviest of the alkali-metal elements, has no stable isotopes and exists only in radioactive form (see table).

In general, the alkali metals are soft, low-melting, reactive metals. This reactivity accounts for the fact that they are never found uncombined in nature but are always in chemical combination with other elements. This reactivity also accounts for the fact that they have no utility as structural metals (with the possible exception of lithium in alloys) and that they are used as chemical reactants in industry rather than as metals in the usual sense. The reactivity in the alkali-metal series increases in general with increase in atomic weight from lithium to cesium. *See* CESIUM; ELECTROCHEMICAL SERIES; FRANCIUM; LITHIUM; PERIODIC TABLE; POTASSIUM; RUBIDIUM; SODIUM. [MARSHALL SITTIG]

Bibliography: American Chemical Society, *Handling and Uses of the Alkali Metals*, Advan. Chem. Ser. no. 19, 1957; I. Fatt and M. Tashima, *Alkali Metal Dispersions*, 1961; R. N. Lyon (ed.), *Liquid-metals Handbook*, 3d ed., Navexos P-733, 1955.

Alkaline-earth metals

Usually calcium, strontium, and barium, the heaviest members of group IIa of the periodic table (excepting radium). Other members of the group are beryllium, magnesium, and radium, sometimes included among the alkaline-earth metals. Beryllium resembles aluminum more than any other element, and magnesium behaves more like zinc and cadmium. The gap between beryllium and magnesium and the remainder of the elements of group IIa makes it desirable to discuss these elements separately. Radium is often treated separately because of its radioactivity.

J. J. Berzelius first reduced the three alkaline-earth metals to the elementary state, but obtained them as amalgams by electrolysis. Sir Humphry Davy in 1808 isolated the metals in the pure state by distillation of amalgams produced electrolytically. Today industrial preparation of these elements involves electrolysis of their molten chlorides or reduction of their oxides with aluminum.

The alkaline earths form a closely related group of highly metallic elements in which there is a regular gradation of properties. The metals, none of which occurs free in nature, are all harder than potassium or sodium, softer than magnesium or beryllium, and about as hard as lead. The metals are somewhat brittle, but are malleable, extrudable, and machinable. They conduct electricity well; the specific conductivity of calcium is 45% of that of silver. The oxidation potentials of the triad are as great as those of the alkali metals.

The alkaline earths exist as large divalent cations in all their compounds, in which the elements are present in the 2+ oxidation state. The metals have a gray-white luster when cut but tarnish readily in air. They burn brilliantly in air when heated, and form the metal monoxide, except for barium, which forms the peroxide. A certain amount of nitride is formed simultaneously, especially with calcium. All the metals dissolve readily in acid. Whereas calcium reacts smoothly with water to yield hydrogen, the heavier members react as violently as sodium does. All the metals are soluble in liquid ammonia, yielding strongly reducing, electrically conducting, blue solutions. The order of solubility in water for most salts is calcium > strontium > barium, except that the order is reversed for the fluorides, hydroxides, and oxalates. All three elements unite directly with hydrogen to form hydrides and with nitrogen to form nitrides, but whereas ease of formation of a nitride increases with atomic number, ease of formation of a hydride decreases.

The elements and their compounds find important industrial uses in low-melting alloys, deoxidizers, and drying agents and as cheap sources of alkalinity. *See* BARIUM; BERYLLIUM; CALCIUM; MAGNESIUM; PERIODIC TABLE; RADIUM; STRONTIUM. [REED F. RILEY]

Bibliography: D. G. Cooper. *The Periodic Table*, 4th ed., 1968; B. Henderson and J. E. Wertz, *Defects in the Alkaline Earth Oxides*: *With Applications to Radiation Damage and Catalysis*, 1977.

Alkaloid

One of a group of nitrogenous bases of plant origin, such as nicotine, cocaine, and morphine. Most of the alkaloids show marked physiological activity, and crude extracts of various alkaloid-bearing plants have been used since antiquity because of their curative or poisonous effects. Well over 1000 alkaloids have since been isolated or have been shown to be present in some 97 families of plants. Alkaloid-bearing plants have been found in virtually every habitat in which vascular plants grow.

Occurrence. Alkaloids are found most frequently in the higher seed-bearing plants, and especially in dicotyledons. Occurrence in the lower non-seed-bearers is rare. As more plants are examined, more families are included in the alkaloid-bearing group, for new genera containing alkaloids are reported regularly. Some families contain very few such genera, and the processes which gave rise to the alkaloid-bearers of these groups appear to have involved mutations.

An extreme of another kind is encountered in the Papaveraceae, in which all species contain alkaloids and no mutations resulting in alkaloid-free plants have yet occurred. The majority of plants occupy an intermediate position. For example, all species of *Aconitum* and *Delphinium* elaborate alkaloids, although most of the other genera (*Ranunculus*, *Trollius*, *Anemone*) do not. In general, alkaloids from plants of closely related genera are similar in structure; those from remotely related genera usually differ markedly.

The elaboration of alkaloids in plants is not localized but appears to be a characteristic of all organs including the seeds. However, not all organs of any one species must have alkaloids; for example, the seeds of the tobacco plant and of the opium poppy are devoid of alkaloids. In perennials during the first year of growth, alkaloids seem to be quite evenly distributed among the various organs, but with increased age there appears to be some localization in a few organs. For instance, the bark of arborescent plants is generally richer in alkaloids than are the leaves or shoots; this may be attributed to their accumulation in the bark year after year. While localization of the alkaloids in various organs does not appear to occur in the annuals, there is a marked fluctuation of alkaloid content throughout the growing season. The period of maximum output of alkaloids appears to coincide with the early flowering stage.

When plants elaborate more than one alkaloid, their ratio in the plant is not necessarily the same in all stages of growth. Some alkaloids are virtually absent in the young plant, but increase to isolable amounts as the plant approaches maturity. This may explain why different investigators frequently report the isolation of different alkaloids from the same species. Furthermore, cultural and climatic conditions exert an effect on the alkaloid content of plants. Thus, the amount of alkaloids in opium varies with the source, and the variation is attributable both to varietal differences among the poppies and to climatic conditions. Planters in Java have been very successful in increasing the quinine yield from the cinchona tree by selection and cultivation. Strains of tobacco and of lupines can be selected and cultivated to yield greater or lesser amounts of alkaloids.

The function of alkaloids in the plant is still a subject of speculation. The alkaloids are generally concentrated in the living tissue at points of intense cell activity, from which they are often cast aside and stored in such dead structures as the seed hulls or outer bark. These facts have led to the view that the alkaloids are end products, or byproducts, of amino acid metabolism in plants. Other theories regard alkaloids as reserve materials stored for protein synthesis, protective substances which discourage animal or insect attack, plant stimulants or regulators similar to hormones, or detoxication products rendered harmless by the plant's defense mechanisms.

Isolation, separation, identification. Chemical work on the alkaloids starts with their isolation from plant materials. Because very few plants elaborate a single alkaloid, the main problem is usually the separation of mixtures. The alkaloid mixture is usually isolated by extraction of powdered plant parts with water, alcohol, or dilute acids; or vegetable material may be treated with alkali, and the alkaloid extracted by organic solvents. The crude mixture obtained after preliminary processing is purified by solution in dilute acid and filtration from insoluble material. The alkaloids are then reprecipitated with alkali or extracted with an immiscible solvent from an alkaline solution.

The individual alkaloids are sometimes separated from each other through differences in the solubilities of the bases and their salts. Fractional crystallization is based on these differences. Fractional extraction by countercurrent distribution utilizes differences in both basicities and solubilities of the alkaloids. Other useful processes are absorption chromatography and partition chromatography.

The nomenclature of the individual alkaloids has not been systematized and they draw their names from a variety of sources. A great many important alkaloids have received names derived from those of plants, such as papaverine, quinine, and berberine. A few are named from their physiological action, such as morphine and emetine. Some are named from their physical characteristics, such as hygrine. Only one, pelletierine, has been named for an alkaloid chemist.

The alkaloids are usually crystalline, colorless substances; only a few are liquid. Colored alkaloids are rare. Nearly all alkaloids from crystalline salts with acids, and these salts are frequently used in characterizing the base. Alkaloid reagents are solutions used to detect or identify minute quantities of the natural bases or their derivatives and can be divided roughly into precipitants and color reagents. The precipitating reagents combine with alkaloids to give highly insoluble complexes. The color reagents are usually dehydrating or oxidizing reagents, and these give characteristically colored solutions upon reaction with alkaloids.

The determination of the structures of alkaloids has been one of the most fascinating chapters of organic chemistry. Almost all the known nitrogen-containing ring systems, both saturated and unsaturated, are encountered among the alkaloids. The molecules vary widely in complexity, ranging from the relatively simple systems such as nicotine to the very complex polycyclic structures of morphine and strychnine.

Uses. Many alkaloids have great value in medical practice because of specific pharmacological actions. Thus, morphine and some of its related compounds are the best-known agents for the relief of pain. The curare alkaloids produce paralysis of voluntary muscle and hence are used as an adjunct to anesthesia in surgery. The ergot alkaloids are used clinically to induce motility of the uterus in the last stages of pregnancy. The belladonna alkaloids prevent the normal response of smooth muscle to nervous impulses; they are used particularly to control excess activity of the gastrointestinal tract and to paralyze the accommodation muscle of the eye in ophthalmic practice. Until World War II quinine was the standard antimalarial drug in medical practice.

The isolation of pure individual alkaloids on large scale is sometimes prohibitively expensive. Furthermore, the synthesis of many of the more useful alkaloids is frequently impractical. These facts, together with the undesirable pharmacological side effects accompanying the favorable action of some alkaloids, have resulted in the expenditure of considerable effort on the synthesis of related substitute compounds. In the treatment of malaria, quinine has been largely replaced by synthetic compounds. Synthetic procaine and similar drugs have supplanted naturally occurring cocaine in local anesthesia. *See* NICOTINE; QUININE; STRYCHNINE. [S. MORRIS KUPCHAN]

Bibliography: K. W. Bentley (ed.), *The Alkaloids*, pt. 1, 1957, pt. 2, 1965; T. A. Henry, *The Plant Alkaloids*, 4th ed., 1949; R. H. F. Manske and H. L. Holmes (eds.), *The Alkaloids*, 5 vols., 1950–1955.

Alkane

A member of the series of saturated aliphatic hydrocarbons having the empirical formula C_nH_{2n+2}. The members of this series are also called paraffinic hydrocarbons or simply paraffins. Alkanes are relatively unreactive compared to unsaturated and aromatic hydrocarbons, and are little used in laboratory synthesis. On the other hand, reactions of alkanes are carried out on a huge scale industrially to convert the hydrocarbons found in natural gas and petroleum to the products needed in modern life.

Nomenclature. The alkanes are usually named in accordance with the rules of the International Union of Pure and Applied Chemistry (IUPAC) which adopted the ending "-ane" for saturated hydrocarbons. Branched-chain paraffins are named as derivatives of the longest straight chain in the compound, the location and name of the alkyl radicals in the branches being indicated as prefixes. The carbon atoms in the longest, or parent chain are numbered so that the positions of the branches have the lower of the two possible values. By this system the two isomeric heptanes having a single branch containing a single carbon atom are named 2-methylhexane and 3-methylhexane, as shown in formulas I–III.

$CH_3CH_2CH_2CH_2CH_2CH_2CH_3$
Heptane or *n*-heptane (I)

$CH_3CH(CH_3)CH_2CH_2CH_2CH_3$
2-Methylhexane
or isoheptane (II)

$CH_3CH_2CH(CH_3)CH_2CH_2CH_3$
3-Methylhexane
(no trivial name) (III)

In an alternative, trivial system of naming alkanes, the straight-chain compound is designated with the prefix "*n*-" (an abreviation for normal) and the isomer having a methyl branch on the second carbon atom is given the prefix "iso-" (for instance, isobutane, isopentane, isohexane). In this connection it should be noted that, although the name isooctane correctly designates 2-methylheptane, it should be avoided because of the unfortunate use of the misnomer "isooctane" in the petroleum industry to represent 2,2,4-trimethylpentane.

The prefix "neo-" indicates the *gem*-dimethyl isomer (two methyl branches on the same carbon atom), but can be used unambiguously only in naming neopentane (dimethylpropane) and neohexane (2,2-dimethylbutane).

A selected list of *n*-alkanes is shown in the table.

For convenience in later discussion, the symbol R is used to represent the alkyl group, and X represents a halogen. Thus RH is the general representation of any alkane, and RX is an alkyl halide.

Physical properties. Alkanes containing fewer than 5 carbon atoms are gases at room temperature and atmospheric pressure. The pentanes through the hexadecanes are mostly liquids (some exceptions: dimethylpropane is gaseous; tetramethylbutane, crystalline). Straight-chain alkanes having more than 16 carbon atoms are waxy solids; paraffin wax consists largely of *n*-alkanes containing 20–30 carbon atoms. The boiling points and the melting points of the normal paraffins increase with increasing molecular weight, and the boiling point decreases with increased branching within a family of isomeric alkanes.

The alkanes have lower refractive indices and densities than do the other types of hydrocarbons having the same number of carbon atoms.

The alkanes are soluble in many organic solvents such as alcohol and ether. They are practically insoluble in water.

Occurrence. Natural gas and petroleum contain literally hundreds of alkanes, both straight-chain and branched, and a number of pure alkanes can be isolated from petroleum by careful fractionation. As shown in the table, the number of possible isomers increases rapidly with the number of carbon atoms. Beyond the nonanes and decanes, relatively few branched isomers have been isolated from natural sources or prepared synthetically and characterized, since there has been no particular reason for doing so.

Laboratory synthesis. A number of laboratory procedures are available for the preparation of those alkanes which are not obtainable in pure form from natural sources. The classical Wurtz reaction, represented by reaction (1), often gives poor yields and is of limited utility.

$$\underset{\text{Alkyl halide}}{2RX} + 2Na \rightarrow \underset{\text{Alkane}}{R{-}R} + 2NaX \quad (1)$$

The Kolbe synthesis, consisting of the anodic oxidation of a carboxylic acid, is useful for the synthesis of symmetrical alkanes having an even number of carbon atoms [reaction (2)]. *See* KOLBE HYDROCARBON SYNTHESIS.

$$2RCOOH{-}2e^- \rightarrow R{-}R + 2CO_2 + 2H^+ \quad (2)$$

Branched-chain alkanes may be synthesized by hydrogenation of the corresponding alkene. Alkenes are prepared directly or via the synthesis of

Some *n*-alkanes and their characteristics*

No. of carbons	Formula	Name	Total isomers possible	bp, °C	mp, °C
1	CH_4	Methane	1	−162	−183
2	C_2H_6	Ethane	1	−89	−172
3	C_3H_8	Propane	1	−42	−187
4	C_4H_{10}	Butane	2	0	−138
5	C_5H_{12}	Pentane	3	36	−130
6	C_6H_{14}	Hexane	5	69	−95
7	C_7H_{16}	Heptane	9	98	−91
8	C_8H_{18}	Octane	18	126	−57
9	C_9H_{20}	Nonane	35	151	−54
10	$C_{10}H_{22}$	Decane	75	174	−30
11	$C_{11}H_{24}$	Undecane	—	196	−26
12	$C_{12}H_{26}$	Dodecane	—	216	−10
20	$C_{20}H_{42}$	Eicosane	366,319	334	+36
30	$C_{30}H_{62}$	Tricontane	4.11×10^9	446	+66

*From N. L. Allinger et al., *Organic Chemistry*, 2d ed., Worth Publishers, 1976.

a simple compound (usually an alcohol) having the desired carbon skeleton. Thus, the reaction of the Grignard reagent, RMgX, with an allylic halide, such as $CHR'{=}CHCH_2X$, will yield an alkene, $RCH_2CH{=}CHR'$ [reaction (3)]. Alternatively,

$$CH_3-\underset{\underset{CH_3}{|}}{\overset{\overset{CH_3}{|}}{C}}-MgCl + CH_2{=}CH-CH_2Cl \rightarrow$$

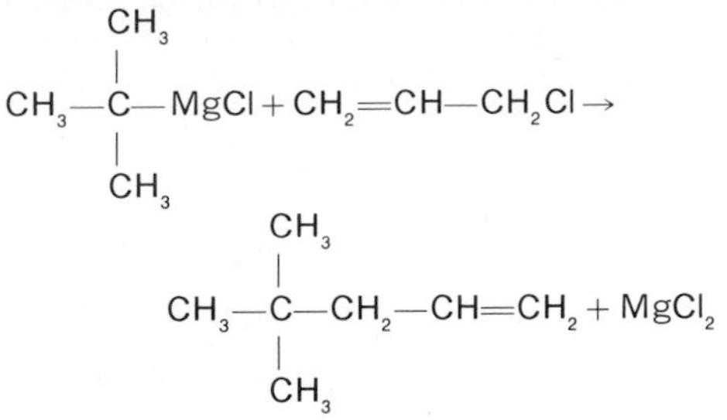

$$CH_3-\underset{\underset{CH_3}{|}}{\overset{\overset{CH_3}{|}}{C}}-CH_2-CH{=}CH_2 + MgCl_2 \qquad (3)$$

the reaction of the Grignard reagent with an aldehyde, ketone, ester, or epoxide yields an alcohol, which is then dehydrated to the alkene.

Industrial synthesis. The petroleum industry has been primarily responsible for the development of reactions for the conversion of the hydrocarbons naturally present in petroleum to other, more useful alkanes.

Alkylation. Alkanes may be condensed with alkenes to yield higher-molecular-weight branched-chain alkanes under both thermal and catalytic conditions. Thermal alkylation takes place at high temperatures (about 500°C) and pressures (about 150–300 atm or 15–30 MPa). Catalytic alkylation proceeds at much lower temperatures and pressures. The temperature is usually in the range from about −30 to about 100°C, depending on the catalyst used; the pressure need only be sufficient to keep the reactants in the liquid phase.

The typical difference between thermal and catalytic alkylation is illustrated by reaction (4).

$$CH_3-\underset{\underset{CH_3}{|}}{\overset{\overset{CH_3}{|}}{CH}} + CH_2 = CH_2$$

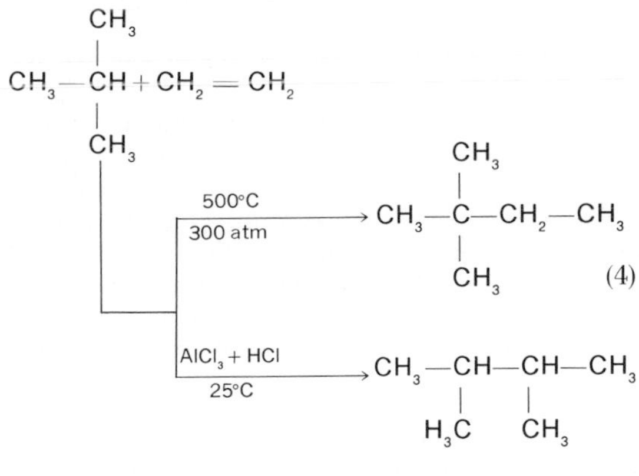

(4)

500°C, 300 atm → $CH_3-\underset{\underset{CH_3}{|}}{\overset{\overset{CH_3}{|}}{C}}-CH_2-CH_3$

$AlCl_3 + HCl$, 25°C → $CH_3-\underset{\underset{H_3C}{|}}{CH}-\underset{\underset{CH_3}{|}}{CH}-CH_3$

The catalytic alkylation of isobutane with gaseous olefins offers an economical means for the production of high-octane motor fuel from refinery gases.

Isomerization. Alkanes undergo conversion into isomeric, more highly branched alkanes when treated with certain catalysts such as aluminum chloride or certain group VIII metals on an oxide support. Sulfuric acid catalyzes the isomerization of alkanes containing tertiary carbon atoms, but not of those containing only primary, secondary, and quaternary carbon atoms. The principal reaction is the shift of an alkyl group (usually methyl) along the carbon chain.

Isomerization is a commercially important reaction for the production of isobutane from butane

for the catalytic alkylation process and for the conversion of low-octane-number normal paraffins to high-octane-number isoparaffins.

Fischer-Tropsch synthesis. The reaction of carbon monoxide and hydrogen in the presence of a cobalt, iron, or ruthenium catalyst at temperatures in the range 170–325°C and pressure of 1–200 atm (100 kpa–20 MPa), depending on the catalyst, yields predominantly straight-chain alkanes containing 3–20 carbon atoms. *See* FISCHER-TROPSCH PROCESS.

Chemical properties. The alkanes, particularly the straight-chain isomers, unlike the alkenes and aromatic hydrocarbons, are not readily affected at mild temperature conditions by acids and oxidizing agents, such as sulfuric acid, nitric acid, and potassium permanganate. Because they are saturated hydrocarbons, they react chiefly by substitution of other atoms or groups of atoms for hydrogen atoms. They also undergo reactions involving scission, or splitting, of carbon-carbon or carbon-hydrogen bonds.

Halogenation. A mixture of mono- and polyhalogenated alkanes is formed by the action of chlorine and bromine on alkanes when the reaction is carried out at high temperatures (thermal reaction) or under irradiation with light of short wavelength (photochemical reaction). Hydrogen halide is formed as a by-product. The halogens react very slowly with alkanes in the dark at room temperature.

The reaction of chlorine with methane is carried out at 400°C industrially to produce a mixture of chloromethane, dichloromethane, trichloromethane, and carbon tetrachloride [reaction (5)], where

$$CH_4 + Cl_2 \rightarrow$$
$$CH_3Cl + CH_2Cl_2 + CHCl_3 + CCl_4 + HCl \qquad (5)$$

the yield of products depends on the ratio of chlorine to methane.

The direct fluorination of alkanes is a highly exothermic reaction, and special apparatus and procedures are necessary to control the reaction. Alkanes are not iodinated when treated with iodine.

Nitration. Alkanes containing tertiary carbon atoms react with dilute nitric acid at 105–110°C (liquid-phase nitration) to yield nitroalkanes [reaction (6)].

$$CH_3-\underset{\underset{CH_3}{|}}{\overset{\overset{H}{|}}{C}}-CH_2-CH_3 + HNO_3 \rightarrow$$

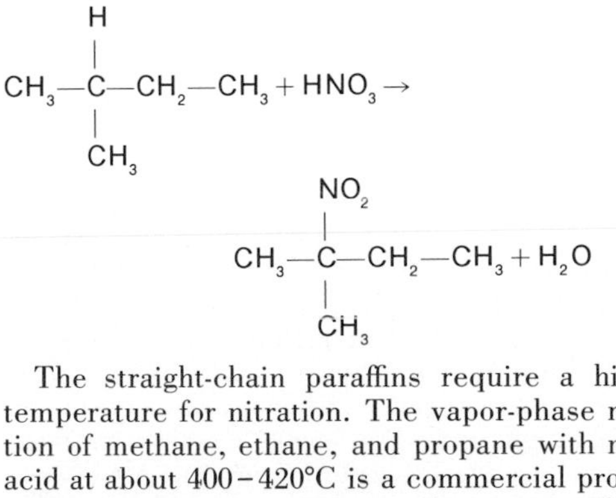

$$CH_3-\underset{\underset{CH_3}{|}}{\overset{\overset{NO_2}{|}}{C}}-CH_2-CH_3 + H_2O \qquad (6)$$

The straight-chain paraffins require a higher temperature for nitration. The vapor-phase nitration of methane, ethane, and propane with nitric acid at about 400–420°C is a commercial process for the manufacture of nitromethane, nitroethane, and 1-nitropropane and 2-nitropropane, which are useful solvents and chemical intermediates. All four compounds are obtained when propane is nitrated. The average composition of the nitration product is about 25% nitromethane, 10% nitroethane, 25% l-nitropropane, and 40% 2-nitropropane, and these compounds are readily separated by distillation.

Sulfonation. Concentrated sulfuric acid does not react appreciably with alkanes at ordinary temperature. On the other hand, fuming sulfuric acid (oleum) reacts with the higher-molecular-weight alkanes, particularly those containing tertiary carbon atoms, yielding sulfonic acids [reaction (7)].

$$RH + H_2SO_4 \cdot SO_3 \rightarrow RSO_3H + H_2SO_4 \qquad (7)$$

A preferred method for obtaining alkanesulfonic acids from alkanes consists in treating the hydrocarbon with a mixture of chlorine and sulfur dioxide [reaction (8)].

$$RH + Cl_2 + SO_2 \rightarrow RSO_2Cl + HCl \qquad (8a)$$

$$RSO_2Cl + H_2O \rightarrow RSO_3H + HCl \qquad (8b)$$

The alkanesulfonic acids are water-soluble strong acids. The sodium salts of the higher-molecular-weight compounds are used as wetting agents and detergents.

Oxidation. The reaction of alkanes with oxygen, yielding carbon dioxide and water, is highly exothermic, an important property in their use as fuels. The heat of combustion of methane (the amount of energy evolved per mole of methane burned) is 212.8 kcal/mole (890 kJ/mole). The heat of combustion increases with increase in molecular weight of the alkane, the increase for each additional CH_2 group being fairly constant at about 156 kcal (653 kJ).

If the supply of oxygen is deficient, the oxidation of the alkane will produce carbon monoxide or even carbon (in the form of carbon black). The oxidation can also be controlled to produce mixtures of alcohols, aldehydes, ketones, and acids; thus, methanol, ethanol, propanols, formaldehyde, acetaldehyde, acetone, acetic acid, and propionic acid are obtained by the air-oxidation of the propane-butane fraction of natural gas. Oxidation of paraffin wax produces (besides some aldehydes and ketones) higher-molecular-weight acids which react with alkali to produce soap.

Decomposition. All alkanes decompose (crack) when heated to high temperatures. Methane is converted to carbon and hydrogen when heated above about 900°C. Ethane yields ethylene and hydrogen at temperatures above about 600°C. Similarly, propane is dehydrogenated to yield propene and hydrogen; however, the chief reaction is one involving scission of a carbon-carbon bond, producing methane and ethylene. The higher alkanes also give products of both dehydrogenation and carbon-carbon scission, ethylene being a major product of their thermal decomposition in the range 500–650°C.

The decomposition of higher alkanes may be carried out in the presence of a catalyst, such as a silica-alumina composite. Then cracking occurs at a lower temperature (400–500°C), and the product contains less ethylene and larger amounts of propene and higher alkenes, as well as more branched-chain hydrocarbons, than are formed in thermal cracking.

Metals such as vanadium, nickel, cobalt, and iron catalyze decomposition of alkanes to carbon and hydrogen.

Dehydrogenation. The thermal dehydrogenation of alkanes of less than six carbon atoms, accompanied by little if any cracking, can be accomplished by use of catalysts, such as the oxides of chromium, molybdenum, or vanadium on alumina at temperatures in the range 450–550°C. Higher alkanes undergo cyclization and dehydrogenation, yielding aromatic hydrocarbons. This sequence of reactions is the basis of the commercial method for the preparation of the important solvent and chemical intermediate toluene [reaction (9)].

$CH_3CH_2CH_2CH_2CH_2CH_2CH_3 \rightarrow$
Heptane

 $—CH_3 + 4H_2$ (9)

Toluene

See ALIPHATIC HYDROCARBON; ALKYLATION; CRACKING; ETHANE; HALOGENATION; HYDROGENATION; ISOMERIZATION; METHANE; NITRATION; OCTANE; PROPANE.

[PAUL E. FANTA]

Bibliography: N. L. Allinger et al., *Organic Chemistry*, 2d ed., 1976.

Alkanolamine

One of a group of viscous, water-soluble amino alcohols of the aliphatic series. They are commercially available in the low molecular weights containing two to six carbons. Ethylene oxide adds to ammonia to yield:

Ethanolamine, $HO—CH_2—CH_2—NH_2$
Diethanolamine, $(HO—CH_2—CH_2)_2NH$
Triethanolamine, $(HO—CH_2—CH_2)_3N$

Propylene oxide gives a similar series. Reduction of nitroalcohols yields homologs of ethanolamine itself. Long-chain fatty acid salts of alkanolamines, soluble in both water and hydrocarbons, act as soaps, wetting agents, or emulsifying agents for insecticide sprays, lubricating coolants, polishes, and waxes. Triethanolamine is an excellent corrosion inhibitor for antifreeze solutions. Alkanolamines readily absorb acidic gases such as CO_2 and H_2S, and these gases may be recovered by heating. The trimethylammonium salt of ethanolamine (choline) is a representative constituent of a class of biological substances known as phospholipids, including the lecithins. *See* AMINE; PHOSPHATIDE; SURFACE-ACTIVE AGENT.

[LEALLYN B. CLAPP]

Alkene

One of the class of acyclic hydrocarbons containing one or more carbon-to-carbon double bonds. Alkenes (also called olefins) and alkynes (also called acetylenes) together constitute the family of organic compounds called unsaturated hydrocarbons, since they contain less than the number of hydrogens found in the corresponding saturated compound, alkane. When the double bond is present in a nonaromatic ring (alicyclic hydrocarbon), the compound is termed a cycloalkene. Hydrocarbons containing more than one double bond are termed dienes, trienes, and so forth, or collectively, polyenes. *See* ALICYCLIC HYDROCARBON; ALKANE; ALKYNE.

Nomenclature. In naming alkenes by the system of the International Union of Pure and Applied Chemistry (IUPAC), the longest chain containing the double bond is identified. The presence of the

Alkenes, dienes, and halogen derivatives

Name	Formula
Ethene (ethylene)	$CH_2{=}CH_2$
Propene (propylene)	$CH_2{=}CHCH_3$
1-Butene	$CH_2{=}CHCH_2CH_3$
2-Butene	$CH_3CH{=}CHCH_3$
2-Methylpropene (isobutylene)	$CH_2{=}\overset{CH_3}{\overset{\vert}{C}}CH_3$
1, 3-Butadiene	$CH_2{=}CHCH{=}CH_2$
2-Methyl-1, 3-butadiene (isoprene)	$CH_2{=}\overset{CH_3}{\overset{\vert}{C}}CH{=}CH_2$
(Chloroprene)	$CH_2{=}\overset{Cl}{\overset{\vert}{C}}CH{=}CH_2$
(Vinyl chloride)	$CHCl{=}CH_2$
Trichloroethylene (Trilene)	$CHCl{=}CCl_2$
Tetrachloroethylene (Perclene)	$CCl_2{=}CCl_2$
Tetrafluoroethylene	$CF_2{=}CF_2$

double bond is indicated by changing the "-ane" ending of the alkane having the same number of carbon atoms to "-ene," and the position of the double bond is indicated by a prefixed number. Examples are given in the table, with common or nonsystematic names which are still frequently used given in parentheses.

Physical properties. The lower alkenes and dienes which have up to five carbon atoms are gases at room temperature and pressure. Higher alkenes are colorless liquids or solids. Like other hydrocarbons, alkenes are insoluble in water. Liquid alkenes have specific gravities well below 1.0.

Occurrence and preparation. The commercially important alkenes are produced on a large scale in the petroleum industry by thermal or catalytic cracking processes.

In the laboratory, most methods for the preparation of alkenes involve some type of elimination reaction, in which atoms or groups on adjacent carbon atoms are removed with concomitant formation of the carbon-carbon double bond [reaction (1)].

$$-\underset{X}{\overset{|}{\underset{|}{C}}}-\underset{Y}{\overset{|}{\underset{|}{C}}}- \rightarrow \;>C{=}C<\; + XY \qquad (1)$$

Specific examples of elimination reactions leading to the formation of alkenes include dehydration of alcohols, pyrolysis of esters, dehydrohalogenation of alkyl halides, dehalogenation of *vic*-dihalides, thermal decomposition of xanthate esters (Chugaev reaction), and pyrolysis of quaternary bases, as well as the large-scale industrial cracking and dehydrogenation processes.

An elimination reaction of special industrial importance is the production of vinyl chloride, a principal component of polyvinyl resins, by the dehydrochlorination of 1,2-dichloroethane. Chlorine is added to ethylene, followed by elimination of hydrogen chloride in a single operation [reaction (2)]. The by-product, hydrogen chloride, is recycled by a catalytic oxychlorination process [reaction (3)].

$$CH_2{=}CH_2 + Cl_2 \rightarrow [ClCH_2CH_2Cl] \rightarrow CH_2{=}CHCl + HCl \qquad (2)$$

$$CH_2{=}CH_2 + 2HCl + \tfrac{1}{2}O_2 \xrightarrow{\text{catalyst}} ClCH_2CH_2Cl + H_2O \qquad (3)$$

Another commercially important monomer, vinylidene chloride, is prepared by dehydrochlorination of 1,1,2-trichloroethane [reaction (4)].

$$ClCH_2CHCl_2 \xrightarrow{Ca(OH)_2} CH_2{=}CCl_2 \qquad (4)$$

The solvents Trilene (trichloroethylene) and Perclene (tetrachloroethylene), much used in the dry cleaning industry, are prepared in a similar way.

Chemical properties. Alkenes may undergo polymerization, cyclization, and addition reactions. A major share of structural and elastic polymers are based on homopolymers or copolymers of alkenes and dienes. Thus, polyethylene is made today by means of high-pressure peroxide-catalyzed processes or by low-pressure methods using Ziegler-type catalysts. Similar to polyethylene is polypropylene made with Ziegler-type catalysts. Polystyrene is made by emulsion polymerization of styrene, and a similar process is used for the production of GR-S synthetic rubber from styrene and butadiene. Butyl rubber from isobutylene and butadiene or isoprene is made by means of aluminum chloride at low temperature. Polybutenes made by aluminum chloride or boron fluoride polymerization of isobutylene or isobutylene-butene mixtures vary in consistency from oils to rubbery polymers. An exact duplicate of the stereochemical structure of polyisoprene in natural rubber has been made with lithium catalysts. Synthetic drying oils have been made from butadiene and butadiene-styrene copolymers with metallic sodium catalysts.

Alkenes and dienes cyclize readily under various conditions. *See* DIELS-ALDER REACTION.

Addition reactions of alkenes are among the most important in the entire field of organic chemistry. Industrially, high-octane gasoline is made by the acid-catalyzed alkylation of the three- and four-carbon alkenes. A variety of alkylated aromatics are made by the alkylation of benzene with olefins, with use of aluminum chloride, hydrogen fluoride at low or moderate temperature, or phosphoric acid on inert carriers at high temperature and pressure. Thus, ethylbenzene is made on a tonnage basis and converted to styrene (I) by catalytic dehydrogenation processes. Cumene (II) for the production of phenol and acetone through the hydroperoxide is made by the catalytic alkylation of benzene with propylene.

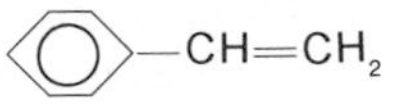

(I)

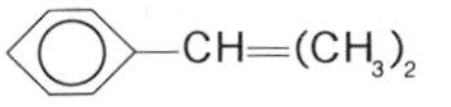

(II)

Oxidation of olefins with air, oxygen, or peracids may occur at the double bond to give epoxides or with ozone to give ozonides which cleave under oxidizing or reducing conditions. With selected reagents such as selenium dioxide and lead tetraacetate, oxidation may take place at a methylene group adjacent to the double bond to give unsaturated ketones or unsaturated hydroxy esters. *See* EPOXIDATION; OZONOLYSIS.

Sulfur, sulfur dioxide, sulfur trioxide, sulfuric acid, hydrogen sulfide, bisulfites, thiocyanogen, and sulfur monochloride add to olefins with varying degrees of ease to give a variety of useful products.

Halogens add to olefins across the double bond at low temperatures, but at high temperatures they may give substitution products. Thus, propylene when chlorinated at 500–600°C yields mainly allyl chloride. Halogenated paraffins add to the double bond in the presence of Friedel-Crafts catalysts. Perhalomethanes may add under the influence of free-radical reagents. Hypochlorous acid adds to olefins to yield chlorohydrins, which may be converted to epoxides or glycols. Hydrogen adds to monoolefins under the influence of catalysts. The oxo reaction for the synthesis of aldehydes and alcohols from olefins, carbon monoxide, and hydrogen is commercially important.

Reactions of conjugated dienes. Conjugated dienes are unique in that both ethylenic linkages in general react as a single unit, in contradistinction to the cumulative or isolated dienes which can react in an independent manner. This property is best illustrated by addition reactions, in which the two components of a reagent add to the terminal carbons of the conjugated diene with the formation of a double bond at the site of the former single bond. A generalized reaction is represented by reaction (5). While 1,4 additions of this type

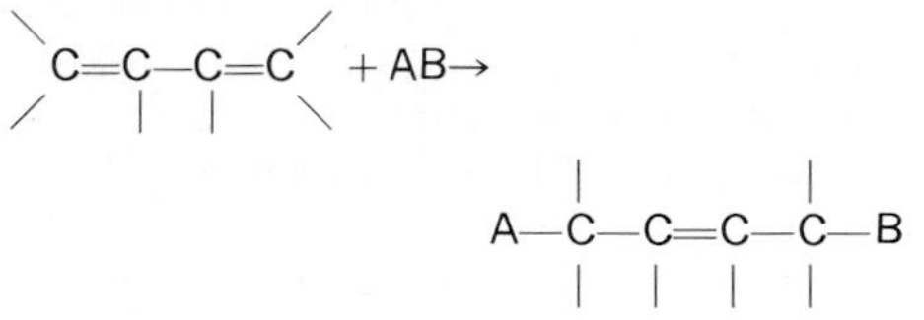

(5)

(termed conjugate additions) are the most common for a conjugated diene, they are not exclusive since many 1,2 additions are known.

Cyclizations involving 1,4 addition occur readily with a variety of reagents. For example, butadiene yields pyrrole with nitrous oxide in the presence of a cadmium-aluminum catalyst, thiophene with hydrogen sulfide, and 3-sulfolene with sulfur dioxide. The Diels-Alder reaction is a unique reaction of conjugated dienes involving 1,4 addition. *See* ALIPHATIC HYDROCARBON; ALKYLATION; CRACKING; HALOGENATION; HYDROGENATION.

[PAUL E. FANTA]

Bibliography: N. L. Allinger et al., *Organic Chemistry*, 2d ed., 1976.

Alkylation

A chemical process in which an alkyl radical is introduced into an organic compound by substitution or addition. By the use of alkylating agents, such as olefins, alcohols, or alkyl halides, alkyl groups are bonded together or to other compounds through carbon, oxygen, nitrogen, sulfur, or metals. Though the general terms addition and substitution cover all alkylation processes, the detailed reactions can be quite dissimilar. Some reactions classified as alkylations are the Williamson ether synthesis, the Friedel-Crafts reaction, and the Wurtz reaction. Some alkylation reactions are shown in reactions (1)–(6).

Alkyl bound to carbon:

$$\underset{\text{Ethylene}}{C_2H_4} + \underset{\text{Benzene}}{C_6H_6} \rightarrow \underset{\text{Ethylbenzene}}{C_6H_5C_2H_5} \quad (1)$$

Alkyl bound to oxygen:

$$\underset{\text{Ethyl alcohol}}{2C_2H_5OH} \rightarrow \underset{\text{Ethyl ether}}{(C_2H_5)_2O} + \underset{\text{Water}}{H_2O} \quad (2)$$

$$\underset{\text{Methyl iodide}}{CH_3I} + \underset{\text{Sodium ethylate}}{C_2H_5ONa} \rightarrow \underset{\text{Methyl ethyl ether}}{CH_3OC_2H_5} + \underset{\text{Sodium iodide}}{NaI} \quad (3)$$

Alkyl bound to nitrogen:

$$\underset{\text{Ethyl chloride}}{2C_2H_5Cl} + \underset{\text{Ammonia}}{3NH_3} \rightarrow \underset{\text{Diethylamine}}{(C_2H_5)_2NH} + \underset{\text{Ammonium chloride}}{2NH_4Cl} \quad (4)$$

Alkyl bound to sulfur:

$$\underset{\text{Ethylene oxide}}{2C_2H_4O} + \underset{\text{Hydrogen sulfide}}{H_2S} \rightarrow \underset{\text{Thiodiglycol}}{S(C_2H_4OH)_2} \quad (5)$$

Alkyl bound to metal:

$$\underset{\text{Ethyl chloride}}{4C_2H_5Cl} + \underset{\text{Sodium-lead alloy}}{4NaPb} \rightarrow \underset{\text{Tetraethyllead}}{(C_2H_5)_4Pb} + \underset{\text{Sodium chloride}}{4NaCl} + \underset{\text{Lead}}{3Pb} \quad (6)$$

Many important chemicals are made by alkylation, some being used directly, others as intermediates for the manufacture of other chemical compounds. Ethylbenzene, for example, is dehydrogenated to make styrene, important in rubber and plastic manufacture. Ethyl ether is the most widely known anesthetic. Tetraethyllead is an important antiknock additive for gasoline. Many other applications could be mentioned, including some of the synthetic detergents (the alkylaryl sulfonates). *See* FRIEDEL-CRAFTS REACTION; ORGANIC REACTION MECHANISM; SUBSTITUTION REACTION; WURTZ-FITTIG REACTION.

[HAROLD C. RIES]

Bibliography: L. F. Albright and A. R. Goldsby (eds.) *Industrial and Laboratory Alkylations*. 1977; R. N. Shreve, *Chemical Process Industries*, 4th ed., 1977.

Alkyne

One of a group of organic compounds containing a carbon-to-carbon triple-bond linkage (—C≡C—). They are termed acetylenes or alkynes. While exhibiting many of the characteristics of alkenes as regards unsaturation, the acetylenes have many unique properties. Since the bonding in alkyne molecules is linear, R—C≡C—R, cis-trans isomerism is not possible. In the simplest alkyne, acetylene (HC≡CH), or in monosubstituted acetylenes, the hydrogen attached to triply bonded carbon is acidic to such a degree that it is replaceable with metals such as sodium. Structural formulas of several alkynes are as follows:

$CH{\equiv}CH$ Ethyne or acetylene	$CH{\equiv}CH_2{-}CH_2{-}CH_3$ 1-Butyne
$CH{\equiv}C{-}CH_3$ Propyne	$CH_3{-}C{\equiv}C{-}CH_3$ 2-Butyne

Synthesis. General methods for the preparation of alkynes depend on dehydrohalogenation of α,β-dihaloparaffins, reaction (1); conversion of

$$\mathrm{R{-}\underset{\displaystyle X}{\overset{\displaystyle H}{C}}{-}\underset{\displaystyle X}{\overset{\displaystyle H}{C}}{-}H \rightarrow R{-}C{\equiv}C{-}H + 2HX} \quad (1)$$

aldehydes or ketones to dihaloparaffins with subsequent dehydrohalogenation, reaction (2); and

$$\mathrm{R{-}CH_2CHO \rightarrow R{-}CH_2CHX_2 \rightarrow R{-}C{\equiv}CH} \quad (2a)$$

$$\mathrm{R{-}COCH_3 \rightarrow R{-}CX_2CH_3 \rightarrow R{-}C{\equiv}CH} \quad (2b)$$

alkylation of metallic acetylides with alkyl halides in liquid ammonia, reaction (3). Grignard reagents of 1-alkynes behave similarly.

$$\mathrm{R{-}C{\equiv}C{-}M + R'X \rightarrow R{-}C{\equiv}C{-}R' + MX} \quad (3)$$

Reactions. The reactions of triply bonded carbon compounds are in general similar to those of compounds containing ethylenic bonds. Addition reactions proceed in two stages to form first a vinyl compound or substituted ethylene, and second the substituted paraffin for example, 1,1-dichloroethane in reaction (5). Thus, hydrogen chloride and hydrogen cyanide add to acetylene to form the technically important intermediates, vinyl chloride, reaction (4), and acrylonitrile, reaction (6). In the

$$\underset{\text{Acetylene}}{\mathrm{HC{\equiv}CH}} + \underset{\text{Hydrogen chloride}}{\mathrm{HCl}} \rightarrow \underset{\text{Vinyl chloride}}{\mathrm{H_2C{=}CHCl}} \quad (4)$$

$$\mathrm{H_2C{=}CHCl + HCl} \rightarrow \underset{\text{1,1-Dichloroethane}}{\mathrm{H_3C{-}CHCl_2}} \quad (5)$$

$$\mathrm{HC{\equiv}CH} + \underset{\text{Hydrogen cyanide}}{\mathrm{HC{\equiv}N}} \rightarrow \underset{\text{Acrylonitrile}}{\mathrm{H_2C{=}CHC{\equiv}N}} \quad (6)$$

presence of catalytic quantities of alkoxides, acetylene adds to alcohols and phenols to give vinyl ethers. Reactions of this type, in which addition takes place by replacement of hydrogen with a vinyl group, are termed vinylation. With a zinc naphthenate catalyst, nuclear vinylation of phenol occurs. Koresin is a polymer made from the nuclear vinyl derivative of *p-tert*-butylphenol. In the presence of zinc or mercuric salts, carboxylic acids add to acetylene to give either vinyl esters or alkylidene dicarboxylates. Alkyl and aryl mercaptans vinylate readily as do compounds having the >NH structure. An interesting development of the latter reaction was the synthesis of a polyvinyl pyrrolidone known as Persiston, which was used as a blood-plasma substitute in Germany during World War II. In the presence of mercuric salts, acetylene hydrates to acetaldehyde, and substituted acetylenes yield ketones.

Another form of addition reaction known as ethynylation involves the addition of acetylene to unsaturated compounds. Addition of acetylene to aldehydes or ketones under certain conditions yields acetylenic carbinols or glycols, as in reactions (7) and (8).

$$\mathrm{R{-}\underset{\displaystyle H}{C}{=}O + HC{\equiv}CH \rightarrow R{-}\underset{\displaystyle OH}{CH}{-}C{\equiv}CH} \quad (7)$$

$$\mathrm{2\,{>}C{=}O + HC{\equiv}CH \rightarrow {>}\underset{\displaystyle OH}{C}{-}C{\equiv}C{-}\underset{\displaystyle OH}{C}{<}} \quad (8)$$

A general reaction of alkynes involves the addition of carbon monoxide and water, or other compounds having an active hydrogen, in the presence of nickel carbonyl. Hydrogen donors such as alcohols, carboxylic acids, amines, and mercaptans may add to yield acrylic derivatives.

Polymerization of alkynes may yield acyclic, aromatic, or alicyclic derivatives. In the presence of cuprous chloride, acetylene dimerizes to vinyl acetylene, which adds hydrogen chloride to give chloroprene (2-chloro-1,3-butadiene). Polymerization of chloroprene in the presence of free radical initiators gives the commercially important synthetic rubber, neoprene. Thermal polymerization of acetylene yields benzene as the major product and also a wide variety of polynuclear aromatic compounds. Of great theoretical and practical importance is the polymerization of acetylene to the cyclic tetramer, cyclooctatetraene. *See* ACETYLENE; ALKENE.

[CHARLES A. COHEN/PAUL E. FANTA]

Bibliography: N. L. Allinger et al., *Organic Chemistry*, 2d ed., 1976.

Alum

A colorless to white crystalline substance which occurs naturally as the mineral kalunite and is a constituent of the mineral alunite. Alum is produced as aluminum sulfate by treating bauxite with sulfuric acid to yield alum cake or by treating the bauxite with caustic soda to yield papermaker's alum. Other industrial alums are potash alum, ammonium alum, sodium alum, and chrom alum (potassium chromium sulfate). Major uses of alum are as an astringent, styptic, and emetic. For water purification alum is dissolved: it then crystallizes out into positively charged crystals that attract negatively charged organic impurities to form an aggregate sufficiently heavy to settle out. Alum is also used in sizing paper, dyeing fabrics, and tanning leather. With sodium bicarbonate it is used in baking powder and fire extinguishers. *See* ALUMINUM; COLLOID. [FRANK H. ROCKETT]

Aluminate

A negative ion usually given the formula AlO_2^- and derived from aluminum hydroxide. Aluminum hydroxide is an ampheteric substance and thus can react with a strong base, such as sodium hydroxide, to form sodium aluminate as in the equation. It

$$\mathrm{Al(OH)_3 + NaOH \rightleftharpoons NaAlO_2 + 2H_2O}$$

also reacts in the more usual manner with acids to form aluminum salts. Solutions of aluminates are strongly basic and the reaction in the equation is easily reversed even by weak acids to form the insoluble aluminum hydroxide. This is the basis for the commercial use of sodium aluminate in the clarification of water. *See* ALUMINUM; AMPHOTERISM. [E. EUGENE WEAVER]

Aluminum

A metallic chemical element, symbol Al, atomic number 13, atomic weight 26.98154, in group IIIA of the periodic system. Pure aluminum is

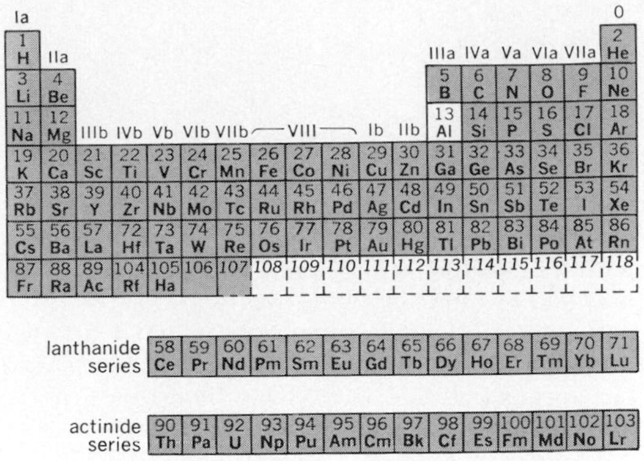
Ia
0
1 H
IIa
IIIa IVa Va VIa VIIa
2 He
3 Li 4 Be
5 B 6 C 7 N 8 O 9 F 10 Ne
11 Na 12 Mg
IIIb IVb Vb VIb VIIb — VIII — Ib IIb
13 Al 14 Si 15 P 16 S 17 Cl 18 Ar
19 K 20 Ca 21 Sc 22 Ti 23 V 24 Cr 25 Mn 26 Fe 27 Co 28 Ni 29 Cu 30 Zn 31 Ga 32 Ge 33 As 34 Se 35 Br 36 Kr
37 Rb 38 Sr 39 Y 40 Zr 41 Nb 42 Mo 43 Tc 44 Ru 45 Rh 46 Pd 47 Ag 48 Cd 49 In 50 Sn 51 Sb 52 Te 53 I 54 Xe
55 Cs 56 Ba 57 La 72 Hf 73 Ta 74 W 75 Re 76 Os 77 Ir 78 Pt 79 Au 80 Hg 81 Tl 82 Pb 83 Bi 84 Po 85 At 86 Rn
87 Fr 88 Ra 89 Ac 104 Rf 105 Ha 106 107 108 109 110 111 112 113 114 115 116 117 118
lanthanide series 58 Ce 59 Pr 60 Nd 61 Pm 62 Sm 63 Eu 64 Gd 65 Tb 66 Dy 67 Ho 68 Er 69 Tm 70 Yb 71 Lu
actinide series 90 Th 91 Pa 92 U 93 Np 94 Pu 95 Am 96 Cm 97 Bk 98 Cf 99 Es 100 Fm 101 Md 102 No 103 Lr

soft and lacks strength, but it can be alloyed with other elements to increase strength and impart a number of useful properties. Alloys of aluminum are light, strong, and readily formable by many metalworking processes; they can be easily joined, cast, or machined, and accept a wide variety of finishes. With a density about one-third that of ferrous alloys, aluminum is used in many structural applications and in transportation where weight saving is an important consideration. The usefulness of the metal is enhanced by its tendency to form an adherent oxide surface that resists corrosion. Because of high electrical conductivity and light weight, aluminum alloys are used extensively in electrical transmission lines. Aluminum reflects radiant energy throughout the spectrum and is odorless, tasteless, nontoxic, nonsparking, and nonmagnetic. Because of its many desirable physical, chemical, and metallurgical properties, aluminum has become the most widely used nonferrous metal.

Natural occurrence. Aluminum is the most abundant metallic element on the Earth and Moon but is never found free in nature. It makes up more than 8% of the solid portion of the Earth's surface. Sea water contains an average of only 0.5 ppm aluminum. The element is widely distributed in plants, where it may be present in significant concentrations, particularly in vegetation in marshy places and acid soils.

Nearly all rocks, particularly igneous rocks, contain aluminum in the form of aluminum silicate minerals. When subjected to weathering under the right conditions, these minerals go into solution; then, depending upon the chemical conditions, aluminum can be precipitated out of the solution as clay minerals or aluminum hydroxides, or both. Any quartz in the parent rock is relatively unattacked by the weathering processes and remains as sand in the deposit. Under conditions which precipitate clay minerals and leach out iron and the alkali and alkaline earth metals, deposits of high-grade kaolin clay can result. Sometimes natural elutriation or mechanical separation by washing occurs which separates the clay from the sand and enhances the purity of the deposit. Under conditions which precipitate aluminum principally as hydroxide minerals, and leach out alkali and alkaline earth metals completely and iron to varying degrees, bauxites are formed. It is also possible for clay minerals to be desilicified to aluminum hydroxides and for aluminum hydroxides to be resilicified to clay minerals.

Bauxites usually consist of mixtures of the following minerals in varying proportions: gibbsite, also known as hydrargillite [$Al(OH)_3$]; boehmite and diaspore [$AlO(OH)$]; clay minerals such as kaolinite [$Al_2Si_2O_5(OH)_4$]; quartz [SiO_2]; and anatase, rutile, and brookite [TiO_2]. Small amounts of magnetite [Fe_3O_4], ilmenite [$FeTiO_3$], and corundum [Al_2O_3] are sometimes present. In addition to the foregoing minerals, bauxites usually contain traces of other insoluble oxides such as those of zirconium, vanadium, gallium, chromium, and manganese. The term alumina trihydrate is often applied to the mineral gibbsite, and alumina monohydrate to the minerals boehmite and diaspore. These terms are misnomers because the minerals are not hydrates in the true sense of the word. However, for processing purposes, bauxites are often classified as trihydrate bauxites and monohydrate bauxites, depending upon their content of the principal mineral or minerals containing the extractable alumina. In general, European bauxites are of the monohydrate type and are geologically older than those in tropical countries, which occur generally as the trihydrate type. This, however, is not an absolute rule. Some European bauxites contain gibbsite along with boehmite and diaspore, and Caribbean bauxites contain small to medium amounts of boehmite along with the gibbsite. Typical bauxites which are used for aluminum production contain 40–60% total alumina (Al_2O_3), 1–15% total silica (SiO_2), 7–30% total Fe_2O_3, 1–5% titania (TiO_2), and 12–30% combined H_2O.

The suitability of a bauxite as a raw material for aluminum production depends not only on its alumina content but also on its content of combined silica, which is usually in the form of the mineral kaolinite. Kaolinite not only contains aluminum which cannot be extracted in the Bayer process, but it also reacts with the caustic-aluminate solution to cause a loss of caustic soda. In some bauxites the clay minerals are concentrated in the fine-particle size range. In this case, if the economics are favorable, the bauxite can be beneficiated by washing it on screens to remove substantial amounts of the clay and thus produce a product containing higher alumina and lower combined silica contents than the original bauxite (see Fig. 1).

History. H. C. Oersted probably prepared the first metallic aluminum (impure) in 1824 by reducing aluminum chloride with potassium amalgam. F. Wöhler is generally credited with the discovery of the metal by virtue of the first isolation of less impure aluminum in 1827 and the first description of a number of its properties. The metal remained a laboratory curiosity until 1854, when Henri Sainte-Claire Deville improved the earlier method of preparation by employing sodium as reductant and was successful in producing larger quantities of relatively pure metal. The first electrolytic preparation of aluminum, also in 1854, was accomplished by electrolysis of fused sodium aluminum chloride independently in the laboratories of Deville in France and R. Bunsen in Germany. *See* ELECTROLYSIS.

The present industrial electrolytic method of production was discovered simultaneously and independently by Charles Martin Hall in the United States and Paul-Louis Héroult of France in 1886. The essentials of their discovery remain the basis

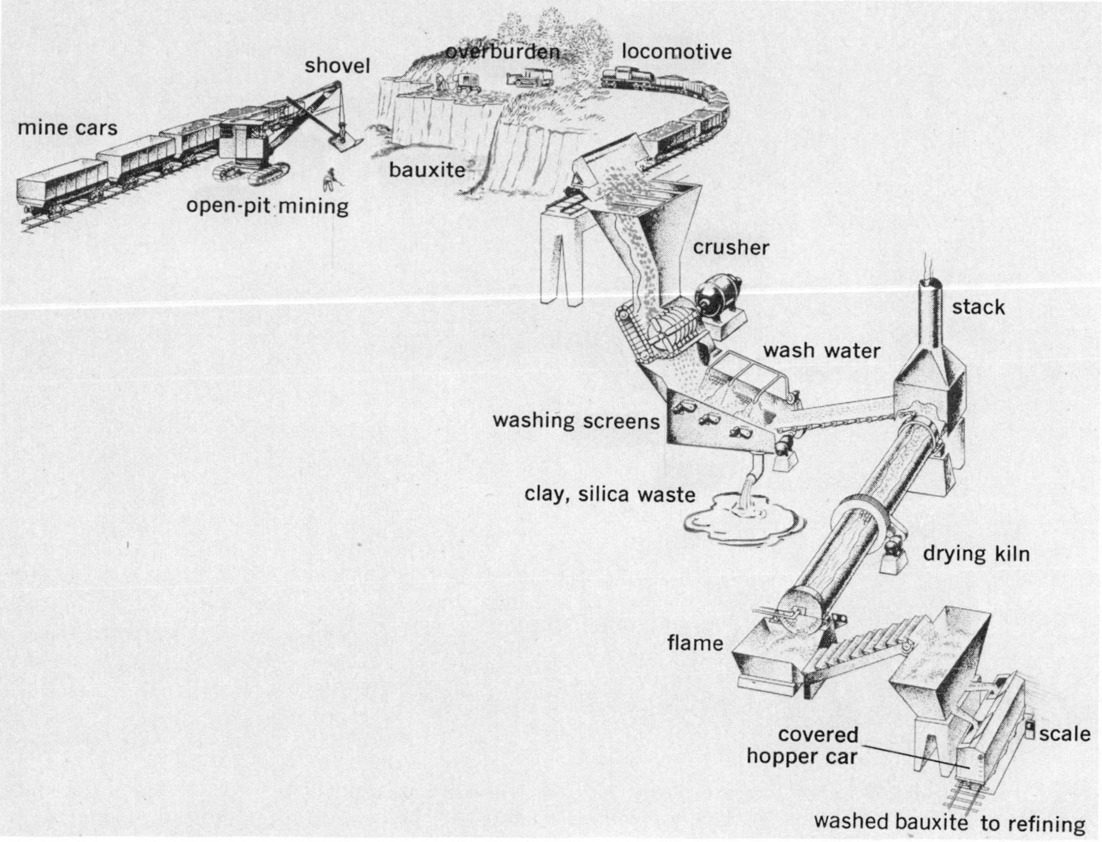

Fig. 1. Bauxite washing. (*Aluminum Company of America*)

for today's aluminum industry. Alumina dissolved in a fluoride fusion (largely cryolite, Na_3AlF_6) is electrolyzed with direct current. Carbon dioxide is discharged at the anode, while the metal is deposited on molten aluminum lying on the carbon lining at the bottom of the cell. The technology for extracting aluminum from its ores was further improved in 1888, when Karl Josef Bayer patented in Germany a method for making pure alumina from bauxite.

Production. The most widely used technology for producing aluminum involves two steps: extraction and purification of alumina from ores, and electrolysis of the oxide after it has been dissolved in fused cryolite. Bauxite is by far the most used raw material at the present time. Technically feasible processes exist for the extraction of alumina from other raw materials such as clays, anorthosite, nepheline syenite, and alunite; however, these processes have not been competitive with the Bayer process in the Western world. With the present rapidly changing economics of bauxite production, some of these processes may well become competitive in the near future. A few of these alternate raw materials are used as a source for alumina in Europe and Asia.

Alumina extraction. A general flow sheet for the Bayer process is shown in Fig. 2. In the process, bauxite is crushed and ground, then digested at elevated temperature (140–230°C) and pressure in a strong solution of caustic soda (80–110 g Na_2O/liter). For monohydrate-type bauxites, in which the alumina occurs in forms which are more difficult to dissolve than in trihydrate-type bauxites, stronger solutions (up to 220 g Na_2O/liter), higher temperatures (up to 300°C) and pressures (as high as 150 atm, or 15.2×10^6 N/m^2), and sometimes longer digestion times are required. The gibbsite, boehmite, or diaspore in the bauxites reacts with the caustic soda to form soluble sodium aluminate. The residue, known as red mud, contains the insoluble impurities and the sodium aluminum silicate compound, referred to as desilication product, formed by the reaction of clay minerals with the sodium aluminate–caustic soda solution. The red mud is separated from the solution by countercurrent decantation and filtration. After cooling, the solution is supersaturated with respect to alumina. It is seeded with recycled synthetic gibbsite (alumina trihydrate) and agitated. A large part of the alumina in solution thus crystallizes out as gibbsite. This gibbsite is classified into product and seed, the seed being recycled and the product washed. Wash water and spent liquor, after concentration by evaporation, are recycled to the digestion system. For metal production, the product is calcined at temperatures up to 1300°C to produce alumina containing about 0.3–0.8% soda, less than 0.1% iron oxide plus silica, and trace amounts of other oxides. Some of the precipitated gibbsite is sold for use in the chemicals industry. Specially precipitated pigment-grade gibbsite is used for rubber, plastic and paper fillers, and paper coating. Activated aluminas having high internal surface area are made from the gibbsite by calcination at low to moderate temperatures. They are used as desiccants and catalysts. Fully calcined aluminas are used in the production of ceramics, abrasives, and refractories.

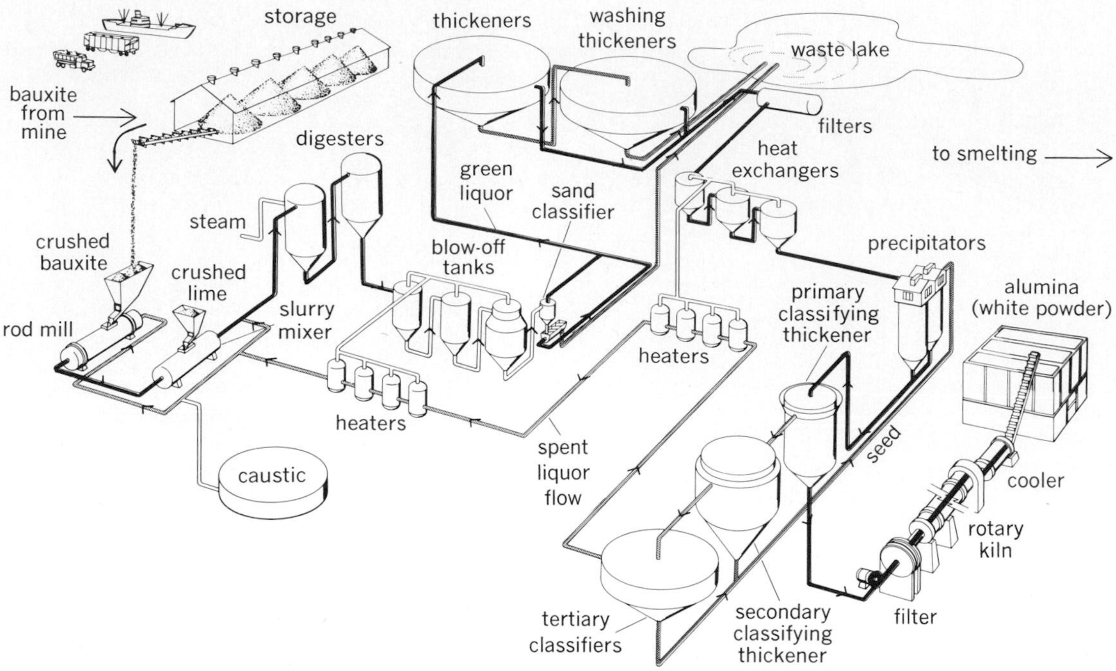

Fig. 2. Processing of alumina from bauxite. (*Aluminum Company of America*)

Alumina can be recovered from low-grade bauxites containing high concentrations of combined silica by the so-called combination or lime-soda sinter process. The process consists of treatment by the Bayer process, following which the red mud is mixed with calcium oxide and sodium carbonate and calcined. This treatment forms sodium aluminate from the aluminous phases in the red mud. The sodium aluminate is then leached out with water, and alumina recovered from the resulting solution.

Processes for treating other ores have been developed which consist of calcination or fusion of the ore with limestone to make calcium aluminate, from which alumina is leached out with sodium carbonate solution and recovered. Such a process is used in the Soviet Union for treating nepheline syenite to yield alumina, and cement as a by-product.

High-iron bauxites have been smelted with limestone in the Pedersen process to yield pig iron and a calcium aluminate slag which can be leached with sodium carbonate solution to recover alumina.

Electrolytic reduction (smelting). Although unchanged in principle, the smelting process of today differs in detail and in scale from the original process discovered by Hall and Héroult. Modern technology has effected substantial improvements in equipment, materials, and control of the process and has lowered the energy and labor requirements and the final cost of primary metal.

In a modern smelter (Fig. 3), alumina is dissolved in cells (pots)—rectangular steel shells lined with carbon—containing a molten electrolyte (bath) consisting mostly of cryolite. The bath usually contains 2–8% alumina. Excess aluminum fluoride and calcium fluoride are added to lower the melting point and to improve operation. Carbon anodes are hung from above the cells with their lower ends extending to within about 1.5 in. (3.8 cm) of the molten metal, which forms a layer under the molten bath. The heat required to keep the bath molten is supplied by the electrical resistance of the bath as current passes through it. The amount of heat developed with a given current depends on the length of the current path through the electrolyte, that is, anode-cathode distance, which is adjusted to maintain the desired operating temperature, usually 960–970°C. A crust of frozen bath, 1–3 in. (2.5–7.6 cm) thick, forms on the top surface of the bath and on the sidewalls of the cell. Alumina is added to the bath or on the crust, where its sorbed moisture is driven off by heat from the cell. While preheating on the crust, the alumina charge serves as thermal insulation. Periodically, the crust is broken and the alumina is stirred into the bath to maintain proper concentration.

The passage of direct current through the electrolyte decomposes the dissolved alumina. Metal is deposited on the cathode, and oxygen on the gradually consumed anode. About 0.5 lb (0.23 kg) of carbon is consumed for every pound of aluminum produced. The smelting process is continuous. Alumina is added, anodes replaced, and molten aluminum is periodically siphoned off without interrupting current to the cells.

Current efficiencies in the industrial electrolytic process are about 85–92%, and the energy efficiency is about 40%. The voltage at the cell terminals is 4–6 V, depending on the size and condition of the cell. Voltage is required to force the current through the entire cell, and the corresponding power (voltage × amperage) is largely converted into heat in the bath. The amount of power required to maintain the temperature is a smaller proportion of the total power input in large cells than in small ones because of the lower ratio of surface to volume. Thus, power consumed per pound of metal is somewhat less in large than in small cells. Consumption of 6–8 kWhr/lb (13–18 kWhr/kg) of aluminum produced includes bus-bar, transformer, and rectifier losses.

A potline may consist of 50–200 cells with a total line voltage of up to 1000 V at current loads of 50,000–225,000 A. Electric power is one of the most costly raw materials in aluminum production. Aluminum producers have continually searched for sources of cheap hydroelectric power, but have also had to construct facilities that produce power from fossil fuels. In the past half century, technological advances have significantly reduced the amount of electrical energy necessary to produce a pound of aluminum. In 1930 the requirement was 12 kWh. In the 1970s a single cell can produce about 1.5 tons (1400 kg) of metal per day and uses only 6 kWh/lb (13 kWh/kg).

Current is led out of the cell to the anode bus bar by a number of carbon block anodes suspended in parallel rows on vertical conducting rods of copper or aluminum. Because impurities in the anodes dissolve in the bath as they are consumed, pure carbon (calcined petroleum coke or pitch coke) is used as raw material. The ground coke is mixed hot with enough coal tar or petroleum pitch to bond it into a block when pressed in a mold to form the "green" anode. This is then baked slowly at temperatures up to 1100–1200°C. In a cavity molded in the top of each block, a steel stub is embedded by casting molten iron around it or by using a carbonaceous paste; the conducting bar is bolted to this stub. Such an electrode is termed a prebaked anode to distinguish it from the Soderberg anode, in which the electrode (single large anode to a cell) is formed in place from a carbonaceous paste which is baked by heat from the pot as it gradually descends into the electrolyte.

The steel shell of the pot is thermally insulated and lined with carbon. The carbon bottom, covered with molten aluminum, serves as cathode of the cell. Electrical connection is made to the carbon cathode by steel bars running through the base of the cell and embedded in the carbon lining.

Molten aluminum is siphoned from the smelting cells into large crucibles. From there the metal may be poured directly into molds to produce foundry ingot, or transferred to holding furnaces for further refining or alloying with other metals to form fabricating ingot. As it comes from the cell, primary aluminum averages about 99.8% purity.

Melting. In plants not adjacent to a smelter, it is necessary to remelt charges, usually consisting of mill scrap or returned scrap with enough primary metal to provide composition control. For casting ingot for fabrication, gas- or oil-fired reverberatory furnaces are most commonly used. Molten metal is generally transferred from the melting furnace to a reverberatory holding furnace or to a ladle for subsequent casting. Metal for foundry use in produc-

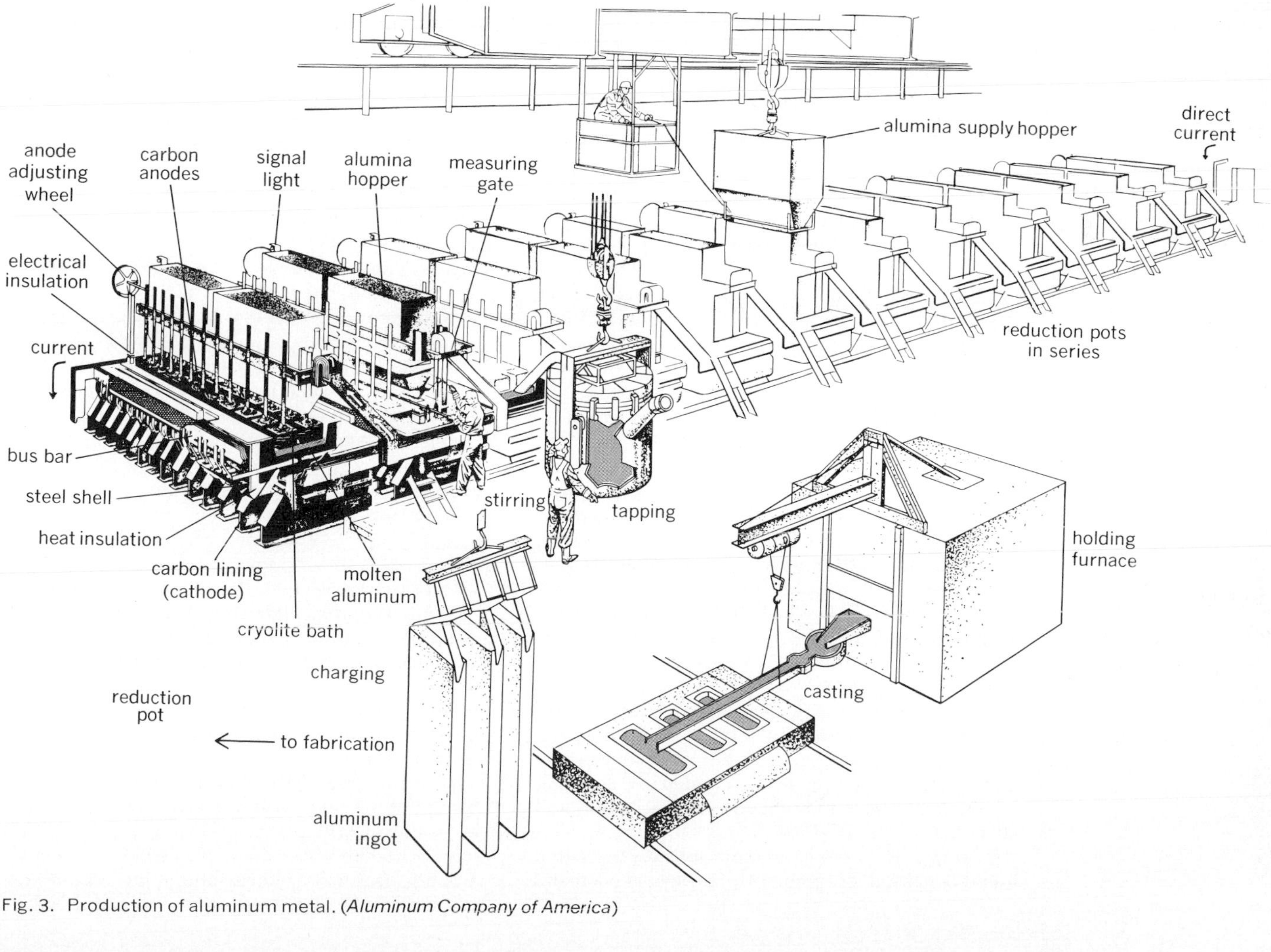

Fig. 3. Production of aluminum metal. (*Aluminum Company of America*)

ing sand, permanent mold, and die castings is usually remelted in gas- or oil-fired crucible furnaces, although in large-scale operations reverberatory furnaces may be used.

Scrap recycling has grown through the years to considerable economic importance and provides a valuable source of aluminum at a much lower energy expenditure than for primary metal. For many years, secondary producers have purchased scrap and reclaimed it, generally into foundry ingot for remelting. Primary producers have also purchased large amounts of mill scrap, as well as discarded cans, and recycled it into ingot used to fabricate sheet for more cans. This closed-circuit type of recycling saves energy, helps to preserve the environment, and conserves metal to the greatest degree.

After remelting, the molten aluminum alloy must be treated to remove dissolved hydrogen, inclusions such as oxides formed during remelting or from the surface of the charge material, and undesirable trace elements, such as sodium, in some alloys. Classically, the practice has been to bubble chlorine or a mixture of nitrogen and chlorine through the melt by means of graphite tubes. Modern practice makes use of a wide variety of in-line systems in which the metal is treated by filtering or fluxing, or both, during transfer between the holding furnace and the casting unit. These processes are more efficient than furnace fluxing, and their use minimizes air pollution, improves metal quality, and saves production time.

Although many types of casting methods for ingot have been used historically, today most ingot for fabricating is cast by the DC (direct chill) process or some modification thereof. This is accomplished on a semicontinuous basis in vertical casting and on a continuous basis in horizontal casting. The DC process consists of pouring the metal into a short mold. The base of the mold is a separate platform that is lowered gradually into a pit while the metal is solidifying. The frozen outer shell of metal retains the still-liquid portion of the metal when the shell is past the mold wall. Water is used to cool the mold, and by impingement on the ingot shell also cools the ingot as it is lowered. Ingot length is limited only by the depth to which the platform can be lowered. In horizontal casting, the same principles are applied, but the mold and base are turned so that the base moves horizontally, removing the constraint of pit depth. Much longer ingot can be practically cast by the horizontal method. By sawing the ingot while casting is progressing, the process can be made truly continuous.

Fabrication methods. Aluminum alloys are fabricated by all the methods known to metalworking and are commerically available in the form of castings; in wrought forms produced by rolling, extruding, forging, or drawing; in compacted and sintered shapes made from powder; and in other monolithic forms made by a wide variety of processes. Some of the common fabrication processes are represented in Fig. 4. In addition, pieces produced separately can be assembled by common joining procedures to build up complex shapes and structures.

Casting. In casting aluminum alloys, molten metal is poured, or forced by pressure, into a mold, where it solidifies in the shape determined by the mold. The temperature is usually 1150–1400°F (620–760°C) depending on the alloy composition and on the type of casting. For die casting and permanent mold casting, the molds are metal and are used repeatedly. Other processes use expendable molds made of sand or plaster.

Rolling. Rolling is the process of reducing material by passing it between pairs of rolls. Aluminum alloy rolled products include plate (0.250 in. or more thick: 1 in. = 2.54 cm), sheet (0.006–0.249 in. thick), and foil (less than 0.006 in. thick). The starting product is ingot, which may be up to 360 in. long, 72 in. wide, and 26 in. thick. Initial reduction is hot (800–1100°F, or 426–593°C, depending on alloy) and is done in a reversing mill. Subsequent rolling operations may be in multistand continuous mills or single-stand mills, and may be hot or cold depending on the thickness and properties required in the final product. Other aluminum alloy rolled products include rod and bar, which are passed through grooved rolls.

Extruding. In extruding aluminum alloys, an ingot or billet is forced by high pressure to flow from a container through a die opening to form an elongated shape or tube, usually at metal temperatures between 550 and 1050°F (287 and 565°C). Hydraulic presses having capacities of 500–14,000 tons (1 ton force = 8896 N) are used. The extrusion process is capable of producing sections with weights of a few ounces to more than 200 lb/ft (300 kg/m), with thicknesses from a few tenths of an inch to about 10 in., with circumscribing circle diameters of 0.25 in. to about 3 ft (1 ft = 0.3 m), and lengths in excess of 100 ft. Tubing in diameters of 0.25–33 in. and pipe up to 20 in. are also produced. Stock that is produced by extrusion is also fabricated to tubing by drawing and to forgings by subsequent metalworking operations.

Forging. Forgings are produced by pressing or hammering aluminum alloy ingots or billets into simple rectangular or round shapes on flat dies and into complex forms in cavity dies. Hydraulic presses capable of forces up to 75,000 tons and mechanical presses with capacities up to 16,000 tons are used. Forging hammers have ram weights of 500–110,000 lb (227–50,000 kg). Metal temperatures of 600–880°F (315–471°C) are generally used, depending on alloy and type of forging.

Drawing. Aluminum alloys can be made into deep-drawn shapes, of which pots and pans and beverage cans are common examples, by forcing sheet or foil into deep holes or cavities of the desired shape. Other drawn products—for example, wire and tubing—are produced by pulling rolled or extruded stock through dies.

Compacting and sintering. Aluminum powders produced by atomizing can be compacted, either alone or mixed with powders of other elements, in dies of complex shape. The compact is then heated to a high temperature to promote additional bonding between the particles to result in products having desired shapes.

Other processes. Aluminum products are machined to final shape and dimension on lathes, joiners, routers, drill presses, shears, grinders, and many other high-speed metal-cutting and metalworking machines. Other fabricating techniques

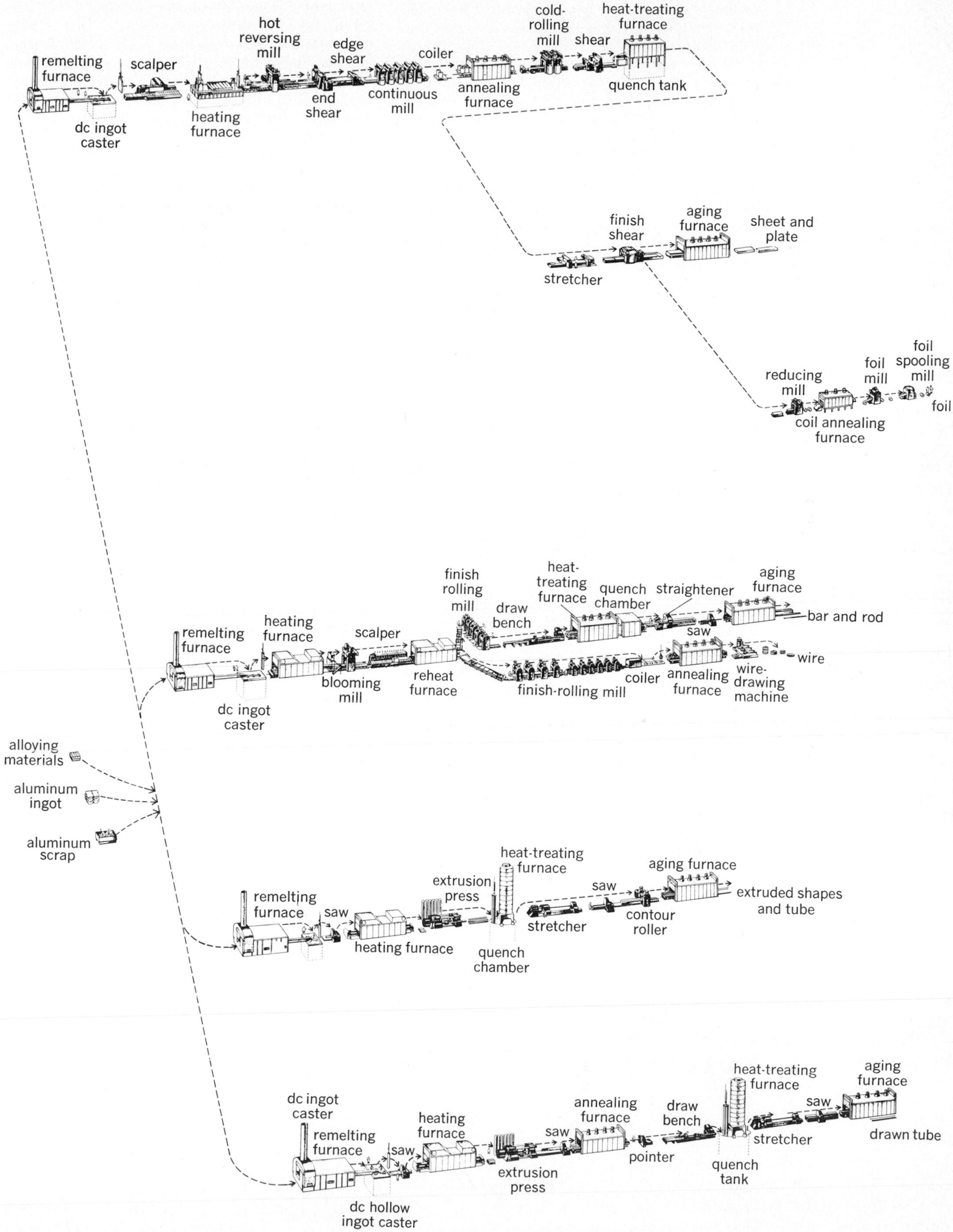

Fig. 4. Fabrication operations. (*Aluminum Company of America*)

include impact extrusion, forming, stamping, embossing, coining, bending, rotary swaging, and cold heading.

Joining. Aluminum products formed into shapes by any of the above processes can be joined by welding, brazing, soldering, and adhesive bonding, and by mechanical means such as crimping, seaming, screwing, bolting, and riveting.

Aluminum alloys. The principal alloying elements in aluminum-base alloys are magnesium, silicon, copper, zinc, and manganese. In wrought products, which constitute the greatest use of aluminum, the alloys are identified by four-digit numbers of the form NXXX, where the value of N denotes the alloy type and the principal alloying element(s) as follows: 1 (Al; at least 99% aluminum by weight), 2 (Cu), 3 (Mn), 4 (Si), 5 (Mg), 6(Mg + Si), 7 (Zn), 8 (other).

Iron and silicon are commonly present as impurities in aluminum alloys, although the amounts may be controlled to achieve specific mechanical or physical properties. Minor amounts of other elements, such as Cr, Zr, V, Pb, and Bi, are added to specific alloys for special purposes. Titanium additions are frequently employed to produce a refined cast structure.

Aluminum-base alloys are generally prepared by making the alloying additions to molten aluminum, forming a liquid solution. As the alloy freezes, phase separation occurs to satisfy phase equilibria requirements and the decrease in solubility as the temperature is lowered. The resultant solidified structure consists of grains of aluminum-rich solid solution and crystals of intermetallic compounds. Elements which lower the freezing point of aluminum, such as Cu, Mg, and Si, tend to segregate to the portions of the grains which freeze last, such as the cell boundaries. Elements which raise the freezing point, such as Cr or Ti, segregate in the opposite manner.

A decrease in solubility with falling temperature also provides the basis for heat treatment of solid aluminum alloys. In this operation, the alloy is held for some time at a high temperature to promote dissolution of soluble phases and homogenization of the alloy by diffusion processes. The limiting temperature is the melting point of the lowest melting phase present. The time required depends both on temperature and on the distances over which diffusion must occur to achieve the desired degree of homogenization. Times of several hours can be necessary with coarse structures such as sand castings. Only a few minutes may be adequate, however, for rapidly heated thin sheet.

The solution heat treatment is followed by a quenching operation in which the article is rapidly cooled, for example, by plunging it into cold or hot water or by the use of an air blast. This produces a supersaturated metallic solid solution that is thermodynamically unstable at room temperature. In several important alloy classes, such as 2XXX, 6XXX, and 7XXX, the supersaturated solution decomposes at room temperature to form fine, submicroscopic segregates or precipitates that are precursors to the equilibrium phases predicted by phase diagrams. The precipitation phenomenon, occurring over periods of days to years, produces substantial increases in strength. Additional precipitation strengthening can be obtained by heating the alloy at temperatures in the range 250–450°F (121–232°C), the time and temperature varying with alloy composition and the objectives with respect to mechanical properties and other characteristics such as corrosion resistance.

Table 1 lists the nominal compositions of a number of commercially important wrought alloys and the type of products for which they are used. The alloys are generally classified in two broad categories depending upon their response to heat treatment as described in the preceding paragraph. Those having no or minor response are identified as non-heat-treatable alloys and include the 1XXX, 3XXX, and 5XXX compositions. Those that do respond are known as heat-treatable alloys and include the 2XXX, 6XXX, and 7XXX compositions.

Included in the non-heat-treatable group is 1350, a special grade used for electrical conductor prod-

Table 1. Nominal composition and forming processes for common wrought aluminum alloys

Alloy	Form†	Si	Cu	Mg	Mn	Zn	Other
1350	b–d	–	–	–	–	–	99.5 Al (min.)
1100	b–d	–	0.1	–	–	–	99.00 Al (min.)
2011	b–d	–	5.5	–	–	–	0.5 Pb, 0.5 Bi
2014	b–e	0.8	4.4	0.4	0.8	–	–
2219	b, e	–	6.3	–	0.3	–	0.1 V, 0.1 Zr
2024	b–e	–	4.4	1.5	0.6	–	–
2036	b	–	2.6	0.45	0.25	–	–
3003	b–d	–	–	–	1.2	–	–
3004	b–d	–	–	1.0	1.2	–	–
5052	b–d	–	–	2.5	–	–	0.25 Cr
5657	b, c	–	–	0.8	–	–	–
5083	b–e	–	–	4.5	0.7	–	0.15 Cr
5086	b–d	–	–	4.0	0.5	–	0.15 Cr
5182	b	–	–	4.5	0.35	–	-
6061	b–e	0.6	0.25	1.0	–	–	0.25 Cr
6063	b–e	0.4	–	0.7	–	–	–
7005	b–e	–	–	1.5	0.5	4.5	0.15 Cr, 0.14 Zr
7050	b, e	–	2.4	2.3	–	6.2	0.12 Zr
7075	b–e	–	1.6	2.5	–	5.6	0.25 Cr

†b = rolled; c = drawn; d = extruded; and e = forged.

ucts. Alloy 1100 is a grade of 99.0% minimum aluminum content with particular controls on Fe, Si, and Cu contents, available in a variety of product forms such as sheet, foil, wire, rod, and tube, used for packaging, fin stock, and a variety of sheet metal applications. The Mn-containing alloy 3003 is a moderate-strength, very workable alloy for cooking utensils, tube, packaging, and lithographic sheet applications. The stronger Al-Mn-Mg alloy 3004 is used for architectural applications, for storage tanks, and especially for drawn and ironed beer and beverage containers.

Alloy 5052 is a workable, corrosion-resistant Al-Mg alloy for many metalworking and metal-forming purposes and marine applications. Where higher strength is required, the higher-Mg-content, weldable 5086 or 5083 alloys may be used. The latter is employed in construction of welded tanks for liquefied gas (cryogenic) transport and storage and for armor plate in military vehicles. Alloy 5182 is also a high-strength Al-Mg alloy that is employed primarily in a highly strain-hardened condition for beverage can ends. Alloy 5657 is a lower-strength material produced with a bright anodized finish for automobile trim and other decorative applications.

The heat-treatable alloys 2014, 2024, and 7075 have high strengths and are employed in aircraft and other transportation applications. Modifications of these basic alloys, such as 2124 and 7475, have been developed to provide increased fracture toughness. High toughness at high strength levels is achieved in thick-section products with 7050. Where elevated temperatures are involved, the 2XXX alloys are preferred. One such alloy, 2219, also has good toughness at cryogenic temperatures and is weldable. This alloy was prominently employed in the fuel and oxidizer tanks serving as the primary structure of the Saturn space vehicle boosters.

Alloy 6061 is widely used for structural applications where somewhat lower strengths are acceptable. For example, 6061 may be used for trailers, trucks, and other transportation applications. Alloy 6063 is a still-lower-strength, extrudable, heat-treatable alloy for furniture and architectural applications. The use of Pb and Bi in 2011 produces a heat-treatable, free-machining alloy for screw-machine products. Alloy 2036 is a moderate-strength alloy with good workability and formability, employed as body sheet in automotive applications.

Table 2 shows the nominal compositions of several important casting alloys. The major alloying addition is silicon, which improves the castability of aluminum and provides moderate strength. Other elements are added primarily to increase the tensile strength. Most die castings are made of alloy 413.0 or 380.0. Alloy 443.0 has been very popular in architectural work, while 355.0 and 356.0 are the principal alloys for sand castings. Number 390.0 is employed for die-cast automotive engine cylinder blocks, while alloy F332.0 is used for pistons for internal combustion engines.

Table 2. Nominal composition and casting procedure for common aluminum casting alloys

Alloy	Form*	Si	Cu	Mg
413.0	D	12	—	—
B†443.0	B, C	5.3	.15 max.	—
F†332.0	C	9.5	3.0	1.0
355.0	B, C	5.0	1.3	0.5
356.0	B, C	7.0	—	0.3
380.0	D	8.5	3.5	—
390.0	C, D	17	4.5	0.55

*B = sand casting; C = permanent mold casting; and D = die casting.

†The letter indicates modifications of alloys of the same general composition or differences in impurity limits, from alloys having the same four-digit numerical designations.

Table 3. Physiochemical properties of pure aluminum

Property	Value
Atomic number	13
Atomic weight	26.98154
Crystal structure	Face-centered cubic
Density, g/cm³ at 25°C	2.698
Melting point, °C	660.37
Boiling point, °C	2494
Latent heat of fusion, cal/g	94.9
Latent heat of vaporization, cal/g	2576
Heat of combustion, cal/g at 25°C	7420
Specific heat, cal/g/°C at 25°C	0.215
Thermal conductivity, cal/cm/°C/s at 25°C	0.566
Coefficient of thermal expansion, 10^{-6}/°C	23
Thermal diffusivity, cm²/s at 25°C	0.969
Electrical resistivity, microhm-cm at 25°C	2.7
Hardness, Mohs	2–2.9
Reflectivity	85–90%

Casting alloys are significant users of secondary metal (recovered from scrap for reuse). Thus, casting alloys usually contain minor amounts of a variety of elements; these do no harm as long as they are kept within certain limits. The use of secondary metal is also of increasing importance in wrought alloy manufacturing as producers take steps to reduce the energy required in producing fabricated aluminum products.

Physical properties. Aluminum is a silvery metal having a density of 2.70 g/cm³ at 20°C. Naturally occurring aluminum consists of a single isotope $_{13}Al^{27}$, which has a cross section for thermal neutrons of 0.21 barn (1 barn = 10^{-28} m²). Aluminum crystallizes in the face-centered cubic structure with edge of the unit lattice cube of 4.0495 A (0.40495 nm). Aluminum is known for its high electrical and thermal conductivities and its high reflectivity. The electrical conductivity of very pure aluminum at room temperature is over 64% (by volume) of the International Annealed Copper Standard. At temperatures below 50 K, the electrical resistivity of pure aluminum is less than that of copper and silver of very high purity. Aluminum becomes superconducting below 1.2 K. A summary of some of the important physical properties of pure aluminum is given in Table 3.

Chemical properties. The electronic configuration of the element is $1s^2\ 2s^2 2p^6\ 3s^2\ 3p^1$. Aluminum exhibits a valence of +3 in all compounds with the exception of a few high-temperature monovalent and divalent gaseous species.

Aluminum is stable in air and resistant to corrosion by sea water and many aqueous solutions and other chemical agents. This is due to protection of the metal by a tough, impervious film of oxide. Growth of this natural oxide film is self-limiting. The thickness (about 50 A or 5 nm, in dry air at room temperature) depends upon conditions and time of exposure. Molten metal is protected by a much thicker oxide skin so that oxidation of the liquid also proceeds very slowly with no agitation. In finely divided form, aluminum reacts with water to form hydrogen and aluminum hydroxide.

At a purity greater than 99.95%, aluminum resists attack by most acids but dissolves in aqua regia. It is used in the storage of nitric, concentrated sulfuric, and organic acids and many other chemical reagents. Its oxide film dissolves in alkaline solutions and corrosion is rapid to give soluble alkali metal aluminate and hydrogen, as in reaction (1).

$$2Al + 2OH^- + 6H_2O \rightarrow 2Al(OH)_4^- + 3H_2 \quad (1)$$

Aluminum is amphoteric and can react with mineral acids to form soluble salts and evolve hydrogen, as in reaction (2).

$$2Al + 6H_3O^+ + 6H_2O \rightarrow 2Al(H_2O)_6^{3+} + 3H_2 \quad (2)$$

Molten aluminum can react explosively with water. The molten metal should not be allowed to contact damp tools or containers.

At high temperatures aluminum reduces many compounds containing oxygen, particularly metal oxides. These reactions are used in the manufacture of certain metals and alloys of the type (thermite reaction) shown in reaction (3).

$$3MO + 2Al \rightarrow Al_2O_3 + 3M \quad (3)$$

Aluminum reduces silicates, especially glass. The reaction may start well below the melting point of aluminum.

Aluminum reacts with the halogens to form the trihalides ($AlCl_3$, $AlBr_3$, and AlF_3). By passing these gases over aluminum at elevated temperatures, the gaseous monohalides can be formed, as in reaction (4).

$$AlCl_3 + 2Al \rightarrow 3AlCl \quad (4)$$

At high temperatures aluminum combines with carbon, nitrogen, sulfur, and phosphorus to form Al_4C_3, AlN, Al_2S_3, and AlP respectively.

Aluminum is attacked by salts of more noble metals. In particular, aluminum and its alloys should not be used in contact with mercury or its compounds.

Products and uses. Applications in building and construction represent the largest single market of the aluminum industry. Millions of homes use aluminum doors, siding, windows, screening, and downspouts and gutters, which require little maintenance and provide a long life. Aluminum is also a major industrial building product. Excellent weather resistance makes it ideal for all climates and locations. Selected aluminum alloys are suitable near the seacoast, where salt spray may be deleterious to other metals. Colored and given additional protection by an electrolytic process called anodizing, aluminum appears in the curtain wall construction of many of the world's tallest buildings.

Transportation is the second largest market. Many commercial and military aircraft have become virtually all-aluminum. In automobiles, aluminum is apparent in interior and exterior trim, grilles, wheels, and air conditioners. Another major use is in automatic transmissions. Some auto radiators, engine blocks, and body panels are made of aluminum to hold down weight and improve fuel economy. Aluminum is also found in rapid transit car bodies, engine parts for diesel locomotives, rail freight and tank cars, bus and truck engines, forged truck wheels, cargo containers, and in highway signs, divider rails, and lighting standards. Aluminum's light weight and corrosion resistance are responsible for applications in pleasure boat hulls, fishing boats, tanks for ships transporting liquified natural gas, and deck houses for naval vessels.

In aerospace, aluminum was in the first plane flown by the Wright brothers, and is found today in aircraft engines, frames, skins, landing gear, and interiors, often making up 80% of a plane's weight. Toughness, lightness, and heat reflective characteristics have made it the preferred material for satellites, Moon rockets, and equipment such as the lunar rover.

The packaging industry is the fastest-growing market. Aluminum helps in the preparation of foods and beverages and keeps them pure during distribution and storage. The fast-cooling, easy-opening, lightweight, and recyclable features of the all-aluminum can have resulted in the use of billions of aluminum beverage containers. Foil pouches and bags, twist-off closures, and easy-open ends have revolutionized the food and beverage packaging industries.

In electrical applications, aluminum wire and cable are major products. The trend toward placing electrical cables underground requires large amounts of aluminum. Aboveground transmission cable is much in demand to satisfy growing power needs. The use of aluminum wiring in residential, commercial, and industrial buildings is on the increase.

Consumer use began before the turn of the century with aluminum cooking utensils. Today, aluminum also appears in the home as cooking foil, hardware, tools, portable appliances, air conditioners, freezers, and refrigerators, and in sporting equipment such as skis, ball bats, and tennis rackets.

There are hundreds of chemical uses of aluminum and aluminum compounds. Aluminum powder is used in paints, rocket fuels, and explosives, and as a chemical reductant. *See* BORON; GALLIUM; INDIUM; THALLIUM. [ALLEN S. RUSSELL]

Bibliography: H. Ginsberg and K. Wefers, *Aluminum and Magnesium*, Ferdinand Enke Verlag, 1971; V. N. Tikhonov, *Analytical Chemistry of Aluminum*, 1973; A. F. Trotman-Dickenson (exec. ed.), *Comprehensive Inorganic Chemistry*, vol. 1, 1973; K. R. Van Horn (ed.), *Aluminum*, 3 vols., American Society for Metals, 1967; P. C. Varley, *The Technology of Aluminum and Its Alloys*, 1970.

Americium

A chemical element, symbol Am, atomic number 95. Americium was discovered by G. T. Seaborg, R. A. James, and L. O. Morgan in plutonium that had been irradiated by reactor neutrons. The nuclear reactions are shown by the reaction below.

$$^{239}Pu(n,\gamma)^{240}Pu(n,\gamma)^{241}Pu \xrightarrow[14\,yr]{\beta^-} {}^{241}Am$$

The isotope ^{241}Am is an alpha emitter with a half-life of 433 years. Other isotopes of americium range in mass from 232 to 247, but only the isotopes of mass 241 and 243 are important. Both have been produced in kilogram amounts. The isotope ^{241}Am is routinely separated from "old" plutonium and sold for a variety of industrial uses such as 59-keV gamma sources and as a component in neutron sources. The longer-lived ^{243}Am (half-life 7400 years) is a precursor in ^{244}Cm production: $^{243}Am(n,\gamma)^{244}Am(\beta^-)^{244}Cm$.

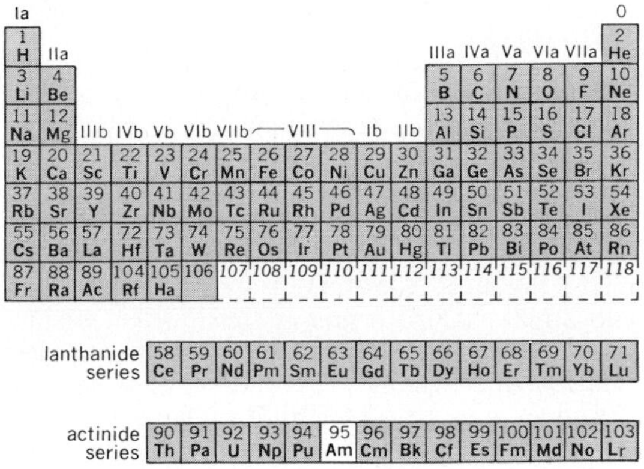

Early studies of americium chemistry in aqueous solution demonstrated that in its most prominent aqueous oxidation state, 3+, it closely resembles the tripositive rare earths. The formal analogy to the rare earths is also marked in anhydrous compounds of both tripositive and tetrapositive americium. Americium is different in that it is possible to oxidize Am^{3+} to both the 5+ and 6+ states.

Oxidation states higher than 3+ (4, 5, and 6) are all reduced in aqueous solution to Am^{3+} by effects of alpha radiation. The formal oxidation states Am^{5+} and Am^{6+} occur in dilute aqueous acid as dioxo monocations AmO_2^{+} and AmO_2^{2+} respectively. In contrast, Am^{4+} is a more powerful oxidant and has been stabilized only in high concentrations of fluoride or phosphate. Both the tetra- and pentapositive states of americium undergo acid disproportionation (self-oxidation to the next higher oxidation state and corresponding reduction to lower oxidation states).

Americium metal has a vapor pressure markedly higher than that of its neighboring elements and can be purified by distillation. The metal has double hexagonal close-packed (dhcp) and face-centered cubic (fcc) phases with a density of 13.67 g/cm³, possibly a body-centered cubic (bcc) phase, and melts at 1175°C. The metal is nonmagnetic and superconducting at 0.79 K. Under high pressure the metal has been compressed to 80% of its room-temperature volume and displays the α-uranium structure.

Dozens of compounds and alloys of americium are now known. Most of the early ones were identified from their x-ray diffraction patterns. *See* ACTINIDE ELEMENTS; BERKELIUM; CURIUM; TRANSURANIUM ELEMENTS.

[ROBERT A. PENNEMAN]

Bibliography: K. W. Bagnall, *The Actinide Elements*, 1972; J. J. Katz and G. T. Seaborg, *The Chemistry of the Actinide Elements*, 1958; C. Keller, *Chemistry of the Transuranium Elements*, Verlag Chemie GmbH, Weinheim, Germany, 1971; W. W. Schulz, *The Chemistry of Americium*, ERDA Crit. Rev. Ser. Rep. no. TID-26971, 1976.

Amide

A derivative of a carboxylic acid with general formula (I), where R is hydrogen or an alkyl or aryl radical. Amides are divided into subclasses, depending on the number of substituents on nitrogen. The simple, or primary, amides are considered to be derivatives formed by replacement of the carboxylic hydroxyl group by the amino group, NH_2. They are named by dropping the "-ic acid" or "-oic acid" from the name of the parent carboxylic acid and replacing it with the suffix "amide," as shown in examples (II)–(V). In the secondary and tertiary amides, one or both hydrogens are replaced by other groups. The presence of such groups is designated by the prefix capital *N* (for nitrogen), as shown in the examples.

$$R{-}\overset{O}{\overset{\|}{C}}{-}NH_2 \qquad (I)$$

$$H\overset{O}{\overset{\|}{C}}{-}NH_2$$

Formamide (II)

$$CH_3\overset{O}{\overset{\|}{C}}{-}NH_2$$

Acetamide (III)

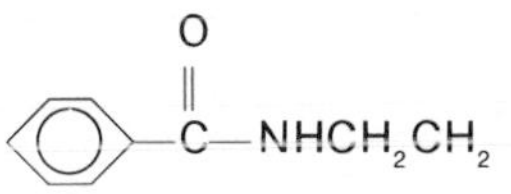

N-ethylbenzamide (IV)

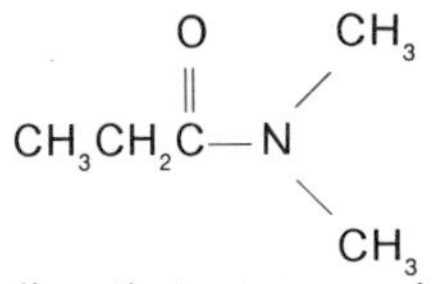

N,N-dimethylpropanamide (V)

Except for formamide, all simple amides are relatively low-melting solids, stable, and weakly acidic. They are strongly associated through hydrogen bonding, and hence soluble in hydroxylic solvents, such as water and alcohol. Because of ease of formation and sharp melting points, amides are frequently used for the identification of organic acids and, conversely, for the identification of amines.

Formation and properties. Commercial preparation of amides involves thermal dehydration of ammonium salts of carboxylic acids. Thus, slow pyrolysis of ammonium acetate, $CH_3COO^-NH_4^+$, forms water and acetamide. *N,N*-dimethylacetamide may be similarly prepared from dimethylammonium acetate,

$$CH_3COO^-\overset{+}{N}H_2(CH_3)_2$$

Acid anhydrides react with ammonia or with primary or secondary amines to form amides. The preparation of acetanilide [reaction (1)] is an example.

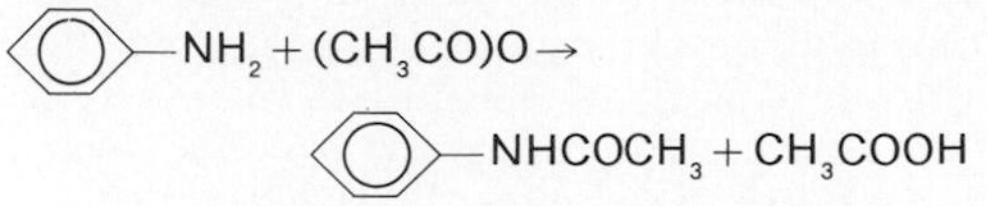

(1)

The reaction of amines with acid chlorides in the presence of aqueous sodium hydroxide (Schotten-Baumann reaction) is often used [reaction (2)].

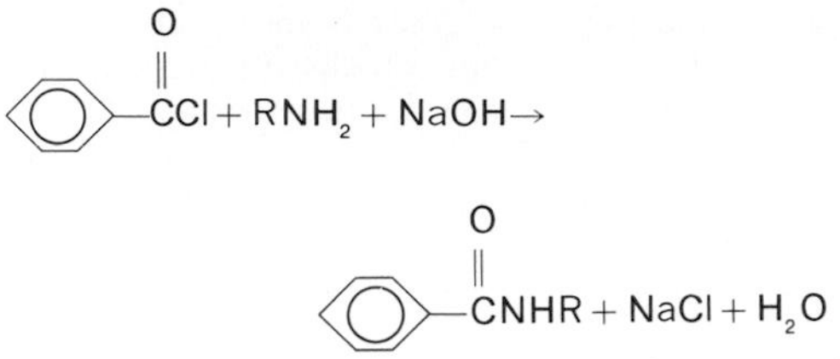

(2)

When urea, the diamide of carbonic acid, is heated with a carboxylic acid, a new amide is formed by an exchange reaction. The other product of the reaction, carbamic acid, decomposes to permit the reaction to go to completion [reaction (3)].

$$CO(NH_2)_2 + CH_3COOH \rightleftharpoons CH_3CONH_2 + H_2NCOOH \quad (3)$$
$$H_2NCOOH \rightarrow CO_2\uparrow + NH_3\uparrow$$

The partial hydrolysis of nitriles also affords a convenient synthesis of a variety of amides [reaction (4)]. The reaction is catalyzed by both acid and base.

$$CH_3CH_2CH_2CH_2C{\equiv}N + H_2O \rightarrow CH_3CH_2CH_2CH_2CONH_2 \quad (4)$$

Uses. Amides are important chemical intermediates since they can be hydrolyzed to acids, dehydrated to nitriles, and degraded to amines containing one less carbon atom by the Hofmann reaction. In pharmacology, acetophenetidin (VI) is a popular analgesic. *N,N*-Dimethylformamide (DMF; VII) is a useful aprotic, highly polar solvent.

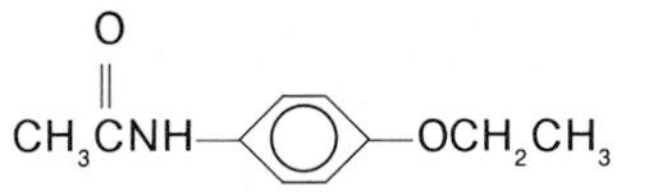

Acetophenetidin (VI)

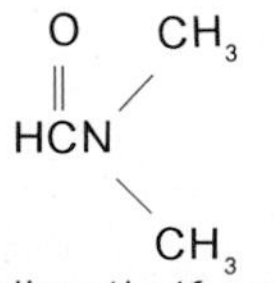

N,N-dimethylformamide (VII)

However, the most important commercial application of amides is in the preparation of polyamide resins, also called nylons. *See* ACID ANHYDRIDE; ACID HALIDE; AMINE; CARBOXYLIC ACID; NITRILE; POLYAMIDE RESINS; UREA.

[PAUL E. FANTA]

Bibliography: N. L. Allinger et al., *Organic Chemistry*, 2d ed., 1976.

Amidine

A compound with the general formula

$$R{-}\overset{\displaystyle NH}{\overset{\|}{C}}{-}NH_2$$

or corresponding *N*-substituted derivatives. Amidines may be considered as derivatives of amides formed by the replacement of the divalent oxygen atom by an imino (>NH) group. They are usually prepared by the reaction of the imide chloride or alkyl imidate and ammonia or amines. The amidines are colorless, crystalline solids. They are strong bases as contrasted with amides. Although somewhat unstable, they form stable salts with acids. Amidines are readily hydrolyzed by hot acids and bases, and they condense with certain reagents to give heterocyclic systems. The pharmacological properties of many have been investigated. *See* AMINE. [ELBERT H. HADLEY]

Bibliography: N. L. Allinger et al., *Organic Chemistry*, 2d ed., 1976.

Amination

The preparation of amines. Amines may be considered derivatives of ammonia, NH_3, where one or more of the hydrogens are replaced by an alkyl (CH_3—), aryl (C_6H_5—), cycloalkyl (C_6H_{11}—), or heterocyclic group. Several processes are employed for the preparation of amines, which are among the most important of organic compounds. They are used for the preparation of dyes, medicinals, surfactants, plastics, rocket fuels, and emulsifiers. *See* AMINE.

Amines can be prepared by reducing a compound that already has a carbon-nitrogen bond, such as a nitro, nitroso, azoxy, or azo compound. They can also be prepared by treating compounds containing (1) a labile group (—Cl, —OH, —SO_3H), (2) a carbonyl (>C═O) group, or (3) a highly reactive structure

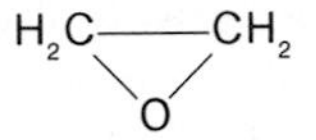

with ammonia. The preparation of amines from compounds containing a C-N bond is termed amination by reduction; the production of amines by reaction with ammonia or a mixture of ammonia and hydrogen is termed ammonolysis or hydroammonolysis. *See* CHEMICAL CONVERSION.

Amination by reduction. A variety of reduction methods have been used for the preparation of amines. The Béchamp reduction involving iron and an acid is the principal example of a commonly used metal and acid combination, but other metals, notably zinc and tin, have been employed. Generally, hydrochloric acid or its acidic salts are preferred, but sulfuric, acetic, and other acids have also served.

Catalytic reduction involves hydrogen and a catalyst such as nickel, copper, platinum, or molybdenum sulfides. Sulfide reductions are used mainly for the partial reduction of polynitro aromatic compounds to nitroamines, for example, *m*-dinitrobenzene to *m*-nitroaniline.

Metal and acid reductions are the most vigorous and as a rule yield primary amines as end products. Catalytic reductions are employed when the nitro compound is a liquid and the production requirements are large and steady, as in the preparation of aniline from nitrobenzene. By a proper selection of reducing agent and careful regulation of the process, reductions may often be stopped at an intermediate stage (azoxy, azo, or hydrazo) to obtain valuable products other than amines.

Iron-and-acid reductions can be carried out with

far less than the theoretical quantity of acid required by the reaction

$$\begin{array}{llll} & C_6H_5NO_2 + & 2Fe + & 7HCl \rightarrow \\ \text{Formula wt.} & 123 & 2(55.84) & 7(36.45) \end{array}$$

$$C_6H_5NH_3^+Cl^- + 2H_2O + 2FeCl_3$$

In industrial practice only about 2% of the amount of acid indicated above is used. A simplified explanation of this phenomenon depends on the hydrolysis of salts of aromatic amines (aniline hydrochloride) which provides the acidic medium.

Amination by ammonolysis. Aqueous ammonia (20–60% NH_3) is used for most aminations in the liquid phase. In such reactions NH_3, and not the addition product, NH_4OH, is the reactant. Examples of the preparation of amines from several classes of compounds by reaction with ammonia follow:

1. Replacement of halogen: aniline from chlorobenzene; mono-, di-, and trialkylamines from alkyl chlorides.
2. Replacement of $—SO_3H$ and $—OSO_3H$ groups: aminoanthraquinone from anthraquinonesulfonic acid; ethylenediamine from aminoethyl hydrogen sulfate.
3. Conversion of alcohols: mono-, di-, and trialkylamines from low-molecular-weight alcohols.
4. Hydroammonolysis of aldehydes and ketones: *n*-propylamine from acrolein; isopropylamine from acetone.
5. Addition reactions: mono-, di-, and triethanolamines from ethylene oxide.

Catalysts. In the ammonolysis of halogen compounds, it is frequently advantageous to use a copper or a silver catalyst. In the production of amines from alcohols, dehydrating catalysts such as aluminum phosphate and aluminum silicate are effective. For the hydroammonolysis of carbonyl compounds in the vapor phase, cobalt and copper-based catalysts are generally employed.

Kinetics. A substantial excess of ammonia is generally used in the conversion of reactive organic compounds to amines. Under such conditions, there is no appreciable change in the NH_3 concentration of the medium, and the rate expression for the process is pseudo first-order. The free energy change for most aminations is favorable at moderate temperatures.

Equipment. Aminations by ammonolysis are generally carried out at elevated temperatures and pressures. Autoclaves are commonly used for both batch and continuous processes up to 700–1000 psi, whereas tubes are preferred for higher pressures and particularly for continuous processes. Indirect heating of autoclaves is accomplished by steam for reactions up to 190°C and a circulating-oil system for higher temperatures. Because the walls of the autoclave are thick, improved heat transfer may be obtained with internal coils instead of with a jacket. Mechanical agitation is essential when the reactants are immiscible. Tubular reactors containing seamless tubes or bored forgings are used for ammonolyses involving pressures of 700–5000 psi.

[P. H. GROGGINS/ROBERT S. KAPNER]

Bibliography: E. C. Herrick et al., *Unit Process Guide to Organic Chemical Industries*, 1979; R. N. Shreve and J. Brink, *Chemical Process Industries*, 4th ed., 1977.

Amine

A member of a group of organic compounds which can be considered as derived from ammonia by replacement of one or more hydrogens by organic radicals. Generally amines are bases of widely varying strengths, but a few which are actually acidic are known.

Amines constitute one of the most important classes of organic compounds. The lone pair of electrons on the amine nitrogen enables amines to participate in a large variety of reactions as a base or a nucleophile. Amines play prominent roles in biochemical systems; they are widely distributed in nature in the form of amino acids, alkaloids, and vitamins. Many complex amines have pronounced physiological activity, for example, epinephrine (adrenalin), thiamin or vitamin B_1, and Novocaine. The odor of decaying fish is due to simple amines produced by bacterial action. Amines are used to manufacture many medicinal chemicals, such as sulfa drugs and anesthetics. The important synthetic fiber nylon is an amine derivative.

Amines are classified according to the number of hydrogens of ammonia which are replaced by radicals. Replacement of one hydrogen results in a primary amine (RNH_2), replacement of two hydrogens results in a secondary amine (R_2NH), and replacement of all three hydrogens results in a tertiary amine (R_3N). The substituent groups (R) may be alkyl, aryl, or aralkyl. Another group of amines are those in which the nitrogen forms part of a ring (heterocyclic amines).

Examples of such compounds are nicotine (I), which is obtained commercially from tobacco for use as an insecticide, and serotonin (II), which plays a key role as a chemical mediator in the central nervous system.

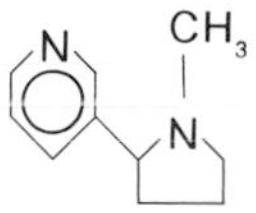

(I)

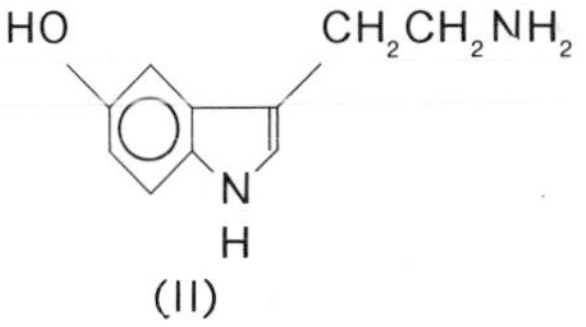

(II)

Many aromatic and heterocyclic amines are known by trivial names, and derivatives are named as substitution products of the parent amine. Thus, $C_6H_5NH_2$ is aniline and $C_6H_5NHC_2H_5$ is N-ethylaniline. For a discussion of definitive rules for naming amines *see* ORGANIC CHEMISTRY.

According to the Brönsted-Lowry theory of acids and bases, amines are basic because they accept protons from acids. In water the equilibrium shown in reaction (1) lies predominantly to the left.

$$RNH_2 + H_2O \rightleftharpoons RNH_3^+ + OH^- \quad (1)$$

The extent to which the amine is successful in picking up a proton from water is given by Eq. (2),

$$K_b = \frac{[RNH_3]^+ [OH]^-}{RNH_2} \quad (2)$$

where the quantities in brackets signify concentrations of the species given. For short-chain aliphatic amines the basic ionization constant K_b lies near 10^{-4}; for aromatic amines $K_b < 10^{-9}$; for ammonia $K_b = 1.8 \times 10^{-5}$. Stable salts suitable for the indentification of amines are in general formed only with strong acids, such as hydrochloric, sulfuric, oxalic, chloroplatinic, or picric.

Reactions and identification. Several test-tube reactions for recognition and characterization of amines are known: the Schotten-Baumann reaction, the Hinsberg test, the carbylamine reaction, and the action of nitrous acid.

The Schotten-Baumann reaction involves treatment of an amine with benzoyl chloride in basic solution. It serves to distinguish tertiary amines from primary and secondary amines by the formation of substituted benzamides from the primary and secondary amines [reactions (3) and (4)]. Tertiary amines do not react with benzoyl chloride. The substituted benzamides are generally insoluble in water, solid, and easily purified, and they have characteristic melting points. Thus they serve to identify the amines.

$$RNH_2 + C_6H_5COCl + OH^- \rightarrow C_6H_5CONHR + Cl^- + H_2O \qquad (3)$$

$$R_2NH + C_6H_5COCl + OH^- \rightarrow C_6H_5CONR_2 + Cl^- + H_2O \qquad (4)$$

The more reactive acylating agents, acetic anhydride and acetyl chloride, give substituted acetamides without added base. This reaction gives the same type of information as the Schotten-Baumann reaction.

Closely related to the Schotten-Baumann reaction is the Hinsberg test. This has the added advantage of distinguishing between primary and secondary amines. It involves reaction of an amine with benzenesulfonyl chloride in alkaline solution. Both it and the Schotten-Baumann test are applicable to both aliphatic and aromatic amines with the exception of those amines which are substantially nonbasic in character. Primary amines give sulfonamides that are soluble in basic solutions; secondary amines give insoluble derivatives; and tertiary amines with no replaceable hydrogen do not react with the reagent. In general, the sulfonamides are solids and are useful for identification of the amines.

Carbylamines (isocyanides) possess a very unpleasant, nauseating odor and are formed by the reaction of any primary amine with chloroform in basic solution [reaction (5)].

$$RNH_2 + CHCl_3 + 3OH^- \rightarrow R{-}N{\equiv}C + 3Cl^- + 3H_2O \qquad (5)$$

Reaction with nitrous acid serves as a further method for distinguishing between various classes of amines. Primary aliphatic amines evolve nitrogen, whereas primary aromatic amines give diazonium compounds, which may be recognized by dye formation on coupling with a suitable second component. Secondary amines of both series give nitrosamines, generally as yellow oils. Tertiary aliphatic amines do not react with nitrous acid, and mixed aliphatic aromatic tertiary amines undergo nuclear nitrosation. Nitrosamines have been identified as potent carcinogenic substances.

In the infrared absorption spectrum, amines exhibit characteristic bands due to N-H stretching and bending, as well as C-N stretching vibrations. In the proton magnetic resonance spectrum, the appearance of the proton on nitrogen is complicated by the rate of exchange and the electrical quadrupole of the ^{14}N nucleus.

In alkaline solutions tertiary amines are oxidized by hydrogen peroxide to amine oxides which, although still basic, are not strong bases as are the quaternary ammonium types [reaction (6)]. *See* ANILINE.

$$R_3N + H_2O_2 \rightarrow R_3NO + H_2O \qquad (6)$$

Aromatic amines undergo halogenation and sulfonation on the ring. However, because of their susceptibility to oxidation, nitration cannot be accomplished without prior protection of the labilizing amino group. This is commonly done by acetylation.

Exhaustive methylation is one of a number of reactions of amines associated with the name of A. W. Hofmann. It involves a sequence of reactions terminating in the thermal decomposition of a quaternary ammonium hydroxide to yield a tertiary amine, usually trimethylamine, water, and an olefin. It has been widely used as a tool in the determination of the structures of complex compounds and, to a lesser extent, for the synthesis of olefins. The sequence of reactions is shown in (7). *See* ALKALOID.

$$RCH_2CH_2NH_2 \xrightarrow{3CH_3I} RCH_2CH_2\overset{+}{N}(CH_3)_3I^- \xrightarrow{AgOH} RCH_2CH_2\overset{+}{N}(CH_3)_3OH^- \xrightarrow{heat} RCH{=}CH_2 + (CH_3)_3N + H_2O \qquad (7)$$

Preparation. Commercial preparation of aliphatic amines can be accomplished by direct alkylation of ammonia (Hofmann method, 1849) or by catalytic alkylation of amines with alcohols at elevated temperatures. Reduction of various nitrogen functions carrying the nitrogen in a higher state of oxidation also leads to amines. Such functions are nitro, oximino, nitroso, and cyano. For the preparation of pure primary amines, Gabriel's synthesis and Hofmann's hypohalite reaction are preferred methods. The Bucherer reaction is satisfactory for the preparation of polynuclear primary aromatic amines. *See* NAPHTHYLAMINE.

Gabriel's synthesis. This is a method for the synthesis of pure primary aliphatic amines by the hydrolysis of an *N*-alkyl phthalimide. The *N*-alkyl phthalimides are prepared by reaction of potassium phthalimide with an alkyl halide (preferably a bromide) [reaction (8)].

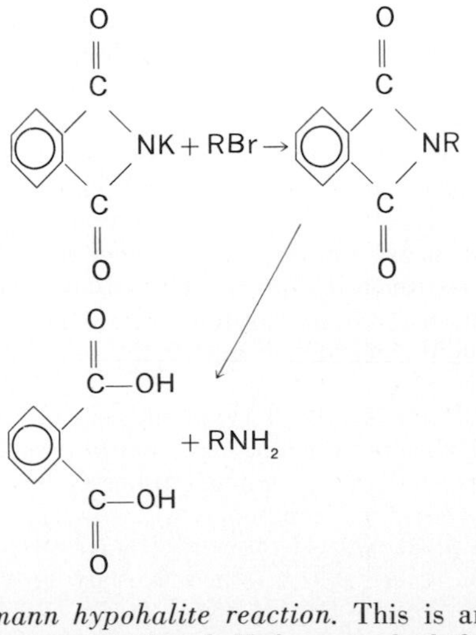

(8)

Hofmann hypohalite reaction. This is another reaction associated with Hofmann, which furnishes a convenient method for the preparation of pure primary amines of either the aliphatic or aromatic series. In the overall sense, it involves conversion of an acid amide to an amine with loss of one carbon atom. The reaction proceeds through the stages shown in (9)–(11). The amine arises by hy-

$$RCONH_2 + NaOX \rightarrow RCONHX + NaOH \quad (9)$$

$$RCONHX + NaOH \rightarrow R{-}N{=}C{=}O + NaX + H_2O \quad (10)$$

$$R{-}N{=}C{=}O + 2NaOH \rightarrow RNH_2 + Na_2CO_3 \quad (11)$$

drolysis of the isocyanate formed by migration of R from carbon to nitrogen in reaction (10). Sodium hypobromite is the common laboratory reagent, but the cheaper calcium hypochlorite is used in commercial applications, such as the manufacture of anthranilic acid from phthalimide. *See* ALKANOLAMINE; QUATERNARY AMMONIUM SALT.

[PAUL E. FANTA]

Bibliography: N. L. Allinger et al., *Organic Chemistry*, 2d ed., 1976.

Ammine

One of a group of complex compounds formed by the coordination of ammonia molecules with metal ions and, in a few instances, such as calcium, strontium, and barium, with metal atoms. Some typical examples of ammines include $[Co(NH_3)_6]Cl_2$ (rose), $[Cu(NH_3)_4]Cl_2$ (blue), $[Cr(NH_3)_6]Cl_3$ (yellow), $[Cr(NH_3)_4Cl_2]Cl$ (*cis*-violet, *trans*-green), $[Ni(NH_3)_6]Cl_2$ (blue), $[Pt(NH_3)_4]Cl_2 \cdot H_2O$ (colorless), and $[Hg(NH_3)_2]Br_2$ (colorless). Although these ammines are formally analogous to many salt hydrates, the general characteristics of the group of ammines differ considerably from those of the hydrates. For example, hydrated Co(III) salts are strong oxidizing agents whereas Co(II) ammines are strong reducing agents. The ammines of principal interest are those of the transition metals and of the zinc family, but even here there is wide variation in stability or rate of decomposition. For example, iron ammines are unstable in aqueous solution; Cu(II) and Co(II) ammines exist in aqueous solution but are decomposed by aqueous acids; Co(III) and Pt(IV) ammines can be recrystallized from strong acids. Ammines are prepared by treating aqueous solutions of the metal salt with ammonia or, in some instances, by the action of dry gaseous or liquid ammonia on the anhydrous salt. These, and similar, differences have motivated much theoretical study of the bonding in such compounds. *See* AMMONIA; COORDINATION CHEMISTRY.

[HARRY H. SISLER]

Ammonia

The most familiar compound composed of the elements nitrogen and hydrogen, NH_3. It is formed as a result of the decomposition of most nitrogenous organic material, and its presence is indicated by its pungent and irritating odor.

Uses. Because of the wide range of industrial and agricultural applications, ammonia is produced in tremendous quantities. Examples of its use are the production of nitric acid and ammonium salts, particularly the sulfate, nitrate, carbonate, and chloride, and the synthesis of hundreds of organic compounds including many drugs, plastics, and dyes. Its dilute aqueous solution finds use as a household cleansing agent. Anhydrous ammonia and ammonium salts are used as fertilizers, and anhydrous ammonia also serves as a refrigerant, because of its high heat of vaporization and relative ease of liquefaction.

Molecular structure. The NH_3 molecule has a pyramidal structure of the type illustrated in the diagram, in which the nitrogen atom has achieved

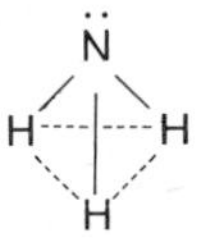

a stable electronic configuration by forming three electron pair bonds with the three hydrogen atoms. The HNH bond angle in the pyramid is 106.75°, which is best explained as resulting from the use of sp^3 hybrid bonding orbitals by the nitrogen atom. This should yield tetrahedral bond angles with the result that one sp^3 orbital is occupied by the unshared pair of electrons. The repulsive effect of this unshared pair, which is concentrated relatively near to the nitrogen nucleus on the shared pairs of electrons forming the N-H bonds, produces a slight compression of the HNH bond angles, thus accounting for the fact that they are slightly less than tetrahedral (109.5°). The dipole moment of the ammonia molecule, 1.5 debyes, is a resultant of the combined polarities of the three N-H bonds and of the unshared electron pair in the highly directional sp^3 orbital. The pyramidal ammonia molecule turns inside out readily, and it oscillates between the two extreme positions at the precisely determined frequency of 2.387013×10^{10} Hz. This property has been used in the highly accurate time-measuring device known as the ammonia clock.

Physical characteristics. The physical properties of ammonia are analogous to those of water and hydrogen fluoride in that the physical constants are abnormal with respect to those of the binary hydrogen compounds of the other members of the respective periodic families. This is particularly true of the boiling point, freezing point, heat of fusion, heat of vaporization, and dielectric constant. These abnormalities may be related to the association of molecules through intermolecular hydrogen bonding. The principal physical constants for ammonia are summarized in Table 1.

Table 1. Physical properties of ammonia

Melting point	−77.74°C
Boiling point	−33.35°C
$\Delta H_{fus.}$ at mp	5657 J $mole^{-1}$
$\Delta H_{vap.}$ at bp	23,350 J $mole^{-1}$
Critical temperature	132.4°C
Critical pressure	113.1°C
Dielectric constant (−77.7°C)	25
Density (−70°C)	0.7253 g cm^{-3}
Density (−30°C)	0.6777 g cm^{-3}
$\Delta H°_{form.}$ (25°C)	46.19 kJ $mole^{-1}$
$\Delta G°_{form.}$ (25°C)	16.64 kJ $mole^{-1}$
$S°$ (298.1 K) (exptl.)	192.5 J deg^{-1} $mole^{-1}$
$C°_p$	35.66 J deg^{-1} $mole^{-1}$
Viscosity (liquid, 25°C)	0.01350 Pa s
Vapor pressure (−20°C)	1426.8 torr (190.22 kPa)
Vapor pressure (0°C)	3221.0 torr (429.43 kPa)
Vapor pressure (20°C)	6428.5 torr (857.06 kPa)
Solubility in H_2O	
(1 atm,* 0°C)	42.8% by wt
(1 atm, 20°C)	33.1% by wt
(1 atm, 40°C)	23.4% by wt

*1 atm = 101.325 kPa.

Ammonia is highly mobile in the liquid state and has a high thermal coefficient of expansion.

Chemical properties. Most of the chemical reactions of ammonia may be classified under three chief groups: (1) addition reactions, commonly called ammonation; (2) substitution reactions, commonly called ammonolysis; and (3) oxidation-reduction reactions.

Ammonation. Ammonation reactions include those in which ammonia molecules add to other molecules or ions either through the mechanism of covalent bond formation using the unshared pair of electrons on the nitrogen atom, through ion-dipole electrostatic interactions, or through hydrogen bonding. Most familiar of the ammonation reactions is the reaction with water, which may be represented schematically as reaction (1). The strong

$$H:N: + H:\ddot{O}: \rightarrow H:N:H:\ddot{O}: \rightarrow$$

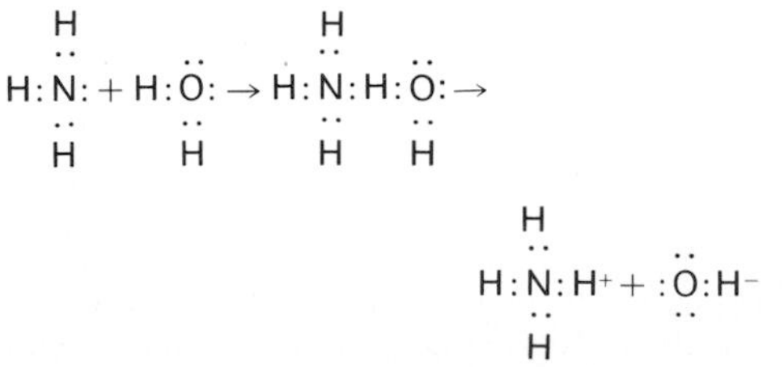

$$H:N:H^+ + :\ddot{O}:H^- \quad (1)$$

tendency of water and ammonia to combine is evidenced by the very high solubility of ammonia in water (700 vol of ammonia gas in 1 vol of water at 20°C and 1 atm ammonia pressure). The ammonia hydrate (ammonium hydroxide) is a weak electrolyte in aqueous solution as indicated by an ionization constant of 1.77×10^{-5} at 25°C. Phase diagrams for the NH_3—H_2O system indicate the existence of $NH_3 \cdot H_2O$ (mp −79.0°C) in the solid state at low temperatures. Under these conditions, the compound $2NH_3 \cdot H_2O$ (mp −78.8°C) also exists. Ammonia reacts readily with strong acids to form ammonium salts, reaction (2). Am-

$$H:N: + H:X \rightarrow H:N:H^+ + :X^- \quad (2)$$

$$(:X^- = \text{an anion})$$

monium salts of weak acids in the solid state dissociate readily into ammonia and the free acid.

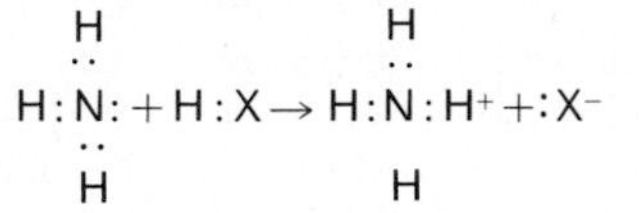

Included among ammonation reactions is the formation of complexes (called ammines) with many metal ions, particularly transition metal ions, such as $Hg(NH_3)_2^+$, $Cr(NH_3)_6^{3+}$, $Zn(NH_3)_4^{2+}$, and $Co(NH_3)_6^{3+}$. *See* AMMINE.

Ammonation occurs with a variety of molecules capable of acting as electron acceptors (Lewis acids). The reaction of ammonia with such substances as sulfur trioxide, sulfur dioxide, silicon tetrafluoride, and boron trifluoride are typical and illustrated by reaction (3).

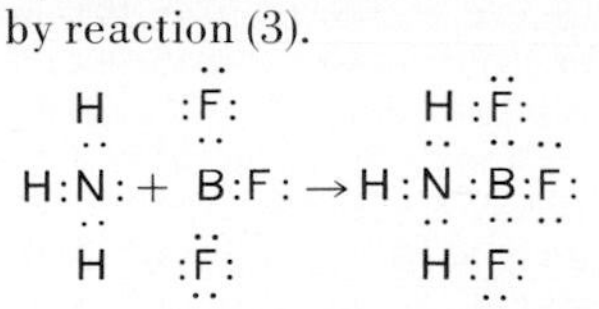

$$H:N: + B:\ddot{F}: \rightarrow H:N:B:\ddot{F}: \quad (3)$$

Ammonolysis. Ammonolytic reactions include reactions of ammonia in which an amide group ($-NH_2$), an imide group ($=NH$), or a nitride group ($\equiv N$) replaces one or more atoms or groups in the reacting molecule. Examples include reactions (4) to (8).

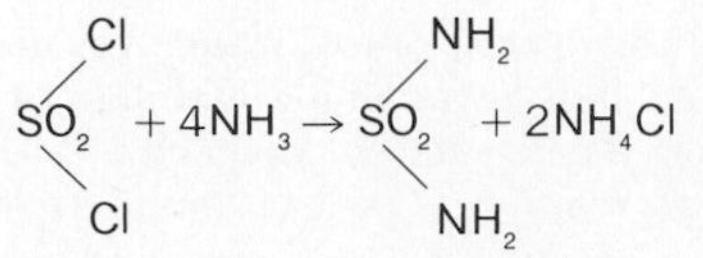

$$SO_2\begin{matrix}Cl\\Cl\end{matrix} + 4NH_3 \rightarrow SO_2\begin{matrix}NH_2\\NH_2\end{matrix} + 2NH_4Cl \quad (4)$$

$$C=O\begin{matrix}Cl\\Cl\end{matrix} + 4NH_3 \rightarrow C=O\begin{matrix}NH_2\\NH_2\end{matrix} + 2NH_4Cl \quad (5)$$

$$CH_3CH_2C\begin{matrix}O\\OCH_3\end{matrix} + NH_3 \rightarrow$$

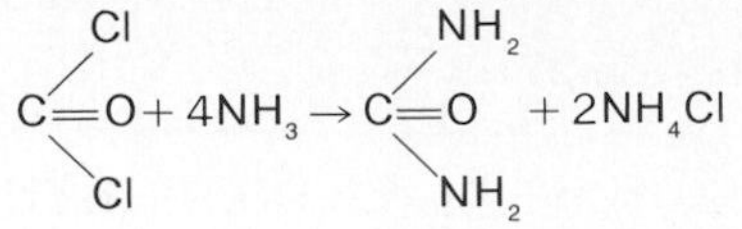

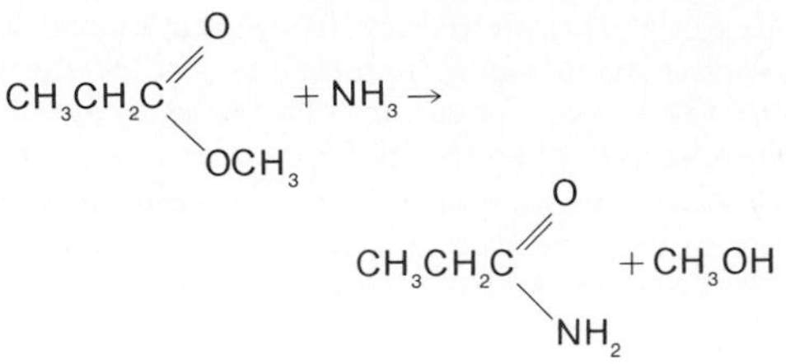

$$CH_3CH_2C\begin{matrix}O\\NH_2\end{matrix} + CH_3OH \quad (6)$$

$$HgCl_2 + 2NH_3 \rightarrow Hg(NH_2)Cl + NH_4Cl \quad (7)$$

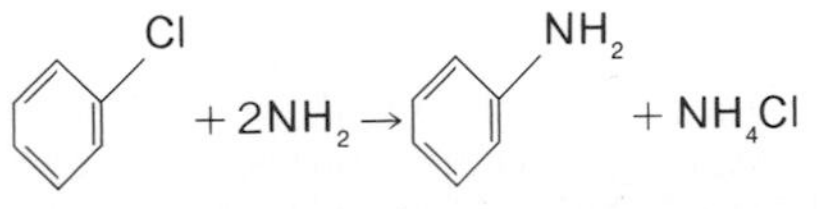

$$\underset{Cl}{\bigcirc} + 2NH_2 \rightarrow \underset{NH_2}{\bigcirc} + NH_4Cl \quad (8)$$

Oxidation-reduction. These reactions may be subdivided into those which involve a change in the oxidation state of the nitrogen atom and those in which elemental hydrogen is liberated. An example of the first group is the catalytic oxidation of ammonia in air to form nitric oxide, shown in reaction (9). In the absence of a catalyst, ammonia burns in oxygen to yield nitrogen. This is shown by reaction (10). Another example, reaction (11), is the reduction with ammonia of hot metal oxides such as cupric oxide. Still another example is the reaction of ammonia with chlorine (12).

$$4NH_3 + 5O_2 \xrightarrow{Pt} 4NO + 6H_2O \quad (9)$$

$$4NH_3 + 3O_2 \rightarrow N_2 + 6H_2O \quad (10)$$

$$3CuO + 2NH_3 \rightarrow 3Cu + 3H_2O + N_2 \quad (11)$$

$$2NH_3 + Cl_2 \rightarrow NH_2Cl + NH_4Cl \quad (12)$$

Oxidation-reduction reactions of ammonia of the second type are exemplified by reactions of active metals with ammonia, shown in reactions (13) and (14).

$$2Na + 2NH_3 \rightarrow 2NaNH_2 + H_2 \quad (13)$$

$$3Mg + 2NH_3 \xrightarrow{\Delta} Mg_3N_2 + 3H_2 \quad (14)$$

Liquid ammonia as a solvent. The physical and chemical properties of liquid ammonia make it appropriate for use as a solvent in certain types of chemical reactions. The solvent properties of liquid ammonia are, in many ways, qualitatively intermediate between those of water and of ethyl alcohol. This is particularly true with respect to dielectric constant; therefore, ammonia is generally superior to ethyl alcohol as a solvent for ionic substances but is inferior to water in this respect. On the other hand, ammonia is generally a better solvent for covalent substances than is water. The chemical properties of ammonia, for example, its ability to undergo ammonation, ammonolysis, and oxidation-reduction reactions, are roughly analogous to reactions of water (hydration, hydrolysis, and oxidation-reduction). Both water and liquid ammonia undergo autoionization; liquid ammonia, reaction (15), undergoes the process to a lesser extent, as its very low ion product constant indicates, reaction (16).

$$2NH_3 \rightleftharpoons NH_4^+ + NH_2^- \qquad (15)$$

$$[NH_4^+][NH_2^-] = 1.9 \times 10^{-33} \text{ at} -50°C \qquad (16)$$

Among the particularly interesting aspects of chemistry in liquid ammonia is the fact that metals of sufficiently low lattice energy, high cation solvation energy, and low ionization energy are reversibly soluble in liquid ammonia. The solutions obtained are highly colored—blue if dilute, bronze if concentrated. Metals which form such solutions include the alkali metals, the heavier alkaline earth metals, and the divalent lanthanides. Concentrations as high as 10 to 20 molal are obtainable for the alkali metals. Dilute solutions exhibit the phenomenon of strong electrolytic conduction, whereas the concentrated solutions act as metallic conductors. At very high dilutions magnetic measurements indicate the presence of 1 mole of unpaired electrons per mole of dissolved metal, but as the concentrations of the solutions are increased, their paramagnetism decreases. Much remains to be learned concerning the nature of these solutions, but presently available data may be interpreted in terms of the five solute species M, $M_{2(am)}$, $M^-_{(am)}$, $M^+_{(am)}$, and $e^-_{(am)}$ participating in equilibria (17)–(19). The ammoniated electron, $e^-_{(am)}$, appears to consist of an electron trapped in a cavity (approximately 3 angstroms or 0.3 nm in diameter) of NH_3 molecules. These solutions are unstable but, in the absence of catalysts, decompose only very slowly to yield the metal amide plus hydrogen gas [reaction (20)]. Solutions of metals in liquid ammonia are excellent reducing agents and are particularly useful for the reduction of organic compounds which are miscible with liquid ammonia.

$$M \rightleftharpoons M^+_{(am)} + e^-_{(am)} \qquad (17)$$

$$2M \rightleftharpoons M_{2(am)} \qquad (18)$$

$$e^-_{(am)} + M \rightleftharpoons M^-_{(am)} \qquad (19)$$

$$e^-_{(am)} + NH_3 \rightarrow \tfrac{1}{2}H_2 + NH_2^- \qquad (20)$$

The usefulness of liquid ammonia as a solvent is based on the differences in the chemical properties of liquid ammonia and of other common solvents, notably water. Principal among these differences (compared with water) are (1) the lesser tendency of ammonia to release protons, (2) the greater electron donor tendency (or proton affinity) of ammonia, and (3) the stronger reducing character of ammonia. Because of the first difference, liquid ammonia may be used as a solvent for very strong bases (such as NH_2^- or $C_2H_5O^-$) which would undergo complete protolysis in aqueous solution. Liquid ammonia may also be used as a solvent for very strong reducing agents (such as solvated electrons) which would immediately displace hydrogen from water. Because of the second difference, liquid ammonia solutions do not provide very strong acids, since all strong acids are converted immediately to ammonium ion, NH_4^+, which is a much weaker acid than hydronium ion, H_3O^+, its counterpart in aqueous systems. In summary, it may be said that as a solvent for chemical reactions, liquid ammonia affords much stronger bases and stronger reducing agents but much weaker acids and weaker oxidizing agents than does water. By application of these differences, a number of interesting synthetic procedures may be carried out in liquid ammonia.

Ammonia system of compounds. Many of the familiar compounds which contain oxygen may be considered to be derived from water as the parent solvent. In an analogous way it is sometimes useful to consider many nitrogen-containing compounds to be derived from ammonia as parent solvent. These latter compounds are sometimes considered to constitute the nitrogen or ammonia system of compounds. This analogy is sometimes very useful in understanding the chemistry of various nitrogen compounds (Tables 2 and 3).

Synthesis of ammonia. The Haber-Bosch synthesis is the major source of industrial ammonia. In a typical process, water gas (CO, H_2, CO_2) mixed with nitrogen is passed through a scrubber cooler to remove dust and undecomposed material. The CO_2 and CO are removed by a CO_2 purifier and ammoniacal cuprous solution, respectively. The remaining H_2 and N_2 gases are passed over a catalyst at high pressures (up to 1000 atm) and high temperatures (approx 700°C) [reaction (21)].

$$3H_{2(g)} + N_{2(g)} \leftrightarrows 2NH_{3(g)} \qquad (21)$$

The ammonia is separated by absorption in water. Processes used vary widely in the sources of N_2 and H_2, treatment of the catalysts, temperature, pressure, and methods of ammonia separation.

Other industrial sources of ammonia include its formation as a by-product of the destructive distillation of coal, and its synthesis through the cyanamide process, which is indicated by reactions (22) and (23).

$$CaC_2 + N_2 \xrightarrow{1000°C} CaCN_2 + C \qquad (22)$$

$$CaCN_2 + 3H_2O\,(\text{steam}) \rightarrow CaCO_3 + 2NH_3 \qquad (23)$$

In the laboratory, ammonia is usually formed by its displacement from ammonium salts (either dry or in solution) by strong bases, as indicated by reaction (24). Another source is the hydrolysis of

Table 2. Analogous compounds in the water and ammonia systems

Aquo	Ammono
$(H_3O)Cl$	$(NH_4)Cl$
$(HO)_2C{=}O$	$(H_2N)_2C{=}NH$
KOH	KNH_2
$Na_2[Zn(OH)_4]$	$Na_2[Zn(NH_2)_4]$
$Cu(H_2O)_4^{2+}$	$Cu(NH_3)_4^{2+}$
MgO	MgNH, Mg_3N_2
$CH_3C(=O)OH$	$CH_3C(=NH)NH_2$
C_2H_5OH	$C_2H_5NH_2$
$(CH_3)_2O$	$(CH_3)_2NH$, $(CH_3)_3N$
$Hg(OH)Cl$	$HgNH_2Cl$
HOCl	H_2NCl

Table 3. Analogous reactions in the water and ammonia systems

Aquo	Ammono
$KOH + (H_3O)Cl \rightarrow KCl + 2H_2O$	$KNH_2 + (NH_4)Cl \rightarrow KCl + 2NH_3$
$Zn + 2H_3O^+ \rightarrow Zn^2 + H_2 + 2H_2O$	$Zn + 2NH_4^+ \rightarrow Zn^{2+} + H_2 + 2NH_3$
$CH_3C(=O)OC_2H_5 + H_2O \xrightarrow{H_3O^+} CH_3C(=O)OH + C_2H_5OH$	$CH_3C(=NH)NHC_2H_5 + NH_3 \xrightarrow{NH_4^+} CH_3C(=NH)NH_2 + C_2H_5NH_2$
$(OH)_2C{=}O + 2OH^- \rightarrow (O^-)_2C{=}O + 2H_2O$	$(NH_2)_2C{=}NH + 2NH_2^- \rightarrow (NH^-)_2C{=}NH + 2NH_3$
$Zn(OH)_2 + 2OH^- \rightarrow Zn(OH)_4^{2-}$	$Zn(NH_2)_2 + 2NH_2^- \rightarrow Zn(NH_2)_4^{2-}$

metal nitrides, as indicated by reaction (25).

$$NH_4^- + OH^- \xrightarrow{Heat} NH_3 + H_2O \quad (24)$$

$$Mg_3N_2 + 6H_2O \rightarrow 3Mg(OH)_2 + 2NH_3 \quad (25)$$

See AMIDE; AMINE; HIGH-PRESSURE PROCESSES; HYDRAZINE; NITROGEN. [HARRY H. SISLER]

Bibliography: H. H. Sisler et al., *College Chemistry: A Systematic Approach*, 4th ed., 1980.

Ammonium salt

A product of a reaction between ammonia, NH_3, and various acids. The general reaction for formation is (1). Examples of ammonium salts are am-

$$NH_3 + HX \rightarrow NH_2X \quad (1)$$

monium chloride, NH_4Cl, ammonium nitrate, NH_4NO_3, and ammonium carbonate, $(NH_4)_2CO_3$. These compounds are addition products of ammonia and the acid. For this reason, their formulas are sometimes written as $[H(NH_3)]X$.

All ammonium salts decompose into ammonia and the acid when heated. Their stability, however, varies according to the nature of the acid. Salts of weak acids decompose at lower temperatures than do salts of strong acids. Ammonium chloride, the salt of the strong acid, hydrogen chloride, HCl, decomposes at 320°C, whereas ammonium sulfide, $(NH_4)_2S$, the salt of the weak acid, hydrogen sulfide, H_2S, decomposes at 32°C. If the salt is heated in a closed vessel, a definite pressure of ammonia is established in the presence of the solid salt. This pressure is determined solely by the temperature and, if the acid is nonvolatile, is called the dissociation pressure at that temperature. For a detailed discussion of such equilibria *see* CHEMICAL EQUILIBRIUM.

If anhydrous ammonia is added to many of the ammonium salts at very low temperatures, salts containing several molecules of ammonia are formed. Ammonium chloride, for example, can add three or six molecules of ammonia to form complex salts. The reaction for the formation of the tetra compound is (2). The number of such

$$[H(NH_3)]Cl + 3(NH_3) \rightarrow [H(NH_3)_4]Cl \quad (2)$$

complexes which may be formed depends upon the nature of the acid radical. When warmed, they lose ammonia; all are unstable above 0°C.

Ammonium chloride is made by absorbing ammonia in hydrochloric acid. It crystallizes from the solution in feathery crystals of the regular crystal system. It is a colorless solid with a density of 1.52. This salt, sometimes called sal ammoniac, is used in galvanizing iron, in textile dyeing, and in manufacturing dry cell batteries.

Ammonium nitrate, NH_4NO_3, a colorless salt with a density of 1.73, is prepared from ammonia and nitric acid. The solid salt deliquesces, or absorbs water from moist air, thus appearing to melt. It is used as a source of nitrous oxide, N_2O, or laughing gas, and in the manufacture of explosives. A mixture of ammonium nitrate and trinitrotoluene is known as amatol.

Ammonium sulfate, $(NH_4)_2SO_4$, obtained from ammonia and sulfuric acid, is a colorless solid with a density of 1.77. It is prepared commercially by passing ammonia and carbon dioxide, CO_2, into a suspension of finely ground calcium sulfate, $CaSO_4$, as shown in reaction (3). Large quantities

$$CaSO_4 + 2NH_3 + CO_2 + H_2O \rightarrow CaCO_3 + (NH_4)_2 SO_4 \quad (3)$$

are also produced as a by-product of coke ovens and coal-gas works. The chief use of ammonium sulfate is as a fertilizer.

Ammonium carbonate, $(NH_4)_2CO_3$, may be prepared by bringing ammonia and carbon dioxide together in aqueous solution, shown in reaction (4).

$$2NH_3 + CO_2 + H_2O \rightarrow (NH_4)_2CO_3 \quad (4)$$

It is also obtained by heating a mixture of ammonium sulfate and a fine suspension of calcium carbonate, shown in reaction (5).

$$(NH_4)_2SO_4 + CaCO_3 \rightarrow (NH_4)_2CO_3 + CaSO_4 \quad (5)$$

Ammonium thiocyanate, NH_4SCN, a colorless solid with a density of 1.31, is prepared by the reaction of ammonia and carbon disulfide, CS_2, as in reaction (6). It is used as a protective agent in dyeing.

$$2NH_3 + CS_2 \rightarrow NH_4SCN + H_2S \quad (6)$$

Except for several complex species such as ammonium chloroplatinate, $(NH_4)_2PtCl_6$, ammonium salts are very soluble in water. In aqueous solutions, they ionize to produce the ammonium ion, NH_4^+, and an acid anion. Solutions of ammonium salts of strong or moderately strong acids are acidic as a result of hydrolysis of the ammonium ion.

The reaction involving a molecule of water and producing a hydrogen ion, H^+, is shown as reaction (7).

$$NH_4^+ + H_2O \rightleftharpoons NH_3 + H_3O^+ \qquad (7)$$

When a strong base is added to a solution of an ammonium salt, ammonia is evolved. This is a test for the presence of NH_4^+ ion and occurs because the reaction of reaction (8) is driven toward the left.

$$NH_3 + H_2O \rightleftharpoons NH_4^+ + OH^- \qquad (8)$$

See AMMONIA; HYDROLYSIS.

[FRANCIS J. JOHNSON]

Amorphous solid

A rigid material whose structure lacks crystalline periodicity; that is, the pattern of its constituent atoms or molecules does not repeat periodically in three dimensions. In the present terminology amorphous and noncrystalline are synonymous. A solid is distinguished from its other amorphous counterparts (liquids and gases) by its viscosity: a material is considered solid (rigid) if its shear viscosity exceeds $10^{14.6}$ poise ($10^{13.6}$ Pa·s). *See* CRYSTAL; VISCOSITY.

Preparation. Techniques commonly used to prepare amorphous solids include vapor deposition, electrodeposition, anodization, evaporation of a solvent (gel, glue), and chemical reaction (often oxidation) of a crystalline solid. None of these techniques involves the liquid state of the material. A distinctive class of amorphous solids consists of glasses, which are defined as amorphous solids obtained by cooling of the melt. Upon continued cooling below the crystalline melting point, a liquid either crystallizes with a discontinuous change in volume, viscosity, entropy, and internal energy, or (if the crystallization kinetics are slow enough and the quenching rate is fast enough) forms a glass with a continuous change in these properties. The glass transition temperature is defined as the temperature at which the fluid becomes solid (that is, the viscosity $= 10^{14.6}$ poise $= 10^{13.6}$ Pa·s) and is generally marked by a change in the thermal expansion coefficient and heat capacity. [Silicon dioxide (SiO_2) and germanium dioxide (GeO_2) are exceptions.] It is intuitively appealing to consider a glass to be both structurally and thermodynamically related to its liquid; such a connection is more tenuous for amorphous solids prepared by the other techniques.

Types of solids. Oxide glasses, generally the silicates, are the most familiar amorphous solids. However, as a state of matter, amorphous solids are much more widespread than just the oxide glasses. There are both organic (for example, polyethylene and some hard candies) and inorganic (for example, the silicates) amorphous solids. Examples of glass formers exist for each of the bonding types: covalent [As_2S_3], ionic [$KNO_3 - Ca(NO_3)_2$], metallic [Pd_4Si], van der Waal's [*o*-terphenyl], and hydrogen [$KHSO_4$]. Glasses can be prepared which span a broad range of physical properties. Dielectrics (for example, SiO_2) have very low electrical conductivity and are optically transparent, hard, and brittle. Semiconductors (for example, As_2SeTe_2) have intermediate electrical conductivities and are optically opaque and brittle. Metallic glasses (for example, Pd_4Si) have high electrical and thermal conductivities, have metallic luster, and are ductile and strong.

Uses. The obvious uses for amorphous solids are as window glass, container glass, and the glassy polymers (plastics). Less widely recognized but nevertheless established technological uses include the dielectrics and protective coatings used in integrated circuits, and the active element in photocopying by xerography, which depends for its action upon photoconduction in an amorphous semiconductor. In optical communications a highly transparent dielectric glass in the form of a fiber is used as the transmission medium. In addition, metallic amorphous solids have been considered for uses that take advantage of their high strength, excellent corrosion resistance, extreme hardness and wear resistance, and unique magnetic properties.

Semiconductors. It is the changes in short-range order (on the scale of a localized electron), rather than the loss of long-range order alone, that have a profound effect on the properties of amorphous semiconductors. For example, the difference in resistivity between the crystalline and amorphous states for dielectrics and metals is always less than an order of magnitude and is generally less than a factor of 3. For semiconductors, however, resistivity changes of 10 orders of magnitude between the crystalline and amorphous states are not uncommon, and accompanying changes in optical properties can also be large.

Electronic structure. The model that has evolved for the electronic structure of an amorphous semiconductor is that the forbidden energy gap characteristic of the electronic states of a crystalline material is replaced in an amorphous semiconductor by a pseudogap. Within this pseudogap the density of states of the valence and conduction bands is sharply lower but tails off gradually and remains finite due to structural disorder (Fig. 1). The states in the tail region are localized; that is, their wave functions extend over small distances in contrast to the extended states that exist elsewhere in the energy spectrum. Because the localized states have low mobility (velocity per unit electric field), the extended states are separated by a mobility gap (Fig. 1) within which charge transport is markedly impeded. In each band, the energy at which the extended states meet the localized states is called the mobility edge.

An ideal amorphous solid can be conceptually defined as having no unsatisfied bonds, a minimum of bond distortions (bond angles and lengths), and no internal surfaces associated with voids. Deviations from this ideality introduce localized states in the gap in addition to those in the band edge tails

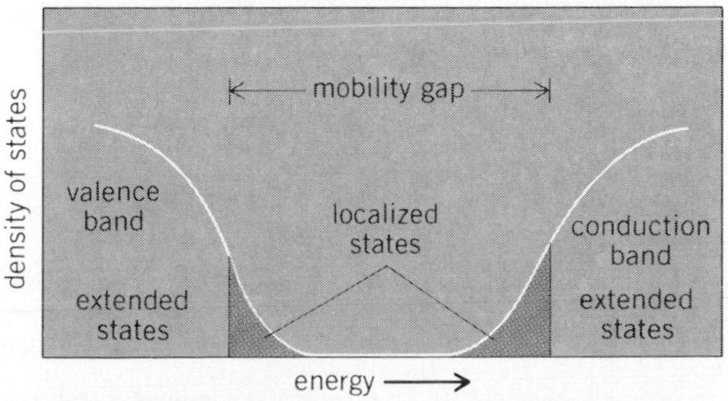

Fig. 1. Density of states versus energy for an amorphous semiconductor.

due to disorder alone. One important defect is called an unsatisfied, broken, or dangling bond. These dangling bonds create states deep in the gap which can act as recombination centers and markedly limit carrier lifetime and mobility. A large number of such states introduced, for example, during the deposition process will dominate the electrical properties.

Charge transport can occur by two mechanisms. The first is conduction of mobile extended-state carriers (analogous to that which occurs in crystalline semiconductors), for which the conductivity is proportional to exp $(-E_g/2kT)$, where E_g is the gap width, T is the absolute temperature, and k is Boltzmann's constant. The second mechanism is hopping of the localized carriers, for which the conductivity is proportional to exp $[-(T_0/T)^{1/4}]$, where T_0 is a constant (Mott's law). At low temperatures carriers hop from one localized trap to another, whereas at high temperatures they can be excited to the mobility edge.

Glassy chalcogenides. One class of amorphous semiconductors is the glassy chalcogenides, which contain one (or more) of the chalcogens sulfur, selenium, or tellurium as major constituents. These amorphous solids behave like intrinsic semiconductors, show no detectable unpaired spin states, and exhibit no doping effects. It is thought that essentially all atoms in these glasses assume a bonding configuration such that bonding requirements are satisfied; that is, the structure accommodates the coordination of any atom. These materials have application in switching and memory devices.

Tetrahedrally bonded solids. Another group is the tetrahedrally bonded amorphous solids, such as amorphous silicon and germanium. These materials cannot be formed by quenching from the melt (that is, as glasses) but must be prepared by one of the deposition techniques mentioned above. An amorphous to crystalline transformation in these materials is irreversible.

When amorphous silicon (or germanium) is prepared by evaporation, not all bonding requirements are satisfied, so a large number of dangling bonds are introduced into the material. These dangling bonds are easily detected by spin resonance or low-temperature magnetic susceptibility and create states deep in the gap which limit the transport properties. The number of dangling bonds can be reduced by a thermal anneal below the crystallization temperature, but the number cannot be reduced sufficiently to permit doping.

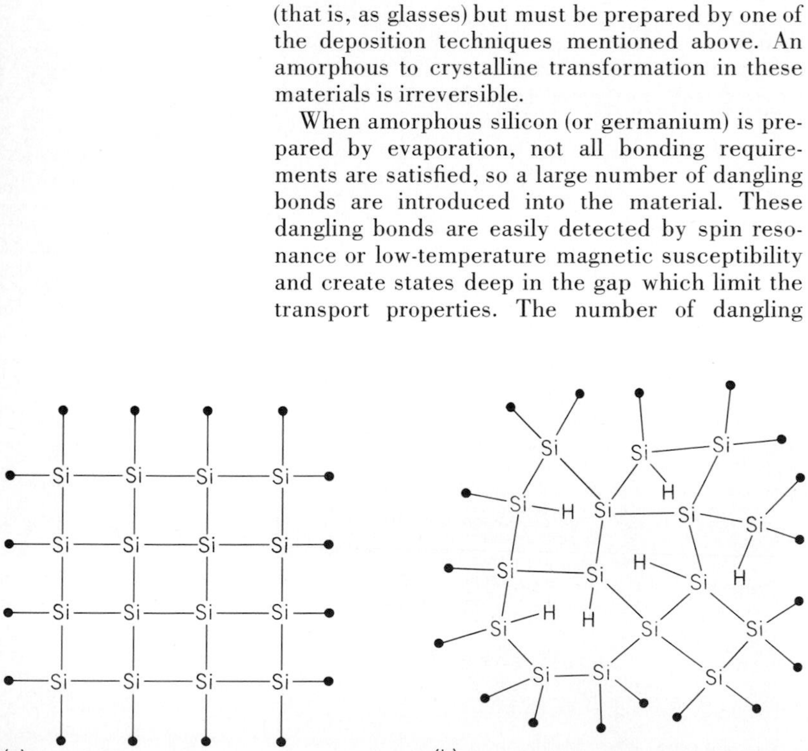

Fig. 2. Bonding of silicon. Bonds which continue the network are shown terminated by a dot. *(a)* Crystalline arrangement. *(b)* Amorphous structure with dangling bonds terminated by hydrogen.

Amorphous silicon prepared by the decomposition of silane (SiH_4) in a plasma has been found to have a significantly lower density of defect states within the gap, and consequently the carrier lifetimes are expected to be longer. This material can be doped *p*- or *n*-type with boron or phosphorus (as examples) by the addition of B_2H_6 or PH_3 to the SiH_4 during deposition. This permits exploration of possible devices based on doping, which are analogous to devices based on doping of crystalline silicon.

One reason plasma-deposited silicon has a significantly lower density of defect states within the gap is that the process codeposits large amounts of hydrogen (typically 5–30% of the atoms, depending upon deposition conditions), and this hydrogen is very effective at terminating dangling bonds (Fig. 2). Other possible dangling-bond terminators (for example, fluorine) have been explored.

The ability to reduce the number of states deep in the gap and to dope amorphous silicon led directly to the development of an amorphous silicon photovoltaic solar cell. Intense effort has been devoted to improving the efficiency of these cells to the 8% level thought to be required for large-scale application. The appeal of amorphous silicon is that it holds promise for low-cost, easily fabricated, large-area cells.

Amorphous silicon solar cells have been constructed in heterojunction, *p-i-n*-junction, and Schottky-barrier device configurations and have been introduced for use in calculators and watches. The optical properties of amorphous silicon provide a better match to the solar spectrum than do those for crystalline silicon, but the transport properties of the crystalline material are better. Experiments indicate that hole transport in the amorphous material is the limiting factor in the conversion efficiency.

[BRIAN G. BAGLEY]

Bibliography: R. H. Doremus, *Glass Science*, 1973; N. F. Mott and E. A. Davis, *Electronic Processes in Non-Crystalline Materials*, 2d ed., 1979; J. Tauc (ed.), *Amorphous and Liquid Semi-conductors*, 1974.

Amphoterism

The manifestation of both acidic and basic properties by one chemical substance. The hydroxides or oxides of aluminum, trivalent chromium, zinc, divalent tin, lead, antimony, arsenic, gold, and platinum are familiar examples of amphoteric substances. Consider the behavior of crystalline $Zn(OH)_2$, which is soluble in both acids and bases. An often-performed laboratory demonstration is the following: To an aqueous solution containing *n* moles of zinc chloride, 2*n* moles of NaOH is added. A white precipitate of $Zn(OH)_2$ (possibly initially hydrated) is produced. Another 2*n* moles of NaOH plus a small excess is now added, and all of the zinc is again in solution. It is now in zincate ions. The zincate ion is the result of the loss of two protons (H^+) by the zinc hydroxide. These are neutralized by 2 moles of OH^-. The formula of the zincate ion is sometimes written as ZnO_2^{--}, but recent studies indicate that the ion is $Zn(OH)_4^{--}$ (possibly containing additional water of hydration). Because zinc hydroxide acts either as an acid or a base, it is said to be an amphoteric substance.

Stepwise phenomena may be detectable if conditions can be sufficiently well controlled to reveal the presence of an intermediate such as $ZnOH^+$ or $HZnO_2^-$. This subject is too specialized for discussion here.

Many organic molecules contain both carboxyl groups (acidic) and amino groups (basic). As a result, they are amphoteric. A very simple example is glycine, $H_2N—CH_2—CO_2H$. In aqueous solution the carboxyl group dissociates to produce a hydrogen ion. In so doing, it leaves a negative charge on one end of the molecule. At the other end of the same molecule the amino group accepts a proton and thereby acquires a positive charge. The result is the hybrid ion (zwitterion), $^+NH_3—CH_2—CO_2^-$.

A solution containing the ion resists either an increase in $[H_3O^+]$ or a decrease because the ion (and glycine itself) are amphoteric. In this example no precipitation is involved. The phenomena are easily observed, however, with a pH meter and in a well-equipped laboratory with electrical apparatus which can detect the large dipole moment of the zwitterion. *See* ACID AND BASE; ELECTROPHORESIS; IONIC EQUILIBRIUM.

[THOMAS F. YOUNG]

Analytical chemistry

The science of chemical characterization and measurement. Qualitative analysis is concerned with the description of chemical composition in terms of elements, compounds, or structural units, whereas quantitative analysis is concerned with the measurement of amount.

Originally, chemical analysis conformed closely to its literal meaning and consisted of the separation of a sample into its components, which were weighed. Thus a rock could be regarded as being composed of metallic or alkaline oxides and nonmetallic or acidic oxides, the amounts of which were determined by weighing the oxides as such or by weighing other forms which could be calculated as oxides.

Later, gravimetric analysis was supplemented by volumetric or (to use a better term) titrimetric analysis, which consists of measurement by the addition of a standardized solution of reagent which reacts with a measured portion of sample to an end point, usually detected by means of an indicator or substance which changes color when a given amount of reagent has been added. Today these forms of chemical analysis are commonly called classical or wet methods, to distinguish them from the so-called instrumental methods, which have greatly enlarged the scope of analytical chemistry since the 1930s. *See* GRAVIMETRIC ANALYSIS; TITRATION; VOLUMETRIC ANALYSIS.

Instrumental methods, which might better be termed physicochemical methods, make up the vast bulk of both qualitative and quantitative analysis today. They include measurements of purely physical characteristics, as well as the use of physical measurements to follow the course of chemical reactions. In the latter category are titrations in which the end points are detected by the measurement of electrical quantities such as electrode potential, conductance, current, impedance, and capacitance; optical quantities such as transmittance, absorbance, emittance, scattering, and reflectance; or other physical quantities such as temperature, density, and surface tension.

In recent years the scope of analytical chemistry has been enlarged in comparison to its original meaning, the determination of chemical composition in terms of the relative amounts of elements or compounds in a sample. It now may involve the spatial distribution of elements or compounds within a sample, the distinction between different crystalline forms of a given element or compound, the distinction between different chemical forms (such as the oxidation states of an element), the distinction between a component on the surface or in the interior of particles, and so forth. To permit these more detailed questions to be answered, as well as to improve the speed, accuracy, sensitivity, and selectivity of traditional analysis, a large variety of physical measurements are used. These include methods based on spectroscopic, electrochemical, chromatographic, chemical, and nuclear principles, which will be discussed in turn.

Spectroscopy. The spectroscopic area includes the measurement of emission, absorption, reflection, and scattering phenomena in various regions of the spectrum ranging from gamma rays and x-rays through the ultraviolet, visible, infrared, and microwave. Coupled with these measurements are a wide variety of sample preparation and excitation techniques. Thus the introduction of a sample as either a solid or a solution into a flame, arc, spark, or plasma may be the source of excitation for the measurement of atomic emission, atomic absorption, or atomic fluorescence spectra.

Molecular spectroscopy. Lower-energy forms of excitation such as ultraviolet, visible, or infrared radiation are used in molecular spectroscopy. X-rays are used either through emission of characteristic radiation, absorption, or diffraction. In the last case, characteristic diffraction patterns reveal information about specific structural entities, such as a particular crystalline form of a chemical compound. Gamma-ray spectroscopy, used to sort out and measure gamma rays according to energy, is usually coupled with nuclear techniques such as neutron activation analysis or photon activation analysis. *See* ACTIVATION ANALYSIS.

Mass spectrometry. Mass spectrometry is an important method for both inorganic and organic analysis. In the inorganic field, spark-source mass spectrometry permits the detection and estimation of some 70 elements. Secondary-ion mass spectrometry (SIMS) involves the bombardment of a surface with an ion beam, causing the ejection of ions characteristic of the sample. These secondary ions are sorted out by mass number for analysis. Ion-probe mass spectrometry permits the scanning of a sample to reveal the point-by-point elemental composition of a surface; by gradual erosion of the surface, it determines the composition in the third dimension as well. In the organic field, important structural information is gleaned through high-resolution mass spectrometry, in which molecular fragments are identified through fractional mass differences in atoms of the same principal mass number. *See* MASS SPECTROMETRY.

Nuclear magnetic resonance. More subtle nuclear phenomena are employed in nuclear magnetic resonance (NMR) spectroscopy. In this technique the chemical environment of atoms possessing nuclear spins can be probed through the interaction between the nuclear spin and an external magnetic field. *See* NUCLEAR MAGNETIC RESONANCE (NMR).

Electron spectroscopy. Several forms of spectroscopy are especially useful for surface analysis. The scanning electron microscope (SEM) involves a finely collimated electron beam which sweeps across the surface to produce an image. At the same time, the surface atoms are excited to emit characteristic x-rays which are sorted out according to their energy to reveal point-by-point elemental composition. *See* ELECTRON SPECTROSCOPY.

The same basic technique is used in the electron microprobe, a specialized instrument designed for high resolution. Low-energy electron diffraction (LEED) involves the observation of the diffraction pattern of a reflected electron beam to give structural information about the arrangement of surface atoms.

Several techniques are used to determine compositional and structural information by measuring the energies of electrons emitted from surface atoms. These include ultraviolet photoelectron spectroscopy (UPS) and x-ray photoelectron spectroscopy (XPS), which differ in the energy of the photon beam striking the surface. The latter technique is sometimes called electron spectroscopy for chemical analysis (ESCA). A related technique, Auger spectroscopy, involves the bombardment of a surface with an electron beam and the observation of the energies of electrons secondarily emitted from surface atoms. Once again, the surface composition and some structural details are revealed. *See* SPECTROSCOPY.

Electrochemical analysis. Methods based on electrochemical principles began with the detection of titration end points through the measurement of electrical quantities, as mentioned above. Coulometry utilizes the faraday (96,500 coulombs) as the basis for quantitative measurement, either via a titration with electrochemically generated reagent or via quantitative electrolysis.

Potentiometry, the measurement of electrode potentials, is used with various ion-selective electrodes. The earliest and most important example is the glass electrode for pH. Later innovations include glass electrodes selective for sodium or potassium ions, solid-state sensors for certain anions and heavy-metal cations, liquid membrane electrodes for anions and cations, gas-permeable membrane electrodes for NH_3, CO_2, and H_2S, composite electrodes incorporating enzymes to detect the substrate, or composite electrodes incorporating the substrate to detect an enzyme system.

Various forms of electroanalysis are carried out by applying an electrical signal to an indicator electrode and observing a readout signal. The applied signal may be a constant or other linear potential or current, a pulse, or a sinusoidal or more complex waveform. The readout may be a current, a potential, or a quantity of electricity measured as a function of time. These methods are named according to the measurements; they include cyclic voltammetry, chronopotentiometry, chronoamperometry, and chronocoulometry. It is characteristic that electrolysis proceeds for short time intervals with diffusion in solution being a rate-controlling process, sometimes complicated by chemical reactions in solution coupled with the primary electrode process. Polarography, a special form of diffusion-controlled electrolysis with the dropping mercury electrode, is useful for a wide variety of inorganic and organic analytical applications. *See* COULOMETRIC ANALYSIS; ELECTROCHEMISTRY; POLARAGRAPHIC ANALYSIS.

Chromatography. Chromatographic methods are based on the separation of mixtures by a phase distribution using multiple stages. The stages are not discrete physical entities but are functionally achieved by using a mobile phase, either liquid or gaseous, flowing past a stationary phase, either solid or liquid (in the form of a film on a solid support). In column chromatography the stationary phase is contained in a cylindrical column, whereas in thin-layer chromatography a flat layer is used. In gas chromatography the mobile phase is gaseous, while in liquid chromatography it is liquid. A special form of liquid chromatography, high-performance liquid chromatography (HPLC), has become especially important for the separation of complex mixtures of nonvolatile materials. Essentially, it involves the use of very small particle size in column packing, with high pressures to increase the rate of flow and shorten the separation time to a few minutes.

Elution is a process by which the sample is caused to move along the column or layer by the addition of a mobile phase. Separation is achieved through differences in the rate of elution of various components. To monitor the elution process, a detection method is used either on the column or thin-layer plate or on the mobile phase as it exists. The detection method may be either nonselective, simply detecting the presence of solute, or selective, identifying each fraction. Thermal conductance in the gas phase is an example of a nonselective detection method, while mass spectrometry is an example of a selective detection method. One- or two-dimensional elution may be used in thin-layer chromatography, and the process may be modified by the application of an electric field to separate charged species, as in electrophoresis. Gel-permeation chromatography is a special form of liquid chromatography in which large molecules move faster than small ones because they fail to permeate the structure of the stationary phase. *See* CHEMICAL SEPARATION TECHNIQUES; CHROMATOGRAPHY.

Chemical methods. Chemical reactions form the basis of many analytical methods besides the classical forms of gravimetric and titrimetric analysis. Thermal methods are a group of methods based on the heating of a sample over a range of temperatures. Thermogravimetry involves the measurement of mass; differential thermal analysis involves a detection of chemical or physical processes through a measurement of the difference in temperature between the sample and a stable reference material; differential thermal coulometry evaluates the heat involved in such processes. In enthalpimetric analysis the heat involved in a solution reaction is measured either during a titration or following the sudden addition of an excess of reagent. In kinetic analysis the dependence of the rate of a chemical reaction on concentration is used to determine the concentration.

Handling and processing methods. Besides this larger array of chemical and physical methods of characterization and measurement, the field of analytical chemistry also includes the areas of sampling, data processing, automation, and research in instrumentation.

Sampling. As mentioned above, the sampling operation sometimes is incorporated into the measurement operation, but more frequently a separate step of procuring samples is involved. Statistical methods are employed to ensure that the sample analyzed is representative of the material under study.

Computers. The digital computer has greatly influenced modern analytical chemistry. In its simplest form, the microprocessor, it is being incorporated as an integral part of an increasing variety of analytical instruments to permit detailed programming of their operation. More complex programming and automation are possible through the microcomputer, which is being used not only to operate a single instrument but to tie various analytical operations together. Still more complex computer operations, such as pattern recognition, are being developed to draw conclusions through comparison of arrays of data.

Automation. Automation of analysis originally was developed as a mechanized version of classical analytical methods, but it has increasingly taken the form of a computer-controlled operation. In principle, optimization of data acquisition is achieved through rapid processing of the initial instrument output and feeding the information back to control the instrument. Increased speed as well as improved sensitivity and accuracy can thus be achieved.

Instrumentation. Finally, instrumentation, the development and perfection of instruments rather than just their application, is a recognized area of analytical chemistry.

Errors in analytical chemistry. Every quantitative measurement is subject to errors of two types, determinate errors and indeterminate errors. Determinate errors are those that in principle can be traced to a cause, and that operate in one direction only. Examples of such causes are incorrect calibration of a weight or volumetric apparatus or loss of a precipitate owing to solubility. Indeterminate errors are those inherent in a given repetitive measurement, beyond the direct control of the observer, that operate to give scatter in both directions. Examples are the scatter due to background noise in electronic measurements or the inherent limit in readability of a meter scale.

Determinate errors can be estimated by comparison between analytical methods operating on different principles, and they can be reduced by seeking and eliminating the causes. Indeterminate errors can be estimated on statistical grounds and reduced by increasing the number of measurements. Accuracy is the deviation of a measurement or set of measurements from the correct result, while precision is a measure of the scatter in a set of measurements. Accuracy is a statement of correctness, while precision is a statement of reproducibility.

The reliability of a quantitative measurement is estimated statistically by assuming a bell-shaped, or gaussian, distribution of indeterminate errors, which frequently is applicable to a set of presumably identical measurements. Thus, it is common practice to express a result as $X \pm a$, where X is the most probable value, namely the average of a number of measurements, and $\pm a$ is the range within which some specified percentage, say 95 or 99%, of measurements can be statistically expected to fall. It is common practice to calculate the standard deviation S, defined as the square root of the sum of the squares of the deviations divided by the number of measurements minus one, as a measure of precision. The value $\pm a$ can then be expressed as a multiple of S, using statistical tables to represent the desired confidence level, say 95 or 99%, for the final result. It should be kept in mind that the confidence interval thus estimated relates only to the indeterminate errors of the measurement.

A special problem of confidence level is involved in expressing the detection limit of an analytical method. Strictly speaking, an analysis can never reveal zero, but only "less than X," where X is related to the precision of the method. Although no uniform procedure is obeyed in the estimation of X, it is common practice to estimate the noise level of a measurement, that is, the random background reading at zero level of analyte or desired constituent, and to express X as a small multiple, say two or three times the noise level. The confidence level of the result "less than X" is statistically related to the multiple used in its estimation, as well as to the number of measurements made.

A distinction should be made between the limit of detection and the limit of measurement, or quantitative estimation. To establish the limit of measurement, a working curve, or plot of response versus analyte concentration, is drawn. The limit of measurement is taken as the minimum concentration at which the signal exceeds background by a factor which is specified. Once again, the confidence level, usually chosen to be at least 95%, is statistically related to the factor chosen as well as to the number of measurements made. *See also* QUANTITATIVE CHEMICAL ANALYSIS.

[HERBERT A. LAITINEN]

Bibliography: G. W. Ewing, *Instrumental Analysis*, 3d ed., 1975; H. A. Laitinen and W. E. Harris, *Chemical Analysis*, 2d ed., 1975; L. Meites (ed.), *Handbook of Analytical Chemistry*, 1963.

Aniline

An aromatic primary amine used in the manufacture of urethane polymers (50%), rubber chemicals (antioxidants, accelerators), agricultural chemicals (herbicides, insecticides, fungicides), and dyes, in that order of volume. Its properties are: boiling point 184°C; melting point −6°C; density 1.0215; refractive index 1.5863; solubility about 3 g/100 g of water; basic dissociation constant $K_b = 3.8 \times 10^{-10}$.

Aniline is prepared by the catalytic reduction of nitrobenzene in the presence of hydrogen. Formerly the reagent for reduction was iron and water in the presence of hydrochloric acid.

Aniline readily undergoes the reactions characteristic of primary aromatic amines. The amine group is a strong ortho-, para-directing group so that chlorination or bromination of aniline leads to a 2,4,6-trihalogen derivative at a rapid rate. Treatment with hypochlorous acid (an oxidizing agent) chlorinates the nitrogen, yielding *N,N*-dichloroaniline. *See* AMINE.

Oxidation of aniline may lead to phenylhydroxylamine, nitrosobenzene, or bimolecular products with certain reagents such as hydrogen peroxide. With other reagents, such as chromic acid or sodium chlorate, the products are azobenzene, *N*-phenylquinonimine, aniline black, and compounds of bimolecular and polymolecular character.

Reduction with hydrogen and a Raney nickel catalyst gives cyclohexylamine, resulting from addition of hydrogen to the aromatic ring.

Carbonic acid derivatives of aniline result from reaction of aniline with phosgene. Diphenylurea, $(C_6H_5NH)_2CO$ (also called carbanilide), is obtained with excess aniline, whereas phenylisocyanate, C_6H_5NCO, is best prepared by heating aniline hydrochloride with excess phosgene. The corresponding thio derivatives result from reaction with the sulfur analog of carbon dioxide, carbon disulfide. Diphenylthiourea (thiocarbanilide), $(C_6H_5NH)_2CS$, precipitates when carbon disulfide is refluxed with aniline. Thiocarbanilide is the starting point for making 2-mercaptobenzothiazole, a rubber accelerator. Phenylisothiocyanate, C_6H_5NCS, is formed by the action of hydrochloric acid on thiocarbanilide. *See* ANTHRANILIC ACID; AROMATIC HYDROCARBON; DIAZOTIZATION; NAPHTHYLAMINE; NITROBENZENE; SULFANILIC ACID; TOLUIDINE. [LEALLYN B. CLAPP]

Anthracene

A colorless crystalline hydrocarbon, $C_{14}H_{10}$, which melts at 216.2°C and boils at 340°C. When pure, anthracene shows a blue-violet fluorescence. It is obtained from coal tar, which usually contains about 1% of the hydrocarbon.

Mild reduction of anthracene yields 9,10-dihydroanthracene, while oxidation yields anthraquinone. Irradiation with ultraviolet light yields a photodimer in which two molecules of the hydrocarbon are linked together through the 9,10, or meso, positions.

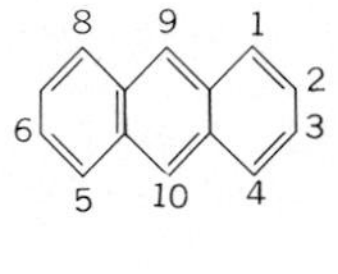

Anthracene

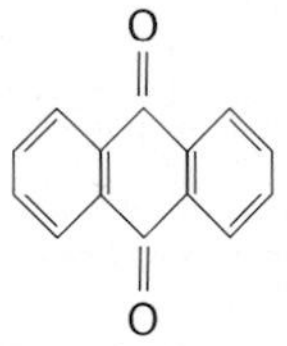

Anthraquinone

Anthracene has had little commercial importance since a direct method for the synthesis of anthraquinone from benzene derivatives was discovered. Anthracene has been used as a light-screening additive for plastics and finds some application in the manufacture of pesticides. *See* AROMATIC HYDROCARBON; POLYNUCLEAR HYDROCARBON. [CHARLES K. BRADSHER]

Anthranilic acid

Ortho-aminobenzoic acid, $NH_2—CO—C_6H_4—COO—Na^+$, a white, crystalline acid with a sweet taste, melting at 146°C, conveniently crystallized from hot water. It is prepared by hypochlorite oxidation of sodium phthalamate. *See* AMINE.

Anthranilic acid is used in the synthesis of diphenic acid and as an intermediate in the preparation of dyes such as thioindigo. The ester, methyl anthranilate, is present in, and is partly responsible for the flavor and odor of, grape juice. It is also an important constituent of the essential oil of orange blossoms. The synthetic ester is used as a grape flavor agent and also as an effective ultraviolet radiation barrier in antisunburn preparations. Anthranilic acid occurs in the blood and urine under various pathological conditions. *See* CARBOXYLIC ACID; DYE.

[EVANS B. REID]

Anthraquinone pigments

Coloring materials which occur in plants, fungi, lichens, and insects. About 50 different derivatives of the parent compound, anthraquinone, have been isolated. Several of the pigments have been used as dyes; others have been utilized as cathartic drugs.

The best-known anthraquinone dye is alizarin, a natural dye known to the ancient Egyptians and Persians. It occurs in the root of the madder

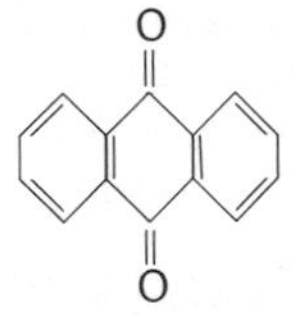

Anthraquinone

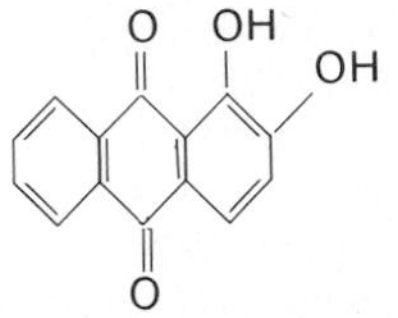

Alizarin

(*Rubia tinctorum*), native to Asia. Madder became an article of commerce in Europe as early as the year 700, and cultivation in Europe increased to an estimated 70,000 tons per year by 1868. Although alizarin was first isolated in 1826, all early attempts to determine its structure failed. The structure was ultimately established in 1868 by C. Graebe and C. Liebermann, who also synthesized it. The synthetic alizarin soon drove the natural product from the market, and a considerable disturbance to the agricultural economy of western Europe resulted.

Another natural anthraquinone pigment is carminic acid. It is the dyeing principle of the cochineal dye, obtained from the dried bodies of the female of the insect *Coccus cacti*, which lives on cactus native to Mexico. This and other natural anthraquinone dyes have largely given way to better and cheaper synthetic dyes.

Emodin is an anthraquinone pigment with strong cathartic properties and is an active principle of cascara, senna, aloe, and rhubarb. *See* DYE. [S. MORRIS KUPCHAN]

Antimonate

The negative radical in salts derived from antimony pentoxide, Sb_4O_{10}, and bases. Most antimonates contain the $[Sb(OH)_6]^-$ ion in spite of the fact that they were called pyro- or meta-antimonates by analogy with phosphates and arsenates. There is very little similarity between the phosphates and antimonates. The hypothetical acid, $HSb(OH)_6$, is not isolated; instead, a gelatinous precipitate of hydrated antimony pentoxide is obtained when antimony solutions are acidified. Sodium salt, $NaSb(OH)_6$, is one of the very few insoluble sodium salts and can be used as a test for the sodium ion. *See* ANTIMONY; ARSENATE; PHOSPHATE.

[E. EUGENE WEAVER]

Antimony

A chemical element, symbol Sb, atomic number 51. Antimony is not a naturally abundant element; it is occasionally found native, often in isomor-

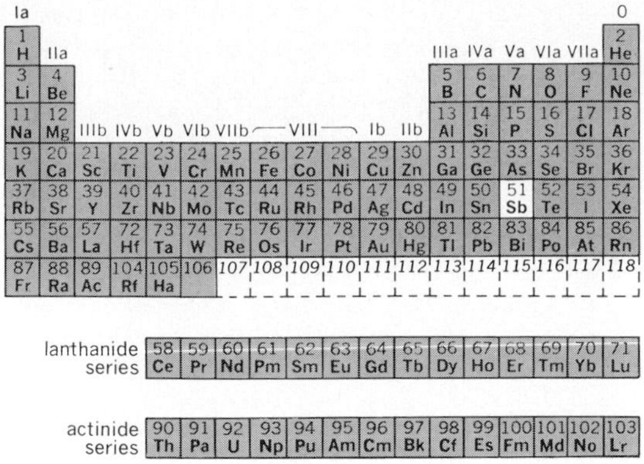

phous mixture with arsenic, as allemonite. The symbol Sb is derived from the Latin name stibium.

General characteristics. The element is composed of two stable nuclides, ${}_{51}Sb^{121}$ (57.25%) and ${}_{51}Sb^{123}$ (42.75%); of the 31 other known nuclides, most of which are short-lived β-emitters, only ${}_{51}Sb^{125}$ has a substantial half-life (2.7 years). The element is available commercially in 99.999+% purity and is finding increasing use in semiconductor technology.

The element is dimorphic, existing as a yellow, metastable form (I), composed of Sb_4 molecules, as in antimony vapor and the structural unit in yellow antimony; and a gray, metallic form (II), which crystallizes with a layered rhombohedral structure, having three nearest-neighboring atoms at 291 picometers and another three at 336 pm. Antimony differs from normal metals in having a lower electrical conductivity as a solid than as a liquid (as does its congenor, bismuth). Metallic antimony is quite brittle, bluish-white with a typical metallic luster, but a flaky appearance. Although stable in air at normal temperatures, it burns brilliantly when heated, with the formation of a white smoke of Sb_2O_3. There are two polymorphic modifications of Sb_2O_3, the high-temperature, cubic form (senarmontite) transforming to the low-temperature, orthorhombic variety (valentinite) at 879 K. Vaporization of the metal gives molecules of Sb_4O_6, which break down to Sb_2O_3 above the transition temperature. Valentinite does not contain Sb_4O_6 molecules, but infinite double chains (III). Some of the more significant properties of atomic and bulk antimony are given in the table.

Some atomic and bulk properties of antimony

Atomic weight	121.75
Electron configuration (^4S ground state)	$[Kr]4d^{10}5s^25p^3$
Covalent radius	141 pm
Ionic radius (Sb^{3+})	90 pm
Metallic radius	159 pm
Ionization energies, 1st–6th in kJ mole^{-1}	834, 1592, 2450, 4255, 5400, 10420
Electrode potential, Sb^{3+}/Sb	0.21 V
Electronegativity (Allred-Rochow)	1.82
Oxidation numbers	−3, 0, +3, +5
Specific gravity	6.691
Melting point	903.9 K
Boiling point	2026.16 K
Electrical resistivity	39.0 μohm cm (273 K)
Toxicity level	0.5 mg m^{-3} of air

Occurrence, extraction, and uses. Antimony occurs in nature mainly as Sb_2S_3 (stibnite, antimonite); Sb_2O_3 (valentinite) occurs as a decomposition product of stibnite. Antimony is commonly found in ores of copper, silver, and lead. The metal antimonides NiSb (breithaupite), NiSbS (ullmannite), and Ag_2Sb (dicrasite) also are found naturally; there are numerous thioantimonates such as Ag_3SbS_3 (pyrargyrite).

Antimony is produced either by roasting the sulfide with iron, or by roasting the sulfide and reducing the sublimate of Sb_4O_6 thus produced with carbon; high-purity antimony is produced by electrolytic refining.

Commercial-grade antimony is used in many alloys (1–20%), especially lead alloys, which are much harder and mechanically stronger than pure lead; batteries, cable sheathing, antifriction bearings, and type metal consume almost half of all the antimony produced. The manufacturing of flameproofing compounds, ceramic enamels, glasses, glazes, and paints utilizes about as much in the form of Sb_2O_3, Sb_2S_5, $SbCl_3$, and $Na[Sb(OH)_6]$. The valuable property of Sn-Sb-Pb alloys, that they expand on cooling from the melt, thus enabling the production of sharp castings, makes them especially useful as type metal (common type metal is composed of 58% Pb, 26% Sn, 15% Sb, 1% Cu); other such alloys are white metal (15% Sb), britannia metal (3.7–15% Sb), pewter (0–7.6% Sb), and tinfoil (0.5% Sb). The main consumer of Sb_2S_5 is the rubber industry. $SbCl_3$ is used as a caustic in medicine and for etching metals. Potassium antimono-tartrate (tarter emetic), $K[C_4H_2O_5Sb(OH)_2] \cdot 1/2\ H_2O$ whose anion is shown in structure (IV), has been used in medicine since the 1700s.

Stereochemistry. Antimony, like arsenic, is essentially a tervalent element, but *d* hybridization enables Sb atoms to form pyramidal, bipyramidal, square-pyramidal, and octahedral sets of bonds. Simple molecules containing multiple bonds are not formed. The trigonal-pyramidal shape of SbH_3, all the Sb trihalides, and of $Sb(CF_3)_3$ has been demonstrated by electron diffraction and molecular mass studies of the vapors and by x-ray studies of the solids. Sb^{5+} forms many compounds containing tetrahedral bonds, as for example GaSb and InSb, which crystallize with the zincblende structure. $SbCl_5$ has the trigonal-bipyramidal configuration in both the vapor and crystalline state. $Sb(C_6H_5)_5$ has a tetragonal-pyramidal shape, in contrast to the P and As analogs, and is the only known example of an M^{5+} species (N, P, As, Bi) forming such a molecule which is not trigonal-bipyramidal. Many substances which used to be formulated as hydrated meta- and pyroantimonates, in analogy with P compounds, are now known to be quite dissimilar in structure. Some of these compounds are $Na[Sb(OH)_6]$, formerly described as $Na_2H_2Sb_2O_7 \cdot 5H_2O$; $Li[Sb(OH)_6]$, formerly referred to as $LiSbO_3 \cdot 3H_2O$; $[Mg(OH_2)_6][Sb(OH)_6]_2$, formerly $Mg(SbO_3)_2 \cdot 12H_2O$; and $[Cu(NH_3)_3(OH_2)_3][Sb(OH)_6]_2$, previously $Cu(NH_3)_3(SbO_3)_2 \cdot 9H_2O$. In the Na and K hexahydroxyantimonates, the alkali metal ions and the Sb atoms occupy the same sites as the Na^+ and Cl^- ions in NaCl, with the Sb atoms at the octahedral centers of $[Sb(OH)_6]^-$ ions. Octahedrally coordinated Sb^{5+} ions occur in $[XeF][Sb_2F_{11}]$ and $[BrF_4][Sb_2F_{11}]$ (V), each of which contains two SbF_6 groups with a common

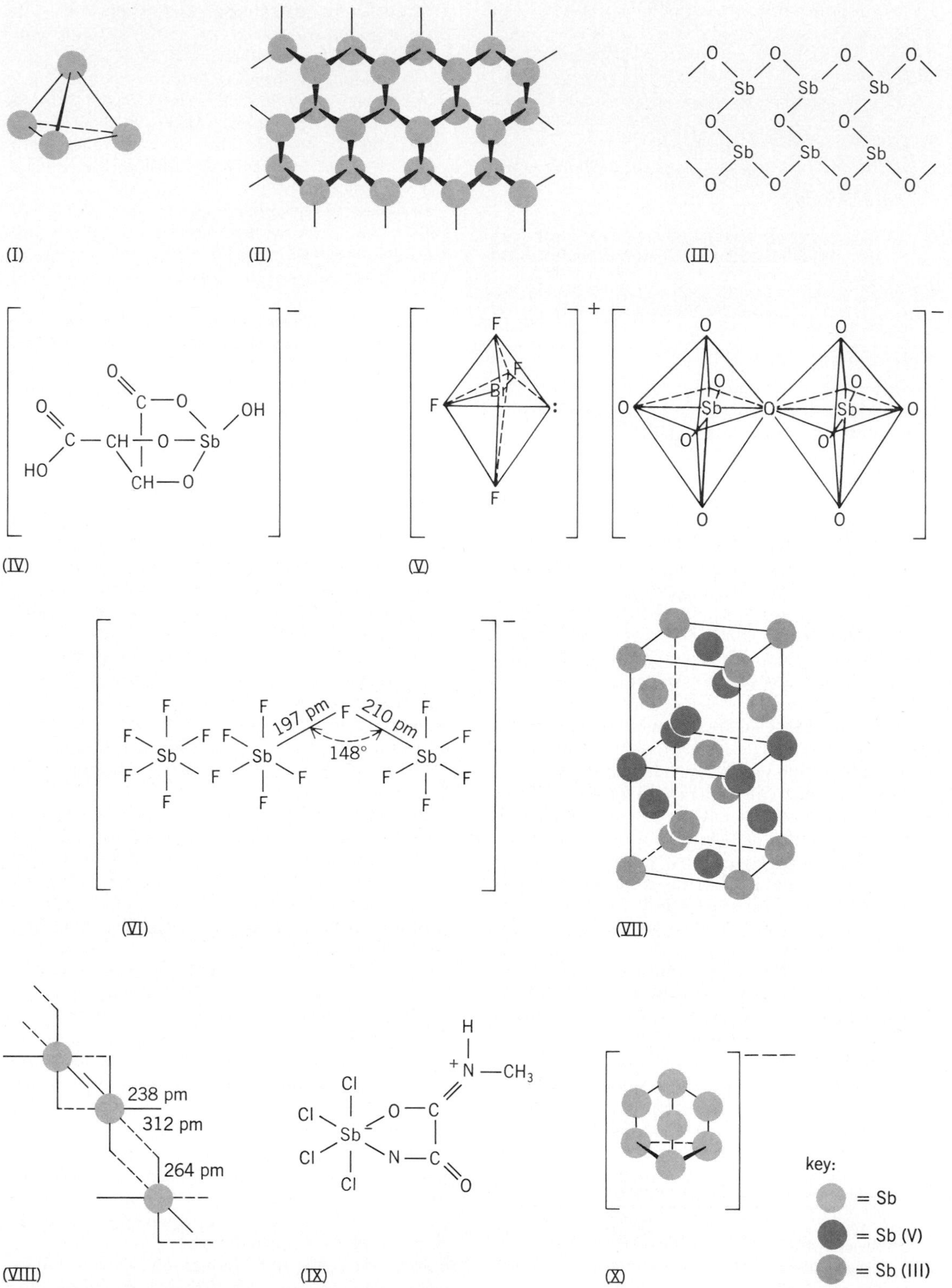

Structure of antimony (I, II) and some antimony compounds (III–X).

vertex; the interesting scarlet, paramagnetic compound (μ = 1.6 Bohr magnetons, or BM), $[Br_2][Sb_3F_{16}]$ (VI), consists of Br_2^+ ions and anions in which the central SbF_6 group shares two bridging F^- ions with similar alternate units. The black compound with empirical composition $(NH_4)_2SbBr_6$ (VII) is actually found to contain both Sb^{5+} and Sb^{3+} ions, each octahedrally coordinated, so that the correct formulation of this substance is $[NH_4]_4[Sb^{3+}Br_6][Sb^{5+}Br_6]$.

There are no known Sb compounds which contain discrete SbX_4^- ions (X = F, Cl, Br, or I), and they most probably do not exist; the compound with the composition $[SbCl_4][C_5H_5NH]$ is known to contain an anion which is composed of infinite "skew" chains of $SbCl_6$ groups sharing edges

(VIII). Many other complex Sb halides contain anions in which octahedral SbX_6 groups share vertices or edges, sometimes with considerable distortion; one such compound is $Sb_{11}F_{43}$, which consists of SbF_6^- anions and a polymeric chain cation, $[Sb_6F_{13}]_n^{5+}$.

Solution chemistry. Soluble Sb compounds produce SbH_3, stibine, when treated with zinc and sulfuric acid, and the resulting mixture of SbH_3 and H_2 burns with a green flame; stibine is less thermally stable than arsine and decomposes to metallic, black antimony if passed through a heated glass tube; this deposit is not oxidized by NaOCl solution as in the case of arsenic. SbH_3 as well as all the trihalides, SbX_3, simple and mixed, are rapidly hydrolyzed by water and are quite volatile. SbF_5 dissolves in liquid AsF_3, with a substantial increase in the conductance of the solution, from which $[AsF_2][SbF_6]$ can be isolated. $SbCl_3$ (mp 346.3 K) has a high dielectric constant, and many metal chlorides dissolve in the melt to give conducting solutions. $SbCl_3$ behaves differently from its P and As analogs in that it is soluble in water, but on further dilution insoluble oxochlorides such as SbOCl and $Sb_4O_5Cl_2$ separate out.

Sb_2O_3, which in the vapor phase exists as Sb_4O_6 molecules, is insoluble in water and in dilute nitric and sulfuric acids, but is soluble in hydrochloric acid and certain organic acids. It dissolves in bases to give solutions of antimonates. No free acids containing either Sb^{3+} or Sb^{5+} can be isolated.

The most widely used method for determining antimony is by precipitation with H_2S from acid solutions of its compounds as Sb_2S_5, which may be then dried in a stream of CO_2 and weighed as such. Titrimetric methods using $KBrO_3$ or $KMnO_4$ solutions as oxidants are also used; this requires the prior reduction of Sb^{5+} to Sb^{3+}, usually with SO_2 or Na_2SO_3 solution.

Organic and cyclic compounds. The organic stibines SbR_3 (R═alkyl) are colorless liquids, soluble in organic solvents but not in water, made by reactions of $SbCl_3$ with zinc alkyls or Grignard reagents. $Sb(CH_3)_3$ is a spontaneously flammable liquid, which reacts with hydrochloric acid to liberate H_2 in the same manner as a strongly electropositive metal; with CH_3I it forms tetramethylstibonium iodide, $[Sb(CH_3)_4]I$, which is essentially completely ionized in solution. Cleavage of trialkylstibines with sodium in liquid ammonia yields dialkylstibylsodium compounds that are useful synthetic reagents in the preparation of new organostibine compounds. $SbCl_5$ and $[CH_3HNCO^-]_2$ react with one another to give a cyclic compound in which the Sb atom is octahedrally coordinated and the *NN'*-dimethyloximido ligand is NO-bonded (IX).

Alloys of antimony with the alkali metals (such as NaSb, KSb_2, Cs_2Sb_3) are remarkably soluble in liquid ammonia, and it has long been suspected that the colored solutions thus formed contain cluster-type ions. Confirmation of this has been difficult to obtain, since on evaporation of the ammonia the alkali metal ion and the cluster ion revert to the metallic alloys. However, the Sb_7^{3-} cage-type cluster ion (X) has been identified by treating small alloy samples with crown ethers in ethylenediamine and isolating the crystalline complex formed by the alkali metal and the crown either combined with the cluster Sb anion. The cyclic stibocan compound

$$\overline{ClSb-S-(CH_2)_2-S-(CH_2)_2-S}$$

has been shown to contain an eight-membered ring with a rather deformed boat conformation. *See* BISMUTH; PHOSPHORUS.

[JOHN L. T. WAUGH]

Bibliography: Chemical Society, London, *Annual Reports*, 1973–1977; J. D. Smith, in J. C. Bailor et al. (eds.), *Comprehensive Inorganic Chemistry*, 1975.

Antioxidant

An inhibitor which is effective in preventing oxidation by molecular oxygen (autoxidation). Such inhibitors have great commercial significance in the preservation of food and food products and in the prevention of deterioration of petroleum products, rubber, and plastics.

Autoxidations are free-radical chain reactions characterized by the interaction of the radicals with oxygen to yield peroxy radicals, organic peroxides, and a broad spectrum of stable oxygenated products. The latter, in the case of foods, are usually of unpleasant taste and odor and render the food unpalatable. It is of interest that antioxidants for foods were known and used long before their function was appreciated. Spices from the Orient served not only to mask unpleasant tastes and odors, but also to prevent the reactions which led to their formation. Modern studies have shown that sage, cloves, oregano, rosemary, and thyme, to name but a few, prevent peroxide development and increase the stability of fats toward oxidation. The active constituents in these spices are phenolic compounds. The autoxidation of gasoline yields gums which foul internal combustion engine fuel systems and which increase combustion chamber deposits. These results increase the octane requirement for the engine. The autoxidation of lubricating oils yields acidic products which accelerate corrosion and engine wear. Oxidation of rubber and plastic products causes chain fission with resultant loss of strength. Discoloration often accompanies this degradation. Hydrocarbon polymers such as polyethylene and polypropylene are particularly subject to such attack and require the addition of appropriate antioxidants to give satisfactory performance.

Mechanism of autoxidation. The reaction of hydrocarbons and their oxidized derivatives with oxygen at low temperatures can be summarized by reactions (1), (2), and (3).

$$RH + O_2 \rightarrow R\bullet + HO_2 \quad (1)$$
$$R\bullet + O_2 \rightarrow ROO\bullet \quad (2)$$
$$ROO\bullet + RH \rightarrow ROOH + R\bullet \quad (3)$$

The initiation reaction (1) is uncertain, but it is definite that some such process must intervene to produce alkyl radicals (R·). Attack is directed at the most labile C—H linkage. In order of reactivity these linkages are allyl > benzyl > tertiary alkyl > secondary alkyl > primary alkyl > aryl. Ample proof exists for the succeeding steps (2) and (3). The free radical R· is regenerated by reaction (3) and repeats the cycle (2) and (3) indefinitely, giving rise to a chain reaction. It is stopped when R· is consumed in some competing reaction. Both

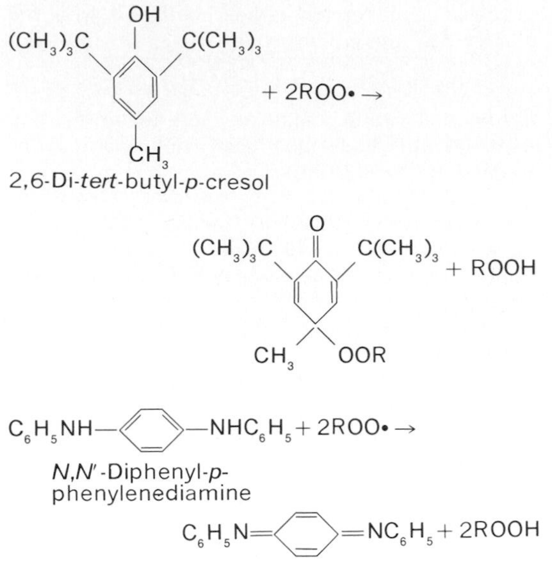

Fig. 1. Oxidation products of typical antioxidants.

the peroxy radical (ROO·) and the hydroperoxide (ROOH) may undergo further reaction to yield more stable oxidized products. These may be alcohols, aldehydes, ketones, acids, and esters. The peroxides themselves, though often used as catalysts for autoxidations and other free-radical chain reactions, can be inert reaction products at relatively low temperatures. Under these conditions, hydroperoxides may function as nonchain oxidants by reaction with aldehydes and olefins present to yield acids and epoxides.

Autoxidation chains are often long, so that a single initiating event may produce many stable product molecules. Thus only a very small amount of an effective antioxidant need be employed for the protection of a large quantity of a substrate. The role of the antioxidant is to provide an alternate path for oxidation which does not involve the substrate. The antioxidant is destroyed in the process and thus does not function indefinitely.

Interruption of the autoxidation chain by an antioxidant takes place at the peroxy radical stage of the chain. Proof of this mechanism was the demonstration that inhibitor (In) efficiency is independent of oxygen partial pressure. The inhibited oxidation process then becomes the reactions shown in (1) and (2), followed by that of reaction (4) in place of (3). In this way, the chain-carrying radical (R·) is not regenerated as long as antioxidant remains.

$$ROO\cdot + In \rightarrow \text{stabilized radical or stable product} \qquad (4)$$

The critical features of oxidation inhibition are the relative reactivity of the antioxidant and the substrate toward the peroxy radical, the stability of the initial product of the radical antioxidant reaction, and the number of radicals with which a given quantity of inhibitor will interact. The first two features determine the efficiency of the inhibition; the third feature determines how long a given quantity of antioxidant will be effective.

Many naturally occurring substances contain antioxidants in their crude states. These inhibitors produce an induction period in autoxidations. During this induction period, absorption of oxygen by the substrate may be so slow as to escape observation. Upon exhaustion of the inhibitor, however, the rate of oxidation quickly increases to a steady level. This level is the same as that for steady-state oxidation of the purified substrate. The length of the induction period observed when an antioxidant is added to a purified substrate has been used extensively as a criterion of antioxidant effectiveness. A quantitative comparison of the relative efficiencies of a series of inhibitors involves a determination of the rate of oxygen absorption during the initial stages of the inhibition period. The efficiency, when measured by this technique, varies considerably with the nature of the antioxidant, as shown in the table.

Types and action of antioxidants. The study of the kinetics of inhibitor action, together with isolation and identification of the products of oxidation of antioxidants has led to the scheme for inhibition shown in reactions (5) and (6).

$$ROO\cdot + InH \rightleftarrows ROOH + In \qquad (5)$$
$$In\cdot + ROO\cdot \rightarrow InOOR \qquad (6)$$

Products of the oxidation of some typical antioxidants are shown in Fig. 1.

The major types of antioxidants now in use are the phenols, the aromatic amines, sulfur compounds, and a variety of naturally occurring materials. The latter find particular use in the protection of foods and cosmetics from oxidation. The

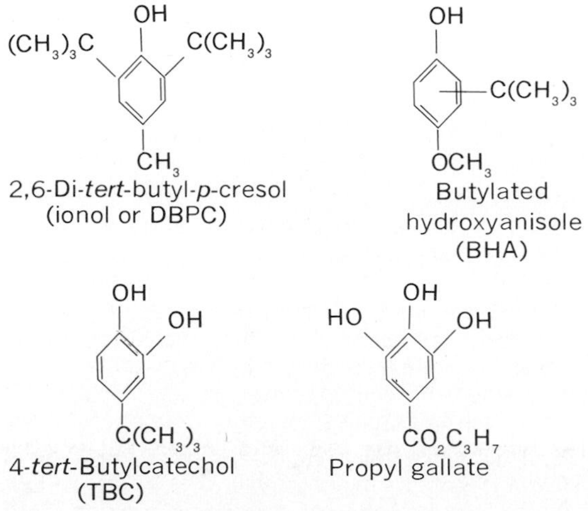

Fig. 2. Phenolic antioxidants.

Inhibition of the oxidation of cumene at 62.5°C and 1 atm O_2 pressure*

Inhibitor	Relative efficiency
Phenol	1.00
2,6-Di-*tert*-butyl-*p*-cresol	3.3
Diphenylpicrylhydrazyl	1.6
4-*tert*-Butylcatechol	14.
N-Methylaniline	1.2
p-Methoxydiphenylamine	6.1
Diphenylamine	2.1
N,N'-Diphenyl-*p*-phenylenediamine	16.
p-Hydroxydiphenylamine	5.6

*G. S. Hammond et al., *J. Amer. Chem. Soc.*, 77:3238, 1955.

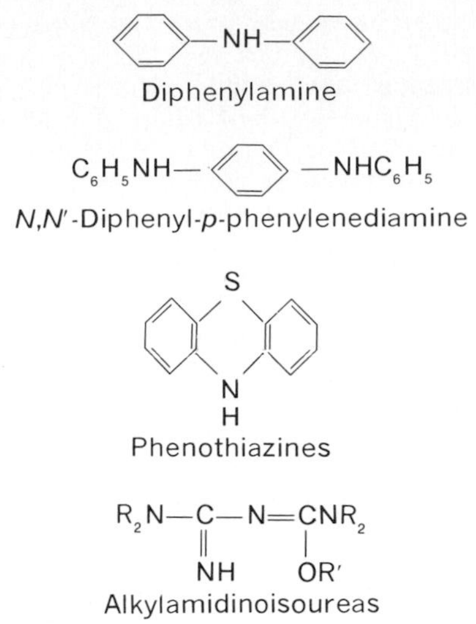

Fig. 3. Nitrogen- and sulfur-containing antioxidants.

structural formulas of some phenolic antioxidants are shown in Fig. 2, and of some nitrogen- and sulfur-containing antioxidants in Fig. 3.

Naturally occurring antioxidants include raw seed oils, wheat germ oil, tocopherols, and gums. The activity of the last-named category may often be increased by the use of synergists. These are substances which have little or no activity alone but which enhance the activity of stronger antioxidants. Some effective synergists are phosphoric, citric, and ascorbic acids.

The wide variety of antioxidants now available is necessitated by the extreme range of conditions under which protection from oxidation is required. An antioxidant which can delay the development of rancidity in stored butter will seldom prove to be suitable for the protection of hot lubricating oil in the crankcase of an automobile. The design of antioxidants for specific uses, however, is still more of an art than a science. Even the qualitative behavior of certain classes of compounds is not always predictable in this connection. *See* CATALYSIS; FREE RADICAL; INHIBITOR.

[LEE R. MAHONEY]

Bibliography: W. O. Lundberg, *Autooxidation and Antioxidants*, 2 vols., 1961–1962.

Aqua regia

A mixture of one part by volume of concentrated nitric acid and three parts of concentrated hydrochloric acid. Aqua regia was so named by the alchemists because of its ability to dissolve platinum and gold. Either acid alone will not dissolve these noble metals. Although chlorine gas and nitrosyl chloride are formed as indicated in Eq. (1),

$$HNO_3 + 3HCl \rightarrow Cl_2 + NOCl + 2H_2O \quad (1)$$

the oxidizing properties of aqua regia are not believed to be increased. Instead, the metal is easier to oxidize because of the high concentration of chloride ions which form a stable complex ion as illustrated with gold in Eq. (2).

$$Au + 4H^+ + NO_3^- + 4Cl^- \rightarrow AuCl_4^- + NO + 2H_2O \quad (2)$$

See CHLORINE; GOLD; NITRIC ACID; PLATINUM.

[E. EUGENE WEAVER]

Arabinose

A plant product which is usually a constituent of polysaccharides such as various gums, hemicelluloses, pectic substances, and some glycosides. It exists in these products mostly in the L form, and usually possesses the furanose modification when it is present in combined form as a constituent of an oligosaccharide, polysaccharide, or glycoside. This pentose sugar is most easily prepared by hydrolysis of the gum of the plant *Prosopis juliflora* (mesquite), *Prunus cerasus* (English cherry), and *Prunus virginiana* (eastern wild cherry). L-Arabinose is obtained in crystalline form as β-L-arabinopyranose (see illustration), mp 160°C, $[\alpha]_D + 191 \rightarrow +105.5°$ (in water).

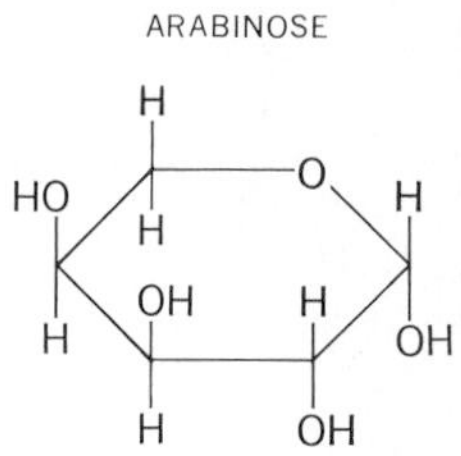

β-L-Arabinopyranose.

The enantiomorph of this sugar, D-arabinose, is relatively rare; it has been found in the aloin glycosides and in the polysaccharides elaborated by *Mycobacter tuberculosis*. *See* POLYSACCHARIDE.

[WILLIAM Z. HASSID]

Argon

A chemical element, Ar, atomic number 18, and atomic weight 39.948. Argon is the third member of group 0 in the periodic table. The gaseous elements in this group are called the noble, inert, or rare gases, although argon is not actually rare. The

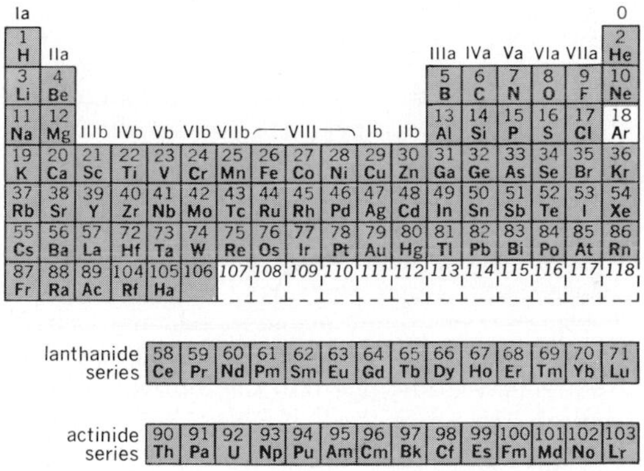

Ia	IIa	IIIb	IVb	Vb	VIb	VIIb	VIII	VIII	VIII	Ib	IIb	IIIa	IVa	Va	VIa	VIIa	0
1 H																	2 He
3 Li	4 Be											5 B	6 C	7 N	8 O	9 F	10 Ne
11 Na	12 Mg											13 Al	14 Si	15 P	16 S	17 Cl	18 Ar
19 K	20 Ca	21 Sc	22 Ti	23 V	24 Cr	25 Mn	26 Fe	27 Co	28 Ni	29 Cu	30 Zn	31 Ga	32 Ge	33 As	34 Se	35 Br	36 Kr
37 Rb	38 Sr	39 Y	40 Zr	41 Nb	42 Mo	43 Tc	44 Ru	45 Rh	46 Pd	47 Ag	48 Cd	49 In	50 Sn	51 Sb	52 Te	53 I	54 Xe
55 Cs	56 Ba	57 La	72 Hf	73 Ta	74 W	75 Re	76 Os	77 Ir	78 Pt	79 Au	80 Hg	81 Tl	82 Pb	83 Bi	84 Po	85 At	86 Rn
87 Fr	88 Ra	89 Ac	104 Rf	105 Ha	106	*107*	*108*	*109*	*110*	*111*	*112*	*113*	*114*	*115*	*116*	*117*	*118*

lanthanide series	58 Ce	59 Pr	60 Nd	61 Pm	62 Sm	63 Eu	64 Gd	65 Tb	66 Dy	67 Ho	68 Er	69 Tm	70 Yb	71 Lu
actinide series	90 Th	91 Pa	92 U	93 Np	94 Pu	95 Am	96 Cm	97 Bk	98 Cf	99 Es	100 Fm	101 Md	102 No	103 Lr

Earth's atmosphere is the only commercial argon source; however, traces of this gas are found in minerals and meteorites. *See* INERT GASES.

Uses. The oldest large-scale use for argon is in filling electric light bulbs. Welding and cutting metal consumes the largest amount of argon. Metallurgical processing constitutes the most rapidly growing application.

Argon and argon-krypton mixtures are used, along with a little mercury vapor, to fill fluorescent lamps. The inert gases make the lamps easier to start, help to regulate the voltage, and supplement the radiation produced by the excited mercury vapor.

Argon mixed with a little neon is used to fill luminous electric-discharge tubes employed in advertising signs (similar to neon signs) when a blue or green color is desired instead of the red color of neon.

Argon is also used in gas-filled thyratrons, Geiger-Müller radiation counters, ionization chambers which measure cosmic radiation, and electron tubes of various kinds. Argon atmospheres are used in dry boxes during manipulation of very reactive chemicals in the laboratory and in sealed-package shipments of such materials.

Welding and cutting metal. The largest single use of argon is in welding and cutting of metals. In the TIG (tungsten–inert gas) and MIG (metal–inert gas) welding processes the electrode is surrounded by a gas cup through which argon flows to blanket the arc and the pool of molten weld metal. The argon blanket persists long enough for the metal to solidify and cool sufficiently to avoid oxidation caused by the normal atmosphere.

In plasma-arc processes an argon-hydrogen mixture flows out of the gas cup and is constricted as it passes through the arc. A high-velocity stream of ionized argon-hydrogen plasma at extremely high temperatures is thus produced. The plasma can be used for cutting metals. Clean cuts through ¼-in. aluminum plate are made at speeds up to 600 in./min (25 cm/s). The plasma-arc process can also be used for welding when very high heat-transfer rates are desired. These inert gas–arc techniques can be used in cutting or welding any metal and thus have become the major welding process, especially for production line work.

Metallurgical processes. Argon may be used in metallurgy whenever nitrogen is detrimental to the metal being processed. Silicon crystals for various solid-state electronic uses are grown in argon atmospheres. In titanium and zirconium production argon atmospheres protect the metal sponge as it is formed from titanium chloride and as it is melted to ingots and also in reheat furnaces in rolling mills. As the final step in aluminum refining it has become the practice to contact the molten metal with a small quantity of chlorine gas. This removes hydrogen and metallic oxides from the melt prior to casting ingots and shapes. The SNIF (spinning nozzle inert flotation) process replaces chlorine with argon, improving the purity of the aluminum while simultaneously reducing environmental pollution. The spinning nozzles introduce the argon sparging gas into the aluminum in the form of extremely small bubbles well dispersed through the melt. The resulting turbulence and mixing of the metal cause the small nonmetallic particles to agglomerate into larger aggregates which are floated to the surface by the gas bubbles. *See* ALUMINUM; ZIRCONIUM.

Steel production. Stainless steel is now produced predominantly by the AOD (argon-oxygen decarburization) process. In this process argon acts both as a temperature-controlling diluent of oxygen and as a gas-scavenging sparging medium. Dilution of oxygen is necessary to hold the temperature resulting from carbon oxidation to a level at which the refractory lining will have a reasonable life. Argon sparging action promotes removal of carbon monoxide, hydrogen, and other gases from the melt. The product is of exceptionally good quality, especially in the case of extra low-carbon alloys, with a high recovery of alloy ingredients and consequent low production costs. Carbon steel can be degassed by sparging with argon. This process eliminates hydrogen and significantly reduces oxygen and oxide solid inclusions. Argon purging of ingot molds and pouring nozzles maintains the cleanliness of the degassed steel. By proper combination of argon treatment and mold protection, a finished quality roughly equivalent to that of vacuum-processed steel can be produced.

Occurrence and origin. Argon constitutes 0.934% by volume of the Earth's atmosphere. Of this argon, 99.6% is the argon-40 isotope; the remainder is argon-36 (0.337%) and argon-38 (0.063%). There is good evidence that all the argon-40 in the air was produced by the radioactive decay of the radioisotope potassium-40.

It is estimated that about 0.0004% by weight of the Earth's crust is argon. Argon also occurs outside the Earth; the best estimate is that there are about 150,000 atoms of argon per 1,000,000 atoms of silicon in the visible universe, as compared with only about 9 atoms of argon per 1,000,000 atoms of silicon on the Earth (including the Earth's atmosphere).

Argon was discovered by Sir William Ramsay in 1894 as the result of Lord Rayleigh's observation that samples of nitrogen prepared from ammonia were always lighter than "nitrogen" prepared by absorbing the oxygen, carbon dioxide, and moisture from air. Ramsay prepared argon by passing this atmospheric "nitrogen" over hot magnesium until no more gas was absorbed. The inert remainder was identified as a new element by the previously unobserved lines in its spectrum.

Radioactive isotopes. The following radioactive isotopes of argon are known: Ar^{35}, Ar^{37}, Ar^{39}, Ar^{41}, and Ar^{42}. None of these occur in nature; they are produced to a small extent in particle accelerators such as the cyclotron or by the neutron bombardment of the appropriate atomic species. In air-cooled or air-ventilated atomic reactors, some of the stable Ar^{40} in the air is converted by neutron absorption to radioactive Ar^{41}, which has a half-life of 110 min. Because of the short half-life, the production of Ar^{41} does not represent a long-term hazard, but the discharge of this radioisotope to the atmosphere must be carried out in such a way that living organisms will not be endangered. This is done by greatly diluting the radioactive air and by discharging it from high stacks.

Properties. Argon is colorless, odorless, and tasteless. The element is a gas under ordinary conditions, but it can be liquefied and solidified readily. Some salient properties of the gas are listed in the table.

Argon does not form any chemical compounds in the ordinary sense of the word, although it does form some weakly bonded clathrate compounds with water, hydroquinone, and phenol. There is one atom in each molecule of gaseous argon. *See* CLATHRATE COMPOUNDS.

Production and distribution. Most argon is produced in air-separation plants. Air is liquefied and

Properties of argon

Atomic number	18
Atomic weight (atmospheric argon)	39.948
Melting point (triple point), °C	−189.4
Boiling point at 1 atm pressure, °C	−185.9
Gas density at 0°C and 1 atm pressure, g/liter	1.7840
Liquid density at normal boiling point, g/ml	1.3998
Solubility in water at 20°C, ml argon (STP) per 1000 g water at 1 atm partial pressure of argon	33.6

subjected to fractional distillation. Because the boiling point of argon is between that of nitrogen and oxygen, an argon-rich mixture can be taken from a tray near the center of the upper distillation column. The argon-rich mixture is further distilled and then warmed and catalytically burned with hydrogen to remove oxygen. A final distillation removes hydrogen and nitrogen, yielding a very-high-purity argon containing only a few parts per million of impurities.

From a minor by-product after World War II, argon has become a major industrial gas. Production has strained air-separation capacity. Some additional argon is also produced as a by-product of ammonia production. Nitrogen for Haber process ammonia synthesis comes from the atmosphere and is usually accompanied by 1% or more of argon. As hydrogen and nitrogen combine to form ammonia, the argon concentration increases in the reactor. A purge stream containing up to 15% argon is removed from the reactor, and, because of the 60% hydrogen content, it is burned as boiler fuel. A few plants have been equipped to separate this purge stream, producing argon for sale and recycling the hydrogen content to the ammonia synthesis reactor. *See* AMMONIA.

For primary shipment of argon in liquid form, insulated railroad tank cars with capacities of 50 to 80 tons of argon and highway trailers of 20–25-ton capacity are used. For smaller users cylinders easily handled by a single workman contain 300–500 lb (1 lb=0.45 kg) of argon in liquid form. High-pressure gas cylinders containing up to 30 lb of argon are also used.

Analytical determination. The principal modern methods of detecting and quantitatively determining the argon content in gases are mass spectrometry and gas chromatography. Until these methods were developed, it was necessary to separate argon from other inert gases by selective low-temperature adsorption on activated carbon in order to determine how much argon was present in a mixture. The older method of identifying argon is by its characteristic emission spectrum, obtained by passing a gas sample through an electric discharge tube at low pressure and analyzing the emitted light with a spectrometer.

[A. W. FRANCIS]

Bibliography: I. Asimov, *The Noble Gases*, 1966; H. L. Clever, *Argon Solubilities*, 1979; G. A. Cook (ed.), *Argon, Helium, and the Rare Gases*, 1961; R. A. Jones and B. S. Kirk, *Helium-Group Gases*, in R. E. Kirk and D. F. Othmer (eds.), *Encyclopedia of Chemical Technology*, 2d ed., vol. 10, 1966.

Aromatic

A term which describes those organic compounds having physical and chemical properties resembling those of benzene. This quality, aromaticity, is reflected in the stability of these compounds. Although the carbons in aromatic compounds are bonded to three atoms rather than four, aromatic compounds do not behave chemically like unsaturated compounds. Their stability is considered to be the result of a resonating electron structure in which all carbon-to-carbon bonds have the same length. Aromatic compounds show a marked tendency to undergo substitution rather than addition reactions.

There are four main classes of aromatic compounds: benzenoid aromatic compounds, containing one benzene ring; polynuclear benzenoid, containing two or more fused benzene rings; nonbenzenoid, containing planar cyclic carbon rings other than benzene; and those heterocyclic compounds whose properties resemble benzene.

Aromatic compounds owe their name to their original derivation from fragrant natural compounds. *See* AROMATIC HYDROCARBON; BENZENE; RESONANCE.

[BETTY RICHMAN]

Aromatic hydrocarbon

An organic compound with a chemistry similar to that of benzene. Usually aromatic hydrocarbons have structures that contain at least one benzene ring. Their properties are so different from those of other hydrocarbons that they are considered to be a separate class of compounds, the arenes. Several important aromatic hydrocarbons are benzene, naphthalene, toluene, styrene, biphenyl, and anthracene.

Aromatic character. Certain of the properties of benzene and its derivatives are regarded as typical, and any compound possessing most of these characteristics is said to have aromatic character.

Diminished unsaturation. Even prior to the invention of structural formulas, it had been observed that a low hydrogen-carbon ratio was characteristic of compounds which react rapidly by addition. For example, acetylene, C_2H_2, reacts almost instantaneously with bromine yielding additon products, reaction (1). Although under special

$$HC{\equiv}CH + 2Br_2 \rightarrow CHBr_2{-}CHBr_2 \qquad (1)$$

conditions benzene can be made to yield addition products, at room temperature and in the absence of a catalyst the reaction with bromine is quite slow and proceeds by substitution rather than by addition, reaction (2). Sulfuric acid, which adds to

$$C_6H_6 + Br_2 \rightarrow C_6H_5Br + HBr \qquad (2)$$

alkenes, or olefins, at room temperature, reacts slowly with benzene on heating to give a substitution product.

Ease of formation. Benzene and its derivatives are easily formed form a variety of nonaromatic compounds. Acetylene when passed through a hot tube yields benzene, reaction (3); under much the

$$3HC{\equiv}CH \rightarrow C_6H_6 \qquad (3)$$

same conditions even methane yields some benzene. The action of phosphorus pentoxide on camphor gives *p*-cymene, reaction (4), while the de-

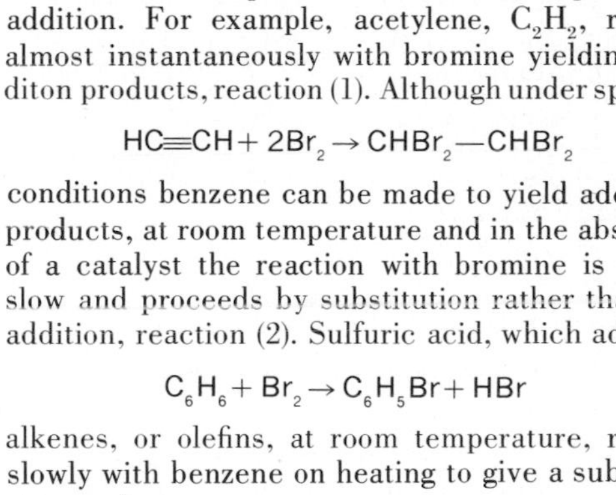

(4)

hydrogenation of cyclohexane by means of platinum, palladium, sulfur, or selenium results in the formation of benzene, reaction (5). The peculiar

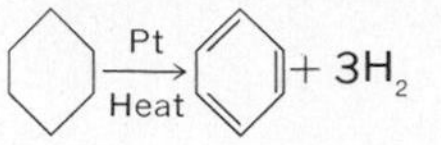

(5)

stability of the six-membered ring is seen in the dehydrogenation of cyclooctane by the action of selenium to yield *p*-xylene, reaction (6).

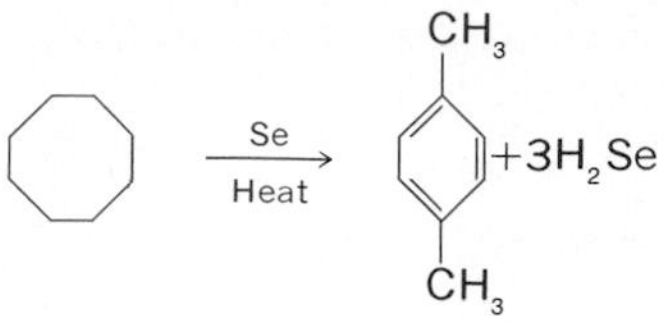

(6)

Stability of nucleus. Another characteristic of benzene derivatives is stability of the benzene nucleus. Whereas alkenes and alkynes react even at 0°C with alkaline solutions of potassium permanganate, an oxidizing agent, benzene is not attacked at 100°C. Besides its resistance to oxidation the benzene ring possesses good thermal stability and shows little tendency to rearrange during reaction.

Modification of attached groups. Some groups when attached to a benzene nucleus undergo reactions which are different from those observed when the groups are attached to an alkane chain. One such group is the hydroxyl, —OH. The aliphatic alcohols, with a hydroxyl replacing a hydrogen atom of an alkane, are very weak acids; even less acidic than water. The phenols, having a hydroxyl group attached to an aromatic ring, are sufficiently acidic to dissolve in sodium hydroxide solution with the formation of a salt, reaction (7).

$$C_6H_5OH + NaOH \rightarrow C_6H_5ONa + H_2O \quad (7)$$

The amino group, $—NH_2$, when attached to the benzene ring is less basic than ammonia or aliphatic amines, although it is still sufficiently basic to form salts with dilute mineral acids, reaction (8).

$$C_6H_5NH_2 + HCl \rightarrow C_6H_5NH_3Cl \quad (8)$$

These salts when treated with nitrous acid at 0–5°C behave quite differently from those of aliphatic amines. Instead of the immediate liberation of nitrogen, a diazonium salt is formed, reaction (9).

$$C_6H_5NH_3^+Cl^- + HONO \rightarrow C_6H_5N_2^+Cl^- + H_2O \quad (9)$$

Compounds in which a halogen atom is bound directly to an aromatic ring are, in the absence of activating groups, almost inert to the action of boiling aqueous alkali. Under the same conditions most alkyl halides yield alcohols and alkenes.

Benzene problem. Since benzene is the most important aromatic compound, and since the whole concept of aromatic character has been developed in terms of benzene and its derivatives, benzene has been the focus of attention for everyone interested in the chemistry of aromatic compounds. In 1869 F. A. Kekule made his classic proposal that the six carbon atoms of benzene were arranged in a hexagon (I) with a hydrogen attached to each carbon. In order to meet objections which were offered concerning the failure to isolate two different ortho derivatives (III) and (IV), Kekule extended his theory in 1872 to provide for a dynamic equilibrium between (I) and (II). More troublesome was the problem of the three double bonds used to account for the fourth val-

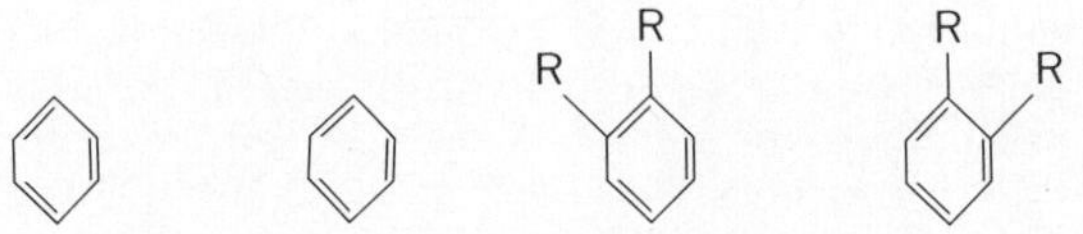

(I) (II) (III) (IV)

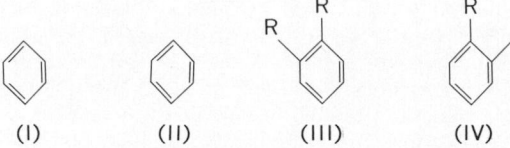

ence of each of the six carbon atoms. The double bond, as it appears in nonaromatic organic compounds, signifies unsaturation and might lead one to think that the characteristic mode of reaction for benzene is addition rather than substitution. Only comparatively recently has the double-bond problem been resolved. From x-ray and electron diffraction studies, as well as from investigations of the Raman and infrared spectra, it is now clear that the six carbon atoms of benzene lie in a plane and are arranged in the regular hexagon, the carbon-carbon bond distance being 1.39 A. By measurements made on other compounds, it has been found that the carbon-carbon bond distance for a single bond (ethane) is 1.54 A, while for a double bond (ethylene) it is only 1.34 A. All this evidence indicates conclusively that all six bonds of benzene are alike and that all are intermediate in nature between double and single bonds.

Resonance. An explanation for the hybrid nature of the carbon-carbon bonds of benzene can be found in the concept of resonance. A covalent single bond is formed by the sharing of a single pair of electrons between adjacent atoms, while the double bond involves the sharing of two pairs of electrons. Formulas (I) and (II) for benzene, as proposed by Kekule, differ from each other only in the arrangement of the electrons; there is no difference in the arrangement of the atoms. According to the resonance theory, the larger the number of alternative arrangements of the electrons, the greater is the stability of the molecule. Since benzene is a resonance hybrid of formulas (I) and (II) plus three other less important forms, the stability of benzene is easy to understand. This stabilization is shown by the measurements of the heats of hydrogenation of some hydrocarbons (G. B. Kistiakowsky, 1936); although the hydrogenation of alkenes is usually exothermic (28–33 kcal/mole), the hydrogenation of benzene to yield cyclohexadiene-1,3 is slightly endothermic, Eq. (10),

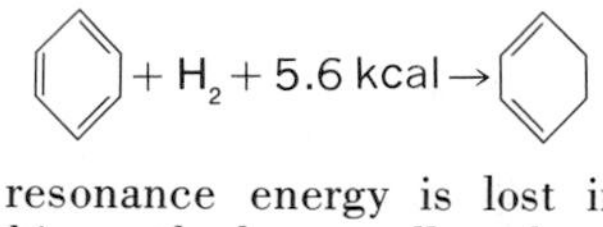

(10)

because resonance energy is lost in the reaction. By this method, as well as by calculations based upon heats of combustion, it has been shown that the resonance energy of benzene is 35–39 kcal/mole. *See* RESONANCE.

Sextet theory. Molecular orbital theory gives a picture of the benzene molecule in which each carbon is, in effect, singly bonded (sharing one pair of electrons) with both adjacent carbon atoms as well as with the hydrogen. This leaves six electrons for which the regions of highest probability lie in two doughnut-shaped regions above and below but parallel to the plane of the benzene ring. From the nature of the orbital involved, these six electrons are referred to as π-electrons.

Nonbenzenoid aromatic systems. Since benzene can be thought of as representing a cyclic conjugated system of alternating double and single bonds, the question was naturally raised as to whether a four-carbon or an eight-carbon cyclic system of alternating double and single bonds would prove as aromatic as that with six. Although cyclobutadiene (V) is known in the form of metallic complexes (Schröder, 1958), the failure of all attempts to liberate the free hydrocarbon would certainly indicate the lack of significant resonance energy in the cyclobutadiene system. Cyclooctatetraene (VI), which can be synthesized from acet-

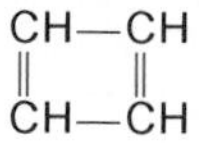

(V)

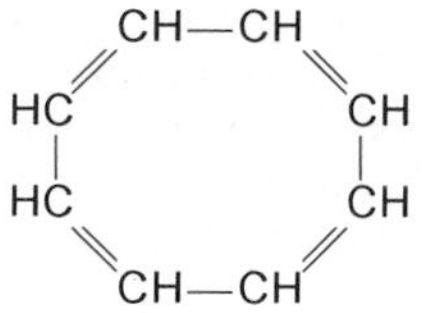

(VI)

ylene, behaves like a cycloalkene with four double bonds. It reacts by addition rather than substitution and has a resonance energy of less than 4.8 kcal/mole, compared with 35–39 kcal/mole for benzene. It is known that cyclooctatetraene is nonplanar, and it is believed that the lack of planarity inhibits resonance. From molecular orbital theory, E. Hückel (1936) derived the rule that completely conjugated planar cyclic systems having $(4n+2)$ π-electrons would have substantial resonance energy. Thus the electron sextet of benzene is merely the simplest example $(n=1)$ of a series of aromatic systems.

Franz Sondheimer (1959–1960) synthesized cyclotetradecaheptaene $(n=3)$ and cyclooctadecanonaene $(n=4)$. He found that the nonplanar C_{14} system is much less stable than the planar, or nearly planar, C_{18} system. The C_{18} system $(n=4)$ is more stable than linear analogs, but it does not show the stability characteristic of benzene and its derivatives.

Aromatic ions. In addition to unchanged cyclic systems having an electron sextet, there exist aromatic systems which are aromatic only because of the gain or loss of an electron. The cyclopentadienyl ion (VII) is an example of an aromatic negative ion. The circle in the middle of the formula implies a sextet of π electrons and the minus sign in the circle indicates that there is an unlocalized negative charge on the system. The inherent stability of the system is seen in ferrocene (VIII), in

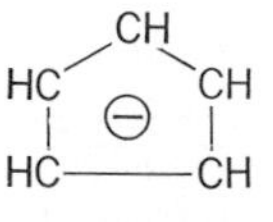

(VII)

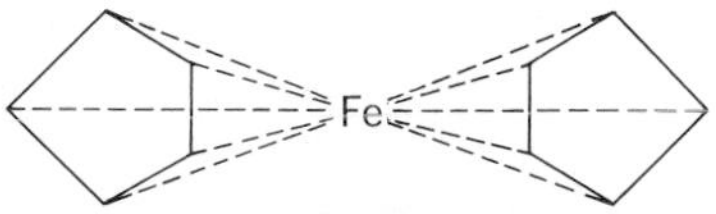

(VIII)

which a ferrous ion is combined with two cyclopentadienyl ions. Although cyclopentadiene (IX) undergoes addition reactions very readily, ferrocene is much less reactive and reacts by substitution rather than addition.

The cycloheptatrienylium ion (X) is stabilized by an electron sextet made possible by the removal of an electron from the system. Tropone (XI) must be a resonance hybrid to which the dipolar ion (XII) makes an important contribution, for it shows little carbonyl activity. *See* FERROCENE.

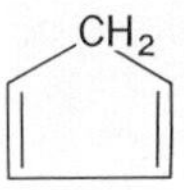

(IX)

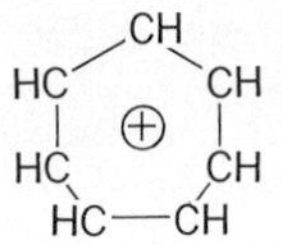

(X)

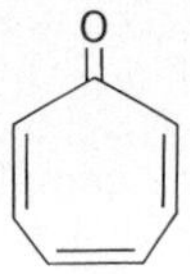

(XI)

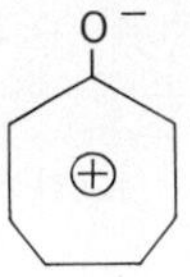

(XII)

Polycyclic systems. Benzologs of benzene are usually classed as aromatic hydrocarbons regardless of their mode of reaction. Examples of benzologs are naphthalene (XIII), anthracene (XIV), and phenanthrene (XV). With increasing numbers of fused rings, there is an increasing tendency for addition rather than substitution to occur. This is particularly true if the fused rings are arranged in a linear fashion as in anthracene (XIV), but both anthracene and phenanthrene will add 1 mole of bromine at room temperature.

Double bonds. Although polycyclic benzenoid hydrocarbons are usually indicated as having double bonds (XIII–XV), actual double bonds are no

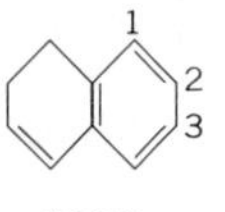

(XIII)

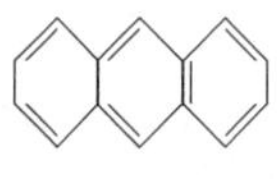
(XIV)

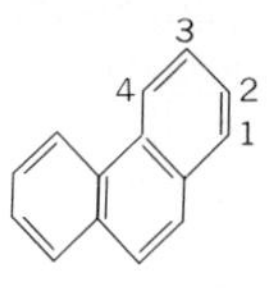

(XV)

more present in polycyclic hydrocarbons than they are in benzene. Certain advantages arise from the use of the double-bond notation provided the bonds are drawn as specified by Fries' rule. According to this rule, double bonds should be drawn so that there is a maximum number of benzenoid rings (three conjugated double bonds). Note that this can be done so that all rings are benzenoid in naphthalene (XIII) and phenanthrene (XV), but it is impossible with anthracene (XIV) where one of the terminal rings must be drawn without the full quota of three double bonds. These formulas usually make it possible to predict those peripheral bonds which have highest "double-bond character" and in this way provide a key to the prediction of the course of a number of reactions. For instance, if an allyloxy group were at position 2 of naphthalene (XIII), it is reasonably certain that the allyl group would rearrange on heating to position 1 rather than position 3. In the same way, if a hydroxyl group were at position 2 in phenanthrene (XV), diazo coupling would be effected at position 1 rather than position 3. Several other types of reactions can be predicted similarly.

Azulene. The only nonbenzenoid polycyclic aromatic hydrocarbon of any importance is azulene (XVI). Most of the azulenes known were pre-

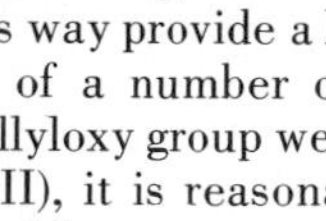
(XVI)

pared by dehydrogenation of a reduced azulene over a platinum or palladium catalyst. Azulene, like naphthalene, has 10 π-electrons, but the resonance energy is only 46 kcal/mole as compared with 77 kcal/mole for naphthalene. Being less stable, azulene may be isomerized to naphthalene under suitable conditions.

Molecular complexes. Aromatic hydrocarbons, and particularly polycyclic aromatic hydrocarbons, have the ability to form stable molecular complexes with some polynitro compounds. Some of these molecular complexes have proved of value in the isolation and identification of aromatic derivatives. Among the most useful nitro compounds for this are picric acid, trinitrobenzene, and trinitrofluorenone. *See* ANTHRACENE; BENZENE; CARBORANE; HYDROCARBON; NAPHTHALENE; PHENANTHRENE; POLYNUCLEAR HYDROCARBON.

[CHARLES K. BRADSHER]

Bibliography: G. M. Badger, *Aromatic Character and Aromaticity*, 1969; S. Coffey (ed.), *Rodd's Chemistry of Carbon Compounds*, vol. 3, pt. C, 1973, pt. G, 1978.

Aromatization

The conversion of any nonaromatic hydrocarbon structures, especially those found in petroleum, to aromatic hydrocarbons. There are numerous routes and means to accomplish this transformation, the simplest and most important of which are (1) direct dehydrogenation of naphthenes to aromatics, reaction (1); (2) dehydroisomerization of naphthenes to aromatics, reaction (2); (3) dehydrocyclization of aliphatics to aromatics, reaction (3); and (4) high-temperature condensation of hydrocarbons to aromatics, reaction (4).

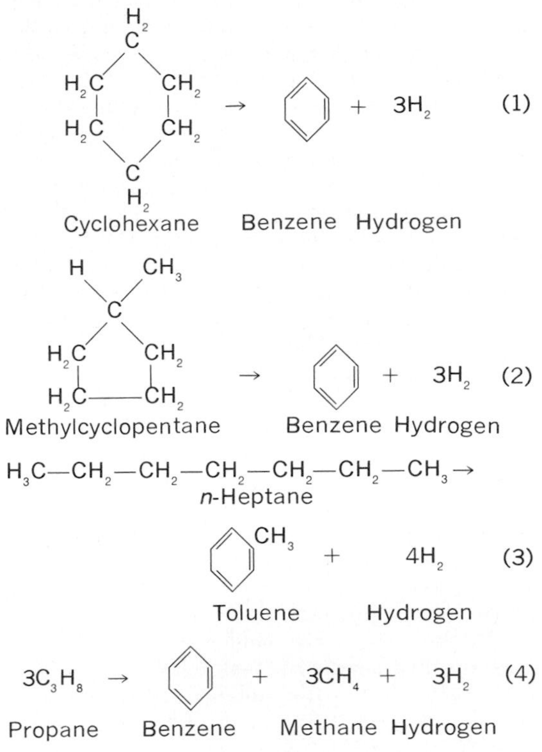

Reaction (1) was performed on a huge scale by the petroleum industry during World War II for the production of toluene for TNT, starting from methylcyclohexane and utilizing molybdenum oxide on porous activated alumina or mixed tungsten-nickel sulfides as catalysts. At that time, reaction (3), employing chromia on alumina catalysts, was investigated extensively but was found impracticable commercially.

Beginning in 1948, reforming of naphthas with catalysts comprising small amounts of platinum on an acidified alumina support has provided a means of aromatization that has rapidly displaced the earlier processes, since it accomplishes reactions (1), (2), and (3) readily and simultaneously. It is now a major process for making benzene, toluene, and other aromatics from petroleum sources.

Reaction (4) merely illustrates one type of reaction that may occur in the high-temperature (600–800°C) thermal cracking of petroleum fractions. Small hydrocarbons from the cracking of heavier components, as well as any originally present, may condense to aromatics, usually with considerable splitting to methane as well as hydrogen. Thermal cracking processes emphasizing this type of reaction have achieved only limited use in the petroleum industry.

[BERNARD S. GREENSFELDER/MOTT SOUDERS]

Bibliography: B. T. Brooks et al. (eds.), *The Chemistry of Petroleum Hydrocarbons*, vol. 2, 1955; W. A. Gruse and D. R. Stevens, *Chemical Technology of Petroleum*, 3d ed., 1960.

Arsenate

A negative ion having the formula AsO_4^{3-}. Arsenates are derived from orthoarsenic acid, H_3AsO_4. Arsenates are quite similar to the phosphates because the term arsenates also includes meta-arsenates and pyroarsenates as well as three series of salts from orthoarsenic acid.

The various arsenates are also very similar to the phosphates in their solubilities and crystal form. The alkali metal salts are the only soluble tertiary orthoarsenates. Lead and calcium tertiary salts, $Pb_3(AsO_4)_2$ and $Ca_3(AsO_4)_2$, are among those arsenates used in insecticides, all of which are poisonous to man.

Because of their similarities, it is difficult to distinguish between the PO_4^{3-} and AsO_4^{3-} ion. However, silver nitrate gives a chocolate-brown precipitate, Ag_3AsO_4, with arsenate, whereas the corresponding phosphate is yellow. *See* ARSENIC; ARSENITE; PHOSPHATE. [E. EUGENE WEAVER]

Arsenic

A chemical element, symbol As, atomic number 33. Arsenic is found widely distributed in nature (approx $5 \times 10^{-4}\%$ of the Earth's crust). It is one of the 22 known elements composed of only one stable nuclide, $_{33}As^{75}$ (nuclear charge number 33, mass number 75); the atomic weight is 74.92158. There are 17 other radioactive arsenic nuclides

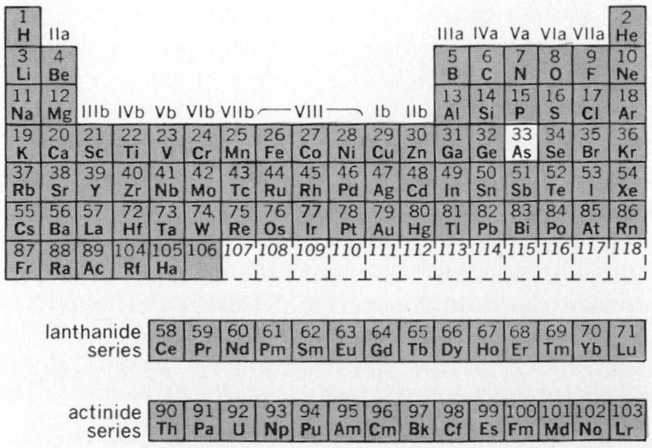

known, with mass numbers ranging from 68 to 85 and half-lives from 0.017 s to 76 days.

General characteristics. Elemental arsenic is one of the few materials now available in 99.9999+% purity, which is largely used in the laser material, GaAs, and as a doping agent in the manufacture of various solid-state devices. Arsenic is found native as the mineral scherbenkobalt, but generally occurs among surface rocks combined with sulfur or metals such as Mn, Fe, Co, Ni, Ag, or Sn.

There are three polymorphic modifications of arsenic. The yellow cubic α-form is made by condensing the vapor at very low temperatures, is metastable, is soluble in CS_2, and consists of tetrahedral As_4 units (I). The black β-polymorph is isostructural with black phosphorus (II), also metastable, and both these modifications revert to the stable γ-form, gray or metallic, rhombohedral arsenic, on heating or exposure to light. The metallic form is a moderately good thermal and electrical conductor and is brittle, easily fractured, and of low ductility. Gray arsenic has a layer structure (III) in which each atom has three equidistant nearest neighbors at 251 picometers and another set of three neighbors at 315 pm; atoms within each layer are covalently bonded to three others, and the puckered layers are held together by metallic bonding. Arsenic vapor contains As_4 molecules at temperatures up to 1100 K; at higher temperatures dissociation to As_2 occurs but is not complete even at 1870 K. Arsenic combines on heating with O_2, S_8, and halogens, and dissolves in hot concentrated H_2SO_4 with evolution of SO_2 to produce As_4O_6. Some of the important atomic and bulk properties of arsenic are given in the table.

Occurrence, extraction, and uses. The principal arsenic mineral is FeAsS (arsenopyrite, mispickel); other metal arsenide ores are $FeAs_2$ (löllingite), NiAs (nicolite), CoAsS (cobalt glance), NiAsS (gersdorffite), and $CoAs_2$ (smaltite). Naturally occurring arsenates and thioarsenates are common, and most sulfide ores contain arsenic. As_4S_4 (realgar; IV) and As_4S_6 (orpiment) are the most important sulfur-containing minerals; on exposure to light realgar is easily converted to orpiment. As_4S_3 and As_4S_{10} are also known. An isomer of realgar has been made by heating the elements together to 770–875 K and rapidly cooling the melt. In each form, As_4S_4 molecules can be distinguished containing As_4 tetrahedra with S atoms in four of the six possible positions between pairs of As atoms, but the four positions for S bridges are different; both modifications grow together from the same melt.

Some atomic and bulk properties of arsenic

Electron configuration (^4S ground state)	[Ar] $3d^{10}4s^24p^3$
Covalent radius	121 pm
Ionic radius (As^{3+})	69 pm
Metallic radius	139 pm
Ionization energies 1st–6th in kJ mole^{-1}	947, 1798, 2734, 4834, 6040, 12300
Electrode potential, As^{3+}/As	0.25 V
Electronegativity (Allred-Rochow)	2.20
Oxidation numbers	−3, 0, +3, +5
Specific gravity (α, β, γ)	2.026, 4.7, 5.727
Melting point (γ)	1090 K (3.6 MPa)
Boiling point (γ)	889 K (sublimes)
Electrical resistivity	33.3 μohm cm (273 K)
Toxicity level	0.5 mg m^{-3} of air

The oxide, arsenolite, As_4O_6 (VI), is found as the product of the weathering of other arsenical minerals, and is also recovered from flue dusts collected during the extraction of Ni, Cu, and Sn from their ores; it also results when the arsenides of Fe, Co, or Ni are roasted in air or oxygen. The element may be obtained by roasting FeAsS or $FeAs_2$ in the absence of air or by reduction of As_4O_6 with carbon, when As_4 may be sublimed away. It is not usual to produce the element by reduction, since the oxide finds many more direct applications.

Elemental arsenic has few uses; about 0.5% added to lead increases the surface tension of the molten metal and thus allows spherical lead-shot to be produced, which is harder than pure lead. As_4O_6 and the lead and calcium arsenates find some use in forestry and agriculture as insecticides, pesticides, and preservatives. The oxide is also used in glass manufacture. The arsenic sulfides are used as pigments and in pyrotechnics. $NaH_2AsO_4{\cdot}H_2O$ on heating forms a variety of condensed oxo-anions, such as $Na_2H_2As_2O_7$, $Na_2H_2As_3O_{10}$, and $(NaAsO_3)_n$. The dihydrogen arsenate itself is used in medicine as are several other arsenic compounds. Most of the medicinal uses of arsenic compounds depend on their toxic nature.

Stereochemistry. Arsenic is essentially a tervalent element and tends to form three electron-pair bonds approximately at right angles, leaving the $4s^2$ electrons unpaired, but hybridization also occurs, resulting in some distortion toward more nearly tetrahedral bonds. The 3+ state is stable with respect to disproportionation to the 5+ and 0 states. All five electrons may participate in bonding (such as AsF_5), d-hybridization giving bipyramidal, square-pyramidal, and octahedral sets of bonds. Simple molecules containing multiple bonds are not formed. For example, $(AsCH_3)_n$ is not structurally similar to azomethane, $CH_3{-}N{=}N{-}CH_3$. There is no evidence for the existence of $[AsCl_4][AsCl_6]$ or of halide-bridged structures, although $[NF_4][AsF_6]$, prepared by heating NF_3, F_2, and AsF_5 under pressure, is reported to be structurally similar to $[PCl_4][PCl_6]$.

In $CoAs_3$, which occurs as the mineral skutterudite (VII), a nearly square group of As_4 atoms is found (sides 246 and 257 pm) such that each As atom has a nearly regular tetrahedral environment of two Co and two As neighbors and each Co atom has a slightly distorted octahedral coordination of six As atoms. An important distinction between phosphorus and arsenic is that only one pure arsenyl halide is known, $AsOF_3$. There are no As analogs of the NH_4^+ and PH_4^+ ions; however, the substituted arsonium ions $[As(CH_3)_4]Br$ and $[As(C_6H_5)_4]I$ are known. Tetrahedral bonds are formed by As^{5+} in the AsO_4^{3-} ion and in compounds such as GaAs, BAs, and InAs, all of which crystallize with the zincblende structure. In LiAs, the As atoms form helical chains, the Li atoms having approximately octahedral coordination. Trigonal-bipyramidal structures have been demonstrated for $As(C_6H_5)_5$ in the solid state and for AsF_5 in the vapor phase. Octahedral bonds have been confirmed in $KAsF_6$ and $[NO][AsF_6]$ and in the trimeric ion $[AsF_4O]_3^{3-}$ (VIII). An interesting arsenic cluster type ion, $[As_7]^{3-}$ (IX), has been reported as

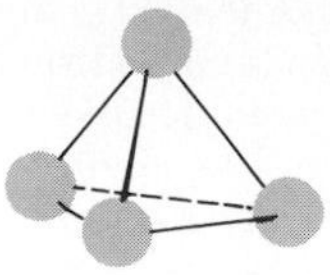

(I)

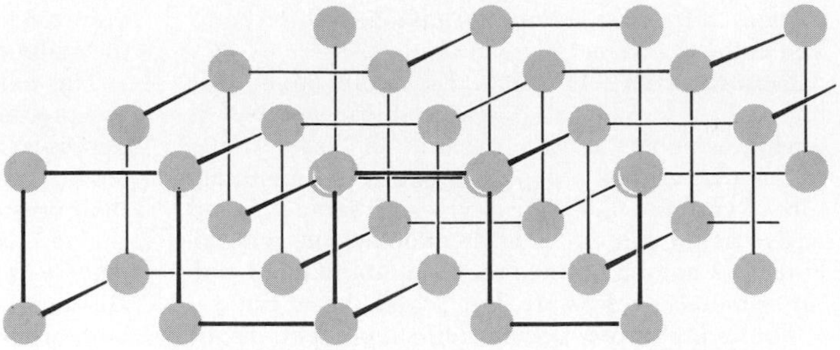

(II)

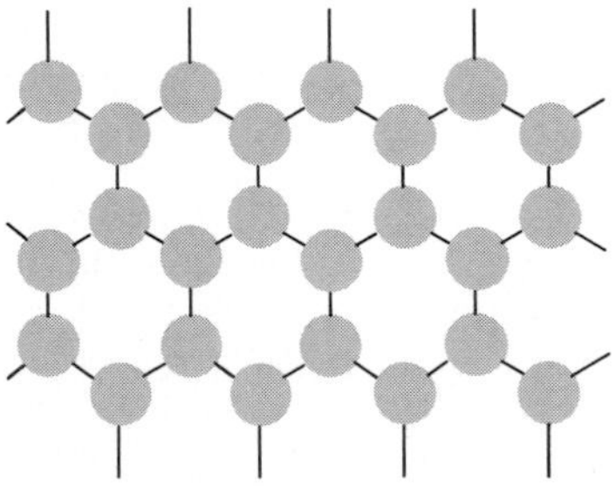

(III)

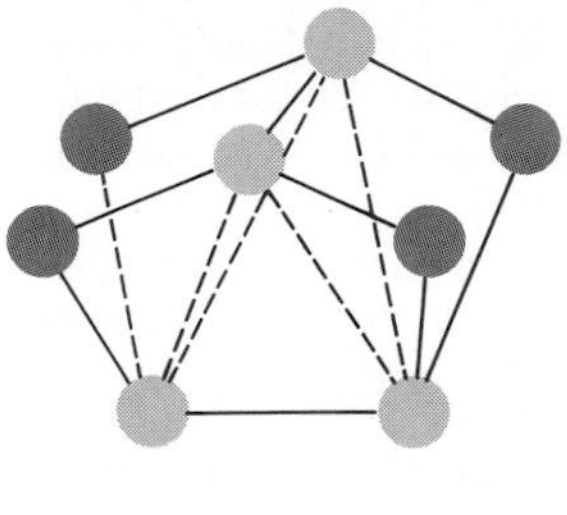

(IV)

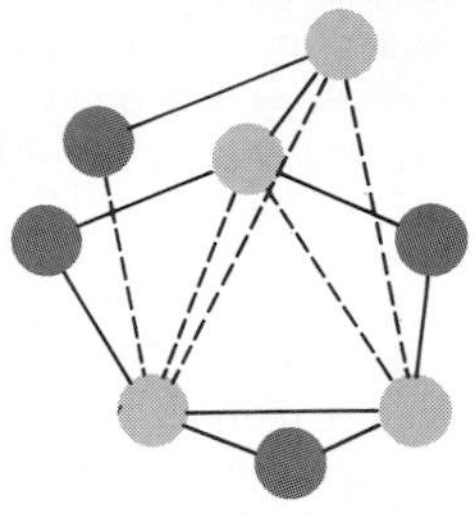

(V)

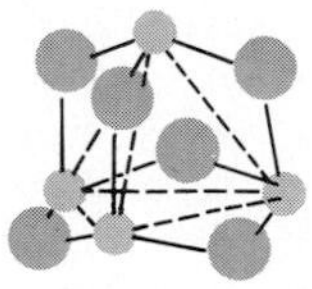

(VI)

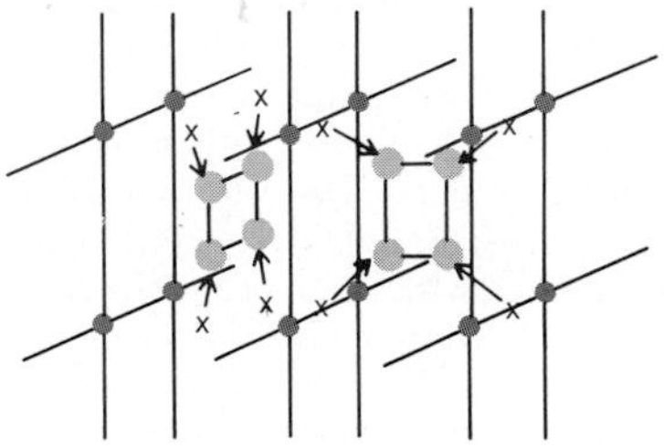

(VII)

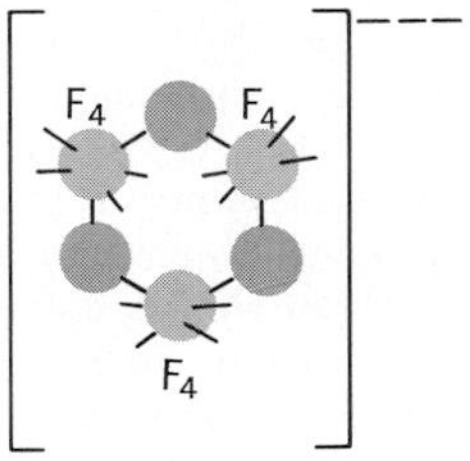

(VIII)

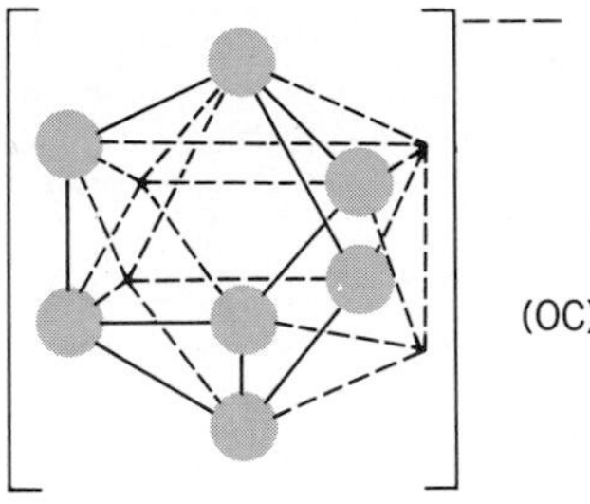

(IX)

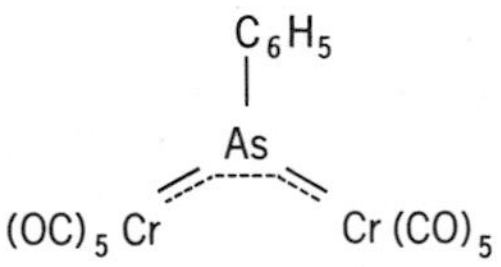

(X)

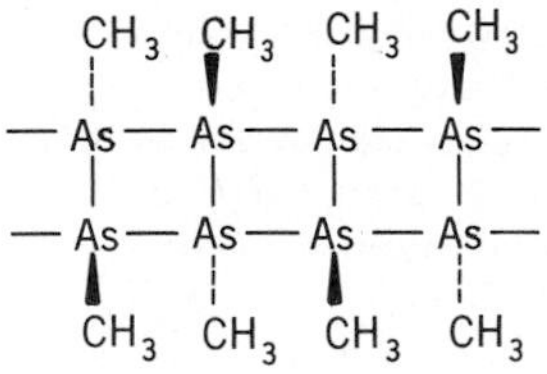

(XI)

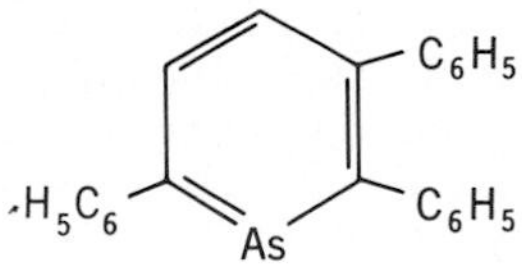

(XII)

key: – As – S – O • – Co

Structure of arsenic (I–III) and some arsenic compounds (IV–XII).

a constituent of Ba_3As_{14}, obtained by fusing the elements together at about 1050 K; related cluster ions have also been found in $[Te_2Se_2]$ $[AsF_6]$ and $[Te_3Se_4]$ $[AsF_6]_2$, and if an additional Se^{2+} is added to $Te_2Se_4^{2+}$, the resulting species, $Se_5Te_2^{4+}$, becomes isostructural with $[As_7]^{3-}$ and with the intermetallic compound, α-As_4Se_3. The bonding in these species is intermediate between ionic and metallic.

Solution chemistry. The oxide As_4O_6 is moderately soluble in water to give solutions of "arsenious acid," not known in the free state, although there are many arsenites; it is most probably a hydroxy acid, $As(OH)_3$ or $H_3[As(OH)_6]$ with $K_a \simeq 10^{-10}$. Arsenic acid may be obtained by oxidizing As_4O_6 with nitric acid as $[H_3AsO_4]_2 \cdot H_2O$, composed of AsO_4 tetrahedra linked by hydrogen bonds. Arsenic acid is widely used as the starting material for the preparation of organic arsenic compounds. The redox potential of As^{5+}/As^{3+} is strongly pH-dependent [reactions (1) and (2)], so that in strongly

$$H_3AsO^4 + 2H^+ + 2e^- \rightarrow As(OH)_3 + H_2O \quad (1)$$
$$E^0 = +0.569 \text{ V}$$
$$AsO_4^{3-} + 4H_2O + 2e^- \rightarrow As(OH)_4^- + 4OH^- \quad (2)$$
$$E^0 = -0.67 \text{ V}$$

acid solution, arsenic acid oxidizes I^- to I_2 (E^0, I_2/I^- = 0.54 V), but in neutral and alkaline solution I_2 oxidizes arsenites to arsenates, which provides a means of determining As^{3+} in solution.

As_4S_4 is precipitated by H_2S from acidified arsenite solutions and dissolves in alkali sulfides to form thioarsenites such as Na_3AsS_3. The fluoarsenates such as $K_2As_2F_8O_2$ contain As atoms octahedrally coordinated by four F^- and two bridging oxygen atoms. The compound $MgNH_4AsO_4 \cdot 6H_2O$, precipitated from arsenates by a mixture of $MgCl_2$, NH_4Cl, and ammonia, is quantitatively converted to $Mg_2As_2O_7$ on ignition and serves as a means of determining arsenic gravimetrically. With $AgNO_3$ solution, arsenites produce yellow silver arsenite, and arsenates produce brown silver arsenate, serving as qualitative tests for soluble arsenic compounds. The extremely toxic gas arsine, AsH_3, is formed by the reduction of soluble arsenic compounds by Zn and H_2SO_4, or by decomposing metallic arsenides with dilute H_2SO_4. AsH_3 burns with a blue flame, which if allowed to impinge on a cool surface deposits a black film of arsenic—the so-called Marsh test, which is capable of detecting as little as 1 μg of arsenic.

Organic and cyclic compounds. The trialkyl derivatives of AsH_3 are colorless liquids obtained by reaction of arsenic trihalides with zinc dialkyls or with Grignard reagents. They are oxidized by O_2, S_8, or halogens to compounds such as $(CH_3)_3AsS$ and $(CH_3)_3AsCl_2$ and react with alkyl halides to form tetraalkylarsonium salts. The cacodyl radical, $(CH_3)_2As$—, exists in compounds such as $[(CH_3)_2As]_2O$, $(CH_3)_2AsCl$, and $(CH_3)_2AsCN$. All of these compounds have a repulsive odor.

Numerous organoarsenic compounds are now known. Research on these substances was greatly accelerated by the early discovery that some of them were trypanocides. The trialkylarsines behave as electron donors in forming coordination compounds such as $(R_3As)_2PdCl_2$, and a large number of such complexes with transition metals have been isolated. The intensely colored arsinidine complex, $C_6H_5As[Cr(CO)_5]_2$ (X), contains a planar tricoordinated As atom. Of the various cyclopolyarsanes, $(AsCH_3)_5$ and $[As(C_6H_5)]_6$ consist of puckered five- and six-membered rings respectively, while the unusual organometallic semiconductor $(CH_3As)_n$ (XI), which is a purple polymer, has a ladder structure. The arsenin ring in 2,3,6-triphenylarsenin (XII) has been shown to contain four equal C-C bond lengths. *See* ANTIMONY; PHOSPHORUS. [JOHN L. T. WAUGH]

Bibliography: Chemical Society, London, *Annual Reports*, 1973–1977; J. D. Smith, in J. C. Bailor et al. (eds.), *Comprehensive Inorganic Chemistry*, 1975.

Arsenite

A negative ion having the formula AsO_3^{3-}. Arsenites are derived from arsenious acid. Arsenious acid is the name given to aqueous solutions of As_4O_6 although the pure acid has never been isolated. These solutions are weakly acidic and can be neutralized by bases to give a variety of ortho-and meta-arsenites. Paris green, $Cu_2(C_2H_3O_2)(AsO_3)$, is used as an insecticide and Scheele's green, $CuHAsO_3$, is used as a pigment. *See* ARSENATE; ARSENIC. [E. EUGENE WEAVER]

Astatine

A chemical element, At, atomic number 85. Astatine is the heaviest of the halogen groups, filling the place immediately below iodine in group VII of the periodic table. Astatine is a highly unstable element existing only in short-lived radioactive

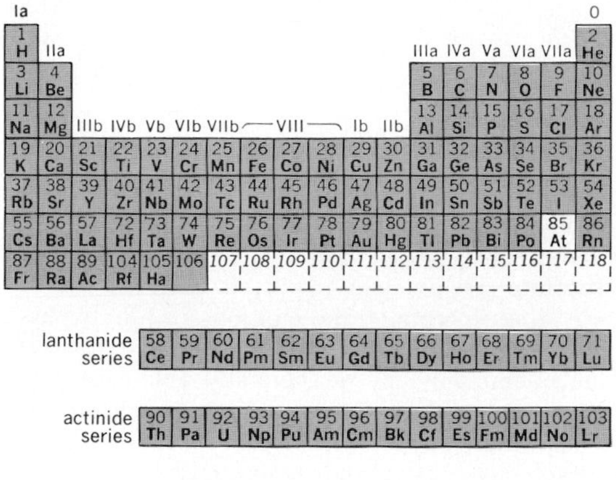

forms. About 25 isotopes have been prepared by nuclear reactions of artificial transmutation, but the longest-lived of these is At^{210}, which decays with a half-life of only 8.3 hr. There is no hope that a stable or long-lived form will be found in nature or prepared artificially. The most important isotope, used for tracer studies, is At^{211}.

The first identification of an isotope of element 85 was made in 1940 by D. R. Corson, K. R. MacKenzie, and E. G. Segré, who bombarded bismuth with helium ions in a cyclotron and identified an isotope of element 85 with mass number 211 among the products of the reaction. This isotope decays 60% by capture of an orbital electron and 40% by emission of an α-particle. The net half-life is 7.5 hr. The name astatine is from the Greek word for unstable.

Astatine does exist in nature in uranium minerals, but only in the form of trace amounts of short-lived isotopes, continuously replenished by the slow decay of uranium. Radium A in the uranium family of radioactive isotopes undergoes very slight branching decay to produce a 2-sec isotope, At^{218}. Actinium K in the actinouranium series undergoes very slight branching decay to produce a 0.9-min isotope, At^{219}. The total amount of astatine in the Earth's crust is less than 1 oz.

In aqueous solution, astatine resembles iodine except for differences attributable to the fact that astatine solutions are of necessity extremely dilute. Like the halogen iodine, when astatine exists as a free element in solution, it is extracted by benzene. The element in solution is reduced by agents such as sulfur dioxide and is oxidized by bromine. It is more electropositive than the other halogens. It has oxidation states with coprecipitation characteristics similar to those of the iodide ion, free iodine, and the iodate ion. Powerful oxidizing agents produce an astatate ion, but not a perastatate ion. Intermediate positive oxidation states exist, but are poorly characterized. The free state is most readily obtained and is characterized by high volatility and high extractability into organic solvents. Astatine resembles polonium in its ready deposition on copper, bismuth, and silver surfaces, and in its coprecipitation with insoluble sulfides and on freshly precipitated tellurium metal.

Animal experiments show that At^{211} is similar to iodine in that it is readily taken up by the thyroid gland, and in that it causes the selective destruction of thyroid tissue by the heavily ionizing, short-range α-particles. *See* HALOGEN ELEMENTS; RADIOACTIVITY.

[EARL K. HYDE]

Bibliography: I. M. Kolthoff, P. J. Elving, and E. B. Sandell (eds.), *Treatise on Analytical Chemistry*, vol. 6, 1964.

Asymmetric synthesis

The chemical synthesis of a pure enantiomer or of an enantiomorphic mixture in which one enantiomer predominates, without the use of resolution. Absolute asymmetric synthesis is achieved by providing dissymmetric conditions by purely physical means such as circularly polarized light or, more commonly, by an optically active reagent. Where a new asymmetric center is produced, the process may be called asymmetric induction, and the selective destruction of one member of an enantiomorphic pair may be called asymmetric decomposition.

Few absolute asymmetric syntheses have been accomplished experimentally, although the natural occurrence of optically active substances presupposes significant numbers of such syntheses early in the evolutionary process. Experimentally, the photochemical decomposition of *N,N*-dimethyl-1-azidopropionamide by circularly polarized light of wavelength ~3000 A is a striking example. Right circularly polarized light destroys the levo enantiomorph more rapidly, leaving the residual amide dextrorotatory. The reverse is true for left circularly polarized light. However, only about 0.5% of the optical activity of the pure enantiomorphs is achieved in this manner.

Asymmetric synthesis generally involves introduction of a new asymmetric center into a pure optically active compound; since the original substance is not symmetrical, the probability of producing exactly equal quantities of the new diastereoisomers (epimers) is small. Thus if an α-keto acid is reduced to an α-hydroxy acid, the product is racemic or optically inactive. However, if the acid is first esterified with a pure optically active alcohol, the reduction of the α-keto ester affords a mixture of epimers in which either the dextro or levo form predominates. If the mixture of the α-hydroxy esters is separated into its pure components and each epimer is then hydrolyzed, resolution of the racemic hydroxy acid is accomplished. However, if the epimeric mixture is hydrolyzed directly to give a partially optically active hydroxy acid, asymmetric synthesis results.

Many variations of this technique are possible, and it may not be necessary to prepare and isolate an optically active intermediate. For example, it is possible to produce mandelonitrile enriched with either enantiomer by appropriate choice of an optically active base to catalyze the reaction, as shown in the following reaction.

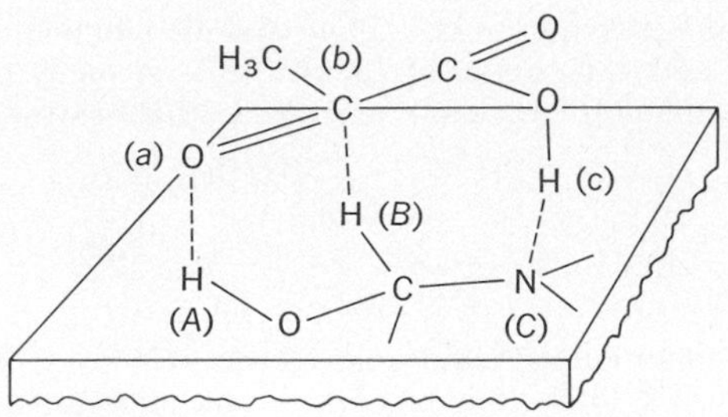

Hypothesized surface and points of contact in an enzymatic reaction. Letters defined in text.

$$C_6H_5CHO + HCN \rightarrow C_6H_5C^*HOHCN$$

Enzymatic reduction. The completely asymmetric synthesis of pure optically active lactic acid by enzymatic reduction of pyruvic acid is a limiting example of asymmetric synthesis. It has been proposed (Ogston's hypothesis) that the symmetrical precursor in an enzymatic reaction makes contact with the enzyme at three specific points; thus as shown in the figure, the centers *a*, *b*, and *c* in the substrate are chemically held to points *A*, *B*, and *C* in the enzyme, which for simplicity may be pictured as a flat surface. Since the substrate has a plane of symmetry, it can make better contacts *aA*, *bB*, and *cC* from one side of this plane than from the other.

The following hypothetical mechanism serves to illustrate. Let points *a*, *b*, and *c* be respectively the ketonic carbon, the ketonic oxygen, and the acid group of pyruvic acid, $CH_3COCOOH$, and let points *A*, *B*, and *C* in the enzyme respectively be the hydrogen attached to a secondary carbinol carbon, the hydroxyl hydrogen of the same alcohol, and a neighboring basic center. Arrangement (I) then represents the best three-point contact, since approach of pyruvic acid from the opposite side of its plane of symmetry would line up the contacts *aB*, *bA*, and *cC*; and the carbon-carbon interchange of hydrogens would not be possible. One need only assume that the primary bonding is the salt formation (*cC*) and that it is essential for the reduction to occur by transfer of hydrogen from carbon to carbon. The enzymatic destruction of one member of an enantiomorphic pair (enzymatic resolution) follows similar principles.

Asymmetric reagents. The use of an asymmetric reagent for the introduction of an asymmetric center into a symmetrical molecule is a later development in asymmetric synthesis. Thus the asymmetric reduction of an unsymmetrical ketone, A—CO—B, by an optically active aluminum alkoxide is characteristic. Such reactions depend upon the principle that a more favorable reaction intermediate obtains as a result of a better fit from one side of the ketone (I) than from the other (II)

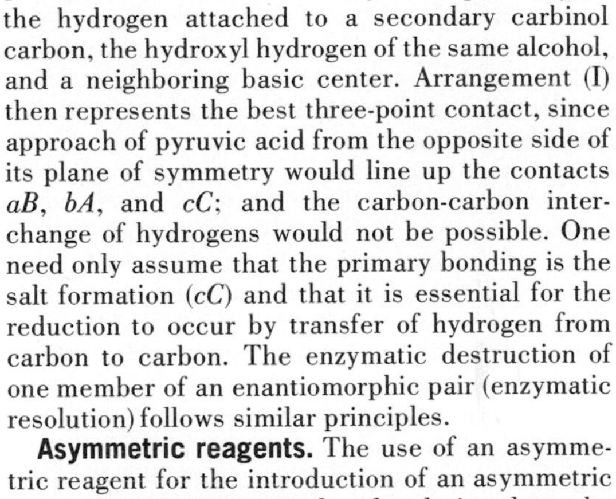

(I) (II)

because of steric interactions of the ketone substituents (A being larger than B) with those of the reducing agent (D being larger than E). The lack of absolute stereospecificity is not surprising since such a complex does not possess the third point of contact specified for enzymatic reduction.

Asymmetric synthesis in the laboratory serves principally for the determination of configurations and for study of reaction mechanisms rather than practical synthesis of optically active products. *See* CONFORMATIONAL ANALYSIS; OPTICAL ACTIVITY; RACEMIZATION; STEREOCHEMISTRY.

[WYMAN R. VAUGHAN]

Atom

The individual structure which constitutes the basic unit of any chemical element. This structure, consisting of a positively charged nucleus surrounded by a number of electrons of total negative charge equal to the positive charge on the nucleus, is essentially identical for all atoms of any one element. The nuclear charge, measured in units of the electronic charge, is called the atomic number and specifies the element. Masses of the atoms of stable elements range from 1.67×10^{-24} g to 3.95×10^{-22} g. Diameters of atoms are on the order of 10^{-8} cm; nuclear diameters are approximately 10^{-12} cm. *See* ISOTOPE.

[F. A. JENKINS/W. W. WATSON]

Atomic beams

Unidirectional streams of neutral atoms passing through a vacuum. These atoms are virtually free from the influence of neighboring atoms but may be subjected to electric and magnetic fields so that their properties may be studied. The technique of atomic beams is identical to that of molecular beams. For historical reasons the latter term is most generally used to describe the method as applied to either atoms or molecules.

The method of atomic beams yields extremely accurate spectroscopic data about the energy levels of atoms, and hence detailed information about the interaction of electrons in the atom with each other and with the atomic nucleus, as well as information about the interaction of all components of the atom with external fields.

[POLYKARP KUSCH]

Atomic constants

That group of physical constants which play a fundamental role in the basic theories of physics. These include the speed of light in vacuum, c; the magnitude of the charge on the electron, e, which is the fundamental unit of electric charge; the mass of the electron, m_e; Planck's constant, h; and the fine-structure constant, α.

These five quantities typify the different origins of the fundamental constants: c and h are examples of quantities which appear naturally in the mathematical formulation of certain physical theories—Einstein's theories of relativity, and quantum theory, respectively; e and m_e are examples of quantities which characterize the elementary particles of which all matter is constituted; and α, the fundamental constant of quantum electrodynamics (QED), is an example of quantities which are combinations of other fundamental constants, but are actually constants in their own right since the same combination always appears together in the basic equations of physics. (In the International System of Units, or SI, which is the unit system used throughout this article, $\alpha = \mu_0 ce^2/2h$, where μ_0, the so-called permeability of vacuum, is exactly $4\pi \times 10^{-7}$ henry/meter.)

Reasons for measurement. Reliable numerical values for the fundamental physical constants are required for two main reasons. First, they are necessary if quantitative predictions from physical theory are to be obtained. Second, and even more important, the self-consistency of the basic theories of physics can be critically tested by a careful intercomparison of the numerical values of fundamental constants obtained from experiments in the different fields of physics.

History of measurement. Although measurements of the fundamental constants date back to the 17th- and 18th-century determinations of c and G (the Newtonian gravitational constant), the field began to blossom only after 1900 with the onset of the modern era of physics. Great progress has occurred since World War II as a direct result of the technological advances made during the war in the fields of electronics and microwaves. In addition to c and G, other constants measured between 1900 and World War II include e by means of R. A. Millikan's oil drop experiment; the Faraday constant, F, using iodine- and silver-based coulometry; the Avogadro constant, N_A, by means of an x-ray technique; and the ratios e/m_e and h/e. Important postwar determinations, mostly related to atomic physics, include the proton gyromagnetic ratio, γ_p; the proton magnetic moment in nuclear magnetons, μ_p/μ_N; the free electron g-factor, g_e; and the fine structure and ground state hyperfine splitting of atomic hydrogen which, in combination with theory, leads to values of α.

In general, the accuracy of fundamental constants determinations has continually improved over the years. (By accuracy is meant the relative size of the uncertainty which must be assigned to the numerical value of the measured constant to indicate how far from the true but unknown value it may be. The uncertainty arises primarily from experimental limitations.) Whereas in the past, 100 ppm (0.01%) and even 1000 ppm (0.1%) measurements were commonplace, today 0.01 ppm and better determinations are not unusual (ppm = parts per million).

Impact of Josephson effect measurement. The fundamental constants field was given a new impetus in 1967 with the determination by W. H. Parker, B. N. Taylor, and D. N. Langenberg of the ratio $2e/h$ using the so-called ac Josephson effect in superconductors. The impact has largely been in the area of QED. To compare the theoretical predictions of QED with experiment requires an accurate value of α. Before the $2e/h$ measurement, the most accurate α values were obtained from experiment with the aid of theoretical equations containing significant contributions from QED. It was thus difficult to compare QED theory and experiment unambiguously since the theory had to be evaluated by using direct values of α derived from the experiments themselves. Such comparisons were therefore limited to the testing of internal consistency. Now, however, by combining the value of $2e/h$ with the measured values of certain other constants, a highly accurate indirect value of α can be obtained without any essential use of QED theory. Consequently, unambiguous comparisons can be made between QED theory and experiment.

Hyperfine splitting of hydrogen. Among the quantities for which such a comparison was of critical importance in 1967 was the hyperfine splitting (hfs) in hydrogen. It can be measured experimentally to the phenomenal accuracy of 1 part in 10^{12} by using the hydrogen maser. In contrast, the theoretical QED equation for the hfs, which involves only well-known constants and α, is limited to an accuracy of a few parts per million because of the difficulty in calculating some of the terms in the equation from theory.

One such term, called the proton polarizability correction, δ_N, arises from the fact that the proton in the hydrogen atom has an internal structure of its own. However, all calculations of δ_N show it to be rather small, 1 or 2 ppm at most. The small size of the theoretical value for δ_N was in marked contrast to that implied by the value of α accepted in 1967 (obtained from a measurement of the fine structure splitting in deuterium). When this value was used to calculate a theoretical value for the hydrogen hfs, when this was then compared with the experimental hydrogen maser value of the hfs, and when their difference was assumed to arise solely from the existence of a polarizability correction, it was found that $\delta_N = (43 \pm 9)$ ppm. This meant that the probability for δ_N to be as small as predicted by theory was only 1 in 20,000, a clear discrepancy. On the other hand, when the new value of α obtained from the Josephson effect measurement of $2e/h$ was used, it was found that $\delta_N = (2.5 \pm 4.0)$ ppm, consistent with the theoretical predictions. Thus the Josephson effect value of α removed the discrepancy and resulting challenge to QED.

This case is an excellent example of how fundamental constants experiments carried out in one field of physics can have important implications for other fields—a low-temperature solid-state physics experiment has given information about the excited states of the proton, a subject usually associated with high-energy physics. It is therefore a good example of the unity of physics as well as the role played by measurements of the fundamental constants.

Methods of obtaining α. In practice, there are several methods of obtaining indirect α values from $2e/h$. One involves F; μ_p/μ_N; the atomic mass of the proton, M_p; and the conversion factor relating the so-called as-maintained ampere in terms of which F was measured to the absolute or SI ampere. Another involves the Rydberg constant, R_∞; c; γ_p; the proton magnetic moment in units of the Bohr magneton, μ_p/μ_B; and the conversion factor relating the so-called as-maintained ohm associated with the electrical units in terms of which both $2e/h$ and γ_p are measured to the SI ohm. The ampere conversion factor and other similar conversion factors play an important role in the fundamental constants field because knowledge of their numerical values is necessary in order to express all measured quantities in the same system of units, that is, SI. Since the uncertainty in some of these factors is relatively large, they must in many cases be considered equal in importance to the particular fundamental constants with which they are associated, and separate experiments must be undertaken for their determination.

Least-squares method. This discussion illustrates the complex relationships which can exist among groups of constants and conversion factors, and that a particular constant may be determined either directly by measurement or indirectly by an appropriate combination of other directly measured constants. If the direct and indirect values have comparable accuracy, then both must be taken into account in order to arrive at a best value for that quantity. (By best value is meant that value believed to be closest to the true but unknown value.) Generally, each of the several routes which can be followed to a particular constant, both direct and indirect, will give a slightly different numerical value. Such a situation may be satisfactorily handled by the mathematical method known as least-squares. This technique provides a self-consistent procedure for calculating best "compromise" values of the constants from all of the available data. It automatically takes into account all possible routes and determines a single final value for each constant being calculated. It does this by weighting the different routes according to their relative uncertainties. The appropriate weights follow from the uncertainties assigned the individual measurements constituting the original set of data.

Least-squares studies of the constants were begun by R. T. Birge in the late 1920s, and continued by others, notably J. W. M. DuMond and E. R. Cohen. The two most recent studies were those of Taylor, Parker, and Langenberg in 1969, based on their Josephson effect determination of $2e/h$; and of Cohen and Taylor in 1973, carried out under the auspices of the CODATA Task Group on Fundamental Constants. (CODATA, the Committee on Data for Science and Technology, is under the jurisdiction of the International Council of Scientific Unions, ICSU.) The recommended values of Cohen and Taylor's 1973 adjustment were officially adopted for international use by the 8th CODATA General Assembly in September 1973. Recommended values of selected constants from this adjustment are given in Table 1 and were those still in general use as of 1980.

To carry out a least-squares adjustment, the data are first divided into two groups: (1) the more precise data, or auxiliary constants, which have uncertainties sufficiently small that they can be considered as exactly known—for example, the speed of light, which has an uncertainty of only 0.004 ppm; and (2) the less precise, stochastic input data, which in the 1973 adjustment of Cohen and Taylor had uncertainties larger than about 0.2 ppm. A subset of constants is then chosen in terms of which all of the stochastic input data can be individually expressed, if necessary, with the aid of the auxiliary constants. It is actually the constants composing this subset which are directly subject to adjustment. In the 1973 work of Cohen and Taylor, these so-called adjustable constants were taken to be α, K, N_A, $\overline{R}$, Λ, and μ_μ/μ_p. Here, K is the conversion factor relating the ampere as maintained by the International Bureau of Weights and Measures (BIPM) in France to the SI ampere; $\overline{R}$ is the conversion factor relating the BIPM as-maintained ohm to the SI ohm; Λ is the conversion factor relating the kilo-X-unit, a unit of length used in the field of x-rays, to the SI unit of length, the meter; and μ_μ/μ_p is the ratio of the magnetic moment

Table 1. Recommended values of selected fundamental constants as taken from the 1973 least-squares adjustment of Cohen and Taylor

Quantity	Symbol	Numerical value*	Uncertainty, ppm	Units†
Speed of light in vacuum	c	299792458(1.2)	0.004	$m \cdot s^{-1}$
Fine-structure constant,	α	7.2973506(60)	0.82	10^{-3}
$\mu_0 ce^2/2h$	α^{-1}	137.03604(11)	0.82	
Elementary charge	e	1.6021892(46)	2.9	10^{-19} C
Planck's constant	h	6.626176(36)	5.4	$10^{-34}\ J \cdot s$
Avogadro constant	N_A	6.022045(31)	5.1	$10^{23}\ mol^{-1}$
Electron rest mass	m_e	9.109534(47)	5.1	10^{-31} kg
Proton rest mass	m_p	1.6726485(86)	5.1	10^{-27} kg
Ratio of proton mass to electron mass	m_p/m_e	1836.15152(70)	0.38	
Faraday constant, $N_A e$	F	9.648456(27)	2.8	$10^4\ C \cdot mol^{-1}$
Rydberg constant, $\mu_0^2 c^3 e^4 m_e/8h^3$	R_∞	1.097373177(83)	0.075	$10^7\ m^{-1}$
Bohr radius, $\alpha/4\pi R_\infty$	a_0	5.2917706(44)	0.82	10^{-11} m
Free electron g-factor, or electron magnetic moment in Bohr magnetons	$g_e/2 = \mu_e/\mu_B$	1.0011596567(35)	0.0035	
Free muon g-factor, or muon magnetic moment in units of $eh/4\pi m_\mu$	$g_\mu/2$	1.00116616(31)	0.31	
Bohr magneton, $eh/4\pi m_e$	μ_B	9.274078(36)	3.9	$10^{-24}\ J \cdot T^{-1}$
Proton gyromagnetic ratio	γ_p	2.6751987(75)	2.8	$10^8\ s^{-1} \cdot T^{-1}$
Proton magnetic moment in Bohr magnetons	μ_p/μ_B	1.521032209(16)	0.011	10^{-3}
Ratio of electron and proton magnetic moments	μ_e/μ_p	658.2106880(66)	0.010	
Proton magnetic moment in nuclear magnetons	μ_p/μ_N	2.7928456(11)	0.38	
Nuclear magneton, $eh/4\pi m_p$	μ_N	5.050824(20)	3.9	$10^{-27}\ J \cdot T^{-1}$
Ratio of muon and proton magnetic moments	μ_μ/μ_p	3.1833402(72)	2.3	
Ratio of muon mass to electron mass	m_μ/m_e	206.76865(47)	2.3	
Muon rest mass	m_μ	1.883566(11)	5.6	10^{-28} kg
Compton wavelength of the electron, $h/m_e c$	λ_C	2.4263089(40)	1.6	10^{-12} m
Gravitational constant	G	6.6720(41)	615	$10^{-11}\ m^3 \cdot s^{-2} \cdot kg^{-1}$

*The numbers in parentheses are the one-standard-deviation uncertainties in the last digits of the quoted value, and the unified atomic mass scale $^{12}C = 12$ has been used throughout.

†C = coulomb, J = joule, kg = kilogram, m = meter, mol = mole, s = second, T = tesla.

of the muon to that of the proton. With just these six quantities and the aid of selected auxiliary constants, a series of equations, generally known as the observational equations, were formed for all 27 separately available items of stochastic data. These 27 observational equations were then solved (with the aid of a computer) for the least-squares adjusted values of the six adjustable constants, and their uncertainties. Optimum values in the least-squares sense for other constants not directly subject to adjustment were then calculated from the six adjustable constants. (But this does not apply to the auxiliary constants since, for the purpose of the least-squares adjustment, they are taken to be exactly known.)

Critical analysis. Critical analysis of the input data, and deciding what uncertainty should be assigned each measurement, is the main problem in adjusting constants—the weight a particular stochastic datum carries in an adjustment is proportional to the reciprocal of the square of its uncertainty. Another important task is deciding how to handle "discrepant" data, that is, measurements which differ from each other by statistically significant amounts in comparison with their assigned uncertainties. Such data cannot be included in an adjustment uncritically because the inconsistencies imply either incorrect uncertainty estimates or the presence of unknown measurement errors.

When confronted with such a situation, the constants adjuster can in general either (1) include the inconsistent data, but only after expanding (increasing) their assigned uncertainties so that they are no longer discrepant; or (2) decide on as sound a theoretical and experimental basis as feasible which of the inconsistent data are least reliable, and discard them, but expand no uncertainties. Thus, there are subjective factors in adjusting constants, and different reviewers may treat the same data differently, obtaining a somewhat different set of best values.

Comparison of 1969 and 1973 values. Some of the pitfalls of discarding data may be seen by comparing the 1969 adjustment of Taylor, Parker, and Langenberg with the 1973 adjustment of Cohen and Taylor. The most critical problem facing Taylor and colleagues was the internal inconsistency among the five available values of μ_p/μ_N, and an inconsistency between two of these values and the one available measurement of F. After much thought and analysis, Taylor and colleagues decided to discard the two "high" values of μ_p/μ_N and to

Table 2. Comparison of the recommended values of selected fundamental constants as taken from the 1973 least-squares adjustment of Cohen and Taylor; the 1969 adjustment of Taylor, Parker, and Langenberg; and the 1963 adjustment of Cohen and DuMond

Quantity*	Value, 1973 adjustment, and ppm uncertainty		Value, 1969 adjustment, and ppm uncertainty		Change 1973–1969, ppm	Value, 1963 adjustment, and ppm uncertainty		Change 1973–1963, ppm
α^{-1}†	137.03604(11)	0.82	137.03602(21)	1.5	+0.15	137.0388(6)	4.4	−20
e	1.6021892(46)	2.9	1.6021917(70)	4.4	−1.6	1.60210(2)	12	+56
h	6.626176(36)	5.4	6.626196(50)	7.6	−3.0	6.62559(16)	24	+88
m_e	9.109534(47)	5.1	9.109558(54)	6.0	−2.6	9.10908(13)	14	+50
N_A	6.022045(31)	5.0	6.022169(40)	6.6	−21	6.02252(9)	15	−79
μ_p/μ_N	2.7928456(11)	0.38	2.792782(17)	6.2	+23	2.79276(2)	7.2	+34
F	9.648456(27)	2.8	9.648670(54)	5.5	−22	9.64870(5)	5.2	−25

*The units for e are 10^{-19} C; for h, 10^{-34} J · s; for m_e, 10^{-31} kg; for N_A, 10^{23} mol^{-1}; for F, 10^4 C · mol^{-1}.

†α^{-1}, the reciprocal of the fine structure constant, is given rather than α because it is a simpler number.

retain the remaining three and F. The most difficult problem facing Cohen and Taylor also had to do with μ_p/μ_N and F. However, in the intervening 4 years, two new, very accurate (sub-ppm) μ_p/μ_N determinations were completed, which showed that the high values of μ_p/μ_N discarded in 1969 were more nearly correct than the three retained low values, and that it was F which was probably in error and should be the quantity discarded. This shift in outlook accounts for the large changes in the recommended values for certain constants as given in the 1969 and 1973 adjustments. These changes are readily apparent in Table 2, where the values of selected constants resulting from the 1973 adjustment are compared with their counterparts from the 1969 adjustment and the 1963 adjustment. (The changes occurring between the 1963 and 1969 values are primarily due to the change in α resulting from the measurement of $2e/h$ using the Josephson effect.) The table shows that (1) knowledge of the numerical values of the fundamental constants continually improves as new measurements become available; (2) the constants are so intimately related to one another that a significant shift in the value of one will usually give rise to large shifts in the values of others; and (3) no set of recommended constants, such as is given in Table 1, should be taken as final and unalterable. Indeed, determinations of several constants since the 1973 adjustment, including N_A, R_∞, γ_p, and F, seem to indicate that there may be important changes in some of the 1973 recommended values when the next least-squares adjustment is carried out. *See* AVOGADRO NUMBER; ELECTROCHEMICAL EQUIVALENT; ELECTRON SPIN; FINE STRUCTURE (SPECTRAL LINES).

[BARRY N. TAYLOR]

Bibliography: E. R. Cohen and B. N. Taylor, The 1973 least-squares adjustment of the fundamental constants, *J. Phys. Chem. Ref. Data*, 2(4): 663–734, 1973; D. N. Langenberg and B. N. Taylor (eds.), *Precision Measurement and Fundamental Constants*, Proceedings of the International Conference on Precision Measurement and Fundamental Constants, Washington, DC, August 1970, NBS SP-343, 1971; F. D. Rossini, *Fundamental Measures and Constants for Science and Technology*, 1974; J. H. Sanders and A. H. Wapstra (eds.), *Atomic Masses and Fundamental Constants 4*, Proceedings of the 4th International Conference on Atomic Masses and Fundamental Constants, Teddington, England, September 1971, published 1972; J. H. Sanders and A. H. Wapstra (eds.), *Atomic Masses and Fundamental Constants 5*, Proceedings of the 5th International Conference on Atomic Masses and Fundamental Constants, Paris, June 1975; B. N. Taylor, W. H. Parker, and D. N. Langenberg, *The Fundamental Constants and Quantum Electrodynamics*, 1969, also published in *Rev. Mod. Phys.*, 41(3):375–496, July 1969.

Atomic mass unit

An arbitrarily defined unit in terms of which the masses of individual atoms are expressed. One atomic mass unit is defined as exactly 1/12 of the mass of an atom of the nuclide ^{12}C, the predominant isotope of carbon. The unit, also known as the dalton, is often abbreviated amu, and is designated by the symbol u. The relative atomic mass of a chemical element is the average mass of its atoms expressed in atomic mass units. *See* RELATIVE ATOMIC MASS.

Before 1961, two versions of the atomic mass unit were in use. The unit used by physicists was defined as 1/16 of the mass of an atom of ^{16}O, the predominant isotope of oxygen. The unit used by chemists was defined as 1/16 of the average mass of the atoms in naturally occurring oxygen, a mixture of the isotopes ^{16}O, ^{17}O, and ^{18}O. In 1961, by international agreement, the standard based on ^{12}C superseded both these older units. It is related to them by: 1 amu (international)≅1.000 318 amu (physical)≅1.000 043 amu (chemical).

[JONATHAN F. WEIL]

Atomic number

The number of elementary positive charges (protons) contained within the nucleus of an atom. It is denoted by the letter Z. For an electrically neutral atom, the number of planetary electrons is also given by the atomic number. Atoms with the same Z (isotopes) belong to the same element. The lightest element, hydrogen, has $Z=1$. The heaviest naturally occurring element, uranium, has $Z=92$. All elements up to and including $Z=103$ (lawrencium) either occur in nature or have been created artificially. The atomic number of an atom is altered during radioactive decay: For α-emission, $Z \rightarrow Z-2$; for β^--emission, $Z \rightarrow Z+1$; for β^+-emission or electron capture, $Z \rightarrow Z-1$. When specifically written, the atomic number is usually placed before and below the elemental symbol, for example, $_1H$, $_{92}U$. *See* MASS NUMBER; RADIOACTIVITY.

[HENRY E. DUCKWORTH]

Atomic spectroscopy

Analytical atomic spectroscopy utilizing the emission, absorption, or fluorescence of light at discrete wavelengths by atoms in a vaporized sample for the determination of the elemental composition of the sample. The sample, usually an aqueous solution, is introduced into a high-temperature device, such as an electrical discharge, flame, or furnace, where it is dissociated into its component atoms. In atomic emission spectroscopy, the sample atoms absorb energy by collisions with high-kinetic-energy atoms and molecules in the hot gas, promoting electrons to higher energy levels within the atom (excitation). A fraction of these excited atoms then return to their normal or "ground" state by emission of electromagnetic radiation at discrete energies (and, thus, discrete wavelengths) corresponding to the energy differences between the electronic energy levels involved. Just as excited atoms can emit radiation, ground-state atoms can absorb radiation of the appropriate wavelength at these same discrete energies, and this reverse of the emission process forms the basis of atomic absorption spectroscopy. A third technique, atomic fluorescence spectroscopy, is basically a combination of the first two: the sample atoms first absorb radiation from an external source, becoming excited atoms, and the reradiation (fluorescence) is measured.

Atomic absorption. In its usual configuration, an atomic absorption spectrometer consists of four basic components: (1) a light source, emitting the spectrum of the desired analyte element; (2) a sample atomization cell, such as a flame or graphite tube furnace; (3) a monochromator, to isolate the desired source emission line; and (4) a detector/readout system, to allow measurement of the change in source line intensity by sample atom absorption.

The light source is usually a hollow cathode discharge lamp, prepared with the desired element. Radiation from the source passes through the atomization cell, where it is absorbed by the same atoms (as the source) produced when the sample is introduced. Calibration is achieved with standard solutions containing known concentrations of the element being determined.

In the case of flame atomization, fuel/oxidant mixtures like acetylene/air are used for most easily atomized elements. Many other elements exist mostly as monoxides (that is, molecules, and thus not available for atomic absorption) at the relatively low temperature of an acetylene/air flame. For these, the analytical sensitivity can be greatly enhanced by going to a higher-temperature flame, such as acetylene/nitrous oxide, shifting the exothermic equilibrium $M + O \rightleftarrows MO$ toward the metal (M).

A significant development was the nonflame atomization systems, particularly the graphite tube furnaces. These devices consist basically of a graphite tube, into which a small amount of sample is placed, which is then heated electrically to a high temperature, vaporizing and atomizing the sample. These devices improve the analytical sensitivity for many elements by about 100-fold, compared to chemical flames. This improvement is due to several factors, such as an oxygen-free inert-gas atmosphere sheathing the tube, the reducing power of the hot carbon, and a relatively long residence time of sample atoms in the small-volume optical path. These devices are particularly useful when the amount of sample available is limited, since sample volumes are usually in the microliter range. Not without limitations, these devices often suffer from matrix effects and exhibit relatively poor precision because of the small sample size.

As an analytical technique for trace elemental determinations, atomic absorption spectrometry has several important advantages: it has high sensitivity for many elements, with perhaps 60 or so elements measurable in the parts-per-million range ($\sim 0.1 - 10\ \mu g/ml$, or $\sim 10^{-7} - 10^{-5}$ g on an absolute basis) with flame atomization, and perhaps half of these measurable at 10–1000 times lower concentrations with nonflame atomization; it is simple, rapid, relatively low-cost, and highly specific (few interferences).

The principal limitations of the technique are: a separate line source is needed for each element to be determined (for many applications, where only one or a few elements are to be determined in a sample, this will not be an important limitation; the technique does not lend itself well to simultaneous multielement determinations); and it has limited dynamic range (for a given determination, all samples must be within about a one-decade concentration interval).

Atomic fluorescence. Atomic fluorescence spectrometry utilizes the same basic components as atomic absorption in a different optical arrangement. The light source(s) is placed at right angles to the optical axis. The fluorescence intensity is directly proportional to the source intensity (over the atom's absorption line width), and so intense sources such as simple electrodeless discharge lamps are used, rather than the hollow cathode lamps used in atomic absorption. The source is not on the optical axis, and so it need not produce narrow lines. In fact, a single, broadband continuum source can be used quite successfully, eliminating the need to change the source for each element determined. Tuneable dye lasers show considerable promise as sources because of their extremely high intensity.

Atomic fluorescence spectrometry combines some of the advantages of emission and absorption and has a few of its own. It has the simplicity, speed, cost, and specificity advantages of atomic absorption, the wide dynamic range and adaptability to multielement determinations of atomic emission, and higher demonstrated and potential sensitivity for many elements. The nonflame atomizers are equally advantageous in this technique. Its principal limitation is scattering of source radiation (common to fluorescence techniques), and so caution must be used in the design of atomization systems to eliminate particulate matter from the optical path. [CLAUDE VEILLON]

Atomic emission. Atomic emission spectroscopy relies on the use of an energetic medium to promote the electronic excitation of atoms in the gas phase and the subsequent emission of light at wavelengths characteristic of the atoms involved. A typical system consists of the following: the energy medium for vaporization, atomization, and excitation of the material to be analyzed; a dispersive unit which isolates the various wavelengths of light emitted; and a measurement or detection device. Three-component systems of this type are widely used to determine what elements are present in a variety of materials (qualitative analysis) and how much of each element is present (quantitative analysis). The latter relies on the fact that the intensity of light emitted at a particular wavelength is proportional to the concentration of atoms which are capable of emitting at the wavelength.

Energy media which may be used for vaporization, atomization, and excitation include electrical discharges, combustion flames, plasmas, and laser beams. Electrical discharges are produced between two conductive electrodes, at least one of which contains the sample material of interest or is composed of that material. High-purity graphite is commonly used as at least one of the electrodes because it is inexpensive and easy to prepare. Common discharges used include direct-current (dc) arcs, alternating-current (ac) arcs, high-voltage ac sparks, and various combinations of these. Although electrical discharges are most often used for the analysis of solid materials, they may also be utilized for the analysis of gases and liquids. Combustion flames are used almost exclusively for the analysis of liquids. Plasmas produced by the interaction of electrical fields alternating in the radio or microwave frequency range with inert gases such as argon or helium are becoming very popular for the excitation of elements contained in solution samples. High-energy laser beams capable of vaporizing and exciting components of solid samples are often used by focusing the beam through a microscope to permit analysis of very small samples or highly localized sections of larger samples. The choice of the excitation method to be used must be based on the type of material to be analyzed, the elements to be determined, and other requirements imposed by the purpose of the particular analysis.

The dispersion system used to isolate the wavelengths of light emitted by the sample almost exclusively relies on the use of diffraction gratings. If the dispersion system is designed to permit observation of only a single wavelength at any one time by scanning the wavelength region of interest, it is referred to as a monochromator. If it permits the simultaneous observation of several wavelengths without scanning, it may be classed as a polychromator. The monochromator is used to determine different elements in a sequential mode of operation, while the polychromator permits simultaneous measurements. The higher cost of a polychromator may consequently be offset by the reduction in time required to determine several elements.

Photomultiplier tubes or photographic films and plates are widely used to observe the dispersed radiation and measure its intensity at any wavelength(s) characteristic of the element(s) to be determined. Photomultipliers instantaneously provide a photocurrent proportional to the light intensity incident on them. Sensitive ammeters measure the photocurrent, which may be related to the concentrations to be determined. Such readout systems in combination with a wavelength dispersion unit are referred to as spectrophotometers if their wavelength resolution capability is low, or spectrometers if it is high. Direct-reading spectrometers have 10 to 60 photomultiplier tubes, each dedicated to measuring the light intensity emitted by a particular element at a characteristic wavelength. Such units are quite expensive and complex to operate but provide a means of rapidly obtaining analysis results. Dispersion units which rely on photographic detection are known as spectrographs. Such detection takes advantage of the well-known effect of light on photographic emulsions deposited on films or glass plates. All wavelengths covered by the dispersion system are permanently recorded in the form of a negative print; the concentration of an element may be determined by measuring the degree of blackness (optical density) of the negative at the appropriate wavelength with a densitometer. Advantages include the ability to determine a wider range of elements simultaneously (up to 70–80) and lower equipment acquisition costs. Longer times are required, however, to measure the photo image densities and convert these to concentrations for the elements of interest. Both the direct and photographic readout systems must be calibrated by analysis of standard materials for which the concentrations of the elements to be determined are known.

Atomic emission spectroscopy has proven to be a very useful means of analysis; its scope of application is diverse. The approach is widely used in industry, government, and university research laboratories for the identification and determination of many elements as means of solving very complex problems. [RODNEY K. SKOGERBOE]

Bibliography: C. L. Chajrabarti (ed.), *Progress in Analytical Atomic Spectroscopy*, 1980; S. A. Clyburn et al., *Anal. Biochem.*, 63:231–240, 1975; M. Slavin, *Emission Spectrochemical Analysis*, 1971; J. D. Winefordner (ed.). *Trace Analysis*, Chemical Analysis Series, vol. 46, pp. 123–181, 1976; J. D. Winefordner and T. J. Vickers, *Anal. Chem.*, 46(5):192R–227R, 1974.

Auger effect

An internal photoelectric process in an atom in which, for example, instead of the emission of a single characteristic x-ray from the filling of a vacancy in the K shell of electrons by an electron from a higher shell, an additional electron from the $L, M, \ldots$ shell is emitted with a kinetic energy equal to the energy of this x-ray minus the binding energy of this ejected electron. Such an electron is called an Auger electron. If the K-shell vacancy is in an atom of high atomic number Z, the return to the normal state may involve the emission of several x-rays and Auger electrons. That fraction of the vacancies in any electron shell that is filled by this Auger process is called the Auger yield, just as the fraction filled with accompanying x-ray emission is the fluorescence yield. The Auger effect is sometimes called autoionization. *See* X-RAY FLUORESCENCE ANALYSIS.

[WILLIAM W. WATSON]

Avogadro number

The number of molecules N_0 in one mole of a substance, 6.02×10^{23}. Equal numbers of moles of all substances contain the same numbers of molecules. For a perfect gas, the molar volume contains a number of molecules equal to the Avogadro number. *See* MOLE.

Determinations of the Avogadro number by a

variety of independent methods give results that agree very closely. The earliest estimates of its value were made during the latter half of the 19th century from the kinetic theory of gases. The viscosity of a gas can be shown by the kinetic theory to be proportional to $(1/N_0)\sigma^2$, where σ is the diameter of the molecules of the gas. If the molecules of a gas are considered as a number of elastic spheres of finite volume, then the expression relating the pressure p, temperature T, and volume V of a gram-molecule is $p(V-b)=RT$, where b is a constant representing the volume occupied by the molecules. It can be shown that b is proportional to $N_0\sigma^3$. If b and the viscosity are found experimentally, a rough value of N_0 can be calculated. The value of N_0 obtained by this method is 5×10^{23}, remarkably close to the accepted value. *See* VISCOSITY.

In 1909, J. Perrin determined values of N_0 from studies of the Brownian movement of colloidal particles and from the effect of gravity on their distribution with height. Albert Einstein showed that, for a particle moving completely randomly, as does a colloidal particle, the mean square of its displacement $\bar{x}^2$ in a given direction over a time t is related to the diffusion coefficient D by Eq. (1).

$$\bar{x}^2=2Dt \tag{1}$$

If the particle obeys Stokes' law, Eq. (2) holds, where η is the viscosity of the water and a is the diameter of the particle.

$$D=\frac{RT}{6\pi N_0\eta a} \tag{2}$$

Using these formulas and studying the mean square displacements of a very large number of colloidal particles over varying times through a microscope, Perrin obtained his N_0 value. *See* COLLOID.

The numbers of particles n_1 and n_2, with energies E_1 and E_2, can be shown by Boltzmann's law to be related by Eq. (3). Now the potential energies

$$\frac{n_1}{n_2}=e^{[(E_2-E_1)N_0]/RT} \tag{3}$$

of particles at heights h_1 and h_2 in a colloidal suspension are given by $E_1=Wh_1$ and $E_2=Wh_2$, where W is the effective weight of the particles, allowing for the buoyancy of the water, calculated from their radius and density. Thus Eq. (4) follows.

$$N_0=\frac{RT}{W(h_1-h_2)}\ln\left(\frac{n_2}{n_1}\right) \tag{4}$$

Perrin obtained particles of uniform size by fractional centrifuging and measured their radius microscopically. He measured the numbers of particles at different heights by direct counting of the number of particles in the field of view of the microscope. By this method, he obtained a value of 7.2×10^{23} for N_0.

By radioactivity measurements, a more accurate value of N_0 was obtained by B. B. Boltwood and E. Rutherford in 1911. They separated some radium salt from its decomposition products and measured the volume of helium produced after a known time and the rate of emission of α-particles per second per gram of radium. They also measured the decrease in the amount of radium during the course of the experiment. They obtained a value of 6.1×10^{23} for N_0. *See* RADIOACTIVITY; RADIUM.

In 1917, R. A. Millikan determined N_0 by a direct measurement of the charge on the electron e. He measured the rate of fall of electrically charged oil drops under gravity, and the rate of rise of the same drops when a vertical electric field was applied. From these measurements, a value for e was obtained. Now e and the Avogadro number N_0 are related by the expression $F=N_0e$, where F is the faraday, the amount of electricity that will release 1 gram-equivalent on electrolysis. From Millikan's value of e and a knowledge of F, a value of 6.07×10^{23} was obtained for N_0. *See* ELECTROCHEMICAL EQUIVALENT.

The most accurate value of N_0 is obtained from x-ray measurements and density data. The wavelength of the x-rays is measured with a ruled diffraction grating, and the lattice spacing of a crystal d is determined by Bragg's relation, Eq. (5),

$$\lambda=2d\sin\theta \tag{5}$$

using x-rays of the same wavelength. The volume v per molecule is related to d by Eq. (6), where Φ is a

$$v=\Phi\, d^3/n \tag{6}$$

geometrical factor and n is the number of molecules in the unit cell. The density of crystal ρ must also be measured very accurately. Then the Avogadro number N_0 is equal to the ratio of the molecular weight M to the weight of a molecule m, as shown in Eq. (7), and so N_0 can be found. Crystals

$$N_0=\frac{M}{m}=\frac{M}{\rho v}=\frac{nM}{\rho\Phi\, d^3} \tag{7}$$

of calcite, $CaCO_3$, have been used in these determinations, and R. T. Birge gave a value of $N_0=(6.0228_3\pm0.00011)\times10^{23}$. T. Batuecas has pointed out that there is an uncertainty of 0.01 in the atomic weight of calcium, and he has used diamond instead of calcite to give a value of $(6.0236\pm0.00007)\times10^{23}$ for N_0.

The most recent x-ray measurements give $6.0231_6\times10^{23}$ for N_0. The constants e, the charge on the electron; h, Planck's constant; F, the faraday; and N_0 are of course related, and in 1963 a committee of the National Academy of Sciences and the National Research Council of the United States recommended that the following values be adopted: $e=1.60210\times10^{19}$ coulomb, $h=6.6256\times10^{-34}$ joule sec, $F=9.64870\times10^4$ coulomb mol^{-1}, and $N_0=6.02252\times10^{23}$ mol^{-1}. These values have been recommended internationally by the Twenty-third Conference of the International Union of Pure and Applied Chemistry. *See* ATOMIC CONSTANTS. [THOMAS C. WADDINGTON]

Avogadro's law

The principle that equal volumes of all gases and vapors, under the same conditions of temperature and pressure, contain identical number of molecules; also known as Avogadro's hypothesis. From Avogadro's law the converse follows that equal numbers of molecules of any gases under identical conditions occupy equal volumes. Therefore, under identical physical conditions the gram-molecular weights of all gases occupy equal volumes.

Avogadro's law is not strictly obeyed by real gases at ordinary temperatures and pressures, although the deviations are only slight. At high pressures the deviations may be large. Avogadro's law can be shown to follow theoretically from the simple kinetic theory of gases. *See* GAS.

[THOMAS C. WADDINGTON]

Azeotropic mixture

A solution of two or more liquids, the composition of which does not change upon distillation. The composition of the liquid phase at the boiling point is identical to that of the vapor in equilibrium with it, and such mixtures or azeotropes form constant-boiling solutions. The exact composition of the azeotrope changes if the boiling point is altered by a change in the external pressure. A solution of two components which form an azeotrope may be separated by distillation into one pure component and the azeotrope, but not into two pure components. Standard solutions are often prepared by distillation of aqueous solutions until the azeotropic composition is reached. At 760 mm pressure, hydrogen chloride and water form an azeotrope containing 20.24% by weight of HCl. *See* DISTILLATION; SOLUTION.

[FRANCIS J. JOHNSTON]

Azide

One of several types of compounds containing the $—N_3$ group and derived from hydrazoic acid, HN_3. The organic azides are represented by two groups of compounds, the alkyl and aryl azides, RN_3, and the acid azides, $RCON_3$. The inorganic azides are important commercially because of the use of lead azide, $Pb(N_3)_2$, in priming compositions and initial detonating agents.

Sodium azide, NaN_3, is produced commercially by passing nitrous oxide over fused sodium amide, according to the equation below.

$$2NaNH_2 + N_2O \rightarrow NaN_3 + NaOH + NH_3$$

The alkali and alkaline-earth metal azides are water-soluble; hence NaN_3 is usually the starting material for the production of lead azide or other desired salts. *See* NITROGEN.

[E. EUGENE WEAVER]

Azole

A suffix designating organic compounds with a five-membered *N*-heterocycle containing two double bonds. *See* HETEROCYCLIC COMPOUNDS.

Formally the term may be, but seldom is, applied to pyrrole. In practice, azole is reserved for diunsaturated rings containing two or more heteroatoms, one of which is nitrogen. Familiar azole systems are given in the table. Examples of other monocyclic systems are formulated as (I) to (III): some benzo derivatives are shown as (IV) to (VI). *See* IMIDAZOLE; OXAZOLE; PYRAZOLE; THIAZOLE.

[WALTER J. GENSLER]

Azole systems

X	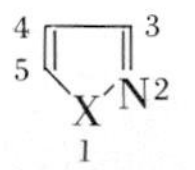	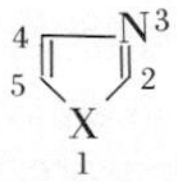
NH	Pyrazole or 1,2-diazole	Imidazole, glyoxaline, or 1,3-diazole
O	Isoxazole or 1,2-oxazole	Oxazole or 1,3-oxazole
S	Isothiazole or 1,2-thiazole	Thiazole or 1,3-thiazole

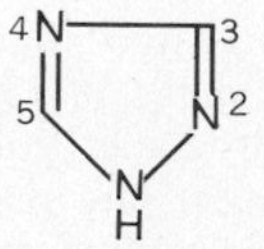

1,2,4-Triazole (I)

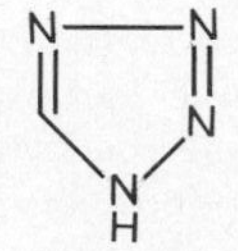

Tetrazole (II)

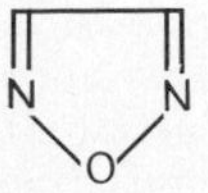

1,2,5-Oxadiazole or furazan (III)

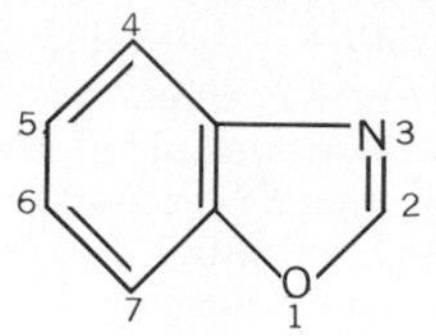

Benzoxazole (IV)

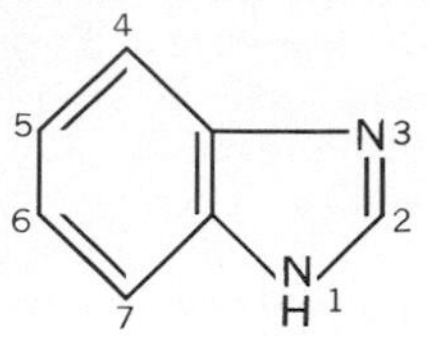

Benzimidazole (V)

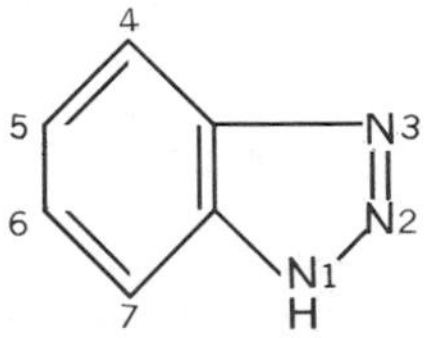

Benzotriazole (VI)

Band spectrum

A spectrum consisting of groups or bands of closely spaced lines. Band spectra are characteristic of molecular gases or chemical compounds. When the light emitted or absorbed by molecules is viewed through a spectroscope with small dispersion, the spectrum appears to consist of very wide asymmetrical lines called bands. These bands usually have a maximum intensity near one edge, called a band head, and a gradually decreasing intensity on the other side. In some band systems the intensity shading is toward shorter waves, in others toward longer waves. Each band system consists of a series of nearly equally spaced bands called progressions; corresponding bands of different progressions form groups called sequences.

Six spectra of diatomic molecular fragments are illustrated in the photograph. The spectrum of a discharge tube containing air at low pressure is shown in *a*. It has four band systems: the γ-bands of nitrogen oxide (NO, 2300–2700 A), negative nitrogen bands (N_2^+, 2900–3500 A), second-positive nitrogen bands (N_2, 2900–5000 A), and first-positive nitrogen bands (N_2, 5500–7000 A). The spectrum of high-frequency discharge in lead fluoride vapor in *b* has bands in prominent sequences. The spectrum in *c* shows part of one band system of SbF, and was obtained by vaporizing SbF into active nitrogen. Emission from a carbon arc cored with barium fluoride (BaF_2) and absorption of BaF vapor in an evacuated steel furnace are illustrated in *d*. These spectra were obtained in the second order of a diffraction grating, as were the spectra in *e* and *f*. The photograph *e* is that of the CN band at 3883 A from an argon dis-

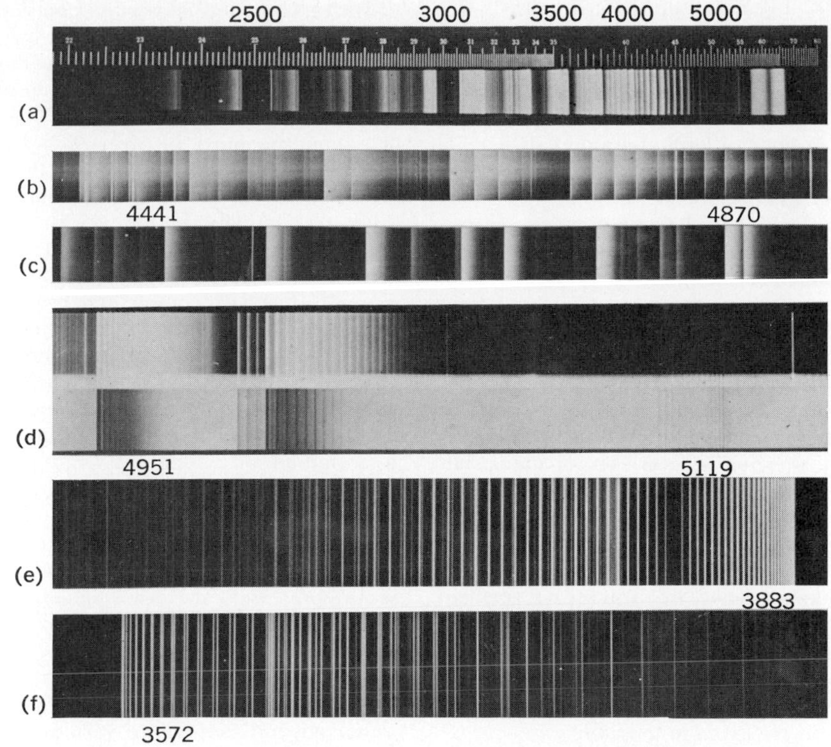

Photographs of band spectra of (*a*) a discharge tube containing air at low pressure; (*b*) high-frequency discharge in lead fluoride vapor; (*c*) SbF (*b* and *c* taken with large quartz spectrograph, after Rochester); (*d*) BaF emission and absorption; (*e*) CN; and (*f*) NO. The measurements are in angstroms. (*From F. A. Jenkins and H. E. White, Fundamentals of Optics, 3d ed., McGraw-Hill, 1957*)

charge tube containing carbon and nitrogen impurities, and *f* is a band in ultraviolet spectrum of NO, obtained from glowing active nitrogen containing a small amount of oxygen.

When spectroscopes with adequate dispersion and resolving power are used, it is seen that most of the bands obtained from gaseous molecules actually consist of a very large number of lines whose spacing and relative intensities, if unresolved, explain the appearance of bands of continua (parts *e* and *f* of the figure). For the quantum mechanical explanations of the details of band spectra *see* MOLECULAR STRUCTURE AND SPECTRA.

[W. F. MEGGERS/W. W. WATSON]

Barium

A chemical element, Ba, with atomic number 56 and atomic weight of 137.34. Barium is eighteenth in abundance in the Earth's crust, where it is found to the extent of 0.04%, making it intermediate in amount between calcium and strontium, the other alkaline-earth metals.

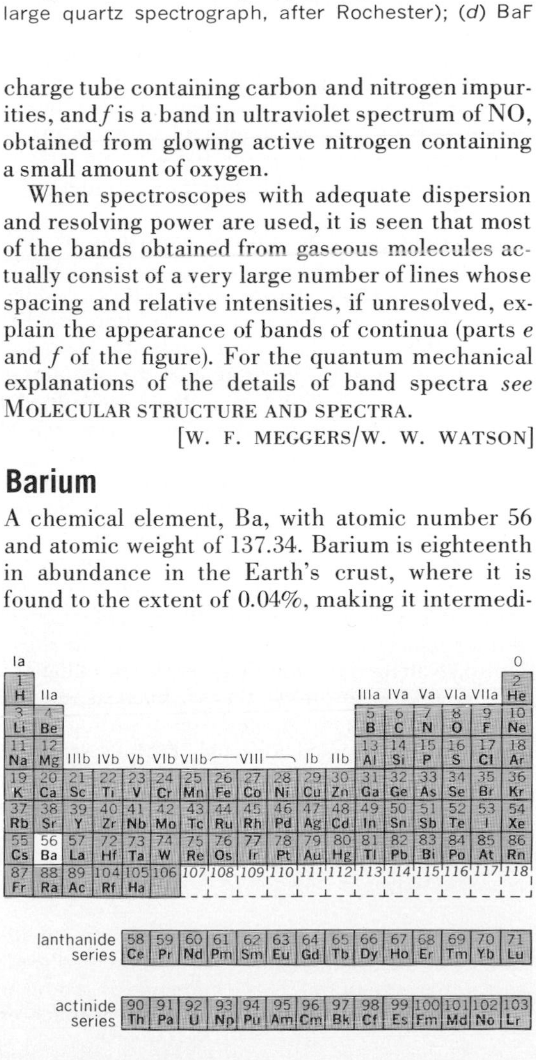

Natural occurrence. Barium compounds are obtained from the mining and conversion of two barium minerals. Barite, barium sulfate, is the principal ore and contains 65.79% barium oxide. Other names for this mineral include baryta, barytes, and heavy spar. Ordinarily, the naturally occurring substance is a heavy, whitish, translucent mineral, but may be any color and transparent to opaque. The crystallinity is not always evident, and granular and fibrous ore bodies are common. Earthy deposits have been described, but are uncommon. Impurities are common, and include calcium sulfate, calcium carbonate, calcium fluoride, silica, alumina, and strontium sulfate. The largest deposits in the United States are in Colorado, California, and Nevada.

Witherite, sometimes called heavy spar, is barium carbonate and is 72% barium oxide. It crystallizes in the orthorhombic system, and because the crystals are nearly all twinned, they resemble hexogonal pyramids. The mineral may be colorless, white, yellowish, or grayish. Witherite is much less common than barite, and in its usual mode of occurrence, it is associated in veins with galena. It is found in northeastern England, Germany, and El Portal, Calif.

The first reference to barite in the literature was in the 17th century. Heavy spar was found near Bologna, Italy, and was experimented with by a shoemaker and alchemist named Casciarola. He noticed that, when mixed with combustibles and heated to redness, it emitted a phosphorescent glow. Barite thus received the name Bologna stone. In 1750 A. S. Marggraf produced sulfuric acid from barite, but believed the base to be lime. K. W. Scheel discovered baryta and found that a white material was obtained with sulfuric acid. Scheele in 1779 first distinguished the oxide, which he obtained from barite, from lime. Barite mining began in the United States about 1845.

Extraction. The metal was first isolated by Sir Humphry Davy in 1808 by electrolysis. The great affinity of this element for halogens, oxygen, nitrogen, hydrogen, and sulfur has made the metal expensive and difficult to produce. Industrially, only small amounts are prepared; these are used in barium-nickel alloys for spark-plug wire (the barium increases the emissivity of the alloy) and in frary metal, which is an alloy of lead, barium, and calcium used in place of babbitt metal because it can be cast. The fused-salt electrolysis to yield the metal has not been satisfactory on an industrial scale, so aluminum reduction of barium oxide in large retorts is used.

The metal reacts with water more readily than do strontium and calcium, but less readily than sodium; it oxidizes quickly in air to form a surface film that inhibits further reaction, but in moist air it may inflame. Freshly cut pieces have a lustrous gray-white appearance, and the metal is both ductile and malleable. The physical properties of the elementary form are given in the table.

Properties of barium

Atomic number	56
Atomic weight	137.34
Isotopes (stable)	130, 132, 134, 135, 136, 137, 138
Atomic volume, cm^3/g-atom	36.2
Crystal structure	Face-centered cubic
Electron configuration	2 8 18 18 8 2
Valence	2+
Ionic radius (A)	1.35
Boiling point, °C	1140 (?)
Melting point, °C	850 (?)
Density, g/cm^3 at 20°C	3.75
Latent heat of vaporization at boiling point, kj/g-atom	374

The metal is sufficiently active chemically to react with most nonmetals. It burns in oxygen at elevated temperatures to form the peroxide, BaO_2 (erroneously called the dioxide); it forms a carbide (acetylide), BaC_2, and a boride, BaB_6, which shows metallic conductivity. The absorption of hydrogen by barium metal becomes noticeable at 120°C. At 180°C the absorption is vigorous, and the reaction goes rapidly to completion to form the hydride, BaH_2, which is a more powerful reducing agent than the corresponding compounds of calcium and strontium, but otherwise similar to calcium hydride. The organometallic compounds $(C_6H_5)_2Ba$ and $BaZn\,(CH_2CH_3)_4$ are isolable as white, reactive solids. $(CH_3CH_2)_2Ba$, prepared in solution, behaves as a Grignard reagent, as shown in the reaction below.

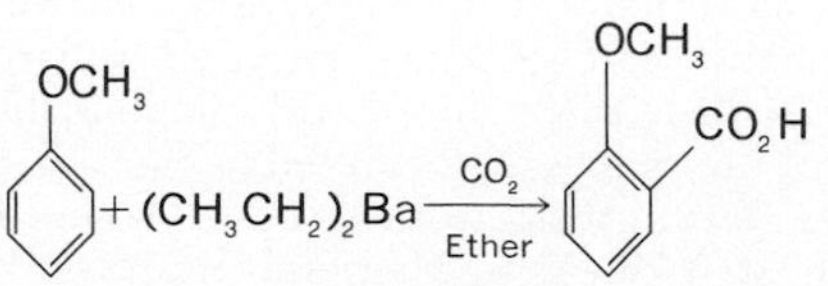

Barite. Commercial barite is marketed in several forms: as crude barite washed free of clay and picked to remove adhering impurities, as ground barite obtained by grinding and purifying crude ore, and as a barium compound. The crude barite of commerce must be over 95% barium sulfate with a ferric oxide content of less than 1%. For the glass trade, however, the amounts are 96 and 0.1%, respectively. Flotation with soap or cationic reagents can be applied satisfactorily to barite ores. Most barite is leached before being finely ground; this is accomplished by sulfuric acid in lead-lined tanks, with the occasional addition of sodium chloride and other chemical reagents. Crude barite is ground to a fine powder to be used as a filler for rubber, oilcloth, linoleum, plastics, and resins. Bleached barite, prime white, is used as a pigment or extender in white paint, but is less satisfactory than blanc fixe (precipitated barium sulfate) or lithopone for this application. Granular barite is applied in glass manufacture to improve the properties of molded materials and as a scavenging flux in brass manufacture. Finely ground barite is used as a thixotropic mud in oil-well drilling.

Principal compounds. For the manufacture of barium compounds, soft (easily crushable) barite is preferred, but crystalline varieties may be used. Crude barite is crushed to pea size and then mixed with pulverized coal in the proportion 1 part coal to 4 parts barite by weight. The mixture is roasted 4 hr in a rotary reduction furnace, and the barium sulfate is thus reduced to barium sulfide or black ash. Black ash is roughly 70% barium sulfide and is treated with hot water to make a solution used as the starting material for the manufacture of many compounds. When, in compound manufacture, a sodium salt is added to the barium sulfide solution, sodium sulfide is produced as a valuable by-product.

Lithopone, a white powder consisting of 20% barium sulfate, 30% zinc sulfide, and less than 3% zinc oxide, is produced when zinc sulfate is added to hot barium sulfide solution. The precipitate is dried, calcined at 500°C in a muffle furnace, and quenched in water. The product is then ground to a pulp, washed and filtered, dried, and passed through a disintegrator. Lithopone is widely used as a pigment in white paints.

Blanc fixe is chemically precipitated barium sulfate obtained by adding sodium sulfate to the barium solfide solution. It is a white, insoluble, amorphous powder having chemical properties identical with, but physical properties different from, those of barite. It is used in the manufacture of brilliant coloring compounds. It is the best grade of barium sulfate for paint pigments, excelling the best ground barite in whiteness, fineness, and oil adsorption. Because of the large absorption of x-crease the contrast. Although barium salts are highly poisonous, the insolubility of barium sulfate rays by barium, the sulfate is used to coat the alimentary tract for x-ray photographs in order to in-makes its use acceptable. The commercial product is marketed as a powder or as a pulp with 30% water.

Barium carbonate is obtained as a white precipitate by adding sodium carbonate or carbon dioxide to the base solution. This compound is useful in the ceramic industry to prevent efflorescence on claywares such as brick or tile. It is used also as a pottery glaze, in optical glass, and in rat poisons.

Barium chloride is obtained by roasting barite with coal and calcium chloride or by adding chloride to the sulfide solution. From witherite, it may be obtained by treatment with hydrochloric acid. It is sold as a monohydrate and is used in purifying salt brines, in chlorine and sodium hydroxide manufacture, as a flux for magnesium alloys, as a water softener in boiler compounds, and in medicinal preparations.

Barium nitrate, or the so-called baryta saltpeter, is obtained commercially by adding sodium nitrate to a solution of barium chloride or barium sulfide. It is also obtained by dissolving witherite, barium oxide, or barium hydroxide in nitric acid. It finds use in pyrotechnics and signal flares (to produce a green color), and to a small extent in medicinal preparations.

Barium oxide, known as baryta or calcined baryta, is available as a white or yellowish-white powder or as grayish-white amorphous lumps. It reacts at elevated temperatures with oxygen to form the peroxide, and it absorbs carbon dioxide and water strongly, even at room temperature. It is formed from either barium nitrate or barium carbonate by calcination. It finds use both as an industrial drying agent and in the case-hardening of steels.

Barium peroxide, BaO_2, is made as described above, and is sometimes used as a bleaching agent because it releases hydrogen peroxide when treated with cold water.

Barium chromate, lemon chrome or chrome yellow, is sold as a yellow insoluble powder, and is made by addition of potassium chromate to various soluble barium compounds. It is used in yellow pigments and safety matches.

Barium chlorate, $Ba(ClO_3)_2$, is produced by the electrolysis of barium chloride, and is sold as a white, soluble monohydrate in the form of monoclinic plates or prisms. It finds use in the manufacture of pyrotechnics.

Barium acetate and cyanide are used industrially as a chemical reagent and in metallury, respectively.

Analytical methods. Qualitatively, barium minerals moistened with hydrochloric acid give a green flame test, but boron and phosphorus will interfere if not previously separated. Barium may be determined quantitatively by precipitating the sulfate and weighing it. *See* RADIUM.

[REED F. RILEY]

Bibliography: J. Donohue, *Structure of the Elements*, 1974; M. F. Lappert, *Main Group Elements: Hydrogen and Groups 1–3*, 1975; D. M. Ritter, *First Four Groups*, 1965.

Base

Any chemical species, ionic or molecular, capable of accepting or receiving a proton (hydrogen ion) from another substance. The other substance acts as an acid in giving up the proton. A substance may act as a base, then, only in the presence of an acid. The greater the tendency to accept a proton, the stronger the base. The hydroxyl ion acts as a strong base. Water acts as a weaker base in the reaction shown in reaction (1).

$$H_2O + H^+ \rightleftharpoons H_3O^+ \qquad (1)$$

This broader definition replaces the older concept of a base as a substance yielding hydroxyl ions in solution. A more general concept of a base is that of a substance which donates a pair of electrons to form a coordinate bond as shown in reaction (2). The SO_3^{--} ion thus acts as a base in the

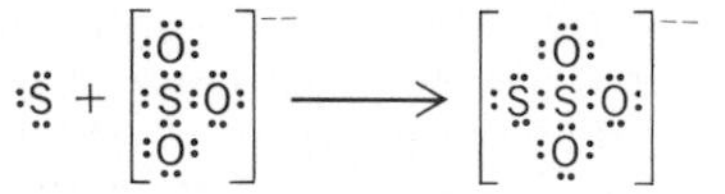

(2)

donation of an electron pair when the $S_2O_3^{--}$ ion is formed by reaction with sulfur. *See* ACID AND BASE.

[FRANCIS J. JOHNSTON]

Bibliography: B. W. Jensen, *The Lewis Acid-Base Concepts: An Overview*, 1979.

Beam-foil spectroscopy

A method of determining the energies and mean lives of excited electronic levels in monatomic ions and in atoms. Beams of particles (ions) of any element are energized in a particle accelerator and then sent in vacuum through a thin foil, usually of carbon. The particles emerge from the foil in various stages of ionization. Numerous energy levels are excited in each of those stages of ionization. The spontaneous loss of the energy of excitation takes place as the beam moves downstream from the foil. That loss is detected by means of the electromagnetic radiation or the electrons which the ions emit. Only experiments dealing with the light emission will be considered, since they represent the great majority of beam-foil researches.

Optical spectra. For optical radiations (4 to 700 nm), the wavelength is determined by a spectrometer employing a diffraction grating. More energetic forms of light (x-rays) are studied with energy-sensitive silicon crystals to which some lithium has been added. In both techniques, the radiation is transformed into an electrical pulse which is then amplified and counted, and the number of counts is plotted against the wavelength. Such a graph (a

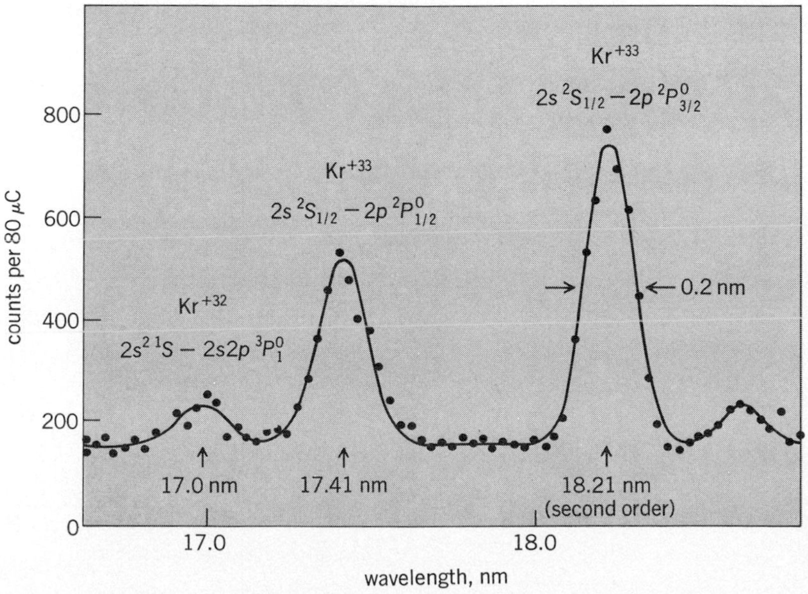

Fig. 1. Spectrum of krypton, excited by sending 714-MeV ions through a carbon foil with a thickness of 600 μg/cm². (*Data courtesy of J. A. Leavitt et al.*)

spectrum) contains peaks of various intensity occurring at wavelengths characteristic of the electronic structure of the excited ion. In the example shown in Fig. 1, three spectral lines are identified; the quantum-mechanical notation above each peak describes the origin of the lines.

Mean lives. The mean life of a level provides sensitive information about the quantum nature of the level itself and the mechanism whereby it connects, by either the absorption or emission of light, to some other level. In beam-foil spectroscopy, this mean life is found from observations on the intensity of a particular spectral line as a function of the separation of the emitter and the exciter foil. Beam-foil spectroscopy is presently the only known way of measuring mean lives for highly ionized systems. An example of the technique is shown in Fig. 2, in which the excitation mechanism is the same as in Fig. 1. The symbol τ stands for the mean life, the subscript identifying which electronic level is decaying. The 9.11-nm line in Fig. 2 is in first order, while the corresponding second-order line at 18.21 nm appears in Fig. 1.

Degree of ionization. The degree of ionization which can be achieved depends on the atomic number of the ions and its energy. Early work was restricted to a few elements at the low end of the periodic table and to energies of a few million electronvolts. Subsequently, more than 60 of the 94 naturally occurring elements were used, at energies extending to nearly 900 MeV. In the 900-MeV work, ions of rhodium, element 45, were accelerated in the heavy-ion linear accelerator (HILAC) at the Lawrence Berkeley Laboratory of the University of California. Figures 1 and 2 are taken from a similar experiment on krypton, element 36, at 714 MeV. In both cases, the incident particles lost so many electrons that only three-electron (lithiumlike) or four-electron (berylliumlike) ions remained. Studies were made of the spectra (Fig. 1) and mean lives (Fig. 2) reflecting transitions of one of these remaining electrons. To reach the same degree of ionization in a hot plasma would require a temperature of 2.5×10^7°C.

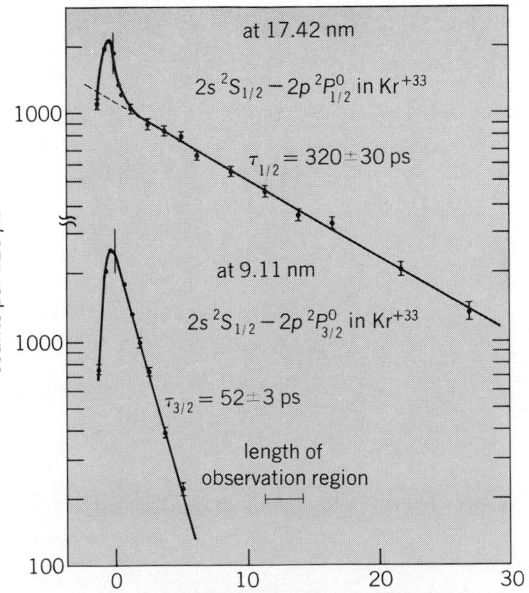

Fig. 2. Lifetime data for two levels in three-electron krypton ions, excited by the same mechanism as in Fig. 1. (*Data courtesy of J. A. Leavitt et al.*)

Relativistic effects. The Coulomb force acting on the electrons of highly ionized atoms is so large that they move with speeds that are appreciable fractions of the speed of light; this means that the electronic motions must be treated relativistically. Since such theories are complicated, there is some uncertainty as to which kind of calculation is the best representation of the data; the beam-foil spectroscopy work on spectra and lifetimes is done partly to clarify the theory. Thus the krypton experiment and earlier work on iron, element 26, showed that the so-called dipole-length calculations are to be preferred over the dipole-velocity approach to the level lifetimes. An interesting relativistic effect is that electronic transitions which are forbidden under ordinary circumstances can become dominant over the usual allowed transitions. Such effects have been seen in the work at HILAC.

Lamb shift. A major advance in atomic theory was made when W. E. Lamb, Jr., showed that the Dirac treatment of the hydrogen atom had to be revised. The revision consisted of a small change in the energy levels of a one-electron system, the Lamb shift. Among other things, it is strongly dependent on the atomic number of the ion, and beam-foil spectroscopy experiments have been done to see if current calculations are applicable to one-electron ions up to chlorine and argon, elements 17 and 18. Preliminary results suggest that the theory, quantum electrodynamics, is satisfactory.

The Lamb shift also occurs in two-electron ions, and that has been investigated up to oxygen, element 8.

Beam-foil interaction. The fundamental process which gives rise to the excitation phenomenon is still poorly understood. Efforts have been made to develop a semiempirical theory of beam-foil spectroscopy, the basic data being the relative populations of the excited levels as a function of particle energy and element. *See* SPECTROSCOPY.

[STANLEY BASHKIN]

Bibliography: D. D. Dietrich et al., Oscillator strengths of the $2s\,^2S_{112} - 2p\,^2P^0_{1/2,\,3/2}$ transitions in Fe XXIV and the $2s\,^1S_0 - 2s2p\,^3P_1^{\,0}$ transition in Fe XXIII, *Phys. Rev. A*, 18:208–211, 1978; N. A. Jelley et al., Lamb shift and fine structure in $1s\,3s\,^3S$–$1s\,3p\,^3P$ transitions in helium-like oxygen, *J. Phys. B: Atom. Molec. Phys.*, 12:2605–2611, 1979; Proceedings of the 5th International Conference on Beam-foil Spectroscopy, *J. Phys.*, C1:1–367, 1979.

Benzaldehyde

A colorless, liquid aldehyde, C_6H_5CHO, boiling at 179°C and possessing a characteristic aromatic odor resembling that of bitter almonds. It is produced industrially at the rate of about 5,000,000 lb annually. Principal uses of benzaldehyde are as a flavoring agent and as an intermediate in chemical syntheses. For discussions of chemical properties and methods of manufacture *see* ALDEHYDE.

[HARRY A. STANSBURY, JR.]

Benzene

A colorless, liquid, inflammable, aromatic hydrocarbon of chemical formula C_6H_6 which boils at 80.1°C and freezes at 5.4–5.5°C. In the older

American and British technical literature benzene is designated by the German name benzol. In current usage the term benzol is commonly reserved for the less pure grades of benzene.

Benzene is used as a solvent and particularly in Europe as a constituent of motor fuel. In the United States the largest uses of benzene are for the manufacture of styrene and phenol. Other important outlets are in the production of dodecylbenzene, aniline, maleic anhydride, chlorinated benzenes (used in making DDT and as moth flakes), and benzene hexachloride, an insecticide.

X-ray and electron diffraction studies indicate that the six carbon atoms of benzene, each with a hydrogen atom attached, are arranged symmetrically in a plane, forming a regular hexagon. The hexagon symbol (I), commonly used to represent the structural formula for benzene, implies the presence of a carbon atom at each of the six angles and, unless substituents are attached, a hydrogen

(I)

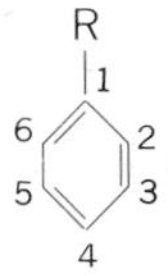

(II)

at each carbon atom. Whereas the three double bonds usually included in the formula (I) are convenient in accounting for the addition reactions of benzene, present evidence is that all the carbon-to-carbon bonds are identical.

Prior to World War II most of the benzene commercially produced was a by-product of the steel industry. High-temperature pyrolysis of coal to produce metallurgical coke for the steel industry yields gases and tars from which benzene may be recovered.

In the present era nearly all commercial benzene is a product of petroleum technology. The gasoline fractions obtained by reforming or steam cracking of feedstocks from petroleum contain benzene and toluene which can be separated economically. Unfortunately the presence of benzene and toluene are important for the high octane number of these same gasoline fractions, with the result that the prices of benzene and toluene are affected by that of lead-free gasoline. Benzene may also be produced by the dealkylation of toluene.

Despite its low hydrogen-to-carbon ratio, benzene reacts chiefly by substitution. Because of the symmetry of the molecule, a given reagent will react to afford only one monosubstituted benzene. Some typical substitution products are included in the table. *See* SUBSTITUTION REACTION.

A large number of disubstitution products of benzene are known. For any two substituents there are only three possible arrangements. The groups may be adjacent to each other (ortho, *o*-). They may be located on carbons separated from each other by a single CH, that is, at positions 1 and 3 of formula (II) (meta, *m*-), or they may be at any two opposite corners of the hexagon (para, *p*-). The three possible dichlorobenzenes are shown below.

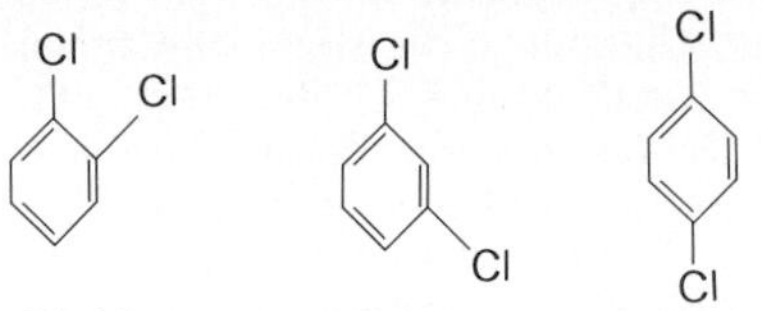

o-Dichlorobenzene *m*-Dichlorobenzene *p*-Dichlorobenzene

The position of substituents in trisubstituted or polysubstituted benzenes is usually indicated by numbers. For instance, the trichlorobenzene in which all three chlorines are adjacent is called 1, 2, 3-trichlorobenzene.

Benzene may also react by addition. Hydrogenation in the presence of a catalyst yields cyclohexane, C_6H_{12}, an intermediate in the manufacture of fibers, while addition of chlorine (usually in the presence of ultraviolet light) yields benzene hexachloride, $C_6H_6Cl_6$, an insecticide.

The largest commercial uses for benzene are in the manufacture of ethylbenzene (a styrene intermediate), cumene (an intermediate in the manufacture of phenol), and cyclohexane (an intermediate in the manufacture of fibers).

Benzene is a toxic substance, and prolonged exposure to concentrations in excess of 35–100 parts per million in air may lead to symptoms ranging from nausea and excess fatigue to anemia and leukopenia. In addition, the Occupational Safety and Health Administration has listed benzene as a potential carcinogen. *See* AROMATIC HYDROCARBON. [CHARLES K. BRADSHER]

Bibliography: T. W. G. Solomons, *Organic Chemistry*, 1978; K. Weissermel and H.-J. Arpe, *Industrial Organic Chemistry, 1978.*

Substitution products of benzene

Reagents	Formula (II), R=	Name of product
H_2SO_4	$-SO_3H$	Benzenesulfonic acid
$HNO_3+H_2SO_4$	$-NO_2$	Nitrobenzene
Cl_2+Fe	$-Cl$	Chlorobenzene
$C_2H_4+AlCl_3$	$-C_2H_5$	Ethylbenzene

Benzoate

A salt or ester of benzoic acid with the formula shown below, formed by replacing the acidic hy-

$$C_6H_5\overset{\displaystyle O}{\overset{\|}{C}}-OH$$

drogen of the carboxyl group by a metal or an organic radical. In nature, benzoates occur chiefly in resins and balsams. Sodium benzoate, one of the few food preservatives permitted by law, is used extensively to inhibit the growth of microorganisms in food, pharmaceuticals, and cosmetics. The esters, characterized by agreeable odors, are used in perfumes, medicines, and plasticizers. Benzyl benzoate is an important one and is used as an antispasmodic, as an antiseptic, and in perfumes. *See* BENZOIC ACID; CARBOXYLIC ACID; ESTER. [ELBERT H. HADLEY]

Bibliography: N. W. Desrosier, *The Technology of Food Preservation*, 4th ed., 1977.

Benzoic acid

The aromatic carboxylic acid with formula C_6H_5COOH. Benzoic acid melts at 122.4°C, boils at 250°C, is slightly soluble in water, and is relatively soluble in alcohol and ether. It is slightly stronger than simple aliphatic acids (K_a at 25°C = 6.4×10^{-5}).

Benzoic acid is found free and combined (gum

benzoin; many berries) in nature. The acid of commerce is prepared by hydrolysis of benzotrichloride $C_6H_5CCl_3$, using lime and iron powder catalyst; by monodecarboxylation of phthalic acid, using phthalic anhydride, steam and zinc oxide or chromate catalyst; and by oxidation of toluene, using manganese dioxide and sulfuric acid or using air in the presence of a cobalt catalyst.

Many derivatives of benzoic acid possess important properties rendering them valuable to industry, commerce, or medicine. Thus tertiary butyl peroxybenzoate, on decomposition, functions as an initiator of free-radical reactions. *See* FREE RADICAL.

The ester benzyl benzoate is used as a miticide. Iodinated benzoic derivatives, especially 2,4,6-triiodo-3-acetamidobenzoic and 2,4,6-triiodo-3,5-diacetamidobenzoic acids, are opaque to x-rays and, since they are readily excreted by the kidneys, are used in urological and cholecystographic examinations as contrast media. Esters of *para*-hydroxybenzoic acid are bacteriostatic agents. The ester *ortho*-hydroxybenzoic acid (salicylic acid) is of extreme importance: acetylation furnishes acetylsalicylic acid, or aspirin. The ester methyl salicylate is present in, and responsible for, the properties of oil of wintergreen; the synthetic material is used as a flavoring agent, and is added to rubbing compounds for athletes. The phenyl ester, salol, is not readily hydrolyzed except under basic conditions. It is therefore used as a protective, enteric coating for pills designed for intestinal infections. The ester *para*-aminobenzoic acid (PABA) has been shown to be necessary for growth in certain bacteria (pneumococci, staphylococci, and streptococci); it is an antagonist for sulfonamides. *See* SULFONAMIDE.

Chemically, the acid is relatively inert because of the meta-directing carboxyl group. Among its limited uses in direct synthesis are chlorination, using chlorine and ferric chloride (catalyst) at room temperature, to give mainly *m*-chlorobenzoic acid, together with some 2,5- and 3,4-dichlorobenzoic acids; and nitration with sulfuric and nitric acids, to form *m*-nitrobenzoic acid (the same reaction at elevated temperature gives 3,5-dinitrobenzoic acid).

Benzoic acid is used in curing tobacco; as a preservative (sodium benzoate) of foods (juices and catsup); in dyes; as a mordant in cloth printing (calico); in many esters (benzoates); and as a reference standard in volumetric analysis.

Ingestion of sodium benzoate (from preserved foods) results in excretion of hippuric acid, $C_6H_5CONHCH_2COOH$, formed via enzymatically induced benzoylation of the amino acid glycine. *See* BENZOATE; CARBOXYLIC ACID; SUBSTITUTION REACTION.

[EVANS B. REID]

Bibliography: E. Costa and P. Greengard (eds.), *Mechanism of Action of Benzodiazepines*, 1975; Merck and Co., *The Merck Index*, 9th ed., 1976.

Berkelium

Element number 97, symbol Bk, the eighth member of the actinide series of elements. In this series the 5*f* electron shell is being filled, just as the 4*f* shell is being filled in the lanthanide (rare-earth) elements. These two series of elements are very

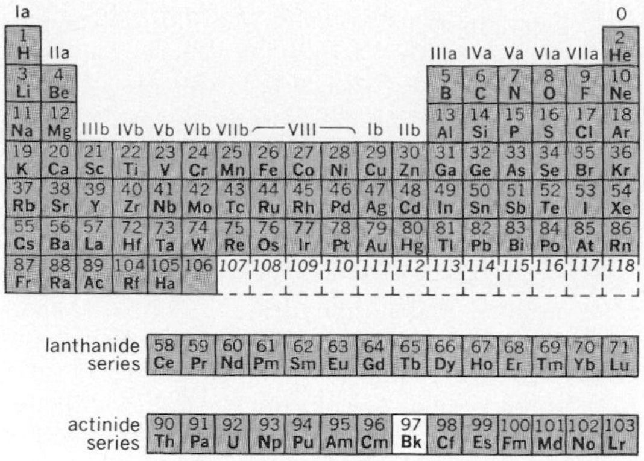

similar in their chemical properties and berkelium, aside from small differences in ionic radius, is especially similar to its homolog terbium.

Berkelium does not occur in the Earth's crust because it has no stable isotopes. It must be prepared by means of nuclear reactions using more abundant target elements. These reactions usually involve bombardments with charged particles, irradiations with neutrons from high-flux reactors, or production in a thermonuclear device. In the latter device the targets are exposed to an extremely high, instantaneous neutron flux.

Nine isotopes of berkelium are known, ranging in mass from 243 to 251 and in half-life from 1 hr to 1380 years. The most easily produced isotope is ^{249}Bk, which undergoes beta decay with a half-life of 314 days and is therefore a valuable source for the preparation of the isotope ^{249}Cf; the latter isotope is especially useful for the investigation of the chemical properties of element number 98, californium. The isotope ^{249}Bk is produced in amounts of tens of milligrams by successive neutron capture reactions beginning with ^{244}Cm and beta decay of ^{249}Cm. The berkelium isotope with longest half-life is ^{247}Bk (1380 years), but it is difficult to produce in sufficient amounts to be applied to the studies of berkelium chemistry, where it would be most valuable.

Before 1958, studies of the chemical properties of the element were limited to a tracer scale. These experiments indicated that berkelium possesses properties characteristic of the actinide elements: a 3+ oxidation state (yellow-green) in aqueous solution and a water-soluble nitrate, chloride, sulfate, perchlorate, and sulfide. Berkelium coprecipitrates with lanthanum fluoride or hydroxide. In contrast to neighboring actinide elements, berkelium may be oxidized in aqueous solution to the 4+ state (yellow) by strong oxidizing agents, such as bromate or ceric ions. This behavior is interpreted as an enhanced ease of removal of the eighth electron in the 5*f* shell to attain the especially stable configuration of a half-filled shell of seven 5*f* electrons.

Much knowledge of the chemical properties of berkelium is derived from investigations on the micro scale by using microgram to milligram amounts of the pure element in the form of the isotope ^{249}Bk. These studies are subject to certain difficulties and limitations resulting from the intense radioactivity of ^{249}Bk and its rapid decay into the neighboring element californium. The first isolation of weighable amounts of the pure element was accomplished by B. B. Cunningham and S. G. Thompson in 1958, when they measured its magnetic susceptibility and obtained certain spectral information. These measurements confirmed the electronic structure expected on the basis of G. T. Seaborg's actinide hypothesis. Cunningham and his coworkers prepared the trivalent compounds Bk_2O_3, BkF_3, $BkCl_3$, and BkOCl. They examined the structures by x-ray diffraction methods and found them to be isostructural with the preceding actinides, as expected. The only tetravalent compound characterized was the dioxide BkO_2. Subsequently $BkBr_3$, BkI_3, and BkF_4 were prepared. Cunningham also observed 16 absorption peaks of Bk(III) ions in the wavelength region 320–700 mμ. Many lines in the emission spectrum of Bk, as well as several absorption bands of Bk^{+4} in solution, have been observed by R. Gutmacher, K. Hulet, and their coworkers.

Solvent extraction methods involving the use of organic phosphates, for example, hydrogen di(2-ethylhexyl)orthophosphoric acid (HDEHP) in heptane, have been especially useful in separating Bk(IV) from other actinides and from lanthanide elements.

Berkelium metal is chemically reactive, exists in two crystal modifications, and melts at 986°C. Berkelium was discovered in 1949 by Thompson, A. Ghiorso, and Seaborg at the University of California in Berkeley and was named in honor of that city. The new element was produced by bombarding americium-241 with helium ions accelerated in a cyclotron. The discovery of berkelium had a special significance in that it depended on the solution of many problems which were also essential to the discovery of many heavier elements. For example, it was necessary to synthesize the highly radioactive target elements used for its preparation and to work out methods for safely handling the high levels of radioactivity involved. Similarly, new systematics of the radioactive properties of the isotopes were developed in order that the modes of decay and half-lives of the undiscovered isotopes could be predicted with reliability. For the first time many new methods of separation of the actinide elements were developed which were both rapid and efficient. *See* ACTINIDE ELEMENTS; TRANSURANIUM ELEMENTS.

[GLENN T. SEABORG]

Bibliography: S. Ahrland et al., *The Chemistry of the Actinides*, 1975; B. B. Cunningham, Berkelium and californium, *J. Chem. Educ.*, 36:32–37, 1959; B. B. Cunningham, Chemistry of the actinide elements, *Annu. Rev. Nucl. Sci.*, 14:323–346, 1964; E. K. Hyde et al., *The Nuclear Properties of the Heavy Elements*, 3 vols., rev. ed., 1971; G. T. Seaborg, History of the synthetic actinide elements, *Actinides Rev.*, 1:3–38, 1967; G. T. Seaborg, *The Transuranium Elements*, 1958.

Beryllium

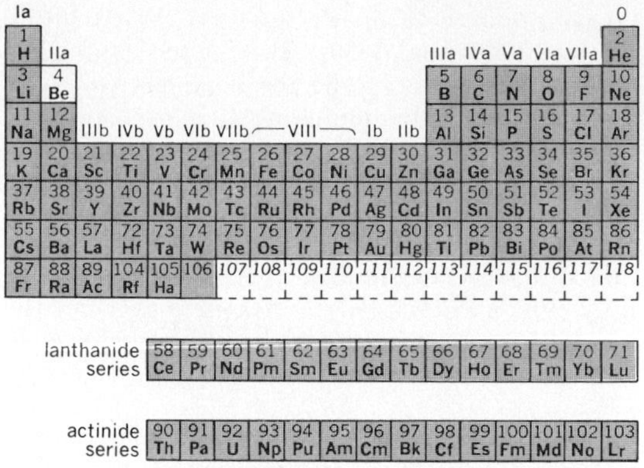

A chemical element, Be, atomic number 4, with an atomic weight of 9.0122. Beryllium, a rare metal, is one of the lightest structural metals, having a density about one-third that of aluminum. Beryllium has a number of unusual and even unique properties. An indication of the broad impact of beryllium in modern science is given by the variety of topics relating to it in the literature: cosmological theories, isotropic dating, meteorological mechanisms, rocket construction, neutron reflection and moderation for nuclear reactors, enzyme inhibition, and toxicology.

The largest volume uses of beryllium metal are in the manufacture of beryllium-copper alloys and in the development of Be-containing moderator and reflector materials for nuclear reactors. Addition of 2% beryllium to copper forms a nonmagnetic alloy which is six times stronger than copper. These beryllium-copper alloys find numerous applications in industry as nonsparking tools, as critical moving parts in aircraft engines, and in the key components of precision instruments, mechanical computers, electrical relays, and camera shutters. Beryllium-copper hammers, wrenches, and other tools are employed in petroleum refineries and other plants in which a spark from steel against steel might lead to an explosion or fire.

Beryllium has found many special uses in nuclear energy because it is one of the most efficient materials for slowing down the speed of neutrons and acting as a neutron reflector. Consequently, much beryllium is used in the construction of nuclear reactors as a moderator and as a support or alloy with the fuel elements.

Commercial production. The discovery of beryllium in 1798 stemmed from the observation by the French mineralogist R. J. Haüy that the optical properties of the emerald were identical with those of the commoner mineral beryl. At the request of Haüy, L. N. Vauquelin chemically analyzed beryl and emerald and proved that both substances had the same composition and contained a new element. Because the salts of beryllium are sweet-tasting, the new element was first called glucinium. The name beryllium was given by F. Wöhler in Germany in 1828 who, simultaneously with W. Bussy in France, first succeeded in isolating the pure metal by fusing beryllium chloride with metallic potassium. In 1898 the French chemist P. Lebeau prepared beryllium of 99.5–99.8% purity by an electrolytic process. Within 20 years a commercial process for beryllium production was developed by A. Stock and H. Goldschmidt in Germany. The process involved electrolysis of the fluoride salt at temperatures near the melting point of beryllium. The important ability of beryllium to age-harden copper with nickel additions was discovered in 1926 by M. G. Corson, an American metallurgist.

Small-scale production of beryllium and its al-

Table 1. Physical properties of beryllium*

Atomic and mass properties	
Mass number of stable isotopes	9
Atomic number	4
Outer electronic configuration	$1s^2 2s^2$
Atomic weight	9.0122
Atomic diameter	2.221 A
Atomic volume	4.96 cm^3/mole
Crystal structure	Hexagonal close-packed
Lattice parameters	$a = 2.285$ A $c = 3.583$ A
Axial ratio	$c/a = 1.568$
Field of cation (charge/radius^2)	17
Density†, 25°, x-ray (theoretical)	1.8477 ± 0.0007 g/cm^3
Density, 1000°, x-ray	1.756 g/cm^3
Radius of atom (Be^0)	1.11 A
Radius of ion, Be^{++}	0.34 A
Ionization energy ($Be^0 \rightarrow Be^{++}$)	27.4 eV
Thermal properties	
Melting point	1285°C (2345°F)
Boiling point‡	2970°C (5378°F)
Vapor pressure (T = K°)	$\log P$ (atm) $= 6.186 + 1.454 \times 10^{-4} T - (16{,}700/T)$
Heat of fusion	250–275 cal/g
Heat of vaporization	53,490 cal/mole
Specific heat (20–100°)	0.43–0.52 cal/(g)(°C)
Thermal conductivity (20°)	0.355 cal/(cm^2)(cm)(sec)(°C) (42% of copper)
Heat of oxidation	140.15 cal
Electrical properties	
Electrical conductivity	40–44% of copper
Electrical resistivity	4 microhms/cm (0°C) 6 microhms/cm (100°C)
Electrolytic solution potential, Be/Be^{++}	$E^0 = -1.69$ volts
Electrochemical equivalent	0.04674 mg/coulomb

*The data are from three principal sources: D. W. White and J. E. Burke (eds.), *The Metal Beryllium*, 1955; U.S. Bureau of Mines, *Materials Survey: Beryllium*, 1953; and M. C. Udy, H. L. Shaw, and F. W. Boulger, Properties of beryllium, *Nucleonics*, 11:52–59, 1953.

†Measured values vary from 1.79 to 1.86, depending on purity and method of fabrication.

‡Obtained by extrapolation of vapor pressure data, not considered very reliable.

loys began in the United States and abroad during the late 1920s and 1930s. The United States has been and continues to be the dominant producer and consumer of beryllium metal, alloys, and compounds.

When beryllium atoms are bombarded with α-particles from radium, the nucleus is broken down to yield a profusion of neutrons. It was this reaction which led to the discovery of the neutron in the 1930s. A radium-beryllium source provided the neutrons for the studies which led Enrico Fermi to the construction of the first nuclear reactor in 1942.

The possible use of beryllium as a moderator in nuclear reactors was first recognized in 1940 by C. V. Wilson at the University of Chicago. The ability of beryllium to withstand high temperatures and corrosion makes it particularly attractive for nuclear reactors; however, the brittleness of the metal and oxide presents many difficulties.

Certain health problems involving beryllium were recognized during the period 1933–1942. At this time a puzzling new occupational disease was reported from Germany, Italy, and the Soviet Union. Many workers employed in the extraction of beryllium from beryl became acutely ill, and about 10–20% of those affected died. Many of the workers had inhaled beryllium in the form of corrosive salts, such as the fluoride. This acute form of beryllium poisoning produced symptoms such as chills, fever, painful cough, and fluid accumulation in the lungs, symptoms similar to those of poisoning by irritating gases such as chlorine.

In 1946 a delayed form of beryllium poisoning appeared in patients who had been engaged in the manufacture of fluorescent lamps in which beryllium compounds were incorporated in the phosphors used to coat the inside of the lamp tubes. In these cases workers were exposed to insoluble forms of beryllium, and their illness became noticeable only after 6 months to 3 years. Cases are now known in which the delayed form of beryllium poisoning, called berylliosis, did not appear until 15 years or longer following exposure. One of the characteristic aspects of berylliosis is the appearance of lesions in the lung called granuloma. The disease is also known as chronic pulmonary granulomatosis.

Procedures have been developed to protect those working with beryllium so that its toxicity no longer need hinder the manufacture of beryllium products.

Physical and chemical properties. Some of the important physical and chemical properties of beryllium are given in Table 1. One of the most striking differences between beryllium and the other main group II elements in the periodic table lies in the very small ionic radius r, which is less than one-half that of the next smallest cation, Mg^{++}. Further, Be^{++} possesses an exceptional deforming power, as shown by the value of e/r^2 (e = charge), which is 17, compared to 3.3 for Mg^{++}, and 1.8 for Ca^{++}, and 9.2 for Al^{3+}. Beryllium shows a remarkable resemblance to aluminum in its chemical behavior.

Like aluminum, beryllium forms a protective oxide skin, and its stability in air even at red heat is due to this protective oxide layer. The protective influence of the film may be disrupted when the metal is placed in contact with more noble metals or alloys which may result in current flow and cause galvanic corrosion. Beryllium metal is appreciably resistant to attack by oxygen-free sodium metal at elevated temperatures. It is difficult, however, to obtain and maintain sodium metal completely free of oxygen. The attack rate of liquid sodium on beryllium can be reduced by the incorporation into the system of an element, such as calcium, which forms a more stable oxide than beryllium. The resistance of beryllium metal to liquid metals is important in the design of heat-transfer units in nuclear reactors.

Beryllium has a very strong tendency to form covalent compounds with a maximum covalency of 4. Its salts so uniformly have four molecules of water of crystallization for every atom of beryllium that any compound which has not may be assumed to be a complex ion formed by displacement of coordinated H_2O or to be a purely covalent substance such as beryllium acetylacetonate. Compounds of beryllium are readily hydrolyzed in water. Beryllium complexes containing donor nitrogen or sulfur atoms are decomposed by water because of the strength of the Be-O bond. In aqueous solution the chelates of beryllium with salicylic acid analogs are among the most stable.

Beryllium compounds form tetrahedrons which are characteristic of many elements having the coordination number 4. This is the result of the directional characteristics of the orbital electronic distribution and intensity. Hybrid bonds of the sp^3 type giving tetrahedral symmetry, such as those of carbon and silicon, are found with compounds of beryllium. Because beryllium lacks d orbitals, only tetrahedral compounds are formed.

The tetrahedral arrangement of the valencies of beryllium leads to optical activity of many of its compounds just as with carbon compounds. The classical example is that found by W. H. Mills and R. A. Gotts, who in 1926 separated the two optical isomers of the beryllium salt of benzoylpyruvic acid.

It is of analytical importance that beryllium does not react strongly with the universal complexing agent, ethylenediaminetetraacetic acid (EDTA). At pH 5–5.5, 500 g of the tetrasodium salt of EDTA will completely bind 1 mole of magnesium or calcium, but only 0.002 mole of beryllium. Also of importance in this connection is the fact that aluminum interference in the analysis of beryllium can be greatly minimized because aluminum reacts quite strongly with EDTA under conditions where beryllium does not.

Solutions of beryllium salts are acidic. In 1815 J. J. Berzelius noted that a solution of beryllium sulfate dissolves up to several molecular proportions of beryllium oxide or hydroxide. This behavior indicates the existence of beryllium hydroxo or oxo complexes. During hydrolysis, when the average number of OH^- bound per Be^{++} exceeds unity, precipitation begins. At least one polynuclear hydroxo complex is formed.

The main product of the hydrolysis is the polynuclear complex, $Be_3(OH)_3^{3+}$, which is presumed to form a six-membered ring, three tetrahedrons being linked by sharing OH corners:

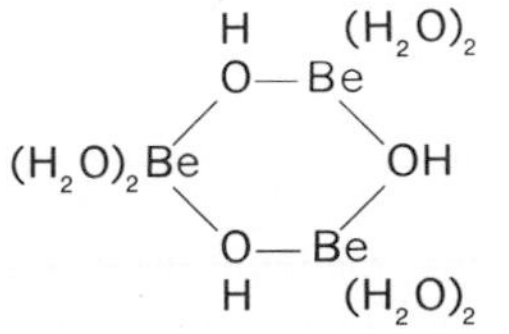

If the tetrahedrons were filled with water, the complex formula would be $Be_3(OH)_3(H_2O)_6^{3+}$, as shown here.

Beryllium can form polymeric compounds as well as a class of covalent compounds which show remarkable thermal stability. Some can be distilled at temperatures above 300°C at atmospheric pressure without decomposition. Examples of these compounds include the basic beryllium carboxylates, $Be_4O(RCO_2)_6$, and neutral chelate compounds of beryllium such as the well-known beryllium acetylacetonate which was discovered in 1894 and which is volatile, melts at 108°C, and boils undecomposed at 270°C; it is insoluble in water, but is hydrolyzed on boiling, and it is readily soluble in organic solvents.

Beryllium alkyls are prepared, as shown by reaction (1), by reaction with Grignard reagents or

$$2RMgX + BeX_2 \rightarrow R_2Be + 2MgX_2 \qquad (1)$$

organolithium compounds with beryllium chloride. Another preparative procedure, shown in reaction (2), involves the disproportionation of alkylberyllium halides on heating.

$$2RBeX \rightarrow R_2Be + BeX_2 \qquad (2)$$

X-ray diffraction studies show that solid dimethylberyllium forms a polymeric chain in which each beryllium atom is attached tetrahedrally to four carbon atoms and each carbon atom to two beryllium atoms. The structure is electron deficient as are the aluminum aryls:

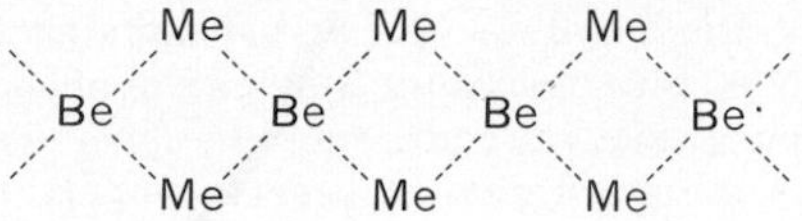

Each dotted line represents a half bond because there are only half enough electron pairs for the number of attachments shown.

Principal compounds. The following list shows some of the principal compounds of beryllium.

Acetylacetonate	Fluoride, BeF_2
Ammonium beryllium fluoride	Hydroxide
Aurintricarboxylate	Nitrate, $Be(NO_3)_2 \cdot 4H_2O$
Basic acetate, $BeO \cdot Be_3(CH_3COO)_6$	Nitride, Be_3N_2
Basic beryllium carbonate	Oxide, BeO
Beryllate, BeO_2^{--}	Perchlorate, $Be(ClO_4)_2 \cdot 4H_2O$
Beryllium ammonium phosphate	Plutonium-beryllium, $PuBe_{13}$
Bromide, $BeBr_2$	Salicylate
Carbide, Be_2C	Silicates (emerald)
Chloride, $BeCl_2$	Sulfate, $BeSO_4 \cdot 4H_2O$
Dimethyl, $Be(CH_3)_2$	Uranium-beryllium, UBe_{13}

Many of the compounds listed are useful as intermediates in processes for the preparation of ceramics, beryllium oxide, and beryllium metal. Other compounds are useful in analysis and organic synthesis.

Nuclear properties. Beryllium is the only element in the periodic table with both an even atomic number and but a single stable isotope. Several radioactive isotopes can be prepared. Two radioactive isotopes of beryllium, Be^7 and Be^{10}, are formed in the upper atmosphere as a result of intense cosmic-ray activity causing spallation of the nuclei of nitrogen and oxygen. A short-lived isotope, Be^{11}, with a half-life of 14.1 sec, has been produced by neutron bombardment of boron.

Neutrons are produced in many nuclear reactions involving beryllium. The mixed radium-beryllium source is often used as a neutron source, yielding 460 neutrons per 10^6 disintegrations resulting from the reaction, $Be^9(\alpha,n)C^{12}$. Photoneutrons are produced from the action on beryllium of γ-rays of energy greater than 1.66 MeV, according to the reaction $Be^9(\gamma,n)Be^8$. The Be^8 isotope decomposes into two α-particles in less than 10^{-14} sec.

Beryllium is an excellent moderator for neutrons and compares favorably with hydrogen, deuterium, helium, carbon, and oxygen. Beryllium and beryllium oxide are useful as reflecting materials for nuclear reactors, and the metal is useful as an alloying element and container for fuel elements.

Analytical techniques. Quantitative analyses for beryllium can be divided into those methods used for the assay of macro amounts (multimilligram quantities) and those used for the assay of micro or trace amounts (multimicrogram or less).

It is often necessary to remove interfering substances from a beryllium sample. Numerous methods are available. The use of selective complexing or chelating agents is common. Metals such as copper, nickel, iron, and calcium, but not beryllium, are chelated by EDTA at pH 5.5. Consequently, the beryllium can be precipitated as the hydroxide, as the beryllium ammonium phosphate, or as one of many organic salts. In specific cases, such as the separation of excess copper, excess ammonium hydroxide dissolves copper as a complex ion, whereas beryllium hydroxide remains insoluble.

Selective precipitation of foreign elements often utilizes 8-hydroxyquinoline. At suitable pH in buffered media, the quinolates of many metals are insoluble, whereas beryllium remains in solution.

Ion-exchange methods, especially in combination with complexing agents, find many uses for separation of beryllium. Beryllium absorbed on organic cation exchanger is eluted completely, for example, by sulfosalicylic acid at a suitable pH, whereas calcium and copper remain on the cation exchanger. At pH 3.5, EDTA converts most metals to anionic complexes, whereas beryllium remains cationic. Passage of the EDTA solution through a cation exchanger results in absorption of beryllium, whereas the anionic complexes pass through.

From concentrated hydrochloric acid solution, numerous cations are retained by a column of an anion-exchange resin while beryllium passes through unabsorbed. The reverse situation is observed when the beryllium and other cations in concentrated lithium chloride are passed through an anion exchanger.

Other methods of removing impurities from beryllium in analytical operations include solvent extraction, volatilization, and electrolysis with a mercury cathode.

Analysis of beryl ore for beryllium involves a fusion with sodium carbonate followed by silicate removal. In some cases it saves time to fuse beryl with sodium fluoride, followed by fuming with sulfuric acid to remove the fluoride.

Occurrence. Beryllium is surprisingly rare for a light element, constituting about 0.005% of the Earth's crust. It is about thirty-second in order of abundance, occurring in concentrations approximating those of cesium, scandium, and arsenic. Actually, the abundance of beryllium and its neighbors, lithium and boron, are about 10^{-5} times those of the next heavier elements, carbon, nitrogen, and oxygen. As a light element, beryllium should be a great deal more abundant on Earth. It is believed that the beryllium nucleus is destroyed by collision with high-energy protons, as in the Sun and other stars. The scarcity of beryllium on Earth is interpreted by some to indicate that Earth passed through a stage of stellar temperatures.

A summary of the natural occurrence of beryllium in various materials is shown in Table 2. More beryllium is present in the air of the larger urban centers, especially coal-burning communities, than in rural areas.

At least 50 different beryllium-bearing minerals are known, but in only about 30 is it a regular constituent. The only beryllium-bearing mineral of industrial importance is beryl, a hexagonal beryllium aluminum silicate having the ideal composition $Be_3Al_2Si_6O_{18}$, equivalent to 5% Be or 14.0% BeO, 19% Al_2O_3, and 67% SiO_2. The precious forms of beryl, emerald and aquamarine, approach the ideal composition. Commercial grades of beryl contain about 10–12% BeO, 17–19% Al_2O_3, 64–70% SiO_2, 1–2% alkali metal ores, and 1–2% iron and other ores.

Table 2. Natural occurrence of beryllium*

Material	Concentration, ppm	
	Average	Range
Earth's crust		3–6
Soil	0.37	0.13–0.88
Shales		0.13–6.00
Bauxite		0.13–0.60
Coal	1.55	0.07–3.0
Coal ash	4.0	8000†
Air	0.0002‡	0.0001–0.003‡
Biological		
Lung		0.00–1.98§
Urine and blood	0.33§	0.00

*From data summarized by J. Cholak, The analysis of traces of beryllium, *A.M.A. Arch. Ind. Health*, 19:205–210, 1959. †Upper limit. ‡Measured in μg/m³. §Measured in μg/100 g.

The emerald is a transparent, intensely green variety of beryl. The hexagonal crystals of beryl itself may be greenish, bluish, or rosy, but in the field they tend to assume the color of the granitic rocks with which they are generally associated. Occasionally, they attain gigantic size, up to 2–3 ft in diameter and several feet in length, and weigh several tons.

World beryllium deposits are divided into two groups, the beryl-bearing pegmatite varieties of granitic rocks and the beryl-bearing nonpegmatite rocks. However, nearly all beryllium produced has been extracted from beryl-bearing pegmatites, which are found in many countries. The most important or potentially important areas are found in Brazil, Argentina, Canada, the United States, the Congo, Rhodesia, South-West Africa, the Union of South Africa, and India.

Beryllium is found in numerous minerals because the very small diameter of the beryllium ion favors its limited entry into other silicate minerals so that the formation of high-concentration beryllium minerals is restricted. In nonpegmatite rocks beryllium is mainly a minor constituent, and only rarely do these minerals contain as much as 0.2% BeO as is found with the helvite group, $(Mn,Fe,Zn)_4Be_3Si_3O_{12}S$.

Extraction and mining. Most beryl is recovered from surface mines by handpicking, to prevent reduction of beryl size. Crystals smaller than 1 in. in diameter cannot be economically processed. About 100 tons of rock must be combed by hand to pick 1/2–1 ton of beryl crystals. In ore deposits where the beryl is fine-grained, milling processes are required. No rock benefaction process has been commercialized to date. However, flotation methods have been proved on a semiworks scale which yield 80–90% recovery of beryl from ores containing 0.5% beryl or more, but seldom give more than 70% recovery from 0.2% beryl feed.

The total known world resources of beryl, including deposits containing as little as 0.1% of beryl, has been estimated at about 4,000,000 tons. The estimated world's resource of beryl in deposits containing 1% beryl or the equivalent is about 200,000 tons. South American and African nations produce 80% of the total used. Because of its beryllium-consuming industries, the United States is the largest beryllium consumer.

Uses. One of the largest uses of beryllium metal is in nuclear reactors as a moderator to lessen the speed of fission neutrons and as a reflector to reduce leakage of neutrons from the reactor core. Beryllium is useful in nuclear applications because of its relatively high neutron-scattering cross section, low neutron-absorption cross section, and low atomic weight.

Another large-scale use of beryllium is in the manufacture of beryllium bronze, which has high tensile strength and a capacity for being hardened by heat treatment. Beryllium-copper molds are used in manufacturing plastic furniture with the appearance of wood-grain surfaces.

A small but important use of beryllium is in sheet or foil form as window material in x-ray tubes. Beryllium transmits x-rays 17 times better than an equivalent thickness of aluminum and 6–10 times better than Lindemann glass. This, together with its high melting point, makes possible the use of x-ray beams of greater intensity.

Beryllium oxide is used in the manufacture of high-temperature refractory material and high-quality electrical porcelains, such as aircraft sparkplugs and ultrahigh-frequency radar insulators. Its high thermal conductivity and its high-frequency electrical insulating properties find application in electrical and electronic fields.

Another use of beryllium oxide is as a slurry for coating of graphite crucibles to insulate the graphite and to avoid contaminations of melted alloys with carbon. Beryllium oxide crucibles are used where exceptionally high-purity or reactive metals are being melted. In the field of beryllium-oxide ceramics, a type of beryllia has been developed that can be formed into custom shapes for electronic and microelectronic circuits. Beryllium oxide has a high thermal conductivity, equal to that of aluminum, and excellent insulating properties, which permits closer packing of semiconductor functions in silicon integrated circuits.

Several salts of beryllium find limited uses. Beryllium nitrate is used in combination with thorium nitrate for fabricating incandescent gas mantles. Sodium-beryllium fluoride is useful for the preparation of glass with a high permeability to ultraviolet rays, as a flux in certain porcelain enamels, and in coating of rods for welding light metals. Beryllium halides and other salts find limited use as catalysts in organic chemical reactions.

Light weight, very high elastic modulus, and heat stability of beryllium make it an attractive material for use as construction material in aircraft and missiles. However, its lack of ductility is a drawback. Were it not for its toxicity and scarcity, beryllium would find use as a rocket fuel because it produces more heat energy per unit volume than any other element. In multistage missiles a small weight reduction in the final stage, such as might be achieved by using beryllium in place of steel, permits a much larger weight reduction in the earlier stages in terms of fuel and structure. Research in the utilization of beryllium metal and beryllium-containing materials for aircraft and missile uses is carried out very actively. These and other still-developing applications together with the continuing uses of beryllium in nuclear technology sustain the ever-mounting production levels of beryllium.

The two radioactive isotopes of beryllium, Be^7 with a half-life of 53 days and Be^{10} with a half-life of 2.5×10^6 years, are useful in geological age determinations. Beryllium-10 especially fills an important gap for dating purposes, falling between C^{14} and the long-lived nuclides (5×10^4 years to at least 10^7 years). The isotopes of beryllium are found in ground-level air, rain, snow, and sea sediments. *See* ALKALINE-EARTH METALS.

[JACK SCHUBERT]

Bibliography: *Beryllium: Physico-Chemical Properties of Its Compounds and Alloys*, Atomic Energy Review Ser., Spec. Issue no. 4, 1974; D. A. Everest, *Chemistry of Beryllium*, 1964; J. Schubert, Some aspects of the chemistry and biochemistry of beryllium, *Chimia (Switz.)*, 1959; U.S. Bureau of Mines, Beryllium, *Mineral Industry Surveys*, issued annually; D. Webster and G. J. London (eds.), *Beryllium Science and Technology*, 2 vols., 1979.

Biphenyl

A colorless hydrocarbon (melting point 70.0°C, boiling point 255.9°C, and density 1.9896), also called diphenyl or phenylbenzene, which crystallizes as small plates (see formula). Although small amounts are present in coal tar, the biphenyl of commerce is obtained by the pyrolysis of benzene at about 750°C or as a by-product in the thermal dealkylation of toluene.

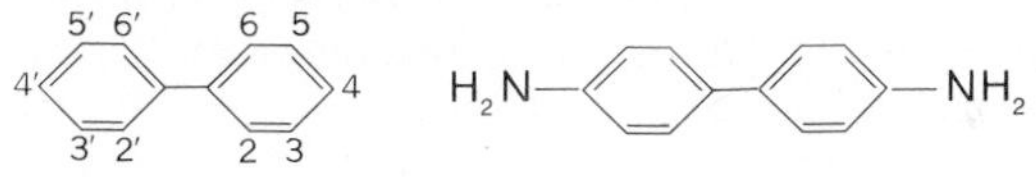

Biphenyl   Benzidine

Biphenyl undergoes substitution in the 2 and 4 positions in the nitration reaction and in position 4 in the Friedel-Crafts or sulfonation reaction. Hydrogenation of biphenyl yields cyclohexylbenzene and bicyclohexyl.

The most important derivative having the biphenyl nucleus is benzidine, with the formula given here, an intermediate in the manufacture of dyes, but benzidine is not usually prepared from biphenyl. Certain biphenyl derivatives having large groups at the 2,2′,6.6′ positions are restricted in rotation about the central bond and have been resolved into optical antipodes. *See* OPTICAL ACTIVITY.

The high thermal stability and low vapor pressure of biphenyl make it a valuable heat-transfer medium, which can be used without high-pressure equipment. Biphenyl is used for this purpose either alone or mixed with diphenyl ether. In certain applications it is chlorinated or alkylated to produce agents in some ways superior to the parent hydrocarbon. Unfortunately, polychlorinated biphenyls (PCBs), despite their much prized thermal and chemical stability, are definitely hazardous to health if ingested and have been listed by the Occupational Safety and Health Administration as potential carcinogens. Biphenyl possesses some antifungal activity, and paper for wrapping oranges is sometimes impregnated with it. Biphenyl is inflammable, flash point (open cup) 124°C. *See* AROMATIC HYDROCARBON; POLYNUCLEAR HYDROCARBON.

[CHARLES K. BRADSHER]

Bibliography: G. Egloff, *Physical Constants of Hydrocarbons*, vol. 3, 1946; K. Higuchi (ed.), *PCB Poisoning and Pollution*, 1976; E. H. Rodd (ed.), *Chemistry of Carbon Compounds*, vol. 3, pt. B, 1964.

Bismuth

The metallic element, Bi, of atomic number 83 and atomic weight 208.980 belonging in the periodic table to group Vb, which also includes nitrogen, phosphorus, arsenic, and antimony. Bismuth is the most metallic element in this group in both physical and chemical properties. The only stable isotope is that of mass 209. The electron configuration of the incomplete shells (*O* and *P*) is: $5s^2$, $5p^6$, $5d^{10}$, $6s^2$, $6p^3$.

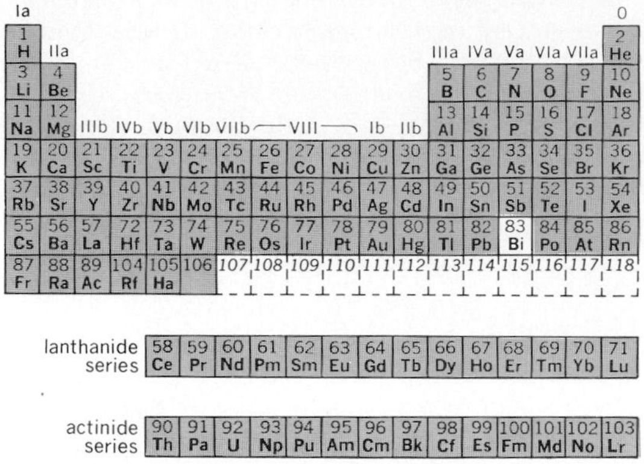

Uses. The main use of bismuth is in the manufacture of low-melting alloys, many of them melting below 100°C. These low-melting alloys are used in fusible elements in automatic sprinklers, special solders, safety plugs in compressed gas cylinders, and automatic shutoffs for gas and electric water-heating systems. Some bismuth alloys, which expand on freezing, are used in castings and in type metal. Bismuth and its alloys are often added to molten iron, steels, and aluminum alloys to produce castings and forgings which can be machined more readily. Because bismuth has low neutron-absorption cross section, low melting point, and high boiling point, tests were begun on its application to reactor technology as a reactor coolant. Another important use of bismuth is in the manufacture of pharmaceutical compounds. Various bismuth preparations have been employed in the treatment of skin injuries, alimentary diseases, such as diarrhea and ulcers, and syphilis. The oxide and basic nitrate are perhaps the most widely used compounds of bismuth. The trioxide is used in the manufacture of glass and ceramic products, while the basic nitrate is used in porcelain painting to fire on gilt decoration. Bismuth telluride is used extensively for thermoelectric cooling and for low-temperature thermoelectric power production.

Occurrence. It is estimated that the Earth's crust contains about 0.00002% bismuth. It occurs in nature as the free metal and in ores such as the sulfide, bismuthinite (also known as bismuth glance), and as the oxide, bismite. The principal ore deposits are in South America. Bismuth also occurs as a minor constituent in a number of lead, copper, and tin ores.

Production. Because its melting point is low, bismuth may be obtained from its native ores simply by heating them; this treatment melts the bismuth, which flows off the ores. The procedure of liquation was formerly used, as was also reduction of oxide ores by carbon or reduction of sulfide ores with iron. At present, however, the primary source of bismuth in the United States is as a by-product in the refining of copper and lead ores. Bismuth remains with the copper or lead after smelting. It is removed from lead by electrolytic refining, crystallization, and by the Betterton-Kroll process. During the electrolytic refining, the bismuth remains in the anode sludge. In the crystallization process, lead crystallizes first from the alloy as it is cooled, leaving a molten alloy enriched in bismuth. The Betterton-Kroll process depends upon the formation of calcium-bismuth and magnesium-bismuth compounds whose high melting points and lower densities permit them to be removed from the lead bath. The enriched bismuth dross is freed of calcium, magnesium, and lead by chlorination.

Physical properties. Bismuth is a gray-white, lustrous, hard, brittle, coarsely crystalline metal. It is one of the few metals which expand on solidification, the expansion being 3.3%. The thermal conductivity of bismuth is lower than that of any metal, with the exception of mercury. Of all metals, bismuth is the most diamagnetic and has the greatest Hall effect. The vapor consists of a mixture of monomer, Bi, and dimer, Bi_2. Table 1 cites the chief physical and mechanical properties of bismuth.

Chemical properties. Bismuth is inert in dry air at room temperature, although it oxidizes slightly in moist air. It rapidly forms an oxide film at temperatures above its melting point, and it burns at red heat, forming the yellow oxide, Bi_2O_3. The metal combines directly with halogens and with sulfur, selenium, and tellurium; however, it does

Table 1. Physical and mechanical properties of bismuth

Melting point, °C	271.4	
Boiling point, °C	1559	
Heat of fusion, kcal/mole	2.60	
Heat of vaporization, kcal/mole	36.2	
Vapor pressure, mm Hg	1	917°C
	10	1067°C
	100	1257°C
Density, g/cm³	9.80	20°(solid)
	10.03	300°(liquid)
	9.91	400°(liquid)
	9.66	600°(liquid)
Mean specific heat, cal/g	0.0294	0–270°C
	0.0373	300–1000°C
Coefficient of linear expansion	13.45×10^{-6}/°C	
Thermal conductivity, cal/(sec)(cm²)(°C)	0.018	100°(solid)
	0.041	300°(liquid)
	0.037	400°(liquid)
Electrical resistivity, μohm-cm	106.5	0°(solid)
	160.2	100°(solid)
	267.0	269°(solid)
	128.9	300°(liquid)
	134.2	400°(liquid)
	145.3	600°(liquid)
Surface tension, dynes/cm	376	300°C
	370	400°C
	363	500°C
Viscosity, centipoise	1.662	300°C
	1.280	450°C
	0.996	600°C
Magnetic susceptibility, cgs units	-1.35×10^{-6}	
Crystallography	Rhombohedral, $a_0 = 4.7457$ A	
Thermal-neutron absorption cross section, barns	0.032 ± 0.003	
Modulus of elasticity, lb/cm²	4.6×10^6	
Shear modulus, lb/cm²	1.8×10^6	
Poisson's ratio	0.33	
Hardness, Brinell	4–8	

not combine directly with nitrogen or phosphorus.

Bismuth is not attacked at ordinary temperatures by air-free water, but it is slowly oxidized at red heat by water vapor. Bismuth does not dissolve in nonoxidizing acids, but does dissolve in nitric acid and in hot concentrated sulfuric acid. The formation of intermetallic compounds involves mainly the strongly electropositive metals.

Principal compounds. Almost all compounds of bismuth contain trivalent bismuth. However, bismuth can occasionally be pentavalent or monovalent. Compounds of bismuth, previously regarded as being in the 2+ or 4+ state, have not been verified. Sodium bismuthate and bismuth pentafluoride are perhaps the most important compounds of Bi(V). The former is a powerful oxidizing agent and the latter a useful fluorinating agent for organic compounds. Bismuth trihalides can be prepared by the reaction of the corresponding hydrogen halide on Bi_2O_3, or by direct union of the elements. All the trihalides except the trifluorides are rapidly hydrolyzed in contact with water, the corresponding oxyhalide being formed. Bismuth trioxide is obtained by burning the metal or by heating the basic carbonate, sulfate, or nitrate. It dissolves in acids, but is insoluble in dilute alkali hydroxides. At red heat, the oxide can be reduced easily to the metal. Bismuth hydroxide, $Bi(OH)_3$, is precipitated by hydroxyl ion from bismuth salt solutions. It is difficult to obtain in a pure state, since it is colloidal in nature. The sulfide, Bi_2S_3, is formed when hydrogen sulfide is passed through a bismuth salt solution and is soluble only in a concentrated acid. It is not very stable, and decomposition to the elements takes place slightly above 100°C. Bismuth nitrate, $Bi(NO_3)_3 \cdot 5H_2O$, is obtained by dissolving bismuth in nitric acid and evaporating the solution. When the crystals are heated, bismuthyl nitrate, BiO (NO_3), is formed. The sulfate, $Bi_2(SO_4)_3$, is prepared by heating the metal or its chloride or nitrate in concentrated sulfuric acid, evaporating the solution, and heating the resulting crystals to 350°C. The hydride, bismuthine, BiH_3, is not stable and decomposes slowly even at ordinary temperatures or instantaneously on heating. Properties of some compounds are given in Table 2.

Analytical. In the conventional qualitative analysis scheme, bismuth, a member of the copper series, is precipitated with H_2S, and the precipitate is dissolved in nitric acid. Addition of base yields a new precipitate which, upon addition of sodium stannite solution, is converted to metallic bismuth. For quantitative determinations, bismuth may be precipitated and weighed as the sulfide or oxychloride, or it may be precipitated as the hydroxide and weighted as the trioxide. *See* ANTIMONY; ARSENIC.

[SAMUEL J. YOSIM]

Bibliography: D. G. Cooper, *The Periodic Table*, 4th ed., 1968; F. A. Cotton and G. Wilkinson, *Advanced Inorganic Chemistry*, 3d ed., 1972; M. Dub (ed.), *Organometallic Compounds: Methods of Synthesis, Physical Constants and Chemical Reactions*, vol. 3: *Compounds of Arsenic, Antimony and Bismuth*, 2d ed., 1968; J. D. Smith, *The Chemistry of Arsenic, Antimony and Bismuth*, 1975.

Table 2. Properties of some bismuth compounds

Compound	Color	Density, g/cm³	Melting point, °C	Boiling point, °C	Heat of formation, kcal/mole
BiF_3	Gray-white	5.32	727	>700	−216
$BiCl_3$	White	4.75	232	447	−90.6
$BiBr_3$	Yellow	5.69	218	461	−71
BiI_3	Gray-black	5.7	408		−46
Bi_2O_3	Yellow	8.9	817		−138
$Bi(OH)_3$	White	4.4	100 (decomposes)		
Bi_2S_3	Black	7.39	685 (decomposes)		

Boiling point

The temperature at which the transition from the liquid to the gaseous phase occurs. For pure substances at a fixed pressure, the boiling or vaporization process occurs at a single temperature; as heat is added, the temperature remains constant until all the liquid has boiled.

The normal boiling point is defined as the boiling point at a total applied pressure of 1 atm, that is, the temperature at which the vapor pressure of the liquid equals 1 atm (760 mm Hg).

The pressure dependence of the boiling point (expressed as absolute temperature T_b) is given by the Clapeyron equation, which is shown below. In

$$\frac{dT_b}{dP} = \frac{T_b \Delta V_v}{\Delta H_v}$$

the equation ΔV_v and ΔH_v are the volume change and the heat absorbed during the vaporization process, and P is the pressure exerted on the liquid. All are necessarily positive, so the boiling point increases with applied pressure. For substances boiling in the region of room temperature, the rate of change of boiling point with temperature is approximately 0.04°/mm Hg (where the pressure is approximately 1 atm). For example, for water at 100°C and 1 atm pressure, $dT_b/dP =$ 0.0369°/mm Hg.

The boiling point cannot be raised indefinitely, however. As the pressure is increased, the density of the gas phase increases until it finally becomes indistinguishable from the liquid phase with which it is in equilibrium; this is the critical temperature, above which no distinct liquid phase exists. Helium has the lowest normal boiling point (4.2 K) of any substance, and tungsten carbide has one of the highest (6300 K).

The boiling point is a rough measure of the intermolecular potential energy of the system, although the heat of vaporization is a better measure. A liquid whose molecules are held together weakly boils at a low temperature; substances with stronger intermolecular forces boil at higher temperatures.

For solutions of two or more components, the boiling process normally occurs over a range of temperatures. A distinction is made between the boiling point (or bubble point), the temperature at which the first bubbles of vapor appear, and the condensation point (or dew point), the temperature at which the last trace of liquid disappears, or alternatively at which the first droplets of liquid appear. Measurement of the boiling point of a solution and the difference between it and the boiling point of the pure solvent provides a convenient method of determining the molecular weight of a dissolved nonvolatile solute. *See* PHASE EQUILIBRIUM.

The well-known process by which liquids evaporate into air at temperatures far below their normal boiling points (for example, the evaporation of water at room temperature) can be regarded as an extreme case of "boiling" in a system of several components. At equilibrium the liquid phase (essentially one substance since the solubility of the air is small) coexists with a gas phase, which is a mixture of air with the vapor of the liquid at a partial pressure almost equivalent to its vapor pressure. For this mixture the dew point is virtually the same as the boiling point of the pure liquid, whereas the bubble point is a temperature as low as the freezing point of the liquid (below which the solid can also evaporate). Similar principles are involved in steam distillation.

Precision measurement of boiling points is known as ebulliometry. The principal experimental difficulty is to be sure that true equilibrium is reached and that the liquid is not superheated into a metastable state. *See* CRITICAL CONSTANT; DISTILLATION; LIQUID; SOLUTION; VAPOR PRESSURE.

[ROBERT L. SCOTT]

Bibliography: G. M. Barrow, *Physical Chemistry*, 2d ed., 1966; W. J. Moore, *Physical Chemistry*, 3d ed., 1962.

Bond angle and distance

In molecules two atoms are said to be bonded to each other when there are strong attractive forces between them. These two atoms constitute a chemical bond. One important characteristic of a chemical bond is the separation (called the bond length or distance) of the two nuclei of the atoms involved. A given atom may be involved in more than one chemical bond, and another important characteristic is the angle between bonds sharing a common atom. It is also often important to know the angle between two bonds which are connected by a third bond. This angle is called the dihedral or torsional angle. *See* CHEMICAL BONDING.

Physical properties. An obvious reason for the importance of bond angles and lengths is that they determine the shape and size of a molecule, and thus determine many physical properties. For example, molecules which are spherical in shape will pack in a crystal in a different manner from long, rodlike molecules, and will likely have a different density and crystal structure. Similarly, the molar volumes in the liquid are likely to be different as well as the diffusion coefficients. Another example is the observation that the melting points for normal alkanes with an even number of carbons behave differently from those with an odd number. This arises because of their different shapes and the way in which they interact in the solid.

Bonding information. Bond angles and distances are functions of the type of chemical binding present and give information about the electronic structure of the molecule. For example, bond lengths shorten considerably in going from a single bond to a triple bond (Table 1), and are thus dependent on the hybridization present. This dependence on hybridization also manifests itself in the variation of the length of C-C single bonds as the hybridization of the C atoms is changed (Table 2). *See* CHEMICAL STRUCTURES.

Since the strength of a bond increases with its multiplicity, there is an inverse relation between the strength of a bond and its length. Related to the strength of a bond is its stretching-force constant, and in fact general relations exist between the stretching-force constant for a bond and its length. One of the simplest is Badger's rule, shown in the equation below, where k is the force

$$k = 1.86\,(r - d_{ij})^{-3}$$

constant, r is the bond length, and d_{ij} is a constant for atoms from any two rows i and j of the periodic table. This relationship was first enunciated by Richard Badger in 1934 and was applied only to diatomics, but it has since been found to hold for polyatomics as well.

Bond angles are also dependent on electronic structure. Table 3 shows the nominal values for different hybridization. These values for the hypothetical ideal cases can be compared with those given in Table 4. If s and p atomic orbitals are assumed to be the only ones involved in the bonding, the bond angles around a central atom can be used to determine the hybridization of the central atom.

Biological effects. Dimensions and structure are exceedingly important in biological molecules. For example, the functioning of a molecule such as

Table 1. Variation of bond length with bond multiplicity

Bond	Molecule	Length, A	Multiplicity
C-C	CH_3CH_3	1.53	Single
C-C	CH_2CH_2	1.33	Double
C-C	CHCH	1.21	Triple
C-O	CH_3OH	1.43	Single
C-O	CH_3COCH_3	1.21	Double
C-N	CH_3NH_2	1.47	Single
C-N	CH_3CN	1.16	Triple

Table 2. Variation of C-C single-bond length with hybridization

Molecule	Length, A	Nominal hybridization
CH_3CH_3	1.53	sp^3-sp^3
$CH_3CH{=}CH_2$	1.51	sp^3-sp^2
$CH_3C{\equiv}CH$	1.48	sp^3-sp

Table 3. Variation of bond angle with hybridization

Hybridization	Angle
sp	180°
sp^2	120°
sp^3	109.5°
p	90°

Table 4. Comparison of observed bond angles with nominal hybridization

Angle	Molecule	Value	Nominal hybridization
CCH	$HC{\equiv}CH$	180°	sp
HCH	$H_2C{=}CH_2$	117°	sp^2
CCH	H_3CCH_3	110°	sp^3
ClPCl	PCl_3	100°	p

deoxyribonucleic acid depends very precisely on the structure of the individual, small components which make up this very large molecule. Another example is the great specificity of enzymes, which is believed to be associated with the exact shape and size of the enzyme. *See* HYDROGEN BOND.

Determination. Since bond angles and lengths are important molecular parameters, considerable experimental effort is devoted to their determination, and extensive compilations of values for individual molecules have been made. The experimental methods which are used to determine molecular structure can be divided into two groups. One group is made up of diffraction methods, and the other is composed of spectroscopic methods.

Diffraction methods make use of the fact that not only photons but also particles can behave as waves and can be diffracted. If the de Broglie wavelength of a particle or the wavelength of a photon is of the same order of magnitude as interatomic distances, then they will be diffracted by molecules. Either electrons or neutrons may be used as particles, and x-ray photons are of the appropriate wavelength. X-ray and neutron diffraction are applied to solids, whereas electron diffraction is used on gases. *See* X-RAY DIFFRACTION.

Spectroscopic methods, with a few exceptions, make use of the quantization of the rotational energies of molecules in the gas phase. The rotational energy of a molecule is determined by its moments of inertia, and these are in turn functions of bond angles and lengths. Rotational transitions are observed either as fine structure in infrared, visible, and ultraviolet spectroscopy, or directly in microwave and far-infrared spectroscopy. *See* SPECTROSCOPY. [VICTOR W. LAURIE]

Bibliography: J. C. D. Brand and J. C. Speakman, *Molecular Structure*, 2d ed., 1975; C. A. Coulson, *Valence*, 2d ed., 1961; J. N. Murrell et al., *Valence Theory*, 2d ed., 1970; L. C. Pauling, *The Nature of the Chemical Bond*, 3d ed., 1960.

Borane

One of a class of binary compounds of boron and hydrogen, often referred to as boron hydrides. The term borane is sometimes used to denote substances which may be considered to be derivatives of the boron-hydrogen compounds, such as BCl_3 and $B_{10}H_{12}I_2$.

The simplest borane is B_2H_6; other boranes of increasingly higher molecular weight are known, one of the least volatile of which is an apparently polymeric solid of composition $(BH)_x$. Certain boranes, such as BH_3 and B_3H_7, are not known as such, but can be prepared in the form of adducts with electron-donor molecules. The formulas, names, melting points, and boiling points of some of the commoner boranes are listed in the table.

The most spectacular projected large-scale use of the boranes and their derivatives is in the field of high-energy fuels for jet planes and rockets. The heats of combustion of the boranes per unit weight are such as to permit specific impulses approaching 300 sec, a distinct improvement over hydrocarbon fuels. The thermal decomposition of B_2H_6 has been used to produce coatings of pure elementary boron for neutron-detecting devices and for applications requiring hard, corrosion-resistant surfaces. Boranes can be used as vulcanizing agents

Properties of boranes

Formula	Name*	Melting point, °C	Boiling point, °C
B_2H_6	Diborane (6)	−165	−92.5
B_4H_{10}	Tetraborane (10)	−120	16
B_5H_9	Pentaborane (9)	−47	58.4
B_5H_{11}	Pentaborane (11)	−123	65†
$B_{10}H_{14}$	Decaborane (14)	99.5	213

*The nomenclature used is that of G. W. Schaeffer and T. Wartik, as presented at the March, 1954, meetings of the American Chemical Society in Kansas City. The prefix designates the number of boron atoms in the molecule, and the numeral suffix indicates the number of hydrogen atoms.

†Extrapolated boiling point.

for natural and synthetic rubbers, and are especially effective in the preparation of silicone rubbers.

Structure. The molecular structures possessed by the boranes are exhibited by no other class of substances. Because of the lack of sufficient electrons for the formation of the requisite number of covalent bonds, normal covalently bonded structures of the hydrocarbon type are not possible. The boranes are sometimes referred to as electron-deficient substances. The hydrogen bridge bonding in B_2H_6,

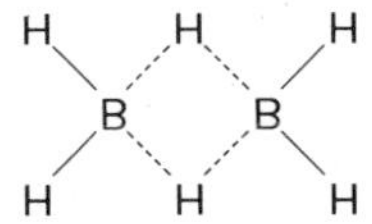

in which only four electrons are utilized to link the two central hydrogen atoms to the two boron atoms, is illustrative of the type of bonding which prevails in the boranes. In no case are the simple chain and ring configurations of carbon chemistry encountered in the more complex boranes. Instead, the boron atoms are situated at the corners of polyhedrons. An example of such a structure is that of pentaborane (9), illustrated in Fig. 1.

Preparation. The first important work on the boranes was that of Alfred Stock and coworkers, whose investigations in this field began about 1912. Their method of preparation, which utilized the decomposition of metallic borides (mainly magnesium boride) by mineral acids as a first step, yielded exceedingly small amounts of the boranes. The discouragingly small yields reported by Stock were significantly improved upon by H. I. Schlesinger and A. Burg, who developed an electrical discharge method based on the overall reaction

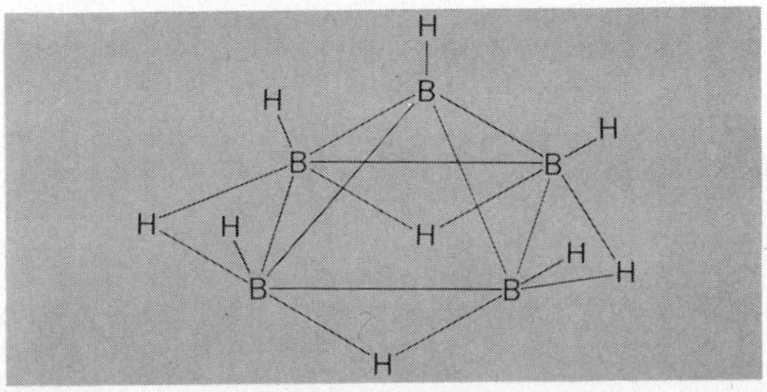

Fig. 1. Structure of pentaborane (9).

shown by reaction (1). Later a number of chemical

$$6H_2 + 2BBr_3 \rightarrow B_2H_6 + 6HBr \quad (1)$$

methods involving the metathetical exchange of hydride ion for halide or alkoxy groups were evolved by Schlesinger, H. C. Brown, and coworkers. One example of the latter is reaction (2). Reac-

$$6LiH + 2BF_3 \xrightarrow{Et_2O} 6LiF + B_2H_6 \quad (2)$$

tions of this type have been utilized to prepare B_2H_6 on a commercial scale. *See* HYDROBORATION.

No practical direct routes to the higher-molecular-weight boranes are known. Instead, substances such as B_4H_{10}, B_5H_9, and $B_{10}H_{14}$ are generally prepared by the thermal decomposition of B_2H_6, a process in some ways analogous to the cracking of hydrocarbons.

BORANE

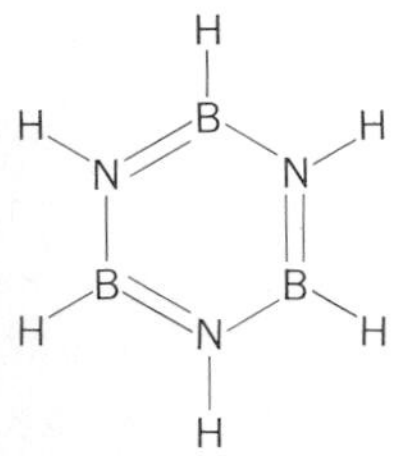

Fig. 2. Molecular structure of borazine.

Reactions. As a class, the boranes are quite reactive substances and are generally decomposed, at times explosively, on contact with air. Their reactivities with air and water decrease with increasing molecular weight; B_2H_6 reacts instantaneously, whereas $B_{10}H_{14}$ must be heated with air or water to obtain an appreciable reaction rate. Because boranes react readily with air, laboratory investigations are almost invariably carried out in all-glass vacuum apparatus or in inert-atmosphere dry boxes. Prolonged exposure to hydrocarbon greases, in which the boranes dissolve and with which they slowly react, must be avoided. Animal tests have indicated that the boranes are quite toxic, and severe physical discomfort can result from the inhalation of their vapors. However the characteristic odor of boranes, which has been compared with that of chocolate but which is distinctly disagreeable, is so intense that it provides adequate warning before a dangerous level of exposure is reached.

With the possible exception of $B_{10}H_{14}$, the known boranes are not indefinitely stable at room temperature. They decompose more or less rapidly to yield elementary hydrogen and boranes richer in boron. When stored as a gas at approximately one atmosphere pressure, a sample of B_2H_6 at room temperature will normally decompose to the extent of several percent per month. The order of stability for the well-characterized boranes is $B_{10}H_{14} > B_5H_9 > B_2H_6 > B_5H_{11} > B_4H_{10}$. B_2H_6 can be completely decomposed to its elements at 500–600°C, whereas temperatures several hundred degrees higher are required for the quantitative conversion of $B_{10}H_{14}$ to its elements.

Because of its ready availability, the chemistry of B_2H_6 has been studied much more extensively than that of the other boranes. The previously mentioned reaction of B_2H_6 with oxygen proceeds according to reaction (3), and that with water yields hydrogen and boric acid, as shown by reaction (4).

$$B_2H_6 + 3O_2 \rightarrow B_2O_3 + 3H_2O \rightleftharpoons 3B(OH)_3 \quad (3)$$

$$B_2H_6 + 6H_2O \rightarrow 6H_2 + 2B(OH)_3 \quad (4)$$

Both of these reactions, particularly the former, are highly exothermic. Stable partial hydrolysis or partial oxidation products have not been isolated. Alcoholysis yields alkoxyboranes and hydrogen.

With electron-donor molecules, B_2H_6 often forms simple addition compounds of BH_3, as shown by reaction (5). More complex boranes can

$$B_2H_6 + 2NMe_3 \rightarrow 2Me_3NBH_3 \quad (5)$$

sometimes be split into more complex component boranes; reaction (6) represents the splitting.

$$B_4H_{10} + OR_2 \rightarrow \tfrac{1}{2}B_2H_6 + B_3H_7 \cdot OR_2 \quad (6)$$

With ammonia at low temperatures, B_2H_6 forms a solid product commonly called the diammoniate or diborane; this is shown by reaction (7). The

$$B_2H_6 + 2NH_3 \rightarrow B_2H_6 \cdot 2NH_3 \quad (7)$$

structure of this solid has been extensively investigated, and the evidence is strongly in favor of the formulation $[BH_2(NH_3)_2]^+[BH_4]^-$. When the diammoniate of diborane is heated at elevated temperatures, hydrogen elimination occurs, as in reaction (8). The reaction product, borazine, has physical

$$3B_2H_6 + 6NH_3 \rightarrow 2B_3N_3H_6 + 12H_2 \quad (8)$$

(but not chemical) properties closely resembling those of benzene, and its molecular structure (Fig. 2) is formally similar to that of the carbon compound.

An important class of compounds, the borohydrides, results from the reaction of B_2H_6 with metal hydrides. With lithium hydride, for instance, lithium borohydride is formed, as in reaction (9). A similar reaction, which gives rise to a triborohydride, is indicated by reaction (10).

$$2LiH + B_2H_6 \xrightarrow{Et_2O} 2LiBH_4 \quad (9)$$

$$NaH + B_4H_{10} \xrightarrow{Et_2O} \tfrac{1}{2}B_2H_6 + NaB_3H_8 \quad (10)$$

Treatment of B_2H_6 with ethylenic compounds produces alkyl boranes, as shown by reaction (11).

$$B_2H_6 + C_2H_4 \rightarrow B_2H_5C_2H_5 \quad (11)$$

A large excess of ethylene leads to the formation of triethylborane. *See* CARBORANE.

The known derivatives of the boranes (other than BH_3) are relatively few in number. Several halo, alkyl, and amino boranes have been reported but, in general, these have not been extensively characterized. *See* BORON; METAL HYDRIDES.

[THOMAS WARTIK]

Bibliography: H. C. Brown, *Boranes in Organic Chemistry*, 1972; E. L. Muetterties (ed.), *Borron Hydride Chemistry*, 1975; H. I. Schlesinger and A. B. Burg, Recent developments in the chemistry of the boron hydrides, *Chem. Rev.*, 31(1):1–41, 1942; A. E. Stock, *Hydrides of Boron and Silicon*, 1957.

Borate

A generic term referring to salts related to boric oxide, B_2O_3, or commonly to only the salts of orthoboric acid, H_3BO_3.

Only orthoboric acid is important in its free state. Some of the others are known only in the form of salts (see table).

The naming of these compounds is additionally complicated by the fact that other polyborates are known and several systems for naming the salts are used. For example, borax is also known as pyroborate, diborate, or sodium (1:2) borate, this latter name based on the formula $Na_2O \cdot 2B_2O_3$.

A dilute solution of any sodium borate contains

Boric acids and their salts

Formula	Name	Salt	Name
H_3BO_3 or $B_2O_3{\cdot}3H_2O$	Boric or orthoboric acid	Na_3BO_3	Sodium borate, sodium orthoborate
HBO_2 or $B_2O_3{\cdot}H_2O$	Metaboric acid	$Na_2BO_2{\cdot}4H_2O$	Sodium metaborate
$H_4B_4O_7$ or $2B_2O_3{\cdot}H_2O$	Tetra- or pyroboric acid	$Na_2B_4O_7{\cdot}10H_2O$	Borax, sodium tetraborate

mainly $H_2BO_3^-$ and BO_2^- ions. The borates are used in manufacture of borosilicate glasses, ceramic glazes, vitreous enamels, detergents and water-softening agents, flameproofing materials, preservatives, and fluxes.

Minerals related to borax are the main source of borates. The largest deposits are found in California. Recent work has produced boron nitride, an abrasive harder than diamond, and boron hydrides, which are used as rocket fuels. *See* BORON.

[E. EUGENE WEAVER]

Bibliography: N. N. Greenwood, *The Chemistry of Boron*, 1975; L. Y. Mayelev, *Borate Glasses*, 1960; E. L. Muetterties (ed.), *Boron Hydride Chemistry*, 1975.

Boride

A binary compound of boron and a metal, formed by heating a mixture of the two elements. Boron does not conform to the usual rules of valence in forming these compounds. The borides, along with the carbides and the nitrides, are called interstitial compounds. These compounds have crystal structures and properties very similar to those of the original metal; hence, it is suggested that the boron is located in the cavities of the metallic lattice.

The borides are very hard and refractory and show high electrical conductivity. Typical borides are Ni_2B, NiB_2, MnB, MnB_2, and Cu_3B_2.

A compound of boron and nitrogen, boron nitride, is an abrasive harder than diamond. *See* BORON; SOLID-STATE CHEMISTRY.

[E. EUGENE WEAVER]

Boron

A chemical element, B, atomic number 5, atomic weight 10.811, in group III of the periodic table. It has three valence electrons and is nonmetallic in behavior. It is classified as a metalloid and is the only nonmetallic element which has fewer than four electrons in its outer shell. The free element is prepared in crystalline or amorphous form. The

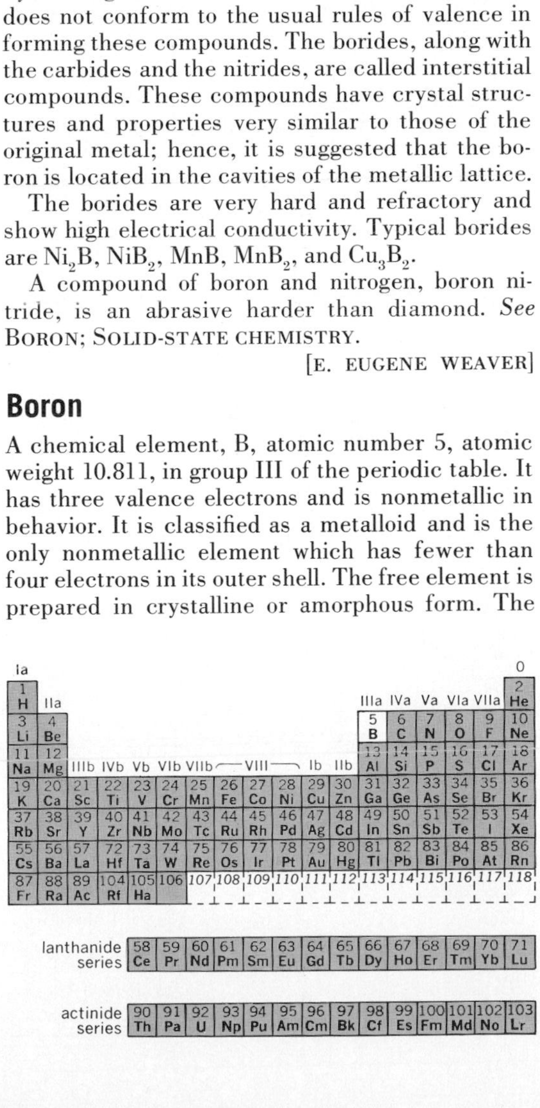

crystalline form is an extremely hard, brittle solid. It is of jet-black to silvery-gray color with a metallic luster. One form of crystalline boron is bright red. The amorphous form is less dense than the crystalline and is a dark-brown to black powder.

In the naturally occurring compounds, boron exists as a mixture of two stable isotopes with atomic weights of 10 and 11. The B^{10} isotope, which normally makes up 18.8% of the total boron, is an effective absorber for thermal neutrons. The large relative difference between the atomic masses of the isotopes B^{10} and B^{11} permits their separation and concentration by fractional distillation and diffusion processes. Such processes have been operated on a large scale using boron trifluoride etherate as the reaction medium.

Sir Humphry Davy and Joseph Gay-Lussac with Louis Jacques Thénard obtained the element from boric acid almost concurrently in 1807 and 1808, respectively. Davy used electrolysis, whereas Gay-Lussac and Thénard used potassium to reduce the boric acid.

Very little in the way of analyses accompanied the work of these original investigators. The first serious investigation pertaining to boron was carried out at the end of the 19th century by Henri Moissan, who in 1892 first obtained boron of better than 98% purity. He also repeated the work of earlier investigators and reported for the first time analyses which showed large amounts of impurities, usually oxygen and the reducing metal.

Uses. Boron and boron compounds have numerous uses in many fields, although elemental boron is employed chiefly in the metal industry. Its extreme reactivity at high temperatures, particularly with oxygen and nitrogen, makes it a suitable metallurgical degasifying agent. It is used to refine the grain of aluminum castings and to facilitate the heat treatment of malleable iron. Boron considerably increases the high-temperature strength characteristics of alloy steels. Elemental boron is used in the atomic reactor and in high-temperature technologies.

In combination with plastics or aluminum, boron provides an effective lightweight neutron-shielding material. Boron-containing shields are valuable because of their satisfactory mechanical properties and because boron absorbs neutrons without producing high-energy γ-rays. Rods and strips of boron steel have been used extensively as control rods in atomic reactors.

The physical properties that make boron attractive as a construction material in missile and rocket technology are its low density (15% lighter than aluminum), extreme hardness, high melting point, and remarkable tensile strength in filament form. Production of boron fibers by vapor deposition methods has been developed on a commercial scale. These fibers are being used in an epoxy (or other plastic) carrier material or matrix. The resulting composite is stronger and stiffer than steel and 25% lighter than aluminum. The composite's balance of strength and stiffness makes it ideal for aircraft applications, where high performance is of primary importance. Another development in this area is the incorporation of boron filaments in metal matrices.

Boron can be produced in various shapes by conventional hot-pressing methods. Temperatures of the order of 2000°C are required and the operation must be carried out in an inert atmosphere. The use of boron has been patented for such diversified applications as in motor-starting devices, phonograph needles, lightning arresters, thermal cutouts for transformers, igniters in rectifier and control tubes, alloys resistant to high-temperature abrasion and scaling, constant-potential controllers, thermoelectric couples, and resistance thermometers.

Certain compounds of boron, such as borax and boric acid, have been known and used for a long time in glass, enamel, ceramic, and mining industries. Refined borax, $Na_2B_4O_7 \cdot 10H_2O$, is an important ingredient of a variety of detergents, soaps, water-softening compounds, laundry starches, adhesives, toilet preparations, cosmetics, talcum powder, and glazed paper. It is also used in fireproofing, disinfecting of fruit and lumber, weed control, and insecticides, as well as in the manufacture of leather, paper, and plastics.

Preparation. The literature describes a variety of methods for the preparation of elemental boron from boron oxide, halides, hydrides, and other boron-containing compounds. These methods involve electrothermic, electrochemical, and direct pyrolysis procedures, and may be classified as follows: thermoreduction of boron-oxygen compounds with an active metal, reaction (1); alkali metal reduction of boron halides, reaction (2); hydrogen reduction of the halides, reaction (3); carbothermic reduction of borates, reaction (4); electrolysis of fused borates or other boron-containing compounds, reaction (5); and thermal decomposition of boron hydrides, reaction (6). Of

$$B_2O_3 + 3Mg \rightarrow 2B + 3MgO \quad (1)$$
$$BCl_3 + 3Na \rightarrow B + 3NaCl \quad (2)$$
$$2BCl_3 + 3H_2 \rightarrow 2B + 6HCl \quad (3)$$
$$Na_2B_4O_7 + 7C \rightarrow 2Na + 7CO + 4B \quad (4)$$
$$2KBF_4 + 6KCl \rightarrow 2B + 8KF + 3Cl_2 \quad (5)$$
$$B_2H_6 \rightarrow 2B + 3H_2 \quad (6)$$

these methods, reactions (1), (3), and (5) are used for the commercial production of boron.

In general, production of high-purity (above 99%) boron involves secondary treatment (such as vacuum degassing, or controlled halogenation) of the crude products as obtained by any of the above methods.

Properties. Many properties of boron have not been sufficiently established experimentally. The literature shows considerable inconsistency in physical properties; this is the result of the questionable purity of some sources of boron, as well as of the variations in the methods and temperatures of preparation. A summary of the physical properties is shown in the table. Samples of boron having the highest purity, whether crystalline or amorphous, are black in color. The samples of so-called Moissan (amorphous) boron exhibit various shades of brown, which are associated with the presence of various amounts of boron suboxide impurities. Crystalline boron, although relatively brittle compared to diamond, is second only to diamond in hardness. Facets on the minute single crystals are clearly visible when viewed through a microscope of at least 30 power.

Three crystalline modifications of boron have been established. A tetragonal form contains 50 boron atoms in its unit cell. Its density is 2.31 g/cm³. Two rhombohedral modifications (α and β) have densities of 2.46 and 2.35 g/cm³, respectively. The α-rhombohedral modification is said to form in the 800–1100°C temperature range. The tetragonal modification is prepared in the 1100–1300°C range, while the β-rhombohedral form may be obtained above 1300°C. Boron possesses a large, negative temperature coefficient of resistivity, which means that its electrical conductivity rises rapidly with increase in temperature. E. Weintraub's data show an increase by a factor of almost 2×10^6 over the 100–1000°C temperature range. W. C. Shaw found that the resistivity of boron decreases by a factor of almost 10^{10} between the temperatures of −70 and 700°C.

The chemical properties of boron depend largely on the physical form as well as on the purity of samples. Amorphous boron oxidizes slowly in the air even at room temperature, and is spontaneously flammable at about 800°C. A special reactive form of amorphous boron may be prepared by the magnesiothermic reduction of anhydrous alkali borates. This form may catch fire in air at temperatures below 100°C, and is generally characterized by an extremely small particle size (smaller than 1 μ) and high alkali-metal impurity (5–10%) content.

Boron is not affected by either hydrochloric or hydrofluoric acids, even on prolonged boiling. Crystalline boron is quite stable to heat and oxidation even at relatively high temperatures. It is slowly attacked and oxidized by hot concentrated nitric acid and by mixtures of sodium dichromate and sulfuric acid. Hydrogen peroxide and ammonium persulfate also slowly oxidize crystalline boron. These reagents are found to act violently on samples of amorphous boron. All varieties of boron are completely oxidized by molten mixtures of alkali carbonates and hydroxides. Chlorine, bromine, and fluorine act easily on boron with formation of the corresponding boron halides (bromine acts at 700°C, chlorine at 410°C, and fluorine at room temperature). Boron reacts vigorously with sulfur at about 600°C to form a mixture of boron sulfides. Boron nitride is formed when boron is

Physical properties of boron

Property	Temp., °C	Value
Density		
Crystalline	25–27	2.31 g/cm³
Amorphous	25–27	2.3 g/cm³
Mohs hardness		
Crystalline		9.3
Melting point		2100 °C
Boiling point		2500 °C
Resistivity	25	1.7×10^6 ohm-cm
Coefficient of thermal expansion	20–750	8.3×10^{-6} cm/°C
Heat of combustion	25	302.0 ± 3.4 kcal/mole
Entropy		
Crystalline	25	1.403 cal/(mole)(deg)
Amorphous	25	1.564 cal/(mole)(deg)
Heat capacity		
Gas	25	4.97 cal/(mole)(deg)
Crystalline	25	2.65 cal/(mole)(deg)
Amorphous	25	2.86 cal/(mole)(deg)

heated in a nitrogen or ammonia atmosphere above 1000°C. At high temperatures, boron combines with phosphorus and with arsenic to form a phosphide, BP, and an arsenide, BAs. On further heating, these materials lose phosphorus and arsenic, respectively, and form more stable phosphorus and arsenic hexaborides, B_6P and $B_{5-7}As$. Boron is not reactive with hydrogen until temperatures of 1800–2000°C are reached. It reacts with silicon to form silicon borides at temperatures above 2000°C. Boron reacts with a majority of metals and metallic oxides at high temperatures to form metallic borides.

Natural occurrence. Boron makes up 0.001% of the Earth's crust. It is never found in the uncombined or elementary state in nature. Besides being present to the extent of a few parts per million in sea water, it occurs as a trace element in most soils and is an essential constituent of several rock-forming silicate minerals, such as tourmaline and datolite. The presence of boron in extremely small amounts seems to be necessary in nearly all forms of plant life, but in larger concentrations, it becomes quite toxic to vegetation. Only in a very limited number of localities are high concentrations of boron or large deposits of boron minerals to be found in nature; the more important of these seem to be primarily of volcanic origin.

As a result, the world's important deposits of borates are to be found only in such localities as the barren wastes of south-central and southwest Asia, portions of Asia Minor, the pampas of South America, and the desert areas of California and Nevada, all of which are immediately adjacent to regions characterized by evidence of former intensive volcanic activity.

The United States accounts for about 95% of the world's reported output of borax. Argentina, Italy, and Turkey are also producers of borates. It is believed that there are considerable deposits of borate ores in the Soviet Union, but the extent of present exploitation is unknown.

Extraction and refining. Borax is produced from bedded deposits and from natural brines found in the United States. The entire output comes from California.

The deposit of kernite (rasorite) and borax (tincal) in the Kramer district of California is the world's principal source of boron compounds. The deposit consists of two beds of borate minerals approximately 200–250 ft (60–75 m) thick interspersed with shale and covered by 150–750 ft (45–230 m) of overburden. Until the late 1950s the ore was mined by shrinkage-stoping and room-and-pillar methods. The overburden has now been stripped off so that open-pit mining operations can be carried out. The crude ore is crushed and concentrated by magnetic separators, calciners, and various air-classifying units to yield a borate concentrate of the desired composition and density. The ore concentrates are then further refined at the mine site by specially adapted crystallization techniques.

A variety of refined borate products, such as sodium metaborate, potassium pentaborate, and ammonium borate, and several grades of boric acid, obtained from acidification of borax with sulfuric acid, are routinely produced.

Another important source of borate minerals is a dry lake containing an alkaline brine permeating the mineral bed that carries, in addition to borax, various other alkaline and miscellaneous mineral salts. This dry lake, known as Searles Lake, is located in northwestern San Bernardino County, Calif.

In the Trona process, brine is obtained from wells drilled in the salt body, approximately 60–70 ft (18–21 m) below the surface of the lake. This brine is transported to the plant, where it is stirred and blended with various end liquors resulting from previous operations of the process. It is then evaporated to a desired concentration. During this evaporation, carbonate, sulfate, and chloride of sodium are removed by means of salt traps and filters. At the end of the evaporation process, the brine is essentially saturated with potassium chloride. It is then cooled rapidly under conditions that cause potassium chloride to crystallize, whereas borax and other salts remain in solution. The crude borax of the supersaturated solution is crystallized upon further cooling and is further refined by recrystallization techniques. Various grades of sodium tetraborate and of boric acid are manufactured by this process.

In a new process, boron values are directly recovered from the Searles Lake brine with the aid of an aliphatic polyol dissolved in kerosine. Sodium, potassium, and borate ions form a chelate complex with the polyol. The kerosine solution is then contacted with dilute sulfuric acid, giving a solution containing boric acid and sodium and potassium sulfates. After a combined evaporation and crystallization operation, high-purity boric acid and mixed sodium-potassium sulfates are recovered as separate products.

Another process for the recovery of borax consists of carbonating the alkaline brine with CO_2 gas. Sodium bicarbonate is precipitated and this procedure serves to recover sodium carbonate. The resulting reduction in alkalinity increases the borax solubility by permitting the formation of the more soluble acid borates. After filtering off the bicarbonate crop, a portion of fresh raw brine is added to adjust the alkalinity back to that of the tetraborate. The mixture is then cooled to crystallize a substantial crop of borax.

In the Tuscan region of Italy, boric acid, together with ammonium salts and carbon dioxide, is recovered from hot springs and fumaroles. In addition to supplying considerable amounts of borates, sufficient to satisfy most of Italy's domestic needs, these hot springs generate appreciable quantities of electrical power at low cost.

Inorganic compounds. Refined boric acid, H_3BO_3, is used as a raw material to make other boron compounds. It is used in leather manufacturing, electroplating, antiseptics, cosmetics, electrolytic condensers, and hydrogen-ion control.

Sodium perborate, $NaBO_3 \cdot 4H_2O$, usually prepared electrolytically from borax and caustic, is used as a bleach, disinfectant, in toothpastes and mouthwashes, and in electroplating.

Anhydrous sodium tetraborate, $Na_2B_4O_7$, prepared by high-temperature dehydration of borax, is used extensively as a replacement for the parent material.

Boric oxide or boric anhydride, B_2O_3, is produced by the dehydration of boric acid. In addition to its wide use as an intermediate in the production of boron halides and metallic borides, it is used in the atomic energy industry as a thermal neutron absorber (boraxal). It is used in glass manufacture, welding fluxes, and ore processing.

Boron carbide, B_4C, is produced by the high-temperature (about 2500°C) interaction of boric oxide, B_2O_3, and carbon in an electrical resistance-type furnace. It is a black, lustrous solid. It is used extensively as an abrasive, because its hardness approaches that of the diamond. It is also used as an alloying agent, particularly in molybdenum steels.

The use of boron carbide in nuclear engineering, both as a construction material and as a neutron absorber, is believed to be potentially of considerable magnitude.

Metallic borides are usually prepared by hot sintering of the elements. They may also be prepared by carbothermic and aluminothermic reduction of metal oxide–boron oxide mixtures. A number of metallic borides of high purity have been prepared on a semicommercial scale by vapor-deposition methods, in which the volatile halides of the elements are deposited on a hot substrate, such as tungsten or tantalum wire, in an atmosphere of hydrogen. Borides have much in common with true metals. They have a high electrical conductivity, high melting points, and extreme hardness. Many metal borides are now used as components of cermet compositions in the ever-growing technology or high-temperature refractory materials. *See* BORIDE.

Nitrogen compounds. Boron forms a large number of compounds with nitrogen. At low temperatures, diborane reacts with ammonia or amines to form ammoniates or the amine boranes. When diborane and ammonia are heated above 200°C, hydrogen is eliminated and a six-membered boron-nitrogen ring compound, borazine, is formed (Fig. 1). By analogous methods, substituted borazines

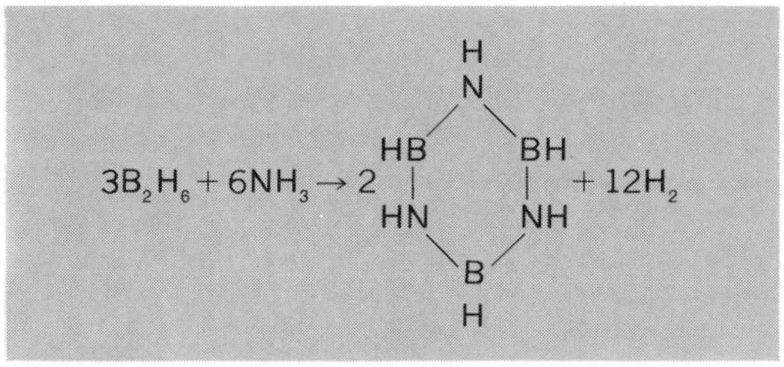

Fig. 1. Formation of borazine.

may be obtained from amines and diborane. The reaction of boron trichloride with ammonium chloride at elevated temperatures produces a substituted borazene, in which a chlorine atom is attached to each boron atom (Fig. 2). Borazines are of interest because of their similarity in physical properties to aromatic organic compounds. Borazine is structurally very similar to benzene, and has been referred to as "inorganic benzene."

In addition to the ring compounds of borazine type, boron halides, boron hydrides, and borate esters form a number of addition compounds with ammonia and amines, in which boron acts as a Lewis acid. Some of these compounds, particular-

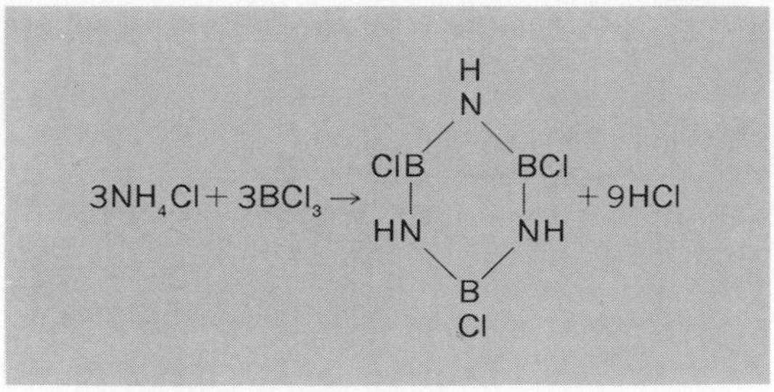

Fig. 2. Formation of a substituted borazine.

ly the amine-boranes, are useful as reducing agents in organic synthesis and have been suggested as polymerization catalysts and petroleum additives. *See* ACID AND BASE.

Because of the unusual stability of the boron-nitrogen chemical bond, the compounds containing this bond are converted to boron nitride on heating to high temperatures. These and similar compounds that boron forms with phosphorus and arsenic have been extensively investigated as possible building blocks in the field of polymer chemistry.

Boron nitride, BN, has many potential commercial applications. It is a white, fluffy powder with a greasy feel. It has an x-ray diffraction pattern almost identical with that of graphite, indicating a close similarity in structure, but it does not conduct electricity. Boron nitride may be prepared in a variety of ways, for example, by the reaction of boric oxide with ammonia, alkali cyanides, and ammonium chloride, or of boron halides with ammonia. The unusually high chemical and thermal stability, combined with the high electrical resistance of boron nitride, suggests numerous uses for this compound in the field of high-temperature technology. Boron nitride can be hot-pressed into molds and worked into desired shapes. A cubic form of boron nitride, called borazon, has been prepared at pressures near 65,000 atm and temperatures near 1500°C. It is comparable to diamond in hardness and apparently has properties superior to those of diamond with regard to oxidation, electrical resistance, and thermal stability. This material will probably take its place with diamond for industrial grinding.

A tough, coherent, hard abrasive compact, consisting of a cubic form of boron nitride bonded with boron carbide, has been developed.

Halides. Boron trihalides are colorless, volatile compounds that are extremely susceptible to moisture. Boron trifluoride is a gas (bp −101°C) and has been an article of commerce for a considerable period. Boron trichloride (bp 12.4°C) and boron tribromide (bp 90.4°C) have also become available on a commercial scale. Boron triiodide is a white crystalline solid (mp 49.6°C). Boron halides, with the exception of the iodide, can be made by heating elemental boron or borides with halogens. The iodide can be prepared only indirectly, for example, from boron trichloride and hydriodic acid or from iodine and sodium borohydride. Boron trifluoride is most readily obtained from the reaction of boric acid, hydrogen (or a metallic) fluoride, and sulfuric acid.

Boron halides, and in particular the trifluoride and trichloride, are valuable catalysts and are used extensively in the chemical industry. They are valuable intermediates in the production of metal borohydrides and various boron-hydrogen compounds. Boron trichloride is used in the refining of aluminum, magnesium, zinc, and copper, and as a fire-extinguishing agent for magnesium and other metal fires.

Boron trifluoride combines with fluoride ions to form fluoborate (borofluoride) ions, BF_4^-. Several of the metallic fluoborates and fluoboric acid have important uses in the electroplating industry.

Subhalides. A number of boron subhalides, in which the halogen–boron ratio is less than three, have been prepared and studied. They are relatively unstable substances and include the diboron tetrahalides B_2Cl_4, B_2F_4, B_2Br_4, and B_2I_4.

Organic compounds. Organic compounds of boron are numerous and varied. They may be broadly subdivided into two classes: alkyl and aryl derivatives of boron, in which boron is directly bonded to carbon; and the boron-oxygen-carbon compounds, such as borate esters and boroxines. These latter substances may be regarded as derivatives of boric acid or boric oxide.

Boron alkyls, such as trimethyl boron (a gas, bp −21.8°C) and triethyl boron (liquid, bp 95°C), are extremely reactive materials. Most of the boron alkyls are spontaneously inflammable in air. Controlled oxidation yields alkyl boron oxides, R·BO, which dissolve in water to form alkyl boric acids, $R{\cdot}B(OH)_2$. Boron alkyls are usually prepared from boron halides and metal alkyls, such as zinc, or Grignard reagents, RMgX, or by the reaction of diborane with unsaturated hydrocarbons.

Boron aryls, such as triphenyl boron (mp 142°C), are usually solids. The aryl compounds are also prepared by the Grignard method from boron halides. The boron aryls are somewhat less sensitive to oxygen than their alkyl analogs.

The aryl and the alkyl compounds of boron are potentially useful intermediates for the synthesis of other boron compounds. Sodium tetraphenyl borohydride, an analytical reagent, is used for the estimation of potassium, rubidium, and cesium.

Borate esters are derivatives of boric acid (or oxide) and alcohols or phenols. They may be prepared either directly from parent materials or indirectly from the reaction of boron halides with alcohols or phenols, or by ester interchange reactions. In general borate esters are extremely sensitive to hydrolysis, but if the alcohol is sufficiently hindered sterically, this hydrolytic tendency can be appreciably reduced. The simplest borate ester, trimethyl borate, is a valuable intermediate for the preparation of metal borohydrides. In addition to synthetic possibilities for borate esters, a number of miscellaneous uses have been described, principally in the patent literature. Some of these include the use of borate esters in cosmetic preparations, as antioxidants for rubber and alcohols, as curing agents for epoxy resins, as petroleum additives, in pharmaceutical preparations, in plasticizers, and as surface-action agents.

Boric oxide dissolves readily in borate esters to form materials known as boroxines. Although most of these boroxines will burn in the air, their combustion product is glassy boric oxide, and for these reasons they are being employed as extinguishing agents for metal fires. Boroxines can also be used as polymerization catalysts.

Analysis. The detection of boron is generally based upon the green coloration which is imparted by its volatile compounds to an alcohol flame or a nonluminous gas flame. Minerals are tested for boric acid by mixing them with calcium fluoride and sulfuric acid and introducing them close to the lower margin of a flame. If boric acid is present, volatile boron trifluoride is formed; this gives a green coloration to the flame.

With yellow turmeric paper, free boric acid or acidified solutions of borates give a red-brown coloration, which appears only after the paper is dried. If the paper is then moistened with ammonia, it is temporarily colored blue-black.

Boric acid can be determined quantitatively by conversion to methyl borate and subsequent saponification with calcium hydroxide. If a weighed amount of calcium oxide is used, the increase in weight after ignition represents directly the amount of boric oxide taken up.

Boric acid (or soluble borates) are conveniently determined volumetrically. For this purpose, the alkali content is first determined by titration with standard acid (usually hydrochloric); methyl orange serves as indicator. Then, by addition of glycerol, or better, mannitol or invert sugar (fructose), the weak boric acid is converted into a stronger monobasic acid; this can then be titrated directly with standard sodium hydroxide, using phenolphthalein as indicator.

For the estimation of boron in metallic borides, boron carbide, boron nitride, and elemental boron, the sample is first digested to a soluble sodium borate by fusion in a mixture of sodium carbonate and sodium nitrate. The boron in the resultant melt is converted to boric acid by treatment with hydrochloric acid, and the resulting solution is titrated with standard sodium hydroxide in the presence of mannitol. *See* BORANE.

[F. H. MAY; V. V. LEVASHEFF]

Bibliography: I. E. Campbell and E. M. Sherwood, *High Temperature Materials and Technology*, 1967; W. Gerrard, *The Organic Chemistry of Boron*, 1961; C. A. Hampel (ed.), *Rare Metals Handbook*, 2d ed., 1971; R. E. Kirk and D. F. Othmer (eds.), *Encyclopedia of Chemical Technology*, vol. 4, 3d ed., 1978; E. L. Muetterties (ed.), *Boron Hydride Chemistry*, 1975; K. Niedenzu (ed.), *Boron-Nitrogen Chemistry*, 1964; H. Steinberg (ed.), *Progress in Boron Chemistry*, vol. 1, 1964, vols. 2–3, 1970.

Boyle's law

A law of gases which states that at constant temperature the volume of a gas varies inversely with its pressure. This law, formulated by Robert Boyle (1627–1691), can also be stated thus: The product of the volume of a gas times the pressure exerted on it is a constant at a fixed temperature. The relation is approximately true for most gases, but is not followed at high pressures. The phenomenon was discovered independently by Edme Mariotte about 1650 and is known in Europe as Mariotte's law. *See* GAS.

[FRANK H. ROCKETT]

Bromate

A negative ion having the formula BrO_3^- and derived from bromic acid, $HBrO_3$.

Bromic acid exists only in aqueous solution. The

solid bromates are stable and can be prepared, as shown in the reaction below, by passing bromine into a warm solution of sodium carbonate.

$$3Br_2 + 3Na_2CO_3 \rightarrow 5NaBr + NaBrO_3 + 3CO_2$$

Since bromine has the oxidation state of 5+ in bromates and these bromates are strong oxidizing agents, these compounds are useful in analytical chemistry. Commercially, they are used in permanent-wave preparations and in flour treatment. *See* BROMINE.

[E. EUGENE WEAVER]

Bromide

A compound which is derived from hydrobromic acid, HBr, and contains the bromine atom in the 1− oxidation state. The chemistry of the bromine atom, and its ability to form covalent and ionic bromides, is similar to that of chlorine. The properties of bromides are intermediate between those of chlorides and iodides. The bromide ion is a better reducing agent and forms less stable complexes than chloride. *See* CHLORIDE.

The solubilities of bromides are very similar to those of chlorides, and $PbBr_2$, CuBr, AgBr, and Hg_2Br_2 are the common insoluble salts. The bromides are usually produced from bromine, which in turn is obtained from salt brines or sea water.

Bromides are used in medicine as sedatives in treatment of nervous disorders. Silver bromide is used in photographic films and paper.

Bromide ion can be detected in aqueous solution by oxidizing it to the free element with chlorine. When extracted into carbon tetrachloride, it imparts an amber color to the solvent. *See* BROMINE; COMPLEX COMPOUNDS; IODIDE.

[E. EUGENE WEAVER]

Bromine

A chemical element, Br, atomic number 35, atomic weight 79.909, which normally exists as Br_2, a dark-red, low-boiling but high-density liquid of intensely irritating odor. The name is derived from the Greek *bromos*, meaning stench. This is the only nonmetallic element that is liquid at normal temperature and pressure. Bromine is very reactive chemically; one of the halogen group of elements, it has properties which are intermediate between those of chlorine and iodine. *See* HALOGEN ELEMENTS.

The most stable valence states of bromine in its salts are −1 and +5, although +1, +3, and +7 are known. Within wide limits of temperature and pressure, molecules of the liquid and vapor are diatomic, Br_2, with a formula weight of 159.818. There are two stable isotopes (^{79}Br and ^{81}Br) that occur naturally in nearly equal proportion, so that the atomic weight is 79.909. A number of radioisotopes are also known. A. J. Balard is recognized as the discoverer of this element in 1826, even though it appears that other workers made the discovery simultaneously. The most notable of these was J. von Liebig, who appropriately called it "chloro-iodide."

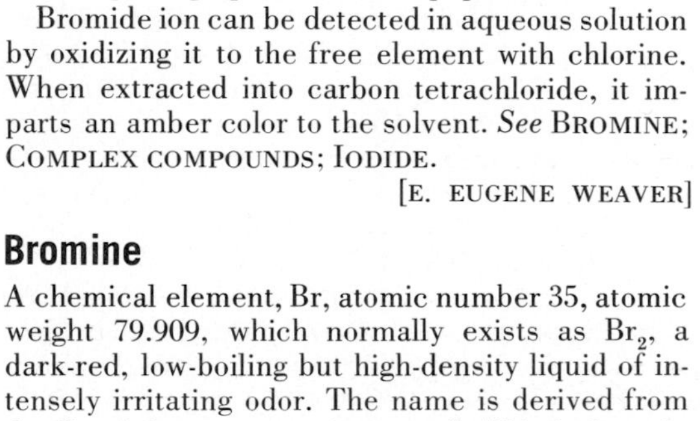

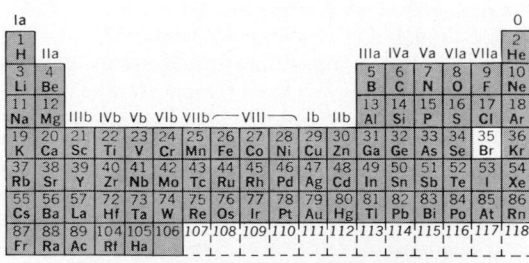

Table 1. Physical properties of bromine

Flash point			None
Fire point			None
Freezing point, °C			−7.27
Density, 20°C			3.1226
Pounds per gallon, 25°C			25.8
Boiling point, 760 mm Hg, °C			58.8
Refractive index, 20°C			1.6083
Latent heat of fusion, cal/g			15.8
Latent heat of vaporization, cal/g, bp			44.9
Vapor density, g/liter, standard conditions (0°C, 1 atm)			7.139
Viscosity, centistokes, 20°C			0.314
Surface tension, dynes/cm, 20°C			49.5
30°C			47.3
40°C			45.2
Dielectric constant, 10^5 freq, 25°C			3.33
Compressibility, vapors, 25°C			0.998
Thermodynamic data, cal/(mole K)			
	T,K	*Entropy*	*Heat capacity*
Solid	265.9	24.786	14.732
Liquid	265.9	34.290	18.579

Natural occurrence. Although it is estimated that from 10^{15} to 10^{16} tons of bromine are contained in the Earth's crust, the element is widely distributed and found only in low concentrations in the form of its salts. The bulk of the recoverable bromine, however, is found in the hydrosphere. Sea water contains an average of 65 parts per million (ppm) of bromine, which means that 308,000 tons of bromine are held in a cubic mile of ocean. Though 15,000 tons of sea water must be processed to obtain 1 ton of bromine, sea water is a major commercial source of bromine.

The other major sources of bromine in the United States are underground brines and salt lakes, with commercial production in Michigan, Arkansas, and California. Brines from wells contain 1000–6000 ppm of bromine, but they must be obtained by drilling to depths as great as a mile and more. Chlorine is 300 times more abundant in both brines and sea water. This same ratio is maintained generally in the Earth's crust, which contains an average of 1.6 ppm bromine.

Physical properties. The solubility of bromine in water at 20°C is 3.38 g/100 g solution, but its solubility is increased tremendously in the presence of its salts and in hydrobromic acid. Thus, in a solution of 359 g/liter of sodium bromide, the solubility of bromine is 641.6 g/liter. Bromine is completely miscible with 48% hydrobromic acid and with common solvents, such as carbon disulfide, carbon tetrachloride, chloroform, ether, and glacial acetic acid. The ability of this inorganic element to dissolve in organic solvents is of considerable importance in its reactions. Table 1 summarizes the physical properties of bromine.

Manufacture. The discovery of potash in the salt deposits at Strassfurt in 1856 opened the way for the initial recovery of bromine as a byproduct in Germany in 1858. Bromine was first produced in the United States in 1845 by D. Altes from salt brine.

Commercial recovery of bromine from brines and from sea water involves the oxidation of the bromide ions in solution to free elemental bromine, which is then vaporized from the solution either by air or by steam. Although oxidation was accomplished at one time by electrolysis, all current processes utilize chlorine, according to reaction (1). Vaporization by steam results in the

$$Cl_2 + 2Br^- \rightarrow 2Cl^- + Br_2 \qquad (1)$$

recovery of free bromine directly.

With sea water, bromine must first be concentrated before the steaming-out process becomes economically practical. Consequently, air is used to vaporize the bromine from chlorinated sea water, and then SO_2 is introduced into the dilute bromine-laden air. Subsequent absorption of the hydrogen bromide in a controlled amount of water produces a much more concentrated bromide solution, as shown in reaction (2). Elemental bromine

$$SO_2 + Br_2 + 2H_2O \rightarrow 2HBr + H_2SO_4 \qquad (2)$$

must then be released again by chlorine and steamed from the solution. The by-product hydrochloric and sulfuric acids are recycled to acidify the incoming sea water to the pH necessary for efficient chlorination. The steps in the process are shown in the illustration.

The air-blowing process can also be used when an alkali bromide is the desired end product. Absorption of the bromine from the air stream by sodium carbonate, for instance, produces sodium bromide and sodium bromate, as in reaction (3).

$$3Br_2 + 3Na_2CO_3 \rightarrow 5NaBr + NaBrO_3 + 3CO_2 \qquad (3)$$

The sodium bromate may then be crystallized from solution, or can be reduced with iron, if NaBr is the only desired product. One process utilizes NaBr solution as the absorbing medium for bromine from bromine-laden air.

Inorganic bromides and bromates. Major intermediates in the manufacture of bromides, second only to bromine as a means of introducing bromine into a molecule, are gaseous hydrogen bromide, HBr, and aqueous hydrobromic acid, usually as the constant-boiling solution containing approximately 48% HBr. The reaction of hydrogen and bromine at high temperature in the absence of air is the principal method of manufacture; HBr as the by-product of an organic substitution reaction may also be utilized. Liquefied hydrogen bromide is available commercially in cylinders, and various strengths of aqueous hydrobromic acid, usually 48%, are available in carboys, drums, or tank cars. The physical properties of anhydrous hydrogen bromide are listed in Table 2. With water at 760 mm Hg pressure, the constant-boiling solution contains 47.63% HBr and boils at 124.3°C.

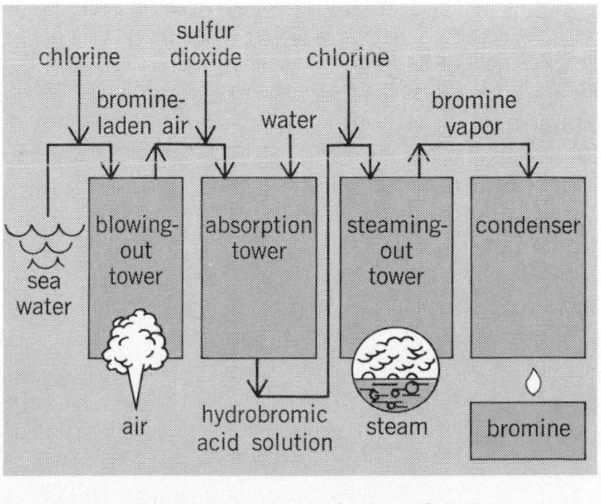

Steps in the extraction process for bromine.

Table 2. Physical properties of hydrogen bromide

Formula weight	80.9
Melting point, °C	−86.8
Boiling point, °C	−66.7
Specific heat, liquid, cal/g	0.176
Specific heat, vapor, 27°C, cal/g	0.085
Heat of fusion at mp, cal/g	71.1
Heat of vaporization at 66.7°C, cal/g	52.0
Critical temperature, °C	89.8
Critical pressure, atm	84

Other valuable inorganic brominating agents include phosphorus tribromide, bromine chloride, aluminum bromide, barium bromide, thionyl bromide, and sulfur bromide.

The alkali bromides can be manufactured by several processes. Thus, ammonia, bromine, and an alkali hydroxide or carbonate will react according to the type reaction in reaction (4). Also, the

$$2NH_3 + 6NaOH + 3Br_2 \rightarrow 6NaBr + N_2 + 6H_2O \qquad (4)$$

reaction of HBr with an alkali hydroxide or carbonate gives the desired product, shown in reaction (5).

$$2HBr + Na_2CO_3 \rightarrow 2NaBr + CO_2 + H_2O \qquad (5)$$

As shown previously, absorption of bromine by alkali carbonates or hydroxides in the blowing-out process of manufacture produces alkali bromide and alkali bromate in a 5:1 mole ratio. If bromate is the desired product, it may be formed from the bromide electrolytically, or by the oxidative action of chlorine in reaction (6). The bromides of

$$NaBr + 6NaOH + 3Cl_2 \rightarrow NaBrO_3 + 6NaCl + 3H_2O \qquad (6)$$

calcium, strontium, barium, magnesium, cesium, and lithium are of commercial importance.

Ammonium bromide is most easily formed by the reaction of ammonia and bromine in water solution with the liberation of nitrogen, as in reaction (7). Aluminum bromide made by direct reaction of

$$3Br_2 + 8NH_4OH \rightarrow 6NH_4Br + N_2 + 8H_2O \qquad (7)$$

aluminum and bromine is a valuable catalyst in alkylation and isomerization reactions. It is preferable to aluminum chloride in some cases because it can be liquefied easily (mp 97.5°C), and it is more soluble and more reactive in organic liquids.

Organic bromides. As numerous as are the inorganic bromides that have found industrial use, the organic bromides have even wider application. Because of the ease of reaction of bromine with organic compounds and the ease of its subsequent removal or replacement, organic bromides have been much studied and used as chemical intermediates. In addition, many of the bromine reactions are so clean-cut that they can be used for the study of reaction mechanisms without complication of side reactions. The ability of bromine to add into unusual places on organic molecules has added to its value as a research tool.

There are several reactions by which bromine and its bromides may be added to an organic molecule. Many of these reactions will give essentially 100% yield of the desired product. The ability of bromine to dissolve both in many common organic solvents and in inorganic bromide solutions permits easy control of the reactions.

The addition of bromine to a double or triple bond to saturate the molecule is usually very easily accomplished, as shown in reaction (8). One exception is perchloroethylene (1,1 2,2-tetrachloroethylene), which resists bromination and requires a catalyst to speed the reaction.

$$\mathrm{RCH{=}CHR} + Br_2 \rightarrow \mathrm{RCHBr{-}CHBrR} \qquad (8)$$

Substitution reactions of bromine on the ring in aromatic molecules take place readily with the aid of a catalyst, such as iron, iron bromide, or aluminum bromide. Direct bromination of saturated aliphatic hydrocarbons requires higher temperatures, with the result that a multitude of products is formed. In the presence of catalysts, the substitution of bromine to the α position of aliphatic acids, aldehydes, and ketones takes place readily. Hydrogen bromide is a mole-for-mole by-product of substitution reactions, as in reaction (9).

$$RCH_2CO_2H + Br_2 \rightarrow RCHBrCO_2H + HBr \qquad (9)$$

Bromine compounds such as phosphorus tribromide, *N*-bromosuccinimide, 1,3-dibromo-5,5-dimethylhydantoin, and 1,2-dibromotetrachloroethane are valuable as highly selective and mild brominating agents. Thus 1,2-dibromotetrachloroethane will react with olefins under the influence of peroxides or ultraviolet light (uv) to yield the corresponding allylic bromide. The reaction proceeds with cyclohexene, as in reaction (10).

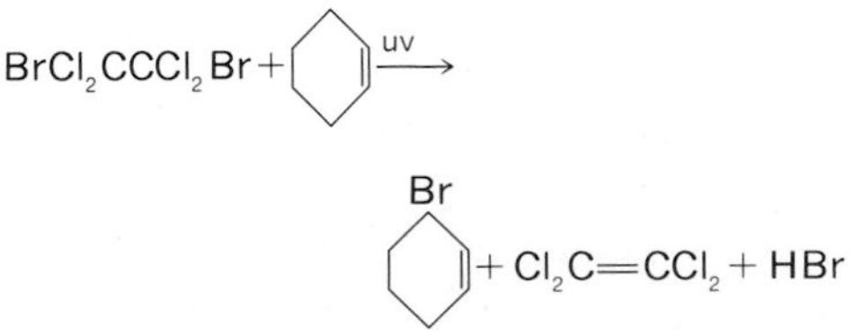

(10)

Phosphorus tribromide is a mild brominating agent that can be used when bromine itself acts too vigorously on a molecule.

A commercially useful reaction for the replacement of hydroxyl groups on organic molecules utilizes sulfur bromide, as shown in reaction (11).

$$S + 3Br_2 + 6ROH \rightarrow 6RBr + H_2SO_4 + 2H_2O \qquad (11)$$

Alkyl bromides may also be prepared by the action of hydrogen bromide or hydrobromic acid on alcohols. Sulfuric acid may be added to take up water and drive the reaction to completion.

Another method for the manufacture of alkyl bromides illustrates a reaction unique to hydrogen bromide. In it, the addition of hydrogen bromide to an olefin, such as allyl chloride, gives the 1,2-chlorobromo compound, as expected from Markownikoff's rule. However, if the reaction is carried out in the presence of free-radical-type catalysts, such as peroxides, or in ultraviolet light, the addition is reversed and the 1,3 compound formed, as shown in reactions (12) and (13). This reverse reaction does not occur with hydrochloric acid, HCl. The reaction is of particular value in placing a reactive bromine on a terminal carbon of a compound for use as an intermediate. Thus, the 1,3 product above finds use in the production of cyclopropane, an anesthetic.

$$CH_2{=}CHCH_2Cl + HBr \rightarrow CH_3CHBrCH_2Cl \qquad (12)$$

$$CH_2{=}CHCH_2Cl + HBr \xrightarrow[\text{catalyst}]{\text{Free radical}} CH_2BrCH_2CH_2Cl \qquad (13)$$

The preparation of alkyl bromides by replacement of chlorine with $AlBr_3$ has been used commercially, particularly for the methane family. However, this preparation is more easily and economically accomplished by the action of hydrogen bromide in the presence of only catalytic amounts of aluminum bromide or aluminum chloride, as shown in reaction (14). Replacement of the hydrogen in such a molecule can be accomplished by high-temperature, vapor-phase, bromination.

$$CHCl_3 + 3HBr \xrightarrow{AlCl_3} CHBr_3 + 3HCl \qquad (14)$$

Both chlorine and bromine are ortho-para directing in their influence on aromatic molecules. Thus, the second bromine that will substitute in benzene will be found predominantly in the para position with some *o*-dibromobenzene, but little or no *m*-dibromobenzene. Each subsequent bromine atom becomes more difficult to add, but with more reactive compounds, such as phenol or aniline, it is difficult to stop short of the symmetrical 2,4,6-tribromo compound. In the case of phenol or aniline, the OH and NH_2 groups are stronger ortho-para directors than halogen is, so these groups will determine the position of substitution. With aromatic compounds having saturated side chains, the presence of heat and light and the absence of metals such as iron favor side-chain substitution, whereas lower temperatures, darkness, and iron favor ring brominations.

The usefulness of organic bromine compounds as chemical intermediates is illustrated by their reactions with alkalies to form either the corresponding alcohols or olefins, with ammonia to form amines, and with phenols to form the corresponding ethers. The carbon-to-carbon chain lengths of organic molecules may be increased by the action of alkyl bromides, utilizing the Grignard or Friedel-Crafts actions. An interesting alkylating agent is bromotrichloromethane, which adds to olefins as shown in reaction (15). Many other methods for using bromine compounds as building blocks for

$$CBrCl_3 + CH_2{=}CH_2 \rightarrow CCl_3CH_2CH_2Br \qquad (15)$$

more complex molecules are available. *See* CHEMICAL DYNAMICS; HALOGENATED HYDROCARBON.

Bromine compounds have a number of properties that make them useful. High specific gravity is a general characteristic, and some bromine compounds are among the most dense organic liquids known. The compound 1,1,2,2-tetrabromoethane has a specific gravity of 2.96, for example. Many of the compounds possess bacteriological and fungicidal activity, and the ability of some, such as aluminum bromide, to act as catalysts has been utilized. The ease of the addition of bromine to a molecule and its subsequent replacement by a more complex molecule makes bromine a useful tool in organic synthesis.

Uses. Historically, a substantial part of bromine production in the United States is used to form ethylene dibromide, an ingredient in leaded gasoline. The ethylene dibromide is used as a scavenger to remove residual lead in internal combustion engines by combining with the lead to form volatile lead bromide which is expelled from the engine via the exhaust gases. The discovery of this phenomenon by T. Midgely and coworkers in 1921 made bromine a chemical commodity of major significance. As leaded gasoline is phased out of the United States market, the use of bromine via ethylene dibromide is expected to diminish with time.

Other chemicals containing bromine have been produced in increased volumes. A major and fast-growing use for bromine is in completion fluids for oil and gas wells, specifically calcium bromide and zinc bromide. These products are useful because of the high specific gravity which can be reached by aqueous solutions of these salts: as high as 1.8 g/cm^3 for calcium bromide and upward of 2.4 g/cm^3 for zinc bromide. Where abnormally high bottom hole pressures are expected, the high density of these fluids controls the oil or gas formation pressure because of the hydrostatic head of the fluid in the well bore. These fluids, being solid-free in nature, are less damaging to the oil-producing formation than fluids that obtain their density by the use of suspended weighting materials, such as barium sulfate. Such solid-weight materials tend to reduce the permeability of the oil-producing formation, and thus restrict production. This use for bromine is expected to grow rapidly with the continued global search for oil and gas that results in deeper and more highly pressured exploratory and producing wells.

Of prime importance for many years has been the use of bromine in the production of sodium and potassium bromides that are used in the photographic industry. Also, several organic bromides have found utility as chemical intermediates in the production of complex chemicals used as herbicides and insecticides. Alkali bromates are also used as oxidizing agents in the production of hair-waving preparations and also in the oxidation of sulfur dyes in the textile industry. Methyl bromide and ethylene dibromide have also found wide use in agriculture as soil and space fumigants.

Another use for bromine compounds is in the area of flame retardants. Compounds that are high in halogen content, for example, chlorine or bromine, are resistant to ignition and burning. Incorporation of these halogen-contining flame retardants into flammable material often reduces the ease of ignition. The addition of these compounds to the base material can be accomplished in two ways. The easier is to physically blend or disperse the compound into the flammable material. Such compounds are called additive-type flame retardants. The second approach is to chemically react the bromine compound into the substrate. Such compounds are known as reactive-type flame retardants. The reactive type are usually more durable than the additive type. *See* FLAMEPROOFING.

Bromine and its compounds have found acceptance as disinfection and sanitizing agents in swimming pools and potable water. Certain bromine-containing compounds are safer to use than the analogous chlorine compounds due to certain persistent residuals found in the chlorine-containing materials. Other bromine chemicals are used as a working fluid in gages, as hydraulic fluids, as chemical intermediates in the manufacture of organic dyes, in storage batteries, and in explosion-suppressant and fire-extinguishing systems. Bromine compounds, because of their density, also find use in the gradation of coal and other minerals where separations are effected by density gradients. The versatility of bromine compounds is illustrated by the commercial use of over 100 compounds that contain bromine.

Handling and safety. Bromine is almost instantaneously injurious to the skin, and it is difficult to remove quickly enough to prevent a painful burn that heals slowly. Extreme precaution, including protective clothing, must be used when handling bromine. Bromine vapor is extremely toxic, but its odor gives good warning; it is difficult to remain in an area of sufficient concentration to be permanently damaging. Bromine can be handled safely, but the recommendations of the manufacturers should be respected.

Bromine is available in 6½-lb (3-kg) bottles, in 225-lb (102-kg) Monel alloy drums, and in nickel- and lead-lined tank cars and trucks. Wet bromine is corrosive to most metals except tantalum, but dry bromine (containing 20 ppm or less water) can be handled at room temperatures with essentially no corrosion in Monel, nickel, and lead. Steel and stainless steel are not usually satisfactory, even for dry bromine. This is because dry bromine can absorb moisture rapidly, and bromine containing more than 50 ppm water is usually very corrosive to steel and stainless steel.

[RANDY C. STAUFFER]

Bibliography: F. A. Cotton (ed.), *Inorganic Syntheses*, vol. 13, 1972; A. J. Downs and C. J. Adams, *The Chemistry of Chlorine, Bromine, Iodine, and Astatine*, vol. 7, 1975; J. Kleinberg (ed.), *Inorganic Syntheses*, vol. 7, 1963, reprint 1978.

Buffers

A solution selected or prepared to minimize changes in hydrogen ion concentration which would otherwise tend to occur as a result of a chemical reaction. In general, chemical buffers are systems which, once constituted, tend to resist further change due to external influences. Thus it is possible, for example, to make buffers resistant to changes in temperature, pressure, volume, redox potential, or acidity. The commonest buffer in chemical solution systems is the acid-base buffer.

Chemical reactions known or suspected to be dependent on the acidity of the solution, as well as on other variables, are frequently studied by measurements in comixture with an appropriate buffer. For example, it may be desirable to investigate how the rate of a chemical reaction depends upon the hydrogen ion activity (pH). This is accomplished by measurements in several buffer systems, each of which provides a nearly constant, different pH. Alternatively, it may be desirable to measure the effects of other variables on a pH-sensitive system, by stabilizing the pH at a convenient value with a particular buffer.

Effectiveness. Buffer action depends upon the fact that, if two or more reactions coexist in a solution, then the chemical potential of any species is common to all reactions in which it takes part, and may be defined by specification of the chemical potentials of all other species in any one of the reactions. To be effective, a buffer must be able to respond to an increase as well as a decrease of the species to be buffered. In order to do so, it is necessary that the proton transfer step of the buffer be reversible with respect to the species involved, in the reaction to be buffered. In aqueous solution the proton transfer between most acids, their conjugate bases and water, is so rapid and reversible that the dominant direct source of protons for a chemical reaction is H_3O^+, the hydronium ion.

An acid-base buffer reaction in water is defined by the reversible reaction in Eq. (1), and the equilibrium constant K_a shown by Eq. (2).

$$BH^+ + H_2O \rightleftharpoons B + H_3O^+ \qquad (1)$$

$$K_a = \frac{[H_3O^+][B]}{[BH^+][H_2O]} \qquad (2)$$

In Eq. (2) the square brackets designate the activity of the species involved. In normal concentrations of buffer (0.1 mole/liter) the activity of the solvent water is essentially constant and approximately that of pure water (55.5 *M*). Thus the position of the equilibrium may be defined by specifying the activity of any two of the three variable species in Eq. (1). Normally this is by means of the equilibrium expression shown as Eq. (3) which,

$$[H_3O^+] = K_a \frac{[BH^+]}{[B]} \qquad (3)$$

upon converting to a logarithmic form, can be reduced to Eq. (4). Here f is the fraction of the total

$$pH = pK_a - \log\frac{(1-f)}{f} - \log\frac{\gamma_{BH^+}}{\gamma_B} \qquad (4)$$

buffer concentration, $(BH^+) + (B)$ existing as B, and γ is the activity coefficient relating activity a to concentration X. This relation is shown by Eq. (5).

$$a_x = \gamma_x(X) \qquad (5)$$

Thus a buffer pH is approximately defined by the dissociation constant K_a of the weak acid system and the ratio of acid to conjugate base concentrations. However, the third term in Eq. (4) indicates that the pH is dependent on the change in activity coefficients with concentration. Effects of this dependency may be eliminated in practice by providing a high and essentially invariant ionic environment in the form of an added pH-neutral strong electrolyte such as KNO_3 or NaCl.

Buffer capacity π is defined as the change in added H_3O^+ necessary to produce a given change in pH, $d[H_3O^+]/d$pH. Since the buffer comes to equilibrium with added H_3O^+, $1/\pi$ may also be defined as $d\mathrm{pH}/df$. Inspection of Eq. (4) shows $1/\pi$ to be a minimum when $f = 1-f$; hence a given buffer system has its highest capacity in a solution composed of equal parts BH^+ and B, and the capacity is directly proportional to the concentrations of BH^+ and B. For these reasons buffers are normally used at concentrations 10–100 times higher than the system to be controlled and, if possible, are selected so that the desired pH is approximately equal to pK_a for the buffer system. As a general rule, weak acid systems are not used to stabilize solutions whose pH is more then 2 pH units removed from pK_a, to ensure that the ratio of BH^+ to B will fall in the range 100–0.01.

Water as solvent. Buffers are particularly effective in water, because of the unusual properties of water as a solvent. Its high dielectric constant (80) tends to promote the existence of formally charged ions (ionization). Because it has both an acidic (H) and a basic (O) group, it may form bonds with ionic species leading to an organized sheath of solvent surrounding an ion (solvation). Water also tends to self-ionize to form its own conjugate acid-base system as shown by Eq. (6), in

$$2H_2O \rightleftharpoons H_3O^+ + OH^- \qquad (6)$$
$$K_{ap} = [H_3O^+][OH^-] = 10^{-14}$$

which K_{ap} is the autoprotolysis constant. The strength of an acid (or base) in solvent water cannot be separated from the reaction of Eq. (6), and the familiar acid (or base) "dissociation constants" are actually measures of the competitive equilibrium reaction of Eq. (1). Strong acids are those for which the K_a of Eq. (2) is very large; weak acids do not completely transfer the proton to water. The strongest acid which may exist in water is H_3O^+; the strongest base is OH^-. Thus, the maximum range of acid level which a solvent can support is governed by its own acid-base properties. In water this range is 14 pH units, or 14 orders of magnitude change in activity of H_3O^+. *See* SUPERACIDS.

The mechanism of buffer action may be regarded as a sequence of the proton transfer steps implied in Eq. (1) coupled with Eq. (6). For example, the result of the chemical production or deliberate addition of an acid, HA, is to cause the water autoprotolysis reaction and the buffer acid reaction to respond to the change shown by Eqs. (7) and (8).

$$HA + H_2O \rightleftharpoons H_3O^+ + A^- \qquad (7)$$

$$H_3O^+ + B \rightarrow BH^+ + H_2O \qquad (8)$$

Addition of a base would be accommodated by the reverse of the reactions of Eqs. (7) and (8). The effect of adding HA depends on the position of the equilibrium shown in Eq. (7); buffer capacity π is usually defined in terms of H_3O^+ added because H_3O^+ is the strongest possible acid in aqueous solution, and would tend to create the maximum possible change in solution pH per mole of added acid. If HA is relatively weak so that its degree of dissociation, in Eq. (1), is small, its effective H_3O^+ addition may be calculated through Eq. (9), where C_a is

$$[H_3O^+] \cong \sqrt{K_a C_a} \qquad (9)$$

the concentration of added HA. A simple calculation using Eq. (9) shows that a given buffer solution will undergo the same change in pH for the addition of 0.1 mole/liter of a weak acid such as acetic acid ($K_a = 10^{-5}$) as for the addition of 0.01 mole/liter of strong acids such as HCl, $HClO_4$, or HNO_3.

In studies of rates of chemical reactions at constant pH, it is necessary that the proton transfer processes of the buffer acid and base and the solvent be rapid with respect to the primary reaction. The phosphate (HPO_4^{2-}-PO_4^{3-}) and carbonate (HCO_3^--CO_3^{2-}) systems, among others, sometimes give anomalous effects because this condition may not be obtained. Buffer rate effects are manifested in different reaction rates for a chemical system in two different buffers of otherwise identical ionic strength and nominal (equilibrium) pH. Later evidence seems to suggest that buffers of low-charge type, for example, NH_3-NH_4^+, react more rapidly than high-charge types such as HPO_4^{2-}-PO_4^{3-}. *See* ACID AND BASE; ACID-BASE INDICATOR; IONIC EQUILIBRIUM.

[A. M. HARTLEY]

Bibliography: R. G. Bates, *Determination of pH: Theory and Practice*, 2d ed., 1973; H. A. Laitinen, *Chemical Analysis*, 2d ed., 1975; E. J. Margolis, *Chemical Principles in Calculations of Ionic Equilibria*, 1966. D. D. Perrin and B. Dempsey, *Buffers for pH and Metal Ion Control*, 1974.

Butanol

Any one of the four isomeric alcohols having the formula C_4H_9OH and molecular weight 74.12. The butanols are colorless, toxic, flammable materials which are soluble in most organic liquids; their solubilities in water depend upon their respective structures. The chemical properties of the butanols depend on the position of the hydroxyl group in the molecule. *n*-Butyl alcohol and isobutyl alcohol are primary alcohols and, therefore, may be oxidized to the respective aldehydes or carboxylic acids, whereas *sec*-butyl alcohol oxidizes to methyl ethyl ketone; *tert*-butyl alcohol cannot be oxidized without degradation. Physical properties of the butanols are shown in the table.

n-Butyl alcohol is produced by condensation of acetaldehyde to crotonaldehyde, followed by catalytic hydrogenation. *n*-Butyl and isobutyl alcohols are also obtained by the reaction of propylene with carbon monoxide and hydrogen, the oxo reaction. Normal butyraldehyde and isobutyraldehyde are produced, which are hydrogenated to the alcohols. *sec*-Butyl alcohol and *tert*-butyl alcohol are produced by hydration of butenes. Small quantities of *n*-butyl alcohol are produced as a by-product of the Ziegler higher alcohol process. Butanols are also produced as by-products of air oxidation of hydrocarbons and of fermentation of sugars.

The major use of the butanols is as chemical intermediates, although significant quantities are used directly as solvents. Important derivatives include plasticizers and acetate ester solvents for plastics and paints. [JOHN W. LYNN]

Bibliography: R. E. Kirk and D. F. Othmer (eds.), *Encyclopedia of Chemical Technology*, vol 4, 3d ed., 1978; J. D. Roberts and M. C. Caserio, *Basic Principles of Organic Chemistry*, 2d ed., 1977.

Cadmium

A relatively rare chemical element, symbol Cd, atomic number 48, closely related to zinc, with which it is usually associated in nature. Cadmium was discovered by F. Stromeyer in Germany in 1817 in a sample of zinc carbonate. It does not occur uncombined in nature, and the one true cadmium mineral, greenockite (cadmium sulfide), is not a commercial source of the metal. Almost all of the cadmium produced is obtained as a by-product of the smelting and refining of zinc ores, which usually contain 0.2–0.4% cadmium.

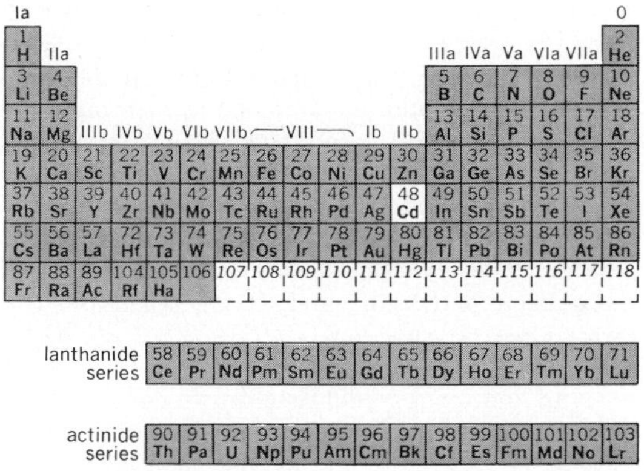

The United States, Canada, Mexico, Australia, Belgium-Luxembourg, and the Republic of Korea are principal sources, although not all are producers. In the United States, the largest producer and consumer of cadmium, domestic production is declining. Average metal purity is better than 99.9%.

Uses. The commercial use of cadmium as an electrodeposited coating on iron and steel for corrosion protection started in 1919 and by 1941 had become by far the largest application of cadmium. Due to its cost and the waste disposal problems associated with its toxicity, however, the use of cadmium in electroplating has decreased to less than half of the present total consumption. Zinc electrodeposited coatings have replaced much of the cadmium. Nickel-cadmium batteries are the second-largest application, with pigment and chemical uses third. Sizable amounts are used in low-melting-point alloys, similar to Wood's metal, and in automatic fire sprinklers, and relatively smaller uses are in brazing alloys, solders, and bearings. Cadmium compounds are used as stabi-

Characteristics of the butanols

Name	Synonym	Structure	bp, °C	fp, °C	Specific gravity
n-Butyl alcohol	1-Butanol	$CH_3CH_2CH_2CH_2OH$	117.7	−90.2	0.810 at 20°C
Isobutyl alcohol	2-Methyl-1-propanol	$(CH_3)_2CHCH_2OH$	108.1	−108	0.805 at 17.5°C
sec-Butyl alcohol	2-Butanol	$CH_3CH_2CHOHCH_3$	99.5	−114.7	0.808 at 20°C
tert-Butyl alcohol	2-Methyl-2-propanol	$(CH_3)_3COH$	82.5	25.55	0.779 at 26°C

lizers in plastics and the production of cadmium phosphors. Because of its great neutron-absorbing capacity, especially the isotope 113, cadmium is used in control rods and shielding for nuclear reactors.

Physical properties. Cadmium is a silvery-white ductile metal with a faint bluish tinge. It is softer and more malleable than zinc, but slightly harder than tin. It has an atomic weight of 112.40, a specific gravity (sp gr) of 8.65 at 20°C, and its metallic radius is 1.54 A (0.154 nm). Its melting point of 321°C and boiling point of 765°C are lower than those of zinc. There are eight naturally occurring stable isotopes, and eleven artificial unstable radio isotopes have been reported. *See* TIN; ZINC.

Chemical properties. Cadmium is the middle member of group IIb (zinc, cadmium, and mercury) in the periodic table, and its chemical properties generally are intermediate between zinc and mercury. Like zinc, cadmium loses its luster in moist air and is rapidly corroded by moist ammonia and moist sulfur dioxide. Most acids will dissolve cadmium, but not as rapidly as zinc. Cadmium is very soluble in concentrated ammonium nitrate solution, and such solutions are used by electroplaters to strip cadmium from plated steel or iron objects. Cadmium oxide and hydroxide, unlike those of zinc, are insoluble in excess sodium hydroxide. The cadmium ion is displaced by zinc metal in acidic sulfate solutions. Cadmium is bivalent in all its stable compounds, and its ion is colorless. It forms the complex ions $Cd(NH_3)_4^{++}$, $Cd(CN)_4^{--}$, and CdI_4^{--}. *See* MERCURY.

Cadmium oxide, CdO, is a brown powder (sp gr 6.95 at 20°C) used mainly to prepare cadmium electroplating baths, and can be made by heating the hydroxide or burning the metal. It is more easily reduced to a metal by heating with carbon than zinc oxide. This property, as well as the fact that cadmium is more volatile than zinc, is important in the recovery of cadmium in zinc ore smelting operations and from lead and copper smelting where zinc and cadmium are present.

Cadmium sulfide, CdS, is obtained as a bright yellow precipitate by passing hydrogen sulfide through a solution containing cadmium ions. The precipitate is insoluble in cold dilute acid, and use is made of this property in the detection of the presence of cadmium by qualitative analysis procedures. CdS (cadmium yellow) and cadmium sulfoselenide (cadmium red) are permanent useful pigments. These compounds exhibit a strong photovoltaic response to visible light, and are becoming of interest in solar energy applications.

The Weston standard cell, which gives a nearly constant potential (1.0186 V), uses cadmium sulfate, $3CdSO_4 \cdot 8H_2O$, as an electrolyte.

Cadmium halide compounds are generally similar to the corresponding zinc halide compounds, except that the cadmium halides are less ionized and do not hydrolyze in water as much as the corresponding zinc compounds. Except for the fluoride, cadmium halides are considerably less soluble in water than the zinc compounds. Cadmium iodide is the least ionized of the cadmium and zinc halides and, like cadmium fluoride, does not form hydrates.

Electroplating of cadmium. The main use of cadmium is in the plating of iron and steel, and to a much less extent of other alloys, such as copper and brass, to protect them from corrosion. Although cadmium can be deposited from acidic baths such as the sulfate (used in the electrowinning of cadmium) and the fluoborate, nearly all commercial plating for protective coatings is done in cyanide baths (composed of 20–25 g/liter cadmium oxide dissolved in 100–125 g/liter concentration sodium cyanide solution) because of its ease of operation and its good throwing power (good distribution of deposit over a complex shape). Cadmium anodes of 99.95% purity are used to replenish the metal as it is deposited on the cathode. The plating process is usually done at temperatures of 20–30°C, voltages of 2–12 V, and current densities of about 1–6 A/dm^2 (approximately 10–60 A/ft^2), depending mainly on metal concentration, stirring, and temperature.

Certain organic compounds (glue, cellulose compounds, wetting agents) are used in the baths to improve deposit brightness and covering power, and to decrease grain size. To further brighten and to improve tarnish resistance, or to form visible passive films which help improve corrosion protection, plated coatings are often given brief dips in chromic acid solutions. Thin cadmium deposits (0.0002 in.; 0.005 mm) offer good protection to iron and steel, especially in marine or rural exposure.

Cadmium solders easily and, unlike zinc, is resistant to alkali. It does not form voluminous corrosion products like zinc does, and therefore is preferred for plating bolts, nuts, and fasteners. It plates much more readily on malleable cast iron than zinc. For all these various reasons it is used extensively. Within the past few years proprietary noncyanide cadmium plating baths have been introduced, but their use is very small.

Toxicity. The fumes of cadmium, its compounds, and solutions of its compounds are very toxic, and cadmium-plated articles should not be used in food, nor should cadmium-coated articles be welded or used in ovens.

Analytical methods. Cadmium can be determined gravimetrically by the sulfate method, precipitation as the beta-napthoquinoline complex, or by electrodeposition from faintly alkaline cyanide solution. Volumetrically, the cadmium ion can be titrated with the chelating agent ethylenediaminetetraacetate (EDTA), using Eriochrome Black T indicator. It is also easily determined polarographically, a method particularly suitable when zinc is present, and also by emission spectroscopy, by atomic absorption spectroscopy, and by electron microprobe analysis.

[WILBUR HAGUE]

Bibliography: J. C. Bailar et al. (eds.), *Comprehensive Inorganic Chemistry*, vol. 3, 1973; C. F. Baker, Cadmium, *Eng. Min. J.*, March 1978; C. A. Hampel (ed.), *Rare Metals Handbook*, 1954; F. A. Lowenheim (ed.), *Modern Electroplating*, 3d ed., 1974; U. S. Bureau of Mines, *Mineral Industry Surveys*, Cadmium, 4th quarter, 1978, Cadmium, 2d quarter, 1979; R. C. Weast (ed.), *Handbook of Chemistry and Physics*, 60th ed., 1979.

Calcium

A chemical element, Ca, of atomic number 20, fifth among elements and third among metals in abundance in the Earth's crust. Calcium compounds make up 3.64% of the Earth's crust. Occurrence of

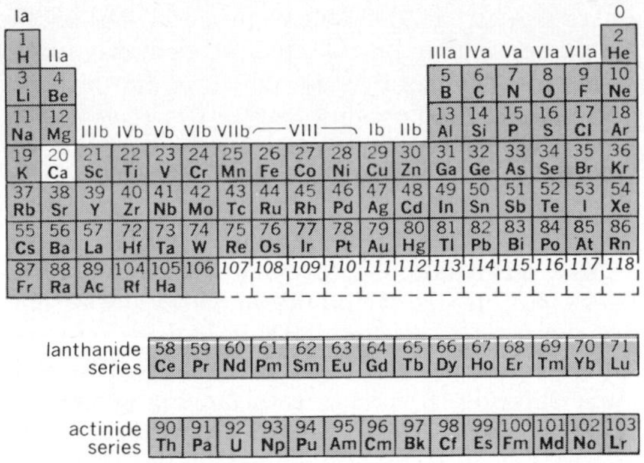

calcium is very widespread; it is found in every major land area of the world. This element is essential to plant and animal life, and is present in bones, teeth, eggshell, coral, and many soils. Calcium chloride is present in sea water to the extent of 0.15%. In Table 1 are listed the major calcium minerals and their formulas.

Lime, calcium oxide, was known and used by the ancient Greeks and Egyptians in the manufacture of mortar.

Table 1. Major calcium minerals

Mineral	Formula
Marble	$CaCO_3$
Limestone	$CaCO_3$
Iceland spar	$CaCO_3$
Calcite	$CaCO_3$
Dolomite	$MgCO_3 \cdot CaCO_3$
Fluorspar	CaF_2
Anhydrite	$CaSO_4$
Gypsum	$CaSO_4 \cdot 2H_2O$
Apatite	$Ca_5(PO_4)_3F$
Asbestos	$CaMg_3(SiO_3)_4$

Extraction of the metal. Calcium metal is prepared industrially by the electrolysis of molten calcium chloride, although the reduction of lime with aluminum metal in vacuum followed by distillation of the calcium metal has been used in times of short supply. Calcium chloride, the raw material, is obtained either by treatment of a carbonate ore with hydrochloric acid or as a waste product from the Solvay carbonate process.

The physical properties of calcium metal are given in Table 2. The metal is trimorphous and is harder than sodium, but softer than aluminum.

Table 2. Properties of calcium metal

Atomic number	20
Atomic weight	40.08
Isotopes (stable)	40, 42, 43, 44, 46, 48
Atomic volume, cm³/g-atom	25.9
Crystal form	Face-centered cubic
Valence	2+
Ionic radius, A	0.99
Electron configuration	2 8 8 2
Boiling point, °C	1487 (?)
Melting point, °C	810 (?)
Density, g/cm³ at 20°C	1.55
Latent heat of vaporization at boiling point, kilojoules/g-atom	399

Like beryllium and aluminum, but unlike the alkali metals, it will not cause burns on the skin. It is less reactive chemically than the alkali metals and the other alkaline-earth metals. The pure metal may be machined in a lathe, threaded, sawed, extruded, drawn into wire, pressed, and hammered into plates. The wire may be extruded at temperatures as low as 420–460°C. Calcium has a tensile strength of 8700 psi, and may be cast if one of the suitable protective fluxes is employed.

In air, calcium forms a thin film of oxide and nitride, which protects it from further attack. At elevated temperatures, it burns in air to form largely the nitride. The commercially produced metal reacts easily with water and acids, yielding hydrogen that contains noticeable amounts of ammonia and hydrocarbons as impurities.

The metal is employed as an alloying agent for aluminum-bearing metal, as an aid in removing bismuth from lead, and as a controller for graphitic carbon in cast iron. It is also used as a deoxidizer in the manufacture of many steels, as a reducing agent in preparation of such metals as chromium, thorium, zirconium, and uranium, and as a separating material for gaseous mixtures of nitrogen and argon. When added to magnesium alloys (0.25%), it refines the grain structure, reduces their tendency to take fire, and modifies the strengthening heat treatment. It finds use also in the precipitation-hardening lead-calcium alloys. Alloys of lead, calcium, and barium are produced by the electrolysis of fused salts in which the fused chlorides of barium and calcium are electrolyzed over a bath of molten lead as a cathode. The ability of calcium to fix the residual gases nitrogen and oxygen in vacuum tubes makes it important as a getter. The phase diagrams of the systems composed of calcium with aluminum, copper, hydrogen, gold, lead, magnesium, nickel, silicon, silver, tin, and zinc have been investigated.

Principal compounds. In general, sodium compounds are preferable to calcium compounds, if they may be substituted for a particular application, because the cost of sodium is approximately one-fifth or one-sixth that of calcium. However, lime, CaO, is a cheaper alkali than is sodium hydroxide. Calcium metal reacts directly with hydrogen at 400°C to form calcium hydride, CaH_2. The product is stable in dry oxygen to 400–500°C, and at elevated temperatures it becomes an excellent reducing agent. It will reduce many inorganic oxides, such as rutile and baddeleyite, ZrO_2, to their metals (Hydrimet process). Sodium chloride may be reduced to sodium, and carbon monoxide to formaldehyde. The hydride also acts as a condensing agent in converting acetone to mesitylene oxide and as a hydrogenation catalyst in converting ethylene to ethane. It has also served as a portable source of hydrogen for cooking and meteorological balloons; the trade names are Hydrolith and Hydrogenate. In this capacity the reaction of 1 lb of the hydride with water gives 16 ft³ (STP) of hydrogen at pressures as high as 1800 psi at rates as high as 1.5 cfm/lb.

Oxide and hydroxide. Calcium oxide is made by the thermal decomposition of carbonate minerals in tall kilns using a continuous-feed process. Care must be taken during the heating to decompose the limestone at a low enough temperature so that the oxide will slake freely with water. If too high a temperature is used, the so-called dead-burnt lime is formed. The oxide is used in high-intensity arc lights (limelights) because of its unusual spectral features and as an industrial dehydrating agent. At high temperatures, lime combines with sand and other siliceous material to form fusible slags; hence the metallurgical industry makes wide use of it during the reduction of ferrous alloys.

The slaking or hydrolysis process for lime produces the slightly soluble [0.219 g $Ca(OH)_2$/100 g H_2O] calcium hydroxide, whose basicity is limited only by its insolubility. Because of the low cost of calcium hydroxide, it is used in many applications where hydroxide ion is needed. During the slaking process, the volume of the slaked lime produced expands to twice that of quicklime, and because of this, it can be used for the splitting of rock or wood. Slaked lime is an excellent absorbent for carbon dioxide to produce the carbonate. Because of the great insolubility of the carbonate, gases are easily tested qualitatively for carbon dioxide by passing them through a saturated lime-water solution and watching for a carbonate cloudiness. The hydroxide is also used in the formation of mortar, which is composed of slaked lime (1 volume), sand (3–4 volumes), and enough water to make a thick paste. The mortar gradually hardens because of the evaporation of the water and the cementing action of the deposition of calcium hydroxide, and because of the absorption of carbon dioxide.

Silicide and carbide. Calcium silicide, CaSi, is an electric-furnace product made from lime, silica, and a carbonaceous reducing agent. This material is useful as a steel deoxidizer because of its ability to form calcium silicate, which has a low melting point. Calcium carbide is produced in enormous quantities each year by heating a mixture of lime and carbon to 3000°C in an electric furnace. The compound, CaC_2, is an acetylide which contains the ionic structure $[C{\equiv}C]^{2-}$, and it yields acetylene upon hydrolysis. Acetylene is the starting material for a great number of chemicals important in the organic chemicals industry. The carbide absorbs atmospheric nitrogen at red heat to form calcium cyanamide, and this reaction is the basis for the cyanamide process for the fixation of atmospheric nitrogen.

Carbonate. Pure calcium carbonate exists in two crystalline forms: calcite, the hexagonal form, which possesses the property of birefringence, and aragonite, the rhombohedral form. Naturally occurring carbonates are the most abundant of the calcium minerals, as is indicated in Table 1. Iceland spar and calcite are essentially pure carbonate forms, whereas marble is a somewhat impure and much more compact variety which, because it may be given a high polish, is much in demand as a construction stone. Although calcium carbonate is quite insoluble in water, it has considerable solubility in water containing dissolved carbon dioxide, because in these solutions it dissolves to form the bicarbonate. This fact accounts for cave formation in which limestone deposits have been leached away by the acidic ground waters.

Halides. The halides of calcium include the phosphorescent fluoride, which is the most widely distributed calcium compound and which, because of its transparency to ultraviolet and infrared radiation, has important applications in spectroscopy. Calcium chloride, obtained as a waste product in the Solvay process, has in the anhydrous form important deliquescent properties which make it useful as an industrial drying agent and as a dust quieter on roads. The equilibrium vapor pressure of water vapor in contact with the systems shown in reaction (1) is 0.34 mm at 25°C.

$$CaCl_2 + 2H_2O \rightleftharpoons CaCl_2 \cdot 2H_2O \qquad (1)$$

Also, this halide finds use in refrigeration plants because of its high solubility in water at low temperatures. The calcium chloride hexahydrate and water system has a eutectic point at −54.9°C. Calcium chloride hypochlorite (bleaching powder) is produced industrially by passing chlorine into slaked lime, and has been used as a bleaching agent and a water purifier. *See* CHLORINE.

Sulfate. Calcium sulfate dihydrate is called gypsum; this mineral is mined in New York, Michigan, Texas, Iowa, and Ohio. It constitutes the major portion of portland cement, and has been used to help reduce soil alkalinity. A hemihydrate of calcium sulfate, produced by heating gypsum at elevated temperatures, is sold under the commercial name plaster of paris. When mixed with water, the hemihydrate reforms the dihydrate, evolving considerable heat and expanding in the process, so that a very sharp imprint of the mold is formed. Thus, plaster of paris finds use in the casting of small art objects and mold testing.

Miscellaneous compounds. Calcium sulfide is formed industrially by the reduction of the sulfate with carbon and, because it dissolves hair, has found some use as a depilatory. Calcium forms a mixed alkyl with diethyl zinc, $CaZn(CH_2CH_3)_4$, which is exceedingly reactive. The reaction of calcium metal with phenyl iodide, reaction (2), shows the formation of a calcium Grignard reagent.

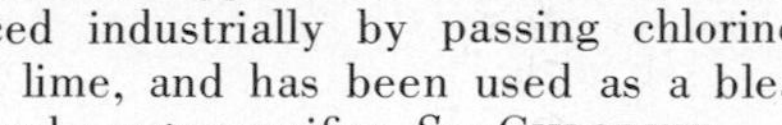

(2)

Calcium boride, CaB_6, has a metallic lattice with interstitial boron atoms. It has been shown to have metallic conductivity.

Analytical methods. The presence of calcium may be detected qualitatively either by forming its insoluble carbonate or by igniting the unknown in a burner flame, to which the calcium imparts a brilliant crimson color. Quantitatively, calcium is determined by precipitation as the oxalate after separation of the strontium and barium. The oxalate is then either ignited to the oxide or oxidized with a standardized solution of potassium permanganate.

Calcium in the biosphere. Calcium is an invariable constituent of all plants because it is essential for their growth. It is contained both as a structural constituent and as a physiological ion. The calcium ion is able to counteract the toxic effects of potassium, sodium, and magnesium ions. Calcium may also affect the growth of plants because its presence in soil affects the alkalinity of the latter.

Calcium is found in all animals in the soft tissues, in tissue fluid, and in the skeletal structures. The bones of vertebrates contain calcium as calcium fluoride, as calcium carbonate, and as calcium phosphate. In some lower animals, magnesium replaces either totally or partially the skeletal calcium. The importance of calcium in animals as a structural constituent is based on its abundance and on the low solubility of the three calcium salts just listed. Calcium is also essential in many biological functions of the vertebrates. *See* METAL HYDRIDES. [REED F. RILEY]

Bibliography: C. L. Mantell, *Industrial Electrochemistry*, 3d ed., 1950; D. M. Ritter, *First Four Groups*, 1965; N. V. Sidgwick, *The Chemical Elements and Their Compounds*, 1950.

Californium

A chemical element, Cf, atomic number 98, the ninth member of the actinide series of elements. Its discovery and production have been based upon artificial nuclear transmutation of radioactive isotopes of lighter elements. All isotopes of californium are radioactive, with half-lives ranging from a minute to about 1000 years. Because of its nuclear instability, californium does not exist in the Earth's crust. The world's supply of californium in 1975 was about a gram. *See* ACTINIDE ELEMENTS; BERKELIUM; RADIOACTIVITY.

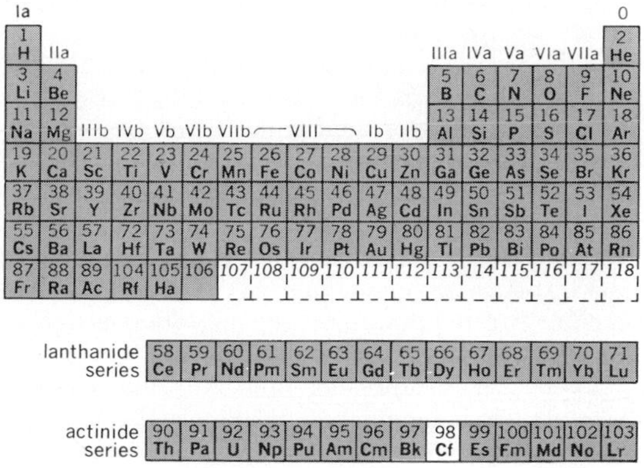

Californium was discovered in 1950 by S. G. Thompson, K. Street, A. Ghiorso, and G. T. Seaborg. The experiments were carried out at the University of California Radiation Laboratory, and the new element was named in honor of the university and the state. The discoverers bombarded a target, consisting of a few millionths of a gram of ^{242}Cm (curium-242), with helium ions accelerated in a cyclotron. Combination of helium nuclei with nuclei of ^{242}Cm, followed by loss of a neutron, gave the mass-245 isotope of element 98. Californium-245 was found to decay partly by capture of an orbital electron and partly by emission of an alpha particle, with a half-life of 45 min.

Several other isotopes of californium (mass numbers 240–245) have been discovered by using a variety of combinations of target isotopes and bombarding particles. The production of ^{246}Cf by bombardment of ^{238}U with carbon ions is an example of the possibility of using comparatively heavy bombarding particles and more abundant target isotopes for synthesis of new elements. For production of amounts of californium isotopes ranging up to a few grams, kilogram amounts of ^{239}Pu have been irradiated with neutrons in nuclear reactors. Successive capture of neutrons by newly formed isotopes and subsequent β-decay to next heavier elements lead eventually to formation of the californium isotopes. Mixtures of californium isotopes with considerably different isotopic composition have been produced by multiple neutron capture in the high, instantaneous neutron flux from thermonuclear explosions (fast time scale). In this latter case compositions especially rich in ^{254}Cf are obtained.

An interesting property of the californium isotopes is that spontaneous fission becomes a very important mode of decay. The isotope ^{254}C decays predominantly (99%) by spontaneous fission and has a half-life of about 60 days. This property has resulted in speculation that this isotope might be responsible for the decay of the light intensity of supernovae, which have a similar half-life, but this hypothesis is not generally accepted. It is also in the region of californium isotopes that a nuclear subshell at neutron number 152 has an important influence on the stabilities of nuclei and in their alpha decay and spontaneous fission half-lives.

The most easily produced isotope for many purposes is ^{252}Cf, which is obtained in gram quantities in nuclear reactors and has a half-life of 2.6 years. It decays partially by spontaneous fission, and has been very useful for the study of fission. It has also had an important influence on the development of counters and electronic systems with applications not only in nuclear physics but in medical research as well. There are also important potential applications of ^{252}Cf for high-intensity neutron sources. The most useful isotope for chemical investigations is ^{249}Cf, which has a half-life of 323 years and is obtained from the beta decay of ^{249}Bk. This californium isotope was first used to measure the magnetic susceptibility of Cf(III) adsorbed on a single resin bead. The susceptibility is similar to that of its lanthanide analog, Dy(III). There is no evidence for an oxidation state of californium other than 3+ in aqueous solution, but +4 and +2 solid compounds exist. Solid compounds include Cf_2O_3, CfO_2 (and intermediate oxides), CfF_3, CfF_4, $CfCl_3$, $CfBr_2$, $CfBr_3$, CfI_2, and CfI_3. Solutions of the tripositive ions are green in color. The color arises from distinct absorption bands in the visible region of the spectrum, the strongest of which appear at 4890–4930, 5900–6180, 6700–6800, and 7200–7600 A.

The chemical properties are similar to those observed for other 3+ actinide elements: a water-soluble nitrate, sulfate, chloride, and perchlorate. Californium is precipitated as the fluoride, oxalate, or hydroxide. Ion-exchange chromatography can be used for the isolation and identification of californium in the presence of other actinide elements. A solution of a complexing agent (α-hydroxyisobutyric acid, for example) desorbs actinides from a cation-exchange resin at different rates; the elements leave the column of resin in distinct fractions and can be identified by their characteristic radioactivities. Solvent extraction with hydrogen di(2-ethylhexyl)phosphoric acid (HDEHP) or quaternary ammonium salts is also used.

Californium metal is quite volatile and can be distilled at temperatures of the order of 1100–1200°C. Chemically reactive, it appears to exist in three crystalline modifications between room temperature and its melting point, 900°C. *See* TRANSURANIUM ELEMENTS. [GLENN T. SEABORG]

Bibliography: S. Ahrland et al., *The Chemistry of the Actinides*, 1975; B. B. Cunningham, Berkelium and californium, *J. Chem. Educ.*, 36:32–37, 1959; B. B. Cunningham, Chemistry of the actinide elements, *Annu. Rev. Nucl. Sci.*, 14:323–46, 1964; E. K. Hyde, I. Perlman, and G. T. Seaborg,

The Nuclear Properties of the Heavy Elements, vols. 1 and 2, 1964; F. L. Oetting et al., *The Chemical Thermodynamics of Actinide Elements and Compounds*, 2 vols., 1977; G. T. Seaborg, History of the synthetic actinide elements, *Actinides Rev.*, 1:3–38, 1967; G. T. Seaborg, *The Transuranium Elements*, 1958.

Calomel

Mercury(I) chloride, Hg_2Cl_2, a covalent compound which is insoluble in water. The substance sublimes when heated. The formula weight is 472.086 and the specific gravity is 7.16 at 20°C. The material is a white, impalpable powder consisting of fine tetragonal crystals. Calomel is manufactured by precipitation when sodium chloride is added to a solution of mercury(I) nitrate or by direct combination of the elements.

Calomel is used in preparing insecticides and medicines. It is well known in the laboratory as the constituent of the calomel reference electrodes which are commonly used in conjunction with a glass electrode to measure pH. *See* CALOMEL ELECTRODE; MERCURY. [E. EUGENE WEAVER]

Calomel electrode

An electrode used widely as a reference electrode of known potential, for example, in electrometric measurements of acidity and alkalinity, in corrosion studies, and in the measurements of the potentials of other electrodes. The calomel electrode consists of mercury covered with a layer of paste of mercury (I) chloride in an aqueous solution of potassium chloride. A small amount of mercury is dispersed in the paste of mercury(I) chloride. It is made in three well-known types: the saturated, the normal, and the one-tenth normal. These designations refer to the concentrations of the potassium chloride solutions used in the preparation. The first type is more widely used than the other two because of the ease with which it may be assembled; no analysis of the concentration is required. Crystals of potassium chloride are placed above the layer of mercury(I) chloride in the saturated type. The calomel electrode is commonly referred to as a half-cell because it includes an aqueous phase of potassium chloride as one arm of the cell to be used in the electric circuit. In some types, a reservoir for the solution of potassium chloride is provided for ease in flushing and replacing the solution in the side arm with new solution.

The potentials of the three types in terms of the standard hydrogen electrode are:

Saturated type:	$-0.2440 + 0.00076(t - 25°C)$ volts
1.0 *N* type:	$-0.2834 + 0.00024(t - 25°C)$ volts
0.1 *N* type:	$-0.3371 + 0.00007(t - 25°C)$ volts

Frequently, calomel electrodes are used in assemblies such as pH meters, in which it is not essential to know their actual potential. In this case the overall system is calibrated with a solution of known pH. *See* ELECTRODE POTENTIAL; HYDROGEN ELECTRODE; HYDROGEN ION; SILVER CHLORIDE ELECTRODE. [WALTER J. HAMER]

Calorimetry

The measurement of the quantity of heat involved in various processes, such as chemical reactions, changes of state, and formation of solutions, or in the determination of the heat capacities of substances. The fundamental unit for these heat measurements is the joule, but many workers prefer to use the calorie, defined as 4.184 absolute joules.

Types of calorimeters. A calorimeter is an apparatus for measuring these heat quantities. Since the problem involves various kinds of processes and experimental work over a range of temperatures, calorimeters have been developed in great variety.

Many calorimeters use a liquid as a heat reservoir, since uniformity of temperature thus can be realized easily by proper stirring, and the temperature change for the system can be evaluated readily. Water is the substance most commonly employed; it is especially satisfactory in measurements near room temperature. However, various organic liquids also have been used. For example, G. B. Kistiakowsky and coworkers, in their measurements of heats of hydrogenation, selected diethylene glycol, which permitted them to operate at temperatures up to 150°C.

In contrast, the aneroid type depends on the use of a metal of high thermal conductivity, such as copper, to promote rapid equalization of temperature. In the Nernst method of measuring heat capacities to low temperatures, the calorimeter is usually constructed so that none of the material under study is more than a few millimeters from a metal surface. Such an aneroid calorimeter can be used over a temperature range of 300 centrigrade degrees or more.

These types represent nonisothermal calorimeters, in which the heat quantities are evaluated by the temperature changes they produce. Various isothermal calorimeters also have been developed in which the heat reservoir contains a liquid in equilibrium with its crystalline solid at the melting point or with its vapor at the boiling point. The Bunsen ice calorimeter, which operates at 0°C with a mixture of ice and water, is a good example. It can receive or give up heat, as the case requires, and the magnitude of this heat quantity can be evaluated by a dilatometric measurement of the volume change in the ice-water mixture. The isothermal calorimeter is simple, and yet it can yield excellent data. Its disadvantage is that it can be operated only at the equilibrium temperature of the two-phase system involved.

All these types of calorimetric apparatus contain (1) the calorimeter proper, or heat reservoir for measuring the amount of heat in the process under study, and (2) a jacket which can be used to control thermal exchange between the calorimeter and its environment. In many cases this jacket is a hollow vessel which may be filled with a liquid, such as water, that can be stirred and accurately regulated in temperature. In the aneroid type the jacket usually is a hollow, cylindrical block of metal with a top and bottom, electrical heating coils, and a suitable thermometer system.

In both the liquid and aneroid types of calorimeters the jacket is frequently kept at constant temperature by a thermostat. Thus, there is some heat exchange between the calorimeter and the jacket during an experiment, and this requires that a correction be made to the observed temperature change within the calorimeter. With the ordinary method, this correction is determined by making a series of temperature readings on the calorimeter before and after the thermal process, with the assumption of Newton's law of cooling, which states that the rate of heat flow is proportional to the temperature difference between an object and its environment. A second method avoids such a correction by keeping the temperature of the jacket during the thermal process equal to that of the calorimeter. Thus, there should be no heat flow between the two. This is adiabatic calorimetry. At first glance it appears more elegant than the ordinary method, but in practice the two methods have been found to yield equally accurate results.

Thermometry. The measurement of temperature is an important factor of calorimetry which has been greatly refined since 1920. Prior to that time, reliance was placed mainly on mercury and other expansion thermometers. The Beckmann thermometer represents the acme among these. It is a differential instrument with a temperature range of about 5°. It can be read to 0.001C°.

Today, resistance thermometers and thermocouples are the instruments most used in high-grade precision measurements. One may have advantages over the other in particular situations. Both types can be used from about −260° to above 700°C. However, the platinum resistance type is the standard. Such an instrument with a resistance of 25 ohms at 0° can be read to better than 0.0001C°.

Bomb calorimeter. This calorimeter is widely used in determining the heats of combustion of organic compounds and has been developed to yield measurements reproducible to within ±0.01%. Its operation is fairly typical of calorimetric processes carried out near room temperature.

Some of the essential features of the apparatus are illustrated in the accompanying schematic figure. The bomb, a strong-walled metal container of 0.3–0.4-liter capacity, is constructed of a corrosion-resistant alloy. It is fitted with a screw head, which contains a valve for introducing oxygen to a pressure of about 30 atm. The weighed sample, liquid or solid, is placed in the combustion crucible and ignited by electrical connections which pass through the head.

The bomb is immersed in about 2.5 liters of water in the calorimeter proper, a metal container which is equipped with a platinum resistance thermometer, a stirrer, and a tightly fitting cover to prevent evaporation of its water. It rests on small ivory pegs within the jacket, the two being separated by a 1-cm air space.

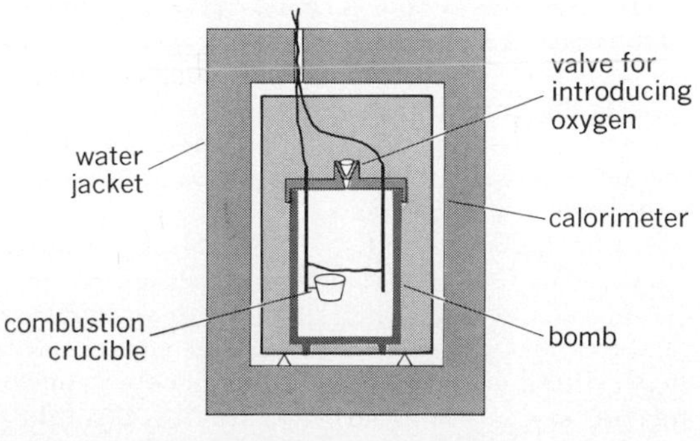

Schematic diagram of bomb calorimeter.

This bomb calorimeter is calibrated either electrically or, more frequently, by burning a sample of standard benzoic acid, supplied by the U.S. National Bureau of Standards. After the heat capacity of the calorimeter has been evaluated, the unknown heat value can be computed from the measured temperature rise. The result so obtained, corrected to 1 atm pressure and computed for 1 mole of the substance, represents ΔE, the heat of combustion at constant volume. The corresponding heat effect at constant pressure can be derived by the relation $\Delta H = \Delta E + P\Delta V$. *See* CHEMICAL THERMODYNAMICS; THERMOCHEMISTRY.

[BRUNO J. ZWOLINSKI]

Bibliography: J. Barthel, *Thermometric Titrations*, 1975; R. S. Porter and J. F. Johnson (eds.), *Analytical Calorimetry*, vols. 2–4, 1970–1977; R. C. Wilhoit, *J. Chem. Educ.*, 44:A571, A629, A685, A853, 1967.

Carbazole

One of a group of organic heterocyclic compounds (also called 9-azafluorenes) containing a dibenzopyrrole system as in the parent compound, carbazole (III). *See* HETEROCYCLIC COMPOUNDS.

The carbazole skeleton occurs in certain dyes, and in the *Strychnos* alkaloids. The parent compound was first found in coal tar. Carbazoles can be synthesized by the Graebe-Ullmann reaction. By this method an *o*-aminodiphenylamine (I) is transformed to a 1-phenylbenzotriazole (II), which on heating loses nitrogen and gives a carbazole (III). Another general scheme, an adaptation of the

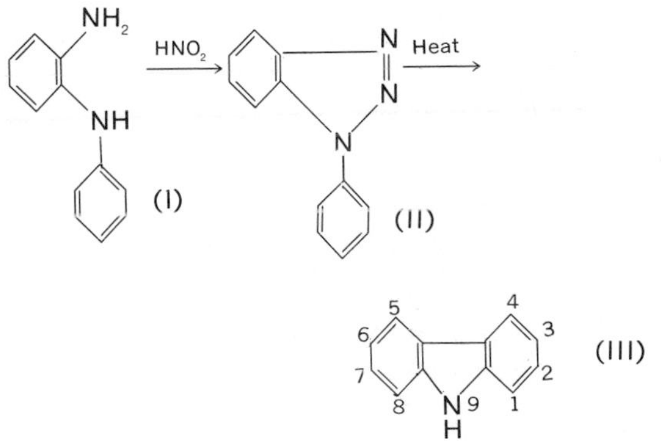

Fischer indole synthesis, converts cyclohexanone phenylhydrazones to 1,2,3,4-tetrahydrocarbazoles, which can then be dehydrogenated. Carbazoles can be obtained also by cyclizing *o*-aminobiphenyls, as well as *o,o'*-diaminobiphenyls.

Carbazole (III) is a colorless, water-insoluble solid, mp 246°C. Substitution occurs at the 3 and 6 positions on nitration, sulfonation, halogenation, and Friedel-Crafts acylation. Carbazole, although sensitive to oxidation, is markedly resistant to heat, acid, or alkali.

The hydrogen at the 9 position is very weakly acidic, but it can be replaced with metals under proper conditions. The 9 position may be substituted with the help of familiar reagents to give 9-acyl- or 9-alkylcarbazoles. 9-Hydroxyethyl-, 9-nitroso-, and 9-vinylcarbazole are formed by reaction of carbazole with ethylene oxide, nitrous acid, and acetylene, respectively. 9-Vinylcarbazole is also formed by pyrolyzing the acetate of 9-hydroxyethylcarbazole. The polymer obtained by peroxidic emulsion polymerization or solvent polymerization of 9-vinylcarbazole has useful electrical insulation properties similar to those of mica.

Hydron Blue R (V), a blue cotton dye, is produced by treating the condensation product (IV) of carbazole and *p*-nitrosophenol with sodium sulfide and sodium hydrosulfite. Hydron Yellow (VI) is

(IV)

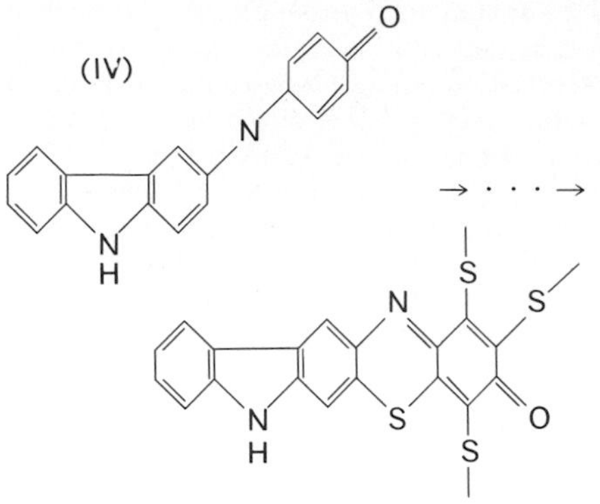

(V)

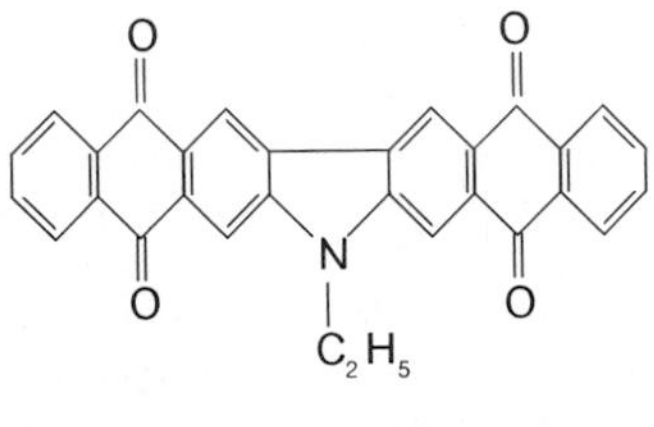

(VI)

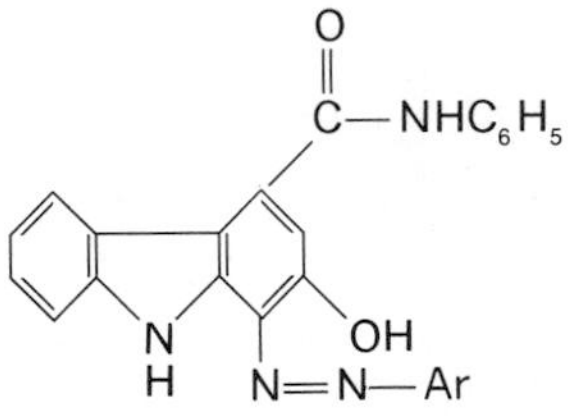

(VII)

representative of a number of carbazole vat dyes. Compound (VII) is a carbazole azo dye. See DYE.

[WALTER J. GENSLER]

Bibliography: R. M. Acheson, *An Introduction to the Chemistry of Heterocyclic Compounds*, 3d ed., 1976; H. A. Lubs (ed.), *The Chemistry of Synthetic Dyes and Pigments*, 1955, reprint 1972; D. M. Young, *Heterocyclic Chemistry*, 1976.

Carbide

A binary compound of carbon with an element less electronegative than carbon. Carbon-hydrogen compounds are excluded. Effectively then, carbides are composed of metal-carbon compounds if boron and silicon are included among the normal metals. Essentially no volatile compounds (except AlC or its dimer) are known, because decomposition sets in at higher temperatures before volatilization of the carbide as such.

Most carbides can be prepared by heating a mixture of the powdered metal and carbon, usually to high temperatures, but not necessarily as high as the melting point. Generally, the same result is possible by heating a mixture of the oxide of the metal with carbon ($CaO + 3C \rightarrow CaC_2 + CO$). Some may be prepared by passing a hydrocarbon vapor over the hot metal in the form of an electrically heated filament. The alkali metal carbides (Li_2C_2, Na_2C_2, K_2C_2, Rb_2C_2, and Cs_2C_2) are best prepared by passing acetylene (C_2H_2) into solutions of the metals in liquid ammonia ($2Na + 2C_2H_2 \rightarrow 2NaHC_2 + H_2$) followed by heating ($2NaHC_2 \rightarrow Na_2C_2 + C_2H_2$). The very unstable carbides (Cu_2C_2, Ag_2C_2, Au_2C_2, ZnC_2, CdC_2, and HgC_2) are prepared by passing acetylene into solutions of the metal salts ($2CuCl + C_2H_2 \rightarrow Cu_2C_2 + 2HCl$). The carbides of these two series of elements are often regarded as acetylides partly in view of the preparative method, but primarily because they react with water to give off acetylene gas ($Na_2C_2 + 2H_2O \rightarrow C_2H_2 + 2NaOH$).

It is useful to classify carbides as ionic (saltlike), metallic (interstitial), and covalent. The more electropositive elements (groups I, II, and III, and to some extent, members of the lanthanide and actinide series of the periodic system) form ionic carbides with transparent (or saltlike) crystals. Belonging to this class are the acetylides listed above and BeC_2, MgC_2, CaC_2, SrC_2, BaC_2, Al_2C_6, and Ce_2C_6. Others yielding predominantly acetylene on reaction with water, but also producing additional hydrocarbons are YC_2, LaC_2, CeC_2, PrC_2, NdC_2, SmC_2, ThC_2, UC_2, and NpC_2. The acetylides contain pairs of carbon atoms in ion units, C_2^{--}. Other ionic carbides (methanides) which yield with water primarily methane but also other hydrocarbons and hydrogen (an example is given in the reaction below) are Be_2C and Al_4C_3,

$$Al_4C_3 + 12H_2O \rightarrow 4Al(OH)_3 + 3CH_4$$

and probably also Sc_4C_3. The action of water on ScC is not known.

The carbide Mg_2C_3 yields allylene (propyne) on treatment with water.

Highly refractory covalent carbides are formed with Si and B. In the electric-arc furnace SiC (diamond structure) is formed from a mixture of SiO_2 and coke. The very hard B_4C can be made similarly from its oxide; it is unusual both structurally and in having a fairly high electrical conductivity.

The group TiC, ZrC, HfC, V_2C, VC, Cb_2C, CbC, Ta_2C, TaC, Mo_2C, MoC, W_2C, and WC has exceedingly high melting temperatures, is very hard, is metallic in appearance and in electrical conductivity, and is classified as metallic. Important refractory units can be prepared by mixing the metal powder with carbon and sintering under pressure. These carbides are reactive neither toward water nor in general toward acids or other chemicals. The relatively smaller carbon atoms fit into the interstices between the metal atoms, and the compounds are therefore sometimes classified as interstitial. Compositions may vary over a considerable range from the simple formulas given. Probably ThC, U_2C_3, UC, CeC, NpC, PuC, and possibly Ce_2C_3, Np_2C_3, and Pu_2C_3 should be included with the metallic carbides.

The carbides of Cr, Mn, Fe, Co, and Ni are intermediate between the interstitial and covalent types, but are much nearer the former in properties. $Cr_{23}C_6$, Cr_7C_3, Cr_3C_2, $Mn_{23}C_6$, Mn_7C_3, Mn_5C_2, Fe_3C, Fe_2C, Co_3C, Co_2C, and Ni_3C react with water or acids to give either simple hydrocarbons or mixtures of hydrocarbon gases. The presence of Fe_3C in iron is an important factor in the properties of steel. See BORIDE; CARBON; HYDRIDE; NITRIDE.

[RUSSELL K. EDWARDS]

Bibliography: R. P. Elliott, *Constitution of Binary Alloys: First Supplement*, 1971; E. K. Storms, *Refractory Carbides*, 1967.

Carbon

A chemical element, C, with an atomic number of 6 and an atomic weight of 12.01115. Carbon is unique in chemistry because it forms a vast number of compounds, larger than the sum total of all other elements combined. By far the largest group of these compounds are those composed of carbon and hydrogen. It has been estimated that there are at least 1,000,000 known organic compounds, and this number is increasing rapidly each year. Although the classification is not rigorous, carbon forms another series of compounds, classified as inorganic, comprising a much smaller number than the organic compounds. *See* ORGANIC CHEMISTRY.

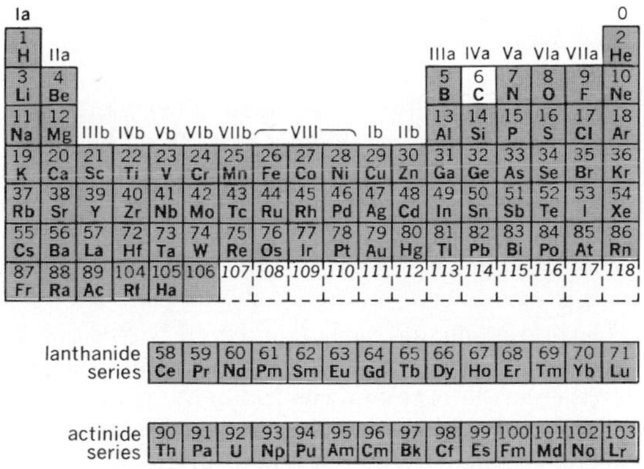

Uses. The free element has many uses, ranging from ornamental applications of the diamond in jewelry to the black-colored pigment of carbon black in automobile tires and printing inks. Another form of carbon, graphite, is used for high-temperature crucibles, arc-light and dry-cell electrodes, lead pencils, and as a lubricant. Charcoal, an amorphous form of carbon, is used as an absorbent for gases and as a decolorizing agent.

The compounds of carbon find many uses. Carbon dioxide is used for the carbonation of beverages, for fire extinguishers, and in the solid state as a refrigerant. Another oxide of carbon, carbon monoxide, finds use as a reducing agent for many metallurgical processes. Carbon tetrachloride and carbon disulfide are important solvents for industrial uses. Gaseous dichlorodifluoromethane, commonly known as Freon, is used in refrigeration devices. Calcium carbide is used to prepare acetylene, which is used for the welding and cutting of metals as well as for the preparation of other organic compounds. Other metal carbides find important uses as refractories and metal cutters.

Occurrence. Carbon and its compounds are found widely distributed in nature. It is estimated that carbon makes up 0.032% of the Earth's crust. Free carbon is found in large deposits as coal, an amorphous form of the element which contains additional complex carbon-hydrogen-nitrogen compounds. Pure crystalline carbon is found as graphite and in small amounts as diamonds.

Extensive amounts of carbon are found in the form of its compounds. In the atmosphere, carbon is present in amounts up to 0.03% by volume as carbon dioxide. Various minerals such as limestone, dolomite, marble, and chalk all contain carbon in the form of carbonate. All plant and animal life is composed of complex organic compounds containing carbon combined with hydrogen, oxygen, nitrogen, and other elements. The remains of past plant and animal life are found as deposits of petroleum, asphalt, and bitumen. Deposits of natural gas contain compounds that are composed of carbon and hydrogen.

Physical and chemical properties. Elemental carbon exists in two well-defined crystalline allotropic forms, diamond and graphite. Other forms, which are poorly developed in crystallinity, are charcoal, coke, and carbon black.

Charcoal is prepared by the ignition of wood, sugar, blood, and other carbon-containing compounds in the absence of air. X-ray diffraction studies reveal that it has a graphite structure but is not very well developed in crystallinity. The lack of crystallinity is the result of defects in the crystal structure and the high surface area. In the activated state, charcoal adsorbs gases, liquids, and solids. Activation of the charcoal is accomplished by treatment of the substance with steam; this tends to remove the adsorbed hydrocarbons from the surface. The adsorbing property of charcoal is related to the large surface area present. It is stated that one cubic centimeter of charcoal has a surface area of 1000 m^2. *See* ADSORPTION.

Coke, another form of amorphous carbon, is prepared by heating coal in the absence of air. It is used primarily for the reduction of metal oxides to the free metals. *See* COKE.

Chemically pure carbon is prepared by the thermal decomposition of sugar (sucrose) in the absence of air. Impurities in the carbon are removed by treatment with chlorine gas at red heat. The substance is then washed with water and the residual chlorine is removed by heating in an atmosphere of hydrogen gas.

The physical and chemical properties of carbon are dependent on the crystal structure of the element. The density varies from 2.25 g/cm^3 for graphite to 3.51 g/cm^3 for diamond. For graphite, the melting point is 3500°C and the extrapolated boiling point is 4830°C. This is abnormally high because a large amount of energy is required to break the three-dimensional covalent bonding in the crystal.

Elemental carbon is a fairly inert substance. It is insoluble in water, dilute acids and bases, and organic solvents. At elevated temperatures, it combines with oxygen to form carbon monoxide or carbon dioxide. With hot oxidizing agents, such as nitric acid and potassium nitrate, mellitic acid, $C_6(CO_2H)_6$, is obtained. Of the halogens, only fluorine reacts with elemental carbon. A number of metals combine with the element at elevated temperatures to form carbides.

Principal compounds. Carbon forms three compounds with oxygen: carbon monoxide, CO; carbon dioxide, CO_2; and carbon suboxide, C_3O_2. The first two oxides are the more important from an industrial standpoint.

Oxides of carbon. Carbon monoxide is prepared in the laboratory, as in reaction (1), by the dehy-

$$HCOOH + H_2SO_4 \rightarrow CO + H_2SO_4{\cdot}H_2O \quad (1)$$

dration of formic acid with concentrated sulfuric

acid. Similarly, other organic acids can be used, for example, oxalic acid, $H_2C_2O_4$. However, in this case, carbon dioxide is obtained along with the carbon monoxide. The highest-purity carbon monoxide is prepared by the thermal decomposition of nickel carbonyl at 200°C, as in reaction (2).

$$Ni(CO)_4 \rightarrow Ni + 4CO \qquad (2)$$

The industrial preparation of carbon monoxide is by the water gas reaction. Water, in the form of steam, is passed over hot coke or coal at 600–1000°C, as represented by reaction (3). The tem-

$$C + H_2O(g) \rightleftharpoons CO + H_2 \qquad (3)$$

perature of the reaction must be carefully controlled because at 500°C carbon dioxide is obtained instead of carbon monoxide. Also, because the reaction is endothermic, air must be passed over the heated coke or coal at regular intervals to maintain the minimum 600°C temperature.

Carbon monoxide is a highly poisonous, colorless, odorless, tasteless gas having a melting point of −205.1°C and a boiling point of −190°C. The gas is stable at room temperature but at elevated temperatures it disproportionates into carbon and carbon dioxide, as shown by reaction (4). This reac-

$$2CO \rightleftharpoons C + CO_2 \qquad (4)$$

tion is accelerated by the presence of certain metal catalysts. At elevated temperatures, the gas is an excellent reducing agent. With metal oxides, such as copper (II) oxide, it gives copper metal and carbon dioxide, as shown by reaction (5). Many other

$$CuO + CO \rightarrow Cu + CO_2 \qquad (5)$$

applications of this reducing property are known, and perhaps the most important is the blast furnace reduction of iron (III) oxide to pig iron.

Carbon monoxide combines with many metals and nonmetals. With chlorine, in the presence of sunlight, it forms highly poisonous phosgene, $COCl_2$; with sulfur, carbonyl sulfide, COS, is obtained. A number of other carbonyl compounds are known, such as carbonyl fluoride, COF_2; carbonyl bromide, $COBr_2$; and carbonyl selenide, COSe.

With metals, a class of compounds known as metal carbonyls are formed. These metals are from groups Ib, IIb, VIa, VIIa, and VIII of the periodic table. The metal carbonyls can be prepared by the direct combination of the metal with carbon monoxide, although several of the compounds require fairly high pressures. With finely divided nickel, the compound $Ni(CO)_4$ is formed at 50°C and normal atmospheric pressure, as shown by reaction (6). With others such as iron, cobalt,

$$Ni + 4CO \rightarrow Ni(CO)_4 \qquad (6)$$

molybdenum, and tungsten, pressures of 200–450 atm and temperatures above 200°C are necessary to prepare $Fe(CO)_5$, $CO_2(CO)_8$, $Mo(CO)_5$, and $W(CO)_6$, respectively.

The metal carbonyls react with the halogens to produce metal carbonyl halides. With hydrogen, a similar reaction takes place to form metal carbonyl hydrides.

Nickel carbonyl finds application in the purification and separation of nickel from other metals. Iron carbonyl has been used in antiknock gasoline preparations and to prepare high-purity iron metal. *See* METAL CARBONYL; NICKEL.

Carbon monoxide may be considered to be the anhydride of formic acid. It does not react with water, however, to form an acid solution. It reacts with a sodium hydroxide solution, at 140°C and under pressure, to give sodium formate, $NaCO_2H$. This solution, on acidification, gives formic acid.

Carbon monoxide has assumed great importance in the field of synthetic organic chemistry. By use of pressures of 100–200 atm, temperatures of 300–600°C, and various mixed metal oxide catalysts, direct combination reactions with hydrogen can be carried out to produce methyl alcohol and benzene, as seen by reactions (7) and (8). Many other similar reactions utilize carbon monoxide.

$$CO + 2H_2 \rightarrow CH_3OH \qquad (7)$$

$$12CO + 3H_2 \rightarrow C_6H_6 + 6CO_2 \qquad (8)$$

As stated previously, carbon monoxide is extremely poisonous. As little as 9 parts of the gas in 10,000 parts of air will cause nausea and headache, and slightly larger amounts will cause death. The physiological action of carbon monoxide poisoning is based upon the formation of a stable compound with the hemoglobin of the blood. This compound is more stable than oxyhemoglobin; thus, body tissues are prevented from receiving the necessary oxygen. Caution should be exercised in handling carbon monoxide at all times. Automobile exhaust gas is a common source of carbon monoxide poisoning.

Carbon monoxide can be detected by the green can be absorbed in an aqueous solution of copper(I) chloride containing hydrochloric acid.

Carbon dioxide is prepared by the combustion of carbon in air or by the decomposition of a metal carbonate with an acid or by heat. *See* CARBON DIOXIDE.

Carbon suboxide, C_3O_2, is a less known oxide of carbon. It is prepared by the dehydration of malonic acid, $HOOCCH_2COOH$, with phosphorus(V) color produced with a mixture of iodine pentoxide and fuming sulfuric acid adsorbed on pumice. Another test is the black color of finely divided palladium metal produced by bubbling the gas through a palladium(II) chloride solution. The gas oxide in a vacuum at 140–150°C, as represented by reaction (9). Carbon suboxide is a gas with an

$$HOOCCH_2COOH \xrightarrow{P_4O_{10}} C_3O_2 + 2H_2O \qquad (9)$$

obnoxious odor and a boiling point of −6.8°C. It combines readily with water to form malonic acid, thus reversing the above reaction.

Carbon tetrahalides. Carbon forms compounds with the halogens which have the general formula CX_4, where X is fluorine, chlorine, bromine, or iodine. At room temperature, carbon tetrafluoride is a gas, carbon tetrachloride is a liquid, and the other two compounds are solids. All are covalent compounds with the usual tetrahedral structure.

By far the most important carbon halide is carbon tetrachloride, CCl_4. It is prepared, as shown by reaction (10), by passing chlorine gas into carbon

$$CS_2 + 3Cl_2 \rightarrow CCl_4 + S_2Cl_2 \qquad (10)$$

disulfide, CS_2, containing a small amount of iodine or antimony (III) chloride. The carbon tetrachloride is separated from the reaction mixture by fractional distillation. Carbon tetrachloride is a colorless, pleasant-smelling, nonflammable liquid with a melting point of −22.9°C, a boiling point of 76.4°C, and a density of 1.595 g/ml at 20°C. The compound is an excellent solvent and finds much use as a solvent for fats, greases, waxes, and many other organic compounds. It is commonly used as a dry-cleaning fluid and also as a spot remover for clothing.

The heavy, nonflammable vapor finds use in fire extinguishers. The dense vapors exclude air and oxygen from the combustion area. It should not be used to extinguish fires where hot metal surfaces are present. In the presence of hot metals as catalysts, water vapor reacts with carbon tetrachloride to produce, among other products, poisonous phosgene.

Carbon tetrabromide and tetraiodide are prepared from the tetrachloride by reaction with aluminum bromide and aluminum iodide, respectively. Both of these substances are thermally unstable at elevated temperatures.

Mixed carbon tetrahalides are also known. Perhaps the most important of them is dichlorodifluoromethane, CCl_2F_2, commonly called Freon. It is prepared by the reaction of hydrogen fluoride with carbon tetrachloride in the presence of antimony(III) chloride. Freon is used as a refrigerant. *See* FREON; HALOGENATED HYDROCARBON.

Metal carbides. Metal carbides are prepared by heating the metal, metal oxide, or metal hydride with carbon or other carbon compounds. The most important of these carbides is calcium carbide, CaC_2. This compound is prepared by heating calcium oxide, CaO, and coke in an electric arc furnace. The most important use for calcium carbide is in the preparation of acetylene, C_2H_2. Acetylene is obtained by the reaction of calcium carbide with water. This is shown by reaction (11). *See* ACETYLENE.

$$CaC_2 + 2H_2O \rightarrow Ca(OH)_2 + C_2H_2 \qquad (11)$$

Many other metals react to form carbides, for example, titanium, hafnium, tantalum, and tungsten. This latter group of metals forms refractory-type carbides; they are noted for their extreme hardness (almost equal to that of diamond) and have very high melting points. For example, tantalum carbide, TaC, has a melting point of about 3900°C. All of these refractory carbides are chemically inert and are important for use as cutting tools in industry. *See* CARBIDE.

Analytical methods. Carbon is usually determined as carbon dioxide. The free element can be burned in air; metal carbonates can be decomposed by acids or heat; carbon monoxide can be burned in air; and organic compounds containing carbon can be oxidized to give carbon dioxide. The evolved carbon dioxide is weighed directly by adsorption in a mixture of sodium hydroxide and calcium hydroxide contained in a glass tube. A qualitative test for carbon dioxide is to pass the gas through limewater, a $Ca(OH)_2$ solution, and observe the presence of a white precipitate of calcium carbonate, $CaCO_3$. *See* GERMANIUM; SILICON. [E. EUGENE WEAVER]

Bibliography: J. W. Mellor and G. D. Parkes, *Comprehensive Treatise on Inorganic and Theoretical Chemistry*, 16 vols., 1922–1937; P. L. Walker and P. A. Thrower (eds.), *Chemistry and Physics of Carbon: A Series of Advances*, vols. 1–15, 1966–1979.

Carbon dioxide

A colorless, odorless, tasteless gas, formula CO_2, about 1.5 times as heavy as air. The specific volume at atmospheric pressure (101,325 N/m^2) and 70°F (21°C) is 8.74 ft^3/lb (0.546 m^3/kg). Under normal conditions, it is stable, inert, and nontoxic.

The decay (slow oxidation) of all organic materials produces CO_2. Fresh air contains approximately 0.033% CO_2 by volume. In the respiratory action (breathing) of all animals and humans, CO_2 is exhaled.

Carbon dioxide gas may be liquefied or solidified. For example, if the gas is compressed to 300 pounds per square inch gage (psig; 1 psi = 6895 N/m^2) and cooled to 0°F (−27°C) it becomes a liquid; or it may be liquefied at 70°F by being compressed to 838 psig. Above the critical temperature of 87.9°F (31.1°C), CO_2 exists only as a gas, regardless of the pressure applied.

If liquid CO_2 is cooled to −69.9°F (−29.5°C) the pressure drops to 60.4 psig, and dry ice snow is formed. This condition of pressure and temperature is known as the triple point of CO_2, at which all three phases, solid, liquid, and gas, may exist in equilibrium with one another. Carbon dioxide cannot exist as a liquid below the triple-point temperature and pressure. If the pressure on dry-ice snow is reduced to atmospheric, its temperature drops to −109.3°F (−41.6°C). As solid CO_2 absorbs heat from its surroundings, it transforms directly to a gas (sublimes), hence the name dry ice. In still air, a film of gaseous CO_2 surrounds the dry ice, and the sublimation temperature is −109.3°F; but in a vacuum or in rapidly moving air in which the CO_2 gas film is stripped away, the sublimation temperature drops to −130°F (−90°C) or lower.

Carbon dioxide is obtained commercially from four sources: gas wells, fermentation, combustion of carbonaceous fuels, and as a by-product of chemical processing.

Applications. Carbon dioxide may be used as a refrigerant, inerting medium, chemical reactant, neutralizing agent for alkalies, and pressurizing agent.

Refrigeration. Solid CO_2 has a greater refrigeration effect than water ice; the latent heat of sublimation of dry ice is 253.8 Btu/lb, whereas the latent heat of fusion of water ice is 143.4 Btu/lb (1 Btu = 1055 J). Furthermore, it is much colder than water ice and sublimes to a gas as it absorbs heat. Solid CO_2 may be furnished to the user in pressed blocks weighing 50–60 lb (1 lb = 0.45 kg) each, or as extruded pellets. It may also be made as snow at the point of use by expanding liquid CO_2 to atmospheric pressure; upon release to the atmosphere, liquid CO_2 at a pressure of 285 psig produces about 47% dry ice snow and 53% gas by weight. However, of the total refrigerating effect originally available in the liquid, 85% remains in the ice and 15% in the vapor when the liquid CO_2 is warmed to 0°F.

Dry ice blocks and pellets are used to chill, firm, and freeze meats, vegetables, and other perishable foods for preservation during transport. Dry ice snow produced from the expansion of liquid CO_2 may be applied directly to the surface of foods in a freezing chamber or tunnel, or it may be deposited on the contents of a container. Refrigeration may be achieved in trucks and other transport modes by vaporization of either solid or liquid CO_2.

Liquid CO_2 is injected into pneumatic conveyor systems to cool the contents during transport. Such items as flour, sugar, plastics, and core sand are quickly chilled by the dry ice snow and cold gas produced in the pneumatic line as liquid CO_2 is introduced through a thermostatically controlled orifice.

In low-temperature testing of aircraft and electronic parts to meet military or manufacturing specifications at −65°F and below, liquid CO_2 is expanded through an orifice to form dry-ice snow, which is either blown directly on the part or circulated in a test chamber; solid CO_2 is used in certain kinds of testing.

Carbon dioxide liquid injected into a hollow plastic shape while it is still in the mold reduces the temperature of the internal surface and, in turn, decreases the time required for the plastic to become sufficiently hardened to be self-supporting, allowing the part to be removed from the mold more quickly. Thus blow-molding equipment can operate on a shorter cycle time, and production can be speeded up.

Carbon dioxide liquid or ice may be used to advantage for removing the flash (mold marks) from molded rubber parts. This is done in an insulated tumbling barrel in which the mechanical action easily removes the flash that has been embrittled by contact with solid CO_2.

Carbon dioxide snow is added directly to choppers and mixers used in the preparation of hamburger, sausage, and prepared meat products such as bologna. The quick chilling reduces meat temperature rapidly, retarding bacterial growth.

Carbon dioxide is used to stiffen shortening for homogeneous blending with the dry ingredients used in piecrust mixes; to chill spices, chemicals, sugar, and rubber during high-speed grinding; and to prevent softening of thermoplastic materials during pulverizing.

Inerting. Carbon dioxide does not react with oxygen, nor does it normally function as an oxidizing agent. At high temperatures, it dissociates into carbon monoxide and oxygen (about 1% at 2800°F, or 1538°C).

Liquid CO_2 is an effective fire-extinguishing agent because it rapidly reduces the temperature of burning materials below the ignition point, and the gas, being heavy and inert, displaces air, usually the source of oxygen, and blankets the flames. It is particularly effective in extinguishing fires when water cannot be used, for example, in oil and electrical fires.

Oil tankers, barges, storage vessels, and pipelines are quickly and safely inerted with CO_2 to allow welding and other repairs. In the automatic electric–tungsten–inert gas welding of steel using a filler wire, CO_2 is used to blanket the arc, thus preventing oxidation of the molten metal.

Carbon dioxide is useful in many other applications in which an inert gas is required to prevent oxidation, as in packaging foodstuffs, spray-drying eggs and other solids, blanketing paint ingredients during manufacture to prevent the formation of "skin," and protecting grain and other bulk foods stored in silos.

Chemical applications. The largest single market for CO_2 for chemical purposes is in the preparation of carbonated beverages, in which the weak carbonic acid formed by the CO_2 acts as a taste enhancer and preservative. Other uses include the hardening of foundry cores in the core box by the reaction of CO_2 with the sand binder, the neutralization of excess alkalinity in water or industrial wastes, the manufacture of salicylic acid for aspirin, the production of pure carbonates and bicarbonates, as an intermediate in the preparation of titanium dioxide, and the stimulation of oil and gas wells.

Pressure medium. Life rafts and life preservers are packaged to include small CO_2 cartridges which permit rapid inflation upon operation of a quick-opening valve. Carbon dioxide in a cartridge is used as the propellent in certain pistols. It is also used as a propellent in pressure packaging, for example, in aerosol cans, and in many instances can replace fluorinated hydrocarbons at a fraction of the cost.

Manufacture. Most CO_2 is obtained as a by-product from steam-hydrocarbon reformers used in the production of ammonia, gasoline, and other chemicals; other sources include fermentation, deep gas wells, and direct production from carbonaceous fuels. Whatever the source, the crude CO_2 (containing at least 90% CO_2) is compressed in either two or three stages, cooled, purified, condensed to the liquid phase, and placed in insulated storage vessels. Commercial liquid CO_2 is usually stored at a pressure of 225–325 psig and is maintained in this range by refrigeration. When the liquid is placed into high-pressure cylinders and stored at ambient temperature, however, the pressure in the liquid rises to 1071 psig when the temperature is 87.8°F (the initial point), or higher, if the cylinder is completely filled with liquid.

Distribution. Carbon dioxide is distributed in three ways: in high-pressure uninsulated steel cylinders; as a low-pressure liquid in insulated truck trailers or rail tank cars; and as dry ice in insulated boxes, trucks, or boxcars.

The size of the high-pressure cylinders is limited because of the weight involved. Most commercial cylinders contain either 20 or 50 lb of CO_2 and are designed to deliver CO_2 gas, which can be reduced to the pressure required by the user through the action of a pressure regulator. However, some cylinders are equipped with a tube that reaches to the bottom of the cylinder and siphons liquid CO_2 to the outlet, provided the temperature of the contents is below the critical temperature (87.9°F).

Truck trailer capacity is restricted by laws which govern the gross vehicle weight. At present, up to 20 tons (1 ton = 0.9 metric ton) of liquid CO_2 may be shipped in an insulated trailer. Suppliers of CO_2 provide their customers with storage vessels varying in capacity from 4 to 150 tons.

When CO_2 is manufactured and transported as dry ice, losses occur because of sublimation. By storing the dry ice in a well-insulated container, losses may be kept within economical limits.

Solid CO_2 can be converted to the liquid form by placing it in a heavy-walled steel vessel known as a converter. After the vessel is sealed, it is allowed to warm to room temperature. As the temperature of the CO_2 rises past the triple point, it is converted to liquid.

[J. S. LINDSEY]

Bibliography: American Society of Heating, Refrigerating, and Air-Conditioning Engineers,

ASHRAE Handbook of Fundamentals, 1977; Compressed Gas Association, *Carbon Dioxide*, Publ. G-6, rev. ed., 1978; Y. S. Touloukian et al., *Thermal Conductivity: Nonmetallic Liquids and Gases*, 1970; M. P. Vukalovich and V. V. Altunin, *Thermophysical Properties of Carbon Dioxide*, 1965, transl. by D. S. Gaunt, 1968; C. L. Yaws, K. Y. Li, and C. H. Kuo, Carbon oxides: CO and CO_2, *Chem. Eng.*, Sept. 30, 1974.

Carbon tetrachloride

A colorless, dense liquid (specific gravity 1.595 at 20°C), formula CCl_4, manufactured by treating carbon disulfide with dry chlorine gas in the presence of a catalyst. This is shown by the reaction below. It is widely used in dry cleaning, as a sol-

$$CS_2 + 3Cl_2 \xrightarrow{SbCl_5} CCl_4 + S_2Cl_2$$

vent for waxes, oils, and greases, in fire extinguishers, particularly for oil and electrical fires, and in the manufacture of Freon refrigerants. Carbon tetrachloride fumes may form phosgene gas, a deadly poison; hence great caution is needed in its use as a fire extinguisher. It was once commonly used as an insecticide and in the treatment of hookworm, but because it irritates the skin, depresses the central nervous system, and, if inhaled, damages the liver and kidneys, it has been supplanted by other chemicals in these uses. During dry cleaning, operators should not breathe air containing more than 100 ppm. After dry cleaning, all carbon tetrachloride should be ventilated from the garment. *See* HALOGENATED HYDROCARBON.

[FRANK H. ROCKETT]

Carbonate

The CO_3^{2-} ion derived from carbonic acid, H_2CO_3, which is a solution of carbon dioxide, CO_2, in water. Because there are two replaceable hydrogen ions in carbonic acid, the salts obtained by neutralizing only one hydrogen are called bicarbonates, for example, sodium bicarbonate, $NaHCO_3$.

Carbonates can be converted into bicarbonates by adding excess carbon dioxide, reaction (1). The

$$Na_2CO_3 + CO_2 + H_2O \rightleftharpoons 2NaHCO_3 \qquad (1)$$

reaction can be reversed by heating. At even higher temperatures a second reaction takes place; for example, limestone is converted to lime, reaction (2). This reaction is commercially important in

$$CaCO_3 \xrightarrow{\Delta} CaO + CO_2 \qquad (2)$$

production of lime and portland cement.

Reaction (1) is the mechanism whereby the ground water in limestone areas dissolves Ca_2CO_3 to give hard water. Bicarbonates in general are more soluble than carbonates. All the carbonates except those of the alkali metals are insoluble.

Sodium carbonate is produced by the Solvay process from the raw materials ammonia, carbon dioxide, and sodium chloride. Sodium carbonate is used in the manufacture of glass, paper, and textiles. *See* CARBON.

[E. EUGENE WEAVER]

Carborane

A generic term for a class of chemical compounds composed of boron, carbon, and hydrogen. The term is also the common name for an important member of this class of compounds, $C_2B_{10}H_{12}$. The study of carboranes and the closely related boranes has led to an expanded understanding of chemical bonding and structure. Carborane derivatives have been used to develop new types of polymers which withstand high temperatures. Carboranes have been reacted with transition metals to produce new types of complexes, some of which exhibit catalytic activity.

Structure. The carboranes and the boranes share a common structural feature. Both have structures which can be rationalized on the basis of a knowledge of triangulated polyhedrons, or fragments of triangulated polyhedrons. Triangulated polyhedrons with 4 to 12 vertexes are especially important in understanding the structures of boranes and carboranes. Three polyhedrons, those with 4, 6, and 12 vertexes, are very symmetrical, and their vertexes are all geometrically equivalent. The polyhedron with 4 vertexes is the tetrahedron; the polyhedron with 6 vertexes is the hexahedron (Fig. 1*a*); and the polyhedron with 12 vertexes is the icosahedron (Fig. 1*b*). Polyhedrons with 5 vertexes (trigonal bipyramid), 7 vertexes (pentagonal bipyramid), 8 vertexes (dodecahedron), 9 vertexes (tricapped trigonal prism), and 10 vertexes (bicapped square antiprism) are not as symmetrical, and all have two different kinds of vertexes (Fig. 1*c*–*f*). The polyhedron with 11 vertexes (octadecahedron) has little symmetry and five different kinds of vertexes (Fig. 1*g*).

The structures of most boranes and carboranes are derived by placing a B-H unit or a C-H unit at the vertexes of the polyhedrons. For example, the borane anion $B_{12}H_{12}^{2-}$ has the icosahedral structure with a B-H unit at each of the 12 vertexes. A carborane exists when a C-H unit replaces a B-H unit at a vertex. Two known icosahedral carboranes are $CB_{10}H_{12}^{-}$ and $C_2B_{10}H_{12}$, in which one and two C-H units reside at the vertexes of the icosahedron. There is only a single isomer of $CB_{10}H_{12}^{-}$, but all three possible isomers of $C_2B_{10}H_{12}$ have been prepared.

Not all compounds consisting of boron, carbon, and hydrogen are carboranes. To qualify as a carborane, a compound must have at least one carbon atom in the three-dimensional skeleton of the structure. If all the carbon atoms are in side groups, replacing a hydrogen atom in a borane, the compound is not a carborane. For example, $C_2H_5B_{12}H_{11}^{2-}$ is a derivative of $B_{12}H_{12}^{2-}$ in which a hydrogen atom has been replaced by an ethyl group; it is thus not a carborane even though it is composed of boron, carbon, and hydrogen.

Derivatives. The most studied set of carborane derivatives are those with two carbon atoms. The formula for this set can be expressed as $C_2B_{n-2}H_n$, for example, $C_2B_{10}H_{12}$, $C_2B_6H_8$, and $C_2B_8H_{10}$. Derivatives are known for values of n from 5 through 12. The three-dimensional structure for each value of n is the polyhedron with n number of vertexes. For example the $C_2B_3H_5$ derivative has the trigonal bipyramidal structure (five vertexes) shown in Fig. 1*c*. There are three isomers for this system, one with the carbon atoms opposite one another at the apex positions, one with a carbon atom in the apex position and a carbon atom in the equatorial position, and one with both carbon atoms in the equatorial positions. Besides the set with two carbon atoms, carboranes with one carbon atom

CARBORANE

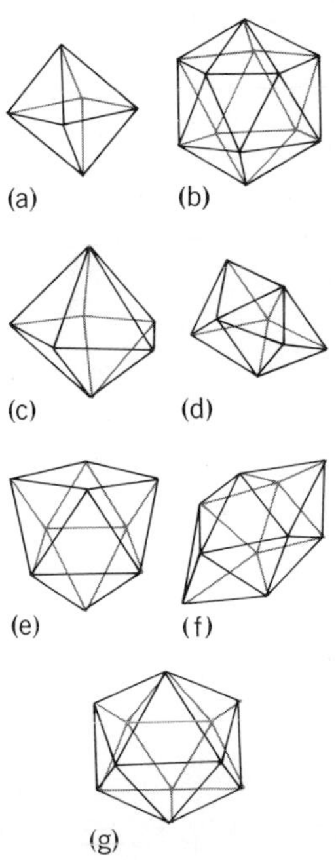

Fig. 1. Skeletal structures of polyhedrons. (*a*) Hexahedron. (*b*) Icosahedron. In figures *c*-*g* the $n = 7-11$ polyhedrons: (*c*) Pentagonal bipyramid. (*d*) Dodecahedron. (*e*) Tricapped trigonal prism. (*f*) Bicapped square antiprism. (*g*) Octadecahedron.

are quite common. They have the general formula $CB_{n-1}H_n^-$; the derivatives with $n = 12$, 11, 10, and 6 have been the most studied.

When the structure of the carborane or borane derivative is represented by one of the regular polyhedrons, the molecule is called a *closo* derivative. When the structure of the carborane or borane is best represented by a fragment of one of the polyhedrons, the molecule is called a *nido* derivative. Most carboranes contain only one or two carbon atoms and are *closo* derivatives. An example of a four-carbon-atom carborane is $C_4B_2H_6$, which has a structure derived by removing an apex position of the pentagonal bipyramid. This structure is called a pentagonal pyramid and is a polyhedral fragment. The arrangement of atoms for $C_4B_2H_6$ is shown in Fig. 2*e*.

Core electrons. The key feature of the polyhedral structure is the number of core electrons (the electrons which hold the polyhedron together). Determination of the number of core electrons in $C_2B_{10}H_{12}$ is as follows. The molecule contains a total of 50 electrons; 24 electrons are used to form the boron-hydrogen bonds and the two carbon-hydrogen bonds. The 26 electrons left are the core electrons, which hold the boron-hydrogen and carbon-hydrogen units in the icosahedral array. The number of core electrons for some other polyhedrons are octahedron, 14; pentagonal bipyramid, 16; dodechedron, 18; tricapped trigonal prism, 20; bicapped square antiprism, 22; and octadecahedron, 24. The core-electron requirements determine the charge on the polyhedral species. For example, $B_{12}H_{12}$ has only 24 core electrons and requires two more to form the stable icosahedral structure; therefore, $B_{12}H_{12}^{2-}$ is the species which exists. Analogously, $CB_{11}H_{12}^-$, but not $CB_{11}H_{12}$,

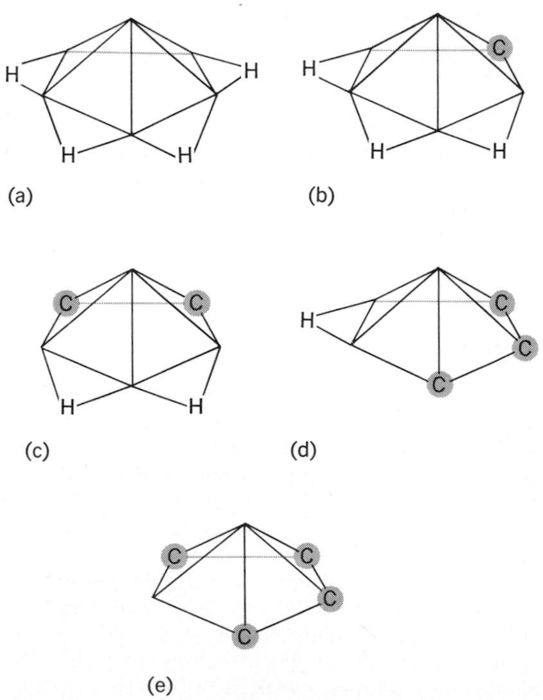

Fig. 2. The series depicting the stepwise loss of the bridging hydrogen atoms. The nonbridging hydrogen atoms are not shown. (*a*) B_6H_{10}, (*b*) CB_5H_9, (*c*) $C_2B_4H_8$, (*d*) $C_3B_3H_7$, and (*e*) $C_4B_2H_6$.

forms. The $C_2B_{10}H_{12}$ molecule has 26 core electrons and is neutral. Other series with the same numbers of core electrons (called isoelectronic series) are known, for example, $B_{10}H_{10}^{2-}$, $CB_9H_{10}^-$, and $C_2B_8H_{10}$. Generally, only neutral and anionic species are observed, and cationic species such as $C_3B_9H_{12}^+$ are not.

When additional core electrons are present in a molecule, the structure of the molecule may change from *closo* to *nido*. This is exemplified by the series $C_2B_4H_6$, $C_4B_2H_6$, and C_6H_6. These three molecules have 14, 16, and 18 core electrons respectively. In the case of benzene, the core electrons are distributed as six carbon-carbon bonds and six π-electrons. The structures of these three molecules change in an interesting manner as the number of core electrons is increased. The $C_2B_4H_6$ molecule is octahedral (*closo*), the $C_4B_2H_6$ molecule is pentagonal bipyramidal (*nido*), and the C_6H_6 molecule is planar. These structural results imply that electron addition to a *closo* polyhedron can lead to an opening of the polyhedron. This type of reactivity has been observed in a number of *closo* $C_2B_{n-2}H_n$ systems when electrons are added with sodium metal used as the source of electrons.

Bridging hydrogen atoms. In many *nido* carboranes and boranes, bridge hydrogen atoms exist. In these cases, an isoelectronic and structurally similar series of molecules frequently is observed. This series of molecules can be understood when it is recognized that replacement of a boron hydrogen unit and a bridge hydrogen unit with a carbon hydrogen unit does not change the total number of core electrons in the molecule. The basic structure consequently remains constant. This is well illustrated for the series B_6H_{10}, CB_5H_9, $C_2B_4H_8$, $C_3B_3H_7$, and $C_4B_2H_6$ in Fig. 2.

Preparation. The most thoroughly studied carborane derivatives contain one and two carbon atoms; consequently the preparative routes to these derivatives will be emphasized. The two carbon carboranes from $C_2B_3H_5$ to $C_2B_5H_7$, as well as $C_2B_{10}H_{12}$, are generally prepared by the reaction of acetylene or a substituted acetylene with a boron hydride. With the exception of $C_2B_{10}H_{12}$, the yields are frequently low, and a mixture of products is usually obtained. The preparation of $C_2B_{10}H_{12}$ by the reaction of decarborane (14) with acetylene in the presence of a Lewis base, shown in reaction (1), has permitted a

$$B_{10}H_{14} + HC{\equiv}CH \xrightarrow[(C_2H_5)_2S]{(C_3H_7)_2O} C_2B_{10}H_{12} \qquad (1)$$

greater examination of this carborane than of any other. By the use of substituted acetylenes, $C_2B_{10}H_{12}$ derivatives with a variety of functional groups on carbon, such as olefin, halomethyl, alkoxy, acetoxy, and dimethylaminomethyl, have been prepared. In all these compounds the two skeletal carbon atoms are next to one another in the icosahedral structure (1,2 isomer). Thermal rearrangements of these compounds are possible in certain cases to give the isomeric 1,7 and 1,12 families (Fig. 3). The changes are shown in reaction (2).

$$1,2\text{-}C_2B_{10}H_{12} \xrightarrow{400^\circ C} 1,7\text{-}C_2B_{10}H_{12} \xrightarrow{600^\circ C} 1,12\text{-}C_2B_{10}H_{12} \qquad (2)$$

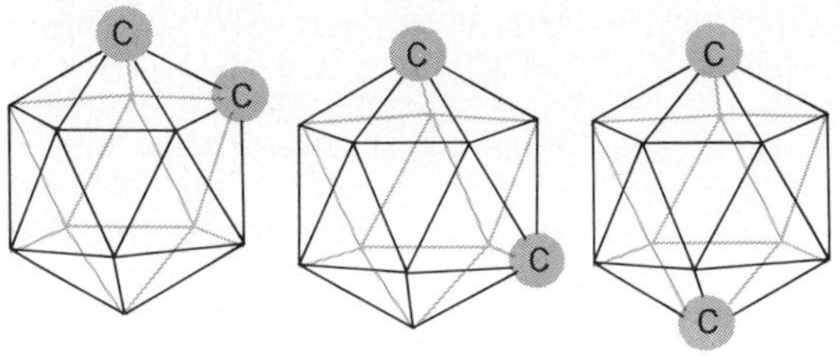
(a) (b) (c)

Fig. 3. Skeletal structures of the *(a)* 1,2 *(b)* 1,7 and *(c)* 1,12 isomers of $C_2B_{10}H_{12}$.

The $C_2B_6H_8$, $C_2B_7H_9$, and $C_2B_8H_{10}$ carboranes, as well as several others, can be prepared from $C_2B_{10}H_{12}$ as shown in reaction (3). Many of these

$$C_2B_{10}H_{12} \xrightarrow{CH_3O^-} C_2B_9H_{12}^- \xrightarrow{H^+} C_2B_9H_{13} \xrightarrow[-H_2]{100°C}$$

$$C_2B_9H_{11} \xrightarrow{H_2O_2} C_2B_7H_{13} \xrightarrow{200°C}$$

$$C_2B_6H_8 + C_2B_7H_9 + C_2B_6H_{10} \quad (3)$$

transformations were performed on derivatives of these carboranes which had methyl or phenyl groups attached to the skeletal carbon atoms. For clarity, these substituents are omitted in the equations.

The preparations of the monocarboranes $CB_9H_{10}^-$ and $CB_{11}H_{12}^-$ are quite unrelated to the $C_2B_{10}H_{12}$ preparation. Reactions (4) and (5)

$$B_{10}H_{14} + 2NaCN \xrightarrow{-HCN} Na_2B_{10}H_{13}CN \xrightarrow{H^+}$$

$$B_{10}C(NH_3)H_{12} \xrightarrow{(CH_3)_2SO_4}$$

$$B_{10}C(N(CH_3)_3)H_{12} \xrightarrow[2.\ H_2O]{1.\ Na} CB_{10}H_{13}^- \quad (4)$$

$$2CsB_{10}CH_{13} \xrightarrow{300°}$$

$$CsB_9CH_{10} + CsB_{11}CH_{12} + 2H_2 \quad (5)$$

describe the synthesis of the monocarboranes. The key to this synthesis is the interesting transformation of $Na_2B_{10}H_{13}CN$ to $B_{10}C(NH_3)H_{12}$ (Fig. 4). In this step a borane with an attached nitrile group is converted smoothly in high yield by reaction with acid to a carborane with an attached NH_3 group. The carbon of the nitrile group assumes a skeletal position, and hydrogen migration to the nitrogen completes the transformation.

Chemistry. The most studied carborane is the icosahedral derivative $C_2B_{10}H_{12}$, and comments

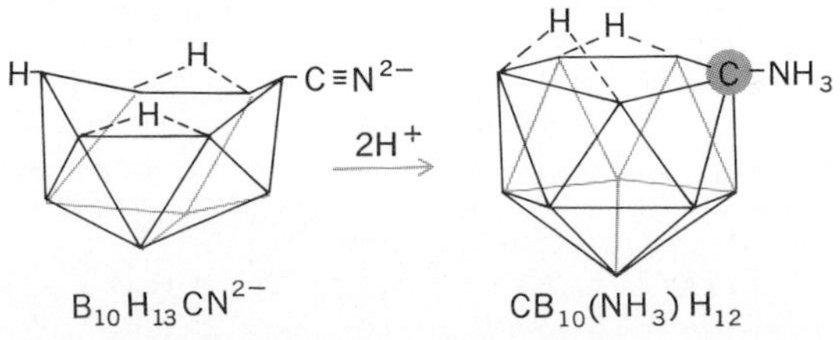

Fig. 4. Rearrangement accompanying the conversion of $B_{10}H_{13}CN^{2-}$ to $CB_{10}(NH_3)H_{12}$. There is one hydrogen atom on each boron in addition to those shown.

concerning the chemistry of carboranes will be limited to this species. The $C_2B_{10}H_{12}$ carboranes are very stable thermally; they are not decomposed at temperatures of 500°C, although the 1,2 isomer rearranges to the 1,7 isomer at 450°C, and to the 1,12 form at 600°C. The stabilities of derivatives are comparable except when substituents of lower stability are present. These carboranes are also very resistant to degradation by acid. For example, C-isopropenyl-carborane can be recrystallized from hot 100% sulfuric acid.

Besides C-substituted derivatives prepared directly by use of substituted acetylenes, C-substituted derivatives of $C_2B_{10}H_{12}$ have been prepared by use of a dilithio derivative, $Li_2C_2B_{10}H_{10}$, in which the two hydrogen atoms on carbon in $C_2B_{10}H_{12}$ have been replaced by lithium. This derivative is prepared by the reaction of $C_2B_{10}H_{12}$ with an alkyl-lithium reagent and undergoes virtually any reaction that is typical of organolithium compounds, thus giving rise to a large number of carbon-substituted derivatives of $C_2B_{10}H_{12}$. Some examples are shown in reactions (6) and (7). The reactions of $Li_2C_2B_{10}H_{10}$, par-

$$Li_2C_2B_{10}H_{10} \xrightarrow[2.\ H^+, H_2O]{1.\ CO_2} (HOOCC)_2B_{10}H_{10} \quad (6)$$

$$Li_2C_2B_{10}H_{10} \xrightarrow{I_2} (IC)_2B_{10}H_{10} \quad (7)$$

ticularly the 1,7 isomer, with difunctional organic and inorganic compounds have led to organo-inorganic polymers of very impressive thermal stabilities. A typical polymer-forming reaction is shown in reaction (8).

$$1{,}7\text{-}Li_2C_2B_{10}H_{10} + (C_6H_5)_2SiCl_2 \xrightarrow{-LiCl} \left[-CB_{10}H_{10}C-\overset{\displaystyle C_6H_5}{\underset{\displaystyle C_6H_5}{\overset{|}{\underset{|}{Si}}}}- \right]_x \quad (8)$$

In contrast to the large number of carbon-substituted derivatives of $C_2B_{10}H_{12}$, the number of boron-substituted derivatives is small. Bromination, chlorination, and fluorination of $C_2B_{10}H_{12}$ have been reported. Mono-, di-, tri-, and tetrabrominated derivatives have been reported, as have chlorinated derivatives containing from 2 to 12 chlorine atoms.

As noted in the preparation of the lower $C_2B_{n-2}H_n$ carboranes, bases such as the methoxide ion remove a B-H unit from both the 1,2 and 1,7 isomers to generate the isomeric 1,2- and 1,7-$C_2B_9H_{12}^-$ ions. The 1,2- and 1,7-$C_2B_9H_{12}^-$ ions react with sodium hydride to produce the 1,2- and 1,7-$C_2B_9H_{11}^{2-}$ ions, respectively.

Both the 1,2- and 1,7-$C_2B_{10}H_{12}$ derived isomers of $C_2B_9H_{11}^{2-}$ are 11-particle icosahedral fragments (Fig. 5) with 26 core electrons, the favored number for a stable icosahedral structure. They therefore react readily with various reagents to regenerate the icosahedral structure. The isomer of $C_2B_9H_{11}^{2-}$ derived from 1,2-$C_2B_{10}H_{12}$ has received the most study. It reacts readily with phenylboron dichloride to form an icosahedral $C_2B_{10}H_{12}$ derivative, as in reaction (9).

$$C_2B_9H_{11}^{2-} + C_6H_5BCl_2 \rightarrow C_6H_5C_2B_{10}H_{11} + 2Cl^- \quad (9)$$

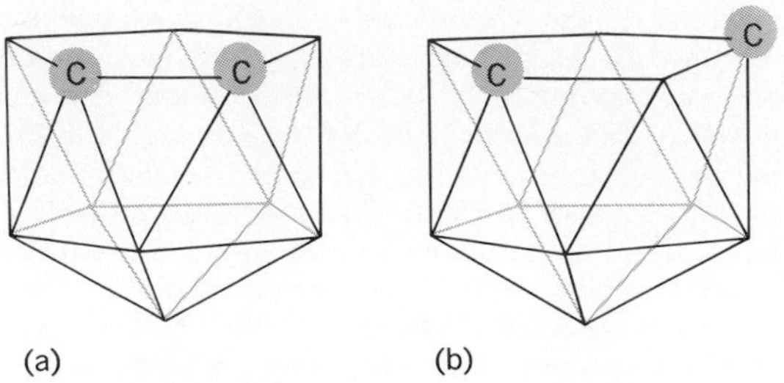

Fig. 5. Skeletal structures of $C_2B_9H_{11}^{2-}$ derived (*a*) from 1,2-$C_2B_{10}H_{12}$ and (*b*) from 1,7-$C_2B_{10}H_{12}$.

Perhaps the major factor which gives carborane and boron hydride chemistry in general its broad versatility is the ability to replace a polyhedral unit with a new unit which can be composed of virtually any other element in the periodic table. This chemistry was first noted for the $C_2B_9H_{11}^{2-}$ ion. A metal was inserted into two $C_2B_9H_{11}^{2-}$ ions to regenerate the icosahedral geometry. The metal also served as the fusion point for the two icosahedrons (Fig. 6). Complexes of the composition $(C_2B_9H_{11})_2M^{N-4}$ (*N* is the oxidation state of the metal, usually II or III) are known for iron, cobalt, nickel, copper, and chromium of the first transition-metal series and for palladium, platinum, and gold of the second and third transition-metal series. Even more numerous are complexes in which the metal replaces a unit in only one icosahedron. Various other ligands complete the coordination requirements of the metal in these cases. Examples of this type of system include $C_5H_5MC_2B_9H_{11}$ (M = Fe, Co, Ni, Cr), $C_2B_9H_{11}Rh(CO)_3$, $C_2B_9H_{11}M(CO)_3^{2-}$ (M = Cr, Mo, W), and $[(C_6H_5)_3P]_2HRhC_2B_9H_{11}$. In the last molecule the rhodium atom binds two triphenylphosphine molecules and a hydrogen atom to complete its coordination requirement. The rhodium complex is the first carborane complex to exhibit high reactivity as a homogeneous catalyst. Besides transition-metal atoms, atoms of the main group can also replace polyhedral units. Icosahedral molecules of this type include $MC_2B_9H_{11}$ (M = Sn, Pb), $C_2H_5AlC_2B_9H_{11}$, and $(CH_3)_3NBeC_2B_9H_{11}$. The ability of other elements to replace polyhedral units appears to be quite general.

It is important to recognize that the chemistry discussed above emphasizes the icosahedral $C_2B_{10}H_{12}$ system. Similar chemistry has been observed in other $C_2B_{n-2}H_n$ systems and in the $C_1B_{n-1}H_n^-$ system. More important, much of the chemistry of these systems and other carborane systems varies from that observed in the icosahedral derivatives. The chemistry of carboranes and related borane derivatives is by no means completely understood, nor have its limits been reached. New and unanticipated chemistry continues to be discovered, and this situation will most certainly obtain for the foreseeable future. *See* BORANE.

[PATRICK A. WEGNER]

Bibliography: R. N. Grimes, *Carboranes*, 1970; R. Köster and M. A. Grassberger, *Angew. Chem. Int. Ed.*, 6:1218, 1967; E. L. Muetterties (ed.), *Boron Hydride Chemistry*, 1975.

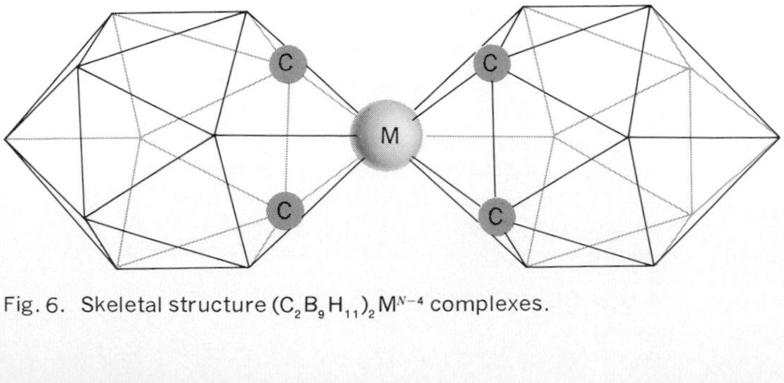

Fig. 6. Skeletal structure $(C_2B_9H_{11})_2M^{N-4}$ complexes.

Carboxylic acid

One of a large family of organic substances widely distributed in nature, and characterized by the presence of one or more carboxyl groups (—COOH). These groups typically yield protons in aqueous solution. *See* ACID AND BASE.

In the type formula, $R(CXY)_nCOOH$, and symbols R, X, and Y can be hydrogen, saturated, or unsaturated groups, carboxyl, alicyclic, or aromatic groups, halogens, or other substituents, and *n* may vary from zero (formic acid, HCOOH) to more than 100, provided that the normal carbon covalence of four is maintained.

Nomenclature and general properties. According to the International Union of Chemistry, an acid is named by its relation to the parent hydrocarbon, but in common practice trivial names are used often to indicate the origin of the substance. Thus HCOOH, formic acid (from Latin *formica*, ant), is properly methanoic acid, and $CH_3(CH_2)_7COOH$, pelargonic acid, from the leaves of *Pelargonium roseum*, is nonanoic acid. Substituents are located numerically on the carbon chain, counting the carboxylic carbon as 1, but an older, common method uses Greek letters, beginning with the carbon adjacent to the carboxyl. The following exemplifies both systems:

$$\overset{\epsilon}{\underset{6}{CH_2}}OH\overset{\delta}{\underset{5}{CH_2}}\overset{\gamma}{\underset{4}{CH}}Br\overset{\beta}{\underset{3}{CH}}Br\overset{\alpha}{\underset{2}{CH_2}}\underset{1}{COOH}$$

is 3,4-dibromo-6-hydroxyhexanoic acid, or β,γ-dibromo-ϵ-hydroxycaproic acid. *See* ORGANIC CHEMISTRY.

Physical and chemical properties of carboxylic acids are represented, grossly, by the resultant of the various chemical groupings present in the molecule. A short-chain aliphatic acid, wherein the carboxyl is dominant, is a pungent, corrosive, water-soluble liquid of abnormally high boiling point (because of molecular association), with specific gravity close to 1 (higher for formic and acetic acids). With increasing molecular weight, the hydrocarbon grouping over-balances the carboxyl; sharpness of odor diminishes, boiling and melting points rise, the specific gravity falls toward that of the parent hydrocarbon, and the water-solubility decreases. Thus the typical high-molecular-weight saturated acid is a bland, waxlike solid.

Ionization and acidity. Ionization occurs on exposure to a proton acceptor (base, in the Brønsted-Lowry sense); in water the opposing equations shown in reaction (1) describe the funda-

$$R(CXY)_nCOOH + H_2O \rightleftharpoons R(CXY)_nCOO^- + H_3O^+ \quad (1)$$

ental equilibrium encountered with any feebly ionized, or weak, acid. Hydration of the proton to form H_3O^+ is virtually complete; thus, the process of ionization is analogous to that of neutralization, the degree to which ionization proceeds varying directly with the basicity of the solvent (proton acceptor), and inversely with the O-H bond strength of the carboxylic hydroxyl group.

The above equilibrium leads to the dissociation constant K_a defined as reaction (2), wherein the

$$K_a = \frac{[H_3O^+][R(CXY)_nCOO^-]}{[R(CXY)_nCOOH]} \qquad (2)$$

square brackets signify concentration, measured in moles per liter of solution. With simple, saturated acids, wherein X and Y are hydrogen, the value of K_a is always small (acetic acid, 1.75×10^{-5}), which means that simple organic acids are only slightly ionized in aqueous solution. However, the degree of dissociation is sensitive to electronegative (electron-attracting) substituents, particularly when located on the α- or 2-carbon (for example, K_a for trichloroacetic acid, 1.3×10^{-1}). *See* IONIC EQUILIBRIUM.

Table 1 presents a list of 30 saturated, monocarboxylic acids, together with their important constants. Also indicated is the main source of each acid; the letter S stands for synthetic, meaning that the acid is manufactured commercially by one or more methods. The fatty acids comprise a group of carboxylic acids, both saturated and unsaturated, which may be obtained by hydrolysis of animal or vegetable fats and oils; these are indicated in the table by the letter F. Acids which are produced by fermentation processes are designated Ferm.

Structure and properties. The unionized carboxyl group has been shown via many classical studies to have the structure

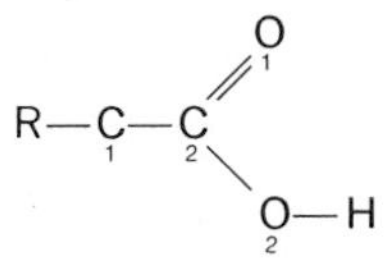

in which carbon 2 is doubly bonded to oxygen 1 and singly bonded to oxygen 2. Geometrical considerations lead to the conclusions that distance $\underset{2}{C}=\underset{1}{O}$ should be shorter than $\underset{2}{C}-\underset{2}{O}$ and the angle $\underset{1}{C}-\underset{2}{C}=\underset{1}{O}$ should be greater than angle $\underset{1}{C}-\underset{2}{C}-\underset{2}{O}$. This is confirmed by measurements on crystalline acids, although the fact that the single bond $\underset{2}{C}-\underset{2}{O}$ never reaches the normal length (about

Table 1. Saturated monocarboxylic acids*

Name	Formula	Main source†	Melting point, °C	Boiling point, °C
Formic	$HCOOH$	S.	8.4	100.5
Acetic	CH_3COOH	S. and Ferm.	16.7	118.1
Propionic	CH_3CH_2COOH	S.	−22.0	141.0
Butyric	$CH_3(CH_2)_2COOH$	F. S. and Ferm.	−4.7	164.1
Valeric	$CH_3(CH_2)_3COOH$	S. and *Valeriana officinalis*	−34.5	186.4
Isovaleric	$(CH_3)_2CHCH_2COOH$	F. and S.	−29.3	176.5
Caproic	$CH_3(CH_2)_4COOH$	F. and S. and by-product in butyric Ferm.	−1.5	205.8
Heptanoic	$CH_3(CH_2)_5COOH$	S. from heptaldehyde	−7.5	223.0
Caprylic	$CH_3(CH_2)_6COOH$	F. and S. Goat's butter	16.3	239.7
Pelargonic	$CH_3(CH_2)_7COOH$	S. Oxidation of oleic and *Pelargonium roseum*	12.5	255.6
Capric	$CH_3(CH_2)_8COOH$	F. Coconut; palm kernel oils	31.5	268.7
Undecylic	$CH_3(CH_2)_9COOH$	S. Reduction of undecenoic acid	29.3	228/160 mm
Lauric	$CH_3(CH_2)_{10}COOH$	F. Coconut and laurel oils	44.1	225/100 mm
Tridecylic	$CH_3(CH_2)_{11}COOH$	S.	41.0	236/100 mm
Myristic	$CH_3(CH_2)_{12}COOH$	F. Nutmeg oil	58.0	250/100 mm
Pentadecylic	$CH_3(CH_2)_{13}COOH$	S.	54.0	257/100 mm
Palmitic	$CH_3(CH_2)_{14}COOH$	F. Palm and olive oils	62.8	271.5/100 mm
Margaric	$CH_3(CH_2)_{15}COOH$	S.	59.9	227/100 mm
Stearic	$CH_3(CH_2)_{16}COOH$	F. Tallow and reduction of oleic acid	69.9	291/110 mm
Nonadecylic	$CH_3(CH_2)_{17}COOH$	S.	66.5	298/100 mm
Arachidic	$CH_3(CH_2)_{18}COOH$	F. Peanut oil	77.0	205/1 mm
Heneicosanoic	$CH_3(CH_2)_{19}COOH$	S.	75.2	
Behenic	$CH_3(CH_2)_{20}COOH$	F. Reduction of erucic acid	80.2	306/60 mm
Tricosanoic	$CH_3(CH_2)_{21}COOH$	S.	79.1	
Lignoceric	$CH_3(CH_2)_{22}COOH$	F. *Adenanthera pavonina* seed; beechwood tar; rotten oak	84.2	
Pentacosanoic	$CH_3(CH_2)_{23}COOH$	S.	83.0	
Cerotic	$CH_3(CH_2)_{24}COOH$	Beeswax; carnauba wax	87.7	
Heptacosanoic	$CH_3(CH_2)_{25}COOH$	S.	87.6	
Montanic	$CH_3(CH_2)_{26}COOH$	Montan wax from lignite	90.9	
Nonacosanoic	$CH_3(CH_2)_{27}COOH$	S.	90.3	
Melissic	$CH_3(CH_2)_{28}COOH$	Beeswax	93.6	

*Based on A. N. Lange, *Handbook of Chemistry*, 10th ed., McGraw-Hill, 1967.
†Abbreviations explained in text.

0.144 nm) means that some resonance is involved between the two forms

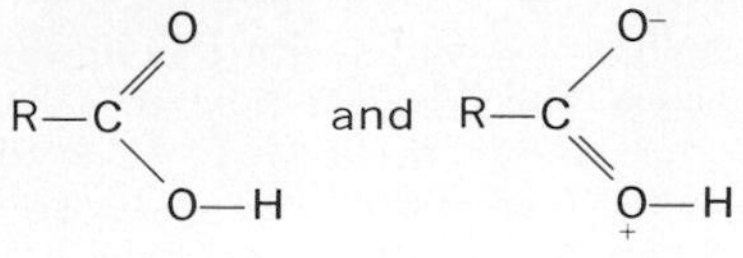

In the ionized state, however, the two resonating forms of the anion are electronically equivalent,

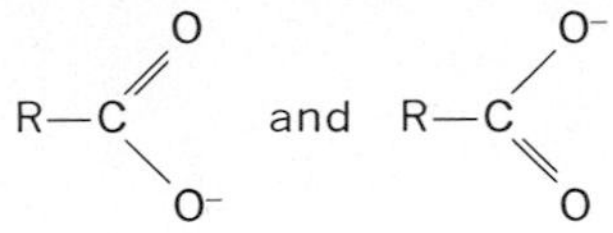

and the actual structure of the carboxylate ion is a stable hybrid of these. *See* RESONANCE.

In the free state (solid, liquid, or gas), the lower acids tend to exist as dimers, in which two molecules are associated through relatively weak hydrogen bonds (energy about 5 kcal/mole):

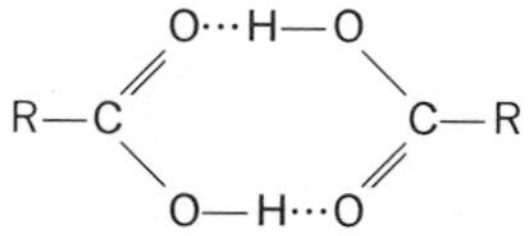

The hydrogen bonds (indicated by dotted valences) account for the abnormally high boiling points of the smaller acids. With longer-chain acids (from C_8) there is a tendency toward polymorphism, ascribable to hydrogen bonding of the following type:

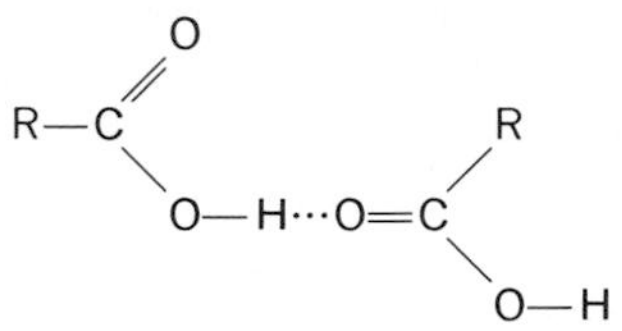

As determined by x-ray diffraction studies, the carboxyl and terminal methyl groups of acids containing an odd number of carbons are on the same side of the molecule; with acids containing an even number of carbons, these groups are on opposite sides, permitting closer packing in the crystal lattice and consequent increased Van der Waals forces. Thus, the melting point of any even-numbered acid lies above that of the preceding and following odd-numbered acid.

Structural variations. These are of two types, (1) variations not involving the carboxyl group, and (2) variations involving the carboxyl, leading to the acid derivatives.

In the first type are such distinctive classes as branched-chain acids, for example, pivalic acid, $(CH_3)_3CCOOH$; alicyclic acids, for example, cyclopropane carboxylic acid

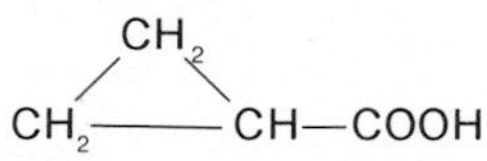

halogenated acids, fluoro, chloro, or bromo; rarely, iodo (*see* TRICHLOROACETIC ACID); hydroxy acids (*see* LACTIDE AND LACTONE; TARTARIC ACID); dicarboxylic and polycarboxylic acids (*see* OXALIC ACID); aromatic acids (*see* BENZOIC ACID; PHTHALIC ACID); β-keto acids; or amino acids (*see* LACTAM AND LACTIM).

An important class of acids contains unsaturated groups, for example, acrylic acid ($CH_2{=}CHCOOH$). Acids of this type are unstable and polymorize readily; hence they and their esters are much studied. *See* POLYACRYLATE RESIN.

Among the variations involving the carboxyl group are found many important classes of substances, derived from acids by suitable substitution within the carboxylic group. Thus, replacement of the hydroxyl group with alkoxyl gives esters (*see* ESTER); replacement of OH by halogen forms acid halides (*see* ACID HALIDE; ACYLATION); subsitution of OH by amino (NH_2) group gives amides (*see* AMIDE; UREA; UREID); intermolecular dehydration of two acids generates the anhydride structure

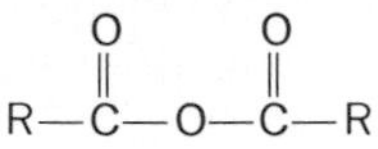

Table 2. Dicarboxylic, aromatic, and unsaturated acids*

Name	Formula	Melting point, °C	Boiling point, °C†
Dicarboxylic acids			
Oxalic	$HOOC{-}COOH$	189.5 (dec)	
Malonic	$CH_2(COOH)_2$	135 (dec)	
Succinic	$(CH_2)_2(COOH)_2$	185–187	235
Glutaric	$(CH_2)_3(COOH)_2$	98.0	302–304
Adipic	$(CH_2)_4(COOH)_2$	152	337.5
Pimelic	$(CH_2)_5(COOH)_2$	105–107 (subl)	272/100 mm
Suberic	$(CH_2)_6(COOH)_2$	142	279/100 mm
Azelaic	$(CH_2)_7(COOH)_2$	106.5	286/100 mm
Sebacic	$(CH_2)_8(COOH)_2$	134	295/100 mm
Aromatic acids			
Benzoic	C_6H_5COOH	121.7	249
o-Toluic	*o*-$CH_3C_6H_4COOH$	104	259/751 mm
m-Toluic	*m*-$CH_3C_6H_4COOH$	111	263
p-Toluic	*p*-$CH_3C_6H_4COOH$	180	275
o-Ethylbenzoic	*o*-$C_2H_5C_6H_4COOH$	68	259
m-Ethylbenzoic	*m*-$C_2H_5C_6H_4COOH$	47	
p-Ethylbenzoic	*p*-$C_2H_5C_6H_4COOH$	113	
o-Fluorobenzoic	*o*-FC_6H_4COOH	122	
m-Fluorobenzoic	*m*-FC_6H_4COOH	124	
p-Fluorobenzoic	*p*-FC_6H_4COOH	182–184	
o-Chlorobenzoic	*o*-ClC_6H_4COOH	140.2	
m-Chlorobenzoic	*m*-ClC_6H_4COOH	154.3	
p-Chlorobenzoic	*p*-ClC_6H_4COOH	239.7	Subl
o-Bromobenzoic	*o*-BrC_6H_4COOH	148–150	Subl
m-Bromobenzoic	*m*-BrC_6H_4COOH	154–155	
p-Bromobenzoic	*p*-BrC_6H_4COOH	251–253	
o-Iodobenzoic	*o*-IC_6H_4COOH	162	
m-Iodobenzoic	*m*-C_6H_4COOH	187	Subl
p-Iodobenzoic	*p*-IC_6H_4COOH	270	Subl
Salicylic	*o*-HOC_6H_4COOH	158.3	211/20 mm
m-Hydroxybenzoic	*m*-HOC_6H_4COOH	201	
p-Hydroxybenzoic	*p*-HOC_6H_4COOH	215	
Gallic	3,4,5-$(HO)_3C_6H_2COOH$	235 (dec)	
Anthranilic	*o*-$H_2NC_6H_4COOH$	144–145	Subl
m-Anthranilic	*m*-$H_2NC_6H_4COOH$	173–174	
p-Anthranilic	*p*-$H_2NC_6H_4COOH$	187–188	
α-Naphthoic	α-$C_{10}H_7COOH$	160–161	300
β-Naphthoic	β-$C_{10}H_7COOH$	184	>300
Unsaturated acids			
Acrylic	$CH_2{=}CH{-}COOH$	14	141
Crotonic (trans)	$CH_3CH{=}CHCOOH$	71.6	185
i-Crotonic (cis)		15.5	
Mesaconic (trans)	$HOOCC(CH_3){=}CHCOOH$	202	250 (dec)
Citraconic (cis)		91	
Tiglic (trans)	$CH_3CH{=}C(CH_3)COOH$	64	198
Angelic (cis)		45	185
Cinnamic (trans)	$C_6H_5CH{=}CHCOOH$	133	300
Allocinnamic (cis)		68	125/18 mm
Fumaric (trans)	$HOOCCH{=}CHCOOH$	287	
Maleic (cis)		130	
Elaidic (trans)	$CH_3(CH_2)_7CH$ ‖ $HOOC(CH_2)_7CH$	44	288/100 mm
Oleic (cis)		13	288/100 mm

*Based on N. A. Lange (ed.), *Handbook of Chemistry*, 10th ed., McGraw-Hill, 1967.

†dec = decomposes, subl = sublimes.

(*see* ACID ANHYDRIDE); the hydrolysis of long-chain esters by alkali hydroxides forms soaps.

Table 2 lists some important dicarboxylic acids, aromatic acids, and unsaturated acids, together with their important physical constants.

Reactions and uses. Acids are used in large quantities in the production of esters, acid halides, acid amides, and acid anhydrides. Decarboxylation of acids to form hydrocarbons containing one less carbon is accomplished by pyrolyzing the sodium or barium salt with soda lime. If the potassium salt is electrolyzed at a platinum anode (Kolbe electrolysis), the hydrocarbon RCH_2CH_2R is produced by the acid $RCH_2COO^-K^+$. Ethylenic and acetylenic acids containing unsaturation α,β to the carboxyl are easily decarboxylated by heat. Aromatic acids, such as benzoic and toluic, are frequently used as sources of hydrocarbons; the decarboxylation is effected by treating the acid in boiling quinoline with a copper powder catalyst.

Many acids obtained by acid hydrolysis of fats or waxes are reduced to the corresponding alcohols, for example, lauric acid to lauryl alcohol. The reaction is carried out in the laboratory by means of lithium aluminum hydride; the industrial procedure utilizes hydrogen at elevated temperatures, over a mixed catalyst of the oxides of copper and chromium.

Acids find wide use in the manufacture of soaps and detergents, in thickening lubricating greases (stearate soaps), in modifying rigidity in plastics, in compounding buffing bricks and abrasives, and in the manufacture of crayons, dictaphone cylinders, and phonograph records. The solvent action of acids finds use in manufacture of carbon paper, inks, and in the compounding of synthetic and natural rubber. Because of the stability of saturated fatty acids toward oxidation, these are often used as solvents for carrying out oxidation reactions upon sensitive compounds.

[EVANS B. REID]

Catalysis

The phenomenon in which a relatively small amount of foreign material, called a catalyst, augments the rate of a chemical reaction without itself being consumed. A catalyst is material, and not light or heat. It augments a rate. The term negative catalyst is obsolete. *See* ANTIOXIDANT; INHIBITOR.

If the reaction $A + B \rightarrow D$ occurs very slowly but is catalyzed by some catalyst, Cat, the addition of Cat must open new channels for the reaction. In a very simple case (1), the two propagation processes, which are fast compared to the uncatalyzed reaction, $A + B \rightarrow D$, provide the new channel for reaction. The catalyst reacts in the first step, but is regenerated in the second step to commence a new cycle. A catalytic reaction is thus a kind of chain reaction.

$$\left.\begin{array}{r} A + \text{Cat} \rightarrow \text{ACat} \\ \text{ACat} + B \rightarrow D + \text{Cat} \end{array}\right\} \text{Chain propagation} \qquad (1)$$
$$A + B \rightarrow D \qquad \text{Overall reaction}$$

If a reaction is in chemical equilibrium under some fixed conditions, the addition of a catalyst cannot change the position of equilibrium without violating the second law of thermodynamics. Therefore, if a catalyst augments the rate of $A + B \rightarrow D$, it must also augment the reverse rate, $D \rightarrow A + B$.

Categories. Catalysis is conventionally divided into three categories: homogeneous, heterogeneous, and enzyme. Heterogeneous catalysis plays a dominant role in chemical processes in the petroleum, petrochemical, and chemical industries. Homogeneous catalysis is important in the petrochemical and chemical industries. Enzyme catalysis plays a key role in all metabolic processes and in some industries, such as the fermentation industry.

Homogeneous. In homogeneous catalysis, reactants, products, and catalyst are all present molecularly in one phase, usually liquid. Examples are the hydrogenation of 1-hexene in a hydrocarbon solvent catalyzed by dissolved $[(C_6H_5)_3P]_3RhH$ [reaction (2)] and the hydrolysis of an ester catalyzed by acid [reaction (3)].

$$CH_2{=}CHCH_2CH_2CH_2CH_3 + H_2 \rightarrow CH_3CH_2CH_2CH_2CH_2CH_3 \qquad (2)$$

$$CH_3COOC_2H_5 + H_2O \xrightarrow{H^+} CH_3COOH + HOC_2H_5 \qquad (3)$$

Heterogeneous. In heterogeneous catalysis, the catalyst is in a separate phase. Usually the reactants and products are in gaseous or liquid phases and the catalyst is a solid. The catalytic reaction occurs on the surface of the solid. Examples are the dehydration and the dehydrogenation of isopropyl alcohol [reactions (4) and (5)]. Reaction (4)

$$CH_3CHOHCH_3 \rightarrow CH_3CH{=}CH_2 + H_2O \qquad \text{Dehydration} \qquad (4)$$

$$CH_3CHOHCH_3 \rightarrow CH_3COCH_3 + H_2 \qquad \text{Dehydrogenation} \qquad (5)$$

can be effected by passing the vapors of the alcohol over alumina at about 300°C, and reaction (5) over copper at 200°C.

Enzyme. Transformations of matter in living organisms occur by an elaborate sequence of reactions, most of which are catalyzed by biocatalysts called enzymes. Enzymes are proteins and therefore of colloidal dimensions. Although studies of interrelations between homogeneous catalysis and heterogeneous catalysis have been developing, enzyme catalysis remains a rather separate area both in the nature of the catalyst and in the type of reactions catalyzed.

Selectivity. In most cases, a given set of reactants could react in two or more ways, as exemplified by reactions (4) and (5). The degree to which just one of the possible reactions is favored over the other is called selectivity. Selectivity is a key property of a catalyst in any practical application of the catalyst.

Since most reactions which proceed readily upon mere mixing of the reactants are likely already to have been discovered, catalysis will be critical in the future of preparative chemistry. *See* HETEROGENEOUS CATALYSIS; HOMOGENEOUS CATALYSIS.

[ROBERT L. BURWELL, JR.]

Catalytic reforming

A process used for upgrading gasoline by improving its antiknock characteristics (increasing the octane number). It is also widely used for the production of aromatic hydrocarbons for the petrochemical industry.

Process reactions. The process utilizes a supported platinum catalyst and involves a number of different reactions. These reactions convert, or reform, the low-octane-number feed components, such as the paraffins, to components with increased octane number. Equations (1)–(5) show typical examples.

Reaction (1) illustrates isomerization (chain branching) of paraffins. Hydrocracking, shown in reaction (2), is usually held to a minimum because it entails formation of some butane and propane, which boil below gasoline and hence represent a yield loss. The most important reactions in catalytic reforming are those leading to the formation of aromatic hydrocarbons because these are high-octane components as well as valuable petrochemical intermediates. Aromatics are formed by dehydrogenation of six-membered ring cycloparaffins, reaction (4); rearrangement and dehydrogenation of five-membered ring cycloparaffins, reaction (3); and cyclization (dehydrocyclization) of paraffins, reaction (5).

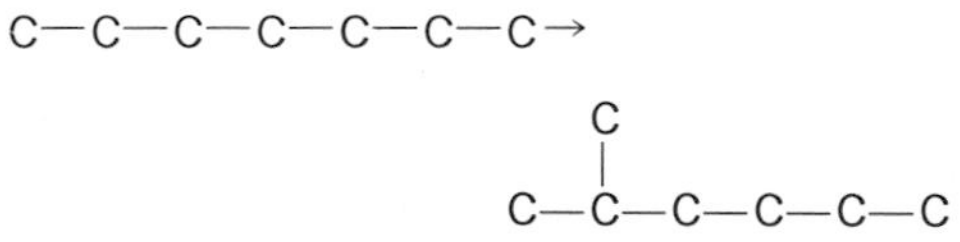
(1)

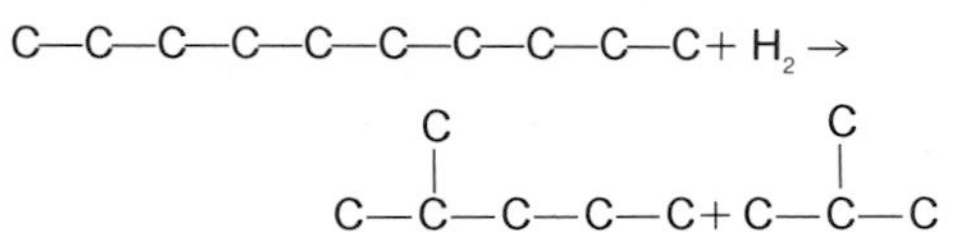
(2)

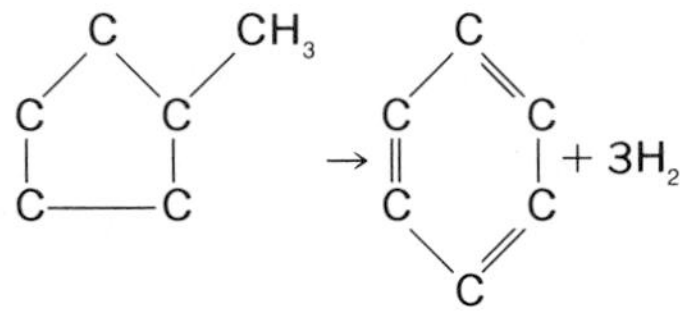
(3)

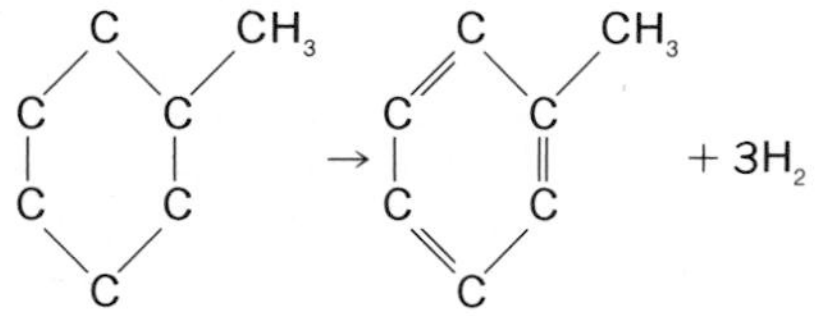
(4)

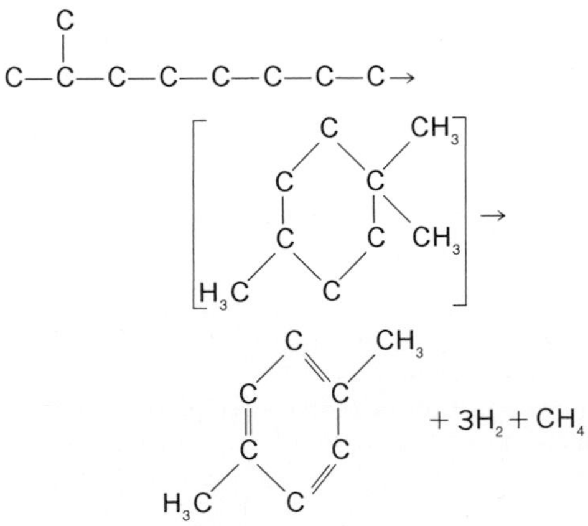
(5)

The product from catalytic reforming consists essentially of aromatics and branched paraffins. Higher-boiling fractions are richer in aromatics, whereas paraffins, particularly in severe reforming, are concentrated in lower-boiling fractions.

The catalysts commonly employed consist essentially of a platinum dehydrogenation component on an acidic support. The dehydrogenation activity and acid activity of such catalysts are carefully balanced to achieve the highest possible yield at a given product octane number or aromatics concentration. A number of bimetallic catalysts have been introduced in which one or more additional elements, such as rhenium, are used to modify and stabilize the platinum component. This results in greatly improved catalyst stability and, in some cases, improvements in activity and liquid-product yield. Such catalysts have found wide acceptance in the refining industry, and the use of platinum-containing catalysts for the reforming of gasoline has grown rapidly since introduction in 1949.

Catalytic reforming units usually have a fractionation section, where the fresh feed is distilled to remove overhead pentane and lower-boiling hydrocarbons and to reject material boiling above the gasoline range (> 400°F or 204°C). Most modern installations also include a pretreatment section in which sulfur compounds and other impurities are removed by reaction with hydrogen over a hydrotreating catalyst, such as cobalt-molybdena-alumina. This pretreating step also removes from the feed arsenic compounds which otherwise poison the platinum catalyst.

The pretreated C_6 to 400°F (204°C) fraction is mixed with hydrogen and preheated to the desired temperature, and the reforming is carried out in three or four reactors in series. Intermediate reheating (between stages) is necessary since the overall reaction is endothermic. The reactions are carried out at temperatures of 800–1050°F (427–566°C) and pressures of 100–500 psig (700–3400 kPa).

The effluent from the reactors goes through heat exchangers to a separator where the liquid product is separated from the hydrogen and other light gases. It is then sent to a stabilizer to produce a finished gasoline as the bottoms product while removing, as overhead, the propane and butane produced during the reaction. Part of the separator gas (consisting mostly of hydrogen) is withdrawn from the system, and the remainder is recycled.

This recycle of a hydrogen-rich gas is an important feature of catalytic reforming because it acts to suppress those side reactions which tend to form carbonaceous deposits on the catalyst. The usual hydrogen recycle (3–6 moles per mole of hydrocarbon) is sufficient to maintain an active catalyst surface for 6–24 months in normal operation. When a catalyst becomes fouled, the carbonaceous deposits are burned off and the catalyst is then returned to service.

Modern requirements for higher-octane gasoline and higher aromatics yields demand higher-severity reforming, which leads to increased fouling rates despite the use of modern bimetallic catalysts. Many reformers are therefore designed with swing reactors which permit regeneration of the catalyst while the unit remains on-stream. In the most recent version, the catalyst is withdrawn continuously from an operating unit, treated in a separate but integral regenerator, and returned to the reaction section. A large percent of the more recently installed units utilize this continuous catalyst regeneration technology.

Aromatics production. Although catalytic reforming is primarily used to upgrade gasoline, the products have also become an exceedingly important source of aromatic hydrocarbons—in fact, the single most important source. These aromatics are used as intermediates in the manufacture of plastics, explosives, detergents, phenols, and other chemicals. The processing schemes are the same, except that somewhat lower pressures are used. The charge stock is a much narrower boiling cut, in the range 200–300°F (93–149°C). The main products include benzene, toluene, and xylenes. The aromatics are concentrated by extraction of the product with a solvent, or by absorption techniques.

Liquid petroleum gas (LPG). LPG consists of propane and butanes and is usually derived from natural gas. In locations where there is no natural gas and LPG is more important than gasoline, or in refineries that require additional isobutane for alkylation, naphtha can be converted to LPG (primarily isobutane) by catalytic reforming. The catalyst is modified in the direction of higher acidity, thus promoting the hydrocracking reactions. Under suitable conditions it is possible to convert 40% of the naphtha to LPG, the by-product being high-octane gasoline. *See* AROMATIC; AROMATIZATION; DEHYDROGENATION; HYDROCARBON; ISOMERIZATION;

[ERNEST L. POLLITZER; VLADIMIR HAENSEL]

Bibliography: G. D. Hobson (ed.), *Modern Petroleum Technology*, 4th ed., 1973; Oil and Gas Journal, *Handbook on Catalytic Reforming*, 1966.

Catechol

A naturally occurring dihydric phenol, also known as pyrocatechol, pyrocatechuic acid, and 1,2-dihydroxybenzene, with the structural formula shown.

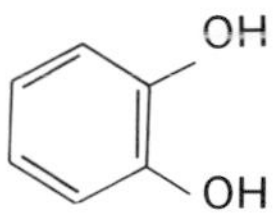

Catechol is a colorless crystalline substance (melting point 105°C) which dissolves readily in water. The aqueous solutions tend to darken on exposure to air. Catechol is only a slightly stronger acid than phenol itself, and it also resembles phenol in being a severe irritant to skin and in possessing systemic toxicity. It forms complexes with many metal ions, including arsenic, tin, and lead.

Although catechol and its derivatives are widely distributed among plants, the substance is produced synthetically for use in the manufacture of antioxidants (stabilizers) for rubber and plastics, and of a feed additive for poultry. Catechol also is used in medicinal and photographic preparations.

The synthetic production of catechol may proceed by alkaline hydrolysis of *o*-dichlorobenzene or *o*-dichlorophenol and also by treatment of phenol-2,4-disulfonic acid with alkali, followed by hydrolysis. Oxidation of salicylaldehyde by hydrogen peroxide (Dakin oxidation) constitutes a useful laboratory preparation of catechol. *See* HYDROQUINONE; PHENOL; RESORCINOL.

[MARTIN STILES]

Cerium

A chemical element, Ce, atomic number 58, atomic weight 140.12. It is the most abundant metallic element of the rare-earth group in the periodic table. The naturally occurring element is made up

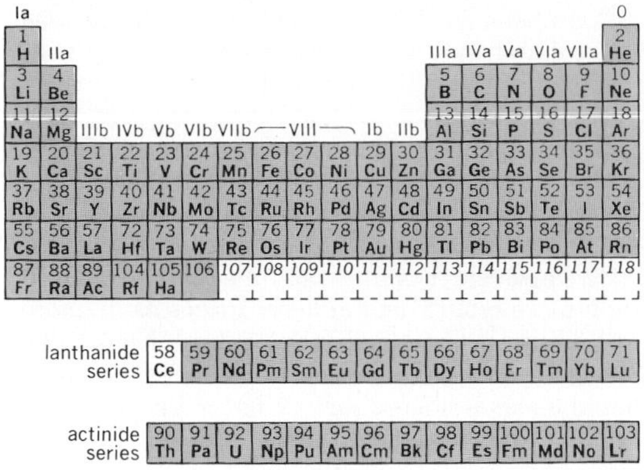

of the isotopes Ce^{136} 0.193%, Ce^{138} 0.250%, Ce^{140} 88.48%, and Ce^{142} 11.07%. A radioactive α-emitter, Ce^{142} has a half-life of 5×10^{15} years. The oxide of the element was discovered in 1803 by M. H. Klaproth, and independently by J. J. Berzelius and W. Hisinger. Cerium occurs mixed with other rare earths in many minerals, particularly monazite and blastnasite, and is found among the products of the fission of uranium, thorium, and plutonium.

Although the common valence of cerium is 3, it also forms a series of quadrivalent compounds and is the only rare earth which occurs as a quadrivalent ion in aqueous solution. Although it can be separated from the other rare earths in high purity by ion-exchange methods, it is usually separated chemically by taking advantage of its quadrivalent state. Ceric oxide, CeO_2, is the oxide usually obtained when cerium salts of volatile acids are heated. Ceric oxide is an almost white powder which is insoluble in most acids, although it can be dissolved in sulfuric acid or other acids when a reducing agent is present. The trivalent salts are white and the quadrivalent salts are usually yellow. The metal is an iron-gray color and it oxidizes readily in air, forming a gray crust of oxide. In the pure state it is not very pyrophoric, but when slightly oxidized or alloyed with iron, it becomes extremely pyrophoric. Mischmetall, an alloy of cerium, is used in the manufacture of lighter flints. Cerium has the interesting property that, at very low temperatures or when subjected to high pressures, it exhibits a fourth allotropic form, face-centered cubic, which is diamagnetic and 18% denser than the common form.

Crude mixtures of cerium with other rare earths have found extensive use in industry. It is used as a "getter" in the metal industry and as an opacifier and polisher in the glass industry. It is one of the main ingredients in Welsbach gas mantles and cored carbon arcs. Cerium metal, in common with the other rare earths, is almost immiscible in molten uranium, and can, therefore, be used as a liquid-liquid extraction agent to remove fission products from spent uranium fuel. The ceric salts are used in analytical chemistry as oxidation agents.

[FRANK H. SPEDDING]

Cesium

A chemical element, Cs, with an atomic number of 55 and an atomic weight of 132.905, the heaviest of the alkali metals in group Ia of the periodic table (except for francium, the radioactive member of the alkali metal family). Cesium was discovered by R. Bunsen in 1860 by spectroanalysis, and was first isolated by electrolysis in 1881. It is a soft, light, very low-melting metal. It is the most reactive of the alkali metals and indeed is the most electropositive and the most reactive of all the elements. Little was known of its properties because of the relative unavailability and high cost of cesium compounds in the chemical market. In 1958, however, cesium salts became available much more readily and at lower prices as by-products of lithium chemicals manufacture. Knowledge of the properties and the reactions of cesium metal should develop accordingly. What is known of the chemical behavior of this most-reactive metal is intriguing, and it seems to offer unusual research possibilities.

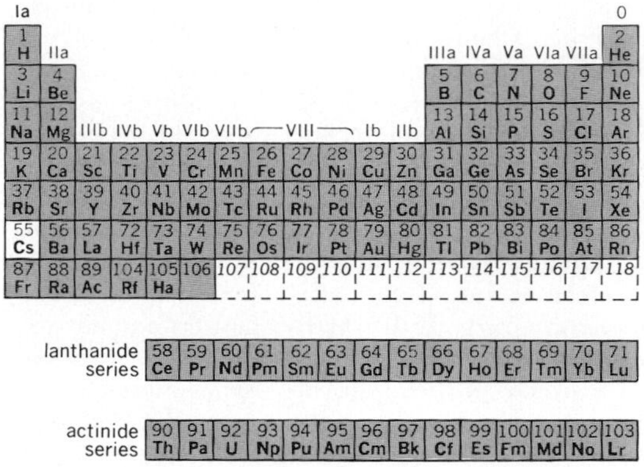

Chemical properties. Cesium reacts vigorously with oxygen to form a mixture of oxides in the same manner as do rubidium and potassium. In moist air, the heat of oxidation may be sufficient to melt and ignite the metal.

Cesium does not appear to react with nitrogen to form a nitride, but does react with hydrogen at high temperatures to form a fairly stable hydride.

Cesium reacts violently with water and even with ice at temperatures as low as −116°C. Little is known about the reaction of cesium with carbon, but acetylides can be formed from cesium and acetylene.

The reaction with halogens is vigorous, and cesium is distinguished among the alkali metals for its ability, presumably due to its large ionic radius, to form stable polyhalides, such as CsI_3.

Cesium reacts with ammonia to form cesium amide and with carbon monoxide in the cold at low pressures to give a crystalline compound of indeterminate composition.

Cesium, in general, undergoes some of the same type of reactions with organic compounds as do the other alkali metals, but it is much more reactive. Thus, it adds directly to ethylene to form a brown solid, $C_2H_4Cs_2$, whereas the other alkali metals will add only to dienes or to olefins activated by aromatic groups. For a discussion of handling techniques *see* SODIUM.

Physical properties. The physical properties of cesium metal are summarized in the table.

Physical properties of cesium metal

Property	Temp., °C	Metric units
Density	20	1.9 g/cm³
Melting point	28.5	
Boiling point	705	
Heat of fusion	28.5	3.8 cal/g
Heat of vaporization	705	146 cal/g
Viscosity	100	4.75 millipoises
Vapor pressure	278	1 mm
	635	400 mm
Thermal conductivity	28.5	0.044 cal/(sec)(cm²)(°C)
Heat capacity	28.5	0.06 cal/(g)(°C)
Electrical resistivity	30	36.6 microhm-cm

Occurrence. Cesium is not very abundant in the Earth's crust, there being only 7 parts per million (ppm) present. However, this concentration still places cesium above beryllium, arsenic, uranium, and boron in abundance. There is only about 0.002 ppm of cesium in solution in sea water. Detectable amounts are found in plant and animal organisms, mineral waters, and soils.

Like lithium and rubidium, cesium is found as a constituent of complex minerals and not in relatively pure halide form as are sodium and potassium. Indeed, lithium, rubidium, and cesium frequently occur together in lepidolite ores, such as those from Rhodesia. Cesium is unlike lithium and rubidium, however, in that a cesium-rich mineral, pollucite, essentially $2Cs_2O \cdot 2Al_2O_3 \cdot 9SiO_2 \cdot H_2O$, does occur. It is found on the island of Elba, in South-West Africa, and in Manitoba. In the United States there are deposits in Maine and South Dakota.

Metallurgical extraction. Cesium metal is not produced on a commercial scale. In the limestone process for the conversion of lepidolite ore to lithium chemicals, however, a mixed alkali carbonate liquor is obtained by carbonation in a submerged combustion evaporator after the separation of the bulk of the lithium values as the hydroxide. Filtration after carbonation removes lithium carbonate and gives a filtrate containing the carbonates of potassium, rubidium, and cesium. The separation of the mixed alkali salts is described in the article on rubidium. *See* RUBIDIUM.

Cesium metal is generally made by thermochemical processes. The carbonate can be reduced by metallic magnesium, or the chloride can be reduced by calcium carbide. Metallic cesium volatilizes from the reaction mixture and is collected by cooling the vapor.

Uses. Cesium metal is used in photoelectric cells, spectrographic instruments, scintillation counters, radio tubes, military infrared signaling lamps, and various optical and detecting devices. Cesium compounds are used in glass and ceramic production, as absorbents in carbon dioxide purification plants, as components of getters in radio tubes, and in microchemistry. Cesium salts have been used medicinally as antishock agents after administration of arsenic drugs. The isotope cesium-137 is supplanting cobalt-60 in the treatment of cancer.

Availability. Since the advent of cesium production as a lithium by-product in 1958, cesium sales and consumption have steadily increased. Requirements for cesium as metal and in compounds range from 1 to 5 tons annually. Commercialization of current research may spur increased production and bring lower unit costs.

Principal compounds. Cesium chloride is the most important cesium compound. The metal is made from the chloride, and the chloride is used as a constituent of getter mixtures for vacuum tubes. The fluoride, carbonate, and sulfate are also available and used in small commercial quantities.

Analytical methods. Cesium can be identified qualitatively by its blue flame. The fact that cesium forms an extremely insoluble alum may be used in the quantitative determination of cesium. *See* ALKALI METALS.

[MARSHALL SITTIG]

Bibliography: W. A. Hart et al., *The Chemistry of Lithium, Sodium, Potassium, Rubidium, Cesium, and Francium*, 1975.

Cetyl alcohol

A straight-chain primary aliphatic alcohol of 16 carbon atoms found free in feces, with the formula $CH_3(CH_2)_{14}CH_2OH$. Its ester with palmitic acid comprises 90% of spermaceti, the wax from marine animals. It is also present in esterified form in wool fat.

[HERBERT E. CARTER; ROY H. GIGG]

Chain reaction

A chemical reaction in which many molecules undergo chemical reaction after one molecule becomes activated. In ordinary chemical reactions, every molecule that reacts must first become activated by collision with other rapidly moving molecules. The number of these violent collisions per second is so small that the reaction is slow. After a chain reaction is once started, it is not necessary to wait for more collisions with activated molecules to accelerate the reaction because the reaction now proceeds spontaneously.

Photochemical reactions. A typical chain reaction is the photochemical reaction between hydrogen and chlorine as described by reactions (1).

$$\begin{aligned} Cl_2 + \text{light} &\rightarrow Cl + Cl \\ Cl + H_2 &\rightarrow HCl + H \\ H + Cl_2 &\rightarrow HCl + Cl \\ Cl + H_2 &\rightarrow HCl + H \end{aligned} \qquad (1)$$

The light absorbed by a chlorine molecule dissociates the molecule into chlorine atoms; these in turn react rapidly with hydrogen molecules to give hydrogen chloride and hydrogen atoms. The hydrogen atoms react with chlorine molecules to give hydrogen chloride and chlorine atoms. The chlorine atoms react further with hydrogen and continue the chain until some other reaction uses up the free atoms of chlorine or hydrogen. The chain-stopping reaction may be the reaction between two chlorine atoms to give chlorine molecules, or between two hydrogen atoms to give hydrogen molecules. Again the atoms may collide with the walls of the containing vessel, or they may react with some impurity which is present in the vessel only as a trace.

The length of the chain, that is, the number of molecules reacting per molecule activated, is determined by the relative rates of the competing reactions, namely, the chain-propagating reaction and the chain-stopping reactions. In the chain reaction just described, 10^6 molecules of hydrogen chloride may be formed by the photodissociation of 1 chlorine molecule.

In photochemical chain reactions, the length of the chain can be determined by measuring the number of photons of light absorbed, that is, the number of molecules activated, and dividing by the number of molecules which react chemically.

Thermal reactions. In thermal reactions, the length of the chain may sometimes be estimated from a knowledge of the intermediate steps and the kinetics involved. The presence of a chain reaction can often be proved by adding a trace of an inhibitor, such as nitric oxide. If the reaction is slowed down greatly by a very small amount of a substance which reacts with the chain-propagating units, the reaction involves a chain. While the inhibitor is being consumed in this way, the reaction is slow. After an induction period, the inhibitor is consumed and the rapid chain reaction then takes place.

Chain reactions are erratic and are reproduced with difficulty in different laboratories because they depend so much on the presence and concentration of accidental impurities which act as inhibitors.

In many chemical reactions, particularly organic reactions at elevated temperatures, the chains are carried by free radicals which are very reactive fragments of molecules that have unshared electrons, such as $\bullet CH_3$, $\bullet C_2H_5$, $\bullet H$, and $\bullet OH$. The thermal decomposition of propane is a typical free-radical chain which follows reactions (2). One

$$\begin{aligned} C_3H_8 &\rightarrow \bullet CH_3 + \bullet C_2H_5 \\ \bullet CH_3 + C_3H_8 &\rightarrow CH_4 + \bullet C_3H_7 \\ \bullet C_3H_7 &\rightarrow \bullet CH_3 + C_2H_4 \\ \bullet CH_3 + C_3H_8 &\rightarrow CH_4 + \bullet C_3H_7 \end{aligned} \qquad (2)$$

molecule of propane is decomposed into free radicals, $\bullet CH_3$ and $\bullet C_2H_5$, which then react with more propane to give the product methane and a free radical, $\bullet C_3H_7$, which decomposes into $\bullet CH_3$ and the product ethylene. The $\bullet CH_3$ reacts with more propane and continues the chain. The chain is terminated by collision of the free radicals with the wall or with each other, in reactions such as (3). Thus it is possible to obtain products of

$$\bullet CH_3 + \bullet C_3H_7 \rightarrow C_4H_{10} \qquad (3)$$

higher molecular weight as well as products of lower molecular weight. The finding of these higher-molecular-weight products supports the theory of free-radical formation and chain reactions.

Certain oxidations in the gas phase are known to be chain reactions. The carbon knock which occurs at times in internal combustion engines is caused by a too-rapid combustion rate caused by chain reactions. This chain reaction is reduced by adding tetraethyllead which acts as an inhibitor.

The polymerization of styrene to give polystyrene and the polymerization of other organic materials to give industrial plastics involve chain reactions. The spoilage of foods, the precipitation of insoluble gums in gasoline, and the deterioration of certain plastics in sunlight involve chain reactions, which can be minimized with inhibitors. *See* ANTIOXIDANT; CATALYSIS; CHEMICAL DYNAMICS; INHIBITOR; PHOTOCHEMISTRY.

[FARRINGTON DANIELS]

Bibliography: K. J. Laidler, *Chemical Kinetics*, 2d ed., 1965; K. J. Laidler, *Reaction Kinetics*, 2 vols., 1963; E. W. R. Steacie, *Atomic and Free Radical Reactions*, 1954.

Chelation

A chemical reaction or process involving chelate ring formation and characterized by multiple coordinate bonding between two or more of the electron-pair-donor groups of a multidentate ligand and an electron-pair-acceptor metal ion. The multidentate ligand is usually called a chelating agent, and the product is known as a metal chelate compound or metal chelate complex. Metal chelate chemistry is a subdivision of coordination chemistry and is characterized by the special properties resulting from the utilization of ligands possessing bridged donor groups, two or more of which coordinate simultaneously to a metal ion. *See* COORDINATION CHEMISTRY.

Ethylenediamine, $H_2NCH_2CH_2NH_2$, is a good example of a bidentate chelating agent consisting of two amino donor groups joined to each other by a two-carbon bridge. The coordination of both nitrogen atoms to the same metal ion would result in the formation of a five-membered chelate ring. If a chelating agent has three groups capable of attaching to a metal ion, as in diethylenetriamine, $NH_2CH_2CH_2NHCH_2CH_2NH_2$, it is tridentate (terdentate); if four, tetradentate (quadridentate); five,

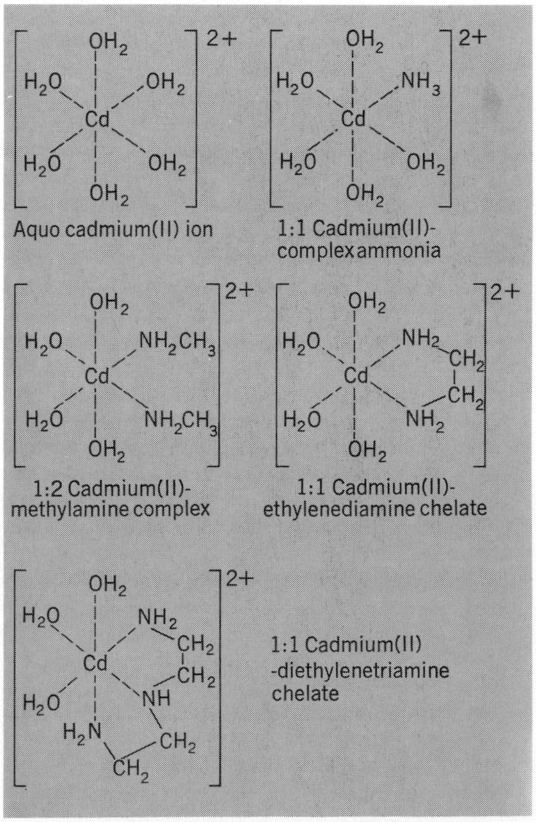

Fig. 1. Formulas of hydrated cadmium(II) ion, two cadmium(II) complexes, and two cadmium(II) chelates.

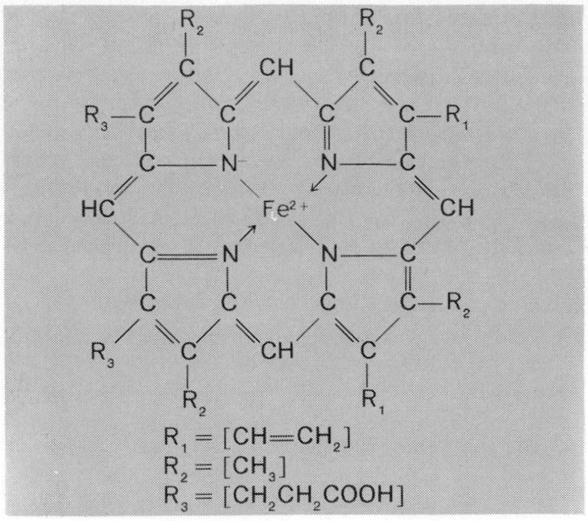

Fig. 2. Heme, a tetradentate chelate of iron(II).

pentadentate (quinquidentate); six, hexadentate (sexadentate); and so on. In general, these chelating agents may be designated as multidentate or polydentate ligands. All chelating agents must be at least bidentate; tridentate agents give fused rings. An example of a nonchelated complex and two chelated complexes of the cadmium(II) ion are illustrated in Fig. 1. In aqueous solution, metal ions are completely solvated, or hydrated, to give an aquo complex, as indicated for the Cd(II) ion in Fig. 1. Thus formation of a complex of a unidentate ligand involves the replacement of a water molecule by the unidentate donor group. Two water molecules are replaced by a single molecule of a bidentate ligand; three water molecules are replaced by a molecule of a terdentate ligand, and so on.

Many of the functional groups of both synthetic and naturally occurring organic compounds can form coordinate bonds to metal ions, producing metal-organic complexes or chelates, many of which are biologically active. Thus chelate compounds are frequently found in an interdisciplinary field of science called bioinorganic chemistry. The biological significance of chelates can be readily recognized if one notes that a large number of biologically important compounds are either metal chelates or chelating agents. This list includes the alpha amino acids, peptides, proteins, enzymes, porphyrins (such as hemoglobin), corrins (such as vitamin B_{12}), catechols, hydroxypolycarboxylic acids (such as citric acid), ascorbic acid (vitamin C), polyphosphates, nucleosides and other genetic compounds, pyridoxal phosphate (vitamin B_6), and sugars. The ubiquitous green plant pigment, chlorophyll, is a magnesium chelate of a tetradentate ligand formed from a modified porphin compound, and similarly the oxygen transport heme of red blood cells contains an Fe(II) chelate of the type illustrated in Fig. 2. *See* ORGANOMETALLIC COMPOUND.

A rapidly growing body of experimental evidence indicates that chelation may be important in the pharmacological action of many drugs. The use of chelating agents to remove certain toxic cations such as lead and plutonium from the body is now widely recognized in medical practice. Intensive

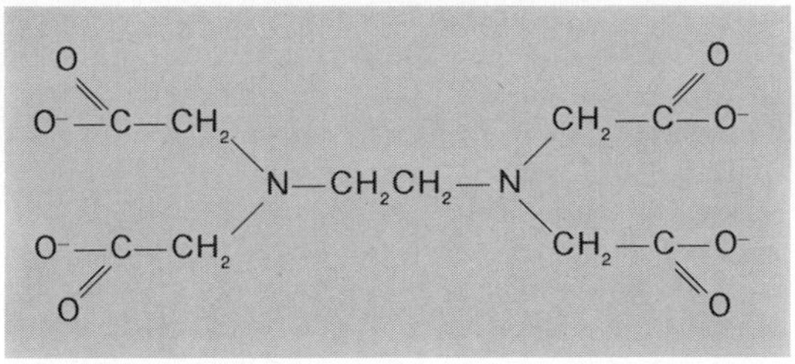

Fig. 3. Structural formula of anion of EDTA, a sexadentate ligand.

research has been undertaken to either develop or discover an effective synthetic or natural chelating agent for the removal of iron deposits in the body that result from certain hereditary metabolic disorders.

Applications. The ability of chelating agents to reduce the chemical activity of metal ions has found extensive application in many areas of science and industry. Ethylenediaminetetraacetic acid (EDTA), a hexadentate chelating agent (Fig. 3), has been employed commercially for water softening, boiler scale removal, industrial cleaning, soil metal micronutrient transport, and food preservation. Nitrilotriacetic acid (NTA) is a tetradentate chelating agent (Fig. 4) which, because of lower cost, has taken over some of the commercial applications of EDTA. Chelating agents of related type have been used in biological systems to produce metal-ion buffers. By selection of the appropriate chelating agent, the free metal-ion concentration can be maintained at a very low and constant concentration level, just as a relatively constant pH can be maintained through the use of a conventional hydrogen-ion buffer system. Furthermore, the colors of certain chelating agents are sensitive to metal-ion concentration in a manner completely analogous to the pH-dependent color changes observed with acid-base indicators; such chelating agents serve as metal-ion indicators in analytical chemistry. The solubility of many chelating agents and metal chelates in organic solvents permits their use in solvent extraction of aqueous solutions for the separation or analysis of metal ions.

Many commercially important dyes and pigments, such as copper phthalocyanines, are chelate compounds. Humic and fulvic acids are plant degradation products in lake and sea-water sediments that have been suggested as important chelating agents which regulate metal-ion balance in natural waters. By virtue of its abundance, low toxicity, low cost, and good chelating tendencies for metal ions that produce water hardness, the tripolyphosphate ion (as its sodium salt) is used in large quantities as a builder in synthetic detergents. Both synthetic ion exchangers and the mineral zeolites are chelating ion-exchange resins which are used in analytical and water-softening applications. As final examples, less conventional chelating agents are the multidentate, cyclic ligands, termed collectively crown ethers, which are particularly suited for the complexation of the alkali and alkaline-earth metals.

Stabilities of metal chelates in solution. One of the most striking properties of chelate ring compounds is their unusual thermodynamic and thermal stability. In this respect, they resemble the aromatic rings of organic chemistry. For example, in reaction (1), β-diketones in the enol form can lose a hydrogen ion and coordinate with a metal cation to give a six-membered ring of unusual thermal stability.

Beryllium acetylacetonate boils without decomposition at 270°C, and G. T. Morgan reported that

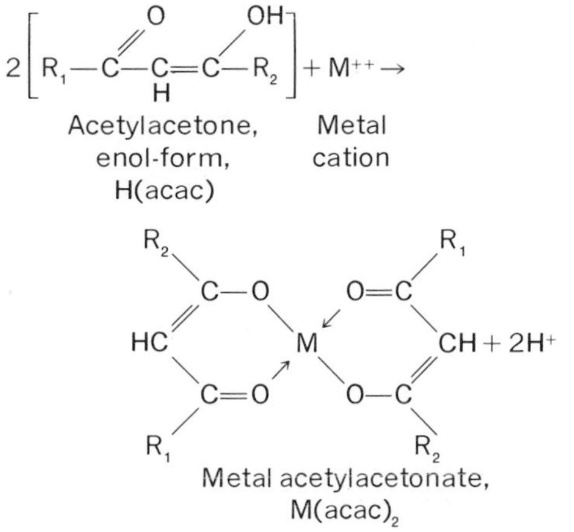

(1)

scandium acetylacetonate, $Sc(acac)_3$, shows very little decomposition at 370°C. This remarkable stability contrasts sharply with the very low stability of coordination compounds containing simpler monodentate ketones, such as acetone.

Because of enhanced thermodynamic stability in solution, chelating agents may greatly alter the behavior of metal ions. The very insoluble compound, ferric hydroxide, will dissolve in a strongly alkaline solution of triethanol amine, $N(CH_2CH_2OH)_3$. Alternatively, the concentration of the ferric ion can be made vanishingly small at pH 2 by the addition of an equimolar amount of bis-(orthohydroxybenzyl)ethylenediamine-*N,N'*-diacetic acid (HBED, Fig. 5).

Since about 1940, both theoretical and practical considerations have focused attention on the factors which contribute to chelate stability. It is convenient to list such factors under three headings: (1), the nature of the metal cation; (2) the nature of the ligand: and (3) the formation of the chelate ring. It should be emphasized that these factors operate together, and their separation is somewhat artificial, but helpful for discussion purposes.

Role of metal in chelate stability. Since nitrogen, oxygen, and sulfur serve as the electron donor atoms in a majority of chelating agents, it is of interest to seek a relationship between the donor

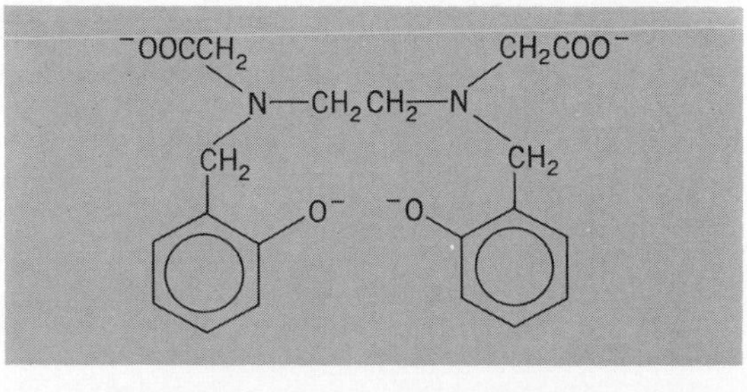

Fig. 5. Structural formula of anion of HBED, a sexadentate ligand.

CHELATION

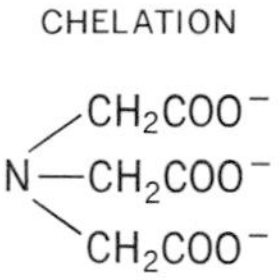

Fig. 4. Structural formula of NTA anion, a quadridentate ligand.

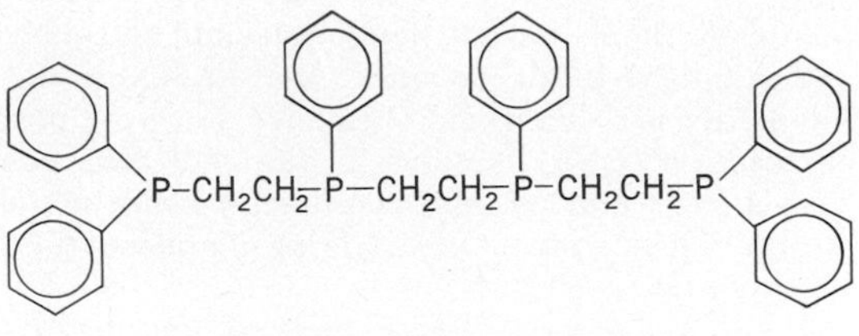

Fig. 6. "Tetraphos," a quadridentate ligand.

atom and the type of metal acceptor atom with which it combines. A large majority of the chelates of Li^+, Na^+, K^+, Rb^+, Cs^+, Mg^{2+}, Ca^{2+}, Sr^{2+}, Ba^{2+}, Al^{3+}, Ga^{3+}, In^{3+}, Tl^{3+}, Ti^{4+}, Zr^{4+}, Th^{4+}, Si^{4+}, Ge^{4+}, and Sn^{4+} contain oxygen as at least one of the donor atoms. It may be furnished as an acid, alcohol, ether, ketone, or other group. These ions coordinate less frequently through two nitrogen or sulfur atoms. Cations of other metals such as vanadium, niobium, tantalum, molybdenum, and uranium, and the cations Be^{2+}, Al^{3+}, and Fe^{3+} show a preference for oxygen as the donor atom, but they may coordinate through nitrogen, sulfur, or phosphorus under special conditions. Cr^{3+}, Fe^{2+}, and the platinum metals show increasing preference for coordination through nitrogen as opposed to oxygen, while Cu^+, Zn^{2+}, Ag^+, Au^+, Cu^{2+}, Cd^{2+}, Hg^{2+}, V^{3+}, Co^{3+}, and Ni^{2+} show a marked preference for nitrogen and sulfur as the donor atoms. The ions of the last group retain the ability to coordinate with oxygen in even greater degree than do the ions of the first group, but their tendency to form bonds through nitrogen is so great that it exceeds their oxygen-binding tendency.

It must be recognized that broad generalizations such as these have many exceptions, particularly in intermediate regions. On the other hand, such generalizations indicate clearly that attempts to arrange elements in the order of their chelating ability can be of significance only when cations of comparable type are selected. Thus, the stabilities of the alkali-metal, the alkaline-earth, and the rare-earth chelates decrease as the charge on the cation decreases or as the size of the cation increases. For example, the chelates of the alkaline-earth metal ions become less stable as the metal ion becomes larger (if the number of chelate rings remains the same), in the order Mg^{2+}, Ca^{2+}, Sr^{2+}, Ba^{2+}, Ra^{2+}. The relationship between ion size and chelate stability is of major importance in the separation of the rare-earth and transuranium elements by ion-exchange processes. The selectivity of the ion-exchange column is increased by the use of appropriate chelating agents in the eluting solution.

Stability sequences established for other metal ions are somewhat less satisfactory. Metal chelates of several substituted β-diketones decrease in stability in the order: $Hg^{++} > (Cu^{++}, Be^{++}) > Ni^{++} > Co^{++} > Zn^{++} > Pb^{++} > Mn^{++} > Cd^{++} > Mg^{++} > Ca^{++} > Sr^{++} > Ba^{++}$. If one restricts stability comparisons to bivalent metals of the first transition series, the following order is obtained: $Zn^{++} < Cu^{++} > Ni^{++} > Co^{++} > Fe^{++} > Mn^{++}$. This latter listing of stabilities appears to be valid for a large variety of chelating ligands.

Lower valence states such as Ir(I) and Rh(I) are effectively coordinated by chelating ligands containing trivalent phosphorus or trivalent arsenic donor groups. For example, recently synthetic coordination chemists have synthesized a series of unusual chelating agents containing trivalent phosphorus, analogous to polyamines, an example of which is illustrated in Fig. 6.

General principles of selective coordination of metal ions by various types of donor atoms have now been worked out, and are described through concepts such as the principles of hard and soft acids and bases, and type A and B character of metal ions. These qualitative principles are based on well-known laws of ionic attraction and polarizabilities of atoms, ions, or molecular groups, as well as on general principles of molecular orbital theory.

Role of ligand in chelate stability. Two general types of groups give rise to coordinate bonds between ligands and metal. These are (1) primary acid groups in which the metal ion can replace hydrogen ions and (2) neutral groups which contain an atom with a free electron pair suitable for bond formation. If two groups from either class 1 or 2, or from both classes, are present in the same molecule in such positions that both groups can form bonds with the same metal ion, a chelate ring may be formed. For example, as shown in Fig. 7, in oxalic acid two groups of type 1 are present; in ethylenediamine two groups of type 2 are present; glycine possesses one of each type; and pyridoxylidine-glycine (the Schiff base of glycine and vitamin B_6) possesses a carboxylate (type 1), an aromatic phenolate (type 1), and an imino group (type 2) available for chelate formation around a metal ion.

In general, anything which increases the localization of negative charge on the donor atom increases its ability to coordinate to a metal atom. Since a hydrogen ion is bound to a ligand by an electron pair in groups of type 1, an increase in electron density on the donor atom will increase

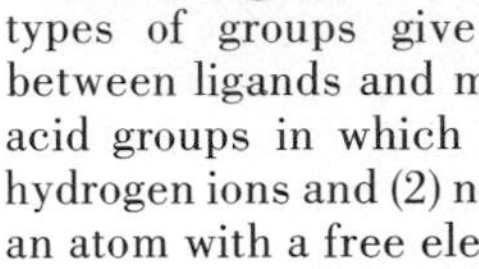

Oxalic acid; –2 type 1 (negative carboxylate) donor groups

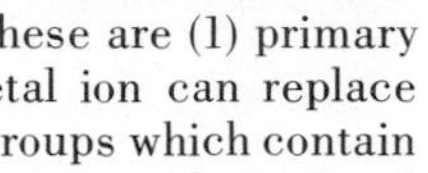

Ethylenediamine; –2 type 2 (neutral amino) donor groups

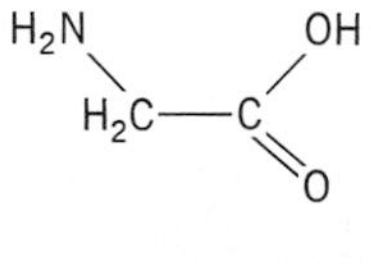

Glycine; –1 type 1 and 1 type 2 donor groups

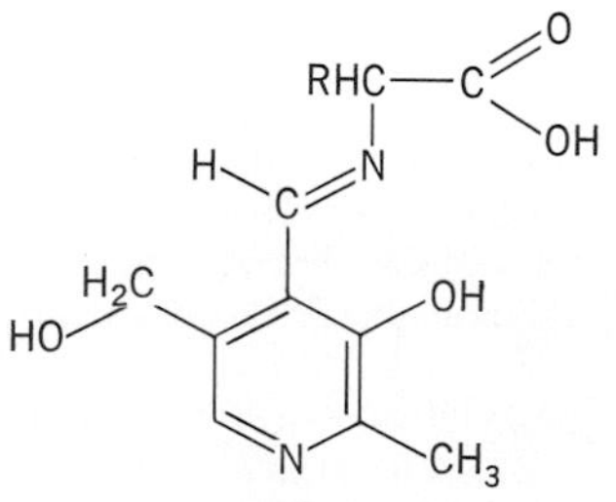

Amino acid Schiff base of pyridoxal (vitamin B_6); –2 type 1 (carboxylate and phenolate) and 1 type 2 (azomethine nitrogen) donor groups

Fig. 7. Chelating ligands indicating the types of donor groups available for coordination.

the ability of the donor to bind either hydrogen ion or metal cation. The ability of a ligand to bind hydrogen ion is frequently referred to as its basic strength. It is not surprising then that, for ligands of rather similar type, an increase in ligand basic strength implies an increase in its metal-chelating ability. A number of researchers have indicated this relationship by plotting the log of the equilibrium constant for the process represented by Eq. (2) against the log of the equilibrium constant for the process of chelate dissocation shown by Eq. (3).

$$H^+ + L^- \rightleftharpoons HL \qquad K_1 = \frac{[HL]}{[H^+][L^-]} \tag{2}$$

$$M^{n+} + L^- \rightleftharpoons M^{(n-1)+} \qquad K_2 = \frac{[ML^{(n-1)+}]}{[M^{n+}][L^-]} \tag{3}$$

L^- is the chelating anion, such as the acetylacetonate anion

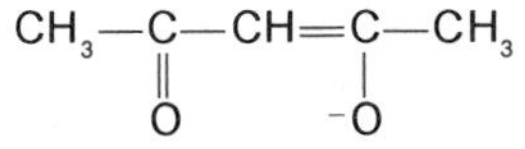

The constant K_1 is the protonation constant (its reciprocal is the familiar acid dissociation constant), and K_2 is designated the stability constant of the metal chelate.

For the organic chemist, the analogy between the hydrogen cation and the metal cation is even more clearly drawn. Rings formed by hydrogen bonding are referred to as chelates; thus, formic acid dimerizes through hydrogen-bond chelation as shown in Fig. 8. The high volatility of *o*-nitrophenol compared with the much lower volatility of its meta or para isomers (Fig. 8) can only be explained in terms of intramolecular versus intermolecular hydrogen bonding. The properties of salicylaldehyde, enhanced enolization in acetoacetic ester, and many other organic compounds are altered by internal chelate-ring formation involving hydrogen bonds. *See* HYDROGEN BOND.

The role of ligand structure on chelate stability

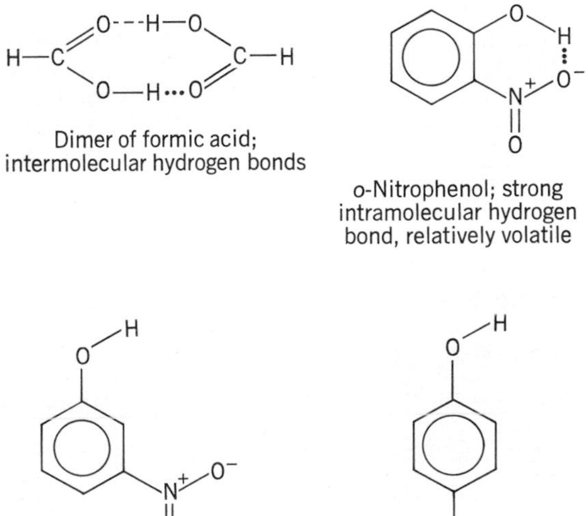

Fig. 8. Compounds forming intramolecular and intermolecular hydrogen bonds.

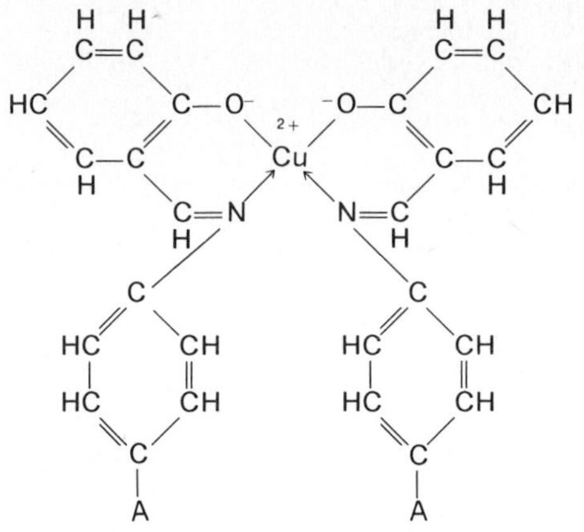

Fig. 9. Structure of a chelated copper compound.

is illustrated by studies which have been conducted on compounds of the type shown by Fig. 9, where A represents an electron-attracting group. In general, it was found that with an increase in the electron-attracting power of group A, electrons were pulled away from the nitrogen atom, resulting in both lower base strength and lower chelating ability for the ligand. As A is changed successively through the groups listed below, chelate stability increases in the order:

$—NO_2 < —SO_3Na <$ 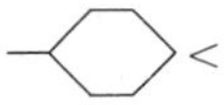 $< —H < —CH_3 < —OH < —OCH_3$

Least stable — Most stable

Similar studies on substituted β-diketone chelates of the type

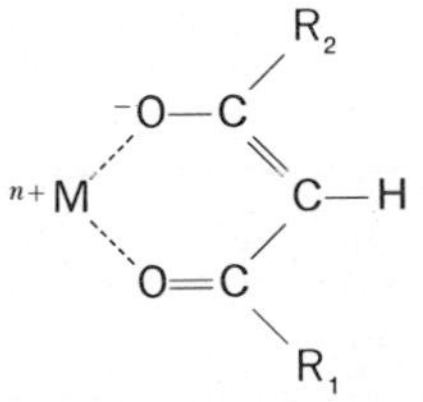

showed that, if R_1 were changed from a methyl group to an electron-withdrawing trifluoromethyl group, the stability of the resulting chelate is greatly decreased. In general, all molecular charge effects which can be invoked to shift charge in an organic molecule, such as inductive and resonance effects, will influence chelate stability. Thus the principles governing charge distribution and bond hybridization in organic chemistry are useful in working out the relationships between ligand structure and chelate stability.

Role of ring closure in chelate stability. The stability factors discussed above are applicable to coordination compounds generally, not to chelates alone. On the other hand, a number of stability factors may be considered to apply uniquely to chelates because of their ring structure. The most obvious variable in this category is ring size, a factor which is uniquely determined by the position of the donor atoms in the chelating ligand. When the groups are present in such a position as to form a five or six-membered ring, the resulting complex is the most stable although four-, seven-, eight-membered, and even larger rings are known. Examples

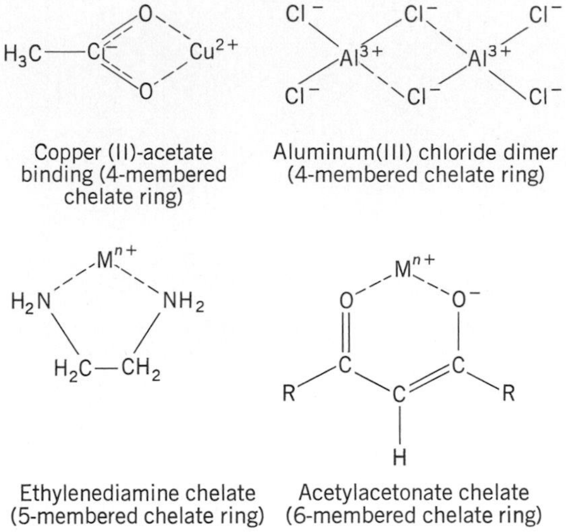

Fig. 10. Metal chelates containing four-, five-, and six-membered rings.

CHELATION

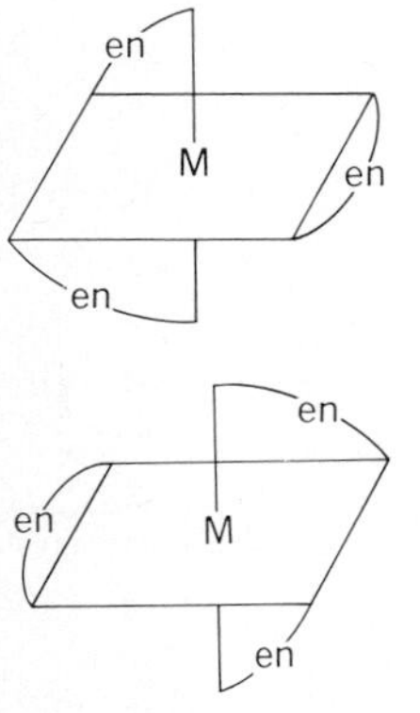

Fig. 11. Optical isomers of chelate complexes.

CHELATION

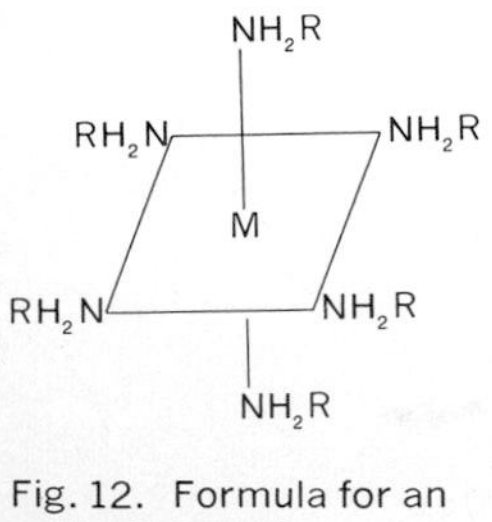

Fig. 12. Formula for an optically inactive nonchelate.

of these would be found among biological ligands as well as in ion exchange resins. The existence of three-membered rings has not been established. Hydrazine, H_2NNH_2, which might in theory form a three-membered chelate ring, appears to be monodentate. There is some evidence for intermediate structures of metal-oxygen complexes

which may be considered three-membered chelate rings. Such compounds may be important reaction intermediates but are generally unstable and present in relatively low concentrations. The four-membered chelate rings are frequently strained. Examples of four-membered chelate rings are copper(II)-carboxylate complexes, and the aluminum chloride dimer (Fig. 10). While five-membered rings are very common and are formed preferentially by saturated organic ligands, ligands containing two double bonds tend to form six-membered structures. If only one double bond is present, five- or six-membered rings may form; five-membered saturated rings are illustrated in Fig. 10 by ethylene-diamine-metal chelates, and the conjugated six-membered rings by the metal acetylacetonates.

Rings of seven or more members are comparatively uncommon, but their existence is well established. As the length of the chain between the two donor atoms increases, so does the tendency of the ligand to form polymetallic complexes. Under such circumstances, the two donor atoms on the same chelate molecule coordinate with different metal atoms rather than with one; thus, a polymeric chain, $M^+—NH_2CH_2CH_2CH_2CH_2NH_2—M^+$, may result instead of a ring structure.

The fact that chelate complexes are usually more stable than comparable nonchelate structures has been called the chelate effect. This effect is partly attributable to the fact that simultaneous rupture of both bonds holding the ligand to the metal is highly improbable, and if only one bond breaks there is a high probability that the broken bond will reform before the second bond is ruptured. As the chain length between the two donor atoms increases, the chance for reformation of the broken bond declines; thus, large rings usually show a decrease in stability. On the other hand, in fused ring systems, formed by polydentate ligands such as ethylenediaminetetraacetic acid, the probability that at least one bond will reform before all bonds are ruptured results in increased complex stability. When the bonding atoms are rigidly positioned around the metal by the organic framework, as in the prophins, the resulting increase in stability is extremely high. It has been reported that the copper phthalocyanine complex, with a completely interlocked ring system, is stable in the vapor phase near 500°C.

Chelate compounds are differentiated from their nonchelate analogs by several properties besides high stability. Although not all chelates are volatile, the existence of low-boiling metal acetylacetonates and related structures is noteworthy. If the coordination number of the metal cation for the oxygen atoms of the acetylacetone (that is, the number of nearest oxygens around the cation) is equal to twice the ionic charge of the cation, the resulting acetylacetonate is volatile; thus, beryllium with a charge of 2 and a coordination number of 4 forms an acetylacetonate which boils at 270°C; aluminum with a charge of 3 and a coordination number of 6 forms an acetylacetonate which boils at 314°C. If the coordination number of the central cation is less than twice the ionic charge, less volatile saltlike complexes are formed.

Isomerism of all types, so important in organic chemistry, is of major concern in chelation, particularly to the biochemist, since desired biological properties are frequently restricted to a particular chelate isomer. Ring formation may result in optical activity where analogous nonchelate structures are inactive. Thus, the ethylenediamine chelate structures shown in Fig. 11 are optical isomers, whereas the analogous methylamine complexes represented in Fig. 12 are optically inactive. *See* CHEMICAL BONDING; CHEMICAL EQUILIBRIUM; COMPLEX COMPOUNDS; STEREOCHEMISTRY; STERIC EFFECT.

[A. E. MARTELL; R. J. MOTEKAITIS]

Bibliography: S. Ahrland in T. D. Dunitz et al. (eds.), *Structure and Bonding*, vol. 5, 1968; S. Chaberek and A. E. Martell, *Organic Sequestering Agents*, 1959; F. P. Dwyer and D. P. Mellor (eds.), *Chelating Agents and Metal Chelates*, 1964; A. E. Martell and M. Calvin, *Chemistry of the Metal Chelate Compounds*, 1952; A. E. Martell and R. M. Smith, *Critical Stability Constants*, vols. 1–4, 1974–1976; A. E. Martell, The chelate effect, *Advances in Chemistry Series*, no. 62, American Chemical Society, pp. 272–294, 1967.

Chemical bond theory

Chemical bonds are the forces that hold atoms together in molecules and solids. Chemical bond theory is the explanation of the physical basis of chemical bonds and of the relationship between chemical bonds and the properties of substances.

The simplest chemical bonds to describe are those resulting from direct coulombic attractions between ions of opposite charge, as in most crystalline salts. These are termed ionic bonds. *See* CHEMICAL STRUCTURES.

Other chemical bonds are of a wide variety of types, ranging from the very weak van der Waals attractions, which bind Ne atoms together in solid neon, to metallic bonds or metallike bonds, in which very many electrons are spread over a lattice of positively charged atom cores and give rise to a stable configuration for those cores. The theory of many of these bond types has been well developed by theoretical chemists. *See* CHEMICAL BONDING; CRYSTAL FIELD THEORY; MOLECULAR ORBITAL THEORY.

Covalent bond. Since the normal covalent bond, in which two electrons bind two atoms together, as in

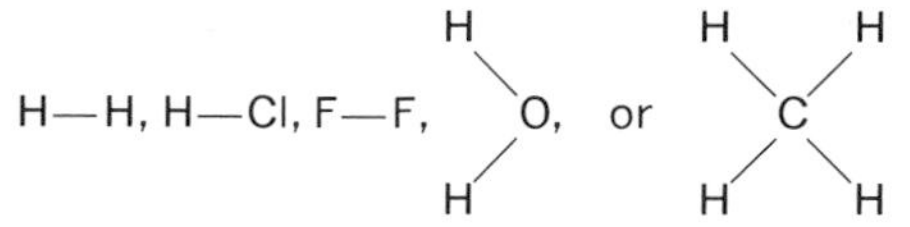

is the most characteristic link in chemistry, an adequate theory to account for it is the central problem in chemical bond theory. The characteristic physical and chemical properties of any molecule are direct consequences of its particular detailed electronic structure. Yet the theory of any one covalent chemical bond, for example, the O-H bond in the water molecule, has much in common with the theory of any other covalent bond, for example, the C-H bond in the methane molecule. An accurate theory of covalent bonds now exists, capable both of treating their qualitative features and of quantitatively accounting for the molecular properties which are a consequence of those features. The theory is a branch of quantum theory. *See* QUANTUM CHEMISTRY.

Hydrogen molecule. A brief outline of the application of quantum theory to the bond in the hydrogen molecule H—H follows. Here two electrons, each of charge $-e$, bind together two protons, each of charge $+e$, with the electrons much lighter than the protons. What must be explained, above all else, is that these particles form an entity with the protons 0.74×10^{-8} cm apart, more stable by $D=$ 109 kcal per mole than two separate hydrogen atoms, where D is the binding energy. In more detail, a molecular energy is involved (ignoring nuclear kinetic energy) that depends on internuclear distance, as shown in the figure. This curve can be determined experimentally, and it can be used to interpret the characteristic spectroscopic properties of hydrogen gas.

The quantum theory accounts for the properties of isolated atoms by assigning atomic orbitals for individual electrons to move in, not more than two electrons at a time. For the hydrogen atom, the orbitals are labeled $1s$, $2s$, $2p_x$, $2p_y$, $2p_z$, and so on, with $1s$ the one having the lowest energy. For the molecule H_2 one electron, say electron 1, might be assigned to a $1s$ orbital on proton A, with $1s_A(1)$ written to signify this; similarly electron 2 might be assigned to the same kind of orbital on proton B, written as $1s_B(2)$. Since independent probabilities multiply and orbitals represent probability amplitudes, the description for the combined system shown in Eq. (1) is arrived at. Unfortunately, this

$$\phi(1,2) = 1s_A(1)1s_B(2) \qquad (1)$$

fails to account for the bond properties; it gives a binding energy of only 10 kcal per mole.

An essential defect of Eq. (1) is the numbering of the electrons; it puts electron 1 on proton A, electron 2 on proton B. Electrons cannot be distinguished experimentally, so they should not be given unique numbers; the function $1s_A(2)1s_B(1)$ would be just as good as the foregoing. It is necessary to use a description that is not affected by interchange of electron labels, as in the additive combination of Eq. (2). (The difference combina-

$$\phi(1,2) = 1s_A(1)1s_B(2) + 1s_A(2)1s_B(1) \qquad (2)$$

tion also is an acceptable description, but it represents an excited state of the molecule.)

Any complete molecular electronic wave function should include electron spin. Symmetric space wave functions like Eq. (2) must be multiplied by antisymmetric spin wave functions to give total wave functions that are antisymmetrical with respect to interchange of electrons. For the ground state of hydrogen, and for the normal covalent bond elsewhere, this requirement means that the electron spins must be paired to give a net electron spin of zero.

The simple relationship described by Eq. (2) qualitatively accounts for the existence of the covalent bond; the predicted binding energy is $D=74$ kcal per mole; and the shape of the curve, with the minimum appearing at 0.80×10^{-8} cm, is right.

The description of Eq. (2) can be systematically improved. The charge acting on the electron may be changed from $^{+}1e$ to the larger value, ^{+}Ze, which is more realistic for the actual molecule. With $Z=1.17$ this gives $D=87$ kcal per mole. Polarization effects may be introduced by taking Eq. (3), where $1\sigma_A = 1s_A + \lambda 2pz_A$ and $1\sigma_B = 1s_B +$

$$\phi(1,2) = 1\sigma_A(1)1\sigma_B(2) + 1\sigma_A(2)1\sigma_B(1) \qquad (3)$$

$\lambda 2pz_B$. This gives $D=93$ kcal. Ionic terms may be introduced, acknowledging the possibility that both electrons may be on one atom, by taking Eq. (4). This also gives (with $Z=1.19$) $D=93$ kcal.

$$\phi(1,2) = \sigma_1[1s_A(1)1s_B(2) + 1s_A(2)\,1s_B(1)]$$
$$+ \sigma_2[1s_A(1)1s_A(2) + 1s_B(1)1s_B(2)] \qquad (4)$$

CHEMICAL BOND THEORY

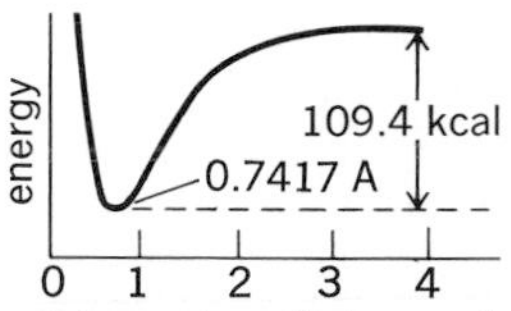

Potential energy of the hydrogen molecule.

Another possible approach is to include both ionic terms and polarization effects, and other terms involving $2s$, $3d$, $4f$, and other orbitals. If this is done, eventually one obtains the observed D value and a potential curve that is in excellent agreement with experiment.

The linear mixing of terms such as $1s_A(1)1s_B(2)$ with terms such as $1s_A(1)1s_A(2)$ is called resonance; the method of mixing covalent and ionic structures is called the valence bond (VB) method. The particular mixing coefficients can be found by the variational principle: The best values for such parameters are those that make the total energy of the molecule, properly computed from quantum mechanics, a minimum. The energy expression only contains terms that have a direct classical interpretation: the kinetic energy of the electrons, their energy of repulsion for one another, their energy of attraction for the nuclei, and the nuclear-nuclear repulsion energy. The breakdown of the binding energy is in accord with the virial theorem: As the molecule is formed from the atoms, the kinetic energy increases by amount D and the potential energy decreases by $2D$. *See* RESONANCE.

Alternative descriptions of H_2 are possible, of which the most important is provided by the molecular orbital (MO) method. Here one puts electrons one at a time into orbitals which are spread over the whole molecule, usually approximating these orbitals by linear combinations of atomic orbitals (LCAO). For H_2 the lowest molecular orbital is $\phi_1 \approx 1s_A + 1s_B$, the next $\phi_2 \approx 1s_A - 1s_B$. The simplest molecular orbital description is displayed in Eq. (5), which represents an equal weighting of

$$\phi(1,2) = \phi_1(1)\phi_2(2) \quad (5)$$

covalent and ionic structures; it gives $D=61$ kcal for $Z=1.00$ and $D=80$ kcal for $Z=1.20$. More suitable is a mixture of this function with the function obtained by promoting both electrons from ϕ_1 to ϕ_2. The result of this configuration interaction process has the form of Eq. (6), and it is identical

$$\phi(1,2) = D_1\phi_1(1)\phi_1(2) + D_2\phi_2(1)\phi_2(2) \quad (6)$$

with the valence bond function of Eq. (4). In this manner more terms can be added, using more orbitals, until, again, the accurate potential energy curve is obtained.

The most accurate description known for the chemical bond in H_2 is a very complicated electronic wave function, formulated by H. James, A. Coolidge, W. Kolos, and C. C. J. Roothaan. It accounts for all known properties of hydrogen to high accuracy and confirms that the nonrelativistic quantum mechanics of E. Schrödinger will suffice for most chemical purposes. The calculated and observed values of D agree absolutely.

Complex molecules. The development of a quantitative treatment of chemical bonds in molecules that are more complicated than H_2 has many inherent difficulties. It constitutes, however, an active and useful field of research which has been stimulated by the development of large and fast electronic computers and advanced numerical and analytical techniques for handling atomic orbitals on many different nuclei. The qualitative and semiquantitative theory preserves the use of many chemical concepts that predate quantum chemistry itself; among these are electrostatic and steric factors, tautomerism, and electronegativity. The quantitative theory provides a physical basis for chemical concepts if they are valid and a tool for weeding them out if they are invalid.

In molecules containing many chemical bonds, it should be possible in many instances to construct accurate descriptions of the whole electronic system from descriptions of the separate electron pairs. If $\phi_A(1,2)$, $\phi_B(3,4)$, . . . , are wave functions for pairs A, B, . . . , including spin, then it may be that the function shown in Eq. (7), where $\mathscr{A}$ represents antisymmetrization with respect to electron interchange, is an accurate wave function for the molecule.

$$\phi = \mathscr{A}[\phi_A(1,2)\,\phi_B(3,4)\ \cdot\cdot\cdot] \quad (7)$$

If this is the case and if certain auxiliary conditions are satisfied, the description is one in which the traditional concept of separate chemical bonds has been preserved. The functions ϕ_A, ϕ_B, . . . , are called geminals; they are important entities in the theory of chemical bonding in large molecules.

To illustrate the level of accuracy of contemporary quantum-chemical calculations, the table gives observed and calculated values for certain spectroscopic properties of the carbon monoxide molecule. In the future, theoretical calculation of properties of small molecules, to a greater accuracy than this, will become routine. See MOLECULAR ORBITAL THEORY. [ROBERT G. PARR]

Spectroscopic properties of CO

Property*	Observed value	Calculated value†
$r_e\ 10^8$ cm	1.128	1.119
ω_e(cm^{-1})	2170.	2357.
$\omega_e x_e$(cm^{-1})	13.5	11.1
B_e(cm^{-1})	1.93	1.97

*r_e = internuclear distance at which potential energy is minimum.
ω_e = a quantity whose square is proportional to the curvature of the potential energy at its minimum.
x_e = anharmonic constant.
B_e = rotational constant.
†According to R. K. Nesbet.

Bibliography: A. L. Companion, *Chemical Bonding*, 2d ed., 1979; H. F. Hameka, *Quantum Theory of the Chemical Bond*, 1975; J. Murrell et al., *The Chemical Bond*, 1978; L. Pauling and E. B. Wilson, Jr., *Introduction to Quantum Mechanics*, 1935.

Chemical bonding

The fundamental fact of chemistry is that elements, such as oxygen, hydrogen, carbon, and iron, can combine to form compounds, such as water, methane, and iron oxide, with properties completely different from those of their components. The description of this in terms of atomic theory is that atoms of the elements attach themselves to each other to form molecules. This property that atoms possess, of joining together to form molecules, is known as chemical bonding.

Atoms contain electric charges—a small, positively charged nucleus surrounded by a cloud of moving, negatively charged electrons. All chemical bonding is caused by the mutual attractions and repulsions of these electric charges. Other types of forces—gravitational, magnetic, nuclear—have negligible direct effect. The electric forces are governed by Coulomb's law. On the other hand, the motions and distributions of the electrons in the atoms are controlled by the laws of quantum mechanics.

Chemical bonds are very strong. To break one bond in each molecule in a gram mole of material will typically require an energy of many tens of kilocalories per mole. Most of the energy used by man is chemical energy, derived from changing chemical bonds in food or fuel.

It is convenient to classify chemical bonding into several types, although all real cases are mixtures of these idealized, purely electrical cases.

Ionic bonding. This is the simplest type of bonding, in which one or more electrons are transferred completely from one atom to another, thus converting the neutral atoms into electrically charged ions. These ions are approximately spherical in shape and attract one another because of their opposite charges. The ions are drawn together until their spherical electron clouds sufficiently interpenetrate and repel one another to balance the force of attraction. Molecules can consist of two or more such ions. Many inorganic crystals can be considered as giant molecules made up of ions. Thus common salt, sodium chloride, consists of a lattice of Na^+ ions (positive sodium) each surrounded by six Cl^- ions (negative chlorine) and vice versa.

In a purely ionic compound or crystal, the ions pack together in a geometrical arrangement which minimizes the total energy. Since unlike charges attract, and like charges repel each other, the positive ions will normally be next to negative ions and vice versa. The radii of ions are characteristic of the element and of the charge. Negative ions (anions) are usually larger than positive ions. The number of negative ions surrounding a positive ion (cation) is mainly determined by the ratio of the radii of the two ions. Thus, a small cation will not be surrounded by as many negative ions as will a larger cation. The number of positive ions surrounding a negative ion is largely governed by the importance of maintaining a local balance of positive and negative charge. Ions act as if the attractive and repulsive forces have no specific directional properties. The number of charges on an ion is called the electrovalence of the element. Sodium, for example, normally carries a single positive charge in its ionic compounds and thus has an electrovalence of +1. Chlorine in most inorganic compounds is an ion with one negative charge, hence it has an electrovalence of −1. Many elements can form several different ions. Thus iron commonly occurs as ferrous iron, Fe^{++}, or as ferric iron, Fe^{+++}, and consequently displays electrovalencies of either +2 or +3. Since completely filled electron shells are especially stable, the ionic state which yields a completed shell is for many elements the most stable one.

Covalent bonding. This is another limiting type of chemical bonding. Here each atom of a bonded pair contributes one electron to form a pair of electrons which move in such a manner as to increase the density of electric charge in the space between the two atoms. The negative charge in the region between the two atoms attracts the two positive nuclei. This type of bond is also called an electron pair bond because its essential feature is the formation of a pair of electrons which spend much time in the region between the two atoms. These two electrons have their spins pointing in opposite directions; in other words, they are paired. If this were not the case, the two electrons could not both have a low kinetic energy (which is important for the stability of the molecule) and also spend much time in the region between the two atoms. This requirement of pairing is due to the basic principles of quantum mechanics, particularly the Pauli exclusion principle.

The number of covalent bonds which an atom can form is called the covalence and is governed by the detailed electron configuration of the atom. The meaning of the historical word valence has been gradually modified so that it now usually refers to the electrovalence or the covalence of an atom. An extremely important case is that of carbon. In most of its compounds, carbon forms four covalent bonds. When these connect it to four other atoms, the directions of the bonds to these other atoms will normally make angles of about 109° to one another, unless the attached atoms are crowded or constrained by other bonds. In other words, covalent bonds have preferred directions. However, to preserve the idea that carbon forms four bonds, it is necessary to introduce the notion of double and triple bonds. Thus in the structural formula of ethylene, C_2H_4, all lines denote covalent

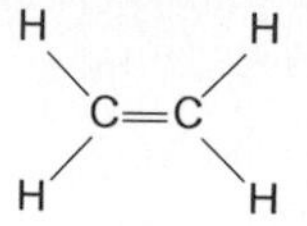

Ethylene

bonds, the double line connecting the carbon atoms being a double bond. This idea has physical reality because such double bonds are distinctly shorter, almost twice as stiff, and require considerably more energy to break completely than do single bonds. However, they do not require twice as much energy to break as a single bond, so it is energetically advantageous for a molecule to open one component of a double bond and add two atoms; for example, H_2 adds to C_2H_4 to form ethane. Similarly, acetylene is written with a triple bond,

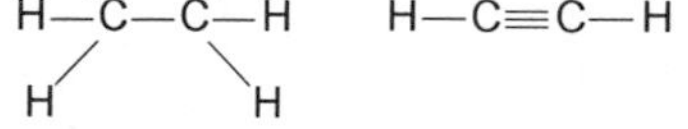

Ethane Acetylene

which is still shorter than a double bond. A carbon-carbon single bond has a length close to 1.54×10^{-8} cm, whereas the triple bond is about 1.21×10^{-8} cm long.

In many compounds the rules for writing bond formulas are not unique. For example, benzene, C_6H_6, can be written in the two forms

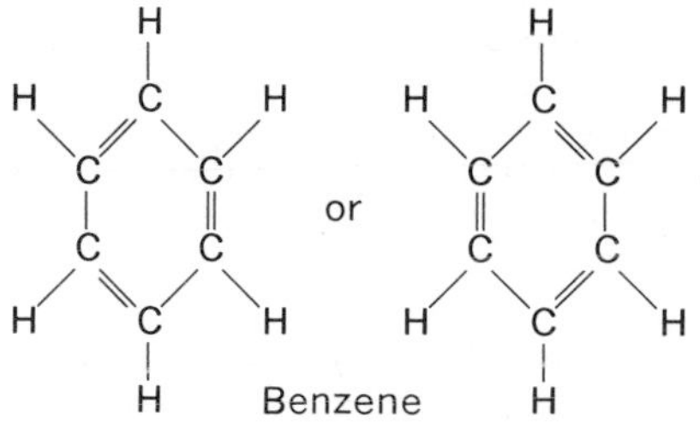

Benzene

Evidence proves that all six C-C bonds are equivalent, so neither formula can be correct. Quantum mechanical arguments show that the correct picture is a blend of the two, in which the bonds have many properties intermediate between those of double and single bonds but in which the whole molecule displays an additional stability. This phenomenon, called resonance, occurs whenever the structure is such that two or more different bond formulas can legitimately be drawn for the same geometry. *See* ELECTRON CONFIGURATION; RESONANCE.

Many substances have some bonds which are covalent and others which are ionic. Thus in crystalline ammonium chloride, NH_4Cl, the hydrogens are bound to nitrogen by electron pairs, but the NH_4 group is a positive ion and the chlorine is a negative ion. On solution in water, ionic crystals undergo dissolution into their separate ions.

Both electrons of a covalent bond may come from one of the atoms. Such a bond is called a coordinate or dative covalent bond or semipolar double bond and is one example of the combination of ionic and covalent bonding. Actually, the electron pair does not have to be symmetrically located in any case; so all degrees of mixing of ionic and covalent character may occur.

The hydrogen bond is a special type of a bond in

which a hydrogen atom links a pair of other atoms. The linked atoms are normally oxygen, fluorine, chlorine, or nitrogen. These four elements are all quite electronegative, a fact which favors a partially ionic interpretation of this kind of bonding. *See* HYDROGEN BOND.

Metallic bonding. This is a third type of chemical bonding, which is exemplified in the common metals. There are several ways of looking at this bonding, but perhaps the simplest is to consider the crystal as consisting of positive ions of the metallic element immersed in a sea of electrons. The attraction of the positive ions for the electrons holds the crystal together. Some of the electrons are free to move about the whole crystal of the metal, and this is what makes the metal an electrical conductor. For further details on metallic bonding *see* CRYSTAL STRUCTURE.

Van der Waals forces. Although many crystals are giant molecules whose atoms are completely linked together by strong ionic or covalent bonds, others consist of discrete molecules, strongly bonded internally, but held to each other by much weaker forces. Most organic crystals are soft and have low melting points. Their discrete molecules are held together by what are called van der Waals or dispersion forces. Although these forces are electrical in nature, they are too weak to be considered as true chemical bonding. The picture is as follows: The electrons in the cloud of negative electricity in a molecule are in rapid motion, and the charge distribution may therefore be thought of as fluctuating in time. At any instant, the distribution in a given molecule may be quite unsymmetrical so that an electric dipole moment results, even though there need not be any average dipole moment. This dipole moment, however, may reverse rapidly or change in direction or magnitude as the electrons move. At any instant, the dipole in one molecule will cause an electric field to act on a neighboring molecule. This electric field will disturb the motion of the electrons in the second molecule in such a way as to produce an instantaneous induced dipole moment in the second molecule. The interaction of the two dipole moments, namely, the original instantaneous one in the first molecule and the induced one in the second, will result always in a net but weak attractive force between the two molecules. This force will then bring the molecules together until the repulsion of their electron clouds for one another keeps them at an equilibrium distance of separation. *See* CHEMICAL STRUCTURES; INTERMOLECULAR FORCES; MOLECULAR STRUCTURE AND SPECTRA; QUANTUM CHEMISTRY; VALENCE.

[E. BRIGHT WILSON, JR.]

Bibliography: A. L. Companion, *Chemical Bonding*, 2d ed., 1979; H. Gray, *Chemical Bonds: An Introduction to Atomic and Molecular Structure*, 1975; J. N. Murrell et al., *The Chemical Bond*, 1978; L. Pauling, *Nature of the Chemical Bond and the Structure of Molecules and Crystals*, 3d ed., 1960.

Chemical compounds

Substances composed of two or more elements which do not vary in composition from sample to sample, and which have fixed and definite physical properties, such as density and refractive index. The elements in compounds cannot be separated by simple physical or mechanical means, but only by chemical treatment. When compounds are formed from their elements, heat is generated or absorbed. These properties distinguish them from mixtures.

For example, if iron filings and sulfur powder are mixed together, the two elements can be separated, either by removing the iron filings with a magnet or by dissolving the sulfur in an appropriate solvent, and the individual particles of each element are distinguishable under a magnifying glass. If iron filings and sulfur are heated together, a chemical reaction takes place and a new substance is formed, iron sulfide, with properties totally different from iron or sulfur. *See* DENSITY; ELEMENTS; MIXTURE.

Most chemical compounds are formed in fixed and definite proportions by weight from their elements, and they obey the laws of chemical combination. However, there are a number of solid compounds, known as nonstoichiometric compounds, that exhibit departures from the law of definite proportions. *See* DEFINITE COMPOSITION, LAW OF; EQUIVALENT WEIGHT; INORGANIC CHEMISTRY; MULTIPLE PROPORTIONS, LAW OF; NONSTOICHIOMETRIC COMPOUNDS.

[THOMAS C. WADDINGTON]

Chemical conversion

The term for the chemical change from reactants to products of a chemical industrial process. Preferred for decades in the petroleum industry, the term is now used to an increasing extent for the entire chemical process industries in place of unit processes, since it is a more exact definition of industrial chemical reactions. The following are important chemical conversions:

Alkylation	Hydrogenation, hydrogenolysis
Amination	Hydrolysis, hydration
Ammonolysis	Ion exchange
Aromatization	Isomerization
Calcination	Neutralization
Combustion	Nitration
Condensation	Oxidation
Dehydration	Polymerization
Diazotization and coupling	Pyrolysis, cracking
Esterification (sulfation)	Reduction
Fermentation	Sulfation
Halogenation	Sulfonation
Hydroformylation (oxo)	

All these terms refer to the classes of chemical changes, each embracing many individual reactions or chemical conversions into which chemical industrial processes can be divided.

Basic principles. Chemical conversions are defined quantitatively as the percentage of reactant converted to a product per mole of reactant charged per single pass through a chemical reactor. By separating the product from any unreacted reactant and by recycling the unreacted chemicals back to the chemical reactor, one or more times, the total conversion or yield may approach 100%. Industrial chemical processes usually contain both chemical conversions and unit operations or unit physical changes. The combination of these may be represented by a flow sheet to depict graphically the complete process from raw material to finished product. Certain unifying aspects of these chemical changes exist simply because they all embody chemical reaction. The most important from a chemical engineering viewpoint involves the laws for chemical equilibria and kinetics. Although both the chemical equilibria and kinetics vary greatly from one chemical conversion to another, the general principles are the same for all processes.

In many cases catalysts are employed to speed up the reaction in order to reach equilibrium within a practical time. Other common factors among industrial chemical reactions are the percent of chemical conversion and the percent of yield. The aim is to raise the chemical conversion to the theoretical chemical equilibrium under the conditions chosen, and to do this within a practical time. Chemical equilibria can be raised in specific cases by changing the proportions of reactants and by altering time, temperature, and pressure. Yield and conversion are illustrated in the reaction $N_2 + 3H_2 \rightleftharpoons 2NH_3$. At about 150 atm pressure and 500°C the yield will be 99+% but the conversion will be only about 14%. In other words, only 14% of the moles charged will be converted into ammonia per pass, but in this case there are practically no losses or by-products, so the yield or net change is theoretically high. On the other hand, in many reactions there are various by-products which reduce the yields and conversions. This is true in many oxidations and nitrations, especially in processes using the aliphatic series of hydrocarbons. *See* NITRATION.

The equipment needed to carry out reactions is of vital significance in any such considerations, from the aspect of corrosion, wall catalysis, and structural strength.

Most technical chemical reactions are exothermic, evolving heat of reaction. The temperature used often affects the chemical equilibrium. The transfer of molecules to and from the reaction site significantly affects the kinetics of conversion. Thus, the applicational concepts of heat, mass, and momentum transfer are critical in evaluation of the chemical conversion process.

Significance of concept. Each chemical conversion emphasizes the unitary or common aspects, such as yield, equilibrium, kinetics, and catalysis. Increases in yield are by and large the most important factors in reducing costs of chemicals. The cost of raw materials represents 50–80% of the processing or manufacturing expense for most chemicals. No distinction need be made between inorganic and organic procedures because remarkably similar conditions in a given chemical conversion prevail for both, for example, in the hydrogenation of N_2 to NH_3 or of CO to CH_3OH. In design of equipment, the engineer is greatly aided by the generalizations arising from the chemical conversion classification rather than by considerations of reactions separately. For additional details see the separate articles on the individual chemical conversions (unit processes). *See* CHEMICAL ENGINEERING; CHEMICAL PROCESS INDUSTRY.

[JOSEPH A. BRINK, JR.]

Bibliography: S. W. Churchill, *The Interpretation and Use of Rate Data*, 1974; M. Modell and R. C. Reid, *Thermodynamics*, 1974; R. N. Shreve, *Chemical Process Industries*, 3d ed., 1967; R. N. Shreve and J. A. Brink, Jr., *Chemical Process Industries*, 4th ed., 1977; J. M. Smith and H. C. Van Ness, *Introduction to Chemical Engineering Thermodynamics*, 3d ed., 1975.

Chemical dynamics

That branch of physical chemistry which seeks to explain time-dependent phenomena, such as energy transfer and chemical reaction, in terms of the detailed motion of the nuclei and electrons which constitute the system.

REACTION KINETICS

Although the ultimate state of a chemical system is specified by thermodynamics, the time required to reach that equilibrium state is highly variable. As a consequence, determining the rate of chemical reactions has proved to be important for practical reasons. For example, diamonds are thermodynamically unstable with respect to graphite, but the rate of transformation of diamonds to graphite is negligible. Rate studies have also yielded fundamental information about the details of the nuclear rearrangements which constitute the chemical reaction.

Traditional chemical kinetic investigations of the reaction between species X and Y to form Z and W, reaction (1), sought a rate of the form given

$$X + Y \rightarrow Z + W \tag{1}$$

in Eq. (2), where $d[Z]/dt$ is the rate of appearance

$$d[Z]/dt = kf([X], [Y], [Z], [W]) \tag{2}$$

of product Z, f is some function of concentrations of X, Y, Z, and W which are themselves functions of time, and k is the rate constant. Chemical reactions are incredibly diverse, and often the function f is quite complicated, even for seemingly simple reactions such as that in which hydrogen and bromine combine directly to form hydrogen bromide. These are actually complex reactions which proceed through a sequence of simpler reactions, called elementary reactions. For reaction (3*d*), the sequence of elementary reactions is a chain mechanism known to involve a series of steps, reactions (3*a*)–(3*c*). This sequence of elementary reactions

$$Br_2 \rightarrow 2Br \tag{3a}$$
$$Br + H_2 \rightarrow HBr + H \tag{3b}$$
$$H + Br_2 \rightarrow HBr + Br \tag{3c}$$
$$H_2 + Br_2 \rightarrow 2HBr \tag{3d}$$

was formerly known as the reaction mechanism, but in the chemical dynamical sense the word mechanism is reserved to mean the detailed motion of the nuclei during a collision.

Bimolecular processes. An elementary reaction is considered to occur exactly as written. Reaction (3*b*) is assumed to occur when a bromine atom hits a hydrogen molecule. The products of the collision are a hydrogen bromide molecule and a hydrogen atom. On the other hand, the overall reaction is a sequence of these elementary steps and on a molecular basis does not occur as reaction (3*d*) is written. With few exceptions, the rate law for an elementary reaction $A + B \rightarrow C + D$ is given by $d[C]/dt = k[A][B]$. The order (sum of the exponents of the concentrations) is two, which is expected if the reaction is bimolecular (requires only species A to collide with species B). The rate

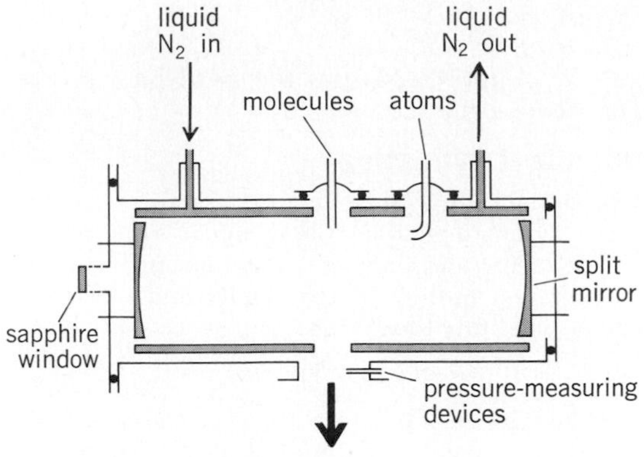

Fig. 1. Reaction vessel for studying infrared chemiluminescence between atoms and molecules at low pressures. (*From D. H. Maylotte, J. C. Polanyi, and K. B. Woodall, J. Chem. Phys., 57:1547–1561, 1972*)

constant k for such a reaction depends very strongly on temperature, and is usually expressed as $k = Z_{AB}\, \rho \exp(-E_a/RT)$. Z_{AB} is the frequency of collision between A and B calculated from molecular diameters and temperature; ρ is an empirically determined steric factor whch arises because only collisions with the proper orientation of reagents will be effective; and E_a, the experimentally determined activation energy, apparently reflects the need to overcome repulsive forces before the reagents can get close enough to react.

Unimolecular processes. In some instances, especially for decompositions, AB → A + B, the elementary reaction step is first-order, Eq. (4),

$$d[A]/dt = d[B]/dt = k[AB] \qquad (4)$$

which means that the reaction is unimolecular. The species AB does not spontaneously dissociate; it must first be given some critical amount of energy, usually through collisions, to form an excited species AB*. It is the species AB* which decomposes unimolecularly.

MOLECULAR DYNAMICS

In principle, it is possible to prepare two reagents in specific quantum states and to determine the quantum-state distribution of the products. In practice, this is much too difficult, and experiments have been limited to preparing one reagent or to determining some aspect of the product distribution. This approach yields data concerning the gross aspects of the dynamics rather than the fine details.

Bimolecular reactions. Molecular-beam and luminescence techniques have played a major role in the development of chemical dynamics. Since these techniques largely complement each other, they are illustrated by discussing the results of a single reaction, the formation of deuterium chloride (DCl), reaction (5).

$$D + Cl_2 \rightarrow DCl + Cl \qquad (5)$$

Chemiluminescence. Reaction (5) is 46 kcal/mol (192 kJ/mol) exoergic. If all of the exoergicity were to go into vibration of the newly formed DCl molecule, enough energy is available so that vibrational levels (v') up to $v' = 9$ could be formed. In a gas-phase reaction at high pressure, much of this energy would be dissipated at the walls of the container, but at very low pressures DCl molecules in excited vibrational states, DCl†, will emit infrared emission prior to undergoing a collision.

An apparatus to study this infrared chemiluminescence is shown in Fig. 1. Deuterium atoms are made by dissociating D_2 gas in an electrical discharge, and are injected into the observation cell at the top. The reagents mix and react inside a vessel with walls at 77 K, which freeze out species hitting the wall. The pressure is kept low to minimize vibrationally deactivating collisions. Infrared (IR) emission from the products is gathered by mirrors at the end, taken out through the sapphire window, and analyzed with an IR spectrometer. By analyzing the spectrum, it is possible to determine which vibration-rotation states are emitting and, as a consequence, which vibration-rotation states are formed in the reaction. Figure 2 shows the relative distribution of vibrational states from the reaction, and also shows, for comparison, the relative distribution of vibrational states calculated from the Boltzmann equation for hot DCl (6000 K). Thermal distributions at any other temperatures would still show a monotonic decline. (The population for $v' = 0$ is not determined by the chemiluminescence experiments, because that state does not emit.) The DCl formed in the reaction clearly has different properties than hot DCl: the vibrational population displays an inversion, and this system (the hydrogen isotope) was the active medium for the first chemically pumped laser. *See* CHEMILUMINESCENCE.

Molecular-beam experiments. Molecules can be isolated in molecular beams, and collisions between these isolated molecules can be observed by

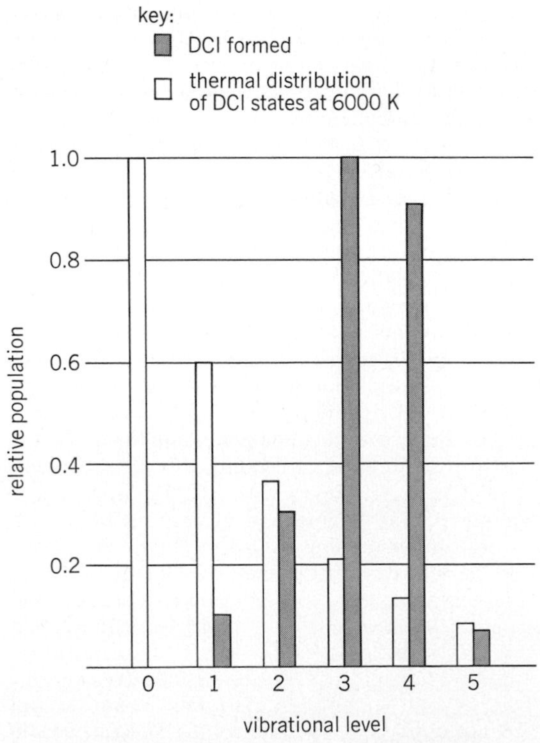

Fig. 2. Relative populations of DCl molecules in different vibrational states formed in the reaction $D + Cl_2 \rightarrow DCl + Cl$.

crossing two tenuous molecular beams in a region of otherwise high vacuum. Figure 3 shows such an experiment. Gaseous atoms or molecules emerge from the ovens, and collimating slits select the particles which are all going in the same direction. The molecular beams cross at the center of rotation of a large platform which can be rotated under vacuum relative to the two beams. Large vacuum pumps maintain a high vacuum, which ensures that collisions take place only at the intersection of the two beams. Product molecules are ionized by electron bombardment, and detected with a quadrupole mass spectrometer housed within a region of ultrahigh vacuum.

Measurements are made of the scattered product intensity and speed at various scattering angles. For ease of interpretation, these data are transformed into the center-of-mass system in which the two reagents approach each other with equal and opposite momenta. Figure 4 shows a contour map of the DCl intensity in the center-of-mass coordinate system in which the D atom is incident from the left and the Cl_2 molecule is incident from the right. The product DCl recoils backward (in the direction from which the D came) in a broad but nonetheless anisotropic distribution. The speed of the product is high, and corresponds to about half of the reaction exoergicity appearing in translational recoil of the products, with the balance appearing in vibration and rotation of the DCl consistent with the chemiluminescence results.

The anisotropic product distribution shows that reaction occurs in a time less than a molecular rotation, ~ 1 picosecond. The partitioning of energy roughly equally between vibration and translation suggests that the major amount of energy is released in repulsion between the DCl and Cl. This repulsion is similar to that experienced by a Cl_2 molecule in photodissociation is shown in Fig. 4. Because the deuterium atom is so light, the direction in which the product is expelled is a measure of the orientation of the Cl_2 molecule during reaction. For this reaction the collinear arrangement D-Cl-Cl is preferred.

Other investigations. Molecular-beam machines (and to some extent chemiluminescence machines) have been modified in various ways to explore

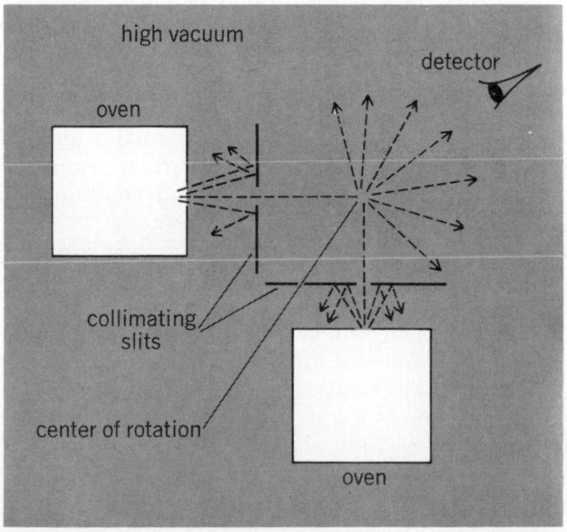

Fig. 3. Schematic diagram of a molecular beam experiment.

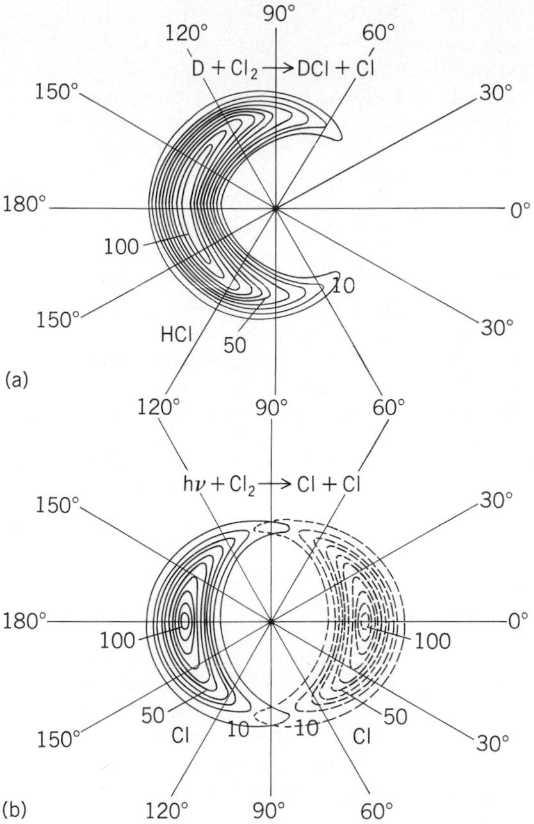

Fig. 4. Contour map of the DCl intensity. (*a*) Center-of-mass angular distribution of DCl from the reaction D+ Cl_2. (*b*) Angular distribution of Cl atoms from photodissociation of Cl_2 molecules. Broken line corresponds to the second Cl atom. (*From D. R. Herschbach. Pure Appl. Chem.. 47:61–73. 1976*)

even finer details of specific reactions. For example, electric and magnetic fields have been used to prepare reagents in various orientations or to determine the magnitude and polarization of the product rotational angular momentum; lasers have been used to prepare reagents in initial vibrational states and to determine the final vibration-rotation state of products by inducing fluorescence from the products; and reactions may be conducted at hyperthermal energies to produce dissociation or even ionization. Much work is concerned with the interaction of light with isolated molecules.

Unimolecular reactions. Unimolecular reactions of collisionally activated reaction complexes have been studied using both crossed-beam and chemiluminescence techniques. The availability of lasers opened new vistas in chemical dynamics, not only because lasers facilitate conventional measurements, as in detecting reaction products by laser-induced fluorescence, but also because they have uncovered new phenomena. Absorption of photons by molecules has been known for many years, and it has become apparent that under the right circumstances a molecule can absorb not one photon, but so many as to cause the molecule to dissociate or even to ionize.

These multiphoton dissociation (MPD) processes can yield different products from thermal reactions, and much confusion surrounded their discovery. The dissociation occurs after a sequence of single-photon absorptions. Absorption of the first photon is a resonant process occurring at a wavelength corresponding to a normal infrared absorption. For polyatomic molecules, a variety of normal vibrations and various combinations and differences exist, and the second photon can be absorbed to give not necessarily the $v = 2$ state of the original normal mode, but possibly some new combination or difference state. Sequential photons are absorbed in this fashion to climb the vibrational ladder both vertically and horizontally. The density of available rotation-vibration levels grows exponentially to become a quasicontinuum where succeeding photons are always resonant. Evidence suggests that an isolated molecule absorbs photons until it is sufficiently activated, whereupon it decomposes randomly.

[PHILIP R. BROOKS]

THEORETICAL METHODS

The goal of chemical dynamics is to understand kinetic phenomena from the basic laws of molecular mechanics, and it is thus a field which sees close interplay between experimental and theoretical research.

Energy distribution. An important question regarding the dynamics of chemical reactions has to do with the product energy distribution in exothermic reactions. For example, because the HF molecule is more strongly bound than the H_2 molecule, reaction (6) releases a considerable amount of

$$F + H_2 \rightarrow HF + H \quad (6)$$

energy (more than 30 kcal/mole or 126 kJ/mol). The two possible paths for this energy release to follow are into translation, that is, with HF and H speeding away from each other, or into vibrational motion of HF.

In this case it is vibration, and this has rather dramatic consequences: the reaction creates a population inversion among the vibrational energy levels of HF—that is, the higher vibrational levels have more population than the lower levels—and the emission of infrared light from these excited vibrational levels can be made to form a chemical laser. A number of other reactions such as reaction (5) also give a population inversion among the vibrational energy levels, and can thus be used to make lasers.

Most effective energy. The rates of most chemical reactions are increased if they are given more energy. In macroscopic kinetics this corresponds to increasing the temperature, and most reactions are faster at higher temperatures. It seems reasonable, though, that some types of energy will be more effective in accelerating the reaction than others. For example, in reaction (7), which has

$$K + HCl \rightarrow KCl + H \quad (7)$$

been studied in a molecular beam, if HCl is vibrationally excited (by using a laser), this reaction is found to proceed approximately 100 times faster, while the same amount of energy in translational kinetic energy has a smaller effect. Here, therefore, vibrational energy is much more effective than translational energy in accelerating the reaction.

For reaction (6), however, translational energy is more effective than vibration in accelerating the reaction. The general rule of thumb is that vibrational energy is more effective for endothermic reactions (those for which the new molecule is less stable than the original molecule), while translational energy is most effective for exothermic reactions.

Lasers. As seen from above, lasers are also an important supplement to molecular-beam techniques for probing the dynamics of chemical reactions. Because they are light sources with a very narrow wavelength, they are able to excite molecules to specific quantum states (and also to detect what states molecules are in), an example of which is reaction (7). For polyatomic molecules—that is, those with more than two atoms—there is the even more interesting question of how the rate of reaction depends on which vibration is excited.

For example, when the molecule allyl isocyanide, $CH_2{=}CH{-}CH_2{-}NC$, is given sufficient vibrational energy, the isocyanide part (—NC) will rearrange to the cyanide (—CN) configuration. A laser can be used to excite a C-H bond vibrationally. An interesting question is whether the rate of the rearrangement process depends on which C-H bond is excited. Only with a laser is it possible to excite different C-H bonds and begin to answer such questions. This question of mode-specific chemistry, that is, the question of whether excitation of specific modes of a molecule causes specific chemistry to result, is a subject of great interest. (For the example above, the reaction is fastest if the C-H bond closest to the NC group is excited.) Mode-specific chemistry would allow much greater control over the course of chemical reactions, and it would be possible to accelerate the rate of some reactions (or reactions at one part of a molecule) and not others.

Models and methods. Many different theoretical models and methods have been useful in understanding and analyzing all of the phenomena described above. Probably the single most useful approach has been the calculation of classical trajectories. Assuming that the potential energy function or a reasonable approximation is known for the three atoms in reaction (6), for example, it is possible by use of electronic computers to calculate the classical motion of the three atoms. It is thus an easy matter to give the initial molecule more or less vibrational or translational energy, and then compute the probability of reaction. Similarly, the final molecule and atom can be studied to see where the energy appears, that is, as translation or as vibration.

It is thus a relatively straightforward matter theoretically to answer the questions and to see whether or not mode-specific excitation leads to significantly different chemistry than simply increasing the temperature under bulk conditions.

The most crucial step in carrying out these calculations is obtaining the potential energy surface—the potential energy as a function of the positions of the atoms—for the system. Figure 5 shows a plot of the contours of the potential energy surface for reaction (6). Even without carrying out classical trajectory calculations, it is possible to deduce some of the dynamical features of this reaction; for example, the motion of the system first surmounts a small potential barrier, and then slides down a steep hill, turning the corner at the bottom of the hill. It is clear that such motion will cause much of the energy released in going down the hill to appear in vibrational motion of HF.

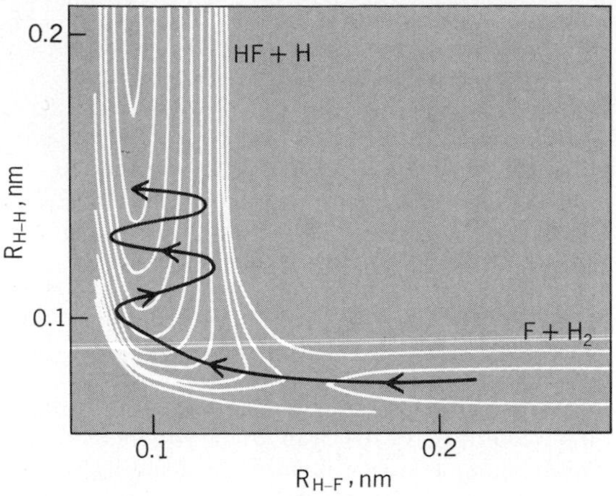

Fig. 5. Contour plot of the potential energy surface for the reaction F + H_2 = HF + H, with a typical reactive trajectory indicated.

This and other theoretical methods are interacting strongly with experimental research in helping to understand the dynamics of chemical reactions. *See* INORGANIC PHOTOCHEMISTRY; LASER PHOTOCHEMISTRY; PHOTOCHEMISTRY.

[WILLIAM H. MILLER]

RELAXATION METHODS

Since about 1950, considerable use has been made of perturbation techniques to measure rates and determine mechanisms of rapid chemical reactions. These methods provide measurements of chemical reaction rates by displacing equilibria. In situations where the reaction of interest occurs in a system at equilibrium, perturbation techniques called relaxation methods have been found most effective for determining reaction rate constants.

A chemical system at equilibrium is one in which the rate of a forward reaction is exactly balanced by the rate of the corresponding back reaction. Examples are chemical reactions occurring in liquid solutions, such as the familiar equilibrium in pure water, shown in reaction (8). The molar equi-

$$H_2O \underset{k_b}{\overset{k_f}{\rightleftharpoons}} H^+(aq) + OH^-(aq) \quad (8)$$

librium constant at 25°C is given by Eq. (9), where

$$K_{eq} = \frac{[H^+][OH^-]}{[H_2O]} = \frac{10^{-14}}{55.5} = 1.8 \times 10^{-16} \quad (9)$$

bracketed quantities indicate molar concentrations. It arises naturally from the equality of forward and backward reaction rates, Eq. (10). Here

$$k_f[H_2O] = k_b[H^+][OH^-] \quad (10)$$

k_f and k_b are the respective rate constants that depend on temperature but not concentrations. Furthermore, the combination of Eqs. (9) and (10) gives rise to Eq. (11). Thus a reasonable question

$$K_{eq} = k_f/k_b = 1.8 \times 10^{-16} \quad (11)$$

might be what the numerical values of k_f in units of s^{-1} and k_b in units of $dm^3\ mol^{-1}\ s^{-1}$ must be to satisfy Eqs. (9) through (11) in water at room temperature. Stated another way, when a liter of 1 *M* hydrochloric acid is poured into a liter of 1 *M* sodium hydroxide (with considerable hazardous sputtering), how rapidly do the hydronium ions, $H^+(aq)$, react with hydroxide ions, $OH^-(aq)$, to produce a warm 0.5 *M* aqueous solution of sodium chloride? In the early 1950s it was asserted that such a reaction is instantaneous. Turbulent mixing techniques were (and still are) insufficiently fast (mixing time of the order of 1 ms) for this particular reaction to occur outside the mixing chamber. The relaxation techniques were conceived by M. Eigen, who accepted the implied challenge of measuring the rates of seemingly immeasurably fast reactions.

The essence of any of the relaxation methods is the perturbation of a chemical equilibrium (by a small change in temperature, pressure, electric-field intensity, or solvent composition) in so sudden a fashion that the chemical system, in seeking to reachieve equilibrium, is forced by the comparative slowness of the chemical reactions to lag behind the perturbation (Fig. 6).

Temperature jump. Reaction (8) has a nonzero standard enthalpy change, ΔH^0, associated with it, so that a small increase in the temperature of the water requires the concentrations $[H^+]$ and $[OH^-]$ to increase slightly, and $[H_2O]$ to decrease correspondingly, for chemical equilibrium to be restored at the new higher temperature. Thus a small sample cell containing a very pure sample of water may be made one arm of a Wheatstone conductance bridge, and further configured so that a pulse of energy from a microwave source (or infrared laser of appropriate wavelength) is dissipated in the sample liquid. The resulting ~2° rise in temperature will produce a small increase in conductance that will have an exponential shape and a time constant or relaxation time $\tau \simeq 27\ \mu s$; τ is the time required for the signal amplitude to drop to $1/e = 1/2.718$ of its initial value, where *e* is the base of natural logarithms.

In pure water at 25°C, $[H^+] = [OH^-] = 10^{-7}\ M$, and for small perturbations, the value for τ is given by Eq. (12), from which it follows that $k_b \simeq 1.8 \times 10^{11}\ dm^3\ mol^{-1}\ s^{-1}$. This is an exceptionally large rate constant for a bimolecular reaction between oppositely charged ions in aqueous solution and is, in fact, larger than that for any other diffusive encounter between ions in water. Eigen and L. De-

$$\tau^{-1} = k_b([H^+] + [OH^-]) + k_f = k_b([H^+] + [OH^-] + K_{eq}) \quad (12)$$

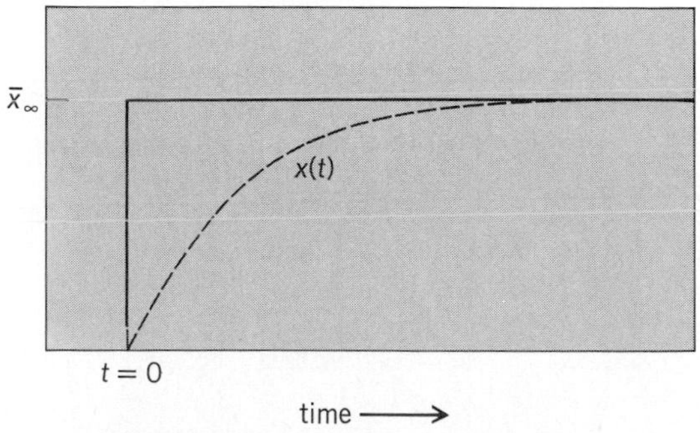

Fig. 6. Relaxational response to a rectangular step function in an external parameter such as temperature or pressure. The broken line represents the time course of the adjustment (relaxation) of the chemical equilibrium to the new temperature or pressure. *(From C. Bernasconi, Relaxation Kinetics, Academic Press, 1976)*

Maeyer, who first determined this rate constant (using another relaxation method called the electric-field jump method), attributed the great speed of the back reaction of the equilibrium, reaction (8), to the exceptionally rapid motion of a proton through water, accomplished by the successive rotations of a long string of neighboring water molecules (Grotthuss mechanism). Since sample solutions can be heated by a mode-locked laser on a picosecond time scale or by a bunsen burner on a time scale of minutes, the temperature jump (T-jump) relaxation method just described is very versatile. The choice of the particular means of effecting the temperature perturbation is dictated only by the requirement that the temperature rise somewhat more rapidly than the time constant of the chemical reaction to be explored, so that a tedious deconvolution can be avoided. The discharge of a high-voltage (15- to 30-kV) capacitor through the sample liquid containing sufficient inert electrolyte to make it a good electrical conductor is the now classic Joule heating T-jump method used by Eigen and coworkers in their pioneering studies. A schematic of such an apparatus is shown in Fig. 7. The 30-kV voltage generator charges the 0.1-microfarad condenser to the voltage at which the spark gap breaks down. The condenser then discharges across the spark gap and through the sample cell, containing an aqueous 0.1 *M* ionic strength solution, to ground. The sample cell is a ~50 ml Plexiglas cell containing two platinum electrodes spaced 1 cm apart and immersed in an aqueous 0.1 *M* ionic strength solution. The surge of current raises the temperature of the 1-ml volume of solution between the electrodes by 10°C in a few microseconds.

Electric-field jump. In a situation, such as reaction (8), in which electrically neutral reactant species dissociate into oppositely charged ions, an especially sensitive tool for measuring rate constants of forward and backward reactions is the electric-field jump (E-jump) technique with conductometric detection. In a strong electric field (of the order of $4 \times 10^6 \text{V m}^{-1}$), a weak acid in solution is caused to dissociate to a greater degree than it would in the absence of the electric field. For weak electrolytes, such as aqueous acetic acid or ammonia, the effect is of the order of 10% or less of the total normal dissociation, even at very high electric-field strengths. However, with a sensitive, high-voltage Wheatstone bridge, the exponential increase with time in the concentration of ions following a precipitous increase in electric-field strength is readily detected. The measured relaxation time (τ) is clearly that corresponding to the high-electric-field environment, but since the rate constants for these reactions differ little in and out of the electric field, no serious problem is posed.

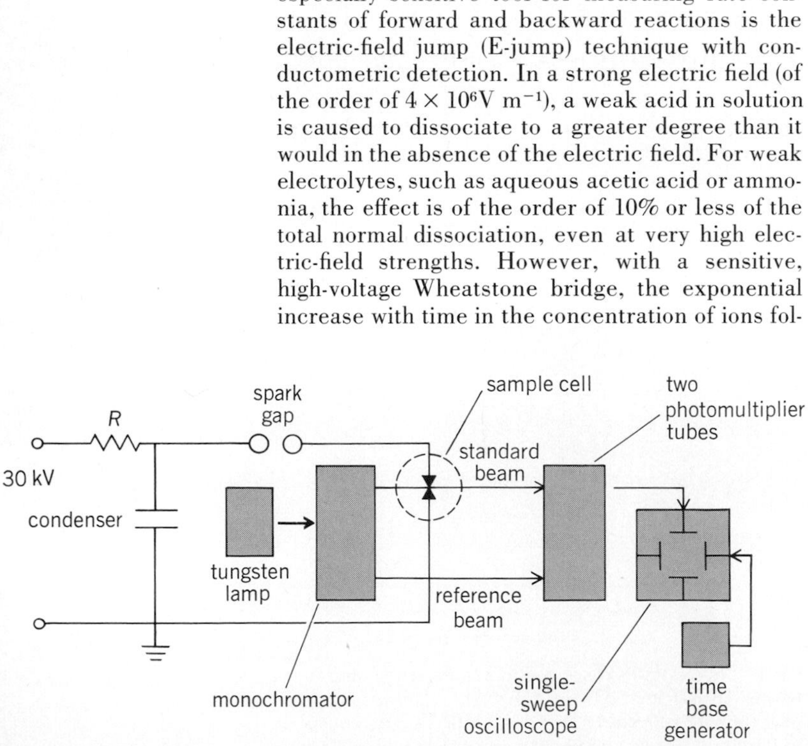

Fig. 7. Schematic of a Joule heating temperature-jump apparatus. (*From H. Eyring and E. M. Eyring, Modern Chemical Kinetics, Reinhold, 1963*)

A more serious concern is that the sample solution may have a very high electrical resistance, so that the supposedly square step function in the electric-field strength is not distorted by a significant voltage drop with concomitant heating of the sample liquid. Problems of working with high voltages, balancing capacitive and inductive effects in a very sensitive conductance bridge (now often circumvented by spectrophotometric detection), and the comparative difficulty of evaluating amplitudes of relaxations (as opposed to their readily determined time constants) are all factors that have worked against the wide use of the E-jump technique. There are many more ways of achieving a T-jump than an E-jump, and ΔH^0 values for chemical equilibria are readily available in the thermodynamic literature, whereas the extent to which a chemical equilibrium is displaced by an electric-field increment is rarely already known and is difficult to determine. Thus the commercialization of the T-jump method and the comparative neglect of the E-jump relaxation technique are readily understood.

Notwithstanding these difficulties, the E-jump technique is without peer for the investigation of the kinetics of solvent autoionization or for the exploration of the properties of weak electrolyte solutes in exotic solvents such as acetonitrile or xenon (the latter liquefied under a pressure of ~50 atm or 5 MPa), so long as the relaxation time to be measured lies in the range 30 ns $< \tau <$ 100 μs.

Ultrasonic absorption. Two other relaxation methods more widely used than the E-jump technique are pressure jump (P-jump) and ultrasonic absorption. Each relies for its effectiveness on a volume change, ΔV^0, occurring in an aqueous sample equilibrium undergoing kinetic investigation. (In a nonaqueous solvent it will frequently be more important that ΔH^0 be large than that ΔV^0 be so for the equilibrium to be susceptible to study by these two relaxation techniques.) As electrically neutral, weak electrolyte solute species dissociate into ions in aqueous solution, there is an increase in the number of solvent molecules drawn into a highly ordered solvation sheath. The higher the charge density of the ion, the more water molecule dipoles are bound and the greater the change in V^0 as reactants become products. Thus the dissociation of an aqueous neodymium(III) sulfate complex is particularly susceptible to study by one or more of the four or five ultrasonic absorption methods that cover the $f \approx 100$ kHz – 1 GHz sound frequency range. Unlike the T-jump and E-jump relaxation methods, which usually employ step function perturbations, the ultrasonic absorption techniques are continuous-wave experiments in which the sample chemical equilibrium absorbs a measurable amount of the sound wave's energy when the frequency of the sound wave (*f*) and the relaxation time of the chemical equilibrium bear the relation to one another given by Eq. (13).

$$\tau^{-1} = 2\pi f \qquad (13)$$

A particularly easy ultrasonic absorption experiment to understand and perform is the laser Debye-Sears technique. A continuously variable frequency sound wave is introduced by a quartz piezoelectric transducer into a 30-ml sample cell that has entrance and exit windows for a visible laser light beam that passes through the cell at about 90° to the direction of travel of the planar sound wave. The regions of compression and rarefaction in the sound wave act as a diffraction grating for the laser light beam. If a chemical equilibrium in the sample strongly absorbs a particular frequency of sound (f), the definition of the "diffraction grating" will deteriorate and the measured intensity of the first-order diffracted laser light will diminish. The frequency of minimum diffracted light intensity will be that of Eq. (12). Figure 8

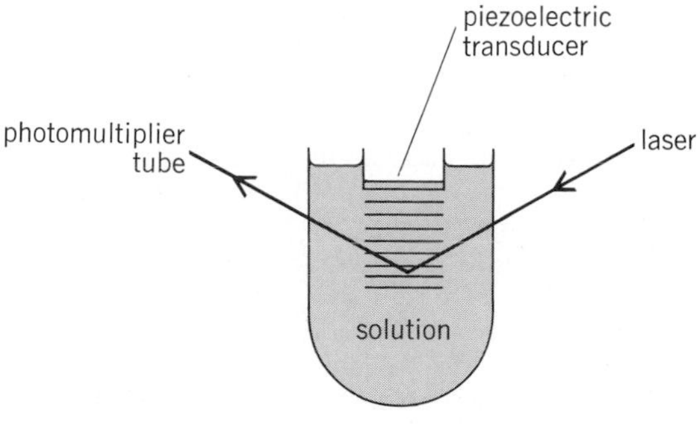

Fig. 8. Schematic of laser Debye-Sears apparatus for measuring ultrasonic absorption (~15–300 MHz) in a sample liquid. *(From W. J. Gettins and E. Wyn-Jones. Techniques and Applications of Fast Reactions in Solution. D. Reidel. Dordrecht. 1979)*

shows a diagram of the apparatus. The piezoelectric (quartz) transducer is cemented to the bottom of a plastic rod that is driven up and down by a stepping motor that rotates a micrometer. The stepping motor is controlled by a mini-computer. The angle of diffraction of the laser beam by the alternating regions of compression and rarefaction in the liquid (suggested by the horizontal lines) is exaggerated in the diagram.

Ultrasonic absorption techniques have been used in kinetic investigations of quite complicated biophysical systems such as the order-disorder transitions that occur in liquid crystalline phospholipid membranes. While the ultrasonic techniques look through a conveniently broad time window at kinetic processes in solutions, this picture window is difficult to "see through" in the sense that many equilibrium processes in solution can absorb sound energy and the responsible process is not instantly identified by a characteristic absorption of electromagnetic radiation as in a spectrophotometric T-jump or E-jump experiment. A further disadvantage arises from the great breadth of the ultrasonic absorption "peaks" in a plot of normalized sound absorption versus sound frequency. Unless multiple relaxation times in a chemical system are quite widely separated in time, they are difficulty to resolve in an ultrasonic absorption spectrum.

Pressure jump. The typical pressure-jump (P-jump) experiment is one in which a liquid sample under about 200 atm (20 MPa) pressure is suddenly brought to atmospheric pressure by the bursting of a metal membrane in the sample cell autoclave. Relaxation times measured spectrophotometrically or conductometrically are thus accessible if $\tau > 100\ \mu s$. This technique has proven particularly useful in the elucidation of micellar systems of great interest for catalysis and for petroleum recovery from apparently depleted oil fields.

The continuous- and stopped-flow techniques antedate somewhat the relaxation techniques described above, and have the sometimes important advantage of permitting kinetic measurements in chemical systems far from equilibrium. The stopped-flow experiment is one in which two different liquids in separate syringes are mixed rapidly in a tangential jet mixing chamber and then the rapid flow of mixed reactants is almost immediately brought to a halt in a spectrophotometric, conductometric, or calorimetric observation chamber. Reaction half-lives exceeding 2 ms are easily accessible.

Other relaxation methods. Stopped-flow equipment has been used in concentration-jump and solvent-jump relaxation kinetic studies. An example of an application of the solvent-jump technique to a system insensitive to concentration-jump is a kinetic study of reaction (14) in mixed CCl_4 – acetic

$$NOCl + n\text{-}BuOH \rightarrow n\text{-}BuONO + HCl \qquad (14)$$

acid solvents of varying composition (Bu = butyl). The thermodynamic treatment of the solvent jump is just about the only aspect of the presently known relaxation techniques that was not described in exhaustive detail by the earliest publications of Eigen and DeMaeyer. *See* CHEMICAL THERMODYNAMICS.

[EDWARD M. EYRING]

Bibliography: C. F. Bernasconi, *Relaxation Kinetics*, 1976; R. B. Bernstein (ed.), *Atom Molecule Collision Theory*, 1979; P. R. Brooks and E. F. Hayes, *State-to-State Chemistry, ACS Symp. Ser. no. 56*, 1977; E. F. Caldin, *Fast Reactions in Solution*, 1964; G. H. Czerlinski, *Chemical Relaxation*, 1966; M. Eigen, *Nobel Lecture: Immeasurably Fast Reactions*, 1967; M. Eigen and L. DeMaeyer, in *Technique of Organic Chemistry*, vol. 8, pt. 2, 1963; W. J. Gettins and E. Wyn-Jones, *Techniques and Applications of Fast Reactions in Solution*, 1979; D. N. Hague, *Fast Reactions*, 1971; G. G. Hammes (ed.), *Investigation of Rates and Mechanisms of Reactions*, 1974; D. R. Herschbach, Molecular dynamics of chemical reactions, *Pure Appl. Chem.*, 47:61–73, 1976; K. Kustin (ed.), Fast reactions, *Methods in Enzymology*, vol. 16, 1969; R. D. Levine and R. B. Bernstein, *Molecular Reaction Dynamics*, 1974; W. H. Miller (ed.), *Dynamics of Molecular Collisions*, 1976.

Chemical engineering

The branch of engineering serving those industries that chemically convert basic raw materials into a variety of products. Starting with ores, salt, sulfur, limestone, coal, natural gas, petroleum, air, water, and so forth, these industries, by chemical processing techniques, produce widely diversified products, such as aluminum, magnesium, and titanium metals; refined petroleum fractions, such as fuels and solvents; synthetic fertilizers; synthetic fibers, resins, and plastics; antibiotics; wood pulp and paper; and petrochemicals.

All products from this family of industries are

formed in chemical processes involving chemical reactions carried out under a wide range of conditions and frequently accompanied by changes in physical state or form. Many steps prior to, between, and after the chemical reactions may be carried out to purify the raw materials and to separate and purify intermediate and final products. These steps involve a number of physical or unit operations that heat and cool, distill, crystallize, filter, mix, and so forth. Along with the chemical reactions they make up a chemical process. *See* CHEMICAL CONVERSION; CHEMICAL PROCESS INDUSTRY.

Functions of engineer. To bring about the conversion of raw materials into such a broad spectrum of products, the chemical engineer starts with basic scientific data on chemical reactions discovered by research chemists. Working first on a small laboratory scale, the engineer develops engineering data for various steps in the process of changing the starting material into a product. Before completing this small-scale investigation, he may actually operate various parts of the process in mini- or pilot-scale equipment, simulating what may happen in the commercial-scale process.

In carrying out these development functions, the chemical engineer relies on his training in scientific and engineering fundamentals and in advanced mathematics to plan the experimental programs and to analyze the resulting data. In his concern for designing the overall process, the chemical engineer may evaluate many alternate schemes before reaching an optimum design. Economic evaluation techniques play an important role in achieving optimum design.

In many cases the engineer will evolve a mathematical model that interrelates all the process variables in mathematical terms. Then by using electronic computing equipment, the design engineer can attain the optimum process design. Ultimately, when the plant is built, he will rely on an on-line compute and sophisticated instrumentation to control the commercial unit.

The plant installation, designed and constructed under the overall direction of a chemical engineer serving as the project manager, usually requires millions of dollars in capital investment. Such a plant, in turn, may be part of a larger complex that is anywhere up to 10 or 15 times as large as the largest of these single plants.

After completion of the plant, chemical engineering continues to play a vital role in the start-up and operation of the facility. Once in operation, the manufacturing unit becomes a continuing problem in optimization to achieve desired product output and specified quality at lowest cost in the face of constantly shifting variables of raw material supply, availability and quality of personnel, condition of plant equipment, and market conditions.

Professional background. There were some 55,000 practicing chemical engineers in the United States, according to the Engineering Manpower Commission. While a large portion of these engineers are engaged in the activities outlined above, others are teaching, consulting, working for engineering firms and equipment suppliers that serve the chemical process industries, or carrying out business functions associated with marketing and selling the industries' products. Still others are applying chemical engineering technology in new fields, such as atomic energy, space, and oceanography.

For their professional group activities, such as technical meetings and seminars devoted to continuing education, chemical engineers work to a large extent through the American Institute of Chemical Engineers. Headquartered in New York, the A.I.Ch.E. has more than 100 local sections throughout the United States. It also supports an extensive technical publishing program.

Until 1968 young men and women trained for the chemical engineering profession by completing a four-year bachelor's degree program at one of 112 engineering colleges accredited by A.I.Ch.E. and by the Engineer's Council for Professional Development. The basic chemical engineering curriculum was established as a uniquely American development when William H. Walker and Arthur D. Little introduced the unit operations concept at Massachusetts Institute of Technology in 1915. Unit operations involve the application or removal of some form of energy in the contacting, separating, transporting, and conditioning of materials by physical means, with or without any chemical changes taking place.

Since that time, however, chemical engineering schools have increased the proportion of science and mathematics in the curriculum. And at the graduate level the schools have diversified the training to include options in design, administration, and management as alternatives to the conventional research-oriented programs.

In 1968 the Goals Committee of the American Society of Engineering Education recommended that professional engineering status should require completing a five-year master's degree program, relegating the four-year program to subprofessional status. However, the A.I.Ch.E. later adopted the view that suitably trained four-year bachelors in chemical engineering are ready to assume professional responsibilities.

For some important related engineering fields *see* ELECTROCHEMICAL PROCESS.

[CALVIN S. CRONAN]

Bibliography: C. A. Clausen and G. C. Mattson, *Principles of Industrial Chemistry*, 1978; T. B. Drew and J. W. Hooper, Jr. (eds.), *Advances in Chemical Engineering*, vols. 1–10, 1956–1978; O. A. Hougen et al., *Chemical Process Principles*, 2d ed., 1954–1960; R. E. Kirk and D. F. Othmer, *Encyclopedia of Chemical Technology*, 3d ed., 1978–1980; R. H. Perry and C. H. Chilton, *Chemical Engineers Handbook*, 5th ed., 1973, J. R. Richardson and D. G. Peacock (eds.), *Chemical Engineering*, 1979.

Chemical equilibrium

In a dynamic or kinetic sense, chemical equilibrium is a condition in which a chemical reaction is occurring at equal rates in its forward and reverse directions, so that the concentrations of the reacting substances do not change with time. In a thermodynamic sense, it is the condition in which there is no tendency for the composition of the system to change; no change can occur in the system without the expenditure of some form of work upon it. From the viewpoint of statistical mechanics, the equilibrium state places the system in a condition of maximum freedom (or minimum restraint) compatible with the energy, volume, and composition of the system. The statistical approach has been merged with thermodynamics into a field called statistical thermodynamics; this merger has been of immense value for its intellectual stimulus, as well as for its practical contributions to the study of equilibria. *See* CHEMICAL THERMODYNAMICS.

Of the three viewpoints, the thermodynamic approach is by far the most powerful and fruitful in treating the quantitative relationships between the position of equilibrium and the factors which govern it. Since thermodynamics is concerned with relationships among observable properties, such as temperature, pressure, concentration, heat, and work, the relationships possess general validity, independent of theories of molecular behavior. Because of the simplicity of the concepts involved, this article will utilize that approach.

Chemical potential. Thermodynamics attributes to each chemical substance a property called the chemical potential, which may be thought of as the tendency of the substance to enter into chemical (or physical) change. Although the chemical potential of a substance cannot be directly measured (except on a relative basis), differences in chemical potential are measurable. (The units are those of energy per mole.)

The importance of the chemical potential lies in its relation to the affinity or driving force of a chemical reaction. Consider the general reaction shown in (1). Let μ_A be the chemical potential per

$$aA + bB \rightleftharpoons gG + hH \qquad (1)$$

mole of substance A, μ_B be the chemical potential per mole of B, and so on. Then, according to one of the fundamental principles of thermodynamics (the second law), the reaction will be spontaneous when the total chemical potential of the reactants is greater than that of the products. Thus, for spontaneous change (naturally occurring processes), (2) applies. When equilibrium is reached, the

$$[g\mu_G + h\mu_H] - [a\mu_A + b\mu_B] < 0 \qquad (2)$$

total chemical potentials of products and reactants become equal; thus Eq. (3) holds at equilibrium.

$$[g\mu_G + h\mu_H] - [a\mu_A + b\mu_B] = 0 \qquad (3)$$

The difference in chemical potentials in Eqs. (2) and (3) is called the driving force or affinity of the process or reaction; naturally, it is zero when the chemical system is in chemical equilibrium.

For reactions at constant temperature and pressure (the usual restraints in a chemical laboratory), the difference in chemical potentials becomes equal to the free energy change ΔG for the process in Eq. (4). The decrease in free energy represents

$$\Delta G = [g\mu_G + h\mu_H] - [a\mu_A + b\mu_B] \qquad (4)$$

the maximum net work obtainable from the process. When no more work is obtainable, the system is at equilibrium. Conversely, if the value of ΔG for a process is positive, some useful work will have to be expended upon the process, or reaction, in order to make it proceed; the process cannot proceed naturally or spontaneously. (The term spontaneously as used here implies only that a process can occur. It does not imply that the reaction will be rapid or instantaneous. Thus, the reaction between hydrogen and oxygen is a spontaneous process in the sense of the term as used here, even though a mixture of hydrogen and oxygen can remain unchanged for years unless ignited or exposed to a catalyst.)

Since by definition a catalyst remains unchanged chemically throughout a reaction, its chemical potential does not appear in Eqs. (2), (3), and (4). A catalyst, therefore, can contribute nothing to the driving force of a reaction, nor can it, in consequence, alter the position of the chemical equilibrium in a system. *See* CATALYSIS.

In addition to furnishing a criterion for the equilibrium state of a chemical system, the thermodynamic method goes much further. In many cases, it yields a relation between the change in chemical potentials (or change in free energy) and the equilibrium concentrations of the substances involved in the reaction. To do this, the chemical potential must be expressed as a function of concentration (and other properties of the substance).

The chemical potential μ is usually represented by Eq. (5), where R is the ideal gas constant, T is

$$\mu = \mu^0 + RT \ln x + RT \ln f \qquad (5)$$

the absolute temperature, $\ln x = \log_e x = 2.3026 \log_{10} x$, x is the concentration of the substance, f is the activity coefficient of the substance, and μ^0 is the chemical potential of the substance in its standard state. *See* CONCENTRATION SCALES.

For substances obeying the laws of ideal solutions (or ideal gases), the last term, $RT \ln f$, is zero, since it is a measure of the deviation from ideal behavior caused by intermolecular or interionic forces. An ideal solution would then be a solution for which $RT \ln f$ is zero over the whole concentration range. For real solutions, the ideal or reference state, where f is unity, is generally chosen as a state of ideal purity (mole fraction = 1) for solids and solvents, and as a state of infinite dispersion (concentration = 0) for gases and solutes. Although the activity coefficient f is regarded as dimensionless, its numerical values will depend upon the particular concentration scale x with which it must be associated.

Although the choice of concentration scales is somewhat a matter of convenience, the following are conventionally used: $x = p$ = partial pressure, for gases; $x = c$ or m = molar or molal concentrations, for solutes in electrolytic solutions; x = mole fraction, for solids and solvents. When the choice is not established by convention, the mole fraction scale is to be preferred.

Activity and standard states. It is often convenient to utilize the product, fx, called the activity of the substance and defined by $a = fx$. The activity may be looked upon as an effective concentration of the substance, measured in the same units as the concentration x with which it is associated. The standard state of the substance is then defined as the state of unit activity (where $a = 1$) and is characterized by the standard chemical potential μ^0. Clearly, the terms μ^0, f, and x are not independent; the choice of the activity scale serves to fix the standard state. For example, for an aqueous solution of hydrochloric acid, the standard state for the solute (HCl) would be an (hypothetical) ideal 1 molar (or molal) solution, and for the solvent (H_2O) the standard state would be pure water (mole fraction = 1). The reference state would be an infinitely dilute solution; here the activity coefficients would be unity for both solute and solvent. For the vapor of HCl above the solution, the standard state would be the ideal gaseous state at 1 atm partial pressure; the reference state would be a state of zero pressure. (For gases, the term fugacity is used instead of activity.) It should be noted that the reference state is a limiting state which in many cases can be reached only through an extrapolation from observed behavior.

Equilibrium constant. If the general reaction in Eq. (1) occurs at constant temperature T and pressure P when all of the substances involved are in their standard states of unit activity, Eq. (4) would become Eq. (6). The quantity ΔG^0 is known as the

$$\Delta G^0 = [g\mu^0{}_G + h\mu^0{}_H] - [a\mu^0{}_A + b\mu^0{}_B] \qquad (6)$$

standard free energy change for the reaction at that temperature and pressure for the chosen standard states. (Standard state properties are commonly designated by a superscript, ΔG^0, μ^0.) Since each of the standard chemical potentials (μ^0) is a unique property determined by the temperature, pressure, standard state, and chemical identity of the substance concerned, the standard free energy change ΔG^0 is a constant (parameter) characteristic of the particular reaction for the chosen temperature, pressure, and standard states.

If, in a reaction, Eq. (1), at constant temperature and pressure, the chemical potential of each substance is expressed in terms of Eq. (5), the free energy change for the reaction, from Eq. (4), becomes Eq. (7) in terms of the activities and the

$$\Delta G = \Delta G^0 + RT \ln \frac{a_G{}^g a_H{}^h}{a_A{}^a a_B{}^b} \qquad (7)$$

standard free energy change, Eq. (6). Equation (7) is often written in the form of Eq. (8), where Q^0 is the

$$\Delta G = \Delta G^0 + RT \ln Q^0 \qquad (8)$$

ratio of the activities of products to the activities of reactants, each activity bearing as an exponent the corresponding coefficient in the balanced equation for the reaction. The standard free energy change ΔG^0 serves as a reference point from which the actual free energy change ΔG can be calculated in terms of the activities of the reacting substances.

When the system has come to chemical equilibrium at constant temperature and pressure, $\Delta G = 0$, from Eq. (3). Equation (7) then leads to the very important relation shown in Eq. (9), where the

$$\Delta G^0 = -RT \ln K^0 \qquad (9)$$

value of K^0 is shown as Eq. (10), and the activities

$$K^0 = \left[\frac{a_G{}^g a_H{}^h}{a_A{}^a a_B{}^b}\right] \qquad (10)$$

are the equilibrium values. The ratio of the activities at equilibrium, K^0, is called the equilibrium constant or, more precisely, the thermodynamic equilibrium constant. (The terms K^0 and Q^0 are written with superscripts to emphasize that they represent ratios of activities.) The equilibrium constant is a characteristic property of the reaction system, since it is determined uniquely in terms of the standard free energy change. The term $-\Delta G^0$ represents the maximum net work which the reaction could make available when carried out at constant temperature and pressure with the substances in their standard states. It should be clear from Eq. (6) that the magnitude of ΔG^0 is directly proportional to the amount of material represented in the reaction in Eq. (1). Likewise, Eqs. (7) to (10) denote this same proportionality through the exponents a, b, g, and h in the terms K^0 and Q^0. Naturally, the value of ΔG^0, as well as K^0 and Q^0, will depend upon the particular concentration scales and standard states selected for the system, so it is essential that sufficient information be stated about a system to prevent any ambiguity.

Equations (8) and (10) can be combined in the form of Eq. (11). When Q^0 for a specified set of

$$\Delta G = RT \ln \frac{Q^0}{K^0} \qquad (11)$$

conditions is larger than K^0 (so that ΔG is positive), the proposed reaction cannot occur. On the other hand, when Q^0 is less than K^0, the proposed process or reaction can occur. The equilibrium constant thus serves as a measure of the position of chemical equilibrium for a system. For the proposed process for which Q^0 is greater than K^0, the reaction system would be moving away from its equilibrium state (impossible of its own accord!), and for a proposed process for which Q^0 is less than K^0, the process would bring the system closer to its equilibrium state (as in all naturally occurring processes).

Instead of activities, values of concentrations and activity coefficients at equilibrium may be used to express the form of the equilibrium constant K^0 in Eq. (12). This gives the equilibrium constant

$$K^0 = \left[\frac{x_G{}^g x_H{}^h}{x_A{}^a x_B{}^b}\right] \cdot \left[\frac{f_G{}^g f_H{}^h}{f_A{}^a f_B{}^b}\right] \qquad (12)$$

as a product of two terms, each of the same form as K^0 itself. The first term, involving concentrations, is directly measurable if the system can be analyzed at equilibrium. On the other hand, the activity coefficient term, as seen in Eq. (13), is frequently difficult to evaluate.

$$\Gamma = \frac{f_G{}^g f_H{}^h}{f_A{}^a f_B{}^b} \qquad (13)$$

Intensive studies of activity coefficients made upon a wide variety of chemical systems have led to a number of simplifying principles and some useful theoretical treatments of the subject. For gases, the activity coefficients differ only slightly from unity for pressures up to 10 atm and can be evaluated from equation-of-state data. For mixtures of nonelectrolytes, the values also appear to be close to unity in many cases. For solutions of electrolytes, the activity coefficients vary greatly with concentration, and in many cases approach unity only below a useful or even meaningful concentration. The theoretical treatments of P. Debye, E. Hückel, and others have systematized the patterns of electrolyte behavior, making possible a reasonable estimate of the activity coefficients in many cases. Many tables of experimental data are available for electrolytes. *See* FUGACITY; SOLUTION.

In general, the function Γ, Eq. (13), approaches unity as the composition of the system approaches that of the reference state, so in practice most equilibrium constants K^0 are evaluated through some suitable extrapolation procedure involving Eq. (12). See the discussion following Eq. (5).

For many approximate calculations or when data for Γ are scarce, it is common to express the equilibrium constant as the concentration term only; that is, Eq. (14) holds. Unless Γ is a rather

$$K = \frac{x_G^{\,g} x_H^{\,h}}{x_A^{\,a} x_B^{\,b}} \tag{14}$$

insensitive function of concentration, the so-called constants obtained in this manner will not be constant at all as the composition is varied, and even though approximately constant, may vary considerably from the true value of K^0. Although the practice of assigning Γ a value of unity will often give adequate results and is frequently the only expedient available, the results should be used with caution.

It is appropriate to point out here that the kinetic concept of chemical equilibrium introduced by C. M. Guldberg and P. Waage (1864) led to the formulation of the equilibrium constant in terms of concentrations. Although the concept is correct in terms of the dynamic picture of opposing reactions occurring at equal speeds, it has not been successful in coping with the problems of activity coefficients and cannot lead to the useful relations, Eqs. (4) to (11), in terms of an energetic criterion for the position of equilibrium. Conversely, the thermodynamic approach yields no relationship between the driving force of the reaction and the rate of approach to equilibrium. *See* CHEMICAL DYNAMICS.

The influence of temperature upon the chemical potentials, and hence upon the equilibrium constant, is given by the Gibbs-Helmholtz equation, Eq. (15). The derivative on the left represents the

$$\left[\frac{d \ln K^0}{dT}\right]_P = \frac{\Delta H^0}{RT^2} \tag{15}$$

slope of the curve obtained when values of ln K^0 for a reaction, obtained at different temperatures but always at the same pressure P, are plotted against temperature. The standard heat of reaction ΔH^0 for the temperature T at which the slope is measured, is the heat effect which could also be observed by carrying out the reaction involving the standard states in a calorimeter at the corresponding temperature and pressure.

For endothermic reactions, which absorb heat (ΔH^0 positive), K^0 increases with increasing temperature. For exothermic reactions, which evolve heat (ΔH^0 negative), K^0 decreases with increasing temperature and the yield of products is reduced. A more useful arrangement of Eq. (15) is shown in Eq. (16). In practice, plots of ln K^0 against $1/T$ are

$$\left[\frac{d \ln K^0}{d(1/T)}\right]_P = \frac{-\Delta H^0}{R} \tag{16}$$

nearly linear for many reactions where the value of ΔH^0 changes slowly with temperature. Hence, over small temperature ranges, Eq. (16) becomes, in integrated form, Eq. (17). This relation is much

$$\ln \frac{K_2^0}{K_1^0} = \frac{\Delta H^0}{R}\left[\frac{T_2 - T_1}{T_1 T_2}\right] \tag{17}$$

used for calculating heats of reaction from two equilibrium measurements or for determining a new equilibrium constant K_2^0 from values of K_1^0 and ΔH^0.

For accurate work, or for extending the calculations over a wide range of temperature, ΔH^0 must be known as a function of temperature before Eq. (15) or (16) can be integrated. When sufficient heat capacity data are available, Kirchhoff's equation, involving the difference in heat capacities between the products and reactants, Eq. (18), may be com-

$$\Delta H^0 = \Delta H_0^{\,0} + \alpha T + \frac{\beta T^2}{2} + \frac{\gamma T^3}{3} \tag{18}$$

bined with Eq. (15) to yield Eq. (19). In these equa-

$$R \ln K^0 = -\frac{\Delta H_0^{\,0}}{T} + \alpha \ln T + \frac{\beta T}{2} + \frac{\gamma T^2}{6} + I \tag{19}$$

tions, α, β, and γ are determined from heat capacity data; the constants $\Delta H_0^{\,0}$ and I require knowledge of one value of ΔH^0 and one value of K^0, or values of K^0 at two temperatures. *See* THERMOCHEMISTRY.

When equilibria are studied under conditions of constant temperature and constant volume, and with use of volume concentrations to fix standard states, the preceding treatment will yield ΔE^0, the internal energy change, which is the calorimetric heat of reaction at constant volume.

Homogeneous equilibria. These involve single-phase systems: gaseous, liquid, and solid solutions. In most cases, solid solutions are so far from ideal that equilibrium constants cannot be evaluated, and such systems are treated in terms of the phase rule. *See* PHASE EQUILIBRIUM.

A typical gas-phase equilibrium is the ammonia synthesis shown in (20). The most natural con-

$$N_2 + 3H_2 \rightleftharpoons 2NH_3 \tag{20}$$

centration measures are mole fraction or partial pressure; molar concentrations might be used. The partial pressures p_i are defined in terms of mole fraction N_i and the total pressure P by $p_i = N_i P$; note that partial pressures in general are not directly observable. For low pressures, where activity coefficients are practically unity, Eq. (21) holds. In turn Eq. (21) may be shown as Eq. (22),

$$K_p^{\,0} = \frac{p^2_{NH_3}}{p_{N_2} p^3_{H_2}} = \frac{N^2_{NH_3}}{N_{N_2} N^3_{H_2}} P^{-2} \tag{21}$$

$$K_p^{\,0} = K_N^{\,0} P^{\Delta n} \tag{22}$$

where Δn is the increase in the number of moles of gases (here $\Delta n = 2$). The mole fraction equilibrium constant $K_N^{\,0}$ is pressure dependent, but $K_p^{\,0}$ is independent of pressure because of the difference in standard states. Consequently, an increase in total pressure P must lead to an increase in $K_N^{\,0}$ in this case. If the increase in total pressure is due to a decrease in volume of the system, the result will be an increased yield of products (NH_3). An increase in pressure brought about by injection of an inert gas into a constant volume system would not affect the partial pressures of the reacting gases nor the ultimate yield of products. The value of $K_N^{\,0}$, and thus that of $\Delta G_N^{\,0}$, will be affected by change in total pressure due to a change in the net work of mixing and unmixing the gases. This reaction, Eq. (20), is exothermic, so best yields will be obtained at lower temperatures. See Eq. (15).

A typical liquid-phase equilibrium is the dissociation of acetic acid in water, reaction (23) and Eq. (24). In this particular case, the activity co-

$$HC_2H_3O_2 + H_2O \rightleftharpoons H_3O^+ + C_2H_3O_2^- \qquad (23)$$

$$\frac{K^0}{\Gamma} = \frac{m_{H_3O^+} m_{C_2H_3O_2^-}}{m_{HC_2H_3O_2} N_{H_2O}} \qquad (24)$$

efficient ratio Γ is slightly less than unity in dilute solutions; the reference state will be the infinitely dilute solution. The mole fraction of water in the applicable concentration range is close to unity, and so numerically plays little part in evaluation of K^0. For this reason, it is commonly omitted in the formulation of the equilibrium constant, and many erroneous explanations exist in the literature. Letting α be the fraction of acetic acid dissociated and m be the total concentration, $K^0/\Gamma = m\alpha^2/(1-\alpha)$; inspection shows that α increases with dilution.

The solvent appears to be inert, since its chemical potential remains practically unchanged over the useful concentration range. As a result of this apparent inertness of the solvent, it is not possible to determine the extent of hydration of any dissolved species from equilibrium studies. Thus, whether the actual ion is H^+, H_3O^+, or $H_9O_4^+$, it is the total stoichiometric concentration that is measured and used in Eq. (24). *See* IONIC EQUILIBRIUM.

Heterogeneous equilibria. These are usually studied at constant pressure, since at least one of the phases will be a solid or liquid. The imposed pressure may be that of an equilibrium gaseous phase, or it may be an externally controlled pressure.

In describing such systems, the nature of each phase must be specified. In the examples to follow, *s*, *l*, and *g* identify solid, liquid, and gaseous phases, respectively. For solutions or mixtures, the composition is needed, in addition to the temperature and pressure, to complete the specification of the system. If not obvious, the identity of the solvent must be given.

In the equilibrium shown as (25), the relation-

$$H_2O(l) \rightleftharpoons H_2O(g) \qquad (25)$$

ship of Eq. (26) holds. Here $K^0 = p/N$, the ratio

$$\Delta G^0 = \mu_g{}^0 - \mu_l{}^0 = -RT \ln \frac{p}{N} \qquad (26)$$

of the vapor pressure p to the liquid mole fraction N. For pure water, the equilibrium constant is simply the standard vapor pressure p^0, and the Clausius-Clapeyron equation is just a special case of the Gibbs-Helmholtz equation, Eq. (15). Now when a small amount of solute is added, decreasing the mole fraction of solvent, the vapor pressure p must be lowered to maintain equilibrium (Raoult's law). The effect of the total applied pressure P upon the vapor pressure p of the liquid is given by the Gibbs-Poynting equation, Eq. (27).

$$\left[\frac{dp}{dP}\right]_T = \frac{V_l}{V_g} \qquad (27)$$

Here V_l and V_g are the molar volumes of liquid and vapor. The vapor pressure will increase as external pressure is applied (activity increases with pressure). If the external pressure is applied to a solution by a semipermeable membrane, an applied pressure can be found which will restore the vapor pressure (or activity) of the solvent to its standard state value. *See* OSMOSIS.

Solubility equilibria are of wide variety. For a solid, such as barium sulfate, $BaSO_4$, which dissociates as shown in (28), the equilibrium rela-

$$BaSO_4(s) \rightleftharpoons Ba^{++} + SO_4^{--} \qquad (28)$$

tionship is shown in Eq. (29). When the solid state

$$K^0 = \frac{a_{Ba^{++}} a_{SO_4^{--}}}{a_{BaSO_4}} \qquad (29)$$

is pure, its mole fraction will be unity. If it is extremely finely divided, its activity coefficient will become greater than unity; with this increase in the activity of the solid state, the solubility must increase to maintain equilibrium. On the other hand, inclusion of foreign ions in the crystal lattice (solid solution formation) may lower the activity of the solid state.

When a gas, such as CO_2, is dissolved in a liquid, its equilibrium with the gas phase is shown in (30). Equation (31), Henry's law, represents the

$$CO_2(soln) \rightleftharpoons CO_2(g) \qquad (30)$$

$$K^0 = \frac{p}{N} \qquad (31)$$

equilibrium constant for this situation. When the gas dissociates in the liquid, reaction (32), then Eq. (33) may be utilized. Here m is the molal concentration.

$$(H^+ + Cl^-)(aqueous) \rightleftharpoons HCl(aqueous) \rightleftharpoons HCl(g) \qquad (32)$$

$$K^0 = \frac{p}{m^2} \cdot \Gamma \qquad (33)$$

The equilibrium constant must reflect this behavior through the proper exponents. Equation (33) might correctly imply that the molecules were dimerized in the gaseous phase, if it were not known from a study of vapor densities that in this case they are not.

Similarly, when a solute distributes itself between two immiscible phases, the equilibrium constant takes the form of Eq. (34). Here n is ratio

$$K^0 = \frac{c_1}{c_2{}^n} \cdot \Gamma \qquad (34)$$

of the molecular weight in phase 1 to that in phase 2. The equilibrium concentrations c in Eq. (34) reflect the relative solubilities of the solute in the two phases; since solubilities may vary widely, distribution operations may provide an effective means for concentrating a widely dispersed solute. *See* EXTRACTION.

When two components form two immiscible phases at equilibrium, each condensed phase will be a saturated solution; complete immiscibility is impossible in principle, since the chemical potential of any component must be the same in both phases. The separation of a liquid system into two liquid phases is a manifestation of the nonideality of the solutions. For practical purposes, however, many solids may be regarded as immiscible because of the stringent requirements associated with formation of the crystal lattice.

In reactions involving condensed and immiscible phases, for example, reaction (35), there can be

$$Pb(s) + 2AgCl(s) \rightleftharpoons PbCl_2(s) + 2Ag(s) \quad (35)$$

no change in concentration of any phase during the reaction. Then the Q^0 term in Eq. (10) will be constant, and ΔG can never become zero. Although reaction is possible, there can be no equilibrium until one of the reactants is completely used up. There might, of course, be one condition of temperature and pressure for which ΔG could be zero. Such is the case in transition phenomena, or melting-freezing phenomena. From the phase-rule viewpoint, the system in reaction (35) lacks one degree of freedom if reaction is possible, so one phase must disappear in order to attain equilibrium.

[CECIL E. VANDERZEE]

Bibliography: G. M. Barrow, *Physical Chemistry*, 4th ed., 1979; D. F. Eggers, Jr., et al., *Physical Chemistry*, 1964; I. M. Klotz, *Introduction to Chemical Thermodynamics*, 2d ed., 1972; I. M. Klotz and R. M. Rosenberg, *Chemical Thermodynamics*, 3d ed., 1972; G. N. Lewis and M. Randall, *Thermodynamics*, 2d ed., rev. by K. S. Pitzer and L. Brewer, 1961.

Chemical fuel

The principal fuels used in internal combustion engines (automobiles, diesel, and turbojet) and in the furnaces of stationary power plants are organic fossil fuels. These fuels, and others derived from them by various refining and separation processes, are found in the earth in the solid (coal), liquid (petroleum), and gas (natural gas) phases.

Special fuels to improve the performance of combustion engines are obtained by synthetic chemical procedures. These special fuels serve to increase the fuel specific impulse of the engine (specific impulse is the force produced by the engine multiplied by the time over which it is produced, divided by the mass of the fuel) or to increase the heat of combustion available to the engine per unit mass or per unit volume of the fuel. A special fuel which possesses a very high heat of combustion per unit mass is liquid hydrogen. It has been used along with liquid oxygen in rocket engines. Because of its low liquid density, liquid hydrogen is not too useful in systems requiring high heats of combustion per unit volume of fuel ("volume-limited" systems). In combination with liquid fluorine, liquid hydrogen produces extremely large specific impulses, and rocket engines using this combination are under development.

A special fuel which produces high flame temperatures of the order of 5000°C is gaseous cyanogen, C_2N_2. This is used with gaseous oxygen as the oxidizer. The liquid fuel hydrazine, N_2H_4, and other hydrazine-based fuels, with the liquid oxidizer nitrogen tetroxide, N_2O_4, are used in many space-oriented rocket engines. The boron hydrides, such as diborane, B_2H_6, and pentaborane, B_5H_9, are high-energy fuels which are being used in advanced rocket engines. *See* BORANE.

For air-breathing propulsion engines (turbojets and ramjets), hydrocarbon fuels are most often used. For some applications, metal alkyl fuels which are pyrophoric (that is, ignite spontaneously in the presence of air), and even liquid hydrogen, are being used.

A partial list of additional currently used liquid fuels and their associated oxidizers is shown in the table.

Liquid fuels and their associated oxidizers

Fuel	Oxidizer
Ammonia	Liquid oxygen
95% Ethyl alcohol	Liquid oxygen
Methyl alcohol	87% Hydrogen peroxide
Aniline	Red fuming nitric acid
Furfural alcohol	Red fuming nitric acid

Fuels which liberate heat in the absence of an oxidizer while decomposing either spontaneously or because of the presence of a catalyst are called monopropellants and have been used in rocket engines. Examples of these monopropellants are hydrogen peroxide, H_2O_2, and nitro-methane, CH_3NO_2.

Liquid fuels and oxidizers are used in most large-thrust (large propulsive force) rocket engines. When thrust is not a consideration, solid-propellant fuels and oxidizers are frequently employed because of the lack of moving parts such as valves and pumps, and the consequent simplicity of this type of rocket engine. Solid fuels fall into two broad classes, double-base and composites. Double-base fuels are compounded of nitroglycerin (glycerol trinitrate) and nitrocellulose, with no separate oxidizer required. The nitroglycerin plasticizes and swells the nitrocellulose, leading to a propellant of relatively high strength and low elongation. The double-base propellant is generally formed in a mold into the desired shape (called a grain) required for the rocket case. Composite propellants are made of a fuel and an oxidizer. The latter could be an inorganic perchlorate such as ammonium perchlorate, NH_4ClO_4, or potassium perchlorate, $KClO_4$, or a nitrate such as ammonium nitrate, NH_4NO_3, potassium nitrate, KNO_3, or sodium nitrate, $NaNO_3$. Fuels for composite propellants are generally the asphalt-oil-type, thermosetting plastics (phenol formaldehyde and phenol-furfural resins have been used) or several types of synthetic rubber and gumlike substances. Recently, metal particles such as boron, aluminum, and beryllium have been added to solid propellants to increase their heats of combustion and to eliminate certain types of combustion instability. *See* HYDROGEN PEROXIDE.

[WALLACE CHINITZ]

Bibliography: S. S. Penner, *Rocket Propulsion and Combustion Research*, 1962; G. A. Sutton and D. M. Ross, *Rocket Propulsion Elements: An Introduction to the Engineering of Rockets*, 4th ed., 1976.

Chemical process industry

This term, abbreviated CPI, is now applied to what has long been known as the chemical and allied industries. It includes industries engaged in the manufacture of chemicals both inorganic and organic, petroleum, petrochemicals, rubber, plastics, fertilizers (agrichemicals), sugar and starch, paper, pulp, synthetic fibers, photography, pharmaceuticals, explosives, food additives and flavors, and dyes. *See* CHEMICAL ENGINEERING; PETROCHEMICAL.

[JOSEPH A. BRINK, JR.]

Bibliography: D. M. Considine, *Chemical and*

Process Technology Encyclopedia, 1974; J. A. Kent (ed.), *Riegels Handbook of Industrial Chemistry*, 7th ed., 1974; R. N. Shreve, *Chemical Process Industries*, 3d ed., 1967; R. N. Shreve and J. A. Brink, Jr., *Chemical Process Industries*, 4th ed., 1977; R. W. Thomas and P. J. Fargo, *Industrial Chemistry*, 1973.

Chemical separation techniques

A method used in chemistry to purify substances or to isolate them from other substances, for either preparative or analytical purposes. In industrial applications the ultimate goal is the isolation of a product of given purity, whereas in analysis the primary goal is the determination of the amount or concentration of that substance in a sample. In principle it is always more convenient to carry out quantitative determinations directly on portions of the original sample. In cases where the analytical methods available are not sufficiently selective to permit this direct approach, it is necessary to employ preliminary separations to reduce the concentration of, or to remove completely, those substances which interfere in the final estimation.

Although special considerations arise in a comparison of separation methods for engineering or laboratory analytical purposes because of differences in the scales of operations, the various separation processes are based on the same principles. There are three factors of importance to be considered in all separations: (1) the completeness of recovery of the substance being isolated, (2) the extent of separation from associated substances, and (3) the efficiency of the separation. The recovery factor or yield R_A of a separation of substance A is defined as Eq. (1), where Q_A and $(Q_A)_0$ are the amounts of A after and before the separation.

$$R_A = \frac{Q_A}{(Q_A)_0} \qquad (1)$$

The degree of separation $S_{B/A}$ of two substances A and B is given by the separation factor R_B for B with respect to A, and is defined as Eq. (2). Although complete separations are usually preferred, they are not always necessary in analytical applications. The degree of purity will depend upon the choice of the method of final estimation. Sometimes merely a reduction in the quantity of foreign substance present is enough to simplify the subsequent analytical task.

$$S_{B/A} = \frac{(Q_A)_0 Q_B}{(Q_B)_0 Q_A} \cong \frac{Q_B}{(Q_B)_0} \cong R_B \qquad (2)$$

The third factor, efficiency, is a measure of the amount of work required to obtain a given amount of product with a prescribed purity. This consideration is of much greater consequence in industrial separations in which both the scale and the cost of the operation are important.

There are many types of separations based on a variety of properties of materials. Among the most commonly used properties are those involving solubility, volatility, adsorption, and electrical and magnetic effects, although others have been used to advantage. The most efficient separation will obviously be obtained under conditions for which the differences in properties between two substances undergoing separation are at a maximum.

The common aspect of all separation methods is the need for two phases. The desired substance will partition or distribute between the two phases in a definite manner, and the separation is completed by physically separating the two phases. The ratio of the concentrations of a substance in the two phases is called its partition or distribution coefficient.

In analytical work the original phase is usually a liquid, that is, a solution of the sample, and the separation is brought about through the addition or formation of a solid, a liquid, or a gaseous second phase. Although the actual separation of the phases may be physical in nature, chemical reactions are usually required to convert or modify the substance to a form which permits the formation of the new phase or the partition of the substance to the second phase. In some separation methods this step may also be accomplished by physical means.

If two substances have very similar distribution coefficients, many successive steps may be required for a separation. The resulting process is called a fractionation.

Based on the nature of the second phase, the more commonly used methods of separation are classified as follows:

1. Methods involving a solid second phase include precipitation, electrodeposition, chromatography (adsorption), ion exchange, and crystallization. These methods involve a solid second phase either through the formation of a slightly soluble product, deposition as a metal on the surface of an electrode, or by physical or chemical adsorption on a suitable solid material.

2. The outstanding method involving a liquid second phase is solvent extraction, in which the original solution is placed in contact with another liquid phase immiscible with the first. Separations are achieved as a result of differences in the distribution of solutes between the two phases. Solid materials may also be separated by extraction with organic solvents.

3. Methods involving a gaseous second phase include gas evolution, distillation, sublimation, and gas chromatography. Mixtures of volatile substances can often be separated by fractional distillation. *See* EXTRACTION; MASS-TRANSFER OPERATION.

[GEORGE H. MORRISON]

Bibliography: E. W. Berg, *Physical and Chemical Methods of Separation*, 1963; J. C. King, *Separation Processes*, 2d ed., 1980; E. S. Perry (ed.), *Techniques of Chemistry Separation and Purification*, 1978.

Chemical structures

Much of chemistry is explainable in terms of the structures of chemical compounds. The understanding of the structures of chemical compounds hinges very strongly on the understanding of the electronic configurations of the elements. The union of atoms, and therefore the formation of compounds from the elements, is associated with interactions among the extranuclear electrons of the individual atoms. Electronic interactions among atoms may occur in either of two ways: Electrons may be transferred from one atom to another, or they may be shared by two (or more) atoms. The first type of interaction is called electrovalence and results in the formation of electrically charged monatomic ions. The second, covalence, leads to the formation of molecules and complex ions. *See* CHEMICAL BONDING.

FORMATION OF IONS

A number of relationships among the electronic structures of the elements are important in determining the properties of ions.

1. Only the outermost shell of electrons is of significance in compound formation. This is commonly called the valence shell, and the electrons contained within it are called the valence electrons.

2. The maximum number of electrons encountered in the valence shell of any single isolated atom is eight.

3. In the case of transition elements, the electrons in the *d* subshell of the penultimate (next outermost) shell are often of importance in compound formation, so that special consideration must be given this complicating factor. This subshell of electrons is considered as essentially a part of the valence shell. Consequently, it must be realized that rules 1 and 2 are rigorously true only for the representative elements, not the transition elements.

4. The number of valence electrons and their subshell assignments are given by the arrangement of the elements in the long form of the periodic table. The elements in any vertical family of the periodic table have the same valence-electron configuration.

a. The elements of group Ia, H, Li, Na, K, Rb, Cs, and Fr are characterized by single valence electrons in *s* subshells. The configuration is ns^1, where *n* refers to the quantum number identifying the valence shell ($n=1$ for H, 2 for Li, 3 for Na). Those in group IIa are characterized by two electrons in filled *s* subshells; that is, the configuration is ns^2. The elements of groups IIIa, IVa, Va, VIa, VIIa, and the inert gases contain, successively, from one to six valence electrons in the *p* subshell, in addition to two *s* electrons. Consequently, it is simply true that the electronic structures of the elements of the "a"-subgroups of the periodic table have numbers of valence electrons equal to the numbers of the vertical periodic groups to which they belong. Further, the first two valence electrons must be assigned to an *s* subshell. The configurations of the valence shells of the elements boron through neon are: B, $2s^22p^1$; C, $2s^22p^2$; N, $2s^22p^3$; O, $2s^22p^4$; F, $2s^22p^5$; Ne, $2s^22p^6$.

b. The elements of the "b"-subgroups of the periodic table differ from those of the "a"-subgroups as a result of the significance of *d* electrons in the penultimate electron shells. The elements of groups IIIb through the triads of group VIII are characterized by electronic configurations containing two valence electrons in the *s* subshell (of the valence shell) and progressively more electrons in the *d* subshell of the penultimate electron shell. This is shown as follows: Sc, $3d^14s^2$; Ti, $3d^24s^2$; V, $3d^34s^2$; Cr, $3d^54s^1$; Mn, $3d^54s^2$; Fe, $3d^64s^2$; Co, $3d^74s^2$; Ni, $3d^84s^2$. As a result of the similar energies of *ns* and $(n-1)d$ electronic states, some irregularities occur among the electronic configurations of these elements. Groups Ib and IIb are characterized by a complete *d* subshell (penultimate) and one and two electrons in the *s* subshell of the valence shell; for example, Cu, $3d^{10}4s^1$; Zn, $3d^{10}4s^2$.

[DARYLE H. BUSCH]

Ionization potential. Positive ions are formed by the removal of valence electrons, and negative ions result from the addition of electrons to the neutral atom. The complete analysis of the energy relationships determining the formation of compounds composed of ions is complex; however, a few aspects of the problem are directly related to the electronic structures of the isolated atoms. The energy required for the removal of an electron from a neutral atom is measured by the ionization potential and relates to reaction (1). A second ion-

$$M(g) \rightarrow M^+(g) + e^- \quad (1)$$

ization potential indicates the amount of energy required to convert the unipositive ion into the dipositive ion, and the successive removal of the remaining electrons may similarly be related to ionization potentials.

From the relative values of the ionization potentials, it is concluded that the more active metals of groups Ia, IIa, and IIIa may more easily be converted into positive ions than the succeeding elements. The valence electrons are more easily removed than are the electrons in energy levels below the valence shell. Along with the accumulated evidence of chemical compositions of large numbers of known compounds, these observations lead to the conclusion that typical positive ions will be formed by the removal of all of the valence electrons of the metallic elements of the "a"-subgroups of the periodic table. The resulting ions are called inert-gas-type ions because they have the electronic configurations of the inert gases preceding the respective elements in the periodic system. As a consequence of the increasing amounts of energy (ionization potentials) required to remove successive electrons from ions, monatomic positive ions that are highly charged do not exist among chemical compounds under the usual conditions of temperature and pressure. The most highly charged ions of any apparent significance are tripositive (Table 1). A number of the actinide elements form tetrapositive monatomic ions, U^{4+}, Th^{4+}, and Pu^{4+}. *See* PERIODIC TABLE.

Table 1. Inert-gas-type ions, listed by periodic group*

Ia	IIa	IIIb	IIIa	Va	VIa	VIIa	Inert gas
Li^+	Be^{2+}		B^{3+}			H^-	He
Na^+	Mg^{2+}		Al^{3+}	N^{3-}	O^{2-}	F^-	Ne
K^+	Ca^{2+}	Sc^{3+}		P^{3-}	S^{2-}	Cl^-	Ar
Rb^+	Sr^{2+}	Y^{3+}			Se^{2-}	Br^-	Kr
Cs^+	Ba^{2+}	La^{3+}			Te^{2-}	I^-	Xe

*The horizontal rows contain isoelectronic series of ions and the corresponding inert gas.

Typical ions. In addition to positive ions of the inert-gas type, two other electronic configurations recur with some regularity among chemical compounds. The elements of groups Ib (Cu, Ag, Au) and IIb (Zn, Cd, Hg) may lose their valence electrons, forming ions having 18 electrons (instead of

Table 2. Pseudoinert-gas-type and helium-type ions

Pseudoinert-gas-type ions			Helium-type ions		
Ib	IIb	IIIa	IIIa	IVa	Va
Cu^+	Zn^{2+}	Ga^{3+}	(Ga^+)	(Ge^{2+})	As^{3+}
Ag^+	Cd^{2+}	In^{3+}	(In^+)	Sn^{2+}	Sb^{3+}
Au^+	Hg^{2+}	Tl^{3+}	Tl^+	Pb^{2+}	Bi^{3+}

8) in their outermost electron shells. These ions are generally called pseudoinert-gas-type ions (Table 2).

In addition, elements of the groups IIIa, IVa, and Va give evidence of forming ions in which only the electrons from the *p* subshell are lost, leaving two electrons in the *s* level of the valence shell. These are summarized in Table 2 under the heading "Helium-type ions."

The addition of an electron, as shown in reaction (2), to a neutral atom with the resulting forma-

$$M(g) + e^- \rightarrow M^-(g) \qquad (2)$$

tion of a negative ion may be associated with the evolution of energy (the electron affinity). It is generally true, however, that the addition of a second electron will require the expenditure of energy. This relationship militates against the formation of highly charged negative ions. The known, monatomic negative ions are formed by the addition of the number of electrons necessary for the atom to attain an inert gas electronic configuration (Table 1), and no monatomic negative ions are known to have a charge more negative than 3−. *See* ELECTRONEGATIVITY.

Simple ions are usually associated with the attainment of favorable electronic configurations. A reaction leading to the formation of a compound (composed of simple ions) from the elements involves the transfer, by some unspecified mechanism, of electrons from the metal atom to the nonmetallic atom. This produces positive and negative ions, as shown by reactions (3)–(5).

$$Na\cdot + \cdot\ddot{\underset{..}{Cl}}: \rightarrow Na^+, :\ddot{\underset{..}{Cl}}:^- \qquad (3)$$

$$Ca: + \cdot\dot{\underset{..}{O}}: \rightarrow Ca^{++}, :\ddot{\underset{..}{O}}:^{--} \qquad (4)$$

$$3Li\cdot + \cdot\dot{\underset{\cdot}{N}}\cdot \rightarrow 3Li^+, :\ddot{\underset{..}{N}}:^{3-} \qquad (5)$$

Reactions (4) and (5) require further comment. More energy is needed to remove the electrons from the metal atoms than is regained from the electron affinities of the nonmetals. However, this is more than balanced by the electrostatic interaction of the oppositely charged ions, provided they are formed in close proximity. In a solid salt, each ion is affected by the interaction of all the other ions in the crystal. Because of the geometric arrangement of the ions, the net effect is one of stabilization. The lattice energy is defined as the energy required to separate one mole of an ionic compound into its constituent ions, separated by large distances.

Born-Haber cycle. Two principal limitations to the occurrence of electrovalent substances deserve mention. The actual occurrence of such a reaction depends on a number of factors, among which the overall energy relationships occupy a position of major importance. The energetics of salt formation are simply indicated in the Born-Haber cycle, which summarizes the several transformations which may be considered as a part of the salt-forming reaction. In the example given in Fig. 1, the formation of a halide salt of a univalent metal is involved. Because all the terms contributing to the heat of formation of the solid salt are endothermic except the lattice energy and the electron affinity, it is clear that stable salts will occur in those instances where the sum of these exothermic quantities compensates sufficiently for the endothermic quantities. The alkali metals are associated with relatively small sublimation energies and ionization potentials so that the formation of simple electrovalent salts should be greatly favored in their case. Similarly, the alkaline-earth metals (except beryllium) are associated with low sublimation energies and ionization potentials. It may be concluded that the alkaline-earth elements should also be characterized by the ready formation of simple salts.

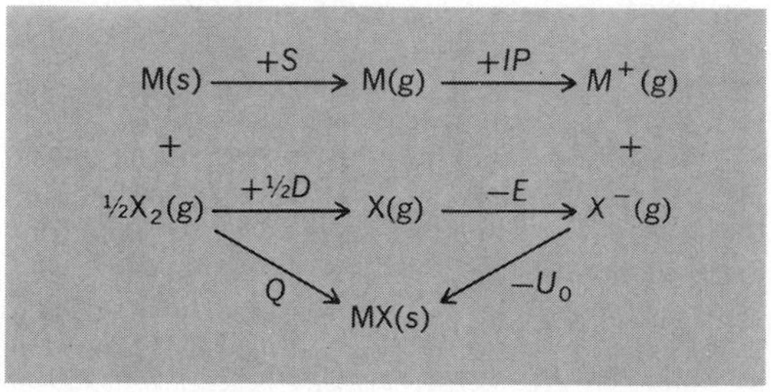

Fig. 1. Born-Haber cycle for formation of solid salt MX; $Q = S + IP + \frac{1}{2}D - E - U_0$, where Q = heat of formation of solid salt MX; S = sublimation energy of the free metal M; IP = ionization potential of the metal; D = dissociation energy of the halogen; E = electron affinity of the halogen; and U_0 = lattice energy of the salt.

The second factor limiting the occurrence of electrovalent salts is best explained in terms of an example. The lattice energy of sodium chloride is 180.4 kcal/mole. Because AgCl has the same arrangement of atoms in the solid and because Ag^+ is only slightly larger than Na^+ (1.13 A compared with 0.95 A), the lattice energy of AgCl should be quite similar to that of NaCl, if both substances are composed of simple spherical ions. Further, the lattice energy of AgCl should be smaller because the oppositely charged ions should be separated at slightly greater distances. A comparison of the lattice energies may be made from the Born-Haber cycle. On the basis of this cycle, the lattice energy of AgCl is calculated to be 211.0 kcal/mole. This result attributes an extra energy equal to 30.6 kcal/mole to the lattice energy of AgCl, in excess of the corresponding value for NaCl. This contradicts the prediction that the lattice energy of the salt containing the larger ion should be smaller. The expected change in lattice energy with cation size is observed in the series of alkali metal halides. The anomaly presented in the case of AgCl requires additional assumptions for its explanation. This example is not unusual. It is commonly observed that the ions in certain salts may not be treated as hard spheres. Two closely related explanations have been utilized to account for these facts.

It may be assumed that the interaction of the charges of the ions on the electron shells of each other results in the polarization or distortion of the ions. Such an interaction would result in induced dipoles or multipoles in the ions. The associated attraction then provides an additional contribution to the lattice energy or, more generally, to the stability of the compound. The relative importance of this mutual polarization depends on the nature of the ions. Eighteen-electron ions, such as Ag^+, are much more polarizable than inert-gas ions, such as Na^+.

The above explanation is based on classical electrostatics, and can be criticized because electrons are not static but are in rapid motion. Furthermore, their motions are governed by the laws of quantum mechanics, not classical mechanics. From this point of view, variations from the behavior expected of simple electrovalent salts may be explained on the assumption that the different atoms, for example, silver and chlorine, are held together by covalent bonds. Since silver atoms are expected to show a greater tendency to form covalent bonds that do sodium atoms, the explanation suggests the correct relationship between NaCl and AgCl. In detail, it would probably be assumed that AgCl exists as Ag^+ and Cl^- ions which are held together by both electrovalent and covalent forces. *See* CHEMICAL BOND THEORY; CHEMICAL BONDING; SOLID-STATE CHEMISTRY.

The preceding example has been discussed at length in order to illustrate a second principal limitation to the occurrence of simple ions in chemical compounds, namely, the formation of covalent bonds. Experimental methods for the determination of the structures of solid compounds are most effective in showing the relative geometric arrangements of atoms or ions, rather than in distinguishing the nature of the bonding forces. In those cases where the geometric arrangement of the atoms is not consistent with expectations based on simple electrostatic arguments, covalent bonding may be inferred. The geometric arrangement of ions in crystals constitutes a large area of chemical knowledge. [R. G. PEARSON]

POLYHEDRAL STRUCTURES

In considering structures more complex than those derived from simple monoatomic ions, the next logical step is to consider single polyhedral aggregates of atoms. At this point it is worth noting that the term "structure" now takes on a much more quantitative meaning. In its most precise sense structure is used to denote a knowledge of the bonding distances and angles between atoms in chemical compounds and, in turn, the geometrical arrangements which they form. These atomic arrangements and the associated distances and angles serve uniquely as "fingerprints" of these atom spatial configurations, and depend very much on understanding the electronic configurations around atoms because, as stated earlier, the union of atoms to form compounds occurs via electronic interactions of one atom with another through their valence electrons. The chemical combination of neutral atoms to produce uncharged species results in molecule formation, whereas the similar combination of atoms or ions possessing a net charge results in the formation of complex ions. A basic understanding of the species formed involves the concept of the coordination polyhedron, which allows a simple classification of the structures of many polyatomic molecules and ions. This type of classification is particularly useful because it conveniently explains the packing together of simple chemical molecules or ions in terms of highly symmetrical polyhedra. It is worth noting that there is an obvious connection between polyhedra and the structures found in crystalline solids formed from them. Crystal formation often involves the linking of convex polyhedra by the sharing of corners, edges, or faces, ultimately forming space-filling assemblies in which all faces of each polyhedron are in contact with faces of other polyhedra. The most important simple polyhedrons are the tetrahedron, the trigonal bipyramid, the octahedron, the pentagonal bipyramid, and the cube (Fig. 2). The most commonly observed of these polyhedral configurations are the tetrahedron (four faces) and the octahedron (six faces).

Coordination polyhedrons. The general properties of a coordination polyhedron are summarized in the identification of a central atom, which in many cases resides at the center of mass of the polyhedron, and the determination of the coordination number of the central atom. The coordination number is equal to the number of groups (atoms, ions) directly attached to the central atom. In the simplest cases, the coordination number is also equal to the number of apexes in the coordination polyhedron. It is therefore concluded that the tetrahedron may occur among molecules and ions having a coordination number of 4; the trigonal bipyramid is associated with a coordination number of 5; the octahedron, with a coordination number of 6; and the pentagonal bipyramid, with a coordination number of 7.

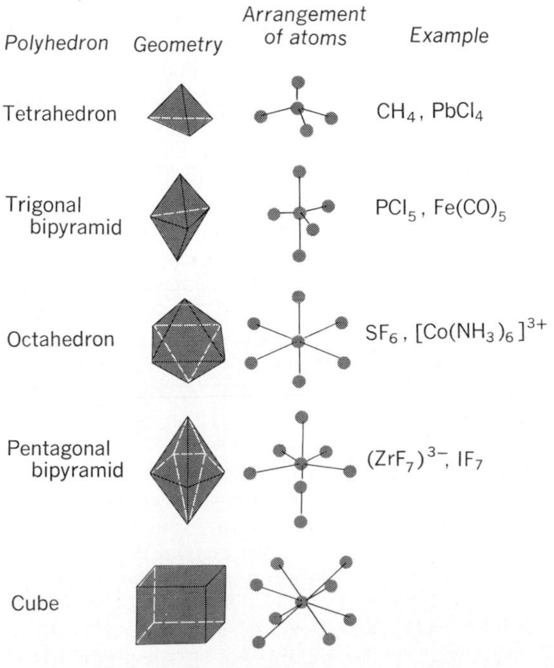

Fig. 2. Coordination polyhedrons.

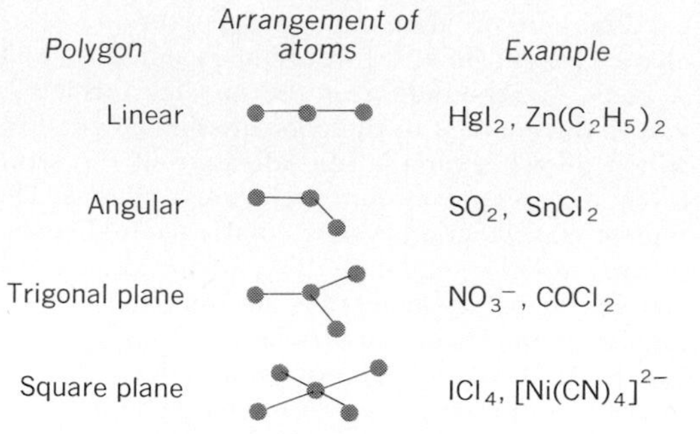

Fig. 3. Coordination polygons.

Lower coordination numbers are associated with plane polygons rather than polyhedrons, as shown in Fig. 3. The coordination number of 4 may be associated with either a polygon (square plane) or a polyhedron (tetrahedron). Similarly, alternative structures exist for the coordination numbers 5 (tetragonal pyramid), 6 (trigonal prism), 7 (irregular structures), and 8 (Archimedean antiprism, dodecahedron); however, these alternative structures are rare in occurrence.

One of the most important problems of modern valence theories has been to explain the formation of and geometrical configurations associated with polyhedral and polygonal structures. With regard to experimental studies of these structures, advances in diffraction studies of crystalline compounds has resulted in the successful "mapping" of valence- or bonding-electron density distributions in numerous chemical systems. X-rays are scattered by the electrons of atoms, and thus it is possible to use diffraction techniques for the location of bonding-electron distributions in chemical systems. This is an extremely exciting area of study being undertaken in many laboratories. *See* COORDINATION CHEMISTRY; MOLECULAR ORBITAL THEORY.

Electrostatic model. The simplest correlative device which accurately summarizes a very large number of such structures and enables the chemist to predict, with a good chance of success, the geometric array of the atoms in a compound of known composition is based on an extreme electrostatic model. This model, or theory, represents the bonds in a purely formal way. The central atom is considered to be a positive ion having a charge equal to its oxidation state. The groups attached to the central atom (the ligands) are then treated either as negative ions or as neutral dipolar molecules. This is illustrated in Table 3 for a series of oxides, oxyanions, halides, and halo complexes. The principal justification for this approach lies in its successful correlation of a vast amount of information.

A number of significant observations related to the formulations of Table 3 should be pointed out. There are series of ions, or ions and molecules, having the same type of composition, differing only in the nature of the central ion and the net charge on the aggregate. Examples are found in the series: NO_3^-, CO_3^{2-}, BO_3^{3-}, ClO_3^-, SO_3^{2-}, PO_3^{3-}; ClO_4^-, SO_4^{2-}, PO_4^{3-}, SiO_4^{4-}; AlF_6^{3-}, SiF_6^{2-}, PF_6^-, SF_6, (ClF_6^+?). The numbers of atomic nuclei and of electrons are the same for all the members of each series; consequently, these are called isoelectronic series. Not only are the several chemical entities in such series isoelectronic, but they are usually identical in geometrical structure (isostructural).

It may also be observed that corresponding ions from a given vertical family of the periodic table commonly vary in coordination number. A useful example is found in N^{5+} and P^{5+} which form NO_3^- and PO_4^{3-}, respectively. In addition, some neutral molecules expand their coordination numbers to form stable anionic halo complexes, whereas others do not. Thus, SiF_4 reacts with fluoride ion to form SiF_6^{--}, whereas CF_4 does not form a similar complex ion. The most satisfactory explanation of these and many related observations is conveniently formulated in terms of the electrostatic model chosen here.

Ionic radii ratios. The necessary condition for stability of the coordination polyhedron MA_n requires that the anions A are each in contact with the central atom M. As a consequence of this condition, the limit of stability of the structure arises in those cases where the anions are also mutually in contact. Larger ligands, or anions, would not be in contact with the central ion. This relationship is usually summarized in terms of the limiting ratio of the radius of the cation, r_M, to that of the anion, r_A,

Table 3. Formal representation of compounds and complex ions as aggregates of ions

Complex ion or molecule	Constituent ions
NO_2^-	N^{3+}, $2O^{2-}$
NO_3^-	N^{5+}, $3O^{2-}$
CO_3^{2-}	C^{4+}, $3O^{2-}$
BO_3^{3-}	B^{3+}, $3O^{2-}$
ClO^-	Cl^+, O^{2-}
ClO_2^-	Cl^{3+}, $2O^{2-}$
ClO_3^-	Cl^{5+}, $3O^{2-}$
ClO_4^-	Cl^{7+}, $4O^{2-}$
SO_3^{2-}	S^{4+}, $3O^{2-}$
SO_4^{2-}	S^{6+}, $4O^{2-}$
PO_3^{3-}	P^{3+}, $3O^{2-}$
PO_4^{3-}	P^{5+}, $3O^{2-}$
SiO_4^{4-}	Si^{4+}, $4O^{2-}$
OF_2	O^{2+}, $2F^-$
CO_2	C^{4+}, $2O^{2-}$
SO_2	S^{4+}, $2O^{2-}$
SO_3	S^{6+}, $3O^{2-}$
NF_3	N^{3+}, $3F^-$
CCl_4	C^{4+}, $4Cl^-$
BCl_3	B^{3+}, $3Cl^-$
ClF	Cl^+, F^-
ClF_3	Cl^{3+}, $3F^-$
BrF_5	Br^{5+}, $5F^-$
IF_7	I^{7+}, $7F^-$
SCl_4	S^{4+}, $4Cl^-$
SF_6	S^{6+}, $6F^-$
PF_3	P^{3+}, $3F^-$
PF_5	P^{5+}, $5F^-$
SiF_4	Si^{4+}, $4F^-$
$AlCl_3$	Al^{3+}, $3Cl^-$
BF_4^-	B^{3+}, $4F^-$
ICl_2^-	I^+, $2Cl^-$
ICl_4^-	I^{3+}, $4Cl^-$
BrF_6^-	Br^{5+}, $6F^-$
PF_6^-	P^{5+}, $6F^-$
SiF_6^{2-}	Si^{4+}, $6F^-$
AlF_6^{3-}	Al^{3+}, $6F^-$
ZnI_4^{2-}	Zn^{2+}, $4I^-$
$CuCl_4^{2-}$	Cu^{2+}, $4Cl^-$
$FeCl_4^-$	Fe^{3+}, $4Cl^-$
FeF_6^{3-}	Fe^{3+}, $6F^-$

below which the anions would no longer be in contact with the cation. Table 4 reports these radius ratios for some of the structures given in Figs. 2 and 3. Regardless of the magnitude of r_M/r_A, the linear structure satisfies the simple condition for stability. However, in those cases where r_M/r_A exceeds 0.155, a third group can be stably attached, and in general, the maximum number of groups which may be associated with a central atom will lead to the most stable chemical species. From this, it follows that all those cases in which r_M/r_A lies between 0.155 and 0.225 should involve a cation with a coordination number of 3, and not of 2. Similarly, in the range $0.225 \leqq r_M/r_A \leqq 0.414$, the tetrahedral structure, with a coordination number of 4, is expected, and an octahedral structure is expected when the radius ratio exceeds 0.414, with the possibility of cubic structures for still larger values of r_M/r_A. As pointed out above, a number of other structures have been observed in addition to those for which limiting values of the radius ratios are given in Table 4. The square planar structure

Table 4. Radius ratios for coordination polyhedrons

Coordination number	Geometric structure	Radius ratios, r_M/r_A
3	Trigonal plane	0.155
4	Tetrahedron	0.225
6	Octahedron	0.414
8	Cube	0.732

has been observed in the case of the coordination number of 4. As a consequence of its geometric relationship to the octahedron, the square planar structure may occur for the same values of the radius ratio as are required by the octahedron. A parallel relationship exists between the trigonal bipyramidal structure (coordination number 5) and the trigonal planar structure (coordination number 3). Because in these cases the radius ratio does not distinguish between the pairs of structures, even though both are known to exist, it is necessary to find additional factors to explain their occurrence. Other complexities could be introduced by extending the treatment properly to include higher coordination numbers.

The majority of the structures of the molecules and complex ions containing central atoms or ions of the inert-gas and pseudoinert-gas types are summarized accurately in terms of the concepts just presented. The structures of compounds derived from central ions having unshared pairs of electrons in their valence electron shells require further consideration.

It is possible to relate the structures of coordination polyhedrons derived from central ions having unshared pairs of electrons by assuming that the sum of the electron pairs plus ligands determines the total geometry, each electron pair being confined to a region in space which might otherwise be occupied by an anion or other ligand. Table 5 illustrates the application of these ideas to coordination numbers from 2 to 6. The relationships summarized in Table 5 are presented in more detail in Fig. 4, where the structures of H_2S, PH_3, and SiH_4 are shown. In this figure, the electron pairs are represented by the volumes in space in which the electrons have the highest probability of being found. Similar treatments are given to molecules and ions derived from the trigonal bipyramidal structure in Fig. 5 and to those derived from the octahedron in Fig. 6.

It may be concluded from the preceding paragraphs that, from a knowledge of the electronic structures of the elements, the sizes of the ions, and the correlations just presented, it is possible to predict the geometric structures of binary compounds and complex ions for given central atoms and ligands. Principal emphasis has been placed on binary halides and oxides and complex oxyanions and haloanions. It should be realized that

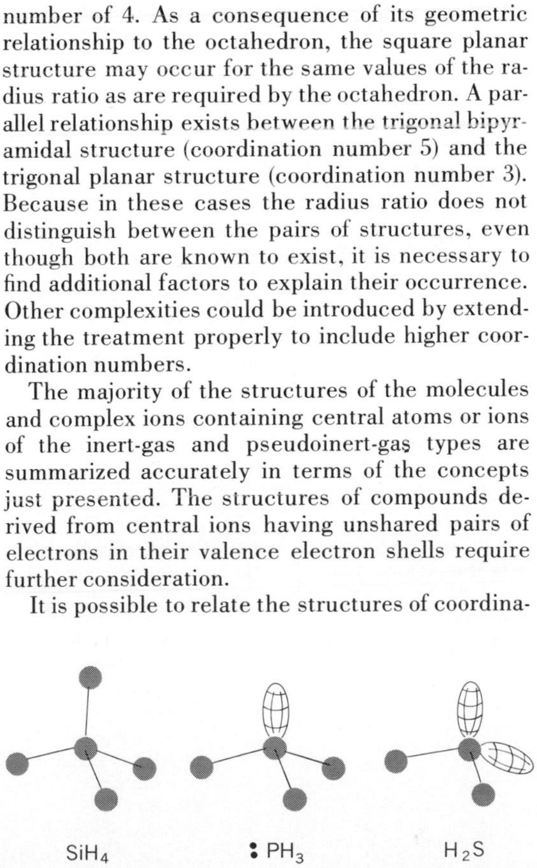

Fig. 4. Structures derived from the tetrahedron.

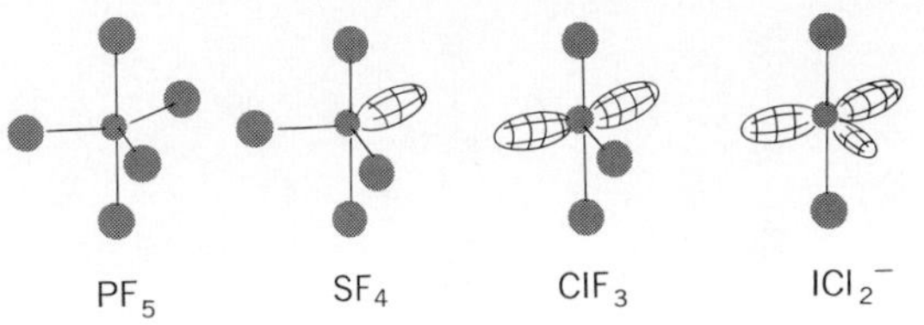

Fig. 5. Structures derived from trigonal bipyramid.

Table 5. Structures of molecules and complex ions derived from central ions having unshared electron pairs

Total number of ligands plus electron pairs	Coordination polyhedron	Number of unshared electron pairs	Geometry of ligands	Examples
2	Linear	0	Linear	$Zn(CH_3)^2$, $HgCl_2$, CO_2
3	Trigonal plane	0	Trigonal plane	BCl_3, SO_3, NO_3^-
		1	Angular	SO_2, NO_2^-, $SnCl_2$
4	Tetrahedron	0	Tetrahedron	CH_4, $SiCl_4$, PCl_4^+
		1	Pyramidal	PCl_3, NF_3, SO_3^{2-}
		2	Angular	OF_2, H_2S, BrF_2^+
5	Trigonal bipyramid	0	Trigonal bipyramid	PCl_5, AsF_5
		1	Irregular tetrahedron	SeF_4, $TeCl_4$, BrF_4^+
		2	T-shaped	ClF_3, BrF_3
		3	Linear	ICl_2^-
6	Octahedron	0	Octahedron	SF_6, PF_6^-, $Sb(OH)_6^-$
		1	Tetragonal pyramid	IF_5, BrF_5
		2	Square planar	ICl_4^-

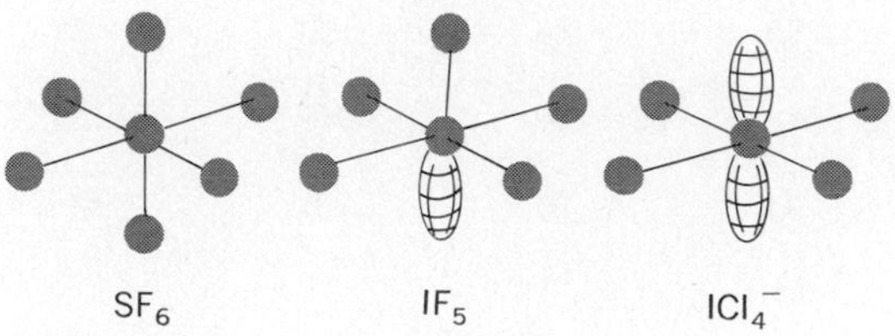

Fig. 6. Structures derived from the octahedron.

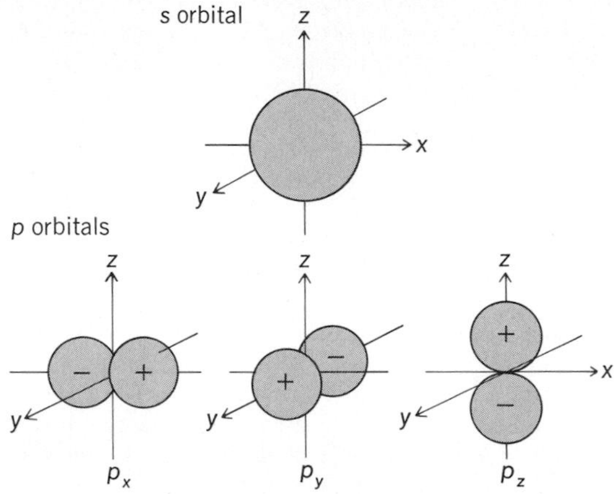

Fig. 7. A pictorial representation of *s* and *p* orbitals.

the same treatment may be applied equally well to most polyatomic aggregates containing nontransition ions as central elements. In addition, the oxyanions and molecular halides of the transition elements may be treated similarly. Thus, VO_4^{3-}, CrO_4^{2-}, and MnO_4^- are tetrahedral in structure, as are $TiCl_4$ and VCl_4. Also, the octahedral fluoro complexes TiF_6^{2-}, VF_6^-, and FeF_6^{3-} are known. Many of the hydrated and ammoniated metallic ions may also be discussed in these terms. Among the particularly useful results of these considerations is the prediction of which compounds have electron pairs that might be shared with molecules or ions having the ability to expand their coordination spheres. In this way the Lewis acids and bases are predicted from their structures. *See* ACID AND BASE.

The most significant limitation of the structural discussion just completed is its complete lack of concern with the detailed nature of the chemical bonds uniting the central atom to its ligands in a coordination polyhedron. It was pointed out in the discussion of monatomic ions that ions of high charge probably do not exist in chemical compounds and that physical properties of many substances are not explainable in terms of simple electrostatic relationships. The quantum-mechanical theories of valence provide the more detailed descriptions of the manner of bonding required to overcome these limitations. Some classes of compounds cannot be systematized, even from the standpoint of their geometric structures, without the use of these more advanced concepts. Two major theories exist for the application of the idea of the covalent bond to the structures of chemical compounds. It is desirable, at this point, very briefly to make use of one of these, the so-called valence-bond concept.

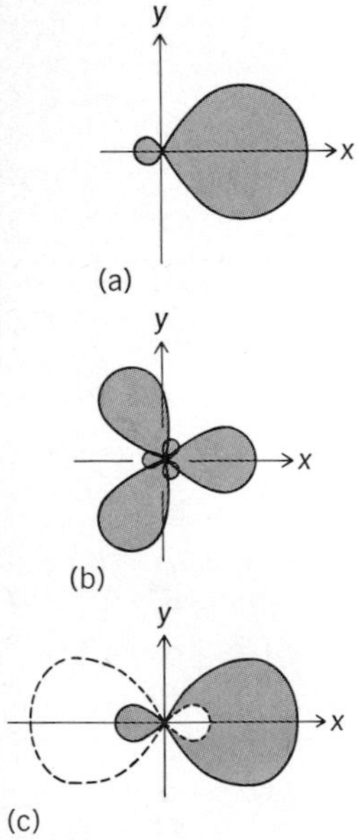

Fig. 8. Pictorial representations of hybrid orbitals formed from *s* and *p* atomic orbitals: (*a*) one tetrahedral (sp^3) orbital, (*b*) trigonal (sp^2) orbitals, and (*c*) collinear (sp) orbitals.

Valence-bond theory. According to the valence-bond theory, the principal requirements for the formation of a covalent bond are a pair of electrons and suitably oriented electron orbitals on each of the atoms being bonded. The geometry of the atoms in the resulting coordination polyhedron is correlated with the orientation of the orbitals on the central atom. The orbitals utilized depend on the energies of the electrons in them. In general, the order of increasing energy of the electron orbitals is $(n-1)d < ns < np < nd$. It is concluded that a nontransition atom having one valence electron will form a covalent bond utilizing an *s* orbital. In those cases where an unshared pair of electrons may be assigned to the *ns* orbital, as many as three equivalent bonds may be formed by utilizing the three *np* orbitals of the central atom. Atomic *s* and *p* orbitals are illustrated in Fig. 7. Because of the orientation of these *p* orbitals with respect to each other, the three resulting bonds should be at 90° to each other. This expectation is nearly realized in PH_3. In order to account for four or six equivalent bonds, or for that matter in order to account for all the remaining polyhedral and polygonal structures, except the angular structure for a coordination number of 2 (with two unshared pairs of electrons on the central atom), an additional assumption is necessary. It is assumed that *s* and *p* orbitals, *s* and *d*, or *s*, *p*, and *d* orbitals may be replaced by new orbitals, called hybridized orbitals. These hybridized orbitals are derived from the original orbitals (mathematically) in such a way that the required number of equivalent bonds may be formed. In the simplest case, it is shown that *s* and *p* may be combined to form two equivalent *sp* hybridized orbitals directed at 180° to each other. Examples of hybrid orbitals formed from atomic *s* and *p* orbitals are shown in Fig. 8. Other sets of hybridized orbitals have been shown to be appropriate to describe the bonding in other structures. These are summarized in Table 6. *See* CRYSTAL FIELD THEORY.

Among inert-gas ions of the first row of eight elements in the periodic table, there are four orbitals available for covalent bond formation, one 2*s* and three 2*p*. Consequently, a maximum of four bonds may be formed. This is in general agreement with the existence of the tetrahedron as the limiting coordination polyhedron among these elements, for example, BeF_4^{2-}, BF_4^-, CCl_4, NH_4^+. Although only Li^+ deviates from this pattern, having a coordination number of 6 in its crystalline halides, these compounds are best treated as simple electrovalent salts. In keeping with the limita-

Table 6. Geometric structure and hybridized bonding orbitals of central atom

Coordination number	Orbitals	Geometric structure
2	p^2	Angular
2	sp	Linear
3	p^3	Trigonal pyramidal
3	sp^2	Trigonal planar
4	sp^3	Tetrahedral
4	d^3s	Tetrahedral
4	dsp^2	Square planar
5	dsp^3	Trigonal bipyramidal
6	d^2sp^3	Octahedral

tion of only four orbitals, the formation of double or triple bonds between atoms of these elements reduces the coordination number of the central atom. Thus, the highest coordination number of a first row element forming one double bond is 3. This is illustrated by

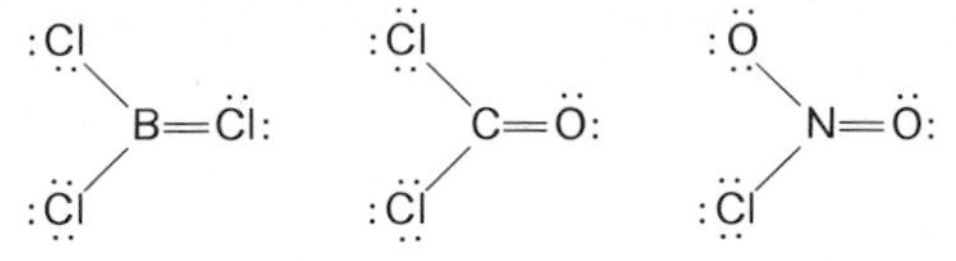

In these and similar examples, the geometric array is determined by the formation of three single bonds utilizing sp^2 hybridized orbitals on the central atom and p orbitals on the ligand. In general, the bonds determining the geometry of a molecule or ion (in this way) are called σ-bonds. The double bond results from the superposition of a second bond, a π-bond, between two atoms. In this example the formation of a π-bond reduces the number of σ-bonds from four to three, thus changing the geometries of the corresponding molecules or ions from tetrahedral to trigonal planar.

The formation of a second π-bond (a triple bond or two double bonds) reduces the coordination number of the atom in question still further, resulting in the linear sp set of hybridized orbitals being utilized in σ-bond formation. *See* VALENCE.

Resonance. With regard to the nature of doubly bonded compounds, another problem arises when such structures are viewed from the standpoint of valence-bond theory. In the species BCl_3, $COCl_2$, NO_2Cl, and many similar substances, nonequivalent bonds are predicted. The doubly bonded oxygen should be closer to the central carbon atom than the singly bonded ones. This is not found to be true experimentally so long as the similar atoms are otherwise equivalent. There is only one observable C—O distance in carbonate, one N—O distance in nitrate, and so on. To account for such facts as these, the concept of resonance must be introduced. If the π-bond exists, it must exist equally between the central atom and all the equivalent oxygen atoms. The resonance method of describing this situation is to say that one of the pictorial structures is inadequate properly to describe the substance, but that enough pictorial structures (resonance structures) should be considered to permute the double bond about all the equivalent bonds. The true structure is assumed to be something intermediate to all the resonance structures and more stable than any of them because it exists in preference to any one of them. The resonance structures for CO_3^{2-} are the following:

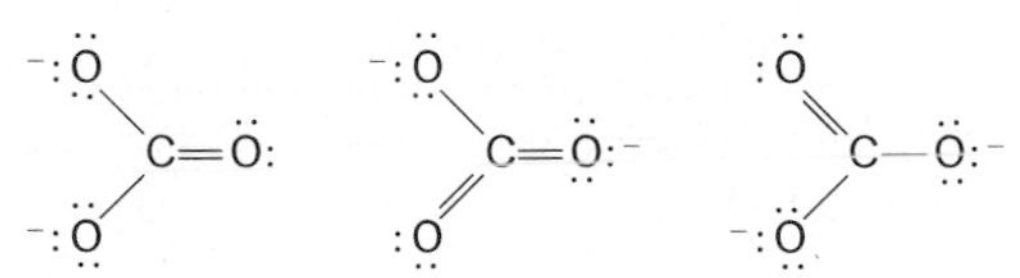

See CONJUGATION AND HYPERCONJUGATION; RESONANCE.

Applications of valence-bond theory. The valence-bond theory may be utilized to systematize structures involving unshared pairs of electrons in a manner very similar to that utilized above (Table 5 and Figs. 4–6). According to this view, the hybridized orbitals of the central atom determine a total polyhedral structure, but either bonds or unshared electron pairs may make use of the hybridized orbitals. Thus, from Fig. 4, the silicon atom in SiH_4 forms bonds directed toward the corners of a tetrahedron by utilizing all four sp^3 hybridized orbitals. In the case of PH_3, the phosphorus atom forms three bonds utilizing three of the same hybridized orbitals; presumably, the fourth orbital is filled by the unshared electron pair.

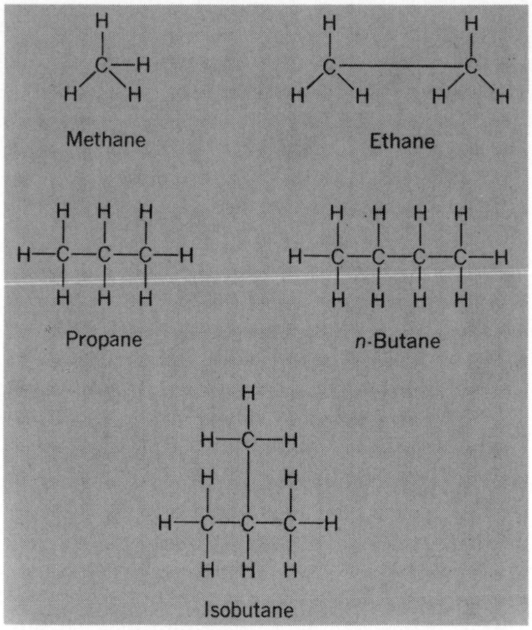

Fig. 9. The saturated hydrocarbons.

The remaining elements of the periodic system (except hydrogen and helium) have $(n-1)d$ or nd orbitals available for bond formation in addition to the ns and np orbitals discussed in connection with the first row of eight elements. Along with the increasing values of the radius ratios for their derivatives, the availability of these orbitals accounts for the increased coordination numbers of compounds. The occurrence of such obviously covalent compounds as PF_5, SF_6, and many others is thereby easily understood.

Many of the compounds of transition-metal ions may also be explained by utilizing the valence-bond concept. The tetrahedral derivatives $TiCl_4$,

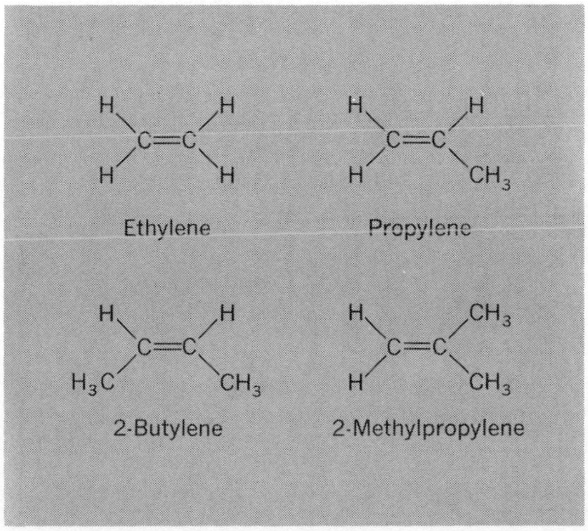

Fig. 11. Structural formulas representing four of the possible olefin compounds.

CHEMICAL STRUCTURES

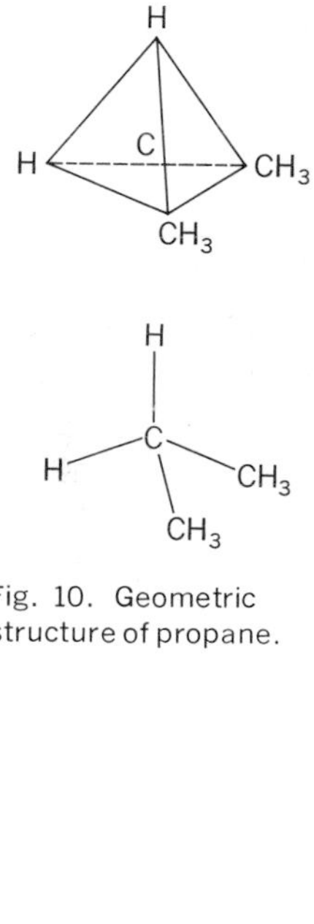

Fig. 10. Geometric structure of propane.

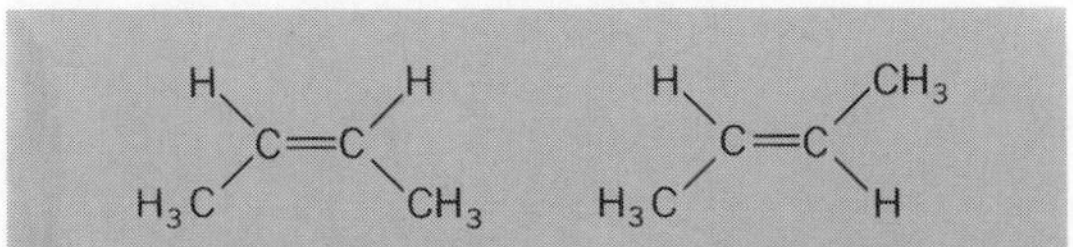

Fig. 12. Cis-trans isomerism among olefins.

VO_4^{3-}, CrO_4^{2-}, MnO_4^-, and FeO_4^{2-}, for example, utilize d^3s hybrid orbitals, whereas the central atoms in $CoCl_4^{2-}$ and $FeCl_4^-$ must be attached through sp^3 orbitals because the d orbitals are not available for bonding in these latter cases. Those central ions having two $(n-1)d$ orbitals may form octahedral structures, such as $[Cr(H_2O)_6]^{3+}$, $[Co(NH_3)_6]^{3+}$, and $[Fe(CN)_6]^{4-}$; whereas those having available a single $(n-1)d$ orbital may form derivatives having square planar structures, for example, $PtCl_4^{2-}$ and $[Ni(CN)_4]^{2-}$. *See* COMPLEX COMPOUNDS.

The relationship between coordination number and the formation of double bonds does not carry over clearly to these elements of atomic number greater than 18. Because none of the coordination polyhedrons commonly encountered would require the utilization of all of the available orbitals on the central atom in σ-bond formation, and because just those d orbitals not involved in hybridization (for σ-bonds) are favorably disposed toward π-bond formation, multiple bonding may at times be present and at other times absent from compounds involving the same coordination number. It follows that the formation of tetrahedral PO_4^{3-}, SO_4^{2-}, and ClO_4^- does not imply the absence of double bonding from the link between the central element and the oxygen.

CONDENSED POLYHEDRAL STRUCTURES

In the preceding discussions of polyatomic molecules and ions, attention has been confined to those species in which single coordination polyhedrons effectively account for the geometric structures. Understanding of these species is central to the problem of the structures of chemical compounds, but does not provide an adequate basis for the descriptions of the structures of all or even a majority of the known compounds. Among the largest numbers of structures not covered by these concepts are those in which two or more coordination polyhedrons are united to form a more complex total structure. Large numbers of the compounds involving such condensed polyhedrons fall naturally into the subject of solid-state chemistry.

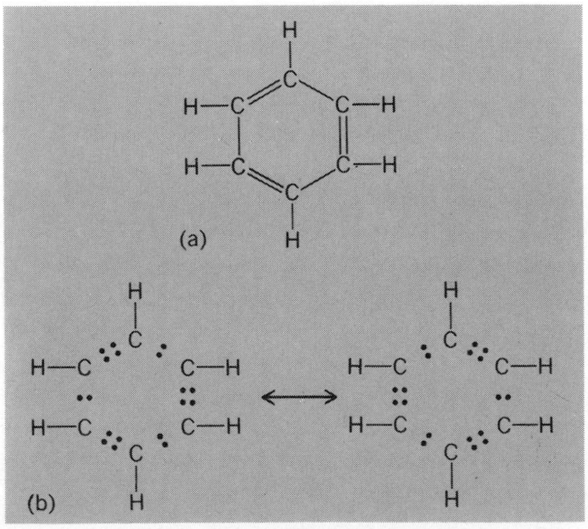

Fig. 13. Benzene molecule. (*a*) Structural formula. (*b*) Two forms in resonance.

Homologous series. The classic homologous series of compounds in organic chemistry provide useful examples involving the condensation of polyhedrons containing the same central element in the individual units. The structures of the first four members of the series of saturated or aliphatic hydrocarbons are shown in Fig. 9. The general formula, C_nH_{2n+2}, represents a large number of compounds extending from the lowest member, methane, CH_4, to polyethylene, a plastic of economic importance in which n is a very large number. The isomerism of the saturated hydrocarbons is easily understood from the example given in Fig. 9. Two ways exist for the linking together of these tetrahedrons. This gives rise to two molecular forms, both of which are stable, well-known compounds. It is an essential part of these structures that each C—C link is linear (because it is merely

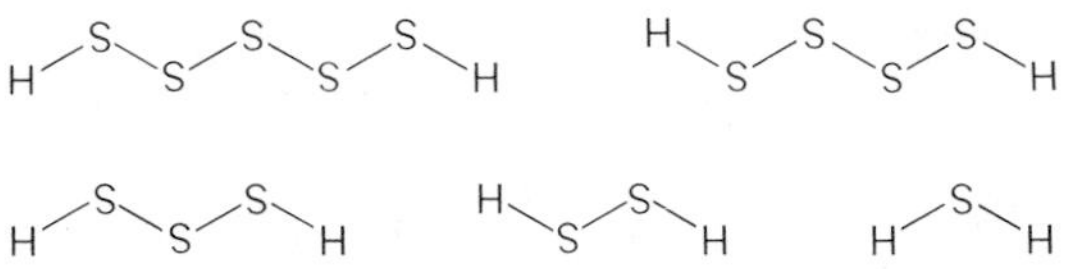

Fig. 14. The sulfanes.

a σ-bond); however, when two carbon atoms are linked to a third, the C—C—C angle is essentially determined by the bond angle of the central carbon atom (that is, the other carbons may be treated as ligands to the first), as shown in Fig. 10. The other familiar homologous series of organic chemistry differ from the saturated hydrocarbons in having at least one unique coordination polyhedron of a different type. The olefins contain two doubly bonded, or unsaturated, carbon atoms, whose polyhedral structures are trigonal planar, and the remainder of the carbons are tetrahedral (Fig. 11). As in the case of the aliphatic hydrocarbons, the olefins exhibit an isomerism which is associated with the branching of the chain structure. In addition, the presence of two linked trigonal planar carbon atoms and the fact that the polyhedrons cannot rotate about the double bond give rise to a different kind of isomerism, called cis-trans isomerism (Fig. 12). *See* BOND ANGLE AND DISTANCE.

The existence of a predicted isomerism provides one of the most important confirmations of theories of chemical structure. In general, the polyhedral view of molecular structure, as described here, has been thoroughly verified by the discovery of the many types of predicted isomerism. The first really convincing proof of the tetrahedral structures of saturated carbon atoms involved optical isomerism. *See* OPTICAL ACTIVITY.

The problem of condensed polyhedrons should be mentioned again. The series of organic compounds shown below with their general formulas

Alkynes	C_nH_{2n-2}
Primary amines	$C_nH_{2n+1}NH_2$
Carboxylic acids	$C_nH_{2n+1}COOH$
Amides	$C_nH_{2n+1}CONH_2$
Nitriles	$C_nH_{2n+1}C{\equiv}N$

and many other such organic series may be explained in a manner similar to that utilized for the saturated hydrocarbons and olefins. In all the substances, the carbon atom dominates the structural relationships. Many additional types of complex polyhedrons are found among organic compounds. The aromatic hydrocarbons are characterized by cyclic arrangements of trigonal planar carbon atoms, as shown for benzene in Fig. 13*a*. The highly symmetrical nature of the benzene molecule is not fully represented by such a structure. The figure indicates the presence of three double and three single bonds in the ring. It has been

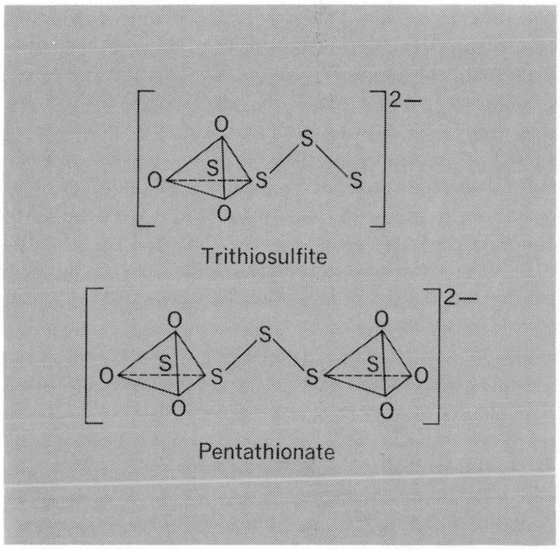

Fig. 15. Polythiosulfite and polythionate structures.

shown that the C-C bonds are all the same and, consequently, the true structure of the substance must be represented by two resonance structures which interchange the single and double bonds. Figure 13*b* illustrates this resonance.

Catenation. The property of an element to link to itself, which is so common for carbon, is called catenation. Other elements exhibit this property to a lesser extent. Nitrogen forms hydrazine, $NH_2{-}NH_2$, the homolog next higher than NH_3; however, the remaining members of the series N_nH_{n+2} are unknown. Similarly, water, H_2O, and hydrogen peroxide, HO—OH, constitute the beginning of a homologous series of which only these two members are known. Sulfur appears to approximate the behavior of carbon most nearly in its proclivity toward catenation. A series of sulfanes, S_xH_2, is known, with x varying from unity to a relatively large number. These substances have the structures shown in Fig. 14. A number of related homologous series exist: the organic sulfides, R_2S_x; the sulfur halides, SCl_2, and S_2Cl_2; the poly-

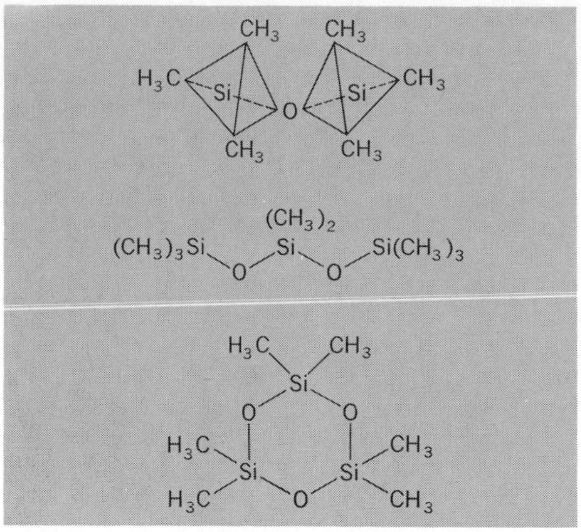

Fig. 16. Structural formulas representing three of the possible siloxane compounds.

thiosulfurous acids, $O_3S_x^{2-}$; and the polythionates, $O_3S{-}S_x{-}SO_3^{2-}$. The structures of representative members of the latter two series are shown in Fig. 15. Catenation occurs to some extent among many other elements, for example, phosphorus, silicon, arsenic, and germanium.

Bridge-forming atoms. A second type of link is commonly involved in joining coordination polyhedrons into more complex structures. This involves the interaction of bridge-forming atoms to link other coordination polyhedrons instead of direct union of the central atoms. The siloxanes provide examples of this behavior (Fig. 16). These interesting and useful substances are related to the silicates in structure.

The polyphosphates also involve the sharing of corners by PO_4 tetrahedrons through oxygen bridges. Linear structures, cyclic structures, and more highly condensed systems are formed (Fig. 17). The structure of P_4O_{10}, the anhydride of these acids, is included for comparison.

In addition to the sharing of corners, coordination polyhedrons may unite by the sharing of edges

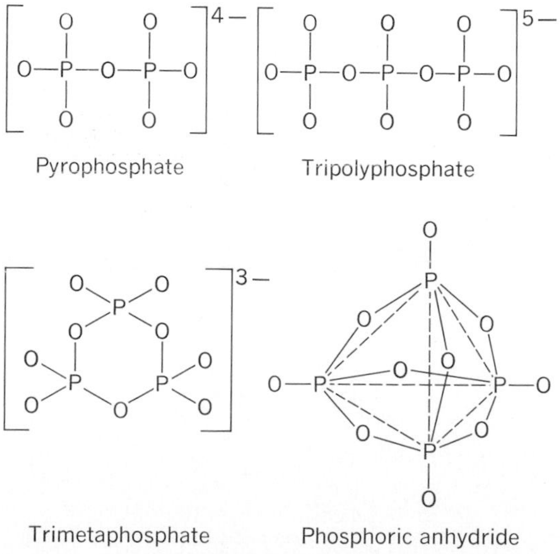

Fig. 17. Polyphosphates and phosphoric anhydride.

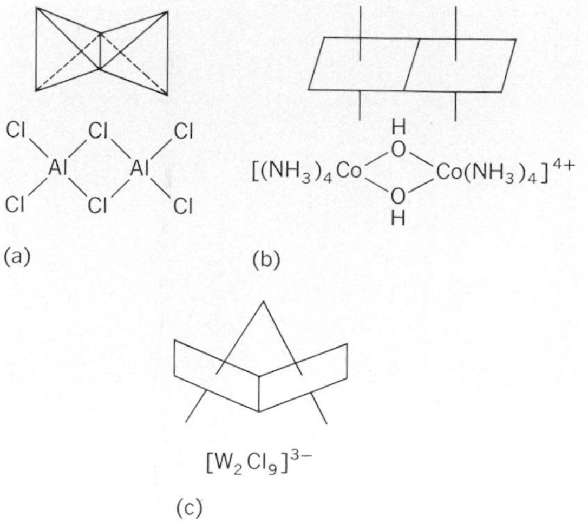

Fig. 18. Condensed polyhedrons. (*a*) Edge-shared tetrahedron. (b) Edge-shared octahedron. (*c*) Face-shared octahedron.

or faces. Of these, edge sharing is the more common (Fig. 18). [JACK M. WILLIAMS]

ELECTRON-DEFICIENT STRUCTURES

A significant number of chemical compounds are now known in which the bonding cannot be accurately described by the utilization of a pair of electrons for each bond between contiguous atoms. Rather, examples exist where the total number of valence-electron pairs is less than the number of bonds. Such molecules are said to have electron-deficient structures which are characterized by the formation of an electron-pair bond between more than two atomic centers. An example is diborane, B_2H_6, the simplest boron hydride. The molecular structure of this compound is shown in Fig. 19*a*. A total of only 12 valence electrons (3 from each boron atom and 1 from each hydrogen atom) is available to form what, under normal circumstances, would be 8 covalent bonds. The bonding in this electron-deficient case can be understood by postulating the existence of a three-center two-

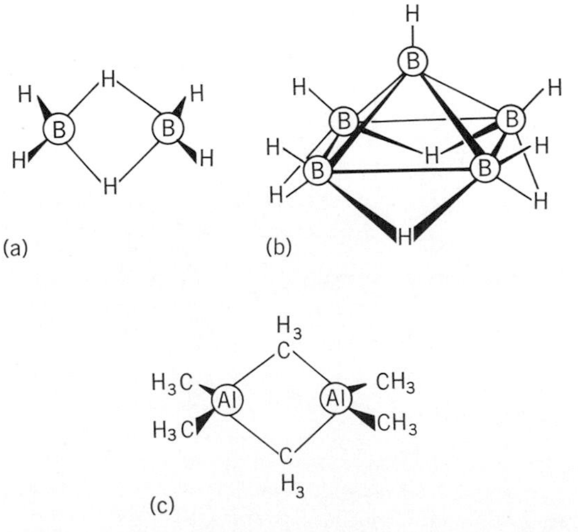

Fig. 19. Structures of some electron-deficient molecules (*a*) Diborane. (*b*) B_5H_9. (*c*) Trimethylaluminum dimer.

electron bond for each B-H-B bridge. That is, the three atoms are joined together by a single electron pair. The formation of such multicenter bonds in response to electron deficiency frequently leads to molecular structures which are cyclic or cage-like. Another, more complex example of this phenomenon in boron chemistry is the polyhedral boron hydride, B_5H_9 (Fig. 19*b*). An example involving metal-to-carbon bonds (an organometallic compound) is trimethylaluminum. Electron deficiency leads to a dimeric structure with bridging methyl groups (Fig. 19*c*). The bridges are joined by three-center two-electron bonds.

[TOBIN J. MARKS]

Bibliography: F. A. Cotton and G. Wilkinson, *Advanced Inorganic Chemistry*, 1972; E. L. Muetterties, *Boron Hydride Chemistry*, 1975.

Chemical thermodynamics

The application of thermodynamic principles to systems involving physical and chemical transformations in order to (1) develop quantitative relationships among the identifiable forms of energy and their conjugate variables, (2) establish the criteria for spontaneous change, for equilibrium, and for thermodynamic stability, and (3) provide the macroscopic base for the statistical-mechanical bridge to atomic and molecular properties. The thermodynamic principles applied are the conservation of energy as embodied in the first law of thermodynamics, the principle of internal entropy production as embodied in the second law of thermodynamics, and the principle of absolute entropy and its statistical thermodynamic formulation as embodied in the third law of thermodynamics.

Basic concepts. The basic goal of thermodynamics is to provide a description of a system of interest in order to investigate the nature and extent of changes in the state of that system as it undergoes spontaneous change toward equilibrium and interacts with its surroundings. This goal implicitly carries with it the concept that there are measurable properties of the system which can be used to adequately describe the state of the system and that the system is enclosed by a boundary or wall which separates the system and its surroundings. Properties that define the state of the system can be classified as extensive and intensive properties. Extensive properties are dependent upon the mass of the system, whereas intensive properties are not. Typical extensive properties are the energy, volume, and numbers of moles of each component in the system, while typical intensive properties are temperature, pressure, density, and the mole fractions or concentrations of the components.

Extensive properties can be expressed as functions of other extensive properties, for instance as in Eq. (1), where the volume V of the system is ex-

$$V = V(U, S, \{n_i\}) \tag{1}$$

pressed in terms of the internal energy U, the entropy S, and $\{n_i\}$, the set of numbers of moles of the various components labeled by the index i. A suitable transformation procedure can be used to replace extensive variables by conjugate intensive variables. For example, the volume can be expressed as in Eq. (2*a*) or (2*b*). Since temperature T

$$V = V'(P, S, \{n_i\}) \tag{2a}$$
$$V = V''(P, T, \{n_i\}) \tag{2b}$$

and pressure P are particularly convenient variables to control and measure in chemical systems, the last form is of great utility. All extensive thermodynamic properties X can be rewritten in this form, namely as Eq. (3), and since all such proper-

$$X = X(T,P,\{n_i\}) \tag{3}$$

ties are linear homogeneous functions of the mass, it can be shown that at a given temperature and pressure Eq. (4) holds, where $\overline{X}_i$ is the partial molar

$$X = \sum_i n_i \overline{X}_i \tag{4}$$

value of the extensive property for the ith component, and is given by Eq. (5), where the notation

$$\overline{X}_i = (\partial X/\partial n_i)_{T,P,\{n_i\}'} \tag{5}$$

$\{n_i\}'$, means that all of the mole numbers are constant except the ith one involved in the derivative. $\overline{X}_i$ is itself intensive.

Specification of boundaries. The concept of a boundary enclosing the system and separating it from the surroundings requires specification of the nature of the boundary and of any constraints the boundary places upon the interaction of the system and its surroundings. Boundaries that restrain a system to a particular value of an extensive property are said to be restrictive with respect to that property. A boundary which restrains the system to a given volume V is a fixed wall. A boundary which is restrictive to one component of a system but not to the other components is a semipermeable wall or membrane. A system whose boundaries are restrictive to energy and to mass or moles of components is said to be an isolated system. A system whose boundaries are restrictive only to mass or moles of components is a closed system, whereas an open system has nonrestrictive walls and hence can exchange energy, volume, and mass with its surroundings. Boundaries can be restrictive with respect to specific forms of energy, and two important types are those restrictive to thermal energy but not work (adiabatic walls) and those restrictive to work but not thermal energy (diathermal walls).

Reversible and irreversible processes. Changes in the state of the system can result from processes taking place within the system and from processes involving exchange of mass or energy with the surroundings. After a process is carried out, if it is possible to restore both the system and the surroundings to their original states, the process is said to be reversible; otherwise the process is irreversible. All naturally occurring spontaneous processes are irreversible. The first law defines the internal energy as a state function or property of the state of a system, and restricts the system and its surroundings to those processes which conserve energy. The second law established which of the permissible processes can occur spontaneously.

First law of thermodynamics. The total energy E of a system is the sum of its kinetic energy T, its potential energy V, and its internal energy U, Eq. (6). If a system has constant mass and its center of

$$E = T + V + U \tag{6}$$

mass is moving with uniform velocity in a uniform potential, then changes in the total energy of the system δE are equal to changes in its internal energy δU. Chemical thermodynamics concentrates on the internal energy of the system, but kinetic and potential energy changes of the system as a whole can be important for chemical systems. The principle of conservation of energy requires that the change in the internal energy of a system be the result of energy transfer between the system and its surroundings. The internal energy U is a function of the set of extensive variables associated with the various forms of internal energy. Each form of internal energy is manifest by the product of an extensive variable and its conjugate intensive variable. Table 1 lists several forms of internal energy, their conjugate pair of variables, and their corresponding work terms.

The internal energy of the system is given by the fundamental equation of state, Eq. (7). Processes

$$U = U(S,V,\{n_i\},\{X_i\}) \tag{7}$$

which give rise to a change in U are then limited to those for which Eq. (8) holds. Since all the exten-

$$dU = (\partial U/\partial S)_{V,\{n_i\},\{X_j\}} dS + (\partial U/\partial V)_{S,\{n_i\},\{X_j\}} dV + \sum_i (\partial U/\partial n_i)_{S,V,\{n_i\}',\{X_j\}} dn_i + \sum_j (\partial U/\partial X_j)_{X,V,\{n_i\},\{X_j\}'} dX_j \tag{8}$$

sive state properties are linear homogeneous functions, the coefficients of the differential terms are themselves intensive and correspond to the conjugate variable of the respective extensive variable.

Table 1. Internal energy and generalized work

Type of energy	Intensive factor	Extensive factor	Element of work
Mechanical			
Expansion	Pressure (P)	Volume (V)	$-PdV$
Stretching	Surface tension (γ)	Area (A)	γdA
Extension	Tensile stretch (F)	Length (l)	Fdl
Thermal	Temperature (T)	Entropy (S)	TdS
Chemical	Chemical potential (μ)	Moles (n)	μdn
Electrical	Electric potential (E)	Charge (Q)	EdQ
Gravitational	Gravitational field strength (mg)	Height (h)	$mgdh$
Polarization			
Electrostatic	Electric field strength (E)	Total electric polarization (P)	EdP
Magnetic	Magnetic field strength (H)	Total magnetic polarization (M)	HdM

Their product is thus the differential work associated with the appropriate form of external energy. Equation (8) then becomes Eq. (9), where μ_i is the

$$dU = TdS - pdV + \sum_i \mu_i dn_i + \sum_j I_j dX_j \qquad (9)$$

chemical potential of the ith component and I_j is the conjugate potential for X_j. The internal energy change given by this expression is dependent only upon the state properties of the system, and hence is independent of the process causing the change. *See* CONSERVATION OF ENERGY; INTERNAL ENERGY.

Heat. Thermal energy exchange or heat (that form of energy transferred as a result of temperature differences between a system and its surroundings) plays a central role in thermodynamics, and is singled out from the other forms of energy or work. This is expressed by Eq. (10), where δq is the

$$dU = \delta q + \delta w \qquad (10)$$

differential thermal energy (heat) absorbed by the system from the surroundings and δw is the differential work performed on the system by the surroundings. Equating Eq. (9) and (10) yields Eq. (11).

$$\delta q = TdS - PdV + \sum_i \mu_i dn_i + \sum_j I_j dX_j - \delta w \qquad (11)$$

Since δw is the work performed by the surroundings on the system, Eq. (11) can be rewritten as Eq. (12), where conservation laws for the nonthermal

$$\delta q = TdS - (P - P_s)dV + \sum_i (\mu_i - \mu_{i_s})dn_i + \sum_j (I_j - I_{j_s})dX_j \qquad (12)$$

extensive properties have been utilized and where the subscript s on the intensive variable identifies it as the value for the surroundings. It is convenient to write this equation as Eqs. (13), where $(-\delta a)$

$$\delta q = TdS - \delta a \qquad (13a)$$

$$dU = TdS + \delta w - \delta a \qquad (13b)$$

is the sum of the nonthermal differential work terms in Eq. (12). The term δa can be either zero or nonzero. If it is zero, the heat absorbed by the system is equal to TdS. In an adiabatic process δq is zero and $TdS = \delta a$, and hence if δa is nonzero, it must correspond to an internally generated thermal energy. This is frequently referred to as the uncompensated heat of a process, since it does not result from the transfer of heat from the surroundings. The first law or energy conservation principle can provide no further insight concerning either the sign or magnitude of δa. This will remain for the statement of the second law to consider. Of course, given the initial and final states of a system, the first law permits thermochemical calculations pertaining to such changes.

Heat capacity. The heat capacity of a system is of particular importance in such thermochemical calculations. The heat capacity is the amount of thermal energy that can be absorbed by a system for a unit rise in temperature. This is defined by Eq. (14), where C_{process} is the heat capacity of a

$$\delta q = C_{\text{process}}\, dT \qquad (14)$$

system for a given type of process. Three commonly considered processes for closed systems and their respective heat capacities are the constant-volume process C_v, the constant-pressure process C_p, and the saturated-vapor process C_s. Further understanding is gained by examination of Eqs. (9) and (10). For a closed system in which only mechanical (PV) work is possible, the internal energy change for a constant volume process is given by Eq. (15), and therefore Eqs. (16) hold. In the same

$$dU = \delta q = C_v dT = TdS \qquad (15)$$

$$C_v = (\partial U/\partial T)_V \qquad (16a)$$

$$\Delta S = \int_{T_1}^{T_2} (C_v/T)dT \qquad (16b)$$

closed system, the internal energy change for a constant pressure process is given by Eq. (17). For

$$dU + PdV = \delta q = C_p dT = TdS \qquad (17)$$

convenience, a new state function called the enthalpy is defined by Eqs. (18), which for a constant-

$$H \equiv U + PV \qquad (18a)$$

$$dH = dU + PdV + VdP \qquad (18b)$$

pressure process yields $dH = dU + PdV$, therefore Eqs. (19) hold. The heat capacity is an exten-

$$C_p = (\partial H/\partial T)_p \qquad (19a)$$

$$\Delta S = \int_{T_1}^{T_2} (C_p/T)dT \qquad (19b)$$

sive property of a system, but it is not a state function since its value is path- or process-dependent. *See* ENTHALPY.

Second law of thermodynamics. There are many possible and essentially equivalent statements of the second law. It will suffice to state the empirical result that in all spontaneous processes the uncompensated heat δa in Eqs. (13) is always positive. Equation (13*a*) can be rewritten as Eq. (20), where the term $\delta q/T$ is the contribu-

$$dS = \delta q/T + \delta a/T \qquad (20)$$

tion to the entropy due to heat exchange with the surrounding (d_eS), while $\delta a/T$ is the contribution to the entropy produced internally as a result of the interconversion of work terms (d_iS). The second law can then be summarized as Eqs. (21),

$$dS = d_eS + d_iS \qquad (21a)$$

$$d_iS \geqslant 0 \qquad (21b)$$

where d_iS greater than zero applies to irreversible processes. When $d_iS = 0$, that is, for a reversible process, Eq. (22) holds. This is the basic

$$dS = \delta q_{\text{rev}}/T \qquad (22)$$

equation for establishing the thermodynamic temperature scale based upon the theoretical limits of reversible cycles. The requirement that $d_iS > 0$ for spontaneous processes provides the criteria for examining the specific conditions for spontaneous paths, and the criteria for establishing the equalibrium states of a system. *See* TEMPERATURE.

In an isolated system, $dU = 0$, for any process that takes place whether spontaneous or not, whereas $dS = d_iS$ since $d_eS = 0$. Any spontaneous process must therefore increase S until S reaches a maximum value at which point only reversible processes can take place. If a random fluctuation perturbs the equilibrium state, the system will spontaneously return to the equilibrium state. Thus states in the vicinity of the equilibrium state are said to be unstable with respect to such perturbations. *See* ENTROPY.

Helmholtz free energy. While isolated systems are of theoretical value, they do not play an important role in practical chemical systems, and consequently criteria for systems undergoing changes at constant temperature and either constant volume or constant pressure are required. Examination of Eq. (11) indicates that if a state function defined by Eqs. (23) is combined with Eq. (13*b*),

$$A \equiv U - TS \qquad (23a)$$

$$dA = dU - TdS - SdT \qquad (23b)$$

then Eq. (24) holds, which for constant-tempera-

$$dA = -\,\delta a + \delta w - SdT \qquad (24)$$

ture processes is Eqs. (25), and where $-\delta w_{\max}$ is

$$dA = -\delta a + \delta w \qquad \text{Spontaneous process} \qquad (25a)$$

$$dA = \delta w_{\max} \qquad \text{Reversible process} \qquad (25b)$$

the maximum work that can be performed by the system under constant temperature conditions. If processes take place at constant temperature and constant extensive variables $(V,\{n_i\},\{X_i\})$, then $dA = -\delta a$, and all spontaneous processes are accompanied by a decrease in A and equilibrium is achieved when A is a minimum. The state function A is the Helmholtz free energy, and its characteristic variables are $A(T,V,\{n_i\},\{X_j\})$. *See* FREE ENERGY.

Gibbs free energy. If a state function is defined by Eqs. (26), combining this with equation (13*b*)

$$G \equiv U + PV - TS = H - TS \qquad (26a)$$

$$dG = dU + PdV + VdP - TdS - SdT \qquad (26b)$$

yields Eq. (27), which for a constant temperature

$$dG = -\delta a + \delta w + PdV + VdP - SdT \qquad (27)$$

and pressure process is Eqs. (28). Here $-\delta w_{\text{net}}$ is

$$dG = -\delta a + \delta w + PdV \qquad \text{Spontaneous process} \qquad (28a)$$

$$dG = \delta w_{\max} + PdV = \delta w_{\text{net}} \qquad \text{Reversible process} \qquad (28b)$$

the net maximum work a system can perform in excess of expansion work under conditions of constant temperature and pressure. If processes take place at constant temperature and pressure and constant extensive variables $(\{n_i\},\{X_j\})$, then $dG = -\delta a$, and all spontaneous processes are accompanied by a decrease in G and equilibrium is achieved when G is a minimum. The state function G is the Gibbs free energy or free enthalpy, and its characteristic variables are $G(T,P,\{n_i\},\{X_j\})$.

Equations of state. Table 2 provides a summary of the fundamental equation of state for the internal energy and entropy, as well as the derived equations of state for practical conditions. A more complete understanding of the relationship of equations involving intensive variables as independent variables is based on the mathematical recognition that intensive variables are Legendre transformations of the corresponding conjugate extensive variables. The properties of these transformations are such that derived equations involving intensive variables only incompletely define the state of the system, even though they do completely define changes in state.

Affinity and chemical equilibrium. Many chemical systems can be considered closed systems in which a single parameter ξ can be defined as a measure of the extent of the reaction or the degree of advancement of a process. If the reaction proceeds or the process advances spontaneously, entropy must be produced according to the second law and δa must be positive. In terms of the advancement parameter ξ, this uncompensated heat δa can be given by Eq. (29), where $\underline{A}$ is the

$$\delta a = \underline{A}d\xi = Td_iS \qquad (29)$$

affinity of the process or reaction. The affinity is related to internal entropy production by Eq. (30).

$$\underline{A} = T\,d_iS/d\xi \geqslant 0 \qquad (30)$$

The condition that the entropy production is zero represents equilibrium, and hence $\underline{A} = 0$ is an equivalent condition for equilibrium in a closed system. For spontaneous processes, since the signs of $\underline{A}$ and $d\xi$ must be the same, for positive $\underline{A}$ the process must advance or go in a forward direction in the usual sense of chemical reactions or physical processes, while for negative $\underline{A}$ the process must proceed in the reverse direction.

Equations (13*b*) and (26*b*) can be combined to give Eqs. (31), which indicates that the affinity is

$$dG = -SdT + VdP - \delta a \qquad (31a)$$

$$= -SdT + VdP - \underline{A}d\xi \qquad (31b)$$

itself a state function, Eq. (32).

$$\underline{A} = -(\partial G/\partial \xi)_{P,T} = \underline{A}(T,P,\xi) \qquad (32)$$

Table 2. Equations of state and characteristic variables

Basic function	Equation of state	Differential form
Internal energy (U)	$\nu(S,V,\{n_i\},\{X_i\})$	$dU = TdS - PdV + \sum_i \mu_i dn_i + \sum_j I_j dX_j$
Entropy (S)	$S(U,V,\{n_i\},\{X_i\})$	$dS = (1/T)dU - (P/T)dV + \sum_i (\mu_i/T)dn_i + \sum_j (I_j/T)dX_j$
Enthalphy (H) ($H = U + PV$)	$H(S,P,\{n_i\},\{X_i\})$	$dH = TdS + VdP + \sum_i \mu_i dn_i + \sum_j I_j dX_j$
Helmholtz free energy (A) ($A = U - TS$)	$A(V,T,\{n_i\},\{X_j\})$	$dA = -\,SdT - PdV + \sum_i \mu_i dn_i + \sum_j I_j dX_j$
Gibbs free energy (G) ($G = U + PV - TS$)	$G(P,T,\{n_i\},\{X_j\})$	$dG = -\,SdT + VdP + \sum_i \mu_i dn_i + \sum_j I_j dX_j$

Consider a closed system in which a chemical reaction can be characterized by the stoichiometry of reaction (33). The stoichiometry requires that at

$$\alpha A + \beta B \rightarrow \gamma C + \delta D \quad (33)$$

each time element t in the reaction the number of moles of the ith component n_i be given by Eq. (34),

$$n_i = n_i^0 + \nu_i \xi \quad (34)$$

where n_i^o is the number of moles of i in the initial or original state ($t = o$), ν_i is the stoichiometric coefficient for the ith component as given in the balanced equation (the convention is that ν_i is positive for products and negative for reactants), and ξ is the degree-of-advancement parameter whose range is normally taken to be zero to unity. In terms of differential changes in advancement Eq. (35) holds. For closed systems (constant temperature and pressure) in which only thermal, expansion, and chemical work terms are included, Eqs. (36) and (37) hold. The condition for equilibrium is $\underline{A} = 0$, and thus for a chemical reaction, equilibrium is achieved when Eq. (38) holds. If electrical work is included in Eq. (36), Eqs. (39) and (40) hold. Since $\underline{A} = 0$ is the equilibrium condition, equilibrium in an electrochemical system is given by Eq. (41), where z is the number of equivalents of charge and F is the Faraday constant (the number of coulombs of charge per equivalent).

$$dn_i = \nu_i d\xi \quad (35)$$

$$dG = +\sum_i \mu_i \, dn_i = -\delta a \quad (36a)$$

$$= (\sum_i \nu_i \mu_i) d\xi = -\underline{A} d\xi \quad (36b)$$

$$\underline{A}(T,P,\xi) = -\sum_i \nu_i \mu_i \quad (37)$$

$$\sum_i \nu_i \mu_i = 0 \quad (38)$$

$$dG = \sum_i \mu_i dn_i + EdQ = -\delta a \quad (39a)$$

$$= (\sum_i \nu_i \mu_i + zFE) d\xi = -\underline{A} d\xi \quad (39b)$$

$$\underline{A}(T,P,E,\xi) = -\sum_i \nu_i \mu_i - zFE \quad (40)$$

$$\sum_i \nu_i \mu_i = -zFE \quad (41)$$

More than one reaction can take place in a chemical system, each characterized by a degree-of-advancement parameter, and thus for r independent reactions, Eqs. (42) hold, where the equilibrium condition is Eq. (43). At equilibrium each of the $\underline{A}_r$ must be zero, but for the spontaneous condition, in equality (44) holds. In a two-reaction system inequality (45) holds, but now $\underline{A}_1$ and $d\xi_1$ do not necessarily have the same sign. If their signs are different, the first reaction can be driven in the nonspontaneous direction by the second reaction. The reactions are then said to be coupled, and this is a common situation in biological systems. *See* CHEMICAL EQUILIBRIUM.

$$dG = \sum_i \mu_i dn_i = -\delta a \quad (42a)$$

$$= \sum_r \underline{A}_r \delta\xi_r \quad (42b)$$

$$= \sum_r (\sum_i \nu_{ir} \mu_{ir} \delta\xi_r) \quad (42c)$$

$$\sum_r \underline{A}_r d\xi_r = 0 \quad (43)$$

$$\sum_r \underline{A}_r d\xi_r > 0 \quad (44)$$

$$\underline{A}_1 d\xi_1 + \underline{A}_2 d\xi_2 > 0 \quad (45)$$

Chemical potential. The affinity of a chemical reaction establishes the spontaneous direction of the reaction, and consequently methods for determining the affinity are important in thermochemical studies. As shown above, the affinity is simply related to the stoichiometric coefficients of the reaction and the chemical potentials of the reactants and products in the reaction. It is necessary therefore to investigate some of the properties of the chemical potential and to develop convenient methods of using it to calculate the affinity.

The chemical potential of a single-phase pure substance can be expressed as $\mu_i = \mu_i(T,P)$, whereas for a component in single-phase solution $\mu_i = \mu_i(T,P,\{x_i\})$, where $\{x_i\}$ is the set of independent mole fractions ($x_i = n_i/\Sigma n_i$). All intensive thermodynamic variables are homogeneous functions of zero degree in mass, and hence Eq. (46) holds, where $\bar{I}_i$ is given by Eq. (47). Equation (46) is a form of the Gibbs-Duhem relationship. Since the chemical potential is intensive, the Gibbs-Duhem equation for a given solution phase can be written as Eq. (48). This is an important result in phase equilibria in heterogeneous systems, and places restraints on the number of independent variables in such systems. *See* PHASE EQUILIBRIUM.

$$\sum_i n_i \bar{I}_i = 0 \quad (46)$$

$$\bar{I}_i = (\partial I_i/\partial n_i)_{T,P,\{n_i\}'} \quad (47)$$

$$\sum_i x_i \bar{\mu}_i = 0 \quad (48)$$

The chemical potential can be represented in several forms, but for chemical studies the form which expresses that it is a partial molar quantity, Eq. (49), is most useful. Since the order of differentiation of exact functions is immaterial, temperature, and pressure derivatives of μ_i are given by Eqs. (50) and (51), where $\bar{S}_i$ and $\bar{V}_i$ are the partial molar entropy and volume of the ith component. Two additional useful differential coefficients are given by Eqs. (52) and (53), where $\bar{H}_i$ and $\bar{C}_{P,i}$ are the partial molar enthalpy and constant-pressure heat capacity.

$$\mu_i = (\partial G/\partial n_i)_{T,P,\{n_i\}'} \quad (49)$$

$$(\partial \mu_i/\partial T)_{P,\{n_i\}} = \partial^2 G/\partial T \partial n_i$$
$$= -(\partial S/\partial n_i)_{T,P,\{n_i\}} = -\bar{S}_i \quad (50)$$

$$(\partial \mu_i/\partial P)_{T,\{n_i\}} = \partial^2 G/\partial P \partial n_i$$
$$= -(\partial V/\partial n_i)_{T,P} = \bar{V}_i \quad (51)$$

$$[\partial(\mu_i/T)/\partial(1/T)]_{P,\{n_i\}} = \bar{H}_i \quad (52)$$

$$(\partial \bar{H}_i/\partial T)_{P,\{n_i\}} = \bar{C}_{P,i} \quad (53)$$

It is quite apparent that a knowledge of the chemical potentials of pure substances and of substances in solution provides the basis of the thermal properties of the substance, as well as the basis for reaction spontaneity. Usually the chemical potential of a substance is expressed in terms of a standard state and a convenient measure of the deviation from that state. Various functional forms could be used, but it is customary to use Eq. (54), where $\mu_i^{\ominus}$ is the chemical potential in a

$$\mu_i(T,P,\{x_i\}) = \mu_i^{\ominus}(T,P,\{x_i\}) + RT \ln a_i^*(T,P,\{x_i\}) \quad (54)$$

designated standard state, R is the gas constant, and a_i^* is the relative activity of the ith substance. The activity is frequently written as Eq. (55), where

$$a_i^* = f_i^* x_i \quad (55)$$

$f_i^*(T,P,\{x_i\})$ is the activity coefficient on the mole fraction basis. Two basic situations arise as the mole fraction approaches its limits of 0 or 1, and it is these limits that determine the functional form of f_i^*. Combining Eq. (54) with (55) gives Eq. (56). If

$$\mu_i = \mu_i^* + RT \ln f_i^* + RT \ln x_i \quad (56)$$

the $\lim_{x_i \to 1} f_i^* = 1$, then Eq. (57), the Raoult law con-

$$\lim_{x_i \to 1} \mu_i = \mu_i^* = \mu_i^*(T,P) \quad (57)$$

vention, holds where μ_i^* is the chemical potential of pure i. The other limit, $\lim_{x_i \to 0} f_i = 1$, implies that $\lim_{x_i \to 1} f_i$ equals some finite value, say f_i^∞, and therefore Eq. (58), the Henry law convention, holds,

$$\lim_{x_i \to 1} \mu_i = \mu_i^{\ominus} + RT \ln f_i^\infty = \mu_i^\infty \quad (58)$$

where f_i^∞ is the activity coefficient at mole fraction unity determined by the limiting behavior of the chemical potential in an infinitely dilute solution of i, and μ_i^∞ is the chemical potential i would have if its dilute solution behavior persisted up to mole fraction one. The two situations can be summarized by Eqs. (59), where f_i^* and f_i^∞ are the activity co-

$$\mu_i = \mu_i^*(T,P) + RT \ln f_i^* x_i \qquad \text{Raoult's law} \quad (59a)$$

$$\mu_i = \mu_i^\infty(T,P,x_i) + RT \ln f_i^\infty x_i \qquad \text{Henry law} \quad (59b)$$

efficients on the Raoult or Henry law basis, respectively.

The activities discussed above are based upon mole fractions. It is frequently more convenient to use other measures of relative amounts, for example, the molality m, defined as the moles of solute per kilogram of solvent, and the molarity c, defined as the moles of solute per liter of solution. If this is done, the chemical potentials are given by Eqs. (60) or (61), where $\mu_i^{\infty,m}$ and $\mu_i^{\infty,c}$ are infinite

$$\mu_i = \mu_i^{\infty,m} + RT \ln f_i^m m_i \quad (60)$$

$$\mu_i = \mu_i^{\infty,c} + RT \ln f_i^c c_i \quad (61)$$

dilution based standard state chemical potentials on the molality and molarity convention, and f_i^m and f_i^c are activity coefficients on the same bases. In each case $\lim_{m_i,c_i \to 0} f_i = 1$.

The chemical potential of gases is frequently discussed in terms of its fugacity or pressure, and Eq. (54) is then written as Eq. (62) or (63), where P_i^*

$$\mu_i = \mu_i^0(T) + RT \ln P_i^* \quad (62)$$

$$\mu_i = \mu_i^0(T) + RT \ln f_i P_i \quad (63)$$

is the fugacity of substance i and is equal of $f_i P_i$, where f_i is the fugacity coefficient and P_i is the partial pressure of i. The fugacity coefficient is defined such that the $\lim_{P \to 0} f_i = 1$. The standard chemical potential $\mu_i^0(T)$ is the value of the chemical potential of a perfect gas, and is a function of temperature only. *See* FUGACITY.

The chemical potential of pure liquids or solids is given by Eq. (64), where $P^{\ominus}$ is the pressure in a

$$\mu_i(T,P) = \mu_i^*(T,P) = \mu_i^{\ominus}(T,P^{\ominus}) + \int_{P^{\ominus}}^{P} \overline{V}_i dP \quad (64)$$

designated standard state. The last term of Eq. (64) is negligible if $(P - P^{\ominus})$ is not very large, since $\overline{V}_i$ for condensed phases is relatively small. Generally the standard chemical potential is taken from tables of Gibbs free energies of formation $\Delta G^{\ominus}$ of pure gases, liquids, or solids or of substances at infinite dilution in particular solvents.

Thermodynamical relationships. The criteria for equilibrium in a chemical system at constant temperature and pressure are given by Eq. (65). If

$$\underline{A} = -\sum \nu_i \mu_i = 0 \quad (65)$$

a reaction takes place in a gaseous mixture, substitution of Eq. (62) for the μ_i gives Eq. (66*a*) or (66*b*), and Eq. (67) holds, where $\Delta G^0_{\text{react}}$ is the Gibbs

$$\sum \nu_i(\mu_i^0(T) + RT \ln P_i^*) = 0 \quad (66a)$$

$$\sum \nu_i \mu_i^0(T) = -RT \ln \prod_i (P_i^*)^{\nu_i} \quad (66b)$$

$$\Delta G^0_{\text{react}}(T) = -RT \ln K_{p^*} \quad (67)$$

free energy change for a reaction going from a standard state of reactants to a standard state of products in accordance with the stoichiometry of the reaction, and K_{p^*} is the equilibrium constant for the reaction in terms of fugacities. At low pressures $P_i^* = P_i$, and the equilibrium constant for reaction (33) is given by Eq. (68). For a reaction in

$$K_p = P_c^\gamma P_d^\delta / P_a^\alpha P_b^\beta \quad (68)$$

solution for which the chemical potential is given on a molarity basis, Eqs. (69)–(71) hold, where K_a^c

$$\sum \nu_i(\mu^{\infty,c} + RT \ln a_i^c) = 0 \quad (69)$$

$$\sum \nu_i \mu_i^{\infty,c} = -RT \ln \prod_i (a_i^c)^{\nu_i} \quad (70)$$

$$\Delta G^{\infty,c}_{\text{react}} = -RT \ln K_a^c \quad (71)$$

is the equilibrium constant for the reaction in terms of molarity-based activities. For dilute solutions, $a_i = c_i$, and for reaction (33), Eq. (72) holds.

$$K_c = c_c^\gamma c_d^\delta / c_a^\alpha c_b^\beta \quad (72)$$

Similar results can be obtained for other concentration-based activities. The activities can be based on whatever measures of activity are most convenient. Clearly the more general expression of the equilibrium is given by Eq. (73). The standard

$$\Delta G^{\ominus}_{\text{react}} = -RT \ln K_a \quad (73)$$

enthalpy, entropy, heat capacity, and volume changes for a chemical reaction can be obtained directly from the appropriate temperature derivatives of the standard free energy of the reaction or its equivalent, the equilibrium constant.

The condition for a reaction to take place spontaneously in the direction from reactants to products requires that $\underline{A}$ be positive, and hence Eqs.

(74) hold, where Q_a is an expression of the form of

$$\underline{A}=-\sum_i \nu_i\mu_i > 0 \quad (74a)$$

$$\underline{A}=-(\sum_i \nu_i\mu_i + RT\ln \prod_i a_i^{\nu_i}) > 0 \quad (74b)$$

$$\underline{A}=RT\ln K_a - RT\ln Q_a > 0 \quad (74c)$$

$$\underline{A}=RT\ln K_a/Q_a > 0 \text{ or } K_a/Q_a > 1 \quad (74d)$$

the equilibrium constant, but involving activities of reactants in their initial state and the activities of the products in their final states. Again, the bases of the activities in Q_a, as in K_a, are arbitrary and selected for convenience.

Third law of thermodynamics. Most understanding of the classical thermodynamics of chemical systems is based upon the first and second laws. As these systems are studied on a molecular rather than on a macroscopic basis, it is apparent that the first and second laws cannot be addressed to this endeavor in a direct manner. Although implicit in the concept of the existence of the fundamental equations of state for U or S is an absolute value of these functions, and therefore an extensive quantity which could be calculated on the basis of molecular properties from quantum mechanics and statistical mechanics, neither the first nor second law considers anything but differences in these state functions. The second law does indeed indicate the existence of an absolute zero for an intensive variable, the temperature, but this is not sufficient to bridge the areas of classical and statistical thermodynamics.

It is found experimentally that for many isothermal processes involving pure phases, Eq. (75)

$$\lim_{T\to 0} \Delta S = 0 \quad (75)$$

holds. This includes phase transitions between different crystalline modifications, solid-state chemical reactions, and even the solid-liquid transition in helium. This, along with Eq. (76), implies

$$S(T,P) - S(0\text{ K}) = \int_0^T (C_p/T)dT \quad (76)$$

that at zero absolute temperature the entropy of pure crystalline phases are equal. If the entropy of pure phases are equal at $T=0$, it is reasonable to take the value $S(0\text{ K})$ to be zero. The statement of the third law then is that the entropy of all pure crystalline phases at 0 K is zero. This makes it possible by using Eq. (76), to calculate the absolute or third-law entropy of a substance from experimental measurements of their heat capacities. Comparison of such experimental values with those calculated by statistical thermodynamic methods has provided evidence for the validity of the third law. In some cases thorough investigation of apparent discrepancies from the third law have led to new conclusions concerning the molecular structure of the substances or new information on the energy level system for the molecules. Calculations of the thermodynamic properties for a gas from the spectroscopic properties of molecules is an important result stemming from the third law.

Thermodynamics of irreversible processes. Classical equilibrium thermodynamics is primarily concerned with calculations for reversible processes, and deals with irreversibility in terms of inequalities. In the case of irreversible processes in systems slightly removed from equilibrium, the rate of internal entropy production d_iS/dt is related to the fluxes J_i associated with thermal, concentration, or other differences in intensive parameters or potentials X_i. This entropy production is then given by Eq. (77). The fluxes include

$$d_iS/dt = \sum_i J_iX_i \geq 0 \quad (77)$$

heat conduction, diffusion, electric conduction, and other direct effects.

In addition, a flux of one type may be coupled to a potential difference of another type. For example, a thermal gradient can result in a mass flux (thermal diffusion), or a concentration gradient in any energy flux. Thermal conductivity, thermoosmosis, and thermoelectric effects are all coupled effects. The fluxes are thus found to be given by Eq. (78),

$$J_i = \sum_j L_{ij}X_j \quad (78)$$

where the L_{ij} are called the phenomenological coefficients. If $L_{ij} \neq 0$, there is a coupling between the flux J_i and the gradient X_j. Microscopic reversibility implies that not far from equilibrium, $L_{ij}=L_{ji}$, and this is known as the Onsager reciprocity relationship. These results, together with the theorem of minimum entropy production, are the basis of investigations of irreversible processes near equilibrium.

Far removed from equilibrium, thermodynamics must be formulated somewhat differently and more cautiously. The interplay of thermodynamic stability and kinetics can give rise to macroscopic structures with both temporal and spatial coherence called dissipative structures. Much theoretical effort is being directed to these studies because of their apparent relevance to biological structures, but it is still too early to assess how far-reaching these theories will be in the future. *See* THERMODYNAMIC PRINCIPLES; THERMODYNAMIC PROCESSES.

[ROBERT A. PIEROTTI]

Chemiluminescence

The type of luminescence wherein a chemical reaction supplies the energy responsible for the emission of light (ultraviolet, visible, or infrared) in excess of that of a blackbody (thermal radiation) at the same temperature and within the same spectral range. Below 500°C, the emission of any light during a chemical reaction is a chemiluminescence. The blue inner cone of a bunsen burner or the Coleman gas lamp are examples.

Many chemical reactions generate energy. Usually this exothermicity appears as heat, that is, translational, rotational, and vibrational energy of the product molecules; whereas, for a visible chemiluminescence to occur, one of the reaction products must be generated in an excited electronic state (designated by an asterisk) from which it can undergo deactivation by emission of a photon. Hence a chemiluminescent reaction, as shown in reactions (1) and (2), can be regarded as the reverse of a photochemical reaction.

$$A + B \rightarrow C^* + D \quad (1)$$

$$C^* \rightarrow C + h\nu \quad (2)$$

The energy of the light quantum $h\nu$ (where h is Planck's constant, and ν is the light frequency) depends on the separation between the ground and the first excited electronic state of C; and the spectrum of the chemiluminescence usually matches the fluorescence spectrum of the emitter. Occasionally, the reaction involves an additional step, the transfer of electronic energy from C* to another molecule, not necessarily otherwise involved in the reaction. Sometimes no discrete excited state can be specified, in which case the chemiluminescence spectrum is a structureless continuum associated with the formation of a molecule, as in the so-called air afterglow: $NO + O \rightarrow NO_2 + h\nu$ (green light). *See* PHOTOCHEMISTRY.

The efficiency of a chemiluminescence is expressed as its quantum yield ϕ, that is, the number of photons emitted per reacted molecule. Many reactions have quantum yields much lower (10^{-8} $h\nu$ per molecules) than the maximum of unity, Einsteins of visible light (1 einstein $= Nh\nu$, where N is Avogadro's number), with wavelengths from 400 to 700 nm, correspond to energies of about 70 to 40 kcal per mole (300 to 170 kilojoules per mole). Thus only very exothermic, or "exergonic," chemical processes can be expected to be chemiluminescent. Partly for this reason, most familiar examples of chemiluminescence involve oxygen and oxidation processes; the most efficient examples of these are the enzyme-mediated bioluminescences. The glow of phosphorus in air is a historically important case, although the mechanism of this complex reaction is not fully understood. The oxidation of many organic substances, such as aldehydes or alcohols, by oxygen, hydrogen peroxide, ozone, and so on, is chemiluminescent. The reaction of heated ether vapor with air results in a bluish "cold" flame, for example. The efficiency of some chemiluminescences in solution, such as the oxidation of luminol (I) and, especially, the reaction of some oxalate esters (II) with hydrogen peroxide, can be very high ($\phi = 30\%$).

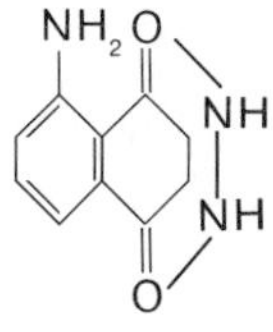

(I)

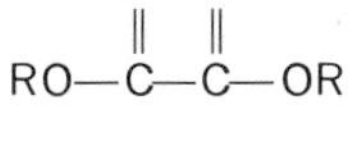

(II)

It is believed that the requirements for chemiluminescence are not only sufficient exothermicity and the presence of a suitable emitter, but also that the chemical process be very fast and involve few geometrical changes, in order to minimize energy dissipation through vibrations. For example, the transfer of one electron from a powerful oxidant to a reductant (often two radical ions of opposite charge generated electrochemically) is a type of process which can result, in some cases, in very effective generation of electronic excitation. An example, with 9,10-diphenylanthracene (DPA), is shown in reaction (3). The same is true of the

$$DPA^{\overline{\cdot}} + DPA^{\overset{+}{\cdot}} \rightarrow DPA^{*} + DPA \qquad (3)$$

decomposition of four-membered cyclic peroxides (III) into carbonyl products, shown in reaction (4),

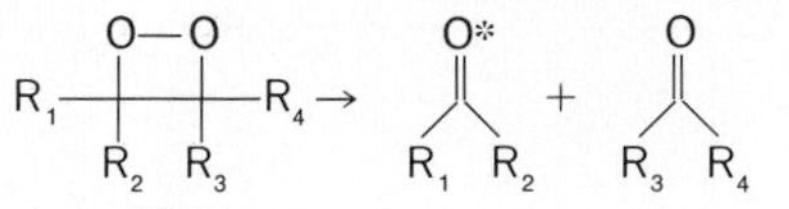

(III) (4)

which may be the prototype of many chemiluminescences.

[THERESE WILSON]

Bibliography: E. J. Bowen (ed.), *Luminescence in Chemistry*, 1968; M. J. Corimier et al. (eds.), *Chemiluminescence and Bioluminescence*, 1973; M. Darlene (ed.), *Methods in Enzymology*, vol. 57, 1978; H. H. Seliger and W. D. McElroy, *Light: Physical and Biological Action*, 1965.

Chemistry

The science of chemistry includes a study of the properties, composition, and structure of matter, the changes in structure and composition which matter undergoes, and the accompanying energy changes. It grew out of the alchemy of the Middle Ages. The exact date of the beginning of chemistry as a science cannot be stated, but frequently given is the time of the correct interpretation of combustion by the great French scientist A. L. Lavoisier, about 1774.

Today the objective of the chemist is to aid in the interpretation of the universe. He has made much progress toward meeting this objective because he knows about not only the structure and composition of many of the materials on the Earth, but also those of the planets, the satellites, the stars, and the materials of interstellar space.

Method of science. The success of chemistry is largely attributed to the use of the scientific method, although not all the discoveries are made by planned research; many of them are made by trial and error and by accident. Nevertheless, the rigorous procedure of observation, classification, theorizing, and experimentation to test the theory runs throughout this entire science.

The huge problem of interpreting the universe is considerably simplified by breaking it down to smaller problems by classifying the great variety of materials in the universe into the two great entities, energy and matter. Energy can be classified as potential or kinetic energy, and can be broken down further into such forms of energy as mechanical, electrical, radiant, chemical, and nuclear. Matter can be classified in a number of different ways. One method is in terms of the physical state – solid, liquid, and gas; but probably the most useful method is in terms of composition – elements, compounds, and mixtures.

Chemical elements. An element is a substance which cannot be broken down to simpler substances by chemical reactions. It is also defined as a substance made of one kind of building block (atom) only. There are only a few more than 100 elements known in the entire universe. A careful study of the elements has indicated that they can be classified into families or groups that further simplify the problem of learning about the universe. This classification is called the periodic table. The elements in each family or group of the periodic table have similar properties. *See* ELEMENTS; PERIODIC TABLE.

The most abundant element on the Earth's sur-

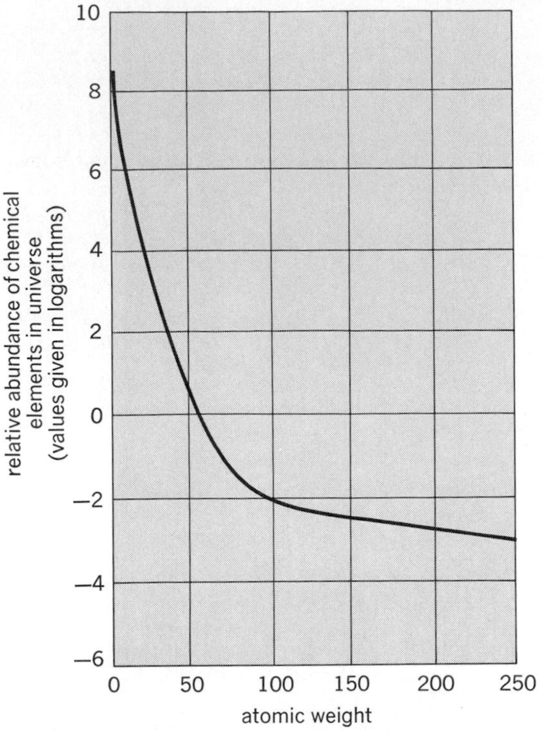

Abundances of the elements in the universe. Elements of low atomic weight are by far the most abundant.

face is oxygen, the next being silicon. The two most abundant elements in the universe are hydrogen and helium (see illustration). Tables 1 and 2 list the most plentiful elements in the Earth's crust and oceans and in the human body.

Laws of chemical combination. The early foundations of chemistry were based on two fundamental laws: the law of conservation of matter and energy and the law of definite composition. The establishment of these laws required much careful research on the quantitative relationship of reactants and products in chemical processes and also of the ratio of weights of the elements in compounds. The chemical balance became the important instrument in this work. After the law of conservation of matter had been established, much research was done on the combining weights of elements. These data not only led to the law of definite composition (the composition of a pure substance is always perfectly definite), but also to the atomic theory, to a system of atomic weights, and to the methods of determining formulas of compounds.

Table 1. Relative abundance of the elements on Earth

	Solid 10-mi crust of Earth, %	Ocean, %	Average, %*
Oxygen	46.71	85.79	49.52
Silicon	27.69		25.75
Aluminum	8.07		7.51
Iron	5.05		4.70
Calcium	3.65	0.05	3.39
Sodium	2.75	1.14	2.64
Potassium	2.58	0.04	2.40
Magnesium	2.08	0.14	1.94
Titanium	0.62		0.58
Hydrogen	0.14	10.67	0.88
Phosphorus	0.130		0.120
Carbon	0.094	0.002	0.087
Chlorine	0.045	2.07	0.188
Other elements	0.391	0.098	0.295
	100.00	100.00	100.00

*Including crust, ocean, and atmosphere.

Table 2. Average composition of the human body

Element	%	Element	%
Oxygen	65.00	Sodium	0.15
Carbon	18.00	Chlorine	0.15
Hydrogen	10.00	Magnesium	0.05
Nitrogen	3.00	Iron	0.004
Calcium	1.50	Iodine	0.00004
Phosphorus	1.00	Fluorine, silicon, and other elements	Very minute amounts
Potassium	0.35		
Sulfur	0.25		

Dalton's atomic theory. To explain several phenomena, including the law of definite composition, which resulted from the observation that elements always combine in the same ratio by weight to form a given compound, John Dalton proposed that matter is made of particles called atoms, that reactions must take place between atoms or groups of atoms, and that atoms of the same element are all alike, but that they differ from atoms of another element. These assumptions are the basis of the accepted modern atomic theory. They are also most useful in explaining another law of chemical combination, called the law of multiple proportions, which states that when two or more elements combine in more than one ratio, they do so in the ratio of small whole numbers.

Atoms, molecules, and ions. Atoms are made up of three fundamental particles: protons and neutrons at the heart or center of the atom, and electrons surrounding the nucleus of protons and neutrons. All the atoms of the same element have the same atomic number (positive charge on the nucleus). Elements whose atoms have the same number of electrons in the outer shell have similar properties; this accounts for the similar properties of the elements in each family of the periodic table. The different atoms tend to react with each other by gaining or losing electrons or sharing pairs of electrons. If the atoms combine by sharing pairs of electrons, molecules are formed. If they combine by gaining and losing electrons, ions are formed. Ions carry positive or negative charges.

Structure of atoms. The present concept of the atom is the result of some of the most significant research in the whole history of chemistry. Some high points of these researches follow:

1896 The discovery of radioactivity by A. H. Becquerel, which opened the atomic era and gave the chemist high-energy atomic projectiles with which to bombard other elements.

1897 The identification of electrons by J. J. Thomson as negative particles of electricity and common to all substances.

1911 The discovery by Sir Ernest Rutherford that the atom is a very porous body and that the nucleus is very tiny and is composed of most of the mass of the atom.

1912 The discovery by H. G. J. Moseley that the positively charged particles are located in the nucleus of the atom.

1913 The invention by Niels Bohr of the "solar system" model of atoms, which was successful in explaining the quantum nature of radiation, the lines in spectra, chemical bonding, and periodic properties of elements.

1919 The discovery by Sir Ernest Rutherford that elements could be transmuted by bombardment with high-energy particles from radium, thus changing their atomic numbers.

1920 to 1930 The discovery of spinning electrons, paired electrons, and the distribution of electrical charge over the atom with higher density in discrete energy levels.

1932 The discovery of the neutron by James Chadwick, which identified the third stable particle in the atoms.

1934 The discovery by J. F. and I. C. Joliot that all elements can be made radioactive by bombardment of their nuclei with high-energy particles.

1939 The discovery of fission by Enrico Fermi, Otto Hahn, and F. Strassmann, and the subsequent interpretation of that process by Lise Meitner.

Work done in the nucleus of atoms indicated they contain a series of energy levels similar to the electron levels on the outside of the atom. Research is ongoing on the forces in the nucleus in an attempt to reach an explanation of the many fleeting, short-lived particles that have been identified in nuclear reactions.

Isotopes. Most of the elements are made of atoms of different atomic weights; such atoms are called isotopes. More than 1000 isotopes, both stable and radioactive, have been identified. Most of the radioactive isotopes of the elements have been produced synthetically. Most of the stable isotopes can be concentrated or separated by electrical means. The main use of isotopes is as tracers; several radioactive isotopes are used as atomic fuel, such as U-235 and Pu-239.

Compounds. Compounds are chemical combinations of elements. They are made of molecules or ions. Several million compounds have been synthesized or have been isolated and identified from products found in plant and animal life or in the minerals of the Earth.

The origin of the compounds in plant tissue is the photosynthesis process which takes place in the green leaves, in which the compounds carbon dioxide and water of the Earth's atmosphere combine in a long series of steps to form such compounds as simple sugar, starch, and cellulose. Animals then use the plant tissue to build animal tissue. Thus, the whole animal kingdom depends upon the plant kingdom, and the plant kingdom, in turn, on the photosynthesis process.

Types of bonds. The six types of chemical bonds that hold atoms to each other are covalent, polar covalent, coordinate covalent, ionic, metallic, and hydrogen bonds. The covalent bonds are formed by sharing pairs of electrons which spin in opposite directions and hence form a strong bond between the atoms. Examples of covalency are H_2, H_2O, NH_3, HCl, and CH_4. The covalent bond is polar if the pair of electrons is located closer to one atom than the other as the result of a difference in attraction for electrons. This gives one end of the molecule a slight positive (+) charge and the other a slight negative (−) charge. The water molecule contains polar covalent bonds; as a result, it has good solvent action for electrovalent compounds. If both electrons are supplied by one atom, the bond is coordinate covalent. This type of bond is formed in acid-base reactions and in the formation of complex ions, for example, $Cu(NH_3)_4^{++}$.

The ionic bond is formed between ions. The metallic bond is present in all metals and is supposed to result from the "sea" of valence electrons free to move through the metal lattice. The hydrogen bond is the result of the polar covalent character of a covalent bond of hydrogen with another element, such as fluorine, oxygen, or sulfur. The boiling point of water is much higher than predicted because the water molecules are polymerized by the formation of hydrogen bonds. It is also important in bonding large organic molecules to each other if they have many OH groups along a chain of carbon atoms. This is particularly true of many large molecules found in plant and animal tissue. *See* CHEMICAL BONDING.

Electrolytes and nonelectrolytes. Compounds are sometimes classified as electrolytes, those which dissolve as ions in solution, and as nonelectrolytes, those which do not. Electrolytes can be further classified as acids, bases, and salts. In water solution, acids have a sour taste, change the colors of indicators, and produce H^+ ions in solution. Bases taste bitter or brackish, change the colors of the indicators, and produce OH^- ions in water. A salt is a product of the reaction of an acid with a base. Generally, an acid is an ion or a molecule that can accept a pair of electrons, that is, H^+ and $AlCl_3$. *See* ACID AND BASE.

A general definition of a base is an ion or a molecule that can donate an electron pair, for example,

$$^{-}{:}\ddot{\underset{\cdot\cdot}{O}}{:}H \quad \text{or} \quad {:}\overset{H}{\ddot{\underset{\underset{H}{\cdot\cdot}}{N}}}{:}H$$

Symbols, formulas, and equations. Abbreviations or symbols are used to designate elements. The symbol of oxygen is O, hydrogen H, helium He, copper Cu, sodium Na, radium Ra, plutonium Pu, and so on. Groups of symbols called formulas are used to designate compounds. The formula for water is H_2O, for carbon dioxide CO_2, for sulfuric acid H_2SO_4, for grain alcohol C_2H_5OH. These symbols and formulas are used as chemists' shorthand to indicate chemical reactions. For example,

Statement:
Water decomposes to form hydrogen and oxygen.

Word equation:

$$\text{Water} \rightarrow \text{hydrogen} + \text{oxygen}$$

Symbol equation:

$$2H_2O \rightarrow 2H_2 + O_2$$

Atomic weight. The nucleus of an atom is made up of neutrons and protons. The total mass of the neutrons and protons along with the electrons gives the mass of the atom. Because the atoms are very tiny, their masses likewise are very small. The weight of an atom of carbon, for example, is 2.0×10^{-23} (or 0.000 000 000 000 000 000 000 020) g. Because this value is so small, a new unit of weight, the atomic weight unit (awu), is often used to express the weights of atoms. It is defined as 1/12 the weight of carbon atom. Carbon therefore weighs 12 awu and is the chemists' standard of reference for atomic weights.

Fields of chemistry. The field of chemistry is a very large one. There are 80 subject sections of chemistry in the chemical publication journal *Chemical Abstracts*. Some of the better-known fields are inorganic chemistry, organic chemistry, physical chemistry, analytical chemistry, biological chemistry, pharmaceutical chemistry, nuclear chemistry, industrial chemistry, colloidal chemistry, and electrochemistry.

Chemical literature. There are many articles published on chemical topics each year in a great variety of journals all over the world. Probably the most important scientific journal is *Chemical Abstracts*, which publishes the abstracts of all the chemistry articles published by any of the journals in the world. This means abstracting articles from some 14,000 scientific journals and publishing over 500,000 abstracts per year.

Chemical industry. The chemical manufacturing industry is the major industry in the United States in terms of value of production. It shows promise of continuing expansion.

Employment for chemists. Chemists perform a number of different and important jobs. In the research laboratory, the chemists work on fundamental research, such as synthesizing new compounds and studying the properties of others. The importance of fundamental research is recognized when it is realized that half of the jobs of the nation can be traced to the research laboratory. In the analytical laboratory, chemists develop methods of analysis of elements and of compounds. In industrial chemistry, one of the important jobs is the synthesis or purification of chemicals on a large scale for production purposes. *See* CHEMICAL ENGINEERING.

There are many other jobs for chemists in industry, such as writing, literature abstracting, promotion of product, and patents. There is also the very important field of teaching in high schools, colleges, and universities.

Chemists work in many different types of laboratories. Many industries, such as oil, plastics, fabrics, rubber, drugs, and iron and steel, have research laboratories and analytical and testing laboratories. There are the government laboratories, such as the Atomic Energy Centers, the Regional Agricultural Research Laboratories, and the National Bureau of Standards. Also, there are the independent research foundations, such as Mellon Institute, Rockefeller Foundation, and Battelle Memorial Institute. The universities have research laboratories in which are trained the graduates who are to become the research chemists for these other laboratories, as well as the teachers for colleges and universities. *See* ANALYTICAL CHEMISTRY; INORGANIC CHEMISTRY; ORGANIC CHEMISTRY; PHYSICAL CHEMISTRY.

[ALFRED B. GARRETT]

Chloramine-T

A derivative of *p*-toluenesulfonamide, prepared by treating the latter with sodium hypochlorite, as shown in the reaction. Dilute solutions have anti-

$$p\text{-}CH_3C_6H_4SO_2NH_2 + NaOCl \rightarrow p\text{-}CH_3C_6H_9SO_2NClNa$$

septic values in dressing and irrigating wounds. *See* ORGANOSULFUR COMPOUND; SULFONAMIDE.

[NORMAN KHARASCH]

Chlorate

A negative ion having the formula ClO_3^-, and derived from chloric acid, $HClO_3$.

Chloric acid exists only in solution. The chlorates are more stable than hypochlorites and chlorites but less stable, and thus better oxidizing agents, than bromates, iodates, and perchlorates. Chlorates are decomposed on heating to liberate oxygen, as in reaction (1).

$$2KClO_3 \rightarrow 2KCl + 3O_2. \qquad (1)$$

Both potassium and sodium chlorates are prepared by electrolysis of a hot chloride solution. Chlorine liberated at the anode is allowed to react with OH^-, as in reaction (2). Hydrogen is liberated at the cathode and the overall reaction is shown in (3).

$$3Cl_2 + 6OH^- \rightarrow ClO_3^- + 5Cl^- + 3H_2O \qquad (2)$$

$$NaCl + 3H_2O \rightarrow NaClO_3 + 3H_2 \qquad (3)$$

Sodium chlorate is used mainly in the production of perchlorates, but also as a weed killer; potassium chlorate is utilized in the manufacture of matches. *See* BROMATE; CHLORINE; IODATE; OXIDIZING AGENT; PERCHLORATE.

[E. EUGENE WEAVER]

Chloride

A compound which is derived from hydrochloric acid, HCl, and which contains the chlorine atom in the 1− oxidation state. Compounds containing this group may be organic or inorganic. The inorganic chlorides can be divided into three classes: covalent chlorides, ionic chlorides, and complexes. The covalent chlorides are recognizable because of their low boiling and melting points. Examples are the nonmetallic chlorides such as boron trichloride, BCl_3, carbon tetrachloride, CCl_4, and phosphorus pentachloride, PCl_5; and metallic chlorides such as aluminum chloride, Al_2Cl_6, stannic chloride, $SnCl_4$, and antimony pentachloride, $SbCl_5$.

The ionic chlorides, the most numerous chlorides, have high melting and boiling temperatures and good electrical conduction in the molten state. Examples are sodium chloride, NaCl, calcium chloride, $CaCl_2$, and magnesium chloride, $MgCl_2$. The dividing line is not clear-cut between these two groups, and beryllium chloride, $BeCl_2$, zinc chloride, $ZnCl_2$, and mercuric chloride, $HgCl_2$, fall in this borderline region. Since formation of a covalent bond is enhanced by a large anion, iodides have the greatest tendency to be covalent, followed by bromides, chlorides, and fluorides in that order.

Most metallic chlorides are water-soluble. Some

of the insoluble or slightly soluble salts include lead chloride, $PbCl_2$, copper(I) chloride, CuCl, silver chloride, AgCl, mercurous chloride, Hg_2Cl_2, thallous chloride, TlCl, and gold(I) chloride, AuCl.

Evidence for the formation of coordination complexes is observed when an insoluble chloride such as silver chloride dissolves in the presence of excess chloride ion to give the complex ion $AgCl_3^{--}$.

Other common chloro complexes are those of copper, cobalt, tin, and platinum, $CuCl_2^-$, $CoCl_4^{--}$, $SnCl_6^{--}$, and $PtCl_6^{--}$, respectively.

Complex formation is a general property of all the halide ions, and complexes are observed with all metal ions, except those of the alkali and alkaline-earth metals. Of the halide complexes, the most stable are the chloro complexes, followed by bromo and iodo complexes. Fluoro complexes are sometimes more stable and at other times less stable than chloro complexes.

Of the chlorides, sodium chloride is most important commercially. Sodium chloride occurs widely in nature and is the starting point for all sodium, chlorine, and compounds containing these two elements. Calcium chloride is obtained as a by-product in the Solvay soda-ash process.

The insoluble silver chloride is used to identify the chloride ion in the absence of bromide ion, Br^-, and iodide ion, I^-. *See* CHEMICAL BONDING; CHLORINE; COMPLEX COMPOUNDS; HALIDE; HALOGENATED HYDROCARBON. [E. EUGENE WEAVER]

Bibliography: F. A. Cotton and G. Wilkinson, *Advanced Inorganic Chemistry*, 3d ed., 1972; J. W. Mellor, *Comprehensive Treatise on Inorganic and Theoretical Chemistry*, 16 vols., 1922–1937.

Chlorine

A chemical element, Cl, of atomic number 17 and atomic weight 35.453. Chlorine exists as a greenish-yellow gas at ordinary temperatures and pressures. It is second in reactivity only to fluorine among the halogen elements, and hence is never

Ia																	0
1 H	IIa											IIIa	IVa	Va	VIa	VIIa	2 He
3 Li	4 Be											5 B	6 C	7 N	8 O	9 F	10 Ne
11 Na	12 Mg	IIIb	IVb	Vb	VIb	VIIb	——VIII——			Ib	IIb	13 Al	14 Si	15 P	16 S	17 Cl	18 Ar
19 K	20 Ca	21 Sc	22 Ti	23 V	24 Cr	25 Mn	26 Fe	27 Co	28 Ni	29 Cu	30 Zn	31 Ga	32 Ge	33 As	34 Se	35 Br	36 Kr
37 Rb	38 Sr	39 Y	40 Zr	41 Nb	42 Mo	43 Tc	44 Ru	45 Rh	46 Pd	47 Ag	48 Cd	49 In	50 Sn	51 Sb	52 Te	53 I	54 Xe
55 Cs	56 Ba	57 La	72 Hf	73 Ta	74 W	75 Re	76 Os	77 Ir	78 Pt	79 Au	80 Hg	81 Tl	82 Pb	83 Bi	84 Po	85 At	86 Rn
87 Fr	88 Ra	89 Ac	104 Rf	105 Ha	106	107	108	109	110	111	112	113	114	115	116	117	118

lanthanide series	58 Ce	59 Pr	60 Nd	61 Pm	62 Sm	63 Eu	64 Gd	65 Tb	66 Dy	67 Ho	68 Er	69 Tm	70 Yb	71 Lu

actinide series	90 Th	91 Pa	92 U	93 Np	94 Pu	95 Am	96 Cm	97 Bk	98 Cf	99 Es	100 Fm	101 Md	102 No	103 Lr

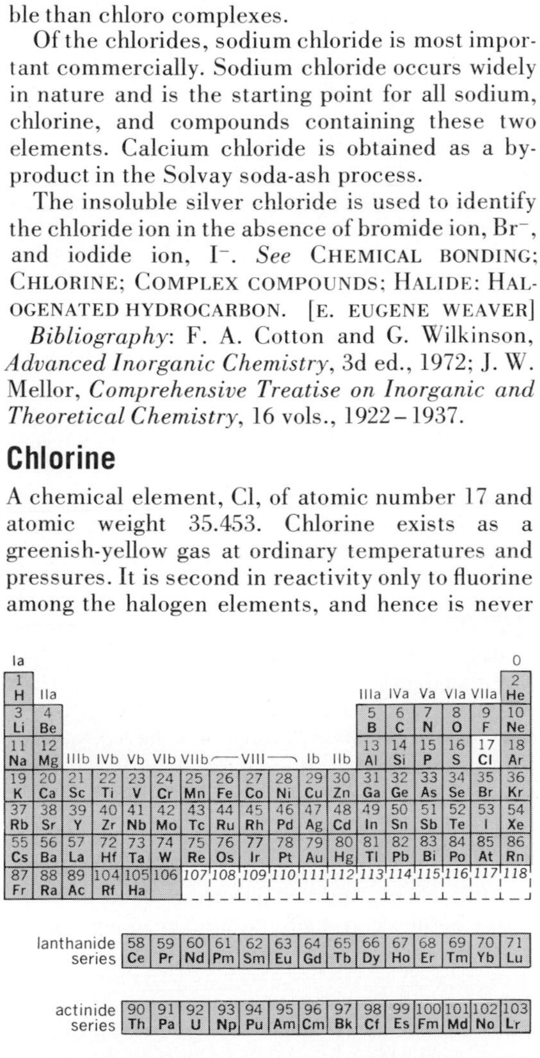

found free in nature, except at the elevated temperatures of volcanic gases. It is estimated that 0.045% of the Earth's crust is chlorine. It combines with metals, nonmetals, and organic materials to form hundreds of chlorine compounds, the most important of which are discussed here. *See* HALOGEN ELEMENTS.

Chlorine and its common acid derivative, hydrochloric (or muriatic) acid, were probably noted by experimental investigators as early as the 13th century. C. W. Scheele identified chlorine as "dephlogisticated muriatic acid" in 1774, and H. Davy proved that a new element had been found in 1810. Extensive production started 100 years later. During the 20th century, the amount of chlorine used has been considered a measure of industrial growth. In 1975 chlorine production ranked seventh on the list of largest-volume chemicals produced in the United States.

The importance of chlorine has changed as new derivatives have been added. Whereas in 1925 paper and pulp used over one-half the chlorine made and chemical products only 10%, by the 1960s paper and pulp use accounted for only 15–17% and chemical uses increased to 75–80%. Sanitation uses have contributed to the growth of large cities, and new textiles, plastics, paints, and miscellaneous uses have raised man's standard of living. Many large chemical companies are based primarily on the manufacture of chlorine and its compounds. For example, 17% of United States production in 1978 went into the production of vinyl chloride monomer. Other chlorinated organics consumed 48% of United States production.

Uses. Chlorine is an excellent oxidizing agent. Historically, the use of chlorine as a bleaching agent in the paper, pulp, and textile industries and as a germicide for drinking water preparation, swimming pool purification, and hospital sanitation has made community living possible.

Chlorine is used to produce bromine from bromides found in brines and sea water. The automotive age increased the production of bromine tremendously for the manufacture of ethylene dibromide for use in gasoline. Of the total bromine produced, such use for ethylene dibromide reached 95% some years ago. Later the use of other bromides increased at a faster rate than ethylene dibromide. Use of bromine has rapidly grown in manufacturing fire retardants for plastics and polymers, as has the use of bromide in high-density fluids for oil recovery. Considering the bromide ions involved as being represented by NaBr in sea water and $CaBr_2$ in brines, the reactions for production of bromine are (1) and (2). Both processes

$$2NaBr + Cl_2 \rightarrow 2NaCl + Br_2 \tag{1}$$

$$CaBr_2 + Cl_2 \rightarrow CaCl_2 + Br_2 \tag{2}$$

are followed by concentration and purification steps. *See* BROMINE.

Compounds of chlorine are used as bleaching agents, oxidizing agents, solvents, and intermediates in the manufacture of other substances. The table lists those industries that are major users of chlorine.

United States chlorine end uses

End use	Percent
Vinyl chloride monomers	20
Pulp and paper	12
Chlorinated ethanes	14
Fluorocarbons	6
Waste and water treatment	5
Miscellaneous organics	11
Methylene chloride	3
Propylene oxide	7
Inorganic chemicals	12
Miscellaneous	10

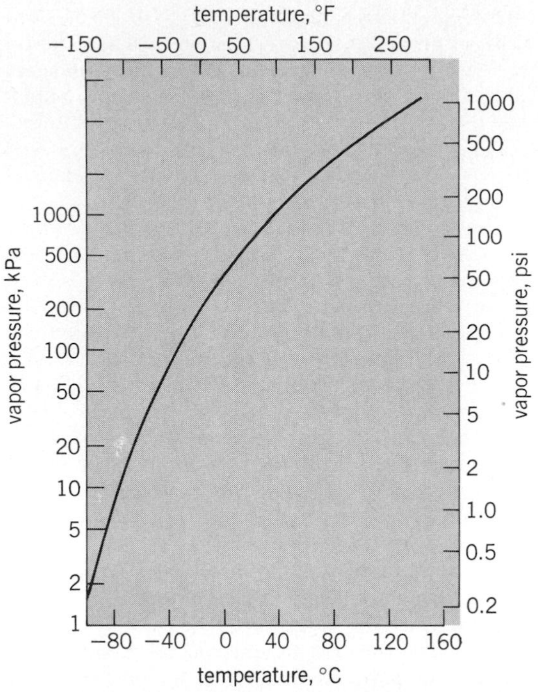

Fig. 1. Vapor pressure of liquid chlorine. (*From Diamond Shamrock Corp., Chlorine Handbook, 1976*)

Natural occurrence. Because many inorganic chlorides are quite soluble in water, they are leached out of land areas by rain and groundwater to accumulate in the sea or in lakes that have no outlets. Sea water contains 18.97 g of chloride ion per kilogram (3% sodium chloride). Solar evaporation produces large deposits of salts in landlocked areas. Similar evaporation in the past is responsible for vast underground deposits of rock salt and brines which may be found in Michigan, central New York, the Gulf Coast of Texas, Stassfurt in Germany, and other places. These deposits are mainly of sodium chloride, the supply of which is unlimited for practical purposes. Other rocks and minerals in the Earth's surface average slightly over 0.03% chloride.

Physical properties. The atomic weight of naturally occurring chlorine is 35.453 (based on carbon at 12). It is formed of stable isotopes of mass 35 and 37; radioactive isotopes have been made artificially. The diatomic gas has a molecular weight of 70.906. The boiling point of liquid chlorine (golden-yellow in color) is −34.05°C at 760 mm, and the melting point of solid chlorine (tetragonal crystals) is −100.98°C. The critical temperature is 144°C; the critical pressure is 76.1 atm; the critical volume is 1.745 ml/g; and density at the critical point is 0.573 g/ml. Thermodynamic properties include heat of sublimation at 7370 ± 10 cal/mole at 0°K, heat of evaporation at 4878 ± 4 cal/mole at −34.05°C, heat of fusion at 1531 cal/mole, heat capacity at 7.99 cal/mole at 1 atm and 0°C, and 8.2 at 100°C. Other properties are shown in Figs. 1–5. Chlorine forms solid hydrates, $Cl_2 \cdot 6H_2O$ (pale-green crystals) and $Cl_2 \cdot 8H_2O$. It hydrolyzes in water as shown in reaction (3).

$$Cl_2 + H_2O \rightarrow HClO + HCl \qquad (3)$$

Chemical properties. Chlorine is one of four closely related chemical elements which have been called the halogen elements. Fluorine is more active chemically, and bromine and iodine are less active. Chlorine replaces iodine and bromine from their salts. It enters into substitution and addition reactions with both organic and inorganic materials. Dry chlorine is somewhat inert, but moist chlorine unites directly with most of the elements.

Safety precautions. Chlorine attacks the tissues of the nose, throat, and lungs. Its pungent odor allows easy detection, which should not be ignored. Concentrations of 15–30 ppm in air for even a short time will irritate the mucous membranes, the respiratory system, and the skin and coughing will usually result. The threshold limit value (TLV) for an 8-h day is one ppm. Higher concentrations are dangerous. In extreme cases the difficulty of breathing may increase to the point where death can occur from suffocation. Chlorine reacts violently with many materials. Explosive mixtures are formed with hydrogen, and metals such as iron will begin to burn in it with only a slight amount of heating. Many organic reactions with chlorine are highly exothermic. Entire industries based on chlorine reactions have excellent safety records, but each use must be carefully considered, and each operation checked for safety hazards. Gas masks should be available.

Manufacture. The first electrolytic process was patented in 1851 by Charles Watt in Great Britain. In 1868 Henry Deacon produced chlorine from hydrochloric acid and oxygen, reaction (4), at

$$4HCl + O_2 \rightarrow 2Cl_2 + 2H_2O \qquad (4)$$

400°C with copper chloride absorbed in pumice stone as a catalyst. The electrolytic cells now used may be classified for the most part as diaphragm and mercury types. Both make caustic (NaOH or KOH), chlorine, and hydrogen. The economics of the chlor-alkali industry mainly involves the balanced marketing or internal use of caustic and chlorine in the same proportions as obtained from the electrolytic cell process.

Diaphragm cells. The Hooker diaphragm cell illustrates the principles of operation with porous asbestos pulp diaphragms. Diaphragms separate the anode compartment, containing the brine feed to the cell and chlorine produced, from the cathode compartment. The cathode compartment con-

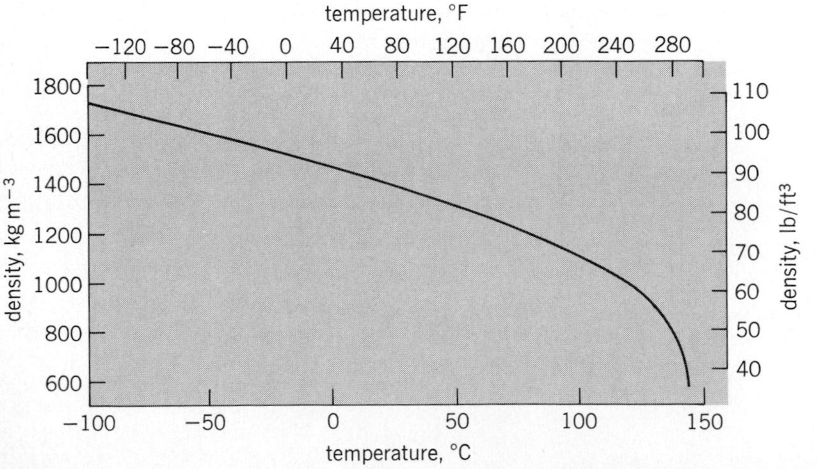

Fig. 2. Density of saturated liquid chlorine. (*From Diamond Shamrock Corp., Chlorine Handbook, 1976*)

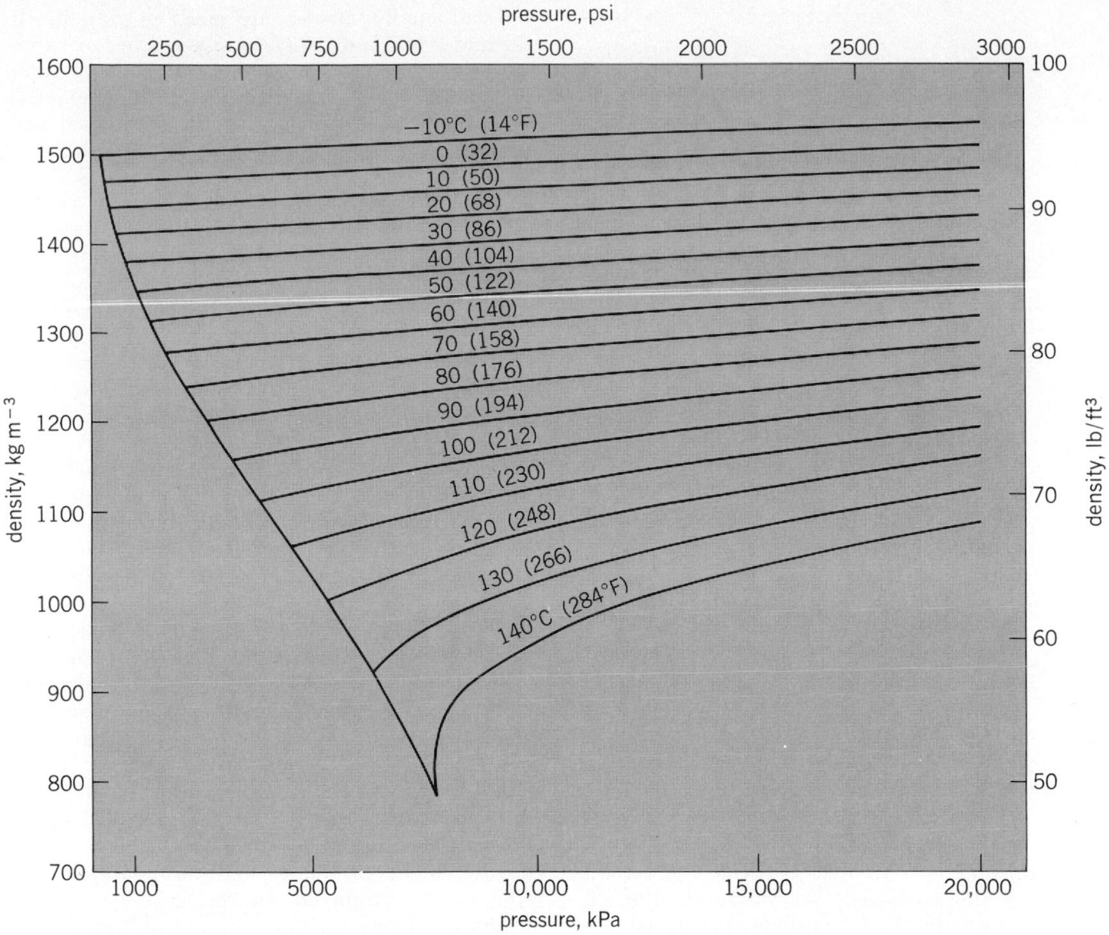

Fig. 3. Density of liquid chlorine. *(From Diamond Shamrock Corp.. Chlorine Handbook. 1976)*

tains the sodium hydroxide (caustic) produced as a solution in undecomposed brine (which has percolated through the diaphragm) and hydrogen gas.

To minimize back diffusion and migration of caustic, the flow rate of brine through the diaphragm is faster than salt can be electrolyzed so that only part of the salt is converted. Diaphragms must be renewed if brine is not sufficiently purified or if oils from graphite and calcium or iron from cell bodies cut down the flow rate. These impurities are all collected by the diaphragms. Diaphragm cells produce solutions containing about 10% sodium hydroxide and 15% sodium chloride. This is used directly in some chemical processes. Usually commercial caustic is evaporated to 50 or 73% solutions or to dry sodium hydroxide, NaOH. Most of the salt then crystallizes out and is returned to the process.

In general, very little unused chlorine capacity exists in the industry. Dow diaphragm cells represent the largest installed capacity, followed by Hooker, Diamond Alkali, and other types of diaphragm cells. Dow diaphragm cells are erected in blocks (traditionally about 50 per unit) which operate as a group in contrast to single-cell units in all other diaphragm installations. In Dow cells the electric current passes through a multiple unit series without electrical cables or bus bars between cells. Abutting frames are pressed together to form a tight block, with each unit connected electrically to the next cell within the frames of the series. Chlorine and hydrogen gases are each collected in inverted trough concrete headers on top of the cells, and caustic at the side through a trap.

The newer diaphragm cell installations operate at currents of 60 to 900 kA/m^2, and in the range of 3.5 V per cell. The voltage is determined by the sum of anode and cathode potentials (dependent on surface characteristics of each electrode, tem-

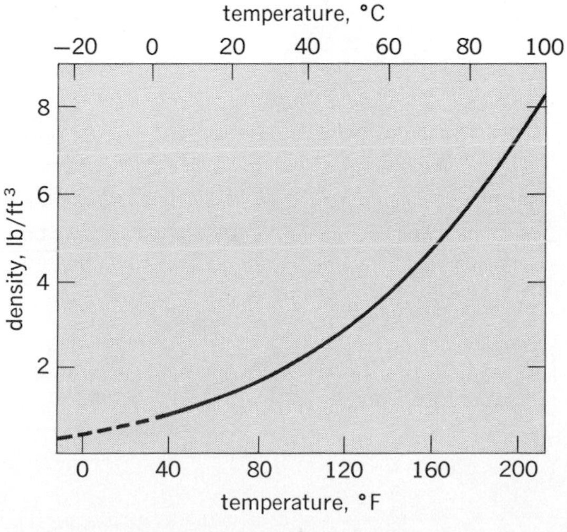

Fig. 4. Relationship between temperature and density of saturated chlorine gas. *(Hooker Chemical Corp.)*

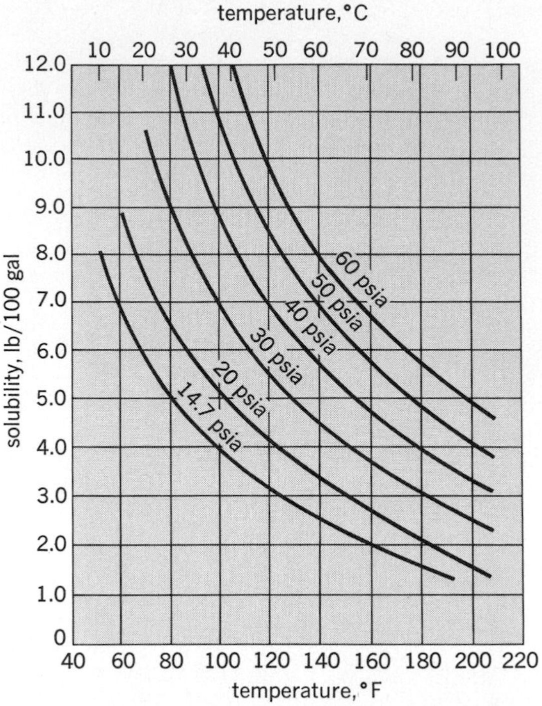

Fig. 5. Equilibrium solubility of chlorine in water. The data were calculated from partial pressure values. (*Chlorine Institute, Inc., Chlorine Manual. 4th ed., 1969*)

perature, and current density, which affect the chlorine and hydrogen overvoltages), the resistance through the electrolyte, the resistance through the diaphragm, the resistance through anode and cathode leads and across contacts, and the diffusion potential.

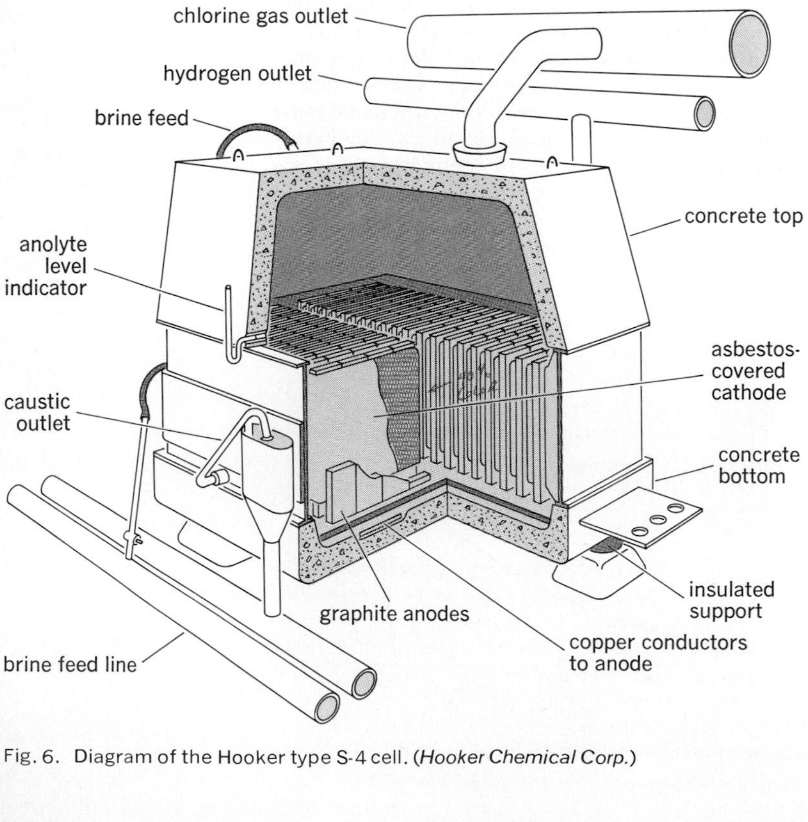

Fig. 6. Diagram of the Hooker type S-4 cell. (*Hooker Chemical Corp.*)

Traditionally, anodes are made of graphite. In the mid-1960s and 1970s, Beer and other researchers developed anodes made of the platinum group metals and having high, near-metallic conductivities. The best results have been obtained with ruthenium dioxide and titanium oxide solid solutions. Diamond Shamrock Technologies, S. A., of Geneva, Switzerland, has licensed these anodes, known as dimensionally stable anodes. These anodes have become the world standard anode material—possessing long life, low operating voltages, and good efficiency and cost. Dow Chemical Company, in its conversion plans for metal anodes, includes the use of its own coating system as well as that of Diamond Shamrock Technologies. Some platinized titanium is also being used in anode production.

Graphite anodes must be replaced periodically, and are therefore an additional item of expense in cell operation. Graphite is oxidized to carbon dioxide gas and trace-chlorinated organics, which leave the cell with the chlorine gas. Iron screen cathodes and concrete cell bodies are smaller items of expense, and need not be replaced as often as the graphite. Some diaphragm cells are cylindrical, such as the historic Vorce, Wheeler, and Gibbs cells. Cylindrical cells usually have less concrete, since the vessels are made of iron, and replacement of these parts is even less frequent. Nelson and Allen-Moore cells are historic rectangular diaphragm cells.

Figure 6 shows a Hooker diaphragm cell, which illustrates the principles of the diaphragm type. Chlorine is produced on the graphite anodes, which are covered with incoming salt brine. Chlorine gas bubbles rise to the top of the anodes and on through a reservoir of incoming brine in the lower portion of the concrete top; they are then collected from the chlorine gas outlet by pipelines connected to each cell. The chlorine contains water vapor, which makes it a very corrosive material. Wet chlorine must be handled in rubber-lined steel, glass, plastic, or similar corrosion-resistant material. However, when cool and dry, chlorine may be handled in equipment made of black iron, steel, copper, nickel, or lead. Processing of chlorine consists of cooling, drying (with sulfuric acid), pumping, and perhaps liquefying for storage or shipment. Liquefaction occurs at increased pressure (usually about 100 psi), perhaps with some cooling or by decreasing temperature to about −40°C at low pressure. Shipments have been common in 100-lb, 150-lb, and 1-ton cylinders; in 16-, 30-, 55-, and 85-ton railroad tank cars, with still larger ones in prospect; in tank trucks; and in tank barges of 550 and 1110 tons capacity. Usually barges contain four or six tanks containing 85 to 185 tons each. In Europe a ship capable of transporting 1600 metric tons is also reported.

Hydrogen gas and caustic are produced inside the asbestos-covered iron screen cathode pockets shown in Fig. 6. The caustic solution drops from the outlet into a funnel (to break the electric current path) and is then collected and pumped to the evaporators. Hydrogen is collected as shown behind the cell, scrubbed with caustic to remove chlorine, and often converted into ammonia or hydrochloric acid or used for hydrogenation of oils.

Mercury cells. This type of cell has been operated during the entire period of commercial production of chlorine, starting with the Castner cell and continuing in the United States with the Mathieson cell. They achieved increased importance in Germany during World War II, and other countries have now renewed their interest. New mercury cell plants are now being erected by many companies in the United States.

In the mercury cell the cathode is metallic mercury to which current is led by the steel body of the cell or other metallic contacts. Sodium amalgam is produced in the cell. It is led away from the cell to a decomposer also containing graphite which, in this case, serves as a cathode. The amalgam serves as an anode in a short-circuited cell.

Pure water is added in the decomposer to produce concentrated sodium hydroxide solutions and hydrogen from the sodium amalgam. Such caustic is usually considered quite pure (depending on the purity of the water and some other factors). It is often sold without further treatment as 50%-strength caustic, or may be evaporated to 73% or solid products. The direct 50% product constitutes the main economy of mercury cell installations compared with diaphragm cell caustic evaporator installations. However, mercury cells operate basically at 0.8 volt higher than diaphragm cells, representing higher power costs. Furthermore, the cost of mercury (of which there are losses in operation) is considerable.

Mercury cells overflow some brine continuously to maintain salt concentration in the electrolyzer. This brine is usually resaturated, purified, and recycled. Sodium amalgam forms in the electrolyzer, rather than caustic and hydrogen. This occurs because the discharge potential of hydrogen is higher than that of sodium because of the high overvoltage of hydrogen on mercury. Cells are usually constructed of steel, and are rubber-lined where necessary for protection against wet chlorine. Graphite anodes or the new metal anodes are parallel to the mercury surface, as shown in Fig. 7, and are suspended from the cover by means of graphite or other conducting pins. Whereas the decomposition voltage in diaphragm cells is about 2.3 V, in mercury cells it is about 3.1 V.

The tremendous interest in mercury cells arises from the large units which have been developed. In Europe mercury cells previously were operated at currents of up to 40,000 A, but since 1950 new designs by all manufacturers of cells have increased the operating range to 100–300 kA. Besides the higher amperage, multilevel layouts and extremely close spacing of cells have cut down

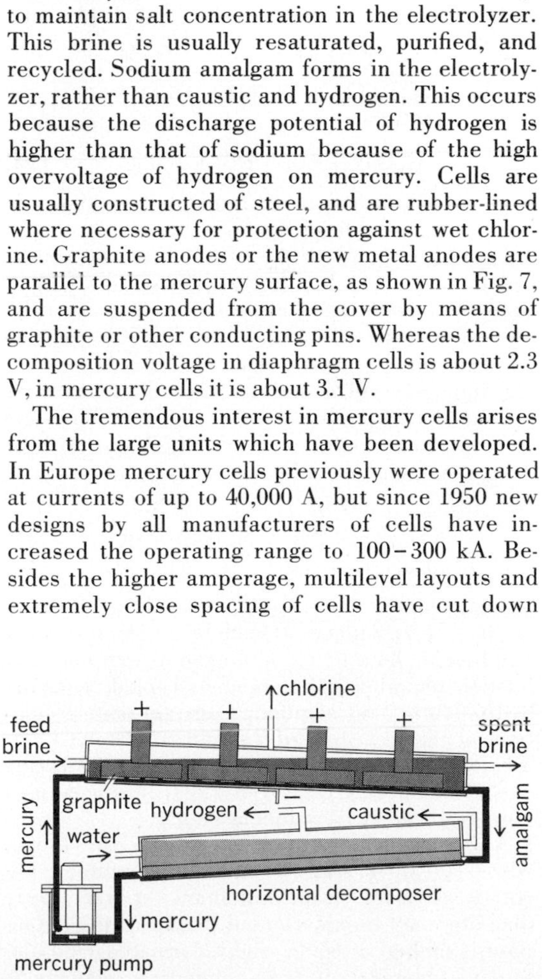

Fig. 7. A mercury cell for producing chlorine.

floor space requirements tremendously. Other new innovations include improvements in cell cover sealing arrangements, automatic anode adjustments, decomposer arrangements, and cleaning devices.

Other methods. Electrolysis of potassium chloride brine in either diaphragm or mercury cells produces potassium hydroxide, KOH, in place of sodium hydroxide. Molten sodium chloride is electrolyzed in cells, such as the Downs cell, operating at 600°C to produce sodium metal rather than sodium hydroxide. Caustic cell liquor can also be converted to the carbonates as alternate products instead of the hydroxides originally produced, resulting in soda ash, Na_2CO_3, and potassium carbonate, K_2CO_3. An interesting new idea related to both the mercury cell and the molten salt cell is the Szechtman cell. Mercury is replaced by molten lead flowing in a closed circuit. Either molten sodium chloride or other molten alkali chlorides could be electrolyzed to produce a variety of coproducts with the chlorine.

Hydrochloric acid is often in oversupply because of its production as a by-product of organic syntheses. The chlorine may be recovered by any one of four available processes: catalytic oxidation of gaseous HCl (Deacon process), direct oxidation of HCl by an inorganic oxidizing agent such as SO_3, electrolysis of HCl, or formation of metal chlorides and oxidation to metal oxide and chlorine.

Small amounts of chlorine are also made as by-products of the electrolysis of fused magnesium chloride, $MgCl_2$, in the production of magnesium metal, and of the electrolysis of fused lithium chloride in lithium metal production. A nonelectrolytic process, the nitrosyl chloride process, has been operating in the United States for many years, and oxidation of HCl with HNO_3 is practiced in some other countries.

West Germany, Great Britain, France, Italy, and Spain have been the major chlorine producers in Europe. Activity in mercury cell development is reflected in an increasing number of new installations in all European countries. While the United States and Canada are the leading Western Hemisphere producers, Latin America has added capacity at an increasing pace. Japan and India have a number of plants and also have developed their chlorine capacity to meet the needs of rising standards of living. A 1975 summary of chlor-alkali producers showed 282 installations around the world.

Natural compounds. Sodium chloride, NaCl, is used directly as mined (rock salt), or as found on the surface, or as brine. It may also be dissolved, purified, and reprecipitated for use in foods or when chemical purity is required. Its main uses are in the production of soda ash and chlorine products. Farm use, refrigeration, dust and ice control, water treatment, food processing, and food preservation are other uses. Calcium chloride, $CaCl_2$, is usually obtained from brines or as a by-product of chemical processing. Its main uses are in road treatment, coal treatment, concrete conditioning, and refrigeration.

Inorganic compounds. Wet chlorine reacts with metals to form chlorides, most of which are soluble in water. It also reacts with sulfur and phosphorus and with other halogens as in reactions (5).

$$
\begin{aligned}
H_2 + Cl_2 &\rightarrow 2HCl \\
2Fe + 3Cl_2 &\rightarrow 2FeCl_3 \\
2S + Cl_2 &\rightarrow S_2Cl_2 \\
S + Cl_2 &\rightarrow SCl_2 \\
S + 2Cl_2 &\rightarrow SCl_4 \\
P_4 + 6Cl_2 &\rightarrow 4PCl_3 \\
P_4 + 10Cl_2 &\rightarrow 4PCl_5 \\
Br_2 + Cl_2 &\rightarrow 2BrCl \\
2F_2 + Cl_2 &\rightarrow ClF + ClF_3
\end{aligned} \tag{5}
$$

Oxygen-containing compounds. The oxides of chlorine, dichlorine monoxide, Cl_2O, chlorine monoxide, ClO, chlorine dioxide, ClO_2, chlorine hexoxide, Cl_2O_6, and chlorine heptoxide, Cl_2O_7, are all made indirectly. Cl_2O is commonly called chlorine monoxide also. Chlorine dioxide, a green gas, has become increasingly important in commercial bleaching of cellulose, water treatment, and waste treatment. It is usually liberated at the plant site by the action of a reducing agent on a chlorate, as shown in reaction (6).

$$2NaClO_3 + H_2SO_4 + SO_2 \rightarrow 2ClO_2 + 2NaHSO_4 \tag{6}$$

Sodium and calcium hypochlorite solutions, made by passing chlorine gas into alkaline solutions, are in general use for their oxidizing power or bleaching power. Various solid hydrates and basic compounds may be prepared and sold, but the most common, called bleaching powder, has been assigned various formulas. The stability of these compounds varies with the water and metallic impurity present. One formula,

$$Ca(OCl)_2 \cdot CaCl_2 \cdot Ca(OH)_2 \cdot 2H_2O$$

would correspond to 35% bleach. $CaOCl_2$ is the simplest formula to use. Hypochlorite production began in earnest in 1823.

Many special processes are used, and mixtures are prepared to stabilize high-available chlorine for bleaching purposes without obtaining detrimental insoluble materials. High-test hypochlorite (HTH) is calcium hypochlorite, $Ca(OCl)_2$, containing 70% available chlorine. Production started in 1923 and increased rapidly. Bleaching and sanitation uses have been increased by swimming pool requirements and account for approximately 90% of the production, the rest being exported. Calcium hypochlorite used in place of chlorine for water treatment is required in some heavily populated areas. *See* HYPOCHLORITE.

The corresponding acid for hypochlorites is hypochlorous acid, HClO, which may be prepared, as shown in reaction (7), by action of carbonic acid on a hypochlorite.

$$NaClO + H_2CO_3 \rightarrow NaHCO_3 + HClO \tag{7}$$

Chlorites such as $NaClO_2$ are produced by reduction of calcium chlorate by hydrochloric acid to chlorine dioxide, followed by absorption and reduction with caustic and peroxide, as shown in reactions (8) and (9). The use of sodium chlorite permits controlled bleaching, and it is stable unless heated. With organic matter present, explosions can result, however. The corresponding acid for chlorites is chlorous acid, $HClO_2$.

$$Ca(ClO_3)_2 + 4HCl \rightarrow 2ClO_2 + Cl_2 + CaCl_2 + 2H_2O \tag{8}$$

$$2ClO_2 + 2NaOH + H_2O_2 \rightarrow 2NaClO_2 + O_2 + 2H_2O \tag{9}$$

Acidified sodium chlorite bleaches are very popular for cellulose and textile processing.

Chlorates may be prepared, reaction (10), by direct electrolysis of a chloride in cells in which the products of electrolysis are allowed to mix. Graphite anodes are used, and cooling coils or jackets are provided to remove heat. The effluent may be cooled to crystallize out sodium chlorate. Calcium chlorate may be prepared by chlorination of lime with heating of the acidified solution to 50–70°C, followed by filtration and removal of insolubles. Sodium chlorate is used as an intermediate in perchlorate production. It is a weed killer, and is also used in dye preparation, textile and fur processing, and metallurgical operations. Potassium chlorate is used in match manufacture and for fireworks. The corresponding acid for chlorates is chloric acid, $HClO_3$, which can exist in water solution up to 30% concentration.

$$NaCl + 3H_2O \rightarrow NaClO_3 + 3H_2 \tag{10}$$

Potassium chlorate, $KClO_3$, is a white powder or transparent, colorless monoclinic crystals. It is poisonous and may explode on contact with combustible substances.

Sodium chlorate, $NaClO_3$, is a colorless, odorless, cubic or trigonal crystal material. *See* CHLORATE.

Perchlorates may be produced by electrolysis of sodium chlorate, reaction (11), using platinum anodes, followed by conversion to any desired salt, reaction (12). The corresponding acid is perchloric acid, $HClO_4$, a colorless, fuming liquid that can be used as an oxidizing agent.

$$4NaClO_3 \rightarrow 3NaClO_4 + NaCl \tag{11}$$

$$NaClO_4 + KCl \rightarrow KClO_4 + NaCl \tag{12}$$

Potassium perchlorate, $KClO_4$, and ammonium perchlorate, NH_4ClO_4, have become important as fuels for rockets and jet propulsion and as explosives. *See* PERCHLORATE.

Chlorides. Hydrogen chloride, HCl, is a colorless, pungent, poisonous gas which liquefies at 82 atm at 51°C. It boils at −85°C at 1 atm. This acid is frequently produced by the LeBlanc process. The reactions are illustrated in reactions (13) and (14).

$$NaCl + H_2SO_4 \rightarrow NaHSO_4 + HCl \tag{13}$$

$$NaHSO_4 + NaCl \rightarrow Na_2SO_4 + HCl \tag{14}$$

Its major production is as the by-product of many organic chlorinations. It can be made by direct reaction of chlorine and hydrogen in an open combustion chamber submerged in cooled, aqueous hydrochloric acid solution. Constant-boiling solutions at 760 mm pressure contain 20.24% HCl, but a saturated solution at 15°C contains 43.4% HCl and has a specific gravity of 1.231. It is used as a strong acid and as a reducing agent.

Aluminum chloride, $AlCl_3$, is an anhydrous, white, deliquescent, hexagonal crystalline substance. Either scrap aluminum or the oxide (bauxite) may be chlorinated. Scrap or pig aluminum is melted at 660°C and chlorinated, and aluminum chloride vapor is condensed. Bauxite is calcined at 980°C, mixed with coal or coke, pulverized, briquetted with asphalt, recalcined to 18% carbon, heated to 870°C in a bauxite-lined vertical kiln, and chlorinated, reaction (15). Aluminum

$$Al_2O_3 + 3C + 3Cl_2 \rightarrow 2AlCl_3 + 3CO \quad (15)$$

chloride is a catalyst for production of cumene, styrene, and isomerized butane. Of the aluminum chloride uses in anhydrous form, ethylbenzene production used 25%, dyes 30%, detergents 15%, ethyl chloride 10%, drugs 8%, and miscellaneous production 12% in 1965. Hydrated and liquid forms are also available, 50% of which are used in drug and cosmetic production.

Titanium tetrachloride, $TiCl_4$, is a clear, colorless liquid. Production of titanium metal became of considerable interest in the 1950s for military uses. This required production of titanium tetrachloride from chlorination of rutile, TiO_2, as in reaction (16), or ilmenite, $FeO \cdot TiO_2$. Usually mix-

$$TiO_2 + 2Cl_2 \rightarrow TiCl_4 + O_2 \quad (16)$$

tures of carbon and rutile are chlorinated. $TiCl_4$ is also used as a polymerization catalyst and as a smoke screen agent and for sky writing.

Zirconium tetrachloride, $ZrCl_4$, is a colorless solid. Titanium technology is applicable to other metals, among which zirconium became of interest in the late 1950s. Zirconium tetrachloride is manufactured by chlorination of zirconium carbide or zirconium cyanonitride. It is also made by action of chlorine and carbon on the ores zircon, $ZrSiO_4$, and zirconia, ZrO_2, or by treating ores with carbon tetrachloride or phosgene.

Phosphorus trichloride, PCl_3, and phosphorus oxychloride, $POCl_3$, are important reagents in organic chemical synthesis. The oxychloride is a clear, colorless, fuming liquid. Chlorine is bubbled through phosphorus pentoxide in phosphorus trichloride, and the oxychloride is removed by distillation. Reaction (17) illustrates this process.

$$P_2O_5 + 3PCl_3 + 3Cl_2 \rightarrow 5POCl_3 \quad (17)$$

Ferric chloride, $FeCl_3$, is a solid composed of dark, hexagonal crystals. Much chlorine from chemical processes is converted to ferric chloride, which is then used for the manufacture of salts, pigments, pharmaceuticals, and dyes and for photoengraving, preparation of catalysts, and waste and sewage treatment. *See* CHLORIDE.

Organic compounds. Methane may be chlorinated to a variable extent, as shown by the sequence in reaction (18). The reaction may be car-

$$CH_4 \xrightarrow{Cl_2} CH_3Cl + HCl \xrightarrow{Cl_2} CH_2Cl_2 + HCl \xrightarrow{Cl_2}$$
$$CHCl_3 + HCl \xrightarrow{Cl_2} CCl_4 + HCl \quad (18)$$

ried out by light activation or through use of catalysts. It may be controlled to produce methyl chloride mainly, with only slight amounts of the other products. Recycling then produces the other chloromethanes. If all products are produced in one reaction, they may be separated by distillation.

Saturated aliphatic compounds. Methyl chloride, CH_3Cl, is a colorless, noncorrosive, liquefiable gas which condenses to a colorless liquid. A small percent of industrial capacity uses methanol according to reaction (19). Yield is 90%. Direct chlo-

$$CH_3OH + HCl \rightarrow CH_3Cl + H_2O \quad (19)$$

rination of methane is the other major process, also with a yield of 90%. Methyl chloride is used in production of silicone, tetramethyllead, and butyl rubber.

Methylene chloride, CH_2Cl_2, is a clear, colorless, volatile liquid. Direct chlorination of methane from CH_3OH also is the predominant process. Other production results from further chlorination of methyl chloride. It also has applications in production of paint remover, plastics, solvent degreasing, aerosol propellants, and other uses.

Chloroform, $CHCL_3$, is a colorless, nonflammable, volatile liquid. Direct chlorination of methane is the predominent process, although 88% yield may be obtained from the reaction shown in reaction (20), and 80% yield may be obtained by that of reaction (21). Fluorocarbon products and resins

$$2CH_3COCH_3 + 6CaOCl_2 \cdot H_2O \rightarrow 2CHCl_3$$
$$+ Ca(OAc)_2 + 2Ca(OH)_2 + 3CaCl_2 + 6H_2O \quad (20)$$

$$CCl_4 + 2[H] \xrightarrow{(Fe,HCl)} CHCl_3 + HCl \quad (21)$$

consume most of the chloroform made, leaving only a small quantity for dye, drug, and other uses.

Carbon tetrachloride, CCl_4, is a colorless, nonflammable liquid. Various processes involving chlorination of methane, natural gas, or aliphatic hydrocarbons in general yield carbon tetrachloride, but considerable amounts are still produced by the carbon disulfide method, reaction (22),

$$\begin{array}{c} CS_2 + 3Cl_2 \xrightarrow{Fe} S_2Cl_2 + CCl_4 \\ CS_2 + 2S_2Cl_2 \rightarrow 6S + CCl_4 \\ 6S + 3C \rightarrow 3CS_2 \end{array} \quad (22)$$

which proceeds with 90% yield. Carbon tectrachloride is used in the production of chlorofluorohydrocarbons, grain fumigation chemicals, and solvents.

Tetrachloroethane, $Cl_2HCCHCl_2$, is produced by chlorination of acetylene with antimony trichloride catalyst, reaction (23), at about 80°C, and

$$C_2H_2 + 2Cl_2 \xrightarrow{SbCl_3} \begin{array}{c} CHCl_2 \\ | \\ CHCl_2 \end{array} \quad (23)$$

is then separated by distillation. The principal use of the compound is in the manufacture of the solvents perchloroethylene and trichloroethylene. Another process used involves the chlorination of propane, reaction (24).

$$C_3H_8 + 8Cl_2 \rightarrow Cl_2CHCHCl_2 + CCl_4 + 8HCl \quad (24)$$

Ethylene dichloride, $ClCH_2CH_2Cl$, is a colorless, oily, sweet-tasting liquid with a chloroformlike odor. Reaction (25) illustrates its production. Eth-

$$CH_2{=}CH_2 + Cl_2 \rightarrow ClCH_2CH_2Cl \quad (25)$$

ylene dichloride is also produced in the chlorohydrin process for making ethylene glycol; by direct chlorination of ethane as a by-product of ethyl chloride manufacture; and, as illustrated by reaction (26), by treatment of ethylene with hydrochlo-

$$2C_2H_4 + 4HCl + O_2 \rightarrow 2ClCH_2CH_2Cl + 2H_2O \quad (26)$$

ric acid and oxygen (oxychlorination). Ethylene dichloride is used in the production of vinyl chloride, ethylenediamine, and solvents.

Ethyl chloride, C_2H_5Cl, is a colorless gas that is

easily compressed to a colorless, flammable liquid. Ethylene may be treated with hydrochloric acid, by using aluminum chloride catalyst, reaction (27),

$$C_2H_4 + HCl \xrightarrow{AlCl_3} C_2H_5Cl \qquad (27)$$

to give a 90% yield. This is the predominant method of production, but ethyl alcohol may also be used as raw material with 95–98% yield. This is illustrated by reaction (28). Most of the ethyl chlo-

$$CH_3CH_2OH + HCl \rightarrow C_2H_5Cl + H_2O \qquad (28)$$

ride produced is used in the manufacture of tetraethyl lead, but it also is used in the production of ethyl cellulose, and for refrigerant, anesthetic, and miscellaneous uses.

Chloral, CCl_3CHO, is a colorless, oily liquid which became important after World War II with the introduction of newer insecticides. About 50% of chloral production starts with chlorination of ethyl alcohol. The reaction steps in the production of chloral are shown in reaction (29). The yield is

$$
\begin{aligned}
&C_2H_5OH + Cl_2 \rightarrow C_2H_5OCl + HCl \\
&C_2H_5OCl + C_2H_5OH \rightarrow CH_3CH(OH)OC_2H_5 + HCl \\
&CH_3CH(OH)OC_2H_5 + 3Cl_2 \rightarrow \qquad\qquad (29)\\
&\qquad\qquad CCl_3CH(OH)OC_2H_5 + 3HCl
\end{aligned}
$$

$$CCl_3CH(OH)OC_2H_5 \xrightarrow{H_2SO_4} CCl_3CHO + C_2H_5OH$$

83%. Chlorination of acetaldehyde, reaction (30), is

$$2CH_3CHO + 3Cl_2 \rightarrow$$

$$\overset{\text{Mixture}}{(CH_2ClCHO + Cl_2CHCHO)} + 3HCl \qquad (30)$$

the other method of chloral manufacture. The monochloro- and dichloroacetaldehyde mixture reacts with the addition of antimony chloride and more chlorine at 80°C to yield chloral in 70% yield, reaction (31). Various mechanisms have been pro-

$$CH_2ClCHO + Cl_2CHCHO + 3Cl_2 \xrightarrow{SbCl_3}$$

$$2Cl_3CCHO + 3HCl \qquad (31)$$

posed for the reaction, one of them involving the formation of ethyl hypochlorite, decomposition to acetaldehyde, and polymerization to paraldehyde and chlorination.

Other saturated chlorine compounds of interest are ethylidene chloride, $C_2H_4Cl_2$, a colorless liquid (an isomer of ethylene dichloride); pentachloroethane, C_2HCl_5, a colorless, liquid chemical intermediate; hexachloroethane, C_2Cl_6, a white rhombic, solid chemical intermediate and insecticide; propylene chloride, $C_3H_6Cl_2$, a colorless, liquid chemical intermediate, fumigant, insecticide, and solvent; 1,2,3-trichloropropane, $C_3H_5Cl_3$, a colorless, liquid solvent and chemical intermediate; *n*-butyl chloride, C_4H_9Cl, a colorless, liquid solvent and alkylating agent; 1,4-dichlorobutane, $C_4H_8Cl_2$, a colorless liquid used in the manufacture of nylon; amylchlorides, $C_5H_{11}Cl$, chemical intermediates; dichloropentanes, $C_5H_{10}Cl_2$, colorless liquids when pure, used as solvents in the rubber, resin, and paint industries; chlorinated paraffins, liquids or solids, used as plasticizers, lubricant additives, and weather- and flameproofing agents.

Unsaturated aliphatic compounds. Perchloroethylene, $Cl_2C{=}CCl_2$, is a colorless, nonflammable, stable, heavy, mildly hazardous liquid. It may be made from tetrachloroethane by chlorination to pentachloroethane. However, trichlorethylene

may be made first by lime treatment as shown by reaction (32). Trichloroethylene is also a colorless,

$$2CHCl_2CHCl_2 + Ca(OH)_2 \rightarrow$$

$$2ClHC{=}CCl_2 + CaCl_2 + 2H_2O \qquad (32)$$

nonflammable, stable, heavy, mildly hazardous liquid which has other end uses beside perchloroethylene manufacture, such as in metal degreasing and as an extraction solvent. Trichloroethylene may be chlorinated, reaction (33), to pentachloro-

$$ClHC{=}CCl_2 + Cl_2 \rightarrow Cl_3CCHCl_2 \qquad (33)$$

ethane, which produces perchloroethylene on treatment with lime or heat, as shown in reaction (34).

$$2Cl_3CCHCl_2 + Ca(OH)_2 \rightarrow$$

$$Cl_2C{=}CCl_2 + CaCl_2 + 2H_2O \qquad (34)$$

Perchloroethylene may also be made by pyrolysis of carbon tetrachloride or by direct chlorination of ethane. Reaction (35) represents the latter

$$C_2H_6 + 5Cl_2 \xrightarrow{\Delta} Cl_2C{=}CCl_2 + 6HCl \qquad (35)$$

method. Perchloroethylene is used for dry cleaning, metal degreasing, compounding, and miscellaneous purposes.

Intermediates containing chlorine are used in the manufacture of the important compound glycerol. Allyl chloride, $CH_2{=}CHCH_2Cl$, is made, reaction (36), by high-temperature chlorination of pro-

$$CH_2{=}CHCH_3 + Cl_2 \rightarrow CH_2{=}CHCH_2Cl + HCl \qquad (36)$$

pylene from petroleum cracking. This will react, with hypochlorous acid to make dichlorohydrins reaction (37), which may in turn be reacted with

$$CH_2{=}CHCH_2Cl + HOCl \rightarrow$$

$$CH_2OHCHClCH_2Cl \qquad (37)$$

lime to produce epichlorohydrin, reaction (38).

$$2CH_2OHCHClCH_2Cl + Ca(OH)_2 \rightarrow$$

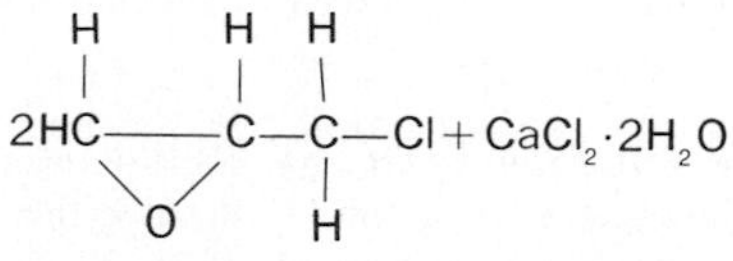

$$2HC\underset{\displaystyle O}{\overset{\displaystyle H}{|}}\!\!-\!\!\overset{\displaystyle H}{\underset{}{C}}\!\!-\!\!\overset{\displaystyle H}{\underset{\displaystyle H}{C}}\!\!-\!\!Cl + CaCl_2 \cdot 2H_2O \qquad (38)$$

This finally may be hydrolyzed to glycerol with sodium hydroxide in 75–80% yield, as in reaction (39). This process (or a variation using sodium hy-

$$HC\underset{\displaystyle O}{\overset{\displaystyle H}{|}}\!\!-\!\!\overset{\displaystyle H}{\underset{}{C}}\!\!-\!\!\overset{\displaystyle H}{\underset{\displaystyle H}{C}}Cl + NaOH + H_2O \rightarrow$$

$$
\begin{aligned}
&CH_2OH + NaCl \qquad (39)\\
&|\\
&CHOH\\
&|\\
&CH_2OH
\end{aligned}
$$

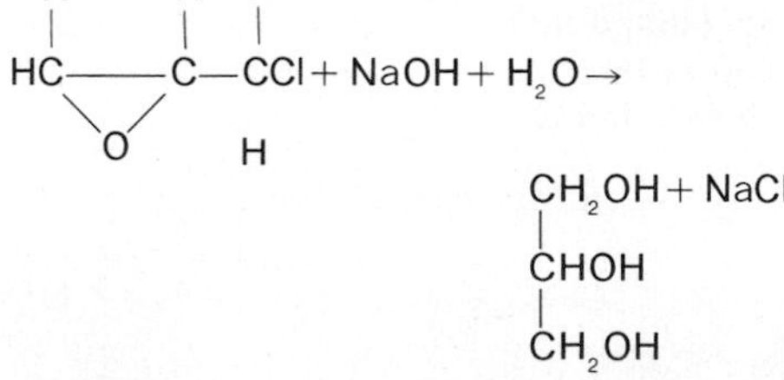

droxide on the allyl chloride to produce $CH_2{=}CHCH_2OH$ before using hypochlorous acid to produce $HOCH_2CHClCH_2OH$, and then treating with sodium hydroxide to make CH_2CHCH_2OH and glycerol) accounts for manufacture of glycerol except that resulting from saponification of fats and oils.

Vinylidene chloride, $C_2H_2Cl_2$, is a low-boiling liquid made by the elimination of hydrogen chloride from 1,1,2-trichloroethane by alkali. It polymerizes easily and hence is used to make copolymers, such as Saran, Geon, and Velon.

Vinyl chloride, C_2H_3Cl, is a colorless and toxic gas made from the elimination of hydrogen chloride from ethylene dichloride. It also polymerizes to make polymers, such as Koroseal, Vinylite, Geon, Tygon, Velon, and Saran.

Most propylene glycol production involves reaction of propylene and hypochlorous acid, as shown by reaction (40). Lime and water treatments then produce the oxide and glycol.

$$CH_2{=}CHCH_3 + HOCl \rightarrow HOCH_2CHClCH_3 \quad (40)$$

Other unsaturated compounds of interest are chloroacetylene, C_2HCl, an unstable gas; dichloroacetylene, C_2Cl_2, a colorless gas which is quite explosive; methallyl chloride, C_4H_7Cl, a colorless liquid which has been used as a fumigant; and chloroprene, C_4H_5Cl, a colorless liquid used to make synthetic rubbers and plastics.

Alicyclic compounds. Benzene hexachloride, $C_6H_6Cl_6$, forms colorless-to-yellow crystals or flakes. Direct chlorination of benzene reaction (41),

$$C_6H_6 + 3Cl_2 \xrightarrow{\text{Light}} C_6H_6Cl_6 \quad (41)$$

using the light from a mercury vapor lamp to activate the reaction, produces a 95% yield of a mixture of geometric isomers. The γ-isomer is toxic to insects and is concentrated by crystallization, steam distillation, and further fractional crystallization. It is used as an insecticide.

Chlordan, $C_{10}H_6Cl_8$, is a viscous, colorless, odorless liquid used as an insecticide.

Aromatic compounds. Chlorobenzene, C_6H_5Cl, is a colorless, mobile, volatile liquid with an almondlike odor. Chlorination of dry benzene at 40°C using ferric chloride or other catalysts in iron tanks produces monochlorobenzene at 70–75% yield, reaction (42). At one time chlorobenzene was

$$C_6H_6 + Cl_2 \rightarrow C_6H_5Cl + HCl \quad (42)$$

important for phenol production. However, since cumene has become the major chemical for phenol manufacture, production of monochlorobenzene has been greatly reduced. Its use as a new material for DDT has also been virtually eliminated with the demise of this product. Small uses still exist to make diphenyl oxide and its derivatives.

Dichlorobenzene, $C_6H_4Cl_2$, exists in two isomeric forms. The ortho form is a colorless, volatile liquid, whereas the para isomer consists of white, volatile crystals. It is obtained by chlorination of monochlorobenzene, as shown in reaction (43).

$$C_6H_5Cl + Cl_2 \rightarrow C_6H_4Cl_2 + HCl \quad (43)$$

Mixtures of mono- and dichlorobenzene may be made slightly alkaline, allowed to settle, and then separated by fractional crystallization. The para form usually is present in three times the amount of the ortho isomer. Small amounts of polychlorobenzenes are simultaneously produced. In 1966 moth repellents and deodorants took most of the *p*-dichlorobenzene made, with dye intermediates and miscellaneous other items accounting for the rest.

Other aromatic chlorine compounds of interest are benzyl chloride, $C_6H_5CH_2Cl$, benzal chloride, $C_6H_5CHCl_2$, and benzotrichloride, $C_6H_5CCl_3$. These three colorless liquids are starting materials for the synthesis of pharmaceuticals (benzedrine, demerol), and are used as constituents of perfumes and as starting materials for the preparation of insecticides and insect repellents. They are also used as lubricants and as intermediates in dye manufacture.

The chlorinated biphenyls find use in electrical insulation, hydraulic mediums, lubricants, and as constituents of paints, varnishes, and plastics. Because these compounds are chemically inert, they have been found to persist in the environment and are being phased out of industrial uses. Chlorinated naphthalenes are used in chemical, lubricant, and electrical insulation applications.

Phenol, C_6H_5OH, may be made by alkaline hydrolysis of the chlorine compound monochlorobenzene. The chlorination of phenol, reaction (44),

$$C_6H_5OH + 5Cl_2 \rightarrow C_6Cl_5OH + 5HCl \quad (44)$$

forms pentachlorophenol, C_6Cl_5OH, which is used as a wood preservative.

Dichlorophenol, $Cl_2C_6H_3OH$, is important in the production of another chlorine compound, 2,4-dichlorophenoxyacetic acid (2,4-D). 2,4-D is produced by reaction of the sodium salt of dichlorophenol with the sodium salt of chloroacetic acid. The compound 2,4-D is widely used as a weed killer.

Miscellaneous compounds. Phosgene, $COCl_2$, is a very poisonous gas. The pure product is a colorless liquid that boils at 7.48°C. It is produced by the reaction of dry carbon monoxide and chlorine gases, conducted over activated charcoal, reaction (45). Phosgene is used primarily in the manufac-

$$CO + Cl_2 \rightarrow COCl_2 \quad (45)$$

ture of toluene diisocyanate, which is becoming important in making urethane foams and special rubbers. Other isocyanates are being used in making adhesives.

Monochloracetic acid, $CH_2ClCOOH$, white deliquescent crystals, is made by chlorination of acetic acid with a catalyst in acid-resistant equipment. It is used as a chemical intermediate for making dyes, coumarin, pharmaceuticals, cosmetics, weed killers, and insecticides. Dichloroacetic acid, $CHCl_2COOH$, results from further chlorination of acetic acid. Trichloroacetic acid, CCl_3COOH, is a white crystalline material which results from still further chlorination of acetic acid. It is used medicinally, and the sodium salt is used as a weed killer.

Other compounds, such as chloramines, have been promoted as bactericides, germicides, and surgical antiseptics. *See* ELECTROCHEMICAL PROCESS; HALOGENATED HYDROCARBON.

[FREDERICK W. KOERKER; ROBERT W. BELFIT, JR.]

Bibliography: Business and Defense Services Administration, *Quart. Ind. Rep. Chem.*, U.S. Department of Commerce, June 1967; Chlorine Institute, Inc., *Chlorine Inst. Bull.*, various dates; Chlorine Institute, Inc., *Chlorine Manual*, 4th ed., 1968; Diamond Shamrock Corp., *Chlorine Handbook*, 1976; Dow Chemical, U.S.A., *Marketing Research*, 1979; Electrochemical Society, Inc., *Annu. Rep. Chlor-Alkali Comm. Ind. Electrolytic Div.*, annual; W. L. Faith, D. B. Keyes, and R. L. Clark, *Industrial Chemicals*, 3d ed., 1965; R. E. Kirk and D. F. Othmer (eds.), *Encyclopedia of Chemical Technology*, vol. 1, 3d ed., pp. 799–865, 1978; C. L. Mantell, *Electrochemical Engineering*, 4th ed., 1960.

Chloroform

A colorless, sweet-smelling, nonflammable liquid, of formula weight 119.39 and formula as shown below. Chloroform has a specific gravity of 1.489 at

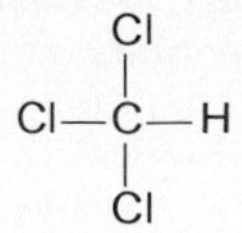

20°C, a boiling point of 61.2°C, and a refractive index, n_D^{25}, of 1.4426. The substance is called trichloromethane also; it is manufactured by the chlorination of ethyl alcohol or acetone in alkaline solution (usually $CaOCl_2$), by the reduction of carbon tetrachloride with iron and steam, or by the direct chlorination of methane. It is insoluble in water but soluble in organic solvents. In the presence of ultraviolet light, or more slowly in the dark, chloroform tends to react with oxygen of the air to produce poisonous phosgene. Therefore, commercial chloroform contains inhibitors, such as ethanol, thymol, or phenolic compounds. It is a powerful anesthetic, but prolonged use causes metabolic disturbances and damage to the heart, liver, and kidney, so that it commonly has been replaced by less toxic anesthetics. It is widely used as an extractant and solvent, a chemical intermediate in the production of dyes and drugs, and in pharmaceuticals as an antispasmodic, sedative (particularly in cough medicines), analgesic liniment, and anthelmintic. *See* HALOGENATED HYDROCARBON. [ELBERT H. HADLEY]

Bibliography: N. L. Allinger et al., *Organic Chemistry*, 2d ed., 1976.

Chloroplatinate

One of a group of compounds containing the anion $PtCl_6^{2-}$, which is derived from chloroplatinic acid. This acid is produced when platinum is dissolved in aqua regia. Of the many salts known, the ammonium and potassium chloroplatinates are the most useful because of their low solubility. The formation of a precipitate is used as a qualitative test for the ammonium and potassium ions. *See* AQUA REGIA; PLATINUM.

[E. EUGENE WEAVER]

Chromate

The CrO_4^{--} ion derived from the unstable acid H_2CrO_4. Chromic oxide, CrO_3, gives, instead of the expected H_2CrO_4, condensed or polyacids such as $H_2Cr_2O_7$ and $H_2Cr_3O_{10}$ in acid solutions. However, the solid chromates are stable and are the most important salts of the series. The dichromates are also well known. *See* DICHROMATE.

The chromates are easily converted to dichromates by the addition of acid. This is accompanied by a color change from yellow to orange and can be represented by the reaction below. Most of the sol-

$$\underset{\text{Yellow}}{2CrO_4^{--}} + 2H^+ \rightleftharpoons \underset{\text{Orange}}{Cr_2O_7^{--}} + H_2O$$

id chromate salts are also yellow, an exception being silver chromate, which is red.

The chromates of calcium, strontium, mercury (I), silver, barium, and lead are insoluble and are used to identify the chromate ion. However, these salts dissolve in strong acid to form dichromates.

Chromates are good oxidizing agents. They are used in paints as pigments and in the tanning of leather. *See* CHROMIUM; OXIDIZING AGENT; PIGMENT. [E. EUGENE WEAVER]

Chromatography

A separation method whereby individual chemical compounds which were originally present in a mixture are resolved from each other by the selective process of distribution between two heterogeneous (immiscible) phases. The distribution of chemical species to be separated occurs in a dynamic process between the mobile phase and the stationary phase. The stationary phase is a dispersed medium, which usually has a relatively large surface area, through which the mobile phase is allowed to flow. The chemical nature of the stationary phase exercises the primary control over the separation process. The greater the affinity of a particular chemical compound (referred to as the solute) for the stationary medium, the longer it will be retained in the system. The mobile phase can be either gas or liquid; correspondingly, the methods are referred to as gas chromatography and liquid chromatography.

Methods. There are four combinations of heterogeneous phase systems, which give rise to four different chromatographic methods: gas-solid, liquid-solid, gas-liquid, and liquid-liquid chromatography. In gas-solid and liquid-solid chromatography, sample molecules are caused to interact physically with the surface of a porous solid by means of a phenomenon called adsorption. Hence, these two methods are also generally referred to as adsorption chromatography. The adsorptive effect of the chromatographic medium for different solutes determines their rates of migration through the medium. In gas-liquid and liquid-liquid chromatography, the liquid stationary phase is held on the surface of an inert solid which serves merely as its support and, ideally, does not participate in the separation process. Primarily, then, the components of a mixture having different solubilities in the stationary phase separate by migrating at different rates. Since the partitioning of sample molecules between the two phases is the basis of these methods, they are also generally referred to as partition chromatography. The rate of migration of a solute can be related to its thermodynamic partition coefficient in a given two-phase system. *See* ADSORPTION.

Uses. Chromatography has been used primarily as a separation and isolation method. Unlike classical chemical separation methods (for example, precipitation or crystalization), chromatography is intended to separate many mixture components in a single-step procedure. The use of chromatography in general analytical procedures, in which the separation of mixture constituents can be followed by their direct identifications and quantitative measurements, has been dramatically increasing. Many analytical tasks cannot be adequately dealt with by any other available methods. Chromatographic methods can also be automated for routine analyses. *See* ANALYTICAL CHEMISTRY.

Because of the wide selection of combinations of mobile and stationary phases, different techniques, and solute measurement principles, chromatography has been applied to the separation of a variety of chemical compounds, both organic and inorganic. These applications cover the entire range of molecular weight, from light gases to macromolecules such as synthetic polymers, proteins, nucleic acids, and even whole subcellular and cellular units. Chromatographic methods can be applied to an incredible concentration range; although some separations can be performed in commercial-scale quantities (grams), it is also feasible to carry out certain highly sensitive analytical determinations on the order of 10^{-15} g (femtogram). Chromatographic separations are based on the physicochemical principles of adsorption and partition; and, conversely, these and related fundamental phenomena can be studied by high-precision chromatography of model systems.

BASIC TECHNIQUES

The most important mode of operation in chromatography is elution. In elution chromatography, the sample (a mixture of components to be separated) is placed as a narrow concentration impulse at the beginning of the chromatographic medium. The mobile phase is then introduced, and the sample components migrate with it through the chromatographic bed. In gas chromatography, in which the mobile phase is most typically an inert inorganic gas, sample molecules are merely transported through the sorption medium. However, in liquid chromatography, the mobile fluid may have an appreciable "extraction" effect on a sample molecule, causing it to migrate more slowly or more quickly; or, alternatively, the mobile-phase molecules can compete with the solute molecules for the available sites on the adsorbent (solid stationary phase).

Although the physical principles on which chromatography is based can be widely demonstrated in many natural phenomena, a Russian botanist, Mikhail Tswett, is credited with discovery of adsorption chromatography as an analytical tool at the beginning of the 20th century. Tswett employed the method to separate various pigments from plant extracts. He correctly interpreted the physical phenomena involved and gave the method its present name. Although the word "chromatography" could be interpreted as referring to a method restricted to colored substances, the term is now used in a more general sense. Tswett himself was aware that chromatography is applicable to both colored and colorless substances.

Column chromatography. A simple chromatographic procedure similar to that developed by Tswett is shown in Fig. 1. In this experiment, an extract of pigments from fresh green leaves, dissolved in a small amount of solvent, is placed on the top of a chromatographic column (Fig. 1*a*). The column consists of powdered sugar packed tightly into a constricted glass tube fitted at the bottom with a cotton plug. When petroleum ether with 0.5%*n*-propanol is applied to the top part of the column, the pigments start migrating down and begin to develop discrete color bands (Fig. 1*b*). By this time, six color bands can be seen, but as the elution process continues, more distinct separation is obtained (Fig. 1*c*). Two important phenomena can be observed in Fig. 1. First, the chemical nature of both the sorbent and the eluting solvent exert primary control over the degree to which individual mixture components are retained on the column. Thus, nonpolar compounds such

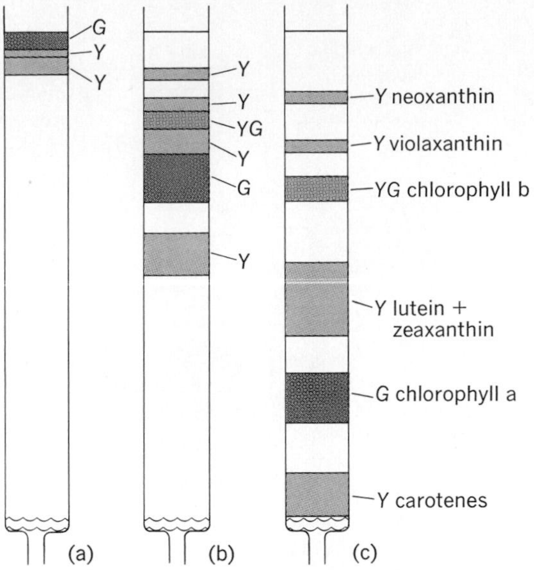

Fig. 1. Column chromatography. Successive steps in formation of chromatogram of green leaf pigments. G signifies green; *Y*, yellow; *YG*, yellow-green. (*a*) Sorption of mixture on powdered sugar from petroleum ether. (*b*) Partial separation after washing with petroleum ether plus *n*-propanol. (*c*) Further separation produced by continued washing.

as carotenes travel more rapidly with the solvent than more polar mixture components do. The rule that "like dissolves (or attracts) like," in chemical terms, is quite generally applicable to chromatography. Second, although four molecular entities are completely separated from the green-leaf extract, lutein and zeaxanthin are not resolved from each other; and, indeed, the fastest-traveling zone of carotenes is also still a mixture. In order to achieve the separation of these components, either a different mobile phase or a different stationary phase, or both, should be tried.

The separated compounds can be further isolated in a pure state (if desired, for instance, for additional chemical studies) by drying the column content and subsequently isolating the individual zones mechanically and extracting with an appropriate solvent. Alternatively, the column can be excessively flushed with the mobile phase, and the separated zones recovered into test tubes inserted under the column.

Modern instrumentation. Although both gas chromatography and modern liquid chromatography utilize the principle of the above experiment (that is, the separation takes place in a column filled with the sorptive medium), the term column chromatography is now universally reserved for the classical system, in which the mobile-phase flow is achieved by means of solvent gravity. A more modern version of the chromatographic system that is now typically used for analytical work is shown schematically in Fig. 2. Such a system, generally applicable to both gas and liquid chromatography, facilitates the introduction of the sample into the controlled stream of mobile phase at the column top and the continuous monitoring of the column effluent by a suitable detector (a device transducing chemical information into electrical signals), followed by a signal recording or, alterna-

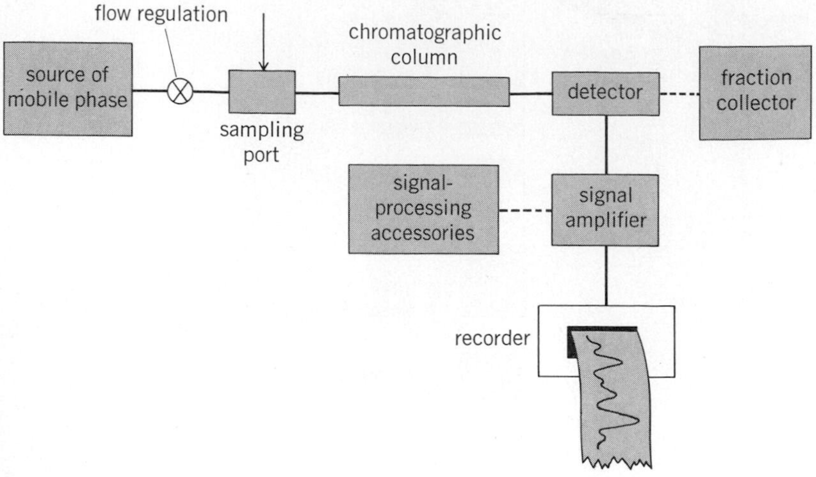

Fig. 2. Schematic diagram of a modern chromatographic instrument. The source of the mobile phase in gas chromatography is a gas cylinder; in liquid chromatography, a high-pressure pump. Interconnecting broken lines indicate the parts often used, but not essential to the basic function.

tively, further processing of recorded information.

Elution chromatogram. A typical recording, called a chromatogram, is shown in Fig. 3. In this experiment, a mixture of naphthoquinone and *ortho-*, *meta-*, and *para*-nitroaniline was separated and detected by means of a liquid chromatograph. The system employed a high-pressure pump, which maintained the mobile-phase flow (0.25% isopropanol in heptane) at 1.0 ml/min through the column, and a flow-cell ultraviolet (uv) photometric detector. The detector was set to monitor changes in absorption of uv light, at 254 nm, during the chromatographic run. Although the mobile phase

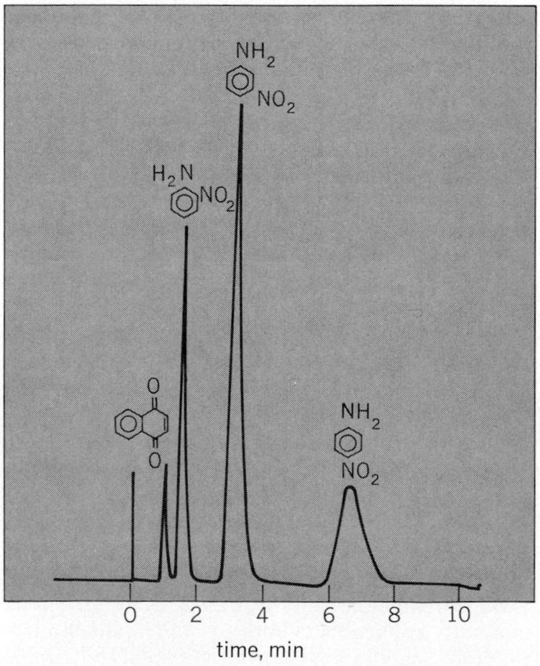

Fig. 3. Typical elution chromatogram obtained by a high-pressure liquid chromatograph. (*From M. Novotny et al., Polar silicone-based chemically bonded stationary phases for liquid chromatography, Anal. Chem., 45:971, 1973*)

alone absorbs uv light to a negligible degree, all the sample constituents absorb strongly owing to their aromatic nature and are consequently detected as peaks as they emerge from the column. The chromatographic column used was a siliceous material modified chemically to incorporate polar hydroxy groups into its surface. The distinct resolution of isomeric nitroanilines reflects their ability to interact with the column material through hydrogen bonding.

Flat-bed chromatography. Other common forms of liquid chromatography take place in flat chromatographic beds instead of columns. Two methods which belong in this category are paper chromatography and thin-layer chromatography.

Paper chromatography. Paper chromatography was developed in the early 1940s by the English biochemists A. J. P. Martin and R. L. M. Synge, in the first reported application of partition chromatography. A typical simple arrangement for a paper chromatography experiment is shown in Fig. 4.

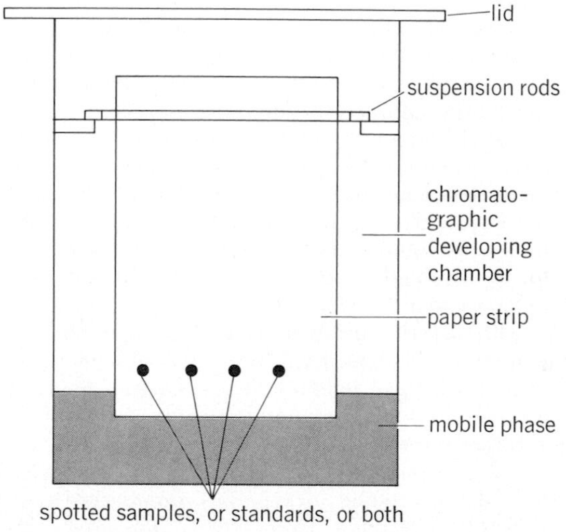

Fig. 4. Apparatus for paper chromatography.

A strip of filter paper that has been equilibrated in a moist atmosphere is spotted near the bottom with a sample (mixture) and placed inside the developing chamber. The water in the moist atmosphere that has been adsorbed by paper is most commonly used as the stationary phase (the paper is a support).

Papers can also be impregnated by other liquids. The mobile phase, usually an organic solvent or a mixture of solvents, ascends the paper strip by means of capillary forces, and the components of the sample mixture are thus separated by the partition effect. Such a process is shown schematically in Fig. 5, in which standard compounds are cochromatographed with the unknown mixture. The analyzed sample migrated from a single spot into five chromatographic zones that traveled at different rates on the paper sheet. In this case, the position of the standard compounds is coincident with that of the two sample spots. It should be emphasized that even though this may indicate that the same compounds might be present in the analyzed mixture, the similarity of chromatographic behavior is not sufficient proof of the identity of the compound.

Thin-layer chromatography. The stationary phase in thin-layer chromatography is a suspension which forms a layer (approximately 250 μm thick) on a metal or glass plate. It is most frequently an adsorbent (with a particle size of several microns) suspended in a suitable solvent, uniformly spread on a plate, and dried. The mobile phase is a liquid that ascends the plate by capillary action. The development of chromatograms is similar to that in paper chromatography. The most frequently used sorption materials are silica gel, alumina, cellulose powder, and various derivatives of cellulose. The selection of the sorption medium depends on the type of sample. Silica gel and alumina are used in combination with nonaqueous organic solvents, whereas cellulose materials are most suitable for chromatography with aqueous mobile phases. Thin-layer chromatography is comparable to paper chromatography in its simplicity. But thin-layer chromatography is superior in terms of rapidity, the variety of stationary phases that can be used, and detection methods.

Retardation factor. The individual separated zones in both paper and thin-layer chromatography are characterized by their R_F values (R_F = retardation factor), defined as the ratio of the distance traversed by the center of the spot to the distance traversed by the leading edge of the solvent. Only colored substances can be seen directly on chromatograms as spots; colorless substances must be made visible. Although certain chemicals can also be detected, after being irradiated with a uv lamp, from either their absorbing or their fluorescent properties, most separated substances are visualized after a selective reaction. A variety of detection agents are used to spray thin-layer plates to produce a suitable formation of colored substances. Quantitative evaluation is based on the measurement of color intensity, either directly by scanning chromatograms through an optical beam, or after quantitative extraction of isolated spots by a suitable solvent.

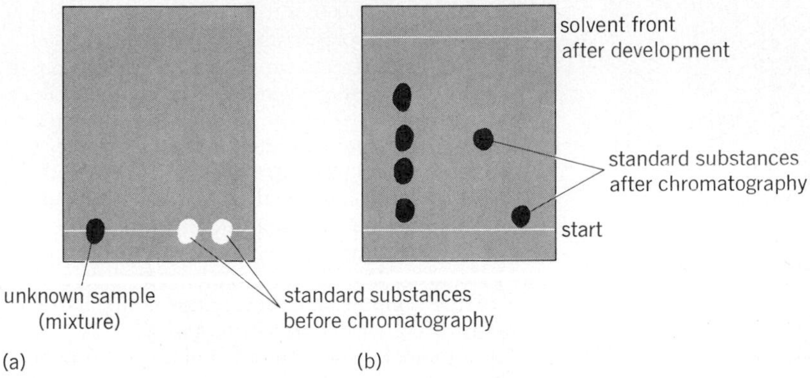

Fig. 5. Paper or thin-layer chromatograms (*a*) before development and (*b*) after development.

FUNDAMENTAL PROCESSES

In any chromatographic process, some components of a given mixture are retained inside the column longer than others. Component retention is primarily a function of the solute distribution between the two phases. If, at equilibrium, a greater amount of solute molecules are found in the mobile phase as compared to those passed into the stationary phase, the solute migration will be fast. However, if the equilibrium favors the stationary phase, the solute band will move more slowly. The selectivity of the retention process is important for chromatographic separations. Various attraction forces between the sample molecules are primarily responsible for retention. They may range in strength from ordinary dispersion (van der Waals) interactions up to hydrogen bonding or reversible complex formation.

Selectivity. Some chromatographic separations can be extremely selective. For example, in the method called affinity chromatography, specific protein molecules (enzymes) are strongly retained by the column material that has attached to its surface a coenzyme or enzyme inhibitor; the other proteins present in the chromatographed mixture are ignored, and pass, unchecked, through the column. It is therefore always necessary to seek the column selectivity that is appropriate to a given separation problem.

Zone spreading. Another crucial parameter for the effectiveness of separations in chromatography is the narrowness of chromatographic zones. Figure 6 shows a comparison of two chromatograms obtained with a model three-component mixture on the columns of identical retention characteristics (selectivity) but of different peak broadening effects. Obviously, the narrower the zones, the greater the number of peaks that can be spaced between the components, and the higher the resolving power of the column.

The mechanism of zone spreading in a chromatographic system is frequently studied. A sample introduced as a narrow concentration plug on the top of a chromatographic column (or as a small compact spot on a thin layer or paper) is subjected to a number of statistical events during the ensuing dynamic flow process. These events, associated to in varying degrees with diffusion spreading, nonuniformity of flow patterns, and the rate at which the sample molecules are redistributed from one phase to the other, are also responsible for broadening the chromatographic zones. The contribution of these individual factors is strongly dependent on the rate of solute transport through the system; an optimum flow rate exists for each

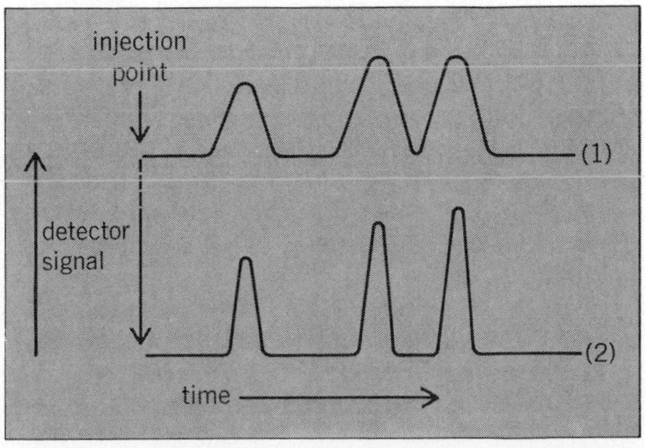

Fig. 6. Chromatograms of a three-component mixture on the columns having identical phase composition, but (1) low efficiency with broad peaks, and (2) high efficiency with narrow peaks.

chromatographic separation. Indirectly, the narrowness of the chromatographic zone may also reflect important column characteristics, such as its geometrical and chemical nature, as well as such factors as size of the packing particles, packing uniformity, and amount of stationary phase.

Retention and peak areas. The most important analytical results derived from a chromatogram are the retention distance for a given solute (its position within the chromatogram) and the detected peak area. The former is somewhat characteristic of the solute's chemical nature; the latter is usually directly related to its quantity.

Migration of a solute band in elution chromatography is described by its retention time (the time elapsed between sample introduction and elution of the peak maximum). The product of the retention time and the flow rate of the mobile phase is called the retention volume. Retention times or volumes of chromatographic peaks are usually related to a standard solute co-chromatographed with the analyzed mixture. For instance, a homologous series of *n*-alkanes (methane, ethane, propane, butane, and so on) is usually employed in gas chromatography for such a purpose.

Quantitation of separated compounds is carried out by comparing peak areas with those obtained from known amounts of identical compounds. Alternatively, calibration curves can be prepared by plotting peak area measurements against corresponding solute concentrations. In order to maintain accuracy of measurements, the so-called internal standard (known amount of a compound that does not interfere with any chromatographic peak within the analyzed mixture) is used and carried through the whole procedure of sample preparation prior to the chromatographic measurement.

GAS CHROMATOGRAPHY

The mobile phase (usually referred to as carrier gas) used in gas chromatography is an inert gas such as helium, nitrogen, hydrogen, or argon. It is supplied from a gas cylinder, and its flow rate through the chromatographic column is regulated. Since separations are carried out in the vapor phase, most parts of a gas chromatograph are temperature-controlled; selection of temperature (or temperature range) is based on the composition of the sample.

The analyzed mixture is injected through a rubber membrane into the stream of preheated carrier gas by means of a small-volume syringe. Liquids or gases of volumes less than microliter size can be reproducibly measured and injected into a gas chromatograph. After the sample is volatilized and carried onto the column, the separation of the components begins. The separated vapor-phase bands are sensed by a detection device situated at the column end. Various properties of separated molecules can give rise to a signal in different detectors.

Separation columns. The three basic types of separation columns used in gas chromatography are packed columns, wall-coated open-tubular columns (also frequently referred to as capillary columns), and porous-layer open-tubular columns. The structure of these columns is illustrated in Fig. 7. Packed columns (Fig. 7*a*) are tubes filled with a sorption material through which the carrier gas flows. Typical dimensions of analytical packed columns are 1–5 m in length and 1–8 mm inside diameter, but much wider columns are employed for preparative-scale separations. Open-tubular columns are prepared from a long, thin tubing (typically 20–100 m in length and 0.2–0.8 mm inside diameter), the inner wall of which is coated with a thin layer of sorptive medium (either a nonvolatile liquid or an adsorbent; see Fig. 7*b*). Because of their high permeabilities and greater lengths, open-tubular columns have considerably higher resolving power than packed columns do. Whereas the typical number of components resolved with a packed column is about 20 to 30, up to several hundred mixture components can be separated in open-tubular columns. Porous-layer open-tubular columns (Fig. 7*c*) are considered "hybrids" which combine some of the advantages of both wall-coated and packed columns.

Detectors. A variety of measurement principles have been used for detection purposes in gas chromatography. The most sensitive detectors are based on gas-phase ionization phenomena. The most common of them, the flame ionization detector, is based on the thermal ionization of solutes in hydrogen flame. An electrode situated above the flame monitors changes in flame conductivity. The hydrogen flame is ordinarily nonconductive, but current can occur, owing to the ionization of solutes emerging from the column. The signal is then amplified and recorded. The flame ionization detector is responsive to all organic compounds.

Separation and measurement of inorganic gases can also be carried out by gas chromatography, but detection principles other than flame ionization must be utilized. Among them, a differential measurement of the thermal conductivity of carrier gas and a solute is the most universally applicable.

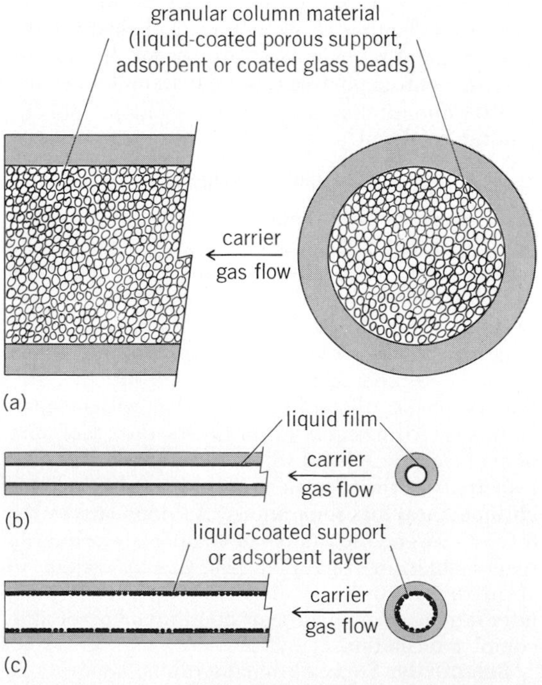

Fig. 7. Structure of main gas chromatographic column types. (*a*) Packed column. (*b*) Conventional (wall-coated) open-tubular column. (c) Porous-layer open-tubular column. (*From C. Horvath, in L. S. Ettre and A. Zlatkis, The Practice of Gas Chromatography, Interscience Publishers, 1967*)

Other gas chromatographic detectors are available that can respond selectively to both organic and inorganic compounds containing sulfur, phosphorus, nitrogen, halogens, or certain electron-absorbing groups.

Chemical nature of solute. Although the position of a chromatographic peak within the recorded mixture (chromatogram) provides some qualitative information, identical retention times of the unknown peak and comparing standard cannot be regarded as proof of structure. Although additional chromatography on a stationary phase of different selectivity may be useful in exercising further judgment on the chemical nature of the solute, a combination of gas chromatography with other physical or chemical methods is a more powerful approach. This is often done after a selected part of column effluent is trapped. Moreover, instrumental methods have been developed in which the structures of separated chemicals are studied "on line" as they emerge from the column exit. For example, the gas chromatograph can be directly coupled with a mass spectrometer or infrared absorption instrument.

Limitations of method. It is not necessary that compounds analyzed by gas chromatography be typically gases or very volatile chemicals. Many substances with molecular weights in excess of 600–800 have appreciable vapor pressures at high temperatures that are nonetheless within the range of gas chromatography. However, the temperatures for most practical applications of this method do not exceed 250°C. Some polar (and consequently nonvolatile) compounds can also be chemically altered, in a defined way, into more volatile derivatives prior to gas chromatographic analysis. Temperature programming of the chromatographic column is frequently employed for the effective resolution of mixture components within a wide range of volatility. In this method, the column temperature is gradually increased at a controlled rate during analysis. *See* GAS CHROMATOGRAPHY.

LIQUID CHROMATOGRAPHY

In liquid chromatography in columns, the mobile phase is percolated through the column by means of either gravity, under pressure generated by a suitable pump, or centrifugal force. High-pressure pumps that generate up to several hundred atmospheres of inlet pressure are used in combination with chromatographic packings of only several microns of particle size. Since the viscosity of a liquid affects its flow at a given pressure, the pressure requirements vary for different liquids achieving identical flow.

Mobile-phase selection. The mobile-phase composition is an important variable in chromatography. Sample solubility, competition between the sample and the mobile-phase molecules for available adsorbent surface, and solvation effects on the structure and function of some chromatographic materials can all influence the separation process. Various types of samples require that solvent systems of different polarities be used, ranging from very nonpolar liquids (such as liquid straight-chain hydrocarbons) to solvents of intermediate polarity (aromatic hydrocarbons, ethers, ketones) to very polar liquids (alcohols and water). Several types of liquid chromatography (ion-exchange, gel-permeation, and affinity chromatography) use aqueous electrolytes as mobile phases.

Successful separation of mixture components having different polarities often requires a change in composition of the mobile phase during a chromatographic run. In liquid chromatography, the "solvent programming" method, commonly referred to as gradient elution, serves a purpose similar to that of temperature programming in gas chromatography. Gradient elution techniques may involve programmed changes of solvent polarity, ionic strength of electrolytes, or pH.

Stationary-phase solubility. In liquid-liquid chromatography, the mobile phase and the stationary phase must be immiscible. Since immiscibility means automatically that the two phases must have significantly different chemical properties, only combinations of either a nonpolar mobile phase and a polar stationary phase or a polar mobile phase and a nonpolar stationary phase are feasible. A system that uses a nonpolar stationary phase and an aqueous mobile phase is frequently called reversed-phase chromatography. For reasons of solubility, gradient elution is not applicable to liquid-liquid systems, and is restricted to adsorption chromatography. However, the problems of stationary-phase solubility in mobile phases of comparable polarity have been solved by the development of a variety of stationary phases which are chemically bonded to their supports and are thus nonextractable with mobile phases.

Ion-exchange chromatography. Ion-exchange chromatography is one of the widely employed forms of liquid chromatography. It is based on selective ionic attractions between variously charged sample constituents and an ionized chromatographic matrix. There are two principal types of ion exchangers: cationic and anionic. The most commonly used ion exchangers consist of an organic polymeric backbone (usually a copolymer of styrene and divinylbenzene) with either acidic or basic exchange sites on its porous surface. The charged resins are capable of exchanging their cations or anions with those ions in the liquid phase which have a greater affinity for the matrix. Exchange interactions that take place during the passage of various ions through the column cause separation into discrete ionic zones.

An application of ion-exchange chromatography is shown in Fig. 8. The separation of four alkali-metal ions is carried out here on a strong cation-exchanger column, with hydrochloric acid solution as the mobile phase. Since the elements in this applications are radioactive (a solution of $^{24}NaCl$, ^{42}KCl, $^{86}RbCl$, and $^{137}CsCl$ was introduced into the column), their elution is monitored by a special radioactivity-measuring detector. Although a +1 charge is common to all chromatographed ions, their different ionic radii give rise to different retention. *See* ION EXCHANGE.

Gel chromatography. The other principal form of liquid chromatography is gel chromatography, also known as gel-permeation chromatography, gel filtration, molecular-sieve chromatography, or steric exclusion chromatography.

Since its introduction in late 1950s, gel chromatography has resulted in tremendous progress in the chemistry of biomacromolecules. Separation in gel chromatography is based on a selective process of penetration of molecules of different sizes and

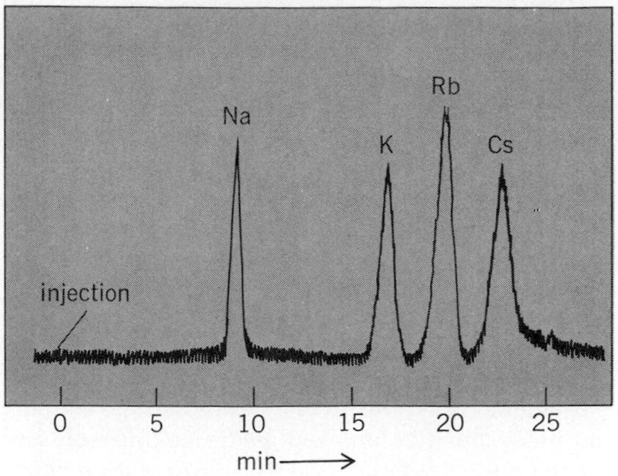

Fig. 8. Ion-exchange separation of alkali-metal cations. (*From J. F. K. Huber and A. M. van Urk-Schoen, Rapid separation of alkali metals by column ion-exchange chromatography, Anal. Chim. Acta, 58:395, 1972*)

shapes through a porous gel medium. This process, in which different molecules are "sized" and thus separated from each other, is schematically shown in Fig. 9*a*. The largest molecules in the mixture are not allowed to penetrate the porous structure at all; the medium-size molecules can penetrate only some pores; and the small molecules can diffuse rather freely inside the medium and can spend a considerably longer time there. Consequently, if the porous material is contained in a column, mixtures of components with differing molecular weights can be effectively resolved (Fig. 9*b*). This "molecular sieve" effect is extremely useful in many important separations. Gel materials are available for applications performed in both aqueous and organic mobile phases. The pore size can be regulated somewhat during the manufacturing process, according to the need for separation problems. Most column materials used in gel chromatography cannot be used at high pressures because of their limited mechanical stability.

Monitoring separation. The extent of separation in liquid chromatography can be monitored either in collected fractions or by a continuous detector.

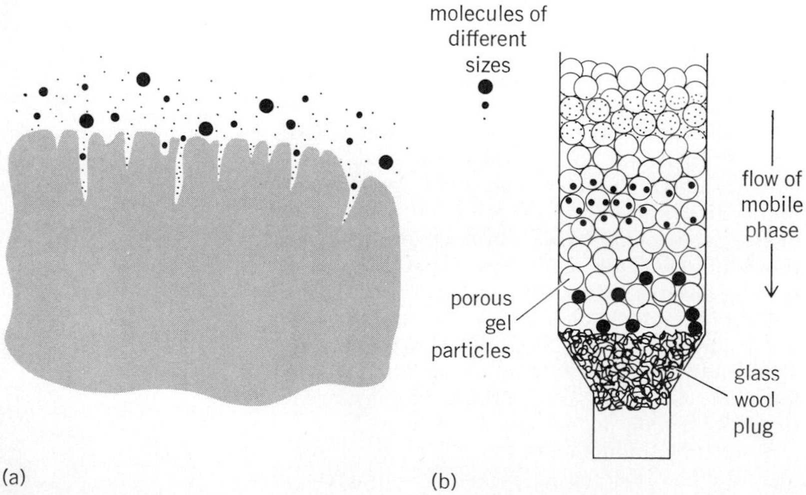

Fig. 9. Penetration of solutes with different molecular weight is schematically shown for (*a*) cross section of enlarged gel particle surface of porous nature and (*b*) get chromatographic column.

The former approach consists of a discontinuous measurement of concentration in automatically collected fractions by a suitable physical method. The fractions are usually collected at selected time intervals or in selected volumes.

Many continuous detectors for measuring the concentration of separated compounds in liquid chromatography have been suggested. Most are based on a measurement of some distinct optical property of solutes as opposed to the mobile phase. Miniature flow cells for spectrophotometry or fluorometry and devices for the differential recording of refractive index are commonly used. In addition, small electrochemical detection cells based on conductometry, coulometry, or polarography can be employed for electrochemically active samples. In general, there is a lack of universal and sensitive detection means comparable to those available in gas chromatography. One approach to achieving this goal is the so-called moving-wire flame ionization detector. In this detector, a portion of the total column effluent is retained on a moving platinum wire, the solvent is evaporated "on line," and the wire with dry residue is passed through a flame ionization detector similar to that used in gas chromatography. However, this detector is much less effective and sensitive than its gas chromatographic version.

Eluent obtained from a liquid chromatographic column can be selectively collected and further studied by chemical or physical methods. Mass spectroscopy, absorption spectroscopy, spectrofluorometry, and nuclear magnetic resonance spectroscopy are commonly used. *See* LIQUID CHROMATOGRAPHY.

APPLICATIONS

There is hardly any major area in the field of chemistry and related sciences that has not been affected by the rapid development of chromatographic methods and their routine applications. Both inorganic and organic compounds have been separated from mixtures. In its various forms chromatography has been succesfully applied to mixtures of small molecules, such as inorganic gases, and mixtures of biological macromolecules with molecular weights up to several million. Gas chromatography is most effectively used for analyses of organic compounds up to the molecular weight of approximately 500, if the requirements of volatility are fulfilled, and if the compounds are not unstable at the column temperatures used. Some of the most spectacular separations ever realized were achieved through this method. For example, Fig. 10 shows the resolution of hydrogen and deuterium spin isomers obtained by high-efficiency gas-solid chromatography at 77.6 K. Similarly, even very fine structural differences within the molecular species, such as different isotopic composition, stereoisomerism, cis-trans isomerism, or optical activity, may frequently give rise to distinguishing chromatographic mobility. Also, it is not unusual for up to several hundred components of a complex natural sample to be resolved by means of high-resolution gas-liquid chromatography.

In all its forms, liquid chromatography is a complementary method to gas chromatography, and often provides results often when gas chromatography fails. It is generally used for separations of more labile (polar) samples and larger (nonvolatile)

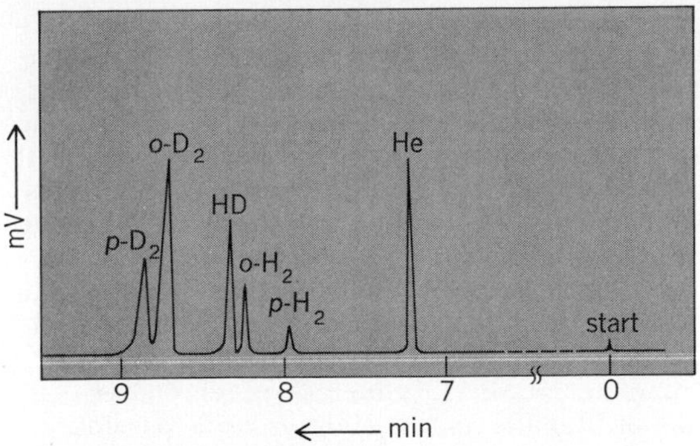

Fig. 10. Separation of hydrogen isotopes and their nuclear spin isomers at cryogenic temperature by high-resolution gas-solid chromatography. *(From M. Mohnke and W. Saffert, in M. van Swaay, ed., Gas Chromatography, Butterworths, London, 1962)*

molecules. The resolving power of liquid chromatography is usually lower than that of gas chromatography, but the simplicity of liquid chromatography is frequently advantageous. Basic biochemical research methods have been particularly advanced by gel chromatography and affinity chromatography.

Both gas and liquid chromatography are becoming indispensable tools in biomedical research, routine clinical determinations, and drug screening programs. Modern chromatographic methods have been developed for forensic "fingerprinting" applications. Gas chromatography has been extensively used in space-related and geochemical research projects.

Both analytical and preparative uses of chromatographic methods are extremely important to food, cosmetic, and pharmaceutical research and industries. These methods are often automated for routine analyses and closed-loop control of industrial streams, or to monitor the efficiency of cracking and distillation processes in the petroleum industry.

Gas chromatography has become the most important analytical method for the determination of various environmental pollutants at extremely low concentrations. *See* CHEMICAL SEPARATION TECHNIQUES. [MILOS V. NOVOTNY]

Bibliography: J. A. Dean, *Chemical Separation Methods*, 1969; J. C. Giddings, *Dynamics of Chromatography*, 1965; E. Heftmann (ed.), *Chromatography*, 2d ed., 1967; B. L. Karger, L. R. Snyder, and C. Horvath (eds.), *An Introduction to Separation Science*, 1973; A. B. Littlewood, *Gas Chromatography*, 2d ed., 1970; L. R. Snyder and J. J. Kirkland, *Introduction to Modern Liquid Chromatography*, 1974.

Chromium

Chemical element, Cr, atomic number 24, atomic weight 51.996, a silver-white metal that is hard and brittle as normally encountered. However, the bulk metal is relatively soft and ductile when unstressed and either effectively scavenged or extremely pure. Chromium was discovered in 1798 by L. N. Vauquelin. Its chief uses are production of noncorrosive, high-strength, heat-resistant characteristics in alloys and as an electroplated coating.

Natural occurrence. Elemental chromium, as such, is not found in nature. The most important mineral occurrence of chromium is in chromite, $FeOCr_2O_3$, which is never found pure. It is a dark-brown to jet-black octahedral mineral containing interstitial impurities, mostly magnesium silicates and other silicates. Some of the FeO is always replaced isomorphically by MgO, and some Cr_2O_3 by Al_2O_3. The chemical composition is usually represented by the formula (Mg,Fe) $(Cr,Al, Fe)_2O_4$ or more simply as $MgO{\cdot}FeO{\cdot}Cr_2O_3{\cdot}Al_2O_3{\cdot}Fe_2O_3$. Some high-grade chromites containing up to 65% Cr_2O_3 are found in nature but are extremely rare.

The usual high-grade ore contains 48% Cr_2O_3, with a Cr-Fe ratio of approximately 3:1. The iron content is found in widely varying amounts. Serpentines, chlorites, and peridotite frequently occur as gangue constituents of the natural chromites. The principal chromite-producing countries are the United States, Canada, Cuba, Guatemala, and Brazil in the Western Hemisphere; Albania, Bulgaria, Cyprus, Finland, Greece, Portugal, and Yugoslavia in Europe; China, Indochina, India, Japan, Pakistan, the Philippines, Turkey, and the Soviet Union in Asia; Sierra Leone, Rhodesia, and the Union of South Africa in Africa; and New Caledonia. Minor occurrences or presently undeveloped reserves have been found in other countries. Up to 4.5 wt % of chromium has been found in some meteorites, and 0.04 to 0.07 μg of chromium are present per liter of sea water. In addition, chromium occurs in a host of less important minerals not listed here.

Of geochemical interest is the fact that basalt found on the Moon contains 0.47 wt % Cr_2O_3, which is 3–20 times as much as that found in representative terrestrial specimens.

Chromite in ore comminuted to 120-μm size can be concentrated rapidly by flotation with either C_{16} or C_{18} amines. Sodium lignin sulfonate is used to suppress the gangue.

Metallurgical extraction. Chromium is produced in the form of an iron alloy, ferrochromium, by the reduction of chromite ores with carbon or silicon in an electric furnace. Ferrochromium is also produced from chromite by a silicothermic reaction in the presence of a suitable oxidizing agent, such as calcium chromate, $CaCrO_4$, sodium nitrate, $NaNO_3$, or manganese dioxide, MnO_2. The latter reaction is exothermic in nature. Chromium metal may also be produced by the exothermic reduction (Goldschmidt) of chemically produced Cr_2O_3, using powdered aluminum as the reductant. Since the use of aluminum powder is associated with explosive hazards and with considerable losses of chromium, and molten aluminum in an

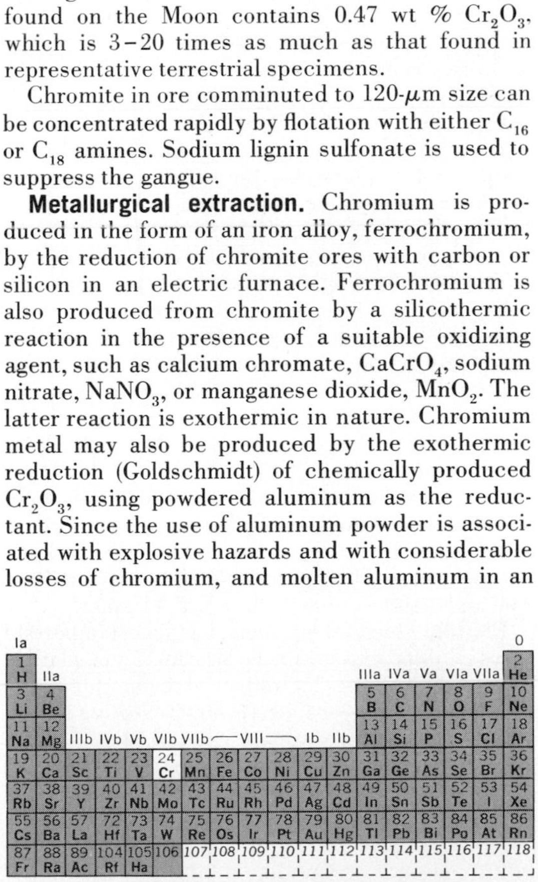

arc furnace at 2720°C reacts too vigorously with Cr_2O_3, molten aluminum is best poured at a lower temperature into a melt of Cr_2O_3 – 44 wt % CaO. With vigorous stirring nearly 94% of the chromium is recovered. The metal quality is good. The silicothermic reaction is generally employed to produce ferrochromium of controlled low-carbon content (0.03 – 0.1% C), although low-carbon ferrochromium is also produced in quantity by the reduction and removal of the carbon of normal high-carbon ferrochrome in a vacuum furnace by iron oxide, chromic oxide, or silica. The ferrochrome produced by this method is usually of very low carbon content (0.01 – 0.03%).

Ferrochromium containing 4% C can be produced in an arc furnace by carbon reduction of chromate ore that has been prereduced at least 50% at 1300 – 1700°C. The ore is smelted with at least a 20% excess of carbon, fluxes and, optionally, with noncarbonaceous reducing agents, such as Fe-Cr-Si alloys. In the prereduction of the ore without fusion, iron oxides are first reduced and the iron carburized.

Chromium metal is also produced on a commercial scale by electrolysis of an ammonium chromium alum solution, prepared either from chromium ore or from high-carbon ferrochromium.

Chromium metal is produced in more limited quantities by the thermal dissociation of chromium (II) iodide in contact with a suitable heated deposition surface under vacuum conditions (the Van Arkel – de Boer process). This is the purest form of chromium presently available. A high-purity product is also produced on a commercial scale by the hydrogen reduction of oxide in electrolytic chromium.

Chromium can be plated electrolytically from chromic acid containing a small amount of sulfuric acid. It is also plated by thermal decomposition of the hexacarbonyl, or from a salt bath containing $CrCl_2$, or from gaseous chromous chloride, $CrCl_2$. It is also produced in so-called sponge form by the hydrogen reduction of $CrCl_2$.

For making chromium chloride, chromium ores and coal are ground to 0.074 and 0.25 mm, respectively, pelleted, dried at 200 – 220°C, and chlorinated at 800 – 950°C in an electric shaft furnace. The iron content is 2 wt %, maximum, and averages $\leqq$ 0.2 wt %, sometimes 0.03 – 0.08 wt %. The chromium loss is 2 – 8 wt %.

Chromium carbide that is dissolved in iron or silicon can be purified by distillation from a graphite hearth and enclosure. A row of electron beams distils the metal. The product contains <0.5 wt % of iron and <100 wt ppm each of S, P, O, and C.

Plating. One of the principal and most important uses of pure chromium is for chromium plating. The advantages of electrodeposited chromium are: (1) attractive appearance, (2) high reflecting power, (3) good adhesion to the base metal, (4) high hardness when desired, (5) good wear resistance, (6) protection against corrosion without sacrifice of wear resistance, (7) possibility of applying thick coatings, (8) possibility of applying porous chromium plating to retain liquid lubricants, (9) possibility of applying crack-free deposits when they are desired, and (10) possibility of rebuilding worn or off-size parts.

The cathode reactions involved in the reduction of chromic acid involve $Cr^{6+} \rightarrow Cr^{3+}$, $2H^+ \rightarrow H_2$, and $Cr^{6+} \rightarrow Cr$. Under improper conditions the electrical reduction of chromic acid may occur without deposition of chromium metal — which deposits by virtue of the third reaction, only.

Since there are different types of electrolytes and deposits, not every one, by itself, will entail all the listed advantages. An unusually uniform distribution of metal on the cathode is obtained at 50°C (current density: 10 – 50 amp/dm^2) when the following electrolyte is used: CrO_3 250 g/liter, H_2SO_4 1.2 g/liter, H_2SiF_6 1.5 g/liter. Direct chromium plating of titanium and its alloys is made possible by breaking down interstitial oxides and hyrides by ultrasonic oscillations during plating.

Under typical conditions, a coating of electrodeposited chromium on iron reduces corrosion attack by about 55%. Chromate can prevent corrosion of a substrate by forming a dense, impenetrable coating. Another mechanism of chromate passivation is by satisfying the free-valence forces of surface metal atoms by chemical bonding with the passivating atoms or ions, without the metal atoms leaving their positions.

In some cases, crack formations in chromium coatings, such as are obtained in the bright-plating range of current density and temperature, interfere with protection against corrosion — although microcracks do not necessarily have this effect. Interruptions of plating can reduce or eliminate macrocracking if their frequency is controlled, along with the current density, temperature, type of substrate, and bath concentration, by preventing the attainment of a critical thickness for crack formation. However, excessive duration or frequency of interruptions of plating may result in loss of luster.

Solid-phase diffusion of chromium coatings on iron or steel forms alloy layers whose properties depend on the temperature and duration of heat treatment. A layer formed below 800°C is chromium-rich and very hard. Layers heat-treated at 600°C contain 85 – 90 wt % Cr.

High bond strengths between chromium and ceramic bodies generally are obtained by chemical interactions or, if any oxides in the body are reduced, by metallurgical interaction. If the chromium deposit is fired in moist ambient hydrogen, the chromium oxidizes and reacts with the substrate ceramic.

A thin Cr-Au film deposited onto TaN-coated Al_2O_3 substrate has acceptable thermocompression bondability in the as-deposited condition, marginal for stabilized films, and acceptable for films that were chemically etched after stabilizing. The thermal stabilization increases chromium oxide formation and thus results in degraded bondability.

Many household appliances are chromium-plated, as are decorative parts of other manufactured items including automobiles. Tools, plug gages, rolls, drum dryers, chemical equipment, dies, mandrels, electric appliances, gears, food machinery, kettles, pans, packing machinery, and many hundreds of other articles are chromium-plated where brightness, beauty, and resistance to wear and rust are necessary or desirable.

Chromizing. Chromium and other electrodeposits, enamels, and phosphate coatings on metal surfaces to be protected do not always provide uniform and reliable protection at high temperatures, particularly in cases where differences in thermal expansion may cause spalling. Recourse is then often made to chromizing. In this process a layer of chromium is applied at a temperature causing inward diffusion to form a composition gradient. An important use of chromizing is to protect nickel against a sulfur-containing atmosphere. Chromizing is commonly done by the thermal diffusion of chromium powder, the reduction of chromium oxide or halide in contact, or the thermal decomposition of a chromium halide or carbonyl.

Diffusion coating. A chromium coating intended for subsequent diffusion treatment can be prepared by immersing an iron base in liquid calcium containing finely dispersed chromium powder. The base loses iron to the liquid while gaining chromium. Another method of creating a diffusion coating of chromium involves first degreasing the steel and pickling it in hydrochloric acid, then embedding the steel in chromium powder in a sealed container, and finally heating it to 900–1100°C for up to 2 hr. Increasing the chromizing temperature within this range sharply reduces the corrosion rate of the product in a moist atmosphere and in kerosine.

Chromium, as well as various other metals, can be codeposited with iron by electroless plating done from an alkaline solution containing a hypophosphite reducing agent and a KNa tartrate complexing agent.

Sputtering. Chromium film about 100 ± 50 A (10 ± 5 nm) thick deposited on the four edges of ground and polished carbon-steel razor blades by conventional diode sputtering adheres strongly to the blade edges, and microscopically rough parts are made smooth.

Vacuum deposition. Chromium to be sublimed for vacuum deposition is commonly first electroplated onto a tungsten filament, which can then be electrically heated to the desired temperature of sublimation. If evaporative deposition of chromium is done in oxygen at a pressure of 10^{-8} torr (vapor pressure of chromium at 1300°C is 1.8×10^{-3} torr), the number of competing oxygen molecules reaching a surface of chromium condensation is great. Since oxygen chemisorbs on chromium, a substantial fraction of the deposited film will be chromium oxide. An impingement ratio of $O_2 : Cr = 10^{-2}$ is the maximum that can result in the deposition of chromium films that are essentially metallic. Since helium at 3×10^{-3} torr is not adsorbed, its presence at this pressure, in place of oxygen, does not interfere with the evaporative deposition of a metallic chromium film of nearly bulk resistivity. On the other hand, increases in oxygen pressure above 10^{-6} torr cause chromium vacuum deposits to contain increasing amounts of Cr_2O_3.

Since pure vapor-deposited chromium adheres much more strongly to glass than do most conductive metals, the prior deposit of chromium can act as a resistive substrate on which conductive metals can then be deposited.

An SiO_2 source at 1100°C and a chromium source at 1500°C can vapor-deposit a high-resistivity Cr-SiO film on a substrate at 400°C. Chromium-silicon monoxide films feature resistivity and stability without large negative temperature coefficients. Increases in SiO content cause increases in resistivity, but they are negatively affected by increases in temperature.

Organometallic compounds for deposition of Cr, Cr-Mo, and Cr-V coatings on carbon steels and on alloy steels include bis-arene complexes of the coating metals. Such compounds decompose on coming into contact with a surface heated to 350–500°C. The surface is then covered with a film of the corresponding metal or alloy. Although the precipitating metal catalyzes further decomposition of the organic radical and, hence, liberates carbon, which passes into the deposit and tends to embrittle it, this effect can be avoided by additions of di-benzene complexes of the same metals to the original compounds.

Aqueous deposition. Conversion coatings, formed by the immersion of a substrate metal in an aqueous solution of chromic acid or of chromate or dichromate salts, are of two types: (1) those that in themselves deposit substantial chromate films on the substrate metal, and (2) those that merely seal or supplement nonmetallic protective coatings, such as oxide or phosphate. For the deposition to occur, activating ions such as sulfate, nitrate, chloride, or fluoride must be added. Hydrogen generated when any of the activators attack the substrate reduces some of the chromium ion to form a hydrated "chromium chromate," $Cr_2O_3 \cdot CrO_3 \cdot xH_2O$, which deposits on the substrate surface if the solution is not too acid. Thus magnesium alloys can be chromated in nearly neutral solutions and aluminum alloys in solutions having either more acidity or more alkalinity. Conversion coatings are decorative, protect cadmium or zinc substrates in rural marine atmospheres, and form a base that bonds readily with many paints. The conversion coat must be removed from zinc prior to soldering but, when rosin flux is employed, need not be removed from cadium, copper, or silver.

Chromium alloys. Consumable-electrode casting, the use of yttrium-stabilized thoria-washed crucibles, the development of high-temperature extrusion techniques, alloys with improved ductility, and methods of protecting against nitride embrittlement make chromium a logical contender for use as jacketing of nuclear fuel elements for operation at temperatures beyond the capabilities of high-nickel alloys. The chromium imparts much of the required high-temperature strength, compatibility with the fuels, and resistance to corrosion in liquid sodium. These alloys are ductile enough for fabrication into thin-wall small-diameter tubing, they can be TIG (tungsten inert gas) welded, and they retain their desirable characteristics even after relatively long-time exposure to neutron irradiation.

Dilute chromium-yttrium alloy with an addition of NbC, TaC, or TiC has finely dispersed carbides and attractive strength characteristics. Resistance to oxidation and nitridation is improved by additional ZrC and HfC dispersants. Tungsten in chromium is superior to molybdenum in strengthening but affects workability adversely. Substitution of cobalt for some of the rhenium in Cr-35 at % Re results in single-phase workable castings, but profuse sigma precipitation occurs in them at 871–982°C. After a breakdown of the as-cast structure—

usually by hot extrusion—a degree of warm working usually is required to make these alloys stronger than stainless steel.

Heating Cr-MgO dispersion alloys at 1000–1200°C causes reaction of all the MgO in the scale layer to form $MgO \cdot Cr_2O_3$. This is a spinel. It is more resistant than tungsten or graphite to the heat flux encountered in rocket reentry.

Addition of 0.5–1.5 wt % of chromium to cast iron containing 3.85 wt % of carbon and 1.81 wt % of silicon markedly increases hardness, shear strength, and compressive strength. The pearlite coarsens, but the graphite flakes are not significantly affected.

Chromium is used extensively for alloying with iron to form stainless steels, special high-strength steels (ferritic), and electrical resistance wires; with nickel and iron to form nickel stainless steels (austenitic); and with manganese and iron to form chromium-manganese stainless steels. In steel it prevents corrosion caused by atmospheric conditions, corrosive waters, acids, and high-temperature conditions. When possible, the alloying addition is made in the form of ferrochrome.

Chromium is alloyed with nickel in varying quantities to form special heat-resistant alloys. The composition of these alloys is usually 15–28% Cr and 5–78% Ni.

Nickel-base alloys of chromium are of particular importance because of their high strength and unique resistance to high temperatures. Other metals can be added to the chromium- or nickel-base alloys, such as iron, Fe; titanium, Ti; columbium (niobium), Cb; cobalt, Co; copper, Cu; molybdenum, Mo; and tungsten, W, to produce modified alloys for specialty uses.

Chromium forms with nickel the well-known electrical resistance alloy Nichrome, which has made possible the large number of modern home appliances such as electric ranges and water heaters.

Chromium is alloyed with cobalt for use in producing cutting tools, and some tungsten, vanadium, and carbon are usually added for these applications. Stellite is probably the best-known commercial form of this type of alloy. These materials also possess desirable high-temperature properties.

Chromium is also used to alloy to a lesser extent with metals such as copper, aluminum, and titanium.

Chromium adds to the durability and adherence of a gold-germanium alloy film used on transparent glass for the selective transmission and reflection of certain parts (visible and invisible) of the solar spectrum. Chromium is also used in solutions for tanning hides, and aluminum chromium phosphate is used as a binder for high-strength ceramics. A powdered chromium-bearing slag is used in place of graphite for dusting molds and cores.

Catalytic applications. Chromium can perhaps compete with more costly platinum when used as a detoxifier of automobile exhausts. Impurities, such as NO_x, SO_2, H_2S, and organic sulfur compounds, can be removed from waste gas by passing it through a glass tube packed with acid absorbent, prepared by immersing dried asbestos in an aqueous CrO_3 solution, and drying it. The service life of the absorbent is 90 hr, compared to 10 hr for activated carbon. The absorbent's relatively low cost favors replacement with reactivated material as required. $CuCr_2O_4$-Al_2O_3 is described as a relatively stable catalyst for the oxidation of CO at high temperatures, and is likewise useful in automotive exhaust-control devices. High catalytic activity and durability is ascribed to a mixture made by heating a chromium-bearing alloy with $Al(OH)_3$ gel. The mixture is suggested for use in decomposing and removing NO_x from automotive exhaust gas. Dipping the steel parts of automobile exhaust converters into molten Al-Cr alloy baths with subsequent heat treatment imparts the required resistance to higher-temperature oxidation and corrosion.

Catalysts containing chromium tris (acetylacetonate) silicic acid, aluminum alkyls, and organometallic-reduced titanium or vanadium compounds have been recommended for the preparation of polyolefins with controlled broad molecular-weight distribution and improved mechanical properties.

A chromium-bearing catalyst has been described for the synthesis of ammonia. $CrCl_2$ is heated with graphite, and the resulting complex is heated with potassium. Passage over the catalyst produced of a 1 : 3 nitrogen-hydrogen atmosphere at 300°C at a pressure of 60 cm mercury and a rate of 15 ml/min gives 6.8 ml condensed NH_3 in 15 hr. The catalyst resists inactivation by oxygen, and therefore does not require reduction during preparation.

Chromium in water. Chromium as a compound in aqueous solution is desirable as a anticorrodent in plant coolant water, and in waste water chromium serves as an important economic source of the metal for a host of applications. In high-temperature recirculating water systems, chromate anticorrodents are superior to all other proven inhibitors.

Wastes resulting from chromate manufacture have had toxic effects on the disposal area, and runoff from the waste has had severe effects on river flora. Chromic ions in soil become less toxic as chromic hydroxide is formed, probably because they are less soluble in that form. This suggests that chemical reduction of chromate in the waste can reclaim the soil. When chromates are used in cooling water, their possible toxicity to aquatic organisms must be considered, as well as environmental quality standards. These standards may involve water temperature, pH, chemical composition, and metallurgical factors. Effluents from chromate-treated waters may require special treatment of the blowdown by chemical reduction, followed by lime precipitation of chromic sulfate.

Chromate ions are removed and recovered from feed water by passage through a bed of basic anion-exchange resin. Then an alkaline solution containing regenerant is passed through to recover the chromate ions. Although Cr_2O_7 present in trace amounts is not readily recovered from waste water by coagulation, sedimentation, or sand filtration, sorption by an activated-carbon column can reduce the chromium content by 94%, and anion-cation exchange can remove 99.7% of the chromium. Since chromium is used as a corrosion inhibitor in the coolant water of many kinds of industrial and nuclear plants, there has been public concern that plant waste water may contain excessive amounts of chromium. Therefore it is usually both economically and ecologically desirable to recover the chromium before the waste water is discharged.

Acidic aqueous electroplating wastes containing 0.4 g/liter of Cr^{6+} as CrO_3 are treated for chromium recovery first in a cation exchanger. The effluent is then treated in an anion-exchange zone to recover Cr^{6+} in the resin and produce demineralized water effluent containing Cr^{6+}. Treating the effluent with a cation-exchange resin produces an aqueous chromic acid solution.

The Cr_2O_3 in a tannery bath can be recovered for reuse by flocculation with Na_2CO_3, decantation, pressing together into filter cake, and redissolving in concentrated H_2SO_4 containing formic or oxalic acid.

Waste water containing cyanide ion, chromic acid, arsenious acid, or cadmium ion can be treated with amines carried in water-insoluble supports (such as crushed red brick) to remove the pollutants from the waste water.

Chromates also have been removed from aqueous solutions, for example, pickling liquors, by adsorption at the bulk of the Cr^{6+} of flocs of hydrous ferric oxide. The Fe^{3+}/Cr^{6+} ratio should be high, but not above 20.

Evaporative recovery of chromium-plating waste has progressed since 1950 when manually operated batches were often processed in jacket kettles. Continually automated systems now incorporate the removal of impurities and the recovery of water by filtering, decantation, and distillation. To minimize water consumption, the chromium waste is rinsed through countercurrent rinse tanks in a closed loop, and the distillate is utilized for rinsing. This is important where water supplies are limited or where the discharge of untreated liquid waste would result in ecological problems.

Medicine and health. Gamma-active radioisotope Cr^{51} is used clinically to label erythrocytes for studying the spleen: to determine its size, function, and blood flow, and even to make possible the diagnosis of splenic rupture with subcapsular hematoma. Intestinal bleeding has been similarly studied, and Cr^{51}-labeled albumin has been used to study the loss of intestinal protein. Radiotherapy of certain types of tumors, implanted with Cr^{51}, had some success in a number of cases involving patients who had received maximum therapy from external-beam radiation and radium.

During production of alumina from bauxite and limestone, the insoluble compounds of Cr^{3+} are transformed to compounds of Cr^{6+} which are present in the dust of sintered products and in alkaline aerosols. Workers exposed to the dust and aerosols develop symptoms of atrophic rhinitis, allergic rhinotomy, and bronchial asthma. Removal of chromium from raw materials in early stages of aluminum production is therefore recommended.

Zinc tetroxychromate, strontium chromate, and lead silicochromate are toxic. Waters containing them can be treated either chemically or by ion exchange.

Physical properties. There are four naturally occurring isotopes of chromium, namely, Cr^{50}, Cr^{52}, Cr^{53}, and Cr^{54}. Their respective percent abundances are 4.31, 83.76, 9.55, and 2.38. The packing fraction of −8.18 applied to these values reduces the atomic weight of natural chromium to 52.0. The neutron capture cross section of chromium, millibarns, is 2900 for thermal neutrons, 10 for 30-kev neutrons, and $2.68\pm^{.82}_{.26}$ for 65-kev and more energetic neutrons. Various unstable isotopes can be produced by radiochemical reactions. The most important is Cr^{51}, which emits soft gamma rays and has a half-life of approximately 27 days. Another tracer isotope is Cr^{56}. The crystal structure of chromium is body-centered cubic, although it transposes to face-centered cubic at 1830°C (3325°F). Other reported structures have turned out to be hydrides, although isolated electrodeposited crystals may be initially hexagonal. Electroplated and polished chromium is bright bluish white. Its reflecting power is 77% that of silver. Chromium has a density at 20°C (68°F) of 7.22 g/cm³ (0.26 lb/in.³) for a single crystal and 7.14 g/cm³ (0.259 lb/in.³) for a polycrystalline solid.

The chromium-lithium ferrite spinel, $Li_2O \cdot 2.5Fe_2O_3 \cdot 2.5CrO_2$, exhibits an adiabatic cooling, negative magnetocaloric effect on magnetic switching near the compensation point.

The Néel point, the temperature at which antiferromagnetic chromium is of maximum magnetic susceptibility, is about 38°C for zero gage pressure and 0°C for a pressure of 8 kilobars. Ordinary paramagnetic chromium has a compressibility of -4.9×10^{-7} bar^{-1}, and the antiferromagnetic phase has a compressibility of -5.9×10^{-7} bar^{-1}.

At 2200°C two molecules per thousand of chromium vapor are dimers.

Nuclear properties. Used in fast nuclear reactors, chromium is superior to all other group VI metals from the standpoint of neutron economy and compares well with vanadium and niobium. Chromium is superior to niobium, molybdenum, and tungsten at the higher flux energies, and superior to tantalum at all energies.

Chromium is an excellent material for supporting foil, such as indium or gadolinium, for the detection of thermal neutrons. The chromium combines high mechanical strength with low scattering cross section.

A patent claims that the treatment of nuclear fuel elements by electrodeposition of chromium plate minimizes corrosion.

Electrical properties. The electrical conductivity is 22.2% that of copper, when measured on the basis of the metal in a very pure and annealed form. The specific resistance at 20°C is 13.0 microhm-cm for annealed and electrodeposited chromium; and 132 microhm-cm at 1800°C.

The electrochemical equivalents are, for valence of 6, 0.08983; valence of 3, 0.17965; and valence of 2, 0.26940.

Mechanical properties. These properties relate to the strength and workability of chromium including, among others, its hardness, tensile strength, and forging, rolling, and drawing characteristics. Chromium has relatively poor forging, rolling, and drawing properties. However, when absolutely free of oxygen, hydrogen, carbon, and nitrogen, it is relatively ductile and can be forged and drawn. It is difficult to keep it free from these elements. Chromium normally demonstrates brittle characteristics which, heretofore, have led many to believe it to be an inherently brittle metal, but it is now known that its brittleness is a function of its purity.

Chromium has a Brinell hardness (as cast) of 110−170; that of electrodeposited chromium is 550−1250, depending on the hydrogen content.

Although the electrodeposited chromium is nearly diamond-hard, a thin deposit is easily damaged when the supporting material is relatively soft. Its hardness is 70–90 when annealed.

Chromium has a low coefficient of friction, being practically one-half that of steel-on-steel and one-third that of babbitt-on-babbitt (static) or three-fourths that of babbit-on-babbit (sliding).

For powder metallurgy the best hot-working temperature is 1250°C (2280°F), and 40,000 psi is the best pressure for compacting.

Sintering is best effected at 1300°C (2370°F) in vacuum until gas-free, followed by heating in an inert atmosphere such as argon at 1600–1700°C (2990–3090°F) to maintain high purity.

Typical compositions of chromium from various processing techniques are as follows: iodide chromium, 99.99+% Cr; electrolytic chromium, 99.00%; thermit chromium, 97.00%; and electric-furnace chromium, 97.00% Cr, 0.03–0.5% C, and 1% Fe.

Thermal properties. The thermal properties of chromium are generally viewed as those which have to do with its melting point, boiling point, heat capacity (in the solid, liquid, and gaseous states), vapor pressure, and heat conductivity.

Chromium has a melting point of 1930 ± 10°C (3505 ± 20°F); its boiling point is 2480°C (4500°F) at 760 mm mercury. Its heat capacity is 5.55 cal/(mole)(°C) at 25°C (77°F) or 0.11 cal/(g)(°C). In the liquid state the heat capacity is 9.40 cal/(mole)(°C).

The latent heat of fusion is 4.20 cal/mole (80.75 cal/g). The latent heat of vaporization is 76,635 cal/mole (1470 cal/g). The vapor pressure at 2097°C (3807°F) is 10^{-1} atm. The thermal conductivity of chromium at 20°C (68°F) is 0.16 cal/(sec)(cm)(°C).

Chemical properties. These properties include the principal chemical characteristics of chromium and its typical reactions with other elements and compounds. Chromium forms three series of compounds with other elements; these can be represented in terms of the chromium oxides: chromium, valence 2, CrO, chromium(II) or chromous oxide; chromium, valence 3, Cr_2O_3, chromium(III) or chromic oxide; and chromium, valence 6, CrO_3, chromium(VI) or chromic acid anhydride.

Chromium metal in acids is normally characterized by surface passivation. Depassivation, which involves brief hydrogen evolution, results from contact in the acid with aluminum, zinc, Cr^{2+}, or other electropositive metals or ions, or from application of a cathodic potential. Once depassivated, chromium dissolves readily in nearly all mineral acids, but not in nitric acid. The dissolution rates decrease in the order of HCl, HBr, H_2SO_4, $HClO_4$. The metal is soluble in sulfuric acid, but insoluble in nitric acid. It is oxidized by water vapor at high temperatures and by carbon monoxide at 1000°C. It reacts at high temperatures with nitrogen and is attacked by fused alkalies. The metal reacts with sulfur vapor and SO_2 at 700°C, and with hydrogen sulfide at 1200°C, and it attacks porcelain at 1600°C. Electrolytic chromium absorbs 250 times its volume of hydrogen.

Chromium and hydrogen form chromium hydride, CrH_2, a black powder having a density of 6.77 g/cm³. The heat of formation is 3800 cal/mole of hydrogen. The hydride decomposes at 350°C (660°F) and at room temperature on long standing. CrH is also formed.

Both CrN and Cr_2N are formed by the interaction of chromium and nitrogen, but there is an appreciable difference in their thermal stabilities. The presence of ammonia in the ambient nitrogen atmosphere enhances the stability of CrN but impairs its homogeneity. The solubility of nitrogen in liquid chromium is approximately 4% by weight.

Chromium and carbon form three positively known carbides of chromium, $Cr_{23}C_6$, Cr_7C_3, and Cr_3C_2. The properties of Cr_3C_2 are well known and are employed commercially. It has a hardness (nominal) of 88_A (equivalent to 72 Rockwell C), density of 7 g/cm³, transverse rupture of 100,000 psi, and an abrasive resistance much higher than steel but not as high as tungsten carbide. The corrosion resistance of Cr_3C_2 is excellent at 1800°F. Its electrical conductivity is 2.2% of that of copper at 20°C. The electrical resistance of the carbide is 80 microhms/(cm)(cm²) at 20°C. It is nonmagnetic. It is resistant to salt spray and sulfuric and nitric acids, and it can be ground and lapped.

Chromium and silicon readily form chromium silicides in the electric furnace. They vary from soft to hard and brittle, depending on the silicon content. They are resistant to all acids except hydrofluoric acid.

The known borides of chromium, Cr_3B_2, CrB, CrB_2, and Cr_3B_4, are attacked by acids.

Chromium is known to alloy with many metals, including aluminum, antimony, beryllium, bismuth, carbon, cobalt, copper, gold, hafnium, iron, lead, manganese, molybdenum, nickel, palladium, platinum, silicon, silver, tin, titanium, tungsten, vanadium, yttrium, zinc, and zirconium.

Principal compounds. Chromium is capable of forming compounds with other elements in oxidation states of II, III, and VI. Table 1 lists the typical compounds of each valence state with oxygen.

Table 1. Valence states of chromium

Compound	Formula	Valence state of Cr
Chromous oxide	CrO	2+
Chromous hydroxide	$Cr(OH)_2$	2+
Chromic tetroxide*	CrO_2	4+
Chromic pentoxide*	CrO_5	5+
Chromic oxide	Cr_2O_3	3+
Hydrous chromic oxide (chromic hydroxide)	$Cr_2O_3 \cdot xH_2O$	3+
Chromites	$(Cr_2O_4)^{2-}$	3+
Chromic anhydride	CrO_3	6+
Chromates	$(CrO_4)^{2-}$ and $(Cr_2O_7)^{2-}$	6+

*Stable only at high pH.

Peroxides, perchromic acid, and perchromates are also known. The most widely known compounds include ferrous chromite ($FeOCr_2O_3$), lead chromate ($PbCrO_4$), barium chromate ($BaCrO_4$), sodium chromate (Na_2CrO_4), sodium dichromate ($Na_2Cr_2O_7$), potassium chromate and potassium dichromate (K_2CrO_4; $K_2Cr_2O_7$), chromic acid (H_2CrO_4), and zinc chromate ($ZnCrO_4$). Almost all metals form chromate compounds with chromium.

Some have extensive use in the production of chromium metal, pigments, tanning agents, textile dyestuffs and dye-fixing agents, chromium-plating preparations, and in numerous organic derivatives.

Sulfides of chromium are known, but are not used extensively by industry. The chromium sulfate, $Cr_2(SO_4)_3$, is a common salt of chromium and finds use in the tanning industry. The sulfate, as well as other chromium salts, forms many coordination compounds with H_2O and NH_3.

The halides (chlorides, fluorides, bromides, and iodides) of chromium are fairly common compounds of this metal. The chloride, for example, has been used to some extent in the production of chromium metal by the hydrogen reduction of chromous chloride, $CrCl_2$. The iodide is used to produce very pure chromium metal by application of a modified Van Arkel–de Boer technique. CrF_2 is the least soluble of the Cr halides.

Nitrogen-containing chromium compounds include the chromic cyanides such as $(NH_4)_3[Cr(CN)_6]$, the azides $Cr(N_3)_3$, chromic amide $Cr(NH_2)_3$, and chromous nitrate $Cr(NO_3)_2$. The industrial uses of these compounds are somewhat limited.

Carbon-containing chromium compounds include the carbonate of chromium ($CrCO_3$), the ammonium carbonate $(NH_4)_2Cr(CO_3)_2$, chromium carbonyl $Cr(CO)_6$, and the thiocarbonate $[Cr_2(CS)_3]_3$, which possess limited industrial usefulness.

The compound VCr_2O_4 occurs in a solid solution prepared from V_2O_5-Cr_2O_3.

Chromium borates of indefinite composition have also been reported. They consist of Cr_2O_3 and B_2O_3 in varying proportions.

Both chromous and chromic salts of acetic acid, formic acid, oxalic acid, and tartaric acid are known and have varying commercial importance.

Thin resistive chromium-cermet films strengthened with silica particles, when laser irradiated, increase in optical density through formation of Cr_3Si crystallites in the irradiated zones. The optical density then falls because of the formation of Cr_3Si aggregates 1000 A (100 nm) in diameter.

Chromium chloride in DMF (dimethylformamide), when complexed with 2-aminobenzoic acid and treated with dicyclohexylamine, gives a dicyclohexylammonium acid chromide that is useful for coloring aluminum yellow, after vapor-deposition of the aluminum on thermoplastic films.

Chromium is used in a variety of electric batteries. Reactants of solid-state electric battery featuring a chromium-fluoride cathode and a calcium anode are reported to give 2.7 volts after charging. Cr_2S_3 or Cr_2O_3 have been used as a cathode in a secondary battery having a lightweight lithium (or lithium alloy) anode. Cathode evaporation losses are decreased by the addition of a molten-salt electrolyte containing Li^+, Li_2S, or Li_2Se_4 to the LiCl-KCl electrolyte. A thermal (fused electrolyte) battery, $Ca[LiCl\text{-}KCl]CaCrO_4$, including a pelletized heat source and depolarizer, has also been developed.

A coloring frit containing MnO_2 and Cr_2O and amounting to 0.7–2% of the glass composition is reported to color glass brown. It is cheaper than the conventional frit containing NiO.

A $PbCrO_4$-Ti system burns at a few milliseconds per centimeter, which makes it useful in igniters. The maximum burning rates are affected by size of fuel particles.

Chromic oxide is used as an indicator of nutrient digestibility in fecal collection apparatus for steers. When added to corn in bovine feed, and recovered in feces, the recovery of chromic oxide is 98.8 to 100.2%, depending on dryness of samples.

The adhesion of diamond grains to the metallic mass in diamond-abrasive stone machining tools is improved by the addition of chromium to a bronze binder. Thus, consumption of diamonds, natural or synthetic, is lowered. In the case of alumina cutting tools used in the machining of alloy steels, the addition of 1–30 mole % of Cr_2O_3 (5 mole % is usually satisfactory) to the alumina before it is sintered significantly reduces the crater wear due to grain pull-out, creep and plastic flow, crack formation, and chemical interaction with the steel base.

An infrared heating device has been designed which consists of a metal tube containing a resistance wire in the middle with magnesia packed around it, and TiO_2 containing 1% Cr_2O_3 in a melt sprayed on the external surface. The chromia improves the resistance to thermal shock, cracking, and peeling.

The variety of characteristic colors exhibited by chromium compounds helps to explain why they are used as pigments and dyes (see Table 2).

Table 2. Characteristics of various chromium pigments and dyes

Compound	Color	Solubility in water
$Ag_2Cr_2O_7$	Red	–
Ag_2CrO_4	Red-brown	–
H_2CrO_4	Red, yellow, or brown	–
Na_2CrO_4	Yellow	+
K_2CrO_4	Yellow	+
$(NH_4)_2CrO_4$	Yellow	+
$CaCrO_4$	Yellow	+
$K_6Cr(CN)_6$	Yellow	Moderate
$BrCrO_4$	Pale yellow	–
$SrCrO_4$	Pale yellow	–
$ZnCrO_4$	Yellow	–
$PbCrO_4$ (chrome yellow)	Yellow	–
$Na_2Cr_2O_7$	Red	+
$K_2Cr_2O_7$	Red	+
$(NH_4)_2Cr_2O_7$	Red	+
$Cr(CO)_6$	White	–

Ammonium chloride reacts with Cr^{3+} to produce such compounds as $[Cr(NH_3)_4Cl]Cl$, violeo-chloride; $[Cr(NH_3)_5Cl]Cl_2$, purpureo-chloride (octahedral, red); and $[Cr(NH_3)_6]Cl_3$, luteo-chloride (yellow crystalline). Cr^{3+} also forms stable complexes with amines, including ethylenediamine, and Cr^{3+} reacts with oxalato groups to form such coordinates as $[Cr(C_2O_4)_3]^{3-}$ and $[Cr(C_2O_4)_2]^-$, which are yellow powders.

Organochromium compounds include dicyclopentadienyl chromium $(C_5H_5)_2Cr$, and also cyano, thiocyano, nitrato, fluoro, bromo, iodo, azido, acetylido, and nitro compounds.

Chromium complexes such as the glycinato complex $[Cr(gly)_2OH]_2$, the phenanthroline complexes $[Cr(phen)_2OH]_2Cl \cdot 6H_2O$ and $[Cr(phen)_2OH]_2I_4 \cdot 4H_2O$, and the oxalato complex $Na_4[Cr(ox)_2 \cdot OH]_2 \cdot 6H_2O$ have structural and magnetic properties whose correlations are explainable in terms of valence bond and superexchange.

Oxides. When a clean layer of chromium is exposed to the atmosphere, some oxygen is adsorbed, but water vapor is responsible for the final layer of oxide.

Cr_2O_3 can be prepared by adding ammonia to chromium salt solutions, or by hydrolysis of urea to form OH^- in chromium salt solutions.

Magnetic CrO_2 can be prepared by pyrolyzing a chromium oxide (Cr_2O_5) in a fused salt bath containing a reductive inorganic compound. The yield of CrO_2 is high and also its purity. CrO_2 is useful in magnetic tapes, especially for recording high frequencies.

Ferromagnetic CrO_2 powder has been prepared by a variety of other procedures. One of them involves dry-treating CrO_3 and Cr_2O_5 at 200°C and 10 kbars pressure. Commercial CrO_3 powder is heated in air at 750°C under a pressure of 30,000 atm for 10 min in order to reduce the CrO_3 powder to granular CrO_2.

Black ferromagnetic CrO_2 powder has also been made by heating a mixture of CrO_3, H_2TeO_4, and CaO with H_2O and 20% NH_3 solution for 2 hr at 390°C in an oxygen atmosphere pressurized at 440 kg/cm^2 in a stainless steel autoclave.

Analytical methods. Chromium can be detected and identified via the chromate ion by precipitation as barium chromate, $BaCrO_4$, or lead chromate, $PbCrO_4$. It is also detected by the change in coloration of the chromate ion from orange to bluish green when a solution of chromate is treated with ferrous ammonium sulfate. More modern methods of detection include the spectrograph x-ray analysis, polarographic analysis, and flame photometry.

There are several principal methods in use today for determining chromium: (1) the gravimetric method, by precipitation and weighing as $BaCrO_4$, or by precipitation and weighing as Cr_2O_3; (2) volumetric methods, such as the ferrous ammonium sulfate titration of a chromate with back titration of the excess ferrous ammonium sulfate with potassium permanganate; it can also be determined volumetrically by the iodometric method in which potassium iodide is added to the chromate solution, and the liberated iodine titrated with a solution of sodium thiosulfate, $Na_2S_2O_3$.

Total chromium, Cr^{6+} and Cr^{3+}, from a sample of a pelletized ion-exchange resin containing as little as 0.50 micromole, can be determined by x-ray fluorescence.

In order to determine Cr^{3+} from the self-color of its aqueous solution, prior determination of the effect of the addition of a known amount of Cr^{3+} is required. It should be oxidized to CrO_4^{--} and then reduced again. By this "addition method" chromium can usually be determined photometrically with a standard deviation of 1%.

In gas chromatography, a flame photometer detector, when equipped with a 425.4-nanometer filter, is linearly sensitive from 0 to 90 nanograms of chromium, and can be used for monitoring chromium in human urine. Conventional gas-chromatographic methods for determining low levels of chromium are sensitive but less specific.

Fresh-water sources have been analyzed for chromium by radiochromatographic analysis following chromatographic separations as volatile complexes (chelates) of metal and organic compounds. Quantitative postirradiation recovery of chromium is possible. Important considerations are possible radiative degradation of the complex as a function of excess reagent, irradiation time, and the neutron flux system.

The chromium chelates of $R^1COCHR^2COR^3$ with fluoralkyl substituents are highly volatile, but those with aromatic substituents are nonvolatile. As the chromium chelates are fairly stable, they are especially suitable for employment in gas chromatography.

As little as 0.01–0.1 mg/ml of chromium in water and solutions can be spectrographically detected when use is made of a porous, hollow carbon electrode filled with a sample solution and excited in a high-voltage condensed spark.

Chromium is determined in various marine substances by spectrophotographic measurement of its diphenylcarbazide complex. This method is applied to solid substances removed from samples of deep and surface sea water by settling and centrifuging.

Chromium in small amounts may be detected by the colorimetric method through use of *s*-diphenylcarbazide as the indicator. *See* TRANSITION ELEMENTS.

[W. D. WILKINSON]

Bibliography: American Society of Metals, *Metals Handbook*, vol. 2, 1979; J. D. Greenwood, *Hard Chromium Plating: A Handbook of Modern Practice*, 2d rev. ed., 1971; F. A. Lowenheim (ed.), *Modern Electroplating*, 3d ed., 1974; National Academy of Sciences, *Chromium*, 1974; C. L. Rollinson, *The Chemistry of Chromium, Molybdenum and Tungsten*, 1975; W. D. Wilkinson, *Fabrication of Refractory Metals*, 1970; W. D. Wilkinson, *Properties of Refractory Metals*, 1969.

Clathrate compounds

Well-defined addition compounds formed by inclusion of molecules in cavities existing in crystal lattices or present in large molecules. The constituents are bound in definite ratios, but these are not necessarily integral. The components are not held together by primary valence forces, but instead are the consequence of a tight fit which prevents the smaller partner, the guest, from escaping from the cavity of the host. Consequently, the geometry of the molecules is the decisive factor.

Clathrate compounds are of theoretical interest because they represent a special type of bonding. In practice they can be used for separation of hydrocarbons, for stabilization of drugs or pesticides, and as enzyme models.

Inclusion compounds can be subdivided into (1) lattice inclusion compounds (inclusion within a lattice which, as such, is built up from smaller single molecules); (2) molecular inclusion compounds (inclusion into larger ring molecules with holes); and (3) inclusion compounds of macromolecules (see table).

The best-known lattice inclusion compounds are the urea and thiourea channel inclusion compounds, which are formed by mixing hydrocarbons, carboxylic acids, or long-chain fatty alcohols with solutions of urea. Urea, in this case the host lattice, crystallizes in the presence of the other molecules in a particular crystal structure which contains long channels or tubes 0.4–0.5 nanome-

Inclusion compounds

Host substance	Shape of cavity	Guest substance examples
Lattice inclusion compounds		
Urea	Channel	Derivatives of straight-chain hydrocarbons
Thiourea	Channel	Branched-chain hydrocarbons
Deoxycholic acid	Channel	Paraffins, fatty acids, aromatic hydrocarbons
Dinitrodiphenyl	Channel	Diphenyl derivatives
Hydroquinone, phenol	Cage	Hydrogen chloride, sulfur dioxide, acetylene
Gas hydrates	Cage	Halogens, inert gases, hydrocarbons
Trithymotide	Cage	Benzene, chloroform
Nickel dicyanobenzene	Cage	Benzene, chloroform
Cyclodextrins	Channel or cage	Hydrocarbons, iodine, aromats
Molecular inclusion compounds		
Cyclodextrins	Cage	Hydrocarbons, iodine, aromats
Bis (*N*,*N'*-alkyl benzidines)	Cage	Benzene, dioxane
Crown ether compounds	Cage	Inorganic cations
Ionophore antibiotics	Cage	Inorganic cations
Inclusion compounds of macromolecular substances		
Clay minerals	Channel or sheet	Hydrophilic substances
Graphite	Sheet	Oxygen, bisulfate ion, alkali metals
Cellulose, starch	Channel	Hydrocarbons, dyes, iodine

ter in diameter. The hydrocarbons are accommodated in these channels, but only unbranched molecules find enough space to fit into the urea channel. The molar ratio of urea to hydrocarbon is simply determined by the lengths of the hydrocarbon molecule, not by any functional group. The longer the paraffin chain, the more urea is necessary in order to envelop the paraffin (Fig. 1). Thiourea has a similar lattice structure, but a larger diameter of the channel. Therefore, only branched paraffins are included; the unbranched ones are too small, and cannot be held tightly (principle of closest packing), as shown in Fig. 2.

Other representatives of lattice inclusion compounds are the choleic acids, which are adducts of deoxycholic acid with fatty acids, and other lipoic substances. Most likely, these compounds play a role in the digestion process by emulsifying or dissolving fat and fat-soluble substances. Hydroquinone and phenol, cyclic anhydrides of phenol carboxylic acids, such as *o*-trithymotide and tetrasalicylide, as well as some other aromatic compounds, form an open crystal lattice which can

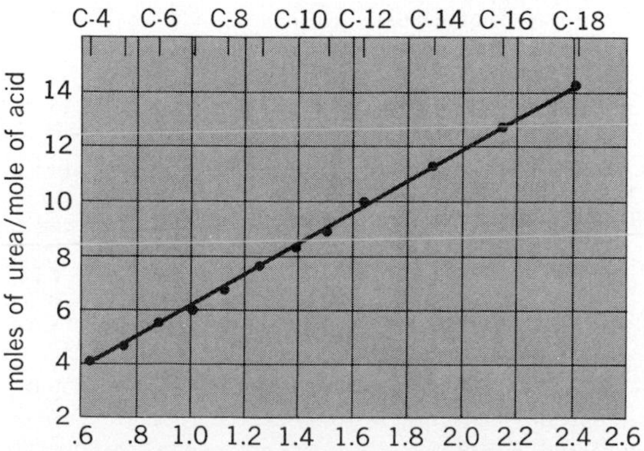

Fig. 1. Dependence of the urea/fatty acid composition of the inclusion compound on the chain length of the fatty acid. (*From W. Schlenk, Jr., Organic occlusion compounds, Fortschr. Chem. Forschg., 2:92–145, 1951*)

accommodate smaller gas and solvent molecules (clathrates in the stricter sense of the word). The gas hydrates, which have been known since about 1810 (for example, $Cl_2 \cdot 6H_2O$), are inclusion compounds of gases in a somewhat expanded ice lattice. The gas or solvent molecules are inserted into definite places within the ice lattice and are surrounded by water molecules on all sides.

The cyclodextrins (cycloamyloses) are cyclic degradation products of starch which contain six, seven, or eight glucose residues (α, β, γ-cyclodextrin) and have the shape of large-ring molecules, the cavity diameters being 0.6–1.0 nm. Here, a great number of inclusion compounds in solution or in the crystalline state can be prepared with various kinds of molecules, the only requirement being the fit of the guest within the cyclodextrin cavity. α-Cyclodextrin forms blue iodine inclusion compounds in which the host molecules are stacked to form a cylinder reminiscent of the starch helix (Fig. 3). Polyiodide, located within the thus formed channellike cavity, consists of I_2, I_3^-, or I_5^- units in α-cyclodextrin, as well as in starch. A similar structure is also likely for the blue iodine·polyvinyl alcohol inclusion compound which exhibits a strong dichroism when stretched. This is used technically to prepare (light) polarizing sheets and spectacles.

Crown ether compounds are cyclic or polycyclic polyether compounds (for example, with the repeating unit . . .O—CH_2—CH_2. . .) capable of including another atom in the center of the ring. In this way, sodium or potassium compounds can be solubilized in organic solvents. Similarly, a series of ionophore antibiotics can complex inorganic cations. *See* CROWN ETHERS AND CRYPTANDS.

Some clay minerals are made up of distinct silicate layers. Between these layers some free space may exist in the shape of channels. Smaller hydrocarbon molecules can be accommodated reversibly within these channels. This phenomenon is used in some technical separation processes for separating hydrocarbons (molecular sieves). Furthermore, ion-exchange processes used for water deionization are based on similar minerals. *See* MOLECULAR SIEVE.

CLATHRATE COMPOUNDS

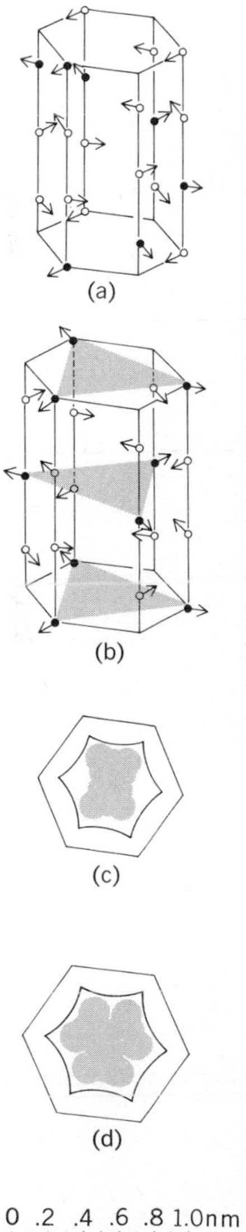

Fig. 2. Schematic drawing of the lattices of (*a, c*) urea and (*b, d*) thiourea; *c* and *d* show accommodation of unbranched and branched paraffin chains, respectively.

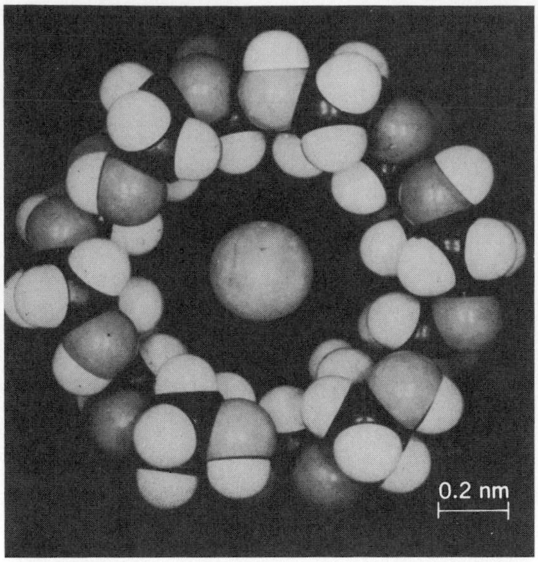

Fig. 3. Molecular model of α-dextrin–iodine compound seen in the direction of the *c* axis, that is, in the direction of the iodine chains, which lie in channels.

Enzymes are believed to accommodate their substrates in active sites, pockets, or clefts prior to the chemical reaction which then changes the chemical structures of the substrates. These binding processes are identical to those of low-molecular-weight inclusion compounds. On the other hand, some low-molecular-weight inclusion compounds can mimic enzyme action, such as ester hydrolysis, but only with modest efficiency.

[FRIEDRICH CRAMER; WOLFRAM SAENGER; D. GAUSS]

Bibliography: M. L. Bender and M. Komiyama, *Cyclodextrin Chemistry*, 1978; V. M. Bhatnagar, *Clathrate Compounds*, 1968; D. J. Cram and J. M. Cram, Design of complexes between synthetic hosts and organic guests, *Accounts Chem. Res.*, 11:8–14, 1978; F. Cramer, *Einschlussverbindungen*, 1954; F. Cramer and H. Hettler, Inclusion compounds of cyclodextrins, *Naturwissenschaften*, 54:625–632, 1967; D. D. MacNicol, J. J. McKendrick and D. R. Wilson, Clathrates and molecular inclusion phenomena, *Chem. Soc. Rev.*, 7:65–87, 1978; W. Saenger, Cyclodextrin inclusion compounds, *Angew. Chem., Int. Ed.*, 19:344–362, 1980; W. Saenger, Einschlussverbindungen, *Umschau*, 74:635–643, 1974.

Coagulation

A separation or precipitation from a dispersed state of suspensoid particles resulting from their growth. In many cases, a suspension of a colloidal crystalline solid may be precipitated by a prolonged heating, during which the larger particles grow at the expense of the smaller.

Certain colloidal suspensions may be precipitated or coagulated by the addition of an electrolyte to the suspending medium.

Solidification of an entire phase may result from a condensation reaction between a solute and solvent, or from a condensation of solute molecules in which the liquid phase is trapped in its meshes. This process is also described by the term coagulation; setting of a gel and clotting of blood are examples. *See* COLLOID; GEL; ISOELECTRIC POINT.

[FRANCIS J. JOHNSTON]

Bibliography: L. Gordon et al., *Precipitation from Homogeneous Solution*, 1959.

Coal chemicals

For about 100 years, chemicals obtained as by-products in the primary processing of coal to metallurgical coke have been the main source of aromatic compounds used as intermediates in the synthesis of dyes, drugs, antiseptics, and solvents. Although some aromatic hydrocarbons, such as toluene and xylene, are now obtained largely from petroleum refineries, the main source of others, such as benzene, naphthalene, anthracene, and phenanthrene, is still the by-product coke oven. Heterocyclic nitrogen compounds, such as pyridines and quinolines, are also obtained largely from coal tar. Although much phenol is produced by hydrolysis of monochlorobenzene and by decomposition of cumene hydroperoxide, much of the phenol, cresols, and xylenols is still obtained from coal tar.

Coke oven by-products, as a percentage of the coal used, are gas 18.5, light oil 1.0, and tar 3.5. Coke oven gas is a mixture of methane, carbon monoxide, hydrogen, small amounts of higher hydrocarbons, ammonia, and hydrogen sulfide. Most of the ammonia is recovered as about 20 lb of ammonium sulfate per ton of coal, but only about 10% of the hydrogen sulfide is recovered as elemental sulfur. Most of the coke oven gas is used as fuel. The composition of a typical coke oven crude light oil is shown in Table 1. The unidentified fractions contain very small amounts of a large number of hydrocarbons, and organic compounds containing oxygen and nitrogen. The uses of coal tar are outlined in Table 2.

Table 1. Analysis of a typical coke oven crude light oil

Component	% by vol
Forerunnings	
Cyclopentadiene	0.5
Carbon disulfide	0.5
Amylenes, unidentified substances	1.0
Crude benzol	
Benzene	57.0
Thiophene	0.2
Saturated nonaromatic hydrocarbons	3.0
Unsaturates, unidentified substances	3.0
Crude toluol	
Toluene	13.0
Saturated nonaromatic hydrocarbons	0.1
Unsaturates, unidentified substances	1.0
Crude light solvent	
Xylenes	5.0
Ethylbenzene	0.4
Styrene	0.8
Saturated nonaromatic hydrocarbons	0.3
Unsaturates, unidentified substances	1.0
Crude heavy solvent	
Coumarone, indene, dicyclopentadiene	5.0
Polyalkylbenzenes, hydrindene, etc.	4.0
Naphthalene	1.0
Unidentified heavy oils	1.0
Wash oil	
(Used to separate light oil from coke oven gas)	5.0

Table 2. Utilization of coal tar in the United States

Use	%
Fuel	21
Benzene	21
Toluene	4
Xylene	1.1
Solvent naphtha	1.2
Other light oil distillates	0.8
Naphthalene	8
Creosote oil	20
Tar acids	5
Tar bases	0.7
Road tars	17
Pitch coke	0.2

Table 3. Coal tar chemicals

Compound	Fraction of whole tar, %	Use
Naphthalene	10.9	Phthalic acid
Monomethylnaphthalenes	2.5	
Acenaphthenes	1.4	Dye intermediates
Fluorene	1.6	Organic syntheses
Phenanthrene	4.0	Dyes, explosives
Anthracene	1.0	Dye intermediates
Carbazole (and other similar compounds)	2.3	Dye intermediates
Phenol	0.7	Plastics
Cresols and xylenols	1.5	Antiseptics, organic syntheses
Pyridine, picolines, lutidines, quinolines, acridine, and other tar bases	2.3	Drugs, dyes, antioxidants

Although several hundred chemical compounds have been isolated from coal tar, a relatively small number are present in appreciable amounts. These may be grouped as in Table 3.

All the compounds in Table 3 except the monomethylnaphthalenes are of some commercial importance. The amounts recovered and sold, however, are only 5–25% of the totals present in the coal tar.

The direct utilization of coal as a source of bulk organic chemicals has been the objective of much research and development. Oxidation of aqueous alkaline slurries of coal with oxygen under pressure yields a mixture of aromatic carboxylic acids. Because of the presence of nitrogen compounds and hydroxy acids, this mixture is difficult to refine. Hydrogenation of coal at elevated temperatures and pressures yields much larger amounts of tar acids and aromatic hydrocarbons of commercial importance than are obtained by carbonization.

Synthetic fuels may be produced from coal by either direct or indirect methods. In direct liquefaction processes, pulverized coal is suspended in a solvent, the resultant slurry is heated and exposed to gaseous hydrogen under pressure, and the liquids produced are separated from the ash and distilled to obtain the fuel fractions. In the indirect process, the coal is decomposed thermally to yield combustible gases, which are synthesized to yield gasoline, alcohols, and waxes. *See* COKE; ORGANIC CHEMISTRY; PYROLYSIS.

[HOWARD W. WAINWRIGHT]

Bibliography: L. L. Anderson and D. A. Tillman, *Synthetic Fuels from Coal*, 1979; H. H. Lowry (ed.), *Chemistry of Coal Utilization*, 2 vols., 1945, suppl. vol., 1963; U.S. Bureau of Mines, *Minerals Yearbook*, vol. 1, 1976.

Cobalt

A metallic chemical element, Co, with an atomic number of 27 and an atomic weight of 58.93. Cobalt is similar to iron and nickel in both its free and combined states. It is widely distributed in nature, making up about 0.001% of the igneous rocks of the Earth's crust, compared with 0.02% for nickel.

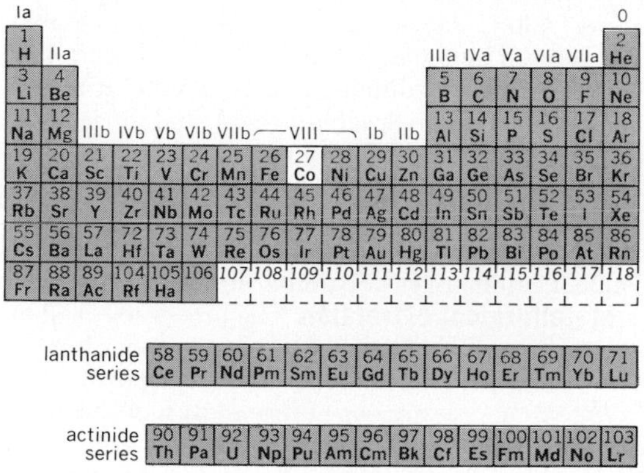

It occurs in meteorites, stars, sea and fresh waters, soils, plants, and animals. Cobalt and its alloys are wear- and corrosion-resistant even at high temperatures. The most important commercial applications are in making alloys for use at high temperatures, magnetic alloys, and alloys for use in machine tools. Cobalt is required in small amounts by plants and animals. The artificially produced radioactive isotope of cobalt, cobalt-60, is widely used in industry, research, and medicine.

Natural occurrence. Traces of cobalt are found in many ores of iron, nickel, copper, silver, manganese, and zinc, but the commercially important cobalt minerals are the arsenides, oxides, and sulfides (Table 1).

Cobalt is mixed with nickel in many of the minerals found near Sudbury, Ontario, Canada, and in

Table 1. Cobalt minerals

Name	Formula	Occurrence
Arsenides		
Smaltite and safflorite	$CoAs_2$	Canada, Morocco, United States
Skutterudite	$CoAs_3$	Canada, Morocco, United States
Cobaltite	CoAsS	Canada, Morocco, United States
Oxidized minerals		
Asbolite	$CoO \cdot 2MnO_2 \cdot 4H_2O$	New Caledonia
Heterogenite	$CoO \cdot 2Co_2O_3 \cdot 6H_2O$	Zaire
Sphaerocobaltite	$CoCO_3$	East Germany, Zaire
Erythrite	$3CoO \cdot As_2O_5 \cdot 8H_2O$	East Germany, Canada, United States
Sulfides		
Carrollite	$CuCo_2S_4$	Zaire, Zambia
Linnaeite	Co_3S_4	United States, Soviet Union

similar copper-nickel sulfide ore bodies. Zaire, where cobalt is associated with copper in the Katanga area, has been the chief producer for many years; other important sources are Zambia, Canada, Morocco, and the Soviet Union. The deposits of nickel-containing laterites found in the Celebes, Cuba, New Caledonia, and many other tropical areas constitute a vast, but as yet virtually untapped, source of cobalt. Another large potential reserve of cobalt, under investigation by several international mining-consortia, exists in the manganese nodules of the ocean floor.

Physical and chemical properties. Cobalt is ferromagnetic and resembles iron and nickel in hardness, tensile strength, machinability, thermal properties, and electrochemical behavior.

At ordinary temperature the stable crystal form of cobalt is close-packed hexagonal, whereas above 400°C face-centered cubic is the stable structure. The metal is unaffected by air or water under normal conditions, and is rapidly attacked by sulfuric acid, hydrochloric acid, and nitric acid, but only slowly by hydrofluoric acid, ammonium hydroxide, and sodium hydroxide. Cobalt exhibits variable valence and forms complex ions and colored compounds, as do all the transition elements. Table 2 summarizes key properties.

Metallurgical extraction. The processes used in extracting cobalt from its ores vary according to the type of ore and location of the ore deposit.

Arsenical ores are concentrated by hand sorting, gravity separation, or froth flotation, and are smelted in a blast furnace with coke and limestone to a speiss (an impure mixture of iron, cobalt, and nickel arsenides). The speiss is ground, roasted with salt, and leached with water. Insoluble chlorides remaining after the leaching process are ground with sulfuric acid, washed, and filtered, and the washings are added to the liquid from the leaching step. The combined solution is oxidized and then neutralized with lime.

Basic ferric arsenate precipitates and is removed, leaving a solution containing cobalt and nickel. The addition of successive portions of sodium hydroxide and sodium hypochlorite precipitates cobalt as the hydroxide, initially pure, and finally admixed with nickel hydroxide. The cobalt precipitate is dried, ground, and formed into pellets, which are reduced by heating with charcoal to cobalt metal.

In the Zambia-Zaire copper belt, cobalt production is an important ancillary operation. Various cobaltiferous materials, such as high-grade ores, concentrates, and slags, are smelted in electric furnaces with coke and a suitable flux to yield a crude alloy of copper, cobalt, and iron containing about 40% cobalt. This alloy is sent to overseas refineries. Treatment with hot 20% sulfuric acid dissolves cobalt and iron. Iron is removed from the solution by treatment with sodium chlorate and carbonate. Cobalt is precipitated as the hydroxide by treatment with sodium hypochlorite and purified in the manner described for arsenical ores.

Table 2. Properties of cobalt

Atomic weight	58.93
Atomic number	27
Melting point	1493°C
Specific gravity	8.9 at 20°C
Electronic configuration	$1s^2\,2s^2\,2p^6\,3s^2\,3p^6\,3d^7\,4s^2$
Oxidation states	2+, 3+

Most of the production from central Africa is in the form of electrolytic cobalt. In Zaire, the residual slimes from the copper-leaching process are dissolved, most of the copper is removed from the solution electrolytically, and the last traces by cementation on cobalt granules. Cobalt in this solution is precipitated as the hydroxide by treatment with lime, removed by filtration, dissolved in sulfuric acid, and electrolytically deposited on steel cathodes, using lead anodes. In Zambia cobalt concentrate is obtained by ore flotation after prior removal of most of the copper minerals in a high-grade copper concentrate. It then is roasted to transform cobalt sulfide to cobalt sulfate and at the same time to leave most of the copper and iron in a water-insoluble state. The roasted concentrate is leached with hot water, and the pH of the solution is adjusted with lime to precipitate all the iron and most of the copper. Residual copper is cemented out on cobalt granules. Cobalt in the solution is then precipitated with lime, dissolved in sulfuric acid, and separated electrolytically.

Recovery of cobalt as a by-product of the nickel industry may take several forms: (1) In electrolytic refining of nickel, cobalt is removed from the anolyte, the solution in the anode compartment of the electrolysis cell, by chlorine and basic nickel carbonate, in the form of an impure slime containing nickel, copper, and iron, as well as cobalt. This slime is dissolved in sulfuric acid and sulfur dioxide, and more fresh slime is added to reoxidize the iron and to precipitate it as ferric hydroxide. Copper is removed by cementation on nickel powder, and cobalt is separated from nickel with sodium hypochlorite. The crude cobaltic hydroxide is redissolved in sulfuric acid and sulfur dioxide, and the hypochlorite separation repeated to reduce the nickel contamination. (2) Where the carbonyl process is employed to refine nickel, cobalt is not volatilized with the nickel at atmospheric pressure but remains in the residue. The residue is roasted and leached, and the copper and iron are removed by precipitation as hydroxide from slightly acid solution. Cobalt is precipitated from the solution as crude cobaltic hydroxide, which is purified to high-grade oxide. (3) In the recovery of nickel by leaching reduced lateritic ores or nickeliferous pyrrhotite with ammonium carbonate, any cobalt present is largely precipitated with the final basic nickel carbonate. (4) Cobalt is also obtained in the pressure leaching of nickel sulfide concentrates in acid or ammonia solution. Metallic nickel and cobalt are precipitated from the solution with hydrogen under elevated pressure and temperature.

Cobalt is also recovered in several industrial countries from iron pyrites, used as a source of sulfuric acid and iron. After roasting for sulfur recovery, the calcine is given a chloridizing roast, leached with water, and the cobalt precipitated with soda ash and chlorine.

Uses. About 80% of the world's cobalt output is used in the metallic state. Alloys that retain their strength and other desirable properties at high temperatures are widely used in jet aircraft, gas turbines, and other equipment that operates at high temperatures. Most of these alloys contain

20–65% cobalt, together with nickel, chromium, molybdenum, tungsten, and other elements. Large quantities of cobalt are used in the production of magnets. The best commercial magnet steel contains 35% cobalt, together with some tungsten and chromium. The Alnico magnet alloys usually contain 6–12% aluminum, 14–30% nickel, 5–35% cobalt, and the balance of iron. In the early 1970s powerful permanent magnets, made from alloys of cobalt and rare earths, became important.

Cobalt is also an important constituent of other permanent magnet alloys, such as Vicalloys, Cunicos, and Remalloy or Comol, and of soft magnet alloys like the Perminvar and Permindur types. It is, of course, found in magnets made from iron-cobalt powder and from cobalt ferrites.

Iron-base and cobalt-base alloys, containing about 6–65% cobalt, together with chromium, tungsten, and other alloying elements, are very hard and resistant to abrasion and corrosion. They are used extensively for cutting tools and hard facing.

Cobalt is employed as the matrix, or binder, for tungsten carbide. Amounts of 3–25% provide the toughness and shock resistance required to make the hard carbide of practical value in drill bits and machine tools. Addition of 5–12% cobalt to standard high-speed steel improves the cutting efficiency at elevated temperatures, and such tools are used in heavy machining.

The artificially produced radioisotope cobalt-60 may be used in place of x-rays or radium in the inspection of materials to reveal internal structure, flaws, or foreign objects. It is usually prepared by neutron irradiation of metallic cobalt. It is employed in cancer therapy and as a radioactive tracer in biology and industry.

Glass-to-metal seals can be made with alloys approximating 18% cobalt, 28% nickel, and 54% iron. An alloy of 36.5% iron, 9.5% chromium, and 54% cobalt has practically zero coefficient of expansion. A dental and surgical alloy, Vitallium, containing essentially 65% cobalt, 30% chromium, and 5% molybdenum or tungsten, is not attacked by body liquids and does not irritate tissues.

Cobalt is also a component in some electrical resistance alloys, beryllium copper and other spring alloys, cathode filaments and cores, and constant-modulus alloys.

Major compounds and their uses. Cobalt(II) chloride, nitrate, and sulfate are all formed by the interaction of the metal, oxide, hydroxide, or carbonate with the corresponding acid. There are three main oxides of cobalt: gray cobaltous oxide, CoO; black cobaltic oxide, Co_2O_3, formed by heating compounds at a low temperature in excess of air; cobaltosic oxide, Co_3O_4, the stable oxide, formed when salts are heated in air at temperatures which do not exceed 850°C. Most common cobalt salts are derivatives of cobalt(II), the higher valence state being encountered only in complex coordination compounds. There is a large number of coordination compounds of cobalt(III), such as the cobalt ammines, $[Co(NH_3)_6]X_3$. An important, naturally occurring cobalt coordination compound is vitamin B_{12}, the antipernicious-anemia factor. Many cobalt salts in solution exhibit remarkable color changes, depending on temperature, concentration, and presence of other ions. These have been ascribed to changes in hydration or solvation, or to the presence of complex ions.

In the glass and ceramic industries, small quantities of cobalt oxide are used to neutralize the yellow tint resulting from the presence of iron in glass, pottery, and enamels. Larger quantities are used to impart a blue color to these products. Cobalt oxide is used in enamel coatings on steel to improve the adherence of the enamel to the metal.

Cobalt linoleates, naphthenates, resinates, and ethylhexoates are excellent driers for paints, varnishes, and inks. Cobalt catalysts have been patented for many industrial reactions, such as Fischer-Tropsch synthesis and the oxo process.

Cobalt sulfate is sometimes added to nickel-plating baths to improve smoothness, brightness, hardness, and ductility of deposits.

Cobalt compounds, such as the chloride, are added in very small amounts to livestock feeds, salt licks, and fertilizers in many parts of the world where a cobalt deficiency exists in the soil and natural vegetation. This prevents serious diseases of cattle and sheep, such as pining, which result from cobalt deficiencies.

Analytical methods. Small quantities of cobalt are readily determined colorimetrically by using various complexing agents or ammonium thiocyanate, or the element may be analyzed spectrographically or polarographically. Larger amounts of cobalt are best determined by electrodeposition analysis. For intermediate quantities, satisfactory procedures are the gravimetric method, employing a separation with nitrosonaphthol and ignition to Co_3O_4, and the potentiometric titration method, wherein cobalt is oxidized in ammoniacal solution by potassium ferricyanide. In many materials, cobalt in the intermediate range can be quickly determined by atomic absorption spectroscopy or x-ray fluorescence. *See* TRANSITION ELEMENTS.

[ROLAND S. YOUNG]

Cobalt in most naturally occurring foodstuffs is largely unabsorbed and eliminated in the feces. Under natural circumstances cobalt does not produce a toxic syndrome in humans. It takes an unusual set of circumstances to produce cobalt toxicity. Intravenously injected, large doses of cobalt have been observed to cause paralysis and enteritis, and sometimes death. When cobalt is injected into the bloodstream, the amount distributed throughout the tissues is very small; higher concentrations are present in the pancreas, liver, spleen, kidney, and bone. Elimination of cobalt is rapid.

Researchers in England have shown that cobalt, like nickel and cadmium, produces a high incidence of cancer, namely rhabdomyosarcoma, when it is injected in powdered form into rats. A number of other metals, such as iron, copper, zinc, manganese, beryllium, and tungsten, are not carcinogenic under the same conditions. Cobalt has an inhibitory effect on cell oxidative metabolism and is toxic to connective-tissue cells grown in culture.

[N. KARLE MOTTET]

Bibliography: D. Nicholls, *The Chemistry of Iron, Cobalt and Nickel*, 1975; I. V. Pyatnitskii, *Analytic Chemistry of Cobalt*, 1972.

Cocaine

The principal alkaloid of coca leaves, a topical anesthetic and stimulant, and popular illicit drug. The molecular formula is given here. Cocaine

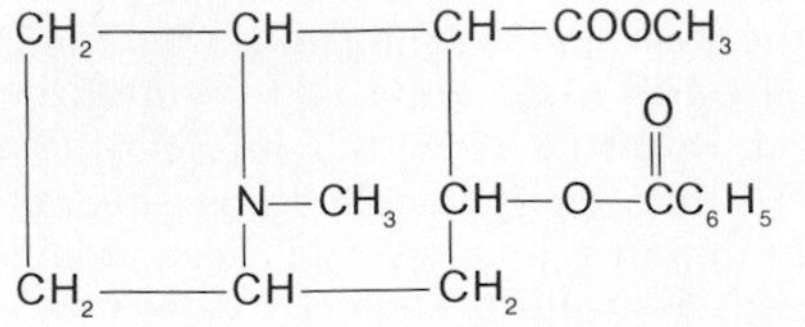

was first isolated by A. Niemann in 1860. In 1884 C. Koller demonstrated its efficacy as an anesthetic in eye surgery, introducing the age of local anesthesia. For the next decade cocaine enjoyed the status of a wonder drug and panacea. It fell into disfavor with increasing reports of acute toxicity and long-term dependence. Today it is used as a topical anesthetic in the eye, nose, mouth, and throat; for injection anesthesia it has been replaced by synthetic drugs with fewer central nervous system effects. *See* COCA.

Cocaine increases heart rate and blood pressure and causes feelings of alertness and euphoria. In animals it can be shown to be strongly reinforcing. It does not produce physical dependence, as alcohol and opiates do, but many people find it hard to use in a stable and moderate fashion if they have access to it in quantity. Although it is quite active orally, most users of illicit cocaine take it intranasally by snuffing; few inject it intravenously. Aside from local irritation of the nasal membranes, moderate users suffer few adverse effects. The soluble hydrochloride salt is the common form. Insoluble cocaine free base may be smoked, a practice that may be more harmful.

Both licit and illicit cocaine are obtained by extraction from coca leaves. Black market production is centered in Bolivia and Peru. Crude cocaine is refined in Colombia and exported from there to North America and Europe. *See* ALKALOID.

[ANDREW T. WEIL]

Bibliography: L. Grinspoon and J. Bakalar, *Cocaine: A Drug and Its Social Evolution*, 1976.

Coke

A coherent, cellular, carbonaceous residue remaining from the dry (destructive) distillation of a coking coal. It contains carbon as its principal constituent, together with mineral matter and residual volatile matter. The residue obtained from the carbonization of a noncoking coal, such as subbituminous coal, lignite, or anthracite, is normally called a char. Coke is produced chiefly in chemical-recovery coke ovens (see figure), but a small amount is also produced in beehive or other types of nonrecovery ovens.

Uses and types. Coke is used predominantly as a fuel reductant in the blast furnace, in which it also serves to support the burden. As the fuel, it supplies the heat as well as the gases required for the reduction of the iron ore. It also finds use in other reduction processes, the foundry cupola, and househeating. About 91% of the coke made is used in the blast furnace, 4% in the foundry, 1% for water gas, 1% for househeating, and 3% for other industries, such as calcium carbide, nonferrous metals, and phosphates. Approximately 1300 lb of coke is consumed per ton of pig iron produced in the modern blast furnace, and about 200 lb of coke is required to melt a ton of pig iron in the cupola.

Coke is classified not only by the oven in which it is made, chemical-recovery or beehive, but also by the temperature at which it is made. High-temperature coke, used mainly for metallurgical purposes, is produced at temperatures of 900–1150°C. Medium-temperature coke is produced at 750–900°C, and low-temperature coke or char is made at 500–750°C. The latter cokes are used chiefly for househeating, particularly in England. The production of these is rather small as compared to high-temperature coke and usually re-

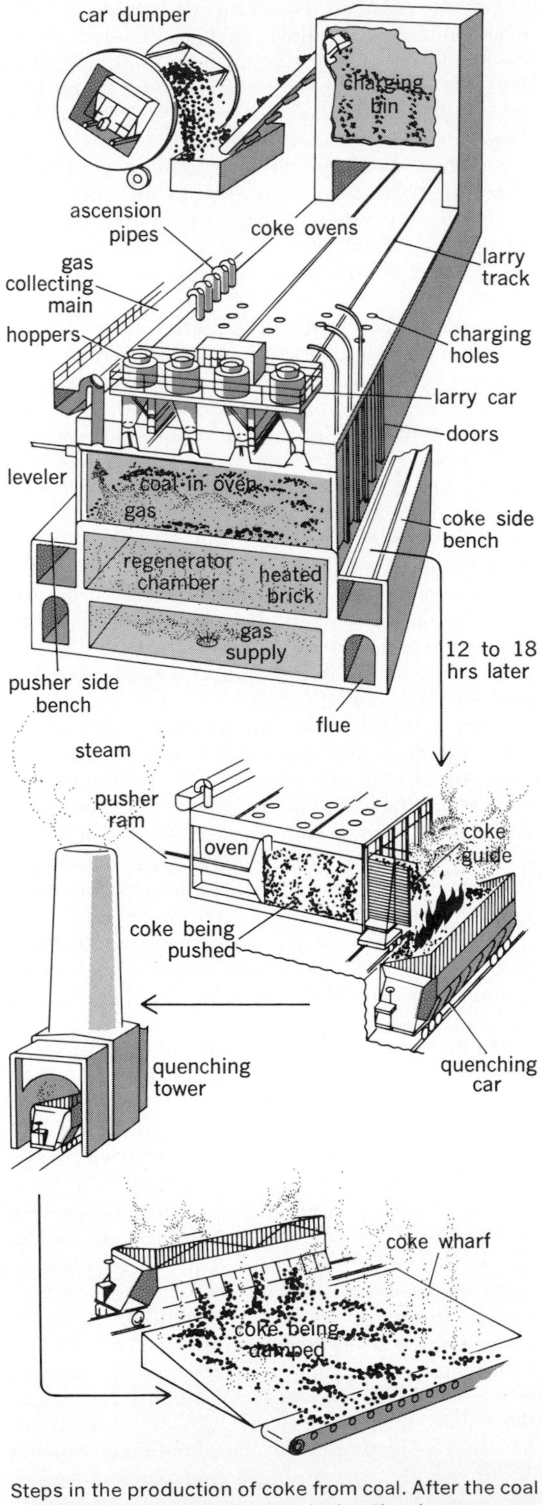

Steps in the production of coke from coal. After the coal has been in the oven for 12–18 hr, the doors are removed, and a ram mounted on the same machine that operates the leveling bar shoves the coke into a quenching car for cooling. (*American Iron and Steel Institute*)

quires special equipment. Coke is also classified according to its intended use, such as blast-furnace coke, foundry coke, water-gas coke, and domestic coke.

Production. Coke is formed when coal is heated in the absence of air. During the heating in the range of 350–500°C, the coal softens and then fuses into a solid mass. The coal is partially devolatilized in this temperature range, and further heating at temperatures up to 1000–1100°C reduces the volatile matter to less than 1% The degree of softening attained during heating determines to a large extent the character of the coke produced.

In order to produce coke having desired properties, two or more coals are blended before charging into the coke oven. Although there are some exceptions, high-volatile coals of 32–38% volatile-matter content are generally blended with low-volatile coals of 15–20% volatile matter in blends containing 20–40% low-volatile coal. In some cases, charges consist of blends of high-, medium-, and low-volatile coal. In this way, the desirable properties of each of the coals, whether it be in impurity content or in its contribution to the character of the coke, are utilized. The low-volatile coal is usually added in order to improve the physical properties of the coke, especially its strength and yield. In localities where low-volatile coals are lacking, the high-volatile coal or blends of high-volatile coals are used without the benefit of the low-volatile coal. In some cases, particularly in the manufacture of foundry coke, a small percentage of so-called inert, such as fine anthracite, coke fines, or petroleum coke, is added.

In addition to the types of coals blended, the carbonizing conditions in the coke oven influence the characteristics of the coke produced. Oven temperature is the most important of these and has a significant effect on the size and the strength of the coke. In general, for a given coal, the size and shatter strength of the coke increase with decrease in carbonization temperature. This principle is utilized in the manufacture of foundry coke, where large-sized coke of high shatter strength is required.

Properties. The important properties of coke that are of concern in metallurgical operations are its chemical composition, such as moisture, volatile-matter, ash, and sulfur contents, and its physical character, such as size, strength, and density. For the blast furnace and the foundry cupola, coke of low moisture, volatile-matter, ash, and sulfur content is desired. The moisture and the volatile-matter contents are a function of manner of oven operation and quenching, whereas ash and sulfur contents depend upon the composition of the coal charged. Blast-furnace coke used in the United States normally contains less than 1% volatile matter, 85–90% fixed carbon, 7–12% ash, and 0.5–1.5% sulfur. If the coke is intended for use in the production of Bessemer or acid open-hearth iron, the phosphorus content becomes important and should contain less than 0.01%.

The requirements for foundry coke in analysis are somewhat more exacting than for the blast furnace. The coke should have more than 92% fixed carbon, less than 8% ash content, and less than 0.60% sulfur.

Blast-furnace coke should be uniform in size, about 2½–5 in., but in order to utilize more of the coke produced in the plant, the practice has been, in some cases, to charge separately into the furnace the smaller sizes after they were closely screened into sizes such as 2½ × 1¾ in., 1¾ × 1 in., down to about ¾ in. in size. There is now a trend, when pelletized ore is used, to crush and screen the coke to a more closely sized material, such as 2½ × ¾ in.

The blast-furnace coke should also be uniformly strong so that it will support the column of layers of iron ore, coke, and stone above it in the furnace without degradation. Standard test methods, such as the tumbler test and the shatter test, have been developed by the American Society for Testing and Materials for measuring strength of coke. Good blast-furnace cokes have tumbler test stability factors in the range of 45 to 65 and shatter indices (2-in. sieve) in the range of 70 to 80%. These are not absolute requirements since good blast-furnace performance is obtained in some plants with weaker coke—but with the ore prepared so as to overcome the weakness.

Foundrymen prefer coke that is large and strong. The coke for this purpose ranges in size from 3 to 10 in. and larger. The size used usually depends upon the size of the cupola and, in general, the maximum size of the coke is about one-twelfth of the diameter of the cupola. High shatter strength is demanded by the cupola operator; shatter indices of at least 97% on 2-in. and 80% on 3-in. sieves have been specified.

Other important requirements of foundry coke are high reactivity toward oxygen so that carbon dioxide is produced with high heat evolution, and low reactivity toward carbon dioxide so that the amount of carbon monoxide formed is minimized.

The requirements for coke intended for house-heating vary in different localities. Usually the coke is of a narrow size range, and the size used depends upon the size and type of appliance. In general, a high ash-softening temperature is desirable (more than 2500°F).

Coke intended for water-gas generators should be over 2 in. in size, and should have an ash content below 10% and a moderately high ash-softening temperature. *See* COAL CHEMICALS; PYROLYSIS. [MICHAEL PERCH]

Bibliography: H. H. Lowry (ed.), *Chemistry of Coal Utilization*, suppl. vol., 1963; P. J. Wilson and J. H. Wells, *Coal, Coke and Coal Chemicals*, 1950.

Colloid

A system of which one phase is made up of particles having dimensions of 10–10,000 A and is dispersed in a different phase. Some examples of colloidal systems are shown in the table.

The unique properties of colloids are due primarily to the large surface areas of the dispersed phase. One of the results of this large surface area is the adsorption of ions and other materials. The adsorbed ions impart an electric charge to the colloidal particles so that they repel each other. Lyophobic colloids exhibit little affinity between the dispersed phase and the solvent; lyophilic colloids are strongly associated with the solvent. *See* ADSORPTION.

Many colloids occur naturally. Others can be prepared by breaking large pieces of material into

Types of colloidal system

Dispersed phase	Dispersing phase	Example
Solid	Solid	Some alloys
Solid	Liquid	Suspensions (muddy water)
Solid	Gas	Smoke, airborne dust
Liquid	Solid	Butter, gels
Liquid	Liquid	Emulsions (milk, mayonnaise)
Liquid	Gas	Fog, aerosols
Gas	Solid	Solid foams (marshmallow)
Gas	Liquid	Foams

smaller pieces, as with a colloid mill, which consists of two accurately machined disks rotating in opposite directions at very high velocities while nearly touching each other. Some colloids can be prepared by condensation methods, in which atoms or molecules are induced to aggregate from solution, or in which atoms or molecules are condensed from the vapor state.

There are three important types of colloids: small solid particles having the same internal structure as the bulk solid phase (sulfur suspensions in water), aggregates formed from smaller molecules (soaps in water), and large molecules of colloidal size (proteins and high polymers).

Optical properties. Colloidal particles are so small that they cannot be seen directly in visible light even with an optical microscope. They can be detected by viewing them at an angle to the incident light path. The electron microscope furnishes a means by which the sizes and shapes of colloidal-size particles can be determined. *See* TYNDALL EFFECT.

Separation. Because colloids are so small in size, they cannot be separated by filtration through paper or porous porcelain crucibles. They can be separated from solvent and dissolved substances by dialysis and ultrafiltration. Airborne colloidal particles can be separated by special techniques. *See* DIALYSIS; ULTRAFILTRATION.

Apparent molecular weights. It is possible to determine the weight-average molecular weight of colloidal particles in solution by light scattering and viscosity measurements. The number-average molecular weight can be determined by osmotic pressure measurements.

Other more specialized aspects of colloidal properties and surface phenomena are discussed in other articles. *See* DONNAN EQUILIBRIUM; ELECTROPHORESIS; EMULSION; GEL; MICELLE.

[QUENTIN VAN WINKLE]

Bibliography: K. J. Mysels, *Introduction to Colloid Chemistry*, reprint 1979; W. J. Popeil, *Introduction to Colloid Science*, 1978.

Combining volumes, law of

The principle that when gases take part in chemical reactions the volumes of the reacting gases and those of the products, if gaseous, are in the ratio of small whole numbers, provided that all measurements are made at the same temperature and pressure. The law is illustrated by the following reactions:

1. One volume of chlorine and one volume of hydrogen combines to give two volumes of hydrogen chloride.
2. Two volumes of hydrogen and one volume of oxygen combine to give two volumes of steam.
3. One volume of ammonia and one volume of hydrogen chloride combine to give solid ammonium chloride.
4. One volume of oxygen when heated with solid carbon gives one volume of carbon dioxide.

It should be noted that the law applies to all reactions in which gases take part, even though solids or liquids are also reactants or products.

The law of combining volumes was put forward on the basis of experimental evidence, and was first explained by Avogadro's hypothesis that equal volumes of all gases and vapors under the same conditions of temperature and pressure contain identical numbers of molecules. *See* AVOGADRO NUMBER.

The law of combining volumes is similar to the other gas laws in that it is strictly true only for an ideal gas, though most gases obey it closely at room temperatures and atmospheric pressure. Under high pressures used in many large-scale industrial operations, such as the manufacture of ammonia from hydrogen and nitrogen, the law ceases to be even approximately true. *See* GAS.

[THOMAS C. WADDINGTON]

Combustion

The burning of any substance, whether it be gaseous, liquid, or solid. In combustion, a fuel is oxidized, evolving heat and often light. The oxidizer need not be oxygen per se. The oxygen may be a part of a chemical compound, such as nitric acid (HNO_3) or ammonium perchlorate (NH_4ClO_4), and become available to burn the fuel during a complex series of chemical steps. The oxidizer may even be a non-oxygen-containing material. Fluorine is such a substance. It combines with the fuel hydrogen, liberating light and heat. In the strictest sense, a single chemical substance can undergo combustion by decomposition, with emission of heat and light. Acetylene, ozone, and hydrogen peroxide are examples. The products of their decomposition are carbon and hydrogen for acetylene, oxygen for ozone, and water and oxygen for hydrogen peroxide.

Solids and liquids. The combustion of solids such as coal and wood occurs in stages. First, volatile matter is driven out of the solid by thermal decomposition of the fuel and burns in the air. At usual combustion temperatures, the burning of the hot, solid residue is controlled by the rate at which oxygen of the air diffuses to its surface. If the residue is cooled by radiation of heat, combustion ceases.

The first product of combustion at the surface of char, or coke, is carbon monoxide. This gas burns to carbon dioxide in the air surrounding the solid, unless it is chilled by some surface. Carbon monoxide is a poison and it is particularly dangerous because it is odorless. Its release from poorly designed, or malfunctioning, open heaters constitutes a serious hazard to human health.

Liquid fuels do not burn as liquids but as vapors above the liquid surface. The heat evolved evaporates more liquid, and the vapor combines with the oxygen of the air.

Spontaneous combustion. This occurs when certain materials are stored in bulk. The oxidizing action of microorganisms often produces the initial heat.

As the temperature increases, the air trapped in the material takes over the oxidation process, liberating more heat. Because the heat cannot be dissipated to the surroundings, the temperature of the material rises still more and the rate of oxidation increases. Eventually the material reaches an ignition point and bursts into flame. Coal is subject to spontaneous combustion and is generally stored in shallow piles to allow the heat of oxidation to dissipate.

Gases. At ordinary temperatures, molecular collisions do not usually cause combustion. At elevated temperatures the collisions of the thermally agitated molecules are more frequent. More important as a cause of chemical reaction is the greater energy involved in the collisions. Moreover, it has been reasonably well established that there is very little combustion attributable to direct reaction between the molecules. Instead, a high-energy collision dissociates a molecule into atoms, or free radicals. These molecular fragments react with greater ease, and the combustion process proceeds generally by a chain reaction involving these fragments. An illustration will make this clear. The combustion of hydrogen and oxygen to form water does not occur in a single step, reaction (1). In this

$$2H + O_2 \rightarrow 2H_2O \qquad (1)$$

seemingly simple case, some fourteen reactions have been identified. A hydrogen atom is first formed by collision; it then reacts with oxygen molecules, reaction (2), forming an OH radical. The lat-

$$H + O_2 \rightarrow OH + O \qquad (2)$$

ter in turn reacts with a hydrogen molecule, reaction (3), forming water and regenerating the H atom

$$OH + H_2 \rightarrow H_2O + H \qquad (3)$$

which repeats the process. This sequence of reactions constitutes a chain reaction. Sometimes the O atom reacts with a hydrogen molecule to form an OH radical and another H atom, reaction (4).

$$O + H_2 \rightarrow OH + H \qquad (4)$$

Thus a single H atom can form a new H atom in addition to regenerating itself. This process constitutes a branched-chain reaction. Atoms and radicals recombine with each other to form a neutral molecule, either in the gas space or at a surface after being adsorbed. Thus, chain reactions may be suppressed by proximity of surfaces; and the number and length of the chains may be controlled by regulating the temperature, the composition of the mixtures, and other conditions.

Under certain conditions, where the rate of chain branching equals or exceeds the rate at which chains are terminated, the combustion process speeds up to explosive proportions; because of the rapidity of molecular events, a large number of chains are formed in a short time so that essentially all of the gas undergoes reaction at the same time; that is, an explosion results. The branched-chain type of explosion is similar in principle to atomic explosions of the fission type, where more than one neutron is generated by the reaction between a neutron and a uranium nucleus. Another cause of explosion in gaseous combustion arises when the rate at which heat is liberated in the reaction is greater than the rate at which the heat dissipates to the surroundings. The temperature increases, accelerates the reaction rate, liberates more heat, and so on, until the entire gas mixture reacts in a very short time. This type of explosion is known as a thermal explosion. There are cases intermediate between branched-chain and thermal explosions which depend upon the type and proportion of gases mixed, the temperature, and the density.

In slow combustion, intermediate products can be isolated. Aldehydes, acids, and peroxides are formed in the slow combustion of hydrocarbons, and hydrogen peroxide in the slow combustion of hydrogen and oxygen. At the relatively low temperature of combustion of paraffin hydrocarbons (propane, butane, ethers) a bluish glow is seen. This light from activated formaldehyde formed in the process is called a cool flame.

In the gaseous combustion and explosive reactions described above, the processes proceed simultaneously throughout the vessel. The gas mixture in a vessel may also be consumed by a combustion wave which, when initiated locally by a spark or a small flame, travels as a narrow intense reaction zone through the explosive mixture. The gasoline engine operates on this principle. Such combustion waves travel with moderate velocity, ranging from 1 ft/sec (0.3 m/sec) in hydrocarbons and air to 20–30 ft/sec (6–9 m/sec) in hydrogen and air. The introduction of turbulence or agitation accelerates the combustion wave. The accelerating wave sends out compression or shock waves which are reflected back and forth in the vessel. Under certain conditions these waves coalesce and change from a slow combustion wave to a high-velocity detonation wave. In hydrogen and oxygen mixtures, the speed is almost 2 mi/sec (3.2 km/sec). The pressure created by detonation can be very high and dangerous.

Combustion mixtures can be made to react at lower temperatures by employing a catalyst. The molecules are adsorbed on the catalyst, where they may be dissociated into atoms or radicals, and thus brought to reaction condition. An example is the catalytic combination of hydrogen and oxygen at ordinary temperatures on the surface of platinum. The platinum glows as a result of the heat liberated in the surface combustion. *See* CHAIN REACTION; CHEMICAL DYNAMICS; FREE RADICAL.

[BERNARD LEWIS]

Spectroscopy. The spectroscopy of combustion is an experimental technique for obtaining data from flames without interferring with the combustion process. The spectrum of light emitted or absorbed in a flame is a physical property of the materials present in the flame. *See* SPECTROSCOPY.

Flame spectra are used in interpreting combustion mechanisms and in determining flame temperatures. Various spectrographic techniques are used depending on the type of measurement desired. The method of line reversal, in which the radiation intensity from thermally excited metal atoms is compared to a blackbody lamp filament of controllable brightness, is one technique that gives a measure of temperature; it can be handled by simple equipment with only filters to isolate radiation. An example is the addition of small amounts of sodium salts to gaseous or liquid fuels—a technique known as sodium D-line reversal.

Band spectra from reactive combustion intermediates are usually studied with a spectroscope of either the prism or grating type; for extremely fine resolution, an interferometer is sometimes used. The dispersed radiation is detected either photographically or by photomultiplier tubes. Photography is of value in recording spectra over a range of wavelengths simultaneously. If only specific lines are of interest, two or more photomultiplier tubes can be used to record relative line brightnesses. A continuous scan of a spectrum can be achieved by moving one phototube and slit along the focus of the spectral radiation and displaying the output on an oscilloscope.

The band spectra from flames has been associated with the quantized energy changes in molecules due to rotation and vibration. Each spectral line in a band spectrum represents a discrete energy level. Under equilibrium temperature conditions, there is ideally a Maxwell-Boltzman distribution of molecules in different energy states. According to the kinetic theory of gases, this is a dynamic equilibrium with molecules having their energy distributed in a specific way over these energy states. The radiation of light is a measure of the number of molecules in the process of changing energy levels. From these measurements, on band spectra, flame temperatures and reaction intermediates can be determined.

[RODERICK S. SPINDT]

Bibliography: N. A. Chigier (ed.), *Progress in Energy and Combustion*, vol. 1, 1976, vol. 2, 1978; R. M. Friston and A. A. Wistenberg, *Flame Structure*, 1965; A. G. Gaydon and H. G. Wolfhard, *Flames: Their Structure, Radiation, and Temperature*, 1979; B. Lewis and G. von Elbe, *Combustion, Flames and Explosions of Gases*, 2d ed., 1961; G. J. Minkoff and C. F. H. Tipper, *Chemistry of Combustion Reaction*, 1962; H. B. Palmer and J. M. Beer (eds.), *Combustion Technology: Some Modern Developments*, 1974; M. L. Parsons et al., *Handbook of Flame Spectroscopy*, 1975; M. L. Parsons and P. M. McElfresh, *Flame Spectroscopy: Atlas of Spectral Lines*, 1971; F. J. Weinberg, *Optics of Flames*, 1963; A. Williams, *The Combustion of Sprays of Liquid Fuels*, 1975; F. A. Williams, *Combustion Theory*, 1965.

Complex compounds

A group of chemical compounds in which a part of the molecular bonding is of the coordinate covalent type.

This article summarizes the different types of compounds that are known and discusses their nomenclature, structure, stereochemistry, and synthesis. For a discussion of the nature of the coordinate bond, and the stability and reactivity of complex compounds *see* CHELATION; COORDINATION CHEMISTRY.

Complex compounds contain a central atom or ion and a group of ions or molecules surrounding it. Many simple hydrates, such as $MgCl_2 \cdot 6H_2O$, are best formulated as $[Mg(H_2O)_6]Cl_2$ because it is known that the 6 molecules of water surround the central magnesium ion. Therefore, $[Mg(H_2O)_6]^{2+}$ is a complex ion, and $[Mg(H_2O)_6]Cl_2$ is a complex compound. The charge on this complex ion is +2, because this is the charge on the magnesium ion and the coordinated water molecules are neutral. However, if the coordinated groups are charged, then the charge on the complex is represented by the sum of the charge on the metal and that of the coordinated ions. See, for example, the progression of charges on the platinum (IV) complexes listed below. Thus the charge is +4 for $[Pt(NH_3)_6]^{4+}$ because Pt is +4 and NH_3 is neutral. But the charge is −2 for $[PtCl_6]^{2-}$ because of the 6 Cl^-, that is, $+4-6=-2$.

COMPLEX COMPOUNDS

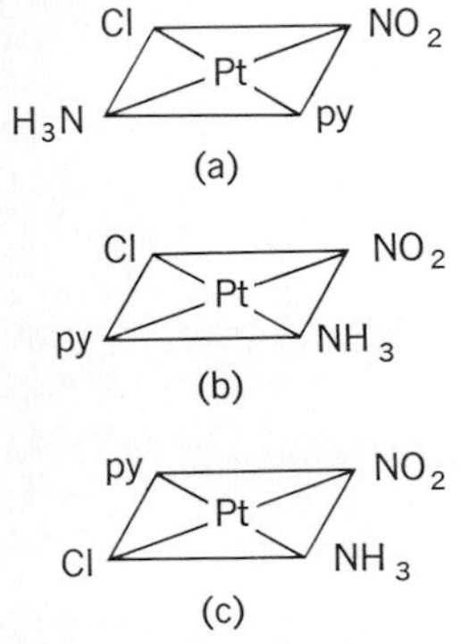

Fig. 3. Isomeric complexes. (*a*) *trans*-Chloropyridine-*trans*-nitroammineplatinum(II). (*b*) *trans*-Chloroammine-*trans*-nitropyridineplatinum(II). (*c*) *trans*-Chloronitro-*trans*-amminepyridineplatinum(II).

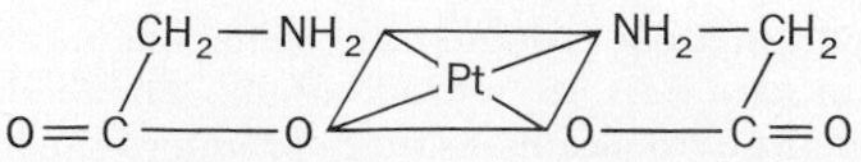

Fig. 1. Structure of *cis*-diglycinatoplatinum(II).

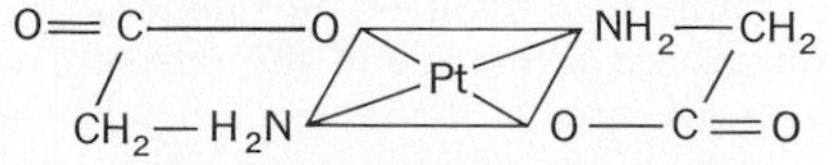

Fig. 2. Structure of *trans*-diglycinatoplatinum(II).

Nomenclature. The metal complex may be a cation, have zero charge, or be an anion, as is exemplified by the following series of complexes:

$[Pt(NH_3)_6]Cl_4$	Hexaammineplatinum(IV) chloride
$[Pt(NH_3)_5Cl]Cl_3$	Chloropentaammine-platinum(IV) chloride
$[Pt(NH_3)_4Cl_2]Cl_2$	Dichlorotetraammine-platinum(IV) chloride
$[Pt(NH_3)_3Cl_3]Cl$	Trichlorotriammine-platinum(IV) chloride
$[Pt(NH_3)_2Cl_4]$	Tetrachlorodiammine-platinum(IV)
$K[Pt(NH_3)Cl_5]$	Potassium pentachloro-ammineplatinate(IV)
$K_2[PtCl_6]$	Potassium hexachloro-platinate(IV)

These compounds are named according to rules set up by the Nomenclature Committee of the In-

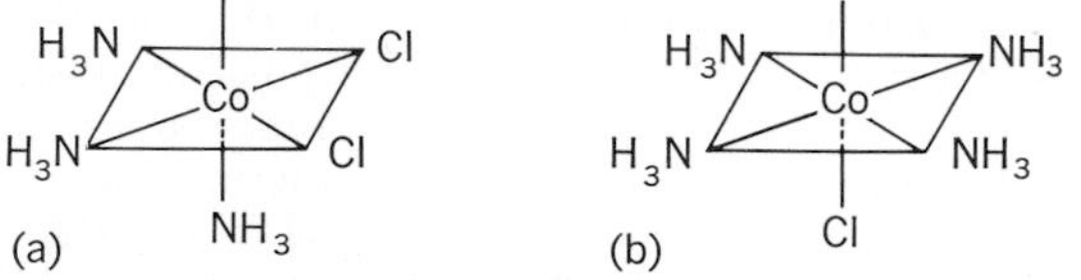

Fig. 4. Structure of (*a*) *cis*-dichlorotetraamminecobalt(III) ion and (*b*) *trans*-dichlorotetraamminecobalt(III) ion.

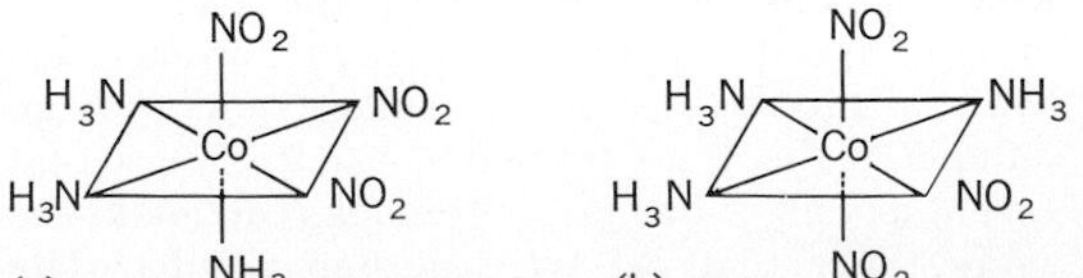

Fig. 5. Structure of (*a*) *cis*-(or facial-)trinitrotriamminecobalt(III) and (*b*) *trans*-(or meridial-)trinitrotriamminecobalt(III).

Table 1. Metal complexes with coordination numbers (C.N.) other than 4 or 6

C.N.	Complex	Structure
2	$[Ag(CN)_2]^-$	Linear
	$[Ag(NH_3)_2]^+$	
3	$[Cu(PR_3)_2I]$	Planar
	$[Au(AsR_3)_2I]$	
5	$[Fe(CO)_5]$	Trigonal bipyramid
	$[Zn(2,2',2''\text{-terpyridine})Cl_2]$	
	$[Ni(PR_3)_2Br_3]$	Tetragonal pyramid
	{Pd[*o*-phenylene-bis-(dimethylarsine)]$_2$I}I	
7	ZrF_7^{3-}	Pentagonal bipyramid
	NbF_7^{2-}	Face-centered trigonal prism
	TaF_7^{2-}	
8	$Mo(CN)_8^{4-}$	Dodecahedral
	TaF_8^{3-}	Archimedean antiprism
9	$Nd(H_2O)_9^{3+}$	Face-centered trigonal prism
10	$Cd_2[Mo(CN)_8(H_2O)_2]$	Unknown structure

ternational Union of Pure and Applied Chemistry. Some of the rules are: (1) Name the cation first as one word followed by the anion as one word. (2) For a cationic complex, name the neutral ligands first, then the negative ligands with an ending of -o, and finally the metal followed by a Roman numeral in parentheses to designate its oxidation state. Neutral ligands are named as the molecule, except that H_2O is aquo and NH_3 is ammine. (3) The prefixes such as di, tri, and tetra are used before simple expressions such as chloro, aquo, and oxalato. Prefixes such as bis, tris, and tetrakis are used before complex expressions such as ethylenediamine, 2,2′-bipyridine, and trialkylphosphine. (4) Neutral complexes are named in the same way except that only one word is required. (5) Anionic complexes are also named according to these same rules except that an -ate ending is used. Additional rules are available to name the more complicated compounds, and some of these will be used in the discussion that follows.

Coordination numbers of 2–10 have been observed for different complex compounds. The most common coordination numbers are 6 and 4. Complexes of a coordination number 6 generally have an octahedral structure, but may also be trigonal prismatic. Complexes of a coordination number 4 are either square planar or tetrahedral. Table 1 gives examples of complexes having coordination numbers other than 4 or 6.

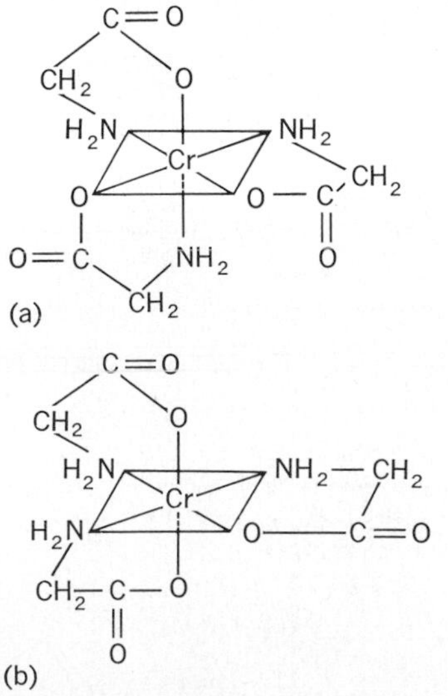

Fig. 6. Structure of (*a*) *fac*-triglycinatochromium(III) and (*b*) *mer*-triglycinatochromium(III).

Structures. Metal complexes exhibit various types of isomerism. In many ways, inorganic stereochemistry is similar to that observed with organic compounds. Geometrical isomers are common among the inert complexes of coordination numbers 4 and 6. Square planar complexes of the type $Pt(NH_3)_2Cl_2$ exist in two forms. Likewise, cis-trans isomers of $Pt(NH_2CH_2COO)_2$ type have been isolated (Figs. 1 and 2). It is apparent from the above examples that the isomer with the same ligands or ligand atoms in adjacent positions is called cis, whereas the trans isomer has its like groups in opposite positions. There are also examples of geometrical isomers of complexes containing four different ligands (Fig. 3). Although geometrical

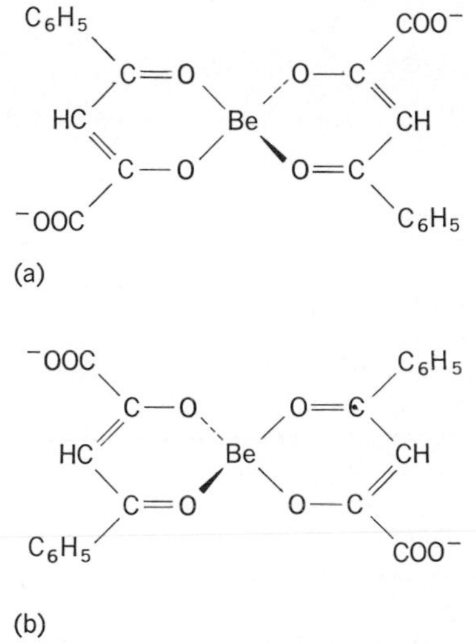

Fig. 7. Optical isomers of bis(benzoylpyruvato) beryllate(II) ion, (*a*) dextro and (*b*) levo.

isomerism of square complexes is most common with platinum(II), it has also been observed with compounds of nickel(II), palladium(II), and gold(III). *See* STEREOCHEMISTRY.

There are many examples of geometrical isomers for 6-coordinated complexes of cobalt(III), chromium(III), and the platinum metals. Most of the examples are of the type $[Co(NH_3)_4Cl_2]^+$ (Fig. 4). If three of the ligands differ from the other three, then only two isomers are possible (Fig. 5). Similar isomers are obtained with an unsymmetrical bidentate ligand (Fig. 6). A complex with 6 different ligands, such as $[Pt(NH_3)(py)(NH_2OH)(Cl)(Br)(NO_2)]$, can exist theoretically in 15 different geometrical forms. In practice, it has not been possible to isolate all of these 15 different isomers.

Optical isomerism is also fairly common among these compounds. For example, 4-coordinate tetrahedral complexes containing unsymmetrical bidentate ligands, such as bis(benzoylpyruvato)beryllate(II) ion, have been resolved (Fig. 7). Optically active complexes of this type are likewise reported for boron(III) and for zinc(II). Most of the examples

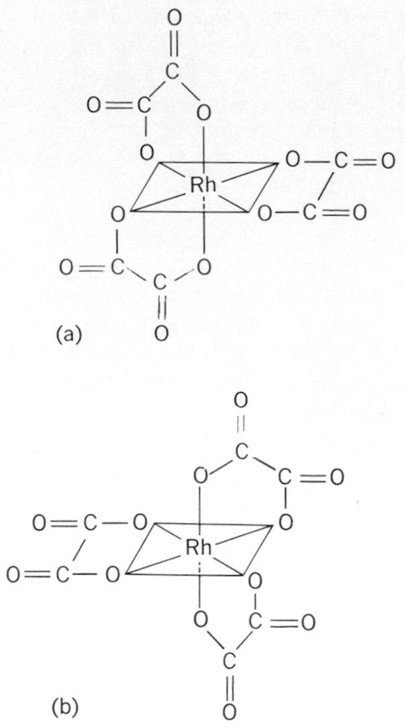

Fig. 8. Optical isomers of trioxalatorhodiate(III) ion, *(a)* dextro and *(b)* levo.

of optical activity occur with 6-coordinated systems containing three bidentate ligands, for example, trioxalatorhodiate(III) ion (Fig. 8). The resolution of 6-coordinated complexes of this type has been reported for the metal ions Al(III), As(V), Cd(II), Co(III), Cr(III), Ga(III), Ge(IV), Ir(III), Fe(II),(III), Ni(II), Os(II),(III), Pt(IV), Rh(III), Ru(II), (III), Ti(IV), and Zn(II).

The cis isomer of a complex containing two bidentate ligands and two monodentate ligands is asymmetric, and therefore exists in the form of mirror-image isomers, whereas the trans form is symmetrical and cannot be optically active (Fig. 9). Since the cis isomer of this type is optically active and the trans isomer is not, the successful resolution of one of the isomers is often used as a proof of its cis structure. Other types of isomerism are known for metal complexes (Table 2).

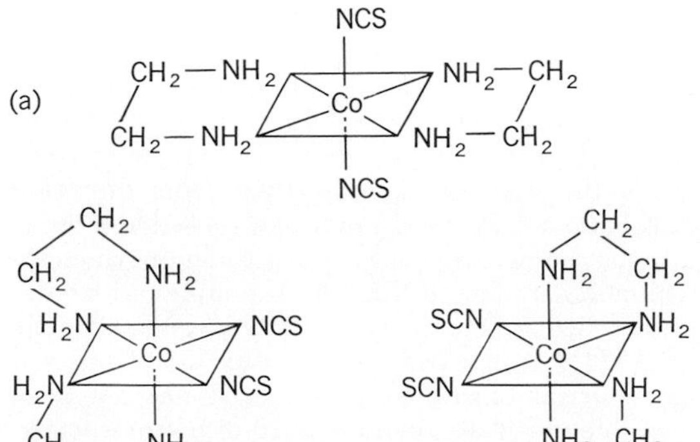

Fig. 9. Symmetric trans form of a complex and asymmetric cis structures. *(a) trans*-Isothiocyanatobis(ethylenediamine)cobalt(III) ion. *(b, c) d,l-cis*-Isothiocyanatobis (ethylenediamine)cobalt(III) ion.

Synthesis. The synthesis of metal complexes containing only one kind of ligand generally involves just the reaction of the metal salt in aqueous solution with an excess of the ligand reagent, reaction (1).

$$[Ni(H_2O)_6](NO_3)_2 + 6NH_3 \rightarrow [Ni(NH_3)_6](NO_3)_2 + 6H_2O \quad (1)$$

The desired complex salt can then be isolated by removal of water until it crystallizes, or by addition of a water-miscible organic solvent to cause it to separate. Many reactions, such as the one cited above, occur readily at room temperature. For the inert complexes (those slow to react), prolonged treatment at more drastic conditions is often necessary.

The preparation of geometrical isomers is much more difficult, and in most cases, the approach used is rather empirical. Generally, reactions yield a mixture of cis-trans products, and these are separated on the basis of their differences in solubility. The trans isomers of platinum(IV) complexes are easily prepared by the oxidation of the appropriate platinum(II) compound (Fig. 10). The cis isomers of cobalt(III) complexes can sometimes be prepared by the reaction of a carbonato complex with the desired acid (Fig. 11).

The easiest geometric isomers to prepare are those of platinum(II). For examples, the cis and the trans isomers of $[Pt(NH_3)_2Cl_2]$ are prepared as shown in reaction (2). The second step in each of

$$K_2[PtCl_4] + 2NH_3 \rightarrow cis\text{-}[Pt(NH_3)_2Cl_2] + 2KCl$$
$$[Pt(NH_3)_4]Cl_2 \xrightarrow{250^\circ} trans\text{-}[Pt(NH_3)_2Cl_2] + 2NH_3 \quad (2)$$

these reactions results in the replacement of the ligand trans to a chloro group (Fig. 12). This phenomenon, that a negative ligand often has a greater labilizing effect on a group in the trans position

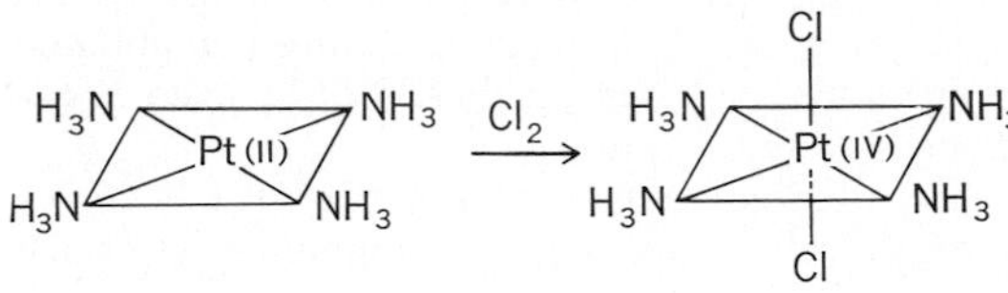

Fig. 10. The reaction in the trans isomer formation of platinum(IV) complexes.

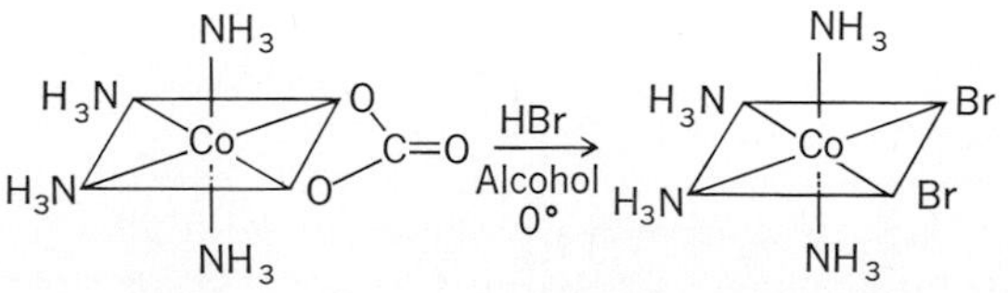

Fig. 11. Cis isomer formation of cobalt(III) complexes.

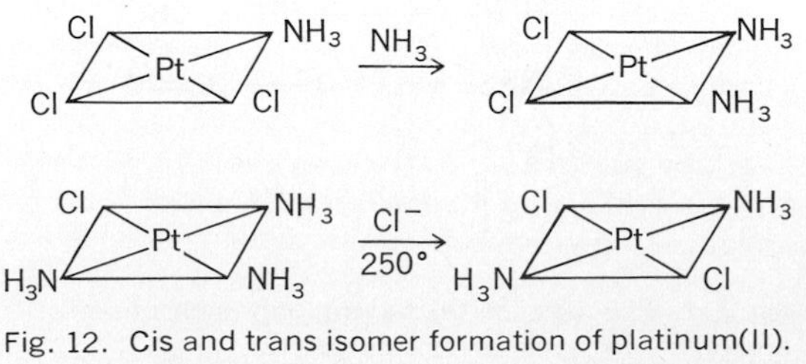

Fig. 12. Cis and trans isomer formation of platinum(II).

Table 2. Types of isomerism for complex compounds

Isomerism	Examples
Geometrical	*cis*- and *trans*-$[Pt(NH_3)_2Br_2]$
Optical	*d*- and *l*-$[Ir(NH_2CH_2CH_2NH_2)_3]Cl_3$
Coordination	$[Co(NH_3)_6][Cr(CN)_6]$ and $[Cr(NH_3)_6][Co(CN)_6]$
Coordination position	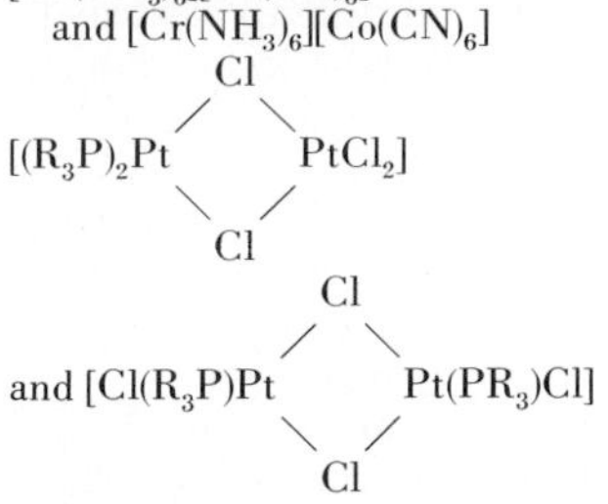
Hydrate	$[Cr(H_2O)_6]Cl_3$ and $[Cr(H_2O)_5Cl]Cl_2 \cdot H_2O$
Ionization	$[Pt(NH_3)_4Cl_2]SO_4$ and $[Pt(NH_3)_4SO_4]Cl_2$
Linkage	$[(NH_3)_5Co—NO_2]SO_4$ and $[(NH_3)_5Co—ONO]SO_4$
Polymerization	$[Co(NH_3)_3(NO_2)_3]$ and $[Co(NH_3)_6][Co(NO_2)_6]$
Conformation	$[Ni(P(C_6H_5)_2CH_2C_6H_5)_2Br_2]$ square-planar and tetrahedral

than does a neutral group, for example, $Cl^- > NH_3$, is called the trans effect. Extensive use has been made of this trans effect in the synthesis of desired platinum(II) complexes. The complex *cis*-$[Pt(NG_3)_2Cl_2]$ (Fig. 12) is used as an antitumor drug for certain types of cancer.

Finally, the separation of optical isomers of metal complexes involves techniques similar to those used for organic compounds. The usual procedure is to convert the racemic mixture into diastereoisomers by means of an optically active resolving agent and then to separate the diastereoisomers by fractional crystallization. Nonionic complexes have been resolved by preferential adsorption on optically active quartz or sugars. *See* AMMINE; HYDRATE.

[FRED BASOLO]

Bibliography: F. Basolo and R. G. Pearson, *Mechanisms of Inorganic Chemistry*, 2d ed., 1967; A. E. Martel (ed.), *Coordination Chemistry*, vol. 1, 1971, vol. 2, 1978.

Concentration scales

Concentration is a very important property of mixtures, because it defines the quantitative relation of the components. In solutions the concentration is expressed as the mass, volume, or number of moles of solute present in proportion to the amount of solvent or of total solution. Each scale of concentration has significant features of experimental simplicity or of theoretically significant relationship.

Percentage. The simplest scale to measure is percentage; hence it is often used for medicinal or household solutions. Weight percent is the number of parts of weight of solute per hundred parts of solution (total). For example, a 10% saline solution contains 10 g of salt in 90 g of water, that is, 100 g total weight. Gaseous mixtures, being difficult to weigh, are often expressed as volume percent. Thus, air is said to contain 78% nitrogen by volume. Solutions of liquids in liquids (say, alcohol in water) may also be expressed in volume percent.

Molarity. To the chemist, the number of moles of solute is of more significance than the number of grams. The molarity (abbreviated *M*) is the number of moles of solute per liter of total solution. Thus, 12 *M* HCl means that the solution contains 12 formula weights (12 × 36.5), or 438 g, of HCl/liter. Concentration is an intensive, rather than an extensive, property of solutions. Thus, 0.5 liter of the acid just mentioned is still 12 *M*, although it contains only 6 moles of solute. As used here, the mole is an amount of substance whose weight in grams is numerically the same as the molecular weight. Chemical engineers sometimes use a mole which is a pound-molecular weight.

Molality. Certain solution properties, for example, the lowering of the freezing point of water by addition of salt, require the use of a concentration scale which relates the number of moles of solute to the weight of solvent rather than to the volume of solution. This scale, called molality (abbreviated *m*), indicates the number of moles of solute/1000 g of solvent. Thus 34.2 g of sucrose ($C_{12}H_{22}O_{11}$, mol wt 342), if dissolved in 200 g of water, has the concentration of 0.5 mole of sucrose/1000 g of water, and hence is 0.5 *m*. For dilute aqueous solutions, the molality is essentially identical with the molarity. In solutions with densities other than unity, the two scales differ. Molality may be computed from molarity, as shown in the equation below, by

$$m = M\left(\frac{1000}{1000d - M \times \text{mol wt}}\right)$$

first subtracting from the weight of 1 liter (1000 × density *d*) the weight of solute, and then scaling the result by proportion to the number of moles/1000 g, where the molecular weight is that of the solute.

Normality. Molarity does not represent reactive capacity for solutes which possess more than one active unit per molecule. Since H_2SO_4 molecules yield twice as many hydrogen ions as do those of HCl, a liter of 0.1 *M* sulfuric acid will neutralize twice as much base as will a liter of 0.1 *M* hydrochloric acid. When it is important to know the reactive capacities of reagents, as in volumetric analysis, the normality scale is used. Normality (abbreviated *N*) is found by multiplying molarity by the number of active units in the formula.

In double-decomposition reactions normality may be an ambiguous concept unless referred to a specific reaction. Phosphoric acid, H_3PO_4, may be reacted with sodium hydroxide to yield NaH_2PO_4, Na_2HPO_4, or Na_3PO_4; thus, H_3PO_4 may have three normalities, depending on whether the reaction involves replacing one, two, or three hydrogen atoms. In oxidation-reduction reactions there is a valence change, and normality must be calculated from the change in the oxidation number—the number of electrons lost or gained. *See* OXIDATION-REDUCTION.

For reaction in a solution, the product of the normality times the volume is the same for both reactants; for a solution of normality N_1, determined by titrating a volume V_1 with volume V_2 of a known solution of normality N_2, this equivalence can be expressed as $V_2N_2 = N_1V_1$. If the volume is expressed in milliliters, the products are in milliequivalents; if in liters, in equivalents.

Formality. In recent years many chemists have sought to avoid the molarity scale lest it imply that electrolyte solutes exist as molecules rather than ions. The formality scale (abbreviated F) represents formula weights per liter. Its values are identical with the molarities of un-ionized solutes.

Mole percent (mole fraction). Many properties of solutions (for example, vapor pressure of one component) are dependent on the ratio of the number of moles of solute to the number of moles of solvent, rather than on the ratios of respective volumes or masses. The mole fraction (abbreviated N_A or X_A for component A) is the ratio of the number of moles of solute to the total number of moles of all components. Thus for 16 g of methanol (0.5 mole) dissolved in 18 g of water (1 mole), the mole fraction of methanol is 0.5/1.5, or 1/3; the mole percent is 33.3. For gases the mole percent is identical with the volume percent. *See* GRAM-EQUIVALENT WEIGHT; GRAM-MOLECULAR WEIGHT; SOLUTION; STOICHIOMETRY; TITRATION.

[ALLEN L. HANSON]

Bibliography: S. W. Benson, *Chemical Calculations*, 2d ed., 1963; E. J. Margolis, *Formulation and Stoichiometry*, 1968; Glen Tilbury, *Problem Solving in Chemistry*, 1962.

Condensation reaction

One of a class of chemical reactions involving a combination between molecules or between parts of the same molecule. A relatively small molecule such as water or alcohol is often eliminated in the process. The conversions of oxygen to ozone and of diethyl pimelate to 2-carbethoxycyclohexanone are examples, as seen in reactions (1) and (2).

$$3O_2 \rightarrow 2O_3 \quad (1)$$

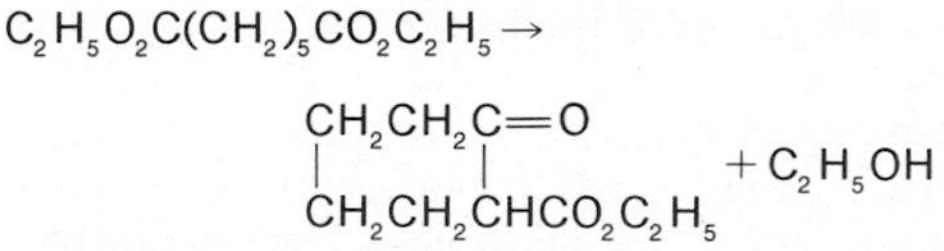
(2)

The term condensation reaction is used very loosely by chemists. Some authors define the term to include all reactions which result in formation of carbon-carbon bonds. General usage, however, does not classify all reactions resulting in formation of carbon-carbon bonds as condensation reactions. Furthermore, the reaction commonly referred to as condensation polymerization involves formation of carbon-oxygen or carbon-nitrogen bonds. *See* POLYMERIZATION.

The following reactions of organic compounds are usually described as condensation reactions.

Aldol condensations. These reactions involve a carbonyl component and an α-hydrogen component, as shown in reaction (3). The carbonyl component is most often an aldehyde (RCHO), but may also be a ketone ($R_2C{=}O$). The α-hydrogen component may be an aldehyde, ketone, simple ester (RCH_2CO_2R), succinate ester, $RO_2CCH_2CH_2CO_2R$ (the Stobbe condensation), carboxylic acid anhydride, $(RCH_2CO)_2O$ (the Perkin reaction), or compounds containing an activated methylene of the type Y—CH_2—Z, where Y and Z may be groups such as —CO_2H, —CO_2R, —CO—R, —CN, —$CONH_2$, or —NO_2 (the Knoevenagel reaction).

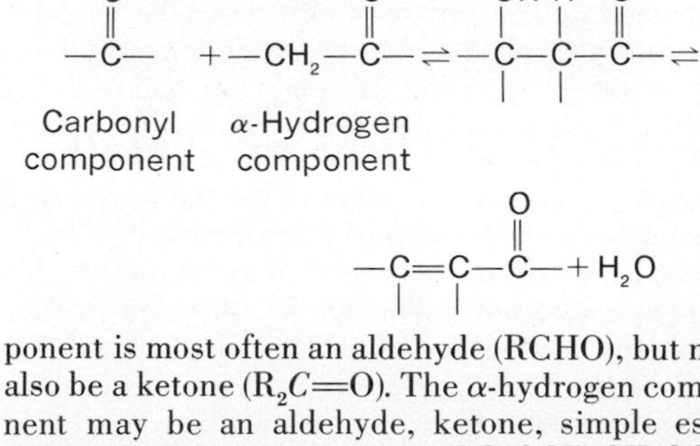
(3)

Aldol condensations are catalyzed by bases such as sodium hydroxide, sodium ethoxide, or amines. Catalysis by acids is feasible in certain cases.

The α-hydroxycarbonyl compounds initially formed in aldol condensations can sometimes be isolated, but they often eliminate water under the experimental conditions and form α,β-unsaturated carbonyl compounds.

At least two α-hydrogen atoms are present in the α-hydrogen component of most aldol condensations, but the first stage of the reaction sometimes involves a single α-hydrogen atom.

The Darzens condensation of aldehydes and ketones with α-haloesters to produce glycidic esters is also an aldol-type condensation, as shown in reactions (4) and (5).

Simple aldol:

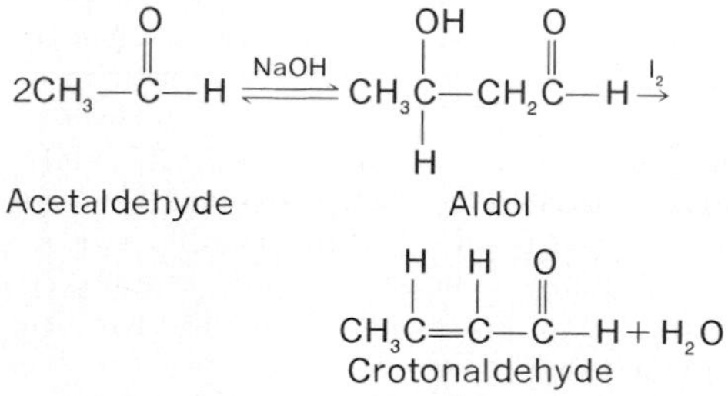
(4)

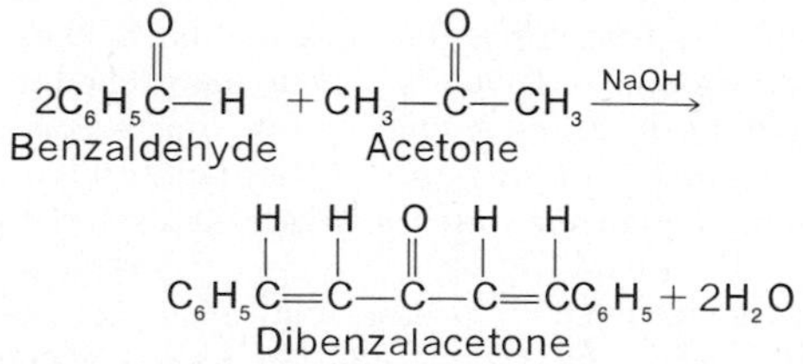
(5)

The Stobbe, Perkin, Knoevenagel, and Darzens reactions are represented by reactions (6)–(9).

Stobbe reaction:

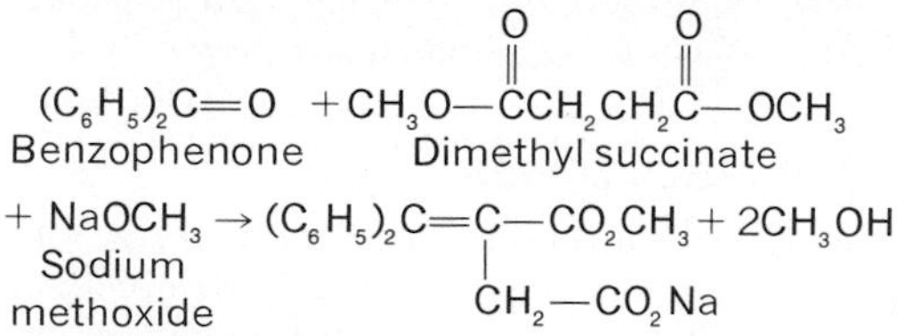
(6)

Perkin reaction:

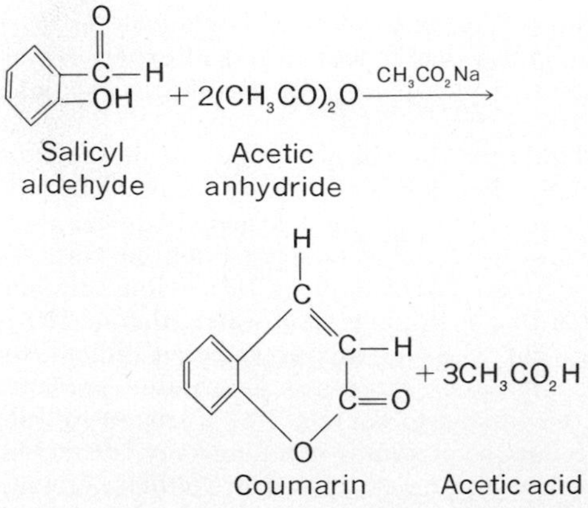
(7)

Knoevenagel reaction:

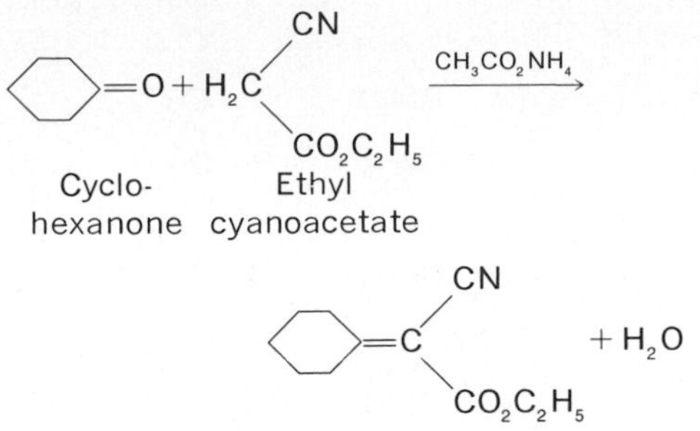

(8)

Darzens reaction:

$$C_6H_5{-}C({=}O){-}CH_3 + ClCH_2CO_2C_2H_5 + NaNH_2 \rightarrow$$

Acetophenone, Ethyl chloroacetate, Sodium amide

$$C_6H_5{-}C(CH_3){-}CHCO_2C_2H_5 \text{ (epoxide, O bridging C and CH)} + NaCl + NH_3 \quad (9)$$

Claisen condensations. The carbonyl component in a Claisen condensation is an ester. Esters of oxalic acid (RO_2CCO_2R) are particularly reactive as carbonyl components. The α-hydrogen component may be an ester or ketone. A generalized reaction is shown by reaction (10).

$$-C({=}O){-}OR + {-}CH_2{-}C({=}O){-} \rightleftharpoons$$

Carbonyl component, α-Hydrogen component

$$-C({=}O){-}CH{-}C({=}O){-} + HOR \quad (10)$$

Claisen condensations involving esters as both components are known as acetoacetic ester condensations and have as products β-ketoesters ($RCO{-}CHRCO_2R$). This type of condensation reaction is represented by reaction (11).

$$2CH_3CO_2C_2H_5 + NaOC_2H_5 \rightarrow$$

Ethyl acetate, Sodium ethoxide

$$Na(CH_3COCHCO_2C_2H_5) + C_2H_5OH \xrightarrow{HCl}$$

$$CH_3{-}C({=}O){-}CH_2{-}C({=}O){-}OC_2H_5 + NaCl \quad (11)$$

Simple esters react with ketones to produce β-diketones ($RCO{-}CHR{-}CO{-}R$), as in reaction (12), and oxalic esters react with ketones, as in reaction (13), to give oxalyl esters ($R{-}CO{-}CHR{-}CO{-}CO_2R$) as products. Condensations of esters

$$C_6H_5CO_2C_2H_5 + C_6H_5{-}C({=}O){-}CH_3 \xrightarrow[(2)\ HCl]{(1)\ NaOC_2H_5}$$

$$C_6H_5{-}C({=}O){-}CH_2{-}C({=}O){-}C_6H_5 \quad (12)$$

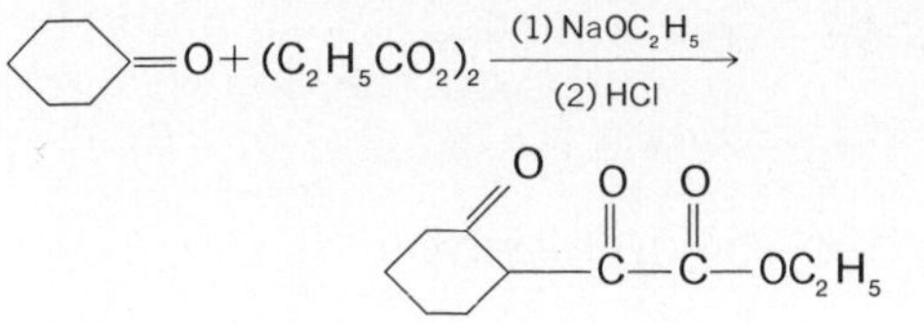

(13)

of dicarboxylic acids which produce cyclic β-ketoesters are known as Dieckman condensations.

Claisen condensations are brought about by strong bases (sodium alkoxides, sodium amide, sodium hydride). At least 1 mole of base for each mole of α-hydrogen component is required.

α-Hydrogen components with only one α-hydrogen atom may be utilized in Claisen condensations if very strong bases such as sodium triphenylmethyl are employed.

Michael condensations. Compounds containing an active methylene group often add to the carbon-carbon double bond of α,β-unsaturated carbonyl compounds in the presence of basic catalysts. This is represented by reactions (14) and (15).

$$-C{=}C{-}C{=}O + X{-}CH_2{-}Y \rightleftharpoons$$

$$X{-}CH(Y){-}C{-}C(H){-}C{=}O \quad (14)$$

$$H_2C{=}CHCO_2C_2H_5 + H_2C(CO_2C_2H_5)_2 \xrightarrow{NaOC_2H_5}$$

Ethyl acrylate, Diethyl malonate

$$CH_2{-}CH_2{-}CO_2C_2H_5 \quad (15)$$
$$|$$
$$CH(CO_2C_2H_5)_2$$

Benzoin condensations. Aromatic aldehydes in the presence of catalytic amounts of potassium cyanide are converted to hydroxy ketones of the type $ArCHOH{-}CO{-}Ar$ (benzoins) reaction (16).

$$2C_6H_5C({=}O){-}H \xrightarrow{KCN} C_6H_5{-}CH(OH){-}C({=}O){-}C_6H_5 \quad (16)$$

Acyloin condensations. Aliphatic esters react with metallic sodium to produce intermediates which are converted by hydrolysis into aliphatic α-hydroxyketones, called acyloins, as shown in reaction (17).

$$2C_3H_7CO_2C_2H_5 \xrightarrow[(2)\ H_2O]{(1)\ Na}$$

$$C_3H_7{-}CH(OH){-}C({=}O){-}C_3H_7 + 2C_2H_5OH + 4NaOH \quad (17)$$

Mannich reaction. The most common application involves a ketone, formaldehyde, and dimethylamine, as in reaction (18). The reaction

$$CH_3{-}C({=}O){-}CH_3 + CH_2O + (CH_3)_2NH_2 \rightarrow$$

$$CH_3{-}C({=}O){-}CH_2CH_2N(CH_3)_2 + H_2O \quad (18)$$

results in replacement of α-hydrogen atoms with one or more dimethylaminomethyl groups. Other secondary amines may be used.

Cyanoethylation. Acrylonitrile adds a variety of weakly acidic compounds. This is shown in reactions (19)–(22).

$$CH_3COCH_3 + 3CH_2{=}CHCN \rightarrow CH_3{-}COC(CH_2CH_2CN)_3 \quad (19)$$

$$(CH_3)_2CHC(H){=}O + CH_2{=}CHCN \rightarrow NCCH_2CH_2C(CH_3)_2{-}C(H){=}O \quad (20)$$

$$CH_3CH_2OH + CH_2{=}CHCN \rightarrow CH_3CH_2OCH_2CH_2CN \quad (21)$$

$$CH_3CH_2NH_2 + 2CH_2{=}CHCN \rightarrow CH_3CH_2N(CH_2CH_2CN)_2 \quad (22)$$

The reaction is catalyzed by strong bases—for example, potassium hydroxide. *See* CYANOETHYLATION.

Pinacol formation. Ketones are converted to tetrasubstituted ethylene glycols (called pinacols) by certain reducing agents, as in reactions (23) and (24).

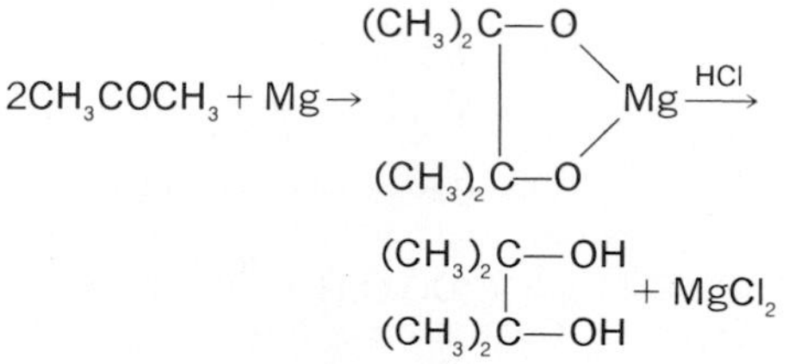

(23)

$$2(C_6H_5)_2CO + (CH_3)_2CHOH \xrightarrow{\text{Light}} (C_6H_5)_2C(OH){-}C(OH)(C_6H_5)_2 + CH_3COCH_3 \quad (24)$$

Reformatsky reaction. Aldehydes and ketones react with α-haloesters in the presence of zinc. Treatment of the reaction mixture with a dilute acid produces β-hydroxy esters, as shown in reactions (25) and (26). *See* REFORMATSKY REACTION.

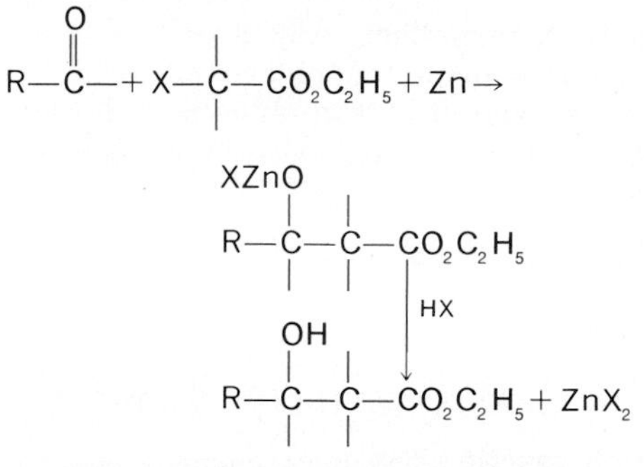

(25)

$$C_6H_5COCH_3 + CH_3CHBrCO_2C_2H_5 \xrightarrow[\text{(2) Acid}]{\text{(1) Zn}} C_6H_5{-}C(CH_3)(OH){-}CH(CH_3){-}CO_2C_2H_5 \quad (26)$$

Additions of Grignard reagents. Grignard reagents (RMgX) add to carbonyl groups and to nitriles. Alcohols, imines, and ketones may be prepared by this method, as shown in reactions (27)–(30).

Organolithium compounds (RLi) and organozinc compounds (R_2Zn) add to carbonyl groups in the same way that Grignard reagents react. *See* GRIGNARD REACTION.

$$R{-}C({=}O){-} + R'{-}MgX \rightarrow R{-}C(OMgX){-}R' \xrightarrow{\text{Acid}} R{-}C(OH){-}R' \quad (27)$$

$$CH_3COCH_3 \xrightarrow[\text{(2) Acid}]{\text{(1) }CH_3CH_2MgBr} CH_3{-}C(OH)(CH_3){-}CH_2CH_3 \quad (28)$$

$$C_6H_5{-}C({=}O){-}OCH_3 \xrightarrow[\text{(2) Acid}]{\text{(1) }C_6H_5MgBr} (C_6H_5)_3COH \quad (29)$$

$$C_6H_5CN \xrightarrow[\text{(2) }NH_4Cl]{\text{(1) }CH_3MgBr} C_6H_5C({=}N{-}H){-}CH_3 \xrightarrow{H_2O} C_6H_5{-}C({=}O){-}CH_3 + NH_3 \quad (30)$$

Friedel-Crafts alkylations and acylations. Alkyl and acyl groups may be substituted for hydrogen atoms in aromatic compounds. Aluminum chloride is the usual catalyst for the reaction, but other strong acids may be used. Alkyl and acyl halides are often the substituting reagents, but alcohols, olefins, anhydrides, organic acids, aldehydes, ketones, ethers, and esters may be employed. Reactions (31)–(34) represent Friedel-Crafts alkylations and acylations. *See* FRIEDEL-CRAFTS REACTION.

$$C_6H_6 + CH_3CH_2Br \xrightarrow{AlCl_3} C_6H_5CH_2CH_3 + HBr \quad (31)$$

$$C_6H_6 + CH_3{-}C({=}O){-}Cl \xrightarrow{AlCl_3} C_6H_5C({=}O){-}CH_3 + HCl \quad (32)$$

$$2C_6H_6 + Cl_3CC({=}O){-}H \xrightarrow{H_2SO_4} (C_6H_5)_2CHCCl_3 + H_2O \quad (33)$$

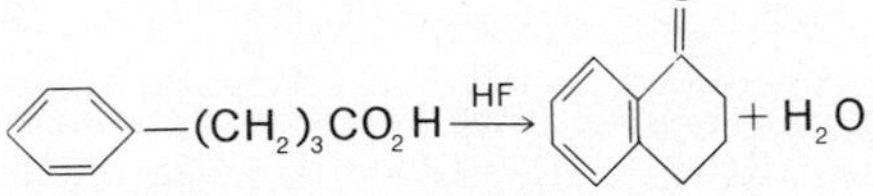

(34)

Alkylations of organometallic compounds. Compounds of the type RMe, where Me represents a metal such as Na, K, Li, or MgX, react with many alkyl halides to eliminate MeX and replace the metal with an alkyl group, as in reaction (35). The

$$RMe + R'X \rightarrow R{-}R' + MeX \quad (35)$$

availability of the compounds RMe and the reactivity of the halides R′X are the limiting factors.

Grignard reagents are readily available from halides and magnesium metal. These reagents are alkylated by reactive halides such as allyl bromide, as in reaction (36). Compounds of the type

$$CH_3CH_2MgBr + BrCH_2CH{=}CH_2 \rightarrow CH_3(CH_2)_2CH{=}CH_2 + MgBr_2 \quad (36)$$

Y—CH_2—Z where Y and Z are activating groups are converted to metallocompounds by treatment with sodium or potassium alkoxides, and may then be alkylated by allyl, primary, or secondary halides. This is shown by reactions (37) and (38).

$$H_2CYZ + MeOR \rightarrow MeHCYZ + HOR \quad (37)$$

$$MeHCYZ + R'X \rightarrow R'HCYZ + MeX \quad (38)$$

A second alkyl group may be introduced by repeating the process, as in reaction (39). Alkylations

$$R'HCYZ \xrightarrow[(2)\ R''X]{(1)\ MeOR} R'R''CYZ \quad (39)$$

of acetoacetic and malonic esters are classical examples of this procedure. These alkylations are represented by reactions (40) and (41).

$$CH_3\overset{\overset{O}{\|}}{C}CH_2CO_2C_2H_5 \xrightarrow[(2)\ CH_3I]{(1)\ NaOC_2H_5} CH_3\overset{\overset{O}{\|}}{C}\underset{\underset{CH_3}{|}}{C}HCO_2C_2H_5 \quad (40)$$

$$CH_3CH(CO_2C_2H_5)_2 \xrightarrow[(2)\ CH_3CH_2Br]{(1)\ NaOC_2H_5} CH_3CH_2\overset{\overset{CH_3}{|}}{C}(CO_2C_2H_5)_2 \quad (41)$$

Ketones with at least one α-hydrogen atom may usually be metallated with sodium amide and subsequently alkylated, as is shown in reaction (42).

$$C_6H_5COCH_3 \xrightarrow{NaNH_2} C_6H_5COCH_2Na \xrightarrow{CH_3I} C_6H_5COCH_2CH_3 \quad (42)$$

See ORGANIC REACTION MECHANISM; ORGANOMETALLIC COMPOUND.

[WILLIAM B. RENFROW, JR.]

Bibliography: R. Adams (ed.), *Organic Reactions*, vols. 1–6, 1942–1951, vol. 8, 1954; J. D. Roberts and M. C. Caserio, *Basic Principles of Organic Chemistry*, 1964.

Conformational analysis

The determination of the arrangement in space of the constituent atoms of a molecule that may rotate about a single bond. These rotations produce changes in conformation. Conformational analysis also includes the study of the chemical and physical differences between the conformations of a molecule. A particular conformation is chemically significant only when capable of finite existence.

Conformations may be treated in a manner analogous to the treatment of geometrical isomers, which are differentiated by the large energy barrier to rotation imposed by the carbon-carbon double bond. However, the barrier to rotation about a carbon-carbon single bond is usually considerably lower, and consequently different conformations are more readily interconvertible.

Ethane theoretically possesses an infinite number of conformations; but if the hydrogen atoms are not differentiated, there are but two significant conformations. Viewed along the axis of the carbon-carbon bond, one of these has the hydrogen atoms of one methyl group exactly lined up behind those of the other methyl group, and the other has the hydrogens of one methyl group falling midway (60° from the former position) between those of the other methyl group. The former is an eclipsed (hydrogens opposed to each other) conformation, and the latter a staggered conformation. The eclipsed conformation represents a free-energy maximum and the staggered conformation a free-energy minimum, presumably due to repulsions, or lack thereof, between the electrons of the carbon-hydrogen bonds.

If the ethane carbons each carry two substituents other than hydrogen, three staggered conformations are possible, and the most stable one is determined by the relative magnitudes of steric repulsions between the substituent groups on the central carbons. The staggered conformations for the (±) configurations (I*a*), (I*b*), and (I*c*), and the meso configurations (II*a*), (II*b*), and (II*c*) of stilbene dibromide are shown here. With reference to the phenyl groups, the conformations shown in (I*c*) and (II*c*) are called trans, whereas the remaining are called skew, or gauche. In general, the energetically most favorable (and therefore most populous) conformation will resemble (I*c*) and (II*c*), the largest groups (phenyl) being trans.

The conversion of either isomer to a stilbene by loss of the bromines requires the bromines to be trans (I*b*) or (II*c*), initially. The molecules must also be in the conformations, with respect to the hydrogens and phenyl groups, resembling the resultant stilbene: cis for (I), phenyls opposed to each other, and trans for (II), phenyls opposed to hydrogens

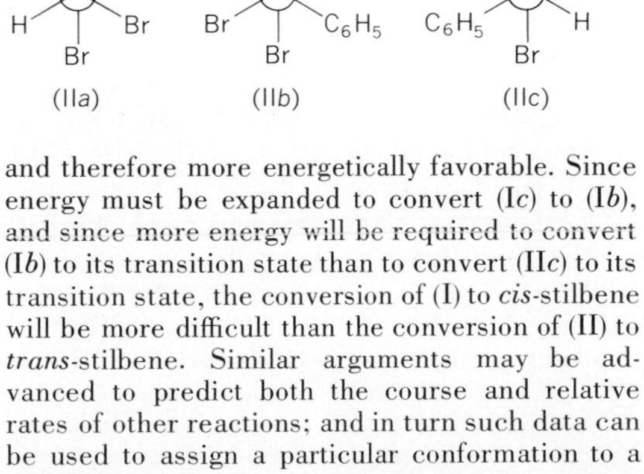

and therefore more energetically favorable. Since energy must be expanded to convert (I*c*) to (I*b*), and since more energy will be required to convert (I*b*) to its transition state than to convert (II*c*) to its transition state, the conversion of (I) to *cis*-stilbene will be more difficult than the conversion of (II) to *trans*-stilbene. Similar arguments may be advanced to predict both the course and relative rates of other reactions; and in turn such data can be used to assign a particular conformation to a molecule.

In cyclic systems containing more than five ring-members, the normal bond angles (carbon = 109°28′) a nonplanar arrangement, and two significant conformations are possible: chair (III*a*) and boat (III*b*). Conformational analysis reveals a

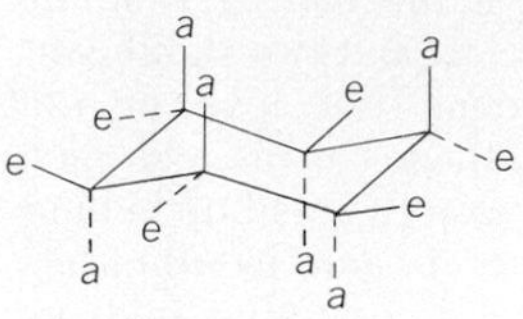

(IIIa)

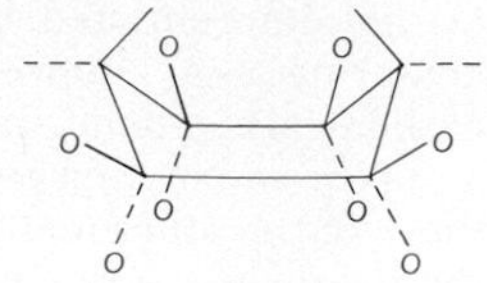

(IIIb)

greater stability for the chair form, in which the hydrogens are completely staggered, whereas the boat form contains four pairs of opposed hydrogens, *o* in (III*b*). In the chair form the hydrogens are directed perpendicularly to the mean plane of the carbons – axial conformation, *a* in (III*a*); or equatorially to the center of the mean plane – equatorial conformation, *e* in (III*a*).

An examination of models suggests that there is less steric interaction if the bonds of a cyclohexane system are rotated to place a single substituent equatorially. When several substituents are present in a given isomer, the most stable conformation is that in which the maximum number of groups are equatorial. When not all groups can be so placed, preference for equatorial placement is given to the largest group.

Similarly the relative thermodynamic stability of a pair of cis-trans disubstituted cyclohexanes can be predicted by conformational analysis: as 1,2-trans isomer can have both substituents equatorial, whereas the cis isomer can have only one substituent equatorial, thereby having greater energy content or lower thermodynamic stability. On the other hand, in the 1,3 isomers both substituents can be equatorial in the cis isomer but not in the trans isomer, and so the cis isomer is thermodynamically more stable.

The extension of conformational analysis to polycyclic systems leads to many useful conclusions concerning the structure, configuration, and mode of reaction of such important natural products as the steroids, terpenes, and alkaloids. *See* STEREOCHEMISTRY.

[WYMAN R. VAUGHAN]

Conjugation and hyperconjugation

CONJUGATION AND HYPERCONJUGATION

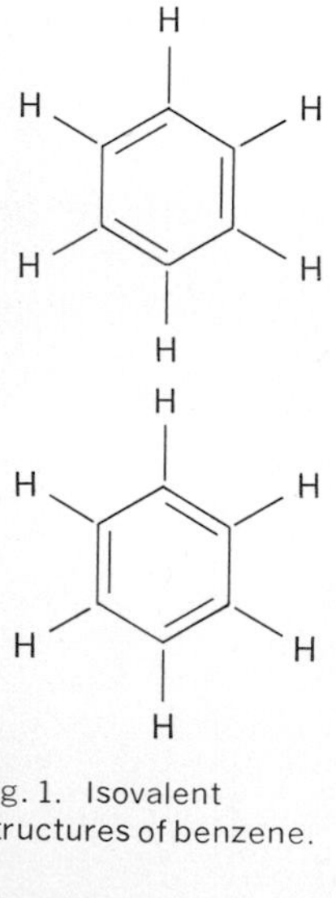

Fig. 1. Isovalent structures of benzene.

An arrangement of bonds in a molecule such that a single bond lies directly between two multiple (that is, double or triple) bonds or between a multiple bond and a group containing a lone π-electron, a π-electron pair or quartet, or a π-electron vacancy makes the molecule show unusual chemical behavior and physical properties. Such molecules and such multiple bonds are called conjugated. Simple examples are 1,3-butadiene ($H_2C{=}CH{-}CH{=}CH_2$) and allylamine ($H_2C{=}CH{-}\ddot{N}H_2$).

The unusual properties of conjugated molecules can be understood theoretically, according to quantum mechanics, in terms of a bond structure in which, in the two examples mentioned, small proportions of $H_2\dot{C}{-}CH{=}CH{-}\dot{C}H_2$ or $H_2C^-{-}CH{=}N^+H_2$, respectively, "resonate with" the main bond structures given above. In these examples one speaks of sacrificial conjugation, since in the minor resonance structures there is one less bond, or else much energy is required to transfer a charge. *See* CHEMICAL BONDING; RESONANCE.

Further examples of sacrificial conjugation include $HC{\equiv}C{-}C{=}CH_2$ and diacetylene ($HC{\equiv}C{-}C{\equiv}CH$). In the latter, both π-bonds of the triple bond are conjugated. In the former, only one π-bond of the triple bond is conjugated (minor resonance structure $H\dot{C}{=}C{=}C{-}\dot{C}H_2$) with the one π-bond of the double bond.

Stronger resonance effects occur in conjugated molecules in which alternative structures with equal numbers of bonds can be written (isovalent conjugation), for example, in benzene (Fig. 1) or the allyl ion ($H_2C{=}CH{-}C^+H_2$ and $H_2C^+{-}CH{=}CH_2$) or radical $H_2C{=}CH{-}\dot{C}H_2$ and $H_2\dot{C}{-}CH{=}CH_2$). Conjugation brings about energy stabilization, but much more so in isovalent than in sacrificial conjugation.

Hyperconjugation. This phenomenon is similar to conjugation in its formulation and manifestations, but the effects are weaker. It occurs when a CH_2 or CH_3 group (or in general an AR_2 or AR_3 group, where A may be any polyvalent atom and R any atom or radical) is adjacent to a multiple bond or to a group containing an atom with a lone π-electron, a π-electron pair or quartet, or a π-electron vacancy. Hyperconjugation can be sacrificial (this is relatively weak) or isovalent (stronger).

Examples of molecules exhibiting first-order sacrificial hyperconjugation are:

$H_3C{-}CH{=}CH_2$
Propylene
(main structure)

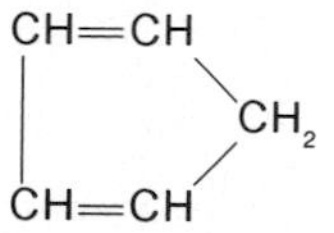

Cyclopentadiene

$H_3C{-}NH_2$
Methylamine

$H_3C{-}CH{=}O$
Acetaldehyde

The analogy to conjugation is brought out by rewriting the main structures as

$H_3{\equiv}C{-}CH{=}CH_2$

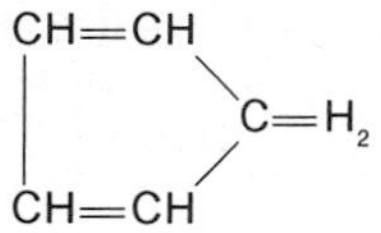

$H_3{\equiv}C{-}CH{=}O$ $H_3{\equiv}C{-}CH{-}\ddot{N}H_2$

with minor resonance structures

$\dot{H}_3{=}C{=}CH{-}\dot{C}H_2$

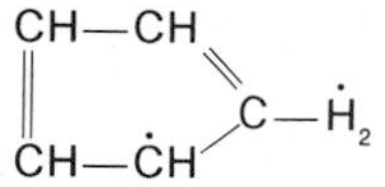

(two structures)

$\dot{H}_3{=}C{=}CH{-}\dot{O}$
(also $H_3^+{=}C{=}CH{-}O^-$)

$H_3C^-{=}C{=}N^+H_2$

A still weaker effect is second-order sacrificial hyperconjugation, as in ethane ($H_3{\equiv}C{-}C{\equiv}H_3$; compare $HC{\equiv}C{-}C{\equiv}CH$) or in ethylene ($H_2{=}C{=}C{=}H_2$; compare $H_2C{=}C{=}C{=}CH_2$).

Examples of isovalent hyperconjugation are $H_3{\equiv}C{-}C^+H_2$ (with hyperisovalent resonance structure $H_3^+{=}C{=}CH_2$) and $H_3{\equiv}C{-}\dot{C}H_2$ (resonance structure $\dot{H}_3{=}C{=}CH_2$). The number of bonds is the same in the two resonance structures, just as in isovalent conjugation, but the second structure is energetically less favorable than the first structure; for this reason the resulting stabilizing resonance energy is less than in isovalent conjugation.

Physical properties. The most notable differences in physical properties for sacrificially conjugated as compared with unconjugated molecules are as follows: (1) Single bonds lying between two multiple bonds are shortened. For example, the lengths of the C-C single bonds in diacetylene and 1,3-butadiene are 1.38 A (1.38×10^{-8} cm) and 1.48 A, as compared with 1.54 A for ethane. The lengths of the double or triple bonds, however, are scarcely affected. According to quantum theory, the observed shortenings result partly from conjugation and partly from differences in the hybridization of the carbon atom in forming single, double, and triple bonds. (2) Electronic absorption spectra begin at considerably longer wavelengths for conjugated than for related unconjugated molecules. (3) Ionization potentials are lower than for isomeric unconjugated molecules. (4) Related to points 2 and 3 is the fact that refractivities and polarizabilities are larger for conjugated than for corresponding unconjugated dienes. (5) As compared with predictions from formulas for unconjugated molecules, conjugated molecules show lower energies, for example, about 6 kcal/mole lower for 1,3-butadiene than for unconjugated dienes; the excess stability is understandable theoretically as resonance energy. *See* BOND ANGLE AND DISTANCE.

More or less similar, but larger, effects occur for isovalent conjugation. Taking benzene as an example, interaction between the two resonance structures removes the distinction between single and double bonds and leaves all bonds of length 1.39 A and yields a resonance energy of about 40 kcal/mole.

In sacrificial and in isovalent hyperconjugation, the effects are similar to those in conjugation, but in general they are smaller. However, in radicals and especially in radical ions, large-energy stabilizations can occur (for example, 16 and 84 kcal/mole in the *t*-butyl radical and cation, respectively).

Observed small dipole moments often can be explained in part by hyperconjugation. Thus for propylene, besides the minor resonance structure $\dot{H}_3{=}C{=}CH{-}\dot{C}H_2$ already mentioned, one has some $H_3^+{=}C^-{-}C^+H{-}C^-H_2$ and $H_3{=}C^+{-}C^-H{-}C^+H_2$, and if the first-named of these two predominates, a dipole moment in the direction $CH_3^+{-}CH{-}CH_2^-$ results, in agreement with experimental indications. However, other factors, including differences in hybridization, must also contribute.

Clear-cut examples of molecules where conjugation or hyperconjugation, respectively, are of primary importance in creating dipole moments are fulvene and cyclopentadiene, with major resonance structures shown in Fig. 2 and minor structures shown in Fig. 3. The minor structures are in addition to the nonpolar structure already mentioned. Fulvene has a dipole moment of 1.2 D, which is explained in terms of the minor structures, and one can safely predict an analogous but smaller dipole moment for cyclopentadiene.

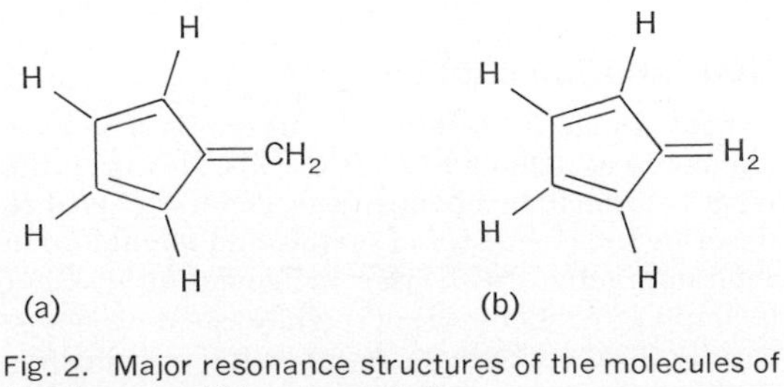

Fig. 2. Major resonance structures of the molecules of (*a*) fulvene and (*b*) cyclopentadiene.

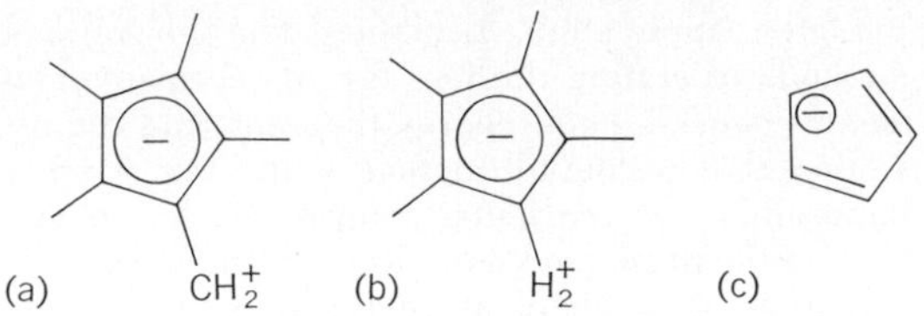

Fig. 3. Minor resonance structures of the molecules of (*a*) fulvene and (*b*) cyclopentadiene. The pentagons in *a* and *b* represent the cyclopentadienyl anion, which is particularly stable because of resonance among five equivalent structures of the type shown by *c*.

[ROBERT S. MULLIKEN]

Bibliography: J. Hinze and D. A. Ramsay (eds.), *Selected Papers of Robert S. Mulliken*, 1975; N. Muller and R. S. Mulliken, Strong or isovalent hyperconjugation in alkyl radicals and their positive ions, *J. Amer. Chem. Soc.*, 80:3489, 1958; R. S. Mulliken, Bond lengths and bond energies in conjugation and hyperconjugation, *Tetrahedron*, 6:68, 1959; R. S. Mulliken, Conjugation and hyperconjugation: A survey with emphasis on isovalent hyperconjugation, *Tetrahedron*, 5:253, 1959; R. S. Mulliken, Discussion, *Tetrahedron*, 17:247, 1962.

Conservation of energy

The principle of conservation of energy states that energy cannot be created or destroyed, although it can be changed from one form to another. Thus in any isolated or closed system, the sum of all forms of energy remains constant. The energy of the system may be interconverted among many different forms—mechanical, electrical, magnetic, thermal, chemical, nuclear, and so on—and as time progresses, it tends to become less and less available; but within the limits of small experimental uncertainty, no change in total amount of energy has been observed in any situation in which it has been possible to ensure that energy has not entered or left the system in the form of work or heat. For a system that is both gaining and losing energy in the form of work and heat, as is true of any machine in operation, the energy principle asserts that the net gain of energy is equal to the total change of the system's internal energy. *See* THERMODYNAMIC PRINCIPLES.

Application to life processes. The energy principle as applied to life processes has also been studied. For instance, the quantity of heat obtained by burning food equivalent to the daily food intake of an animal is found to be equal to the daily amount of energy released by the animal in the forms of heat, work done, and energy in the waste products. (It is assumed that the animal is not gaining or losing weight.) Studies with similar results have also been made of photosynthesis, the process upon which the existence of practically all plant and animal life ultimately depends.

Conservation of mechanical energy. There are many other ways in which the principle of conservation of energy may be stated, depending on the intended application. Examples are the various methods of stating the first law of thermodynamics, the work-kinetic energy theorem, and the assertion that perpetual motion of the first kind is impossible. Of particular interest is the special form of the principle known as the principle of conservation of mechanical energy (kinetic E_k plus potential E_p) of any system of bodies connected together in any way is conserved, provided that the system is free of all frictional forces, including internal friction that could arise during collisions of the bodies of the system. Although frictional or other nonconservative forces are always present in any actual situation, their effects in many cases are so small that the principle of conservation of mechanical energy is a very useful approximation. Thus for a missile or satellite traveling high in space, the dissipative effects arising from such sources as the residual air and meteoric dust are so exceedingly small that the loss of mechanical energy $E_k + E_p$ of the body as it proceeds along it trajectory may, for many purposes, be disregarded.

Mechanical equivalent of heat. The mechanical energy principle is very old, being directly derivable as a theorem from Newton's law of motion. Also very old are the notions that the disappearance of mechanical energy in actual situations is always accompanied by the production of heat and that heat itself is to be ascribed to the random motions of the particles of which matter is composed. But a really clear conception of heat as a form of energy came only near the middle of the 19th century, when J. P. Joule and others demonstrated the equivalence of heat and work by showing experimentally that for every definite amount of work done against friction there always appears a definite quantity of heat. The experiments usually were so arranged that the heat generated was absorbed by a given quantity of water, and it was observed that a given expenditure of mechanical energy always produced the same rise of temperature in the water. The resulting numerical relation between quantities of mechanical energy and heat is called the Joule equivalent, or mechanical equivalent of heat. The present accepted value is one 15° calorie $= 4.1855 \pm 0.0004$ joules.

Conservation of mass-energy. In view of the principle of equivalence of mass and energy in the restricted theory of relativity, the classical principle of conservation of energy must be regarded as a special case of the principle of conservation of mass-energy. However, this more general principle need be invoked only when dealing with certain nuclear phenomena or when speeds comparable with the speed of light (3×10^{10} cm/sec) are involved.

If the mass-energy relation, $E = mc^2$, where c is the speed of light, is considered as providing an equivalence between energy E and mass m in the same sense as the Joule equivalent provides an equivalence between mechanical energy and heat, there results the relation, 1 kg $= 9 \times 10^{16}$ joules.

Laws of motion. The law of conservation of energy has been established by thousands of meticulous measurements of gains and losses of all known forms of energy. It is now known that the total energy of a properly isolated system remains constant. Some parts or particles of the system may gain energy but others must lose just as much. The actual behavior of all the particles, and thus of the whole system, obeys certain laws of motion. These laws of motion must therefore be such that the energy of the total system is not changed by collisions or other interactions of its parts. It is a remarkable fact that one can test for this property of the laws of motion by a simple mathematical manipulation that is the same for all known laws: classical, relativistic, and quantum mechanical.

The mathematical test is as follows. Replace the variable t, which stands for time, by $t + a$, where a is a constant. If the equations of motion are not changed by such a substitution, it can be proved that the energy of any system governed by these equations is conserved. For example, if the only expression containing time is $t_2 - t_1$, changing t_2 to $t_2 + a$ and t to $t_1 + a$ leaves the expression unchanged. Such expressions are said to be invariant under time displacement. When daylight-saving time goes into effect, every t is changed to $t + 1$ hr. It is unnecessary to make this substitution in any known laws of nature, because they are all invariant under time displacement.

Without such invariance laws of nature would change with the passage of time, and repeating an experiment would have no clear-cut meaning. In fact, science, as it is known today, would not exist.

[DUANE E. ROLLER/LEO NEDELSKY]

Bibliography: K. R. Atkins, *Physics*, 3d ed., 1976; D. Halliday and R. Resnick, *Physics*, 3d ed., 1978; G. Laundry et al., *Physics: An Energy Introduction*, 1979; F. W. Sears et al. *University Physics*, 5th ed., 1976; E. P. Wigner, Symmetry and conservation laws, *Phys. Today*, 17(3):34–40, March 1964.

Conservation of mass

The notion that mass, or matter, can be neither created nor destroyed. According to conservation of mass, reactions and interactions which change the properties of substances leave unchanged their total mass; for instance, when charcoal burns, the mass of all of the products of combustion, such as ashes, soot, and gases, equals the original mass of charcoal and the oxygen with which it reacted.

The special theory of relativity of Albert Einstein, which has been verified by experiment, has shown, however, that the mass of a body changes as the energy possessed by the body changes. Such changes in mass are too small to be detected except in subatomic phenomena. Furthermore, matter may be created, for instance, by the materialization of a photon (quantum of electromagnetic energy) into an electron-positron pair; or it may be destroyed, by the annihilation of this pair of elementary particles to produce a pair of photons.

[LEO NEDELSKY]

Coordination chemistry

A field which, in its broadest usage, is acid-base chemistry as defined by G. N. Lewis. However, the term coordination chemistry is generally used to describe the chemistry of metals and metal ions in their interactions with other molecules or ions. For example, reactions (1)–(3) show acid-base-type reactions; the products formed are coordination ions or compounds, and this area of chemistry is known as coordination chemistry.

$$Mg^{2+} + 6H_2O \rightarrow Mg(H_2O)_6^{2+} \quad (1)$$

$$Ni + 4CO \rightarrow Ni(CO)_4 \quad (2)$$

$$Fe^{2+} + 6CN^- \rightarrow Fe(CN)_6^{4-} \quad (3)$$

Thus, it follows that coordination compounds are compounds that contain a central atom or ion and a group of ions or molecules surrounding it. Such a compound tends to retain its identity, even in solution, although partial dissociation may occur. The charge on the coordinated species may be positive, zero, or negative, depending on the charges carried by the central atom and the coordinated groups. These groups are called ligands, and the total number of attachments to the central atom is called the coordination number. Other names commonly used for these compounds include complex compounds, complex ions, Werner complexes, coordinated complexes, chelate compounds, or simply complexes. *See* ACID AND BASE; CHELATION; COMPLEX COMPOUNDS.

Experimental observations as early as the middle of the 18th century reported the isolation of coordination compounds. During that time and for the following 150 years, the valence theory could not adequately account for such materials. As a result, they were referred to as complex compounds, a term which is still in common usage, but not for the same reason. The correct interpretation of these compounds was finally given by Alfred Werner in 1893. He introduced the concept of residual or secondary valence, and suggested that elements have this type of valence in addition to their normal or primary valence. Thus, platinum (IV) has a normal valence of 4 but a secondary valence or coordination number of 6. This then led to the formulation of $PtCl_4 \cdot 6NH_3$ as $[Pt(NH_3)_6]^{4+}$, $4Cl^-$ and of $PtCl_4 \cdot 5NH_3$ as $[Pt(NH_3)_5Cl]^{3+}$, $3Cl^-$. The compound with five ammonias has only three ionic chlorides, the fourth is inside the coordination sphere, and therefore is not readily precipitated upon the addition of silver ion. Although the exact nature of the coordinate bond between metal and ligand remains the subject of considerable discussion, it is agreed that the formulations of Werner are essentially correct.

Coordinate bond. Three theories have been used to explain the nature of the coordinate bond. These are the valence bond theory, the electrostatic theory, including crystal field corrections, and the molecular orbital theory. Currently, the theory used almost exclusively is the molecular orbital theory. The valence bond theory for metal complexes was developed chiefly by Linus Pauling. This theory considers that the pair of electrons on the ligand enter the hybridized atomic orbitals of

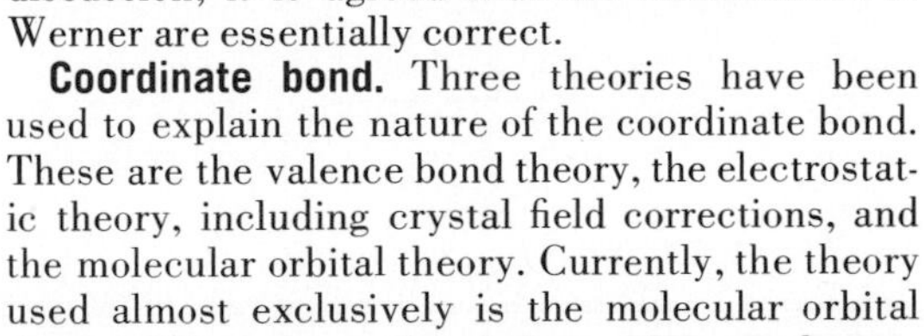

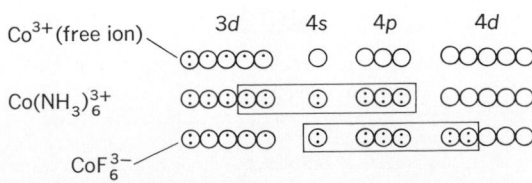

Fig. 1. Representation of the cobalt(III) complexes.

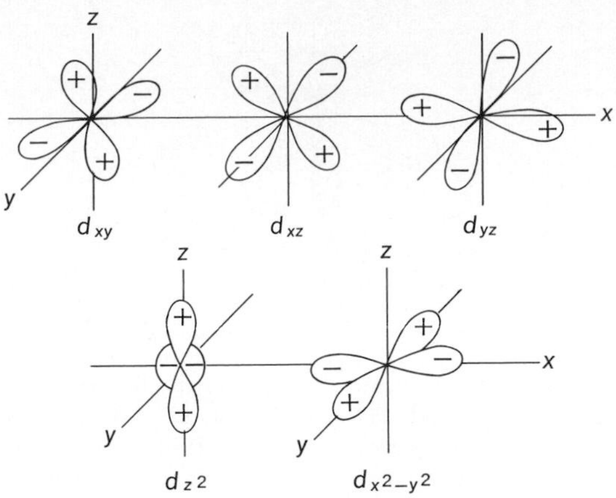

Fig. 2. Orientation of the *d* orbitals in space.

the metal and that the bond is either essentially covalent or essentially ionic. Several of the properties of these substances can be explained on the basis of this theory. For example, cobalt(III) complexes are represented by Fig. 1. The orbital hybridization of $Co(NH_3)_6^{3+}$ is designated as d^2sp^3, and the complex is referred to as an inner orbital complex. Such a representation is consistent with the diamagnetic properties of this cation, for all electrons are paired. The orbital hybridization of CoF_6^{3-} is designated as sp^3d^2, and it is called an outer orbital complex. This ion is known to be paramagnetic, which is in keeping with the four unpaired electrons in the 3*d* orbitals. The term inner orbital is applied if the *d* orbital of a lower energy than the *s* and *p* is used in bonding, whereas outer orbital has reference to systems in which the *d* orbital used is at the same energy level as the *s* and *p*.

The electrostatic theory, plus the crystal field theory for the transition metals, assumes that the metal-ligand bond is caused by electrostatic interactions between point charges and dipoles and that there is no sharing of electrons. Physicists have made good use of this theory to explain the properties of ionic crystalline solids, and it has now been extended to the metal complexes. For the nontransition metals, the parameters needed to determine the strength of the metal ligand bond are the charges and sizes of the central ions and the charges, dipole moments, polarizabilities, and sizes of the ligands. In order to give an adequate explanation of the bonding for transition metals, it is also necessary to consider the orientation of the *d* orbitals in space. The five possible spatial configurations are shown in Fig. 2. For the gaseous ion M, all five of the *d* orbitals are of equal energy. However, as shown in Fig. 3, the $d_{x^2-y^2}$ and d_{z^2} orbitals are pointing directly toward the six ligands at the corners of an octahedron and, because of repulsive interaction with the ligands, are at a higher energy than the d_{xy}, d_{xz}, and d_{yz} orbitals which do not point toward the ligands. *See* CRYSTAL FIELD THEORY.

Thus, on the basis of the crystal field theory, the cobalt(III) complexes referred to above are designated as in Fig. 4. The $Co(NH_3)_6^{3+}$ is called a spin-paired complex, whereas CoF_6^{3-} is called spin-free. This refers to the fact that the electrons are

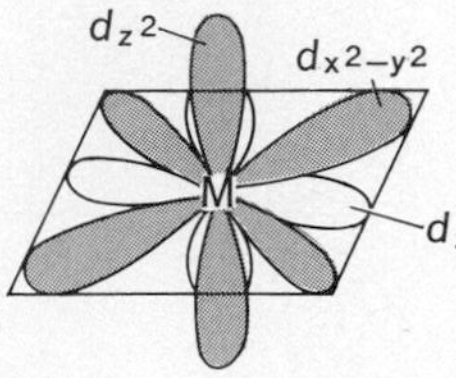

Fig. 3. Diagram of the spatial orientation of *d* orbitals in gaseous ion represented by M.

all paired in the former because of the larger crystal field splitting, Δ_0, whereas in the latter, Δ_0 is small, and the electrons are not paired. In addition to explaining the structure and magnetic properties, this theory affords an adequate interpretation of the visible spectra of metal complexes.

The molecular orbital theory assumes that the electrons move in molecular orbitals which extend over all the nuclei of the metal-ligand system. In this manner, it serves to make use of both the valence bond theory and crystal field theory. The molecular orbital theory is therefore the best approximation to the nature of the coordinate bond because it is sufficiently flexible to permit both covalent and ionic bonding as well as the splitting of *d* orbitals into various energy levels. The complexes CoF_6^{3-} and $Co(NH_3)_6^{3+}$ can be represented by the molecular orbital energy diagrams shown in Fig. 5. *See* MOLECULAR ORBITAL THEORY.

It is apparent from these molecular orbital diagrams that this theory combines the desirable features of both the valence bond and the crystal field theories. The covalent bonding of the valence bond theory appears as the sigma (σ) bonded molecular orbitals, designated here as σ_s, σ_p, and σ_d. Likewise the crystal field splitting (Δ_0) of the crystal field theory now is the energy difference between the nonbonding *d* orbitals d_{xy}, d_{xz}, d_{yz} and the antibonding sigma orbital $\sigma_d{}^*$. More complicated molecular orbital diagrams would include the contribution of π bonding in these systems.

Stabilities of complexes. The stability of metal complexes depends both on the metal ion and the ligand. There is a great deal of quantitative information on the stability of coordination compounds. These data show that in general the stability of metal complexes increases if the central ion increases in charge, decreases in size, and increases in electron affinity. Thus, alkali-metal ions have the least tendency to form complexes, and the highly polarizing transition-metal ions have the greatest tendency. However, even in the least favorable cases of the alkali-metal ions, there is ample evidence that coordination with certain ligands does occur. For the transition-metal ions, it also appears that the electronic configuration of the ion is significant with regard to the stability of its complexes. For example, regardless of the nature of the ligand, the stability of complexes of bivalent transition metals is Mn < Fe < Co < Ni < Cu > Zn. This so-called natural order of stability is explained in terms of the crystal field theory, which indicates that a maximum stabilization effect in spin-free complexes is realized with d^8 systems, such as exists in Ni^{2+}. The platinum metals, because of their large polarizing ability, generally form the most stable metal complexes.

Effect of ligand. Several characteristics of the ligand are known to influence the stability of complexes: (1) basicity of the ligand, (2) the number of metal-chelate rings per ligand, (3) the size of the chelate ring, (4) steric effects, (5) resonance

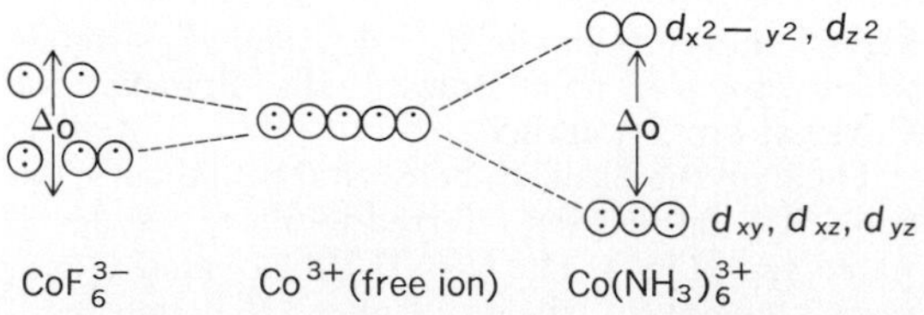

Fig. 4. Representation of the cobalt(III) complexes on the basis of crystal field theory.

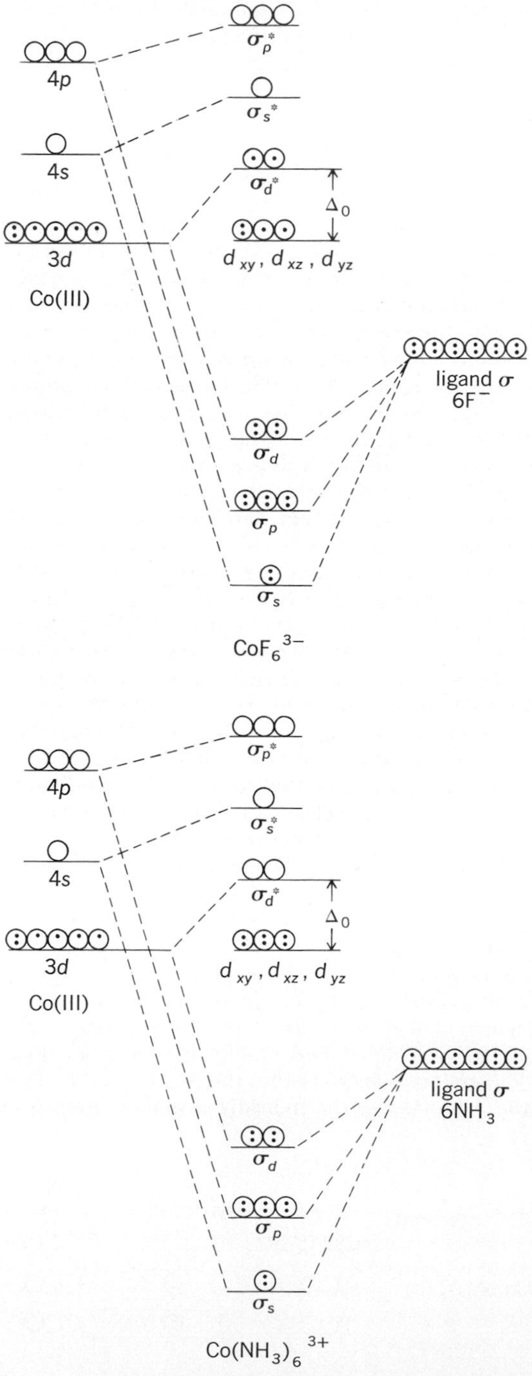

Fig. 5. Representation of the cobalt complexes CoF_6^{3-} and $Co(NH_3)_6^{3+}$ according to molecular orbital theory.

effects, and (6) the ligand atom. Since coordination compounds are formed as a result of acid-base reactions where the metal ion is the acid and the ligand is the base, it follows that generally the more basic ligand will tend to form the more stable complex. Much of the available quantitative data on the formation constants of coordination compounds gives a good linear correlation with the base strength of the ligand. It is also known that a polydentate ligand, one attached to the metal ion at more than one point, forms more stable complexes than does an analogous monodentate ligand. For example, ethylenediamine forms more stable complexes than does ammonia:

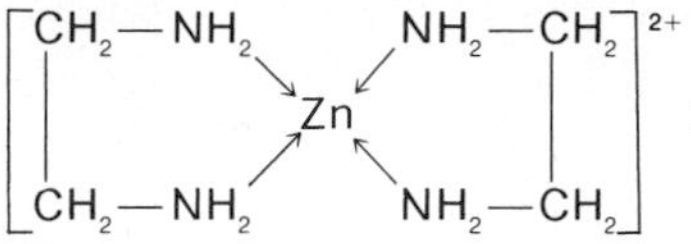

>

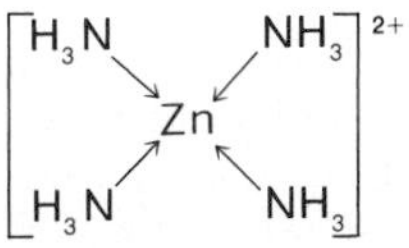

The ethylenediamine complex is referred to as a chelate complex. In general, an increase in the extent of chelation results in an increase in the stability of the complex. Ethylenediaminetetraacetate ion, a sexadentate ligand, is an excellent chelating agent and has found numerous applications as a sequesterant for metal ions. Figure 6 is the structural formula of the chelate complex of calcium ion and ethylenediaminetetraacetic acid.

The size of the chelate ring is likewise an important factor. For saturated ligands such as ethylenediamine, five-membered rings are the most stable, whereas six-membered rings are the most stable for chelates containing one or more double bonds, for example (I) and (II). Examples of smaller and of

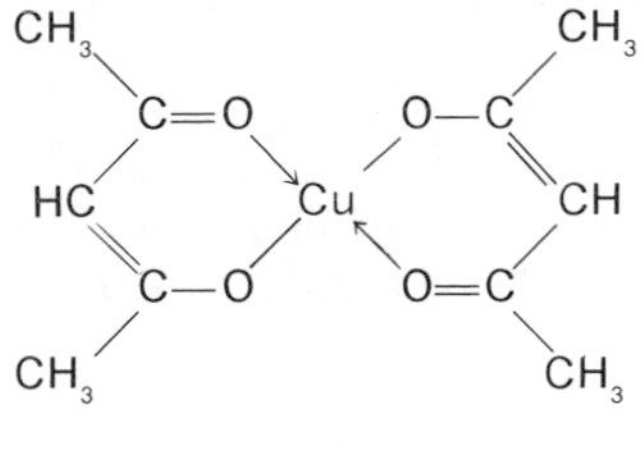

(I)

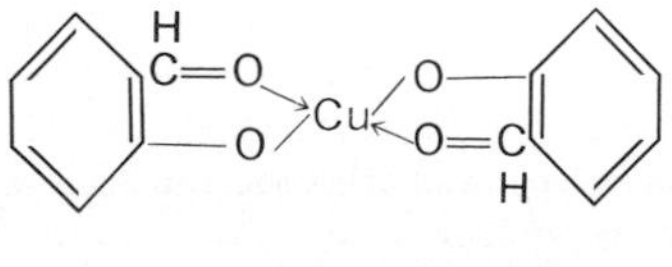

(II)

larger chelate ring systems are known, but these are much less common and less stable. It has been observed that for analogous systems an increase in double bonding in the chelate ring increases the stability of the complex. For ligands of comparable base strengths, the acetylacetonate complexes of type (I) are more stable than the salicylaldehyde compounds of type (II). There are in effect two double bonds per chelate ring in (I) compared to one and a half in (II). This same phenomenon is believed to be responsible for the marked stability of the porphyrin complexes such as chlorophyll, hemoglobin, and the phthalocyanine dyes.

Steric factors often have a very large effect on the stability of metal complexes. This is most frequently observed with ligands having a large group attached to the ligand atom or near it. Thus complexes, of the type shown in Fig. 7, with alkyl groups R in the position designated are much less stable than the parent complex where R = H. This results from the steric strain introduced by the size of the alkyl group on or adjacent to the ligand atom. In contrast to this, alkyl substitution at any other position results in the formation of more stable complexes because the ligand becomes more basic, and the bulky group is now removed from a position near the coordination site.

Finally, the ligand atom itself plays a significant role in controlling the stability of metal complexes. For most of the metal ions, the smallest ligand atom with the largest electron density will form the most stable complex. This means that the second period elements form more stable metal-ligand bonds than do other members of the same group, for example, N > P > As > Sb; O > S > Se > Te; F > Cl > Br > I. It is also known that for this same class of metal ions the stability order is N > O > F. These trends are in accord with the coordinate bond strengths expected on the basis of electrostatic interactions. These metal ions and the ligand atoms they prefer are not very polarizable, and are designated as hard Lewis acids and bases, respectively. A different trend in stability is found for a second class of metal ions, for example, Cu(I), Rh(I), Pd(II), Ag(I), Pt(II), and Hg(II). The stability trends observed for these metal ions are $N \gg P > As \sim Sb$, $O \ll S > Se \sim Te$, and $F \ll Cl < Br < I$. These metal ions and the ligand atoms they prefer are polarizable, and are designated as soft Lewis acids and bases, respectively. Thus, Pearson has proposed the HASAB theory of acids and bases, which states that hard Lewis acids pre-

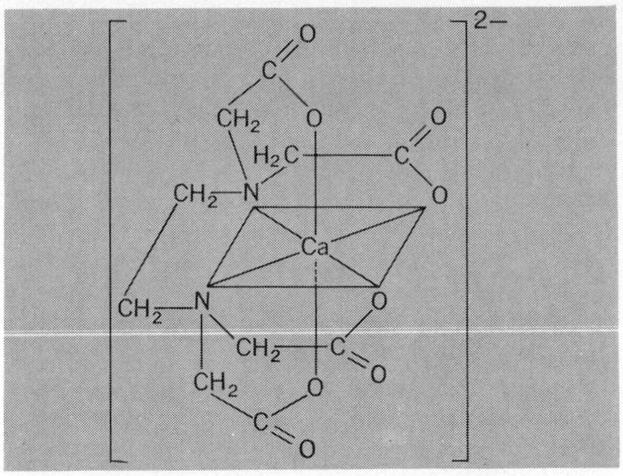

Fig. 6. Structural formula for the chelate complex formed from a calcium ion and ethylenediaminetetraacetic acid.

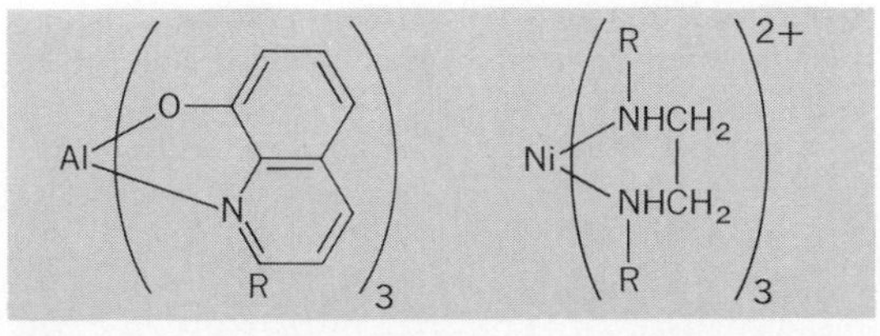

Fig. 7. Structural formulas of metal complexes which are affected by steric factors.

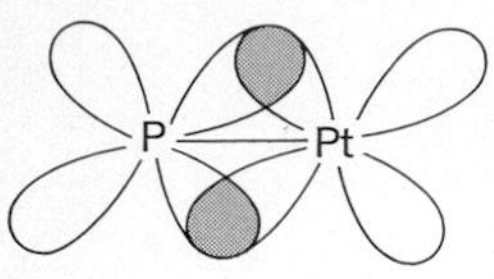

Fig. 8. Overlap of *d* orbitals of platinum with empty *d* orbitals of phosphorus ligand.

fer hard bases and soft Lewis acids prefer soft bases. Since the soft metal ions also have a large number of *d* orbital electrons and the third-period ligand atoms have vacant *d* orbitals, the enhanced stability is attributed to the formation of *dd*-π bonding. This means that the filled *d* orbitals of the metal ion overlap with the empty *d* orbitals of the ligand atom. This is illustrated by Fig. 8. This same class of metal ions is the one that forms olefin complexes such as Zeise's salt, $K[Pt(CH_2{=}CH_2)Cl_3]$. The stability of these compounds is likewise attributed to π-bonding.

Stability and reactivity. Often the most stable complex is also the least reactive or most inert. Therefore, the platinum metal complexes as well as cobalt(III) and chromium(III) complexes are usually very slow to react. However, there are exceptions to this, for instance, the rapid exchange of radiocyanide with the coordinated cyanide in the stable tetracyano complexes $Hg(CN)_4^{2-}$ and $Ni(CN)_4^{2-}$. Several factors, such as the electronic configuration of the central metal ion, its coordination number, and the extent of chelation, all have a marked effect on the rate of reaction of a given compound.

There is sufficient data available to permit a fairly reliable classification for the relative reactivities of six-coordinated systems. Metal complexes are said to be labile if they have reacted completely at room temperature within the time of mixing, such as the instantaneous formation of the deep-blue color of $Cu(NH_3)_4^{2+}$ upon the addition of excess ammonia to $Cu(H_2O)_4^{2+}$. Complexes are designated as being inert if their reactions at room temperature proceed at a detectable rate and have half-lives longer than 2 min. On the basis of such a definition, it is possible to classify the labile complexes as being those of either the outer orbital type, such as CoF_6^{3-}, $Al(C_2O_4)_3^{3-}$, $SnCl_6^{2-}$, $Ni(NH_3)_6^{2+}$, or the inner orbital type with one or more vacant *d* orbital, such as TiF_6^{2-}, $V(CN)_6^{3-}$, $Mo(SCN)_6^{2-}$. The inert complexes are then the inner orbital type with no vacant *d* orbital, for example $Co(NH_3)_6^{3+}$, $Cr(H_2O)_6^{3+}$, $RhCl_6^{3-}$. A similar classification on the basis of the crystal field theory is done in terms of the labile systems being those that are not stabilized greatly by the crystal field.

Reactions of complexes. The reactions referred to above are acid-base reactions, where one basic ligand replaces a less basic ligand coordinated to the acidic metal ion, for example, reaction (4).

$$Co(NH_3)_5Cl^{2+} + H_2O \rightarrow Co(NH_3)_5H_2O^{3+} + Cl^- \quad (4)$$

A reaction of this type is called a nucleophilic substitution reaction and is designated by the symbol S_N. There are at least two fundamentally different pathways conceivable for such a reaction. These are the dissociation and displacement mechanisms which are designated as S_N1 and S_N2, respectively. An S_N1 reaction goes by a two-step process, where the first step is a slow unimolecular heterolytic dissociation, reaction (5), followed by the rapid coordination of the entering group, reaction (6).

$$Co(NH_3)_5Cl^{2+} \xrightarrow{\text{Slow}} Co(NH_3)_5^{3+} + Cl^- \quad (5)$$

$$Co(NH_3)_5^{3+} + H_2O \xrightarrow{\text{Fast}} Co(NH_3)_5H_2O^{3+} \quad (6)$$

An S_N2 reaction is one involving a bimolecular rate-determining step in which one nucleophilic reagent displaces another [reaction (7)]. Examples

$$Co(NH_3)_5Cl^{2+} + H_2O \xrightarrow{\text{Slow}}$$
$$H_2O \cdots Co(NH_3)_5 \cdots Cl^{2+} \xrightarrow{\text{Fast}}$$
$$Co(NH_3)_5H_2O^{3+} + Cl^- \quad (7)$$

of both types of substitution are known for the reaction of metal complexes. The specific reaction just cited, as well as analogous reactions of $Co(NH_3)_4Cl_2^+$ and $Co(NH_3)_4OHCl^+$, appears to proceed by an S_N1 mechanism. However, reactions of platinum(II) complexes are believed to involve an S_N2 process, for example, reaction (8).

$$Pt(NH_3)_3Cl^+ + NO_2^- \xrightarrow{\text{Slow}}$$
$$O_2N \cdots Pt(NH_3)_3 \cdots Cl \xrightarrow{\text{Fast}}$$
$$Pt(NH_3)_3NO_2^+ + Cl^- \quad (8)$$

Most recently the symbols AID were introduced in order to better designate the mechanisms of these reactions. A = association and relates to S_N2, where the entering group becomes associated with the complex [slow step of Eq. (8)]; I = interchange and signifies that a group just outside the complex interchanges positions with a ligand on the metal; D = dissociation and relates to S_N1, where the leaving group dissociates away from the complex [Eq. (5)]. In general, I_d is similar to S_N1 and designates a dissociative interchange mechanism, whereas I_a is similar to S_N2 and designates an associative interchange mechanism.

Finally, coordination may have the effect of stabilizing an unusual oxidation state of the coordinated metal. For example, ligands such as CO, CN^-, CNR, 2,2′-bipyridine, and PR_3 are most effective in stabilizing low-valent states. Thus, compounds of the type $Na_2[Fe(CO)_4]$, Na[V(2,2′-bipyridine)$_3$], $K_4[Ni(CN)_4]$, and $Pt(PR_3)_4$ are known where the assigned oxidation states are Fe (2−), V (1−), Ni (0), and Pt(0). It is believed that the low oxidation states in these systems are stabilized by the formation of π-type molecular orbitals between the metal and ligand atoms. If the ligand orbitals are of low energy and vacant, they can accommodate the *d*-orbital electrons of the metal and permit the addition of other electrons leading to the reduction of the complex. In contrast to this, ligands such as O^{2-}, OH^-, and F^- tend to stabilize high-valence states giving complexes of the type K_3FeO_4, Na_3BiO_4, K_2NiF_6, and K_3CuF_6, where the oxidation states are Fe (6+), Bi (5+), Ni (4+), and Cu (3+). These ligands have no *d* orbitals, but only filled *p* orbitals. One explanation for the stabilization of higher-valent states by such ligands is that in these systems the π-type molecular orbitals cannot be formed so that the metal electrons are forced into higher-energy antibonding orbitals from which they may be readily removed. *See* CHEMICAL BONDING; MAGNETOCHEMISTRY; SOLID-STATE CHEMISTRY; STEREOCHEMISTRY; SUBSTITUTION REACTION.

[FRED BASOLO]

Bibliography: J. C. Bailar, Jr. (ed.), *The Chemistry of the Coordination Compounds*, Amer. Chem. Soc. Monog. no. 131, 1956; F. Basolo and R. G. Pearson, *Mechanisms of Inorganic Reactions*, 2d ed., 1967; R. G. Pearson (ed.), *Hard and Soft Acids and Bases*, 1973.

Coordination number

The number of nearest neighbors of a point in a space lattice, of an atom or an ion in a structure, or of an anion or cation in a complex ion. The figure shows that the coordination number in the three cubic lattices *P*, *I*, and *F* is respectively 6, 8, and 12. In closely packed structures this number is 12. The structures of the subgroup elements are easily described by means of the coordination number of each atom, which is given by the rule of Hume-Rothery, or the 8-*N* rule. Here *N* is the number of the subgroup. Diamond, germanium, and silicon are in subgroup four; the coordination number is 4. Tellurium and selenium are in subgroup six, each atom having two nearest neighbors. Accordingly, the structure is an assembly of strings.

In ionic structures each ion is surrounded by a number of ions of opposite sign. This coordination is respectively 8/8, 6/6, 4/4, 8/4 in the cesium chloride, sodium chloride, zinc sulfide, and calcium fluoride structures. For a discussion of these structures *see* CRYSTAL STRUCTURE.

The anions are, in nearly all cases, larger than the cations. For any given coordination the distribution of the anions is as regular as possible. Each cation is in the center of a coordination polyhedron. The structures can therefore be described in terms of the way in which the anion polyhedrons are packed. As an example, in the rock salt structure, each cation is in the center of an octahedron formed by six anions. Two neighboring octahedrons have a common edge. Along these lines L. Pauling has deduced a number of rules which summarize general features of ionic compounds. The most important are listed here.

1. A coordinated polyhedron of anions is formed about each cation, the cation-anion distance being determined by the radius sum, and the coordination number of the cation by the radius ratio.

2. In a stable coordination structure, the total strength of the valence bonds which reach an anion from all the neighboring cations is equal to the charge of the anion.

3. The existence of edges and particularly of faces common to two anion polyhedrons decreases stability. This effect is large for cations with high valency and small coordination number, and is especially large when the radius ratio approaches the lower limit of stability of the polyhedron.

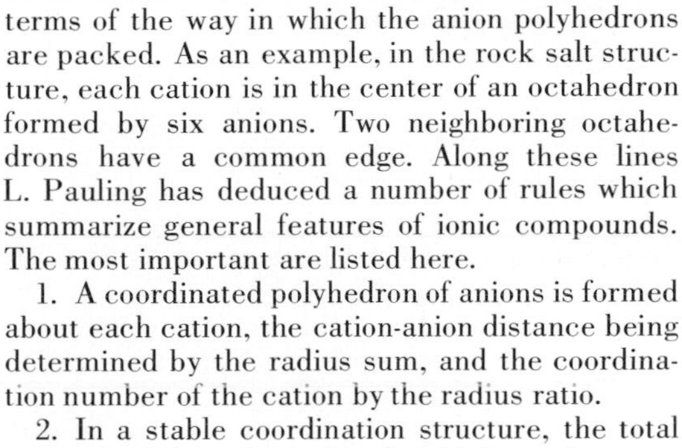

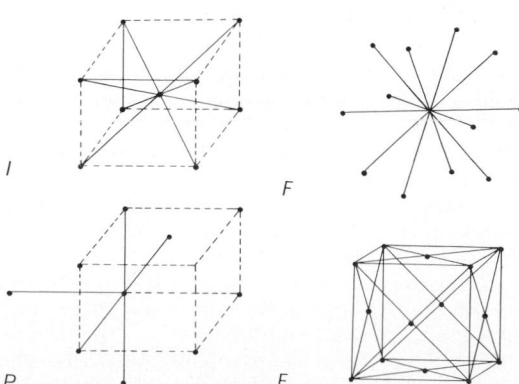

The coordination number in the three cubic lattices *P*, *I*, and *F* is shown to be 6, 8, and 12, respectively.

4. In a crystal containing different cations, those of high valency and small coordination number tend not to share polyhedron elements with each other. [WILLY C. DE KEYSER]

Bibliography: M. J. Buerger, *Crystal Structure Analysis*, 1979; R. C. Evans, *Introduction to Crystal Chemistry*, 2d ed., 1964.

Copper

A chemical element, Cu, atomic number 29, one of the most important nonferrous metals. Its usefulness is accounted for by its combination of chemical, physical, electrical, and mechanical properties and its fairly abundant supply. Copper is one of the first metals to have been used by humans.

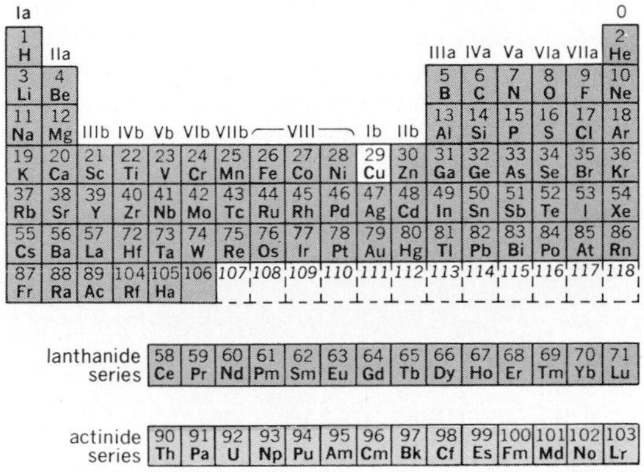

Natural occurrence. By far the greater part of the world's copper is obtained from the sulfide ores chalcocite, Cu_2S; covellite, CuS; chalcopyrite, $CuFeS_2$; bornite, Cu_5FeS_4; and enargite, $Cu_3(As,Sb)S_4$. Oxidized ores include cuprite, Cu_2O; tenorite, CuO; malachite, $CuCO_3Cu(OH)_2$; azurite, $2CuCO_3Cu(OH)_2$; chrysocolla, $CuSiO_3 \cdot 2H_2O$; and brochantite, $Cu_4(OH)_6SO_4$. Native copper, once widespread in the United States, is now mined in quantity only in Michigan. The grade of ore used for copper production has been going steadily downward as the richer ores have become exhausted and the demand for copper has grown. The average ore in the United States contains less than 1% copper, but the average is higher in other countries. There are vast amounts of copper in the ground, available for future use if ores of still lower grades are utilized, and there is no prospect of exhaustion for a long time to come. At the same time, the use of lower-grade ore increases production costs.

Production metallurgy. There are many variations in the production processes, and improvements are constantly being made. Only the most common methods can be indicated here because of space limitations (Fig. 1).

The ores are first concentrated, usually by flotation, to yield concentrates containing 20–40% copper; the grade has been improved in recent years and may exceed 40%.

Roasting the concentrates to lower the sulfur content has usually been the next step. Because the richer concentrates require no roasting, this step is omitted in many plants. Roasted concentrates are mixed with raw concentrates before smelting to obtain a sulfur content that will yield the desired grade of matte in the reverberatory furnace.

Smelting may be done in blast furnaces, but this practice has been replaced, except in a few instances, by reverberatory smelting. The reverberatory furnace has the advantage of being able to smelt finely divided flotation concentrates without sintering, and it is not dependent on coke as fuel. Natural gas, oil, or pulverized coal, whichever is cheapest in the locality of the smelter, can be used. The basis of the process is that reactions between copper oxide and iron sulfide in the molten charge have negative free-energy changes, forming copper sulfide and iron oxide. With excess iron sulfide, the copper sulfide forms a molten solution called copper matte. The iron oxide formed in the reactions, together with that added as flux, unites with silica in the ore to form a slag.

The matte is taken to a converter, in which air is blown through the molten matte to oxidize the iron chiefly to FeO, which is slagged by addition of a siliceous flux. The sulfur in the matte is oxidized to SO_2 and passes off in the gases, leaving copper in the converter. When cast, the copper forms cakes with a surface roughened and blistered by the escape of gases during freezing, and hence is called blister copper (Fig. 2). It is usually 98–99% Cu.

The blister copper is next partly refined in a furnace and cast into anodes. It may be poured in molten form into the anode furnace, or it may be cast into cakes which are charged to the furnace and remelted. An oxidation refining takes place in the furnace, assisted by introduction of a small amount of air below the surface of the molten copper. Little or no flux is required. Impurities whose oxides have a more negative free energy of formation than copper are oxidized and removed as a slag or in the gases. The copper is cast into anodes containing about 99.9–99.3% Cu.

The anodes are hung in electrolytic tanks, spaced alternately with cathodes, in an electrolyte of copper sulfate and free sulfuric acid, contaminated with soluble impurities. Insoluble impurities, in the main those below copper in the electrochemical series of metals, fall to the bottom of the tank; these include gold, silver, selenium, and tellurium. Soluble impurities, in the main those above copper in the electrochemical series, enter the electrolyte and do not plate out at the cathode. These are chiefly nickel and arsenic, together with smaller amounts of other elements. Some of the impurities form insoluble compounds. Thus the copper is freed of nearly all impurities, whereas a furnace-refining process can remove only those more easily oxidized than copper.

The cathodes are usually remelted for casting into shapes such as wirebars, cakes, billets, or ingots. Some of the cathodes are sold without remelting, but the shapes are required for fabrication into such forms as rods, wire, sheet, and tubes. Remelting is done in large reverberatory furnaces, arc furnaces, or induction furnaces. In reverberatory melting, oxygen and sulfur are taken up, and although the cathodes are already 99.98% Cu, the steps carried out in the anode-refining furnace must be largely repeated in order to remove the sulfur and reduce the oxygen content to the amount desired in tough-pitch copper, usually containing 0.02–0.05% oxygen. After being blown with air introduced through small steel pipes, the copper is subjected to poling. This consists of inserting poles of green wood beneath the surface of the molten copper to cause a reaction with the hydrogen and hydrocarbon gases evolved from the wood and to agitate the bath of molten copper, thus exposing it to air. Sulfur in the copper is removed as hydrogen sulfide and sulfur dioxide, and most of the copper oxide is reduced. Poling is now sometimes done by injecting oil or gas into the molten copper. The remaining copper oxide separates in a Cu_2O-Cu eutectic between the grains of copper; a close adjustment of the amount of eutectic is necessary to produce a casting with a nearly level surface. Casting is commonly done in horizontal open molds. Billets and sometimes other shapes are cast in vertical molds.

Continuous casting of oxygen-free copper is practiced on a large scale, the copper being protected from air while molten.

Most copper ores contain gold and silver. The recovery of these and other metals as by-products is an important part of the metallurgy of copper. When the ore is smelted, gold and silver dissolve

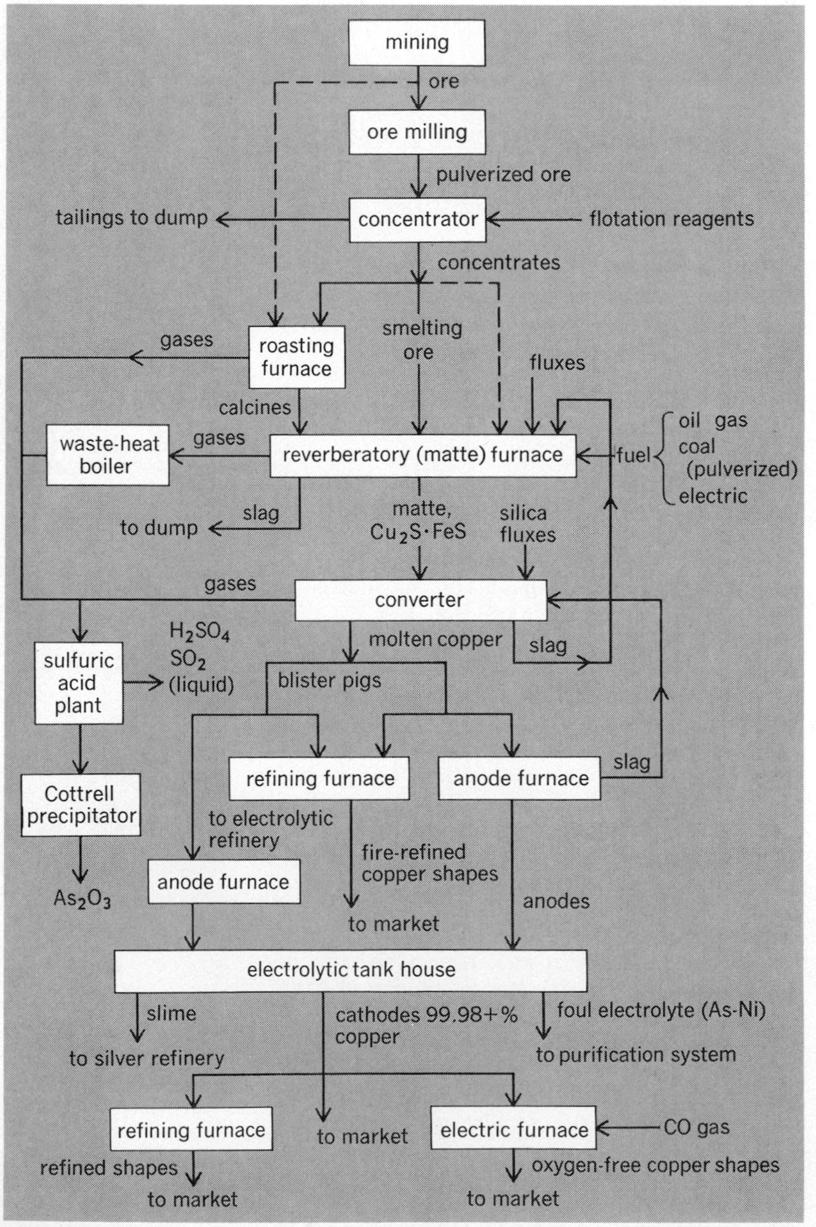

Fig. 1. Flow sheet of copper extraction from sulfide minerals.

in the matte, remaining with the copper through the steps of converting and furnace refining, to be separated in the electrolytic refining tanks.

A considerable amount of copper is produced by hydrometallurgy. The ore is leached without being concentrated, dilute sulfuric acid being the solvent most often used. The process is applied chiefly to oxidized ores. A further limitation of leaching is that the solvents employed do not dissolve gold or silver.

After leaching, it is necessary to recover the copper from the leaching solution. When sulfuric acid is the solvent, the resulting copper sulfate solution may be electrolyzed by use of insoluble anodes. The cathode deposit is pure, comparable with electrolytically refined copper. Another method of precipitating the copper from the leaching solution is to pass it over scrap iron or sponge iron (Fig. 3). Copper is replaced in the solution by iron and precipitates as so-called cement copper. In this case, the precipitated copper is not pure and must be refined.

An important development in the production metallurgy of copper is the use of oxygen, usually in the form of oxygen-enriched air containing commonly 30–50% oxygen. Flash roasting and flash smelting, in which finely ground concentrates are dropped through heated air in a furnace chamber, have grown in use. This may be done in preheated air or by the use of oxygen-enriched air. Another method is the injection of oxygen over the charge banks in the reverberatory furnace. Smelting pelletized concentrates by feeding them into the converters after an initial charge of molten matte is another innovation. Oxygen may also be used in refining furnaces. Oxygen-enriched air yields higher temperatures and accelerates the reactions involved in the roasting, smelting, and refining processes.

In another proposed method of copper recovery, low-grade ore in the ground would be shattered by a nuclear energy explosion, then leached in place with dilute sulfuric acid; the solution would next be pumped to the surface and the copper cemented out on scrap iron.

Chemical properties. Copper is the first element in subgroup Ib of the periodic table, which also includes the other coinage metals, silver and gold. The copper atom has the electronic structure $1s^2 2s^2 2p^6 3s^2 3p^6 3d^{10} 4s^1$. The low ionization potential of the $4s^1$ electron results in easy removal to give the copper (I), or cuprous ion, Cu^+; and the copper (II), or cupric ion, Cu^{++}, is readily formed by removal of one electron from the $3d$ shell.

The copper crystal is face-centered cubic, with lattice dimension $a = 3.6080$ kX units at 18°C, and the closest approach of atoms 2.551 kX units.

The atomic weight of copper is 63.546. It has two stable natural isotopes, Cu^{63}, consisting of 29 protons and 34 neutrons, and Cu^{65}, with 29 protons and 36 neutrons. There are also known nine unstable (radioactive) isotopes having the mass numbers 58, 59, 60, 61, 62, 64, 66, 67, and 68. The half-lives of these are 3.2 sec, 81 sec, 24 min, 3.33 hr, 9.9 min, 12.9 hr, 5.1 min, 61 hr, and 30 sec, respectively.

Copper is characterized by low chemical activity. The standard potential of the ionization $Cu \rightarrow Cu^{++}$ is 0.34 ± 0.01 volt in normal ionic solution at 25°C. The heat of oxidation of copper is $37{,}100 \pm 800$ cal/mole (584 cal/g of Cu) in formation of CuO, and $40{,}000 \pm 700$ cal/mole (315 cal/g of Cu) in formation of Cu_2O.

The element combines chemically in one of three valences. The most common valence is 2+ (cupric), but 1+ (cuprous) is also frequent; the valence 3+ occurs in only a few unstable compounds.

The electrochemical equivalent of bivalent copper is 0.0003294 g/coulomb.

Physical properties. Copper is a comparatively heavy metal. The density of the pure solid is 8.96 g/cm³ at 20°C. The density of commercial copper varies with method of manufacture, averaging 8.90–8.92 in cast refinery shapes, 8.93 for annealed tough-pitch copper, and 8.94 for oxygen-free copper. The density of liquid copper is 8.22 near the freezing point.

The melting point of copper is 1083.0 ± 0.1°C (1981.4 ± 0.2°F). Its normal boiling point is 2595°C (4703°F).

The coefficient of linear expansion of copper is 1.65×10^{-5}/°C at 20°C.

The specific heat of the solid is 0.092 cal/g at 20°C. The change of specific heat with tempera-

COPPER

Fig. 2. Blister copper being poured. (*Anaconda Co.*)

Fig. 3. Mine water containing copper sulfate is passed over scrap iron to precipitate the copper. (*Anaconda Co.*)

ture is given by the expression $0.092 + 0.0000250t$ cal/g, equivalent to $5.85 + 0.00159t$ cal/mole. The specific heat of liquid copper is 0.112 cal/g, and of copper in the vapor state about 0.08 cal/g.

The latent heat of fusion is 48.9 cal/g and of vaporization about 1150 cal/g.

Shrinkage of copper in solidification is 4.92%.

Copper is nonmagnetic; or more precisely, it is slightly paramagnetic, having a mass susceptibility of -0.080×10^6 cgs units/g at 20°C, equivalent to -5.1×10^6 cgs units/mol.

The thermal conductivity of copper, like its electrical conductivity, is very high. It is 0.941 cal/(sec)(cm²)(cm)(°C) at 20°C; this is equivalent to a resistivity of 0.254 thermal ohm/(cm²)(cm).

Electrical properties. The electrical reistivity of copper in the usual volumetric unit, that of a cube measuring 1 cm in each direction, is 1.6730×10^{-6} ohm-cm at 20°C. Only silver has a greater volumetric conductivity than copper. On a relative basis in which silver is rated 100, copper is 94, aluminum 57, and iron 16.

The mass resistivity of pure copper for a length of 1 m weighing 1 g at 20°C is 0.14983 ohm. The conductivity of copper on the mass basis is surpassed by several light metals, notably aluminum. The relative values are 100 for aluminum, 50 for copper, and 44 for silver.

By far the largest use of copper is in the electrical industry, and therefore high electrical conductivity is its most important single property, although for industrial use this property must be accompanied by suitable characteristics in other respects. The conductivity of commercial copper is commonly rated on a percentage basis. This method affords a convenient standard for buying and selling, for comparison of quality for electrical purposes, and for comparison with other metals. The rating is based on percentage of a mass conductivity, adopted as a standard by the International Electrotechnical Commission in 1913 and subsequently by the American Standards Association, American Society for Testing and Materials, and other bodies. The adopted standard mass resistivity is 0.15328 ohm (m, g), and to its reciprocal is assigned the value 100, the conductivities of other specimens being stated in percentages of this conductivity. This standard conductivity is designated the International Annealed Copper Standard (IACS). Commercial electrolytic copper of the highest purity regularly produced has a conductivity of about 102 IACS, and most of the copper sold averages between 100.5 and 101.8.

The volumetric resistivity equivalent to 100 IACS is 1.7241×10^{-6} ohm-cm. Other IACS equivalent resistivity figures used in engineering are 10.371 ohms (mil, ft) and 875.2 ohms (mi,lb).

The resistivity of copper increases with temperature, the amount of change being given by the temperature coefficient 0.00393/°C at 20°C, equivalent to an increase of 6.02×10^{-4} ohm (m, g/°C).

Mechanical properties. Copper is one of the strongest pure metals. It is moderately hard, extremely tough, and wear resistant. The strength of copper is accompanied by high ductility. For example, copper wire of commercial purity with a tensile strength of 42,000 psi and yield point of 36,000 psi has an elongation in 2 in. of 14%. These are typical values after 50% cold reduction. Work-hardened copper is readily softened by annealing. Fully annealed copper wire has a tensile strength of 32,000 psi, yield point of 10,000 psi, and elongation in 2 in. of 45%. The mechanical properties of metals, and also their electrical properties, are strongly dependent on the physical condition of the metal, especially on the amount of strain hardening. They are also affected by temperature and grain size. Figure 4 shows the effect of cold reduction on the electrical conductivity of copper and on its tensile strength for two different grain sizes; the conductivity specimens were wire of 0.204-in. diameter annealed and cold-drawn to several reductions of area; the tensile specimens were strip of 0.040-in. thickness annealed to different grain sizes and cold-rolled to several reductions of area.

The modulus of elasticity of copper in tension (Young's modulus) is 16.5×10^6 psi.

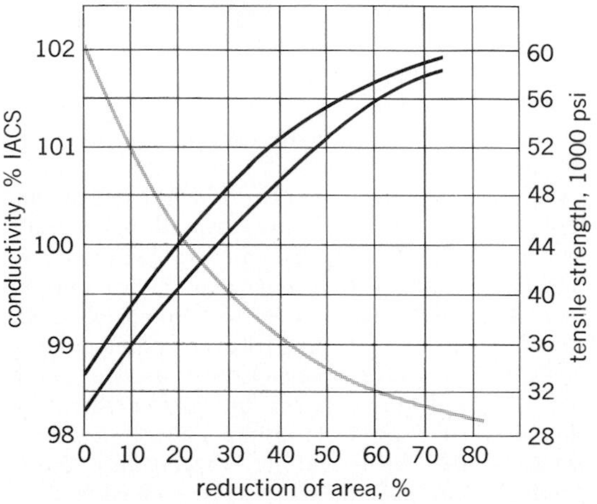

Fig. 4. Effect of cold work on electrical conductivity (light line) and tensile strength (dark lines) of tough-pitch copper. *(After data of D K. Crampton, H. L. Burghoff, and J. L Stacy, in R. A. Wilkins and E. S Bunn, Copper and Copper Base Alloys, McGraw-Hill, 1943)*

Commercial copper. In order to draw a line between marketed grades of copper and copper alloys, the American Society for Testing and Materials has adopted specifications in which metal containing not more than 0.5% of an element or elements other than Cu is designated as copper. These specifications cover the three main commercial varieties, as determined by the method of refining, namely, electrolytic, fire-refined, and Lake copper; also, as determined by the method of processing: tough-pitch, oxygen-free, and deoxidized copper; and a number of specific kinds, which include casting copper, phosphorized copper, high-residual-phosphorus copper, low-residual-phosphorus copper, silver-bearing copper, arsenical copper, selenium-bearing copper, and tellurium-bearing copper.

Electrolytic tough-pitch copper, designated ETP, is the most used type. It has been electrolytically refined, and the cathodes remelted and cast according to procedures which will result in an oxygen content of about 0.02–0.05%. The specifications for ETP copper require a minimum copper content of 99.90% (silver being counted as copper) and electrical conductivity not less than 100.0 IACS.

Oxygen-free copper has been electrolytically

refined and cast under conditions that exclude oxygen, or "coalesced" by heat and pressure into massive form without remelting, or cast with addition of a small amount of a deoxidizing element, usually phosphorus. This last variety is termed deoxidized copper to distinguish it from the other types.

Casting copper is refined in a furnace (fire-refined). It usually contains more oxygen and impurities than electrolytic copper. Little of it is used in the United States, but it is of some importance in England and European countries, especially for nonelectrical purposes. Remelted secondary (scrap) copper is the usual source.

Deoxidized copper, which contains the amount of phosphorus commonly found as residual element after deoxidation, does not have sufficiently high conductivity to meet electrical specifications. The other types of copper free from oxygen have about the same conductivity as ETP. The absence of copper oxide in itself would result in higher conductivity, but this factor may be offset by a slightly lower density which tends to lower conductivity. Another factor tending to lower conductivity is the fact that some impurities present in the copper will not exist as oxides but will be present in solid solution, in which form they have a greater effect in reducing conductivity. The range of variation is relatively small in the first and second types of copper, and electrical specifications are readily met.

Oxygen-free copper costs slightly more than ETP. Any superiority in conductivity is not often the reason for its use; this copper is especially valuable for higher ductility, which enables it to undergo more severe cold work. This ability, of course, is one of the valuable properties of copper in comparison with other structural materials; in the oxygen-free type, this advantage is enhanced. Another valuable characteristic of oxygen-free copper is its ability to be heated in a reducing atmosphere, as in bright annealing, without developing porosity and brittleness because of reduction of copper oxide between the grains.

There has been considerable substitution of other metals (chiefly aluminum) and plastics for copper and copper alloys, because of an increase in the price of copper. Nevertheless, the demand for copper for established uses has continued to grow rapidly; from 1959 to 1966 the world's consumption of copper increased 50% (in the United States, 70%).

Copper alloys. Copper is one of the few metals used to a greater extent in pure form than in alloyed form. However, one of the features of copper is its ability to form a great number and variety of alloys. Brass and bronze in their many varieties are among the oldest and most useful alloys (Fig. 5).

The principal classes of copper alloys are listed below according to composition:

Copper-zinc (binary brasses)
Copper-tin (binary bronzes)
Copper-zinc-tin (special brasses and bronzes)
Copper-zinc-lead and copper-tin-zinc-lead (leaded brasses and bronzes)
Copper-zinc-nickel (nickel silvers)
Copper-zinc-manganese, with or without tin, iron, or aluminum (manganese bronzes)
Copper-tin-phosphorus (phosphor bronze)
Copper-aluminum and copper-aluminum plus iron, nickel, or manganese (aluminum bronzes)
Copper-silicon plus manganese, tin, iron, or zinc (silicon bronzes)
Copper-nickel (cupronickel)
Copper-beryllium and copper-cobalt-beryllium (beryllium copper)

Other elements may be added to some of the above combinations to form other alloy brasses and bronzes.

Fabrication of copper and copper alloys. Pure copper has poor casting qualities caused especially by the release of gases from the molten metal when it solidifies. Machining also is difficult. Castings of intricate or special shapes are seldom made of pure copper, although there are exceptions when high thermal or electrical conductivity in a cast shape is a prime consideration. Wirebars, cakes, billets, and such objects of regular or simple shape intended for working are readily cast if the oxygen content of the molten material is adjusted. In this case, a certain amount of porosity may be tolerated because it will be closed up during working.

Most copper alloys have superior casting qualities, and some of the bronzes and brasses are

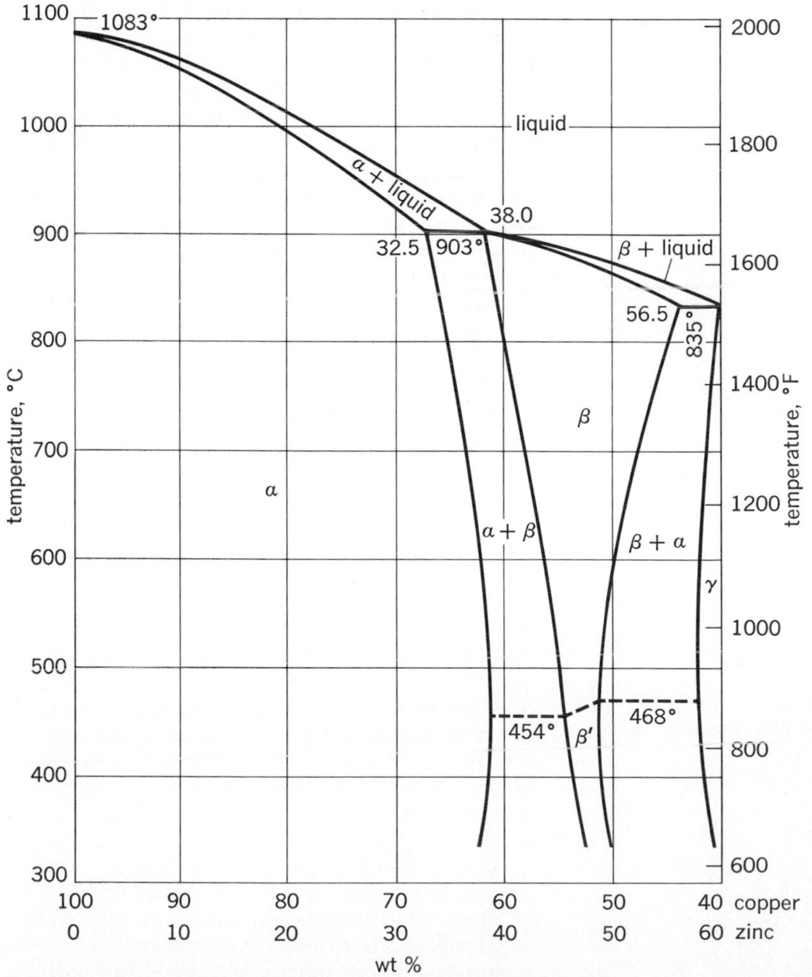

Fig. 5. Constitution of copper-zinc system. After cooling below 300°C, alloys consist of single solid solution α of zinc in copper containing zinc up to about 37%. The constituent β begins to separate from α at 903°C and at room temperature exists with α from 37 to 47% zinc. Amount of β increases as percent of zinc increases.

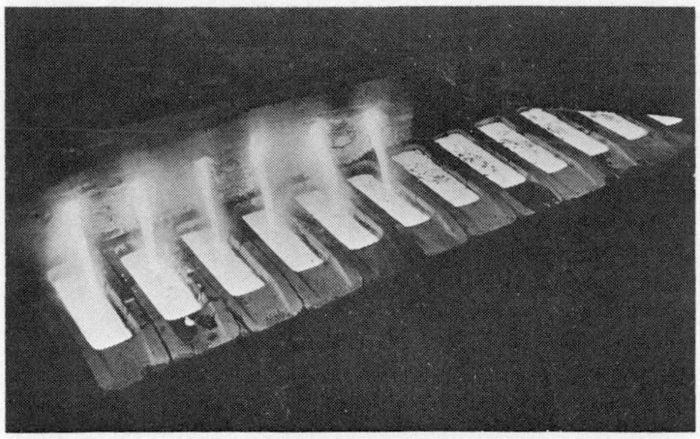

Fig. 6. Pouring bronze melt into ingot molds. (*American Smelting and Refining Co.*)

unexcelled in this respect (Fig. 6). Die castings are made with some compositions of brass.

Pure copper is highly workable, and this is true also of many of the wrought alloys. The most important methods of working are rolling, extrusion, wire drawing, piercing (for billets), and stamping. Other methods employed include forging, pressing, deep drawing, flanging, machining of various kinds, spinning, and cupping.

For manufacture of copper wire, copper wirebars are heated to 700–800°C (1300–1470°F) and rolled to rod 1/4–3/8 in. in diameter. This is drawn cold through a series of dies of tungsten carbide, decreasing the diameter in each draw; diamond dies are used for fine wire. The metal becomes hardened through cold drawing. If not softened by annealing, it is termed hard-drawn wire; this has maximum strength, but its electrical conductivity is relatively low. When annealed at the end of the drawing operation, the wire is soft-drawn and has the maximum conductivity with lower strength. Other tempers, such as quarter-hard and half-hard, are made by annealing after reduction to a certain point and then drawing to final size (Fig. 4).

Copper plate and sheet are rolled from cast copper cakes with annealing to adjust the final temper in a manner similar to that followed in wire drawing. The same type of procedure is used in rolling brass, but an additional factor of considerable importance is grain size. The size of grain is controlled by the temperature and time of annealing after partial reduction. Large grain size yields higher ductility but causes the surface to roughen during working and makes it difficult to polish. Grain sizes below 0.030 mm are used for operations which require relatively little forming and in which a bright surface finish is wanted. Sizes from 0.030 to 0.050 mm permit more severe cold work and yet produce a fairly bright surface; these are most often used. Sizes above 0.050 mm are used for heavy drawing and spinning when the surface finish is unimportant. An orange-peel appearance on the surface may result from drawing (especially deep drawing) materials having too large a grain size.

Copper is one of the principal metals used for electroplating. Not only may the outside coating be of copper, but in chromium plating on steel, the usual method is to plate the steel first with thin undercoatings of copper and nickel. Electroplating is sometimes done with brass or bronze.

Copper is also used for cladding. In this process copper sheet is hot-rolled onto the surface of steel or other metal.

Corrosion resistance. Copper and some of its alloys are very resistant to corrosion. Copper resists corrosion for two main reasons, its relative chemical inertness and the formation of a protective film. Copper lies toward the bottom of the electrochemical series of metals, well below hydrogen. When an electrolytic couple is formed between copper and another metal, copper is in most cases the cathode, so that it is electrochemically protected.

Copper becomes coated with mixtures of oxide, basic carbonate, and basic sulfate. Unlike the protective coating on aluminum and other metals, the film on copper is visible. It is of different colors, most often green, and may be porous and powdery at the surface, while continuous and adherent at the junction of the coating and the metal.

Brass, especially high brass, is subject to two special corrosion processes, season cracking and dezincification. Both processes sometimes occur in other copper alloys and in other metals. Season cracking is a form of stress-corrosion cracking. It is mostly intergranular, occurring even under mildly corrosive conditions when internal stress exists in the brass after cold work. It develops slowly; in addition to stress and time, moisture and the presence of traces of ammonia in the atmosphere are causative factors. It has been troublesome in brass cartridge cases. It is usually avoided by a stress-relieving anneal at a temperature high enough to permit crystalline readjustment, but not so high as to cause softening; 30 min at 300°C (570°F) is often employed.

In dezincification, zinc atoms are removed from the solid solutions more readily than copper atoms. Probably both copper and zinc dissolve initially, copper then being deposited as it is replaced in solution by more zinc. A layer of copper appears on the surface, underlain by porous and weakened alloy. Dezincification occurs in acids, acidic or salt water, and some impure or hard waters. It may also occur in heating thin sections of brass through vaporization of zinc from the surface.

Principal compounds. Of the hundreds of compounds, in addition to those mentioned above only a few are manufactured industrially on a large scale. The most important is copper (II) sulfate pentahydrate or blue vitriol, $CuSO_4 \cdot 5H_2O$. Others include bordeaux mixture, $3Cu(OH)_2CuSO_4$; paris green, a complex of copper metaarsenite and acetate; cuprous cyanide, CuCN; cuprous oxide, Cu_2O; cupric chloride, $CuCl_2$; cupric oxide, CuO; basic cupric carbonate; and copper naphthenate, the most widely used agent for prevention of rotting in wood, fabric, rope, and fishing nets. Leading uses of copper compounds are in agriculture, especially as fungicides and insecticides; as pigments; in electroplating solutions; in primary cells; as mordants in dyeing; and as catalysts. *See* COPPER CHEMISTRY; TRANSITION ELEMENTS.

[ALLISON BUTTS]

Bibliography: American Bureau of Metal Statistics, *Year Book*; American Society for Testing and Materials, *ASTM Standards*, pt. 6: *Copper and Copper Alloys*, 1980; Copper Development Association, *Standards Handbook*: *Copper and Copper Alloys*, 7 pts., 1979; J. H. Mendenhall, *Understanding Copper Alloys*, 1980; J. H. Tatsch, *Copper Deposits*: *Origin, Evolution and Present Characteristics*, 1975.

Copper chemistry

The principal oxidation states of copper in its compounds are 1+ and 2+, with several unstable compounds exhibiting 3+. Compounds in which copper is 1+, copper(I) or cuprous, are unstable in solution with respect to the reaction shown in (1). However, with respect to heat, the copper(I)

$$2Cu^+ \rightleftharpoons Cu^{++} + Cu \qquad (1)$$

compounds are the most stable and are the terminal product of the heating of many copper(II), or cupric, compounds; for example, the reactions (2)–(4).

Two oxides of copper are well known, copper(I) oxide, Cu_2O, and copper(II) oxide, CuO. Red

$$2CuO \xrightarrow{\Delta} Cu_2O + \tfrac{1}{2}O_2 \quad \text{(at about 900°C)} \qquad (2)$$

$$2CuS \xrightarrow{\Delta} Cu_2S + S \quad \text{(at red heat)} \qquad (3)$$

$$CuF_2 \xrightarrow{\Delta} CuF + \tfrac{1}{2}F_2 \quad \text{(at 500°)} \qquad (4)$$

copper(I) oxide can be prepared by the thermal decomposition of copper(II) oxide or by the reduction of a copper(II) salt in basic solution with a weak reducing agent such as glucose. Copper(II) oxide, which is black in color, is prepared by the action of heat on copper metal in air. It is also the terminal product of the thermal decomposition of copper nitrate, copper hydroxide, copper carbonate, and other oxygen-containing compounds.

Two series of copper halide salts are also known, the copper(I) and copper(II) compounds. The copper(II) halides, with the exception of the iodide, can be prepared by the direct combination of the elements, reaction (5). The colors of the anhydrous

$$Cu(s) + X_2(g) \rightarrow CuX_2(s) \qquad (5)$$

compounds vary widely: CuF_2 is colorless; $CuCl_2$ is bright yellow; and $CuBr_2$ is black. However, the 2-hydrates are all blue or blue-green. They are all soluble in water, dilute acids, and ammonia. With ammonia, complexes of the type $Cu(NH_3)_4X_2$ are formed, which explains their solubility in this solvent.

The copper(I) halides, with the exception of CuI, are prepared by the reduction of the copper(II) halides with elemental copper in hydrohalide acid solution according to the reaction shown in (6). All of the copper(I) halides are color-

$$CuX_2 + Cu \xrightarrow{HX} 2CuX \qquad (6)$$

less. Copper(I) iodide can be prepared by the reaction shown in (7), in which brownish-white

$$2CuCl_2 + 4KI \rightarrow 2CuI(s) + 4KCl + I_2 \qquad (7)$$

CuI precipitates out of solution. From vapor-density studies at elevated temperatures, it was concluded that the copper(I) halides were dimeric in structure, or Cu_2X_2. However, from other studies, copper(I) chloride was found to be monomeric, or CuCl. To avoid confusion, the monomeric structure is used throughout this article. Copper(I) cyanide is similar to the halides in that it can be prepared in a manner identical to CuI, reaction (8).

$$2Cu^{++} + 4CN^- \rightarrow 2CuCN(s) + C_2N_2 \qquad (8)$$

Perhaps the most important compound of copper is copper(II) sulfate 5-hydrate. This compound, commonly called blue vitriol, is used as a source of other copper(II) compounds as well as in electroplating and as a fungicide. The 5-hydrate begins to lose water of hydration at about 27°C, to form the 3-hydrate. On further heating, the white anhydrous salt, $CuSO_4$, is produced; this decomposes at temperatures above 702°C to give CuO and SO_3. The addition of the anhydrous salt to water gives the blue 5-hydrate. However, on long standing in aqueous solution, a small amount of basic copper sulfate is formed.

The simple carbonate, $CuCO_3$, is unknown; however, the addition of sodium carbonate to a copper(II) salt solution gives basic salts such as $CuCO_3 \cdot Cu(OH)_2$, commonly called malachite.

The reaction of copper metal with nitric acid produces, among other products, copper nitrate, the 6-hydrate of which is a deep-blue color. Anhydrous copper nitrate sublimes at 150°C in vacuum, and is surprisingly volatile.

Leading uses of copper compounds are in agriculture, especially as fungicides and insecticides; as pigments; in primary electrical cells; as mordants in dyeing; and as catalysts. *See* AMMINE; COPPER; PIGMENT; TRANSITION ELEMENTS.

[WESLEY W. WENDLANDT]

Cotton effect

The characteristic wavelength dependence of the optical rotatory dispersion curve or the circular dichroism curve or both in the vicinity of an absorption band.

When an initially plane-polarized light wave traverses an optically active medium, two principal effects are manifested: a change from planar to elliptic polarization, and a rotation of the major axis of the ellipse through an angle relative to the initial direction of polarization. Both effects are wavelength dependent. The first effect is known as circular dichroism, and a plot of its wavelength (or frequency) dependence is referred to as a circular dichroism (CD) curve. The second effect is called optical rotation and, when plotted as a function of wavelength, is known as an optical rotatory dispersion (ORD) curve. In the vicinity of absorption bands, both curves take on characteristic shapes, and this behavior is known as the Cotton effect, which may be either positive or negative (Fig. 1). There is a Cotton effect associated with each absorption process, and hence a partial CD curve or partial ORD curve is associated with each particular absorption band or process. *See* ROTATORY DISPERSION.

Measurements. Experimental results are commonly reported in either of two sets of units, termed specific and molar (or molecular). The specific rotation $[\alpha]$ is the rotation in degrees produced by a 1-decimeter path length of material containing 1 g/ml of optically active substance, and the specific ellipticity θ is the ellipticity in degrees for the same path length and same concentration. Molar rotation $[\varphi]$ (sometimes $[M]$) and molar ellipticity $[\theta]$ are defined by Eqs. (1) and (2). For comparisons among different compounds,

$$[\varphi] = [\alpha]M/100 \qquad (1)$$

$$[\theta] = \theta M/100 \qquad (2)$$

the molar quantities are more useful, since they allow direct comparison on a mole-for-mole basis.

The ratio of the area under the associated partial

COTTON EFFECT

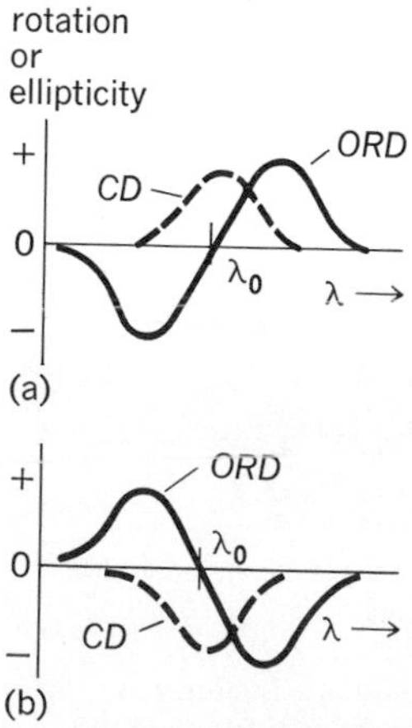

Fig. 1. Behavior of the ORD and CD curves in the vicinity of an absorption band at wavelength λ_0 (idealized). (*a*) Positive Cotton effect. (*b*) Negative Cotton effect.

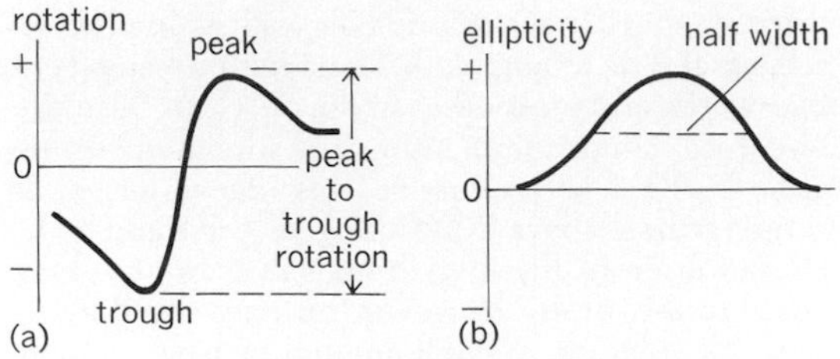

Fig. 2. Curves used to determine relative rotatory intensities. (*a*) Partial ORD curve. (*b*) Partial CD curve.

CD curve to the wavelength of the CD maximum is a measure of the rotatory intensity of the absorption process. Moreover, for bands appearing in roughly the same spectral region and having roughly the same half-width (Fig. 2), the peak-to-trough rotation of the partial ORD curve is roughly proportional to the wavelength-weighted area under the corresponding partial CD curve. In other words, relative rotatory intensities can be gaged from either the pertinent partial ORD curves or pertinent partial CD curves. A convenient quantitative measure of the rotatory intensity of an absorption process is the rotational strength. The rotational strength R_i of the i_{th} transition, whose partial molar CD curve is $[\theta_i(\lambda)]$, is given by Eq. (3).

$$R_i \approx 6.96 \times 10^{-43} \int_0^\infty \frac{[\theta_i(\lambda)]}{\lambda}\, d\lambda \qquad (3)$$

COTTON EFFECT

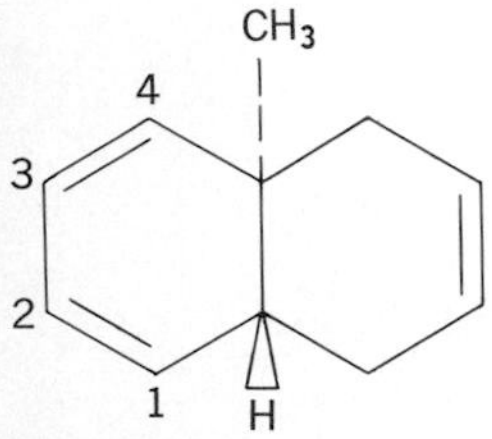

Fig. 3. Structural formula of (+)-*trans*-9-methyl-1,4,9,10-tetrahydronaphthalene.

Molecular structure. The rotational strengths actually observed in practice vary over quite a few orders of magnitude, from $\sim 10^{-38}$ down to 10^{-42} cgs and less; this variation in magnitude is amenable to stereochemical interpretation. In this connection it is useful to classify optically active chromophores, which are necessarily dissymmetric, in terms of two limiting types: the inherently dissymmetric chromophore, and the inherently symmetric but dissymmetrically perturbed chromophore. *See* OPTICAL ACTIVITY.

A symmetric chromophore is one whose inherent geometry has sufficiently high symmetry so that the isolated chromophoric group is superimposable on its mirror image, for example, the carbonyl group $>C{=}O$. The transitions of such a chromophore can become optically active, that is, exhibit a Cotton effect, only when placed in a dissymmetric molecular environment. Thus, in symmetrical formaldehyde, $H_2C{=}O$, the carbonyl transitions are optically inactive, while in ketosteroids, where the extrachromophoric portion of the molecule is dissymmetrically disposed relative to the symmetry planes of the $>C{=}O$ group, the transitions of the carbonyl group exhibit Cotton effects. In such instances the signed magnitude of the rotational strength will depend both upon the chemical nature of the extrachromophoric perturbing atoms and their geometry relative to that of the inherently symmetric chromophore. In a sense, therefore, the chromophore functions as a molecular probe for searching out the chemical dissymmetries in the extrachromophoric portion of the molecule.

The type of optical activity just described is associated with the presence of an asymmetric carbon (or other) atom in a molecule. The asymmetric atom serves notice to the effect that, if an inherently symmetric chromophore is present in the molecule, it is almost assuredly in a dissymmetric environment, and hence it may be anticipated that its erstwhile optically inactive transitions will exhibit Cotton effects. Moreover, the signed magnitude of the associated rotational strengths may be interpreted in terms of the stereochemistry of the extrachromophoric environment, as compared with that of the chromophore. But an asymmetric atom is not essential for the appearance of optical activity. The inherent geometry of the chromophore may be of sufficiently low symmetry so that the isolated chromophore itself is chiral, that is, not superimposable on its mirror image, for example, in hexahelicene.

COTTON EFFECT

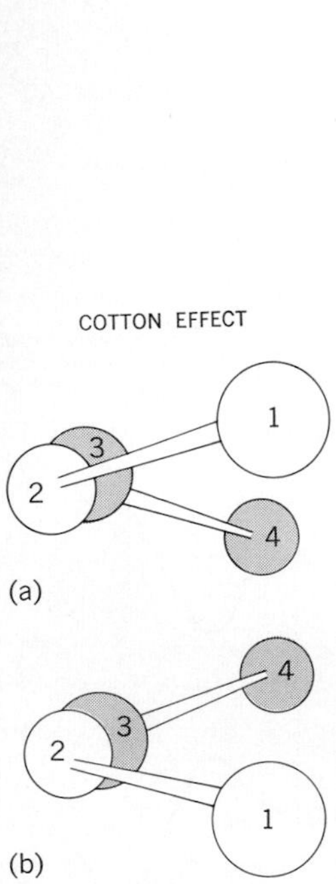

Fig. 4. Schematic representation of the twisted diene chromophore showing the two possible handednesses; the numbering is as indicated in Fig. 3. (*a*) Right-handed. (*b*) Left-handed.

In such instances the transitions of the chromophore can manifest optical activity even in the absence of a dissymmetric environment. In addition, it is very often true that the magnitudes of the rotational strengths associated with inherently dissymmetric chromophores will be one or more orders of magnitude greater ($\sim 10^{-38}$ cgs, as opposed to $< 10^{-39}$ cgs) than those associated with inherently symmetric chromophores. Hence, in the spectral regions of the transitions of the inherently dissymmetric chromophore, it will be the sense of handedness of the inherently dissymmetric chromophore itself that will determine the sign of the rotational strength, rather than perturbations due to any dissymmetric environment in which the inherently dissymmetric chromophore may be situated.

The sense of handedness of an inherently dissymmetric chromophore may be of considerable significance in determining the absolute configuration or conformations of the entire molecule containing that chromophore. Accordingly, the absolute configuration or conformation can often be found by focusing attention solely on the handedness of the chromophore itself. For example, in the chiral molecule shown in Fig. 3 there is a one-to-one correspondence between the sense of helicity of the nonplanar diene chromophore present and the absolute configuration at the asymmetric carbon atoms. Hence there exists a one-to-one correspondence between the handedness of the diene and the absolute configuration of the molecule. Since it is known that a right-handed diene helix (Fig. 4) associates a positive rotational strength with the lowest diene singlet transition in the vicinity of 260 mμ, by examination of the pertinent experimental Cotton effect (positive), the absolute configuration of the molecule is concluded to be as shown.

Other examples of inherently dissymmetric chromophores are provided by the helical secondary structures of proteins and polypeptides. Here the inherent dissymmetry of the chromophoric system arises through a coupling of the inherently symmetric monomers, which are held in a comparatively fixed dissymmetric disposition relative to each other through internal hydrogen bonding. The sense of helicity is then related to the signs of the rotational strengths of the coupled chromophoric system. The destruction of the hydrogen bonding destroys the ordered dissymmetric secondary structure, and there is a concomitant decrease in the magnitude of the observed rotational strengths.

[ALBERT MOSCOWITZ]

Bibliography: *Proc. Roy. Soc. London Ser. A*, 297:1–172, 1976; G. Snatzke (ed.), *Optical Rotatory Dispersion and Circular Dichroism in Organic Chemistry*, 1967; L. Velluz et al., *Optical Circular Dichroism: Principles, Measurements, and Applications*, 1969.

Coulometer

Electrolytic cell for the measurement of quantities of electricity by quantitative determination of chemical changes; sometimes known as a voltameter. Coulometers are used for the determination of quantities of electricity in standardization procedures and some electroanalytical measurements.

Coulometers are based on Faraday's first law, which states that for a current of I amperes flowing for t sec, the coulombs passing are the product of It; consequently, the amount of material deposited on an electrode is $It/96{,}487$ gram-equivalents. If w is the equivalent weight of the substance, the total weight deposited is $Itw/96{,}487$ g. This formula is a mathematical expression of the law of electrolysis, which is used for precise measurements of electrical quantities or current intensity. For this purpose the apparatus used is a coulometer. *See* ELECTROCHEMICAL EQUIVALENT; ELECTROLYSIS.

The quantity of substance produced or consumed in a coulometer is generally determined by weighing, titration, or volume measurements. Many electrode reactions have been proposed, but only a few proceed with the 100% current efficiency that is required for application of Faraday's law. The deposition of silver, copper, or mercury, the electrolysis of water with hydrogen and oxygen evolution, and the oxidation of iodide to iodine are the electrode reactions generally selected for coulometers.

In the silver coulometer (Fig. 1) the quantity of electricity is determined by weighing the amount of silver plated upon the platinum cup cathode during passage of the current. One coulomb of electricity will deposit 0.0011180 g of silver.

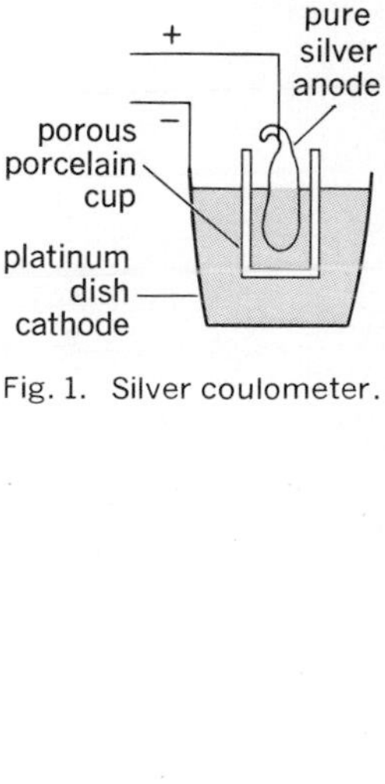

Fig. 1. Silver coulometer.

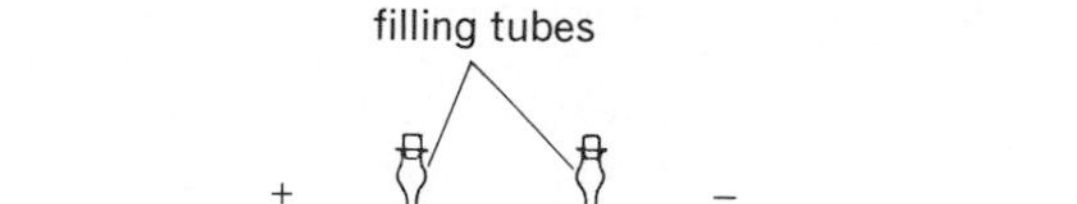

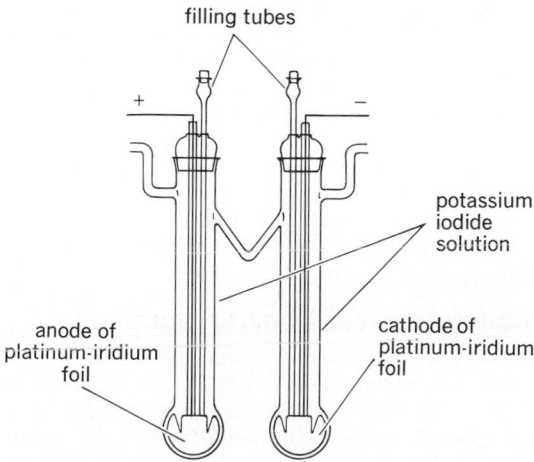

Fig. 2. Iodine coulometer.

The passage of 1 coulomb of electricity through the iodine coulometer (Fig. 2) will liberate 0.001315 g of iodine at the anode.

[PAUL DELAHAY]

Bibliography: H. W. Nuremberg, *Electro-Analytical Chemistry*, vol. 10, 1975; K. J. Vetter, *Electrochemical Kinetics: Theoretical Aspects*, 1967.

Coulometric analysis

An electroanalytical chemistry technique in which the amount of a substance is determined quantitatively by measuring the total amount of electricity required to deplete a solution of this substance. The total electricity or charge consumed is related to the concentration by Faraday's law. The faraday, F, is 96,487 coulombs per equivalent. Coulometric analyses are classified as primary or secondary. Primary coulometric analyses are those in which the substance to be determined is electrolyzed directly. Secondary coulometric analyses are those in which an electrolytically generated intermediate reacts stoichiometrically with the substance to be determined. When the solution is completely depleted, the total charge used is related directly to the amount of starting material. The basic equation for coulometry is stated in Eq. (1). Here Q is

$$Q = \int_0^\infty i_t dt = nFVC^0 \qquad (1)$$

the total charge, i_t is the current at time t, n is the number of electrons exchanged per molecule or equivalents per mole, F is the faraday based on the carbon-12 scale, V is the solution volume, and C^0 is the initial concentration of the substance determined. Secondary coulometric processes are often referred to as coulometric titrations. However, any coulometric analysis is a titration with electrons and is therefore a coulometric titration.

A condition of 100% current efficiency must exist for a coulometric analysis; that is, all the current must be used either directly or indirectly to deplete the desired substance. For a coulometric analysis either the applied current or the applied potential can be controlled. The controlled-current method is seldom used in primary coulometric analysis, for, as the reaction proceeds, the electroactive substance is depleted and is unable to maintain the required current. Another reaction must then provide for part of the current, and there is no longer 100% current efficiency. The controlled-potential method, however, is frequently used for primary coulometric analyses. *See* ELECTROCHEMICAL TECHNIQUES.

Controlled-potential analysis. In controlled-potential coulometry the potential of the working electrode is maintained constant with respect to a reference electrode. The potential can usually be selected from voltametric data, so that only the desired reaction occurs. For an electrode reaction which is diffusion limited and not complicated by chemical side reactions, the current i_t decays according to Eq. (2).

$$i = i_0^{-kt} \qquad (2)$$

In this relation i_0 is the initial current; k is a constant which depends on the geometry of the cell and the electrode, the diffusion coefficient of the electroactive species, and the solution stirring rate; and t is the electrolysis time. The time required for complete depletion is independent of the initial concentration, but it is strongly dependent on the ratio of electrode area to solution volume (cell geometry) and the stirring rate. The end of the reaction is indicated when the current decays to the value of the background current. The total charge is determined with a coulometer. Controlled-potential coulometric analysis is both sensitive and selective. Its sensitivity is, however, limited by the background current. Background currents arise from the charging of the electrode-solution interface whenever there is a change in the potential of the electrode (double-layer charging), and from unwanted faradaic processes usually caused by electroactive impurities. Double-layer charging causes no great errors unless very dilute solutions are being analyzed. The contribution of the unwanted faradaic currents can often be determined by titrating a blank solution.

An increase in the ratio of electrode area to solution volume will result in an increase in the constant k in Eq. (2), and hence a decrease in the time required for complete electrolysis of the substance. Two coulometric techniques which take advantage of this are the flow electrolysis and thin-layer techniques. *See* COULOMETER.

Flow electrolysis. In the flow electrolysis technique the solution to be analyzed is passed at a constant rate through a column packed with electrode material whose potential is controlled. During passage through the column the electroactive substance is depleted from the solution. Under conditions of 100% current efficiency, the concentration of the electroactive substance C can be obtained by Eq. (3), where i is the current, G is the

$$C = \frac{i}{nFG} \tag{3}$$

flow rate of solution, and n and F have their usual significance.

Thin-layer methods. The thin-layer technique is one in which a small volume of solution is entrapped in the vicinity of the working electrode. Electrolysis occurs at the working electrode. The solution thickness is quite small (usually less than 0.1 mm), and all the entrapped electroactive specie is able to diffuse to the electrode surface rapidly and to be reacted upon.

Both controlled-current and controlled-potential coulometry have been used in the thin-layer technique. For controlled current, Eq. (4) holds. Here T

$$\mathrm{T} = \frac{nFAlC^{\circ}}{i} - \frac{l^2}{3D} \tag{4}$$

is the electrolysis time, i is the applied current, A is the electrode area, C^0 is the initial concentration of the electroactive substance, D is the diffusion coefficient, and l is the average solution thickness. Conditions can usually be selected so that $l^2 < 3D$, and $l^2/3D$ in Eq. (4) can be neglected. The thin-layer equation then reduces to Eq. (1). The time required for coulometric analyses can be shortened by this technique.

Controlled-current analysis. For quantitative analysis controlled-current coulometry is used extensively. In this method a constant current is passed between two electrodes. It is useful in primary coulometric analyses only when the substance to be determined is deposited on the electrode surface, such as the reduction of an oxide layer. It has greater usefulness in secondary coulometric analyses. In these processes an intermediate is generated electrolytically and allowed to react stoichiometrically with the substance to be determined. The end point of the reaction can be detected by any method sensitive to the reaction. Electrochemical methods are frequently employed.

Acid-base, complexometric, and redox reactions can be analyzed by this technique. Among the intermediates which can be generated are the ions hydronium, hydroxide, silver(I), silver(II), copper(I), mercury(I), titanium(III), and uranium(V); and the molecules chlorine, bromine, and iodine. The advantages of this method over standard titration techniques are the elimination of the need for storage and standardization of standard solutions, and the ability to generate and use in solution unstable substances, such as silver(II).

Reaction mechanisms. Controlled-potential coulometry can also be used to help elucidate reaction mechanisms. For a system in which the only reaction occurring is the electron transfer at the electrode surface, the number of electrons involved in the reaction can be determined directly from Eq. (1), providing the concentration and the solution volume are known. For systems in which chemical reactions complicate the reaction occurring at the electrode surface, coulometry can often be used in conjunction with other methods to determine the reaction mechanism. For example, when a slow chemical reaction occurs between two electron-transfer reactions, the n value measured by coulometry would correspond to both electron-transfer steps. Voltametry, on the other hand, would give an n value corresponding to the first electron-transfer step. The stirring rate and the concentration of electroactive substance can also affect the n value determined by coulometry.

Reaction kinetics. Controlled-current coulometry often provides a convenient means of measuring the rate at which chemical reactions occur. The reactant can be added as a coulometrically generated intermediate and allowed to react with a substrate in solution. The concentration of the intermediate can be monitored. The rate of generation can be made to equal the rate at which it is depleted by the chemical reaction, and the rate constant for the reaction thus calculated. This method provides for convenient addition of reactant at low concentrations, which allows rather rapid reactions to be measured. *See* ELECTROCHEMICAL EQUIVALENT; ELECTRODEPOSITION ANALYSIS; ELECTROLYSIS; TITRATION.

[GEORGE W. O'DOM]

Bibliography: A. J. Bard, *Anal. Chem.*, vol. 40, no. 5, 64R (1968) and subsequent Fundamental Annual Reviews; L. Meites, *Polarographic Techniques*, 2d ed., 1965; R. W. Murray and C. N. Reilly, *Electroanalytical Principles*, 1978.

Countercurrent transfer operations

Industrial processes in chemical engineering or laboratory operations in which heat or mass or both are transferred from one fluid to another, with the fluids moving continuously in very nearly steady state or constant manner and in opposite directions through the unit. Other geometrical arrangements for transfer operations are the parallel or concurrent flow, where the two fluids enter at the same end of the apparatus and flow in the same direction to the other end, and the cross-flow apparatus, where the two fluids flow at right angles to each other through the apparatus. These two arrangements are ordinarily not as efficient as countercurrent flow, but do find certain applications in industry and the laboratory.

Heat transfer. In heat transfer there can be almost complete transfer in countercurrent operation. The limit is reached when the temperature of the colder fluid becomes equal to that of the hotter fluid at some point in the apparatus. At this condition the heat transfer is zero between the two fluids. The amount of actual heat transfer is determined by economical design, that is, by comparing the value of the transferred heat with the cost of the heat exchange equipment. The economically optimum heat transfer has been studied for many years in engineering and is changing constantly as the costs of basic forms of energy increase.

Most heat transfer equipment has a solid wall between the hot fluid and the cold fluid, so the fluids do not mix. Heat is transferred from the hot fluid through the wall into the cold fluid. Another type of equipment, fewer in number but significant in size and use, does use direct contact between the two fluids—for example, the cooling towers used to remove heat from a circulating water stream. Cooling towers are of the countercurrent type and the cross-flow type.

Mass transfer. This process involves the changing compositions of mixtures, and is done usually by physical means instead of chemical reactions because of the lower costs. Even as heat is transferred from a region of high temperature to one of lower temperature, a material is transferred within a single phase from a region of high concentration to one of lower concentration by processes of molecular diffusion and eddy diffusion. In typical mass transfer processes, at least two phases are in direct contact in some state of dispersion, and mass (of one or more substances) is transferred from one phase across the interface into the second phase. Similar to heat transfer, mass transfer takes place between two immiscible phases until equilibrium between the two phases is attained. In heat transfer, equilibrium denotes an equality of temperature in the two phases, but in mass transfer there is seldom an equality of concentration in the two equilibrium phases. This means that a component may be transferred from a phase at low concentration (but at a concentration higher than that at equilibrium) to a second phase of greater concentration. The approach to equilibrium is controlled by diffusion transport across phase boundaries. Because this is a relatively slow process, the transport rate is increased by increasing the total interfacial area, by decreasing the thickness of the near-stagnant films adjacent to the interface, and by more frequent renewals of the interfacial films.

Although the two phases may be in concurrent flow or cross-flow, usual arrangements have the phases moving in countercurrent directions. The more dense phase enters near the top of a vertical cylinder and moves downward under the influence of gravity. The less dense phase enters near the bottom of the cylinder and moves upward under the influence of a small pressure gradient. In some cases, centrifugal force is used instead of gravity to provide phase separations.

For discussions of specific mass transfer operations *see* ADSORPTION; CHEMICAL SEPARATION TECHNIQUES; CRYSTALLIZATION; DISTILLATION; DRYING; ELECTROPHORESIS; EXTRACTION; LEACHING.

[FRANK J. LOCKHART]

Bibliography: R. H. Perry and C. H. Chilton, *Chemical Engineers' Handbook*, 5th ed., 1973; R. E. Treybal, *Mass Transfer Operations*, 3d ed., 1980.

Cracking

A process used in the petroleum industry to reduce the molecular weight of hydrocarbons by breaking molecular bonds. Cracking is carried out by thermal, catalytic, or hydrocracking methods. Increasing demand for gasoline and other middle distillates relative to demand for heavier fractions makes cracking processes important in balancing the supply of petroleum products.

Thermal cracking depends on a free-radical mechanism to cause scission of hydrocarbon carbon-carbon bonds and a reduction in molecular size, with the formation of olefins, paraffins, and some aromatics. Side reactions such as radical saturation and polymerization are controlled by regulating reaction conditions. In catalytic cracking, carbonium ions are formed on a catalyst surface, where bond scissions, isomerizations, hydrogen exchange, and so on, yield lower olefins, isoparafins, isoolefins, and aromatics. Hydrocracking, a relative newcomer to the industry, is based on catalytic formation of hydrogen radicals to break carbon-carbon bonds and saturate olefinic bonds. Hydrocracking converts intermediate- and high-boiling distillates to middle distillates, high in paraffins and low in cyclics and olefins. Hydrocracking also causes hydrodealkylation of alkyl-aryl components in heavy reformate to produce benzene and naphthalene. *See* HYDROCRACKING.

Thermal cracking. This is a process in which carbon-to-carbon bonds are severed by the action of heat alone. It consists essentially in the heating of any fraction of petroleum to a temperature at which substantial thermal decomposition takes place through a thermal free-radical mechanism followed by cooling, condensation, and physical separation of the reaction products.

There are a number of refinery processes based primarily upon the thermal cracking reaction. They differ primarily in the intensity of the thermal conditions and the feedstock handled.

Visbreaking is a mild thermal cracking operation (850–950°F; 454–510°C) where only 20–25% of the residuum feed is converted to mid-distillate and lighter material. It is practiced to reduce the volume of heavy residuum which must be blended with low-grade fuel oils.

Thermal gas-oil or naphtha cracking is a more severe thermal operation (950–1100°F; 510–593°C) where 45% or more of the feed is converted to lower molecular weight. Attempts to crack residua under these conditions would coke the furnace tubes.

Steam cracking is an extremely severe thermal cracking operation (1100 to 1400°F; 593–760°C) in which steam is used as a diluent to achieve a very low hydrocarbon partial pressure. Primary products desired are olefins such as ethylene and butadiene.

Fluid coking is a thermal operation where the residuum is converted fully to gas-oil products boiling lower than 950°F (510°C) and coke. The thermal conversion is carried out on the surface of a fluidized bed of coke particles.

Delayed coking is a thermal cracking operation wherein a residuum is heated and sent to a coke drum, where the liquid has an infinite residence time to convert to lower-molecular-weight hydrocarbons which distill out of the coke drum, and to coke which remains in the drum and must be periodically removed.

In fluid coking and delayed coking, there is total conversion of the very heavy high-boiling end of the residuum feed.

Although there are many variations of visbreaking and thermal cracking, most commonly a feedstock that boils at higher temperatures than gasoline is pumped at inlet pressures of 75–1000 psig (1 psi = 6895 Pa) through steel tubes so placed in a furnace as to allow gradual heating of the coil to temperatures in the range 850–1100°F (454–593°C). The flow rate is controlled to provide sufficient time for the required cracking to lighter products; the time may be extended by subsequently passing the hot products through a reaction chamber that is maintained at a high temperature. To achieve optimum process efficiency, part of the overhead product ordinarily is returned to the cracking unit for further cracking (Fig. 1).

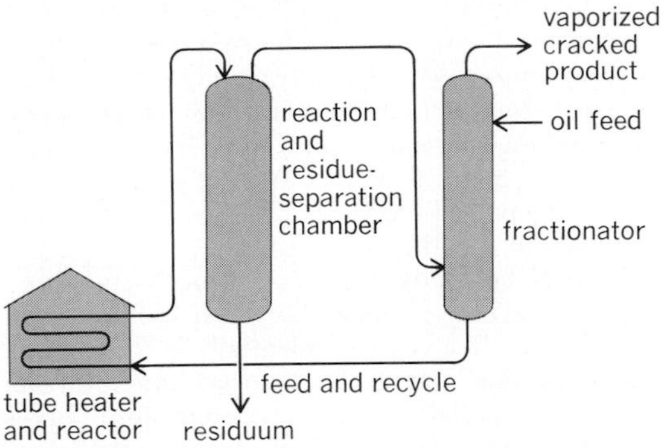

Fig. 1. Thermal cracking unit.

Crude oils differ in their compositions, both in molecular weight and molecular type of hydrocarbon. Since refiners must make products in harmony with market demand, they often need to alter the molecular structure of the hydrocarbons. The cracking of heavy distillates and residual oils increases the yields of gasoline and the light intermediate distillates used as diesel fuels and domestic heating oils, as well as providing low-molecular-weight olefins needed for the manufacture of chemicals and polymers.

Beginning in 1912, thermal cracking proved for many years to be eminently suitable for this purpose. During the period 1920–1940, more efficient automobile engines of higher compression ratios were developed. These engines required higher-octane-number gasolines, and thermal cracking operations in the United States were expanded to meet this need. Advantageously, thermal cracking reactions produce olefins and aromatics, leading to gasolines generally of higher octane number than those obtained by simple distillation of the same crude oils. The general nature of the hydrocarbon products and the basic mechanism of thermal cracking is well described by the free-radical theory of the pyrolysis of hydrocarbons. *See* PYROLYSIS.

In the early 1930s, the petrochemical industry began its growth. Olefinic gases from thermal cracking operations, especially propylene and butylenes, were used as the chief raw materials for the production of aliphatic chemicals. Simultaneously, practical catalytic processes were invented for polymerizing propylene and butylenes to gasoline components, and for dimerizing and hydrogenating isobutylene to isooctane (2,2,4-trimethylpentane), the prototype 100-octane fuel. Just prior to World War II, the alkylation of light olefins with isoparaffins to produce unusually high-octane gasoline components was discovered and extensively applied for military aviation use. These developments resulted in intense engineering efforts to bring thermal cracking to maximum efficiency, as exemplified by a number of commercial processes made available to the industry.

Since World War II, thermal cracking has been largely supplanted by catalytic cracking, both for the manufacture of high-octane gasoline and as a source of light olefins. It is, however, still used widely for the mild cracking of heavy residues to reduce their viscosities and for final cracking of gas oils derived from catalytic cracking.

Since carbon-to-hydrogen bonds are also severed in the course of thermal cracking, two hydrogen atoms can be removed from adjacent carbon atoms in a saturated hydrocarbon, producing molecular hydrogen and an olefin. This reaction prevails in ethane and propane cracking to yield ethylene and propylene, respectively. Methane cracking is a unique case wherein molecular hydrogen is obtained as a primary product and carbon as a coproduct. These processes generally operate at low pressures and high temperatures and in some cases utilize regenerative heating chambers lined with firebrick, or equipment through which preheated refractory pebbles continuously flow. Such conditions also favor the production of aromatics and diolefins from normally liquid feedstocks and are applied commercially to a limited extent despite relatively low yields of the desired products.

Catalytic cracking. This is the major process used throughout most of the world oil industry for the production of high-octane quality gasoline by the conversion of intermediate- and high-boiling petroleum distillates to lower-molecular-weight products. Oil heated to within the lower range of thermal cracking temperatures (850–1025°F; 454–551°C) reacts in the presence of an acidic inorganic catalyst under low pressures (10–35 psig). Gasoline of much higher octane number is obtained than from thermal cracking, a principal reason for the widespread adoption of catalytic cracking. All nonvolatile carbonaceous materials are deposited on the catalyst as coke and are burned off during catalyst regeneration.

In contrast to thermal cracking, residual oils are not generally processed, because excessive amounts of coke are deposited on the catalyst, and inorganic components of these oils contaminate the catalyst. Feedstocks usually are restricted to distillates boiling above gasoline.

Catalytic cracking, as conceived by E. J. Houdry in France, reached commercial status in 1936 after extensive engineering development by American oil companies. In its first form, the process used a series of fixed beds of catalyst in large steel cases. Each of these alternated between oil cracking and catalyst-coke burning at intervals of about 10 min and provided for heat and temperature control.

Successful operation led to major engineering improvements, and the goals of much improved efficiency, enlarged capacity, and ease of operation were achieved by two different systems. One employs a moving bed of small pellets or beads of catalyst traveling continuously through the oil-cracking vessel and subsequently through a regeneration kiln. The beads are lifted mechanically or by air to the top of the structure to flow down through the vessels again. This process has two commercially engineered embodiments, the Thermofor (or TCC) and the Houdriflow processes, which are similar in general process arrangements.

These moving-bed processes are limited in size and are now technically obsolete. They are being replaced by another type of unit, the fluid-solids, as dictated by economic considerations, and no new moving-bed units are being constructed. In the fluid-solids unit, a finely divided powdered catalyst is transported between oil-cracking and air-regeneration vessels in a fluidized state by gaseous streams in a continuous cycle. This system employs the principle of balanced hydrostatic heads of fluidized catalyst between the two vessels. Catalyst is moved by injecting heated oil vapors into the transport line from the regenerator to the reactor, and by injecting air into the transport line from the reactor stripper to the regenerator. Large amounts of catalyst can be moved rapidly; cracking units of total oil intake as great as 180,000 bbl/day (28,800 m^3/day) are in operation (Fig. 2).

In both the moving-bed and fluidized systems, the circulating catalyst provides the cracking heat. Coke deposited on the catalyst during cracking is burned at controlled air rates during regeneration; heat of combustion is converted largely to sensible heat of the catalyst, which supplies the endothermic heat of cracking in the reaction vessel.

Gasoline of 90–95 research octane number without tetraethyllead is rather uniformly produced by catalytic cracking of fractions from a wide variety of crude oils, compared with 65–80 research octane number via thermal cracking, the latter figures varying with crude oil source.

Although the primary objective of catalytic cracking is the production of maximum yields of gasoline concordant with efficient operation of the process, large amounts of normally gaseous hydrocarbons are produced at the same time. The gaseous hydrocarbons include propylene and butylenes, which are in great demand for chemical manufacture. Isobutane and isopentane are also produced in large quantities and are valuable for the alkylation of olefins, as well as for directly blending into gasoline as high-octane components.

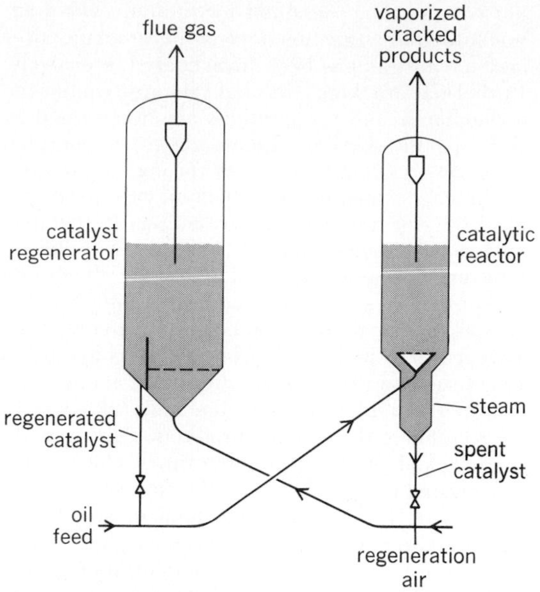

Fig. 2. Fluid catalytic cracking unit.

The other chief product is the material boiling above gasoline, designated as catalytically cracked gas oil. It contains hydrocarbons relatively resistant to further cracking, particularly polycyclic aromatics. The lighter portion may be used directly or blended with straight-run and thermally cracked distillates of the same boiling range for use as diesel and heating oils. Part of the heavier portion is recycled with fresh feedstock to obtain additional conversion to lighter products. The remainder is withdrawn for blending with residual oils to reduce the viscosity of heavy fuel, or else subjected to a final step of thermal cracking.

Thus, the catalytic cracking process is used in refineries to shift the production of products to match swings in market demand. It can process a wide variety of feeds to different product compositions. For example, light gases, gasoline, or diesel oil can be emphasized by varying process conditions, feedstocks, and boiling range of products as shown in the table.

To account for the difference between the prod-

Representative yield structures for three different processing objectives in catalytic cracking

Process variables	Light gases	Gasoline	Light cycle oil
Feed	Light gas oil	Gas oil	Gas oil
Reactor temperature, °F	990	950–990	890–900
Light gases, wt %	4.5	2.8	1.6
Propane/propylene, vol %	15	10.0	7.5
Butane/butylene, vol %	22	16.4	11.2
Gasoline, vol %	46	69.5	32.6
Catalytic diesel oil, vol %	18	10	43.6
Bottom, vol %	5	5	5

uct compositions obtained by catalytic and thermal cracking, the mechanism of cracking over acidic catalysts has been investigated intensively. In thermal cracking, free radicals are reaction intermediates, and the products are determined by their specific decomposition patterns. In contrast, catalytic cracking takes place through ionic intermediates, designated as carbonium ions (positively charged free radicals) generated at the catalyst surface. Although there is a certain parallelism between the modes of cracking of free radicals and carbonium ions, the latter undergo rapid intramolecular rearrangement reactions prior to cracking. This leads to more highly branched hydrocarbon structures than those from thermal cracking, and to important differences in the molecular weight distribution of the cracked products. Furthermore, the cracked products undergo much more extensive secondary reactions with the catalyst.

The catalytic cracking mechanism also favors the production of aromatics in the gasoline boiling range; these reach quite high concentrations in the higher-boiling portion. This characteristic, together with the copious production of branched aliphatic hydrocarbons especially in the lower-boiling portion, is largely responsible for the high octane rating of catalytically cracked gasoline.

Cracking catalysts must have two essential properties: (1) a chemical composition capable of maintaining a high degree of acidity, preferably as readily available hydrogen ions (protons); and (2) a physical structure of high porosity (high surface area). Mechanical durability is also necessary for industrial use.

Cracking catalysts are essentially silica-alumina compositions. A dramatic improvement in catalytic unit performance occurred with the switch from acid-treated clays (montmorillonite or kaolinite) to synthetic silica-aluminas. After 1960, a new group of aluminosilicates, molecular sieve zeolites, have been introduced into the catalyst formulation. These crystalline materials (Fig. 3) have cracking activity 50 to 100 times the previous amorphous catalyst. They permit cracking to greater conversion levels, producing more gasoline, less coke, and less gas.

As the catalyst particles pass through the reactor regenerator system every 3 to 15 min they are gradually deactivated through loss of surface area by the effect of heat and steam and through contamination by the trace metallic components on the feedstocks, mainly nickel, vanadium, and copper. The catalyst particles also undergo mechanical attrition, and fines are lost in the reactor and regenerator gases. To compensate, fresh catalyst is added.

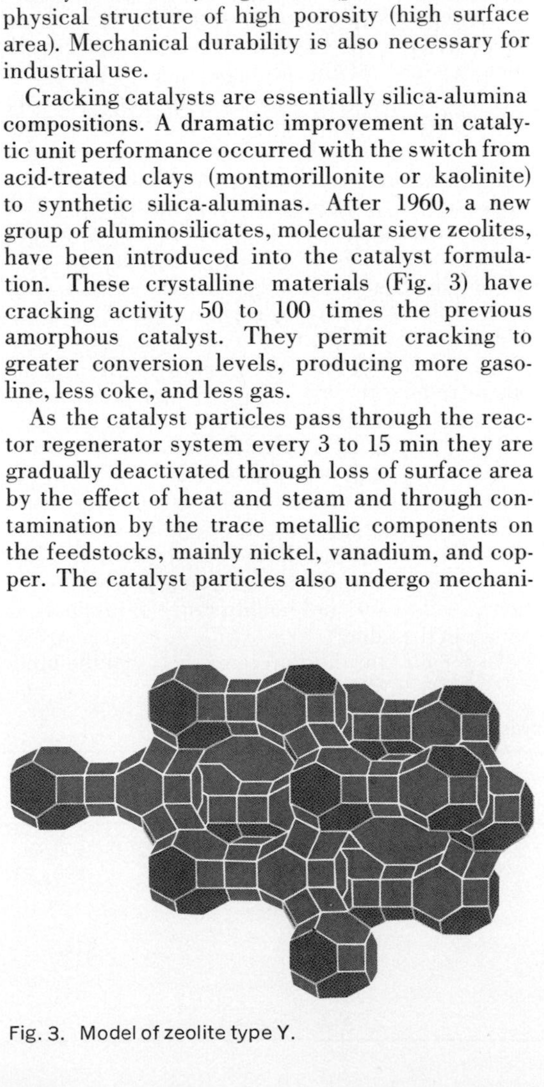

Fig. 3. Model of zeolite type Y.

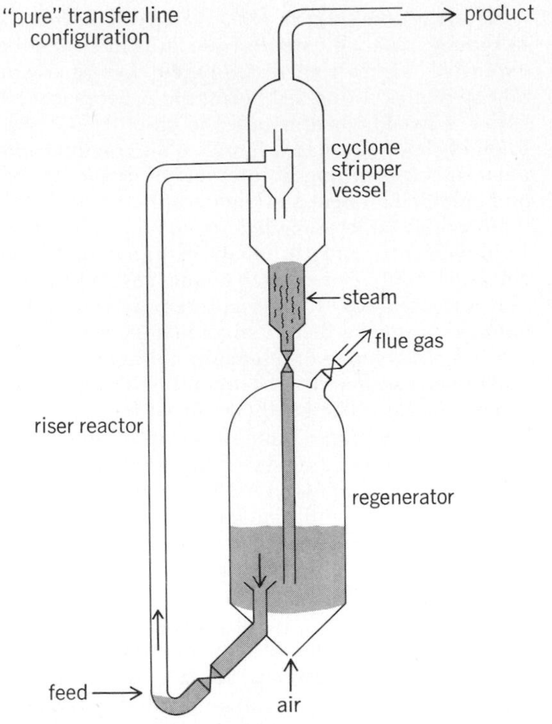

Fig. 4. Riser catalytic cracker.

The new zeolite catalysts have resulted in considerable change in the process itself. The catalysts are more resistant to thermal degradation, and regenerator temperatures can be safely raised to the 1350°F (732°C) level. The carbon on regenerated catalyst is reduced to the 0.05 wt % level resulting in improved gasoline yields. In addition, all the carbon monoxide produced at lower generation temperatures (10% concentration in the regenerator flue gas) can be combusted in the regenerator, making for a more efficient recovery of combustion heat and reduced atmospheric pollution. The effluent CO concentration can be reduced to less than 0.05 vol%.

The high-activity zeolite catalyst has permitted units to be designed with all riser cracking (Fig. 4), wherein all the cracking reaction takes place in a relatively dilute (less than 2 lb/ft^3) catalyst suspension in a 2- or 5-sec residence time. No dense-bed (10 – 15 lb/ft^3) cracking exists in these units.

Many old units are being converted to riser crackers, and virtually all new units feature riser cracking. Some riser crackers are also provided with small dense beds to achieve an optimum yield pattern.

With added emphasis on protection of the environment, complex facilities are provided to remove pollutants from the effluent regeneration gases. Third-stage cyclone collectors, electrostatic precipitators, and scrubbers are used commercially to meet regulations.

To conserve energy in the process, flue gas expanders are used in the flue gas system to provide more than enough energy to compress the regeneration air.

With the coming emphasis on conservation of

petroleum resources, catalytic cracking is assuming a more important role. The heavier products from a refinery will probably be replaced by coal-, shale-, or tar-sand-derived products. Thus, there will be a need for greater conversion of petroleum to gasoline. Emphasis will be on catalytic cracking since it is the cheapest major conversion process. *See* AROMATIZATION; CATALYTIC REFORMING; ISOMERIZATION.

[EDWARD LUCKENBACH]

Bibliography: F. H. Blanding, Reaction rates in catalytic cracking of petroleum, *I and EC*, 45: 1185–1197, June 1953; G. D. Hobson and W. Pohl *Modern Petroleum Technology*, 4th ed., 1973; K. A. Kobe and J. J. McKetta, Jr., *Advances in Petroleum Chemistry and Refining*, vols. 5 and 6, 1962; W. L. Nelson, *Petroleum Refinery Engineering*, pp. 759–818, 1958; M. Sittig, Catalytic cracking techniques in review, *Petrol. Refinery*, 37:263–316, 1952; J. W. Ward and S. A. Quader (eds.), *Hydrocracking and Hydrotreating*, 1975.

Critical constant

If a gas is cooled to a low enough temperature and then compressed, it can be liquefied. For each gas there is a characteristic temperature above which it cannot be liquefied, no matter how great is the applied pressure. This temperature is called the critical temperature T_c. The pressure required to liquefy the gas at this temperature is called the critical pressure p_c and it is, of course, the highest vapor pressure the liquid can exert. The volume occupied by one mole at T_c and p_c is called the critical volume V_c. T_c, p_c, and V_c are usually referred to as the critical constants of the gas. They vary widely from substance to substance, and are well below room temperature for the permanent gases such as oxygen and nitrogen and well above room temperature for compounds such as water and ammonia (see table).

Critical constants of typical gases

Formula	T_c, °K	p_c, atm	V_c, ml/mole
He	5.2	2.26	57.6
H_2	33.3	12.8	65.0
N_2	126.0	33.5	90.2
O_2	154.3	49.7	74.5
CO_2	304.2	73.0	95.7
H_2O	647.2	217.7	45.0
Hg	1823	200	45

Above the critical temperature the distinction between a vapor and a liquid disappears, and the critical temperature may be found by observing the minimum temperature at which the meniscus disappears when a sealed tube containing the liquid is heated. *See* GAS; LIQUID.

[THOMAS C. WADDINGTON]

Crown ethers and cryptands

The crown ethers are macrocyclic polyethers, for example, (I), which are large enough to encircle metal cations. The cryptands are macropolycyclic polyaza-polyethers, where the three-coordinate nitrogen atoms provide the vertices of a defined three-dimensional structure, for example, (II). These compounds possess cavities which can encapsulate or cryptate metal cations.

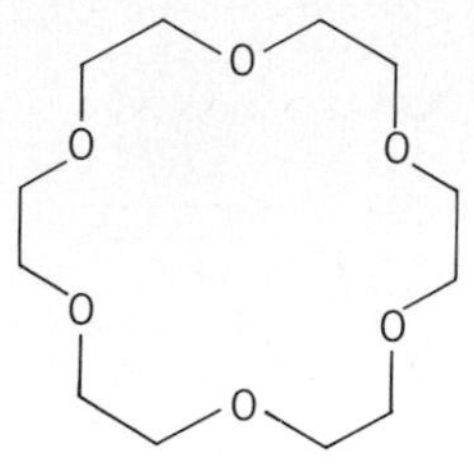

(I) 18-crown-6

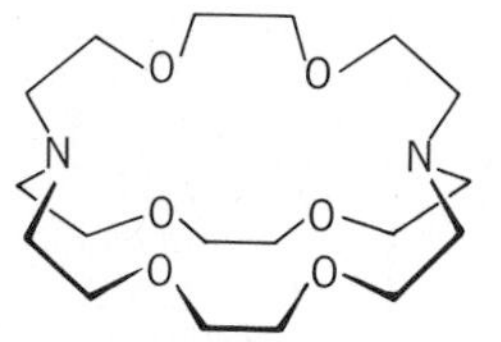

(II) (2.2.2) cryptand

The cryptands and crown ethers have the capability of forming stable and selective complexes with monoatomic or molecular ions and, as such, are potential receptor, carrier or catalyst molecules. *See* COMPLEX COMPOUNDS.

Receptors. A receptor forms a stable and selective complex with a particular substrate, thus exhibiting the phenomenon of recognition. Molecules have been designed, with a suitable array of binding sites, which show stable and selective complexes with various cations and anions.

Metal cation receptors. In the middle 1960s C. J. Pedersen synthesized the first crown ethers and showed that these compounds complexed alkali metal cations much more strongly than any other ligands previously investigated. At the same time it was discovered that certain naturally occurring macrocyclic peptides such as valinomycin and nonactin also formed stable complexes with alkali cations. X-ray structural analysis revealed that the crown ethers encircled the metal atoms, replacing some of the inner-sphere molecules of solvation. The stability of the complexes of different-sized crown ethers with different metal ions showed a good correlation with the fit between the hole defined by the crown ether and the radius of the cation.

Certain macrobicyclic cryptands (II–VIII) are able to form very strong and specific complexes with alkali and alkaline-earth metal cations. These

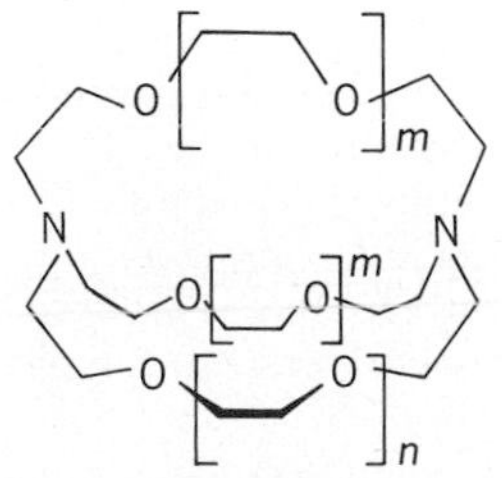

(II)–(VIII)

III (1.1.1) $m = n = 0$
IV (2.1.1) $m = 0, n = 1$
V (2.2.1) $m = 1, n = 0$
II (2.2.2) $m = 1, n = 1$
VI (3.2.2) $m = 1, n = 2$
VII (3.3.2) $m = 2, n = 1$
VIII (3.3.3) $m = n = 2$

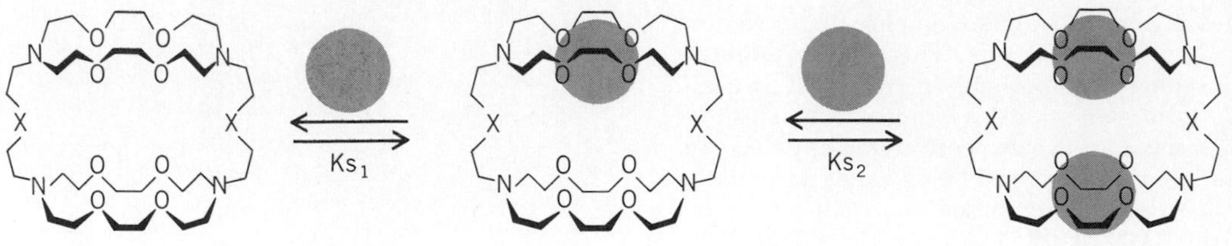

(IX) (1)

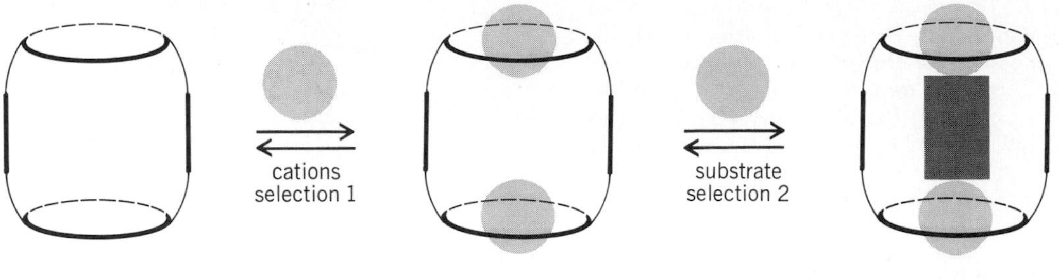

(2)

complexes are several orders of magnitude more stable than those obtained with the crown ethers, and are thus appreciably stable even in water. As in the case of the crown ethers, the cryptands show a selectivity which depends upon the goodness of fit between the cavity size of the ligand and the radius of the cation. The smaller cryptands show peak selectivity toward the alkali metals, discriminating against cations larger and smaller than the ideal size. The larger, more flexible cryptands show plateau selectivity, being able to bind the larger cations equally well, but the smaller cations poorly. In general a high selectivity is observed. The cryptands bind the alkaline-earth metal cations more strongly than alkali metal cations of similar size, in contrast with most natural macrocyclic ligands. The cryptands also complex the lanthanide metal cations. *See* ALKALI METALS; ALKALINE-EARTH METALS; RARE-EARTH ELEMENTS.

Macrotricyclic cryptands. These ligands have been prepared with both cylindrical (IX) and spherical (X) cavities. The cylindrical macrotricycles can bind two alkali metal cations sequentially [reaction (1)], and the possibility exists for the binding of a third substrate between the metal ions, forming a cascade complex [reaction (2)]. In this situation the metal cations mimic the role of the cofactors which are frequently used in a regulatory capacity by enzymes. Substitution of the oxygen atoms by nitrogen or sulfur in macrotricycles like structure (IX) allows the formation of mono- and binuclear complexes with the transition metals. Ligand (X) has a spherical cavity and forms the strongest known complex of Cs^+. *See* CHELATION.

Molecular recognition and anion complexation. As well as binding monoatomic metal ions, the crown ethers form stable complexes with the molecular ammonium ion (XI) and primary alkyl ammonium ions such as ion (XII). A large number of substituted and functionalized crown ethers have been prepared which bind alkyl ammonium salts. Chiral crown ethers containing binaphthyl subunits (for example, XIII) have been obtained, and these compounds show high chiral discrimination between optically active primary alkyl ammonium salts (for example, XIV).

The spherical cryptand (X) complexes the ammonium ion very strongly. The nitrogen atoms in structure (X) are arranged in a tetrahedral array,

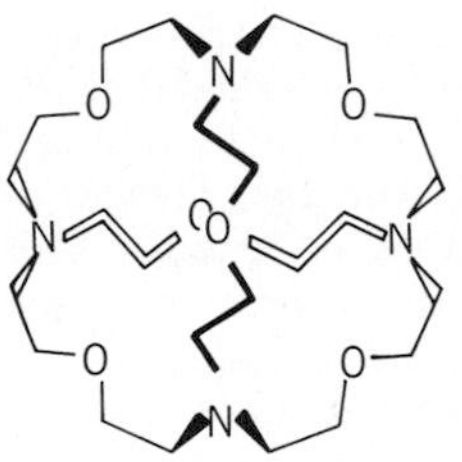

(X)

NH_4^+ (XI)

$CH_3\overset{+}{N}H_3$ (XII)

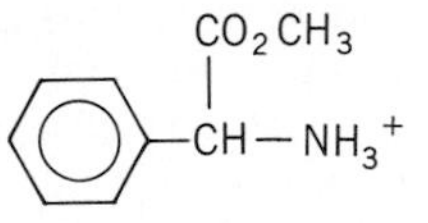

(XIV)

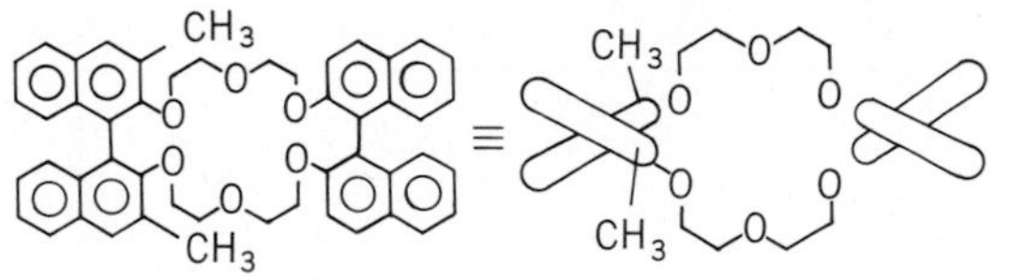

(XIII)

ideally placed to form hydrogen bonds to the tetrahedral NH_4^+ ion (XI) [see illustration *a*]. In its diprotonated form, cryptand (X) complexes a water molecule, again providing an optimum array of hydrogen bonds [illustration *b*]. In its tetraprotonated form, (X) acts as an anion receptor, complexing halide anions [illustration *c*]. The chloride complex is very stable, and the selectivity between chloride and bromide is high.

The hexaprotonated bicyclic ligand (XV) complexes molecular anions [reaction (3)]. The cavity size and shape discriminate for linear triatomic anions, and the azide ion, N_3^-, is particularly strongly bound. In its unprotonated state, ligand (XV) can form binuclear complexes with transition metals. Polyazamacrocycles (crown ethers with the oxygen atoms replaced by nitrogen atoms) in their protonated state are also able to complex

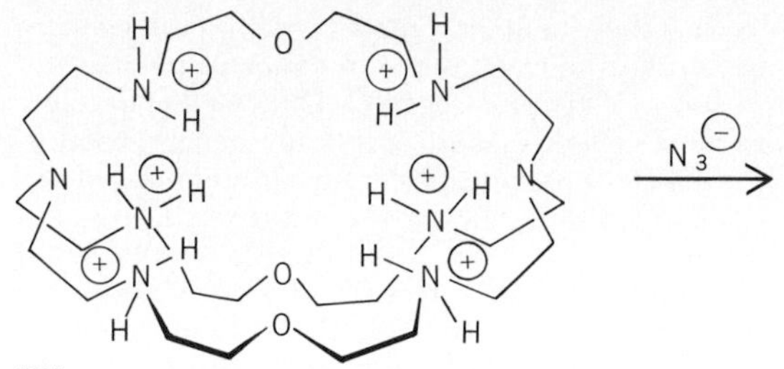

(XV)

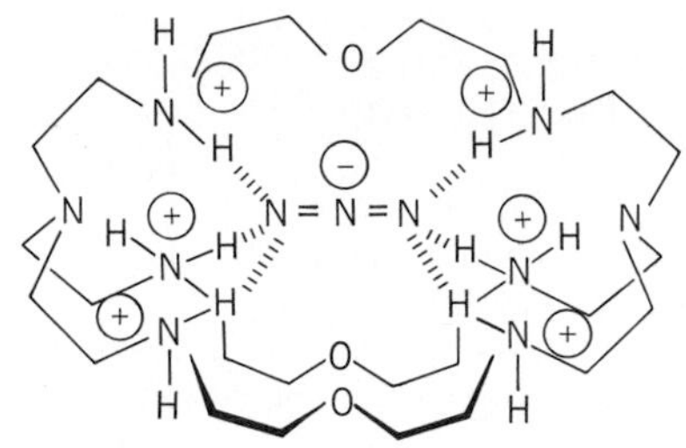

(3)

molecular anions. *See* OPTICAL ACTIVITY; STEREOCHEMISTRY.

Transport. The transport of substrates through membranes is of fundamental importance in biology. The simplest mechanism of transport is that of facilitated diffusion, whereby transport of a substrate is assisted by a carrier molecule. A carrier must possess some of the characteristics of a receptor, in particular showing a high specificity for the substrate. However, it is important that the affinity of the carrier for the substrate is not too high, for if the complex is too stable there will be insufficient free carrier available to diffuse back across the membrane and complete the transport cycle. Also, the carrier and complex must have some solubility in the phase outside the membrane in order to facilitate complexation and decomplexation, and finally the rates of these processes must be rapid. In order to design a carrier to selectively transport one substrate from a mixture, it is necessary that the substrate is not only the most efficiently transported but also the most strongly complexed. *See* CHEMICAL DYNAMICS.

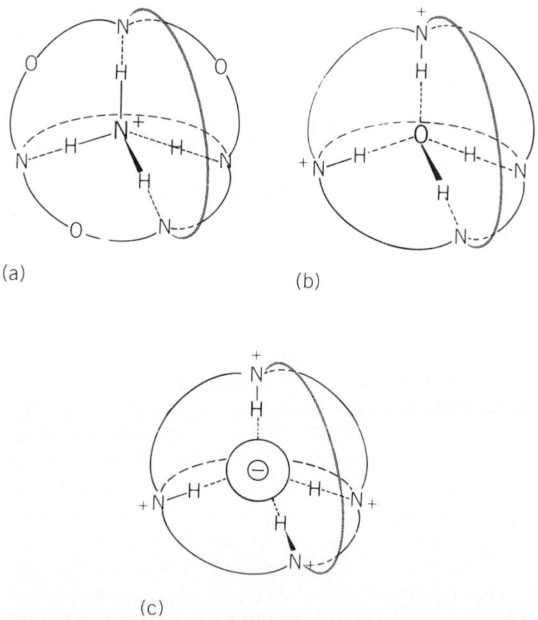

Spherical cryptand complexes. *(a)* Tetrahedral binding of NH_4^+. *(b)* Tetrahedral binding of H_2O. *(c)* Tetrahedral binding of a halide ion.

Applications. Crown ethers and cryptands have numerous applications.

Solubilization of salts. Complexation of the cations of metal salts by crown ethers of cryptands generally increases the solubility of the salts in less polar solvents. Dramatic examples of this effect include the solubilization of potassium permanganate in benzene by dicyclohexyl-18-crown-6 (XVI) [0.02 *M*] and by (2.2.2) cryptand; (2.2.2) cryptand also causes a 10^4 increase in the solubility of barium sulfate in water, dissolving up to 50 g/liter.

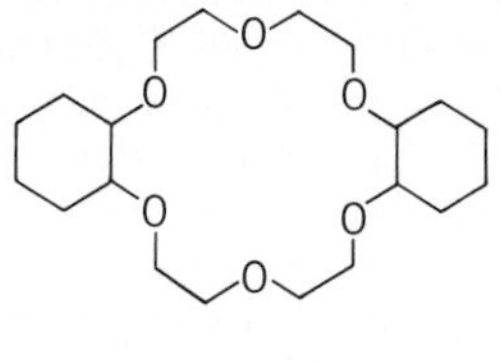

(XVI)

Anion activation. A very important consequence of the solubilization effect of the crown ethers and cryptands is that of anion activation. Complexation of the cation of a salt not only increases its solubility but also reduces ion pairing between the new larger cation and the anion. The poor solvation of anions in polar aprotic solvents gives rise to solutions of so-called naked anions which show a greatly enhanced nucleophilic reactivity. For example, the rate of reaction (4) is increased 280-fold by the addition of 5% dicyclohexyl-18-crown-6. The crown ether acts in a catalytic role and can be recovered at the end of the reaction. On the laboratory scale such catalysis is of interest as it enables reactions to be carried out under milder conditions, perhaps thereby allowing the preservation of other functional groups in the reactant which would otherwise be destroyed. Also, the crown ethers have been shown to initiate novel nucleophilic substitution reactions. From the industrial viewpoint such catalysis is of paramount importance, allowing reactions to be carried out with lower inputs of activation energy, hence conserving energy and reducing the need for capital-intensive high-temperature/pressure reaction plants.

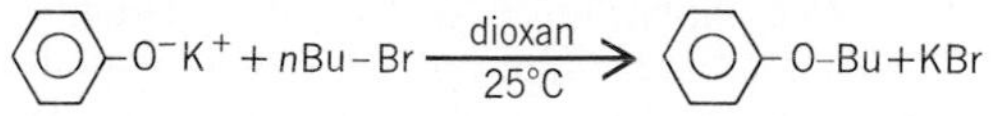

(4)

In general, the cryptands are more effective than the crown ethers for producing naked anions because the metal cation is fully encapsulated, which reduces ion pairing even further. On the debit side, they are more difficult to synthesize and hence more expensive [it should be pointed out that many crown ethers and cryptands, like (I) and (II), are commercially available compounds]. Metal complexes of crown ethers and cryptands are also effective lipophilic cations in the related field of phase transfer catalysis. *See* ORGANIC REACTION MECHANISM; SUBSTITUTION REACTION.

Stabilization of unusual species. The most remarkable example of this phenomenon is the preparation of stable salts of alkali metal anions. Thus the salt $[(2.2.2)Na]^+ Na^-$ formed gold-colored crystals, stable at room temperature, and was characterized by x-ray structural analysis and ^{23}Na nuclear magnetic resonance spectroscopy. Stabilization of metal cluster anions such as Sb_7^{3-}, Pb_5^{2-}, and Sn_9^{4-} by cryptation of the cation (Na^+) has been achieved. In the absence of the cryptand these compounds revert to an alloy of sodium and the heavy metal. *See* NUCLEAR MAGNETIC RESONANCE (NMR); X-RAY CRYSTALLOGRAPHY.

Medicine. The ability of the crowns and cryptands to selectively complex metal ions suggests their use as drugs for the detoxification of organisms contaminated with radioactive or heavy metals. It has been shown that (2.2.2) cryptand facilitates the elimination of radioactive strontium-85 and radium-224 from rats and can also remove lead. This compound shows too many side effects to be used in the medical treatment of human beings, but illustrates a useful approach to these problems.

Isotope separation. The discrimination of the crown ethers and cryptands between metal ions extends, to an appreciable extent, toward the isotopes of a particular metal. When $^{48}Ca^{2+}$ and $^{40}Ca^{2+}$ are partitioned between water and chloroform in the presence of dicyclohexyl-18-crown-6, the organic layer is enriched in $^{40}Ca^{2+}$ by about 1%. Thus considerable enrichments could be achieved by successive liquid/liquid (countercurrent) extractions. Under similar conditions (2.2.1) cryptand (V) causes a 3% isotopic separation between $^7Li^+$ and $^6Li^+$.

Analysis. The receptor properties of the crown ethers and cryptands make these compounds potentially very useful in the area of analytical chemistry. They have been employed in extraction analysis, where the component to be determined is selectively extracted from the unknown mixture. Crown ethers have also been prepared possessing chromophores which change color upon complexation, hence allowing easy detection of the particular metal ion. This technique has been used to measure the concentration of K^+ in blood serum. Exchange resins have been prepared bearing crown ethers and cryptands, and these have been used to extract particular cations from aqueous solution. These resins can also be used in column chromatography where the mixture of metal ions is separated by elution with water. Crown ethers and cryptands can also be used to advantage with ion selective electrodes. *See* CHROMATOGRAPHY; ION EXCHANGE.

Ion selective electrodes are very convenient tools for the determination of ionic concentrations. These devices possess a membrane containing a specific carrier compound. When the electrode is brought into contact with the ion (for which it is specific), the carrier renders the membrane permeable and an electrical potential is set up, proportional to the concentration of the ion present. The naturally occurring macrocyclic antibiotics (valinomycin and such) and the crown ethers have been used as carriers in these electrodes. The synthetic compounds have the advantage that they can (in principle) be designed with particular characteristics. The ion-specific electrodes can be used to determine the stability constants of new crown ethers and cryptands with metal ions by measuring the free metal-ion concentration in the presence of the macrocycle. Thus well-characterized crown ethers and cryptands can be used to investigate the properties of newer members of the group. *See* ION SELECTIVE MEMBRANES AND ELECTRODES. [RICHARD B. SESSIONS]

Bibliography: D. J. Cram and J. M. Cram, *Acc. Chem. Res.*, 11:8–14, 1978; R. M. Izatt and J. J. Christensen (eds.), *Synthetic Multidentate Macrocyclic Compounds*, pp. 1–51, 1978; I. M. Kolthoff, *Anal. Chem. Rev.*, 51:1–22R, 1979; J. M. Lehn, *Acc. Chem. Res.*, 11:49–57, 1978; J. M. Lehn, *Pure Appl. Chem.*, 49:857–870, 1977, and 50:871–892, 1978, and 51:979–997, 1979; C. J. Pedersen, *J. Amer. Chem. Soc.*, 89:2495–2496, 1967.

Crystal

This term, as used in science and technology, usually denotes a single crystal. A single crystal is a solid throughout which the atoms or molecules are arranged in a regularly repeating pattern. In electronics the term crystal is usually restricted to mean a single crystal which is piezoelectric. Examples of single crystals are most gems, piezoelectric quartz crystals used in controlling the frequencies of radio transmitters, and single crystals of galena (lead sulfide) used in crystal radios. *See* SINGLE CRYSTAL.

Most crystalline solids are made up of millions of tiny single crystals called grains and are said to be polycrystalline. These grains are oriented randomly with respect to each other. Any single crystal, however, no matter how large, is a single grain. Single crystals of metals many cubic centimeters in volume are relatively easy to prepare in the laboratory.

Single crystals differ from polycrystalline and amorphous substances in that their properties are anisotropic. Young's modulus, for example, is different for different directions in the crystal. Anisotropy is responsible for the fact that crystals will cleave (split) along very flat planes which are characteristic of the atomic stacking pattern. *See* CRYSTAL STRUCTURE.

[HERMAN H. HOBBS]

Crystal field theory

An essentially ionic approach to chemical bonding which is often used with coordination compounds. These compounds consist of a central transition metal ion that is surrounded by a regular array of coordinated atoms or ligands. Accordingly, the ligands are assumed to be sources of negative charge which perturb the energy levels of the central metal ion. In this respect the ligands subject the metal ion to an electric field which is analogous to the electric or crystal field produced by the regular distribution of nearest neighbors within an ionic crystalline lattice. For example, the crystal field produced by the Cl ion ligand in octahedral $TiCl_6^{3-}$ is considered to be similar to that produced by the octahedral array of the six Cl ions about each Na ion in NaCl. The Na ion with its rare-gas configuration has an electronic charge distribution which is spherically symmetric both within and without the crystal field. The paramagnetic Ti(III) ion, which possesses one $3d$ electron (d^1), has a spherically symmetric charge distribution only in the absence of the crystal field produced by the

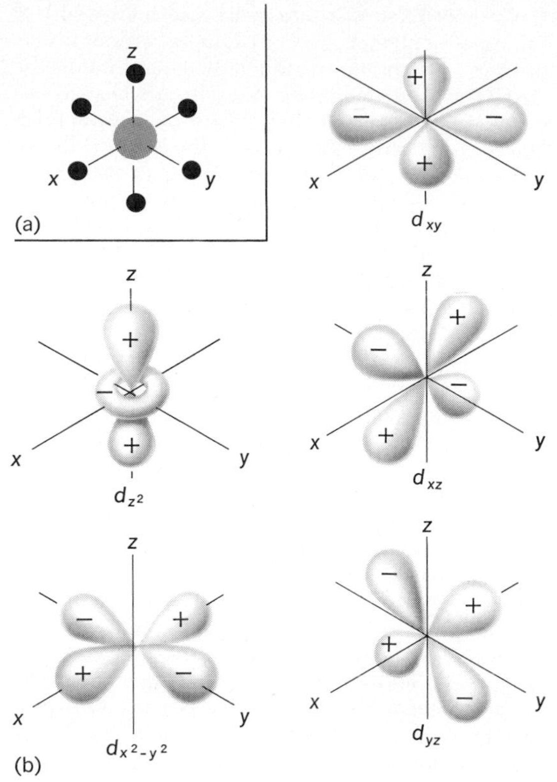

Fig. 1. Spatial orientation of d orbitals with relation to ligands in $TiCl_6^{3-}$. (a) The coordinate system for the octahedral $TiCl_6^{3-}$ ion. (b) The five d orbitals. The d_{z^2} orbital is symmetric with respect to the z axis.

ligands. The presence of the ligands destroys the spherical symmetry and produces a more complex set of energy levels within the central metal ion. The crystal field theory allows the energy levels to be calculated and related to experimental observation. *See* COORDINATION CHEMISTRY.

To illustrate the results of a typical crystal field calculation, assume that the single d electron in the Ti(III) ion will experience a coulombic repulsion with each of the six nearest neighbor Cl ions which are taken as point negative charges. This model for coulombic repulsion may be described mathematically as the summation $eq\sum r^{-1}$, where e and q are the electronic and ligand charges, respectively, and r is the distance between the electron and the ligand. The summation extends over all the ligands. A detailed quantum mechanical calculation can then be made. Fortunately, it is possible to arrive at identical results by a very qualitative procedure. This method considers the spatial orientation of d orbitals with relation to the ligands when both are viewed within the same coordinate system, as shown in Fig. 1.

The d_{z^2} orbital is oriented along the z axis with a ring in the xy plane, while the $d_{x^2-y^2}$ orbital is oriented only along the x and y axes. Although it is not visually obvious, both are equivalent. The d_{xy}, d_{xz}, and d_{yz} orbitals are oriented between the x, y, and z axes and are geometrically equivalent. The ligands are placed along the coordinate x, y, and z axes. An electron in any of the five orbitals will experience a coulombic repulsion due to the crystal field of the ligands. However, since the d_{z^2} and $d_{x^2-y^2}$ orbitals are directed toward the ligands, an electron in these orbitals will undergo far more repulsion than one in either of the three other orbitals. The imposition of six point negative charges, then, will not allow the five d orbitals to be equally energetic (degenerate) as they are in the bare Ti(III) ion, but causes three of these orbitals to be more stable than the remaining two. The difference in energy is termed $10Dq$. Thus, in the ground state of $TiCl_6^{3-}$, the single d electron will be found in the lower set of orbitals, whose energy is $-4Dq$. The crystal field stabilization energy (CFSE) is then said to be $4Dq$ (Fig. 2).

Coordination compounds are often colored. Crystal field theory suggests that in the case of $TiCl_6^{3-}$ its color is a result of an electronic excitation of the electron from the threefold set of orbitals into the twofold set. In the spectrum of $TiCl_6^{3-}$, an absorption band maximum is found at 13,000 cm^{-1}, which is then the energy associated with the transition, or $10Dq$. *See* MOLECULAR STRUCTURE AND SPECTRA.

CRYSTAL FIELD THEORY

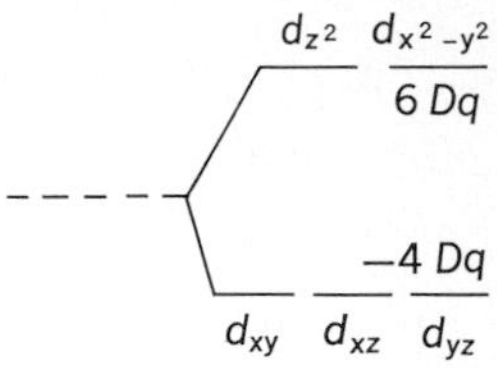

Fig. 2. The splitting of the five d orbitals because of an octahedral crystal field.

Several electrons. When more than one d electron is present, the spectroscopic evaluation of Dq is not as simple, but can generally be accomplished. Nevertheless, the CFSE may be easily and formally obtained for these cases, as shown in the table, since this energy is simply the number of electrons occupying the orbital multiplied by the orbital energy. Thus, the CFSE amounts to $8Dq$ in an octahedral d^2 complex, and $(6\times4)-(2\times6)=12Dq$ in a similar d^8 complex. With d^4, d^5, d^6, and d^7, however, two possibilities exist. If the crystal field is sufficiently strong to overcome the repulsion energy which will result from pairing the electrons in the lower set of orbitals, the maximum number of electrons will be found in the lower set. This situation is termed the strong-field case. In the weak-field case, the electron-pairing (repulsive) energy is greater than the crystal field (attractive) energy and the maximum number of unpaired electrons will result. Thus, in a strong crystal field a d^5 ion should have only one unpaired electron and a CFSE of $20Dq$. This is found in the $Mn(CN)_6^{4-}$ ion, which has been shown experimentally to possess only one unpaired electron. In most complexes of Mn(II), such as $Mn(H_2O)_6^{2+}$, the crystal field is weak so that five unpaired electrons result and the CFSE is zero.

Simple considerations such as these have enabled inorganic chemists to understand why certain coordination compounds containing a given metal ion may exhibit full paramagnetism, while others containing the metal in exactly the same formal oxidation state may show either a much weaker paramagnetism or none at all. A striking example

Crystal field stabilization energy

	Octahedral CFSE (Dq)	
d^n	Weak	Strong
1	4	4
2	8	8
3	12	12
4	6	16
5	0	20
6	4	24
7	8	18
8	12	12
9	6	6

of this is provided by paramagnetic CoF_6^{3-}, which possesses six unpaired *d* electrons and the diamagnetic $Co(NH_3)_6^{3+}$, in which none of the *d* electrons are unpaired. Both contain Co(III).

Tetrahedral array. Tetrahedral complexes may be treated in a similar fashion. The results indicate that the only difference, when compared to the octahedral complex, is in the orbital splitting pattern and the relative magnitude of *Dq*. The tetrahedral array of ligands causes an inversion of the pattern such that the d_{z^2} and $d_{x^2-y^2}$ orbitals lie lowest. The difference in energy between the two sets of orbitals is now found to be 4/9 of that in an octahedral complex, or D_q(tetrahedral) = (4/9) D_q(octahedral). This particularly simple result has led to the understanding of many stereochemical phenomena. An important early example was found in the cation distribution of normal and inverse spinels. The former are double oxides having the general formula $M(II)[M'(III)]_2O_4$ in which the oxygens lie in a close-packed system. The divalent metal ions, M(II), occupy one-eighth of the tetrahedral holes. In the inverse spinel $M'(III)[M(II)M'(III)]O_4$ the divalent metal ions have changed places with one-half of the trivalent ions.

Experimentally, it has been found that $MnCr_2O_4$ [containing Mn(II) with five unpaired electrons] and $NiFe_2O_4$ [containing Fe(III) with five unpaired electrons] have the normal and inverse structures, respectively. A simple application of crystal field theory results in exact agreement with experiment. In $MnCr_2O_4$ the ion at each octahedral site has a stability of 12*Dq*, while the tetrahedral site has no CFSE. The total CFSE is then 24*Dq*. If the structure were inverse, the total CFSE would be only $(4/9) \times 12 + 12 = 17.3Dq$. With $NiFe_2O_4$ the CFSE of the normal spinel is only 5.3*Dq*, but in the inverse structure the CFSE increases to 12*Dq*. In general, agreement between predicted cation distribution and that experimentally observed is good.

The application of these methods to conventional coordination compounds also meets with a fair amount of success. For example, with only one or two exceptions octahedral coordination of Cr(III) and diamagnetic Co(III) prevails in all compounds, as one would predict from crystal field theory. Important exceptions to these rules do exist and point out that other factors, such as ligand-ligand repulsions, sometimes outweigh the CFSE. Octahedral coordination of Ni(II) is favored over the tetrahedral arrangement insofar as CFSE is concerned, yet the tetrahedral $NiCl_4^{2-}$, $NiBr_4^{2-}$, and NiI_4^{2-} ions are well known.

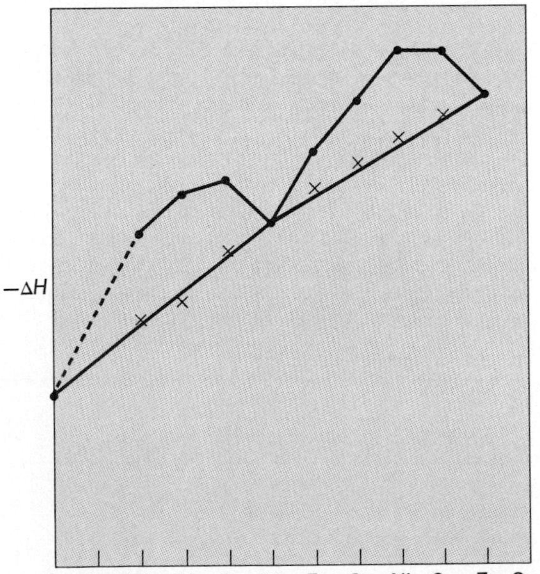

Fig. 3. Variation of heat of hydration $-\Delta H$ of divalent ions of first transition series. Experimental values are given by the filled dots, and the corresponding values, after correction for CFSE, are given by the x's.

Anomalous effects. Heats of hydration, lattice energies, crystal radii, and oxidation potentials of transition metal ions and their complexes contain apparent anomalies which are best explained in terms of an effect due to the crystal field. As an example, when the heats of hydration of the ions of the first transition series are plotted in Fig. 3 with respect to atomic number, a peculiar double-humped curve is obtained. The results for those ions not possessing any CFSE, that is, Ca, Mn, and Zn, lie nearly on a straight line. In the absence of any other effect, it might be expected that with the successive addition of each nuclear charge, a monotonic increase in the heat of hydration would occur. Instead, it was found that the heat of hydration of those metal ions possessing CFSE is far more exothermic than would be predicted by arguments pertaining only to the successive increase in atomic number. In fact, the excess heat is best accounted for in terms of the CFSE. The total heat of hydration may be written as $\Delta H = \Delta H^0 + \text{CFSE}$, where ΔH^0 is the heat of hydration that would be expected for a hypothetical metal ion which ignored the crystal field. The change in ΔH^0 with atomic number would then be expected to be monotonic. When the observed heats of hydration are corrected for the CFSE obtained from spectroscopic data for the resulting $M(H_2O)_6^{2+}$ complexes, the expected monotonic increase is observed. Similar double-humped curves occur with the lattice energies and crystal radii and can be explained by including the effects due to the crystal field.

Stereochemical anomalies can also often be explained through the judicious use of simple arguments. A particularly important example is found in complexes of Cu(II). X-ray crystallography has established that most "octahedral" complexes containing that ion are in fact elongated along one axis. According to a theorem due to H. A. Jahn and E. Teller, this behavior is not unexpected. The theorem states that a system possessing a degenerate ground state will distort in some unspecified manner to remove the degeneracy.

The degeneracy in a regular octahedral complex of Cu(II) is easily illustrated by the possibility of writing the electronic configuration in two distinct, but equally energetic, ways:

$$(d_{xy})^2(d_{xz})^2(d_{yz})^2(d_{z^2})^2(d_{x^2-y^2})^1$$

or $$(d_{xy})^2(d_{xz})^2(d_{yz})^2(d_{x^2-y^2})^2(d_{z^2})^1$$

For each, the CFSE is 6*Dq*. In addition to the twofold degeneracy, there are two separate means by which the degeneracy may be removed. If the ligands along the *z* axis of the octahedron move away from the metal ion while those in the *xy* plane move toward the center of the octahedron, then according to simple crystal field arguments, this movement will result in stabilizing the d_{z^2} orbit-

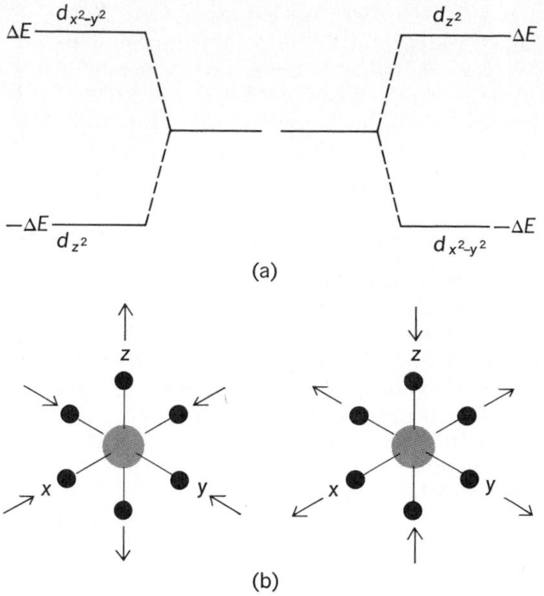

Fig. 4. The splitting of the $d_{x^2-y^2}$ and d_{z^2} orbitals by an axial distortion of the octahedron. (a) Alternate possibilities for writing the electronic configuration. (b) The two modes of distortion.

al with respect to the $d_{x^2-y^2}$ orbital. Alternatively, the completely opposite movement will stabilize the $d_{x^2-y^2}$ orbital with respect to the d_{z^2} orbital (Fig. 4). In either case, an additional increment of energy is added to the CFSE of the Cu(II) complex: $6Dq + \Delta E$. Thus, the driving force for the distortion is the additional stabilization energy ΔE. Crystal field theory is unable to predict which mode of distortion will occur, but it clearly predicts that a distortion should occur. From structure determinations through the use of x-rays, it is found that elongation along one axis is by far the most predominant mode.

Complete theory. The many successes of crystal field theory in the interpretation of the natural phenomena associated with transition metal compounds should not lead one to the conclusion that the bonding within these compounds can be truthfully represented by a strictly ionic model. Crystal field theory is essentially a very specialized form of the more complete molecular orbital theory. The need for the more complete theory becomes obvious when an attempt is made to rationalize the absolute value of Dq, the observation of ligand-to-metal electronic transitions, and certain observations in both nuclear magnetic resonance and electron spin resonance experiments, which can only be interpreted in terms of some covalent bonding. *See* MOLECULAR ORBITAL THEORY.

[R. A. D. WENTWORTH]

Bibliography: T. M. Dunn, D. S. McClure, and R. G. Pearson, *Some Aspects of Crystal Field Theory*, 1965; M. Gerloch and R. C. Slade, *Ligand-Field Parameters*, 1973; H. L. Schlafer and G. Glieman, *Basic Principles of Ligand Field Theory*, 1969.

Crystal structure

The arrangement of atoms or ions in a crystalline solid. Knowledge of the precise ways in which atoms and ions are distributed in crystals is of prime importance in solid-state physics, chemistry, metallurgy, mineralogy, geochemistry, and other fields. In 1912 M. von Laue, W. Friedrich, and P. Knipping first discovered that x-rays could be diffracted by crystals. Prior to this discovery, little was known about the arrangement of atoms and ions in solid materials. Knowledge of crystal structure is and has been obtained mainly from x-ray diffraction data, although electron diffraction and neutron diffraction have become important tools in crystal analysis.

This article discusses the important concepts and terminology involved in the structure of crystalline solids and describes the structure of various metals and some relatively simple crystalline compounds. For related information *see* CHEMICAL BONDING; NEUTRON DIFFRACTION; SOLID-STATE CHEMISTRY; X-RAY CRYSTALLOGRAPHY; X-RAY DIFFRACTION.

CONCEPTS AND TERMINOLOGY

In order to understand and describe crystal structures, several terms and concepts have been developed. Brief explanations of the most important are given in the following paragraphs.

Space lattices. A three-dimensional, indefinitely extended array of points, each of which is surrounded in an identical way by its neighbors, is known as a lattice or space lattice. The space lattice of a crystal is the representation of the periodicity with which matter is distributed in it. It is essential to distinguish a lattice from a crystal structure; a crystal structure is formed by associating with every lattice point an assembly of atoms identical in composition, arrangement, and orientation. The space-lattice concept was introduced by R. J. Haüy as an explanation for the special geometric properties of crystal polyhedrons. It was postulated that an elementary unit, having all the properties of the crystal, should exist, or conversely that a crystal was built up by the juxtaposition of such elementary units. If mathematical points forming the vertices of a parallelepiped *OABC* (defined by three vectors $\overline{OA}$, $\overline{OB}$, $\overline{OC}$) are considered (Fig. 1), a space lattice is obtained by translations parallel to and equal to $\overline{OA}$, $\overline{OB}$, $\overline{OC}$. The parallelepiped is called the unit cell. The vector **r** joining O to any lattice point can be written as $\mathbf{r} = m\overline{OA} + n\overline{OB} + r\overline{OC}$, where m, n, and r are integers.

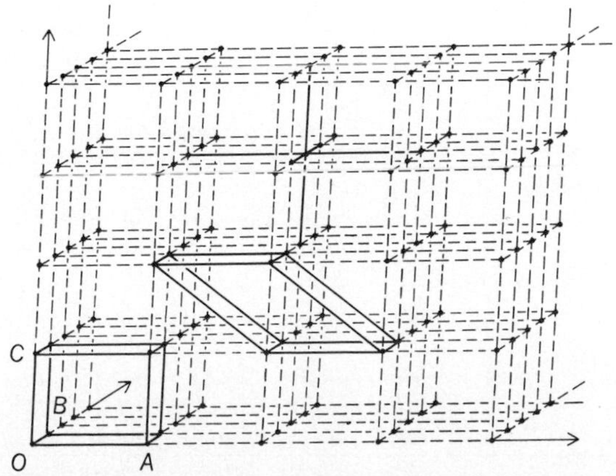

Fig. 1. A space lattice, two possible unit cells, and the environment of a point.

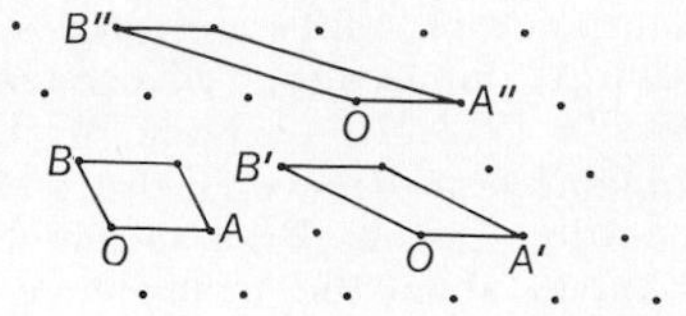

Fig. 2. Two-dimensional lattice, formed by the points in a lattice plane (*hkl*). All possible primitive parallelograms have the same surface area.

In a space lattice two points define a row, and three define a lattice plane. Taking the directions of *OA*, *OB*, *OC* as axes and *ABC* as the unit plane, all planes and rows can be expressed by Miller indices.

In a row [*uvw*] two neighboring points are separated by a distance p_{uvw} which is characteristic for that row and all parallel ones. Two adjacent parallel rows of the family represented by [*uvw*] are a distance r_{uvw} apart. The plane (*hkl*) defines a family of lattice planes in which the points are identically distributed and form a two-dimensional lattice. Figure 2 is an example.

This two-dimensional latrice can be deduced from two vectors such as $\overline{OA}$, $\overline{OB}$. The parallelogram *OAB* is the smallest unit from which the whole assembly can be obtained by parallel translation. This statement applies equally well to the other two parallelograms *OA′B′* and *OA″B″* in Fig. 2 and, in fact, to an infinite number of such parallelograms, all of which have the same surface area $S_{hkl} = p_{uvw} r_{uvw}$. Thus the density of points in a given row $1/p_{uvw}$ is proportional to the distance between rows.

In the same way an infinity of unit cells can be considered in the space lattice. These cells all have the same volume $V = S_{hkl} d_{hkl}$, if d_{hkl} is the distance between two adjacent planes (*hkl*). The distance d_{hkl} is proportional to $1/S_{hkl}$, the latter providing a measure for the density of points or the packing in (*hkl*). This remark, as well as the similar one made for the rows, is important when the properties of the crystals are considered on an atomic scale. The faces and the edges of crystals are, respectively, densely packed lattice planes and rows. It is clear from a mathematical standpoint that from all possible cells the most convenient one should be chosen and that, whenever possible, preference should be given to a parallelepiped with mutually perpendicular edges.

The volume V of the unit cell is given by Eq. (1).

$$V = abc\sqrt{\begin{matrix}1 - \cos^2\alpha - \cos^2\beta - \cos^2\gamma \\ + 2\cos\alpha\cos\beta\cos\gamma\end{matrix}} \qquad (1)$$

where α, β, and γ are the angles defined by the crystal axes, as shown in Fig. 3*a*. The general expression for d_{hkl} (called d for convenience) is given by Eq. (2).

$$\frac{1}{d^2} = \frac{\frac{h}{a}\begin{vmatrix} h/a & \cos\gamma & \cos\beta \\ k/b & 1 & \cos\alpha \\ l/c & \cos\alpha & 1 \end{vmatrix} + \frac{k}{b}\begin{vmatrix} 1 & h/a & \cos\beta \\ \cos\gamma & k/b & \cos\alpha \\ \cos\beta & l/c & 1 \end{vmatrix} + \frac{l}{c}\begin{vmatrix} 1 & \cos\gamma & h/a \\ \cos\gamma & 1 & k/b \\ \cos\beta & \cos\alpha & l/c \end{vmatrix}}{\begin{vmatrix} 1 & \cos\gamma & \cos\beta \\ \cos\gamma & 1 & \cos\alpha \\ \cos\beta & \cos\alpha & 1 \end{vmatrix}} \qquad (2)$$

If $\alpha = \beta = \gamma = 90°$, $\cos\alpha = \cos\beta = \cos\gamma = 0$, and Eq. (2) reduces to Eq. (3), and if furthermore $a = b = c$, as in cubic lattices, this becomes $1/d^2 = (h^2 + k^2 + l^2)/a^2$. When $a = b \neq c$, $\alpha = \beta = 90°$, and $\gamma = 120°$ as in the hexagonal case, Eq. (4) holds.

$$\frac{1}{d^2} = \frac{h^2}{a^2} + \frac{k^2}{b^2} + \frac{l^2}{c^2} \qquad (3)$$

$$1/d^2 = (4/3a^2)(h^2 + k^2 + l^2) + l^2/c^2 \qquad (4)$$

Symmetry can also be considered in a lattice. The lattice being indefinitely extended, there is no longer a point group but an array of regularly repeating symmetry elements. Each lattice point is a center of symmetry. Symmetry planes are parallel to lattice planes. Symmetry axes must be perpendicular to a lattice plane and coincide with, or be parallel to, a row.

Bravais lattices. These are the 14 different possible space lattices obtained on the basis that two lattices are different when the environment of their points is different. The 14 unit cells are represented in Fig. 3. If considered as solids, the combination of symmetry elements they exhibit can be determined. Seven point groups which are respectively the most symmetrical of each system are found. Accordingly the unit cells, and by extension the lattices, can be divided into seven groups corresponding to the seven crystallographic systems. Among the cells shown in Fig. 3, some have points at places other than corners. These cells are not primitive but are multiple cells chosen for convenience. As an example, the three primitive cells of the cubic lattices are, respectively, a cube, a rhombohedron with a plane angle of 109°28′, and a rhombohedron with an angle of 60°. The two rhombohedrons are extremely inconvenient to handle; consequently, the body-centered and face-centered cubes are adopted in their stead. Figure 4 shows the three cubic systems and the relationship of the last two to their primitive rhombohedrons. The letters P, I, and F stand for primitive, body-centered, and face-centered, respectively. A P cell contains one point, since each corner is used eight times in the formation of the complete assembly. An I cell contains two points, and an F cell four (one at the corners and six halves, as each point in the center of the faces belongs to two cells).

The external symmetry of a crystal is due to the periodically repeated arrangement of its atoms or ions. The Bravais lattices give the possible periodicities. A further step must consist in finding the number of possible periodic arrangements, two such arrangements being considered as different when they give rise to a different symmetry.

Two-dimensional structures are very well suited to illustrate this. In Figs. 5 and 6 a numeral 9 is chosen as the object; it is fully asymmetric and no false symmetry can therefore be introduced. It is multiplied by a periodic array of mirror planes and fourfold axes. Several interpenetrating identical lattices are formed. The two arrangements have the same Bravais lattice but not the same symmetry. In three-dimensional space the problem is similar, but other symmetry elements are considered.

Screw axes. These combine the rotation of an ordinary symmetry axis with a translation parallel to it and equal to a fraction of the unit distance in this direction. Figure 7 illustrates such an operation, which is symbolically denoted 3_1 and 3_2. The translation is respectively 1/3 and 2/3. The helices are added to help the visualization, and it is seen that they are respectively right- and left-handed. The projection on a plane perpendicular to the axis shows that the relationship about the axis remains in spite of the displacement. A similar type of arrangement can be considered around the other symmetry axes, and the following possibilities arise: 2_1, 3_1, 3_2, 4_1, 4_2, 4_3, 6_1, 6_2, 6_3, 6_4, and 6_5.

If screw axes are present in crystals, it is clear that the displacements involved are of the order of a few angstroms and that they cannot be distinguished macroscopically from ordinary symmetry axes. The same is true for glide mirror planes.

Glide mirror planes. These combine the mirror image with a translation parallel to the mirror plane over a distance which is half the unit distance in the glide direction. This is illustrated in Fig. 8. Axial glide planes, denoted by *a*, *b*, *c*, have translations which are equal to $a/2$, $b/2$, $c/2$, respectively, where *a*, *b*, *c* are the lattice vectors. Diagonal glide planes, denoted by *n*, have translations of $(a + b)/2$, $(b + c)/2$, or $(c + a)/2$.

These new symmetry elements must be taken into account when the number of possible periodic arrangements is considered. This problem was solved by R. S. Federow (1885), A. M. Schoenflies (1891), and W. Barlow (1894). A total of 230 arrangements or space groups is possible; of these, 32 can be distinguished macroscopically.

Space groups. These are indefinitely extended arrays of symmetry elements disposed on a space lattice. A space group acts as a three-dimensional kaleidoscope: An object submitted to its symmetry operations is multiplied and periodically repeated in such a way that it generates a number of interpenetrating identical space lattices. The fact that 230 space groups are possible means, of course, 230 kinds of periodic arrangement of objects in space. When only two dimensions are considered, 17 space groups are possible.

Space groups are denoted by the Hermann-Mauguin notation preceded by a letter indicating the Bravais lattice on which it is based. For example, P $2_12_12_1$ is an orthorhombic space group; the cell is primitive and three mutually perpendicular screw axes are the symmetry elements. All space groups are listed in advanced textbooks.

J. D. H. Donnay and D. Harker have shown that it is possible to deduce the space group from a detailed study of the external morphology of crystals.

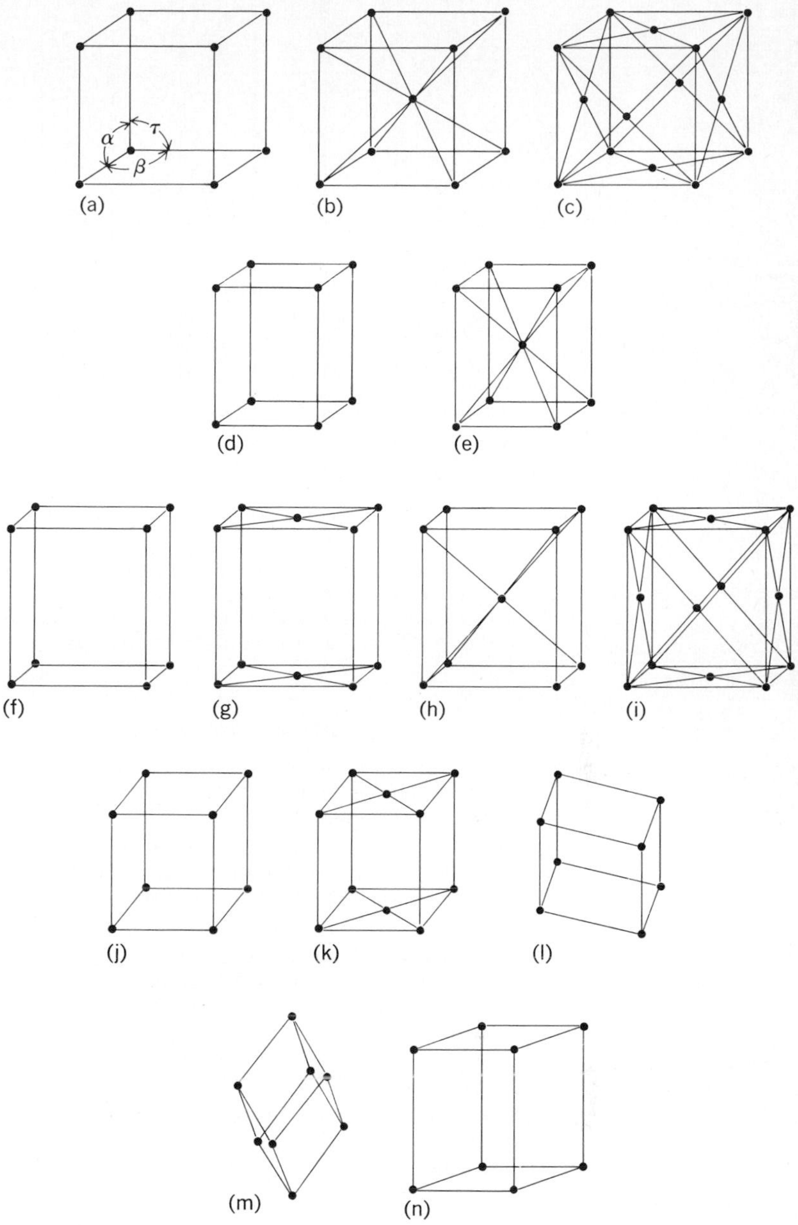

Fig. 3. The 14 Bravais space lattices. (*a*) Simple cubic. (*b*) Body-centered cubic. (*c*) Face-centered cubic. (*d*) Tetragonal. (*e*) Body-centered tetragonal. (*f*) Orthorhombic. (*g*) Base-centered orthorhombic. (*h*) Body-centered orthorhombic. (*i*) Face-centered orthorhombic. (*j*) Monoclinic. (*k*) Base-centered monoclinic. (*l*) Triclinic. (*m*) Trigonal. (*n*) Hexagonal.

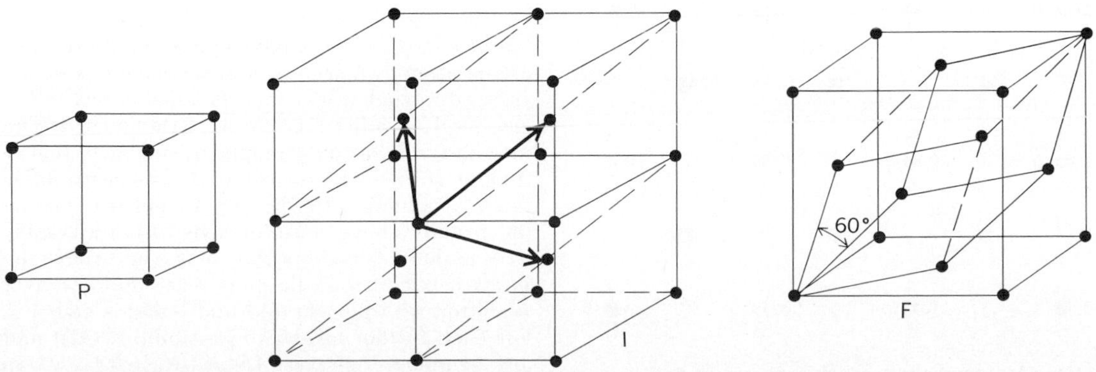

Fig. 4. The three cubic systems. The relationship of the F and I cells to their primitive rhombohedrons is demonstrated. (*After C. Kittel, Introduction to Solid State Physics, 2d ed., copyright © 1956 by John Wiley and Sons, Inc.; reprinted by permission*)

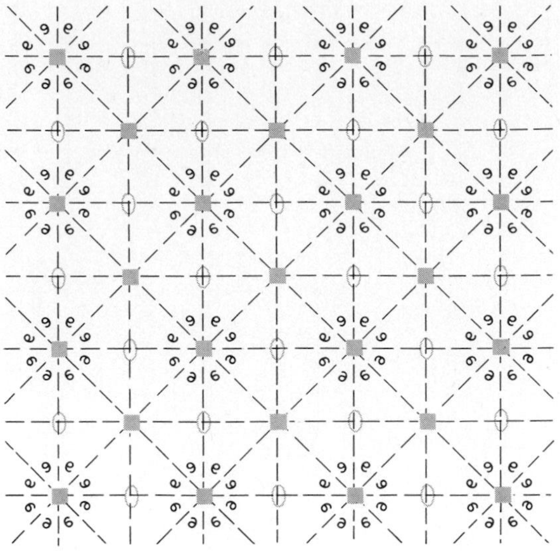

Fig. 5. Periodic pattern obtained by multiplication of asymmetric object by indicated array of symmetry elements. Both pattern and lattice have same symmetry elements. Mirror planes indicated by dotted lines.

COMMON STRUCTURES

A discussion of the three structural systems found in metals and of the five systems found in crystalline compounds is given below.

Metals. In general, metallic structures are relatively simple, characterized by a dense packing and a high degree of symmetry. Manganese, gallium, mercury, and one form of tungsten are exceptions. Metallic elements situated in the subgroups of the periodic table gradually lose their metallic character and simple structure as the number of the subgroup increases. A characteristic of metallic structures is the frequent occurrence of allotropic forms; that is, the same metal can have two or more different structures which are most frequently stable in a different temperature range.

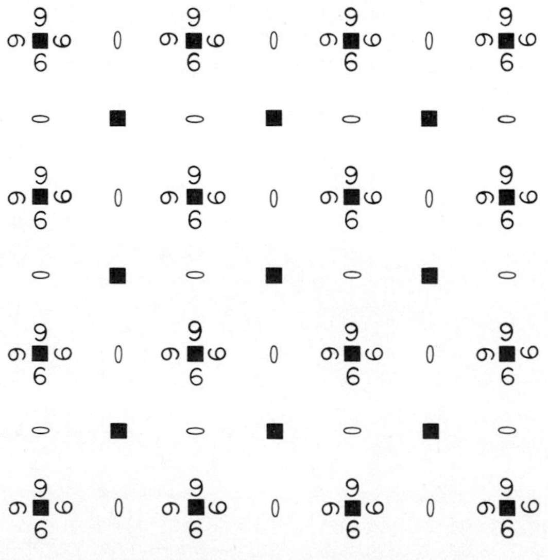

Fig. 6. Arrangement in which the periodic pattern has fewer symmetry elements than the lattice. This shows that the symmetry of the Bravais lattices can be lowered by the structure of the lattice points.

The forces which link the atoms together in metallic crystals are nondirectional. This means that each atom tends to surround itself by as many others as possible. This results in a dense packing, similar to that of spheres of equal radius, and yields three distinct systems; close-packed face-centered) cubic, hexagonal close-packed, and body-centered cubic.

Close packing. For spheres of equal radius, close packing is interesting to consider in detail with respect to metal structures. Close packing is a way of arranging spheres of equal radius in such a manner that the volume of the interstices between the spheres is minimal. The problem has an infinity of solutions. The manner in which the spheres can be most closely packed in a plane *A* is shown in Fig. 9. Each sphere is in contact with six others; the centers form a regular pattern of equi-

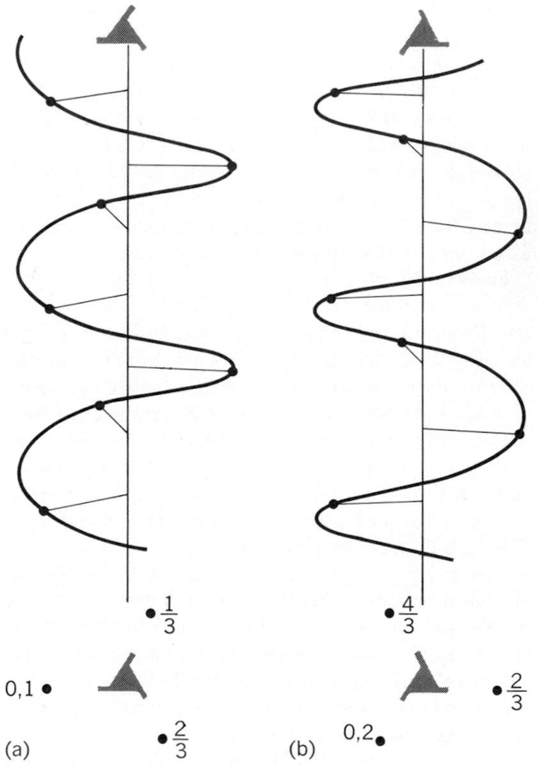

Fig. 7. Screw axes (*a*) 3_1 and (*b*) 3_2. Lower parts of figure are projections on a plane perpendicular to the axis. Numbers indicate heights of points above that plane.

lateral triangles. The cavities between the spheres are numbered. A second, similar plane can be positioned in such a way that its spheres rest in the cavities 1, 3, 5 between those of the layer *A*. The new layer, *B*, has an arrangment similar to that of *A* but is shifted with respect to *A*. Two possibilities exist for adding a third layer. Its spheres can be put exactly above those of layer *A* (an assembly *ABA* is then formed), or they may come above the interstices 2, 4, 6. In the latter case the new layer is shifted with respect to *A* and *B* and is called *C*. For each further layer two possibilities exist and any sequence such as *ABCBABCACBA* . . . in which two successive layers have not the same denomination is a solution of the problem. They are all characterized by the fact that each sphere

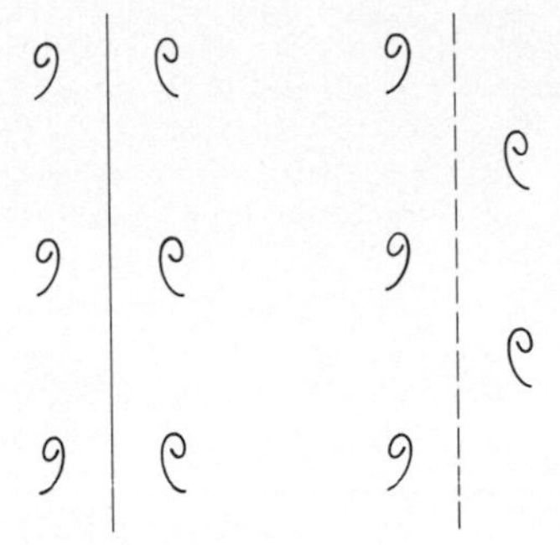

Fig. 8. Symmetry elements involving (*a*) mirror and (*b*) glide plane.

touches 12 others and all are equally densely packed. Such assemblies are rare in crystals. Complicated periodic sequences have, however, been observed, especially in carborundum. One structure is known where 89 layers form a sequence which is regularly repeated. In the vast majority of cases, periodic assemblies with a very short repeat distance occur. The cavities between the spheres, occupying 27% of the total volume, are of two types: tetrahedral cavities between four spheres, and octahedral ones between six spheres. For an assembly of N spheres, $3N$ cavities exist; $2N$ are tetrahedral and N are octahedral.

Face-centered cubic structure. This utilizes close packing characterized by the regular repetition of the sequence *ABC*. The centers of the spheres form a cubic lattice F, as shown in Fig. 10*a*. The densely packed planes of the type *A*, *B*, *C* are perpendicular to the threefold axis and can therefore be written as {111}. This form contains four sets of planes. These being the closest packed planes of the structure, d_{111} is greater than any other d_{hkl} of the lattice. The densest rows in these planes are <110>.

It is relatively easy to calculate the percentage of the volume occupied by the spheres. The unit cell has a volume a^3 and contains four spheres of radius R, their volume being $16\pi R^3/3$. The spheres touch each other along the face diagonal (Fig. 10*a*); $a\sqrt{2}$ is therefore equal to $4R$, or $R = a\sqrt{2}/4$. Substitution gives for the volume of the spheres $\pi\sqrt{2}a^3/6$, which is 73% of the volume of the cube. It is clear that the same percentage of the unit volume is filled in all other close-packed assemblies, that is, assemblies corresponding to other alternations of the planes *A*, *B*, *C*.

Hexagonal close-packed structure. This is a close packing characterized by the regular alternation of two layers, or *ABAB* . . . . The assembly has hexagonal symmetry (Fig. 10*b*); the unit cell is shown in Fig. 11. Six spheres are at the corners of an orthogonal parallelepiped having a parallelogram as its base; another atom has as coordinates (1/3, 1/3, 1/2). The ratio c/a is easily calculated. The length $BD = a$, the edge of the unit cell. The height of the cell is $AE = 2AG = c$, defined in Eqs. (5) and (6).

$$AG = \sqrt{a^2 - \left(\frac{2}{3}\,\frac{a\sqrt{3}}{2}\right)^2} = a\sqrt{2/3} \qquad (5)$$

$$c = 2AG = a\sqrt{8/3} \qquad c/a = \sqrt{8/3} = 1.633 \qquad (6)$$

This latter value is important, for it permits determination of how closely an actual hexagonal structure approaches ideal close packing.

Body-centered cubic structure. This is an assembly of spheres in which each one is in contact with eight others, as shown in Fig. 10*c*.

The spheres of radius R touch each other along the diagonal of the cube, so that measuring from the centers of the two corner cubes, the length of the cube diagonal is $4R$. The length of the diagonal is also equal to $a\sqrt{3}$, if a is the length of the cube edge, and thus $R = a\sqrt{3}/4$. The unit cell contains two spheres (one in the center and 1/8 in each corner) so that the total volume of the spheres in each cube is $8\pi R^3/3$. Substituting $R = a\sqrt{3}/4$ and dividing by a^3, the total volume of the cube, gives the percentage of filled space as 67%. Thus the structure is less dense than the two preceding cases.

The closest packed planes are {110}; this form contains six planes. They are, however, not as

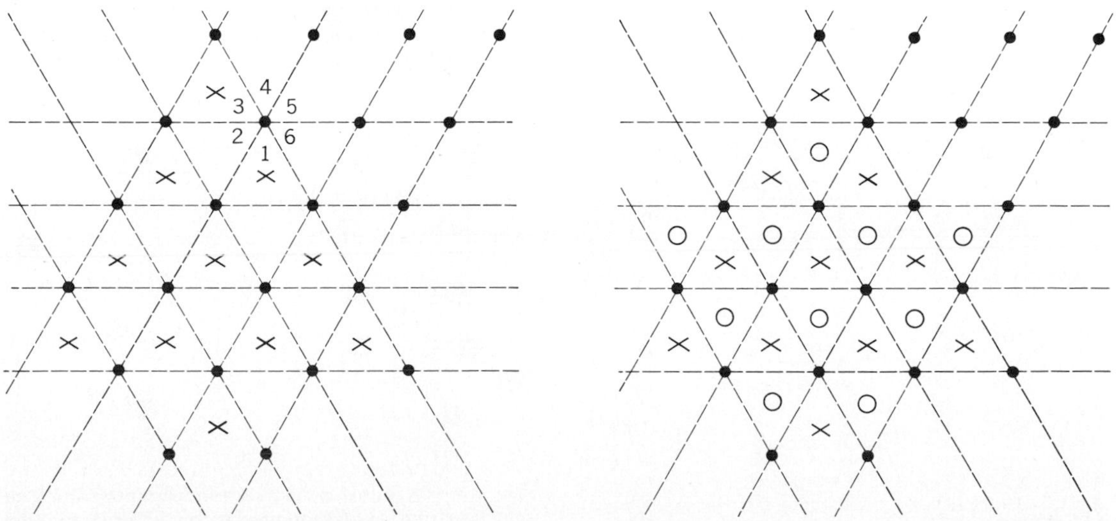

Fig. 9. Close packing of spheres of equal radius in a plane. The centers for the *A*, *B*, and *C* layers are indicated by a dot, a cross, and an open circle, respectively. Cavities between spheres are numbered.

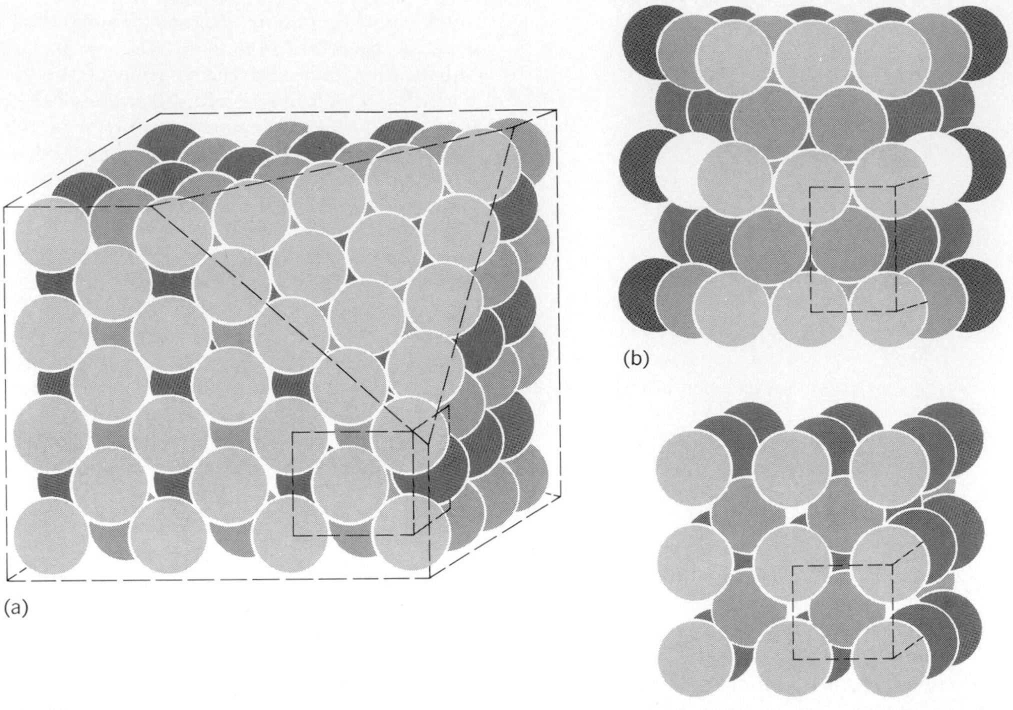

Fig. 10. Close packing of spheres in space. (*a*) Cubic close packing (face-centered cubic). One set of close-packed {111} planes (*A*, *B*, or *C*) is shown. (*b*) Hexagonal close packing. (*c*) Body-centered cubic arrangement.

dense as the *A*, *B*, *C* planes considered in the preceding structures. The densest rows have the four <111> directions.

Tabulation of structures. The structures of various metals are listed in the table, in which the abbreviations fcc, hcp, and bcc, respectively, stand for face-centered cubic, hexagonal close-packed, and body-centered cubic. In this table the hexagonal structures classified as ? are still in doubt. The structures listed hcp are only roughly so; only magnesium has a c/a ratio (equal to 1.62) which is very nearly equal to the ratio (1.63) calculated earlier for the ideal hcp structure. For cadmium and zinc the ratios are respectively 1.89 and 1.86, which are significantly larger than 1.63. Strictly speaking, each zinc or cadmium atom has therefore not twelve nearest neighbors but only six. This departure from the ideal case for subgroup metals follows a general empirical rule, formulated by W. Hume-Rothery and known as the 8-*N* rule. It states that a subgroup metal has a structure in which each atom has 8-*N* nearest neighbors, *N* being the number of the subgroup. *See* COORDINATION NUMBER.

Metal structures

Metal	Modification	Stability range	Structure
Beryllium	α	To melting point	hcp
	?β	630°C to melting point	Hexagonal
Cadmium	α	To melting point	hcp
Calcium	α	To 450°C	fcc
	?β	300–450°C	Hexagonal
	γ	450°C to melting point	bcc
Cerium	α	To melting point	fcc
	?β	50°C	Hexagonal
Chromium	α	To melting point	bcc
	?β	Electrolytic form	Hexagonal
Cobalt	α	To 420°C	hcp mixed with fcc
	β	420°C to melting point	fcc
Gold	α	To melting point	fcc
Iron	α	To 909°C	bcc
	γ	909–1403°C	fcc
	δ	1403°C to melting point	bcc
Lead	α	To melting point	fcc
Magnesium	α	To melting point	Nearly hcp
Nickel	α	To melting point	fcc
	β	Electrolytic form	Hexagonal
Silver	α	To melting point	fcc
Tungsten	α	To melting point	bcc
Zirconium	α	To 862°C	hcp
	β	862°C to melting point	bcc
Zinc	α	To melting point	hcp

Crystalline compounds. Simple crystal structures are usually named after the compounds in which they were first discovered (diamond or zinc sulfide, cesium chloride, sodium chloride, and calcium fluoride). Many compounds of the type A^+X^-,

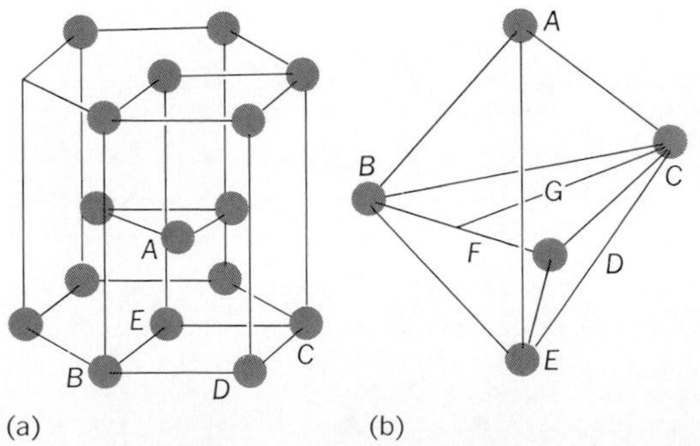

Fig. 11. Hexagonal close-packed structure. (*a*) Three unit cells, showing how the hexagonal axis results. One of the cells is fully outlined. (*b*) Calculation of the ratio c/a. Distance *AE* is equal to height of cell.

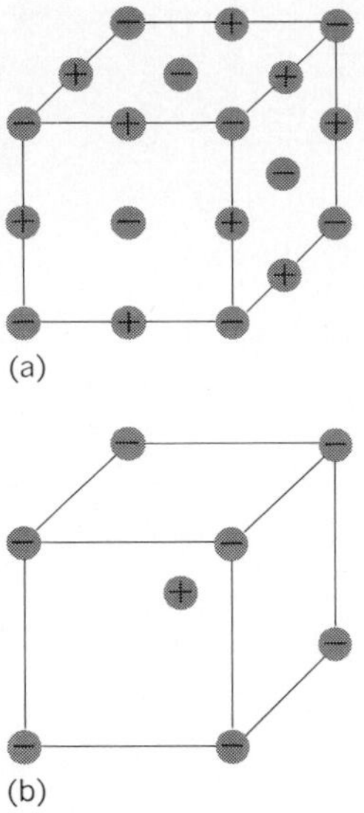

Fig. 12. Simple crystalline compound structures. (*a*) Sodium chloride. (*b*) Cesium chloride.

$A^{++}X_2^{-}$ have such structures. They are highly symmetrical, the unit cell is cubic, and the atoms or ions are disposed at the corners of the unit cell and at points having coordinates which are combinations of 0, 1, 1/2, or 1/4.

Sodium chloride structure. This is an arrangement in which each positive ion is surrounded by six negative ions and vice versa. The arrangement is expressed by stating that the coordination is 6/6. The centers of the positive and the negative ions each form a face-centered cubic lattice. They are shifted one with respect to the other over a distance $a/2$, where a is the repeat distance (Fig. 12*a*). Systematic study of the dimensions of the unit cells of compounds having this structure has revealed that:

1. Each ion can be assigned a definite radius. A positive ion is smaller than the corresponding atom and a negative ion is larger.

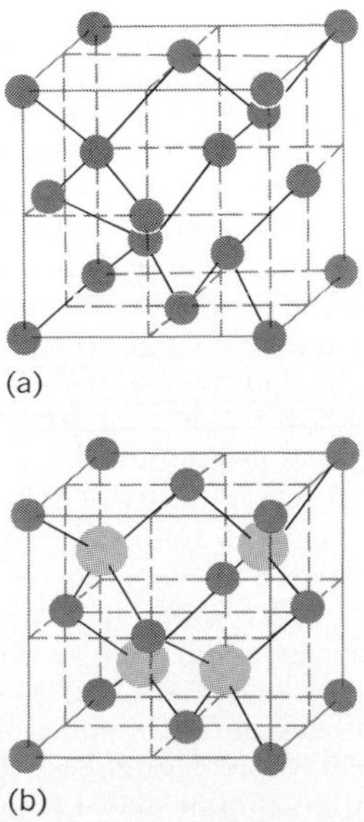

Fig. 13. Tetrahedral crystal compound structures. (*a*) Diamond. (*b*) zinc blende.

2. Each ion tends to surround itself by as many others as possible of the opposite sign because the binding forces are nondirectional.

On this basis the structure is determined by two factors. a geometrical factor involving the size of the two ions which behave in first approximation as hard spheres, and an energetical one involving electrical neutrality in the smallest possible volume. In the ideal case all ions will touch each other; therefore, if r_A and r_X are the radii of the ions, $4r_X = a\sqrt{2}$ and $2(r_A + r_X) = a$. Expressing a as a function of r_X gives $r_A/r_X = \sqrt{2} - 1 = 0.41$. When r_A/r_X becomes smaller than 0.41, the positive and negative ions are no longer in contact and the structure becomes unstable. When r_A/r_X is greater than 0.41, the positive ions are no longer in contact, but ions of different sign still touch each other. The structure is stable up to $r_A/r_X = 0.73$, which occurs in the cesium chloride structure.

Cesium chloride structure. This is characterized by a coordination 8/8 (Fig. 12*b*). Each of the centers of the positive and negative ions forms a primitive cubic lattice; the centers are mutually shifted over a distance $a\sqrt{3}/2$. The stability condition for this structure can be calculated as in the preceding case. Contact of the ions of opposite sign here is along the cube diagonal.

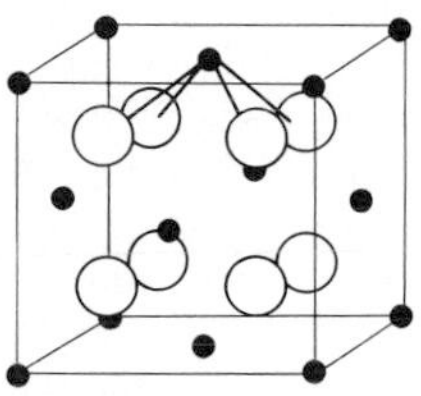

Fig. 14. Calcium fluoride structure.

Diamond structure. In this arrangement each atom is in the center of a tetrahedron formed by its nearest neighbors. The 4-coordination follows from the well-known bonds of the carbon atoms. This structure is illustrated in Fig. 13*a*. The atoms are at the corners of the unit cell, the centers of the faces, and at four points having as coordinates (1/4, 1/4, 1/4), (3/4, 3/4, 1/4), (3/4, 1/4, 3/4), (1/4, 3/4, 3/4). The atoms can be divided into two groups, each forming a face-centered cubic lattice; the mutual shift is $a\sqrt{3}/4$.

Zinc blende structure. This structure, shown in Fig. 13*b*, has coordination 4/4 and is basically similar to the diamond structure. Each zinc atom (small circles in Fig. 13*b*) is in the center of a tetrahedron formed by sulfur atoms (large circles) and vice versa. The zinc atoms form a face-centered cubic lattice, as do the sulfur atoms.

Calcium fluoride structure. Figure 14 shows the calcium fluoride structure. If the unit cell is divided into eight equal cubelets, calcium ions are situated at corners and centers of the faces of the cell. The fluorine ions are at the centers of the eight cubelets. There exist three interpenetrating face-centered cubic lattices, one formed by the Ca ion and two by the F ions, the mutual shifts being (0, 0, 0), (1/4, 1/4, 1/4), (3/4, 3/4, 3/4).

[WILLY C. DEKEYSER]

Bibliography: Akademische Verlagsgesellschaft, Strukturbericht, vols. 1–7, *Z. Krist.*, suppl. vols., 1913–1939, reprint, 1943, and continued by International Union of Crystallography, *Structure Reports*, 1945–; P. J. Brown and J. E. Forsyth, *The Crystal Structure of Solids*, 1978; M. J. Buerger, *Crystal-Structure Analysis*, 1979; M. J. Buerger, *Introduction to Crystal Geometry*, 1977; J. D. H. Donnay and W. Nowacki, *Crystal Data*, Geol. Soc. Amer., Mem., no. 60, 1954; R. C. Evans, *Introduction to Crystal Chemistry*, 2d ed., 1964; N. F. M. Henry and K. Lonsdale, *International Tables for X-ray Crystallography*, vol. 1, 1952.

Crystallization

The formation of a solid from a solution, melt, vapor, or a different solid phase. Crystallization from solution is an important industrial operation because of the large number of materials marketed as crystalline particles. Fractional crystallization is one of the most widely used methods of separating and purifying chemicals. This article discusses crystallization of substances from solutions and melts. Not discussed is biological crystallization, involving formation of teeth and bone, otoconia in the inner ear, renal calculi (kidney stones), biliary calculi (gallstones), crystals in some forms of arthritis, and dental plaque. Polymer crystals obtained from solutions are used to study the properties of these crystals, while crystallization of polymer melts dramatically influences polymer properties; *see* POLYMER. For methods of preparing large crystals *see* SINGLE CRYSTAL. For crystallization from vapors *see* SUBLIMATION. For solubility and other relationships between solid and liquid phases *see* PHASE EQUILIBRIUM; SOLUTION.

Solutions. In order for crystals to nucleate and grow, the solution must be supersaturated; that is, the solute must be present in solution at a concentration above its solubility. Different methods may be used for creating a supersaturated solution from one which is initially undersaturated. The possible methods depend on how the solubility varies with temperature. Two examples of solubility behavior are shown in the illustration. Either evaporation of water or cooling may be used to crystallize potassium nitrate (KNO_3), while only evaporation would be effective for NaCl. An alternative is to add a solvent such as ethanol which greatly lowers the solubility of the salt, or to add a reactant which produces an insoluble product. This causes a rapid crystallization perhaps more properly known as precipitation. *See* PRECIPITATION.

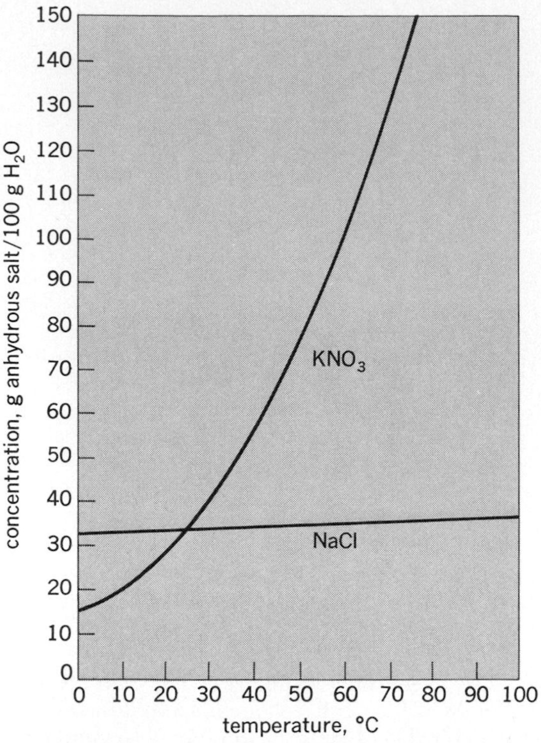

Temperature-solubility curves for two salts.

Nucleation. The formation of new crystals is called nucleation. At the extremely high supersaturations produced by addition of a reactant or a lower-solubility solvent, this nucleation may take place in the bulk of the solution in the absence of any solid surface. This is known as homogeneous nucleation. At more moderate supersaturations new crystals form on solid particles or surfaces already present in the solution (dust, motes, nucleation catalysts, and so on). This is called heterogeneous nucleation. When solutions are well agitated, nucleation is primarily secondary, that is, from crystals already present. Probably this is usually due to minute pieces breaking off the crystals by impact with other crystals, with the impeller, or with the walls of the vessel. *See* NUCLEATION.

Crystal size. Generally, large crystals are considered more desirable than small crystals, probably in the belief that they are more pure. However, crystals sometimes trap (occlude) more solvent as inclusions when they grow larger. Thus there may be an optimal size. The size distribution of crystals is influenced primarily by the supersaturation, the amount of agitation, and the growth time. Generally, the nucleation rate increases faster with increasing supersaturation than does the growth rate. Thus, lower supersaturations, gentle stirring, and long times usually favor large crystals. Low supersaturations require slow evaporation and cooling rates.

Crystal habit. Often the habit (shape) of crystals is also an important commercial characteristic. In growth from solutions, crystals usually display facets along well-defined crystallographic planes, determined by growth kinetics rather than by equilibrium considerations. The slowest-growing facets are the ones that survive and are seen. While habit depends somewhat on the supersaturation during growth, very dramatic changes in habit are usually brought about by additives to the solution. Strong habit modifiers are usually incorporated into the crystal, sometimes preferentially. Thus the additive may finally be an impurity in the crystals.

Fractional crystallization. In fractional crystallization it is desired to separate several solutes present in the same solution. This is generally done by picking crystallization temperatures and solvents such that only one solute is supersaturated and crystallizes out. By changing conditions, other solutes may be crystallized subsequently. Occasionally the solution may be supersaturated with respect to more than one solute, and yet only one may crystallize out because the others do not nucleate. Preferential nucleation inhibitors for the other solutes or seeding with crystals of the desired solute may be helpful. Since solid solubilities are frequently very small, it is often possible to achieve almost complete separation in one step. However, for optimal separation it is necessary to remove the impure mother liquor from the crystals. Rinsing of solution trapped between crystals is more effective for large-faceted crystals. Internally occluded solution cannot be removed by rinsing. Even heating to high temperatures may not burst these inclusions from the crystal. Repeated crystallizations are necessary to achieve desired purities when many inclusions are present or when the solid solubility of other solutes is significant.

Melts. If a solid is melted without adding a solvent, it is called a melt, even though it may be a mixture of many substances. That is the only real distinction between a melt and a solution. Crystallization of a melt is often called solidification, particularly if the process is controlled by heat transfer so as to produce a relatively sharp boundary between the solid and the melt. It is then possible to slowly solidify the melt and bring about a separation, as indicated by the phase diagram. The melt may be stirred to enhance the separation. The resulting solid is usually cut into sections, and the purest portion is subjected to additional fractional solidifications. Alternatively the melt may be poured off after part of it has solidified. Zone melting was invented to permit multiple fractional solidifications without the necessity for handling between each step. Fractionation by the above techniques appears to be limited to purification of small batches of materials already fairly pure, say above 95%.

Fractional crystallization from the melt is also being used for large-scale commercial separation and purification of organic chemicals. It has also been tested for desalination of water. Rather than imposing a sharp temperature gradient, as in the above processes, the melt is relatively isothermal. Small, discrete crystals are formed and forced to move countercurrent to the melt. At the end from which the crystals exit, all or part of the crystals are melted. This melt then flows countercurrent to the crystals, thereby rinsing them of the adhering mother liquor. *See* CRYSTAL STRUCTURE.

[WILLIAM R. WILCOX]

Bibliography: J. W. Mullin, *Crystallization*, 2d ed., 1972; J. Nývlt, *Industrial Crystallization: The Present State of the Art*, 1978; A. D. Randolph and M. A. Larson, *Theory of Particulate Processes: Analysis and Techniques of Continuous Crystallization*, 1971; M. Zief and W. R. Wilcox (eds.), *Fractional Solidification*, 1967.

Curium

A chemical element, Cm, in the actinide series, with an atomic number of 96. The electronic configuration of the neutral atom (beyond the radon core) is $5f^76d7s^2$.

Curium does not exist in the terrestrial environment, but may be produced artificially, as was first done in 1944 by G. T. Seaborg, R. A. James, and Albert Ghiorso, who synthesized and identified the isotope ^{242}Cm. The synthesis was effected by the bombardment of ^{239}Pu (plutonium-239) with helium ions of about 32,000,000 electron volts (32 MeV) energy, $^{239}_{94}\mathrm{Pu}(\alpha,n)^{242}_{96}\mathrm{Cm}$.

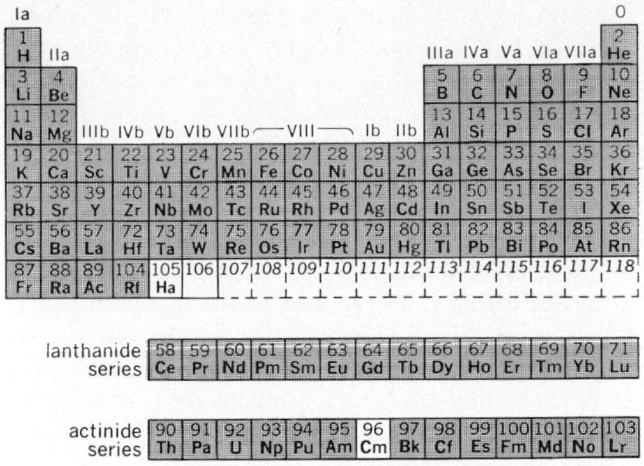

Larger amounts of ^{242}Cm, produced by the prolonged neutron irradiation of ^{241}Am, Eq. (1),

$$^{241}_{95}\mathrm{Am}(n,\gamma)^{242m}_{95}\mathrm{Am}\xrightarrow{\beta^-}{}^{242}_{96}\mathrm{Cm} \qquad (1)$$

provided material for the first isolation of visible amounts of curium by L. B. Werner and I. Perlman in 1947. Higher mass isotopes—^{244}Cm, ^{247}Cm, ^{248}Cm—are now the principal sources of the element.

The known isotopes of curium, along with their half-lives and modes of decay, are listed in the table. Where, for a given mass, separate half-lives and modes of disintegration are listed, the decay processes occur simultaneously.

The isotope ^{244}Cm is of particular interest because of its potential use as a compact, thermoelectric power source, through conversion to electrical power of the heat generated by nuclear decay.

Metallic curium may be produced by the reduction of curium trifluoride with barium vapor at 1350°C. The metal has a silvery luster, tarnishes in air, and has a specific gravity of 13.5. Its room-temperature structure is double-hexagonal close-packed, similar to that observed in lanthanum. A face-centered cubic structure has been found at higher temperatures. The melting point has been determined as 1340±40°C. The metal dissolves

Isotopes of curium

Mass no.	Half-life	Mode of disintegration
238	2.5h	Orbital electron capture, <90%; α-particle emission, >10%
239	2.9h	Orbital electron capture, >99.9%
240	28d	Alpha particle emission
	1.9×10^6y	Spontaneous fission
241	35d	Orbital electron capture, 99+%; α-particle emission, 0.96%
242	162.5d	Alpha particle emission
	6.09×10^6y	Spontaneous fission
243	29y	Alpha particle emission, >99%; orbital electron capture, 0.26%
244	18.12y	Alpha particle emission
	1.35×10^7y	Spontaneous fission
245	8540y	Alpha particle emission
246	4716y	Alpha particle emission
	1.85×10^7y	Spontaneous fission
247	1.56×10^7y	Alpha particle emission
248	3.70×10^5y	Alpha particle emission
	4.11×10^6y	Spontaneous fission
249	65m	Beta particle (electron) emission
250	1.13×10^4y	Spontaneous fission

readily in common mineral acids with the formation of the tripositive ion, Eq. (2). In this process 142 kilocalories of heat are evolved per mole of curium dissolved.

$$Cm + 3H^{+} = Cm^{3+} + 1\tfrac{1}{2}H_2(g) \qquad (2)$$

A number of solid compounds of curium have been prepared and their structures determined by x-ray diffraction. These include CmF_4, CmF_3. $CmCl_3$, $CmBr_3$, CmI_3, Cm_2O_3, and CmO_2. In the trihalides the ionic radius of Cm^{3+} is 0.94 A. Isostructural analogs of the compounds of curium are observed in the lanthanide series of elements. *See* X-RAY DIFFRACTION.

Curium trichloride may be prepared by treating the sesquioxide with HCl(*g*) at 500°C. Synthesis of the tribromide and triiodide may be effected similarly by using HBr(*g*) and HI(*g*). The trifluoride is readily prepared by precipitation of aqueous Cm^{3+} with hydrofluoric acid and drying of the precipitate. This reacts at about 300°C with elemental fluorine to yield the tetrafluoride. The dioxide is obtained by treating curium(III) oxalate with ozone and oxygen at about 300°C. The black oxide has the fluorite structure with crystal lattice parameter, $a = 5.37$ A. The dioxide is relatively unstable, as it may be decomposed in vacuum at 600°C to yield the cubic sesquioxide with $a_0 = 5.50$ A.

In aqueous solution the tetrapositive ion is highly unstable. It may be prepared, however, as a fluoro-complex ion by dissolving solid curium tetrafluoride in 15 *M* CsF at 0°C. The light-yellow solution shows sharp absorption maxima, the strongest being at 4515 and 8640 A. Complete reduction to Cm(III) occurs in about 20 min at room temperature. Cm(III) in aqueous solutions shows little absorption in the visible region of the spectrum, but weak lines are found in the region 3700–4000 A. Strong absorption occurs further into the ultraviolet, particularly near 2800 A.

The magnetic susceptibility of CmF_3, measured near room temperature, gives a calculated effective magnetic moment of 7.9 Bohr magnetons, similar to the value of 7.94 for Gd^{3+}, the lanthanide analog of curium. The spectroscopic g value for Cm^{3+} is 1.99 as compared to 1.93 for Gd^{3+}. The difference is well accounted for by the greater tendency toward *j-j* coupling in the 5*f* as compared with the 4*f* transition series.

Curium is the seventh member of the actinide series of elements. Its half-completed *f* subshell ($5f^7 6d 7s^2$) is analogous to that of gadolinium in the lanthanide series. The chemical properties of curium are so similar to those of the typical rare earths that, if it were not for its radioactivity, it might easily be mistaken for one of these elements. *See* ACTINIDE ELEMENTS; TRANSURANIUM ELEMENTS.

[GLENN T. SEABORG]

Bibliography: *Gmelins Handbuch der anorganischen Chemie: Transurane*, vol. 4, 1972, vols. 7a and 8, 1973; E. K. Hyde, I. Perlman, and G. T. Seaborg, *The Nuclear Properties of the Heavy Elements*, vols. 1 and 2, 1964; J. Katz and G. T. Seaborg, *The Chemistry of the Transuranium Elements*, vol. 3 of *Kernchemie in Einzeldarstellungen*, 1971; G. T. Seaborg, *Man-made Transuranium Elements*, 1963.

Cyanamide

A term used to refer to the free acid, H_2NCN, or commonly to the calcium salt of this acid, $CaCN_2$, which is properly called calcium cyanamide. Calcium cyanamide is manufactured by the cyanamide process, in which nitrogen gas is passed through finely divided calcium carbide at a temperature of 1000°C.

Most plants are equipped to produce the calcium cyanamide from the basic raw materials of air, limestone, and carbon according to reactions (1)–(3).

$$CaCO_3 \xrightarrow{\Delta} CaO + CO_2 \qquad (1)$$

$$CaO + 3C \rightarrow CaC_2 + CO \qquad (2)$$

$$CaC_2 + N_2 \rightleftharpoons CaCN_2 + C \qquad (3)$$

Most calcium cyanamide is used in agriculture as a fertilizer; some is used as a weed killer, as a pesticide, and as a cotton defoliant.

Cyanamide, H_2NCN, is prepared from calcium cyanamide by treating the salt with acid. It is used in the manufacture of dicyandiamide and thiourea. *See* CARBON; NITROGEN.

[E. EUGENE WEAVER]

Cyanate

A compound containing the —OCN group and derived from cyanic acid, HOCN. Cyanates are isomeric with fulminates which have the same atoms arranged thus, —ONC.

Like hydrogen cyanide and thiocyanic acid, cyanic acid may exist in two forms, HO—C≡N and H—N═C═O, the latter of which is called isocyanic acid. The alkali and alkaline-earth metal cyanates contain an anion which may also be written in two forms: $[N{\equiv}C{-}O]^-$ and $[O{=}C{=}N]^-$. These salts are water-soluble.

The heavy metal cyanates are more covalent in nature and are water-insoluble. The main use of cyanates is in the synthesis of organic compounds.

Ammonium cyanate rearranges easily to urea: $NH_4OCN \rightleftharpoons NH_2CONH_2$.

The alkali metal cyanates may be prepared by the direct oxidation of the cyanide with oxygen or another suitable oxidizing agent under controlled conditions. *See* CARBON; CYANIDE; NITROGEN; THIOCYANATE.

[E. EUGENE WEAVER]

Cyanide

A compound containing the —CN group, for example, potassium cyanide, KCN; calcium cyanide, $Ca(CN)_2$; and hydrocyanic (or prussic) acid, HCN. Chemically, the simple inorganic cyanides resemble chlorides in many ways. Organic compounds containing this group are called nitriles. Acrylonitrile, CH_2CHCN, is an important starting material in the manufacture of fabrics, plastics, and synthetic rubber.

HCN is a weak acid, having an ionization constant of 1.3×10^{-9} at 18°C. In the pure state, it is a highly volatile liquid, boiling at 26°C.

HCN and the cyanides are highly toxic to animals, a lethal dose to humans usually considered to be 100 to 200 mg. Death has been attributed, however, to as low a dose as 0.57 mg of HCN per kilogram of body weight. As poisons, they are rapid in their action. When these compounds are inhaled or ingested, their physiological effect involves an inactivation of the cytochrome respiratory enzymes, preventing tissue utilization of the oxygen carried by the blood.

The cyanide ion forms a variety of coordination complexes with transition-metal ions. Representative complexes are those of gold $[Au(CN)_2]^-$, silver $[Ag(CN)_2]^-$, and iron $[Fe(CN)_6]^{4-}$. This property is responsible for several of the commercial uses of cyanides. In the cyanide process, the most widely used method for extracting gold and silver from the ores, the finely divided ore is contacted with a dilute solution of sodium or potassium cyanide. The metal is solubilized as the complex ion, and, following extraction, the pure metal is recovered by reduction with zinc dust. In silver-plating, a smooth adherent deposit is obtained on a metal cathode when electrolysis is carried out in the presence of an excess of cyanide ion.

$Ca(CN)_2$ is extensively used in pest control and as a fumigant in the storage of grain. In finely divided form, it reacts slowly with the moisture in the air to liberate HCN.

In case hardening of metals, an iron or steel article is immersed in a bath of molten sodium or potassium cyanide containing sodium chloride or carbonate. At temperatures of 750°C and above, the cyanide decomposes at the surface, forming a deposit of carbon which combines with and penetrates the metal. The nitrogen also contributes to the increased hardness at the surface by forming nitrides with iron and alloying metals. *See* SURFACE HARDENING OF STEEL.

NaCN and KCN are produced commercially by neutralization of hydrogen cyanide with sodium or potassium hydroxide. The neutralized solution must then be evaporated to dryness in a strictly controlled manner to avoid undue losses. $Ca(CN)_2$ is produced primarily from calcium cyanamide by reaction with carbon in the presence of sodium chloride at 1000°C. *See* COORDINATION CHEMISTRY; FERRICYANIDE; FERROCYANIDE.

[FRANCIS J. JOHNSTON]

Bibliography: M. J. Sienko and R. A. Plane, *Chemical Principles and Properties*, 1974; U.S. Department of Health, Education and Welfare, *The Toxic Substance List*, 1973.

Cyanocarbon

A derivative of hydrocarbon in which all of the hydrogen atoms are replaced by the —C≡N group. Only two cyanocarbons, dicyanoacetylene and dicyanobutadiyne, were known before 1958. Since then hexacyanoethane, tetracyanoethylene, hexacyanobutadiene, and hexacyanobenzene were synthesized. The term cyanocarbon has been applied to compounds which do not strictly follow the above definition: tetracyanoethylene oxide, tetracyanothiophene, tetracyanofuran, tetracyanopyrrole, tetracyanobenzoquinone, tetracyanoquinodimethane, tetracyanodithiin, pentacyanopyridine, diazomalononitrile, and diazotetracyanocyclopentadiene.

Tetracyanoethylene, the simplest olefinic cyanocarbon, is a colorless, thermally stable solid, having a melting range of 198–200°C. It is a strong π-acid, readily forming stable complexes with most aromatic systems. These complexes absorb radiation in or near the visible spectrum.

Reactions of tetracyanoethylene

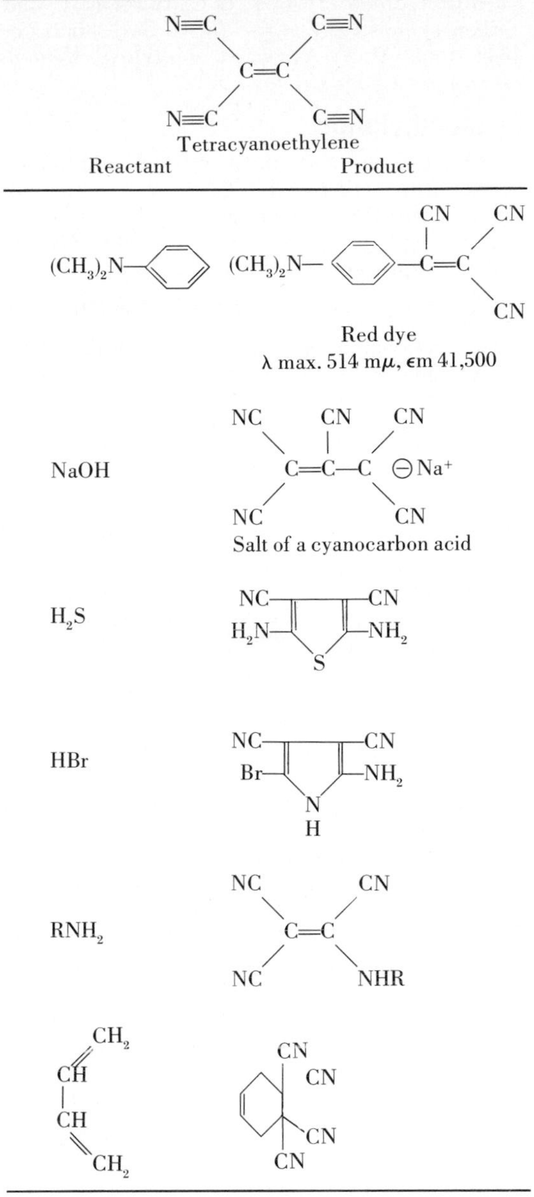

Cyanocarbon acids are among the strongest protonic organic acids known and are usually isolated only as the cyanocarbon anion salt. These salts are usually colored and stabilized by resonance of the type indicated in the reaction below.

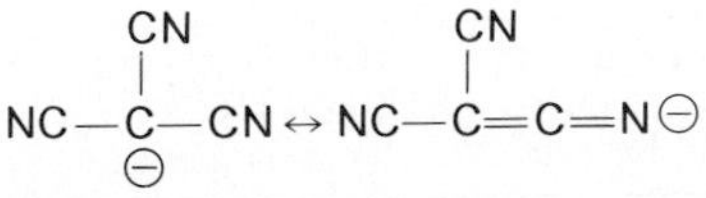

Twenty-five similar forms can be written for the 2-dicyanomethylene-1,1,3,3,-tetracyanopropanediide ion, $(NC)_2C{=}C[C(CN)_2]_2^{--}$, which accounts for the similarity between the acid strength of the free acid and that of sulfuric acid.

The reactivity of cyanocarbons is illustrated by the facile replacement of one of the cyano groups in tetracyanoethylene by nucleophilic attack under very mild conditions and by its addition to dienes. Examples of these reactions are illustrated in the table. *See* ORGANIC CHEMISTRY.

[OWEN W. WEBSTER]

Bibliography: N. L. Allinger et al., *Organic Chemistry*, 2d ed., 1976; T. L. Cairns et al., Cyanocarbon chemistry, *J. Amer. Chem. Soc.*, 80:2775–2844, 1958; M. F. Ashworth, *Analytical Methods for Organic Cyano Groups*, 1971.

Cyanoethylation

A chemical reaction involving the addition of acrylonitrile ($CH_2{=}CH{-}CN$) to compounds carrying a reactive hydrogen, to introduce the β-cyanoethyl grouping ($-CH_2CH_2CN$). The reaction is represented by Eq. (1).

$$ZH + CH_2{=}CH{-}CN \rightleftharpoons ZCH_2CH_2CN \qquad (1)$$

Cyanoethylation can be effected with HBr, HCl, HCN, and with compounds containing the functional groups $-\overset{|}{As}H$, $-\overset{|}{B}H$, $-\overset{|}{C}H$ (when activated by adjacent electron-withdrawing groups), $-\overset{|}{N}H$, $-OH$, $-\overset{|}{P}H$, $-SH$, $-\underset{|}{\overset{|}{Si}}H$, and $-\underset{|}{\overset{|}{Sn}}H$. The products can be subjected further to the usual nitrile reactions, such as hydration, hydrolysis and reduction as in Eqs. (2)–(4).

$$ZCH_2CH_2CN \rightarrow ZCH_2CH_2CONH_2 \qquad (2)$$

$$ZCH_2CH_2CN \rightarrow ZCH_2CH_2COOH \qquad (3)$$

$$ZCH_2CH_2CN \rightarrow ZCH_2CH_2CH_2NH_2 \qquad (4)$$

Because of its versatility, cyanoethylation has assumed great importance in synthesis, as shown by the fact that over 1200 compounds have been used as reactants. Introduction of the cyanoethyl group tends to increase hydrophobic character and resistance to biological attack. Natural products, notably cellulose, have been cyanoethylated to take advantage of these properties.

Substrates for cyanoethylation. The nature of the catalyst is used to classify substrates into three groups:

(1) No catalyst: $-\overset{|}{As}H$, $-\overset{|}{B}H$, $-\overset{|}{N}H$ (aliphatic), $-\overset{|}{P}H$, $-\underset{|}{\overset{|}{Sn}}H$, HBr, and HCl

(2) Basic catalyst (for example, NaOH, $NaOCH_3$, benzyltrimethylammonium hydroxide): $-\underset{|}{\overset{|}{C}}H$, $-\overset{|}{N}H$ (amide and certain heterocycles), $-OH$, $-\underset{|}{P}H$, $-\underset{|}{P}(O)H$, $-SH$, and HCN

(3) Acid catalyst (for example, acetic acid with or without a copper salt): $-\overset{|}{N}H$ (aromatic), $-NH_2$ (*t*-carbinamines)

Yields in cyanoethylation are generally high. The reaction is strongly exothermic and is usually carried out at moderate temperatures in a solvent such as dioxane or *t*-butanol. Acrylonitrile itself has been used as a diluent, but this method is not widely applicable in base-catalyzed cyanoethylation because of competing anionic polymerization of acrylonitrile. High temperatures and high concentrations of base tend to reverse the process.

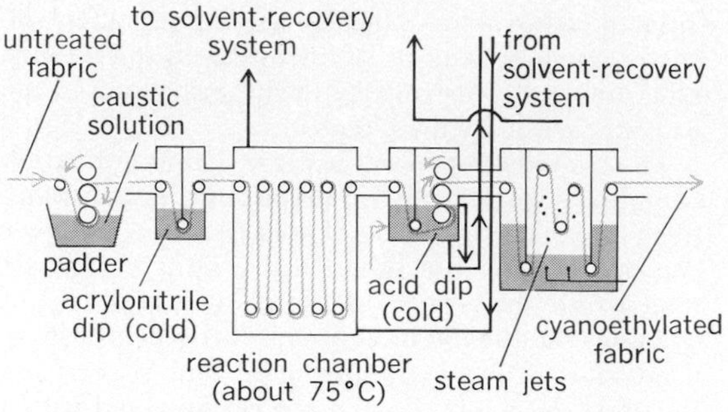

Fig. 1. Flow diagram for cyanoethylation. (*From Cyanoethylation of Cotton, American Cyanamid Co., Monsanto Chemical Co., and the Institute of Textile Technology, September, 1956*)

Cyanoethylation of natural materials. Since the extent of cyanoethylation can be varied by choice of conditions, a variety of products can be obtained from a given material. These may be further modified by reactions of the cyanoethyl group, for example, to give polyelectrolytes.

Partially cyanoethylated cellulose, in which one-sixth of the hydroxyl groups have reacted, is rot- and termiteproof; it has improved resistance to degradation by heat and by acids, as well as altered dyeing characteristics and greater hydrophobicity. Crystallinity and morphology of the cellulose are retained. Utility is in marine nets and ropes, electrical insulation, industrial fabrics, structural wood, and dimensionally stable paper.

In the cyanoethylation of cotton the fabric is first padded with 6% sodium hydroxide solution and with acrylonitrile, and then passed continuously into a chamber heated to about 75°C by acrylonitrile vapor (Fig. 1). After a residence time of about 3.5 min, the caustic-impregnated cotton is neutralized, washed, and dried (Fig. 2). Unreacted acrylonitrile is recovered and recycled.

Highly cyanoethylated cellulose (two-thirds or more of hydroxyl groups reacted) is an amorphous,

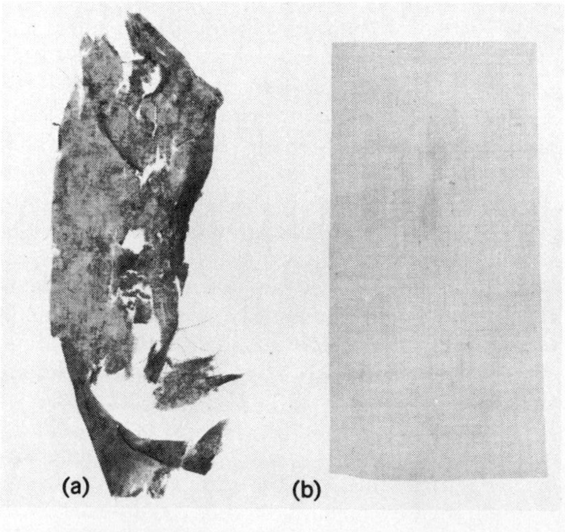

Fig. 2. Appearance of cotton after burial in soil for 2 weeks. (*a*) Untreated. (*b*) Partially cyanoethylated.

thermoplastic material which is soluble in acrylonitrile, dimethylformamide, and pyridine. Unhindered rotation of the polar cyanoethyl segments imparts an unusually high dielectric constant (13.3 at 60 cycles and 25°C), a property useful in capacitors and in matrices for electroluminescent phosphors. Starch, cyanoethylated to retain water solubility, has the ability to prevent soil from redepositing on washed fabrics. More highly cyanoethylated starch is soluble in organic solvents. Cyanoethylsucrose is used as a dielectric fluid.

In the cyanoethylation of proteins the free amino and possibly the amide groups are attacked. Complete resistance to putrefaction results when casein is cyanoethylated with as little as a 5–10% by weight of acrylonitrile. Cyanoethylated wool has increased affinity for both cationic dyes and direct dyes.

Other uses. Cyanoethylation is also used in the preparation of amino acids, chemotherapeutic agents, dyes, blowing agents, long-chain diamines, special-purpose silicones, food stabilizers, and liquids to effect separations by gas chromatography. *See* ACRYLONITRILE.

[NORBERT M. BIKALES]

Bibliography: R. J. Harper, Cyanoethylation, *Kirk-Othmer Encyclopedia of Chemical Technology*, 3d ed., vol. 7, 370–385, 1979; J. D. Roberts and M. C. Caserio, *Basic Principles of Organic Chemistry*, 2d ed., 1977.

Cyanogen

A colorless, highly toxic gas having the molecular formula C_2N_2. Cyanogen belongs to a class of compounds known as pseudohalogens, because of the similarity of their chemical behavior to that of the halogens. Liquid cyanogen boils at −21.17°C and freezes at −27.9°C at 1 atm. The density of the liquid is 0.954 g/ml at the boiling point.

Cyanogen may be prepared by prolonged heating of mercuric cyanide at 400°C, reaction (1),

$$Hg(CN)_2 \rightarrow Hg + C_2N_2 \qquad (1)$$

or by allowing a solution of copper sulfate to flow slowly into a solution of potassium cyanide, reaction (2). The copper (II) cyanide is unsta-

$$2KCN + CuSO_4 \rightarrow Cu(CN)_2 + K_2SO_4 \qquad (2)$$

ble, decomposing to give cyanogen and copper (I) cyanide, reaction (3).

$$2Cu(CN)_2 \rightarrow 2CuCN + C_2N_2 \qquad (3)$$

Cyanogen is also formed by the reaction in the gas phase of hydrogen cyanide and chlorine at 400–700°C in the presence of a catalytic agent such as activated carbon, reaction (4). Cyanogen

$$2HCN + Cl_2 \rightarrow C_2N_2 + 2HCl \qquad (4)$$

chloride, ClCN, is an intermediate in this reaction. When heated to 400°C, cyanogen gas polymerizes to a white solid, paracyanogen $(CN)_x$.

Cyanogen reacts with hydrogen at elevated temperatures in a manner analogous to the halogens, forming hydrogen cyanide, reaction (5).

$$C_2N_2 + H_2 \rightleftharpoons 2HCN \qquad (5)$$

With hydrogen sulfide, H_2S, cyanogen forms thiocyanoformamide or dithiooxamide. These reactions are shown by (6) and (7).

$$C_2N_2 + H_2S \rightarrow NC\text{—}\overset{\overset{\displaystyle S}{\|}}{C}\text{—}NH_2 \qquad (6)$$

$$C_2N_2 + 2H_2S \rightarrow H_2N\overset{\overset{\displaystyle S}{\|}}{C}\text{—}\overset{\overset{\displaystyle S}{\|}}{C}NH_2 \qquad (7)$$

Cyanogen burns in oxygen, producing one of the hottest flames known from a chemical reaction. It is considered to be a promising component of high-energy fuels.

Structurally, cyanogen is written N≡C—C≡N. *See* CYANIDE.

[FRANCIS J. JOHNSTON]

Bibliography: V. Migrdichian, *The Chemistry of Organic Cyanogen Compounds*, 1947; Z. Rappoport (ed.), *The Chemistry of the Cyano Group*, 1971.

Dalton's law

The total pressure that a mixture of gases exerts is equal to the sum of the separate pressures which each of the gases would exert if it alone occupied the whole volume. This law was observed by John Dalton (1766–1844) and is also called the law of additive pressures. It may be stated mathematically as $p_m = p_a + p_b + p_c + \cdots$, where p_m is the pressure produced by the mixture and p_a, p_b, p_c, $\cdots$ are the partial pressures of the several gases in the mixture. The partial pressure is the pressure exerted by a constituent gas that occupies the whole volume occupied by the mixture at the same temperature and pressure in the absence of the other gases. *See* AVOGADRO'S LAW; GAS; THERMODYNAMIC PRINCIPLES.

[GEORGE A. HAWKINS]

Bibliography: T. L. Brown, *General Chemistry*, 2d ed., 1968; M. J. Sienko and R. A. Plane, *Chemistry*, 5th ed., 1975.

Decomposition potential

The electrode potential at which the electrolysis current begins to increase appreciably. Decomposition potentials are used as an approximate characteristic of industrial electrode processes. *See* ELECTROLYSIS.

Decomposition potentials are obtained by extrapolation of current-potential curves, as seen in the illustration. Extrapolation is not precise because there is a progressive increase of current as the electrode potential is varied. The decomposition potential, for a given element, depends on the

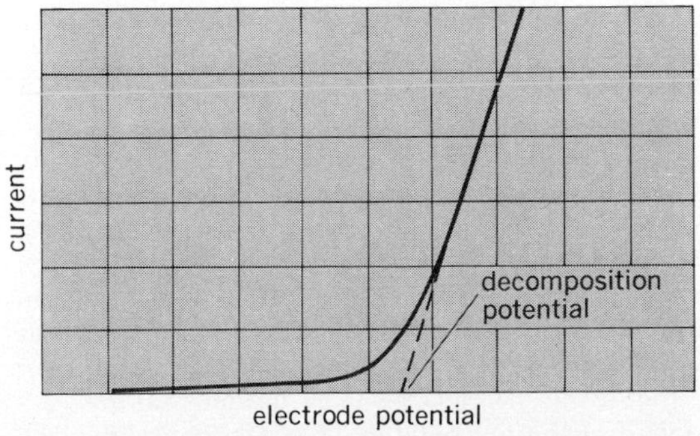

Determination of decomposition potential.

range of currents being considered. The cell voltage at which electrolysis becomes appreciable is approximately equal to the algebraic sum of the decomposition potentials of the reactions at the two electrodes and the ohmic drop, or voltage drop, in the electrolytic cell. The ohmic drop term is quite negligible for electrolytes with high conductance. *See* ELECTROLYTIC CONDUCTANCE.

[PAUL DELAHAY]

Definite composition, law of

The law that a given chemical compound always contains the same elements in the same fixed proportions by weight. Thus, whatever its source, silver chloride always contains 100 g of silver to every 32.85 g of chlorine. If a compound is formed by the union of m atoms of one element, each weighing a, with n atoms of another element, each weighing b, the composition by weight of one molecule of the compound is in the ratio $ma:nb$. This must be the composition of any mass of the compound, provided that all atoms of the same kind have the same weight. It is now known that this is not usually the case but that the atoms of an element may consist of a number of isotopes, having different masses. However, as long as any sample of the element always contains the same relative proportions of the isotopes, the law still holds. *See* ISOTOPE.

This is not the case for lead. Lead is the final product of the decay of three radioactive series, the atomic weights from the three series being 206 from radium, 208 from thorium, and 207 from actinium. Hence both the atomic weight of lead and the proportion of lead in its compounds will vary with the source of the lead. *See* RADIOACTIVITY.

Much more widespread and serious departures from the law of definite composition occur in a large variety of solid compounds (the nonstoichiometric compounds). Nonstoichiometry arises for a number of differing reasons. In the silicate minerals, such as olivine, it occurs as a result of isomorphous replacement. Thus olivine is $(Mg,Fe)_2SiO_4$ and the proportion of magnesium to iron may vary widely from sample to sample. Other solids are simply deficient in metal atoms. Thus ferrous sulfide, FeS, rarely has an atomic ratio Fe/S of precisely unity. Density measurements have shown that the lattice of the sulfur atoms remains intact but that some of the iron atoms are missing. *See* CRYSTAL STRUCTURE; NONSTOICHIOMETRIC COMPOUNDS; STOICHIOMETRY.

[THOMAS C. WADDINGTON]

Dehydrogenation

A reaction in which hydrogen is detached from a molecule. The reaction is strongly endothermic, and therefore heat must be supplied to maintain the reaction temperature. When the detached hydrogen is immediately oxidized, two benefits accrue: (1) the conversion of reactants to products is increased because the equilibrium concentration is shifted toward the products (law of mass action); and (2) the added exothermic oxidation reaction supplies the needed heat of reaction. This process is called oxidative dehydrogenation. On the other hand, excess hydrogen is sometimes added to a dehydrogenation reaction in order to diminish the complete breakup of the molecule into many fragments.

Industry. Most commercial dehydrogenation reactions are catalytic and take place in the vapor phase. However, the noncatalytic, thermal dehydrogenation of isobutane is practiced to produce propylene and isobutylene. Also, alcohols with high boiling points can be dehydrogenated in the liquid phase in the presence of powdered catalysts (for example, active Raney nickel). Air can be bubbled through the liquid mixture to oxidize the hydrogen as it is formed.

Superheated steam (725°C) is often mixed with the dehydrogenation reactor feed to (1) provide the needed heat of reaction, (2) lower the partial pressure and enhance the equilibrium conversion, and (3) decoke the catalyst to extend the reactor onstream time.

Types. The primary types of dehydrogenation reactions are:

1. Vapor-phase conversion of primary alcohols to aldehydes, as in the preparation of acetaldehyde from ethyl alcohol in the presence of a silver catalyst at 550°C, Eq. (1).

$$CH_3CH_2OH + \tfrac{1}{2}O_2 \rightarrow CH_3CHO + H_2O \qquad (1)$$

2. Vapor-phase conversion of secondary alcohols to ketones, as in the preparation of acetone from isopropyl alcohol, Eq. (2), and methyl ethyl ketone from secondary butyl alcohol, Eq. (3), in the presence of copper catalyst at 300°C.

$$CH_3CHOHCH_3 \rightarrow CH_3COCH_3 + H_2 \qquad (2)$$

$$CH_3CHOHCH_2CH_3 \rightarrow CH_3COCH_2CH_3 + H_2 \qquad (3)$$

3. Dehydrogenation of a side chain, as in the preparation of styrene from ethylbenzene, Eq. (4), in the presence of iron oxide catalyst at 600°C.

$$C_6H_5C_2H_5 \rightarrow C_6H_5CH{=}CH_2 + H_2 \qquad (4)$$

4. Catalytic reforming of naphthas and naphthenes in the presence of a platinum catalyst at 500°C and 100–1000 psi (689,576–6,895,757 Pa) for the production of aromatics for high-octane gasoline, toluene for TNT, and *ortho*-, *meta*-, and *para*-xylenes for oxidation to the corresponding phthalic acids.

All four of the listed types of dehydrogenation reactions are of major industrial importance. They account for the production of billions of pounds of organic compounds that enter into the manufacture of lubricants, explosives, plastics, plasticizers, and elastomers. *See* CHEMICAL CONVERSION; HYDROGENATION; OXIDATION PROCESS.

[J. W. FULTON]

Deliquescence

The absorption of atmospheric water vapor by a crystalline solid until the crystal eventually dissolves into a saturated solution. This behavior is well known for certain salts such as hydrated calcium chloride, $CaCl_2 \cdot 6H_2O$, and zinc chloride, $ZnCl_2$, but it is a property of all soluble salts in air of sufficiently high humidity.

Thermodynamically, the condition for deliquescence is that the partial pressure of the water vapor in the air exceed the vapor pressure (aqueous tension) of the water in the saturated solution of the salt. Then the reaction

$$\text{Solid} + \text{water vapor} \rightarrow \text{aqueous solution}$$

will occur spontaneously. The speed at which the

process takes place depends upon the rate of diffusion of water vapor into the crystal lattice, crystal size, and other factors. The process will stop when the water vapor in the atmosphere is depleted to the point at which its partial pressure equals that of the saturated solution.

In general, substances which are highly soluble in water have a greater tendency to deliquesce, since concentrated solutions will have a lower vapor pressure. At 25°C the vapor pressure of pure water is 23.8 mm Hg, whereas that of a saturated solution of $CaCl_2 \cdot 6H_2O$ is only 7.0 mm Hg; this salt then will deliquesce in an atmosphere where the relative humidity exceeds 7.0/23.8 or 30%. For $ZnCl_2$ the situation is much more extreme; it will deliquesce at a relative humidity of 10%. Ordinary sugar (sucrose) at 25°C will deliquesce at humidities above 85%.

Crystalline solids also may absorb water by increasing their water of hydration if the dissociation pressure of the hydrated species to be formed is less than the partial pressure of the water vapor. It is this process, not deliquescence, which is the opposite of efflorescence.

Deliquescent substances can be used to remove water vapor from air, although they have no special advantage over substances which merely add water of hydration and remain crystalline. *See* DESICCANT; DRYING; EFFLORESCENCE; SOLUTION; VAPOR PRESSURE.

[ROBERT L. SCOTT]

Density

The mass per unit volume of a material. The term is applicable to mixtures and pure substances and to matter in the solid, liquid, gaseous, or plasma state. Density of all matter depends on temperature; the density of a mixture may depend on its composition, and the density of a gas on its pressure. Common units of density are grams per cubic centimeter, and slugs or pounds per cubic foot. The specific gravity of a material is defined as the ratio of its density to the density of some standard material, such as water at a specified temperature, for example, 60°F, or, for gases the basis may be air at standard temperature and pressure. Another related concept is weight density, which is defined as the weight of a unit volume of the material. *See* WEIGHT.

[LEO NEDELSKY]

Depolymerization

The reversion of a polymer to a monomer, as shown in the reaction below for a trimer of propylene, 2,4-dimethyl-4-heptene.

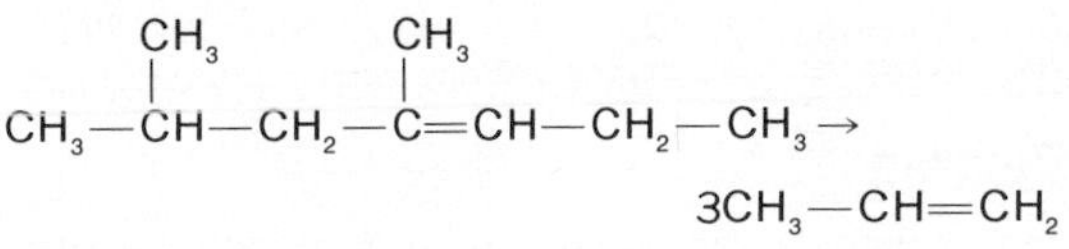

Depolymerization normally is not employed as a distinct process in the petroleum industry, but may occur incidentally during catalytic or thermal cracking operations if the feedstocks contain polymers.

The reaction is accelerated by protonic catalysts, notably inorganic acids placed on porous supports or acidic cracking catalysts such as silica-alumina, at temperatures of 250–500°C, and is then identical in mechanism with the catalytic cracking of olefins. Depolymerization by heat alone will proceed by a thermal cracking mechanism. In both cases the product may contain substantial quantities of polymer fragments differing from the original monomer. *See* CRACKING; POLYMERIZATION.

[BERNARD S. GREENSFELDER/MOTT SOUDERS]

Desiccant

A substance used to withdraw moisture from other materials. Although the removal of large quantities of water is done by evaporation, aided by moving air currents and by elevated temperature, the last traces of moisture are often held very tightly and do not evaporate readily. Furthermore, evaporation ceases when the moisture content of the material is reduced to that of the drying-air current. For final drying, one uses as a desiccant a substance with high affinity for water. It may react with water chemically or retain water through capillarity or adsorption. The drying agent is placed directly into the gas or liquid to be dried; solid materials are placed in a desiccator, a closed vessel in which moisture diffuses to the desiccant through the dry desiccator atmosphere. A desiccant loses potency as it takes on water; often it can be renewed by heating. Desiccants which form hydrates can be selected to maintain certain levels of low humidity in a closed vessel. [ALLEN L. HANSON]

Important types. Silica gel possesses a high adsorptive power because of its extreme capillarity, the capillary pores occupying approximately 50% of its specific volume. The capillaries are probably spine-shaped, and the average pore diameter has been estimated to be 4×10^{-7} cm, which is only about 10 times the diameter of one molecule of adsorbate. The drying efficiency of silica gel depends upon the concentration of water in the gas mixture, the temperature of the gel and gas, the properties of the condensed liquid, its wettability, and the state of the gel itself. Silica gel adsorbs water vapor preferentially in the presence of other vapors. It is readily capable of drying air to a dew point below −94°F.

Activated alumina is prepared from aluminum trihydrate. It is a granular porous adsorbent with properties similar to those of silica gel. Alumina gel has many applications in addition to gas drying. It is used to adsorb gases and vapors from gaseous mixtures and to dry liquids.

Anhydrous calcium sulfate, or Drierite, is prepared from a high grade of gypsum, $CaSO_4 \cdot 2H_2O$, which is dried, crushed, sized, and heated to 450–500°F for 2 hr. This leaves a granular porous form of anhydrous calcium sulfate, with sufficient mechanical strength to support its own weight.

Magnesium perchlorate or Anhydrone, is the equal of any desiccant from the standpoint of drying efficiency. The adsorption rate is rapid, and the first hydrate does not lose water until 275°F is reached, thereby permitting its use for drying gases at higher temperatures than most commercial desiccants. It passes through three hydrate stages, namely, di-, tri-, and hexahydrate. The last, $Mg(ClO_4)_2 \cdot 6H_2O$, represents saturation after adsorption of 48.6% of the dry weight of water, a high capacity. Although drying efficiency tends to decrease after each hydrate formation, even the trihydrate is superior to solid NaOH and $CaCl_2$.

Other types. Other solid desiccants that have been used or studied for gas drying are oxides, such as barium and calcium oxide, and activated carbon. Barium oxide maintains a high drying activity up to 1000°F. Barium oxide also appears to have marked possibilities for the drying of gases at high temperatures.

Calcium oxide has long been used as a desiccant because of its low cost. Although it has high drying efficiency, its capacity is low because of the formation of carbonates on its surface from the carbon dioxide of the air.

Activated carbon is old historically and probably has been as intensively studied as any single adsorbent. Although capable of adsorbing large amounts of water vapor, activated carbon finds its major use in solvent recovery, in odor and taste removal, and as a catalyst and catalyst carrier. Instead of the preferential adsorption of water vapor, as in the case of alumina and silica gel, organic vapors tend to displace any water present on the carbon. *See* ADSORPTION; DELIQUESCENCE; DRYING.

[WILLIAM R. MARSHALL, JR.]

Deuterium

The isotope of the element hydrogen with atomic weight 2.0144 and symbols H^2 or D. Considerations of nuclear stability and a discrepancy between the chemical and physical atomic weights of hydrogen led to the prediction of a stable isotope of hydrogen of mass 2. A successful search for this isotope, deuterium, was made by H. C. Urey, F. G. Brickwedde, and G. M. Murphy in 1931. The terrestrial natural abundance of deuterium is 1 part in 6700 parts of ordinary hydrogen (protium), which has atomic weight 1.0078. Small variations in natural sources are found as a result of fractionation by geological processes. Industrial hydrogen, particularly that generated by the electrolysis of water, may contain significantly less deuterium.

Deuterium is used mainly in the form of heavy water. In the uncombined state it finds uses as a research tool. Liquid deuterium is used in bubble chambers to study the reactions of elementary particles with the deuterium nucleus, the deuteron. Deuterons are frequently accelerated in cyclotrons to study their reactions with other nuclei and also to produce radioactive nuclides. Deuterium gas is used in the direct synthesis of organic compounds for tracer studies. Illustrative of such procedures are exchange reactions with hydrogen-containing substances in the presence of hydrogenation catalysts. If controlled thermonuclear fusion can be achieved, deuterium gas would become an exceedingly important source of power. *See* NUCLEAR FUSION.

Deuterium, D_2, is a gas at room temperature. It is prepared from heavy water, D_2O, either by electrolysis or by reaction of D_2O with metals such as zinc, iron, calcium, and uranium. It is also prepared directly by the fractional distillation of liquid hydrogen. In this process it is necessary to catalyze the disproportionation (unsymmetrical dissociation and recombination) of HD into H_2 and D_2 in order to obtain D_2.

Ordinary deuterium is a molecular mixture of two-thirds *ortho-* and one-third *para-*deuterium. At 20 K the equilibrium composition is 97.8% *o*-deuterium. Deuterium molecules obey the Bose-Einstein statistics, and the ortho species have even rotational quantum numbers, whereas the para species have odd rotational quantum numbers. The analysis of the ortho-para composition of deuterium is most conveniently made by measurement of the thermal conductivity of the gas at 77 K. The physical and chemical properties of *o*- and *p*-deuterium are very similar. In most of its physical and chemical properties, deuterium resembles protium. In most cases, deuterium is slightly less reactive than protium. An intercomparison of some of the physical properties of D_2 with those of H_2 is given in the table.

Selected values of some physical properties of deuterium

Property	n-H_2	n-D_2	97.8% o-D_2
Triple point	13.96 K	18.72 K	18.63 K
Normal boiling point	20.4 K	23.6 K	23.5 K
Critical temperature	33.24 K	38.35 K	
Critical pressure	12.8 atm	16.4 atm	
Heat of fusion	28.0 cal/mole	47.0 cal/mole	
Heat of vaporization (at normal boiling point)	216 cal/mole	293 cal/mole	
Molar volumes of liquid at 20 K	28.3 ml	23.5 ml	

Deuterium gas is usually slightly contaminated with HD. The analysis of the gas for protium is most conveniently carried out by a mass spectrometer. Alternative methods of analysis are by thermal conductivity and the rate of effusion of the gas through an orifice. The gas can be converted to water and the following properties have been used as a basis for analysis: infrared spectrum, density, index of refraction, and nuclear magnetic resonance.

The chemical reactivity of deuterium is less than that of hydrogen because of its lower zero-point energy and smaller collision frequency. At 1000 K deuterium is 32% less dissociated than protium. At room temperature deuterium atoms are electrolyzed out of water in the form of hydrogen gas, at one-eighth the rate of protium atoms. *See* HEAVY WATER; HYDROGEN; TRITIUM.

[JACOB BIGELEISEN]

Bibliography: G. Friedlander, J. W. Kennedy, and J. M. Miller, *Nuclear and Radiochemistry*, 2d ed., 1964; H. W. Woolley, R. B. Scott, and F. G. Brickwedde, Compilation of thermal properties of hydrogen in its various isotopic and ortho-para modifications, *J. Res. Nat. Bur. Std.*, 41:397–475, 1948.

Dextran

A polyglucose biopolymer characterized by preponderance of α-1,6 linkage, and generally produced by enzymes from certain strains of *Leuconostoc* or *Streptococcus*. The chemical properties may be modified chemically for specific usages. While formerly its principal utility was as blood plasma substitute, dextran is also employed as packing material in column chromatography and as pharmaceutical agents; its average molecular

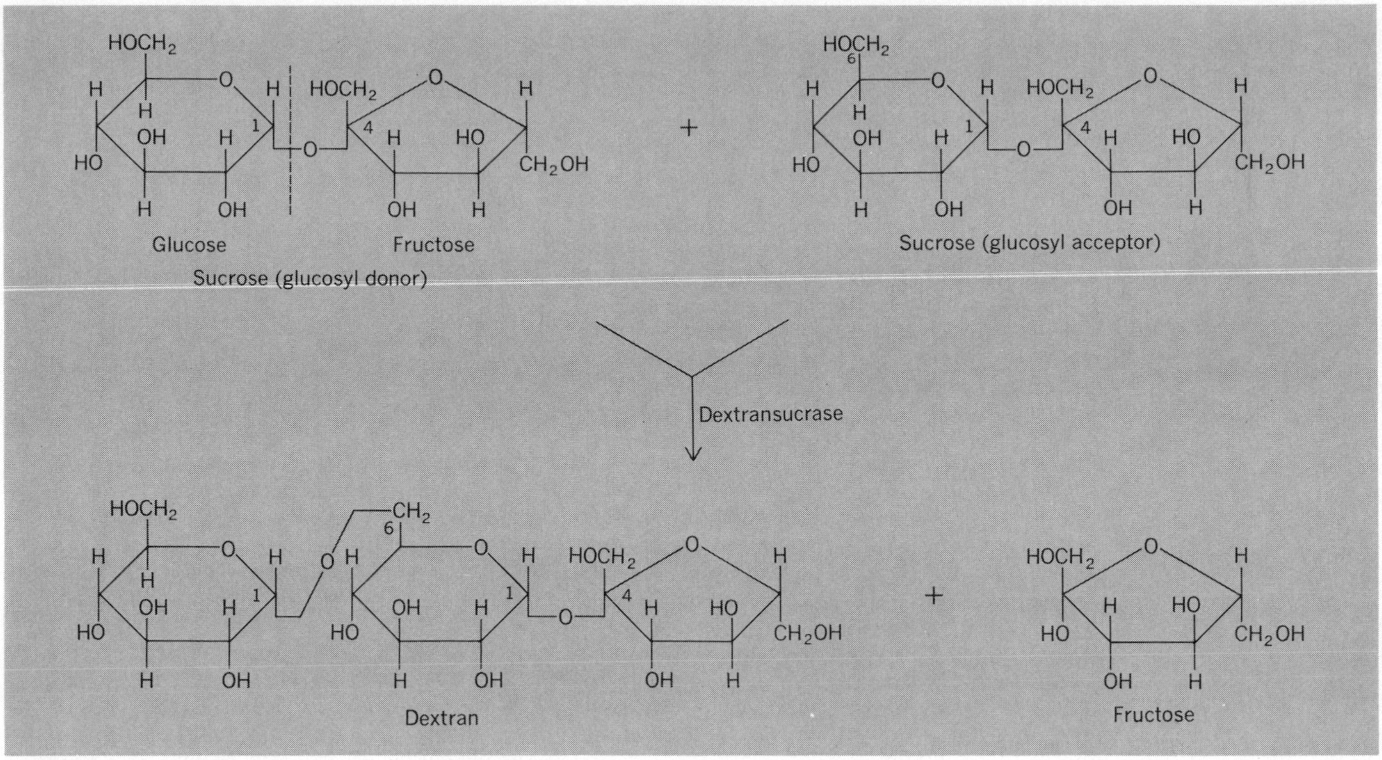

Reaction for formation of dextran from sucrose.

weight determines usage to a great extent. Dextran's chemical and physical properties depend upon the strain of microorganism employed and the environmental conditions imposed upon the bacterium during growth, or the reaction conditions where an enzymatic method of dextran production is employed. *Leuconostoc* and *Streptococcus* species primarily convert sucrose to dextran and fructose. *Acetobacter* species convert dextrin to dextran; the α-1,4 linkage is converted to an α-1,6 linkage. *See* CHROMATOGRAPHY; DEXTRIN; POLYSACCHARIDE.

The conversion of sucrose to dextran and its by-product fructose is a transglucosylation reaction. This transformation is mediated by the enzyme dextransucrase. It is readily obtained extracellularly from suitable strains of *L. mesenteroides* propagated under appropriate conditions. *L. mesenteroides* NRRL B-512F has been adopted by nearly all of the Western nations for the production of dextran.

Dextran, like glycogen and amylopectin, is a branched glucose polymer but differs from amylopectin in that the principal linkage between the anhydroglucose units is of the α-1,6 type; the others are of the 1,4 or 1,3 type. The enzyme system from *L. mesenteroides* NRRL B-512F forms a polymer with about 95% α-1,6 type of linkage. The ratio between the 1,6 and the non-1,6 linkages may vary from 20:1 to almost 1:1, depending on the bacterial strain, and is characteristic of the enzyme obtained from it.

Mechanism. Formation of dextran from sucrose involves the reaction shown in the illustration. The reaction requires two substrates. Dextransucrase is more specific in its requirement for the glucosyl donor substrate (sucrose) than for its acceptor substrates. The acceptor substrate may be sucrose, maltose, isomaltose, α-methylglucoside, and low-molecular-weight dextran. The average molecular weight of the dextran polymerized is controlled by both the type and quantity of the acceptor substrate incorporated either in the fermentation medium or in the enzymatic reaction.

The kinetic constants can be readily varied since dextran is formed from a system where double substrate kinetics apply. Neither the V_{max} nor the Michaelis constant for the polymerization reaction is fixed; they are varied by alteration in the concentration of either cosubstrate (glucosyl donor and glucosyl acceptor). In some respects, the mechanism of the polymerization is similar to that of the condensation type in that water is eliminated. It also displays, however, certain characteristics of the chain-reaction type in that the molecular weight of the dextran produced is essentially that which is desired and requires no further modification.

Production processes. The average molecular weight of dextran produced by most of these organisms is generally on the order of several to hundreds of millions. One strain of *Streptococcus* produces, however, a dextran with a molecular weight of about 100,000 or slightly lower. The high-molecular-weight dextran, as ordinarily produced by *L. mesenteroides*, is hydrolyzed to a product with an average molecular weight of about 75,000 for use as a blood plasma volume expander. More information is available on the utility of dextran from *L. mesenteroides* NRRL B-512F for use as blood plasma substitute than on the dextran from the *Streptococcus* sp. For use as plasma extender, the high-molecular-weight dextran is acid hydrolyzed, fractionated (as to its molecular weight distribution), and purified. The specifications on the dextran for use as a plasma volume extender vary in different countries; the physical, chemical, and biological specifications it must meet are extremely rigid. *See* POLYMERIZATION.

[HENRY M. TSUCHIYA]

Dextrin

A polymer of D-glucose which is intermediate in complexity between starch and maltose. The dextrins are usually obtained by hydrolysis of starch with diastase (amylases). The higher dextrins resemble starch, while the lower dextrins more nearly resemble the sugars. Compared with the original starch, the dextrins produce less viscous solutions. They are soluble in water but insoluble in alcohol. Dextrins may be obtained from starch by controlled hydrolysis with acids. The Lintner method for solubilizing starch consists of subjecting the native starch grains to 7.5% hydrochloric acid at room temperature for 7 days. The degradation product (dextrins) thus produced readily dissolves in water, but still gives a blue color with iodine. More drastic treatment of starch with acid will produce dextrins having purple, red, or no color with iodine. Dextrins are used commercially as adhesives. Tapioca, waxy maize, and sweet potato starch represent the best material for their manufacture.

When a starch is exposed to the action of *Bacillus macerans* or to the bacteria-free filtrate of this microorganism, a mixture of water-soluble dextrins, known as Schardinger dextrins, is produced. From this mixture, three distinct, nonreducing crystalline compounds can be isolated. These dextrins, α, β, and γ, are known to possess cyclic structures consisting of six, seven, and eight 1,4-α-glucosidically linked D-glucose units, respectively.

[WILLIAM Z. HASSID]

Dialysis

A process of selective diffusion through a membrane by dissolved solutes in liquid solution. As dialysis is usually carried out, the membrane permits the diffusion of low-molecular-weight solutes (crystalloids) but prevents the passage of colloidal and high-molecular-weight solutes (macromolecules). Membranes suitable for this purpose include vegetable parchment, animal parchment, goldbeater's skin (peritoneal membranes of cattle), fish bladders, dialyzing cellophane (Visking sausage casing), and collodion (nitrocellulose deposited from alcohol-ether solution).

The solution is contained within such a membrane. The low-molecular-weight solutes are removed by placing pure solvent outside the membrane. This solvent is changed periodically or continuously until the concentration of diffusible solutes in the solution is reduced to near zero. The technique is used extensively in separating and purifying macromolecules of biological origin (see illustration).

DIALYSIS

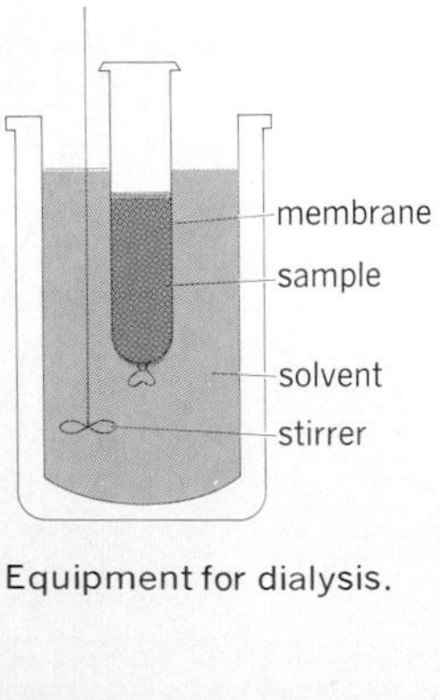

Equipment for dialysis.

Dialysis rates of ionic low-molecular-weight solutes may be increased greatly by applying an electric field to achieve an electrophoretic movement of ions through the dialytic membrane. This combined process of dialysis and electrophoretic transport of solutes through a membrane is known as electrodialysis. *See* ELECTROPHORESIS.

Ions do not migrate readily through membranes that carry in their pores charges of the same sign as the ion; ions of charge opposite to that of the membrane are not prevented from passing through. Such a selective permeability to ions by charged membranes has been used to improve the efficiency of electrodialysis. Membranes of collodion, vegetable parchment, and cellophane carry negative charges in contact with aqueous solutions. Animal membranes show positive charges at low pH and negative charges at high pH.

Membranes of ion-exchange resins are commercially available in cationic and anionic forms. The ion-exchange membranes show a high selectivity for ions of one charge type. *See* ION-SELECTIVE MEMBRANES AND ELECTRODES.

For desalting small quantities of solutions for chromatography and radioisotope tracer studies, micro- and semimicroelectrodialyzers have proved valuable. On a larger scale, electrodialysis combined with ion-exchange membranes has been used to remove salt from sea water. *See* COLLOID; SALINE WATER RECLAMATION.

[QUENTIN VAN WINKLE]

Diastereoisomer

One of a pair of optical isomers which are not mirror images of each other. A given diastereoisomer (or diastereomer) may not be optically active, in which case it is an optically inactive meso form; or it may be optically active, in which case with its nonsuperimposable mirror image it constitutes an enantiomorphic pair of optical isomers. Thus *meso*-tartaric acid is a diastereoisomer of both the (+) and (−) tartaric acids, and it can form two optically active monoesters, each of which is a diastereoisomer of monoesters of the (+) and (−) acids. *See* OPTICAL ACTIVITY.

[WYMAN R. VAUGHAN]

Diazotization

The reaction between a primary aromatic amine and nitrous acid to give a diazo compound. Diazotization is important in organic chemical synthesis. First recognized by Peter Griess in 1858, the reaction is remarkable for the smoothness and completeness with which it can be carried out and for the reactivity of the products formed. Its most striking use is in the large-scale manufacture of the important azo class of dyes, but it has likewise been invaluable in general synthesis in both chemical manufacturing and in research.

Preparation of diazonium salts. The most widely useful method of diazotizing a primary aromatic amine (the direct method) is by slow addition of an aqueous solution of sodium nitrite to a solution of the amine in dilute mineral acid held at 0–10°C, as shown in reaction (1). Excess mineral acid,

$$ArNH_2 + NaNO_2 + 2HX \rightarrow ArN_2X + 2H_2O + NaX \quad (1)$$

usually about 2.5 moles, is used to prevent formation of diazoamino by-products, and since the reaction is exothermic, cooling is required to maintain a temperature at which the diazo product is stable.

To diazotize aminosulfonic acids, such as sulfanilic acid, which are sparingly soluble in the acid solution, the indirect method can be used, in which a solution of sodium nitrite and the sodium salt of the aminosulfonic acid is run into a mineral acid solution containing ice. For weakly basic amines such as 1-aminoanthraquinone, diazotization is carried out in concentrated sulfuric acid, using nitrosylsulfuric acid as the diazotizing agent.

Few primary aromatic amines resist diazotization. Not only can the amino groups on benzene, naphthalene, and their substituted derivatives be diazotized, but heterocyclic amines such as aminothiazoles and aminopyridines will undergo the reaction also. Diamines in which the amino groups are on different aromatic nuclei in the same molecule behave independently with respect to diazotization, although the reaction can be carried out stepwise. When the amino groups are on the same aromatic nucleus, the reactivity of the second amino group to diazotization is lessened by the presence of the diazonium group first formed. If the amino groups are ortho to one another, a triazole ring may result.

In most cases a diazonium salt is used without isolation from the solution in which it was prepared. When isolation of the solid salt is necessary, the amine precursor may be diazotized by reaction with an aklyl nitrite in alcohol solution, followed by the addition of ether to cause precipitation of the desired product. (Solid diazonium salts must be handled with great care, owing to their high explosibility, particularly when dry. Many of them are detonated not only by heat, but by shock or abrasion.)

Structure. Aromatic diazo compounds have been the object of extensive study. Diazotization of aniline in acid solution yields benzenediazonium ion (structure I, where Ar= C_6H_5) as the dominant species. When such a solution is made strongly alkaline, two equivalents of hydroxide ion are consumed, with formation of the isodiazotate anion (III). The elusive diazohydroxide (II) is clearly an intermediate in the interconversion of I and III, reaction (2).

$$\underset{\text{(I)}}{\mathrm{Ar\overset{+}{N}{\equiv}N}} \underset{}{\overset{OH^-}{\rightleftharpoons}} \underset{\text{(II)}}{\mathrm{ArN{=}NOH}} \overset{OH^-}{\rightleftharpoons} \underset{\text{(III)}}{\mathrm{ArN{=}NO^-}} \qquad (2)$$

Diazotization of sulfanilic acid yields the inner salt (IV); diazotized *ortho-* and *para*-aminophenol form the diazo oxides V and VI, respectively, which are highly colored.

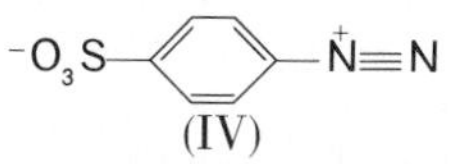
(IV)

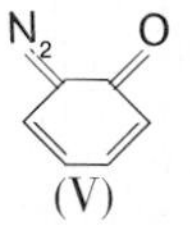
(V)

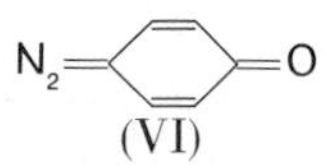
(VI)

Diazotization of typical aliphatic amines does not produce stable diazonium salts. Instead a mixture of alcohol and olefin is produced, which can be recognized as derived by solvolytic decomposition of the corresponding diazonium ion. On the other hand, the related diazoalkanes are stable, though highly reactive, compounds. Both diazomethane (VII) and ethyl diazoacetate (VIII) are useful laboratory reagents.

$$\underset{\text{(VII)}}{\mathrm{\bar{N}{=}\overset{+}{N}{=}CH_2}} \qquad \underset{\text{(VIII)}}{\mathrm{\bar{N}{=}\overset{+}{N}{=}CHCO_2C_2H_5}}$$

Coupling reactions. In coupling reactions between a diazonium salt and an aromatic amine or phenol, the diazonium ion replaces a hydrogen ion located on the ring in a position ortho or para to the activating amino or hydroxy group. The reaction is illustrated by the formation of methyl orange (IX) from diazotized sulfanilic acid and *N,N*-dimethylaniline, reaction (3).

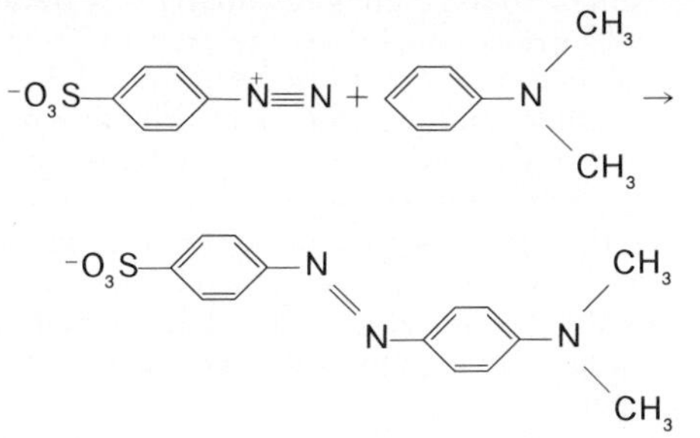
(3)

(IX)

Coupling of a benzenediazonium salt with phenol yields *p*-hydroxyazobenzene (X, where Ar= C_6H_5) by preferential attack at the position para to the hydroxy group. When there is no free para position, as in *β*-naphthol, an ortho position is attacked, yielding structure XI. When treated with a

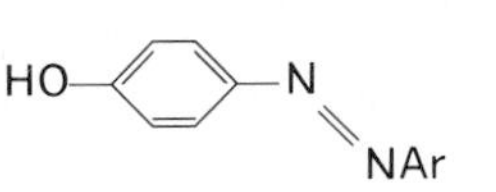
(X)

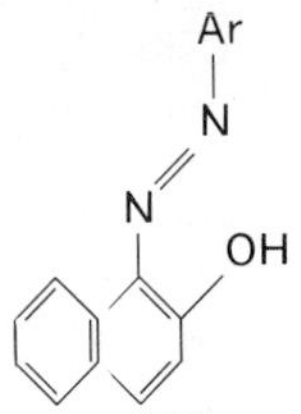
(XI)

primary amine, such as aniline, a diazonium salt couples to the nitrogen atom to form a triazene, Arn-N=N—NHAr.

Coupling of the electrophilic diazonium ion to a phenol occurs more rapidly in mildly alkaline solution, where the phenolate ion confers enhanced electron density on the ring, than in acid solution, where the free phenol is the dominant species. If the solution is too strongly alkaline, the rate of coupling is very slow because of conversion of the reactive diazonium ion (I) to the unreactive species (III) by reaction (2). Coupling to an aromatic amine is usually carried out in neutral to weakly acid medium. In practice it is possible to couple selectively two different diazonium salts with an amino phenol such as "H-acid" (8-amino-l-naphthol-3,6-disulfonic acid) by careful control of the pH. Under slightly acidic conditions, a diazo group can be introduced ortho to the amino substituent, followed by raising the pH and introduction of the second diazo group ortho to the hydroxy substituent. Thus the dye naphthol blue black (XII) is prepared from H-acid and two diazonium salts, derived from *p*-nitroaniline and aniline, respectively.

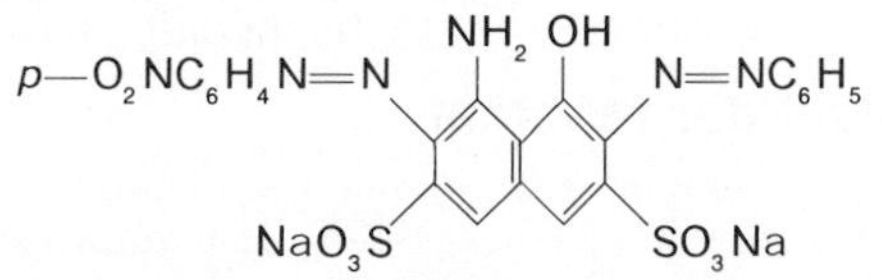
(XII)

Displacement reactions. Replacement of the diazonium group is an important synthesis tactic in aromatic chemistry. Heating an aqueous acidic solution of a diazonium salt causes displacement of the nitrogen by hydroxyl, reaction (4), and this

$$Ar{-}\overset{+}{N}{\equiv}N \xrightarrow[\text{heat}]{H_2O} Ar{-}OH + N_2 + H^+ \qquad (4)$$

reaction is a useful laboratory synthesis of phenolic compounds from amines. Similarly, aromatic halides and nitriles can be synthesized by treatment of the diazonium salt with an appropriate source of halide and cyanide ions (Sandmeyer reaction). The introduction of fluorine by this route (Schiemann reaction) is particularly important because of the difficulties encountered with direct fluorination. Displacement of nitrogen by hydrogen (deamination) can be effected by using ethanol, sodium stannite, or hypophosphorous acid as the reducing agent.

Diazonium salts are also intermediates in the synthesis of biaryls (for example, the Gomberg-Bachmann reaction) and in certain cyclizations, such as the conversion of diazotized 2-amino-α-phenylcinnamic acid to phenanthrene-9-carboxylic acid (Pschorr reaction).

Reduction. Diazonium salts are reduced to aryl-substituted hydrazines by treatment with sodium sulfite.

Uses. The light sensitivity of aromatic diazo compounds has led to their use in photocopying processes. Azo dyes derived by coupling diazonium salts to aromatic amines and phenols remain an important class of dyes, although the use of some members of this class in food has been restricted in recent years because of carcinogenic properties observed in experiments with laboratory animals. *See* ANILINE; DYE. [MARTIN STILES]

Bibliography: S. Patai (ed.), *The Chemistry of the Diazonium and Diazo Groups*, pt. 2, 1978; P. A. S. Smith, *The Chemistry of Open-Chain Nitrogen Compounds*, vol. 2, 1966; H. Zollinger, *Diazo and Azo Chemistry: Aliphatic and Aromatic Compounds*, 1961.

Dichromate

The $Cr_2O_7^{--}$ ion which is derived from dichromic acid, $H_2Cr_2O_7$. This acid and other polyacids are obtained when chromium trioxide, CrO_3, is dissolved in water. The dichromates are converted to chromates under basic conditions as indicated by the reaction below.

$$\underset{\text{Orange}}{Cr_2O_7^{--}} + 2OH^- \rightleftharpoons \underset{\text{Yellow}}{2CrO_4^{--}} + H_2O$$

Most dichromates are water-soluble and give orange-colored solutions. Dichromates are good oxidizing agents, and this property is utilized in chemical processes, analytical chemistry, explosives, and safety matches. Dichromates are also used in electroplating, tanning, photography, and pigments. *See* CHROMATE; CHROMIUM.

[E. EUGENE WEAVER]

Diels-Alder reaction

The 1,4-addition of an alkene (the dienophile) to a conjugated diene. The reaction, also known as the diene synthesis, is one of the most valuable and versatile methods for the preparation of compounds containing a six-membered ring, and it proceeds most rapidly when the dienophile is substituted for by electron-attracting groups, such as —C=O or —C≡N. Two examples are given in reactions (1) and (2).

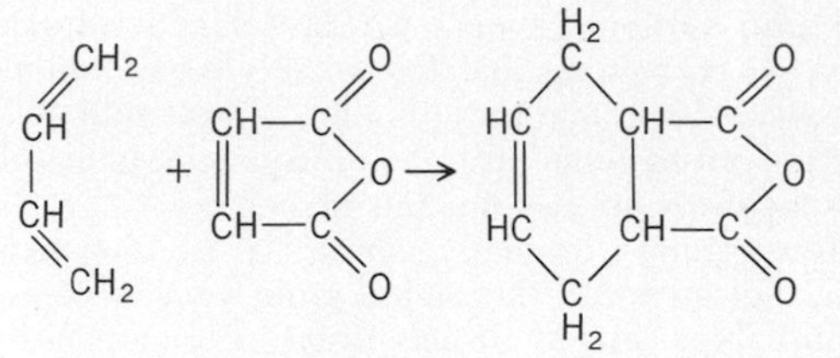

Diene (1,3-butadiene) Dienophile (maleic anhydride) Adduct (1)

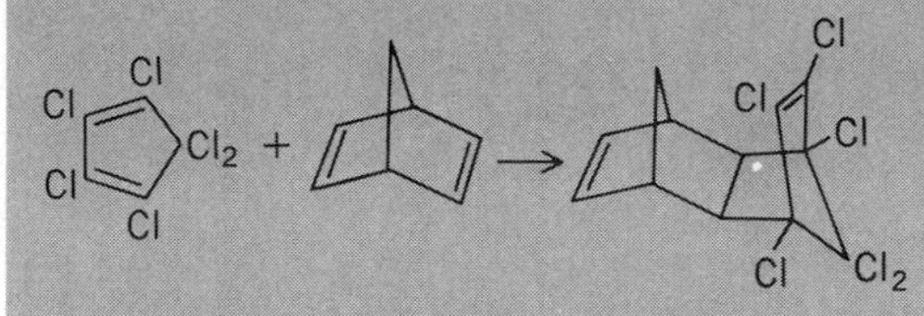

Hexachlorocyclopentadiene Norbornadiene Aldrin (2)

Diene component. With the exception of a few highly substituted diolefins which interfere with the reaction because of steric effects, most alkyl and aryl homologs of butadiene react readily. Alicyclic dienes are especially reactive, except in those cases where the adduct formed would contain a bridged ring having a double bond at the bridgehead. Aromatic compounds such as styrene, in which part of the conjugation is in the ring, react in a normal manner, but heterocyclic dienes may show anomalous behavior. Furan reacts normally to give a six-membered ring containing a bridged oxygen, whereas thiophene and pyrrole do not react in this manner.

Dienophile component. Most commonly, dienophiles consist of compounds containing structure >C=C< or —C≡C—. However, dienophiles are not restricted to unsaturated carbon compounds and adducts form with >C=N—, —C≡N, —N=N—, and —N=O as the dienophile component. Quinones are especially reactive with dienes, requiring only mild conditions.

Reaction conditions. The Diels-Alder or diene-synthesis reaction does not require a catalyst, nor is the reaction retarded by the presence of oxidation inhibitors with which dienes are commonly treated to prevent formation of peroxides. For reactive dienophiles such as maleic anhydride, it is sufficient to mix the reactants in molar proportion, usually in a solvent such as benzene, and reaction takes place at room temperature—often with the evolution of heat. On the other hand, the reaction of ethylene with cyclopentadiene to form bicyclo-[2.2.1]-2-heptene, shown by reaction (3), requires

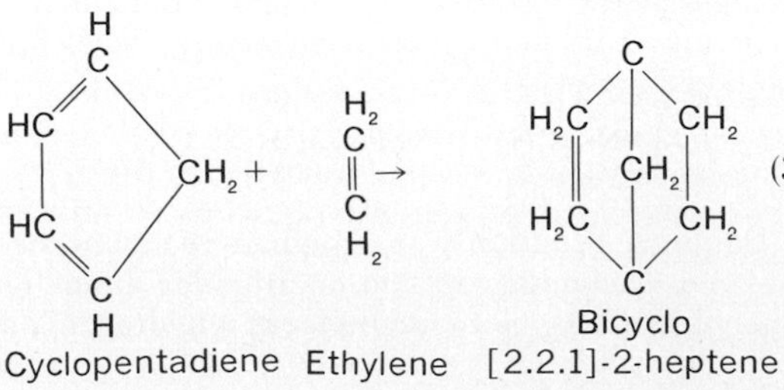

Cyclopentadiene Ethylene Bicyclo [2.2.1]-2-heptene (3)

temperatures in the range of 190–220°C and pressures in the range of 20–80 atm.

Instances are known where the diene synthesis is a reversible reaction. For example, cyclohexene, the adduct of 1,3-butadiene and ethylene, is quantitatively transformed into its components when subjected to high temperature for a short contact time at low pressure. Similarly, dicyclopentadiene (I) yields two moles of cyclopentadiene when maintained at its boiling point of 170°C.

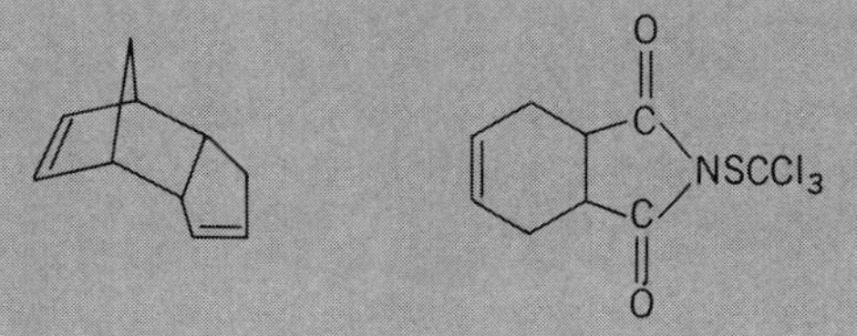

Dicyclopentadiene (I) Captan (II)

Industrially the diene synthesis is used in the production of the insecticides aldrin and dieldrin. The adduct of butadiene and maleic anhydride is used in the synthesis of the important fungicide captan (II).

[PAUL E. FANTA]

Bibliography: N. L. Allinger et al., *Organic Chemistry*, 2d ed., 1976; A. Wassermann, *Diels-Alder Reactions*, 1965.

Digitoxigenin

A steroid isolated from the seeds or leaves of *Digitalis purpurea*. It exists as a glycosidic derivative in the plant (see illustration). In this combined

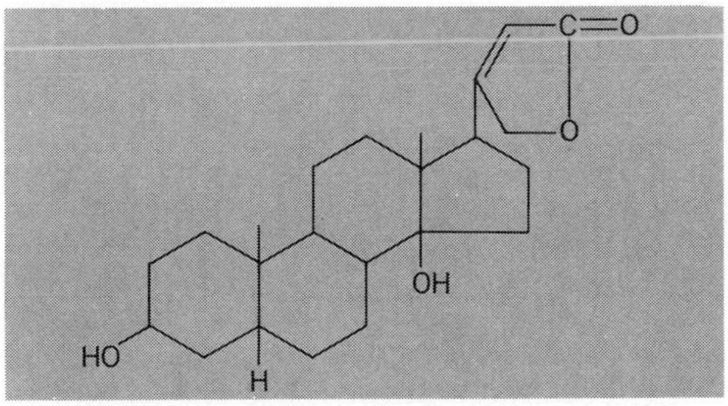

Structural formula for digitoxigenin.

form of glycoside it is highly toxic, but in proper dosage it is of value in the control of arteriosclerotic and hypertensive heart disease, especially when auricular fibrillation is present.

[RALPH I. DORFMAN]

Dimethyl sulfoxide

A versatile solvent, abbreviated DMSO, used industrially and in chemical laboratories as a medium for carrying out chemical reactions. Its uses have been extended to that of a chemical reagent where DMSO itself is involved in a chemical change. It is the simplest member of a class of organic compounds which are typified by the polar sulfur-oxygen bond represented in the resonance hybrid shown in reaction (1). The molecule (see

$$CH_3\overset{\overset{O}{\|}}{S}CH_3 \longleftrightarrow CH_3\overset{\overset{\bar{O}}{|}}{\underset{+}{S}}CH_3 \qquad (1)$$

illustration) is pyramidal in shape with the oxygen and the carbons at the corners.

DIMETHYL SULFOXIDE

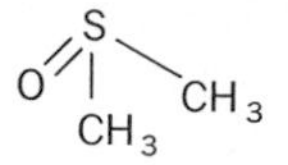

Structural formula for dimethyl sulfoxide.

Because of the polarity of the sulfur-oxygen bond, DMSO has a high dielectric constant (ϵ^{20} = 48.9) and is slightly basic. It forms crystalline salts with strong protic acids, reaction (2), and it coordi-

$$CH_3\overset{\overset{O}{\|}}{S}CH_3 + HCl \rightarrow \left[CH_3\overset{\overset{\overset{H}{|}}{\overset{O}{|}}}{S}CH_3\right]^+ Cl^- \qquad (2)$$

nates with Lewis acids. It is a colorless, odorless (when pure), and very hygroscopic stable liquid (bp 189°C, mp 19.5°C). It is manufactured commercially by reacting the black liquor, from digestion in the kraft pulp process, with molten sulfur to form dimethyl sulfide, CH_3SCH_3. The dimethyl sulfide is oxidized with nitrogen tetroxide to yield DMSO. This highly polar aprotic solvent is water- and alcohol-miscible and will dissolve most polar organic compounds and many inorganic salts.

Solvent. As an aprotic solvent, DMSO strongly solvates cations, leaving a highly reactive anion, reaction (3). Thus in DMSO, basicity and nucleo-

$$CH_3\overset{\overset{O}{\|}}{S}CH_3 + A^+B^- \rightarrow CH_3\overset{\overset{\overset{A}{|}}{\overset{O}{|}}}{\underset{+}{S}}{-}CH_3 + B^- \qquad (3)$$

philicity is enhanced, and it is a superior solvent for many elimination, nucleophilic substitution, and solvolysis reactions in which nucleophile and base strength are important. Nucleophilic substitution reactions in which halogens or sulfonate esters are displaced by anions such as cyanide, alkoxide, thiocyanate, azide, and others are accelerated 1000 to 10,000 times in DMSO over the reaction in aqueous alcohol. Reaction (4) illustrates nucleo-

$$(CH_3)_3CCH_2Cl + NaCN \xrightarrow{DMSO} (CH_3)_3CCH_2CN + NaCl \qquad (4)$$

philic substitution with the cyanide anion. Dimethyl sulfoxide is also superior to protic solvents as a medium for elimination reactions. Reaction (5) illustrates the use of DMSO in the elimination of a tosyl group from a steroid derivative.

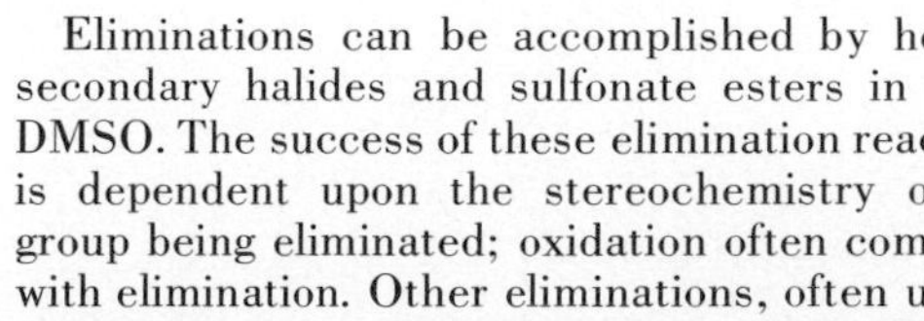

(5)

Eliminations can be accomplished by heating secondary halides and sulfonate esters in 100% DMSO. The success of these elimination reactions is dependent upon the stereochemistry of the group being eliminated; oxidation often competes with elimination. Other eliminations, often unsuccessful in protic solvents, are accomplished by the use of potassium *t*-butoxide ($K^{+-}OC(CH_3)_3$) in DMSO.

The elimination of tertiary and secondary alcohols has been accomplished by refluxing in 100% DMSO. Reactions (6) and (7) illustrate this elimina-

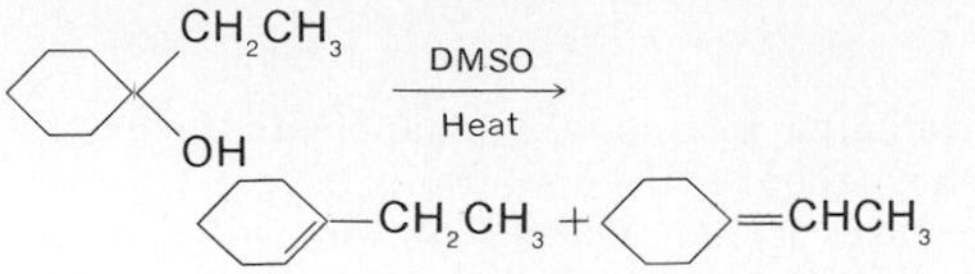

(6)

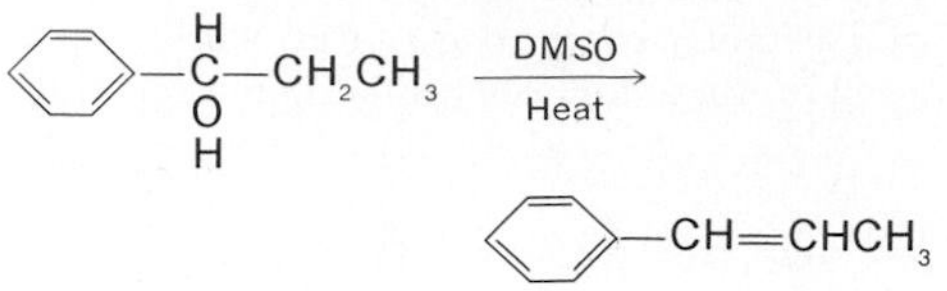

(7)

tion from a tertiary and secondary alcohol respectively.

The distinction between solvent and reactant in this and previous eliminations, to date, has not been clarified. Dimethyl sulfoxide has found use as a suitable solvent for thermal decarboxylations and base-catalyzed double-bond isomerizations. A double-bond isomerization utilizing DMSO and potassium *t*-butoxide is illustrated by reaction (8).

$$CH_2{=}CHCH_2OR \xrightarrow[\text{DMSO}]{K^{+}\,{}^{-}OC(CH_3)_3} CH_3CH{=}CH{-}OR \quad (8)$$

The high dielectric constant makes DMSO a useful solvent for dissolving resins, polymers, and carbohydrates. It is employed commercially as a spinning solvent in the manufacture of synthetic fibers.

Chemical reactions. The compound is capable of taking part in several types of chemical reactions. These reactions are described below.

Coordination compounds. Dimethyl sulfoxide acts as an electron donor to form coordination compounds with a large number of inorganic ions. In most complexes the DMSO ligand is attached through the oxygen atom, and the bond strength is roughly that of the corresponding aquo derivative. In addition to the typical metal-ion complexes, DMSO also forms some complexes with nonmetallic compounds. For example, BF_3 forms a one-to-one complex which is highly hygroscopic and low melting.

Synthetic intermediate. Acting as an acid, DMSO reacts as indicated by reaction (9) with sodium hydride to form the strong base methylsulfinyl carbanion. Methylsulfinyl carbanion, used effectively as a base catalyst in many condensation reactions, itself condenses with carbonyl compounds, particularly esters, as illustrated by reaction (10), to give *β*-keto sulfoxides in high yield. The synthetic importance of the easily prepared *β*-keto-sulfoxides lies in the fact that they can be smoothly reduced by aluminum amalgam to methyl ketones in a large variety of cases, reaction (11).

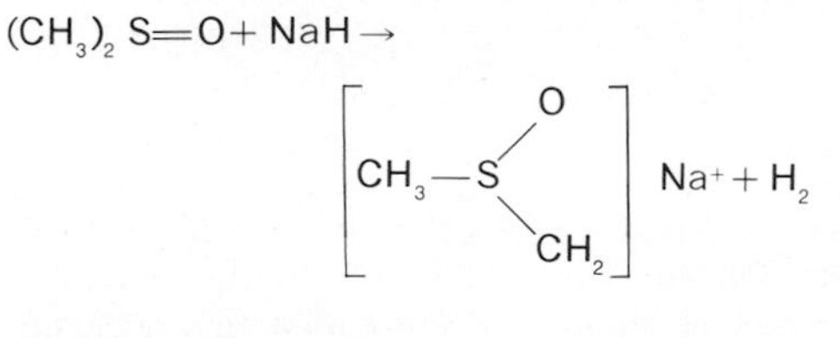

(9)

$$R{-}COOR' + CH_3\underset{\underset{O}{\|}}{S}CH_2^{-} \rightarrow R{-}\underset{\underset{O}{\|}}{C}{-}CH_2\underset{\underset{O}{\|}}{S}{-}CH_3 + R'O^{-} \quad (10)$$

$$R{-}\underset{\underset{O}{\|}}{C}{-}CH_2{-}\underset{\underset{O}{\|}}{S}{-}CH_3 \xrightarrow{\text{Al-Hg}} R{-}\underset{\underset{O}{\|}}{C}{-}CH_3 \quad (11)$$

See CONDENSATION REACTION; ORGANIC CHEMICAL SYNTHESIS.

Biological properties. The apparent low toxicity and high skin permeability have led to extensive studies in numerous biological systems, including humans. Inorganic salts or small-molecular-weight organic compounds dissolved in DMSO can be transported across skin membrane, indicating a potential hazard in commercial use. Although many claims have been made about the potential of DMSO as a medicinal, it is still an investigational drug.

Oxidizing agent. Under the proper experimental conditions many substrates (primary or secondary alcohols, halides, or sulfonates) can be oxidized to the corresponding carbonyl compound by DMSO. Although the evidence is not yet conclusive in every case, there is a strong indication that most of the DMSO oxidations involve the same dimethylalkoxysulfonium salt intermediate, which subsequently reacts with a base to give the observed carbonyl product and dimethyl sulfide. There are two routes, labeled pathways A and B in reaction (12), by which a substrate may be converted into

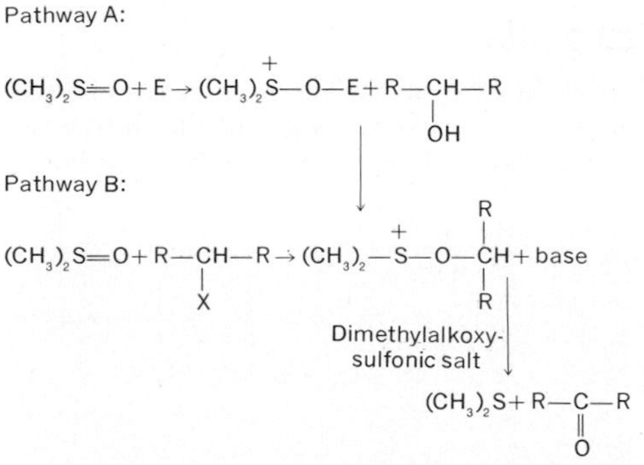

(12)

the dimethylalkoxysulfonium salt intermediate. The route is determined by the structure of the substrate.

Pathway A involves reaction of DMSO with an intermediate activating electrophilic species E, which is subsequently displaced by the substrate to be oxidized, usually an alcohol, to form the dimethylalkoxysulfonium salt.

Pathway B involves a leaving group X($X = Cl^-$, Br^-, I^-, or sulfonate) being displaced by DMSO acting as a nucleophile, and resulting directly in the dimethylalkoxysulfonium salt.

The most common technique involves use of dicyclohexylcarbodiimide as the E group and pyridine as the base and is applicable to primary or secondary alcohol groups in an almost unlimited variety of compounds including alkaloids, steroids, carbohydrates, and other complex substances. The mild conditions, uncomplicated experimental details, and high yields with which most oxidations can be effected have elevated this approach to oxidation into prominence.

[W. W. EPSTEIN; F. W. SWEAT]

Bibliography: E. Block (ed.), *Reactions of Organosulfur Compounds*, 1978; Crown Zellerback Corp., *Dimethyl Sulfoxide as a Reaction Solvent*, 1968, reprint 1978; L. F. Fieser and M. Fieser, *Reagents for Organic Synthesis*, 1967; C. J. M. Stirling (ed.), *Organic Sulphur Chemistry*, 1975.

Diphenylmethane

A colorless hydrocarbon $(C_6H_5)_2CH_2$, melting point 25.9°C, boiling point 263.2°C, described as having an odor resembling that of geraniums or oranges. Its structural formula is shown. It can be pre-

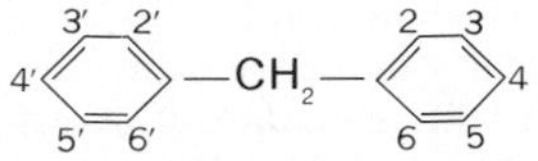

pared by the action of formaldehyde (HCHO) on benzene in the presence of sulfuric acid, by the action of benzyl chloride, $C_6H_5CH_2Cl$, on benzene in the presence of aluminum chloride, or by the reduction of benzophenone, $C_6H_5COC_6H_5$.

Diphenylmethane may be hydrogenated in the presence of nickel to yield dicyclohexylmethane, $(C_6H_{11})_2CH_2$. Oxidation of diphenylmethane yields benzophenone. Substitution reactions usually occur at the 4,4′ and 2,4′ positions.

Although diphenylmethane has been used as a perfume for soaps, it is not an important item of commerce. A derivative, 2,2′-dihydroxy-3,5,6,3′,5′,6′-hexachlorodiphenylmethane, has been used as a bacteriostatic agent in soaps. *See* POLYNUCLEAR HYDROCARBON. [CHARLES K. BRADSHER]

Distillation

The process of producing a gas or vapor from a liquid or solid. However, the term sublimation is used ordinarily to describe the vaporization of a solid. Heat is generally supplied to the liquid during the distillation, although in special cases the latent heat required for the vaporization may be obtained from the internal energy of the liquid.

The main purpose of distillation is either the separation of volatile components from nonvolatile materials, or the separation of a mixture of volatile components. The separation of volatile components from nonvolatile materials is carried out by a simple distillation in which the material is placed in a still and heated, and the vapor removed and condensed. Simple distillation is similar to the process of evaporation, but the former term usually describes the operation in which the volatile material is a desired product, whereas evaporation generally is applied to aqueous solutions of nonvolatile materials in which the nonvolatile material is the desired product. Simple distillation is frequently used for high-boiling organic compounds; to prevent thermal degradation of the product, the operation is usually carried out either at reduced pressure, termed simple vacuum distillation, or with the addition of steam, termed steam distillation. *See* EVAPORATION; EVAPORATOR; SUBLIMATION.

Although simple distillation can be applied to mixtures of volatile components, the separation obtained is usually not complete, particularly if the components have boiling points that are close to each other. To obtain greater separations in such cases, fractional distillation is employed. In this process, the vapors from the still are permitted to come in contact with a portion of the condensate in a countercurrent or stepwise countercurrent operation. Because of its lower operating and capital costs, this countercurrent type of operation has completely replaced the multiple distillation-condensation method formerly employed. This process was originally developed in France for alcoholic beverages, and it has been widely adopted for both laboratory and industrial operations because it is usually the most effective method of separating mixtures of miscible volatile liquids. It is so effective and efficient that it is frequently employed to separate mixtures which are not normally liquids. For example, most industrial oxygen is produced by liquefying air and fractionally distilling the liquid air. In this case, the separation is so effective that not only high-purity oxygen and nitrogen but also argon, neon, krypton, xenon, and other noble gases are recovered in commercial quantities. The isotopes of hydrogen have been separated by fractional distillation at even lower temperatures. In contrast is the high-temperature fractional distillation of zinc and cadmium mixtures. The temperature range between liquid hydrogen and liquid zinc includes the boiling points of most liquids, and as a result fractional distillation has wide application.

Vapor-liquid equilibriums. Separation of two liquids is possible only when the composition of the vapor is different from that of the liquid from which it was produced. The separation will be easier if there is a great difference between the composition of the vapor and that of the liquid, but separations may be practical even when the difference is small. The design of distillation equipment is usually based on vapor-liquid equilibrium compositions.

The vapor-liquid equilibrium data needed for distillation work are either obtained experimentally or estimated from physical chemistry relationships. To obtain experimental data, it is necessary to bring the vapor and liquid to equilibrium with each other. Samples of each are then removed without altering the equilibrium, and each phase is analyzed. Such data are difficult and time-consuming to obtain; consequently, the amount of vapor-liquid equilibrium data is very limited.

Because of the difficulty of determining vapor-liquid equilibrium data experimentally, the calculation or estimation of such data is important. The principles of physical chemistry form the basis of understanding the equilibrium between liquid and vapor phases. The rules of Raoult, Dalton, and Henry are of particular importance. For nonideal solutions, the equations of Margules and Van Laar have been used to estimate the activity coefficients, and although they have worked well in some cases, they have not been satisfactory for many other mixtures. The use of fugacities has been of great assistance in calculating the effect of high pressures on vapor-liquid equilibriums. High pressures are frequently used so that low-boiling materials can be fractionated using available water supplies as the coolant, but the pressures must be lower than the critical pressure of the mixtures to obtain the desired operation. In checking experimental vapor-liquid equilibrium data, the Duhem equation has been particularly useful. *See* BOILING POINT; CHEMICAL THERMODYNAMICS; FUGACITY; GAS; LIQUID; SOLUTION; VAPOR PRESSURE.

A useful method of correlating vapor-liquid data

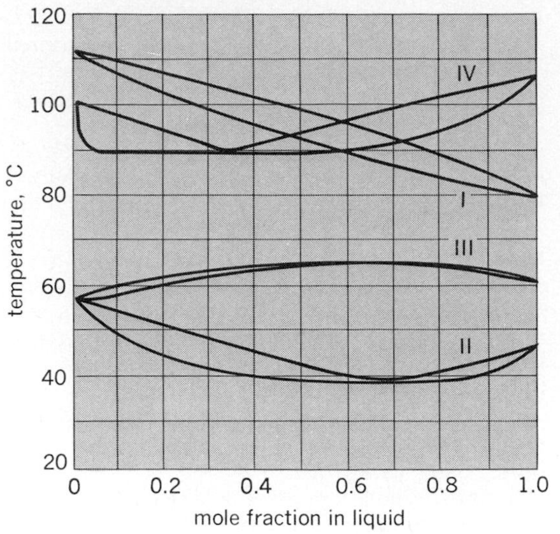

Fig. 1. Temperature-composition diagram. Curve I, benzene-toluene; II, acetone-carbon disulfide; III, acetone-chloroform; IV, isobutyl alcohol-water.

is comparing the volatility of one component in a mixture relative to that of another. This is called the relative volatility and is defined as Eq. (1). Here

$$\alpha_{ab} = (y_a/x_a/y_b/x_b) \qquad (1)$$

α_{ab} is the relative volatility of component A to component B, and y_a, x_a, y_b, and x_b are the mole fractions in the vapor and liquid of A and B, respectively. If $\alpha_{ab} = 1.0$, the ratio of A to B is the same in the vapor and the liquid, and no relative separation has been obtained between them. If $\alpha_{ab} < 1.0$, the vapor contains less A relative to B than the liquid phase; if $\alpha_{ab} > 1.0$, the reverse is true.

Relative volatility changes slowly with temperature and may either increase or decrease with rise in temperature. Because most distillations are carried out at essentially constant pressure (variable temperature), the low dependence of relative volatility on temperature increases its usefulness.

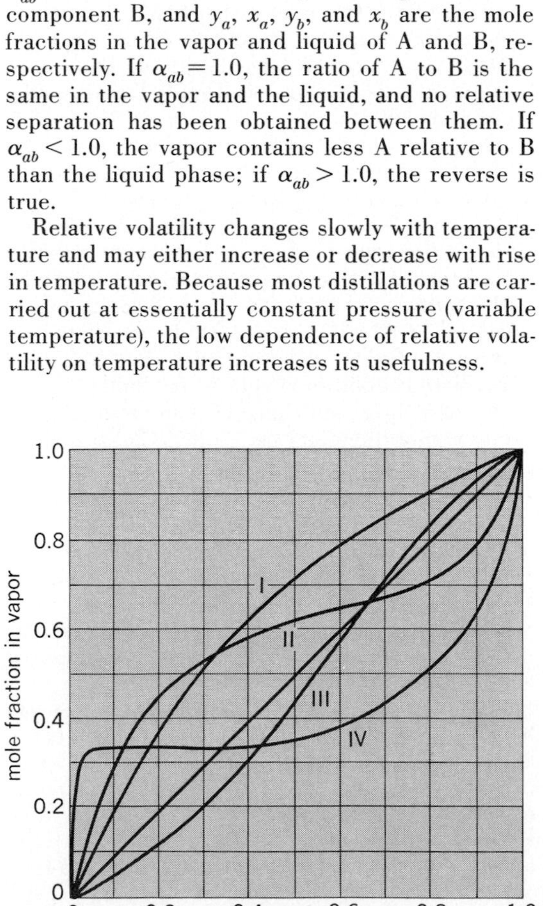

Fig. 2. Composition-composition (y-x) diagram. Curve I, benzene-toluene; II, acetone-carbon disulfide; III, acetone-chloroform; IV, isobutyl alcohol-water.

For distillation purposes, the data for binary mixtures are usually presented for constant total pressure on either a temperature-composition diagram or a y-x diagram that gives the mole fraction of one component in the vapor as a function of its mole fraction in the liquid (Figs. 1 and 2).

The temperature-composition diagram gives more information, but is not as easily used as a y-x diagram, and the latter is most often used. Curve I is for benzene and toluene and is typical of systems that are approximated by Raoult's law. The mole fraction of benzene in the vapor is always greater than its mole fraction in the liquid, and the relative volatility of benzene to toluene shown in Fig. 2 is relatively constant at about 2.4. Curve II is for a system in which the equilibrium curve crosses the $y = x$ line. Thus, at one point the vapor and liquid compositions are equal; this composition is termed an azeotrope, azeotropic mixture, or constant boiling mixture because it will vaporize without any change in composition and, therefore, without any change in temperature during the evaporation. Mixtures having less acetone than the azeotrope have a higher concentration of acetone in the vapor than in the liquid, and the relative volatility of acetone to carbon disulfide is greater than 1.0. The relative volatility is 1.0 at the azeotropic composition and becomes less than 1.0 for mixtures containing more acetone. In this latter region, the concentration of acetone in the vapor is less than in the liquid. The azeotrope, or constant boiling mixture, has a higher vapor pressure than either acetone or carbon disulfide and therefore, at a given total pressure, it boils at a lower temperature. It is termed a minimum boiling azeotrope or a minimum constant boiling mixture. A y-x curve that crosses the $y = x$ line with a slope less than that of the $y = x$ line will have a minimum boiling azeotrope at the composition represented by the point of intersection.

Curve III illustrates a mixture whose y-x curve also crosses the $y = x$ diagonal, but in this case the slope is greater than that of the $y = x$ line, and the mixture corresponding to the intersection of the two curves will be a maximum boiling azeotrope. This mixture will distill without change in composition, and it will have a higher boiling point at a given pressure than either acetone or chloroform.

Curve IV is similar to curve II, except that for a considerable region the vapor composition is constant. This curve is characteristic of partially miscible liquid systems. *See* PHASE EQUILIBRIUM.

Simple distillation. This operation can be carried out either as a continuous steady-state operation or batchwise. In the continuous system, the liquid to be separated is added to the still as the feed at a steady rate, a portion of it is vaporized by the heating coil, the vapor produced is condensed as the distillate or overhead product, and the unvaporized liquid is continually removed as the still or bottom product. In such a system (Fig. 3), the vapor product is usually of a composition close to the value in equilibrium with the liquid leaving the still. For this type of distillation, the material balance is Eq. (2), where V, L, F = moles per unit time

$$Vy + Lx = Fz \qquad (2)$$

of vapor, unvaporized liquid, and feed, respectively, and y, x, z = mole fractions of a component in the corresponding stream.

Combined with the overall material balance, this gives Eq. (3), which together with data relating y and x, such as a y-x equilibrium curve, makes it possible to calculate y and x for any fraction vaporized.

$$\frac{V}{F} = \text{fraction vaporized} = \frac{z-x}{y-x} \qquad (3)$$

In a simple batch distillation, the material to be separated is added to the still before the distillation is begun and no additional feed is added during the cycle. As the distillation is continued, the amount of liquid in the still decreases and the compositions of the vapor leaving and of the liquid remaining in the still continually change with time. Rayleigh developed an equation for a binary distillation of this type. If the relative volatility α is constant during the distillation, the Rayleigh equation becomes Eq. (4), where W_0 is the original number of moles added to the still, W is the number of moles remaining in the still, x_0 is the original mole fraction, and x is the mole fraction corresponding to W.

$$\ln \frac{W}{W_0} = \frac{1}{\alpha - 1} \ln \frac{x(1-x_0)}{x_0(1-x)} + \ln \frac{1-x_0}{1-x} \qquad (4)$$

Steam distillation. In this operation, steam is introduced directly into the liquid in the still. The method is usually limited to those cases in which the solubility of the steam in the liquid is low at the operating temperature and pressure. It is employed with relatively high-boiling organic materials which would decompose if they were distilled directly at atmospheric pressure or with liquids that have such poor heat-transfer characteristics that excessive local overheating would result with indirect heating. By steam distillation, a volatile material can be separated from nonvolatile impurities, or mixtures can be separated with results about equivalent to those obtained with simple distillation. Other gases or vapors could be used instead of steam, but steam usually is the most desirable from the viewpoint of cost and ease of recovering the vaporized materials. The heat required for vaporization must be supplied by indirect heat, or by the steam. For maximum steam economy, the temperature of the still should be as high as possible without undesirable thermal effects, and the total pressure should be as low as is consistent with the condensation of the vapor mixture. When a simple distillation is carried out under a high vacuum, the rate at which molecules leave the surface, rather than equilibrium, determines the composition of the vapor. Such an operation is termed molecular distillation. *See* MOLECULAR DISTILLATION.

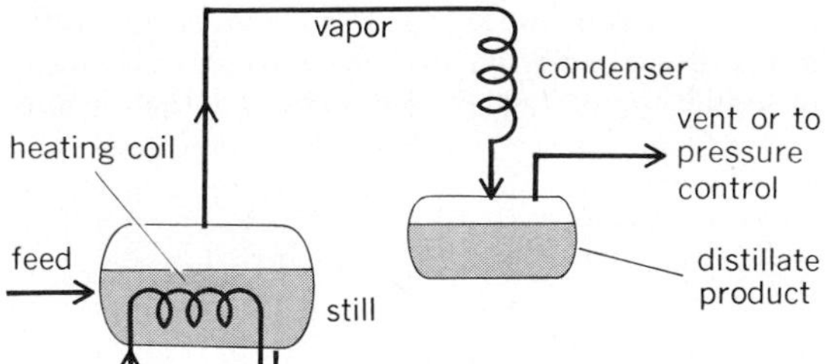

(a)

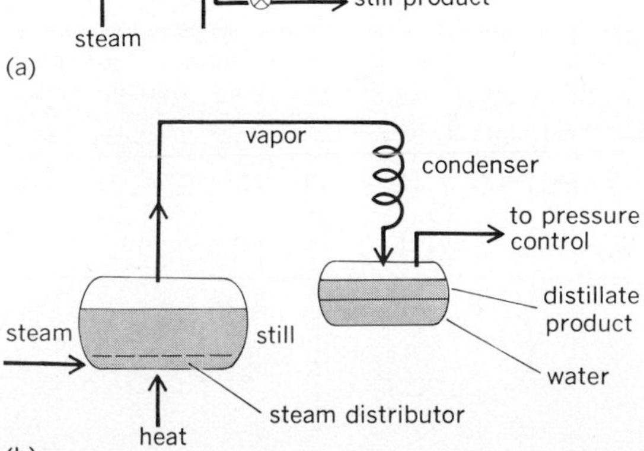

(b)

Fig. 3. Types of distillation. (*a*) Continuous simple distillation. (*b*) Batch steam distillation.

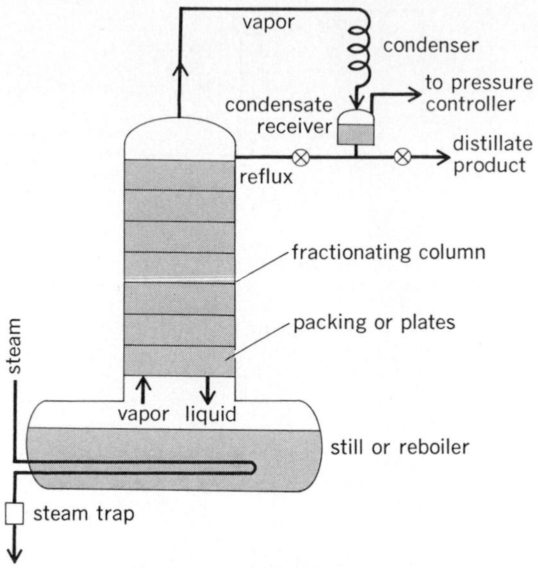

Fig. 4. Batch fractional distillation system.

Fractional distillation. In this operation, the vapor produced in the still is brought into contact with a portion of the condensate in a countercurrent or stepwise countercurrent system. The operation can be batch or continuous. The unit consists of a still to which a vertical column is attached (Fig. 4). This column is filled with some type of packing or plate construction which will permit the descending liquid added at the top to come into contact with the vapor rising from the still. The vapor from the top of the column is condensed and a portion of it is removed as product; the remaining liquid is returned to the top of the column as liquid, called reflux. The flow rates of liquid and vapor are adjusted so that, at every place within the column, the liquid has a higher concentration of the more volatile components than corresponds to equilibrium with the vapor with which it is in contact. As a result, the more volatile components pass from the liquid to the vapor and the less volatile components pass in the reverse direction. The vapor becomes progressively more enriched in the volatile components as it flows up the column to the condenser, and the liquid becomes more concentrated in the less volatile components as it flows down the column to the still.

The separation obtainable by such a system is much greater than that of a simple distillation; it depends on the relative volatility, the effectiveness

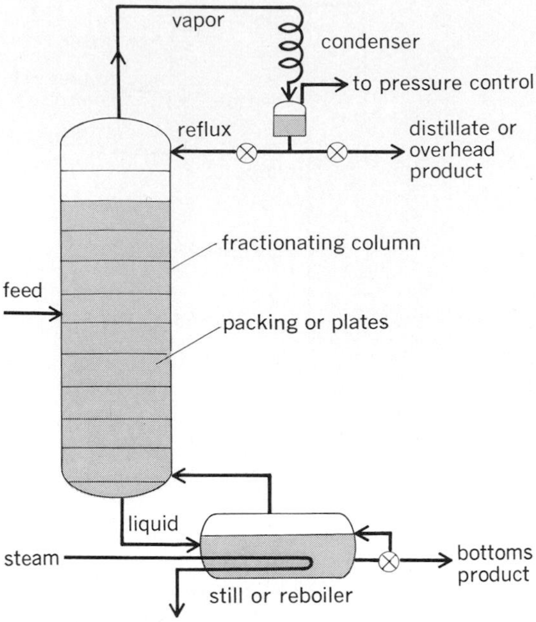

Fig. 5. Continuous fractional distillation system.

of the transfer between the phases, and the ratio of the liquid to vapor in the column (or the fraction of the condensate returned as reflux as compared to that removed as product). The relative volatility is essentially fixed by the components involved, although it can be altered by changing the operating pressure. The effectiveness of the contacting device is determined by its ability to make a rapid exchange of the components between the vapor and the liquid; the exchange is mainly a function of interfacial area produced and the flow characteristics of the vapor and liquid. The interfacial area usually is proportional to the volume of the column, but increased height is more effective than increased cross-sectional area of the column in improving the efficiency of the interphase transfer because of the effects of vapor velocity and liquid distribution.

In a batch fractional distillation column, the overhead product is continually changing in composition, and fractions of increasing boiling point are obtained as the operation proceeds. As a result, considerable labor is required to control the process. It is most commonly used for laboratory fractionations and for small industrial operations. When large quantities of materials are to be separated on a regular basis, it is more economical to operate the system on a continuous basis (Fig. 5). In this system, the feed to be separated is added continuously at some position in the column, vapor is introduced at the bottom, and reflux is introduced at the top. The vapor for the bottom of the column is usually produced by vaporizing a portion of the liquid from the column; the remainder of the liquid from the column is the bottom product. By such a system, it is possible to obtain an overhead product that contains a high concentration of the more volatile components in the feed and a bottom product that contains a high concentration of the less volatile components. High degrees of separation are obtainable, and the factors that determine the effectiveness of the separation are the same as those described for the batch system. The feed for such an operation can be liquid or vapor or a mixture of the two. In cases in which the bottom product is essentially water, as in the fractionation of an ethanol-water mixture, it is possible to use steam directly as the vapor for the column instead of boiling the bottom liquid. In cases in which it is difficult to condense all of the overhead vapors, for example, petroleum fractions containing hydrogen or methane, or in which the overhead product is desired as a vapor, it is possible to condense only that portion needed for reflux and to remove the rest as a vapor product.

A very large number of different contacting arrangements has been used in the columns. Laboratory columns and small industrial columns (1 – 2 ft in diameter or less) are frequently filled with packing which may be almost any small solid having a size from one-tenth to one-fiftieth that of the column diameter. In laboratory columns, many types of packings have been used, and small rings, spheres, wire spirals, and specially shaped and perforated metal pieces are common. Packings may be made of metal, glass, ceramic, or carbon, or any other material that can be suitably shaped and that will stand the operating conditions (Fig. 6). In larger packed columns, it is customary to use rings or specially shaped pieces, although packings such as coke lumps or bricks have been used. For larger-diameter columns, it is cheaper to use some type of plate arrangement instead of packing. These plates may be perforated disks, special grids, or more complicated arrangements such as bubble-cap plates (Fig. 7). Some of these are similar to packing in their action because the liquid flows down from plate to plate while the vapor passes up by it, but most of the plate units are designed to produce a bubbling type action. For example, on a perforated plate there is liquid over the perforations through which the vapor bubbles as it flows up the column, and the liquid flows down through the perforations or through special pipes of channels arranged for this purpose.

For large-diameter columns, the plate construction is not only less expensive, but is generally more effective for the interphase transfer because of liquid channeling encountered in large packed

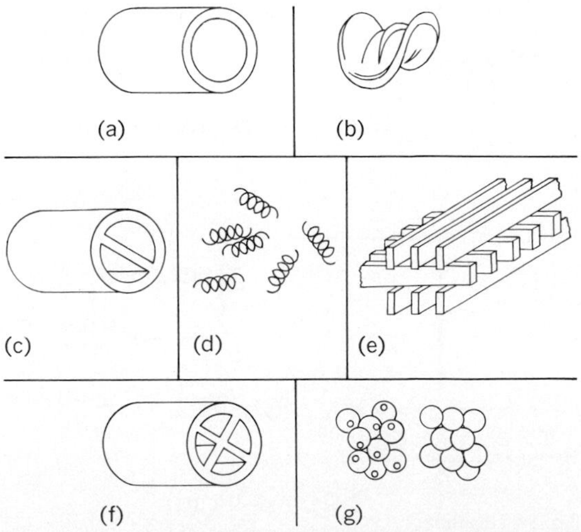

Fig. 6. Typical column packings. (*a*) Raschig ring; (*b*) Berl saddle; (*c*) Lessing ring; (*d*) helices; (*e*) wood grids; (*f*) partitioned ring; (*g*) spheres and bead.

columns. The unsatisfactory liquid-vapor contacting that results from channeling in packed towers can be improved by redistributing the liquid within the column. *See* FRACTIONATING COLUMN.

Design calculations for fractionating columns are usually made on the basis of the theoretical plate, which was defined by E. Sorel as a plate for which the average composition of the vapor leaving the plate was of a composition equal to the vapor in equilibrium with the liquid leaving the plate. In such designs, it is usually desirable to calculate two limiting cases as well as the actual operating conditions. The effectiveness of the column increases as the ratio of the liquid reflux to overhead product, called reflux ratio, is increased; the fewest theoretical plates will be required when this ratio is very large. This limit is called total reflux. The other limit is the lowest reflux ratio that will give the desired separation even if an infinite number of theoretical plates were used. It is termed minimum reflux ratio. Both of these limits require columns of infinite volume to obtain a finite product rate and are therefore not of practical design, but they indicate the minimum number of theoretical plates and the minimum heat consumption that can be used for a given system.

In the case of a steady-state binary distillation for which the relative volatility is a constant, M. R. Fenske has given an equation for the number of theoretical plates at total reflux for a column operating with a reboiler and a condenser that produces no separation between the reflux and distillate product. This is shown as Eq. (5), where N=

$$N+1=\frac{\ln[(x_a/x_b)_D/(x_a/x_b)_B]}{\ln \alpha_{ab}} \quad (5)$$

number of theoretical plates; x_a, x_b = mole fractions of A and B, respectively; D and B refer respectively to overhead and bottom products; and

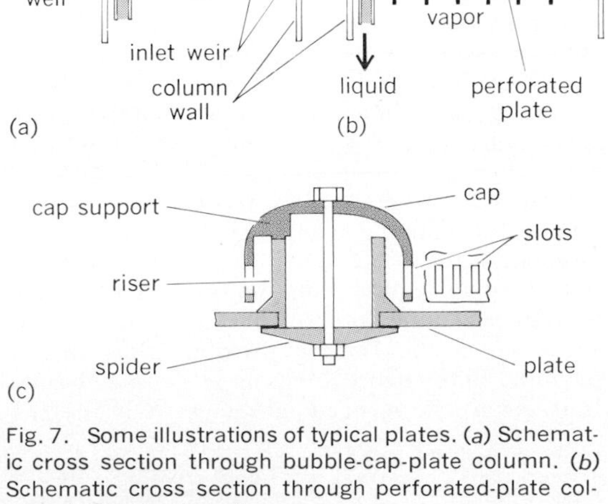

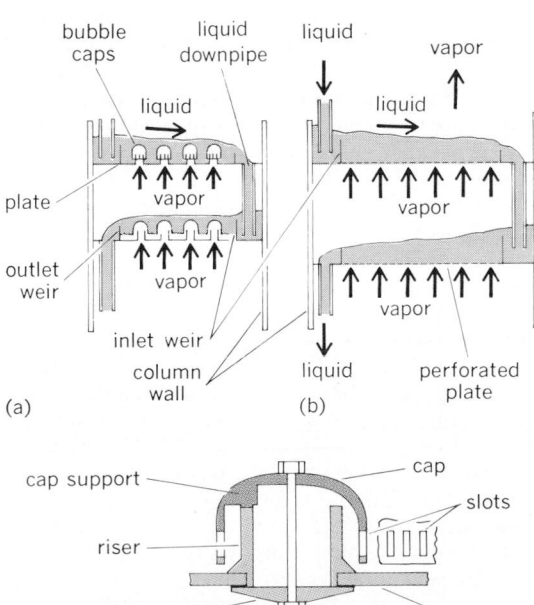

Fig. 7. Some illustrations of typical plates. (*a*) Schematic cross section through bubble-cap-plate column. (*b*) Schematic cross section through perforated-plate column. (*c*) Typical bubble cap.

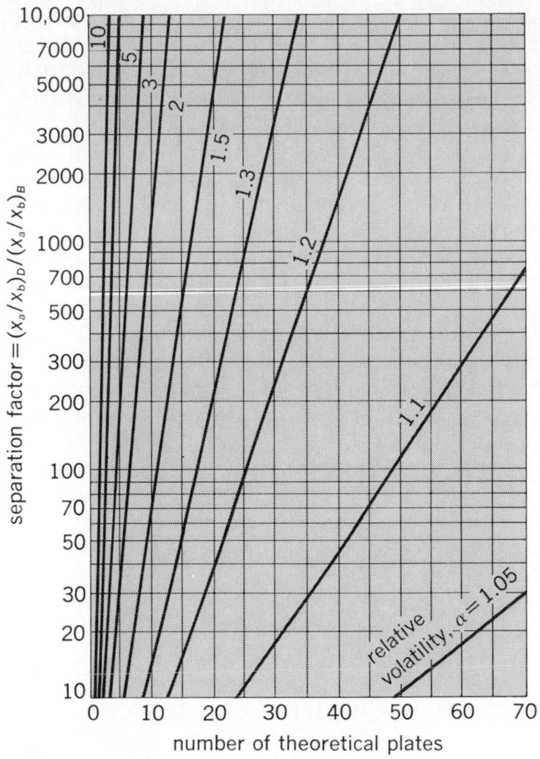

Fig. 8. Number of theoretical plates at total reflux.

α_{ab} = relative volatility of A to B. Figure 8 shows a plot of this relation.

Many graphical and analytical methods for binary mixtures have been proposed for estimating the number of theoretical plates required as a function of reflux ratio and for total reflux and the minimum reflux ratio.

Multicomponent mixtures are more complicated to study than binary systems, and the necessary equilibrium and enthalpy data are usually not available. Although a batch column can separate such a mixture into a number of fractions of relatively high purity, the continuous column illustrated can give only two fractions. In this case, the overhead product could be relatively pure in the lowest boiling component but the bottoms would contain a mixture of the other components; or the bottoms could be relatively pure in the highest boiling components and the overhead product would contain a mixture of the more volatile materials. Many high-purity fractions can be obtained from a multicomponent mixture with a continuous system by using a number of columns in series.

Plate efficiency. This term is used to express the relationship between the performance of an actual plate and a theoretical plate. Many definitions have been given for plate efficiency, but the simplest to use is the overall column efficiency, which is the ratio of the number of theoretical plates required for a given separation divided by the actual number of plates needed. Plate efficiency depends mainly on the characteristic of the mixture being separated and to a lesser degree on the design of the plate. The most important characteristics of the system are the relative volatility and the viscosity of the liquid in the column.

Although packings and other contacting methods do not give the definite stepwise arrangement obtained with perforated and bubble-cap plates, the analysis on the basis of theoretical plates is so convenient that this method is often employed and a height of packing that makes a separation equivalent to that of a theoretical plate is used. This is called the height equivalent of a theoretical plate, H.E.T.P. An alternate method of design uses the number of transfer units, N.T.U., and the height of a transfer unit, H.T.U.

As the relative volatility of the components approaches 1.0 (close-boiling components or mixtures near an azeotropic composition), separation by fractional distillation becomes difficult and large equipment and high heat consumption are required. The relative volatility in such cases can frequently be changed by using a different total pressure which will make the separation easier, but this is generally not very effective. A more common technique is to add another component in the liquid phase that will so alter the volatilities that the separation of the desired components can be made more economically. Depending on the characteristics of the added component, this technique is termed azeotropic distillation or extractive distillation. *See* MASS-TRANSFER OPERATION.

[EDWIN R. GILLILAND]

Bibliography: R Billet, *Distillation Engineering*, 1979; Lord Rayleigh, On the distillation of binary mixtures, *Phil. Mag.*, [6] 4(23):521–537, 1902; D. P. Tassios (ed.), *Extractive and Azeotropic Distillation*, 1972; M. Van Winkle, *Distillation*, 1967.

Disulfide

One of a group of organosulfur compounds, RSSR′, that may be symmetrical (R = R′) or unsymmetrical (R and R′, different). They are of great biochemical interest, since the S-S link occurs in natural products such as cystine, α-lipoic acid, and insulin. The cleavage of the S-S bond in proteins, for example, has important biological and industrial interests in the dehairing of hides, in the preparation of cysteine from wool, in wavesetting of hair, and in petroleum refining. Disulfides are also of interest in the manufacture of polysulfide rubbers and polymers. Higher, linear polysulfides such as trisulfides and tetrasulfides are also known. *See* ORGANOSULFUR COMPOUND; THIOETHER.

[NORMAN KHARASCH]

Donnan equilibrium

The particular equilibrium set up when two coexisting phases are subject to the restriction that one or more of the ionic components cannot pass from one phase into the other. This equilibrium was first recognized by F. G. Donnan in 1911. Commonly, the restriction is caused by a membrane which is permeable to the solvent and small ions but impermeable to colloidal ions or charged particles of colloidal size. The presence of a membrane is not essential, since the restriction of movement on the charged colloid can be provided by a centrifugal or gravitational field or by gel coherence.

An immediate consequence of such a restriction in a system is the uneven distribution of diffusible ions at equilibrium. This is apparent in the following example. Let the initial state of a system be a solution of the ions Na^+ and R^- of concentration c_1 separated from a solution of sodium chloride of concentration c_2 by a membrane freely permeable to all but the R^- ions.

$$\begin{array}{cc|cc} Na^+ & R^- & Na^+ & Cl^- \\ c_1 & c_1 & c_2 & c_2 \end{array}$$

Then at equilibrium, a certain concentration of chloride ions, x, will have diffused through the membrane, accompanied by the same number of sodium ions in order to preserve electrical neutrality on both sides of the membrane, and the final equilibrium will be:

$$\begin{array}{ccc|cc} Na^+ & R^- & Cl^- & Na^+ & Cl^- \\ (c_1+x) & & x & (c_2-x) & (c_2-x) \\ & (1) & & (2) & \end{array}$$

It can be shown thermodynamically that at equilibrium the product of the concentrations, or more strictly, the activities of the sodium and chloride ions, will be the same on both sides of the membrane. Hence may be formed

$$[Na^+]_1[Cl^-]_1 = [Na^+]_2[Cl^-]_2$$

or

$$(c_1+x)x = (c_2-x)^2$$

Obviously, the diffusion of the chloride ions (and an equal number of sodium ions) through the membrane has been hindered by the presence of the nondiffusible ion R^-. Calculations based on this equation, which are confirmed by experimental measurements, show that sodium chloride is almost completely prevented from diffusing through the membrane if it is present in small concentration relative to the concentration of the nondiffusible ion R^-. As the relative concentration of sodium chloride is increased, more of it diffuses through the membrane. Finally, when the salt concentration is very high relative to that of R^-, an even distribution of sodium and chloride ions on either side of the membrane is approached. A similar equilibrium is also attained when the diffusible salt has no ion common to the colloidal electrolyte.

An important example of Donnan equilibrium is the dialysis of a solution of a colloidal electrolyte against pure water. Sodium ions from the colloidal $R^-{\cdot}Na^+$ will diffuse through the membrane and be replaced by an equivalent number of hydrogen ions. This phenomenon is called membrane hydrolysis and is helpful in explaining certain membrane equilibria in biological cells and tissues. Obviously if the water is renewed continuously, complete hydrolysis will ultimately ensue.

Two important consequences arise from the Donnan equilibrium. The first is that the observed osmotic pressure, that is, the difference in hydrostatic pressure on the two sides of the membrane, will always exceed that of R^- except when a large excess of salt is added. An illustration of this effect is the behavior of ionic gels (for example, protein gels) when immersed in water. Ionic groups attached to the structure of the gel cannot diffuse out into the surrounding solution, and osmosis causes swelling of the gel. The gel will behave according to the Donnan equilibrium; the swelling is found to be reduced by addition of salts. The second consequence of the Donnan distribution is that a potential difference E is set up at the membrane. It is given by the equation below; R is the gas

$$E = \frac{RT}{F}\ln\frac{[Na^+]_1}{[Na^+]_2} = \frac{RT}{F}\ln\frac{[Cl^-]_2}{[Cl^-]_1}$$

constant, T is the absolute temperature, and F is Faraday's constant. This is the origin of the difference in potential between a suspension and its intermicellar liquid. In soil chemistry this is known as the suspension effect. *See* COLLOID; DIALYSIS; ION-SELECTIVE MEMBRANES AND ELECTRODES. [GEORGE S. MILL; W. O. MILLIGAN]

Bibliography: F. G. Donnan, The theory of membrane equilibria, *Chem. Rev.*, 1(1):73–90, 1924; R. B. Martin, *Introduction to Biophysical Chemistry*, 1964.

Drying

An operation in which a liquid, usually water, is removed from a wet solid in equipment termed dryers. The use of heat to remove liquids distinguishes drying from mechanical dewatering methods such as centrifugation, decantation or sedimentation, and filtration, in which no change in phase from liquid to vapor is experienced. Drying is preferred to the term dehydration, which is sometimes used in connection with the drying of foods. Dehydration usually implies removal of water accompanied by a chemical change. Drying is a widespread operation in the chemical process industries. It is used for chemicals of all types, pharmaceuticals, biological materials, foods, detergents, wood, minerals, and industrial wastes. Drying processes may evaporate liquids at rates varying from only a few ounces per hour to 10 tons per hour in a single dryer. Drying temperatures may be as high as 1400°F (760°C), or as low −40°F (−40°C) in freeze drying. Dryers range in size from small cabinets to spray dryers with steel towers 100 ft high and 30 ft in diameter. The materials dried may be in the form of thin solutions, suspensions, slurries, pastes, granular materials, bulk objects, fibers, or sheets. Drying may be accomplished by convective heat transfer, by conduction from heated surfaces, by radiation, and by dielectric heating. In general, the removal of moisture from liquids (that is, the drying of liquids) and the drying of gases are classified as distillation processes and adsorption processes, respectively, and they are performed in special equipment usually termed distillation columns (for liquids) and adsorbers (for gases and liquids). Gases also may be dried by compression. *See* ADSORPTION; DISTILLATION.

Drying of solids. In the drying of solids, the desirable end product is in solid form. Thus, even through the solid is initially in solution, the problem of producing this solid in dry form is classed under this heading. Final moisture contents of dry solids are usually less than 10%, and in many instances, less than 1%.

The mechanism of the drying of solids is reasonably simple in concept. When drying is done with heated gases, in the most general case, a wet solid begins to dry as though the water were present alone without any solid, and hence evaporation proceeds as it would from a so-called free water surface, that is, as water standing in an open pan. The period or stage of drying during this initial phase, therefore, is commonly referred to as the constant-rate period because evaporation occurs at a constant rate and is independent of the solid present. The presence of any dissolved salts will cause the evaporation rate to be less than that of pure water. Nevertheless, this lower rate can still be constant during the first stages of drying.

A fundamental theory of drying depends on a knowledge of the forces governing the flow of liquids inside solids. Attempts have been made to develop a general theory of drying on the basis that liquids move inside solids by a diffusional process. However, this is not true in all cases. In fact, only in a limited number of types of solids does true diffusion of liquids occur. In most cases, the internal flow mechanism results from a combination of forces which may include capillarity, internal pressure gradients caused by shrinkage, a vapor-liquid flow sequence caused by temperature gradients, diffusion, and osmosis. Because of the complexities of the internal flow mechanism, it has not been possible to evolve a generalized theory of drying applicable to all materials. Only in the drying of certain bulk objects such as wood, ceramics, and soap has a significant understanding of the internal mechanism been gained which permits control of product quality.

Most investigations of drying have been made from the so-called external viewpoint, wherein the effects of the external drying medium such as air velocity, humidity, temperature, and wet material shape and subdivision are studied with respect to their influence on the drying rate. The results of such investigations are usually presented as drying rate curves, and the natures of these curves are used to interpret the drying mechanism. Figure 1 shows a series of typical drying-rate curves.

The constant-rate period of drying when heat is supplied by convection is susceptible to theoretical and analytical treatment because it is essentially independent of the solid material. When drying is accomplished by heat transfer from hot gases,

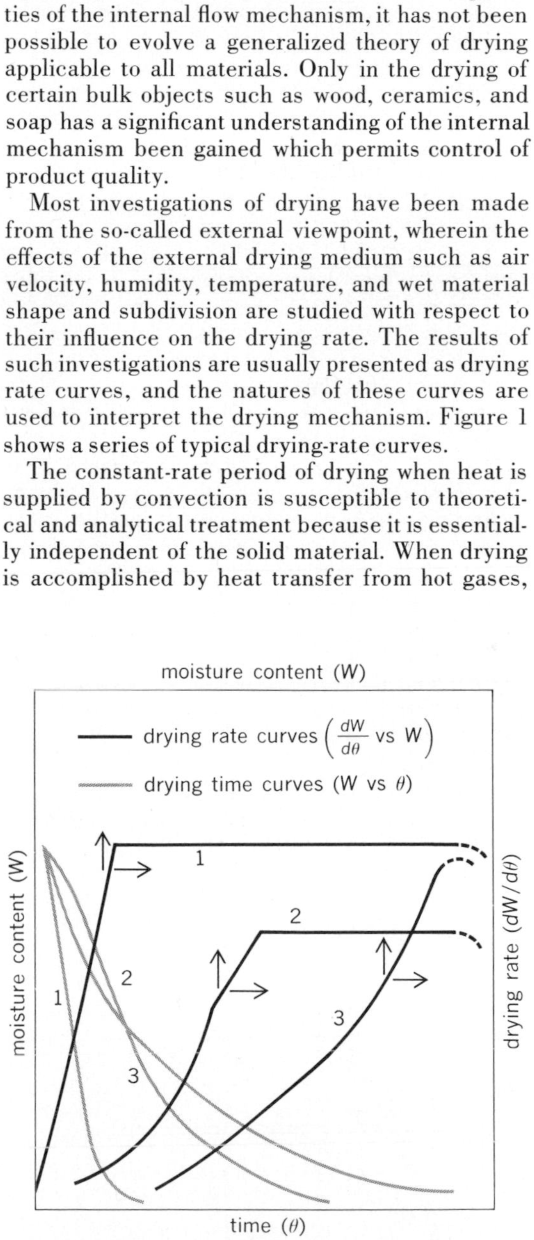

Fig. 1. Drying-time and drying-rate curves illustrating the general problem of drying: (1) Curves typical of a layer of thin material with most of the drying in the constant rate. (2) A more general case in which two stages in the falling-rate period occur. Typical of granular materials. (3) A case in which no constant rate occurs. Typical of homogeneous and colloidal materials such as soap, gelatin, and viscous solutions.

which also remove the evolved vapors, the constant rate may be expressed in terms of heat-transfer rates or mass-transfer rates.

A constant rate of evaporation at the surface of the solid maintains the surface at a constant temperature, which, in the absence of other heat effects, is very nearly the wet-bulb temperature of the air. This temperature may range from 70 to 130°F (21–54°C) for convection drying, depending on the temperature and humidity of the air and on radiation. This so-called wet-bulb cooling effect is one reason why heat-sensitive solids can be dried in air at temperatures well above the decomposition temperature of the solid.

The magnitude of the constant rate can vary widely, depending on the degree of subdivision of the material, that is, the manner in which the material is exposed to the drying air. Thus, the rate of drying in spray dryers can be several hundred-thousand-fold greater than the rates in tray dryers.

A number of empirical expressions based on experimental studies have been developed for estimating the constant rate for different physical configurations of the wet material.

When materials are dried in contact with hot surfaces, termed indirect drying, the air humidity and air velocity may no longer be significant factors controlling the rate. Instead, the "goodness" of contact between the wet material and the heated surfaces, plus the surface temperature, will be controlling. This may involve agitation of the wet material in some cases.

The falling-rate period is not as amenable to treatment as the constant-rate period because the falling rate depends largely on the internal structure of the solid and the mechanism of moisture flow therein. In falling-rate processes, the rate of drying decreases gradually until the moisture content of the material approximates the equilibrium value. The equilibrium moisture content of a material is that moisture content to which a given material can be dried under specific conditions of air temperature and humidity. A typical equilibrium moisture curve is shown in Fig. 2. A unique characteristic of hygroscopic materials is that they hold or retain water at a vapor pressure less than water at the same temperature. Moisture so retained is termed bound moisture. Materials in which water exerts its normal vapor pressure at all moisture contents are termed nonhygroscopic, and in general, are easier to dry.

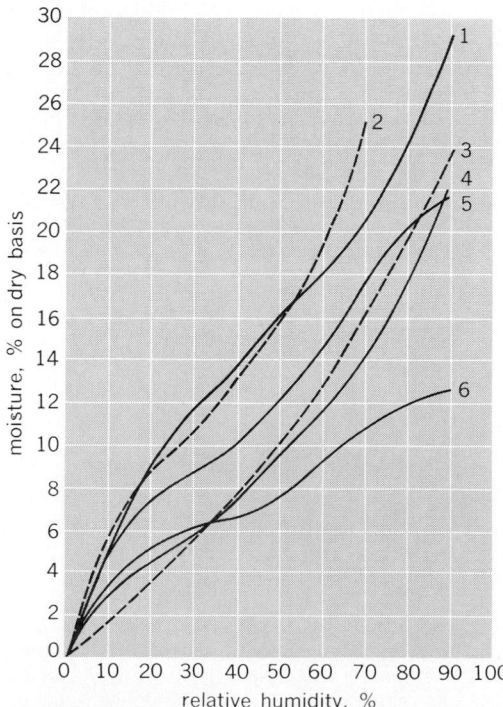

Fig. 2. Equilibrium moisture content of miscellaneous organic materials at 70°F (21°C); 1, leather; 2, tobacco; 3, soap; 4, wood; 5, catgut; 6, glue.

Classification of dryers. Drying equipment for solids may be conveniently grouped into three classes on the basis of the method of transferring heat for evaporation. The first class is termed direct dryers; the second class, indirect dryers; and the third class, radiant heat dryers. In the chart in Fig. 3, each class is subdivided into batch and continuous. Batch dryers are restricted to low capacities and long drying times. Most industrial drying operations are performed in continuous dryers. The large numbers of different types of dryers reflect the efforts to handle the large numbers of wet materials in ways which result in the most efficient contacting with the drying medium. Thus, filter cakes, pastes, and similar materials, when preformed in small pieces, can be dried many times faster in continuous through-circulation dryers than in batch tray dryers. Similarly, materials which are sprayed to form small drops, as in spray drying, dry much faster than in through-circulation drying.

Direct dryers. The general operating characteristics of direct dryers are: (1) drying is accomplished by convection heat transfer between the wet solid and a hot gas, the latter removing the vaporized liquid as well as supplying the heat needed for evaporation; (2) the heating medium may be steam-heated air, gases of combustion, a heated inert atmosphere such as nitrogen, or a superheated vapor such as steam; (3) drying temperatures may range from prevailing atmospheric temperatures to 1400°F (760°C); (4) at drying temperatures below the boiling point of the liquid, increasing amounts of vapor of this liquid in the drying gas will decrease the rate of drying and increase the final liquid content of the solid; (5) when the drying temperatures are above the boiling point throughout the process, an increase in the vapor content of the gas or air, in general, will have no retarding effect on the drying rate, and no effect on the final moisture content; (6) for low-temperature drying, dehumidification of the drying gas is often required when high atmospheric humidities prevail; (7) the efficiency of direct dryers will increase with an increase in the inlet temperature of the drying gas at a fixed exhaust temperature.

Indirect dryers. The general operating characteristics of indirect dryers are as follows: (1) Drying by the transfer of heat by conduction and some radiation to the wet material; conduction usually occurs through a metallic retaining wall. The source of heat is generally condensing steam, but may also be hot water, gases of combustion, mol-

All types of dryers used for producing a dry, solid product from a wet feed.

Direct Dryers

Heat transfer for drying is accomplished by direct contact between the wet solid and hot gases. The vaporized liquid is carried away by the drying medium, that is, the hot gases. Direct dryers might also be termed convection dryers.

Infrared or Radiant Heat Dryers

The operation of radiant heat dryers depends on the generation, transmission, and absorption of infrared rays.

(Dielectric heat dryers operate on the principle of heat generation within the solid by placing the latter in a high-frequency electric field.)

Indirect Dryers

Heat for drying is transferred to the wet solid through a retaining wall. The vaporized liquid is removed independently of the heating medium. Rate of drying depends on the contacting of the wet material with hot surfaces. Indirect dryers might also be termed conduction or contact dryers.

Continuous

Operation is continued without interruption as long as wet feed is supplied. It is apparent that any continuous dryer can be operated intermittently or batchwise if so desired.

Batch

Dryers are designed to operate on a definite size batch of wet feed for given time cycles. In batch dryers the conditions of moisture content and temperature continuously change at any point in the dryer.

Continuous

Drying is accomplished by material passing through the dryer continuously and in contact with a hot surface.

Batch

Batch indirect dryers are generally well adapted to operate under vacuum. They may be divided into agitated and nonagitated types.

Direct Continuous Types

1. Continuous tray dryers such as continuous metal belts, vibrating trays utilizing hot gases, vertical turbo-dryers.
2. Continuous sheeting dryers. A continuous sheet of material passes through the dryer either as festoons or as a taut sheet stretched on a pin frame.
3. Pneumatic conveying dryers. In this type drying is often done in conjunction with grinding. Material conveyed in high temperature, high velocity gases to a cyclone collector.
4. Rotary dryers. Material is conveyed and showered inside a rotating cylinder through which hot gases flow. Certain rotary dryers may be a combination of indirect and direct types; for example, hot gases first heat an inner shell and then pass between on inner and outer shell in contact with the wet solid.
5. Spray dryers. Dryer feed must be capable of atomization either by a centrifugal disk or a nozzle.
6. Through-circulation dryers. Material is held on a continuous conveying screen, and hot air is blown through it.
7. Tunnel dryers. Material on trucks is moved through a tunnel in contact with hot gases.

Direct Batch Types

1. Batch through-circulation dryers. Material held on screen bottom trays through which hot air is blown.
2. Tray and compartment dryers. Material supported on trays which may or may not be on removable trucks. Air blown across material on trays.

Indirect Continuous Types

1. Cylinder dryers for continuous sheets such as paper, cellophane, and textile piece goods. Cylinders are generally steam heated, and rotate.
2. Drum dryers. These may be heated by steam or hot water.
3. Screw conveyor dryers. Although these dryers are continuous, operation under a vacuum is feasible. Solvent recovery with drying is possible.
4. Steam-tube rotary dryers. Steam or hot water can be used. Operation on slight negative pressure is feasible to permit solvent recovery with drying.
5. Vibrating tray dryers. Heating accomplished by steam or hot water.
6. Special types such as a continuous fabric belt moving in close contact with a steam heated platten. Material to be dried lies on the belt and receives heat by contact.

Indirect Batch Types

1. Agitated pan dryers. These may operate atmospherically or under vacuum, and can handle small production of nearly any form of wet solid; that is, liquids, slurries, pastes, or granular solids.
2. Freeze dryers. Material is frozen prior to drying. Drying in frozen state is then done under very high vacuum.
3. Vacuum rotary dryers. Material is agitated in a horizontal, stationary shell. Vacuum may not always be necessary. Agitator may be steam heated in addition to the shell.
4. Vacuum tray dryers. Heating done by contact with steam-heated or hot-water-heated shelves on which the material lies. No agitation involved.

Fig. 3. Classification of dryers, based on methods of heat transfer.

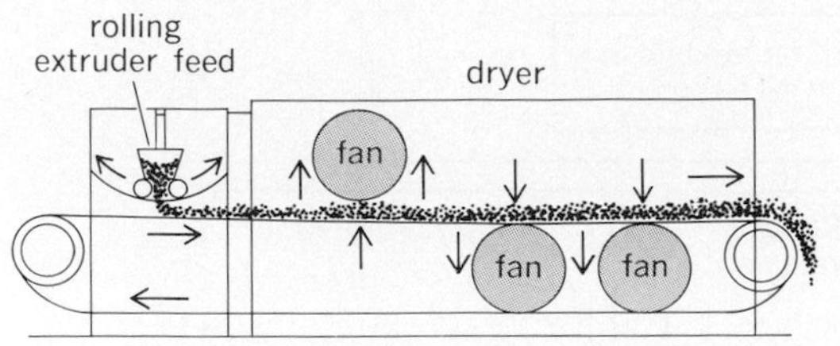

Fig. 4. Diagram of path of travel of permeable bed of wet material through a three-unit through-circulation dryer. (*Proctor and Schwartz, Inc.*)

ten heat-transfer salts, hot oil, or electric heat. (2) The drying temperature of the surface of contact may range from below freezing to 1000°F (537°C). (3) Indirect dryers are especially suited to drying under reduced pressures and with inert atmospheres, and are therefore well adapted to the recovery of solvents. (4) Indirect dryers using condensing steam usually have a high efficiency because heat is supplied according to the demand, but, as in all dryers, the efficiency falls off markedly when very low final moisture contents are required. (5) Indirect dryers can handle dusty materials more readily than direct dryers. (6) The operation of indirect dryers is frequently characterized by some method of agitation to improve the contact between the hot metal surface and wet material. The nature of this contact determines the drying rate of indirect dryers—heavy, granular materials generally give higher heat-transfer coefficients of contact than fluffy, bulky solids.

Radiant-energy dryers. These operate by the transfer of heat from a radiant source to the wet material being dried. The temperature of the radiant source may range from hot water or steam temperatures, 200–350°F (93–176°C), to the temperatures of incandescent surfaces, 1500–2500°F (815–1371°C). The generating medium may be steam, hot liquids, gas flames, or electricity, depending on the temperature desired and the equipment design.

Special types. Dielectric-heat dryers do not fall in any of the above classes inasmuch as their operation depends on the generation of heat by high-frequency fields inside the material being dried so that heat will actually flow out of the interior of the solid. These dryers are used to dry large bulky objects which have a long internal path for moisture flow.

Direct batch dryers. In operation, heated air circulates over the wet material being dried. The wet solid is supported according to its physical form. Lumber, ceramics, and similar massive objects are stacked in piles or on racks; textile skeins, painted objects, and hides are suspended on hangers; and granular materials, pastes, slurries, and liquids are placed in trays, which may be supported on stationary or movable racks. Good performance of this type of dryers depends on uniform, equal air velocities across all the wet material.

In batch–through-circulation dryers, heated air is blown through the wet material on screen-bottom trays instead of across. The material to be dried must be permeable to air flow.

Dryers of this type, but of binlike design, are used extensively in the explosives industry to dry gunpowder, and in food processing to dry and condition certain foodstuffs, such as grains and corn.

Direct continuous dryers. Tunnel and continuous tray dryers usually consist of long, enclosed housing or tunnels through which wet material is moved on trucks. Hot air is blown through the trucks. Air flow may be parallel, counter flow, or at right angles (cross flow) to the movement of the trucks. The trucks may move continuously or semicontinuously through the tunnel. Tunnel dryers may be operated adiabatically, that is, without the addition of heat in the tunnel, or the air may be reheated periodically during its passage through the tunnel.

Wet granular materials are held on the trays of trucks; foodstuffs, rayon cakes, pottery, and large ceramic objects are held on racks; textile skeins are draped over rods; and hides are pasted on glass plates or hung on frames.

In continuous through-circulation dryers, heated air is blown through a permeable bed of wet material as it passes continuously through the dryer. Drying rates are much higher than in the usual tray or tunnel dryers because of the large surface area exposed per unit weight of material and because the smaller particles offer less resistance to internal moisture flow.

The operation of this type of dryer depends on whether or not the wet material is in a state of subdivision suitable for through circulation of the hot air. Some materials are already in such a permeable state. Many materials require special preliminary treatment, termed preforming, to form them into permeable beds. Preforming may include the processes of scoring on rotary filters, granulating, extruding, briqueting, flaking, and predrying on finned drums.

One type of through-circulation dryer consists of a horizontal conveying screen, which moves through a tunnellike housing. A permeable bed of the wet material is supported and conveyed on the screen, and hot air or gas is blown vertically up or down through the bed (Fig. 4). Through-circulation drying may also be performed in rotary-type dryers, which convey the material by a tumbling action imparted by the rotation of the dryer shell.

Conveying-screen through-circulation dryers are widely used in chemical plants to dry fibrous, flaky, and granular materials, such as cotton linters, rayon staple, cellulose acetate, silica gel, sawdust, and minerals, materials that are well suited to through circulation of air without prior treatment. They are also used extensively to dry materials, such as starch, pigments, calcium carbonate, insecticides, dyes, and intermediates, which must

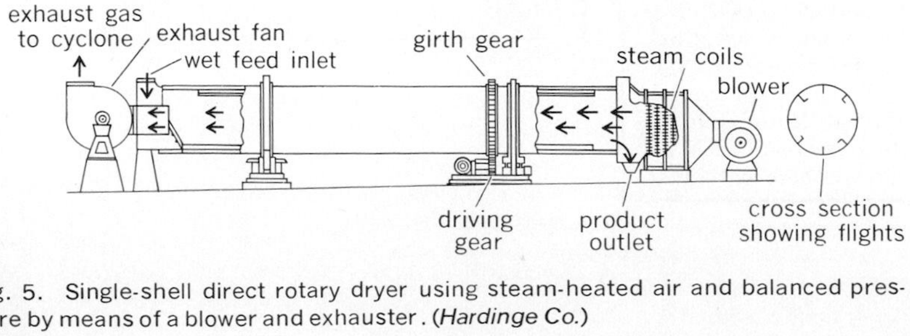

Fig. 5. Single-shell direct rotary dryer using steam-heated air and balanced pressure by means of a blower and exhauster. (*Hardinge Co.*)

be preformed by one of the methods noted above.

Direct rotary dryers are used extensively in the chemical industry. A rotary dryer consists of a cylinder slightly inclined to the horizontal and rotated on suitable bearings. The rotary action of the cylinder serves to convey wet material from one end to the other while passing hot gases axially through the dryer shell. The contact of the solids and gases is further improved by means of flights arranged within the dryer shell so as to shower the wet material through the hot gas stream. A rotary dryer without such lifting flights is usually called a rotary kiln.

Rotary dryers may operate with air flow either parallel or countercurrent to the flow of the wet solid (Fig. 5).

Pneumatic conveying dryers, sometimes termed flash or dispersion dryers, operate on the basis of simultaneously conveying and drying a wet solid in a high-velocity stream of hot gas. Temperatures up to 1400°F are used in these dryers. The short contact times involved permit using gas temperatures above the decomposition temperature of the material. The gas stream acts as both the conveying and heating medium. Gas velocities on the order of 75 ft/sec are used. Frequently, the wet material is in such a form that some disintegrating action is required before it can be conveyed. A schematic diagram of this type of dryer is shown in Fig. 6. It is applicable to granular, free-flowing materials, such as coal, whey, and sodium chloride, and to sludges, filter-press cakes, and similar nongranular solids which require disintegration for proper dispersion. It is common practice to recycle dry product into the wet feed in order to facilitate dispersion and handling.

Spray dryers operate on the principle of creating a highly dispersed liquid state in a high-temperature (up to 1400°F or 760°C) gas zone. The heart of a spray-drying process is the creation of small liquid droplets by spraying. This may be accomplished by means of: (1) high-pressure nozzles, (2) pneumatic nozzles, and (3) high-speed rotating disks. Almost any pumpable liquid, from a thin

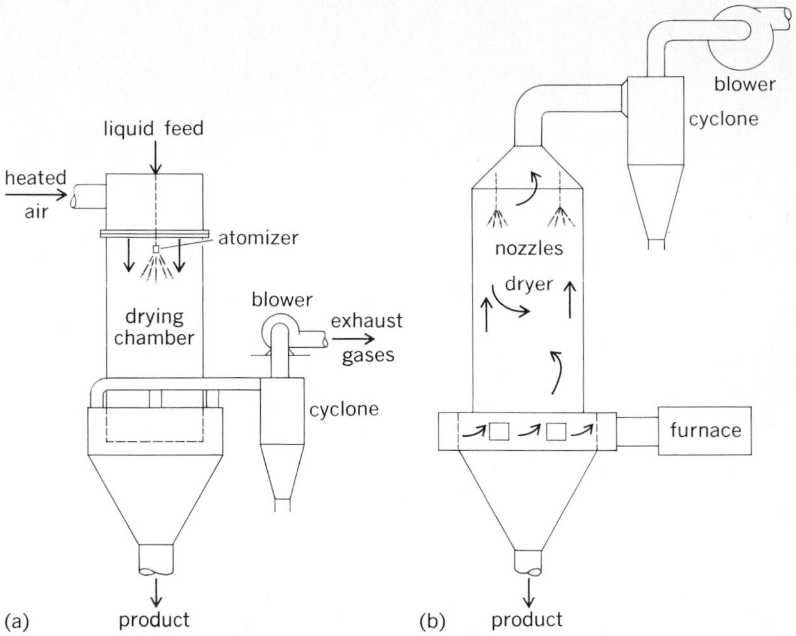

Fig. 7. Spray dryers. (*a*) Cocurrent. (*b*) Countercurrent.

clear liquid to a pasty sludge, can be atomized sufficiently for spray drying. Liquids above a viscosity of 1500 centipoises, however, are very difficult to atomize and spray dry. Generally, fine atomization will not produce a large percentage of droplets less than 5 micrometers (μm) in diameter. The particle size under conditions of so-called coarse atomization will be on the order of 200–600 μm. Because of the high surface-volume ratio of small drops, the actual drying time in spray dryers may be considerably less than 1 sec for high-temperature operation.

Spray dryers are made in a multitude of designs. The drying gases may flow cocurrent with or countercurrent to the spray, and the spray may be directed vertically up or down, or horizontally. Various types of spray dryers are shown in Fig. 7. They are used extensively in the chemical, pharmaceutical, and food industries.

Fluidized-bed dryers operate by having heated air pass upward at sufficient velocity through a column or layer of granular, wet material to cause it to fluidize and become mixed by turbulent action. Wet feed may be introduced at the bottom of the column, and dry product removed from the top, as shown in Fig. 8.

In direct continuous-sheeting dryers, heated air is circulated over or through a continuously moving wet sheet, which is supported by a variety of methods.

Direct continuous-sheeting dryers occur in a variety of types. The festoon or loop dryer permits drying of continuous sheet material in a relaxed state by festooning or looping it over rolls, which in turn are conveyed through the dryer. Tenter dryers are used to dry a continuous sheet of material under tension. Heated air is usually blown perpendicularly from slots or nozzles against both sides of the sheet.

Direct continuous-sheeting dryers are used to dry sheet materials that are sufficiently strong mechanically in the wet and dry states to support

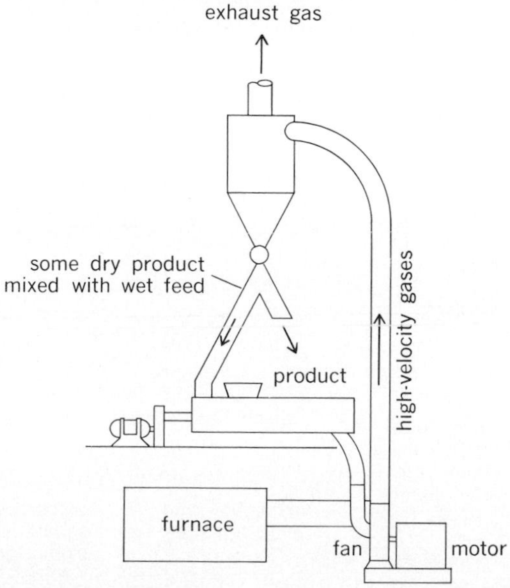

Fig. 6. Dispersion dryer.

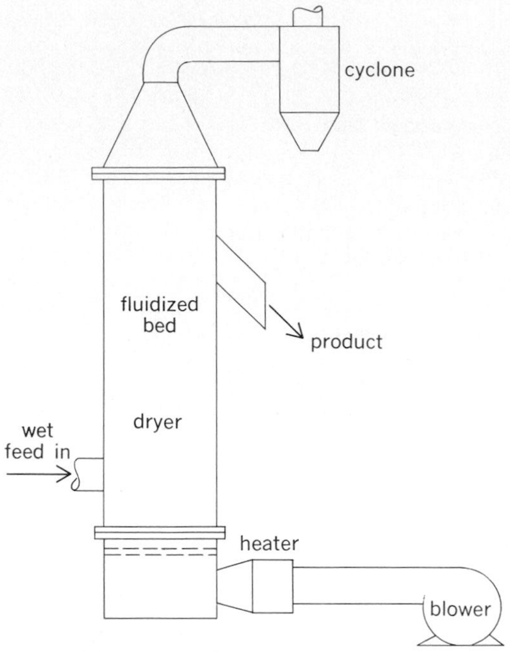

Fig. 8. Fluidized-bed dryer.

their own weight or withstand tension as required. Their widest use is in the textile field, in the manufacture of coated and impregnated fabrics, and in the preparation of films and coated papers.

Indirect batch dryers. Vacuum shelf dryers are indirect batch dryers which generally consist of a vacuum-tight cubical or cylindrical chamber of cast-iron or steel plate, heated supporting shelves inside the chamber, a vacuum source, and a condenser. In operation, heat transfer takes place by conduction through metal surfaces and interfaces to the wet material held on the shelves.

Vacuum shelf dryers are used extensively for drying pharmaceuticals, temperature-sensitive or easily oxidizable materials, and small batches of high-cost products where any product loss must be avoided. This type is particularly useful for recovery of valuable solvents or vapors.

Vacuum freeze dryers are used principally for drying materials that would be destroyed by the loss of volatile ingredients or by drying temperatures above the freezing point. The material is dried in the frozen state, so that a process of sublimation is involved, that is, ice sublimes directly to water vapor. Because the material dries in a rigid frozen condition, shrinkage is minimized, and the resulting structure of the dry solid is usually porous and readily soluble. This led to the term lyophilization in the early development of freeze drying.

The equipment required for this method of drying consists of a heated drying chamber where sublimation occurs, a piping system for the transport of vapors, and a vapor-removal system for condensable and noncondensable gases. Removal of the condensable vapors may be accomplished by freezing on cold surfaces, absorption in liquid desiccants, or adsorption on solid desiccants.

One application of freeze drying (the Cryochem process) involves conduction heat transfer to the frozen solid held on a metallic surface. However, should the metal surface temperature rise above the freezing point of the solid, melting will occur. A second method of freeze drying utilizes heat transfer by radiation.

Agitated-pan dryers consist of a bowl- or pan-shaped receptacle in which wet material is stirred or agitated in contact with hot surfaces. The operation may be atmospheric or vacuum. Agitated-pan dryers are used to handle small batches of materials that can withstand agitation during drying. They are suitable for pastes and slurries, and for materials containing valuable solvents which must be recovered.

The vacuum rotary dryer is another type of batch indirect dryer with agitator. It consists of a horizontal shell in which agitator blades attached to a horizontal rotating shaft revolve and agitate the material being dried (Fig. 9). Heat is supplied by condensing steam in a jacket surrounding the shell, by hot water, or by other heat-transfer fluids. Vacuum rotary dryers are used for large batches of materials that must be dried in the absence of air or where the recovery of solvents is required.

Indirect continuous dryers. Indirect rotary dryers are similar mechanically and in appearance to the direct rotary dryers discussed above. They differ primarily in that heat is transferred to the material through the metal shell or from steam tubes located around the dryer shell, rather than from hot gases as in the case of direct rotary dryers.

A screw-conveyor dryer is essentially a jacketed conveyor in which material is heated and dried as it is conveyed. In one type, the jacket may extend only to the top of the conveyor, which is left open to the atmosphere. This is termed a trough dryer. When the jacket encloses the conveyor completely, a slight negative pressure is required to sweep out the evaporated moisture.

Vibrating-conveyor dryers consist of a vibrating, heated, solid deck over which the wet solid moves. This deck can be heated by hot gases, steam, methanol vapors, or Dowtherm vapors, which pass through a jacket fastened to the deck upon which the material is conveyed. Direct gas flames can also be used as the source of heat. A hood equipped with an exhaust fan and placed over the

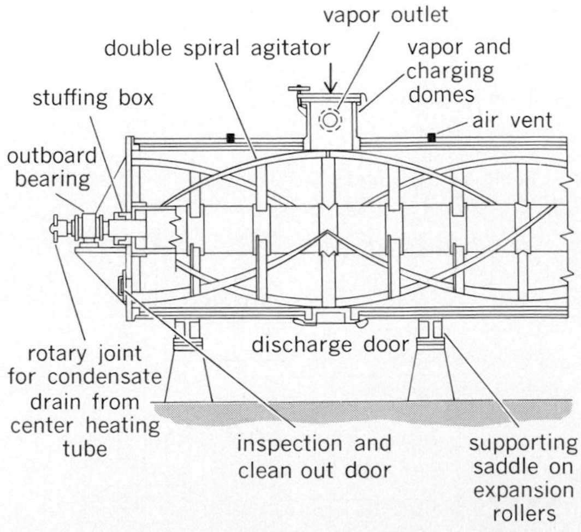

Fig. 9. Typical vacuum rotary dryer. (*Blaw-Knox Co.*)

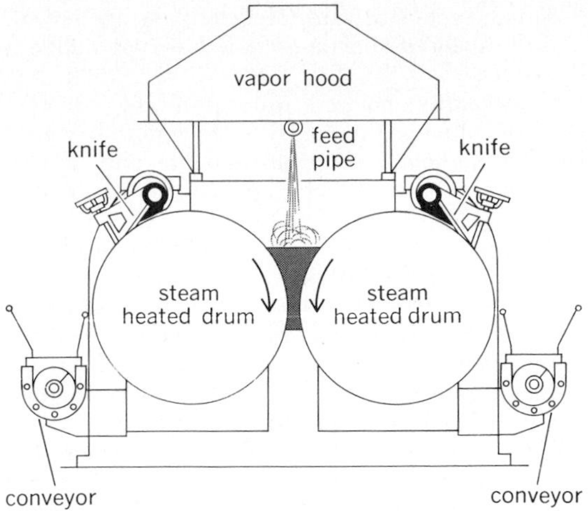

Fig. 10. Double-drum dryer with pipe feed. (*Buflovak Equipment Division, Blaw-Knox Co.*)

deck removes the evaporated liquid. Infrared lamps may be mounted above the deck to increase drying rates.

Drum drying consists of applying a liquid material, solution, slurry, or paste to a revolving heated metal drum, which conducts heat to the wet film to evaporate the water during a partial revolution of the drum (Fig. 10). The dry material is scraped from the drum by a stationary knife. Drum dryers may be designated by type as atmospheric double-drum, atmospheric single-drum, atmospheric twin-drum, vacuum single-drum, and vacuum double-drum dryer.

In drum drying, the product is exposed to heat for only short periods of time. This has the advantage that, although the product may approach the temperature of the drum surface, there usually is no adverse effect from overheating.

Cylinder dryers, sometimes called can dryers or drying rolls, differ from drum dryers in that they are used for materials in a continuous-sheet form. Cylinder dryers may consist of one large cylindrical drum, such as the so-called Yankee dryer; more often they comprise a number of drums arranged so that a continuous sheet of material can pass over them in series. Typical of this arrangement are Fourdrinier paper-machine dryers, cellophane dryers, and slashers for textile piece goods and fibers.

The size of commerical cylinder dryers covers a wide range. The individual rolls may be 2–6 ft in diameter and up to 20 ft in width. In some cases, the width of the rolls decreases throughout the dryer in order to conform to the shrinkage of the sheet.

Drying of gases. The removal of 95–100% of the water vapor in air or other gases is frequently necessary. Gases having a dew point of −40°F (−40°C) are considered commercially dry. The more important reasons for the removal of water vapor from air are: (1) comfort, as in air conditioning; (2) control of the humidity of manufacturing atmospheres; (3) protection of electrical equipment against corrosion, short circuits, and electrostatic discharges; (4) requirement of dry air for use in chemical processes where moisture present in air adversely affects the economy of the process; (5) prevention of water adsorption in pneumatic conveying; and (6) as a prerequisite to liquefaction.

Gases may be dried by the following processes: (1) absorption by use of spray chambers with such organic liquids as glycerin, or aqueous solutions of salts such as lithium chloride, and by use of packed columns with countercurrent flow of sulfuric acid, phosphoric acid, or organic liquids; (2) adsorption by use of solid adsorbents such as activated alumina, silica gel, or molecular sieves; (3) compression to a partial pressure of water vapor greater than the saturation pressure to effect condensation of liquid water; (4) cooling below dew point of the gas with surface condensers or cold-water sprays; and (5) compression and cooling, in which liquid desiccants are used in continuous processes in spray chambers and packed towers—solid desiccants are generally used in an intermittent operation that requires periodic interruption for regeneration of the spent desiccant.

Desiccants are classified as solid adsorbents, which remove water vapor by the phenomena of surface adsorption and capillary condensation (silica gel and activated alumina); solid absorbents, which remove water vapor by chemical reaction (fused anhydrous calcium sulfate, lime, and magnesium perchlorate); deliquescent absorbents, which remove water vapor by chemical reaction and dissolution (calcium chloride and potassium hydroxide); or liquid absorbents, which remove water vapor by absorption (sulfuric acid, lithium chloride solutions, and ethylene glycol).

The mechanical methods of drying gases, compression and cooling and refrigeration, are used in large-scale operations, and generally are more expensive methods than those using desiccants. Such mechanical methods are used when compression or cooling of the gas is required.

Liquid desiccants (concentrated acids and organic liquids) are generally liquid at all stages of a drying process. Soluble desiccants (calcium chloride and sodium hydroxide) include those solids which are deliquescent in the presence of high concentrations of water vapor.

Deliquescent salts and hydrates are generally used as concentrated solutions because of the practical difficulties in handling, replacing, and regenerating the wet corrosive solids. The degree of drying possible with solutions is much less than with the corresponding solids; but, where only moderately low humidities are required and where large volumes of air are dried, solutions are satisfactory. *See* DESICCANT; EVAPORATION; FILTRATION; GAS ABSORPTION OPERATIONS; VAPOR PRESSURE. [WILLIAM R. MARSHALL]

Bibliography: W. R. Marshall, Jr., *Atomization and Spray Drying*, 1954: R. H. Perry (ed.), *Chemical Engineers' Handbook*, 4th ed., 1963.

Drying oil

Oils are classified as nondrying, semidrying, or drying, according to their ease of autoxidation and polymerization to form a hard, dry film on exposure to air. More than 800,000,000 lb (360,000,000 kg) of drying oils is used annually in the United States in paints and varnishes. Drying oils are relatively highly unsaturated; that is, they are composed of triglycerides constructed from unsaturated fatty acids. The best drying oils contain sev-

Glycerides present in drying oils, %*

Name	Saturated	Oleate	Linoleate	Linolenate	Eleostearate	Licanate
Cottonseed	25	40	35			
Soybean	14	26	52	8		
Dehydrated castor oil	5	10	85			
Linseed	10	18	17	55		
Perilla	7	14	16	63		
Tung	5	7	3		85	
Oiticica	10	6	10			74

*Based on N. A. Lange (ed.), *Handbook of Chemistry*, 10th ed., McGraw-Hill, 1967.

eral nonconjugated double bonds per molecule; thus, a good drying oil should have a high iodine value (about 130), should on hydrolysis yield only a small percentage of saturated acids (palmitic and stearic), and should furnish on hydrolysis large percentages (above 65%) of combined unsaturated acids, such as oleic, linoleic, linolenic, licanic, and eleosteric acids. *See* VINYLOGY.

The table gives the percentage content, as saturated or unsaturated glycerides, of the common drying oils.

Raw drying oils, that is, untreated drying oils, are not suitable for paints and varnishes because they polymerize too slowly, and various methods have been introduced to improve the polymerization process. One method involves boiling the oil after addition of soluble resin-acid salts of cobalt, manganese, or lead (such salts are known as driers); the product, called boiled oil, dries in approximately one-fifth the time in which raw oil dries. Boiled oil is considerably more viscous than raw oil, and is used in the production of paints, varnishes, and enamels. Blown oil is produced by blowing air through the oil (to which driers have been added) at about 120°C (248°F); it is said to have superior wetting or surface-covering properties. Stand oil has been partially polymerized, with admixture of driers, by heating to 260–280°C (500–536°F). This material is used extensively in antifouling paints, printing inks, and linoleum, as well as in varnishes and enamels. Linseed oil is the most widely used drying oil in paints and varnishes. [EVANS B. REID]

Dye

A colored substance which imparts more or less permanent color to other materials.

Not all colored substances are dyes, however. If red iron rust is ground with white sugar, the resulting mixture has an over-all reddish appearance. The sugar has been colored by pigmentation, and the iron rust has been used as a pigment. Examination of the mixture under a microscope shows distinct white and red particles, and a separation can be made by dissolving the sugar out of the mixture with water. If the red iron rust is added to white cloth in water, some of the red particles may cling to the cloth, but they can be removed by rubbing or by washing with soap. Some colored substances (usually organic chemical compounds) may be added to cloth in water, and after a period of soaking, usually accompanied by heat and agitation, the cloth will be colored; no separate colored particles can be seen under ordinary microscopic examination, and the color cannot be removed by washing, even with soap. The cloth has been dyed, and the colored substance is a dye (also called a dyestuff).

Customarily, colored water-insoluble substances are called pigments. Dyes are generally water-soluble, although some are soluble only during application, after which they become insoluble. *See* PIGMENT.

The mechanism by which soluble colored substances enter the internal structure of fibers and there become fixed has been variously explained in terms of the physical and chemical concepts of the times when the explanations were given. It is said to be an adsorption phenomenon, a salt formation, a quasi-chemical union caused by hydrogen bonding, or an ether linkage, and in some cases it is considered to be a true solution effect. The end result, however, is that the dye has imparted a color (not necessarily that of the solid dye itself) to the fiber which is more or less resistant to washing or removal by similar mechanical operations. The dye is said to be fixed on and to have affinity, or substantivity, for the material it has colored. The material is designated as the substrate. If the color is quite resistant to washing and light, it is called a fast color; if the color is easily removed or fades quickly, it is a fugitive dye.

Because not all water-soluble colored substances are dyes, various attempts have been made to relate chemical constitution with color and substantivity. One of the earlier and still very useful explanations was given by O. N. Witt, who stated in 1876 that all colored organic compounds (called chromogens) contain certain unsaturated chromophoric groups which are responsible for the color, and if these compounds also contain certain auxochromic groups, they possess dyeing properties. Examples of chromophores are the groups $—NO_2$, —N═N—, and ═CO, and of auxochromes, $—NH_2$, —OH. The auxochromes also influence hue according to their nature, number, and position on the chromogen molecule. An elaboration of this theory in 1888 by H. E. Armstrong regarded all chromophores as being quinoid (═R═) in structure. Many later studies have added explanations of the nature and variation of color in organic compounds, but the Witt theory provides a frame of reference for most dyes which is simple, practical, and satisfactory for all but the specialist in dye chemistry. *See* SPECTROPHOTOMETRIC ANALYSIS.

Dyestuffs may be classified in various ways, according to color (blue, red, and so on), origin (natural—from vegetable and animal matter—or synthetic), chemical structure, kinds of material to which they are applied, and method of application.

Color or hue. Commercial dyes are usually named to indicate the hue imparted to the dyed article. This color is not necessarily the same as that of the solid dyestuff, nor is it the same as the color of the dye solution from which the dyeing is made.

Origin. The dyes used earliest were natural dyestuffs (saffron, henna, cochineal, and logwood) derived from plants or animals. Water extracts of various plants gave solutions of yellows and browns; the extract of the Mediterranean mollusk (*Murex brandaris*) gave Tyrian purple, so expensive that it was the mark of a king; and indigo came from plants of the genus *Indigofera*. Most natural dyes are of the mordant type that requires a fixing agent. With the advent of the synthetic dyes, which are far more varied in shades and properties, and generally more brilliant, the production of the natural colors has decreased to a very small portion of the total market. Their greatest use now is in the dyeing of leather.

Synthetic dyes were an early result of the development of chemistry as a science, and the first commercial production in 1857 in England of a synthetic organic chemical was that of the first commercially produced synthetic dye, mauve, discovered in 1856 by Sir William Henry Perkin. The early history of organic chemistry was primarily a record of investigations of material and methods for making synthetic dyes. Because one of the early and important materials used in these syntheses was aniline (derived from coal tar), the synthetic dyes have been known as coal-tar dyes and aniline dyes. Their manufacture was at first so vigorously pursued in Germany that it became almost exclusively a German industry. Only in World War I (1914–1918) did the loss of dye imports from Germany lead to the development of their manufacture in other countries.

The manufacture of dyes proceeds from simple raw materials, mostly aromatic hydrocarbons such as benzene (benzol), toluene, and naphthalene, with introduction of other chemical groups, such as nitro, amino, halogen, and sulfonic acid. These intermediate compounds are then further processed by many special operations in organic chemistry such as diazotization, coupling, condensation, and fusion to give the final dyestuff. Following the chemical manufacture, the dye may be treated to give it special physical properties, and is standardized for strength. It may be sold as a dry powder, as a paste, or in solution. *See* DIAZOTIZATION.

Of the many thousands of dyes which have been synthesized in laboratories, about 3500 have had actual commercial use. These have been indexed in several compilations, the latest and most complete being the *Colour Index*, 3d ed., 1971, published by both the Society of Dyers and Colourists (England), and the American Association of Textile Chemists and Colorists. This encyclopedic work is the major reference for the dye chemist. Each dye listed is given an individual five-digit number as identification regardless of the specific manufacturer. Although some of the earliest dyes have common names with general usage, commercial dyes are sold under a large number of names, many of which relate to the same material. Each dyestuff manufacturer has his own nomenclature, which usually consists of three or four parts. First comes his own trademarked name for the class, then the hue, then words, letters, or numbers which describe the shade and other characteristics, and finally designation of the strength or physical form (powder, paste, and double-strength powder). Thus, Erie Black RX Conc. Paste is a trade name for the color (CI30235), which is a direct cotton dye, black with a reddish shade, of extra quality (improved over earlier manufacture), and standardized in strength much stronger than the ordinary type.

Chemical structure. The most precise and scientific classification of dyes is based upon their chemical structure. This is the classification of interest to the research chemist and the manufacturer of dyes.

The table shows the 22 classes that are listed in the *Colour Index*, with typical structures and examples.

Utilization. This classification is based on the materials to which the dyes are applied.

Cloth of natural fibers. Coloring cotton, wool, linen, and silk is by far the largest use for dyes.

Cloth of synthetic fibers. These include regenerated cellulose (viscose), cellulose acetate, nylon, polyacrylonitrile, and polyester. Regenerated cellulose fiber can be dyed with dyes for cotton; dyeing other synthetic fibers with the colors commonly used for natural fibers is often less satisfactory or impossible. Special classes of dyestuffs were created for these newer polymeric materials.

Paper. This is colored both by dyeing and by pigmentation with finely divided colored materials. The color is usually added to the raw stock in the beater before the sheets are formed, or on the calender. Dyes are also applied in the coatings used on paper, such as the wax-dye coating used for duplicating (carbon) paper.

Leather. One of the earliest materials to be colored, leather has retained the use of natural dyes to a greater extent than most other materials, but there are many synthetic textile and specific dyes for leather. Because of the difficulty of penetrating the closely knit structure of leather, surface coloring by spraying or brushing is widely practiced.

Wood. Dyeing or tinting of wood is done with dyes in water, alcohol, or other solvents which evaporate after the wood has been painted with or soaked in the dye solution.

Pigments. Many soluble dyes are converted into pigments by forming insoluble salts (that is, replacing sodium in a dye salt with calcium) for use in lacquers, paints, and printing inks. This insoluble salt, called a toner, may also be deposited on inorganic fillers, such as aluminum hydroxide, to form a lake.

Food. The appearance of many foods is enhanced by artificial coloring. Butter and margarine are colored yellow, fruits and sauces which are often dulled by the process of preserving are brightened by addition of dye, and large amounts of dyestuffs are used in soft drinks and candies. In most countries, only certain colors are permitted to be used in this way.

In the United States, food must be colored by a natural dye or by choosing a synthetic from a list of certified food colors, of which the harmlessness to the human system is under constant check and review by a Federal agency. Elaborate studies of the toxicity and pharmacology of any new color must be provided by the manufacturer before it is admitted to the list.

Oils. Lubricating oils and gasolines are colored for improvement in appearance and for identification of grade, type of use, or merely as a trademark of the manufacturer. Waxes, shoe polishes, and candles are also colored.

Plastics and rubbers. Coloration of resins, plastics, and elastomers may be done by solution of a dyestuff or by dispersion of a pigment in the sub-

Some important dyes

Class; basic chemical structure; chromophore	Example	Remarks
(1) Nitroso (quinone oxime) *o*-Nitrosophenol (or *o*-nitroso-naphthol); —N=O	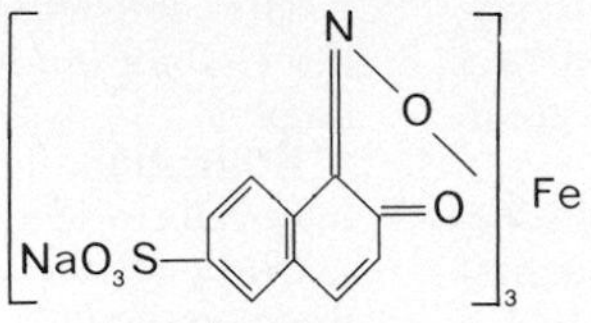 CI 10020 CI Acid Green 1 Naphthol Green B	Mordant dyes, used only as lakes of metals; prepared by action of nitrous acid on phenols or naphthols
(2) Nitro *o*-and *p*-Nitrophenols and *o*- and *p*-nitroamines; 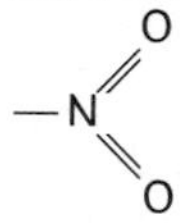	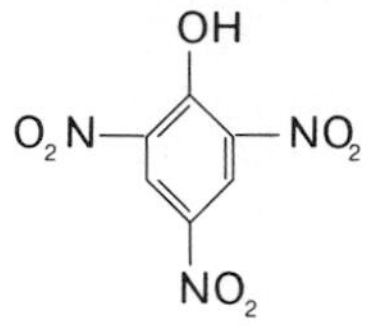 CI 10305 Picric acid	Prepared by action of nitric acid on phenols, naphthols, and amines
(3) Azo Aromatic and heterocyclic azo compounds; —N=N—		Most numerous class of dyes
(a) Monazo R—N=N—R′	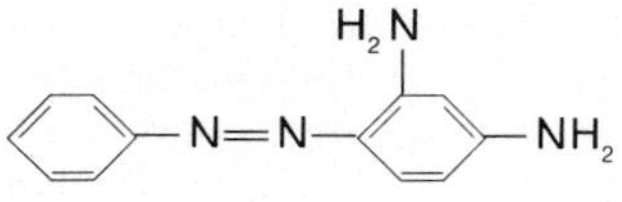 CI 11270 CI Basic Orange 2 Chrysoidine	Prepared by action of nitrous acid on primary arylamine to give diazo compound, which is coupled with aromatic amino or hydroxy compound
(b) Disazo R—N=N—R′— N=N—R″	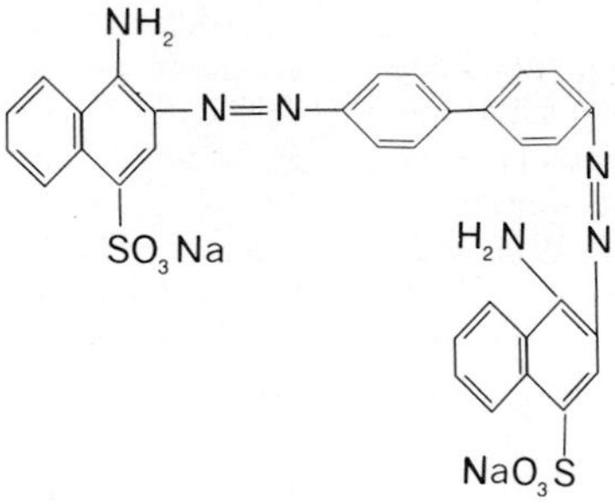 CI 22120 CI Direct Red 28 Congo Red	Prepared by nitrous acid on primary diamine, with coupling of the resulting tetrazo compound with two mols of aromatic amino or hydroxy compound
(c) Trisazo R—N=N—R′— N=N—R″— N=N—R‴	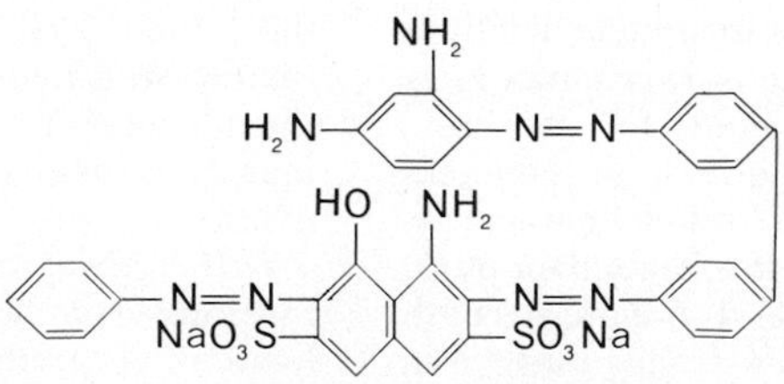 CI 30235 CI Direct Black 38 Direct Black EW	Prepared from four intermediates by rediazotizations and further couplings

Some important dyes (cont.)

Class; basic chemical structure; chromophore	Example	Remarks
(d) Polyazo With four or more azo groups	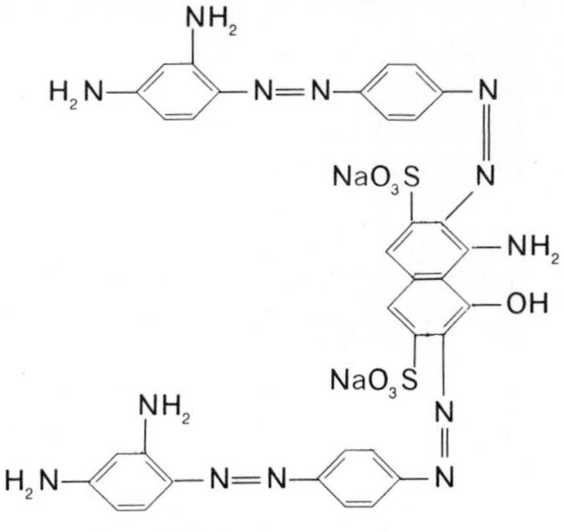 CI 35255 CI Direct Black 19 Columbia Fast Black G	
(4) Azoic Azo compounds with —N=N— as chromophore formed on fiber from components:		
(a) Fast-color base, RNH_2	 *p*-Nitroaniline	Primary amine which may be diazotized
(b) Fast-color salt, $RN\equiv N$ (with Cl on N)	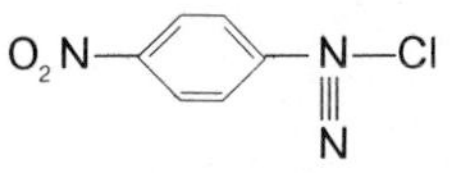 CI 37035 CI Azoic Diazo Compound 37 Diazotized *p*-nitroaniline Fast Red GG Salt	Diazo salt stabilized with additives, or by formation of complex salts
(c) Nitrosamine	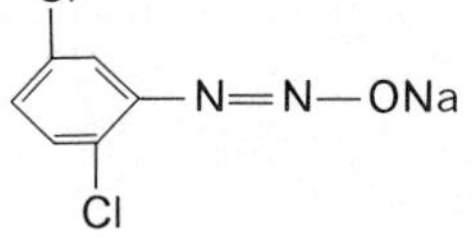 CI 37015	Special form of stabilized diazo
(d) Diazo amino	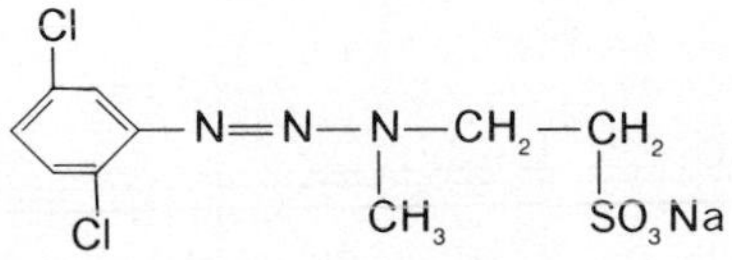	Special form of stabilized diazo
(e) Coupling component	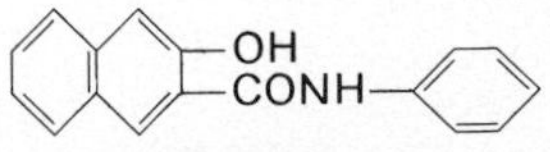 CI 37505 CI Azoic Coupling Component 2 Naphthol AS	Condensation products of 3-hydroxy-2-naphthoic acid with amines

Some important dyes (cont.)

Class; basic chemical structure; chromophore	Example	Remarks
(5) Stilbene 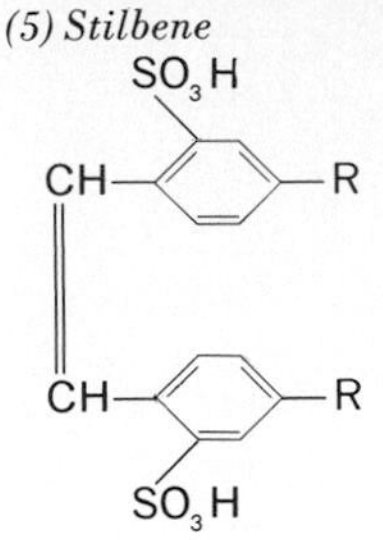 stilbene plus azo or azoxy 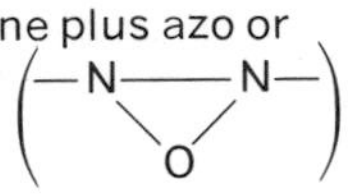	Not known CI 40003 CI Direct Orange 15 Diamine Orange D	Condensation products of 5-nitro-*o*-toluene sulfonic acid with alkali, forming first dinitrostilbene disulfonate, which condenses with itself or with hydroxy and amino compounds
(6) Diphenylmethane (ketone imine) 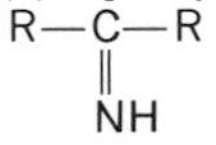 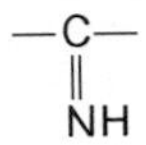	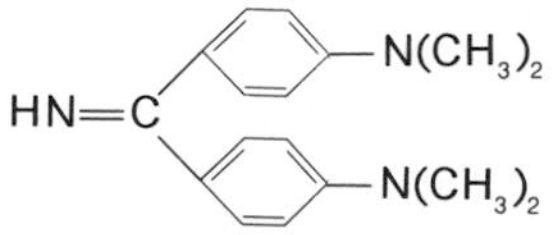 CI 41000 CI Basic Yellow 2 Auramine O	From Michler's ketone, heated with ammonium and zinc chlorides
(7) Triarylmethane $R_2C{=}Ar{=}NH$(or $=O$) $>C{=}Ar{=}NH$(or $=O$)		Condensation products of aromatic aldehydes with arylamines or phenols
(a) Diamino	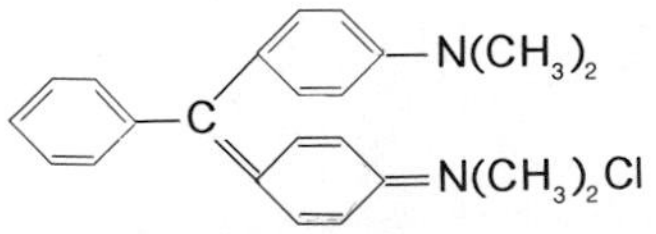 CI 42000 CI Basic Green 4 Malachite Green	
(b) Triamino	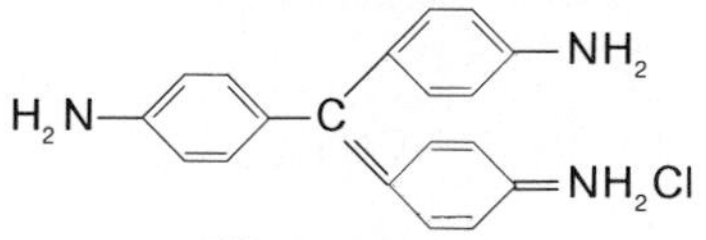 CI 42500 CI Basic Red 9 Pararosaniline	
(c) Aminohydroxy	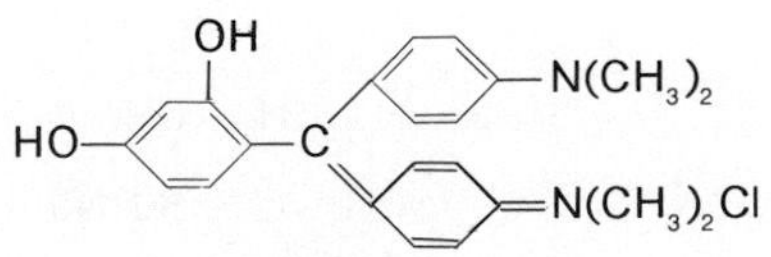 CI 43520 Resorcine Violet	
(d) Hydroxy	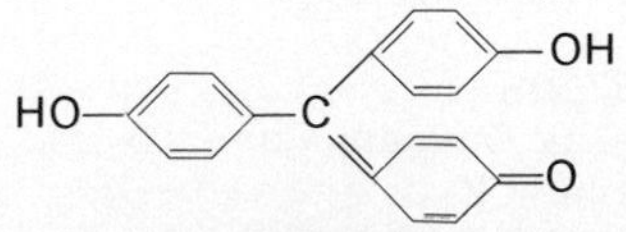 CI 43800 Aurine	

Some important dyes (cont.)

Class; basic chemical structure; chromophore	Example	Remarks
(e) Diphenyl naphthyl methane	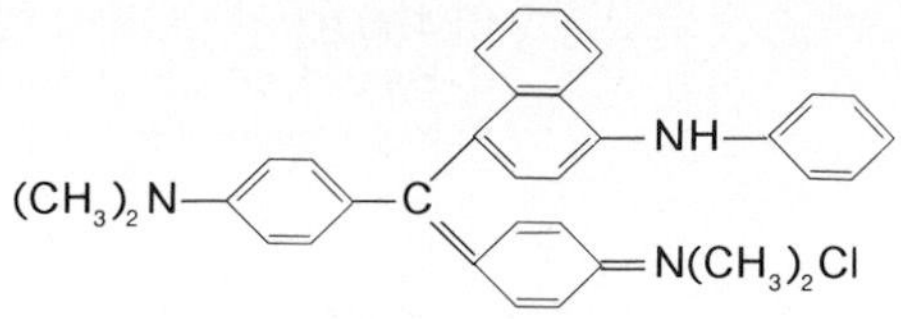 CI 44045 CI Basic Blue 26 Victoria Blue B	
(8) Xanthene 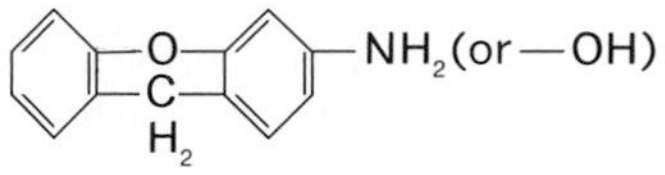 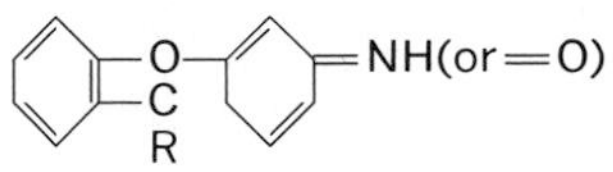		Basic mordant dyes; subclasses are pyronines, succineins, saccha-reins, rosamines, rhodamines, rhodols, fluorones (hydroxy- and anthrahydroxy-phthaleins)
(a) Amino	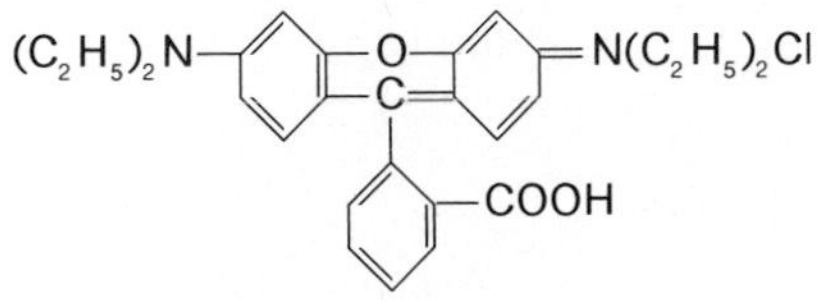 CI 45170 CI Basic Violet 10 Rhodamine B	
(b) Hydroxy	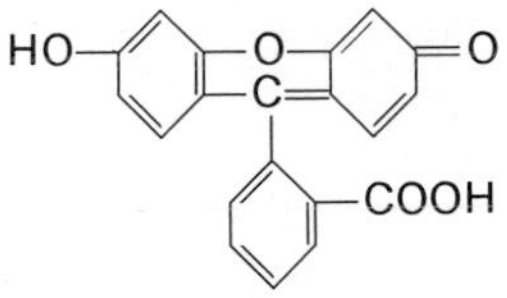 CI 45350 CI Acid Yellow 73 Fluorescein	Condensation of phthalic anhydride with hydroxy compounds
(9) Acridine 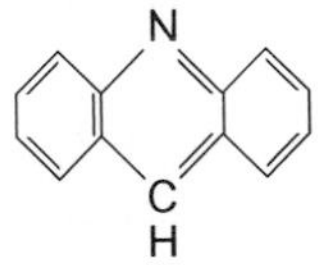 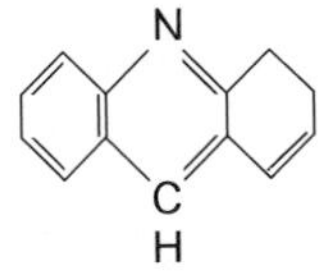	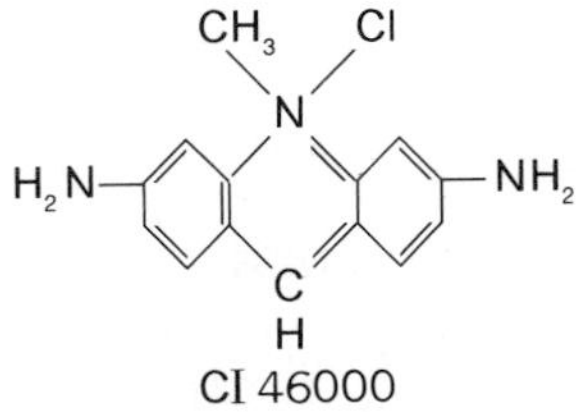 CI 46000 Acriflavine	Condensation of *m*-diamine with aldehyde, cyclize and oxidize
(10) Quinoline 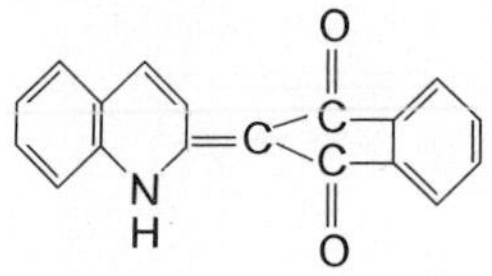 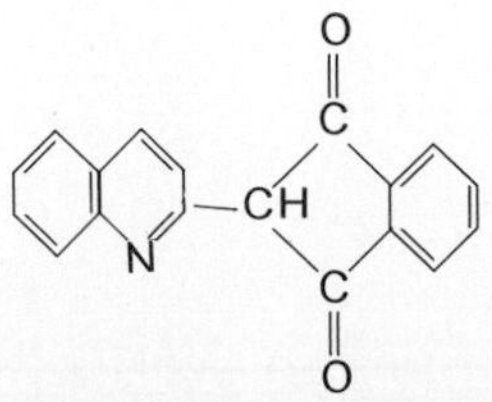	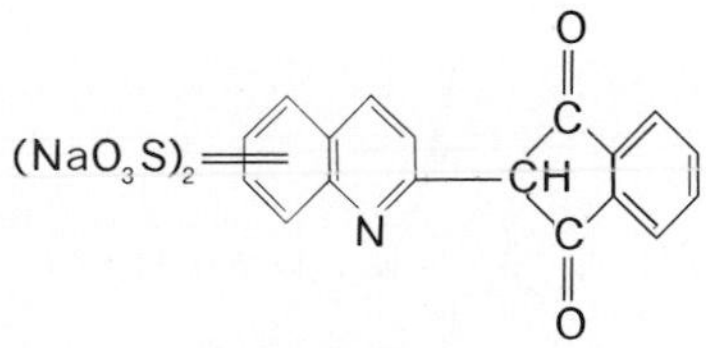 CI 47005 CI Acid Yellow 3 Quinoline Yellow	Condensation of quinolines with phthalic anhydride; solvent and basic dyes for paper and wool; when sulfonated, give acid wool dyes

Some important dyes (cont.)

Class; basic chemical structure; chromophore	Example	Remarks
(11) Methine and polymethine 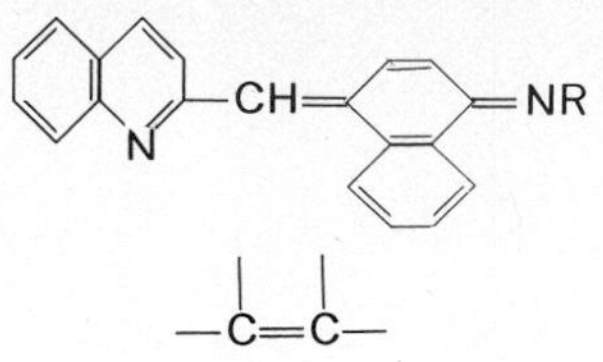	CI 48070 CI Basic Red 12 Astraphloxine FF	Quinoline, benzothiazole, or trimethyl indoline nuclei linked together with methine chains; main uses in photography
(12) Thiazole 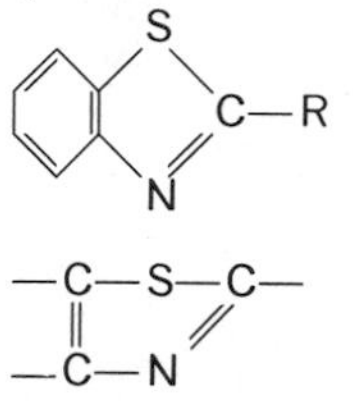	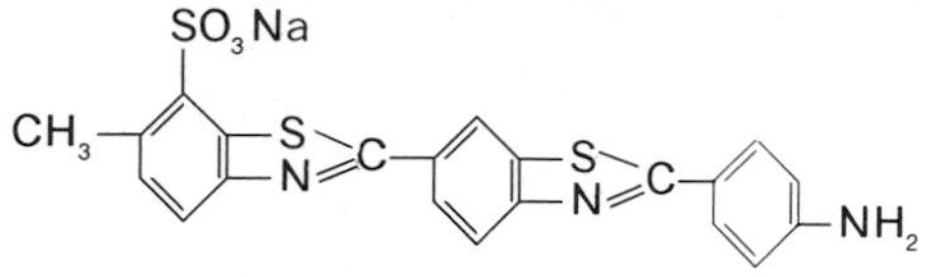 CI 49000 CI Direct Yellow 59 Primuline	Heat *p*-toluidine with sulfur, then sulfonate This structure is valuable if incorporated into other classes of dyes, enhancing substantivity
(13) Indamine and indophenol 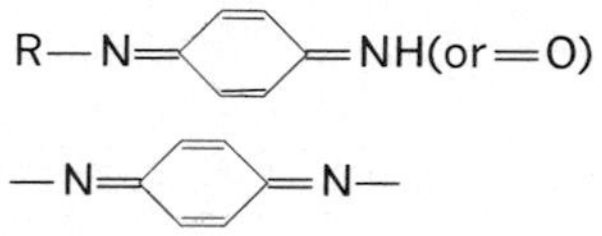	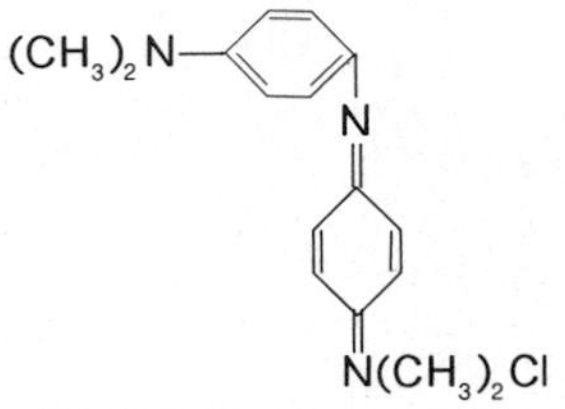 CI 49405 Bindschedler's Green	Oxidation of a *p*-diamine or *p*-aminophenol in presence of amine or phenol No use in textile dyeing; intermediates for sulfur colors; used in color photography
(14) Azine 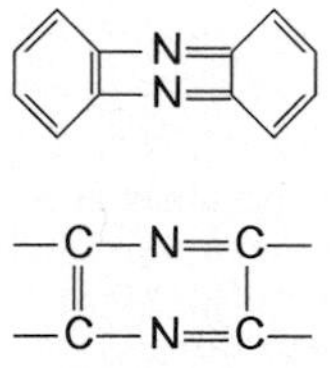	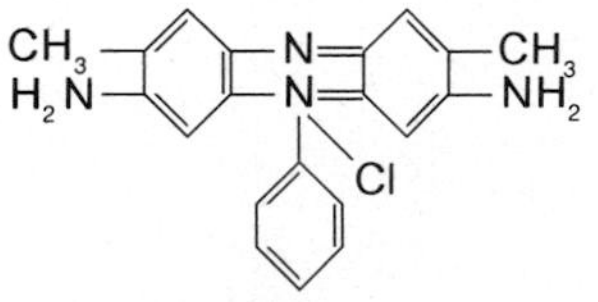 CI 50240 CI Basic Red 2 Safranine T	Basic dyes; subclasses are quinoxalines, eurhodines and eurhodols, aposafranines, safranines, indulines, nigrosines
(15) Oxazine 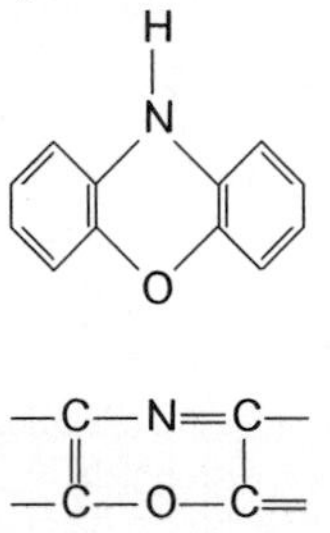	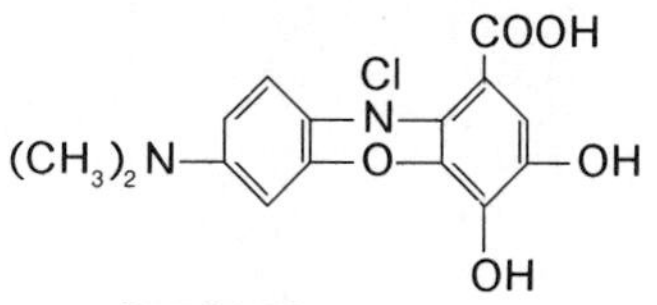 CI 51030 CI Mordant Blue 10 Gallocyanine	Basic dyes; subclasses are monooxazines, dioxazines; oxazones
(16) Thiazine 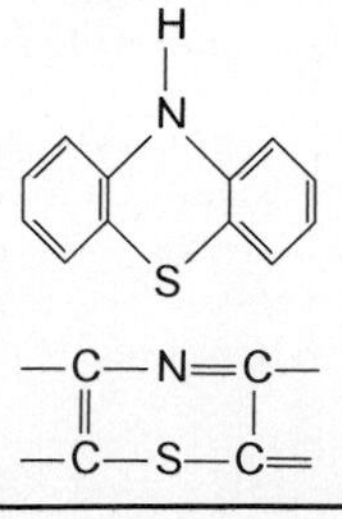	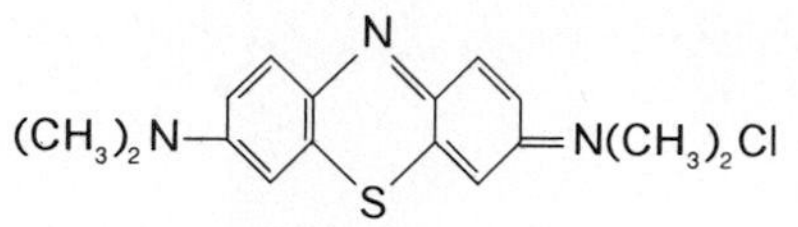 CI 52015 CI Basic Blue 9 Methylene Blue	Basic dyes

Some important dyes (cont.)

Class; basic chemical structure; chromophore	Example	Remarks
(17) Sulfur Structure not known, contains thiazole, thiazin, thianthrene rings with mercapto and polysulfide links; chromophore not known	CI 53185 CI Sulfur Black 1 Sulfur Black	Made by heating a variety of organic substances with sulfur and alkali polysulfides
(18) Lactone 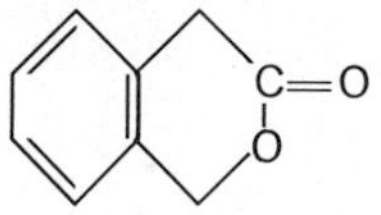 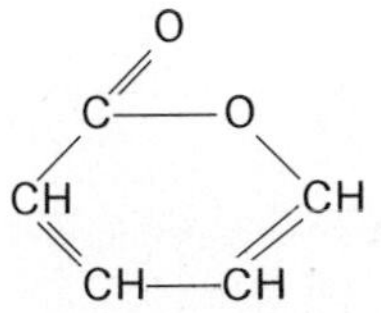	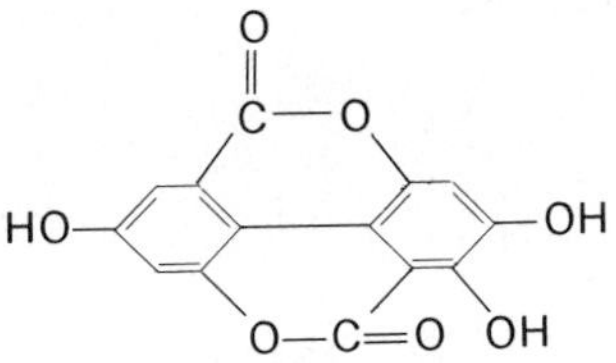 CI 55000 Resoflavine W	Oxidation of polyhydroxy aromatic compounds; for chrome-mordanted wool; of little importance
(19) Amino ketone and hydroxy ketone 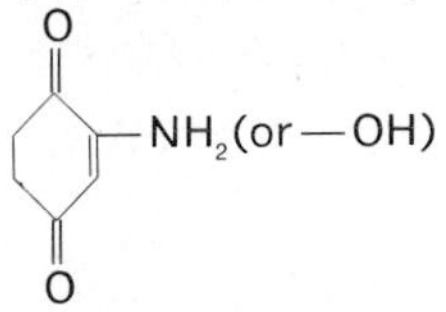 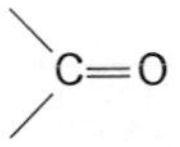	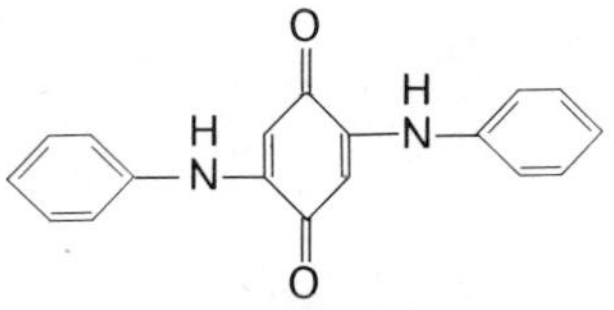 CI 56000 Helindone Yellow CA	Of little importance
(20) Anthraquinone 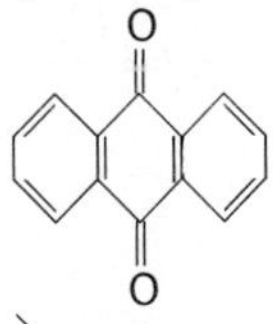 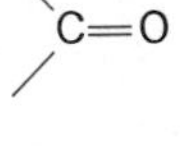		
(a) Acid	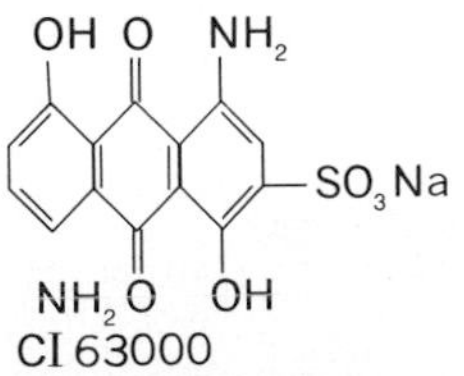 CI 63000 CI Acid Blue 43 Alizarin Sapphire SE	Important wool dyes; sulfonated amino- or hydroxyanthraquinones
(b) Mordant	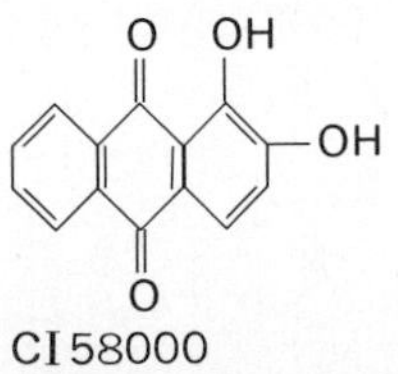 CI 58000 CI Mordant Red 14 Alizarine	

Some important dyes (cont.)

Class; basic chemical structure; chromophore	Example	Remarks
(c) Disperse	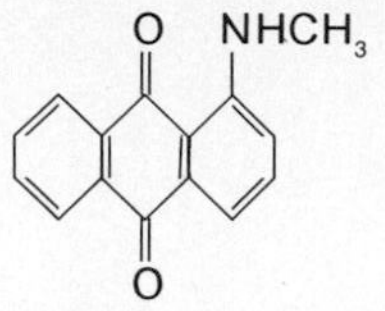 CI 60505 CI Disperse Red 9	Containing no water-solubilizing groups; for acetate silk and other synthetic fibers
(d) Vat	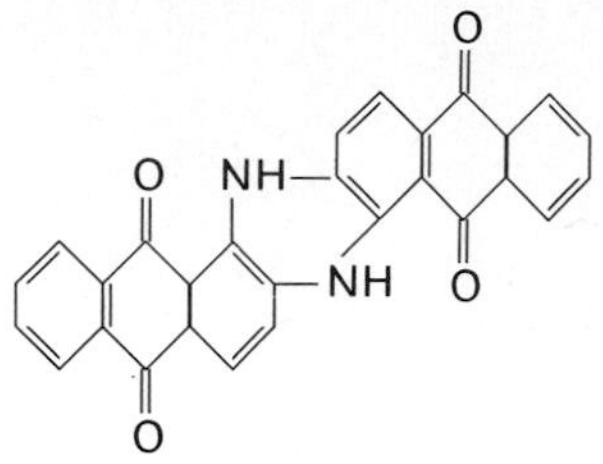 CI 69800 CI Vat Blue 4 Indanthrone	Most important class of dyes
(e) Esters of leuco vat dyes	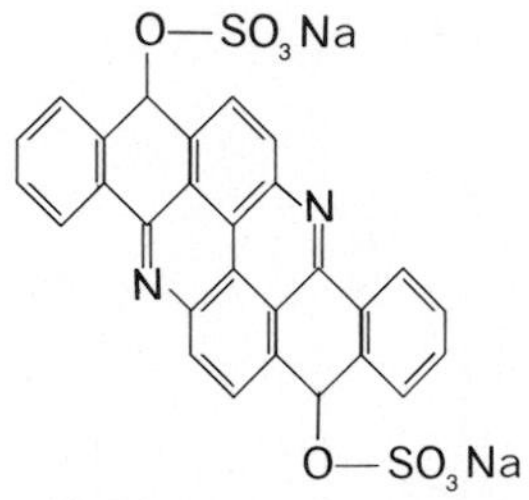 CI 70601 CI Solubilized Vat Yellow 1 Solubilized Flavanthrone	Water-soluble forms of vat dyes
(21) Indigoid 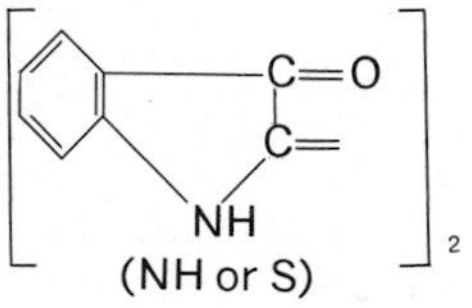 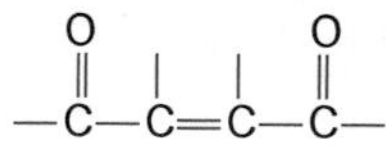		Vat dyes; solubilized forms may also be made as in (20*e*)
(a) Indigoid	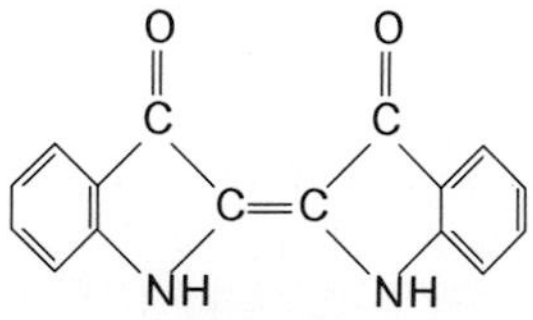 CI 73000 CI Vat Blue 1 Indigo	
(b) Thioindigoid	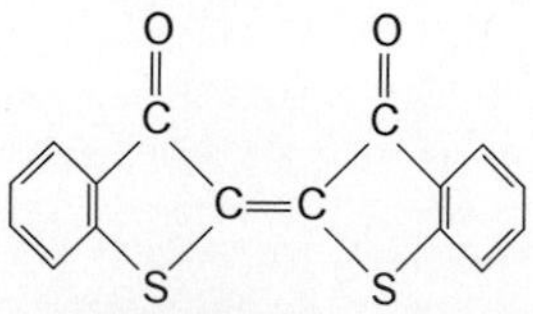 CI 73300 CI Vat Red 41 Thioindigo Red B	

Some important dyes (cont.)

Class; basic chemical structure; chromophore	Example	Remarks
(22) Phthalocyanine	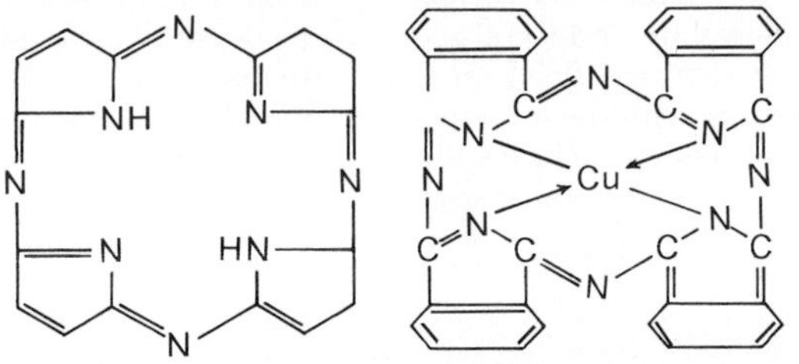 CI 74160 CI Pigment Blue 15 Phthalocyanine Blue	Of great importance in pigments

strate. In this coloring operation, there is no need for substantivity, and choice of color is determined solely by the solubility, shade, and fastness properties desired.

Biological samples. The dyeing or staining of tissues of animal and vegetable matter is an important technique in physiological and medical studies. Microorganisms and cell structure are made more visible and differentiated under the microscope by selective staining with dyes of varying affinity for protoplasm.

Photography. Certain dyes (cyanines, of the polymethine class) are added to photographic emulsions to increase sensitivity to light in special regions of the spectrum. Color photographs are produced by formation of dyes from their components within the emulsion layer.

Indicators. Some property of the environment may bring about a change in hue in certain dyes, thus giving information to the observer about this property. Indicators usually respond to acidity, alkalinity, and oxidation or reduction, but may respond to heat, humidity, electricity, and water hardness. *See* ACID-BASE INDICATOR.

Miscellaneous. Soap, synthetic detergents, cosmetics, ink, hair, fur, metals, anodized aluminum, and many other materials are colored with dyes of various types. Dyes are also used to produce colored smokes, particularly for military identification.

Methods of application. This classification is used most frequently by the practical dyer.

Acid dyes. These are salts of organic acids (sulfonic and carboxylic) and are usually marketed as the sodium salts. The acid groups confer water solubility on the dyestuff molecule. When dissolved, the dye ionizes (separates into particles with opposite electric charges), with the dye structure in the anionic (carrying the negative charge) part.

These dyes are used principally for wool, natural silk, synthetic fibers of polyamide and polyacrylic nature, leather, and paper. They are normally applied to the fiber in a solution containing some sulfuric or acetic acid, although some acid dyes will be fixed on the fiber from a neutral bath.

Because of improved fastness resulting from treatment of the dyed material with metal salts, especially chromium, a large number of acid dyes are available which contain metal atoms as part of the anion, as distinguished from the cationic metal atoms which form the salts of the organic acid. These anionically bound metal atoms do not exhibit the usual reactions of metal ions and are said to be chelated. These metalized (or premetalized) dyes have markedly superior fastness, approaching the vat dyes in this property.

Some organic chemical compounds (mainly stilbene derivatives) have substantivity for fibers, but do not absorb light in the visible spectrum and hence show no color. They do transform some of the ultraviolet light which they absorb into visible light, thereby increasing the amount of white light reflected from them. This gives a bluing effect to yellowed materials and makes them appear whiter. Such products, used on cloth and paper, are called whites dyes, optical bleaches, or optical brighteners.

Basic dyes. These, too, are salts, but of organic bases containing amino and imino groups. The colored base is combined with a colorless acid, such as hydrochloric or sulfuric, and in solution, the dye structure is in the cation (carrying the positive charge). These dyes have exceptional brightness, but generally have only fair to poor fastness except when applied to acrylic fibers. They have wide utility for coloring wool, silk, leather, acrylic fibers, and paper. They can also be applied to cotton if the cotton has been mordanted (treated with tannin, tannin and antimony salts, and alum).

Mordant dyes. Dyes which have little or no affinity for certain substrates may yet be fixed thereon if a mordant has first been applied. The fixation of the color is principally the result of reaction with the mordant material. These colors are also called adjective colors, as compared to the direct-dyeing substantive colors. Basic dyes which are applied to mordanted cotton are commonly still called basic colors, so that the designation mordant practically covers only acid-dye types and alizarin. The most common mordants are chromium salts for chrome dyeing processes. Often the chromium compound is applied during the dyeing, along with the dye, or it can be applied after the dyeing. These are called respectively meta-, mono-, or auto-chroming, and after- or top-chroming operations. These treatments may not only change the shade of the original dyeing (the self-shade), but they also may improve fastness of the dyeing to both water and light. *See* MORDANT.

Direct dyes. These dyes are normally sodium salts of sulfonic acids, and the colored part of the molecule is the anion. They differ from the acid dyes in that they are so substantive to cotton or other cellulosic fibers that they are fixed on the fiber from an aqueous solution with the assistance only of additions of common salt or sodium sulfate to the dyebath. Some authorities limit the designation substantive to these direct colors because of their outstanding affinity. These colors are also of importance in coloring paper, leather, and silk and have many miscellaneous uses. Because of the ease of application, they constitute the bulk of the package dyes used by the housewife. This is an important class of dyestuffs, exceeded in numbers only by the acid dyes and in quantity used (in the United States) only by vat dyes.

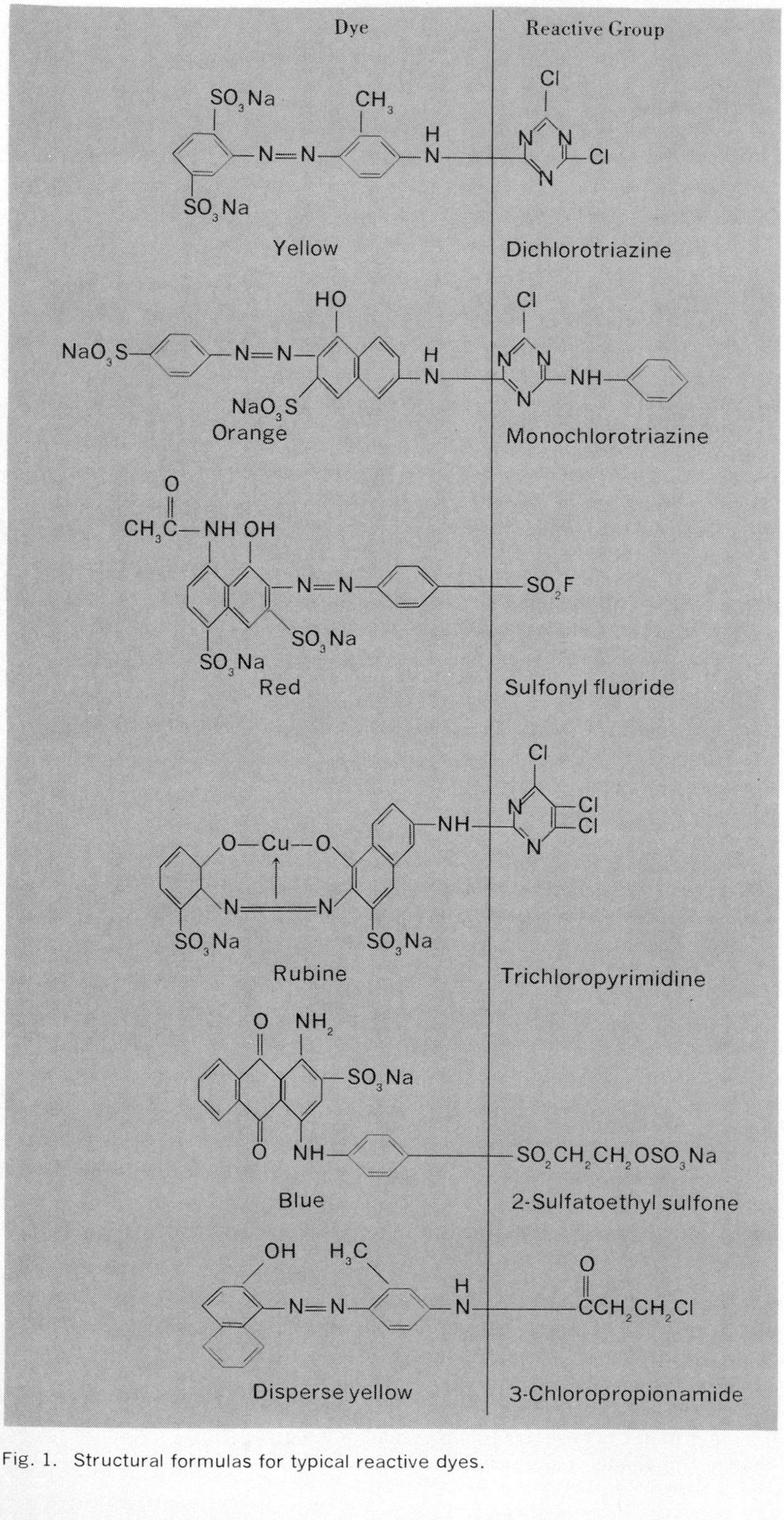

Fig. 1. Structural formulas for typical reactive dyes.

Many direct dyeings can be improved in wetfastness by an after treatment such as with copper or chromium salts, formaldehyde, resins, and cationic fixing agents. It is also possible to modify dyeings made by dyes of certain structures by a further chemical treatment (diazotization of a free amino group in the dye molecule, followed by coupling with a developer) of the dye on the fiber. Dyes suitable for such treatments are called developed dyes. These treatments may or may not change the self-shade.

Ingrain dyes. These are dyes which are formed directly on the substrate by some type of chemical action. The principal subclasses are the azoic dyes and the oxidation dyes. In a few instances, dyes of the phthalocyanine type are developed on the fiber by special treatments.

Azoic dyes are water-insoluble azo compounds which have been formed within the substrate by chemical reaction of the intermediate components. Generally, the dyeing operation proceeds by dipping the cloth into a solution of one of the components (a hydroxy or amino component), drying the cloth without rinsing, and then treating it with a solution of the other component (usually a diazo component which must be kept cold to prevent decomposition; hence, these colors are often called ice colors). Because the product of the reaction, which will then be deposited throughout the fiber, is a water-insoluble color, it is actually a pigment instead of a dyestuff. These colors are used for cotton and rayon, and in great quantity in printing.

Oxidation dyes are produced directly in the fiber or other substrates by a chemical oxidation following the impregnation of the substrate with certain aromatic amines. The final product is probably a pigment instead of a dyestuff. These colors are used principally for the dyeing of hair and fur.

Disperse dyes. These colors were originally developed for use on cellulose acetate, a synthetic fiber. This fiber has little affinity for the older known classes of dyestuffs. Some very slightly water-soluble colored materials transfer to this fiber from a water suspension if the colors are extremely finely divided particles. This results in a solution of the dye in the solid fiber. Use of these dispersed colors has extended to many of the newer synthetic fibers developed after cellulose acetate.

Vat dyes. This class of colors is distinguished by the special method of application needed—a vatting operation, wherein the water-insoluble color is made soluble by a chemical reduction of the chromophore, the ketonic =C=O group, to the ≡C—OH group, the leuco compound, which in the presence of alkali forms the water-soluble leuco salt, ≡C—OAlk. This vat solution, which is often a different color from the original insoluble material, has affinity for cotton, dyeing it with the shade of the vat solution. Upon oxidation with air or oxidizing agents, the reaction reverses to form the original water-insoluble color, leaving it deposited in the fiber as a pigment.

One group of vat colors includes stabilized leuco salts, which are salts of leuco esters (products of the leuco compound with sulfuric acid) and which do not oxidize in the air. The dyer is saved the trouble of the vatting operation; he need only apply to the cloth and then acidify and oxidize.

The two major subclasses of vat colors are the indigos and the anthraquinones. The latter are outstanding in their fastness to water, to light, and to chemicals. Since discovery of the first of the anthraquinones in 1901, this class has become the most important in dyestuffs.

Sulfur dyes. Dyes of this class are also applied by a vatting technique which makes the solubilized color substantive to cellulosic fibers. These colors are made by treating a wide variety of organic compounds with sulfur and sodium sulfides. With a few exceptions, the final reaction products are not well-identified chemical compounds with known structures. These water-insoluble dyes are dissolved in alkaline sodium sulfide solution, which serves both as reducing agent and as source of alkali. After the dye is applied in the soluble leuco form, oxidation produces the insoluble dye on the fiber.

These colors are mostly used on cotton and viscose rayon. They have moderate all-around fastness and are relatively cheap. Hence, they are used in large quantities.

Solvent dyes. These colors are soluble in organic solvents such as benzene, gasoline, alcohol, acetone, oils, fats, and waxes. Solvent dyes color merely by solution of the dye in the substrate. They may be subclassified as spirit-soluble colors with principal solubility in alcohol and as oil-soluble colors which are soluble in benzene and vegetable and mineral oils. Uses include wood stains and varnishes, lacquers, printing and writing inks, butter and margarine, and plastics and resins.

[WESLEY MINNIS]

Reactive dyes. Dyes that during the process of dyeing form a covalent bond with the substrate are known as reactive dyes. They are of particular interest on cellulosic fibers, where good washfastness is obtained without resorting to the production of large insoluble agglomerates within the fiber such as occur with vat and azoic dyes. Reactive dyes were introduced in 1955, and it is now possible to use almost any chromophore on cellulosic fibers and thus achieve a combination of brilliance and fastness not previously possible.

Structure. A reactive dye (Fig. 1) is a soluble color that contains a group capable of reacting with the hydroxyl or amino groups in cellulose, wool, silk, and other substrates. The reaction must be rapid and be carried out in the presence of water. Generally, an acid acceptor and heat are used to promote the reaction. Reaction efficiency varies from 60 to 95%, and unreacted dye must be removed by scouring to ensure washfastness.

Two types of reactive groups have been found successful in dyes which combine chemically with cellulose: active-halogen compounds which form ester or amide links, and activated vinyl groups which form ether links (Fig. 2). Other reactions

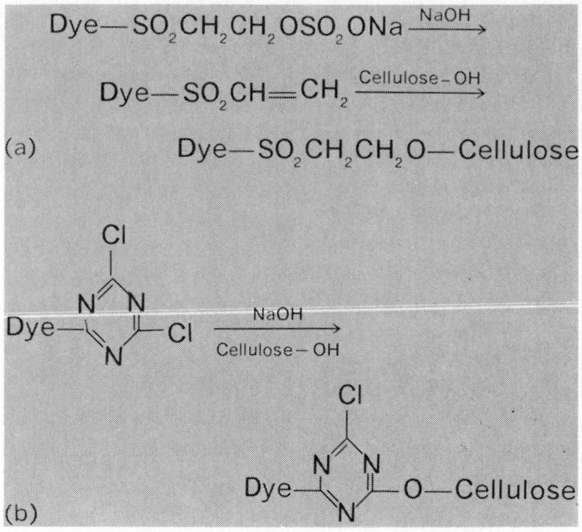

Fig. 2. Constitutional formulas depicting the reactive dye reaction mechanisms with cellulose. (*a*) Addition across double bond. (*b*) Esterification.

have been investigated but either have not given efficient reaction or have not produced a bond that is adequately stable under alkaline washing treatment.

Application. Reactive dyes can be applied by either continuous or batch techniques. The most popular continuous methods are the pad, dry chemical pad, steam and scour, or pad thermofix and scour procedures (Fig. 3). Batch application involves exhausting the dye with salt and then introducing the acid acceptor to promote reaction with the fiber. The method of application depends on the type of reactive groups and on the material being dyed. Nitrogenous fibers generally react at neutral or slightly acid conditions, and hydroxy-containing fibers require an alkali to promote reaction. When the reactive group is stable in the presence of the acid binder at room temperature the process becomes simply pad, heat, and scour.

Use and potential. The major penetration of reactive dyes has been into the cellulosic fiber dye market, as they offer improved washfastness over direct dyes, improved brightness over vat dyes, and improved rubbing fastness and shade range over azoic dyes. This penetration will deepen as greater production makes them cheaper. Reactive dyes are one of the more important single types of cellulosic dye. To a lesser degree, they are used for polyamide fibers.

The simplicity of continuous application and the brightness of reactive dyes make them useful in printing as well as dyeing. Both prints and dyed

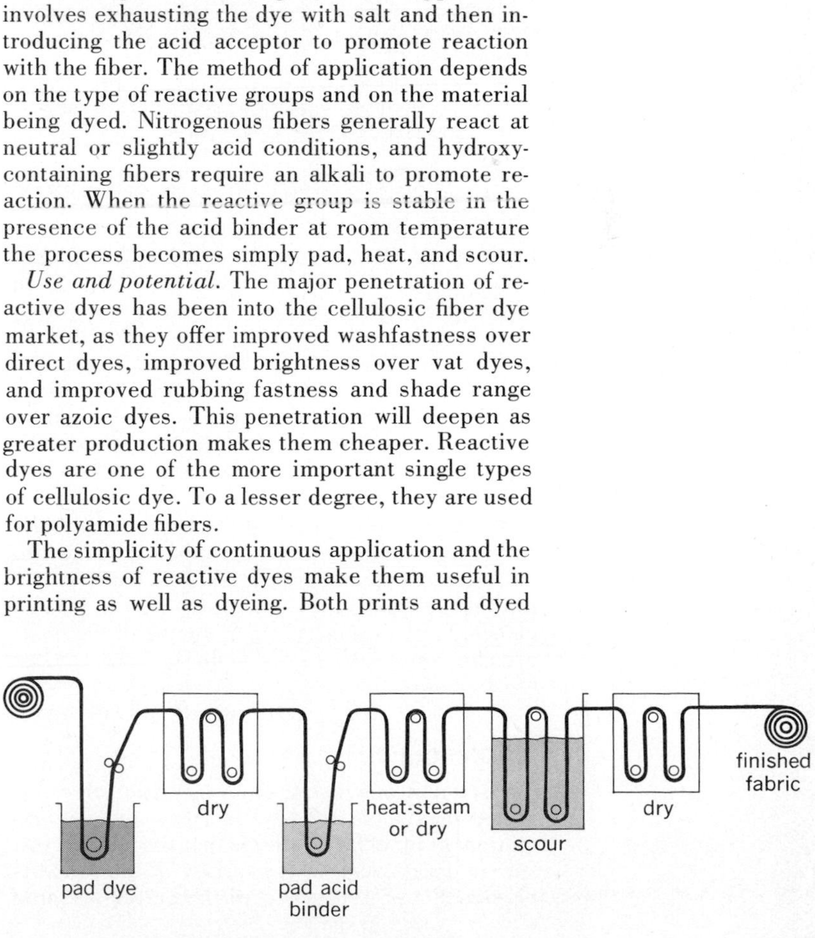

Fig. 3. Continuous application of reactive dyes.

fabrics may be resin treated without significantly affecting the dye properties when selected resins are used. A one-step process for the simultaneous dyeing and resin finishing of cellulosic textiles has been developed. *See* HETEROCYCLIC COMPOUNDS; QUINONE.

[DONALD R. BAER/DAVID H. ABRAHAMS]

Bibliography: M. Bogle, *Textile Dyes, Finishes, and Auxiliaries*, 1977; T. S. Gore et al. (eds.), *Recent Progress in the Chemistry of Natural and Synthetic Colouring Matters and Related Fields*, 1962; R. M. Johnston and M. Saltzman (eds.), *Industrial Color Technology*, 1977; H. A. Lubs (ed.), *The Chemistry of Synthetic Dyes and pigments*, 1955, reprint 1972; M. Simmons, *Dyes and Dyeing*, 1978; K. Venkataraman (ed.), *The Chemistry of Synthetic Dyes*, vols. 1–8, 1952–1978.

Dysprosium

A metallic rare-earth element, Dy, atomic number 66 and atomic weight 162.50. The naturally occurring element is composed of the following stable isotopes: Dy^{156} 0.052%, Dy^{158} 0.090%, Dy^{160} 2.294%, Dy^{161} 18.88%, Dy^{162} 25.53%, Dy^{163} 24.97%, Dy^{164} 28.18%. Dysprosium was discovered by L. de Boisbaudran in 1886. It forms a white oxide, Dy_2O_3, which dissolves in acid to give a yellowish-green solution. For properties of the metal *see* RARE-EARTH ELEMENTS.

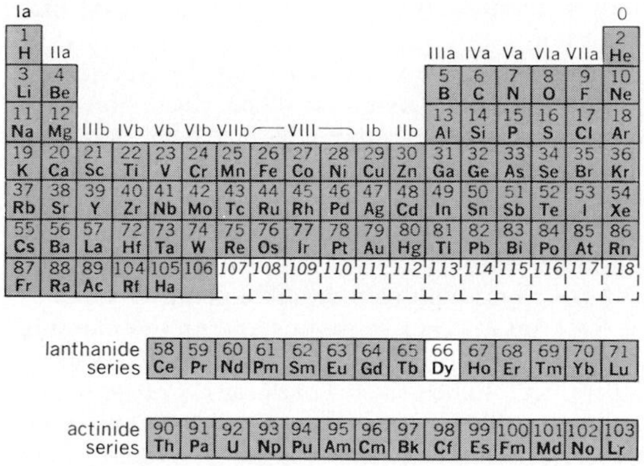

The metal is attacked readily by air at high temperatures, but at room temperatures, in massive blocks, it is fairly stable in the atmosphere and remains shiny for long periods of time. Dysprosium is paramagnetic, but as the temperature is lowered, it becomes antiferromagnetic at the Néel point (178K) and ferromagnetic at the Curie point (85K). At very low temperatures, the metal shows strong anisotropic magnetic properties. It is easy to saturate dysprosium in the direction of the hexagonal planes, but it is almost impossible, with the fields available in the laboratory, to do so at right angles to the plane.

[FRANK H. SPEDDING]

Efflorescence

The spontaneous loss of water (as vapor) from hydrated crystalline solids. The thermodynamic requirement for efflorescence is that the partial pressure of water vapor at the surface of the solid (its dissociation pressure) exceed the partial pressure of water vapor in the air.

A typical efflorescent substance is Glauber's salt, $Na_2SO_4 \cdot 10H_2O$. At 25°C the dissociation pressure for the process in reaction (1) is 19.4 mm Hg,

$$Na_2SO_4 \cdot 10H_2O\,(s) \rightarrow Na_2SO_4\,(s) + 10H_2O\,(g) \quad (1)$$

81% of the saturation vapor pressure of pure water at this temperature. In a sufficiently humid atmosphere Glauber's salt also can deliquesce by the process shown in reaction (2). The vapor pressure

$$Na_2SO_4 \cdot 10H_2O\,(s) + H_2O\,(g) \rightarrow Na_2SO_4 \text{ (saturated aqueous solution)} \quad (2)$$

of the saturated solution is 21.9 mm Hg, 92% of the vapor pressure of pure water. Thus, Glauber's salt at 25°C is stable in atmospheres having relative humidities of 81–92%; below 81% it effloresces; above 92% it deliquesces.

The spontaneous loss of water normally requires that the crystal structure be rearranged, and consequently, efflorescent salts usually go to microcrystalline powders when they lose their water of hydration. *See* DELIQUESCENCE; PHASE EQUILIBRIUM; VAPOR PRESSURE.

[ROBERT L. SCOTT]

Einsteinium

A chemical element, Es, atomic number 99, a member of the actinide series in the periodic table. It is not found in nature but is produced by artificial nuclear transmutation of lighter elements. All isotopes of einsteinium are radioactive, decaying with half-lives ranging from a few seconds to about 1 year. The world's supply in 1975, produced primarily as ^{253}Es, consisted of a few milligrams from nuclear reactors. *See* ACTINIDE ELEMENTS; RADIOACTIVITY.

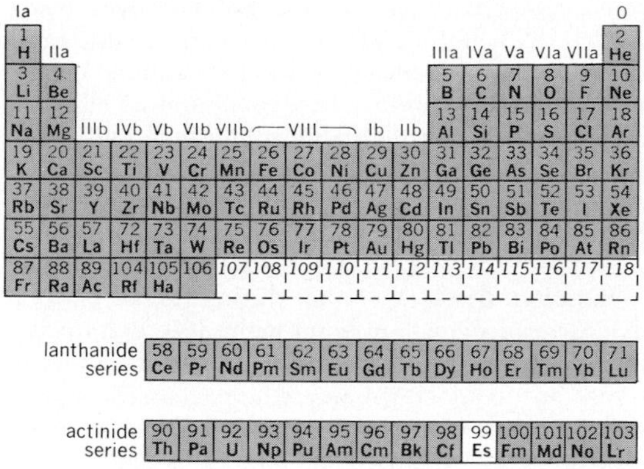

Einsteinium was discovered in 1952 in the debris of a hydrogen bomb by scientists of the Los Alamos, Argonne, and University of California laboratories of the U.S. Atomic Energy Commission. A very high flux of neutrons at the instant of explosion resulted in multiple neutron capture by uranium present in the bomb to give the mass-253 isotope of uranium. A subsequent chain of β-decays through mass-253 isotopes of intervening elements finally gave ^{253}Es. The isotope ^{253}Es was separated from other actinide elements by the ion-exchange method and was found to decay by α-emission with a half-life of 20 days. A number of other einsteinium isotopes, ranging in mass from 243 to 255 and in half-life from about 20 seconds to 400 days, have since been discovered by using the more conventional techniques of bombardment of heavy-element isotopes with deuterons, helium ions, or heavy ions, or by neutron bombardment in reactors. *See* CALIFORNIUM.

Tracer scale studies of the chemical properties of einsteinium first indicated its existence in the 3+ oxidation state characteristic of actinide elements. As other tripositive actinides do, einsteinium coprecipitates with lanthanum fluoride or hydroxide. The principal method used for the separation and identification of einsteinium in the presence of other actinides is ion-exchange chromatography. Actinides are desorbed from cation-exchange resin at different rates when a complexing agent is passed through the column of resin. As a result, the elements leave the column in distinct fractions and can be identified by their characteristic radioactivity.

The magnetic susceptibility of submicrogram amounts of einsteinium has been measured by using ^{253}Es (half-life, 20 days). The existence of the longer-lived ^{254}Es (half-life, 276 days), as larger amounts become available, will allow extension of chemical investigations. Einsteinium is the heaviest actinide element to be isolated in weighable form.

In 1967 B. B. Cunningham and coworkers measured seven optical absorption peaks from Es(III) in hydrochloric acid solution. The peaks ranged in wavelength from 3600 to 6100 A, and they explain the green color observed in solid compounds and concentrated solutions of einsteinium.

Einsteinium exists in normal aqueous solution essentially as Es^{3+} (green), although Es^{2+} can be produced under strong reducing conditions. Solid compounds such as Es_2O_3, $EsCl_3$, EsOCl, $EsBr_2$, $EsBr_3$, EsI_2, and EsI_3 have been made.

Einsteinium metal is chemically reactive, is quite volatile, and melts at 860°C; one crystal structure is known. *See* TRANSURANIUM ELEMENTS.

[STANLEY G. THOMPSON; GLENN T. SEABORG]

Bibliography: B. B. Cunningham, Chemistry of the actinide elements, *Ann. Rev. Nucl. Sci.*, 14: 323–346, 1964; *Gmelins Handbuch der anorganischen Chemie: Transurance*, vol. 4, 1972, vols. 7a and 8, 1973; E. K. Hyde, I. Perlman, and G. T. Seaborg, *The Nuclear Properties of the Heavy Elements*, vols. 1 and 2, 1964; J. J. Katz and G. T. Seaborg, *The Chemistry of the Actinide Elements*, 1958; C. Keller, *The Chemistry of the Transuranium Elements, Kernchemie in Einzeldarstellungen*, vol. 3, 1971; G. T. Seaborg, History of the synthetic actinide elements, *Actinides Rev.*, 1:3–38, 1967; G. T. Seaborg, *Man-Made Transuranium Elements*, 1963; G. T. Seaborg, *The Transuranium Elements*, 1958.

Electrochemical equivalent

The weight of a substance, according to Faraday's law, produced or consumed by electrolysis with 100% current efficiency during the flow of a quantity of electricity equal to 1 faraday or 96,487 coulombs (1 coulomb corresponds to a current of 1 amp during 1 sec). Electrochemical equivalents are essential in the calculation of the current efficiency of an electrode process.

The electrochemical equivalent of a substance is equal to the gram-atomic or gram-molecular weight of this substance divided by the number of electrons involved in the electrode reaction. For example, the electrochemical equivalent of zinc, for which two electrons are required in order to deposit one atom, is Zn/2 or 65.37/2 g. Thus, the faraday is equal to the product of the charge of the electron times the number of electrons (the Avogadro number) required to react with 1 atom- or molecule-equivalent of substance. The value of the faraday computed in this manner agrees with values obtained from electrochemical determinations. The relative error on the value 96,487 coulombs is smaller than ±0.01%. *See* COULOMETER; ELECTROLYSIS. [PAUL DELAHAY]

Electrochemical process

The principles of electrochemistry may be adapted for use in the preparation of commercially important quantities of certain substances, both inorganic and organic in nature.

INORGANIC PROCESSES

Inorganic chemical processes can be classified as electrolytic, electrothermic, and miscellaneous processes including electric discharge through gases and separation by electrical means. In electrolytic processes, chemical and electrical energy are interchanged. Current passed through an electrolytic cell causes chemical reactions at the electrodes. Voltaic cells convert chemicals into electricity. Electrothermic processes use electricity to attain the necessary temperature for reaction. For related information *see* ELECTROCHEMISTRY; ELECTROLYSIS; ELECTROLYTIC CONDUCTANCE; ELECTROMOTIVE FORCE (CELLS).

Voltaic cells are used for the intermittent production of small amounts of electricity. When the chemicals involved are exhausted and must be replaced, the unit is called a primary cell. A special case of the primary cell is the fuel cell in which the fuel and an oxidizer are fed continuously to the cell, converted to electricity, and the products removed. If exhausted components can be revived by passing electricity backward through the unit, it is called a secondary cell, storage battery, or

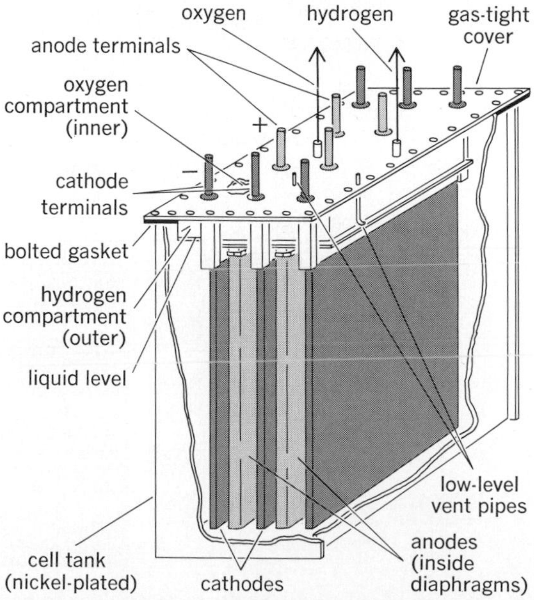

Fig. 1. Diagram of Stuart hydrogen-oxygen cell. (*Electrolyser Corp. Ltd., Toronto, Canada*)

accumulator. Cells may be connected in parallel or in series to form a battery.

Electrolysis in aqueous solutions. The electrolysis of water to form hydrogen and oxygen, according to the reaction $2H_2O \rightarrow 2H_2 + O_2$, may be considered as the simplest process for aqueous electrolytes. It does not compete with hydrogen from propane or from natural gas and with oxygen from liquid air, except in small installations. While simplicity, high hydrogen purity requirement, and lower capital cost (in small plants) have justified electrolytic plants, severely rising energy costs have limited such applications. The electrolyte is 18–30% NaOH or KOH, the latter having a lower resistance but higher cost. The cathode is steel and the anode is nickel-plated steel separated by a diaphragm, usually of asbestos. Figure 1 is a diagram of this type of cell. A cell voltage of 2.0–2.5 volts is the summation of the decomposition voltage of water (1.23 volts at room temperature) and the oxygen and hydrogen overvoltages, plus the *IR* drop through the electrolyte, electrode contacts, and bus bars. The raw material is distilled or demineralized water. There are monopolar cells operating up to 20,000 amperes, bipolar or filter-press cells using 2000–5000 amperes, and 150 cells in series at 300–400 volts overall. Cells are avaliable which operate at 600 psi (4.1 MPa). A. T. Kuh described 11 cells using 25–30% KOH and indicated designs operating at pressures up to 200 atm (20 MPa). *See* HYDROGEN; OXYGEN.

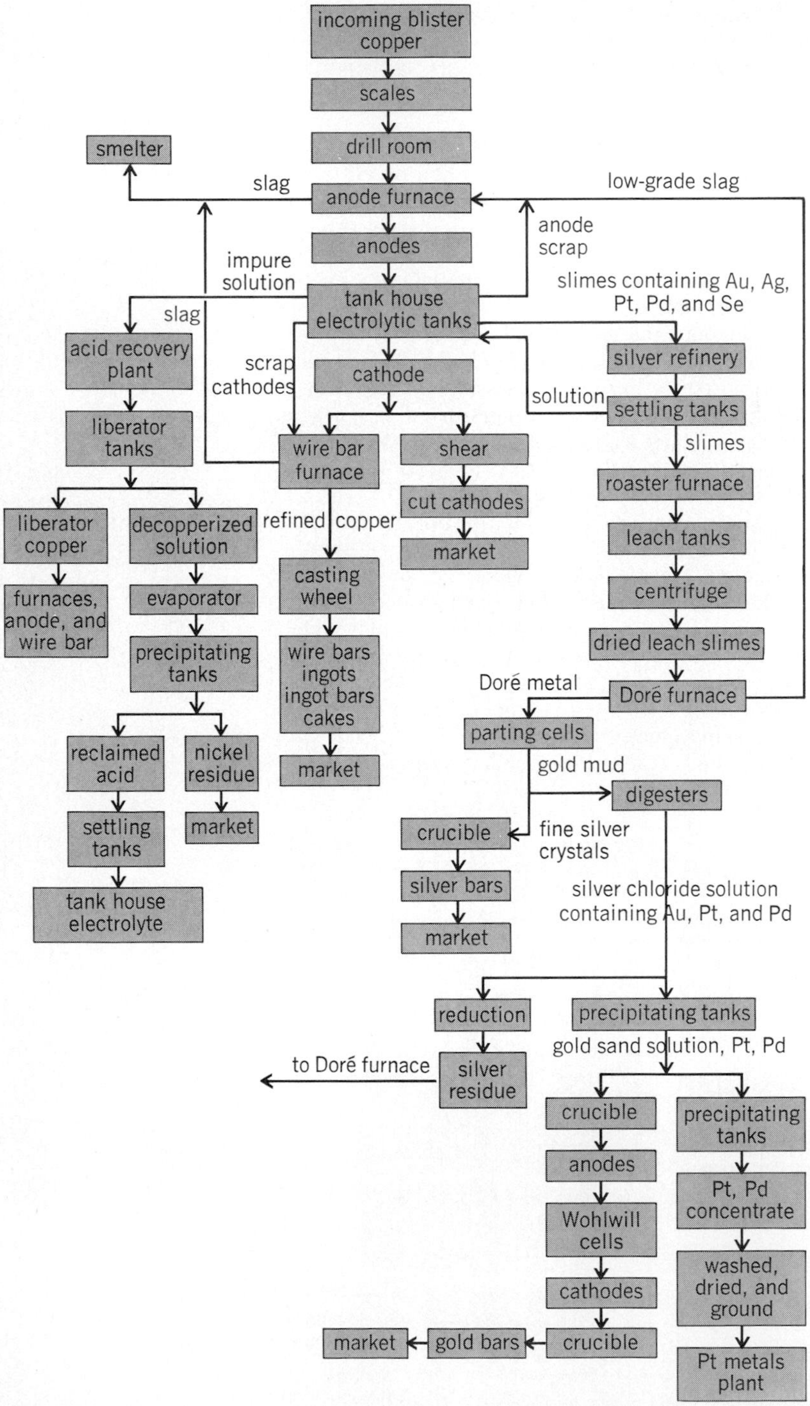

Fig. 2. Flowsheet for a copper plant and the refinery by-products resulting from a product of 99.97% conductor copper. (*Ontario Refining Co., International Nickel Co. of Canada, Ltd.*)

Heavy water, or deuterium oxide, used in moderating nuclear reactors is also a by-product of the electrolysis of water. Protium (H^1) is preferentially discharged, so that the electrolyte becomes richer in deuterium. Electrolysis must be combined with catalytic exchange or distillation processes when used in primary or earlier stages of concentration; it is also used in final stages to produce 99–100% concentration. *See* DEUTERIUM; HEAVY WATER.

Metallurgical applications. Protective or decorative coatings on a base metal such as steel are obtained by electroplating. Plating may also be used to replace worn metal or to provide a wear-resistant surface. The final surface may require several layers of different metals or even layers of the same metal deposited under varying conditions. The metals plated are: copper, cadmium, chromium, cobalt, gold, iron, lead, nickel, the platinum metals, silver, tin, and zinc, and many alloys. Electrogalvanizing is preferred over hot dipping for applying zinc to steel. Tin plate for containers is electrolytic. Factors affecting the resulting plate include pretreatment and cleaning of the metal surface, current density, concentration of metal ions, agitation, temperature, conductance of solution, pH, and addition agents.

Electroforming is a method of forming or reproducing articles by electrodeposition. In contrast to electroplating, the product is removed from the base surface or mold. A nonconducting surface can be made conductive by metallizing or with graphite powder. A low-melting-point metal mold can be used, or a mold can be plated with a metal from which the final metal can be removed. The electrodeposits may be up to 0.5 in. (13 mm) thick, and after completion are removed from the mold. Phonograph records, electrotype, textile replicas, and brass instrument bells are common products of electroforming, which is also widely used in the automotive and electronic industries.

Electrodeposition of metal powders is used to produce particles in the 1- to 1000-μm range for use in powder metallurgy and metallic pigments. Powdered iron made electrolytically makes stronger parts than other types, and electrolytic powdered copper makes parts that are easier to machine and have improved wear.

In the anodizing of aluminum articles, a coating

Table 1. Electrolytics for various individual metals

Metal	Source	Method	Electrolyte	Application
Antimony	Crude metal	Refining	Sulfates	Alloys, batteries, chemicals
Antimony	Antimony ores	Winning	Sulfides	
Bismuth	Lead refining slimes	Refining	Chlorides	Alloying agent, medicine
Cadmium	Zinc residues and slimes	Winning	Sulfates	Electroplating, alloys
Chromium	Chromium ores	Winning	Sulfates	High-temperature, alloys
Cobalt	Complex cobalt ores	Winning	Sulfates	Alloying agent, electronics
Cobalt	Complex nickel ores	Refining	Sulfates and chlorides	Alloying agent, electronics
Copper	Copper ores, complex ores	Winning	Sulfates	Electrical conductors, wire, brass alloys
Copper	Crude metal, secondary materials, waste	Refining	Sulfates	Electrical conductors, wire, brass alloys
Copper powder	Refined metal	Plating	Sulfates	Powders for powder metallurgy, oilless bearings
Copper sheet	Purified electrolyte	Plating	Sulfates	Sheet for printed circuits, electronics
Gallium	Sodium aluminate liquors	Winning	Caustic	Low-melting-point metals and alloys
Gold	Copper refining slimes	Refining	Chlorides	Jewelry, dentistry, plating
Indium	Residues, wastes, ores	Refining	Chlroides	Silver alloys, jewelry, television
Iron	Low-carbon steel	Refining	Sulfates	Powder metallurgy
Lead	Crude lead	Refining	Fluosilicates	Separation of bismuth, et al., alloys, fittings
Manganese	Manganese ores	Winning	Sulfates	Stainless steels, carbon-free metal, alloys
Nickel	Crude metal, nickel matte	Refining	Sulfate-chloride	Plating, alloys, stainless steel
Silver	Copper refinery slimes, lead residues, crude metal, ore	Refining	Nitrates	Jewelry, electrical applications, alloys
Silver	Silver alloys, photographic wastes	Winning	Nitrates	Jewelry, electrical applications, alloys
Solder	Waste and crudes	Refining	Fluosilicate	Metal joining, electronic components
Tin	Crude metal	Refining	Cresol-sulfonates	Tin plate, solder, alloys
Zinc	Zinc ores, complex ores	Winning	Sulfates	Die-casting alloys, battery cups, brass, galvanizing
Zinc	Ores and residues	Winning	Caustic	Chemicals, paint

approximately 0.001 in. (25 μm) thick is applied, which can be dyed or made impervious. The article to be anodized is cleaned and made anodic in sulfuric acid solution.

Electrolytic polishing of metals is accomplished by making the article anodic in an electrolyte of mixed acids, such as phosphoric-chromic acids or phosphoric-sulfuric acids. High points on the surface apparently dissolve, to give a unique polish not achievable by mechanical polishing.

Electrolytic machining of metals is accomplished by making the metal part anodic in a suitable electrolyte. Metal dissolves at the anode and hydrogen discharges at the cathode. The cathode is contoured as a negative of the shape to be developed in the workpiece. By circulating the elec-

Table 2. Anodes, cathodes, and diaphragms used on commercial processes

Metal	Method	Anode	Anolyte-Catholyte		Diaphragm	Cathode	Voltage
Antimony	Winning	Insoluble	Same		No	Steel	2.5–3
Cadmium	Winning	Insoluble	Same		No	Aluminum	4.0
Chromium	Winning	Insoluble	More acid	Acid	Yes	Hastelloy	4.2
Cobalt	Refining	Soluble	Same		No	Cobalt–stainless steel	2.5
Copper	Refining	Soluble	Same		No	Copper	0.2–0.3
Copper	Winning	Insoluble	Same		No	Copper	2–2.1
Gold	Refining	Soluble	Same		No	Gold	0.5–2.8
Lead	Refining	Soluble	Same		No	Lead	0.35–0.45
Manganese	Winning	Insoluble	Acid	Alkaline	Yes	Stainless	
Nickel	Refining	Soluble	Same	Pure	Yes	Nickel	2.4
Silver	Refining	Soluble	Same		Yes	Stainless steel or carbon	1.3–5.4
Tin	Refining	Soluble	Same		No	Tin	0.3
Zinc	Winning	Insoluble	Same		No	Aluminum	3.25–3.7

trolyte under very high pressure and using high current densities (30–2000 amperes/in.^{2} or 5–300 A/cm^2), practical machining rates are achieved. Very hard and very thin metal surfaces can be machined without changing the heat treatment.

Electrorefining is a process for purifying metals and recovering their impurities, which at times are more valuable than the original metal. Copper from its ore or scrap is purified of volatilizable impurities, cast into an anode, dissolved in an electrolyte in a cell, and deposited at a cathode as a very pure, highly conductive metal for electrical engineering purposes (Fig. 2). Gold, silver, platinum, selenium, and tellurium are recovered as by-products. Nickel is freed of copper by using a diaphragm between anode and cathode, with cobalt and the platinum metals as by-products. Lead is freed from bismuth; tin from lead, antimony, and bismuth with silver as a by-product; and zinc from copper and lead, with cadmium as a by-product, to make commercial a 99.99% pure metal for die casting. The last operation is usually re-

Table 3. Energy consumption of electrochemical products

Industry	kWhr/lb	lb/kWhr	Voltage/tank, cell, or furnace	Range, amperes/unit	Line voltage	Range, kVA/cell or furnace	kVA/line
Electrolytic refining							
Copper: multiple system,	0.09–0.2	5–13.3	0.2–0.3	6000–15000	80–200	1.2–4.5	480–3,000
series system	0.074	13.5	120–200				
Gold (troy lb)	0.15	6.6	0.5–2.8	150–500	3–10	0.1–1.4	0.5–5
Lead	0.07–0.09	11.1–14.3	0.35–0.5	5000–6000	100–185	1.75–3.0	500–1200
Nickel	1.1	0.9	2.4–2.6	5000–6000	220–230	12–16	1300–1500
Silver, Moebius (troy lb)	0.27–0.3	3.3–3.7	2.3–2.8	400–500	45–250	1–1.4	18–125
Silver, Thum (troy lb)	0.33–0.6	1.67–3.0	1.3–5.4	150–200	200–220	0.3–1.2	30–50
Solder	0.08	12.5	0.34				
Tin	0.085	11.8	0.3–0.35	5000	100–200	1.5–2.0	500–1000
Electrowinning							
Antimony			2.5–3	1500	300	4	450
Cadmium	0.65–0.97	1.03–1.54	2.5–3.1				
Chromium	5.0–8.4	0.12–0.2	4.2–4.3	10,000	250–300	40–45	2500–3000
Cobalt	1.2–1.56	0.64–0.83	3.7–4.5				
Copper	0.89–1.34	0.77–1.12	2.0–2.12	10,000–25,000	150–200	20–30	1500–5000
Manganese	4–4.5 dc	0.22–0.25	5–5.4	6000	600	30–40	3600
	5–5.3 dc	0.19–0.2					
Silver	0.58–1.0	1–1.73	1–1.5	300		0.3–0.5	0.3–0.5
Zinc	1.4–1.6	0.61–0.71	3–3.7	5000–10,000	600–800	15–40	3000–8000
Metal melting							
Copper	0.12–0.15	6.67–8.33	85–225	12,000–30,000		6000	
Copper alloys	0.15–0.3	3.33–6.67					
Steel: cold charge,	0.237–0.35	2.86–4.22	80–250	30,000–100,000		25,000–33,000	
hot charge	0.05–0.2	5–20	80–250	30,000–100,000		25,000–33,000	
Zinc	0.045	22.2					
Metal powders							
Copper	0.3–0.5	2–3.3	0.3–0.5	10,000–15,000			
Iron	1.20	0.83	2.5				
Nickel	1.13	0.83	1.4–1.5				
Zinc	1.37	0.73	3.4				

*1 kWhr/lb = 7.937 MJ/kg; 1 lb/kWhr = 0.1260 kg/MJ.

ferred to as electrowinning. This has rapidly replaced pyrometallurgy or fire processing because it can eliminate sulfur dioxide atmospheric pollution and contamination by particulates.

Electrowinning, sometimes termed aqueous electrometallurgy, involves processing of metallic ores, usually of very low metal content but large in volume, by leaching solutions, usually sulfates, to obtain metal-containing electrolytes which can be processed with insoluble anodes and metal cathodes. Examples are found in cadmium, copper, cobalt, manganese, and zinc.

Electrorefining, electrowinning, and electroforming are summarized as to the individual metals in Table 1, while Table 2 gives anodes, cathodes, and diaphragms for commercially successful operation. Table 3 lists energy consumption of aqueous electrochemical operations.

Electrolytic corrosion of metals. This occurs because some parts of the surface of metals act as anodes and corrode, whereas other parts act as cathodes and do not corrode.

Cathodic protection is provided if the whole surface is made cathodic to a separate anode and sufficient voltage is available between the two electrodes. This type of protection is used to inhibit corrosion of boilers, condensers, underground pipelines, ships, and water tanks. Sacrificial anodes of zinc, magnesium, or aluminum alloy may provide the potential, or inert anodes such as graphite, stainless steel, or platinum-plated titanium may be used with power supplied from a rectifier.

Anodic protection can be used to create a passive layer on the surface of some metals, such as steel and stainless steel in some environments. It is a practical method of controlling corrosion of tanks in the chemical industry, but is not feasible for copper or brass vessels. The tank is made anodic. An inert cathode, such as platinum-clad metal, is installed in the liquid in the tank. Current is applied so as to maintain a predetermined voltage between the anodic surface and a reference electrode, such as silver–silver chloride, in the liquid. Equipment to maintain precise potential control at high current output has made anodic protection practical. For example, a current of 0.0015 ampere/ft^2 (0.016 A/m^2) at +0.900 volt to a silver–silver chloride electrode will maintain passivity of a carbon-steel tank holding 93% sulfuric acid at 80°F (27°C).

Alkali-chlorine processes. Electrolysis of alkali halides is the basis of the alkali-chlorine and chlorate industries. Chlorine, Cl_2, and caustic soda, NaOH (or caustic potash, KOH), are made by electrolysis of brine, a solution of sodium chloride, NaCl, in water. This is represented by reaction (1).

$$2NaCl + 2H_2O \xrightarrow{\text{Electricity}} 2NaOH + Cl_2 + H_2 \quad (1)$$

Two processes are used to prevent the products from mixing: the diaphragm cell and the mercury cell. In the diaphragm cell process (Fig. 3) an asbestos diaphragm is interposed between a graphite anode and an iron screen cathode. Saturated purified brine fed around the anode passes through the diaphragm to the cathode. Chlorine is formed at the anode. Hydrogen is released at the cathode, leaving NaOH as a 10–15% solution and 10–15% residual NaCl in the cell liquor. By evaporation to 50% NaOH, the NaCl crystallizes out and is recycled. The decomposition voltage of brine to form chlorine and hydrogen is 2.3 volts. At 0.75 ampere/in.^{2} (11.6 A/dm^2) the average voltage components of a diaphragm cell are as follows:

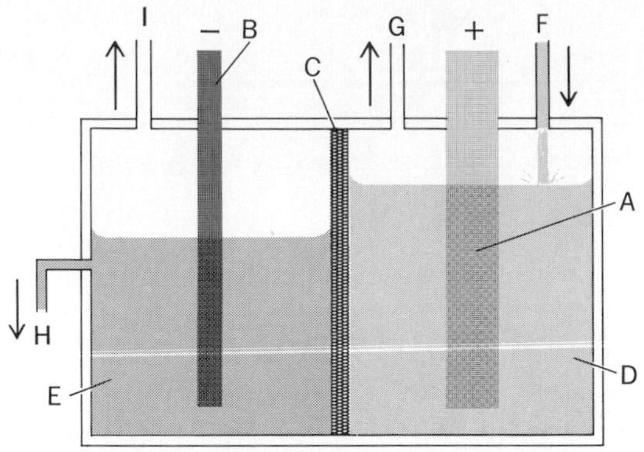

Fig. 3. Diagram of diaphragm cell for chlorine and caustic soda. A = graphite anode, B = iron screen cathode, C = asbestos diaphragm, D = anode compartment for brine and chlorine, E = cathode compartment for NaOH-NaCl cell liquor, F = brine inlet, G = chlorine outlet, H = cell liquor (caustic soda) outlet, I = hydrogen outlet.

Anode potential	1.50 volts
Cathode potential	1.25
Anolyte	0.47
Diaphragm	0.30
Conductors	0.18
Total	3.70 volts

The current efficiency is 95.5–96.5%, due to some oxygen discharge at the anode and some chlorine being carried through the diaphragm. An installation is pictured in Fig. 4. Dow, Diamond, and Hooker are popular designs of diaphragm cells in use in the United States.

In the mercury cell process brine is electrolyzed between graphite anodes and a flowing mercury cathode, forming a dilute (0.2–0.4%) sodium amal-

Fig. 4. Photograph of an installation of Hooker S-4 diaphragm alkali-chorine cells. (*Hooker Chemical Corp.*)

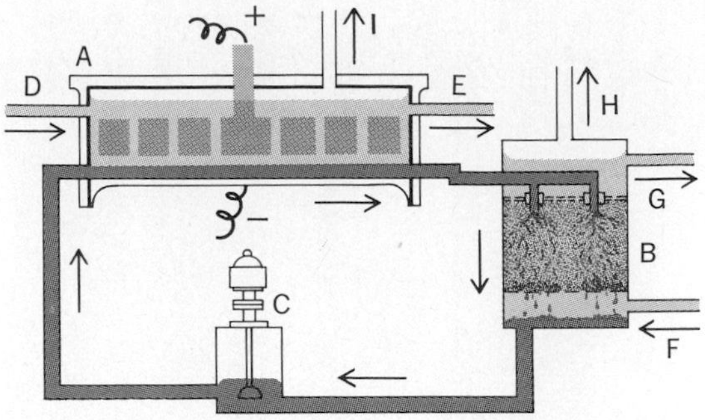

Fig. 5. Diagram of a mercury alkali-chlorine cell. A= electrolyzer, B= decomposer with graphite packing, C= mercury pump, D= feed brine, E= spent brine, F= water, G= 50% caustic soda, H= hydrogen, I= chlorine.

gam which is decomposed in another compartment by water in contact with graphite surfaces to form H_2 and NaOH. Figure 5 is a diagram of the mercury cell. The products of the mercury cell are purer than those of the diaphragm cell. To offset the cost of mercury, a much higher current density is used in mercury cells. A longitudinal section of a mercury cell is seen in Fig. 6, and a photograph of another is shown in Fig. 7. Typical components of voltage in a mercury cell at 5.12 amp/in.^{2} (80 amp/dm^2) are as follows:

Anode potential, reversible	1.34 volts
Cathode potential, reversible	1.76
Decomposition voltage	3.10 volts
Anode polarization	0.35 volt
Cathode polarization	0.06
Electrolyte	0.60
Conductors and contacts	0.29
Total cell voltage	4.40 volts

Economic factors dictate the use of higher current densities, equal to or exceeding 6.5 amperes/in.^{2} (100 A/dm^2). Cell voltage at these higher current densities can be calculated for good cell designs on the market from the equation: $V = 3.20 + 0.015C$, where C is the cathode current density in amperes per square decimeter. The current efficiency is approximately 95%. Inefficiency reactions are demonstrated in reactions (2) through (6).

$$2OH^- \longrightarrow \tfrac{1}{2}O_2 + H_2O + 2e \quad \text{(at anode)} \quad (2)$$

$$H^+ + e \longrightarrow \tfrac{1}{2}H_2 \quad \text{(at cathode)} \quad (3)$$

$$Cl_2 + 2Na(Hg) \longrightarrow 2NaCl + (Hg) \quad \text{(at cathode)} \quad (4)$$

$$Cl_2 + H_2O \longrightarrow Cl^- + ClO^- + 2H^+ \quad (5)$$

$$3ClO^- \longrightarrow ClO_3^- + 2Cl^- \quad (6)$$

Adverse conditions can increase these inefficiencies. *See* CHLORINE.

Two important factors have developed in chlorine technology: (1) The Nafion diaphragm of the synthetic resin type is replacing the deposited asbestos diaphragm. (2) The dimensionally stable anode (DSA) is replacing the graphite anode. The DSA is a titanium substrate with a platinum-group coating of metals and oxides. Its use eliminates the continuously necessary voltage increase or anode-cathode spacing adjustment needed because of graphite-anode wear.

In addition, there has been a swing away from mercury cells because of widespread publicity of so-called mercury poisoning by mercury discharges. These discharges have been reduced by better hydrogen cooling and recycle of metal dross, cutting mercury losses as well as permitting better "housekeeping."

The Japanese government pressured firms to eliminate mercury cells by March 1978 and to replace them by diaphragm units. Mercury-cell discharges caused deaths before 1975, resulting from the release of discharges into coastal waters and the ingestion of fish containing mercury. The chemical companies involved paid heavy fines and damages.

In the United States the trend away from Mercury cells has been evident since 1970, and all added capacity has been through diaphragm cells with dimensionally stable metallic anodes.

Sodium hypochlorite is formed when the products of the electrolysis of brine are mixed. Electrolytic cells have been built for this purpose, but have limited or special use, such as for sterilization of swimming pools and algae control in power plant condensers. Sodium hypochlorite is usually made chemically.

Sodium chlorate is made in cells with graphite or lead peroxide anodes and steel cathodes. When mixing is encouraged, changes take place accord-

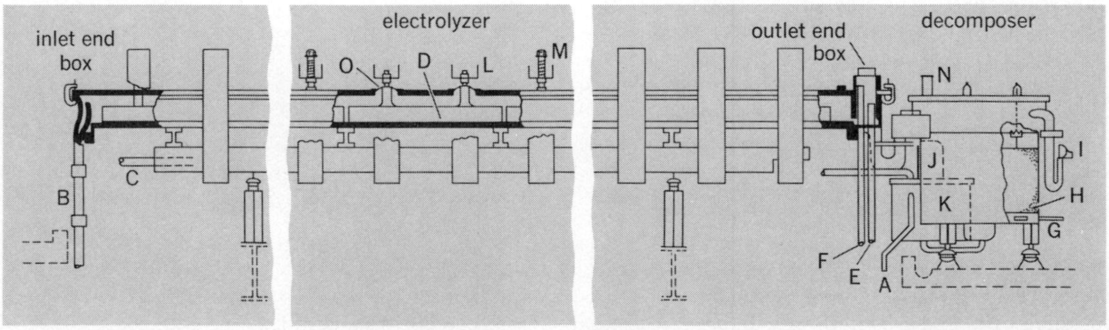

Fig. 6. Longitudinal section of Olin Mathieson E-11 mercury cell. A= dilute caustic outlet, B= brine inlet, C= mercury return, D= anode, E= brine-chlorine outlet, F= outlet end box vent, G= water inlet, H= graphite packing, I= caustic outlet, J= mercury pump, K= mercury pump sump, L= anode support bus, M= lifting screws, N= hydrogen outlet, O= anode seal. (*Olin Mathieson Chemical Corp.*)

ing to reactions (7) through (9). The overall reaction is labeled reaction (10). The temperature is kept below 40°C in cells using graphite anodes to prevent excessive attack. The optimum efficiency is at pH 6.8; hydrochloric acid is added as required. Sodium dichromate prevents reduction of chlorate at the steel cathode. The conversion of hypochlorite to chlorate is a somewhat slow chemical reaction, occurring partly in the cells and partly in a rundown tank. Salt is added and electrolysis is continued until the sodium chloride is down to about 100 g/liter and the chlorate has reached the desired concentration. It is then recovered by crystallization. Cells operate at 3–3.5 volts, 1–3 A/dm², and 80–85% current efficiency. Energy consumption is about 2.5 kWhr/lb (20 MJ/kg).

$$6NaCl + 6H_2O \xrightarrow{\text{Electricity}} 6NaOH + 3Cl_2 + 3H_2 \qquad (7)$$

$$6NaOH + 3Cl_2 \longrightarrow 3NaClO + 3NaCl + 3H_2O \qquad (8)$$

$$3NaClO \longrightarrow NaClO_3 + 2NaCl \qquad (9)$$

$$NaCl + 3H_2O \xrightarrow{\text{Electricity}} 3H_2 + NaClO_3 \qquad (10)$$

Hydrochloric acid electrolysis is of interest for recovery of chlorine from HCl resulting as a by-product from organic chlorinations. A filter-press electrolyzer is used which has 30–50 unit cells with polyvinyl-cloth diaphragms and graphite electrodes. Hydrochloric acid of 30–33% concentration is fed to the anode compartment. Weak acid is withdrawn at about 20% and reconcentrated by absorbtion of HCl gas. The graphite anode is not attacked as long as the concentration is kept at 20% HCl or higher. The current efficiency is 92–96%, the loss being due to electrical leakage. Energy consumption is 1800 kWhr/2000 lb (7.1 MJ/kg) chlorine (direct current). The voltage balance of a unit cell is as follows:

Anode potential	1.02 volts
Cathode potential	0.28
Anode polarization	0.2
Cathode polarization	0.5
Electrolyte, diaphragm	0.3
Total	2.30 volts

Oxidations and reductions. These reactions occur in all cells, but in a narrower sense oxidation reactions are those in which oxygen or chlorine at the anode oxidizes some material to form a new compound; reduction reactions are those in which hydrogen, liberated at the cathode, reduces a material to a new product. There are no commercial applications of inorganic electrochemical reductions by this narrow definition.

Sodium perchlorate is made by oxidation of a solution containing $NaClO_3$ at pH 6.1–6.4 by use of a platinum anode and an iron cathode, with chromate in the electrolyte. A lead dioxide anode may be used with a stainless-steel or nickel cathode, with no chromate in the electrolyte. Energy consumption is 1.4–1.6 kWhr/lb (11–13 MJ/kg) $NaClO_4$ (direct current). Other perchlorates are made by metathesis with $NaClO_4$.

Persulfuric acid, $H_2S_2O_8$, is made by oxidizing sulfuric acid as an intermediate in the production of hydrogen peroxide. The reactions for the process are shown in reactions (11) and (12). Alkali

Fig. 7. Olin Mathieson chlor-alkali E-812 mercury cell; 300,000-ampere capacity.

$$2HSO_4^- \xrightarrow{\text{Electricity}} H_2S_2O_8 + 2e \qquad (11)$$

$$H_2S_2O_8 + 2H_2O \longrightarrow 2H_2SO_4 + H_2O_2 \qquad (12)$$

persulfates can be made in the same way. The cell has smooth platinum anodes, a porous stoneware diaphragm, and a lead cathode cooled to 30°C. The reversible potential is 2.18 volts and the operating voltage 5.0–5.5 volts. The energy released as heat in the cell must be removed by cooling with stoneware or glass coils in the cell. Hydrogen peroxide is recovered by distillation. *See* PEROXIDE.

Lead dioxide anodes are used in sodium chlorate, sodium perchlorate, sodium bromate, and periodate or periodic acid regeneration cells. The material is dense and is made from a lead nitrate solution. For some uses it is produced on a steel base, from which it is removed mechanically and chemically. It is also applied to tantalum, platinum-clad tantalum, or graphite.

Periodic acid is used in producing dialdehyde starch, and the spent solution from the oxidation can be regenerated in a cell by use of a lead dioxide anode, a porous ceramic diaphragm, and an iron cathode.

Electrolytic manganese dioxide for batteries is made by electrolyzing hot $MnSO_4$ solutions at pH 6.5–7.5 by use of graphite electrodes. The MnO_2 deposited on the anode is pulverized with the graphite and separated mechanically. Energy consumption is 1 kWhr/lb (8 MJ/kg) MnO_2. The quality of the MnO_2 for battery use depends on cell temperature and anode current density.

Ion-permeable membrane cells. These utilize diaphragms made of ion-exchange resins. Cation-permeable membranes permit cations to pass through but not anions, whereas the reverse holds for anion-permeable membranes. A diagram of the

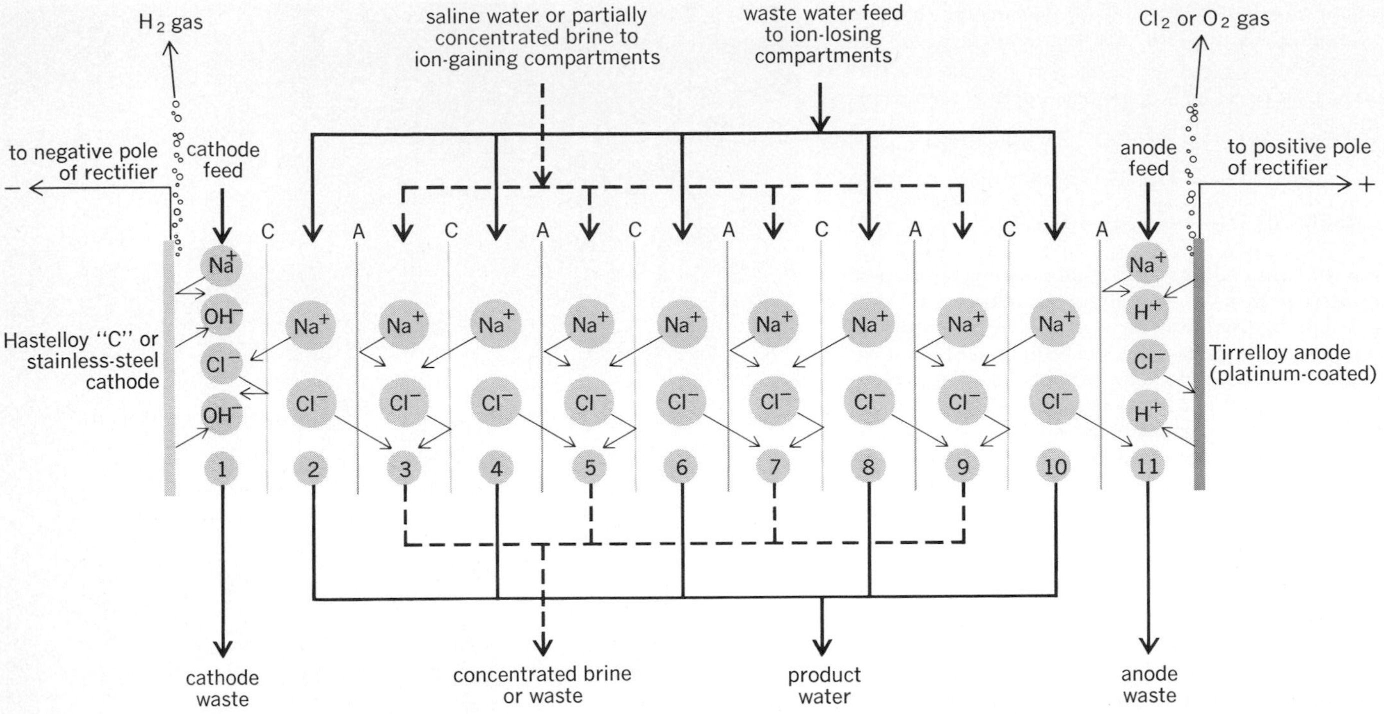

Fig. 8. Diagram of the basic ion and water flow in electric membrane stack. C = cation membrane; A = anion membrane; Na^+ = any cation, such as sodium; Cl^- = any anion, such as chloride. Numbers = compartments.

movement of ion and water in an electric membrane stack is shown in Fig. 8. Purification of sea water is the most important application. Salt has been recovered from sea water which has been concentrated in this way. *See* ION-SELECTIVE MEMBRANES AND ELECTRODES.

Fused-salt electrolysis. Aluminum, barium, beryllium, cerium and misch metal, fluorine, lithium, magnesium, sodium, molybdenum, thorium, titanium, uranium, and zirconium are obtained by electrolysis of fused salts, because water interferes with the desired reaction. Raw materials must all be purified before addition to fused-salt economical as in aqueous electrolytes. Metallizing is a process of depositing a metal as an alloy on a substrate from a fused complex metal salt.

Aluminum is produced in steel pots, lined with carbon, graphite, or silicon carbide, containing an electrolyte of alumina dissolved in fused cryolite, $AlF_3 \cdot 3NaF$, at 950–1000°C. The pool of aluminum in the bottom of the pot is the cathode. Contact is made by iron bars buried in the carbon lining or by titanium diboride contacts. Anodes are prebaked carbon blocks or the Söderberg type made from carbon paste baked in place. The arrangement of the apparatus for aluminum production is shown in Fig. 9. Aluminum is siphoned out periodically. Oxygen released at the carbon anode forms carbon monoxide. *See* ALUMINUM.

Barium, used in the electronics industry and in alloys, is obtained by aluminum reduction at low pressures and high temperatures.

Beryllium is made by batch electrolysis of fused salt, starting with 25% $BeCl_2$ and 75% NaCl in a chrome-iron pot, which acts as cathode to a graphite anode. Beryllium is deposited on the wall of the pot and is cleaned out and broken up when cold. Salt is washed out with water. The metal is in the form of bright crystalline flakes. A beryllium-copper eutectic can be made by using a copper cathode. Beryllium alloys containing copper and nickel are made from BeO in arc furnaces.

Calcium has been made by electrolysis of pure fused calcium chloride at about 800°C. In this case the cathode is solid calcium, and is mechanically withdrawn from the cell as a "carrot." Calcium is now made by aluminum reduction at low pressures and high temperatures.

Cerium and misch metal are made from $CeCl_3$ or mixtures of chlorides of cerium, lanthanum, and neodymium in fused-salt electrolysis with NaCl. Misch metal is used for lighter flints. *See* RARE-EARTH ELEMENTS.

Fluorine for separation of uranium isotopes is produced by electrolysis of 40% HF in KF between carbon anodes and steel cathodes at 88–100°C. A diaphragm of Monel screen keeps the products H_2 and F_2 separated. Dry HF gas is bubbled continuously into the electrolyte. At a current density of 1 ampere/in.^{2} (15 A/dm^2) the cell operates at 9–12 volts and 96% current efficiency. Energy consumption is 3.0 kWhr/lb (24 MJ/kg) fluorine. The theoretical decomposition potential is 2.85 volts.

Lithium is made by electrolysis of fused 60% LiCl and 40% KCl at 450–500°C in a cell similar to the Downs sodium cell.

Magnesium is produced by electrolysis of fused 25% $MgCl_2$ and 75% NaCl at around 700°C. The Dow process for making magnesium from sea water uses material approximating $MgCl_2 \cdot 2H_2O$, which is fed around the graphite anodes, where dehydration occurs. A diagram of the cell is shown in Fig. 10. Gas from the anode compartment is wet chlorine, air, and hydrogen chloride; the latter is used to make fresh magnesium chloride from magnesium hydroxide. Magnesium metal is deposited on steel cathodes, which direct the metal to a collecting zone. The cell is a cast-steel pot in a furnace setting. Other cells use molten anhydrous magnesium chloride feed. They have brick-lined steel bodies with graphite anodes. A diagram of this cell type is illustrated in Fig. 11. Magnesium chloride, with other chlorides, is separated in vacuum crystallizers from brines, dehydrated, and melted in electric resistance cells from which molten $MgCl_2$ is tapped periodically to feed the cells. Molten magnesium is ladled from the cells and cast into molds. The cells operate at 6–7 volts and 80–88% current efficiency and use 8–8.5 kWhr/lb (63–67 MJ/kg) metal.

Sodium was once made by electrolysis of fused NaOH, but since 1929 it has been made by electrolysis of NaCl in the Downs cell. The electrolyte is 40% NaCl and 60% $CaCl_2$ at 590°C. The cell consists of a brick-lined steel vessel. Four graphite anodes project upward from the bottom. The cathode is made of steel cylinders concentric with the anodes and supported from iron arms extending through the sides of the cell, which also conduct current. A diaphragm of 26-mesh-per-inch (10 mesh-per-centimeter) iron screen directs the sodium into an inverted trough leading to a riser pipe, which cools the metal and conducts it to a collecting tank beside the cell. Chlorine is collected in a nickel cone inverted over the anode. Pure dry salt is fed to the cell. The reactions in the cell are Eqs. (13) and (14). The metal at cell temperature is

$$2NaCl + 2e \longrightarrow 2Na + Cl_2 \qquad (13)$$

$$2Na + CaCl_2 \rightleftharpoons Ca + 2NaCl \qquad (14)$$

5% Ca, but as it cools, the second equation is reversed so that the cool metal is about 1% Ca. It is filtered just above the melting point of the sodium, and the final product is under 0.04% Ca. Cells of 38,000 amp operate at 7 volts and 83% current efficiency. Energy consumption is about 4 kWhr/lb (32 MJ/kg) metal.

Molybdenum, thorium, titanium, uranium, and zirconium can all be made by electrolysis of their complex halides, K_3MoCl_6, $ThF_4{\cdot}KF$, K_2TiF_6,

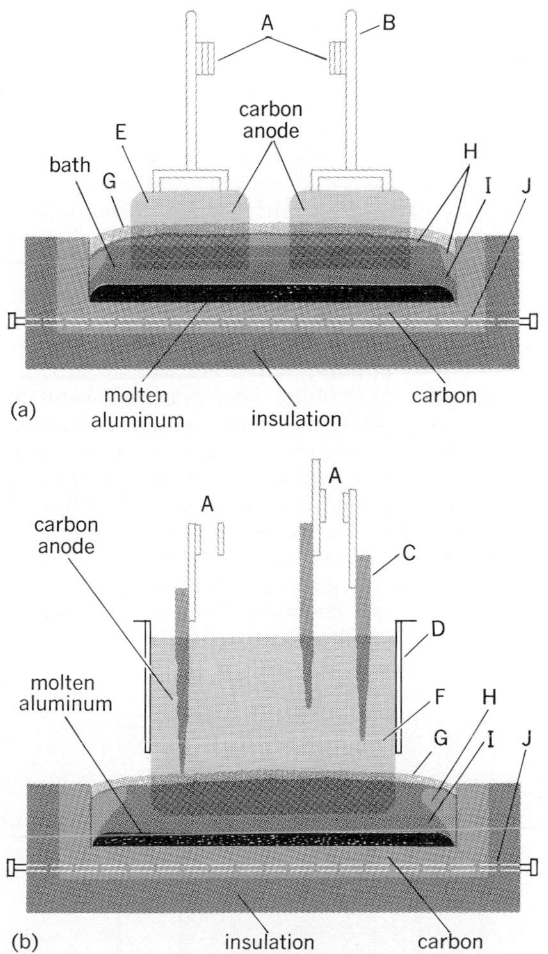

Fig. 9. Two types of aluminum cells, (*a*) utilizing prebaked carbon anode and (*b*) utilizing Söderberg carbon anode. A = anode bus, B = anode rod, C = anode stub, D = anode casing, E = prebaked carbon anode, F = Söderberg carbon anode baked in place, G = crust of frozen bath and alumina, H = frozen bath, I = bath (electrolyte), J = steel pin cathode collector.

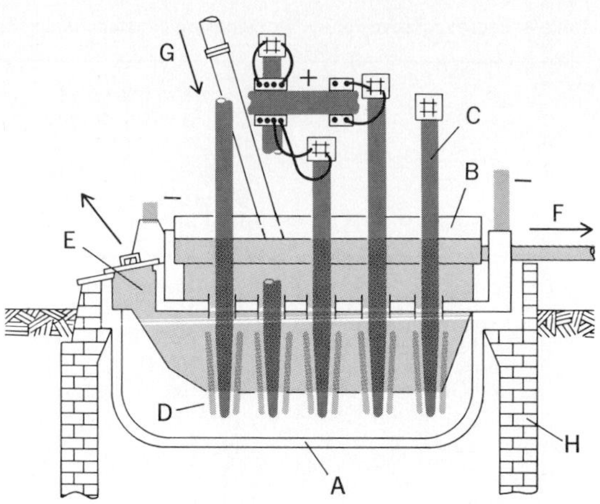

Fig. 10. Cross-sectional diagram of the Dow Chemical Company magnesium cell. A = steel container, B = ceramic cover, C = graphite anodes, D = steel cathodes, E = magnesium collecting well, F = chlorine outlet, G = magnesium chloride feed, H = furnace setting.

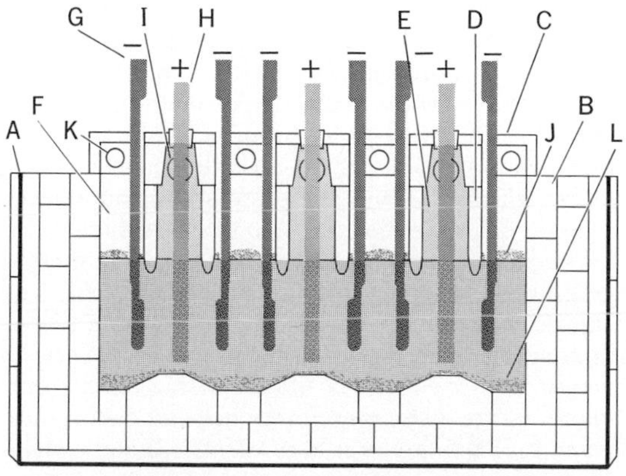

Fig. 11. Cross-sectional diagram of anhydrous magnesium chloride cell. A = steel box, B = ceramic lining, C = ceramic cover, D = ceramic separators, E = anode (chlorine) compartment, F = cathode (metal) compartment, G = iron cathode, H = graphite anode, I = chlorine outlet, J = magnesium, K = cathode compartment vent, L = mud.

Table 4. Representative values for electrode consumption, voltage, current, and current density for electric furnace products

Product	Electrode consumption: Approx. lb/ton of product*	Electrode consumption: Approx. lb/1000 kWhr†	Voltage range, phase to phase	Current range, secondary amperes	Current density, amperes/in.^{2}
Ferrosilicon, 50%	30–40	7	130–195	35,000–55,000	35–55
Ferrosilicon, 65%	60–90	10	125–170	35,000–50,000	35–60
Ferrosilicon, 75%	90–130	12	120–170	35,000–50,000	35–60
High-carbon ferrochrome	30–50	10	110–200	25,000–40,000	25–40
Silicomanganese	80–100	23	110–140	35,000–70,000	35–70
Ferromanganese	30–50	16	85–100	30,000–60,000	35–65
Calcium carbide	40–60	17	120–200	30,000–80,000	40–70
Phosphorus	35–70	3	250–400	13,000–30,000	20–30
Refined copper	4–6	19	80–190	12,000–30,000	80–110
Nonferrous castings	4–5	15	100–110	800–5000	110–300

*1 lb/ton = 0.5 kg/metric ton. †1 lb/kWhr = 0.1260 kg/MJ.

ELECTROCHEMICAL PROCESS

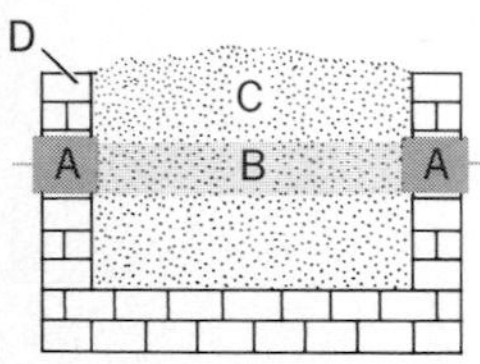

Fig. 12. Diagram of resistance furnace. A = current conductors, B = conducting core of granulated carbon, C = granular charge, D = furnace wall of brick and refractory.

ELECTROCHEMICAL PROCESS

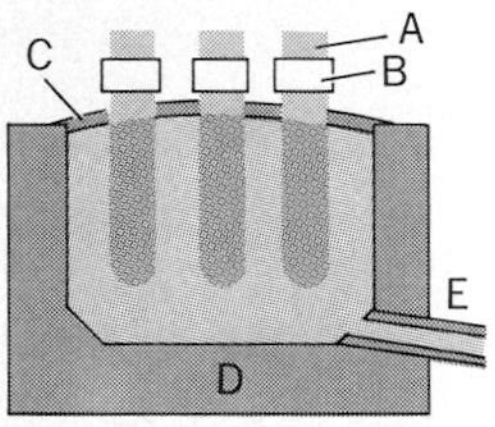

Fig. 13. Diagram of three-phase arc furnace. A = carbon or graphite electrodes, B = current clamps, C = refractory roof, D = refractory shell, E = tap hole.

KUF_5, and K_2ZrF_6, in molten NaCl. Inert atmospheres are required, and the cell is usually a graphite crucible acting as anode with a graphite or molybdenum cathode, on which the metal deposits as crystals or powder. The batch of metal on the cathode is cooled in an inert atmosphere, then broken off, pulverized, and leached with water. The metal powder is used as such or melted in a vacuum arc furnace.

In the case of tantalum, pure K_2TaF_7 is heated to 900°C in a graphite pot acting as anode with a removable metal cathode. When the cathode is loaded with deposited metal, it is removed and quickly replaced. The bath is replenished with K_2TaF_7. The cathode deposit is pulverized and washed with acid. The metal powder can be compacted by sintering.

Electrothermics. The manufacture of many products requires temperatures higher than can be obtained by combustion methods. Electric heat can usually be developed at, or close to, the point where it is required, so that it is relatively quick. It permits easy control of the atmosphere for oxidizing, reducing, or neutral conditions.

Products of the electric furnace include iron and steel; ferralloys; nonferrous metals and alloys; the exotic metals titanium, zirconium, hafnium, thorium, and uranium; and nonmetallic products such as calcium carbide, calcium cyanamide, sodium cyanide, silicon carbide, boron carbide, graphite, fused alumina, magnesium oxide, quartz, silica, thoria, zirconia, lime, spinel, kyanite, sodium aluminate, dolomite, boric acid and borides, carbides of zirconium and titanium and related metals, graphite, phosphorus and phosphoric acid, and chlorides of magnesium, boron, zirconium, and titanium. An electric smelting process converts ilmenite into iron and a titanium slag for pigment manufacture. Zinc metallurgy uses an electric furnace. Steam is generated in electric boilers where economically feasible.

Increasing amounts of United States steels and all of its stainless steels and special alloys are made in electric furnaces (17% in 1970, 25% in 1980, with further increases expected in the future). Furnace electrodes are increasing in size to handle the increasing furnace amperages. Table 4 gives values for electrode consumption, furnace voltage, secondary amperage (from transformer to electrodes), and current density. Furnaces of 24-ft (7.2-m) diameter are no longer rare, nor are 50,000-ampere units.

Methods of electric heating utilize resistance, arcs, or induction. Resistance furnaces, as shown in Fig. 12, may use the substance being heated as a resistor, or auxiliary resistors may be used. Arc furnaces (Fig. 13) may have a direct arc between an electrode and the material being heated, for example, steel scrap or the arc may be between two or more electrodes. The arc may be indirect or it may be submerged in the material being melted. Induction furnaces (Fig. 14) use the crucible or its charge as the closed secondary circuit of a transformer with low- or high-frequency alternating current. The temperature in the carbon arc is approximately 3100°C. Graphite or amorphous carbon are used for electrodes, depending on economics. Metals such as titanium, zirconium, and hafnium may be compacted into electrodes which are consumed by resistance or arc melting or combinations of both in a nonreactive-atmosphere or vacuum furnace. Furnaces as large as 50,000 kW have been developed for the calcium carbide and phosphorus industries with use of 60-in. (1.5-m) diameter electrodes in a 40-ft (12-m) furnace. Some types of furnaces are adaptable to several products, but usually the design is developed for an individual product.

Zone refining of metals for the electronics indus-

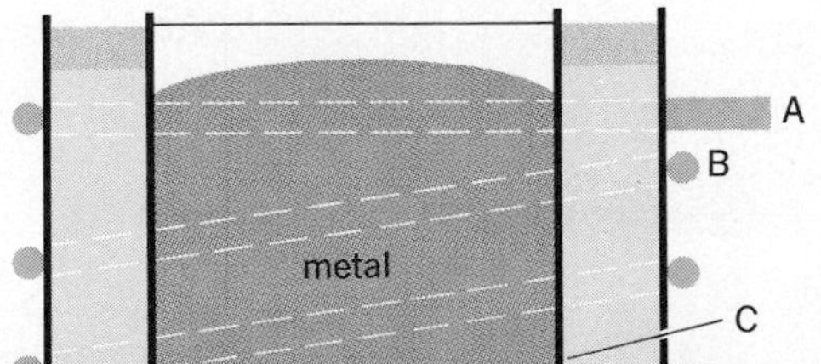

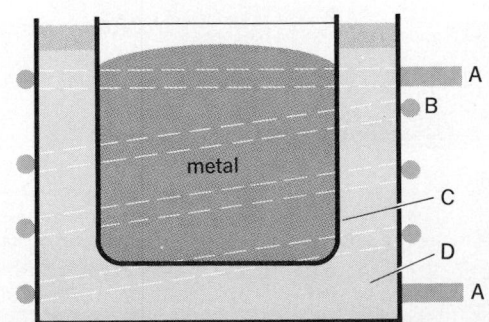

Fig. 14. Induction furnace. A = terminals of high frequency, B = copper coil, C = crucible, D = refractory.

try, such as silicon for diodes and transistors, is accomplished by induction melting of the metal in a narrow zone and slow movement of the molten zone in the metal ingot from one end to the other in an evacuated or inert gas–filled enclosure. Impurities move toward the end of the ingot. The operation is repeated until the desired purity is obtained.

High-melting-point metals, carbides, oxides, and nitrides are melted with a plasma-arc torch by coating objects with molten droplets carried in a jet of inert gas passed through the torch. Raw material, such as a powder or wire, is fed through a dc arc between a tungsten cathode and water-cooled copper anode. A strong bond with the base material is obtained.

Processes in gases. Electrical discharge through gases has industrial application in ozone production and nitrogen fixation.

Ozonizers consist of two metal electrodes with an air gap and a dielectric, such as gas, between them. One commercial unit operates at 15,000 volts, 60 Hz, and 35–40 watts/ft^2 (375–430 W/m^2). Very dry air passed through an air gap will then contain 10–12 mg ozone per liter. Fixation of nitrogen by passing air through an arc furnace, thus forming oxides of nitrogen, was practiced when power was cheap in Norway, France, and Italy, but has been replaced by conventional processes. *See* OZONE.

Electromagnetic separation. Magnetic separation removes tramp iron from mixtures of granular solids, suspensions, and solutions. Magnetic separation is also used to separate solids of various magnetic susceptibilities, such as mineral fractions.

Electrodialysis. This is the separation of low-molecular-weight electrolytes from aqueous solutions by migration of the electrolyte through semipermeable membranes in an electric field. It is used on an industrial scale for deashing starch hydrolyzates and whey, and in many municipalities for producing potable water from saline water. Its uses also include the concentration of liquid foods such as dairy products and citrus juices, the recovery of sulfite pulp waste and pickling acid, and the isolation of proteins. *See* COLLOID; DIALYSIS; SALINE WATER RECLAMATION.

Electrophoretic deposition. This is the deposition of a nonconductive material in a finely divided state from a suspension in an inert medium. Electrophoresis is the migration of colloidal particles, which acquire positive or negative charges in an electric field. The process is useful in electropainting; for instance, electropainting of automobile bodies and other objects has now been adopted on a large scale. Rubber latex is an example of a negatively charged colloid which can be plated on an anode. Electronic components can be coated with inorganic salts, oxides, and ceramics suspended in organic media. Bitumen can be electrodeposited out of an aqueous dispersion onto the anodic surface of steel pipe by use of an axially placed cathode. *See* ELECTROPHORESIS.

Electroendosmosis. This is the movement of a liquid with respect to an immobilized colloid in an electric field. The process is used in the dehydration of peat, dye pastes, and clay. Dies in a clay extrusion press can be "lubricated" by making the die cathodic, which attracts a film of water to it from the wet clay. It is also used commercially for dewatering soils in mining, road building construction, and other civil engineering works.

Electrostatic technique. The deposition of charged particles from suspension in gases has many useful applications. The Cottrell electrostatic precipitator removes dusts and mists from gases. In one form, a fine wire is axial in a pipe and insulated from it. The pipe is grounded, and the wire is negative in a high-voltage dc circuit. Particles in the gas that is passing through the pipe become electrically charged and move to the pipe. Liquid particles drain off, while solids are periodically vibrated off.

Spray painting with a high voltage between the spray gun and the work is particularly effective in providing an even coating with an economical use of paint on irregular and open surfaces, such as a screen. Complete coverage is achieved because an uncovered spot or pore has a higher field strength and attracts the charged droplets to it.

In xerography a sheet of plain paper is electrically sensitized in those areas corresponding to an original so that colored resin particles carrying an opposite charge are attracted and retained only on the sensitized areas, thus producing a visible image corresponding to the original.

Abrasive paper and cloth are coated with an adhesive and abrasive powders attracted to the base material in an electrostatic field. Pile fabrics can be produced in a similar manner, with the short fibers oriented by the electric field. Powdered adhesive can be applied to paper.

Plant design for electro process. The greatest changes have taken place in the electrical distribution system portion of electrolytic plants. Copper bus-bars have been taken for granted for years. They show good corrosion resistance, low contact voltages, and contacts can be kept clean readily. Aluminum is not as corrosion-resistant, and contacts are poor unless specifically designed. Aluminum and copper are competitive conductors when the ratio of the cost of a given amount of aluminum to that of copper is about 1.7. Aluminum is below the equality price, copper is above. Therefore, the newest electrolytic plants cannot afford the luxury of copper bus-bars, but still want the advantages of copper contacts—namely, the copper-surface aluminum bus-bar and the copper-shoe aluminum crossbar, achieved by plating, machined sections, explosive forming, and duplex preparation.

Electrolytic engineering encompasses cell design; electrical isolation; use of sumps instead of sewers; economic selection of bus-bars; solution circuit isolation; under-floor distribution of lines, supports, conductors, crystallizing liquors, and plastic pipe and conduits; lined and unlined cells; contacts and their inspection; and avoidance of pollution.

Early in plant development, the problems of impurities in solutions, and the effects of traces on the electrolytic deposit were evident, appreciated, and solved, largely by Edisonian methods. Maximum tolerances were set up, and purification procedures were developed to a high level in copper and zinc. Crystallizing solutions were a problem for manganese. In this area, control of deposition-potential, voltage-amperage and amperage-time recording instruments gave the concept of impurity concentrations affecting the deposition potential and the current density needed to reach this potential. There was an explanation for receding or resolution of deposits. In chromium the problem was solved, in effect, by converting the ore during purification into a chemical compound purified by crystallization. Purification of electrowinning solutions of magnesium, sodium, potassium, barium, calcium, iron, aluminum, antimony, arsenic, tin, cadmiun, silica, lead, nickel, and cobalt by chemical methods is a common occurrence.

The reconditioning of cathodes for metal deposition and ease of product removal in the electrowinning of manganese and chromium is a repetitive, demanding, and skilled-labor operation. Lines have been designed and put in operation resembling the continuous, conditioning, cleaning, pickling, and preparation lines of the electroplating phases of the automotive industry, a long-delayed cross-fertilization. When electrolytic copper sheet needed separate plants for printed electronic circuits, auxiliary electrodes were developed to control edges and ensure more uniform thickness. Perhaps manufacture of starting sheets can be mechanized by adopting continuous electroplating lines.

The concept of the electrolytic cell as an insulated mechanism, supported on structures and insulated from them, with separate support for conductors, supply lines, circulating systems, all insulated and electrically broken so that they are not shunt circuits, has been well established for decades. Tanks were wood, but are now unreinforced concrete, lined or unlined, and in few cases glass fiber or cloth-reinforced plastics, such as polyvinyl chloride or its competitors. In general the copper industries studies have shown that the investment cost for a lead-lined concrete cell is a little more than half that of a ribbed, equivalent-strength plastic. Because of local availability, tin refining cells are polyvinyl-chloride-lined concrete units, but chromium metal is made in plastic cells. A few foreign plants employ "painted" or thin-lined concrete.

Selenium rectifiers are widely used, to a greater extent than coppper oxide, for small power supplies, control and signaling circuits, and for applications in which their relative immunity to voltage surges is desired.

ITE built many mechanical rectifiers for electrochemical and other industrial uses up to about 1958. The largest single installation consisted of four 4000-kW machines in parallel for a chlorine-caustic cell line. There was no other American manufacturer except General Electric which installed only one machine.

Since about 1960, there have been no mercury-arc rectifiers built in the United States and very few in Europe for electrolytic or industrial applications, except for high-voltage direct current, welding, and transmission. In the United States and in Europe, monocrystalline diode rectifiers, first germanium and then silicon, have completely replaced mercury-arc rectifiers in industrial applications. Manufacturers have converted to silicon in many multianode mercury-arc rectifiers and excitron and ignition units. All new installations and expansions utilize silicon rectifiers.

Silicon rectifiers range in sizes up to 90,000 amperes and up to nearly 1000 volts, with a maximum rating of 30,000 kVA, shipable fully assembled except for the walk-in structures.

Since the early 1960s, the only type of power rectifiers for electrolytic loads, such as aluminum reduction, metal refining, electrowinning, chlorine and chlorate cells, and so on, are silicon rectifiers. General Electric has not furnished mercury-arc rectifiers for this type of application since the late 1950s. General Electric does not make any mechanical rectifiers and has not since the mid-1960s; this is true for all other domestic rectifier manufacturers.

ORGANIC PROCESSES

Organic electrochemistry was once regarded as a tantalizing area with many important laboratory achievements but few successes in commercial practice. This situation is changing, however, in that electroorganic processes are likely to prove commercially advantageous if they can fulfill either of two conditions: (1) performance under conditions of voltage corresponding thermodynamically to the conversion of an organic group to a reduced or oxidized group, with the cell products relatively easy to isolate and purify; (2) performance of a highly selective, specific technique to make an addition at a double bond, or to split a particular bond (for example, between carbon atoms 17 and 18 of a complex molecule having 25 carbon atoms).

Selectivity and specificity are highly important in electroorganic processes for the manufacture of complicated molecules of vitamins and hormones—as well as for the medicinal products whose action on pathogenic organisms is a function of their spatial arrangement, steric forms, and resonance.

The electrolytic approach can also be competitive for some low-cost, tonnage products. Here continuous processing is important, and only a single phase should be present, that is, a solution rather than an emulsion, dispersion, or mechanical mixture. Only for fairly valuable products is it practical to find a conducting solvent and then to engineer around it.

The electrolytic oxidation and reduction of organic compounds differ from the corresponding and more familiar inorganic reactions only in that organic reactions tend to be complex and have low yields. The electrochemical principles are precisely those of inorganic reactions, while the procedures for handling the chemicals are precisely those of organic chemistry.

Most organic molecules are insoluble in the aqueous solutions that are the best electrical conductors. Unless solubility can be increased, the only other approach is to use organic solvents. These make relatively poor conductors; hence are encountered power loss, heat build-up, chemical inversion, and often, stepwise, complex reactions. (Alkali-metal aromatic sulfonates have been suggested as electrolytes, but even here the solubility of some organic materials is limited and the rate of reaction can be too slow.)

Oxidations. Commercial success in organic electrochemistry has come about by well-engineered combinations of organic and inorganic techniques in areas where strictly chemical methods are either impossible or inefficient, for example, in catalytic hydrogenation or oxidation.

The conventional oxidation reagents of the organic chemist are expensive. There is no market for the oxidant once it is reduced, and chemical regeneration is prohibitive in cost. Accordingly, these reagents are avoided. Electrolytic regeneration, however, can be relatively inexpensive if linked to a carefully controlled organic operation. Typical of this approach is manganese dioxide oxidation of anthraquinone with electrolytic regeneration of the oxidant. In the electrolytic oxidation of anthracene there is a cost-efficient process that utilizes a 20% sulfuric acid suspension with a small quantity of ceric sulfate as a catalyst. Other examples include chromic acid oxidation of oleic acid to perlargonic and azelaic acids, where the oxidant is regenerated electrolytically, and the electrolytic regeneration of periodic acid (a costly oxidizing reagent) in the dialdehyde starch process. These involve savings not only in the purchase price of the oxidants but also in the disposal cost of products that cannot be marketed.

Illustrating a different type of reaction, in which selectivity gives electrolysis its great advantage, A. Cooper and C. L. Mantell reported on the oxidation of cholesteryl acetate dibromide. This starting material was dissolved in carbon tetrachloride, suspended by mixing in a 4.5 *M* solution of sulfuric acid, and electrolyzed at a lead dioxide anode. The anodically prepared film of lead dioxide, when maintained at potentials (relative to a calomel electrode) above 1.5 volts, was a catalyst for the oxidation. A new and mild oxidation at the catalytic electrode surface resulted in attack on the tertiary hydrogen atom at the 25 carbon. The mildness of the oxidation prevents decomposition of the unconverted raw material and permits recovery of up to 96% of the cholesteryl acetate dibromide and derivatives. Conversion of starting material ranges from 30 to 55%. The molar yield of products ranges from 85 to 93%; the best chemical methods have given 5–7% weight yields. The process is being scaled up for commercial applications in hormone preparation. An electrolytic step with a mercury electrode is important in the successful synthesis of vitamin A.

In electrolytic reactions the cell surface is the only source of reductants or oxidants, and these must be produced at highest efficiencies. The mass action effects, concentrations, temperature of reaction, reaction velocities, diffusion, and equilibria between the initial and final products of the reaction all apply to electrochemical reduction and oxidation in the same manner as do the corresponding reactions carried on outside the electrolytic cell.

Materials that are reduced absorb hydrogen at the cathode, and may be considered cathodic depolarizers; those that are oxidized absorb oxygen at the anode, and are anodic depolarizers. Oxidation reactions may involve substances other than oxygen, such as chlorine.

Anodes are selected with high oxygen or halogen overvoltage, and cathodes are selected with a high hydrogen overvoltage. (Because the accumulation of electrolysis products at anode and cathode causes polarization, the overvoltages are needed to move the products away and keep the process going.)

Reductions. Substances that are easy to reduce may be acted on, at the interface of cathodes, with low hydrogen overvoltage. Hard-to-reduce materials may require much higher overvoltages, which are reached through either the cathode composition or the current density.

The aromatic nitrogen-containing organic compounds have been extensively studied by many researchers. As early as 1900, F. Haber and K. Elbs showed that the nitro compounds of the type RNO_2, where R is an organic radical, could either be directly reduced to the amine RNH_2 (that is, aniline), or successively reduced to the nitroso product RNO, the beta aryl hydroxylamine RNHOH, the azoxy product RN—O—NR, and the material containing an azo group RN═NR, which in turn is reduced to the hydrazo form RNH═NHR or the idene type as $H_2NR{=}RNH_2$.

Although commercial processes based on these reactions were successful, they were eventually replaced by improved nonelectrolytic processes. The older work is now being reviewed to see if the use of constant-potential equipment, for example, Anatrol instrument system (developed by Continental Oil Co. but sold to Magna Corp., which marketed it), will open up new commercial applications.

Oxidation of starch. The process for the oxidation of starch to dialdehyde starch by periodic acid, continuously regenerated at the anode of an electrochemical cell, was first developed by the U.S. Department of Agriculture in Peoria, Ill. Dialdehyde starch is an important polymeric aldehyde used in paper manufacture and leather tanning. In this process a lead anode, coated with lead dioxide, is immersed in a suspension of starch; the cathodes are steel, enclosed in porous aluminum oxide diaphragms. One difficulty with this process is that free iodine, formed by the migration of iodide ions to the anode, tends to react with starch, despite the presence of the diaphragm.

In 1960 an improved two-stage process was devised, where just enough periodic acid is added to oxidize the starch in a straightforward chemical reaction. The iodic acid produced during the reaction is then separated from the dialdehyde starch and sent to an electrolytic cell for reconversion to periodic acid. In the Mantell-Peoria cell, this reconversion is achieved with current efficiencies of 80–90%. Materials of construction are structural plastics that are unaffected by the reagents or reactants; only the electrodes are metal. Continuous operation is achieved, with the instrumentation permitting good control of temperature, concentration, coversions per pass, gas venting or collection, external recirculation rates, and other rate factors. The wide range over which such control can be exercised allows the cell to function like a reactor or catalytic converter. In fact, the cell has many of the characteristics of the catalytic reactors, the main difference being that in the cell the variables are related to the type and surface of the electrodes, while in the catalytic reactor they are related to the surface area of the catalyst and its support.

All piping is manifolded exterior to the cell (unlike the hydrogen-oxygen cell), and the feeds are valve controlled externally. The electrode spacing is prearranged, and all reaction chambers are in parallel both relative to flow and electrically. Because valve closure can shut off a fraction of the parallel streams through the cell, shutdowns because of mechanical or other failures are minimized. The electrolytes are simple solutions, not emulsions or suspensions, and do not contain stabilizers or salting-in components. Gas evolution, separation, and collection are separately handled in anolyte and catholyte streams. These two streams may be blended in any proportion desired for maintenance of pH and control of concentration.

Adiponitrile production. In 1965 the Monsanto Co. developed and commercialized an electroorganic process for adiponitrile production. Monsanto used power as a reagent in the electrolytic reductive coupling (or hydrodimerization) of acrylonitrile to adiponitrile, a component of nylon-6,6. The adiponitrile molecule is a six-carbon dimer, essentially made up of two molecules of three-carbon acrylonitrile.

In the multimillion-dollar plant that Monsanto's Chemstrand Division put on stream, each cell consists of a lead cathode and an alloy anode separated by a sulfonated polystyrene resin membrane. Spacers support the membrane and prevent it from touching the electrodes. The cells are made of polypropylene with neoprene gaskets and are grouped in banks of 24. The reactants, the catholyte and anolyte, circulate separately in an upflow direction through a bank of cells in parallel; the direct current passes through the bank in series. As mentioned earlier, the catholyte contains a quaternary ammonium salt, that is, a McKee salt. This provides ions for electrical conductivity and increases the acrylonitrile's solubility above 10%; lesser concentrations would lead to the formation of proprionitrile. The catholyte is circulated to remove the film of product at the cathode surface.

There are some related patents in this area, and also a du Pont Belgian patent for a rather similar approach.

Lead alkyl production. The Nalco Corp. also developed an electroorganic process. The process is for the electrolytic production of lead alkyls at a cost competitive with chemical processes employing lead-sodium alloys. Electrochemical synthesis of tetramethyllead or tetraethyllead is achieved by employing a Grignard reagent and lead shot. The electrolyte consists of alkyl magnesium halides, which ionize in ether solutions. The solution is fed to 8000-gal (30,280-liter) electrolytic cells, with lead pellet anodes, the walls of the cell as cathode, and a nonconducting low-permeability diaphragam. The reaction is shown in reaction (15),

$$4R^- + Pb - 4e \rightarrow R_4Pb \qquad (15)$$

where R represents the methyl or ethyl groups.

The $MgCl^+$ ions migrate to the cathode cell walls. Magnesium chloride and metallic magnesium are formed according to reaction (16).

$$4MgCl^+ + 4e \rightarrow 2Mg + 2MgCl_2 \qquad (16)$$

Nalco uses a variable-voltage approach. As the rate of reaction drops and the resistance of the cell increases, more and more voltage is applied to the cell, thus maintaining a constant rate of reaction per unit of time.

Mixed alkyl lead compounds may be made by electrolyzing ethylmagnesium chloride and adding methyl chloride in the cells; or, conversely, by starting with methylmagnesium chloride and then adding ethyl chloride.

The electrolysis cell bank is made up of 10 cells, each with an associated recirculation surge drum. In the cell the copper bus handles the low-voltage, high-amperage current loads. The lead pellet storage hopper and tubular conveyor equipment are located directly above the cell dome. The upper third of the recirculating drum is an insulated refrigerant-storage vessel that supplies coolant to carry off heat developed during the course of electrolysis. Refrigerant vapors are returned to centrifugal compressors.

Other oxidations. Gluconic acid and oxidation products of various sugars are in production in various parts of the world. Reduction products such as sorbitol and mannitol, initially products of an electrolytic cell, are made by catalytic hydrogenation as a more economic process. Electrolytic oxidation of wastes in sewage and water is being intensely studied, reviving work done in the 1920s.

[CHARLES L. MANTELL]

Bibliography: M. M. Baizer (ed.), *Organic Electrochemistry: An Introduction and a Guide*, 1973; D. R. Crow, *Principles and Applications of Electrochemistry*, 2d ed., 1979; H. Gerisher and C. W. Tobias (eds.), *Advances in Electrochemistry and Electrochemical Engineering*, 1978; A. Kuhn (ed.), *Industrial Electrochemical Processes*, 1971; C. L. Mantell, *Batteries and Energy Systems*, 1970; C. L. Mantell, *Solid Wastes: Origin, Collection, Processing, and Disposal*, 1975; D. J. Pickett, *Electrochemical Reactor Design*, 1977.

Electrochemical series

A series in which the metals are listed in the order of their chemical reactivity, the most active at the top and the less reactive or more "noble" metals at the bottom. In a broader sense such an activity series need not be limited to the metals but may be carried on through the electronegative (nonmetallic) elements as well. See the table for a list of common elements.

The electrochemical series as it applies to metals was first established by laboratory experiments in which the purpose was to determine which met-

Electrochemical series of the elements*

Lithium	Li	Aluminum	Al	Molybdenum	Mo
Potassium	K	Titanium	Ti	Tin	Sn
Rubidium	Rb	Zirconium	Zr	Lead	Pb
Cesium	Cs	Manganese	Mn	Germanium	Ge
Radium	Ra	Vanadium	V	Tungsten	W
Barium	Ba	Niobium	Nb	**Hydrogen**	**H**
Strontium	Sr	Boron	B		
Calcium	Ca	Silicon	Si	Copper	Cu
Sodium	Na	Tantalum	Ta	Mercury	Hg
Lanthanum	La	Zinc	Zn	Silver	Ag
Cerium	Ce	Chromium	Cr	Gold	Au
Magnesium	Mg	Gallium	Ga	Rhodium	Rh
Scandium	Sc	Iron	Fe	Platinum	Pt
Plutonium	Pu	Cadmium	Cd	Palladium	Pd
Thorium	Th	Indium	In	Bromine	Br
Beryllium	Be	Thallium	Tl	Chlorine	Cl
Uranium	U	Cobalt	Co	Oxygen	O
Hafnium	Hf	Nickel	Ni	Fluorine	F

*According to standard oxidation potentials E^0 at 25°C.

als would displace others from solutions of their salts. Thus a clean strip of zinc immersed in a solution of copper sulfate is soon found to be covered by a deposit of copper, while zinc in turn goes into solution from the strip as zinc ions. By definition, then, zinc is a more reactive metal than copper, since it will displace copper from a solution of Cu^{++} ions. The reaction is readily seen to be an oxidation-reduction transfer of electrons, which can be summarized by the equation below. Simi-

$$Cu^{++} + Zn = Cu + Zn^{++}$$

larly, copper will displace silver from a solution containing Ag^{+} ions, depositing crystals of metallic silver and coloring the solution with Cu^{++} ions. From these observations an activity series may be set up in the order Zn, Cu, Ag. By exhaustive experiments with other metals it becomes possible to draw up a complete list in the order of chemical activity, in which the metals at the top of the list are those which are found to give up their electrons most readily (that is, are the most electropositive elements). Such a list is shown in the table, where lithium exhibits the most reactivity as a metal.

The ease with which an isolated atom of an element gives up an electron, known as the first ionization potential, is a precise physical quantity which can be measured by electrical experiments on gases or vapors at low pressure. The replacement experiments which determine the order of the electrochemical series take place in a very different environment, since they involve solid phases and also aqueous solutions with their consequent hydration effects. Moreover, it might well be expected that displacement reactions in solution would depend upon the concentrations of the reagents used, and also upon the presence or absence of other dissolved substances.

To obtain a more accurate and reproducible activity series, it is best to turn to the more exact quantity called electrode potential, or oxidation-reduction potential, which is defined as the voltage developed by a sample of pure metal immersed in a solution of one of its salts (at unit activity and at 25°C) versus a hydrogen electrode immersed in hydrochloric or sulfuric acid of equivalent concentration. For further details about this measurement *see* ELECTRODE POTENTIAL.

It is evident that by confining the experimental conditions to a standard concentration and temperature the hydration and concentration effects can be kept quite constant, making possible a more exact listing of metals according to their activity. Hence any present-day electrochemical series must rely on the measurements of oxidation potential and should be in agreement with the accepted values determined from such electrochemical cells. Such reliance has the further advantage that the series need not then be confined to metals but may be extended to the nonmetals, or electronegative elements. As before, those metals which will liberate hydrogen from dilute acids (such as hydrochloric or sulfuric) will stand above hydrogen in the series, while those metals and nonmetallic elements which will not liberate hydrogen from such dilute acids will stand below hydrogen in the list. Since the oxidation potentials are also related to the equilibrium constants for reversible reactions, it becomes possible to calculate oxidation potentials from other information when direct experiments are inconvenient, as in the case of the alkali metals versus aqueous solutions of their salts. *See* ELECTROCHEMISTRY; ELECTRONEGATIVITY; OXIDATION-REDUCTION.

[EUGENE G. ROCHOW]

Bibliography: F. A. Cotton and G. Wilkinson, *Advanced Inorganic Chemistry*, 3d ed., 1972; C. A. Hampel (ed.), *Encyclopedia of Electrochemistry*, 1964, reprint 1972.

Electrochemical techniques

Experimental methods developed to study the physical and chemical phenomena associated with electron transfer at the interface of an electrode and solution. The objective is to obtain either analytical or fundamental information regarding electroactive species in solution. Fundamental electrode characteristics may be investigated also.

The physical and chemical phenomena important in electrode processes generally occur very close to the electrode surface (usually within a few microns). Mass transfer of species involved in an electrode process to and from the bulk of solution is one important aspect. Inclusion of a large excess of inert electrolyte in most electrochemical systems eliminates electrical migration as an important means of mass transfer for electroactive species, and only convection and diffusion are considered.

Important chemical aspects of electrode processes include the oxidation or reduction occurring as a result of electron transfer, and coupled chemical reactions. Coupled reactions are initiated by production or depletion of the primary products or reactants at the electrode surface.

The primary experimental variables involved in electrochemical techniques are the potential E, the current I, and the time t. Either the potential or current at the working electrode is controlled and the other observed as a function of time. The many ways in which either may be controlled give rise to the wide variety of controlled-potential or controlled-current techniques.

The general scheme for electron transfer at an electrode in solution is shown in Eq. (1), where O is

$$O + ne^- \underset{k_b}{\overset{k_f}{\rightleftarrows}} R \qquad (1)$$

the oxidized form, R is the reduced form, and n is the number of electrons transferred. When k_f and k_b, the rate constants for the forward and back reaction, respectively, are very fast, the system is called reversible and the Nernst equation, Eq. (2),

$$E = E^{0\prime} + (0.059/n)\log(C_O/C_R) \qquad (2)$$

holds. In this relation E is the electrode potential, $E^{0\prime}$ is the formal standard potential for the redox couple, and C_O and C_R are concentrations at the electrode surface. In the following discussions, only reversible reduction processes will be considered, although oxidations are equally applicable. *See* ELECTROCHEMICAL PROCESS; ELECTRODE; ELECTRODE POTENTIAL; ELECTROLYSIS.

Controlled potential. Two methods have been developed in this area and are described.

Constant potential with convection. The electrode potential is held constant or varied slowly

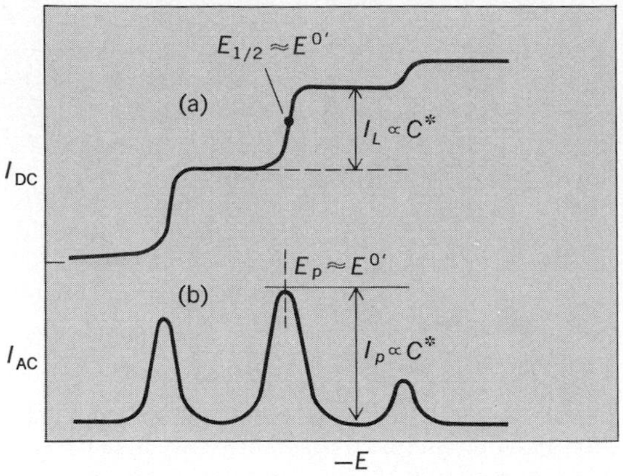

Fig. 1. Plots of the variation of current with constant applied potential in reduction processes. (*a*) Direct current versus voltage at a rotated working electrode, where three successive reductions of electroactive species occur. (*b*) Alternating current versus direct-current voltage applied at a working electrode in an alternating-current polarographic experiment.

with time (about 5 mV/sec) as the solution is stirred or the electrode rotated at a constant rate. The current is measured as a function of potential, and it increases sharply whenever the potential passes through the region of $E^{0'}$ for the particular electroactive species involved. Between reduction steps, current plateaus are established, the heights of which are proportional to the concentrations of each electroactive species. This is plotted in Fig. 1*a*, where I_L is the limiting or plateau current which is proportional to the bulk concentration of electroactive species C^*. The point designated $E_{1/2}$ is the half-wave potential, about equal to the formal standard potential $E^{0'}$. In Fig. 1*b* an alternating current potential is superimposed on the direct current one of Fig. 1*a*. E_p is peak voltage and I_p is alternating current.

Potentiostatic chronoamperometry. This technique consists of maintaining a constant potential without convection. When a reducing potential is imposed instantaneously on a stationary working electrode in quiescent solution, current will rise sharply and then decay as the electroactive species in the electrode vicinity is depleted by electrolysis. The magnitude of the current is proportional to the bulk concentration of electroactive species and is related to the electrode potential through the Nernst equation, Eq. (2). If the potential is sufficiently beyond $E^{0'}$, the Nernst equation demands complete conversion to the reduced form. Thus, under these conditions the current is diffusion-controlled and decays with $1/t^{1/2}$. This is shown by Eq. (3), where F is the faraday, A is the

$$I = nFAC^*(D/\pi t)^{1/2} \qquad (3)$$

electrode area, D is the diffusion coefficient for the electroactive species, and C^* is the bulk concentration of electroactive species.

One important variation of potentiostatic chronoamperometry is to apply a double (or cyclic) potential step. During the initial step, electrolysis occurs, depleting the oxidized form O, but producing the reduced form R in the immediate vicinity of the electrode. If the potential is instantaneously switched back to the initial value, species R will be reoxidized, and an inverted I-t curve will be obtained with well-defined diffusion-controlled decay characteristics (Fig. 2). If species R is chemically reactive in the solution (coupled chemical reaction), its contribution to the reverse current will be diminished not only because of mass transfer limitations, but also because of chemical decay. The ratio of reverse to forward currents (i_a/i_c) will be smaller for larger chemical rates and a specific switching time τ. Thus, within electronic limitations, one can adjust τ to fit the reactivity of the chemical system in order to evaluate the kinetics.

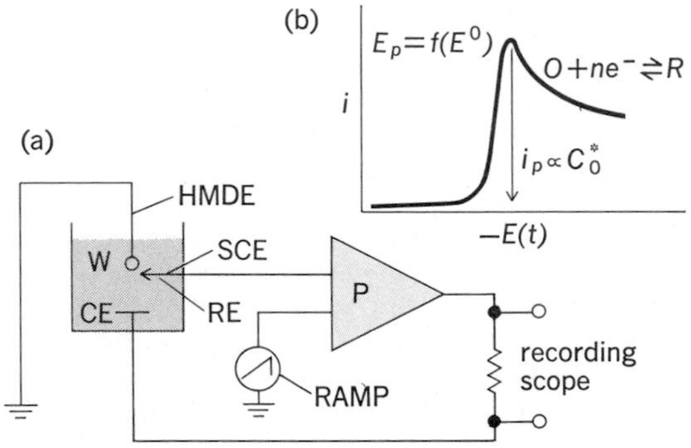

Fig. 3. Stationary electrode polarography. (*a*) Generalized instrumentation and (*b*) the current-voltage curve obtained on the scope. HMDE = hanging mercury drop electrode; SCE = saturated calomel reference electrode; CE = auxiliary counter electrode; P = potentiostate; W = working electrode; RE = reference electrode; and RAMP = linearly increasing voltage signal.

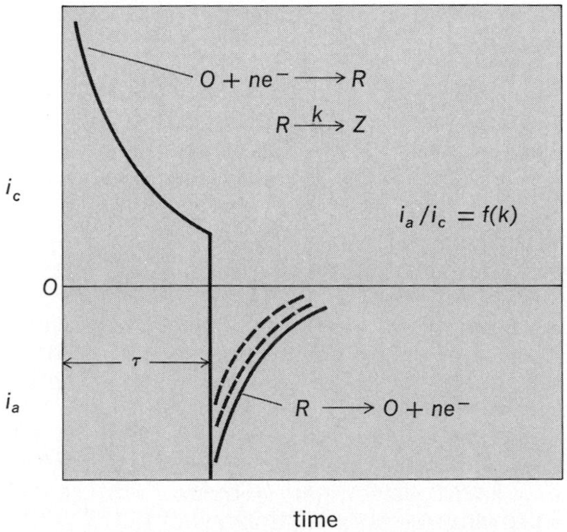

Fig. 2. Current-time behavior for double potential-step experiment at a stationary working electrode. Dashed lines in reverse current-time curve represent cases where the rate of chemical reaction is fast compared to the time scale of the experiment.

Variable potential. Several procedures that have been developed in this area are described.

Linearly varying potential (LVP). The term chronoamperometry with LVP is applied here. When the potential of a stationary electrode in quiescent solution is varied in a linear fashion, as in conventional polarography, but without restrictions on the sweep rate, a peak-shaped stationary electrode polarogram is obtained. The generalized system and a current-voltage curve are given in Fig. 3. Peak current I_p is proportional to bulk concentration of electroactive species, and the peak potential E_p is related to $E^{0'}$. Analytically, the approach is applicable over the range of 10^{-3} to 10^{-6} *M* electroactive species.

A variation, called cyclic voltammetry, involves application of a triangular potential sweep, allowing one to sweep back through the potential region just covered. A reverse current I_r is obtained with characteristics related to the chemical reactivity of the initial electrode product. This is analogous to cyclic potential-step chronoamperometry. Current-voltage curves are displayed in Fig. 4.

An enhancement in analytical sensitivity is obtained by another variation called stripping analysis. Here one applies a constant reducing potential for a period of time (about 1 to 60 min), sufficient to collect substantial amounts of the reduced species as an amalgam or deposit at the working electrode. Then the concentrated product is stripped by applying a linear oxidizing potential sweep. The stripping current is considerably larger than would be obtained from a direct measurement. The stripping current is proportional to the amount accumulated, which in turn is proportional to bulk concentration and prior electrolysis time. Sensitivity to 10^{-9} *M* is achieved.

Alternating potential. Three variations of this method are outlined below.

Alternating-current polarography. This involves an approach identical to conventional polarography at a dropping mercury electrode, except that a small-amplitude (about 1–15 mV), sinusoidal alternating voltage (frequency about 10–100 Hz) is superimposed on the direct-current controlling potential. Only the ac component of the electrolysis current is measured. A plot of I_{ac} versus E_{dc} results in a derivative-type curve with symmetrical triangular peaks (Fig. 1*b*). The primary analytical advantage is that the interference from succeeding or preceding reductions is considerably reduced. Sensitivity to 10^{-6} *M* is possible.

Square-wave polarography. This also involves the basic polarographic approach, except that a small-amplitude (about 1–25 mV) square wave (frequency about 225 Hz) is imposed on the controlling dc potential. The current measured is an alternating component obtained by measuring the current at times near the end of each half-cycle. This avoids the large capacitive background current associated with instantaneous electrode potential changes and allows greater sensitivity. Analyses over the range of 10^{-3} to 10^{-7} *M* are possible.

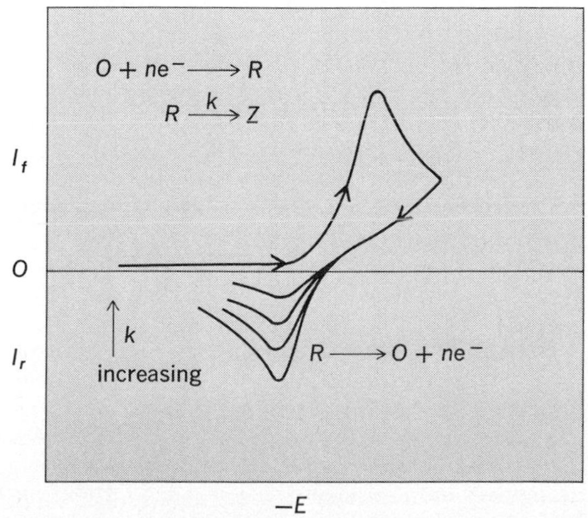

Fig. 4. Current-voltage curves in cyclic voltammetry.

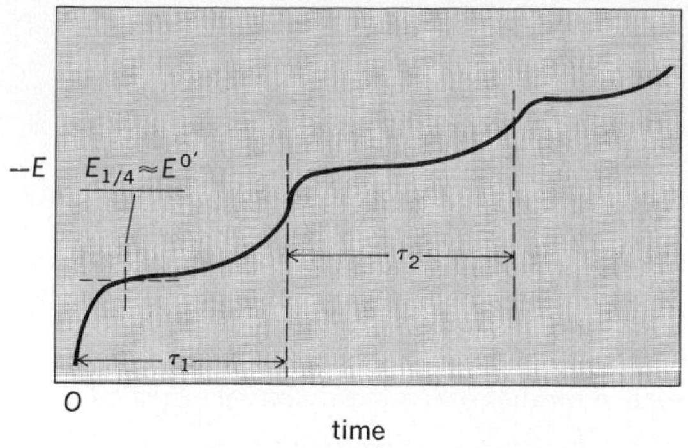

Fig. 5. Potential-time curves for chronopotentiometric experiments. Curve represents successive reduction of three different electroactive species in solution.

Pulse polarography. This is a variation of square-wave polarography, where a single small potential step is imposed on the controlling dc potential during the life of each mercury drop. The change in current with each step is measured and plotted as a function of the dc potential. Like ac and square-wave polarography, a derivative-type curve is obtained. Sensitivity is about the same as square-wave polarography.

Controlled current. Several analytical methods based on controlled current have evolved.

Chronopotentiometry. A constant current is imposed at the working electrode, and its potential is monitored with time. The electrode must assume a potential which will cause electrolysis sufficient to maintain the imposed current. Thus, the electrode adopts first the potential of the most easily reduced species. As this species is depleted near the electrode, the potential shifts to that of the next most easily reduced species. This continues until reduction of solvent or electrolyte occurs. Characteristic curves are plotted in Fig. 5. The transition time τ is related to the bulk concentration of the electroactive species giving rise to the transition. For the first transition, C^* is proportional to $\tau^{1/2}$. However, for the reduction of a second species the transition time τ_2 depends on the first transition time. Thus, the term $[(\tau_1+\tau_2)^{1/2}-\tau_1^{1/2}]$ is proportional to the bulk concentration of the second reducible substance. Further steps are similarly complicated.

Cyclic chronopotentiometry. This involves following a normal chronopotentiometric experiment with an instantaneous current reversal. The product is reoxidized and a reverse transition is seen. Results obtained are analogous to the cyclic voltammetric and double potential-step experiments. The ratio of the reverse transition time τ_r to the forward transition time τ_f reflects the kinetics of a succeeding chemical reaction of species *R*. For negligible chemical reactivity $\tau_r/\tau_f = 1/3$.

Coulostatic analysis. This does not involve controlled current, but the application of a very short, large pulse of current to the electrode. This pulse serves to charge up the capacitive electrode–solution interface to some new potential. The cell circuit is then opened, and the return of the working electrode potential to its initial value is monitored. The open-circuit condition requires that the current necessary to discharge the electrode interface comes from electrolysis of electroactive species in solution. The change in electrode potential ΔE versus $t^{1/2}$ results in a straight-line plot, the slope of which is proportional to concentration, as in Eq. (4), where C is the electrical capacitance associated with the electrode-solution interface.

$$\pm \Delta E = \frac{2nFD^{1/2}C^*}{\pi^{1/2}C} t^{1/2} \qquad (4)$$

Thin-layer electrochemistry. Electrochemical techniques can be applied in cells where only a 10–100-μm-thick solution is electrolyzed. The principal advantage is that, for experiments lasting more than about 1 sec, mass transfer limitations can be ignored. Thus, correlations between observed parameters and system characteristics are more straightforward. For example, for a chronopotentiometric experiment the transition time represents complete depletion of the electroactive species. Thus, C^* is proportional to τ for each of a sequence of reduction steps. Also, if a linear potential sweep is applied, a symmetrical current peak is obtained, the height or area of which is related to the concentration of electroactive species.

[SAM P. PERONE]

Bibliography: B. E. Conway, *Theory and Principles of Electrode Processes*, 1965; P. Delahay, *Double Layer and Electrode Kinetics*, 1965; P. Delahay, *New Instrumental Methods in Electrochemistry*, 1954; C. N. Reilley, in I. M. Kolthoff and P. J. Elving (eds.), *Treatise on Analytical Chemistry*, 1963; H. Schmidt and M. Von Stackelberg, *Modern Polarographic Methods*, 1963; D. E. Smith, in A. J. Bard (ed.), *Electroanalytical Chemistry*, vol. 1, 1966; P. Zuman and I. M. Kolthoff, *Progress in Polarography*, vols. 1 and 2, 1962.

Electrochemistry

The science dealing with the chemical changes accompanying the passage of an electric current, or the reverse process in which a chemical reaction is used as the source of energy to produce an electric current, as in a battery. Electric conduction occurs through the motion of charged particles. The charged particles may be electrons (as in metals or semiconductors) or ions, which are electrically charged atoms, molecules, or molecular aggregates. Ionic conduction in electrolytes (liquid solutions, molten salts, and certain ionically conductive solids) is a phase of electrochemistry. Conduction in metals, semiconductors, and gases is generally considered a portion of physics. Other aspects of electrochemistry are described below. *See* ELECTROLYTIC CONDUCTANCE.

Galvanic cells. These are better known as electric batteries. Many chemical reactions can be arranged to produce electrical energy by physically separating the reaction into two half-reactions, one supplying electrons to an electrode forming the negative terminal of the cell, and the other removing the electrons from the positive terminal. In the lead storage battery, for example, electrons are supplied to the negative terminal by the half-reaction shown in (1), which represents the oxida-

$$Pb + SO_4^{--} \rightarrow PbSO_4 + 2e^- \qquad (1)$$

tion of metallic lead to form lead sulfate. At the positive terminal, lead dioxide is reduced to lead sulfate by the half-reaction shown in (2).

$$2e^- + PbO_2 + 4H^+ + SO_4^{--} \rightarrow PbSO_4 + H_2O \quad (2)$$

The electrons flowing in the external circuit from the negative to the positive terminal constitute the desired electric current. Charging of the lead storage battery by forcing a current to flow in the reverse direction results in the reversal of both half-reactions, and the storage of electric energy in the form of lead and lead dioxide. Such a cell is called a secondary cell, in contrast to a primary cell, such as the Leclanché cell or familiar dry cell, which is not designed to be recharged. In the Leclanché cell, electric energy is produced by the oxidation of zinc and the reduction of manganese dioxide at a carbon electrode. The electrolyte is a moist mixture of zinc chloride, ammonium chloride, and powdered carbon. The fuel cell is designed for the continuous production of electric current through the consumption of oxidant and reductant at separate electrodes. The most common fuel cell is the hydrogen-oxygen (or air) cell with alkaline electrolyte. Many other systems, including hydrocarbon-oxygen, carbon monoxide–oxygen, lithium-chlorine, and sodium-sulfur have been proposed.

Electrodeposition. The most important type of chemical reaction brought about by the passage of electric current is the deposition of a metal at a cathode from a solution of its ions. Electroplating of many metals, such as silver, cadmium, nickel, and chromium is used for protective and decorative coatings. Electroforming is a variety of electrodeposition in which an article to be reproduced is rendered conductive by spraying a thin metallic coating, then electroplated with a metallic deposit that is stripped from its substrate and filled with backing to reproduce the original article. Electrowinning is used for the commercial production of active metals, such as aluminum, magnesium, and sodium, from molten salts and others, such as copper, manganese, and antimony, from aqueous solution. Electrorefining is commonly used to purify metals such as silver, lead, and copper. The impure metal is used as the anode, and purified metal is deposited at the cathode.

Electrolytic processes. Many electrode reactions other than metal deposition are of commercial or scientific use. Electrolysis of brine to yield chlorine at the anode, hydrogen at the cathode, and sodium hydroxide in the electrolyte is an important industrial process. Many organic compounds can be prepared electrolytically.

Electrothermics. While not strictly electrochemical, electrothermics is generally recognized as a part of the field. It includes high-temperature processes involving electric arc or resistance furnaces.

Electroanalytical chemistry. Many electrochemical measurements are useful for analytical purposes. Electrodes that are commonly used for analytical purposes through measurement of their potentials include the glass electrode for pH measurements, and ion-selective electrodes for certain ions, such as sodium or potassium ion (special glass compositions), calcium ion (liquid membrane), and fluoride ion (doped lanthanum fluoride single crystals). Polarography involves the use of a dropping mercury electrode as one electrode of an electrolytic cell. Qualitative analysis is carried out by measurement of characteristic potentials (half-wave potentials) for electrode processes, and quantitative analysis by measurement of diffusion-controlled currents. Coulometry involves the application of Faraday's law for analytical purposes.

Several methods involving electrolysis during short periods of electrolysis permit the application of diffusion theory (in the absence of convection) to calculate mass transport rates. These include chronopotentiometry (measurement of potential-time transients under constant current conditions), and linear sweep and cyclic voltammetry (measurement of currents with linear voltage scan). Several titration methods involve electrochemical measurements, for example, conductometric, potentiometric, and amperometric titrations.

Electrode kinetics. Studies of kinetics of electrode processes are valuable not only for the understanding of mechanisms of electrode reactions, but also for the study of homogeneous reactions occurring in solutions either preceding or following the charge-transfer step. Such studies are made by pulse or transient techniques, or by steady-state methods involving dynamic systems, such as rotating disc electrodes for flowing solutions. *See* ELECTROCHEMICAL TECHNIQUES.

Miscellaneous phenomena. Electrochemical transport of ions through synthetic or natural membranes is important for processes, such as desalination of water and electrodialysis. In biological systems, the transmittal of nerve impulses and the generation of electrical signals, such as brain waves, are basically of electrochemical origin. A set of related phenomena can be grouped together under electrokinetic behavior, including the motion of colloidal particles in an electric field (electrophoresis), the motion of the liquid phase relative to a stationary solid under the influence of a potential gradient (electroosmosis), and the inverse generation of a potential gradient caused by a flowing liquid (streaming potential). Alternating-current phenomena, such as dielectric behavior, double-layer charging, and faradaic rectification, may also be included in a general definition of electrochemistry. Corrosion and passivation of metals are electrochemical in nature.

[HERBERT A. LAITINEN]

Bibliography: J. O. Bockris, *Electrochemistry*, 1976; C. A. Hampel (ed.), *Encyclopedia of Electrochemistry*, 1964, reprint 1972; K. Vetter, *Electrochemical Kinetics: Theoretical Aspects, 1967.*

Electrode

An electrical conductor through which an electric current enters or leaves a conducting medium, whether it be an electrolytic solution, solid, molten mass, gas, or vacuum. For electrolytic solutions, many solids, and molten masses, an electrode is an electric conductor at the surface of which a change occurs from conduction by electrons to conduction by ions. For gases and vacuum, the electrodes merely serve to conduct electricity to and from the medium. *See* ELECTRODE POTENTIAL; ELECTRODEPOSITION ANALYSIS; ELECTROLYSIS; ELECTROMOTIVE FORCE (CELLS).

[WALTER J. HAMER]

Electrode potential

The potential which a metal or gas electrode takes up relative to a solution of ions.

The various electrodes encountered in electrochemical work may be grouped into seven types: (1) metal – metal ion, (2) amalgam, (3) nonmetal nongas, (4) gas, (5) metal – insoluble salt, (6) metal – insoluble oxide, and (7) oxidation-reduction. Any of these electrodes may be combined with any other to give a cell, the emf of which is equal to the algebraic sum of the potentials of the two electrodes.

Metal–metal ion electrodes. When a metal is immersed in an electrolyte, an equilibrium tends to be established in which a steady difference of electric potential exists across the region of the interface between metal and solution. This equilibrium electrode potential results from the ionization of the atoms of the metal until the displacement of electric charges produced thereby exactly balances the tendency for additional metallic atoms to ionize. If the metal electrode M has a valence of n, the reaction which takes place is shown in (1),

$$M \rightleftharpoons M^{n} + ne^{-} \qquad (1)$$

where e^- indicates an electron and M^{n+} an ion in solution. Examples of metal – metal ion electrodes are zinc, copper, and sodium electrodes.

Because the potential of such an electrode changes with the concentration of ions, it is necessary to adopt some standard concentration at which to compare the potentials of various electrodes. The standard electrode potential E^0, expressed in volts, is defined as the potential of an element immersed in a solution of its ions at unit activity, that is, the effective concentration of 1 mole/1000 g of water. The electrode potential E at other concentrations is given by Eq. (2), where T is

$$E_M = E^0_M - \frac{RT}{nF} \ln a_{M^{n+}} \qquad (2)$$

the absolute temperature, F is the faraday (96,487 coulombs), R the gas constant, and $a_{M^{n+}}$ the effective concentration (activity) of M^{n+} ions in the solution. At 25°C Eq. (2) can be written as Eq. (3).

$$E_M = E^0_M - \frac{0.05916}{n} \log a_{M^{n+}} \qquad (3)$$

Because the single electrode potential E involves the activity of an individual ionic species, it has no strict thermodynamic significance. This difficulty is overcome by defining the standard hydrogen electrode as an arbitrary zero of potential. Electrode potentials based on this zero are thus said to refer to the hydrogen scale. Such a potential is actually the electromotive force (emf) of a cell obtained by combining the given electrode with a standard hydrogen electrode. *See* ELECTROCHEMICAL SERIES; HYDROGEN ELECTRODE.

Gas electrodes. A gas electrode is formed by partially immersing an inert metal (usually platinized platinum) in a solution of the ions of the gas. The gas must establish a reversible equilibrium with the ions in solution in the presence of the metal. The metal wire or foil helps establish an equilibrium between the gas and its ions and serves as the electric contact for the electrode.

The potential of such an electrode is determined by the pressure of the gas and the activity of its ions in solution. Thus, for the chlorine electrode, the reaction is shown in (4), and the equation

$$\tfrac{1}{2}Cl_2(g) + e^- \rightleftharpoons Cl^- \quad (4)$$

for the electrode potential E_{Cl_2} is shown in Eq. (5),

$$E_{Cl_2} = E^0{}_{Cl_2} - \frac{RT}{F} \ln \frac{a_{Cl^-}}{P^{1/2}{}_{Cl_2}} \quad (5)$$

where P_{Cl_2} is the pressure of chlorine in atmospheres, a_{Cl^-} the effective concentration (activity) of chloride ions in the solution, and $E^0{}_{Cl_2}$ the standard electrode potential for the chlorine electrode, which is equal to 1.3583 volts at 25°C. The standard electrode potential of a gas electrode is defined as the potential of the electrode when the gases involved in the reaction are at a fugacity of 1 atm, that is, an effective pressure of 1 atm, and all dissolved substances are at an effective concentration (activity) of 1 molal, that is, 1 mole/1000 g of water. *See* FUGACITY.

The most important gas electrode is the hydrogen electrode, which is reversible to hydrogen ions. The reaction for this electrode is shown in (6), and the electrode potential is given by Eq.

$$\tfrac{1}{2}H_2(g) \rightleftharpoons H^+ + e^- \quad (6)$$

(7). But $E^0{}_{H_2}$, the standard electrode potential of

$$E_{H_2} = E^0{}_{H_2} - \frac{RT}{F} \ln \frac{a_{H^+}}{P^{1/2}{}_{H_2}} \quad (7)$$

hydrogen, is the reference of all emf measurements and is taken by definition to be zero at all temperatures. Thus Eq. (7) becomes Eq. (8).

$$E_{H_2} = -\frac{RT}{F} \ln \frac{a_{H^+}}{P^{1/2}{}_{H_2}} \quad (8)$$

Another gas electrode which has received considerable attention is the oxygen electrode, whose potential depends on the activity of hydroxyl ions. However, unlike the hydrogen and chlorine electrodes, the oxygen electrode cannot be made reversible because no suitable electrode material has been found which can catalyze the establishment of the equilibrium between oxygen and hydroxyl ions. This equilibrium is shown in (9).

$$\tfrac{1}{2}O_2 + H_2O + 2e^- \rightleftharpoons 2OH^- \quad (9)$$

The standard potential of the oxygen electrode cannot be determined directly from emf measurements because of the irreversible behavior of this electrode. It is possible, however, to derive the value in an indirect manner, and it has been found to be +0.401 volt.

Measurements. In order to measure the potential of any electrode, it is necessary, in principle, to combine the electrode with a reference electrode such as the hydrogen electrode, which has an arbitrarily assigned potential, and to measure the total voltage across the two half-cells. The potential of the reference electrode is then subtracted to give the required electrode potential. For various reasons, such as the difficulty in setting up a hydrogen gas electrode, several subsidiary reference electrodes, whose potentials are known on the hydrogen scale, have been devised. The most common of these is the calomel electrode, which consists of mercury in contact with a solution of potassium chloride saturated with mercurous chloride. *See* CALOMEL ELECTRODE; SILVER CHLORIDE ELECTRODE.

A simple voltmeter cannot be used alone for measuring the emf of a small cell because the operation of the voltmeter draws some current, which causes chemical changes at the electrodes and produces a different voltage. To avoid these difficulties, the emf is measured by balancing against the cell a known voltage under conditions such that practically no current flows.

The constructional details of a potentiometer, as the apparatus for measuring emf is called, is shown schematically in the illustration. In this

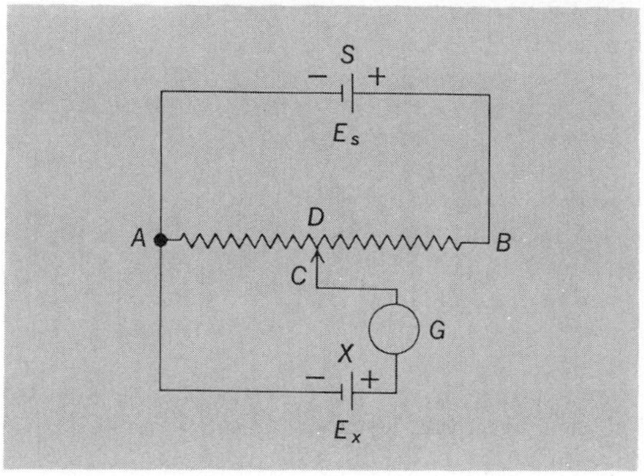

Poggendorff compensation method for measuring emf.

diagram S is a cell of known emf E_S, whose potential is impressed across a uniform resistance AB. Connected with S in such a way that the two emfs oppose each other is the source X of unknown potential E_X. To find E_X the sliding contact C is moved along AB until a position D is found at which the galvanometer G gives no deflection. From E_S and the distances AB and AD, the unknown emf E_X is found as follows. Since E_S is impressed across the full length AB, E_S for any given current passing through the resistance must be proportional to AB. Again, because E_X is impressed only across the distance AD, it must be proportional to this length. Consequently, dividing E_X by E_S gives Eqs. (10) and (11).

$$\frac{E_X}{E_S} = \frac{AD}{AB} \quad (10)$$

$$E_X = \left(\frac{AD}{AB}\right) E_S \quad (11)$$

At the present time the majority of emf measurements are made by means of special potentiometers, available commercially, which operate on the foregoing principle. In these potentiometers, the conductor AB consists of a number of resistance coils with a movable contact, together with a slide wire for fine adjustment. A standard cell is used for calibration purposes, and the emf of the cell being studied can be read with an accuracy of 0.1 millivolt or better.

The standard cell that is widely employed for emf measurements is some form of the Weston cell. It is highly reproducible, its emf remains constant over long periods of time, and it has a small temperature coefficient. In order to retain constancy of emf during use, only very minute currents should be drawn from the cell, as is actually done if the potentiometer is operated properly. *See* ELECTROCHEMISTRY; ELECTROMOTIVE FORCE (CELLS); ELECTRONEGATIVITY; OXIDATION-REDUCTION. [RICHARD GLICKSMAN]

Bibliography: J. Bockris et al., *An Introduction to Electrochemical Science*, 1974; B. E. Conway, *Theory and Principles of Electrode Processes*, 1965; D. R. Crow, *Principles and Applications of Electrochemistry*, 2d ed., 1979.

Electrodeposition analysis

An electroanalytical chemistry technique in which the product of the electrode reaction is insoluble and is either deposited upon or dissolved into the electrode. This method usually results in the complete depletion in solution of the deposited substance. Electrodeposition is applied most commonly to the separation of metals from solution and to their subsequent quantitative determination. However, it may be used simply as a separation technique. *See* ELECTROLYSIS.

The quantities of metals deposited may be determined by weighing a suitable electrode before and after deposition (electrogravimetry). Among the elements that may be determined by deposition are cadmium, nickel, zinc, copper, cobalt, antimony, tin, silver, and gold. At the anode the oxides of lead, PbO_2, and manganese, MnO_2, may be deposited. Halides may be deposited on a silver anode as the respective silver halides.

For a strictly reversible reaction, the potential of an electrode in equilibrium (no net electrodeposition occurring) with a solution of its ions of given concentration may be determined by using the Nernst equation, Eq. (1), which relates the formal

$$E = E^{0\prime} + \frac{RT}{nF} \ln[M] \qquad (1)$$

potential of an electrode to the equilibrium potential in the presence of a solution of its ions. In Eq. (1) E is the equilibrium potential (versus a reference electrode) of the electrode, $E^{0\prime}$ is the formal potential, n is the number of electrons needed for the reduction of one metal ion, F is the faraday (96,487 coulombs/equivalent), R is the gas constant, T is the absolute temperature, and $[M]$ is the molar concentration of metal ions in equilibrium with the electrode. Variation of equilibrium electrode potentials with concentration of several ions is shown in Fig. 1. *See* ELECTRODE POTENTIAL.

When the electrode potential is made more cathodic than the equilibrium value, electrodeposition occurs until the metal ion concentration is lowered to that value which is in equilibrium with the electrode at the applied potential. A tenfold change in concentration at 30°C may be effected by a $0.060/n$ volt change in potential. By making the electrode sufficiently cathodic, the metal ions remaining in the solution may be reduced to a negligible concentration. Two metallic species may be separated by adjusting the potential of the cathode so that it is less cathodic than the equilibrium potential of the metal to be left in solution, and more cathodic than the initial equilibrium potential of the metal to be removed. Practically all of the latter metal will be deposited as equilibrium is reestablished. In cases where the formal potentials of the simple metal ions are not sufficiently far apart to allow complete separations, it is often possible to lower the concentration of the interfering free metal ions in solution by adding a reagent to convert most of them to a complex-ion species. After selectively complexing the metal to remain in solution, the potential required to deposit the low concentration of free ions of this metal is sufficiently more cathodic than that required for the other metal to make complete separation possible. Figure 2 is a diagram of the electrodeposition apparatus.

Equation (1) cannot be used to select a deposition potential for a reaction which proceeds irreversibly at the electrode surface. For such a reaction the potential must be selected from current-potential curves.

The electrode material most commonly used for electroseparations is platinum. This metal is used as anode for most oxidations, and as cathode for depositions of the more readily reduced metals. Because of its low overvoltage on platinum, hydrogen is evolved and thereby interferes with the deposition of metals requiring very cathodic potentials for reduction.

Mercury is frequently used as a cathode material because its extremely high hydrogen overvoltage permits most metals to be deposited without interference from hydrogen evolution. As an anode, mercury finds very limited use because it is oxidized quite readily. Original procedures required removal of mercury from the deposited metals by distillation. However, coulometry is of-

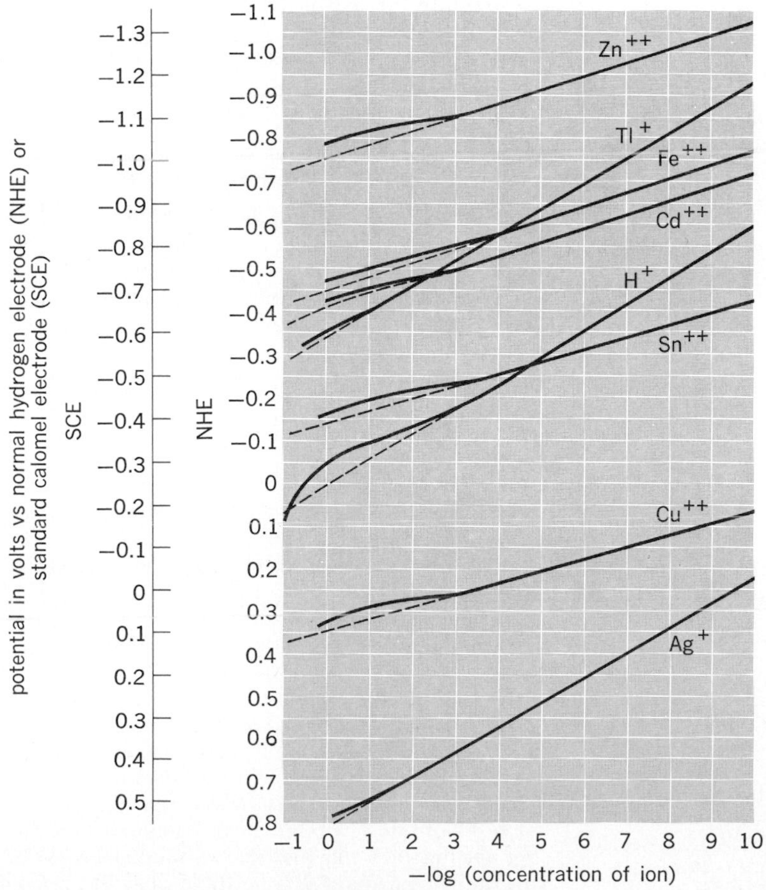

Fig. 1. Equilibrium electrode potentials. (*G. W. Ewing, Instrumental Methods of Chemical Analysis, 2d ed., McGraw-Hill, 1960*)

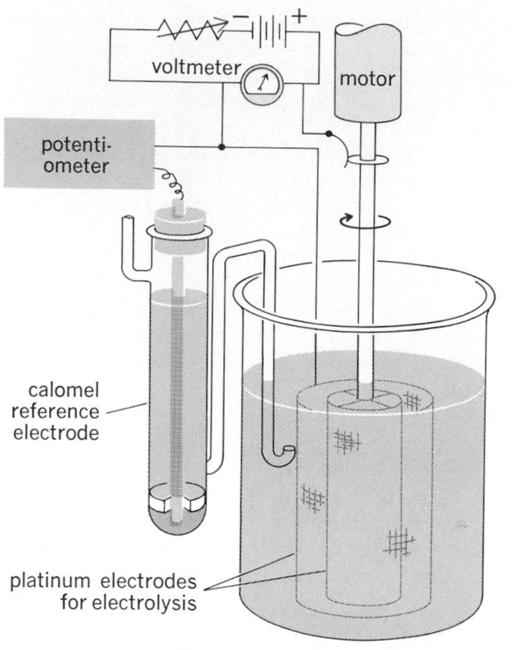

Fig. 2. Diagram of the electrodeposition apparatus. (*G. W. Ewing, Instrumental Methods of Chemical Analysis, 2d ed., McGraw-Hill, 1960*)

ten effective for removing the metal from the mercury and obtaining a measurement of it.

Constant-potential electrolysis. The potential of the working electrode (the cathode when a metal is being deposited) is the most critical factor in obtaining a desired separation. It is not possible, however, to obtain constant potential at an electrode by applying a constant voltage to the cell. The applied voltage is expended in accordance with Eq. (2). In this expression, E_{ap} is the voltage

$$E_{ap} = E_{an} - E_c + \omega_{an} + \omega_c + IR \qquad (2)$$

applied to the cell, E_{an} is the equilibrium potential of the anode, E_c is the equilibrium potential of the cathode, ω_{an} is the anodic overvoltage or increase in anode potential beyond the equilibrium value needed to pass the current of I amperes, ω_c is the corresponding cathodic overvoltage, and R is the ohmic resistance of the solution. The overvoltages for metal deposition and dissolution are usually less than 0.1 volt, with notable exceptions for iron, cobalt, and nickel. Because they depend upon solution conditions and electrode form in a manner which is not fully understood, overvoltages cannot be predicted with certainty, thus limiting the accuracy with which the current in a given electrolysis may be calculated.

In electrolysis at constant applied voltage, the potential of a working cathode is less than the total applied voltage. When the applied voltage is adjusted so that the critical potential for a separation cannot be exceeded, the potential of the electrode is significantly less than this critical value throughout most of the deposition due to IR drop, and the electrolysis proceeds very slowly. If a stable reference electrode is placed in the solution and the voltage between this and the working electrode is maintained constant at the critical value by periodic adjustment of the voltage applied to the cell, much more rapid deposition results. Such manual control, however, is tedious. Instruments which automatically perform this task are known as potentiostats, and several types have been described in the literature. When the cathode potential is to be controlled, the desired voltage between the cathode and reference is maintained constant by an electronic or electromechanical servo system which changes the total voltage applied between the anode and cathode. When the anode potential is controlled, the voltage between the anode and the reference is kept constant by the same means. Anode and cathode potentials cannot both be controlled at the same time.

Internal electrolysis. Electrodeposition without externally applied voltage was an early but simple method of approximating a controlled cathode potential. An active metal, such as magnesium, which dissolves spontaneously is made the anode of a cell, and an inert electrode, such as platinum, is made the cathode. When the electrodes are shorted together, the potential of the cathode is equal to, and created by, the potential of the anode. By judicious choice of the anode metal and the concentration of the reagent in which it dissolves, the cathode potential may be made to assume predetermined values over most of the useful range. Deposition at the cathode occurs at the expense of dissolution of the anode. The current flow is limited by the magnitude of the spontaneous voltage of the cell, and by its internal resistance. Internal electrolysis has been used relatively little in recent years.

Constant-current electrolysis. This process precludes the possibility of electrode-potential control by electrical means. While the concentration of metal ions remains large, the potential of the cathode stays near the equilibrium potential. However, as the ions in the solution are depleted, the desired cathode reaction is not able to use all of the current forced through the cell, so the potential becomes more cathodic until an additional reaction, such as the deposition of another metal or the evolution of hydrogen, takes place to maintain the current. In most electroseparation devices a large voltage is applied to the cell. In order to effect a clean separation of two metals under these conditions, it is necessary to interpose a harmless reaction to limit the potential of the cathode before it exceeds the equilibrium potential of the metal to be left in solution. The reaction most often employed is hydrogen evolution, which may be made to occur at various potentials by proper adjustment of the pH. Through use of selective complexing agents to change the relative effective concentrations of the ions (and indirectly their equilibrium potentials), and also through control of pH to limit the cathode potential, many simple combinations of metals can be separated by this form of electrolysis. *See* COULOMETRIC ANALYSIS.

[GEORGE W. O'DOM]

Bibliography: A. J. Bard, *Electroanalytical Chemistry: A Series of Advances*, vols. 1–9, 1966–1976; A. J. Bard and L. R. Faulkner, *Electrochemical Methods: Fundamentals and Applications*, 1980; I. M. Kolthoff and P. J. Elving (eds.), *Treatise on Analytical Chemistry*, pt. 1, vol. 4, 1963; L. Meites and H. C. Thomas, *Advanced Analytical Chemistry*, 1958.

Electrokinetic phenomena

Phenomena associated with the movement of charged particles through a continuous medium or with the movement of a continuous medium over a charged surface. The four principal electrokinetic phenomena are electrophoresis, electroosmosis, streaming potential, and sedimentation potential, or Dorn effect. These phenomena are related to one another through the zeta potential ζ of the electrical double layer which exists in the neighborhood of the charged surface.

Electrically charged layers. The distribution of electrolyte ions in the neighborhood of a negatively charged surface and the variation of potential ψ with distance from the surface are shown in Fig. 1. According to O. Stern, two different layers of ions are associated with the charged surface. The layer of ions immediately adjacent to the surface is called the Stern layer. The ions of this layer are held to the charged surface by a combination of electrostatic attraction and specific adsorption forces, such as short-range van der Waals interactions and chemical bonds. The thickness δ of this layer is assumed to be equal to the ionic radius of the adsorbed ion species. The second layer of ions in the Gouy layer. The boundary between the two layers is the limiting Gouy plane. The ions in the Gouy layer are acted upon only by electrostatic forces and thermal motions of the liquid environment (Brownian motion), and they form a diffuse atmosphere of opposite charge (positive charge in Fig. 1) to the net charge at the limiting Gouy plane. The net charge density of the diffuse ion atmosphere of the Gouy layer decreases exponentially with distance from the limiting Gouy plane. The Gouy layer forms the positive half of an electrical double layer, and the charged surface plus the Stern layer form the negative half. The effective distance of separation $1/\kappa$ between the two halves of the double layer is determined by the concentration of electrolyte (ionic strength). For an electrolyte of univalent ions in water at 25°C, the relationship for $1/\kappa$ from the Debye-Hückel theory is Eq. (1), where c is the concentration of electrolyte (moles/liter).

$$\frac{1}{\kappa}=\frac{3\times 10^{-8}}{\sqrt{c}} \tag{1}$$

Variation of potential ψ with distance x from the charged surface is shown by a solid curve in Fig. 1. Here ψ_0 represents the thermodynamic reversible electrode potential which is independent of the properties of the electrical double layer and dependent only on the activity of the ion which is in reversible electrochemical equilibrium with the substance of the charged surface. The potential ψ decreases linearly with increasing distance x in the region of the Stern layer. In the region of the Gouy layer, ψ decreases exponentially with increasing distance x, as shown by G. Gouy and W. Chapman.

Displacement of charged layers. In the four listed electrokinetic phenomena, a displacement occurs at some plane (plane of shear) between the charged surface and its atmosphere of ions. The position of the slipping plane in Fig. 1 is shown to be located in the Gouy layer. The potential of the plane of shear is the ζ-potential. From the theories of Gouy and Chapman, for spherical particles Eq. (2) holds. Here $1/\kappa$ is the effective thickness of the

$$\zeta=\frac{q}{Da}\left(\frac{1}{1+\kappa a}\right) \tag{2}$$

double layer, q the net charge of the particle inside the plane of shear, D the dielectric constant of the liquid, and a the particles radius at the plane of shear. For flat surfaces Eq. (3) holds where e is the

$$\zeta=\frac{4\pi e}{D\kappa} \tag{3}$$

charge per unit area of surface. Equations (2) and (3) show that ζ-potential is determined by the net charge at the plane of shear and $1/\kappa$, the effective thickness of the ion atmosphere. In turn, ζ-potential controls the rate of transport between the charged surface and the adjacent liquid. The relationship between rate of transport v_E and ζ-potential which is valid for all four electrokinetic phenomena is Eq. (4), where v_E is the velocity of the

$$v_E=\frac{D\zeta E}{4\pi\eta} \tag{4}$$

liquid at a large distance from the charged surface, E is the field strength (volts/cm), and η is the viscosity of the liquid. The conditions for validity of Eq. (4) are that the double layer thickness ($1/\kappa$) must be small compared to the radius of curvature of the surface; the substance of the surface must be nonconducting; and the surface conductance of the interface must be negligible. The equations which relate ζ-potential to electroosmotic flow rate and streaming potential may be obtained from Eq. (4) by use of Poiseuille's law for laminar flow through a capillary. For electrophoresis and sedimentation potential (Dorn effect), v_E is the velocity of the particles. E is the applied field strength for

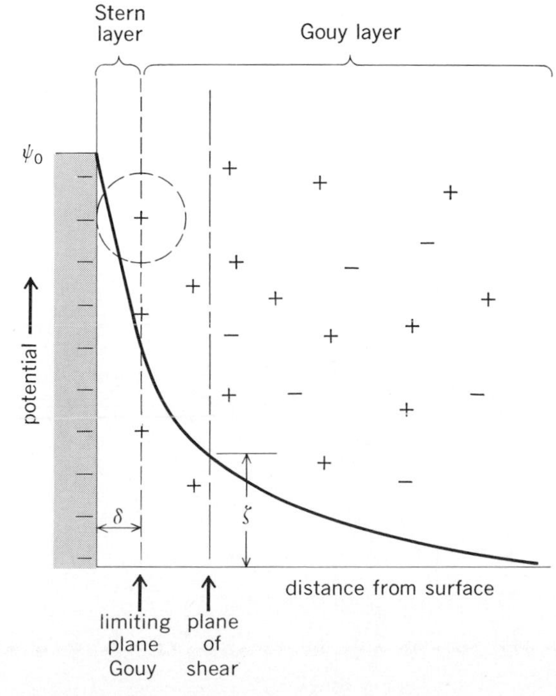

Fig. 1. Electrical double layer.

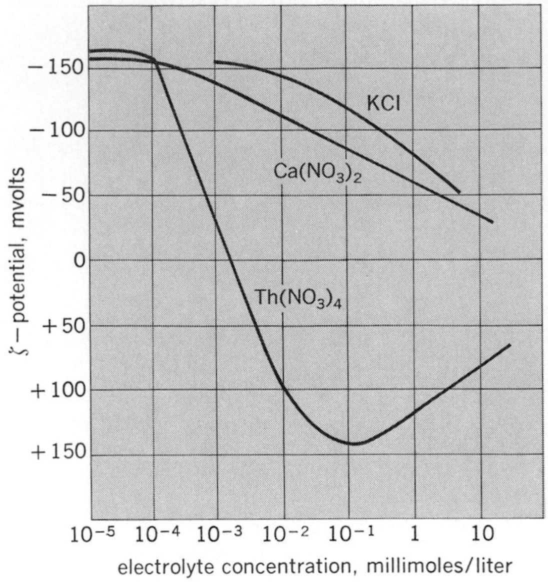

Fig. 2. Effect of electrolyte concentration on the ζ-potential of glass-water interfaces.

electrophoresis, whereas it is the gradient of potential developed by the sedimentation of charged particles in the Dorn effect.

Electrophoresis, electroosmosis, and streaming potential experiments have been shown to yield identical ζ-potentials for several different interfaces, particularly glass-water and protein-water systems. The sedimentation potential has not been significantly studied.

The effect of electrolytes on the ζ-potential of glass-water interfaces is shown in Figs. 2 and 3. As

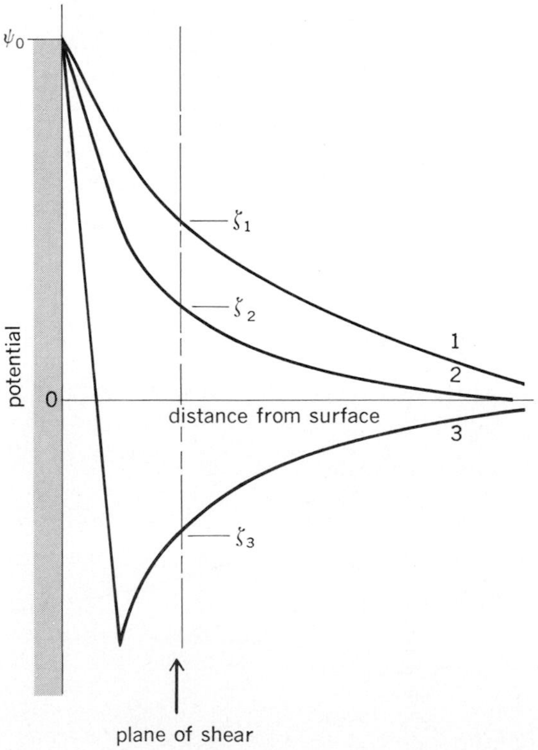

Fig. 3. Reversal of ζ-potential by ion adsorption.

shown in Fig. 2, an increase in electrolyte concentration produces a decrease in ζ-potential, and ions of high charge of opposite sign to that of the surface can completely reverse the sign of the ζ-potential. The explanations for these two effects are given in Fig. 3, where the variation in ψ with distance from the surface is shown for low concentration of electrolyte in curve 1; moderate concentration of electrolyte in curve 2; and charge reversal by adsorption of ions (Th^{4+} on glass) in curve 3. Curves 1 and 2 show that an increase in electolyte concentration reduces ζ-potential by reducing $1/\kappa$, as indicated by Eqs. (1), (2), and (3). Curve 3 shows that reversal of charge by ion adsorption occurs in the Stern layer and that this gives rise to a ζ-potential of opposite sign to the original value. *See* COLLOID; ELECTROPHORESIS; STREAMING POTENTIAL.

[QUENTIN VAN WINKLE]

Electrolysis

A method by which reactions are carried out in solutions of electrolytes or in molten salts by use of electricity. As shown in the diagram, the elec-

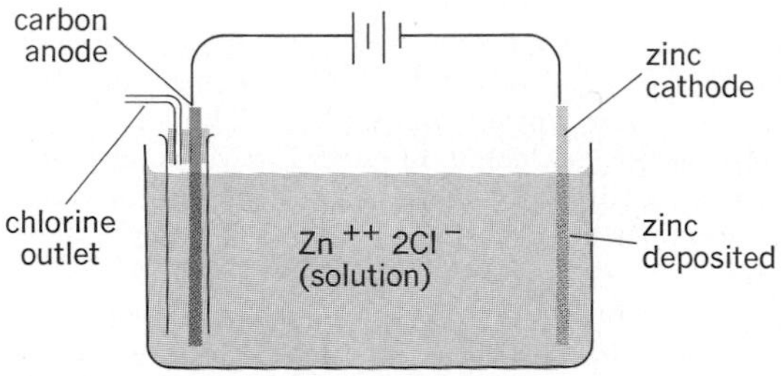

Electrolysis of zinc chloride solution.

trodes of an electrolytic cell are immersed in an electrolyte solution or in molten salts and are connected to a direct-current power supply. One or several reactions occur at each electrode when current flows through the cell. Reduction, a reaction in which electrons are consumed, occurs at the electrode called the cathode; oxidation occurs at the anode. For instance, sodium is produced at the cathode by reduction, and chlorine at the anode by oxidation in the electrolysis of molten sodium chloride.

Applications are important and varied: industrial production of chemicals, metallurgical extraction of metals, electroplating of metals, metal finishing, and production of electricity in batteries. Metallic corrosion often involves electrolytic processes. For application to analytical chemistry *see* ELECTRODEPOSITION ANALYSIS; POLAROGRAPHIC ANALYSIS.

Theory. The quantity of electrolysis products, their rate of production, and quite often their nature depend on electrolysis conditions. According to Faraday's law, the quantity of substance being consumed or produced by a single electrode reaction is proportional to the quantity of electricity consumed in electrolysis. This quantity of electricity is equal to the product of the current multiplied by the duration of electrolysis for a constant current or to the integral of the current over the duration of electrolysis for a variable current. *See* COULOMETER; ELECTROCHEMICAL EQUIVALENT.

The nature and the relative abundance of electrolysis products at each electrode generally depend on the electrode potential. Direct control of potential is rarely used, and electrode potentials in industrial cells are controlled indirectly by adjustment of the current density (current per unit area) at each electrode. Control is achieved because the current density depends on the electrode potential. Control of the electrolysis current has the advantage of allowing connection of several identical electrolytic cells in series in industrial installations. *See* DECOMPOSITION POTENTIAL; ELECTROMOTIVE FORCE (CELLS).

Electrolytic cells are characterized by their current efficiency and their power consumption efficiency. The current efficiency is the ratio of the quantity of a substance being consumed or produced to the theoretical quantity of this substance as calculated from Faraday's law. The current efficiency of a single electrode reaction occurring without losses, such as side reaction, electrolysis of the solvent, and evaporation, is 100%.

The second efficiency characteristic of an electrolytic cell, the power consumption efficiency, is the ratio of the theoretical electrical power to the actual electrical power that is consumed in the production or consumption of a given quantity of substance. The power consumption efficiency is smaller than 100% because of overvoltage phenomena, losses of products by side reactions, and ohmic drop (voltage drop) in the cell. Power efficiencies as low as 50% or even lower are not uncommon.

Applications. Industrial applications for inorganic substances are many. Such important chemicals as hydrogen, oxygen, hydrogen peroxide, chlorine, and sodium hydroxide are produced by electrolysis. Water is enriched in deuterium oxide (heavy water) by electrolysis. (There is isotopic separation because the overvoltage for discharge of deuterium ions is larger than for hydrogen ions.) Certain metals such as aluminum, magnesium, and sodium are produced by electrolysis of molten salts. Deposition of these metals from aqueous solution is impossible because this reaction requires higher cathodic potentials than hydrogen evolution. Likewise, fluorine is produced by oxidation of fluoride ion in anhydrous hydrofluoric acid; electrolysis of aqueous solution of fluoride produces oxygen because this reaction occurs at lower anodic potentials than fluorine evolution.

Electroplating of thin layers of a corrosion-resistant metal on fabricated objects is an important technique. Chromium and nickel are most commonly used, but electroplating of other metals such as gold, silver, and copper also has applications. The composition of the electrolytic bath influences the structure and surface finish of the metallic coating; numerous formulas involving metallic complexes and organic additives have been developed empirically. Electroplating is also applied to the industrial refining of copper, silver, gold, and nickel. Certain metals (copper, zinc, and cadmium) are extracted from low-grade ores by electrolysis of a solution of their ores (electrowinning). The opposite reaction of electroplating —anodic oxidation—is applied to metal finishing in electropolishing.

Electrolysis of organic compounds has found only a few industrial applications, although numerous electrode reactions have been studied. Purely chemical preparative methods are more economical and often simpler than electrolysis. In some cases, however, electrolysis involves reactions which are more easily controlled than purely chemical methods. Important reactions include reduction of nitro compounds, aldehydes, ketones, carboxylic acids, unsaturated compounds, and halogenated substances; oxidation of fatty acids (Kolbe reaction), alcohols, aldehydes, ketones, and sugars; and halogenation by anodic oxidation. *See* ELECTROCHEMISTRY.

[PAUL DELAHAY]

Bibliography: B. E. Conway, *Theory and Principles of Electrode Processes*, 1965; A. Kuhn (ed.), *Industrial Electrochemical Processes*, 1971; F. A. Lowenstein (ed.), *Modern Electroplating*, 3d ed., 1974; C. L. Mantell, *Industrial Electrochemistry*, 3d ed., 1950; K. J. Vetter, *Electrochemical Kinetics: Theoretical Aspects*, 1967.

Electrolyte

A chemical compound which when fused or dissolved in certain solvents, usually water, will conduct an electric current. The passage of the current is always accompanied by decomposition of the electrolyte, called electrolysis, which takes place at the electrodes. All electrolytes in the fused state or in solution give rise to ions which conduct the electric current. The phenomena of electrolysis are summarized by Faraday's laws of electrolysis. All acids, bases, and salts are electrolytes. *See* ELECTRODE; ELECTROLYSIS; ION; SOLVENT.

Electrolytes are divided into strong and weak electrolytes. Strong electrolytes usually contain a stable ionic bond and are wholly ionized in solution and usually in the crystalline state. Weak electrolytes are only partially ionized in solution. Metallic hydroxides and salts are usually strong electrolytes, for example, potassium hydroxide and sodium chloride. A weak electrolyte contains a covalent bond which on dissolving in a solvent such as water may be transformed into an ionic bond. A solution of a weak electrolyte contains both the ionic and the covalent forms in equilibrium; for example, acetic acid in water consists of a mixture of undissociated molecules, CH_3COOH, and of the ions CH_3COO^- and H^+. *See* CHEMICAL BONDING; IONIC EQUILIBRIUM.

[THOMAS C. WADDINGTON]

Electrolytic conductance

The transport of electric charges, under electric potential differences, by particles of atomic or larger size. This phenomenon is distinguished from metallic conductance, which is due to the movement of electrons. The charged particles that carry the electricity are called ions.

Positively charged ions are termed cations; the sodium ion, Na^+, is an example. The negatively charged chloride ion, Cl^-, is typical of anions. The negative charges are identical with those of electrons or integral multiples thereof. The unit positive charges have the same magnitude as those of electrons but are of opposite sign. Colloidal particles, which may have relatively large weights, may be ions, and may carry many positive or negative charges. Electrolytic conductors may be solids, liquids, or gases. Semiconductors have properties that are intermediate between the metallic and electrolytic types.

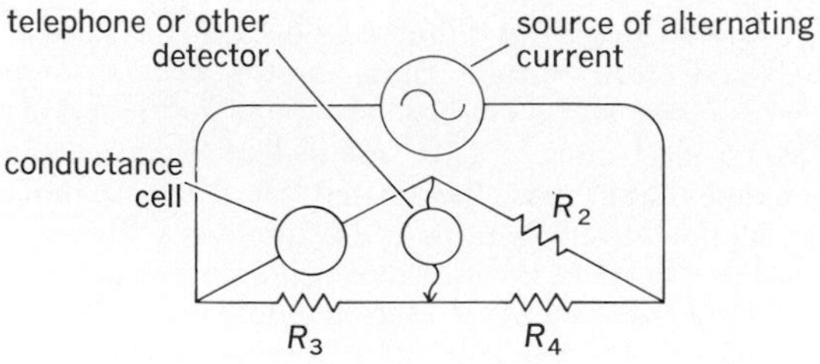

Fig. 1. Wheatstone bridge circuit for the measurement of electrolytic conductance. R_3 and R_4 are fixed resistances; R_2 is a variable resistance.

Measurement. Conductances are usually reported as specific conductances κ, which are the reciprocals of the resistances of cubes of the materials, 1 centimeter (cm) in each dimension, placed between electrodes 1 cm square, on opposite sides. These units are sometimes called mhos, that is, ohms spelled backward. Conductances of solutions are usually measured by Friedrich Kohlrausch's method, in which a Wheatstone bridge is employed. Such a bridge is shown diagrammatically in Fig. 1. The resistances R_3 and R_4 (usually of the same value) form two arms of the bridge. Resistance R_2 is adjustable, and the remaining arm is the cell holding the electrolytic conductor, or as is usually stated, solution of electrolyte. Direct current and the usual galvanometers cannot be used because of an apparent failure of Ohm's law. Passage of direct current produces chemical reactions and a back electromotive force (emf) is generated by the galvanic action of the products. By using an alternating current, the electrochemical reactions occurring when the current is briefly passed in one direction may be reversed when the direction of the current is changed. When a small alternating-current (ac) input signal is used, practically all of the electric charge passed during each half cycle is stored in the electric double layer, which acts as a capacitor. The electrodes are usually made of platinum and are platinized, that is, coated with finely divided platinum. The surface area, and hence the electrode capacitance, is thereby greatly increased. By making measurements at several frequencies and extrapolating to infinite frequency, the effect of electrode reactions can be eliminated. For less exact measurements, a fixed frequency of 60–1000 Hz is commonly used.

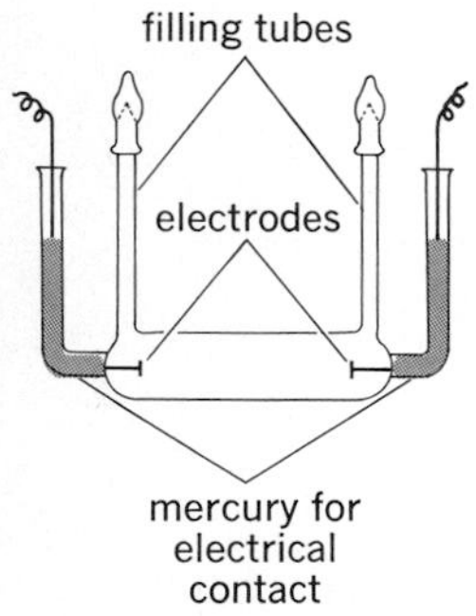

Fig. 2. Conductance cell

To determine the conductance C, that is, the reciprocal resistance $1/R$ of the cell of Fig. 1, the resistance R_2 of the bridge is adjusted until a minimum of sound is heard in the telephone.

Greater sensitivity may be obtained by electronic amplification of the off-balance signal and by using a "tuning eye" or an oscilloscope to detect the point of balance. When the bridge is in balance, the conductance is given by the relation $C = R_4/R_2R_3$. From this the specific conductance κ may be obtained from the equation $\kappa = KC$, in which K is the cell constant. Occasionally, this constant can be computed from the dimensions of the cell. Usually, however, it is determined by using a solution whose κ value is accurately known from measurements in such a cell, or, as was done by G. Jones and B. C. Bradshaw, by comparison with the specific conductance of mercury.

For precision work, care must be taken to avoid errors due to electrical reactances. This has been done in bridges designed by Jones and R. C. Josephs, and by Theodore Shedlovsky. A typical, properly designed conductance cell is shown in Fig. 2. The cell is filled with solution through the center tubes. Electrical contact is made with the electrodes by platinum wires sealed through the glass wall. These connect the mercury in the outside tubes, which are widely spread to avoid errors due to electrical capacity.

Equivalent conductance. Although many substances and mixtures show electrolytic conductance, the greater part of the research on the subject has been on aqueous solutions of salts, acids, and bases. There have been considerable data accumulated for solutions of such electrolytes in nonaqueous solvents, such as alcohols. The data are usually given in terms of equivalent conductance Λ, which is defined by Eq. (1), in which κ is the

$$\Lambda = \frac{1000\kappa}{c} \qquad (1)$$

specific conductance and c is the concentration in equivalents per liter. Values of Λ change with the concentration and, in general, increase as the solutions measured are made more dilute, that is, as c is decreased. A plot of values of the equivalent conductance Λ against $\sqrt{c}$ for some typical electrolytes is shown in Fig. 3. Svante Arrhenius, who was the first to assume that electrolytic conductance is due to freely moving charged ions, explained the decrease of Λ with increasing c by assuming that the number of ionic carriers gets smaller as the concentration increases, and he computed a degree of dissociation α by the formula shown in Eq. (2). The term Λ_0 is obtained by determining Λ at a

$$\alpha = \frac{\Lambda}{\Lambda_0} \qquad (2)$$

series of low concentrations and extrapolating to a limiting value, termed the equivalent conductance at infinite dilution. Though Eq. (2) has been shown by later work to give nearly the right values of α

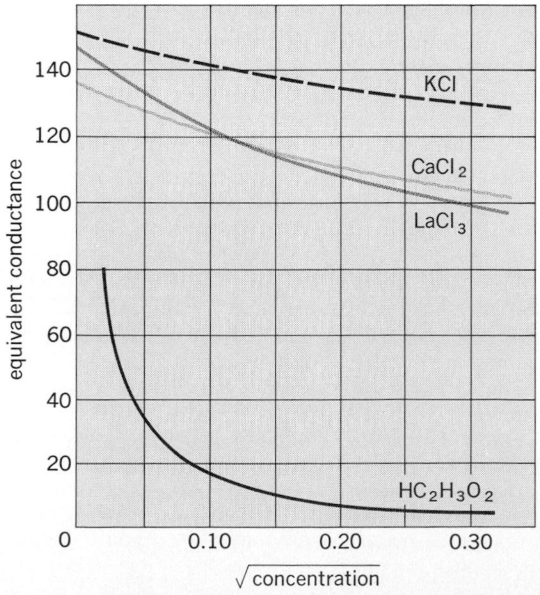

Fig. 3. Equivalent conductance at infinite dilution for some typical electrolytes.

for certain poorly conducting solutions, it is now considered to be much in error for the so-called strong electrolytes. These include most salts, such as potassium chloride, KCl, and sodium sulfate, Na_2SO_4, and inorganic acids and bases, such as hydrochloric acid, HCl, and sodium hydroxide, NaOH. For an electrolyte which yields two types of ion, it can be shown that Eq. (3) holds, in which F

$$\Lambda = F\alpha(U^+ + U^-) \tag{3}$$

is the faraday and U^+ and U^- are the mobilities, or speeds, under unit potential difference of the positive and negative ions, respectively. For Eq. (2) to hold, these mobilities must be constant from the equivalent concentration c at which Λ is measured to infinite dilution. This requires that the transference numbers of the ions be constant in the same range, which is seldom the case.

Since the advent of the Debye-Hückel theory of interionic attractions, strong electrolytes have been considered to be completely dissociated, that is, the term α of Eq. (3) is equal to unity for these substances. The decreases observed in the values of the equivalent conductances Λ, with increases in concentration, are assumed to be due to reductions in the values of the ionic mobilities U^+ and U^-. According to the theory of P. Debye and E. Hückel, the ion possesses an ionic atmosphere distributed with radial symmetry around the ion as center. This is due to the fact that interionic attractions and repulsions, together with thermal vibrations, tend to produce a slight preponderance of negative ions around a positive ion, and vice versa. The presence of this atmosphere leads to the lowering of ionic mobilities with increasing ion concentrations. The adaptation of the Debye-Hückel theory for conducting solutions is due to Lars Onsager. His equation for very dilute uni-univalent electrolytes, such as sodium chloride, NaCl, is displayed in Eq. (4), in which θ and σ are given

$$\Lambda = \Lambda_0 - (\theta\Lambda_0 + \sigma)\sqrt{c} \tag{4}$$

for uni-univalent electrolytes by Eqs. (5) and (6), in

$$\theta = \frac{8.16 \times 10^5}{(DT)^{3/2}} \tag{5}$$

$$\sigma = \frac{8.28}{\eta(DT)^{1/2}} \tag{6}$$

which D is the dielectric constant at the absolute temperature T and η is the viscosity.

Equation (4) yields accurate values of the data for strong electrolytes up to concentrations of about 0.001 M, above which there are small deviations. Modifications of Eq. (4) for solutions of salts of higher valence types, such as calcium chloride, $CaCl_2$, and lanthanum chloride, $LaCl_3$, are available and have also been found to agree with the data for dilute solutions.

Onsager, in his derivation of Eq. (4), treated the ions as point charges. Later Raymond Fuoss and Onsager extended the theory to include the radii of the ions and also the effects of higher concentrations. The ion sizes obtained from conductance data agree closely with those calculated from activity measurements.

Equation (4), or its empirical and theoretical extensions, can be used to obtain values of Λ_0 from the data on equivalent conductances of dilute solutions. Some typical figures for limiting equivalent conductances Λ_0 of typical strong electrolytes in aqueous solution at 25°C are listed below.

HCl	426.16	NH_4Cl	149.86	$MgCl_2$	129.40
NaOH	247.8	KNO_3	144.96	NaI	126.94
KBr	151.9	$CaCl_2$	135.84	NaCl	126.45
KI	150.38	Na_2SO_4	129.9	LiCl	115.03

Ion conductances. Values of the limiting equivalent conductance Λ_0 may be assigned to each of the ions of an electrolyte. Thus, for potassium chloride, $\Lambda_{0,KCl} = \lambda_{0,K^+} + \lambda_{0,Cl^-}$, and the value of λ_{0,Cl^-} is the same whether it is derived from measurements on HCl, NaCl, or KCl solutions. This additive relation is known as Kohlrausch's law of the independent mobility of ions. However, it is necessary to obtain the value of λ_0 for at least one ion constituent independently in order to establish the ion conductances of the other ions. The relation used is shown in Eqs. (7), in which Λ_0 is the lim-

$$t_0^+\Lambda_0 = \lambda_0^+ \quad \text{or} \quad t_0^-\Lambda_0 = \lambda_0^- \tag{7}$$

iting equivalent conductance of an electrolyte and t_0^+ and t_0^- are the limiting transference numbers of the positive and negative ion constituents, respectively.

The same value of λ_{0,Cl^-}, within 0.02%, is obtained from precision conductance and transference measurements on solutions of hydrogen, lithium, sodium, and potassium chlorides. Values of the limiting ionic conductance at 25°C are given below for some ions.

H^+	349.82	$1/2Ba^{++}$	63.64
OH^-	198	Ag^+	61.92
SO_4^{--}	79.8	$1/2Ca^{++}$	59.50
Br^-	78.4	$1/2Mg^{++}$	53.06
I^-	76.8	Na^+	50.11
Cl^-	76.34	HCO_3^-	44.48
K^+	73.52	$CH_3CO_2^-$	40.9
NH_4^+	73.4	$CH_2ClCO_2^-$	39.7
NO_3^-	71.44	Li^+	38.69

Nonaqueous systems. In addition to the study of water solutions of electrolytes, considerable study has been given to electrolytes in nonaqueous and mixed solvents. In general, the same principles as those outlined above apply to the interpretation of the results. However, fewer of the electrolytes are completely dissociated, and the degrees of dissociation of the weaker acids and bases are lower. This is due to the fact that, in general, the dielectric constants of nonaqueous solvents are smaller than those of water, so that the attractions between positive and negative ions are greater.

It will be observed that in this article discussion is confined to quite dilute solutions of electrolytes. For concentrated solutions few generalizations of any value can be given.

Molten salts exhibit a wide range of conductivities, depending upon their structures. Salts of alkali and alkaline earth metals usually are largely ionic in character and are highly conductive in the molten state, whereas heavy metal salts may be essentially covalent and exhibit little or no conductivity. Thus the conductivities of liquid $AsCl_3$ and $BiCl_3$ near their melting points are approximately 10^{-6} and 0.44 ohm^{-1}-cm^{-1}, respectively, reflecting the more ionic structure of $BiCl_3$.

If, instead of using quite low potentials in the measurement of electrolytic conductances, voltages of the order of 100,000 are employed, the conductances observed are no longer constant but tend to increase with the potential used. Under these conditions Ohm's law evidently is not valid. This increase of conductance with high potentials is called the Wien effect. This effect is in accord with the interionic attraction theory. When the velocity of the ions becomes sufficiently great, the ion atmospheres do not have time to form to their full extent, so that both the electrophoretic and time of relaxation effects exert less influence on the conductance. However, a large Wien effect is also found for weak acids and bases. It would appear that the high potentials produce, temporarily, additional ionization of these substances. This explanation has been proposed and discussed theoretically by Onsager. If very high frequencies are used in the measurements, an increase in the conductance, termed the Debye-Falkenhagen effect, is observed. This can also be explained by the interionic attraction theory. *See* ELECTROCHEMISTRY; ELECTROMOTIVE FORCE (CELLS).

[HERBERT A. LAITINEN]

Bibliography: H. S. Harned and B. B. Owen, *Physical Chemistry of Electrolytic Solutions*, 3d ed., 1958; R. A. Robinson and R. H. Stokes, *Electrolyte Solutions*, 2d ed., 1966.

Electromotive force (cells)

When two dissimilar electrodes are connected through an external conducting circuit, a difference in electric potential exists between them. Although this difference in potential is sometimes called the potential difference of the electrode couple, it is customary to say that the galvanic cell composed of the two dissimilar electrodes exhibits an electromotive (driving) force. This electromotive force (emf) is the resultant of the relative potential forces of the two dissimilar electrodes at which electrochemical reactions occur during cell operation. The two dissimilar electrodes need not be of unlike metals; for example, the metals may both be copper, but with the two coppers immersed in solutions of different concentration or composition. Likewise, both electrodes may be of the same gas, but with the pressure of the gas different at the two electrodes. *See* ELECTRODE POTENTIAL.

Cell types. Galvanic cells are of two general types, reversible and irreversible. If the chemical reactions at the electrodes can be exactly reversed by reversals in the direction of the current flow at the electrodes, the cell is said to be reversible; if the chemical reactions cannot be reversed, or if entirely different reactions occur on current reversal, the cell is then of the irreversible type. An example of a reversible cell is

$$(-)Pb(s)|PbSO_4(s)|H_2SO_4(aq)|PbO_2(s)|PbSO_4(s)|Pb(s)\,(+)$$

where s = solid, aq = aqueous solution, and the vertical lines represent the interfaces between different phases. This cell is the familiar lead-acid storage cell (or battery), widely used for a variety of purposes, including starting, ignition, and lighting for automobiles. When the cell is discharged, that is, when electric energy is being drawn from it, reactions (1) and (2) occur at the negative and

$$Pb(s) + SO_4^{--}(aq) \rightarrow PbSO_4(s) + 2e^- \qquad (1)$$

$$PbO_2(s) + 4H^+(aq) + SO_4^{--}(aq) + 2e^- \xrightarrow{Pb} PbSO_4(s) + 2H_2O(l) \qquad (2)$$

positive electrodes, respectively. Here l = liquid. The overall cell reaction is (3), whereby lead sul-

$$Pb(s) + PbO_2(s) + 2H_2SO_4(aq) \rightarrow 2PbSO_4(s) + 2H_2O(l) \qquad (3)$$

fate and water are formed in the cell reaction. The lead in the reaction at the positive electrode merely serves as an electronic conductor.

Now when the cell is charged, the reverse of the above reactions occurs, and lead sulfate is converted back to the initial state. The cell is, therefore, called a reversible one.

If, however, a cell is prepared by immersing zinc and platinum electrodes in perchloric acid, an irreversible cell results. This cell may be represented as

$$(-)\,Zn(s)|HClO_4(aq)|Pt(s)\,(+)$$

When the cell is discharged, reactions (4) and (5) occur at the negative and positive electrodes, respectively.

$$Zn(s) \rightarrow Zn^{++}(aq) + 2e^- \qquad (4)$$

$$2H^+(aq) + 2e^- \rightarrow H_2(\text{gas, on platinum surface}) \qquad (5)$$

The overall cell reaction is (6). Here g = gas.

$$Zn(s) + 2H^+(aq) \rightarrow Zn^{++}(aq) + H_2\ (g) \qquad (6)$$

Now, if the cell is charged, that is, subjected to an electrolyzing current, instead of discharged, reactions (7) and (8) occur at the negative and positive electrodes, respectively.

$$2H^+(aq) + 2e^- \rightarrow H_2(g, \text{on zinc surface}) \qquad (7)$$

$$2H_2O(l) \rightarrow O_2(g, \text{on platinum surface}) + 4H^+(aq) + 4e^- \qquad (8)$$

The overall cell reaction is then (9), or the

$$2H_2O(l) \rightarrow 2H_2(g) + O_2(g) \qquad (9)$$

simple electrolysis of water. Since the electrode reactions at each electrode for the charge differ from those obtained for the discharge, the cell is of the irreversible type.

In many cases the reversibility of a cell, or of the electrodes composing the cell, can be determined only by means of precise electrical measurements. Reversibility in the strict sense is determined by the response of a cell to very small (infinitesimal) discharging or charging current. In practice, measurements are made simultaneously of the current and the electromotive force of the cell for different values of the current. The slope of the electromotive force of the cell versus the current is ascertained for both the charging and the discharging currents. For reversible cells, the two slopes should be identical; furthermore, these slopes should be reproducible for repeated reversals in the direction of the flow of current through the cell. Also, for reversible cells, the internal resistance is small, and the magnitude of the slopes will be nearly zero. This means that the electromotive force of a reversible cell is insignificantly affected by the passage of very small currents through it in either direction. The reversibility of a galvanic cell is primarily a function of the reversibility of the two electrodes composing the cell, and the same criteria may be used to establish the reversibility of the electrodes.

Energy relations. Conditions may be chosen whereby practically no current flows through a cell in either direction. These conditions may be achieved by balancing the electromotive force of the cell against the electromotive force of another cell of known and steady value by using a highly sensitive galvanometer. When this state is achieved, the measured electromotive force of the cell E represents its reversible or maximum value. This value represents the maximum driving force of the cell. When it is multiplied by the total number of coulombs corresponding to the cell reaction, it gives the maximum electrical work, or free energy, which the cell is capable of producing. Thus, $E \times nF = -\Delta G$, where n is the number of electrons (or valence change) involved in the cell reaction, F is the faraday, and ΔG is the change in (Gibbs) free energy associated with the cell reaction. Furthermore, if the variation of the electromotive force of the cell with temperature is measured, the heat of reaction ΔH for the cell may be computed by the Gibbs-Helmholtz relation shown in Eq. (10),

$$\Delta H = -nFE + nFT\frac{dE}{dT} \qquad (10)$$

which may also be written as $\Delta H = \Delta G + T\Delta S$, where ΔS is the entropy change for the cell reaction since $nF(dE/dT) = \Delta S$.

If a galvanic cell has a negative emf-temperature coefficient, the heat of the reaction exceeds the free-energy change. Thus, the available electrical work is less than that corresponding to the heat of the reaction for the cell; the cell warms in operation, and heat is lost to the surroundings. Conversely, if the emf-temperature coefficient is positive, the free energy exceeds the heat of reaction, and the cell tends to cool when in operation; this cooling is overcome if the cell is maintained under isothermal conditions by absorption of heat from the surroundings. Thus, the total available electrical energy from such a cell is a resultant of the inherent changes in the heat content of the cell and the heat absorbed from the surroundings. *See* FREE ENERGY.

Electromotive force measurements. As stated above, in measuring the emf of a cell, a reference cell of known emf must be available to effect a comparison. This reference cell should have an emf that is known in terms of physical laws and units, and not one chosen arbitrarily. The emf of reference cells (and then, through comparisons, the emf of all cells) is known in terms of the mks (meter-kilogram-second) system of electromagnetic units through Ohm's law, $E = IR$, where E is emf in volts, I is current in amperes, and R is resistance in ohms. In practice, then, the emf of a cell is equal to IR, and it may be balanced against the IR drop across a resistor of known value through which a known current is flowing. If the values of the resistor and the current are known in mks electromagnetic units, then the emf of a cell is given in like units.

Standard cells. The standard cells for this purpose are the cadmium amalgam standard cells of the saturated type proposed by Edward Weston in 1892. This type of cell is the most reversible galvanic cell known and retains a constant emf to within a few microvolts for many decades. This cell consists of a cadmium amalgam anode (negative element), a mercury–mercurous sulfate cathode (positive element), and a saturated solution of cadmium sulfate containing crystals of $CdSO_4 \cdot \frac{8}{3}H_2O$. This cell may be represented by

$$\text{Cd-Hg}(10\%)(2p)|CdSO_4\cdot{}^8\!/_3H_2O(c)|\;CdSO_4(sat.\,aq)|CdSO_4\cdot{}^8\!/_3H_2O(c)|\;Hg_2SO_4(s)|Hg(l)$$

where $2p$ = two phase, c = crystals, *sat. aq* = saturated aqueous solution, and the other symbols have the meaning given above. This cell has an emf, when freshly made, of 1.018636 volts at 20°C, and for precise measurements must be maintained at a constant temperature. Its emf at other temperatures between 0 and 43.5°C may be calculated from Eq. (11), where t is in degrees Celsius. Saturated standard cells should not be used above 43.5°C; at this temperature, the crystals of $CdSO_4 \cdot \frac{8}{3}H_2O$ are converted to $CdSO_4 \cdot H_2O$. Although standard cells prepared with the monohydrate are stable, they can be used with confidence only at temperatures above 43.5°C; when such cells are cooled, the monohydrate reverts to $CdSO_4 \cdot \frac{8}{3}H_2O$, but the rate of conversion is slow and the emf of such cells is erratic for indefinite periods.

$$E_t = 1.018636 - 0.0000406(t-20) - 0.00000095(t-20)^2 + 0.00000001(t-20)^3 \qquad (11)$$

The cadmium standard cell is also made in the unsaturated type, that is, with no crystals of $CdSO_4 \cdot \frac{8}{3}H_2O$, and is the type widely used in recording instruments, with potentiometers, and in pH meters. A solution of cadmium sulfate that is saturated at 4°C is used in its preparation; the solution is then unsaturated at higher and normal temperatures. It is made portable by placing cork or plastic septa over the positive and negative elements. This cell has a very low emf-temperature coefficient (0.000005 volt/°C), about one-tenth that of the saturated type. On the average, this cell decreases in emf at the rate of 20 μvolts per year, and its ultimate life is about 10 years.

For the saturated type of cell, the overall reaction of the cell is (12), where x moles of Cd are associated with y moles of Hg in the amalgam, and m is the number of moles of water associated with 1 mole of $CdSO_4$ in the saturated solution. For the unsaturated cell the cell reaction is simply (13).

$$xCd(yHg)(l+s) + Hg_2SO_4(s) + \frac{8/3}{m-8/3}CdSO_4\cdot mH_2O(l) = (CdSO_4\cdot{}^8\!/_3H_2O) + 2Hg(l) + (x-1)Cd(yHg)(l+s) \qquad (12)$$

$$xCd(yHg)(l+s) + Hg_2SO_4(s) = CdSO_4(aq) + 2Hg(l) + (x-1)Cd(yHg)(l+s) \qquad (13)$$

A cross-sectional sketch of the saturated type of

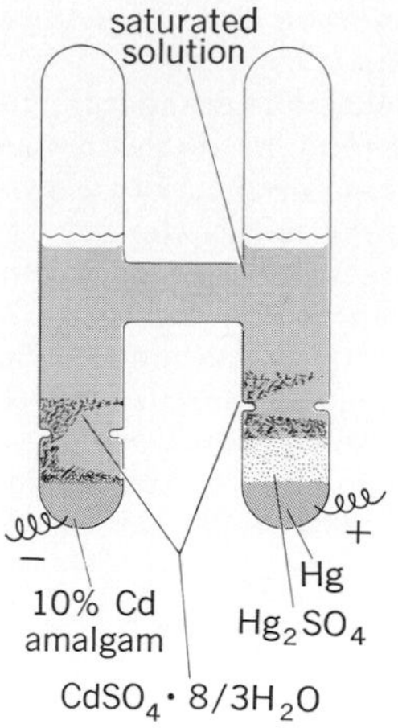

Cross-sectional diagram of Weston cell.

cell is shown in the illustration. The unsaturated type is made similarly except that no crystals are used and cork or plastic septa are placed above the amalgam and the mercurous sulfate paste. The H-form container is made of Kimball glass with platinum wires sealed in the bottom of each limb. Mercury purified by several vacuum distillations is placed at the bottom of one limb and is used to prepare the amalgam used in the bottom of the other limb. The cadmium is purified by sublimation. A 10% (sometimes 12.5%) amalgam of cadmium is added while warm and in a single phase; on cooling, it becomes two-phased, with the solid phase being an isomorphous mixture of cadmium and mercury. Mercury(I) sulfate, prepared electrolytically, is then placed over the mercury, and crystals of $CdSO_4 \cdot \frac{8}{3}H_2O$ are added to both limbs. Then a saturated solution of cadmium sulfate is added to a level slightly above the crossarm, and the cell is sealed. The unsaturated cell is usually mounted in a nontransparent case, and the saturated cell is housed in oil baths or in thermoregulated air baths. After proper aging, these cells serve as reliable standards of emf with which the emf of all other cells are compared. *See* CALOMEL ELECTRODE; ELECTROCHEMISTRY; HYDROGEN ELECTRODE; SALT BRIDGE.

[WALTER J. HAMER]

Bibliography: P. Delahay, *New Instrumental Methods in Electrochemistry*, 1954; W. J. Hamer, *Standard Cells: Their Construction, Maintenance, and Characteristics*, 1965; D. J. G. Ives and G. J. Janz, *Reference Electrodes*, 1960; W. M. Latimer, *Oxidation Potentials*, 2d ed., 1962; G. W. Vinal, *Primary Batteries*, 1950.

Electron affinity

The amount of energy release when an electron at rest is captured by a species M, producing the negative ion M^-. The electron affinity of a species M can also be thought of as the ionization potential of the negative ion M^-. Stated in terms of a chemical equation, the electron affinity of a species M is equal to the exothermocity of the reaction $e + M \rightarrow M^-$, where the negative ion M^- is left in its lowest electronic, vibrational, and rotational state.

If the electron affinity of M is negative, the M^- ion is unstable with respect to decomposition into $M + e$. Most atoms have positive electron affinities, even though there is no net Coulomb attraction between the electron and the atom until the electron is close enough to be "a part of the atom." The simple rules of chemical valency provide a qualitative guide to the magnitude of electron affinities. Thus the noble gases, which have a filled outer electronic shell and are chemically inert, are not capable of binding an additional electron to form a negative ion. The largest electron affinities are possessed by the halogens, atoms which require only one additional electron to fill the valence shell.

The major exception to this concept is that multiply charged negative ions—for example, O^{--}, one of many multiply charged negative ions which are stable in solution—are not stable in the gas phase. The ability to place more than one additional electron in the valence shell of a neutral atom or molecule appears to come from the medium; the solvent shell surrounding the ion in liquid solutions and the amorphous or crystalline region surrounding the ion in solids.

Experimental methods. While accurate ionization potentials of the elements have been known for a number of years, comparable data for electron affinities of the elements have become available only more recently. In order to determine the ionization potential of an element, one can simply make a vapor of the element, place it in an optical spectrometer, and look for the onset of absorption corresponding to the photoionization process $h\nu + A \rightarrow A^+ + e$. The energy of the photon corresponding to the threshold wavelength for this process is the ionization energy of the species A. The analogous method for determination of an electron affinity is through observation of the threshold of the very similar photodetachment reaction $h\nu + A^- \rightarrow A + e$. Again, the threshold wavelength for this process corresponds to the electron affinity of the species A. Unfortunately, it has not proved possible to produce sufficiently large densities of negative ions to be able to observe directly the threshold for the photodetachment process in a photoabsorption measurement; consequently, determination of accurate electron affinities has lagged far behind the determination of accurate ionization potentials.

The major experimental advances which now enable accurate electron affinity determinations have been the development of ion-beam techniques and the availability of intense light sources in the form of lasers. In modern experiments to determine electron affinities, negative ions are formed in an electrical discharge, extracted through a small aperture into a high-vacuum region, formed into a negative ion beam, mass-analyzed, and intersected by an intense laser beam. The laser-beam–negative-ion-beam intersection takes place in a high-vacuum region where no collisions are likely. The occurrence of a photodetachment event is determined by detection of the photodetached electron.

Two experimental methods have evolved which can produce accurate electron affinities. In the first method the laser is a tunable laser, and one searches for the wavelength corresponding to the threshold for the photodetachment process. In this case the electron affinity is given directly by the threshold wavelength, and is in principle determinable to accuracies of 10^{-6} eV. In the second type of

1 H 0.7542							2 He <0
3 Li 0.620	4 Be <0	5 B 0.282	6 C 1.268	7 N ⩽0	8 O 1.462	9 F 3.399	10 Ne <0
11 Na 0.546	12 Mg <0	13 Al 0.442	14 Si 1.385	15 P 0.743	16 S 2.0772	17 Cl 3.615	18 Ar <0
19 K 0.5012	20 Ca <0	31 Ga 0.3	32 Ge 1.2	33 As 0.80	34 Se 2.0206	35 Br 3.364	36 Kr <0
37 Rb 0.4860	38 Sr <0	49 In 0.3	50 Sn 1.25	51 Sb 1.05	52 Te 1.9708	53 I 3.061	54 Xe <0
55 Cs 0.4715	56 Ba <0	81 Tl 0.3	82 Pb 0.349	83 Bi 0.947	84 Po 1.9	85 At 2.8	86 Rn <0

21 Sc <0	22 Ti 0.2	23 V 0.526	24 Cr 0.667	25 Mn <0	26 Fe 0.164	27 Co 0.667	28 Ni 1.157	29 Cu 1.226	30 Zn <0
39 Y ≈0	40 Zr 0.429	41 Nb 0.886	42 Mo 0.747	43 Tc 0.7	44 Ru 1.1	45 Rh 1.138	46 Pd 0.558	47 Ag 1.303	48 Cd <0
57 La 0.5	72 Hf <0	73 Ta 0.323	74 W 0.816	75 Re 0.15	76 Os 1.1	77 Ir 1.566	78 Pt 2.128	79 Au 2.3086	80 Hg <0

Periodic table showing the best values for the electron affinities of the elements. All values are reported in electronvolts. The value <0 implies that the negative ion is unstable with respect to decomposition to an electron and a neutral atom. The solid bar below represents the relative uncertainty in the electron affinity. (*From H. Hotop and W. C. Lineberger, Binding energies in atomic negative ions, J. Phys. Chem. Ref. Data, 4:539–576, 1975*)

experiment, called photoelectron spectroscopy, a fixed-frequency laser (of known photon energy) is employed, and electrostatic fields are used to determine the kinetic energy of the ejected electron. From simple energy conservation arguments, the electron affinity is then given by the photon energy less the kinetic energy of the ejected electron. This latter technique is quite general, but is limited in accuracy by the resolution of the electron energy analyzer (typically 10^{-2} eV). *See* MASS SPECTROMETRY.

Periodic trends. These laser photodetachment studies dramatically improved knowledge of the electron affinities of the elements. The illustration is a periodic table showing the current best values of the electron affinities of the elements. Most of the data shown here were obtained by using laser photodetachment methods. The periodic trends in electron affinities and the qualitative effects described earlier are immediately apparent. In addition, a number of more subtle trends are observable. For example, while one expects that filled-shell species such as the rare gases will not be capable of binding an additional electron, the illustration shows that half-filled shells (for example, N and P) also exhibit small or negative electron affinities. Again, this effect is the result of the fact that a half-filled valence shell is spherically symmetric and behaves somewhat as though it were a filled shell. A similar situation is also seen for half-filled *d*-shells, as in the transition metals.

These same techniques are beginning to provide a number of accurate electron affinity determinations for molecules and free radicals, and new insight into the structural and chemical properties of ions in the gas phase. [W. C. LINEBERGER]

Bibliography: R. R. Corderman and W. C. Lineberger, Negative ion spectroscopy, *Annu. Rev. Phys. Chem.*, 30:347–376, 1979; H. Hotop and W. C. Lineberger, Binding energies in atomic negative ions, *J. Phys. Chem. Ref. Data*, 4:539–576, 1975; B. K. Janousek and J. I. Brauman, Electron affinities, in M. T. Bowers (ed.), *Gas Phase Ion Chemistry*, 1980; H. S. W. Massey, *Negative Ions*, 3d ed., 1976.

Electron configuration

The orbital arrangement of an atom's electrons. Negatively charged electrons are attracted to a positively charged nucleus to form an atom or ion. Although such bound electrons exhibit a high degree of quantum-mechanical wavelike behavior, there still remain particle aspects to their motion. Bound electrons occupy orbitals that are somewhat concentrated in spatial shells lying at different distances from the nucleus. As the set of electron energies allowed by quantum mechanics is discrete, so is the set of mean shell radii. Both these quantized physical quantities are primarily specified by integral values of the principal, or total, quantum number n. The full electron configuration of an atom is correlated with a set of values for all the quantum numbers of each and every electron. In addition to n, another important quantum number is l, an integer representing the orbital angular momentum of an electron in units of $h/2\pi$, where h is Planck's constant. The values 1, 2, 3, 4, 5, 6, 7 for n and 0, 1, 2, 3 for l together suffice to describe the electron configurations of all known normal atoms and ions, that is, those that have their lowest possible values of total electronic energy. The first seven shells are also given the letter designations *K, L, M, N, O, P*, and *Q* respectively. Electrons with l equal to 0, 1, 2, and 3 are designated *s, p, d*, and *f*, respectively.

In any configuration the number of equivalent electrons (same n and l) is indicated by an integral exponent (not a quantum number) attached to the letters *s, p, d*, and *f*. According to the Pauli exclusion principle, the maximum is s^2, p^6, d^{10}, and f^{14}. For example, the configuration $1s^2 2s^2 2p^6 3s$ of the ground or lowest energy state of sodium means that there are two electrons with $n = 1$, $l = 0$; two with $n=2$, $l=0$; six with $n=2$, $l=1$; and one with $n = 3$, $l = 0$. Higher values of n and l can be achieved by excitation of normal atoms, either through photon absorption or by particle impact, temporarily moving one or more electrons to unfilled shells of larger radii and increasing the atom's electronic energy. Such excited electronic configurations are inherently unstable. Excited electrons ultimately fall back down to their normal locations closer to the nucleus, a process accompanied by the emission of one or more quanta of electromagnetic radiation (photons). While the atom is excited, the partially depleted shells are said to possess vacancies.

An electron configuration is categorized as having even or odd parity, according to whether the sum of p and f electrons is even or odd. Strong spectral lines result only from transitions between configurations of unlike parity.

Insofar as they are known from spectroscopic investigations, the electron configurations characteristic of the normal or ground states of the first 103 chemical elements are shown in the table. In the next-to-last column of the table, the spectral term of the energy level with lowest total electronic energy is shown. The main part of the term

Distribution of electrons in the atoms

Element and atomic number		K	L		M			N				O				Ground term	Ionization potential, eV
		1, 0 1s	2, 0 2s	2, 1 2p	3, 0 3s	3, 1 3p	3, 2 3d	4, 0 4s	4, 1 4p	4, 2 4d	4, 3 4f	5, 0 5s	5, 1 5p	5, 2 5d	5, 3 5f		
H	1	1	—	—	—	—	—	—	—	—	—	—	—	—	—	$^2S_{1/2}$	13.5981
He	2	2	—	—	—	—	—	—	—	—	—	—	—	—	—	$^1S_0$	24.5868
Li	3	2	1	—	—	—	—	—	—	—	—	—	—	—	—	$^2S_{1/2}$	5.3916
Be	4	2	2	—	—	—	—	—	—	—	—	—	—	—	—	$^1S_0$	9.322
B	5	2	2	1	—	—	—	—	—	—	—	—	—	—	—	$^2P^0_{1/2}$	8.298
C	6	2	2	2	—	—	—	—	—	—	—	—	—	—	—	$^3P_0$	11.260
N	7	2	2	3	—	—	—	—	—	—	—	—	—	—	—	$^4S^0_{3/2}$	14.534
O	8	2	2	4	—	—	—	—	—	—	—	—	—	—	—	$^3P_2$	13.618
F	9	2	2	5	—	—	—	—	—	—	—	—	—	—	—	$^2P^0_{3/2}$	17.422
Ne	10	2	2	6	—	—	—	—	—	—	—	—	—	—	—	$^1S_0$	21.564
Na	11	Neon configuration			1	—	—	—	—	—	—	—	—	—	—	$^2S_{1/2}$	5.139
Mg	12				2	—	—	—	—	—	—	—	—	—	—	$^1S_0$	7.646
Al	13				2	1	—	—	—	—	—	—	—	—	—	$^2P^0_{1/2}$	5.986
Si	14				2	2	—	—	—	—	—	—	—	—	—	$^3P_0$	8.151
P	15				2	3	—	—	—	—	—	—	—	—	—	$^4S^0_{3/2}$	10.486
S	16				2	4	—	—	—	—	—	—	—	—	—	$^3P_2$	10.360
Cl	17				2	5	—	—	—	—	—	—	—	—	—	$^2P^0_{3/2}$	12.967
Ar	18				2	6	—	—	—	—	—	—	—	—	—	$^1S_0$	15.759
K	19	Argon configuration					—	1	—	—	—	—	—	—	—	$^2S_{1/2}$	4.341
Ca	20						—	2	—	—	—	—	—	—	—	$^1S_0$	6.113
Sc	21						1	2	—	—	—	—	—	—	—	$^2D_{3/2}$	6.54
Ti	22						2	2	—	—	—	—	—	—	—	$^3F_2$	6.82
V	23						3	2	—	—	—	—	—	—	—	$^4F_{3/2}$	6.74
Cr	24						5	1	—	—	—	—	—	—	—	$^7S_3$	6.765
Mn	25						5	2	—	—	—	—	—	—	—	$^6S_{5/2}$	7.432
Fe	26						6	2	—	—	—	—	—	—	—	$^5D_4$	7.870
Co	27						7	2	—	—	—	—	—	—	—	$^4F_{9/2}$	7.86
Ni	28						8	2	—	—	—	—	—	—	—	$^3F_4$	7.635
Cu	29						10	1	—	—	—	—	—	—	—	$^2S_{1/2}$	7.726
Zn	30						10	2	—	—	—	—	—	—	—	$^1S_0$	9.394
Ga	31						10	2	1	—	—	—	—	—	—	$^2P^0_{1/2}$	5.999
Ge	32						10	2	2	—	—	—	—	—	—	$^3P_0$	7.899
As	33						10	2	3	—	—	—	—	—	—	$^4S^0_{3/2}$	9.81
Se	34						10	2	4	—	—	—	—	—	—	$^3P_2$	9.752
Br	35						10	2	5	—	—	—	—	—	—	$^2P^0_{3/2}$	11.814
Kr	36						10	2	6	—	—	—	—	—	—	$^1S_0$	13.999
Rb	37	Krypton configuration								—	—	1	—	—	—	$^2S_{1/2}$	4.177
Sr	38									—	—	2	—	—	—	$^1S_0$	5.693
Y	39									1	—	2	—	—	—	$^2D_{3/2}$	6.38
Zr	40									2	—	2	—	—	—	$^3F_2$	6.84
Nb	41									4	—	1	—	—	—	$^6D_{1/2}$	6.88
Mo	42									5	—	1	—	—	—	$^7S_3$	7.10
Tc	43									5	—	2	—	—	—	$^6S_{5/2}$	7.28
Ru	44									7	—	1	—	—	—	$^5F_5$	7.366
Rh	45									8	—	1	—	—	—	$^4F_{9/2}$	7.46
Pd	46									10	—	—	—	—	—	$^1S_0$	8.33

symbol is a capital letter, S, P, D, F, and so on, that represents the total electronic orbital angular momentum. Attached to this is a superior prefix, 1, 2, 3, 4, and so on, that indicates the multiplicity, and an anterior suffix, 0, ½, 1, 3/2, 2, 5/2, and so on, that shows the total angular momentum, or J value, of the atom in the given state. A sign ° above the J value signifies that the spectral term and electron configuration have odd parity.

The last column of the table presents the first ionization potential of the atom when this has been derived from spectroscopic observations. In any atomic spectrum, two or more spectral lines with certain similar properties may form a series such that the reciprocal wavelengths $1/\lambda$ (number of waves per centimeter $= \sigma$) can be closely represented by a formula of the Rydberg type, $\sigma = L - R/(n + \mu)^2$, in which L is the limit of the series. R is called the Rydberg constant, and the principal quantum number n has successive integral values to which a constant fractional part μ is added. The second term vanishes when n approaches infinity, and the series limit is thus evaluated. This limit is usually coincident with the ground state of the ion, and is thus a measure (in wave-number units) of the energy required to remove from an atom its least firmly bound electron and transform a neutral atom into a singly charged ion. The energy re-

Distribution of electrons in the atoms (cont.)

Element and atomic number	Configuration of inner shells	*N* 4, 3 4*f*	*O* 5, 0 5*s*	*O* 5, 1 5*p*	*O* 5, 2 5*d*	*O* 5, 3 5*f*	*P* 6, 0 6*s*	*P* 6, 1 6*p*	*P* 6, 2 6*d*	*Q* 7, 0 7*s*	Ground term	Ionization potential, eV
Ag 47	Palladium configuration	–	1	–	–	–	–	–	–	–	$^2S_{1/2}$	7.576
Cd 48		–	2	–	–	–	–	–	–	–	$^1S_0$	8.993
In 49		–	2	1	–	–	–	–	–	–	$^2P^0_{1/2}$	5.786
Sn 50		–	2	2	–	–	–	–	–	–	$^3P_0$	7.344
Sb 51		–	2	3	–	–	–	–	–	–	$^4S^0_{3/2}$	8.641
Te 52		–	2	4	–	–	–	–	–	–	$^3P_2$	9.01
I 53		–	2	5	–	–	–	–	–	–	$^2P^0_{3/2}$	10.457
Xe 54		–	2	6	–	–	–	–	–	–	$^1S_0$	12.130
Cs 55	The shells 1*s* to 4*d* contain 46 electrons	–	The shells 5*s* to 5*p* contain 8 electrons		–	–	1	–	–	–	$^2S_{1/2}$	3.894
Ba 56		–			–	–	2	–	–	–	$^1S_0$	5.211
La 57		–			1	–	2	–	–	–	$^2D_{3/2}$	5.5770
Ce 58		1			1	–	2	–	–	–	$^1G^0_4$	5.466
Pr 59		3			–	–	2	–	–	–	$^4I^0_{9/2}$	5.422
Nd 60		4			–	–	2	–	–	–	$^5I_4$	5.489
Pm 61		5			–	–	2	–	–	–	$^6H^0_{5/2}$	5.554
Sm 62		6			–	–	2	–	–	–	$^7F_0$	5.631
Eu 63		7			–	–	2	–	–	–	$^8S^0_{7/2}$	5.666
Gd 64		7			1	–	2	–	–	–	$^9D^0_2$	6.141
Tb 65		(8)			(1)	–	(2)	–	–	–	($^8G_{13/2}$)	5.852
Dy 66		10			–	–	2	–	–	–	$^5I_8$	5.927
Ho 67		11			–	–	2	–	–	–	$^4I^0_{15/2}$	6.018
Er 68		12			–	–	2	–	–	–	$^3H_6$	6.101
Tm 69		13			–	–	2	–	–	–	$^2F^0_{7/2}$	6.184
Yb 70		14			–	–	2	–	–	–	$^1S_0$	6.254
Lu 71		14			1	–	2	–	–	–	$^2D_{3/2}$	5.426
Hf 72	The shells 1*s* to 5*p* contain 68 electrons				2	–	2	–	–	–	$^3F_2$	6.865
Ta 73					3	–	2	–	–	–	$^4F_{3/2}$	7.88
W 74					4	–	2	–	–	–	$^5D_0$	7.98
Re 75					5	–	2	–	–	–	$^6S_{5/2}$	7.87
Os 76					6	–	2	–	–	–	$^5D_4$	8.5
Ir 77					7	–	2	–	–	–	$^4F_{9/2}$	9.1
Pt 78					9	–	1	–	–	–	$^3D_3$	9.0
Au 79	The shells 1*s* to 5*d* contain 78 electrons					–	1	–	–	–	$^2S_{1/2}$	9.22
Hg 80						–	2	–	–	–	$^1S_0$	10.43
Tl 81						–	2	1	–	–	$^2P^0_{1/2}$	6.108
Pb 82						–	2	2	–	–	$^3P_0$	7.417
Bi 83						–	2	3	–	–	$^4S^0_{3/2}$	7.289
Po 84						–	2	4	–	–	$^3P_2$	8.43
At 85						–	2	5	–	–	$^2P^0_{3/2}$	
Rn 86						–	2	6	–	–	$^1S_0$	10.749
Fr 87						–	2	6	–	(1)		
Ra 88						–	2	6	–	2	$^1S_0$	5.278
Ac 89						–	2	6	1	2	$^2D_{3/2}$	5.17
Th 90						–	2	6	2	2	$^3F_2$	6.08
Pa 91						2	2	6	1	2	$^4K_{11/2}$	5.89
U 92						3	2	6	1	2	$^5L_6$	6.05
Np 93						4	2	6	1	2	$^6L_{11/2}$	6.19
Pu 94						6	2	6	–	2	$^7F_0$	6.06
Am 95						7	2	6	–	2	$^8S^0_{7/2}$	5.993
Cm 96						7	2	6	1	2	$^9D^0_2$	6.02
Bk 97						(9)	2	6	(0)	(2)	$^6He^0_{5/2}$	6.23
Cf 98						(10)	2	6	(0)	(2)	$^5I_8$	6.30
Es 99						(11)	2	6	(0)	(2)	$^4I^0_{15/2}$	6.42
Fm 100						(12)	2	6	(0)	(2)	$^3H_6$	6.50
Md 101						(13)	2	6	(0)	(2)	$^2F^0_{7/2}$	6.58
No 102						(14)	2	6	(0)	(2)	$^1S_0$	6.65
Lw 103						(14)	2	6	(1)	(2)		

quired to ionize an atom is usually expressed in electronvolts (1 eV = 8065.48 wave numbers) and is called its first ionization potential.

Molecules consist of two or more atoms that at least partially share one or more electrons with one another. Sets of quantum numbers specify molecular electron configurations in much the same way as for atoms. However, the physical meaning of some molecular quantum numbers is different than that for the atomic case, as is the term notation. *See* MOLECULAR STRUCTURE AND SPECTRA.

[JAMES E. BAYFIELD]

Electron paramagnetic resonance (EPR) spectroscopy

The study of magnetic resonance spectra of materials which show paramagnetism because of the magnetic moment of unpaired electrons. EPR spectra are usually presented as plots of the absorption or dispersion of the energy of an oscillating magnetic field of fixed radio frequency versus the intensity of an applied static magnetic field. *See* ELECTRON SPIN.

EPR spectroscopy has been used for detection and identification of paramagnetic materials, for determinations of electronic structure, for studies of interactions between molecules, and for measurements of nuclear spins and moments. Among the wide variety of paramagnetic substances to which EPR spectroscopy has been applied are free radicals (including free atoms), impurity centers, and compounds of the transition elements, rare earths, and actinides. EPR spectra have been obtained from gases, liquids, and solids. Much of the work has been done with the oscillating magnetic field either in the vicinity of 9×10^9 Hz (X-band) or 24×10^9 Hz (K-band). Measurements at other frequencies lying between 10^6 and 10^{11} Hz have been performed.

Spectra characteristic of individual paramagnetic molecules, uncomplicated by magnetic interactions with neighboring paramagnetic molecules, may be obtained only from dilute solutions in diamagnetic solvents. The required degree of dilution depends on the nature of the magnetic molecules. Concentrations lower than 10^{16} molecules per cubic centimeter frequently must be used for solutions of organic free radicals, whereas concentrations as high as 10^{20} molecules per cubic centimeter may sometimes be tolerated for solutions of inorganic ions.

In some cases spin-lattice relaxation (exchange of magnetic energy with thermal motions of the environment) obscures the spectra. The effects are especially pronounced in inorganic magnetic ions and frequently require the use of low temperatures. The spectra of organic free radicals, on the other hand, are not severely affected and may usually be obtained at ordinary temperatures.

Solids. Maximum information is yielded by EPR spectra of solid solutions in single crystals. The spectrum of a single species may contain scores of

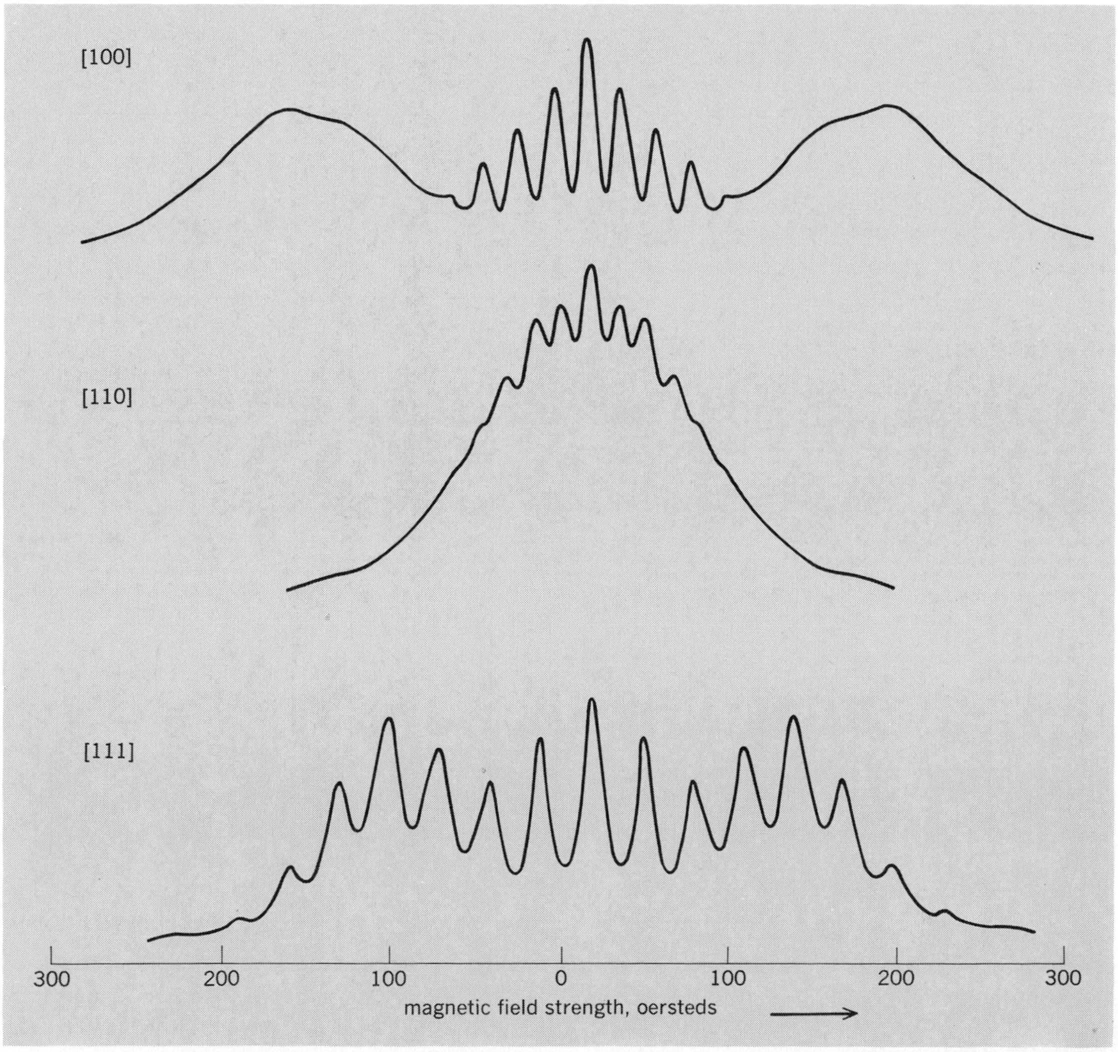

Fig. 1. Electron paramagnetic resonance spectra showing energy absorption versus magnetic field strength for a dilute solution of FeF_6^{3-} in a single crystal of Na_2KGaF_6. The indices [111], [110], and [100] give the direction of the magnetic field relative to the crystal axes. (*Courtesy of L. Helmholz*)

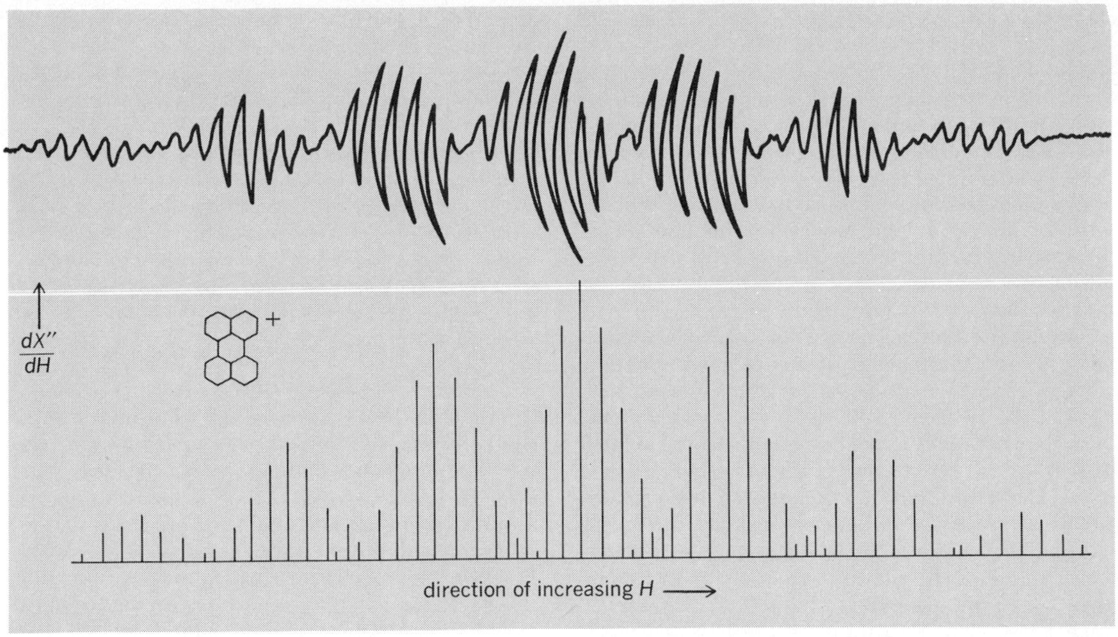

Fig. 2. Spectrum of dX''/dH versus H (X'' is the energy absorption, H is the magnetic field strength) for a liquid solution of the free radical perylene positive ion. Vertical lines give the positions and intensities of the spectral lines as calculated from molecular orbital theory. (*Courtesy of E. de Boer and S. I. Weissman*)

lines, their positions and intensities varying with orientation of the specimen relative to the static magnetic field. The many-lined structure results in part from interactions of the orbital motion of the electrons with the electric fields of the environment and from interactions of the magnetic moments of the electrons with nuclear magnetic moments. The latter effect (hyperfine interaction) is the sole cause of the splittings in the EPR spectra of most organic free radicals. Most magnetic ions exhibit pronounced anisotropy in their EPR spectra (Fig. 1) because of the anisotropic nature of the orbitals (wave functions) which give rise to their magnetism. Most organic free radicals and a few ions with highly symmetrical charge distributions or with highly quenched orbital magnetism exhibit little anisotropy in their EPR spectra.

From analysis of hyperfine interactions with nuclei whose spins and magnetic moments are known, details of the distribution of electrons about the nuclei may be determined. The average value of the cube of the reciprocal of the distances between electrons and nuclei, the orientation of the orbits relative to the crystal axes, and the density of unpaired electrons about the various nuclei in a free radical may be evaluated from the hyperfine structure.

Relative values of nuclear moments of isotopes may be found from the relative splittings produced by them in the same chemical environment. If only one isotope is available, its nuclear magnetic moment may be obtained from EPR spectra only if suitable properties of the electronic orbits are known. Nuclear spins, on the other hand, may be found simply by counting hyperfine components.

Liquids. Much work has been done, particularly with organic free radicals, in liquid solutions, where less information is obtainable by EPR spectroscopy than in crystals. Only averages of orientation-dependent properties can be observed in motions in liquids. The nature of the average is dependent on the rapidity of the motions. Well-resolved EPR spectra may be obtained in liquids only if the variations of positions of lines with orientation of molecular axes are not great. Many organic free radicals fulfill this requirement and yield lines only a few tenths of an oersted (1 Oe = 79.6 A/m) broad in liquid solutions. Highly characteristic spectra ranging from those containing only one line (such as the semiquinone of chloranil) to others containing more than 100 lines (triphenylmethyl) have been recorded. The spectra of organic free radicals are symmetrical about a center (Fig. 2). At fields of 3200 oersteds (25 kA/m), the centers usually lie within a few oersteds of the position of the resonance of the spin of a free electron, that is, one which has no spin-orbit interaction.

Hyperfine interactions are responsible for the complexity of the EPR spectra of most free radicals. Most of the splittings observed thus far have been produced by ^{1}H, which has a nuclear spin quantum number of $\frac{1}{2}$ ($I = \frac{1}{2}$). Splittings by ^{2}H, ^{11}B, ^{13}C, ^{14}N, and ^{15}N have also been studied. The contribution to the splitting by each kind of proton (in a free radical the protons differ in chemical environment) is determined by observation of the EPR spectra of radicals with appropriate substitutions of ^{1}H by ^{2}H ($I = 1$).

Rates of electron transfer. Migration of electrons among different molecules may produce measurable effects on the EPR spectra. In favorable cases, electron spin rates and mechanisms may be detected. In a stable free radical with resolved hyperfine structure, each line is associated with the frequency of precession of the electron spin in the presence of a particular arrangement of nuclear spins. When the electron jumps to a molecule with a different arrangement of nuclear spins, its spin precesses at a different frequency. Jumps occurring with mean time $1/\bar{\nu}$ between them add breadth $\bar{\nu}$ to the spectral lines as long as $\bar{\nu}$ is small compared with the separation of the various frequencies of precession. The method has been applied to measurements of electron-transfer reactions with second-order rate constants in the range $10^6 - 10^9$ liters/(mole)(s).

When $\bar{\nu}$ becomes large compared with separation of lines, a new spectrum appears. The hyperfine structure of the new spectrum reveals the nature of the groups of atoms which accompany the electron in its migrations.

Motional effects. The effects described above arise from modulation of the hyperfine interaction accompanying electron transfer from one molecule to another. Reorientation of a single paramagnetic molecule produces similar effects through modulation of the anisotropic part of the hyperfine interaction. The effect has been exploited for probing motional processes in large molecular structures, particularly ones of biological importance. Radicals containing the nitroxide group (NO) have been inserted into hemoglobins, enzymes, membranes, and other complex systems with little impairment of biological function. Owing to the large anisotropy in the nitrogen hyperfine interaction, the EPR spectra are sensitive to reorientation of the radical. Characteristic times for reorientation in the range 10^{-9} to 10^{-5} s may be measured.

Multiple resonance methods. Greatly enhanced resolution may be obtained in many cases through simultaneous irradiation with several frequencies. The most useful of these methods is electron nuclear double resonance (ENDOR). The material is simultaneously irradiated at one of its EPR resonant frequencies and by a second oscillatory field whose frequency is swept over the range of nuclear frequencies. The intensity of the EPR response is measured as a function of the second frequency. Changes in EPR intensity accompany passage through nuclear reasonance. Resonances as narrow as 10 kHz are observed. The spectra are far more easily interpreted than are conventional EPR ones, owing to the fact that each variety of nucleus yields only a doublet in ENDOR, independent of the number of such nuclei. The triphenylmethyl radical which contains 15 protons (3 para, 6 ortho, 6 meta) has 196 lines in its EPR spectrum, and only 3 doublets in its ENDOR spectrum. In crystalline materials, proton ENDOR has permitted precise location of hydrogen atoms.

Transient methods. The experimental methods described above rely on steady-state responses; that is, absorption or dispersion at each part of the spectrum is recorded only after all transients have subsided. Just as in nuclear magnetic resonance, the transients often contain more information than the steady-state responses. Echoes associated with electronic paramagnetism have now been observed in a variety of systems. They are particularly useful in materials with inhomogeneously broadened lines. A succession of two microwave pulses evokes a conventional echo; and a succession of three pulses, stimulated echoes. From the way in which the echo intensities vary with time intervals between pulses, relaxation parameters and hyperfine splittings may be obtained. Developments in solid-state technology, such as wide-banded microwave amplifiers and very rapid digitizers promise enhanced sensitivity of detection, as well as richer information than is now available. *See* NUCLEAR MAGNETIC RESONANCE (NMR).

[S. I. WEISSMAN]

Bibliography: N. M. Atherton, *Electron Spin Resonance: Theory and Applications*, 1973; A. Carrington and A. D. McLachlan, *Introduction to Magnetic Resonance with Applications to Chemistry and Chemical Physics*, 1967; D. J. E. Ingram, *Free Radicals as Studied by Electron Spin Resonance*, 1958; T. L. Squires, *An Introduction to Electron Spin Resonance*, 1964; J. E. Wertz, Nuclear and electronic spin magnetic resonance, *Chem. Rev.*, 55:829–955, 1955.

Electron spectroscopy

A form of spectroscopy which deals with the emission and recording of the electrons which constitute matter—solids, liquids, or gases. The usual form of spectroscopy concerns the emission or absorption of photons (x-rays, ultraviolet (uv) rays, visible or microwave wavelengths, and so on). Electron spectra can be excited by x-rays, which is the basis for electron spectroscopy for chemical analysis (ESCA), or by uv photons, or by ions (electrons; see Fig. 1). By means of x-ray or uv photons with energy $E_{h\nu}$ photoelectron spectra (PES) are produced when electrons with binding energies E_b are emitted with energy E_{kinetic} from bound molecular states, according to the equation below.

$$E_{\text{kinetic}} \times E_{h\nu} - E_b - E_r - \varphi$$

E_r is a usually negligible small recoil energy, and φ a small work function correction which can be attributed to the spectrometer material and to contact potentials. For insulating solid materials, precautions are taken for stabilizing the surface potential. For solids, the binding energies are preferably referred to the Fermi level, and for gases, to the vacuum level.

Modes of excitation. By means of ESCA, complete sets of photoelectron lines can be excited from the internal (core) levels as well as from the external (valence) region (Fig. 2). Also, complete sequences of the Auger electron lines are automatically obtained in this mode. A convenient source of excitation is the x-radiation from Al at 1487 eV. With the use of spherically bent quartz crystals this radiation can be further monochromatized to an inherent width of 0.2 eV. The commonly used light source for uv excitation of photoelectron spectra within the valence electron region is an He lamp that produces highly monochromatic radiation at 21.2 eV (internal width less than 10 meV) and (with less intensity) at 40.8 eV. Intermediate in energy for excitation are ultrasoft x-rays (Y Mξ at 132 eV). Another source of excitation in the intermediate region is provided by synchrotron radiation, which can be continually varied by means of a suitable monochromator. Excitation by means of an electron beam is an alternative mode to obtain Auger electron lines that has the advantage of ease of production. Ordinary photoelectron lines are not produced by that radiation. In order to compensate for the low-signal-to-background ratio when Auger electron spectra are excited by an electron beam impinging on a solid surface, the spectral distribution is frequently differentiated.

Applications. In ESCA, only electrons which are expelled from a surface layer of less than 50 A (5 nm) of a solid material contribute to the electron

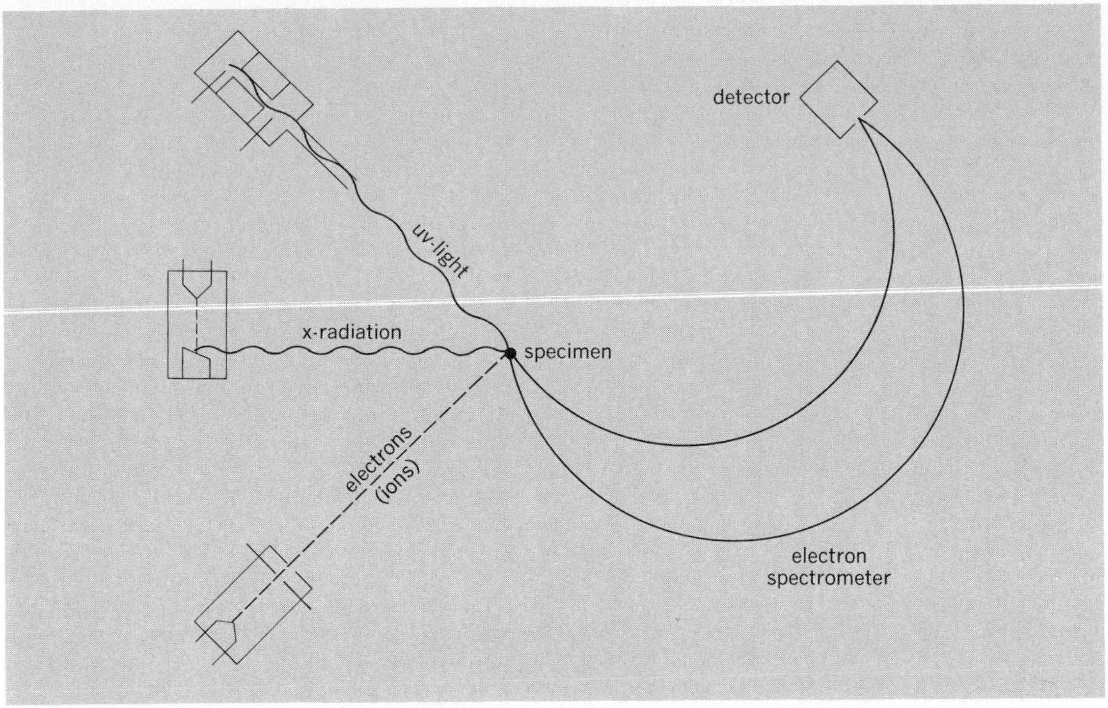

Fig. 1. Excitation of electron spectra recorded in high-resolution instruments.

line with the kinetic energy given above. Electrons from the interior of the material are scattered out from the line, and form a low background which does not interfere with the line character of the ESCA spectrum. The electron lines are extremely sharp and well suited for precision measurements. With a high-resolving ESCA spectrometer which has a magnetic or electrostatic focusing dispersive system, the electron lines have widths which are set by the limit caused by the uncertainty principle (the "inherent" widths of atomic levels). With a suitable choice of radiation, electron spec-

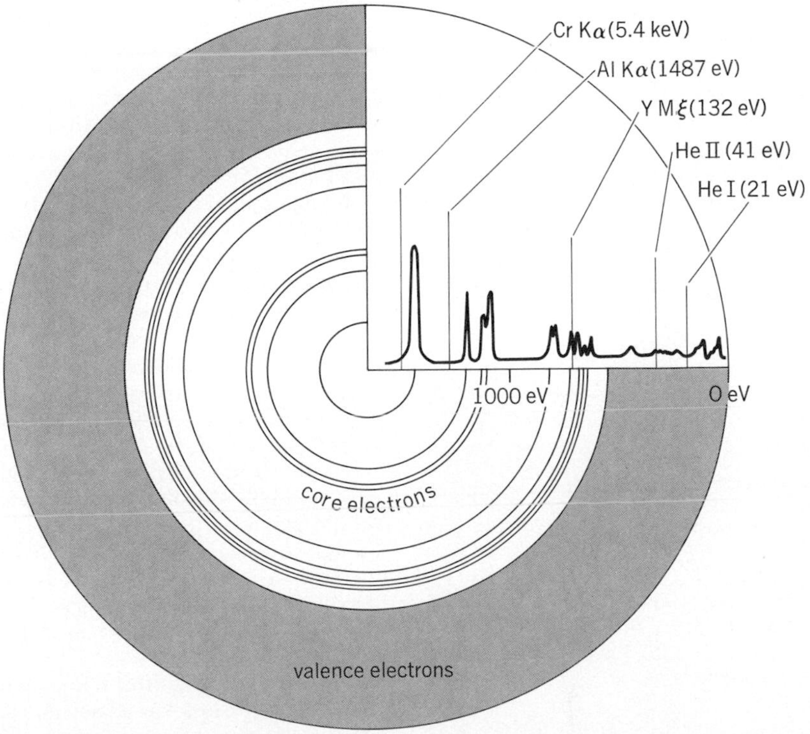

Fig. 2. The system of levels of an atom in a molecule can be divided into a valence electron region and an atomic-core region. Excitation of electron lines from the various regions can be made at different photon energies.

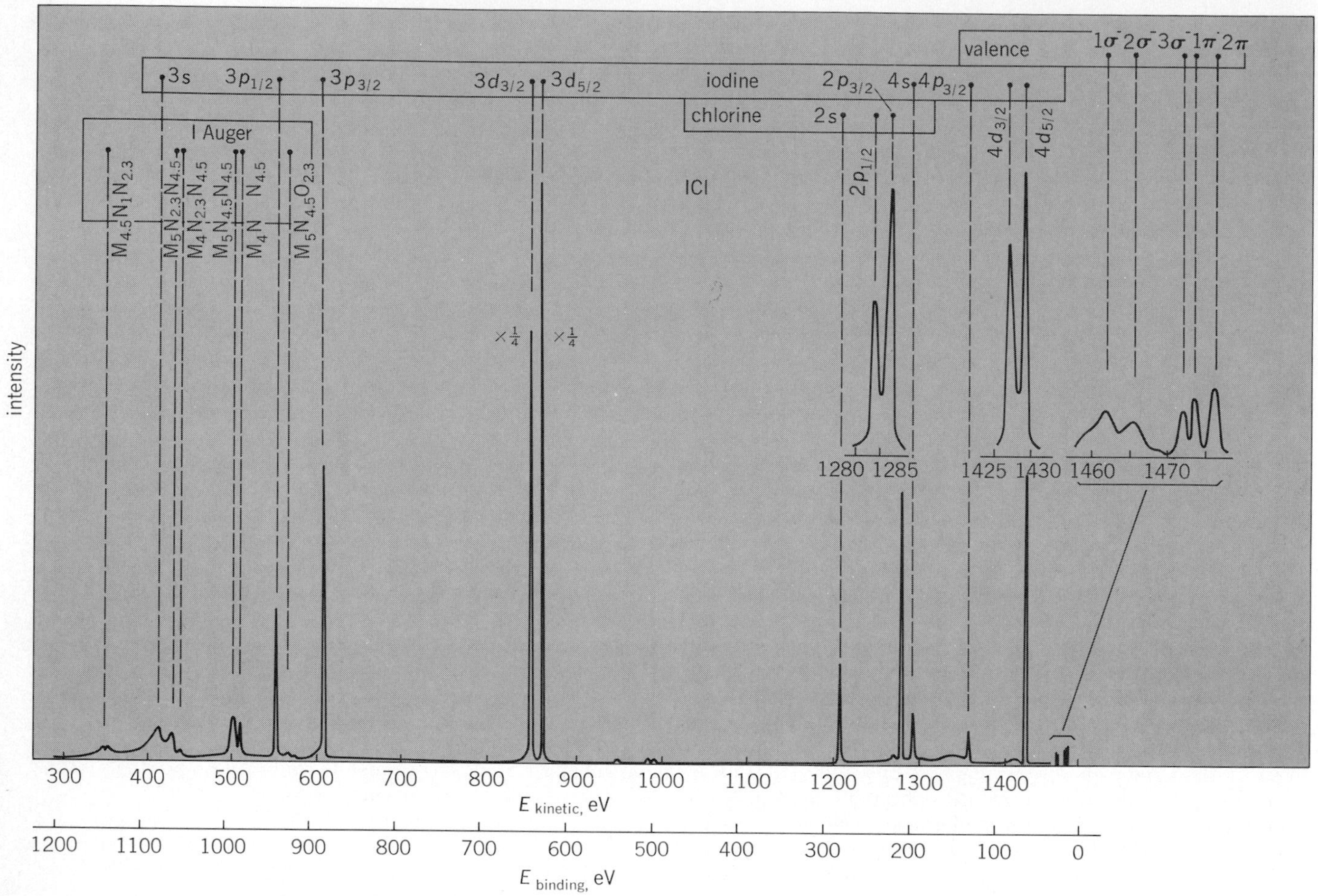

Fig. 3. Electron spectrum of the vapor of iodine chloride at a pressure of 0.04 torr, excited by monochromatic Al $K\alpha$ radiation with the ESCA instrument (scan spectrum). The *M*- and *N*-shell levels are excited in iodine, and the *L* levels in chlorine. At low binding energies (high kinetic energies), the valence spectrum of the molecule is recorded. The *MNN* Auger electron lines of iodine can also be seen.

troscopy reproduces directly the electronic level structure from the innermost shells (core electrons) to the atomic surface (valence or conduction band; see Fig. 3). Furthermore, all elements from hydrogen to the heaviest ones can be studied

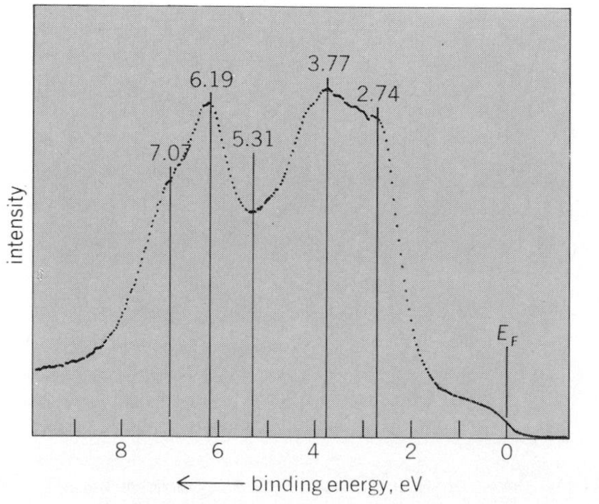

Fig. 4. Conduction electron spectrum of gold excited by monochromatized Al $K\alpha$ radiation. E_F = Fermi edge.

even if the element occurs together with several other elements and even if the element represents only a small part of the chemical compound. *See* LINE SPECTRUM.

When applied to solid materials, ESCA is a typical surface spectroscopy with applications to problems such as chemical surface reactions, for example, corrosion or heterogeneous catalysis. In such cases, ultrahigh vacuum conditions are usually required, with a vacuum less than 10^{-9} torr. ESCA also reproduces bulk matter properties such as valence electron band structures (Fig. 4). With uv excitation, the resolution is sufficient to resolve vibrational structures in valence electron spectra for gases (Fig. 5). In fact, electron spectroscopy can supply a detailed knowledge of the valence orbital structure for all molecules which can be brought into gaseous form with pressures of 10^{-5} torr or more. A convenient pressure region is 10^{-2} torr. Under certain conditions, liquids and solutions of various compositions can be studied by ESCA techniques.

ESCA chemical shifts. When atoms are brought close together to form a molecule, the electronic orbitals of each atom are perturbed. Inner orbitals, that is, those with higher binding energies, may still be regarded as atomic and belonging to speci-

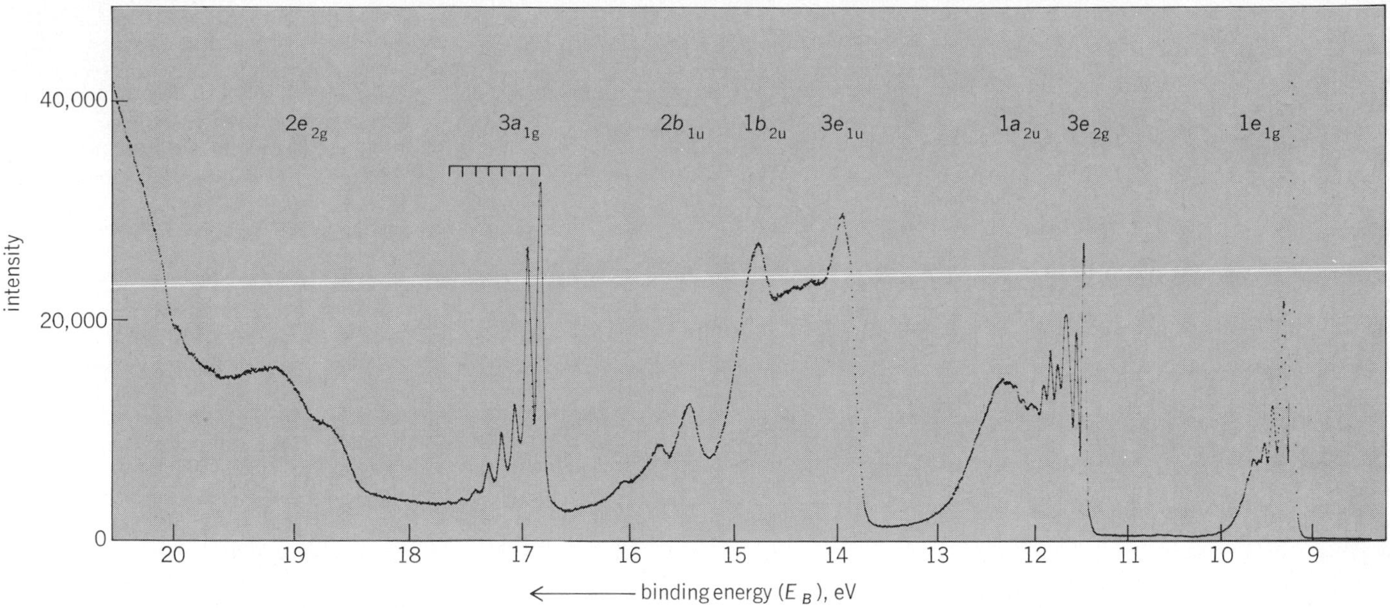

Fig. 5. Benzene valence electron spectrum excited by HeI radiation at 21.22 eV, showing vibrational structures.

fied atoms within the molecule, whereas the outer orbitals combine to form the valence-level system of the molecule. These orbitals play a more or less active part in the chemical bonds which are formed between the atoms in the molecule and which define the chemical properties. The chemical bonds affect the charge distribution so that the original neutral atoms can be regarded as charged to various degrees; a neutral molecule has a net charge of zero. The individual atoms in the molecule can be regarded as spheres with different charges set up by the transfer of certain small charges from one atomic sphere to the neighboring atoms taking part in the chemical bond. Inside each charged sphere the atomic potential is constant, in accordance with classical electrostatic theory. The result of this atomic potential is to shift the whole system of inner levels in an atom by a small amount, the same amount for each level. Levels belonging to different atoms in the molecule are generally shifted differently, however, and if the ESCA chemical shifts for individual atoms in the molecule are measured, a mapping can be made of the distribution of charge or potential in the molecule. This, then, is a reflection of the chemical bondings between the atoms, which in turn can be described by the orbitals in the valence-level system.

A unique feature of ESCA is that, if the exact position of the electron lines characteristic of the various elements in the molecule is measured, the area of inspection can be moved from one atomic species to another in the molecular structure. Figure 6 shows the electron spectrum from the 1*s* level of the carbon in ethyl trifluoroacetate. All four carbon atoms in this molecule are distinguished in the spectrum. The lines appear in the same order from left to right, as do the corresponding carbon atoms in the structure shown in the figure. If the structure of the molecule is known, the charge distribution can be estimated in a simple way by using, for example, the electronegativity concept and assuming certain resonance structures. More sophisticated quantum-chemical treatments can also be applied. Conversely, if, by means of ESCA, the approximate charge distribution is known, conclusions concerning the structure of the molecule can be drawn.

Experimental evidence obtained so far for various elements in a large number of molecules indicates strong correlations between chemical shifts and calculated atomic charges. A typical correlation curve for the 2*p* level in sulfur obtained from more than 100 compounds containing this element is shown in Fig. 7. Similar curves are obtained for

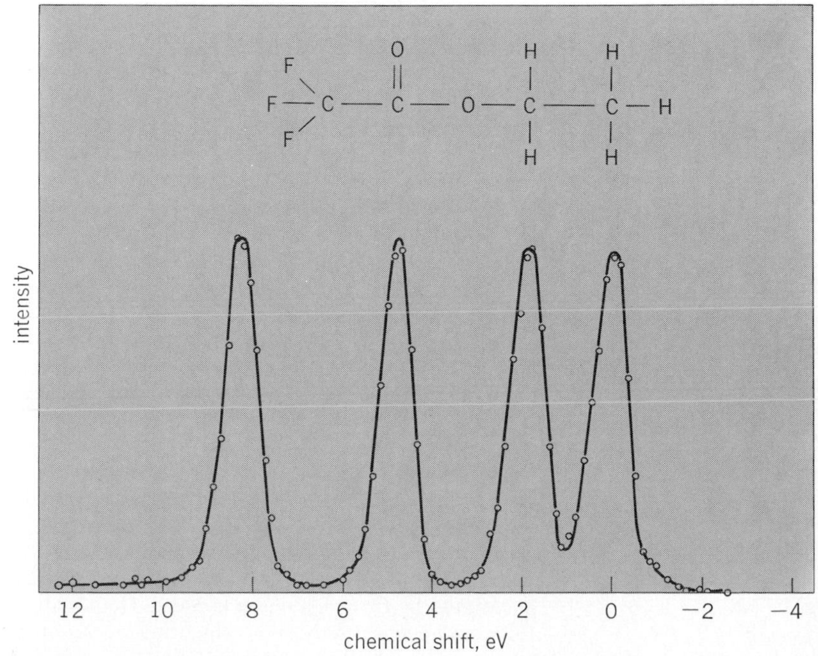

Fig. 6. Electron spectrum from the 1*s* level of carbon in ethyl trifluoroacetate; binding energy = 291.2 eV.

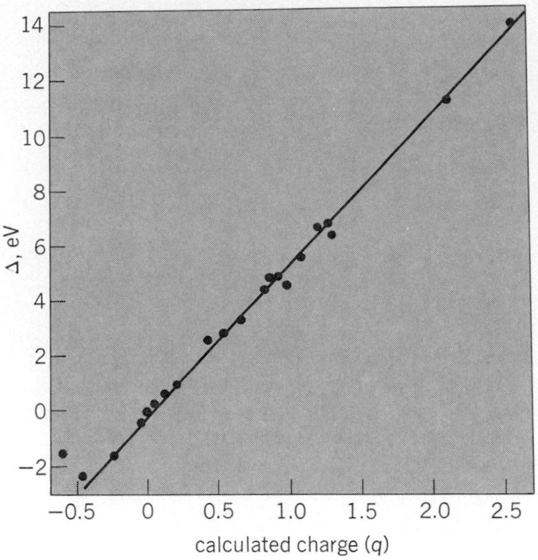

Fig. 7. Binding energies for the sulfur 2p electrons versus the calculated charge. The points indicate averages from more than 100 compounds.

other elements such as carbon, nitrogen, oxygen, and phosphorus. Chemical shifts are also observed in the electron lines due to the Auger effect. Second-order chemical shifts front groups situated farther away in the molecule (inductive effects) are also observed.

The theoretical importance of the chemical shift effect lies in the study of molecular electronic structure. In other contexts, its usefulness lies precisely in its ability to specify chemical composition, in particular, at surfaces. For example, a metal and its oxide give two distinctly different lines because of the chemical shift effect. It is often easy to follow the rate of oxidation at a surface, or the removal of an adsorbed gas layer from it. The chemical composition of a surface or the changes due to various chemical or physical treatments are amenable to detailed examination. *See* ELECTRON CONFIGURATION; ELECTRONEGATIVITY; MOLECULAR ORBITAL THEORY; SPECTROSCOPY.

[KAI SIEGBAHN]

Bibliography: C. R. Brundle and A. D. Baker (eds.), *Electron Spectroscopy: Theory, Techniques, and Applications*, vols. 2 and 3, 1979; T. A. Carlson, *Photoelectron and Auger Spectroscopy*, 1975; W. Dekeyser et al. (eds.), *Electron Emission Spectroscopy*, 1973; K. Siegbahn et al., *ESCA: Atomic, Molecular and Solid State Structure Studied by Means of Electron Spectroscopy*, 1967; K. Siegbahn et al., *ESCA Applied to Free Molecules*, 1969.

Electron spin

That property of an electron which gives rise to its angular momentum about an axis within the electron. Spin is one of the permanent and basic properties of the electron. Both the spin and the associated magnetic dipole moment of the electron were postulated by G. E. Uhlenbeck and S. Goudsmit in 1925 as necessary to allow the interpretation of many observed effects, among them the so-called anomalous Zeeman effect, the existence of doublets (pairs of closely spaced lines) in the spectra of the alkali atoms, and certain features of x-ray spectra.

All theory that concerns itself with electronic, nuclear, atomic, and molecular phenomena includes the electron spin in its formulation to obtain a theoretical structure consistent with experimental observation. The electron thus possesses the intrinsic property of spin angular momentum (rotational motion about an axis), in addition to the intrinsic properties of charge and mass.

The spin quantum number is s, where s is always 1/2. This means that the component of spin angular momentum along a preferred direction, such as the direction of a magnetic field, is $\pm\frac{1}{2}\hbar$ where $\hbar = h/2\pi$ and h is Planck's constant. The spin angular momentum of the electron is not to be confused with the orbital angular momentum of the electron associated with its motion about the nucleus. In the latter case the maximum component of angular momentum along a preferred direction is $l\hbar$, where l is the angular momentum quantum number and may be any positive integer or zero. The total orbital angular momentum is $\sqrt{l(l+1)}\,\hbar$. In this discussion the terms angular momentum or magnetic dipole moment describe the maximum component of these quantities along a field direction.

Electron magnetic moment. The electron has a magnetic dipole moment by virtue of its spin. The approximate value of the dipole moment is the Bohr magneton μ_0 which is equal to $eh/4\pi mc = 9.27 \times 10^{-21}$ erg/oersted, where e is the electron charge measured in electrostatic units, m is the mass of the electron, and c is the velocity of light. (In SI units, $\mu_0 = 9.27 \times 10^{-24}$ J/T.) The orbital motion of the electron also gives rise to a magnetic dipole moment μ_l that is equal to μ_0 when $l = 1$ (Fig. 1). In Fig. 1 is shown the simple case of an electron in circular motion in a magnetic field. The electron is shown with a spin angular momentum parallel to the orbital angular momentum. Physically it could equally well be in the opposite direction. The direction of the dipole angular momentum (both are vectors), and the magnetic moments are therefore negative. For a positron (a positively charged particle having the same mass and magnitude of charge as the negatively charged electron) the magnetic moments are positive, that is, in the same direction as the angular momentum.

The orbital magnetic moment of an electron can

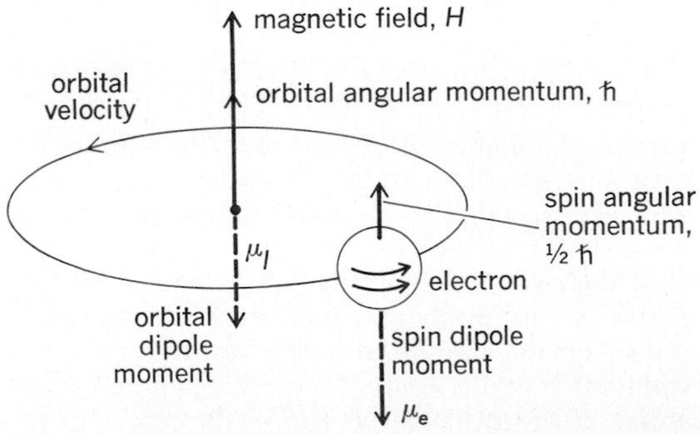

Fig. 1. Diagram of orbital angular momentum G_l and orbital magnetic moment μ_l due to a negative charge revolving in circle of radius a.

readily be deduced with the use of the classical statements of electromagnetic theory in quantum-mechanical theory; the simple classical analog of a current flowing in a loop of wire describes the magnetic effects of an electron moving in an orbit. The spin of an electron and the magnetic properties associated with it are, however, not possible to understand from a classical point of view. The classical radius of the electron is $e^2/(2mc^2) = 1.41 \times 10^{-13}$ cm; a reasonable distribution of mass and electric charge within this radius which would lead to a magnetic moment μ_0 leads to the calculation of a peripheral velocity of the electron far greater than the velocity of light. This is, of course, wholly precluded by the special theory of relativity. No theory of the structure of the electron has been formulated which makes the spin of the electron amenable to simple pictorial understanding. Nevertheless, the interpretation of the spectra of atoms and molecules, the magnetic properties of materials, and other phenomena on an atomic scale unambiguously require that the electron have the property of spin. To the extent that physical theory assumes these properties and does not concern itself with questions about the structure of the electron, it is quite adequate to deal with a large range of physical phenomena in a highly quantitative way.

In the Landé g-factor, g is defined as the negative ratio of the magnetic moment, in units of μ_0, to the angular momentum, in units of $\hbar$. For the orbital motion of an electron, $g_l = 1$. For the spin of the electron the appropriate g-value is $g_s \cong 2$; that is, unit spin angular momentum produces twice the magnetic moment that unit orbital angular momentum produces.

The following discussion is limited to atoms which have a single electron outside of closed electron shells. Both the orbital and spin angular momenta of the electrons within closed shells add up in such a way that their net angular momentum is zero. The single electron outside closed shells may have $l = 0, 1, 2, \ldots$ By the usual rules developed from quantum mechanics the total angular momentum quantum number of the electron, which is called j, is $l \pm s$ (Fig. 1) except when $l = 0$, in which case the total angular momentum is s. For instance, when $l = 1$, $j = \frac{1}{2}$ or $\frac{3}{2}$. These relations may be represented by vector diagrams, as shown in Fig. 2. Since the revolution of the electron about the nucleus produces a magnetic field at the electron, and since the electron has a magnetic dipole moment, the energy of the atom is different when the vector s is parallel to l $(j = l + \frac{1}{2})$ and when s is antiparallel (180°) to l $(j = l - \frac{1}{2})$. Thus the spin of the electron causes a doubling of the energy levels in all single-electron atoms except when $l = 0$, in which case level is single. In the case of sodium, the familiar yellow lines (the D lines) comprise a closely spaced doublet that arises from a transition from a state of $l = 1$ to one of $l = 0$. The doubling of the lines is a direct consequence of the existence of the electron spin.

Energy-level splitting. The total electronic magnetic moment of an atom depends on the state of coupling between the orbital and spin angular momenta of the electron (Fig. 2). In the single-electron case, an atom in the state for which $l = 0$ has only the magnetic moment associated with the spin, and this moment may be oriented in either of two directions with respect to an externally applied magnetic field. However, when the atom is in a state for which $l = 1$, j can be $\frac{1}{2}$ or $\frac{3}{2}$ and the Landé g-factors for these two states are $\frac{2}{3}$ and $\frac{4}{3}$ respectively. That is, the magnetic moment per unit angular momentum is equal neither to that which characterizes the orbital motion nor to that which characterizes the spin motion. In a magnetic field, a single energy level characterized by l and j is split into several energy levels, each described by the component of j, m_j, along a magnetic field, where $m_j = j, j-1, \ldots, -(j-1), -j$. The energy of the level is the zero field energy plus the term $m_j g_j \mu_0 H$. In a transition between two energy levels that gives rise to a spectral line, the energy of the emitted or absorbed photon is the zero field energy difference between the two levels plus the difference between two magnetic energy terms as given previously. The resultant splitting of the line, which is single at zero magnetic field, into a line of two or more components is called the Zeeman effect. The Zeeman effect has been called anomalous when it is not explicable purely in terms of the orbital motion of the electron. The introduction of the electron spin allows the interpretation of all observed Zeeman effects.

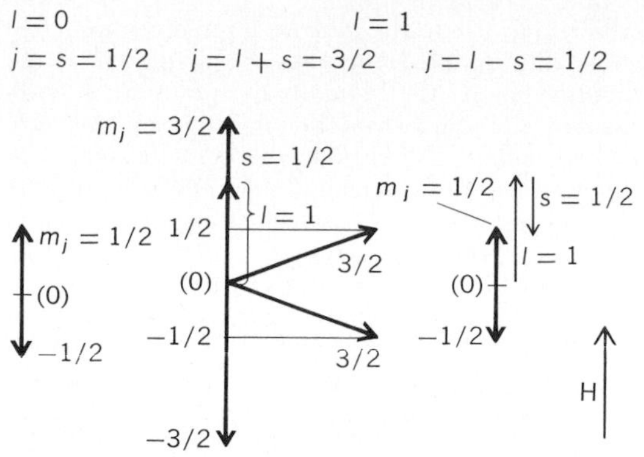

Fig. 2. Diagram illustrating addition of l and s vectors into a resultant j vector. The allowed values of m_j are also shown. In no case does the (0), in parentheses, give an allowed value.

Atomic beam measurements. Prior to 1940 spectroscopic measurements had been made only on optical spectra, and Zeeman effects were small effects superimposed on lines of considerable natural width. Exact measurements of the splitting of lines in a magnetic field, therefore, could not be made on such lines, and all data on the Zeeman effect were consistent with the statement that $g_s = 2g_l$. With the development of spectroscopy by the atomic beam method, a new order of precision in the measurement of the frequencies of spectral lines became possible. By using the atomic-beam techniques, it became possible to measure g_s/g_l directly, with the result $g_s/g_l = 2$ (1.001168 ± 0.000005). The magnetic moment of the electron therefore is not μ_0 but $1.001168\mu_0$, or equivalently the g factor of the electron departs from 2 by the so-called g-factor anomaly defined as $a = (g_2 - 2)/2$ so that $\mu = (1 + a)\mu_0$. Thus the first molecular beam work gave $a = 0.001168$.

Calculation of g-factor anomaly. It is not possible to give a qualitative description of the effects which give rise to the g-factor anomaly of the electron. The detailed theoretical calculation of the quantity is in the domain of quantum electrodynamics, and involves the interaction of the zero-point oscillation of the electromagnetic field with the electron. Comparison of theoretical determination of a with its experimental measurement constitutes the most accurate and direct existing test of the theory of quantum electrodynamics.

Theoretical work on the g-factor anomaly, based on the principles of quantum electrodynamics, began simultaneously with its experimental discovery. The initial prediction of the value of the anomaly was made by J. Schwinger in 1948, who showed that the anomaly could be written as $a = 0.5\ (\alpha/\pi)$, where α, the fine-structure constant, is given by $\alpha = e^2/\hbar c \simeq 1/137$. Shortly thereafter it was shown that a could be expressed as a power series in α (more customarily, α/π), that is, one can write $a = A(\alpha/\pi) + B(\alpha/\pi)^2 + C(\alpha/\pi)^3 + D(\alpha/\pi)^4 + \ldots$. Calculation of B and C from quantum electrodynamics has proven to be very difficult, with errors occurring at various stages of the work. However, it is now generally agreed that $A = 0.500$, $B = -0.328478$, and $C = 1.184 \pm 0.007$. In order to find a theoretical value for a, one also needs an accurate value of α. Such a value is given by work at the National Bureau of Standards where, from a number of precise measurements of quantities which enter the definition of α, it has been found that $\alpha^{-1} = 137.035963 \pm 0.000015$. The result of substituting this value of α into the expression for a gives the most recent theoretical value of a as a(theoretical) = 0.001 159 652 570 ± 0.000 000 000 150. The uncertainty quoted for this number is due primarily to error in α and also to error in the coefficient. It is possible that both of these sources of error will be reduced dramatically. A calculation of D has been undertaken. This is of importance since, if $D = 1$, this term contributes 0.000 000 000 027 to a(theoretical), and such a contribution, as small as it is, must be considered in future research. *See* ATOMIC CONSTANTS.

Measurement of g-factor anomaly. The initial molecular-beam method for measuring a was based on measurement of g_s/g_l, which yields $1 + a$ rather than a itself. Such experiments have been superseded by two new types of experiments which have the major advantages of measuring a directly, and of measuring a for electrons trapped in electric and magnetic fields but not bound to atoms. The second feature means that various corrections due to the presence of an atom are now not necessary.

The new experiments may be readily understood, qualitatively, by noting that if an electron (spin angular momentum $\hbar/2$) moves at a low velocity v ($v/c \ll 1$) perpendicular to a magnetic field B, the particle will rotate in the field at a frequency (the cyclotron frequency) given by $f_C = (1/2\pi)(eB/mc)$, while its magnetic moment (and spin) will precess about the field at a frequency (the spin precession frequency) given by $f_S = (1 + a)f_C$. The difference between these frequencies, often called the g-2 or difference frequency, is given by $f_D = f_S - f_C = af_C$. It represents the rate at which the spin of the particle rotates relative to the particle's velocity.

One technique for measuring a consists of trapping very-low-energy electrons in a magnetic field and measuring f_D and f_C simultaneously, with a being given by the relation $a = f_D/f_C$.

A slightly different version of the same technique consists of trapping electrons whose velocity is about half the speed of light (electrons accelerated to several hundred thousand volts) in a magnetic field and determining the same ratio. Relativistic effects manifest themselves in both the expressions for f_S and f_C, but cancel exactly when the ratio f_D/f_C is formed. Consequently comparison of f_D/f_C, using both the high- and low-energy techniques, constitutes an excellent check of the separate measurements, as well as one of the most precise tests extant of the special theory of relativity, which predicts that the frequency ratio should be independent of velocity.

The most accurate experimental result has been obtained by using the low-energy technique. This result is a(experimental) = 0.001 159 652 200 ± 0.000 000 000 040, and it is in agreement, within the limits of error, with a(theoretical). The comparison of a(experimental) with a(theoretical) presented above constitutes the most precise confrontation of any experiment with a theoretical prediction in the history of science, a confrontation at the level of 150 parts per billion. Even more accurate experiments at both high and low energy and, as mentioned above, improved theoretical predictions have been undertaken.

[ARTHUR RICH]

Bibliography: E. Merzbacher, *Quantum Mechanics*, 1970; A. Rich and J. C. Wesley, The current status of the lepton g-factors, *Rev. Mod. Phys.*, 44(2):250–283, 1972; F. K. Richtmyer, E. H. Kennard, and T. Lauritsen, *Introduction to Modern Physics*, 5th ed., 1969; V. F. Weisskopf, Recent developments in the theory of the electron, *Rev. Mod. Phys.*, 21(2):305–315, 1949.

Electronegativity

Electronegativity, according to L. Pauling, is "the power of an atom in a molecule to attract electrons to itself." With the concept of electronegativity, a vast number of observations of chemical and physical properties have been either correlated or predicted. Quantitative definitions and scales of electronegativity have been based not on electron distribution itself but on properties which were assumed to reflect electronegativity.

The electronegativity of an element depends upon its valence state and thus is not an invariant atomic property. As an example, the electron-withdrawing ability of an sp^n hybrid orbital centered on carbon and directed toward hydrogen increases as the percentage of s character in the orbital increases in the series ethane < ethylene < acetylene. Thus, according to this concept of orbital electronegativity, each element exhibits a range of electronegativity values. In the following paragraphs, a few scales of electronegativity will be discussed.

The original scale, proposed by Pauling in 1932, is based upon the difference Δ between the energy of the A-B bond in the compound AB_n and the mean of the energies of the homopolar bonds A-A and B-B, as in Eq. (1). The A-B bond energy ex-

$$\Delta = E(\text{A-B}) - \frac{E(\text{A-A}) + E(\text{B-B})}{2} \qquad (1)$$

ceeds the arithmetic mean of the A-A and B-B bonds to an increasing extent as the elements A and B diverge in electron-attracting ability. The difference in electronegativities of bonded atoms is proportional to the square root of the energy difference Δ, as in Eq. (2). The proportionality factor

$$\chi_A - \chi_B = 0.208\sqrt{\Delta} \quad (2)$$

0.208 converts the units of energy from kilocalories to electron volts. After the electronegativity of one element is arbitrarily assigned, other values of electronegativity can be calculated from thermochemical data. Values for selected elements in common oxidation states are presented in the table. Electronegativity increases with increasing oxidation state. For example, $\chi_{Sn(II)} = 1.80$ and $\chi_{Sn(IV)} = 1.96$.

R. S. Mulliken proposed that the electronegativity of an element is given by the average of the valence-state ionization potential and electron affinity: $\chi_M = (IP_V + EA_V)/2$. The quantities IP_V and EA_V are not observable properties of the ground state of an atom, but are energies for a hypothetical state of the isolated atom having the same electronic configuration (hybridization, electron-electron interaction, and so forth) as the atom in the molecule. The Mulliken approach has a sound theoretical basis, is consistent with Pauling's original definition, and gives orbital electronegativities, not invariant atomic electronegativities. Valence-state ionization potentials and electron affinities have been calculated from the equations $IP_V = IP_g + P^+ - P^0$ and $EA_V = EA_g + P^0 - P^-$, where IP_g and EA_g are ground-state potentials and affinities, respectively, and P^+, P^0, and P^- are promotion energies of the positive ion, atom, and negative ion, respectively. The calculation of *d*-orbital electronegativity by the Mulliken method has not been accomplished for nontransition elements due to the lack of spectroscopic data. Since electronegativity is a sensitive function of *d*-orbital hybridization and since the extent of *d*-orbital participation generally cannot be ascertained quantitatively, the calculations of electronegativities for the heavier elements are limited.

The energy of an ion relative to the neutral atom can be expressed as a power series $E = aq + bq^2 + cq^3 + dq^4$, where q is the formal oxidation state or ionic charge for a particular state of ionization. Z. R. P. Iczkowaski and J. L. Margrave defined the electronegativity of a neutral atom as the derivative, $\chi_{IM} = (dE/dq)_{q=0}$. As a fairly good approximation, the last two terms in the above power series can be dropped, giving Eq. (3). The units of χ_{IM} are energy/

$$\chi_{IM} = \frac{dE}{dq} = \frac{d(aq + bq^2)}{dq} = a + 2bq \quad (3)$$

electron and the magnitudes are the same, in accordance with theory, as those of Mulliken if E is evaluated only from the electron affinity and the first ionization potential. The quantity a is the Mulliken electronegativity for a neutral atom, and electronegativity is shown by Eq. (3) to increase linearly with increasing positive charge. By using IP_V and EA_V values, H. H. Jaffé and coworkers calculated the electronegativities of certain vacant, singly occupied, and doubly occupied orbitals.

Electronegativity was defined by A. L. Allred and E. G. Rochow as the force of attraction between a nucleus and an electron from a bonded atom. The electrostatic force was calculated simply from the effective nuclear charge and the atomic radius, as in Eq. (4).

$$\chi = \frac{0.359 Z_{eff}}{r^2} + 0.744 \quad (4)$$

A quantum-defect electronegativity scale has been developed from potentials based on atomic spectral data, and a nonempirical scale has been calculated by an ab initio method using floating gaussian orbitals (FSGO).

Other methods for calculating electronegativities utilize such observables as bond-stretching force constants, electrostatic potentials, spectra, and covalent radii. The fact that the various scales of electronegativity have different dimensions (energy$^{1/2}$, energy/electron, force, potential, and so forth) or no dimension reflects the widespread results of differences in electron-attracting ability. The measurement of electronegativities involves observations of properties dependent upon electron distribution. Close agreement of electronegativity values obtained from measurements of several diverse properties lends confidence and utility to the concept. [A. LOUIS ALLRED]

Average electronegativities from thermochemical data

Element	Value	Element	Value
H	2.20	Al	1.61
Li	0.98	Ga	1.81
Na	0.93	In	1.78
K	0.82	Tl	2.04
Rb	0.82	C	2.55
Cs	0.79	Si	1.90
Be	1.57	Ge	2.01
Mg	1.31	Sn	1.96
Ca	1.00	Pb	2.33
Sr	0.95	N	3.04
Ba	0.89	P	2.19
Sc	1.36	As	2.18
Ti	1.54	Sb	2.05
V	1.63	Bi	2.02
Cr	1.66	O	3.44
Mn	1.55	S	2.58
Fe	1.83	Se	2.55
Co	1.88	F	3.98
Ni	1.91	Cl	3.16
Cu	1.90	Br	2.96
Zn	1.65	I	2.66
B	2.04		

Bibliography: A. L. Allred and E. G. Rochow, A scale of electronegativity based on electrostatic force, *J. Inorg. Nucl. Chem.*, 5(4):269–288, 1958; A. L. Allred and E. G. Rochow, Electronegativity values from thermal data, *J. Inorg. Nucl. Chem.*, 17:215–221, 1961; W. Gordy and W. J. Orville-Thomas, Electronegativities of the elements, *J. Chem. Phys.*, 24:439, 1956; J. Hinze and H. H. Jaffé, Electronegativity: I. Orbital electronegativity of neutral atoms, *J. Amer. Chem. Soc.*, 84: 540, 1962; J. Hinze and H. H. Jaffé, Electronegativity: II. Bond and orbital electronegativities, *J. Amer. Chem. Soc.*, 85:148, 1963; J. E. Huheey, *Inorganic Chemistry*, 2d ed., 1978; L. Pauling, *Nature of the Chemical Bond*, 3d ed., 1960; J. St. John and A. N. Bloch, Quantum-defect electronegativity scale for nontransition elements, *Phys. Rev. Lett.*, 33:1095–1098, 1974; G. Simons, M. E. Zandler, and E. R. Talaty, Nonempirical electronegativity scale, *J. Amer. Chem. Soc.*, 98:7869–7870, 1976.

Electrophilic and nucleophilic reagents

Electrophilic reagents are chemical species which, in the course of chemical reactions, acquire electrons, or a share in electrons, from other molecules or ions. Although this definition embraces all oxidizing agents and all Lewis acids, electrophilic reagents are ordinarily thought of as cationic species, such as H^+, NO_2^+, Br^+, or SO_3 (or carriers of these species such as HCl, CH_3COONO_2, or Br_2), which can form stable covalent bonds with carbon atoms. Electrophilic reagents frequently are positively charged ions (cations). *See* ACID AND BASE.

Nucleophilic reagents are the opposite of electrophilic reagents. Nucleophilic reagents give up electrons, or a share in electrons, to other molecules or ions in the course of chemical reactions. Nucleophilic reagents frequently are negatively charged ions (anions). Typical nucleophilic reagents are hydroxide ion (OH^-), halide ions (F^-, Cl^-, Br^-, and I^-), cyanide ion (CN^-), ammonia (NH_3), amines, alkoxide ions (such as CH_3O^-) and mercaptide ions (such as $C_6H_5S^-$). *See* SUBSTITUTION REACTION. [JOSEPH F. BUNNETT]

Electrophoresis

The migration of electrically charged particles in solution or suspension in the presence of an applied electric field. Each particle moves toward the electrode of opposite electrical polarity. For a given set of solution conditions, the velocity with which a particle moves divided by the magnitude of the electric field is a characteristic number called the electrophoretic mobility. The electrophoretic mobility is directly proportional to the magnitude of the charge on the particle, and is inversely proportional to the size of the particle. An electrophoresis experiment may be either analytical, in which case the objective is to measure the magnitude of the electrophoretic mobility, or preparative, in which case the objective is to separate various species which differ in their electrophoretic mobilities under the experimental solution conditions.

Tiselius cell. The phenomenon of electrophoresis was first observed in 1807 by the Russian physicist F. F. Reuss, but electrophoresis was not employed as an experimental technique until the introduction of a new electrophoresis apparatus by Arne Tiselius in 1937. The apparatus of Tiselius detected electrophoretic motion by the moving-boundary method, in which a boundary is created between the solution of particles to be examined and a sample of pure solvent. As the particles migrate in an electric field, the boundary between solution and solvent can be observed to move, and if there are a number of species in the solution with different electrophoretic mobilities, a series of boundaries of various shapes and magnitudes can be detected. An electrophoresis experiment in a Tiselius cell is depicted in Fig. 1. Using his apparatus, Tiselius demonstrated the heterogeneity of human blood plasma, and showed for the first time that the globulin molecules could be separated into different classes, which were designated alpha, beta, and gamma globulin. The moving-boundary method was used for 3 decades to separate complex mixtures of charged macromolecules in solution and to study the physical characteristics of

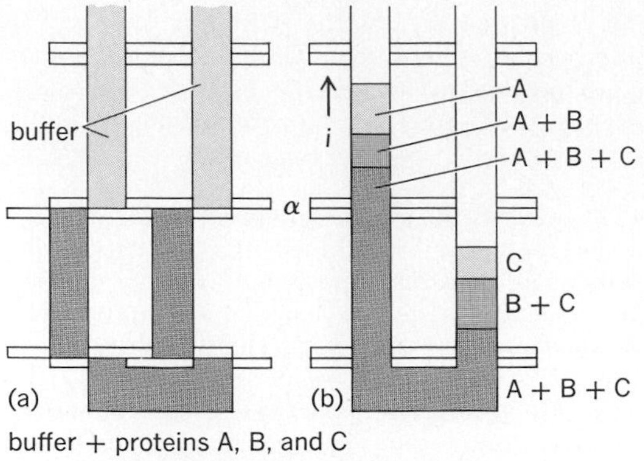

Fig. 1. Diagrammatic representation of Tiselius electrophoresis cell, showing (*a*) formation and (*b*) motion of electrophoretic boundaries of a mixture of proteins.

solutions of proteins and other macromolecules of biological and industrial importance.

Gel techniques. The resolving power of electrophoresis has been greatly improved by the introduction of the use of gel supporting media. The gel matrix prevents thermal convection caused by the heat which results from the passage of electric current through the sample. The absence of convection reduces greatly the mixing of the various parts of the sample, and therefore allows for more stable separation. The dimensions of the cross-links of the gel may also provide a molecular sieving effect, which increases the resolving power of the electrophoretic separation of molecules of different size. In addition, the gel media may support a gradient of a separate reagent, which assists in the separation of macromolecules. Gradients of pH and of reagents of various types may be combined in two-dimensional arrays for even greater resolving power. A very successful derivative of the gel technique has been the determination of the molecular weights of protein molecules by electrophoresis of the molecules in a gel medium which contains substantial amounts of detergent. The detergent denatures the protein molecules, changing them from globular, compact structures to long, flexible polymers which are coated with detergent molecules. These polymers move in the electric field through the gel medium with a velocity which is determined by the length of the polymer, and therefore by the molecular weight of the protein unit. This method is now the most common technique for the determination of molecular weights of proteins in biochemical studies.

Isoelectric focusing. An important variation of the electrophoresis technique is isoelectric focusing. In this technique the medium supports a pH gradient which includes the isoelectric pH of the species being studied. Many charged macromolecules have both positive and negative charges on their surfaces, and the electrophoretic mobility is related to the net excess of charge of one type or the other. As the pH becomes more acidic, the number of positive charges increases, and as the pH becomes more basic, the number of negative charges increases. For each molecule of this type, there is one pH at which the net charge on the surface is zero, so that the molecule does not move when an electric field is applied and thus has an electrophoretic mobility of zero. This pH is called the isoelectric pH. If the molecule is introduced into a pH gradient which includes its isoelectric pH, it will migrate to the position of the isoelectric pH and then become stationary. In this way, all molecules of a given isoelectric pH will migrate to the same region—hence the term isoelectric focusing. The method of isoelectric focusing is particularly good for the analysis of microheterogeneity of protein species and other species which may differ slightly in their chemical content. *See* ISOELECTRIC POINT.

Isotachophoresis. One important variant of electrophoresis is the phenomenon of isotachophoresis, in which ionic species move with equal velocity in the presence of an electric field. The isotachophoretic condition is maintained when particles of different mobilities form boundaries within the solution so that the most mobile ions form the leading edge of the moving sample, followed by the less mobile ions in the order of their mobility. The reason that the ions with different mobility can move with the same velocity is that the electric field in each of the regions is inversely proportional to the mobility of the ionic species in order to maintain a constant current throughout the sample. Isotachophoresis has a number of important analytical and preparative applications which are featured by its advantages of high resolving power, sensitivity, speed, and the ability to concentrate rather than to dilute the components which are being analyzed.

Particle electrophoresis. The methodology of electrophoresis may be modified considerably when the particles undergoing analysis are of sufficient size to be viewed either with the naked eye or with the assistance of an optical microscope. This general area of particle electrophoresis has its most important applications in the analysis of the surface charge of living cells and in the study of various types of particles used in industrial coating processes. The most straightforward technique, called optical cytopherometry or microelectrophoresis, is that in which a human experimenter views the particles in an electric field under an optical microscope and determines manually the amount of time necessary to traverse a given distance. Although time-consuming and tedious, this technique has had many important applications since the 1940s. Attempts to modernize this method have included the introduction of high-speed photography, television technology, and the study of particle electrophoresis by the laser Doppler effect. Particle electrophoresis can be conducted under many of the conditions which were originally developed for smaller macromolecules. The use of gradients of density and pH and the methods of isoelectric focusing and isotachophoresis are commonly applied to particles which are even large enough to be viewed with the naked eye.

A common interference effect in the performance of particle electrophoresis is the sedimentation of the particles in the field of the Earth's gravity. This effect can be minimized in importance by performing the experiments in a medium of equal density, by performing the electrophoresis in a vertical direction and accounting for the effect of sedimentation, or by performing the experiment in such a short period of time that the degree of sedimentation in the Earth's gravity is not a significant interfering effect.

Laser applications. Application of the optical laser to electrophoretic detection has resulted in the development of a technique which can be used for analytical electrophoresis experiments on particles of all sizes. The basic principle is that the highly monochromatic (single-frequency) laser light impinges upon the particles, and is scattered from the particles in all directions. When observing the laser light which has been scattered from a moving particle, one can detect that there is a slight shift in the frequency of the light as a result of the motion of the particle. This is the Doppler effect, which causes the change in the apparent tone of passing trains or cars and which is the operating principle of other familiar techniques such as radar. The application of the laser Doppler principle to electrophoresis experiments, often called electrophoretic light scattering (ELS) has become an important method for the rapid determination of electrophoretic velocities. The complete electrophoretic mobility distribution of a sample of many particles can be determined in a time as short as 1 s with a precision heretofore unobtainable by standard technology.

An electrophoretic light-scattering apparatus with a sample Doppler spectrum is shown in Fig. 2. The very slight Doppler shifts caused by the electrophoretic motions are detected by the electronic "beats" which result when the scattered light is incident simultaneously with an unshifted reference beam, or "local oscillator," on a photodetector. An important application of electrophoretic light scattering is the characterization of leukemic

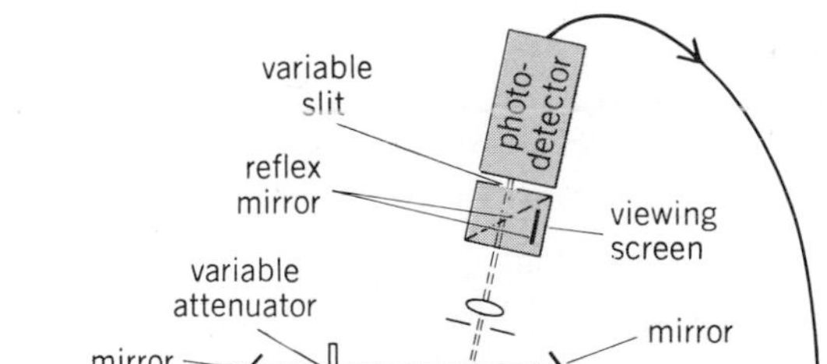

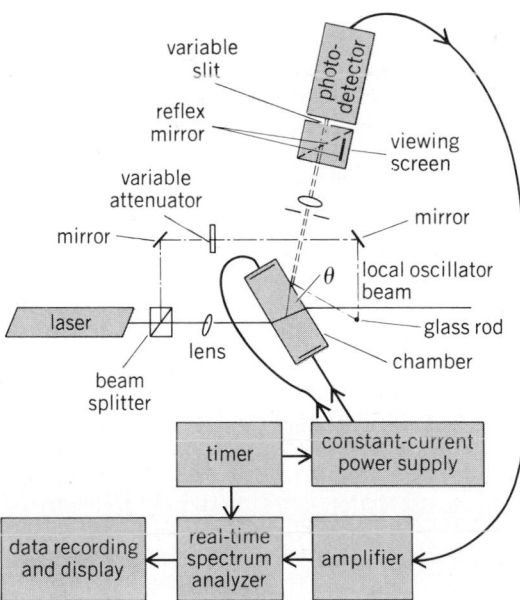

Fig. 2. Diagram of an electrophoretic light-scattering apparatus. Doppler-shifted light from the particles moving in the chamber is mixed with an unshifted reference beam (the local oscillator). The beat frequencies from the photodetector are spectrum-analyzed to produce the electrophoretic mobility histogram. (*From B. R. Ware, The study of biological surfaces by laser electrophoretic light scattering, in G. M. Hieftje, ed., New applications of lasers to chemistry, ACS Symposium Ser. no. 85, 1978*)

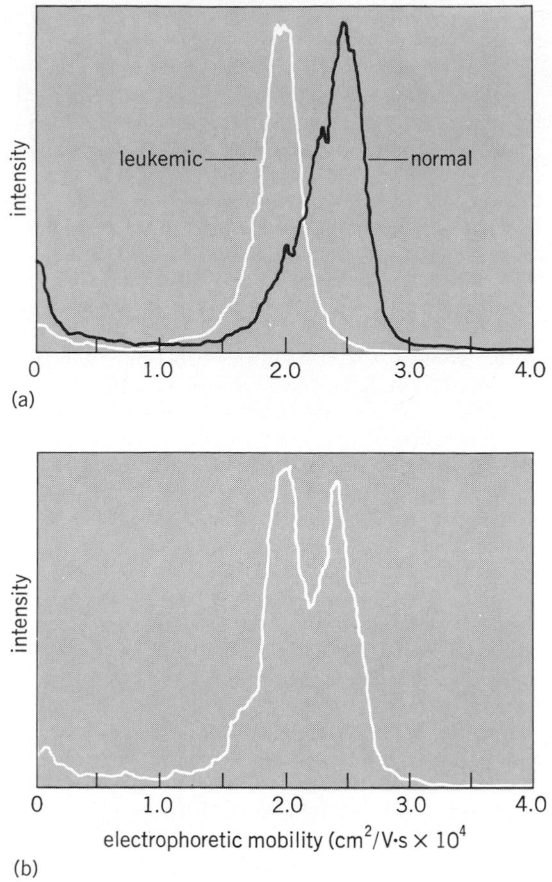

Fig. 3. Characterization of leukemic cells with electrophoretic light scattering. (*a*) Spectra of leukemic cells and normal cells. (*b*) Spectrum for mixture of the two cell types. (*From B. A. Smith, B. R. Ware, and R. S. Weiner, Electrophoretic distributions of human peripheral blood molecular white cells from normal subjects and from patients with acute lymphocytic leukemia, Proc. Nat. Acad. Sci. U.S.A., 73:2388, 1976*)

cells (Fig. 3). The normal lymphocytes and leukemic cells isolated in the same way have distinctly different electrophoretic mobilities and mobility distributions, as shown by the two spectra in Fig. 3*a*. The spectrum in Fig. 3*b* shows the simultaneous detection of leukemic and normal cells in a mixture, based solely on their Doppler-detected electrophoretic mobilities. Electrophoretic light scattering has been used for the study of many types of living cells, cell organelles, viruses, proteins, nucleic acids, and synthetic polymers. It is anticipated that the continued refinement and application of this extremely precise and rapid technique will result in an even broader scope and significance of the many forms of electrophoresis. *See* ELECTROLYTIC CONDUCTANCE.

[B. R. WARE]

Bibliography: P. J. Karol and M. H. Karol, Isotachophoresis, *J. Chem. Ed.*, 55:626–630, 1978; D. H. Leaback, Electrophoresis in protein analysis, *Chem. Brit.*, 10(10):376–383, 1974; B. R. Ware, The study of biological surfaces by laser electrophoretic light scattering, in G. M. Hieftje (ed.), *New Applications of Lasers to Chemistry*, Amer. Chem. Soc. Symp. Ser. no. 85, 1978.

Element 104

The first element beyond the actinide series, and the twelfth transuranium element. In 1964 G. N. Flerov and coworkers at the Dubna Laboratories in the Soviet Union claimed the first identification of element 104 and a little later suggested the name kurchatovium (symbol Ku). *See* ACTINIDE ELEMENTS; TRANSURANIUM ELEMENTS.

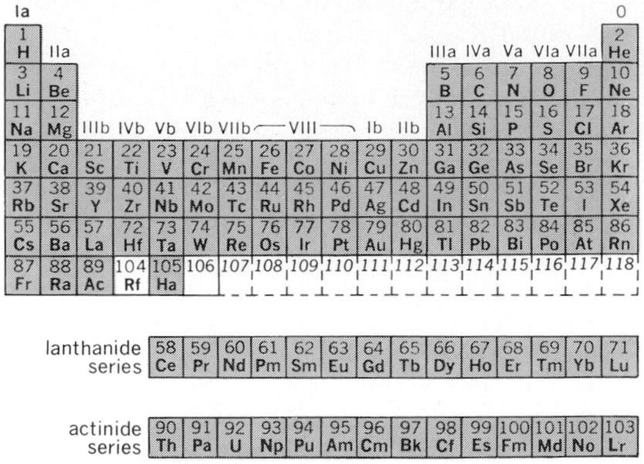

The Dubna group claimed the preparation of element 104, mass number 260, by irradiating plutonium-242 with neon-22 ions in their 130-cm heavy-ion cyclotron. The postulated nuclear reaction was $^{242}Pu + ^{22}Ne \rightarrow ^{260}104 + 4$ neutrons. Using 113–115-MeV incident neon particles, they observed an apparent spontaneous fission (SF) activity with a half-life of 0.3 ± 0.1 s. Atoms produced in the bombardments recoiled out of the plutonium target to be stopped in a moving belt of nickel ribbon which, in turn, passed in front of passive fission detectors made of mica or a special glass; after bombardment, chemical etching of the detector plates made fission tracks visible under a microscope. The spatial distribution of the tracks was determined by the half-life of the activity. The number of tracks per bombardment was very small, only about one per 5 h of irradiation being observed on the average, so that it was extremely difficult to do any definitive experiments that would prove the validity of their assignment of the activity to an isotope of element 104. *See* NOBELIUM.

The Dubna team used the only tools available to them at the time to try to define their activity: an excitation function and cross bombardments using other targets and heavy-ion projectiles. While their observed excitation function with $^{242}Pu + ^{22}Ne$ was consistent with the presumed reaction, it is now known that such a function does not exclude other reactions to make lower nuclides of lower atomic number.

On the other hand, cross bombardments did not seem to make the 0.3-s SF activity. On the basis of this rather meager evidence, they announced their discovery of element 104.

In 1966 I. Zvara and coworkers at Dubna announced the results of chemical experiments which they felt definitely identified the 0.3-s activity as an isotope of element 104. According to G. T. Seaborg's actinide concept, element 103 should have a completed shell of fourteen 5*f* electrons. Element 104 should, therefore, be a member of the 6*d* transition series of elements and thus resemble its homolog, hafnium. Its chemical properties should be quite different from the actinide elements. *See* PERIODIC TABLE.

The chemical identification experiments were conducted in the following manner. Recoil fragments from the bombardments were stopped in a stream of nitrogen gas, which was then combined with a mixture of $NbCl_5$ and $HfCl_4$ (3:1) in a duct heated to 350°C. After transversing a hot filter, the gas was passed between mica detector plates. Under these conditions it had already been shown that the formation of the volatile chlorides of group IV elements (hafnium and zirconium) takes place rapidly and this makes possible a separation by a factor of 50 from the less volatile group III elements (scandium) and the actinide elements. The Soviet scientists concluded that the distribution of the fission fragment tracks in the mica detectors was consistent with the 0.3-s half-life and thus indicated that a volatile chloride had been formed as expected for element 104. They published this work as a confirmation of their 0.3-s SF activity being due to that element.

Attempts were soon made at the then Lawrence Radiation Laboratory, University of California at Berkeley, by A. Ghiorso and coworkers to achieve additional confirmation of the Dubna discovery, since the work at Dubna seemed as if it might be correct in spite of the pragmatic SF systematics of the Ghiorso group, which predicted a much shorter half-life for $^{260}104$. Their procedure was to utilize the larger formation cross sections that become available when higher-Z targets and lower-Z projectiles are used to make the same product nucleus. In a long series of bombardments carried out over the period of a year they bombarded einsteinium, californium, and curium targets with ions of mass from ^{11}B to ^{18}O, but no 0.3-s SF activity at all was observed, although they did discover a new activity with a half-life in the range of 10–30 ms.

An atom that decays by SF provides little in the way of a fingerprint, since one fission differs little from another. Half-life can be useful to some extent. However, if two species are present with similar half-lives, there is little that can be done to distinguish them, even if thousands of atoms are made, unless a chemical separation is achieved. On the other hand, an atom that decays by the emission of an alpha particle with a distinctive energy and half-life to form a daughter, which in turn decays by the emission of an alpha particle with a different alpha energy and half-life, can be uniquely identified even though only a few atoms are produced. This basic philosophy became the ultimate goal in the search by Ghiorso and coworkers for element 104. Over the next couple of years they developed the vertical wheel (VW) alpha detection system. *See* ELEMENT 105; ELEMENT 106.

By 1969 the Berkeley group had succeeded in unambiguously discovering two alpha-emitting isotopes of element 104 with mass numbers 257 and 259 by bombarding ^{249}Cf with ^{12}C and ^{13}C projectiles from the Berkeley heavy-ion linear accelerator (HILAC). The respective alpha-emitting daughters ^{253}No and ^{255}No, which are well-known nuclides, were identified as arising from mothers with respective half-lives of 4.5 and 3 s. The $^{257}104$ nucleus was found to emit a complex alpha-particle spectrum with many groups ranging in energy from 8.70 to 9.00 MeV, whereas the $^{259}104$ spectrum showed two principal lines at 8.77 and 8.86 meV. In a separate experiment an SF emitter with a half-life of about 11 ms was produced which they tentatively assigned to $^{258}104$ although there was no way to be sure of such an assignment. Since the Berkeley group felt they had demonstrated conclusively that the Dubna discovery was not valid, they suggested that element 104 be named rutherfordium with the symbol Rf in honor of Lord Rutherford.

A long-lived isotope of element 104 was then discovered at Berkeley in bombardments of ^{248}Cm with ^{18}O ions. The isotope $^{261}104$ was found to be an alpha emitter with a half-life of 65 s and an alpha energy of 8.3 MeV. It was possible to perform some simple ion-exchange experiments with this isotope which showed that element 104 behaved similarly to hafnium in aqueous solution. These experiments were unambiguous in showing that element 104 was not a member of the actinide family. *See* ION EXCHANGE.

When the Dubna group endeavored to repeat their work on element 104, they discovered that their old 0.3-s half-life was seriously in error, finding only 0.1-s decay activity in experiments performed outside of their cyclotron. Apparently about 75% of their original activity was due to a neutron-induced background. This new shorter half-life meant that their thermal chromatography experiments could not have made use of this nuclide since the transit time of the gas through their column was about 1 s. The Dubna workers decided that they must have been observing SF activity from a small fission branch of the 3-s $^{259}104$ alpha-emitting isotope discovered by the Berkeley group. There is no firm evidence that this isotope does indeed decay part of the time by SF. The alternate explanation for the fissioning atoms that transited through the Dubna hot column is that they were due to ^{256}Fm atoms which attached themselves in tiny quantities to aerosols formed in the target chamber.

Since the Dubna group persisted in their claim that the isotope $^{260}104$ decayed by SF with the relatively long half-life of 100 ms (which later became 80 ms), Ghiorso and coworkers made another attempt to find this activity by using a metal tape system to spread out the long-lived background from the 2.63-h SF emitter ^{256}Fm. In bombardments of ^{249}Bk with ^{15}N and ^{248}Cm with ^{16}O, there were no observations of any 80-ms SF activity down to a limit more than order of magnitude less than the value claimed by Dubna. In these same bombardments, however, the Berkeley group found a 21-ms SF activity, apparently the same one they had seen as a 10–30-ms activity in 1968. It is quite possible that this activity is due to $^{260}104$, but it seems more likely due to a nuclide of lower Z.

A number of years after the original discovery at Berkeley, a team at Oak Ridge National Laboratory confirmed discovery of the isotope ^{257}Rf by detecting the characteristic nobelium x-rays following alpha decay. However, the International Union of Pure and Applied Chemistry has not yet adopted an official name for element 104 because of the conflicting American-Soviet claims.

[ALBERT GHIORSO]

Element 105

A chemical element. It was synthesized and identified unambiguously for the first time in March, 1970, at the heavy-ion linear accelerator (HILAC) in the Lawrence Radiation Laboratory, Berkeley, Calif. The team that made the discovery consisted

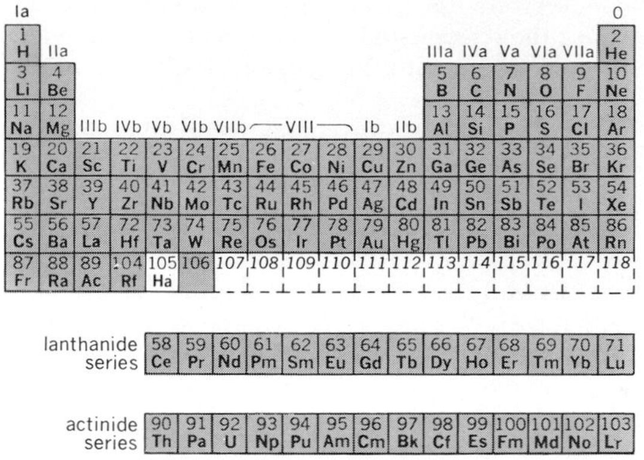
Ia
0
1 H
IIa
IIIa IVa Va VIa VIIa
2 He
3 Li 4 Be
5 B 6 C 7 N 8 O 9 F 10 Ne
11 Na 12 Mg
IIIb IVb Vb VIb VIIb VIII Ib IIb
13 Al 14 Si 15 P 16 S 17 Cl 18 Ar
19 K 20 Ca 21 Sc 22 Ti 23 V 24 Cr 25 Mn 26 Fe 27 Co 28 Ni 29 Cu 30 Zn 31 Ga 32 Ge 33 As 34 Se 35 Br 36 Kr
37 Rb 38 Sr 39 Y 40 Zr 41 Nb 42 Mo 43 Tc 44 Ru 45 Rh 46 Pd 47 Ag 48 Cd 49 In 50 Sn 51 Sb 52 Te 53 I 54 Xe
55 Cs 56 Ba 57 La 72 Hf 73 Ta 74 W 75 Re 76 Os 77 Ir 78 Pt 79 Au 80 Hg 81 Tl 82 Pb 83 Bi 84 Po 85 At 86 Rn
87 Fr 88 Ra 89 Ac 104 Rf 105 Ha 106 107 108 109 110 111 112 113 114 115 116 117 118
lanthanide series
58 Ce 59 Pr 60 Nd 61 Pm 62 Sm 63 Eu 64 Gd 65 Tb 66 Dy 67 Ho 68 Er 69 Tm 70 Yb 71 Lu
actinide series
90 Th 91 Pa 92 U 93 Np 94 Pu 95 Am 96 Cm 97 Bk 98 Cf 99 Es 100 Fm 101 Md 102 No 103 Lr

of A. Ghiorso, M. J. Nurmia, J. A. Harris, K. A. Y. Eskola, and P. Eskola. They suggested that the new element be called hahnium, symbol Ha, in honor of Otto Hahn, the discoverer of nuclear fission. This name has not been officially accepted in the international committee on nomenclature because of a conflicting claim by a Soviet group. A review committee with international membership has now been set up by the International Union of Pure and Applied Chemistry (IUPAC) to establish the matter of priority of discovery so as to decide whether element 105 will be called hahnium or nielsbohrium, the Soviet choice.

The hahnium isotope had a half-life of 1.6 sec and decayed by emitting alpha particles with energies 9.06 (55%), 9.10 (25%), and 9.14 (20%) MeV. It was shown to be of mass 260 by identifying lawrencium-256 as its daughter by two different methods.

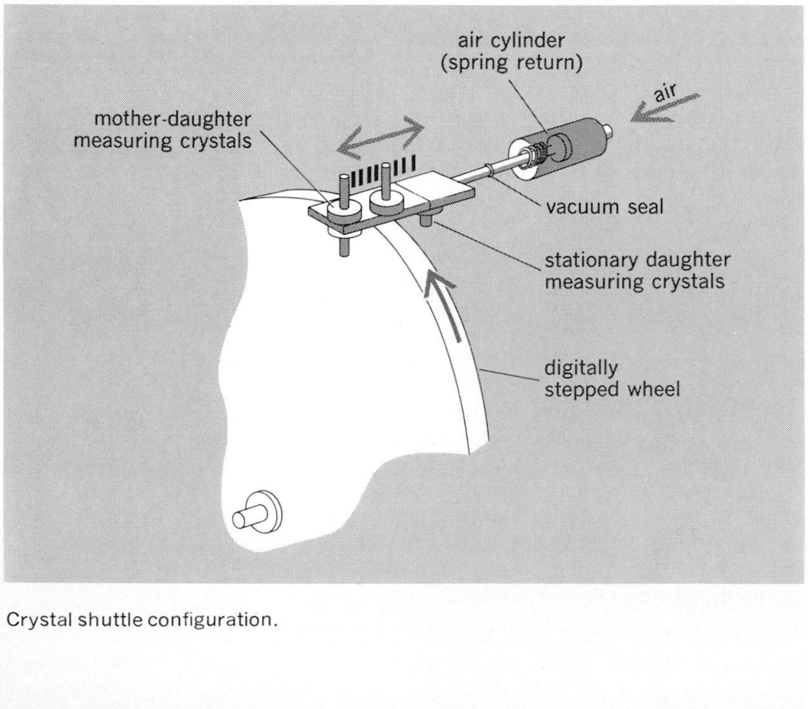

Crystal shuttle configuration.

Characteristic alpha emissions. The method used to produce the new alpha-particle activity of hahnium was similar to the technique of identifying the alpha-particle-emitting isotopes of element 104, except that the new process was considerably more elaborate and sophisticated. The hahnium nuclide was formed by bombarding a 60-microgram (μg) target of californium-249 with 84-MeV nitrogen-15 ions. The transmutation recoil atoms which were knocked out of the 300-$\mu g/cm^2$ target by the intense nitrogen beam (approximately 4×10^{12} ions/s) were stopped in helium gas at a pressure of 600 torr. They were then carried by the helium stream through a small orifice into a rough vacuum, where they impinged upon the periphery of a 45-cm-diameter magnesium wheel. Most of the transmutation products were adsorbed in a small area on the surface. The wheel was rotated periodically to place these collection spots next to solid-state detectors arranged at high geometry so that the alpha-particle radiations could be examined with good resolution. *See* ELEMENT 104.

Since the rate of production of the element 105 atoms was very low (only about 10 atoms per hour were detected), it was necessary to make many measurements simultaneously. To this end, seven detecting stations were used, and four surface-barrier alpha-particle detectors were employed at each station. As shown in the illustration, these four detectors were arranged in pairs, two crystals alternately being placed (shuttled quickly) next to the flange of the wheel so that they could record the alpha particles emitted by the "mother" element 105 atoms and two crystals fixed in position so that they would be opposite two similar mother crystals and thus aid in the detection of the "daughter" element 103 (lawrencium-256) atoms. The daughter atoms could have reached the mother detectors only as a result of the recoil energy imparted to them by the alpha-particle decay of their mother atoms. The known daughter activity, 30-s 8.4-MeV lawrencium-256, was thereupon identified as a unique recoil product and shown to be transferred to the mother crystals with the same half-life as that observed for the complex 9.1-MeV peak. This half-life was determined by the relative number of alpha particles of this energy which were detected at each station.

Further proof. An additional proof that ^{256}Lr was the daughter of ^{260}Ha was afforded by an experiment in which there was shown to be a time correlation between the disappearance of ^{260}Ha and the appearance of ^{256}Lr as measured by certain characteristic alpha emissions. The interval from the time of emission of an approximately 9.1-MeV alpha particle to that of an approximately 8.4-MeV alpha particle was shown to correspond with the expected sequence.

A rough excitation function (relative cross section as ordinate, energy in millions of electronvolts as abscissa) for the ^{249}Cf (^{15}N,4n) ^{260}Ha reaction was found to peak at 84 MeV. In addition, the peak production rate of the 9.1-MeV activity was found to be about 1.5 alpha counts per microamperehour, which corresponds to a cross section of 3×10^{-33} cm^2. Both of these quantities agree with the values predicted for this reaction.

Soviet experiments. Previous work on element 105 was reported in 1968 by G. N. Flerov and colleagues at Dubna Laboratories in the Soviet Union. They claimed to have discovered two isotopes of element 105 produced by the bombardment of ^{243}Am by ^{22}Ne ions. The transmutation products were carried by a helium gas stream through an annular semiconductor detector which viewed the collecting surface. In the gross spectrum alpha-particle peaks were observed at 8.3, 8.7, 9.0, and 11.6MeV, which were attributed mostly to a lead impurity in the target. In the high-energy region between 8.8 and 10.3 MeV the Soviet scientists looked for delayed coincidences with alpha particles in the range from 8.35 to 8.6 MeV. They knew that this region was occupied partly by ^{256}Lr, and assumed that ^{257}Lr also emitted similar alpha groups with the same half-life. They found peaks at 9.4 and 9.7 MeV which seemed to give a statistically valid correlation with the lower-energy alphas and came to the conclusion that they were detecting ^{261}105 with $E_\alpha = 9.4 \pm 0.1$ MeV and $0.1 < T_{1/2} < 3$ s, and ^{260}105 with $E_\alpha = 9.7 \pm 0.1$ MeV and $T_{1/2} > 0.01$ s. The rate of production of these events was extremely low, with only 10 delayed coincidences observed in 400 microampere-hours of heavy ion bombardment.

The Lawrence Radiation Laboratory work did not confirm the above findings for the following reasons: (1) The isotope ^{260}105 has an energy of only 9.1 MeV; and (2) the daughter of ^{261}105 does not have similar decay properties to ^{256}Lr, as assumed by the Dubna group, and thus could not give a valid alpha-alpha correlation (the Lawrence group found that ^{257}Lr has an energy of 8.87 MeV and $T_{1/2} = 0.7$ s). *See* NUCLEAR CHEMISTRY.

At about the same time as the Berkeley discovery of ^{260}Ha, the Dubna group reported finding a 2-s spontaneous-fission emitter in other bombardments of americium-243 by neon-22. They thought it could be due to element 105, and gas chromatography experiments of the same type used in their element 104 work were cited to show that the 2-s activity behaved like the element 105 homolog Ta. Their evidence was far from conclusive, but it is possible that the spontaneous-fission activity was due to a branching decay of ^{261}Ha, since in later experiments the Berkeley group identified this isotope as predominantly an alpha emitter with energy of 8.93 MeV and a half-life of 1.7 ± 0.8 s. The Dubna group suggested the name nielsbohrium for element 105 on the basis of this work.

A third isotope of hahnium with mass 262 was later discovered by the Berkeley group. It was found to have an 8.45-MeV alpha energy and a half-life of 42 s. Again, its daughter, ^{258}Lr, was identified by recoil-milking and time-correlation techniques.

The isotopes ^{255}Ha and ^{257}Ha have been reported as spontaneous fission (SF) emitters with half-lives of about 2 s and about 6 s, respectively. The Dubna group bombarded Pb and Bi isotopes with ^{51}V ions and ^{50}Ti, using a rotating wheel device to observe these activities with passive dielectric track detector sheets. They made their assignments by assuming that the only reactions which could be observed by low bombarding energies would be of the compound-nucleus type. This assumption has now been called into question by several other groups as a result of experiments which demonstrate that deep-inelastic-transfer reactions also occur at low energies so that it is possible the SF activities could be due to elements of lower atomic number. Because of this possibility all of the new SF activities made by bombardments of Pb and Bi to make isotopes of elements from $Z = 93$ to 107 must be considered as unproved.

[ALBERT GHIORSO]

Bibliography: G. N. Flerov, *Proceedings of the International Conference on Nuclear Structure*, Tokyo, 1967; A. Ghiorso, *The Transuranium Elements: The Mendeleev Centennial*, the 13th Robert A. Welch Foundation Conference on Chemical Research, November 1969; A. Ghiorso et al., *Phys. Rev. Lett.*, 24:1498, 1970; J. Sanada (ed.), *Suppl. J. Phys. Soc. Jap.*, 24:237, 1968.

Element 106

The element with the atomic number 106 was synthesized and identified in 1974. This is the fourteenth of the synthetic transuranium elements, the laboratory-made elements heavier than uranium. Uranium is the heaviest commonly occurring element in nature (traces of plutonium, atomic number 94, have been detected in nature), having the atomic number 92. The lightest transuranium element, neptunium, with atomic number 93, was synthesized first in 1940, and the heaviest previous transuranium element, atomic number 105, was synthesized in 1970. *See* ELEMENT 105.

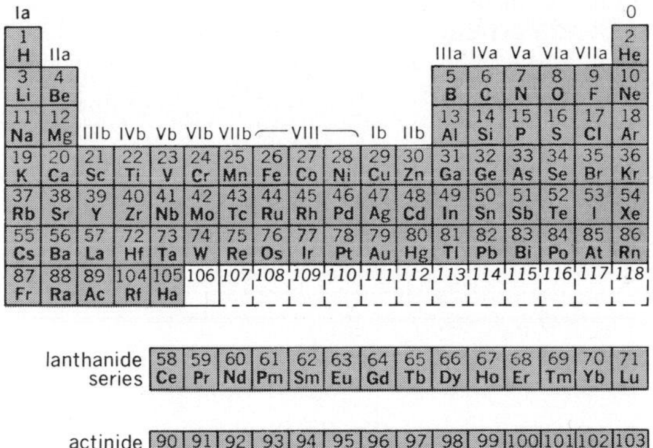

This discovery of element 106 took place nearly simultaneously in two widely separated nuclear laboratories, the Lawrence Berkeley Laboratory at the University of California and the Joint Institute for Nuclear Research at Dubna (near Moscow). Two independent and different approaches, using bombardment with heavy ions, were used in this difficult accomplishment. In view of the near simultaneity of the experiments at the two laboratories, and their very different nature, neither group has suggested a name for the element. Consequently, for the time being, it will have the simple designation "element 106."

On the basis of its projected position in the periodic table, element 106 is expected to have chemical properties similar to those of tungsten (atomic number 74).

Berkeley group. The Berkeley group, under the leadership of A. Ghiorso, used as its source of heavy ions the Super-Heavy Ion Linear Accelerator, the SuperHILAC. They bombarded a 259-μg target of californium (the isotope ^{249}Cf, with atomic number 98 and mass number 249) with ^{18}O ions (of energy 95 MeV and intensity 3×10^{12} ions per second). This resulted in the production and positive identification of the isotope of element 106 with the mass number 263, $^{263}106$, which decays with a half-life of 0.9 ± 0.2 s by the emission of alpha particles of principal energy 9.06 ± 0.04 MeV. This isotope is produced in the reaction in which four neutrons are emitted: $^{249}Cf(^{18}O,4n)$. The definite identification consists of the establishment of the genetic link between the element 106 alpha-particle-emitting isotope ($^{263}106$) and the previously identified alpha-particle-emitting daughter $^{259}104$ (3-s half-life) and granddaughter $^{255}102$ (3-min half-life) nuclides; that is, the demonstration of the decay sequence: $^{263}106 \xrightarrow{\alpha} {}^{259}Rf \xrightarrow{\alpha} {}^{255}No \xrightarrow{\alpha}$. (Rf is rutherfordium, element 104.) *See* ELEMENT 104.

In the Berkeley experiments the genetic linkage was established by use of specially designed apparatus in which the recoiling $^{263}106$ atoms were swept in a stream of helium gas to the surface of a wheel that could be rotated close to solid-state detectors in such a manner that the genetically related alpha particles of $^{263}106$, ^{259}Rf, and ^{255}No were recorded in their natural time sequence. The operation of the apparatus was controlled by a computer, which also recorded on magnetic tape all of the experimental data. A total of 73 $^{263}106$ alpha particles and approximately the expected corresponding number of ^{259}Rf daughter and ^{255}No granddaughter alpha particles were recorded.

Dubna group. The Dubna group, under the leadership of G. N. Flerov and Yu. Ts. Oganessian, produced its heavy ions with their 3.10-cm heavy-ion cyclotron. They chose lead (atomic number 82) as the target because, they believe, its closed shells of protons and neutrons and consequent small relative mass results in minimum excitation energy for the compound nucleus and therefore an enhancement in the cross section for the production of the desired product nuclide. They bombarded ^{207}Pb and ^{208}Pb with ^{54}Cr ions (atomic number 54, energy 280 MeV, 2×10^{11} ions per second) to find a product that decays by the spontaneous fission mechanism with the very short half-life of 7 ms. They assign this to the isotope $^{259}106$, suggesting reactions in which three or two neutrons are emitted: $^{207}Pb(^{54}Cr,2n)$ and $^{208}Pb(^{54}Cr,3n)$.

In the Dubna experiments the ^{54}Cr ions impinged on the lead target on the surface of a rotating disk (see illustration). Dielectric (mica) detectors were placed in the position of a surrounding sleeve at a distance of 3 mm from the rotating lead target. The product nuclei, decaying by spontaneous fission, projected their recoiling fragments into the mica detectors, which were examined after the bombardment to determine the number and location, and hence the half-life, of these decaying nuclei. A total of 51 spontaneous fission events were ascribed to $^{259}106$. The identification depends on rather uncertain deductions concerning the nature of the new type of nuclear reactions that the Dubna scientists postulate and new correlations on the dependence of spontaneous fission half-lives on atomic and mass numbers. *See* NOBELIUM; NUCLEAR CHEMISTRY; PERIODIC TABLE; TRANSURANIUM ELEMENTS.

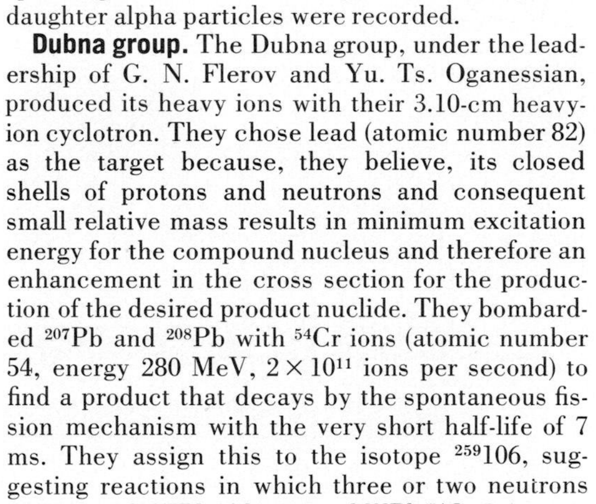

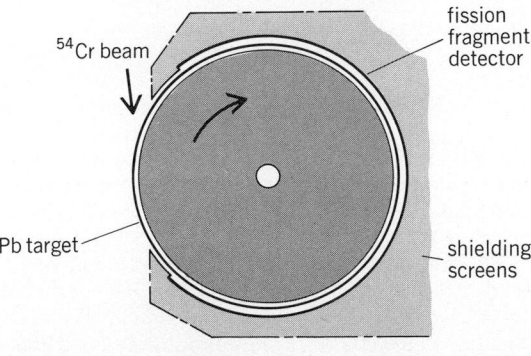

Schematic view of device for detecting short-lived spontaneous fission nuclei; Flerov group.

[GLENN T. SEABORG]

Bibliography: G. N. Flerov, in *Proceedings of the International Conference on Reactions between Complex Nuclei*, Nashville, vol. 2, p. 459, 1974; A. Ghiorso et al., *Phys. Rev. Lett.*, 33:1490, 1974.

Elements

An element is a substance made up of atoms with the same atomic number. Some common elements are oxygen, hydrogen, iron, copper, gold, silver, nitrogen, chlorine, and uranium. Approximately 75% of the elements are metals and the others are nonmetals. Most of the elements are solids at room temperature, two of them (mercury and bromine are liquids, and the rest are gases.

Occurrence and classification. A few of the elements are found in nature in the free (uncombined) state. Some of these are oxygen, nitrogen, the noble gases (helium, neon, argon, krypton, xenon, and radon), sulfur, copper, silver, and gold. Most of the elements in nature are combined with other elements in the form of compounds. The most abundant element on the earth is oxygen; the next most abundant is silicon. The most abundant element in the universe is hydrogen and the next most abundant is helium.

The elements are classified in families or groups in the periodic table. Elements are also frequently classified as metals and nonmetals. A metallic element is one whose atoms form positive ions in solution, and a nonmetallic element is one whose atoms form negative ions in solution. *See* PERIODIC TABLE.

Atoms of a given element have the same atomic number, but may not all have the same atomic weight (see table). Atoms with identical atomic numbers but different atomic weights are called isotopes. Oxygen, for example, is made up of atoms whose atomic weights are 16, 17, and 18. Hydrogen is made up of isotopes 1, 2, and 3; the isotopes of masses 2 and 3 are called deuterium and tritium, respectively. Carbon is made up of isotopes 11, 12, 13, and 14. Carbon-14 is radioactive and is used as a tracer in many chemical experiments.

All the elements have isotopes, although in certain cases only synthetic isotopes are known. Thus, fluorine exists in nature as ^{19}F, but the artificial radioactive isotope ^{18}F can be prepared. Many of the isotopes of the different elements are unstable, or radioactive, and hence disintegrate to form stable atoms either of that element or of some other element. *See* RADIOACTIVITY.

Origin and uses. The origin of the chemical elements is believed to be the result of the synthesis by fusion processes at very high temperatures

Table of atomic weights, 1973[h]

Name	Symbol	Atomic number	Atomic weight	Footnotes
Actinium	Ac	89	(227)	
Aluminium	Al	13	26.98154	*a*
Americium	Am	95	(243)	
Antimony	Sb	51	121.75*	
Argon	Ar	18	39.948*	*b, c, d, g*
Arsenic	As	33	74.9216	*a*
Astatine	At	85	(210)	
Barium	Ba	56	137.33	
Berkelium	Bk	97	(247)	
Beryllium	Be	4	9.01218	*a*
Bismuth	Bi	83	208.9804	*a*
Boron	B	5	10.81	*c, d, e*
Bromine	Br	35	79.904	*c*
Cadmium	Cd	48	112.41	
Caesium	Cs	55	132.9054	*a*
Calcium	Ca	20	40.08	*g*
Californium	Cf	98	(251)	
Carbon	C	6	12.011	*b, d*
Cerium	Ce	58	140.12	
Chlorine	Cl	17	35.453	*c*
Chromium	Cr	24	51.996	*c*
Cobalt	Co	27	58.9332	*a*
Copper	Cu	29	63.546*	*c, d*
Curium	Cm	96	(247)	
Dysprosium	Dy	66	162.50*	
Einsteinium	Es	99	(254)	
Erbium	Er	68	167.26*	
Europium	Eu	63	151.96	
Fermium	Fm	100	(257)	
Fluorine	F	9	18.998403	*a*
Francium	Fr	87	(223)	
Gadolinium	Gd	64	157.25*	
Gallium	Ga	31	69.72	
Germanium	Ge	32	72.59*	
Gold	Au	79	196.9665	*a*
Hafnium	Hf	72	178.49*	
Helium	He	2	4.00260	*b, c*
Holmium	Ho	67	164.9304	*a*
Hydrogen	H	1	1.0079	*b, d*
Indium	In	49	114.82	
Iodine	I	53	126.9045	*a*
Iridium	Ir	77	192.22*	
Iron	Fe	26	55.847*	
Krypton	Kr	36	83.80	*e*
Lanthanum	La	57	138.9055*	*b*
Lawrencium	Lr	103	(260)	
Lead	Pb	82	207.2	*d, g*
Lithium	Li	3	6.941*	*c, d, e, g*
Lutetium	Lu	71	174.97	
Magnesium	Mg	12	24.305	*c, g*
Manganese	Mn	25	54.9380	*a*
Mendelevium	Md	101	(258)	
Mercury	Hg	80	200.59*	
Molybdenum	Mo	42	95.94	
Neodymium	Nd	60	144.24*	
Neon	Ne	10	20.179*	*c, e*
Neptunium	Np	93	237.0482	*f*
Nickel	Ni	28	58.70	
Niobium	Nb	41	92.9064	*a*
Nitrogen	N	7	14.0067	*b, c*
Nobelium	No	102	(255)	
Osmium	Os	76	190.2	*g*
Oxygen	O	8	15.9994*	*b, c, d*
Palladium	Pd	46	106.4	
Phosphorus	P	15	30.97376	*a*
Platinum	Pt	78	195.09*	
Plutonium	Pu	94	(244)	
Polonium	Po	84	(209)	
Potassium	K	19	39.0983*	
Praseodymium	Pr	59	140.9077	*a*
Promethium	Pm	61	(145)	
Protactinium	Pa	91	231.0359	*f*
Radium	Ra	88	226.0254	*f, g*
Radon	Rn	86	(222)	
Rhenium	Re	75	186.207	*c*
Rhodium	Rh	45	102.9055	*a*
Rubidium	Rb	37	85.4678*	*c*
Ruthenium	Ru	44	101.07*	
Samarium	Sm	62	150.4	
Scandium	Sc	21	44.9559	*a*
Selenium	Se	34	78.96*	
Silicon	Si	14	28.0855*	*d*
Silver	Ag	47	107.868	*c*
Sodium	Na	11	22.98977	*a*
Strontium	Sr	38	87.62	*g*
Sulfur	S	16	32.06	*d*
Tantalum	Ta	73	180.9479*	*b*
Technetium	Tc	43	(97)	
Tellurium	Te	52	127.60*	
Terbium	Tb	65	158.9254	*a*
Thallium	Tl	81	204.37*	
Thorium	Th	90	232.0381	*f, g*
Thulium	Tm	69	168.9342	*a*
Tin	Sn	50	118.69*	
Titanium	Ti	22	47.90*	
Tungsten (Wolfram)	W	74	183.85*	
Uranium	U	92	238.029	*b, c, e, g*
Vanadium	V	23	50.9414*	*b, c*
Xenon	Xe	54	131.30	*e*
Ytterbium	Yb	70	173.04*	
Yttrium	Y	39	88.9059	*a*
Zinc	Zn	30	65.38	
Zirconium	Zr	40	91.22	

SOURCE: Inorganic Chemistry Division, Commission on Atomic Weights, International Union of Pure and Applied Chemistry. Incorporates changes approved at 28th IUPAC Conference, Madrid, 1975. Published in *Pure Appl. Chem.*, 37(4):591–603, 1974.

[a]Element with only one stable nuclide.

[b]Element with one predominant isotope (about 99–100% abundance); variations in the isotopic composition or errors in its determination have a correspondingly small effect on the value of A_r(E).

[c]Element for which the value of A_r(E) derives its reliability from calibrated measurements (that is, from comparisons with synthetic mixtures of known isotopic composition).

[d]Element for which known variations in isotopic composition in terrestrial material prevent a more precise atomic weight being given: A_r(E) values should be applicable to any "normal" material.

[e]Element for which substantial variations in A_r from the value given can occur in commercially available material because of inadvertent or undisclosed change of isotopic composition.

[f]Element for which value of A_r is that of most commonly available long-lived isotope.

[g]Element for which geological specimens are known in which element has anomalous isotopic composition.

[h]Scaled to the relative atomic mass. $A_r(^{12}C) = 12$. The atomic weights of many elements are not invariant but depend on the origin and treatment of the material. The footnotes to this table elaborate the types of variation to be expected for individual elements. The values of A_r(E) given here apply to elements as they exist naturally on Earth and to certain artificial elements. When used with due regard to the footnotes, they are considered reliable to ±1 in the last digit or ±3 when followed by an asterisk. Values in parentheses are used for certain radioactive elements whose atomic weights cannot be quoted precisely without knowledge of origin; the value given is the atomic mass number of the isotope of that element of longest known half-life.

(in the order of 100,000,000°C and higher) of the simple nuclear particles (protons and neutrons) first to heavier atomic nuclei such as those of helium and then on to the heavier and more complex nuclei of the light elements (lithium, boron, beryllium, and so on). The helium atoms bombard the atoms of the light elements and produce neutrons. The neutrons are captured by the nuclei of elements and produce heavier elements. These two processes—fusion of protons and neutron capture—are the main processes forming the chemical elements. Furthermore, the energy of the Sun and the stars is derived primarily from the fusion of hydrogen nuclei and electrons to form helium nuclei. It is believed that this element-producing fusion process is occurring even today in the hot stars.

The elements form the raw materials for the great chemicals industry today. Various metals are used for structural materials, protective coatings, ornamental devices, jewelry, and tableware. Such nonmetals as chlorine, bromine, hydrogen, sulfur, and nitrogen are important for the manufacture of many of the common chemicals of commerce. Helium is used to inflate dirigibles, neon is used to make neon light signs, and radon is used as a source of radioactive rays for therapy.

A number of elements, found in only very slight traces or not at all in nature, have been synthesized. Those elements are technetium, promethium, astatine, francium, and all the elements with atomic numbers above 92. These elements have been synthesized by a variety of nuclear reactions that involve transmuting atoms of one element into atoms of another by bombarding that element with neutrons or fast-moving particles (protons, deuterons, and α-particles) which will change the atomic number to that of the new element. Not only have these elements been synthesized, but isotopes of all the other elements also have been synthesized. Today the hope of the alchemist has been realized in that all the elements have been transmuted. *See* CHEMISTRY; ISOTOPE.

[ALFRED B. GARRETT]

Emission spectrochemical analysis

A technique used in qualitative or quantitative chemical analysis and conducted by monitoring and measuring the spectrum of light caused to be emitted by the material to be analyzed. In general, there are many ways in which to conduct an emission spectrometric measurement; the differences among approaches result mainly from the choice of location within the electromagnetic spectrum at which to observe emitted radiation. However, emission spectrochemical analysis traditionally refers to those analytical determinations based on radiation in the visible through vacuum ultraviolet region of the electromagnetic spectrum (wavelengths about 800 to 100 nm). The technique is used principally to detect (qualitative analysis) and determine (quantitative analysis) metals and some nonmetals. Under optimum conditions, as little as 10^{-10} g of an element per gram of sample can be determined. Routine concentration ranges in which emission spectrometry is used are from approximately 10^{-8} g per gram of sample to 10^{-2} g per gram of sample (1% by weight).

The steps in emission spectrochemical analysis are: vaporization and atomization of sample; excitation of atomic vapor; resolution of emitted radiation; and observation and measurement of resolved radiation.

Vaporization and atomization. A number of approaches to emission spectrochemical analysis have been developed, all of which follow the four steps listed, but which differ fundamentally in the first two steps. These steps are involved with the transfer of energy to the sample so that the sample will emit characteristic radiation. The application, control, and transfer of different forms of energy pose significantly different equipment requirements. Therefore, different versions of emission spectrometry have developed. The principal versions of vaporization, atomization, and excitation are by: ac spark, dc arc, electrical plasmas (dc plasma jet, radio-frequency-induced plasma, microwave-induced plasma), and chemical flame.

It is necessary first to produce a vapor (vaporization) in which all the compounds of the sample are broken down into their constituent atoms (atomization). It is these individual atoms which are detected and quantitated. If the sample is originally a solid, it can be made part of an anode-cathode pair. When an electric discharge is struck across these electrodes, for example producing an ac spark or a dc arc, the sample is vaporized into the electrode gap. There are numerous ways to configure an electrode such that the sample is an integral part of it; the desirable ones make it possible to reproduce the procedure simply and easily so that the technique is routinely applicable.

A more exotic way to produce a cloud of atoms from a solid sample is to fire a high-power pulse of laser radiation at the sample. The material vaporized when the small laser crater is blasted out is in just the state required for an atomic emission experiment. This approach has not yet been used extensively because high-pulsed power lasers are extremely expensive, and sample vapor cannot be reproduced in quantity. Consequently, it is not a cost-effective approach, and it is restricted to qualitative or semiquantitative analyses at best.

A solid sample may also be dissolved to bring it into an accessible solution state. The liquid solution sample can then be made part of an electrode. However, there are simpler ways in which to use liquid samples.

Dissolved solids or liquid samples can be broken down into microscopic-sized droplets, in a process called nebulization, and a stream of these micrometer-sized droplets can be injected directly into a discharge of hot gas or a flame. The hot gas then evaporates the solvent in the droplet, vaporizes the resulting solid particle, and atomizes the compounds of which the particle is made.

The hot gas can be produced in many ways. The complex mixture of atmospheric constituents and the electrode material in a dc arc discharge form a gas which reaches about 7500 K. Nebulized sample has been sprayed into dc arcs, but this approach tends to destabilize the discharge, resulting in poor signal precision.

A variety of techniques, in which sample nebulization systems and components for generation of a stable hot gas work together, have been developed. The earliest and most successful of these techniques is that of nebulizing the sample, mixing it homogeneously with the fuel and oxidant in a burner manifold, so that as the flame burns, the sample is vaporized and atomized. Many approaches for nebulizing the sample, mixing the droplets and gases, and configuring the burner manifold and burner head have proved successful. However, chemical flames (at least the types which are easily managed in the laboratory) reach temperatures of only 3000–4000 K. As a consequence, many compounds, for example refractory oxides and carbides, are not atomized because they are capable of existing at these temperatures. *See* FLAME PHOTOMETRY.

The general term electrical plasmas is used to refer to a group of discharges developed for the generation of temperatures close to 10,000 K. No known chemical compounds can exist at this temperature. These electrical plasmas, which will be discussed below, represent nearly optimal devices to use for vaporization and atomization, for they have been developed in parallel with sample injection systems so that signal stability is retained.

Excitation of atomic vapor. In the majority of spectrochemical instruments, the same process is used to produce the atomic vapor and to excite the atoms to cause emission of characteristic radiation. Thermal energy is imparted through collisional processes and is then released in the form of radiation. Mechanisms other than collisional mechanisms contribute to transfer of energy to analyte atoms. These might be: resonance with an applied electromagnetic field, excited-state interlevel transfer, radiative transfer, or other suprathermal processes. The extent of contribution from steps such as these, particularly when the experiment is conducted at atmospheric pressure, is not clear, and also not likely to be large.

In any case, the energy which is used resides in a hot gas. The hot gas is produced in a number of different ways. A chemical flame produces a hot gas from the combustion of some sort of fuel, for example acetylene or nitrous oxide (laughing gas), with the release of many thousands of calories of heat. Chemical flames normally have temperatures between 1000 and 4000 K.

An electric discharge which is caused by applied potentials or applied currents across an electrode gap (ac spark or dc arc) produces a hot gas of atmospheric species, electrode material, and electrons. Temperatures up to about 7500 K are produced.

Newer kinds of electric discharges, termed electrical plasmas, produce hot gases of much higher temperature. A plasma is a cloud of vapor in which more than 1% of the atoms are ionized, that is, have their outer electrons stripped away from them. In general, the greater the extent of ionization, the greater the amount of energy resident in the hot gas.

Modern electrical plasmas are produced in two principal ways. DC plasma jets are electrode-based plasmas in which a conventional dc arc is ignited in an electrode gap by applying a current. However, unlike the dc arc mentioned above, a dc plasma jet also has a highly controlled flow of cooling gas surrounding the arc. This cooling flow decreases the electrical conductivity of the outer regions of the arc, causing the applied current to be carried in a path of reduced volume. This is called

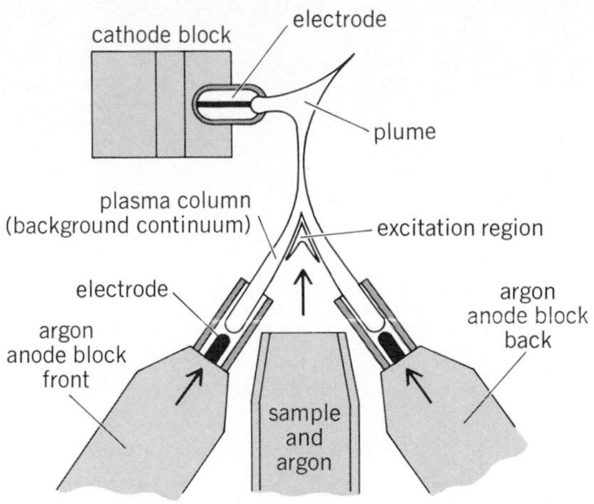

Fig. 1. Direct-current plasma jet device.

the thermal pinch effect. As a result, the current density of the arc rises dramatically and the temperature of the gas through which the arc is carried increases. Temperatures of up to 50,000 K have been recorded with dc plasma jets, but routine laboratory devices regularly generate plasmas of between 4000 and 10,000 K (Fig. 1). In a dc plasma jet device, each electrode block consists of an electrode, either pyrolytic graphite or thoriated tungsten, and an electrode sleeve through which the cooling gas causing the thermal pinch flows. Consequently, this device actually produces three plasma jets. This is done to ensure stability of operation and produce a large excitation region.

Inductively coupled plasmas are non-electrode based plasmas in which a carrier gas, usually argon or helium, is caused to reach a plasma state through the interaction of electrons, carrier gas atoms, and an applied radio-frequency or microwave-frequency field. Electrons will resonate with the magnetic vector impressed through an induction coil surrounding a flow of carrier gas, or with the electrical vector impressed into a resonant cavity through which the carrier gas flows. The electrons eventually gain extremely high kinetic energies through this resonance. In spite of the fact that the number of electrons that gain this high energy is small, the result is that they impart this energy through collisions to the much more numerous and massive carrier gas atoms. The carrier gas atoms eventually obtain a sufficient amount of energy to reach excited states. At this point the plasma has ignited, and the carrier gas can release its energy by emitting radiation, losing electrons (ionizing), or colliding with other atoms (thermal transfer). Temperatures routinely attained in induced plasmas are also between 4000 and 10,000 K, although much higher temperatures have been observed (Fig. 2).

Once the hot gas is produced, the vaporized and atomized sample is injected into it, or alternatively the hot gas functions as the vaporizer and atomizer. In any case, the plasma is used to impart energy to the analyte atoms so that they may reach any number of excited states. While more than one mechanism is followed as the analyte releases this energy, the route which is principally desired is that of radiational deactivation. That is, the

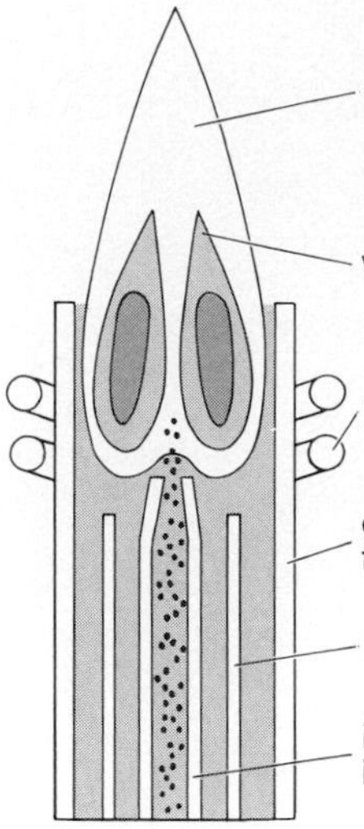

Fig. 2. Inductively coupled radio-frequency plasma.

analyte loses the energy by ejecting photons (emitting light).

Because the analyte atoms are allowed to attain a large number of excited states, a wide range of radiations (different wavelengths of light) are emitted. The energy gained and lost relates directly, through Planck's equation, $E = hc/\lambda$, to wavelengths between the infrared and vacuum ultraviolet regions.

Resolution. Identification of the emitting element and quantitative measurement of the radiation emitted requires that the light emitted be resolved into its component wavelengths; that is, a spectrum of the light must be produced. This is done with a spectrograph or spectrometer using photographic or photoelectric detection, respectively. The instrument dispersion must be on the order of 0.1 nm/mm to 1.0 nm/mm, so that individual wavelengths are adequately separated from each other at the detector. Prism spectrometers are no longer used very frequently, since they have inadequate dispersion over the entire wavelength range. Diffraction-grating instruments have uniform wavelength dispersion and can be produced so as to give quite high resolution without wasting much input radiation. In general, emission spectrochemical analysis requires higher spectral resolution than other types of spectrometric experiments because of the larger number of spectral lines produced.

Observation and measurement. Visual observation is rarely used in routine spectrochemical analysis. The human eye does not possess high enough resolving power to sort out the complex spectra usually produced.

Photographic observation has long been the method of choice for spectrochemical detection. It is used mainly for qualitative analysis and quantitative analysis at the minor- and high-trace-constituent levels. Glass plates or film may be employed. Spectrograms usually are viewed with a projection comparator, which projects sample and standard plates on a split-field, ground-glass screen, permitting any spectrum on one plate to be brought adjacent to and in register with any spectrum on the other. Many comparators also incorporate a microphotometer. Any emulsion is suitable for the 230–430-nm range; at higher wavelengths, special sensitization is needed. Emulsion sensitivity varies directly with grain size; thus, high sensitivity is obtained at the cost of resolving power, and vice versa. Sensitivity and contrast vary with wavelength and age for all emulsions. Aging conditions can be retarded by storage at 32°C.

Different production batches of a given emulsion are seldom identical; if quantitative work is to be done, a 6 to 12 months supply of plates or film should be bought from a single batch and stored at 32°C.

Measurement of line intensity ratios requires calibration of the emulsion, that is, construction of a curve showing blackening of the emulsion as a function of the intensity of incident light of the wavelength used. Lines whose intensities are to be compared photographically should be as close as possible in wavelength.

Densitometers or microphotometers measure spectral-line blackening by scanning the illuminated spectrum with a fine receiving slit or by projecting a fine illuminated line onto and through the spectrogram and onto a photocell.

The major advantage of photographic observation is that it is both the least expensive and the simplest method for obtaining the entire spectrum simultaneously (multielement analysis) and storing it nearly indefinitely. It is not, however, the most sensitive method, and thus is not useful for the important realm of trace-element and ultratrace-element analysis.

Photoelectric observation affords extremely sensitive and rapid detection. But one photomultiplier tube (PMT) must be used for each and every wavelength to be monitored. The photocurrent output is directly proportional to the rate at which photons strike the photoanode. This current is retrieved by various electronic means, and is amplified to give an electric signal which is then proportional to the amount of analyte present. Photoelectric detection is best suited to trace and ultratrace determinations because of its high sensitivity. But it presents significant problems for simultaneous multielement analyses due to its one tube per wavelength limitation. However, arrays of photomultiplier tubes have been constructed for the simultaneous observation of up to 60 wavelengths.

Another approach to multielement detection is to use only one photomultiplier tube and to adjust the spectrometer precisely so that successive wavelengths pass by the tube. This is called spectrometer scanning. Its principal drawback is that each wavelength is viewed sequentially, and so it takes longer to monitor more wavelengths. Its principal advantage is that there is no limit to which wavelengths can be viewed, all of them being accessible.

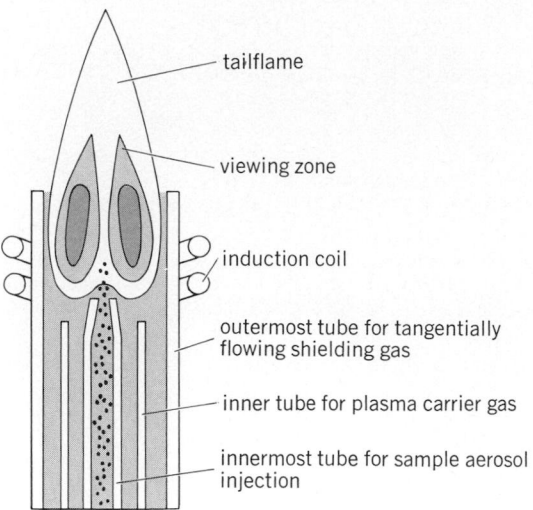

A developing approach to multielement, simultaneous detection is the use of imaging electronic devices, such as tv tubes, photodiode arrays, and charge-coupled and charge-injection devices (CCDs and CIDs). These solid-state devices can all be considered to be electronic versions of photographic plates. Consequently, all the simultaneity and simplicity of photographic detection and the high sensitivity and speed of photoelectric detection are obtained at once. The drawbacks of these devices are that they have low resolution and are quite expensive. Thus they require even higher-resolution spectrometers. However, they have increasing applications in emission spectrochemical analysis. *See* SPECTROCHEMICAL ANALYSIS; SPECTROSCOPY.

[ANDREW T. ZANDER]

Bibliography: L. H. Ahrens, *Spectrochemical Analysis*, 2d ed., 1961; J. A. Dean and T. C. Rains, *Flame Emission and Atomic Absorption Spectrometry*, vol. 1, 1969; G. W. Ewing, *Instrumental Methods of Chemical Analysis*, 1957; J. D. Winefordner, *Trace Analysis: Spectroscopic Methods for Elements*, 1976.

Emulsion

A dispersion of one liquid in a second immiscible liquid. Since the majority of emulsions contain water as one of the phases, it is customary to classify emulsions into two types: the oil-in-water (O/W) type consisting of droplets of oil dispersed in water, and the water-in-oil (W/O) type in which the phases are reversed. The continuous liquid is referred to as the dispersion medium, and the liquid which is in the form of droplets is called the disperse phase.

A stable emulsion consisting of two pure liquids cannot be prepared; to achieve stability, a third component, an emulsifying agent, must be present. Generally, the introduction of an emulsifying agent will lower the interfacial tension of the two phases.

Classification. A large number of emulsifying agents are known; they can be classified broadly into several groups. The largest group is that of the soaps, detergents, and other compounds whose basic structure is a paraffin chain terminating in a polar group. Water-soluble soaps (for example, sodium or potassium soaps) are effective in stabilizing oil-in-water emulsions; in this case, the paraffin chains of the soap molecules are concentrated in the oil droplets, and their polar groups are oriented toward the continuous water medium. Dissociation of the polar groups leaves the oil droplets with a charge situated at the interface, and the counter ions form a diffuse double layer in the water, thus preventing coalescence. Water-insoluble soaps (for example, calcium soaps) are effective in stabilizing water-in-oil emulsions. In this case the stabilizing action is believed to be similar to that exhibited by certain solid powders. For a powder to act as an emulsifier, it must be wetted more by one phase than by the other. Whichever phase shows the greater wetting power will become the dispersion medium, because such powders congregate at the interfaces and present the greater portions of their surfaces to the liquid which wets them preferentially. For example, precipitated sulfur, which is wetted preferentially by water, stabilizes oil-in-water emulsions; lampblack, which is wetted preferentially by oil, stabilizes water-in-oil emulsions. Many naturally occurring emulsions, such as milk or rubber latex, are stabilized by proteins. Egg yolk proteins stabilize mayonnaise and salad dressing. In these and similar types of emulsions, stability results from the formation of a protective coating of the protein around each droplet of the disperse phase. Certain hydrophilic colloids such as gum arabic or gelatin also stabilize water-in-oil emulsions by a similar mode of action.

Geometrically, the maximum amount of one liquid which can be dispersed in another in the form of spheres of equal size is about 74% of the total available space, independent of the diameter of the spheres. However, considerably more concentrated emulsions of either type can be obtained because droplets are not necessarily uniform in size, and the emulsifying agent permits distortion of the droplets without coalescence. The creams used in cosmetics are examples of high-concentration emulsions.

Properties. Various methods are available for determining the type of a particular emulsion, but recognition of the following three characteristics is usually sufficient for most purposes. (1) The electrical conductivity of an oil-in-water emulsion is much greater than that of a water-in-oil emulsion. (2) A water-soluble dye such as methyl orange will color an oil-in-water system easily, but will not color a water-in-oil system. For an oil-soluble dye such as fuchsin, the reverse is true. (3) An emulsion will mix perfectly with more of its continuous phase when this is added in pure form.

Emulsions may be prepared readily by shaking together the two liquids or by adding one phase drop by drop to the other phase with some form of agitation, such as irradiation by ultrasonic waves of high intensity. In industry, emulsification is accomplished by means of emulsifying machines. In a typical machine, a mixture of the two liquids containing an emulsifying agent is forced through a narrow slit between a rapidly rotating rotor and a stator. The preparation of stable emulsions must be controlled carefully, since emulsions are sensitive to such variations as the mode of agitation, the nature and amount of the emulsifying agent, and temperature changes.

The breaking of emulsions is necessary in many industrial operations, for example, in the separation of water-in-oil emulsions in the petroleum industry and in product recovery from emulsions produced by the steam distillation of organic liquids. Emulsions may be broken by (1) addition of multivalent ions of charge opposite to the emulsion droplet, (2) chemical action (addition of acids to emulsions stabilized by soaps), (3) freezing, (4) heating, (5) aging, (6) centrifuging, (7) application of high-potential alternating electric fields, and (8) treatment with ultrasonic waves of low intensity. *See* COLLOID.

[GEORGE S. MILL; W. O. MILLIGAN]

Enantiomorph

One of an isomeric pair of chemical compounds whose molecules are nonsuperimposable mirror images. One molecular configuration of such dissymmetric substances is capable of rotating plane-polarized light to the right, dextro or (+) form, while the mirror image rotates the light equally to the left, levo or (−) form. Each member of such an enantiomorphic pair (optical isomers) possesses identical chemical and physical properties except for interaction with other dissymmetric systems, that is, other optically active substances or plane-polarized or circularly polarized light. With the dextro form of another dissymmetric system, a given (+) form behaves exactly as the corresponding (−) form does with the levo form. *See* OPTICAL ACTIVITY.

[WYMAN R. VAUGHAN]

Enthalpy

For any system, that is, the volume of substance under discussion, enthalpy is the sum of the internal energy of the system plus the system's volume multiplied by the pressure exerted by the system on its surroundings. This may be expressed as $U + PV = H$, where U is the system's internal energy, P the pressure of the system, V the system's volume, and H the enthalpy of the system. The sum of $U + PV$ is given the special symbol H primarily as a matter of convenience because this sum appears repeatedly in thermodynamic discussion. Consistent units must, of course, be used in expressing the terms in the above equation. Previously, enthalpy was referred to as total heat or heat content, but these terms are misleading and should be avoided. Enthalpy is, from the viewpoint of mathematics, a point function, as contrasted with heat and work, which are path functions. Point functions depend only on the initial and final states of the system undergoing a change; they are independent of the paths or character of the change. Mathematically, the differential of a point function is a complete or perfect differential.

Because the absolute value of internal energy of even a simple system is usually unknown, recorded values of enthalpy are relative values measured above some convenient but arbitrarily chosen datum. Thus in the steam tables of Keenan and Keyes, the datum is liquid water at 32°F (0°C) and under its own vapor pressure. At this state water is assumed to have an enthalpy equal to zero. Under this assumption the internal energy of water in this state is a negative quantity equal to PV. No complication is introduced by this fact, although visualization of negative energies of this kind may be disturbing to some. There is limited utility for absolute enthalpies because only the changes in enthalpy are measurable. It is instructive to examine the utility of the enthalpy function in terms of some simple but important thermodynamic processes.

The first law of thermodynamics is merely a statement of the law of conservation of energy. The first law alone indicates that:

1. For a chemical reaction carried out at constant pressure and temperature with no work performed except that resulting from keeping the internal and external pressure equal to each other as the volume changes, the change in enthalpy of the system (the material taking part in the chemical reaction) is numerically equal to the heat that must be transferred to maintain the above-mentioned conditions. This heat is often loosely referred to as the heat of reaction. More properly, it is the enthalpy change for the reaction.

2. So-called heat balances on heat exchangers, furnaces, and similar industrial equipment that operate under steady flow conditions are really enthalpy balances.

3. The work developed in a steadily running adiabatic engine or turbine is equivalent to the enthalpy change of the fluid passing through the engine.

4. The adiabatic, irreversible, steady flow of a stream of materials through a porous plug or a partially opened valve under circumstances where the change in kinetic energy of flow is negligible (a Joule-Thomson process) results in no change in enthalpy of the flowing stream. Although no change in enthalpy results from this process, there is a loss in the energy available for doing work as a result of the pressure drop across the plug or valve. For change in enthalpy with pressure or temperature *see* THERMODYNAMIC PRINCIPLES. *See also* ENTROPY; THERMODYNAMIC PROCESSES.

[HAROLD C. WEBER; WILLIAM A. STEELE]

Entropy

A function first introduced in classical thermodynamics to provide a quantitative basis for the common observation that naturally occurring processes have a particular direction. Subsequently, in statistical thermodynamics, entropy was shown to be a measure of the number of microstates a system could assume. Finally, in communication theory, entropy is a measure of information. Each of these aspects will be considered in turn. Before the entropy function is introduced, it is necessary to discuss reversible processes.

Reversible processes. Any system under constant external conditions is observed to change in such a way as to approach a particularly simple final state called an equilibrium state. For example, two bodies initially at different temperatures are connected by a metal wire. Heat flows from the hot to the cold body until the temperatures of both bodies are the same. As another example, a vessel containing a gas is connected through a stopcock to an evacuated vessel. When the stopcock is opened, the gas expands to fill the whole of the available space uniformly. It is common experience that the reverse processes never occur if the systems are left to themselves; that is, heat is never observed to flow from the cold to the hot body, nor will the gas compress itself into one of the vessels. Max Planck classified all elementary processes into three categories: natural, unnatural, and reversible.

Natural processes do occur, and proceed in a direction toward equilibrium. Unnatural processes move away from equilibrium and never occur. If $A \rightarrow B$ is a natural process between states A and B, then $B \rightarrow A$ is an unnatural process. A reversible process is an idealized natural process that passes through a continuous sequence of equilibrium states. Consider the evaporation of a liquid in the presence of its vapor at a pressure P. Let the equilibrium vapor pressure of the liquid be p. If $P < p$, liquid evaporates as a natural process. If $P > p$, evaporation is an unnatural process and will not occur; indeed, the opposite process—condensation—will take place. Finally, if $P = p$, both processes of condensation and evaporation are reversible and can be initiated by a very slight increase or decrease in the external pressure P.

A useful idea is that a reversible process may be exactly reversed by an infinitesimal change in the external conditions. If a hot object is placed adjacent to a much colder object, the heat-flow direction cannot be reversed by small changes in the temperature of either object. In reversible processes, work is accomplished through small pressure differences, and heat transfer occurs through small temperature differences.

Entropy function. The state function entropy S puts the foregoing discussion on a quantitative basis. The function is not derived in this article; but, rather, some of its properties are stated, and its implications are discussed mainly by example. Entropy is related to q, the heat flowing into the system from its surroundings and to T, the absolute temperature of the system. The important properties for this discussion are:

1. $dS > q/T$ for a natural change.
   $dS = q/T$ for a reversible change.

It is necessary to introduce both S and T together. A formal derivation would show T^{-1} as an integrating factor leading to the complete differential dS.

2. The entropy of the system S is made up of the sum of all the parts of the system so that $S = S_1 + S_2 + S_3 \cdots$. *See* TEMPERATURE; THERMODYNAMIC PRINCIPLES.

Heat flow. Consider two bodies, α and β, at different temperatures separated by an adiabatic (no heat transfer) wall. If the two bodies are connected by a fine wire that allows a small heat flow q from α to β, then $dS_\alpha = -q/T_\alpha$ and $dS_\beta = q/T_\beta$.

For the whole system, Eq. (1) holds. If $T_\alpha > T_\beta$,

$$dS = dS_\alpha + dS_\beta = q\left(\frac{1}{T_\beta} - \frac{1}{T_\alpha}\right) \qquad (1)$$

$dS > 0$, and heat flows from α to β as a natural process. The process could be continued until $T_\alpha = T_\beta$ and $dS = 0$.

Once the constraint of the adiabatic wall is abrogated, the entropy increases to a maximum value, and T_α becomes equal to T_β. This is a special case of the most important notion in thermodynamics; that is, the system will assume that equilibrium state which maximizes the entropy at constant energy, consistent with the constraints.

Nonconservation of entropy. In his study of the first law of thermodynamics, J. P. Joule caused work to be expended by rubbing metal blocks together in a large mass of water. By this and similar experiments, he established numerical relationships between heat and work. When the experiment was completed, the apparatus remained unchanged except for a slight increase in the water temperature. Work (W) had been converted into heat (Q) with 100% efficiency. Provided the process was carried out slowly, the temperature difference between the blocks and the water would be small, and heat transfer could be considered a reversible process. The entropy increase of the water at its temperature T is $\Delta S = (Q/T) = (W/T)$.

Since everything but the water is unchanged, this equation also represents the total entropy increase. The entropy has been created from the work input, and this process could be continued indefinitely, creating more and more entropy. Unlike energy, entropy is not conserved. *See* CONSERVATION OF ENERGY.

Although the heat transfer is considered to be reversible in order to calculate the entropy increase, the overall process of converting work into heat is irreversible. The frictional process that converts kinetic energy into the heat of the metal blocks is a natural process. In fact, the impossibility of the reverse process is Lord Kelvin's statement of the second law of thermodynamics. Heat cannot be completely converted into work without other changes occurring in the surroundings. For example, a gas in a cylinder can be expanded reversibly by extracting heat from a large constant-temperature bath. All of the heat extracted from the bath is converted into work, but eventually the pressure of the gas system would be reduced to an unusable level. The system has changed, and the process cannot continue indefinitely. If one tries to convert heat into work through a system undergoing a cycle so that the system will return to its initial state, one finds that only a portion of the heat input does work and that the remainder must be rejected to a lower temperature; this is just the process which takes place in a heat engine. *See* THERMODYNAMIC PROCESSES.

Degradation of energy. Energy is never destroyed. But in the Joule friction experiment and in heat transfer between bodies, as in any natural process, something is lost. In the Joule experiment, the energy expended in work now resides in the water bath. But if this energy is reused, less useful work is obtained than was originally put in. The original energy input has been degraded to a less useful form. The energy transferred from a high-temperature body to a lower-temperature body is also in a less useful form. If another system is used to restore this degraded energy to its original form, it is found that the restoring system has degraded the energy even more than the original system had. Thus, every process occurring in the world results in an overall increase in entropy and a corresponding degradation in energy. R. Clausius stated the first two laws of thermodynamics as: "The energy of the world is constant. The entropy of the world tends toward a maximum."

Increasing entropy and mixing. Once the atomic theory of matter is accepted, the entropy concept can be made much clearer. It is then found through statistical thermodynamics that the increase of entropy toward its maximum value at equilibrium corresponds to the change of the system toward its most probable state consistent with the constraints. The most probable state represents the most mixed or most random state. Mixing must be given a broad interpretation which includes particle or configurational mixing, and spreading of energy over the particles or thermal mixing. Diffusion of one gas into another represents obvious configurational mixing and increased entropy. Irreversible expansion of a gas represents configurational mixing of the molecules over the available space. Heat flow represents spreading of the kinetic energy between the particles. Friction spreads the kinetic energy of the body over the constituent particles. Sometimes the energy-spread entropy increase and the configurational entropy increase are not compatible, and a compromise is struck. A subcooled liquid adiabatically crystallizes to a lower configurational entropy but gains even more entropy through the additional energy levels made available. The same sort of behavior occurs in partially miscible liquids—some configurational entropy is sacrificed in order to gain a large amount of energy-spread entropy.

Absolute entropy. The third law of thermodynamics (Nernst's heat theorem) refers to the vanishing of entropy at zero temperature. In 1912 Planck proposed that the theorem applied to pure crystalline solids. However, the theorem is now known to be applicable to gases and, by all reasonable expectation, is applicable to any system. Thus, any substance at finite temperatures has an absolute entropy, the value of which can be determined from either calorimetric or spectroscopic data. Absolute entropies, together with thermochemical data, are very useful in the calculation of equilibrium compositions of reaction systems.

The statistical viewpoint is that a thermodynamic state at finite temperatures corresponds to many microstates. During an observation the microstates of a system undergo continuous rapid transitions. Since entropy is proportional to the logarithm of the number of available microstates, the Nernst theorem implies that the thermodynamic state at zero temperature corresponds to a single microstate. Thus, at zero temperature, even a ferromagnetic material should exist in a single state, fully magnetized in a direction determined by its inevitable interactions with the environment.

[WILLIAM F. JAEP]

Measure of information. The probability characteristic of entropy leads to its use in communication theory as a measure of information. The absence of information about a situation is equivalent to an uncertainty associated with the nature of the situation. This uncertainty, designated H, is the entropy of the information about the particular situation, Eq. (2), where $p_1, p_2, \ldots, p_n$ are the proba-

$$H(p_1, p_2, \ldots, p_n) = -\sum_{k=1}^{n} p_k \log p_k \qquad (2)$$

bilities of mutually exclusive events, the logarithms are taken to an arbitrary but fixed base, and $p_k \log p_k$ always equals zero if $p_k = 0$. For example, if $p_1 = 1$ and all others ps are zero, the situation is completely predictable beforehand; there is no uncertainty and so the entropy is zero. In all other cases the entropy is positive.

In introducing entropy of an information space, C. E. Shannon described a source of information by its entropy H in bits per symbol. The ratio of the entropy of a source to the maximum rate of signaling that it could achieve with the same symbols is its relative entropy. One minus relative entropy is the redundancy of the source.

[FRANK H. ROCKETT]

Bibliography: J. Aczel and Z. Daroczy, *Measures of Information and Their Characterizations*, 1975; H. B. Callen, *Thermodynamics*, 1960; K. G. Denbigh, *Principles of Chemical Equilibrium*, 3d ed., 1971; J. D. Fast, *Entropy: The Significance of the Concept of Entropy and Its Applications in Science and Technology*, 2d ed., 1968; A. I. Khinchin, *Mathematical Foundations of Information Theory*, 1957; C. E. Shannon, A mathematical theory of communication, *Bell Syst. Tech. J.*, 27(3): 379–423, 27(4):623–656, 1948; R. C. Tolman, *The Principles of Statistical Mechanics*, 1980; K. Wark, *Thermodynamics*, 3d ed., 1977.

Epoxidation

The conversion of olefins (or other substances containing carbon-carbon double bonds) into epoxy (oxirane) compounds. Direct oxidation with oxygen to oxiranes has been used industrially only with ethylene, where oxidation over silver yields ethylene oxide [reaction (1)]. Epoxidation with peroxy-

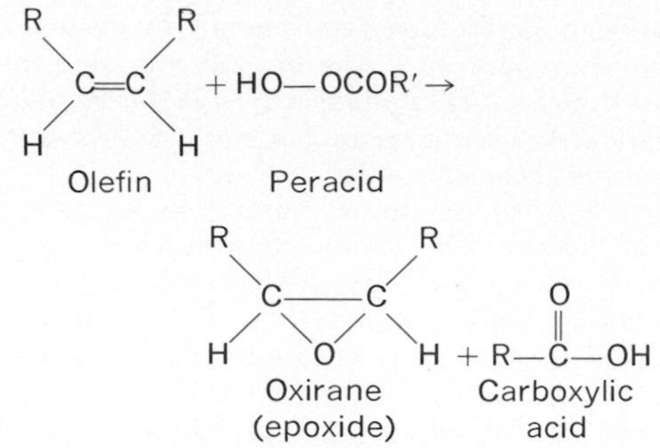

(1)

acids is the most generally applied method [reaction (2)]. Other epoxidation procedures involve oxidation of the olefin with organic peroxides, permanganates, chromates, or dehydrochlorination of chlorohydrins with caustic alkalies. *See* ETHYLENE OXIDE.

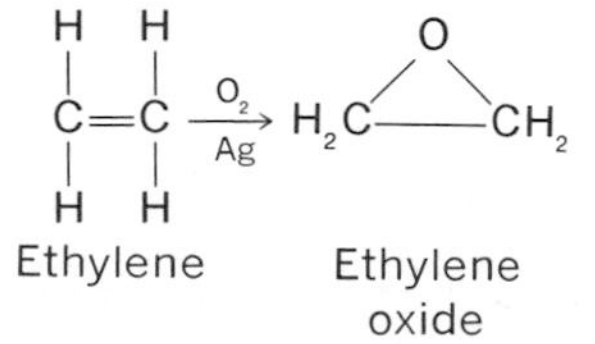

(2)

Peracetic acid or performic acid are readily prepared peroxyacids and are convenient to use for epoxidations. Acetic acid or formic acid is combined with 30% hydrogen peroxide in the presence of an acid catalyst such as cation exchange resin to yield the corresponding peroxyacid. Peracetic acid may also be prepared by vapor-phase oxidation of acetaldehyde with oxygen in suitably passivated, inert reactors, then absorbing the vapors in an unreactive solvent such as ethyl acetate or octane.

The molecular structure of the unsaturated material determines the readiness of epoxidation, the ratio of reactants needed, the temperature used, and the catalyst required. Methyl oleate and similar esters undergo epoxidation faster than terminally unsaturated olefins, for example, 1-tetradecene. Perbenzoic acid and monoperphthalic acid also give good yields of epoxides with olefins. Certain organic hydroperoxides in the presence of molybdenum compounds produce epoxides through a reaction similar to peracid epoxidation. *See* ALKENE; SUBSTITUTION REACTION.

[FRANK WAGNER]

Bibliography: R. Adams (ed.), *Organic Reactions*, vol. 7, 1954.

Epoxy resin

A polyether resin formed by the polymerization of bisphenol A and epichlorohydrin. Epoxy resins are used as coatings, adhesives, castings, and foams. Laminates of epoxy resin and glass cloth have been used to make pipes, to repair damaged automobile bodies, and to make small-boat hulls. *See* POLYETHER RESINS. [MARVIN YELLES]

Bibliography: H. Lee and K. Neville, *Epoxy Resins: Their Applications and Technology*, 1957.

Equivalent weight

The number of parts by weight of an element or compound which will combine with or replace, directly or indirectly, 1.008 parts by weight of hydrogen, 8.00 parts of oxygen, or the equivalent weight of any other element or compound. The term equivalent weight comes from the law of equivalent proportions, which states that the weights of two elements A and B which combine separately with identical weights of another element C are either the weights in which A and B combine together, or are related to them in the ratio of small whole numbers. A standard weight of 8.000 parts is chosen for oxygen. For all elements, the atomic weight is equal to the equivalent weight times a small whole number, called the valence of the element. *See* CHEMICAL COMPOUNDS; ELEMENTS; VALENCE.

An element can have more than one valence and therefore more than one equivalent weight. The use of the terms is explained below.

1. Ammonia, NH_3, contains 1 atom of nitrogen combined with 3 atoms of hydrogen. Since the equivalent weight of hydrogen is equal to its atomic weight, the equivalent weight of nitrogen is 1/3 its atomic weight and its valence is 3.

2. Magnesium oxide, MgO, contains 1 atom of magnesium combined with 1 atom of oxygen. Since the equivalent weight of oxygen is 1/2 its atomic weight, the equivalent weight of magnesium is also 1/2 its atomic weight and its valence is 2.

3. Phosphorus forms two chlorides, phosphorus trichloride, PCl_3, and phosphorus pentachloride, PCl_5. Since the equivalent weight of chlorine is equal to its atomic weight, in the trichloride the equivalent weight of phosphorus is 1/3 its atomic weight and its valence is 3, and in the pentachloride the equivalent weight is 1/5 its atomic weight and its valence is 5.

The equivalent weight of a compound depends on the reaction in which it takes part. Thus:

1. In the reaction between potassium iodate, KIO_3, and silver nitrate, $AgNO_3$, one molecule of silver iodate, $AgIO_3$, is precipitated for every molecule of silver nitrate. This is represented by reaction (1). Since the equivalent weight of silver

$$KIO_3 + AgNO_3 \rightarrow AgIO_3(\text{ppt}) + KNO_3 \qquad (1)$$

is its atomic weight, the equivalent weight of potassium iodate, in this reaction, is its molecular weight.

2. When potassium iodate, KIO_3, is reduced to iodine, I_2, by potassium iodide, KI, three molecules of iodine are produced per molecule of potassium iodate. This is represented by reaction (2).

$$KIO_3 + 5KI + 6HCl \rightarrow 3I_2 + 3H_2O + 6KCl \qquad (2)$$

Since the equivalent weight of iodine is 1/2 its molecular weight, the equivalent weight of potassium iodate, in this reaction, is 1/6 its molecular weight.

This concept, together with that of gram-equivalent weight, tends to have been abandoned, and relations are expressed in terms of balanced stoichiometric chemical equations and relative numbers of moles reacting. *See* ELECTROCHEMICAL EQUIVALENT; MOLE; OXIDATION-REDUCTION; STOICHIOMETRY. [THOMAS C. WADDINGTON]

Erbium

A chemical element, Er, atomic number 68, atomic weight 167.26, belonging to the rare-earth group. The naturally occurring element is made up of the stable isotopes Er^{162} 0.136%, Er^{164} 1.56%, Er^{166}

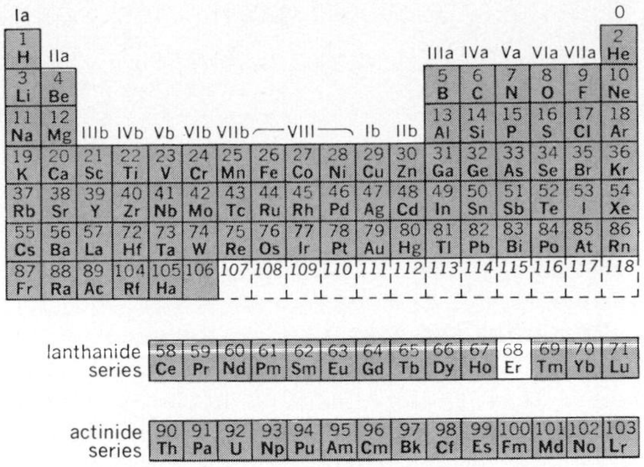

33.41%, Er^{167} 22.94%, Er^{168} 27.07%, Er^{170} 14.88%. This metallic element was discovered in 1843 by C. G. Mosander, who originally named the oxide terbia. The names terbia and erbia became confused in the early literature, so by general consent after 1860 this oxide was called erbia. The rose-pink oxide, Er_2O_3, dissolves in mineral acids to give rose-colored solutions. The salts are paramagnetic and the ions are trivalent. At low temperatures the metal is antiferromagnetic and at still lower temperatures becomes strongly ferromagnetic. The Néel point is 78 K and the Curie point about 20 K. In the antiferromagnetic region between 20 and 78 K, the magnetism shows an interesting but complicated periodic arrangement of the magnetic moments along the *c* axis of the hexagonal metal crystals. For properties of the metal *see* RARE-EARTH ELEMENTS. [FRANK H. SPEDDING]

Ester

The product of a condensation reaction (esterification) in which a molecule of an acid unites with a molecule of alcohol with elimination of a molecule of water, reaction (1).

$$\underset{\text{Acid}}{\mathrm{RC(=O)OH}} + \underset{\text{Alcohol}}{\mathrm{HOR'}} \rightarrow \underset{\text{Ester}}{\mathrm{RC(=O)OR'}} + H_2O \qquad (1)$$

At one time it was thought that esterification was analogous to neutralization, and esters are still named as though they are "alkyl salts" of carboxylic acids, as shown in the following examples:

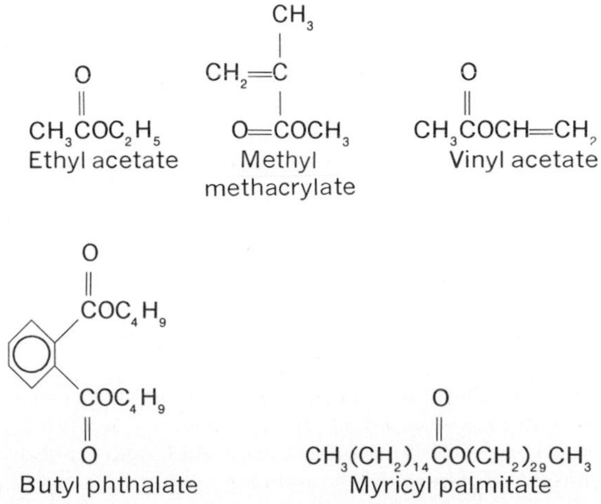

CH_2ONO_2
|
$CHONO_2$
|
CH_2ONO_2
Glyceryl trinitrate
(nitroglycerin)

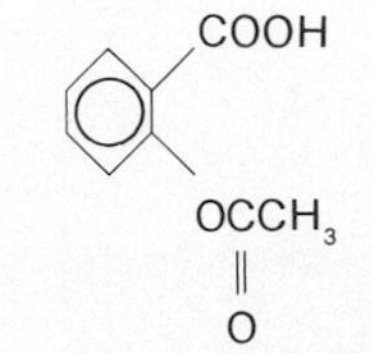

Acetylsalicylic
acid (aspirin)

$CH_3OSO_2OCH_3$
Dimethyl
sulfate

Properties and uses. Esters are generally insoluble in water and have boiling points slightly higher than hydrocarbons of similar molecular weight. An ester may often be characterized by its infrared absorption spectrum. For example, saturated aliphatic esters have a distinctive, strong C=O absorption band at 1750–1735 cm^{-1}, and a second band assigned to C—O stretching at 1300–1000 cm^{-1}.

Ethyl and butyl acetates are volatile industrial solvents, used particularly in the formulation of lacquers. The ethyl acetate produced in the United States is primarily used as a solvent. Higher-boiling esters such as butyl phthalate are used as softening agents (plasticizers) in the compounding of plastics. The natural waxes of biological origin are largely simple esters. For example, a principal component of beeswax is myricyl palmitate. *See* SOLVENT.

Esters of cellulose (cellulose triacetate) are used in photographic film and as a textile fiber (acetate rayon). Cellulose acetatepropionate and cellulose acetatebutyrate have become important as thermoplastic materials. Cellulose nitrate, containing 10.5–11% nitrogen, is called celluloid pyroxylin; with alcohol and camphor (a plasticizer), it forms celluloid. Dynamite cotton is cellulose nitrate of 11.5–12.3% nitrogen content, and gun cotton is cellulose nitrate of 12.5–13.5% nitrogen. Cordite and ballistite are made from gun cotton, which is plasticized with glyceryl trinitrate (nitroglycerin). Dimethyl and diethyl sulfates (esters of sulfuric acid) are excellent agents for alkylating organic molecules that contain labile hydrogen atoms, for example, starch and cellulose.

Esters of unsaturated acids, for example, acrylic or methacrylic acid, are reactive and polymerize rapidly, yielding resins; thus, methyl methacrylate yields a polymethyl methacrylate resin (Lucite). Analogously, esters of unsaturated alcohols are reactive and readily react with themselves; thus, vinyl acetate polymerizes to polyvinyl acetate. The polyester resins known as glyptals result from the polyesterification of glycerol with phthalic anhydride; the process can be controlled to yield either a fusible or an infusible resin. When the polyesterification is carried out in the presence of a long-chain, unsaturated acid of the drying oil type, the oxidative polymerization of the latter is superimposed upon the polyesterification, resulting in hard, synthetic, weather-resistant enamels, suitable for automobile finishes. Polyesterification of ethylene glycol with terephthalic acid results in a polyester fiber. If the material is formed in sheets, it is a useful photographic film.

Many low-molecular-weight esters have characteristic, fruitlike odors: banana (isoamyl acetate), rum (isobutyl propionate), and pineapple (butyl butyrate). These esters are used to some extent in compounding synthetic flavors and perfumes. *See* CARBOXYLIC ACID: DRYING OIL; POLYESTER RESINS; SOLVENT.

Esterification. In the broadest sense, esterification is any reaction in which at least one of the products is an ester. There are many routes to the formation of esters. Some of the more important reactions for preparing esters take place between the following pairs of compounds: (1) an acid and an alcohol, (2) an acid anhydride and an alcohol, (3) an acid chloride and an alcohol, (4) an acid and an unsaturated hydrocarbon such as an olefin or an acetylene, (5) an ester and an alcohol, (6) an ester and an acid, and (7) two different esters. This article treates esterification in only a limited sense—the reaction between a carboxylic acid (RCOOH) and an alcohol (R′OH) to give the ester and water. For discussions of reactions of an ester with an alcohol, an acid, or another ester *see* TRANSESTERIFICATION.

Esterification reactions are generally reversible and accompanied by relatively small heat effects of the order of a few kilocalories per mole of ester. Although the reactions generally take place in a single liquid phase in the presence of a catalyst, a limited number of esters have been prepared by passing the reactant vapors over a solid catalyst. In the presence of a catalyst, the reaction is commonly conducted at a temperature of about 100°C; in the absence of a catalyst, a temperature of about 250°C is used to give a reasonable reaction rate. The pressure at which the reaction is conducted is determined only by the volatility of the components of the system. It is usually atmospheric pressure. In order to produce most esters economically, some means must be provided for completing the reaction by removing one or more of the products. *See* ACID ANHYDRIDE; ACID HALIDE.

In a typical industrial procedure for the preparation of ethyl acetate, a mixture of acetic acid, excess ethanol, and sulfuric acid is passed into an esterifying column heated to reflux. A ternary azeotrope containing 70% ethanol, 20% ester, and 10% water separates into layers, one of which contains 85% ethyl acetate. The ester may be purified by fractional distillation, and the recovered starting materials are recycled.

Other commercially important esters are prepared as follows. Dibutyl phthalate is prepared from phthalic anhydride and butanol in a stepwise reaction to form first the monoester and then the diester; cellulose acetate from purified α-cellulose and a mixture of acetic anhydride and acetic acid; alkyd resins from phthalic anhydride, unsaturated fatty acids, and glycerol; nitroglycerine (glycerol trinitrate) from glycerol and the proper mixture of nitric acid and sulfuric acid. Aspirin, the world's most used analgesic, is prepared by the reaction of salicylic acid with acetic anhydride below 90°C, and is purified by recrystallization. *See* AZEOTROPIC MIXTURE.

The fact that esterification involves an equilibrium was established in 1862 by M. P. E. Berthelot in his study of the ethyl alcohol–acetic acid system. If 1 mole each of acetic acid and ethyl alcohol react, it is found at equilibrium that ⅔ mole each of ethyl acetate and water is present at room temperature, along with ⅓ mole each of alcohol and acid. This can be applied to the equilibrium equation shown as Eq. (2). where K_E is the equilibrium

$$K_E = \frac{[CH_3COOC_2H_5] \times [H_2O]}{[CH_3COOH] \times [C_2H_5OH]} \quad (2)$$

constant and the square brackets signify concentrations in moles per liter of the enclosed reagent. This gives for K_E the value 4. Unless the temperature is deliberately changed, this value is fixed for ethyl alcohol and acetic acid (different alcohol-acid systems have different though characteristic equilibrium constants); indeed, regardless of the starting concentrations of acid and alcohol, the value 4 is maintained.

The mechanism of direct esterification has been much studied. The use of isotopic oxygen (^{18}O) shows that in the reaction of an acid with an alcohol of primary or secondary type, the ester oxygen comes from the alcohol and the acid oxygen goes to form water. Moreover, in an ordinary acid-catalyzed esterification, the rate of reaction is dependent upon the concentrations of both the carboxylic acid and the alcohol. These observations are accommodated by the mechanistic picture shown as reactions (3) through (6).

$$R{-}\overset{\overset{\displaystyle O}{\|}}{C}{-}OH + H^+ \text{ (from catalyst)} \rightleftharpoons RC(OH)_2^+ \quad (3)$$

$$RC(OH)_2^+ + R'OH \rightleftharpoons RC(OH)_2(OR') + H^+ \quad (4)$$

$$RC(OH)_2(OR') + H^+ \rightleftharpoons RC(OH)(OR')^+ + H_2O \quad (5)$$

$$RC(OH)(OR')^+ \rightleftharpoons R\overset{\overset{\displaystyle O}{\|}}{C}{-}OR' + H^+ \quad (6)$$

In the case of tertiary alcohols, isotopic oxygen studies show that the ester oxygen comes from the carboxylic acid, and the hydroxyl from the alcohol goes to form water, implying the modified picture represented as reactions (7) and (8).

$$R_3C{-}O{-}H + H^+ \text{ (from catalyst)} \rightleftharpoons R_3C^+ + H_2O \quad (7)$$

$$R_3C^+ + HO{-}\overset{\overset{\displaystyle O}{\|}}{C}{-}R \rightleftharpoons R_3C{-}O{-}\overset{\overset{\displaystyle O}{\|}}{C}{-}R + H^+ \quad (8)$$

Aromatic acids having substituents in both ortho positions are so hindered in their reaction with alcohols that direct esterification is impracticable. When such acids are dissolved in 100% sulfuric acid and the resulting solution is poured into an alcohol, a good yield of the ester is obtained in a few minutes. The mechanism of this reaction involves an intermediate acyl ion, as shown in reactions (9) and (10).

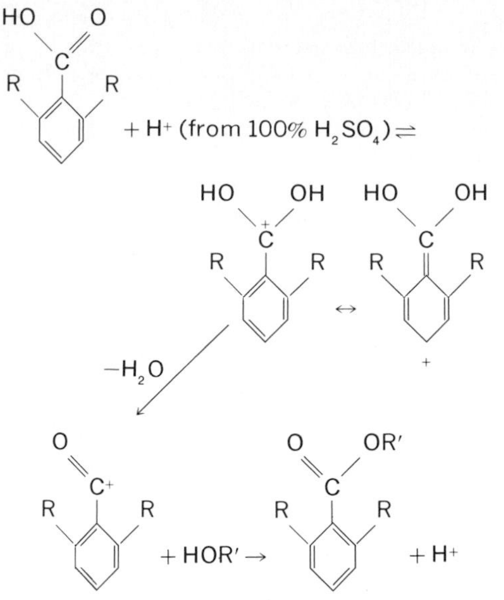

(9)

(10)

Hydrolysis. The splitting of esters in such a way as to regenerate the parent acid and alcohol is an example of hydrolysis. It is important, especially in dealing with naturally occurring esters such as those found in animal and vegetable fats, oils, and waxes. In the presence of dilute mineral acid, hydrolysis of an ester is the reverse of acid-catalyzed esterification; an excess of water is used to ensure complete splitting, and the reaction is carried out at elevated temperatures to speed up the process. Often alcohol is added to solubilize the reactants. Esters formed from glycerol and long-chain carboxylic acids (fats and oils), from long-chain acids and long-chain alcohols (waxes), and simple esters of mono-, di-, or polycarboxylic acids with primary, secondary, or tertiary alcohols, are hydrolyzable under acid conditions, using dilute hydrochloric or sulfuric acids or Twitchell's reagent (prepared from benzene or naphthalene, oleic acid, and concentrated sulfuric acid). However, esters of di-ortho-substituted aromatic carboxylic acids (for example, 2,6-dimethylbenzoic acid) are hindered with respect to hydrolysis, and must be treated according to the Newman technique, which involves first solution in 100% sulfuric acid, and then addition to excess cold water.

The reaction of an ester with a base to form an alcohol and salt of the acid is a type of hydrolysis historically called saponification. Ordinary household soaps are made by the saponification of natural fats and oils of plant or animal origin. Such soaps are typically mixtures of the sodium salts of C_{12} and higher fatty acids, and a by-product of the soapmaking industry is glycerol, reaction (11).

$$\begin{matrix} CH_2OCR \ (\text{C=O}) \\ | \\ CHOCR^1 \ (\text{C=O}) \\ | \\ CH_2OCR^2 \ (\text{C=O}) \\ \text{Glyceride} \end{matrix} + 3NaOH \rightarrow \begin{matrix} CH_2OH \\ | \\ CHOH \\ | \\ CH_2OH \\ \text{Glycerol} \end{matrix} + \begin{matrix} RCOONa \\ R^1COONa \\ R^2COONa \\ \text{Soaps} \end{matrix} \quad (11)$$

Catalytic reduction of esters can be effected at elevated temperature (250°C) and pressure (15–20 atm or 1.5–2.0 MPa), using molecular hydrogen and a copper chromite catalyst; this furnishes a convenient means for the preparation of long-chain mono- or dihydroxy alcohols from esters of the corresponding mono- or dicarboxylic acids. Thus, diethyl succinate ($C_2H_5OOC-CH_2CH_2-COOC_2H_5$) is reduced to form ethyl alcohol (C_2H_5OH) and butylene glycol ($HO-CH_2CH_2CH_2CH_2-OH$). For laboratory reductions, either sodium dissolving in alcohol (Bouveault-Blanc method), or lithium aluminum hydride is preferred.

Esters usually react well with Grignard reagents to yield tertiary alcohols. *See* GRIGNARD REACTION.

Acetoacetic ester synthesis. An ester of special importance in laboratory synthesis is ethyl acetoacetate, often called acetoacetic ester. Upon treatment with sodium ethoxide, acetoacetic ester forms the salt, ethyl sodioacetoacetate, which reacts with a primary alkyl bromide or iodide (RX) to form the alkyacetoacetate. Hydrolysis and decarboxylation of the latter product provides a general route to the synthesis of methyl ketones, reaction (12). *See* KETONE.

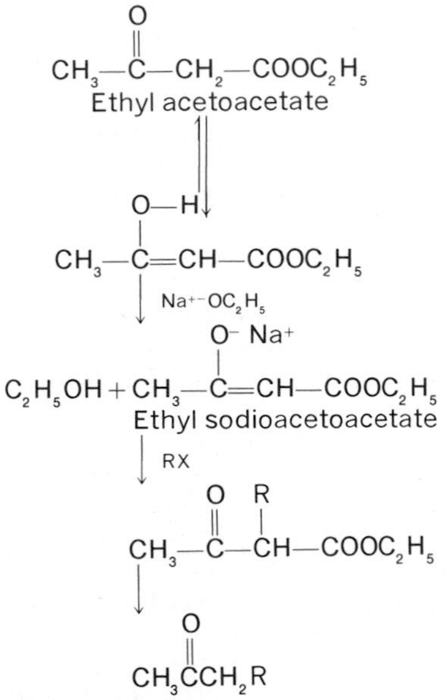

(12)

Malonic ester synthesis. Ethyl malonate reacts with a strong organic base such as sodium ethoxide to form the sodiomalonate. Upon treatment with a primary alkyl halide (RX), a substituted malonic ester is formed, which upon hydrolysis and decarboxylation yields a carboxylic acid. A variety of carboxylic acids have been prepared by reaction sequence (13).

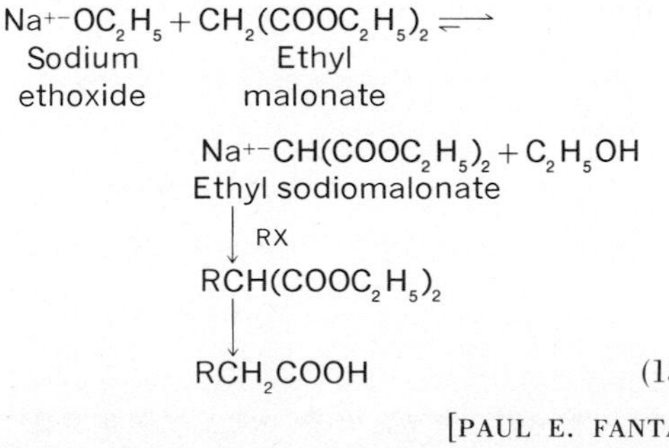

(13)

[PAUL E. FANTA]

Bibliography: N. L. Allinger et al., *Organic Chemistry*, 2d ed., 1976.

Ethane

A member of the alkane or paraffin series of hydrocarbons, formula CH_3CH_3. It occurs in natural gas, but in much smaller quantities (5–20%) than methane, the principal component (50–90%). It is a colorless, odorless, normally gaseous hydrocarbon having a freezing point of −183.3°C and a boiling point of −88.6°C.

Ethane undergoes thermal reactions more readily than does methane. It begins to undergo dehydrogenation to ethylene and hydrogen at about 485°C; commercial operations are usually conducted at 800°C. As the temperature is raised, the reaction is accompanied by the formation of carbon, methane, acetylene, butadiene, and aromatic hydrocarbons. Pyrolysis of the ethane and propane portion of natural gas is used as an industrial method for production of ethylene. *See* ALKANE; CRACKING. [LOUIS SCHMERLING]

Ether

One of a class of organic compounds characterized by the structural feature of an oxygen atom linking two hydrocarbon groups, R—O—R′. Ethers are used widely as solvents, both in chemical manufacture and in the research laboratory. The most important ether is ethyl ether, $C_2H_5OC_2H_5$.

The hydrocarbon radicals R and R′ may be identical (simple ether) or different (mixed ether). They may be aromatic or aliphatic, and the names of the ethers correspond to the hydrocarbon groups present. Thus, CH_3-O-CH_3 is methyl ether, rarely dimethyl ether, and $C_6H_5-O-CH_3$ is phenyl methyl ether.

Manufacture and preparation. Simple ethers may be considered to be the anhydrides of alcohols and are manufactured from alcohols by catalytic dehydration, as in reaction (1), or from olefins by

$$2ROH \rightarrow ROR + H_2O \qquad (1)$$

controlled catalytic hydration, as in reaction (2).

$$2CH_3CH{=}CH_2 + HOH \rightarrow (CH_3)_2CH-O-CH(CH_3)_2 \qquad (2)$$

Mixed ethers of definite structure may be prepared by the Williamson synthesis, shown by reaction (3). This synthesis was of considerable significance

$$C_3H_7ONa + C_2H_5Br \rightarrow C_3H_7OC_2H_5 + NaBr \qquad (3)$$

historically because a knowledge of the structure of ethers was important in developing the radical theory, a stepping-stone to the present extensive knowledge concerning the arrangements of atoms in the molecules of organic compounds. A closely related reaction is that which takes place between cellulose, alkali, and ethyl chloride to yield an important plastic, the polyethyl ether of cellulose known as ethyl cellulose.

Properties. Ethers are less soluble in water than are the corresponding alcohols, but are miscible with most organic solvents. Low-molecular-weight ethers have a lower boiling point than the corresponding alcohols, but for those ethers containing radicals larger than butyl, the reverse is true. The boiling points approximate those of hydrocarbons of the same molecular weight and geometry, indicating that association of ether molecules in the liquid state is negligible. Inertness at moderate temperatures, an outstanding chemical characteristic of the saturated alkyl ethers, leads to their wide use as reaction media. The organic magnesium compounds known as the Grignard reagents, RMgX, perhaps the most used reagents in organic synthesis, are almost always prepared in ether solutions, and suspensions of alkali metals in ethers are often employed. At higher temperatures, however, ethers are split by the alkali metals and by the halide salts of metalloids.

Ethers may also be split by hydrogen halides. Hydrogen iodide, HI, for example, often reacts at room temperature to form an alcohol and an alkyl iodide. Ethers react with chlorine and bromine considerably more readily than do the corresponding hydrocarbons. The initial reaction involves the formation of hydrogen halide and the substitution of a halogen atom for one of the hydrogens on a carbon adjacent (alpha) to oxygen, as shown by reaction (4). Because such α-halogens are reactive,

$$CH_3CH_2—O—CH_2CH_3 + Cl_2 \rightarrow CH_3CHCl—O—CH_2CH_3 + HCl \quad (4)$$

the halogenated ethers are convenient intermediates for synthesis. Halogenated ethers are known in which the halogen is on a carbon other than that adjacent to the oxygen, but they are relatively inert.

On standing, ethers react with the oxygen of the air to form peroxides. Before distillation, it is essential that any considerable accumulation of peroxides be destroyed, by alkaline hydrolysis or by treatment with a reducing agent, such as ferrous hydroxide. On concentration and heating, ether peroxides detonate with dangerous violence. Some ethers form saltlike addition compounds with Lewis acids, the halogens, or picric acid. These addition compounds are theoretically related to, but usually much less stable than, the corresponding derivatives of amines. This property permits the separation of ethers from inert hydrocarbons by extraction with concentrated sulfuric acid.

Identification of ethers is difficult. Often the more reactive components of a mixture are removed by chemical reagents, and the residual ethers are identified by a combination of their failure to react and their specific physical properties. The inertness of ethers is utilized in the syntheses of complicated organic molecules, an objectionably reactive alcohol group being protected by converting it to an unreactive ether. Hydrogen iodide may be used to regenerate the alcohol from the ether when the need for protection has passed.

Unsaturated ethers undergo the reactions usually associated with the double bond. Vinyl ethers, in which the double bond is adjacent to the oxygen, are readily polymerized or copolymerized with such monomers as vinyl acetate to yield useful polymers. Vinyl ethers also react in the presence of acid catalysts with compounds that possess active hydrogens. Thus, with alcohols, they form acetals.

Ethyl ether. The best known of the ethers is ethyl ether, sometimes called diethyl ether or simply ether, $CH_3CH_2OCH_2CH_3$. It is used in industry as a solvent and in medicine as an anesthetic.

The older process of manufacture involved heating ethyl (grain) alcohol to moderate temperatures with catalytic quantities of sulfuric acid, H_2SO_4. Both ethyl ether and ethyl alcohol are manufactured by the controlled catalytic hydration of ethylene, a by-product from the production of gasoline. Solid acid catalysts and a flow process are generally used instead of the sulfuric acid method, and the proportion of alcohol to ether is controlled by variation of temperature and reactant concentrations.

The anesthetic properties were first noticed by Paracelsus (1490–1541) and were rediscovered by Michael Faraday in 1818, but it was not until 1846 that its potential as a surgical anesthetic was demonstrated, by W. T. G. Morton. The ethyl ether intended for anesthetic use differs from the ordinary variety in that possibly injurious impurities are removed. Peroxides are particularly harmful, and storage conditions must inhibit their formation.

When ether is used as a solvent, its high volatility can cause loss. However this volatility is advantageous in that ether can be readily removed from the concentrated or crystallized product. The toxicity to humans is low, and recovery from overexposure is rapid and complete. It readily forms explosive mixtures with air, and on standing in containers which have been opened, it forms dangerous peroxides. Its freezing point is −117.4°C; boiling point, 34.6°C; density, 0.7146; and refractive index, 1.35424. The solubility of ether in water is 6.18%, and of water in ether, 1.2%.

Cyclic ethers. Several cyclic ethers are of special importance and interest. The simplest of these is ethylene oxide or oxirane (I), made industrially by the oxidation of ethylene with air over a silver catalyst. The major portion is used as an intermediate in the hydrolytic manufacture of ethylene glycol. Ethylene oxide is also used in the preparation of nonionic emulsifying agents, plastics, plasticizers, one type of synthetic rubber, and several important synthetic textiles. Another important use is as a gaseous sterilizing agent.

Dioxane or 1,4-dioxane (II) is prepared by the catalytic dimerization of ethylene oxide. It is unusual among substances of low dielectric constant (2.21) in that it is soluble in water in all proportions. Extensively used as a solvent industrially, it readily dissolves fats, waxes, natural and synthetic resins, cellulose ethers, and lacquers, and it is employed by biologists to prepare paraffin-impregnated tissue sections.

Furan (III), made by the decarbonylation of furfural, is the most important ether obtained from an agricultural source. Most of it is hydrogenated to form the useful solvent tetrahydrofuran (IV).

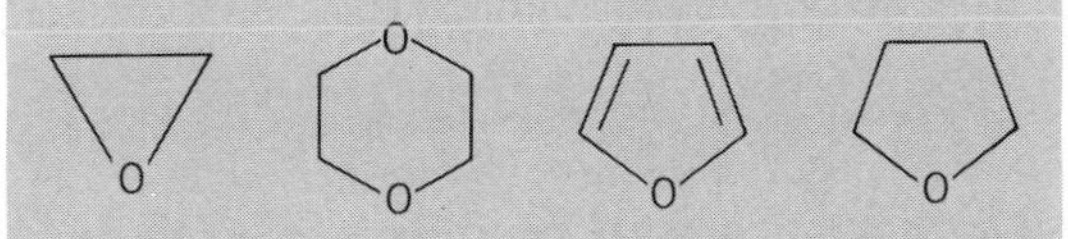

(I) (II) (III) (IV)

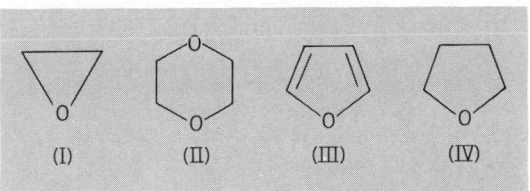

Crown ethers. Certain large-ring polyethers, the crown ethers, are able to increase the solubility of alkali metal salts in nonpolar organic solvents. Specific metal complexes are formed by these crown ethers with alkali metal cations, the specificity for a given cation depending upon the hole in the middle of the crown ether structure. The 18-membrane cyclic polyether containing six oxygen atoms is known as 18-crown-6 (V). It readily forms

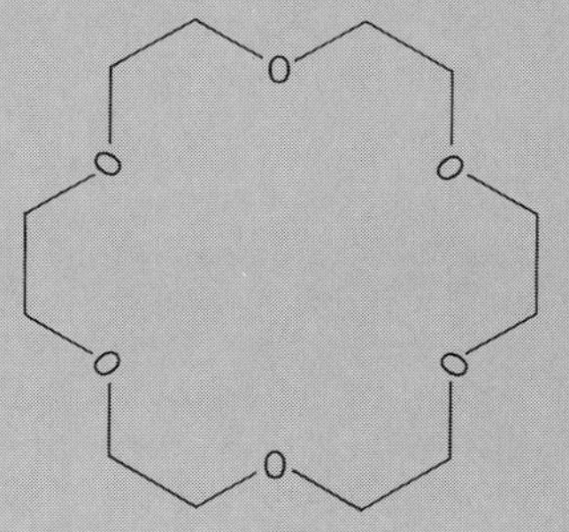

(V)

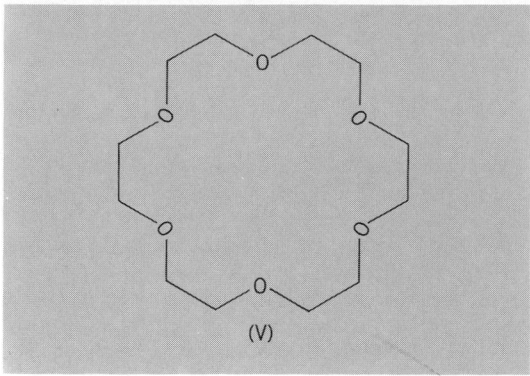

a complex with the potassium cation. Potassium permanganate dissolves in benzene in the presence of 18-crown-6, forming a purple solution that can oxidize alcohols, alkenes, and alkylbenzenes under neutral conditions. *See* CROWN ETHERS AND CRYPTANDS; FURAN; HETEROCYCLIC COMPOUNDS.

[PAUL E. FANTA]

Bibliography: N. L. Allinger et al., *Organic Chemistry*, 2d ed., 1976.

Ethyl alcohol

Probably the best known of the alcohols, ethyl alcohol is also called alcohol, ethanol, grain alcohol, industrial alcohol, fermentation alcohol, cologne spirits, ethyl hydroxide, and methylcarbinol. Pure ethyl alcohol is a colorless, limpid, volatile liquid which is flammable and toxic and has a pungent taste. It boils at 78.4°C and melts at −112.3°C, has a specific gravity of 0.7851 at 20°C, and is soluble in water and most organic liquids. It is one of the most important industrial organic chemicals. Billions of pounds of it are produced annually. Ethyl alcohol is produced by chemical synthesis and by fermentation or biosynthetic processes.

Uses. Ethyl alcohol is used as a solvent, extractant, antifreeze, and intermediate in the synthesis of innumerable organic chemicals. It is also an essential ingredient of alcoholic beverages.

Various grades of ethyl alcohol are produced, depending on their intended use. U.S. Pharmaceutical (U.S.P. XV) grade is the water azeotrope of ethyl alcohol and is 95% ethyl alcohol by volume. National Formulary (N.F.X) grade is 99+% ethyl alcohol by weight; it is also called absolute, or anhydrous, alcohol. This grade is generally prepared by azeotropic dehydration with benzene and therefore usually contains about 0.5% benzene. Denatured alcohol contains a small amount of a malodorous or obnoxious material to prevent the use of this nontaxed grade of ethyl alcohol for beverage purposes. Ethyl alcohol which is employed in the manufacture of beverages, medicine, and flavoring is taxed and all ethyl alcohol production is closely supervised by the government.

The concentration of ethyl alcohol is often expressed as proof, which is simply twice the volume percent of ethyl alcohol. Thus, a 100-proof whiskey contains 50% by volume of ethyl alcohol.

The major use of ethyl alcohol is as a starting material for various organic syntheses. Bimolecular dehydration of ethyl alcohol gives diethyl ether, which is employed as a solvent, extractant, and anesthetic. Dehydrogenation of ethyl alcohol yields acetaldehyde, which is the precursor of a vast number of organic chemicals, such as acetic acid, acetic anhydride, chloral, butanol, crotonaldehyde, and ethylhexanol. Reaction with carboxylic acids or anhydrides yields esters which are useful in many applications. The hydroxyl group of ethyl alcohol may be replaced by halogen to give the ethyl halides. Treatment with sulfuric acid gives ethyl hydrogen sulfate and diethyl sulfate, a useful ethylating agent. Reaction of ethyl alcohol with aldehydes gives the respective diethyl acetals, and reaction with acetylene produces the acetals, as well as ethyl vinyl ether. Treatment of ethyl alcohol with ammonia produces acetonitrile, which may be reduced to ethylamine. These and other ethyl alcohol—derived chemicals are used in dyes, drugs, synthetic rubber, solvents, extractants, detergents, plasticizers, lubricants, surface coatings, adhesives, moldings, cosmetics, explosives, pesticides, and synthetic fiber resins.

Synthetic production. Ethyl alcohol is produced commercially by several synthetic processes.

Catalytic hydration of ethylene. Ethylene, produced by petroleum cracking and from natural gas, is treated with water at high temperatures in the presence of acidic catalysts to produce ethyl alcohol.

Sulfuric acid hydration of ethylene. Ethylene is treated with concentrated sulfuric acid to produce ethyl hydrogen sulfate and diethylsulfate, which are then hydrolyzed to ethyl alcohol and dilute sulfuric acid.

Fischer-Tropsch process. Ethyl alcohol is a major by-product in the synthesis of methanol by the reaction of carbon monoxide and hydrogen over iron catalysts.

[JOHN W. LYNN]

Industrial fermentation. The chemical produced in largest volume by industrial fermentation is ethyl alcohol (ethanol, C_2H_5OH). Ethyl alcohol is produced only from hexose sugars by yeast according to the overall reaction shown below. The

$$\underset{\text{Hexose}}{C_6H_{12}O_6} \rightarrow \underset{\text{Ethanol}}{2C_2H_5OH} + \underset{\text{Carbon dioxide}}{2CO_2}$$

highest yield of alcohol obtainable is about 96% of the theoretical. The theoretical yield is about 51% by weight, based on the sugar fermented. Disaccharides, such as sucrose and maltose, are first split by hydrolytic enzymes of the yeast cell into monosaccharides, the hexose sugars. Polysaccharides, such as starch and cellulose, cannot be fermented directly by yeast, but must be hydrolyzed to simple sugars. This is usually done by the use of enzymes and sometimes by heating with a mineral acid. The most common substrates for the production of industrial alcohol are molasses, grain, sulfite waste liquor, and wood waste. The fermentation is carried out for the most part in the absence of air. This decreases the yield of cells and the amount of carbon dioxide produced from respiration, and it increases the yield of alcohol. The alcohol is finally purified and concentrated by distillation (Fig. 1).

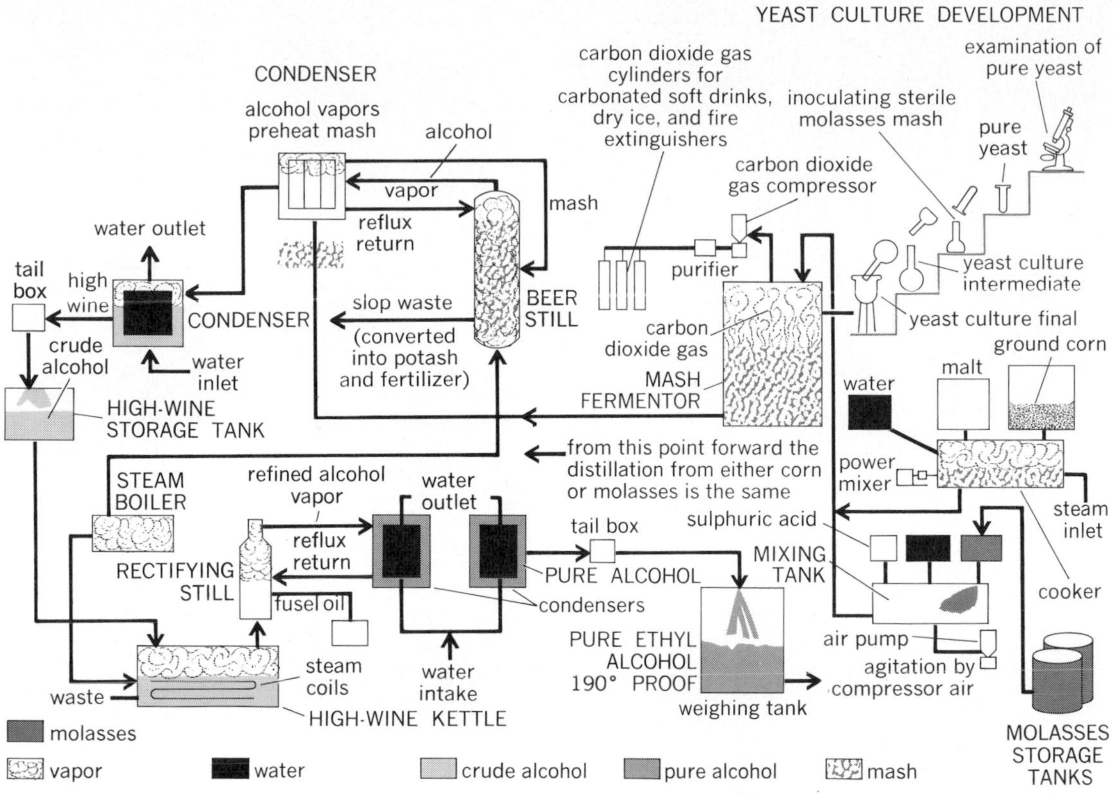

Fig. 1. Flow sheet of industrial ethyl alcohol manufacture. Fermentation is usually carried out in the absence of air. *(From S. C. Prescott and C. G. Dunn, Industrial Microbiology, 2d ed., McGraw-Hill, 1949)*

Molasses alcohol. This is a relatively simple process, since about 90% of the sugars present in blackstrap molasses is in fermentable form as sucrose and invert sugar. In so-called high-test, or invert, molasses, about 95% of the sugars is fermentable by yeast. Molasses is diluted with water to a sugar content of 14–18% and pumped directly into the fermentor. During filling 2–4% of an actively fermenting yeast mash is also added. The filling may require about 8 hr, during which time the yeast multiplies actively. The pH is adjusted with sulfuric, lactic, or hydrochloric acid to 4–5. Sometimes a little ammonium sulfate is added to supply nitrogen for yeast growth, but generally molasses contains sufficient minerals to allow ample yeast growth.

Fermentation produces 26 kcal of heat per mole of hexose fermented, and the fermentors are therefore cooled to a temperature of 80–90° F. If this is not done, alcohol production is impaired. Fermentation is complete after 36–72 hr, depending on the type of molasses used. Usually 2.3–2.7 gal of blackstrap molasses is required to produce 1 gal of 190-proof alcohol. The principal by-products of the fermentation are carbon dioxide, produced in yields almost as high as alcohol, and fusel oil, the high-boiling fraction of the distillation. The evaporated and concentrated molasses residue can be used as stock feed. The nonconcentrated stillage residue can be used directly as a soil fertilizer. Continuous processes, rather than the batch method, are also used for industrial alcohol production.

Grain alcohol. Most distilleries use distiller's barley malt as a source of the enzyme amylase to convert the starch to fermentable sugars. After the grain is milled, the flour is cooked to hydrate and gelatinize the starch. Cooking is often done in pressure cookers, with a steam pressure of 100 psi. Both batch and continuous processes are used. After partial cooking, distiller's malt of high diastatic power is added so that the mixture reaches a temperature of about 145°F. The starch is liquefied rapidly and about 75–80% is converted to maltose and other fermentable sugars and the remainder to residual dextrins, called limit dextrins. In some countries the conversion process on the starch is accomplished by the use of a diastatic mold, *Rhizopus delemar.*

The mold is grown in the mash for 24 hr prior to inoculation with yeast. The mash is cooled from 145°F to 70–85°F by the use of heat exchangers. Yeast nutrients are supplied by adding 20–25% stillage to the volume of cooled mash. Fermentation and distillation take place essentially as described previously. The residual dextrins are slowly converted to maltose during the 2–3 day fermentation period.

Sulfite waste liquor. This by-product is also a suitable source of carbohydrates for industrial alcohol fermentation by yeast (Fig. 2). Sulfite waste liquor from softwoods, or gymnosperms, is preferred over that from hardwood, or angiosperms, since the former has a much lower content of pentoses which are not fermentable by yeast. Since the content of fermentable sugars is only about 1–2% in the liquor, this solution is used without dilution.

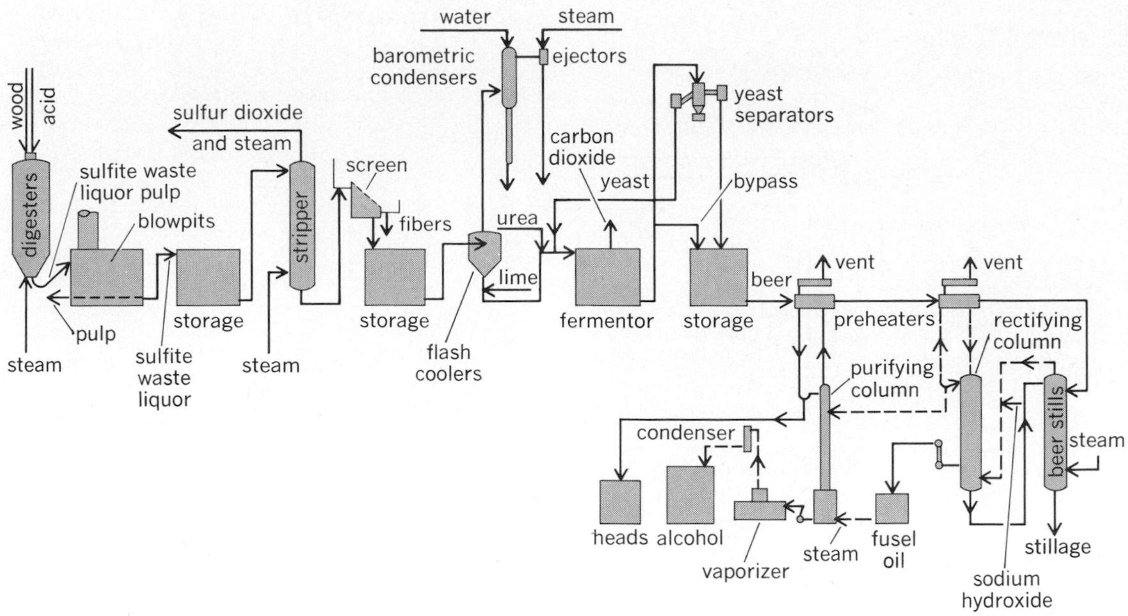

Fig. 2. Generalized flow diagram for alcohol production from sulfite waste liquor plant. (*From L. A. Underkofler and R. J. Hickey, eds., Industrial Fermentations, Chemical Publishing, 1954*)

Preparation of the liquor for fermentation is similar to the process described for the production of fodder yeast from this source. Yields varying from 10–20 gal of 95% alcohol per ton of air-dried pulp can be obtained, depending on the type of wood and the extent of hydrolysis of the hemicelluloses in the wood.

Wood waste as a source of alcohol. The polysaccharides of wood are converted to fermentable sugars by acid hydrolysis. Since use is made of the cellulose fraction of wood after hydrolysis to glucose, alcohol yields are much higher than those from sulfite waste liquor. Yields of 10–22 liters of alcohol per 100 kg of wood have been reported.

[HERMAN J. PHAFF; EMIL M. MRAK]

Bibliography: H. J. Peppler (ed.), *Microbial Technology*, 1967; F. Reiff et al. (eds.), *Die Hefen*, vol. 2: *Technologie der Hefen*, 1962; A. H. Rose, *Industrial Microbiology*, 1961; L. A. Underkofler and R. J. Hickey (eds.), *Industrial Fermentations*, vol. 1, 1954.

Ethylene

A colorless gas, formula CH_2=CH_2, with a boiling point of −103.8°C and a melting point of −169.4°C. Ethylene is the most important synthetic organic chemical in terms of volume, sales value, and number of derivatives. About half of the ethylene produced is used in the manufacture of polyethylene; ethylene dichloride and vinyl chloride production uses about 20%, synthesis of ethylene oxide and derivatives account for about 12%, and styrene production consumes about 8% of the ethylene. Other important derivatives are ethanol, vinyl acetate, and acetaldehyde.

Production. Thermal cracking of hydrocarbons in the presence of steam is the most important and widely used process for producing ethylene. Cracking is done at about 1600°C and 30 psia (21 kPa absolute) pressure, followed by rapid quenching to below 1000°C. Ethylene is recovered by low-temperature fractionation at 500–550 psia (340–380 kPa) and purified by very-low-temperature (approximately −65°C) gas-separation procedures to remove hydrogen, methane, and ethane.

In the United States, hydrocarbon gases—ethane through the butanes—account for about 65% of the feedstocks used to produce ethylene. With these clean raw materials, yields are good (about 50%), and efficiencies and conversion are high (about 80%). Liquid refinery products such as naphthas, kerosines, and gas oils constitute about 25% of process raw materials in the United States. While with these more complex feedstocks the yields and efficiencies are about 30%, the trend has been toward the use of the more available, heavier feedstocks. In some countries where petroleum raw materials are expensive, ethylene is produced by the dehydration of fermentation ethanol.

Derivatives. Polyethylene, the most important derivative of ethylene, is produced by both high- and low-pressure processes to make high- and low-density, high-molecular-weight thermoplastic polymers. Aluminum alkyl catalysts (Ziegler polymerization) are used to polymerize ethylene to relatively low-molecular-weight, straight-chain hydrocarbon derivatives which are convertible to even-numbered-carbon linear olefins, alcohols, and acids. Commercial processes use palladium-catalyzed oxidation of ethylene to produce acetaldehyde, or if acetic acid is used as the solvent, vinyl acetate. Chlorination and oxychlorination processes are used to make vinyl chloride. Ethylene oxide is produced by silver-catalyzed oxidation of ethylene. Acid-catalyzed hydration of ethylene produces ethanol competitively with fermentation processes. *See* ACETYLENE; ALKENE; ETHYL ALCOHOL; ETHYLENE OXIDE; POLYMER.

[ROBERT K. BARNES]

Bibliography: S. A. Miller (ed.), *Ethylene and Its Industrial Derivatives*, 1969.

Ethylene chlorohydrin

A colorless, mobile liquid, $HOCH_2CH_2Cl$, of characteristic ethereal order. It boils at 128.7°C at 760 mm Hg and melts at −62.6°C. The compound forms a constant-boiling mixture with water boiling at 97.8°C; the mixture contains 42.3% by weight of chlorohydrin. Commercial production is by simultaneous addition of chlorine and ethylene into a packed tower, through which a countercurrent stream of water is circulated. A dilute solution (5–8%) of the chlorohydrin is continuously withdrawn and concentrated to the constant-boiling mixture by distillation. Approximately 5–10% of the ethylene is converted to ethylene dichloride, which is sold as a solvent and fumigant or converted to vinyl chloride.

The major use for the chlorohydrin is in the production of ethylene oxide, which is made on a large scale for the manufacture of synthetic detergents. *See* ALKENE; ETHYLENE.

[CHARLES A. COHEN]

Ethylene glycol

A colorless, nearly odorless, sweet-tasting, hygroscopic liquid, formula $HOCH_2CH_2OH$. It is relatively nonvolatile and viscous and is the simplest member of the glycol family. Ethylene glycol freezes at −13°C, boils at 197.6°C, and is completely soluble in water, common alcohols, and phenol. Low molecular weight, low volatility, water solubility, and low solvent action on automobile finishes make ethylene glycol ideal as a radiator antifreeze and coolant; water mixtures (58–80% glycol) freeze below −46°C (−50°F). *See* GLYCOL.

Worldwide, about half of all ethylene glycol is used in polyester resins, and about a third goes into antifreeze. Other uses for this commodity chemical are in explosives, brake and shock-absorber fluids, and alkyl-type resins. *See* POLYESTER RESINS.

The original commercial process for ethylene glycol involved hydrolysis of ethylene chlorohydrin derived from chlorine and ethylene. By 1979 all commercial ethylene glycol was produced by the vapor-phase oxidation of ethylene. A new commercial process has been developed to produce ethylene glycol directly from ethylene. This process uses a tellurium oxide/bromide ion catalyst to oxidize ethylene in acetic acid solution. Another patented process describes production of ethylene glycol from synthesis gas (carbon monoxide and hydrogen).

Ethylene glycol undergoes the simple reactions of alcohols such as etherification, condensation, oxidation, and esterification. Mono- and dialkyl ethers of ethylene glycol are formed by its reaction with sodium hydroxide and dialkyl sulfates or alkyl halides. Dioxane, a cyclic diether, as well as diethylene glycol, can be prepared by acid-catalyzed dehydration of ethylene glycol. Di-, tri-, and polyethylene glycols are best produced by the acid- or base-catalyzed reaction of ethylene oxide with ethylene glycol. Commercially di-, tri-, and some tetraethylene glycols are produced as by-products of industrial ethylene glycol processes. Acid-catalyzed reactions of aldehydes and ketones with ethylene glycol produce five-membered cyclic acetals and ketals called 1,3-dioxolanes. Vapor-phase catalytic dehydrogenation (oxidation) of ethylene glycol can produce either 2-hydroxymethyl-1,3-dioxolane or glyoxal. Liquid-phase nitric acid or air oxidation produces oxalic, glycolic, and formic acids, formaldehyde, and glycol aldehyde and glyoxal. With nitric acid under dehydrating conditions, ethylene glycol forms the dinitrate ester, which is employed in conjunction with nitroglycerin to produce explosives that have low freezing points.

Dibasic acids or anhydrides react with ethylene glycol to form polyester condensation polymers. Reaction with terephthalic acid or its esters produces polyester resins which can be spun to fibers that find wide use in clothing and general fabrics applications. The polymer also has important film applications. Condensation of glycol with unsaturated diacids, for example maleic acid, followed by free-radical cross-linking reactions with polymerizable olefins such as styrene produces another commercially important polymer class identified as unsaturated polyester resins. Saturated and unsaturated polyester resins constitute the major use of ethylene glycol in the world today. *See* ETHER; ETHYLENE OXIDE; POLYETHYLENE GLYCOL.

[ROBERT K. BARNES]

Ethylene oxide

The simplest cyclic ether or epoxide, with the formula C_2H_4O, and the structure

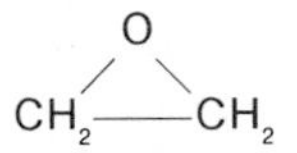

It is also called epoxyethane and oxirane. Ethylene oxide was discovered in 1859 by C. A. Wurtz. His preparation from ethylene chlorohydrin and aqueous base became the first commercial process, practiced during the early 1920s and until the mid-1940s. Thereafter the direct, silver-catalyzed vapor-phase oxidation of ethylene was the process of choice. Commercial processes use either air or oxygen to oxidize ethylene at low conversion and high selectivity to ethylene oxide. *See* ETHYLENE; HETEROCYCLIC COMPOUNDS.

Ethylene oxide is a colorless gas boiling at 10.4°C and melting at −112°C, with refractive index 1.3597 at 7°C, and density 0.8969 kg/liter at 0°C. Its vapors are flammable and explosive, and it is considered a relatively toxic liquid and gas. It is miscible in all proportions with water, alcohols, ethers, and other organic solvents. *See* ETHER.

Ethylene oxide reacts, usually by acid or base catalysis, with most active hydrogen compounds such as water, alcohols, acids, phenols, amides, hydrogen sulfide, and hydrogen cyanide; it also reacts with other organic compounds such as carbon dioxide, amines, ammonia, and Grignard reagents and with itself. Ethylene oxide's reactivity can be ascribed to the strained ring structure.

About 50% of the ethylene oxide produced is converted to ethylene glycol by reaction with water. Primary uses for ethylene glycol are in the manufacture of polyester resins and as an automobile antifreeze. About 30% of the ethylene oxide produced is used in the manufacture of nonionic surfactants, glycol ethers, and ethanolamines. Gaseous ethylene oxide in CO_2 or difluoromethane is used as a fumigant and a sterilizing agent for medical equipment. *See* ETHYLENE GLYCOL; POLYETHYLENE GLYCOL; POLYOXYETHYLATION OF ALCOHOL. [ROBERT K. BARNES]

Bibliography: S. A. Miller (ed.), *Ethylene and Its Derivatives*, 1969.

Ethylenediaminetetraacetic acid

A chelating agent for metallic ions, abbreviated EDTA. Tetrasodium EDTA is the most common form in commerce, but other metallic chelates are marketed, for example, iron, zinc, and calcium. Tetrasodium EDTA is a white solid, very soluble in water and forming a basic solution. Prepared from ethylenediamine, formaldehyde, and sodium cyanide in basic solution, or from ethylenediamine and sodium chloroacetate, EDTA is a strong complexing and chelating agent. It reacts with many metallic ions to form soluble chelates. As such, it is widely used in analysis to retain alkaline earths and heavy metals in solution. The iron chelate is useful in lawn management and gardening as a replacement for ferrous sulfate (copperas). Calcium EDTA is used in controlling the deterioration of natural sea water in salt-water aquariums. Calcium disodium salt of EDTA is used in pharmaceuticals to prevent calcium depletion of the body during therapy. *See* CHELATION; COMPLEX COMPOUNDS. [FRANK WAGNER]

Europium

A chemical element, Eu, atomic number 63, atomic weight 151.96, a member of the rare-earth group. The stable isotopes, Eu^{151} 47.82% and Eu^{153} 52.18%, make up the naturally occurring element.

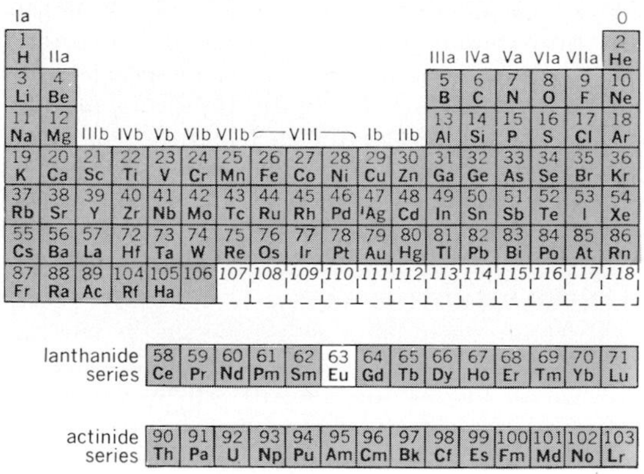

The existence of this metallic element was first suspected in 1889, when Sir William Crookes found an absorption band in the spectra of samarskite. He attributed this band to a new element, which he named "S." In 1896 the element was found by E. Demarcay, and he named it europium after Europe. "S" and europium were later shown to be the same element. Its compounds were first prepared in the fairly pure state by G. Urbain and H. Lacombe in 1904. The trivalent salts have a very pale pink color. A divalent series of compounds is known. Europium is one of the rarer of the naturally occurring rare earths. During the period 1936–1941, H. N. McCoy published a series of papers giving methods for preparing divalent europium salts. He succeeded in separating several hundred grams of europium salts from rare-earth concentrates which he had accumulated over a number of years. He gave or loaned this material to a large number of research workers, and as a result the properties of europium are now better known than the properties of many other rare earths. The metal is the second most volatile of the rare earths and has a considerable vapor pressure at its melting point. Reduction of europium salts with calcium or alkali metals usually produces divalent europium salts. The metal is best prepared by vacuum distillation of a mixture of europium oxide and lanthanum metal. The metal is very soft, it is rapidly attacked by air, and really belongs more to the calcium-strontium-barium series than to the rare-earth series. It probably has only two electrons in its conduction bands instead of three. Europium not only has a high absorption cross section for neutrons, but so also do a number of the isotopes formed when neutrons are added. This property makes the element attractive to the atomic industry, since the elements can be used in control rods and as nuclear poisons. These poisons are materials added to a nuclear reactor to balance the excess reactivity at start-up, and are so chosen that the poisons burn out at the same rate as the excess activity decreases.

Since 1964 europium oxide has attained considerable industrial importance. The television industry uses considerable quantities of phosphors, such as europium-activated yttrium orthovanadates, and other europium-activated yttrium phosphors have been patented. These phosphors give a brilliant red color and are used in the manufacture of television screens. *See* RARE-EARTH ELEMENTS. [FRANK H. SPEDDING]

Evaporation

The process by which a substance in the liquid state is converted into the vapor state. The molecules of substances in a condensed state are held to one another by strong forces of attraction, which are balanced by equally strong repulsive forces. Tending to overcome the potential energy of attraction is the escaping tendency of molecules, which arises from their kinetic energy. The kinetic energy, and therefore the escaping tendency of molecules, is a function of temperature. At each temperature a certain fraction of the molecules possesses enough kinetic energy to overcome the forces of attraction of surrounding molecules and to escape from the surface of the liquid. If the process occurs at constant volume, the equation below will hold. Here n_v is the num-

$$\frac{n_v}{n_l} = e^{-\Delta E/RT}$$

ber of molecules per milliliter in the vapor, n_l is the number of molecules per milliliter in the liquid, ΔE is the difference in molar internal energy of the gas and liquid, R is the gas constant, and T is the absolute temperature. As the molecules which possess excess kinetic energy evaporate from the liquid, the average kinetic energy of the remaining molecules decreases and the temperature drops. In order to maintain the temperature constant, heat must be furnished to the liquid. Ordinarily, evaporation occurs, not at constant volume, but at constant pressure. The quantity of energy required to evaporate 1 mole of liquid at constant pressure is called the molar latent heat of vaporization, ΔH, and is related to the internal energy by the first law of thermodynamics, $\Delta H = \Delta E + P\Delta V$, where $P\Delta V$

represents the work done by the vapor in expanding to a volume $\Delta V = V_{gas} - V_{liq}$ against the atmospheric pressure P. The molar volume of the liquid is ordinarily negligible by comparison with that of the vapor, and the gas obeys the ideal gas law to a first approximation ($PV = RT$). Hence, the latent heat of vaporization is given by $\Delta H = \Delta E + RT$. It is a function of temperature and is normally measured calorimetrically at the normal boiling point. For nonassociated liquids the latent heat of vaporization is given approximately by Trouton's rule, $\Delta H/T_b = 22$, where ΔH has the units of calories, and T_b is the normal boiling point on the absolute temperature scale (K).

The following factors affect the rate of evaporation of a liquid: the rate at which heat is supplied to the liquid to furnish the latent heat of vaporization; the rate at which the liquid is stirred to bring to the surface molecules having sufficient kinetic energy to escape; and the rate at which the vapor above the liquid is changed to provide the best conditions for escape of molecules from the surface of the liquid.

For the conversion of a substance from the solid to the liquid state *see* SUBLIMATION. *See also* EVAPORATOR; LIQUID; VAPOR PRESSURE.

[NORMAN H. NACHTRIEB]

Evaporator

A device used to vaporize part or all of the solvent from a solution. The valuable product is usually either a solid or a concentrated solution of the solute. If a solid, the heat required for evaporation of the solvent must have been supplied to a suspension of the solid in the solution, otherwise the device would be classed as a drier. The vaporized solvent may be made up of several volatile components, but if any separation of these components is effected, the device is properly classed as a still or distillation column. When the valuable product is the vaporized solvent, an evaporator is sometimes mislabeled a still, such as water still, and sometimes is properly labeled, such as boiler-feedwater evaporator. In the great majority of evaporator installations, water is the solvent that is removed.

Uses. Evaporators are used primarily in the chemical industry. Common salt is made by boiling a saturated brine in an evaporator. The salt precipitates as a solid in suspension in the brine. This slurry is pumped continuously to a filter, from which the solids are recovered and the liquid portion returned for further evaporation. In the manufacture of pulp and paper, the waste liquor from cooking the wood is a dilute solution of inorganic cooking chemicals and soluble, organic wood constituents. The liquor is disposed of by concentrating it in evaporators to a strength at which the liquor will burn in a boiler. This avoids a water-pollution problem, recovers the inorganic chemicals for reuse, and provides sufficient heat energy not only to operate the evaporator but also to supply other needs of the mill as well. In the alkali industry, salt brine passed through a diaphragm-type electrolytic cell yields a dilute solution of salt and caustic soda. Evaporation to a strength of about 50% NaOH purifies the caustic by causing precipitation of practically all of the salt. It also simplifies shipment by bringing about a more than fivefold reduction in volume. Further evaporation of the caustic solution to a final temperature of about 700°F (371°C) produces a practically anhydrous product that freezes at about 600°F (316°C) and is shipped as a solid. Evaporators are widely used in the food industry, usually as a means of reducing volume to permit easier storage and shipment. Evaporators are also the most commonly used means of producing potable water from sea water or other contaminated sources.

Classification. The vaporization of solvent requires large amounts of heat. Provisions for transferring this heat to the solution constitute the largest element of evaporator cost and the principal means of distinguishing between types of evaporators. Practically all evaporators fall into one of the following categories:

1. Those heated by a flame that burns below the liquid surface, and in which the hot combustion gases are bubbled through the liquid.
2. Those in which the flame and combustion gases are separated from the boiling liquid by a metal wall, or heating surface.
3. Those in which steam or other condensable vapor is the source of heat, and in which the steam condenses on one side of the heating surface and the heat is transmitted through the wall to the boiling liquid.

Submerged-combustion evaporators (type 1) are used primarily to concentrate solutions that would deposit a heat-insulating blanket of scale on the solid heating surfaces of other types of evaporators. Since the evolved vapor is mixed with the combustion gases, neither the solvent vapor nor its heat content can be recovered easily.

Direct-fired evaporators (type 2) are typified by the steam boiler and the old maple-syrup kettle. They are not commonly used for the concentration of solutions, primarily because local overheating can cause formation of insulating deposits on the heating surface, which then becomes overheated and may melt or burn through. Also, a large heating surface is needed to recover the heat in the combustion gases, and there is usually no cheap metal that will resist attack by both the combustion gases and the boiling liquid.

Steam-heated evaporators (type 3) are by far the most common, primarily because condensing steam gives up its heat so readily—condensing steam film coefficients are usually in excess of 1000 Btu/(hr)(ft^2)(°F)[5.7 kW/(m^2)(°C)]. Thus the design of the evaporator and materials of construction can be suited to the solution being concentrated instead of being dictated by the problem of getting heat to the heating surface. The heating surface is usually in the form of metal tubes, since this represents the most economical method of putting the largest heating surface in the smallest volume. The tubes may be vertical or horizontal, and the boiling liquid may be either inside or outside, depending on the characteristics of the solution, such as viscosity, ratio of feed to evaporation, and whether or not a salt is deposited. *See* DISTILLATION.

Operation. The vaporization capacity of an evaporator is determined by the usual rules of heat transfer, and is directly proportional to the area of heating surface, to the difference between condensing steam and boiling liquid temperatures, and to the coefficient of heat transfer. The heat-transfer coefficient is usually limited by conditions on the boiling liquid side, although the condensing steam film and the resistance of the metal wall have some influence. Various means are employed to increase the boiling-film coefficient and all involve movement of the liquid relative to the heating surface. A great many evaporators use only natural convection to accomplish this circulation. Typical is the long-tube vertical type shown in the illustration. Feed liquid enters the bottom of a nest of vertical tubes and begins to boil as it passes up the tubes. The boiling causes a large increase in volume, which accelerates the liquid to high velocities and gives good heat-transfer performance. The vapor-liquid mixture is separated in the chamber at the top of the tubes. The liquor may all be discharged as product, or part may be recirculated to the feed inlet.

A pump or agitator may be used to force the liquid past the heating surface if even higher heat-transfer coefficients are needed, for example, when corrosive conditions dictate use of the smallest possible area of an expensive alloy. Forced circulation evaporators are also used for scaling liquids or those from which a salt is to be crystallized, since there is less tendency for solids to form on the heating surfaces. Such evaporators usually consist of a flash chamber or crystallizing chamber, a conventional shell-and-tube heat exchanger, and a pump to circulate fluid from the chamber to the exchanger and back to the flash chamber. Another type of forced-circulation evaporator for extremely viscous, heat-sensitive, or foamy materials either rotates the heating surfaces or employs wipers that sweep the material around the walls of a stationary surface. *See* MOLECULAR DISTILLATION.

The water vapor evolved in an evaporator is usually about the same in quality and quantity as the steam used to heat the evaporator—the only difference being that the vapor has a lower pressure and hence lower condensing temperature. It is possible to compress the vapor so that it can be

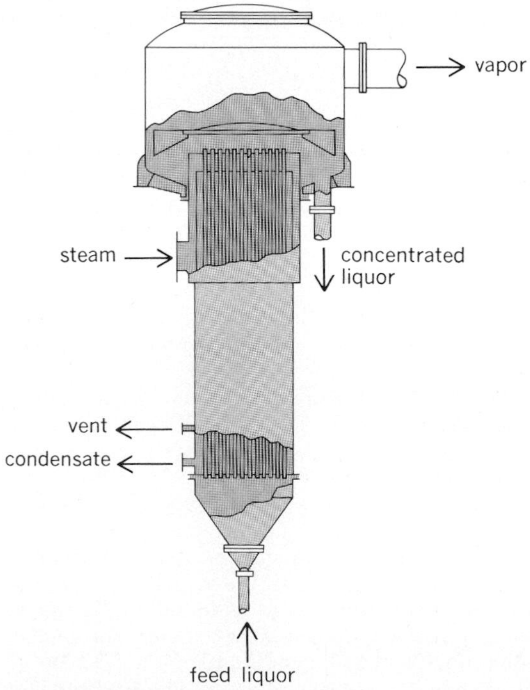

Long-tube vertical evaporator.

used as the heating steam in the same evaporator. Such thermocompression evaporators require far less energy for the compressor than would be needed to generate fresh steam. However, to keep the compressor cost and power consumption within reason, it is necessary to use a narrow compression ratio and this requires a large and expensive evaporator.

Assuming a perfect compressor, the power requirement is given by the Carnot equation: $W = Q\Delta T/T$, where W and Q are work required and amount of heat pumped in the same units, ΔT is the difference between saturated steam temperatures at compressor discharge and suction pressures, and T is the absolute suction temperature. For pure water boiling at atmospheric pressure (212°F, 672° Rankine or 100°C or 373 K), the latent heat is 970.3 Btu/lb (2,257 kJ/kg). At a temperature difference of 10°F (5.6°C), the ideal work required is only (970.3)(10)/(672) = 14.4 Btu/lb (33.6 kJ/kg), or 35.3 kWhr per 1000 gal of water evaporated (33.6 MJ per 1000 liters). This is only 1.5% of the heat energy required for a simple evaporator, but the energy is expensive mechanical energy rather than low-grade heat energy.

If pure water is being boiled, this 10°F (5.6°C) temperature difference established by the compressor is available as the driving force to transfer heat from the compressed steam through the heating surface to the boiling liquid. If the boiling liquid is an aqueous solution, it has a boiling temperature higher than that of pure water at the same pressure and the difference in these temperatures is called the boiling point elevation (bpe). This bpe cannot be utilized as a part of the driving force. Thus, if the solution being evaporated had a bpe of 30°F (16.7°C), the compressor would have to work across a 40°F (22.2°C) compression range and would require four times the above power to establish the same 10°F (5.6°C) driving force in the evaporator.

An alternative method of reducing the energy requirement of an evaporator is to use the water vapor evolved in one part to heat another part in which liquid is boiled at a lower temperature. The water vapor evolved in this part, termed an effect, can then be used to heat another effect boiling at still lower temperature (and pressure), and so on. The ultimate limit is determined by the need to discharge to a heat sink the heat contained in the vapor from the last effect, which amounts to most of the heat supplied by steam to the first effect (the balance of the entering heat leaves with the condensate and concentrate, which are generally hotter than the feed solution). The heat sink is usually water from a river or other source, which limits the boiling temperature in the last effect to about 100°F (38°C).

If the condensate of last-effect vapors is valuable, the vapors are condensed in a shell-and-tube heat exchanger by the cooling water. If the solution is the desired product, the vapors are brought into direct contact with a shower of cooling water. This is usually done in a barometric condenser, which is elevated to such a height (about 34 ft or 10.3 m) that the high vacuum cannot prevent the water from draining out by gravity. To maintain the vacuum in the condenser, it also is necessary to remove noncondensible gases with a vacuum pump. These gases originate in the feed, as air dissolved in the condenser water and as air leakage into the evaporator. Only in the rarest of circumstances, as when a very-low-boiling temperature is needed to avoid degradation of the solution, is it feasible to compress the entire last-effect vapor for discharge at a higher temperature to the heat sink.

Such multiple-effect evaporators are more expensive than single-effect units because each effect can operate at only a fraction of the total difference, ΔT, between the temperature at which heat is accepted from the prime steam and the temperature at which it is rejected to the heat sink. For a single effect, the amount of heat transfer surface A required is given by the equation: $A = Q/U\Delta T$, where Q is the rate at which heat must be transferred to achieve the desired evaporation rate and U is the coefficient of heat transfer.

In a triple effect, each effect evaporates only about one-third of the water and must transfer only about one-third of the total heat, but the total temperature difference must also be divided between the three effects. Thus the heating surface in each effect may be written: $A_n = (Q/3)/U(\Delta T/3) = Q/U\Delta T$. Each effect must have substantially the same amount of surface as a single effect evaporator. The steam required, however, goes down in inverse proportion to the number of effects. The choice of the proper number of effects involves an economic balance between the first cost of equipment and the continuing costs of steam and cooling water. The great majority of evaporator installations employ the multiple-effect principle and as many as a dozen effects have been used. Another means of achieving multiple-effect steam economy is the multistage flash cycle, which has been developed primarily for producing fresh water from sea water. *See* SALINE WATER RECLAMATION.

[FERRIS C. STANDIFORD]

Bibliography: W. L. Badger and J. T. Banchero, *Introduction to Chemical Engineering*, 1955; R. Kirk and D. Othmer (eds.), *Encyclopedia of Chemical Technology*, vol. 8, 1965; J. H. Perry, *Chemical Engineers' Handbook*, 4th ed., 1963; F. C. Standiford, *Chem. Eng.*, pp. 158–176, Dec. 9, 1963.

Excited state

In quantum mechanics, a stationary state of higher energy than the lowest stationary state or ground state of a particle or a system of particles. Customarily, only bound stationary states, which generally are at most denumerably infinite in number, are spoken of as excited, although the formal quantum theory often treats the noncountable unbound stationary states on an equal footing with the bound states. Conventionally, the excited states are ranked in order of increasing energy; that is, the second excited state has higher energy than the first, which lies higher than the zeroth or ground state. Unlike the ground state, excited states frequently are degenerate. *See* GROUND STATE.

[EDWARD GERJUOY]

Extraction

A method of separating the constituents of a mixture utilizing preferential solubility of one or more components in a second phase. Commonly, this added second phase is a liquid, while the mixture to be separated may be either solid or liquid. As a mundane example, the preparation of tea or coffee is a process of liquid/solid extraction whereby water selectivity dissolves certain components of the mixture, leaving behind the insoluble residue (as tea leaves or coffee grounds). If the starting mixture is a liquid, then the added solvent must be immiscible or only partially miscible with the original and of such a nature that the components to be separated have different relative solubilities in the two liquid phases.

Solvent extraction processes can be divided into two broad categories according to the origins of the differential solubility. On the one hand, it arises from purely physical differences between the two solutes, such as polarity, while in other cases it can be traced to definite chemical interaction between solute and solvent.

Principles. The ratio of the concentrations of a particular dissolved substance (solute) in two coexisting liquid phases at equilibrium is shown in Eq. (1), where D is the distribution coefficient, and x_A

$$D_A = \frac{x_A}{y_A} \qquad (1)$$

and y_A are the concentrations of A in the two phases. When the separation of two components in a mixture is under consideration, the ease of separation is conveniently measured by the separation factor α, which is the ratio of the distribution coefficients of the two components between the two solvents (the equivalent of relative volatility in distillation), as in Eq. (2).

$$\alpha_{AB} = \frac{D_A}{D_B} \qquad (2)$$

Although the basic concepts of equilibrium outlined above define the potential for separation, they give no indication of the rate of the process. When two phases that are not at equilibrium are contacted together, the rate of transfer of solute between them depends on the extent to which the concentrations of the solute in the two phases differ from the equilibrium value as fixed by the distribution coefficient. In the classical two-film theory, it is assumed that the two phases are actually in equilibrium at the interface and that the resistance to mass transfer is concentrated in thin films on either side of the interface. Mass transfer through these films takes place by molecular diffusion. Thus, Eq. (3) applies where k is the film

$$\text{Rate} = k_x I(x_A - x'_A) = k_y I(y'_A - y_A) \qquad (3)$$

mass-transfer coefficient for a particular phase; I is the interfacial area over which mass transfer takes place; x'_A and y'_A are the solute concentrations in the bulk of a phase; and x_A and y_A are the concentrations in a phase adjacent to the interface.

Important variables affecting rate of mass transfer are temperature and agitation. When two immiscible liquids are mixed together, one will break into droplets, forming the dispersed phase while the other remains coherent as the continuous phase. Increasing the degree of agitation gives smaller droplets, hence greater total surface area, enhancing the mass-transfer rate.

Having achieved dispersion and mass transfer, it is necessary to segregate the phases for separate removal. Coalescence of droplets then becomes important, and various chemical and mechanical aids are available for speeding this step.

Chemistry of extraction. Many applications of extraction, particularly those concerned with the separation of metals, depend upon chemical interaction between solute and solvent. Categories of major importance are ion-association systems and chelate compounds.

Ion association systems. Ion association systems involve the pairing of oppositely charged ions under the influence of electrostatic attraction. Metals may be incorporated as either the cationic or anionic species in an aqueous phase; by association with an oppositely charged ion, an uncharged molecule is formed which can be extracted into the organic phase. Examples of metals in the anionic form include halides, thiocyanates, nitrates, and perchlorates, while the organic solvent is typically an oxygenated compound (for example, ethers, alcohols, ketones). An example is the extraction of uranium by tributyl phosphate (TBP), shown in reaction (4). The position of equilibrium in metals

$$\begin{array}{ll} \text{Aqueous} & UO_2^{++} + 2NO_3^- \rightleftharpoons UO_2(NO_3)_2 \\ \text{Organic} & \qquad\qquad + 2TBP \\ & \qquad\qquad \Downarrow\Uparrow \\ & \qquad UO_2(NO_3)_2\,2TBP \end{array} \quad (4)$$

extraction reactions of this type depends upon the pH of the system. Since different metals exhibit different dependencies, control of the pH of extraction allows sharp separations between different metals.

Chelate systems. Metal chelates are cyclic coordination compounds containing a metal atom in the ring. A reagent that has found widespread application in analytical chemistry is 8-quinolinol (I) which contains two atoms (N and O) that will coordinate with metal ions to form a five-membered ring as shown in reaction (5). Thus aluminum 8-

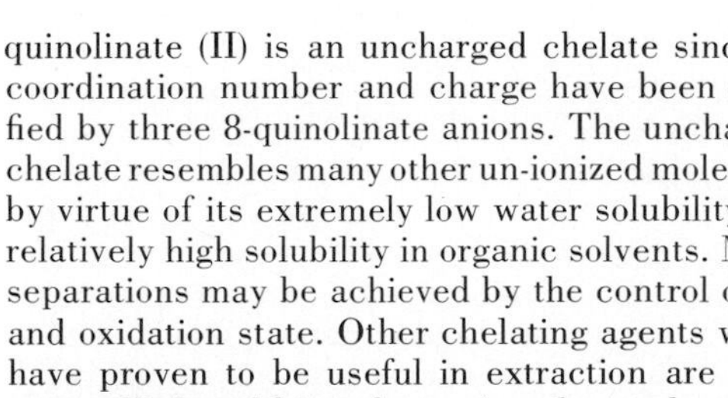

(I) (II) (5)

quinolinate (II) is an uncharged chelate since its coordination number and charge have been satisfied by three 8-quinolinate anions. The uncharged chelate resembles many other un-ionized molecules by virtue of its extremely low water solubility and relatively high solubility in organic solvents. Metal separations may be achieved by the control of pH and oxidation state. Other chelating agents which have proven to be useful in extraction are dithiozone (diphenylthiocarbazone) and acetylacetone. *See* CHELATION.

Contacting. The three basic methods for contacting solvent (S) with the feed mixture (F) are shown in the illustration.

In single-stage contacting, the two phases are

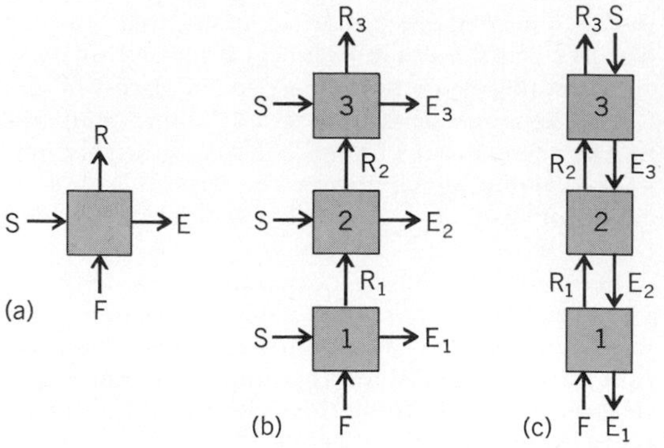

Contacting methods used in extraction: (*a*) single stage; (*b*) crosscurrent; and (*c*) countercurrent.

mixed to equilibrium and allowed to settle, after which the extract phase, E (solvent plus dissolved solute), and raffinate phase R (residual stream depleted in solute) are separately withdrawn. The separation attainable by this method is limited by the equilibrium prevailing under the extraction conditions. In order to further improve the degree of separation, multistage contacting is necessary. Within each component stage, equilibrium is established, but the terminal raffinate and extract phases are not in equilibrium with each other.

In crosscurrent extraction, fresh solvent is added at each stage; enhanced separation is obtained, but solvent requirement is high.

Countercurrent extraction is the most efficient and is the choice for commercial operation whenever possible. In this latter approach, raffinate and extract phases flow countercurrently and emerge at opposite ends of the contactor. Many stages can be used to give a high degree of separation while maintaining a modest solvent requirement.

Liquid/solid extraction. Liquid/solid extraction may be considered as the dissolving of one or more components in a solid matrix by simple solution, or by the formation of a soluble form by chemical reaction. Today the largest use of liquid/solid extraction is in the extractive metallurgical, vegetable oil, and sugar industries. The field may be subdivided into the following categories: leaching, washing extraction, and diffusional extraction. Leaching involves the contacting of a liquid and a solid (usually an ore) and the imposing of a chemical reaction upon one or more substances in the solid matrix so as to render them soluble. In washing extraction the solid is crushed to break the cell walls, permitting the valuable soluble product to be washed from the matrix. Sugar recovery from cane is a prime example of this method. In diffusional extraction the soluble product diffuses across the denatured cell walls (no crushing involved) and is washed out of the solid. The recovery of beet sugar is an excellent case in point.

Particle size is significant in all cases since it is a direct function of the total surface area that will be available for either reactions or diffusion. It is probably of greatest importance in extracting cellular materials because a reduction in particle size also results in an increase in the number of cells being ruptured. In ores, porosity and pore-size distribution greatly affect the rate of extraction because the solvent must flow or diffuse in and out of the pores and, in many cases, the solute moves through the pores to the particle surface by diffusion.

Liquid/liquid extraction. Liquid/liquid extraction separates the components of a homogeneous liquid mixture on the basis of differing solubility in another liquid phase. Because it depends on differences in chemical potential, liquid/liquid extraction is more sensitive to chemical type than to molecular size. This makes it complementary to distillation as a separation technique. One of the first large-scale uses was in the petroleum industry for the separation of aromatic from aliphatic compounds. The original process employed liquid sulfur dioxide as solvent. More recently, sulfolane (thio-cyclopentane-1,1-dioxide) has replaced sulfur dioxide for extraction of lighter aromatics due to its greater selectivity and ease of recovery. For the selective separation of higher-molecular-weight aromatics and aliphatics as in lubricating oil manufacture, phenol and furfural are the most widely used solvents.

Liquid/liquid extraction has found application for many years in the coal tar industry. The recovery of tar acids from crude tar oil by washing with an aqeous solution of alkali is an example where chemical interaction between solute and solvent determines differential solubility.

On a smaller scale, extraction is a key process in the pharmaceutical industry for recovery of antibiotics from fermentation broths. Penicillin is obtained by extraction into solvents such as amyl acetate at relatively low pH values (2 to 2.5) and is then stripped from the organic phase by treatment with a buffered aqueous solution at about pH 7 to 7.5. Other examples in this field are the recovery and separation of vitamins and the production of alkaloids from natural products.

Equipment. On a laboratory scale, the separatory funnel is the simplest device for achieving mixing and subsequent phase separation. To obtain continuous extraction of a small feed sample, the Soxhlet extractor is employed.

If the starting mixture is a solid, it must first be ground down to the required particle size. On a commercial scale, the equipment which is used for achieving efficient contacting with the liquid solvent tends to be specific to the type of industry and may employ screw-type conveyors, successive mixing/settling tanks, or a variety of proprietary designs.

Liquid/liquid extraction using countercurrent processing is preferably carried out in a vertical column. The internals are designed to disperse one phase into droplets, and to accomplish mass transfer between this phase and the continuous phase in a series of stages. These stages may be physical partitions comprising plates or trays, or theoretical concepts in the case of differential contactors such as a packed column. See CHEMICAL EQUILIBRIUM; CHEMICAL SEPARATION TECHNIQUES; COUNTERCURRENT TRANSFER OPERATIONS.

[BRUCE M. SANKEY]

Bibliography: D. Dyrssen et al. (eds.), *Solvent Extraction Chemistry*, 1967; C. Hanson (ed.), *Recent Advances in Liquid-Liquid Extraction*, 1971; C. Hanson, *Chem. Eng.*, 75:76–98, Aug. 26, 1968; R. H. Perry et al. (eds.), *Chemical Engineers' Handbook*, 5th ed., 1973; R. N. Rickles, *Chem. Eng.*, 72:157–172, Mar. 15, 1965; R. E. Treybal, *Liquid Extraction*, 2d ed., 1963.

Fermium

A chemical element, Fm, atomic number 100, the eleventh element in the actinide series. Fermium does not occur in nature; its discovery and production have been accomplished by artificial nuclear transmutation of lighter elements. Isotopes of

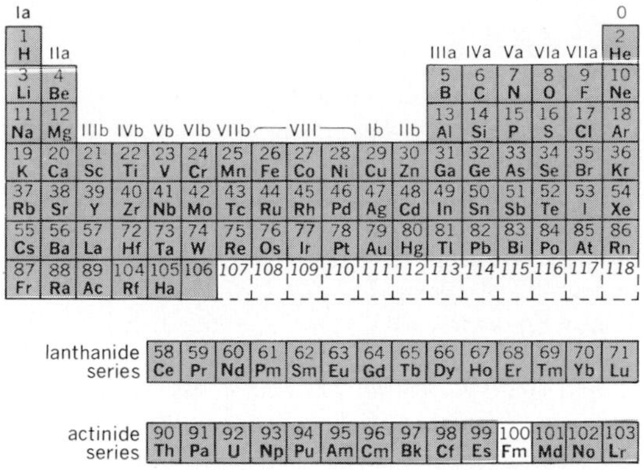

mass number 244–259 have been discovered; all are radioactive, decaying with half-lives ranging from a few milliseconds to 100 days. The total weight of fermium which has been synthesized is much less than one-millionth of a gram. See ACTINIDE ELEMENTS; RADIOACTIVITY.

The first isotope of fermium to be discovered was ^{255}Fm. In 1952 scientists of the Los Alamos, Argonne, and University of California laboratories of the U.S. Atomic Energy Commission isolated ^{255}Fm from the debris of a hydrogen bomb explosion. Uranium present in the weapon underwent multiple neutron capture to give ^{255}U; a subsequent chain of β-decays yielded finally the mass 255 isotope of fermium. Other fermium isotopes have been produced by bombardment of heavy element isotopes with helium, beryllium, carbon, oxygen, and other heavy ions accelerated in cyclotrons or linear accelerators. For production of larger amounts of fermium isotopes, gram quantities of ^{239}Pu or heavier isotopes are irradiated with neutrons for periods of several years in nuclear reactors. Isotopes of fermium and the other higher actinide elements are isolated as products of successive neutron capture and β-decay. See CALIFORNIUM.

Spontaneous fission, which becomes increasingly important in the heavier elements, is the major mode of decay for ^{244}Fm, ^{256}Fm, and ^{258}Fm. The longest-lived isotope is ^{257}Fm, which has a half-life of about 100 days. This isotope is produced in high-flux nuclear reactors and by thermonuclear devices as a result of successive neutron capture and beta decay processes. The amounts produced, however, are extremely small. Fermium-258 decays by spontaneous fission with a half-life of 0.38 msec. This suggests the existence of an abnormality at this point in the nuclear periodic table.

Fermium exists predominantly in the 3+ oxidation state in aqueous solution, but strong reducing conditions produce Fm^{2+}, which has greater stability than Es^{2+} and less stability than Md^{2+}. Ion-exchange chromatography has played a significant part in the separation and identification of fermium. Fermium is separated from other actinide elements when a complexing agent is passed through a column of cation-exchange resin onto which the actinides have been adsorbed; extraction chromatography with hydrogen di(2-ethylhexyl) phosphoric acid (HDEHP) is also used.

Fermium in the form of the 3.24-hr ^{254}Fm has been identified in the "metallic" zero valent state in an atomic-beam magnetic resonance experiment. *See* NUCLEAR CHEMISTRY; TRANSURANIUM ELEMENTS. [GLENN T. SEABORG]

Bibliography: S. Ahrland et al., *The Chemistry of the Actinides*, 1975; L. B. Asprey and R. A. Penneman, The chemistry of the actinides, *Chem. Eng. News*, 45:75–91, 1967; B. B. Cunningham, Chemistry of the actinide elements, *Annu. Rev. Nucl. Sci.*, 14:323–346, 1964; *Gmelins Handbuch der anorganischen Chemie*, 8th ed., new supplement, vol. 20d, pt. 2, 1975; E. K. Hyde, I. Perlman, and G. T. Seaborg, *The Nuclear Properties of the Heavy Elements*, vols. 1 and 2, 1964; J. J. Katz and G. T. Seaborg, *The Chemistry of the Actinide Elements*, 1958; C. Keller, *The Chemistry of the Transuranium Elements, Kernchemie in Einzeldarstellungen*, vol. 3, 1971; G. T. Seaborg, *The Transuranium Elements*, 1958; V. I. Spitsyn and J. J. Katz (eds.), *Moscow Symposium on the Chemistry of Transuranium Elements*: *Proceedings*, 1976.

Ferric compound

A compound containing iron in the 3+ oxidation state. The ferric, iron(III), state is obtained when iron is dissolved in oxidizing acids or when ferrous solutions are oxidized.

Although ferrous hydroxide is a fairly strong base, ferric hydroxide is much weaker. In fact, the hydrated ions acts as an acid [reaction (1)].

$$Fe(H_2O)_6^{3+} + H_2O \rightleftharpoons Fe(H_2O)_5(OH)^{2+} + H_3O^+ \quad (1)$$

Ferric salts are easily hydrolyzed in this manner to produce $Fe(H_2O)_3(OH)_3$ or $Fe(OH)_3 \cdot 3H_2O$, after two more steps similar to Eq. (1).

Because the composition of this red-brown gelatinous precipitate varies, it also is called a hydrous ferric oxide and written $Fe_2O_3 \cdot xH_2O$. The presence of this hydrous oxide in colloidal form gives ferric solutions their characteristic yellow and often muddy appearance.

The oxide, Fe_2O_3, known under many descriptive names, is used as a paint pigment, polishing compound, and rouge.

Ferric chloride, $FeCl_3$, is covalent in nature when anhydrous, and gas density measurements indicate that the formula of the gas is Fe_2Cl_6. It is used as a coagulant and as a mordant and etching compound. A double salt, $(NH_4)_2SO_4 \cdot Fe_2(SO_4)_3 \cdot 24H_2O$, is one of a series of compounds called alums. Ferric ion is detected by the blood-red reaction product with thiocyanate ion which forms ferric thiocyanate [reaction (2)].

$$FeCl_3 + 3NH_4CNS \rightarrow Fe(CNS)_3 + 3NH_4Cl \quad (2)$$

See FERROUS COMPOUND; IRON.

[E. EUGENE WEAVER]

Ferricyanide

A compound containing the complex ion $[Fe(CN)_6]^{3+}$, where iron has an oxidation state of 3+. The coordination complex can be considered a derivative of ferricyanic acid, $H_3[Fe(CN)_6]$, which has been isolated as a red-brown solid.

In the nomenclature of coordination compounds, $Na_3[Fe(CN)_6]$, for example, should be called sodium hexacyanoferrate(III).

Although the ferricyanides of the alkali and alkaline-earth metals are water-soluble, the remainder are water-insoluble.

Ferricyanides are prepared from ferrocyanides by oxidation of the iron with chlorine. The uses are much the same as for ferrocyanides. When a solution of ferrous salt is mixed with a solution of an alkali ferricyanide, a precipitate is obtained which is called Turnbull's blue and is not $Fe_3[Fe(CN)_6]_2$. It is thought to be the same as prussian blue and may be represented as $K[Fe^{II}(CN)_6Fe^{III}]$. *See* FERROCYANIDE; IRON.

[E. EUGENE WEAVER]

Ferrocene

Dicyclopentadienyl iron, $(C_5H_5)_2Fe$, is an orange crystalline solid with a melting point of 174°C. The compound sublimes at 100°C, is diamagnetic, and has a dipole moment of zero. The structure of ferrocene, in which an atom of iron is sandwiched between two parallel cyclopentadiene rings, is shown in the illustration.

FERROCENE

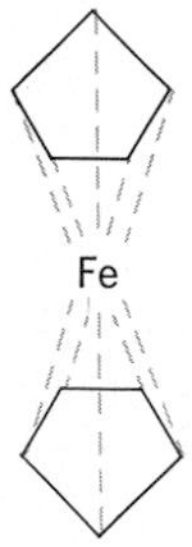

Structural diagram of ferrocene.

Ferrocene was first prepared in 1951 by two research groups. Two preparation routes were used: first, the oxidation of cyclopentadienylmagnesium bromide with anhydrous ferric chloride, and second, the heating of powdered iron with cyclopentadiene at 300°C in a nitrogen atmosphere. The compound is prepared conveniently in the laboratory by the action of anhydrous ferrous chloride on the sodio derivative of cyclopentadiene in tetrahydrofuran. A further simplification in procedure consists in treating a solution of cyclopentadiene in diethylamine with either anhydrous ferric or ferrous chloride.

The mean iron-carbonyl distance is approximately 2.04 A and the carbon-carbon distance is near 1.40 A. Ferrocene has been shown to possess a pentagonal antiprismatic structure in the crystal state. The precise nature of the ring metal bond is still in dispute, largely in regard to the degree in which the π-electron participates in bonding with metal orbitals.

Ferrocene is quite stable to heat and to certain kinds of chemical attack. It resists pyrolysis at 470°C. The compound is recovered unchanged from boiling water, boiling 10% aqueous caustic, and from boiling concentrated hydrochloric acid. In contrast to cyclopentadiene, ferrocene resists hydrogenation at 200°C and 2000-psi pressure, and it shows no reactivity toward a typical dienophile, such as maleic anhydride. Ferrocene is oxidized to a water-soluble trivalent iron cation.

The most outstanding feature of the chemistry of ferrocene is its aromatic character. The aromaticity of ferrocene is evidenced by the manner in which it behaves toward electrophilic reagents. The molecule undergoes typical Friedel-Crafts acylation to yield either the mono- or diacetylated product. Ferrocene also undergoes other electrophilic substitution reactions typical of a reactive aromatic compound. Some reactions, such as nitration, are not observed with ferrocene under the usual conditions because of the ease of oxidation of ferrocene under acidic conditions. Such derivatives can, however, be prepared by modified methods.

Ferrocene is metalated with *n*-butyllithium to produce mono- and bislithioferrocene, which can then undergo further reactions typical of such an organolithium compound.

Only relatively minor efforts appear to have been expended to date with regard to commercial uses of ferrocene and its derivatives.

[FRANK D. POPP]

Bibliography: K. Plesske, Ring substitutions and secondary reactions of aromatic-metal π-complexes, *Angew. Chem. Int. Ed.*, 1:312–327, 394–399, 1962; M. D. Rausch, Metallocene chemistry: A decade of progress, *Can. J. Chem.*, 41: 1289–1314, 1963; M. D. Rausch, *The History and Development of Organotransition Metal Chemistry*, Advances in Chemistry Series, no. 62, Werner Centennial, American Chemical Society, 1966; M. Rosenblum, *Chemistry of the Iron Group Metallocenes*, pt. 1, 1965.

Ferrocyanide

A compound that contains the complex ion $[Fe(CN)_6]^{4+}$, where iron has an oxidation state of 2+. The coordination complex can be considered a derivative of ferrous and cyanide salts, as it is manufactured from these compounds, or a derivative of ferrocyanic acid, $H_4[Fe(CN)_6]$, which has been isolated as a white solid.

In the nomenclature of coordination compounds, $K_4[Fe(CN)_6]$, for example, should be called potassium hexacyanoferrate (II).

The alkali and alkaline-earth metal hexacyanoferrates (II), with the exception of barium, are easily soluble in water. Most of the heavy metal salts are insoluble; hence, they are used in analytical chemistry.

The sodium and potassium salts are the most important and are used in making blue pigments, dyes, blueprint paper, and ferricyanide.

When a solution of ferric salts is added to a ferrocyanide solution, a precipitate with a complicated structure is obtained which is called prussian blue. This precipitate is not $Fe_4[Fe(CN)_6]_3$, which can be prepared under controlled conditions. Prussian blue is now thought to be the same as Turnbull's blue. *See* CYANIDE; FERRICYANIDE; FERROUS COMPOUND; IRON. [E. EUGENE WEAVER]

Ferrous compound

A compound containing iron in the 2+ oxidation state. This oxidation state, iron(II), is obtained when iron metal is dissolved in nonoxidizing acids such as hydrochloric or dilute sulfuric. Insoluble ferrous hydroxide, $Fe(OH)_2$, is a white, gelatinous solid which turns green, then reddish-brown, as it is oxidized to the ferric state on exposure to air. Solutions of ferrous salts have a pale-green color and are also oxidized by air to ferric salts.

Ferrous sulfate is obtained as a by-product from the pickling of steel and by air oxidation of pyrites, FeS_2, and is crystallized from water as greenish crystals, $FeSO_4 \cdot 7H_2O$. It is used in water purification, writing inks, pigments, and medicine.

Mohr's salt or ferrous ammonium sulfate, $Fe(SO_4) \cdot (NH_4)_2SO_4 \cdot 6H_2O$, crystallizes as light-green crystals from solutions containing ferrous sulfate and ammonium sulfate. Because the solid is not easily oxidized by air, it is used in the analytical laboratory as a source of ferrous ion. *See* FERRIC COMPOUND; IRON. [E. EUGENE WEAVER]

Filtration

The separation of solid particles from a fluid-solids suspension of which they are a part by passage of most of the fluid through a septum or membrane that retains most of the solids on or within itself. The septum is called a filter medium, and the equipment assembly that holds the medium and provides space for the accumulated solids is called a filter. The fluid may be a gas or a liquid. The solid particles may be coarse or very fine, and their concentration in the suspension may be extremely low (a few parts per million) or quite high (>50%).

The object of filtration may be to purify the fluid by clarification or to recover clean, fluid-free particles, or both. In most filtrations the solids-fluid separation is not perfect. In general, the closer the approach to perfection, the more costly the filtration; thus the operator of the process cannot justify a more thorough separation than is required.

Gas filtration. It is frequently necessary to remove solids (called dust) from a gas-solids mixture because: (1) the dust is a contaminant rendering the gas unsafe or unfit for its intended use; (2) the dust particles will ultimately separate themselves from the suspension and create a nuisance; or (3) the solids are themselves a valuable product that in the course of its manufacture has been mixed with the gas. One method of separating solids from a gas is by gas filtration. Among other methods are centrifugal separation, infringement, scrubbing, and electrostatic precipitation.

Three kinds of gas filters are in common use. Granular-bed separators consist of beds of sand, carbon, or other particles which will trap the solids in a gas suspension that is passed through the bed. The bed depth can range from less than an inch to several feet. Bag filters are bags of woven fabric, felt, or paper through which the gas is forced; the solids are deposited on the wall of the bag. A simple example is the dust collector of a household vacuum cleaner. Air filters are light webs of fibers, often coated with a viscous liquid, through which air containing a low concentration of dust, normally < 5 grains/1000 ft³ (approx. 28 m³), can be passed to cause entrapment of the dust particles. The filter installed in the air-intake lines of most household furnaces is an air filter. Gas filters can be used to remove particles in the size range from 1 nm to 100,000 nm.

Liquid filtration. Liquid-solids separations are important in the manufacture of chemicals, polymer products, medicinals, beverages, and foods; in mineral processing; in water purification; in sewage disposal; in the chemistry laboratory; and in the operation of machines such as internal combustion engines. Filtration is the most versatile and widely used operation for such applications. In some cases, however, gravity or centrifugal sedimentation is preferred to filtration.

Liquid filters are of two major classes, cake filters and clarifying filters. The former are so called because they separate slurries carrying relatively large amounts of solids. They build up on the filter medium as a visible, removable cake which normally is discharged "dry" (that is, as a moist mass), frequently after being washed in the filter. It is on the surface of this cake that filtration takes place after the first layer is formed on the medium. The feed to cake filters normally contains at least 1%

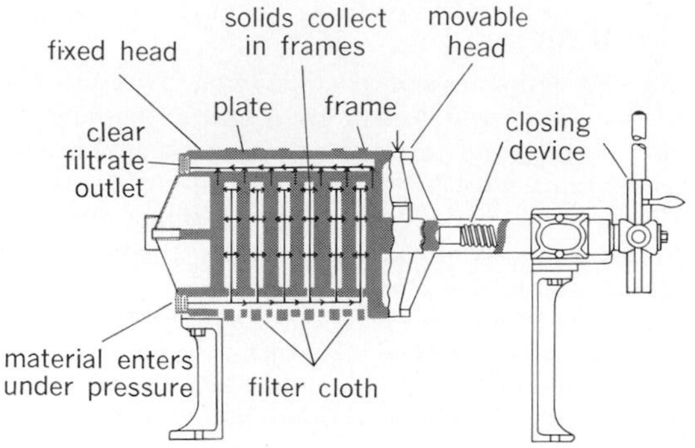

Fig. 1. Corner-feed, closed-discharge filter press, a type of plate-and-frame press. (*From T. Shriver and Co.*)

solids. Clarifying filters, on the other hand, normally receive suspensions containing less than 0.1% solids, which they remove by entrapment on or within the filter medium without any visible formation of cake. The solids are normally discharged by backwash or by being discarded with the medium when it is replaced.

Cake filters. There are three classes of cake filters, depending on the driving force producing the separation. In pressure filters, superatmospheric pressure is applied by pump or compressed gas to the feed slurry, and filtrate is discharged at atmospheric pressure, or higher. In vacuum filters, the pressure of the atmosphere is the driving force, a vacuum being applied to the filtrate collection chamber. In centrifugal filters, centrifugal force causes the filtrate to flow through the cake and filter medium.

Cake filters are also classified as batch-operating or continuous. In general, pressure filters are batch or intermittent filters, vacuum filters are continuous machines, and centrifugal filters may be either.

Pressure filters. There are three principal industrial classes of pressure filters: filter presses, horizontal plate filters, and leaf filters. They may be operated at 250 psi (1,723,750 Pa) or higher feed pressure.

Filter presses are the simplest and most versatile pressure filters. One form, the plate-and-frame press, is shown in Fig. 1. This filter consists of a number of rectangular or circular vertical plates, alternating with empty frames, assembled on horizontal rails. One fixed head and one movable head hold the plates and frames together. Both sides of each plate and the inner surfaces of the heads constitute the filtration area. The medium consists of filter cloths hung over the plates and covering all filter areas. The plates and frames are compressed by a powerful screw, so that the cloths also act as gaskets to prevent the filtrate from leaking between the plates. Filtrate collects on the plate surfaces and leaves the press through drainage channels; cake collects in the frames and is dumped out after the filtration cycle is finished.

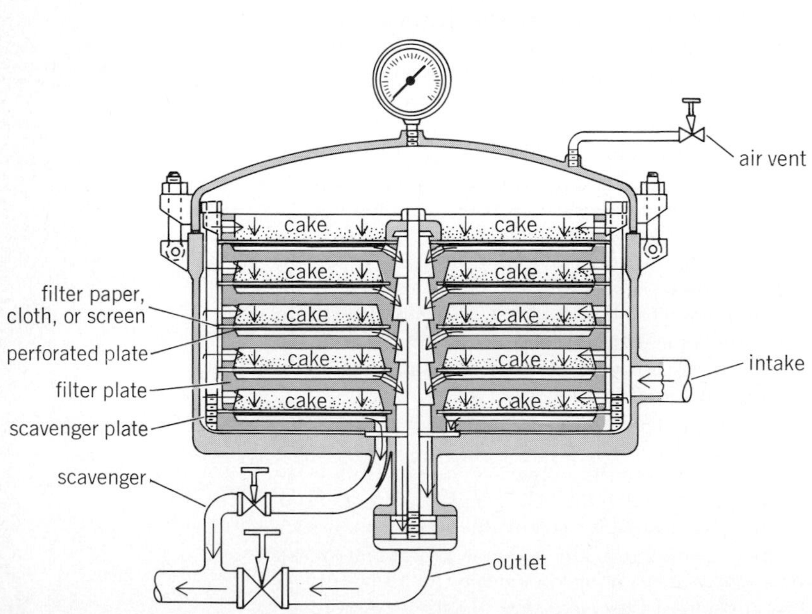

Fig. 2. Sparkler horizontal plate filter. (*From R. H. Perry and G. H. Chilton, eds., Chemical Engineers' Handbook, 5th ed., McGraw-Hill, 1973*)

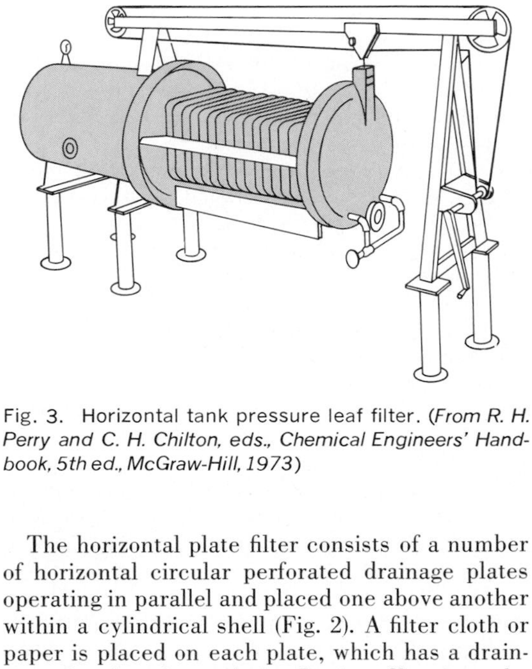

Fig. 3. Horizontal tank pressure leaf filter. (*From R. H. Perry and C. H. Chilton, eds., Chemical Engineers' Handbook, 5th ed., McGraw-Hill, 1973*)

The horizontal plate filter consists of a number of horizontal circular perforated drainage plates operating in parallel and placed one above another within a cylindrical shell (Fig. 2). A filter cloth or paper is placed on each plate, which has a drainage chamber beneath it. During filtration, the whole shell is filled with slurry under filtering pressure. At the end of the cycle, the shell is drained and the filter cake is removed.

Pressure leaf filters consist of pressure tanks filled with slurry into which a number of filter leaves connected to drainage pipes are immersed. The leaves may be grids covered with fitted cloth bags, shallow boxes of screen cloth, or porous cylindrical tubes of metal or porcelain. They may be horizontal or vertical. A horizontal tank filter with vertical leaves is shown open for cleaning in Fig. 3. When the filter cake can be flushed away instead of recovered dry, a pressure leaf filter can be operated for many cycles over a period of months without being opened. Such filters are adaptable to automatic programmed operation.

Continuous-vacuum filters. These filters are used principally for free-filtering solids. Their most important forms are rotary-drum, rotary-disk, horizontal belt, and horizontal pan or table filters.

The horizontal drum filter is a cloth-covered horizontal cylinder partially immersed in the slurry

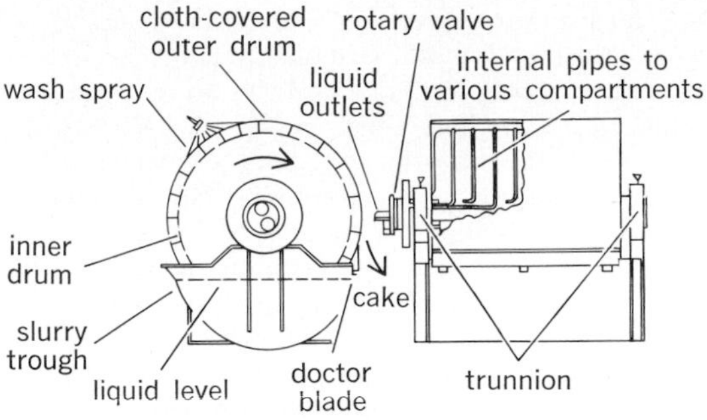

Fig. 4. Continuous vacuum rotary filter. (*From W. L. McCabe and J. C. Smith, Unit Operations of Chemical Engineering, McGraw-Hill,* 2d ed., 1967)

(Fig. 4). The periphery of the drum consists of a number of individual filter elements, each connected to one of the trunnions (drum supports) by pipes. The trunnion contains a special valve which applies suction to each element from the time it enters the slurry until just before it reaches the doctor blade. It then directs compressed air to the element until the element again submerges. Suction is then reestablished. Filtrate flows to the trunnion under the action of the vacuum and then out of the filter, as does wash water sprayed on the cake. The cake is sucked dry of wash and doctored off the filter.

In a disk filter, the rotating drum is replaced by an array of parallel disks, the faces of which are the cloth-covered filtering surface (Fig. 5). Each disk is divided into trapezoidal filter elements controlled by a valve, as in the drum filter.

Horizontal filters are useful when heavy solids are to be filtered or when thorough washing is required. They consist either of an endless belt or a rotating annular table, the top surface of each being the filtering surface. The belt passes over a suction box; the table is divided into filter elements connected to a valve, as in the drum filter. Instead of a table being used the filter elements can be separated into individual trays, annularly arrayed, that can be tipped to assist in the cake dumping.

Clarifying filters. When the object of filtration is to produce a solids-free liquid from a dilute suspension, a clarifying filter may be used. Alternatively, sedimentary clarifiers or centrifuges may be chosen. *See* SEDIMENTATION.

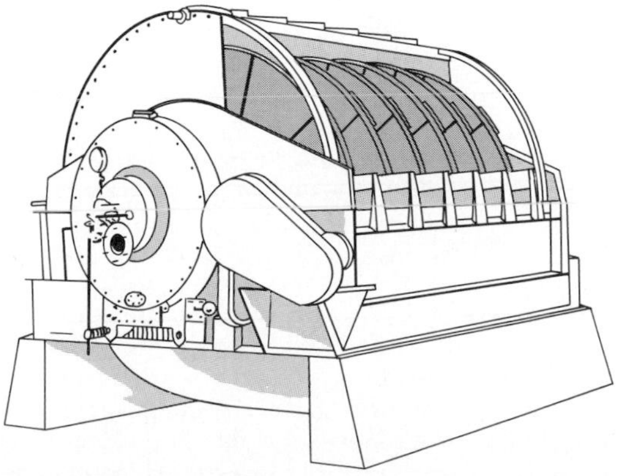

Fig. 5. American disk filter. (*From R. H. Perry and C. H. Chilton, eds., Chemical Engineers' Handbook, 5th ed., McGraw-Hill, 1973*)

Clarifying filters are pressure filters that employ a very fine filter medium, the pores of which are sufficiently small to prevent passage of particles of the size that must be removed. In cake filtration, cake particles bridge relatively large openings in the medium and create a new filtering surface; in clarifying filtration, each particle is collected on or within the filter medium.

Clarifying filters may be in the form of plate filters (analogous to shallow-frame filter presses), disk filters (analogous to horizontal plate filters), cartridges (analogous to leaf filters), and granular beds. The last-mentioned type is a special case; the principal example of the use of granular beds is in water filtration, commonly carried out by sand or coal filters operated by gravity instead of under pressure. A cartridge filter is shown in Fig. 6.

Although the filter medium is important in every filtration, it assumes paramount importance in clarifications. Aside from granular beds, there are four general types of clarifying mediums: spaced wires that strain out solids (lower limit, 25 μm); depth mediums, masses of fibers or interconnected pores that trap particles by interception (lower limit, 5 μm); micronic sheets or pads of carefully sized fibers (lower limit, 0.01 μm); and hyperfiltration membranes (lower limit, 1 nm). A special case of granular-bed clarification occurs when a cake of diatomaceous earth is laid down as a precoat on a coarse filter, thereby converting it to a clarifying filter.

[S. A. MILLER]

Bibliography: B. Anderson and J. B. Moore, *Optimal Filtering*, 1979; C. O. Bennett and J. E. Myers, *Momentum, Heat, and Mass Transfer*, 2d ed., 1974; W. L. McCabe and J. C. Smith, *Unit Operations of Chemical Engineering*, 3d ed., 1975; C. Orr, *Filtration: Principles and Practices*, pt. 1, 1979; R. H. Perry and C. H. Chilton (eds.), *Chemical Engineers' Handbook*, 5th ed., 1973.

FILTRATION

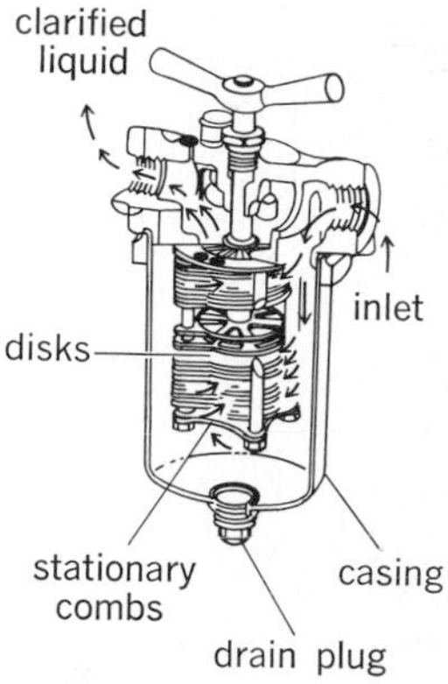

Fig. 6. Cartridge filter. (*W. L. McCabe and J. C. Smith, Unit Operations of Chemical Engineering, McGraw-Hill, 1956*)

Fine structure (spectral lines)

A term referring to the closely spaced groups of lines observed in the spectra of the lightest elements, notably hydrogen and helium. The components of any one such group are characterized by identical values of the principal quantum number n, but different values of the azimuthal quantum number l and the angular momentum quantum number j.

According to P. A. M. Dirac's relativistic quantum mechanics, those energy levels of a one-electron atom that have the same n and j coincide exactly, but are displaced from the values predicted by the simple Bohr theory by an amount proportional to the square of the fine-structure constant α. The constant α is dimensionless, and nearly equal to 1/137. Its value is actually $0.007297351 \pm 0.000000006$. In 1947 deviations from Dirac's theory were found, indicating that the level having $l = 0$ does not coincide with that having $l = 1$, but is shifted appreciably upward. This is the celebrated Lamb shift named for its discoverer, Willis Lamb, Jr. Modern quantum electrodynamics accounts for this shift as being due to the interaction of the electron with the zero-point fluctuations of the electromagnetic field.

In atoms having several electrons, this fine

structure becomes the multiplet structure resulting from spin-orbit coupling. This gives splittings of the terms and the spectral lines that are "fine" for the lightest elements but that are very large, of the order of an electronvolt, for the heavy elements.

[F. A. JENKINS/W. W. WATSON]

First-order transition

A change in state of aggregation of a system accompanied by a discontinuous change in enthalpy, entropy, and volume at a single temperature and pressure. This transition may be between liquid and gas, between solid and gas, between solid and liquid, or between two solid phases. For the differences between first- and second-order transition *see* SECOND-ORDER TRANSITION. *See also* BOILING POINT; MELTING POINT; PHASE EQUILIBRIUM; SUBLIMATION; TRANSITION POINT.

[ROBERT L. SCOTT]

Fischer-Tropsch process

The synthesis of hydrocarbons and, to a lesser extent, of aliphatic oxygenated compounds by the catalytic hydrogenation of carbon monoxide. The synthesis was discovered in 1923 by F. Fischer and H. Tropsch at the Kaiser Wilhelm Institute for Coal Research in Mülheim, Germany. The reaction is highly exothermic, and the reactor must be designed for adequate heat removal to control the temperature and avoid catalyst deterioration and carbon formation. The sulfur content of the synthesis gas must be extremely low to avoid poisoning the catalyst. The first commercial plant was built in Germany in 1935 with cobalt catalyst, and at the start of World War II there were six plants in Germany producing more than 4,000,000 bbl/year of primary products. Iron catalysts later replaced the cobalt.

Following World War II, considerable research was conducted in the United States on the iron catalysts. One commercial plant was erected at Brownsville, Tex., in 1948, which used a fluidized bed of mill scale promoted by potash. Because synthetic oil was not competitive with petroleum, the plant was shut down within a few years. A Fischer-Tropsch plant constructed in South Africa (SASOL) about the same time has continued to produce gasoline, waxes, and oxygenated aliphatics. The SASOL plant gasifies inexpensive coal in Lurgi generators at elevated pressure. After purification, the gas is sent to entrained and fixed-bed synthesis reactors containing iron catalysts.

Research at the U.S. Bureau of Mines Pittsburgh Coal Research Center has resulted in the development of active and low-cost iron catalysts of steel lathe turnings and flame-sprayed magnetite. With these catalysts, improved reactors have been tested by use of gas recycle to control the operating temperature while maintaining a low pressure drop.

Methanation. Since about 1960, interest has grown in the United States in catalytic methanation to produce high-Btu gas from coal. Synthesis gas containing three volumes of hydrogen to one volume of carbon monoxide is reacted principally to methane and water, as shown in Eq. (1). Nickel catalysts, first discovered by P. Sabatier and J. B. Senderens in 1902, are still the principal catalysts for methanation. Active precipitated catalysts were developed at the British Fuel Research Station, and Raney nickel catalysts were tested in fluid and fixed-bed reactors by the Bureau of Mines and the Institute of Gas Technology at the Illinois Institute of Technology. The Bureau of Mines has developed a technique for applying a thin layer of Raney nickel on plates and tubes by flame-spraying the powder. This has led to development of efficient gas recycle and tube-wall reactors. Commercial production of high-Btu gas from coal in the United States is expected to start about 1980 to supplement natural gas supplies.

Reactions. Typical Fischer-Tropsch reactions for the synthesis of paraffins, olefins, and alcohols are the pairs of equations labeled (1), (2), and (3), respectively.

$$(2n+1)H_2 + nCO \xrightarrow{\text{Co catalysts}} C_nH_{2n+2} + nH_2O$$
$$(n+1)H_2 + 2nCO \xrightarrow{\text{Fe catalysts}} C_nH_{2n+2} + nCO_2 \qquad (1)$$

$$2nH_2 + nCO \xrightarrow{\text{Co catalysts}} C_nH_{2n} + nH_2O$$
$$nH_2 + 2nCO \xrightarrow{\text{Fe catalysts}} C_nH_{2n} + nCO_2 \qquad (2)$$

$$2nH_2 + nCO \xrightarrow{\text{Co catalysts}} C_nH_{2n+1}OH + (n-1)H_2O$$
$$(n+1)H_2 + (2n-1)CO \xrightarrow{\text{Fe catalysts}} C_nH_{2n+1}OH + (n-1)CO_2 \qquad (3)$$

The primary reaction on both the cobalt and the iron catalysts yields steam, which reacts further on iron catalysts with carbon monoxide to give hydrogen and carbon dioxide. On cobalt catalysts, at synthesis temperatures (about 200°C, or 392°F), the reaction $H_2O + CO = CO_2 + H_2$ is much slower than on iron catalysts at synthesis temperatures (250–320°C, or 482–617°F). All these synthesis reactions are exothermic, yielding 37–51 kcal/mole of carbon in the products or 4700–6100 Btu/lb of product.

The hydrocarbons formed in the presence of iron catalysts contain more olefins than those formed in cobalt catalyst systems. The products from both catalysts are largely straight-chain aliphatics; branching is about 10% for C_4, 19% for C_5, 21% for C_6, and 34% for C_7. Aromatics appear in small amounts in the C_7 and in larger amounts in the higher boiling fractions. Operating conditions and special catalysts required, such as nitrides and carbonitrides of iron, have been ascertained for the production of higher proportions of alcohols.

[JOSEPH H. FIELD]

Bibliography: H. A. Dirksen and H. R. Linden, *Inst. Gas Technol. Res. Bull.* no. 31, July, 1963; J. H. Field et al., *Ind. Eng. Chem. Prod. Res. Develop.*, 3:150–153, June, 1964; J. C. Hoogendoorn and J. M. Salomon, *Brit. Chem. Eng.*, pp. 238–243, May, 1957; L. A. Moignard and F. J. Dent, *Gas Times*, pp. 40–41, May 11, 1946; H. H. Storch, N. Golumbic, and R. B. Anderson, *The Fischer-Tropsch and Related Syntheses*, 1951.

Flame photometry

A branch of spectrochemical analysis in which samples in solution are excited to luminescence by introduction into a flame. Flame photometry is particularly useful in determining small amounts of lithium, sodium, and potassium in solution. Only small samples are required, and an analysis usually can be carried out in relatively short time.

Emission flame photometry. In the Lundegardh vaporizer, a compressed-air aspirator vaporizes the solution within a chamber. The smaller droplets are carried along into the fuel-gas stream and then to the burner orifice. As each droplet passes through the flame, the solvent is evaporated and the solutes are vaporized, dissociated, and optically excited. This system usually is used with slow-burning fuel mixtures, such as illuminating gas and air. The burner may be similar to a meker burner. Most flame photometry in the United States, however, now is done with the Beckman burner, which aspirates solution directly into a hydrogen-oxygen or acetylene-oxygen flame. A diagram of this aspirator-burner is shown in Fig. 1.

Since flame temperatures are comaratively low, only the more volatile compounds are vaporized from the residue. Also, only the less stable molecules are dissociated, and only the more easily excited spectral lines and bands are emitted. Flame line-emission spectra are thus relatively simple, although molecular band spectra caused by the fuel or metallic oxides sometimes cause spectral interference.

The line or band of interest is isolated with a monochromator, and its intensity measured photoelectrically. The size and type of monochromator required depend on the spectral region studied and the complexity of the spectrum. For the 4000–7700 A region, a small or medium prism instrument often is adequate, although a grating monochromator would be preferable. In the 7700–8600 A range, which is used in rubidium and cesium analyses, the grating gives much superior performance.

Flame photometry is more accurate, precise, and sensitive than wet chemical methods for the determination of small amounts of alkali metals, and it rivals wet methods in the determination of some other elements. Standard deviations are usually 1–5% of the amount present. Sensitivity limits for various elements go as low as fractions of a microgram per liter to several hundred milligrams per liter, depending on the element, the solvent, the solute composition, the portion of flame used, the resolving power of the monochromator, and the sensitivity and signal-to-noise ratio of the detection system. Since flames are steady, internal standards are not needed to compensate for fluctuations. Matrix effects are usually compensated for by comparing the sample with synthetic standards prepared in a matric simulating that of the sample or by using the standard-addition technique. *See* SPECTROCHEMICAL ANALYSIS.

Atomic absorption analysis. This is a method of determining the concentration of a metal, usually in solution. The solution is converted to a vapor in which at least some of the metal is in the form of individual neutral atoms. The absorbance of this vapor is measured at a resonant absorption wavelength of the metal in question. This absorbance is then compared with that given by standard solutions. For some elements (for example, zinc, cadmium, antimony, lead, bismuth, mercury, and tellurium) detection limits are much better than those achieved by emission flame photometry; for most elements the two methods are comparable in sensitivity. The atomic absorption method has fewer spectral line interference problems than does emission flame photometry. It is finding increasing use in many fields because of its high sensitivity and versatility.

The basic components of an atomic absorption spectrophotometer (Fig. 2) are the light source, the vaporizer, the monochromator, and the detector. The light source most commonly used is a hollow cathode discharge tube containing the element to be determined. The lines emitted by this tube are approximately of the same spectral width (~0.03 A) as the absorption lines of the metallic vapor, and are thus absorbed very efficiently. A continuous light source could be used instead, but higher quality monochromators and detection systems would then be needed. On the other hand, discharge tubes have been made which emit only the resonance radiation of the element in question. These light sources would often make it possible to do without the monochromator entirely.

The absorbing metallic vapor is usually created by nebulizing the solution and passing the resulting aerosol through a flame. If the gases used are hydrogen and oxygen, the aerosol is injected into the flame at the orifice of the burner. The flame is then either directed vertically through the optic axis or injected into an open-ended tube which is coaxial with that axis. If a slow-burning fuel mixture is used (for example, acetylene-air, acetylene-oxygen, or acetylene–nitrous oxide), the aerosol is added to the fuel mixture in a mixing chamber. The flame then emerges from a narrow slot, usually about 10 cm long, parallel to and immediately below the optic axis.

The monochromator must have resolving power sufficient to separate the absorbed spectral line from the closest nonabsorbed line emitted by the discharge tube.

The light signal, after passing through the atomic vapor and the monochromator, is detected by a photomultiplier tube. To prevent light emitted by the flame from "diluting" the absorption effect, the radiation from the discharge tube can be periodically interrupted at some fixed frequency before it enters the flame. The signal from the photomulti-

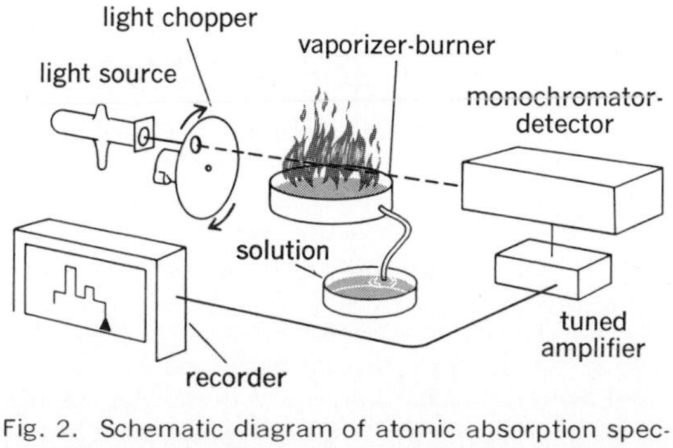

Fig. 2. Schematic diagram of atomic absorption spectrophotometer for determining metal concentrations.

FLAME PHOTOMETRY

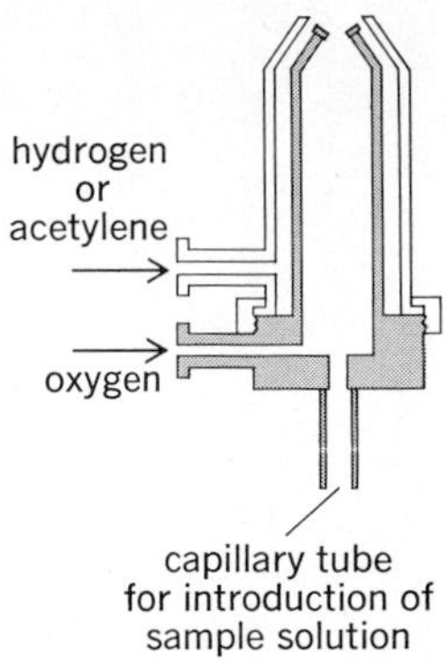

Fig. 1. Beckman aspirator-burner for flame photometry.

plier is then passed through an ac amplifier tuned to the frequency in question. Signals which are unmodulated (for example, from the flame) or not modulated at the frequency in question are filtered out at this point.

Fluctuations in the intensity of the light source may create spurious variations in the absorption effect being measured. This difficulty can be avoided by observing the ratio of the intensity of light which has passed through the flame to that of light which has not. [CYRUS FELDMAN]

Bibliography: F. Burriel-Martí and J. Ramírez-Muñoz, *Flame Photometry*, 1957; J. A. Dean, *Flame Photometry*, 1960; R. Herrmann and C. T. J. Alkemade, *Chemical Analysis by Flame Photometry*, 1963; R. Mavrodineanu and H. Boiteux, *Flame Spectroscopy*, 2d ed., 1965; N. S. Poluektov, *Techniques in Flame Photometric Analysis*, 1961; E. Pungor, *Flame Photometry Theory*, 1967.

Flameproofing

The process of treating materials so that they will not support combustion. Although cellulosic materials such as paper, fiberboard, textiles, and wood products cannot be treated so that they will not be destroyed by long exposure to fire, they can be treated to retard the spreading of fire and to be self-extinguishing after the igniting condition has been removed.

Treatments. Numerous methods have been proposed for flameproofing cellulosic products. One of the simplest and most commonly used for paper and wood products is impregnation with various soluble salts, such as ammonium sulfate, ammonium phosphate, ammonium sulfamate, borax, and boric acid. Special formulations are often used to minimize the effects of these treatments on the color, softness, strength, permanence, or other qualities of the paper.

For some applications, these treatments are not suitable because the salts remain soluble and leach out easily on exposure to water. A limited degree of resistance to leaching can be achieved by the addition of latex, lacquers, or waterproofing agents. In some cases the flameproofing agent can be given some resistance to leaching by causing it to react with the cellulose fiber (for example, urea and ammonium phosphate).

Leach-resistant flameproofing may also be obtained by incorporating insoluble retardants in the paper during manufacture, by application of insoluble materials in the form of emulsions, dispersions, or organic solutions, or by precipitation on, or reaction with, the fibers in multiple-bath treatments. The materials involved are of the same general types as those used for flameproofing textiles and include metallic oxides or sulfides and halogenated organic compounds. Because of the higher material cost and more elaborate equipment required, such treatments are used on paper only when unusual conditions justify their cost.

Action of retardants. Various theories have been proposed to explain the flameproofing action. In general, these are quite similar to those considered in the flameproofing of textiles. Two types of combustion may be involved. In one, the volatile decomposition products burn with a flame; in the other, the solid material undergoes flameless combustion, or afterglow. In general, the alkaline types of retardants are effective in preventing afterflame, and the acid types control afterglow. A few, such as ammonium phosphates and halogenated products, reduce both flame and glow.

The theory for prevention of afterflame having the most support is a chemical one, which holds that effective flame retardants direct the decomposition of the cellulose when heated so as to minimize the formation of volatile flammable products and increase the amount of water and solid char formed. The prevention of afterglow is usually attributed to a modification of the flameless combustion to make it less exothermic and therefore incapable of maintaining itself, for example, by formation of carbon monoxide instead of carbon dioxide. Other theories include the formation of a coating or froth which excludes oxygen and smothers the combustion, the formation of nonflammable gases which dilute the flammable products and exclude oxygen, and the thermal theories which hold that heat is dissipated by endothermic reactions or by conduction. It is probable that these actions do contribute in some cases.

The susceptibility of paper or fiberboard to combustion may also be modified by the introduction of noncombustible but otherwise inert materials, such as mineral fillers, asbestos, and glass or ceramic fibers, into the paper or board during manufacture. The high proportions required result in products with properties determined largely by noncombustible components rather than by cellulosic fibers.

The behavior of paper, fiberboard, or wood products exposed to fire depends also on structural factors. The density of the material and the ratio of mass to surface exposed affect the ease of ignition and the rate of spreading or penetration by the fire. Fire-resistant coatings minimize the spread of flame. Heat-reflective coatings or insulating coatings prevent or retard the material beneath them from reaching the ignition temperature. Intumescent coatings are formulated to form a thick insulating foam when exposed to fire.

A wide variation exists in methods for evaluating the efficiency of a flameproofing treatment or of suitability for a given use. Paper is usually tested by igniting with a flame of specified character for a predetermined period of time and then noting duration of afterflame and of afterglow and the amount of char length or char area. Structural materials are tested by other methods which involve features of construction in addition to those of composition. Methods have been standardized by the Technical Association of the Pulp and Paper Industry (TAPPI), American Society for Testing and Materials (ASTM), and various governmental agencies. *See* COMBUSTION.

[T. A. HOWELLS]

Flash point

The temperature at which a flash of flame is first observed when a small lighted flame is passed over a sample of a flammable substance in a cup. The flash point is an important test for fire and safety regulations. So also is the fire point, or temperature at which sustained burning first occurs. These tests are indicative of the lowest-boiling components in the oil and are applied to solvents, distillate fuels, lubricating oils, and many other petroleum products. [MOTT SOUDERS]

Flocculation

The formation of larger particles of a solid phase dispersed in a solution by the gathering together of smaller particles. The process whereby initial aggregates (having dimensions of a few unit cells) in a solution develop spontaneously into particles of a new stable phase is known as nucleation. When these particles grow to size sufficient to scatter visible light, they are known as colloids. Dispersions of particles of colloidal size (colloidal solutions) are frequently stable, coalescence of the particles into larger aggregates settling under gravitational or centrifugal forces being prevented by their similar residual and mutually repelling charges. By changing the ionic environment in which colloidal particles exist, for example, by adding salts containing multicharged ions, the colloidal particles can be made to undergo further aggregation, or flocculation. Silver chloride is a good example of a compound which can be precipitated as a flocculated colloid. *See* COLLOID; NUCLEATION; PRECIPITATION; SOLUBILITY PRODUCT CONSTANT.

[LOUIS GORDON/ROYCE W. MURRAY]

Fluidization

The processing technique employing a suspension or fluidization of small solid particles in a vertically rising stream of fluid—usually gas—so that fluid and solid come into intimate contact. This is a tool with many applications in the petroleum and chemical process industries. There are major possibilities of even greater growth and utility in the future. Suspensions of solid particles by vertically rising liquid streams are of lesser interest in modern processing, but have been shown to be of use, particularly in liquid contacting of ion-exchange resins. However, they come in this same classification and their use involves techniques of liquid settling, both free and hindered (sedimentation), classification, and density flotation. *See* ION EXCHANGE.

Properties of fluidized systems. The interrelations of hydromechanics, heat transfer, and mass transfer in the gas-fluidized bed involve a very large number of factors which must be understood for mathematical and physical analysis and process design. Because of the excellent contacting under these conditions, numerous chemical reactions are also possible—either between solid and gas, two fluidized solids with each other or with the gas, or most important, one or more gases in a mixture with the solid as a catalyst. In the usual case, the practical applications in multimillion-dollar plants have far outrun the exact understanding of the physical, and often chemical, interplay of variables within the minute ranges of each of the small particles and the surrounding gas phase. Accordingly, much of the continuing development of processes and plants is of an empirical nature, rather than being based on exact and calculated design.

The fluidized bed results when a fluid, usually a gas, flows upward through a bed of suitably sized, solid particles at a velocity sufficiently high to buoy the particles, to overcome the influence of gravity, and to impart to them an appearance of great turbulence, similar to that of a violently boiling liquid. Fluid velocities must be intermediate between that which would lift the particles to maintain a uniform suspension and that which would sweep the particles out of the container. The fluidized bed is in a relatively stable condition of vigorous contacting of fluid and solids, with a lower boundary at the point of fluid inlet and an upper definite and clearly marked boundary surface at which the gas disengages itself from the system. The upper boundary usually goes up or down with increase or decrease of the gas velocity.

Particle sizes are often in the range of 30–125 μm, but they may vary to 1/4 in. (6 mm) or more. Superficial gas velocities may range 0.02–1.0 ft/s (0.006–0.3 m/s), depending on factors such as the relative densities of the gas and solid, the size and shape of solid particles, and the number of particles per cubic foot (bed density) desired.

Conditions within the bed are intermediate between that of a packed column and that of a pneumatic transporter of the particles. In pneumatic transport, the slippage of the gas past the solid particles is often relatively small, and the components of friction and inertia carry the solid particle along. In the fluidized bed, the slippage of gas past the solid particle is sufficient to balance the gravitational effect or weight, but is not sufficient to supply a continuing velocity.

An increase in fluid velocity increases the lift or upward drag on the particles. This causes them to rise, thereby increasing the bed voidage and the effective cross section available for gas passage and also decreasing the number of particles per unit volume and the density of the bed as a whole. This decreases the interstitial velocity to reduce the drag on the individual particles, which then settle somewhat until the gravitational and the fluid-dynamic forces on the particles ultimately come into balance. Further increase in fluid velocity causes further bed expansion until the bed passes beyond the limits of the containing vessel and the particles are transported.

There is excellent contact between the solid particles themselves and the particles with the gas. If there is a temperature difference, there is an excellent flow of heat between the particles themselves and between the particles and the gas. Particularly if heat is absorbed, as in drying of a solid by a hot gas, or if heat is given off because of chemical reaction, heat transfers to equate the temperatures of the gas and the solid very quickly.

Similarly, there is a major opportunity for mass transfer by molecular diffusion to or from the solid particles and between the solid particles and the gas phase. Examples are the movement of reactants in the gaseous phase toward the surface of solid particulate catalysts, and the reverse movement of products away from the catalytic surface back to the body of the gas phase.

Principal applications. With such excellent opportunities for heat and mass transfer to or from solids and fluids, fluidization has become a major tool in such fields as drying, roasting, and other processes involving chemical decomposition of solid particles by heat. An important application has been in the catalysis of gas reactions, wherein the excellent opportunity of heat transfer and mass transfer between the catalytic surface and the gas stream gives performance unequaled by any other system. In the petroleum industry, for example, the cracking of hydrocarbon vapors in the presence of solid catalysts affords many advantages. Among these are a minimum tendency for overheating of one particle or one part of the catalytic bed as compared to another, and the immediate removal of the reactant products, resulting in greatly increased efficiency of the catalyst. The catalyst itself often becomes carbonized on the surface because of the coking tendency of the reacting hydrocarbons. This catalyst, as shown in the diagram, may then be removed by pneumatic transport or otherwise to a second fluidized reactor wherein an oxidation is accomplished, again with maintenance of relatively uniform conditions. The rejuvenated catalyst may be returned to the prime reactor by pneumatic or other transfer lines, and the process can be continued indefinitely.

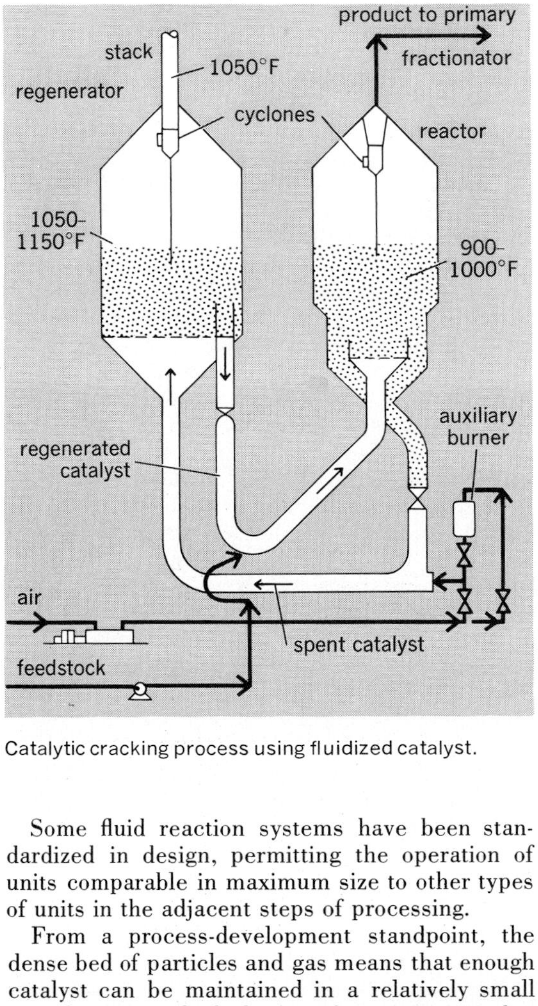

Catalytic cracking process using fluidized catalyst.

Some fluid reaction systems have been standardized in design, permitting the operation of units comparable in maximum size to other types of units in the adjacent steps of processing.

From a process-development standpoint, the dense bed of particles and gas means that enough catalyst can be maintained in a relatively small vessel to give a high degree of conversion with a low pressure drop causing gas flow. The mechanical design of a fluidization system and accessories is comparable, in many respects, to that of a body of liquid. However, special devices are needed to control fluid and solid handling, to introduce gas to the fluid bed, and to control the level of the solid particles, which usually are allowed to overflow the top of vertical pipes, as in maintaining the level of a true liquid. Vapor-disengaging units, filters, and cyclones, for removal of minute amounts of fines which might otherwise escape, blowers, pumps, pneumatic transport lines, and bucket elevators are also needed.

One important component in many units, besides the fluidizing chamber—often the reactor—is the standpipe, which allows a recycling of solids and builds up pressure on the solids for transfer purposes. In a large catalytic cracking unit, the rate of circulation of solids between vessels may be as much as 50 tons/min (0.75 metric ton/s).

Fluidized versus fixed beds. While catalytic particle–gas reactions have long been done in fixed beds, these cannot all be converted to fluidized operations. There are notable advantages and disadvantages of each. The advantages of the fluidized bed have been evidenced in the much better mixing, which results in greatly improved heat transfer, mass transfer, and uniformity. Thus there may often be better heat and material recovery, that is, increased yields, lower catalyst losses, and sometimes lower pressure drops. Fluidized beds allow ready additions of solids, combustion products, or direct combustion within the bed to give a much greater flexibility of process design and operation.

On the other hand, the concurrent flow in a fluidized reactor is disadvantageous, and multibed reactors are necessary to secure the advantages of countercurrent flow. There may be mixing of products with raw materials throughout the bed, and particularly when the solid is a reactant, it is difficult to obtain a high conversion and almost impossible to obtain a 100% conversion, except in a multibed system. Some attrition and thus loss of catalyst particles and changing fluidization properties with changing average size results. Recovery equipment for fines may be necessary, either because of their value or their nuisance as air pollutants if lost to the atmosphere. Similarly, there is considerable erosion of the walls, and especially of transfer lines. Because of the nature of fluidization, many reactions requiring a catalyst bed cannot be so operated, for example, those that give waxy or gummy materials which would agglomerate on the catalyst particles and also those wherein properties of the solids or of the gas stream, the necessary gas velocities, and so on, are outside of the mechanical possibilities of the fluidization operation. Thus, fixed beds are in many cases either required or most advantageous.

Individual industries. The simplest use of a fluidized bed is as a simple heat exchanger, with direct contact between solids and fluid—either liquid or gas—possibly to recover heat from a hot gas or from a hot solid by contacting with a cold solid or a cold gas, respectively. Such a heat exchanger, with no chemical reactions, is often used as a supplementary fluidized bed to one wherein there is a chemical reaction. Within a single bed, there is no possible countercurrent action, although multiple beds can be used and their particle flow arranged countercurrently to that of the fluid.

A basic example of diffusion is in the use of particles of an adsorbent, such as activated carbon, to remove vapors of solvent from a fluidizing stream of carrier gas, such as air. The air loses the vapor molecules by their diffusion first through a gas film surrounding the particles and then into the solid mass of the particles. The solvent-charged particles overflow the top of the bed to form a second bed, fluidized usually with steam to desorb the solvent. The steam-solvent mixture is separated by distillation, and the solvent-free particles are recycled to the first bed. Obviously, this example of mass transfer of molecules by diffusion to and from the particles also includes a transfer of heat in both beds.

A similar bed without chemical reaction is used as a dryer of solid particles. In this example, the moisture must diffuse to the surface of the particles, from which it is evaporated into the gas stream. It must again be noted that, because the flows in the bed are not countercurrent, a single unit cannot achieve complete dryness, except with an excess in heat consumption.

Very similar is the removal of chemically combined water and carbon dioxide, or both, in what is usually called calcination, at temperatures about 1400–2200°F (760–1200°C). Multiple beds in countercurrent are used to minimize heat requirement. A necessary added expense is the grinding of limestone to a size which is practical to fluidize, usually not over 1/4 in. maximum dimension. The crushed limestone descends from a top-drying fluidized bed through a series of beds to one where combustion of fuel oil is also taking place to supply the heat for calcination both in this bed and in the higher beds. The lowest bed receives and cools the burnt lime with the incoming air for combustion, which is thus preheated.

Roasting of pyrites to give sulfur dioxide is practiced throughout the world in possibly hundreds of fluidized beds; the combustion operation may be controlled better than in other types of roasting furnaces. Similarly, additional roasters are used to convert the iron oxide from hematite to magnetite, which then is heated to give pellets for steel manufacture.

Other sulfide ores are roasted in closely controlled fluidized beds to give metal sulfates, which may then be leached to recover their metal values from the gangue.

There are other examples of fluidized beds where there are more complicated chemical reactions between gases and solid. Included are the carbonization of wood and of coal to form charcoal and coke with concomitant volatile products, the combustion of pulverized coal and other solid fuels, and the production of portland cement clinker. More modern applications have been the production of titanium chloride in a bed of a ground titanium mineral, rutile, and coke fluidized with chlorine, and the production of uranium compounds for nuclear fuels.

Fluid coking is an interesting cracking process, in which the undistillable residues of petroleum towers are coked in a fluidized reactor at about 1000°F (540°C) to give coke and a distillate. Coke deposits on the fluidized particles of seed coke, prepared by fine grinding of some of the product, and these particles are built up to the desired product sizes for withdrawal and sale.

The greatest use in number and size of fluidized reactors and in volume of throughput has been in catalytic cracking and reforming of petroleum and its fractions. Scores of variations have developed the processing by particles of catalyst petroleum vapor streams. The modern petroleum industry, could not otherwise operate under present conditions and costs. *See* CATALYSIS; CRACKING; GAS ABSORPTION OPERATIONS; MASS-TRANSFER OPERATION.

[DONALD F. OTHMER]

Bibliography: J. F. Davidson (ed.), *Fluidization*, 1978; J. R. Grace and J. M. Matsen (eds.), *Fluidization*, 1980; D. Kunii and O. Levenspiel, *Fluidization Engineering*, 1969, reprint 1977; F. A. Zenz, Fluidization, in *Encyclopedia of Chemical Technology*, 3d ed., vol. 10, pp. 548–581, 1980.

Fluids

Substances with no reference configuration of permanent significance. Aggregates of matter in which the molecules are able to flow past each other without limit and without fracture planes forming are usually classified as fluids. The subdivisions of fluids known as gases, vapors, and liquids, each of which exhibits successively closer association of the molecules and is distinguished by different thermodynamic and mechanical properties, compose a group of easily distinguishable fluids, some having mainly Newtonian and the rest mainly non-Newtonian flow properties.

Solid aggregates of matter, whether continuous or divided in the form of powders, can usually be distinguished from fluids on the molecular scale in that long-term and long-range order are apparent in the structure.

There are some substances, the semisolids, which appear to be able to flow without fracture and yet which also have some of the attributes of solids such as the ability to form free-standing figures. Butter in a temperate environment is an excellent example of a material with this dual nature.

Similarly, some fluids possess elastic properties, the so-called viscoelastic fluids. Such substances are not easy to classify in one or another category, and their properties are not well understood in depth.

Gases. Gases at low density are quite amenable to the theoretical treatment of J. C. Maxwell in which the molecules are visualized as rigid, nonattracting spheres in relative motion, as shown in Fig. 1.

The theoretical expression for the viscosity μ of a gas, derived from considerations of momentum exchange between the spheres in different planes, is given by Eq. (1), where m is the mass of

$$\mu = \frac{2}{3\pi^{3/2}} \frac{\sqrt{mkT}}{d} \qquad (1)$$

one molecular sphere, k is the Boltzmann constant, T is the absolute temperature, and d is the sphere diameter.

An important consequence of Eq. (1) is that the viscosity of a low-density gas is proportional to the square root of the absolute temperature but is in-

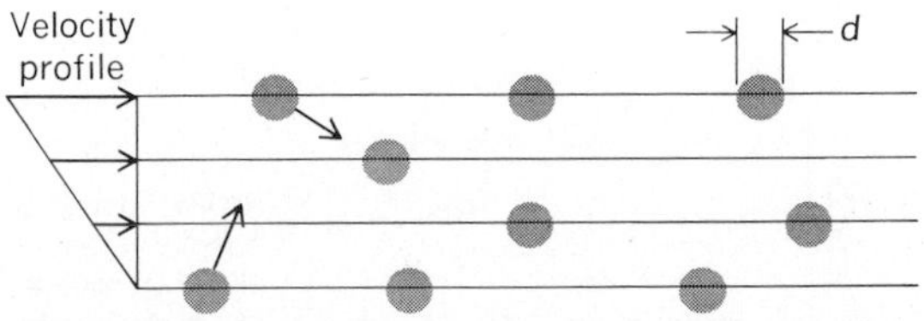

Fig. 1. Maxwell's molecular system for gases.

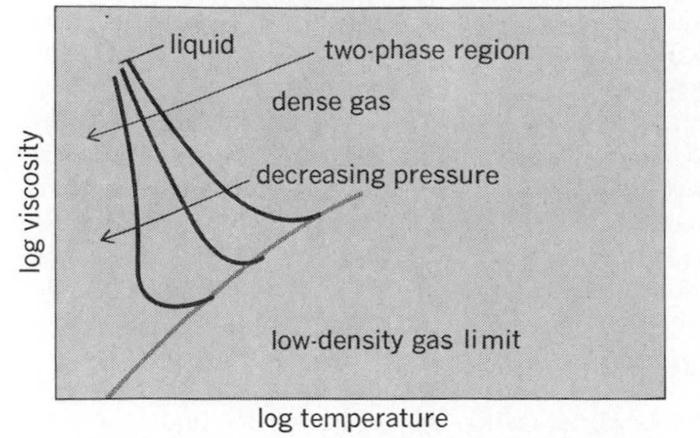

Fig. 2. The effect of varying temperature and pressure upon the fluid viscosity.

dependent of pressure. This is radically different from the behavior of liquids.

As condensation conditions are approached, the viscosity starts to differ significantly from that predicted by Eq. (1). The temperature exponent of 0.5 increases to between 0.6 and 1.0, and the gas viscosity is *no longer independent of pressure.* The general behavior of the viscosity as the temperature and pressure are varied is shown in Fig. 2. The viscosity of gas mixtures is generally not simply related to the viscosity of the components and the mole fractions present. *See* GAS.

Liquids. An approximate theory which enables the viscosity of liquids to be estimated from other physical properties is based on an application of the theory of rate processes developed by E. Eyring and coworkers. In this theory it is assumed that migration of liquid molecules occurs by their achieving enough vibrational energy to surmount the barrier surrounding their current site. *See* LIQUID.

In a state of rest the energy barrier is assumed symmetrical, but under the action of applied stress the barrier is distorted so that the molecule is biased toward jumping in the forward direction (Fig. 3). The relationship between the shear stress τ and the shear rate $\dot{\gamma}$ derived by Eyring and coworkers is given by Eq. (2), where ΔG_0 is a free

$$\dot{\gamma}=\frac{a}{\delta}\left(\frac{kT}{h}\exp-\frac{\Delta G_0}{RT}\right)\left(2\sinh\frac{a\tau\bar{v}}{2\delta RT}\right) \qquad (2)$$

energy of activation indicated in Fig. 3, h is Planck's constant, k is the Boltzmann constant, $\bar{v}$ is the molar volume, R is the gas constant, T is the absolute temperature, and a and δ are molar spacings defined in Fig. 3.

Note that generally the shear stress–shear rate of relation (2) is non-Newtonian. For fluids of *small molecular weight* it is known from the amassed experimental data that Newton's viscous law, Eq. (3), is generally appropriate for all isothermal

$$\tau=\mu\dot{\gamma} \qquad (3)$$

flows. The exceptional condition of periodic stresses of megahertz frequency may find Eq. (3) inadequate even for small molecules.

The Eyring formula, Eq. (2), predicts that the stress will be in phase with the shear rate, *but this is not generally found in non-Newtonian flow.*

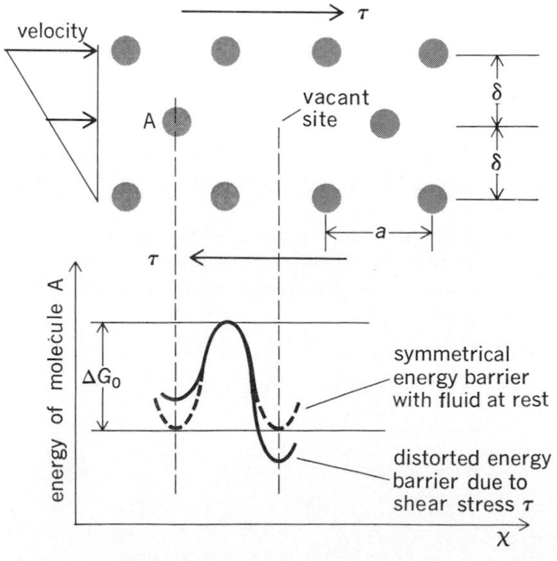

Fig. 3. Eyring molecular system for liquids.

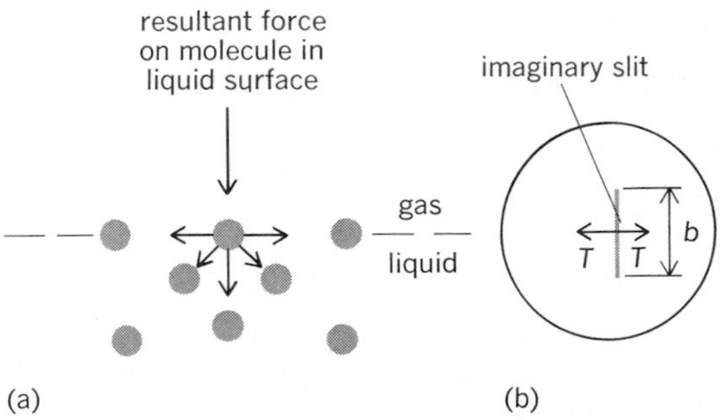

Fig. 4. Fluid-surface phenomena. (*a*) Forces (attractive on one hemisphere only) on a liquid molecule at a gas-liquid interface. (*b*) Surface tension which is due to molecular forces.

The temperature dependence of the viscosity can often be described over limited temperature ranges by the exponential relation shown in Eq. (4).

$$\mu=A\exp\frac{E}{RT} \qquad (4)$$

The effect of high pressure can also often be described by the same type of relation, with pressure replacing temperature in Eq. (4). This is an important consideration in the field of lubrication.

Particle suspensions. The addition of rigid, noninteracting spheres to a Newtonian fluid increases the viscosity according to the Einstein equation, Eq. (5), where c is the volume concentration of the

$$\mu=\mu'(1+2.5c) \qquad (5)$$

spheres, μ' is the viscosity of the suspending liquid, and μ is the viscosity of the suspension. For concentrated solutions a power series in the volume concentration is required.

Fluid surfaces. A molecule of liquid in a gas-liquid interface finds itself in a different force field from those in the core of the gas or the liquid. This results from the surface molecule having attractive forces only on one hemisphere, as shown in Fig. 4*a*. The resultant effect is that the liquid surface at any interface appears to be in a state of tension, like the surface of an inflated balloon.

An imaginary slit in the surface of Fig. 4*b* would have a tensile force per unit length, given by Eq. (6).

$$t = \frac{T}{b} \tag{6}$$

The well-known capillary rise in a fine bore tube (Fig. 5) results from surface tension effects and is a common method for measuring surface tension from Eq. (7), where ρ is the liquid density, g is the

$$t = \frac{\rho g H r}{2 \cos \theta} \tag{7}$$

gravitational acceleration, H is the capillary meniscus elevation, and r is the capillary tube radius.

Equation (7) yields a static surface tension, but if there is shearing in the surface, then surface viscosity will be exhibited as a two-dimensional analog of bulk viscosity. Furthermore, if macromolecules are present in the surface, shear elasticity will exhibit itself as a surface phenomenon.

Diffusion. The tendency for a substance to achieve uniform distribution throughout the space available to it is known as diffusion. It is exhibited by all classes of matter, whether gas, liquid, or solid, but is most vigorous in the case of gases. The basic demonstration is the placing together of the mouths of two jars, one containing a heavy gas, say nitrogen which is placed in the bottom jar, and the other a light gas, say hydrogen which is placed in the top jar. After a time has elapsed, the hydrogen and nitrogen will be found to be evenly distributed throughout the two jars.

Graham's law of diffusion, which is an approximation, states that the rate of diffusion of a gas is inversely proportional to the square root of the density, as in expression (8). Diffusion separation is

$$D \propto \left(\frac{1}{\rho}\right) \tag{8}$$

an important process in the field of chemical engineering.

[J. HARRIS; W. L. WILKINSON]

Bibliography: R. A. Alberty and F. Daniels, *Physical Chemistry*, 5th ed., 1979; G. Barrow, *Physical Chemistry*, 4th ed., 1979.

Fluoborate

In the broadest sense, the term fluoborate refers to a group of compounds related to the borates in which one or more oxygens have been replaced by fluorines. In a strict sense, fluoborate refers to the BF_4^- ion which is derived from tetrafluoroboric acid, HBF_4. Many other fluoborates are known only in the form of salts. Some of the postulated parent acids are H_2BF_5, $H_2B_2F_6$, $H_4B_4F_{10}$, H_2BOF_3, H_2BO_2F, and H_3BO_2F. The lithium, sodium, ammonium, alkaline-earth, and heavy-metal salts of these acids are soluble. The fluoborates are used as fluxes and in electroplating. *See* BORATE; BORON; FLUORIDE; FLUORINE. [E. EUGENE WEAVER]

Fluoride

A compound derived from hydrofluoric acid, HF, in which the fluorine atom is in the 1− oxidation state.

The fluorides are different in many respects from the other halides. *See* CHLORIDE; HALIDE.

The small size of the fluoride ion means that the extra electron is held tightly; it is difficult to convert the fluoride ion to free fluorine. The small size also means that there is less tendency to form covalent bonds in simple salts such as mercuric fluoride, HgF_2, which has an ionic lattice, whereas its chlorine analog, mercuric chloride, $HgCl_2$, has a molecular crystal. In direct contrast to this, many fluorides are known in which the central atom is in a higher oxidation state than that observed for other halides and these compounds are often covalent, for example, vanadium pentafluoride, VF_5, chromium tetrafluoride, CrF_4, bismuth pentafluoride, BiF_5, and sulfur hexafluoride, SF_6.

A group of particular interest is the hexafluorides. They are very volatile for their relatively high molecular weight of about 300 g/mole. Hexafluorides of sulfur, selenium, tellurium, molybdenum, tungsten, rhenium, osmium, iridium, platinum, uranium, neptunium, and plutonium are known. Xenon hexafluoride, XeF_6, has been prepared only recently and demonstrates the ability of the noble gas family elements to form compounds.

The fluoride ion is involved in many stable coordination complex ions such as fluoborate, BF_4^-, fluoaluminate, AlF_6^{3-}, fluosilicate, SiF_6^{--}, hexafluophosphate, PF_6^-, and hexafluoroferrate, FeF_6^{3-}. Fluorine also replaces oxygen in anions to give fluorophosphates, fluoborates, fluosilicates, and fluosulfonates. Certain esters of fluophosphoric acid are known as nerve gases, and others have been developed as effective insecticides.

The solubilities of the inorganic fluorides are somewhat different from those of other halides; notably, silver fluoride, AgF, thallous fluoride, TIF, and mercurous fluoride, Hg_2F_2, are much more soluble whereas magnesium fluoride, MgF_2, calcium fluoride, CaF_2, strontium fluoride, SiF_2, barium fluoride, BaF_2, and lithium fluoride, LiF, are much less soluble than the corresponding chlorides.

Of the naturally occurring fluorine compounds, cryolite, Na_3AlF_6, is important in aluminum metallurgy and fluorspar, CaF_2, as a flux in metallurgy.

Sodium fluoride is used as an insecticide and rodenticide. Since a small amount of F^- ion (0.8–1.6 ppm) helps to prevent cavities in teeth, the fluoridation of municipal water is widespread.

Fluoride may be detected analytically by the addition of sulfuric acid. The hydrogen fluoride evolved will etch glass. *See* FLUOBORATE; FLUOROCARBON; FLUOSILICATE; HALOGENATED HYDROCARBON; INERT GASES; ORGANOPHOSPHORUS COMPOUND; POLYFLUOROOLEFIN RESINS.

[E. EUGENE WEAVER]

Fluorine

A chemical element, symbol F, atomic number 9, the member of the halogen family that has the lowest atomic number and atomic weight. Although only the isotope with atomic weight 19 is stable, the other, radioactive isotopes between atomic weight 17 and 22 have been artificially prepared and have half-lives between 4 seconds for ^{22}F and 110 min for ^{18}F. Fluorine is the most electronegative element, and by a substantial margin the most chemically energetic of the nonmetallic elements.

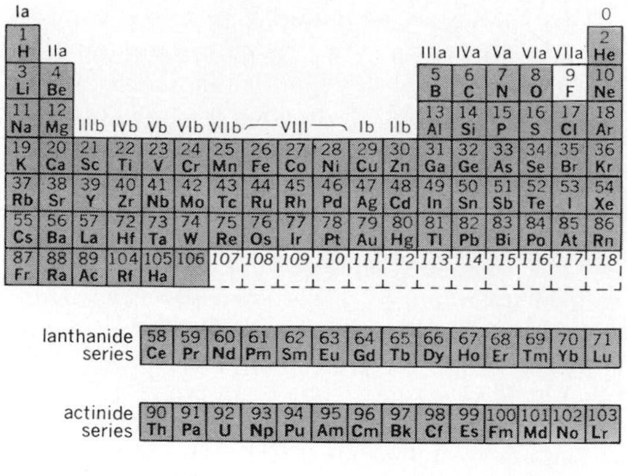

Consequently, it enters into combination with most other chemical elements, and is found in nature only in such combination. Binary fluorides have been demonstrated for all the available elements except helium, neon, and argon. *See* HALOGEN ELEMENTS.

Properties. The element fluorine is a pale yellow gas at ordinary temperatures. The odor of the element is somewhat in doubt. A very distinctive musty odor is detected whenever the element is present in the atmosphere. Since this odor is the same as that of oxygen difluoride, which results from the reaction of fluorine with water, there is a question as to whether the element itself has ever been smelled. Some physical properties are listed in the table. J. Berkowitz and A. C. Wahl have discussed the dissociation energy of fluorine and resolved the discrepancy by reinterpreting the earlier experimental approaches using the best available experimental and theoretical data and bringing them into conformity with the more recent photoionization measurements. The current best value is given in the table.

Physical properties of fluorine

Property	Value
Atomic weight	18.998403
Boiling point, °C	−188.13
Freezing point, °C	−219.61
Critical temperature, °C	−129.2
Critical pressure, atm	55
Density of liquid at b.p., g/ml	1.505
Density of gas at 0°C + 1 atm, g/l	1.696
Dissociation energy, kcal/mol	36.8
Heat of vaporization, cal/mol	1510
Heat of fusion, cal/mol	121.98
Transition temperature (solid), °C	−227.61

The electron configuration of the fluorine atom is $1s^2 2s^2 2p^5$. The tendency to complete the outer shell with eight electrons is the driving force for the extreme chemical activity of fluorine. The valence of fluorine in compounds is always −1.

The reactivity of the element is so great that it will react readily at ordinary temperatures with many other elementary substances, such as sulfur, iodine, phosphorus, bromine, and most metals. Since the products of the reactions with the nonmetals are in the liquid or gaseous state, the reactions continue to the complete consumption of the fluorine, frequently with the evolution of considerable heat and light. Reactions with the metals usually form a protective metallic fluoride which blocks further reaction, unless the temperature is raised. Aluminum, nickel, magnesium, and copper form such protective fluoride coatings.

On steel the film is less protective and may grow thicker, but fluorine can be stored even in ordinary steel gas cylinders. The fluorine molecule is readily dissociated, and therefore is even more reactive. In addition to heat and the electric discharge mechanism, electromagnetic radiation well up into the visible range in wavelength is energetic enough to initiate reaction, even with the noble gas xenon. Fluorine reacts with considerable violence with most hydrogen-containing compounds, such as water, ammonia, and all organic chemical substances whether liquids, solids, or gases. The reaction of fluorine with water is very complex, yielding mainly hydrogen fluoride and oxygen with less amounts of hydrogen peroxide, oxygen difluoride, and ozone. Fluorine displaces other nonmetallic elements from their compounds, even those nearest fluorine in chemical activity. It displaces chlorine from sodium chloride, and oxygen from silica, glass, and some ceramic materials. In the absence of hydrofluoric acid, however, fluorine does not significantly etch quartz or glass even after several hours at temperatures as high as 200°C.

Fluorine is a very toxic and reactive element. Many of its compounds, especially inorganic, are also toxic and can cause severe and deep burns. Care must be taken to prevent liquids or vapors from coming in contact with the skin or eyes.

Production of the element. The preparation of the element can be accomplished satisfactorily only by means of an electric current. No chemical procedure has been found to produce the element in more than very small amounts. The essential feature of all electrical methods is the electrolysis of liquid hydrogen fluoride that has been made conducting by addition of an alkali fluoride. Potassium fluoride is the salt used for this purpose, and there are three temperature ranges in which potassium fluoride—hydrogen fluoride mixtures are liquid. The first isolation of the element by H. Moissan in 1886 employed electrolysis of liquid hydrogen fluoride at a temperature below 0°C in a platinum vessel with platinum electrodes. In the range 230–320°C an electrolyte of the approximate composition of 1 mole of hydrogen fluoride per mole of potassium fluoride can be used to produce fluorine in iron or copper vessels, using a graphite electrode for the anode. A United States team of scientists discovered this procedure during World War I. Great difficulty resulted from the failure of the electrical insulator at the high temperature in the presence of both elementary fluorine and hydrogen fluoride. Two hours was about as long as this operation could be continued before the insulation failed. In 1923 at the University of California, it was found that portland cement made an excellent insulator for use at this high temperature. This permitted the high-temperature fluorine-producing cells to be operated continuously.

A typical modern plant producing more than 3 tons (2.7 metric tons) of fluorine per day might use 50 or more cells with 32 carbon anodes and steel cathodes, and a current density of 140 amperes/ft^2. The electrolyte is nominally $2HF \cdot KF$, and is used at a temperature of about 100°C. Large users, such as the producer of uranium hexafluoride, pipe fluorine directly to their process reactors. Where this is impractical, as in the rocket industry, shipments of liquid fluorine may be made by insulated tank truck, while for laboratory use fluorine is most conveniently shipped in mild-steel (carbon steel with a maximum of about 0.25% carbon) tanks as a low pressure (300 psi; 1 psi = 6895 Pa) gas.

While many fluorides, for example, halogen fluorides and xenon fluorides, are still made from the element, many compounds can be more conveniently synthesized by way of reactive intermediates.

Natural occurrence. At an estimated 0.065% of the Earth's crust, fluorine is roughly as plentiful as carbon, nitrogen, or chlorine, and much more plentiful than copper or lead, though much less abundant than iron, aluminum, or magnesium. Compounds whose molecules contain atoms of fluorine are widely distributed in nature. Many minerals contain small amounts of the element, and it is found in both sedimentary and igneous rocks. Deposits of fluorspar (calcium fluoride), the chief ore, are found in many parts of the surface of the Earth. Large deposits of high-grade ore are not plentiful, but there are many low-grade or small deposits. Cryolite, Na_2AlF_6, which is used both in the ceramic industries and the metallurgy of aluminum, is found in quantities in only a few places. Fluorapatite, $Ca_5F(PO_4)_3$, frequently called rock phosphate, is the most plentiful mineral. In many minerals the fluoride ion (size 1.36 A, or 0.136 nm) is randomly inserted, replacing the hydroxyl ion (1.40 A, or 0.140 nm), or O^{--} ion with an appropriate cation balancing charge.

Uses. The compounds of fluorine have been used since humans' earliest attempts to chemically modify the materials of the environment. Since early metallurgy, ceramics and glass working were limited to materials that could be melted at the relatively low temperatures available, and the discovery of a mineral that made melts and slags more fluid at low temperatures was an important technological advance. This fluorine-containing mineral, which was called flores and is now called fluorspar or fluorite, is the source of the name of the element.

Fluorine-containing compounds are used to increase the fluidity of melts and slags in the glass and ceramic industries. Fluorspar is introduced into the blast furnace to reduce the viscosity of the slag in the metallurgy of iron. Cryolite is used to form the electrolyte in the metallurgy of aluminum. Aluminum oxide is dissolved in this electrolyte, and the metal is reduced electrically from the melt. *See* ALUMINUM.

The use of halocarbons containing fluorine as refrigerants was patented in 1930, and these volatile and stable compounds have found a growing market in aerosol propellants as well as in refrigeration and air-conditioning systems. Although halogens naturally exist in the atmosphere, the extreme chemical stability of chlorofluorocarbons erupted in a major controversy in 1974 concerning their use as aerosol propellants and their effect on atmospheric pollution. When released into the atmosphere, they rise undecomposed to the stratosphere, where the increased intensity of ultraviolet radiation causes decomposition to chlorine atoms and other products. The chlorine is reported to deplete the ozone layer, which limits the intensity of the Sun's ultraviolet radiation reaching the Earth's surface. Greater ultraviolet penetration to the Earth could result in increased skin cancer, weather modification, and agricultural damage.

Newer uses for fluorine compounds became prominent during World War II. In the enrichment of the fissionable isotope ^{235}U, the most important process employed uranium hexafluoride. This stable, volatile compound was by far the most suitable material for isotope separation by gaseous diffusion. Enrichment plants using uranium hexafluoride remain a key factor in economical production of nuclear energy.

J. H. Simons and coworkers demonstrated the catalytic use of hydrogen fluoride in organic reactions. The catalytic alkylation of isoparaffins found wide use in the production of high-octane aviation fuel needed for World War II planes.

While consumers are mostly unaware of the fluorine compounds used in industry, some compounds have become familiar to the general public through minor but important uses, such as additive to toothpaste and nonsticking surfaces on frying pans and razor blades. With Teflon, the family of fluorocarbon polymers has become a familiar household word.

Because of the strong affinity of fluorine for bone tissue, the 110 minute half-life radioactive isotope, ^{18}F, has found ever-increasing use in medical research and clinical diagnosis. The short half-life compared to ^{85}Sr, also used for bone scans, permits use of larger doses with little radiation exposure. In addition, ^{18}F accumulates in the bone much more rapidly than either strontium or calcium, making it very useful as a tracer for bone scintigraphy. ^{18}F-labeled chemicals are used for early diagnosis of bone diseases, to test the effectiveness of toothpaste, and for the development of new radiopharmaceuticals. ^{18}F is conveniently prepared by bombardment of normal ^{19}F with energetic neutrons of greater than 10 MeV, the binding energy of a neutron in fluorine, in a cyclotron or linear accelerator. The nuclear reaction $^{19}F(n, 2n)$-^{18}F produces ^{18}F.

In rocket propellant systems, fluorine provides more energy per pound than oxygen in burning fuels. While pure fluorine is most energetic, mixtures with liquid oxygen or reactive fluorine compounds may have engineering advantages.

Inorganic compounds. In all fluorine compounds the high electronegativity of this element suggests that the fluorine atom has an excess of negative charge. It is convenient, however, to divide the inorganic binary fluorides into saltlike (ionic lattice) nonvolatile metallic fluorides and volatile fluorides, mostly of the nonmetals. Some metal hexafluorides and the noble-gas fluorides show volatility that is frequently associated with a molecular compound. Volatility is often associated with a high oxidation number for the positive element.

The extreme chemical properties of fluorine are expected from its position in the upper-right corner of the periodic table. While in some ways fluorine is a typical halogen, the difference between fluorine and the next halogen, chlorine, is so large that the properties of fluorine-containing compounds cannot be predicted by analogy with the compounds of chlorine. A measure of the metallic or nonmetallic properties of an element is indicated by its molal electrode potential. When this experimentally determined number is negative, it can be associated with the nonmetallic properties. On this basis fluorine (−2.85 V) is farther away from chlorine (−1.36 V) than chlorine is from hydrogen (0.00 V). Also, its low dissociation energy relative to that for chlorine helps explain the greater reactivity of fluorine. Differences between fluorine and chlorine take many forms. The fluoride ion is most difficult to polarize, and differing trends in the lattice stability and solvation energy for chloride and fluoride explain the high solubility of silver fluoride and the low solubility of silver chloride, while the corresponding chlorides and fluorides of barium, calcium, and aluminum have reverse solubility behavior. More important is the striking difference in behavior between the chlorine-substituted hydrocarbons, with a relatively weak carbon chlorine bond and relatively few examples of stable substitution products, and the unlimited variety of compounds with very stable carbon fluorine bonds. The latter compounds are discussed below.

The metals characteristically form nonvolatile ionic fluorides where electron transfer is substantial and the crystal lattice is determined by ionic size and the predictable electrostatic interactions, for example, NaF, BaF_2, LaF_3, and ZrF_4. When the coordination number and valence are the same, for example, BF_3, SiF_4, and WF_6, the binding between metal and fluoride is not unusual, but the resulting compounds are very volatile, and the solids show molecular lattices rather than ionic lattice structures. For higher oxidation numbers, simple ionic lattices are less common and, while the bond between the central atom and fluorine usually still involves transfer of some charge to the fluorine, molecular structures are identifiable in the condensed phases. If the symmetry is high, as for hexafluorides in particular, the compound is volatile with a short liquid range (often the solid sublimes at atmospheric pressure). For pentafluorides, on the other hand, there is usually a long liquid range, a tendency to form polymeric viscous liquids through fluorine bridges, and a tendency to form complexes by adding fluoride ions to form symmetrical hexafluoroanions.

In addition to the binary fluorides, a very large number of complex fluorides have been isolated, often with a fluoroanion containing a central atom of high oxidation number. The binary saltlike fluorides show a great tendency to combine with other binary fluorides to form a large number of complex or double salts. Cryolite, $3NaF.AlF_3$, is one such compound, best considered as a sodium fluoroaluminate.

The long-sought hypofluorous acid, HOF, was recently prepared by circulating fluorine through wet Teflon rings and condensing the HOF as a white solid. It has the smallest known unconstrained oxygen bond angle, 97°, which is explained by the measured $-0.5\ e$ charge on the fluorine and the $+0.5\ e$ charge on the hydrogen.

The extremely electronegative behavior of fluorine is best illustrated in the volatile compounds it forms with nonmetals: the halogen fluorides, ClF, ClF_3, ClF_5, BrF, BrF_3, BrF_5, IF_5, and IF_7; the oxygen fluorides, OF_2, O_2F_2, and O_4F_2; and the noble-gas fluorides, XeF_2, XeF_4, XeF_6, and KrF_2. Volatile compounds are also formed with the other nonmetals, nitrogen, sulfur, and phosphorus. Most of these fluorides are rather reactive, substituting fluorine into oxides and halides, fluorinating metals, and in many cases proving more reactive than fluorine itself.

As water is to oxygen, hydrogen fluoride is to fluorine: a solvent, an intermediate in synthesis, and in many respects, a most interesting and important compound. Boron trifluoride is a gas with unusual properties as a catalyst for organic chemical condensation reactions. The silicon fluorides are volatile. The best-known one, SiF_4, is the parent compound of the fluorosilicates in which it is combined with metallic fluorides.

Nearly all the volatile fluorides react with water to produce the oxyfluorides or oxides of the elements in addition to hydrogen fluoride. To this, however, there are several exceptions. If the fluorines fit tightly around the central atom to prevent access by other atoms, the compound is resistant not only to hydrolysis but to many other chemical reactions. Sulfur hexafluoride is an example of such a compound. It is highly unreactive, although sulfur tetrafluoride hydrolyzes and undergoes other reactions readily. Carbon tetrafluoride is another example of a compound that is resistant to chemical reactions for the same reason. Its related compound, silicon tetrafluoride, reacts readily with water. In this compound the central atom is larger, and therefore, the four fluorine atoms surrounding it do not protect it as well.

Organic compounds of fluorine. The fluorine-containing compounds of carbon can be divided into fluorine-containing hydrocarbons and hydrocarbon derivatives (organic fluorine compounds) and the fluorocarbons and their derivatives. The fluorine atom attached to the aromatic ring, as in fluorobenzene, is quite unreactive. In addition, it reduces the reactivity of the molecule as a whole. Dyes, for example, that contain fluorine (or the CF_3 group which is equivalent to a fluorine atom) attached to the aromatic ring are more resistant to oxidation and are more light-fast than dyes that do not contain fluorine. Most aliphatic compounds, such as the alkyl fluorides, are unstable and lose hydrogen fluoride readily. These compounds are difficult to make and to keep and are not likely to become very important.

Fluorocarbons. Compounds that contain only carbon and fluorine are called fluorocarbons. If they contain other elements, they are fluorocarbon derivatives. They are not considered organic compounds in that they do not participate in the chemical reactions that have become the basis of organic chemistry. The replacement of hydrogen by fluorine greatly increases the chemical and thermal stability of the molecule because of the high bond energy of the C-F bond and the greater shielding effect of the fluorine on the carbon backbone. Together with the ability of the carbon atom to form a large variety of structural chains in combination with other carbon atoms, this increased stability gives rise to a large number and variety of fluorocarbons and fluorocarbon derivatives.

There is a still more numerous class of fluorine-containing compounds of carbon. Its molecules contain parts which are essentially fluorocarbon in nature and also parts which are organic or hydrocarbon in nature. There are potentially many millions of such hybrid molecules. The fluorocarbon parts retain their physical properties and chemical inertness, whereas the organic parts retain their chemical activity for organic reactions. Such compounds are useful for purposes demanding properties intermediate between the fluorocarbon and the organic. They are particularly useful in cases in which the addition of an organic part to a fluorocarbon molecule can provide reactivity toward, or compatibility with, organic substances. The organic part provides means by which this molecule can be attached to some organic material, such as the fibers of cloth or paper. This is advantageous since it provides the organic material with the desirable surface properties of the fluorocarbons.

Fluorocarbons, like hydrocarbons, do not mix with polar liquids, such as water; in addition, fluorocarbons do not mix with hydrocarbons either. As a third nonmiscible phase, not found in nature, they provide a system which excludes most natural products. Fluorocarbon components and, particularly, the hybrids with fluorocarbon fragments therefore give stain and soil resistance to many solids, reduce the surface tension of liquids, reduce evaporation and flammability of volatile organic liquids, and improve the leveling of emulsions. They are, therefore, being used in a growing number of commercial products, such as soil repellants in textiles and paper, as surfactants in application of waxes, and as components in chromium-plating solutions.

The fluorocarbons and their derivatives are a relatively recent addition to the classes of chemical compounds. Despite earlier studies of the reaction of carbon and carbon-containing compounds, the simplest compound of the elements, CF_4, was obtained pure only in 1926; the second, C_2F_6, was identified in 1930; and the third, $CF_2{=}CF_2$, was prepared and identified in 1933. The demonstration of the fluorocarbons as an entire field of chemical substances by the isolation and identification of members of homologous series of substances came only in 1937.

Fluorocarbons were first prepared by the reaction of the elements fluorine and carbon with the help of mercury as a catalyst. Since this method produces a mixture of a very large number of gaseous, liquid, and solid substances which are difficult to separate, other means of producing individual compounds have been sought. Numerous complex chemical procedures have been devised. In one of these the element fluorine is caused to react with organic compounds with the aid of a catalyst and under exacting physical conditions. In another, the element fluorine is first used to produce metallic fluorides, such as silver difluoride or cobalt trifluoride, and these are than caused to react with the organic substance. These two methods have produced many compounds in the laboratories but seem to be unsuited for large-scale industrial use. Xenon difluoride has been shown to be a specific fluorinating agent. It is reported to be a useful reagent which effects substitution of fluorine for hydrogen in aromatics and substituted aromatics for good yields of fluoroaromatics.

One method of producing fluorocarbons that is currently used industrially starts with the production of C_2F_4. This is obtained from chloroform, CCl_3H, by a replacement reaction which produces CF_2ClH. This compound loses HCl upon heating, and $CF_2{=}CF_2$ is formed. This tetrafluoroethylene can combine with itself to form a useful polymer, called Teflon TFE, which is a chemically inert plastic. Since Teflon TFE is a thermal setting plastic, products are produced by heating under pressure through a powder-sinter-gel stage; coating may be applied as emulsions. The final plastic surface has a low coefficient of friction. Since few foreign objects stick to such surfaces, they are reasonably easy to clean, and Teflon surfaces find increasing use in kitchen utensils, as well as bearings, skis, and razor blades. The copolymer of the perfluorinated propylene and ethylene, Teflon FEP, is thermoplastic and can be softened and worked at elevated temperatures. These polymers tend to be more transparent than Teflon TFE, and are essentially similar in surface and chemical resistance properties.

Another useful plastic is polychlorotrifluoroethylene which is transparent and thermoplastic at lower temperatures, and quite chemically resistant, although much less so than the fully fluorinated polymers. Although not as low in surface friction, it has greater dimensional stability and does not "cold flow" under pressure as does Teflon TFE.

A more flexible industrial method of producing fluorocarbons and many derivatives is the electrochemical process. In this process an electric current is passed through a liquid mixture of hydrogen fluoride and some compound of carbon in a metal container in which there are metallic electrodes. The anode is nickel. Hydrogen is produced; and if the fluorocarbon product is gaseous, it escapes with the hydrogen and is separated therefrom. If it is liquid, it sinks to the bottom of the vessel and is readily removed, since it is insoluble in the hydrogen fluoride liquid. The fluorocarbon product frequently does not have the structure of the organic starting material. The material to be used for a specified product is known only from experience. This process produces not only the fluorocarbons proper, whose molecules contain only carbon and fluorine atoms, but substances containing oxygen, nitrogen, and sulfur in addition to carbon and fluorine. There are also compounds containing small percentages of hydrogen in the products from the process.

The saturated fluorocarbons are resistant to chemical attack by most other chemical reagents, whether these be acidic or basic, oxidizing or reducing up to relatively high (about 600°C) temperatures. Compounds in which the number of fluorine atoms is insufficient to satisfy all the valence bonds of the carbon atoms, and which would be called unsaturated in organic chemistry, sometimes exhibit the properties of organic olefins and sometimes do not.

Fluorocarbon derivatives. The oxygen-containing fluorocarbons can be considered to be derived from OF_2, in which one or both fluorine atoms are replaced by fluorocarbon radicals. CF_3COF is a very reactive chemical, but CF_3OCF_2 or higher-molecular-weight fluorocarbon oxides, such as $C_3F_7OC_2F_5$, are almost as unreactive chemically as the parent fluorocarbons. Although of similar chemical formula, these compounds are chemically unlike the organic ethers. Compounds of the type $R_1R_2C{=}O$ are much more reactive than the fluorocarbon oxides. (R is used an an abbreviation of a fluorocarbon radical, such as CF_3, C_2F_5, or C_3F_7.) On reaction with water, compounds of type RCOF produce fluorocarbon carboxylic acids, RCO_2H, which are nearly as strong as common mineral acid and which can be reduced to aldehydes and alcohols and thus become the starting point for the synthesis of the hybrid fluorocarbon-organic type of substance.

The nitrogen-containing fluorocarbons are derived from NF_3. Compounds of the class $R_1R_2R_3N$ are almost as resistant to chemical attack as the fluorocarbon oxides. They are not amines, since they do not react with acids to form salts. They are properly called fluorocarbon nitrides. Compounds of the group R_1R_2NF and RNF_2 are also known. Compounds such as RSF_5 and $R_1R_2SF_4$ are somewhat less resistant to chemical attack than the fluorocarbon nitrides.

The physical properties of the fluorocarbons and their inert oxygen, nitrogen, and sulfur derivatives combined with their resistance to chemical attack up to relatively high temperatures make them potentially superior substances for resins, plastics, oils, waxes, greases, elastomers, and fibers.

[IRVING SHEFT]

Bibliography: R. E. Banks, *Fluorocarbons and Their Derivatives*, 1970; R. D. Chambers, *Fluorine in Organic Chemistry*, 1973; W. G. Dauben (ed.), *Organic Reactions*, vol. 21, 1974; H. J. Emeleus, *The Chemistry of Fluorine and Its Compounds*, 1969; M. Hudlicky, *Organic Fluorine Chemistry*, 1971; Paul Tarrant (ed.), *Fluorine Chemistry Reviews*, vols. 1–7, 1967–1974; J. C. Tatlow et al. (eds.), *Advances in Fluorine Chemistry*, vols. 1–7, 1960–1973.

Fluorocarbon

An organic compound containing fluorine directly bonded to carbon. Simple fluorocarbons are made by the liquid-phase reaction of carbon tetrachloride with hydrogen fluoride catalyzed by antimony trifluoride, according to the equation below. Temperatures and pressures may be adjusted to vary the product ratio. CCl_3F is employed mainly as an aerosol propellant, and CCl_2F_2 is used as a refrigerant. They are low-boiling, odorless, nonflammable gases.

$$2CCl_4 + 3HF \rightarrow CCl_3F + CCl_2F_2 + 3\,HCl$$

Higher fluorocarbons are prepared by passing an olefin through a bed of cobalt fluoride, regenerated with fluorine. The more fully fluorinated products are nontoxic. Tetrafluoroethylene, CF_2CF_2, is formed from chloroform and SbF_3 at high temperatures. It undergoes polymerization in the presence of free-radical initiators to form the highly inert polytetrafluoroethylene. Aromatic monofluorocarbons are made from amines by a diazonium reaction with hydrogen fluoride, or by substitution of fluorine upon chloroaromatic compounds. Successive fluorines are added with increasing difficulty. The perfluoroaromatics are moderately reactive.

Consumption of the fluorocarbon aerosol propellants has been declining because of concern over their possible effect on the stratospheric ozone layer, and a government prohibition for certain applications, such as deodorants, hairsprays, perfumes, polishes, some insecticides, and food products. The largest use for these products is as a refrigerant. Some fluorocarbons are used as blowing agents in the manufacture of both flexible and rigid polyurethane foams and in polystyrene and polyethylene foams. Almost all the remaining fluorocarbons are used as grease-removing solvents, especially in electronic and aerospace industries, and for the production of fluoropolymers.

Polymers of tetrafluoroethylene (PTFE) were the first fluoropolymers marketed, and remain a significant force in the industry. They have superb chemical resistance, excellent nonstick properties, and good electrical insulating characteristics, all the result of being completely fluorinated. PTFE cannot be processed by conventional thermoplastic techniques because it is so highly viscous above its crystalline melting point. Its fabrication demands methods similar to those used for ceramics or refractory metals. Copolymers of tetrafluoroethylene with hexafluoropropene, ethylene, vinylidene fluoride, vinyl chloride, and related monomers give fluoropolymers that are more readily processed than PTFE but have inferior chemical and physical properties. *See* FREON; HALOGENATED HYDROCARBON; POLYFLUOROOLEFIN RESINS.

[FRANK WAGNER]

Fluorometric analysis

A method of chemical analysis in which a sample, exposed to radiation of one wavelength, absorbs this radiation and reemits radiation of the same or longer wavelength. If this reemission occurs in about 10^{-9} s, it is called fluorescence. Reemission after about 10^{-6} s or more is called phosphorescence. Fluorescence analysis utilizes the reemitted light to determine the material which is the source of the fluorescence.

Fluorometer. The radiation source is usually a mercury arc, and one line of its spectrum is isolated by the primary filter. The light then passes through the sample, and the fluorescence is measured at an angle, usually a right angle, to the light beam. Figure 1 is a diagram of the components

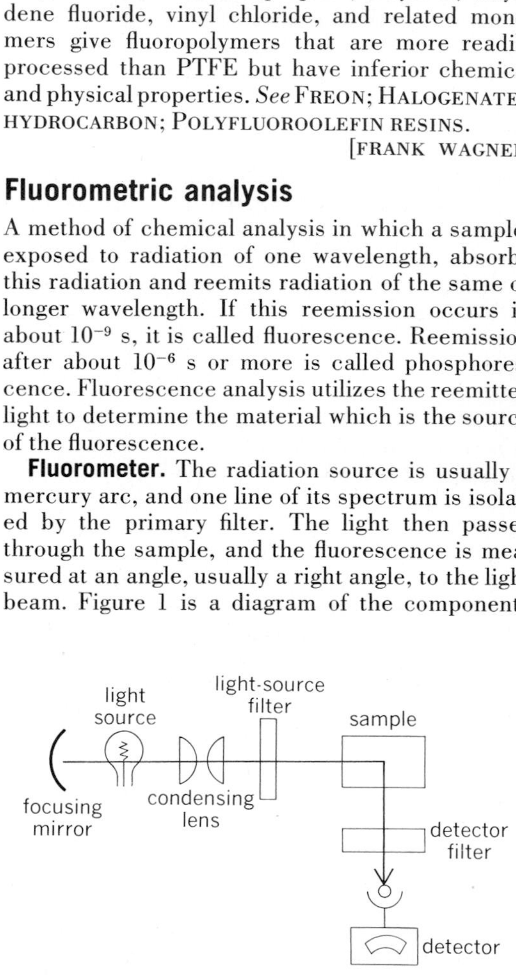

Fig. 1. Diagram of a simple fluorometer.

involved in the technique. The secondary filter before the detector eliminates scattered radiation of the wavelength of the source. The arrangement of a fluorometer is very similar to that of a nephelometer, and it has the same advantage of sensitivity—being able to detect very small amounts of reemitted radiation. Fluorometric analysis can also be employed in the x-ray region by using a molybdenum or tungsten source, appropriate metal filters, and, usually, a solid sample.

For quantitative work at very low fluorescent intensities, the intensity is almost directly propor-

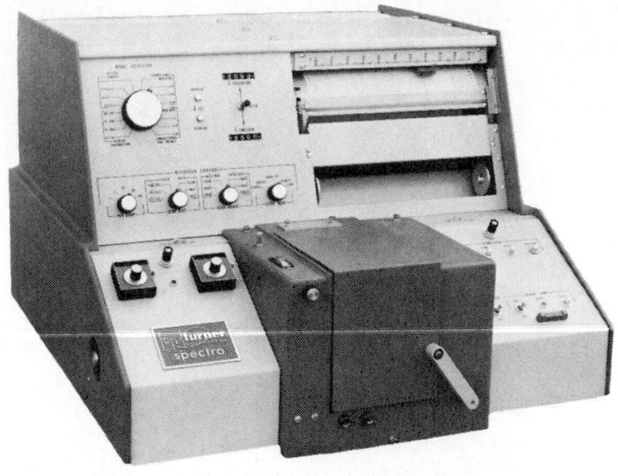
Fig. 2. Turner 210 "Spectro." (*G.K. Turner Associates*)

tional to the concentration of the fluorescing material. At high fluorescent intensities there are usually appreciable deviations from any linear function of fluorescent intensity and concentration. For this reason, fluorescence analysis is usually carried out in a very dilute solution, and a calibration curve of fluorescent intensity versus concentration of emitter is carefully prepared.

Variables. Among the many variables which must be closely watched in quantitative fluorescence analysis are scattering of the incident light by colloidal particles in solution, absorption of the fluorescent light by colored materials in solution, either quenching or intensification of the fluorescence of the compound in question by other ions or compounds present in the solution, fluorescence of other compounds present in the solution, and often the extreme dependence of fluorescence intensity on the temperature of the solution.

Applications. In spite of the drawbacks mentioned above, fluorescence is widely used in analytical work because of its sensitivity and selectivity in many systems. For example, vitamin B_2 (riboflavin) is usually determined by its fluorescence in solutions as dilute as 0.001 μg/ml. Other compounds or ions commonly determined by fluorescence in solution are β-phenylnaphthylamine, an organic antioxidant; thiamin (vitamin B_1), after reaction with ferricyanide; aluminum, after complexing by 8-hydroxyquinoline; and zirconium, after reaction with morin (an organic compound). The fluorescence technique has been extended to the determination of amino acids, proteins, and nucleic acids. Solids are also often analyzed by fluorescence. In particular, uranium compounds fluoresce with a yellow color, and uranium in ores is often detected after fusion of the ore with sodium fluoride. Many sensitive qualitative tests for the presence of various inorganic ions have been developed, based on the selective formation of a fluorescing compound.

Atomic fluorescence method. A later atomic fluorescence method is based upon the intensity of the fluorescent emission when atoms in a flame are excited by absorption of radiation. With this technique metal ions can be determined directly and with remarkable sensitivity, for example, Zn 0.001 ppm, Cd 0.002 ppm, and Hg 0.1 ppm.

Fluorescence spectroscopy. Using spectrofluorometers, fluorescence spectroscopy has been developed to increase the selectivity of fluorometry. In this technique the emitted fluorescent light is passed through a monochromator so that the fluorescence emission spectrum can be recorded. It is possible to measure several fluorescing compounds in the same solution, provided that they have sufficiently different fluorescence emission spectra. In obtaining this selectivity, some sensitivity is lost since some fluorescent light is lost in the monochromator, and only a small portion of the total fluorescent energy emitted is measured at any one wavelength.

This loss in sensitivity can be minimized by the use of a more powerful xenon arc lamp as the radiation source. A further increase in selectivity is effected by using different wavelengths of incident radiation to excite the fluorescence. Advanced spectrofluorometers, which give absolute excitation and emission spectra by correcting for any nonlinearity in the radiation source and detector, are commercially available (Fig. 2).

[ROBERT F. GODDU; JAMES N. LITTLE]

Flash-photolysis resonance fluorescence. The use of fluorescence analysis has extended beyond standard analytical applications. The sensitivity and selectivity of the technique, coupled with the easily characterized relationship between intensity and concentration, make it well suited for following the chemical reactivity of selected species. More specifically, recognition of these analysis traits has led to the development of a novel experimental technique for measuring absolute rate constants for gas-phase reactions involving free radicals (principally atoms and diatomic radicals). The method combines pulsed ($\sim 10^{-5}$ s duration) photolytic production of the radical with real-time resonance fluorescence monitoring of its subsequent temporal behavior. This technique, flash-photolysis resonance fluorescence, revolutionized gas-phase chemical kinetics research in the 1970s.

The typical experimental apparatus (Fig. 3) is by no means unique, and there are numerous variations. The basic components include a vacuum reaction chamber, a pulsed photolysis source (flash lamp), an analytical light source (reasonance lamp), and a detection system. The latter three components are situated on mutually perpendicular axes of the cell. In a typical experiment a gas mixture consisting of a photolytic radical source, molecular reactant, and inert diluent are flash-photolyzed in the cell under preselected conditions

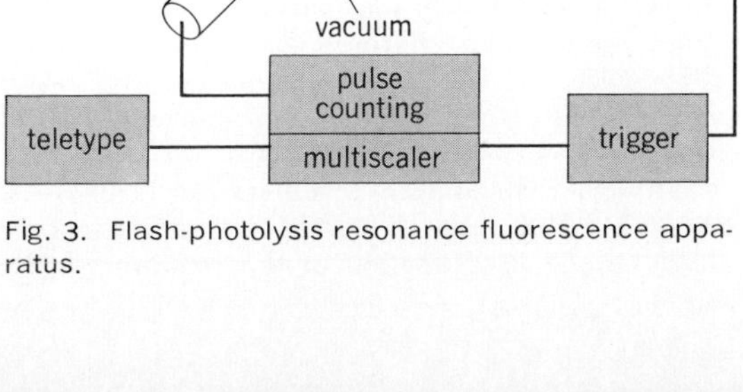

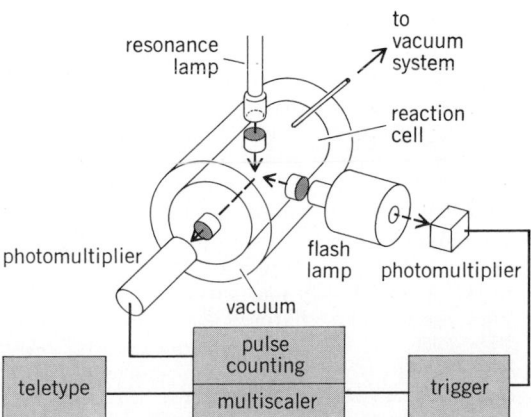

Fig. 3. Flash-photolysis resonance fluorescence apparatus.

of light intensity and wavelength distribution. In this way the desired uniform concentration of transient reactive species is produced across the central region of the cell. The analytical light source probes an intersection of this central region continuously, thereby generating a fluorescence emission from the radical species. Conditions are generally chosen such that the reaction is first-order in radical concentration. In order to achieve these conditions as well as to minimize secondary reactions between the radical and either reaction or photolytic products, radical concentrations less than 10^{11} cm^{-3} must frequently be used. The linearity of fluorescence intensity with concentration under these conditions permits real-time tracking of the reactive behavior of the transient species.

The versatility of this technique for probing atomic reactions is a result of the development of intense vacuum ultraviolet atomic line sources. The intensity and spectral purity of these electrodeless discharge lamps make them ideally suited for measuring very low atom concentrations. Such resonance lamps, when operated under optically thin conditions, are an extremely sensitive analytical tool (particularly for the very strong atomic resonance transitions in the vacuum ultraviolet spectral region). The use of rotational-vibrational bands within the electronic absorption spectra of a free radical has shown similar utility for resonance fluorescence detection of diatomic radical species. Due to difficulties in matching lamp spectral outputs with the absorption band, the technique is less sensitive for such radicals than for atoms. Consequently, laser sources tuned to specific rotational-vibrational transition of the radical have proved more effective due to the increased intensity and accompanying reduction in scattered light.

Other modifications of the flash-photolysis resonance fluorescence technique have been concerned with variations in the radical production system. The replacement of the spark discharge flash lamp (designed originally for vacuum ultraviolet operation) with a near-ultraviolet or visible laser has increased the versatility of reactant production. By selecting an appropriate absorption band of the precursor, reactive transients can be prepared in specific energy states.

The application of resonance fluorescence analysis for real-time kinetics investigations is limited only by the fluorescing ability of the free radical and the availability of a suitable resonance light source. This versatility has led to the adaptation of this measurement technique to other kinetics experiments utilizing entirely different production and analysis methodologies. Discharge flow reactors are a prime example of such expanded use. *See* CHEMICAL DYNAMICS.

[MICHAEL J. KURYLO]

Bibliography: W. Braun and M. Lenzi, Resonance fluorescence method for kinetics of atomic reactions, *Disc. Faraday Soc.*, 44:252–262, 1967; D. Davis and W. Braun, Intense vacuum ultraviolet atomic line sources, *Appl. Opt.*, 7:2071–2074, 1968; A. H. Gunn, *Introduction to Fluorimetry*, 1963; D. M. Hercules, *Fluorescence and Phosphorescence Analysis*, 1966; M. A. Konstantinova-Schlezinger, *Fluorimetric Analysis*, 1965; M. J. Kurylo, Flash photolysis resonance fluorescence investigation of the reaction of OH radicals with dimethyl sulfide, *Chem. Phys. Lett.*, 58:233–237, 1978; L. Meites, *Handbook of Analytical Chemistry*, 1963; H. H. Willard, L. L. Merrit, and J. A. Dean, *Instrumental Methods of Analysis*, 5th ed., 1974.

Fluosilicate

The SiF_6^{--} ion which is derived from fluosilicic acid, H_2SiF_6.

The fluosilicates are prepared from solutions of fluosilicic acid. The salts of barium, potassium, rubidium, cesium, and sodium are only sparingly soluble, whereas many of the heavy-metal salts are quite soluble.

Sodium fluosilicate, Na_2SiF_6, is used in the fluoridation of water supplies and in laundry scours, enamel frits, insecticides, and wood preservatives. *See* FLUORINE; SILICATE.

[E. EUGENE WEAVER]

Fluxional compounds

Molecules that undergo rapid intramolecular rearrangements among equivalent structures in which the component atoms are interchanged. The rearrangement process is usually detected by nuclear magnetic resonance (NMR) spectroscopy, which can measure rearrangement rates from 0.5 to 10,000 s^{-1}. With sufficiently rapid rates, a single resonance is observed in the NMR spectrum for a molecule that might be expected to have several nonequivalent nuclei on the basis of its instantaneous structure.

Organic structures. Within organic chemistry, degenerate Cope rearrangements represented some of the first examples of interconversions between equivalent structures, but these were relatively slow (Fig. 1*a*). The rate of this rearrangement is rapid in more complex molecules. The epitome of degeneracy is reached in bullvalene, which has more than 1,200,000 equivalent structures (Fig. 1*b*) and rapidly interconverts among them.

Fluxional molecules are frequently encountered in organometallic chemistry, and rapid rearrangements which involve migrations about unsaturated organic rings are commonly observed. The best known (called ring-whizzers) are cyclopentadienyl and cyclooctatetraene complexes of iron (Fig. 2).

Inorganic structures. Inorganic structures also exhibit fluxional phenomena, and five-coordinate complexes provide the greatest number of well-

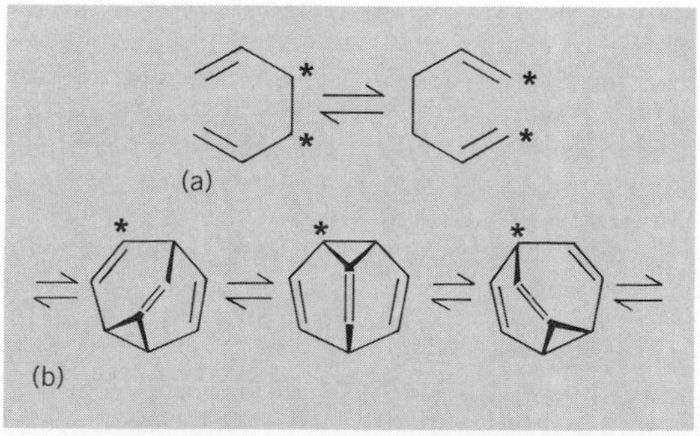

Fig. 1. Degenerate rearrangements. (*a*) Cope rearrangement. (*b*) Rearrangement of bullvalene. The asterisks indicate labels for the atoms so that they can be distinguished from one another.

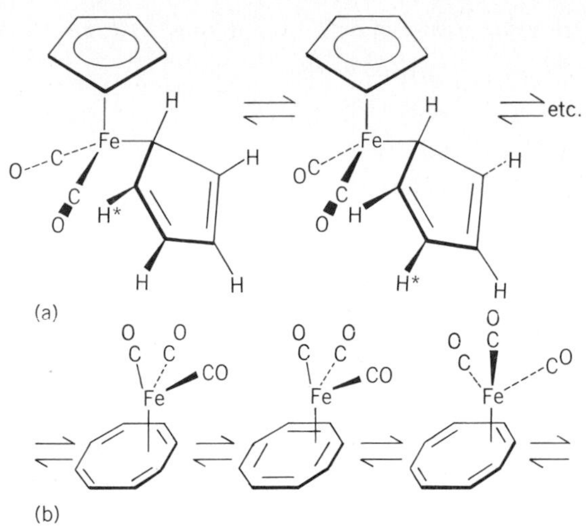

Fig. 2. Rearrangements about unsaturated organic rings. *(a)* Cyclopentadienyl iron complex. *(b)* Cyclooctatetraene iron complex.

known examples. Although the static structure would suggest a significant difference between equatorial and axial fluorine nuclei, phosphorus pentafluoride rearranges rapidly and shows only a single ^{19}F resonance split by P-F spin-spin coupling even though the static structure would suggest a significant difference between equatorial and axial fluorines. The retention of this coupling during the rearrangement demands that the process be intramolecular. This distinguishes a fluxional process from other intermolecular exchange processes which might give rise to dynamic NMR spectra. Detailed line-shape analysis of the NMR spectra can also provide insight into the possible paths of the rearrangements. Thus, the ring-whizzers have been shown to rearrange preferentially by moving to an adjacent position rather than across the ring.

The rearrangement of PF_5 involves interconversion of the trigonal bipyramidal molecule (I) to a square pyramidal configuration (II) and back (Fig. 3). The square pyramidal structure is of sufficiently high energy that none of that structure is observed by spectroscopic techniques that operate on a faster time scale than does NMR. If two such nonequivalent structures are present in observable concentrations but interconvert rapidly to cause averaging in the NMR experiments, they are said to be stereochemically nonrigid. This term is generally taken to embrace all compounds that undergo rapid reversible intramolecular rearrangements. Thus, fluxional compounds are a subset of nonrigid compounds with equivalent structures. Nonequivalent structures, that is, tautomers, might be stereochemically nonrigid if they rearranged rapidly, but would not be considered fluxional.

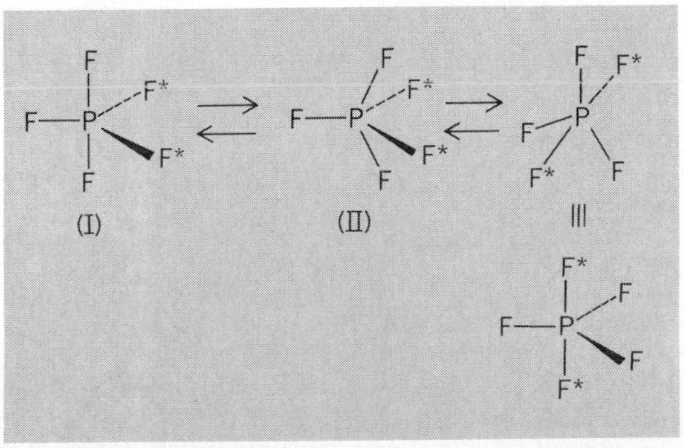

Fig. 3. Rearrangement of PF_5.

Some workers prefer to reserve the term fluxional for molecules in which bonds are broken and reformed in the rearrangement process. Hence, of the examples above, only bullvalene and the iron complexes would be termed fluxional, whereas all would be considered stereochemically nonrigid. *See* NUCLEAR MAGNETIC RESONANCE (NMR); RESONANCE; TAUTOMERISM.

[J. W. FALLER]

Bibliography: F. A. Cotton, Fluxional organometallic molecules, *Acc. Chem. Res.*, 1:257–265, 1968; W. von E. Doering and W. R. Roth, Thermal rearrangements, *Angew. Chem.* (Internat. Edit.) 2(3):115–122, 1963; J. W. Faller, Fluxional and nonrigid behavior of transition metal organometallic π-complexes, *Adv. Organometal. Chem.*, 16: 211–239, 1977; L. M. Jackman and F. A. Cotton, *Dynamic Nuclear Magnetic Resonance Spectroscopy*, 1975.

Formaldehyde

The simplest aldehyde, formula HCH=O. Because of its extreme reactivity, even with itself, it cannot be readily isolated or handled in the pure state. Therefore, it is produced and marketed as an aqueous solution (usually 37–50% formaldehyde by weight), sometimes known as Formalin. It is also sold as the solid hydrated polymer known as paraformaldehyde or paraform. Approximately 60% of all the formaldehyde produced is for captive use.

Uses. Formaldehyde is used principally to produce synthetic resins and adhesives by reaction with phenols, urea, and melamine. This use accounts for about 75% of the total production. Approximately 15% is used in the manufacture of textiles, dyes, drugs, paper, leather, photographic materials, embalming agents, disinfectants, and insecticides. Increasing amounts are used for the industrial production of pentaerythritol and hexamethylenetetramine, reactions (1) and (2).

$4HCH{=}O + CH_3CHO + Ca(OH)_2 + H_2O \rightarrow$
Formaldehyde, Acetaldehyde, Calcium hydroxide

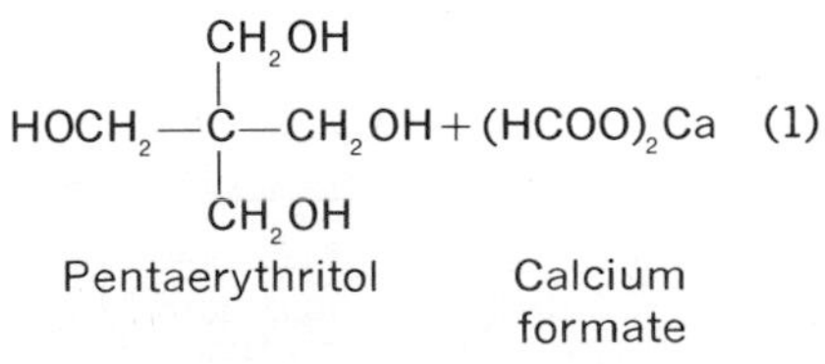

$6\,HCH{=}O + 4\,NH_3 \rightarrow$
Formaldehyde, Ammonia

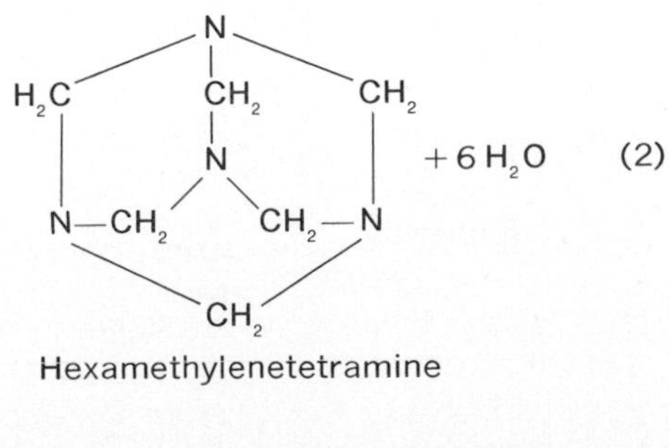

Production. Most of the formaldehyde is manufactured by the oxidation of methanol with air over a metal catalyst in the temperature range of 450–650°C. Catalysts are either silver, copper, or an iron-molybdenum mixture. The reaction is shown as (3).

$$2CH_3OH + O_2 \rightarrow 2HCH{=}O + 2H_2O \quad (3)$$
Methanol

Increasing quantities of formaldehyde are produced by the partial oxidation of hydrocarbons from natural gas. Air is used as the source of oxygen, excess hydrocarbon to minimize complete oxidation, and steam to quench the reaction. The yields of formaldehyde are low, but the raw materials are plentiful and low in cost. Since other products, such as alcohols, acids, ketones, and other aldehydes, are formed, the purification of formaldehyde produced by this process may involve fractionation, liquid-liquid extraction, azeotropic distillation, and extractive distillation.

Properties. Formaldehyde undergoes many of the general reactions which are typical of aldehydes. Some interesting reactions which are somewhat specific for formaldehyde are shown in reactions (4) and (5). Bis-chloromethyl ether has been

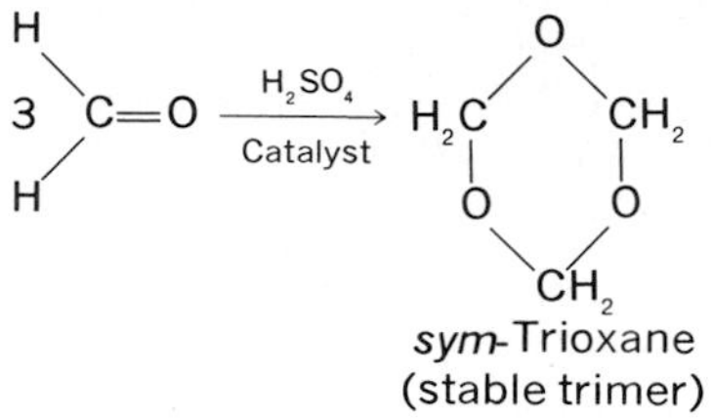

$$3 \begin{matrix} H \\ C{=}O \\ H \end{matrix} \xrightarrow[\text{Catalyst}]{H_2SO_4} \quad (4)$$

sym-Trioxane
(stable trimer)

$$2 \begin{matrix} H \\ C{=}O \\ H \end{matrix} + 2HCl \rightarrow ClCH_2OCH_2Cl + H_2O \quad (5)$$
Hydrogen   Bio-chloromethyl
chloride   ether

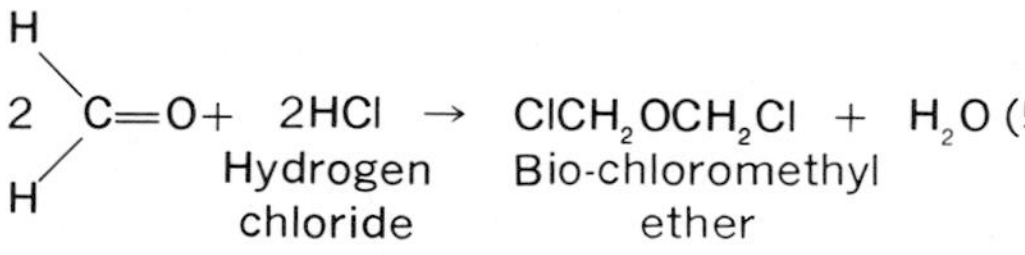

classified by the U.S. Occupational Safety and Health Administration as a carcinogen. *See* ALDEHYDE.

The Mannich reaction (6) involves the reaction

$$(CH_3)_2NH{\cdot}HCl + \begin{matrix} H \\ C{=}O \\ H \end{matrix} + CH_3\overset{O}{\overset{\|}{C}}CH_3 \rightarrow$$

Dimethylamine   Formal-   Acetone
hydrochloride   dehyde

$$CH_3\overset{O}{\overset{\|}{C}}CH_2CH_2N(CH_3)_2{\cdot}HCl + H_2O \quad (6)$$
4-(Dimethylamino)-2-
butanone hydrochloride

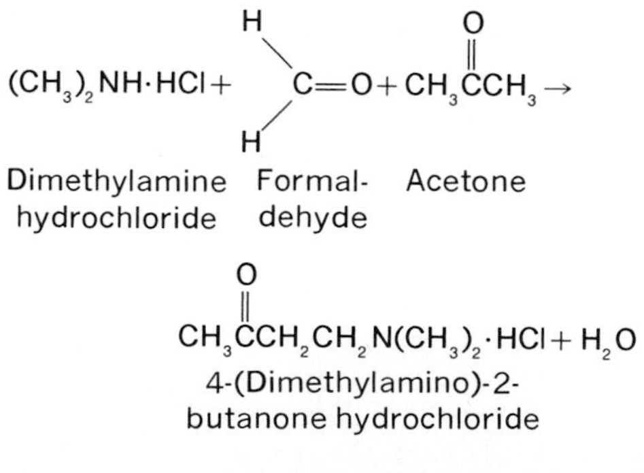

of an amine hydrochloride with formaldehyde and a ketone. Reaction of formaldehyde with carbon monoxide is shown in reaction (7). Reac-

$$\begin{matrix} H \\ C{=}O \\ H \end{matrix} + CO + H_2O \xrightarrow[\text{Catalyst}]{\text{High pressure}}$$
Carbon
monoxide

$$HOCH_2COOH \quad (7)$$
Hydroxyacetic
acid
(glycolic acid)

tion (8) represents the Reppe ethynylation reaction.

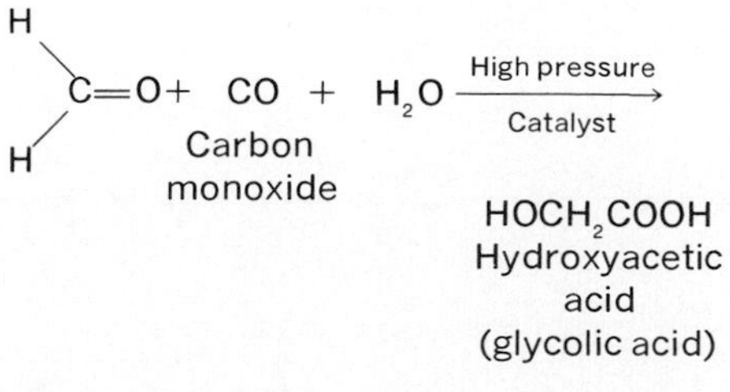

$$2 \begin{matrix} H \\ C{=}O \\ H \end{matrix} + HC{\equiv}CH \xrightarrow[\text{catalyst}]{\text{Copper}}$$
Acetylene

$$HOCH_2C{\equiv}CCH_2OH \quad (8)$$
2-Butyne-1,4-diol

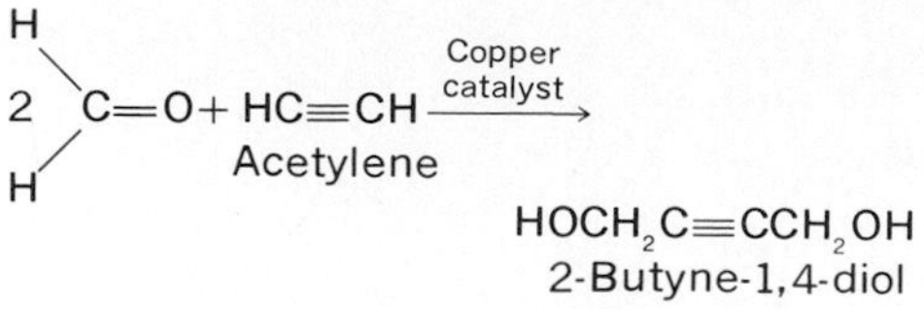

The Prins reaction (9) involves the reaction of formaldehyde with olefinic aromatic hydrocarbons in the presence of an organic acid and a mineral acid catalyst.

$$\begin{matrix} H \\ C{=}O \\ H \end{matrix} + C_6H_5CH{=}CH_2 + 2CH_3COOH \xrightarrow{H_2SO_4}$$
Styrene   Acetic acid

$$C_6H_5\overset{OOCCH_3}{\overset{|}{CH}}{-}CH_2CH_2OOCCH_3 + H_2O \quad (9)$$
1-Phenyl-1,3-propanediol
diacetate

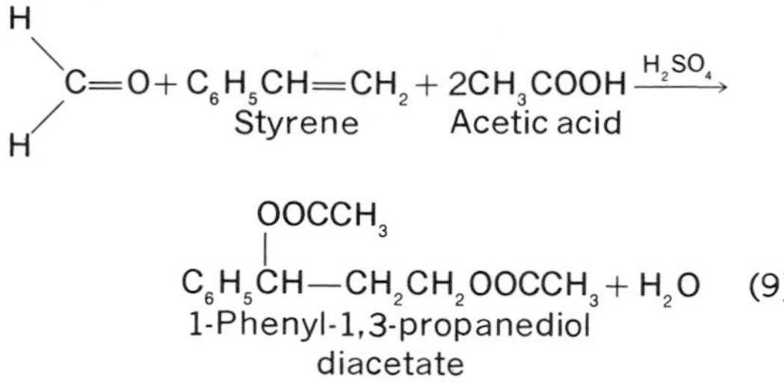

The reaction of formaldehyde with phenols, urea, and melamine to form resins proceeds first with formation of methylol (—CH_2OH) derivatives followed by intermolecular dehydration to produce methylene (CH_2) linkages. This reaction is catalyzed by either alkali or acid, and the extent of resinification is controlled by temperature, reaction time, and concentration. *See* PHENOLIC RESIN.

[L. MADESTAU]

Formic acid

A colorless, pungent liquid having a melting point of 8.4°C and the formula HCOOH. Formic acid is miscible with water, alcohol, ether, and glycerol. It is toxic, and causes painful skin wounds. It is a constituent of the stings of ants, stinging caterpillars, and stinging nettles.

Sodium formate is produced by heating carbon monoxide with sodium hydroxide under pressure at about 210°C. Treatment with sodium acid sulfate or dilute sulfuric acid liberates free formic acid.

Formic acid is stronger than acetic acid. With carbonates, oxides, and hydroxides of alkali or alkaline-earth metals, it gives formates (salts); with alcohols it forms esters (formates). *See* ESTER.

It is used as a reducing agent for metallic ions and dyes, in dehairing and tanning, as a latex coagulant, in the preparation of a large number of esters, as a source of allyl alcohol via its reaction with glycerol, and in the preparation of oxalic acid (by heating of sodium formate). With concentrated sulfuric acid, it decomposes, liberating carbon monoxide. *See* CARBOXYLIC ACID; OXALIC ACID.

[EVANS B. REID]

Fractionating column

An apparatus used widely for countercurrent contacting of vapor and liquid to effect separations by distillation or absorption. In general, the apparatus consists of a cylindrical vessel with internals designed to obtain multiple contacting of ascending vapor and descending liquid, together with means for introducing or generating liquid at the top and vapor at the bottom.

Figure 1 shows a column that can be applied to distillation. A vapor condenser is used to produce

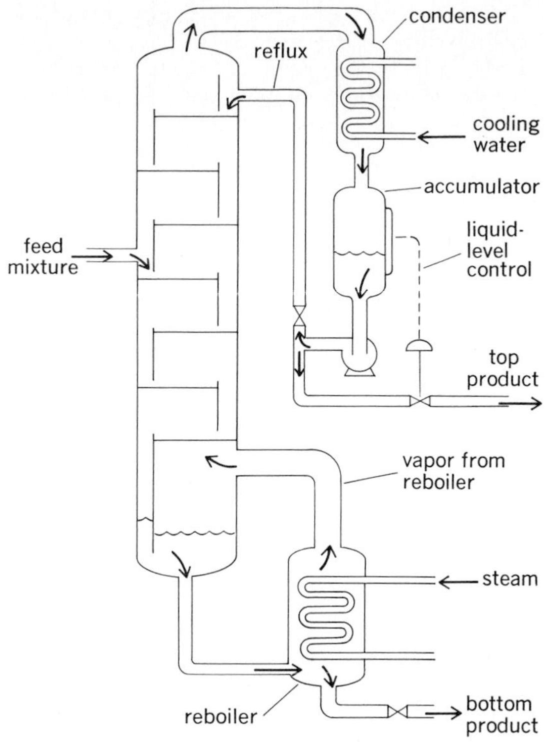

Fig. 1. Elements of a fractionating column.

liquid (reflex) which is returned to the top, and a liquid heater (reboiler) is used to generate vapor for introduction at the bottom. In a simple absorber, the absorption oil is the top liquid and the feed gas is the bottom vapor. In all cases, changes in composition produce heat effects and volume changes, so that there is a temperature gradient and a variation in vapor, and liquid flows from top to bottom of the column. These changes affect the internal flow rates from point to point throughout the column and must be considered in its design.

Fractionating columns used in industrial plants range in diameter from a few inches to 40 ft and in height from 10 to 200 ft. They operate at pressures as low as a few millimeters of mercury and as high as 3000 psi; at temperatures from −300 to 700°F. They are made of steel and other metals, of ceramics and glass, and even of such materials as bonded carbon and plastics.

A variety of internal devices has been used to obtain more efficient contacting of vapor and liquid. The most widely used devices are the bubble-cap plate, the perforated or sieve plate, and the packed column (Fig. 2).

Plates. The bubble-cap plate is a horizontal deck with a large number of chimneys over which circular or rectangular caps are mounted to channel and distribute the vapor through the liquid. Liquid flows by gravity downward from plate to plate through separate passages known as downcomers. The number of caps and the size of downcomers are designed to fit the expected internal flow rates of liquid and vapor.

The bubble-cap plate is the most common type of contacting device used in fractionating columns. It can handle a wide range of vapor and liquid flows, is not obstructed by small amounts of rust and other solids, and is manufactured by a large number of suppliers.

The perforated or sieve plate is a horizontal deck with a multiplicity of round holes or rectangular slots for distribution of vapor through the liquid. The sieve plate can be designed with downcomers similar to those used for bubble-cap trays, or it can be made without downcomers so that both liquid and vapor flow through the perforations in the deck. The simple sieve tray has the disadvantage of a restricted operating range so that operation at vapor flows much less than the design rate results in poor contacting.

A number of variations and improvements have been made in the sieve tray to improve the operating range. The most successful of these is the valve tray, in which a check valve is placed over each perforation to adjust for the variation in vapor flow. The valve usually consists of a disk over each perforation which is held within a cage attached to the deck. The disk moves up and down and adjusts to the vapor flow.

Although it is older than the bubble-cap plate, the sieve tray did not obtain wide use until stainless steel sheets became readily available to overcome difficulties with corrosion and fouling. With the newer improvements, such as the valve tray, this type of plate promises to supplant the bubble-cap plate in many services.

Packings. The packed column is a bed or succession of beds made up of small solid shapes over

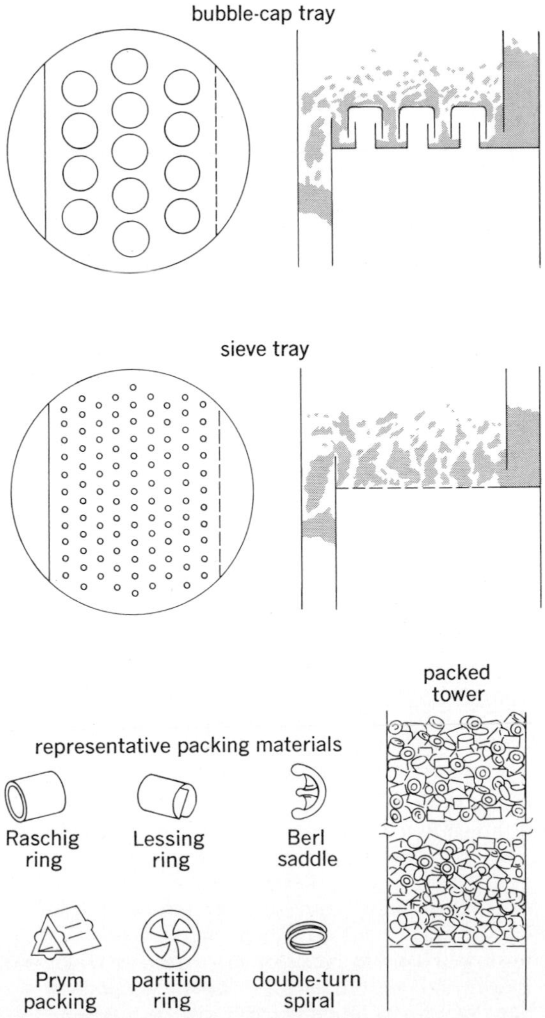

Fig. 2. Contacting devices for fractionating column.

which liquid and vapor flow in tortuous countercurrent paths. Some of the more common types of packing are shown in Fig. 2. Expanded metal or woven mats are also used as packing. The packed column is used without downcomers, but in larger sizes usually has horizontal redistribution decks to collect and redistribute the liquid over the bed at successive intervals of height. The packed column is widely used in laboratories. It is often used in small industrial plants, especially where corrosion is severe and ceramic or glass materials must be chosen.

Other contacting means sometimes used in fractionating columns are the disk-and-doughnut tray for very viscous or dirty liquids, the spiral ribbon for very small columns, the spray column, and the welled-wall column. *See* DISTILLATION; GAS ABSORPTION OPERATIONS.

[MOTT SOUDERS]

Bibliography: R. H. Perry (ed.), *Chemical Engineers' Handbook*, 5th ed., 1973; A. H. Skelland, *Diffusional Mass Transfer*, 1974; B. D. Smith, *Design of Equilibrium Stage Processes*, 1963.

Francium

A chemical element, Fr, atomic number 87, an alkali metal element falling below cesium in group Ia of the periodic table. Distinguished by nuclear instability, francium exists only in short-lived radioactive forms, the most durable of which has a half-life of 21 min. The chief isotope of francium is

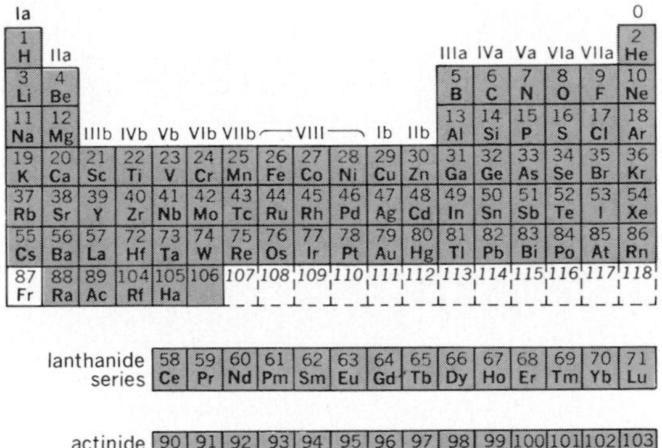

actinium-K, an isotope of mass 223, which arises from the radioactive decay of the element actinium. Actinium in turn is a decay product of the rare uranium isotope U^{235} and hence is a normal component of uranium minerals. Actinium decays chiefly by β-decay, but the French scientist Marguerite Perey discovered that 1% of the disintegrating actinium atoms emit α-particles and are thus converted to atoms of element 87 of mass number 223. Actinium-K emits β-particles with a half-life of 21 min. Perey's experiments with this isotope constituted the first reliable identification of an isotope of element 87; she was thus credited with its discovery and named it francium.

Other isotopes of francium, none as long-lived as AcK, have been synthesized by the bombardment of thorium targets with high-energy protons, deuterons, or helium ions. From the properties of the known isotopes, it is reasonably certain that no long-lived form of element 87 will ever be found in nature or synthesized artificially.

The chemical properties of francium can be studied only on the tracer scale. The element has all the properties expected of the heaviest alkali element. With few exceptions, all the salts of francium are water-soluble. Francium remains in aqueous solution when other elements are precipitated as insoluble hydroxides, fluorides, sulfides, or chromates. Radiochemical purification of francium can be achieved simply, by removing actinium and other foreign radioactivity by successive precipitation of a series of insoluble scavenger precipitates, following which the radiations of francium can be sought in the supernatant solution. Coprecipitation with cesium carrier in conjunction with perchlorate, chlorophatinate, or silicotungstate anions is a useful method for tracer amounts of francium. Francium can also be separated by use of cation-exchange resins. *See* ALKALI METALS.

[EARL K. HYDE]

Bibliography: K. W. Bagnall, *Chemistry of the Rare Radioelements*, 1957; E. K. Hyde, *Radiochemistry of Francium*, U.S. Atomic Energy Commission, 1960; I. M. Kolthoff, P. J. Elving, and E. B. Sandell (eds.), *Treatise on Analytical Chemistry*, vol. 6, pt. 2, 1964.

Free energy

A term in thermodynamics which in different treatments may designate either of two functions defined in terms of the internal energy E or enthalpy H, and the temperature-entropy product TS.

The function $(E-TS)$ is the Helmholtz free energy and is the function ordinarily meant by free energy in European references. The Gibbs free energy is the function $(H-TS)$. For the Lewis and Randall school of American chemical thermodynamics, this is the function meant by the free energy F. To avoid confusion with the symbol F as applied elsewhere to the Helmholtz free energy, the symbol G has also been used. A recent development was the introduction of the name free enthalpy, with symbol G, for the Gibbs function. *See* WORK FUNCTION.

Theory. For a closed system (no transfer of matter across its boundaries), the work which can be done in a reversible isothermal process is given by the series shown in Eq. (1). For these conditions,

$$W_{\text{rev}} = -\Delta A = -\Delta(E-TS) \quad = -(\Delta E - T\,\Delta S) \qquad (1)$$

$T\,\Delta S$ represents the heat given up to the surroundings. Should the process be exothermal, $T\Delta S < 0$, then actual work done on the surroundings is less than the decrease in the internal energy of the system. The quantity $(\Delta E - T\,\Delta S)$ can then be thought of as a change in free energy, that is, as that part of the internal energy change which can be converted into work under the specified conditions. This then is the origin of the name free energy. Such an interpretation of thermodynamic quantities can be misleading, however; for the case in which $T\,\Delta S$ is positive, Eq. (1) shows that the decrease in "free" energy is greater than the decrease in internal energy. *See* CHEMICAL THERMODYNAMICS.

For constant temperature and pressure in a reversible process the decrease in the Gibbs function G for the system again corresponds to a free-energy change in the above sense, since it is equal to the work which can be done by the closed system other than that associated with its change in volume ΔV under the given constant pressure P. The relations shown in Eq. (2) can be formed since $\Delta H = \Delta E + P\Delta V$.

$$\Delta G = -(\Delta H - T\,\Delta S) = W_{\mathrm{net}} = W_{\mathrm{rev}} - P\,\Delta V \quad (2)$$

Each of these free-energy functions is an extensive property of the state of the thermodynamic system. For a specified change in state, both ΔA and ΔG are independent of the path by which the change is accomplished. Only changes in these functions can be measured, not values for a single state.

The thermodynamic criteria for reversibility, irreversibility, and equilibrium for processes in closed systems at constant temperature and pressure are expressed naturally in terms of the function G. For any infinitesimal process at constant temperature and pressure, $-dG \geqq \delta w_{\mathrm{net}}$. If δw_{net} is never negative, that is, if the surroundings do no net work on the system, then the change dG must be negative or zero. For a reversible differential process, $-dG > \delta w_{\mathrm{net}}$; for an irreversible process, $-dG > \delta w_{\mathrm{net}}$. The free energy G thus decreases to a minimum value characteristic of the equilibrium state at the given temperature and pressure. At equilibrium, $dG = 0$ for any differential process taking place, for example, an infinitesimal change in the degree of completion of a chemical reaction. A parallel role is played by the work function A for conditions of constant temperature and volume. Because temperature and pressure constitute more convenient working variables than temperature and volume, it is the Gibbs free energy which is the more commonly used in thermodynamics.

Partial molal quantities. For a particular homogeneous phase in the absence of surface, gravitational, and magnetic forces, the free energy G depends on the numbers of moles of the constituents present, the temperature T, and the pressure P. Let Ω represent the total number of constituents, n_i the number of moles of typical constituent i, and designate by subscript n constant composition, by subscript n_j constancy of the number of moles of all constituents except n_i; then Eq. (3) is formed.

$$dG(T,P,n_1,\ldots,n_\Omega) = \left(\frac{\partial G}{\partial T}\right)_{P,n} dt + \left(\frac{\partial G}{\partial p}\right)_{T,n} + \sum_{i=1}^{\Omega} \left(\frac{\partial G}{\partial n_i}\right)_{T,P,n_j} dn_i \quad (3)$$

In Eq. (3) the term $\left(\frac{\partial G}{\partial n_i}\right)_{T,P,n_j}$ is the chemical potential μ_i of the ith constituent. It is identical to the partial molal free energy $\overline{G}_i$ of Lewis and Randall. It then follows that Eq. (4) holds. *See* SOLUTION.

$$dG = -S\,dT + V\,dP + \sum_{i=1}^{\Omega} \mu_i\,dn_i \quad (4)$$

Because the chemical potentials at constant T,P are intensive variables whose values are fixed, like that of the density, by the relative number of moles of the various constituents present, and are independent of the total mass of the phase, this equation can be integrated for constant T,P and relative composition starting from $n_i = 0$ to obtain Eq. (5).

$$G(T,P,n_1,\ldots,n_\Omega) = \sum_{i=1}^{\Omega} n_i\mu_i \quad (5)$$

This yields Eq. (6). Consistency with the expression for dG in Eq. (4) requires that Eq. (7) hold.

$$dG = \sum_{i=1}^{\Omega} \mu_i\,dn_i + \sum_{i=1}^{\Omega} n_i\,d\mu_i \quad (6)$$

$$S\,dT - V\,dP + \sum_{i=1}^{\Omega} n_i\,d\mu_i = 0 \quad (7)$$

This is the Gibbs-Duhem equation. For constant temperature and pressure, this relation imposes a condition on the composition variation of the set of chemical potentials.

Heterogeneous systems. The free energy of a closed, heterogeneous system is the sum of the free energies of its various phases. In the absence of such a constraint as provided by the subdivision of the system by a rigid, semipermeable membrane, the general thermodynamic criterion of equilibrium requires that the temperature and pressure be uniform throughout the system and that the chemical potential of each constituent have a common value for all phases in which it is present. Further, if any of the constituents can be formed from others, the chemical potentials of the reactants and products are related in accordance with the stoichiometry of the reaction equation. Thus, for the reaction in Eq. (8), at equilibrium Eq. (9) can be formed. Expressing each chemical potential μ_i in terms of the standard value μ_i^0 and its associated activity term $RT \ln a_i$ results in Eq. (10).

$$\mathrm{A} + 2\mathrm{B} \rightleftharpoons 3\mathrm{C} + 4\mathrm{D} \quad (8)$$

$$\mu_{\mathrm{A}} + 2\mu_{\mathrm{B}} = 3\mu_{\mathrm{C}} + 4\mu_{\mathrm{D}} \quad (9)$$

$$RT \ln \left(\frac{a_{\mathrm{C}}^3 a_{\mathrm{D}}^4}{a_{\mathrm{A}} a_{\mathrm{B}}^2}\right)_{\mathrm{equil}} = -(3\mu_{\mathrm{C}}^0 + 4\mu_{\mathrm{D}}^0 - \mu_{\mathrm{A}}^0 - 2\mu_{\mathrm{B}}^0) = -\Delta G^0 \quad (10)$$

In Eq. (10) ΔG^0 is called the standard free-energy change for the reaction. Its value depends on the standard states chosen, but for a given temperature and pressure, it is a constant characteristic of the reaction involved. A true equilibrium constant K then results as shown by Eqs. (11) and (12). If the pressure for each standard state is fixed and independent of the pressure of the reaction system, ΔG^0 and hence K are functions of temperature only. This is the conventional approach in treating gas-phase equilibria, but not ordinarily for condensed phases.

$$K = \left(\frac{a_{\mathrm{C}}^3 a_{\mathrm{D}}^4}{a_{\mathrm{A}} a_{\mathrm{B}}^2}\right)_{\mathrm{equil}} \quad (11)$$

$$RT \ln K = -\Delta G^0 \quad (12)$$

Since the activities can be correlated with partial pressures or concentrations through fugacity coefficients or activity coefficients, this thermodynamic approach eliminates the uncertainties otherwise associated with equilibrium calculations based on the law of mass action. *See* CHEMICAL EQUILIBRIUM; FUGACITY.

The prediction of an equilibrium constant then requires the calculation of ΔG^0 for the reaction. The so-called third-law method involves calculation for the reaction at 25°C of the value of ΔH^0, the standard heat of reaction, from tabulated standard heat of formation data and of ΔS^0 from tabulated third-law entropies. These are combined in the sense of $\Delta G^0 = \Delta H^0 - T\,\Delta S^0$ to permit calculation of the equilibrium constant for 25°C. This in turn is used for evaluation of the integration constant in the integration of the relation in Eq. (13).

$$\frac{d \ln K}{dT} = \frac{\Delta H^0}{RT^2} \quad (13)$$

The integration requires expression of ΔH^0 as a function of temperature, which necessitates a knowledge of the heat capacities $C^0_{P(i)}$ for the various reactants over the temperature range involved.

Alternatively, if values of the free-energy function $(G^0 - H^0_{298}/T)$ are available, either from experimental measurement or from statistical thermodynamical computations, they can be combined with the standard heat of reaction at 25°C to give the desired result, Eq. (14).

$$\Delta G^0 = \Delta\left(\frac{G^0 - H^0_{298}}{T}\right) + \frac{\Delta H^0_{298}}{T} \quad (14)$$

See ENTROPY; THERMOCHEMISTRY.

[PAUL BENDER]

Bibliography: K. G. Denbigh, *Principles of Chemical Equilibrium*, 2d ed., 1968; E. Fermi, *Thermodynamics*, 1956; J. W. Gibbs, *Collected Works*, vol. 1, 1948; K. S. Pitzer and L. Brewer, *Thermodynamics*, rev. ed., 1961; F. T. Wall, *Chemical Thermodynamics*, 2d ed., 1965.

Free radical

Any molecule or atom which possesses one unpaired electron. There are some molecules which contain more than one unpaired electron (for example, oxygen); they normally are not considered as free radicals. Free radicals can be chemically very reactive (for example, the methyl radical) or they can be very stable entities (for example, nitric oxide).

Free radicals were frequently, and incorrectly, postulated throughout the 19th century. Avogadro's hypothesis was not taken too seriously by the early organic chemists, and substances such as C_2H_6 and C_4H_{10} were frequently described as CH_3 and C_2H_5. By the end of the 19th century this situation had been cleared up, and the impossibility of the independent existence of free radicals appeared to be well established. This situation was completely upset by Moses Gomberg's discovery of the triphenyl methyl radical. Its reactivity toward O_2, I_2, and NO at room temperature furnished strong evidence that hexaphenylethane in solution is dissociated according to the equilibrium reaction (1). Since then many similar radicals

$$(C_6H_5)_3C{-}C(C_6H_5)_3 \rightleftharpoons 2(C_6H_5)_3C \quad (1)$$

have been discovered, and such compounds are now freely postulated in organic mechanisms.

Free radicals can be grouped into three major classes: atoms (for example, H, F, and Cl), inorganic radicals (for example, OH, CN, NO_2, and ClO_3), and organic radicals (for example, CH_3, CH_3CH_2, and $C_6H_6^-$). Such radicals are of great importance since they often appear as intermediates in thermal and photochemical reactions. Radicals are also known to initiate and propagate polymerization and combustion reactions.

Production of atoms and radicals. In general, free radicals are formed by the rupture of a bond in a stable molecule with the production of two fragments, each with an unpaired electron. The resulting free radicals may participate in further reactions or may combine to reform the original compound. In the gas phase, equilibria such as reaction (2) may be established, especially at elevated tem-

$$R_1{-}R_2 \rightleftharpoons R_1 + R_2 \quad (2)$$

peratures. Since in most cases recombination occurs at nearly every collision of R_1 and R_2, the composition of the equilibrium mixture under ordinary circumstances indicates only a minute amount of decomposition into radicals. Even though radical partial pressures may be less than 10^{-6} mm of mercury and their lifetime less than 10^{-3} sec, these radicals do play an important role in reaction kinetics. The transitory existence of such radicals has been verified by spectroscopic study.

There are many ways in which radicals can be generated—among these are thermal decomposition, electric discharge photochemical reactions, electrolysis at an electrode such as mercury or platinum, rapid mixing of two reactants, and gamma- or x-ray irradiation. In thermal methods a stable molecule is decomposed at an elevated temperature. In exceptional circumstances the dissociation into radicals at equilibrium may be considerable. Thus hydrogen atoms may be produced by heating hydrogen to a very high temperature, reaction (3). At 1900 K this equilibrium

$$H_2 \rightleftharpoons 2H \quad (3)$$

corresponds to 1% dissociation into atoms when the pressure is 1 atm. In a few cases substances in solution are considerably dissociated into radicals at room temperature. It is then possible to obtain radicals of apparent long life in high concentrations. Examples are hexaphenylethane, $(C_6H_5)_3C{-}C(C_6H_5)_3$, which is about 3% dissociated into triphenylmethyl radicals, $C(C_6H_5)_3$, in benzene solution at 5°C at a concentration of 2–3%, and hexa-(*p*-phenyl)-biphenylethane, $(C_6H_5{-}C_6H_4)_3C{-}C(C_6H_5{-}C_6H_4)_3$, which is virtually 100% dissociated under similar circumstances.

Usually, however, thermal decompositions are essentially irreversible under the conditions employed. Most gaseous organic substances decompose wholly or in part by a mechanism involving an initial split into radicals. For example, in reaction (4), ethane forms two methyl radicals. *See* HYDROCARBON; PYROLYSIS.

$$C_2H_6 \rightarrow 2CH_3 \quad (4)$$

Radicals may also be produced electrically by passing gas through an electrical discharge at high speed. Atomic spectra are produced in this way, and the method is frequently used for the investigation of the chemical reactions of hydrogen, oxygen, and nitrogen atoms.

Photochemical methods of producing free radicals are very common. Almost all gaseous organic compounds decompose photochemically via free radicals, and the method is thus of very wide applicability. Two of the most widely used substances are chlorine and acetone. Chlorine decomposes on irradiation with light in the continuous region of its absorption spectrum to give chlorine atoms, reaction (5). Many reactions of chlorine atoms have

$$Cl_2 + h\nu \rightarrow 2Cl \quad (5)$$

been investigated by this method. The acetone

photolysis has also been widely investigated. There is no doubt that the primary split, using radiation in the range 2537 to 3130 A, is reaction (6). This

$$CH_3COCH_3 + h\nu \rightarrow CH_3 + CH_3CO \qquad (6)$$

reaction has been one of the most frequently used sources of methyl and acetyl radicals. Some of the subsequent reaction steps will be mentioned later. A photochemical production of free radicals appears to be the primary chemical step in photosynthesis by plants and certain bacteria. *See* PHOTOCHEMISTRY.

Electrolysis often produces radicals at one of the electrodes. For instance, the electrolysis of anthracene in dimethylformamide causes the anthracene negative ion to be generated at the cathode.

Rapid mixing of two suitable reactants can lead to moderate steady-state concentrations of unstable radical intermediates. For instance, rapid mixing of aqueous solutions of $TiCl_3$ and of H_2O_2 results in the formation of the OH radical, presumably by reaction (7).

$$Ti^{3+} + H_2O_2 \rightarrow Ti^{4+} + OH + OH^- \qquad (7)$$

High-energy radiation such as gamma-rays and x-rays, as well as high-energy particles given off by radioactive nuclei, can cause extensive disruption of molecules leading to radical and ion production. Such systems are important in cancer research and in research on the mechanisms of radiation damage. However, the reactions are usually very complex.

Detection and estimation. The earliest methods of detection involved the chemical properties of the radicals. Later and more reliable methods use absorption spectroscopy, mass spectrometry, and principally electron spin resonance spectroscopy.

The methyl radical was first detected by Paneth by the so-called mirror removal method. Radicals were prepared by thermal decomposition in a very fast flow system employing an inert carrier gas. The resulting stream of radicals plus carrier passed down a tube and over a thin "mirror" of lead deposited on the wall. In the presence of the radicals the lead mirror was removed, and it was proved conclusively that this was due to the formation of $Pb(CH_3)_4$, presumably by the reaction of CH_3 radicals with the lead. Many other radicals were detected in this way, but the results were erratic, and the method has been largely superseded.

Another chemical method of detecting radicals uses traps. If iodine is added to a photochemical system, the rapid reactions (8) and (9) occur. Under

$$R + I \rightarrow RI \qquad (8)$$

$$R + I_2 \rightarrow RI + I \qquad (9)$$

appropriate conditions all radicals present may be removed from the system and trapped as iodides. The method has proved very useful in conjunction with other detailed information, but it has many pitfalls if used without caution.

Absorption spectroscopy is a simple method of detecting free atoms and has been used successfully for many years. More recently the method has been successfully employed for the detection of OH, NH, NH_2, CN, CF_2, CF, C_2, CH, CHO, C_3, CH_3O, C_2H_5O, NCO, NCS, CH_3S, HNO, PH_2, CH_3, and many other radicals with varying degrees of certainty. The development of flash photolysis, by which large amounts of energy may be absorbed by a gas in a short time, has greatly advanced the detection of radicals by absorption spectroscopy. In an interesting variation of the method, an organic compound in the form of a glass is photodecomposed at very low temperatures, and the radicals formed are thus trapped in a frozen matrix. They have an indefinitely long life, and their spectrum can be determined at will.

An ingenious method for the detection and the estimation of the concentration of radicals is mass spectrometry. A rapid flow system is used in which radicals are produced thermally or photochemically. The gas stream is sampled through the leak of the mass spectrometer. An electron energy is used such that radicals are ionized but stable molecules are not dissociated. Any radical ions detected in the mass spectrometer must come from radicals already present in the gas, and a direct and unequivocal detection of radicals is possible. Also, under favorable circumstances the method can be used for the quantitative estimation of the concentration of radicals.

Electron spin resonance (ESR) spectroscopy is by far the best technique presently available for the detection and characterization of free radicals in solids, liquids, or gases. The technique is very sensitive (10^{-9} *M* free radicals can be detected) and, furthermore, it is specific to molecules which contain one or more unpaired electrons. Each radical generally has a characteristic ESR spectrum which can be used both for identification and to deduce structural information about the radical. For instance, the radical CF_3 was shown to have a pyramidal structure by analyzing the ESR spectrum. Analysis of the ESR spectra of radicals such as triphenylmethyl yields information concerning the distribution of the unpaired electron over the molecule.

Free-radical mechanisms. In well-investigated photochemical reactions it is frequently possible to establish a mechanism involving a set of free-radical steps beyond any reasonable doubt by the techniques listed above. A typical example is the thermal decomposition of ethane, reactions (10)–(15). General mechanisms of this type, the so-

$$C_2H_6 \rightarrow 2CH_3 \qquad (10)$$

$$CH_3 + C_2H_6 \rightarrow CH_4 + C_2H_5 \qquad (11)$$

$$C_2H_5 \rightarrow C_2H_4 + H \qquad (12)$$

$$H + C_2H_6 \rightarrow C_2H_5 + H_2 \qquad (13)$$

$$H + C_2H_5 \rightarrow C_2H_6 \qquad (14)$$

$$2C_2H_5 \rightarrow C_4H_{10} \qquad (15a)$$

$$\rightarrow C_2H_4 + C_2H_6 \qquad (15b)$$

called Rice mechanisms, have been well established in many cases by detailed investigation of the kinetics. Several types of free-radical reaction are known to be involved:

1. The rate of the bond-splitting reaction (10) is governed mainly by the dissociation energy of the C–C bond. *See* THERMOCHEMISTRY.

2. The activation energy of the radical decomposition reaction (12) is usually in the range of 10–40 kcal/mole, and the reaction is a typical unimolecular decomposition.

3. Much detailed information on metathesis reactions, such as (11) and (13), has been derived from photochemical investigations. For reac-

tions of this type which involve CH_3, the activation energy is usually in the neighborhood of 7–12 kcal/mole, and the collision theory steric factor is about 10^{-3} to 10^{-4}. Results from reactions of C_2H_5 or C_3H_7 are similar (that is, only about one collision in 10^3-10^4 which meet the energy requirements results in chemical reaction). Data from H atom reactions such as (13) are much less certain, but the reactions are much faster than those of methyl radicals. Information on this type of reaction now exists for a large number of reactions of alkyl radicals and, to a limited extent, for radicals containing chlorine, fluorine, oxygen, and nitrogen.

4. Reaction (15*a*) is an example of radical recombination, and (15*b*) of disproportionation. Such radical-radical reactions are rapid and occur at a high percentage of collisions between the reacting species, that is, with a low activation energy and a high steric factor.

The overall Rice mechanisms are chain reactions, reaction (10) initiating the chain, (12) and (13) propagating it, and (14) and (15) terminating it. Similar mechanisms are involved in a large number of other thermal and photochemical reactions. Gradually a body of information is being built up about the rates of the individual free-radical steps. When such information is widely available, it will become possible to make detailed predictions about the mechanisms of many organic reactions. *See* CHEMICAL DYNAMICS; ELECTRON PARAMAGNETIC RESONANCE (EPR) SPECTROSCOPY.

[JAMES R. BOLTON]

Bibliography: A. M. Bass and H. P. Broida (eds.), *Formation and Trapping of Free Radicals*, 1960; A. Carrington, *Microwave Spectroscopy of Free Radicals*, 1974; A. Carrington and A. D. McLachlan, *Introduction to Magnetic Resonance*, 1979; W. A. Pryor, *Frontiers of Free Radical Chemistry*, 1980; W. A. Pryor, *Introduction to Free Radical Chemistry*, 1966; C. J. M. Stirling, *Radicals in Organic Chemistry*, 1965.

Freon

One of a group of polyhalogenated derivatives of methane and ethane containing fluorine and, in most cases, chlorine or bromine. These include such compounds as trichlorofluoromethane and dichlorodifluoromethane, also known as Freon-11 and Freon-12, respectively. The freons possess such noteworthy characteristics as nonflammability, excellent chemical and thermal stability, and low toxicity. Additional properties include high density, low boiling point, low viscosity, and low surface tension. This unique combination of properties has made the freons particularly suitable for use as refrigerants.

The freons are usually chemically and thermally stable. This stability results directly from the presence of fluorine atoms in the molecule; generally, the more fluorine present, the greater the stability of the compound. Although freon compounds are related, each series member has a different degree of stability and chemical structure.

The most popular freons are Freon-11 and Freon-12. These compounds are prepared by the reaction of hydrogen fluoride with carbon tetrachloride in the presence of a catalyst. This catalyst ($SbCl_4F$) is obtained by a reaction between antimony pentachloride and hydrogen fluoride, as shown in reaction (1). Chlorine in carbon tetrachloride is

$$SbCl_5 + HF \rightarrow SbCl_4F + HCl \quad (1)$$

replaced by fluorine, as in reactions (2) and (3).

$$CCl_4 + HF \xrightarrow{SbCl_4F} CCl_3F + HCl \quad (2)$$

$$CCl_3F + HF \xrightarrow{SbCl_4F} CCl_2F_2 + HCl \quad (3)$$

Freon-11 (CCL_3F), boiling point 23.8°C (74.8°F), is widely used in commercial and industrial air conditioning systems and water coolers. Freon-12 (CCl_2F_2), boiling point −29.8°C (−21.6°F), is the most commonly used of the freon refrigerants. Applications include household and commercial refrigeration and air conditioning.

Other freons are chlorodifluoromethane, or Freon-22 ($CHClF_2$); chlorotrifluoromethane, or Freon-13 ($CClF_3$); 1,2-dichloro-1,1,2,2-tetrafluoroethane, or Freon-113 ($CClF_2$-$CClF_2$); and 1,1,2-trichloro-1,2,2-trifluoroethane, or Freon-114 ($CClF_2$-CCl_2F). Some bromine-containing freons are bromotrifluoromethane, or Freon-13B1 ($CBrF_3$) and 1,2-dibromo-1,1,2,2-tetrafluoroethane, or Freon-114B2 ($CBrF_2$-$CBrF_2$).

In addition to their use as refrigerants, the freons serve as propellants in aerosol products, as solvents, and as intermediates in the synthesis of other fluorine compounds. The most common propellant is Freon-12, alone or in combination with another freon. Freon-13B1 and other freons containing bromine are used as fire-extinguishing agents. Lack of odor, in addition to the properties previously mentioned, renders the freons especially desirable for these purposes. As solvents, the freons rank above the hydrocarbons and below chlorinated solvents. Further application has been in the field of polymers and plastics, where several of the freons are important intermediates. *See* FLUORINE; FLUOROCARBON; HALOGENATED HYDROCARBON.

[ELLEN CLARKE; MAX G. GERGEL]

Friedel-Crafts reaction

A substitution reaction, catalyzed by aluminum chloride, in which an alkyl (R—) group or an acyl (RCO—) group replaces a hydrogen atom of an aromatic nucleus. This general reaction is the most important member of a larger group of aromatic substitution reactions known to be catalyzed by conventional or Lewis acids.

Alkylation. In the classical alkylation reaction, an alkyl halide (RX) serves as the alkylating agent. In reaction (1), an excess of the hydrocarbon

$$R{-}Cl + C_6H_6 \xrightarrow{AlCl_3} R{-}C_6H_5 + HCl \quad (1)$$

benzene (C_6H_6) is the solvent. In case the aromatic hydrocarbon is not so readily available, a smaller quantity may be used in an inert solvent such as carbon disulfide or petroleum ether.

Alkylations usually need less than 1 mole of aluminum chloride ($AlCl_3$) per mole of alkyl halide, and a molar ratio of 1:10 is common. Tertiary halides (R_3CX) are superior to secondary (R_2CHX) or primary (RCH_2X) halides for the alkylation reaction. Rearrangement of the product may occur if the alkyl group is larger than ethyl (C_2H_5). For example, with *n*-propyl chloride ($CH_3CH_2CH_2Cl$), a large part of the product would be isopropylbenzene, $C_6H_5CH(CH_3)_2$, rather than the expected *n*-propylbenzene, $C_6H_5CH_2CH_2CH_3$. Since alkyl groups attached to an aromatic nucleus facilitate further substitution, it is difficult to prevent di- or polysubstitution during the reaction. The structure of the dialkylation and trialkylation products cannot be predicted simply from a knowledge of the rules of substitution, since aluminum chloride is a catalyst for the intermolecular and intramolecular migration of alkyl groups. Alkenes (C_nH_{2n}) may be substituted for alkyl halides in the Friedel-Crafts reaction, and the industrial synthesis of ethylbenzene and isopropylbenzene (cumene) has been carried out in this way.

Acylation. For acylation of aromatic hydrocarbons, acyl halides, reaction (2), have proved most valuable although acid anhydrides, reaction (3),

$$R—COCl + C_6H_6 \xrightarrow{AlCl_3} R—COC_6H_5 + HCl \qquad (2)$$

$$(RCO)_2O + C_6H_6 \xrightarrow{AlCl_3} RCOC_6H_5 + RCOOAlCl_2 + HCl \qquad (3)$$

have also been used. For each mole of acid chloride present, slightly more than 1 mole of $AlCl_3$ is commonly required, whereas with acid anhydrides slightly more than 2 moles of $AlCl_3$ are necessary. The acyl group deactivates the benzene ring toward further substitution. Therefore, it is usually easy to obtain good yields of monosubstitution products. In contrast to the alkyl group, the substituted acyl group does not undergo rearrangement, making acylation a more reliable synthetic tool than the alkylation reaction.

Modifications of the Friedel-Crafts reaction are of value in the acylation of certain heterocyclic compounds such as thiophene and furan. Modifications have also been used in bringing about cyclization reactions through intramolecular acylation. *See* AROMATIC HYDROCARBON; BENZENE; SUBSTITUTION REACTION.

[CHARLES K. BRADSHER]

Bibliography: G. A. Olah (ed.), *Friedel-Crafts and Related Reactions*, 4 vols., 1963–1965.

Fuel cell

An electric cell that converts the chemical energy of a fuel directly into electric energy in a continuous process. The efficiency of this conversion can be made much greater than that obtainable by thermal-power conversion. In the latter the chemical reaction is made to produce heat by combustion. The heat is then transformed partially into mechanical energy by a heat engine, which drives a generator to produce electric energy. Further loss is involved if the direct current generated is converted into alternating current.

Although, in principle, the nature of the reactants is not limited, the fuel-cell reaction almost always involves the combination of hydrogen with oxygen, reaction (1). At 25°C and 1 atm pres-

$$H_2(g) + \tfrac{1}{2}O_2(g) \rightarrow H_2O(l) \qquad (1)$$

sure, that is, standard temperature and pressure (STP), the reaction takes place with a free energy change (ΔG) of $\Delta G = -56.69$ kcal/mole, that is, 237,000 joules/mole water.

If the reaction is harnessed in a galvanic cell working at 100% efficiency, a cell voltage of 1.23 volts results. In actual service such cells have shown steady-state potentials in the range 0.9–1.1 volts, with reported coulombic efficiencies of the order 73–90%.

Fuel cells are of 200–500 watts capacity and 50–100 mA/cm² current density. Larger prototypes have been produced, some as large as 15 kW capacity, while a system under study is expected to provide 100 kW.

In the present stage of development, it is difficult to make a classification of the fuel-cell types. The most popular and successful type remains the classical H_2-O_2 fuel cell of the direct or indirect type. In the direct type, hydrogen and oxygen are used as such, the fuel being produced in independent installations. The indirect type, employs a hydrogen-generating unit which can use as raw material a wide variety of fuel. The reaction taking place at the anode is as in reaction (2), and at the cathode as in reaction (3).

$$2H_2 + 4OH^- \rightarrow 4H_2O + 4e^- \qquad (2)$$

$$O_2 + 2H_2O + 4e^- \rightarrow 4OH^- \qquad (3)$$

Because of the low solubility of H_2 and O_2 in electrolytes, the reactions take place at the interface electrode-electrolyte, requiring a large area of contact. This is obtained with porous materials called upon to fulfill the following main duties: The materials must provide contact between electrolyte and gas over a large area, catalyze the reaction, maintain the electrolyte in a very thin layer on the surface of the electrode, and act as leads for the transmission of electrons.

The porosity is obtained by the Raney technique or by sintering. When flooding of the pores is feared, the electrode is made with double porosity, fine at the electrolyte side and coarse at the gas side. The catalytic effect is obtained with noble metals, mainly platinum, silver, nickel, cobalt, and palladium.

The thickness of the electrolyte layer, on which depends the internal resistance of the cell, is controlled by pore size, wetting properties, and pressure of the fuel gas. When pressure is used, care must be taken not to increase it to the extent that gas is allowed to bubble through the electrolyte because of the danger of forming an explosive hydrogen-oxygen mixture. The fuel cells may work with acid or alkaline electrolytes.

The acid electrolytes require costly corrosion-resistant construction materials but are not sensitive to CO and CO_2 in the fuel, which may lead to the buildup of carbonates. Some models using phosphoric acid proved quite successful. The alkaline electrolytes are more practical, and they are found in most fuel cells produced industrially at present.

Some fuel cells are designed to work with molten carbonates as electrolyte, at temperatures as high as 800°C. These cells are attractive because they can use reformed hydrocarbon fuels, require a small investment, and can be made as large units. They are insensitive to carbon oxides but, at least in the present situation, are affected by important shortcomings, such as excessive size, rapid corrosion of metallic parts, and long periods of heating required before useful service.

The cells may use any alkali metal carbonate or eutectic mixtures of the same. The reactions taking place are those known for the systems involving hydrogen and oxygen or carbon monoxide and oxygen. In the case of carbon monoxide, the reaction step at the anode is as in reaction (4) and at the cathode as in reaction (5).

$$CO + CO_3^{--} \rightarrow 2CO_2 + 2e^- \quad (4)$$

$$CO_2 + \tfrac{1}{2}O_2 + 2e^- \rightarrow CO_3^{--} \quad (5)$$

The total cell reaction is given by Eq. (6), with a

$$2CO + O_2 \rightarrow 2CO_2 \quad (6)$$

free energy change of $\Delta G = -61.45$ kcal/mole of CO. At 100% efficiency, this gives a theoretical cell voltage of 1.34 volts at STP.

Principal fuel-cell reactions. The principal overall reactions which have been employed in fuel-cell work are summarized in Tables 1 and 2.

Table 1. Theoretical cell potentials at various temperatures

Reaction	Cell potential, volts					
	25°C	100°C	250°C	500°C	750°C	1000°C
$C + O_2 \rightarrow CO_2$	1.02	1.02	1.02	1.02	1.02	1.01
$2C + O_2 \rightarrow 2CO$	0.71	0.75	0.82	0.93	1.04	1.15
$2CO + O_2 \rightarrow 2CO_2$	1.33	1.30	1.23	1.11	1.00	0.88
$2H_2 + O_2 \rightarrow 2H_2O$	1.23	1.18	1.12	1.05	0.97	0.90

Table 2. Theoretical material consumption

Reaction	Temperature, °C	Consumption, lb/kWh		
		Anode	Cathode	Total
$C + O_2 \rightarrow CO_2$	750	0.344	0.918	1.262
$2C + O_2 \rightarrow 2CO$	750	0.474	0.632	1.106
$2CO + O_2 \rightarrow 2CO_2$	750	1.15	0.656	1.816
$2H_2 + O_2 \rightarrow 2H_2O$	100	0.070	0.56	0.63
	750	0.085	0.68	0.763

The direct anodic use of carbon has been practically abandoned in modern fuel-cell work. Carbon potentials seem entirely due either to carbon monoxide, CO, or to hydrogen, H_2, formed at high temperature by direct reaction between the carbon and the electrolyte. For example, in the Jacques cell, which consists of carbon electrodes and iron (air) electrodes in molten sodium hydroxide, H_2 is liberated at the carbon by reaction with the electrolyte. It is this H_2 which is responsible for the observed potential.

Modern fuel cells use gaseous fuels, either H_2 or CO or mixtures of these gases. The oxidizer is normally oxygen or air. Hydrocarbons have not been made to function anodically. Where potentials have been measured, they are attributed to decomposition of the hydrocarbon to liberate H_2. For example, methane decomposes at high temperatures, as shown by reaction (7).

$$2CH_4 \rightarrow 2C + 4H_2 \quad (7)$$

For technical reasons, it is simpler to use the carbon or hydrocarbon fuel in a chemical reactor to produce the active gases, H_2 and CO, than to attempt to operate a cell under the conditions best suited for the chemical reaction. Typical chemical production of the active gases might be as in reaction (8).

$$2C + O_2 \rightarrow 2CO \quad (8)$$

Theoretically, CO has a requirement of 1.15 lb/kwhr when reacted anodically against an oxygen cathode. To produce 1.15 lb CO, 0.493 lb carbon is needed. Hence a perfect process, starting with 0.493 lb carbon, would yield 1 kWh. A mixture of H_2 and CO can also be produced by reacting carbon with steam, as shown by reaction (9).

$$C + H_2O \rightarrow CO + H_2 \quad (9)$$

The complete engineering design of the chemical reactor in conjunction with the fuel cell has been extensively studied. The main difficulties in the past were due to the large concentration of CO in the reformed fuel, which poisons the Pt catalyst often used in the fuel cell proper. At present, various new catalysts have been developed, such as Pt-Rh, Pt-Ir, and Pt-Ru, well capable of processing H_2-CO fuel mixtures.

The present fuel generators are capable of covering most of the needs of established fuel cells from almost pure hydrogen to 1:1 mixtures of H_2 and CO. One project, sponsored by the Office of Coal Research, U.S. Department of the Interior, has as its object a 100-kW coal reactor–solid electrolyte fuel-cell system.

Hydrogen-oxygen fuel cell. Work with H_2 has established that it can operate efficiently at moderate temperatures, with polarization decreasing as the temperature is increased. This permits the use of aqueous solutions. One has been reported to have, at 25°C, the following characteristics:

Current density, A/ft^2	0	1	10	50
Cell voltage	1.12	1.01	0.95	0.70

From the point of view of the working principle, three well-developed systems should be mentioned: those of General Electric, Allis-Chalmers, and Bacon. Only the General Electric system had been tested in service in 1968, but semiindustrial tests conducted on the Allis-Chalmers and Bacon systems show that these also can be considered as ready for practical use. Some of their general characteristics are listed in Table 3.

Table 3. Characteristics of three fuel-cell systems

Type	Principle	Temperature, °C	Power, watts	Application
General Electric	Ion-exchange membrane	25–35	100–1000	Gemini
Allis-Chalmers	Porous Ni electrodes and porous electrolyte vehicle	90–100	2000	Space flight
Bacon (Pratt and Whitney)	Porous Ni electrodes	200–220	500–1500	Apollo

In the same category should be mentioned the hydrazine-air fuel cell under investigation by Monsanto Research Corp. and Allis-Chalmers, together with some research centers of the U.S. Army.

This cell is based on the reaction shown in (10). It is a medium-size system intended to

$$N_2H_4 + O_2 \rightarrow N_2 + 2H_2O \quad (10)$$

produce 60–300 watts for the Monsanto project and up to 3 kw for the Allis-Chalmers project. The unit cell voltage is 0.6–0.7 volt, and its greatest advantage is that it uses a condensed fuel, convenient in some applications. A different concept is used in the alkaline metal–oxygen fuel cells developed by the M. W. Kellogg Co. In an actual unit the electrochemical process involves oxidation of Na with oxygen from air, with NaOH as electrolyte. The reaction taking place at the anode is as in reaction (11) and at the cathode as in reaction (12), with the cell reaction given as (13).

$$4Na \rightarrow 4Na^+ + 4e^- \quad (11)$$

$$O_2 + 4e^- + 2H_2O \rightarrow 4OH^- \quad (12)$$

$$4Na + O_2 + 2H_2O \rightarrow 4NaOH \quad (13)$$

Because Na as such is too reactive, the cell uses a sodium amalgam which is quite stable in concentrated NaOH. The sodium amalgam–oxygen fuel cell possesses some remarkable features. It provides almost 1.5 volts at steady state and very high current densities of the order of 200 mA/cm^2, is insensitive to water quality, and requires a small gas consumption. It is expected to play an important role in some applications.

Ion-exchange membrane types. An interesting fuel cell in the laboratory stage is the one which uses inorganic ion-exchange membranes. This type of cell offers two potential advantages: It tolerates higher temperature operation and has a higher ionic conductivity because of the higher density of ion-exchange sites.

An advanced laboratory fuel-cell model developed by Armour Research Foundation uses zirconyl phosphate as an ion-exchange electrolyte, hydrogen and oxygen as fuels, and platinum black as catalyst. The fuel cell is capable of delivering almost 1 volt, but the current density reached so far is too small. The inorganic ion-exchange membrane requires water for its operation.

In the cells using solid electrolytes of the ionic conductive type, the need for water is eliminated. The electrolyte currently studied by Westinghouse Electric is a mixture of zirconium oxide and calcium oxide (0.85:0.15) and is expected to work at about 1000°C. Another promising electrolyte is the mixture zirconium oxide and yttrium oxide (0.9:0.1).

Closed cycle types. The last fuel-cell type to be considered is the closed-cycle type, in the frame of which the reactants are recovered by an auxiliary process. Some systems, such as those developed by Electro-Optical Systems Division of Xerox, are simply made of two converse cycles operating one at a time. The regeneration consists in producing hydrogen and oxygen by electrolysis when electrical energy is available and shifting to cell performance on stored fuel when the production of electrical energy becomes necessary. In the regenerative system developed by the United Aircraft, the electrolysis current is provided by a solar-energy converter made of regular silicon solar cells.

Another possibility is that of radiochemical regeneration. A model under development by Union Carbide is that proposed by J. A. Ghormley, using ferrous sulfate irradiated by gamma radiation. The electrical efficiency reported to the gamma radiation absorbed had been evaluated at 3%. The unit cell voltage is about 0.6 volt, and the module containing 6 units should deliver 5 watts over a period of 2 years.

Problem areas. In the development of fuel cells, there are some general difficulties which must be solved before they become mature industrial propositions. The main problems are as follows.

Catalyst. Its importance increases while the temperature decreases. With few exceptions the catalyst is a very expensive constituent of the cell, and not only will its price increase when industrial production requires increased amounts, but it may even become unavailable at any price. This suggests that future models will tend to work at higher temperatures and pressures (like Bacon's fuel cell) at which cheaper catalysts (Ni, NiO, and so on) can be used.

Capital cost. At an estimated $1000/kw investment the fuel cell is not commercially attractive. By various standard improvements, its capital cost should be reduced to one-tenth this value before becoming competitive with other electrical energy sources.

Heat transfer. In most fuel cells the reaction product is water in gaseous or liquid form. At 100% efficiency there are 421 g of water per kilowatt-hour to dispose of. In addition, about 30% of the heat of reaction, that is, about 260 kcal/kW, must be disposed of in order to maintain the working temperature at a level at which the electrolyte is not decomposed and the material does not become too susceptible to corrosion.

High-temperature fuel cells. The use of carbon monoxide has been limited to high-temperature cells. One carbon monoxide cell using air as the cathodic material operated at about 700°C to yield 0.75–0.85 volt at 32 A/ft^2. It has been demonstrated that molten-salt electrolyte cells can operate at 550–800°C on inexpensive hydrocarbons if steam is admitted to prevent carbon deposition. *See* ELECTRODE POTENTIAL.

[J. DAVIS; L. ROZEANU]

Bibliography: E. Findl and M. Klein, Electrolytic regenerative hydrogen-oxygen fuel-cell battery, *Proceedings of the 20th Annual Power Sources Conferences*, 1966; M. I. Gillibrand and G. B. Lomax, Factors affecting the life of fuel cells, *Proceedings of the 20th Annual Power Sources Conferences*, 1966; D. W. McKee and A. G. Scarpellino, Electrocatalysts for hydrogen/carbon monoxide fuel cell anodes, *Electrochem. Technol.*, 6:101, 1968; R. Noyes, *Fuel Cells for Public Utility and Industrial Power*, 1978; L. Oniciv, *Fuel Cells*, transl. by J. Hammel. 1976.

Fugacity

A function introduced by G. N. Lewis to facilitate the application of thermodynamics to real systems. Thus, when fugacities are substituted for partial pressures in the mass action equilibrium constant expression, which applies strictly only to the ideal case, a true equilibrium constant results for real systems as well.

The fugacity f_i of a constituent i of a thermodynamic system is defined by Eq. (1) [where μ_i is

$$\mu_i = \mu_i^* + RT \ln f_i \qquad (1)$$

the chemical potential and μ_i^* is a function of temperature only], in combination with the requirement that the fugacity approach the partial pressure as the total pressure of the gas phase approaches zero. At a given temperature, this is possible only for a particular value for μ_i^*, which may be shown to correspond to the chemical potential the constituent would have as the pure gas in the ideal gas state at 1 atm pressure. This definition makes the fugacity identical to the partial pressure in the ideal gas case. For real gases, the ratio of fugacity to partial pressure, called the fugacity coefficient, will be close to unity for moderate temperatures and pressures. At low temperatures and appropriate pressures, it may be as small as 0.2 or less, whereas at high pressures at any temperature it can become very large.

The fugacity concept is not restricted to gaseous systems, however. Because of its relation to the chemical potential, the basic thermodynamic criterion of equilibrium requires that the fugacity of a constituent have the same value at equilibrium for every phase in which it is present. This permits the indirect determination of the fugacity for a condensed phase through the calculation of the value for the equilibrium vapor phase, for which the fugacity may be computed routinely if the dependence of its volume on temperature, pressure, and composition is known. Thus, results are readily obtained for a pure gas, but because of the more extensive data required, accurate calculations have been made for very few mixtures.

For an ideal solution, the fugacity is given by the mole fraction of the constituent multiplied by the fugacity of the pure constituent at the temperature and pressure of the solution. For liquid solutions, this is the thermodynamic counterpart of Raoult's law, but the relation applies also to ideal gaseous solutions and can serve for the prediction of the properties of real gas mixtures. Where no equation of state data are available for pure gases, their fugacities can be estimated by means of the generalized fugacity coefficient chart. At sufficiently high dilution in liquid solution, the fugacity of a nondissociating solute will become proportional to its concentration, the proportionality constant depending on the concentration scale used; this is the thermodynamic statement of Henry's law.

The dependence of the fugacity on temperature at constant pressure and composition is given by Eq. (2). Here H_i^* is the enthalpy per mole for the

$$\left(\frac{\partial \ln f_i}{\partial T}\right)_{P,\,\text{comp}} = \frac{H_i^* - \overline{H}_i}{RT^2} \qquad (2)$$

constituent in the gas phase at very low pressure, and $\overline{H}_i$ its contribution per mole to the enthalpy of the system for the state under consideration. *See* CHEMICAL EQUILIBRIUM; CHEMICAL THERMODYNAMICS; GAS; PHASE EQUILIBRIUM; SOLUTION.

[PAUL BENDER]

Bibliography: K. S. Pitzer and L. Brewer, *Thermodynamics*, rev. ed., 1961; F. T. Wall, *Chemical Thermodynamics*, 3d ed., 1974.

Fulminate

A compound containing the —ONC group and derived from fulminic acid, HONC. Fulminates are isomeric with cyanates; that is, cyanates have the same atoms but in different arrangement, —OCN.

The fulminates are commercially important because of the use of mercury fulminate, $Hg(ONC)_2$, in priming compositions and as an initial detonating agent. Mercury fulminate is very sensitive to impact, friction, and heat, but lead azide is replacing mercury fulminate as a detonating agent. *See* AZIDE; CYANATE.

[E. EUGENE WEAVER]

Furan

One of a group of organic heterocyclic compounds containing a diunsaturated ring of four carbon atoms and one oxygen atom. Furan (I) is a typical member of the group. Furfural (II) and some of its

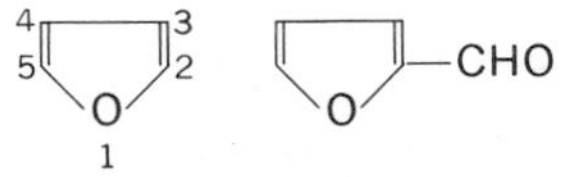

(I) (II)

close relatives, such as furfuryl alcohol, tetrahydrofurfuryl alcohol, and tetrahydrofuran, are important chemicals of commerce. *See* FURFURAL; HETEROCYCLIC COMPOUNDS.

Properties and preparation. Furan (I) is a colorless, volatile liquid, bp 31.4°C, n_D^{20} 1.42140, density (20/4) 0.9378, which is stable to alkali but not to mineral acid. Its water solubility is approximately 1% at room temperature. On exposure to air, furan decomposes very slowly by autoxidation. Substituted furans, particularly negatively substituted furans, are much less sensitive. The furan system is aromatic, with resonance energies of 17–25 kcal/mole reported. Nitration, halogenation, acylation, mercuration, and sulfonation reactions occur with relative ease. The extreme sensitivity of furan itself to strong acid precludes many direct electrophilic substitutions. The presence of negative groups stabilizes the ring. Unoccupied 2 or 5 positions are favored as the site of substitution.

Addition reactions, in contradistinction to substitution reactions, have also been observed. Furan, for example, condenses with dienophiles in the Diels-Alder reaction.

The most versatile general furan synthesis is the acid-catalyzed cyclization of 1,4-dicarbonyl compounds (III). The R groups in (III) may be hydro-

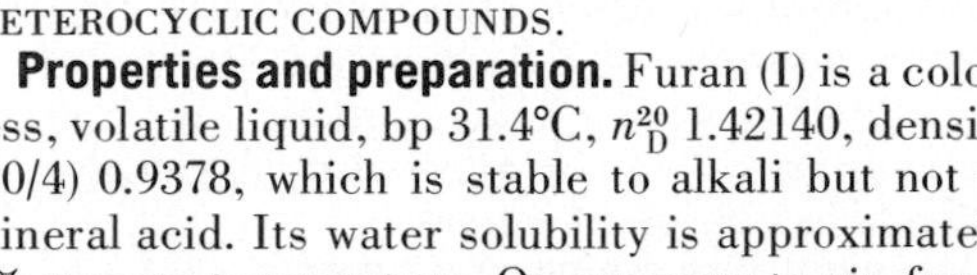

(III)

gen, aryl, alkyl, or carbethoxyl. Another synthesis (Treibs) transforms an α,β-unsaturated ketone (IV) to an unsaturated δ-sultone (V), which on pyrolysis loses sulfur dioxide and forms the furan. Furans are also formed in the acid-catalyzed dehydration of carbohydrates, the most important example being the production of furfural from the polysaccharides present in annually renewable agricultural waste materials (VI). Furan is obtained from

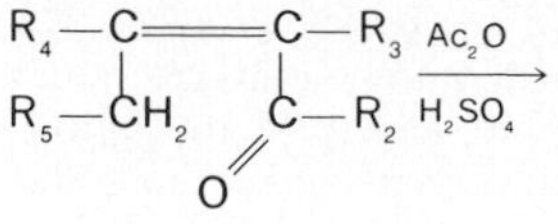

(IV)

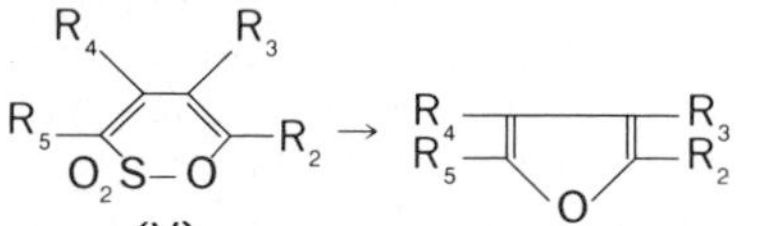

(V)

furfural either directly by catalytic decarbonylation or by oxidation to furoic acid (VII) followed by decarboxylation.

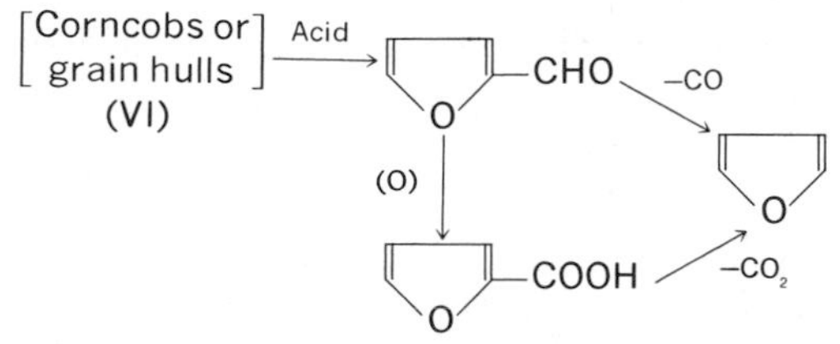

(VII)

Important derivatives. Oxidation of furans generally disrupts the system, although some ring oxidations give useful products. Ring reduction is possible, the standard method being catalytic rather than chemical. Several products derived from furfural directly or indirectly by hydrogenation are of considerable industrial importance.

Reaction of furans with halogens, although not altogether free of complications, can be used for the synthesis of halofurans. Furansulfonic acids are prepared by sulfonation with sulfuric acid, or with a sulfur trioxide–pyridine complex. Nitration of furan compounds generally makes use of fuming nitric acid in acetic anhydride solvent. The synthesis of the bacteriostatic and bacteriocidal agent nitrofurazone, Furacin (IX), involves such a nitration of derivative (VIII).

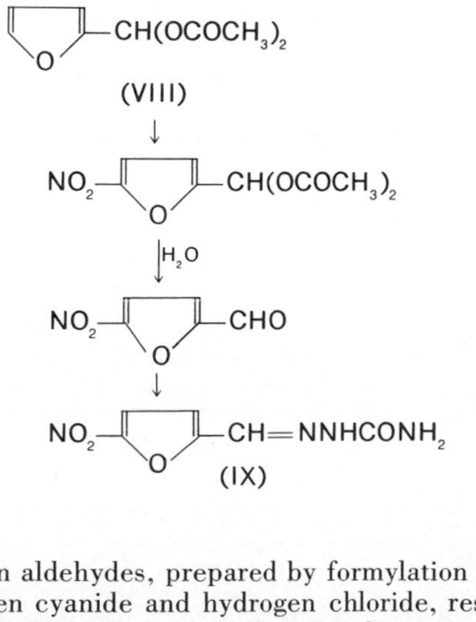

Furan aldehydes, prepared by formylation with hydrogen cyanide and hydrogen chloride, resemble benzaldehyde in their functional group reactions. Furyl ketones, prepared by Friedel-Crafts acylation, react normally. All the furan mono- and polycarboxylic acids are known. Removal of the carboxyl group from the 2 or 5 positions by decarboxylation is easier than from the 3 or 4 positions.

Tetrahydrofuran (XI) is prepared either by decarbonylation of furfural to furan followed by catalytic hydrogenation, or by the Reppe process from butynediol (X). Tetrahydrofuran is a water-miscible,

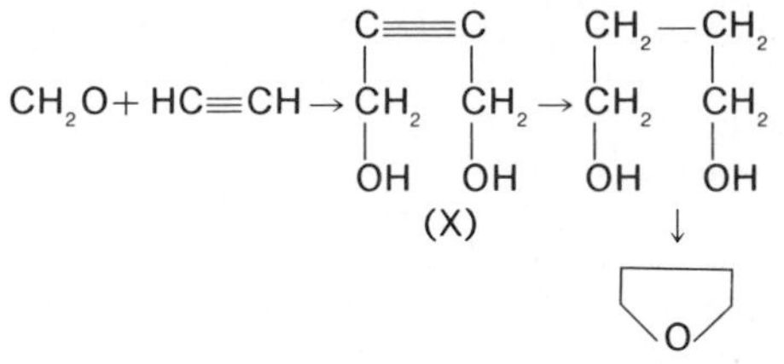

(XI)

colorless liquid, bp 66°C, density (20/4) 0.8880, and n_D^{20} 1.4073. It is an industrial solvent, and also has served as an intermediate in the synthesis of butadiene for polymerization, adiponitrile (XII) for nylon,

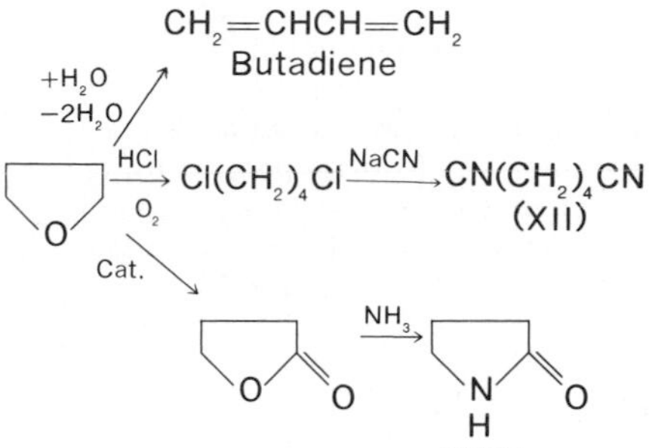

(XIII)

lon, and pyrrolidone (XIII) for the blood extender polyvinylpyrrolidone. *See* PYRAN.

[WALTER J. GENSLER]

Bibliography: G. M. Badger, *The Chemistry of Heterocyclic Compounds*, 1961; A. P. Dunlop and F. N. Peters, *The Furans*, Amer. Chem. Soc. Monogr. no. 119, 1953; L. A. Paquette, *Principles of Modern Heterocyclic Chemistry*, 1968; M. V. Sargent and T. M. Cresp, Furans, in P. G. Sammes (ed.), *Comprehensive Organic Chemistry*, vol. 4, 1979.

Furfural

When pure, a colorless liquid aldehyde boiling at 161.7°C and also called furfuraldehyde, fural, 2-furaldehyde, or 2-furancarboxaldehyde. It is obtained by the digestion of corn cobs or oat hulls with dilute mineral acid. A wide variety of agricultural by-products also yield furfural if treated in a similar manner; thus, the potential supply of this highly reactive aldehyde is virtually limitless.

Uses. Furfural was first produced on a commercial scale in 1922, and the subsequent reductions in price have aroused increasing interest in this chemical. It is used as a general synthetic intermediate in the preparation of chemicals, many of which compete with the same chemicals derived from coal and petroleum. Large amounts are also used in the preparation of molding resins and other polymers of value to the plastics industry. Its unusual solvent properties make it useful in the refining of vegetable and lubricating oils and in extracting certain components, such as butadiene, from cracked refinery gases. Other uses to which furfural has been put are as an insecticide, herbicide, fungicide, and embalming fluid.

FURFURAL

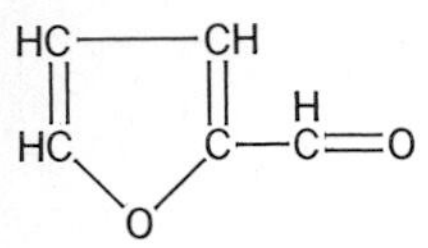

Structural formula of furfural.

Chemical properties. Furfural has the structural formula shown in the figure. It contains not only an aldehyde group but also an ether linkage (C—O—C) and a system of alternating single and double bonds (diene structure). These provide several sites in the molecule for reaction with other compounds. It is similar to benzaldehyde in most of its reactions as an aldehyde and in its ability to undergo substitution reactions for hydrogen atoms on the ring. Resin formation can take place by reaction at the aldehyde group or through the diene-aldehyde system (also known as a resinophore grouping). The ether linkage provides a point at which the ring can be opened to give straight-chain compounds.

Three general routes to furfural derivatives are employed in the chemical industry. It can react as an aldehyde with phenols, ketones, esters, and other materials to produce resins and compounds valuable as plasticizers. Catalytic removal of the aldehyde group from furfural gives furan, C_4H_4O, which can be hydrogenated to tetrahydrofuran. The latter is a widely used solvent and also an intermediate in the preparation of polytetramethylene ether glycol, an important ingredient of a variety of polyurethane elastomers and spandex-type fibers. Catalytic hydrogenation of furfural gives furfuryl alcohol and tetrahydrofurfuryl alcohol. The former is used in the preparation of resins, and the latter can be converted to dihydropyran. Dihydropyran is either hydrogenated to give tetrahydropyran, which resembles tetrahydrofuran in its solvent properties, or converted directly to polymeric materials and chemicals important to the industry. *See* ALDEHYDE; BENZALDEHYDE; FURAN; POLYMERIZATION.

[ALEX E. BRODHAG, JR.]

Fused-salt solution

A nonaqueous solvent system particularly useful in coordination chemistry. Fused salts are a large class of liquids which are composed largely of ions. Many simple inorganic salts melt at rather high temperatures (greater than 600°C), forming liquids which have high specific conductivities, 1–6 $ohm^{-1} \cdot cm^{-1}$ (see Table 1). There is, however, a number of exceptions to this generalization; for example, the electrical conductivity of $AlCl_3$ decreases sharply upon melting due to the formation of a molecular liquid (Al_2Cl_6). *See* HIGH-TEMPERATURE CHEMISTRY; SOLVENT.

The use of binary or ternary melt compositions results in liquids which typically have much lower melting temperatures and somewhat lower specific conductivities than do pure salts (see Table 2). The choice of melts for use as solvents is frequently based on such considerations as the availability and cost, the lowest melting temperature attainable, and the ease of purification of the solvent, as well as the width of the electrochemical span and the spectroscopic transparency of the melt. Several molten-salt solvents, such as the LiCl–KCl eutectic and the equimolar $NaNO_3$–KNO_3 melt, have been extensively studied, and many of their physical and chemical properties are well known. For example, an abbreviated electromative force (emf) series in the LiCl–KCl eutectic at 450°C is shown in Table 3. Many melt systems, particularly ternary and more complex compositions, have been only partially characterized, and their physical properties have to be estimated by using the information available for less complex systems, such as the binary component melts.

Chloroaluminate systems. For some melt systems, such as those containing aluminum halides, the melt composition can be adjusted to change the acid-base properties of the solvent which, in turn, determine the chemistry of solutes dissolved in these melts. The $AlCl_3$–NaCl system has been studied extensively and will be used to illustrate the concepts involved. The modified Lewis acidity (an acid in this system is defined as a chloride ion acceptor) and $pCl^- \equiv -\log [Cl^-]$ (this concept is similar to pH) are frequently used in discussing these systems. Raman spectral measurements have shown that the $AlCl_4^-$ ion is the predominant anion in the equimolar $AlCl_3$–NaCl melt. As the $AlCl_3$/NaCl molar ratio is made greater than one, the melt becomes acidic and the concentration of the $Al_2Cl_7^-$ ion increases. Near the equimolar point the dominant equilibrium is that shown in reaction (1). The mole fraction equilibrium constant for this equilibrium has been determined from potentiometric measurements as 1.1×10^{-7} at 175°C. The most basic sodium chloroaluminate system corresponds to a NaCl-saturated melt in which the pCl^- is constant in the presence of solid NaCl and is equal to 1.1 at 175°C. Two other equilibria [reactions (2) and (3)] become important

$$2\,AlCl_4^- \rightleftharpoons Al_2Cl_7^- + Cl^- \qquad (1)$$

$$3\,Al_2Cl_7^- \rightleftharpoons 2\,Al_3Cl_{10}^- + Cl^- \qquad (2)$$

$$2\,Al_3Cl_{10}^- \rightleftharpoons 3\,Al_2Cl_6 + 2Cl^- \qquad (3)$$

as the melt acidity is increased by increasing the $AlCl_3$ content. Reaction (1) is analogous to the dissociation of water [reaction (4)]. Similar equilibria have been proposed for other melt systems containing a covalent halide such as $FeCl_3$ or $SbCl_3$.

$$2\,H_2O \rightleftharpoons H_3O^+ + OH^- \qquad (4)$$

A major difference between water (and related

Table 1. Some physical properties of single-component melts

Property	NaCl	LiCl	KCl	NaF	NaBr	NaOH	Na_2S	$NaNO_3$	Na_2CO_3	Na_2SO_4	$MgCl_2$	$CaCl_2$	$AlCl_3$
Melting point, °C	800±2	610±2	770±1	995±3	747	318	1170±10	307±1	858±1	884±2	714±2	782±5	192±1 (@ 2.3 atm*)
Density, g/cm³	1.542 (@827°C)	1.490 (@637°C)	1.512 (@797°C)	1.928 (@1027°C)	2.320 (@774°C)	1.772 (@345°C)	–	1.866 (@337°C)	1.959 (@887°C)	2.058 (@907°C)	1.668 (@747°C)	2.070 (@807°C)	1.231 (@217°C)
Specific conductivity, $ohm^{-1} \cdot cm^{-1}$	3.687 (@827°C)	5.854 (@637°C)	2.229 (@797°C)	5.019 (@1027°C)	2.976 (@774°C)	2.332 (@345°C)	–	1.107 (@337°C)	2.978 (@887°C)	2.333 (@907°C)	1.077 (@747°C)	2.147 (@807°C)	7.45×10^{-7} (@217°C)
Viscosity, cp†	0.986 (@827°C)	1.377 (@637°C)	1.020 (@797°C)	1.78 (@1027°C)	1.26 (@800°C)	2.83 (@400°C)	–	2.57 (@337°C)	3.40 (@887°C)	8.53 (@967°C)	2.04 (@747°C)	2.90 (@807°C)	0.271 (@217°C)

*2.3 atm = 233 kPa.
†1 centipoise = 10^{-3} Pa · s.
SOURCE: G. Mamantov, Molten salt electrolytes in secondary batteries, in D. W. Murphy et al. (eds.), *Materials for Advanced Batteries*, NATO Conf. Ser. VI: Materials Science, vol. 2, pp. 111–122, Plenum, 1980.

Table 2. Some physical properties of several molten-salt solvents

Property	LiCl – KCl	NaCl – KCl	$CaCl_2$ – NaCl	$AlCl_3$ – NaCl	Na_2S_x	LiF – LiCl – LiBr	Li_2CO_3 – Na_2CO_3 – K_2CO_3
Liquidus temperature, °C	355 (eutectic, 58.5 mole % LiCl)	685 (equimolar)	500 (eutectic, 48 mole % NaCl)	107 (eutectic, 36.8 mole % NaCl) ~155 (equimolar)	230 (Na_2S_3) 255 (Na_2S_5)	445 (eutectic, 22 mole % LiF, 21 mole % LiCl, 47 mole % LiBr)	397 (eutectic, 43.5 mole % Li_2CO_3, 31.5 mole % Na_2CO_3, 25.0 mole % K_2CO_3)
Density, g/cm³	1.646 (for 60 mole % LiCl @ 447°C)	1.571 (@ 717°C)	1.855 (for 49.1 mole % NaCl @ 787°C)	1.691 (for 48 mole % NaCl @ 177°C); 1.644 (for 38.2 mole % NaCl @ 177°C)	1.888 (for Na_2S_3 @ 327°C); 1.876 (for $Na_2S_{4.8}$ @ 327°C)	2.19 (for eutectic @ 500°C)	2.110 (for eutectic @ 467°C)
Specific conductivity, ohm⁻¹ · cm⁻¹	1.615 (for 58.8 mole % LiCl @ 457°C)	2.396 (for 51.2 mole % NaCl @ 717°C)	1.183 (for 51.8 mole % NaCl @ 557°C)	0.462 (for 50 mole % NaCl @ 187°C); 0.197 (for 31 mole % NaCl @ 187°C)	0.561 (for Na_2S_3 @ 327°C); 0.296 (for $Na_2S_{5.1}$ @ 327°C)	—	0.516 (for eutectic @ 457°C)
Viscosity, cp*	1.46 (for 60 mole % LiCl @ 617°C)	1.58 (for 51.2 mole % NaCl @ 727°C)	4.43 (for 50 mole % NaCl @ 657°C)	2.645 (for 50 mole % NaCl @ 187°C); 3.246 (for 31.1 mole % NaCl @ 187°C)	24.87 (for $Na_2S_{3.1}$ @ 327°C); 19.37 (for $Na_2S_{5.2}$ @ 347°C)	—	5.402 (for eutectic @ 487°C)

*1CP = 10^{-3} Pa · s.
SOURCE: G. Mamantov, Molten salt electrolytes in secondary batteries, in D. W. Murphy et al. (eds.), *Materials for Advanced Batteries*, NATO Conf. Ser. VI: Materials Science, vol. 2, pp. 111 – 122, Plenum, 1980.

solvents) and the chloroaluminate solvent system is the ability of the latter to stabilize unusually low oxidation states of metals in acidic melts. Thus ions such as Cd_2^{2+}, Bi^+, and Hg_3^{2+} are quite stable in $AlCl_3$-rich melts. This unusual chemistry has been attributed to the small tendency for disproportionation of the low oxidation state; such disproportionation would be favored by large concentrations of halide ions which complex preferentially the higher oxidation state. For example, reaction (5) is favored by the presence

$$Cd_2^{2+} \rightleftharpoons Cd + Cd^{2+} \qquad (5)$$

of chloride ions; the chloride concentration can be made small by adding a Lewis acid such as $AlCl_3$. The chemistry in molten chloroaluminates bears considerable similarity to that in the so-called superacid systems, such as $HSO_3F – SbF_5$. Thus sulfur, selenium, and tellurium in both media form homopolyatomic cations such as S_8^{2+} and Se_4^{2+}, as well as yet poorly characterized radical species. *See* CHEMICAL EQUILIBRIUM; SUPERACIDS.

It should be stressed that small or moderate changes in melt composition can have a drastic effect on the chemistry of solutes in the chloroaluminate melts. For example, Ti(II) is a stable oxidation state in melts quite rich in $AlCl_3$, such as $AlCl_3$ – NaCl (65 – 35 mole %). In less acidic melts, Ti(II) forms a precipitate, while in melts which contain an excess of NaCl, no evidence for the formation of divalent titanium has been obtained. Similarly in alkali fluoride melts, only Ti(IV) and Ti(III) exist; the reduction of trivalent titanium leads directly to titanium metal. The use of fluoride media results in greater stability of the high oxidation states because many metal ions in these oxidation states form very stable complexes with fluoride ions.

It is interesting to contrast the reduction of refractory metal ions, such as those of niobium, tantalum, and tungsten, in the basic alkali fluoride media, for example, the LiF – NaF – KF eutectic (Flinak) and in the acidic chloroaluminate melts. In Flinak the reduction proceeds through several steps (some of which are irreversible) to the metal, which may be obtained in coherent nondendritic form. In $AlCl_3$-rich chloroaluminate melts, the metal clusters, such as W_6Cl_{12}, are readily formed and the formation or plating of the metal may not be possible. On the other hand, the cluster species formed may be of interest in other applications, such as catalysis. For example, it has been reported that $Ir_4(CO)_{12}$ in acidic $AlCl_3$ – NaCl melts catalyzes the methanation reaction (formation of CH_4 from CO and H_2).

Applications. Fused-salt solutions are being investigated for use as catalyst systems, in rechargeable batteries, and in fuel cells.

Table 3. The emf series in LiCl – KCl eutectic at 450°C*

Couple	E^0_M(Pt), V	E^0_m(Pt), V	E^0_X(Pt), V	Precision, V
Li(I)/Li(0)	−3.304	−3.320	−3.410	0.002
Na(I)/Na(0)	−3.25	−3.23	−3.14	0.008
Mg(II)/Mg(0)	−2.580	−2.580	−2.580	0.002
Al(III)/Al(0)	−1.762	−1.767	−1.797	0.009
Zn(II)/Zn(0)	−1.566	−1.566	−1.566	0.002
Cr(II)/Cr(0)	−1.425	−1.425	−1.425	0.003
Fe(II)/Fe(0)	−1.172	−1.172	−1.172	0.005
Se(I),C/Se_x^{2-}	−1.141	−1.172	−1.252	0.002†
S(I),C/S_x^{2-}	−1.008	−1.039	−1.219	0.002†
Cu(I)/Cu(0)	−0.957	−0.941	−0.851	0.004
Ni(II)/Ni(0)	−0.795	−0.795	−0.795	0.002
Ag(I)/Ag(0)	−0.743	−0.727	−0.637	0.002
HCl(g)/H_2(g), Pt	−0.694	−0.710	−0.800	0.005
Pt(II)/Pt(0)	0.000	0.000	0.000	0.002
Cu(II)/Cu(I)	+0.061	+0.045	−0.045	0.002
Fe(III)/Fe(II)	+0.086	+0.070	−0.020	0.003
Au(I)/Au(0)	+0.205	+0.221	+0.311	0.008
	+0.322	+0.306	+0.216	0.002

*E^0_M, E^0_m, and E^0_X are standard potentials on the molarity scale, molality scale, and mole fraction scale, respectively.
†Extrapolated.
SOURCE: G. Mamantov, Molten salt electrolytes in secondary batteries, in D. W. Murphy et al. (eds.), *Materials for Advanced Batteries*, NATO Conf. Ser. VI: Materials Science, vol. 2, pp. 111 – 122, Plenum, 1980.

Catalysts. The use of fused salts as media for organic reactions is of increasing interest. Not only does the fused-salt environment provide for a better thermal control of the reaction (heat dissipation is readily possible), but the fused salt may serve as a catalyst for the reaction. For example, molten $SbCl_3$ and $ZnCl_2$ have been found to be effective hydrocracking catalysts for coal. It has been found that polycyclic hydrocarbons, such as anthracene, undergo several types of reactions in $SbCl_3$ melts, including formation of radical cations and protonated species which react further to form condensed systems, such as anthra[2.1-*a*]aceanthrylene. The unraveling of this complex chemistry may lead to more effective utilization of molten salts as catalytic media.

Batteries and fuel cells. Another technological area of great interest in which molten salts play a key role is that of advanced batteries and fuel cells. Thus, LiCl – KCl eutectic is the solvent in the rechargeable Li(Al) – FeS (or FeS_2) battery, and sodium polysulfide melts are employed in the promising sodium/sulfur battery which operates at about 350°C. Also under development is a rechargeable cell Na/Na^+ conductor/S(IV) in $AlCl_3$ – NaCl. This cell at 200 – 250°C has an open circuit voltage of 4.2 V as well as high energy densities; it operates at 200 – 250°C.

Molten carbonates, such as the ternary Li_2CO_3 – NA_2CO_3-K_2CO_3 melt, are used in fuel cells which employ H_2/CO mixtures and oxygen as electrochemical fuels. *See* FUEL CELL.

[GLEB MAMANTOV]

Bibliography: M. Blander (ed.), *Molten Salt Chemistry*, 1964; G. J. Janz, *Molten Salts Handbook*, 1967; G. Mamantov (ed.), *Characterization of Solutes in Non-Aqueous Solutes*, 1978; B. R. Sundheim (ed.), *Fused Salts*, 1964.

Gadolinium

A metallic chemical element, Gd, atomic number 64 and atomic weight 157.25, belonging to the rare-earth group. The naturally occurring element is composed of the following isotopes: Gd^{152} 0.20%, Gd^{154} 2.15%, Gd^{155} 14.73%, Gd^{156} 20.47%, Gd^{157} 15.68%, Gd^{158} 24.87%, Gd^{160} 21.90%. Gd^{152} is the radioactive α-emitter with a half-life of 1.1×10^{14} years. In 1880 J. C. G. Marignac obtained a new rare earth in an impure condition and termed it "Y." In 1886 he gave it the name gadolinium in honor of the Swedish scientist J. Gadolin. The oxide, Gd_2O_3 in powdered form is white, and solutions of the salt are colorless. For properties of the metals *see* RARE-EARTH ELEMENTS.

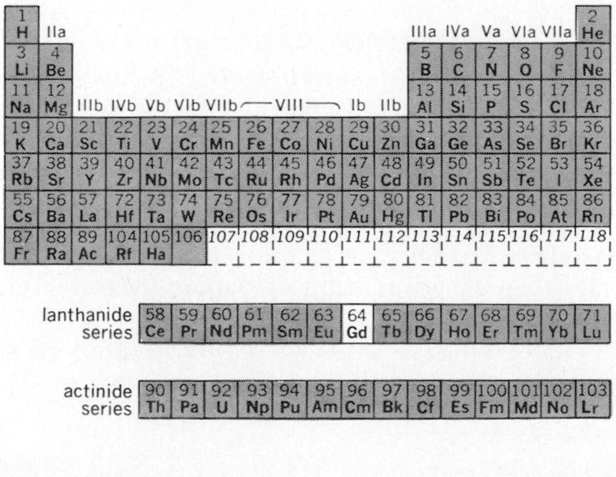

Gadolinium salts are of particular interest because they were the first salts used to obtain temperatures below 1 K by means of magnetic cooling. Gadolinium is of interest in nuclear fission because isotopes 155 and 157 have high thermal neutron-absorption cross sections, 58,000 and 240,000 barns, respectively. They may be used as burnable poison materials in controlling nuclear chain reactions. Gadolinium metal is paramagnetic and becomes strongly ferromagnetic below room temperatures. The Curie point, where this transition occurs, is about 16 K.

[FRANK H. SPEDDING]

Bibliography: J. E. Powell, F. H. Spedding, and D. B. James, The separation of rare earths, *J. Chem. Ed.*, 37(12):629 – 633, 1960.

Gallium

A chemical element, Ga, atomic number 31 and atomic weight 69.72. Gallium was discovered by Lecoq de Boisbaudran in France in 1875. The discovery followed an extended spectroscopic search, instigated because Boisbaudran had noted that a gap occurred in the regularity of the spectral lines between aluminum and indium in the third group of the periodic arrangement of the elements. After

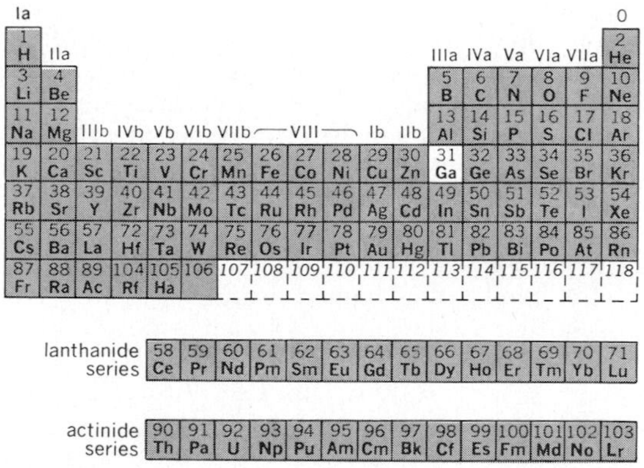

examining hundreds of mineral samples, Boisbaudran found the missing spectral lines in a sample of zincblende from the Pyrenees. In the same year, he succeeded in isolating a few grams of the metallic element and named it gallium in honor of France.

Uses. The uses for gallium were mostly experimental until the 1970s. Its great temperature range in the liquid state has prompted its use to some extent in high-temperature thermometers and manometers. In an alloy with silver and tin, gallium helps to form a suitable replacement for amalgam in dental fillings. Gallium is also used to solder nonmetallic materials, including gems, to metals. Gallium has received considerable attention as a heat-exchange medium for nuclear reactors. However, its highly corrosive effects on most metals, except tungsten and rhenium, at elevated temperatures, and its high price have discouraged this use.

The addition of gallium to cadmium for arc lamps does not greatly modify the cadmium arc spectrum but prevents the cadmium from adhering to the quartz or glass and from breaking it on cooling. Gallium can also be used in vapor arc lamps which are intended for analysis because of its wide frequency range. Addition of gallium iodide to a high-pressure mercury lamp increases the radiation intensity in the radiation region of 400–420 nm.

Possible applications in the preparation of dyes, as well as in chemical analysis and purification, may result from the ability of gallium to form acidoaminate complexes with certain organic acids.

Gallium arsenide can be used in a system for translating mechanical motion into electrical impulses. The properties of this compound are given in Table 1. A movable grating is interposed between a light-emitting gallium arsenide diode (forward-biased) and a photosensitive gallium arsenide diode (reverse-biased). If the grating is attached to a diaphragm or other mechanical system, its motion can modulate the light as it impinges on the junction of the reverse-biased diode. Low-temperature resistance thermometers are made of *n*-type gallium arsenide.

Synthetic superconducting articles are prepared by making a porous matrix of vanadium or tantalum, impregnating it with gallium hydride (liquid at room temperature), and heating to form V_3Ga or Ta_4Ga. Superconductors can also be made from composites of unalloyed copper with hollow vanadium cores filled with V_2Ga_5 powder. Mechanical reduction of the composite to fine wire, followed by heat treatment, produces the superconductors, while the V_2Ga_5 reacts with vanadium to form V_3Ga.

Gallium ions are used (as an alternative to chromium) to replace some of the smaller aluminum ions in a lanthanum-calcium aluminate crystal used as the solid electrolyte in a fuel cell. The oxygen electrode is a surface of the crystal in contact with oxygen gas; and the hydrogen electrode is another surface, one that is in contact with hydrogen gas.

Gallium can also be used as the positive electrode in a regenerative battery having a molten salt of an alkali or alkaline-earth metal as the electrolyte, and a molten metal (corresponding to the positive ion of the salt) as the negative electrode. The electrolyte is isolated from the molten electrodes by a porous separator. The two liquid electrodes interact by diffusion, the voltage varying inversely with increasing dilution of the gallium. Accordingly, depleted or diluted electrode metals are continuously replaced with pure metals and the two interdiffused electrode metals can be separated and recovered by distillation. The cell is therefore capable of continuous operation, has a high power output, and is low in weight. Because of its low density, gallium has an advantage over denser metals, such as bismuth or mercury, that might be used for the anodes.

Table 1. Properties of gallium arsenide

Melting point	1240°C
Band gap	
room temperature	1.43 eV
90 K	1.48 eV
0 K	1.52 eV
Electron mobility	12,000 cm^2/volt/sec
Hole mobility	450 cm^2/volt/sec
Dielectric constant	11.1
Resistivity	6.2×10^8 ohm-cm
Intrinsic resistivity	3.7×10^8 ohm-cm
Intrinsic electron concentration	$10^7/cm^3$
Effective mass ratio (m^*/m_0)†	0.068
Effective dielectric constant ratio (E^*/E_o)†	10.9
Density	
electrons	5.3 g/cm^3
holes	0.5 g/cm^3
Hall mobility at 300 K	8500 cm^2/volt/sec
Thermal conductivity (300 K)	
impure	0.44 watt/cm K
high-purity	0.58 watt/cm K
Heat content (298 K)	8.85 ± 0.50 kcal/mole
Entropy (298 K)	
solid	1.42 ± 0.30 kcal/deg mole
liquid	1.3 kcal/deg mole
Free energy (ΔG°_{298})	8.43 ± 1.59 kcal/g-mole

†Here the asterisk indicates effective superlattice, and the subscript zero indicates standard or intrinsic superlattice.

Toxicologically, gallium is a safe metal, and its association with other metals sometimes makes them less hazardous. For example, nickel implants in rats have induced sarcoma formation, but tumors are never induced by gallium-containing subcutaneous implants such as GaSNiSn or GaNiSn.

Some gallium-containing catalysts have been developed for various purposes. One formulation, SiO_2(33 wt %)-Al_2O_3(5 wt %)-Ga_2O_3, has a high activity as a cracking catalyst for hydrocarbon oil and a low yield of coke and gas. As little gallium as 0.001 wt % of a diester can catalyze the polymerization of the latter. In the process a gallium halide with ammonia, zinc acetate, and glycol are heated with a diester, such as a dialkyl terephthalate, to give a colorless polymer. Zinc oxide doped with gallium is another catalyst used for organic reactions carried out at 300–400°C. Dibutylphosphinic acid and $GaCl_3$ at 100–250°C can catalyze the polymerization of epoxides, and a catalyst containing gallium, iridium, and platinum is effective in converting heptane into an aromatic compound.

As the cation of an arsenide, a phosphide, a telluride, a selenide or of an antimonide or as a dopant in germanium, silicon, or silicon carbide, gallium has given excellent results in a semiconductor for use in rectifiers, transistors, photoconductors, light sources, laser or maser diodes, and refrigerating devices. The excellent semiconductive properties of the gallium-base compounds, particularly at higher temperatures and at both lower and higher frequencies than are usable with germanium and silicon, are attested by the hundreds of reports of investigations and patents that have appeared annually in the 10 years following 1958, and the thousands produced each year in the 1970s. To meet the need for semiconductor-grade gallium, the metal is being produced with a purity of 99.9999%, and its use in electronic devices is being internally studied all over the world. A brief summary of some of these developments is given in Tables 2 and 3.

Interest in low-energy-gap materials centers about thermoelectricity and infrared detection, and in high-energy-gap materials it centers about high-temperature electronic devices. Matters of importance are purification (zone refining, segregation coefficients, vapor pressures of impurity and host, and reactivities); analysis (neutron activation, mass spectroscopy, and so on); shaping and deformation (cutting, grinding, working, cleavage, diffusion, and dislocations); stoichiometry; phase stability (transformations and nucleation and growth); and thermodynamic parameters (lattice energies, energies of formation, work functions, ionization potentials, contact with leads, surface

Table 2. Typical gallium-containing energy-gap compounds having gallium arsenide as a base compound

Construction	Use and behavior
pn Type, doped with In-10Cu-(S,Se,Te)	Diodes
2 GaAs wafers in parallel	Microwave oscillators (peak power at 205 W at 1540 MHz); efficiency, 3–9%
p-Type, doped with Fe and S; deep acceptor (Fe) conc: 10^{16} to 5×10^{17} cm^{-3}	Electroluminescent transistors, with improved high-frequency performance and current-gain stability at high temperatures
n-Type; active region only a few micrometers thick	Injection lasers whose delay time t_d between current I and light emission at 77 K is given by $t_d = \ln[I/(I - I_{th})]$, where I_{th} is the threshold current, and is the spontaneous lifetime (about 2 nanoseconds); (activated by superimposed current pulse)
$GaAs_{1-x}P_x (x = 50\%)$ *pn* type (proportions correspond to desired wavelength, which can thus be extended into visible spectrum)	Laser or maser; transmits intelligence or pumps optical masers
np Type, stacked (since power dissipation arises primarily across the bulk region of a diode and GaAs has relatively low thermal conductivity, minimizing the thickness of the *p* and *n* junctions and stacking them increases the peak output power)	Laser diode; an 845-nm luminous area 0.05×0.05 cm can emit 200 peak power pulses per second of greater than 300 W into a cone of 0.01 steradians when operated at 77 K; current density can be as low as 300 A/cm^2
Forward-biased diode	Refrigeration; more than 90% of the photons emitted have energies $h\nu$ higher than the applied voltage; this excess comes from lattice heat, removal of which, in the form of photon energy, should lead to refrigeration

Table 3. Typical gallium-containing energy-gap compounds having gallium phosphide as a base compound

Construction	Use and behavior
n-Type	Rectifiers; operate at −55 to +500°C; devices 0.2 in. in diameter can rectify 15 A at room temperature; units with a junction diameter of 5 mils can carry forward currents of 0.5 A at 3–4 V and reverse currents of 10 mA at 10–20 V; at room temperature the ratio of forward to reverse current at a given voltage is 10^6–10^7; at 300°C the ratio is 10^3; at 500°C the ratio is 10
n-Type or *p*-type ($5 \times 4 \times 0.3$ mm); some Cu adsorbed on surface; intrinsic resistivity, *n*-type or *p*-type, depends on amount; 10^{17}/cm^3 free electrons	Light sources and photocells; resistivity in the dark is 10^9 ohm-cm; entropy at 298 K = 12.4 eu
Polycrystal doped with Zn (Si and Cu impurities); impact excitation of the carrier concentration with Si impurity is 10^{15}–10^{17} cm^{-3}; that with Cu impurity is 10^{-8}	Capacitor plates and light sources; a capacitor plate consists of a metal electrode with a 0.3-mm coating of GaP powder mixed with a melanine aldehyde resin; for light sources, brightness of electroluminescence is a 3d power function of electric field strength and is greater at 77 K than at room temperature
p-Type; dopants: O, Cu, Zn	Light sources and photocells; a red (Zn-O) band photoluminescent time decay at 1.82 eV (20 K) is nonexponential with time and shortens with increasing Zn; switching time is 10^{-6} sec and quantum efficiency at room temperature is 1%
Diffused *pn* junction (reverse-biased)	Electroluminescent transistors; at 298 K light emission occurs at defects due to dislocations and precipitates; at 77 K it occurs at a different bias range
pn Heterojunction, with epitaxial film of GaS	Photoconductor; two maxima in the spectral distribution of photosensitivity—one in the shortwave region and one near the red limit of photosensitivity
Contains Ge, with or without shallow acceptors (Zn or Cd) or donors (Te or Se); deep donor O	Photoemitter; emission bands at 77 K and 1.942, 1.907, 1.823, 1.815, 1.38, and 1.59 eV

preparation, precipitation, and structure).

The large energy gap of gallium arsenide (GaAs) favors its use in transistors. Even when GaAs is heated to 450°C, its intrinsic carriers compare in number with those produced by silicon at 250°C or by germanium at 100°C. In GaAs transistors, electron mobility is very much greater than hole mobility, and maximum hole concentration is very much greater than maximum electron concentration. Doping with tin can increase the carrier concentration from about $1.3 \times 10^{14}/cm^3$ to $1 \times 10^{15}/cm^3$ at 300 K. A high uniform mobility can be retained. GaAs also undergoes an important band-to-band direct phototransition in which a photon accounts for the energy of a hole-electron pair, and there is no change in momentum.

Photoluminescence of GaAs has bands at 1.23 eV and 1.5 eV. The intensity of the former is increased by a layer of SiO_2 (or of SiO_2/P_2O_5 or Si_3N_4) applied at 350°C, and that of the latter is decreased. But at 650°C the band at 1.23 eV is absent and a band, perhaps due to residual copper, is present at 1.31 eV. Other bands, some of them phonon bands, have been detected and used in computing the number of emitted phonons and the Stokes shift (13 meV) in luminescence.

Scores of GaAs-$Ga_{1-x}Al_x$ systems have been employed as heterostructure injection lasers. Some degraded double lasers of this type have exhibited a local heating effect at the dark lines. Multilayer heterostructures consisting of alternating layers of GaAs and $Al_{0.2}Ga_{0.8}As$ have exhibited optically pumped laser oscillation.

High-purity GaP can be prepared by heating gallium with phosphorus in a quartz ampul which is in a partially covered surrounding crucible in a furnace at 1300°C. The high efficiencies reported in Table 3 for the green-light emission of GaP mesa diodes were obtained partly through the use of a large interval of gradual cooling (900–700°C), during which the *pn* junction formed at 850°C. Accomplishment of the liquid-phase epitaxial growth at the lowest possible temperature improves the efficiency and service lifetime.

Because gallium production involves many chemical operations on large quantities of raw material, its price is relatively high. Annual production is only a few hundred pounds.

Occurrence and extraction. Gallium occurs in nature in only very low concentrations, usually less than 0.01%. It is widely distributed, its abundance having been estimated at 15 g/ton in the Earth's crust. This is about equal to the abundance of lead and 30 times that of mercury, but there are no ores sufficiently rich in gallium to make its direct extraction from an ore economically feasible. Germanite, the richest ore, contains an average of only 0.6% gallium. Consequently, gallium is obtained as a by-product from the production of other metals or from waste products.

Following Boisbaudran's discovery, 40 years elapsed before further interest in the recovery of gallium occurred. In 1915 F. G. McCutcheon, in the United States, observed that the lead-containing residues from the refining of zinc smelted from the tri-state ores would exude drops of a liquid metal following exposure to rain. An analysis of these drops showed that the liquid metal was an alloy containing about 94% gallium and 6% indium. He concluded that sodium and calcium in the residue reacted with the rainwater and released the rare metals. Working with these residues, McCutcheon produced the world's first pound of gallium and pioneered in the development of a commercial process for producing gallium from zinc smelter residues. Following chemical concentration of the gallium, the metal is recovered electrolytically from a strongly alkaline solution and finally purified by means of recrystallization.

In addition to the recovery from zinc smelter residues, gallium is also produced in a state of high purity from the sodium aluminate liquor obtained in the refining of bauxite by the aluminum industry.

Bauxite, the primary source of aluminum, contains gallium in the amount of a few thousandths of a percent. In the regular Bayer process of refining the bauxite prior to smelting to aluminum, the bauxite is dissolved in caustic soda, and pure aluminum oxide trihydrate is precipitated by seeding and cooling the aluminate liquor. After repeated dissolution of bauxite and precipitation of alumina, gallium builds up in the recycled liquor to an equilibrium value of about 0.1 g/liter. This liquor is therefore a relatively rich source of gallium.

Gallium is separated from alumina in the liquor by utilizing the fact that gallium is more acid than alumina. If the aluminate liquor taken at the end of the alumina trihydrate precipitation cycle is treated with lime, most of its alumina can be precipitated, whereas the gallium remains in solution. A carbon dioxide treatment of the liquor will then precipitate the gallium along with the fraction of alumina remaining. An alternate procedure (believed to be practiced in Europe) is to precipitate the bulk of the alumina by a slow gassing with carbon dioxide and then to precipitate the gallium and the remaining alumina by a second gassing with carbon dioxide. In either case the final precipitate, a gallia-rich alumina, is dissolved in caustic soda and, following a sodium sulfide treatment to remove heavy metals, is electrolyzed to obtain metallic gallium.

Traces of gallium in the Bayer-process type of sodium aluminate liquors can be removed by electrolysis with the use of cathodes of Sn, In, Pb, Zn, Sn-Ga, or Pb-Sn-Ga alloy. For a Bayer-type liquor containing 320 g/liter Na_2O, 160 g/liter Al_2O_3, and 0.35 g/liter gallium, a tin cathode 0.3 mm in thickness may be used for electrolysis at 41°C and 0.22 A/cm² (0.022 nm/cm²). This extracts 33% of the gallium and produces on the surface of the tin cathode a 30-μm-thick layer containing 1.4% Ga. This gallium may be quantitatively removed by treatment with fused NaOH.

In Great Britain and Europe gallium is recovered from coal ashes. The ash (flue dust) is fused with caustic soda (or perhaps with Na_2CO_3, K_2CO_3, and KNO_3), the cake is dissolved in water, and most of the alumina and silica is removed by controlled precipitation with hydrochloric acid. Gallium to be quantitatively assayed can then be separated from interfering elements in a 6 *N* hydrochloric acid solution by repeated extractions with amyl acetate. The extracted gallium can be stripped into water and be determined colorimetrically with Rhodamine B.

The caustic soda, and so on, need not be used for the decomposition of gallium-bearing residuals from which germanium has been recovered as a volatile halide. The gallium can be extracted from the residual wastes in a 4–5 *M* hydrochloric acid solution by treatment with 0.1–0.5 *M* trioctyl amine, or *N*-octylamine in kerosine, benzene, or carbon tetrachloride. The gallium is then recovered in a 2–5% caustic solution fed countercurrently to the amine solution in an extraction column.

Gallium ion, along with aluminum and calcium ions, can be separated from many other cations adsorbed on the H^+ form of a cation-exchange resin (for example, 100–200 mesh AG-50 W-X8) by selective elution with suitable amounts and concentrations of hydrochloric acid.

Gallium can be "salted out" from solution in molten zinc by the addition of zinc sulfide (sphalerite). Hence, if sphalerite containing traces of gallium is dissolved in zinc, the gallium will be precipitated and can be recovered as impure metal. *Thiobacillus ferroxidans*, an oxidizing bacillus in an acidic solution, acting over a period of 1.5 months, enables gallium to be leached from sphalerite twice as rapidly as it is leached in a sterile solution.

Prior to World War II, most of the world's gallium was produced in Germany. Intermediate products from the processing of the copper-bearing schists of the Mansfeld district were used.

In many processes, the ultimate step is electrolysis of either an alkaline or an acid solution of the gallium. In most cases, sodium gallate is the solution electrolyzed. Alumina is present in this solution in substantial amount, but very little enters the gallium. Gallium is deposited at the cathode, which may be platinum, nickel, or the liquid gallium itself. The anode may be carbon, nickel, or platinum. The current density at the cathode is relatively high and the potential is about 5 volts. The yield can often be improved by using liquid Ga(5 wt %)-Zn as the cathode, depositing the gallium at 70°C, and increasing the current density to 150 mA/cm^2.

If vanadium is present with the aluminate and gallate, a separation by direct electrodeposition is impracticable. The gallium may then be recovered electrolytically as a mercury amalgam by a patented method. In the process a thin layer of mercury at the bottom of a rubber-lined or plastic tank (with an overflow at one end) serves as a cathode and nickel as the anode. An endless chain of vanes immersed 1–2 min in the mercury stirs it with a horizontal movement about 1 ft/sec. The entering electrolyte should contain only 0.22 g of Na_2O per liter. The electrolyte-volume/cathode-surface ratio is 40–50 gal/m^2. For an 8-ft^2 cathode the electrolyte flow is 200 gal in 24 hr. Electrolysis is at 4.5 volts, 5.4 A/ft^2 cathode current density, and 40–50°C. The composition of the amalgam is controlled at Ga 1.0 max and Na 0.4–0.5%. Discharged electrolyte contains about 80 mg of gallium per liter, as well as a suspension of reduced vanadium compounds, and should be filtered for reuse.

If an aluminate-gallate solution is contacted with a sodium-mercury amalgam in a rotating open vessel, amalgam enriched in gallium collects at the bottom of the vessel.

Gallium chloride prepared by the action of chlorine or HCl on impure liquid gallium may be separated from traces of other metals by a twofold distillation. The $GaCl_3$ is then reduced at 200°C with high-purity aluminum. Either $GaCl_3$ or HCl can then reduce the aluminum content in the gallium product to 1 wt ppm. All other impurities will be below the limits of spectrophotometric or polarographic detection.

France is one of the largest producers of gallium in Western Europe. The French production and that of the Soviet Union are believed to be obtained either from sodium aluminate liquor or from other intermediate products in the refining of bauxite.

Physical properties. Solid gallium looks bluish gray when exposed to the atmosphere. Liquid gallium is silver white with a bright mirror surface. The freezing point of gallium is lower than that of any metal except mercury (−39°C) and cesium (−28.5°C). Because of the complex structure of the stable solid (orthorhombic, space group D_{2h}^{18}), the pure liquid has a marked tendency to supercool and may be held in an ice bath for days without crystallizing. However, addition of a single crystal of solid metal, or even of ice, causes rapid crystallization. Cavitation, for example, that caused by the passage of ultrasound, nucleates crystallization. Gallium that normally supercools at 25°C can be nucleated in an ultrasound field with but a few degrees of supercooling.

Pure gallium, when undercooled below −17°C at 1 atm pressure, can crystallize to either a stable Ga(I), that is, α-Ga, or to a metastable phase, Ga(II), that is, β-Ga, which also forms at high pressures (for example, greater than 12 kilobars). The temperature of metastable transformation from the liquid is −15.8°C at atmospheric pressure (large samples at −16.3°C). The densities of α-Ga and β-Ga at −22°C are 5.92 and 6.23 g/cm^3, respectively. Since the density of undercooled liquid gallium is 6.13 g/cm^3, the melt must expand when it crystallizes to α-Ga and contract when it transforms to β-Ga. The formation of β-Ga is thus aided by pressure. Both α-Ga and β-Ga are orthorhombic. A second metastable form of gallium, Ga(III), that is, γ-Ga, which appears at −35.6°C, also is orthorhombic. The cell parameters of these forms are given in Table 4. Between −55 and −100°C γ-Ga is more probable than β-Ga. Below −115°C β-Ga is always obtained.

The interatomic distances in the alpha-gallium orthorhombic structure, which is complex, are d_1, the distance separating a given atom from a first nearest neighbor; d_2, the distance of the given

Table 4. Parameters of α-Ga, β-Ga, and γ-Ga

Form	Cell parameters, angstroms*	Atoms per unit cell
I (α-Ga)	a = 4.524	8
	b = 4.523	
	c = 7.661	
II (β-Ga)	a = 3.17	4
	b = 2.90	
	c = 8.13	
III (γ-Ga)	a = 10.10 ± 0.07	4
	b = 13.56 ± 0.04	
	c = 5.19 ± 0.04	

*10 angstroms = 1 nanometer.

Table 5. Thermal and mechanical properties of gallium

Melting point	29.78°C (85.60°F)
Boiling point	2237°C (4059°F)
Fusion temperature T, K, as a function of pressure P, kilobars	
for 1 atm to 30 kbars	$T = 318(P + 4/34)^{0.246} + 69e^{(-0.145P)}$
for 30–70 kbars	$T = 318(P + 5/35)^{0.317} + 69e^{(-0.115P)}$
Hardness, Mohs scale	1.5
Density	
Solid	
at 20°C (68°F)	5.907 g/cm³*
at 29.67°C (85.35°F)	5.9037 g/cm³
Liquid	
at −35.6°C (−32.1°F)	6.153 ± 0.008 g/cm³
at −16.3°C (−2.66°F)	6.136 g/cm³
at 29.8°C (85.65°F)	6.0948 g/cm³
at 1100°C (2012°F)	5.445 g/cm³
Atomic volume	
solid	11.81 cm³/g-atom
liquid	11.44 cm³/g-atom
Coefficient of linear thermal expansion, 0–30°C (32–86°F)	18×10^{-6}/°C
Latent heat of fusion	
α-Ga	9.16 cal/g
γ-Ga	8.35 ± 0.15 cal/g
Coefficient of thermal conductivity (liquid) at melting point	0.07–0.09 cal/cm²/cm/°C/sec
Electrical resistivity (liquid) at melting point	24.85 μ-ohm-cm
Undercooled	
before α-Ga crystallization, −40 to +60°C	25.4–27.3 μ-ohm-cm
before β-Ga crystallization, −20 to +60°C	$24.2–25.9 \times 10^{-6}$ μ-ohm-cm
solid	−60°C 0°C 20°C
a axis†	38.9 50.5 54.3
b axis†	11.9 16.1 17.4
c axis†	5.65 7.5 8.1
Entropy	
ΔS_γ	2.45 cal/(mole)(K)
ΔS_β	2.47 cal/(mole)(K)
Total heat capacity at 2 K	7.2×10^{-4} cal/K
Vapor pressure at 1130°C	10^{-2} torr
Surface free energy, α-Ga	~77 erg/cm²
$\sigma = 708 - 0.0031\,[T(\text{K}) - 302.93] - 0.000067[T(\text{K}) - 302.93]^2$	
Volume changes	
for $L \rightarrow$ α-Ga	$\Delta V_L = 0.00529$ cm³/g
for $L \rightarrow$ β-Ga	$\Delta V_L = 0.0123 \pm 0.0001$ cm³/g
for $L \rightarrow$ γ-Ga	$\Delta V_L = 0.00136$
Critical-state parameters	
density	1.58 g/cm³
temperature	5410 K
pressure	235 meganewtons/m²

*(g/cm³) = 6.08 − 0.006[T K) − 302.93]
†All values in μ-ohm-cm.

atom from a next nearest neighbor; d_3, the distance of the atom from the second next nearest neighbor; and so on.

Gallium, when amorphous, has a strong electron-phonon interaction that enhances the bare density of states by a factor of 3; but between 0 and 20 K this enhancement decreases by about 30%. Being temperature-independent, the Hall constant is not affected by the temperature dependence of electron-phonon enhancement.

The structures of undercooled and normal liquid gallium are the same. The aggregation of atoms into nuclei for the initiation of crystallization has a sharp temperature dependence, occurring as an abrupt process at the temperature of heterogeneous nucleation. The crystallization of one droplet causes "cascade" crystallization. The fast growth of gallium crystals is attributed, in part, to screw dislocations and other defects. Gallium oxide promotes nucleation.

The time t for a sample to crystallize is related to the free surface energy σ, the volume V, the temperature T, and the fusion temperature T_f, by the relation shown in Eq. (1), where K and B are mate-

$$1/t = KVe^{-B\sigma^3/T(T_f - T)^2} \quad (1)$$

rials constants. Small volumes, as in emulsions, should therefore freeze most slowly. Drops of emulsified gallium 120–300 μm diameter heated at 100°C and cooled 0.3°C per minute solidify at −60 to −75°C.

The solidification temperature T^- is a function of the temperature T^+ at which the liquid is heated above T_f. If the liquid is heated at 50°C, for example, solidification is likely at −17°C. Removal of oxide by hydrochloric acid treatment of liquid gallium may lower T^- to −40°C. Heating has relatively little influence on gallium free of oxide film. Surface ionization is practicable for gallium because it has a low ionization potential. Ion beams of high purity are therefore readily produced from gallium.

Other physical properties of gallium are given in Table 5.

Wetting of ceramics. With a constant heating rate of 5–10°C per minute, liquid gallium on graphite at 100°C forms a wetting angle of 138° and adhesion work of 180 erg; at 600°C the angle is 144° and the adhesion work is 150 erg. Gallium at high temperatures (600–800°C) wets magnesia but does not wet beryllia or alumina. The spark spectrum of gallium consists of two violet lines of wavelengths 417.1 and 403.1 nm.

Nuclear and radiation properties. The energies of the most energetic γ-rays from the decay of Ga^{66} are 4832 ± 7, 4485 ± 10, 4313 ± 60, and 4132 ± 11 keV. A listing of resonance energies and neutron cross sections for Ga^{69} and Ga^{71} is given in Table 6.

The reaction $Ga^{71}(\nu,e^-)$ Ga^{71} can be used for recording solar neutrinos from the reactions $H^1(p,e^+\nu)$ H^2 and $Be^7(e^-,\nu)$ Li^7.

For Ga^{72} the relative intensities of gamma lines in the spectrum are 100, 25.4, and 50.5 for energies of 2201, 2490, and 2508 keV, respectively. Gamma

Table 6. Resonance energies and neutron cross sections for Ga^{69} and Ga^{71}

Isotope	Resonance energy (E_o), eV	Neutron cross section, barns
Ga^{69}	112 ± 2	13.5
	340 ± 10	13
	710 ± 30	10
Ga^{71}	95 ± 2	15.5
	290 ± 9	50
	380 ± 10	40
	710 ± 30	10

spectrum Ga^{72} enriched up to 99 at. % and possessing a half-life of 14.2 ± 0.3 hr is obtained by irradiation of normal Ga_2O_3 with thermal neutrons in a reactor for 1 hr.

For 14.8-MeV neutron activation, the cross section for the reaction $As^{75}(n.He^3)$ Ga^{73} is 0.5 mb.

Deformation. Gallium deforms by slip on (011) and (0$\bar{1}$1) planes in the [0$\bar{1}$1] and [0.11] directions, respectively, and on the (100) and (001) planes in the [010] direction. Slip on noncrystallographic planes in the ⟨010⟩ zone is a composite of alternating slip on the (001) and (100) planes. Twinning occurs when the applied stress tends to equalize the *a* and *b* axes. Twins have (110)[110] and (111)-[110] systems. Work hardening is low under all conditions. At low temperatures gallium becomes extremely brittle and fractures by cleavage, invariably near the grips. Twinning is associated with cleavage fracture on the (110) planes.

Gallium is different from most metals (but similar to water) in that it expands on solidifying. This property, coupled with its freezing point of 29.8°C, makes it necessary to use elastic containers, such as rubber bulbs or flexible plastic bottles, for shipment, since alternate freezing and melting of the gallium would cause breakage of a rigid container and loss or contamination of the metal.

Chemical properties. Gallium is chemically similar to aluminum. It is amphoteric but slightly more acid than aluminum; for example, a solution of sodium gallate is more stable than a solution of sodium aluminate. This property is used in the commercial preparation of gallium, as previously described. The normal valence of both aluminum and gallium is 3+, and the two metals form corresponding hydroxides, oxides, and salts. However, it appears that gallia (gallium oxide) forms only a monohydrate, whereas alumina (aluminum oxide) forms a monohydrate and a trihydrate. Both aluminum and gallium form alums and a long list of organometallic compounds.

Gallium burns in air when heated to 500°C. It reacts vigorously with boiling water, but only slightly in water at room temperature.

The salts of gallium are colorless; they are prepared directly from the metal, since purification of the metal is simpler than purification of the salts. Gallium trichloride is soluble in many organic liquids, including ether, benzene, carbon tetrachloride, and carbon disulfide. Ether extraction from a water solution is used commercially to purify gallium chloride, and its solubility in many organic liquids is the basis for its use as a catalyst in organic synthesis.

Micro amounts of gallium in steel can be separated by using organic compounds, complexed with quarternary ammonium salts and determined spectrophotometrically. A preliminary separation of gallium from the specimen in acid solution by the addition of 4-(2-pyridilazo)-resorcine (PAR) or phenyl florone is followed by the complexing with zephiramine to improve color sensitivity.

In the extraction of gallium, when present in micro amounts in chloride solutions, the dielectric constants of such organic solvent extractants as BuOAc, MeCOBu-iso($ClCH_2CH_2$)O, and $MeNO_2$, affect the degree to which extraction is suppressed by the presence of Sb^{3+} and Sb^{5+}.

Although beryllia and alumina are used as support materials for gallium at high temperatures, some chemical interaction occurs between these ceramics and gallium at temperatures above 1000°C.

Gallium acetyltrifluoro-β-diketone and gallium benzoyltrifluoroacetonate-β-diketone are chelates that sublime at 1 mm Hg pressure and 76 ± 2°C and 105 ± 10°C, respectively. Purification of gallium by fractional sublimation is thus possible.

Trivalent gallium undergoes incipient extraction from 2 *M* HCl solution into β,β′-dichlorodiethyl ether, and is effectively transferred to the organic phase from 6 *M* HCl. The gallium is accompanied by trivalent gold and thallium, but most of the other elements are only weakly extracted or remain completely in the aqueous phase.

Indium, iron, and antimony can be extracted from gallium in chloride solutions with diisopropyl ether, and the gallium can then be extracted from HCl-LiCl solutions. Also, gallium, as a low-concentration impurity in aluminum and its compounds, can be extracted as $H(GaCl_4)$ by BuOAc from a hydrochloric acid solution.

Gallium can also be determined by atomic absorption spectrometry in premixed inert-gas, entrained air-hydrogen flames at 287.4 nm. Interference from other elements and acids is nearly eliminated by the addition of one part per thousand of MgAs chloride.

Gallium metal can be dissolved by heating with either the caustic alkalies or the common mineral acids. However, the ease of attack decreases with increasing purity of the metal. Most metal impurities can be removed from gallium metal by alternate treatment at ordinary temperatures with nitric acid and hydrochloric acid, with intermediate, thorough washing with water. Nitric acid disperses the metal into small droplets, and hydrochloric acid causes the droplets to coalesce.

Alloys and compounds. Gallium forms low-melting eutectic alloys with several metals, and intermetallic compounds with many others. The unusually low-melting eutectic gallium-base alloys (containing substantial alloying additions) and their melting temperatures include those listed below.

Ga(~1 wt %)-Al, mp 26.3°C
Ga(5 wt %)-Zn, mp 25°C
Ga(8 wt %)-Sn, mp 20°C
Ga(12 wt %)-Sn(6 wt %)-Zn, mp 17°C
Ga(24 wt %)-In, mp 15.7°C
Ga(21.5 wt %)-In(16 wt %)-Sn, mp 10.7°C

All aluminum contains small amounts of gallium as a harmless impurity, but intergranular penetration by larger amounts of gallium at 30°C causes catastrophic failure on loading.

A negative temperature coefficient of ultrasound velocity in gallium at 600–1000°C becomes less pronounced with the addition of antimony until, for 50 wt % Sb, the coefficient at 750–900°C is practically zero.

Internal fields at gallium nuclei in iron were 110 kgauss when measured by means of the spin-echo technique at 4.2 K.

Electrically deposited or vacuum-deposited gallium on high-purity polycrystalline aluminum penetrates the aluminum at an accelerated rate if the temperature is above 30°C, and the deposit is thicker than is required for coverage. This rapid penetration occurs through the grain boundary etching by liquid gallium. This behavior can be prevented by limiting the thickness of the newly deposited film to that required for coverage, or, if possible, by keeping the ambient temperature below the melting point of gallium.

Ga(25 wt %)-In attacks austenitic chromium-nickel stainless steel at 320°C. Reaction products include $FeGa_4$, which is very brittle, and FeIn. Titanium, however, resists the Ga-In alloy up to 350°C but is attacked at 400°C.

Gallium in magnesium increases hardness and forms a protective surface oxide film. Up to 5 atom % Ga also increases the tensile and yield strengths of magnesium and its ductility.

A lens of liquid gallium on mercury dissolves faster at 20°C than a solid lens under identical experimental conditions. The halo of a liquid-gallium lens spreads approximately six times faster than the halo of a solid lens. The diffusion coefficient of gallium in mercury at 25°C is 1.57×10^5 cm²/sec. The solubility of mercury in gallium is 6.5 wt % at 35°C, and 8.6 wt % at 100°C. The solubility of gallium in mercury is 0.13 wt % at −38.9°C, 0.24 wt % at 19°C, 1.3 wt % at 35°C, and 1.4 wt % at 100°C.

Nickel can contain up to 28 at. % (31.59 wt %) of gallium in solid solution. A slight excess, corresponding to Ni(30 at. %)-Ga, results in a hardness of 440 Bhn.

The solubility of potassium in gallium at 32°C is 4×10^{-6} %.

Delta Pu(0.85 at. %)-Ga is face-centered cubic and should be more ductile and easy to fabricate than α-Pu, β-Pu, or even γ-Pu, provided the delta alloy does not transform during deformation. Uniaxial compression at 23–55°C results in progressive transformation to α-, β-, and γ-phases accompanied by increases in density. At 75–130°C the density increases linearly with strains to 13% and then more slowly. The α- and γ-phases have been detected. At 155°C, however, no density change occurs up to 43% strain. The delta phase presumably remains stable and workable at that temperature. In order to ensure stability of the plutonium delta phase under room-temperature rolling, the presence of 3.4% of gallium is necessary. Even then deformation by grinding can still transform a thin surface layer to alpha.

Gallium-uranium alloys have interesting electrical properties. UGa is antiferromagnetic with an anisotropy constant of $K = 18^8$ erg cm^{-3}. U_3Ga can be superconducting, and it can be made into multifilamentary wire. Critical currents excited at speeds of more than 200 kilo-oersteds/sec do not cause small coils of this wire to degrade.

The addition of 1 part by volume of liquid gallium to 3–4 parts of molten tin forms an alloy that, on cooling, is about three times as hard as pure tin. Unlike tin, the alloy cannot be etched by aqueous $FeCl_3$. The lattice parameters are reduced slightly by the alloying.

Gallium additions up to 20 wt % reduce the Knoop hardness of silicon from 1000 to 175 H_K. Further additions have a more gradual effect. The heat of solution of silicon in gallium is 16.2 kcal/mole. The diffusion of gallium in silicon is retarded by phosphorus and arsenic, provided the donor concentration is in excess of a threshold value.

Table 7. Thermodynamic properties of Ga_2O_2

Temperature K	$H_T - H_{298.15}$, cal/mole	$S_T - S_{298.15}$, cal/deg-mole	$\Delta H_{formulation}$, kcal/mole	$\Delta F_{formulation}$, kcal/mole
400	2,390	6.88	−263.75	−231.1
800	13,360	25.70	−262.7	−198.7
1200	25,430	37.91	−260.95	−167.05
1600	37,910	46.88	−259.0	−136.05
1800	—	—	−256.85	−105.55
2000	50,710	54.02	—	—

Compounds. Known Ga^{2+} compounds include GaS, GaSe, GaTe, and $GaBr_2$. GaTe and $GaBr_2$ are unstable and readily undergo oxidation or disproportionate. For example, $GaCl_2$ reacts with water to produce $GaOCl_2$ plus hydrogen; and two molecules of $GaCl_2$, or of other gallium dihalides, can disproportionate to $Ga^+(GaX_4)^-$, where X is the halide. The resulting structure is diamagnetic. Ga^+ is ordinarily unstable, but it may be stabilized in the presence of large anions, forming $Ga(AlCl_4)$, for example. Ga_2S and Ga_2Se have been formed, but Ga_2O, if it exists, is extremely unstable, $GaSnTe_2$, Ga_2SnTe_3, and $Ga_3Sn_3Te_{10}$, either as liquids or solids, are diamagnetic.

Gallium monoxide is insoluble in cold water but readily soluble in acids. The monoxide has been produced by reaction (2).

$$Ga_2O_3 + H \xrightarrow{730-830°C} 2GaO + H_2O \qquad (2)$$

At 830–900°C hydrogen reduces Ga_2O_3 completely to gallium and steam. An intermediate oxide, Ga_2O, results from reaction (3).

$$4Ga + Ga_2O_3 \xrightarrow{850°C} 3Ga_2O \qquad \text{(vacuum cool)} \qquad (3)$$

The specific heat of gallium sesquioxide is given by $C_p(Ga_2O_3) = 0.083 - 22.03$ cal/(mole)(K) at 16–300 K, and $C_p(Ga_2O_3) = 11.77 + 25.2 \times 10^{-3}T$ cal/(mole)(K) at 298–923 K. Other thermodynamic properties of gallium sesquioxide are given in Table 7. *See* ALUMINUM; INDIUM; THALLIUM.

[W. D. WILKINSON]

Bibliography: R. Evans (ed.), *Liquid Metals*, 1976; R. E. Kirk and D. F. Othmer (eds.), *Encyclopedia of Chemical Technology*, 3d ed., vol. 11, 1980; A. M. Oymov and A. P. Savostin, *Analytical Chemistry of Gallium*, 1971; E. N. Simons, *Guide to Uncommon Metals*, 1967.

Gas

A phase of matter characterized by relatively low density, high fluidity, and lack of rigidity. A gas expands readily to fill any containing vessel. Usually a small change of pressure or temperature produces a large change in the volume of the gas. The equation of state describes the relation between the pressure, volume, and temperature of the gas. In contrast to a crystal, the molecules in a gas have no long-range order.

At sufficiently high temperatures and sufficiently low pressures, all substances obey the ideal gas, or perfect gas, equation of state, shown as Eq. (1), where p is the pressure, T is the abso-

$$pv = RT \quad (1)$$

lute temperature, v is the molar volume, and R is the gas constant. Absolute temperature T expressed on the Kelvin scale is related to temperature t expressed on the Celsius scale as in Eq. (2).

$$T = t + 273.16 \quad (2)$$

The gas constant is

$$R = 82.0567 \text{ cm}^3\text{-atm/(mole)(K)}$$
$$= 82.0544 \text{ ml-atm/(mole)(K)}$$

The molar volume is the molecular weight divided by the gas density.

Empirical equations of state. At lower temperatures and higher pressures, the equation of state of a real gas deviates from that of a perfect gas. Various empirical relations have been proposed to explain the behavior of real gases. The equations of J. van der Waals (1899), Eq. (3), of P. E. M. Berthelot (1907), Eq. (4), and F. Dieterici (1899), Eq. (5),

$$\left(p + \frac{a}{v^2}\right)(v - b) = RT \quad (3)$$

$$\left(p + \frac{a}{Tv_2}\right)(v - b) = RT \quad (4)$$

$$pe^{a/vRT}(v - b) = RT \quad (5)$$

are frequently used. In these equations, a and b are constants characteristic of the particular substance under consideration. In a qualitative sense, b is the excluded volume due to the finite size of the molecules and roughly equal to four times the volume of 1 mole of molecules. The constant a represents the effect of the forces of attraction between the molecules. In particular, the internal energy of a van der Waals gas is $-a/v$. None of these relations gives a good representation of the compressibility of real gases over a wide range of temperature and pressure. However, they reproduce qualitatively the leading features of experimental pressure-volume-temperature surfaces.

Schematic isotherms of a real gas, or curves showing the pressure as a function of the volume for fixed values of the temperature, are shown in Fig. 1. Here T_1 is a very high temperature and its isotherm deviates only slightly from that of a perfect gas; T_2 is a somewhat lower temperature where the deviations from the perfect gas equation are quite large; and T_c is the critical temperature. The critical temperature is the highest temperature at which a liquid can exist. That is, at temperatures equal to or greater than the critical temperature, the gas phase is the only phase that can exist (at equilibrium) regardless of the pressure. Along the isotherm for T_c lies the critical point, C, which is characterized by zero first and second partial derivatives of the pressure with respect to the volume. This is expressed as Eq. (6). At temperatures

$$(\partial p/\partial v)_c = (\partial^2 p/\partial v^2)_c = 0 \quad (6)$$

lower than the critical, such as T_3 or T_4, the equilibrium isotherms have a discontinuous slope at the vapor pressure. At pressures less than the vapor pressure, the substance is gaseous; at pressures greater than the vapor pressure, the substance is liquid; at the vapor pressure, the gas and liquid phases (separated by an interface) exist in equilibrium.

Along one of the isotherms of the empirical equations of state discussed above, the first and second derivatives of the pressure with respect to the volume are zero. The location of this critical point in terms of the constants a and b is shown below; p_c and v_c are the pressure and volume at the critical temperature.

	Van der Waals	*Berthelot*	*Dieterici*
p_c	$\frac{a}{27b^2}$	$\left(\frac{aR}{216b^3}\right)^{1/2}$	$\frac{a}{4e^2b^2}$
v_c	$3b$	$3b$	$2b$
T_c	$\frac{8a}{27Rb}$	$\left(\frac{8a}{27Rb}\right)^{1/2}$	$\frac{a}{4Rb}$
$\frac{p_cv_c}{RT_c}$	0.3750	0.3750	0.2706

Some typical values of p_cv_c/RT_c for real gases are as follows: 0.30 for the noble gases, 0.27 for most of the hydrocarbons, 0.243 for ammonia, and 0.232 for water. The van der Waals and Berthelot equations of state, Eqs. (3) and (4), cannot quantitatively reproduce the critical behavior of real gases because no substance has a value of p_cv_c/RT_c as large as 0.375. The Dieterici equation, Eq. (5), gives a good representation of the critical region for the light hydrocarbons but does not represent well the noble gases or water.

At temperatures lower than the critical point, the analytical equations of state, such as the van der Waals, Berthelot, or Dieterici equations, give S-shaped isotherms as shown in Fig. 2. From thermodynamic considerations, the vapor pressure is determined by the requirement that the cross-hatched area DEO be equal to the cross-hatched area AOB. Under equilibrium conditions, the portion of this isotherm lying between A and D cannot occur. However, if a gas is suddenly compressed, points along the segment AB may be realized for a short period until enough condensation nuclei form to create the liquid phase. Similarly, if a liquid is suddenly overexpanded, points along DE may occur for a short time. For low temperatures, the point E may represent a negative pressure corresponding to the tensile strength of the liquid. How-

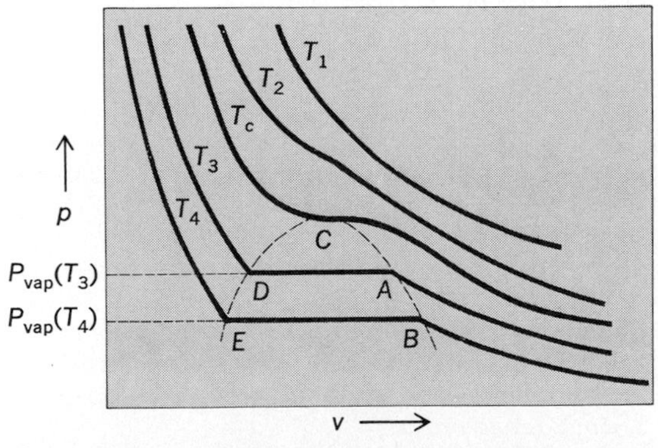

Fig. 1. Schematic isotherms of a real gas. C is the critical point. Points A and B give the volume of gas in equilibrium with the liquid phase at their respective vapor pressures. Similarly, D and E are the volumes of liquid in equilibrium with the gas phase.

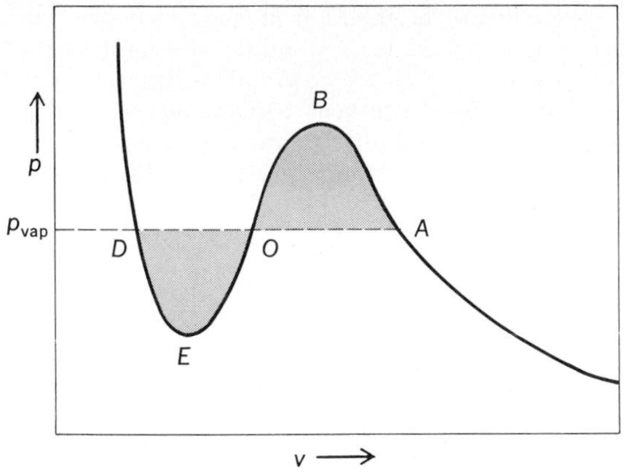

Fig. 2. Schematic low-temperature isotherm as given by van der Waals, Berthelot, or Dieterici equations of state. Here the line *DOA* corresponds to the vapor pressure. The point *A* gives the volume of the gas in equilibrium with the liquid phase, and *D* gives the volume of the liquid. The segment of the curve *DE* represents overexpansion of the liquid. The segment *AB* corresponds to supersaturation of the vapor. However, the segment *EOB* could not be attained experimentally.

ever, the simple analytical equations of state cannot be used for a quantitative estimate of these transient phenomena. Actually, it is easy to show that the van der Waals, Berthelot, and Dieterici equations give poor representations of the liquid phase since the volume of most liquids (near their freezing point) is considerably less than the constant b.

Principle of corresponding states. In the early studies, it was observed that the equations of state of many substances are qualitatively similar and can be correlated by a simple scaling of the variables. To describe this result, the reduced or dimensionless variables, indicated by a subscript r, are defined by dividing each variable by its value at the critical point. These variables are given in Eqs. (7)–(9).

$$p_r = p/p_c \qquad (7)$$
$$T_r = T/T_c \qquad (8)$$
$$v_r = v/v_c \qquad (9)$$

In its most elementary form, the principle of corresponding states asserts that the reduced pressure, p_r, is the same function of the reduced volume and temperature, v_r and T_r, for all substances. An immediate consequence of this statement is the statement that the compressibility factor, expressed as Eq. (10), is a universal function of

$$z = pv/RT \qquad (10)$$

the reduced pressure and the reduced temperature. This principle is the basis of the generalized compressibility chart of O. A. Hougen and K. M. Watson (Fig. 3). This chart was derived from data on the equation-of-state behavior of a number of common gases.

It follows directly from the principle of corresponding states that the compressibility factor at the critical point z_c should be a universal constant. It is found experimentally that this constant varies somewhat from one substance to another. On this account, A. L. Lydersen, R. A. Greenkorn, and Hougen (see Hougen in bibliography) have developed empirical tables of the compressibility factor and other thermodynamic properties of gases as functions of the reduced pressure and reduced temperature for a range of values of z_c. Such generalized corresponding-states treatments are very useful in predicting the behavior of a substance on the basis of scant experimental data.

Theoretical considerations. The equation-of-state behavior of a substance is closely related to the manner in which the constituent molecules interact. Through statistical mechanical considerations, it is possible to obtain some information about this relationship. If the molecules are spherically symmetrical, the force acting between a pair of molecules depends only on r, the distance between them. It is then convenient to describe this interaction by means of the intermolecular potential $\varphi(r)$ defined so that the force is the negative of the derivative of $\varphi(r)$ with respect to r. *See* INTERMOLECULAR FORCES.

Two theoretical approaches to the equation of state have been developed. In one of these approaches, the pressure is expressed in terms of the partition function Z and the total volume V of the container in the manner of Eq. (11). Here k is the

$$p = kT(\partial \ln Z/\partial V) \qquad (11)$$

Boltzmann constant or the gas constant divided by the Avogadro number N_0, $k = R/N_0$. For a gas made up of spherical molecules or atoms with no internal structure, the partition function is given as Eq. (12). In this expression, φ_{ij} is the energy of

$$Z = \frac{1}{N!}\left(\frac{2\pi mkT}{h^2}\right)^{3N/2} \times \int \exp\left(-\sum_{i>j}\frac{\varphi_{ij}}{kT}\right) dv_1\, dv_2 \cdots dv_N \qquad (12)$$

interaction of molecules i and j and the summation is over all pairs of molecules, h is Planck's constant, N is the total number of molecules, and the integration is over the three cartesian coordinates of each of the N molecules. The expression for the partition function may easily be generalized to include the effects of the structure of the molecules and the effects of quantum mechanics.

In another theoretical approach to the equation

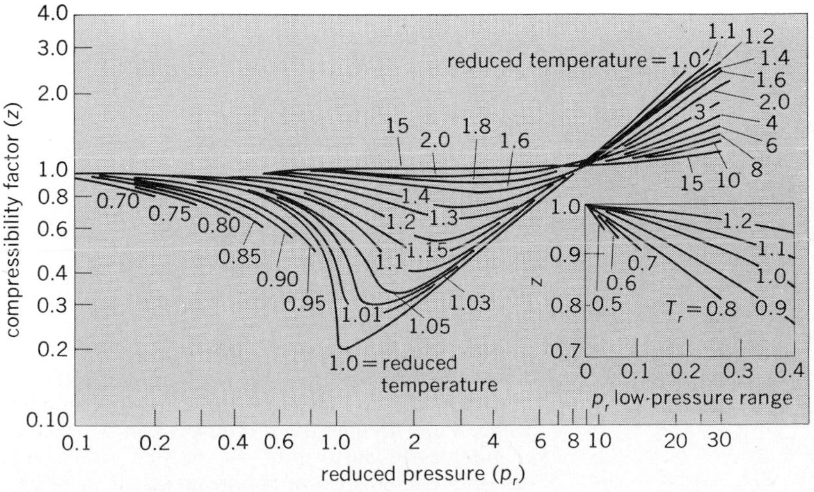

Fig. 3. The compressibility factor pv/RT as a function of the reduced pressure $p_r = p/p_c$, and reduced temperature $T_r = T/T_c$. (*From O. A. Hougen, K. M. Watson, and R. A. Ragatz, Chemical Process Principles, pt. 2, Wiley, 1959*)

of state, the pressure may be written as Eq. (13),

$$p=\frac{NkT}{V}-\frac{2\pi N^2}{3V^2}\int g(r)\frac{d\varphi}{dr}r^3\,dr \qquad (13)$$

where $g(r)$ is the radial distribution function. This function is defined by the statement that $2\pi(N^2/V)$ $g(r)r^2\,dr$ is the number of pairs of molecules in the gas for which the separation distance lies between r and $r+dr$. The radial distribution function may be determined experimentally by the scattering of x-rays. Theoretical expressions for $g(r)$ are being developed.

The compressibility factor $z=pV/NkT$ may be considered as a function of the temperature, T, and the molar volume, v. In the virial form of the equation of state, z is expressed as a series expansion in inverse powers of v, as in Eq. (14). The

$$z=1+B(T)/v+C(T)/v^2+\cdots \qquad (14)$$

coefficients $B(T)$, $C(T)$, . . . , which are functions of the temperature, are referred to as the second, third, . . . , virial coefficients. This expansion is an important method of representing the deviations from ideal gas behavior. From statistical mechanics, the virial coefficients can be expressed in terms of the intermolecular potential. In particular, the second virial coefficient is Eq. (15). If the

$$B(T)=2\pi N_0\int(1-e^{-\varphi/kT})\,r^2\,dr \qquad (15)$$

intermolecular potential is known, Eq. (15) provides a convenient method of predicting the first-order deviation of the gas from perfect gas behavior. The relation has often been used in the reverse manner to obtain information about the intermolecular potential. Often $\varphi(r)$ is expressed in the Lennard-Jones (6–12) form, Eq. (16), where ϵ and

$$\varphi(r)=4\epsilon\left[\left(\frac{\sigma}{r}\right)^{12}-\left(\frac{\sigma}{r}\right)^{6}\right] \qquad (16)$$

σ are constants characteristic of a particular substance. Values of these constants for many substances have been tabulated. In terms of these constants, the second virial coefficient has the form of Eq. (17), where $B^*(kT/\epsilon)$ is a universal func-

$$B(T)=(2/3)\pi N_0\sigma^3B^*(kT/\epsilon) \qquad (17)$$

tion. If all substances obeyed this Lennard-Jones (6–12) potential, the simple form of the law of corresponding states would be rigorously correct. *See* CHEMICAL THERMODYNAMICS.

[C. F. CURTISS; J. O. HIRSCHFELDER]

Bibliography: J. O. Hirshfelder et al., *Molecular Theory of Gases and Liquids*, 1964; R. Holub and P. Vonka, *The Chemical Equilibrium of Gaseous Systems*, 1976; O. A. Hougen, K. M. Watson, and R. A. Ragatz, *Chemical Process Principles*, pt. 2, 1959.

Gas absorption operations

The separation of solute gases from gaseous mixtures of noncondensables by transfer into a liquid solvent. This recovery is achieved by contacting the gas stream with a liquid that offers specific or selective solubility for the solute gas or gases to be recovered. The operation of absorption is applied in industry to purify process streams or recover valuable components of the stream. It is used extensively to remove toxic or noxious components (pollutants) from effluent gas streams.

Examples of application of gas absorption are the removal of hydrogen sulfide and mercaptans from natural gas, the recovery of carbon monoxide in petrochemical syntheses, the removal of sulfur oxides from power plant stack gases, and the recovery of silicon tetrafluoride and hydrogen fluoride from fertilizer production stack gases.

The absorption process requires the following steps: (1) diffusion of the solute gas molecules through the host gas to the liquid boundary layer based on a concentration gradient, (2) solvation of the solute gas in the host liquid based on gas-liquid solubility, and (3) diffusion of the solute gas based on concentration gradient, thus depleting the liquid boundary layer and permitting further solvation. The removal of the solute gas from the boundary layer is often accomplished by adding neutralizing agents to the host liquid to change the molecular form of the solute gas. This process is called absorption accompanied by chemical reaction.

The factors affecting the rate of absorption or the kinetics of transfer from the host gas to the host liquid are indicated in the table.

Historically, the kinetics of gas absorption have been represented by statistical average overall coefficients in the form of Eq. (1), where N_A is the

$$N_A=K_{G\alpha}V(P-P^*) \qquad (1)$$

rate of mass-transfer quantity of solute gas transferred per unit time; $K_{G\alpha}$ is the mass-transfer coefficient based on partial pressure gradient from the host-liquid bulk concentrations, quantity per unit time per volume of the transfer system per unit partial pressure gradient; V is volume of the transfer system; P is partial pressure of the solute gas in the bulk of the host gas; and P^* is partial pressure of the host gas in equilibrium with the bulk of the host liquid.

An alternative method of sizing the system is the transfer unit concept expressed in Eqs. (2)–(4),

$$N_{oG}=\int\frac{dp}{P-f(P^*)}\,\frac{[1-f(P_f)]}{1-P} \qquad (2)$$

$$H_{oG}=\frac{Gm}{K_{G\alpha\pi}}\,\frac{[1-f(P_f)]}{1-P} \qquad (3)$$

$$H=N_{oG}H_{oG} \qquad (4)$$

Factors affecting the kinetics of the gas absorption process

Zone	Factor
Gas	Gas velocity
	Molecular weights and sizes of host and solute gases
	Temperature
	Concentration gradient
Solvation	Solubility in host liquid
Liquid	Molecular weights and sizes of host liquid and solute gas
	Temperature
	Viscosity of liquid
	Liquid surface velocity
	Surface renewal rate
	Concentration gradient of solute gas
	Concentration gradient of neutralizing reagent

where H is height of the absorber; N_{oG} is number of transfer units; H_{oG} is height of a transfer unit; $K_{G\alpha}$, P, and P^* are as in Eq. (1); P_f is partial pressure of the solute gas in the interfacial film between gas and liquid; π is total pressure of operation; and Gm is molar gas flow rate.

The effect of the relative resistances in the liquid and gas phases are established by the relationship shown in Eq. (5), where $K_{G\alpha}$ is as in Eq. (1); $k_{G\alpha}$

$$K_{G\alpha} = \frac{1}{\dfrac{1}{k_{G\alpha}} + \dfrac{m}{k_{L\alpha}}} \tag{5}$$

is mass-transfer coefficient in the gas phase only, quantity per unit per volume of transfer system per unit partial pressure gradient; $k_{L\alpha}$ is mass-transfer coefficient in liquid phase only, quantity per unit time per unit volume per unit concentration gradient; and m is partial pressure–liquid phase concentration relationship of solute gas.

The gas-phase and liquid-phase mass-transfer coefficients are established as Eqs. (6) and (7),

$$k_G = \frac{D_G \pi}{Rt\chi P_{BM}} \tag{6}$$

$$k_L = Z\left(\frac{D_L \phi}{\pi t}\right)^{1/2} \tag{7}$$

where D_G and D_L are diffusivities in the gas and liquid phases, respectively; R is the universal gas constant; T is the absolute temperature; χ is the effective thickness of the gas boundary layer; P_{BM} is the partial pressure of the inert gases; ϕ is the surface renewal rate; and t is time of contact.

It has been established that except for unique situations of very high solubility, the transfer in the liquid phase tends to become the controlling factor in rate of absorption.

The emphasis in design of systems has been to increase the surface renewal rate and to increase the effective liquid diffusivity by neutralizing the solute gas in the liquid phase as rapidly as possible. In the case of neutralization or total removal, the mass-transfer coefficient in the liquid phase follows the relationship shown in Eq. (8), where k_L' is

$$\frac{k_L'}{k_L} = 1 + \left(\frac{q}{Cai}\frac{D_B}{D_L}\right)^n \tag{8}$$

liquid-phase mass-transfer coefficient when chemical reaction occurs; q is concentration of unreacted reagent in the bulk of the host liquid; Cai is concentration of solute gas in the liquid phase (physically absorbed) in equilibrium with P; D_B and D_L are diffusivities of the solute gas and reagent in the host liquid, respectively; and n is an exponent, generally less than 1. The transfer systems require effective contact of liquid and gas, with maximum surface contact and surface renewal, and minimum consumption of energy. The types of equipment used for this service are: venturis (high surface, no surface renewal); spray towers (high surface, no surface renewal); tray towers (high surface renewal); and packed towers (high surface, high surface renewal).

The venturi system (Fig. 1), requiring high energy input, is often used where large quantities of particulates are present in the gas or the scrubbing liquid has large concentrations of suspended solids. The spray tower (Fig. 2) is used where the transfer requirements are low. The tray tower (Fig. 3) is not used extensively because of poor efficiency in low-concentration gas systems. The packed tower (Fig. 4) is the predominant system used in gas absorption because of its inherent differential mechanism, low power consumption, and capability for handling low concentrations of particulates

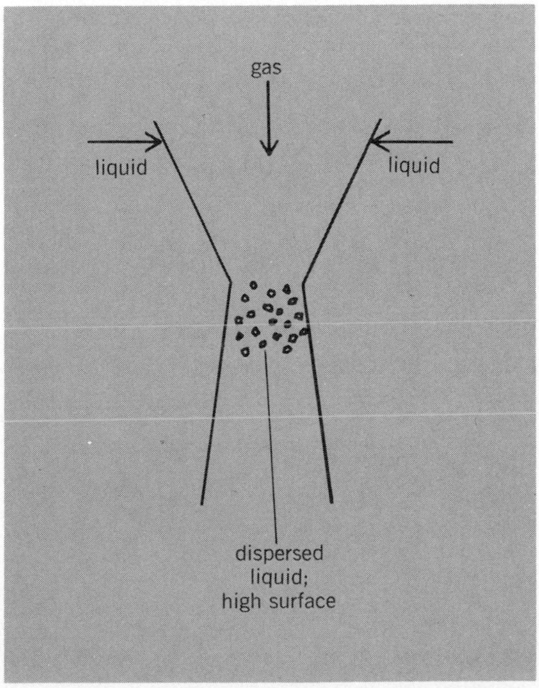

Fig. 1. Diagram of the operation of a venturi system.

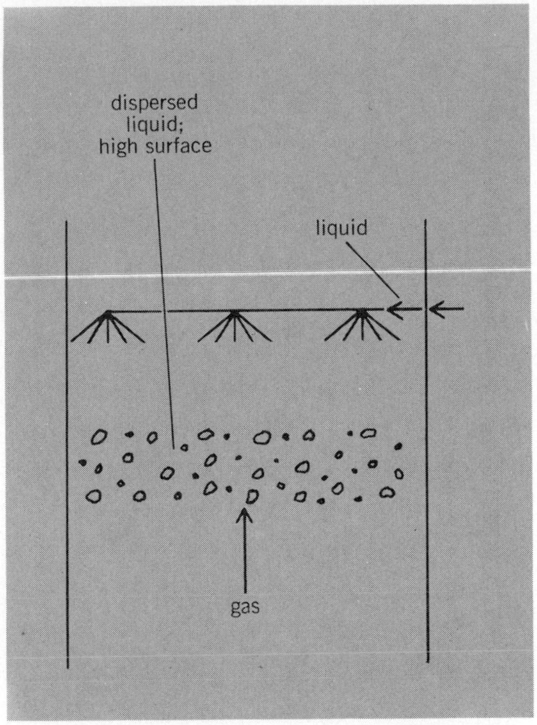

Fig. 2. Diagram of the operation of a spray tower.

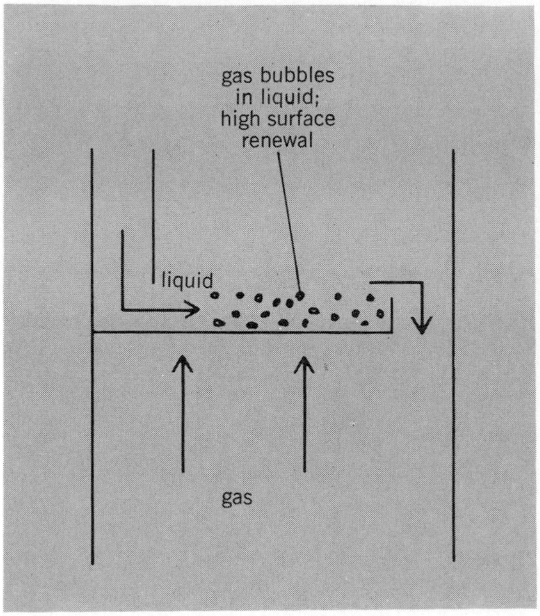

Fig. 3. Diagram of the operation of a tray tower.

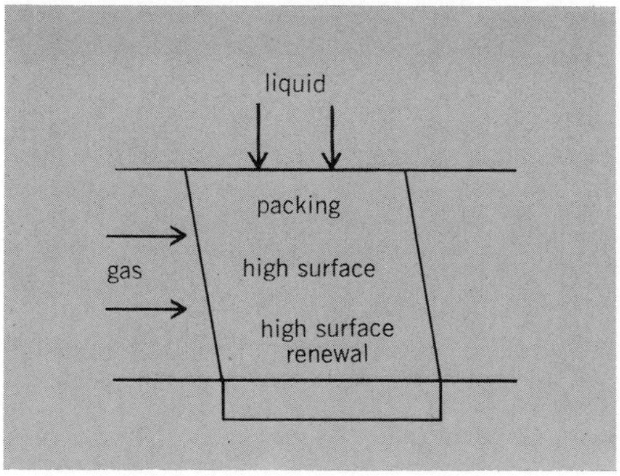

Fig. 4. Diagram of the operation of a packed tower.

in gas streams. It is often used in conjunction with a solids recovery unit such as a venturi.

The packed column volúme generally is randomly filled with a commercially available geometric shape, ranging from 1 to 4 in. (2.5 to 6 cm) in the largest dimension, forms that provide either large surfaces or capability for surface renewal, or both.

The gas absorption equipment, usually called scrubbers, generally use water or solutions of reagents in the water as the host liquid, although hydrocarbon liquids are used for specific applications in the chemical industry. The efficiency of absorption can be uniquely high, especially in the case of absorption accompanied by chemical reaction. Reduction of the solute gas concentration to approximately 1–5 ppm is often required for air-pollution control. These effluent concentrations have been achieved in air-pollution control in the steel, fertilizer, glass, and pulp and paper industries. The gas absorption equipment range in capacity from 100 cfm (cubic foot per minute) or 4.7×10^{-2} m^3/s to 2,000,000 cfm (940 m^3/s). *See* DISTILLATION; MASS-TRANSFER OPERATION.

[AARON J. TELLER]

Gas and atmosphere analysis

Qualitative identifications and quantitative determinations of gaseous substances essential for the evaluation of the air quality in the ambient air and in the industrial workplace. The conventional methods of gas analysis used in research, process control, and medical technology were not adequate to meet the demands for techniques to analyze air contaminants, some of which are recognized as having adverse effects on health, welfare, and property. In the United States two important laws, the Clean Air Act of 1963, along with the subsequent amendments, and the Occupational Safety and Health Act of 1970, stimulated interest in new methods capable of analyzing gases from the ambient air, the breathing zone of workers, and emission sources. Requirements for gas analyses of air pollutants are demanding, and range from the qualitative identification of a few microliters of an unknown gas to the quantitation of a trace contaminant at the parts-per-billion level.

The terms air pollution and gas analysis are used here in the broad sense. Air pollution refers to unwanted air contaminants in both the ambient air and the work environment, since many of the same pollutants are found at both places and are assessed using similar procedures. Gas analysis refers to the analysis of both gases and vapors. The term vapor is used for the gaseous phase of substances which are liquids or solids at standard temperature and pressure. Thus the gaseous states of gasoline, mercury, and water are examples of vapors, whereas methane, hydrogen, and ozone are gases. Many important pollutants are vapors that have arisen from the volatilization or decomposition of organic materials.

QUALITATIVE IDENTIFICATION

The qualitative identification of air pollutants can be extremely complex, and may require the use of several instruments which provide complementary information about composition and structure. Since the entire sample is often limited to milligram or microgram quantities, the classical identification methods, such as boiling point and refractive index determinations, functional group tests, combustion analyses, and derivative preparations, have been largely replaced by instrumental methods. Information for identification purposes is now generally obtained from instruments such as mass, nuclear magnetic resonance, infrared, and ultraviolet spectrometers that rely upon the response of a molecule to an energy probe.

Mass spectroscopy. This is probably the single most powerful technique for the qualitative identification of volatile organic compounds, and has been particularly useful in the identification of many environmental contaminants. When a sample is introduced into the mass spectrometer, electron bombardment causes the parent molecule to lose an electron and form a positive ion. Some of the parent ions are also fragmented into characteristic daughter ions, while other ions remain intact. All of the ions are accelerated, separated, and focused on an ion detector by means of either a magnetic field or a quadrupole mass analyzer. Using microgram quantities of pure materials, the mass spectrometer yields information about the molecular weight and the presence of other atoms, such as nitrogen, oxygen, and halogens, within the molecule. In addition, the fragmentation pattern often provides a unique so-called fingerprint of a molecule, allowing positive identification. If the gas is a mixture, interpretation of the mass spectral data is difficult since the fragmentation patterns are superimposed. However, interfacing the mass spectrometer to a gas chromatograph provides an elegant solution to this problem. *See* MASS SPECTROMETRY.

Gas chromatography. A gas chromatograph is essentially a highly efficient apparatus for separating a complex mixture into individual components. When a mixture of components is injected into a gas chromatograph equipped with an appropriate column, the components travel down the column at different rates and therefore reach the end of the column at different times. The mass spectrometer located at the end of the column can then analyze each component separately as it leaves the column. In essence, the gas chromatograph allows the mass spectrometer to analyze a complex mixture as a series of pure components. More than 100 compounds have been identified and quantified in automobile exhaust by using a gas chromatograph–mass spectrometer combination. *See* GAS CHROMATOGRAPHY.

QUANTITATIVE ANALYSIS

Once a qualitative identification of an important pollutant has been established, further interest often centers on quantifying the levels of the pollutant as a function of time at various sites.

Methods. The methods employed chiefly for quantification can be classified for convenience into direct and indirect procedures. Direct-reading instruments may analyze and display their results in a few seconds or minutes, and can operate in a continuous or semicontinuous mode. Indirect methods are those involving collection and storage of a sample for subsequent analysis. Both direct and indirect methods have inherent advantages and disadvantages. By using indirect methods, samples with several pollutants can be simultaneously collected from a number of different sites with relatively inexpensive collection devices and later analyzed at a central laboratory. On the other hand, direct methods may require one instrument for each pollutant at each sampling site, and thus may become prohibitively expensive. However, reduction of the delay before the results are available may be the basis for selecting a direct over an indirect method, or the pollutant in question may not be stable under the conditions of storage. For example, if a worker needs to enter an enclosure with an atmosphere potentially hazardous to life because of an oxygen deficiency, the presence of an explosive mixture, or the presence of a high concentration of a toxic gas, it is essential to have the analytical results available in a few minutes.

Direct methods. These consist of methods utilizing colorimetric indicating devices and instrumental methods.

Colorimetric indicators. Three types of direct-reading colorimetric indicators have been utilized: liquid reagents, chemically treated papers, and glass tubes containing solid chemicals (detector tubes). The simplest of these methods is the detector tube. Detector tubes are constructed by filling a glass tube with silica gel coated with color-forming chemicals. For use, the ends of the sealed tube are broken and a specific volume of air, typically 100 cm^3, is drawn through the tube at a controlled rate. The first detector tube, which appeared in 1917, estimated the concentration of carbon monoxide by observing the intensity of the color which developed when carbon monoxide was present. A detector tube was developed in 1935 which depended on the length rather than the intensity of the stain to indicate the concentration of hydrogen sulfide. These tubes rely on the fact that the capacity of the packing to absorb a pollutant is limited and that, after saturation, successive molecules penetrate farther down the tube; hence the length of the stain is concentration-dependent. Detector tubes often utilize the same color-forming chemical to detect several different gases, and therefore may be nonspecific for mixtures of these gases. Temperature, humidity, age, and uniformity of packing also influence the performance. Detector tubes for analyzing approximately 200 different gases are commercially available. Accuracy is sometimes low, and detector tubes for only 38 gases meet the National Institute for Occupational Safety and Health (NIOSH) accuracy requirement of $\pm 25\%$. For some gases, semicontinuous analyzers have been developed which operate by pulling a fixed volume of air through a paper tape impregnated with a color-forming reagent. The intensity of the color is then measured for quantification. Phosgene, arsine, hydrogen sulfide, nitric oxide, chlorine, and toluene diisocyanate have been analyzed by indicating tapes.

Direct-reading instruments. With the availability of stable and sensitive electronics, direct-reading instruments capable of measuring gases directly at the parts-per-billion range were developed. Most direct-reading instruments contain a sampling system, electronics for processing signals, a portable power supply, a display system, and a detector. The detector or sensor is a component that is capable of converting some characteristic property of the gas into an electrical signal. While there are dozens of properties for the bases of operation of these detectors, the most sensitive and popular detectors are based on electrical or thermal conductivity, ultraviolet or infrared absorption, mass spectrometry, electron capture, flame ionization, flame photometry, heat of combustion, and chemiluminescence. Many of these detectors respond to the presence of nanogram (10^{-9} g) quantities, and even to picogram (10^{-12} g) levels. In addition to improved accuracy, precision, and analysis time, another advantage is that most instruments produce an electrical signal which can be fed into a computer for process control, averaging, and record keeping. Rapid fluctuations and hourly, daily, and yearly averages are readily obtained. Other instruments are essentially manual methods which have been automated. For example, the "manual" method for nitrogen dioxide employs a bubbler containing an azo dye that forms a pink color. The absorbance of the color is then measured on a spectrophotometer and related to concentration by means of a calibration curve. An instrument is available which performs all of these functions. *See* SPECTROPHOTOMETRIC ANALYSIS.

1. Heat of combustion. Many portable direct-

reading meters for explosive atmospheres are based on the principle of catalytic or controlled combustion on a heated filament. The filament is usually one arm of a Wheatstone bridge circuit. The resulting heat of combustion changes the resistance of the filament (usually platinum), and the resulting imbalance is related to the concentration of the gas. A meter displays the results. This method is nonspecific, and gives an indication of all combustible gases present in the range from about 100 ppm to a few percent.

2. Chemiluminescence. The phenomenon of chemiluminescence is employed for the determination of levels of ozone, oxides of nitrogen, and sulfur compounds. Chemiluminescence is an emissive process which occurs when all or part of the energy of a chemical reaction is released as light rather than heat. A familiar example is the "cold light" of fireflies. Ozone levels in the range from 1 to 100 ppb can be determined by measuring the emission at 585 nm which occurs when ozone is mixed with excess ethylene. Similarly, nitric oxide (NO) levels from 10 ppb to 5000 ppm can be measured by a chemiluminescence method. The analysis of nitrogen dioxide is performed by reducing NO_2 to NO in a catalytic converter and measuring the NO emission. Mixtures of NO and NO_2 can be analyzed by measuring the NO level by chemiluminescence, reducing the NO_2 to NO, and again measuring the chemiluminescence level. The NO_2 level is obtained by difference. Measurement of the chemiluminescence produced by thermal means in a hydrogen-rich flame allows the detection of sulfur compounds from 1 to 1000 ppb. With the use of appropriate light filters, interferences by other gases can be reduced. *See* CHEMILUMINESCENCE.

3. Infrared absorption. Measurement of carbon monoxide utilizes a variety of other principles, including infrared absorption at a fixed wavelength, nondispersive infrared absorption, measurement of the heat of reaction in oxidizing carbon monoxide to carbon dioxide, and gas chromatography.

4. Gas chromatography. The gas chromatograph is widely used both for the analysis of collected samples and as a semicontinuous direct-reading instrument. The availability of many different detectors capable of measuring the effluent from the gas-chromatograph column and the development of valves for automatic injection of samples and for directing sample flow have further extended the versatility of this instrument. Accessories are available which can collect samples from about 12 different distant sites through sampling lines for sequential injection directly into a gas chromatograph for analysis. Another application involves the use of valves to inject an air sample into a gas chromatograph equipped with a flame ionization detector to measure the methane content. A second air sample is injected directly into the detector to measure the total hydrocarbon content. The methane signal is subtracted from the total hydrocarbon content to provide an indication of the smog potential.

Indirect methods. For indirect methods, the main collection devices are freeze traps, bubblers, evacuated bulbs, plastic bags, and solid sorbents. Because of their convenience. solid sorbents now dominate collection procedures. NIOSH has developed a versatile method for over 200 industrially important vapors, based on the sorption of the vapors on activated charcoal or, to a lesser extent, on other solid sorbents such as silica gel and porous polymers. Typically, in this technique a few liters of air are pulled through a glass tube containing about 100 mg of charcoal. The charcoal tube is only 7 cm × 6 mm, and has the advantage that it can be placed on the worker's lapel. A battery-operated pump small enough to fit into a shirt pocket is connected by a plastic tube to the collecting device, so that the contaminants are continuously collected from the breathing zone of the worker. Many solvent vapors and gases are efficiently trapped and held on the charcoal. The ends of the sample tube are then capped, and the tube is returned to a laboratory for analysis. In the laboratory the tube is broken open, and the charcoal poured into carbon disulfide to desorb the trapped vapors. Following desorption, a sample of the solution is injected into a gas chromatograph for quanfication.

This technique has been highly successful for several classes of compounds, such as aromatics, aliphatics, alcohols, esters, aldehydes, and chlorinated compounds. Sulfur- and nitrogen-containing compounds can also be analyzed by using a gas chromatograph which is equipped with a sulfur- or nitrogen-sensitive detector.

Considerable efforts continue to be directed toward the development of new analytical principles and miniaturization of existing instruments.

[WILLIAM R. BURG]

Bibliography: American Conference of Governmental Industrial Hygienists, *Air Sampling Instruments*, 5th ed., 1978; M. Katz, (ed.), *Methods for Air Sampling and Analysis*, Intersociety Committee, American Public Health Association, 1977; National Institute for Occupational Safety and Health, U.S. Department of Health, Education and Welfare, Public Health Service, *NIOSH Manual of Analytical Methods*, vols. 1–4, 1978; A. C. Stern, (ed.), *Air Pollution*, 3d ed., vol. 3, 1977.

Gas chromatography

A method for the separation and analysis of complex mixtures of volatile organic and inorganic compounds. Most compounds with boiling points less than about 250°C can be readily analyzed by this technique. A complex mixture is separated into its components by eluting the components from a heated column packed with sorbent by means of a moving-gas phase. A typical chromatogram is shown in Fig. 1.

Classification. Gas chromatography may be classified into two major divisions: gas-liquid chromatography, where the sorbent is a nonvolatile liquid called the stationary-liquid phase, coated as a thin layer on an inert, granular solid support, and gas-solid chromatography, where the sorbent is a granular solid of large surface area. The moving-gas phase, called the carrier gas, is an inert gas such as nitrogen or helium which flows through the chromatographic column packed with the sorbent. The solute partitions, or divides, itself between the moving-gas phase and the sorbent and moves through the column at a rate dependent upon its partition coefficient, or solubility, in the liquid phase (gas-liquid chromatography) or upon its adsorption coefficient on the packing (gas-solid chromatography) and the carrier-gas flow rate. Open

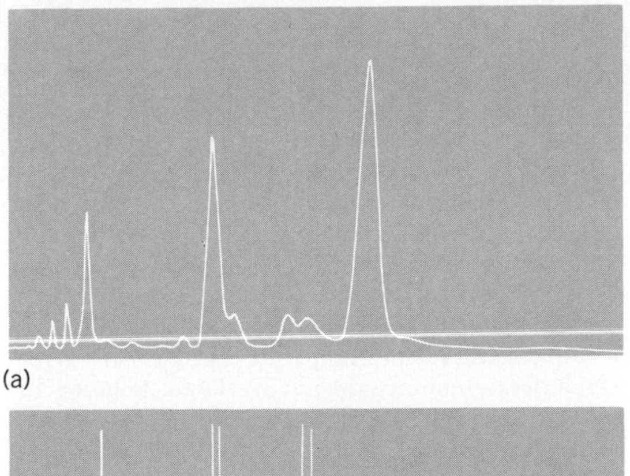

(a)

(b)

(c)

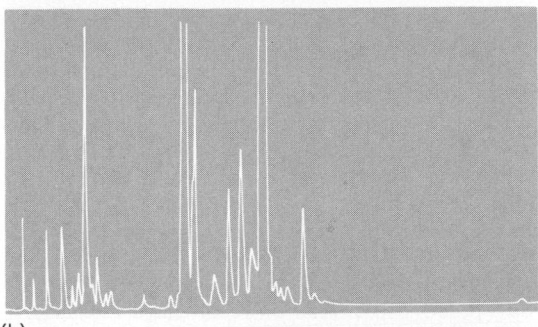

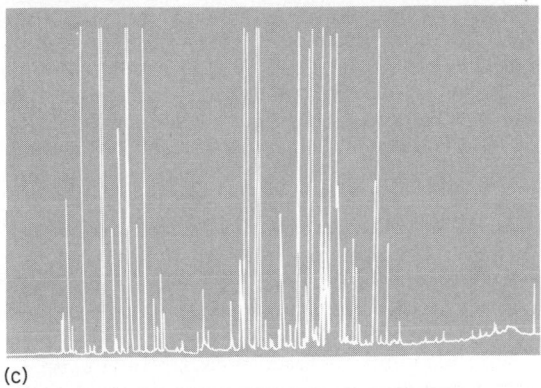

Fig. 1. Three generations in gas chromatography. Peppermint oil separated on (*a*) a 1/4 in. × 6 ft (6 mm × 1.8 m) packed column; (*b*) a 0.03 in. × 500 ft (0.8 mm × 150 m) stainless steel capillary column; and (*c*) a 0.25 mm × 50 m glass capillary column. (*From W. Jennings, The use of glass capillary columns for food and essential oil analysis, J. Chromatogr. Sci., 17:636–639, 1979*)

tubular glass or stainless steel capillary tubes of 0.005–0.02 in. (0.13–0.5 mm) inside diameter and length often as great as 300 ft (91 m), coated on the inside walls with a nonvolatile stationary-liquid phase, are also widely used in the separation of complex mixtures.

Gas-solid chromatography is historically the older of the two forms of gas chromatography but was formerly limited primarily to the separation of permanent gases or relatively nonpolar solutes of low molecular weight, whereas gas-liquid chromatography has been the more popular and more versatile approach to the separation of a wide range of higher-molecular-weight compounds, owing to the large choice of liquid phases available. Research in gas-solid adsorption chromatography, however, has shown considerable progress in decreasing objectionable peak tailing, thus providing higher separation efficiencies than were formerly possible. It is entirely possible that gas-solid chromatography will achieve greater status in future years. *See* ADSORPTION.

Basic apparatus. The apparatus used in gas chromatography consists of four basic components: a carrier-gas supply and flow controller, a sample inlet system providing a means for introduction of the sample, the chromatographic column and associated column oven, and the detector system. A schematic diagram of a working instrument is shown in Fig. 2.

Carrier-gas and flow monitor. Although the carrier gas is most commonly nitrogen or helium, other gases such as carbon dioxide, argon, xenon, and hydrogen are occasionally used. If the detector is a thermal conductivity detector, then helium is the carrier gas of choice. This is because helium has the largest thermal conductivity of all gases with the exception of hydrogen, a flammable and explosive gas. Since the sensitivity of the thermal conductivity detector is a function of the difference in thermal conductivity between the pure carrier gas and that of a mixture of the carrier gas with solute, helium gives high sensitivity. Use of a carrier gas of higher molecular weight will improve column efficiency. Therefore nitrogen or perhaps a gas of even higher molecular weight is preferred to helium if a detector other than thermal conductivity is being used.

A rotometer may be used in the carrier-gas system to give an approximate indication of flow rate. A rotometer consists of a graduated tube with slowly increasing inside diameter and a glass or metal ball that is suspended in the gas flow within the tube at a height dependent upon the flow rate. Since the position of the ball is a function of both the flow rate and the column back pressure when positioned at the column inlet, a rotometer can be used only for rough approximations of flow rate. A soap-bubble flowmeter is used for more accurate measurements.

Sample inlets. These are of two general types depending upon whether the sample is gaseous, liquid, or solid. Liquid samples are generally injected by means of a calibrated hypodermic syringe through a silicon rubber septum into a metal or glass-lined metal injection port, while gaseous samples are introduced by means of a valve and sample loop system. Liquid sample sizes commonly range from 0.5 to 5 μl. Thus, only very small amounts of material are required for gas chroma-

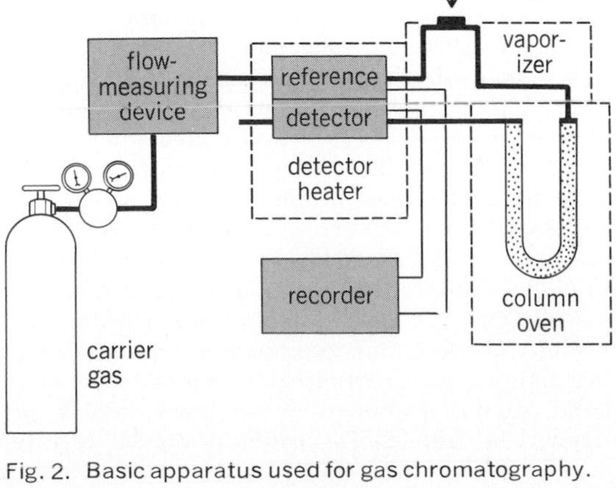

Fig. 2. Basic apparatus used for gas chromatography.

tographic analysis, 1 μl being approximately equal to 1/50 of a normal-sized drop. Gaseous samples of 0.5–1 ml are often used. Injection techniques somewhat similar to those used for liquid samples are used for solids. In order to produce sharp chromatographic peaks with minimum peak overlap, solid and liquid samples must be vaporized rapidly upon injection by maintaining the injection port at a temperature greater than the boiling point of the sample. With some gas chromatographic detectors, the sensitivity is so great that an inlet splitter must be used to reduce the sample amount to 1/10 or 1/100 of the amount injected. Sample splitters are commonly used with capillary columns, owing to the relatively small amount of liquid phase used with these columns and the much reduced sample capacity which results.

Column. This is the heart of the gas chromatograph, and separation of components on packed columns depends more on the choice of liquid phase than on any other factor. Typically the column is a glass or metal tube of 1/8 or 1/4 in. (6 or 13 mm) in diameter and 4–6 ft (1.2–1.8 m) in length, packed with an inert diatomaceous earth support coated with a nonvolatile liquid to 3–20% by weight. In open tubular or capillary column technology, the support for the thin film of liquid phase is the wall of the capillary itself. Support-coated open tubular columns are also sometimes used, the sample capacity of the columns being increased by the presence of very loosely packed support or by a roughening of the capillary walls.

In choosing a liquid phase for a particular separation, several factors are considered, including: solute-solvent interactions (for example, hydrogen bonding), which if present gives special selectivity for a particular solute; temperature limitations of the liquid phase; and the possibility of rare irreversible reactions on the column. Common liquid phases include high-molecular-weight hydrocarbons such as Apiezon L and the natural product, squalane, silicone gum rubbers such as SE-30 and methyl phenyl silicones, polyethylene glycols such as Carbowax 20M, and certain esters of low volatility such as triethylene glycol succinate. The general rule "like dissolves like" is useful in liquid-phase selection, since it is generally true that polar liquid phases make good solvents for polar solutes while nonpolar liquid phases are often the choice for the separation of nonpolar solutes.

In packed column gas chromatography, the liquid phase is adsorbed as a uniform, thin film over the surface of an inert solid support. A good solid support should have the following characteristics: chemical inertness, mechanical strength, relatively large surface area per unit volume, low pressure drop, and thermal stability. Materials with surface areas of 1–6 m^2/g are available which fulfill most of these requirements. The most commonly used support in gas chromatography is diatomaceous earth (kieselguhr), which consists of agglomerates of the siliceous skeletons of diatoms treated to produce materials known variously as Chromosorb-P and -W, Celite, C-22 firebrick, Dicalite, Sil-O-Cel, Embacel, and Sterchamol. Chromosorb-G is a diatomaceous earth treated to produce an inert solid, similar in properties to Chromosorb-W but much less friable. To reduce chemical reactivity, the solid support is frequently acid-washed, rinsed, dried, and treated with a silanizing reagent such as trimethylchlorosilane or dimethyldichlorosilane. The supports are generally size-graded to fall in the range 60–80 or 80–100 mesh.

Detectors. The two most popular detectors are the flame ionization detector and the thermal conductivity detector. The flame ionization detector is the more sensitive, with a limit of detection of 0.1–0.001 that of the thermal conductivity detector. The flame ionization detector has a large linear dynamic range, the range in sample size over which the detector responds in a linear manner. In this detector the sample is burned in a small hydrogen flame as it is eluted from the column. Upon combustion of the organic material, charged species are formed in the flame, and these species are collected by a polarized collector electrode. The signal is then fed to an electrometer, a signal being observable for samples as small as 10^{-11} g. Although the detector is sensitive to most carbon compounds, it gives little or no signal for CO_2, CO, CS_2, H_2O, O_2, N_2, NH_3, the rare gases, or oxides of nitrogen and sulfur. The detector has enjoyed widespread use in gas chromatographic analysis and is obviously attractive in pollution studies involving trace organic compounds in air and water owing to its insensitivity toward the air and the water.

The thermal conductivity cell was one of the first gas chromatographic detectors developed. It consists of a metal filament or a thermistor, heated by an electric current from a power supply, and positioned in the gas stream at the end of the column. The filament increases in temperature when organic materials are eluted, since organic compounds have a thermal conductivity less than that of the helium carrier gas, and the heat is therefore conducted away less rapidly.

Wire filaments increase their resistance about 0.4% per degree while thermistors decrease their resistance by about −4% per degree. Thus thermistor thermal conductivity detectors are more sensitive than filaments at low temperatures but become less sensitive at higher detector temperatures. The change in resistance is sensed by making the detector filament a part of a Wheatstone bridge. The advantages of the thermal conductivity detector are that it is sensitive toward all components except the carrier gas being used, and it is nondestructive. It also has a large linear dynamic range, but the limit of detection is less than that of the flame ionization detector.

Other popular gas chromatographic detectors include the flame photometric detector, the electron capture detector, and the thermionic detector. The flame photometric detector may be made selective toward individual metals in organometallic compounds, volatile metal chelates, and volatile metal halides, as well as toward compounds of sulfur and phosphorus. An inexpensive monochromator system or a filter system may be used to isolate the emission wavelength of a particular element, thus making the detector highly selective. The electron capture detector is one of the most sensitive detectors for compounds with electron-capturing properties, such as halogenated compounds and highly conjugated polynuclear aromatic hydrocarbons. It is insensitive to compounds with little or no electron-capturing properties. This detector has proved of particular value in the analysis of pesticides, many of which are chlorinated compounds. If certain alkali metal salts (such as Rb_2SO_4 and CsBr) are placed in the flame of the flame ionization detector and the flow of hydrogen and air is properly adjusted, it is found that the detector becomes particularly sensitive to nitrogen and phosphorus compounds. This orientation is called the alkali flame or thermionic detector.

Quantitative and qualitative analysis. Qualitative information in gas-liquid chromatography is obtained from retention measurements of the peak, and quantitative information is obtained from the area under a peak or, with control of certain variables, from the peak height. Retention time is constant for a given substance on a given column at a given column temperature and carrier-gas flow rate. Thus, a comparison of the retention time of an authentic sample with a component of an unknown mixture gives useful qualitative evidence. Combination of gas chromatography with mass spectrometry provides the ultimate in qualitative information and has been used extensively in research in recent years.

The area under a chromatographic peak is measured by means of an electromechanical integrator, an on-line computer system, a planimeter, triangulation approximations (such as multiplication of peak height by the peak width at half height), or sometimes crudely by cutting out the recorded chromatogram and weighing on a balance. Calibration curves may be made by plotting the area or peak height as a function of concentration or amount of sample.

Nonanalytical applications of gas chromatography include measurement of the heat of solution, adsorption, and other thermodynamic quantities; the surface area of powders; the vapor pressure of materials; chemical kinetics; and equilibrium constants. *See* MASS SPECTROMETRY.

Important equations. A 4-ft (1.2-m) packed gas chromatography column typically has the equivalent of 600–2000 theoretical plates, N. Capillary columns of 1,000,000 plates have been prepared. N is a measure of the column efficiency and is given by Eq. (1) where x is the distance to the peak maxi-

$$N = 16(x/y)^2 \qquad (1)$$

mum from the point of injection, and y is the peak width measured at the base.

Column efficiency and peak broadening may be evaluated by Eq. (2) where H is the average height

$$H = 2\lambda d_p + 2\gamma D_g/u + (8/\pi^2)(k/(1+k)^2)(d_l^2/D_l)u \qquad (2)$$

equivalent to a theoretical plate ($=L/N$); L is the column length; λ and γ are related to the uniformity in size and shape of the particles in the column; d_p is the average particle diameter; D_g and D_l are the diffusivity of the solute in the gas and liquid phases, respectively; u is the average linear gas velocity; k is the partition ratio, related to the thermodynamic partition coefficient; and d_l is the average thickness of the liquid layer adsorbed on the solid support. This equation shows that optimum column efficiency (smallest value of H) is obtained by using: particles of small diameter, uniform in size; a carrier gas of high molecular weight at an optimum velocity of generally 3–10 cm/s; a nonvolatile liquid phase of low viscosity which will retain the solute well on the column; and a uniform distribution of the liquid phase in a thin layer over the surface of a solid support possessing a large specific surface area. *See* CHROMATOGRAPHY; GAS AND ATMOSPHERIC ANALYSIS.

[RICHARD S. JUVET, JR.]

Bibliography: J. R. Conder and C. L. Young, *Physicochemical Measurement by Gas Chromatography*, 1979; S. Dal Nogare and R. S. Juvet, Jr., *Gas-Liquid Chromatography: Theory and Practice*, 1962; R. S. Juvet and F. M. Zado, A new selective–non-selective flame photometric detector for inorganic and organic gas chromatography, *Anal. Chem.*, 38:569–573, 1966.

Gas constant

Boyle's law and Charles' law may be combined into a single expression, $pV/T =$ a constant, showing how the volume V of a given mass of gas depends upon its temperature T and pressure p. If the mass of gas chosen is 1 g-mole, then the constant, known as the gas constant, is written as R. Hence, $pV = RT$ for 1 g-mole. *See* GAS.

The numerical value of the gas constant R is obtained by dividing the molar or gram-molecular volume of a perfect gas at the melting point of ice and a pressure of 760 mm of mercury at the same temperature (the standard atmosphere) by the absolute temperature at the ice point, $R = V_0/T_0$. The best values available for R, in various units, are given below.

$$\begin{aligned} R &= 8.31433 \pm 0.00034 \times 10^7 \text{ ergs/(deg)(mole)} \\ &= 1.98717 \text{ cal/(deg)(mole)} \\ &= 8.20544 \pm 0.00037 \times 10^{-2} \text{ liter-atm/(deg)(mole)} \end{aligned}$$

The kinetic theory of gases relates R to the specific heats of a perfect monatomic gas at constant pressure C_p and constant volume C_v. C_p is equal to $\frac{5}{2}R$ and C_v to $\frac{3}{2}R$. Maxwell's law of the equipartition of energy shows that any degree of freedom of a system possesses an energy of $\frac{1}{2}RT$ per mole, making a specific heat contribution of $\frac{1}{2}R$. A perfect monatomic gas has three degrees of freedom, three independent directions of molecular motion, and a specific heat C_v of $\frac{3}{2}R$. In a perfect diatomic gas molecules rotate in two independent directions; their specific heat C_v is $\frac{5}{2}R$.

[THOMAS C. WADDINGTON]

Bibliography: R. H. Perry and C. H. Chilton (eds.), *Chemical Engineers' Handbook*, 5th ed., 1973.

Gel

A two-phase colloidal system consisting of a solid and a liquid. Gels behave as elastic solids and retain their characteristic shape, whereas sols (colloidal dispersions) possess the shape of the container. Commonly, gels have a low solid content, for example, 2–5% for ferric oxide, and as little as 0.1% for coagulated blood. Gels include jellies or transparent elastic gels rich in liquid, and gelatinous precipitates which are believed to consist of minute particles of jelly. Gels or jellies which have dried until apparently solid may be called zerogels, and sometimes will swell or redisperse to form a sol when treated with a suitable solvent. The lyophobic gels (usually inorganic) may be prepared by double decomposition reactions employing high reagent concentrations. Rapid mixing favors the formation of clear, transparent gels. Other methods include slow coagulation of sols by electrolytes and concentration of sols by slow evaporation of the dispersion medium.

The lyophilic gels (usually organic), such as gelatin, agar-agar, and certain soaps, may be prepared by cooling a sol prepared at an elevated temperature or by allowing air-dry gels to swell in a solvent. The setting or gelation of a sol is characterized by (1) time of set, (2) gelation temperature, (3) critical concentration of setting, and (4) rate of viscosity increase.

Setting of a sol to a gel occurs with negligible heat effects; this phenomenon is designated as the isothermal sol-gel transformation. Gels such as clays, which contain platelike particles, may be transformed into sols by shaking; these sols revert to gels on standing. This phenomenon is called thixotropy. Rheopexy refers to the converse process, in which gelation occurs more rapidly when the sol is stirred or vibrated.

The classical explanations of gel structure were the so-called honeycomb and brush-heap theories. In the former, the continuous phase was the solid, with the liquid contained in the connected holes or pores. In the latter theory, the liquid was the continuous phase, and the pores or capillaries were the interstices between the solid particles. The brush-heap theory appears to be correct for many common gels. For example, one sample of silica gel, as viewed in the electron microscope, consisted of small primary particles about 100 A in diameter. The continuous pore structure (the interstices between the loosely packed primary particles) had a most frequent pore radius of 30–35 A. These considerations do not preclude the possibility that some types of gels, such as sintered or partially sintered gels, may possess the honeycomb structure. *See* COLLOID; PRECIPITATION; RHEOLOGY.

[W. O. MILLIGAN]

Bibliography: S. Voyutsky, *Colloid Chemistry*, 2d ed., 1978.

Gel permeation chromatography

A separation technique involving passage of a liquid moving phase through a column containing the stationary phase which consists of a porous material. Gel permeation chromatography (GPC) affords a rapid method for separating high-molelecular-weight materials, with separation based on differences in molecular size. It is of particular importance for research in biological systems and synthetic polymers. Figure 1 is a schematic diagram of a chromatograph designed for GPC.

Procedure. The stationary phase is a porous inorganic solid or cross-linked gel containing a range of pore sizes. The mobile phase is a liquid, usually water or a buffered solution for biological systems. Organic solvents are used for synthetic polymers. The mobile phase is circulated through the column either by pumping or by gravity flow. The sample to be separated is introduced at the head of the column. As it progresses through the column, small molecules can enter all pores larger than the molecule, while larger molecules can fit into a smaller number of pores, again only those larger than the molecule. Thus the larger the molecules, the smaller the amount of pore volume available into which the molecules can enter. The sample emerges from the end of the column in inverse order of molecular size, with the larger molecules eluting first followed by the smaller ones. In order to determine the amount of sample emerging, a detector is used which produces a signal proportional to sample concentration. The volume of the mobile phase is also recorded to provide a means of characterizing the molecular size of the emerging molecule.

Figure 2 shows a typical gel permeation chromatogram; the emerging solute gives an increasing detector signal which passes through a maximum and returns to zero. This signal in terms of relative amount of emerging solute is normally recorded as a function of elution volume. Some sort of volume recording device, typically a siphon, places a blip on the curve at each interval when it discharges. This affords an internal calibration of volume. If a constant volume pump is used, a time axis with a known flow rate per unit time may be used.

Calibration curve. In order to convert elution volume to molecular weight, a calibration curve is required. This is normally prepared by injecting samples of polymer with a narrow molecular weight distribution of a known molecular weight. The molecular weight is determined by an independent method such as viscometry, light scattering, and osmometry. If a wide molecular weight range is covered, then a plot such as is shown in Fig. 2 may be established of molecular weight versus elution volume.

The procedure outlined above requires fractions for each specific polymer or high-molecular-weight substance of known molecular weight. There has been an extensive search for a "universal" calibration curve, that is, for one on which a wide variety of substances could be represented by a single line. It has been established that a plot of the log of the intrinsic viscosity times the molecular weight versus elution volume gives such a curve. Many high-molecular-weight materials can be measured from a single curve established for one type of material. This quantity, intrinsic viscosity times molecular weight, is proportional to the hydrodynamic radius. It appears therefore that separation is primarily based on molecular size. There are other contributing factors such as diffusion that affect separation, but these seem to be relatively unimportant compared to the size factor.

Once a calibration curve relating elution volume and molecular weight is available, it is comparatively simple to convert the raw data from the chromatogram, that is, relative concentration versus elution volume, to a molecular weight distribution. These computations can be performed readily on a desk calculator, although computer programs are available that do this automatically. The usual parameters calculated are the cumulative molecular weight distribution and the differential molecular weight distribution. Typical plots of these two distributions are shown in Fig. 3. Additionally, the various molecular weight averages are computed. These are the number, viscosity weight, Z, and $Z+1$ averages. The location of these is indicated on Fig. 3 for the two distribution curves. Other parameters such as inhomogeneity factor and breadth of distribution may be computed if desired. Thus the formerly laborious process of determining precise molecular weight distributions becomes very simple with the use of GPC.

Column packings. A wide variety of materials have been used for column packing; aqueous solution, cross-linked dextrans (Sephadex), and polyacrylamide have found widespread use. For organic solvents, polymer gels such as cross-linked polystyrene and various other cross-linked copolymers are frequently employed. Rigid porous materials, principally silica gel and glass, also give good separations and have the advantage of high mechanical strength.

Detectors. In the simplest systems fractions are collected on a time or volume basis, the solvent removed, and the remaining fractionated material characterized by any of the standard chemical or physical tests. The more sophisticated automated systems employ a detector to continuously monitor the concentration of the emerging material in the eluent. Detectors that have been used widely include differential refractometers, infrared and ultraviolet spectrometers, colorimetric detectors, and various combinations of these. For the analytical determination of molecular weight distribution or for the separation of materials that are difficult to resolve, sample sizes are small. This requires a detector with high sensitivity. Therefore, the most frequently employed detectors operate in a differential mode. Temperature control is required, and flow rate variations must be kept to a minimum.

Associated equipment. Pumping systems employed range from columns using gravity flow to high-pressure precision pumps that can be regulated to give a constant flow rate with very good accuracy. The choice of systems depends on the resolution required. There is a wide variety of sample injection systems, ranging from a hypodermic needle to multiport valves. The columns themselves may be metal or glass. Diameters range from a few millimeters to several centimeters, and length from a few centimeters to many meters. In general, column efficiency increases directly with column length, while band spreading increases as the square root of column length. Thus, in principle, with a long enough column it should be possible to separate very difficult resolvable materials. The time for analysis, however, also increases linearly with column length for a given flow rate.

Flow rate. The effect of flow rate on efficiency in GPC has been the subject of a number of studies. Column performance may be defined in terms of the number of plates per unit length as measured on monodispersed substances. In general, it is found that efficiency decreases with increasing flow rate. The decrease, however, is not large, and in typical flow rates for analytical columns of approximately 5 mm inside diameter where flow rates are of the order of 1 to 5 ml per minute, only a slight decrease of a few percent is noted in the flow range from 1 to 5 ml. At higher rates the decrease in efficiency becomes more pronounced. As the analysis time is inversely proportional to the flow rate, it is of advantage to use the highest flow permissible that will give the required resolution.

Sample size. Normally the sample sizes used in analytical GPC are in the 2- to 4-mg range. For higher resolution, samples in the microgram range may be required.

Solvent and temperature. Selection of the solvent and temperature is normally based on two factors. The first is that the sample must be soluble. The second is compatibility with the detector

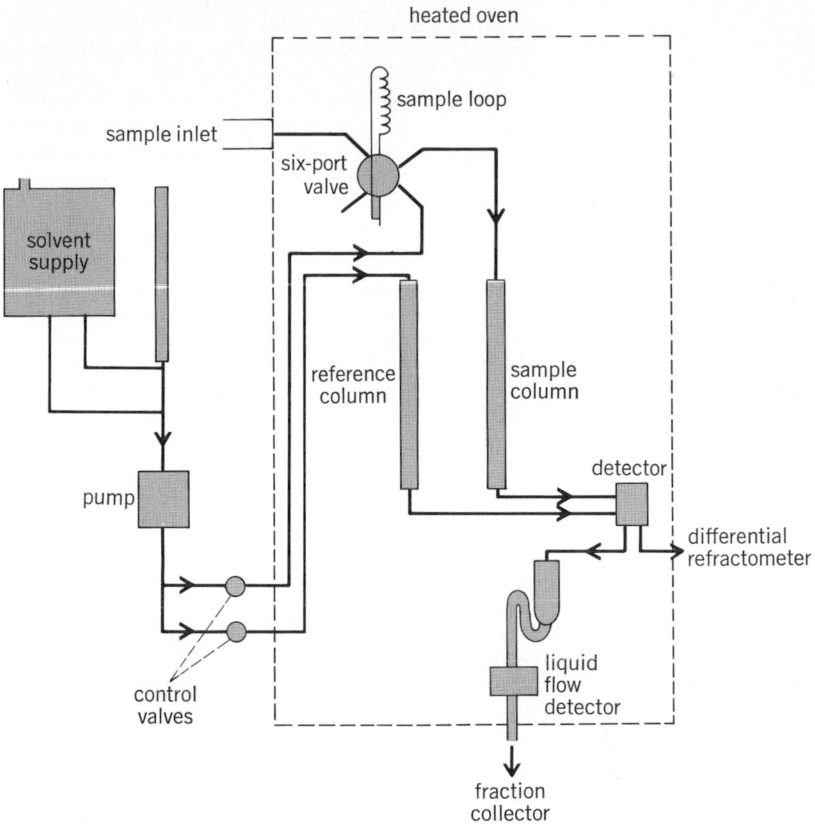

Fig. 1. Schematic diagram of a gel permeation chromatograph.

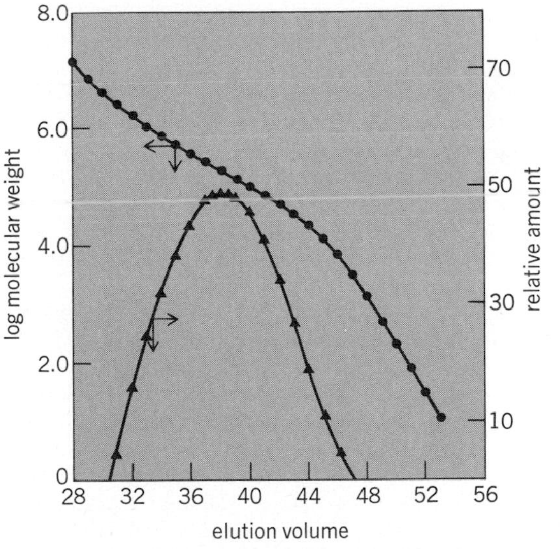

Fig. 2. A typical calibration curve and gel permeation chromatogram.

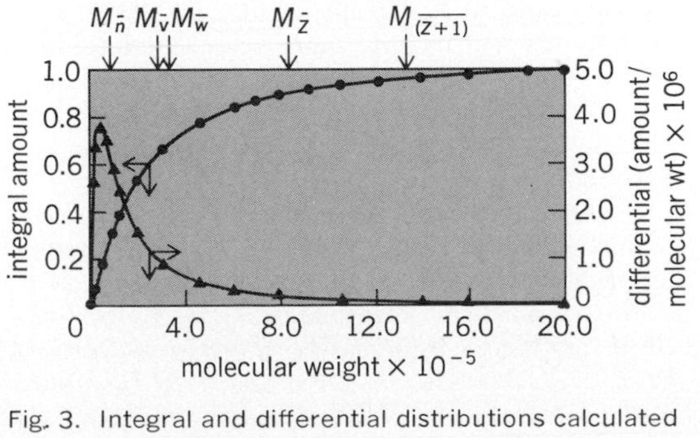

Fig. 3. Integral and differential distributions calculated from the data shown in Fig. 2.

system. In the differential refractometer, for example, one prefers a solvent with as big a difference in refractive index from the solute as possible, since the signal from the refractometer is directly proportional to the difference in refractive index. For a spectrometric detector, however, a solvent that does not absorb in the frequency range used is required.

Unlike most chromatographic separations, temperature has very little effect on the separability, as the separation is based on the hydrodynamic radius of the molecule. There are minor effects on this radius with changes in temperature. Normally, however, all the species will change in the same direction and approximately the same amount so that there is little overall effect on the relative separation. There are minor changes in efficiency due principally to changes in viscosity. Lower viscosity in the solvent phase improves mass transfer and gives a faster approach to equilibrium. The changes due to viscosity are usually also minor; therefore, the temperature chosen is normally ambient or near ambient simply for experimental convenience. For some synthetic polymers, for example, the polyolefins, that are difficultly soluble, it is necessary to operate at elevated temperatures. Temperature ranges from 0 to 150°C have been reported.

Applications. Gel permeation chromatography has had widespread applications. In the field of determining molecular weight distribution of synthetic polymers, at least 50 types of polymers have been studied. These range over alkyd resins, butyl rubbers, cellulose esters, elastomers, polyethylene, polypropylene, polystyrene, and many others. Additionally, the ability to determine the molecular weight distribution and changes in distribution has had many applications in areas such as blending distributions, chain lengths in crystals, interaction in solutions, radiation studies, mechanical degradation studies, research on the mechanism of polymerization, polymerization reactor control, and evaluation of the possibility of processing polymers.

In the field of natural and biological polymers, again numerous systems have been separated and analyzed. Among these are acid phosphatases, adrenalin, albumin, amino acids and their derivatives, enzymes, blood group antibodies, collagen and related compounds, peptides, and proteins.

Additionally, although not widely used for this purpose, GPC is capable of making separations of low-molecular-weight compounds. This is particularly valuable when both high- and low-molecular-weight compounds are present in the same sample.

Gel permeation chromatography is a rapidly growing field. It is safe to predict that future developments will include more sensitive detectors, columns with higher resolutions, more rapid analyses, and an ever-increasing number of applications to a wide and diverse number of systems. *See* CHROMATOGRAPHY.

[ANTHONY R. COOPER; JULIAN F. JOHNSON]

Bibliography: K. H. Altgelt and L. Segal (eds.), *Gel Permeation Chromatography*, 1971; M. J. R. Cantow and J. F. Johnson, in T. H. Gouw (ed.), *Guide to Modern Methods of Instrumental Analysis*, 1972; J. V. Dawkins, *Gel Permeation Chromatography of Polymers*, 1978; T. Kremmer and L. Boros, *Gel Chromatography*, 1979.

Germanium

A brittle, silvery-gray, metallic chemical element, Ge, atomic number 32 and atomic weight 72.59, with properties between silicon and tin. Germanium is distributed widely in the Earth's crust in an abundance of 6.7 parts per million (ppm). Germanium is found as the sulfide or is associated with sulfide ores of other elements, particularly those of copper, zinc, lead, tin, and antimony.

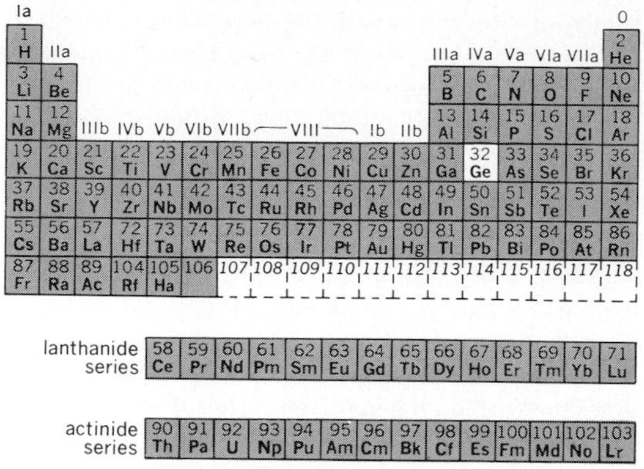

In 1864 John A. R. Newlands noted that there was a missing element between silicon and tin in the periodic table. In 1871 D. I. Mendeleev predicted on the basis of his newly formulated periodic table, not only the existence of eka-silicon, but also many of the properties of the element germanium from a study of the properties of adjacent elements in the table. The predictions were verified by Clemens A. Winkler in 1886, when he isolated the element from a Freiberg mineral, argyrodite, $4Ag_2S \cdot GeS_2$. Winkler named the element after his native country, Germany.

The classification of germanium as a rare metal probably accounted for the fact that little interest was shown in finding a use for it until more than 50 years after its discovery. The first account of any work on the uses of germanium was in 1922, when the therapeutic value of germanium oxide in treating anemia was reported. Industrial development of the metal was first sought by the Eagle-Picher Co. (1935) in the tri-state district of Missouri, Kansas, and Oklahoma. The company had erected a germanium recovery facility at Henryetta, Okla., by 1941. Fundamental research programs at Purdue University and Bell Telephone Laboratories, using germanium dioxide and germanium produced by the Eagle-Picher Co., paved the way to a variety of new concepts in solid-state physics, electronics, and ultrahigh-purity metallurgy. Germanium was officially introduced as an industrial material during World War II and achieved its initial growth in the years after the war. First to make use of the new product, and still a consumer, was the semiconductor industry.

Occurrence and production. The metal does not occur as a native ore in sufficient quantity to justify extraction by itself. However, it is easily obtained as a by-product from the treatment of other metal-containing ores. The minerals in Table 1 are known sources of germanium from such ores.

The world reserve base of germanium is estimated at 1,800,000 kg. A process for the recovery of

Table 1. Minerals containing Ge

Name	Composition	Ge content, wt %
Argyrodite	$4Ag_2S \cdot GeS_2$	6.7
Canfieldite	Ag_8SnS_6	1.8
Germanite	$7CuS \cdot FeS \cdot GeS_2$	8.7
Renierite	$(Cu, Fe, Ge, As)_x S_y$	7.8
Ultrabasite	$(Pb,Ag,Ge, Sb)_x S_y$	4.0

this element from ash and flue dusts resulting from the burning of certain coals for power generation was utilized to a small extent during the 1970s. If this were increased significantly by switching from gas to coal, the potential worldwide germanium resources would become several billion kilograms.

Modern industrial production of germanium is based primarily on the intermediate products and wastes of the processing of pyrometallic and zinc sulfide ores and coal. Germanium is recovered as a by-product from plants using germanium-containing raw materials.

In the treatment of zinc concentrates from the central United States, the germanium tends to be found in certain cadmium-rich dusts or in the residues from horizontal retort furnaces. The cadmium and zinc plants then yield products suitable for treatment in hydrochloric acid and subsequent distillation. The germanium is recovered as a crude tetrachloride, which is purified very carefully by solvent extraction and redistillation. The refined tetrachloride is hydrolized in demineralized water, yielding germanium dioxide. The latter is filtered off, washed, dried, and packaged for shipment or is processed further by hydrogen reduction to the metal state.

Some electrolytic zinc producers in Belgium and Italy obtain, during the lixiviation and purification operation of their ores, a concentrate rich in germanium. This concentrate is suitable for the distillation and refining treatment by hydrochloric acid.

Studies concerned with coal as a source of germanium have indicated another potential source of the metal, that is, living plants which are known to concentrate germanium. According to G. W. Monier-Williams, germanium is taken up from soils by cereals, especially oats. Oak and beech humus from Germany contain as much as 70 ppm germanium.

The germanium is recovered from flue dusts obtained from the manufacture of producer gas. The germanium (from germaniferous coals) remains in the coke produced from the coal. Combustion of the coke in a limited supply of air, as in the manufacture of producer gas, drives off the germanium with the gas. When the hot gas is burned, metallic constituents are converted into oxides which are deposited in the cooler parts of the flue system. These flue dust deposits may contain as high as 2% germanium, and in England they constitute the richest sources of germanium, as well as of gallium. Production of germanium was started in 1950 using flue dust as the primary raw material.

Germanium metal production (Fig. 1), involves the reduction of dioxide to metal powder in graphite boats at 650°C. An atmosphere of hydrogen or cracked ammonia (nitrogen and hydrogen) is employed. After reduction, the temperature is increased to 1100°C to melt the germanium powder. The molten metal may be cooled uniformly to a solid ingot, or the boat may be cooled gradually from end to end, causing a segregation of impurities in the last portion to solidify. The surface of the first-reduction ingot is then usually cleaned in an etching solution, and the end with the high impurity level is cropped off and reprocessed chemically.

The ingot is then purified by zone refining in a graphite boat. It is usually passed through a series of induction coil heaters, each of which causes a molten zone, or a single induction coil with a mechanism for multiple passes can be used. Impurities tend to remain in the molten zone because of their higher solubility in the liquid state and become concentrated at the freeze-out end of the ingot. The latter is cut off and returned to the cycle for additional purification, either by zone refining or by complete recycling.

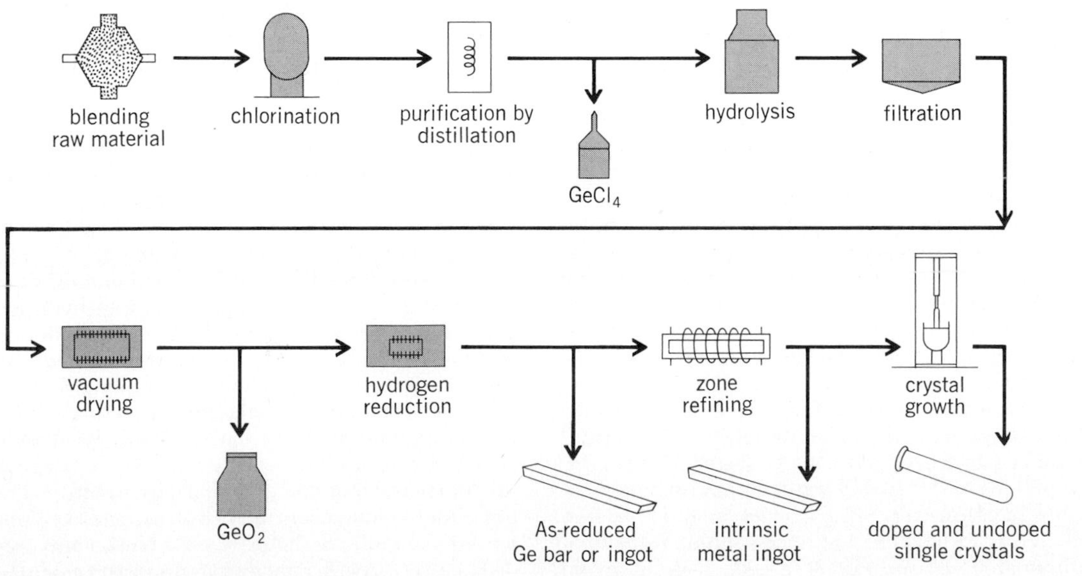

Fig. 1. Refining flow chart for germanium.

The most generally accepted standard measurement of purity for germanium is resistivity. Resistivity measurements are made by using a two-point or a four-point probe on an abraded surface at 25°C. The two-point probe gives more reliable readings on polycrystalline germanium. Although specifications for the metal have changed, typical supplier specifications for resistivity can be given as follows:

First reduction metal	5 ohm-cm minimum, *n*-type at 25°C
Intrinsic metal	40 ohm-cm minimum, *n*-type at 25°C
Undoped monocrystals	40 ohm-cm minimum, *n*-type at 25°C
Doped monocrystals	To customer's specification

The physical shape, size, and weight of the intrinsic polycrystalline ingots or bars vary widely. The cross-sectional shape is usually hemispherical, trapezoidal, or round and varies in cross-sectional area from 0.3 to 3.0 in.^{2} (2 to 20 cm^2). Ingot or bar length is from 2.0 to 30 in. (5 to 75 cm), and the weights vary from approximately 30 to 250 g per linear inch (12 to 100 g per linear centimeter).

Physical and chemical properties. Germanium is in group IVB of the periodic table with carbon, silicon, tin, and lead. It has a metallic appearance but exhibits the physical and chemical properties of a metal only under special conditions since it is located in the periodic table where the transition from nonmetal to metal occurs. Germanium melts at 937.4°C and crystallizes in the diamond lattice with a coordination number of 4, a much smaller value than that for the atoms in a metal. There are 4.42×10^{22} atoms/cm^3, the lattice constant a at 20°C is 0.56575 nm, and the expansion coefficient is $5.9-6.6 \times 10^{-6}$/°C. Germanium undergoes an approximate 5% expansion in volume upon solidification and has a hardness of 6 on the Mohs scale and 190 on the Brinell. At room temperature there is little indication of plastic flow and consequently it behaves like a brittle material. Sawing, lapping, and polishing are done by techniques similar to those that are usually used for glass and quartz.

For pure material (intrinsic germanium) the resistivity is 47 ohm-cm at 27°C. Table 2 summarizes the physical properties of germanium.

Since the oxide (GeO_2) of germanium is somewhat soluble (0.45 g/100 ml water at 25°C) in water, it provides little protection for germanium in oxidizing aqueous solutions. Silicon, on the other hand, does form an insoluble protective oxide (SiO_2) and thus is less reactive than germanium. Germanium is quite insoluble in water and nonoxidizing electrolytes in the absence of dissolved oxygen at temperatures up to 100°C. It dissolves in concentrated nitric or sulfuric acid. Nitric acid–hydrofluoric acid mixtures and molten alkalies, such as sodium or potassium hydroxide, readily dissolve germanium. The dissolution of germanium in a hydrogen peroxide concentration of 20% at pH of 4–5 gives a rate of approximately 1 μm/min. However, if a certain amount of free alkali (pH 7–8) is present, a dissolution rate of approximately 3 μm/min is noted. Sodium hypochlorite solution is a strong oxidant of germanium and is used as a etchant in the semiconductor industry for polishing germanium by a chemical mechanical process.

Germanium compounds. As it exists in compounds, germanium is either divalent or tetravalent. The divalent compounds (oxide, sulfide, and all four halides) are easily reduced or oxidized. The tetravalent compounds are more stable.

Oxygen compounds. Two oxides of germanium are known, GeO and GeO_2. A mass spectrometric study of the oxides of germanium has shown that GeO exists in polymeric forms.

Germanium monoxide is made by the action of water on germanium dichloride, $GeCl_2$, or by the action of water or ammonia on germanium chloroform. Germanium monoxide is a brown to black solid and sublimes at 710°C and is insoluble in cold or hot water, acids, and alkalies. It is soluble in chlorine water as well as hydrogen peroxide–ammonium hydroxide solutions.

Two crystalline forms of GeO_2 have been observed. The hexagonal form of GeO_2, a white powder, is obtained by the hydrolysis of germanium tetrachloride, $GeCl_4$. This form is soluble in water, 0.45 g/100 ml water at 25°C, 0.51 g/100 ml water at 30°C, and 1.00 g/100 ml water at 100°C. It reacts with hydrochloric acid to yield $GeCl_4$ and with hydrofluoric acid to yield fluorogermanic acid, H_2GeF_6. Ignition of the hexagonal form of GeO_2 at 380°C alters part of the oxide to the tetragonal form, which is insoluble in water and does not react with hydrochloric or hydrofluoric acid. The tetragonal form of GeO_2 is quite inert; however, it is slightly soluble in sodium hydroxide.

An amorphous, transparent, glassy form of GeO_2 is formed by rapid cooling of melted GeO_2. This form does not have a definite melting point and is soluble in water. It devitrifies into an opaque white glass.

Halogen compounds. Germanium forms both the dihalides and tetrahalides of fluorine, chlorine, bromine, and iodine (symbols GeX_2 and GeX_4). GeF_2 is obtained as a white sublimate from the reaction of germanium tetrafluoride vapors and germanium. The germanium dichloride, a white powder, can be obtained in a similar manner.

Germanium dibromide is formed as a by-product in the preparation of tribromogermane by the action of hydrogen bromide on heated germanium powder. Germanium diiodide is obtained from germanium tetraiodide by reduction with hypophosphorous acid. Decomposition of GeI_2 to form tetraiodide and germanium occurs above 300°C and has been investigated as an intermediate reaction to deposit thin films of germanium.

Germanium tetrafluoride may be obtained by heating barium hexafluorogermanate, $BaGeF_6$. Germanium tetrachloride is obtained by the action of chlorine on metallic germanium or of hydrogen chloride on germanium dioxide. Every commercial method for the extraction of germanium from its ores or concentrates involves the preparation of $GeCl_4$. Germanium tetrachloride is a colorless liquid which boils at 84.0°C and solidfies at −49.5°C. Its density at 25°C is 1.874 g/cm^3. The tetrachloride fumes in air and is hydrolized by water. It does not dissolve in concentrated hydrochloric acid but is very soluble in dilute hydrochloric acid and does not react with concentrated sulfuric acid. It is soluble in absolute ethanol, carbon disulfide, ben-

Table 2. Properties of germanium

Properties	Value or description
General	
Atomic weight	72.59 (based on carbon-12)
Atomic number	32
Atomic density/cm³	4.42×10^{22}
Crystal structure	diamond
Lattice constant A	5.6575
Density at 25°C, g/cm³	5.323
Melting point, °C	937.4
Boiling point, °C	2830
Vapor pressure at 27°C, atm	1.1×10^{-9}
Ionic radius (M^{4+})	0.53
Color	silvery gray
Critical temperature, °C	4170
Critical pressure, atm	910
Electrical	
Energy gap at 300 K, eV	0.67
Intrinsic ρ at 27°C, ohm-cm	47
Dielectric constant, ϵ	16.3 ± 0.2
Electron mobility drift, cm²/volt-sec	3900 ± 100
Hole mobility drift, cm²/volt-sec	1900 ± 50
Density of intrinsic electrons, n_i, at 300 K/cm³	2.4×10^{13}
np (where k = Boltzmann's constant); n = electrons and p = holes	$3.1 \times 10^{32} T^3 \exp(-0.785/kT)$
Electron diffusion constant (cm²/sec)	100
Hole diffusion constant (cm²/sec)	48.7
Critical electric field-electrons (volt/cm)	900
Critical electric field-holes (volt/cm)	1400
Thermal	
Debye temperature, K	362–371
Thermal conductivity, cal/sec/cm²/°C	0.15
Specific heat (0–100°C), cal/g/°C	0.074
Thermal expansion coefficient (linear) at 25°C (0–300°C)	6.1×10^{-6}
Latent heat of fusion, cal/g atom	8100
Heat of vaporization, cal/g atom	79,900
Entropy (cal/mole/K) at 298 K	10.14
Mechanical	
Hardness, Mohs scale	6
Brinell scale	190
Bulk rigidity modulus, dyne/cm²	1.3×10^{12}
Elastic constants,* dyne/cm²,	
c_{11}	1.292×10^{12}
c_{12}	0.479×10^{12}
c_{44}	0.670×10^{12}
Young's modulus, dynes/cm²,	
E_{111} at 300 K	1.400×10^{12}
Bulk modulus, dynes/cm², B	0.750×10^{12}
Shear modulus, dynes/cm², G_{111}	0.494×10^{12}
Volume compressibility, cm²/dyne	1.3×10^{-12}
Surface tension, dynes/cm at 937°C	600–630

Properties	Value or description
Optical	
Refractive index, n_0	3.994
Refractive index at 2.0μm	4.125
Refractive index at 12.0μm	4.002
Refractive index, temperature dependence at 2.25μm, dn/dT	5.25×10^{-4}
Emissivity, 1000 K	0.56
1200 K	0.53
1000 K	
1200 K	
Magnetic	
Specific susceptibility, $\times 10^6$	−0.122 at 20°C −0.146 at −183°C
Diamagnetic susceptibility, $\times 10^6$	(Ge^{+2}) − 16.8 (Ge^{+4}) − 10 (Ge^{-4}) − 100

Piezoelectric

Piezoresistance coefficients, 25°C

Material	Resistivity (ohms)	Π_{44}(10^{-12}cm²/dyne)
n-Ge	~1	−138
p-Ge	~1	+97

Temperature dependence of piezoresistance:

$$n\text{-Ge}\ \frac{T\Pi_{44}}{2} = 2 \times 10^{-8}\ \text{cm}^2\ \text{K/dyne}$$

Pressure-induced phase transition at 25°C

$P_{app} = 1.2 \times 10^8$ dyne/cm²

Electrochemical	
$Ge = Ge^{2+} + 2e^-$	E° = 0.0 volt
$Ge + 50H^- = HGeO_5^- + 2H_2O + 4e^-$	E° = −1.0 volt
$Ge + 2H_2O = GeO_2 + 4H^+ + 4e^-$	E° = −0.15 volt
$Ge + 3H_2O = H_2GeO_3 + 4H^+ + 4e^-$	E° = 0.131 volt
Electronegativity	1.8–2.0

Radioisotopes

Isotopes	*Half-life*	*Radiations*
Ge^{65}	1.5 min	β^+, 3.7 MeV; γ, 0.67–1.72 MeV
Ge^{66}	2.7 hr	β^+, 1.3–2.0 MeV; K; γ, 0.38–0.183 MeV
Ge^{67}	19 min	β^+, 3.1–2.3–1.6 MeV; K; γ, 0.17–0.92–3.4 MeV
Ge^{68}	280 days	β^+, 1.89 MeV; K; γ, 1.07 MeV
Ge^{69}	40 hr	β^+, 1.21 MeV; K; γ, 1.12–2.0 MeV
Ge^{71}	11 days	K, 0.0104 MeV
Ge^{75}	82 min	β^-, 1.18–0.92 MeV; γ, 0.265–0.63 MeV
Ge^{77}	11 hr	β^-, 2.20–1.38–0.71 MeV; γ, 0.215–0.265 MeV
Ge^{78}	2.1 hr	β^-, 0.9 MeV

*c_{11} = stiffness (or elastic) coefficient in direction normal to a given plane; c_{12} = stiffness coefficient parallel to said plane above; c_{44} = shear strain compound between above two planes.

zene, chloroform, ethyl ether, and other nonpolar organic solvents.

In preparing films of pure germanium, $GeCl_4$ is reduced in the gas phase, followed by crystallization of the germanium. The reduction is carried out in hydrogen and produces germanium and hydrogen chloride. The substrates used for the deposition of the film can be graphite, quartz, tungsten, or molybdenum. Detailed studies of the reaction $GeCl_4 + 2H_2 \rightarrow Ge + 4HCl$ have been carried out in the semiconductor industry because of its use in epitaxial technology. The reaction between $GeCl_4$ and a 20·80 copper silicon alloy at 400–500°C produces silicon tetrachloride and metallic germanium. Germanium tetrachloride can be extracted almost completely with carbon tetrachloride from 9 to 12 N solutions of hydrochloric acid. The reextraction of germanium with water is quite easily accomplished. Hydrocarbons extract $GeCl_4$ equally well. The boiling points of $GeCl_4$ and arsenic trichloride are close to each other. Therefore, in the process of distillation of $GeCl_4$, use is made of oxidizing agents which make it possible to convert the arsenic to the pentavalent state. Repeated distillations are necessary to obtain high-purity $GeCl_4$, used for the vapor deposition of high-purity germanium thin films.

Germanium tetrabromide is formed when bromine vapors are passed over heated germanium (burns with a yellowish flame) or when hydrobromic acid reacts with germanium dioxide. Germanium tetraiodide is obtained by treating germaniun dioxide with 57% hydroiodic acid. The tetraiodide will form by heating germanium to approximately 600°C in iodine vapor.

Hydrogen compounds. Until 1961 only three volatile binary germanium hydrogen compounds had been positively identified. These were germane, GeH_4, digermane, Ge_2H_6, and trigermane, Ge_3H_8. However, some work on the hydrolysis of magnesium germanide and the effect of an electrical discharge on GeH_4 has shown that several higher germanes are capable of existence. A number of hitherto unknown halogen derivatives of monogermane have also been characterized. In addition to these volatile germanes, solid nonvolatile materials containing germanium and hydrogen have long been known. A high-yield germane reaction (73%) is obtained when an aqueous solution of sodium borohydride is added to an aqueous acidic solution of germanium dioxide.

Monogermanes decompose measurably at about 280°C into germanium and hydrogen. This reaction has been investigated extensively by the semiconductor industry as a technique for depositing thin films of germanium. The hydrides GeH_4 and Ge_2H_6 are much less easily oxidized than SiH_4 and Si_2H_6. Monogermane slowly reacts with oxygen at 160°C, whereas digermane is rapidly oxidized at about 100°C. Monogermane is soluble in sodium hypochlorite and slightly soluble in hot hydrochloric acid. Alkali metals in liquid ammonia with germane produce H_3GeK or H_3GeNa. These compounds are white air-sensitive solids, unstable at room temperature.

Sulfur compounds. Germanium monosulfide is precipitated by hydrogen sulfide from aqueous solutions containing compounds of divalent germanium. Germanium disulfide is insoluble in strongly acidic media. Thus GeS_2 can be made by the action of hydrogen sulfide on a strongly acidic solution of GeO_2. The melting point of GeS_2 is approximately 800°C. Molten germanium disulfide is a transparent, light-brown, mobile liquid, which on cooling forms an amber glass with a density of 5.81 g/cm^3. Multiple treatment of germanium disulfide with nitric acid or hydrogen peroxide results in the formation of germanium dioxide.

Organogermanium compounds. Organogermanium compounds are many in number and, in this respect, germanium resembles silicon. Organogermanium compounds can be grouped as (1) alkyl compounds of germanium hydrides of the type R_4Ge, R_3Ge-GeR_3, and so on; these can be made from the tetrahalides by action of a Grignard reagent or zinc alkyl; (2) compounds with halogenated germanium hydrides of the type R_nGeX_{4-n} (where R is an organic base and X is a halogen); these can be made by treating the tetraalkyls with halogens; (3) oxidized compounds of the type $(R_3Ge)_2O$; the simpler compounds are formed by the hydrolysis of the bromide; (4) compounds of the type $R_2Ge(OH)_2$; little is known about the diols, and diethyldibromogermane is converted to a diol but loses water to a polymeric form; (5) esters of germanic acid $Ge(OR)_4$ (where the R is a methyl, ethyl, propyl, or butyl group); these compounds can be obtained by reaction (1); and (6) nitrogen compounds of germanium containing organic bases; for example, germanium tetrachloride reacts with ethylamine to form $Ge(NC_2H_5)_2$.

$$GeCl_4 + 4NaOCH_3 \rightarrow Ge(OCH_3)_4 + 4NaCl \quad (1)$$

Some of the compounds synthesized are dimethyldichlorogermane, diphenyldibromogermane, triphenylcyanoethylgermane, triphenylcyclohexylaminegermane, triphenyl(5-hydroxypentyl) germane, triphenyl(cyanoethyl)germane, triphenyl-(2-carbamylethyl)germane, triphenyl(2-acetoxyethyl)germane, triphenyloctylgermane, tetraphenylaminegermane, triphenyl(5-hydroxypentyl)germane, triphenyl(cyanoethyl)germane, triphenylpropylgermanium, tetra-*n*-hexylgermanium, hexaphenyldigermane, triphenylgermanium sulfide, triphenyl(thiomethyl)germane, triphenyl(thiobenzyl)-germane, triphenylgermanium thiocyanate, triphenylgermanium thio,bis-(triphenylgermanium) disulfide, triphenylthiobutylgermane, diethyldichlorogermane, tetraethoxygermane, tri-*n*-butylchlorogermane, and diphenyldichlorogermane.

Studies are being conducted on a number of these compounds as to their physical and chemical properties. Interest in organogermanium compounds has centered around their biological action. Germanium in its derivatives appears to have a lower mammalian toxicity than tin or lead compounds.

Alloys. Binary, ternary, and quaternary systems of most metals with germanium have been investigated. Figure 2 illustrates most of the binary systems reported.

Two alloys of importance to the semiconductor industry are germanium-aluminum and germanium-gold. At 55 wt % germanium, the germanium-aluminum system forms a eutectic which melts at 423°C. At 12 wt % germanium, the germanium-gold system forms a eutectic which melts at 356°C. The two alloys are very useful in forming electrical contact systems for germanium transistors, diodes, and rectifiers. The Ge-Au system has been evaluated as to its use for dental alloys because of its good dimensional stability upon cooling, thus allowing precision castings to be made. The germanium-silicon alloy system forms a continuous series of solid solutions. It has been explored quite extensively from the standpoint of its semiconductor characteristics.

Germanium has been investigated as a possible alloying agent for zirconium in the development of corrosion-resistant, high-strength zirconium alloys. Small additions of germanium are known to give increased hardness and strength to copper, aluminum, and magnesium. In addition, small quantities of germanium improve the rolling properties of

Li		Be			B		C		N		O				
Na		Mg			Al		Si		P		S				
K		Ca		(Sc)		Ti		V		Cr		Mn	Fe	Co	Ni
	Cu		Zn		Ga		**Ge**		As		Se				
Rb		(Sr)		Y		Zr		Nb		Mo			Ru	Rh	Pd
	Ag		Cd		In		Sn		Sb		Te				
Cs		Ba		La		Hf		Ta		W		Re	Os	Ir	Pt
	Au		Hg		Ti		Pb		Bi		Po				

rare earths: Rn, Fr, Ra, Ac
actinides: Th, Pa, U, Np, Pu, Am, Cm, Bk, Cf, E_S

Fig. 2. Chart of the binary germanium alloy systems. The elements are arranged in the form of the periodic table. The literature contains questionable data on strontium and scandium as to their being true alloys.

some alloys and do not create any appreciable increase in production cost. Fundamental studies on the magnetic properties of iron-germanium and manganese-germanium alloys have been made. These materials are ferromagnetic, as are the alloys UGe_2, PuGe, and CrGe.

The solid solubilities of tin, gallium, aluminum, arsenic, antimony, lithium, indium, zinc, silver, lead, copper, gold, nickel, and iron in liquid and solid germanium have been determined. The solubility of these elements in germanium is very important in the preparation of germanium for semiconductor and optoelectronic applications, especially in the areas of crystal growth, diffusion, metallization technology, and optoelectronic technology.

Uses. The properties of germanium are such that there are several important applications for this element.

Semiconductor. During World War II germanium was intensively investigated for its use in the rectification of microwaves for radar applications. Several types of diodes were developed.

A major development in electronics occurred in 1948 with the invention of the transistor. This solid-state device, which was first made of germanium, has had a profound influence on all electronic applications. The transistor captured the hearing-aid and radio markets, then moved into industrial applications, such as computers and guidance control systems for missile and antimissile systems. Transistors cover signal processing from direct current to gigahertz frequencies, with power-handling capabilities from microwatts to hundreds of watts. The sale of both germanium diodes and transistors peaked in 1966 and then declined because of inroads made by discrete silicon devices and integrated circuits. The United States trend is expected to continue downward as more and more of the electronic functions are designed around integrated circuits. Little technical or scientific effort has been expended on germanium integrated circuits. However, when ultrahigh-speed switching circuitry is required, germanium inherently is better than silicon since it has mobility values for electrons and holes greater by a factor of 2 than those for silicon.

The development of germanium semiconductor technology has quite closely paralleled the development of the germanium transistor. In fabricating a transistor, a need arose for high-purity polycrystalline germanium to be used in the growth of single-crystal germanium, since it was soon discovered that not only purity but also good crystal structure was required for good device performance. Technology for purifying $GeCl_4$, GeO_2, and polycrystalline germanium was developed. One such process, developed by W. G. Pfann, was zone refining, which enables movement of a molten zone down an impure polycrystalline rod, with the impurities being carried to the end of the rod. Knowledge of the solubilities of impurity elements in the liquid and solid states of germanium (C_l and C_s, respectively) had to be obtained, as well as their segregation constants ($K_{seg}=C_s/C_l$). The growth processes for single crystals of germanium had to be developed. Several techniques have been extensively studied. Two of the most common techniques are the Czochralski or Teal-Little method and zone leveling. The general method, shown in Fig. 3, proceeds by allowing germanium to slowly freeze or crystallize onto a single crystal seed which may be rotated and withdrawn from the melt. The growth (freeze) rate is controlled by a combination of the temperature of the melt and the amount of heat lost from the crystal by conduction up the seed and by radiation from its surfaces. Crystals from 1/16 in. to 6 in. in diameter have been grown by this process. Growth rates vary with diameter and desired crystal properties but are usually in the inches-per-hour range.

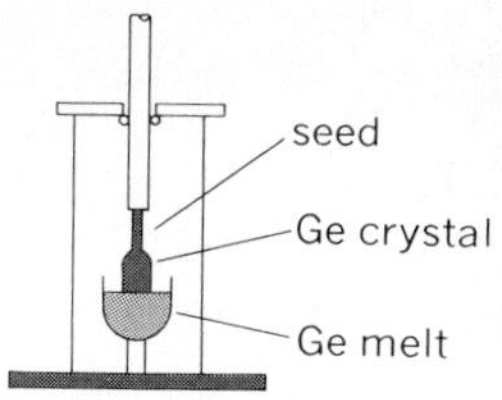

Fig. 3. Czochralski crystal-growing technique.

The zone-leveling technique, or horizontal method, is shown in Fig. 4. Here the seed and polycrystalline charge or ingot are usually loaded into a quartz boat. The boat and contents are then placed into the quartz tube of the zone leveler. With this technique, melting occurs on the leading edge of the molten zone, and freezing occurs on the trailing edge as the furnace is moved. Growth rates vary with crystal properties desired but are usually in the inches-per-hour range. The advantage of this process over the Teal-Little for germanium is the extremely uniform resistivity profile of the crystal due to the uniform mass of molten germanium in the growth process. The constant mass of molten germanium enables the ratio of dopant to germanium to remain the same, thus producing uniform resistivity.

In the fabrication of germanium devices, the single-crystal material must be sawed, lapped, or ground and then polished into thin-slice forms with flat, damage-free surfaces. Various techniques have been developed to produce slices with these characteristics. Some of these are slicing with diamond wheels, grinding with diamond, or lapping with an abrasive slurry. These techniques induce damage in the form of microcracks and fissures, which must be removed by chemical, electrochemical, or chemical-mechanical polishing methods.

It must be pointed out that germanium has a salvage value that makes it very economical for the semiconductor user to collect it at various stages in materials processing. The user pays a reclamation fee based on contained germanium in the salvage material. Broad salvage categories are:

Scrap metal	100% contained Ge
Saw sludge	100% contained Ge
Lapping sludge	5–15% contained Ge
Spent etchants	<1% contained Ge

A large percentage of the single-crystal germanium ends up as salvage material for reprocessing.

In attempts to develop better utilization of

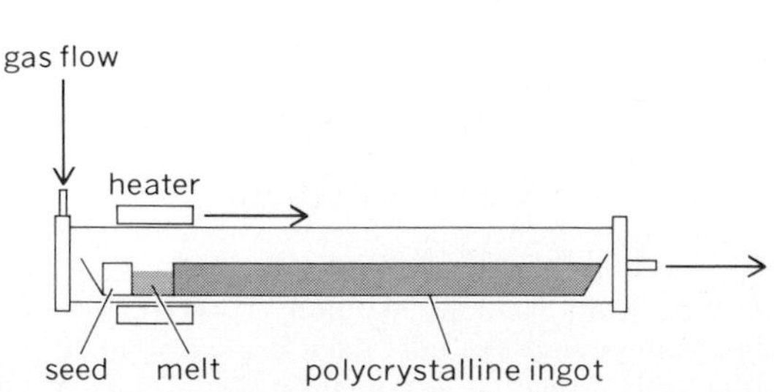

Fig. 4. Zone-leveling technique.

single-crystal material and to circumvent all the above processing steps, two other growth techniques (modifications of melt growth) have been explored. One is dendritic growth, which depends on the growth characteristics of twinned crystals, and the other is shaped crystal growth, which is dependent on a shaped heat zone. Both techniques can produce thin ribbons of germanium with desired characteristics for some type of devices.

Vapor-phase crystal growth at temperatures well below the melting point is now widely used for growing thin films or layers onto slices and is usually referred to as epitaxial growth. The primary advantage of gas-phase growth is that it allows doping impurities to be changed quite rapidly so that thin (micrometer range) layers of quite different resistivities can be grown sequentially. Direct evaporation or a chemical means may be used as the method of transport. Chemical transport reactions usually used are the thermal decomposition of germane and the hydrogen reduction of $GeCl_4$. Disproportionation reactions of the type shown by series (2) may be used to transport ger-

$$\underbrace{\mathrm{Ge+GeI_4 \rightarrow GeI_2}}_{T_1}\overbrace{\mathrm{\rightarrow Ge+GeI_4}}^{T_2} \qquad T_1 > T_2 \qquad (2)$$

manium from polycrystalline feed stock to the seed or slice. The latter technique has been used for extremely low temperatures (300–400°C range). All these techniques have been used to grow epitaxial films with the desired crystallographic and electrical characteristics.

Optoelectronics. Another use that has surfaced due to the cost of material such as gallium arsenide (GaAs) and the inability to control its growth is the use of Czochralski- or Teal-Little-grown single-crystal germanium as a substrate for vapor phase growth of GaAs and gallium arsenide phosphide (GaAsP) thin films used in some light-emitting diodes (LEDs). These devices have been used in digital displays for calculators, watches, and so on.

Infrared optical materials. Germanium lenses and filters are being used in instruments which operate in the infrared region of the spectrum. Windows and lenses of germanium are vital components of some laser and infrared guidance or detection systems. Glasses prepared with germanium dioxide have a higher refractivity and dispersion than do comparable silicate glasses and may be used in wide-angle camera lenses and microscopes. The GeO_2-TiO_2-P_2O_5-type glasses have excellent infrared transmission characteristics that make them ideal for use as windows for the protection of ultrasensitive infrared detectors used in the space programs. A 60% GeO_2, 37% PbO, and 3% TiO_2 glass has a 74% transmittance in the 2.5–7.0 μm region. The germanate glasses have a softening temperature above 400°C, can withstand thermal shock, have sufficient hardness, and have chemical durability.

Another area with considerable activity and importance is the study of nonoxide chalcogenide glasses as potential infrared optical materials, several of which use germanium. Two-, three-, and four-component systems have been evaluated; the qualitative results obtained are summarized in Table 3 in terms of suitability for application in the 3–5 and 8–14 μm regions. Strong absorption means the glass is not useful over the entire wavelength region. As can be seen from Table 3, some of the germanium-bearing glasses have very good infrared characteristics. The newer chalcogenide glasses have superior physical properties to As_2S_3 glass and are easier to polish and are not as brittle. When compared to oxide optical glasses, chalcogenide glasses must be classified as soft weak materials with low softening points.

Table 3. Qualitative evaluation of glasses from nonoxide chalcogenide systems

System	Softening point, °C	Refractive index at ~5 μm	Absorption* 3–5	Absorption* 8–14
As-S	200	2.4	W	M
As-Se	200	2.8	—	W
Ge-S	420	2.3	W	S
Ge-Se	360	2.6	—	S
Si-P-Te	180	3.4	—	M
Si-Sb-Se	270	3.3	S	S
Si-Sb-S	280	—	S	S
Ge-P-Se	420	2.4–2.6	M	S
Ge-P-S	520	2.0–2.3	W	S
Si-As-Te	475	2.9–3.1	—	M
Ge-As-Te	270	3.5	—	—
Ge-P-Te	380	3.5	—	—
Ge-As-Se	450	—	—	M
Ge-Sb-Se	325	2.6	—	—
As-Se-Te	200	2.6–3.1	—	M
As-S-Se-Te	195	2.1–2.9	—	M
Si-Ge-As-Te	325	3.1	—	M

*Absorption symbols: — = no appreciable absorption; W = weak absorption; M = medium absorption; S = strong absorption.

Gallium-doped single-crystal germanium has been used for low-temperature thermometers and bolometers. The sensitivity of these devices in the important temperature range 1–10 K has been employed to provide an accurate (0.00X K) secondary temperature standard in many laboratories. The thermometer elements are encapsulated to prevent photovoltaic effects. Their principle of operation is the large change in resistance of gallium-doped germanium at low temperatures. A germanium thermometer with a room temperature resistivity of approximately 0.05 Ω-cm has, for example, a resistance of 0.5 ohm at 77 K and a resistance of 1000 ohms at 4.2 K. The germanium bolometer can be used as a far-infrared detector and is operated in a liquid helium cryostat. The photon energy entering the optics port is absorbed by the germanium chip and thus causes heating and a change in its resistance.

Mercury-doped and copper-doped germanium are used as infrared detectors. These are single-crystal, extrinsic, photoconductive devices. The mercury-doped germanium detector is ideally suited for the detection of infrared radiation in the 8–14 μm spectral window of the atmosphere but will operate at 2–14 μm. The copper-doped germanium device is capable of detection of infrared radiation of 2–27 μm. Development work on optical telecommunications systems which employ germanium-cored glass fibers as the transmission medium looks promising and is continuing.

General uses. Synthetic garnet materials containing germanium have magnetic properties which may be of interest in high-power microwave devices and magnetic bubble memories. The magnesium germanate phosphor is used in some fluorescent lights. A striking thermoluminescence is observed in vitreous germanium oxide, containing aluminum as an impurity. Germanium devices have been used to convert the energy of radioactive decay into electrical energy, for example, the germanium-lithium drift detector.

The catalytic properties of atomically clean germanium surfaces have been investigated. Their use in chemical reactions, such as the hydrogenation of coals and chemical nickel plating, have been reported. Some petroleum refining operations and processes for the production of polyester fibers (especially those produced in Europe and Japan) employ germanium catalysts. Germanium dioxide has been used as a catalyst in oxidative polymerization reactions.

Sulfuric acid with small amounts of germanium lowers the internal resistance to current flow in a storage battery. This reduces heating and results in concurrent faster current drain, thus more usable ampere-hours. The germanium may be introduced directly into the sulfuric acid electrolyte or into the lead-oxide paste which coats the battery grids. Germanium films have been applied on reflectors because of their high reflectivity and good corrosion resistance. Crystal analyzers of single-crystal germanium are used in x-ray spectroscopy. Copper-germanium and platinum-germanium thermocouples are very sensitive. Germanium-silver film resistors have been developed with the temperature coefficients less than 0.0001/°C.

The pharmacological uses of germanium compounds are still being investigated. The trialkylgermanium compounds, in particular, are being investigated as to their antimicrobial activity.

Analytical chemistry. Samples are usually taken into solution by a sodium carbonate fusion followed by a cold-water extraction. Germanium is separated by precipitation with hydrogen sulfide in 6 *N* sulfuric acid from all the elements except arsenic, antimony, tin, and molybdenum. Of these, arsenic usually is the only problem element. The germanium sulfide is dissolved in dilute ammonium hydroxide, transferred to a distillation flask, and distilled from a 1·1 HCl solution under 1 atm (101 kPa) of chlorine. The chlorine oxidizes the trivalent arsenic to its pentavalent form whose chloride is involatile. This allows the germanium to distill over, leaving the arsenic behind.

Since $GeCl_4$ is very soluble in such organic solvents as carbon tetrachloride, chloroform, benzene, and aliphatic hydrocarbons, an extraction method is very useful in separating germanium from interfering elements. The germanium tetrachloride is extracted from solutions in concentrated hydrochloric acid (9 *N*) by using carbon tetrachloride.

Identification. Germanium may be detected in solution by using hydrogen sulfide, which precipitates GeS_2; potassium ferrocyanide, which precipitates $(GeO_2)Fe(CN)_6 2H_2O$; hydrogen selenide, which in an aqueous solution of formaldehyde ties up germanium as a yellow precipitate; phenylfluorone, which gives a very sensitive reaction and is used as a quantitative technique; and other organic reagents, such as alizarin, hydroxyazobenzene, glycerol, and mannitol.

Determination. Germanium is commonly determined gravimetrically by precipitation with hydrogen sulfide, a 5% tannin solution, and orthohydroxyquinoline or dibromoorthohydroxyquinoline followed by conversion to GeO_2.

A volumetric method of determination also exists. Germanium in aqueous solution reacts with mannitol to form a strong monoprotic acid complex, which can be titrated with sodium hydroxide. In the presence of strong electrolytes, the germanium-mannitol complex liberates iodine quantitatively from potassium iodide–potassium iodate solution. The liberated iodine is titrated with sodium thiosulfate. Another method involves precipitation of the salt of molybdogermanic acid and pyridine. The precipitate is dissolved in a mixture of HCl and ethylalcohol, which is treated with an excess of bromide-bromate solution and then potassium iodide, thus liberating free iodine, which is titrated with sodium thiosulfate. Another method, and probably the best, involves the use of fructose or inverted sucrose. This is added to a neutral solution of GeO_2; the pH of the solution drops because of the formation of a complex germanic acid, which is titrated with an alkali.

Microgram quantities of germanium are determined spectrophotometrically with phenylfluorone as the color-developing reagent. Germanium reacts with ammonium molybdate to form a soluble heteropoly germanomolybdate complex, which can be used to determine germanium photometrically. The complex can be reduced with ferrous sulfate to yield molybdenum blue, which permits a more sensitive measurement of the germanium content.

Germanium is separated from the interfering elements by distillation of the tetrachloride or by extraction of the $GeCl_4$ with carbon tetrachloride prior to polarography. The reduction of the tetravalent germanium in alkaline and neutral buffer solutions is irreversible. The half-wave potential of the reduction of Ge^{4+} to the metal on a mercury electrode in an ammonium chloride medium is 1.4 volts. Divalent germanium in 6*N* hydrochloric acid can be reduced to metal on a dropping mercury electrode with a half-wave potential of 0.45–0.50 volt at a Ge concentration of 1×10^{-4} mole/liter.

In an emission spectrographic method the lines usually employed for the determination of germanium are 270.96, 265.115 and 303.908 nm. This method is quite simple and is suitable for mass analyses of similar products.

There are no satisfactory radiochemistry techniques for germanium. Most of the isotopes have too short a half-life to be useful. Germanium-71 has an 11-day half-life but has a very weak emission. Neutron activation analysis for germanium is also not very satisfactory because of the generation of arsenic-77 and other interfering elements.

[PAUL S. GLEIM]

Bibliography: V. I. Davydov, *Germanium*, 1966; Germanium, *Ann. Mining Rev.*, July, 1967; Germanium Information Center, *Germanium*, Midwest Research Institute, Kansas City, Mo., 1964; M.

Hansen, *Constitution of Binary Alloys*, 1958; L. P. Hunter, *Handbook of Semiconductor Electronics*, 3d ed., 1970; M. S. Kiver, *Transistors*, 1962; J. W. Mellor, *Inorganic and Theoretical Chemistry*, vol. 7, 1957; M. Neuberger, *Germanium Data Sheets AD610828*, Defense Documentation Center, U.S. Department of Commerce, February, 1965; U.S. Bureau of Mines, *Mineral Commodity Summaries*, issued annually; U.S. Department of the Interior, *Minerals Yearbook*, issued annually.

Gibbs function

The Gibbs function G, also known as Gibbs free energy or free enthalpy, is defined in the equation shown, where E is the internal energy, p is the pressure, v is the volume, T is the absolute temper-

$$G = E + pv - TS$$

ature and S is the entropy. The Gibbs function is most useful in analyzing systems held at constant temperature and pressure. Under these conditions, the change in the Gibbs function ΔG of a system is a measure of the maximum attainable work, not including the work of displacing the environment. For example, ΔG represents the maximum electrical work obtainable from a galvanic cell. When the only work done by the system at constant temperature and pressure is displacing the environment, the equilibrium state is characterized by G having reached its minimum value. Since chemical processes frequently occur at constant temperature and pressure, the Gibbs function is extensively used in chemical engineering for calculating phase equilibrium and reaction equilibrium. *See* CHEMICAL THERMODYNAMICS; FREE ENERGY. [WILLIAM F. JAEP]

Bibliography: R. W. Haywood, *Equilibrium Thermodynamics for Engineers and Scientists*, 1980; J. B. Jones and G. A. Hawkins, *Engineering Thermodynamics*, 1960; S. L. Kittsley, *Physical Chemistry*, 3d ed., 1969; M. Mark and A. R. Foster, *Thermodynamic Principles and Applications*, 1979; W. J. Moore, *Physical Chemistry*, 3d ed., 1962; K. Wark, *Thermodynamics*, 3d ed., 1977.

Glycerol

The simplest trihydric alcohol, with the formula $CH_2OHCHOHCH_2OH$. The name glycerol is preferred for the pure chemical, but the commercial product is usually called glycerin. It is widely distributed in nature in the form of its esters, called glycerides. The glycerides are the principal constituents of the class of natural products known as fats and oils.

Uses. Glycerin is used in nearly every industry. With dibasic acids, such as phthalic acid, it reacts to make the important class of products known as alkyd resins, which are widely used as coatings and in paints. Because of its valuable emollient and demulcent properties, it is used in innumerable pharmaceutical and cosmetic preparations. It is an ingredient of many tinctures, elixirs, cough medicines, and anesthetics. It is a basic medium for toothpaste.

In foods, it is an important moistening agent for baked goods and is added to candies and icings to prevent crystallization. It is used as a solvent and carrier for extracts and flavoring agents and as a solvent for food colors.

Because of its humectant properties, it is sprayed on tobacco before it is processed to prevent crumbling and is added to adhesives and glues to keep them from drying too fast. Many specialized lubrication problems have been solved by using glycerin or glycerin mixtures.

The production of glycerin has increased at a steady rate. Many millions of pounds are used each year to plasticize various materials. As much as 15% is added to cellophane to render it pliable. It is included in meat casings and special types of paper for the same purpose. Sheets and gaskets made from ground cork are plasticized with glycerin.

Of its chemical derivatives, the esters of glycerin are the most important. Nitroglycerin (glyceryl trinitrate) is used in the manufacture of dynamites and propellants. The mono- and diesters of higher fatty acids can be formed by direct reaction of glycerin with the acid or by a transesterification reaction of a glyceride with glycerin. These esters are used as emulsifiers in foods and preparation of baked goods and for modification of alkyd resins.

Properties. When pure, glycerin is a colorless, odorless, viscous liquid with a sweet taste. It is completely soluble in water and alcohol but is only slightly soluble in many common solvents, such as ether, ethyl acetate, and dioxane. Glycerin is insoluble in hydrocarbons. It boils at 290°C at atmospheric pressure and melts at 17.9°C. Its specific gravity is 1.262 at 25°C referred to water at 25°C, and its molecular weight is 92.09. It has a very low mammalian toxicity.

Production. Glycerin was first discovered in 1779 by Carl W. Scheele, who made it by heating olive oil with litharge. Until after World War II, nearly all the glycerin of commerce was produced as a by-product in the manufacture of soap or from the hydrolysis (splitting) of fats and oils. However, at present a substantial portion of the material made in the United States is prepared synthetically from propylene.

In the process of soapmaking, called saponification, fat reacts with aqueous sodium hydroxide. The crude product is coagulated by the addition of salt. The acid portion of the fat combines with the sodium hydroxide to form solid soap, and the glycerin liberated in the reaction remains in the salt solution, which is called spent lye.

In the process involving the hydrolysis of fats, they react with water to give the component acids and glycerin. An aqueous solution of the latter is produced, called glycerin sweet water. From this liquid and from spent lye from soapmaking, glycerin is obtained in much the same way. After a preliminary treatment to remove impurities, the water of solution is evaporated under reduced pressure. The residual glycerin is filtered while hot to remove precipitated salts. For most applications, it is necessary to refine it further by fractional distillation under reduced pressure.

Three synthetic routes from propylene to glycerin are used on a large scale. In the first of these, propylene is converted to allyl chloride, which is then treated with aqueous chlorine to make glycerin dichlorohydrins which may be directly hydrolyzed to glycerin. Alternately, the dichlorohydrins are hydrolyzed to epichlorohydrin with a further hydrolysis to glycerin. In a second process, propylene is oxidized in the vapor phase to acrolein. The acrolein is converted to glycerin by successive reactions with hydrogen peroxide, water, and hydrogen. In the third route, propylene is reacted with aqueous chlorine to make propylene oxide. This is isomerized to allyl alcohol, which in turn is reacted with peracetic acid and then with water to make glycerin.

Several grades of glycerin are marketed, including high gravity, dynamite, yellow distilled, USP (U.S. Pharmacopoeia), and CP (chemically pure). USP grade is water-white and suitable for use in foods, pharmaceuticals, and cosmetics, or for any purpose where the product is designed for human consumption. *See* POLYHYDRIC ALCOHOL.

[PHILIP H. COOK]

Bibliography: W. L. Faith et al., *Industrial Chemicals*, 3d ed., 1965; C. S. Minor and N. N. Dalton (eds.), *Glycerol*, ACS Monogr. no. 117, 1953.

Glycol

Any of a class of compounds characterized by having two hydroxyl (—OH) groups on separate carbons of an organic structure, usually linear and aliphatic. Products in this class are also called diols, dihydric alcohols, or in special cases, polyglycols. Glycols are generally water-soluble and hygroscopic, viscous liquids, even up through hexamethyleneglycol (1,6-hexanediol). Polyethylene glycols, that is, polyglycols derived by polymerization of ethylene oxide, are water-soluble at all molecular weights. Polypropylene glycols from propylene oxide become increasingly less water-soluble at molecular weights above 425.

Reactions. Glycols undergo typical reactions of alcohols such as esterification, etherification, salt formation, and additions to activated double bonds. They also react with diacids or diesters to produce long-chain polyester polymers, an important family of products which find broad application in the manufacture of fibers, films, and molded plastic products. Reaction of glycols, especially polypropylene glycols, with diisocyanates produces another important family of products, polyurethanes, used in rigid and flexible foams and elastomers. *See* POLYESTER RESINS; POLYMER; POLYVINYL RESINS.

1,2-Diols. The most common subclassification of glycols is the 1,2-diols, of which ethylene glycol and 1,2-propylene glycol are the important members. Generally, 1,2-diols may be produced by hydrolysis of epoxides, chlorohydrins, or 1,2-dichlorides, or by catalytic reduction of α-keto-or α-hydroxyaldehydes or ketones. Ethylene glycol is produced by the hydrolysis of ethylene oxide. In another commercial process, ethylene glycol is produced directly from ethylene and oxygen in a liquid-phase reaction using tellurium oxide as a catalyst. Ethylene glycol is used mainly in polyester resin manufacture and as an antifreeze. *See* ETHYLENE GLYCOL.

1,2-Propylene glycol (1,2-propanediol) is produced by the reaction of propylene oxide with water. Propylene oxide for this purpose is manufactured mainly by the chlorohydrin process based on propylene and chlorine. Another, more economical route to propylene oxide uses *t*-butylbenzene or ethylbenzene hydroperoxides to epoxidize propylene catalytically at about 100°C and 600 psia (4.2 MPa absolute pressure). Propylene glycol is used as a solvent for flavors, extracts, and food colors; as a humectant in tobacco, baked goods, and packaged foods; and as a preservative.

Other glycols. Other glycols of commercial importance are prepared by special reactions leading to the specific diol substitution desired. For example, 1,3-butanediol is prepared by the reduction of acetaldol, and 1,4-butanediol is derived from the reaction of acetylene and formaldehyde which produces 1,4-butynediol. The diol is then hydrogenated to 1,4-butanediol. 1,5-Pentanediol is derived by reduction of glutaraldehyde. These diols find use in polyester and polyurethane resins, as plasticizer intermediates, and as solvents. Hexylene glycol, 2-methyl-2,4-pentanediol, is derived by reduction of the acetone condensation product, mesityl oxide. This glycol, because of its special solvent and wetting properties, finds use as a blending and coupling agent for immiscible liquid systems, such as cutting oils and emulsions. An excellent insect repellent, 2-ethyl-1,3-hexanediol, is obtained by catalytic hydrogenation of butyraldol. This diol is also a blending and coupling agent. Thiodiethylene glycol, obtained by the reaction of ethylene oxide with hydrogen sulfide, is an excellent solvent for dyestuffs, has antioxidant and lubricant properties, and is used as an intermediate in the manufacture of wash-and-wear fabrics and mustard gas. *See* POLYETHYLENE GLYCOL; POLYHYDRIC ALCOHOL.

[R. K. BARNES]

Bibliography: G. O. Curme and F. Johnston, *Glycols*, ACS Monogr. no. 114, 1952; Make propylene oxide direct, *Hydrocarbon Proc.*, 46(4): 141–143, 1967.

Gold

A chemical element, Au, atomic number 79 and atomic weight 196.967, a deep yellow, soft, and very dense metal. Gold is classed as a heavy metal and as a noble metal; commercially, it is the most familiar of the precious metals. Copper, silver, and gold make up group Ib of the periodic table of elements.

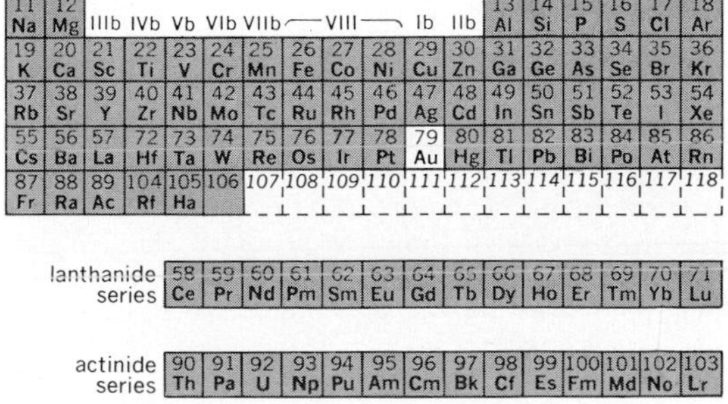

There is only one stable isotope of gold, that of mass number 197. Some 24 radioactive isotopes, 5 of which have additional metastable nuclear isomers, have mass numbers from 177 to 204. The half-lives of the radioactive species range from 1.35 sec for ^{177}Au to 183 days for ^{195}Au.

Throughout history gold has been known and prized. The Latin name for gold, *aurum* (glowing dawn), is the source of the chemical symbol Au. In 353 B.C. Xenophon wrote that the earliest records of Grecian gold mines were lost in antiquity. Gold has been found in Stone Age Egyptian tombs.

Uses. Consumption of gold in jewelry accounts for about three-fourths of the world's production of gold. Industrial applications, especially electronic, consume another 10–15%. The remainder is divided among medical and dental uses, coinage, and bar stock for governmental and private holdings. Gold coins and most decorative gold objects are actually gold alloys, because the metal itself is too soft (2.5–3 on the Mohs scale) to be useful with frequent handling. Some architectural gold leaf is almost pure gold. Although gold has long been used in noncorroding electrical contacts and wires, its uses in semiconductor electronics are much more diverse. Gold components are used in diodes and transistors, and removable gold layers are applied to certain electronic parts to provide chemical and thermal resistance during fabrication. Gold is commonly used in the beam-lead contacts of interlocking semiconductor components. Another use of gold is as a "killer," which speeds the reestablishment of charge equilibrium after a signal pulse and thus permits faster switching or more faithful reproduction of high-frequency signals.

Radioactive ^{198}Au, which has a half-life of 2.7 days, is used in medical irradiation, in diagnosis, and in a number of industrial applications as a tracer. Another tracer use is in the study of movement of sediment on the ocean floor in and around harbors. The properties of gold toward radiant energy have led to development of efficient energy reflectors for infrared heaters and cookers and for focusing and retention of heat in industrial processes. Gold-coated window glass has been used in office buildings to achieve large reductions in summer heat gain and winter heat loss, while retaining satisfactory light transmission. Combinations of gold films with dielectric light-interference films can increase the light transmission still further without loss of the insulating effect. Heated gold coatings are used to defrost windshields in some aircraft, ships, and locomotives. A number of gold complexes of phosphines and phosphites, usually also containing organic sulfur compounds, have been reported to be effective in treating arthritis.

Gold's potential for hydrogenation and dehydrogenation catalysis has been much studied but little used in actual processing. In general, gold does not take up hydrogen directly by adsorption at ordinary temperatures, and its activity as a catalyst depends on induction of the adsorption process by other means. Mixtures of gold and palladium have been found to be more efficient catalysts than palladium alone in very restricted instances, but gold is still considered inferior as a catalyst to the platinum-group metals. The shaping of small articles of solid gold by electroforming has come into more common use. The gold is electroplated onto a negative master or mandrel, then removed for use. Electrodes, dental inlay models, and grids for electron microscopy have been made by this process. Commemorative gold models of the Moon were made after the first manned lunar landing. Proof gold, which is 99.999+% pure, is used for certain chemical and physical standardizations. Private citizens in the United States may now own gold bars, since the restriction of many years' duration has now been lifted. Federal law prescribes that part of the currency of the United States is backed by the value of gold bullion stored at Fort Knox, KY.

Occurrence. Gold occurs widely throughout the world, but usually very sparsely, so that it is quite a rare element. It has been estimated that all the gold metal ever refined would fit into a cube about 50 ft on a side. Sea water contains low concentrations of gold, on the order of 10 μg per ton (10 parts of gold per trillion parts of water). Somewhat higher concentrations accumulate on plankton or on the ocean bottom. At present, no economically feasible process is visualized for extracting gold from the sea. Native, or metallic, gold and various telluride minerals are the only forms of gold found on land. Native gold may occur in veins among rocks and ores of other metals, especially quartz or pyrite, or it may be scattered in sands and gravel (alluvial gold). The gold nuggets found in the western United States in the 19th century were typical alluvial deposits. One alluvial nugget found in Australia weighed about 600 lb. Most native gold contains silver in amounts ranging from 1 to 50%. Telluride minerals are less common than native gold; among them are calaverite, $AuTe_2$, sylvanite, $AuAgTe_4$, and petzite, $(Au,Ag)_2Te$. South Africa leads the world in gold production. Other principal areas of production are in the western United States (Utah, California, South Dakota, as well as Alaska), Australia, Mexico, Canada, China, India, and the Soviet Union.

Gold metal. The density of gold is 19.3 times that of water at 20°C, so that 1 ft³ of gold weighs about 1200 lb. Masses of gold, like those of other precious metals, are measured on the troy scale, which counts 12 oz to the pound. A troy ounce, however, is larger than the common avoirdupois ounce, as follows:

$$\begin{aligned} 1 \text{ oz (troy)} &= 1.097 \text{ oz (avoir)} \\ 12 \text{ oz (troy)} &= 1 \text{ lb (troy)} \\ &= 13.17 \text{ oz (avoir)} \\ &= 0.823 \text{ lb (avoir)} \end{aligned}$$

Gold melts at 1064.43°C and boils at 2860°C. It is somewhat volatile well below its boiling point; mixtures of platinum and gold can be separated by keeping them melted for a long time to drive off the gold. Gold is a good conductor of heat and electricity; its electrical resistivity is 2.35 microhms/cm at 20°C. Gold is the most malleable and ductile metal. It can easily be made into translucent sheets 0.00001 mm thick or drawn into wire weighing only 0.5 mg/m. The quality of gold is expressed on the fineness scale as parts of pure gold per thousand parts of total metal, or on the karat scale as parts of pure gold per 24 parts of total metal. Gold may be obtained commercially in the form of sheets, tubes, foil, wire, powder, leaf, or granules. Gold and silver form true solutions (alloys) in any proportions. Gold readily dissolves in mercury to form amalgams. This property of gold has been made the basis of a sensitive instrument which detects mercury vapor in air by the change in electrical reistance of a thin gold film.

Chemical properties. Gold is one of the least-active metals chemically. It does not tarnish or burn in air. It is inert to strong alkaline solutions and to all pure acids except selenic acid. In order to dissolve gold chemically, it is best to use a combination of oxidizing power and complexing ability such as is found in a mixture of nitric and hydrochloric acids, called aqua regia because it can dissolve the royal metal. A similar effect occurs when gold dissolves in solutions of cyanide ion (a complexing agent) containing oxygen (the oxidizing agent), forming $[Au(CN)_2]^-$ complex ions. Gold also reacts with bromine at room temperature and with fluorine, chlorine, iodine, and tellurium at higher temperatures. One of the interesting characteristics of gold is its ability to exist in a sol, or colloid. Depending upon the particle size of the gold, aqueous gold sols may be red, blue, or purple. Tannin, formaldehyde, and phenylhydrazine are some of the common reducing agents added to solutions of gold compounds to generate the sols. The beautiful sol known as purple of Cassius is generated by addition of tin(II) chloride. This sol may possibly contain complexes of gold instead of, or in addition to, colloidal gold. Some of the atomic and ionic properties of gold are listed.

Electronic configuration	$1s^2, 2s^2, 2p^6, 3s^2, 3p^6, 3d^{10}, 4s^2, 4p^6, 4d^{10}, 4f^{14}, 5s^2, 5p^6, 5d^{10}, 6s^1$
Ionization potentials	
1st electron loss	9.23 eV
2d electron loss	20.0 eV
Ionic radius (Au^+); (Au^{3+})	0.137 nm; 0.085 nm
Covalent radius (tetrahedral)	0.150 nm
Crystal structure	Face-centered cubic
Oxidation potentials (acid solution)	$Au \rightleftarrows Au(I) + e^-$ $E^0 \approx -1.7$ volts; $Au \rightleftarrows Au(III) + 3e^-$ $E^0 = 1.5$ volts
Electronegativity	142, where F = 4.10

Compounds. Gold may be either unipositive or tripositive in its compounds, as indicated by the table. So strong is the tendency for gold to form complexes that all the compounds of the 3+ oxidation state are complex. Gold(III) fluoride, the most stable of the gold halides, most closely approaches ionic character among the compounds of gold, but it nevertheless has a polymeric, three-dimensional structure. The compounds of the 1+ oxidation state are not very stable and tend to be oxidized to the 3+ state or reduced to metallic gold. All compounds of either oxidation state are easy to reduce to the metal. This situation follows the usual rule that compounds of inactive (noble) metals readily revert to the metals, whereas the reduction of compounds of active metals is difficult.

In its complex compounds gold forms bonds most readily and stably with halogens and sulfur, less stably with oxygen and phosphorus, and only weakly with nitrogen. Bonds between gold and carbon are fairly stable, as in the cyanide complexes and a variety of organogold compounds. The complexes of 1+ gold are usually linear, having two groups attached to the metal atom in diametrically opposite positions. Complexes of 3+ gold are usually square planar, having four groups attached to the metal. Coordination of five or six groups is known, but rare. Gold is unlike silver and copper in

Various compounds of gold

Name	Formula	Properties, uses, remarks
Tetrachloroauric acid 4-hydrate; chloroauric acid; "gold chloride"	$HAuCl_4 \cdot 4H_2O$	Soluble complex compound; solutions are a common commercial form of combined gold; used to prepare other gold compounds and gold sols
Sodium tetrachloroaurate 2-hydrate	$NaAuCl_4 \cdot 2H_2O$	Made from $HAuCl_4$ solutions; soluble compound; used in gold plating, photographic tinting, and some medicines
Gold(I) cyanide	$(AuCN)_x$	Does not exist as single AuCN molecules, but as long chains of alternate gold atoms and CN groups
Cesium hexachlorodiaurate	$Cs_2Au_2Cl_6$	Contains gold in both 1+ and 3+ oxidation states; originally thought to be a compound of gold in 2+ state
Gold(III) chloride	$(AuCl_3)_2$	Red soluble compound, made by reaction of gold and chlorine, or by reaction of $HAuCl_4$ with chlorine; forms $HAuCl_4$ in hydrochloric acid solution; with ammonia, forms explosive fulminating gold; structure consists of two square-planar $AuCl_4$ groups sharing an edge
Gold(III) hydroxide	$Au(OH)_3$	Yellow-brown insoluble compound; formed by addition of hydroxide ion to $AuCl_3$ or $HAuCl_4$ solution; probably a hydrous form of the oxide Au_2O_3; dissolves in most acids; easily reduced to metallic gold; is amphoteric, dissolving in excess hydroxide ion to form complex hydroxoaurate ions; used in medicines, gilding liquids, gold-plating, coloring rubber and porcelain
Sodium dicyanoaurate	$NaAu(CN)_2$	Used to gold-plate radar and other electronic parts, clocks, and jewelry
Alkyl gold mercaptides	$t\text{-}C_4H_9$, $t\text{-}C_8H_{17}$, $t\text{-}C_{12}H_{25}$, and $t\text{-}C_{16}H_{33}$ Au mercaptides	Decompose at 200–260°C; used in thermal gold-plating of stainless steel and plastics
Gold t-dodecyl mercaptide	$Au(t\text{-}C_{12}H_{25}S)$	Decomposes at 150°C; highly soluble in toluene; used in thermal gold-coating of plastics and other heat-sensitive nonconductors.
Gold cluster compound	$Au_{11}[P(p\text{-}C_6H_4F)_3]_7I_3$	Compound in which a central gold atom is surrounded by 10 other gold atoms, each bonded to one of the other ligands, one of a class of compounds with structures similar to those of some gold alloys

its formation of true (sigma-bond) organometallic compounds that are stable. The best-known of such compounds contain gold bonded to manganese, as in the compound $(C_6H_5O)_3PAuMn(CO)_5$. In this compound the triphenoxyphosphine gold group behaves similarly to a halogen (Cl, Br, or I) in the manganese carbonyl halides. *See* ORGANOMETALLIC COMPOUND.

Analytical methods. The fire-assay method with a lead collector is the traditional method of choice for gold in native, mineral, or alloyed form. It requires special high-temperature equipment. Qualitative detection of gold in solution is usually accomplished by addition of reducing agents to liberate metallic gold, which is identifiable either by the yellow color of the compact metal or by a characteristic color of a gold sol. Two color reactions of gold that are particularly sensitive in solution are those with dithizone and benzidine. Gold may be separated from the platinum-group metals by precipitation with hydroquinone in dilute hydrochloric acid. Quantitative analysis of gold can be done by means of an indirect titration with ethylenediamine-tetraacetate ion (EDTA). The nickel in standard $Ni(CN)_4^{2-}$ solution is displaced by gold ion, then determined by addition of an excess of EDTA reagent and back titration with $MnSO_4$ solution, using Eriochrome Black T as an indicator. Another volumetric procedure uses the reaction between tetrachloroaurate ion and iodide ion. The gold is reduced from the 3+ state to the 1+ state, and the iodine released is titrated with standard sodium thiosulfate solution. There is a special fire-assay method for the evaluation of gold ores.

Instrumentally, gold in alloys has been determined by flame photometry. Gold in solution can be determined polarographically in 0.1 *M* KOH or KCN electrolyte. Atomic absorption spectrometry can detect to about 0.3 ppm in solution, and neutron-activation analysis can detect as little as 0.00002 μg of gold in minerals, alloys, semiconductors, and biological samples.

Toxicity. Gold salts are not considered to be among the more poisonous compounds of metals, but some toxicity has been observed. Most of the toxic reactions to gold are on the skin, which may develop a rash or other dermatitis. Some of these cases, however, may be caused by residual radioactivity of minor contaminants in the gold. Gastrointestinal upset has occasionally been observed after excessive ingestion of gold compounds, and more serious internal disorders, although rare, have occurred. The usual antidote for poisoning by gold is dimercaprol, $HSCH_2CHSHCH_2OH$, also called BAL (British Anti-Lewisite), which forms a stable complex with gold and deactivates its physiological effect, permitting rapid elimination from the body.

[WILLIAM E. COOLEY]

Bibliography: J. A. Dean (ed.), *Lange's Handbook of Chemistry*, 12th ed., 1978; B. F. G. Johnson and R. Davis, Gold, in J. C. Bailer et al. (eds.), *Comprehensive Inorganic Chemistry*, vol. 3, 1973; G. G. Hawley, *The Condensed Chemical Dictionary*, 9th ed., 1977; R. J. Puddephatt, *The Chemistry of Gold*, 1978; J. H. Tatsch, *Gold Deposits: Origin, Evolution, and Present Characteristics*, 1975; R. C. Weast (ed.), *Handbook of Chemistry and Physics*, 57th ed., 1976.

Gram-equivalent weight

The equivalent weight of an element or compound, expressed in grams (g), on a scale in which C^{12} (the isotope of carbon of nuclear mass 12) has an equivalent weight of 3 g in those compounds in which its formal valency is 4. This replaces the older definition in which the fixed point on the scale was the equivalent weight of oxygen, 8.000 g. Because variable valency is much more common in carbon than in oxygen, it is now necessary to specify the formal valency in defining the equivalent weight concept. The gram-equivalent weight of an electrolyte is usually the weight in grams which is associated with 1 faraday of electricity. For example, the gram-equivalent weight of sodium chloride, Na^+Cl^-, is equal to the gram-molecular weight; that of calcium chloride, $Ca^{2+}Cl_2^-$, is one-half the gram-molecular weight; that of aluminum sulfate, $Al_2^{3+}(SO_4)_3^{2-}$, is one-sixth the gram-molecular weight. This concept, together with that of equivalent weight, tends to be abandoned, and relations are expressed in terms of balanced stoichiometric chemical equations and relative numbers of moles reacting. *See* ELECTROCHEMICAL EQUIVALENT; ELECTROLYSIS; ELECTROLYTE; EQUIVALENT WEIGHT; GRAM-MOLECULAR WEIGHT; MOLE; STOICHIOMETRY; VALENCE.

[THOMAS C. WADDINGTON]

Gram-molecular weight

The molecular weight of an element or compound expressed in grams (g), that is, the molecular weight on a scale on which the atomic weight of the C^{12} isotope of carbon is taken as 12 exactly. This replaces the earlier scale on which the atomic weight of oxygen was taken as 16.00 g. In the International System of Units, gram-molecular weight is replaced by the mole.

The ratio of the gram-molecular weights of any two elements or compounds must be identical with the ratio of the absolute weights of their individual molecules. Therefore, the gram-molecular weights of all elements or compounds contain the same number of molecules. This number, called the Avogadro number, *N*, is 6.022×10^{23}.

Since they contain the same number of molecules, the gram-molecular weights of all gases occupy the same volume at the same temperature and pressure. At 0°C and 1 atm (101,325 Pa) pressure this volume, called the gram-molecular volume, is 22.4 liters. *See* AVOGADRO NUMBER; GAS; MOLE; MOLECULAR WEIGHT; RELATIVE MOLECULAR MASS.

[THOMAS C. WADDINGTON]

Gravimetric analysis

That branch of quantitative chemical analysis in which a desired constituent is converted (usually by precipitation) to a pure compound or element of definite, known composition, and is weighed. In a few cases, a compound or element is formed which does not contain the constituent but bears a definite mathematical relationship to it. In either case, the amount of desired constituent can be determined from the weight and composition of the precipitate. The following are the essential steps in a conventional gravimetric analysis.

Dissolution. A sample is weighed and dissolved in a suitable solvent. Water and dilute mineral acids dissolve most inorganic substances, but occasionally concentrated acids or the specific effect of hydrofluoric acid are required. Some refractories require fusions to convert them to acid-soluble products. *See* SOLUBILIZING OF SAMPLES.

Precipitation. After removal of any interfering substances, the desired constituent is precipitated by the addition of the appropriate reactant to the properly prepared solution. Conditions are regulated so that coprecipitation of foreign substances is minimized. In most cases, favorable conditions are attained by precipitating the constituent from a well-stirred, highly diluted, hot solution by the drop-by-drop addition of a dilute reagent in only slight excess over the amount theoretically required. In order to obtain a more nearly pure product, it is usually desirable to dissolve the precipitate in a suitable solvent and to form it a second time. *See* PRECIPITATION.

Digestion. In order to be readily filterable, a suspension of a precipitate is ordinarily allowed to stand at high temperature for the period of time necessary to permit amorphous particles to clot, or crystalline particles to increase in size. This digestion is usually carried out on an electric hot plate so regulated as to maintain a temperature just below the boiling point of the liquid. The addition of paper pulp is helpful in aiding subsequent filtration, but may be used only in cases where the precipitate is to be ignited at a temperature that causes the pulp to burn off.

Filtration. Filtration is accomplished by pouring the suspension of a precipitate through a suitable filtering medium. Whatever the medium, as much of the supernatant liquid as possible is decanted through the filter, and the transference of the precipitate is delayed as long as possible. The common filters are as follows.

Filter paper for gravimetric use is specially prepared and has undergone a treatment with hydrofluoric acid and with other acids so that on ignition it gives an ash of known, and usually negligible, weight. Papers of different degrees of porosity are available, and in a given filtration that grade is chosen which gives as rapid a filtration as possible and yet retains the precipitate completely. In general, gelatinous precipitates require a coarse-mesh paper; fine crystalline precipitates, a fine-mesh paper. Suction is almost never used in paper filtration.

A Gooch crucible is a perforated-base procelain crucible, and the filtering medium is a pad of asbestos produced by pouring into the crucible a suspension of asbestos fibers which are matted by applying suction. A perforated plate on top of the pad holds it in place. Fiber-glass disks may be used instead of asbestos. Suction is applied during filtration; the crucible is dried to constant weight and is weighed before and after the filtration. Gooch crucibles are usually used for precipitates that attain a definite composition at the moderate temperature of a drying oven.

A Munroe crucible differs from a Gooch crucible in that it is made of platinum and uses a platinum sponge (produced by igniting alcohol-moistened ammonium chloroplatinate) as the filtering medium. It retains even fine precipitates and can be heated to a high temperature. It is too expensive for routine use.

An Alundum crucible is made from aluminum oxide and is porous throughout. It therefore needs no additional filtering medium and can be heated to a high temperature.

A Selas crucible has glazed porcelain sides and an unglazed porcelain base which serves as the filtering medium.

A glass filtering crucible has a sintered glass base for the filtering medium and is favored in cases where the precipitate needs to be heated only to the moderate temperatures of a drying oven.

Washing. Precipitates are washed (usually with hot water) until essentially free from soluble foreign matter. Occasionally an aqueous solution of an electrolyte is used to prevent peptization of the precipitate, with resulting conversion to a colloidal solution. Ammonium nitrate is favored for this purpose since it is removed by volatilization when the precipitate is subsequently ignited. Washing is more efficient if the wash water is added in several portions with intermediate drainage.

Drying and ignition. Some substances can be dried to constant weight by heating to relatively low temperatures (110–275°C) in drying ovens. High-temperature ignition is usually carried out on a precipitate that has been filtered on paper. First the paper is smoked off at a low temperature, and the residue is ignited either in an electric muffle furnace or over a free flame. In the latter case, one of the following types of burners is used:

A bunsen burner, a simple tube with an opening at the base to permit air to be mixed with the illuminating gas used.

A tirrill burner, a modification of the bunsen burner which allows greater flexibility in the adjustment of the air-gas mixture.

A meker burner, which is larger in diameter than the tirrill burner and has a grid at the top to give broad, hot flame.

A blast lamp which supplies air (or oxygen) under pressure to the illuminating gas used.

The temperature delivered to the contents of a platinum crucible by a bunsen or tirrill burner is about 850°C; that by a grid-top burner is about 1000°C; that by a gas-air blast lamp about 1150°C.

Cooling. A dried or ignited precipitate is cooled in a desiccator, which is a jarlike receptacle containing a desiccant, and which, in analytical chemistry, is used principally to allow a heated crucible and its contents to come to room temperature without taking on moisture from the air. Anhydrous calcium chloride, although not a very efficient desiccant, is commonly used for this purpose because of its low cost. Other desiccants in the order of increasing effectiveness are anhydrous barium perchlorate, sodium hydroxide sticks, silica gel, aluminum oxide, anhydrous calcium sulfate, calcium oxide, anhydrous magnesium perchlorate, barium oxide, and phosphorus pentoxide.

Calculations. At least two weighings are required for each analysis – the original sample, and the dried or ignited residue. From these weights, the percentage or proportion of the desired constituent may be calculated from the equation below.

$$\%A = \frac{\text{wt of residue} \times \text{factor} \times 100}{\text{wt of sample}}$$

The factor is determined from a knowledge of the chemical relationships between the weight of substance A contained in, or equivalent to, a fixed weight of residue of known composition. *See* ANALYTICAL CHEMISTRY; ELECTRODEPOSITION ANALYSIS; QUANTITATIVE CHEMICAL ANALYSIS; STOICHIOMETRY. [STEPHEN G. SIMPSON]

Bibliography: A. I. Vogel, *Vogel's Textbook of Quantitative Inorganic Analysis*, 4th ed., 1978.

Grignard reaction

A reaction between an alkyl or aryl halide and magnesium metal in a suitable solvent, usually absolute ether. This is represented by reaction (1),

$$RX + Mg \xrightarrow{\text{Ether}} RMgX \qquad (1)$$

where R stands for alkyl or aryl, and X for chlorine, bromine, or iodine. The organomagnesium halides produced by this reaction are known as Grignard reagents and are useful in many chemical syntheses. They are named after Victor Grignard, who discovered them and developed their use as synthetic reagents, for which he received a Nobel prize in 1912.

The scope of the Grignard reaction is extremely broad, and Grignard reagents have been prepared from many kinds of alkyl and aryl halides. In general, alkyl chlorides, bromides, and iodides and aryl bromides and iodides react readily. A few halides, such as aryl chlorides, react very sluggishly, and require specially activated magnesium, modified reaction techniques, the use of high-boiling solvents, and long reaction periods. Vinylmagnesium halides can be prepared by using tetrahydrofuran as the solvent.

The structure of a Grignard reagent is usually written RMgX, where X represents a halogen. However, the actual structure in ether solution is more complex. Equilibrium is rapidly established between the organomagnesium halide (RMgX) and the corresponding dialkylmagnesium (RMgR), as in reaction (2). These two species (RMgX and

$$2RMgX \rightleftharpoons R_2Mg + MgX_2 \qquad (2)$$

R_2Mg) are reactive, and are solvated in ether solvents by coordination of the ether oxygen to magnesium. In solution they further associate as dimers or higher polymers. Thus, while oversimplified, a Grignard reagent can be considered as RMgX.

Probably the most common synthetic use of the Grignard reaction involves the reaction of Grignard reagents with carbonyl compounds, followed by hydrolysis to produce a variety of products containing new carbon-carbon bonds. For example, in reaction (3), a Grignard reagent with carbon dioxide at low temperatures produces salts of carboxylic acids, from which the acids can be readily isolated. Addition of Grignard reagents to the carbonyl group of many aldehydes and ketones produces primary, secondary, or tertiary alcohols. The production of these alcohols is shown in (4)–(6).

$$RMgX + CO_2 \rightarrow RCOOMgX \xrightarrow{H_2O} RCOOH + MgXOH \qquad (3)$$

$$RMgX + CH_2{=}O \rightarrow RCH_2OMgX \xrightarrow{H_2O} RCH_2OH + MgXOH \qquad (4)$$

A primary alcohol

$$RMgX + R'CH{=}O \rightarrow RR'CHOMgX \xrightarrow{H_2O} RR'CHOH + MgXOH \qquad (5)$$

A secondary alcohol

$$RMgX + R'R''C{=}O \rightarrow RR'R''COMgX \xrightarrow{H_2O} RR'R''COH + MgXOH \qquad (6)$$

A tertiary alcohol

The reaction of Grignard reagents with carbon dioxide, aldehydes, and ketones serves as a convenient means of lengthening the carbon chain. Further, a hydrocarbon chain can be increased by two carbon atoms by the reaction of a Grignard reagent with ethylene oxide, as in reaction (7).

$$RMgX + \underset{\backslash\,O\,/}{CH_2CH_2} \rightarrow RCH_2CH_2OMgX \xrightarrow{H_2O} RCH_2CH_2OH + MgXOH \qquad (7)$$

The resulting alcohol can be converted to the corresponding halide, and the process repeated.

Grignard reagents react with esters to produce ketones initially. The reaction as a rule does not stop at this stage, however, since additional Grignard reagent reacts with the ketone to yield a tertiary alcohol.

Acid chlorides and anhydrides react in an analogous manner to produce ketones, which react further to give tertiary alcohols. A valuable modification of this procedure involves prior reaction of the Grignard reagent with cadmium or zinc chloride. The organocadmium or organozinc compounds are less reactive than these of magnesium and react so slowly with ketones that the conversion of the ketone to the tertiary alcohol can usually be avoided.

When aliphatic Grignard reagents are exposed to air or oxygen, they are oxidized to alkoxides, which on hydrolysis produce alcohols. Aromatic Grignard reagents react poorly in this way, and only unsatisfactory yields of phenols can be obtained.

The reaction of Grignard reagents with alkyl halides which contain reactive halogen atoms produces hydrocarbons by a process of alkylation, as in reaction (8). In a similar manner, the

$$RMgX + XR \rightarrow R{-}R + MgX_2 \qquad (8)$$

addition of a variety of metal halides, such as silver bromide or cobaltous chloride, tends to couple Grignard reagents, as in reaction (9).

$$2RMgX + 2AgBr \rightarrow R{-}R + 2Ag + 2MgXBr \qquad (9)$$

Compounds containing active hydrogen atoms react readily with Grignard reagents. Reaction with water brings about immediate decomposition and formation of the corresponding hydrocarbon, as in reaction (10). In certain instances, hydrogen

$$RMgX + H_2O \rightarrow RH + MgXOH \qquad (10)$$

attached to carbon is sufficiently acidic to react with Grignard reagents. The reactions of ethylmagnesium bromide with acetylene, reaction (11),

$$2C_2H_5MgBr + HC{\equiv}CH \rightarrow 2C_2H_6 + BrMgC{\equiv}CMgBr \qquad (11)$$

and cyclopentadiene, reaction (12), are examples. These reactions enable formation of additional Grignard reagents which could otherwise not be prepared by conventional methods.

Grignard reagents react with a variety of nitrogen- and sulfur-containing compounds, such as imines, nitriles, and sulfoxides. Furthermore, the

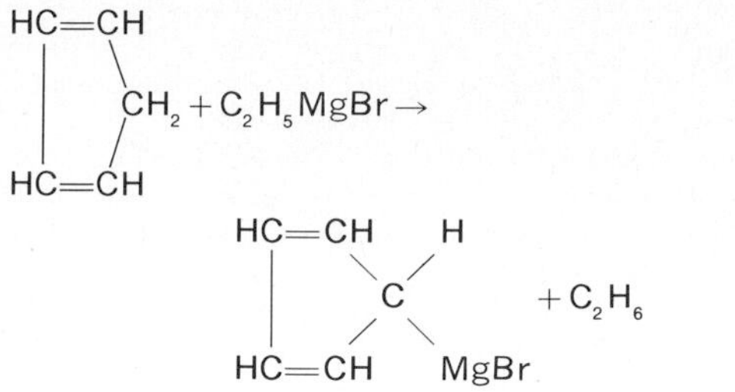

(12)

reaction of Grignard reagents with halides of boron, phosphorus, silicon, and tin is a very convenient procedure for producing organometallic compounds of these elements. An example, reaction (13), is the preparation of tetraphenylsilane

$$4C_6H_5MgBr + SiCl_4 \rightarrow (C_6H_5)_4Si + 4MgBrCl \quad (13)$$

from phenylmagnesium bromide and silicon tetrachloride. *See* ORGANOMETALLIC COMPOUND.

[PAUL E. FANTA]

Bibliography: J. D. Roberts and M. C. Caserio, *Basic Principles of Organic Chemistry*, 2d ed., 1977.

Ground state

In quantum mechanics, the stationary state of lowest energy of a particle or a system of particles. The ground state may be bound or unbound; when bound, its energy generally is a finite amount less than the energy of the next higher or first excited state. In the typical circumstance that the potential energy is zero at infinite separation, the magnitude of the negative ground-state energy is the binding energy, that is, the energy required to seperate all the particles infinitely. *See* EXCITED STATE.

[EDWARD GERJUOY]

Hafnium

A metallic element, symbol Hf, atomic number 72, and atomic weight 178.49. The naturally occurring isotopes are 174 (0.18%), 176 (5.2%), 177 (18.47%), 178 (27.10%), 179 (13.75%), and 180 (35.25%). It is

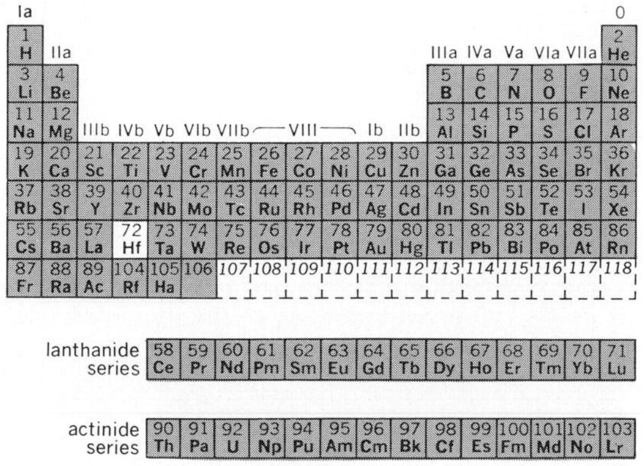

one of the less abundant elements in the Earth's crust. Hafnium always occurs with zirconium, often in small proportions, but it may equal or slightly exceed the zirconium content in alvite, $(Be,Hf,Th,Y,Zr)O_2 \cdot SiO_2 \cdot xH_2O$, and cyrtolite, $3(Zr,Hf)O_2 \cdot 2SiO_2 \cdot 3H_2O$, and frequently exceeds the zirconium content in thortveitite, $(Sc,Y)_2Si_2O_7$, containing hafnia and zirconia. Hafnium makes up 2% of the combined weight of hafnium and zirconium in the zircon of commerce, $(Zr,Hf)O_2 \cdot SiO_2$, which is recovered from certain beach sands. Most of the zirconium and hafnium of industry are derived from these sands, but some is also derived from baddeleyite, $(Zr,Hf)O_2$, in which the hafnium content may be as low as 0.5% or as high as 2% of the zirconium.

Hafnium is a lustrous, silvery metal that melts at about 2222°C. Reported values of the boiling point vary greatly, from about 2500 to about 5100°C. The crystal structure is hexagonal close-packed below 1300°C (alpha form) and body-centered cubic above (beta form). There are virtually no uses of the metal other than in control rods for nuclear reactors, where its high cross section of 105 barns to thermal neutrons is the most significant property. Hafnium must be separated from zirconium that is to be used for cladding fuel elements used in reactors, where the low cross section of zirconium, 0.18 barn, is important. Many studies of hafnium alloys have been reported. The alloy Ta_4HfC_5 has the highest melting point of any substance known, about 4215°C.

The chemistry of hafnium is almost identical with that of zirconium, being comparable in similarity to that of deuterium with hydrogen. Both pairs of elements have atomic-weight ratios of 2:1. The similarity of hafnium to zirconium is a consequence of the lanthanide contraction, which brings the ionic radii to very nearly identical values, 0.78 and 0.79, respectively. More than a century elapsed (1789 to 1923) between the discovery of zirconium and the announcement by D. Coster and G. von Hevesy of the presence of hafnium in zirconium. Before (and since) the discovery of hafnium, this element was extracted with zirconium from its ores and passed with zirconium into all derivatives. Since the chemical properties are so similar, there has been no incentive to separate the hafnium except for making nuclear studies and components of nuclear reactors.

The separation processes for hafnium and zirconium have depended upon taking advantage of differences in equilibrium constants of the same reaction, rather than upon different reactions of the two elements. For a summary of these methods *see* ZIRCONIUM.

The few qualitative differences in chemical behavior between hafnium and zirconium follow.

1. Rufigallic acid gives a red solution with either hafnium or zirconium in a solution moderately acidic with hydrochloric acid, but on increasing the acidity, only the hafnium solution turns yellow.

2. Hafnium and zirconium both form 1:1 complexes with β-hydronaphthazarin, but only zirconium also forms a 1:2 complex.

3. The compound $Hf(NO_3)_4 \cdot N_2O_5$ has been sublimed from preparations in which both zirconium and hafnium were present in the reagents, but no zirconium was present in the sublimate. A zirconium compound similar in structure to the hafnium compound did not form.

4. In strong hydrochloric acid solution, Neothorin gives a red-violet solution or precipitate with hafnium, but a blue-violet one with zirconium. Hydrogen peroxide alters the color of the zirconium complex to orange and that of the hafnium complex to pink.

5. The addition of diarsine to a solution of zirconium tetrachloride in tetrahydrofuran leads to

the rapid precipitation of an adduct. The hafnium analog precipitates only after a delay.

[WARREN B. BLUMENTHAL]

Bibliography: R. Clark et al., *The Chemistry of Titanium, Zirconium, and Hafnium*, 1975; J. H. Schemel, *Manual on Zirconium and Hafnium*, 1977.

Half-life

The time required for one-half of a given material to undergo chemical reactions; also, the average time interval required for one-half of any quantity of identical radioactive atoms to undergo radioactive decay.

Chemical reactions. The concept of the time required for all of the material to react is meaningless, because the reaction goes very slowly when only a small amount of the reacting material is left and theoretically an infinite time would be required. The time for half completion of the reaction is a definite and useful way of describing the rate of a reaction.

The specific rate constant k provides another way of describing the rate of a chemical reaction. This is shown in a first-order reaction, Eq. (1),

$$k=\frac{2.303}{t}\log\frac{c_0}{c} \qquad (1)$$

where c_0 is the initial concentration and c is the concentration at time t. The relation between specific rate constant and period of half-life, $t_{1/2}$, in a first-order reaction is given by Eq. (2). In a

$$t_{1/2}=\frac{2.303}{k}\log\frac{1}{1/2}=\frac{0.693}{k} \qquad (2)$$

first-order reaction, the period of half-life is independent of the initial concentration, but in a second-order reaction it does depend on the initial concentration according to Eq. (3).

$$t_{1/2}=\frac{1}{kc_0} \qquad (3)$$

See CHEMICAL DYNAMICS. [FARRINGTON DANIELS]

Radioactive decay. The activity of a source of any single radioactive substance decreases to one-half in 1 half-period, because the activity is always proportional to the number of radioactive atoms present. For example, the half-period of Co^{60} (cobalt-60) is $t_{1/2}=5.3$ years. Then a Co^{60} source whose initial activity was 100 curies will decrease to 50 curies in 5.3 years. The activity of any radioactive source decreases exponentially with time t, in proportion to exp $-0.693\ t/t_{1/2}$. After 1 half-period (when $t=t_{1/2}$) the activity will be reduced by the factor $e^{-0.693}=1/2$. In 1 additional half-period this activity will be further reduced by the factor 1/2. Thus, the fraction of the initial activity which remains is 1/2 after 1 half-period, 1/4 after 2 half-periods, 1/8 after 3 half-periods, 1/16 after 4 half-periods, and so on.

The half-period is sometimes also called the half-value time or, with less justification, but frequently, the half-life. The half-period is 0.693 times the mean life or average life of a group of identical radioactive atoms. The probability is exactly 1/2 that the actual life-span of one individual radioactive atom will exceed its half-period. *See* RADIOACTIVITY. [ROBLEY D. EVANS]

Halide

A compound of the type MX, where X may be fluorine, chlorine, bromine, iodine, or astatine, and M may be another element or organic radical. The compounds are considered derivatives of the hydrohalic acids, HX, where the halogen X has an oxidation state of −1. All the halides form both covalent and ionic inorganic salts, coordination compounds, and many organic compounds.

Many properties show a gradual change as one compares the halides. The size of the ions increases from F^- to I^-. The strength of the halide as a reducing agent also increases from F^- to I^-. It becomes more difficult to produce the element from the halide as one goes from I^- to F^-, electrolysis being required to produce elemental fluorine. The stability of complexes formed generally increases from I^- to F^-. The solubilities of metal halides in water increase going from I^- to F^-. As a group the halides are very soluble. *See* BROMIDE; CHLORIDE; COMPLEX COMPOUNDS; FLUORIDE; HALOGEN ELEMENTS; HALOGENATED HYDROCARBON; HALOGENATION; IODIDE.

[E. EUGENE WEAVER]

Haloform reaction

The halogenation of acetaldehyde or a methyl ketone in aqueous basic solution. The reaction is characteristic of compounds that contain a CH_3CO— group linked to a hydrogen or to another carbon atom. The halogen is dissolved in aqueous base (sodium hydroxide or carbonate) to form an alkali hypohalite, which is the halogenating agent. This is added to the ketone. The reaction occurs in two phases: the methyl group is fully substituted by halogen to give an intermediate trihaloketone, reaction (1); the trihaloketone breaks down in the

$$RCOCH_3 + 3NaOX \rightarrow RCOCX_3 + 3NaOH \qquad (1)$$

presence of alkali to produce the haloform and the sodium salt of a monocarboxylic acid, reaction (2).

$$RCOCX_3 + NaOH \rightarrow RCOONa + HCX_3 \qquad (2)$$

Sufficient alkali is produced in the initial halogenation to cleave the trihaloketone so that the reactions proceed simultaneously. The haloforms which can be produced by this method are chloroform, bromoform, and iodoform.

Since the solution of hypohalite is mildly oxidizing, compounds which are easily oxidized to acetaldehyde or to a methyl ketone also give the haloform reaction. Examples are ethyl, isopropyl, and *sec*-butyl alcohols. Curiously, the compound CF_3-$CCl=CCl_2$, on oxidation, is routinely cleaved by potassium permanganate and sodium hydroxide to yield sodium and potassium salts of trifluoroacetic acid, without undergoing the haloform reaction. This attests to the stability of the CF_3— group. *See* TRIFLUOROACETATE.

Acetone or ethyl alcohol reacts with calcium hypochlorite to produce chloroform. Sodium hypobromite and sodium hypoiodite react in the same manner to yield bromoform and iodoform, respectively. *See* CHLOROFORM; HALOGENATED HYDROCARBON.

The haloform reaction is so characteristic that it may be used as a dependable qualitative test in distinguishing methyl ketones and methyl carbinols. Iodoform, for example, is an easily identifiable halide which precipitates from an aqueous solution as a yellow solid. Similarly, the test may be used to distinguish between ethyl and methyl alcohols, since ethyl alcohol forms a precipitate of iodoform, whereas methyl alcohol does not. All methyl ketones, however, give the iodoform test. Secondary alcohols which can be oxidized to methyl ketones likewise undergo the haloform reaction. *See* IODOFORM.

In addition to use as a test reaction and as a method of synthesis for haloforms, the haloform reaction may be used to prepare acids from methyl ketones when conventional methods of synthesis are not feasible.

Treatment of the methyl ketone with a basic solution of iodine yields iodoform together with a solution of the acid salt. After filtration of the iodoform and subsequent acidification of the basic solution, the acid is obtained routinely.

The haloform reaction was formerly applied industrially, specifically in the manufacture of chloroform, but it has been virtually displaced by more economical methods.

[ELLEN CLARKE; MAX G. GERGEL]

Halogen elements

The halogen family consists of the elements fluorine, F; chlorine, Cl; bromine, Br; iodine, I; and astatine, At. Estimates of the relative abundances of these elements in sea water in parts per million, and in the lithosphere (Earth's crust to a depth of 10 mi or 16 km) are given in Table 1. From direct measurements it is also known that all the halogen elements except astatine exist in the Earth's atmosphere.

Table 1. Relative abundances of the halogens, in ppm

Element	Sea water	Lithosphere
Fluorine	1.4	770
Chlorine	18,980	550
Bromine	65	1.6
Iodine	0.05	0.3

PROPERTIES

The halogens are the best-defined family of elements. They have an almost perfect gradation of physical properties. The increase in atomic weight from fluorine through iodine is paralleled by increases in density, melting and boiling points, critical temperature and pressure, heats of fusion and vaporization, and even in progressively deeper color (fluorine is pale yellow; chlorine, yellow-green; bromine, dark red; and iodine, deep violet).

Chemically, fluorine is the most powerful oxidizing agent know. The heavier halogens have progressively less oxidizing ability. Each forms an acid with hydrogen, and salts with metals. The properties of these acids and salts show as consistent a relationship as the elements themselves. Organic halogen compounds generally show progressively increased stability in the order iodine, bromine, chlorine, fluorine.

Although all halogens generally undergo the same types of reactions, the extent and ease with which these reactions occur vary markedly. Fluorine in particular has the usual tendency of the lightest member of a family of elements to exhibit reactions not comparable to the other members. Each halogen must be considered individually, both in its preparation and in its reaction. *See* ASTATINE; BROMINE; CHLORINE; FLUORINE; HALIDE; HALOGENATION; IODINE; PERIODIC TABLE.

[ALBERT A. GUNKLER]

Table 2. Atmospheric halogen concentrations

Source	F	Cl	Br	I
Particles (ng/scm) in marine air	0.2	5000	7	2
Organic gases (ppt)	1000	2500	20	3
Inorganic gases (ppt)	?	1000	1–10	1–5
Precipitation (μg/liter)	5–150	100–10,000	10	5

ATMOSPHERIC HALOGENS

Despite the chemical similarities of the halogen elements, a rich variety of chemical forms and processes are involved in the injection of halogen-containing material into the atmosphere, the chemical and physical transformations that occur in the air, and the eventual removal of the halogen atoms from the atmosphere. Although their concentrations are known to vary with time and location, the values listed in Table 2 are typical.

Particulates. In airborne particles whose diameters range from 0.1 μm to over 10 μm in the marine atmosphere, there is typically 5000 nanograms of chlorine per standard cubic meter (scm) of air, and lesser amounts of F, Br, and I as shown in Table 2. In continental air there is normally less chlorine, bromine, and iodine in airborne particles (aerosols) than near the sea, while the opposite is true for fluorine. In polluted areas there are often much higher levels of halogens in particles; specific examples include fluoride from aluminum and steel mills and phosphate-fertilizer plants and bromide from the burning of gasoline additives. Generally, the halogen form in particles and in precipitation is the halide: fluoride (F^-), chloride (Cl^-), bromide (Br^-), and iodide (I^-), although iodate (IO_3^-) may also occur. A particularly interesting anomaly is that the amount of iodine in the marine aerosol is much larger compared to that for Cl, Br, and Na than that expected from the composition of sea water. Part of the explanation for the iodine richness of marine aerosols could be a correspondingly high iodine content of organic surface films on sea water that enter the air on bursting bubbles and sea spray.

Gases. In gaseous form the halogens exist in the atmosphere in both organic and inorganic molecules. While there is at least one atmospherically significant naturally occurring organohalogen, methyl chloride (CH_3Cl), much of the atmospheric burden is anthropogenic. Table 2 shows that a typical concentration of organic F compounds is 1000 parts per trillion by volume. Corresponding values for organochlorine, -bromine, and -iodine compounds are 2500 ppt, 20 ppt, and 3ppt. The dominant individual species in this class of compounds are anthropogenic. They include the chlorofluorocarbons CF_2Cl_2, $CFCl_3$, $C_2F_4Cl_2$, $C_2F_3Cl_3$; the chlorocarbons CH_3CCl_3 and CCl_4 (which might also have a natural source), CH_2Cl_2, and C_2Cl_4; and perfluoromethane, CF_4. Generally, the environmental stability or inertness of these organohalogens increases with the number of halogen atoms substituted for hydrogen atoms and from the heaviest to the lightest halogen elements. The unusual stability of chlorofluorocarbons in sea-level air allows accumulation in the air; their levels are increasing each year.

Dissolved halides. Most of the measurements of halides in precipitation are of the halide ion dissolved in rainwater but some studies have been performed on melted snow. Table 2 shows typical concentrations measured in micrograms per liter of liquid precipitation. For fluoride the lowest levels, 5 μg/liter. are seen far from areas influenced by continental dust or industrial pollution. For chloride the lowest values, 100 μg/liter, are measured over continents, away from the influence of chloride-rich sea-salt aerosol. However, in polluted urban areas hydrochloric acid, HCl, is often a major component in acidic rain. The low values in Table 2 for Br^- and I^- in rainfall represent marine precipitation.

Chlorofluorocarbons. One class of halogen compounds, the anthropogenic chlorofluorocarbons mentioned above, despite being very useful as aerosol-spray-can propellants and as refrigerants, causes serious side effects. The same property of chemical inertness that makes CF_2Cl_2 and $CFCl_3$ useful also allows their concentrations to grow. Once in the air, there appears to be only one means of destroying these tightly constructed artificial molecules: upward air motions eventually carry these molecules into the upper atmosphere (stratosphere, 8–35 mi or 13–56 km above the surface) where they are attacked and decomposed by harsh ultraviolet (uv) light and energetic oxygen atoms. Chlorine atoms are released as the chlorofluorocarbons decompose, and the following chain reaction ensues:

$$\begin{array}{c} Cl + O_3 \rightarrow ClO + O_2 \\ ClO + O \rightarrow Cl + O_2 \\ \hline \text{net: } O + O_3 \rightarrow 2O_2 \end{array}$$

Ozone (O_3) and its precursor O atoms are thus destroyed in the high stratosphere. Stratospheric ozone is very important as a natural shield against biologically damaging ultraviolet rays from the Sun. Also, the absorption of solar energy by stratospheric O_3 is a major source of heat for the upper atmosphere. Natural wind systems are largely generated in this way. In addition to the chain reaction above, the chlorine species participate in many other photochemical reactions in the stratosphere. It is clear that human usage of CF_2Cl_2, $CFCl_3$, and similar chemicals is having a large impact on the chemistry of the high-altitude air, but it is not clear how much the atmospheric O_3 is being depleted by these pollutants. Continued usage of CF_2Cl_2 and $CFCl_3$ is predicted to be capable of causing a 12% loss of atmospheric O_3 and to cause a large redistribution of the ozone with altitude. The latter disturbance could affect global wind systems and climate. The facts that these chlorofluorocarbons are present in air everywhere on Earth and have reached high altitudes are well established by reliable measurements. A second and perhaps equally important side effect of the chlorofluorocarbons is that their increasing concentrations are causing more of the Earth's infrared energy to be trapped. An appreciable global warming through a "Greenhouse effect" is indicated. *See* PHOTOCHEMISTRY.

The projected stratospheric consequences of chlorofluorocarbon usage are largely due to the fact that each chlorofluorocarbon molecule can reside in the atmosphere for 50–100 years, thus allowing accumulation to occur. A great deal of elemental chlorine, Cl_2, is used in water treatment and in bleaching with no apparent long-lasting or global environmental effects. Chlorinated solvents are also used widely, but most of them are similarly short-lived in the open air. Potential problems due to accidental release of radioactive iodine (the 129 and 131 isotopes) from uranium fission include the possibility of accumulation in the human thyroid. *See* FLUOROCARBON.

[RALPH J. CICERONE]

Halogenated hydrocarbon

One of a group of halogen derivatives of organic hydrogen- and carbon-containing compounds. The group includes monohalogen compounds (alkyl or aryl halides) and polyhalogen compounds that contain the same or different halogen atoms. The halogenated hydrocarbons may be classified under the general headings of alkyl halides, unsaturated (vinyl and allyl) halides, polyhalogenated compounds, and halogenated aromatic hydrocarbons.

Uses. The alkyl halides, especially bromides and iodides, are used extensively as research intermediates. Several are essential pharmaceutical intermediates. Alkyl chlorides are important industrially in the manufacture of other organic compounds, especially alcohols and esters.

The polyhalogenated hydrocarbons find many uses. They are generally chemically inert, and are widely used in many industrial processes as solvents, degreasers, and dry-cleaning agents. The polybromides, more reactive than the polychlorides, are frequently used in the synthesis of other compounds. The polyfluorides, often with attached chorine and bromine, are called freons and are widely used as refrigerants. The unsaturated polyhalogen compounds are also valuable in a variety of organic syntheses.

The halogenated aryl hydrocarbons are used in organic syntheses as solvents and as reactants. Generally, the chlorine compounds are more inert than the bromine and iodine compounds.

The aralkyl- or benzyl-type halides are more reactive than the aryl halides. They are used chiefly as chemical intermediates.

Alkyl halides. These are monohalogenated derivatives, RX, of the alkanes, or saturated hydrocarbons. Typical of these are methyl iodide (CH_3I), ethyl chloride (CH_3CH_2Cl), and *n*-propyl bromide ($CH_3CH_2CH_2Br$). The alkyl chlorides and bromides may be obtained by direct halogenation of the hydrocarbons. Thus, alkyl chlorides are prepared by direct chlorination of alkanes, particularly methane, ethane, propane, butane, and pentane. Chlorine gas is used alone, with light or heat to promote reaction, or with catalysts such as aluminum chloride, ferric chloride, and other metallic chlorides. At lower temperatures, the relative rates of formation of chlorides are in the order tertiary > secondary > primary. At higher temperatures, this tendency is less significant. Chlorination may proceed so vigorously that special techniques are necessary to avoid excessive decomposition that causes low yields of desired products. The process may be controlled frequently by using an excess of the hydrocarbon, an inert solvent, or a diluent inert gas. In all cases, mixtures of monochloro, dichloro, and polychloro compounds result; these can be separated, usually, by fractional distillation.

Direct bromination of the alkanes is restricted to the simple paraffins and olefins (unsaturated hydrocarbons). The reactions are generally conducted either in the vapor or in the liquid phase, and heat and light are used as initiators. Because oxygen decreases the rate of reaction (this indicates a free-radical mechanism), the reactions are run in an inert atmosphere. Higher temperatures favor substitutive bromination; on the other hand, bromine adds best to olefins at lower temperatures.

Iodine does not react with hydrocarbons. Fluorine reacts violently and causes extensive decomposition unless the reaction is carefully controlled.

Alkyl halides (chlorides, bromides, iodides, fluorides) may also be prepared by the addition of hydrogen halide to olefins. The order of reactivity is HI > HBr > HCl. With an unsymmetrical olefin, the addition of the hydrogen halide follows Markownikoff's rule. This rule states that the negative radical (F, Br, Cl, I) attaches to the carbon atom which has the smaller number of hydrogen atoms. Thus, propylene ($CH_3CH{=}CH_2$) adds hydrogen iodide to form isopropyl iodide (CH_3-$CHICH_3$) and not *n*-propyl iodide ($CH_3CH_2CH_2I$), as shown in reaction (1). Exceptions to Markownikoff's rule are those reactions involving the addition of hydrogen bromide. In the absence of oxygen, normal addition is effected. However, when peroxides and oxygen are present, the addition may be abnormal. For example, propylene, in the presence of peroxides and oxygen, adds hydrogen bromide to produce *n*-propyl bromide and not the expected isopropyl bromide, as in reaction (2). This is explained on a basis of the difference in reaction mechanisms. Normal addition is ionic in character, whereas peroxide-catalyzed addition is a free-radical process.

$$CH_3CH{=}CH_2 + HI \rightarrow CH_3CHICH_3 \qquad (1)$$

$$CH_3CH{=}CH_2 + HBr \xrightarrow[\text{Oxygen}]{\text{Peroxides}} CH_3CH_2CH_2Br \qquad (2)$$

Alkyl fluorides can be prepared by the addition of hydrogen fluoride to an olefin, as in reaction (3).

$$CH_2CH_2 + HF \rightarrow CH_3CH_2F \qquad (3)$$

However, the monofluorides are markedly unstable. This is in direct contrast to the polyfluorides in which more than one fluorine atom is attached to the same carbon atom. These are not only stable, but chemically inert as well. It should be noted that monofluoro compounds are frequently quite toxic.

Alkyl halides (iodides, bromides, chlorides) may be made conveniently on a laboratory scale from the corresponding alcohol by the action of hydrogen halides (anhydrous or concentrated aqueous solution) or of inorganic acid halides, particularly the phosphorus halides. The halogen atom replaces the hydroxyl group of the alcohol, and water is formed as a by-product, as in reaction (4). Anhydrous hydrogen halide may be passed directly into the alcohol, or the alcohol may be heated with a concentrated aqueous solution of the acid, usually in the presence of a catalyst. The relative rates of reaction of the hydrogen halides are HI > HBr > HCl; and of the alcohols, tertiary > secondary > primary alcohol. Anhydrous zinc chloride increases the rate of reaction of hydrochloric acid with primary and secondary alcohols. Sulfuric acid catalyzes the reaction of hydrobromic acid with alcohols.

$$ROH + HBr \rightarrow RBr + H_2O \qquad (4)$$

Other halogenating agents include thionyl chloride and the phosphorus halides. The latter are used especially for preparing bromides and iodides, particularly of the higher alkanes, from the respective alcohols, as shown in reaction (5).

$$3ROH + PBr_3 \rightarrow 3RBr + P(OH)_3 \qquad (5)$$

Alkyl iodides can be produced by the reactions of metallic iodides (sodium or potassium) with other alkyl halides (bromides or chlorides) in a solvent such as acetone, as in reaction (6). The metallic iodides exchange iodine for bromine or chlorine where the halogen to be exchanged is relatively reactive. Other excellent methods for the preparation of alkyl iodides are by the reaction of phosphoric acid, a metallic iodide, and an alcohol; and by the gradual addition of red phosphorus to free iodine in the appropriate alcohol. The preparation of the lower-molecular-weight iodides by this method is hazardous.

$$RBr + NaI \xrightarrow{\text{Acetone}} RI + NaBr \qquad (6)$$

Alkyl fluorides may be made by the interchange of fluorine for bromine in an organic bromide. Mercury(II) fluoride or antimony trifluoride is used as the fluorinating agent.

Unsaturated halides. Halogenated unsaturated hydrocarbons are classified in two groups, the vinyl halides and the allyl halides. The vinyl-type compounds are those in which the halogen atom is attached to an unsaturated carbon atom. A typical example, vinyl chloride ($CH_2{=}CHCl$), is the most important of the unsaturated halogen compounds. It is prepared by the reaction of hydrogen chloride with acetylene, using a catalyst, as in reaction (7); by dehydrohalogenation of ethylene dichloride, as in reaction (8); and in the high-temperature chlorination of ethane, as in reaction (9).

$$HC{\equiv}CH + HCl \xrightarrow{\text{Catalyst}} CH_2{=}CHCl \qquad (7)$$

$$ClCH_2CH_2Cl + NaOH \xrightarrow{\text{Heat}} CH_2CHCl + NaCl + H_2O \qquad (8)$$

$$H_3CCH_3 + Cl_2 \xrightarrow{\text{High temp.}} ClCH{=}CH_2 \qquad (9)$$

Vinyl chloride is widely used as a monomer in the manufacture of polymeric substances, such as plastics and synthetic fibers. Methods of preparation similar to those for vinyl chloride are used for most vinyl-type halides. The vinyl compounds are noted for the stability of the carbon-halogen bond. *See* VINYL RESIN.

Allyl halides possess a halogen on a carbon atom adjacent to an unsaturated carbon atom. Typical of the allylics are allyl chloride ($CH_2{=}CHCH_2Cl$) and crotyl chloride ($CH_3CH{=}CHCH_2Cl$). These halides (chlorides, bromides, iodides) are prepared from the respective unsaturated alcohols by the action of the corresponding hydrohalogen acid, as in reaction (10). Industrially, allyl chloride is pro-

$$CH_2{=}CHCH_2OH + HCl \rightarrow CH_2{=}CHCH_2Cl \quad (10)$$

duced by the high-temperature chlorination of propylene. The iodide may be obtained from glycerol by treatment with either hydriodic acid or a mixture of phosphorus and iodine. The allyl halides are quite reactive and are used extensively as organic intermediates. They are powerful lacrimators.

Polyhalogen compounds constitute those halogen-containing hydrocarbons that have two or more atoms of halogen per molecule. These are the polybromo, polychloro, and polyfluoro compounds, together with the mixed halogeno hydrocarbons. The polyiodides are, with few exceptions, difficult to synthesize and generally unstable. The polychlorides and polybromides are produced by the high-temperature chlorination and bromination of the alkanes. Such polychloro compounds as methylene chloride (CH_2Cl_2), chloroform ($CHCl_3$), and carbon tetrachloride (CCl_4) are produced on a commercial basis by the controlled chlorination of methane. At high temperatures, both substitutive chlorination and combined chlorination and fission of the carbon chain produce polychlorinated alkanes and olefins. Reaction products are generally carbon tetrachloride, tetrachloroethylene, and hexachloroethane. Ethylene dichloride and propylene dichloride are made from the olefin and gaseous chlorine. Many other polychlorides and polybromides may be made by direct halogenation of olefins. This method is used chiefly for bromination, because the reaction with chlorine is so vigorous that it is sometimes explosive. Bromination with liquid bromine goes smoothly at room temperature or lower to yield dibromo products with olefins, and tetrabromo products with acetylenes. This reaction is shown in reactions (11) and (12).

$$RCH{=}CH_2 + Br_2 \xrightarrow{\text{Room temp.}} RCHBrCH_2Br \quad (11)$$

$$HC{\equiv}CH + 2Br_2 \rightarrow Br_2HCCHBr_2 \quad (12)$$

Mixed polyhalogen compounds result from halogenation of the unsaturated halides, as in reaction (13). Olefinic polyhalogen derivatives are

$$CH_2{=}CHCH_2Cl + Br_2 \rightarrow BrCH_2CHBrCH_2Cl \quad (13)$$

made by careful dehydrohalogenation of other polyhalogen compounds by treatment with concentrated sodium or potassium hydroxide solution (aqueous or alcoholic), according to reaction (14).

$$CHCl_2CHCl_2 + NaOH \xrightarrow{\text{Heat}} CHCl{=}CCl_2 \quad (14)$$

Polyfluorinated hydrocarbons are made by heating chloro or bromo compounds with antimony trifluoride or mercuric fluoride, as in reaction (15).

$$RCX_2R + SbF_3 \xrightarrow{\text{Heat}} RCF_2R + SbFCl_2 \quad (15)$$

The fluorine in the metallic salt exchanges for the halogen atom in the organic compound. The relative ease of replacement follows the general order: $I > Br > Cl$. Organic polyfluoro compounds, in which the fluorine atoms are attached to the same carbon atom, are more stable than those in which the fluorine atoms are attached to adjacent carbons. Antimony trifluoride is an effective fluorinating agent, particularly if chlorine gas or antimony pentachloride is added to the reaction mixture. About 10% by weight of the antimony trifluoride is recommended. *See* FREON.

The most important polyiodides are methylene iodide (CH_2I_2) and iodoform (CHI_3). Methylene iodide has the highest density of any liquid except mercury. It is used by mineralogists to determine the density of minerals, and to separate them from each other. It is prepared by the reduction of iodoform with sodium arsenite solution, as in reaction (16). Iodoform is produced by the iodination of

$$CHI_3 + Na_3AsO_3 + NaOH \rightarrow CH_2I_2 ; NaI + Na_2AsO_4 \quad (16)$$

methyl ketones in aqueous basic solution or by the electrolytic oxidation of ethyl alcohol in aqueous carbonate and iodine solution. Iodoform is used as an antiseptic. *See* HALOFORM REACTION; IODOFORM.

Halogenated aromatic hydrocarbons. These include the halogen derivatives of the benzenoid carbon-hydrogen compounds, and are divided into two types, halides in which halogen is attached to a ring carbon, and halides in which halogen is attached to a carbon atom of a side chain. The former are called aryl halides; the latter are aralkyl derivatives (benzyl halides), which are similar to the alkyl halides.

Aryl halides, such as chlorobenzene, and the halogen compounds of naphthalene and anthracene, can be made from the hydrocarbons themselves, or from a derivative of the hydrocarbons such as the nitro or amino. Commercially, benzene is chlorinated or brominated in the presence of a halogen carrier (iron, aluminum, or iodine) at 50°C, as shown in reaction (17). Aluminum chloride, a

$$C_6H_6 + Br_2 \rightarrow C_6H_5Br + HBr \quad (17)$$

very active catalyst, produces di- and polyhalogenated products. These can be separated by fractional distillation or crystallization. Chloro derivatives may also be made from benzene and hydrogen chloride by treatment with metallic oxides in air at high temperatures. Iodination with iodine requires the use of a suitable oxidizing agent, such as nitric acid, iodic acid, or persulfate, as in reaction (18). If other alkyl groups are

$$5C_6H_6 + 2I_2 + HIO_3 \rightarrow 5C_6H_5I + 3H_2O \quad (18)$$

present on the ring, reactions must be carried out in the dark to avoid side-chain halogenation. The halogenation of compounds containing a side chain, such as toluene, results in a mixture of the ortho and para isomers, as in reactions (19). These

$$CH_3C_6H_5 + Br_2 \xrightarrow{FeBr_3} 1,2\text{-}CH_3C_6H_4Br \quad (19a)$$

$$CH_3C_6H_5 + Br_2 \xrightarrow{FeBr_3} 1,4\text{-}CH_3C_6H_4Br \quad (19b)$$

can be separated by fractional crystallization. Halogenated aromatic hydrocarbons can be prepared from amino derivatives of the hydrocarbon (amines are made by reduction of nitro compounds). This method involves making a diazonium salt from the amine and replacing the salt with halogen using a metallic halide at 0°C, as in reaction (20). The reaction has the advantage of

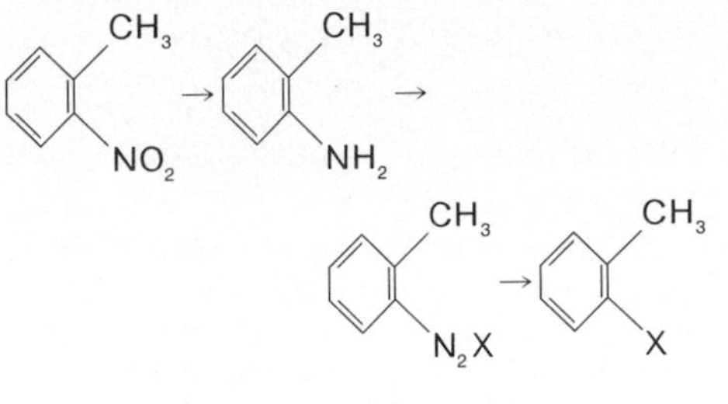

(20)

producing a pure halogen compound, because substitution of halogen occurs only in the position of the amino group. This process is used when there are other alkyl groups present on a ring, and a particular isomer is desired. Fluorine, chlorine, bromine, and iodine may be substituted on the ring for an amino group, using the diazonium reaction. *See* DIAZOTIZATION.

Aralkyl halides, containing halogen in the side chain, are similar to the alkyl halides. As in the alkane series, the iodides, bromides, and chlorides can be prepared from the corresponding alcohols by treatment with a suitable halogenating agent (hydrogen halide, phosphorus halide, thionyl chloride, sulfuryl chloride) or by direct substitution in the case of the bromides and chlorides. Side-chain substitution is helped by heating and by use of sunlight or ultraviolet light. With a longer side chain, substitution at low temperature (0°C) in the presence of ultraviolet radiation occurs on the carbon atom adjacent to the ring (α position), as in reaction (21). At higher temperatures, it may occur on the second carbon away from the ring (β position), as shown in reaction (22). To avoid nuclear sub-

$$C_6H_5CH_2CH_3 + Cl_2 \rightarrow C_6H_5CHClCH_3 + HCl \quad (21)$$

$$C_6H_5CH_2CH_3 + Cl_2 \rightarrow C_6H_5CH_2CH_2Cl + HCl \quad (22)$$

stitution, the system must be kept entirely free of iron or iron chloride. Mono-, di-, and trichloro derivatives are obtained. With long side chains, complicated mixtures result, and other methods must be used to obtain the desired product. The phosphorus halides are particularly useful for preparing the monohalides from alcohols. Fluorine compounds are made by treating halides with antimony trifluoride. *See* BROMINE; CHLORINE; FLUORINE; HALOGENATION; IODINE; SUBSTITUTION REACTION. [ELLEN CLARKE; MAX G. GERGEL]

Halogenation

A chemical reaction or process which results in the formation of a chemical bond between a halogen atom and another atom. Reactions resulting in the formation of halogen-carbon bonds are especially important, and the emphasis in this article is primarily on such reactions. The halogenated compounds produced are employed in many ways, for example, as solvents, intermediates for numerous chemicals, plastic and polymer intermediates, insecticides, fumigants, sterilants, refrigerants, additives for gasoline, and materials used in fire extinguishers. *See* HALOGEN ELEMENTS.

Numerous compounds are halogenated by a wide variety of commercial and laboratory processes. Halogenation reactions can be subdivided in several ways, for example, according to the type of halogen (fluorine, chlorine, bromine, or iodine), type of material to be halogenated (paraffin, olefin, aromatic, hydrogen, and so on), and operating conditions and methods of catalyzing or initiating the reaction.

Halogenation reactions with elemental chlorine, bromine, and iodine are of considerable importance. Because of high exothermocities, fluorinations with elemental fluorine tend to have high levels of side reactions. Consequently, elemental fluorine is generally not suitable for direct fluorination. Two types of reactions are possible with these halogen elements, substitution and addition.

Substitution halogenation. This type of halogenation is characterized by the substitution of a halogen atom for another atom (often a hydrogen atom) or group of atoms (or functional group) on paraffinic, olefinic, aromatic, and other hydrocarbons. A chlorination reaction of importance that involves substitution is that between methane and chlorine. The overall reaction between methane and chlorine is shown in reaction (1).

$$CH_4 + Cl_2 \rightarrow CH_3Cl + HCl \quad (1)$$

Substitution reactions with paraffins involve several free-radical reaction steps; only the most important steps are shown below. In the initiating step, chlorine-free radicals are formed as follows: Cl_2 (elemental chlorine) $\rightarrow$ 2 Cl· (chlorine-free radical). Breaking elemental chlorine into free radicals requires energy, and can be accomplished thermally at high temperatures (about 250°C or higher) or at lower, temperatures (including ambient temperatures) by means of radiation (gamma radiation, sunlight, and so on).

The chlorine-free radical starts a chain mechanism with methane, as in reactions (2) and (3).

$$CH_4 + Cl\cdot \rightarrow CH_3\cdot + HCl \quad (2)$$

$$CH_3\cdot + Cl_2 \rightarrow CH_3Cl + Cl\cdot \quad (3)$$

Similar reaction steps also occur with other paraffins or hydrogen. The relative rates of halogenation vary according to the type of carbon-hydrogen bond, as follows: tertiary > secondary > primary. Free-radical reactions such as those shown above are sometimes affected to a considerable extent by the surface of the reactor employed. The kinetics of chlorination can vary by factors of at least four or five, depending on the material of construction of the reactor or the past history of the reactor.

The table gives approximate heats of reaction

Heats of reaction in kilocalories per mole in radical halogenation

Reaction	F_2	Cl_2	Br_2
$RH + X\cdot \rightarrow R\cdot + HX$	−33	−3	+13
$R\cdot + X_2 \rightarrow RX + X\cdot$	−71	−24	−21

for the two steps in the chain mechanism, such as those shown in reactions (2) and (3) for substitution halogenation with fluorine, chlorine, and bromine. Halogenation reactions are exothermic; the level of exothermicity is: $F_2 > Cl_2 > Br_2 > I_2$. *See* SUBSTITUTION REACTION.

Addition halogenation. Chlorine, bromine, and iodine react readily with most olefins; the reaction between ethylene and chlorine, to form 1,2-dichloroethane, reaction (4), is one of considerable

$$CH_2{=}CH_2 + Cl_2 \rightarrow CH_2Cl{-}CH_2Cl \quad (4)$$

commercial importance, since it is used in the manufacture of vinyl chloride. This reaction occurs readily in the gas phase at temperatures in the range of 80 to 130°C when the inner surface of the reactor is coated with calcium or lead chlorides.

Reaction (4) also occurs readily in the liquid phase by means of ionic reactions at temperatures from 30 to 70°C. Ferric chloride, often formed by reaction of chlorine with iron in the walls of the reactor, is an excellent catalyst.

Addition reactions with bromine or iodine are frequently used to measure quantitatively the number of —CH=CH— (or ethylenic-type) bonds in organic compounds. Bromine numbers or iodine values are measures of the degree of unsaturation of the hydrocarbons.

Halogenation of aromatics. Substitution halogenation on the aromatic ring can be made to occur by means of ionic reactions. The chlorination reactions with elemental chlorine occur under conditions similar to those used for addition chlorination of olefins; temperatures of 20–50°C are suitable, and ferric chloride is an effective catalyst. Polychlorination of benzene results in various isomers of dichloro-, trichloro-, and tetrachlorobenzenes; even higher levels of chlorination sometimes occur.

When liquid benzene is contacted with chlorine at essentially ambient temperature in the presence of radiation (light radiation, gamma radiation, and so on), but in the absence of catalysts, the chlorine adds to the benzene by means of a free-radical chain mechanism to form benzene hexachloride, an effective insecticide. The overall reaction is shown as reaction (5).

$$C_6H_6 + 3Cl_2 \xrightarrow[\text{Radiation}]{30-40°C} C_6H_6Cl_6 \qquad (5)$$

Bromine and benzene also react in the presence of sunlight; benzene hexabromide is formed, however, with difficulty.

When alkyl benzenes are halogenated, the halogen can react either on the aromatic nucleus or on the alkyl group. Free-radical (and gas-phase) chlorinations generally result in substitution, primarily on the alkyl groups; such reactions occur readily at 120–130°C. Ionic, low-temperature reactions, however, favor substitution on the nucleus.

At 400–500°C, chlorobenzenes are produced from benzene by free-radical chlorination steps. The resulting products have an isomer distribution quite different from that obtained by means of ionic chlorinations at essentially ambient temperatures.

Halogenations with hydrogen halides. Numerous methods are available for halogenating with hydrogen halides. Oxychlorination reactions can be used with olefinic, paraffinic, and aromatic hydrocarbons. The reaction with ethylene, reaction (6), is of major industrial importance for the pro-

$$CH_2{=}CH_2 + 2HCl + {}^1\!/_2O_2 \xrightarrow[250-400°C]{CuCl_2} CH_2Cl{-}CH_2Cl + H_2O \qquad (6)$$

duction of 1,2-dichloroethane in modern vinyl chloride plants.

Hydrogen halides can also be made to react with many ethylenic and acetylenic compounds. Examples are shown in reactions (7) and (8).

$$CH_2{=}CH_2 + HCl \xrightarrow[100°c]{ZnCl_2} CH_3CH_2Cl \qquad (7)$$

$$CH{\equiv}CH + HCl \xrightarrow[90-140°C]{HgCl_2} CH_2{=}CHCl \qquad (8)$$

Before 1965 reaction (8) was the main method used in producing vinyl chloride. Fluorocarbons widely used as refrigerants are produced from carbon tetrachloride; the production of trichlorofluoromethane is shown in reaction (9). Alcohols

$$CCl_4 + HF \xrightarrow[100°F]{SbF_5} CCl_3F + HCl \qquad (9)$$

react with hydrogen halides to form alkyl halides, as shown in reactions (10) and (11).

$$CH_3OH + HBr \rightarrow CH_3Br + H_2O \qquad (10)$$

$$CH_3CH_2OH + HCl \xrightarrow[110-140°C]{ZnCl_2} C_2H_5Cl + H_2O \qquad (11)$$

Electrochemical techniques are employed commercially for substituting fluorine atoms for hydrogen on the alkyl group of organic acids, amines, nitriles, and alcohols. In an electrolytic cell operated at low voltages hydrogen fluoride is used as the fluorinating agent; hydrogen is also a product.

Miscellaneous halogenating agents. Many compounds have been employed as halogenating agents. Several examples are: sodium hypochlorite, phosgene ($COCl_2$), thionyl chloride, sulfuryl chloride, ferric chloride, antimony chlorides, and phosphorus chlorides, which can be used for certain chlorination reactions; hypobromites, *N*-bromosuccinimide, and *N*-bromoacetamide, which have proved to be brominating agents; antimony pentafluoride, silver difluoride, and lead tetrafluoride, metal fluorides which are used as fluorinating agents; and iodine monochloride and alkali hypoiodites, which are iodination agents.

[LYLE F. ALBRIGHT]

Bibliography: P. B. DeLamare, *Electrophilic Halogenation*, 1976; V. Gutmann, *Main Group Elements: Groups 6–7*, 1975; R. T. Morrison and R. N. Boyd, *Organic Chemistry*, 3d ed., 1973; M. Grayson (ed.), *Kirk-Othmer Encyclopedia of Chemical Technology*, 3d ed., vol. 8, 1979.

Heavy water

A form of water in which the hydrogen atoms of mass 1 (^1H) ordinarily present in water are replaced by deuterium (symbol D or ^2H), the heavy stable isotope of hydrogen of mass 2. The molecular formula of heavy water is D_2O (or ^2H_2O).

Properties. Because the mass difference between ^1H and ^2H is the largest for any pair of stable (nonradioactive) isotopes in the periodic table, many of the physical and chemical properties of the pure isotopic species and their respective compounds differ to a significant extent. Selected physical properties of ^1H_2O and ^2H_2O are compared in Table 1.

Heavy water, judging from its higher melting and boiling points, its higher viscosity, and its surprisingly high temperature of maximum density, is a distinctly more structured liquid than is ordinary water. Heavy water is more extensively hydrogen-bonded, and the hydrogen bonds formed by ^2H are somewhat stronger than are those of ^1H.

Table 1. Physical properties of ordinary and heavy water

Property	^1H_2O	$^2H_2O(D_2O)$
Molecular weight, ^{12}C scale	18.015	20.028
Melting point, °C	0.00	3.81
Normal boiling point, °C	100.00	101.42
Temperature of maximum density, °C	3.98	11.23
Density at 25°C, g/cm^3	0.99701	1.1044
Critical constants		
Temperature, °C	374.1	371.1
Pressure, mPa	22.12	21.88
Volume, cm^3/mol	55.3	55.0
Viscosity at 55°C, mPa·s	0.8903	1.107
Refractive index, n_D^{20}	1.3330	1.3283

Nonpolar solutes induce structure in ^2H_2O to a greater extent than in ^1H_2O, and structure-breaking salts are more disruptive in ^2H_2O, largely because there is more structure to break. With increasing temperature, structure is broken down more rapidly in ^2H_2O than in ^1H_2O, and as a result ^2H_2O may be more structured or less structured than is ^1H_2O, depending on the temperature. Many ionic solutes are distinctly less soluble in D_2O than in H_2O.

The nuclear properties of ^1H and ^2H are very different. The capture cross section (Σ_a) of deuterium for low energy of neutrons is much smaller than that of ordinary hydrogen. For ^1H_2O, Σ_a is 2.2×10^{-2} cm^{-1} as compared with the value of 8.5×10^{-6} cm^{-1} for ^2H_2O. The nuclear spin of ^1H is 1/2, but that of ^2H is 1. Thus, both isotopes can be used in nuclear magnetic resonance spectroscopy. The resonance frequency for ^2H is much lower than that of ^1H at a given magnetic field, and the relative sensitivity for detection is much higher for ^1H than for ^2H.

The principal difference in chemical behavior between ^1H and ^2H derives from the generally greater stability of chemical bonds formed by ^2H. The most important factor contributing to the difference in bond energy is the lower zero-point vibrational energy (of the order of 5.021–5.275 kJ/mol) for chemical bonds formed by ^2H. Both kinetic and static isotope effects result. The ion product constant of ^2H_2O is 1.11×10^{-15} at 25°C compared to 10^{-14} for ^1H_2O. The difference by a factor of 10 in ionization constant causes values of pH and pD for solutions of identical composition in the two media to differ significantly. The nonidentity of pH and pD for solutions of a given composition can be critical for phenomena sensitive to hydrogen ion activity, and can make for serious problems in the interpretation of experiments in many biological systems. In the pH range 2–9, it has been established that pD = pH + 0.41 (molar scale; 0.45 molal scale) as measured by an ordinary glass electrode.

Replacement of more than one-third of the ^1H by ^2H in the body fluids of animals by administration of heavy water, or two-thirds of the hydrogen in higher plants, is lethal. However, numerous green and blue-green algae have been successfully cultured in >99.5% ^2H_2O on carbon dioxide and inorganic nutrients. Fully deuterated organisms can, in turn, be used to start a food chain for the growth of nutritionally more demanding bacteria, molds, and even protozoa in fully deuterated form. These organisms contain >99.5% D, and cultures of these organisms of unnatural isotopic composition have been successfully maintained for years in a fully deuterated form.

Preparation. Deuterium is present in ordinary water to the extent of 0.0145%. Water is the only practical source of deuterium, and very large amounts of water must be processed to produce relatively small amounts of heavy water. Deuterium can be extracted from water (as highly enriched ^2H_2O) by electrolysis, by distillation, or by chemical exchange. Electrolysis of ordinary water is very costly in energy, and has been used only for small-scale production or to enrich partially concentrated material. Distillation of water has been successfully used to separate highly enriched D_2O from ordinary water, but again the energy costs are high per unit of separative work performed. More efficient is the distillation of liquid hydrogen at cryogenic temperatures. Excellent fractionation factors can be achieved in the distillation of liquid hydrogen. The power requirements are modest, but the requirements for very large hydrogen feeds of very high purity (traces of oxygen, nitrogen, carbon monoxide, and so on, are solids at liquid hydrogen temperatures and may clog the apparatus) have mitigated against large-scale use. As the production of hydrogen for synthetic ammonia or coal liquefaction expands, however, by-product extraction of deuterium from the hydrogen stream by distillation of liquid hydrogen may well become important.

Various chemical exchange reactions have been considered for concentrating deuterium from natural water (Table 2). Isotope exchange reactions between hydrogen gas and water or ammonia, or between hydrogen sulfide and water, provide the best point of departure for the large-scale manufacture of heavy water. The H_2O/H_2S process is the chemical exchange process of principal commercial interest. Because the equilibrium constant for the distribution of deuterium between H_2S gas and liquid H_2O is temperature-dependent, the process can be carried out in the form of a dual-temperature exchange process. In the dual-temperature H_2O/H_2S process, exchange of ^2H between H_2S (gas) and H_2O (liquid) is carried out at elevated pressure (~2MPa or 20 atm) at 120–

Table 2. Chemical exchange reactions for concentrating deuterium from water

Reaction	Equilibrium constant: Low temp. (°C)	Equilibrium constant: High temp. (°C)
$H_2O(l) + HDS(g) \rightleftarrows HDO(l) + H_2S(g)$	2.18 (20°)	1.83 (130°)
$NH_3(l) + HD(g) \xrightleftharpoons{\text{catalyst}} NH_2D(l) + H_2(g)$	6.60 (−50°)	4.42 (0°)
$H_2O(g) + HD(g) \xrightleftharpoons{\text{catalyst}} HDO(g) + H_2(g)$	3.62 (25°)	2.43 (125°)

140°C; ^2H displaces ^1H in the H_2S and becomes concentrated in the gas. At 30°C the equilibrium is shifted and reversed, and ^2H is stripped from the gas into the liquid water. The dual-temperature exchange process is carried out in a pair of gas-liquid contacting towers; the cold tower operates at 30°C, the hot tower at 120–140°C. Water fed to the system flows downward through the cold tower and then through the hot tower countercurrent to a stream of hydrogen sulfide. The water is progressively enriched in deuterium as it passes through the cold tower and progressively depleted in transit through the hot tower. Deuterium-enriched water is drawn from the bottom of the cold tower, and H_2S enriched in deuterium is removed from the top of the hot tower. Further enrichment then occurs in a second stage. To produce 1 metric ton of D_2O, the plant must process 41,000 tons of water and must cycle 135,000 tons of H_2S. The product from the exchange plant is enriched to around 15% deuterium, and enrichment to 99.5% is accomplished by vacuum distillation. The dual-temperature GS H_2S/H_2O exchange process is used by all heavy-water plants producing more than 20 metric tons of D_2O per year. Heavy water produced in the United States by the GS process is priced at $220/kg in bulk quantities.

Uses. The only large-scale use of heavy water in industry is as a moderator in nuclear reactors. The Canadian heavy-water-moderated natural-uranium-fueled CANDU reactor uses 0.85 metric ton of heavy water per electrical megawatt of installed capacity. The heavy water is not consumed by reactor operations, and the demand for heavy water is determined by the rate at which new nuclear power stations are built. About 3500 metric tons of heavy water were contained in operating CANDU nuclear reactors as of early 1981.

Deuterium may become important in energy production by nuclear fusion by the reaction $^2_1H + ^2_1H \rightarrow ^3_1H + ^1_1H + 4.0$ MeV. It has been estimated that the entire annual energy requirements for the United States in the year 2020 could be supplied by the D-D nuclear fusion reaction by the amount of deuterium contained in 5000 metric tons of D_2O.

Small amounts of heavy water are used to grow fully deuterated organisms, which serve as a source of fully deuterated compounds of biological importance. These are finding increasing use in research techniques such as small-angle neutron scattering, in high-resolution nuclear magnetic resonance spectroscopy of immobilized samples, and in the study of isotope effects. The applications of deuterium in biological research at present require only small amounts of heavy water, but large-scale requirements for heavy water may develop from such research.

Analysis. The principal methods for determining the deuterium content of heavy water are density determinations, infrared spectroscopy in the ~3-μm region, and absorption spectrophotometry in the near-infrared region of the spectrum between 1 and 2 μm. Mass spectroscopy, interferometry, falling-drop methods, and nuclear magnetic resonance spectroscopy have also been used for determining the $^1H/^2H$ ratio in heavy water. Absorption spectrophotometry in the near infrared is a particularly useful procedure because it can be carried out in conventional recording spectrophotometers. *See* DEUTERIUM; LIQUID; TRITIUM; WATER.

[JOSEPH J. KATZ]

Bibliography: I. Kirshenbaum, *Physical Properties and Analysis of Heavy Water*, 1951; J. J. Katz and H. L. Crespi, in *Isotope Effects in Chemical Reactions*, ACS Monogr. no. 167, 1971; G. M. Murphy (ed.), *Production of Heavy Water*, 1955; H. K. Rae (ed.), *Separation of Hydrogen Isotopes*, ACS Symp. Ser. no. 68, 1978; J. Spevack, U.S. Patents 2,787,526, 2,895,803, and 4,008,046.

Helium

A gaseous chemical element, He, atomic number 2 and atomic weight 4.0026. Helium is one of the noble gases in group 0 of the periodic table. It is the second lightest element. The world's chief source of helium is a group of natural gas fields in the United States. *See* INERT GASES.

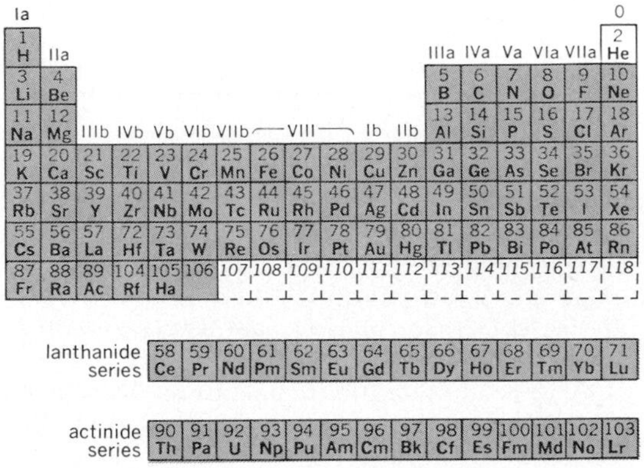

Uses. Helium was first used as a lifting gas in balloons and dirigibiles. This use continues for high-altitude research and for weather balloons.

Welding. The principal use of helium is in inert gas–shielded arc welding. Using helium instead of argon permits a greater heat release, which is useful in welding very heavy sections or in high-speed machine welding of long seams. By mixing helium and argon, the optimum heat release can be obtained for different welding jobs.

Superconductive devices. The greatest potential for helium use continues to emerge from extreme-low-temperature applications. Helium is the only refrigerant capable of reaching temperatures below 14 K. In the laboratory many fundamental properties of matter are studied at temperatures near absolute zero with helium refrigeration. Infrared detectors and masers operate with exceptionally low noise distortion at these low temperatures. The chief value of ultra-low temperature is the development of the state of superconductivity, in which there is virtually zero resistance to the flow of electricity. Very large currents are carried by even small conductors with little loss of voltage. Electromagnets producing immensely powerful magnetic fields can be made small and light and are energized with modest amounts of electric power through the use of superconducting windings. These magnets are already used in particle accelerators, bubble chambers, and plasma confinement for nuclear physics research. Thermonuclear and magnetohydrodynamic (MHD) power plants are expected to use superconducting magnets. Additional applications are developing in electric motors and generators. Superconductive devices make highly sensitive detectors of electric voltage and frequency, magnetic field strength, and temperature, especially at low temperature levels.

Lasers. Helium use in gas-discharge lasers is also increasing. Energy is transferred by helium to the lasing gas, carbon dioxide or neon, for example.

Rockets. Consumption of helium as a pressurizing gas in liquid-fueled rockets has declined with the completion of the Apollo space program. Because it is light, inert, and relatively insoluble in the fuel and oxidizer fluids, helium is an ideal material to fill the tankage as the liquids are consumed.

Breathing mixtures. Use of helium-oxygen breathing mixtures for divers at great depths is required to eliminate the narcotic effects of nitrogen. The low density and low viscosity of helium also reduce the work of breathing. Similarly, helium-oxygen breathing mixtures promote both intake of oxygen and removal of carbon dioxide for patients whose breathing passages are constricted.

Nuclear reactors. Inertness and heat-transfer capability make helium an excellent working fluid for gas-cooled nuclear power reactors. Because the reactor core is composed of graphite and ceramic materials, very high temperatures can be attained without damage. Helium-cooled reactors operate with the highest efficiency of all reactor types. In addition to electric power generation, it appears likely that with helium working fluid nuclear reactors may provide the process heat for coal gasification, steel making, and various chemical processes.

Chemical analysis. Helium is the most frequently used carrier gas for chemical analysis by gas chromatography. It is the most sensitive leak detection fluid and can be used at extremes of high and low temperature.

Occurrence and origin. Terrestrial helium is believed to be formed in natural radioactive decay of heavy elements. Most of this helium migrates to the surface and enters the atmosphere. The atmospheric concentration of helium (5.25 ppm at sea level) could be expected to be higher. However, its low molecular weight permits helium to escape into space from the upper atmosphere at a rate roughly equal to its formation.

Natural gases contain helium at concentrations higher than in the atmosphere. Higher helium contents are found in deposits originating in older strata and are trapped by very impermeable cap rock. More than half the natural gas discovered to date worldwide contains noncommercial quantities of helium ($<0.1\%$). Less than one-tenth of natural-gas deposits have economically attractive helium content ($>0.4\%$). These rich deposits are concentrated in the Rocky Mountains and High Plains areas in the United States.

Because helium is a minor constituent, most helium-bearing gases are exploited for their hydrocarbon content. About 2/3 of the known accumulations of helium has already been produced from the ground. Only a few percent of this was put to use. The rest was simply dissipated to the atmosphere as the associated hydrocarbons were consumed. The remaining known fields in the United States will be exhausted in 10–15 years. New gas fields are being discovered every year, and this trend will continue for some time. However, few of these fields are expected to contain high concentrations of helium because the geology of the newer fields tends to be unfavorable to helium accumulation.

Most uses of helium are for advanced technology with further development likely. High-efficiency production, distribution, and utilization of electric power appears to be dependent on helium in a variety of ways. Demand for helium should continue to grow for many decades. As a result, a shortage could develop around the turn of the century.

The United States government was conserving helium as a safeguard. After extraction from the fuel gas, helium was injected into a government-owned gas field. However, in 1973, following a ruling by the Court of Appeals, the Department of the Interior halted the purchase of helium from private contractors for storage in the Bureau of Mines conservation system. By the end of 1973 over 38×10^9 ft^3 (1.08×10^9 m^3) had been placed in storage. Private producers also have been storing helium, but in smaller quantities. By the end of 1977, 1.7×10^6 ft^3 (4.8×10^4 m^3) was in private storage for future redelivery.

Helium is far more abundant outside the Earth than on the Earth; the best estimates are that only 0.000001% by weight of the Earth's crust, including the atmosphere, is helium, whereas about 23% by weight of the visible universe (stars, nebulae, and interstellar space) is helium.

Discovery. The existence of a bright yellow line in the spectrum of the Sun's prominences was discovered during a solar eclipse on Aug. 18, 1868, by six different observers at different places on the Earth. On Oct. 20, 1868, another observer, J. N. Lockyer, observed three yellow emission lines in the Sun's chromosphere and recognized that one of these did not correspond to any of the known dark absorption lines in the Sun's spectrum. In 1869 G. Rayet first ascribed the new line to some element other than hydrogen or sodium. In 1871 Lockyer recognized that the yellow line corresponded to a new element, which he and E. Frankland called helium. In 1895 Sir William Ramsay, in England, examined the spectrum of a gas liberated from a Norwegian uranium-thorium-lead mineral called cleveite and found the spectrum to contain a brilliant yellow line which W. Crookes identified as the helium line. Thus, helium was found for the first time on Earth.

Properties. Helium is a colorless, odorless, and tasteless gas. It has the lowest solubility in water of any known gas. It is the least reactive element and forms essentially no chemical compounds. The density and the viscosity of helium vapor is very low. Thermal conductivity and heat content are exceptionally high. Helium can be liquefied, but its condensation temperature is the lowest of any known substance. At pressures below 25 atm (1 atm = 101,325 N/m^2) helium remains liquid even at absolute zero. Above 25 atm helium solidifies at about 1.1 K (see Table 1). The properties of liquid helium change radically below 2.19 K. Viscosity virtually disappears and heat transfer rate increases beyond that of any other known substance.

Table 1. Properties of helium

Property	Value
Atomic number	2
Atomic weight	4.0026
Melting point* at 25.2 atm pressure	−272.1°C (1.1 K)
Triple point (solid, helium I, helium II)	−271.37°C (1.78 K)
Triple point = λ-point (helium gas, helium I, helium II)	−270.96°C (2.19 K)
Boiling point at 1 atm pressure	−268.94°C (4.22 K)
Gas density at 0°C and 1 atm pressure, g/liter	0.17847
Liquid density at its boiling point, g/ml	0.1249
Solubility in water at 20°C, ml helium (STP)/1000 g water at 1 atm partial pressure of helium	8.61

*The melting point varies with the pressure.

Production and distribution. Helium is extracted from natural gas by successively liquefying and removing all the other components of the gas stream. The remaining vapor stream contains 70–85% helium. The balance is primarily nitrogen with small concentrations of hydrogen, methane, and neon. Crude helium gas of this type is sent to storage in gas fields owned by the Bureau of Mines.

In purifying the crude helium for current use, the gas is compressed to about 3000 psi (1 psi = 6895 N/m^2) and cooled successively to about −340°F (−207° C).

The small remaining impurities are adsorbed on activated charcoal at this low temperature. Refined helium contains less than 50 ppm total impurities.

Helium must be transported long distances from the extraction plants in western Texas, Oklahoma, and Kansas to the main use areas on the East and West coasts. The older method employs heavy steel pressure vessels mounted on rail cars to carry the helium compressed to 4000 psi.

Private plants have been built to liquefy the helium for transport in large highway trailers. Each trailer carries 10,000 gal (37.9 m^3) of liquid helium (more than three times the capacity of the rail tube cars). This system reduces transport costs and makes available liquid helium for use in cryogenic applications.

Analytical determination. The principal methods of detecting the presence of helium are mass spectrometry, gas chromatography, and emission spectroscopy. The first two of these methods are also used for the quantitative determination of helium. A third quantitative method involves adsorbing all gases present in a sample except the helium on activated charcoal cooled by liquid nitrogen, measuring the volume or pressure of the residue, and then checking the purity of the residue by its emission spectrum. The only gases that are likely to interfere with this method are hydrogen and neon. If hydrogen is present, it can be removed over hot copper oxide. If neon is present, a second adsorption step on freshly activated charcoal at the temperature of liquid nitrogen will probably remove it, leaving only helium in the gaseous state. [ARTHUR W. FRANCIS]

Helium-3. This rare stable isotope of helium was discovered by L. W. Alvarez and R. Cornog in 1939. Its concentration in nature is so low, approximately one part per hundred million in well helium, that it was 1951 before sufficient quantities of pure gas became available for experimentation. The gas was then, and continues to be, obtained as a by-product from the decay of tritium, the heavy isotope of hydrogen. Tritium is produced in a nuclear reactor from the reaction between lithium and a neutron [reaction (1)]. The tritium is separated from the helium and stored. Helium-3 gradually builds as the tritium decays with a 12-year half-life [reaction (2)]. The ^{3}He nucleus is composed of two protons and one neutron, one fewer than for ^{4}He; as a consequence, ^{3}He is a fermion whereas ^{4}He is a boson. The two isotopes are the exemplars of Fermi-Dirac and Bose-Einstein systems, respectively. It is principally for this reason that helium, an apparently featureless chemical element, has been studied intensively.

$$^{6}Li_{3} + {}^{1}n_{0} \rightarrow {}^{4}He_{2} + {}^{3}H_{1} \qquad (1)$$

$$^{3}H_{1} \rightarrow {}^{3}He_{2} + \beta\text{-particle} \qquad (2)$$

Both isotopes remain liquid to the absolute zero of temperature when cooled under their own vapor pressures, but the similarity ends there. The physical properties of the two isotopes are so dissimilar that they can be considered different materials. In fact, below 0.85 K, solutions of ^{3}He in ^{4}He will separate into two phases if the mole fraction of ^{3}He exceeds 0.65. At absolute zero, a dense phase of ^{4}He containing 6.6% ^{3}He will be in equilibrium with a light phase of pure ^{3}He. A powerful, widely used technique of refrigeration between 1 K and 10 mK is based on this phenomenon.

The important physical parameters of ^{3}He, along with those of ^{4}He for comparison, are presented in Table 2; the phase diagram for solid and liquid is shown in the illustration. The minimum in the solidification pressure, at 318 mK, implies that the entropy of liquid ^{3}He is less than that of the solid. This conclusion follows directly from the Clausius-Clapeyron equation (3), where dP/dT is the rate of change of pressure with temperature along the melting line, S_{liq} and S_{sol} are the entropies of liquid and solid along the melting line, and V_{liq} and V_{sol} are their respective volumes. As ($V_{liq} - V_{sol}$) is always a positive quantity, the quantity ($S_{liq} - S_{sol}$) must be negative for the solidification line to exhibit a negative slope. The entropy of the liquid is less than the entropy of the solid. This behavior has been used for refrigeration and is called Pomeranchuk cooling after the Russian physicist who predicted the phenomenon. The transitions from solid

$$\frac{dP}{dT} = \frac{S_{liq} - S_{sol}}{V_{liq} - V_{sol}} < 0 \qquad (3)$$

Table 2. Physical parameters for liquids ^{3}He and ^{4}He

Parameter	^{3}He	^{4}He
Critical temperature, K	3.32	5.20
Critical pressure, atm*	1.15	2.26
Boiling temperature, K	3.19	4.21
Minimum solidification pressure, atm*	29	25
Density at 1 K, g/cm^3	0.0816	0.1450

*1 atm = 101, 325 N/m^2.

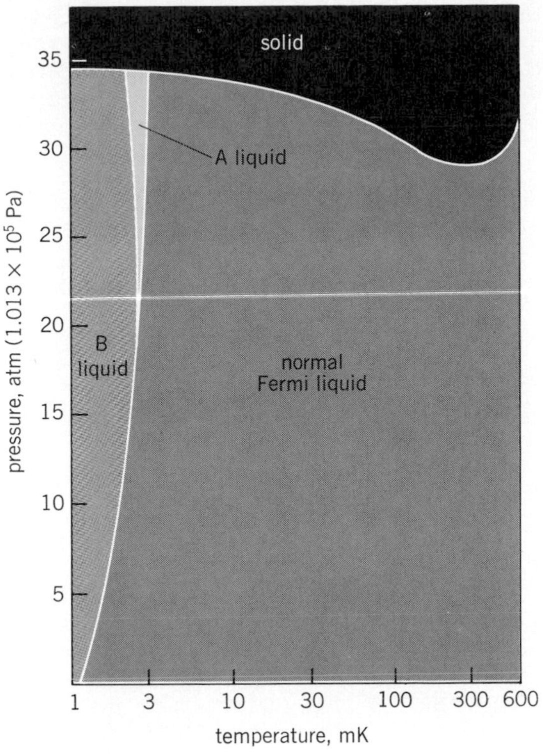

Phase diagram of ^{3}He.

to A and B liquids were first discovered in a Pomeranchuk refrigeration cell.

Below 100 mK, the normal Fermi liquid region is characterized by a specific heat linear in temperature, a viscosity proportional to the inverse square of the temperature, a temperature-independent nuclear susceptibility, a negative coefficient of thermal expansion, and the appearance of a unique mode of sound propagation termed zero sound. The damping of this mode increases as the square of the temperature and is frequency independent, in contrast to normal hydrodynamic sound, which is attenuated as the inverse square of the temperature and the square of the frequency.

The A and B phases, which make their appearance below 3 mK, are remarkable manifestations of quantum behavior on a macroscopic scale. Whereas superconducting electrons, which are also fermions, pair in a singlet state, spin $S=0$, according to the Bardeen-Cooper-Schrieffer theory, ^{3}He atoms are paired in a triplet state, $S=1$ ($S_z=0, \pm 1$, where S_z is the component of the spin along a magnetic field). The A phase is distinguished from the B phase by the absence of the $S_z=0$ state. If a preferred spatial direction is defined by a magnetic field, many properties of ^{3}He-A are anisotropic. For example, both the velocity and the attenuation of zero sound are strongly dependent on the angle between the field direction and the propagation direction. The B phase, on the other hand, is isotropic. Both phases can be described by a two-fluid model: a normal fluid of density ρ_n and superfluid of density ρ_s, wherein the density of the liquid $\rho=\rho_n+\rho_s$. The superfluid has a vanishing viscosity, so the total viscosity decreases as ρ_s increases.

The transition from normal Fermi liquid to the A phase is second order; that is, there is neither a latent heat nor a volume change. The transition from A → B is first-order, with the latent heat equal to 15.4 ergs/mol (1.54 microjoules/mol) at 2.2 mK.

[BERNARD M. ABRAHAM]

Bibliography: L. W. Alvarez and R. Cornog, *Phys. Rev.*, 56:379, 1939; K. H. Benneman and J. B. Ketterson (eds.), *The Physics of Liquid and Solid Helium*, pt. II, 1978; *Bureau of Mines' Mineral Yearbook*, annual; *The Energy Related Applications of Helium*, U.S. Energy Research and Development Administration Report, Apr. 11, 1975; D. D. Osherhoff, R. C. Richardson, and D. M. Lee, *Phys. Rev. Lett.*, 28:885–888, 1972; I. Pomeranchuk, *Zh. Eksp. Teor. Fiz.*, 20:919–924, 1950; J. Wilks, *The Properties of Liquid and Solid Helium*, 1967.

Heterocyclic compounds

Cyclic compounds in which the rings include at least one atom of an element different from the rest. Homocyclic compounds are cyclic compounds in which all the ring atoms are the same. In organic homocyclic compounds the annular atoms are all carbons. If the molecule contains carbon atoms, then it is organic (most types of heterocyclic compounds studied to date are organic compounds). An example of an organic heterocyclic compound is oxazoline (I), which is composed of five annular atoms, two of which are not carbon. An example of an inorganic heterocyclic compound is the phosphonitrilic chloride (II); the six-membered ring is composed of alternating atoms of nitrogen and phosphorus.

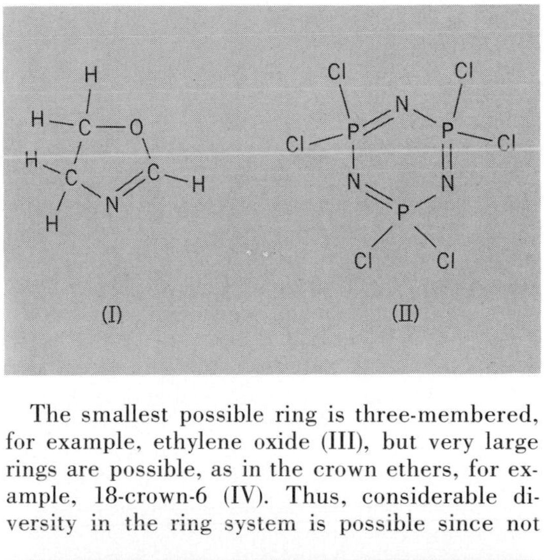

(I) (II)

The smallest possible ring is three-membered, for example, ethylene oxide (III), but very large rings are possible, as in the crown ethers, for example, 18-crown-6 (IV). Thus, considerable diversity in the ring system is possible since not

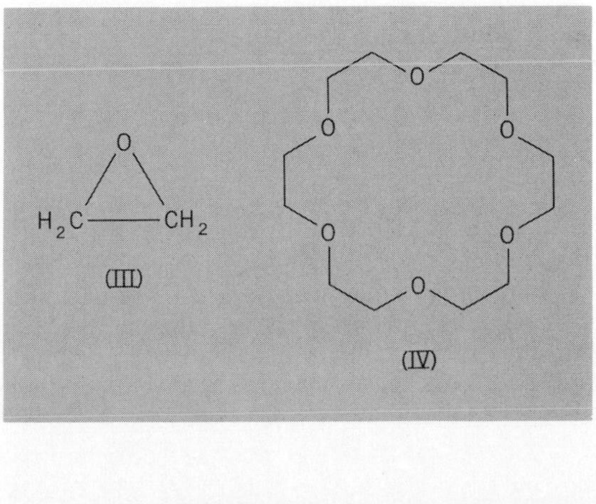

(III) (IV)

only may the size of the ring vary, but also the nature and number of heteroatoms in the ring can vary. The cycle may contain only single bonds and is thus saturated; it may include one or more double bonds; or it may possess aromatic unsaturation characteristics of benzene, that is, it is heteroaromatic. As with the homocyclic compounds, heterocyclic compounds can contain more than one ring, either heterocyclic or homocyclic.

The more familiar heteroelements in an organic compound are oxygen, nitrogen, and sulfur, but many elements have been or could be incorporated into a heterocyclic system. Compounds containing a ring formed by internal hydrogen bonding are not classed as heterocyclic. There is no limitation as to what kind of heteroatom may participate in ring formation, provided it has the appropriate bond angles and geometry.

Natural and synthetic compounds. Heterocyclic compounds are encountered in a very large number of groups of organic compounds. Inorganic heterocyclic compounds are also very common, but have not been studied systematically in terms of their heterocyclic characteristics in most cases. Naturally occurring heterocyclic compounds are extremely common as, for example, most alkaloids, sugars, vitamins, DNA and RNA, enzymic cofactors, plant pigments, many of the components of coal tar, many natural pigments (such as indigo, chlorophyll, hemoglobin, and the anthocyanins), antibiotics (such as penicillin and streptomycin), and some of the essential amino acids, (for example, tryptophan), and many of the peptides (such as oxytocin). Some of the most important naturally occurring high polymers are heterocyclic, including starch and cellulose. The major groups of natural products that are not mainly heterocyclic are the fats and most of the terpenes, steroids, and essential α-amino acids, though exceptions do exist.

Important synthetic dyes, such as phthalocyanins and phthaleins, as well as a large number of drugs, poisons, and medicinal products (both natural and synthetic) are heterocyclic. These include sulfothiazole, pyrethrin, strychnine, most of the antihistamines, the ergot alkaloids, morphine, the barbiturates, the tranquilizers Librium and Valium, and dihydrocannabinol (the main active constituent in marijuana). Many synthetic polymers (such as polyvinylpyridine, polyvinylpyrrolidone, and melamineformaldehyde), a number of industrial solvents (such as pyridine, dioxane, and tetrahydrofuran), and certain bulk chemicals (ethylene oxide, propylene oxide, and several aromatic nitrogen heterocycles obtained by synthesis from petroleum hydrocarbons or from coal tar) are heterocyclic.

Properties. The chemical properties of saturated heterocyclic compounds are usually very similar to those of appropriate open-chain analogs, provided account is taken of conformational differences. For example, the chemistry of *N*-methylpiperidine (V) is not very different from that of ethylmethylpropylamine (VI). Differences in properties originate in stereochemical factors. One such factor concerns four- and three-membered heterocycles in which the ring bonds are distorted with respect to bond angle and bond length, just as are those in cyclobutane and cyclopropane. These strained heterocyclic systems show enhanced activity in processes involving ring opening. Thus ethylene oxide (III), although formally an ether, is far more reactive than its open-chain analog, dimethyl ether (VII). On the other hand, some poly-

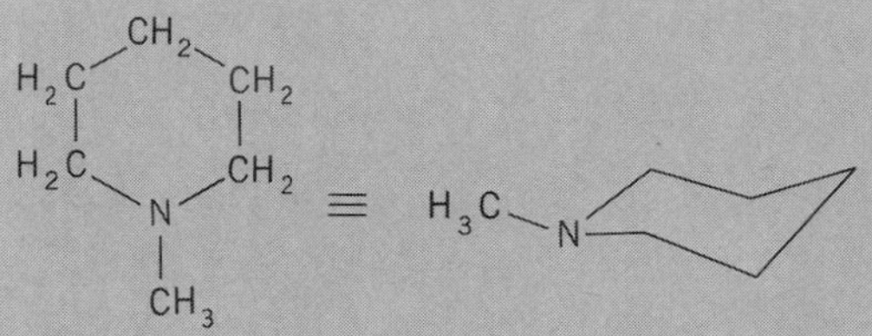

(V)

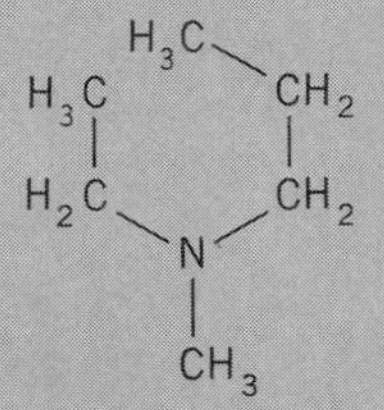

(VI)

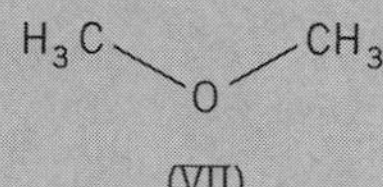

(VII)

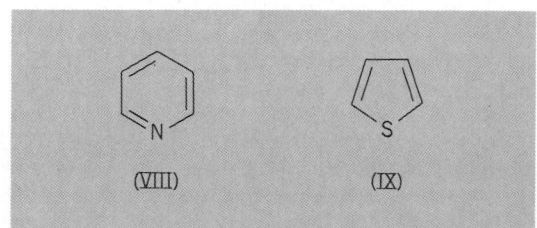

unsaturated heterocyclic systems exhibit chemical properties implied in the term aromatic. For example, pyridine (VIII), the parent of six-membered heteroaromatic nitrogen compounds, is a thermally stable material resisting oxidation, undergoing aromatic substitution (particularly nucleophilic substitution—the ring nitrogen atom makes electrophilic substitution difficult and the electrophile usually attacks the nonbonded pair of electrons on the ring nitrogen atom), and possessing appreciable resonance energy. Thiophene (IX) is a five-membered sulfur heterocycle, but it has aromatic

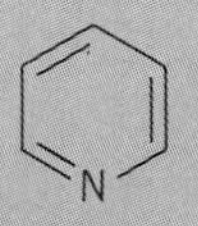

(VIII)

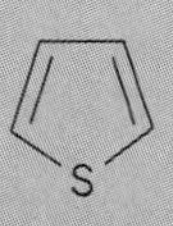

(IX)

properties similar to those of benzene. Such aromatic character may be expected in planar, unsaturated rings in which six electrons derived from ring unsaturation and from available electron pairs of the heteroatoms are delocalized over the whole ring system.

Nomenclature. Heterocyclic compounds may be named systematically. Many heterocycles, however, have nonsystematic names that are usually preferred by practicing chemists over the systematic ones. In the systematic approach to no-

Suffixes for degree of unsaturation

	Number of double bonds in ring							
	With nitrogen				Without nitrogen			
Ring size	Max.	2	1	0	Max.	2	1	0
3			irine	iridine			irine	irane
4		ete	etine	etidine		ete	etene	etane
5	ole		oline	olidine	ole		olene	olane
6	ine	*	*	*	in	*	*	inane or ane
7	epine	*	*	*	epin	*	*	epane

*Add prefix, such as dihydro and tetrahydro, to the name of the ring with maximum unsaturation.

menclature the ring size is denoted by the appropriate stem. For example, three-membered saturated rings without nitrogen would have a name ending in -irane. The nature of the heteroatom is denoted by such prefixes as oxa-, thia-, or aza-, for oxygen, sulfur, or nitrogen, respectively. Thus, ethylene oxide (III) becomes oxirane. A five-membered unsaturated ring would have a name ending in -ole. Thiophene (IX) thus becomes thiole. A six-membered unsaturated ring containing nitrogen would have a name ending in -ine according to this scheme. Pyridine (VIII) then becomes azine. Actually, the trivial names for these three systems are commonly accepted, and the systematic names are not often used. The degree of unsaturation is specified in the suffix as indicated in the table. It is important to note that the suffix is slightly modified when nitrogen is absent from the heterocyclic ring. The numbering of the ring begins with the heteroatom of highest priority and proceeds around the ring so as to give other heteroatoms or substituents the lowest number possible. The illustration shows

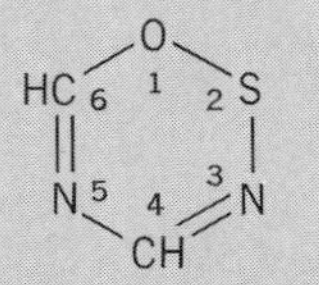

1,2,3,5-Oxathiadiazine

1,2-Oxazole, or preferably, isoxazole

(X)

Azete or azacyclobutadiene

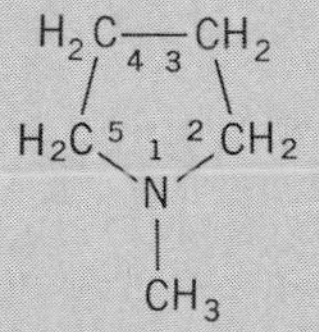

1-Methylazolidine, or preferably, 1-methylpyrrolidine

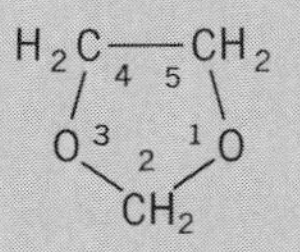

1,3-Dioxolane

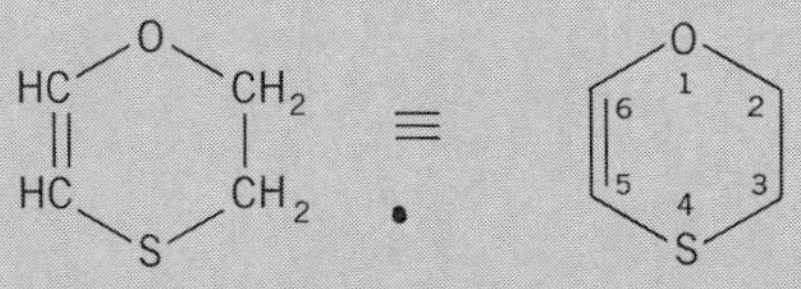

2,3-Dihydro-1,4-oxathine

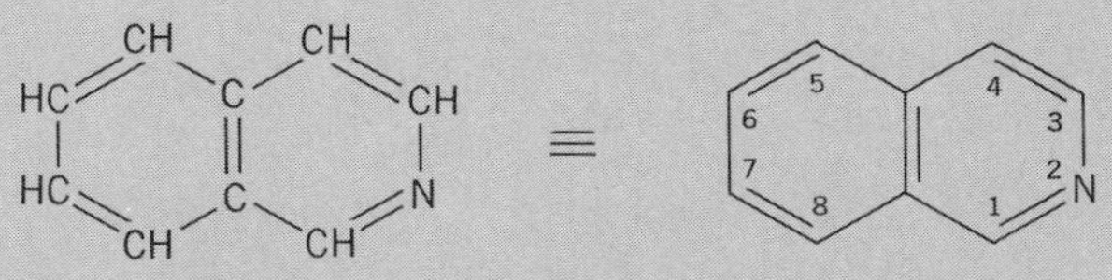

2-Azanaphthalene, or preferably, isoquinoline

Structural formulas of heterocyclic compounds. The ring positions are numbered.

how the systematic naming rules are applied and also how the ring atoms are numbered. An apex at which no specific atom is shown [for example, the equivalent structure (X) for isoxazole] represents =CH—.

For details about the specific heterocyclic systems *see* ACRIDINE; AZOLE; CARBAZOLE; FURAN; IMIDAZOLE; INDOLE; ISOQUINOLINE; OXAZOLE; PYRAN; PYRAZOLE; PYRIDINE; PYRROLE; QUINOLINE; THIAZOLE; THIOPHENE.

[R. A. ABRAMOVITCH]

Bibliography: R. M. Acheson, *An Introduction to the Chemistry of Heterocyclic Compounds*, 3d ed., 1976; *Aromatic and Heteroaromatic Chemistry*, vols. 1–4, *Specialist Periodical Reports*, Chemical Society, London 1973–1976; J. A. Joule and G. F. Smith, *Heterocyclic Chemistry*, 1972; A. R. Katritzky, *Advances in Heterocyclic Chemistry*, vols. 1–24, 1963–1980; A. R. Katritzky, *Physical Methods in Heterocyclic Chemistry*, vols. 1–2, 1963; M. H. Palmer, *The Structure and Reactions of Heterocyclic Compounds*, 1967; L. A. Paquette, *Principles of Modern Heterocyclic Chemistry*, 1968; A. M. Patterson and L. T. Capell, *The Ring Index*, ACS Monogr. no. 84, 1940; T. E. Peacock, *Electronic Properties of Aromatic and Heterocyclic Molecules*, 1965; Q. N. Porter and J. Baldas, *Mass Spectrometry of Heterocyclic Compounds*, 1971; E. H. Rodd, *Chemistry of Carbon Compounds*, vol. 44 and B, 1957; E. C. Taylor, *Principles of Heterocyclic Chemistry*, ACS Film Courses, 1974; A. Weissberger and E. C. Taylor, *The Chemistry of Heterocyclic Compounds*, a continuing series.

Heterocyclic polymer

Essentially, linear high polymers comprising heterocyclic rings, or groups of rings, linked together by one or more covalent bonds. As the search has continued for polymeric materials having useful properties at high temperatures (500°C or higher), much attention has been given to heterocyclic polymers. The possibility of forming rigid molecules that can be ordered into anisotropic arrays having exceptional stiffness may become of even greater interest. As a group such polymers are often both mechanically rigid and inherently resistant to thermal degradation.

Some of these polymers form molecules in which the rings are fused together, as shown symbolically in the illustration (ladder polymers), and some form molecules in which fused rings are joined by single bonds (stepladder polymers). Similar considerations hold for simple aromatic systems, for example, linear polymers of benzene, but the heterocyclic systems have, in general, been more useful in application.

HETEROCYCLIC POLYMER

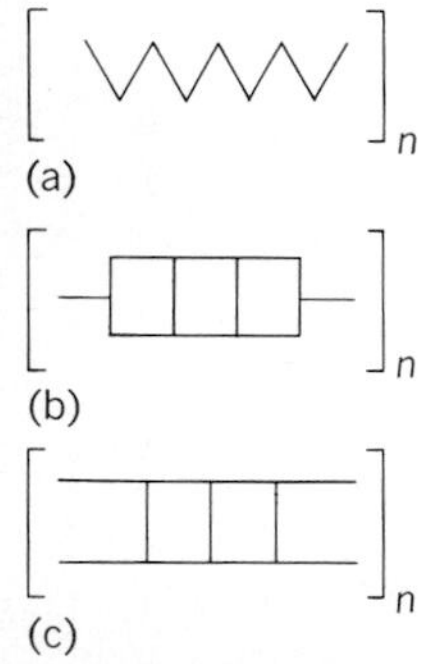

Structural units in linear polymers. (*a*) Simple linear polymer. (*b*) Stepladder polymer. (*c*) Ladder polymer.

Three heterocyclic polymers have been developed to the point of commercial availability: polyimides, polybenzimidazoles, and polybenzothiazoles. At least the first two appear to have established specialized markets. These three resins are discussed in this article, along with an amide-modified polymide. The development of additional members of the family is anticipated; information about many polymers of potential interest can be found in the bibliographical material.

In practice, the heterocyclic resins have rather high molecular weights and high glass temperatures. Some cross-linking may also be introduced during curing. Because of the insolubility and infusibility of the unmodified polymers, processing and fabrication was originally accomplished in stages. First, soluble prepolymers were prepared and fabricated into the final form desired (film, molding, coating, impregnated glass cloth, and so on). In this stage, the heterocyclic rings are not yet closed. Closure of the rings by condensation reactions was then effected by heating, and volatile by-products were eliminated. In contrast, poly(amide-imide) and the newer unmodified resins can be processed by more conventional techniques.

Major applications for these rather expensive polymers are as metal-to-metal adhesives and as laminating resins for fibrous composites for structural applications in the aerospace industry. Other applications requiring both strength and resistance to oxidation at elevated temperatures have developed, including valve seats, bearings, and turbine blades. *See* POLYMERIC COMPOSITE.

Polyimide resins. The basic synthetic reaction to form the prepolymer is the condensation of an aromatic dianhydride with an aromatic diamine. Thus, pyromellitic dianhydride may be added to 4,4′-diaminodiphenyl ether in an anhydrous medium to give a high-molecular-weight polyamic acid, as shown in reaction (1). Solvents such as

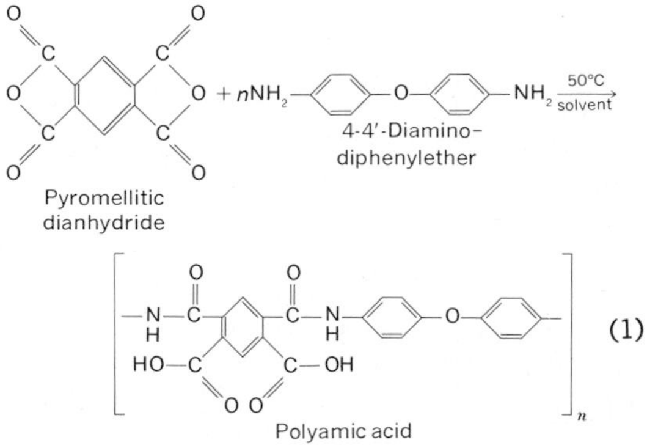

(1)

N,N-dimethylformamide are suitable media. The ingredients must be extremely pure, and must be present in equal molar amounts. Other dianhydrides and diamines may be used.

The prepolymer solution may then be used to impregnate fiber-glass cloth or other reinforcement, or applied to other substrates. Then, the solvent must be driven off and the rings closed by condensation, usually in stages, reaction (2). Fre-

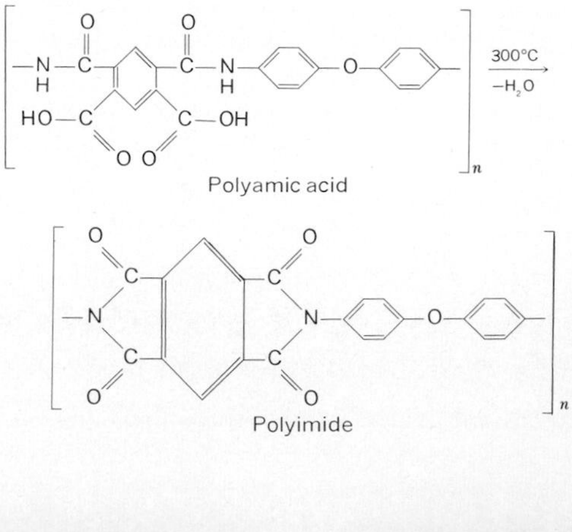

(2)

quently, most of the solvent is removed in a precuring stage and the major condensation effected later, at temperatures in the range 180–380°C. A subsequent postcuring step, also at elevated temperatures, may be used. The structure shown is of the stepladder type; ladder structures may be obtained by use of phenyl or fused-ring diamines.

Properties depend on the structures of the ingredients and on the reaction and curing conditions. Fiber-glass composites that are made using typical resins retain considerable strength after aging for 100 hr at 500–600°C. Adhesive bonds also show good resistance to aging at high temperatures. Certain polyimides are available as films and as molding powders, and some can be spun into fibers.

By introducing amide groups into a polyimide, some thermal stability is sacrificed, but improved processibility is gained. A typical structure is given in notation (3). Applications have developed in high-performance electrical connectors, engine parts, pumps, valves, and turbines.

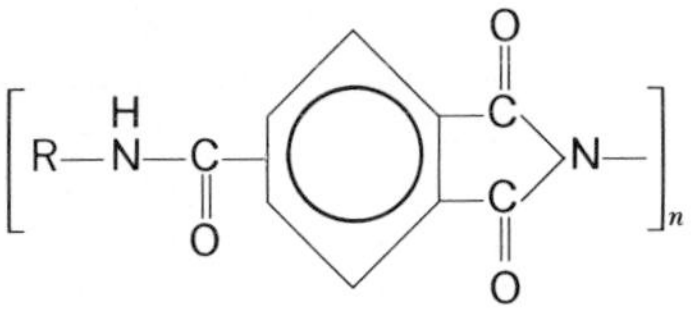
(3)

Polybenzimidazoles. The basic reaction for the synthesis of aromatic polybenzimidazoles is the reaction of an aromatic tetramine with an aromatic diacid or diester. Although the details of the intermediate steps are not completely understood, the overall reaction for a typical example, the condensation of 3,3′-diaminobenzidine and the diphenylester of isophthalic acid, is as shown in reaction (4).

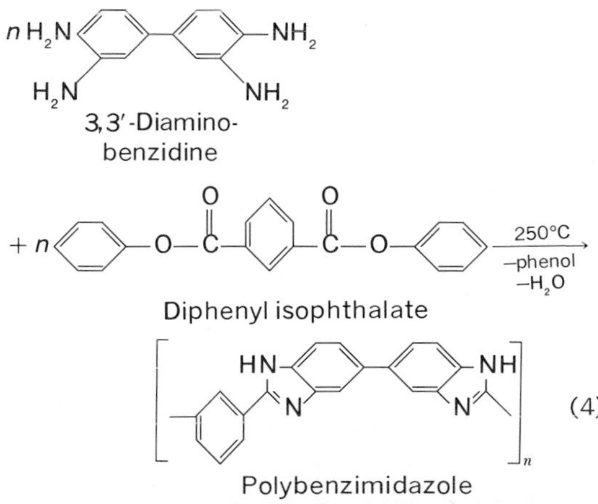

(4)

Polymerizations may be conducted in the melt, and to varying degrees of reaction. Partially polymerized resins may be dissolved in solvents such as *N,N′*-dimethylformamide and applied as a solution for laminating and adhesive applications. As with other heterocyclic polymers, final curing is effected at elevated temperatures (up to about 400°C) under pressure.

Properties depend on the structures of the ingredients, and on reaction and curing conditions. In general, polybenziomidazoles are somewhat less stable in air at high temperatures than the polyimides. Applications are mainly as laminating and adhesive resins for composites and metals.

Polybenzothiazoles. Polybenzothiazoles are typically prepared by the reaction of a dimercaptobenzidine with an aromatic diacid, diester, or diacyl chloride, as shown in reaction (5).

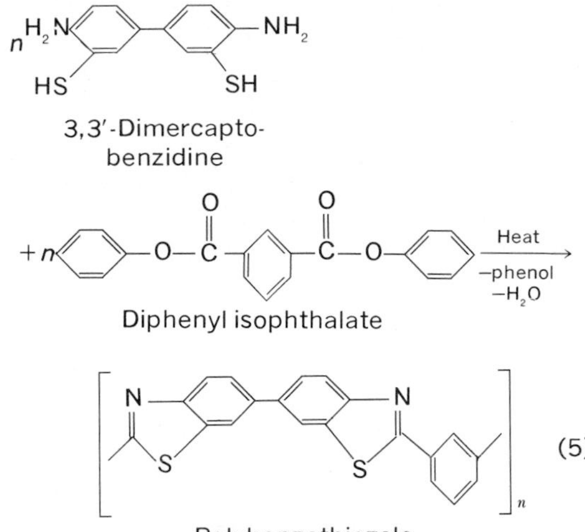

(5)

A mainly soluble prepolymer is prepared by carrying the reaction forward to only a limited extent. As with the other resins in this family, the solvent is removed and the reaction is completed by heating after application to the substrate desired. Polybenzothiazoles find the same applications as other heterocyclic polymers and are intermediate in stability between polyimides and polybenzimidazoles. *See* POLYESTER RESINS; POLYETHER RESINS; POLYMER; POLY-P-XYLYLENE RESINS; POLYSULFONE RESINS.

[JOHN A. MANSON]

Bibliography: H. Lee, D. Stoffey, and K. Neville, *New Linear Polymers*, 1967; *Modern Plastics Encyclopedia*, 1979–1980; G. Odian, *Principles of Polymerization*, 1970.

Heterogeneous catalysis

A chemical process in which the catalyst is present in a separate phase. In the usual case, the catalyst is a solid and the reactants and products are in gaseous or liquid phases. *See* CATALYSIS.

Heterogeneous catalysis proceeds by the formation and subsequent reaction of chemisorbed complexes which can be considered to be surface chemical compounds. In the very simple case where A → B is slow in the absence of catalyst, one might have reaction (1).

$$\left.\begin{array}{l} * + A \rightarrow *A \\ A* \rightarrow B* \\ B* \rightarrow B + * \end{array}\right\} \text{chain propagating steps} \qquad (1)$$
$$A \rightarrow B$$

Reaction A → B is fast if the three preceding steps are fast. Here, * represents a catalytic site on the surface of the catalyst, A* → B* is called a surface reaction, * + A → *A represents the chemisorption of A, and B* → B + * represents desorption of B. *See* ADSORPTION.

With most sets of reactants, more than one reaction will be thermodynamically possible. The degree to which a given catalyst favors one reaction compared with other possible reactions is called the selectivity of the catalyst for reaction (1). Two aspects of a catalyst are of particular importance: its selectivity, and its activity, which can be taken as the rate of conversion of reactants by a given amount of catalyst under specified conditions. Ideally, the rate will be proportional to the amount of catalyst.

History. Heterogeneous catalytic processes were discovered in the 1810s by Sir Humphry Davy and L. J. Thénard. Indeed, the present use of noble metal catalysts to oxidize the hydrocarbon vapors in automobile exhaust goes back to Davy's discovery of the oxidation of hydrocarbon vapors on platinum, and one could say that the Haber process for the synthesis of ammonia goes back to Thénard's observation that ammonia decomposed into its elements when passed over hot iron.

The first plant for the manufacture of sulfuric acid by the contact process, reactions (2), went

$$\begin{array}{c} 2SO_2 + O_2 \rightarrow 2SO_3 \\ SO_3 + H_2O \rightarrow H_2SO_4 \end{array} \quad (2)$$

on-stream in 1875. This was the first important heterogeneous catalytic process to be used in the chemical industry. For many years, the oxidation catalyst was platinum, but it was replaced by vanadium pentoxide (V_2O_5) on silica gel some years ago. However, major development of heterogeneous catalysis started only about 1900. It could be noted that the first industrial process for the manufacture of nitric acid, reaction (3), began in 1906, al-

$$NH_3 \xrightarrow{O_2} NO \xrightarrow{O_2} NO_2 \xrightarrow{H_2O,\ O_2} HNO_3 \quad (3)$$

though in 1838, C. F. Kuhlmann had discovered that platinum catalyzed the oxidation of ammonia to nitric oxide. In current usage, the catalyst is Pt-Rh gauze at 900°C.

By the 1950s, heterogeneous catalytic processes had come to dominate the petroleum, petrochemical, and chemical industries. As examples, in 1979 the United States production of sulfuric acid in reactions (2) was 40,000,000 tons (1 ton = 0.9 metric ton), and 15,000,000 tons of ammonia was synthesized by the Haber process, reaction (12) in the table. Of this ammonia, 4,000,000 tons was converted to nitric acid by reaction (3). About 70% of each of the 15,000,000 barrels (2.4×10^6 m^3) of crude oil refined each day in the United States is exposed to at least one heterogeneous catalytic process. Heterogeneous catalysis is a critical feature in energy conservation and interconversion, and is a key feature in the production of synthetic fuels from coal and oil shale. *See* FISCHER-TROPSCH PROCESS.

Reactors. Both industrially and in the laboratory, catalytic reactions are usually effected in one of three kinds of reactors: batch, flow, and gradientless. Examples are shown in the illustration. In the flow reactors, there is usually a large change in concentrations of products and reactants between the entrance and exit of the bed. In the gradientless reactor, the changes are kept very small, either by recirculation of 99% of the exit gases from the catalyst bed in *b* back to the entrance to the bed, or by use of a stirred-flow reactor, of which the fluidized reactor in *c* is one form.

Catalysts. A selection of heterogeneous catalytic processes of scientific or industrial interest is shown in the table.

Since catalytic activity will ordinarily be proportional to surface area, most catalysts are used in forms with large specific areas, a_s. The low-area catalyst of reaction (3) is a rather unusual case. Higher-area metal powders are often used for liquid-phase reactions in batch reactors like that shown in *a*. For example, finely divided nickel, $a_s = 25-40$ m^2 g^{-1}, is used for the hydrogenation of unsaturated glycerides in the manufacture of margarine from vegetable oils. *See* HYDROGENATION.

Supported catalysts are widely used. In these, the catalytic ingredient is dispersed in the internal porosity of such supports as silica gel, γ-alumina, and charcoals. These supports have large areas in the internal porosity, 100–1000 m^2 g^{-1}, and their average pore diameters are 2–20 nm. Tiny crystallites of such metals as platinum, palladium, rhodium, and nickel can be formed in the pore structure. Supported oxides and sulfides of transition metals are also used. Supported catalysts have the advantage that the area of the catalytic ingredient can be very large. Since, however, the support granules can have diameters in the 1–5 mm range, large flows of gas produce only moderate pressure drops across a catalyst bed. Supported catalysts are much more resistant to coalescence of the catalytic ingredient than are powders. Further, deposition of carbonaceous residues accompanies many catalytic reactions. In some cases, the catalyst can be regenerated by burning off the residues. Such regeneration would result in drastic losses of area with metal powders.

Supported catalysts are particularly prone to problems with diffusion. Reaction (1*a*) must be preceded by the diffusion of the reactant through the pore structure of the support to a catalytic site. Reaction (1*c*) must be followed by the diffusion of the product out of the support. Heat must also flow in and out of the granule of support. Such matters receive particular attention in the chemical engineering aspects of catalysis.

One important type of catalyst exposes strongly acidic sites in its internal porosity. Such catalysts are used to crack larger molecules of hydrocarbon into smaller ones in petroleum refining. Other cat-

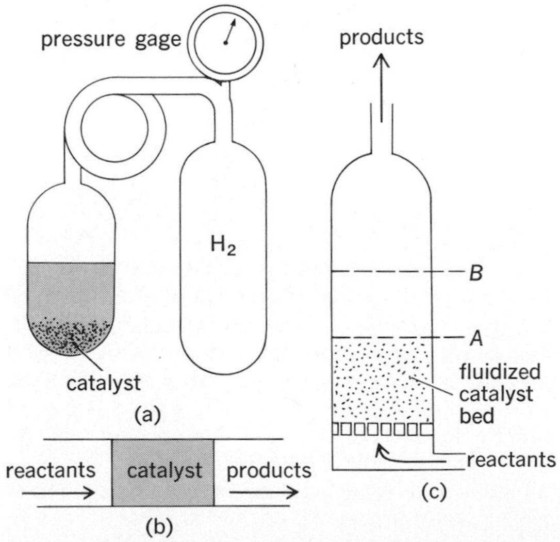

Catalytic reactors. (*a*) Batch. (*b*) Continuous fixed-bed. (*c*) Continuous fluidized reactor.

Some typical heterogeneous catalytic reactions

Catalyst	Reaction	T,°C*	Reaction number
	Hydrogenation		
Pt, Pd, Rh, Ni, as powders or supported on SiO_2, Al_2O_3, or C	$H_2+C_2H_4 \rightarrow C_2H_6$	−100	(4)
	$H_2+D_2 \rightarrow 2HD$	−180	(5)
	$C_3H_8+D_2 \rightarrow C_3H_7D$–$C_3D_8$+HD	50	(6)
	Cyclopropane+$H_2 \rightarrow C_3H_8$	40	(7)
	$C_2H_6+H_2 \rightarrow 2CH_4$	250	(8)
Cr_2O_3 activated to generate coordinatively unsaturated Cr^{3+}	$H_2+C_2H_4 \rightarrow C_2H_6$	−100	(9)
	$H_2+D_2 \rightarrow 2HD$	−180	(10)
	$D_2+C_3H_8 \rightarrow C_3H_7D$+HD	200	(11)
Fe	$3H_2+N_2 \rightarrow 2NH_3$	400†	(12)
Cu^+/ZnO	$2H_2+CO \rightarrow CH_3OH$	350†	(13)
Pt, Cr_2O_3	Heptane→toluene+$4H_2$	450	(14)
Pt, Cu	Acetone+H_2→2-propanol	75	(15)
	Polymerization		
Cr^{2+}/SiO_2	C_2H_4→linear polyethylene	50	(16)
	Olefin metathesis		
Mo^{4+}/Al_2O_3	$2C_3H_6 \rightarrow C_2H_4+CH_3CH{=}CHCH_3$	50	(17)
	Oxidation		
Pt, many oxides of transition metals	$2H_2+O_2 \rightarrow 2H_2O$	0–200	(18)
	$2CO+O_2 \rightarrow 2CO_2$	50–200	(19)
	$CH_4+2O_2 \rightarrow CO_2+2H_2O$	200–350	(20)
Ag	$2C_2H_4+O_2$→ethylene oxide	250	(21)
Bismuth molybdate	$C_3H_6+NH_3+3/2O_2 \rightarrow CH_2{=}CHCN+3H_2O$	450	(22)
V_2O_5/SiO_2	Naphthalene+O_2→ phthalic anhydride	350	(23)
Fe_3O_4	$H_2O+CO \rightarrow H_2+CO_2$	450	(24)

*The lowest temperature at which significant yields are obtained in a flow reactor.

†Because of the small value of the equilibrium constants at the operating temperatures, these reactions are run at about 100 atm (10 MPa) in order to get adequate conversions.

alysts, called dual-functional catalysts, have a hydrogenating catalytic ingredient on an acidic support. These are also of major importance in processing petroleum. *See* CATALYTIC REFORMING; CRACKING; HYDROCRACKING.

Another type of catalyst consists of an organometallic complex deposited on such supports as silica or alumina. These catalysts have been called heterogenized homogeneous catalysts, and they accompany a recent development in which the nature of surface sites on heterogeneous catalysts has been interpreted in terms of coordination chemistry and homogeneous catalysis.

Most catalysts exhibit coordinatively unsaturated surface sites *(cus)* which are capable of reacting with molecules in the gas or liquid phases to form chemisorbed intermediates, as in reaction (1*a*). For example, the atoms at the surface of a crystallite of platinum must be coordinatively unsaturated. There has been interest in applying the studies of surface chemistry and physics on particular crystal surfaces—(111), (110), or (100)— of such metals as platinum to the interpretation of catalytic reactions and chemisorption on metals.

High catalytic activity for a given reaction requires adsorption of the reactants at *cus* sites to be of intermediate strength. If the adsorption is too weak, the reactants are unlikely to be activated; if too strong, A* will be unreactive. With good catalysts, then, it is likely that there will be other compounds which adsorb more strongly than the reactants, with partial or complete blockage of the catalytic reaction. Such compounds are said to be poisons. For example, transition metals are poisoned by bases like R_2S, R_3P, and CO. Poisoning is common in all types of catalysis.

Mechanism. The kinetics of heterogeneous catalytic reactions are often rather complicated. In general, Rate = (amount of catalyst)$f(T,P_i,$ or $C_i)$, where T is the temperature of the catalyst, P_i is the partial pressure of component i, and C_i is the concentration of component i. The index i covers reactants and products. A reaction having a rate-limiting process, $A^* + B^* \rightarrow C^*$, will have a rate proportional to $\theta_A\theta_B$ where θ_A is the fraction of surface sites represented by * which have been converted to A*, and θ_B, that converted to B*. The rate in terms of the θ_i's (the fraction of the surface covered by A and B) can be related to the P_i's through adsorption isotherms for the components. *See* ADSORPTION.

Considerable work has been aimed at the determination of the chemical mechanism of particular catalytic reactions. Such work has depended upon kinetics, isotopic tracers, stereochemistry, and structure variation in reactants. A simple example is that which is commonly accepted for the hydrogenations of olefins, the Horiuti-Polanyi mechanism as illustrated for ethylene in reaction (4). It

$$H_2C{=}CH_2 + 2* \rightarrow H_2\overset{*}{C}{-}\overset{*}{C}H_2$$

or (4)

$$H_2C{=}CH_2 + * \rightarrow H_2C{=}CH_2 \downarrow *$$

is as yet unclear which formulation for the chemisorption of ethylene is most appropriate. The geometries of the two forms would be nearly the same. The adsorption of ethylene is accompanied by the dissociative chemisorption of hydrogen reaction (5), followed by reactions (6) and (7).

$$H_2 + 2* \rightarrow 2H* \qquad (5)$$

$$*H_2C{-}CH_2* + H* \rightarrow *CH_2CH_3 + 2* \qquad (6)$$

$$*CH_2CH_3 + H* \rightarrow CH_3CH_3(g) + 2* \qquad (7)$$

Much still remains to be learned about the details of catalytic mechanisms. *See* HOMOGENEOUS CATALYSIS. [ROBERT L. BURWELL, JR.]

Bibliography: M. Boudart and R. L. Burwell, Jr., in E. S. Lewis (ed.), *Techniques of Chemistry*, vol. 6, 1974; R. L. Burwell, Jr., Heterogeneous catalysis, *Surv. Progr. Chem.*, 8:1, 1977; C. N. Satterfield, *Heterogeneous Catalysis in Practice*, 1980; C. L. Thomas, *Catalytic Processes and Proven Catalysts*, 1970.

Hexamethylenetetramine

A white, odorless, water-soluble compound having a cage-structure of four 6-membered rings of three carbons and three nitrogens, $(CH_2)_6N_4$. The structural formula is shown below. Formaldehyde and

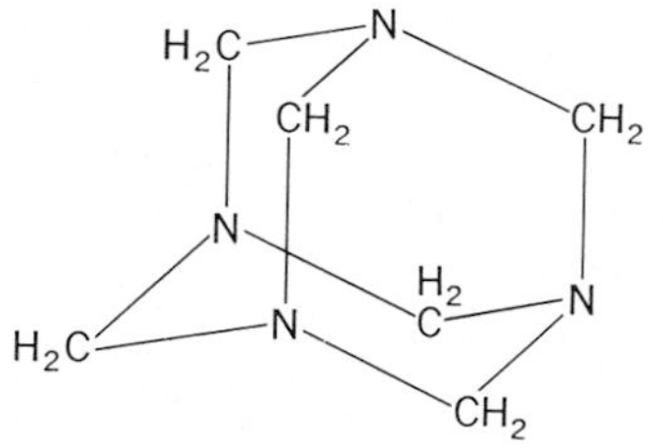

ammonia condense to hexamethylenetetramine and water. Slow reversibility of this reaction allows its use as a source of either ammonia or formaldehyde, the latter as disinfectant or urinary antiseptic. Nitration in the presence of ammonium nitrate and acetic anhydride gives very efficient conversion to trimethylenetrinitramine (cyclonite, hexogen, RDX), an explosive of higher brisance than trinitrotoluene (TNT). It was manufactured for use with TNT in large bombs during World War II. Alkyl halides slowly ammonolyze in acid to primary amines in the presence of the compound (Delepine reaction). *See* FORMALDEHYDE.

[LEALLYN B. CLAPP]

Hexaphenylethane

A colorless, crystalline hydrocarbon which melts with decomposition at 145–147°C. Moses Gomberg first prepared the hydrocarbon from triphenylchloromethane by the action of silver, zinc, or mercury metal in the absence of air.

If hexaphenylethane, protected from air or oxygen, is dissolved in benzene or carbon disulfide, it gives a yellow solution. Colorless hexaphenylethane may be obtained again by evaporation of the solvent. If a solution of hexaphenylethane is exposed to air, it becomes colorless, yielding a peroxide $(C_6H_5)_3COOC(C_6H_5)_3$, and if it is treated with iodine, it yields triphenylmethyl iodide.

Gomberg correctly concluded that the hydrocarbon (I) dissociates in solution yielding triphenylmethyl radicals (II). The extent of dissociation de-

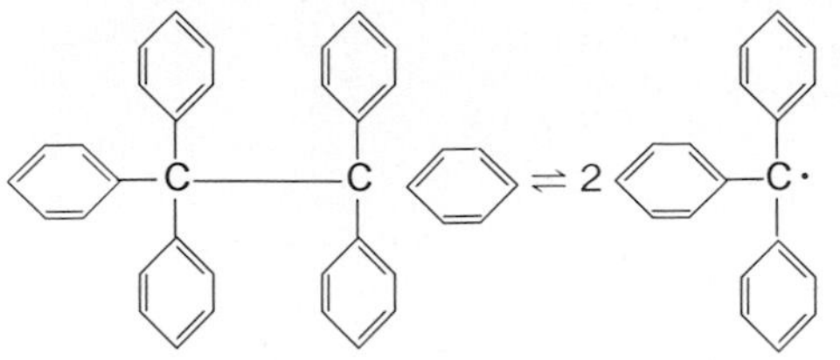

(I) (II)

pends upon concentration and is greatest at higher dilutions. Dissociation also increases with higher temperature. The degree of dissociation in solution is measured by boiling-point elevation, freezing-point lowering, or colorimetric methods, or by measurement of paramagnetic susceptibility. The last method is possible since free radicals have an unpaired electron and are attracted to a magnet; undissociated hexaphenylethane is not.

Other hexaarylethanes which are even more highly dissociated than hexaphenylethane have been synthesized. For example, hexa-(*p*-biphenylyl)-ethane is 100% dissociated in 2–3% benzene solution at 5°C, while under the same conditions hexaphenylethane is only 1–3% dissociated into radicals. *See* FREE RADICAL; TRIPHENYLMETHANE.

[CHARLES KILGO BRADSHER]

High-pressure chemistry

Chemistry at very high pressures, arbitrarily chosen to be above 10,000 bars, and mainly concerned with solid and liquid states. A bar is 10^6 dynes/cm^2, or 1.0197 kg/cm^2, or 0.9869 atm, or 10^5 newtons/m^2. Multiples of the bar are the kilobar (1 kbar = 10^3 bars) and the megabar (1 Mbar = 10^6 bars). At 25°C and 10 kbar, nearly all ordinary gases are liquid or solid, and only a few liquids are not frozen; thus most high-pressure chemistry involves either higher temperatures, at which chemical reactions can occur at appreciable rates, or studies of internal arrangements in solids.

Figure 1 illustrates the range of high pressures which exist in nature as well as those which can be attained in the laboratory. Three broad ranges of pressures are also shown. In the lowest range, from 1 bar to about 10^5 bars, normal low-pressure chemical behavior prevails, and only minor departures from the usual valence and coordination rules are found. However, as discussed later, many interesting changes in materials can be effected in this pressure range as atoms are forced into new bonding arrangements. In the second range, from 10^5 to 10^9 bars, the energy added by compression becomes comparable with chemical bond energies, so that outer-shell electronic orbits are distorted and atoms and molecules change in character. A general tendency toward more metallic behavior is observed as the electrons become less strongly fixed to particular atoms, and chemical bonds may be broken. In the third region, upward of about 10^9 bars, the delocalization of electrons is extensive, and the material consists of a mixture of ions and electrons, so that chemical bonds are of little importance. The boundaries on these three pressure ranges are, of course, only approximate, and show some variation according to the temperature and the atoms involved.

Equipment. High-pressure chemical phenomena can be studied in a wide variety of types of equipment, depending on the pressure and temperature range and the object of the study. The highest laboratory pressures, several megabars, are achieved, albeit but for a few microseconds, by the accelerative forces generated by high explosives or high-velocity impacts in the so-called shock-wave techniques. Static presures which may be exerted for minutes or hours have reached a maximum of about 400 kbars between diamond faces on specially supported anvils of the type shown in Fig. 2*a*, but the specimens are quite thin, with typical diameters of 1 mm, and the temperature range is limited. Larger specimens, 1 cm^3 or more in size, may be studied up to about 100 kbar and 2000°C or higher in cylindrical, tetrahedral, or cubical apparatus of the types that are indicated in Fig. 2*a*, *c*, and *d*.

There is no theoretical upper limit to the static pressures which may be achieved, but practical limits are imposed by the magnitude of the forces required, the materials of construction available, and the stress gradients in them. The changes occurring in the compressed material may be monitored in place, for example, by optical, x-ray, or electrical techniques, for in many cases the material reverts to its original state upon release of pressure. Pressure apparatus is conveniently calibrated by observing definite changes in certain substances, for example, resistance changes in bismuth at about 25 kbar or in lead of 130 kbar.

High-pressure effects. The simplest effect of high pressure is the closer compression of atoms. The noble gases and alkali metals are quite compressible (potassium shrinks to half its original volume under a pressure of 100 kbars), whereas most oxides and the stronger metals are considerably stiffer. However, at a pressure exceeding about 100 kbars, most of the easily compressed electronic clouds are tightened up, and the compressibilities of most substances approach each other. Usually, only minor amounts of energy compared with chemical bond energies can be added by compression to 100 kbars. Nevertheless, the atoms of the substance can thereby be forced much closer together than they would be by cooling to low temperatures. An immediate consequence is an increase in melting temperature with pressure for most substances, since nearly all solids, with the exception of unusual ones like ice and bismuth, expand when they melt, and the process of melting absorbs heat as the entropy of the substance increases. The thermodynamic relationship shown in Eq. (1) expresses the change in melting

$$\frac{dT}{dP}=\frac{T\Delta V}{\Delta H} \qquad (1)$$

temperature T with pressure P as a function of the volume increase ΔV and the heat absorbed in melting ΔH for a given quantity of the substance. Thus, pressures of 50–100 kbar increase melting points about 100°C for substances such as iron, whose ΔV of melting is small, to several hundred degrees for substances such as NaCl, whose ΔV of melting is large. At 1 bar, NaCl melts in an iron crucible, but at 100 kbar iron can be melted in an NaCl crucible.

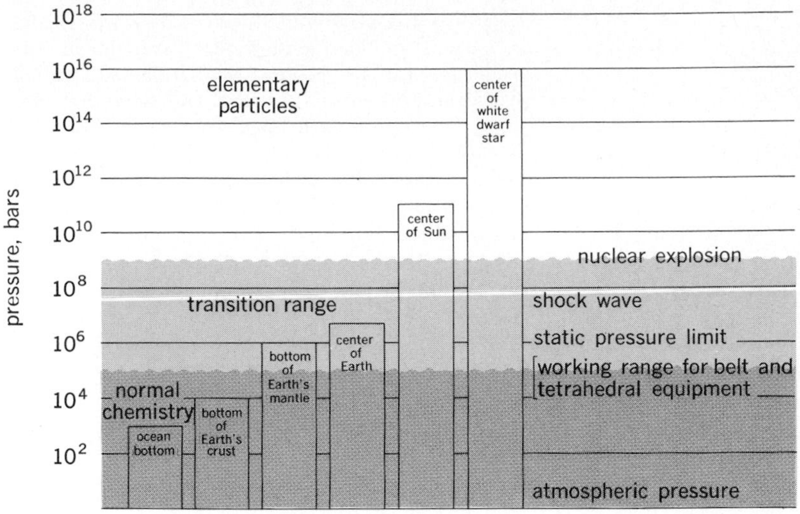

Fig. 1. Range of existing natural high pressures.

Substances which consist of large molecules are easily stiffened or frozen by high pressures. The mobility of the molecules is sharply decreased by a sort of interlocking and tangling effect; thus for the substance to be sheared, chemical bonds must be broken, a process which requires considerable energy. For example, ordinary oils become so stiff at a few dozen kilobars that they are useless for transmitting pressure, and a droplet of pressure-frozen oil is capable of denting a steel plate. This stiffening phenomenon limits the study of most reactions of organic molecules to low pressures because they are rather large and "freeze" easily, but yet are usually not stable enough to withstand the temperatures necessary for liquefaction or intermolecular reactions.

The thermodynamic relationship shown in Eq. (1) applies not only to melting but also to phase changes or internal structure changes in solids. In general, the higher-density forms are favored at high pressures, but since ΔH can be positive or negative, the effects of pressure and temperature may oppose or reinforce each other in determining the stability of a particular high-pressure phase. Phase changes between solids rarely run freely to follow the theoretical equilibrium line, but instead tend to be sluggish or exhibit a region of indifference. The stronger or more refractory the solid, or the greater the atomic displacements involved in the change, the broader the region of indifference. An outstanding example is diamond, which at room conditions persists at nearly 20 kbar out of its stability field. The existence of a region of indifference can hamper the study of some high-pressure phenomena, but on the other hand, it may permit the recovery of high-pressure phases for more detailed analysis at room conditions. The region of indifference may shrink drastically in the presence of solvents or catalysts which can promote the phase change. One of the more effective solvents, especially for systems containing oxides,

HIGH-PRESSURE CHEMISTRY

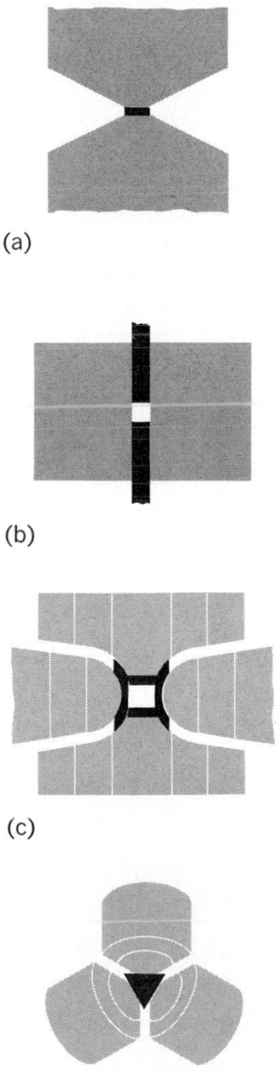

Fig. 2. Basic types of static high-pressure equipment. (*a*) Bridgman anvil. (*b*) Piston and cylinder. (c) Belt design. (*d*) Tetrahedral design.

is hot water. For chemically reducing systems such as carbon, the diamond-graphite transformations are assisted by the group VIII metals, iron, nickel, platinum, and so on. Other carbon solvents, such as AgCl or CdO, exist but do not permit diamond to form, apparently because they do not favor carbon cations. For transformations in the BN system, nitrides, either molten or in metallic solution, are effective catalysts. The rule "like dissolves like" is a useful, though not infallible, guide in these matters.

Some interesting chemical effects of pressure are related to shifts in chemical equilibria. The free-energy change ΔG determines chemical equilibria at temperature T through an expression of the form of Eq. (2), where R is the gas constant,

$$\Delta G = RT \ln (a_m a_n \cdots / a_x a_y \cdots) \quad (2)$$

and a_m, a_n, etc., are terms related to the concentrations of reactants and products in a reaction of the type $m + n + \cdots = x + y + \cdots$. For most reactions, ΔG ranges about 50 kcal (1 kcal = 4184 J) each side of zero. At 43 kbar, a change in volume of 1 cm^3 corresponds to a change in energy of 10 kcal, and a change of 1 cm^3 per gram mole of reacting substance is not large, so that it is easy to find large shifts in chemical equilibria produced by pressure. Phase changes can also be regarded as special kinds of intramolecular reactions whose equilibria are shifted by pressure when interatomic bonds are broken and reformed.

Some general rules exist regarding phase changes or electronic bonding changes in solids produced by pressure.

The first rule recognizes that open structures stabilized by relatively weak ionic or Van der Waals forces can easily collapse under pressure to denser structures. For example, eight crystalline forms of water ice have been identified. KCl collapses from the NaCl structure to the more closely packed CsCl structure at about 25 kbar. White phosphorus, made up of separate rings of four P atoms, is permanently pushed into the denser, more stable, semiconducting black form at pressures of a few tens of kilobars.

The second rule results from the compression of the outermost electronic shells and the delocalization of the electrons so that they are not as firmly fixed to particular atoms, and so that greater numbers of atoms can cluster around a given atom as the strongly directed valence forces are weakened. This is the same behavior observed as atomic number increases: Most of the lighter elements, with a few valence bonds per atom, are nonmetallic, whereas most of the heavier elements are metallic with higher valence numbers. Thus, pressure favors not only more metallic behavior but also behavior more like that of elements of higher atomic number. For example, upon compression to 100–130 kbar, silicon and germanium, normally semiconductors with the relatively open diamond structure, collapse to the white tin structure and become good metallic electrical conductors, equivalent to aluminum in that respect. As another example, silica, with four oxygen atoms about each silicon atom, changes at pressures of about 100 kbar and moderate temperature in the presence of water to stishovite, which has the rutile structure of TiO_2, with six oxygen atoms about each silicon. Many other examples of similar behavior in silicate systems are known wherein pressures of 100 kbar or so force common crustal minerals into crystalline forms embodying higher coordination numbers, similar to those found in oxide compounds of heavier atoms. These effects are important in considerations of the nature of the deeper layers of the Earth's crust.

The approach to the metallic state with increasing pressure may take different forms. Substances such as phosphorus, iodine, and selenium become substantially metallic at 100–150 kbar. On the other hand, when normally insulating organic compounds such as pentacene or hexacene, which consist of five or six aromatic rings fused togethers, are compressed to 150–200 kbar, they become semiconductors, since electrons are able to travel between the large molecules as the electronic clouds overlap. Certain complex compounds containing linear chains of metal atoms, such as Magnus's green salt, $Pt_2Cl_4(NH_4)_4$, increase in electrical conductivity up to 160 kbar and then show a decrease; evidently, conductivity is favored only in particular ranges of interatomic spacing. Mössbauer studies have shown that ferric ion is reversibly reduced to ferrous ion in many compounds at pressures in the 100–200 kbar range. Theoretical calculations indicate that even hydrogen could become metallic at pressures variously estimated to be 2–18 Mbars, but such pressures would be extremely difficult to generate and use, even with shock-wave techniques.

By changing the environment of atoms, high pressures can have strong effects on cooperative phenomena such as magnetism and superconductivity. For example, the high-pressure form of iron found above about 110 kbar at 25°C is not ferromagnetic. Superconducting transition temperatures are generally lowered by the application of high pressures.

Many chemical reactions proceed through an intermediate state whose volume may be larger or smaller than that of the reactants. When the volume is smaller, the rate of reaction may be increased by pressure. Usually the intermediate state is more voluminous, especially in solids, and the rate of reaction is reduced by pressure. Here, as with chemical equilibria, the energy change due to pressure in the 30–100 kbar range can be comparable with ordinary chemical bond or reaction-activation energies, and large pressure effects are possible. Certain reaction pathways may be effectively blocked so that new paths are followed. For example, at 130 kbar, where diamond is stable at temperatures up to 3000°C or more, the pyrolysis of some organic compounds, especially those consisting mainly of aromatic rings, produces graphite as the initial product whereas paraffin or polyethylene loses hydrogen to form waxy, dense solids of increasing microcrystalline diamond content as the pyrolysis temperature increases. A hexagonal form of diamond can be preapared by subjecting highly crystalline graphite to pressure of 130 kbar and 1500°C. The reaction proceeds to some extent at 25°C but proceeds much further on heating. (Poorly crystalline or partly amorphous graphite yields ordinary cubic diamond.) The hexagonal diamond can be recovered at room pressure and temperature if it has been heated at high pressure. Hexagonal graphitic BN can be converted to a wurtzite form by the application of 130 kbar at 30°C, but at high temperatures, or in the presence of a molten catalyst-solvent, the cubic form of BN, which is slightly more stable than the wurtzite form, is obtained. [ROBERT H. WENTORF, JR.]

Bibliography: R. S. Bradley (ed.), *Advances in High Pressure Research*, vols. 1–4, 1966–1974; H. G. Drickamer and C. W. Frank, *Electronic Transitions and the High Pressure Chemistry and Physics of Solids*, 1973.

High-pressure processes

Changes in the chemical or physical state of matter subjected to high pressure. The earliest high-pressure chemical process of commercial importance was the Haber synthesis of ammonia from hydrogen and nitrogen developed in Germany prior to World War I. The synthesis of diamonds from graphite developed in the early 1950s is a high-pressure physical process. Raising the pressure on a system may result in several kinds of change. It causes a gas or vapor to become a liquid, a liquid to become a solid, a solid to change from one molecular arrangement to another, and a gas to dissolve to a greater extent in a liquid or solid. These are physical changes. A chemical reaction under pressure may proceed in such a fashion that at equilibrium more of the product forms than at atmospheric pressure; it may also take place more rapidly under pressure; and it may proceed selectively, forming more of the desired product among multiple possible products.

Pressures higher than that of the atmosphere are expressed in bars and kilobars (kb) as well as in other units. A bar is 10^5 pascals (Pa), or 10^5 newtons per square meter (N/m^2), which are the units for pressure in the International System of units. These units are too small for convenient use in high-pressure processes, hence the bar is used. The bar is approximately 1 atm (1 bar = 0.9869 standard atmosphere, 760 mm Hg).

Physical processes. Increasing the pressure on a gas or vapor compresses it to a higher density and so to a smaller volume. If the pressure exceeds the vapor pressure, the vapor will condense to a liquid which occupies a still smaller volume. A vapor may be condensed at a higher temperature when it is under pressure; this permits the use of cooling water to remove the latent heat instead of more costly refrigeration.

Solids also change from a less dense phase to a more dense phase under the influence of increases in pressure. The density of diamond is about 1.6 times greater than that of graphite because of a change in the spatial arrangement of the carbon atoms. The temperatures and pressures used in the commercial synthesis of diamond range up to 3000 K and 100,000 atm. A molten metal is required as a catalyst to permit the atomic rearrangement to take place at economical rates of conversion. Metals such as tantalum, chromium, and iron form a film between graphite and diamond.

The highest static pressure attained in the laboratory is about 600,000 atm. At still higher pressures, the electrons are stripped from the atomic nuclei and matter loses its identity as recognizable atoms. This situation apparently exists in the centers of the white dwarf stars, where pressures of the order of 10^{16} atm prevail.

Chemical processes. In a manner similar to its effect during a physical change in which the volume of a system decreases, pressure also favors a chemical change where the volume of the products is less than the volume of the reactants. This is Le Chatelier's principle, which applies to systems in equilibrium. This general principle may be derived more precisely by thermodynamic reasoning, and thermodynamics is used to predict the effect of pressure on physical and chemical changes which lead to an equilibrium state.

Ammonia production. Ammonia is formed according to the reaction shown below. The ammo-

$$N_2 + 3H_2 \rightleftharpoons 2NH_3$$

nia content at equilibrium in a mixture which contains initially a ratio of 3 moles of hydrogen to 1 mole of nitrogen is shown in Fig. 1. At 1 atm only a fraction of 1% ammonia is formed. The ammonia content increases greatly when the pressure is raised. At 100 atm and 200°C there would be about 80% ammonia at equilibrium. However, a very long time is required to form ammonia under these conditions, and consequently commercial processes operate at higher temperatures and pressures and use a catalyst to obtain higher rates of reaction. Many combinations of pressure and temperature have been used. The largest number of plants now operate in the region of 300 atm and 450–500°C. A higher pressure process is carried out at about 1000 atm and 500–650°C.

A catalyst must be used or the reaction is too slow. Thus the process does not operate at 100 atm and 200°C in spite of the favorable conversion at equilibrium because no catalyst has been found to provide an adequate rate of reaction under these conditions. Iron is the catalyst used at higher temperatures and pressures. The physical form of the catalyst is very important. Iron shavings or lumps are not effective. A suitable catalyst has been made from a fused mixture containing 66% Fe_2O_3, 31% FeO, 1.0% K_2O, and 1.8% Al_2O_3. The last two components are promoters and significantly increase the activity of the catalyst, but they alone are not catalysts. The mixture must not be contaminated with other impurities or the catalyst may be

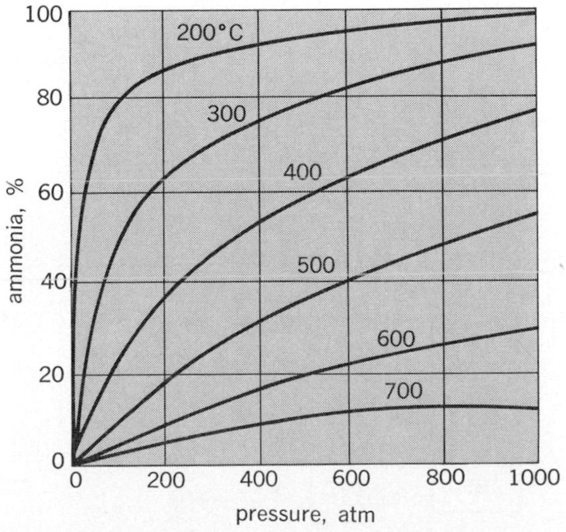

Fig. 1. Effect of combined temperature and pressure on the formation of ammonia. (*From E. W. Comings, High Pressure Technology, McGraw-Hill, 1956*)

poisoned. Sulfur, phosphorus, and arsenic poison the catalyst permanently, whereas gases such as oxygen, water vapor, and carbon oxides reduce the catalyst's activity only while they are in contact with it and are temporary poisons. Before the catalyst is used, the iron oxides are first reduced to iron by the hydrogen of the reacting mixture. The catalyst is activated by this treatment and must not be exposed to gases, which may contain poisons. Its life may be from several months to several years. Increasing the pressure increases the rate of reaction of the mixture in contact with a suitable catalyst.

A flow sheet for an ammonia process operating at 1000 atm is shown in Fig. 2. Nitrogen is obtained from the air, and hydrogen comes from natural gas. The process involves preparation of the nitrogen-hydrogen mixture, purification to exclude catalyst poisons, compression to high pressure, and circulation of the synthesis gas through a closed loop which contains two synthesis converters charged with catalyst. Because of the incomplete conversion in one converter, the unreacted gas is passed on to the next converter after ammonia has been condensed from the partially reacted gas mixture.

The temperature of the reaction is carefully controlled. Higher temperatures shorten the life of the catalyst and reduce the degree of conversion at equilibrium, whereas lower temperatures reduce the rate of reaction. The reaction mixture entering the converter is relatively cool. It is heated to the reaction temperature inside the converter by heat exchange with the reacting gas and with the gas that has passed through the catalyst. The reaction is exothermic, and the converter is designed to remove some of the heat of reaction while the gas is passing through the catalyst.

The converters are large pressure vessels containing a heat exchanger and a catalyst basket. The converter for the 1000-atm process contains 16.5 ft^3 of catalyst and produces 250–400 lb of ammonia per hour in each cubic foot of catalyst. This is comparable to 144 ft^3 of catalyst in a converter for a 300-atm process, which produces 50 lb of ammonia per hour per cubic foot of catalyst. Molecular sieves (refractory compounds with controlled porosity) are coming into use as a base for catalysts to reduce the volume and increase the life of the catalysts. The high reaction temperature is confined to the catalyst basket inside the vessel, and the thick walls of the vessel are held at a lower temperature.

Methanol production. Methanol is synthesized from hydrogen and carbon monoxide at 200 atm and 600°F in a similar manner. The catalyst contains aluminum oxide, zinc oxide, chromium oxide, and copper. Higher alcohols are produced at pressures of 200–1000 atm and temperatures up to 1000°F with a similar catalyst to which potassium carbonate or chromate has been added.

Polyethylene production. Polyethylene has been produced at pressures in the ranges 3–4, 20–30, 40–60, and 1000–3000 atm. This latter is probably the highest pressure yet used in the commercial synthesis of an organic chemical product. The ethylene is polymerized in a stainless steel tubular reactor at 375°F with small amounts of oxygen as a catalyst.

Phenol production. Phenol can be formed from chlorobenzene mixed with 18% sodium hydroxide solution at a pressure of 330 atm. Pressure is employed in this instance to maintain the mixture in the liquid phase at a temperature high enough for the hydrolysis reaction to proceed at an acceptable rate. A tubular reactor installed in a furnace provides for heating the mixture to 680°F.

Hydrogenation. Hydrocracking and hydrodesulfurization in the refining of gasoline and fuel oils are carried out at pressures up to 200 atm (3000 lb/in.^{2}) and temperatures of 800°F and higher. The large volumes processed require very large reaction vessels, 8 ft in diameter by 80 ft long with 8-in.-thick walls, for example.

Other chemical reactions are carried out at elevated pressure. A reaction which has been observed to take place in the laboratory under high pressure may later be found to take place at lower pressure with a suitable catalyst.

Apparatus. High-pressure apparatus is carefully designed to provide for the measurement of pressure and temperature, a vessel with sufficient strength, a closure to prevent leakage, easy access to the interior, for the compression of gases

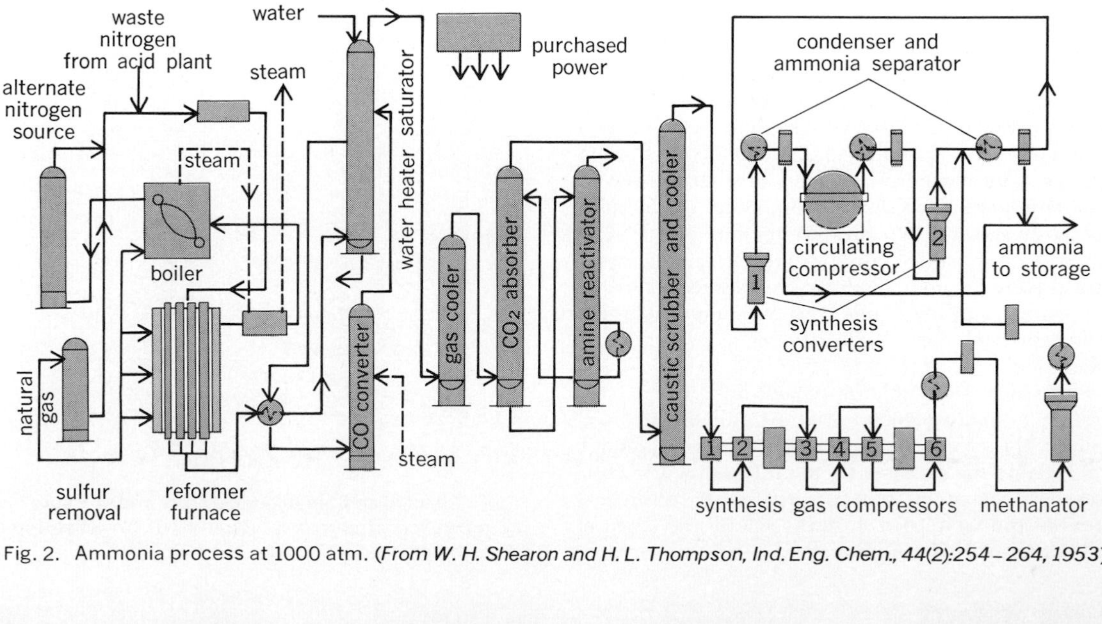

Fig. 2. Ammonia process at 1000 atm. (*From W. H. Shearon and H. L. Thompson, Ind. Eng. Chem., 44(2):254–264, 1953*)

and liquids at high pressures, and for the safety of people and equipment in case of ruptures or explosions. The deadweight gage is the principal primary gage for calibrating other pressure-measuring instruments. The deadweight gage consists of a free piston balanced between the force of gravity acting on weights at one end and the pressure of oil contained in the system at the other end.

P. W. Bridgman received the Nobel Prize in physics in 1946 for his pioneering work in the physics of high pressure. He was the originator of a self-sealing closure based on the unsupported area principle. This principle is illustrated in Fig. 3. The sealing gasket is initially under a low or moderate pressure. As pressure rises in the vessel, the pressure in the gasket is automatically maintained at a level higher than in the vessel. This ensures against leakage of the contents. Pressure, contained in the vessel *a*, acts upward on the steel disk *c*, which in turn acts on the gasket *e*. The gasket is supported by the steel ring *d*, which has a smaller area than *c* because of the unsupported area in the center. A balance of forces on the gasket requires that the forces acting downward balance those acting upward. The upward force is the product of the pressure in the vessel and the area of the lower face of *c*. This force is balanced by an equal force downward, which is the product of the pressure in the gasket and the area of the gasket. Because the area of the gasket is smaller than the area of the lower face of *c*, the pressure in the gasket must always be higher than the pressure in the vessel. *See* CHEMICAL EQUILIBRIUM; PHASE EQUILIBRIUM.

[EDWARD W. COMINGS]

Bibliography: R. S. Bradley, *High Pressure Physics and Chemistry*, 2 vols., 1963; P. W. Bridgman, *The Physics of High Pressure*, reprint 1949; J. F. Harvey, *Theory and Design of Pressure Vessels*, 2d ed., 1974; W. R. D. Manning and S. Labrow, *High Pressure Engineering*, 1972; I. L. Spain and J. Paauwe, *High Pressure Technology*: *Applications and Processes*, vol. 2, 1977.

HIGH-PRESSURE PROCESSES

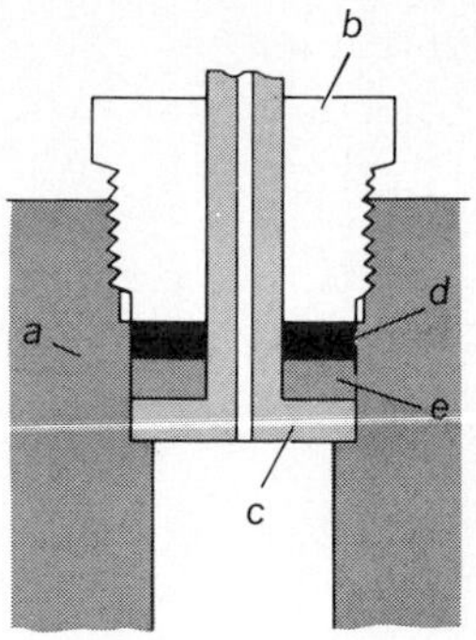

Fig. 3. Bridgman unsupported area gasket. (*From E. W. Comings, High Pressure Technology, McGraw-Hill, 1956*)

High-temperature chemistry

The study of chemical phenomena occurring above 500 K. High temperatures represent one of the important variables available to scientists for increasing the variety of possible chemical reactions over that expected for classical ground-state atoms and molecules. One can enhance the relative population of excited rotational, vibrational, and electronic states by increasing the temperature and thus effectively create new species and new mechanisms for reaction. The potentialities of this approach are well illustrated by the three laws of high-temperature chemistry: (1) At high temperatures everything reacts with everything. (2) The higher the temperature, the faster the reaction. (3) The products may be anything. With an infinity of species available at high temperatures, the "golden age" of chemical synthesis is probably still in the future.

High temperatures also provide a common tie among the various options for energy production, conversion, or storage. For maximum thermodynamic efficiency, an energy production cycle should operate with a working fluid at as high a temperature as possible, and exhaust the spent fluid at as low a temperature as possible. Thus, in the combustion of coal to produce electric power or in the combustion of gasoline or diesel fuel to propel a car or an airplane, there is a need for materials of construction which allow operation of such devices at higher temperatures. In the evaluation of new fuels or propellants, higher flame temperatures are among the desirable properties often sought.

It is convenient to discuss temperatures in terms of energy and to note that 11,500 K corresponds to one electron volt. In this sense, the particles emitted by radioactive nuclei or accelerated in cyclotrons and synchrotrons, which have energies in the keV, MeV, and BeV ranges, are effectively at temperatures of $\sim 10^7$ K, $\sim 10^{10}$ K, and $\sim 10^{13}$ K, respectively, and "high-energy physics" is synonymous with "ultrahigh temperature chemistry".

Traditional high-temperature chemistry of the past 25 years has been mainly concerned with phenomena in the range of 500–3000 K, although exotic flames can produce temperatures up to ~ 6000 K, shock waves can generate temperatures up to $\sim 25{,}000$ K, electric arcs can be operated in constricted modes to produce temperatures of $\sim 50{,}000$ K, and nuclear processes begin to occur at temperatures in the millions of degrees range. Laser-excitation of selected energy states can produce species with effective temperatures in the range of 10^8 K. The major goals of high-temperature scientists include (1) the characterization of all important gaseous molecules, ions, and condensed phases—molecular formulas and structures, energy levels, thermodynamic properties, and the details of chemical bonding; (2) the establishment of reaction rate parameters and correlations with molecular properties; (3) the development of unique approaches to chemical syntheses and the preparation of new materials; and (4) the development of new techniques for generating or containing or utilizing high-temperature fluids in connection with energy production, conversion, and storage.

Gaseous species. Studies of thermal decomposition, vaporization, and/or sublimation of inorganic compounds have shown that complex gaseous species are much more common than usually realized. For example, condensed alkali metal sulfates vaporize appreciably as M_2SO_4 (gas); alkali metal carbonates vaporize as M_2CO_3 (gas); and even NH_4X(solid) yields NH_4X(gas). At high temperatures the relative fraction of complex species in the vapor actually increases with the increasing temperature for many systems. Vapors at high pressures and high temperatures can thus be complex. Rapid mass spectrometric sampling techniques have demonstrated unequivocally the existence of polymers such as $(Ar)_m$, $(CO_2)_n$, and $(H_2O)_q$ where m, n, and q are 2,3,4, . . . , 8,9, 10, ... As a further complication, various oxidation states can exist in high-temperature systems and thereby a still greater variety of possible molecular species. Typical systems which have been characterized by a combination of mass spectrometric, optical, infrared, nuclear magnetic resonance, and electron spin resonance spectroscopy and chemical studies are the difluorides of group IV (CF_2, SiF_2, GeF_2, SnF_2, PbF_2) and their polymers.

In particular, the technique of generating high-temperature gaseous species and then reacting them with low-temperature molecules on a surface which is maintained at low temperatures (~−190°C) has been very productive. P. L. Timms, P. S. Skell, J. L. Margrave, and others have developed practical approaches to a great variety of organometallic syntheses through the use of molecules such as SiF_2, C_2, C_3, BF, SiO, and SiS, and various atoms—such as C, Si, Li, Mg, Fe, Cu, Ni, Pd, and Pt. Matrix-isolation spectroscopy at 4–100 K provides the opportunity for gaining detailed information about atomic/molecular parameters—energy levels, frequencies, bond angles—which are necessary for calculation of thermodynamic functions and the prediction of chemical reactions.

Condensed systems. Condensed systems at high temperatures provide additional versatility in the chemical world since one fails to observe almost uniquely in high-temperature systems that atoms do combine in the ratio of small whole numbers. Such observations on gases led to the formulation of the law of definite proportions and other basic rules of early chemistry, but these rules no longer hold exactly for condensed systems at temperatures in the 1000–3000 K range. Experiments have established melting points of refractory solids (binary carbides, borides, silicides, and so forth) at 1 atm pressure in the 2000–4000 K range, and unless one goes to higher pressures than 1 atm there really is no sold-state chemistry at temperatures greater than ~4000 K. Obviously, there is a chemistry of liquid systems which might be of some technological importance, and ingenious techniques for generating and maintaining reactive high-temperature liquids have been described by A. V. Grosse and associates.

Phase diagrams have demonstrated conclusively that one can prepare crystalline solids in an almost infinite number of compositions. For example, there are "pure" compounds fitting classical valences, and then one can substitute, as in $MgAl_2O_4$, where Al^{3+} ions can occupy some of the Mg^{2+} sites or Mg^{2+} ions can occupy some of the Al^{3+} sites so that there is an almost continual variation in properties. One can prepare Ta-O-C phases in which the gross stoichiometry is $Ta_1O_xC_{1-x}$ and, further, find it extremely difficult to detect the fact that oxygen is even present since the sizes of the oxygen and carbon atoms are so nearly the same in this lattice. Discovery of oxynitride phases and the establishment of essentially continuous solid solutions over wide ranges of compositions for many oxide, sulfide, carbide, boride, and other systems have been reported. The variations in compositions of alloys are similarly extremely complex; the variety possible is only beginning to be appreciated. Techniques for presenting phase diagrams have been described in which an attempt is made to present a single diagram which summarizes the behavior of whole classes of alloy systems rather than that of individual binary or ternary combinations. *See* NONSTOICHIOMETRIC COMPOUNDS.

Properties of high-temperature solids are, of course, not reliably interpreted until the exact stoichiometric and structural data have been obtained. The current status of high-temperature thermodynamic properties is acceptably characterized by saying that no material is described thermodynamically to better than ±0.5% up to 1500 K, or better than to ±1–2% from 1500–2000 K. There are practically no reliable data at temperatures greater than 2000 K except for a handful of basic materials—tungsten, molybdenum, tantalum, and aluminum oxide.

There is a great need for experimental techniques by which thermodynamic measurements can be reliably extended into this high-temperature region. There is already experimental evidence which is not explainable by current theoretical viewpoints on solids and liquids at high temperatures. Vaporization, sublimation, and other heterogeneous equilibrium studies have been made for many systems, and these are, of course, sensitive to the exact nature of the surfaces, to variations in stoichiometry as gases are evolved from a condensed phase, and to other factors which may accompany deviations from equilibrium.

Electrochemical techniques. One of the unique ways in which high-temperature thermodynamic data have been obtained involves the use of electrochemical techniques of the same sort as those which were widely applied during the 1920–1940 period for the establishment of reliable thermodynamic reference data for aqueous systems and for simple solids at temperatures near 25°C. Carl Wagner and associates have devised various ingenious approaches to high-temperature electrochemistry. From their data high-precision free energies of formation for nonstoichiometric phases can be derived over fairly wide ranges of temperature. However, eventually at 1500 K and higher, one begins to have difficulty with vaporization, melting, and high diffusion rates.

Calorimetry. Calorimetry at temperatures up to 3000 K has become routine through the use of electron bombardment heating and levitation heating. In electron bombardment heating, one boils electrons out of a thermionic emitter, such as thoriated tungsten, and accelerates them across a potential drop of a few thousand volts to strike the conducting sample and raise its temperature by means of the kinetic energy transferred and also by the heating caused by the electric resistance of the material. Levitation heating of conducting samples is accomplished with standard radio-frequency induction heaters and a pair of oppositely wound (left-handed and right-handed) coils. Samples weighing up to 1000 g can be levitated, melted, and cast in a containerless, controlled atmosphere. Copper, gold, platinum, tantalum, graphite, and many other materials have been levitated and, in all cases except graphite, raised to temperatures above their melting points (3000 K and higher). Levitation calorimetry has provided heats of fusion and heat capacities for many metals, alloys, and conducting compounds in both solid and liquid states.

Chemical kinetics. From the viewpoint of heterogeneous kinetics, the high-temperature area has been studied extensively, with the practical concerns in the literature involving (1) the rates of interaction between various corrosive gases (oxygen, nitrogen, sulfides, halogens, and so on) and surfaces of various pure metals, alloys, and ceramic structural materials, and (2) the catalytic effects of solids of various stoichiometries on various gas reactions as in hydrocarbon refining, in the preparation of SO_3 or NH_3, or in the catalytic conversion of $CO + H_2$ to methanol, of CO to CO_2, or of NO_x to N_2 and O_2.

Homogeneous gas kinetics in the region above 1000 K has been mainly concerned with reactions of neutral atomic and molecular species which were stable at high temperatures, as in typical flames. The roles of H, O, OH, and of intermediate combustion products like HO_2, MO, and CHO have been established by direct mass-spectrometer probing of flames. The importance of electrons and atomic or molecular ions in flame kinetics has also been recognized. This knowledge is of interest in identifying the most efficient combustion systems for energy generation and in elucidating the chemical parameters which are crucial to the development of flame-retardant materials.

Another fertile area for high-temperature kinetics research has been the study of the reaction rates of atoms or molecules which have been produced by either photochemical, laser-pulse, or electric-arc excitation or by pulsed thermal dissociation processes and then allowed to react either in systems at relatively higher pressures by random collisions or in systems at low pressures by molecular-beam techniques. Thus, a hydrogen atom created from the H_2 molecule in an arc or by thermal dissociation, requires 2.24 eV/atom, and therefore is a chemical species of the sort that might be expected in a very high temperature system. The thermodynamic potential of such atoms for reaction is much greater than that for most molecular species, and reaction rates are usually measurable although relatively fast. Selective excitation to specific energy levels is possible with the tunable dye lasers now available.

Several types of monitoring techniques, rapid-scan infrared spectroscopy, electron spin resonance spectroscopy, and mass spectrometry, are typically used in these studies. Gas reaction rates in mass spectrometers also are being widely explored by scientists interested in ion-molecule reactions. Of course, since ions are created endothermally by the impartation of several electron volts of energy, they are high-temperature species. Many new types of high-temperature reactants can be created. It is possible that a synthetic chemistry making use of ion-molecule reactions may someday be of economic significance. Ion sputtering is already a widely used technique for the preparation of electronic circuitry.

Summation. Techniques for generation of temperatures to ~10^8 K and for the measurement of molecular parameters, thermodynamic properties, and kinetic parameters have been developed and used in high-temperature chemistry. Since this field concerns all of the elements of the periodic table and all of the various types of measurements and observations that can be made, almost every kind of experimental and theoretical work is being done in high-temperature chemistry. The factor binding the field together is simply the condition of high temperature, or high energy, as it might be called. The range of high-temperature chemistry makes it attractive for those whose interests in science are broad and whose imagination operates in the areas of physics, engineering, and chemistry. The potentialities of this field in the technology of the future seem almost unpredictable.

In the last 200 years scientists have done almost all of their work within a fairly narrow temperature range—only a few hundred degrees Celsius. When one considers the elements of the periodic table with their infinite number of excited electronic states and all of the possible ways in which they can combine or interact, it is clear that no textbook has really begun to describe chemistry at high temperatures. Almost any combination of atoms is stable at some temperature and pressure, and the properties of each member of this infinity of chemically complex systems cannot be predicted by currently available theories.

[JOHN L. MARGRAVE]

Bibliography: L. Eyring (ed.), *Advances in High Temperature Chemistry*, vols. 1–4, 1967–1975; J. W. Hastie, *High Temperature Vapors*, 1975; R. Hauge, J. Hastie, and J. L. Margrave, *Annu. Rev. Phys. Chem.*, 21:475, 1970; J. L. Margrave (ed.), *High Temperature Science*, vols. 1–8, 1969–1976.

Holmium

A chemical element, Ho, atomic number 67, atomic weight 164.93, a metallic element belonging to the rare-earth group. The stable isotope Ho^{165} makes up 100% of the naturally occurring element.

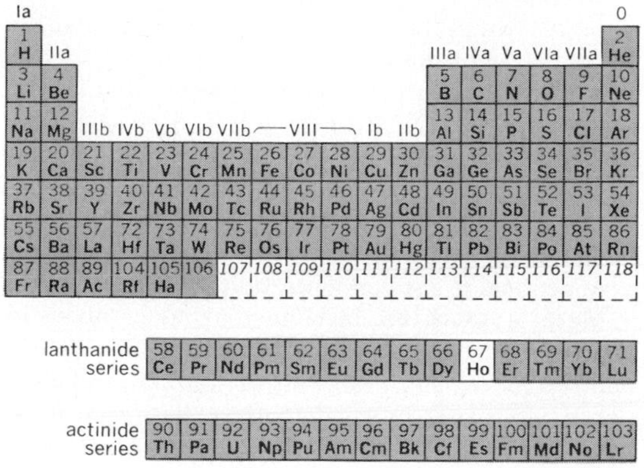

It was discovered in 1878 by J. L. Soret, and independently by P. T. Cleve in 1879. The oxide, Ho_2O_3, has a pale-green color, and it is soluble in mineral acids to give the trivalent ion. It forms yellow salts. For properties of the metal *see* RARE-EARTH ELEMENTS.

The metal is paramagnetic, but as the temperature is lowered, it changes to antiferromagnetic and then to the ferromagnetic system. The Néel point occurs about 132 K and the Curie point is in the neighborhood of 20 K.

[FRANK H. SPEDDING]

Homogeneous catalysis

A process in which a catalyst is in the same phase as the reactant. A homogeneous catalyst is molecularly dispersed (dissolved) in the reactants, which are most commonly in the liquid state. Catalysis of the transformation of organic molecules by acids or bases represents one of the most widespread types of homogeneous catalysis. In addition, the catalysis of organic reactions by metal complexes in solution has grown rapidly in both scientific and industrial importance. *See* CATALYSIS.

Acid-base catalysis. The two principal areas are specific acid (or base) catalysis and general acid (or base) catalysis. Specific acid catalysis refers to reactions in which only the oxonium ion (H_3O^+) can act as the catalyst. A common example is the hydrolysis of simple acetals, reaction (1).

$$CH_3-\underset{\underset{OR}{|}}{\overset{\overset{CH_3}{|}}{C}}-OR \underset{}{\overset{H_3O^+}{\rightleftharpoons}} CH_3-\underset{\underset{OR}{|}}{\overset{\overset{CH_3}{|}}{C}}-\overset{H}{\underset{\oplus}{OR}} \rightleftharpoons$$

$$CH_3-\underset{\underset{OR}{|}}{\overset{\overset{CH_3}{|}}{C}}\oplus + ROH \underset{}{\overset{H_2O}{\rightleftharpoons}} CH_3-\underset{\underset{OR}{|}}{\overset{\overset{CH_3}{|}}{C}}-OH$$

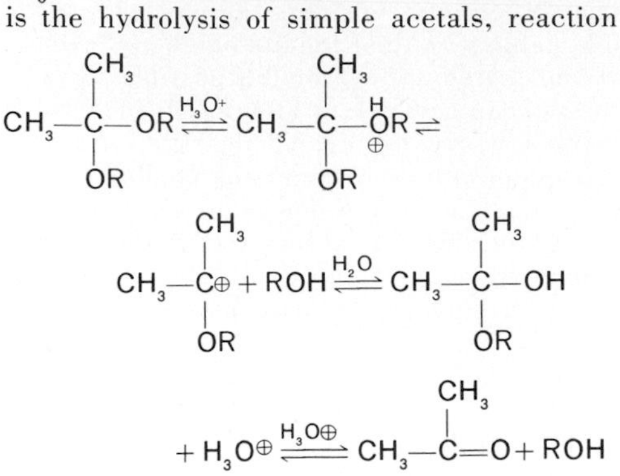

$$+ H_3O\oplus \overset{H_3O\oplus}{\rightleftharpoons} CH_3-\underset{}{\overset{\overset{CH_3}{|}}{C}}{=}O + ROH \qquad (1)$$

Specific acid catalysis is found to be characteristic of reactions in which there is rapid, reversible protonation of the substrate before the slow, rate-limiting step.

Reactions which are catalyzed by proton donors in general are considered to be subject to general acid catalysis. General acid catalysis often becomes important only at higher acidity levels. The proton is a convenient and powerful agent for the distortion of the electronic configuration of a substrate in order to facilitate reaction. The mechanism by which this occurs has many variants. For example, a covalent bond may be more easily broken after protonation of one of the bonded atoms; the reaction, $ROH_2^\oplus \to R^\oplus + H_2O$ is easier than $ROH \to R^\oplus + OH^\ominus$.

Exactly the same distinction can be made in catalysis by bases as was made above for acids. Thus, in specific base catalysis the reaction rate is proportional to the concentration of $OH^\ominus$.

Metal complexes. In homogeneous catalysis by coordination compounds of transition metals, the catalyst is usually deployed in solution and most commonly exists in a molecularly dispersed form. Thus, all sites are potentially active for catalysis, and in many cases catalysis is observed under much milder reaction conditions than found with heterogeneous catalysis by metals and metal oxides.

The catalysis of the incorporation of carbon monoxide into organic substrates by transition metal complexes is technologically important. The hydroformylation or oxo reaction [reaction (2)] in

$$RCH{=}CH_2 + CO + H_2 \to$$
$$RCH_2-CH_2CHO + RCH-CH_3 \qquad (2)$$
$$\underset{\underset{CHO}{|}}{}$$

which an olefin is reacted with carbon monoxide and hydrogen to generate a mixture of linear and branched aldehydes, was discovered in 1938. The first catalyst found was dicobalt octacarbonyl, $Co_2(CO)_8$, and this is still used extensively today in commercial operations. The steps involved in this reaction are summarized in reactions (3).

$$Co_2(CO)_8 + H_2 \rightleftharpoons 2HCo(CO)_4 \qquad (3a)$$

$$HCo(CO)_4 + olefin \rightleftharpoons$$
$$HCo(CO)_3(olefin) + CO \qquad (3b)$$

$$HCo(CO)_3(olefin) \rightleftharpoons RCo(CO)_3 \qquad (3c)$$

$$RCo(CO)_3 + CO \rightleftharpoons RCo(CO)_4 \qquad (3d)$$

$$RCo(CO)_4 + CO \rightleftharpoons$$
$$(R-CO)Co(CO)_4 \qquad (3e)$$

$$(RCO)Co(CO)_4 + HCo(CO)_4 \to$$
$$RCHO + Co_2(CO)_8 \qquad (3f)$$

Some of these steps represent transformations which are common to many sequences found in homogeneous catalysis. Thus, step (3c), the insertion of a coordinated olefin into the metal-hydride bond to generate a metal-alkyl bond, is a frequently encountered method of metal-carbon bond formation. Step (3e), the formation of a metal-acyl bond by alkyl migration to a coordinated carbon monoxide, is a key step in most catalytic (and stoichiometric) syntheses involving the incorporation of carbon monoxide into organic molecules.

While the reaction steps above can be conducted in a stoichiometric manner under very mild reaction conditions, in order for the system to function catalytically at rates which are desirable for industrial processes, the reaction temperature is maintained at greater than 120° C and the reaction pressures are usually in excess of 200 atm (20 MPa).

Other catalyst systems have been discovered which can perform hydroformylation reactions under much milder reaction conditions than cobalt. In particular, rhodium complexes containing triarylphosphine ligands can catalyze hydroformylation reactions at very rapid rates at ~100°C and 30 atm (3 MPa) pressure of synthesis gas (CO + H_2). Another important difference between the rhodium and cobalt catalysts is that the rhodium system can generate a much higher proportion of linear aldehyde product [reaction (2)]. This effect is related to the greater steric crowding around the metal when triarylphosphines are present in the coordination sphere. This is an example of the influence of stereochemistry around the metal on the stereochemical course of the catalytic reaction, and this phenomenon is an important feature of many homogeneously catalyzed reactions. *See* STERIC EFFECT.

Another reaction involving the catalysis of the incorporation of carbon monoxide which has assumed considerable commercial importance is the synthesis of acetic acid from methanol, reaction (4). The reaction is catalyzed by both cobalt

$$CH_3OH + CO \to CH_3CO_2H \qquad (4)$$

and rhodium complexes in the presence of an iodide cocatalyst or promoter. The mechanism of the rhodium-catalyzed reaction is reasonably well understood, as shown in reactions (5). The reaction which generates the metal carbon bond, that is, step (5b), is rate-determining in the catalytic cycle.

$$CH_3OH + HI \rightleftharpoons CH_3I + H_2O \qquad (5a)$$

$$[Rh(CO)_2 I_2]^- + CH_3I \rightleftharpoons$$
$$[Rh(CO)_2(CH_3) I_3]^- \qquad (5b)$$

$$[Rh(CO)_2(CH_3) I_3]^- \to$$
$$[Rh(CO)(COCH_3) I_3]^- \qquad (5c)$$

$$[Rh(CO)(COCH_3) I_3]^- + CO \rightleftharpoons$$
$$[Rh(CO)_2(COCH_3) I_3]^- \qquad (5d)$$

$$[Rh(CO)_2(COCH_3)I_3]^- \rightarrow [Rh(CO)_2I_2]^- + CH_3COI \quad (5e)$$

$$CH_3COI + H_2O \rightarrow CH_3CO_2H + HI \quad (5f)$$

The commercial reactors utilizing rhodium catalysts are operated at temperatures in the range of 150–200°C and pressures of less than 40 atm (4 MPa). The rate of the reaction is sufficiently rapid that the amount of the very expensive rhodium catalyst required is very small.

A wide range of olefin transformation reactions are catalyzed by transition metal complexes. Some of the more important reactions are isomerization, dimerization, polymerization, and metathesis. Nickel catalysts convert olefins into a mixture of dimers, trimers, and higher oligomers. The rate is particularly rapid with ethylene. The catalytic species are typically generated in place from various nickel complexes by reaction with an alkyl-aluminum compound. Uncharacterized hydridonickel species are postulated to be the active catalysts. The mechanism for dimerization and polymerization can be visualized as a series of sequential olefin insertions into Ni-H and Ni-C bonds followed by β-hydride elimination. The dimerization of propylene and higher olefins can give rise to linear, monobranched and dibranched olefin products. The relative amount of each type of product depends on the ligands coordinated to the nickel. *See* HYDRIDO COMPLEXES.

Migratory insertion of alkenes into alkyl-metal bonds is also represented in the polymerizations catalyzed by a variety of transition metal species. For example, ethylene is polymerized by a catalyst prepared in place by the reduction of $TiCl_4$ with alkyl-aluminum compounds. This is the basis for the commercial Ziegler-Natta process for the preparation of high-density polyethylene, which is practiced on a large scale throughout the world.

Oxidation. Transition metal complexes act as homogeneous catalysts in many different types of oxidation process. Two main categories of reaction can be recognized, involving either one-electron or two-electron processes.

Autooxidation. The involvement of transition metal complexes in one-electron, radical processes is most evident in the so-called autooxidation reactions whereby hydrocarbons are oxidized to various oxygen-containing compounds by radical chain processes. The general scheme is shown in reactions (6). While metal species can enhance the rate

$$\text{Initiation:} \quad \text{Initiator} + RH \rightarrow inH + R^\bullet \quad (6a)$$

$$\text{Propagation:} \quad R^\bullet + O_2 \rightarrow RO_2^\bullet \quad (6b)$$

$$RO_2^\bullet + RH \rightarrow RO_2H + R^\bullet \quad (6c)$$

$$\text{Termination:} \quad R^\bullet + RO_2^\bullet \rightarrow RO_2R \quad (6d)$$

$$2RO_2^\bullet \rightarrow RO_4R \quad (6e)$$

$$RO_4R \rightarrow \text{nonradical products} + O_2 \quad (6f)$$

of several of the above steps, the most common pathway for catalysis of liquid-phase autooxidations involves the metal-catalyzed decomposition of alkyl hydroperoxides, of which reactions (7) and (8) are examples. Cobalt and manganese salts are

$$RO_2H + Co^{II} \rightarrow RO^\bullet + Co^{III}(OH) \quad (7)$$

$$(RO_2)Co^{III} \rightleftharpoons RO_2^\bullet + Co^{II} \quad (8)$$

particularly effective in promoting autooxidation processes. The oxidation of *p*-xylene to terephthalic acid (the key monomer involved in the manufacture of polyester) is carried out on a very large scale using a cobalt-bromide catalyst.

Indirect oxidation. Transition metal complexes find utility in the catalysis of various types of indirect, two-electron oxidations. Examples of these indirect processes are the so-called Wacker reaction, in which olefins are oxidized to aldehydes or ketones by palladium (II) compounds, with concomitant reduction of the palladium. The palladium is then reoxidized in a separate reaction by a combination of a copper salt and oxygen. The best-known example of the Wacker reaction is the oxidation of ethylene to acetaldehyde, reaction (9).

$$CH_2{=}CH_2 + \tfrac{1}{2}O_2 \rightarrow CH_3CHO \quad (9)$$

The reaction is conducted in an aqueous medium in the presence of palladium and copper chlorides as the catalyst system. The generally accepted mechanism is shown in reactions (10).

$$PdCL_4^{2-} + C_2H_4 \rightleftharpoons [PdCl_3(C_2H_4)]^- + Cl^- \quad (10a)$$

$$[PdCl_3(C_2H_4)]^- + H_2O \rightleftharpoons [PdCl_2(H_2O)(C_2H_4)] + Cl^- \quad (10b)$$

$$[PdCl_2(H_2O)(C_2H_4)] + H_2O \rightleftharpoons [PdCl_2(OH)(C_2H_4)]^- + H_3O^+ \quad (10c)$$

$$[PdCl_2(OH)(C_2H_4)]^- + H_2O \rightleftharpoons [HOCH_2CH_2PdCl(H_2O)] + Cl^- \quad (10d)$$

$$[HOCH_2CH_2PdCl(H_2O)] \rightarrow CH_3CHO + Pd + HCl + H_2O \quad (10e)$$

$$Pd + 2CuCl_2 \rightarrow PdCl_2 + 2CuCl \quad (10f)$$

$$CuCl + \tfrac{1}{2}O_2 + HCl \rightarrow CuCl_2 + \tfrac{1}{2}H_2O \quad (10g)$$

Another important indirect oxidation process was developed in the 1970s. This is the metal-catalyzed epoxidation of olefins with alkyl hydroperoxides, reaction (11). Various molybdenum, vanadium, and chromium complexes act as catalysts for this reaction, by pathways which are still rather poorly understood.

$$C{=}C + RO_2H \rightarrow C{-}C \text{ (bridged by O)} + ROH \quad (11)$$

Adiponitrile, $NC(CH_2)_4CN$, is produced as a precursor of hexamethylenediamine, one of the building blocks of Nylon 66. The selective addition of two moles of hydrogen cyanide to butadiene has been developed into a valuable new synthesis of adiponitrile. In one variant of this process, a zerovalent nickel catalyst, $Ni[P(OAryl)_3]_4$, can be used to bring about a series of reactions including HCN addition to butadiene, isomerization of cyanolefins, and hydrocyanation of 4-pentene-nitrile, reactions (12)-(14). A key step in reactions (12) and (14) appears to be generation of a nickel hydride species capable of reacting with an olefin.

$$CH_2{=}CH{-}CH{=}CH_2 + HCN \rightarrow CH_3{-}CH{=}CH{-}CH_2{-}CN \quad (12)$$

$$CH_3{-}CH{=}CH{-}CH_2{-}CN \rightleftharpoons CH_2{=}CH{-}CH_2{-}CH_2CN \quad (13)$$

$$CH_2{=}CH{-}CH_2{-}CH_2CN + HCN \rightarrow NC(CH_2)_4CN \quad (14)$$

An outline of the mechanism is shown in reactions (15)–(19), where L represents phosphite or phosphine.

$$NiL_4 + HCN \rightleftharpoons HNi(CN)L_3 + L \quad (15)$$

$$HNi(CN)L_3 + RCH{=}CH_2 \rightarrow HNi(CN)(olefin)L_2 + L \quad (16)$$

$$HNi(CN)(olefin)L_2 \rightarrow RCH_2CH_2Ni(CN)L_2 \quad (17)$$

$$RCH_2CH_2Ni(CN)L_2 \rightarrow NiL_2 + RCH_2CH_2CN \quad (18)$$

$$NiL_2 + 2L \rightarrow NiL_4 \quad (19)$$

Perhaps the most elegant illustration of the selectivity achievable with homogeneous catalysts is found in the asymmetric hydrogenation of unsymmetrical olefins in the presence of rhodium complexes containing optically active phosphine ligands. Through this process, it is possible to prepare a number of optically active α-amino acids from the corresponding unsaturated precursors, reaction (20). (The carbon marked with an asterisk

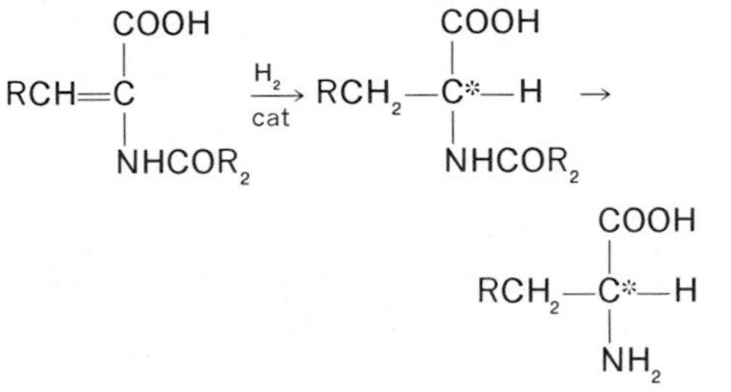

(20)

is an asymmetric center.) The energy difference between the optical enantiomers is very small, but nevertheless, with suitable optically active phosphine ligands, one isomer can be produced with greater than 90% selectivity. This approach is used commercially in the synthesis of L-DOPA, the drug that is used in the treatment of Parkinson's disease. *See* ASYMMETRIC SYNTHESIS; OPTICAL ACTIVITY; ORGANIC CHEMICAL SYNTHESIS; STEREOSPECIFIC CATALYST. [DENIS FORSTER]

Hydrate

A particular form of a solid compound which has water in the form of H_2O molecules associated with it. For example, anhydrous copper sulfate is a white solid with the formula $CuSO_4$. When crystallized from water, a blue crystalline solid which contains water molecules as part of the crystals is formed. Analysis shows that the water is present in a definite amount, and the hydrate may be given the formula $CuSO_4 \cdot 5H_2O$. Four of the water molecules are attached to the copper ion in the manner of coordination complexes, and the fifth water molecule is related to the sulfate and presumably held by hydrogen bonding.

Water can also be present in definite proportions in the crystal without being associated directly with the anion or cation. The water occupies a definite place in the crystal lattice. Alums, with their 12 molecules of water, are examples of this.

Gas hydrates are compounds with a definite composition obtained when water and certain gases are chilled sufficiently. Chlorine, carbon dioxide, and acetylene all form such icy hydrates. Similarly, methane, ethane, and propane yield solid crystalline hydrates. Methane hydrate hampers natural-gas transmission in pipelines, and much research has been devoted to its formation and decomposition. Less heat is involved in forming the lower hydrocarbon hydrates than in forming pure ice. Because these hydrates selectively reject the salts present in sea water, these hydrocarbon hydrates have been investigated as low-energy ways to desalt saline waters.

Some crystalline hydrates do not have definite proportions of water. Zeolites and similar silicate minerals, certain clays, and metallic oxides have variable proportions of water in their hydrated forms. *See* COORDINATION CHEMISTRY; HYDRATION. [FRANK WAGNER]

Bibliography: F. Franks (ed.), *Water: A Comprehensive Treatise*, vol. 2: *Water in Crystalline Hydrates*, 1973.

Hydration

The incorporation of molecular water into a complex with the molecules or units of another species. The complex may be held together by relatively weak forces or may exist as a definite compound. Many salts form solid hydrates when exposed to water vapor under certain conditions of temperature and pressure. Copper sulfate, for example, forms a monohydrate ($CuSO_4 \cdot H_2O$) when exposed at 25°C to water vapor at a pressure of 0.8 mm of mercury. At higher pressures other hydrates are formed. Water is lost from these compounds when they are heated or when the water vapor pressure falls below a minimum value. Solids forming hydrates at low pressures are used as drying agents. *See* DELIQUESCENCE; DESICCANT; EFFLORESCENCE; SOLUTION; SOLVATION.

[FRANCIS J. JOHNSTON]

Hydrazine

A colorless liquid, H_2NNH_2 (boiling point 114°C), with a musty, ammonialike odor. Physically it is similar to water, but chemically it is reducing, decomposable, basic, and bifunctional. Its derivatives range from simple salts to ring compounds, polymers, and coordination complexes. Major uses of hydrazine include such diverse applications as rocket fuels (since combustion of hydrazine is highly exothermic), corrosion inhibition in boilers, and syntheses of biologically active materials.

Hydrazine is manufactured by two routes: the reaction of chloramine with ammonia and the reaction of sodium hypochlorite with urea. Both processes require the presence of glue or gelatin to inhibit catalytic decomposition of the product by unreacted oxidants. Because hydrazine forms an azeotrope containing 31% water (boiling point 121°C), anhydrous hydrazine is isolated from aqueous process streams by extractive distillation with aniline.

Added to feedwater, hydrazine reduces rust in

boilers to a hard film of magnetic iron oxide and reduces oxygen to water on catalytic metal surfaces. Metal ions such as Cu^{++} and Ni^{++} are reduced to free metals, and organic nitro compounds are reduced to amines by hydrazine. The energetic reaction of hydrazine with strong oxidants, such as nitric acid, is utilized in rocket propulsion. The thermal decomposition of hydrazine produces free radicals and gases, useful in rubber curing and foam-rubber production.

A slightly weaker base than ammonia, hydrazine forms most of the analogs of ammonia derivatives as well as distinctive hydrazine derivatives in which both nitrogen atoms are involved. For example, hydrazine forms two series of salts, such as $N_2H_4 \cdot HCl$ and $N_2H_4 \cdot 2HCl$, and forms not only hydrazones, $RCH{=}NNH_2$, but also azines, $RCH{=}NN{=}CHR$, by reaction with aldehydes. In a manner similar to ammonia, hydrazine attacks polar bonds, in one case displacing ammonia from urea to form semicarbazide, a reaction which can be reversed by an excess of ammonia.

Prominent uses of hydrazine derivatives include rocket fuels [1,1-dimethylhydrazine, $(CH_3)_2NNH_2$]; antituberculin drugs (isonicotinic hydrazide, $C_5H_4N \cdot CO \cdot NHNH_2$); plant-growth regulators (maleic hydrazide, $NH \cdot CO \cdot CH{=}CH \cdot CO \cdot NH$, and β-hydroxyethyl hydrazine, $HOCH_2CH_2NHNH_2$); dye and explosive intermediates [aminoguanidine, $NH_2 \cdot C(NH) \cdot NHNH_2$]; algaecides and fungicides [copper dihydrazinium sulfate, $CuSO_4 \cdot (N_2H_5)_2SO_4$]; soldering fluxes (hydrazine hydrobromide, $N_2H_4 \cdot HBr$); blowing agents for foam rubber (azides, $R \cdot CO \cdot N_3$, and sulfonyl hydrazides, $R \cdot SO_2 \cdot NHNH_2$); insecticides (1,4-diphenylsemicarbazide, $C_6H_5 \cdot NH \cdot CO \cdot NHNH \cdot C_6H_5$); heterocycle syntheses (thiosemicarbazide, $NH_2 \cdot CS \cdot NHNH_2$); and polymers (dihydrazide-formaldehyde resins). *See* AMMONIA; NITROGEN.

[THEODORE H. DEXTER]

Bibliography: R. E. Kirk and D. F. Othmer, *Encyclopedia of Chemical Technology*, vol. 12, 3d ed., 1980.

Hydride

A compound containing hydrogen and another element. Many of the elements combine with hydrogen to form compounds, and even though hydrogen is less negative than the second element in the compound, they are called hydrides. For example, H_2S is called a hydride in the discussion here, although it is usually called hydrogen sulfide and properly so. Hydrides can be divided into three classes of compounds: covalent hydrides, saltlike hydrides, and metallic hydrides.

Covalent hydrides refer to compounds such as H_2O, H_2S, and NH_3, which are volatile. Covalent hydrides are formed from the nonmetals. Carbon hydrides would refer to natural gas, CH_4, and other hydrocarbons containing chains of carbons. Boron hydrides also may contain more than one boron atom and have some importance as rocket fuels. Saltlike hydrides are ionic in nature and nonvolatile. Lithium hydride, LiH, is a typical example. These compounds contain negative hydrogen ions which migrate to the positive pole when molten compounds are electrolyzed. These compounds are used as sources of hydrogen, as drying agents, and in organic syntheses.

Metallic hydrides include compounds with simple formulas such as uranium hydride, UH_3, or palladium hydride, PdH_x. Palladium metal will absorb various amounts of hydrogen, depending upon the temperature. Some of these compounds are thought to contain the hydrogen in the holes in the metallic lattice. They retain many metallic characteristics. *See* HYDROCARBON; HYDROGEN; METAL HYDRIDES.

[E. EUGENE WEAVER]

Hydrido complexes

Complex hydrides containing a hydride ligand bonded to a central atom. The prefix hydro instead of hydrido is sometimes used. Sodium tetrahydridoborate, $NaBH_4$, (or sodium tetrahydroborate, originally called sodium borohydride), and lithium tetrahydridoaluminate, $LiAlH_4$ (originally called lithium aluminum hydride), are important reducing agents in synthetic and industrial reactions. $NaBH_4$ is employed in aqueous or alcoholic solutions, and $LiAlH_4$ is employed in ethers. Sodium cyanotrihydridoborate, $NaBH_3CN$, can be used in acidic medium. A family of aluminum-based reducing agents is now available commercially, including sodium diethyldihydridoaluminate, $NaAlH_2(C_2H_5)_2$; sodium tri-*tert*-butoxohydridoaluminate, $NaAlH(O\text{-}t\text{-}C_4H_9)_3$; and sodium bis(2-methoxyethoxo)dihydridoaluminate, $NaAlH_2(OCH_2CH_2OCH_3)_2$. All are soluble in aromatic hydrocarbons. They are rather expensive, but their specific reducing powers make them attractive for synthesizing high-value products, such as pharmaceuticals, flavorings, fragrances, dyes, and insecticides.

Similar hydrides are C_6H_5MgH and $C_6H_5Mg_2H_3$. If zinc is considered a nontransition metal, then complex hydrides such as Li_3ZnH_5, Li_2ZnH_4, $LiZnH_3$, $NaZn_2H_5$, $LiZnH(CH_3)_2$, and $LiZn(CH_3)_2AlH_4$ may be grouped with the above compounds. *See* HYDROBORATION; METAL HYDRIDES.

Transition metals. More than a thousand hydrido complexes of transition metals have been prepared. Excepting Sc, Y, and La, hydrido com-

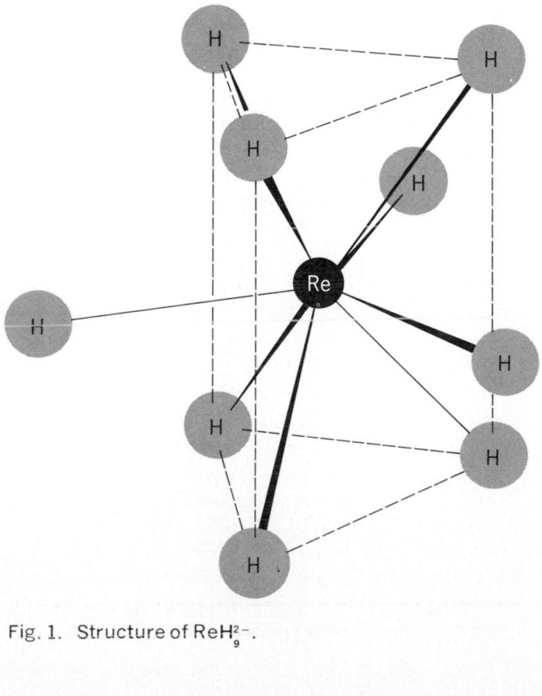

Fig. 1. Structure of ReH_9^{2-}.

HYDRIDO COMPLEXES

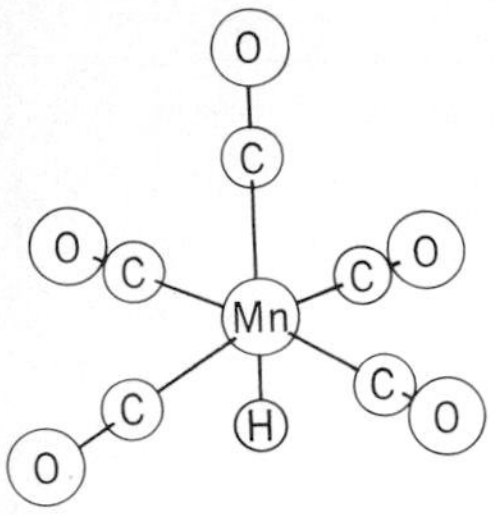

Fig. 2. Structure of $[HMn(CO)_5]$.

pounds of all three transition series metals are known. In only two cases do the complex anions contain the central atom bonded to hydride ions and no other ligand; these are K_2ReH_9 and K_2TcH_9 (see below). The rest contain H-M-L linkages, where hydrogen is bonded to a transition metal M which is also bonded to one or more pi-bonded ligands L [for example, CO, CN^-, $C_5H_5^-$, PF_3, or $P(C_6H_5)_3$]. The metal-hydrogen bonds are stabilized by the pi bonding between the metal and other ligands. The M-H bond lengths are generally 0.16 to 0.17 nm. These compounds are exceedingly diverse in nature and undergo a bewildering multitude of reactions. A few industrial processes involve hydrido complexes. Some chemists strongly suspect that the enzymes which convert atmospheric nitrogen to ammonia (fixation) depend on Fe-H or Mo-H functions. *See* COMPLEX COMPOUNDS; COORDINATION CHEMISTRY.

Enneahydridorhenate ion. A compound of formula K_2ReH_9 was isolated in 1960. Brackets are frequently employed to identify complex ions or molecules, for example, $K_2[ReH_9]$. The structure of the enneahydridorhenate ion, ReH_9^{2-}, is that of a triangular prism with three extra hydrogen atoms bonded through three faces, as shown in Fig. 1. One of these facial hydrogen atoms can be replaced by a phosphine molecule, the compound $K[ReH_8PH_3]$ being formed. The technetium compound analogous to $K_2[ReH_9]$, namely $K_2[TcH_9]$, is the second transition metal hydrido complex known which has only hydride ligands. Both compounds are fairly stable in alkaline aqueous solution.

Carbonyl hydrides. Hydrido complexes of metals also bonded to carbonyl ligands were the first such compounds discovered in 1931. The iron compounds were prepared by the action of an alkali on iron pentacarbonyl as shown in reaction (1).

$$[Fe(CO)_5] + 3OH^- \rightarrow [HFe(CO)_4]^- + CO_3^{2-} + H_2O \quad (1)$$

Acidification of the product from reaction (1) yields dihydridotetracarbonyliron, $[H_2Fe(CO)_4]$. The corresponding cobalt compound is prepared by hydrogenation of dicobalt octacarbonyl as shown in reaction (2).

$$[Co_2(CO)_8] + H_2 \rightleftarrows 2[HCo(CO)_4] \quad (2)$$

Some other hydridocarbonyls are $[HMn(CO)_5]$ (Fig. 2), $[HRe(CO)_5]$, and $[HV(CO)_6]$. The effective atomic number rule is valid in these cases. These compounds are acidic, and many derivatives are known, such as $[(CO)_4Co—Hg—Co(CO)_4]$, a substance containing metal-metal bonds. The hydridocarbonyls $[HCo(CO)_4]$ and $[HV(CO)_6]$ are strong acids, while $[H_2Fe(CO)_4]$, with $pK_1 = 4.4$, is about as acidic as acetic acid. All of these hydridocarbonyls are volatile, toxic compounds with obnoxious odors. Several hydridocarbonyls are listed in Table 1.

Knowledge of the hydrogen-metal bond in these complexes derives primarily from neutron and x-ray diffraction studies, nuclear magnetic resonance studies, and infrared absorption spectra. Proton (hydride) nuclear magnetic resonance shifts are exceptionally large, and to high field. *See* INFRARED SPECTROSCOPY; NEUTRON DIFFRACTION; NUCLEAR MAGNETIC RESONANCE (NMR); X-RAY DIFFRACTION.

The hydridocarbonyls are chemically reactive substances, as illustrated by olefin and carbon monoxide insertion reactions of the cobalt compound, reaction (3). The last compound reacts with

$$[HCo(CO)_4] \xrightarrow{H_2C=CH_2} [CH_3CH_2Co(CO)_4] \xrightarrow{CO} [CH_3CH_2COCo(CO)_4] \quad (3)$$

HYDRIDO COMPLEXES

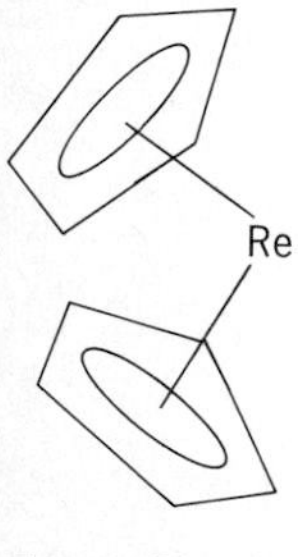

Fig. 3. Structure of $[HRe(C_5H_5)_2]$. The aromatic C_5H_5 groups are represented by pentagons.

Table 1. Some mononuclear hydrido complexes

Group	Formula
IVB	$[HZrCl(C_5H_5)_2]$, $[H_2Zr(C_5\{CH_3\}_5)_2]$
VB	$[H_3Ta(C_5H_5)_2]$
VIB	$[HCr(C_5H_5)(CO)_3]$, $[H_2Mo(C_5H_5)_2]$, $[H_3Mo(C_5H_5)_2]^+$, $[H_2W(C_5H_5)_2]$, $[HW(C_5H_5)(CO)_3]$
VIIB	$[HMn(CO)_5]$, $[HMn(PF_3)_5]$, $[HRe(C_5H_5)_2]$, $[H_2Re(C_5H_5)_2]^+$
VIII	$[HFe(CO)_4]^-$, $[HFe(C_5H_5)(CO)_2]^+$, *cis*-$[H_2Fe(PF_3)_4]$, $[HCo(CO)_4]$, $[HCo(PF_3)_4]$, $[HCo(CO)_3P(C_6H_5)_3]$, $[H_2Co(P\{C_6H_5\}_3)_3]$, $K_3[HCo(CN)_5]$, $[HRu(C_5H_5)_2]^+$, $[HRu(C_5H_5)(CO)_2]$, $[HRuCl(P\{C_6H_5\}_3)_3]$, $[HPtCl(P\{C_2H_5\}_3)_2]$

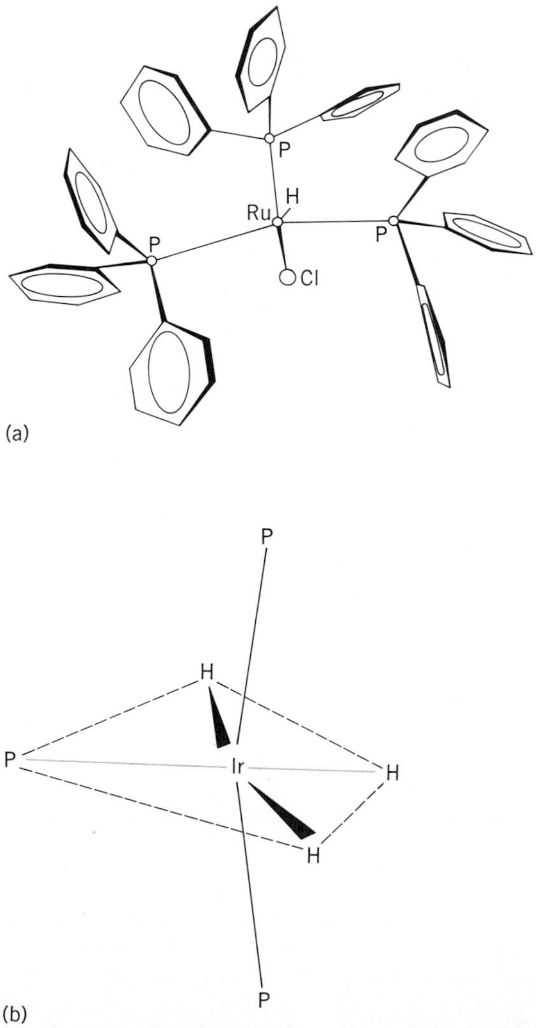

Fig. 4. Hydridophosphine structures. (*a*) $[HRuCl(P\{C_6H_5\}_3)_3]$. (*b*) $[H_3Ir(P\{C_6H_5\}_3)_3]$, with the C_6H_5 groups not shown.

hydrogen to regenerate $[HCo(CO)_4]$, forming an aldehyde, as shown in reaction (4). Thus hydrido-

$$[CH_3CH_2COCo(CO)_4] \xrightarrow{H_2} [HCo(CO)_4] + CH_3CH_2CHO \quad (4)$$

tetracarbonylcobalt functions as a catalyst. Further hydrogenation of the aldehyde leads to an alcohol. The above sequence of reactions is known as hydroformylation or the oxo reaction, and is employed industrially to produce long-chain alcohols. *See* HYDROFORMYLATION; METAL CARBONYL.

Hydridocyclopentadienyl complexes. The hydrocarbon cyclopentadiene, C_5H_6, readily loses one hydrogen ion to form the cyclopentadienide ion, $C_5H_5^-$. The iron(II) derivative, $[Fe(C_5H_5)_2]$, is bis (cyclopentadienyl)iron, the famous sandwich compound, ferrocene. *See* FERROCENE.

In 1955 hydrido-bis(cyclopentadienyl)rhenium, $[HRe(C_5H_5)_2]$, was synthesized, and nuclear magnetic resonance studies proved that the hydrogen atom is bonded to the metal (Fig. 3). The compound is a Brönsted base, and neutralizes dilute acids to form $[H_2Re(C_5H_5)_2]^+$. In 60% dioxane the value of pK_b for $[HRe(C_5H_5)_2]$ is 8.5; pK_b for am-

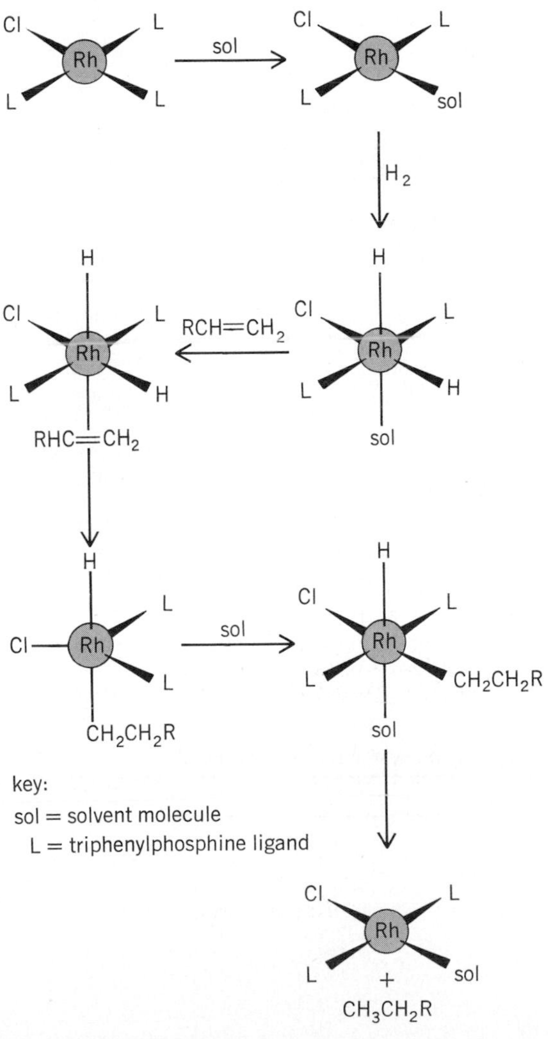

Fig. 5. Reaction scheme for hydrogenation of an olefin using Wilkinson's catalyst.

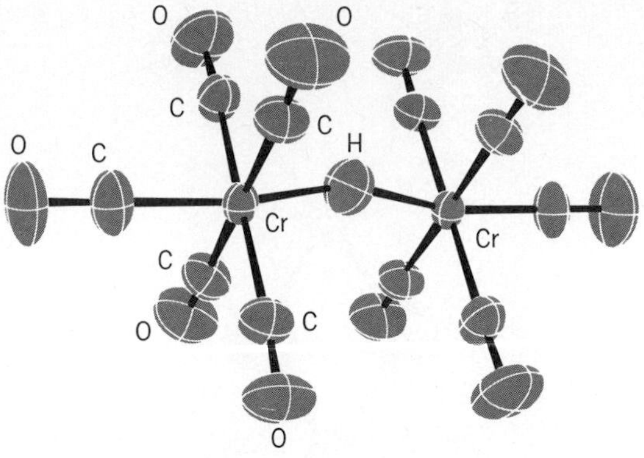

Fig. 6. Structure of $[HCr_2(CO)_{10}]^-$.

monia under the same conditions is 8.85. It is believed that in $[HRe(C_5H_5)_2]$ and similar compounds there are three orbitals directed from the metal atom; electrons in these orbitals cause the two C_5H_5 rings to be angular rather than parallel. In $[H_2Re(C_5H_5)_2]^+$, hydrogen atoms are bonded through two of these orbitals, and all three are occupied in $[H_3Ta(C_5H_5)_2]$ (which is not a base) and in $[H_3W(C_5H_5)_2]^+$. Mixed ligand complexes such as $[HCr(C_5H_5)(CO)_3]$ were also discovered in 1955. Several hydridocyclopentadienyl complexes are listed in Table 1.

When attempts were made to prepare $[HM(C_5H_5)_2]$, where M is Co, Rh, or Ir, it was found that the hydrogen atom bonds, not to the metal atom, but to one of the five-carbon rings, resulting in $[M(C_5H_5)(C_5H_6)]$, which is not a hydrido complex.

Hydridophosphine and hydridocyano complexes. Trifluorophosphine (PF_3), triethylphosphine ($P(C_2H_5)_3$) and triphenylphosphine ($P(C_6H_5)_3$) are important ligands in hydrido-transition metal complexes. Examples of the first type are the colorless liquids $[HMn(PF_3)_5]$, *cis*-$[H_2Fe(PF_3)_4]$, and $[HCo(PF_3)_4]$, which resemble the corresponding carbonyls. The structures of $[HRuCl(P\{C_6H_5\}_3)_3]$ and of $[H_3Ir(P\{C_6H_5\}_3)_3]$ are sketched in Fig. 4. Within a given periodic family the stability of a given type of hydride increases, as shown by the following trend: $[HNiCl(P\{C_2H_5\}_3)_2]$ can be detected but not isolated. $[HPdCl(P\{C_2H_5\}_3)_2]$ can be isolated, but is unstable. $[HPtCl(P\{C_2H_5\}_3)_2]$ is stable in air even at 100° C, and can be vacuum-distilled. Certain hydrido complexes of the platinum metals function as hydrogenation catalysts at room temperature and 1 atm (100 kPa) pressure. The catalysis is attributed to the ability of the metal atom to coordinate, via a vacant site, the reactant molecules, thereby orienting them and lowering the activation energy for bond making and breaking. The transition metal atom can supply or accept electrons as necessary. The hydrogen atom is facile, and under these circumstances is inserted into the substrate molecule. One well-known compound of this type is chloro-tris(triphenylphosphine)rhodium(I), Wilkinson's catalyst. Its role in catalyzing the hydrogenation of an olefin ($RCH{=}CH_2$) is shown in Fig. 5.

The mobility of the hydrido ligand is further shown by the reversible insertion of ethylene into an H-Pt bond. This is shown in reaction (5),

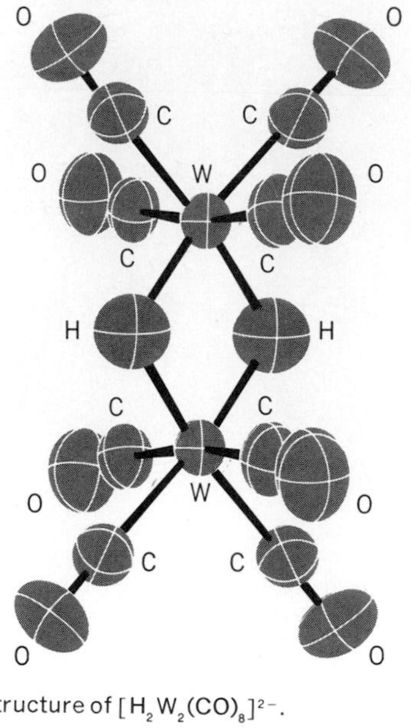

Fig. 7. Structure of $[H_2W_2(CO)_8]^{2-}$.

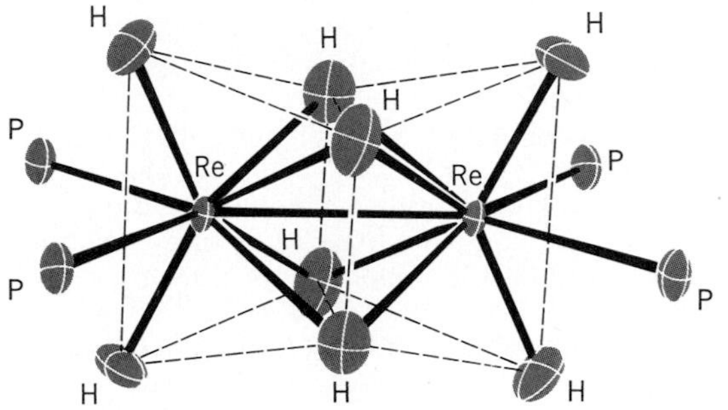

Fig. 8. Core of the $[H_8Re_2(P\{C_2H_5\}_2C_6H_5)_4]$ molecule. All organic groups are omitted for clarity.

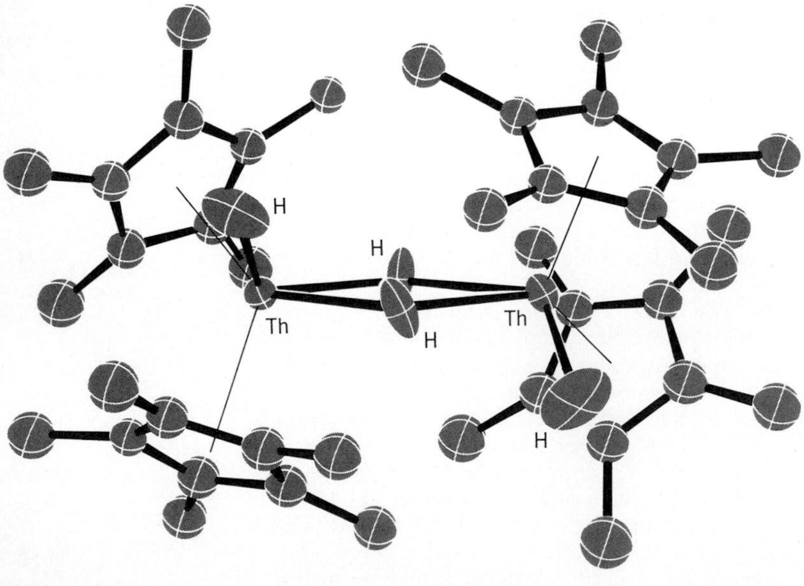

Fig. 9. Structure of $[H_2Th(C_5\{CH_3\}_5)_2]_2$. The hydrogen atoms of the methyl groups are omitted for clarity.

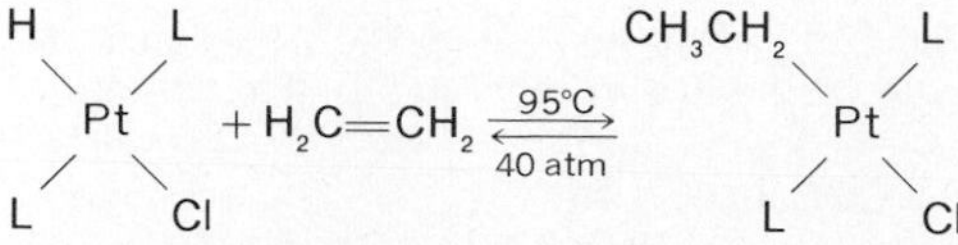

(5)

where the ligand L is $P(C_6H_5)_3$. The conversion of orange chloro-tris(triphenylphosphine)iridium(I) to its colorless hydrido isomer is shown in reaction (6). Hydrido(tri-*n*-butylphosphine)copper(I),

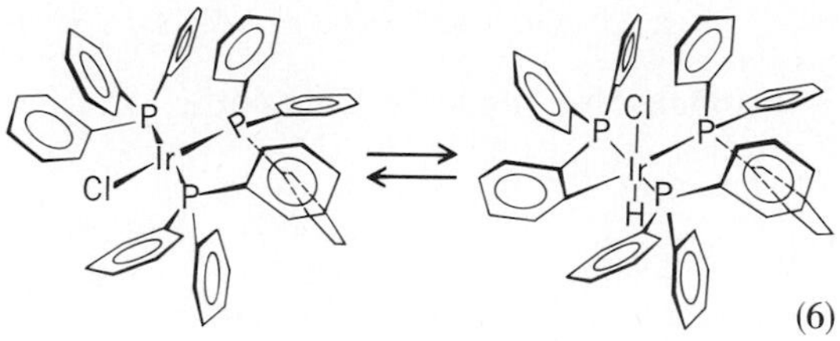

(6)

$[HCuP(C_4H_9)_3]$ (which might be a polymer), is a mild, selective reducing agent employed in organic syntheses. It and similar complexes are very unstable and must be employed below −20°C.

Reversible hydrogenation of the cyanocobalt(II) complex yields the hydridocyanocobalt(III) derivative, shown in reaction (7). Hydrido complexes of

$$2[H_2OCo(CN)_5]^{3-} + H_2 \rightleftarrows 2[HCo(CN)_5]^{3-} + 2H_2O \quad (7)$$

this type are catalysts in selectively reducing diene hydrocarbons such as butadiene, C_4H_6, to the monoene stage, C_4H_8. Some hydridophosphine and hydridocyano complexes are listed in Table 1.

Polynuclear hydrido complexes. Reduction of $Cr(CO)_6$ with $NaBH_4$ yields the yellow binuclear (two-chromium) anion, $[HCr_2(CO)_{10}]^-$. The earliest structural studies of this ion indicated that the hydrogen atom lies on the Cr-Cr axis, bonding the two $Cr(CO)_5$ units together. More recently it was demonstrated that the Cr-H-Cr linkage is inherently bent, as indicated in Fig. 6. The hydrogen atom is about 0.03 nm from the center of the Cr-Cr bond. In $[HW_2(CO)_{10}]^-$ the hydrogen atom is about 0.07 nm off-axis. In these compounds there is an electron-deficient three-center linkage. Only one electron pair is involved, as in the case of diborane. When $[C_5H_5Mo(CO)_3]_2$ is dissolved in concentrated sulfuric acid, it is protonated to $[C_5H_5Mo(CO)_3]_2H^+$, which also has a hydrogen bridge bond. In $[H_2W_2(CO)_8]^{2-}$ there are two such bridging atoms, as sketched in Fig. 7. The molecule $[H_8Re_2(P\{C_2H_5\}_2C_6H_5)_4]$ has an Re-Re bond and no less than four bridging H atoms, as well as four terminal H atoms. The core of this structure is

Table 2. Some polynuclear hydrido complexes

Group	Formula
VB	$[HNb_6I_{11}]$
VIB	$[HCr_2(CO)_{10}]^-$, $[C_5H_5Mo(CO)_3]_2H^+$, $[HW_2(CO)_{10}]^-$, $[H_2W_2(CO)_8]^{2-}$
VIIB	$[H_2Mn_2(CO)_9]$, $[H_8Re_2(PR_3)_4]$*, $[H_2Re_3(CO)_{12}]^-$
VIII	$[HFe_3(CO)_{11}]^-$, $[HCo_6(CO)_{15}]^-$, $[H_2Ni_2(CO)_6]$, $[H_3Ni_4(C_5H_5)_4]$, $[HNi_{12}(CO)_{21}]^{3-}$
IB	$[HCuP(C_6H_5)_3]_6$

*R = ethyl or phenyl.

shown in Fig. 8. Each hydrogen atom in $[H_3Ni_4(C_5H_5)_4]$ is bonded to three nickel atoms. The hydrogen atom of $[HCo_6(CO)_{15}]^-$, a cluster ion, is situated right in the center of the octahedron formed by the six cobalt atoms. A recently discovered hydrido complex of thorium is the dimer of $[H_2Th(C_5\{CH_3\}_5)_2]$. The organic ligands are pentamethylcyclopentadienyl groups. As shown in Fig. 9, two of the hydrogen atoms are bridging and two are terminal. Some polynuclear hydrido complexes are listed in Table 2. [JAMES C. WARF]

Bibliography: R. Bau (ed.), *Transition Metal Hydrides*, 1978; A. P. Ginsberg, *Transition Metal Chemistry*, vol. 1, 1965; E. L. Muetterties, *Transition Metal Hydrides*, 1971.

Hydroboration

The process of producing organoboranes by the addition of diborane to unsaturated organic compounds. In ether solvents the addition of diborane to such molecules is exceedingly rapid and essentially quantitative. This reaction therefore makes the organoboranes readily available. Such organoboranes are finding increasing application as intermediates for organic synthesis.

Procedures. Diborane is highly soluble in tetrahydrofuran, where it exists as the addition compound tetrahydrofuran-borane. Such solutions are often used for hydroboration, and merely involve bringing the two reactants together as indicated by reaction (1).

$$3RCH{=}CH_2 + C_4H_8O{:}BH_3 \rightarrow (RCH_2CH_2)_3B + C_4H_8O \qquad (1)$$

Alternatively, sodium borohydride may be utilized to achieve hydroboration by the addition of boron trifluoride etherate. This is shown by reaction (2). Usually the organoborane is not isolated

$$12RCH{=}CH_2 + 3NaBH_4 + 4(C_2H_5)_2O{:}BF_3 \rightarrow 4(RCH_2CH_2)_3B + 3NaBF_4 + 4(C_2H_5)_2O \qquad (2)$$

but is utilized in place, similar to applications of the Grignard reagent in synthesis.

Scope and stoichiometry. Essentially all molecules containing one or more double or triple bonds undergo rapid conversion to organoboranes by this procedure. In general, disubstituted olefins react to give trialkylboranes, trisubstituted olefins react rapidly to give dialkylboranes and only slowly beyond, and tetrasubstituted olefins react rapidly to the monoalkylborane stage. The respective reactions are indicated by (3), (4), and (5).

$$3\,CH(CH_3){=}CH(CH_3) + BH_3 \rightarrow (CH_3CH_2CH(CH_3){-})_3B \qquad (3)$$

$$2\,(H_3C)_2C{=}CH(CH_3) + BH_3 \rightarrow ((H_3C)_2HC{-}CH(CH_3){-})_2BH \qquad (4)$$

Disiamylborane

$$(H_3C)_2C{=}C(CH_3)_2 + BH_3 \rightarrow H{-}C(CH_3)_2{-}C(CH_3)_2{-}BH_2 \qquad (5)$$

t-Hexylborane

Such substituted organoboranes are often very useful for controlled hydroborations, as indicated by reactions (6) and (7).

$$R_2BH + CH_2{=}CHCH_2Cl \rightarrow R_2BCH_2CH_2CH_2Cl \qquad (6)$$

$$RBH_2 + CH_2{=}C(CH_3){-}CH{=}CH_2 \rightarrow RB\langle CH_2CHCH_3,\ CH_2CH_2 \rangle \text{ (ring)} \qquad (7)$$

The hydroboration of unsaturated molecules containing functional groups is usually easily accomplished. Reaction (8) illustrates this technique.

$$R_2BH + CH_2{=}CHCH_2CO_2C_2H_5 \rightarrow R_2BCH_2CH_2CH_2CO_2C_2H_5 \qquad (8)$$

In this way organoboranes containing such functional groups become readily available for organic synthesis.

Asymmetric syntheses. The hydroboration of optically active α-pinene yields an optically active dialkylborane. This can be utilized to introduce an asymmetric center, as illustrated in reaction (9),

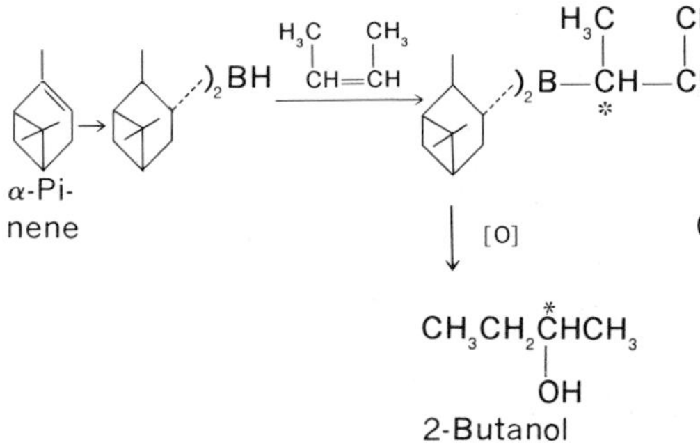

(9)

the synthesis of optically active 2-butanol in high optical purity. In reaction (9) the asymmetric center is indicated by an asterisk and [O] indicates an oxidation step.

Directive effects. Hydroboration generally proceeds to place the boron atom at the less substituted of the two carbon atoms of a double bond (anti-Markovnikov addition). Since the boron atom may readily be replaced by many functional groups, such as hydroxyl and amino, this makes possible the anti-Markovnikov hydration and amination of double bonds, as shown by reaction (10).

$$RCH{=}CH_2 \rightarrow RCH_2CH_2{-}B\!< \rightarrow RCH_2CH_2OH \qquad (10)$$

Cis addition. The hydroboration reaction appears to involve a simple four-center cis addition of the hydrogen-boron bond to the carbon-carbon double bond. Reaction (11) indicates this addition.

$$RCH{=}CH_2 + H{-}B\!< \rightarrow [RCH{\cdots}CH_2,\ H{\cdots}B{-}] \rightarrow RCH_2CH_2{-}B{-} \qquad (11)$$

The oxidation proceeds with retention of configuration. Thus the hydroboration-oxidation of cyclic olefins, such as of 1-methylcyclohexene in reaction (12), provides the pure trans alcohol.

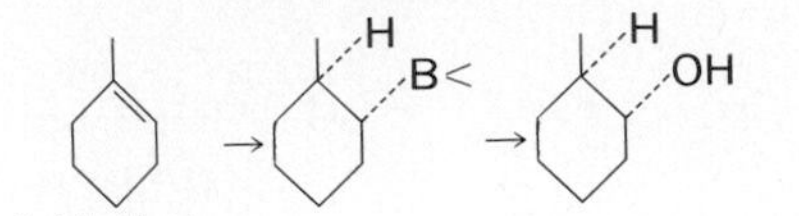

(12)

1-Methyl-cyclo-hexene

Steric effects. The hydroboration reaction is quite sensitive to steric influences. Thus it hydroborates bicyclic olefins, such as norbornene in reaction (13), predominantly from the less hindered

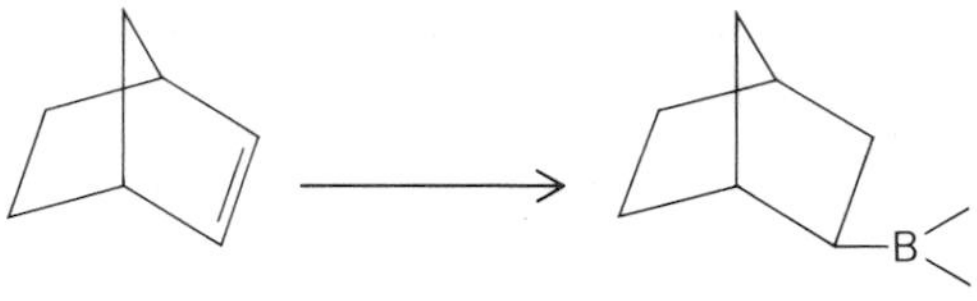

(13)

Norbornene 99.5% *exo*-

side. Moreover, the reaction does not cause rearrangements even in labile systems.

Stereospecific syntheses. These unusual characteristics—(1) anti-Markovnikov addition, (2) freedom from rearrangement, (3) cis addition, and (4) high degree of steric control—give the hydroboration reaction major importance for achieving stereospecific syntheses. This is indicated in the reaction sequence (14), which is the synthesis of isopinocampheol from α-pinene.

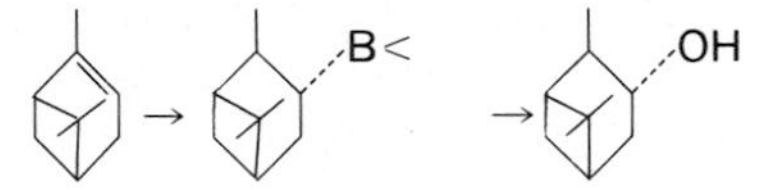

(14)

α-Pinene Isopinocampheol

Isomerization. Although the hydroboration reaction is remarkably free of rearrangements of the carbon structure, even in labile systems, the boron atom is capable of facile migration around the carbon skeleton at temperatures of 100–150°C. The reaction achieves a thermodynamic equilibrium among the possible organoboranes with that particular carbon skeleton. The preferred isomer is the one in which the boron atom is in the least crowded position. Reaction (15) illustrates this isomerization.

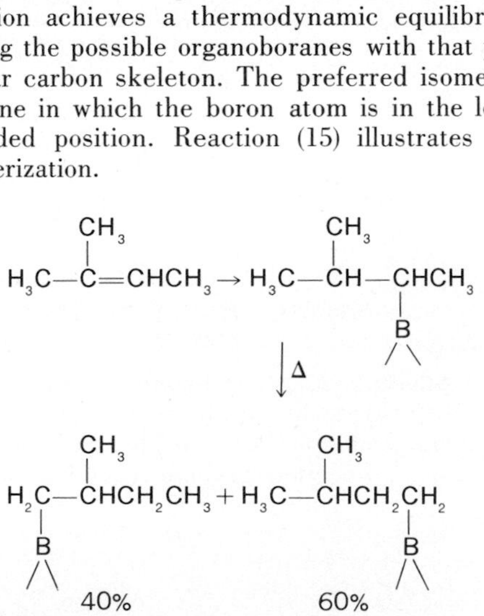

(15)

Displacement reaction. Heating an organoborane with another olefin transfers the boron atom to the new olefin, liberating the original alkyl group as an olefin, as in reaction (16). The reaction can be

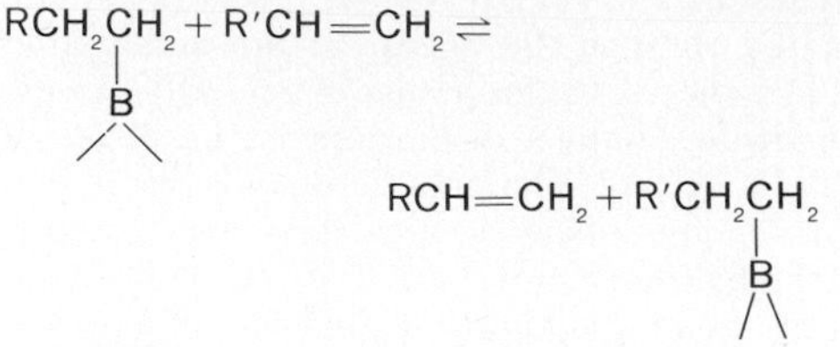

(16)

made to proceed to essential completion by taking advantage of differences in volatility of the olefins, by using an excess of the displacing olefin, or by using an olefin, such as ethylene, which forms a very stable organoborane.

Contrathermodynamic isomerization. A combination of isomerization of the organoborane and the displacement reaction makes it possible to move double bonds from the more stable position at an alkyl branch or internal position to the less stable, unbranched terminal position. This is shown by the sequence in reaction (17).

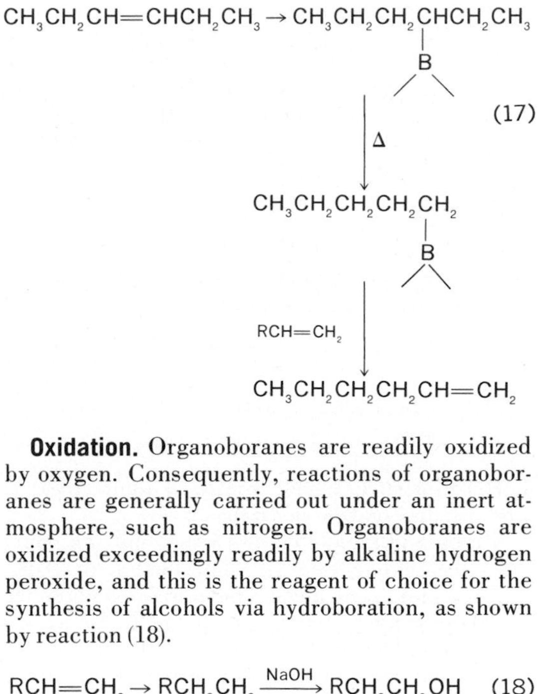

(17)

Oxidation. Organoboranes are readily oxidized by oxygen. Consequently, reactions of organoboranes are generally carried out under an inert atmosphere, such as nitrogen. Organoboranes are oxidized exceedingly readily by alkaline hydrogen peroxide, and this is the reagent of choice for the synthesis of alcohols via hydroboration, as shown by reaction (18).

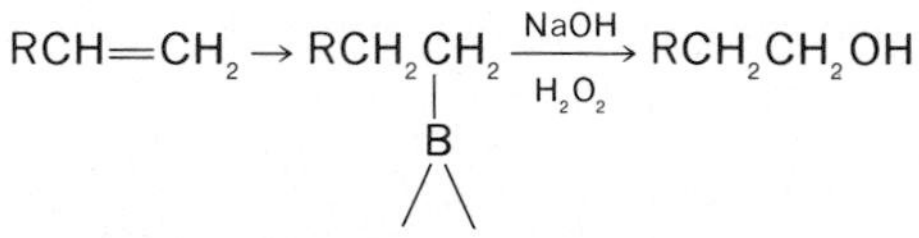

(18)

Amination. Chloramine and *o*-hydroxylaminesulfonic acid convert organoboranes into the corresponding amine. Amination with *o*-hydroxylaminesulfonic acid is shown by reaction (19).

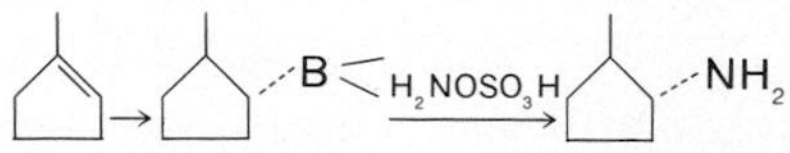

(19)

Coupling. Treatment of an olefin after hydroboration with alkaline silver nitrate produces a coupled product, reactions (20) and (21), that is a new compound composed of two molecules of the original.

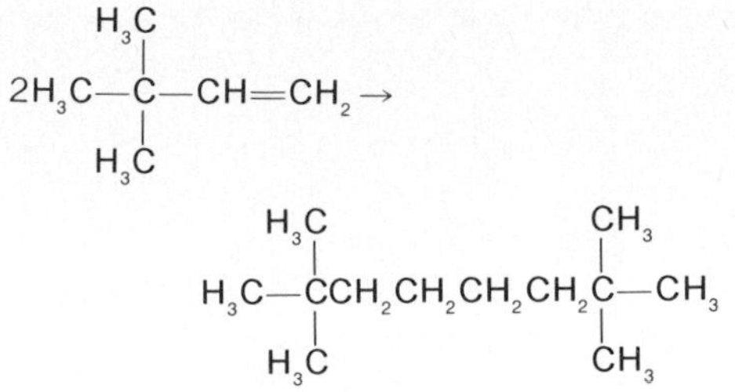

$$2H_3C-\underset{\underset{H_3C}{|}}{\overset{\overset{H_3C}{|}}{C}}-CH=CH_2 \rightarrow$$

$$H_3C-\underset{\underset{H_3C}{|}}{C}CH_2CH_2CH_2CH_2\underset{\underset{CH_3}{|}}{C}-CH_3 \quad (20)$$

$$2CH_2=CH(CH_2)_8CO_2C_2H_5 \rightarrow$$

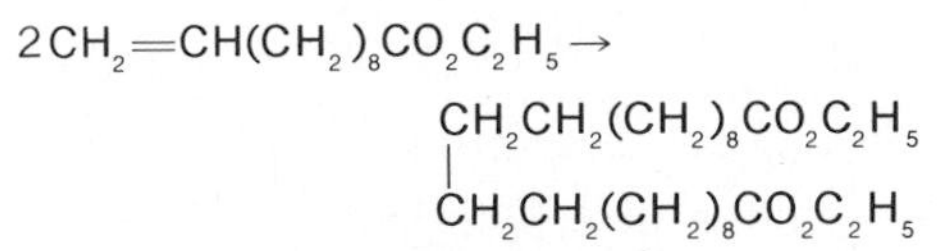

$$\begin{array}{l} CH_2CH_2(CH_2)_8CO_2C_2H_5 \\ CH_2CH_2(CH_2)_8CO_2C_2H_5 \end{array} \quad (21)$$

1,4-Additions. Organoboranes react very rapidly with certain α,β-unsaturated aldehydes and ketones to give saturated products. This is illustrated by reaction (22).

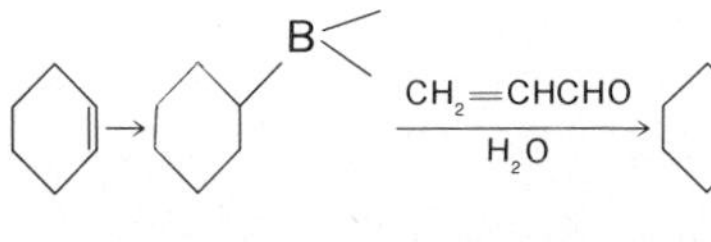

$$(22)$$

Condensations. Organoboranes react with α-halo-substituted carbanions to transfer alkyl groups from boron to carbon. This is shown by reactions (23) and (24), where the symbol KOt-Bu represents potassium tertbutoxide.

$$R_3B+CH_2BrCO_2C_2H_5 \xrightarrow{KOt\text{-}Bu} RCH_2CO_2C_2H_5 \quad (23)$$

$$R_3B+CH_2Br_2CO_2C_2H_5 \xrightarrow{KOt\text{-}Bu} RCHBrCO_2C_2H_5 \quad (24)$$

Carbonylation. Organoboranes react with carbon monoxide to produce intermediates which can be converted to tertiary alcohols, as in reaction (25),

$$R_3B+CO \rightarrow R_3CBO \xrightarrow{[O]} R_3COH \quad (25)$$
$$\text{Tertiary alcohol}$$

secondary alcohols, ketones, aldehydes, methylol derivatives, ring ketones, as in reaction (26), and polycyclics, as in reaction (27).

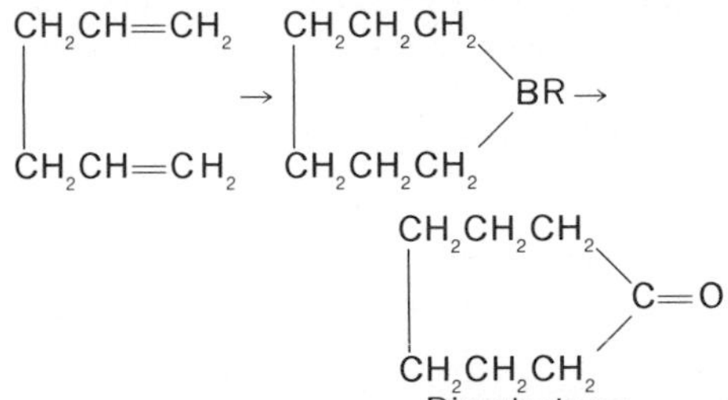

$$(26)$$
Ring ketone

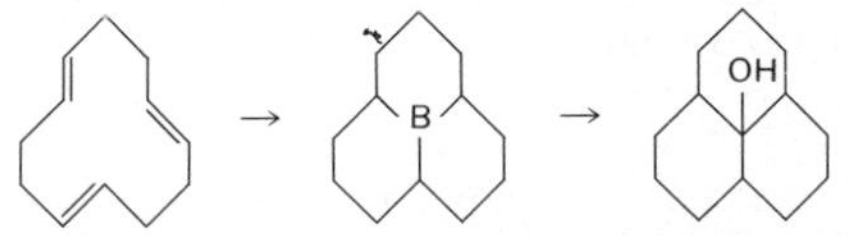

$$(27)$$
Polycyclic
compound

Future developments. It is already apparent that the organoboranes are among the most versatile synthetic intermediates that are available to the organic chemist. The hydroboration reaction

has made these intermediates readily available. *See* BORANE; CARBORANE; METAL HYDRIDES; ORGANIC CHEMICAL SYNTHESIS; ORGANOMETALLIC COMPOUND.

[HERBERT C. BROWN]

Bibliography: H. C. Brown, *Hydroboration*, 1962; H. C. Brown, *Organic Syntheses via Boranes*, 1975; G. E. Coates et al., *Principles of Organometallic Chemistry*, 1968; G. Cragg, *Organoboranes in Organic Chemistry*, 1973; T. Onak, *Organoborane Chemistry*, 1975.

Hydrocarbon

One of a group of chemical compounds composed only of hydrogen and carbon. The very large variety of hydrocarbons can best be divided into three classes (aliphatic, alicyclic, and aromatic), each of which may be further divided into a number of subclasses.

Aliphatic hydrocarbons. These are open-chain compounds which may be saturated or unsaturated. The saturated compounds, known as paraffin hydrocarbons or alkanes, include methane and its homologs having the empirical formula C_nH_{2n+2}. The unsaturated compounds fall into a number of homologous series: (1) those containing one double bond (ethylene and its homologs) and having the formula C_nH_{2n} are known as olefins or alkenes; (2) those containing one triple bond (acetylene and its homologs) are called acetylenes or alkynes and have the formula C_nH_{2n-2}; (3) those having two double bonds (allene, 1,3-butadiene, and 1,4-pentadiene represent three types) are diolefins or alkadienes and also have the formula C_nH_{2n-2}; (4) those having a larger number of double or triple bonds or of both double and triple bonds are named in analogous fashion as alkatrienes, alkatetraenes, alkadiynes, alkenynes, and alkadienynes (Fig. 1). *See* ALIPHATIC HYDROCARBON.

Alicyclic hydrocarbons. These nonaromatic cyclic (ring) compounds fall into a larger number of classes than do the aliphatic hydrocarbons because the rings may be of various sizes, they may

$$\underset{\text{Ethane}}{CH_3-CH_3} \quad \underset{\text{Ethylene}}{CH_2=CH_2} \quad \underset{\text{Acetylene}}{CH\equiv CH}$$

$$\underset{\text{1,2-Butadiene}}{CH_2=C=CH-CH_3} \quad CH_2=\underset{\underset{CH_3}{|}}{C}-CH=CH_2$$
2-Methyl-1,3-butadiene
(isoprene)

$$CH_2=CH-CH_2-CH=CH_2$$
1,4-Pentadiene

$$CH\equiv C-C\equiv C-CH_3$$
1,3-Pentadiyne

$$CH_3-\underset{\underset{CH_3}{|}}{C}=CH-CH=CH-\underset{\underset{CH_3}{|}}{C}=CH-CH_3$$
2,6-Dimethyl-2,4,6-octatriene
(alloocimene)

$$CH_2=CH-C\equiv CH$$
1-Buten-3-yne

$$CH_2=CH-CH=CH-C\equiv CH$$
1,3-Hexadien-5-yne

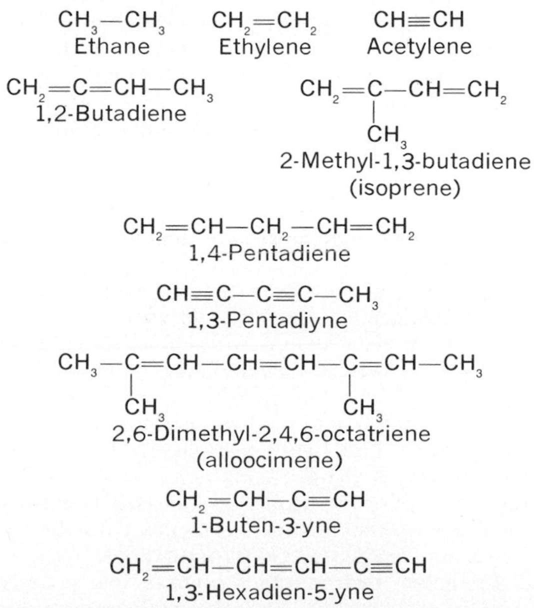

Fig. 1. Formulas of some aliphatic hydrocarbons.

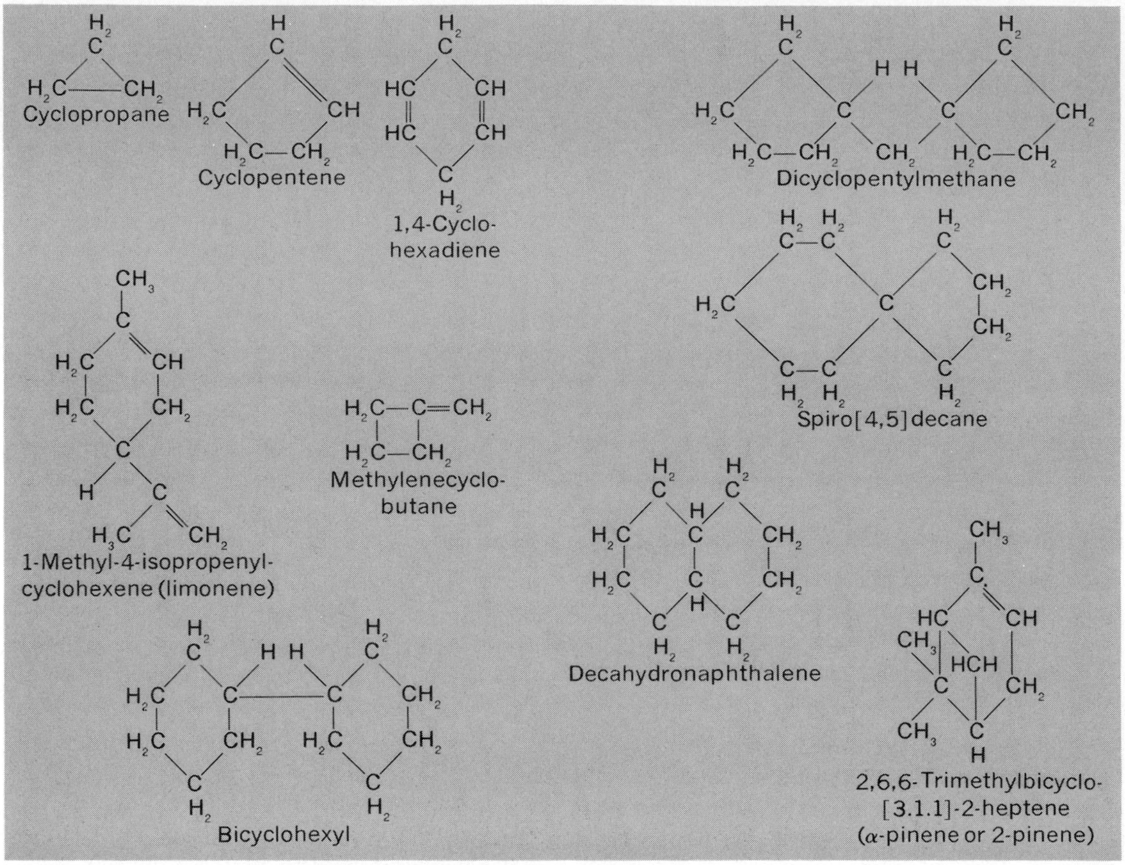

Fig. 2. Typical alicyclic hydrocarbons.

be saturated or unsaturated, and the individual members may contain one or more rings. Typical alicyclic hydrocarbons are shown in Fig. 2. The saturated monocyclic compounds, C_nH_{2n}, known as cycloparaffins or cycloalkanes, include a large number of homologous series of which those containing five or six carbon atoms in the ring (cyclopentanes and cyclohexanes) are the most stable and common. The unsaturated monocyclic hydrocarbons include the cycloolefins, or cycloalkenes (C_nH_{2n-2}), and cyclodioıefins, or cycloalkadienes (C_nH_{2n-4}). There are very few examples of cycloacetylenes, or cycloalkynes, since a ring containing a triple bond would be under strain and therefore unstable.

Olefin is sometimes used as a generic term to include not only alkenes but also cycloalkenes and hydrocarbons containing more than one ethylenic double bond.

The rings in polycyclic alicyclic hydrocarbons may be connected in a variety of ways. These are typified by bicyclohexyl, dicyclopentylmethane, spiro[4.5]decane, decahydronaphthalene, and 2-pinene.

The saturated alicyclic hydrocarbons are sometimes called naphthenes, particularly by petroleum chemists. The word naphthene is related to the fact that cyclopentane and cyclohexane homologs have been isolated from the naphtha fraction of petroleum. *See* ALICYCLIC HYDROCARBON.

Aromatic hydrocarbons. These compounds contain at least one 6-membered benzene ring, incorporating what appear to be three conjugated double bonds (Fig. 3). However, none of these bonds actually have the olefinic character associated with alkenes and cycloalkenes. The aromatic hydrocarbons fall into a number of types: (1) benzene, alkylbenzenes (ethylbenzene), alkenylbenzenes (styrene), and alkynylbenzenes (phenylacetylene); (2) cycloalkylbenzenes (cyclopentylbenzene) and cycloalkenylbenzenes (phenylcyclohexenes); (3) fused ring polynuclear hydrocarbons (indene, indan, naphthalene, tetrahydronaphthalene, fluorene, anthracene); (4) polynuclear aromatic hydrocarbons having directly united rings (biphenyl, binaphthyl) or rings united through aliphatic carbon (diphenylmethane, 1,2-diphenylethylene, diphenylacetylene). *See* AROMATIC HYDROCARBON.

Natural sources of hydrocarbons. Of these sources, the largest are natural gas and petroleum. The composition of natural gas varies greatly depending on the area from which it is obtained. The methane content usually ranges from about 50 to 90%; ethane, about 5 to 20%; propane, about 3 to 18%; and butanes about 1 to 7%.

Petroleum is a complex mixture of liquid and solid hydrocarbons, its composition also varying with the source. Its chief components are paraffinic hydrocarbons, saturated five- and six-carbon atom alicyclic hydrocarbons, and aromatic hydrocarbons. Olefins are usually absent. Petroleum is distilled into a number of commercial fractions: gasoline, boiling at about 20–200°C; kerosine, 175–275°C; and heating oils, 250–400°C. Refining of higher boiling fractions by treatment

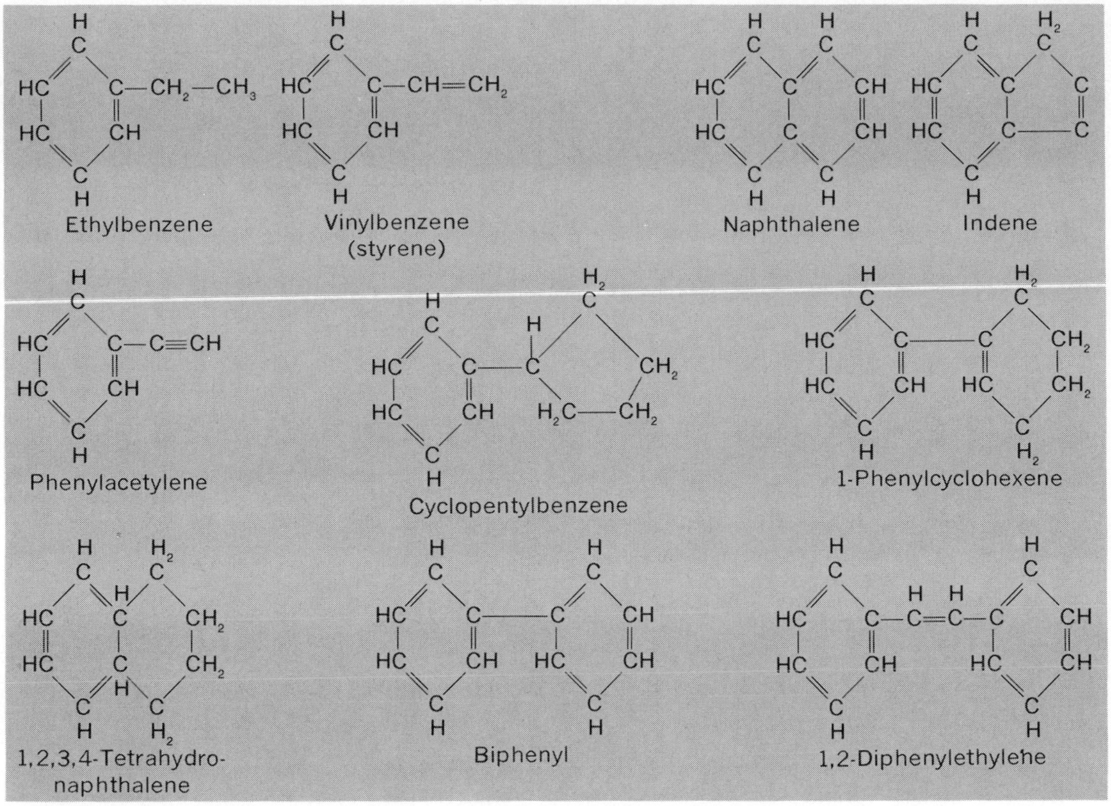

Fig. 3. A few of the aromatic hydrocarbons.

with acids or clays and by crystallization yields lubricating oils, white mineral oils, petrolatums (petroleum jelly), and paraffin wax, composed predominantly of straight-chain alkanes.

Some normal paraffins have been isolated from vegetable products. For example, *n*-heptane, used as a standard fuel in determination of the knock rating (octane number) of gasoline, was formerly obtained by distillation and chemical treatment of the oil obtained from the resin of the Jeffrey pine; it is now prepared by distillation from straight-run gasoline.

Hydrocarbons are obtained by the destructive distillation or carbonization of coal (heating in the absence of air) at 350–1000°C, producing coal gas and coal tar. Coal gas contains methane together with smaller amounts of ethane, ethylene, benzene, toluene, cyclopentadiene, naphthalene, and nonhydrocarbon products such as hydrogen, ammonia, carbon monoxide, carbon dioxide, hydrogen sulfide, hydrogen cyanide, cyanogen, and nitric oxide. The normally liquid or solid hydrocarbons in the gas (namely the cyclopentadiene, benzene, toluene, and napthalene) can be recovered. The gas is further treated to remove toxic components and then used as a domestic fuel.

The volatile oils obtained from certain plants (caraway or dill), trees (pine), and citrus fruits (lemon or orange) contain members of a large group of hydrocarbons known as terpenes, having the formula $C_{10}H_{16}$. These exist both as aliphatic compounds (alkatrienes such as alloocimene, or 2,6-dimethyl-2,4,6-octatriene) or more usually, as alicyclic compounds (limonene or 1-methyl-4-isopropenylcyclohexene, α-pinene, or 2,6,6-trimethylbicyclo[3.1.1]-2-heptene). Related hydrocarbons include the sesquiterpenes, $C_{15}H_{24}$; diterpenes, $C_{20}H_{32}$; triterpenes, $C_{30}H_{48}$; and polyterpenes, $C_{10n}H_{16n}$.

Rubber, which like the terpenes is a polymer of isoprene, may be considered an open-chain polyterpene. The rubber molecule contains one double bond per isoprene unit. *See* TERPENE.

[LOUIS SCHMERLING]

Bibliography: E. Clar, *Polycyclic Hydrocarbons*, 2 vols., 1964; P. W. Jones and P. Leber (eds.), *Polynuclear Aromatic Hydrocarbons*, 1979; E. L. Kugler and F. W. Steffgen (eds.), *Hydrocarbon Synthesis from Carbon Monoxide and Hydrogen*, 1979; A. Oblad et al. (eds.), *Hydrocarbon Chemistry*, 1979.

Hydrocarbon resin

Brittle or gummy materials prepared by the polymerization of several unsaturated constituents of coal tar, rosin, or petroleum. The hydrocarbon resins are inexpensive and find many uses in rubber and asphalt formulations and in coating and calking compositions. Brief descriptions will be given of the coumarone-indene resins, the petroleum resins, and the polyterpene resins.

Coumarone-indene resins. Coumarone and indene, structural formulas of which are shown below, occur together in coal tar fractions. Indene

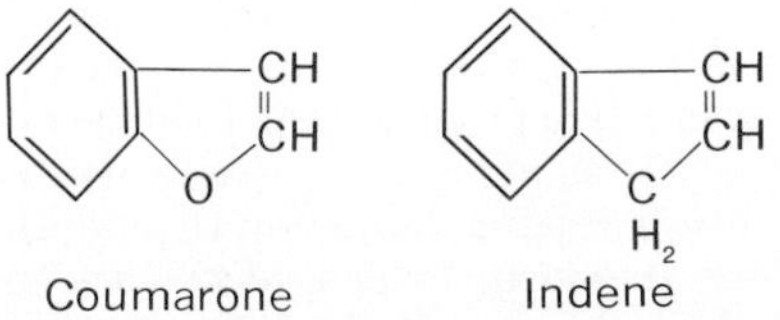

Coumarone Indene

usually predominates, and the mixture may contain small amounts of cyclopentadiene and styrene. Polymerization, which occurs through the carbon-carbon double bond, may be readily effected in the presence of sulfuric acid to yield soft to hard, brittle polymers. The molecular weights vary from about 500 to 4000, and the colors of the products range from yellow to dark brown. The soft products are used as plasticizers and tackifiers in calking formulations and rubber compositions. The hard products are used to stiffen and strengthen synthetic rubber, and they also are used in floor coverings, roofing materials, varnishes, and metal paints, and in combination with other resins.

Indene without coumarone can be obtained from petroleum naphtha, and it polymerizes to yield products which are generally similar to the coumarone-indene resins.

Petroleum resins. Mixtures of polymerizable dienes, for example, cyclopentadiene, and unsaturated compounds are obtained from gasoline-refining operations. The properties and uses of the resins are generally similar to those of the coumarone-indene resins. The resins are recommended for use in paper coatings and hot-melt adhesive formulations. Aqueous suspensions can be used in making water-based paint formulations.

Polyterpenes. The α- and β-pinene obtained from turpentine polymerize in the presence of acid catalysts with the formation of soft to hard polymers of molecular weights up to about 2000. The hard products have softening points up to about 100°C.

The properties and uses of the polyterpenes are generally similar to those of the other hydrocarbon resins described above. They are used in oil paints and adhesive formulations. *See* POLYMERIZATION.

[JOHN A. MANSON]

Hydrocracking

A catalytic, high-pressure process flexible enough to produce either of the two major light fuels—high octane gasoline or aviation jet fuel. It proceeds by two main reactions: adding hydrogen to molecules too massive and complex for gasoline and then cracking them to the required fuels. The process is carried out by passing oil feed together with hydrogen at high pressure (1000–2500 psig) and moderate temperatures (500–750°F) into contact with a bifunctional catalyst, comprising an acidic solid and a hydrogenating metal component. Gasoline of high octane number is produced, both directly and through a subsequent step such as catalytic reforming; jet fuels may also be manufactured simply by changing conditions with the same catalysts. The process is characterized by a long catalyst life (2–4 years), though a slow decline in activity occurs, caused by the deposition of carbonaceous material on the catalyst. Regeneration at intervals by burning off these deposits restores the activity, but eventually the catalyst porosity is destroyed and it must be replaced.

Generally, the process is used as an adjunct to catalytic cracking. Oils, which are difficult to convert in the catalytic process because they are highly aromatic and cause rapid catalyst decline, can be easily handled in hydrocracking, because of the low cracking temperature and the high hydrogen pressure, which decreases catalyst fouling. Usually, these oils boil at 400–1000°F, but it is possible to process even higher-boiling feeds if very high hydrogen pressures are used. However, the most important components in any feed are the nitrogen-containing compounds, because these are severe poisons for hydrocracking catalysts and must be removed to a very low level.

Hydrocracking was carried out on a practical scale in Germany and England starting in the 1930s. In this early work, a common hydrocracking catalyst was tungsten disulfide on acid-treated clay; thus, both hydrogenation and acidic components were present. Generally, a light oil from coal or coking products was vaporized and passed over the catalyst at high pressure. After separation of gasoline from the products, the unconverted material was returned to the reactor with a fresh portion of feed. Because this catalyst was not very active, the process had to be carried out at very high pressures and temperatures (4000 psig; 750°F). It was costly and the products were not of high quality.

Research in the United States concentrated on the development of much more active catalysts, a different mode of operation, and the use of heavier oil feeds. As a result, the reaction is carried out in two separate, consecutive stages; in each, oil and hydrogen at high pressure flow downward over fixed beds of catalyst pellets placed in large vertical cylindrical vessels.

First stage. In the first, or pretreating, stage the main purpose is conversion of nitrogen compounds in the feed to hydrocarbons and to ammonia by hydrogenation and mild hydrocracking. Typical conditions are 650–740°F, 150–2500 psig, and a catalyst contact time of 0.5–1.5 hr; up to 1.5 wt% hydrogen is absorbed, partly by conversion of the nitrogen compounds, but chiefly by aromatic compounds which are hydrogenated. It is most important to reduce the nitrogen content of the product oil to less than 0.001 wt% (10 parts per million). This stage is usually carried out with a bifunctional catalyst containing hydrogenation promotors, for example, nickel and tungsten or molybdenum sulfides, on an acidic support, such as silica-alumina. The metal sulfides hydrogenate aromatics and nitrogen compounds and prevent deposition of carbonaceous deposits; the acidic support accelerates nitrogen removal as ammonia by breaking carbon-nitrogen bonds. The catalyst is generally used as 1/8 × 1/8 in. or 1/16 × 1/8 in. pellets, formed by extrusion.

Second stage. Most of the hydrocracking is accomplished in the second stage, which resembles the first but uses a different catalyst. Ammonia and some gasoline are usually removed from the first-stage product, and then the remaining oil, which is low in nitrogen compounds, is passed over the second-stage catalyst. Again, typical conditions are 600–700°F, 1500–2500 psig hydrogen pressure, and 0.5–1.5 hr contact time; 1–1.5 wt% hydrogen may be absorbed. Conversion to gasoline or jet fuel is seldom complete in one contact with the catalyst, so the lighter oils are removed by distillation of the products and the heavier, high-boiling product combined with fresh feed and recycled over the catalyst until it is completely converted.

The catalyst for the second stage is also a bifunctional catalyst containing hydrogenating and acidic components. Metals such as nickel, molybdenum, tungsten, or palladium are used in various combinations, dispersed on solid acidic supports such as synthetic amorphous or crystalline silica-aluminas, such as zeolites. These supports contain strongly acidic sites and sometimes are enhanced by the incorporation of a small amount of fluorine. A long period (for example, 1 year) between regenerations is desirable; this is achieved by keeping a low nitrogen content in the feed and avoiding high temperatures, which lead to excess cracking with consequent deposition of coke on the catalyst. When activity of the catalyst does decrease, it can be restored by carefully controlled burning of the coke.

The catalyst is the key to the success of the hydrocracking process as now practiced, particularly the second-stage catalyst. Its two functions must be most carefully balanced for the product desired; that is, too much hydrogenation gives a poor gasoline but a good jet fuel. The oil feeds are composed of paraffins, other saturates, and aromatics—all complex molecules boiling well above the required gasoline or jet-fuel product. The catalyst starts the breakdown of these components by forming from them carbonium ions, that is, positively charged molecular fragments, via the protons (H^+) in the acidic function. These ions are so reactive that they change their internal molecular structure spontaneously and break down to smaller fragments having excellent gasoline qualities. The hydrogenating function aids in maintaining and controlling the ion reactions and protects the acid function by hydrogenating coke precursors off the catalyst surface, thus maintaining catalyst activity. Any olefins formed in the carbonium ion decomposition are also hydrogenated.

Products. The products from hydrocracking are composed of either saturated or aromatics compounds; no olefins are found. In making gasoline, the lower paraffins formed have high octane numbers; for example, the 5- and 6-carbon number fractions have leaded research octane numbers of 99–100. The remaining gasoline has excellent properties as a feed to catalytic reforming, producing a highly aromatic gasoline which, with added lead, easily attains 100 octane number. Both gasolines are suitable for premium-grade motor gasoline. Another attractive feature of hydrocracking is the low yield of gaseous components, such as methane, ethane, and propane, which are less desirable than gasoline. When making jet fuel, more hydrogenation activity of the catalysts is used, since jet fuel contains more saturates than gasoline.

The hydrocracking process is being applied in other areas, notably, to produce lubricating oils and to convert very asphaltic and high-boiling residues to lower-boiling fuels. Its use will certainly increase greatly in the future, since it accomplishes two needed functions in the petroleum-fuel economy: Large, unwieldly molecules are cracked, and the needed hydrogen is added to produce useful, high-quality fuels. *See* AROMATIZATION; CRACKING; HYDROGENATION; ISOMERIZATION.

[CHARLES P. BREWER]

Bibliography: *Advan. Petrol. Chem. Refining*, 8:168–191, 1964; W. F. Bland and R. L. Davidson (eds.), *Petroleum Processing Handbook*, sec. 3, pp. 16–25, 1967; *Hydrocarbon Process.*, 47(9): 139–144, 1968; Hydroprocesses, *Kirk-Othmer Encyclopedia of Chemical Technology*, 2d ed., vol. 11, 1966.

Hydroformylation

An aldehyde synthesis process that falls under the general classification of a Fischer-Tropsch reaction but is distinguished by the addition of an olefin feed along with the characteristic carbon monoxide and hydrogen. In the oxo process for alcohol manufacture, hydroformylation of olefins to aldehydes is the first step. The second step is the hydrogenation of the aldehydes to alcohols. At times the term "oxo process" is used in reference to the hydroformylation step alone. In the hydroformylation step, olefin, carbon monoxide, and hydrogen are reacted over a cobalt catalyst to produce an aldehyde which has one more carbon atom than the feed olefin. The olefin conversion takes place by the addition of a formyl group (CHO) and a hydrogen atom across the double bond. This is represented by reaction (1). *See* FISCHER-TROPSCH PROCESS.

$$R{-}CH{=}CH_2 + CO + H_2 \xrightarrow{Co} \begin{cases} R{-}CH_2{-}CH_2{-}CHO \\ R{-}\underset{\displaystyle CHO}{\underset{|}{CH}}{-}CH_3 \end{cases} \quad (1)$$

The aldehyde is then treated with hydrogen to form the alcohol. In commercial operations, the hydrogenation step is usually performed immediately after the hydroformylation step in an integrated system.

A wide range of carbon number olefins, C_2–C_{16}, have been used as feeds. Propylene, heptene, and nonene are frequently used as feedstocks to produce normal and isobutyl alcohol, isooctyl alcohol, and primary decyl alcohol, respectively. Feed streams to oxo units may be single-carbon-number or mixed-carbon-number olefins.

The lower-carbon-number alcohols such as butanols are used primarily as solvents, while the higher-carbon-number alcohols go into the manufacture of plasticizers, detergents (surfactants), and lubricants.

Reactants. Reactants are CO, H_2, and olefins. The H_2 and CO are usually fed in a 1-to-1 ratio, as synthesis gas from methanol conversion. The hydroformylation takes place in the liquid phase. If the olefins are normally gaseous at the reactor conditions, a heavier liquid solvent is used as a suspension medium. The synthesis gas is usually fed in considerable excess of the required stoichiometric amount.

The reaction mass is contacted with the cobalt catalyst either by flowing the liquid reaction mass over a fixed bed of supported cobalt catalyst or by pumping a cobalt catalyst slurry into the liquid reaction mass in a continuous stirred tank or plug flow reactor. In the latter case, the catalyst is carried along with the reaction mass in what is frequently called the slurry system.

The olefin reactants are fed in blocked operation as either single-carbon-number C_7, C_8, or C_9 or in multiple-carbon-number feeds, such as C_{11}–C_{13}. For straight-chain olefins the double bond can be

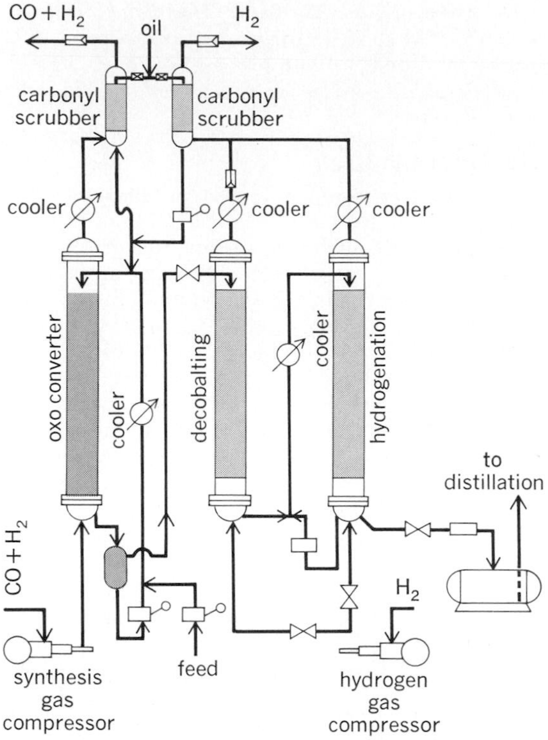

Fig. 1. Oxo process with fixed catalyst bed.

either in the terminal position or internal position. The product distribution is approximately the same in either case, roughly 60–40% normal alcohol and 40–60% alpha branched alcohol. Although the product distribution is nearly the same, the reaction rate when the double bond is in the terminal position is almost 3.5 times faster than the rate with double bond in an internal position.

The reaction rate has been found to be directly proportional to the olefin concentration, the cobalt concentration, and the hydrogen pressure, and inversely proportional to carbon monoxide pressure. The proposed mechanism for the reaction begins with the formation of an olefin-carbonyl complex and carbon monoxide from the reaction of

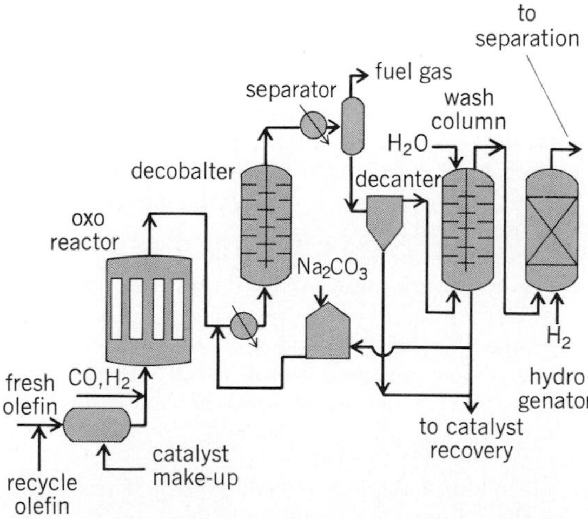

Fig. 2. The slurry process, in which the catalyst is carried with the liquid reactant.

dicobalt octacarbonyl with olefin, as shown in reaction (2). The complex decomposes through a reac-

$$C_0(CO)_8 + RCH{=}CH_2 \rightarrow C_0(CO)_7RCH{=}CH_2 + CO \quad (2)$$

tion with a hydrogen or cobalt hydrocarbonyl to form the aldehyde and a precursor of dicobalt octacarbonyl. The heat of reaction of the hydroformylation step is about 30 kcal/mole (125,500 J/mole). The capability of removing the heat of reaction is often the limiting factor in the reactor capacity.

Commercial operations. Two general types of liquid-phase process have been used. The first process employs a fixed bed of cobalt catalyst with the liquid reactant mixture flowing past the catalyst. The second type of process, the slurry type, carries the catalyst along with the liquid reactant. The slurry type of reactors can be either the mechanically agitated, stirred-tank type or the gas-sparged type. A stirred-tank type with mechanical agitation can be used when the operating pressure is not too high. However, many operations are run at high pressure, 100–300 atm ($1-3 \times 10^6$ Pa). For these processes, a gas-sparged slurry-type reactor can be designed to use the gas rising through the liquid to provide mixing and thus eliminate the mechanical agitator and seals.

The fixed-bed process is shown schematically in Fig. 1. In this process, soluble cobalt salts of fatty acids or naphthenates are pumped with the olefin to the top of the first reactor and flow countercurrent to the synthesis gas. One type of fixed-bed catalyst consists of 2% metallic cobalt on a pumice carrier. Part of the cobalt is converted to carbonyl, leaves the reactor with the overhead product, and is replaced by cobalt salts in the feed. Unreacted synthesis gas leaving the top of the reactor is cooled, passed through a packed tower countercurrent to the olefin feed to remove cobalt carbonyl, and recycled to the reactor.

The second vessel is a decobalting converter in which cobalt carbonyl, dissolved in the product from the first reactor, is decomposed by treatment with hydrogen at about 200–220 atm ($2-2.2 \times 10^6$ Pa) and 125–150°C. The liquid enters at the top and flows countercurrent to the hydrogen. Metallic cobalt is deposited on the packing. Gas flows from the top of the decobalting unit to the carbonyl scrubber, and the liquid leaving the bottom is sent to the hydrogenation reactor for conversion to alcohol.

The slurry type of process shown schematically in Fig. 2 begins with cobalt oxide being mixed with recycle olefins. This slurry is then combined with the main olefin feed stream to give a 3–5 wt % cobalt slurry. This reaction slurry is fed into the bottom of the first reactor along with the synthesis gas. This process may employ up to five reactors in series. In the reactors, the catalyst reacts with the synthesis gas to form cobalt hydrocarbonyl, which is a gas at the reactor conditions of 180°C and 220 atm. The liquid-gas mixture exits the last reactor of the series, and an aqueous solution of sodium carbonate is injected into this stream to form the water-soluble sodium cobalt carbonylate. The aqueous phase is separated from the organic stream in the decanter, and the organic stream is then water-washed to remove the last traces of catalyst. The washed aldehyde stream is then fed to the hydrogenator from which the product alcohol goes to the separation train downstream. The catalyst is recovered from the wash water and recycled to the process.

[D. L. HOLT]

Hydrogen

The first chemical element in the periodic system. Under ordinary conditions it is a colorless, odorless, tasteless gas composed of diatomic molecules, H_2. The hydrogen atom, symbol H, consists of a nucleus of unit positive charge and a single

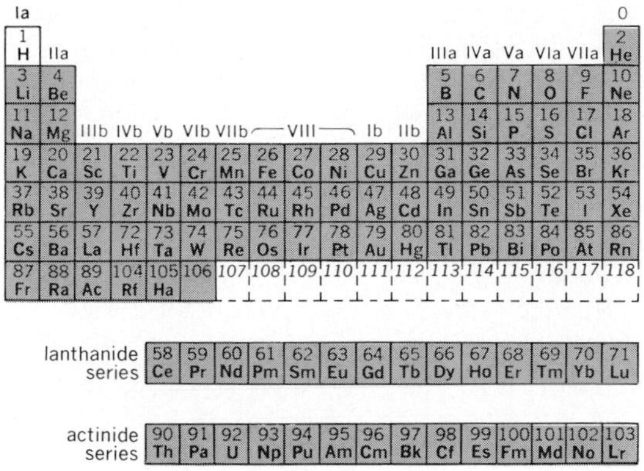

electron. It has atomic number 1 and an atomic weight of 1.00797. The element is a major constituent of water and all organic matter, and is widely distributed not only on the Earth but throughout the universe. There are three isotopes of hydrogen: protium, mass 1, makes up 99.98% of the natural element; deuterium, mass 2, makes up about 0.02%; and tritium, mass 3, occurs in extremely small amounts in nature but may be produced artificially by various nuclear reactions. *See* DEUTERIUM; ISOTOPE; TRITIUM.

Although hydrogen had been produced by earlier workers by the reaction of metals with acid, it was Henry Cavendish (1731–1810) who first distinguished it from other flammable gases.

Uses. The largest single use of hydrogen is in the synthesis of ammonia. Ammonia plants are often built adjacent to petroleum refineries or coking plants to utilize by-product hydrogen which might otherwise be wasted. A rapidly expanding use for hydrogen is in petroleum-refining operations, such as hydrocracking and hydrogen treatment for removal of sulfur. Large quantities of hydrogen are consumed in the catalytic hydrogenation of unsaturated liquid vegetable oils to make solid fats. Hydrogenation is used in the manufacture of organic chemicals, such as alcohols from esters and glycerides, amines from nitriles, and cycloparaffins from aromatic hydrocarbons. Methanol is produced commercially by reaction of hydrogen with carbon monoxide. Reaction of hydrogen with chlorine is a major source of hydrochloric acid.

Large quantities of hydrogen are used in the United States space program. It is used both as a rocket fuel, in conjunction with oxygen or fluorine, and as a propellant for nuclear-powered rockets. These uses have led to rapid development of the technology of the production and handling of liquid hydrogen. Liquid hydrogen is also used in bubble chambers for studying high-energy particles from nuclear accelerators. Hydrogen gas is used in the oxyhydrogen torch and in the atomic hydrogen torch to produce high temperatures for welding and cutting metals. It is used in the metallurgical industries to reduce metal oxides, such as those of tungsten and molybdenum, to provide a reducing atmosphere in the heat treatment of metals, and in the manufacture of metal hydrides. It is used increasingly in the production of iron from its ores. Although once used extensively to inflate dirigible balloons, it has been largely superseded by helium because of a number of disastrous explosions. When not used directly at the site of manufacture, hydrogen is transported and stored as gas in steel cylinders at a pressure of 120–150 atm, or as liquid in large, well-insulated tanks.

Natural occurrence. Hydrogen, in the free state, is only a minor component of the Earth. It constitutes less than 1 ppm of the atmosphere. It is found in the gases from some volcanoes, oil wells, and coal mines. It may be liberated as a product of the decomposition of organic matter and has been observed in the intestinal gases of animals. In the combined state, hydrogen makes up 0.76% of the weight of the Earth's crust to make it the ninth most abundant element; 13.5% of the atoms of the Earth's crust are hydrogen, exceeded in number only by oxygen and silicon. Most of this hydrogen is present in sea water, of which it constitutes 10.82% by weight. Other important occurrences are in minerals, as hydrates, in the hydrocarbons of petroleum deposits, and in the organic constituents of all living organisms.

Hydrogen is believed to constitute approximately 90% of the atoms in the universe. The thermonuclear energy produced by fusion reactions of hydrogen nuclei is the source of most of the energy radiated by the Sun and other stars.

Physical properties. Ordinary hydrogen has an atomic weight of 1.00797, and a molecular weight of 2.01594. The gas has a density at 0°C and 1 atm of 0.08987 g/liter. Its specific gravity, compared to air, is 0.0695. The lightest substance known, it has a buoyancy in air of 1.203 g/liter. Some additional properties of hydrogen are listed in Table 1.

Table 1. Properties of hydrogen

Property	Value
Melting point	−259.2°C
Boiling point at 1 atm	−252.8°C
Density of solid at −259.2°C	0.0866 g/cm^3
Density of liquid at −252.8°C	0.0708 g/cm^3
Critical temperature	−240.0°C
Critical pressure	13.0 atm
Critical density	0.0301 g/cm^3
Specific heat at constant pressure	
Gas at 25°C	3.42 cal/(g)(°C)
Liquid at −256°C	1.93 cal/(g)(°C)
Solid at −259.8°C	0.63 cal/(g)(°C)
Heat of fusion at −259.2°C	14.0 cal/g
Heat of vaporization at −252.8°C	107 cal/g
Thermal conductivity at 25°C	0.000444 cal/(cm)(cm^2)(sec)(°C)
Viscosity at 25°C	0.00892 centipoise

Hydrogen dissolves in water to the extent of 0.0214 volume per volume of water at 0°C, 0.018 volume at 20°C, and 0.016 volume at 50°C. It is somewhat more soluble in organic solvents, and 0.078 volume dissolves in 1 volume of ethanol at 25°C. Many metals adsorb hydrogen. Palladium is particularly notable in this respect, and dissolves about 1000 times its volume of the gas. The adsorption of hydrogen in steel may cause "hydrogen embrittlement," which sometimes leads to the failure of chemical processing equipment.

The hydrogen atom has an ionization potential of 13.54 volts. The hydrogen nucleus (proton, mass 1) has a spin of $1/2\,\hbar$ and a magnetic moment of 2.79270 nuclear magnetons. Its absorption cross section for thermal neutrons is 0.332×10^{-24} cm^2.

The hydrogen molecule may exist in either of two forms, known as ortho- and parahydrogen. These distinct forms are possible because the nucleus of the hydrogen atom is spinning in a toplike manner and two atoms may combine with their nuclei spinning in the same direction (ortho) or in opposite directions (para). These spin isomers, as they are called, are ordinarily fairly stable but may be rapidly interconverted by a suitable catalyst, such as activated charcoal or platinized asbestos. The ratio of the two isomers in an equilibrium mixture varies markedly with the temperature: near the absolute zero, equilibrium hydrogen consists entirely of the para modification; at room temperature and above, parahydrogen constitutes 25% and orthohydrogen 75% of the mixture. Pure parahydrogen may be readily prepared by passing liquid hydrogen over activated charcoal. Pure orthohydrogen has been prepared only in small amounts by separating it from parahydrogen by processes such as thermal diffusion or gas chromatography. Physical properties of parahydrogen are measurably different from those of ordinary hydrogen. For example, parahydrogen melts at −259.34°C and boils at −252.90°C, whereas ordinary hydrogen melts at −259.21°C and boils at −252.77°C. The thermal conductivity of parahydrogen is markedly greater than that of the ortho form; the difference is used in analyzing mixtures of the two. When ordinary hydrogen is liquified, the heat evolved during the slow conversion of the equilibrium mixture to parahydrogen is responsible for the evaporation of large amounts of liquid hydrogen during storage. These losses may be averted by catalytically converting all of the orthohydrogen to the para form during the liquefaction. The orthopara-hydrogen interconversion can be catalyzed by free hydrogen atoms. Measurement of the rate of interconversion permits the determination of hydrogen-atom concentrations in chemical reactions.

Chemical properties. At ordinary temperatures hydrogen is a comparatively unreactive substance unless it has been activated in some manner, for example, by a suitable catalyst. At elevated temperatures it is highly reactive.

Although ordinarily diatomic, molecular hydrogen dissociates at high temperatures into free atoms according to reaction (1). The heat of disso-

$$H_2 \rightleftharpoons 2H \tag{1}$$

ciation at 25°C is 104.2 kcal/mole. The calculated percentage dissociation is 0.08 at 2000 K, 7.8 at 3000 K, 62.2 at 4000 K, and 95.5 at 5000 K. Atomic hydrogen is also produced when an electrical discharge is passed through hydrogen gas at low pressure, or when a mixture of hydrogen and mercury vapor is irradiated with light of wavelength 2537 A from a mercury arc.

Atomic hydrogen is a powerful reducing agent, even at room temperature. It reacts with the oxides and chlorides of many metals, including silver, copper, lead, bismuth, and mercury, to produce the free metals. It reduces some salts, such as nitrates, nitrites, and cyanides of sodium and potassium, to the metallic state. It reacts with a number of elements, both metals and nonmetals, to yield hydrides such as NaH, KH, H_2S, and PH_3. With oxygen atomic hydrogen yields hydrogen peroxide, H_2O_2. With organic compounds atomic hydrogen reacts to produce a complex mixture of products. With ethylene, C_2H_4, for example, the products include ethane, C_2H_6, and butane, C_4H_{10}. The heat liberated when hydrogen atoms recombine to form hydrogen molecules is used to obtain very high temperatures in atomic hydrogen welding.

Hydrogen reacts with oxygen to form water, as shown by reaction (2). The heat of reaction is 57.6

$$2H_2 + O_2 \rightarrow 2H_2O \tag{2}$$

kcal/mole of hydrogen. At room temperature this reaction is immeasurably slow, but is accelerated by catalysts, such as platinum, or by an electric spark, and then may take place with explosive violence. In the absence of catalysts, reaction between hydrogen and oxygen becomes measurable at about 300°C and rapid above 500°C. The reaction is believed to take place in a series of steps constituting chain reaction (3). The intense heat of

$$\begin{aligned} OH + H_2 &\rightarrow H_2O + H \\ H + O_2 &\rightarrow O + OH \\ O + H_2 &\rightarrow OH + H \end{aligned} \tag{3}$$

the reaction is utilized in the oxyhydrogen torch for cutting and welding metals.

Hydrogen reacts less vigorously with the other group VI elements. The reaction (4) with sulfur is exothermic by only 5 kcal/mole; the correspond-

$$H_2 + S \rightarrow H_2S \tag{4}$$

ing reactions with selenium, reaction (5), and tellurium, reaction (6), are endothermic.

$$H_2 + Se \rightarrow H_2Se \tag{5}$$

$$H_2 + Te \rightarrow H_2Te \tag{6}$$

The reactivity of hydrogen with the halogens decreases in the order fluorine, chlorine, bromine, and iodine. The reaction (7) with fluorine is violent,

$$H_2 + F_2 \rightarrow 2HF \tag{7}$$

even in the dark at −252°C. Properly controlled, it has been used to produce the extremely high temperature of 4000°C. The reaction (8) with chlo-

$$H_2 + Cl_2 \rightarrow 2HCl \tag{8}$$

rine is slow under ordinary conditions, but may become explosive under the influence of light or heat. As in the case of oxygen, a chain reaction (9) is involved.

$$\begin{aligned} Cl_2 + h\nu &\rightarrow Cl + Cl \\ Cl + H_2 &\rightarrow HCl + H \\ H + Cl_2 &\rightarrow HCl + Cl, \text{ etc.} \end{aligned} \tag{9}$$

With nitrogen, hydrogen undergoes the important reaction in (10) to give ammonia. *See* AMMONIA.

$$N_2 + 3H_2 \rightarrow 2NH_3 \qquad (10)$$

Hydrogen reacts at elevated temperatures with a number of metals, including lithium, sodium, potassium, calcium, strontium, and barium, to give hydrides. An example is given by reaction (11). *See* METAL HYDRIDES.

$$H_2 + 2Li \rightleftharpoons 2LiH \qquad (11)$$

The oxides of many metals are reduced by hydrogen at elevated temperatures either to the free metal or to lower oxides. Metals which can be produced from their oxides in this way include copper, silver, bismuth, mercury, tungsten, iron, nickel, and cobalt. Similarly, the chlorides of silver, copper, nickel, and cobalt react with hydrogen at 300–750°C to give hydrogen chloride and the free metal. Metallic bromides and iodides are more easily reduced than the chlorides.

Hydrogen reacts at room temperature with the salts of the less electropositive metals, such as gold, silver, copper, or mercury, in aqueous solution, and reduces them to the metallic state. These reductions are ordinarily quite slow, but are markedly accelerated if the hydrogen is adsorbed on platinum or palladium.

In the presence of a suitable catalyst, such as platinum, palladium, or nickel, hydrogen reacts with unsaturated organic compounds and adds to the double bond. With ethylene, the reaction is as shown in (12). Aldehydes and ketones are similarly

$$H_2C{=}CH_2 + H_2 \rightarrow H_3C{-}CH_3 \qquad (12)$$

reduced to alcohols. Acetaldehyde, for example, is reduced to ethyl alcohol, as shown by reaction (13). *See* HYDROGENATION.

$$CH_3CH{=}O + H_2 \rightarrow CH_3CH_2OH \qquad (13)$$

Principal compounds. Hydrogen is a constituent of a very large number of compounds containing one or more other elements. Such compounds include water, acids, bases, most organic compounds, and many minerals. Compounds in which hydrogen is combined with a single other element are commonly referred to as hydrides. These may be divided into three general classes: the ionic or saltlike hydrides, the covalent or molecular hydrides, and the transitional metal hydrides.

With the halogens, fluorine, chlorine, bromine, and iodine, hydrogen forms hydrides containing 1 atom of hydrogen per atom of halogen. These compounds are usually referred to as hydrogen halides, for example, hydrogen chloride, HCl, or as hydrohalic acids, for example, hydrochloric acid. The hydrogen halides are gases at room temperature. Hydrofluoric acid $(HF)_x$ is polymeric; the others are monomeric. Their physical properties are given in Table 2. The hydrogen halides dissolve in water to give strongly acid solutions. Hydrogen fluoride is commonly prepared from calcium fluoride and sulfuric acid, as shown in reaction (14). Hydrogen chloride is similarly prepared

$$CaF_2 + H_2SO_4 \rightarrow CaSO_4 + 2HF \qquad (14)$$

from sodium chloride and sulfuric acid. Hydrogen bromide and hydrogen iodide may be prepared by direct union of the elements or by hydrolysis of the corresponding phosphorus trihalides, PBr_3 and PI_3.

Table 2. Properties of hydrogen halides

Property	Hydrogen fluoride, $(HF)_x$	Hydrogen chloride, HCl	Hydrogen bromide, HBr	Hydrogen iodide, HI
Melting point, °C	−83.1	−114.8	−86.9	−50.7
Boiling point, °C	19.5	−84.9	−66.8	−35.4
Density of liquid, g/ml	0.991	1.194	2.77	2.85
At temperature °C	19.5	−86	−67	−47

Hydrogen forms two compounds with oxygen, water, H_2O, and hydrogen peroxide, H_2O_2. The physical properties of these compounds are compared in Table 3. Hydrogen peroxide may be prepared by the hydrolysis of a metal peroxide, as shown by reaction (15), or by electrolysis reactions.

Table 3. Properties of hydrogen peroxide and water

Property	Water, H_2O	Hydrogen peroxide, H_2O_2
Melting point, °C	0	−0.9
Boiling point, °C	100	151.4
Density, g/ml	1.000 at 4°C	1.465 at 0°C

$$Na_2O_2 + 2H_2O \rightarrow H_2O_2 + 2NaOH \qquad (15)$$

It is unstable, and slowly evolves oxygen on standing. It is a powerful oxidizing agent, and reacts vigorously with many organic compounds.

With sulfur, hydrogen forms the compound hydrogen sulfide, H_2S, a colorless gas with the odor of rotten eggs. It may be prepared by the direct reaction of hydrogen with sulfur, or by treating a metal sulfide with an acid. A series of polysulfides of hydrogen, with the formulas H_2S_2, H_2S_3, H_2S_4, H_2S_5, and H_2S_6, are also known. With selenium and tellurium, hydrogen forms the compounds hydrogen selenide, H_2Se, and hydrogen telluride, H_2Te.

Boron forms a series of covalent hydrides, the best-known member of which is the gas, diborane, B_2H_6. Other members of the series are pentaborane, B_5H_9, a liquid, and decaborane, $B_{10}H_{14}$, a crystalline solid. Because of their high heats of combustion, these compounds and their derivatives have received considerable attention as possible rocket fuels.

The largest class of the covalent hydrides comprises the compounds of hydrogen with carbon, known as hydrocarbons. The first member of this series is methane, CH_4, a gas.

Among the hydrides of nitrogen are ammonia, NH_3, and hydrazine, N_2H_4. Ammonia is a colorless gas which melts at −78°C and boils at −33.3°C. Hydrazine is a liquid with a melting point of 1.4°C and a boiling point of 113.5°C.

Other covalent hydrides are silane, SiH_4; phosphine, PH_3; arsine, AsH_3; and stibine, SbH_3.

For additional details on the compounds of hydrogen *see* ACID AND BASE; HYDRAZINE; HYDROCARBON; HYDROGEN FLUORIDE; HYDROGEN PEROXIDE; HYDROGENATION; PEROXIDE; SILICON; WATER.

Preparation. A large number of methods may be used to prepare hydrogen gas. The choice of method is determined by such factors as the quantity of hydrogen desired, the purity required, and the availability and cost of raw materials. Among the processes frequently used are the reactions of metals with water or acids, the electrolysis of water, the reaction of steam with hydrocarbons or other organic materials, and the thermal decomposition of hydrocarbons.

Small amounts of hydrogen are readily prepared in the laboratory by the reaction of zinc with dilute hydrochloric acid; this reaction is shown in (16).

$$Zn + 2HCl \rightarrow ZnCl_2 + H_2 \quad (16)$$

The gas may be purified by passing it through an acidified solution of potassium permanganate or potassium dichromate and then through a solution of sodium hydroxide. It may be dried by passage through concentrated sulfuric acid or over silica gel. Other metals more electropositive than hydrogen will likewise liberate hydrogen gas upon treatment with water or acid solutions. Thus, metallic sodium reacts violently with water according to reaction (17). The reaction may be moderated by

$$2Na + 2H_2O \rightarrow 2NaOH + H_2 \quad (17)$$

amalgamating the sodium with mercury. The dissolution of iron filings in dilute sulfuric acid has also been frequently used as a source of hydrogen.

For the preparation of somewhat larger quantities of hydrogen, for example, for use in filling balloons, a convenient method is the reaction (18)

$$CaH_2 + 2H_2O \rightarrow Ca(OH)_2 + 2H_2 \quad (18)$$

of calcium hydride with water. In this reaction 1 kg of calcium hydride will produce about 1 m^3 of hydrogen. Another process sometimes used for the production of moderate amounts of hydrogen is the dissolution of aluminum or of silicon in alkali, shown by reactions (19) and (20).

$$2Al + 2NaOH + 6H_2O \rightarrow 2NaAl(OH)_4 + 3H_2 \quad (19)$$

$$Si + 4NaOH \rightarrow Na_4SiO_4 + 2H_2 \quad (20)$$

Hydrogen gas is produced on a large scale industrially. The amount produced annually in the United States is over 100,000,000,000 ft^3, excluding that consumed in the manufacture of ammonia and methanol or in petroleum refining. The principal raw materials for hydrogen production are now hydrocarbons, such as natural gas, oil refinery gas, gasoline, fuel oil, and crude oil. In catalytic steam-hydrocarbon reforming, which has become the dominant production process, volatile hydrocarbons are reacted with steam over a nickel catalyst at 700–1000°C to produce carbon oxides and hydrogen. The carbon monoxide formed, as indicated in reaction (21), using propane as a typical

$$C_3H_8 + 3H_2O \rightarrow 3CO + 7H_2 \quad (21)$$

hydrocarbon, is converted to carbon dioxide according to reaction (22). The latter conversion

$$CO + H_2O \rightarrow CO_2 + H_2 \quad (22)$$

takes place at 350°C over an iron oxide catalyst.

The steam-hydrocarbon processes have largely replaced the once widely used steam–water gas process. Water gas, a mixture of carbon monoxide and hydrogen, is made by treating coke or coal with steam at a temperature of 1000°C or higher, as shown in reaction (23). The carbon monoxide is

$$C + H_2O \rightarrow CO + H_2 \quad (23)$$

converted to carbon dioxide as in the hydrocarbon processes. Hydrogen may be separated from carbon dioxide by scrubbing with an aqueous solution of monoethylamine.

Hydrogen is also produced industrially by the electrolysis of water, containing dissolved potassium hydroxide. Although comparatively expensive, this process generates hydrogen of very high purity (over 99.9%). High purity oxygen is a by-product. The overall reaction is (24).

$$2H_2O \rightarrow 2H_2 + O_2 \quad (24)$$

Other important industrial processes are the reaction of steam with iron at 800°C, reaction (25),

$$Fe + H_2O \rightarrow FeO + H_2 \quad (25)$$

and the thermal dissociation of ammonia, reaction (26), used for the production of comparatively

$$2NH_3 \rightarrow N_2 + 3H_2 \quad (26)$$

small amounts of hydrogen for metal treating or for catalytic hydrogenation.

Analytical methods. There are no fully satisfactory reagents for the direct absorption of hydrogen. Colloidal palladium absorbs large amounts of hydrogen, but the reagent is unstable, and it is necessary to remove carbon dioxide, olefins, and carbon monoxide before absorption of the hydrogen. Silver phosphate and silver borate have also been used to absorb hydrogen from a gas mixture.

The most commonly used methods for determining hydrogen in a gas mixture depend on burning the hydrogen to water and measuring the reduction in volume of the gas. The oxidation may be performed by adding excess oxygen and exploding the mixture either by passing an electrical spark between platinum electrodes immersed in the gas, or by passing the gas over a platinum wire heated to about 600°C. Since 1/2 mole of oxygen is required to burn 1 mole of hydrogen, the volume of hydrogen is equal to two-thirds the reduction in volume. Alternatively, the hydrogen may be oxidized by passing it through a tube filled with copper oxide heated to 290°C. The latter procedure is quite specific for hydrogen, because the only other common gas oxidized by copper oxide at this temperature is carbon monoxide, the oxidation of which results in no volume change. Because no oxygen gas is consumed in this procedure, the volume of hydrogen is equal to the measured reduction in volume.

A number of physical measurements may be used to determine hydrogen, especially in mixtures with a single other gas whose identity is known. The exceptionally high thermal conductivity and low density of hydrogen make the measurement of either of these properties a sensitive indicator of hydrogen content. Other physical methods useful for hydrogen analysis include measurement of the velocity of sound in the gas and measurement of the refractive index. The hydrogen content of quite complex gas mixtures can often be determined by mass spectrometry or by gas chromatography.

The determination of chemically bound hydrogen, particularly in organic compounds, is usually performed by burning the compound in a stream of oxygen at about 700°C. The combustion water is absorbed in a dehydrating agent, such as anhydrous magnesium perchlorate; the amount of hydrogen is calculated from the increase in weight of the absorption tube. Alternatively, the combustion water may be determined by gas chromatography, as is done in some automated systems for simultaneous determination of carbon, hydrogen, and nitrogen in organic compounds. Nuclear magnetic resonance (NMR) spectroscopy may be used, not only for the quantitative determination of total bound hydrogen, but also for measurement of hydrogen attached at specific locations in a molecule.

[LOUIS KAPLAN]

Bibliography: K. E. Cox and K. D. Williamson, Jr., *Hydrogen: Its Technology and Implications*, vols. 1–5, 1977–1979; M. F. Lappert, *Main Group Elements: Hydrogen and Groups 1–3*, 1975; H. Remy, *Treatise on Inorganic Chemistry*, vol. 1, 1956.

Hydrogen bond

The interaction which occurs when a hydrogen atom, covalently bonded to an electronegative atom (as in A—H), interacts with another atom to form the aggregate A—H · · · Y. The shortest and strongest bond is indicated as A—H, while the secondary and weaker interaction is written as H · · · Y. Thus A—H is a proton donor, while (Y) is a proton acceptor which often contains lone pair electrons and can act as a base. The strongest hydrogen bonds are formed between the most electronegative (A) atoms such as fluorine, nitrogen, and oxygen which interact with (Y) atoms having electronegativity greater than that of hydrogen (C, N, O, S, Se, F, Cl, Br, I). The weakest of hydrogen bonds are formed by acidic protons of C—H groups, as in chloroform and acetylene, and by olefinic and aromatic π-electrons acting as (Y).

Bond energies. The majority of hydrogen bonds have energies in the range 4–6 kcal/mole (17–25 kJ/mole) and involving those between O—H functional groups (as in water, alcohols, or acids) or N—H groups (as in amides or amines) and oxygen atoms (as in water, alcohols, carbonyls, or esters). The strongest hydrogen bond known is that found in the hydrogen difluoride ion, $(F—H—F)^-$, which has been variously estimated at 37–55 kcal/mole (155–230 kJ/mole). Therefore, the average hydrogen bond is of much lower energy than a normal chemical bond (>100 kcal/mole or 418 kJ/mole). Although hydrogen bonding gives rise to a specific interaction between atoms, resulting in a complex with characteristic A—H · · · Y distances and angles, especially in the solid state, it is difficult to establish a lower limit for the H—bond enthalpy because experimental methods of detection are becoming increasingly more sensitive and accurate.

The weaker the hydrogen bond, the shorter the lifetime of the complex it forms. The detection of weak hydrogen bonds often amounts to measuring shorter and shorter lifetimes of rapidly associating and dissociating species in equilibrium. This is a difficult problem because proton transfer along hydrogen bonds belongs to the fastest known chemical reactions, and in most experimental studies only mean values and "average" structures are determined.

An important aspect of weak hydrogen bond formation is that the different molecular aggregates which do form can be easily and reversibly transformed. Thus the small energy changes resulting in the rapid making and breaking of hydrogen bonds in biological systems are of great importance; for example, hydrogen bonding determines the configuration of the famous α-helix, and the structures of most proteins, thereby serving an important function in determining the nature of all living things.

Spectroscopy. Even though slight energy changes are usually involved in hydrogen bond formation, the aggregate once formed changes almost every measurable physical property of the original species. When investigating hydrogen bonding, the most frequently used techniques are infrared and nuclear magnetic resonance spectroscopy. In the infrared method the A—H stretching frequency is shifted to lower values and is accompanied by band broadening and increased intensity. Such changes are usually easily discernible, but in the case of very strong hydrogen bonds, such as $(F—H—F)^-$, the shift and broadening is so drastic that it is difficult to assign the new frequencies correctly. In the case of nuclear magnetic resonance, hydrogen bonding usually shifts the proton resonance to lower fields. *See* INFRARED SPECTROSCOPY; NUCLEAR MAGNETIC RESONANCE (NMR).

Neutron diffraction. While infrared and nuclear magnetic resonance techniques can yield considerable information about hydrogen bond formation, far greater information content is derived from diffraction studies of crystalline solids. The method of choice is neutron diffraction (versus x-ray diffraction), because in the former case hydrogen atoms scatter almost as well as any other atom while their scattering is often swamped out in the x-ray case. Neutron diffraction crystal structure analysis has become the best probe available for the study of the geometry of A—H · · · Y bonds. Using this technique, it is generally observed that the A—H separation is ~0.10 ± 0.01 nm and is much less than the H · · · Y separation; that is, hydrogen bonds are usually asymmetric. In the extreme case the hydrogen atom may be equally bonded to both atoms, as in certain O—H—O and $(F—H—F)^-$ containing systems where the atomic environment around A and Y is identical and symmetric. In such systems strong hydrogen bond formation is indicated when the A · · · Y separation is ~0.02–0.03 nm less than the sum of the van der Waals radii. However, contrary to prior belief, even the shortest and strongest hydrogen bond known, $(F—H—F)^-$, may be asymmetric. This was demonstrated in a neutron diffraction study of *p*-toluidinium hydrogen difluoride in which the two terminal F atoms exist in vastly different F · · · H—N hydrogen bonding environments (see illustration). The F—H distances are unequal, and the $(F—H—F)^-$ ion is asymmetric, because of the very different N—H · · · F hydrogen bonding environments around the F atoms. Thus it seems that the H atom of a strong $(X—H—X)^-$ bond is a probe of the X atom environment. Indeed, the hydrogen dichloride ion, $(HCl_2)^-$, with a bond energy of about 12 kcal/mole (49 kJ/mole) appears to be symmetric (centered H atom) in some salts and asymmetric in

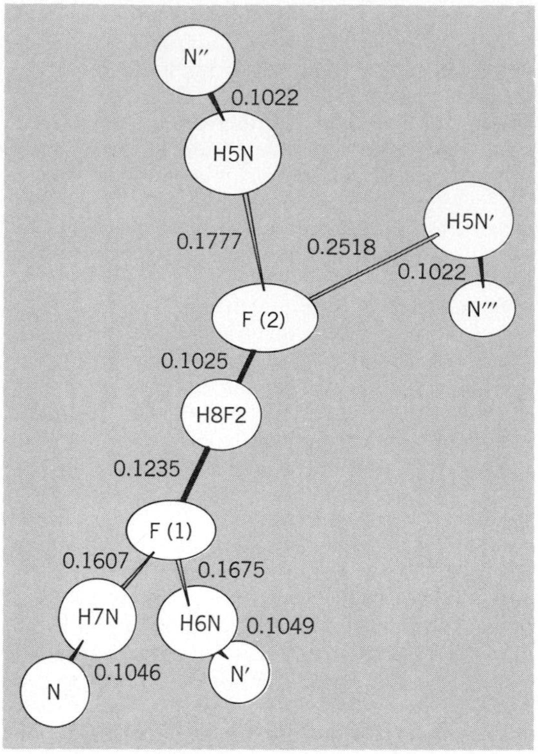

Hydrogen difluoride ion, $(F—H—F)^-$, geometry. Distances are in nanometers, and atoms are represented as ellipsoids of 50% probability. Numbers following symbols for the elements signify that like atoms are not equivalent structurally. The primes on N atoms indicate that they are structurally equivalent and related by symmetry operations.

others. Other types of hydrogen atom interactions such as those of M—H · · · M and C—H · · · M, where M is a metal, are becoming increasingly important in catalysis. *See* NEUTRON DIFFRACTION.

Theory. Developments in theory have made it possible to better define certain contributions to hydrogen bond energies. The relative importance of forces of different origin (dispersion, polarization, exchange, coulomb, and so on) have become possible to estimate by using both molecular orbital methods and perturbation theory. In general, it appears that quantum theory gives reliable descriptions of isolated dimers and trimers, but fails when dealing with large clusters of the type found in condensed phases.

[JACK M. WILLIAMS]

Bibliography: P. Schuster, G. Zundel, and C. Sandorfy (eds.), *The Hydrogen Bond: Recent Developments in Theory and Experiments*, vols. I–III, 1976.

Hydrogen electrode

A noble metal of large surface area covered with hydrogen gas in a solution of hydrogen ion saturated with hydrogen gas. The hydrogen gas may also be bubbled continuously over the noble metal; in this case, the affluent gas must be saturated with water vapor. Platinum is generally used as the noble metal. A large surface area is achieved by plating the surface with platinum sponge in a solution of chloroplatinic acid. In general, the noble metal is used in foil form and is welded to a wire sealed in the bottom of a hollow glass tube which is partially filled with mercury. Electrical contact to an external circuit is made through the mercury usually by a copper wire.

The hydrogen electrode is reversible to hydrogen ion and its electric potential is a logarithmic function of hydrogen-ion activity. Because of this relation, this electrode is used to measure hydrogen-ion activity. It cannot be used in reducing or oxidizing solutions or in solutions that poison the surface of the noble metal. Accordingly, it has limited application in the measurements of hydrogen-ion activity of solutions.

When the hydrogen-ion activity is unity and the hydrogen-gas pressure is at 1 atm, the potential of the hydrogen electrode is conventionally assigned a value of zero at all temperatures, and is used as a reference or standard electrode to which the potentials of all other electrodes are referred. All electrodes having a positive potential relative to the standard hydrogen electrode have greater reducing power than hydrogen; conversely, those of lower potential have a lower reducing power. Elements arranged in a series of their reducing power constitute what is called the electromotive-force series of the elements. *See* ELECTROCHEMICAL SERIES.

The potential of the hydrogen electrode varies with hydrogen-gas pressure p_H and with hydrogen-ion activity a_{H^+} according to Eq. (1), where R is the

$$E=-\frac{RT}{F}\ln a_{H^+}+\frac{RT}{2F}\ln p_{H_2} \tag{1}$$

gas constant, T the absolute temperature, and F the faraday. At 25°C this relation is as given in Eq. (2). Thus, for a 10-fold change in hydrogen-ion ac-

$$E=-0.059158\log a_{H^+}+0.029579\log p_{H_2} \tag{2}$$

tivity and a constant pressure of hydrogen gas, the potential of the hydrogen electrode changes by 0.059158 volt. At constant hydrogen-ion activity, a 10-fold change in hydrogen-gas pressure causes a change in the hydrogen electrode potential of 0.029579 volt. In determining the pressure of the hydrogen gas, the vapor pressure of the solution p should be subtracted from the observed barometric pressure P. The observed potential of the electrode is corrected to 1 atm hydrogen pressure by Eq. (3). For approximate measurements at room

$$E_{1\ \mathrm{atm}}=E_{\mathrm{obs}}+\frac{RT}{2F}\ln\frac{760}{(P-p)} \tag{3}$$

temperature, corrections for the vapor pressure of the solution may be neglected.

The hydrogen electrode may be used at atmospheric pressure from 0 to 100°C and at higher temperatures for pressures exceeding 1 atm. It may also be used in nonaqueous solutions which do not have reducing or oxidizing power on the electrode.

The electrode may be prepared in three general forms: (1) dipping type, (2) stationary bubbling type, and (3) rocking type. In the dipping type, the gas is introduced by a tube adjacent to the hydrogen electrode; both are supported at the top by a stopper or other means. The dipping type may also be of the type shown in the illustration. In the stationary bubbling type, the gas is introduced below the hydrogen electrode (see illustration), and the vessel is provided with a side arm whereby it may

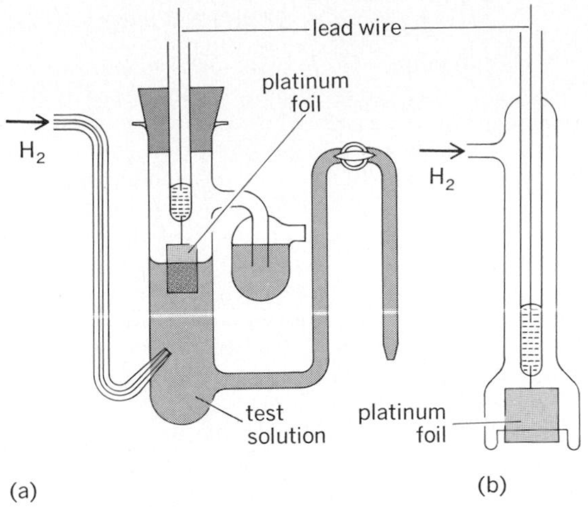

Hydrogen electrodes. (*a*) Stationary bubbling type and (*b*) dipping type.

be joined to another electrode or half-cell. In the rocking type, the solution is agitated and made to move over the hydrogen electrode in a rocking motion; contact to another electrode or half-cell is made through a stopcock and a flexible rubber connection. This type is sealed, and therefore has limited life. *See* ELECTRODE POTENTIAL; HYDROGEN ION.

[WALTER J. HAMER]

Hydrogen fluoride

The hydride of fluorine and the first member of the family of halogen acids. Anhydrous hydrogen fluoride is a mobile, colorless liquid that fumes strongly in air. It has the empirical formula HF, melts at −83°C, and boils at 19.8°C. The vapor is highly aggregated, and gaseous hydrogen fluoride deviates from perfect gas behavior to a greater extent than any other gaseous substance known. Aggregate formation in both the vapor and liquid phase arises from unusually strong hydrogen-bond interactions. Hydrogen fluoride is prepared on the large industrial scale by treating fluorspar (calcium fluoride, CaF_2) with concentrated sulfuric acid. The crude product is purified by fractional distillation to yield a product containing more than 99.5% HF; the remaining impurities are principally water and small amounts of sulfur dioxide, silicon tetrafluoride, and boron trifluoride. Very dry hydrogen fluoride can be obtained by electrolysis or by treatment with reagents such as fluorine or cobaltic fluoride that react with water. *See* HYDROGEN BOND.

Properties. Anhydrous hydrogen fluoride is an extremely powerful acid, exceeded in this respect only by 100% sulfuric acid. Like water, hydrogen fluoride is a liquid of high dielectric constant that undergoes self-ionization and forms conducting solutions with many solutes. Because anhydrous hydrogen fluoride is a superacid, many organic solutes dissolve in it to form stable carbonium ions. Alkali metal fluorides and silver fluoride dissolve readily in hydrogen fluoride to form conducting solutions. The alkali metal fluorides are bases in the hydrogen fluoride system and correspond to solutions of alkali metal hydroxides in water. Conversely, antimony pentafluoride and boron trifluoride act as acids in hydrogen fluoride and accentuate the already strong acid properties of the solvent. *See* SUPERACIDS.

Anhydrous hydrogen fluoride dissolves a wide variety of organic compounds. Oxygen-, nitrogen-, and sulfur-containing compounds usually have high solubility in liquid hydrogen fluoride, generally higher than that found in water. Aromatic hydrocarbons are moderately soluble, and even saturated aliphatic hydrocarbons show appreciable solubility. Despite the fact that hydrogen fluoride is a strong dehydrating agent, many organic solutes can be recovered unchanged from hydrogen fluoride solution. Surprisingly, the enzymes trypsin and lysozyme survive dissolution and recovery from solution in anhydrous liquid hydrogen fluoride with full retention of biological activity.

Uses. Hydrogen fluoride is a widely used industrial chemical. It was formerly used in the petroleum refining industry for the isomerization of aliphatic hydrocarbons to form more desirable automotive fuels, but this application has been superseded by other methods. The largest industrial use of hydrogen fluoride is in making fluorine-containing refrigerants (Freons, Genetrons).

An increasingly important use of hydrogen fluoride is in the preparation of organic fluorocarbon compounds by the Simons electrochemical process. In this procedure, an organic compound is dissolved in hydrogen fluoride, and an electric current passed through the solution, whereupon the hydrogen atoms in the organic solute are replaced by fluorine. Hydrogen fluoride is employed in the electrochemical preparation of fluorine and for the preparation of inorganic fluorides. Thus, hydrogen fluoride is used for the conversion of uranium dioxide to uranium tetrafluoride, an intermediate in the preparation of uranium metal and uranium hexafluoride. With the great increase in nuclear energy-produced electricity, this represents an important use of hydrogen fluoride.

Important organic reactions may in some cases be performed to advantage in hydrogen fluoride solution, and nitration, sulfonation, diazotization, cyclization, and polymerization reactions have been carried out in this medium. Anhydrous hydrogen fluoride can be used for the carboxylation of olefins with carbon monoxide, for the alkylation of isoparaffins with olefins, for the esterification of fatty acids, and for the removal of protective groups in peptide preparation by solid-phase reactions.

Aqueous solutions of hydrogen fluoride (hydrofluoric acid) are relatively weakly acidic as compared to hydrochloric acid. Fluoride salts are formed by reaction of hydrofluoric acid with metal oxides and carbonates. Of particular importance is the rapid reaction of hydrofluoric acid or anhydrous hydrogen fluoride with silica, which leads to the application of these substances as etching agents for glass.

Both hydrogen fluoride and hydrofluoric acid cause unusually severe burns; appropriate precautions must be taken to prevent any contact of the skin or eyes with either the liquid or the vapor. *See* FLUORIDE; FLUORINE; HALOGENATED HYDROCARBON; ISOMERIZATION.

[JOSEPH J. KATZ]

Bibliography: F. A. Cotton and G. Wilkinson, *Advanced Inorganic Chemistry*, 4th ed., 1980; J. H. Simons (ed.), *Fluorine Chemistry*, vol. 1, 1950.

Hydrogen ion

A proton combined with a number of water molecules. It is often written as H_3O^+ and called the hydronium ion. However, this species is best considered as an excess proton on a tetrahedral group of four water molecules and so would be designated as $H_9O_4^+$. For simplicity, it is most commonly written as H^+ (aq).

Chemical activity. Since it is formed by the self-ionization of water according to reaction (1), the

$$H_2O \rightleftharpoons H^+(aq) + OH^-(aq) \quad (1)$$

hydrogen ion is present in all aqueous solutions. This formation also means that H^+(aq) is always found in the company of the hydroxide ion, OH^-(aq). The relationship between the concentrations of these two species is a very important property of water and at 25°C is given by Eq. (2). The

$$K_w = [H^+(aq)][OH^-(aq)] = 10^{-14} \quad (2)$$

equilibrium constant K_w increases rapidly with increasing temperature, and for more precision the concentrations in Eq. (2) should be replaced by the activities of the ions. *See* IONIC EQUILIBRIUM.

Reaction (1) and Eq. (2) indicate that the H^+(aq) and OH^-(aq) concentrations in pure water are equal to each other with a value of 10^{-7} mole/liter. Any aqueous solution with this concentration of H^+(aq) is called a neutral solution. If the H^+(aq) concentration is greater than 10^{-7} mole/liter, the solution is called acidic. Basic solutions are those in which the H^+(aq) concentration is less than 10^{-7} mole/liter. It is clear from Eq. (2) that as the H^+(aq) concentration increases the OH^-(aq) concentration must decrease, and vice versa. In the most straightforward system, acids are substances that can donate an H^+(aq), and bases are substances that can accept one. Therefore, the action of common acids can be written, for example, as reaction (3) or (4), and the action of common bases

$$HCl(aq) \xrightarrow{H_2O} H^+(aq) + Cl^-(aq) \quad (3)$$

$$H_2SO_4(aq) \xrightarrow{H_2O} H^+(aq) + HSO_4^-(aq) \quad (4)$$

can be exemplified by reaction (5). Metal cations

$$NH_3(aq) + H^+(aq) \rightarrow NH_4^+(aq) \quad (5)$$

can also act as acids, particularly when they are small ions with high positive charges. An example of such a hydrolysis reaction is shown in reaction (6). The reaction of an acid and a base is called a

$$Al^{3+}(aq) \rightleftharpoons AlOH^{2+}(aq) + H^+(aq) \quad (6)$$

neutralization reaction and has great importance. A typical example of a neutralization is reaction (7). The products of such a neutralization are al-

$$HCl(aq) + NaOH(aq) \rightarrow NaCl(aq) + H_2O \quad (7)$$

ways a salt, in this case NaCl, and water. *See* ACID AND BASE.

The reaction of H^+(aq) with bicarbonates, carbonates, sulfites, bisulfites, and sulfides produces the volatile gases carbon dioxide, sulfur dioxide, and hydrogen sulfide, respectively. The hydrogen ion also reacts with metals above hydrogen in the electromotive force series to produce hydrogen gas and the cation of the metal [reaction (8)]. *See* ELECTROCHEMICAL SERIES.

$$Zn(s) + 2H^+(aq) \rightarrow H_2(g) + Zn^{2+}(aq) \quad (8)$$

Concentration. Hydrogen ion concentration determines the course of many chemical reactions that occur in living organisms and in the chemical industry. The control of hydrogen ion concentration is achieved in living organisms and in the laboratory by buffer systems. These are chemical mixtures designed to resist change in hydrogen ion concentration. Careful control of acid concentration is crucial in industries such as brewing, pharmaceutical manufacturing, electroplating, and textiles. Water sanitation, whether in a swimming pool or a water treatment plant, demands close attention to acidity. *See* BUFFERS.

Electrical conductivity. Another property of the H^+(aq) ion is important in both theoretical and practical ways. H^+(aq) is the best conductor of electricity of any ion in aqueous solution. Its conductance at 25°C is almost five times as large as the next-most-conducting ion. This abnormally large conductance is due to a unique mechanism of electricity conduction. Ordinary ions must travel physically through the solution to conduct electricity. The H^+(aq) conducts electricity by a chain mechanism whereby an excess proton can be attached to one side of a water molecule cluster, and another excess proton lost from the opposite side of the same cluster. This high conductivity is of great practical importance in batteries such as the common lead-acid cell, where the sulfuric acid electrolyte keeps the internal cell resistance low. *See* ELECTROLYTIC CONDUCTANCE.

pH. The hydrogen ion concentration can vary over fourteen powers of 10. To avoid dealing with such exponentials, the concept of pH was advanced by S. P. L. Sørensen in 1909. It is defined by Eq. (9). In precise work the activity of the hy-

$$pH \equiv -\log[H^+(aq)] \quad (9)$$

drogen ion would be used to replace the concentration in Eq. (9). Since all aqueous solutions contain both hydrogen ion and hydroxide ion, it is possible to define all degrees of acidity and basicity on the pH scale. The table, which gives hydrogen ion concentrations, hydroxide ion concentrates, and pH for aqueous solutions at 25°C, illustrates certain characteristics of the pH scale. Neutral solutions have a pH of 7, while acid solutions have a pH less than 7, and basic solutions have a pH greater than 7. A change of one pH unit corresponds to a tenfold change in acidity. The pH is also useful in pointing out the differences between strong acids such as hydrochloric acid and weak acids such as acetic acid, the main component of vinegar. For a strong acid, reaction (10) is almost

Relation between ion concentrations and pH in water

$[H^+(aq)]$	$[OH^-(aq)]$	pH	$[H^+(aq)]$	$[OH^-(aq)]$	pH
1	1×10^{-14}	0	1×10^{-7}	1×10^{-7}	7
1×10^{-1}	1×10^{-13}	1	$1 + 10^{-8}$	1×10^{-6}	8
1×10^{-2}	1×10^{-12}	2	1×10^{-9}	1×10^{-5}	9
1×10^{-3}	1×10^{-11}	3	1×10^{-10}	1×10^{-4}	10
1×10^{-4}	1×10^{-10}	4	1×10^{-11}	1×10^{-3}	11
1×10^{-5}	1×10^{-9}	5	1×10^{-12}	1×10^{-2}	12
1×10^{-6}	1×10^{-8}	6	1×10^{-14}	1	14

$$HCl(aq) + H_2O \rightarrow H^+(aq) + Cl^-(aq) \quad (10)$$

100% complete. Therefore, a 0.1 mole/liter solution of HCl has a pH of 1.0. For the weak acid, reaction (11) occurs to a very limited extent. A 0.10

$$CH_3COOH(aq) + H_2O \rightleftharpoons H^+(aq) + CH_3COO(aq) \quad (11)$$

mole/liter solution of acetic acid has a pH of 2.9 and thus is almost 100 times less acidic than the hydrochloric acid solution. *See* PH.

Determining concentrations. Two general methods are used for the determination of hydrogen ion concentrations. For relatively crude work, colorimetric methods are commonly used. These methods depend on the fact that certain natural and synthetic dyes have colors that depend on the hydrogen ion concentration. At times, paper is impregnated with such an indicator. By dipping the paper into the aqueous solution, the color of the paper changes due to the acidity of the solution. The paper can then be compared to a standard printed color series, and a rough idea of the pH obtained. In another application, a small amount of such an indicator dye can be added to a solution whose acidity is being determined. Small increments of a base solution are then added in a titration. If the indicator is chosen correctly, the solution will change color when the exact amount of base has been added to completely neutralize the original acid. A simple calculation will then give the amount of acid initially present. *See* ACID-BASE INDICATOR.

In most precise work, a potentiometric method is used for the determination of hydrogen ion concentration. This method depends on an electrode whose potential is sensitive to hydrogen ion concentration. The only electrode commonly in use for practical pH measurements is the glass electrode. A glass electrode, together with a reference electrode, is placed in the solution of unknown acidity. The potential of the electrochemical cell formed is then measured by a high-input-impedance voltmeter.

Many good commercial pH meters are available. Because it is difficult to relate the measured voltage directly to a hydrogen ion concentration, the pH meter–electrode combination is calibrated with buffer solutions of well-defined pH values. The calibrated instrument can then be used to measure the pH of unknown solutions. The importance of pH measurement has led to the creation of very precisely defined pH standards at the National Bureau of Standards. It is now possible to make meaningful pH measurements, even in extreme conditions such as very high pressures (1000 bars or 100 megapascals) or very high temperatures (200°C). The technology of pH measurement has advanced to the point where many industrial processes are automatically controlled through pH measurements on process streams. *See* ION-SELECTIVE MEMBRANES AND ELECTRODES; TITRATION; VOLUMETRIC ANALYSIS. [G. ATKINSON]

Bibliography: R. G. Bates, *Determination of pH: Theory and Practice*, 1964; J. O'M. Bockris and A. K. N. Reddy, *Modern Electrochemistry*, vol. 1, 1970; H. T. S. Britton, *Hydrogen Ions*, 2 vols., 4th ed., 1965; G. Eisenman (ed.), *Glass Electrodes for Hydrogen and Other Cations*, 1967.

Hydrogen peroxide

A binary compound of hydrogen and oxygen, empirical formula H_2O_2, used mostly in dilute aqueous solutions as an oxidizing agent. It was discovered in 1818 by the French chemist Louis-Jacques Thenard, who named it *eau oxygénée*. Its most remarkable feature is its tendency to decompose readily into water and oxygen, the first observed instance of contact catalysis.

Properties. Anhydrous hydrogen peroxide is a clear, colorless liquid, of nearly the same viscosity and dielectric constant as water, but of greater density, 1.442 $g \cdot cm^{-3}$ at 25°C. Like water, it is strongly associated through hydrogen bonds. It boils at 150°C with violent, sometimes explosive decomposition. With considerable supercooling, it can be frozen into needle-shaped crystals which melt at −0.41°C. The decomposition, strongly exothermic (690 $cal \cdot g^{-1}$ or 2.89 $kJ \cdot g^{-1}$), is almost always catalytic, becoming homogeneous only in the vapor above 420°C. The rate varies considerably with the nature of the catalyst, the temperature, and the surface-volume ratio of the sample. Decomposition by light begins only in the near ultraviolet. As a solvent, hydrogen peroxide resembles water, except that acids and bases show much lower electrical conductivity. Although a fairly strong oxidant, it can act as a mild reducing agent, for example, with permanganates and perchromates.

The H_2O_2 molecule has the skew chain configuration shown in the illustration, the simplest example of internal rotation or torsion about a single bond. This is a semirigid structure, because the torsion is hindered by a rather low potential barrier (1 $kcal \cdot mol^{-1}$ or 4 $kJ \cdot mol^{-1}$ for the trans versus 7 $kcal \cdot mol^{-1}$ or 29 $kJ \cdot mol^{-1}$ for the cis configuration).

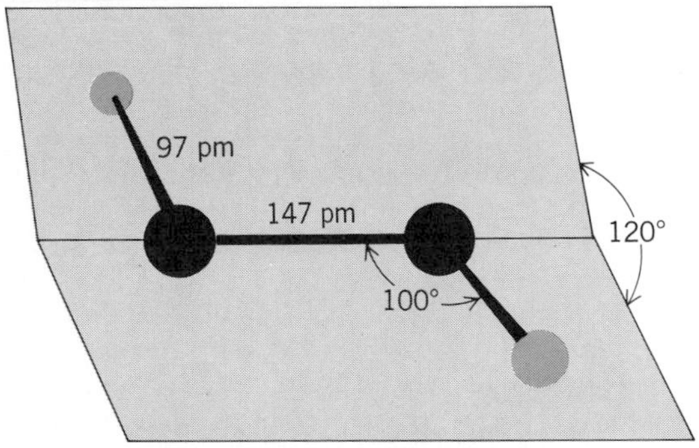

Structural parameters of the H_2O_2 molecule.

Formation. Hydrogen peroxide occurs only in traces in nature, mostly in rain and snow. It has not yet been detected in interstellar space. Its formation, usually in low concentration, has been studied in a variety of systems: (1) from the elements in the products of the oxyhydrogen flame quickly quenched in liquid air, or in explosion of hydrogen-rich mixtures near 550°C; (2) from water, liquid, or vapor irradiated by ultraviolet light of a wavelength shorter than 185 mm, with and without sensitizer, and by ionizing radiation; and (3) from a silent electric discharge through water vapor at atmospheric pressure, or under reduced pressure (below 10^{-3} atm) in electrodeless discharges through a fast-flowing system, followed by quick chilling in a liquid-nitrogen trap. Under optimum conditions, concentrations up to 50% can be reached. This last system, developed for the synthesis of deuterium peroxide, D_2O_2, from heavy water, has led to identification of the long-postulated higher oxides, H_2O_3 and H_2O_4, as metastable intermediates.

Manufacture. The original barium peroxide–sulfuric acid method was superseded long ago by the electrochemical process, which involves oxidation of sulfuric acid, or ammonium, or potassium sulfate in concentrated solutions at high current density on a platinum anode. This also is now largely replaced by the anthraquinone process based on the cyclic oxidation-reduction of some substituted anthraquinone as shown in the reaction below. The anthraquinone derivative (I) dis-

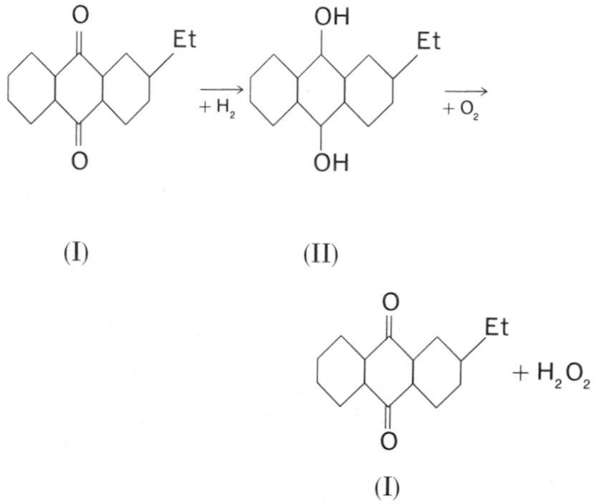

solved in an appropriate solvent is first reduced catalytically by hydrogen at atmospheric pressure. The resulting hydroquinone (II) is then oxidized by air, and the hydrogen peroxide (5 to 15 g per liter) is removed by countercurrent extraction with water, thereby regenerating the working substance. The efficiency of the latter step and the avoidance of side reactions are essential for economical operation.

Uses. Hydrogen peroxide is used mainly for bleaching cotton and other fibers, natural or synthetic. Increasing amounts are used in the pulp and paper industry. Because of its milder action on fibers and the fact that it leaves no undesirable residue, it is preferable to chlorine and its compounds. Its well-known cosmetic use as hair bleach consumes relatively little of the commercial 10% (30 volume) solution. In medicine it is useful for cleansing wounds and cuts, although its antiseptic action is rather slow. A limited but important use of the concentrated peroxide is for energy production in rockets, submarines (during submersion), airplanes (at takeoff), and the steering of space vessels. Fast decomposition is achieved by sudden addition of a catalyst or by blowing the vapor through a porous bed of catalyst-impregnated ceramics. The jet from a 90% solution can reach temperatures of about 750°C. A land vehicle with such a jet engine is claimed to have broken the sound barrier in the California desert. *See* CHEMICAL FUEL.

Handling. Hydrogen peroxide, especially when concentrated, requires great care in handling and storing. When dropped on paper or wood, it can start a fire. Contact with the skin causes blotches that can be painful, but they disappear after a few hours without leaving traces. Appropriate containers are made of Pyrex glass, Teflon, or polyethylene; for industrial storage and shipment, tanks or carboys are made of electropolished stainless steel or pure aluminum. All containers must be fitted with a vent for the escape of oxygen, and be of such design as to prevent spilling.

Slight decomposition may be overcome by distillation, or by addition of some stabilizer (such as zinc salts, phosphates, or ascorbic acid). For special purposes, a 90% product, stabilizer-free, is sometimes available. It can be concentrated to over 99% by fractional distillation in an all-Pyrex-glass still. The last traces of water are easily removed by fractional crystallization, provided that the solid is frozen very slowly into a large single crystal. *See* HYDROGEN; OXYGEN; PEROXIDE.

[PAUL A. GIGUÈRE]

Bibliography: W. C. Schumb, C. N. Satterfield, and R. L. Wentworth, *Hydrogen Peroxide*, ACS Monogr. no. 128, 1955.

Hydrogenation

The chemical reaction of hydrogen with another substance, generally an unsaturated organic compound, and usually under the influence of temperature, pressure, and catalysts. There are several types of hydrogenation reactions. They include: (1) the addition of hydrogen to reactive molecules; (2) the incorporation of hydrogen accompanied by cleavage of the starting molecules (hydrogenolysis); and (3) reactions in which isomerization, cyclization, and so on, result. Other reactions that involve molecular hydrogen and catalysts are reductive amination (hydroammonolysis) and hydroformylation (oxo reaction).

Hydrogenation is synonymous with reduction in which oxygen or some other element (most commonly nitrogen, sulfur, carbon, or halogen) is withdrawn from, or hydrogen is added to, a molecule. When hydrogenation is capable of producing the desired reduction product, it is generally the simplest and most efficient procedure.

Hydrogenation is used extensively in industrial processes. Important examples are the synthesis of methanol, liquid fuels, hydrogenated vegetable oils, fatty alcohols from the corresponding carboxylic acids, alcohols from aldehydes prepared by the aldol reaction, cyclohexanol and cyclohexane from phenol and benzene, respectively, and hexamethylenediamine for the synthesis of nylon from adiponitrile.

Hydrogenolysis. This term refers particularly to cleavages in a molecule associated with the addition of hydrogen. Hydrogenolysis is analogous to hydrolysis and ammonolysis, which involve cleavage of a bond induced by the action of water and ammonia, respectively. Chemical bonds which are broken by hydrogenolysis reactions include carbon-carbon, carbon-oxygen, carbon-sulfur, and carbon-nitrogen. Two examples are hydrodealkylation of toluene to form benzene and methane, reaction (1), and reduction of methyl laurate to form lauryl alcohol and methanol, reaction (2).

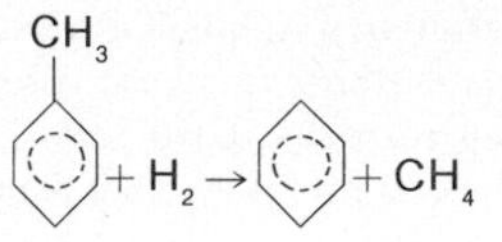
(1)

$$CH_3(CH_2)_{10}COOCH_3 + H_2 \rightarrow CH_3(CH_2)_{10}CH_2OH + CH_3OH \quad (2)$$

Catalytic hydrogenation. A variety of organic compounds can be hydrogenated easily in the presence of a catalyst. Acetylenes readily add two moles of hydrogen giving the saturated derivatives, as shown in reaction (3), where R and R′ are aliphatic, aromatic, or certain other groups. Under proper conditions the hydrogenation can be stopped at the intermediate olefin stage.

$$RC{\equiv}CR' + 2H_2 \rightarrow RCH{=}CHR' + H_2 \rightarrow RCH_2CH_2R' \quad (3)$$

Catalytic hydrogenation of olefins can be carried out either in gas or in liquid phase, depending on their molecular weights. A nickel-containing catalyst and sometimes platinum or palladium catalysts are employed.

Aromatic compounds may be reduced either in the vapor phase at atmospheric pressure or in the liquid phase at hydrogen pressures up to 200 atm ($1\ atm = 10^5$ Pa). In the latter case, aromatics, such as benzene, toluene, and *p*-cymene, can be hydrogenated readily in the presence of a nickel catalyst. In the case of naphthalene or substituted naphthalenes, the product may be the tetra- or decahydronaphthalene derivative.

The reduction of carbonyl compounds, such as aldehydes and ketones, to the corresponding alcohols is represented by reaction (4), where R is an aliphatic or aromatic group, and R′ may be the same group or a hydrogen atom. Frequently when R is an aromatic group, it is difficult to stop the reduction at the alcohol. Instead, it proceeds further to yield a hydrocarbon, RCH_2R'. In general, aldehydes are reduced more rapidly than ketones, although there are numerous examples in which both undergo reduction at room temperature and only a few atmospheres of hydrogen pressure. Often a small amount of water (1–10%) is added to the feed to the hydrogenator to suppress ether formation.

$$RCOR' + H_2 \rightarrow RCHOHR' \quad (4)$$

Other processes. Hardening of various animal fats and vegetable oils (such as soybean, cottonseed, fish, whale, and peanut) is carried out on a large scale by partial hydrogenation. The resultant plastic fats have a consistency and other properties suitable for the manufacture of shortenings, margarine, soaps, and a variety of other edible and industrial products. Chemically, the process involves the conversion of glycerides of unsaturated fatty acids (for example, oleic and linoleic) to saturated ones. Mild conditions (such as 100–750°C, 1–14 atm, and Ni catalyst) are employed to avoid the hydrogenolysis of the ester linkage. The conversion of olein to stearin may be expressed as reaction (5).

$$(C_{17}H_{33}COO)_3C_3H_5 + 3H_2 \rightarrow (C_{17}H_{35}COO)_3C_3H_5 \quad (5)$$

Unlike the hardening of fats, which involves only the hydrogenation of ethylene linkages, the hydrogenolysis of the carboxyl group of acids and esters takes place with the formation of alcohols, as shown by reaction (6). The olefinic bonds in the fatty chain may or may not be reduced. A reduced copper–ammonium chromate catalyst is used at 350–400°C.

$$RCOOH + 2H_2 \rightarrow RCH_2OH + H_2O \quad (6)$$

The synthesis of methanol from carbon monoxide and hydrogen is carried out at high pressures (3000–5000 psi; 1 psi = 6895 Pa), because the reaction (7) involves a decrease in volume. The practical reaction temperature range is small. Below 300°C, the rate is slow; above 400°C, the equilibrium becomes unfavorable. Mixed catalysts consisting of oxides of zinc, chromium, manganese, or aluminum, such as zinc oxide with 10% chromium oxide, are utilized. Carbon monoxide can also be hydrogenated to give various hydrocarbons and higher alcohols.

$$CO + 2H_2 \rightarrow CH_3OH \quad (7)$$

Petroleum, tar, and coal are hydrogenated to: (1) improve existing products; (2) convert low-grade materials such as heavy oils into valuable fuels; and (3) transform solid fuels such as lignites and coal into liquid fuels. Several of these processes are hydrodesulfurization, hydrocracking, and hydrodealkylation. By a proper selection of catalysts and operating conditions, such hydrogenations can be directed to give desired end products and at the same time cause impurities that are common catalytic poisons, such as sulfur, nitrogen, and oxygen, to be detached from their molecular linkages and to be removed as hydrogen sulfide, ammonia, and water.

Catalytic hydrogenation may continue to increase in importance, size, and the variety of processes. Examples are possible commercialization of coal-hydrogenation processes and the conversion of petroleum residues and shale oils to lighter fractions.

Thermodynamics. Hydrogenation reactions are generally reversible. Catalysts affect the rate or speed of reaction, but have nothing to do with the inherent tendency of the reaction to proceed. To know whether or not the reaction is feasible, the free energy change of a reaction, ΔG, can be determined. For example, for the reaction represented by (8), the change in free-energy content per atom of carbon in each molecule at 400°C is shown in reactions (9) and (10).

$$CO(g) + 3H_2(g) \rightleftharpoons CH_4(g) + H_2O(g) \quad (8)$$

$$-40{,}500 + 0 + \Delta G = -3600 - 50{,}000 \quad (9)$$

$$G = -13{,}100\ \text{cal at } 400°C\ (1\ \text{cal} = 4184\ J) \quad (10)$$

The decrease in free energy means that, at this temperature, reduction of carbon monoxide to methane is possible. At 1000°C, the reverse reaction for the production of hydrogen and carbon monoxide from natural gas and steam is feasible. It so happens that both the forward and the reverse reactions are industrially important. *See* FREE ENERGY.

Hydrogenation reactions are exothermic, that is,

heat is released during the reaction. Typically, the heat release per gram mole of hydrogenated material formed is about 28–30 kcal for the hydrogenation of alkenes to alkanes and is 50 kcal for the hydrogenation of benzene to cyclohexane. The heat of reaction must be removed by heat exchangers in the reactor or is utilized to heat the feeds to the reaction temperature.

Effect of temperature. The reaction temperature affects the rate and the extent of hydrogenation as it does any chemical reaction. Practically every hydrogenation reaction can be reversed by increasing temperature. High temperatures often lead to loss of selectivity and, therefore, yield of desired product, if a second functional group is present. As a practical measure, hydrogenation is carried out at as low a temperature as possible compatible with a satisfactory reaction rate. Although the optimum temperature depends on the catalyst type and age, the temperatures for hydrogenation reactions are generally below 500°C.

Effect of pressure. Hydrogenation rates are generally increased by increasing the hydrogen pressure. Pressure also increases the equilibrium yield in hydrogenations where there is a decrease in volume as the reaction proceeds. For economic reasons, many industrial hydrogenation processes are carried out under an imposed pressure but seldom above 300 atm.

Catalysts. For industrial applications, hydrogenation catalysts are generally solids consisting of metals, metal oxides, and some salts. These catalysts may be classified in accordance with their customary use. Vigorous catalysts suitable for the hydrogenation of alkyne and alkene linkages, aldehydes, and ketones include nickel and cobalt, and molybdenum and tungsten oxides or sulfides. Mild catalysts useful for stepwise hydrogenations of aldehydes and ketones include oxides of copper, zinc, and chromium, and metallic platinum and palladium. Molybdenum sulfide and especially tungsten disulfide are active catalysts for operations at 3000 psi. These catalysts are useful for the hydrogenation of unsaturates and to effect the cleavage of C-C, C-O, and C-N bonds.

In recent years, a wide range of metal ions and complexes has been found to catalyze hydrogenation reactions homogeneously in solution. These ions and complexes have been derived from a variety of metals, including platinum, cobalt, rhodium, and copper. Homogeneous catalytic systems are inherently simpler chemically and kinetically, and are often more selective. Judging from the patent activity in this area, the use of homogeneous catalysts to effect hydrogenation shows considerable promise.

Equipment. There are two common types of reaction vessels. The first is used with liquids or solids, as in the hydrogenation of oils and viscous hydrocarbons. Internal agitators bring about an intimate mixing of organic compound, catalyst, and hydrogen. Alternatively, hydrogen is kept dispersed in the oil by recirculating the gas extracted from the head space of the reactor to the bottom by means of a blower. Usually, these are batch processes, although continuous mode of operation is becoming more attractive. The second type of reactor resembles a column or tube containing a fixed bed of catalyst, and is used where the organic compound has sufficient vapor pressure at the reaction temperature, as in the synthesis of methanol from carbon monoxide, to permit gas-phase, continuous operations.

The design and construction of process equipment which can withstand hydrogen gas at high temperature or high pressure, or both, are complicated. Alloy steels are the most common materials of construction. *See* AMINATION; DEHYDROGENATION; FISCHER-TROPSCH PROCESS; HIGH-PRESSURE PROCESSES; HYDROFORMYLATION; HYDROGEN; ORGANIC CHEMICAL SYNTHESIS; OXIDATION-REDUCTION.

[ROBERTO LEE]

Bibliography: M. Freifelder, *Catalytic Hydrogenation in Organic Synthesis*: *Procedures and Commentary*, 1978; B. R. James, *Homogeneous Hydrogenation*, 1973; P. N. Rylander, *Catalytic Hydrogenation over Platinum Metals*, 1967.

Hydrolysis

Literally destruction, decomposition, or alteration of a chemical substance by water. In discussions of aqueous solutions of electrolytes the term hydrolysis is applied especially to reactions of cations (positive ions) with water to produce a weak base, or of anions (negative ions) to produce a weak acid. A salt of a weak acid or of a weak base, or of both a weak acid and weak base, is then said to be hydrolyzed. The degree of hydrolysis is the fraction of the ion which reacts with the water. The term solvolysis is employed for reactions of solutes with solvents in general.

A frequently quoted example is the hydrolysis of sodium acetate, which is highly dissociated in water, to yield sodium ion, Na^+, and acetate ion, $CH_3CO_2^-$. The acetate ion is represented below by the symbol A^-. Because acetic acid is a weak acid (it has a small dissociation constant), the acetate ion, A^-, combines with some of the hydrogen ion present in aqueous solutions, as shown in reaction (1).

$$A^- + H_3O^+ \rightleftharpoons HA + H_2O \qquad (1)$$

The concentration of the hydrogen ion is involved in the equilibrium, reaction (2), between wa-

$$2H_2O \rightleftharpoons H_3O^+ + OH^- \qquad (2)$$

ter and the hydrogen ion (often called the hydronium ion) and the hydroxide ion. Because reaction (1) removes a product of reaction (2), each reaction proceeds to produce more of its products (the substances on the right side of each equation). The net result is reaction (3), which represents the reaction

$$A^- + H_2O \rightleftharpoons HA + OH^- \qquad (3)$$

which is described as the hydrolysis of the acetate ion. When equilibrium is attained, the concentrations are such as to conform to the equilibrium constants of all three reactions (one of which is a mathematical consequence of the other two).

Textbooks contain a simple equation for the calculation of the degree of hydrolysis of an ion in a dilute salt solution.

For reaction (3), the degree of dissociation x is

given by Eq. (4), where the brackets indicate molar

$$x=\frac{[OH^-][HA]}{[A^-]}=\frac{[OH^-][H^+][HA]}{[H^+][A^-]}=\frac{K_w}{K_a} \quad (4)$$

concentrations, K_w is the ion-product constant of water, and K_a is the dissociation constant of the acid, HA.

As the equation implies, the fraction hydrolyzed may be affected in one of several ways.

1. The hydrolyzed fraction may become larger as the dissociation constant of the weak acid or base produced by hydrolysis becomes smaller. For example, sodium cyanide is more highly hydrolyzed than sodium acetate because hydrocyanic acid is a weaker acid than acetic acid; consequently a solution of sodium cyanide dissolved in pure water is slightly more alkaline than one of sodium acetate.

2. The hydrolyzed fraction may become larger as the dissociation constant of the solvent becomes larger. For example, the degree of hydrolysis of sodium acetate in water is larger at 50°C than at 25°C because the dissociation constant of water increases with temperature and because the dissociation constant of acetic acid is smaller at 50°C than at 25°C.

3. The fraction may become larger as the salt concentration becomes smaller. Because of oversimplification, the equation usually given in textbooks implies that the degree of hydrolysis approaches 100% as concentration approaches zero. Actually, the limit approached is only about 0.5% for sodium acetate in carbon dioxide—free water at 25°C.

4. The fraction may become much larger if the electrolyte is the salt of both a weak acid and a weak base. If the dissociation constants are nearly equal at the temperature of the solution, the solution need be neither appreciably acid nor appreciably alkaline.

Other types of electrolytic hydrolysis may occur. Sometimes a solution may become sufficiently acidic or sufficiently alkaline to cause the formation by precipitation of another phase, for example, a solid basic salt. Even the anion of the strong acid HCl may be hydrolyzed if solid sodium chloride is heated in the presence of moisture. Some of the very volatile HCl is lost, and the residual melt when cooled and dissolved in pure water is demonstrably alkaline. In this example the hydrolysis does not occur because of the formation of a weak acid in solution, but because of the formation of a volatile compound. *See* ACID AND BASE; HYDROGEN ION; IONIC EQUILIBRIUM.

[THOMAS F. YOUNG]

Bibliography: H. Freiser and Q. Fernando, *Ionic Equilibria in Analytical Chemistry*, 1963, reprint 1979; G. Scuttchard, *Equilibrium in Solutions and Surface and Colloid Chemistry*, ed. by I. Scheinberg, 1976.

Hydroquinone

A dihydric phenol ($C_6H_6O_2$; colorless crystals, melting point 169°C) in which the two hydroxyl groups are in the para or 1,4 positions. The important relationship between hydroquinone and *p*-benzoquinone (commonly called simply quinone) is illustrated in the reaction below. This rapid and

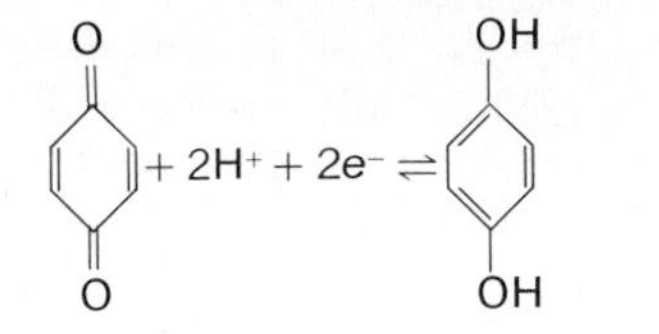

reversible oxidation-reduction equilibrium is characterized by a standard reduction potential (E_0) of 0.699 V at 25°C.

The facile oxidation of hydroquinone is the basis for the most important uses of the compound and its derivatives. As a photographic developer, hydroquinone reacts with light-activated silver bromide crystals to form black metallic silver (reduction product) and quinone (oxidation product). Various alkylated derivatives of hydroquinone and the monomethyl ether (hydroxyanisole) find use as preservatives and antioxidants.

The commercial production of hydroquinone has proceeded by reduction of quinone with iron and water. Quinone has usually been produced by oxidation of aniline, using manganese dioxide. The production of hydroquinone from *p*-diisopropylbenzene by a process analogous to the cumene process for phenol has been reported. *See* CATECHOL; PHENOL; RESORCINOL.

[MARTIN STILES]

Hydroxide

One type of hydroxyl compound (OH-containing) whose solutions have a bitter taste, feel slippery to the touch, neutralize acids, and change the color of red litmus to blue. They are also called bases and have a high concentration of OH^- ions in solution. When an oxide of an element, XO, X_2O, is dissolved in water, a hydroxyl compound of the type XOH is formed. These molecules, XOH, can be ionized to give either OH^- ions or H^+ ions. The former are called bases and the latter acids. Ordinarily, when X is a metal, bases are obtained, for example, NaOH; and when X is a nonmetal, acids are obtained, for example, $S(OH)_6$, which loses two molecules of water and exists as H_2SO_4. Some insoluble oxides, Al_2O_3, for example, have both acid and basic properties. This class of compounds is said to be amphoteric.

Although only the alkali metal hydroxides are extensively soluble in water, the moderately soluble alkaline-earth metal hydroxides, and even quite insoluble oxides, such as Fe_2O_3, have basic properties and react with acids.

Some moderately soluble hydroxides, such as $Ba(OH)_2$ and $Ca(OH)_2$, are strong bases, that is, highly ionized. When solutions of heavy metal salts are made basic, insoluble hydroxides are precipitated. In reality these compounds often do not contain OH^- ions, although they have basic properties. They are hydrates of one form or another.

The compounds containing more water than is required for the hydroxide formula $Mg(OH)_2 \cdot xH_2O$, for example, are called hydrous hydroxides. Those containing less than the required water, for example, $Al_2O_3 \cdot H_2O \cdot xH_2O$, are called hydrous hydrates; and those that contain water not in any fixed amount are called hydrous oxides, for example, $Fe_2O_3 \cdot xH_2O$.

Hydroxides are very important compounds industrially, sodium hydroxide being the most important. It is used in making other chemicals, rayon, petroleum products, and soap.

Hydroxides are not the only compounds used as bases; others are phosphates, carbonates, and oxides. *See* ACID AND BASE; HYDRATE.

[E. EUGENE WEAVER]

Hydroxylation reaction

One of several types of reactions used to introduce one or more hydroxyl groups into organic compounds. Unsaturated compounds are the most common starting materials. Only those reactions having the widest use are discussed.

Monohydroxylation. Reaction (1) is especially useful when isomeric alcohols cannot form, as in the conversion of ethylene to ethyl alcohol. Reactions with longer-chain olefins produce primarily secondary and branched-chain alcohols. Reaction (2) is relatively new, but promises to have wide use when the starting materials and products do not react with boiling formic acid. Lewis acids are preferred catalysts. Reaction (3) is of wide applicability. Hydrogenation can be effected catalytically or by reduction with metal hydrides. Reaction (4) is a classical one, and is rather limited.

Dihydroxylation. Reaction (5) is the most widely used dihydroxylation reaction. Organic peracids need not be separately prepared, but most frequently are formed and utilized in place by the interaction of hydrogen peroxide with formic or acetic acid. Reaction (6) is the classical dihydroxylation reaction, but the large amounts of manganese dioxide produced make the reaction awkward on a large laboratory scale. Reaction (7) is also

$$-CH{=}CH- + H_2SO_4 \rightarrow -CH_2-\underset{\underset{SO_3OH}{|}}{CH}- \quad (1)$$

$$-CH_2-\underset{\underset{SO_3OH}{|}}{CH}- + H_2O \rightarrow -CH_2-\underset{\underset{OH}{|}}{CH}-$$

$$-CH{=}CH- + HCO_2H \xrightarrow{H^+} -CH_2-\underset{\underset{O-C-H}{|}\underset{\|}{}O}{CH}-$$

$$-CH_2-\underset{\underset{O-C-H}{|}}{CH}- + H_2O \rightarrow -CH_2-\underset{\underset{OH}{|}}{CH}- \quad (2)$$

$$-CH{=}CH- + \text{Organic peracids} \rightarrow -CH\underset{\underset{O}{\diagdown\diagup}}{-}CH- \quad (3)$$

$$-CH\underset{\underset{O}{\diagdown\diagup}}{-}CH- + H_2 \rightarrow -CH_2-\underset{\underset{OH}{|}}{CH}-$$

$$-CH{=}CH- + HBr \rightarrow -CH_2-\underset{\underset{Br}{|}}{CH}-$$

$$-CH_2-\underset{\underset{Br}{|}}{CH}- + H_2O \xrightarrow{OH^-} -CH_2-\underset{\underset{OH}{|}}{CH}- \quad (4)$$

$$-CH{=}CH- + \text{Organic peracids} \rightarrow -CH\underset{\underset{O}{\diagdown\diagup}}{-}CH-$$

$$-CH\underset{\underset{O}{\diagdown\diagup}}{-}CH- + H_2O \xrightarrow[\text{or } OH^-]{H^+} -CH-CH-\underset{OH\ \ OH}{} \quad (5)$$

$$-CH{=}CH + KMnO_4 \xrightarrow{OH^-} -\underset{\underset{OH}{|}}{CH}-\underset{\underset{OH}{|}}{CH}- \quad (6)$$

$$-CH{=}CH- + OsO_4 \xrightarrow[\text{solvent}]{\text{Inert}} -CH-CH- \quad (7)$$
$$\underset{OH\ \ \ OH}{}$$

$$-CH{=}CH- + I_2 + RCOOAg \rightarrow -\underset{\underset{\underset{\underset{R}{|}}{C{=}O}}{|}}{CH}{-}{-}\underset{\underset{\underset{\underset{R}{|}}{C{=}O}}{|}}{CH}-$$

$$-\underset{\underset{\underset{\underset{R}{|}}{C{=}O}}{|}}{CH}{-}\underset{\underset{\underset{\underset{R}{|}}{C{=}O}}{|}}{CH}- + H_2O \rightarrow -\underset{\underset{OH}{|}}{CH}-\underset{\underset{OH}{|}}{CH}- \quad (8)$$

$$-CH{=}CH- + HOCl \rightarrow -\underset{\underset{OH}{|}}{CH}-\underset{\underset{Cl}{|}}{CH}-$$

$$-\underset{\underset{OH}{|}}{CH}-\underset{\underset{Cl}{|}}{CH}- + H_2O \xrightarrow{OH^-} -\underset{\underset{OH}{|}}{CH}-\underset{\underset{OH}{|}}{CH}- \quad (9)$$

$$-CH{=}CH- + H_2O_2 \xrightarrow[OsO_4]{\textit{tert}\text{-Butyl alcohol}}$$
$$-\underset{\underset{OH}{|}}{CH}-\underset{\underset{OH}{|}}{CH}- \quad (10)$$

$$-CH{=}CH- + HCHO \xrightarrow{H^+} -\underset{\underset{OH}{|}}{CH}-\underset{\underset{CH_2OH}{|}}{CH}- \quad (11)$$

widely applicable and gives high yields, but osmium tetroxide is expensive and toxic. Reaction (8), known also as the Prévost reaction, is of interest in that isomeric α-glycols can be obtained from the reaction of an olefin, iodine, and silver acetate, depending on whether the reaction is conducted in wet or dry acetic acid. Reaction (9), also a classical one, has limited value. Reaction (10) is analogous to reaction (7) but has the advantage that only catalytic quantities of osmium tetroxide are required.

Reaction (11), to obtain 1,3-glycols, is known as the Prins reaction. Under some conditions and with some olefins, 1,3-glycols predominate, but competing reactions limit its scope and utility.

See EPOXIDATION; ETHYLENE GLYCOL; ETHYLENE OXIDE; GLYCOL; ORGANIC REACTION MECHANISM; SUBSTITUTION REACTION.   [DANIEL SWERN]

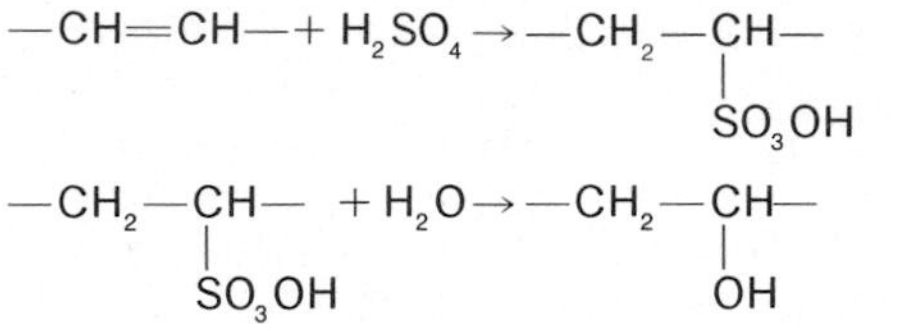

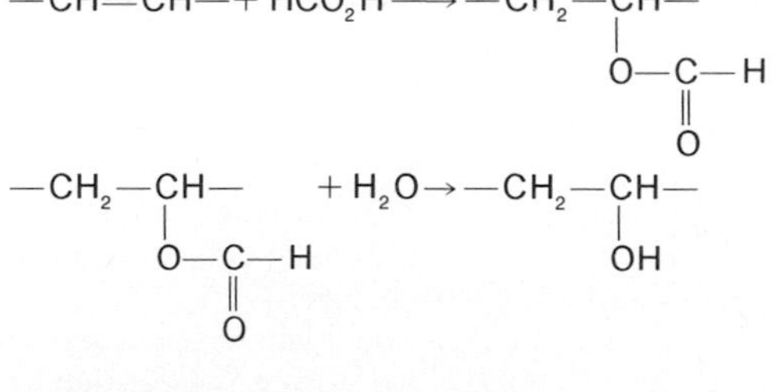

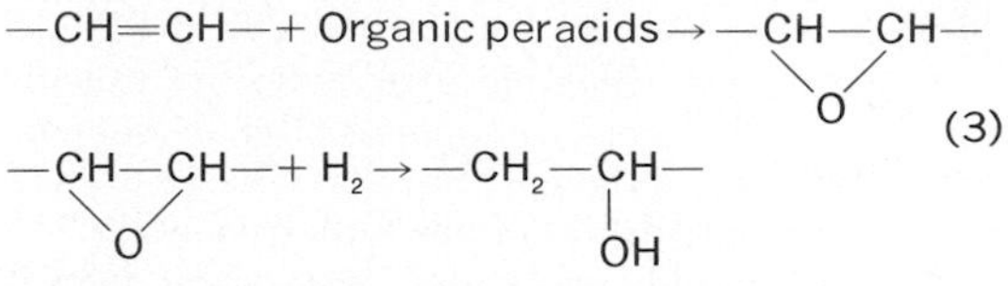

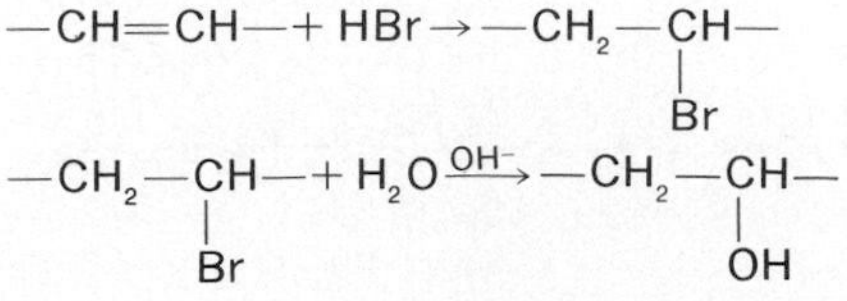

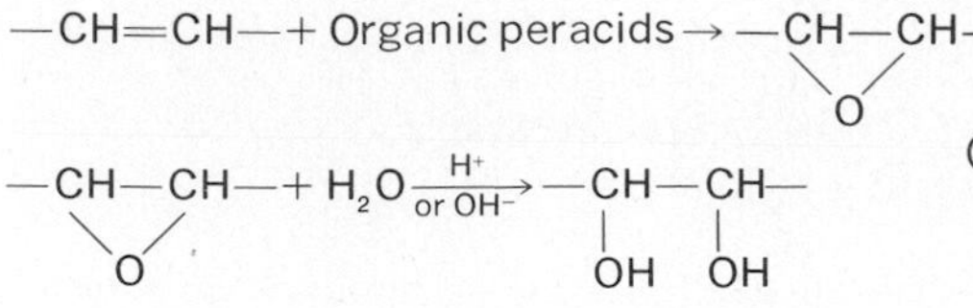

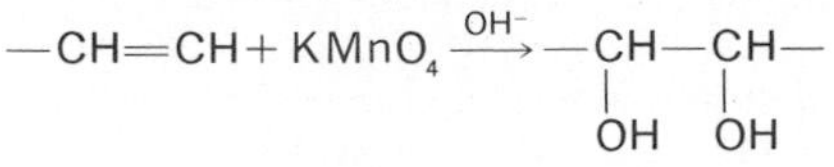

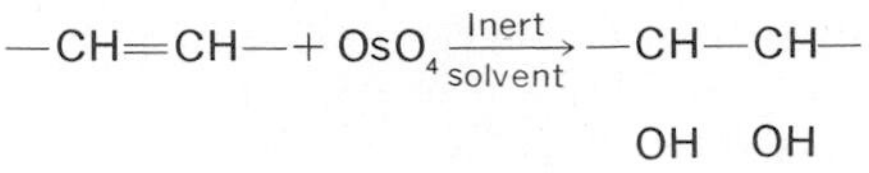

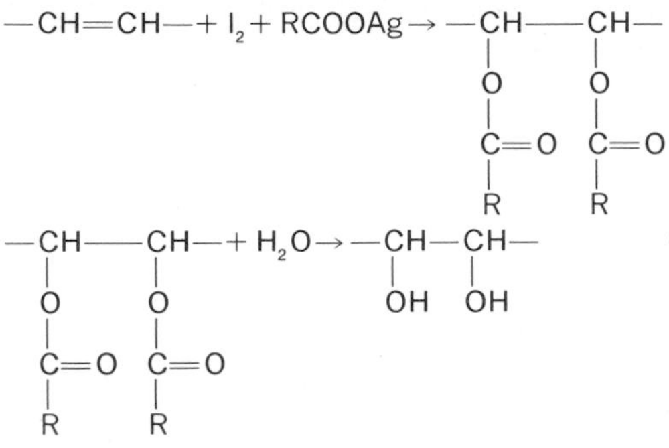

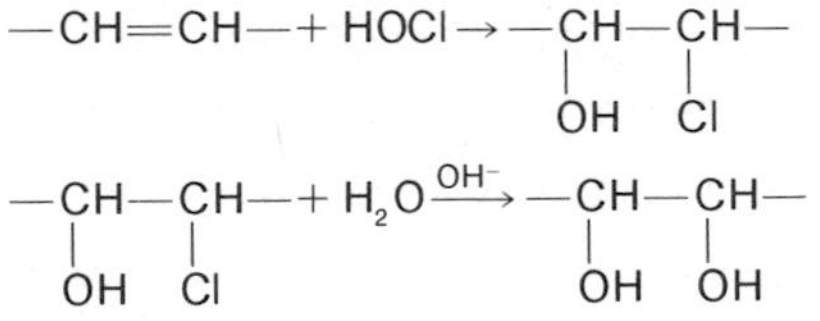

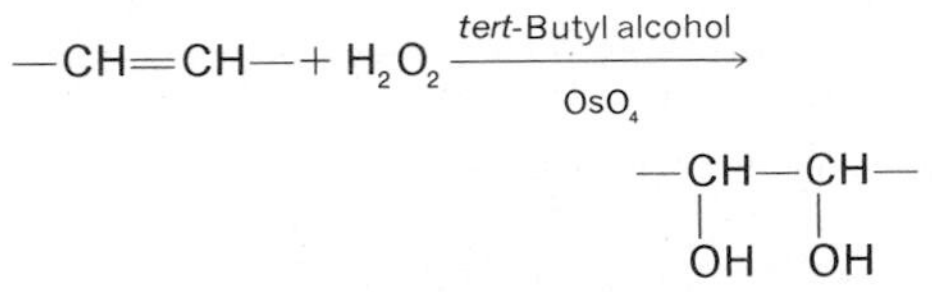

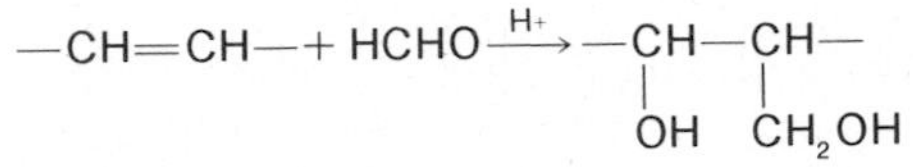

Hypochlorite

The negative ion, ClO^-, derived from hypochlorous acid, $HClO$. The hypochlorite ion is best known as an oxidizing agent and as a constituent of bleaching agents.

Hypochlorites are prepared by passing chlorine gas into alkaline solution and in this procedure the chlorine is simultaneously oxidized and reduced, as in reaction (1).

$$Cl_2 + 2NaOH \rightarrow NaCl + NaClO + H_2O \qquad (1)$$

Sodium hypochlorite is extensively used as a bleach in the textile industry and in laundries and as a disinfectant in the home.

When chlorine is passed over slaked lime, a product called bleaching powder or chlorinated lime is obtained which approximates the formula CaCl(OCl). A related preparation, high-test hypochlorite (HTH), which contains twice as much available chlorine as bleaching powder, may be represented by the formula $Ca(OCl)_2$.

These preparations are shipped in powder form to the point of use and converted to sodium hypochlorite by adding soda lime as indicated by reaction (2).

$$Na_2CO_3 + Ca(ClO)_2 \rightarrow 2NaClO + CaCO_3 \qquad (2)$$

See CHLORATE; CHLORINE; OXIDIZING AGENT; PERCHLORATE.

[E. EUGENE WEAVER]

Hypofluorous acid

An unstable acid with the chemical formula HOF. The preparation of hypofluorous acid was first claimed in 1932, but this claim was almost certainly incorrect. In 1968 HOF was identified after photolysis of a mixture of fluorine and water in a solid N_2 matrix. The compound was prepared and isolated in 1971 by passage of a stream of fluorine over ice. The most efficient preparations involve circulating fluorine over ice at −40°C, removing water from the product stream with traps at −50 and −78°C, and collecting the product in a trap at −183°C.

Hypofluorous acid is a colorless gas at room temperature. At very low temperature it is a white solid, melting at −117°C to a pale yellow liquid that boils somewhat below room temperature. The HOF molecule is bent, with an unusually small bond angle of 97°. The compound has a standard heat of formation of −23.5±1 kcal/mol (−98±4 kJ/mol) and a standard Gibbs free energy of formation of −20.5±1 kcal/mol (−86±4 kJ/mol).

Hypofluorous acid is unstable to the formation of HF and O_2. The gas decomposes with a 30–60 min half-life at room temperature, while the liquid has a tendency to explode. The compound reacts rapidly with water to form H_2O_2 in acid solution and O_2 in base. Most oxidizable substrates are attacked by HOF. Its mode of oxidation appears to be by oxygen transfer, as evidenced by the production of H_2O_2 and by the oxidation of HSO_4^- to HSO_5^-, peroxymonosulfate. In alkaline solution, HOF oxides bromate to perbromate. Oxygen transfer from HOF provides a means of synthesizing a variety of compounds with isotopically labeled oxygen atoms. *See* PERBROMATE.

The fluorine atom in HOF is negative, unlike the chlorine atom in HOCl. Hence, HOF acts as an oxygenating and hydroxylating agent rather than as an oxidative fluorinating agent. Aromatic compounds are oxidized to phenols, and the mechanism resembles that invoked for the oxidation of aromatic compounds in living organisms. Octaethyl porphyrin is converted to its previously unknown N-oxide.

Hypofluorous acid adds to olefins to form α-fluoro alcohols. The sense of the addition is opposite to that observed with HOCl, as would be expected from the polarization of the HOF molecule: $HO^{\delta+}$—$F^{\delta-}$. [EVAN H. APPELMAN]

Bibliography: E. H. Appelman, Non-existent compounds: Two case histories, *Acc. Chem. Res.*, 6:113–117, 1973.

Imidazole

One of a group of organic heterocyclic compounds (also called iminazoles, glyoxalines, and 1,3-diazoles) containing a five-membered diunsaturated ring with two nonadjacent nitrogen atoms as part of the ring. Imidazole (I) is a typical member of the group. *See* AZOLE; HETEROCYCLIC COMPOUNDS.

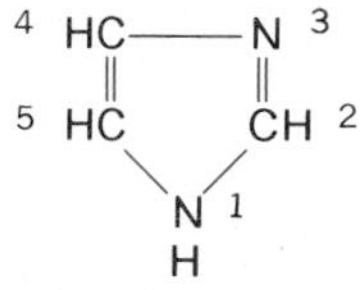

(I)

The imidazole ring system is found in a number of natural products, for example, in the α-amino acid histidine (II) and in the alkaloid pilocarpine (III). Histamine (IV) is associated with allergic

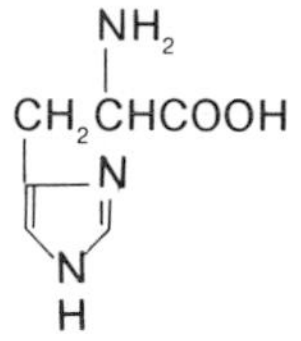

(II)

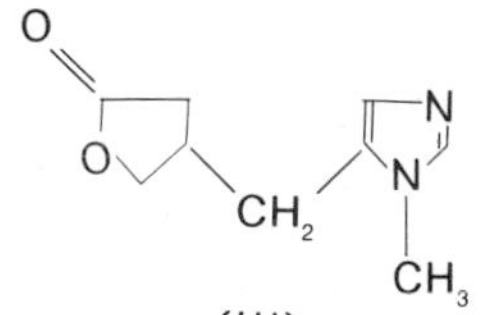

(III)

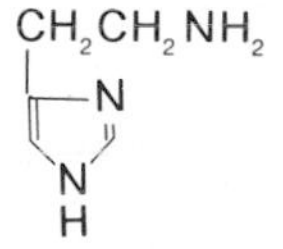

(IV)

response, and ostensibly is the biological target against which the antagonistic action of synthetic antihistaminics is directed. The biologically important purine system contains an imidazole ring fused to pyrimidine. The imidazole ring, present in enzyme proteins as the histidine side chain, is involved in enzyme-catalyzed reactions, presumably by serving as an efficient acyl-transfer agent. The same kind of imidazole ring in the blood protein, globin, holds heme and globin together by coordinating with the iron atom of the heme.

Properties and preparation. Imidazole itself is a water-soluble solid, mp 90°C, which is basic enough (pK_a 6.95 at 25°) to form stable salts with both organic and inorganic acids. Imidazole is also weakly acidic, since the hydrogen at the 1 position may be replaced by metal. The low volatility of imidazole, bp 256°C, is indicative of considerable association by intermolecular hydrogen bonding.

Imidazole is a resonance system showing the chemical behavior of a moderately aromatic ring. The system is stable to oxidation by nitric acid, hexavalent chromium, and alkaline permanganate, and it is, in general, resistant to ring reduction. Amino groups at position 4 may be diazotized normally. Imidazole undergoes electrophilic substitutions such as bromination, nitration, or sulfonation at position 4, and azo coupling at position 2. The imidazole ring is opened by attack of peroxide and peracids. Alkylation on nitrogen is possible to give first 1-alkylimidazoles and then 1,3-dialkylimidazolium cations. Hot alkali disrupts the ring in the quaternary imidazolium compounds.

Several general methods of synthesis are known. Combination of α-halo ketones with amidines (V) gives imidazoles, a process emphasizing

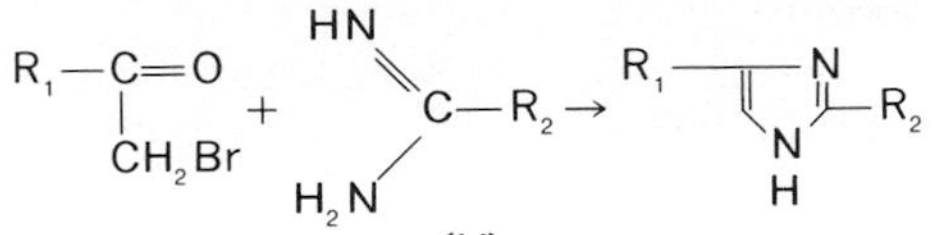

(V)

the amidine structure of imidazole. α-Dicarbonyl compounds react with ammonia and aldehydes to give imidazoles (VI). Further, α-amino ketones or

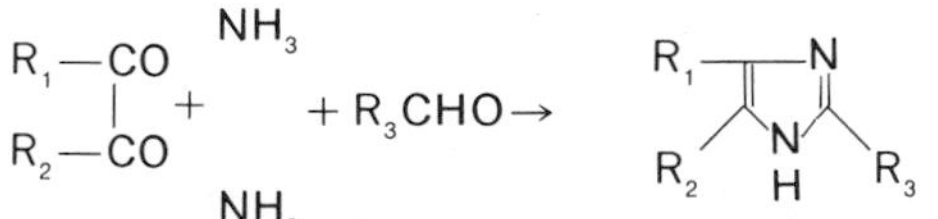

(VI)

aldehydes (VII) condense with thiocyanate to give 2-mercaptoimidazoles (VIII).

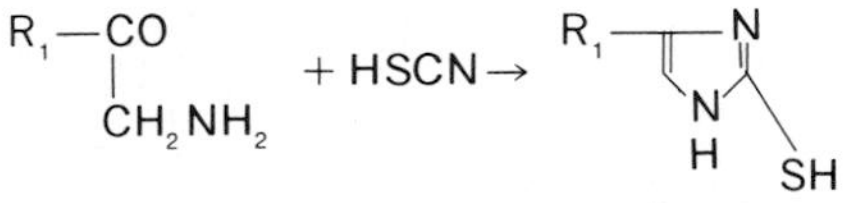

(VII) (VIII)

Important derivatives. Imidazolines and imidazolidines are the names assigned to dihydro- and tetrahydroimidazoles, respectively. Such compounds are generally formed by ring-closure processes involving positions 1, 2, or 3. Ring carbonyl derivatives are known. Biotin (IX), for example, is a condensed 2-imidazolidone.

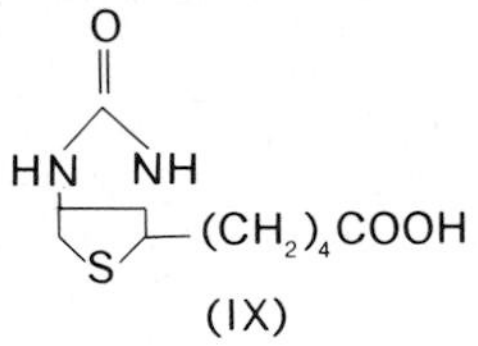

(IX)

Hydantoins or 2,4-diketoimidazolidines (X) are prepared by condensation of a ketone or aldehyde with potassium cyanide and ammonium carbonate.

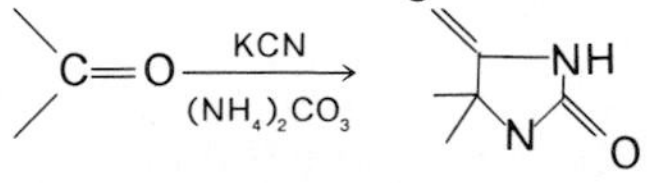

(X)

Hydantoins have been studied extensively in connection with the physiological activity of 5,5-disubstituted derivatives and because of the possibility of converting hydantoins to α-amino acids. The 5,5-diphenyl derivative, Dilantin sodium (XI), is used as an anticonvulsant in treatment of epilepsy. The 5-ethyl-5-phenyl derivative, Nirvanol (XII),

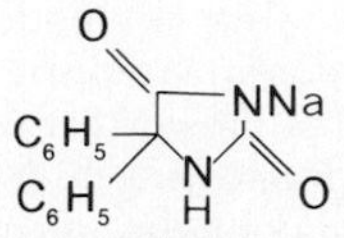

(XI)

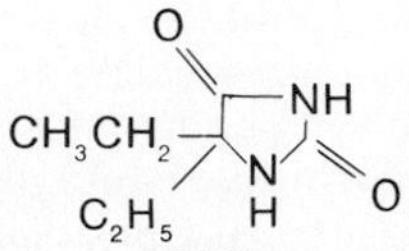

(XII)

is an effective hypnotic. Complete hydrolysis of hydantoins generates an α-amino acid (XIII, XIV).

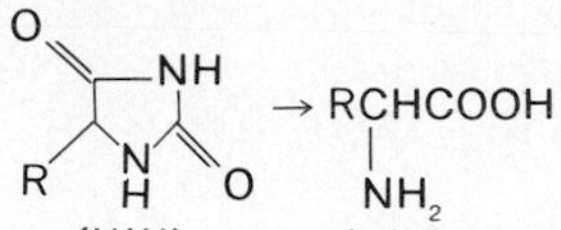

(XIII) (XIV)

This route constitutes one of the standard syntheses for α-amino acids. 1,3-Dibromo- and 1,3-dichloro-5,5-dimethylhydantoin are convenient sources of positive halogen.

Allantoin, or 5-ureidohydantoin (XV), is the end product of purine metabolism in most mammals

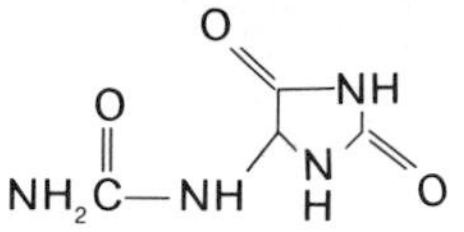

(XV)

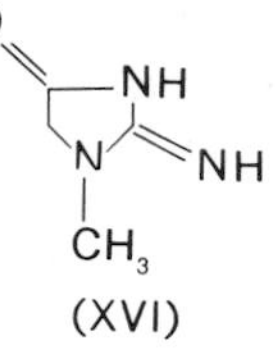

(XVI)

(but not including humans) and is an intermediate in the purine metabolism of crustaceans and amphibia. Creatinine (XVI) is formed irreversibly from creatine and is excreted. [WALTER J. GENSLER]

Bibliography: R. N. Acheson, *An Introduction to the Chemistry of Heterocyclic Compounds*, 1967; R. C. Elderfield (ed.), *Heterocyclic Compounds*, vol. 5, 1957; K. Hofmann, *Imidazole and Its Derivatives*, pt. 1, 1953.

Indene

A colorless, liquid hydrocarbon (C_9H_8, also called benzocyclopentadiene) with the structure shown below, which boils at 181°C and freezes at

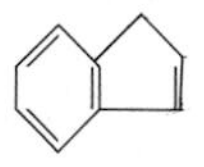

−2°C. It is usually obtained from the light-oil fraction (boiling point 178–182°C) produced in the carbonization of coal but is also obtained by pyrolysis of certain petroleum fractions.

Indene resembles cyclopentadiene in that one hydrogen of the methylene (CH_2) group may be replaced by sodium. Indene may be oxidized to phthalic acid or reduced to indan (C_9H_{10}). In the presence of acid, indene polymerizes. Copolymers with benzofuran have been manufactured on a small scale for use in coatings and floor coverings. *See* POLYNUCLEAR HYDROCARBON.

[CHARLES K. BRADSHER]

Indium

A chemical element, In, atomic number 49, a member of subgroup IIIa and the fifth period of the periodic table. The valence electron notation corresponding to its ground-state term, $5s^25p^1$, accounts for the maximum oxidation state of III in its compounds. Compounds of oxidation state I and apparent oxidation state II are also reported. Indium has a relative atomic weight of 114.82.

Indium occurs in the Earth's crust to the extent of about $1 \times 10^{-5}\%$ and is normally found in concentrations of 0.1% or less. It is widely distributed in many ores and minerals but is largely recovered from the flue dusts and residues of zinc-processing operations.

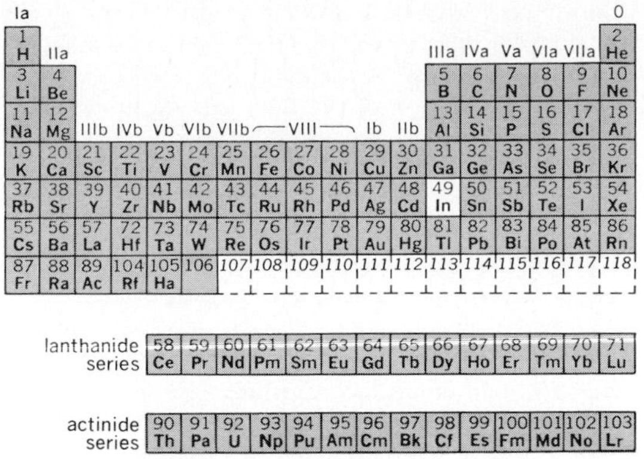

Indium is used in soldering lead wires to germanium transistors and as a component of the intermetallic semiconductor used for germanium transistors. Indium arsenide, antimonide, and phosphide are semiconductors with unique properties. Other uses of indium are sleeve-type bearings to reduce corrosion and wear, glass-sealing alloys, and dental alloys.

Metal. Indium is recovered from the residues of zinc processing by an acid leach followed by chemical separations from the accompanying elemental impurities such as zinc, cadmium, aluminum, arsenic, and antimony. Final purification by aqueous electrolysis at a controlled potential yields a product of 99.9% purity. Indium is a soft metal which can be easily scratched with the fingernail. It adheres to other surfaces when rubbed across them. It has a melting point of 156.4°C, a boiling point of over 2000°C, and a density of 7.31 g/cm³ at 20°C. It crystallizes in the face-centered tetragonal structure and has a metal radius of 1.66 A.

The standard oxidation potential for the reaction, shown below, is +0.34, which accounts for

$$In_{(s)} \rightleftharpoons In^{3+}_{(aq)} + 3e$$

the fact that the metal dissolves in all acids to give solutions of In^{3+}. The metal reacts slowly with oxygen of the air up to its melting point, but at higher temperatures it readily yields yellow In_2O_3. Indium reacts directly with the halogens and other nonmetals such as sulfur, selenium, and phosphorus when warm.

Compounds. The trivalent chloride, bromide, and iodide are dimeric in the vapor state. The iodide is also dimeric in the solid state with approximate tetrahedral symmetry about the metal ions, whereas the bromide and iodide have a layer lattice structure in which the metal ions sit in distorted octahedra of anions. The bromide and iodide each exhibit an alternate form. The trifluoride has a structure in which the metal ion is distributed in one-third of the octahedral holes throughout three dimensions, resulting in a high-melting (1170°C) compound. The radius for the In^{3+} ion in a CN_6 environment is 0.80 A (0.08 nm). The anhydrous fluoride is only slightly soluble in water, whereas the other trihalides are very soluble and are recovered from aqueous solution as hydrates. The soluble indium trihalides hydrolyze, yielding species such as $In(OH)^{2+}$, $In_2(OH)_4^{4+}$ depending upon the metal ion concentration, and $InCl_n^{3-n}$ ($n = 1-7$) depending upon the halide ion concentration, cation, and solvent. A 0.1 *M* solution of InX_3 has a pH of 3; the hydrate hydroxide precipitates from such a solution at pH 3.4. Indium hydroxide dissolves neither in excess hydroxyl ion nor in the presence of ammonia. The carbonate, oxalate, and sulfide are insoluble in water.

The anydrous halides, except the fluoride, react as Lewis acids, forming 1:2 adducts with a variety of Lewis bases. Although the $MX_3 \cdot 2NMe_3$ compounds have a trigonal bipyramidal structure, $InI_3 \cdot 2DMSO$ appears to have a solid-state structure which contains $[InI_2(DMSO)_4^+][InI_4^-]$ species. On the other hand, five-coordinate complexes of InI_3 are formed with olefinic phosphines as in $InI_3 \cdot 2[P(C_6H_4)_3]$. Variable coordination with NCS^- is observed depending upon the cation present. Ph_4As^+ and $Ph_3PCH_2Ph^+$ yield five-coordinate species $[In(NCS)_5^{3-}]$, whereas Me_4N^+, Et_3NH^+, and $Bu_4{}^nN^+$ result in six-coordinate species $[In(NCS)_6^{3-}]$.

Indium(III) forms six-coordinate complexes with bidentate ligands such as 2,2′-dipyridyl, dicarboxylic acids, β-diketones, catechol, 8-hydroxyquinoline, and diethoxythiophosphate, while an eight-coordinate complex $[InT_4^-]$ is obtained with tropone, T^-. The bidentate ligand, *cis*-1,2-$S_2C_2(CN)_2^{2-}$, mnt, yields $In(mnt)_2^-$, $In(mnt)_2X^{2-}$, and $In(mnt)_3^{3-}$. Indium(I) halides form $XIn[1,2,S_2C_2(CF_3)_2]$.

Reduction of the anhydrous halides (except iodide) with hydrogen or a hydrogen-hydrogen halide mixture leads to compounds of composition InX_2. These compounds are known to be $In^I(In^{III}X_4)$. Reduction of the trihalides with indium metal results in products of composition InX. Indium monochloride has a distorted NaCl structure, while the bromide and iodide structures are unknown. Compounds of composition In_2Cl_3, In_2Br_3, and In_7Cl_9 appear to have the compositions $3In^+$, $InCl_6^{3-}$; $2In^+$, $In_2Br_6^{2-}$; and $6In^+$, $InBr_6^{3-}$, $3Br^-$. The reduced-state compounds are unstable in aqueous solution with respect to disproportionation to the metal and oxidation state III. InX and InX_2 compounds are insoluble in most organic solvents but dissolve in aniline and morpholine, from which solutions compounds of composition $In(morpholine)_2X$ and $In(aniline)_4X_2$ are obtained. Morpholine acts as a bidentate ligand, and conductivity studies suggest that the species in solution is a 1:1 electrolyte. Aniline acts as a unidentate ligand, and in solution the species probably is $[In(aniline)_4]^+[InX_4]^-$.

The reaction of indium metal with mercury alkyls or aryls yields organometallic compounds of the R_3In class. Trimethylindium in the solid state has a unique tetrameric structure in which unsymmetrical $CH_3 \cdots In—CH_3$ bridges are present. In solution the trimethylakyl is a monomer. Trimethylindium reacts with primary and secondary aliphatic phosphines to give $(Me_2InPMe_2)_n$ and methane. Trialkyl indium etherates react with triorganosilanols to give compounds of composition $(X_2M—O—M'R_3)_2$, where X = methyl or Cl; M^1 = C, Si, or Ge; and R = methyl or phenyl. These compounds contain a framework of four-membered In—O rings. The trialkyls also react with some carboxylic, phosphinic, thiophosphinic, and sulfuric acids to give eight-membered rings, each containing two In—Me groups. A series of monomeric four-coordinate complexes, $(C_6F_5)_3InL$, where L = py, Ph_3P, Ph_3PO, Ph_3AsO, have been prepared as well as five-coordinate complexes,

$(C_6F_5)_3InL_2$, with L=DMSO or THF. Bipyridyl, $Ph_2P(CH_2)_2PPh_2$, and *N,N,N′,N′*-tetramethylenediamine appear to be bridging ligands in compounds of composition $[(C_6H_5)_3In]_2L$. Insertion into the In—C bond occurs in the reaction of $(CH_3)_3In$ with SO_2 to yield $In(SO_2Me)_3$, while In—C bond rupture occurs when $(C_2H_5)_3In$ reacts with CH_2X_2 (X=Br or I) or CCl_4 to produce $(C_2H_5)_2InCH_2X$ and $(C_2H_5)_2InCCl_3$ respectively. Indium metal reacts with alkylhalides (RX, X=Br, I and R=Me, Et, *n*-Pr, *n*-Bu) to yield products of composition $R_3In_2X_3$, which when heated in the presence of KBr or KI give R_2InX. The Me_2In^+ ion is stable in aqueous solution at 0°C for several days. It appears to possess a linear structure. A series of quinolinato complexes of the type R_2InQ (R=Me, Et, or Bu′) have been prepared. Dialkylindium compounds Et_2In-$[C(NO_2)_2R_2]$, where R=H or Me, $Me_2In(dtc)$, $Et_2In(OX)$, and $R_2InOC_2H_4NMe_2$, where R=Me or Et, are obtained by reaction of the indium trialkyl and α-mononitroalkanes, $SSCNMe_2$(dtc), 8-hydroxyquinoline(OXH), and $HOC_2H_4NMe_2$ respectively. $RInX_2$ compounds (X=Br or I, R=Me, Et, Pr, or Bu) have been prepared by the reaction of InBr or InI with the proper alkyl halide. The iodides appear to be $InMe_2^+InI_4^-$, while the bromides are probably polymeric. Cyclopentadiene forms a composition C_2H_5In, which is known to be a monomer in the gaseous phase but probably is polymeric in the solid. It is not considered to be an ionic compound. Adducts of composition $(Cp)In \cdot BX_3$ (X=F, Cl, Br, or CH_3) have been prepared. Spectroscopic evidence suggests that the cyclopentadienyl group is σ-bonded. Solid Cp_3In consists of infinite polymeric chains of σ-bonded Cp_2In units bridged by Cp groups. The cyclopentadienyl groups form a slightly distorted tetrahedral environment about the indium.

Numerous compounds containing indium-metal bonds have been prepared. The reaction of indium metal with $Mn_2(CO)_{10}$ at 140°C yields $In[Mn(CO)_5]_3$. $In[Co(CO)_4]_3$ is also known. The cobalt and manganese atoms are essentially trigonal about the indium, and the four carbon monoxides and indium atom are trigonal bipyramidal about the cobalt and manganese. The indium-metal bonds are cleaved by halogens or hydrogen halides to form $X_{3-n}[In(Mn(CO)_5)]_n$, where X=Cl or Br and n=1 or 2. Cobalt forms similar compounds. The ions $In[Mn(CO)_5]_2^+$, $In[Co(CO)_4]_2^-$, and $In[Co(CO)_4]_4^-$ have been obtained. The crystal structure of $Br_3In_3Co_4(CO)_{15}$ consists of a six-membered ring of alternate In and Br atoms with each indium atom bonded to separate $Co(CO)_4$ groups and a central cobalt atom possessing three carbon monoxides. Indium(I) bromide reacts with $Co_2(CO)_8$ by insertion to give $Co(CO)_4$—$InBr_2In$-$[Co(CO)_4]$, but with $W(CO)_6$ only the ion $W(CO)_5$-$InBr_3^-$ is obtained. The insertion of InX(Cl,Br) into a variety of compounds has yielded products such as $XIn[Co(CO)_4]_2$, $XIn[CpFe(CO)_2]_2$, and In-$[Mo(CO_3)_2Cp]_3$.

Analysis. Indium may be determined quantitatively with 8-hydroxyquinoline by precipitation of the compound $In(C_9H_6ON)_3$ at 70–80°C from a sodium acetate–acetic acid buffer, followed by drying at 120°C and direct weighing of the precipitate. An alternate to the direct weighing is the bromometric titration procedure after solution of the compound in warm 10–15% hydrochloric acid or the colorimetric determination at 400 mμ of a chloroform solution of the 8-hydroxyquinolate. Indium may also be determined by atomis absorption spectroscopy. *See* GALLIUM; THALLIUM.

[EDWIN M. LARSEN]

Bibliography: F. A. Cotton and G. Wilkinson, *Advanced Inorganic Chemistry*, 3d ed., 1972; R. T. Sanderson, *Inorganic Chemistry*, 1967.

Indole

The parent compound of a group of organic heterocyclic compounds containing the indole nucleus, which is a benzene ring fused to a pyrrole ring as in indole itself (I). *See* CARBAZOLE; PYRROLE.

Indole can exist in two tautomeric forms, the more stable enamine form (I) and the 3-H-indole or imine form (II). Unsubstituted 3-H-indoles (sometimes called indolenines) and a structural isomer of indole, isoindole (III), are not stable, but have been

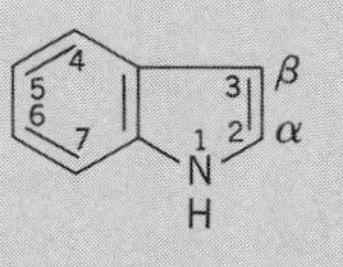

(I)

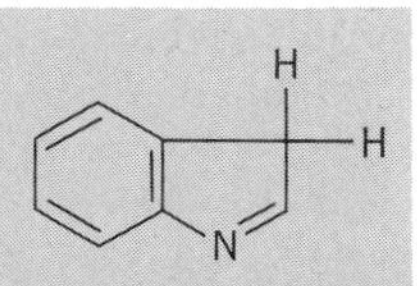

(II)

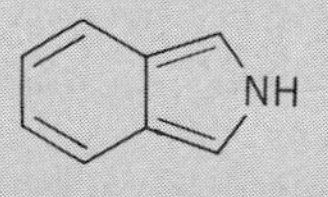

(III)

shown to be reaction intermediates. They are isolable when properly substituted.

The importance of the indole ring lies in its presence in a large number of naturally occurring compounds. Some of these are the plant growth hormone indole-3-acetic acid (IV), the animal tissue constituent serotonin (V), the amino acid tryptophan (VI), the pigment indigo (VII), and the antitumor antibiotic mitomycin A (VIII). In addition,

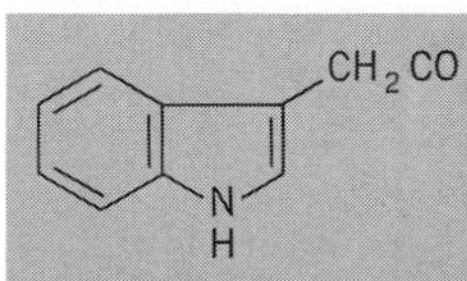

(IV)

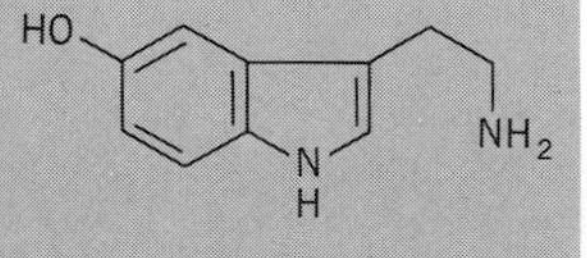

(V)

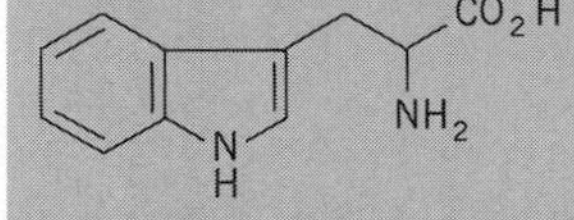

(VI)

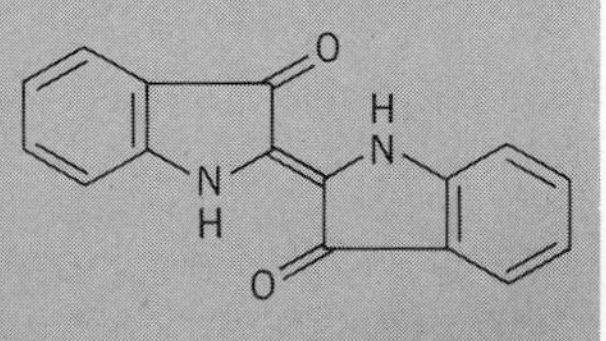

(VII)

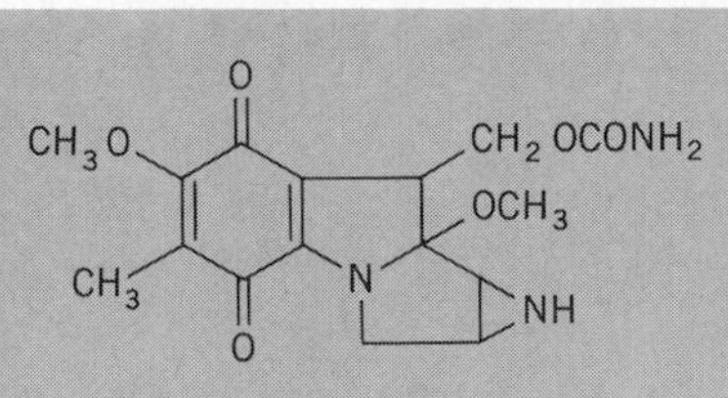

(VIII)

the indole nucleus is found in the indole alkaloids and in a number of synthetic drugs and pigments.

Physical properties. Indole (I) is a steam-volatile, colorless solid, melting point 52.5°C, boiling point 253°C. It is found in small amounts in coal tar, feces, and flower oils. Indole has a characteristic ultraviolet spectrum with absorption maxima at 216, 266, 287, and 276 nm. It fluoresces (shines) when irradiated with ultraviolet light. These, and other spectral properties, vary widely for the many derivatives of the enamine form.

Preparations. Indoles are generally synthesized by constructing a five-membered nitrogen ring on a preformed benzene ring. The Fischer synthesis (IX–XI) is the most common and versatile of many known methods. In this method, the phenylhydrazone (X) of some ketone (IX) is treated with acid to give the substituted indole (XI). The R groups in the ketone may be varied widely, and the phenylhydrazone ring may be substituted. Other syntheses include the Bischler synthesis from anilines (XII) and haloketones (XIII) or the ring closure of *o*-toluides (XIV–XV).

Chemical properties. Despite the presence of nitrogen, indole is not basic in the sense that it dissolves in aqueous acid or turns litmus blue. In fact, the hydrogen on the nitrogen is about as acidic as an aliphatic alcohol hydrogen (pK_a = about 17). Indole is an aromatic compound and undergoes electrophilic substitutions much like benzene, although it is much more reactive than benzene. Its reactivity is comparable to that of phenol, and it undergoes a number of reactions similar to those

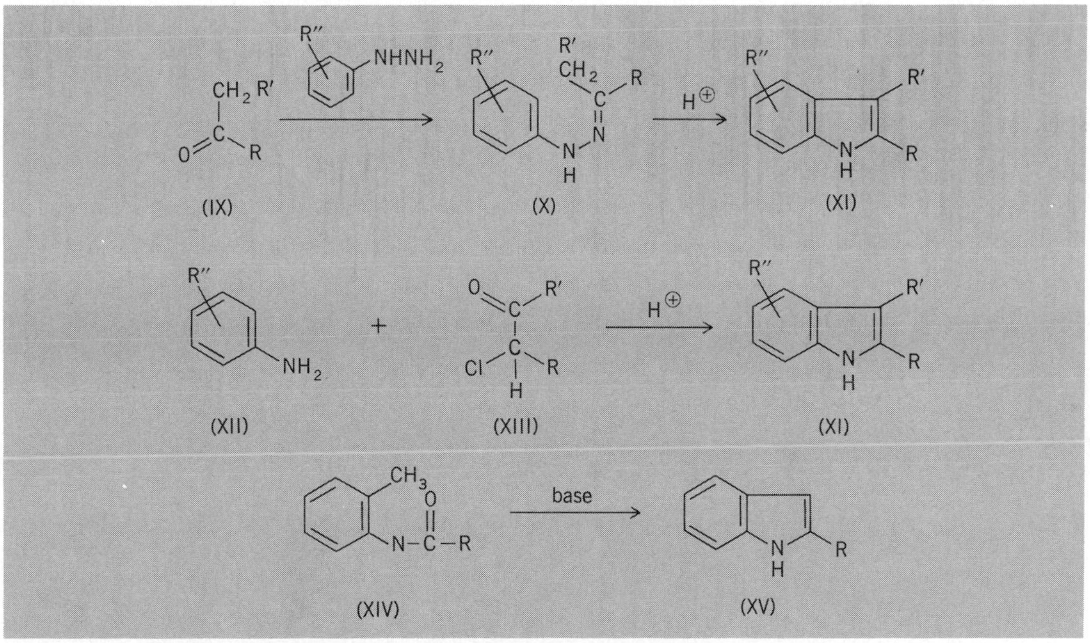

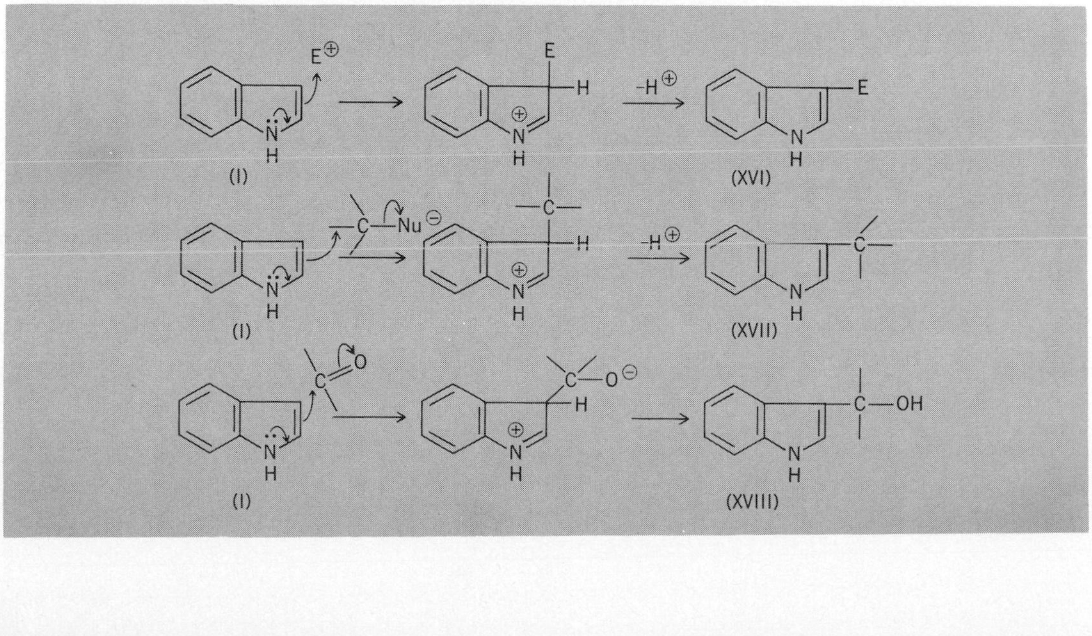

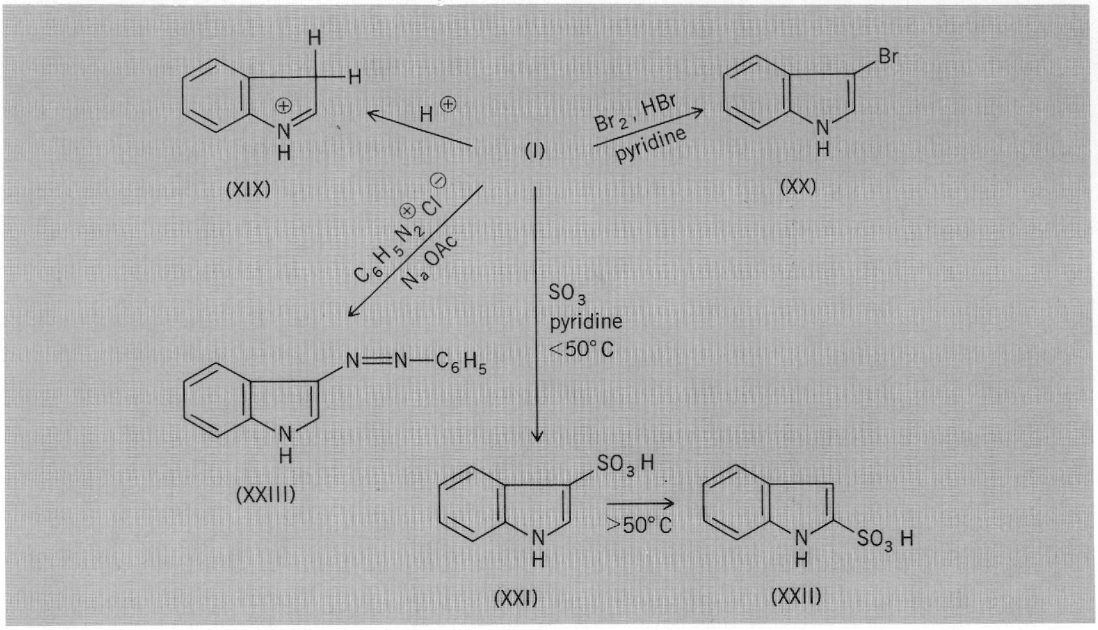
(I)
H⊕
(XIX)
Br2, HBr
pyridine
(XX)
C6H5N2⊕Cl⊖
NaOAc
N=N—C6H5
(XXIII)
SO3
pyridine
<50° C
SO3H
>50° C
(XXI)
(XXII)

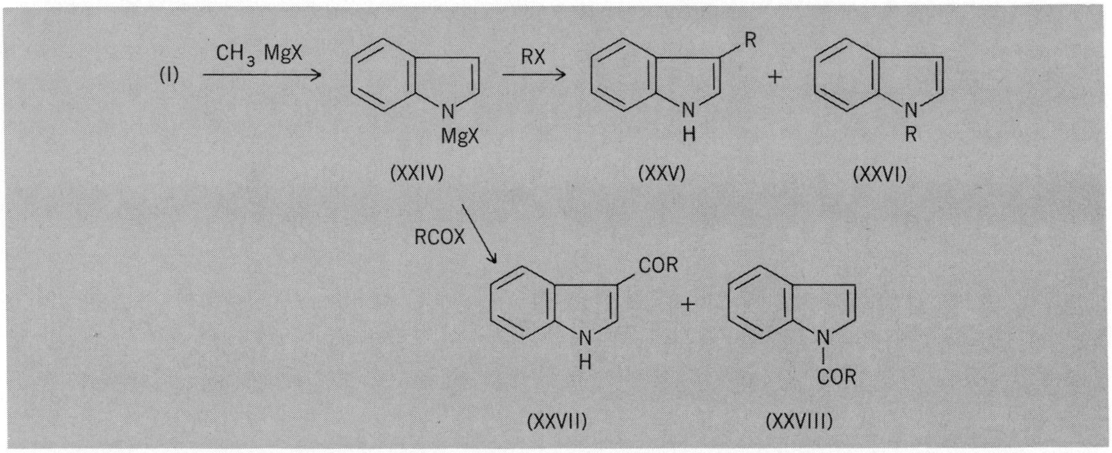
(I)
CH3MgX
MgX
RX
(XXIV)
(XXV)
(XXVI)
RCOX
COR
(XXVII)
(XXVIII)

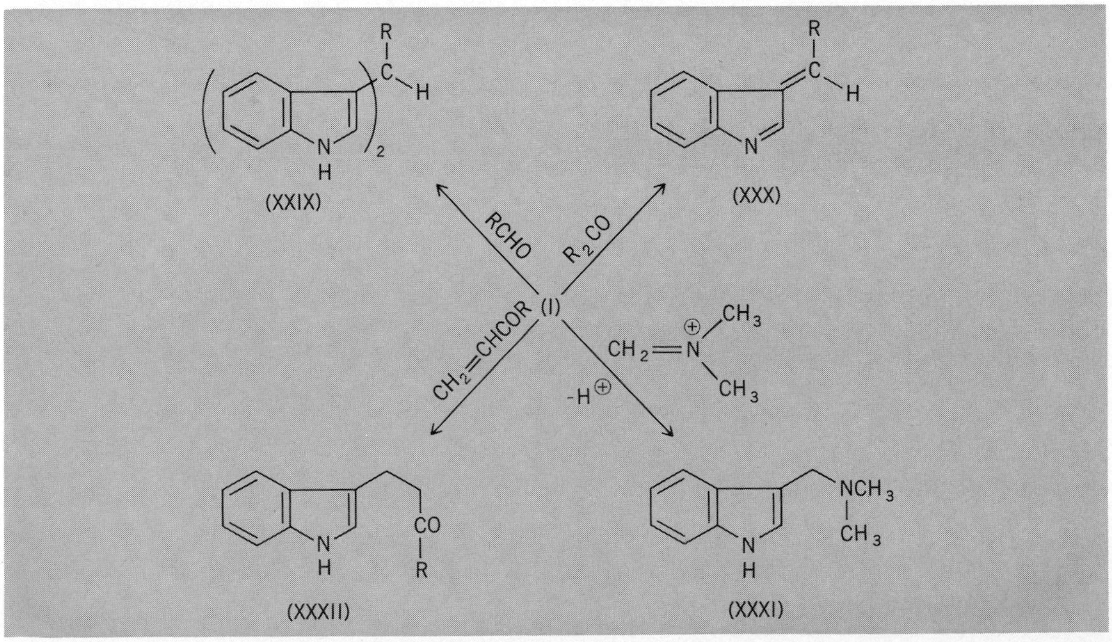
(XXIX)
(XXX)
RCHO
R2CO
(I)
CH2=CHCOR
CH2=N⊕(CH3)2
-H⊕
(XXXII)
(XXXI)

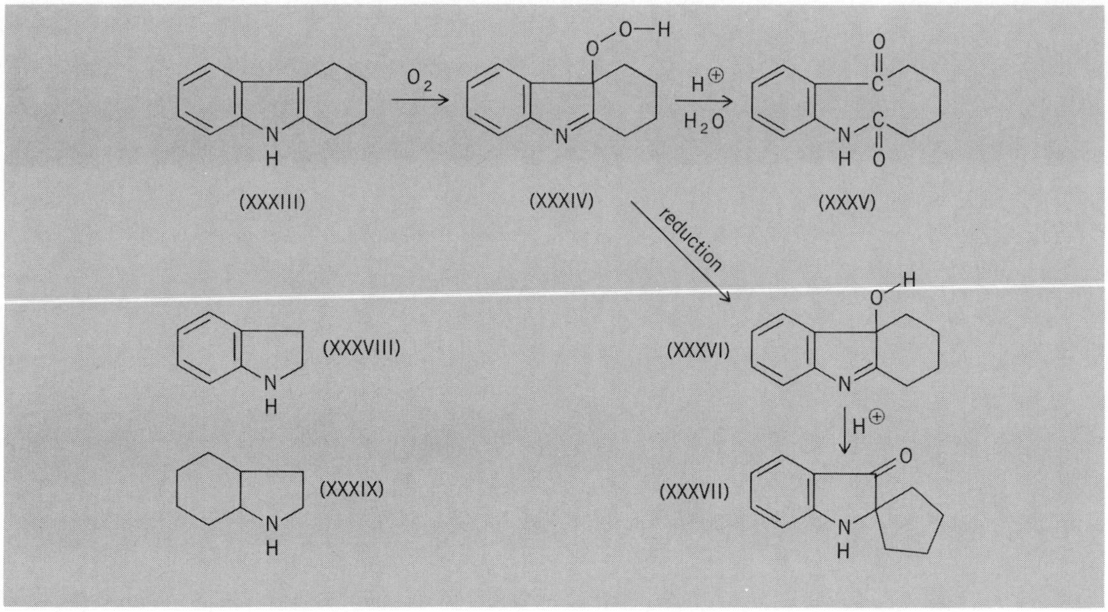

of phenol. Much of the chemistry of indole can be summarized mechanistically by reaction scheme (I–XVI, XVII, XVIII). Essentially, indole is an efficient provider of electrons which can react with an electrophile (E^+) to yield a substituted product (XVI), or which can take part in the displacement of a nucleophile (Nu^-) from carbon to yield an alkylated product such as (XVII). When the electrophile is a polarized double bond such as a carbonyl or an imine, addition reactions take place to give products such as (XVIII).

Thus, indoles react under mild conditions with many electrophiles, and the reactivity of the various positions is in approximately the following order, $3 \gg 2 \gg 6 > 4 > 5 > 7$. Some specific reactions involve protonation to give (XIX), halogenation to give (XX), sulfonation to give (XXI) which rearranges to the thermodynamically more stable (XXII), and diazonium coupling to give (XXIII).

Alkylation and acylation, such as in reaction (5), are more facile when the nucleophilic character of the indole is enhanced by converting it into a metallic salt (using the acidic proton on nitrogen). Thus, the nitrogen Grignard reagent of indole (XXIV) can be prepared from (I) and any simple Grignard reagent. The metallic salt or complex is then alkylated or acylated by a nucleophilic displacement to give such products as (XXV) to (XXVIII). Both the N and the 3-substituted products are formed.

In the third type of reaction (see XVIII above), indole reacts with aldehydes or ketones to give (XXIX) or (XXX) respectively, with imines to give (XXXI), and with activated alkenes to give (XXXII).

Indole itself reacts slowly with air and rapidly with most oxidizing agents to give intractable polymeric tars. However, substituted indoles undergo a number of smooth oxidations to give isolable products. Air oxidations appear to involve a peroxide intermediate such as (XXXIV) formed from 1,2,3,4-tetrahydrocarbazole (XXXIII). The peroxide can then undergo a number of reactions to give products such as (XXXV) to (XXXVII). Indole can be reduced to give 2,3-dihydroindoles (indolines) such as (XXXVIII) or octahydro derivatives such as (XXXIX).

Indole alkaloids. The indole alkaloids are a large group of substances containing the indole nucleus and can be isolated from plants. They contain a number of physiologically active materials,

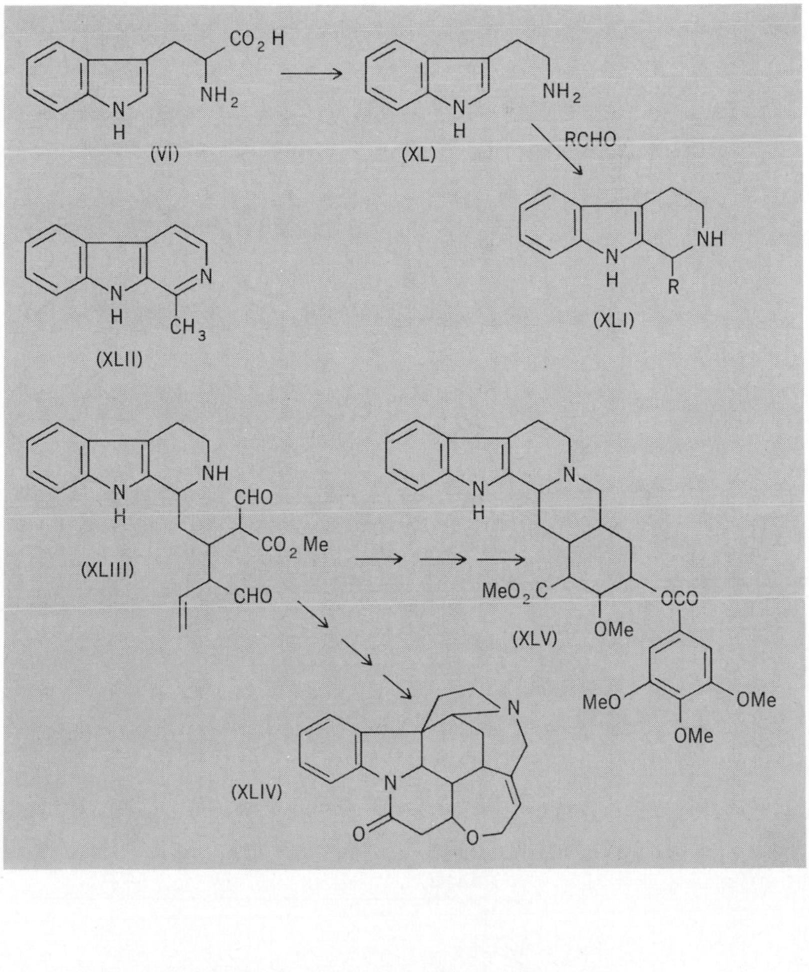

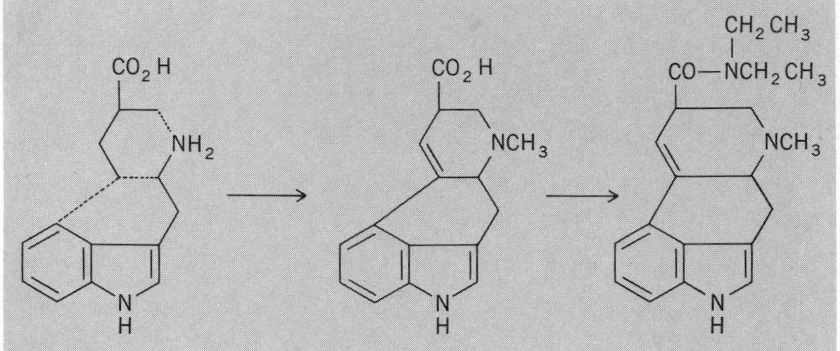

(XLVI) (XLVII) (XLVIII)

such as strychnine (XLIV), reserpine (XLV), some forms of curare poison, and the rye-fungus drug, ergot. The indole alkaloids are synthesized in tissue from the amino acid tryptophane (VI) by a series of complex enzyme-moderated processes. The processes involve the decarboxylation of (VI) to give tryptamine (XL), which reacts with some aldehyde to give a compound containing the β-carboline ring system (XLI). When the R on the aldehyde is methyl, the simple β-carboline alkaloids, such as harman (XLII), result. When R is a more complex fragment, as in (XLIII), the more complex indole alkaloids such as strychnine (XLIV) and reserpine (XLV) result (after a number of other reactions). When tryptamine is combined with a five-carbon unit, as shown in dotted lines in (XLVI), a key component of ergot, lysergic acid (XLVII), arises. When the acid group of (XLVII) is converted to the *N,N*-diethylamide, lysergic acid diethylamide or LSD (XLVIII) is obtained. LSD is a synthetic, and not a naturally occurring, substance. There are many hundreds of other indole alkaloids. *See* ALKALOID; HETEROCYCLIC COMPOUNDS.

[JAMES M. BOBBITT]

Bibliography: R. C. Elderfield (ed.), *Heterocyclic Compounds*, vol. 3, 1952; W. J. Houlihan (ed.), *Indoles*, pts. 1–3, 1972; R. H. Manske, *The Indole Alkaloids*, 1965; R. J. Sundberg, *The Chemistry of Indoles*, 1970.

Inert gases

The inert gases, listed in the table, constitute group 0 of the periodic table of the elements. They are now better known as the noble gases, since stable compounds of xenon have been prepared. The term rare gases is a misnomer, since argon is plentiful and helium is not rare in the United States and some other countries.

All these gases occur to some extent in the Earth's atmosphere, but the concentrations of all but argon are exceedingly low. Argon is plentiful, constituting almost 1% of the air.

All isotopes of radon are radioactive, the longest lived having a half-life of about 4 days. Each of the other inert gases has at least two stable (nonradioactive) isotopes, in addition to one or more radioactive isotopes.

All the gases are colorless, odorless, and tasteless. They are all slightly soluble in water, the solubility increasing with increasing molecular weight. They can be liquefied at low temperatures, the boiling point being proportional to the atomic weight. All but helium can be solidified by reducing the temperature sufficiently, and helium can be solidified at temperatures of 0–1 K by the application of an external pressure of 25 atm or more.

The noble gases are all monatomic. The outer shell of each of the atoms (unless strongly excited by radiation, electron bombardment, or other disturbing effects) is completely filled with electrons, and the gases are generally chemically inert. However, since the outer electrons in xenon are relatively far from the nucleus, they are held rather loosely; and they are capable of interacting with the outer electrons of fluorine atoms and the atoms of some other elements to form fairly stable compounds, such as XeF_2 and XeF_4. Radon can presumably form compounds similar to those of xenon. A fluoride of krypton which is stable at −8°C has been prepared. No compounds of any of the lighter noble gases have been obtained. *See* KRYPTON COMPOUNDS; XENON COMPOUNDS.

In an electric discharge, as in a mass spectrometer, very short-lived ions and molecules of all the noble-gas atoms can be formed; examples are Ar_2^+, HgHe, HgNe, HgAr, $(ArKr)^+$, and $(NeNe)^+$. In the presence of methane, $(XeCH_3)^+$ and other highly unstable ions can be formed.

All the inert gases except helium and radon are produced in concentrated form by the liquefaction and distillation of air, followed by special purification processes. Helium is obtained from certain natural gases containing 0.5% or more He. Radon is obtained by collecting the gas, called radium emanation, given off in the radioactive decay of radium. *See* ARGON; HELIUM; KRYPTON; NEON; PERIODIC TABLE; RADON; XENON.

[A. W. FRANCIS]

Bibliography: V. G. Fastovskii et al., *Inert Gases*, Moscow, 1964 (English translation available from U.S. Department of Commerce, Clearinghouse for Federal Scientific and Technical Information).

The inert gases

Name	Symbol	Atomic number	Atomic weight
Helium	He	2	4.0026
Neon	Ne	10	20.183
Argon	Ar	18	39.948
Krypton	Kr	36	83.80
Xenon	Xe	54	131.30
Radon	Rn	86	(222)

Infrared spectroscopy

The study of the interaction of material systems with electromagnetic radiation in the infrared region of the spectrum. The infrared region is valuable for the study of the structure of matter because the natural vibrational frequencies of atoms in molecules and crystals fall in the infrared range. Some gaseous molecules also have rotational frequencies in the far-infrared range, and certain frequencies corresponding to the energy levels of electrons in solids and in large molecules lie in the near infrared. For a detailed discussion of molecular vibration and rotation *see* MOLECULAR STRUCTURE AND SPECTRA.

The infrared absorption spectrum of a molecule

is highly characteristic, and often has been referred to as a molecular fingerprint. The spectrum can thus be used for molecular identification. Because the absorption of radiation at various infrared frequencies is quantitatively related to the number of absorbing molecules in a system, quantitative analysis is also possible.

The usefulness of an infrared absorption spectrum for identification and chemical analysis was recognized as long ago as 1890. In the early 1900s the American physicist W. W. Coblentz determined the infrared spectra of hundreds of substances and clearly demonstrated the potential value of such spectra. Unfortunately, the instrumentation of that day was cumbersome and necessarily homemade, so that few physicists and chemists were attracted by Coblentz's work. Only after the development of commercial electronic devices for amplification and recording of a continuously scanned spectrum in the 1940s was extensive use made of the technique.

Instrumentation and techniques. The usual arrangement for measurement of an infrared spectrum is shown schematically in Fig. 1. A source Q sends a beam of continuous infrared radiation to a spherical condensing mirror C, which passes the beam through S, the sample to be studied. Some of the infrared frequencies in the beam are absorbed strongly, some weakly. The reduced beam passes on and comes to a focus at the entrance slit of the monochromator M. The latter is an infrared spectrometer which disperses the radiation into a spectrum. One frequency at a time appears at the exit slit of M, from which the radiation of that frequency is passed by a suitable optical system to the detector T. The detector (a thermocouple or other device) converts the radiant energy into an electrical signal, which is amplified electronically at A and recorded by a chart recorder CR.

An infrared spectrum is a record of intensity of infrared radiation as a function of frequency or wavelength. To produce such a record, the chart recorder is driven in synchronism with the dispersing system of the monochromator M by some common driving mechanism D. In this way a given position on the chart corresponds directly to a given frequency setting of M, at which setting radiation of that frequency is emerging from M.

For many basic reasons—atmospheric absorption, variation of source intensity with frequency, changing dispersion in the spectrometer, and the like—the electrical output of the detector would not be constant even if the sample S were completely transparent. To correct for these variations it is necessary to determine two spectra, one with the sample S in the beam and one with S removed from the beam. The absorption of S as a function of frequency can then be computed from these two spectra. The individual spectra on which the computation is based are called single-beam spectra.

The computation is laborious, time consuming, and potentially unreliable because of changes in the entire system between the two determinations of spectra. These difficulties are avoided if a second optical path, shown in dotted lines in Fig. 1, is introduced. The second optical path, called the reference beam, is made as nearly like the first as possible, except for the absence of the sample. In fact, the reference beam may contain an absorption cell R which differs from S only in the absence of the sample itself. For instance, if the sample S were in solution, R would contain the same amount of solvent as S.

The operation of the double-beam spectrometer, often called a spectrophotometer, consists of a rapid switching of the beam (say 10 times per second) back and forth between S and R by alternately placing plane mirrors 1 and 1′ in the optical system. The identical mirrors 2 and 2′ are permanently placed. The spectrum is scanned continuously as for single-beam operation, but the beams through S and R are compared 10 times per second and the chart records the energy passing through S relative to that through R. In this way the variations mentioned cancel out.

Typical spectra. Typical mid-infrared spectra, plotted automatically as percent transmission of the sample on a linear frequency scale (wave number in cm^{-1}), are shown in Fig. 2. Samples of gases, liquids, and solids can be readily measured. Techniques for high and low temperature of sample and for small samples (down to about 1 mg or less in special cases) are in common use.

Percent transmission T, the quantity usually plotted by commercial instruments, is defined in Eq. (1). Here $I_{0,\nu}$ is the intensity of infrared radia-

$$T_\nu = 100\, I_\nu / I_{0,\nu} \qquad (1)$$

tion of frequency ν entering the sample and I_ν is the intensity of the same radiation after passing through the sample. The percent transmission T_ν at frequency ν is different in principle at different values of ν. A quantity of fundamental importance, the absorbance A_ν, is defined in Eq. (2). The ab-

$$A_\nu = \log\,(I_{0,\nu}/I_\nu) = -\log\,(T_\nu/100) \qquad (2)$$

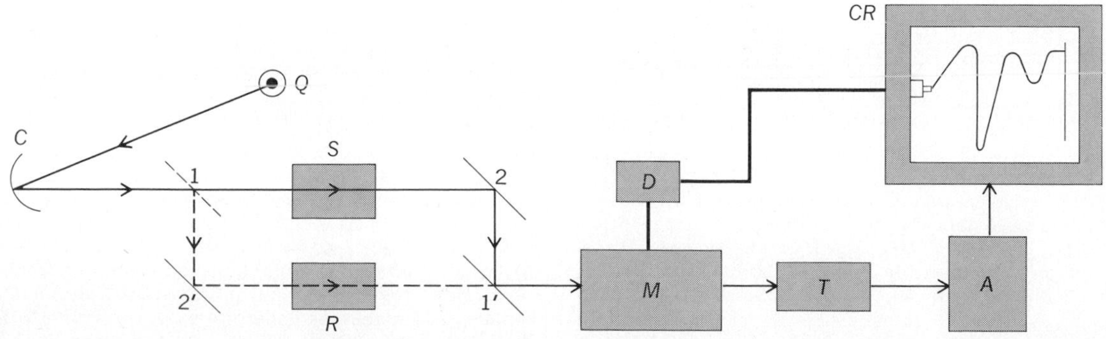

Fig. 1. Recording infrared spectrometer.

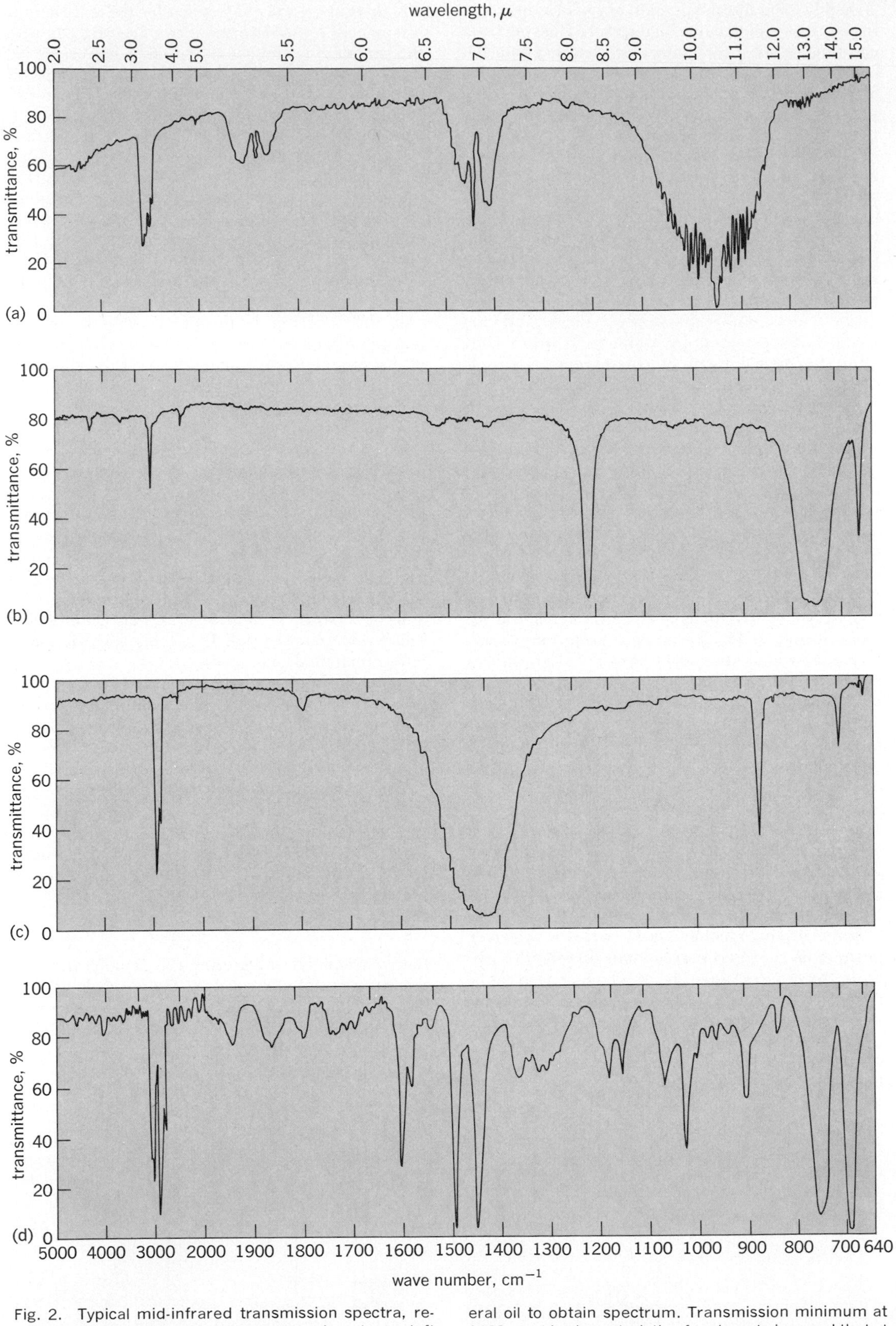

Fig. 2. Typical mid-infrared transmission spectra, recorded automatically. Note compressed scale on left portion of abscissa. (*a*) Spectrum of ethylene gas. Transmission minima in range 850–1050 cm^{-1} result from modulation of ethylene vibrational frequency at 950 cm^{-1} by molecular rotational frequencies. (*b*) Spectrum of liquid chloroform. (*c*) Spectrum of powdered crystalline calcium carbonate. Powder was suspended in mineral oil to obtain spectrum. Transmission minimum at 1430 cm^{-1} is characteristic of carbonate ion, and that at 2900 cm^{-1} is characteristic of CH groups in the oil. (*d*) Spectrum of an amorphous high polymer (polystyrene). Detail here shows why infrared spectra are sometimes called molecular fingerprints by workers in this discipline of science.

sorbance A_ν is proportional to the number of absorbing molecules, and by evaluating the proportionality constant at frequency ν for a given kind of molecule in a particular system, the number of such molecules in other systems of the same kind may be measured quantitatively.

Interferometric methods. Infrared spectra are measured in many applications by the technique of Fourier-transform spectroscopy (FTS). In dispersive (prism or grating) spectroscopy a spectrum is recorded by continuous scanning of the spectrum at successive frequencies (Fig. 1). In FTS the entire frequency range of interest is passed simultaneously through an interferometer, which produces an output signal containing all these frequencies. The quantitative way in which this signal varies as the condition for interference within the interferometer is varied is called an interferogram (Fig. 3). The interferogram can be made to yield the spectrum as a function of frequency by the mathematical procedure known as a Fourier transform. Although this procedure is complicated, small and powerful digital computers are available so that an interferometer and a digital computer can be combined into a single unit which produces the transformed spectrum with negligible delay.

A block diagram of an interferometric spectrometer and auxiliary components is shown in Fig. 4. The source Q sends a beam containing a complete range of infrared frequencies into the interferometer I. The beam is first divided at the semitransparent beam splitter BS, which transmits half to the variable plane mirror M_V and reflects half to the fixed plane mirror M_F. The separate beams are returned by M_V and M_F to the beam splitter, where they are reunited after having traveled distances differing by some amount L which is continuously variable. The reunited beams interfere at BS, where they are partially reflected and sent out of the interferometer to the sample S. After passing through S, the radiation is converted to an electrical signal at the detector *det*. The output of *det* as L changes is shown in Fig. 3. This output is processed electronically at A, stored in the computer memory, and then transformed to the spectrum. The spectrum is read out in suitable form, for example, as a plot on the chart recorder CR of percent transmission versus frequency. Appropriate instructions to the computer are provided from the command post, which may be a teletypewriter. For double-beam operation the optical system at S may resemble the two-beam arrangement through S and R in Fig. 1.

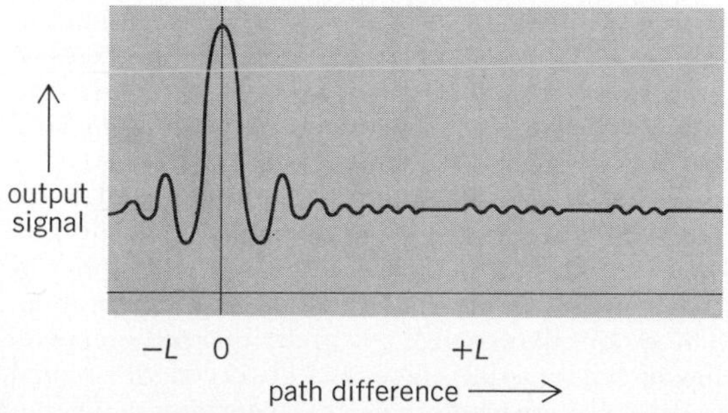

Fig. 3. A representative interferogram.

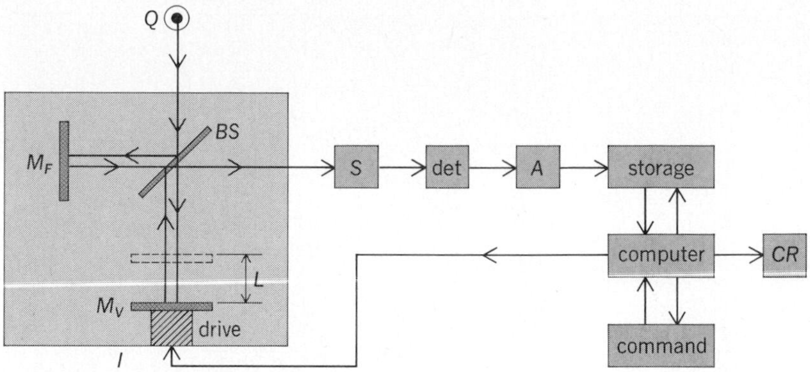

Fig. 4. Block diagram of an FTS system.

The virtues of interferometric spectroscopy as compared to dispersive (that is, grating or prism) spectroscopy are:

1. Enormous superiority in the effective use of the limited radiant power in infrared sources. This superiority, which leads to much larger signal-to-noise ratios in the transformed spectra, arises from two fundamental differences between interferometers and spectrometers: first, the inteferometer processes all frequencies in the input radiant power simultaneously (the multiplex advantage), whereas the dispersive spectrometer processes them one at a time. Second, the spectrometer needs narrow entrant and exit slits to do this processing, and the solid angle of radiant power accepted from the source is correspondingly restricted. The interferometer does not have this restriction and thus can accept a much larger solid angle of input radiation (the through-put advantage). The combination of these two advantages may lead to a superiority of several orders of magnitude in signal-to-noise ratio of the FTS over that of a dispersive spectrometer having the same resolution and scanning rate. Alternatively, the FTS will be able to record at the same signal-to-noise ratio and scanning rate with an increase in the resolution of more than an order of magnitude. This latter superiority is illustrated in Fig. 5, which shows the absorption spectrum of ethylene gas recorded from 940 to 960 cm^{-1} by a Digilab FTS-14 spectrometer. The effective scanning rate in cm^{-1}/s and the signal-to-noise ratio are about the same in Fig. 2*a* (dispersive spectrum of ethylene) and in Fig. 5, but the resolution (as measured by the reciprocal of the spectral bandpass $\Delta\nu$) is 8 times higher in the latter ($\Delta\nu \simeq 2$ cm^{-1} at 950 cm^{-1} in Fig. 2*a*, $\Delta\nu = 0.25$ cm^{-1} in Fig. 5). *See* RESOLVING POWER (OPTICS).

2. The interferometer does not require any filters or other order-sorting devices.

3. The wavelength or wave-number scale of the interferogram is automatically provided by the scanning parameter of the interferometer. This parameter is usually controlled to high precision by an auxiliary laser interferometer.

4. The spectral bandpass $\Delta\nu$ computed from the interferogram is constant throughout the spectrum. It is equal to, or somewhat larger than, $1/2L_{max}$ where L_{max} is the maximum excursion of the moving mirror, depending on how the data are processed.

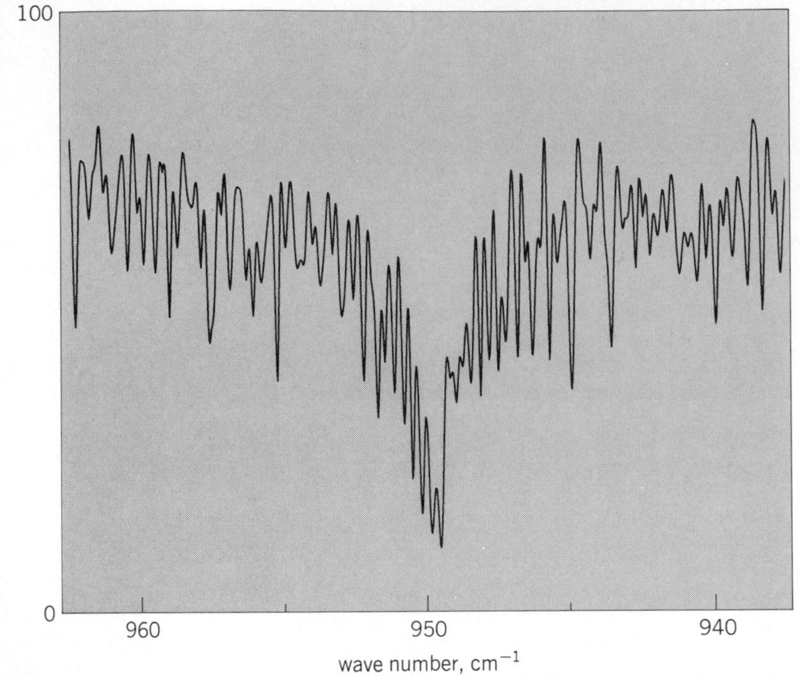

Fig. 5. FTS spectrum of ethylene gas, 940–960 cm^{-1}. Path length of sample, 10 cm; gas pressure, 30 torr (4.0×10^3 Pa); recording time, ~ 30 min; spectral bandpass $\Delta\nu$, 0.25 cm^{-1}.

5. The computer needed to calculate the Fourier transform can be used in addition for the manipulation of the spectroscopic data at any stage. It can also be programmed to control the mechanical operation of the interferometer, the electronics system, and the read-out device (for example, the recorder).

Use of tunable lasers. The sharpness of the frequency and the high power per unit solid angle and unit spectral bandpass in a laser beam make it attractive for infrared spectroscopy at ultrahigh resolution. The main problem is the tuning of the laser frequency, that is, varying some parameter in the laser system so that its frequency may be varied continuously and accurately. The table shows the frequency ranges over which various kinds of infrared lasers can be tuned. It is apparent from the table that their extremely high resolution (small spectral bandpass) makes infrared spectroscopy with tunable lasers a quite different kind of enterprise from dispersive or Fourier transform spectroscopy. An example of this resolution is shown in Fig. 6, where the absorption spectrum of ethylene gas at 950 cm^{-1} measured with a tunable semiconductor diode laser (SDL) is illustrated (compare Figs. 2*a* and 5). The triangular background with maximum at 949.3 cm^{-1} is the slit function of the dispersive spectrometer used to eliminate unwanted SDL modes.

Tunable infrared lasers have been used mainly to measure the energy levels due to the vibration and rotation of molecules in the gas phase at low pressures. By this means, very accurate values of such molecular parameters as internuclear distances, electric dipole moments, vibrational frequencies, internal force fields, and the like may be measured. Other anticipated uses of laser infrared spectroscopy include the monitoring of the composition of the atmosphere. For example, E. D. Hinkley and associates have monitored the amount of carbon monoxide in a horizontal path of 600 m at ground level with an SDL; the sensitivity achieved was of the order of 5 parts per billion.

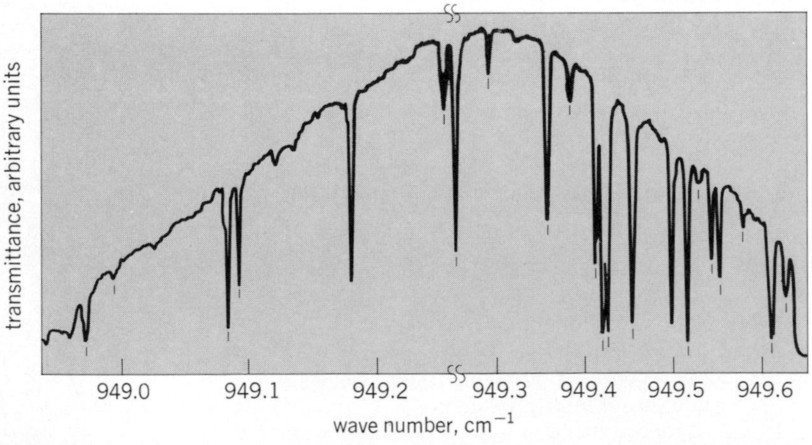

Fig. 6. Absorption spectrum of ethylene gas obtained by tuning a $Pb_x\ Sn_{1-x}Te$ SDL in the range 948.9–949.7 cm^{-1}. Path length, 10 cm; pressure, 0.08 torr (1.1×10^2 Pa); effective spectral bandpass, $\Delta\nu < 0.001$ cm^{-1}. (*From G. P. Montgomery, Jr., and J. C. Hill, High-resolution diode-laser spectroscopy of the 942.9 cm⁻¹ band of ethylene, J. Opt. Soc. Amer., 65:579–585, 1975*)

Frequency ranges of tunable infrared lasers*

Type of laser	Frequency range, cm^{-1}	Minimun spectral bandpass $\Delta\nu$, cm^{-1}
Semiconductor diode laser (SDL)	300–10,000*	3×10^{-6}
Spin-flip Raman laser (SFR)	1600–2000 700–1000	3×10^{-6} 3×10^{-2}
Zeeman-tuned gas laser (ZTG)	1100–3300	3×10^{-3}
High-pressure CO_2 laser (HPG)	900–1100	3×10^{-4}
	Nonlinear devices	
Optical parametric oscillator (OPO)	900–10,000⁺	3×10^{-2}
Difference frequency generator (DFG)	1600–3300	5×10^{-4}
Two-photon mixer (TPM)	900–1100	3×10^{-5}
Four-photon mixer (FPM)	400–5000	1×10^{-1}

*Adapted from K. W. Nill, Tunable infrared lasers, Opt. Eng., 13:516–522, 1974.

Applications. An infrared spectrum consists of a plot of T or A as a function of ν (or of wavelength λ). The basic information provided by the spectrum is a set of ν values at which the substance is absorbing strongly, that is, at which T_ν is a minimum (Fig. 2) or A_ν is a maximum. These frequencies of maximum absorption usually correspond to the actual vibrational frequencies of the absorbing molecules or to some arithmetical combination of such vibrational frequencies. If the molecules are in the vapor phase, absorption maxima may also be observed at frequencies which are combinations of frequencies of molecular rotation and vibration. The qualitative usefulness of an infrared spectrum lies in the fact that the set of observed vibrational frequencies characterizes the absorbing molecule.

Qualitative chemical analysis. Infrared spectra

can be used for the following purposes.

1. To identify pure chemical compounds by comparison of the spectrum of an unknown with previously recorded spectra of pure compounds. Catalogs of spectra are available, and in addition there are practical methods for encoding the information in the spectra and storing it in a computer memory system or on punched cards. The stored data can then be used for fast identification of an unknown spectrum.

2. To identify the constituents of mixtures. When a mixture contains only two or three constituents, it is often possible to identify them directly from the spectrum of the mixture. More commonly it is necessary to fractionate the mixture by gas chromatography or other technique and to identify the individual compounds in the mixture by their infrared spectra. For this purpose Fourier transform spectroscopy is advantageous because of its speed and superior signal-to-noise ratio; combined chromatographic and FTS equipment is commercially available.

3. To show the presence of a group of atoms, a so-called functional group, in a molecule of unknown or doubtful structure. It has been known since the 1890s that certain groups of atoms—for example, a methyl group (CH_3), a carbonyl group (CO), a nitrate ion (NO_3^-)—have characteristic absorption frequencies that are relatively independent of the rest of the molecule or crystal in which the group occurs. Literally hundreds of such group frequencies are known.

Quantitative chemical analysis. There is a linear quantitative relationship between the absorbance A_ν, defined in Eq. (2), and the number of absorbing molecules. Thus the quantitative analysis of mixtures by infrared means is feasible. The infrared method is not particularly sensitive, the limit of detection and measurement of minor constituents being in the range 0.1–1.0%, except in favorable circumstances. An example of the latter is the detection of a component of a gaseous mixture. If the component has strong absorption in a spectral range where the rest of the mixture is transparent, much smaller quantities may be detected. In general, the higher the resolution (the narrower the spectral bandpass) of the infrared instrument, the lower the minimum concentration that may be detected, provided the width of the spectral lines of the absorbing constituent is of the same order as the spectral bandpass. The advantage of narrow spectral bandpass disappears if it is much smaller than the spectral line width. Thus the sensitivity of the SDL measurement of atmospheric carbon monoxide (5 parts per billion over a path of 600 m) would be much lower (that is, much improved) if the spectral line width of CO ($\sim$0.2 cm^{-1} at atmospheric pressure) were reduced to that of the SDL used ($\sim$0.0001 cm^{-1}).

The precision of quantitative measurement is mainly limited by the signal-to-noise ratio, and Fourier transform spectroscopy therefore offers an advantage for quantitative work. With dispersive spectroscopy, precision of measurement is seldom better than 1% of the quantity being measured and may be considerably worse. Infrared methods are especially useful in the quantitative determination of isomeric substances and in measurement of constituents of a chemical equilibrium. *See* ANALYTICAL CHEMISTRY; QUANTITATIVE CHEMICAL ANALYSIS.

Determination of molecular structure. Structures of molecules can be determined to varying degrees of refinement from infrared spectra. If only a few independent parameters (interatomic distances and bond angles) are required to specify the structure, as is the case with a small symmetrical molecule, these can be evaluated from the moments of inertia of the molecule, which can in turn be measured from rotational frequencies, usually observed as fine structure in a vibrational absorption. The structural parameters of carbon dioxide, methane, ethylene (Figs. 2*a* and 5), and ethane, for example, have been evaluated with high precision from their infrared spectra.

If the number of parameters is too large to be determined in this way, it may nevertheless be possible to draw conclusions about the molecule's shape without measuring its size. The number of vibrational frequencies which appear in the infrared spectrum is related to the molecular symmetry, and it is often possible to infer the symmetry from the observed spectrum. Such inferences are more reliable if they are based on combined data from both infrared and Raman spectra. *See* RAMAN EFFECT.

It is still possible to say something about the structure of large molecules of little or no symmetry from their spectra if one is content with a statement about the presence or absence of various functional groups. The organic chemist often finds such statements very valuable. The nature of functional groups in the molecules of high polymers or of natural products such as the steroids can be determined from their infrared spectra, and this permits information about their structure to be obtained.

Study of solids, catalysts, and matrices. The infrared spectra of crystalline solids give information about modes of vibration of crystals, about hydrogen-bond vibrations when such bonds are present in them, and about electronic energy states in semiconductors and superconductors. Fourier transform spectrometers are regularly used by solid-state physicists, particularly for their advantages in the spectral range 1–200 cm^{-1}. At higher frequencies Fourier transform spectrometers are useful in the study of materials adsorbed on the surfaces of catalysts. The FTS signal-to-noise advantage is especially important because of the optical heterogeneity of these samples. A similar heterogeneity in samples trapped at low temperatures in condensed rare gases or other types of matrices likewise makes FTS the technique of choice for the investigation of trapped transient species. *See* SPECTROSCOPY.

[RICHARD C. LORD]

Bibliography: L. J. Bellamy, *Infra-red Spectra of Complex Molecules*, 3d ed., 1975; G. R. Harrison, R. C. Lord, and J. R. Loofbourow, *Practical Spectroscopy*, 1948; G. Herzberg, *Infrared and Raman Spectra of Polyatomic Molecules*, 1945; R. T. Ku, E. D. Hinkley, and J. O. Sample, Long-path monitoring of atmospheric carbon monoxide with a tunable diode laser system, *Appl. Opt.*, 14:854–861, 1975; H. Walther (ed.), *Topics in Applied Physics: Laser Spectroscopy of Atoms and Molecules*, 1974.

Inhibitor

A substance which is capable of stopping or retarding a chemical reaction. To be technically useful, such compounds must be effective in low concentrations, usually under 1%. The type of reaction which is most easily inhibited is the free-radical chain reaction. The study of inhibitor action is often used as a diagnostic test for free-radical chain character of a reaction. Vinyl polymerization and autoxidation are two important examples of the class. Another reaction type for which inhibitors have been found is corrosion, particularly in aqueous systems. The economic importance of corrosion inhibition can scarcely be overestimated. An understanding of inhibitor action depends on an understanding of the processes which are to be interrupted.

Inhibition of vinyl polymerization. This type of inhibitor action must be considered in terms of the accepted mechanism for the polymerization process, which may be summarized as reaction sequence (1). The symbol P represents a catalyst,

$$\begin{array}{l} \left.\begin{array}{r} P \rightarrow 2R\bullet \\ R\bullet + M \rightarrow \sim\!\sim M\bullet \end{array}\right\} \text{initiation} \\ \sim\!\sim M\bullet + M \rightarrow \sim\!\sim M\bullet \ \text{propagation} \\ 2\sim\!\sim M\bullet \rightarrow \text{termination by dimerization or disproportionation} \end{array} \quad (1)$$

often a peroxide, R• is a free radical derived from the catalyst, M is a monomer, and ~~M• is a growing polymer chain. The polymerization will be stopped or inhibited if some added substance (an inhibitor) reacts more readily than does the monomer with R• to yield a product which will not sustain the polymerization. Every reaction chain is stopped until the inhibitor is consumed. If the added substance (a retarder) is somewhat less reactive, the monomer can compete more successfully for the initiating radicals, so that the result is retardation rather than total inhibition. The difference between inhibition and retardation is one of degree rather than of kind.

Phenolic compounds and quinones. These interact with an initiating radical or a growing polymer chain either by hydrogen atom abstraction or by radical addition to an unsaturated linkage. These interactions are represented by sequence (2).

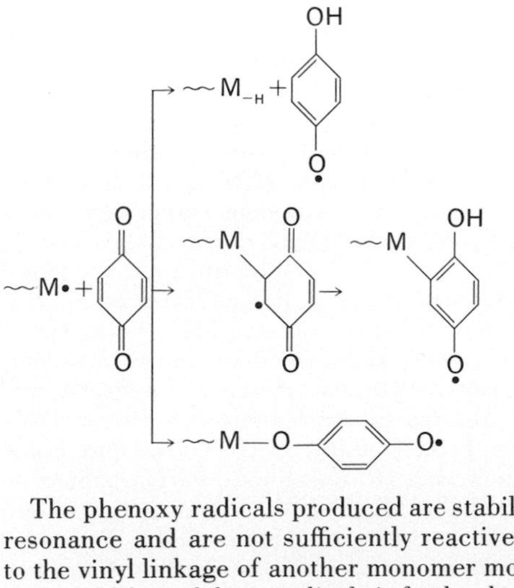
(2)

The phenoxy radicals produced are stabilized by resonance and are not sufficiently reactive to add to the vinyl linkage of another monomer molecule. The usual fate of these radicals is further hydrogen atom loss by reaction with a second polymer radical or by disproportionation with another phenoxy radical to yield a quinone and a hydroquinone, both of which may continue to act as inhibitors. All the reaction possibilities shown have been demonstrated experimentally. Although the efficiencies of phenols and quinones for the interruption of polymerizations vary with the structures of these molecules, they may be classed as inhibitors. Aromatic amines react similarly.

Nitroaromatics. As typified by trinitrobenzene, nitroaromatics function as retarders rather than as inhibitors in most polymerizations. It is necessary, however, to consider the specific reaction involved. Thus polynitroaromatics inhibit the polymerization of vinyl acetate, retard that of styrene, and have no effect on that of methyl methacrylate. No clear-cut mechanism has been established for the interaction of nitro compounds with free radicals.

Monomers. Both monomers and the radicals derived from them differ greatly among themselves in reactivity. Thus, although certain monomers may copolymerize with one another, others may actually function as inhibitors. Styrene and vinyl acetate, for example, both polymerize well when alone. Styrene, however, inhibits the polymerization of vinyl acetate. This occurs because the vinyl acetate radical and the styrene monomer are highly reactive, whereas the styrene radical and the vinyl acetate monomer are not. A small amount of styrene added to vinyl acetate will rapidly react with any vinyl acetate radicals formed when polymerization is initiated. The resulting styrenelike radical will react only very slowly with the vinyl acetate monomer. In the overall process the chain-carrying radical is converted to one which is too unreactive to carry on the chain.

Autoinhibition. This action, sometimes called allylic termination, is exhibited by monomers which contain the highly reactive allylic C—H linkage. Free radicals are capable of hydrogen atom abstraction from copresent molecules as well as of addition to an unsaturated linkage. The ease with which this abstraction reaction is carried out by a given radical is a function of the reactivity of the C—H linkage which is attacked. The reactivity of radicals containing these C—H linkages increases in the order aryl < primary alkyl < secondary alkyl < tertiary alkyl < allyl < benzyl. Because of this high reactivity, a monomer containing an allylic C—H, allyl acetate, for example, functions as its own retarder, as shown by sequence (3).

$$\begin{array}{c} \sim\!\sim M\bullet + CH_2{=}CHCH_2OAc \\ \swarrow \quad \searrow \\ \sim\!\sim MCH_2\dot{C}HCH_2OAc \qquad \sim\!\sim MH + CH_2{=}CH\dot{C}HOAc \end{array} \quad (3)$$

The resonance-stabilized allylic radical will react with the monomer only very slowly. The predominant further reaction is dimerization. Not only is polymerization slowed in this case, but the molecular weight of the polymer formed is low.

Miscellaneous inhibitors. Oxygen, iodine, and nitric oxide interact rapidly with free radicals to yield stable products. Inclusion of these materials in polymerizing systems thus leads to effective inhibition. It is of particular interest that oxygen will copolymerize, under carefully controlled conditions, with certain monomers, styrene, for example, to yield high-molecular-weight polymeric peroxides. The repeating unit is shown in formula (4).

$$[\text{—CH}(C_6H_5)CH_2OO\text{—}]_n \qquad (4)$$

Iodine and nitric oxide have been used extensively to detect and in some cases to identify alkyl free radicals in nonchain as well as in chain reactions. This method has proved to be of great value in defining primary processes in photochemical decompositions of aldehydes and ketones.

Inhibition of corrosion. Metallic corrosion in conducting media is electrochemical in nature. Local electrolytic cells are set up because of the presence of impurities, crystal lattice imperfections, or strains within the metal surface. The result is dissolution of the metal from the anodic regions. Corrosion inhibitors now in use may operate at the anodes or the cathodes or provide physical protection over the entire surface.

Anodic inhibitors. These are mild oxidants which reduce the open-circuit potential difference between local anodes and cathodes and increase the polarization of the former. Sodium chromate and sodium nitrite are most commonly used. The former is used in air conditioners, refrigeration systems, automobile radiators, power plant condensers, and similar equipment. Sodium nitrite finds special use in the protection of petroleum pipelines. It is effective even on rusty, mild steel. An extension of the nitrite type is the use of nitrite salts of secondary amines as vapor-phase inhibitors. The inclusion of a salt such as dicyclohexylammonium nitrite with a packaged steel object provides effective protection against corrosion.

Cathodic inhibitors. Compounds such as calcium bicarbonate and sodium phosphate, in an aqueous medium, deposit on metal surfaces films that provide physical protection against corrosive attack.

Organic inhibitors. These are usually long-chain aliphatic acids and the soaps which are derived from them. Adsorption of these compounds on metal surfaces gives a hydrophobic film which protects the metal from corrosion by many agents. As little as 0.1% of palmitic acid, for example, is sufficient to protect mild steel from attack by nitric acid. *See* ANTIOXIDANT; CATALYSIS; FREE RADICAL; POLYMERIZATION.

[LEE M. MAHONEY]

Bibliography: P. G. Ashmore, *Catalysis and Chemical Inhibition of Reactions*, 1963; R. L. Le Mar, *VCI Bibliography and Abstracts*, U.S. Atomic Energy Commission, 1958.

Inorganic chemistry

The chemical reactions and properties of all the elements in the periodic table and their compounds, with the exception of the element carbon. The chemistry of carbon and its compounds falls in the domain of organic chemistry. The boundaries of inorganic chemistry with the other major areas of chemistry are not precisely defined, and it is often a matter of taste as to whether a particular topic is to be included in the field of inorganic chemistry or is to be considered physical or even organic chemistry. Physical chemistry may be defined as the application of quantitative and theoretical methods to chemical problems, and is a methodology rather than a specific body of knowledge. Investigations into theoretical inorganic chemistry or the study of problems in inorganic chemistry by quantitative and sophisticated physical methods may be considered either inorganic or physical chemistry quite arbitrarily. In similar fashion, organometallic compounds may be considered to be in the sphere of either inorganic or organic chemistry. To an increasing extent, the inorganic chemist is concerned with problems that once were considered the prerogative of physical chemists, organic chemists, or even biochemists. *See* PHYSICAL CHEMISTRY.

Synthetic inorganic chemistry. The reactivity of the elements of the periodic table varies enormously, and over a much wider range than is encountered in organic chemistry. Consequently, the inorganic chemist must frequently employ unusual apparatus and techniques. The elements range from the rare gases, which are unreactive and form very few chemical compounds, to the extremely reactive halogens and alkali metals. Flourine is perhaps the most reactive element known; it forms compounds with all other elements, including the rare gases. Because of the great reactivity of fluorine and the closely related halogen fluorides, special metal and plastic apparatus must be used in experimentation. Both fluorine and the important compound hydrogen fluoride attack glass, and this common material of construction cannot be used for experimentation with these substances. Methods have been developed to effect direct reaction between elemental fluorine and organic compounds without completely destroying the organic reactant, and this procedure greatly increases the number of fully fluorinated organic compounds that can be readily obtained. *See* FLUORINE; FLUOROCARBON; HYDROGEN FLUORIDE; PERIODIC TABLE.

Another element important in synthetic inorganic chemistry since World War II is boron. The hydrides of boron were first obtained by reaction of a metal boride with a solution of an aqueous acid. The procedure was very difficult and tedious, and it required weeks or months of labor to obtain a few cubic centimeters of the gaseous product, which turned out to be the simplest boron hydride, B_2H_6 (diborane). Many years later it was discovered that yields could be increased by passing boron trichloride and hydrogen through an electric discharge, but the yields were still distressingly low. Under the impetus of wartime urgency, chemical syntheses for the boron hydrides were developed. Using the readily available alkali metal hydrides and boron trifluoride as starting materials, boron hydrides were obtained in very good yield. The metal borohydrides, such as $Al(BH_4)_3$ (b.p. 44.5°C), when they can be prepared, are likely to be the most volatile compounds of the metal. *See* BORANE; BORON.

The carboranes are a relatively new class of organoboron compounds with a rich chemistry and many potential applications. Hydroboration, the addition of diborane to organic compounds containing a carbon-carbon double bond, is a preparative reaction that has many important applications in organic synthesis. Many new compounds of elements such as silicon, germanium, gallium, phosphorus, and nitrogen, some of which were at one time considered too unstable to be synthesized, have been prepared by the synthetic inorganic chemist. An outstanding achievement in modern inorganic synthesis is the preparation of the noble gas fluorides. The discovery that the noble gases could form compounds with fluorine and oxygen has resulted in a whole new area of inorganic research. *See* CARBORANE; HYDROBORATION; INERT GASES.

Many boron compounds react violently with air and water. It is necessary, therefore, to use all-glass vacuum systems for carrying out experiments with these substances. Vacuum-line techniques are widely used in inorganic chemistry for the manipulation of volatile, highly reactive compounds such as the hydrides of phosphorus, silicon, and related compounds, and an array of vacuum lines often is the hallmark of a laboratory for synthetic inorganic chemistry.

The synthesis of inorganic compounds as an end in itself, which at one time was perhaps the most characteristic of the activities of the inorganic chemist, has been largely superseded by synthesis directed to specific goals; among these are compounds required to elucidate fundamental laws of chemical binding, reactivity, and structure; inorganic polymers with unusual thermal and mechanical properties; compounds or systems with superconductor or electron conductor properties; and compounds with particular virtues as reagents in organic syntheses.

Coordination chemistry. The transition elements in the periodic table form coordination compounds with a great variety of organic donor molecules. Traditionally, coordination chemistry has been a major interest of inorganic chemists, and it continues to be a prominent feature of the inorganic chemistry landscape. However, interest in maintaining a large flow of new coordination compounds whose principal virtue is that they can be easily prepared has considerably diminished in most quarters. The principal interest in coordination chemistry is in characterizing the physical properties of coordination compounds with unusual properties or uses. Thus, the synthesis of multidentate ligating agents such as the crown ethers and the cryptates, which have high specificity for binding particular ions, has become a very important activity. Macrocyclic polyethers and macrobicyclic polyethers with tertiary amino groups at the bridgeheads mimic many of the properties of ion-transporting agents in living systems. The template properties of transition-metal ions for the synthesis of multidentate ligands such as the porphyrins and corrins that occur in nature have also received increasing attention. These and other initiatives in coordination chemistry have become an important component in the emerging discipline of bioinorganic chemistry. *See* COORDINATION CHEMISTRY; CROWN ETHERS AND CRYPTANDS.

Organometallic compounds. Organometallic compounds constitute a borderline area of study between inorganic and organic chemistry. Until 1950, the organometallic compounds known were entirely compounds of the principal families in the periodic table; no stable derivatives of subgroup metals or transition metals had been prepared. Today a very large number of organometallic compounds of the transition and subgroup elements with metal-carbon bonds to cyclopentadiene, C_5H_6, aromatic hydrocarbons such as benzene, C_6H_6, as well as other organic compounds, have been synthesized. These compounds form interesting derivatives with carbon monoxide, CO, and can also undergo numerous reactions, and the structure and properties of these compounds pose interesting theoretical problems. Homogeneous catalysis by organometallic compounds has become one of the most active areas of catalysis research. Hydride complexes of ruthenium, rhodium, and iridium are catalysts for hydrogenation, olefin isomerization, hydroformylation, and olefin polymerizations. Much current effort is devoted to the development of homogeneous organometallic catalysts for the reduction of carbon monoxide by hydrogen; these catalysts are intended to be more selective and to function under milder conditions than those required in heterogeneous catalysis. Organometallic reagents are also being sought for highly selective syntheses of complicated molecules such as pharmaceuticals employed in therapy. Palladium complex compounds are being used in the synthesis of alkaloids, and the preparation of such important classes of compounds as the insect pheromones and the prostaglandins is being facilitated by the use of nickel and aluminum organometallic compounds. *See* HOMOGENEOUS CATALYSIS; HYDRIDO COMPLEXES; ORGANOMETALLIC COMPOUND.

Solid-state chemistry. The elements and compounds that form the subject matter of inorganic chemistry exhibit a very wide range of physical properties which run the gamut from helium, the substance of lowest known melting and boiling point, to elements such as tungsten, titanium, and carbon, which are among the most refractory high-melting substances known. The study of solid-state reactions and compounds that can be prepared by them has assumed great importance, and is another major research area in inorganic chemistry. Unlike gaseous compounds and most organic compounds, which obey the law of definite proportions, many solid compounds, particularly those of the transition elements, exhibit variability of composition, or as it is frequently designated, nonstoichiometry. When a solid compound deviates from simple stoichiometric relations, it contains an excess of either positively charged metal cations or negatively charged anions. Such solid systems frequently show unusual electronic properties, which are made use of in many solid-state devices. Photovoltaic cells, for example, which convert light to electricity, belong to this group of nonstoichiometric solid-state devices. *See* NONSTOICHIOMETRIC COMPOUNDS.

Transistors, thermistors, phosphors, and light-emitting diodes are also important solid-state devices. Because these materials are frequently prepared by solid-phase reactions, high temperatures may be required for the reaction. High-temperature chemistry has greatly expanded in scope, not only in the preparation of nonstoichiometric compounds, but also in the preparation of refractories useful at very high temperatures in space and nuclear technology. Although many reactions can be forced to proceed in the desired direction by only increasing the temperature, this is not always adequate in all instances. A new dimension in solid-state chemistry has been added by the simultaneous use of high temperature and extremely high pressures. With the equipment now available, it is possible to carry out solid-state reactions at temperatures of approximately 2500°C (4532°F) and pressures of the order 10^5 mPa (10^6 bars). Under conditions of high temperature and pressure, ordinary carbon (graphite) can be converted into diamond, and since the pressure is actually sufficient to distort the electron orbitals, new varieties of matter can be prepared. For example, liquid hydrogen has been converted to a metallic form at these ultrahigh pressures. *See* HIGH-PRESSURE CHEMISTRY; HIGH-TEMPERATURE CHEMISTRY.

The solid-state systems used for semiconductors, photovoltaic devices, and laser photodiodes have traditionally been highly ordered, crystalline materials. Recently, substances in the amorphous or glassy state have been found to possess very interesting properties for similar uses. Amorphous silicon appears to have considerable potential in solar energy conversion cells for direct production of electricity, and amorphous silicon and germanium have semiconductor properties quite analogous to their crystalline counterparts. Amorphous vanadium or molybdenum sulfides can store lithium ions in nonaqueous lithium storage batteries better than crystalline materials can. Amorphous systems are only imperfectly understood, but they possess much scientific and technical interest and can, therefore, be expected to attract continued scientific investigation.

Layered materials have been discovered that have unusual properties, and this discovery has revived interest in the synthesis and study of layered compounds. Graphite intercalation compounds, for example, consist of layers of carbon atoms that have halogen or other atoms or molecules inserted between the carbon layers. Certain of these intercalation compounds have an electrical conductivity comparable to that of metallic copper, and the possibility that such systems could replace metallic electrical conductors has aroused much interest. Intercalated room-temperature superconductors are also being sought. Intercalation compounds derived from graphite or from transition-metal sulfides are effective and highly selective reagents in organic synthesis. Many layered compounds can accommodate layers of ions with high mobility. These systems are in effect solid electrolytes or charge carriers and may contribute significantly to the long-sought lightweight high-capacity electric storage battery.

Still another class of electronic conductors comprises chains and clusters of metal atoms. Linear chains of transition-metal atoms (Krogmann salts) are one-dimensional electron conductors. Metal cluster compounds also have extremely interesting electrical properties. Organometallic compounds containing metal clusters with as many as 28 platinum metal atoms and 44 carbonyl ligands have been made. The metal atoms in the cluster are directly bonded to each other and exhibit many of the properties of solid metal catalysts. The clusters often exhibit superconductivity at higher than customary temperatures, and have, therefore, become a point of departure in the search for new superconductive materials that may be suitable for electric power transmission at liquid hydrogen rather than liquid helium temperatures. Metal cluster chemistry ranks among the most promising of the approaches now being explored in the search for new catalysts. *See* METAL CLUSTER COMPOUNDS; SOLID-STATE CHEMISTRY.

Geochemical aspects. Many of the synthetic procedures in inorganic chemistry carried out at high temperatures and pressures have considerable interest in geochemistry. Some of the earliest inorganic chemistry was practiced in connection with mineralogy. Mineral syntheses, or the preparation of inorganic compounds identical with those found in nature, provide important information to the geochemist. Not only is such research helping to explain the sequence of chemical reactions and conditions responsible for the formation of minerals in nature, but many minerals and gems such as diamond, ruby, sapphire, quartz, and corundum are now manufactured on the large industrial scale. High-temperature reactions at very high pressures or hydrothermal reactions at high temperatures and pressure are employed for this purpose. With the advent of ultra-high-pressure equipment, it has become feasible to study chemical reactions under conditions approximating those many miles below the Earth's surface, and such studies are expected to add greatly to the understanding of geochemical phenomena. *See* HIGH-PRESSURE PROCESSES.

Nuclear science and energy. The development of nuclear energy since World War II has provided a great impetus to inorganic chemistry. The discovery of the transuranium elements is one of the outstanding events in chemical science, and opened up an entirely new area of the periodic table for investigation. The chemistry of the actinide 5*f* elements has revealed many surprises, and despite the great amount of new chemistry that has already been acquired, much still remains to be learned about the very complex phenomena associated with the oxidation-reduction behavior, the ions in solution, and the metallic states of the actinide elements. A particularly active field of actinide element research has been the synthesis of organometallic compounds of the 5*f* elements. The solvent-extraction and ion-exchange procedures developed for separating the actinide elements have had widespread applications in other areas of inorganic chemistry. Nuclear technology has also provided the impetus for the development of other separation procedures, for example, the separation of zirconium and hafnium, and of the rare-earth elements from each other, and has generally served to reinforce the traditional interest of the inorganic chemist in separations procedures. The development of safe, effective processing methods for the intensely radioactive spent nuclear fuels from breeder reactors, and the solution to the problem of safe storage and disposal of nuclear waste, remain paramount challenges to the inorganic chemist. *See* ACTINIDE ELEMENTS; ION EXCHANGE; NUCLEAR CHEMISTRY; RARE-EARTH ELEMENTS; SOLVENT EXTRACTION.

Applications in organic chemistry. Many of the most important advances in organic chemistry since 1900 resulted from the introduction of inorganic substances as reagents. Synthetic reactions based on magnesium metal gave rise to the vast corpus of Grignard chemistry, and the investigation of the metal carbonyls provided the impetus for the development of acetylene chemistry. Hydroboration, a process for producing organoboranes, can be used for stereospecific syntheses, asymmetric syntheses, and isomerizations. Hydroboration must be reckoned as among the most versatile synthetic procedures added to the armory of the organic chemist in recent years. Other inorganic substances that have found important use in inorganic chemistry are selenium for dehydrogenation reactions; lead tetraacetate and thallium compounds for selective oxidations; aluminum chloride as a catalyst for alkylation, acylation, and ring-closure reactions; anhydrous hydrogen fluoride for diazotization, nitration, and sulfonation; and lithium aluminum hydride (and various of its derivatives) and alkali-metal borohydrides for selective reduction reactions. Metal or carbon atoms in the gas phase are remarkable reagents for gas-phase organic reactions. *See* GRIGNARD REACTION; METAL CARBONYL; ORGANIC CHEMICAL SYNTHESIS; ORGANIC CHEMISTRY.

Reaction kinetics and mechanisms. Organic reactions generally proceed with the skeleton of the molecule remaining intact. Inorganic gas-phase reactions, on the contrary, are usually characterized by a complete disruption of molecular structure followed by reorganization to form the products of the reaction. Gas-phase reactions of inorganic compounds are thus in principle more complicated than are the usual organic reactions. Electron transfer in the oxidation-reduction reactions of transition-metal ions and compounds in solution has been a particularly important topic of theoretical interest. As is the case with electron transfer reactions in general, the role of electron tunneling in inorganic redox reactions still has many obscure features. This is particularly true of the very complex redox reactions of the actinide elements. Hydrolytic reactions of highly charged ions in aqueous solutions are also important in the mechanisms of many inorganic reactions and are important aspects of modern transition-element and actinide element chemistry. Increasingly, these subjects have become the concern of physical and theoretical chemists, but they also continue to be an important part of contemporary inorganic chemistry. *See* CHEMICAL DYNAMICS; OXIDATION-REDUCTION.

Bioinorganic chemistry. Inorganic chemistry is under pressure, as are other areas of chemical research, to escalate attention from simple molecules to more complicated systems. Like the organic chemist, the inorganic chemist has turned to biology as a new field to explore, and much of the most interesting contemporary work in organic chemistry is directed to the inorganic chemical aspects associated with living organisms.

Metal ions are very important participants in many biological phenomena. These range from the role of trace elements as essential nutrients to such questions as the function of metal ions in respiratory and photosynthetic pigments. Other important questions relate to the function of the metal ion in metal-containing enzymes involved in oxidation or reduction reactions, and to the transport of metal ions across membranes. Platinum and other metal coordination compounds are being investigated for possible use in the therapy of cancer. The basis for the use of lithium ion to control manic-depressive fluctuations in mood also poses important questions for the inorganic chemist.

A major activity in bioinorganic chemistry is the preparation and study of model systems intended to mimic the behavior of important biological entities. Thus, the construction of model systems for the prosthetic group of respiratory pigments (hemoglobin, myoglobin), oxidative enzymes (peroxidase), electron transfer proteins (cytochrome, ferredoxin, plastocyanin), and enzymes such as nitrogenase (the nitrogen-fixing enzyme) is under very active investigation by inorganic chemists.

The role of zinc ion in such enzymes as carbonic anhydrase and proteolytic peptidases has also focused attention on the inorganic components of these and other important enzyme systems. The application of modern spectroscopic techniques, particularly infrared and magnetic resonance spectroscopy, has suggested a role for the central magnesium atom of chlorophyll in the light-energy conversion step of photosynthesis. *See* SPECTROSCOPY.

Inorganic technology. The production of inorganic chemicals is a basic aspect of the chemical industry, and the heavy inorganic chemicals sulfuric acid, ammonia, chlorine, and phosphoric acid provide indispensable materials for many industries. The production of glass, ceramic, cement, fertilizer, and metals essentially involves reactions of inorganic chemistry carried out on the large scale. These industries have become less traditional in nature because of the contributions from inorganic chemical research, and it is altogether likely that modern inorganic chemistry will continue to make critically important contributions to the continued evolution of inorganic technology.

[JOSEPH J. KATZ]

Bibliography: A. W. Addison et al. (eds.), *Biological Aspects of Inorganic Chemistry*, 1977; F. A. Cotton and G. Wilkinson, *Advanced Inorganic Chemistry*, 4th ed., 1980; G. Eichhorn (ed.), *Inorganic Biochemistry*, 2 vols., 1973.

Inorganic photochemistry

Principally the study of the light-induced behavior of various metal compounds. The physical and chemical properties of substances are generally altered by the absorption of light. Typical metal compounds have a characteristic number (coordination number) of molecules or ions (ligands) directly bonded to the metal center. This article will refer to six-coordinate compounds (for example, ML_6^{n+}). Many of these compounds are colored, and much interest has been aroused by speculation that some metal compounds could mediate the transformation of solar radiation into useful chemical or electrical energy. *See* COORDINATION NUMBER.

The photochemistry of metal compounds has grown in concert with modern theories of the electronic structure of molecules and of chemical bonding in molecules. Photochemical studies are often designed to probe and test these theories. The range of pertinent studies spans most of the subdisciplines of chemistry and includes or bears on such topics as photophysics, the development of laser materials, catalysis, photosynthesis, oxidation-reduction chemistry, acid-base chemistry, organometallic chemistry, metalloenzyme chemistry, solid-state chemistry, and surface chemistry. *See* CHEMICAL BONDING.

Excited states. The absorption of light results in a rearrangement of electrons within a molecule. In many molecules, the new electronic configuration can persist for a significant period of time. It is useful to regard these excited states of molecules as new chemical species with chemical properties distinctly different from those of the ground state. Chemical properties depend on electronic configurations, and the ground-state and excited-state electronic configurations differ. In general, the bond lengths (angles, and so on) of the excited state of a molecule will be different from those of the ground-state (or thermally equilibrated) molecule. As a consequence, some of the energy used to generate an excited state is degraded to heat as the excited molecule relaxes to a bonding arrangement compatible with the new electronic configuration. In addition, the initial excited state (*X) may rapidly convert (or cross) to a lower-energy excited state (*Y) with yet another electronic configuration (see illustration). After light absorption

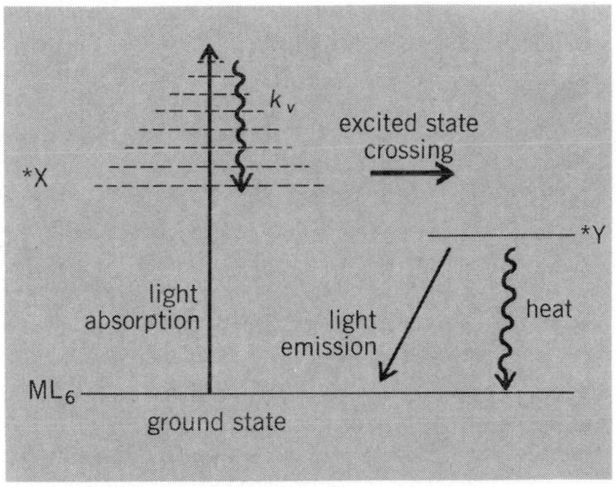

Qualitative electronic energy level scheme for a six-coordinate metal compound (ML_6). Absorption of light by ML_6 results in an excited state $*(ML_6)$ in which electrons are rearranged (either *X or *Y). Any bond length difference between ML_6 and $*(ML_6)$ results in the initial excited state being vibrationally excited (bonds stretched or compressed from the equilibrium position). Vibrational relaxation k_v is rapid in condensed phases.

in metal compounds, the time required for generation of the lowest-energy excited state is usually very short (less than a nanosecond). As a consequence, any chemistry due to the higher-energy excited states (for example, *X in the illustration) must occur very quickly, and either is intramolecular or involves the nearest neighbors to the excited molecule in condensed phases. The lowest-energy excited states of some metal compounds (*Y in the illustration) can exist for nearly a millisecond. This is long enough for many collisions in condensed phases, and such excited-state species can often react with other molecules present in the medium (bimolecular reactions). The lifetimes of excited molecules are limited by: the probability of chemical reaction; the probability of return to the ground state with the emission of light; and the probability of return to the ground state without light emission, but with the generation of heat energy. *See* ELECTRON CONFIGURATION.

Unimolecular reactions. The initial steps of the simplest excited-state chemical reactions involve only the excited molecule or solvent molecules, and may be classified as: excitation [reaction (1)];

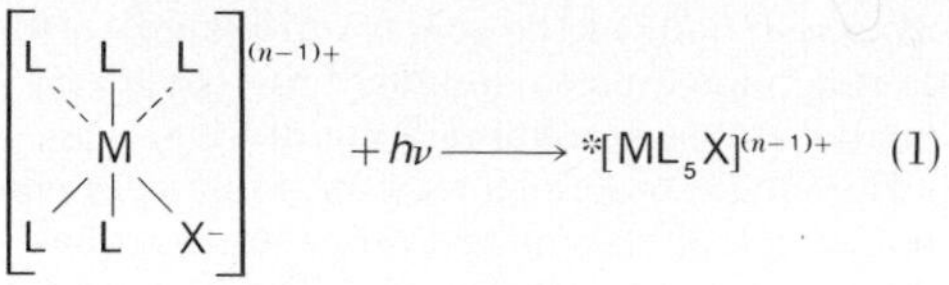

decrease in coordination number [reaction (2)];

$$*[ML_5X]^{(n-1)+} \longrightarrow ML_5^{n+} + X^- \text{ (or L)} \qquad (2)$$

increase in coordination number, for S, a solvent species [reaction (3)]; and homolysis, as in oxida-

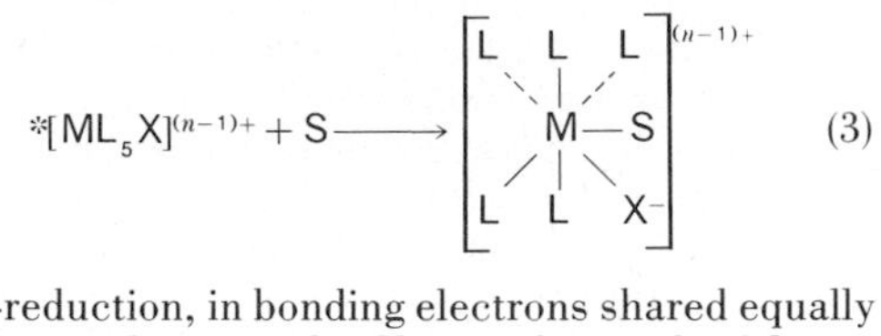

tion-reduction, in bonding electrons shared equally in the products, with $\cdot X$ as a free radical [reaction (4)].

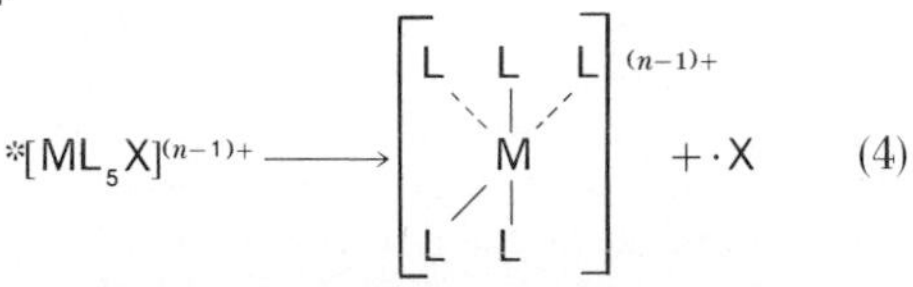

Each of these chemical processes, (2), (3), or (4), gives rise to an unstable, therefore reactive, chemical intermediate. Such intermediates are often useful in the synthesis of new compounds, as in reaction (5). When these intermediates are gener-

$$ML_5^{n+} + Y^- - [ML_5(Y^-)]^{(n-1)+} \qquad (5)$$

ated on a surface or in an inert matrix, their lifetimes may be enhanced, and they may function as reactive catalytic sites. *See* FREE RADICAL.

The changes in the number of ligands coordinated to the central metal [reaction (2) or (3)] can be theoretically related to the relationship between the electronic configuration of the molecular species and the nature of the coordinate covalent bond. The net chemical changes observed depend on the chemistry of the intermediate species [$(ML_5)^{n+}$ or $(ML_5SX)^{(n-1)+}$] as well as the chemistry of the reactive excited state [$*(ML_5X)^{(n-1)+}$].

While excited-state reactions which result in changes in coordination number produce a single reactive species, the intermediate metallofragment, homolytic cleavage of a metal ligand bond results in formation of a pair of very reactive species: a reduced metallofragment and a one-electron oxidized ligand species. These very reactive substances tend to recombine to regenerate the original metal compound, in addition to reacting with themselves or other solution species. The recombination reaction may be very rapid (in less than a nanosecond) when these species are formed in close proximity, or it may be somewhat slower (of the order of 10^{-6} s) if the species manage to separate a few molecular diameters. In order for a simple homolytic process to be observed, the minimum energy difference between the reactive excited state and the ground state must exceed the M^{n+}—(X^-) bond energy. However, there are some systems in which the solvent can assist the homolytic process by means of a concerted displacement of the departing radical fragment. The net quantum yields for photohomolytic processes are reasonably large (that is, greater than 0.1) only when the recombination reactions are slow compared to other reactions of the reactive fragments. Homolytic reactions tend to dominate the chemistry of the excited states of cobalt(III) complexes.

Homolytic reactions turn out to be very efficient for methylcobalamin and other organocobalt complexes related to vitamin B_{12}. These materials are very highly colored (deep red), and about 40% of the light photons absorbed result in homolytic cleavage of the cobalt-carbon bonds. Various recombination reactions decrease the observed product yields to less than 30%. This photosensitivity results in degradation of these natural complexes, and photoinduced reactions are not important to their enzymatic function. However, the small photonic energy required for cobalt-carbon homolysis (80–200 kJ mol^{-1}) is a manifestation of the weakness of this chemical bond. Thermal cleavage of the cobalt-carbon bond has been postulated as a key step in the enzymic functioning of coenzyme-B_{12}.

Inevitably, many of the photoinduced unimolecular processes found for metal compounds cannot be neatly placed in the above categories. Among the most intriguing are processes which involve bond-breaking or bond-making processes on a ligand. For example, irradiations of some rhodium(III) or iridium(III) azide complexes ($M^{III}—N_3^-$) result in cleavage of the $N^-—N_2$ bond forming a coordinated nitrene, $M^{III}—N^-$. Another remarkable example is cleavage of the $O_2C—(CO_2^{2-})$ bond in some oxalate complexes

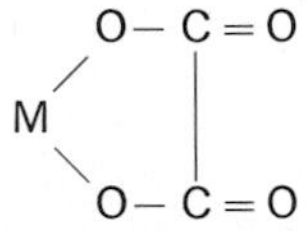

to form a metal-carbon bonded formate complex

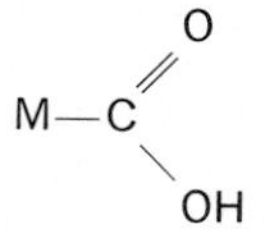

There are a number of photoinduced isomerizations, for example,

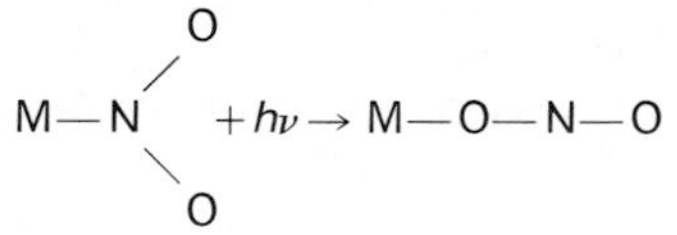

which might appear to fall outside the categories listed above, but in which the primary photoprocess appears to involve metal-ligand homolysis, followed by very rapid recombination of the fragments before they can be separated by one or more solvent molecules. *See* NITROGEN COMPLEXES.

Biomolecular excited-state reactions. Reactions between excited states and molecules other than solvent species are very important when the electronically excited molecules are long-lived. Metal compounds are frequently facile oxidation-reduction reagents, and metal-compound excited states can be employed to displace oxidation-reduction equilibria. The important categories of bimolecular excited-state reactions are electron transfer, shown in reaction (6*a*) or (6*b*), and

$$^*[ML_6{}^{n+}] + A \rightarrow ML_6{}^{(n-1)+} + A^+ \qquad (6a)$$

$$^*[ML_6{}^{n+}] + B \rightarrow ML_6{}^{(n+1)+} + B^- \qquad (6b)$$

quenching by electronic energy transfer, as in reaction (7).

$$[ML_6{}^{n+}]^* + C \rightarrow ML_6{}^{n+} + C^* \qquad (7)$$

The transfer of an electron from one species to another constitutes a net redistribution of electrical charge. As a consequence, there is necessarily some difference in the solvation of reactants and products; in addition, there will often be differences in the bond lengths of reactants and products. The larger these differences in bond length or solvation, the slower will be the rate of the electron transfer process. Excited-state lifetimes are necessarily short, so that electron transfer reactions must occur rapidly if they are to be of any consequence. Very rapid excited-state electron transfer reactions occur: most often between large molecules (thus minimizing solvation energies); most often between molecules in which the electron transfer does not result in large bond-length changes; and in part because the excitation energy stored in the excited molecule helps make the reactants appreciably less stable than the products. Typical of the metal compounds employed in the study of such reactions are *tris*-bipyridyl complexes, $M(LL)_3{}^{n+}$, where

(LL)= 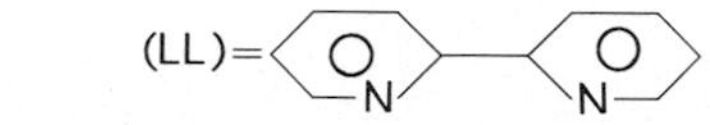

and $M^{n+} = Ru^{2+}$ or Cr^{3+}. Enough energy is released in the electron transfer process to overcome some of the intrinsic factors limiting the rate. Indeed, typical metal compound excited states can store enough energy that even the recombination reactions $ML_6{}^{(n-1)+} + A^+ \rightarrow ML_6{}^{n+} + A$ or $ML_6{}^{(n+1)+} + B^- \rightarrow ML_6{}^{n+} + B$, could ideally produce 1–2 V of electrical energy in a battery. Such excited-state electron transfer reactions afford a convenient means of generating very reactive intermediate species (A^+, B^-).

Quenching by electronic energy transfer is a theoretically complex process since it requires simultaneous relaxation of a donor electron and excitation of an electron in the acceptor molecule. Although such reactions tend to be degradative in most systems, they can be employed to generate chemically interesting acceptor molecule excited states which cannot be easily populated by direct light absorption.

Attempts to utilize metal compounds to mediate the transformation of light energy from the Sun into a useful chemical fuel have focused largely on the cleavage of water, shown in reaction (8). Sever-

$$H_2O + ML_6{}^{n+} + h\nu \rightarrow H_2 + \tfrac{1}{2}\,O_2 + ML_6{}^{n+} \qquad (8)$$

al metal-compound excited states store enough energy to promote this reaction as it is written. However, detailed consideration of reaction (8) reveals that it must involve at least two water molecules and four electrons per product molecule formed, and the energy requirements for this reaction cannot be simply equated to the energy available from single-electron transfer processes of molecular excited states, as in reactions (6*a*) and (6*b*). On the other hand, reactive intermediates generated in electron transfer processes might be capable of transforming more than one equivalent of electrons (for example, by reduction of H^+ to a coordinated hydride, H^-). Appropriate intermediates might be homogeneous (for example, low-valent complexes in solution) or heterogeneous (for example, colloidal metals or metal oxides). By mounting the absorbing metal complex on the surface of an electrode, it might be possible to effect a catalytic cycle for photochemical fuel generation (the electrode could be used to replace any electrons transferred to or from the electrolytic medium). Related applications may involve photocurrents induced at semiconductor electrodes. *See* CHEMICAL DYNAMICS; COORDINATION CHEMISTRY; LASER PHOTOCHEMISTRY; PHOTOCHEMISTRY; REACTIVE INTERMEDIATES.

[JOHN F. ENDICOTT]

Bibliography: A. W. Adamson and P. D. Fleischauer (eds.), *Concepts of Inorganic Photochemistry*, 1975; V. Balzani and V. Carassiti, *Photochemistry of Coordination Compounds*, 1970; J. R. Bolton (ed.), *Solar Power and Fuels*, 1977; M. S. Wrighton (ed.), *Inorganic and Organometallic Photochemistry*, Advances in Chemistry Series, no. 168, 1978.

Inorganic polymer

A giant molecule linked by covalent bonds but with an absence or near-absence of hydrocarbon units in the main molecular backbone; these may be included as pendant side chains. Carbon fibers, graphite, and so forth are considered inorganic polymers. Much of inorganic chemistry is the chemistry of high polymers. This article will be restricted to compounds which, on melting or dissolution, give high polymers, as shown by their viscosity. For compounds that do not melt or dissolve without chemical change, both the absence of an equilibrium vapor pressure and the observation of a dissociation pressure resulting from depolymerization bring them into the framework of the definition.

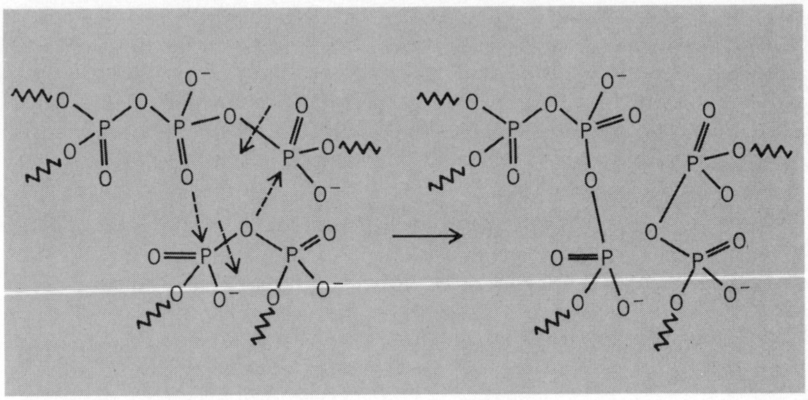

Fig. 1. Bond interchange in a cross-linked polyphosphate. (*From N. H. Ray, Inorganic Polymers, Academic Press, 1978*)

Properties. Some special characteristics of many inorganic polymers are a higher Young's modulus and a lower failure strain compared with organic polymers. Relatively few inorganic polymers dissolve in the true sense, or alternatively, if

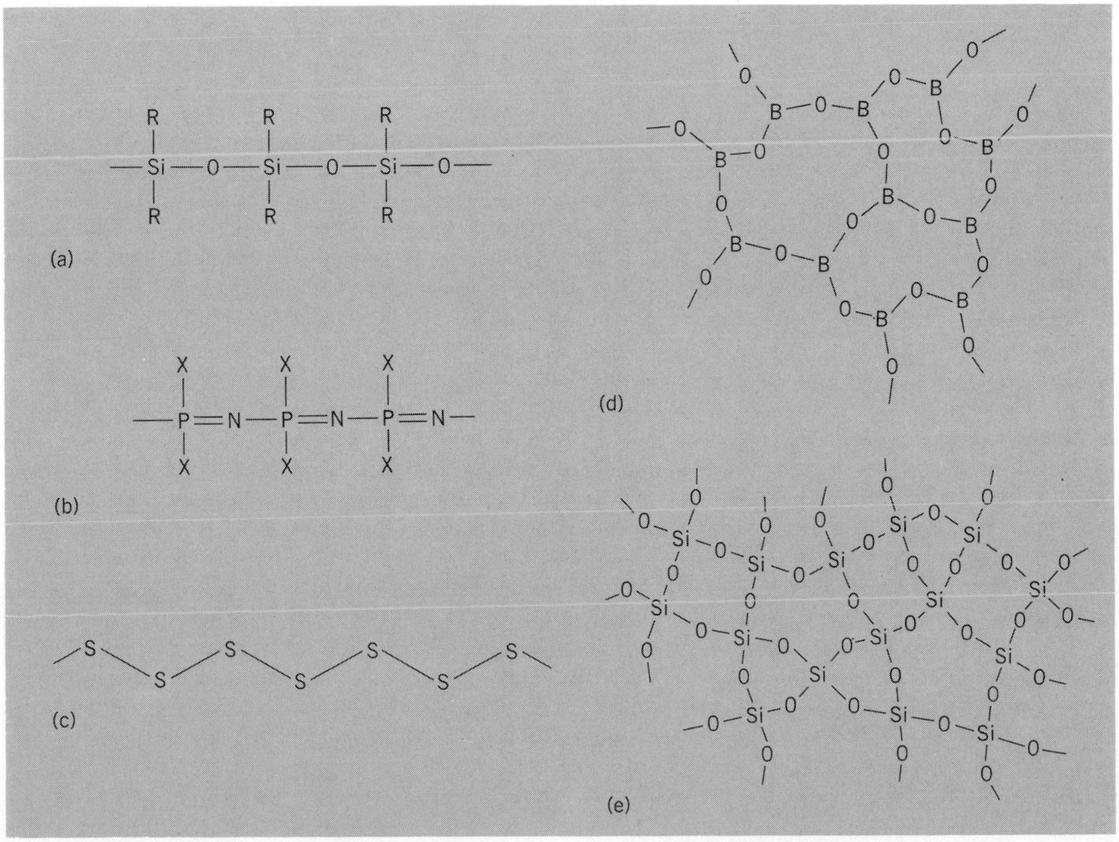

Fig. 2. Polymers with varying connectivities: (*a*) siloxanes, two; (*b*) phosphazenes, two; (*c*) sulfur, two; (*d*) boric oxide, three; (*e*) amorphous silica, four. (*From N. H. Ray, Inorganic Polymers, Academic Press, 1978*)

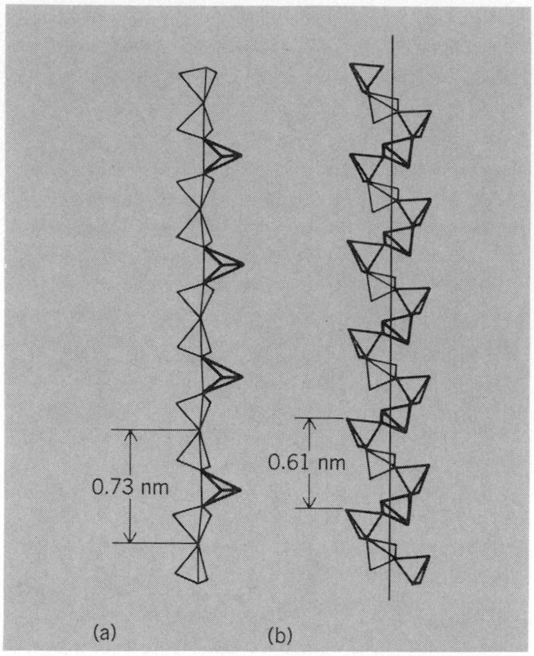

Fig. 3. Anionic polymer chain in (*a*) Madrell's salt and (*b*) Kurrol's salt. (*From E. Thilo, Inorganic Polymers, Chem. Soc. London Spec. Publ. no. 15, 1961*)

INORGANIC POLYMER

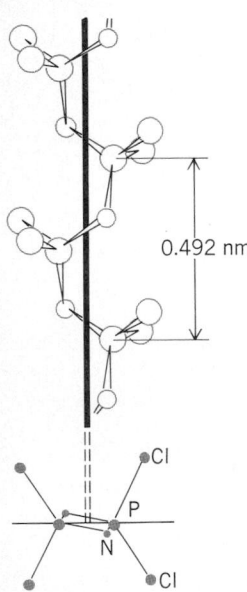

Fig. 4. Structure of poly(dichlorophosphazene). (*From E. Giglio, F. Pompa, and A. Ripamonti, J. Polym. Sci., 59:293, 1962*)

they swell, few can revert. Crystallinity and high glass transition temperatures are also much more common than in organic polymers. In highly cross-linked inorganic polymers, stress relaxation frequently involves bond interchange (Fig. 1). *See* CHEMICAL BOND THEORY.

The properties of inorganic polymers require a different technology from that of their organic counterparts. Such technology is either completely new (such as reconstructive processing—the spinning of an inorganic compound on an organic support or binder subsequently removed by oxidation/volatilization), or it has been adapted from other fields, for example, glass technology. Thus reconstructed vermiculite can give flexible sheets. Yarn, paper, woven cloth, and even textiles can be made from alumina and zirconia fibers by the spinning/volatilization process. A micaforming glass ceramic is resistant to thermal and mechanical shock and can be worked with conventional metal-working tools.

Classification. Inorganic polymers can be classified in a number of ways. Some are based on the composition of the backbone, such as the silicones (Si—O), the phosphazenes (P—N), and polymeric sulfur (S—S). Others are based on their connectivity, that is, the number of network bonds linking the repeating unit into the network. Thus the silicones based on R_2SiO, the phosphazenes based on NPX_2, and polymeric sulfur each have a connectivity of two, while boric oxide based on B_2O_3 has a connectivity of three, and amorphous silica based on SiO_2 has one of four (Fig. 2).

Types. The number of inorganic polymers is very large. Sulfur, selenium, and tellurium all form high polymers. Polymers of sulfur are usually elastomeric, and those of selenium and tellurium are generally crystalline. In the melt at 220°C, the molecular weight of the sulfur polymer is about 12,000,000, that of selenium about 800,000.

Linear polyphosphates, $(MPO_3)_n$, can be obtained in crystalline and glassy forms. Madrell's salt (used as dentifrice polishing agent) and Kurrol's salt, both based on sodium metaphosphates, are examples of linear polyphosphates, and have three and four PO_4 units respectively in the repeating distance (Fig. 3). Magnesium polyphosphate can be prepared as a gum, while quaternary ammonium polyphosphates with one or more long alkyl chains are like greases; the latter are used as boundary lubricants.

Poly(dichlorophosphazenes) (X = Cl; Fig. 4) can be obtained from cyclic precursors either as elastomers or as soluble polymers. Both types are hydrolytically unstable, and both degrade, as do silicones, at high temperatures (250–300°C) to cyclic oligomers. The soluble poly(dichlorophosphazenes) can be converted by a variety of nucleophiles to other polymeric products no longer containing the hydrolytically unstable phosphorus-chlorine bond. The best known of these is a copolymer with mixed $X = OCH_2CF_3$ and $OCH_2C_3F_7$ groups—a rubber with a glass transition temperature at −77°C—reputedly the elastomer which can be used at lower temperatures than any other.

Sulfur nitride and derivatives. Polymeric sulfur nitride, $(SN)_n$ (Fig. 5), forms fiberlike crystals and possesses a metallic conductivity parallel to the fiber axis that increases with decreasing temperature. At 0.26 K the polymer becomes superconducting; it is the first example of a nonmetallic polymer with this property. Polymeric sulfur nitride has been put forward as an electrode material with unusual properties. Derivatives of $(SN)_n$, such as polythiazyl bromides, $(NSBr_x)_n$ ($x = 0.4$ or 0.25), also have metallic properties.

Silicones. Perhaps best known of all the synthetic polymers based on inorganic molecular structure, are the silicones, which are derived from the basic units shown in Fig. 6. These units are put together in various proportions and arrangements and, in conjunction with different R groupings, they give a wide variety of materials, varying from oils to waxes, resins, and elastomers. Silicones have a large number of uses as hydraulic and dielectric fluids, lubricants, antifoaming agents, and mold-release agents, and in addition, have been incorporated in many waxes and polishes. The resins can be used as electric insulators, and the silicone elastomers can be used at both higher and lower temperatures than most natural and synthetic organic rubbers. These examples

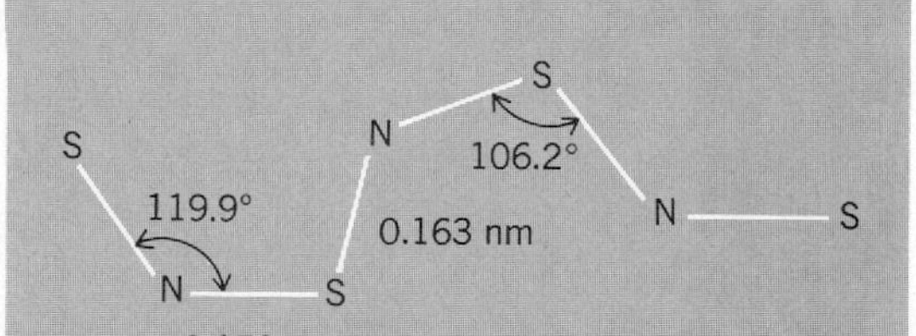

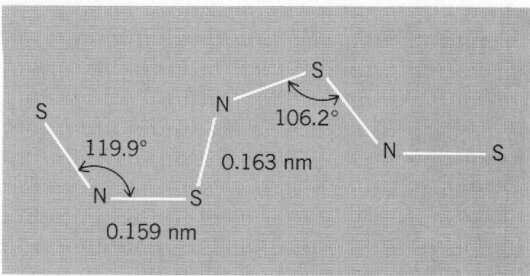

Fig. 5. Structure of polymeric sulfur nitride (phase I). (*From N. H. Ray, Inorganic Polymers, Academic Press, 1978*)

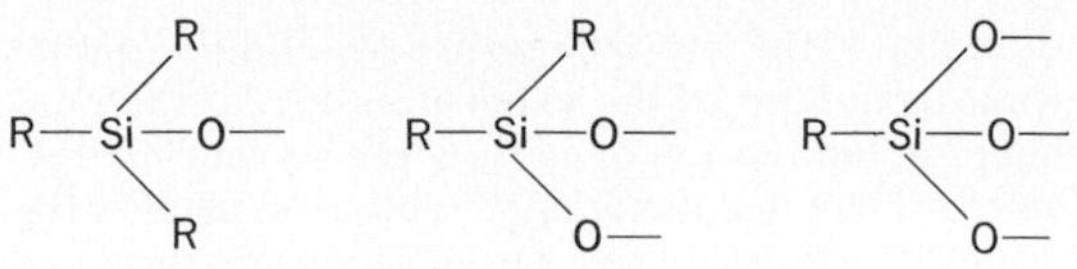

Fig. 6. Building units for silicones.

represent polymers with a connectivity of two. *See* SILICONE RESINS.

Chalcogenide glasses. These are amorphous cross-linked polymers with a connectivity of three. Probably the best known is arsenic sulfide, $(As_2S_3)_n$, which can be used for infrared transparent windows. Threshold and memory switching are also interesting properties possessed by these glasses.

Ultraphosphate glasses resemble glassy organic plastics and can be processed by the same methods, such as extrusion and injection molding. They are used for antifouling surfaces for marine applications and in the manufacture of nonmisting spectacle lenses. They can be considered to be structurally derived from amorphous polymeric phosphorus pentoxide, $(P_2O_5)_n$, in which some P—O—P bridges have been hydrolysed and metal cations introduced.

Graphite. This is a well-known two-dimensional polymer with lubricating and electrical properties. Intercalation compounds of graphite can have supermetallic anisotropic properties.

Boron polymers. Structurally related to graphite is hexagonal boron nitride, $(BN)_n$ (Fig. 7). Like graphite, it has lubricating properties, reflecting the relationship between molecular structure and physical properties, but unlike graphite, it is an electrical insulator. Molybdenum disulfide, $(MoS_2)_n$, with a similar and related structure, is also a solid lubricant. Both graphite and hexagonal boron nitride can be readily machined. Outstanding properties of the latter include high thermal and chemical stability and good dielectric properties. Crucibles and such items as nuts and bolts can be made from this material.

Borate glasses with comparatively low softening points are used as solder and sealing glasses and can be prepared by fusing mixtures of metal oxides with boric oxide, $(B_2O_3)_n$.

Boron phosphate, $(BPO_4)_n$, is a crystalline polymer but when admixed with alkali gives glassy polymers with uses similar to those of ultraphosphates. Gels of $(BPO_4)_n$ in concentrated sulfuric acid have been advocated as nonspillable electrolytes in lead-acid accumulators. In suitable acid solutions aluminum phosphate gives viscous solutions that are used as refractory cement and as binders for abrasives.

Silicate polymers. The silicates, both crystalline and amorphous, supply a very large number of inorganic polymers. Examples include the naturally occurring fiberlike asbestos and sheetlike mica. The industrially important water-soluble alkali metal silicates can give highly viscous polymeric solutions. Borosilicate glasses form another important group of silicate polymers. The Pyrex type is well known for its resistance to thermal shock; the leached Vycor type is porous and can be used for filtering bacteria and viruses. Asbestos occurs as ladder polymers, of which crocidolite is the most important, and as layer polymers exemplified by chrysotile. The zeolites, many of which have been found naturally or have been synthesized, are three-dimensional network polymers. Their uses as molecular sieves are well known. *See* MOLECULAR SIEVE; SILICATE.

Other polymers. Silicon nitride, $(Si_3N_4)_n$, is another macromolecule with interesting properties. Prepared by heating of silicon powder in an atmosphere of nitrogen (nitridation) at above 1200°C, the product is a material which can be machined readily and whose good thermal shock resistance and creep resistance at high temperatures, which is further improved by admixture of another inorganic macromolecule silicon carbide, make it useful for applications in gas turbines, diesel engines, thermocouple sheaths, and a variety of components.

Allotropic forms of carbon and boron nitride are diamond and cubic boron nitride, both preparable by high-temperature and high-pressure syntheses and characterized by extreme hardness, which make them useful industrially in cutting and grinding tools.

The structurally related elemental silicon is, if suitably doped, a well-known semiconductor material.

Phosphorus oxynitride, $(PON)_n$, isoelectronic with silica, $(SiO_2)_n$, and prepared by thermolysis of phosphoryl triamide, $P(O)(NH_2)_3$, can be drawn as fibers from the melt. It has been suggested as a bonding agent for asbestos.

Copolymers of *meta*-carboranes with short-chain polysiloxanes, for example, $—Me_2Si—CB_{10}H_{10}C—(SiMe_2O)_n—(n=1$ to 4), have some use at high

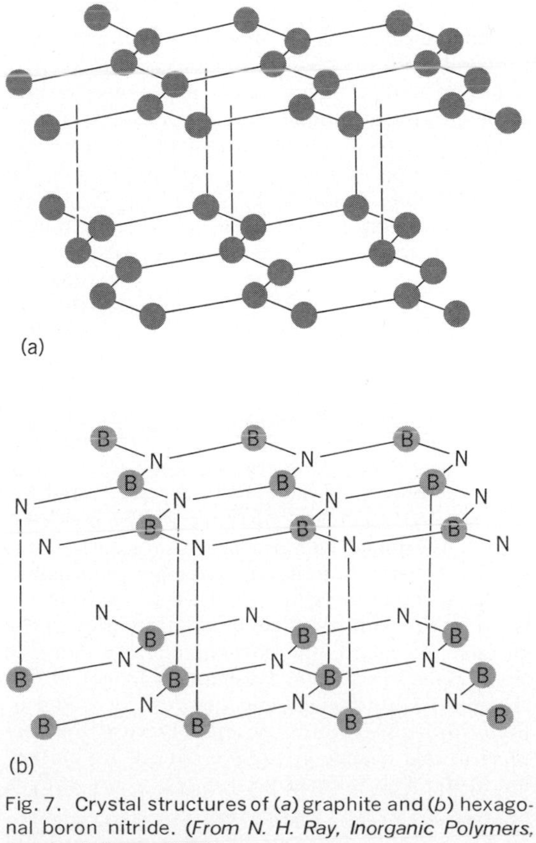

Fig. 7. Crystal structures of (*a*) graphite and (*b*) hexagonal boron nitride. (*From N. H. Ray, Inorganic Polymers, Academic Press, 1978*)

temperatures as expensive specialty polymers.

Carbon fibers are prepared by the controlled oxidation under tension of suitable organic polymers. Although expensive, they have specialized applications, especially as composites.

Importance. Inorganic polymeric materials are growing in importance as a result of a combination of two major factors: the depletion of the world's fossil fuel reserves (the basis of the petrochemical industry) and the ever-increasing demands of modern technology, coupled with environmental and health regulations, such as flame retardancy and nonflammability. *See* INORGANIC CHEMISTRY; POLYMER; POLYMERIZATION. [ROBERT A. SHAW]

Bibliography: P. Popper (ed.), *Special Ceramics*, 1960; H. Rawson, *Inorganic Glass-Forming Systems*, 1967; N. H. Ray, *Inorganic Polymers*, 1978; A. F. Wells, *Structural Inorganic Chemistry*, 4th ed., 1975.

Interface of phases

The boundary between any two phases. Among the three phases, gas, liquid, and solid, five types of interfaces are possible: gas-liquid, gas-solid, liquid-liquid, liquid-solid, and solid-solid. The abrupt transition from one phase to another at these boundaries, even though subject to the kinetic effects of molecular motion, is statistically a surface only one or two molecules thick.

A unique property of the surfaces of the phases that adjoin at an interface is the surface energy which is the result of unbalanced molecular fields existing at the surfaces of the two phases. Within the bulk of a given phase, the intermolecular forces are uniform because each molecule enjoys a statistically homogeneous field produced by neighboring molecules of the same substance. Molecules in the surface of a phase, however, are bounded on one side by an entirely different environment, with the result that there are intermolecular forces that then tend to pull these surface molecules toward the bulk of the phase. A drop of water, as a result, tends to assume a spherical shape in order to reduce the surface area of the droplet to a minimum.

Surface energy. At an interface, there will be a difference in the tendencies for each phase to attract its own molecules. Consequently, there is always a minimum in the free energy of the surfaces at an interface, the net amount of which is called the interfacial energy. At the water-air interface, for example, the difference in molecular fields in the water and air surfaces accounts for the interfacial energy of 72 ergs/cm² of interfacial surface. The interfacial energy between the two liquids, benzene and water, is 35 ergs/cm², and between ethyl ether and mercury is 379 ergs/cm². These interfacial energies are also expressed as surface tension in units of dynes per centimeter.

The surface energy at an interface may be altered by the addition of solutes that migrate to the surface and modify the molecular forces there, or the surface energy may be changed by converting the planar interfacial boundary to a curved surface. Both the theoretical and practical implications of this change in surface energy are embodied in the Kelvin equation, Eq. (1), where P/P_0 is

$$\ln\frac{P}{P_0}=\frac{2M\gamma}{RT\rho r} \qquad (1)$$

the ratio of the vapor pressure of a liquid droplet with diameter r to the vapor pressure of the pure liquid in bulk, ρ the density, γ the surface energy, and M the molecular weight. Thus, the smaller the droplet the greater the relative vapor pressure, and as a consequence, small droplets of liquid evaporate more rapidly than larger ones. The surface energy of solids is also a function of their size, and the Kelvin equation can be modified to describe the greater solubility of small particles compared to that of larger particles of the same solid. *See* ADSORPTION; SURFACE-ACTIVE AGENT.

Contact angle. At liquid-solid interfaces, where the confluence of the two phases is usually termed wetting, a critical factor called the contact angle is involved. A drop of water placed on a paraffin surface, for example, retains a globular shape, whereas the same drop of water placed on a clean glass surface spreads out into a thin layer. In the first instance, the contact angle is practically 180°, and in the second instance, it is practically 0°. The study of contact angles reveals the interplay of interfacial energies at three boundaries. The illustration is a schematic representation of the cross

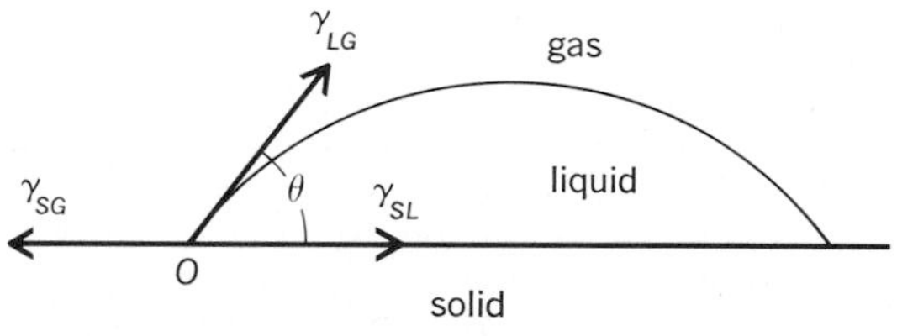

Contact angle θ at interface of three phases.

section of a drop of liquid on a solid. There are solid-liquid, solid-gas, and liquid-gas interfaces that meet in a linear zone at O. The forces about O that determine the equilibrium contact angle are related to each other according to Eq. (2), where the γ

$$\gamma_{SG}=\gamma_{SL}+\gamma_{LG}\cos\theta \qquad (2)$$

terms represent free energies at the interfaces and θ is the contact angle. Since only γ_{LG} and θ can be measured readily, the term adhesion tension is defined by Eq. (3). Adhesion tension, which is the

$$\gamma_{LG}\cos\theta=\gamma_{SG}-\gamma_{SL}=\text{adhesion tension} \qquad (3)$$

free energy of setting, is of critical importance in detergency, dispersion of powders and pigments, lubrication, adhesion, and spreading processes.

The measurement of interfacial energies is made directly only upon liquid-gas and liquid-liquid interfaces. In measuring the liquid-gas interfacial energy (surface tension), the methods of capillary rise, drop weight on pendant drop, bubble pressure, sessile drops, Du Nuoy ring, vibrating jets, and ultrasonic action are among those used. There is a small but appreciable temperature effect upon surface tension, and this property is used to determine small differences in the surface tension of a liquid by placing the two ends of a liquid column in a capillary tube whose two ends are at different temperatures. The determination of interfacial energies at other types of interfaces can be inferred only by indirect methods. *See* FREE ENERGY; PHASE EQUILIBRIUM; SURFACE TENSION.

[WENDELL H. SLABAUGH]

Bibliography: American Chemical Society, *Contact Angle, Wettability, and Adhesion*, Advan. Chem. Ser. no. 43., 1964; J. T. Davies and E. Rideal, *Interfacial Phenomena*, 2d ed., 1963.

Interhalogen compounds

The elements of the halogen family (fluorine, chlorine, bromine, and iodine) possess an ability to react with each other to form a series of binary interhalogen compounds (or halogen halides) of general composition given by XY_n, where n can have the values 1, 3, 5, and 7, and where X is the heavier (less electronegative) of the two elements. All possible diatomic compounds of the first four halogens have been prepared. In other groups a varying number of possible combinations is absent. Although attempts have been made to prepare ternary interhalogens, they have been unsuccessful; there is considerable doubt that such compounds can exist. *See* HALOGEN ELEMENTS.

Formation. In general, interhalogen compounds are formed when the free halogens are mixed as gases, or, as in the case of iodine chlorides and bromides, by reacting solid iodine with liquid chlorine or bromine. Most of the nonfluorinated interhalogens also readily form when solutions of the halogens in inert solvent (for example, carbon tetrachloride) are mixed. It is also possible to form them by the reaction of a halogen with a salt of a more electropositive halogen, such as $KI + Cl_2 \rightarrow KCl + ICl$. Higher polyhalides can also be prepared by reacting more electronegative halogen with a corresponding halogen halide, for example, $ICl + Cl_2 \rightarrow ICl_3$ or $ClF_3 + F_2 \rightarrow ClF_5$. Chlorine pentafluoride can also be prepared by reacting a $MClF_4$ salt with fluorine (M = alkali metal), $MClF_4 + F_2 \rightarrow MF + ClF_5$. A list of known interhalogen compounds and some of their physical properties is given in Table 1. Interhalogen compounds containing astatine have not been isolated as yet, although the existence of AtI and AtBr has been demonstrated by indirect measurements.

Table 1. Known interhalogen compounds

	XY	XY_3	XY_5	XY_7
	ClF	ClF_3	ClF_5	IF_7
mp	−154°C	−76°C	−103°C	
bp	−101°C	12°C	−14°C	4.77 (sublimes)
	BrF	BrF_3	BrF_5	
mp	≈−33°C	8.77°C	−62.5°C	
bp	≈20°C	125°C	40.3°C	
	IF	IF_3	IF_5	
mp	—	−28°C	10°C	
bp	—	—	101°C	
	BrCl	ICl_3*		
mp	≈−54°C	101°C		
bp	—	—		
	ICl†			
mp	27.2°C(α)			
bp	≈100°C			
	IBr			
mp	40°C			
bp	119°C			

*In the solid state the compound forms a dimer.
†Unstable β-modification exists, mp 14°C.

Stability. Thermodynamic stability of the interhalogen compounds varies within rather large limits. In general, for a given group the stability increases with increasing difference in electronegativity between the two halogens. Thus for XY group the free energy of formation of the interhalogens, relative to the elements in their standard conditions, falls in the following order: IF > BrF > ClF > ICl > IBr > BrCl. It should be noted, however, that the fluorides of this series can be obtained only in minute quantities since they readily undergo disproportionation reaction, for example, $5IF \rightarrow 2I_2 + IF_5$. The least-stable compound, bromine chloride, has only recently been isolated in the pure state. Decrease of stability with decreasing difference of electronegativity is readily apparent in the higher interhalogens since compounds such as $BrCl_3$, IBr_3, or ICl_5 are unknown. The only unambiguously prepared interhalogen containing eight halogen atoms is IF_7. *See* ELECTRONEGATIVITY.

Reactivity. The reactivity of the polyhalides reflects the reactivity of the halogens they contain. In general, they behave as strong oxidizing and halogenating agents. Most halogen halides (especially halogen fluorides) readily attack metals, yielding the corresponding halide of the more electronegative halogen. In the case of halogen fluorides the reaction results in the formation of the fluoride, in which the metal is often found in its highest oxidation state, for example, AgF_2, CoF_3, and so on. Noble metals, such as platinum, are resistant to the attack of the interhalogens at room temperature. Halogen fluorides are often handled in nickel vessels, but in this case the resistance to attack is due to the formation of a protective layer of nickel (II) fluoride.

All halogen halides readily react with water. Such reactions can be quite violent and, with halogen fluorides, they may be explosive. Reaction products vary depending on the nature of the interhalogen compound. For example, in the case of chlorine trifluoride, reaction with an excess of water yields HF, Cl_2, and O_2 as reaction products.

The same reactivity is observed with organic compounds. The nonfluorinated interhalogens do not react with completely halogenated hydrocarbons, and solutions of ICl, IBr, and BrCl are quite stable in dry carbon tetrachloride or hexachloroethane, as well as in fluorocarbons, as long as the solvents are very dry. They readily react with aliphatic and aromatic hydrocarbons and with oxygen- or nitrogen-containing compounds. The reaction rates, however, can be rather slow and dilute solutions of ICl can be stable for several hours in solvents such as nitrobenzene. Halogen fluorides usually react vigorously with chlorinated hydrocarbons, although IF_5 can be dissolved in carbon tetrachloride and BrF_3 can be dissolved in Freon 113 without decomposition. All halogen fluorides react explosively with easily oxidizable organic compounds. *See* FLUORINE.

Halogen halides, like halogens, act as Lewis acids and under proper experimental conditions may form a series of stable complexes with various organic electron donors. For example, mixing of carbon tetrachloride solutions of pyridine and of iodine monochloride leads to the formation of a solid complex, $C_5H_5N \cdot ICl$. The same reaction can occur with other heterocyclic amines and ICl, IBr, or ICl_3. Addition compounds of organic electron donors with IF, IF_3, and IF_5 have been reported. In all cases it is the iodine atom which is directly attached to the donor atom.

A number of interhalogen compounds conduct electrical current in the liquid state. Among these are ICl and BrF_3. For example, electrical conductance of molten iodine monochloride is comparable to a concentrated aqueous solution of a strong electrolyte (4.52×10^{-3} ohm^{-1} cm^{-1} at 30.6°C). The conductances, however, are much smaller than those of fused salts and, therefore, it can be concluded that the bonding in these compounds is largely covalent. Electrical conductance is due to self-ionization reactions, as shown in reactions (1) and (2).

$$3ICl \rightleftarrows I_2Cl^+ + ICl_2^- \quad (1)$$

$$2BrF_3 \rightleftarrows BrF_2^+ + BrF_4^- \quad (2)$$

The above behavior leads to the possibility of studying acid-base reactions. In these systems an acid is any compound which generates the solvo-cation, while a base would generate solvo-anions. Thus SbF_5 would be an acid in liquid bromine trifluoride, reaction (3), while an electrovalent fluoride would be a base, reaction (4). *See* SUPERACID.

$$SbF_5 + BrF_3 \rightleftarrows BrF_2^+ + SbF_6^- \quad (3)$$

$$KF + BrF_3 \rightleftarrows K^+ + BrF_4^- \quad (4)$$

Analogy with acid-base reactions in water is obvious, as shown by reactions (5) and (6). Such rela-

$$K^+OH^- + H_3O^+Cl^- \rightarrow K^+Cl^- + 2H_2O \quad (5)$$

$$K^+BrF_4^- + BrF_2^+SbF_6^- \rightarrow KSbF_6 + 2BrF_3 \quad (6)$$

tions have been studied in BrF_3, ClF_3, and IF_5. Numerous salts of the interhalogen acid-base systems have been isolated and studied.

Thus, numerous compounds have been formed containing either interhalogen anions (solvo-anions) or interhalogen cations (solvo-cations) simply by adding the appropriate acid or base to a liquid halogen halide or a halogen halide in an appropriate nonaqueous solvent. In addition, cations derived from previously unknown compounds can be prepared by using powerful oxidizing agents, such as KrF^+ salts. For example, even though BrF_7 has not been unambiguously prepared to date, a compound containing BrF_6^+ has been prepared by R. J. Gillespie and coworkers according to reaction (7).

$$BrF_5 + KrF^+AsF_6^- \rightarrow BrF_6^+AsF_6^- + Kr \quad (7)$$

Table 2. Known interhalogen anions

Three-membered	Five-membered	Seven-membered	Nine-membered
ClF_2^-	ClF_4^-	ClF_6^-	IF_8^-
BrF_2^-	BrF_4^-	BrF_6^-	
ICl_2^-	IF_4^-	IF_6^-	
IBr_2^-	ICl_4^-		
$IBrCl^-$	$I_2Cl_3^-$		
$BrCl_2^-$	$I_2Cl_2Br^-$		
I_2Cl^-	$I_2ClBr_2^-$		
Br_2Cl^-	I_4Cl^-		
I_2Br^-			

Table 3. Known interhalogen cations

Three-membered	Five-membered	Seven-membered
Cl_2F^+	ClF_4^+	ClF_6^+
ClF_2^+	BrF_4^+	BrF_6^+
BrF_2^+	IF_4^+	
ICl_2^+		
I_2Cl^+		
IBr_2^+		
I_2Br^+		
$BrCl_2^+$		
Br_2Cl^+		
$IBrCl^+$		

Pentahalides can also be formed by the addition of an interhalogen compound to a trihalide ion as shown in reaction (8).

$$ICl + ICl_2^- \rightarrow I_2Cl_3^- \quad (8)$$

A compilation of interhalogen cations and anions which have been previously prepared is given in Tables 2 and 3.

[TERRY SURLES]

Bibliography: R. J. Gillespie and G. J. Schrobilgen, The hexafluorobromine (VII) cation, BrF_6^+, *Inorg. Chem.*, 13:1230, 1974, and references therein; A. I. Popov, Interhalogen compounds and polyhalide anions, in V. Gutmann (ed.), *MTP International Review of Science: Inorganic Chemistry*, ser. 1, vol. 3, chap. 2, 1972; A. I. Popov, Polyhalogen complex salts, in V. Gutmann (ed.), *Halogen Chemistry*, vol. 1, 1967; A. I. Popov and T. Surles, Interhalogen compounds and polyhalide anions, in V. Gutmann (ed.), *MTP International Review of Science*: *Inorganic Chemistry*, ser. 1, vol. 3, chap. 6, 1975, and references therein; L. Stein, Physical and chemical properties of halogen fluorides, in V. Gutmann (ed.), *Halogen Chemistry*, vol. 1, 1967; W. W. Wilson, B. Lands, and F. Aubke, The new interhalogen cations $BrCl_2^+$ and Br_2Cl^+, *Inorg. Nucl. Chem. Lett.*, 11:529, 1975, and references therein.

Intermetallic compounds

Materials composed of two or more types of metal atoms, which exist as homogeneous, composite substances and differ discontinuously in structure from that of the constituent metals. They are also called, preferably, intermetallic phases. Their properties cannot be transformed continuously into those of their constituents by changes of composition alone, and they form distinct crystalline species separated by phase boundaries from their metallic components and mixed crystals of these components; it is generally not possible to establish formulas for intermetallic compounds on the sole basis of analytical data, so formulas are determined in conjunction with crystallographic structural information.

The term "alloy" is generally applied to any homogeneous molten mixture of two or more metals, as well as to the solid material that crystallizes from such a homogeneous liquid phase. Alloys may also be formed from solid-state reactions. In the liquid phase, alloys are essentially solutions of metals in one another, although liquid compounds may also be present. Alloys containing mercury are usually referred to as amalgams. Solid

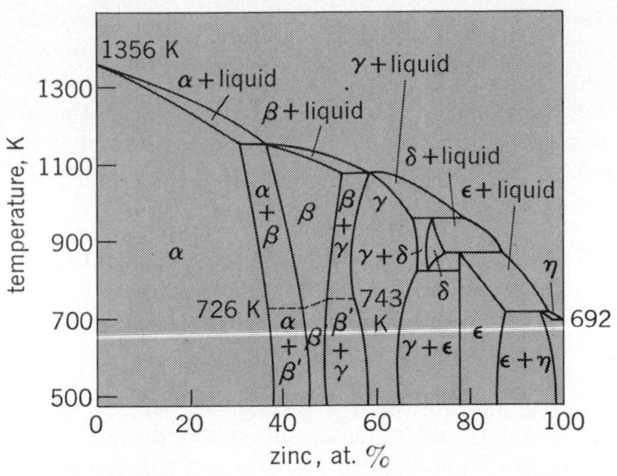

Fig. 1. Equilibrium-phase diagram for the copper-zinc (brass) system, showing the succession of phases which result with increasing concentration of zinc. The ordered phases β′ and γ′ exist only below 726 K and 743 K, respectively. (*J. L. T. Waugh, The Constitution of Inorganic Compounds, pt. M: Metals and Intermetallic Compounds, Wiley-Interscience, 1972*)

alloys may vary greatly in range of composition, structure, properties, and behavior.

Phase transformations. Much of the accumulated experimental information about the nature of the interaction and the phase transformations in systems composed of two or more metals is contained in phase or equilibrium diagrams, such as Fig. 1, which depicts the phase relationships in the copper-zinc (brass) system. Such phase diagrams, even for binary metal systems, may be of all degrees of complexity, ranging from systems showing the formation of simple solid solutions to dozens of intermetallic phases exhibiting structural (polymorphic), order-disorder, magnetic, bond-type, or deformation-type transformations as a function of composition or temperature, or both. Intermetallic compounds are composed of two or more metals. They may be stable over only a very narrow or over a relatively wide range of composition, which may be stoichiometric, as for the compounds GaAs, $PdCu_{13}$, $Zr_{57}Al_{43}$, and $Nb_{48}Ni_{39}Al_{13}$, or nonstoichiometric, as for the compounds $Co_{1-x}Te_x$ (where x extends continuously from 0.5 to 0.67) and $Al_{\sim0.8}Ge_{\sim0.2}Nb_3$ (an important superconducting alloy). *See* PHASE EQUILIBRIUM.

Crystal structure. The crystal structures found for intermetallic compounds may likewise range from the simple rock-salt structure displayed by BaTe to the extremely complex arrangement found for $NaCd_2$, Mg_{32} $(Al,Zn)_{49}$, and Cu_4Cd_3. Thus, a continuous range of solid solutions is observed in the K-Rb system; a simple eutectic (mixture with the minimum melting point) in the Sn-Pb system; a single intermetallic compound, $PbMg_2$, in the Pb-Mg system; and three different intermetallic compounds, CuZn, Cu_5Zn_8, and $CuZn_3$ (Hume-Rothery phases), in the brass system. Examples of disordered and ordered structures (superlattices or superstructures), stable above and below a certain temperature, respectively, are shown in Fig. 2. *See* CRYSTAL STRUCTURE.

All of the metallic elements, with the exception of Mn and Sn, crystallize with at least one polymorphic modification having a face-centered cubic (fcc), body-centered cubic (bcc), or close-packed hexagonal (cph) structure. Substitutional intermetallic compounds are preferentially formed between metal pairs that do not differ in metallic radii by more than about 15%, that adopt the same crystal structure, and that are of similar electronic structure and electronegativity. More stable compounds are formed between metal pairs with substantially different electrochemical characteristics. Many intermetallic compounds have compositions and structures that are determined largely by the relative sizes of the atoms involved. For example, $MgCu_2$, $NaAu_2$, $CaAl_2$, KBi_2, ZrW_2, $AgBe_2$, MgZnNi, and $BiAu_2$ all crystallize with the same structure; bismuth behaves as the smaller atom in KBi_2, and as the larger one in $BiAu_2$.

Another very large class of intermetallic compounds results from one kind of metal atom occupying the interstitial cavities in the close-packed structure of the other, usually in some preferential set or subset of lattice sites. If all the octahedral cavities in an fcc array of metal atoms are occupied by a dissimilar type of metal atom, the rock-salt structure results (CaSe, SrTe, UC); the nickel arsenide structure results when the corresponding set of octahedral holes in the cph structure is similarly occupied (CrSb, FeTe, PtSn); if exactly half of the tetrahedral cavities are occupied by one kind of metal atom, the sphalerite structure is derived from the fcc (GaAs, InSb), and the wurtzite structure from the cph (CdSe, ZnTe), arrangements. Among the more complex intermetallic structures, a large number involve icosahedral

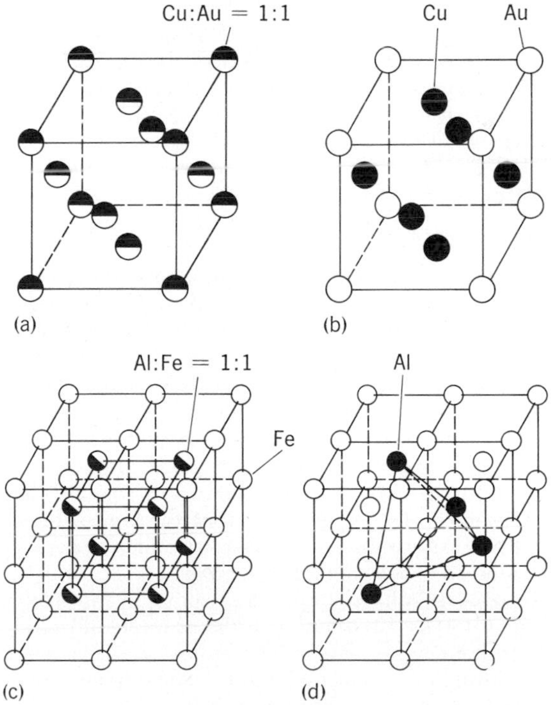

Fig. 2. Superstructure formation in the Cu-Au and the Al-Fe systems. (*a*) Disordered CuAu and (*b*) ordered Cu_3Au structures; (*c*) disordered Fe_3Al and (*d*) ordered Fe_3Al structures. The ordering of the atoms in *b* compared with *a* involves their redistribution over the lattice sites of an fcc unit cell; ordering of atoms in *d* relative to *c* involves their rearrangement over the lattice sites of a bcc structural unit. (*J. L. T. Waugh, The Constitution of Inorganic Compounds, pt. M: Metals and Intermetallic Compounds, Wiley-Interscience, 1972*)

coordination of a larger metal atom about a smaller one ($MoAl_{12}$, $MgCu_2$, $Mg_{32}\{Al,Zn\}_{49}$). Coordination numbers greater than 12 (14, 15, and 16) are common among the more complex structures. The Hume-Rothery electron compounds with closely related structures but apparently unrelated stoichiometry are determined by electron-to-atom ratios. Thus, metal pairs for which the electron-atom ratio is 3:2 crystallize with either bcc (Ag_3Al, NiAl), β-Mn complex cubic (Cu_5Si, $CoZn_3$) or cph (Cu_3Ga, Ag_7Sb) structures; if the ratio is 21:13, the complex γ-brass structure results ($Na_{31}Pb_8$, Cu_5Zn_8); and if the ratio is 7:4, cph structures (Au_5Al_3, Ag_3Sn) are obtained. Several groups of intermetallic compounds have been classified on the basis of the results of x-ray crystallographic determinations, such as the three types of Laves phases represented by $MgZn_2$, $MgCu_2$, and $MgNi_2$, which are derived from fcc, cph, and a combination of these structures.

Other characteristics. In addition to the electronic structure of the constituent metal atoms (which determines electron-atom ratios), electronegativity and metallic radii differences, packing and structural considerations (which determine the symmetry and extent of the so-called Fermi surfaces), and thermodynamic factors (such as the entropy and enthalpy changes on interaction between different metals) must be taken into account in order to understand why some intermetallic compounds exist as ordered or disordered phases, exhibit stoichiometric or nonstoichiometric compositions, or exist over a range of compositions. *See* NONSTOICHIOMETRIC COMPOUNDS; SOLID-STATE CHEMISTRY.

[JOHN L. T. WAUGH]

Bibliography: A. P. Cracknell and K. C. Wong, *The Fermi Surface*, 1973; L. T. Waugh, *The Constitution of Inorganic Compounds*, pt. M: *Metals and Intermetallic Compounds*, 1972; J. H. Westbrook (ed.), *Intermetallic Compounds*, 1967, reprint 1977.

Intermolecular forces

Attractive or repulsive interactions that occur between all atoms and molecules. These forces, which become significant at molecular separations of about 1 nm or less, are much weaker than forces associated with chemical bonds or electrostatic interactions of charged particles. They are important, however, since they are responsible for many of the physical properties of solids, liquids, and pressurized gases. Intermolecular forces also determine to an important extent the three-dimensional arrangement of biological molecules, polymers, and even smaller molecules.

Description. A simple description of intermolecular forces can begin with the example of two interacting argon atoms. The atoms are electrically neutral and do not undergo chemical bond formation.

Figure 1 shows the potential energy of two argon atoms as a function of their separation. At distances of about 1 nm or greater this energy is essentially zero and the atoms exert no forces on each other. (The force is the negative gradient, or slope, of the potential energy.) Between 0.4 and 0.8 nm the potential energy decreases and the atoms experience forces of attraction. For distances less than 0.3 nm the potential energy rises sharply as the atoms repel each other. At a distance of 0.38 nm the forces of attraction and repulsion balance each other. The potential energy (and corresponding intermolecular forces) between other pairs of atoms exhibits the same general shape as shown in Fig. 1, although the quantitative values of energy and separation are somewhat different. For intermolecular forces between molecules the relative orientations as well as distances are important and the description is more complex. In general, for either atoms or molecules at separations of 0.3 nm or less, the intermolecular forces are repulsive. At longer range, usually greater than 0.3 nm, the intermolecular forces are attractive. And at some intermediate distance, usually 0.3–0.4 nm (which depends on orientation in the case of molecules), the intermolecular forces of attraction and repulsion just balance.

Origin. The origin of intermolecular forces is again most simply discussed by considering two interacting atoms. Quantum mechanics indicates that the rapid motion of the electrons causes instantaneous fluctuations in the charge density around the nucleus. For atoms far apart the electrons in one atom move independently of electrons in the other atom, and on the average the charge distribution is symmetric as shown in Fig. 2*a*. At distances where attractive forces become important, the average charge distribution is still symmetric. However, an instantaneous fluctuation in the electron distribution in one atom can now affect its neighbor nearby. A charge separation in one atom occurs when the electron cloud shifts toward one side of the atom, barring its nucleus to a slight extent. In the other atom the electrons have moved in concert toward this barred nucleus, and an electrostatic attraction is set up. This is illustrated schematically in an exaggerated fashion

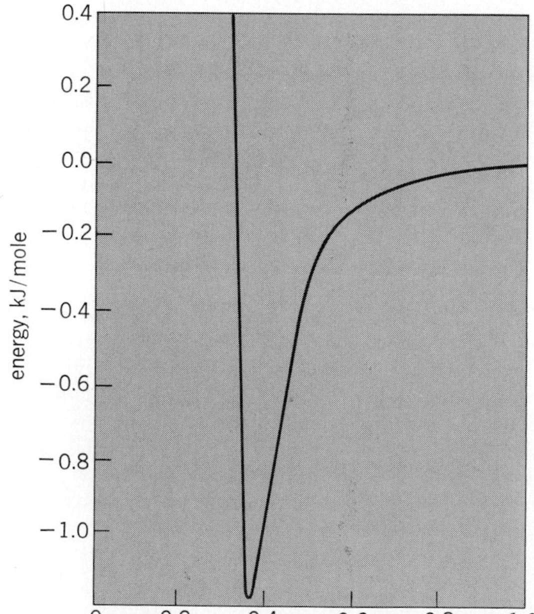

Fig. 1. The intermolecular potential energy of two argon atoms. (*From J. M. Parson, P. E. Siska, and Y. T. Lee, Intermolecular potentials from cross-beam differential elastic scattering measurements, IV:Ar+Ar, J. Chem. Phys., 56: 1511–1516, 1972.*)

in Fig. 2*b*. At another instant the electron clouds may shift in the opposite direction, and the other atom has its nuclear charge partially exposed to the electrons of its neighbor. The electron motions in both atoms are correlated so that an attractive electrostatic force is maintained while the averaged motions assure a symmetric distribution about each atom. These attractive forces are often called London or dispersion forces.

At small separations the electron clouds can overlap, and repulsive forces are set up. These are called Pauli or exchange forces and are also explained by quantum mechanics. They are essentially a consequence of the reluctance of electrons to be confined into the same small region of space. Atoms or molecules brought close together will respond to exchange forces by a permanent distortion of their electron distribution as shown in Fig. 2*c*.

All atoms and molecules experience dispersion and exchange forces, which thus are a common component of intermolecular forces. Neutral molecules, in addition, may interact with each other because they possess permanent electrical polarity expressed as a dipole, quadrupole, or higher multipole moments. The electrostatic forces associated with these interacting multipole moments depend on the orientation of the molecules and may be either attractive or repulsive. The corresponding energies are usually somewhat less than dispersion or exchange energies. The dispersion, exchange, and permanent multipole electrostatic forces taken together are usually called van der Waals forces. Energies associated with the formation of hydrogen bonds (that is, between two HF or H_2O molecules) are somewhat larger than van der Waals energies.

Interactions considerably stronger than those just discussed sometimes occur between atoms or molecules. The energies of chemical bond formation are hundreds of times greater than that shown by the intermolecular potential well of Fig. 1. Electrostatic interactions between charged particles are likewise relatively strong. These interactions are usually not classified as intermolecular forces. *See* CHEMICAL BONDING.

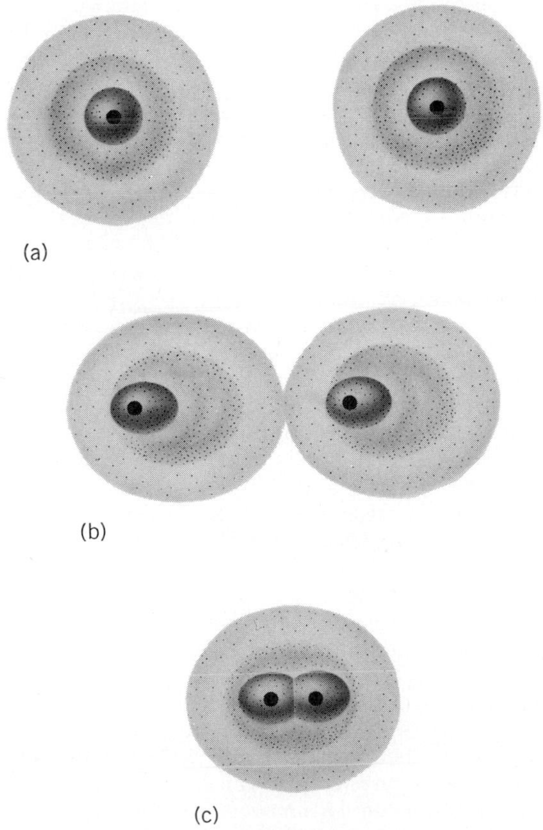

Fig. 2. Schematic illustration of intermolecular interaction. (*a*) There is no interaction between the atoms that are 1 nm or more apart. (*b*) For atoms separated by about 0.8 nm or less, dispersion forces which are attractive result from correlated fluctuations of the electron charge distribution in the atoms. (Distribution shown is greatly exaggerated.) (*c*) For the atoms closer together, 0.3 nm or less, exchange forces which are repulsive cause a permanent distortion of the electron charge distribution. (Distribution shown is greatly exaggerated.)

Occurrence. Intermolecular forces are responsible for many of the bulk properties of matter in all its phases. A realistic description of the relationship among pressure, volume, and temperature of a gas must include the effects of attractive and repulsive forces between molecules. At increased pressures and sufficiently low temperatures the attractive forces between molecules in the gas will cause it to liquefy. The viscosity, surface tension, and diffusion of liquids are examples of physical properties which are a consequence of intermolecular forces. Repulsive forces prevent the molecules from approaching one another too closely and account for the high compressibility of liquids. Intermolecular forces between near and distant neighbors dictate the ordered molecular arrangements in crystalline solids. These forces also account for the elasticity of solids. A detailed accounting of the intermolecular forces in the condensed phase is complex since it must include the interactions of each molecule with many of its neighbors. Nevertheless, the energy of each pair of atom interactions is approximately described by an intermolecular potential of the sort shown in Fig. 1.

Intermolecular forces are also important between atoms within a molecule. Even for a molecule as small as ethane, CH_3CH_3, they direct important details in the molecular structure. Chemical bonds dictate the arrangement of the hydrogen atoms about each carbon atom as well as the distance between carbon atoms. However, intermolecular forces mold the final structure, which keeps the hydrogen atoms on one CH_3 group staggered with respect to those on the other CH_3 group. Thus the staggered rather than the eclipsed configuration for ethane as shown in Fig. 3 is the most stable three-dimensional structure. In an analo-

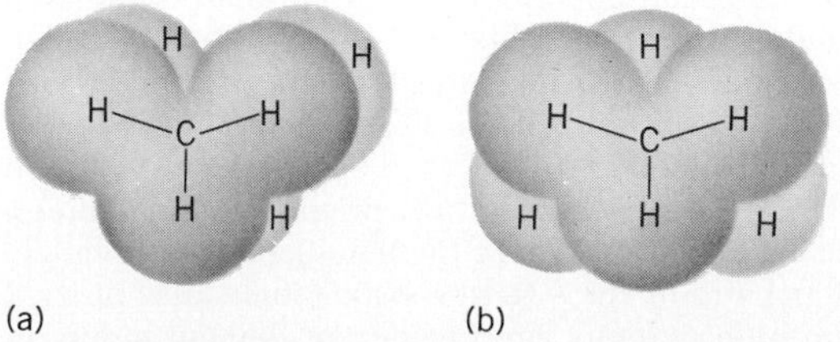

Fig. 3. Intermolecular forces and ethane. (*a*) Eclipsed configuration. (*b*) Staggered configuration.

gous way for larger molecules, proteins, and other biological molecules, the complex spatial arrangement assumed is determined in part by the balance of attractive and repulsive intermolecular forces between atoms that are chemically bonded within the molecule. *See* PROTEINS.

Atoms and molecules may be held to a solid surface by intermolecular or van der Waals forces. This weak bonding, called physisorption, has many important applications. The trapping out of molecules from the gas phase onto cooled surfaces is the basis of pumps for producing vacuums. Undesirable odors or colors in food or water may sometimes be removed by use of filters which capture by physisorption the offending contamination. The selective adsorption of molecules by surfaces is a useful method for separation of mixtures of molecules. *See* ADSORPTION; CHROMATOGRAPHY.

Study methods. The importance of intermolecular forces has been responsible for their extensive study for many decades. In the early 1970s most of the information on intermolecular forces was inferred from the study of matter in bulk. Measurements of the viscosity of gases, or crystal structure of solids, for example, were used to estimate the quantitative nature of the intermolecular interactions that must produce these physical properties. However, it has since been found that studies of individual molecular interactions yield the information more directly.

In molecular beam experiments, low-density streams of atoms or molecules are directed so that individual particles collide. The way in which the molecules rebound as a result of their collision is determined by their initial velocities which can be controlled. Intermolecular forces can be extracted from the experimental data. The intermolecular potential energy curve shown in Fig. 1 was obtained from studies of the collision dynamics of argon atoms. Mappings of the potential energy surfaces of other atoms and molecules are being obtained by this technique.

Another approach is to study van der Waals molecules. In these experiments clusters of atoms or molecules are formed at low temperatures in the gas phase because of their intermolecular attractions for each other. Clusters of two or three atoms or molecules are called van der Waals molecules. For example, gaseous argon at the temperature of the boiling liquid (−186°C) contains about 98% Ar atoms, and the remaining 2% are Ar_2 van der Waals molecules. The ultraviolet spectrum of the gas at low temperatures reveals features due to Ar_2 which can be used to characterize the bond strength and the intermolecular forces between the argon atoms in the van der Waals molecule.

INTERMOLECULAR FORCES

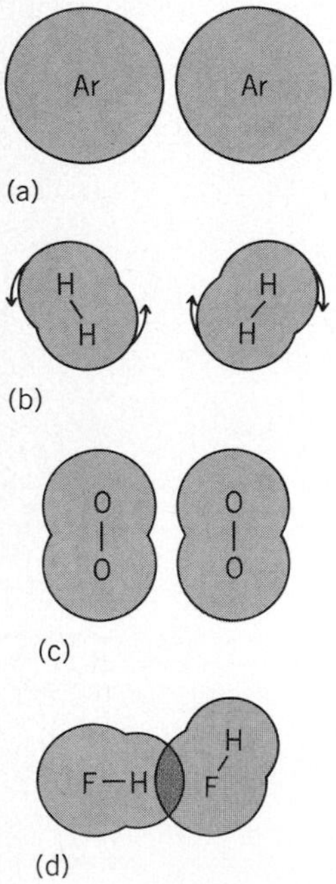

Fig. 4. The structures of some van de Waals molecules. (*a*) Argon. (*b*) Hydrogen. (*c*) Oxygen. (*d*) Hydrogen fluoride. (*From G. Ewing, Structure and properties of van der Waals molecules, Accounts Chem. Res., 8:185–192, 1975*)

Spectroscopy of van der Waals molecules formed by clusters of chemically bonded molecules has also revealed much about intermolecular forces which depend on the orientation of the molecules within the cluster. Gaseous H_2, O_2, or HF contains small concentrations of $(H_2)_2$, $(O_2)_2$, or $(HF)_2$. The structures of these van der Waals molecules are shown in Fig. 4. The chemical bonds in H_2, O_2, or HF are about 0.1 nm long and not affected by the formation of the 0.3–0.4 nm intermolecular bond of the van der Waals molecule. In $(H_2)_2$ the intermolecular forces do not depend much on the orientation of either H_2, and as a consequence each H_2 molecule, while weakly bound to its neighbor, rotates freely within the cluster. The arrows shown in Fig. 4 are meant to represent this freedom of internal rotation. The $(O_2)_2$ van der Waals molecule appears to reside in a rectangular configuration, while $(HF)_2$ exhibits a bent structure characteristic of hydrogen bond formation. While chemical bonds produce rigid molecules with well-defined geometries, intermolecular forces maintain rather floppy structures of the van der Waals molecules. Internal motions in $(O_2)_2$ or $(HF)_2$ produce considerable distortions of the static structure representations in Fig. 4. The structures of several dozen van der Waals molecules are now known. The determination of properties of this new class of compounds promises to provide a deeper insight into the nature of intermolecular forces. *See* MOLECULAR STRUCTURE AND SPECTRA.

Theoretical approaches to intermolecular interactions have taken two directions. Detailed quantum-mechanical calculations have been performed on the interactions of very simple systems, for example, two He atoms. These calculations seek to determine the wave functions, importance of the correlated motions of the electrons, and the precise nature of the energy of the interaction. This theoretical approach then seeks a deeper understanding of the quantum-mechanical origin of intermolecular forces. A more pragmatic approach uses the electron distribution of the isolated molecule from previous calculations. This distribution is treated as an "electron gas" with an associated electric field. It is the response of the interacting molecules to these electric fields that is responsible for intermolecular forces. Calculations of the electron gas model appear to produce reliable intermolecular energies for both interacting atoms and molecules, with a modest amount of computational effort. *See* SOLUTION; VALENCE.

[GEORGE E. EWING]

Bibliography: B. L. Blaney and G. E. Ewing, Van der Waals molecules, *Ann. Rev. Phys. Chem.*, 27:553–586, 1976; S. T. Ceyer and G. A. Somojai, Surface scattering, *Annu. Rev. Phys. Chem.*, 28: 477–499, 1977; J. O. Hirschfelder, C. F. Curtiss, and R. B. Bird, *Molecular Theory of Gases and Liquids*, 1954; T. Kihara, *Intermolecular Forces*, 1978; Y. S. Kim and R. G. Gordon, Unified theory for intermolecular forces between closed shell atoms and ions, *J. Chem. Phys.* 61:1–16, 1974.

Internal energy

A characteristic property of the state of a thermodynamic system, introduced in the first law of thermodynamics. For a static, closed system (no bulk motion, no transfer of matter across its boundaries), the change ΔE in internal energy for a process is equal to the heat Q absorbed by the system from its surroundings minus the work w done by the system on its surroundings. Only a change in internal energy can be measured, not its value for any single state. For a given process, the change in internal energy is fixed by the initial and final states and is independent of the path by which the change in state is accomplished.

The internal energy includes the intrinsic energies of the individual molecules of which the system is composed and contributions from the interactions among them. It does not include contributions from the potential energy or kinetic energy of the system as a whole; these changes must be accounted for explicitly in the treatment of flow systems. Because it is more convenient to use an independent variable (the pressure P for the system instead of its volume V), the working equations of practical thermodynamics are usually written in terms of such functions as the enthalpy $H=E+PV$, instead of the internal energy itself. *See* CHEMICAL THERMODYNAMICS; ENTHALPY.

[PAUL J. BENDER]

Iodate

A negative ion having the formula IO_3^-, and derived from iodic acid, HIO_3. Some salts such as $KH_2I_3O_9$ and $KH(IO_3)_2$ indicate that the acid may exist as polymers of the ion indicated by the empirical formula. Sodium and potassium iodates are the most important salts and are used in medicine.

Iodates occur along with $NaNO_3$ in Chile saltpeter. They are prepared in an electrolytic reaction similar to that for the preparation of chlorates or by oxidation of iodides with chlorine.

The iodates are more stable and are weaker oxidizing agents than bromates and chlorates. *See* BROMATE; CHLORATE; IODINE.

[E. EUGENE WEAVER]

Iodide

A compound which contains the iodine atom in 1− oxidation state and which is derived from hydriodic acid, HI.

The chemistry of iodine and its ability to form covalent and ionic iodides is very similar to the properties described for chloride. *See* CHLORIDE.

In comparing the iodide ion with the other halide ions, it should be pointed out that the iodides are more covalent, are the best reducing agents of the group, and form the least stable complexes. The aqueous solubilities of the metal iodides are much the same as the chlorides but in general a little lower. In organic solvents the order of solubility is frequently reversed. Bismuth and mercuric iodides are only slightly soluble. Iodide ion combines with free iodine to form the triodide ion, I_3^-.

Sodium or potassium iodide is added to table salt to prevent malfunction of the thyroid gland. Silver iodide is used in photographic films and papers.

Iodide can be detected in solution by oxidizing it to the free element with chlorine. It imparts a violet color to the solvent when extracted into carbon tetrachloride. *See* COMPLEX COMPOUNDS; HALIDE; HALOGENATED HYDROCARBON; IODINE.

[E. EUGENE WEAVER]

Iodine

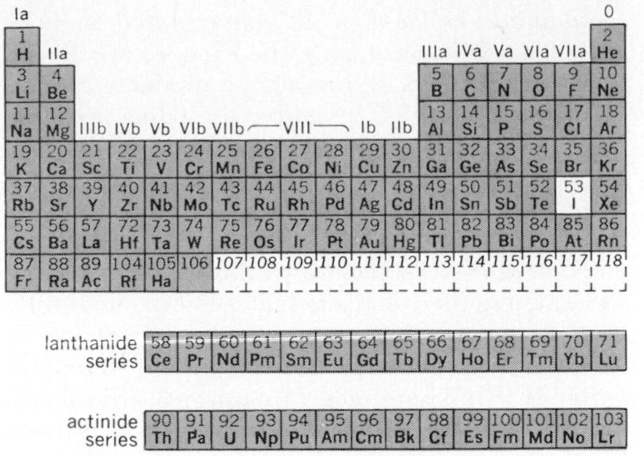

A nonmetallic element, symbol I, atomic number 53, relative atomic mass 126.9045, the heaviest of the naturally occurring halogens, which form group VIIa of the periodic table. Under normal conditions iodine is a black, lustrous, volatile solid; it is named after its violet vapor. *See* HALOGEN ELEMENTS.

The chemistry of iodine, like that of the other halogens, is dominated by the facility with which the atom acquires an electron to form either the iodide ion I^- or a single covalent bond —I, and by the formation, with more electronegative elements, of compounds in which the formal oxidation state of iodine is +1, +3, +5, or +7. Iodine is more electropositive than the other halogens, and its properties are modulated by: the relative weakness of covalent bonds between iodine and more electropositive elements; the large sizes of the iodine atom and iodide ion, which reduce lattice and solvation enthalpies for iodides while increasing the importance of van der Waals forces in iodine compounds; and the relative ease with which iodine is oxidized. Some properties of iodine are listed in the table. *See* ASTATINE; BROMINE; CHLORINE; FLUORINE.

Occurrence, isolation, and production. Iodine occurs widely, although rarely in high concentration and never in elemental form. The average concentration in the Earth's crust is estimated to be 0.3 parts per million (ppm). Despite the low concentration of iodine in sea water (about 0.05 ppm), certain species of seaweed (for example *Laminaria*) can extract and accumulate the element.

In the form of calcium iodate, iodine is found in the caliche beds in Chile, which at one time were the major commercial source of the element; the iodine content is 0.02–1%. In the isolation process, sodium iodate, extracted from the caliche into solution, is treated with the exact amount of sodium bisulfite needed to reduce all of the iodate to iodide, as shown in reaction (1). The resulting

$$2IO_3^- + 6HSO_3^- \rightarrow 2I^- + 6SO_4^{2-} + 6H^+ \qquad (1)$$

acid solution is then mixed with just enough fresh mother liquor to effect precipitation of iodine, according to reaction (2). The product is purified by sublimation.

$$5I^- + IO_3^- + 6H^+ \rightarrow 3I_2 + 3H_2O \qquad (2)$$

Oxidation of iodides is the other way in which iodine is prepared. In 1812 the element was discovered during the treatment of calcined ash of kelp

Some important properties of iodine

Property	Value
Electronic configuration	$[Kr]4d^{10}5s^25p^5$
Relative atomic mass	126.9045
Electronegativity (Pauling scale)	2.66
Electron affinity, eV	3.13
Ionization potential, eV	10.451
Covalent radius, —I, A	1.33
Ionic radius, I^-, A	2.12
Boiling point, °C	184.35
Melting point, °C	113.5
Specific gravity (20/4)	4.940

(containing iodides) with concentrated sulfuric acid (an oxidizing agent). In France, the Soviet Union, and Japan, where alginic products are isolated from seaweed, quantities of iodine are similarly obtained as a valuable by-product. Iodine also occurs as iodide ion in commercially exploitable concentrations (typically 40 ppm) in some oil well brines in California, Michigan, and Japan; the extraction processes generally utilize chlorine as the oxidant, and the volatile iodine is blown out of solution in a stream of air.

Nuclear properties. The sole stable isotope of iodine is ^{127}I (53 protons, 74 neutrons), whose properties allow the investigation of iodine derivatives by nuclear quadrupole resonance spectroscopy and Mössbauer spectroscopy. Of the 22 artificial isotopes (masses between 117 and 139), the most important is ^{131}I, with a half-life of 8 days. It is widely used in radioactive tracer work and certain radiotherapy procedures.

Physical and chemical properties. Iodine exists as diatomic I_2 molecules in solid, liquid, and vapor phases, although at elevated temperatures (> 200°C) dissociation into atoms is appreciable. Short intermolecular I . . . I distances in the crystalline solid indicate strong intermolecular van der Waals forces; consistent with this picture, iodine has a band gap of 1.3 eV (comparable with some forms of phosphorus and selenium, which are semiconductors) and under high pressure exhibits the electrical characteristics of a metal.

Iodine is moderately soluble in nonpolar liquids [for example, CCl_4, aliphatic hydrocarbons], and the violet color of the solutions suggests that I_2 molecules are present, as in iodine vapor; I_4 aggregates have also been detected in some concentrated solutions. Other liquids (aromatic hydrocarbons, alcohols, ethers, amines, organic sulfides) dissolve larger amounts of iodine to give deep red or brown solutions. The increased solubility and change of color result from charge transfer interactions, in which a solvent molecule donates electron density to a vacant orbital of the I_2 molecule. Although the bonding involved is rather feeble, many such crystalline adducts of definite composition have been isolated [for example, $N(CH_3)_3,I_2$], and some complexes even persist in the vapor phase [such as C_6H_6,I_2]. Charge transfer complexes are often important intermediates in halogenation reactions.

The reactions of iodine in water are usually rapid and are governed by the reduction potentials given in Fig. 1. Iodine is only slightly soluble in water, and the brown solution is acid because of a disproportionation reaction, shown in Eq. (3). In

$$I_2 + H_2O = HOI + H^+ + I^- \qquad (3)$$

$$K_{eq} = \frac{[H^+][I^-][HOI]}{[I_2(aq)]} = 2 \times 10^{-13}$$

alkaline solution, extensive conversion of iodine to iodide and hypoiodite ions may be postulated, as in Eq. (4), but the IO^- ion is so unstable as to have

$$I_2 + 2OH^- = I^- + IO^- + H_2O \qquad (4)$$

$$K_{eq} = \frac{[I^-][IO^-]}{[I_2(aq)][OH^-]^2} = 30$$

eluded detection by conventional methods; disproportionation to iodate and iodide, as in Eq. (5), is

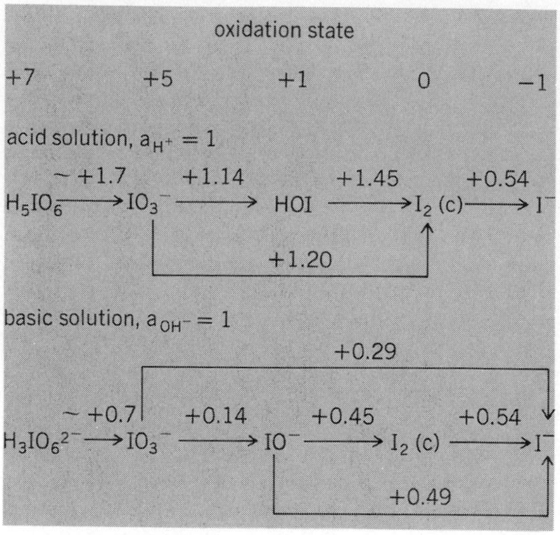

Fig. 1. Standard reduction potentials for iodine species in aqueous solution, E^0 in volts.

$$3IO^- = 2I^- + IO_3^- \qquad (5)$$

$$K_{eq} = \frac{[I^-]^2[IO_3^-]}{[IO^-]^3} = 10^{20}$$

kinetically and thermodynamically very favorable, and the overall alkaline hydrolysis is shown in reaction (6). That the reverse reaction, namely the

$$3I_2 + 6OH^- \rightarrow IO_3^- + 5I^- + 6H_2O \qquad (6)$$

combination of iodate and iodide to form iodine, as in reaction (2), occurs in acid solution demonstrates the importance of pH in aqueous iodine chemistry.

Treatment of iodine in alkaline solution with strong oxidizing agents (for example, hypochlorite) generates iodate, which by an excess of oxidant is converted to the +7 oxidation state as periodate. By contrast, oxidation of iodine in anhydrous, very acidic media (for instance, oleum, fluorsulfuric acid) produces variously the cations IO_2^+, IO^+, I_3^+, and blue paramagnetic I_2^+, all of which have been isolated as salts. While the discrete I^+ cation does not exist, complexes with pyridine and thiourea have been prepared respectively as $(py)_2I^+I^-$ and $[(H_2N)_2CS]_2I^+I^-$; the N−I−N and S−I−S units are linear and symmetrical.

Reduction of iodine (to I^-) in aqueous solution is more difficult than reduction of the other halogens, and strong reducing agents are necessary. Hydrogen sulfide, H_2S, effects the transformation, as does the thiosulfate ion, $S_2O_3^{2-}$, which is oxidized by iodine to tetrathionite $S_4O_6^{2-}$, as in reaction (7);

$$2S_2O_3^{2-} + I_2 \rightarrow S_4O_6^{2-} + 2I^- \qquad (7)$$

this reaction is used in the analytical determination of iodine. Iodide ion is readily oxidized to iodine by moderate oxidants such as bromine; in acid solution it is slowly oxidized by atmospheric oxygen.

Although it is usually less vigorous in its reactions than the other halogens, iodine combines directly with most elements. Important exceptions are the noble gases, carbon, nitrogen, and some noble metals. The readiness with which reaction occurs depends critically on the conditions, especially temperature, phase, and the presence of impurities. Typical organic reactions of iodine include: electrophilic substitution of aromatic compounds to form aryl iodides; iodination of the carbon atom adjacent to a carbonyl function; and addition of I_2 across the multiple bonds of unsaturated hydrocarbons. Hydrogen iodide, a frequent by-product, may inhibit these reactions unless it is continuously removed.

Inorganic compounds. The inorganic derivatives of iodine may be grouped into three classes of compounds: those with more electropositive elements, that is, iodides; those with other halogens; and those with oxygen. The last two classes, especially, show close analogies with the chemistry of neighboring elements tellurium and xenon.

Iodides. The properties of iodides depend on the identity of the more electropositive element. Iodine rarely brings out the highest oxidation state in its partner: thus PI_3 and ReI_4 are known, but not PI_5 or ReI_7. Conversely, low oxidation states may be stabilized as iodides, for example, the +2 oxidation state of the lanthanide elements.

Iodides of the more electropositive metals, groups Ia and IIa, are typical ionic solids with three-dimensional or layer lattices; they dissolve in water with dispersion of iodide ion. Lithium iodide is soluble in ether, a phenomenon often attributed to covalency but actually due to a favorable balance of solvation and lattice energies.

Nonmetals give iodides which exist as covalent molecules in solid, liquid, and vapor, for example, BI_3; similar derivatives are found for metals in high oxidation states, for instance, TiI_4. Typically, they are low-melting solids and are rapidly hydrolyzed by water with evolution of hydrogen iodide, as in reaction (8). The colorless gas HI (bp

$$PI_3 + 3H_2O \rightarrow H_3PO_3 + 3HI \qquad (8)$$

− 35.3°C) can also be prepared by direct combination of the elements over a platinum catalyst; its solution in water (called hydriodic acid) may contain up to 70% by weight of HI.

Iodides of the less electropositive metals, such as HgI_2, or metalloids, such as SbI_3, have layer or chain structures in the solid state, using partially covalent bonds between the atoms. They are usually insoluble in water, but sublime readily on heating and dissolve in organic solvents as molecular species. The decomposition of some iodides into the elements at high temperature has been utilized in the van Arkel process for the production of metals and metalloids of very high purity (for instance, titanium, hafnium, and silicon).

Iodide ion forms complex anions in solution with many metal and metalloid cations, and with other acceptor molecules; the complexes are often deeply colored, and are readily isolated as salts, such as K_2HgI_4, $KAgI_2$, $[(C_6H_5)_4P]_2TeI_6$, and $[(CH_3)_4N]BCl_3I$.

Compounds with other halogens. The simple binary compounds ICl and IBr are low-melting solids with halogenlike properties; at ordinary temperatures they are somewhat dissociated into the parent halogens in the vapor phase. Iodine trichloride crystallizes as a dimer I_2Cl_6, but is apparently completely dissociated (into ICl and Cl_2) in the vapor. Of the fluorides, IF_5 and IF_7 are reactive, moisture-sensitive, thermally stable, and respectively liquid and gas under normal conditions; intermediate fluorides IF and IF_3 disproportionate to iodine and IF_5 well below room temperature, as in reaction (9).

$$5IF \rightarrow 2I_2 + IF_5 \qquad (9)$$

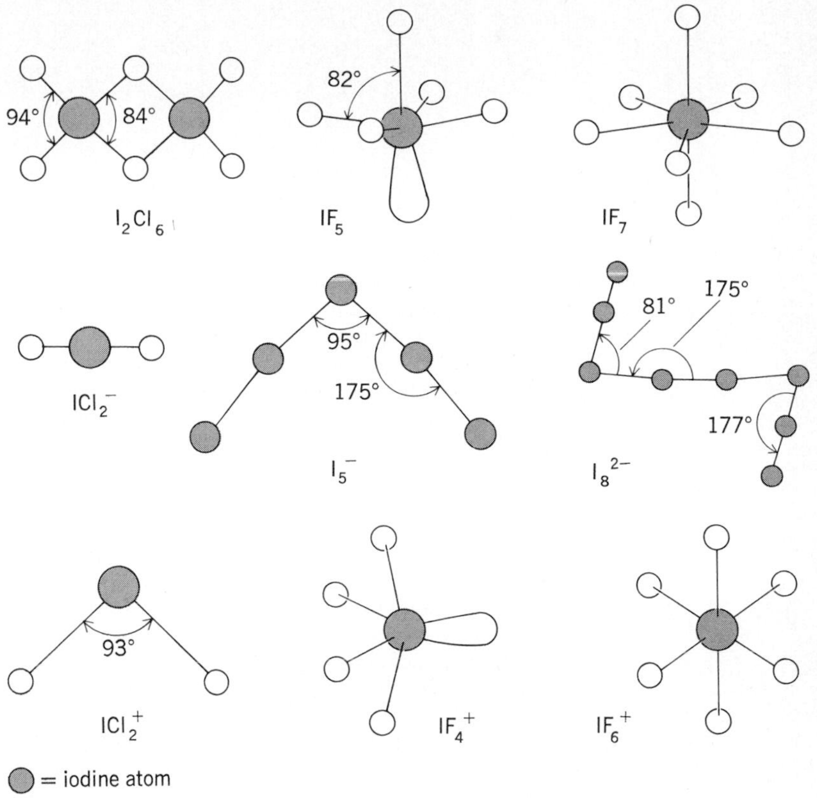

Fig. 2. The shapes of some interhalogen molecules and polyhalide ions.

Most interhalogen compounds react under suitable conditions to gain or lose halide ions, giving a series of anions and cations, shown in reactions (10)−(12). Other species include $ClIBr^-$, ICl_4^-,

$$I_2 + I^- \rightarrow I_3^- \qquad (10)$$

$$IF_7 + NOF \rightarrow [NO]^+[IF_8]^- \qquad (11)$$

$$IF_5 + SbF_5 \rightarrow [IF_4]^+[SbF_6]^- \qquad (12)$$

IF_6^-, ICl_2^+, and IF_6^+. The heaviest halogen always lies at the center of these molecules. Detailed knowledge of their structures, summarized in Fig. 2, has been crucial to the development and refinement of theories of molecular structure and bonding. The prevalence of angles close to 90 and 180° shows the clear relation to noble-gas compounds. *See* MOLECULAR STRUCTURE AND SPECTRA; XENON.

Polyiodide ions such as I_5^- or I_8^{2-}, isolable as black crystalline salts of large cations, are best treated as complexes of I_2 and I^-. A charge transfer complex involving polyiodide anions and starch has an intense blue-black color, and starch-iodide mixtures are used as indicators in titrations involving iodine.

Oxygen compounds. The +5 oxidation state is the most stable. Solid iodic acid contains $(HO)IO_2$ molecules joined by hydrogen bonds. Dehydration of the acid is reversible and occurs in two stages, giving at 100°C a hemihydrate, HI_3O_8, and at 200°C iodine pentoxide, I_2O_5. Strong intermolecu-

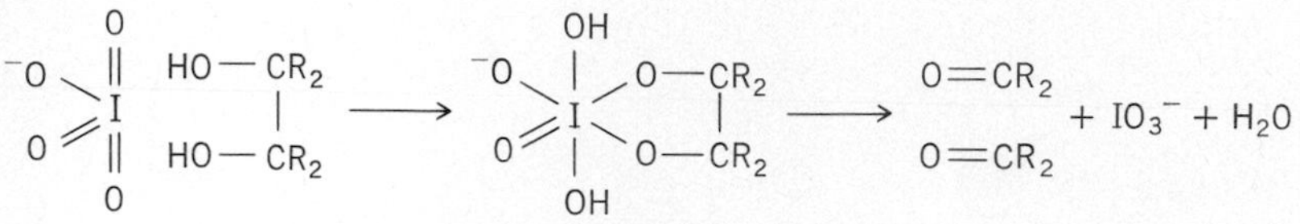

Fig. 3. Mechanism of oxidation of an α-glycol by periodate.

lar I · · O bonding is noted in all these compounds. Acid solutions of iodates are strong oxidizing agents. *See* IODATE.

White crystalline periodic acid is a genuine ortho acid, $(HO)_5IO$. It is a weak acid, and the anions present in its solutions include $H_4IO_6^-$, $H_3IO_6^{2-}$, $H_2IO_6^{3-}$, $H_2I_2O_{10}^{4-}$, and IO_4^-; the latter is found only in strong alkali. Salts of all these anions have been isolated, as well as derivatives such as Ag_3IO_5, Na_5IO_6, and $K_4I_2O_9$. The adoption of six-coordination about iodine in periodates is attributable to the large size of iodine, which allows high coordination numbers.

The periodates are powerful oxidizers in acid solution, converting manganese(II) to permanganate; the cleavage of 1,2-diols by periodic acid (Fig. 3) is stereospecific to cis units, and is a valuable tool in the chemistry of carbohydrates and nucleic acids. Aqueous solutions of periodic acid slowly evolve oxygen and ozone, while the salts lose oxygen on heating to form iodates or, occasionally, novel paramagnetic derivatives of iodine(VI). *See* PERIODATE.

The intermediate acids HIO and HIO_2 are very unstable, but two lower oxides are known: I_2O_4 is characterized as iodosyl iodate $IO^+IO_3^-$, and I_4O_9 as $I(IO_3)_3$. Several unstable covalent compounds have been prepared, as in reactions (13) and (14),

$$I_2 + S_2O_6F_2 \rightarrow 2IOSO_2F \qquad (13)$$

$$I_2 + 6ClOClO_3 \rightarrow 2I(OClO_3)_3 + 3Cl_2 \qquad (14)$$

which are formally iodine(I) or iodine(III) esters of strong oxyacids, and include a nitrate, $IONO_2$, a perchlorate, $I(OClO_3)_3$, and fluorosulfates $IOSO_2F$ and $I(OSO_2F)_3$. The relation to interhalogen compounds is clearly shown by the square planar IO_4 unit of the $[I(OClO_3)_4]^-$ anion (compare ICl_4^-). Iodine triacetate, $I(OCOCH_3)_3$, made from iodine, acetic acid, and fuming nitric acid as oxidant, undergoes electrolysis with the generation of iodine (trapped as silver iodide) at a silvered platinum cathode.

Organic compounds. Organoiodine compounds fall into two categories: the iodides; and the derivatives in which iodine is in a formal positive oxidation state by virtue of bonding to another, more electronegative element. The chemistry of all these compounds is summarized in Fig. 4.

Iodides. The simple organoiodides resemble the other halides in their properties. Because the C-I bond is the weakest of the carbon-halogen bonds, organoiodides are the least stable and most reactive of the organohalides: they are thus useful intermediates. The large mass of the iodine atom makes iodides relatively dense and involatile.

Iodoalkanes are made from the corresponding alcohol with either HI or PI_3. Iodomethane, so prepared, is a colorless liquid [bp 42.5°C, *d*(20/4) 2.28] which decomposes in daylight to iodine and ethane; it reacts (Fig. 4*a*) with, for example, primary amines, dialkyl sulfides, and magnesium (yielding a Grignard reagent). Iodoarenes may be made from the hydrocarbon, iodine, and nitric acid. *See* GRIGNARD REACTION; HALOGENATED HYDROCARBON; HALOGENATION.

Positive oxidation states. An extensive chemistry of such compounds exists for iodine (Fig. 4*b*) but not for chlorine or bromine. Even for iodine, alkyl derivatives are unstable; an electronegative organic group (CF_3, C_6H_5) is necessary for a stable compound.

A simple alkyl iodine(III) compound, dimethyliodonium(III) hexafluoroantimonate, has been prepared from iodomethane and hexafluoroantimonic acid. It is a potent methylating agent and unstable except in very electrophilic media. The related diphenyliodonium cation, $(C_6H_5)_2I^+$, is rather stable, and many of its salts may be made by double decomposition reactions; the hydroxide is a strong base and ionic, but the chloride, $(C_6H_5)_2ICl$, crystallizes as a chlorine-bridged dimer, resembling I_2Cl_6.

The iodoso compounds RIO and iodoxy compounds RIO_2 are white solids, sparingly soluble in water, and powerful oxidizing agents which explode on heating. Iododichlorides $RICl_2$ have found some application as chlorinating agents. The CIF_4 skeleton of $C_6H_5IF_4$ is shaped like IF_5, but with a carbon atom in the unique axial position.

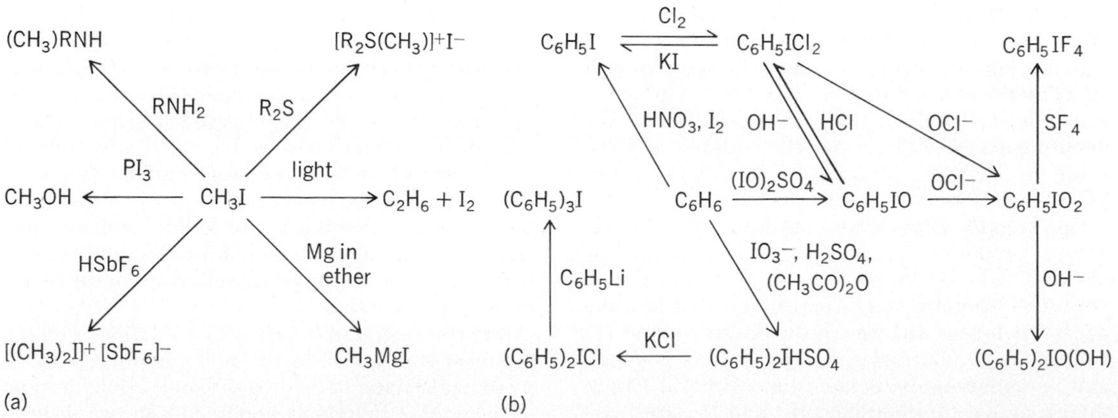

Fig. 4. The chemistry of the organoiodine compounds. (*a*) Reactions of iodomethane. (*b*) Reactions of those organoiodine compounds in which iodine is in a positive oxidation state.

Biological importance. Iodine appears to be a trace element essential to animal and vegetable life. Iodide and iodate in sea water enter into the metabolic cycle of most marine flora and fauna, while in the higher mammals iodine is concentrated in the thyroid gland, being converted there to iodinated amino acids (chiefly thyroxine and iodotyrosines). They are stored in the thyroid as thyroglobulin, and thyroxine is apparently secreted by the gland. Iodine deficiency in mammals leads to goiter, a condition in which the thyroid gland becomes enlarged.

Uses. The bactericidal properties of iodine and its compounds bolster their major uses, whether for treatment of wounds or sterilization of drinking water. Also, iodine compounds are used to treat certain thyroid and heart conditions, as a dietary supplement (in the form of iodized salt), and for x-ray contrast media.

Major industrial uses are in photography, where silver iodide is a constituent of fast photographic film emulsions, and in the dye industry, where iodine-containing dyes are produced for food processing and for color photography. Two important catalysts are NiI_2 (for the addition of CO to organic compounds) and TiI_4 (for the production of stereospecific polymers). *See* DYE.

[CHRIS ADAMS]

Bibliography: F. A. Cotton and G. Wilkinson, *Advanced Inorganic Chemistry*, 3d ed., 1972; A. J. Downs and C. J. Adams, *The Chemistry of Chlorine, Bromine, Iodine, and Astatine*, 1975.

Iodoform

A yellow, hexagonal solid with a penetrating odor also called triiodomethane, CHI_3. Its specific gravity is 4.08 and melting point 119°C. It is prepared by the action of iodine in a basic solution, NaOI, on ethanol or acetone or by the electrolysis of an alkaline I_2-KI solution in the presence of ethanol or acetone. It serves as a qualitative test for the groups

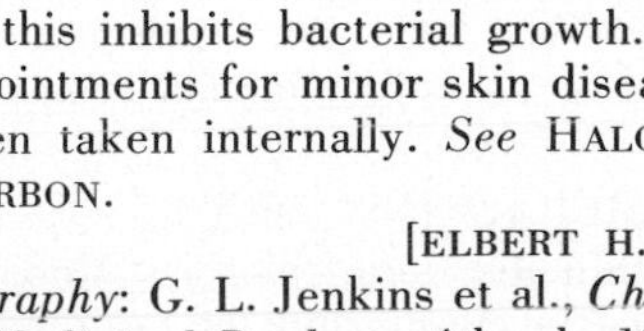

It is soluble in organic solvents and insoluble in water. It has weak bactericidal properties and exerts antiseptic action when applied to raw wounds, because of the liberation of free iodine. Iodoform also acts as an inhibitor for wound secretion, and this inhibits bacterial growth. Its chief use is in ointments for minor skin diseases. It is toxic when taken internally. *See* HALOGENATED HYDROCARBON.

[ELBERT H. HADLEY]

Bibliography: G. L. Jenkins et al., *Chemistry of Organic Medicinal Products*, 4th ed., 1957; C. R. Noller, *Textbook of Organic Chemistry*, 3d ed., 1966.

Ion

An atom, or group of atoms, which by loss or gain of one or more electrons has acquired an electric charge. If the ion is formed from an atom of hydrogen or an atom of a metal, it is usually positively charged; if the ion is formed from an atom of a nonmetal or from a group of atoms, it is usually negatively charged. The number of electronic charges carried by an ion is called its electrovalence. The charges are denoted by superscripts which give their sign and number; for example, a sodium ion, which carries one positive charge, is denoted by Na^+; a sulfate ion, which carries two negative charges, by SO_4^{--}. *See* CHEMICAL BONDING.

Salts are usually composed of orderly arrangements of ions which are not free to move easily in the solid. However, when the salt is fused or dissolved in water, the ions become free, and when an electric field is applied to the salt in solution, the positively charged cations move to the cathode and the negatively charged anions move to the anode. At the electrodes the ions lose their electric charge. This process is called electrolysis. *See* ELECTROLYSIS; SALT; VALENCE.

[THOMAS C. WADDINGTON]

Ion exchange

The reversible exchange of ions of the same charge sign between a solution, usually aqueous, and an insoluble solid in contact with it. The phenomenon was first recognized in soils, where it accounts for the absorption and retention of water-soluble fertilizers. When potash (potassium chloride) is applied to soil, potassium ions are taken into the soil, while chemically equivalent amounts of sodium and calcium ions are released. If the potassium-loaded soil is now washed with a solution of common salt (sodium chloride), some of the potassium is released, while sodium takes its place. Water alone will not remove potassium ions from the soil, because there are no negative ions in the water to balance the charges. Chemically, the exchange is represented by reactions (1) or (2).

$$K^+ + Na\cdot soil \longrightarrow Na^+ + K\cdot soil \quad (1)$$

$$2K^+ + Ca\cdot soil \longrightarrow Ca^{2+} + 2K\cdot soil \quad (2)$$

A solid ion exchanger has an extended, open molecular framework that includes electrically charged, ionic groups. A cation exchanger exchanges positive ions and therefore has negative ions built into its framework. An anion exchanger has positive ions in its framework. The ions of the lattice are called the fixed ions; the smaller ions of opposite charge that can change places with ions in the solution are called counterions; small ions having the same charge as the fixed ions, which may enter the exchanger under conditions to be discussed below, are called co-ions.

Sometimes exchange occurs at the surface, as in glass and some clays, but more commonly the exchanger has a porous structure that may be crystalline, with channels, cavities, or layered spaces that let ions move in and out, or may be amorphous, like a cross-linked organic polymer. Solvent molecules enter these structures when the exchanger is in contact with solutions.

MATERIALS

Ion exchange is common in nature. Many synthetic ion exchangers have been made which have important uses in chemical processing and chemical analysis.

Ion-exchange resins. These materials have the widest range of practical use as ion exchangers. They are organic polymers or condensation products. An important class is made by copolymerizing styrene, $C_6H_5CH{=}CH_2$, with divinylbenzene, $H_2C{=}CH \cdot C_6H_4 \cdot CH{=}CH_2$, to give the cross-linked three-dimensional network shown below. After the polymer is formed, the ionic groups are intro-

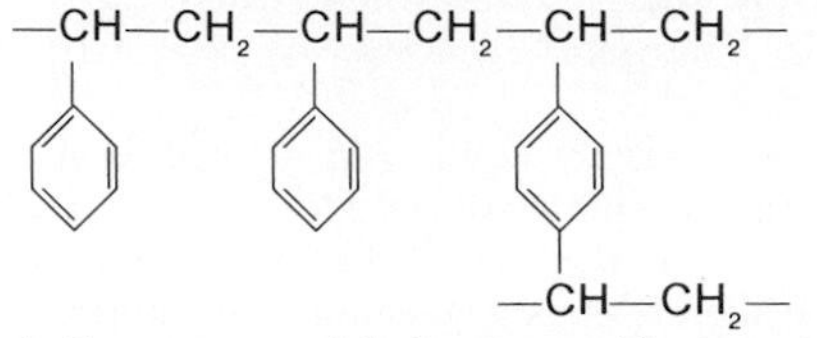

duced. Treatment with fuming sulfuric acid introduces sulfonic acid groups, $-SO_3^-H^+$; chlormethyl ether attaches the groups $-CH_2Cl$, which react with tertiary amines to give $-CH_2{-}NR_3^+Cl^-$, whose hydroxide is a strong base. A great variety of special exchangers with tailor-made functional groups can be made from chlormethylated polystyrene (carrying $-CH_2Cl$). A commercial example is the chelating resin with $-CH_2N(CH_2COOH)_2$ groups. Other exchangers have phosphonate, $-PO_3H_2$, and carboxyl, $-COOH$, groups. Resins with carboxyl groups are made by copolymerizing methyl methacrylate, $CH_3C(COOCH_3){=}CH_2$, with divinylbenzene (cross-linking agent) and then hydrolyzing the product. Resins made by condensation include products from phenol (or substituted phenols, like phenolsulfonic acid) and formaldehyde, and the weak-base resins made from polyamines and epichlorhydrin. Some products, notably the styrene-divinylbenzene polymers, have strong aromatic character, while others are predominantly aliphatic.

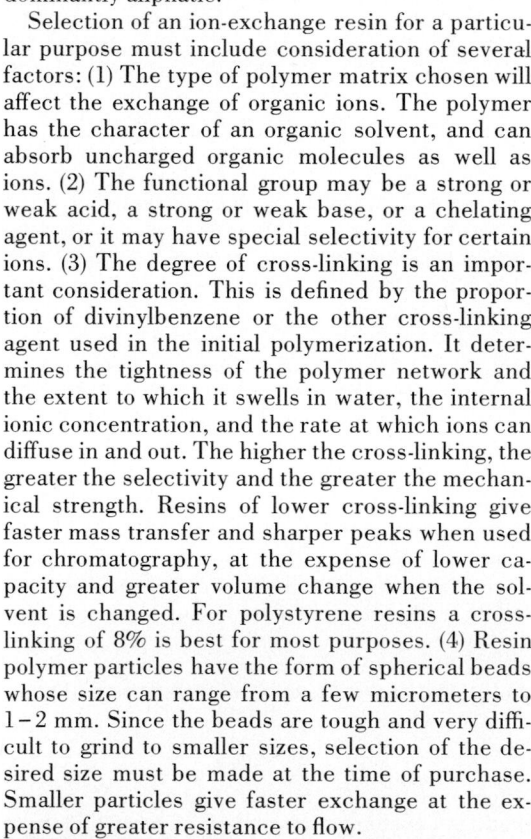

ION EXCHANGE

Fig. 1. Macroporous resin bead.

Selection of an ion-exchange resin for a particular purpose must include consideration of several factors: (1) The type of polymer matrix chosen will affect the exchange of organic ions. The polymer has the character of an organic solvent, and can absorb uncharged organic molecules as well as ions. (2) The functional group may be a strong or weak acid, a strong or weak base, or a chelating agent, or it may have special selectivity for certain ions. (3) The degree of cross-linking is an important consideration. This is defined by the proportion of divinylbenzene or the other cross-linking agent used in the initial polymerization. It determines the tightness of the polymer network and the extent to which it swells in water, the internal ionic concentration, and the rate at which ions can diffuse in and out. The higher the cross-linking, the greater the selectivity and the greater the mechanical strength. Resins of lower cross-linking give faster mass transfer and sharper peaks when used for chromatography, at the expense of lower capacity and greater volume change when the solvent is changed. For polystyrene resins a cross-linking of 8% is best for most purposes. (4) Resin polymer particles have the form of spherical beads whose size can range from a few micrometers to 1–2 mm. Since the beads are tough and very difficult to grind to smaller sizes, selection of the desired size must be made at the time of purchase. Smaller particles give faster exchange at the expense of greater resistance to flow.

Macroporous resins. Also called macroreticular resins, these materials combine fast exchange with low resistance to flow. Seen under the microscope, the beads look rough and opaque, compared with the smooth, transparent beads of ordinary (microporous, homogeneous, or gel-type) polymer-type resins. They are aggregates of very small, highly cross-linked particles some tens of nanometers across, penetrated by channels through which solutions can flow (Fig. 1). Most of the exchange takes place on the surfaces of the tiny microspheres, though some also occurs in depth. Macroporous resins are used extensively in chemical processing. They are robust and rigid, and their volume changes little when one solvent is substituted for another. They are especially suited to nonaqueous solvents.

Other exchangers. Ion exchangers made from cellulose and dextran are used in biochemistry to hold and exchange large ions like those of polypeptides and nucleic acids. Many synthetic inorganic exchangers are available, like the molecular sieves, aluminosilicates that have crystal structures with large cavities holding the exchangeable ions. Molecular sieves are used more as absorbents and catalysts than as ion exchangers. Another class of inorganic ion exchangers is the hydrous oxides of elements of the fourth, fifth, and sixth groups of the periodic table, and combinations of these oxides. Many of these materials show high selectivity because the microcrystalline environment of the fixed ions imposes severe restrictions on the size and charge of the counterions. Thus hydrous antimony pentoxide holds sodium ions very strongly, and is used to remove radioactive ^{24}Na from solutions following irradiation for activation analysis. Hydrous tin dioxide can exchange both anions and cations, and is especially selective for Li and F. The inorganic exchanger that has been most studied is zirconium phosphate, which has high selectivity for Rb, Cs, Sr, and Ba and is used to remove long-lived ^{90}Sr and ^{137}Cs from radioactive wastes. *See* MOLECULAR SIEVE.

Equilibria in ion exchangers. Ion-exchange resins absorb water through the hydration of their ions, and the interior of a microporous or gel-type resin is a concentrated solution, perhaps 5–10 molal. The osmotic pressure of this solution is very large, 100 atm (10 MPa) or more, and it is this pressure that causes the resin to swell. Chromatographic columns are never packed with dry resin; the resin is first stirred with water in an open vessel and then transferred to the column.

Placed in a salt solution, an ion-exchange resin absorbs water but relatively little salt. This fact is a consequence of the Donnan equilibrium, which requires that the chemical potential of a salt or any independent chemical species be the same inside the resin as out. For sodium chloride, using concentrations as an approximation to activities, Eq. (3) applies. (C^2 NaCl represents the square of the

$$[Na^+]_{in}[Cl^-]_{in} = [Na^+]_{out}[Cl^-]_{out} = C^2_{NaCl} \qquad (3)$$

concentration of sodium chloride.) More generally, Eq. (4) gives the equilibrium for a salt A_aB_b. Within

$$[A^{+b}]^a_{in}[B^{-a}]^b_{in} = [A^{+b}]^a_{out}[B^{-a}]^b_{out} \qquad (4)$$

the exchanger the concentration of counterions (Na^+ in a sodium-loaded cation exchanger) is very high, and the co-ion concentration $[(Cl^-)_{in}]$ is very small. Salt is excluded from the resin, yet water can penetrate freely. The Donnan equilibrium is the key to the action of ion-exchange membranes.

The exchange of ions is governed by the familiar law of chemical equilibrium. An exchange of ions of equal charge, like $K^+ + NaRes \rightleftharpoons Na^+ + KRes$, has an $\rightleftharpoons$ equilibrium quotient (K_c), expressed in concentrations, given by Eq. (5). Multiplying K_c by

$$K_c = \frac{[\text{KRes}][\text{Na}^+]}{[\text{NaRes}][\text{K}^+]} \quad (5)$$

activity coefficient ratios gives the true or thermodynamic equilibrium constant. It is easy to generalize the constant for ions of unequal charge. An important equilibrium of this type is: $Ca^{2+} + 2NaRes = 2Na^+ + CaRes_2$. It is evident by inspection that this equilibrium is affected by concentration. If water is added to a flask containing calcium and sodium ions and an ion exchanger, calcium ions are displaced from the solution into the exchanger.

In many applications of ion exchange, one of the ions is present in much lower concentration than the other, and it is necessary to determine how the less abundant species is distributed between the exchanger and the solution. A distribution ratio D is defined by Eq. (6). Its value depends on the quo-

$$D = \frac{\text{concentration of ions A in exchanger}}{\text{concentration of ions A in solution}} \quad (6)$$

tient K_c and on the concentration of the more abundant species B. The units of D are usually (milliliters of solution) (grams of resin)$^{-1}$. *See* DONNAN EQUILIBRIUM.

Selectivity. For the alkali and alkaline-earth metal ions occurring in the common polystyrenesulfonate resins, the selectivity orders are: Li<Na<K<Rb<Cs and Mg<Ca<Sr<Ba. Cesium is held four times as strongly as lithium in an 8% cross-linked resin. The ions held most weakly, Li^+ and Mg^{2+}, are the most hydrated; entering the resin, they cause the greatest swelling. The work of swelling adds to the free energy of exchange and lowers the equilibrium constant.

In anion exchange the halide ions show a parallel selectivity order: F<Cl<Br<I. Here, however, iodide is held 100 times as strongly as fluoride in an 8% cross-linked resin. Hydration is a minor factor with anions, compared to cations; the bare anions are larger, and it seems that a major factor in anion-exchange selectivity is the disturbance of the hydrogen-bonded structure of water. Larger ions disturb the water structure more and are displaced from the water into the resin. Among the oxy-anions, selectivity orders are $NO_2^- < NO_3^-$, $ClO^- < ClO_3^- < ClO_4^-$, $SO_3^{2-} < SO_4^{2-}$, and so on.

Hydrogen and hydroxyl ions are held weakly or strongly, depending on the acid or base strength of the resin's functional groups. Two types of strong-base anion exchanger are in common use. Type I has functional groups $-N(CH_3)_3^+$, and its hydroxide is an extremely strong base, so that hydroxyl ions are held more weakly than fluoride; type II has groups $-N(CH_3)_2C_2H_4OH^+$, and its hydroxide is a weaker base; hydroxyl ions are held nearly as strongly as chloride.

Metal-ion selectivities are enhanced by forming complex ions in solution. A powerful way to separate metal ions is by anion exchange in hydrochloric acid solutions. Most metals form negatively charged chloride complexes which are absorbed by anion-exchange resins. An example is iron(III), whose chloride complex, $FeCl_4^-$, is absorbed from 9 M hydrochloric acid with $D = 10^4$. Below 1 M hydrochloric acid, the complex dissociates, and iron is stripped from the resin. Distribution ratios vary enormously from metal to metal, and Al, Na, Ca, and so on are not absorbed. Two factors affect distribution: stability of the complexes in solution and binding of the complex ions by the resin.

Another way to get selectivity is to use an exchanger with selective functional groups. Iminodiacetate chelating resins were mentioned above; other special chelating groups can be grafted onto resins to make selective absorbents. Sugar-alcohol groups attached to a macroporous polymer give an absorbent selective for borate ions.

APPLICATIONS

Ion exchange has numerous applications for industry and for laboratory research.

Water softening and deionization. Water conditioning provides the largest market for ion exchange. Natural water from rivers and wells is never pure; it is usually hard; that is, it contains calcium and magnesium salts that form curds with soap and leave hard crusts in pipes and boilers. Hard water is softened by contact with a cation exchanger that carries sodium ions. Calcium and magnesium ions enter the exchanger and displace sodium ions, which do not cause hardness.

The exchanger is packed into a bed, or column, through which the water is passed (Fig. 2). In this way the exchange, which is reversible, is forced to completion in one direction. Water flowing down the bed constantly meets fresh, unreacted exchanger that absorbs even low concentrations of calcium ions, so that the water leaving the bed is virtually free from calcium. Eventually the bed becomes saturated, and magnesium and calcium ions break through. Flow is then stopped, and the bed is regenerated by passing concentrated sodium chloride brine. The ion exchange is reversed, calcium and magnesium ions are driven out of the bed, and sodium ions take their place. After rinsing, the bed is ready to soften more water.

To deionize water, that is, to remove all its dissolved salts, two resins are necessary: a strong-acid cation exchanger carrying hydrogen ions and a strong-base anion exchanger carrying hydroxyl ions. First the water passes through the cation

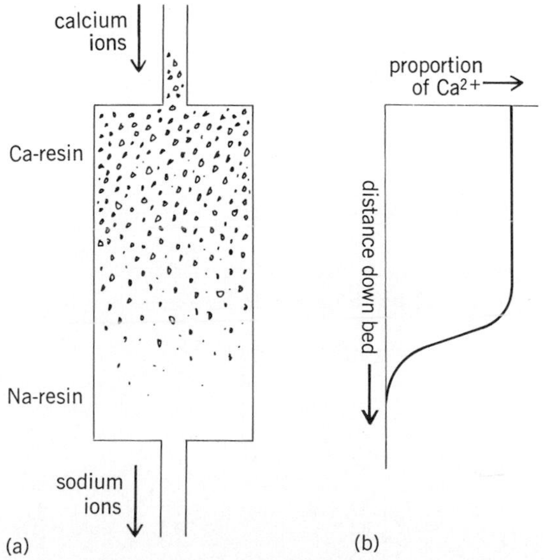

Fig. 2. Water softening by resin bed. (*a*) Ion-exchange bed. (*b*) Graph showing change in calcium ion concentration as water goes down the column.

exchanger (Fig. 3). All its cations are replaced by hydrogen ions, so that the water that emerges carries a mixture of acids. Bicarbonate ions give carbonic acid, which is partially vented as carbon dioxide gas. The acid mixture then enters the bed of anion exchanger, whose hydroxyl ions combine with the hydrogen ions of the acids to produce water, while the anions of the acids are retained. Ion-free water leaves the bed.

When the resins are exhausted, the cation exchanger is regenerated with dilute sulfuric acid, and the anion exchanger with sodium hydroxide. Type II anion exchangers are easier to regenerate than type I, but the stronger type I bases are needed to remove very weak acids, like silicic.

Instead of using two separate beds, the anion and cation exchange resins can be mixed and put in the same column. The mixed bed gives water of higher purity than the separate beds. However, regeneration is more difficult. The two kinds of resin must be separated by an upward current of water; this floats the less dense anion exchanger and leaves the cation exchanger below. Then, by a suitable piping arrangement, the two resins are regenerated and rinsed separately, and mixed by an air jet for reuse.

Deionization by resin beds is practical for waters of low salt content, but not for highly saline waters like sea water. There distillation must be used, or electrolytic desalting with ion-exchange membranes. *See* WATER SOFTENING.

Sugar refining. Ion exchange removes salts but not nonelectrolytes. Salts present in raw sugar juices interfere with the crystallization of sugar. They are removed by ion exchange. Macroporous resins are preferred for this application because they foul less easily and react faster with viscous solutions than the common resins.

Waste treatment (hydrometallurgy). Toxic ions like Cu^{2+}, Pb^{2+}, CrO_4^{2-}, and $Fe(CN)_6^{4-}$ are removed by ion exchange from industrial waste waters and mine drainage, preventing pollution and recovering valuable metals in useful form. Special

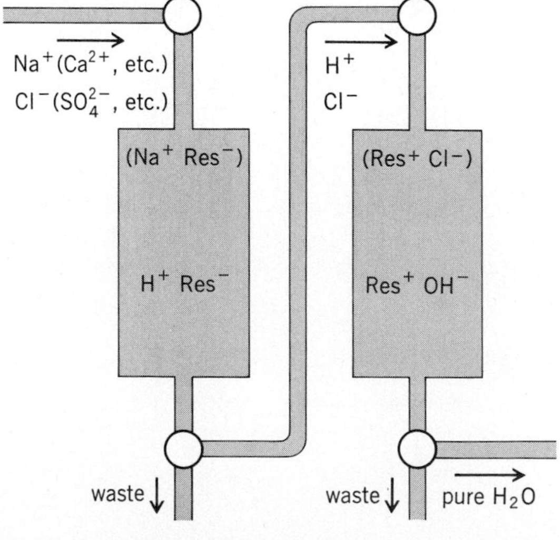

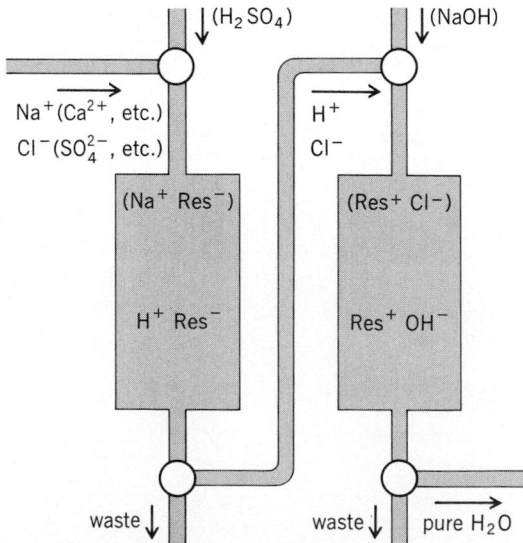

Fig. 3. Two-stage deionization by ion exchange. Valves are for regeneration with H_2SO_4 and NaOH.

selective resins are often used. Uranium is recovered from low-grade ores by leaching with dilute sulfuric acid, then absorbing the uranyl sulfate complex anions on a strong-base anion-exchange resin, which has a high affinity for uranium sulfate; it is so high, in fact, that a batch process may be used. Batch operation avoids having columns become clogged with fine suspended matter. A variant on this process uses sodium carbonate to leach the ore; complex carbonate anions are absorbed by the resin. In yet another development, sodium carbonate solutions are pumped underground to extract uranium in place and bring it to the surface.

Chemical analysis. Ion exchange is used very widely on the laboratory scale for analytical purposes. Many useful separations can be made in short resin columns (5–15 cm) with gravity flow. Trace metals can be collected from large volumes of water, interfering anions and cations can be separated, and metals can be separated through anion exchange of their chloride complexes. This last operation is often used in conjunction with activation analysis. Separation factors are high, and separations are fast and easy. Ion exchange columns can be made very small and are ideally suited for radiochemical use; indeed, a classic feat of ion exchange was the isolation and recognition of five atoms of the element mendelevium. With long resin columns and a succession of eluents (extracting solutions), using simple gravity flow, complex mixtures of elements obtained from rocks and minerals can be separated into individual fractions, each containing one element in concentrated form, which can then be measured by any method one chooses. Such separation schemes are slow and tedious but reliable, and serve as reference methods to check the validity of fast routine methods of analyzing complex materials. *See* ACTIVATION ANALYSIS.

Ion-exchange chromatography. In modern high-performance liquid chromatography, separation is coordinated with measurement. The columns are efficient separators because they are packed with small particles around 10 μm in diameter. Because the particles are small, liquids must be forced through the columns under pressure. At the column exit there is a detector that continuously monitors the concentrations of the separated substances as they emerge.

Two factors have limited the role of ion exchange as a separating mechanism in liquid chromatography. One is the physical properties of the resins: their softness and the slow diffusion of ions in and out. The other is the lack of sensitive detectors for inorganic ions. In spite of these limitations, cation-exchange resin columns have been used for years to analyze mixtures of amino acids. The acids are bound to the resin as their cations and eluted by a series of buffers of increasing pH. They are detected by pumping in a reagent, ninhydrin, that forms colored compounds with amino acids, then passing the liquid through a photometric detector. By present standards the operation is rather slow, taking some 4–6 h to measure the 20 common amino acids.

Ion chromatography has made possible the rapid routine analysis of inorganic ions. It has two distinctive features: first, a kind of resin that allows

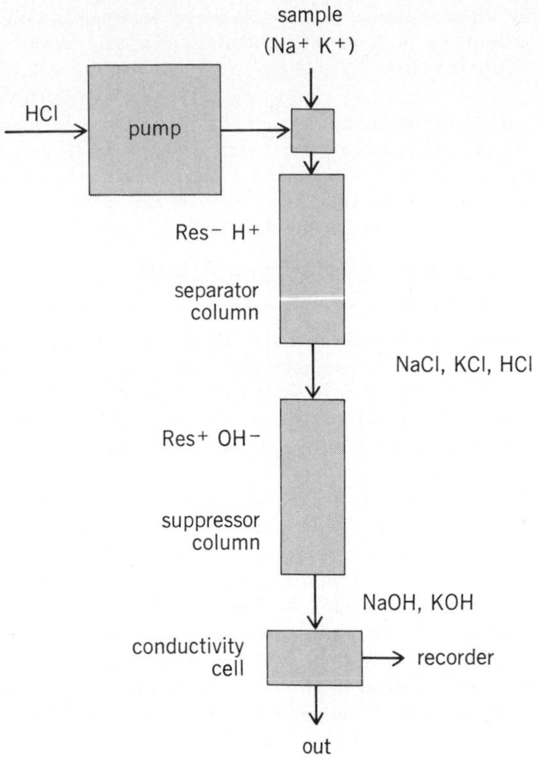

Fig. 4. Ion chromatography process.

rapid ion exchange, and second, a suppressor column that removes excess eluent and makes it possible to use electrical conductivity for detection. The resins are hard polystyrene beads some 20 μm across, coated with a surface layer some 1−2 μm thick that has the ion-exchanging groups. They give fast exchange at the expense of low capacity. The separator column (Fig. 4) is packed with one of these resins; the suppressor column contains an ordinary high-capacity resin. When a mixture of anions is to be analyzed, a sample (250 μl) is introduced into the separator column, which is packed with an anion-exchange resin. It is carried in a flowing solution of sodium carbonate and bicarbonate, 0.003 M in each salt. The anions leave this column as their sodium salts, accompanied by excess eluent. The solution now enters the separator column, which is packed with a strong-acid cation-exchange resin in its hydrogen form. The eluent ions are converted to carbonic acid, which has a low electrical conductivity; the anions to be analyzed give their acids, such as sulfuric and hydrochloric, which conduct very well. Detection limits for the anions of strong acids are 1 nanomole and less; six anions can be separated in 15 min.

Heavy-metal ions and organic ions can generally be detected by means other than electrical conductivity, and the only problem in their analysis by ion-exchange chromatography is to find an efficient, fast-acting ion exchanger. Pellicular resins, glass spheres coated with a thin layer of ion-exchange resin, were introduced in 1967 and were a landmark in modern liquid chromatography. Ion-exchanging groups are bonded to porous silica in a variety of ways. Reactions (7) show a sequence which gives a fast-acting exchanger of high capacity.

$$
\begin{aligned}
&-\text{Si—OH} + CH_3O \cdot SiR_2 \cdot CH{=}CH_2 \rightarrow \\
&\quad -\text{Si—O—}SiR_2 \cdot CH{=}CH_2 + C_6H_5CH{=}CH_2 \rightarrow \\
&\quad\quad -\text{Si—O—}SiR_2 \cdot (CH_2)_4 \cdot C_6H_5 + H_2SO_4 \rightarrow \\
&\quad\quad\quad -\text{Si—O—}SiR_2 \cdot (CH_2)_4 \cdot C_6H_4SO_3H \qquad (7)
\end{aligned}
$$

Paired-ion chromatography. Another way to achieve rapid ion exchange is by dynamic coating of a porous solid. A solution containing a long-chain colloidal electrolyte like cetyl trimethylammonium bromide is passed through a reversed-phase packing like octadecyl-bonded porous silica. The long hydrocarbon tails of the colloidal electrolyte are absorbed into the hydrocarbon surface, turning it into an anion exchanger. In analyzing a mixture of sulfa drugs, compounds which are weak acids that form anions at pH 7, a buffered solution of pH 7, which is also about 0.001 M in cetyl trimethylammonium bromide, is pumped through the column, and the sample is injected. The drugs are retained as ion pairs on the surface of the packing, in equilibrium with their ions and ion pairs in the solution, and come out of the column at different times, when they are "detected" by their absorption of ultraviolet light. Mass transfer is fast, and peaks are as sharp as in conventional reversed-phase chromatography.

Smaller pairing ions like tetrabutylammonium do not coat the surface, but rather form ion pairs in solution, and these ion pairs are partitioned between the absorbent and the moving liquid.

Ion-exchange membranes. The Donnan equilibrium governs the concentrations of co-ions and counterions within an ion-exchange resin that is in contact with an ionic solution. If the concentration of fixed ions in the resin is high and that of the outside solution is low, the concentration of co-ions is very low and, for simplicity, they may be considered to be excluded from the resin altogether. A

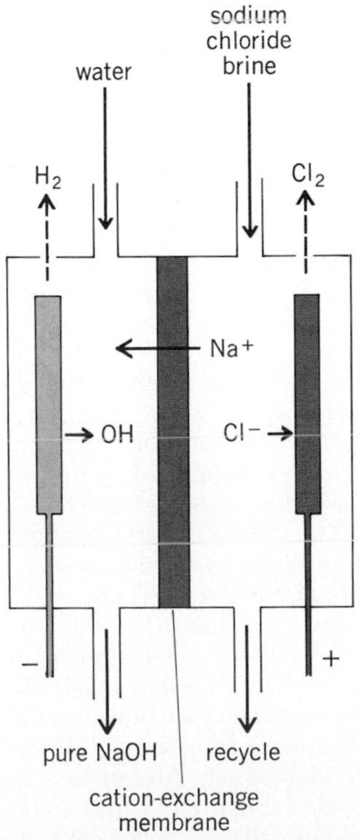

Fig. 5. Alkali-chlorine cell with membrane separator.

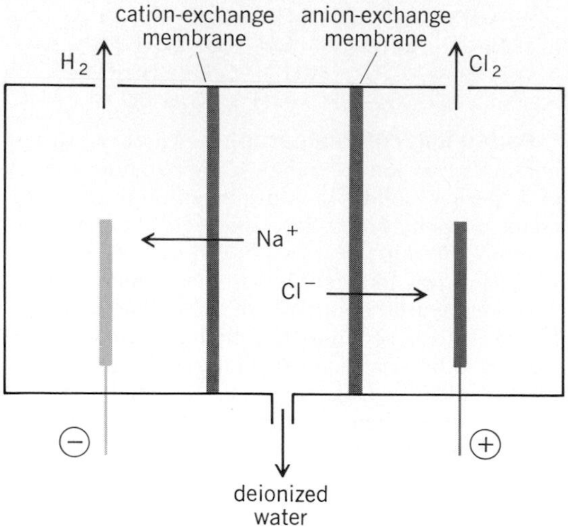

Fig. 6. Ion-exchange membrane cell for desalting by electrodialysis.

membrane made from an ion-exchanging polymer is freely permeable to counterions and to water, but not to co-ions.

Such membranes are used as separators in electrolytic and voltaic cells. Figure 5 shows an alkali-chlorine cell in which the anode and cathode compartments are separated by a cation-exchange membrane. Salt solution is fed into the anode compartment, where chloride ions are discharged to give chlorine gas, and sodium ions cross the membrane to the cathode compartment, taking water with them, but no chloride ions. Hydrogen is formed at the cathode, and pure sodium hydroxide, free from chloride, accumulates in the cathode compartment.

To deionize water by electrolysis, both anion- and cation-exchange membranes are needed. The simplest arrangement is the three-compartment cell shown in Fig. 6. As the current passes, sodium ions leave the middle compartment through the cation-exchange membrane while chloride ions leave through the anion-exchange membrane, until nearly pure water remains in the middle compartment. In practice, many membranes are used, with cation-exchange and anion-exchange membranes alternating. Salt accumulates in alternate compartments, leaving deionized water in the others.

This process (often called electrodialysis) is practical for desalting saline groundwaters but only marginal for sea water. Another way to use ion-exchange membranes in desalting is reverse osmosis. Pressure is applied to the water on one side of the membrane. Water can flow through the membrane, but co-ions cannot, and because, in the absence of an electric current, negative ions cannot flow without positive ions, no salt flows through the membrane. To make the water flow, the applied pressure must exceed the osmotic pressure of the saline water. Reverse osmosis is used widely for purifying water and wastewater, but the membranes used are not the same as the ion-exchange membranes described. Materials like cellulose acetate are used that are cheaper and mechanically stronger. They do, however, carry ion-exchanging groups, and the Donnan equilibrium has a role in their action. *See* DIALYSIS; ION-SELECTIVE MEMBRANES AND ELECTRODES; SALINE WATER RECLAMATION.

[H. F. WALTON]

Bibliography: K. Dorfner, *Ion Exchangers: Properties and Applications*, 2d ed., 1972; F. Helfferich, *Ion Exchange*, 1962; H. F. Walton (ed.), Ion exchange chromatography, *Benchmark Papers in Analytical Chemistry*, vol. 1, 1976.

Ion-selective membranes and electrodes

Membrane-based devices, involving permselective, ion-conducting materials, used for measurement of activities of species in liquids or partial pressures in the gas phase. Permselective means that ions of one sign may enter and pass through a membrane.

Properties. Ion-selective electrodes are generally used in the potentiometric mode, and they superficially resemble the classical redox electrodes of types 0 (inert), 1 (Ag/Ag^+), 2 ($Ag/AgCl/Cl^-$), and 3 ($Pb/PbC_2O_4/CaC_2O_4Ca^{2+}$). The last, while ion-selective, depend on a redox couple (electron exchange) rather than ion exchange as the principal origin of interfacial potential difference. Ion-selective electrodes have the typical shorthand form:

Cu; Ag ;	AgCl ;	Electrolyte; M^+ Cl^-	Membrane permselective to M^+
Lead wire	Inner reference electrode	Inner filling solution	

or Cu; Ag; Membrane permselective to M^+

The former is the ionic-contact "membrane" configuration, and the latter is the "all-solid-state" membrane configuration. In the former, both membrane interfaces are ion-exchange-active, and the potential response depends on M^+ activities in both the test solution and the inner filling solution. In the latter, the membrane must possess sufficient electronic conductivity to provide a reversible, stable electron-exchange potential difference at the inner interface, with ion exchange only at the test solution side.

Potentiometric responses of ion-selective electrodes take the form in Eq. (1) when an ion-selec-

$$V(\text{measured}) = V^0 + \frac{RT}{F} \ln \left[a_i^{1/z_i} + \sum_j (k_{ij}\, a_j)^{1/z_j} \right] \quad (1)$$

R = the universal gas constant
T = the absolute temperature
F = the Faraday constant (96,487 coulombs/equivalent)
V^0 = formal reference potential

tive electrode is used with an external reference electrode, typically a saturated-calomel reference electrode with a salt bridge, to form a complete electrochemical cell. Activities of the principal ion, a_i, and interfering ions, a_j, are in the external "test" solution and correspond to ions M^{z_i} and M^{z_j}, where z_i and z_j are charges with sign. The ion M^{z_j} is written first because it is the principal ion favored in the membrane, for example, high ion-exchange constant and high mobility. The k_{ij}

values are "selectivity coefficients" which are experimentally determined, but can be related to extrathermodynamic quantities such as single-ion partition coefficients and single-ion activity coefficients and mobilities. When only one ion is present in a test solution, or the membrane is ideally permselective for only one ion, this equation simplifies as Eq. (2). V^0 can be written explicitly in terms of

$$V(\text{measured}) = V^0 + \frac{RT}{z_i F}[\ln a_i] \qquad (2)$$

activities of species at the inner interface, or in terms of solid-state activities for the all-solid-state configuration. Equation (1), variously known as the Horovitz, Nicolsky, or Eisenman equation, resembles the Nernst equation, Eq. (2), but the former originates from different factors. Equation (1) cannot be derived from first principles for ions of general charge. However, when $z_i = z_j$, the equation can be derived by various means, including thermodynamic (Scatchard) equations and transport models (Nernst-Planck equations), mainly. The response slope $2.303RT/z_iF$ is $59.14/z_i$ mV/decade of activity at 25°C. Measurements reproducible to +0.6 mV are typically achieved, and activities can be reproducible to ±2% for monovalent ions. A normal slope is considered "nernstian," and can persist over a wide activity range, especially for solid electrodes, for example, 24 decades in the Ag^+,S^{2-} system, and 12 decades of H^+ using Li^+-based glass membrane electrodes. Less than nernstian slopes, and in the limit zero slope, can occur at low activities of sensed species, and can occasionally occur at very high activities. Ultimate low-level response (detection limit) is determined by the solubility of the membrane ion-exchanging material, although impurities may cause premature failure. Because of the logarithmic dependence of potential response on activities, activity measurements using ion-selective electrodes are particularly suited to samples with wide activity variations. Standardization against pure samples or samples of the type to be determined is required. Precise measurements of concentrations over narrow ranges are not favorable, but are possible by elaborate standardization schemes: bracketing, standard additions, and related methods involving sample pretreatment. *See* DONNAN EQUILIBRIUM.

Ion-selective electrodes are most often cylindrical, 6 in. (15 cm) long and 0.25 or 0.5 in. (6 or 13 mm) in diameter, with the lead wire exiting at the top and the membrane sensor at the lower end. However, inverted electrodes for accommodating single-drop samples, and solid electrodes with a drilled hole or a special cap for channeling flowing samples past a supported liquid membrane, are possible configurations. The conventional format is intended for dip measurements with samples large enough to provide space for an external reference electrode. Single combination electrodes are useful for smaller samples, because the external electrode is built-in nearly concentrically about the ion-selective electrode. Drilled and channeled-cap electrodes are intended for use with flowing samples. Microelectrodes with membrane-tip diameters of a few tenths of a micrometer have been constructed for single-cell and other measurements in the living body (see illustration for construction details).

Ion-selective electrodes are intended to be used to monitor and measure activities of flowing, or stirred, solutions because electrodes detect and respond to activities only at their surfaces. The time responses of solid and liquid membrane electrodes to ideal step activity changes of the principal ion (already present in the membrane) can be very rapid: 200 ms for glass to about 30 μs for AgBr. Generally this fast response cannot be observed or used, because sample mixing or diffusion of a fresh sample to an electrode surface is the limiting process. Also, almost any time two ions are simultaneously determining response, interior diffusion potential generation is involved in reaching a new steady-state potential. Similarly, formation of hydrated surface layers or layers of adsorbed matter layers introduces diffusion barriers. Response times from 2 to 20 s can be expected. About 20–30 samples per hour may be analyzed manually by the dip method, and about 60 per hour when samples are injected into a flowing stream of electrolyte. *See* ELECTRODE POTENTIAL; ELECTROLYTIC CONDUCTANCE.

Classification and responsive ions. Ion-selective electrodes are classified mainly according to the physical state of the ion-responsive membrane material, and not with respect to the ions sensed. It has also proved superfluous to distinguish between homogeneous membranes and those that are made from a homogeneous phase supported physically in voids of an inert polymer, or from two homogeneous phases intimately mixed, so-called heterogeneous membranes.

Glass membrane electrodes. Mainly used for hydrogen ion activity measurements, they predate the wider variety of membrane electrodes developed after 1960. Glass electrodes are based on alkali ion silicate compositions. Superior pH-sensing glasses (pH 1 to 13 or 14) result from lithium silicates with addition of di-, tri-, and tetravalent heavy-metal oxides. The latter are not chain formers. Membranes responsive to Na^+, K^+, NH_4^+, and some other cations use additional Al_2O_3 or B_2O_3, or both. The pH glasses are highly selective for H^+ over other monovalent ions. The Na^+-sensing glasses are not intrinsically very selective for Na^+ over H^+, but useful pNa measurements can be made, even in excess K^+ at pH 7 or above. No glasses with high selectivity of K^+ over Na^+ have been found. Chalcogenide glasses containing low contents of Cu^{2+} or Fe^{3+}, while called glasses, are thought to be semiconductor electrodes with a high component of electron exchange, rather than ion exchange, for establishment of interfacial potential responses to Cu^{2+} and Fe^{3+}.

Electrodes based on water-insoluble inorganic salts. These electrodes include sensors for F^-, Cl^-, Br^-, I^-, CN^-, SCN^-, S^{2-}, Ag^+, Cu^{2+}, Cd^{2+}, and Pb^{2+}. The compounds used are silver salts, mercury salts, sulfides of Cu, Pb, and Cd, and rare-earth salts. All of these are so-called white metals whose aqueous cations (except La^{3+}) are labile. The salts themselves are Frenkel-defect solids which possess the necessary ionic conductivity. Ag_i^+ (interstitials) or Ag_v^- (vacancies) are the mobile species in the silver salts, while F^- interstitials are mobile in LaF_3. These materials are ion exchangers, and show no diffusion potential. Single crystals, doped and undoped, may be used as

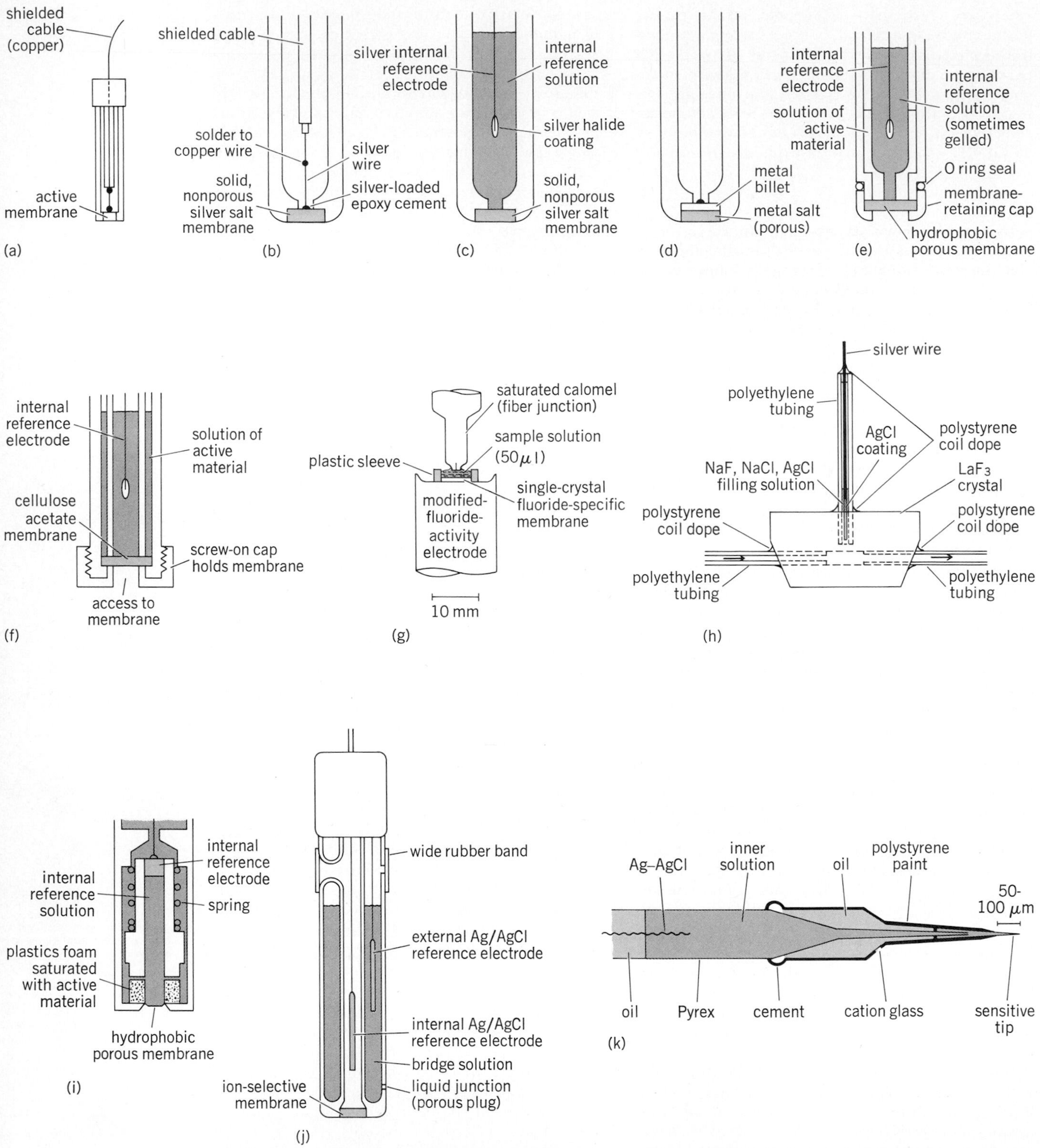

Diagrams of ion-selective electrodes. (*a*) Typical electrode configuration—in this example, an all-solid-state ion-selective electrode. (*b*) Enlarged view of the construction for metal-contacted-membrane ion-selective electrodes. (*c*) Enlarged view of the construction for internal electrolyte-contacted-membrane ion-selective electrodes. (*d*) Enlarged view of the construction for an electrode. (*e–g*) Enlarged views of constructions for liquid ion-exchanger-membrane ion-selective electrodes. (*h*) An inverted electrode microcell using a fluoride-sensing material; the reference electrode (external) is saturated calomel [SCE] (*from R. A. Durst and J. K. Taylor, Anal. Chem., 39:1483, 1967*). (*i*) Construction of a flow-through crystal electrode (*from H. I. Thompson and G. A. Rechnitz, Anal. Chem., 44:300, 1972*). (*j*) A combination electrode illustrating the usual active membrane surrounded by an attached Ag/AgCl external reference electrode. (*k*) An example of a cation-sensing microelectrode used in biological research (*from R. N. Khuri, W. J. Flanagan, and D. E. Oken, J. Appl. Physiol., 21:1568, 1966*).

membranes. Pressed pellets using inert binders such as polyethylene or an insoluble salt such as Ag_2S (for the silver halide electrodes) are popular.

In addition, powdered salts may be suspended in silicone rubber or polyvinyl chloride (about 50:50% by weight) to form heterogeneous flexible membranes. CuS-Ag_2S, CdS-Ag_2S, and PbS-Ag_2S pressed pellets formed at about 250°C are indirectly responsive to the divalent metal ion activities through control of Ag^+ activities at the electrode surface and in leached layers or surface pores by means of the common ion effect.

Electrodes using liquid-ion exchangers. These are electrodes supported in the voids of inert polymers such as cellulose acetate, or in transparent films of polyvinyl chloride, and provide extensive examples of devices for sensing. Fewer-cation-sensing liquid-ion exchanger systems have been found. The principal example (and among the most important) is the Ca^{2+}-responsive electrode based on calcium salts of diesters of oil-soluble phosphonic acids. Anion-sensing electrodes typically use an oil-soluble cation Aliquat (tricapryl methylammonium) or a metal ion-uncharged organic chelating agent (Ni^{2+} or Fe^{2+} phenanthroline or substituted phenanthroline cations) in a support matrix. Sensitivity is virtually assured if the salt is soluble in a mediator solvent, typically a nitro aromatic or esters of difunctional carboxylic acids: adipic, sebacic, or phthalic. Selectivity poses a severe problem since these electrodes, based on hydrophobic materials, tend to respond favorably to many oil-soluble anions. Thus construction of electrodes for the simple inorganic anions F^-, OH^-, HCO_3^-, and HPO_4^{2-} is difficult. Yet many electrodes respond to SCN^-, I^-, Br^-, NO_3^-, ClO_4^-, and BF_4^- in accordance with the Hofmeister lyotropic series. Surfactant anion sensors use salts such as hexadecylpyridinium dodecylsulfate in *o*-dichlorobenzene; surfactant cation sensors use a picrate salt of the species to be measured. Acetylcholine may be measured in the presence of choline, Na^+, and K^+ using the tetra-*p*-chlorophenylborate salt in a phthalate ester in polyvinyl chloride, for example.

Neutral carrier-based sensors for monovalent and divalent cations are closely related to ion-exchanger-based electrodes. Both systems may involve ion-exchange sites, particularly negative mobile sites arising from mediators or negative fixed sites arising from hydrolysis of support materials. All of the available neutral carriers are hydrophobic complex formers with cations, and they may be either cyclic or open-chain species. These compounds permit selective extraction (leading to permselectivity) for ions such as K^+, Na^+, NH_4^+, and Ca^{2+} that would ordinarily not dissolve as simple inorganic salts in the hydrocarbonlike membrane phase. Valinomycin is the best-known example, and its use in supported solvents such as diphenylether and *p*-NO_2 cymene provides an electrode with sensitivity of 10^5 for K^+/Na^+. *See* ION EXCHANGE.

Electrodes with interposed chemical reactions. These electrodes, with chemical reactions between the sample and the sensor surface, permit a new degree of freedom in design of sensors for species which do not directly respond at an electrode surface. Two primary examples are the categories of gas sensors and of electrodes which use enzyme-catalyzed reactions. Gas sensors for CO_2, SO_2, NH_3, H_2S, HCl, and others can be made from electrodes responsive to H^+, S^{2-}, or Cl^-. By enclosing a pH glass membrane in a thin layer of dilute $NaHCO_3$, an electrode for partial pressure of CO_2 is formed, since H^+ increases in a known way with increasing dissolved CO_2. Similarly, immobilized enzymes convert a substrate such as urea or an amino acid to ammonia, which can be sensed and monitored by the underlying electrode. However, increased sensitivity is accompanied by an increased response time. Each diffusion and diffusion-reaction barrier slows the transport and increases the time constant of the overall sensor electrode. *See* CALOMEL ELECTRODE; ELECTRODE.

Applications. Electrodes for species identified above are, for the most part, commercially available. In addition, electrodes have been made and reported that are responsive to many other species. A few of these are: Cs^+, Tl^+, Sr^{2+}, Mg^{2+}, Zn^{2+}, Ni^{2+}, UO_2^{2+}, Hg(II), HSO_3^-, SO_4^{2-}, IO_4^-, ReO_4^-, halide anion complexes of heavy metals (for example, $FeCl_4^-$), pyridinium, pyrocatechol violet, vitamins B_1 and B_6 and many cationic drugs, aromatic sulfonates, salicylate, trifluoroacetate, and many other organic anions. Applications may be batch or continuous. Important batch examples are potentiometric titrations with ion-selective electrode end-point detection, determination of stability constants of complexes and speciation identity, solubility and activity coefficient determinations, and monitoring of reaction kinetics, especially for oscillating reactions. Ion-selective electrodes serve as liquid chromatography detectors and as quality-control monitors in drug manufacture. Applications occur in air and water quality (soil, clay, ore, natural-water, water-treatment, sea-water, and pesticide analyses); medical and clinical laboratories (serum, urine, sweat, gastric-juices, extracellular-fluid, dental-enamel, and milk analyses); and industrial laboratories (heavy-chemical, metallurgical, glass, beverage, and household-product analyses). *See* ANALYTICAL CHEMISTRY; CHROMATOGRAPHY; TITRATION.

[RICHARD P. BUCK]

Bibliography: P. L. Bailey, *Analysis with Ion-Selective Electrodes*, 1976; R. P. Buck, in *Analytical Chemistry Biennial Reviews*, vol. 42, no. 284R, 1970, vol. 44, no. 270R, 1972, vol. 46, no. 28R, 1974, vol. 48, no. 23R, 1976, and vol. 50, no. 17R, 1978; P. W. Cheung et al., *Theory, Design and Applications of Solid State Chemical Sensors*, 1978; A. K. Covington, *Ion-Selective Electrode Methodology*, 2 vols., 1979; H. Freiser (ed.), *Ion-Selective Electrodes in Analytical Chemistry*, 1978; J. Koryta, *Ion-Selective Electrodes*, 1975; N. Lakshminarayanaiah, *Membrane Electrodes*, 1976.

Ionic equilibrium

An equilibrium in a chemical reaction in which at least one ionic species is produced, consumed, or changed from one medium to another.

Types of equilibrium. A few examples can illustrate the wide variety of types of ionic equilibrium which are known.

Dissolution of an unionized substance. The dissolution of hydrogen chloride (a gas) in water (an ionizing solvent) can be used to illustrate this type. Reactions (1), (2), and (3) all represent exactly the

$$HCl(g) \rightleftharpoons H^+ + Cl^- \quad (1)$$

$$HCl(g) + H_2O \rightleftharpoons H_3O^+ + Cl^- \quad (2)$$

$$HCl(g) + 4H_2O \rightleftharpoons H_9O_4^+ + Cl^- \quad (3)$$

same equilibrium. Reaction (1) ignores the hydration of the proton and is preferred for many purposes when the hydration (or solvation) of the proton is irrelevant to a particular discussion. Reaction (2) is written in recognition of the widely held belief that free protons do not exist in aqueous solution. Reaction (3) indicates that another three molecules of water are very firmly bound to the H_3O^+ ion (the hydronium ion). There is no implication in reaction (3), however, that the total number of molecules of water attached to, or weakly affected by, the hydronium ion may not be considerably larger than three.

Not much is known about the solvation of ions, although it has been proved that each chromic ion, Cr^{3+}, in dilute aqueous solution holds at least six water molecules. Only a few other similar data have been clearly established. Hence equations for the other examples cited below are written without regard to solvation, except that the hydrogen ion is usually written in accordance with common practice as H_3O^+. *See* ACID AND BASE.

Dissolution of a crystal in water. The dissociation of solid silver chloride, reaction (4), illus-

$$AgCl(crystal) \rightleftharpoons Ag^+ + Cl^- \quad (4)$$

trates this type of equilibrium. *See* SOLUBILITY PRODUCT CONSTANT.

Dissociation of a strong acid. Nitric acid, HNO_3, dissociates as it dissolves in water, as in reaction (5). At 25°C about one-half the acid is dissociat-

$$HNO_3 + H_2O \rightleftharpoons H_3O^+ + NO_3^- \quad (5)$$

ed in a solution containing 10 (stoichiometric) moles of nitric acid per liter.

Dissociation of an ion in water. The bisulfate ion, HSO_4^-, dissociates in water, as in reaction (6). About one-half the HSO_4^- is dissociated in an

$$HSO_4^- + H_2O \rightleftharpoons H_3O^+ + SO_4^{--} \quad (6)$$

aqueous solution containing about 0.011 mole of sulfuric acid per liter at 25°C.

Dissociation of water itself. In pure water at 25°C the concentration of each ion is about 10^{-7} mole/liter, but increases rapidly as temperature is increased. This equilibrium is represented by reaction (7).

$$2H_2O \rightleftharpoons H_3O^+ + OH^- \quad (7)$$

Formation of a complex ion. In water or in a mixture of fused (chloride) salts, complex ions, such as $ZnCl_4^{--}$, may be formed, as in reaction (8). *See* COMPLEX COMPOUNDS.

$$Zn^{++} + 4Cl^- \rightleftharpoons ZnCl_4^{--} \quad (8)$$

Dissociation of a weak acid. In water acetic acid dissociates to form hydrogen (hydronium) ion and acetate ion, as in reaction (9).

$$CH_3CO_2H + H_2O \rightleftharpoons H_3O^+ + CH_3CO_2^- \quad (9)$$

Electrochemical reaction. Reaction (10) takes place "almost reversibly" when the equilibrium

$$\tfrac{1}{2}H_2(g) + AgCl(s) + H_2O(l) \rightleftharpoons H_3O^+ + Cl^- + Ag(s) \quad (10)$$

shown exists. A small current is allowed to flow through an electric cell consisting of an aqueous solution of HCl saturated with silver chloride, a hydrogen electrode, and a silver electrode. Saturation is maintained by an excess of solid silver chloride which for convenience is sometimes mixed with the silver or plated, as a coating, on the metal. The electrode is then called a silver–silver chloride electrode.

Many additional types of equilibria could be mentioned, including those reactions occurring entirely in the gaseous phase and those reactions occurring between substances dissolved in two immiscible liquids.

Quantitative relationships. Each reaction obeys an equilibrium equation of the type shown as Eq. (11).

$$\frac{[H^+][Cl^-]}{[HCl(g)]}\frac{f_+f_-}{\gamma_g} = Q_cQ_f = K \quad (11)$$

The activity coefficient γ_g can be ignored here because it is very nearly unity. The terms f_+ and f_- are the respective activity coefficients of H^+ and Cl^- but cannot be determined separately. Their product can be determined experimentally and can also be calculated theoretically for very dilute solutions by means of the Debye-Hückel theory of interionic attraction. Because γ_g and f_+f_- are nearly unity, Eq. (11) demands that the pressure of HCl gas above a dilute aqueous solution be proportional to the square of the concentration of the solute. Q_c is called the concentration quotient and Q_f the quotient of activity coefficients.

Similarly, the dissociation of acetic acid obeys Eq. (12). Early work on electrolytes revealed that

$$\frac{[H^+][CH_3CO_2^-]}{[CH_3CO_2H]}\frac{f_+f_-}{f_u} = Q_cQ_f = K \quad (12)$$

Q_c, the concentration quotient, was constant within the limits of accuracy attainable at the time. Later work revealed that the measured concentration quotient Q_c first increases as concentration is increased from very small values and then decreases sharply. The initial increase is due largely to the electrical forces between the ions which reduce the product f_+f_-. There is some evidence that the subsequent decrease in Q_c and the concomitant rise in Q_f are due to removal of some of the monomeric acetic acid by the formation of dimeric acetic acid, $(CH_3CO_2H)_2$. The fact is, however, that knowledge concerning activity coefficients in solutions other than very dilute ones is not yet understood. Even the experimental methods for the measurement of the molecular species involved in some equilibria were not evolved until recently. *See* ELECTROLYTIC CONDUCTANCE; HYDROLYSIS.

[THOMAS F. YOUNG]

Bibliography: G. M. Barrow, *Physical Chemis-*

try, 2d ed., 1966; J. Bjerrum et al., Stability constants of metal-ion complexes, with solubility products of inorganic substances, *Chem. Soc. London*, pts. 1 and 2, 1957–1958; W. M. Latimer, *Oxidation Potentials*, 2d ed., 1952.

Iridium

A chemical element, Ir, atomic number 77, and atomic weight 192.2. Iridium in the free state is a hard, white metallic substance.

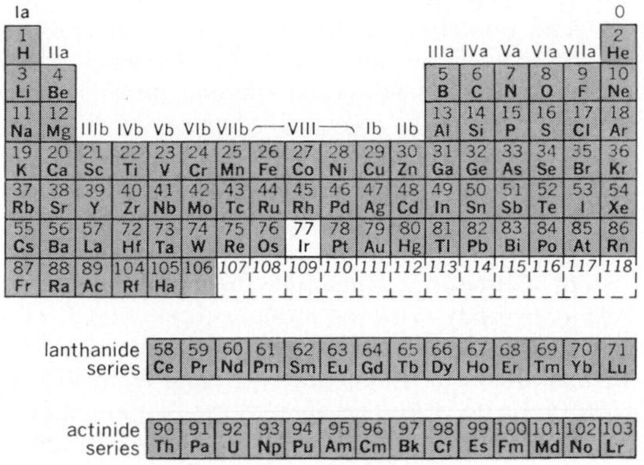

Chemical and physical properties. When hot, iridium is reasonably ductile, and is worked hot for this reason. Iridium retains its ductility in the cold state after hot working as long as it is not annealed. When annealed, it loses its cold ductility and is then rather brittle. It may be fabricated into fine wire and thin sheet if care is taken to avoid annealing.

Iridium has considerably less oxidation resistance than platinum or rhodium, but more so than ruthenium or osmium. A thin, adherent oxide film, IrO_2, forms above about 600°C. This oxide dissociates around 1100°C, and above this temperature a volatile oxide, possibly IrO_3, is responsible for iridium weight loss. Although such weight loss is greater than that of platinum or rhodium, iridium is the only metal which can be used unprotected in air up to 2300°C with any degree of life expectancy. Iridium is resistant to many molten metals such as sodium, potassium, mercury, bismuth, and lithium. It is only slowly attacked by molten lead, cadmium, tin, silver, and gold. It is rapidly attacked by molten copper, aluminum, zinc, and magnesium. Iridium is not attacked by any acid, including aqua regia. A 30% iridium–platinum alloy is practically insoluble in aqua regia. Iridium is slightly attacked by fused sodium and potassium hydroxides and by fused sodium bicarbonate. Fusion with alkaline oxidizing fluxes is necessary to convert iridium to a soluble form. Dissolution of iridium may also be accomplished by fusing it with sodium chloride while treating the melt with chlorine. The extreme inertness of iridium-rich alloys creates problems in their analysis and refining. Iridium has a strong tendency to form coordination compounds. The table gives values for important physical properties of iridium.

Principal compounds. Iridium trichloride, $IrCl_3$, is a green, water-insoluble compound, made by treating iridium powder with chlorine at 500°C. Sodium iridium(IV) chloride, $Na_2IrCl_6 \cdot 6H_2O$, is a black, water-soluble crystalline solid made by heating a fused mixture of iridium and sodium chloride in chlorine, and then dissolving the resulting melt in water. Sodium iridium(III) chloride, $Na_3IrCl_6 \cdot 12H_2O$, is an olive-green, water-soluble crystalline solid made by reducing a solution of sodium iridium(IV) chloride. Ammonium iridium(IV) chloride, $(NH_4)_2IrCl_6$, is a red-black, relatively insoluble crystalline solid made by adding ammonium chloride to a solution of sodium iridium(IV) chloride. Iridium trihydroxide, $Ir(OH)_3 \cdot xH_2O$, is a green-black, insoluble solid made by hydrolyzing a solution of iridium(III) chloride.

Metallurgical extraction. Osmiridium, obtained in the extraction of platinum, is fused with zinc and subsequently digested with hydrochloric acid to convert the material to a fine powder. This powder is then fused with an alkaline oxidizing flux which converts the iridium to an acid-soluble form. The procedure is rarely quantitative, so that the insoluble residues must be recycled. Various hydrolytic separations are available to separate the iridium from other metals. Often the insolubility of iridium in lead is used to separate it from other precious metals. The relative insolubility of ammonium chloroiridate in water may also be used to effect separation. When heated, this compound yields metallic iridium. Separations are not quanti-

Physical properties of iridium

Property	Value
Atomic weight, C^{12} = 12.00000	192.2
Naturally occurring isotopes and percent abundance	191 (37.3), 193 (62.7)
Crystal structure	Face-centered cubic
Lattice constant *a* at 25°C, angstrom units	3.8394
Thermal neutron capture cross section, barns	440
Common chemical valence	3, 4
Density at 25°C, g/cm³	22.55
Melting point, °C	2447
Boiling point, °C	4500
Specific heat at 0°C, cal/g	0.0307
Thermal conductivity 0–100°C, cal cm/cm² sec °C	0.35
Linear coefficient of thermal expansion 20–100°C, micro-in./in./°C	6.8
Electrical resistivity at 0°C, microhm-cm	4.71
Temperature coefficient of electrical resistance 0–100°C/°C	0.00427
Tensile strength (1000 psi)	
Soft	160–180
Hard	300–360
Young's modulus at 20°C	
psi, static	75.0×10^6
psi, dynamic	76.5×10^6
Hardness, DPN	
Soft	200–240
Hard	600–700

tative; thus, lengthy recycling of tailings is required. *See* OSMIUM.

Uses. At ordinary temperatures iridium is the most corrosion-resistant element known. Iridium exhibits catalytic activity for many reactions—hydrogenation, dehydrogenation, oxidation, and others. However, it has not been widely applied since usually another precious metal is found superior. Difficulties in fabrication limit use of iridium in the pure form at ordinary temperatures. For such use it is usually alloyed with platinum as the base, since a platinum–30% iridium alloy, for instance, is almost as corrosion-resistant as iridium and is a great deal easier to fabricate. *See* PLATINUM.

Pure iridium is used in special aircraft spark plugs, where its resistance to lead corrosion is outstanding. Iridium crucibles are used in the high-temperature growth of laser crystals and high-temperature glass melting. This may be done in reducing or inert atmosphere or air. In the latter case, the iridium is often protected from oxidation by a ceramic coating. Iridium with rhodium is used in very-high-temperature thermocouples, the only such thermocouples which can be used in air above 1800°C. Iridium alloyed with 5% tungsten may be used as a spring material up to 800°C. Iridium is also used at elevated temperatures as wires in vacuum gages which may be exposed to air at times. In this and other applications it may be reinforced by a thorium oxide dispersion through the metal. Iridium may be applied over other materials by cladding, plating from aqueous or fused salt baths, and deposition from organometallic media. *See* RHODIUM.

[HENRY J. ALBERT]

Iron

A chemical element, Fe, atomic number 26, and atomic weight 55.847. Iron is the fourth most abundant element in the crust of the Earth (5%). It is a malleable, tough, silver-gray, magnetic metal. It melts at 1540°C, boils at 2800°C, and has a density

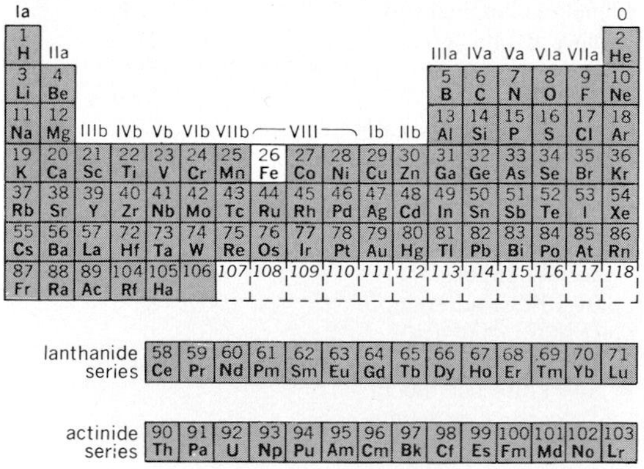

of 7.86 g/cm³. Each atom has 26 electrons, and the four stable, naturally occurring isotopes have masses of 54, 56, 57, and 58. The two main ores are hematite, Fe_2O_3, and limonite, $Fe_2O_3 \cdot 3H_2O$; other ores are magnetite, Fe_3O_4, taconite (an iron silicate), and siderite, $FeCO_3$. Pyrites, FeS_2, and chromite, $Fe(CrO_2)_2$, are mined as ores for sulfur and chromium, respectively. Iron is found in many other minerals, and it occurs in ground waters and in the red hemoglobin of blood.

The greatest use of iron is for structural steels; cast iron and wrought iron are made in quantity, also. Magnets, dyes (inks, blueprint paper, rouge pigments), and abrasives (rouge) are among the other uses of iron and iron compounds.

The free metal is obtained in bulk by reduction of the ore by coke. Pure iron is difficult to obtain because other elements are held tenaciously; a pure material, may, however, be obtained by reduction of the oxide with hydrogen or by electrolysis. The chemistry of the zero oxidation state is chiefly that of the alloys; addition of trace impurities (carbon, phosphorus, silicon, nickel, manganese, chromium, and cobalt) has a marked effect on the properties of the metal.

Properties of the metal. There are several allotropic forms of iron. Ferrite or α-iron is stable up to 760°C. The change to β-iron involves primarily a loss of magnetic permeability because the lattice structure (body-centered cubic) is unchanged. The allotrope called γ-iron has the cubic close-packed arrangements of atoms and is stable from 910 to 1400°C. Little is known about δ-iron except that it is stable above 1400°C and has a lattice similar to that of α-iron.

The metal is a good reducing agent and, depending on conditions, can be oxidized to the 2+, 3+, or 6+ state. In most iron compounds, the ferrous ion, iron(II), or ferric ion, iron(III), is present as a distinct unit. Iron(II) is found in simple compounds such as FeO, FeS, $FeBr_2$, and $FeCO_3$, and in hydrated compounds such as $FeSO_4 \cdot 7H_2O$, $FeF_2 \cdot 8H_2O$, $FeCl_2 \cdot 4H_2O$, and ferrous acetate tetrahydrate. Ferrous sulfate and ferrous ammonium sulfate (Mohr's salt), $(NH_4)_2SO_4 \cdot FeSO_4 \cdot 6H_2O$, have been employed as standards for oxidimetric titrations. The oxide, FeO, and hydroxide, $Fe(OH)_2$, are quite basic; they react with strong and weak acids to form salts. The ferrous compounds are usually light yellow to dark green-brown in color; the hydrated ion, $Fe(H_2O)_6^{2+}$, which is found in many compounds and in solution, is light green. This ion has little tendency to form coordination complexes except with strong reagents such as cyanide ion, polyamines, and porphyrins.

Oxidation of ferrous ion to ferric ion is moderately difficult in acid solution, but occurs readily in basic solution because of the insolubility of ferric hydroxide. The electrode potential for the ferrous-ferric reaction is dependent on the complexing species present in the solution. Anions such as CN^-, F^-, and PO_4^{3-} stabilize the 3+ oxidation state, whereas amines such as phenanthroline stabilize the 2+ state.

Coordination compounds. The ferric ion, because of its high charge (3+) and its small size (0.53 A as compared with 0.75 A for ferrous ion), has a strong tendency to hold anions. Ferric hydroxide is only weakly basic. Salts are formed with strong acids, but they are normally hydrated, as in $Fe(NO_3)_3 \cdot 9H_2O$ and ferric ammonium alum, $NH_4Fe(SO_4)_2 \cdot 12H_2O$. Because of similarities in size and charge, aluminum ion and ferric ion have quite analogous chemistries, although ferric hydroxide is not amphoteric. The anhydrous chlorides and bromides have dimeric vapors, for example, Fe_2Cl_6, and catalyze certain organic reactions.

The hydrated ion, $Fe(H_2O)_6^{3+}$, which is found in solution, combines with OH^-, F^-, Cl^-, CN^-, SCN^-, N_3^-, $C_2O_4^{2-}$, and other anions to form coordination complexes. As the concentration of anion increases, more groups combine with each cation until combinations such as $FeCl_4^-$ and $Fe(CN)_6^{3-}$ are reached.

The ferrate ion FeO_4^{2-}, which contains iron in the 6+ oxidation state, is obtained by reaction with strong oxidants such as hypochlorite in alkaline solution. Salts such as Na_2FeO_4 and $BaFeO_4$ are readily prepared and are fairly stable. Ferrate solutions are reddish purple. They release oxygen slowly in alkaline media, rapidly in acid.

An interesting aspect of iron chemistry is the array of compounds with bonds to carbon. Cementite, Fe_3C, is a component of steel. The cyanide complexes of both ferrous and ferric iron are very stable and are not strongly magnetic in contradistinction to most iron coordination complexes. The cyanide complexes form colored salts, including the famous prussian blue, $KFe_2(CN)_6$, made from ferric ion and potassium ferrocyanide, $K_4Fe(CN)_6$. The compound Turnbull's blue, made from ferrous ion and potassium ferricyanide, $K_3Fe(CN)_6$, is believed to be identical to prussian blue. There are many compounds containing five cyanide groups and one other group (such as NO, CO, SO_3^{2-}, NO_2^-, NH_3, and H_2O) about the iron. The compound $Na_2Fe(NO)(CN)_5{\cdot}2H_2O$ is one such compound; it can be used to test for sulfide ion.

The three carbonyls $Fe(CO)_5$, $Fe_2(CO)_9$, and $Fe_3(CO)_{12}$ also have iron-to-carbon bonds. Iron pentacarbonyl is formed directly by the reaction of iron with carbon monoxide under pressure; the other carbonyls are synthesized by indirect means. The pentacarbonyl compound is used as a source of pure iron. *See* METAL CARBONYL.

About 1950, a new type of iron compound was discovered. Cyclopentadiene can react with iron oxide to give dicyclopentadienyl iron, Fe $(C_5H_5)_2$. Many alternative syntheses of this compound, called ferrocene, are available. The structure is that of an iron atom between two symmetrical pentagonal rings, thus bringing to mind the name sandwich compound. Ferrocene has remarkable thermal stability. *See* ORGANOMETALLIC COMPOUND.

Many derivatives of ferrocene with substituent groups attached to the ring carbons are known; their chemistries are more in the line of organic reactions than reactions of iron. Ferrocene can be oxidized to the ferrocinium ion, $Fe(C_5H_5)_2^+$, and salts containing this ion have been prepared. *See* COBALT; FERROCENE; NICKEL; TRANSITION ELEMENTS.

[JOHN O. EDWARDS]

Isocyanate

One of a group of neutral derivatives of primary amines, formula R—N═C═O, that are obtained by reaction with excess phosgene and loss of hydrogen chloride, reaction (1). Two double bonds on carbon presage a very reactive system, useful in the identification of alcohols and phenols as urethanes (RNHCOOR′), reaction (2); amines as ureas, reaction (3); alkyl halides as anilides (C_6H_5 NHCOR), reaction (4); and carboxylic acids as amides (RNHCOR′), reaction (5). *See* AMINE.

$$RNH_2 + COCl_2 \rightarrow RNHCOCl \rightarrow R{-}N{=}C{=}O + HCl \quad (1)$$

$$RNCO + R'OH \rightarrow RNHCOOR' \quad (2)$$

$$RNCO + R'NH_2 \text{ (or } R'R''NH) \rightarrow RNHCONHR' \text{ (or } RNHCONR'R'') \quad (3)$$

$$RX \rightarrow RMgX + C_6H_5NCO \rightarrow C_6H_5N{=}CR(OMgX) \xrightarrow{H_2O} C_6H_5NHCOR \quad (4)$$

$$RNCO + R'COOH \rightarrow RNHCOOCOR' \rightarrow RNHCOR' + CO_2 \quad (5)$$

Isocyanates hydrolyze in air to the unstable carbamic acids (RNHCOOH) which revert to the original amines by loss of carbon dioxide, reaction (6).

$$RNCO + H_2O \rightarrow RNHCOOH \rightarrow RNH_2 + CO_2 \quad (6)$$

Unhydrolyzed isocyanate then reacts with the amine to give a symmetrical urea, RNHCONHR. Normally an important reaction only because the nuisance must be avoided, controlled hydrolysis of free isocyanate groups to liberate carbon dioxide is the basis of one type of foamed plastic. *See* POLYURETHANE RESINS.

To avoid hydrolysis by atmospheric moisture, isocyanates are best kept in sealed glass containers.

Replacement of oxygen by sulfur, as occurs in isothiocyanates, reduces the activity considerably.

[LEALLYN B. CLAPP]

Isoelectric point

The pH value of the dispersion medium of a colloidal suspension at which the colloidal particles do not move in an electric field; that is, the particles are electrophoretically inert (see illustration). The isoelectric point is often employed to characterize colloidal material such as the proteins. However, a range of values is usually necessary, since the isoelectric point varies detectably with (1) the size of the particles, (2) the purity, and (3) the concentration of other than hydrogen ions.

If the pH value of an extrinsic sol (stability attributed primarily to electrical charge) is adjusted toward the isoelectric point, coagulation will occur at or near the isoelectric point. Intrinsic sols (stability attributed primarily to solvation) may be carried to the isoelectric point without coagulation, but such sols will be in a region of minimum stabil-

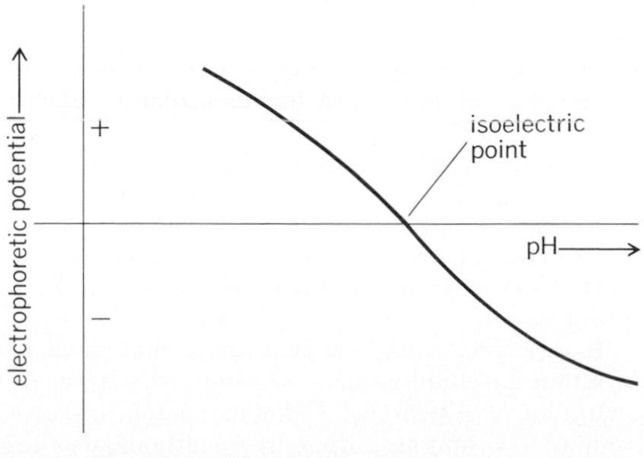

Graph showing the isoelectric point.

ity, so that a minimum concentration of desolvating agent will cause coagulation. Likewise, viscosity changes often reach a minimum at or near the isoelectric point. *See* COLLOID; ELECTROPHORESIS; ION-SELECTIVE MEMBRANES AND ELECTRODES.

[W. O. MILLIGAN]

Isomerization

Rearrangement of the atoms within hydrocarbon molecules. Isomerization processes of practical significance in petroleum chemistry are (1) migration of alkyl groups, (2) shift of a single-carbon bond in naphthenes, and (3) double-bond shift in olefins.

Migration of alkyl groups. An example of alkyl group migration (skeletal isomerization) is reaction (1). Isomerization to more highly branched con-

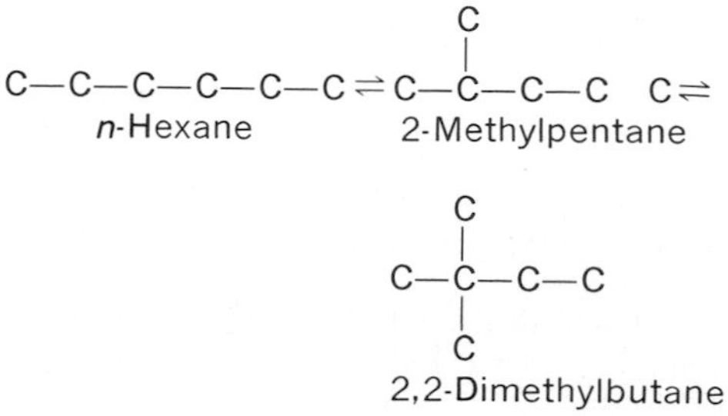

(1)

figurations has commercial importance since it results in improvement in combustion quality in the automobile engine as measured by octane number and increased chemical reactivity because tertiary carbon atoms result. The unleaded, motor-method octane numbers of the hexane isomers shown in reaction (1) are 26.0, 73.5, and 93.4, respectively. Normal butane is converted to isobutane (which has a tertiary carbon atom) to attain chemical reactivity with olefins in alkylation reactions where *n*-butane is inert.

Isomerization of paraffins is a reversible first-order reaction limited by thermodynamic equilibrium which favors increased branching at lower temperatures. Undesirable cracking reactions leading to catalyst deactivation occur at higher temperatures. They are controlled by adding a cracking suppressor such as hydrogen.

Conversion of normal butane to isobutane is the major commercial use of isomerization. Usually, it is carried out in either liquid- or vapor-phase over aluminum chloride catalyst promoted with hydrogen chloride. In the vapor-phase process (250–300°F), the aluminum chloride is often supported on bauxite. In the liquid-phase processes (180°F), it is dissolved in molten antimony trichloride or used in the form of a liquid complex with hydrocarbon. A second type of catalyst for vapor-phase isomerization (300–850°F) is a noble metal, usually platinum, supported on a carrier. This may be alumina with halide added to provide an acidic surface. All the processes are selective (95–98% to isobutane). Approximately 60% of the *n*-butane feed is converted per pass to isobutane in the liquid-phase process.

Isopentane, a high-octane component used in aviation gasoline, is made commercially by isomerization of *n*-pentane. Petroleum naphthas containing five- and six-carbon hydrocarbons also are isomerized commercially for improvement in motor-fuel octane numbers. Noble-metal catalyst is normally used with higher-molecular-weight feeds. Isomerization of paraffins above six carbon atoms is of less importance, since octane improvement is limited by predominance of monomethyl branching at equilibrium. Skeletal isomerization is an important secondary reaction in catalytic cracking and catalytic reforming. Aromatics and olefins undergo skeletal isomerization as do paraffins.

Single-carbon bond shift. This process, in the case of naphthenes, is illustrated by reaction (2).

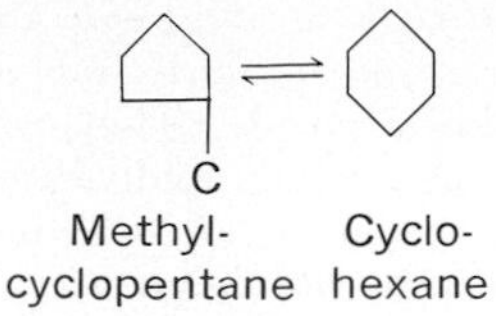

(2)

Cyclohexane and methylcyclohexane have been produced commercially by liquid-phase isomerization of the five-carbon ring isomers over aluminum chloride–hydrocarbon-complex catalyst promoted by hydrogen chloride. Conversion per pass is high, selectivity excellent, and reaction conditions mild (190°F). Cyclohexane is a raw material for making nylon, and it may be dehydrogenated to benzene. Methylcyclohexane has been used to make synthetic nitration-grade toluene.

Shift of a double-bond. This process is usefully applied when a specific olefin is needed for chemical synthesis, as in reaction (3). Double-bond

$$\underset{\text{1-Butene}}{C{-}C{-}C{=}C} \rightleftharpoons \underset{\text{2-Butene}}{C{-}C{=}C{-}C} \qquad (3)$$

shift occurs selectively over acidic catalysts at temperatures below 450°F. However, the proportion undergoing skeletal isomerization increases as temperature is increased until, at temperatures in the range of 600–950°F, equilibrium is approached at fairly high space velocities. Equilibrium favors movement of double bonds to the more stable internal positions (85.6% 2-butene at 400°F), and octane improvement accompanies this shift; however, the increase of octane number normally is insufficient to justify the cost of processing thermally cracked gasolines solely for this purpose. This type of isomerization occurs as a secondary reaction in the catalytic cracking and catalytic polymerization processes, in part accounting for the high octane numbers of the gasolines. *See* ALKYLATION; AROMATIZATION; CRACKING; MOLECULAR ISOMERISM.

[GEORGE E. LIEDHOLM]

Isoprene

A five-carbon, conjugated diolefin, or diene, having the structure shown. It does not occur naturally

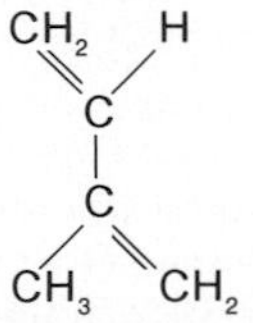

but is obtained by the destructive distillation of gas oil, naphthas, and rubber. It is also prepared by the catalytic decomposition of dipentene. It is commercially available in about 96% purity and is used in the production of butyl rubber.

The terpenes may be regarded as multiples of

the isoprene unit, C_5H_8. Indeed, isoprene may also be the foundation for other important plant products such as phytol, the sterols, and the carotenoids.

Isoprene is a mobile, colorless liquid boiling at 34.1°C, having a specific gravity of 0.862. It exhibits all the characteristic reactions of dienes of this type. Isoprene polymerizes readily to form dimers and high-molecular-weight resins. The principal dimer formed is isopropenyl methyl cyclohexene as shown below. Polymerization of isoprene during

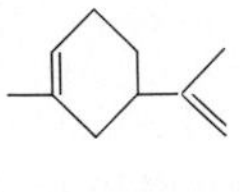

storage can be controlled by avoiding contact with oxygen, by using inhibitors such as *tert*-butyl catechol, and by keeping it at low temperatures. *See* TERPENE. [WILLIAM MOSHER]

Isopropylphenol

One of three isomeric compounds, having structures shown in the figure. The nomenclature of the derivatives of these compounds is directly parallel to that of the cresol analogs.

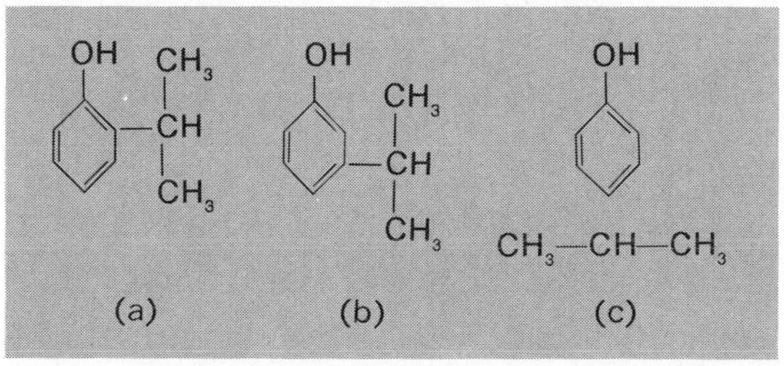

Structural formulas for isopropylphenol. (*a*) Ortho, (*b*) meta, and (*c*) para configurations.

Production. These compounds can be prepared by a number of syntheses. Starting from cumene, the same processes may be used as those used for the production of phenol from benzene. Of little interest or use prior to the introduction of the cumene hydroperoxide process of phenol manufacture, supplies of these compounds are very largely derived as by-products of this process.

Uses. In many of the applications of which cresols have historically been used, one or another of the isopropylphenols can be used. As a by-product of an extremely large industrial manufacturing operation, that is, phenol from cumene, isopropylphenols are potentially plentiful and are free of many troublesome impurities commonly found in the cresols or cresylic acids of commerce, which are largely derived from coal tars or waste products of petroleum-refining operation. These advantages have actually led at least one large European supplier of triarylphosphates to undertake a wholly synthetic route to these popular plasticizers, using isopropylphenol instead of the traditional cresol. The uniformity and improved properties of the end-use derivatives are relied upon to justify the somewhat higher costs. There are as yet no reliable published statistics on prices or production and sales of isopropylphenol, but substantial quantities are known to move in commerce. *See* PHENOL. [ROBERT I. STIRTON]

Isoquinoline

One of a group of organic compounds containing a benzene ring fused to the 3,4 positions of pyridine. Isoquinoline (I) is a representative member of the

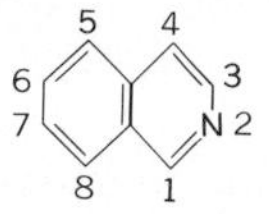

(I)

group. Quinoline produced from coal tar contains approximately 1–4% of isoquinoline, and it is an important source of the latter material. Separation is effected by selective extraction of the more basic isoquinoline with acid, or by selective precipitation and fractional crystallization of salts. Repeated fractional freezing and distillation have furnished pure isoquinoline. Many plant alkaloids, especially those in the cactus, opium, and curare groups, are isoquinoline derivatives. *See* HETEROCYCLIC COMPOUNDS; QUINOLINE.

Properties and preparation. Isoquinoline is a colorless, odorous, liquid with bp 243.3°C, mp 26.5°C, and n_D^{30} 1.62078. Its stability to acid, base, or heat is high. Isoquinoline, which is somewhat more basic than quinoline (the pK_a's are, respectively, 5.14 and 4.51), can be protonated to form simple salts and alkylated to form quaternary salts. With some exception, the reactions of isoquinoline have analogous counterparts in the reactions of quinoline. Substitution reactions such as nitration, sulfonation, bromination, and mercuration are observed.

Phenylethylamines (II) are starting points in two general syntheses. The Bischler-Napieralski method cyclizes an *N*-acylated derivative (III) with phosphorus oxychloride to a 3,4-dihydroisoquinoline (IV). The Pictet-Spengler sequence converts phenylethylamine (II) by reaction with an aldehyde to an imine (V), which cyclizes in the presence of mineral acid to a 1,2,3,4-tetrahydroisoquinoline (VI). Dehydrogenation of either 3,4-dihydro or 1,2,3,4-tetrahydroisoquinoline by chemical or catalytic methods is generally uncomplicated, so that fully unsaturated isoquinolines (VII) are

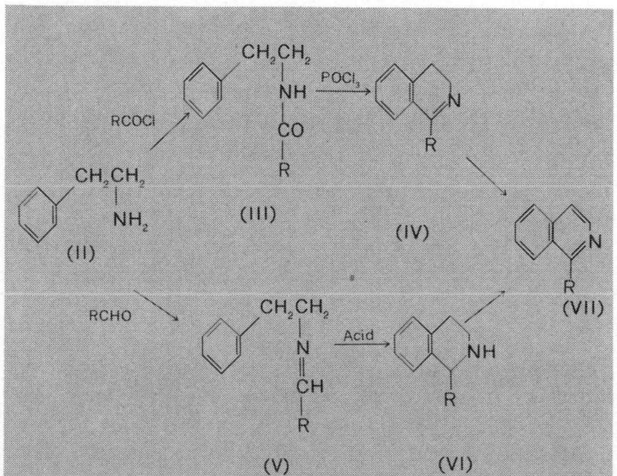

readily obtained. In another kind of synthesis (Pomeranz-Fritsch), sulfuric acid is used to cyclize anils (IX) produced by the reaction of aminoacetal (VIII) and an aromatic aldehyde.

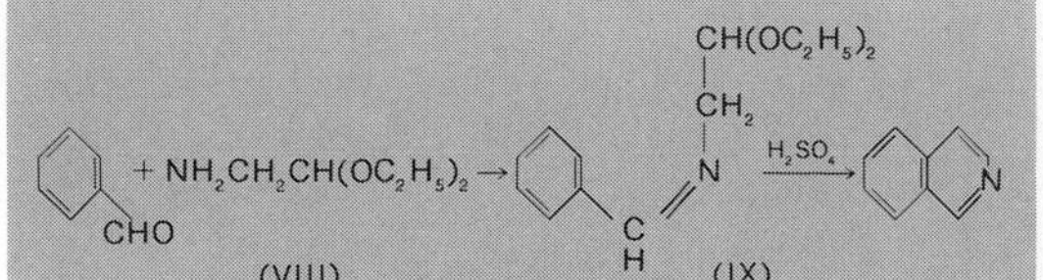

(VIII) (IX)

When the aromatic ring in phenylethylamine (II) is activated by a phenolic hydroxyl group para to the point of cyclization, the Pictet-Spengler reaction (V – VI) occurs under exceptionally mild conditions. Biosynthesis of isoquinoline plant products proceeds in this way. [WALTER J. GENSLER]

Alkaloid derivatives. The isoquinoline skeleton is incorporated in a large number of alkaloids so that over 700 isoquinoline alkaloids are known. Most of these can be broken down into two broad categories depending upon whether they are derived biogenetically from the benzylisoquinolines (±)-coclaurine (X) or (+)-reticuline (XI). Both X

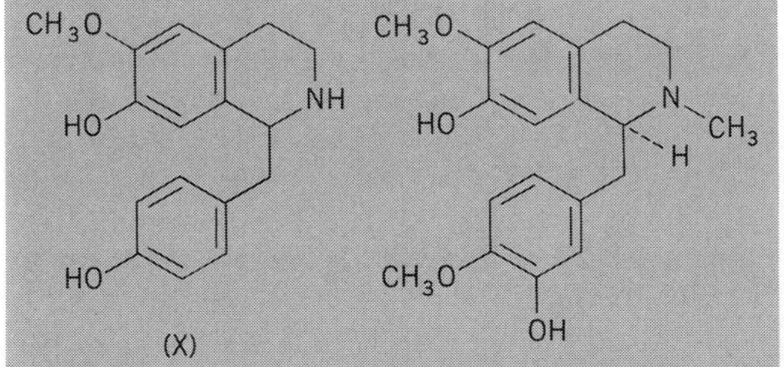

(X) (XI)

and XI are products of the metabolism of the amino acid tyrosine. The oxygenated substituents usually occur in the form of a phenol, a methoxyl, or a methylenedioxy group.

Alkaloids from (±)-coclaurine. (±)-Coclaurine (X) or its *N*-methyl analog tend to dimerize to provide a large variety of bisbenzylisoquinoline alkaloids, a prime example being the powerful neuromuscular blocking agent (+)-tubocurarine.

Alternatively, a dimeric alkaloid such as (+)-berbamunine may undergo phenolic oxidative coupling to give rise to a proaporphine-benzylisoquinoline alkaloid, which can then undergo a dienonephenol rearrangement to supply an alkaloidal aporphine-benzylisoquinoline.

Reactions parallel to the above can occur with a monomeric (±)-coclaurine (X) or *N*-methylcoclaurine unit. The initial phenolic oxidation product is a proaporphine (XII) which can undergo dienone-phenol rearrangement to a trioxygenated aporphine (XIII).

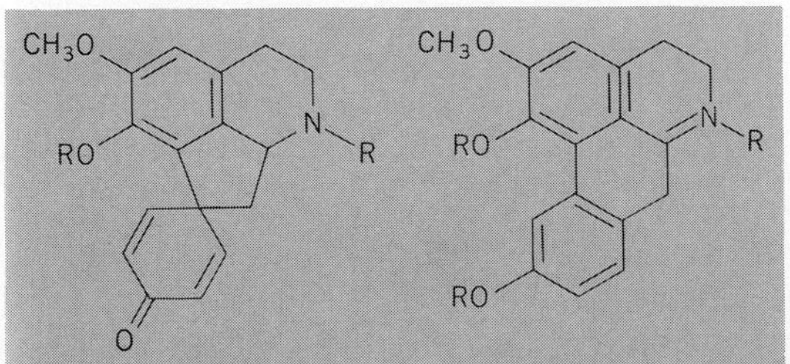

(XII) (XIII)

Alkaloids from (+)-reticuline. The chemistry of (+)-reticuline (XI), which possesses an additional oxygen in the bottom ring, differs in some important respects from that of X. The tendency to dimerize is minimal, and no bisbenzylisoquinolines derived from two reticuline units are known. Rather, there is a proclivity for internal cyclization to generate a pavine, a dibenzopyrrocoline, a berbine, or an aporphine alkaloid.

Berbines can themselves act as precursors to other isoquinoline alkaloids such as the phthalideisoquinolines, the protopines, the spirobenzylisoquinolines, and the rhoeadines.

Emetine bases. Alkaloids of the emetine type have a somewhat different biogenetic history, originating in part from a monoterpenoid precursor.

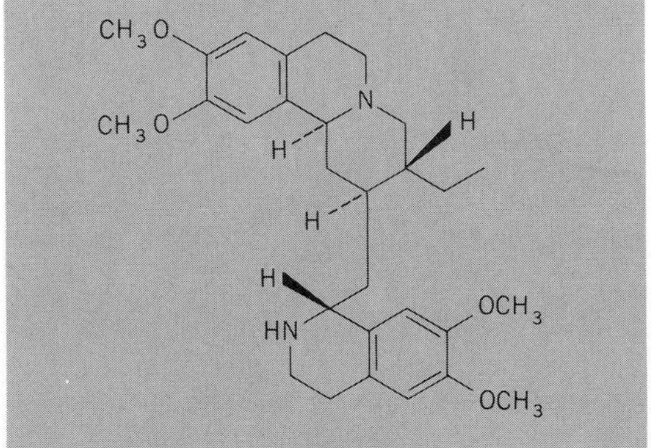

(XIV)

Emetine (XIV) itself is used in the treatment of amebic infections. [MAURICE SHAMMA]

Bibliography: J. A. Joule and G. F. Smith, *Heterocyclic Chemistry*, 1972; T. Kametani, *The Chemistry of the Isoquinoline Alkaloids*, vol. 2, Sendai Institute of Heterocyclic Chemistry, 1974; M. Shamma, *The Isoquinoline Alkaloids, Chemistry and Pharmacology*, 1972; M. Shamma and J. L. Moniot, *Isoquinoline Alkaloids Research, 1972–1977*, 1979; D. W. Young, *Heterocyclic Chemistry*, 1976.

Isotope

One of two or more nuclidic species of an element having identical number of protons (*Z*) in the nucleus but different number of neutrons (*N*). Isotopes differ in mass but chemically are the same element. All naturally occurring elements have radioactive isotopes, and the majority have at least one stable nuclide. Some elements which occur in nature, such as uranium, are radioactive but have isotopes with long half-lives.

The isotopic composition of an element is generally determined by mass spectrometry. Of the 83 elements present on Earth in significant amounts, 20 possess only a single stable nuclide and are referred to as mononuclidic or anisotopic. The others have 2 to 10 stable isotopes. *See* MASS SPECTROSCOPE.

Nuclear stability. Of the 287 nuclidic species listed in the table, 168 have even-even structure (even number of protons and even number of neutrons in the nucleus), 57 have even-odd, 53 have odd-even, and only 9 have odd-odd. This indicates the pairing tendency of the nuclear constituents.

Natural isotopic abundances of the elements

Element		Mass no.	Atom %
1	H*	1	99.985
		2	0.015
		3	0.00013
2	He*	4	≈100.
3	Li*	6	7.5
		7	92.5
4	Be	9	100.
5	B*	10	19.8
		11	80.2
6	C*	12	98.89
		13	1.11
7	N*	14	99.64
		15	0.36
8	O*	16	99.756
		17	0.039
		18	0.205
9	F	19	100.
10	Ne*†	20	90.51
		21	0.27
		22	9.22
11	Na	23	100.
12	Mg	24	78.99
		25	10.00
		26	11.01
13	Al	27	100.
14	Si	28	92.2
		29	4.7
		30	3.1
15	P	31	100.
16	S*	32	95.00
		33	0.76
		34	4.22
		36	0.02
17	Cl	35	75.77
		37	24.23
18	Ar†	36	0.34
		38	0.07
		40	99.59
19	K	39	93.26
		40	0.01
		41	6.73
20	Ca	40	96.937
		42	0.65
		43	0.14
		44	2.08
		46	0.003
		48	0.19
21	Sc	45	100.
22	Ti	46	8.0
		47	7.5
		48	73.7
		49	5.5
		50	5.3
23	V	50	0.25
		51	99.75
24	Cr	50	4.35
		52	83.79
		53	9.50
		54	2.36
25	Mn	55	100.
26	Fe	54	5.85
		56	91.7
		57	2.14
		58	0.31
27	Co	59	100.
28	Ni	58	68.3
		60	26.1
		61	1.1
		62	3.6
		64	0.9
29	Cu	63	69.2
		65	30.8

Element		Mass no.	Atom %
30	Zn	64	48.9
		66	27.8
		67	4.1
		68	18.6
		70	0.6
31	Ga	69	60.0
		71	40.0
32	Ge	70	20.7
		72	27.5
		73	7.7
		74	36.4
		76	7.7
33	As	75	100.
34	Se	74	0.9
		76	9.0
		77	7.6
		78	23.5
		80	49.8
		82	9.2
35	Br	79	50.69
		81	49.31
36	Kr†	78	0.35
		80	2.25
		82	11.6
		83	11.5
		84	57.0
		86	17.3
37	Rb	85	72.17
		87	27.83
38	Sr†	84	0.56
		86	9.84
		87	7.0
		88	82.6
39	Y	89	100.
40	Zr	90	51.4
		91	11.2
		92	17.1
		94	17.5
		96	2.8
41	Nb	93	100.
42	Mo	92	14.8
		94	9.1
		95	15.9
		96	16.7
		97	9.5
		98	24.4
		100	9.6
44	Ru	96	5.5
		98	1.9
		99	12.7
		100	12.6
		101	17.1
		102	31.6
		104	18.6
45	Rh	103	100.
46	Pd	102	1.0
		104	11.0
		105	22.2
		106	27.3
		108	26.7
		110	11.8
47	Ag	107	51.83
		109	48.17
48	Cd	106	1.2
		108	0.9
		110	12.4
		111	12.8
		112	24.0
		113	12.3
		114	28.8
		116	7.6
49	In	113	4.3
		115	95.7

Element		Mass no.	Atom %
50	Sn	112	1.0
		114	0.7
		115	0.4
		116	14.7
		117	7.7
		118	24.3
		119	8.6
		120	32.4
		122	4.6
		124	5.6
51	Sb	121	57.3
		123	42.7
52	Te	120	0.1
		122	2.5
		123	0.9
		124	4.6
		125	7.0
		126	18.7
		128	31.7
		130	34.5
53	I	127	100.
54	Xe†	124	0.1
		126	0.1
		128	1.9
		129	26.4
		130	4.1
		131	21.2
		132	26.9
		134	10.4
		136	8.9
55	Cs	133	100.
56	Ba	130	0.1
		132	0.1
		134	2.4
		135	6.6
		136	7.9
		137	11.2
		138	71.7
57	La	138	0.09
		139	99.91
58	Ce	136	0.2
		138	0.3
		140	88.4
		142	11.1
59	Pr	141	100.
60	Nd	142	27.1
		143	12.2
		144	23.9
		145	8.3
		146	17.2
		148	5.7
		150	5.6
62	Sm	144	3.1
		147	15.0
		148	11.2
		149	13.8
		150	7.4
		152	26.7
		154	22.8
63	Eu	151	47.8
		153	52.2
64	Gd	152	0.2
		154	2.2
		155	14.9
		156	20.6
		157	15.7
		158	24.7
		160	21.7
65	Tb	159	100.

Element		Mass no.	Atom %
66	Dy	156	0.06
		158	0.1
		160	2.34
		161	18.9
		162	25.5
		163	24.9
		164	28.2
67	Ho	165	100.
68	Er	162	0.1
		164	1.6
		166	33.4
		167	22.9
		168	27.0
		170	15.0
69	Tm	169	100.
70	Yb	168	0.1
		170	3.1
		171	14.3
		172	21.9
		173	16.2
		174	31.7
		176	12.7
71	Lu	175	97.4
		176	2.6
72	Hf	174	0.2
		176	5.2
		177	18.5
		178	27.1
		179	13.8
		180	35.2
73	Ta	180	0.012
		181	99.988
74	W	180	0.1
		182	26.3
		183	14.3
		184	30.7
		186	28.6
75	Re	185	37.40
		187	62.60
76	Os†	184	0.02
		186	1.58
		187	1.6
		188	13.3
		189	16.1
		190	26.4
		192	41.0
77	Ir	191	37.4
		193	62.6
78	Pt	190	0.01
		192	0.79
		194	32.9
		195	33.8
		196	25.3
		198	7.2
79	Au	197	100.
80	Hg	196	0.2
		198	10.1
		199	16.9
		200	23.1
		201	13.2
		202	29.7
		204	6.8
81	Tl	203	29.5
		205	70.5
82	Pb†	204	1.4
		206	24.1
		207	22.1
		208	52.4
83	Bi	209	100.
90	Th	232	100.
92	U*	234	0.0054
		235	0.7200
		238	99.2746

*Isotopic composition of the element may be somewhat variable with specific geological or biological origin of the sample. Commercial chemicals may have, in some cases, quite anomalous composition as the result of processes of isotope separation.

†The element may vary in isotopic composition in some samples because one or more of the isotopes result from radioactive decay, or from nuclear processes in nature, such as spontaneous fission of uranium or α,n reactions on light elements.

The extra stability of a 50-proton configuration is indicated by the existence of 10 isotopes of tin.

Isotopic abundance. This refers, unless otherwise specified, to the isotopic composition of the naturally occurring terrestrial element (see table). Some elements are observed to vary in isotopic composition. The variability ranges from a few per mil to a percent or two, although variations greater than this are observed in some samples. This variation occurs for several reasons. In the lighter elements—hydrogen, lithium, and boron, for instance—the isotopes differ enough in mass and to some extent in chemical reactivity that processes of distillation or chemical exchange between different chemical compounds of the element can produce significant differences in isotopic composition. Indeed, exchange reactions are used in the case of hydrogen, lithium, boron, carbon, nitrogen, and oxygen to separate isotopes of these elements on a relatively large scale.

Elements which take part in the life cycle of living organisms will vary somewhat in isotopic composition because of exchange reactions and diffusion through membranes. Slight differences in reaction rates are also important.

The composition of some elements may be variable because one or more of the isotopes are stable products of radioactive decay. Thus three of the four lead isotopes, ^{208}Pb, ^{207}Pb, and ^{206}Pb, are end products of the decay of thorium and of ^{235}U and ^{238}U respectively. The fourth lead isotope, ^{204}Pb, does not come from any known decay chain. The rare potassium isotope, ^{40}K, which is present in only 0.012%, has a half-life of 1.28×10^9 years. It decays by beta-particle emission to stable ^{40}Ar and by electron capture to ^{40}Ca. The argon in the atmosphere, approximately 1.1%, is 99.6% in ^{40}Ar. Argon in potassium-bearing minerals will differ in composition from atmospheric argon just as the composition of lead will depend upon its past association with thorium and uranium. These decays and the decay of rubidium, ^{87}Rb, to ^{87}Sr are the basis of methods used in the determination of geological age.

The three nuclides ^{40}K, ^{40}Ar, and ^{40}Ca are examples of "isobars" in that they have the same mass number, $N+Z$, but differ in the number of protons in the nucleus. The radioactive potassium isotope is an example of an odd-odd nucleus, while the stable ^{40}Ar and ^{40}Ca are both even-even.

Anomalous isotopic compositions in some elements occur because of nuclear processes in nature. The discovery, in 1972, of a "fossil reactor" in the Oklo uranium deposit in Gabon (West Africa) is such a case. As a result of a chain reaction, perhaps 1.7×10^9 years ago, much of the uranium in this formation is depleted in the fissionable isotope, ^{235}U. Fission products are present in the composition found in reactor waste, and some elements in the surrounding rocks have been modified in isotopic composition by neutron absorption and subsequent radioactive decay.

The only nonterrestrial materials which are available for comparison of isotopic composition are meteorites and the lunar materials returned to Earth by the several Apollo manned missions and the Soviet unmanned lunar probes. Isotopic compositions in these are generally identical with terrestrial samples within the precision of the measurements. Differences, when they are found, can be identified as caused by radioactive decay, cosmic-ray bombardment and, in the case of lunar surface materials, bombardment by solar "wind" particles. In iron meteorites the spallation of iron nuclei by very energetic particles from cosmic rays produces helium, neon, and some other light elements with anomalous isotopic composition. The helium $^3He/^4He$ ratio is used as an indicator of the cosmic ray "bombardment" age. *See* RADIOACTIVITY.

Use of separated isotopes. Isotopes of certain elements possess unique or peculiar properties, and their separation or enrichment is desirable. Deuterium, ^2H, which occurs in an abundance of about 0.16% in terrestrial hydrogen, is useful for moderating neutrons in heavy-water reactors and is expected to form the fuel of fusion reactors in the future. Very large quantities of deuterated water have been produced by a variety of distillative, exchange, and electrolysis processes. The desirable fissionable isotope of uranium, ^{235}U, is enriched by gaseous diffusion in uranium hexafluoride gas in very large plants.

Of the biologically important elements, only carbon and hydrogen have radioisotopes of sufficiently long half-lives to be used in tracer studies in living organisms. The use of large amounts of ^3H and ^{14}C is not desirable and, in any case, the mass differences are so great that they do not act precisely like the isotopes ^1H and ^{12}C. Studies of metabolism, drug utilization, and other reactions in living organisms are best done with stable isotopes like ^{13}C, ^{15}N, ^{18}O, and ^2H. Compounds are tagged by introducing a high concentration of the isotope into the molecular structure, and the metabolized products are studied using the mass spectrometer to measure the altered isotopic ratios.

The process of isotope dilution consists of adding a known amount of material containing the tracer isotope, allowing the system to reach chemical equilibrium and then recovering a small sample sufficient in size for a mass spectrometric measurement of the new isotopic composition. This is a method of very wide applicability.

Atomic weight. Atomic weight is the ratio of the average mass per atom of the natural isotopic composition of an element to 1/12 of the mass of an atom of the nuclide ^{12}C. It is a dimensionless number. The atomic weight of a mononuclidic element is simply the atomic mass of that nuclide relative to 1/12 of the mass of $^{12}C = 12$ exactly. These masses can be measured with very high precision. Thus the atomic weight of the mononuclidic element beryllium is 9.01218. Atomic weights of the elements having two or more isotopic species are increasingly based upon calculations from isotopic composition and atomic masses. The mass spectrometer is not an absolute instrument because of certain inherent discriminations. Comparisons must be made with isotopic standards carefully prepared by gravimetric procedures from separated isotopes of high chemical and isotopic purity. This is hardly less exacting than the chemical determination of atomic weights used early in this century by T. W. Richards and coworkers at Harvard University. The "Harvard method" involved the careful precipitation and weighing of silver chloride formed by the stoichiometric reaction of silver nitrate with the chloride of the element being investigated. Thus, one would determine the germanium chloride/silver chloride ratio ($GeCl_4/4AgCl$) and thus the ratio of the atomic weight to that of silver. Even in the 1975 Table of Atomic Weights some of the polynuclidic elements, such as germanium, tin, and mercury, are still based upon chemical ratios measured by the Harvard method. *See* ELEMENTS.

[A. E. CAMERON]

Bibliography: J. F. Duncan and G. B. Cook, *Isotopes in Chemistry*, 1968; S. Glasstone, *Sourcebook on Atomic Energy*, 3d ed., 1967; N. E. Holden and F. W. Walker, (eds.), *Chart of the Nuclides*, Educational Relations, General Electric Co., 1972; J. Robos, *Introduction to Mass Spectrometry*, 1968; A. Romer (ed.), *Radiochemistry and the Discovery of Isotopes*, 1970.

Isotopic irradiation

The subjection of a material to radiation from radioactive isotopes (radioisotopes). Although this article treats irradiation by radioactive isotopes and deals primarily with the gamma rays emitted by the isotopes, many other forms of radiation from many different sources are often used for irradiation purposes. The use of radioactive isotopes to irradiate various materials can be divided into a number of major areas, including industrial, medical, and research. For most applications the radioactive isotope is placed inside a capsule, but for some medical applications the radioisotope is dispersed in the material to be irradiated.

The radiation from radioisotopes produces essentially the same effect as the radiation from electron linear accelerators and other high-voltage particle accelerators, and the choice of which to use is based primarily on convenience and cost. Although the radioisotope radiation source does not require the extensive and complex circuitry necessary for a high-voltage radiation source, its radiation is always present and requires elaborate shielding for health protection and specialized mechanisms for bringing the irradiated objects into and out of the radiation beam. Further, the radiation output decreases with time according to the half-life of the radioisotope, which must therefore be replaced periodically.

The most commonly used radioisotopes and their major applications are listed in the table.

Industrial applications. The two main radioisotopes for industrial processing are cobalt-60 with a half-life of 5.261 years and an average gamma-ray energy of 1.25 MeV, and cesium-137 with a half-life of 30.3 years and a gamma-ray energy of 0.662 MeV. The application of industrial irradiation is increasing and includes cross-linking plastics, curing polymeric insulations on wires and cables, sterilizing medical devices, and food preservation. Further, cobalt-60 sources are used for such diverse applications as making a concrete patching material for subzero climates; making plastic rods and tubes that can be made into bolts, nuts, and rivets; sterilizing medical disposables; irradiating potatos to prevent sprouting; curing of an acrylic polymer after it has impregnated wood flooring; and sludge processing. Cesium-137 is also used for sludge processing to produce fertilizer or cattle feed. Cesium-137 is available in large quantities in the reprocessing of fuel elements from nuclear reactors, and the process helps solve the problem of disposing of cesium from this source.

Medical applications. Radioisotopes are used in the treatment of cancer by radiation. Encapsulated sources are used in two ways: the radioisotopes may be external to the body and the radiation allowed to impinge upon and pass through the patient (teletherapy), or the radiation sources may be placed within the body (brachytherapy).

For teletherapy purposes cobalt-60 is the most commonly used isotope, with source strengths ranging from several thousand curies up to 10^4 curies (1 Ci $= 3.7 \times 10^{10}$ Bq). Some cesium-137 irradiators have been built, but cesium-137 radiation is not as penetrating as that from cobalt-60.

Historically radium-226 has been used as the brachytherapy encapsulated source, but due to serious problems which can arise from a health physics aspect if the capsule breaks, radium is now being replaced with cesium-137. Iridium-192 is also used for brachytherapy applications.

In order to avoid the long half-life of radium, its daughter product radon-222, which is an inert gas, was used in very small glass or gold seeds. Today radon has been replaced by iodine-125 or gold-198 seeds, which are put directly in a tumor and permanently left in place.

For some medical applications the radioisotope

Most commonly used radioisotopes and major applications

Isotope	Half-life	Average gamma-ray energy, MeV	Main uses
Radium-226 in equilibrium (^{226}Ra)	1602 years	0.83	Medical (brachytherapy)
Gold-198 (^{198}Au)	2.697 days	0.412	Medical (brachytherapy—permanent implants)
Iridium-192 (^{192}Ir)	74.2 days	0.38	Medical (brachytherapy)
Cesium-137 (^{137}Cs)	30.3 years	0.662	Industrial, medical (brachytherapy), research irradiators
Cobalt-60 (^{60}Co)	5.261 years	1.25	Industrial, medical (teletherapy), research irradiators
Iodine-125 (^{125}I)	60 days	0.035	Medical (brachytherapy—permanent implants)
Iodine-131 (^{131}I)	8.065 days	0.187, average beta-ray energy	Medical (dispersal technique)

is dispersed in the body; the most commonly used is iodine-131. The specific advantage of internal therapy with iodine-131 is that the thyroid gland concentrates the element iodine. When the radioisotope is administered either orally or intravenously in a highly purified form, it goes to the thyroid, where certain forms of thyroid disorders and cancers can be treated by the radiation. Unlike the other radioisotopes discussed here, the iodine-131 therapeutic effectiveness depends upon the beta rays emitted, not the gamma rays.

Another medical use for radioisotopes is the requirement that all transfusions of blood be irradiated before given to certain patients. This is done so that the irradiation destroys most of the lymphocytes present. For this purpose self-contained irradiators are used. Cobalt-60 or cesium-137 encapsulated sources are permanently fixed and sealed off within the radiation shield, and the irradiation chamber is introduced into the radiation field by rotation. Such units are also extensively used for radiobiological and other research purposes.   [PETER R. ALMOND]

Ketene

A colorless, highly reactive gas, bp $-56°C$, mp $-151°C$. It is soluble in ether and acetone, and decomposes in water and alcohol. The ketene molecule shown below contains a highly reactive carbonyl group $>C=O$.

$$\begin{array}{c} H \\ \diagdown \\ C=C=O \\ \diagup \\ H \end{array}$$

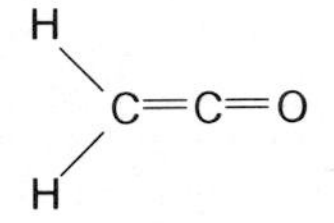

Ketene is produced by the pyrolysis of acetone or acetic acid vapor, as in reactions (1) and (2).

$$CH_3-\overset{\overset{\displaystyle O}{\|}}{C}-CH_3 \xrightarrow{700°C} CH_2=C=O + CH_4 \quad (1)$$

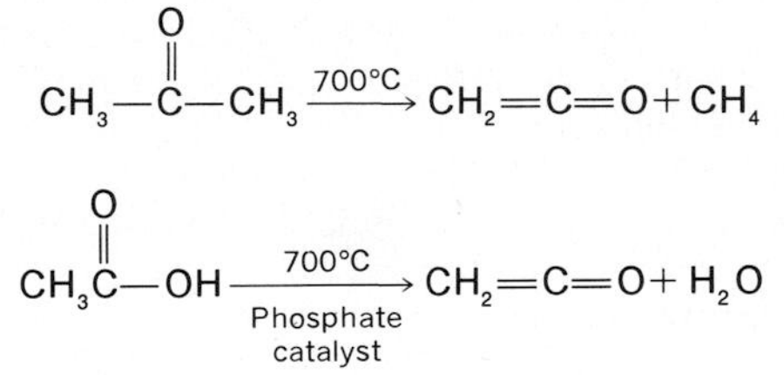

$$CH_3\overset{\overset{\displaystyle O}{\|}}{C}-OH \xrightarrow[\text{Phosphate catalyst}]{700°C} CH_2=C=O + H_2O \quad (2)$$

Carboxylic acids add to ketene to give acid anhydrides, as in reaction (3), and this is the basis of

$$CH_2=C=O + RCOOH \rightarrow$$

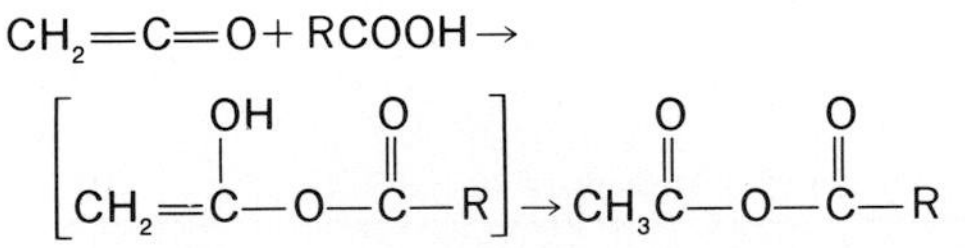

$$\left[CH_2=\overset{\overset{\displaystyle OH}{|}}{C}-O-\overset{\overset{\displaystyle O}{\|}}{C}-R \right] \rightarrow CH_3\overset{\overset{\displaystyle O}{\|}}{C}-O-\overset{\overset{\displaystyle O}{\|}}{C}-R \quad (3)$$

an important commercial process for the manufacture of acetic anhydride. Ketene reacts with the enol form of ketones to produce enol acetates, as in reaction (4). Molecules containing O—H or

$$CH_3\overset{\overset{\displaystyle O}{\|}}{C}-R \rightleftharpoons CH_2=\overset{\overset{\displaystyle OH}{|}}{C}-R \xrightarrow{CH_2=C=O}$$

$$\begin{array}{c} \overset{\overset{\displaystyle O}{\|}}{O-C-CH_3} \\ | \\ CH_2=C-R \end{array} \quad (4)$$

N—H bonds add to the carbonyl group, for example, reaction (5). Dimerization of ketene to diketene

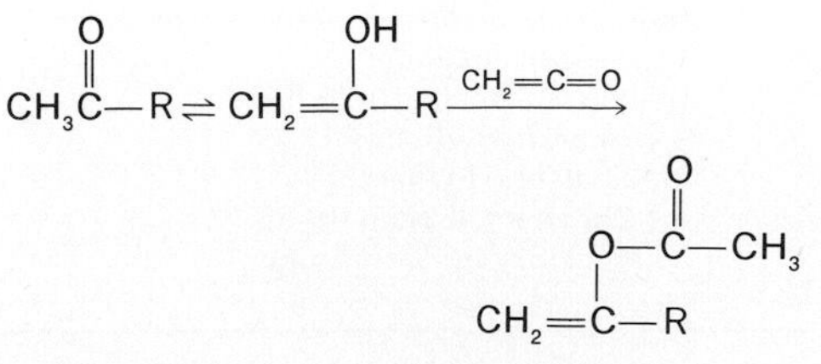

$$CH_2=C=O + R_2NH \rightarrow \left[CH_2=\overset{\overset{\displaystyle OH}{|}}{C}-NR_2 \right] \rightarrow$$

$$CH_3-\overset{\overset{\displaystyle O}{\|}}{C}-NR_2 \quad (5)$$

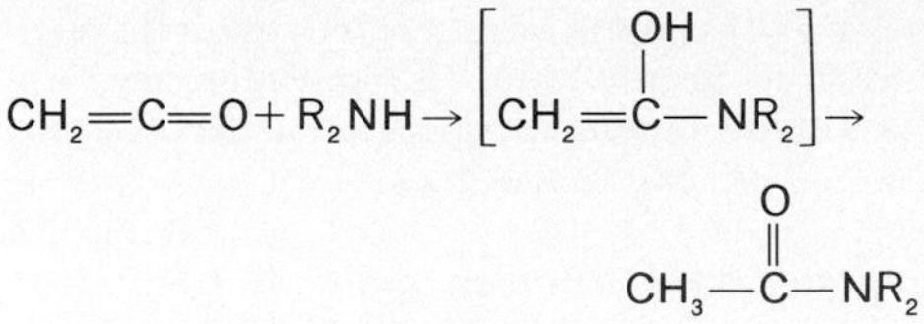

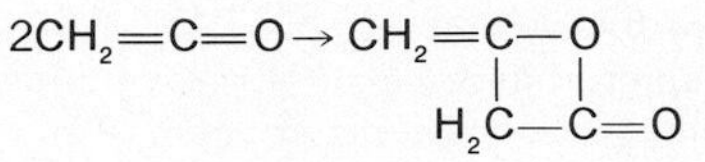

$$2CH_2=C=O \rightarrow \begin{array}{c} CH_2=C-O \\ |\qquad| \\ H_2C-C=O \end{array} \quad (6)$$

occurs on storage of ketene even at low temperature, as in reaction (6). See KETONE.

[DAVID A. SHIRLEY]

Bibliography: J. D. Roberts and M. C. Caserio, *Basic Principles of Organic Chemistry*, 1964.

Ketone

One of a class of chemical compounds of the general formula given below. In the formula,

$$\begin{array}{c} R \\ \diagdown \\ C=O \\ \diagup \\ R' \end{array}$$

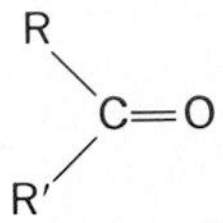

R and R' are alkyl, aryl, or heterocyclic radicals. The groups R and R' may be the same or different or incorporated into a ring as in $CH_2CH_2CH_2CH_2C=O$ (cyclopentanone). The ketones, acetone and methyl ethyl ketone, are used as solvents. Ketones are important intermediates in the syntheses of organic compounds.

Biacetyl, Camphor, Cyclopentadecanone

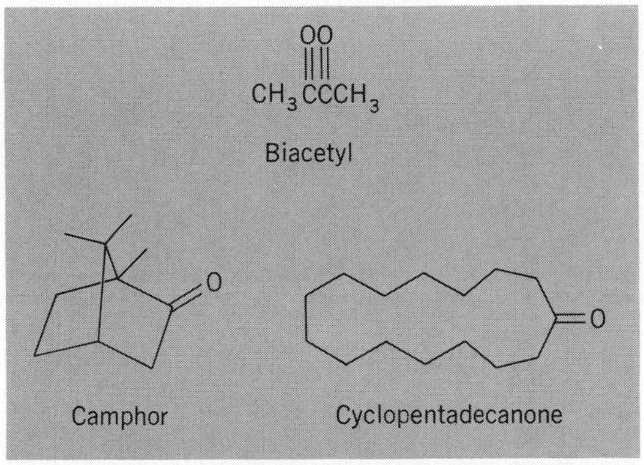

By common nomenclature rules, the R and R' groups are named, followed by the word ketone, for example, $CH_3CH_2COCH_2CH_3$ (diethyl ketone), $CH_3COCH(CH_3)_2$ (methyl isopropyl ketone), and $C_6H_5COC_6H_5$ (diphenyl ketone). The nomenclature of the International Union of Pure and Applied Chemistry uses the hydrocarbon name corresponding to the maximum number of carbon atoms in a continuous chain in the ketone molecule, followed by "one," and preceded by a number designating the position of the carbonyl group in the carbon chain. The first two ketones above are named 3-pentanone and 3-methyl-3-butanone.

Properties. The lower-molecular-weight ketones are colorless liquids. Acetone and methyl ethyl ketone are miscible with water; the water solubility of the higher homologs decreases with increasing number of carbon atoms. In the infrared absorption spectrum, the intense carbonyl stretching band near 1715 cm^{-1} is useful for identification and characterization of ketones. Because of their characteristic odors, various ketones are of use in the flavoring and perfumery industry. Biacetyl is a principal ingredient in the flavoring of margarine, camphor is valued for its medicinal odor, although it appears to have no therapeutic value, and cyclopentadecanone (Exaltone) has a musk odor and is used in perfumes.

Reactions of ketones. Addition to the carbonyl group is the most important type of ketone reaction. Ketones are generally less reactive than aldehydes in addition reactions. Methyl ketones are more reactive than the higher ketones because of steric group effects.

Hydrogen adds catalytically to the carbonyl group, and lithium aluminum hydride gives the same type of product—the secondary alcohol, as shown in reaction (1). The Grignard reagent adds

$$RCOR' \xrightarrow[\text{or } LiAlH_4]{H_2(\text{Ni or Pt})} RCHOHR' \qquad (1)$$

to the carbonyl group, and tertiary alcohols are formed by hydrolysis, reaction (2). Hydrogen cya-

$$RCOR' \xrightarrow{R''MgX} RC(OMgX)(R'')R' \xrightarrow[H^+]{H_2O} R\text{—}C(OH)(R'')\text{—}R' \qquad (2)$$

nide and sodium bisulfite add to methyl ketones, as in reaction (3). Alcohols do not add readily to the

$$CH_3COR \begin{cases} \xrightarrow{NaHSO_3} CH_3C(SO_3Na)(OH)R \\ \xrightarrow{HCN} CH_3C(CN)(OH)R \end{cases} \qquad (3)$$

ketone carbonyl as they do to the aldehyde carbonyl, but ketals may be formed by the action of orthoformates, as in reaction (4). Amine deriva-

$$RCOR' + HC(OC_2H_5)_3 \xrightarrow{H^+} R\text{—}C(OC_2H_5)_2R' + HCOOC_2H_5 \qquad (4)$$

tives such as hydroxylamine (NH_2OH), phenylhydrazine ($C_6H_5NHNH_2$), and semicarbazide ($NH_2CONHNH_2$) add to the carbonyl by breaking an N-H bond and with subsequent loss of water, as in reaction (5). The resulting oximes, phenylhydra-

$$RCOR' + C_6H_5NHNH_2 \xrightarrow{H^+} RC(=NNHC_6H_5)R' + H_2O \qquad (5)$$

zones, and semicarbazones are useful derivatives for the identification and characterization of ketones.

Ketones supply alpha hydrogen in aldol-type condensation reactions, as in reaction (6), but can

$$H_2CO + CH_3COR \xrightarrow{OH^-} RCOCH_2CH_2OH \xrightarrow{H^+} RCOCH{=}CH_2 + H_2O \qquad (6)$$

supply a carbonyl group only to a limited extent because of its lower reactivity. An exception is the self-condensation of acetone to diacetone alcohol represented by reaction (7).

$$2CH_3COCH_3 \xrightarrow{Ba(OH)_2} CH_3C(OH)(CH_3)CH_2COCH_3 \qquad (7)$$

Ketones are oxidized less readily and less selectively than aldehydes, to give oxidation products such as carboxylic acids, by cleavage of the bonds from the carbonyl carbon atom to the adjacent atom.

Methyl ketones give the haloform reaction with solutions of iodine in aqueous potassium hydroxide, reaction (8).

$$CH_3COR \xrightarrow[KOH]{I_2} CHI_3 + RCOOK \qquad (8)$$

Chlorine or bromine will substitute the alpha hydrogen atoms of ketones, as in reaction (9). The

$$RCH_2COR' + Cl_2 \rightarrow RCH(Cl)COR' + HCl \qquad (9)$$

resulting α-haloketones are quite reactive in displacement reactions of the halogen atom.

Preparation. Ketones are formed by the dehydrogenation or oxidation of secondary alcohols, as shown in reaction (10), and this is the method used

$$RCH(OH)R' \rightarrow RCOR' \qquad (10)$$

industrially for the preparation of acetone and methyl ethyl ketone. In the United States, acetone and methyl ethyl ketone are produced primarily by this reaction.

Aromatic ketones may be prepared by the Friedel-Crafts acylation reaction using either acid halides as shown in reaction (11) for the preparation of acetophenone, or anhydrides as shown in reaction (12) for the preparation of *o*-benzoylbenzoic acid.

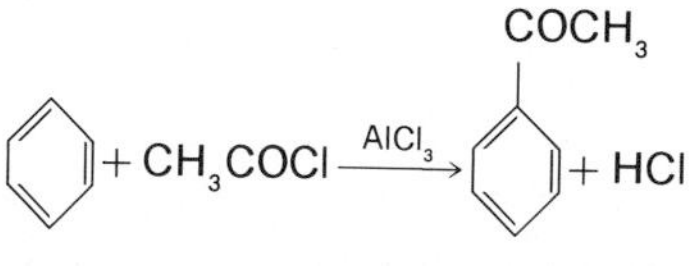

(11)

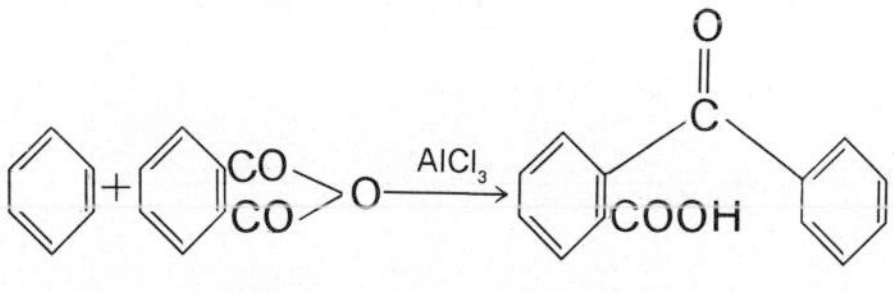

(12)

Organocadmium reagents give ketones on reaction with acid halides, reaction (13).

$$RCOCl + R'CdX \rightarrow RCOR' + CdXCl \qquad (13)$$

Beta-ketoesters formed in the Claisen condensation are cleaved by aqueous sodium hydroxide solution to form ketones, reactions (14) and (15). The

$$2RCH_2COOEt \xrightarrow{EtONa} RCH_2COCHRCOOEt + EtOH \qquad (14)$$

$$RCH_2COCHRCOOEt + H_2O \xrightarrow{NaOH} RCH_2COCH_2R + CO_2 + EtOH \quad (15)$$

Dieckmann condensation of esters of dibasic acids leads in similar fashion to cyclic ketones, reactions (16) and (17).

$$EtOOC(CH_2)_4COOEt \xrightarrow{NaOEt} CH_2CH_2CH_2COCHCOOEt \quad (16)$$

$$CH_2CH_2CH_2COCHCOOEt + H_2O \xrightarrow{NaOH} CH_2CH_2CH_2CH_2CO + CO_2 + EtOH \quad (17)$$

Both cyclic and open chain ketones are formed by pyrolysis of calcium or thorium salts of dibasic acids or monobasic acids, reactions (18) and (19).

$$\tfrac{1}{2}CaOOC(CH_2)_4COO\tfrac{1}{2}Ca \xrightarrow{Heat} CH_2CH_2CH_2CH_2CO + CaCO_3 \quad (18)$$

$$2RCOO\tfrac{1}{2}Ca \xrightarrow{Heat} RCOR + CaCO_3 \quad (19)$$

See ALDEHYDE; CONDENSATION REACTION; FRIEDEL-CRAFTS REACTION; GRIGNARD REACTION; HALOFORM REACTION; REFORMATSKY REACTION; STERIC EFFECT. [PAUL E. FANTA]

Bibliography: N. L. Allinger et al., *Organic Chemistry*, 2d ed., 1976.

Kinetic methods of analysis

The measurement of reaction rates for the analytical determination of the initial concentrations of the species of interest taking part in chemical reactions. This technique can be used since, in most cases, the rates or velocities of chemical reactions are directly proportional to the concentrations of the species taking part in the reactions.

The rate of a chemical reaction is measured by experimentally following the concentration of some reactant or product involved in the reactions as a function of time as the reaction mixture proceeds from a nonequilibrium to an equilibrium or static state (steady state). Thus, kinetic techniques of analysis have the inherent problem of the increased experimental difficulty of making measurements on a dynamic system, as time is now a variable which is not present in making measurements on an equilibrium system. However, kinetic methods often have advantages over equilibrium techniques in spite of the increased experimental difficulty. For example, the equilibrium differentiations or distinctions attainable for the reactions of very closely related compounds are often very small and not sufficiently separated to resolve the individual concentrations of a mixture without prior separation. However, the kinetic differentiations or distinctions obtained when such compounds are reacted with a common reagent are often quite large and permit simultaneous analysis.

A further advantage of kinetic methods is that they permit a larger number of chemical reactions to be used analytically. Many reactions, both inorganic and organic, are not sufficiently well-behaved to be employed analytically by equilibrium or thermodynamic techniques. Many reactions attain equilibrium too slowly; side reactions occur as the reactions proceed to completion, or the reactions are not sufficiently quantitative (do not go to completion) to be applicable. However, a kinetic-based technique can often be employed in these cases simply by measuring the reaction rate of these reactions during the early or initial portion of the reaction period. Also, the measurement of the rates of catalyzed reactions generally is a considerably more sensitive analytical method for the determination of trace amounts of a large number of species than equilibrium methods. *See* CATALYSIS; CHEMICAL DYNAMICS.

Kinetic methods of analysis can be divided into three basic categories: (1) methods employing uncatalyzed reactions, (2) methods employing catalyzed reactions, and (3) methods for the simultaneous in-place determination of mixtures.

Uncatalyzed reactions. In order to determine the concentration of a single species in solution, the rate is measured of an irreversible uncatalyzed reaction of the type shown by (1), where the single

$$A + R \xrightarrow{K_A} P \quad (1)$$

species A reacts with the reagent R to form a product or products P. In general, the types of reaction employed for kinetic analysis are second-order irreversible reactions. The rate of the reaction is most conveniently measured by following the rate of formation of the product P as a function of time (although the change in concentration of either A or R could also be followed). Any method of measuring concentrations, such as titrimetry, spectrophotometry, and polarography, can be employed. The rate of formation of the product, $d[P]_t/dt$, is given by differential equation (2). In this relation $[A]_0$ is the initial or original concentration of

$$\frac{d[P]_t}{dt} = k_A([R]_0 - [P]_t)([A]_0 - [P]_t) \quad (2)$$

the species to be determined, $[R]_0$ is the initial concentration of the reagent (added in known concentration), and $[P]_t$ is the concentration of product formed at any time t; thus, the terms $([R]_0 - [P]_t)$ and $([A]_0 - [P]_t)$ equal the concentration of A and R, respectively, remaining at any time t, and k_A is the

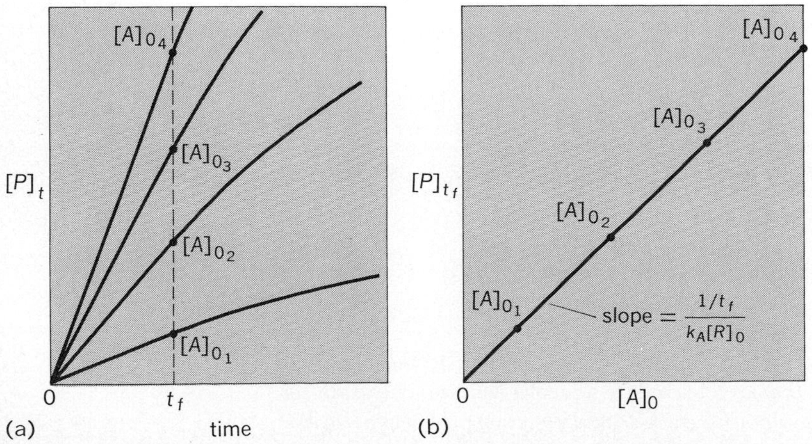

Fig. 1. Fixed-time method for uncatalyzed reactions. (*a*) Variation of $[P]$ as a function of time for different initial concentrations of $[A]_0$. (*b*) Variation of amount of P formed at fixed time t_f with initial concentration of A.

second-order rate constant for the reaction of A with R.

If the rate of the reaction given by Eq. (1) is measured only during the initial portion of the reaction (during a time period chosen so that the reaction is only 2–3% complete), the concentration of product formed $[P]_t$ is small compared to $[A]_0$ and $[R]_0$. Thus, on rearrangement, Eq. (2) becomes Eq. (3). Integration of Eq. (3) between the

$$[A]_0 = \frac{d[P]_t/dt\,(\text{initial})}{k_A[R]_0} \tag{3}$$

time interval $t=0$ to t and for an initial concentration of $P=0$ at $t=0$ gives Eq. (4). Both fixed- and

$$[A]_0 = \frac{[P]_t/t}{k_A[R]_0} \tag{4}$$

variable-time methods may be used to calculate $[A]_0$ when employing Eq. (4).

When using the fixed-time method, the concentration of P formed at a chosen fixed time t_f is measured experimentally. As t_f and $[R]_0$ are thus kept constant for all measurements and k_A is a constant for all conditions, the term $\dfrac{1/t_f}{k_A[R]_0}$ is the proportionately constant relating $[A]_0$ to $[P]_{t_f}$, as shown in Fig. 1. For the variable-time method, with $[R]_0$ constant for all solutions, the time $t_{[P]_f}$ necessary for the reaction to form a fixed amount of product $[P]_f$ is measured. As $[P]_f$ is now a constant, $[A]_0$ is related to $1/t_{[P]_f}$ by the proportionately constant, $P_f/k_A[R]_0$, as shown in Fig. 2.

Catalyzed reactions. A catalyst may be broadly defined as an agent which alters the rate of a chemical reaction without shifting the equilibrium of the reaction. Although the catalyst undoubtedly enters into the reaction mechanism at a critical stage, it does so in a cyclic manner and hence does not undergo a permanent change. Therefore a catalyst speeds up the rate of attainment of the equilibrium of a system, but it does not change the position of the equilibrium and it is not consumed during the reaction.

In many reactions involving a substance which acts as a catalyst, the concentration of the catalyst is directly proportional to the rate of the reaction. Thus the rates of these reactions can be employed for the analytical methods and are extremely sensitive since the catalytic agents are not consumed but participate in the mechanism in a cyclic manner. The amount of the catalyst that can be determined employing catalyzed reactions is several orders of magnitude smaller than can be found by most direct equilibrium methods. Also, in many cases—enzyme-catalyzed (biological catalyst) reactions in particular—the catalyzed reaction of a reactant is extremely specific with respect to the chemical nature of the reactant. Such reactions, therefore, can be employed for the in-place analysis of a particular reactant in the presence of either a large excess or a large number of other species which would interfere with conventional equilibrium techniques, unless separation was performed prior to the analysis reaction.

The general mechanism for catalyzed reactions involves the combination of the catalyst C and the reactant, called the substrate S, to form a "complex" X, which then decomposes to form the product. Thus the catalyst is regenerated and combines

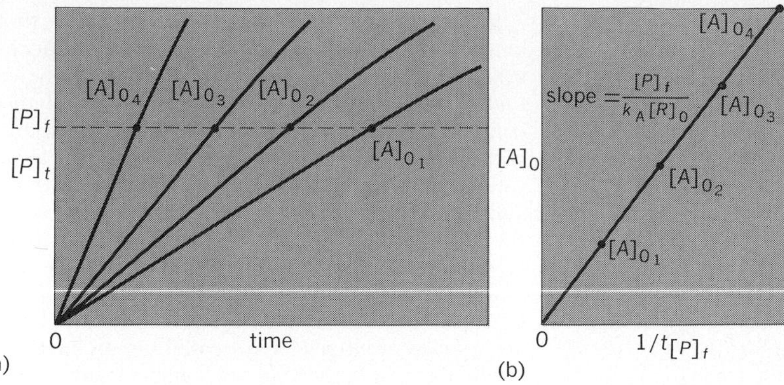

Fig. 2. Variable-time method for uncatalyzed reactions. (a) Variation of $[P]$ as a function of time for various initial concentrations of $[A]_0$. (b) Variation of $1/t_{[P]_f}$ required to reach a fixed value of product formed $[P]_f$ with initial concentration of A.

once more with a substrate molecule. The general mechanism is given by reactions (5) and (6),

$$C + S \underset{k_{-1}}{\overset{k_1}{\rightleftarrows}} X \tag{5}$$

$$X + R \overset{k_2}{\rightarrow} P + C \tag{6}$$

where R is a reagent molecule (which in many catalyzed reactions is not needed) that reacts with the complex to give the product P, and k_1 and k_{-1} are the rate constants for the formation of the complex and for the conversion of the complex back to C and S, respectively; the equilibrium constant K for the reaction given by (5) is equal to k_1/k_{-1}, and k_2 is the rate constant for the decomposition of X to form the product and C, as in (6). In many catalyzed reactions the rate of the decomposition reaction (6) is slow compared to the rate of the conversion back to C and S ($k_2 \ll k_{-1}$). Thus, in this case, (6) is the rate-determining step. Under these conditions, the rate of reaction of S in a catalyzed process is given by Eq. (7).

$$-\frac{d[S]_t}{dt} = \frac{k_2[C]_0[S]_0[R]_t}{[S]_0 + 1/K} \tag{7}$$

When the equilibrium constant K is relatively small, so that $[S]_0$ is small compared to $1/K$, and when R is in a nonlimiting excess (or nonexistent) such that its concentration does not change during the initial 2–3% of the reaction, Eq. (7) becomes Eq. (8), where the constant K' equals $k_2K[R]_0$. Thus

$$-\frac{d[S]_t}{dt} = (K')[C]_0[S]_0 \tag{8}$$

the initial rate, $-(d[S]_t)/(dt)$, is directly proportional to the initial concentration of either the catalyst or the substrate, and the initial concentration of either can be determined. If a catalyst is to be determined, the substrate is added to each sample in nonlimiting excess (so that the concentration of S does not change during the initial 2–3% of the reaction). If a substrate is to be measured, the catalyst concentration added is not critical in any way because the catalyst is not consumed during the reaction. The choice of the catalyst concentration is dictated only by the range of initial rates that are conveniently measured. Either the fixed- or variable-time method of analysis can be employed using catalyzed reactions.

Simultaneous determinations. Often the thermodynamic properties of closely related species are such that conventional equilibrium analytical techniques are unable to resolve the analytical concentrations of the components of the mixture without prior separation. However, in general the rates of reaction of such species in a mixture are sufficiently different to enable one to determine the initial concentration of each species without resorting to separation, which is a considerable saving of time and labor. Such techniques are called differential kinetic analysis methods. There are a large number of differential kinetic methods applicable to both first- and second-order reactions. Each of these methods has special conditions under which it is most applicable. The principles of these methods are complex and an explanation of each is beyond the scope of this section. However, the detailed explanation of the logarithmic extrapolation method given below illustrates the general concepts of differential kinetic methods.

Consider two competing second-order irreversible reactions of the type shown by (9) and (10),

$$A + R \xrightarrow{k_A} P \qquad (9)$$

$$B + R \xrightarrow{k_B} P \qquad (10)$$

where k_A and k_B are the second-order rate constant for the reaction of A and B with R, respectively. If the concentration of R is 50 to 100 times greater than the sum of the concentration of A plus B (the components of interest in the analysis), the reaction is pseudo first order because $[R]_t$ is constant and is equal to the initial concentration $[R]_0$. The differential rate expressions for the reactions given by (9) and (10) then become Eqs. (11) and (12).

$$-\frac{d[A]_t}{dt} = k_A[R]_0[A]_t = k'_A[A]_t \qquad (11)$$

$$-\frac{d[B]_t}{dt} = k_B[R]_0[B]_t = k'_B[B]_t \qquad (12)$$

The sum of the concentrations of A and B which react competitively to form a common product P is given at any time t by the rate expression, as shown by Eq. (13).

$$\frac{d[P]_t}{dt} = -\left[\frac{d[A]_t}{dt} + \frac{d[B]_t}{dt}\right] = k'_A[A]_t + k'_B[B]_t \qquad (13)$$

On integration of Eq. (13) between the limits $t=0$ and $t=\infty$, Eq. (14) is formed. In the case where the rate of reaction of component A is larger than that of B, the term $[A]_0e^{-k'_At}$ eventually becomes very small compared to $[B]_0e^{-k'_Bt}$ at some time t after A has reacted essentially to completion ($[A]_t \cong 0$) and can be considered negligible. Thus, by taking the logarithm of both sides of Eq. (14),

$$[P]_\infty - [P]_t = [A]_t + [B]_t$$
$$= [A]_0e^{-k'_At} + [B]_0e^{-k'_Bt} \qquad (14)$$

Eq. (15) is obtained, which predicts that a plot

$$\ln([A]_t + [B]_t) = \ln([P]_\infty - [P]_t)$$
$$= -k'_Bt + \ln[B]_0 \qquad (15)$$

$\ln([A]_t + [B]_t)$ or $\ln([P]_\infty - [P]_t)$ versus time t will yield a straight line with a slope of $-k'_B$ and an intercept (at $t=0$) equal to $\ln[B]_0$. The value of $[A]_0$ is then obtained by subtracting $[B]_0$ from the total initial concentration of the mixture, $[A]_0 + [B]_0$, which must be determined either by independent methods or from $[P]_\infty$. A typical reaction rate curve of this type is illustrated in Fig. 3. *See* CHAIN REACTION; ELECTROCHEMICAL TECHNIQUES; INHIBITOR.

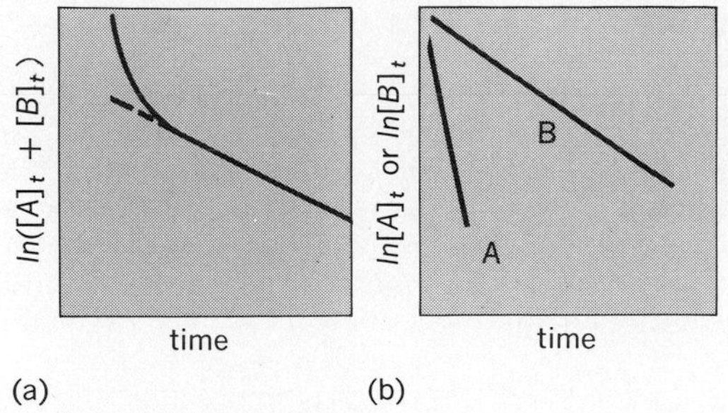

Fig. 3. Logarithmic extrapolation method for a mixture of species A and B reacting by first-order kinetics. (*a*) Rate data obtained for the mixture. (*b*) Rate data obtained for each component separated.

[HARRY B. MARK, JR.]

Bibliography: S. W. Benson, *Foundations of Chemical Kinetics*, 1980; A. A. Frost and R. G. Pearson, *Kinetics and Mechanism: A Study of Homogeneous Reactions*, 2d ed., 1960; K. J. Laidler, *Chemical Kinetics*, 2d ed., 1965; H. B. Mark, Jr., G. A. Rechnitz, and R. A. Greinke, *Kinetics in Analytical Chemistry*, 1968; R. W. Ramette, *Chemical Equilibrium and Analysis*, 1981; C. N. Reilley, *Advances in Analytical Chemistry*, vol. 2, 1963; K. B. Yatsimerskii, *Kinetic Methods of Analysis*, 1966.

Kolbe hydrocarbon synthesis

The production of an alkane by the electrolysis of a water-soluble salt of a carboxylic acid. The carboxylate ions are discharged at the anode, yielding carbon dioxide and alkyl radicals which couple to form the saturated hydrocarbon. Hydrogen is liberated at the cathode, as in the equation below.

$$2RCOONa + 2H_2O \rightarrow RR + 2CO_2 + 2NaOH + H_2$$

Good yields of alkanes are obtained with the straight-chain acids containing 5–18 carbon atoms. Alkenes rather than alkanes are formed if the acid contains an alkyl branch at the carbon atom adjacent to the carboxyl group, that is, at the α position of the acid. *See* ALKANE.

[LOUIS SCHMERLING]

Krypton

A gaseous chemical element, Kr, atomic number 36, and atomic weight 83.80. Krypton is one of the noble gases in group 0 of the periodic table. *See* INERT GASES.

Uses. The principal use for krypton is in filling electric lamps and electronic devices of various types. Krypton-argon mixtures are widely used to fill fluorescent lamps.

The higher the molecular weight of the inert gas used to fill tungsten-filament lamp bulbs, the less the tendency of the tungsten to evaporate when heated; thus, krypton is a better filling gas, although more expensive, than argon. Two different uses are made of this advantage: Krypton-filled lamps can be operated at the same brightness as

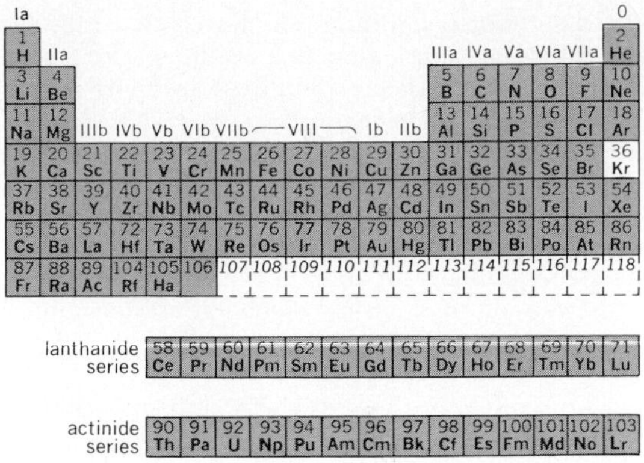

those filled with argon, in which case the life of the lamp is a good deal longer; or they can be operated at higher filament temperatures to give a brighter light at greater efficiency (more lumens per watt), but the latter alternative reduces the life of the lamp somewhat. Brighter lamps are used in slide projectors and projectors for home movies.

Krypton is also used to fill electric-arc lamps. An example is the lamp which will pierce fog for 1000 ft or more. Lamps of this type are arranged in rows to mark airplane runways at night. They flash 40 times a minute; each flash lasts only 17 μsec so as not to blind the pilots.

Radioactive krypton-85 is relatively inexpensive and is finding a number of uses. One is in leak-testing of sealed containers. Another is in the continuous measurements of the thicknesses of materials, such as sheets of metal and plastics. Another use is in lamps which give off light for several years with no source of energy other than the radioactivity of the krypton; in these, the invisible radiation from the krypton activates a phosphor coated on the inner glass walls of the lamp, and the phosphor gives off light continuously until the krypton has decayed to a low level of radioactivity. Krypton-85 can be used to detect abnormal heat openings; the introduction of a small quantity of this gas into the human body is practical because the krypton is not retained by the body; it remains there for only a short time, working its way out by way of the bloodstream and lungs.

Occurrence. The only commercial source of stable krypton is the air, although traces of krypton are found in minerals and meteorites. Krypton constitutes 1.14 ppm by volume of the Earth's atmosphere, and this krypton is almost entirely a mixture of the following isotopes, none of which is radioactive: 78 (0.35%), 80 (2.27%), 82 (11.56%), 83 (11.55%), 84 (56.90%), and 86 (17.37%). The relative abundance of the particular isotope is given in parentheses after each mass number.

A mixture of stable and radioactive isotopes of krypton is produced in nuclear reactors by the slow-neutron fission of uranium.

It is estimated that about 2×10^{-8}% of the weight of the Earth is krypton. Krypton also occurs outside the Earth; the best estimate is that there are about 51 atoms of krypton for each 1,000,000 atoms of silicon in the visible universe, silicon being used as a standard of abundance.

Discovery. Krypton was discovered in England in 1898 by Sir William Ramsay and M. W. Travers; they found it in the less volatile part of the inert-gas mixture left after the oxygen and nitrogen had been removed chemically from a sample of air. The fact that a new element was present was ascertained by the discovery of new lines in the emission spectrum of the residual gas.

Radioactive isotopes. The following radioactive isotopes of krypton are known: Kr^{76}, Kr^{77}, Kr^{79}, Kr^{81}, Kr^{83m} (metastable form of Kr^{83}), Kr^{85}, Kr^{87}, Kr^{88}, Kr^{89}, Kr^{90}, Kr^{91}, Kr^{92}, Kr^{93}, Kr^{94}, Kr^{95}, and Kr^{97}. As mentioned above, they are produced as by-products of the nuclear fission of uranium in nuclear reactors. They can also be formed in particle accelerators, such as the cyclotron, or by the neutron bombardment of the appropriate atomic species. The only radioactive isotope which is produced in more than traces and also has a reasonably long lifetime (half-life about 10 years) is krypton-85.

Properties. Krypton is a colorless, odorless, and tasteless gas. Table 1 gives some physical properties of krypton. The outer shell of the krypton atom is filled with electrons in a stable structure. For this reason there is only one atom in a molecule, and no one has yet been able to prepare a chemical compound of krypton which is stable at room temperature. However, the compound KrF_4 has been prepared; it is stable only at −80°C and below. *See* KRYPTON COMPOUNDS.

Table 1. Physical properties of krypton

Property	Value
Atomic number	36
Atomic weight (atmospheric krypton only)	83.80
Melting point, triple point °C	−157.20
Boiling point at 1 atm pressure, °C	−153.35
Gas density at 0°C and 1 atm pressure, g/liter	3.749
Liquid density at its boiling point, g/ml	2.413
Solubility in water at 20°C , ml krypton (STP) per 1000 g water at 1 atm partial pressure krypton	59.4

Krypton can be trapped in crystals of hydroquinone, phenol, and other host compounds to form clathrates. The clathrate of hydroquinone is stable at room temperature and has found a use as a convenient method of supplying radioactive Kr^{85}. *See* CLATHRATE COMPOUNDS.

Production. Stable krypton is produced in air-separation plants. Air is liquefied and distilled; krypton and xenon remain with the oxygen. The liquid oxygen is redistilled to concentrate the krypton and xenon from a few parts per million to a few percent. The rare gases are then adsorbed from the liquid oxygen onto silica gel, desorbed, separated, and purified. Final purification is carried out by passing the krypton over hot titanium metal, on which all except inert gas impurities are removed.

Of the radioactive krypton isotopes produced in nuclear reactors, the only one that has a half-life of over about 3 hr is krypton-85; this isotope has a half-life of about 10 years. Thus, when krypton from a nuclear reactor has been stored for several days, the only radioactive isotope left is krypton-85. Table 2 gives the approximate isotopic composition of the krypton thus produced. The composition of the krypton mixture varies somewhat with conditions in the reactor. As nuclear reactors come into wide use for power production, a large

Table 2. Isotopes in reactor-produced krypton

Mass number	Percentage by volume	
	Stable	Radioactive
78, 80, 82	1–3	
83	13–14	
84	29–33	
85		5–6.5
86	46–50	

amount of radioactive krypton will become available. Since no cheap method of separating the radioactive from the stable isotopes of krypton is known, the nuclear reactor is not at present a source of stable krypton, even though about 95% produced in it consists of stable isotopes.

Analytical methods. The principal modern methods of detecting and quantitatively determining the krypton content in gases are mass spectrometry and gas chromatography. Until these methods were developed, it was necessary to separate krypton from other gases by selective low-temperature adsorption on activated carbon in order to determine how much krypton was present in a mixture. The older method of detecting krypton makes use of its characteristic emission spectrum, obtained by passing a gas sample through an electric discharge tube at low pressures and analyzing the light with a spectrometer.

[ARTHUR W. FRANCIS]

Bibliography: I. Asimov, *The Noble Gases*, 1966; G. A. Cook, *Argon, Helium and the Rare Gases*, 2 vols., 1961; B. L. Smith and J. P. Webb, *The Inert Gases: Model Systems for Science*, 1971.

Krypton compounds

Compounds in which krypton is one of the combining elements. Until 1962, krypton, like the other noble gases, was thought to be chemically inert. Even now the only well-characterized krypton compounds are the difluoride, KrF_2, and its complexes. An early report of a tetrafluoride, KrF_4, has not been confirmed, and there is no convincing evidence for the existence of krypton compounds in aqueous solution.

Krypton difluoride can be prepared by passing an electric discharge through a 1:1 mixture of gaseous fluorine and krypton at −188°C. After the reaction is complete, any excess krypton and fluorine are pumped away, the reactor is slowly warmed, and the KrF_2 is sublimed into a glass vessel. KrF_2 can also be prepared by the action of ionizing radiation on a gaseous mixture of Kr and F_2 at −60 to −150°C and by the action of near-ultraviolet light on a solution of krypton in liquid fluorine. *See* FLUORINE.

Krypton difluoride is a volatile solid at room temperature and is easily sublimed to give clear, colorless crystals. Spectroscopic studies show that the KrF_2 molecule is linear and symmetrical. The compound decomposes slowly at room temperature to krypton and fluorine, but it is stable at −78°C. Reaction of KrF_2 with water is very rapid, yielding krypton, oxygen, and aqueous HF. Reaction with chlorine compounds produces ClF_5. Krypton difluoride reacts with strong fluoride acceptors, such as antimony pentafluoride, to form fluorine-bridged addition complexes that may be regarded as containing the cations KrF^+ and $Kr_2F_3^+$. Some of these complexes are more stable than KrF_2 itself. However, they are also very powerful fluorinating agents and have been used to synthesize other novel compounds, such as AuF_5.

The short-lived excited molecule KrF^* is formed when a mixture of krypton and a gaseous fluorine-containing compound is excited with an electron beam or an electrical discharge. It decomposes with the emission of ultraviolet light at 248 nm and is of interest as a laser.

[EVAN H. APPELMAN: JOHN G. MALM]

Bibliography: N. Bartlett and F. O. Sladky, The chemistry of krypton, xenon, and radon, *Comprehensive Inorganic Chemistry*, vol. 1, 1973; H. H. Claassen, *The Noble Gases*, 1966; J. G. Malm and E. H. Appelman, The chemical compounds of xenon and other noble gases, *At. Energ. Rev.*, 7(3), 1969.

Lactam and lactim

Cyclic amides, the nitrogen analogs of lactones, are called lactams. For example, γ-aminobutyric acid readily forms γ-butyrolactam (also known as pyrrolidone) upon heating, as in reaction (1). The tautomeric enol form of a lactam is known as a lactim.

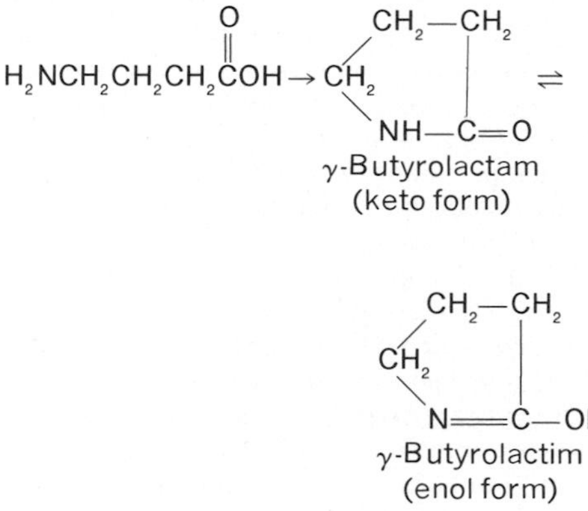

(1)

The δ-amino acids similarly form δ (six-membered-ring) lactams upon heating, but larger- and smaller-ring lactams must be made by indirect methods, such as addition of ketenes to imines for β-lactams and ring expansion via Beckmann rearrangement of cyclic ketoximes for ε- and higher lactams.

Several lactams are of considerable industrial importance. Pyrrolidone is made by the treatment of γ-butyrolactone with ammonia and is a useful specialty solvent. Upon addition to acetylene, it forms vinylpyrrolidone, which is polymerized to polyvinylpyrrolidone, as in reaction (2), and is used in aerosol hair sprays.

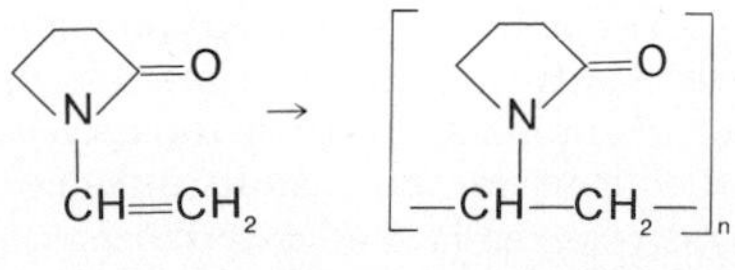
(2)

ε-Caprolactam is made commercially by the Beckmann rearrangement of cyclohexanone oxime, and polymerizes when subjected to heat and pressure to form the fibrous polymer known as nylon-6, as in reaction (3).

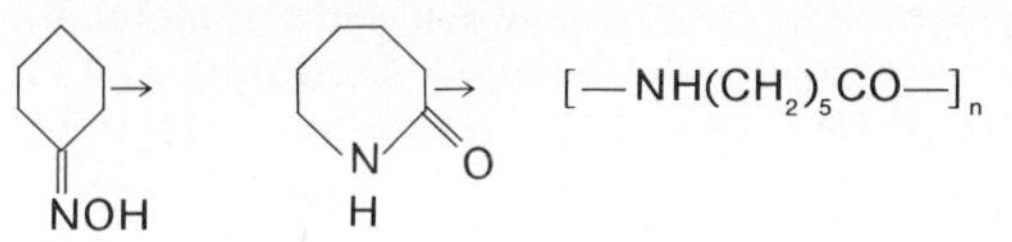
(3)

See LACTIDE AND LACTONE.

[PAUL E. FANTA]

Bibliography: N. L. Allinger et al., *Organic Chemistry*, 2d ed., 1976.

Lactate

A salt or ester of lactic acid (2-hydroxypropanoic acid, α-hydroxypropionic acid) in which the acidic hydrogen of the carboxyl group has been replaced by a metal or an organic radical (symbol X). The structural formulas of the two optical isomers are

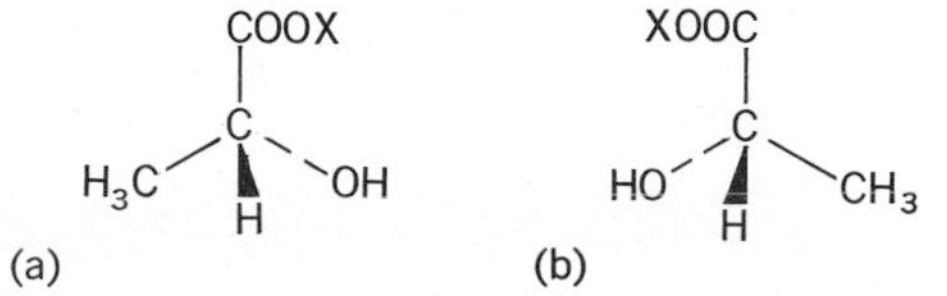

Optical isomers of lactic acid. (*a*) D(−) Form. (*b*) L(+) Form.

given in the illustration. Alkali metal salts have practical applications, for example, as blood coagulants in calcium therapy (calcium), as antiperspirants (aluminum), in treatment of anemia (iron), and as plasticizers (sodium). Lower-molecular-weight esters are water-soluble liquids. Esters are used as plasticizers and as solvents for lacquers, cellulose acetate, and cellulose nitrate. *See* CARBOXYLIC ACID; ESTER; SOLVENT.

[ELBERT H. HADLEY]

Lactide and lactone

Lactides are cyclic, intermolecular double esters formed from α-hydroxy acids. Thus, lactic acid, $CH_3CHOHCOOH$, on heating forms the lactide shown in (1). Glycolic acid, $HOCH_2COOH$, be-

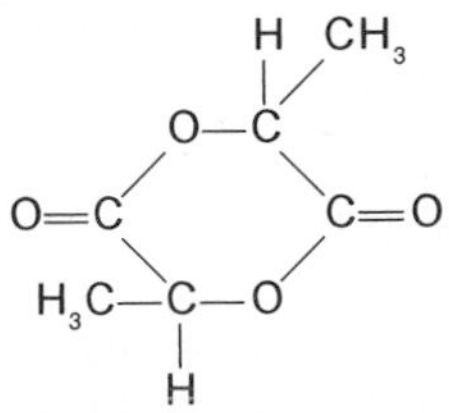
(1)

haves analogously. In each case bimolecular interaction occurs, forming the strain-free 6-membered ring. Most lactides are relatively low-melting solids and are easily hydrolyzed by base to form salts of the parent acids, for example, sodium lactate.

Lactones are internal cyclic mono esters formed by γ- or δ-hydroxy acids spontaneously; thus, γ-hydroxybutyric acid, $HOCH_2CH_2CH_2COOH$, forms γ-butyrolactone, with structural formula (2).

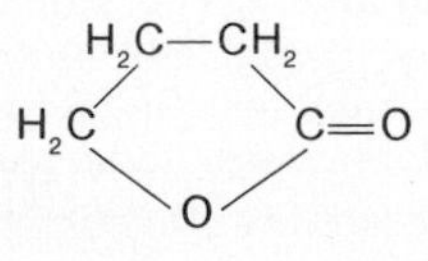
(2)

Other lactones of smaller or greater ring size are prepared specially.

The γ- and δ-lactones are commonly prepared by either hydrolysis or distillation of γ- or δ-halo acids, by treatment of unsaturated acids with aqueous hydrobromic or sulfuric acids, or by partial reduction of cyclic acid anhydrides. β-Lactones result from the reaction of ketene with aldehydes or ketones. Reaction of ketene with formaldehyde is shown in reaction (3). Large-ring lactones

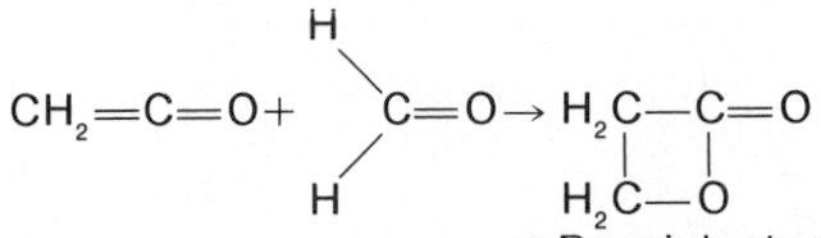
(3)

β-Propiolactone

can be made by oxidation of cyclic ketones with Caro's acid; thus, cyclohexanone yields ε-caprolactone.

The lower lactones are neutral liquids that react with bases (alkalies, ammonia, and amines) to give open-chain derivatives of the parent hydroxy acids. Very-large-ring (macrocyclic) lactones, for example 15 or 16 carbons, have pronounced musk odors (perfumes).

Unsaturated lactones (butenolides) are widely distributed: for example, the angelica lactones; penicillic acid and protoanemonine (mold metabolites); coumarin (tonka bean); and dicoumarol (spoiled sweet clover), used medicinally as a hemorrhaging or anticlotting agent in coronary thrombosis. *See* ESTER.

[EVANS B. REID]

Lanolin

The hydrous sheep's-wool wax (primarily cholesterol esters of higher fatty acids) derived as a by-product from the preparation of raw wool for the spinner. The crude wool wax is purified, melted, mixed with about 30% water, and allowed to harden into a soft, pale-yellow, unguentlike substance.

Lanolin is widely used as a base for emollients in cosmetics and shampoos. Perhaps because of its slight antiseptic effect and high resistance to rancidity it has continued wide use as a base for skin ointments. Actually, lanolin hinders absorption of the medication through the skin, as compared to other substances such as olive oil or lard. Lanolin is soluble in alcohol, ether, chloroform, and benzene.

[FRANK H. ROCKETT]

Bibliography: D. R. Crump, *The Composition of Wool Wax and Some Notes on the Chemistry of Lanosterol*, N. Z. Dept. Sci. Ind. Res. Chem. Div. Rep. 2203, 1975; D. Swern (ed.), *Bailey's Industrial Oil and Fat Products*, 4th ed., 1979.

Lanthanide contraction

The name given to an unusual phenomenon encountered in the rare-earth series of elements. The radii of the atoms of the members of this series decrease slightly as the atomic number increases. Starting with element 58 in the periodic table, the balancing electron fills in an inner incomplete 4*f* shell as the charge on the nucleus increases. According to the theory of atomic structure, this shell can hold 14 electrons; so starting with element 58, cerium, there are 14 true rare earths. Lanthanum has no electrons in the 4*f* shell, cerium has 1, and

Atomic and ionic radii of rare-earth metals

Element	Radius, A 3+ion[a]	Metal crystal structure[b]	Metallic radii, A	
			c	*d*
Sc		hcp	1.6545	1.6280
Y		hcp	1.8237	1.7780
La	1.061	hcp	1.8852	1.8694
Ce	1.034	fcc	1.8248	
Pr	1.013	hcp	1.8363	1.8201
Nd	0.995	hcp	1.8290	1.8139
Pm	0.979			
Sm	0.964	rhom-hcp	1.8105	1.7943
Eu	0.950	bcc	1.994	
Gd	0.938	hcp	1.8180	1.7865
Tb	0.923	hcp	1.8005	1.7626
Dy	0.908	hcp	1.7952	1.7515
Ho	0.894	hcp	1.7887	1.7428
Er	0.881	hcp	1.7794	1.7340
Tm	0.869	hcp	1.7688	1.7237
Yb	0.858	fcc	1.9397	
Lu	0.848	hcp	1.7516	1.7171

[a]Data from D. H. Templeton and Carol H. Dauben, *J. Amer. Chem. Soc.*, 76:5237–5239, 1954.

[b]Data from F. H. Spedding, A. H. Daane, and K. W. Herrmann, *Acta Cryst.*, 9(7):559–563, 1956; hcp, hexagonal close-packed; fcc, face-centered cubic; rhom, rhombic; bcc, body-centered cubic.

[c]Data from K. W. Herrmann, Doctoral thesis; radii calculated from atoms in basal plane.

[d]Data from K. W. Herrmann, Doctoral thesis; radii between layers.

lutetium, 14. The 4*f* electrons play almost no role in chemical valence; therefore, all rare earths can have three electrons in their valence shell and they all exist as trivalent ions in solution. As the charge on the nucleus increases across the rare-earth series, all electrons are pulled in closer to the nucleus so that the radii of the rare-earth ions decrease slightly as the compounds go across the rare-earth series. Any given compound of the rare earths is very likely to crystallize with the same structure as any other rare earth. However, the lattice parameters become smaller and the crystal denser as the compounds proceed across the series. This contraction of the lattice parameters is known as the lanthanide contraction. For many compounds the lattice parameters decrease only partway across the series, and when the contraction has progressed to that point, a new crystalline form develops. Frequently, both crystalline forms can be observed for a number of the elements. For this reason, the rare-earth series is of particular interest to scientists because many of the parameters determining the properties of a substance can be kept constant while the lattice spacings can be varied in small increments across the series.

The atomic and ionic radii of atoms are not clearly defined. The atoms can be polarized by the neighboring atoms and there is no clear-cut boundary between the electrons associated with one atom and another. Therefore, the atomic radii will vary somewhat from compound to compound, and the absolute values depend on the method of calculation. However, if most of the parameters are assumed constant, and the difference in lattice parameters in the rare-earth crystalline series is attributed to the rare-earth ion or atom, then the lanthanide contraction becomes clearly evident. Although scandium and yttrium are not members of this series, the information is usually wanted at the same time and is given for completeness. The atomic radii of the trivalent ion and the metal atoms are given in the table. *See* PERIODIC TABLE; RARE-EARTH ELEMENTS. [FRANK H. SPEDDING]

Lanthanum

A chemical element, La, atomic number 57, atomic weight 138.91. Lanthanum, the second most abundant element in the rare-earth group, is a metal. The naturally occurring element is made up of the isotopes La^{138}, 0.089%, and La^{139}, 99.91%.

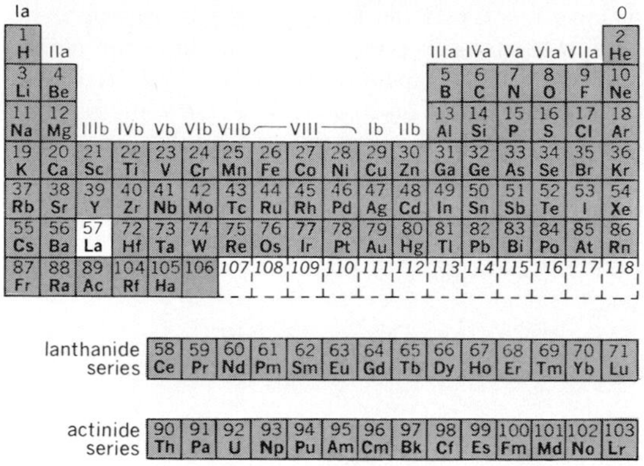

La^{138} is a radioactive positron emitter with a half-life of 1.1×10^{11} years. The element was discovered in 1839 by C. G. Mosander and occurs associated with other rare earths in monazite, bastnasite, and other minerals. It is one of the radioactive products of the fission of uranium, thorium, or plutonium. Lanthanum is the most basic of the rare earths and can be separated rapidly from other members of the rare-earth series by fractional crystallization. Considerable quantities of it are separated commercially, since it is an important ingredient in glass manufacture. Lanthanum imparts a high refractive index to the glass and is used in the manufacture of expensive lenses. The metal is readily attacked in air and is rapidly converted to a white powder. For other properties of the metal *see* RARE-EARTH ELEMENTS.

Lanthanum becomes a superconductor below about 6 K in both the hexagonal and face-centered crystal forms. [FRANK H. SPEDDING]

Laser photochemistry

The branch of physical chemistry in which chemical reactions are induced, altered, or monitored by laser light. Lasers have had an immense impact on the field of photochemistry by providing scientists with an extremely intense, nearly monochromatic source of light. There are lasers that extend from wavelengths of 157 nm (ultraviolet) to 10,600 nm (infrared) [1 nm = 10^{-9} m]; for comparison, the entire visible spectrum extends from only 400 nm (violet) to 700 nm (red). Figure 1 displays some of the more common laser systems and their operating wavelengths.

Advantages of lasers. The main advantages of a laser over a conventional light source (such as discharge or arc lamps) for study of photochemistry are threefold. First, laser light has a very small divergence angle, typically less than 1/100 of a degree; over the length of a football field the beam diameter grows by less than 2 cm. The high degree of collimation of laser light permits very efficient illumination far from the source.

Second, laser light is exceptionally pure in color. The spectral purity of light is measured in reciprocal centimeters or wave numbers, cm^{-1}, defined as the frequency width divided by the speed of light, $\Delta\nu/c$. While the full visible spectrum is 10,000 cm^{-1} wide, a typical laser may be as narrow as a few cm^{-1}, and with care, frequency widths as small as several billionths of a wave number have been obtained. This is of central importance in photochemistry, since molecules may absorb over only an extremely small range of frequencies ($\sim$0.01 cm^{-1}); only a miniscule fraction of the photons from a spectrally broad light source is actually utilized. Additionally, the spectral purity of a laser permits specific excitation of one out of thousands of closely spaced absorption features, a property which makes state-selected chemistry and isotope separation possible. This near monochromaticity can be disadvantageous as well, if the available laser frequencies do not overlap with absorption features in the molecule of interest. This difficulty has been largely solved for the visible and near ultraviolet, where organic dye lasers and nonlinear frequency doubling techniques have been developed that permit continuous tuning over a significant portion of the spectrum. Optical parametric oscillators and F-center lasers can tune over a sizable portion of the near infrared.

Third, lasers are generally more powerful than conventional light sources. A continuous wave (cw) argon ion laser will produce 10 W at 514.5 nm. Pulsed lasers, which compress the light energy into very short time periods (10^{-6} to 10^{-12} s), can generate correspondingly higher peak powers, typically from 10^3 to 10^9 W.

All in all, the laser provides the photochemist with a source of light between 10 and 20 orders of magnitude more spectrally bright than previously available.

Laser-initiated chemistry. One area of laser photochemistry is the study of single-collision chemical reaction dynamics. Ideally, the scientist would like to generate reagent molecules in a specific state and monitor the progress of the reaction as the product states are formed. This state-to-state picture of reaction dynamics requires extremely sensitive and selective species generation and detection, a task that typically demands the intensity and spectral purity of a laser light source. Additionally, the reagent state preparation must occur on a time scale significantly shorter than that of the reaction studied; fast pulsed lasers are therefore very convenient. Selective reagent generation may proceed by direct photolysis to form highly reactive radicals, as in the laser-initiated chain reactions (1).

$$Cl_2 \xrightarrow[300\ nm]{laser} 2Cl^{\cdot} \qquad (1a)$$

$$Cl^{\cdot} + H_2 \longrightarrow HCl + H^{\cdot} \qquad (1b)$$

$$H^{\cdot} + Cl_2 \longrightarrow HCl + Cl^{\cdot} \qquad (1c)$$

Since the chain reaction continues to propagate, a single laser photon in such a system can trigger the generation of thousands of product molecules.

Alternatively, the photons need not dissociate the molecule, but rather provide it with specific internal excitation, as in reactions (2), where the

$$HBr \xrightarrow[3.46\,\mu m]{laser} HBr^{*} \qquad (2a)$$

$$HBr^{*} + I \longrightarrow HI + Br \qquad (2b)$$

asterisk represents vibrational excitation of the molecule. Without this vibrational excitation, the reaction is endothermic and proceeds extremely slowly at room temperature. The extra vibrational energy has been observed to accelerate the reaction by factors of 10^9. In this fashion, lasers can selectively pump energy into a chemical system; monitoring the subsequent reaction provides highly detailed information on the dynamics of the reactive collisions. This is in sharp contrast with early temperature-dependence studies of reactions, which could never isolate the specific effects of vibrational versus translational excitation of reagents, for example.

Provided that the reaction is exothermic, analyses of the final states can often be obtained from the light emitted by the excited products, or chemiluminescence. For typical reaction exothermicities, however, the product chemiluminescence is in the infrared, where photon detection is not very sensitive. An often-used alternative technique for molecules that absorb in the visible or near ultraviolet is laser-induced fluorescence. Here a separate laser illuminates the reaction zone and is tuned over the wavelength region where the products absorb. Molecules in different states absorb light at slightly different frequencies and can then emit via a process known as fluorescence. The fluorescence is typically in the visible, where very sensitive photon detectors can be used. Its intensity provides a measure of the concentrations of product molecules in a particular final state. This technique is limited to molecules that exhibit discrete absorption in the range of a tunable laser. *See* CHEMILUMINESCENCE.

A natural and practical application of this research is the use of lasers to drive specific chemical reactions. Successful examples of laser-in-

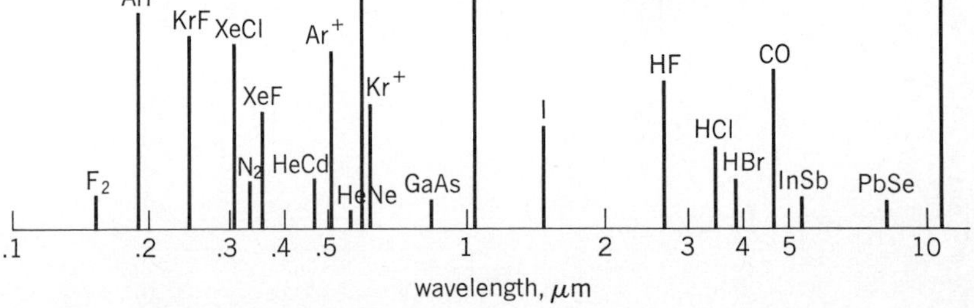

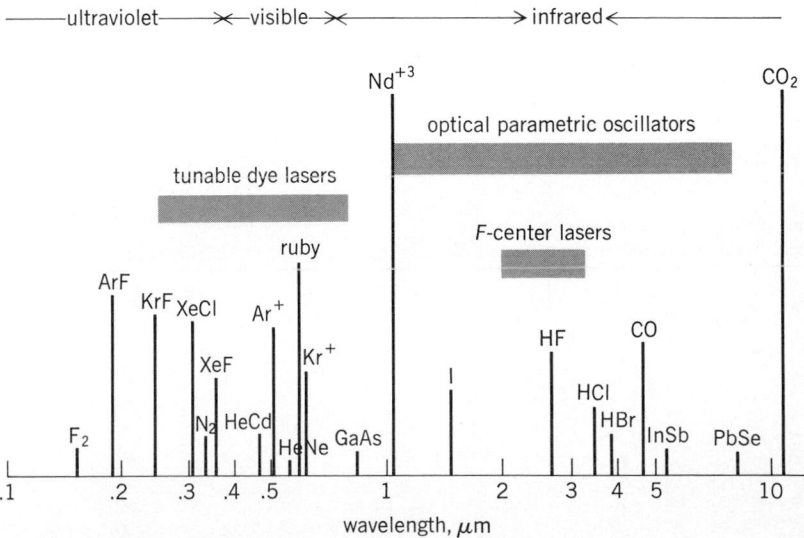

Fig. 1. Laser systems and their operating wavelengths.

duced chemistry are readily found and represent potential directions for industrial use. Carbon dioxide laser irradiation of diborane (B_2H_6) produces icosaborane ($B_{20}H_{16}$) in high yield, whereas conventional pyrolysis generates a mixture of products in which $B_{20}H_{16}$ is not found. Lasers have been used to generate catalysts in a gas-phase mixture, which in turn accelerate a selected chemical reaction. For example, photolytic ejection of CO can serve to catalyze the hydrogenation process [reactions (3)]. In some chemical syntheses, the

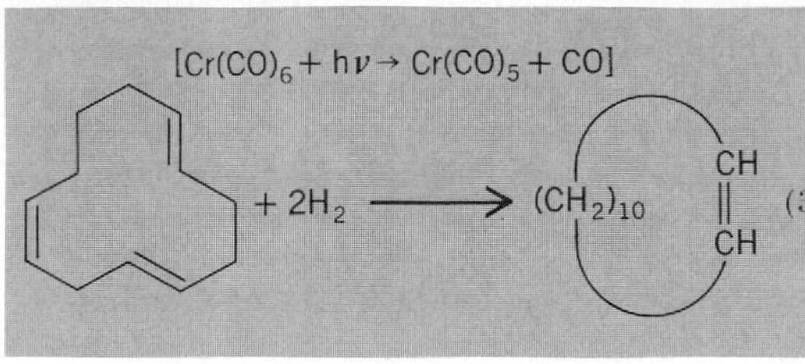

(3)

laser can be used to convert unwanted products into a desired product, as in the photoisomerization of *cis*-1–4-hexadiene [reaction (4)]; or the

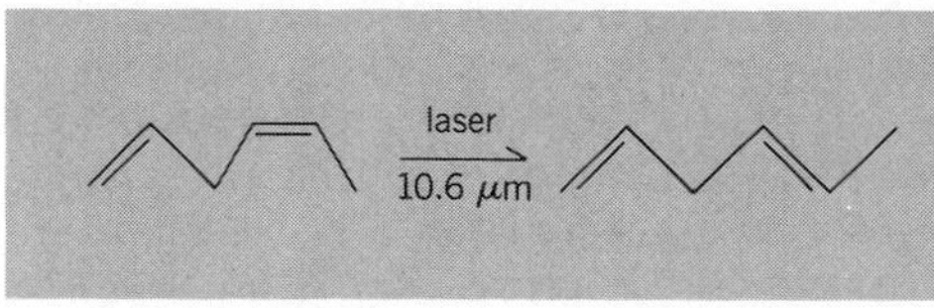

(4)

laser can be used to purify a final product by selective photodegradation of trace impurities. An interesting pharmaceutical application of lasers is in the key, light-induced reaction in vitamin D_3 synthesis, where the KrF laser may prove to be an economical alternative to the mercury-arc light source.

A particularly promising direction for laser-induced chemistry utilizes intense vibrational excitation by powerful infrared lasers to switch on a reaction pathway that normally would not proceed at all. For example, boron trichloride ordinarily reacts with benzene to form $C_6H_5BCl_2$ only at temperatures in excess of 600°C even in the presence of a palladium catalyst. On the other hand, a brief exposure of the same reagents to pulsed CO_2 laser light at 10.6 μm yields this product at room temperature without a catalyst. Bromination of pentafluorobenzene has also been induced by CO_2 laser light [reaction (5)]. On exposure to a 50-W cw CO_2

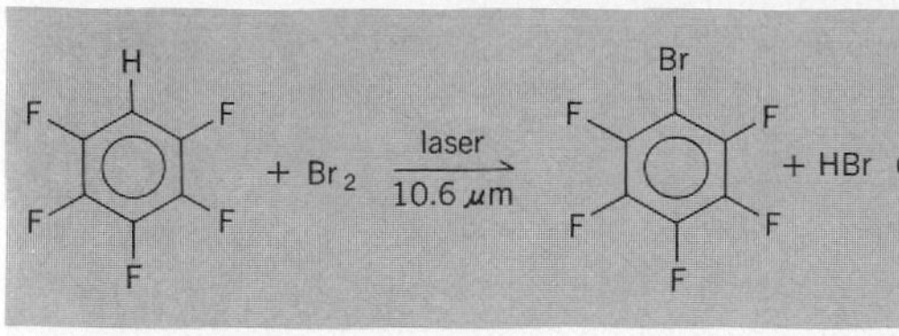

(5)

laser, the reaction is carried out to 50% completion in several minutes, yielding no side products. The mechanism in both of these examples has been shown to involve highly vibrationally excited intermediates.

The potential large-scale industrial market for laser-initiated chemistry is limited by the significant cost of generating laser photons. A mole (6.023×10^{23}) of photons from a CO_2 laser costs considerably less than one from a visible Ar^+ laser. Consequently, only very efficient, and preferably nonstochiometric (that is, catalytic), laser-assisted photochemical systems are considered to be economically feasible.

Picosecond photochemistry. Picosecond studies involve the use of state-of-the-art, ultrashort (10^{-12}-s) laser pulses to probe the reaction dynamics of very fast processes. One picosecond is approximately the time between collisions in a liquid or the time required for a full rotation of a small molecule. These picosecond pulses are formed by mixing together several pulses of very slightly different frequencies, a method known as mode locking. For the majority of the pulse duration, the individual frequencies are out of phase, and destructively interfere with each other. But for a very short time, all the frequencies are in phase, resulting in a brief but intense burst of light. This pulse is then split into a pump and a probe pulse. The intense pump pulse passes directly through the sample, while the weaker probe pulse is first expanded into many subpulses and then delayed optically by a series of glass plates before it is passed through the sample (Fig. 2). Since light travels 50% more slowly in glass than in air, these pulses emerge staggered in time. As a result, the sample is illuminated by the initial pump pulse followed by a staccato burst of probe pulses. By measurement of the absorption of the delayed probe pulses, the fast chemical processes initiated by the pump pulse can be monitored on an unprecedented time scale. Some early applications of this technique investigated the recombination dynamics of photo-dissociated iodine in solution. Scientists have also been able to study the rapid light-induced structural changes in the large biological molecules responsible for vision.

Photon dissociation. A long-standing goal of the laser photochemist has been to manipulate the pathway of a photochemical process. One vigorously pursued approach to this problem involves using extremely intense lasers to induce dissociation of a specific bond in a molecule—in effect, molecular photosurgery. The electrons of a molecule in the ground state arrange themselves to shield the positively charged nuclei from one another, which tends to hold the molecular fragments together in a chemical bond. In many electronically excited states, the electrons cannot shield the

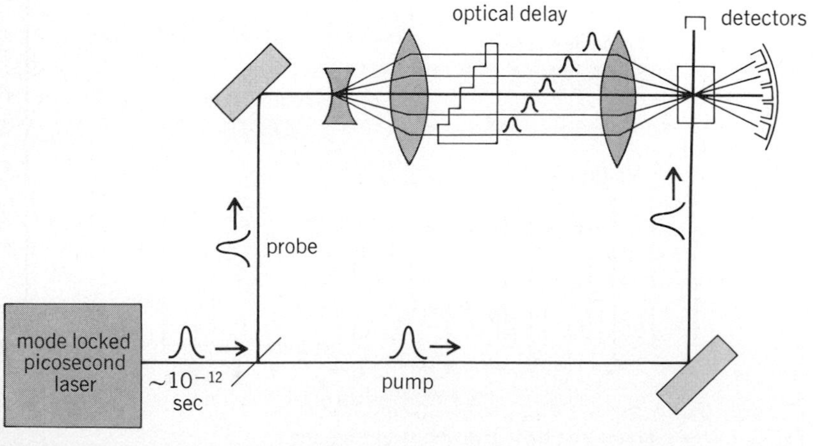

Fig. 2. Laser system for picosecond photochemistry.

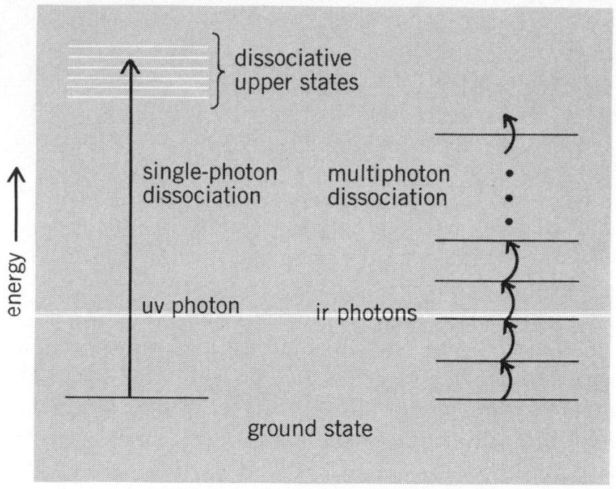

Fig. 3. Energy levels of excited molecules in single-photon and multiphoton dissociation.

nuclei effectively; a molecule in one of these states rapidly dissociates.

Single photon. This type of dissociation is simply the excitation of the electrons in a bond from their stable ground state to a repulsive upper state. In principle, therefore, selective bond rupture requires determining the ground- and upper-state energy separation and illuminating the molecule with a photon of the appropriate wavelength (typically in the ultraviolet; Fig. 3). In practice, however, the dissociation occurs so rapidly (less than a picosecond) that the molecule is somewhat insensitive to the precise photon energy. As a result, several different excited states can result from the absorption of the same photon, which in turn can cause the molecule to dissociate into many different undesired products.

Multiphoton. An alternative approach is to supply the requisite energy via a large number of less energetic photons. In the ground electronic state there are a number of individual energy levels corresponding to different amounts of excitation of a particular vibrational mode of the molecule. The energy spacing of these levels is approximately constant, typically between 1 and 10 kcal/mole (4 and 40 kJ/mole). An appropriate-wavelength infrared laser can therefore excite molecules sequentially up this vibrational manifold by using the individual states as one would use rungs on a ladder. Eventually the molecule becomes so vibrationally excited that it dissociates. The advantage of this multiphoton approach is that the individual vibrational states are relatively long-lived, and thus absorb over a very narrow range of frequencies. This permits very selective excitation of the molecules and thus greater control of the products formed.

The practicality of this technique is limited by the difficulty of obtaining a laser that generates the correct infrared frequency. Additionally, a typical bond strength of 50–100 kcal/mole (200–400 kJ/mole) requires consecutive absorption of over 20 infrared photons, which demands extremely high laser powers. This latter difficulty has been overcome with pulsed CO_2 lasers, which can generate up to 3×10^{11} W in a 10^{-9}-s pulse. (Over the duration of these short pulses, such a laser generates more energy than Niagara Falls.) The former difficulty is still largely unsolved, and thus experimentation has been limited mainly to molecular systems with infrared absorptions that happen to coincide with frequencies of the CO_2 laser.

In principle, infrared multiphoton dissociation could be an extremely selective, photosurgical tool, provided that the vibrational energy remains in the desired bond on the time scale of the excitation. In actuality, the initially localized vibrational energy probably extends throughout the molecule on a picosecond time scale. The multiphoton results most likely do not represent selective dissociation, but rather a collision-free molecular heating that ultimately leads to fragmentation. *See* PICOSECOND MOLECULAR PROCESSES.

Laser isotope separation. An early hope for large-scale practical application of laser photochemistry was to achieve the separation of isotopes. In typical chemical reactions, which involve only the electrons surrounding the nucleus, the isotopes of a given element behave nearly identically. (Fundamentally, this is why they are difficult to separate using standard chemical techniques.) The nuclear chemistry of the isotopes, however, can be vastly different. For example, ^{235}U is a fissionable isotope used in atomic reactors, whereas ^{238}U, with three extra neutrons, not only does not sustain the fission process but actually inhibits it. The elements occur naturally as a mixture of isotopes (the $^{235}U/^{238}U$ natural abundance ratio is less than 1%), and the difficulty and expense of any purification scheme is considerable. The conventional UF_6 gas diffusion separation method, which relies on fractions of a percent difference in diffusion rates of the two isotopic species, consumes 10^5kWh per mole of ^{235}U recovered. (Compare this, however, with the 5×10^6 kWh/mole generated by the product in a nuclear reactor.) Consequently, an immense effort has been invested in developing cost-efficient techniques for laser-induced separation of isotopes.

In principle, this is a straightforward process. Because of the small differences in mass, compounds of different isotopes of the same element exhibit absorption spectra that are slightly shifted with respect to one another. If a laser can pump energy preferentially into one isotope and thereby induce a permanent change, then an efficient separation is feasible. One separation technique involves crossing a beam of molecules with an intense laser tuned to an absorption feature in one of the desired isotopes. Photons possess a small but finite momentum; molecules that absorb light from the laser will be deflected away from the rest of the beam. The magnitude of the deflection is very small unless multiple absorptions occur, significantly limiting this technique. Another method uses lasers to accelerate a chemical reaction by selective vibrational excitation of a particular isotope. This desired isotope is predominant among the reaction products, and can be readily separated from the original natural-abundance mixture. By far the most promising isotope separation scheme uses intense pulsed lasers for selective multiphoton dissociation. With this technique, the laser preferentially dissociates molecules of a particular isotope; the isotopically enriched radical fragments then react with added scavengers, and can be removed. Application of this technique has achieved spectacular isotopic separations for a number of elements, most notably sulfur, boron, nitrogen, and hydrogen.

For example, photodissociation of SF_6 by pulsed CO_2 laser excitation of the ν_3 mode has produced dissociation yields in the beam near unity with isotopic enrichments of the $^{34}S/^{32}S$ ratio by a factor of 10. A far more dramatic demonstration is observed in the hydrogen/deuterium isotope separation in trifluoromethane, CF_3H (or CF_3D). In this system, a pulsed CO_2 laser selectively excites the ν_5 mode of the deuterium-bearing molecules up to dissociation [reactions (6) and (7)]. The DF product is then

$$CF_3D \xrightarrow[10.2\,\mu m]{laser} :CF_2 + DF \quad (6)$$

$$CF_3H \xrightarrow[10.2\,\mu m]{laser} \text{little or no photodissociation} \quad (7)$$

extracted as the source of enriched deuterium. The isotopic selectivity (that is, the D/H ratio of products divided by the natural abundance ratio) has been shown to be 11,000 under optimum conditions, with essentially 100% dissociation of the sample per laser pulse, and a photon efficiency on the order of 10%. Estimated energy consumption per mole of deuterium generated is less than 28 kWh, which compares rather favorably with a cost of 30 kWh/mole by the conventional gas diffusion process.

Unfortunately, separation of the uranium isotopes has proven substantially more elusive. A major difficulty has been finding an intense laser source that emits at a wavelength absorbed by the few gaseous compounds of uranium. Much of the research thrust has been with uranium hexafluoride, UF_6, which absorbs predominantly in the 16-μm region, but has much weaker overtone and combination bonds (sums of the fundamental absorption frequencies) which can be accessed by a CO_2 laser. The development of a powerful 16-μm laser for use in uranium isotope separation is the focus of intense research effort. *See* CHEMICAL DYNAMICS; PHOTOCHEMISTRY.

[DAVID J. NESBITT]

Bibliography: *Laser Photochemistry and Diagnostics: Recent Advances and Future Prospects*, NSF/DoE Seminar Rep., June 4–5, 1979; B. A. Lengyel, *Lasers*, 1971; C. B. Moore (ed.), *Chemical and Biochemical Applications of Lasers*, vols. 1–4, 1974; R. P. Wayne, *Photochemistry*, 1970.

Laser spectroscopy

Spectroscopy with laser light or, more generally, studies of the interaction between laser radiation and matter. Lasers have led to a rejuvenescence of classical spectroscopy, because laser light can far surpass the light from other sources in brightness, spectral purity, and directionality, and if required, laser light can be produced in extremely intense and short pulses. The use of lasers can greatly increase the resolution and sensitivity of conventional spectroscopic techniques, such as absorption spectroscopy, fluorescence spectroscopy, or Raman spectroscopy. Moreover, interesting new phenomena have become observable in the resonant interaction of intense coherent laser light with matter. Some of these effects have become the basis for powerful new spectroscopic methods, which offer unprecedented spectral resolution, or which permit the investigation of properties of matter that could not be observed previously. Laser spectroscopy has become a wide and diverse field, with applications in numerous areas of physics, chemistry, and biology.

Tunable sources. Early lasers, such as ruby lasers or helium-neon lasers, worked only at a few discrete wavelengths, determined by narrow spectral lines of the active medium. The vigorous advances of laser spectroscopy since about 1970 are largely due to the development of laser sources which are highly monochromatic, and in which the wavelength can be tuned continuously over a wide spectral range.

Tunable coherent sources are available for any wavelength region from the far infrared into the near-vacuum ultraviolet. Tunable infrared lasers include high-pressure molecular gas lasers, semiconductor diode lasers, spin-flip Raman lasers, and color-center lasers. In the visible region, organic dye lasers have proved particularly versatile and powerful spectroscopic sources. Both pulsed and continuous-wave dye lasers cover the entire visible spectrum, including the bordering near-infrared and near-ultraviolet regions. Intense shorter-wave ultraviolet radiation can be generated with excimer lasers over limited regions.

The generation of harmonic frequencies or sum frequencies in nonlinear optical crystals or gases provides a valuable method to produce tunable coherent ultraviolet radiation from laser light of longer wavelengths. Optical parametric oscillators and difference-frequency crystal mixers offer corresponding alternatives for the generation of coherent infrared radiation. Stimulated Raman scattering can be used to shift the frequency of a tunable laser by some integer multiple of a molecular vibration frequency and thus to extend the tuning range. These nonlinear frequency-mixing and shifting techniques work best with the intense radiation from pulsed lasers.

Although pulsed tunable lasers can be highly monochromatic by conventional standards, their line width can never be narrower than the inverse pulse duration. The highest resolution is obtained with continuous-wave lasers. Considerable engineering efforts have been devoted to the active frequency stabilization of such lasers. By electronically locking the frequency of a continuous-wave dye laser to some reference interferometer, for instance, a line width of less than 1 MHz or a resolution on the order of 1 part in 10^9 is achieved routinely, and line widths down to a few hundred hertz have been produced with more sophisticated feedback controls.

Much engineering effort is devoted to the development of highly accurate wavelength and frequency meters, which can do justice to the narrow line width of tunable lasers.

Absorption spectroscopy. Lasers can replace conventional light sources and spectrographs or monochromators for absorption spectroscopy. The spectral purity of laser light can eliminate instrumental resolution limits, the high intensity helps to overcome detector noise, and the good directionality permits long or folded absorption paths. All these factors contribute to improved sensitivity.

Rather than by measuring the attenuation of a laser beam in the sample, the absorption can often be monitored indirectly with still higher sensitivity. For instance, absorption of a modulated laser beam will produce a sound wave in the sample, because some of the light is converted to heat. This sound wave can be picked up by a microphone (optoacoustic detection). Resonant absorption of laser light by atoms or molecules in a gas discharge can change the ionization probability. The resulting changes in discharge current or voltage across the tube are easily measured (optogalvanic detection). The highest sensitivities at visible and ultraviolet wavelengths have been obtained by monitoring the fluorescence or photoionization of laser-excited atoms or molecules in a gas or molecular beam. Single atoms have been selectively detected in this way.

Intracavity absorption. Very high sensitivity has also been obtained by placing an absorbing sample inside the resonator of a broad-band dye laser without any optical tuning elements. Any absorption lines can be detected as dips in the laser emission spectrum, when analyzed by a spectrograph. This intracavity spectroscopy can surpass the sensitivity of a conventional single-pass absorption measurement by a large factor (on the order of 10^5) because the sample is effectively traversed by the light many times, and the competition between many simultaneously oscillating modes strongly disfavors modes with slightly increased losses.

Multiphoton absorption. It has long been known that atoms can be excited to a higher quantum state by simultaneously absorbing two or more photons which together provide the necessary energy. The probability for an N-photon transition grows initially with the Nth power of the intensity. Only the high intensity of laser light has made it possible to observe two- and multi-photon absorption in the optical region. Two-photon spectroscopy permits the study of states with the same parity as the absorbing level, which are not normally reached by single-photon transitions, and it requires photons of less energy, which are sometimes more readily produced.

Fluorescence spectroscopy. Intense laser light is a very effective means to pump a large fraction of an absorbing species to some excited quantum level. Hence, lasers can greatly increase the sensitivity of such classical spectroscopic methods as fluorescence spectroscopy, optical pumping, level-crossing spectroscopy, or double-resonance spectroscopy. Moreover, lasers make it possible to apply these techniques to atomic and molecular transitions at wavelengths where intense spectral lamps are not available, and stepwise excitation permits studies even of highly excited states, including autoionizing levels and very high Rydberg states of atoms.

Studies of the line shape of fluorescent emission at high intensities have permitted interesting tests of the predictions of quantum-electrodynamic theory.

Raman spectroscopy. Lasers have revolutionized Raman spectroscopy, that is, the observation of scattered light at wavelengths other than that of the exciting light. The high intensity of laser light has greatly increased the sensitivity of this form of two-photon spectroscopy, where the energy difference between incident photon and scattered photon corresponds to a resonant transition between states of equal parity. Tunable lasers have enhanced the sensitivity, resolution, and versatility of Raman spectroscopy even more, by making it possible to excite close to some intermediate resonant transition (resonance Raman spectroscopy), or to observe stimulated Raman scattering with the help of a second, tunable probe laser beam. Polarization anisotropies associated with Raman transitions provide additional information and can further increase the detection sensitivity. *See* RAMAN EFFECT.

Frequency mixing. The high intensity of laser light has also made possible a new class of spectroscopic methods which rely on nonlinear frequency mixing in the sample. For instance, if a sample is irradiated simultaneously by two strong laser beams of frequencies ω_1 and ω_2, the nonlinear response of the driven dipoles leads to the generation of coherent light at a new frequency, $2\omega_1 - \omega_2$. This frequency mixing can be described in terms of a third-order nonlinear susceptibility of the sample.

If the frequency ω_2 is tuned to the (red-shifted) Stokes line of a Raman transition, excited by ω_1, a resonance is observed in the intensity of the new wave at the (blue-shifted) anti-Stokes frequency. This type of frequency mixing is sometimes referred to as coherent anti-Stokes Raman spectroscopy (CARS). It can be used to study Raman transitions even in the presence of strong incoherent background radiation, such as in the observation of flames.

High-resolution laser spectroscopy. Lasers have led to particularly noteworthy progress in the field of very-high-resolution spectroscopy. They have become powerful tools to investigate the structure of atoms, molecules, and ions. They can be used to study fine and hyperfine splittings, Zeeman and Stark splittings, light shifts, collision broadening, collision shifts, and other attributes of spectral lines. Moreover, lasers make it possible to measure the wavelengths of spectral lines with unprecedented accuracy. The laser wavelength can be locked to an atomic or molecular transition, providing an accurate standard of length or frequency. Such lasers have become important tools for precision metrology, making possible accurate measurements of fundamental constants and stringent tests of basic physics laws.

Doppler-free spectroscopy. The sharpest spectral lines are generally those in atoms or molecules which are relatively free and undisturbed. Such molecules, however, are almost inevitably moving with high thermal velocity. Molecules which are moving toward an observer appear to emit or absorb light at higher frequencies than molecules at rest, and molecules moving away appear to absorb at lower frequencies. In a gas, with molecules moving at random in all directions, the lines appear blurred, with typical Doppler widths on the order of 1 part in 1,000,000. To take advantage of the very narrow instrumental line width of laser sources, it is necessary to overcome this Doppler broadening of spectral lines.

The oldest method of Doppler-free spectroscopy is the transverse observation of a well-collimated molecular beam, so that the range of velocities along the line of sight is much restricted. However, molecular-beam equipment tends to be expensive and cumbersome, and it is difficult to observe rare species or molecules in short-living excited states in this way. Fortunately, a number of clever schemes have been devised which permit Doppler-free spectroscopy of simple gas samples. These include, in particular, techniques of saturation spectroscopy which use the laser light itself to label molecules of slow velocity.

Saturation spectroscopy. Laser light can easily be intense enough to partly saturate the absorption of a spectral line. That is, those molecules which have absorbed a light quantum are temporarily removed from the initial state. The absorption from that level is reduced or saturated, at least until some relaxation process can replenish the supply of absorbing molecules. In a gas, a monochromatic laser beam will resonantly interact only with those molecules which have the right axial velocity to be Doppler-shifted into resonance. The resulting velocity-selective saturation can be used for Doppler-free spectroscopy.

The first step was the realization by Willis E. Lamb, Jr., in 1963 that the two waves traveling in opposite directions inside a gas laser could work together to saturate the emission (rather than absorption) of those atoms which happen to have a zero component of velocity along the laser axis. Thus the power output would decrease when the laser length was adjusted to produce the light wavelength that would interact with those stationary atoms. This Lamb dip was soon observed and was used for high-resolution spectroscopy. However, for some time it was limited to studying the laser transitions themselves, or those few molecular lines which happened to coincide with gas laser wavelengths.

In 1970 C. Borde and T. W. Hansch introduced independently a now commonly used form of saturation spectroscopy which achieves good sensitivity with samples outside the laser resonator and which is particularly well suited for use with broadly tunable laser sources. As illustrated in Fig. 1, the output of a tunable laser is divided by a beam splitter into a stronger saturating beam and a weaker probe beam that are traversing an absorbing gas sample along the same path but in nearly opposite directions. When the saturating beam is on, it bleaches a path through the cell; that is, it depletes those molecules which are Doppler-shifted into resonance, and a stronger probe signal is received at the detector. As the saturating beam is alternately stopped and transmitted by a chopper, the probe signal is modulated. However, that happens only when both beams interact with the same molecules, and those can only be molecules which are standing still or at most moving transversely. Thus the method picks out those molecules which have near zero component of velocity along the laser beams and ignores others. Figure 2 illustrates the power of the technique by comparing a Doppler-broadened absorption profile of the well-known Balmer-alpha line of atomic hydrogen with one of the first saturation spectra, recorded with a pulsed dye laser in a glow discharge.

The described methods work well only if the sample has noticeable absorption and if the laser is strong enough to excite a substantial fraction of the resonant molecules. Higher sensitivity can sometimes be achieved by observing the absorption of laser light indirectly, via the emitted fluorescence or via acoustooptic or optogalvanic detection. In this case, it is advantageous to monitor the nonlinear interaction of two counterpropagating laser beams via intermodulation; that is, the two laser beams are chopped at two different frequencies f_1 and f_2, and the spectrum is recorded as a modulation in the signal at the sum or difference frequency $f_1 + f_2$ or $f_1 - f_2$.

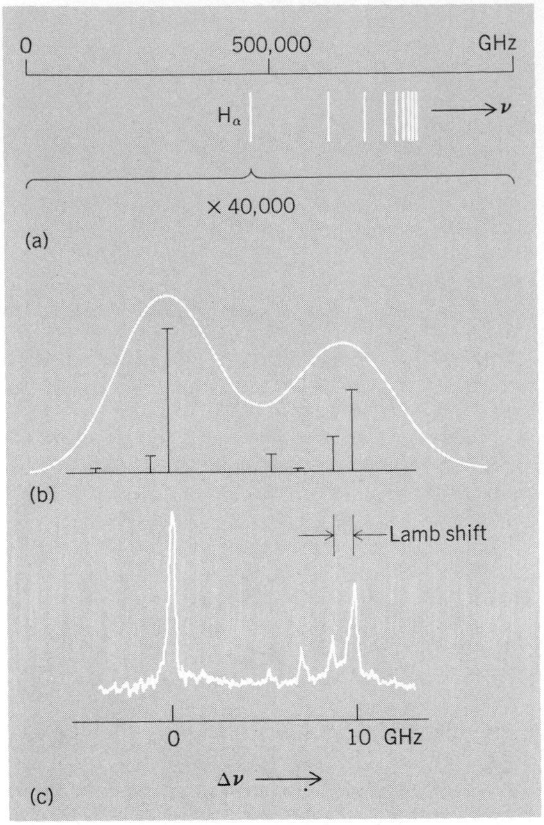

Fig. 2. Laser spectroscopy of atomic hydrogen. (*a*) Balmer series. (*b*) Doppler-broadened absorption profile (300 K) of the Balmer-alpha line with theoretical fine structure. (*c*) Saturation spectrum of Balmer-alpha line with resolved 2*S*-2*P* Lamb shift. (*Recorded by T. Hänsch and collaborators*)

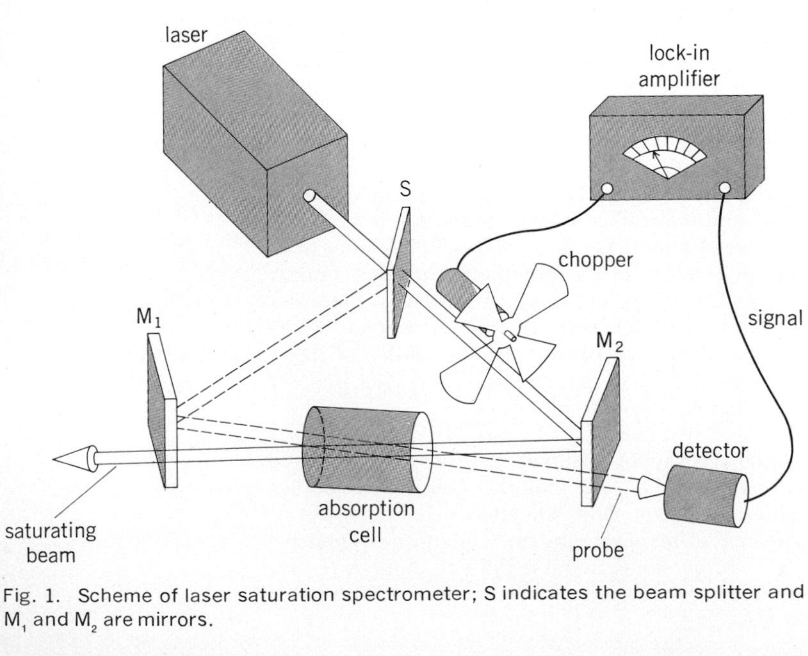

Fig. 1. Scheme of laser saturation spectrometer; S indicates the beam splitter and M_1 and M_2 are mirrors.

Polarization spectroscopy. There are also purely optical methods, which can be considerably more sensitive than the older saturated absorption method by suppressing the fluctuating background of probe light on which the signal has to be detected. These include, in particular, the technique of polarization spectroscopy introduced in 1976 by C. Wieman and Hansch. A polarization spectrometer looks similar to a saturation spectrometer but takes advantage of the fact that small changes in light polarization can be detected more easily than changes in light intensity. The probe beam "sees" the sample placed between nearly crossed linear polarizers so that only very little light arrives at the photodetector. The saturating beam is made circularly polarized by a birefringent plate. Alternatively, a linearly polarized beam is used with its polarization axis rotated at 45°.

Normally, in a gas, molecules have their rotation axes distributed at random in all directions. But the probability for absorbing polarized light depends on the molecular orientation. Thus the saturating beam depletes preferentially molecules with a particular orientation, leaving the remaining ones polarized. These can then be detected with high sensitivity because they can change the polarization of the probe beam. The probe acquires a component that can pass through the crossed polarizer into the detector, but again this happens only near the center of a Doppler-broadened line where both beams are interacting with the same molecules.

Polarization spectroscopy makes it possible to observe fewer molecules with lower light intensity so that external causes of line broadening and shifts are more easily avoided. A promising related technique, saturated interference spectroscopy, can work even in spectral regions where good polarizers are not available.

Doppler-free two-photon spectroscopy. There is a completely different approach to Doppler-free laser spectroscopy which does not rely on velocity selection, but which works only for two- or multiphoton spectroscopy, where the frequencies and directions of the exciting photons can be chosen so that the momenta of the absorbed photons add to zero.

Doppler broadening in two-photon excitation can be eliminated to first order simply by reflecting the output of an intense monochromatic laser back onto itself and placing a gas sample in the resulting standing wave field, so that the molecules can absorb two photons of equal energy coming from opposite directions (Fig. 3). From a moving molecule, one beam will appear Doppler-shifted toward the blue, and the other will appear shifted toward the red by an equal amount. The sum frequency is hence constant, independent of the molecular velocity. If the number of excited molecules is observed during a laser scan, a sharp resonance appears on a low, Doppler-broadened background, produced by each traveling wave separately. The Doppler-free signal is strongly enhanced because all molecules, regardless of their velocity, contribute.

Resolution limits. The resolution of any of the described methods is ultimately limited by the natural line width of the observed transition. However, in practice, a large number of additional causes of line broadening demand attention, including collision effects, power broadening and light shifts, relativistic "transverse" Doppler shifts, and imperfections in the laser wave fronts. Another important limitation is transit time broadening, in which the number of field oscillations a molecule can see, and hence the resolution, is limited by the finite time during which a molecule traverses the laser beam. Transit time broadening can sometimes be alleviated with the help of two or more spatially separate laser fields. In this case, narrow-band interference fringes appear in the spectrum, similar to the Ramsey fringes commonly observed in molecular-beam spectroscopy with separate radio-frequency fields.

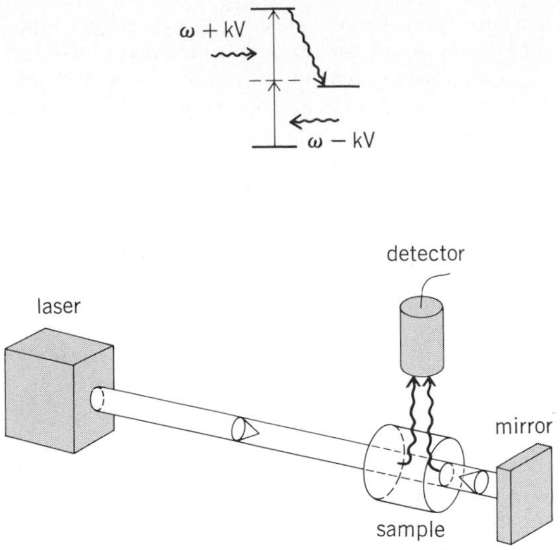

Fig. 3. Setup for Doppler-free two-photon spectroscopy.

Other approaches are being explored which promise to overcome some of the resolution limits. The resonant radiation pressure of intense laser light can be used to cool gases rapidly to very low temperature, far below the equilibrium condensation point. Such radiation cooling has been demonstrated for ions in a radio-frequency quadrupole trap or a Penning trap. It has also been suggested as a means to trap slow atoms with laser light fields, taking advantage of dielectric forces.

Hole-burning spectroscopy. Although most solid samples show only rather broad spectral features, some extremely narrow natural line widths have been observed in certain ions and molecules in host crystals near liquid helium temperature. Generally, such lines appear broadened because crystal-site-dependent statistical field variations cause varying line shifts. The true natural line width can be observed by a method closely related to saturation spectroscopy in gases. A monochromatic laser interacts resonantly only with a group of absorbing ions or molecules at selected crystal sites. By temporarily removing these from their absorbing level, it produces a narrow-band "hole-burning" in the absorption profile, which can be observed with a second tunable probe laser beam.

Time-resolved laser spectroscopy. Lasers permit the generation of extremely short and intense light pulses, down to subnanosecond or subpicosecond duration, which can be powerful tools for studies of transient phenomena in the interaction of light with matter.

Short laser pulses permit, in particular, measurements of excited-state lifetimes and studies of relaxation processes in atoms, molecules, and ions in the gaseous, liquid, and solid phase with unprecedented temporal resolution.

Pulsed lasers can be used to excite atoms or molecules to a superposition of two or more closely spaced levels. Interference effects lead to a modulation of the subsequent spontaneous emission. Such quantum beats provide information about the detailed level structure.

Lasers have also made it possible to observe interesting coherent transient phenomena in the interaction of light with resonant transitions. Such effects include free induction decay, optical nutations, photon echoes, and self-induced transparency. They often have analogs in phenomena previously observed in the microwave region in studies of nuclear magnetic resonance, and they are useful not only for measurements of phase relaxation processes, but also for understanding the intricacies in the interaction of light with matter. *See* NUCLEAR MAGNETIC RESONANCE (NMR); SPECTROSCOPY. [THEO W. HANSCH]

Bibliography: N. Bloembergen (ed.), Nonlinear spectroscopy, *Proceedings of the International School of Physics, "Enrico Fermi"*, 1977; J. Hall and J. L. Carlsten (eds.), *Laser Spectroscopy III*, Springer Series in Optical Sciences, vol. 7, 1977; V. S. Lethokhov and V. P. Chebotaev, *Nonlinear Laser Spectroscopy*, Springer Series in Optical Sciences, vol. 4, 1977; K. Shimoda (ed.), *High Resolution Laser Spectroscopy*, Topics in Applied Physics, vol. 132, 1976; H. Walther (ed.), *Laser Spectroscopy of Atoms and Molecules*, Topics in Applied Physics, vol. 2, 1976; H. Walther and K. W. Rothe (eds.), *Laser Spectroscopy IV*, Springer Series in Optical Sciences, vol. 21, 1979.

Lawrencium

A chemical element, symbol Lr, atomic number 103. Lawrencium, named after E. O. Lawrence, is the eleventh transuranium element; it completes the actinide series of elements. The symbol Lw was originally proposed by the discoverers: it was changed to Lr in 1963. *See* ACTINIDE ELEMENTS.

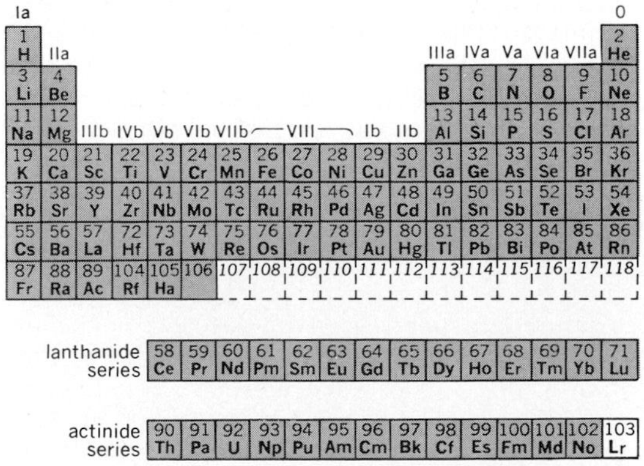

The element was discovered in February 1961 at the Heavy-Ion Linear Accelerator (HILAC) in the Lawrence Radiation Laboratory in Berkeley, CA, by A. Ghiorso, T. Sikkeland, A. E. Larsh, and R. M. Latimer, when a 2-μg target including the californium isotopes 249, 250, 251, and 252 was bombarded with boron-11 ions. After being stopped in helium, the recoiling atoms of the new element were attracted electrically to a metallized mylar tape which was pulled successively in front of four semiconductor alpha detectors. In many runs, a total of about 100 events of an alpha activity with energy 8.6 MeV and half-life about 8 sec was detected. By means of excitation functions and cross-bombardment techniques (that is, the use of other heavy ions and other targets) as well as by the use of α-systematics, it was concluded that an isotope of element 103 with a probable mass number of 257 or 258, or both together, was the source of the α-particle activity. (Later Berkeley work showed that the activity was due to the isotope 258, and a better half-life of 4.2 sec was determined). *See* CALIFORNIUM.

In 1965 at the Dubna laboratories near Moscow, teams under the general direction of G. N. Flerov discovered another isotope, ^{256}Lr, produced by the reaction of oxygen-18 ions with americium-243. By recoil-milking a known daughter product from alpha decay, they measured a half-life of about 45 sec. Later work in their laboratory measured a complex α-spectrum (8.35–8.60 MeV) and a 25–35-sec half-life for this nuclide.

In 1968 at Berkeley, R. J. Silva and coworkers used this isotope in a tour de force of chemistry to make a crucial observation of the behavior of element 103 in aqueous solution. Whereas nobelium, element 102, had been found to have a stable +2 state and was indeed very difficult to oxidize to a +3 valency in aqueous solution, lawrencium was shown to have a stable +3 state just as the other heavy actinide elements do. This finding is what was expected for the last element in the actinide transition series.

By 1971 work in the Berkeley laboratory had established the nuclear properties of all the isotopes of lawrencium from mass 255 to mass 260. ^{260}Lr is an alpha emitter with a half-life of 3 min and consequently is the longest-lived isotope known.

The very neutron-deficient isotopes ^{251}Lr and ^{253}Lr have been reported by the Dubna group from bombardments of ^{203}Tl and ^{205}Tl by ^{50}Ti projectiles. Spontaneous fission half-lives of about 2 and 5 sec were observed, but no atomic number identification was possible. The same possibility for error exists in these experiments as for the neutron-deficient element 105 isotopes made in similar reactions. *See* ELEMENT 105. [ALBERT GHIORSO]

Bibliography: E. D. Donets et al., *Sov. J. At. Energy*, 19:995–999, 1965; G. N. Flerov et al., *Study of Alpha Decay of ^{256}103 and ^{257}103*, Dubna Preprint no. E7-3257, 1967; A. Ghiorso et al., New element: Lawrencium, atomic number 103, *Phys. Rev. Lett.*, 6(9):473–475, 1961.

Leaching

The removal of a soluble fraction, in the form of a solution, from an insoluble, permeable solid with which it is associated. The separation usually involves selective dissolving, with or without diffusion, but in the extreme case of simple washing it consists merely of the displacement (with some mixing) of one interstitial liquid by another with which it is miscible. The soluble constituent may be solid (as the metal leached from ore) or liquid (as the oil leached from soybeans).

Leaching is closely related to solvent extraction,

in which a soluble substance is dissolved from one liquid by a second liquid immiscible with the first. Both leaching and solvent extraction are often called extraction. Because of its variety of applications and its importance to several ancient industries, leaching is known by a number of other names: solid-liquid extraction, lixiviation, percolation, infusion, washing, and decantation-settling. The liquid used to leach away the soluble material (the solute) is termed the solvent. The resulting solution is called the extract or sometimes the miscella.

The mechanism of leaching may involve simple physical solution, or dissolution made possible by chemical reaction. The rate of transportation of solvent into the mass to be leached, or of soluble fraction into the solvent, or of extract solution out of the insoluble material, or some combination of these rates may be significant. A membranous resistance may be involved. A chemical reaction rate

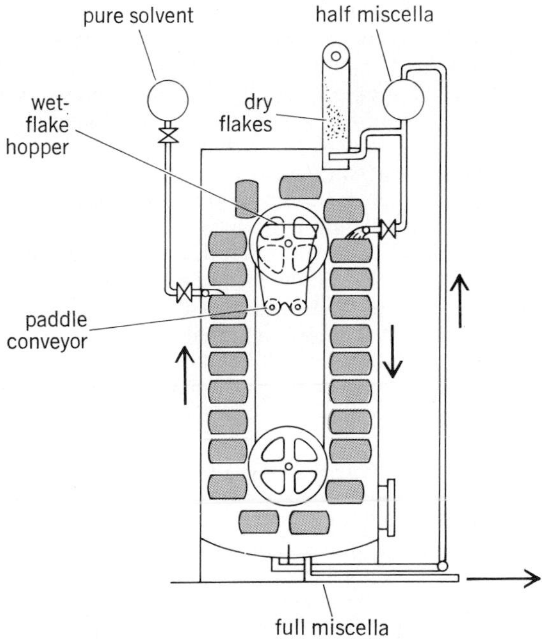

Fig. 1. Bollman extractor. (*From W. L. McCabe and J. C. Smith, Unit Operations of Chemical Engineering, 2d ed., McGraw-Hill, 1967*)

may also affect the rate of leaching. The general complication of this simple-appearing process results in design by chiefly empirical methods. Whatever the mechanism, however, it is clear that the leaching process is favored by increased surface per unit volume of solids to be leached and by decreased radial distances that must be traversed within the solids, both of which are favored by decreased particle size. Fine solids, on the other hand, cause mechanical operating problems during leaching, slow filtration and drying rates, and possible poor quality of solid product. The basis for an optimum particle size is established by these characteristics.

Leaching processes fall into two principal classes: those in which the leaching is accomplished by percolation (seeping of solvent through a bed of solids), and those in which particulate solids are dispersed into the extracting liquid and subse-

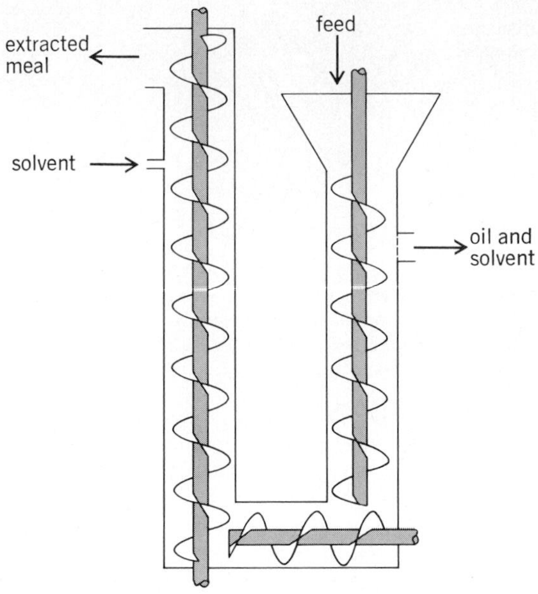

Fig. 2. Hildebrandt screw-conveyor extractor. (*From W. L. McCabe and J. C. Smith, Unit Operations of Chemical Engineering, 2d ed., McGraw-Hill, 1967*)

quently separated from it. In either case, the operation may be a batch process or continuous. *See* EXTRACTION; MASS-TRANSFER OPERATION; SOLVENT EXTRACTION.

Percolation. In addition to being applied to ores and rock in place and by the simple technique of heap leaching, percolation is carried out in batch tanks and in several designs of continuous extractors.

The batch percolator is a large circular or rectangular tank with a false bottom. The solids to be leached are dumped into the tank to a uniform depth. The are sprayed with solvent until their solute content is reduced to an economic minimum and are then excavated. A simple example is the brewing of coffee in a percolator (repeated extraction) or a drip pot (once-through). Countercurrent flow of the solvent through a series of tanks is common, with fresh solvent entering the tank containing the most nearly exhausted material. Some leach tanks operate under pressure, to contain volatile solvents or increase the percolation rate.

Continuous percolators employ the moving-bed principle, implemented by moving baskets that carry the solids past solvent sprays, belt or screw conveyors that move them through streams or showers of solvent, or rakes that transport them

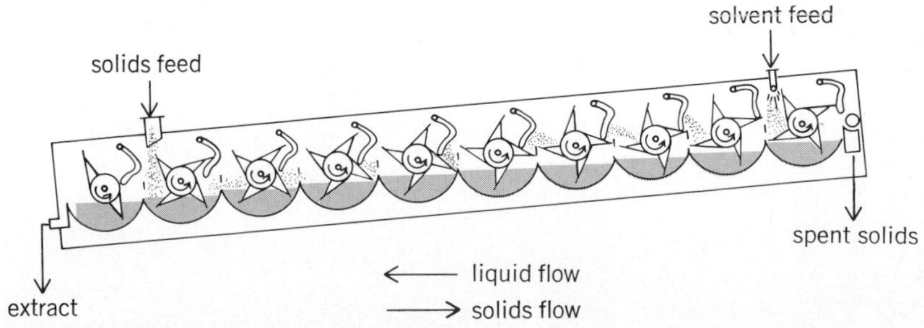

Fig. 3. Kennedy extractor. (*From R. H. Perry and C. H. Chilton, eds., Chemical Engineer's Handbook, 5th ed., McGraw-Hill, 1973*)

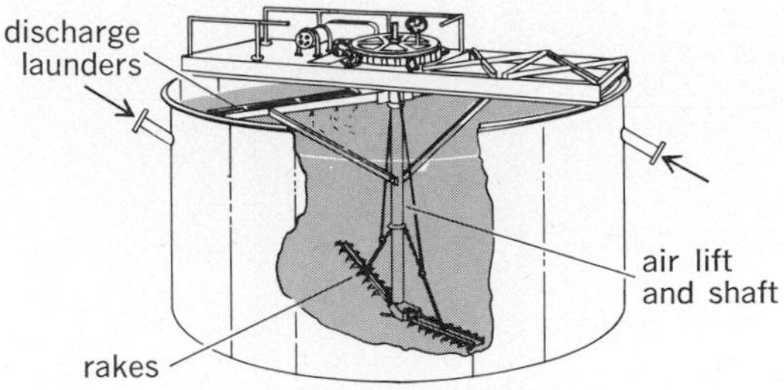

Fig. 4. Dorr agitator for batch washing of precipitates. (*From W. L. Badger and J. T. Banchero, Introduction to Chemical Engineering, McGraw-Hill, 1955*)

along a solvent-filled trough. In a revolving-basket type like the Rotocel extractor, bottomless compartments move in a circular path over a stationary perforated annular disk. They are successively filled with solids, passed under solvent sprays connected by pumps so as to provide countercurrent flow of the extracting liquid, and emptied through a large opening in the disk. Alternating perforated-bottom extraction baskets may be arranged in a bucket-elevator configuration, as in the Bollman extractor (Fig. 1). On the up cycle, the partially extracted solids are percolated by fresh solvent sprayed at the top; on the down cycle, fresh solids are sprayed with the extract from the up cycle. A screw-conveyor extractor like the Hildebrandt extractor (Fig. 2) moves solids through a V-shaped line in a direction opposite to the flow of solution. The solids may be conveyed instead by rakes or paddles along a horizontal or inclined trough counter to the direction of solvent flow, as in the Kennedy extractor (Fig. 3). In the last two types, the action is predominantly percolation but involves some solids dispersal because of agitation by the conveyors. Horizontal continuous vacuum filters of the belt, tray, or table type sometimes are used as leaching equipment. *See* FILTRATION.

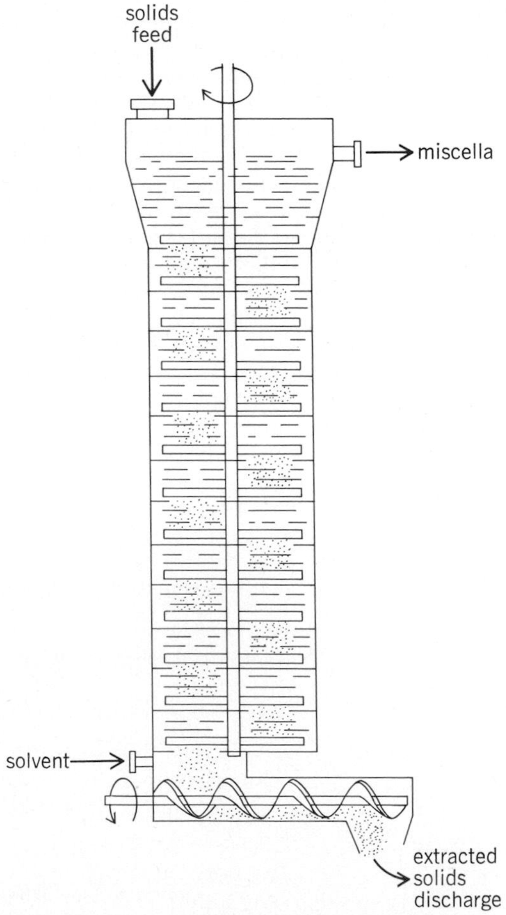

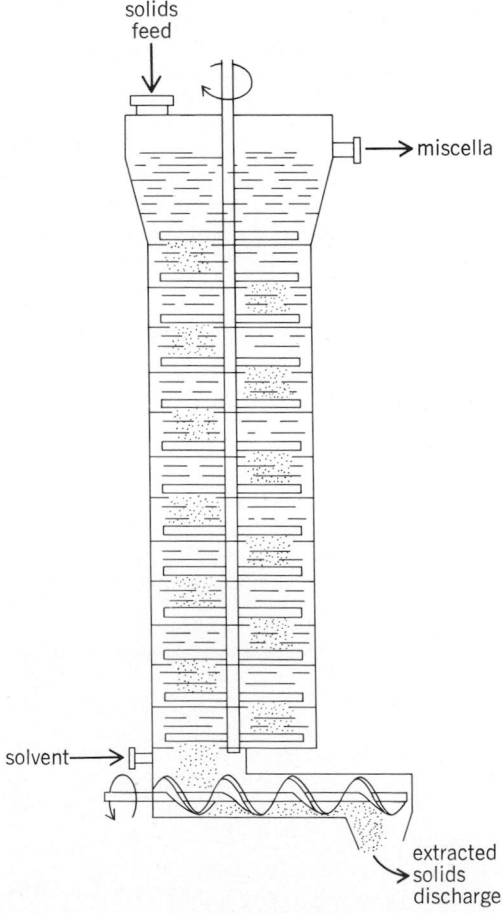

Fig. 5. Bonotto extractor. (*From R. H. Perry and C. H. Chilton, eds., Chemical Engineer's Handbook, 5th ed., McGraw-Hill, 1973*)

Dispersed-solids leaching. Equipment for leaching fine solids by dispersion and separation, a particularly useful technique for solids which disintegrate during leaching, includes batch tanks and continuous extractors.

Inasmuch as the purpose of the dispersion is usually only to permit exposure of the particles to unsaturated solvent, the agitation in a batch-stirred extractor need not be intense. Air agitation is often used. Examples are Pachuca tanks, large cylinders with conical bottoms and an axial air nozzle or air-lift tube. If mechanical agitation is employed, a slow paddle is sufficient. The Dorr agitator (Fig. 4) combines a rake with an air lift. In all cases, the mixture of solids and liquid is stirred until the maximum economical degree of leaching has occurred. The solids are then allowed to settle. The extract is decanted, and the solids, sometimes after successive treatments with fresh solvent, are removed by shoveling or flushing.

Continuous dispersed-solids leaching is accomplished in gravity sedimentation tanks or in vertical plate-extractors. An example of the latter is the Bonotto extractor shown in Fig. 5. Staggered openings in the plates allow the solids, moved around each plate by a wiping radial blade, to cascade downward from plate to plate through upward-flowing solvent.

Gravity sedimentation thickeners can serve as effective continuous contacting and separating devices for leaching fine solids. A series of such units properly connected provides true continuous countercurrent washing (known as CCD for continuous countercurrent decantation) of the solids. *See* COUNTERCURRENT TRANSFER OPERATION.

[SHELBY A. MILLER]

Bibliography: C. J. King, *Separation Processes*, 2d ed., 1980; *Kirk-Othmer Encyclopedia of Chemical Technology*, 3d ed., vol. 9, 1980; R. H. Perry and C. H. Chilton (eds.), *Chemical Engineers' Handbook*, 5th ed., 1973; J. R. Richardson and D. G. Peacock (eds.), *Chemical Engineering*, 1979; R. E. Treybal, *Mass-Transfer Operations: Chemical Engineering*, 1980.

Lead

A chemical element, Pb, atomic number 82 and atomic weight 207.19. Lead is a heavy metal (specific gravity 11.34 at 16°C), of bluish color, which tarnishes to dull gray. It is pliable, inelastic, easily fusible, melts at 327.4°C, and boils at 1740°C. The normal chemical valences are 2 and 4. It is relatively resistant to attack by sulfuric and hydrochloric acids but dissolves slowly in nitric acid. Lead is amphoteric, forming lead salts of acids as well as metal salts of plumbic acid. Lead forms many salts, oxides, and organometallic compounds.

Industrially, the most important lead com-

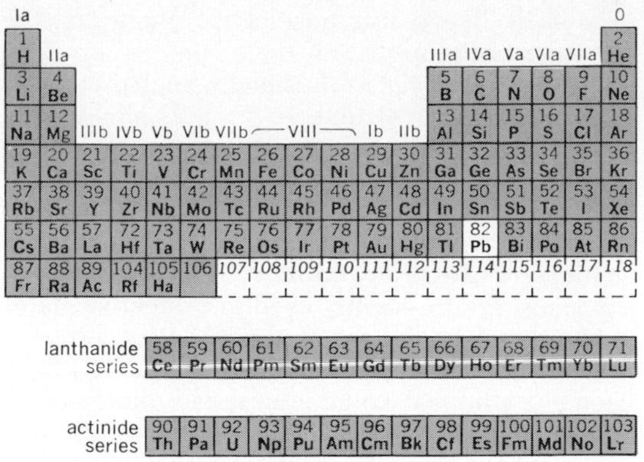
Ia 0
1 H IIa IIIa IVa Va VIa VIIa 2 He
3 Li 4 Be 5 B 6 C 7 N 8 O 9 F 10 Ne
11 Na 12 Mg IIIb IVb Vb VIb VIIb VIII Ib IIb 13 Al 14 Si 15 P 16 S 17 Cl 18 Ar
19 K 20 Ca 21 Sc 22 Ti 23 V 24 Cr 25 Mn 26 Fe 27 Co 28 Ni 29 Cu 30 Zn 31 Ga 32 Ge 33 As 34 Se 35 Br 36 Kr
37 Rb 38 Sr 39 Y 40 Zr 41 Nb 42 Mo 43 Tc 44 Ru 45 Rh 46 Pd 47 Ag 48 Cd 49 In 50 Sn 51 Sb 52 Te 53 I 54 Xe
55 Cs 56 Ba 57 La 72 Hf 73 Ta 74 W 75 Re 76 Os 77 Ir 78 Pt 79 Au 80 Hg 81 Tl 82 Pb 83 Bi 84 Po 85 At 86 Rn
87 Fr 88 Ra 89 Ac 104 Rf 105 Ha 106 107 108 109 110 111 112 113 114 115 116 117 118
lanthanide series 58 Ce 59 Pr 60 Nd 61 Pm 62 Sm 63 Eu 64 Gd 65 Tb 66 Dy 67 Ho 68 Er 69 Tm 70 Yb 71 Lu
actinide series 90 Th 91 Pa 92 U 93 Np 94 Pu 95 Am 96 Cm 97 Bk 98 Cf 99 Es 100 Fm 101 Md 102 No 103 Lr

pounds are the lead oxides and tetraethyllead. Lead forms alloys with many metals and is generally employed in the form of alloys in most applications. Alloys formed with tin, copper, arsenic, antimony, bismuth, cadmium, and sodium are all of industrial importance. *See* TETRAETHYLLEAD.

Lead compounds are toxic and have resulted in poisoning of workers from misuse and overexposure. However, lead poisoning is presently rare because of the industrial application of modern hygienic and engineering controls. The greatest hazard arises from the inhalation of vapor or dust. In the case of organolead compounds, absorption through the skin may become significant. Some of the symptoms of lead poisoning are headaches, dizziness, and insomnia. In acute cases there is usually stupor, which progresses to coma and terminates in death. The medical control of employees engaged in lead usage involves precise clinical tests of lead levels in blood and urine. With such control and the proper application of engineering control, industrial lead poisoning may be entirely prevented.

Lead is one of the oldest known metals, and the earliest archeological specimens date from about 3000 B.C. It is mentioned several times in the Old Testament, and in ancient Egypt it was used to glaze pottery and make ornamental objects. Lead was used extensively by the Romans for water pipes, even to the extent of being standardized by size and length. Some ancient Roman pipe is still intact and in serviceable condition, attesting to lead's outstanding corrosion resistance.

Natural occurrence. Yearly world production of lead is about 2,500,000 tons. The leading countries in lead mining and production are Australia, the United States, Canada, and the Soviet Union. However, the United States is unique in that it consumes much more lead than it produces (about 50% of the world's consumption). Other important sources of lead are Mexico, Peru, Yugoslavia, Germany, Morocco, South-West Africa, and Spain. Most of the lead produced in the United States is mined in Missouri, with additional lead from Idaho, Utah, Arizona, Colorado, and Montana.

Lead rarely occurs in its elemental state. The most common ore is the sulfide, galena. The other minerals of commercial importance are the carbonate, cerussite, and the sulfate, anglesite, which are much more rare.

Lead also occurs in various uranium and thorium minerals, arising directly from radioactive decay. Because certain isotopes are concentrated in lead derivatives from such sources, both the atomic weight and density of such samples vary significantly from normal lead.

Lead ores generally occur in nature associated with silver and zinc. Other metals commonly occurring with lead ores are copper, arsenic, antimony, and bismuth. Most of the world production of arsenic, antimony, and bismuth arises from their separation from lead ores.

Commercial lead ores may contain as little as 3% lead, but a lead content of about 10% is most common. The ores are concentrated to 40% or greater lead content before smelting. A variety of mechanical separation processes may be employed for the concentration of lead ores, but the sulfide ores are generally concentrated by flotation processes, for which they are particularly suitable.

More than one-third of the lead produced in the United States is derived from reclaimed lead or lead alloy. The chief source of this secondary lead is scrapped automobile storage batteries. The secondary lead is obtained by melting of scrap and is generally not purified but rather reused as lead alloy. Secondary lead from storage batteries contains antimony and is reused in battery manufacture. The lead scrap which contains tin is reused in the manufacture of solder.

Uses. As implied in the discussion of scrap lead above, the largest single use of lead is for the manufacture of storage batteries. Other important applications are for the manufacture of tetraethyllead, cable covering, construction, pigments, solder, and ammunition.

The lead storage battery consists of a negative plate of porous lead, a positive plate of lead peroxide, and an electrolyte of sulfuric acid solution. The plates are made by casting perforated grids from an alloy of lead containing 7–12% antimony and small amounts of tin. The antimony is added to give the grids better hardness and corrosion resistance, whereas the tin is added to give better casting properties. The negative grid is coated with a paste of litharge (PbO) or a mixture of litharge and finely divided metallic lead. The positive grid is coated with a mixture of litharge and finely divided lead and in some cases red lead (Pb_3O_4). After the batteries are assembled and charged with sulfuric acid, the plates are "formed" by passing an electric current through them to convert the lead oxide at the negative plate into finely divided lead sponge and convert the lead oxide at the positive plate to lead peroxide. The chemical reactions which take place on charging or discharging a battery are very complex but may be indicated in simple form by reaction (1). Because sulfuric acid

$$PbO_2 + 2H_2SO_4 + Pb \underset{\text{Charge}}{\overset{\text{Discharge}}{\rightleftharpoons}} 2PbSO_4 + 2H_2O \qquad (1)$$

is consumed on discharge of the battery, the sulfuric acid concentration is used as a measure of the state of charge of lead batteries.

In 1957 an important addition to the family of lead batteries was the introduction of a chargeable dry cell. The battery is composed of an alkali–lead oxide–silver system and is capable of repeated discharge and recharge without affecting its capacity. Also noteworthy are its operability at both low and high temperatures and its long-term shelf life.

The manufacture of tetraethyllead is the second largest use of lead. When 2–3 ml of this antiknock agent is added to 1 gal of gasoline, the octane rating of the fuel is raised about 10 octane numbers. This increase in octane number permits the use of higher compression ratios in automobile engines, thereby improving engine efficiency. For example, it has been shown that an average decrease in gasoline consumption of about 45% can be obtained in going from 5.25:1 to 10:1 compression ratio under conditions of constant performance.

Tetraethyllead is manufactured by the reaction of ethyl chloride with an alloy of sodium and lead. The chemical reaction is given by reaction (2).

$$4PbNa + 4C_2H_5Cl \rightarrow (C_2H_5)_4Pb + 4NaCl + 3Pb \quad (2)$$

Lead is melted and mixed with sodium to form an alloy containing about 10% sodium. The solid alloy is treated with ethyl chloride in autoclaves at moderate temperature and pressure in the absence of solvent. The crude product is separated by steam distillation and is purified by air-blowing to remove organic bismuth compounds arising from bismuth impurities in the original lead metal. The tetraethyllead is then blended with ethylene dichloride and ethylene dibromide, which serve as scavengers to aid in the removal of lead deposits from engines. It is sold in this form to petroleum refiners as antiknock fluid for addition to their automotive gasoline stocks. In aviation gasoline, ethylene dibromide alone is used as the scavenger.

Other commercial organolead antiknock agents which have been introduced include tetramethyllead, mixtures of tetramethyllead and tetraethyllead, and redistributed mixtures containing all the possible methylethyllead compounds as well as the original tetraethyllead and tetramethyllead. Tetramethyllead is manufactured both by a process based on methyl chloride and sodium-lead alloy, analogous to the manufacture of tetraethyllead, and by a new electrochemical process using a lead anode and an ether–methylmagnesium chloride electrolyte.

A much smaller commercial use for tetraethyllead arises from its ability to alkylate mercury compounds. The resulting organomercury compounds are used as fungicides.

In addition to these traditional uses of organolead compounds, some high-value applications are being developed. These applications are in such diverse fields as catalysts for polyurethane foams, marine antifouling paint toxicants, biocidal agents against gram-positive bacteria, protection of wood against marine borers and fungal attack, preservatives for cotton against rot and mildew, molluskicidal agents to kill snails which act as intermediate hosts for the serious tropical disease "snail fever" (bilharzia, or schistosomiasis), anthelmintic agents against tapeworms in domestic animals, wear-reducing agents in lubricants, and corrosion inhibitors for steel.

Because of its excellent resistance to corrosion, lead finds extensive use in construction, particularly in the chemical industry. It is resistant to attack by many acids because it forms its own protective oxide coating. Because of this advantageous characteristic, lead is used widely in the manufacture and handling of sulfuric acid. The corrosion resistance of lead is also utilized in the cathodic protection of metal structures. This is accomplished by impressing an electric charge on the structure to be protected; the structure serves as the cathode and lead as the anode. Such protection is useful for large structures such as ships, pipelines, and bridges.

Because of the poor structural strength of lead, it is generally used in the form of its alloys, particularly in combination with antimony, or is used as coatings or plates on stronger structural metals. Tin alloys are frequently used in protective plates to impart mechanical strength and better corrosion resistance. To some extent, the lead that has been used in traditional architectural and plumbing applications in the past, such as roofing and flashings, for example is now being displaced by other materials. However, lead is being used increasingly in construction because of its excellent sound attenuation properties. Changes in building codes have established maximum permissible noise levels for multistoried structures; these have accelerated the trend toward the use of lead sheet as either an integral part of the wall partition or as a barrier in the air space above suspended ceilings. Lead has also been used in construction for many years because of its vibration-damping properties. This use also is growing. Heavy machinery and even large buildings are isolated from vibration by placing them on lead and asbestos pads. A splendid example in antivibration building techniques is the Pan American Building built over Grand Central Terminal in New York. In inertial guidance research, the experimental chambers are isolated from adjacent ground and structures by the use of massive lead pads.

Lead has long been used as protective shielding for x-ray machines. Because of the expanded applications of atomic energy, radiation-shielding applications of lead have become increasingly important. Basically, shielding effectiveness against radiation depends on density, and lead has the highest density of the commonly available materials. Wherever glass windows are required in radiation equipment, a type of glass containing large amounts of lead is used.

Lead sheathing for telephone and television cables continues to be a sizable outlet for lead. The unique ductility of lead makes it particularly suitable for this application because it can be extruded in a continuous sheath around the internal conductors. The lead used in this application is generally alloyed with small amounts of arsenic and bismuth.

The use of lead in pigments has been a major outlet for lead but is decreasing in volume. White lead, $2PbCO_3 \cdot Pb(OH)_2$, is the most extensively used lead pigment. It is prepared from metallic lead by treatment with acetic acid, air, and carbon dioxide. It is an excellent pigment because of its outstanding chemical affinity for paint vehicles and its great hiding power. Pigments such as red lead (Pb_3O_4) and blue lead (a combination of basic lead sulfate, $PbSO_4 \cdot PbO$, zinc oxide, and carbon) are used as metal protective pigments in paints. Other lead pigments of importance are basic lead sulfate and lead chromates. The lead chromates are frequently used as dyes in formulating yellow, orange, red, and green paints. *See* PIGMENT.

The high density of lead, which permits a maximum of striking power with a minimum of air resistance, has made lead the ideal metal for bullets and shot. Lead shot is manufactured by a unique process which involves dropping molten lead into water from heights up to 125 ft, thus freezing lead in the spherical form assumed by its droplets.

Principal compounds. From a commercial standpoint, the most important lead compounds have been mentioned above. In addition, a considerable variety of lead compounds, such as silicates, carbonates, and salts of organic acids, are used as heat and light stabilizers for polyvinyl chloride plastics. These compounds are both inexpensive and effective. They function as hydrogen chloride acid acceptors, thereby preventing the autocatalytic breakdown of the plastic by the acid.

Certain inorganic lead compounds find specialized use. Lead silicates are used for the manufacture of glass and ceramic frits, which are useful in introducing lead into glass and ceramic finishes. Lead azide, $Pb(N_3)_2$, is the standard detonator for explosives. Lead arsenates are used in large quantities as insecticides for crop protection. Organic insecticides have displaced lead arsenate to some extent, but not completely because they have not been found as effective in certain applications.

Among the newer uses for inorganic lead compounds, litharge (lead oxide) is widely employed at a level of approximately 2% lead to improve the magnetic properties of barium ferrite ceramic magnets. High lead ferrites, containing approximately 20% lead, have been developed with superior magnetic properties. Also, a calcined mixture of lead zirconate and lead titanate, known as PZT in the trade, is finding increasing markets as a piezoelectric material. The most important application is in ultrasonic cleaning equipment. Finally, lead telluride is finding some use as the active component of thermoelectric generators.

[HYMIN SHAPIRO; JAMES D. JOHNSTON]

Bibliography: American Society for Metals, *Metals Handbook, Properties and Selection: Nonferrous Alloys and Pure Metals*, 1979; J. Donohue, *The Structures of Elements*, 1974, reprint 1981; Lead Industries Association, *Modern Uses of Lead in the Construction Industry*, 1980; R. E. Kirk and D. F. Othmer (eds.), *Encyclopedia of Chemical Technology*, vol. 14, 3d ed., 1981; D. O. Rausch et al (eds.), *Lead-Zinc Update*, 1977; E. G. Rochow and E. W. Abel, *The Chemistry of Germanium, Tin, and Lead*, 1975.

Line spectrum

A discontinuous spectrum characteristic of excited atoms, ions, and certain molecules in the gaseous phase at low pressures, to be distinguished

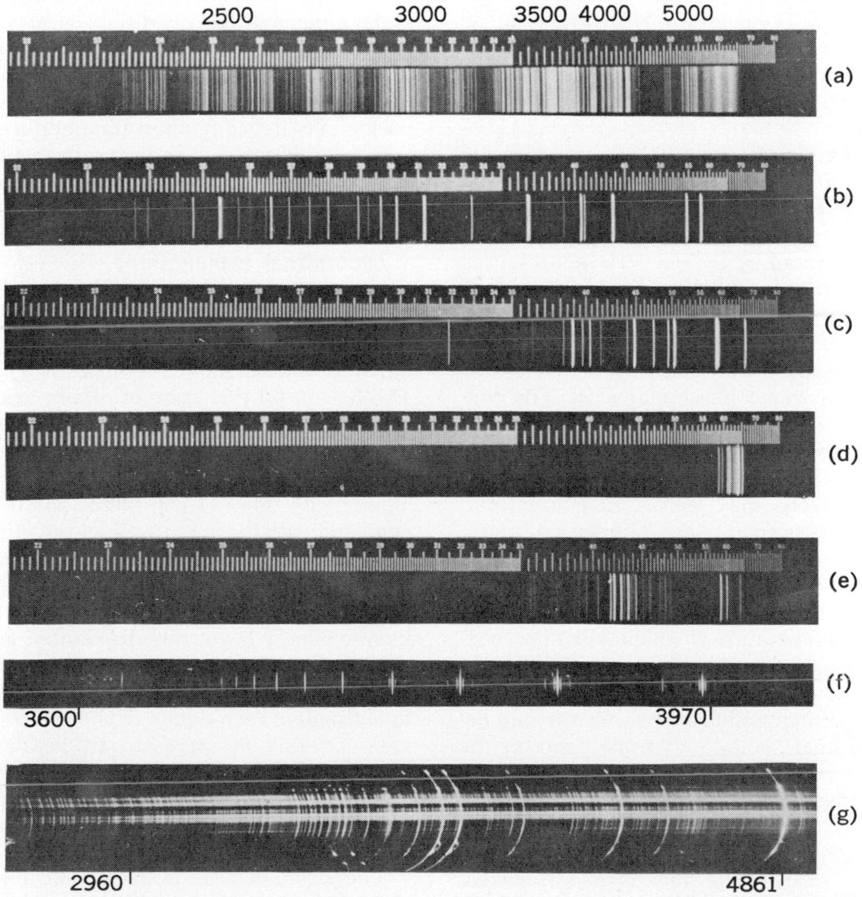

Photographs of common line spectra, wavelengths given in angstroms; emission spectra (*a–e*) all taken with the same quartz spectrograph. (*a*) Spectrum of iron arc. (*b*) Mercury spectrum from an arc enclosed in quartz. (*c*) Helium in a glass discharge tube. (*d*) Neon in a glass discharge tube. (*e*) Argon in a glass discharge tube. (*f*) Balmer series of hydrogen in the ultraviolet, photographed with a grating spectrograph. (*g*) Emission spectrum from gaseous chromosphere of the Sun, a grating spectrum taken without a slit at the instant immediately preceding a total eclipse, when the rest of the Sun is covered by the Moon's disk. Two strongest images, H and K lines of calcium, show marked prominences, or clouds, of calcium vapor. Other strong lines are caused by hydrogen and helium. (*From F. A. Jenkins and H. E. White, Fundamentals of Optics, 3d ed., McGraw-Hill, 1957*)

from band spectra, emitted by most free molecules, and continuous spectra, emitted by matter in the solid, liquid, and sometimes gaseous phase. If an electric arc or spark between metallic electrodes, or an electric discharge through a low-pressure gas, is viewed through a spectroscope, images of the spectroscope slit are seen in the characteristic colors emitted by the atoms or ions present.

To avoid the overlapping of close spectral images, the slit illuminated by the light source is made very narrow. The spectrum then appears as an array of bright line slit images on a dark background (see illustration). Under certain conditions spectra show dark absorption lines against a bright background. *See* SPECTROSCOPY.

[GEORGE R. HARRISON]

Liquid

A state of matter intermediate between that of crystalline solids and gases. Macroscopically, liquids are distinguished from crystalline solids in their capacity to flow under the action of extremely small shear stresses and to conform to the shape of a confining vessel. Liquids differ from gases in possessing a free surface and in lacking the capacity to expand without limit. On the scale of molecular dimensions liquids lack the long-range order that characterizes the crystalline state, but nevertheless they possess a degree of structural regularity that extends over distances of a few molecular diameters. In this respect, liquids are wholly unlike gases, whose molecular organization is completely random.

LIQUID

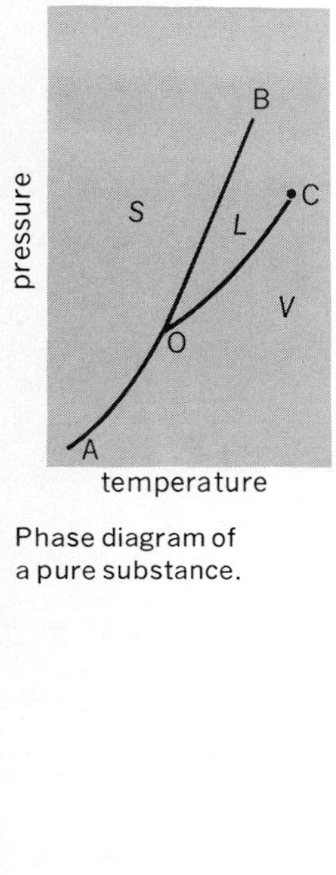

Phase diagram of a pure substance.

Thermodynamic relations. The thermodynamic conditions under which a substance may exist indefinitely in the liquid state are described by its phase diagram, shown schematically in the figure. The area designated by L depicts those pressures and temperatures for which the liquid state is energetically the lowest and therefore the stable state. The areas denoted by S and V similarly indicate those pressures and temperatures for which only the solid or vapor phase may exist. The connecting lines OC, OB, and OA define pressures and temperatures for which the liquid and its vapor, the solid and its liquid, and the solid and its vapor, respectively, may coexist in equilibrium. They are usually termed phase boundary or phase coexistence lines. The intersection of the three lines at O defines a triple point which, for the three states of matter under discussion, is the unique pressure and temperature at which they may coexist at equilibrium. Other triple points exist in the phase diagram of a substance that possesses two or more crystalline modifications, but the one depicted in the figure is the only triple point for the coexistence of the vapor, liquid, and solid. Line OA has its origin at the absolute zero of temperature and OB, the melting line, has no upper limit. The liquid-vapor pressure line OC is different from OB, however, in that it terminates at a precisely reproducible point C, called the critical point. Above the critical temperature no pressure, however large, will liquefy a gas. *See* TRIPLE POINT.

Along any of the coexistence curves the relationship between pressure and temperature is given by the Clausius-Clapeyron equation: $dP/dT = \Delta H/T\Delta V$, where ΔV is the difference in molar volume of the corresponding phases (gas-liquid, gas-solid, or liquid-solid) and ΔH is the molar heat of transition at the temperature in question. By means of this equation the change in the melting point of the solid or the boiling point of the liquid as a function of pressure may be calculated. When a liquid in equilibrium with its vapor is heated in a closed vessel, its vapor pressure and temperature increase along the line OC. ΔH and ΔV both decrease and become zero at the critical point, where all distinction between the two phases vanishes. *See* PHASE EQUILIBRIUM.

Transport properties. Liquids possess important transport properties, notably their capacity to transmit heat (thermal conductivity), to transfer momentum under shear stresses (viscosity), and to attain a state of homogeneous composition when mixed with other miscible liquids (diffusion). These nonequilibrium properties of liquids are well understood in macroscopic terms and are exploited in large-scale engineering and chemical-process operations. Thus, the rate of flow of heat across a layer of liquid is given by $\dot{Q} = \kappa\, dT/dx$, where $\dot{Q}$ is the heat flux, dT/dx is the thermal gradient, and κ is the coefficient of thermal conductivity. Similarly, the shearing of one liquid layer against another is resisted by a force equal to the momentum transfer: $F = \dot{p} = \eta\, dv/dx$, where dv/dx is the velocity gradient and η is the coefficient of viscosity. Likewise, the rate of transport of matter under nonconvective conditions is governed by the gradient of concentration of the diffusing species: $J = -D\, dC/dx$, where J is the matter flux and D is the coefficient of diffusion. Each of these transport coefficients depends upon temperature, pressure, and composition and may be determined experimentally. An a priori calculation of κ, η, and D is a very difficult problem, however, and only approximate theories exist.

Theoretical explanations. In fact, although a great deal of effort has been expended, there still exists no satisfactory theory of the liquid state. Even so commonplace a phenomenon as the melting of ice has no adequate theoretical explanation. The reason for this state of affairs lies in the tremendous structural and dynamical complexity of the liquid state. To understand this, it is useful to compare the structural and kinetic properties of liquids with those of crystalline solids on the one hand and with those of gases on the other.

In crystals, atoms or molecules occupy well-defined positions on a three-dimensional lattice, oscillating about them with small amplitudes; their kinetic energy is entirely distributed among these quantized vibrational states up to the melting point. This nearly perfect spatial order is revealed by diffraction techniques, which utilize the coherent scattering of x-rays or particles having wavelengths comparable with interatomic spacings. The structural and dynamical properties are sufficiently tractable mathematically so that the theory of solids is quite well understood.

The theory of gases is also simple, but for quite a different reason. No vestige of positional regularity of atoms remains in gases, and their energy resides entirely in high-speed translational motion. Except for collisions, which deflect their motions into new straight-line trajectories, atoms in gases do not interact with one another; vibrational modes in monatomic gases are absent.

Liquids, by contrast, lie intermediate between

gases and crystals from both a structural and dynamic point of view. Kinetic energy is partitioned among translational and vibrational modes, and diffraction studies reveal a degree of short-range order that extends over several molecular diameters. Moreover, this "structure" is continually changing under the influence of vibrational and translational displacements of atoms. Physical reality may be attributed to this short-range structure, nevertheless, in the sense that a time average over the huge number of possible configurations of atoms may show that a fairly definite number of neighboring atoms lie close to any arbitrary atom. At a somewhat greater distance from this reference atom, the density of neighbors oscillates above and below the average density of atoms in the liquid as a whole.

Information about the degree of local order is contained in the radial distribution function, a mathematical property which may be deduced from diffraction measurements. This is the starting point for a theory of the liquid state, and although efforts by J. G. Kirkwood and his students, by H. Eyring and his collaborators, and by J. E. Lennard-Jones and A. F. Devonshire have yielded partial successes, prodigious mathematical difficulties lie in the path of an entirely satisfactory solution. *See* BOILING POINT; MELTING POINT; VISCOSITY; X-RAY DIFFRACTION.

[NORMAN H. NACHTRIEB]

Bibliography: P. A. Egelstaff (ed.), *An Introduction to the Liquid State*, 1967; J. P. Hansen and I. R. McDonald, *The Theory of Simple Liquids*, 1976; P. Kruus, *Liquids and Solutions: Structure and Dynamics*, 1977; H. N. Temperley and D. H. Trevena, *Liquids and Their Properties: A Molecular and Microscopic Treatise*, 1978.

Liquid chromatography

A separation technique involving passage of a liquid moving phase through a solid or a liquid stationary phase. Interest in modern liquid chromatography derives from the fact that with this approach separation problems can be attacked that are difficult or even impossible to solve by other techniques. The major separation method up to the present time has been gas chromatography; yet this method is limited to volatile components (mol. wt < 300) that are thermally stable. Most biochemical mixtures are therefore chromatographed only with great difficulty (in most cases after the preparation of a volatile derivative). Liquid chromatography, however, is compatible with the samples commonly encountered in biochemical and organic mixtures (that is, mol. wt > 300, ionic, heat-sensitive, and so on), and this is its greatest attribute. *See* GAS CHROMATOGRAPHY.

Historically, column liquid chromatography was the first chromatographic method. For the most part, however, it has remained as a preparative tool to obtain milligram-to-gram quantities of purified material. It is only since about 1970 that major advances in speed and efficiency have occurred such that performance characteristics similar to those of gas chromatography have been achieved. Another major development has been the introduction of new instrumentation, especially high-sensitivity detectors and high-pressure pumps. Advances in selectivity and flexibility of liquid chromatographic systems also have taken place. Finally, many new applications have broadened the scope of the method.

High performance. From theoretical considerations developed first with reference to gas chromatography, it became clear in the late 1960s that liquid chromatographic conditions were not optimized for speed and efficiency. Typically at that time, columns were packed with support particles of 100–200-μm diameter, and mobile phase flow was controlled by gravity feed of the eluent (velocity about 0.01 cm/sec). A 50-cm column might generate 100 theoretical plates and 0.01 effective plate per second.

To improve performance, much smaller particle diameters (about 5 μm), velocities a hundredfold greater (about 1 cm/sec), and columns with narrower diameters (about 2–3 mm) are now employed. Performance characteristics of 10,000 plates per meter and 35 effective plates per second are possible. These modifications have resulted in the need for high-pressure pumps to drive the mobile phase through the column. Pressures of 1000 to 6000 psi (where 1 psi is about 6890 N/m^2) are now typically employed. Of course, the low compressibility of liquids minimizes the safety hazards with this equipment. Figure 1 shows the separation of a series of alcohols by liquid-liquid chromatography on 5–6-μm porous silica.

Ion-exchange resin particles in the diameter range of 10–15 μm have been used since about 1963 for the separation of amino acid and peptide mixtures, as well as complex body fluid mixtures (for example, urine). Indeed, major advances have been made in the application of this method to clinical diagnosis. However, workers found diffi-

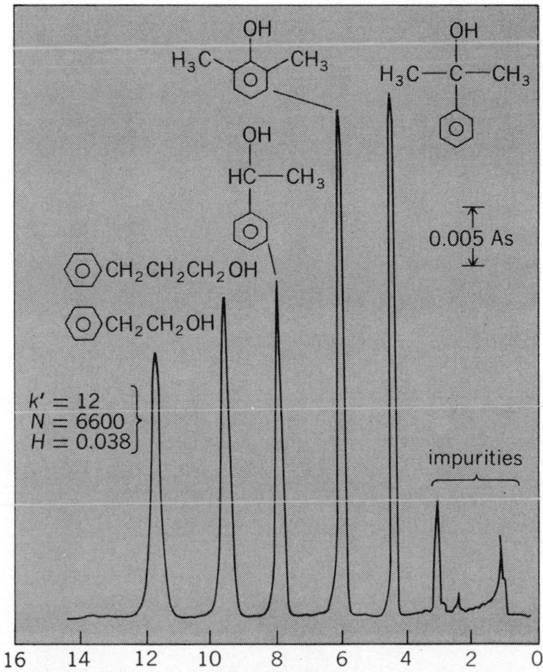

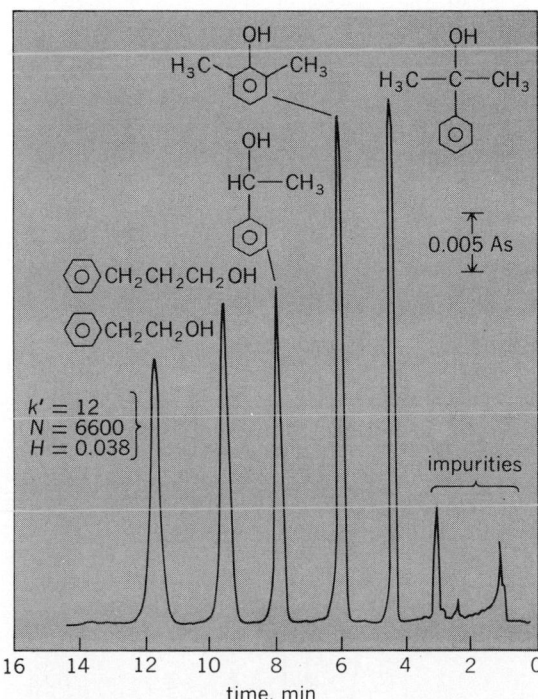

Fig. 1. Liquid-liquid chromatography with porous microspheres. Capacity factor is represented by k', N is the number of theoretical plates, H the height equivalent to a theoretical plate, and As the absorbance unit. (*From J. J. Kirkland, J. Chromatogr. Sci., 10:583, 1972*)

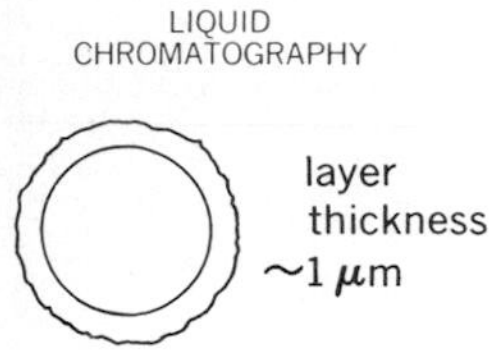

Fig. 2. Diagrammatic illustration of porous layer beads. Particle diameter about 30 μm, porous shell about 1 μm.

culty in efficiently and reproducibly packing porous silica or alumina particles less than 20 μm in diameter. In 1972 slurry packing methods were introduced which produce excellent columns with these materials down to 5 μm. Because of the high pressures involved, column lengths are typically no greater than 30 cm; however, more than 10,000 theoretical plates can be generated with these short lengths. These columns have been used for the separation and analysis of a variety of mixtures, with both liquid-liquid and liquid-solid chromatography employed. *See* ION EXCHANGE.

Although support columns packed with small-diameter particles will be used with increasing frequency in the future, at present porous layer bead supports provide the major load in liquid chromatography. As shown in Fig. 2, these materials consist of a solid spherical core (about 30 μm diameter) coated with a thin porous layer of adsorbent (about 1 μm thick). The larger particle diameter and high density allow easy and reproducible packing characteristics. The specially designed supports result from the need to reduce pore depths and thus enhance mass transfer into and out of the particles. The quite slow diffusion of components in liquids would severely limit the velocity of the mobile phase by the particle, if fully porous materials of > 30 μm were to be used. The drawback to the exclusive application of porous layer beads is the low sample capacity (and peak capacity), a result of the small amount of stationary phase. Yet, since their introduction in 1969–1970, porous layer beads have been used quite extensively, and, as noted above, they represent the most popular support materials at the present time for liquid-liquid, liquid-solid, and ion-exchange chromatography. Small-diameter support (about 5 μm) columns provide greater speed characteristics by a factor of 10 and somewhat greater sample capacity; however, porous layer bead columns make fewer demands on equipment and are easier to manufacture.

Instrumentation. Detectors are available which fulfill the demands of high sensitivity, low dead volume, rapid response, and selectivity. The ultraviolet detector is most frequently selected because of its high sensitivity. Previously, commercial devices used only the 254-nm line of a mercury discharge tube. In 1972 it became possible to use other Hg lines (for example, 280 nm), and indeed several instruments now allow the simultaneous detection of the chromatogram at two wavelengths. This is valuable when complex mixtures are employed. The differential refractive index is the second most widely used detector. Although its sensitivity is generally less than that of the ultraviolet detector, it is a more universal sensor. More interest is being shown in the moving wire detector, in which the effluent from the column is deposited onto a moving platinum wire. The mobile phase is evaporated, leaving a residue of the solute which is then pyrolyzed into a flame ionization detector. This device finds increasing use with sophisticated multisolvent gradients, as indicated by Fig. 3. Other detectors are valuable in specialized areas, for example, polarographic and radiometric applications.

The development of pumps specifically designed for high-performance liquid chromatography has also played an important role in the advancement of the field. It is possible to obtain relatively pulse-free pumps operating a constant flow rate up to 6000 psi. Moreover, these devices permit simple digitization of the flow rate and with simultaneous operation of two pumps allow the convenient dialing of a variety of solvent gradient programs.

Modes of liquid chromatography. Much activity occurred during the early 1970s in the use of chemically bonded phases in liquid chromatography. Here, organic substances are bonded in either a monomeric or a polymeric fashion to a solid support such as silica or alumina. One advantage of these phases is that selectivity can be "tailor-made" for a particular application. Moreover, certain of these phases act as bonded liquids, and as such they require no precolumn to maintain saturation of the mobile phase, and they can be use in gradient elution.

The development of techniques to enhance the peak capacity (that is, number of solutes that can be resolved) of the liquid chromatographic system has also been important. Coupled columns are finding increasing use in this regard. Here, two columns with different capacities or selectivities are placed in series. By proper use of switching valves the sample components are resolved on the column best suited to their separation.

Finally, R. P. W. Scott developed a 15-solvent series for gradient elution which overcomes the problem of displacement effects with silica or alumina as support. A particularly striking example of the separation of a complex mixture is shown in Fig. 3.

Applications. The application of modern liquid chromatography to various separation problems is increasing at a rapid pace. Pesticides, drugs of abuse, antioxidants, and nonionic surfactants are only a few of the substances to which the method can be applied. Liquid chromatography, moreover, played a central role in the successful synthesis of vitamin B_{12}. Finally, it is interesting to note that 20 or more phenylthiohydantoin derivatives of amino acids can now be separated in less than 1 hr. This

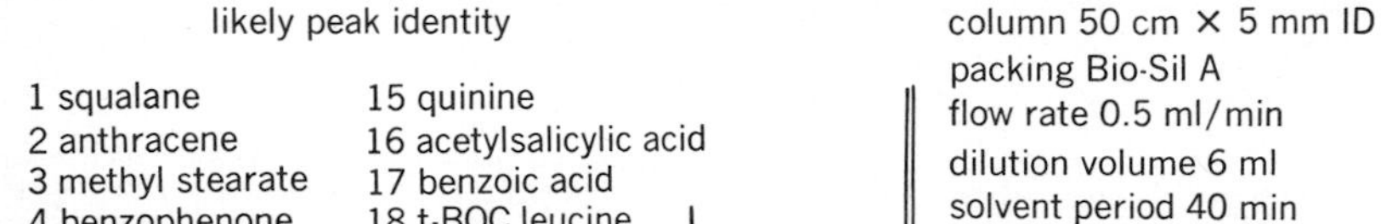

Fig. 3. Chromatogram of a mixture containing solutes of widely differing polarities using incremental gradient elution. (*From R. P. W. Scott and P. Kucera, Anal. Chem., 45:749, 1973*)

development will prove of great value to workers in protein sequencing. *See* ADSORPTION; CHEMICAL SEPARATION TECHNIQUES; CHROMATOGRAPHY.

[BARRY L. KARGER]

Bibliography: N. Hadden et al., *Basic Liquid Chromatography*, 1972; J. J. Kirkland (ed.), *Modern Practice of Liquid Chromatography*, 1971; A. Pryde and M. T. G. Gilbert, *Application of High Performance Liquid Chromatography*, 1970; R. Scott (ed.), *Contemporary Liquid Chromatography*, 1976; R. Snyder and J. J. Kirkland, *Introduction to Modern Liquid Chromatography*, 2d ed., 1979.

Liquid crystals

A state of matter that mixes the properties of both the liquid and solid states. Liquid crystals may be described as condensed fluid states with spontaneous anisotropy. They are categorized in two ways: thermotropic liquid crystals, prepared by heating the substance, and lyotropic liquid crystals, prepared by mixing two or more components, one of which is rather polar in character (for example, water). Thermotropic liquid crystals are divided, according to structural characteristics, into two classes, nematic and smectic. Nematics are further subdivided into ordinary and cholesterics.

When the solid which forms a liquid crystal is heated it undergoes transformation into a turbid system that is both fluid and birefringent. The consistency of the fluid varies with different compounds from a paste to a free-flowing liquid. When the turbid system is heated, it is converted into an isotropic liquid (optical properties are the same regardless of the direction of the measurement). These changes in phases can be represented schematically as follows:

$$\text{Solid} \underset{\text{cool}}{\overset{\text{heat}}{\rightleftharpoons}} \text{liquid crystal} \underset{\text{cool}}{\overset{\text{heat}}{\rightleftharpoons}} \text{liquid}$$

On cooling the system, the process reverses and goes from isotropic liquid to liquid crystal and finally to the solid. *See* LIQUID.

Lyotropic liquid crystals often have an amphiphilic component, a compound with a polar head attached to a long hydrophobic tail. Sodium stearate and lecithin (a phospholipid) are typical examples of amphiphiles. Starting with a solid amphiphile and adding water, the lamellar structure (molecular packing in layers) is formed. By stepwise addition of water, the molecular packing may take on a cubic structure, then hexagonal, then micellar, followed by true solution. The process is reversed by withdrawing water. Thousands of compounds will form liquid crystals on heating, and still more will do so if two or more components are mixed. A few representative compounds are listed in the table.

Classification and structure. Conventionally, matter is considered to have only three states: solid, liquid, and gas. In the solid state, the molecules or atoms show small vibrations about rigidly fixed lattice positions, but they cannot rotate. A liquid will take the shape of its container and will bound itself at the top with its own free surface. The liquid state is characterized by relatively unhindered rotation and no long-range order. The space in a system constituting a gas is sparsely occupied. The molecules are free to occupy the entire volume of their container.

Liquid crystals are a state of matter that combines a kind of long-range order (in the sense of a solid) with the ability to form droplets and to pour (in the sense of waterlike liquids). They also exhibit properties of their own such as the ability to form monocrystals with the application of a normal magnetic or electric field; an optical activity of a magnitude without parallel in either solids or liquids; and a temperature sensitivity which results in a color change in certain liquid crystals. Thermotropic liquid crystals are either nematic or smectic.

Nematic structures. Nematic liquid crystals are subdivided into the ordinary nematic and the cholesteric-nematic. The molecules in the ordinary nematic structure maintain a parallel or nearly parallel arrangement to each other along the long

Name, formula, and liquid crystalline range of selected compounds

THERMOTROPIC LIQUID CRYSTALS

1. Nematic liquid crystals

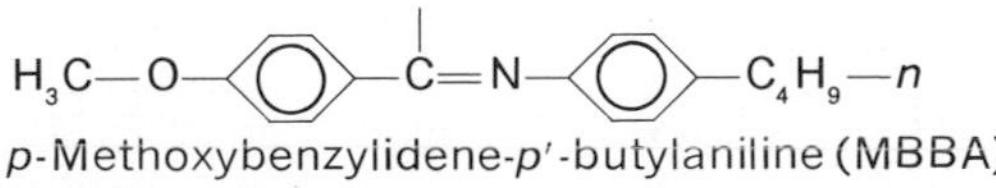

p-Methoxybenzylidene-*p*′-butylaniline (MBBA) — 21–47°C

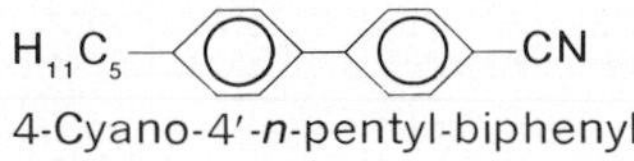

4-Cyano-4′-*n*-pentyl-biphenyl — 24–35°C

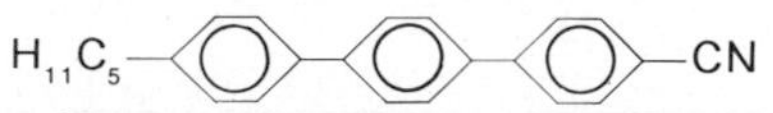

4-Cyano-4″-*n*-pentyl-*p*-terphenyl — 131–240°C

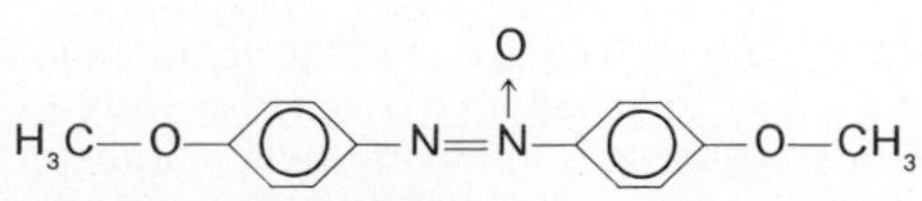

p-Azoxyanisole (PAA) — 117–137°C

(*continued*)

Name, formula, and liquid crystalline range of selected compounds (cont.)

THERMOTROPIC LIQUID CRYSTALS (cont.)

2. Cholesteric esters

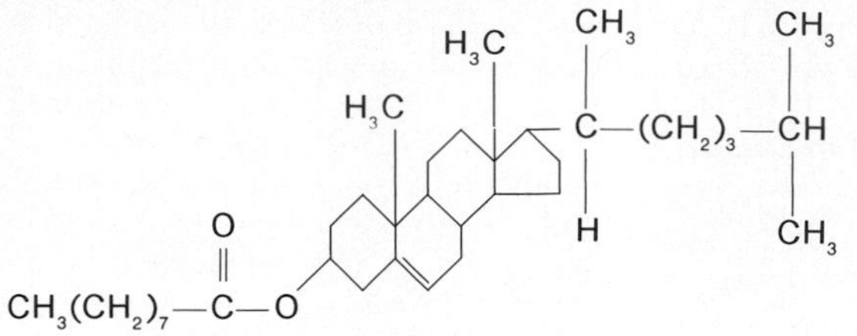

145–179°C

$CH_3(CH_2)_7$—C—O
Cholesteryl nonanoate

3. Noncholesteryl, chiral-type compound

H_3C—O—⟨⟩—C=N—⟨⟩—C=C—C—O—CH_2—C—C_2H_5

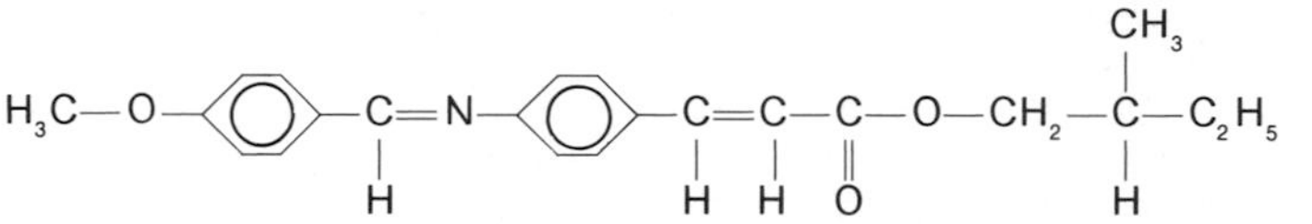

76–125°C

(–)-2-Methylbutyl-*p*-(*p'*-methoxy-benzylideneamino) cinnamate

4. Smectic A

⟨⟩—⟨⟩—C=N—⟨⟩—$COOC_2H_5$

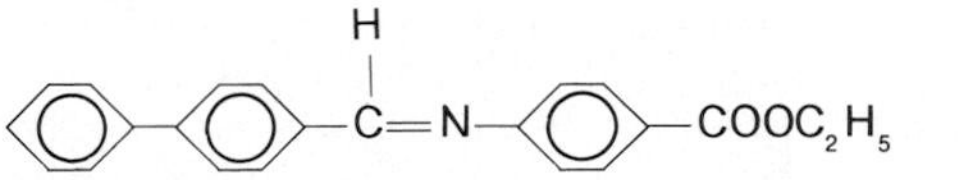

121–131°C

Ethyl *p*(*p'*-phenylbenzalamino) benzoate

5. Smectic B

H_5C_2O—⟨⟩—C=N—⟨⟩—CH=CH—$COOC_2H_5$

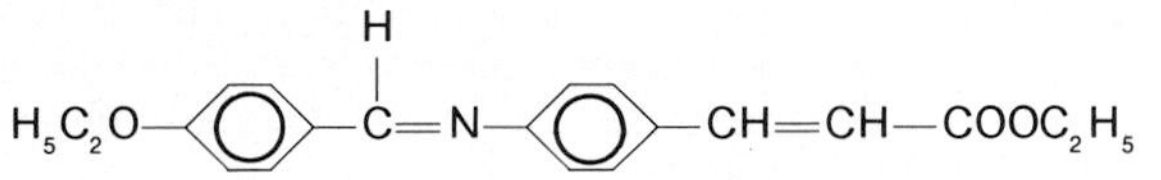

77–116°C

Ethyl *p*-ethoxybenzal-*p'*-aminocinnamate

6. Smectic C

n—$H_{17}C_8$—O—⟨⟩—COOH

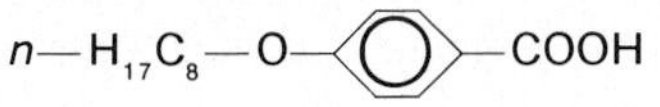

108–147°C

p-n-Octyloxybenzoic acid

LYOTROPIC LIQUID CRYSTALLINE COMPOUNDS

1. Sodium stearate

$CH_3(CH_2)_{16}COO^- Na^+$

2. α-Lecithin

CH_2—O—CO—$(CH_2)_{16}$—CH_3
CH—O—C—O—$(CH_2)_{16}$—CH_3
         O^-
CH_2—O—P—O—CH_2—CH_2—$N^+(CH_3)_3$
         ‖
         O

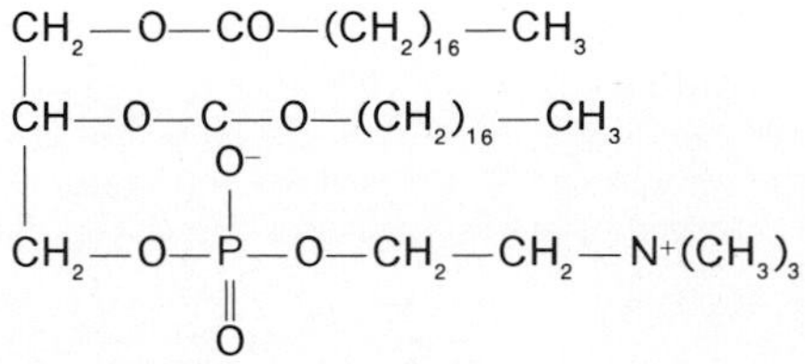

(Stearic acid derivative)

molecular axes (Fig. 1*a*). They are mobile in three directions and can rotate about one axis. This structure is one-dimensional.

When the nematic structure is heated, it is generally transformed into the isotropic liquid (Fig 1*b*). The nematic structure is the highest-temperature mesophase in thermotropic liquid crystals. The energy required to deform a nematic liquid crystal is so small that even the slightest perturbation caused by a dust particle can distort the structure considerably.

In the cholesteric-nematic structure (Fig. 1*c*), the direction of the long axis of the molecule in a given layer is slightly displaced from the direction of the molecular axes of the molecules in an adjacent layer. If a twist is applied to this molecular packing, a helical structure is formed. The helix has a pitch which is temperature-sensitive. The helical structure serves as a diffraction grating for visible light. Chiral compounds show the cholesteric-nematic structure (twisted nematic), for example, the cholesteric esters.

Smectic structures. The term smectic covers all thermotropic liquid crystals that are not nematics. At least seven smectic structures have been described. Indications are that two more can be added, making a total of nine from smectic A(S_A) to smectic I (S_I). The alphabetic subscripts only indicate the order in which the smectic structures were first recognized and identified. In most smectic structures, molecules are arranged in strata. The molecules (except in smectic D) are arranged in layers with their long axes parallel to each other. They can move in two directions in the plane and can rotate about one axis. Those within layers can be in neat rows or randomly distributed.

Smectic liquid crystals may have structured or unstructured strata. Structured smectic liquid crystals have long-range order in the arrangement of molecules in layers to form a regular two-dimensional lattice. The most common of the structured liquid crystals is smectic B. Molecular layers are in well-defined order, and the arrangement of the molecules within the strata is also well ordered. The long axes of the molecules lie perpendicular to the plane of the layers. In the smectic A (Fig. 1*d*) structure, molecules are also packed in strata, but the molecules in a stratum are randomly arranged. The long axes of the molecules in the smectic A structure lie perpendicular to the plane of the layers. Molecular packing in a smectic C (Fig. 1*e*) is the same as that in smectic A, except

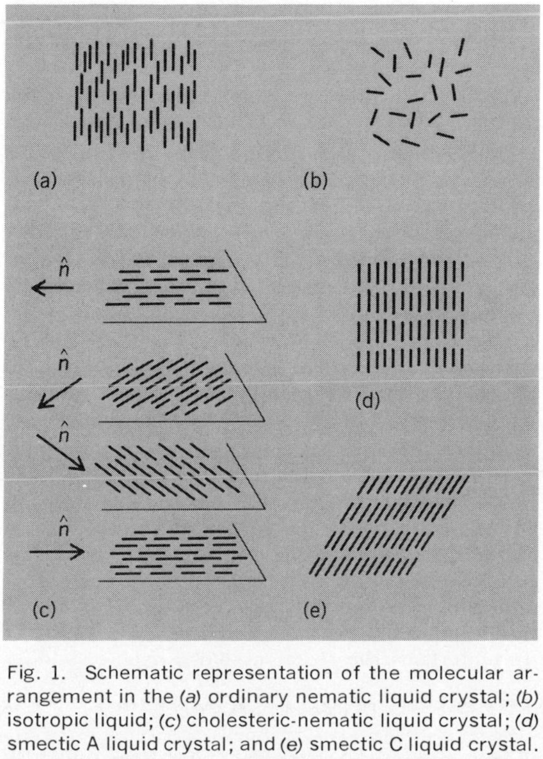

Fig. 1. Schematic representation of the molecular arrangement in the *(a)* ordinary nematic liquid crystal; *(b)* isotropic liquid; *(c)* cholesteric-nematic liquid crystal; *(d)* smectic A liquid crystal; and *(e)* smectic C liquid crystal. *(From G. H. Brown and J. J. Wolken, Liquid Crystals and Biological Structures, Academic Press, 1979)*

the molecules in the stratum are tilted at an angle to the plane of the stratum.

There is also a unique kind of liquid crystal known as the smectic D which is isotropic, but nevertheless shows three-dimensional order in the molecular packing of the structure.

Applications. Liquid crystals have many applications. They are used as displays in digital wristwatches, calculators, panel meters, and industrial products. They can be used to record, store, and display images which can be projected onto a large screen. They also have potential use as television displays.

Displays. The two features which make them more desirable for displays than other materials are lower power consumption and the clarity of display in the presence of bright light. The power requirements are often so low that a digital display on a wristwatch requires about the same power as does the mechanism which runs the watch. The two modes most widely used in liquid crystal displays are dynamic-scattering and field-effect.

In displays, the liquid crystal cell design usually begins with a thin film of a room-temperature liquid crystal sandwiched between two transparent electrodes (glass coated with a metal or metal oxide film). The thickness of the liquid crystal film is 6–25 μm and is controlled by a spacer which is chemically inert. The cell is hermetically sealed to eliminate oxygen and moisture, both of which may chemically attack the liquid crystalline material.

In dynamic scattering, if no electric field is applied, the cell is transparent. However, on addition of electric field to the liquid crystal, the cell becomes opaque. The field-effect display utilizes twisted nematic liquid crystals. The cell is prepared by rubbing the glass surface directionally or by chemically treating the surface and by adding a chiral compound to nematic liquid crystals. Digital displays are made by photoetching a seven-segment pattern onto one or both of the indium–tin oxide–coated glass plates. Reflection displays have one of the plates coated with a reflecting layer. Transmissive displays have etched surfaces on both plates.

The field-effect display is more widely used in watch and pocket-calculator displays. In a liquid crystal watch that displays hours and minutes (Fig. 2), the quartz crystal accurately controls the oscillating circuit. The oscillator frequency is typically 32,768 Hz. Each of the time pulses is decoded to give outputs that are needed for seven-segment displays. All electronic watches operate on the same principles, regardless of the display technique (for example liquid crystals and light-emitting diodes). New display techniques, such as multiplexing, will reduce the number of electrical leads to the segments in digital and bar-graph displays.

The electronics for the wristwatch display is complex, and the dividers, driver circuits, decoders, and counters contain about 1500 transistors. All of these transistors can be collected on a silicon area of approximately 0.25 by 0.40 cm. The display for field-effect requires low voltages (1.5 to 5.0 V), and little power.

The optics of the liquid crystal display are shown in Fig. 3. The cell with the nematic liquid crystal is placed between two crossed polarizers. Polarized light entering the cell follows the twist of

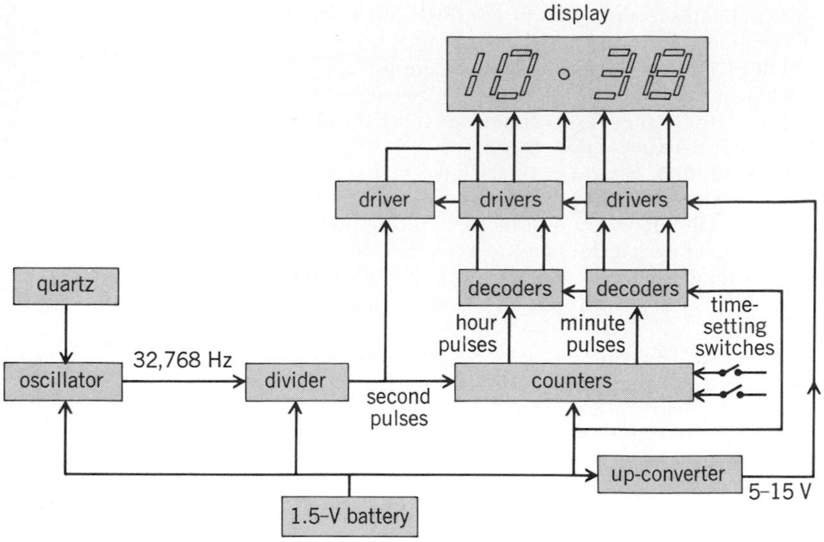

Fig. 2. Block diagram of a liquid crystal watch. *(From F. D. Saeva, ed., Liquid Crystals: The Fourth State of Matter, Marcel Dekker, 1979)*

the nematic liquid crystal, is rotated 90°, and as such can allow passage of the light through the second polarizer (Fig. 3*a*). Application of an electric field changes the molecular alignment in the liquid crystal such that the polarization is not altered in the cell and no light is transmitted. If a mirror is used behind the second polarizer, the display will appear black (Fig. 3*b*) when voltage is applied. If, in addition, one of the electrodes is shaped in the pattern of segments of digits, then a numeric display will appear when the voltage is on. By changing the direction of the polarizers, the digit can be made to appear white on a black background.

Nondestructive testing. Cholesteric-nematic liquid crystals undergo a change in color with a small change in temperature, a property that can be used for nondestructive testing.

Thermometers. Desk thermometers are available which use cholesteric-nematic liquid crystals. Observation of skin temperature changes following blockage of the sympathetic nervous system enables the physician to determine if neurological and vascular pathways are open. Monitoring skin temperature over extended areas provides a more detailed and readily interpreted indication of circulatory patterns than point measurements with thermocouples and thermistors.

[GLENN H. BROWN]

Bibliography: G. H. Brown and J. J. Wolken, *Liquid Crystals and Biological Structures*, 1979; J. D. Margerum and L. J. Miller, Electro-optical applications of liquid crystals, *J. Coll. Int. Sci.* 58: 559–580, 1977; F. D. Saeva (ed), *Liquid Crystals: The Fourth State of Matter*, 1979; A. Skoulios, Amphiphiles: Organization et diagrammes de phases, *Ann. Phys.*, 3:421–450, 1978.

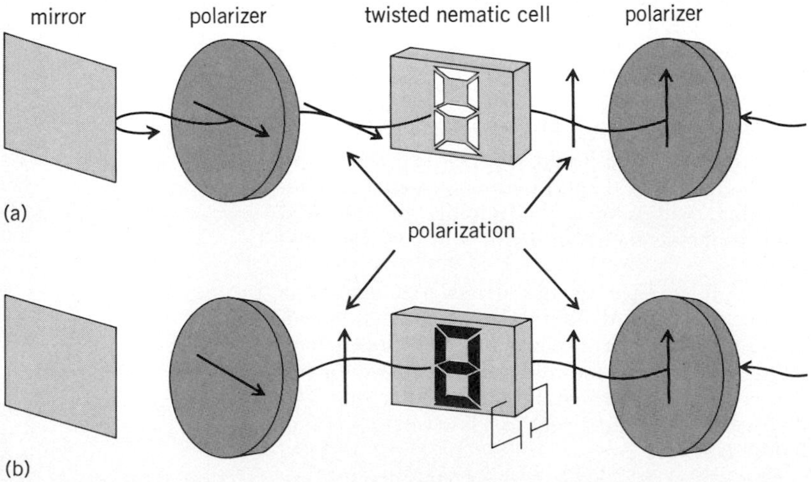

Fig. 3. Optics of field-effect display device when a twisted nematic cell is utilized. *(a)* Optics without electric field applied to liquid crystal cell. *(b)* Optics with electric field applied.

Lithium

A chemical element, Li, atomic number 3, and atomic weight 6.939. Lithium heads the alkali metal family in group Ia in the periodic table. In nature it is a mixture of the isotopes Li^6 and Li^7. Much of the lithium processed in the United States has been subjected to isotope separation to obtain the pure lighter isotope, Li^6, important in thermonuclear processes; lithium available in the market may thus have a different atomic weight. Lithium,

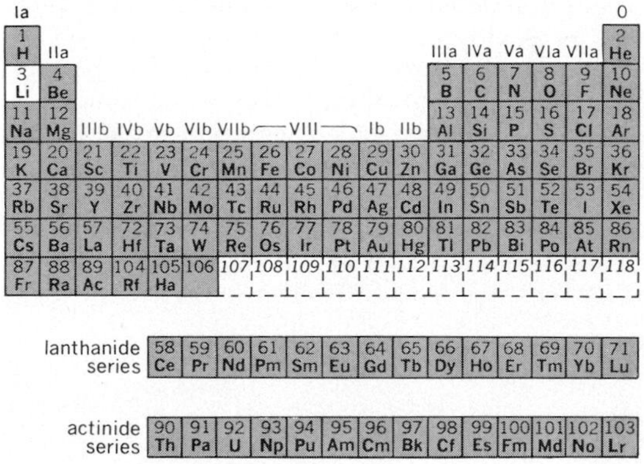

the lightest solid element, is a soft, low-melting, reactive metal. In many physical and chemical properties it resembles the alkaline-earth metals in group IIa as much as, or more than, it does the alkali metals. First discovered by J. A. Arfvedson in Sweden during an analysis of petalite ore, it was first isolated by Sir Humphry Davy in 1818 by electrolysis.

Uses. The major industrial use of lithium is in the form of lithium stearate as a thickener for lubricating greases. Lithium-base greases combine high water resistance with high-temperature resistance and good low-temperature properties. These all-purpose greases replace a multitude of specialized greases, and they have captured about one-third of the total automotive grease market.

Another important use of lithium compounds is in ceramics, specifically in porcelain enamel formulation. The major action of the lithium is as a flux. In ceramic mixtures, lithium carbonate readily undergoes the reaction shown below, yielding

$$Li_2CO_3 \rightarrow Li_2O + CO_2$$

the Li_2O as one constituent of the oxide mixtures of which most ceramics are composed. Alternatively, lithium aluminum silicate ores can be used directly.

There are a number of lesser uses for lithium and its compounds. Lithium hydroxide is used as an additive to give longer life and higher output in the alkaline storage batteries known as Edison cells. Lithium chloride and fluoride are used in welding and brazing fluxes; lithium-copper and lithium-silver alloys are used as self-fluxing brazing alloys. Lithium perchlorate has been suggested as an oxidizer for solid-propellant rocket mixtures; it offers a higher percentage of available oxygen than any other perchlorate due to the low atomic weight of the lithium.

Occurrence. Lithium is a moderately abundant element and is present in the Earth's crust to the extent of 65 parts per million (ppm). This places lithium a little below nickel, copper, and tungsten, and a little above cerium and tin in abundance. In sea water there is about 0.1 ppm of lithium. It is found in human and animal organisms, in soil, in mineral waters, and in plants such as cocoa, tobacco, and seaweed. Its presence in the Sun's atmosphere has been established.

Although lithium occurs widely in nature, the concentrations in ores are generally quite low, unlike sodium and potassium whose relatively pure halides may be mined or pumped from brine wells. Lithium resembles rubidium and cesium in its manner of occurrence as part of complex silicate minerals. Lithium ore assays are usually expressed in terms of percent of lithium oxide, Li_2O, and 1.0% of Li_2O affords a commercially processable ore. Most commercial ores run from 1 to 3% Li_2O; virgin ores rarely contain as much as 6%. In North America the most common lithium ore is spodumene which generally occurs in pegmatite formations in association with feldspar. The largest known reserves are in the Kings Mountain area of North Carolina, in Quebec, and in Manitoba. Substantial quantities of lepidolite and petalite occur in Africa, particularly in Rhodesia. Amblygonite, the other commercially important lithium mineral, is found in Europe, Africa, and South America. A nonmineral source of lithium is the brine in Searles Lake, Calif. Table 1 lists the four commercially important lithium ores.

Table 1. Commercially important lithium minerals

Mineral	Simplified formula	% Li_2O, theoretical
Spodumene	$LiAlSi_2O_6$	8.03
Lepidolite	$LiKAl_2F_2Si_3O_9$	4.09
Petalite	$LiAlSi_4O_{10}$	4.89
Amblygonite	$LiAlFPO_4$	10.10

Metallurgical extraction. Lithium ores are concentrated from 1–3% Li_2O to 4–6% Li_2O by heavy media separation using dense, nonaqueous liquids, and by froth flotation. The silicate ores, which are those processed most widely, are then chemically cleaved by acid or alkaline processes.

In the acid process, spodumene ore is first heated in a kiln to about 1000–1100°C to change the naturally occurring alpha spodumene to beta spodumene, a more friable, less dense crystal form which is readily attacked by the acid. The ore is then roasted in a second kiln with an excess of sulfuric acid. Water leaching of the kiln product yields lithium sulfate which is treated with sodium carbonate to give lithium carbonate. The carbonate is converted to the chloride by reaction with hydrochloric acid.

In the alkaline process, spodumene or lepidolite ores are ground and calcined with about 3 parts limestone to 1 part lithium ore at 900–1000°C. Water leaching of the kiln product yields lithium hydroxide which is converted to the chloride by reaction with hydrochloric acid.

Dry lithium chloride is the feed material for the manufacture of lithium metal by electrolysis. The cell bath is actually a molten mixture of lithium chloride and potassium chloride. The cell is operated at 400–420°C with a voltage across the cell of 8–9 volts. The current consumption in electrolysis is about 18 kwhr per pound of lithium produced. The illustration shows a typical lithium cell.

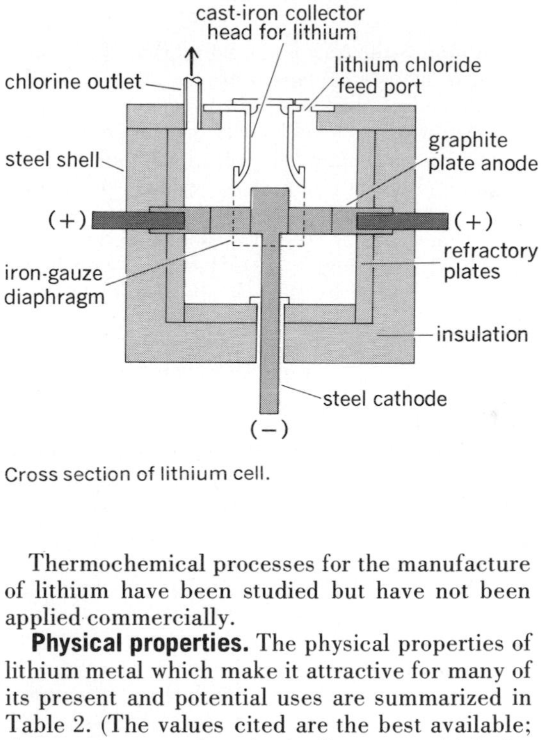

Cross section of lithium cell.

Thermochemical processes for the manufacture of lithium have been studied but have not been applied commercially.

Physical properties. The physical properties of lithium metal which make it attractive for many of its present and potential uses are summarized in Table 2. (The values cited are the best available; there are many values in the literature which disagree with them.)

Noteworthy among the physical properties are the high specific heat (heat capacity), large temperature range of the liquid phase, high thermal conductivity, low viscosity, and very low density.

Lithium metal is soluble in liquid ammonia and is slightly soluble in the lower aliphatic amines, such as ethylamine. It is insoluble in hydrocarbons. There are no published data on the solubility of lithium metal in molten inorganic salts.

Chemical properties. Lithium undergoes a large number of reactions with both organic, and inorganic, reagents.

Inorganic reactions. Lithium reacts with oxygen to form the monoxide, Li_2O, and the peroxide, Li_2O_2. The superoxide, LiO_2, is not known and may not exist because of insufficient room for two oxygen atoms around the small lithium atom.

Table 2. Physical properties of lithium metal

Property	Temperature °C	Temperature °F	Metric (scientific) units	British (engineering) units
Density	20	68	0.534 g/cm³	33.7 lb/ft³
	400	752	0.490 g/cm³	30.6 lb/ft³
	800	1472	0.457 g/cm³	28.6 lb/ft³
Melting point	179	354		
Boiling point	1317	2403		
Heat of fusion	179	354	103.2 cal/g	186 Btu/lb
Heat of vaporization	1317	2403	4680 cal/g	8420 Btu/lb
Viscosity	200	392	5.62 millipoises	11.1 kinetic units
	400	752	4.02 millipoises	8.2 kinetic units
	600	1112	3.17 millipoises	6.7 kinetic units
Vapor pressure	727		0.78 mm	0.015 lb/in.²
	1077		91.0 mm	1.76 lb/in.²
Thermal conductivity	216	420	0.109 cal/(sec)(cm²)(cm)(°C)	26.5 Btu/(hr)(ft²)(°F)
	539	1002	0.073 cal/(sec)(cm²)(cm)(°C)	17.6 Btu/(hr((ft²)(°F)
Heat capacity	100	212	0.90 cal/(g)(°C)	0.90 Btu/(lb)(°F)
	300	573	1.02 cal/(g)(°C)	1.02 Btu/(lb)(°F)
	800	1472	0.99 cal/(g)(°C)	0.99 Btu/(lb)(°F)
Electrical resistivity	0	32	1.34 microhm-cm	
	100	212	12.7 microhm-cm	
Surface tension	200–500	392–932	About 400 dynes/cm	

Lithium is the only alkali metal that reacts with nitrogen at room temperature to form a nitride, Li_3N, which is black.

Lithium reacts readily with hydrogen at about 500°C to form lithium hydride, LiH. Of all the alkali metals, lithium reacts most readily with hydrogen, and is the only alkali metal forming a hydride stable enough to be melted without decomposition. An important reaction of LiH is that with boron fluoride diethyletherate, $BF_3 \cdot (C_2H_5)_2O$, to yield diborane, B_2H_6, or lithium borohydride, $LiBH_4$.

The reaction of lithium metal with water is exceedingly vigorous. Lithium reacts directly with carbon to form the carbide, Li_2C_2. Lithium combines readily with the halogens at elevated temperatures, forming halides with the emission of light. Lithium does not burn in chlorine unless heated, and it reacts only superficially with liquid bromine, presumably because of the formation of protective halide coatings.

With ammonia, lithium reacts to form the amide, $LiNH_2$. The reaction may be carried out by heating lithium in a stream of ammonia at 400°C, or by reaction with liquid ammonia in the presence of a catalyst such as iron. On heating, the amide loses ammonia to form the imide, Li_2NH; lithium is the only alkali metal which forms an imide.

Lithium and carbon monoxide react to give lithium carbonyl, LiCO, a rather unstable product.

Organic reactions. While lithium does not react with paraffin hydrocarbons, it does undergo addition reactions with arylated alkenes and with dienes. The addition of lithium to diene systems is the basis for the catalytic polymerization of dienes to synthetic elastomers having natural rubber properties. One recently developed synthetic rubber employs lithium metal dispersed in petroleum jelly as the catalyst to polymerize isoprene to *cis*-polyisoprene, the exact structural counterpart of natural hevea rubber from *Hevea brasiliensis* (rubber trees). Lithium also reacts with acetylenic compounds, replacing the acetylenic hydrogen atom, and forming lithium acetylides, which are important in the synthesis of vitamin A.

Lithium metal reacts with alcohols forming lithium alkoxides, LiOR, and liberating hydrogen. Alkoxides form less readily with lithium than they do with sodium, and their commercial utility is much less than that of the sodium alkoxides.

Lithium adds to carbonyl compounds such as aldehydes and ketones giving a lithium addition compound as an intermediate which can be hydrolyzed to an alcohol. Similarly, organolithium compounds add to aldehydes and ketones giving addition products which yield alcohols upon hydrolysis. Organolithium compounds react with organic acids to give addition products which can be hydrolyzed to ketones; this reaction is used in one step of the vitamin A synthesis.

Lithium can also be used in conjunction with ammonia or amines to effect unique reductions of various organic compounds, as for example the reduction of aromatic hydrocarbons to monoolefins.

Not only lithium metal but also compounds derived directly from lithium metal are important in organic chemistry. Thus, lithium amide, $LiNH_2$, made from lithium and ammonia, is used to introduce amino groups, as a dehalogenating agent, and as a catalyst. Lithium aluminum hydride, $LiAlH_4$, made by the reaction of lithium hydride and aluminum chloride, is a powerful reducing agent for specific linkages in complex molecules. This reagent has become important in many organic syntheses.

Availability. Lithium is available in two grades, a regular grade containing 99.8% lithium and a low-sodium grade containing 0.005% sodium or less.

The metal is available commercially in the form of shot, wire, ribbon, and in cylindrical bricks. The shot are shipped under hydrocarbon and the wire and ribbon are coated with petrolatum. The bricks (usually 1 lb in weight) are shipped in hermetically sealed metal cans.

Lithium salts are many times more expensive than the corresponding sodium or potassium compounds.

Handling. Handling of lithium metal is much the same as handling sodium metal with a few specific exceptions:

1. Because lithium reacts with nitrogen, argon or helium must be used instead of nitrogen to blanket molten lithium.

2. Because molten lithium reacts with glass and ceramics, metal equipment must be used.

3. Lithium fires are most easily extinguished by the use of graphite powder, in contrast to the use of the metal halide or carbonate powders for the extinguishing of other alkali metal fires.

4. Lithium behaves differently in the presence of other materials. *See* SODIUM.

Principal compounds. The most important lithium compound in terms of major uses and pounds produced is lithium hydroxide. As discussed earlier, it is the end product resulting from water leaching of the clinker produced by limestone roasting of lithium ores. It is also produced from the lithium sulfate from the sulfuric acid roasting process. It is a white powder, and the material of commerce is actually lithium hydroxide monohydrate, $LiOH \cdot H_2O$. Tallow, or another natural fat, is cooked with the lithium hydroxide, producing lithium stearate, which is used as a thickener or gelling agent to transform oil into lithium-base lubricating grease. The U. S. Atomic Energy Commission also makes substantial purchases of the hydroxide.

Lithium carbonate, $LiCO_3$, finds application in the ceramic industries, particularly in the manufacture of frits (powdered glass) for porcelain enamel formulation.

Both lithium halides, lithium chloride and lithium bromide, form concentrated brines having the ability to absorb moisture over a wide temperature range; these brines are used in commercial air conditioning systems. The chloride and fluoride have low melting points, high boiling points, and high solvent power for metal oxides; these properties lead to commercial applications in welding and brazing fluxes for aluminum, magnesium, and titanium.

Lithium hydride, discussed earlier under inorganic reactions of the metal, is also of major importance. It was used in large amounts during World War II as a compact source of hydrogen for military balloon inflation. The deuteride formed by lithium-6 can be converted in a thermonuclear reaction to two atoms of helium with the evolution of large amounts of energy; this energy may provide the key to thermonuclear power production.

Analytical methods. In most quantitative analytical separations, all other elements are removed leaving the alkali metal ions in solution. Lithium chloride is separated from the other alkali chlorides by virtue of its greater solubility in organic solvents, and then converted to lithium sulfate, dried, and weighed.

For qualitative identification of lithium, the characteristic crimson color produced when lithium compounds are introduced into a hot flame is very satisfactory. A flame spectrophotometer may be used to measure the intensity of the characteristic lithium color at 670.8 millimicrons wavelength. This technique is the most suitable for routine determination of lithium because it is fast and accurate. *See* ALKALI METALS. [MARSHALL SITTIG]

Bibliography: S. Beer (ed.), *Liquid Metals: Chemistry and Physics*, 1972; F. A. Cotton and G. Wilkinson, *Advanced Inorganic Chemistry*, 3d ed., 1972.

Lutetium

A chemical element, Lu, atomic number 71, atomic weight 174.97, a very rare metal and the heaviest member of the rare-earth group. The naturally occurring element is made up of the stable isotope ^{175}Lu, 97.41%, and the long-life β-emitter ^{176}Lu

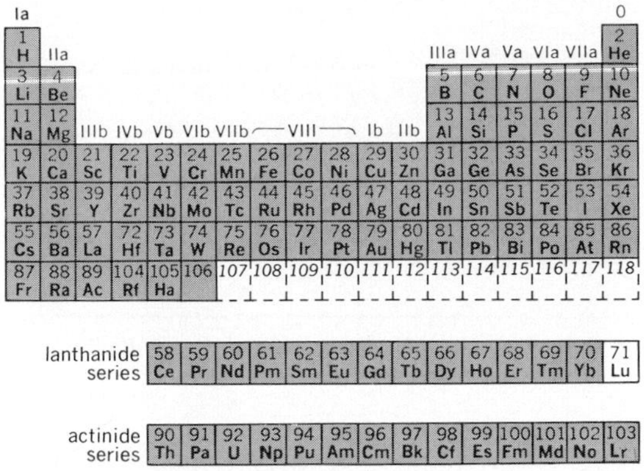

with a half-life of 2.1×10^{10} years. It was discovered in 1906 by G. Urbain, who named it lutecium, after an ancient name for Paris. It was also discovered independently by C. F. Auer von Welsbach, who named the element cassiopeium, but lutetium is the accepted form at present. The salts are colorless and give a trivalent ion in solution.

Lutetium, along with yttrium and lanthanum, is of interest to scientists studying magnetism. All of these elements form trivalent ions with only subshells which have been completed, so they have no unpaired electrons to contribute to the magnetism. Their radii with regard to the other rare-earth ions or metals are very similar so they form at almost all compositions either solid solutions or mixed crystals with the strongly magnetic rare-earth elements. Therefore, the scientist can dilute the magnetically active rare earths in a continuous manner without changing appreciably the crystal environment. *See* MAGNETOCHEMISTRY; RARE-EARTH ELEMENTS. [FRANK H. SPEDDING]

Magnesium

A metallic chemical element, Mg, in Group IIa of the periodic system, atomic number 12, atomic weight 24.312. Magnesium is silvery white and extremely light in weight. The specific gravity is 1.74, and the density is 1740 kg/m³ (0.063 lb/in.³ or 108.6 lb/ft³). Because of this lightness combined with alloy strength suitable for many structural

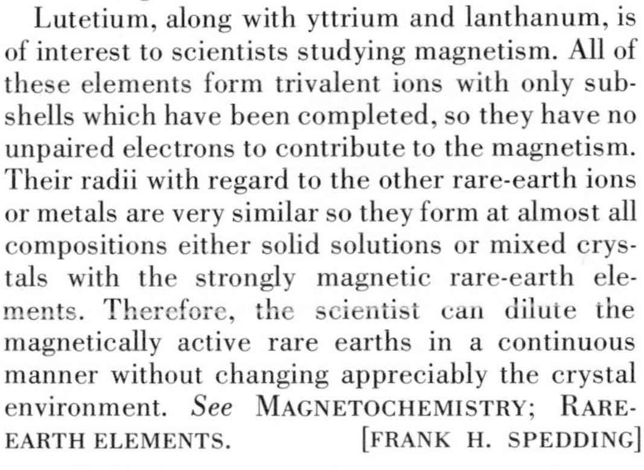

Table 1. World producers of magnesium*

Country	Producer	Raw material	Process
Canada	Dominion Magnesium	Dolomite	Pidgeon ferrosilicon
France	Pechiney-Ugine Kuhlmann	Dolomite	Magnetherm ferrosilicon
Italy	Cromodora	Dolomite	Ferrosilicon
Japan	Furukawa Magnesium Co.	Dolomite	Ferrosilicon
Norway	Norsk Hydro-Elektrisk	Dolomite and sea water	Electrolytic
United States	Alamet Div. of Calumet & Hecla	Dolomite	Pidgeon ferrosilicon
	American Magnesium Co.	Underground brine	Electrolytic
	The Dow Chemical Co.	Sea water	Electrolytic
	Oregon Metallurgical	Underground brine	Electrolytic
	Titanium Metals Corp.	Recycled $MgCl_2$	Electrolytic
China	Data not available		
Soviet Union	Data not available	Dolomite and carnallite	Electrolytic

*Based on information from *Metals Week*, 40(25):21, June 23, 1969.
†Estimated.

uses, magnesium has long been known as industry's lightest structural metal.

With a density only two-thirds that of aluminum, magnesium is used in countless applications where weight saving is an important consideration. The metal also has, however, many desirable chemical and metallurgical properties which account for its extensive use in a variety of nonstructural applications.

Natural occurrence. Magnesium is very abundant in nature, occurring in substantial amounts in many rock-forming minerals such as dolomite, magnesite, olivine, and serpentine. In addition, magnesium is also found in sea water, subterranean brines, and salt beds. It has been estimated that magnesium constitutes 2.5% of the Earth's crust, making it the eighth most abundant chemical element and the sixth most abundant metallic element. It is the third most abundant structural metal in the Earth's crust, exceeded only by aluminum and iron.

More than 60 minerals occurring in nature contain the element magnesium, but only a few are important commercially for the production of magnesium and its compounds. In the United States the important sources are brucite, dolomite, magnesite, natural brines, sea water, and sea-water bitterns. Olivine and serpentine offer a huge potential, but these minerals are not being used extensively for the production of magnesium compounds. Although California, Nevada, and Washington are major sources for brucite, dolomite, and magnesite, huge deposits of magnesium ores are widespread throughout the United States. Large reserves are found in Austria, Brazil, Canada, Czechoslovakia, Greece, India, Manchuria, the Soviet Union, Venezuela, and Yugoslavia.

History. Sir Humphry Davy is generally credited with the discovery of magnesium in 1808, when he established that the compound magnesia alba (magnesium oxide) was the oxide of a new metal. Davy passed potassium vapors over hot magnesium oxide and extracted the reduced magnesium with mercury. He also electrolyzed magnesium sulfate, using mercury as a cathode. In both cases he obtained magnesium in the form of an amalgam, but it is not known definitely to what extent he actually obtained metallic magnesium. The first isolation of magnesium is attributed to the French scientist A. A-B. Bussy, who in 1828 fused anhydrous magnesium chloride with metallic potassium and obtained magnesium that was substantially pure. Michael Faraday in 1833 was the first to produce magnesium electrolytically by the electrolysis of molten magnesium chloride, using a voltaic cell. In 1852 R. Bunsen designed an electrolytic cell in which hollow carbon cathodes were used to collect the molten magnesium to prevent burning when it came into contact with air.

Production of the metal. Two major methods of producing magnesium are used throughout the world, as shown in Table 1. They are the electrolytic and the silicothermic processes. Magnesium is produced from sea water by the electrolytic process. The silicothermic, or ferrosilicon, process uses dolomite as the raw material.

Electrolytic process. The electrolysis of magnesium chloride to yield chlorine and metallic magnesium is the basis of this process. Although magnesite, dolomite, and natural brines have been used as raw materials, the principal source is sea water, which contains about 0.13% magnesium. Because of this content of magnesium and the fact that an efficient and economical method for its extraction has been developed, the world's supply of magnesium is considered limitless. As an example, it has been calculated that if magnesium were extracted from sea water at the rate of 100,000,000 tons/year (1 ton = 0.9 metric ton) for 1,000,000 years, the magnesium content of sea water would drop only to 0.12%. The method of extracting magnesium from sea water is shown in Fig. 1. The sea water is pumped into large settling tanks where it is mixed with lime obtained by roasting oyster shells dredged from the ocean bottom. The lime converts the magnesium into insoluble magnesium hydroxide (milk of magnesia) which is filtered out. This hydroxide is then treated with hydrochloric acid, obtained from chlorine by reaction with natural gas, to produce magnesium chloride solution. The water is evaporated, and the dry magnesium chloride is fed to the electrolytic cells which break up the compound into metallic magnesium and chlorine. The chlorine gas is recycled to make hydrochloric acid, and the magnesium metal is poured into ingots.

Silicothermic process. The ferrosilicon process, although first originated experimentally in Germany, was developed commercially during World War II by L. M. Pidgeon of Canada. The flow diagram in Fig. 2 shows the actual steps. Ferrosilicon, an alloy of silicon and iron, is mixed with calcined dolomite ore and pressed into small briquets. These are charged into a steel retort, put under vacuum, and heated to about 2200°F. (The silicon reduces the magnesium oxide formed by calcining of the dolomite) to form a vapor of metallic magnesium which condenses in the cool end of the retort extending from the furnace. In this process, the magnesium is removed from the retort in the form of crystals, which are subsequently melted and cast into ingots.

Properties. The properties of magnesium in metallic form are best divided into two categories, physical and chemical. The former are listed in Table 2. Many of these properties provide magnesium with distinctive qualities that determine some of its uses.

Magnesium is very active chemically, as indicated by its high position in the electromotive force series of metals. It will actually displace hydrogen from boiling water, and a large number of metals can be prepared by thermal reduction of their salts and oxides with magnesium. The metal will com-

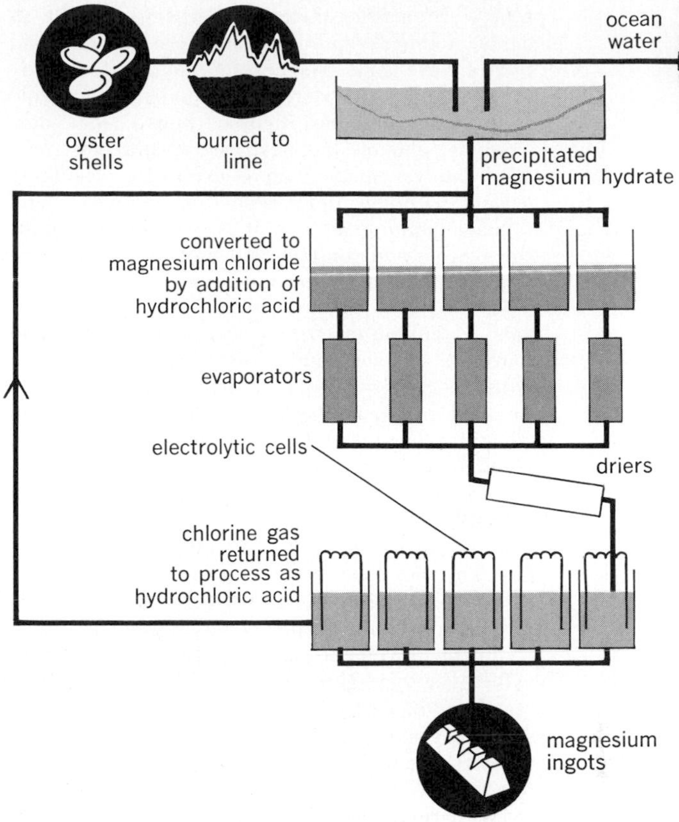

Fig. 1. Diagram showing the electrolytic extraction of magnesium from sea water.

Table 2. Physical properties of primary magnesium (99.9% pure)

Property	Value
Atomic number	12
Atomic weight	24.312
Atomic volume, cm³/g-atom	14.0
Crystal structure	Close-packed hexagonal
Lattice parameters	
A_0	3.203A
C_0	5.199A
Axial ratio, c/a	1.624
Electron arrangement in free atoms	(2) (8) 2
Mass numbers of the isotopes	24, 25, 26
Percent relative abundances of Mg^{24}, Mg^{25}, Mg^{26}	77, 11.5, 11.5
Density, g/cm³ at 20°C	1.738
Specific heat, cal/g/°C at 20°C	0.245
Thermal conductivity, cal/cm²/°C/sec at 20–100°C	0.367
Thermal diffusivity, cm²/sec	0.87
Coefficient of thermal expansion, in./°C at 20–100°C	0.0000261
Melting point, °C	650
Boiling point, °C	1110±10
Latent heat of fusion, cal/g	88±2
Latent heat of vaporization cal/g mole	1260±30
Heat of combustion, cal/g mole	145,000
Electrical resistivity, microhm-cm at 20°C	4.46
Electrical conductivity at 20°C	
Vol % annealed copper standard	38.6
Mass % annealed copper standard	198.0

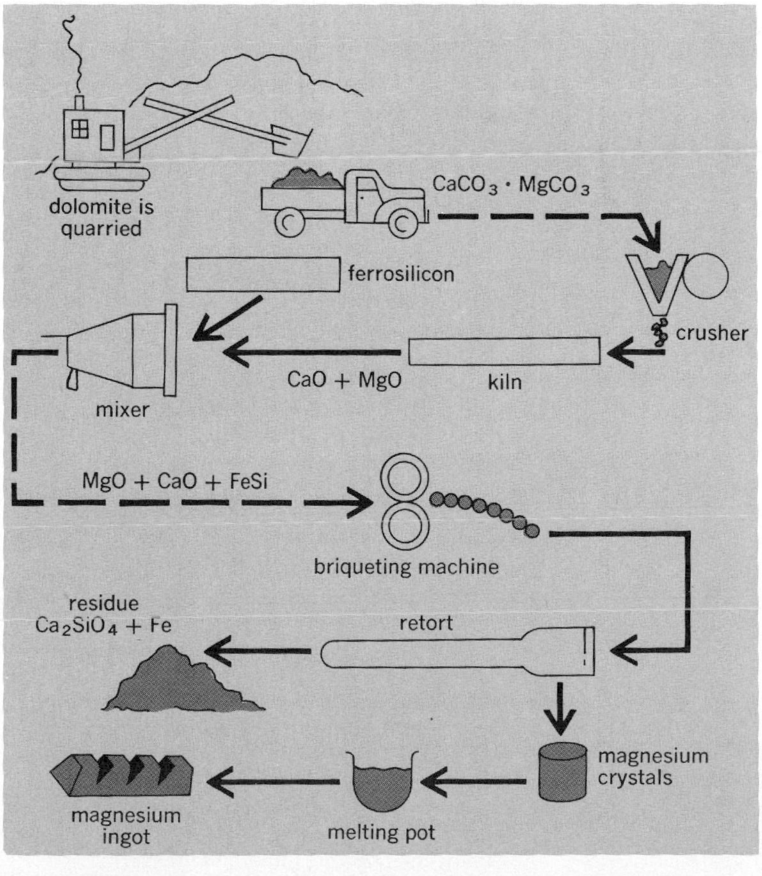

Fig. 2. Diagram showing extraction of magnesium metal from dolomite ore by ferrosilicon process.

bine with most nonmetals and with practically all acids, noted exceptions being pure chromic and hydrofluoric acids. Magnesium reacts only slightly or not at all with most alkalies and many organic chemicals, including hydrocarbons, aldehydes, alcohols, phenols, amines, esters, and most oils. As a catalyst, magnesium is useful for promoting organic condensation, reduction, addition, and dehalogenation reactions. It has long been used for the synthesis of complex and special organic compounds by the well-known Grignard reaction. Principal alloying ingredients include aluminum, manganese, zirconium, zinc, rare-earth metals, and thorium. Certain combinations of these alloying constituents produce magnesium alloys suitable for sand, permanent mold, and die castings; extrusions; forgings; and sheet and plate with excellent mechanical properties at room and elevated temperatures. *See* GRIGNARD REACTION.

Workability. The excellent machinability of magnesium is one of its outstanding characteristics. The metal can be machined at higher speeds and with larger feeds and depths of cut than is possible with most other commonly used metals.

Magnesium can be cast and fabricated by practically every metal-forming method known. The metal can be cast in sand or permanent molds and also by the die-casting process, which is used when the quantities desired are sufficiently great. Magnesium can also be cast by some of the less common methods, including plaster-mold, centrifugal, shell-mold, and investment processes.

Magnesium is rolled into sheet and plate, and can be extruded into rods, bars, tubing, and an almost endless variety of structural and special shapes.

Deep and shallow drawing, stretch forming, bending, spinning, and impact extrusion are representative of the types of forming operation which can be used to fabricate magnesium.

The forging of magnesium is accomplished by methods much the same as those used for forging other metals. Both press and hammer equipment are used, but the former is most commonly employed because the physical structure of magnesium makes the metal better adapted to the squeezing action of the forging press.

Magnesium parts can be joined by any of the common methods, such as gas-shielded arc welding, electric resistance spot and seam welding, adhesive bonding, and riveting. Brazing and gas welding, although not so frequently used as the other methods, are also suitable for joining magnesium.

Uses. Modern developments in the magnesium industry have greatly extended the fields of usefulness for this lightweight metal. There are new alloys with interesting and useful properties. Also, there are new and improved fabricating techniques. Magnesium alloys have structural and nonstructural applications. Most of the structural uses were at first in the aircraft industry, for fuselages, engine parts, and landing wheels. Later, with a better price position and supply, use of the alloys broadened into materials handling equipment, luggage, photo and optical instruments, lawnmowers, portable tools, and others.

Nonstructural uses of magnesium include its application as an alloying element in aluminum, zinc, and certain other nonferrous alloys. Magnesium is also used as an oxygen scavenger in the manufacture of nickel and copper alloys. In very finely divided form, magnesium finds some use in pyrotechnics, both as pure magnesium and when alloyed with 30% and more of aluminum. The cathodic protection of other metals from corrosion is a nonstructural use for magnesium that has become well known, largely in connection with domestic water heaters and pipelines. It is used in dry-cell batteries for longer service life. Gray-iron foundries use magnesium alloys as addition agents to the ladle, just prior to pouring a casting, with the result that the graphite particles become nodular and the properties of the cast iron are greatly improved. Magnesium, a powerful reducing agent, is used for the production of titanium, zirconium, beryllium, uranium, and hafnium. Because of its rapid yet controlled etching characteristics, as well as its lightness, magnesium finds expanding usage in the photoengraving field.

Table 3. Principal magnesium compounds and uses

Compound	Uses
Magnesium carbonate	Refractories, production of other magnesium compounds, water treatment, fertilizers
Magnesium chloride	Cell feed for production of metallic magnesium, oxychloride cements, refrigerating brines, catalyst in organic chemistry, production of other magnesium compounds, flocculating agent, treatment of foliage to prevent fire and resist fire, magnesium melting and welding fluxes
Magnesium hydroxide	Chemical intermediate, alkali, medicinal
Magnesium oxide	Insulation, refractories, oxychloride and oxysulfate cements, fertilizers, rayon-textile processing, water treatment, papermaking, household cleaners, alkali, pharmaceuticals, rubber filler catalyst
Magnesium sulfate	Leather tanning, paper sizing, oxychloride and oxysulfate cements, rayon delustrant, textile dyeing and printing, medicinal, fertilizer ingredient, livestock-food additive, ceramics, explosives, match manufacture

Principal compounds. Magnesium compounds are used extensively in industry and agriculture. The compounds of major importance are the carbonate, chloride, hydroxide, oxide, and sulfate, which are produced and used in considerable tonnage. Other magnesium compounds, including the bromide, nitrate, phosphate, acetate, silicate, and trisilicate, also find considerable usage in industry.

Magnesium compounds are very important to industry. Uses include production of magnesium metal, refractories, insulation, fertilizers, textile processing, leather tanning, papermaking, ceramics, explosives, and medicinals. Table 3 lists the major magnesium compounds and indicates some of their more significant applications.

Several other magnesium compounds, both organic and inorganic, are used for a variety of industrial purposes—as chemical reagents, medicinals, catalysts, and mild abrasives.

[WILLIAM H. GROSS; STEPHEN C. ENCLESON]

Bibliography: A. Beck (ed.), *The Technology of Magnesium and Its Alloys*, 3d ed., 1943; E. F. Emley, *Principles of Magnesium Technology*, 1966; W. H. Gross, *The Story of Magnesium*, 1949; C. S. Roberts, *Magnesium and Its Alloys*, 1960.

Magnetic separation methods

All materials possess magnetic properties. Substances that have a greater permeability than air are classified as paramagnetic; those with a lower permeability are called diamagnetic. Paramagnetic materials are attracted to a magnet; diamagnetic substances are repelled. Very strongly paramagnetic materials are classified as ferromagnetic and include such metals as iron, nickel, and cobalt, and such minerals as magnetite, pyrrhotite, and ilmenite. Such substances can be separated from weakly or nonmagnetic materials by the use of low-intensity magnetic separators. Minerals such as hematite, limonite, and garnet are weakly magnetic and can be separated from nonmagnetics by the use of high-intensity separators.

Magnetic separators are widely used to remove tramp iron from ores being crushed, thereby protecting the crushers, to remove contaminating magnetics from food and industrial products, to recover magnetite and ferrosilicon in the float-sink methods of ore concentration, and to upgrade or concentrate ores. Tramp iron may be removed by a stationary magnet suspended over a conveyor belt or the material may be passed over a magnetic pulley; in both cases, magnetics are attracted by the magnetic field and separated.

Magnetic separators are extensively used to concentrate ores, particularly iron ores, when one of the principal constituents is magnetic. When the chief economic mineral is magnetite, iron ores can be cheaply and effectively separated by low-intensity separators. Such separators may be dry or wet. If an ore can be crushed to give substantial liberation of minerals at sizes coarser than 1/4 in., separations can be made on the +1/4-in. sizes on a belt-type magnetic cobber (see illustration). Such a machine is usually used to cob out or reject waste materials. If clean mineral occurs at this size, it can be recovered by splitting the discharge or by retreatment on a lower strength machine. Wet magnetic separators are usually used to treat ore finer than 1/8 in. These separators may be of the belt type or of the more common rotating-drum type. Drum-type separators consist of one or more rotatable drums having inner nonrotatable magnet elements with 3–7 poles. The magnets may be either electromagnets or permanent magnets. After the feed enters the machine as a slurry, the magnetics are attracted to the pole pieces and are carried to a discharge point on the surface of the drum. Many types of box designs are in use. The concurrent type is frequently used on relatively coarse material to reject a clean waste product. The countercurrent type is employed on fine ore to give a clean concentrate. Magnets may be of either the electromagnetic or permanent type. Electromagnets were formerly used almost exclusively but are now used mainly when exceptionally high field strengths are required or when it is desirable to vary magnet strength. Permanent magnets are now widely used since modern materials permit charging and retaining high field strengths permanently. Most permanent magnets are of the alnico type but the ceramic types containing barium ferrites are coming into use. Several types of magnetic separators employing alternating current have been devised but have found little commercial use.

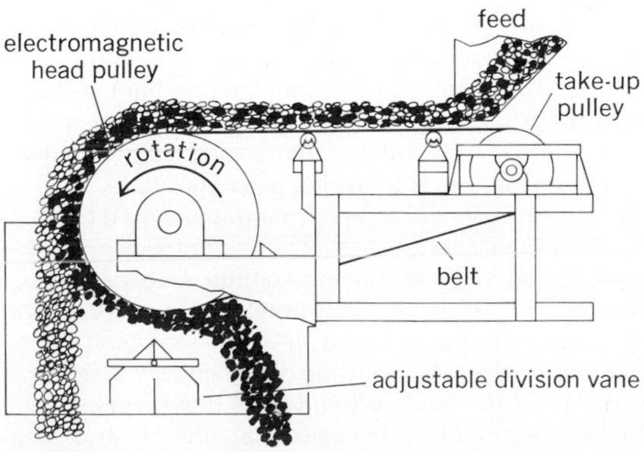

Belt-type magnetic cobber. (*Stearns Magnetic Products*)

High-intensity separators for the separation of weakly magnetic minerals are usually of the dry type. Surface-tension effects usually rule out wet separations. Since magnetic attraction varies inversely as the square of the distance, weakly magnetic minerals must be brought close to the magnets if they are to be separated. Both belt-type and induced-roll machines are used. The ore must be fully dried and closely sized for best results. *See* CHEMICAL SEPARATION TECHNIQUES.

[FRED D. DE VANEY]

Bibliography: C. J. King, *Separation Processes*, 2d ed., 1980; R. H. Perry and C. H. Chilton, *Chemical Engineers' Handbook*, 5th ed., 1973; R. E. Treybal, *Mass Transfer Operations*, 3d ed., 1980.

Magnetochemistry

The branch of chemistry which studies the interrelationship between a magnetic field and atomic and molecular structures.

When a substance is placed in a magnetic field of strength H, the magnetic induction B is given by Eq. (1). The quantity I is the intensity of magnetization, and $I/H = \kappa$ is the magnetic susceptibility per unit volume. The magnetic susceptibility per unit mass is $\kappa/D = \chi$, where D is the density.

$$B = H + 4\pi I \qquad (1)$$

A substance in a magnetic field acquires an intensity of magnetization which may be either smaller or larger than that induced in a vacuum by the same field. In the first case, the substance is said to be diamagnetic. In the second case, the substance may be paramagnetic, ferromagnetic, or antiferromagnetic.

Diamagnetism, a universal property of matter, is usually of the order of magnitude 10^{-6} to 10^{-5}. Temperature-dependent paramagnetism, on the other hand, arises only when an atom, ion, or molecule possesses a permanent magnetic moment either in the ground state or in an excited state. A permanent magnetic moment is the result of the presence of one or more unpaired electrons. Paramagnetic susceptibilities are of the order of magnitude 10^{-4} to 10^{-3}.

A substance composed of atoms with permanent magnetic moments which are very near to one another (for example, iron metal) may display ferromagnetism. This phenomenon occurs when large numbers of the atoms with permanent magnetic moments interact so that their individual moments align in a parallel fashion, giving rise to a large resultant moment.

On the other hand, a similar substance (for example, manganese metal) may display antiferromagnetism. Here, the magnetic moments align in an antiparallel fashion, thus largely canceling the individual magnetic moments of the atoms. Parallel versus antiparallel alignment depends, among other factors, upon interatomic distances. Magnetic theories are not yet refined enough to enable one to predict whether a given substance will be ferromagnetic or antiferromagnetic.

In general, the susceptibility of diamagnetic substances is independent of temperature and of field strength. The susceptibility of paramagnetic substances is often inversely proportional to the absolute temperature but independent of field strength. The susceptibility of ferromagnetic and antiferromagnetic substances is dependent on both temperature and field strength in a rather complicated way.

There are many methods available for the measurement of magnetic susceptibility. Most of these methods involve measuring the force exerted on a sample by a magnetic field.

Atomic diamagnetism. The only important application of diamagnetic ionic susceptibilities is their use as correction factors for measured susceptibilities. All substances, even though paramagnetic, have an underlying diamagnetism. A precise determination of the paramagnetic susceptibility of sodium neptunyl acetate, for example, should include subtraction from the measured molar susceptibility of the diamagnetic ionic susceptibilities of sodium, acetate, and neptunyl ions.

The diamagnetism of atoms and ions can be calculated theoretically by considering electron density distributions summed for each electronic shell. In addition, a large number of diamagnetic susceptibilities have been determined empirically. There is, in general, agreement between the measured values and those calculated theoretically.

Molecular diamagnetism. Estimates of the diamagnetism of organic compounds are based primarily on empirical methods. From measurements on a large number of compounds, B. Pascal concluded that diamagnetic susceptibilities could be represented by Eq. (2), where n_A is the number

$$\chi_M = \Sigma n_A \chi_A + \lambda \quad (2)$$

of atoms of susceptibility χ_A in the molecule, and λ is a constitutive correction depending on the nature of the bonds between the atoms. In this expression, χ_A is not the theoretical atomic susceptibility referred to in the previous section, but is a purely empirical constant derived from the measured susceptibilities. This procedure, when applied to organic compounds, often gives results that are within 1% of the experimentally determined values.

To illustrate the Pascal method, a simple example will suffice. According to this method, the molar susceptibility of ethyl bromide (C_2H_5Br) is given by $2\chi_C + 5\chi_H + \chi_{Br} + \lambda$. In this case, λ is the constitutive correction for the C-Br band. The magnitude of these quantities is shown in computation (3). The experimentally observed molar susceptibility is -53.3×10^{-6}.

$$\{-[(2\times 6.00) + (5\times 2.93) + 30.6] + 4.1\} \times 10^{-6} = -53.1\times 10^{-6} \quad (3)$$

The magnetic susceptibility of a noncubic substance varies along different crystal axes. Susceptibilities measured along different crystal axes are called principal susceptibilities, and the difference between principal susceptibilities is called the magnetic anisotropy of the substance. A large amount of structural information can be obtained from measurements of principal susceptibilities. For example, the principal susceptibility of graphite perpendicular to the hexagonal axis of a single graphite crystal is about -0.5×10^{-6}, or nearly the same as the powder susceptibility of diamond. However, along the hexagonal axis, the susceptibility of graphite is -21.5×10^{-6} at room temperature. It is thought that the large diamagnetic susceptibility along the hexagonal axis is a result of the diamagnetism of conduction electrons, which are present in graphite but not in diamond.

Atomic paramagnetism. As stated previously, an atom with unpaired electrons has a permanent magnetic moment and is therefore paramagnetic. The present discussion will restrict itself to the paramagnetism exhibited by transition element compounds of the iron group and the lanthanide and actinide series. The magnetic properties of the palladium and platinum group compounds and the magnetochemistry of coordination compounds are not discussed here.

The modern quantum-mechanical theory of magnetism was, to a great extent, developed by J. H. Van Vleck. One of the triumphs of this theory is the remarkable agreement between the theoretically calculated magnetic moments of the lanthanide ions and those experimentally determined. The paramagnetism of the lanthanides arises from unpaired electrons in the 4*f* shell which are unique because they are but little affected by the electric fields of the surrounding anions. *See* RARE-EARTH ELEMENTS.

It is, therefore, a good first approximation near room temperature to calculate the magnetic moments of the lanthanide group ions on the assumption that their behavior is that of free gaseous ions.

A somewhat similar situation obtains with respect to the actinide series of compounds. Here, the paramagnetism arises from unpaired electrons in the incomplete 5*f* shell. These electrons are less well shielded from the crystalline electric fields than are the 4*f* electrons. Therefore, in calculating the magnetic moments of actinide ions one must take account of the splitting of the energy levels by the crystalline field. When this is done, very satisfactory agreement between calculated and observed moments is obtained in most cases.

The effect of the crystalline electric fields on the magnetic properties of ions is even more striking in the case of the iron group compounds with their unpaired 3*d* electrons. In general, the magnetic moment of an unpaired electron is proportional to the vector sum of the orbital and the spin angular momentum vectors. The electric fields in a crystal or a solution containing an iron group ion interact so strongly with the orbital part of the moment that they quench its contribution to the magnetic moment almost entirely. Therefore, the observed moments for these compounds are often in very close agreement with moments calculated using the so-called "spin-only" formula.

Molecular paramagnetism. Oxygen has two unpaired electrons in its normal state and is therefore paramagnetic. The molar susceptibility of oxygen over a wide range of pressures and temperatures is given by the simple equation $\chi_M = 0.993/T$ where T is the absolute temperature. Other paramagnetic gases are NO, NO_2, ClO_2, and ClO_3.

There are a large number of organic compounds which possess one or two unpaired electrons. These compounds are known as free radicals. One of the most famous examples is hexaphenylethane which dissociates in benzene solution to give the free radical, triphenylmethyl. Other compounds, such as α,α-diphenyl-β-picrylhydrazyl are stable free radicals even in the solid state.

Many organic compounds, of which fluorescein and naphthalene are examples, are normally diamagnetic. These materials, however, become paramagnetic when exposed to ultraviolet light. The reason for this is that light excites the molecules to triplet or phosphorescent states which are characterized by two unpaired electrons. Magnetic susceptibility measurements on such materials during irradiation have yielded important information on the mechanism of phosphorescence. Paramagnetic resonance measurements on naphthalene have proved earlier static susceptibility measurements.

Ferro- and antiferromagnetism. Ferromagnetic substances are distinguished chiefly by their large susceptibilities at low magnetic fields and by the fact that their specific magnetization is a function of field, up to the field at which the substance is saturated. Above a certain temperature, the Curie point, all ferromagnetic substances lose their ferromagnetism and become paramagnetic.

Antiferromagnetic substances also undergo a transition at a temperature which is characteristic for each material. Above this temperature, known as the Néel point, the substance is paramagnetic. The susceptibility of the material in the antiferromagnetic state is field dependent and is smaller than the susceptibility above the Néel point.

Some examples of ferromagnetic materials are iron, cobalt, nickel, gadolinium, uranium hydride, and nickel disulfide.

Examples of antiferromagnetic substances are manganese, titanium trichloride, uranium trichloride, and neptunium dioxide. *See* ELECTRON PARAMAGNETIC RESONANCE (EPR) SPECTROSCOPY; MOLECULAR STRUCTURE AND SPECTRA.

[DIETER M. GRUEN]

Bibliography: R. Carlin and A. J. Duynevelt, *Magnetic Properties of Transition Metal Compounds*, 1977; J. B. Goodenough, *Magnetism and the Chemical Bond*, 1963, reprint 1976; P. W. Selwood, *Chemisorption and Magnetization*, 1967; E. C. Stoner, *Magnetism and Matter*, 1934; J. H. Van Vleck, *The Theory of Electric and Magnetic Susceptibilities*, 1932.

Manganese

A chemical element, Mn, of atomic number 25 and atomic weight 54.9380. Manganese is one of the transition elements of the first long period of the periodic table, falling between chromium and iron. It has certain properties in common with both of these metals. Although relatively little known or used in its pure form, it has great practical importance in the manufacture of steel.

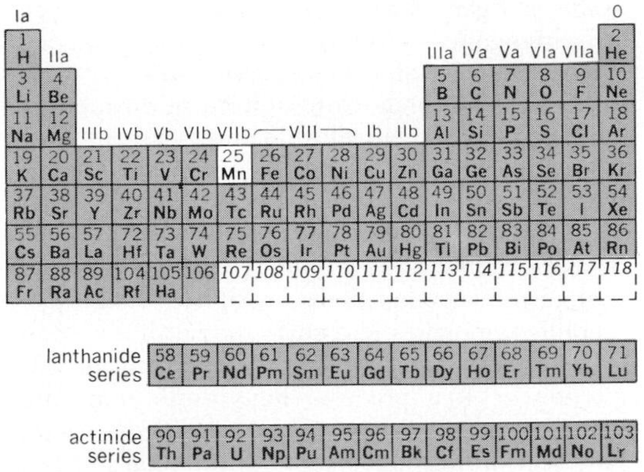

Ores and occurrence. The major ores of manganese are the oxides in hydrated and dehydrated forms, and, to a lesser extent, the silicates and carbonates. Ores suitable for metallurgical purposes contain more than 40% manganese and are at least nine parts manganese to one of ore. However, a high-intensity arc process has been developed to exploit ores of lower content, such as rhodonite. The principal producing countries, in decreasing order of output, are the Soviet Union, India, the Republic of South Africa, Ghana, and Morocco. The typical percentage analysis of a high-grade ore of Soviet origin is MnO, 18.8; MnO_2, 62.4; Fe_2O_3, 1.3; SiO_2, 6.9; P_2O_5, 0.33; Al_2O_3, 1.1; CaO, 1.4; and H_2O, 5.5. This is equivalent to 53.97% Mn, 0.89% Fe, and 0.15% P.

Use in steelmaking. Manganese is an essential ingredient in the large-scale manufacture of wrought steel and there is no effective substitute for it. It assists in the deoxidation of the steel bath and also, by combining with sulfur, markedly improves the hot-working properties of the steel. For this purpose, manganese is used in the form of ferro-manganese, which contains approximately 80% manganese. It is used in high- and low-carbon grades, the former containing 6–7% carbon, the latter, only 0.1% carbon or less. High-carbon ferro-manganese is made by the direct reduction of manganese ore with coke in a blast furnace. Ores of low silicon content containing at least 40% manganese and with a manganese-to-iron ratio of 9:10 are required for this purpose. The required ratio of manganese to iron is higher than might be thought necessary to produce an 80% manganese alloy because manganese has a high vapor pressure and consequently only 80% of it is recovered from the blast furnace, whereas the recovery of iron is practically 100%. Low-carbon ferro-manganese is normally produced by the reduction of manganese ore with silico-manganese in an electric-arc furnace. Silico-manganese, which contains 69–71% manganese, 18–22% silicon, 4–6% iron, and less than 1% carbon, is also made in an arc furnace by the treatment of ores of medium-iron and low-phosphorus content with carbon and silicon. Low-carbon ferro-manganese naturally costs much more to produce than the high-carbon alloy but its use is essential to produce steels in which carbon must be kept at a low level, such as certain austenitic and heat-resistant steels.

Production of pure manganese. Manganese of 97–98% purity is readily made on a commercial scale by the reduction with aluminum of high-grade manganese ore having a very low iron content. Manganese of even higher purity is made by the electrolysis of manganese(II) sulfate solution buffered with ammonium sulfate in a diaphragm cell employing lead-alloy anodes and alloy-steel cathodes. The manganese(II) sulfate is produced by leaching roasted high-purity ore with used catholyte solution from the electrolytic cell. This process, which is now operated on a commercial scale, produces manganese with less than 0.1% of metallic impurities at a cost comparable with that of low-carbon ferro-manganese. Prior to the introduction of this process, high-purity manganese was a metallurgical rarity produced from the aluminothermic manganese of commercial purity by distillation and condensation in a vacuum.

Properties. Manganese melts at 1244±3°C and boils at 2095°C. In the solid state it exists in four allotropic modifications, with the following transformation temperatures:

$$\alpha \rightleftharpoons \beta \quad 700°C$$
$$\beta \rightleftharpoons \gamma \quad 1079°C$$
$$\gamma \rightleftharpoons \delta \quad 1143°C$$

The $\alpha \rightleftharpoons \beta$ transformation is extremely sluggish and for this reason the transformation on heating appears to occur at appreciably higher temperatures than during cooling. The β modification can be retained at room temperature by quenching from temperatures in the β range. The γ modification cannot, however, be completely retained on quenching. Electrodeposited manganese commonly assumes the γ form in the as-deposited condition but transforms to the α modification on standing at room temperature. δ-Manganese is stable only at high temperatures. The physical properties of manganese are summarized in the table.

Manganese in either the α or β modifications at room temperature is both hard and brittle and cannot be plastically deformed to any significant extent. Ductility is, however, associated with the simpler structure of γ-manganese. Manganese can be electrodeposited in this form in plating baths free from reduced sulfur compounds, and although it rapidly transforms to the α form at room temperature, it can be stored by refrigeration for long periods of time without transformation. The ductility of the electrodeposited γ-manganese is such that sheet steel plated in this manner can be fabricated like tin or zinc-plated steel without damage to the manganese coating. Subsequent transformation to α-manganese is associated with a very considerable hardening to a level of 800–900 Brinell.

Manganese readily oxidizes in air to form a brown oxide coating. It also readily oxidizes at elevated temperatures. In this respect its behavior is more akin to its neighbor in the periodic table of higher atomic number, iron, than its neighbor with lower atomic number, chromium. A layered oxide scale is formed on oxidation in air, the layer nearest to the metal being MnO, and the outer layer, Mn_3O_4. The thickness of the MnO layer progressively increases, whereas that of the Mn_3O_4 layer decreases on oxidation at temperatures above 800°C. Below this temperature a third oxide layer, Mn_2O_3, occurs, and below about 450°C a fourth outermost layer of MnO_2 is stable.

Alloys. These can be classed conveniently as either ferrous or nonferrous alloys.

Ferrous alloys. Reference has already been made to the use of manganese as a deoxidant and desulfurizing additive in steelmaking. All commercial grades of steel, therefore, contain manganese as a minor hut essential alloying constituent whose presence influences the cleanliness of the steel and its ability to be hot-worked. Small quantities of manganese also confer on the steel an improved resistance to impact, especially at low temperatures. Plain carbon, unalloyed steels normally contain about 0.3–0.7% manganese, resulting from its use in normal steelmaking procedures. Manganese is also used in somewhat larger quantities for low-alloy steels for constructional purposes and in high-duty cast irons. Rail steels commonly contain 0.9–1.2% manganese and there are other steels containing 1.3–1.6% manganese. These steels have an increased tensile strength compared with steels of comparable carbon content and normal manganese contents and are produced in considerable tonnages.

The addition of manganese to iron lowers the temperature of the $\alpha \rightleftharpoons \gamma$ transformation of iron and with more than about 12% manganese the transformation is suppressed to below room temperature. These austenitic, nonmagnetic manganese steels have extremely important properties. As normally manufactured they contain 11–14% manganese and 1.0–1.4% carbon. They are used in the austenitic condition resulting from quenching from about 1050°C, at which temperature the

Physical properties of manganese

Property	α-Manganese	β-Manganese	γ-Manganese
Specific heat at 25°C, cal/(g)(°C)	0.114	0.155	0.120
Specific gravity at 20°C	7.21	7.29	7.21
Linear coefficient of expansion at 20°C	22.3×10^{-6}	24.9×10^{-6}	14.8×10^{-6}
Electrical resistivity, ohm-cm	$150-260 \times 10^{-6}$	90.0×10^{-6}	40.0×10^{-6}
Temperature coefficient of resistivity/°C	$2-3 \times 10^{-4}$	12.0×10^{-4}	60.0×10^{-4}
Crystal structure	Cubic	Cubic	Face-centered cubic; on cooling, it forms face-centered tetragonal crystals

carbides go into solution. In this condition they have a hardness of 230 Brinell and very unusual mechanical properties. With an ultimate tensile strength of 65–70 tons/in.^{2}, the elongation and reduction of area are 60 and 45%, respectively. Unlike most strong metals, they show very little necking in a tensile test, the reduction of area being much the same all along the gage portion of the test piece. This is due to the fact that these steels show a very marked degree of work-hardening and this is the basis of their very successful use for parts, such as dredger buckets, track links, and rail crossings, which must withstand a high degree of shock and abrasion.

Manganese is also used in certain of the austenitic nickel-chromium steels, which have a wide use for corrosion- and heat-resistant applications. Steels of the 18% chromium, 8% nickel; and 25% chromium, 20% nickel types are very well known. Manganese can be substituted for some of the nickel in steels of this type without loss of their austenitic structure and their corrosion- and heat-resistant properties. Compositions of this type are Cr, 17%; Mn, 6%; Ni, 4%; Cr, 18%; Mn, 10%; Ni, 4%; and Cr, 16%; Ni, 15%; Mn, 6%; and Mo, 6%.

Nonferrous alloys. Manganese has a limited but nevertheless important use in nonferrous alloys. It is an effective deoxidizer of copper-base alloys and improves their mechanical properties. When added to brass containing 40–45% zinc, it increases the tensile strength at the rate of 0.7 ton/in.^{2} per 1% of manganese, and 1% of manganese increases the elongation by 5%. Further additions decrease the ductility but continue to increase the strength. A brass containing 39% zinc and 1% manganese with small quantities of about 0.25% of iron and aluminum is known as manganese bronze. It has a tensile strength of about 26.5 tons/in.^{2} with an elongation of 47% and, because of its resistance to corrosion, is commonly used for marine propellers and low-temperature steam-turbine blades.

Both copper-manganese and copper-manganese-nickel alloys have very interesting properties. An alloy of the latter system containing approximately 84% copper, 4% nickel, and 12% manganese is a well-known alloy for electrical purposes (manganin). It has a resistivity of about 75×10^{-6} ohm-cm and a temperature coefficient of only 1×10^{-5} at 20°C. Other alloys in the system have even more remarkable properties; for example, an alloy of 60% manganese and 20% each of copper and nickel has a resistivity of about 190×10^{-6} ohm-cm combined with a temperature coefficient of resistance at least as low as that of manganin.

This alloy system also contains alloys with very high coefficients of thermal expansion. An alloy containing 72% manganese, 10% nickel, and 18% copper has a maximum expansion coefficient of 27×10^{-6}/°C and is made commercially as a high-expansion element of bimetals for thermostats.

Both aluminum and magnesium alloys commonly contain small quantities of manganese. Binary aluminum-manganese alloys containing 1.25–2.0% manganese have limited applications, but manganese is also present to the extent of 0.5–0.6% in the age-hardening duralumin type of alloys and other precipitation-hardening alloys where it has important effects in modifying the aging behavior and the corrosion resistance.

[ARTHUR H. SULLY]

Compounds. Manganese is a fairly reactive metal. Although the massive metal is somewhat slow to react, the powdered metal reacts with ease and, in some cases, quite vigorously. When it is heated in air or oxygen, powdered manganese forms a red oxide, Mn_3O_4. With water, at room temperature, hydrogen and manganese(II) hydroxide, $Mn(OH)_2$, are formed. In the case of acids, because manganese is such a reactive metal, hydrogen is liberated along with the formation of a manganese(II) salt. Manganese reacts at elevated temperatures with the halogens, sulfur, nitrogen, carbon, silicon, phosphorus, and boron.

In its many different compounds, manganese has oxidation states of 1+ through 7+. The most common oxidation states are 2+, 4+, and 7+. All the compounds, except those containing Mn^{II}, are deeply colored. For example, potassium permanganate, $KMnO_4$, forms aqueous solutions that are reddish-purple; potassium manganate, K_2MnO_4, forms deep-green solutions.

Oxides. Manganese forms six oxides: MnO, Mn_3O_4, Mn_2O_3, MnO_2, MnO_3, and Mn_2O_7. The most common and best known are MnO, Mn_3O_4, MnO_2, and Mn_2O_7. The oxides that occur naturally are pyrolusite, MnO_2, braunite, Mn_2O_3, and hausmannite, Mn_3O_4. All the oxides are difficult to reduce to the free metal with carbon monoxide. They can be reduced only under pressure with hydrogen. However, lower oxides can be obtained, in some cases, by controlled thermal decomposition of more oxygen-rich oxides.

Manganese dioxide, MnO_2, which has a gray to gray-black color, is by far the most important oxide. Although it is usually found in the free state, manganese dioxide can be prepared by heating manganese nitrate, $Mn(NO_3)_2$, at 180–200°C, or by heating a mixture of manganese carbonate, $MnCO_3$, and potassium chlorate at 300°C. There is some question about the purity of the product obtained; it has been stated that the purest product contained only 98% MnO_2. This was attributed to the formation of a mixture of MnO_2 and MnO.

This oxide is insoluble in water and in weak acids and bases. However, it does react with concentrated sulfuric acid, reaction (1), and with a reducing acid, such as concentrated hydrochloric acid, at room temperature or slightly higher, reaction (2).

$$2MnO_2 + 2H_2SO_4 \rightarrow 2MnSO_4 + O_2 + 2H_2O \quad (1)$$

$$MnO_2 + 4HCl \rightarrow MnCl_2 + Cl_2 + 2H_2O \quad (2)$$

With ice-cold concentrated hydrochloric acid, it gives a red to brown solution which probably contains manganese(III) chloride, $MnCl_3$, and manganese(IV) chloride, $MnCl_4$.

In the presence of air or other oxidizing agents, MnO_2 reacts with fused potassium hydroxide to form potassium manganate, K_2MnO_4, reaction (3).

$$2MnO_2 + 4KOH + O_2 \rightarrow 2K_2MnO_4 + 2H_2O \quad (3)$$

In weakly acid solution, potassium manganate disproportionates to form, among other products, potassium permanganate, $KMnO_4$, reaction (4). Potassium permanganate is a powerful oxidizing agent, both in organic and inorganic chemistry. In

$$3K_2MnO_4 + H_2O \rightarrow 2KMnO_4 + MnO_2 + 4KOH \quad (4)$$

acid solution, five oxygen atoms are liberated by each two formula weights that react, as in reaction (5), whereas in basic solution, only three oxygen atoms are liberated, as in reaction (6). The compound is also useful as an oxidizing agent in the analytical chemistry of certain metal ions.

$$2KMnO_4 + 3H_2SO_4 \rightarrow K_2SO_4 + 2MnSO_4 + 3H_2O + 5O \quad (5)$$

$$2KMnO_4 + H_2O \rightarrow 2MnO_2 + 2KOH + 3O \quad (6)$$

Halides. The most common manganese compounds with fluorine, chlorine, bromine, and iodine are those containing manganese in a 2+ oxidation state. These compounds, having the formulas MnF_2, $MnCl_2$, $MnBr_2$, and MnI_2, respectively, can generally be prepared by the direct combination of manganese metal with the halogen.

The most useful of the halides are manganese (II) fluoride, MnF_2, and manganese(II) chloride, $MnCl_2$. The fluoride is best prepared by heating NH_4MnF_3 in an atmosphere of carbon dioxide. It is a colorless compound which is slightly soluble in water. The pink-colored chloride can be prepared by the reaction of manganese dioxide with warm concentrated hydrochloric acid or by the action of dry hydrogen chloride on MnO, $MnCO_3$, or manganese metal.

Other compounds. In the presence of strong bases, solutions of manganese(II) salts form the pink-colored, insoluble manganese(II) hydroxide $Mn(OH)_2$. In the presence of air or oxygen, $Mn(OH)_2$ gradually oxidizes to form dark-brown products which contain $Mn(OH)_3$ or possibly $MnO \cdot MnO_2 \cdot nH_2O$. Manganese(II) hydroxide, because of its insolubility in water, is a fairly weak base. When heated in the absence of air, it forms MnO, whereas in air the main product is Mn_3O_4.

Ammonium sulfide or alkali metal sulfides precipitate pink-colored manganese(II) sulfide, MnS, from solutions of manganese(II) salts. The sulfide, which is useful in the analytical chemistry of manganese, is readily soluble in dilute acids.

Manganese(II) carbonate, $MnCO_3$, is precipitated from solution on the addition of sodium or potassium hydrogen carbonate. It is easily decomposed on heating or by treatment with dilute acids.

Other important manganese(II) salts are the pink-colored manganese(II) nitrate 6-hydrate, pink-colored manganese(II) sulfate 4-hydrate, and colorless manganese(II) phosphate.

Uses. The compounds of manganese have many uses in industry. Manganese dioxide is used as a drying agent or catalyst in paints and varnishes, as a decolorizer in glass manufacturing, and in dry cells. Potassium permanganate is used for bleaching purposes, for decolorization of oils, and as an oxidizing agent in preparative and analytical chemistry. *See* OXIDIZING AGENT.

Analytical methods. Manganese(II) can be determined gravimetrically by precipitation as manganese ammonium phosphate, $MnNH_4PO_4 \cdot H_2O$, followed by ignition to manganese pyrophosphate, $Mn_2P_2O_7$, which is weighed. Ammonium sulfide, which forms insoluble manganese sulfide, can be used to separate manganese from certain other ions. A sensitive qualitative test for manganese is to fuse a small quantity of the compound with potassium nitrate. Green-colored manganate ion, MnO_4^{--}, is formed in the reaction and can be easily identified.

[W. W. WENDLANDT]

Bibliography: F. A. Cotton and G. Wilkinson, *Advanced Inorganic Chemistry*, 4th ed., 1980; G. P. Glasby (ed.), *Marine Manganese Deposits*, 1977; C. A. Hampel (ed.), *Rare Metals Handbook*, 2d ed., 1961, reprint 1971; Max Planck Society for the Advancement of Science, Gmelin Institute for Inorganic Chemistry, *Manganese, Pt. C, The Compounds: Section 2*, 1975.

Mass defect

The difference between the mass of an atom and the sum of the masses of its individual components in the free (unbound) state. The mass of an atom is always less than the total mass of its constituent particles; this means, according to Albert Einstein's well-known formula, that an energy of $E = mc^2$ has been released in the process of combination, where m is the difference between the total mass of the constituent particles and the mass of the atom, and c is the velocity of light.

The mass defect, when expressed in energy units, is called the binding energy, a term which is perhaps more commonly used.

[W. W. WATSON]

Mass number

The mass number A of an atom is the total number of its nuclear constituents, or nucleons, as the protons and neutrons are collectively called. The mass number is placed (by North American practice) following and above the elemental symbol, thus U^{238}, or (by international agreement) before and above it, thus ^{238}U. Because of the approximate equality of the proton and neutron masses, and the relative insignificance of that of the electron, the mass number gives a useful rough figure for the atomic mass; for example, $H^1 = 1.00814$ atomic mass units (amu), $U^{238} = 238.124$ amu, and so on. The mass number is reduced by four during α-emission, but it is not altered during β-decay or electron capture. *See* ATOMIC NUMBER.

[HENRY E. DUCKWORTH]

Mass spectrometry

An analytical technique for identification of chemical structures, determination of mixtures, and quantitative elemental analysis, based on application of the mass spectrometer. Organic and inorganic molecular structure determination is based on the fragmentation pattern of the ion formed when the molecule is ionized; further, because such patterns are distinctive, reproducible, and additive, mixtures of known compounds may be quantitatively analyzed. Quantitative elemental analysis of organic compounds requires exact mass values from a high-resolution mass spectrometer; trace analysis of inorganic solids requires a measure of ion intensity as well.

Methods of ion production. For analysis of organic compounds the principal methods are electron impact, chemical ionization, field ionization, and field desorption.

Electron impact. When a gaseous sample of a molecular compound is ionized with a beam of energetic (commonly 70-V) electrons, part of the energy is transferred to the ion formed by the collision, as shown in reaction (1). For most molecules

$$A—B—C + e \rightarrow A—B—C^{+} + e + e \qquad (1)$$

the production of cations is favored over the production of anions by a factor of about 10^4, and the following discussion pertains to cations. The ion corresponding to the simple removal of the electron is commonly called the molecular ion and normally will be the ion of greatest *m*/*e* ratio in the spectrum. In the ratio *m* is the mass of the ion in atomic mass units and *e* is the charge of the ion measured in terms of the number of electrons removed (or added) during ionization. Occasionally the ion is of vanishing intensity, and sometimes it collides with another molecule to abstract a hydrogen or another group. In these cases an incorrect assignment of the molecular ion may be made unless further tests are applied. Proper identification gives the molecular weight of the sample.

The remaining techniques were devised generally to circumvent the problem of the weak or vanishingly small intensity of a molecular ion.

Chemical ionization. Here the ions to be analyzed are produced by transfer of a heavy particle (H^+, H^-, or heavier) to the sample from ions produced from a reactant gas. Frequently the reactant gas is methane at pressures of 0.2–2.0 torr (1 torr = 133.32 Pa). As above, the initial process in methane upon electron impact is ionization to yield CH_4^+ ions; some of these have enough energy to fragment to CH_3^++H. These ions in turn react with neutral methane as in reactions (2) and (3). The

$$CH_4^+ + CH_4 \rightarrow CH_5^+ + CH_3 \qquad (2)$$

$$CH_3^+ + CH_4 \rightarrow C_2H_5^+ + H_2 \qquad (3)$$

resulting ions are strong Brönsted acids and react by proton transfer as in reactions (4) and (5) to ionize the molecule of interest. The $C_2H_5^+$ ion also reacts by hydride abstraction, as in reaction (6). Other

$$CH_5^+ + A—B—C \rightarrow H—A—B—C^+ + CH_4 \qquad (4)$$

$$C_2H_5^+ + A—B—C \rightarrow H—A—B—C^+ + C_2H_4 \qquad (5)$$

$$C_2H_5^+ + A—B—C—H \rightarrow A—B—C^+ + C_2H_6 \qquad (6)$$

gases may also be used for ionization: H_2, H_2O, NH_3, and isobutane are common. In these the reactive ions are H_3^+, H_3O^+, HN_4^+, and $C_4H_9^+$ [the conjugate acid of $(CH_3)_2C{=}CH_2$], respectively. The thermochemistry of these proton transfer reactions and many others is summarized by reference to Fig. 1. The proton affinity is the negative of the enthalpy change for the solvation of a proton by the compound, as in reaction (7). It follows that

$$A—B—C + H^+ \rightarrow A—B—C—H^+ \qquad (7)$$

A—B—C—H^+ protonates molecules with greater proton affinities in the absence of kinetic complications; such reactions are exothermic. Recently, proton affinities have been determined both by high-pressure techniques akin to chemical ionization and by the ion cyclotron resonance technique. Thus H_3^+ transfers the most energy when it protonates a molecule; NH_4^+ transfers the least of the four examples. If the energy transferred (typically 10–50 kcal/mole or 40–200 kJ/mole) is great enough, fragmentation can occur, but there is

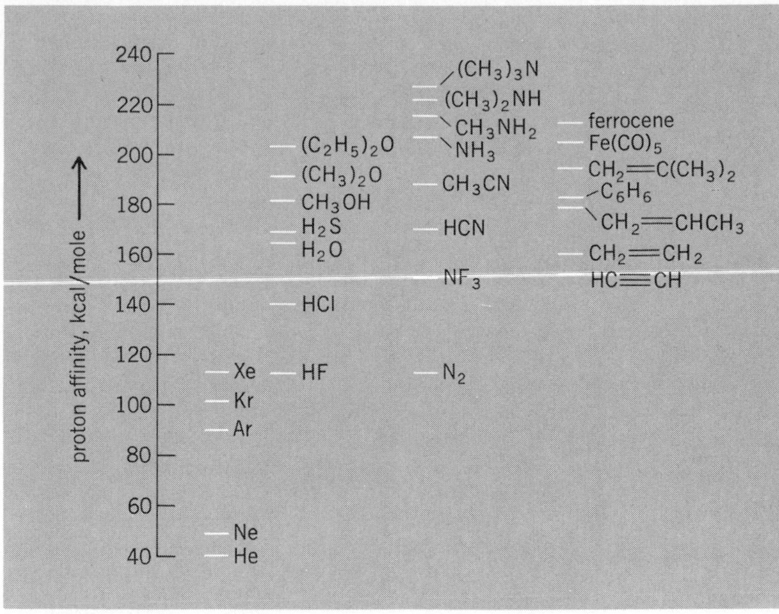

Fig. 1. Representative gas-phase proton affinities of molecules.

much less in chemical ionization than in electron impact mass spectra.

In addition to heavy-particle transfer of other types (transfer of CH_3^+, $C_2H_5^+$, Cl^-), another important method is charge exchange. For a gas with an ionization potential greater than that of the molecule of interest, reaction (8) is exothermic,

$$G^+ + A—B—C \rightarrow A—B—C^+ + G \qquad (8)$$

and the excess energy is transferred as internal energy. He^+, Ne^+, and Ar^+ transfer large amounts of internal energy and molecular ions are weak; CO and NO are useful gases for this process. Occasionally a combination of chemical ionization and charge exchange is used, with mixtures of Ar and H_2O yielding information about both the molecular weight and important fragments, for example.

The term negative chemical ionization is usually intended to include all ionization processes yielding negative ions under source pressure conditions characteristic of chemical ionization. In some cases ions may be formed by a true chemical ionization process like proton transfer, as in reaction (9), in which OH^- is the reagent ion, produced from

$$RCOOH + OH^- \rightarrow RCOO^- + H_2O \qquad (9)$$

H_2O or $N_2O + CH_4$. In others the reagent gas only mediates the energy of free electrons to low values compatible with capture by neutral molecules to yield negative molecular ions. Molecules with several electronegative atoms have large cross sections for this latter process. Thus the direct analysis of polychlorinated aromatics, and the analysis of small peptides after derivatization by pentafluorobenzoylation, are possible in the femtogram region.

New applications of ions formed by chemical ionization (atmospheric pressure ionization; plasma chromatography, in which the drift time of the ion through a flowing gas is measured) also have femtogram sensitivities. *See* PLASMA CHROMATOGRAPHY.

Field ionization and field desorption. For less volatile material, the sample is ionized when it is in a very high field gradient (several volts per angstrom) near an electrode surface. Figure 2 illustrates the distortion of the molecular potential well so that an electron tunnels from the molecule to the anode. The ion thus formed is repelled by the anode. Typically, the lifetime of the ion in the mass spectrometer source is much less (10^{-12} to 10^{-9} s) than in electron impact. Because little energy is transferred as internal energy and the ion is removed rapidly, little fragmentation occurs, and the molecular weight is more easily determined. Field ionization is also used in the time-resolved study of ion fragmentation and rearrangement called field ionization kinetics. This technique permits determination of the fragmentation products at specific times from about 10^{-11} to 10^{-8} s after ionization by energy-focusing of ions at different points in the field ionization source. In this way simple fragmentations uncomplicated by rearrangements of hydrogen or other atoms in the molecular ion may be observed at the shortest times yet used for sampling ions, and complex rearrangements of ions may be defined from studies over a range of times.

The electrode may be a wire, razor blade, point, or collection of points; all of these have a small radius of curvature of the tip or edge. In field ionization the sample is admitted as a gas and ionized within a few angstroms of the surface; a fraction may actually be adsorbed. Minimal heating is required. In field desorption, the sample is coated on the metal surface, for example, by dipping it into a solution of the sample or by coating the solution from a microsyringe; then the solvent is evaporated. Field desorption is useful for compounds which cannot be evaporated easily; ionic or thermally labile compounds such as phospholipids, water-soluble vitamins, peptides, and sugars, may be studied, with heating sufficient to cause sample mobility on the surface so that the sample can migrate to areas of highest field gradient, but not enough to evaporate it. The above samples require internal heating by passing a current through the wire, for example. External heating, especially by a focused Ar^+ ion laser beam, can produce spectra of vitamin B_{12} (molecular ion of 1354 daltons) and overcome lattice energies of Ba^{2+} salts to yield Ba^{2+} ions. Uniform internal heating to the Curie point of the prepared electrode (called an emitter) produces reproducible pyrolysis spectra of bacteria which differ sufficiently to have possible taxonomic application.

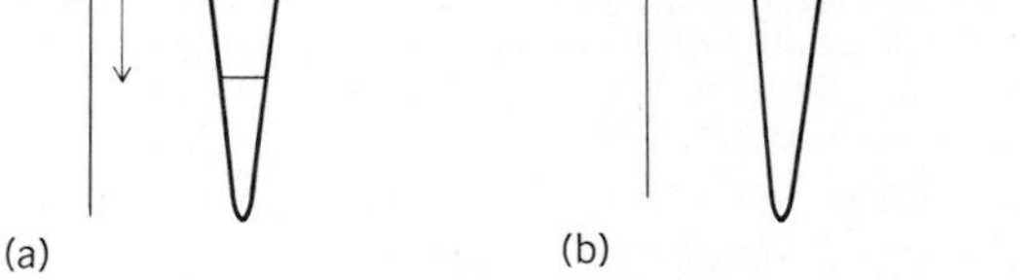

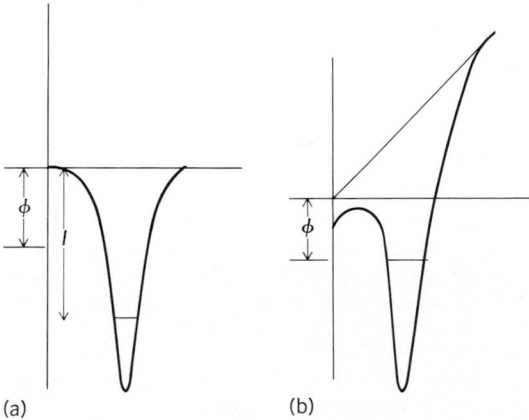

Fig. 2. Potential well of a molecule: (*a*) undistorted, with ionization potential *I* near a metal with work function ϕ in the absence of a field gradient; and (*b*) in a strong field gradient. Near the metal surface the external field raises the most weakly bonded electron to the Fermi level so that tunneling through the small barrier may occur.

The generation of surfaces for these techniques is varied. In field ionization, conditioning of a razor blade by heating in the mass spectrometer source in the presence of an organic sample improves sensitivity. In field desorption, creation of more suitable multipoint surfaces has led to techniques for the growth of microneedles perpendicular to the wire axis by deposition from benzonitrile vapor at very high (approximately 1200°C) temperatures, room-temperature deposition of Si whiskers from a SiH_4/Ar mixture, or electrodeposition of metal dendrites from salt solutions. *See* LASER SPECTROSCOPY.

Cationization. This is a form of field desorption in which cations, most commonly Li^+ but often other alkali metals, are field-desorbed from a salt coating on a wire and pass through a gaseous or adsorbed sample M. MLi^+ ions are produced in some cases where simple field desorption fails.

Electrohydrodynamic ionization. A high electric-field gradient induces ion emission from a droplet of a liquid solution, that is, the sample and a salt dissolved in a solvent of low volatility. An example would be the sample plus NaI dissolved in glycerol. Spectra include peaks due to cationized molecules of MNa^+, $MNa(C_3H_8O_3)_n^+$, and $Na(C_3H_8O_3)_n^+$.

Rapid heating methods. A sample heated very rapidly may vaporize before it pyrolyzes. Techniques for heating by raising the temperature of a source probe on which the sample is coated by 200 K/s, or by bombardment of the solid sample with energetic electrons or ions or light (laser desorption), are now developed. Bombardment by electrons is achieved simply by inserting a probe with the sample directly into the electron beam of an electron impact source (in-beam electron ionization). The energetic ions may likewise be the plasma in a chemical ionization source (in-beam chemical ionization) or, alternatively, ions in a beam of 2.5 kV Ar^+ directed to a surface coated with the sample (without the coating, this is the inorganic surface analysis technique of secondary ion mass spectrometry). The most desirable method of this variety is the californium-252 plasma desorption source. It must be noted that nomenclature is not unified in this area; some authors also refer to ion-beam chemical ionization as plasma desorption.

The ^{252}Cf source produces fission fragments which penetrate a thin foil on which a film of sample has been coated. The fission fragment deposits energy by electronic excitation in a cylindrical track with a 20-nm diameter. Molecular, $(M + 1)^+$, and $(M - 1)^-$ ion ejection from the surface then occurs as the energy is dissipated through the remaining lattice in a shock wave. Molecular ions of mass up to 4000 daltons have been observed with this technique.

Assignment of empirical formula. If the spectrum is a high-resolution spectrum, the deviation of the molecular weight from an integral value is used to determine the elemental composition (for example, $^{12}C_{12}{}^{1}H_{10}$ has molecular weight 154.0782; $^{12}C_9{}^{1}H_{14}{}^{16}O_2$, 154.0994; $^{12}C_{10}{}^{1}H_{16}{}^{16}O$, 154.1358; and $^{12}C_{11}{}^{1}H_{22}$, 154.1721). If it is low-resolution, then the worker takes advantage of the natural abundances of isotopes (for example, ^{12}C, 98.9%; ^{13}C, 1.1%; ^{35}Cl, 75.4%; ^{37}Cl, 24.6%; ^{79}Br, 50.6%; and ^{81}Br, 49.4%) to calculate at least part of the empirical formula by application of the binomial coefficients. In Fig. 3, the intensity of the (M + 1) peak is 11% that of the M peak. Now only ions containing one ^{13}C ion contribute to the (M + 1) peak, and the probability that any C chosen randomly is ^{13}C is 1.1%. If one examines 10 C atoms as a group, the probability that one is ^{13}C is 10 times greater, or 11%; hence the molecule contained 10 carbons.

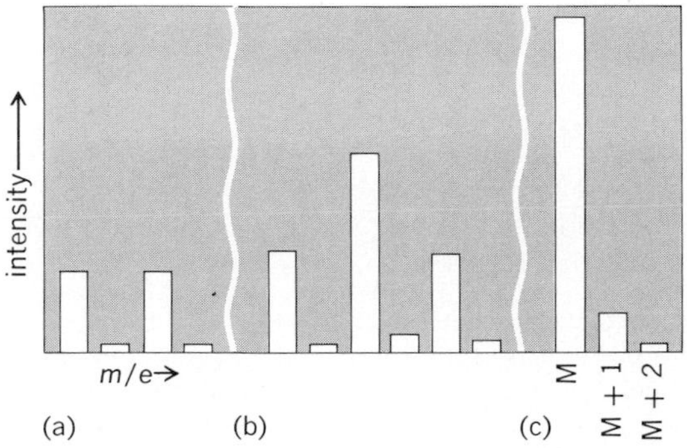

Fig. 3. Use of isotope distribution to determine elemental composition. (*a*) Molecular ion for a compound containing one Br atom. (*b*) Molecular ion for a compound containing two Br atoms. (c) Molecular ion for a compound containing neither Br nor Cl.

Fragmentation patterns. Since the amount of energy transferred to the molecule is much more than that required simply to ionize it or to break some of the bonds in the remaining ion, some of the molecules fragment after ionization, by competing consecutive decompositions, as indicated in reactions (10)–(15). These ions are separated by

$$A{-}B{-}C^{+\cdot} \rightarrow A{-}B^+ + C\cdot \quad (10)$$

$$A{-}B{-}C^{+\cdot} \rightarrow A{-}B\cdot + C^+ \quad (11)$$

$$A{-}B{-}C^{+\cdot} \rightarrow A\cdot + B{-}C^+ \quad (12)$$

$$A{-}B^+ \rightarrow A + B^+ \quad (13)$$

$$B{-}C^+ \rightarrow B^+ + C \quad (14)$$

$$A{-}B{-}C^{+\cdot} \rightarrow A{-}C^{+\cdot} + B \quad (15)$$

mass in the spectrometer and produce the mass spectrum. Although fragments at almost every mass are usually produced, the most favorable routes for decomposition, which give the most intense peaks in the spectrum, are characteristic of the functional groups in the molecule. Therefore, the intense ions, particularly those at the high-mass end of the spectrum, are used in the assignment of structure to compounds. The common mechanistic rationalizations used for interpretation assume that the charge is localized at the functional group, because the electron lost is most likely to be a π-electron or a nonbonding electron. Other less empirical explanations, notably the statistical approach of the quasi-equilibrium theory of mass spectra, have been outlined in the literature.

The reactions of functional groups bearing electronegative atoms are typically loss of the electronegative atom or the group containing it [X = Cl, Br, I, OR, SR, acyl, in reaction (16)] or loss of a substituent group on the α-carbon [R = H or alkyl, X = OR, NR_2, SR, Cl, Br, in reaction (17); R= H, alkyl, or aryl, X = H, OH, OR, NR_2, Cl, Br, I, alkyl, or aryl, in reaction (18)]. These are simple fragmen-

$$R{-}\overset{+\cdot}{X} \rightarrow R^+ + X\cdot \quad (16)$$

$$R{-}CH_2{-}\overset{+\cdot}{X} \rightarrow R\cdot + CH_2{=}X^+ \quad (17)$$

$$R{-}\overset{O^{+\cdot}}{\overset{\|}{C}}{-}X \rightarrow R{-}C{\equiv}O^+ + X\cdot \quad (18)$$

tations [reactions (10)–(12)] corresponding to losses of a radical; the remaining ion is an even-electron ion. Except for the molecular ion, which is an odd-electron ion, the principal ions in a spectrum are even-electron ions unless special structural requirements are met. In these less common cases, prominent odd-electron ions unless special structural requirements are met. In these less common cases, prominent odd-electron ions, formed by the loss of a molecule, are found.

The most general case of odd-electron ion formation is illustrated in reaction (19), where in general

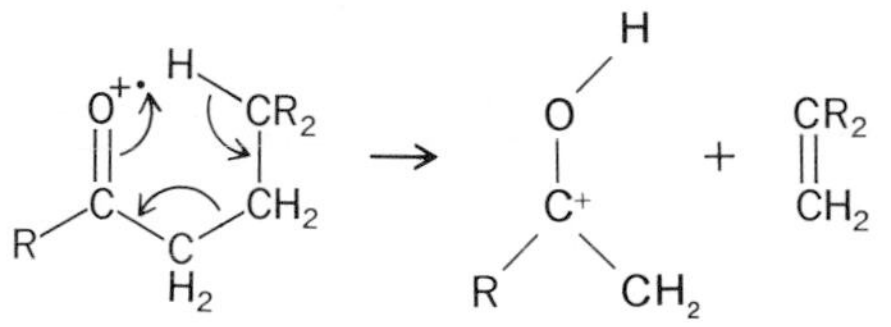

(19)

a γ-hydrogen is abstracted by a multiply bonded electronegative atom with rupture of the β-bond; this reaction is called the McLafferty rearrangement. Other cases of odd-electron ion formation involve interactions of ortho substituents on aromatic systems and certain decompositions of cyclic molecules, which may or may not be rearrangements [reaction (15)]. Typical electron-impact mass spectra of volatile compounds are shown in Fig. 4. The spectrum of acetophenone is characterized only by simple bond-cleavage reactions, but there is an intense peak in the spectrum of valerophenone corresponding to a McLafferty rearrangement in which C_4H_8 is lost from the molecule.

Even-electron ions, such as many fragment ions found in electron-impact spectra and also (M + 1) ions from chemical ionization, decompose most often to smaller even-electron ions by loss of a small molecule. Illustrations are given by reactions (20) and (21), from the chemical ionization of small

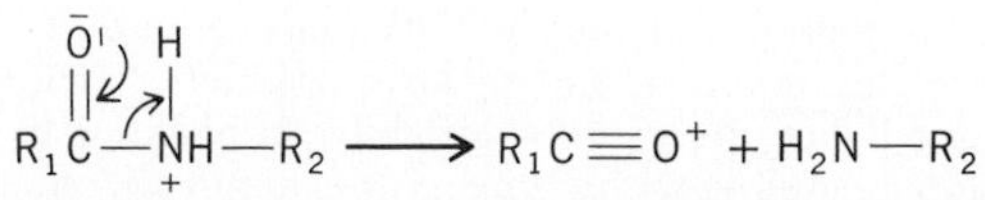

(20)

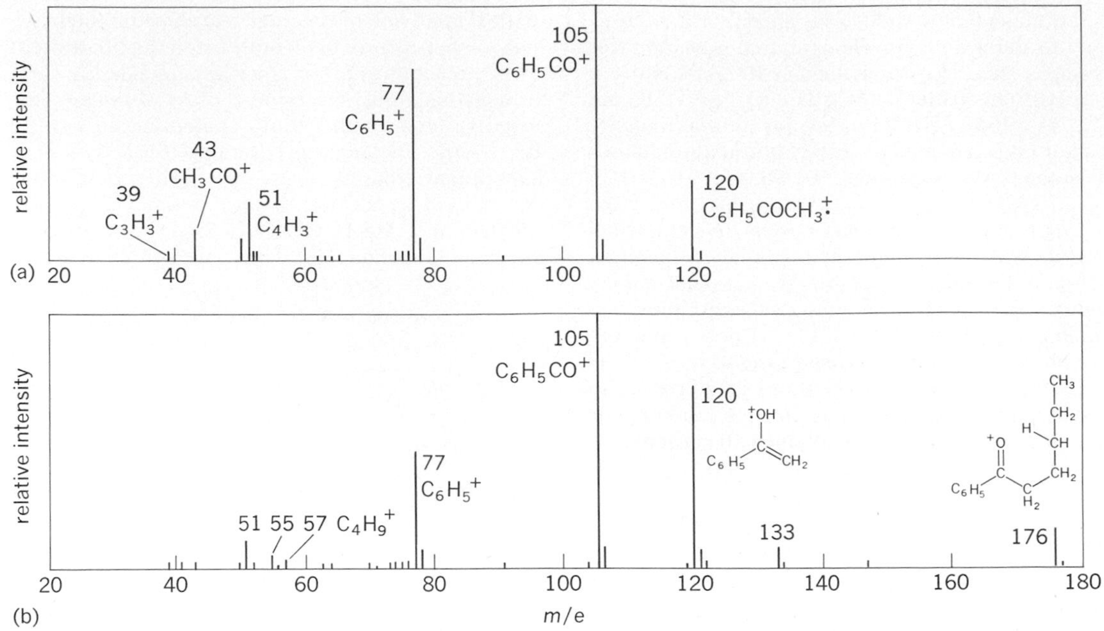

Fig. 4. Mass spectra of (*a*) acetophenone and (*b*) valerophenone, using electron impact.

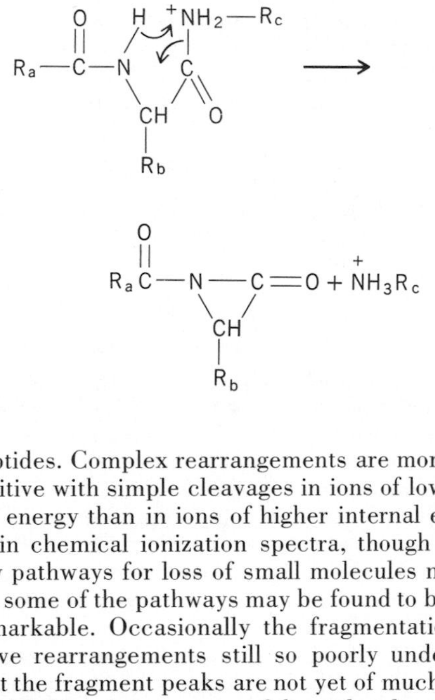

(21)

peptides. Complex rearrangements are more competitive with simple cleavages in ions of low internal energy than in ions of higher internal energy; so in chemical ionization spectra, though only a few pathways for loss of small molecules may exist, some of the pathways may be found to be quite remarkable. Occasionally the fragmentations involve rearrangements still so poorly understood that the fragment peaks are not yet of much use in establishing the structure of the molecule.

Typical spectra for the various techniques discussed are shown in Fig. 5 for the involatile compound creatine; both the molecular ion in Fig. 5*a* and the (M + 1) ions in Fig. 5*b*–*d* lose water by a rearrangement process. It is not universally true that the chance of finding the molecular ion increases as one progresses from electron impact to chemical ionization, field ionization, and field desorption, but these results are typical.

Further useful information about fragments is obtained, first, from the potentials required to ionize the molecule and to form fragments and, second, from observations of shifts of peaks with isotopic labeling techniques. For example, in reaction (19), substitution of D for H at the β position does not increase the mass of the product ion, but when H at the γ position is replaced by D, the mass of the product is shifted to higher mass by one unit. Finally, correlations with trends observed in ordinary chemical reactions yield information. For example, in both aromatic and aliphatic systems, orders of reactivities depending on substitution patterns show correspondence; in some aromatic systems the correlation is quantitative. If an unknown compound is suspected of belonging to a certain group of compounds, then its structure may often be established by comparison of its spectrum with spectra of compounds of known structure. This technique is particularly valuable in the study of natural products. The shift technique, the interpretation of substitution patterns when the molecular ion and some fragments are shifted by the same amount, for example, 30 units by a methoxy group, is helpful in studying alkaloid spectra. In addition to correlation with typical solution phenomena, relation of fragmentations to pyrolytic reactions and to photochemical processes is studied. The McLafferty reaction is analogous to a well-known photochemical reaction of ketones, the Norrish type II rearrangement, in which an olefin molecule is lost from the alkyl ketone while the enol form of a smaller ketone is produced by the influence of light. Other ionization techniques (by large electric-field gradients, by photons, or by other ions) produce different types of spectra useful in structural analysis. *See* PHOTOCHEMISTRY.

Metastable ions. Correlation of spectra with structure requires the additional use of further decomposition products beyond the first fragmentation steps, but it is difficult to define the origin of products possibly arising by more than one path, for example, B^+ in reactions (13) and (14). A clue to origin of fragments is given by metastable ions: If in reaction (13) some AB^+ ions decompose after they are accelerated but before they are magnetically deflected, a broad peak (Fig. 6) appears at an m/e value numerically equal to m_B^2/m_{AB}. In Fig. 6

this broad peak is seen at $m/e = 56.5$. The value 56.5 equals $77^2/105$; therefore, the metastable peak corresponds to the reaction $105 \rightarrow 77$ and indicates that at least part of the $C_6H_5^+$ ion is formed by the loss of CO from $C_6H_5CO^+$. The order of the decomposition steps is suggested by the collection of metastable ions and bears on the organization of the parts of the molecule into the whole.

Study of metastable ions and other ions which can be made to decompose between acceleration and mass analysis has been carried out through ion kinetic energy spectroscopy (IKES) and mass-analyzed ion kinetic energy spectroscopy (MIKES), also called direct analysis of daughter ions (DADI). In the IKES technique, which may be studied with a conventional double-focusing mass spectrometer, those ions which decompose between the accelerating region and the electric sector will form ionic fragments with only a fraction of the kinetic energy imparted to the original ions. The fraction is the ratio of the mass of the fragment to that of the original ion. Scanning the voltage of the electric sector and detecting the ions which leave it provides a scan of ions according to their kinetic energy. Frequently intensities of ions differ substantially for even closely related isomers; thus this technique is useful for fingerprinting difficult compounds.

In MIKES, the electric sector and the magnetic sector in the double-focusing mass spectrometer are reversed, so that the ions, after acceleration, enter the magnetic sector and are analyzed according to mass before they enter the electric sector. By setting the magnetic field, one chooses ions of a certain mass for study; those which decompose before entering the electric sector have only the fraction of the kinetic energy according to the ratio of fragment and original mass noted above, and by sweeping the electric sector, the amounts of daughter ions from this one particular precursor may be observed. The intensities of ions are useful for structure determination, although they are somewhat affected by the distribution of energies within the fragmenting ion. Much information about energetics of decomposing ions has been gained from studies of the width of peaks due to their products in MIKES, for these are always broadened because kinetic energy T is released in the fragmentation. It has been shown that T is a characteristic of molecular structure; for example, the McLafferty product ion formed by the loss of C_2H_4 from the molecular ion of butyrophenone, as in reaction (19), fragments by loss of CH_3, releasing 48 meV (4.6 kJ/mole) of T. Since the molecular ion of acetophenone releases 7 meV (0.7 kJ/mole) energy, the McLafferty product does not have the acetophenone structure. Further studies of rearrangements have led to analyses of the partitioning of energy between internal and kinetic energy in fragmentation, as a function of geometry of the activated complex and other parameters.

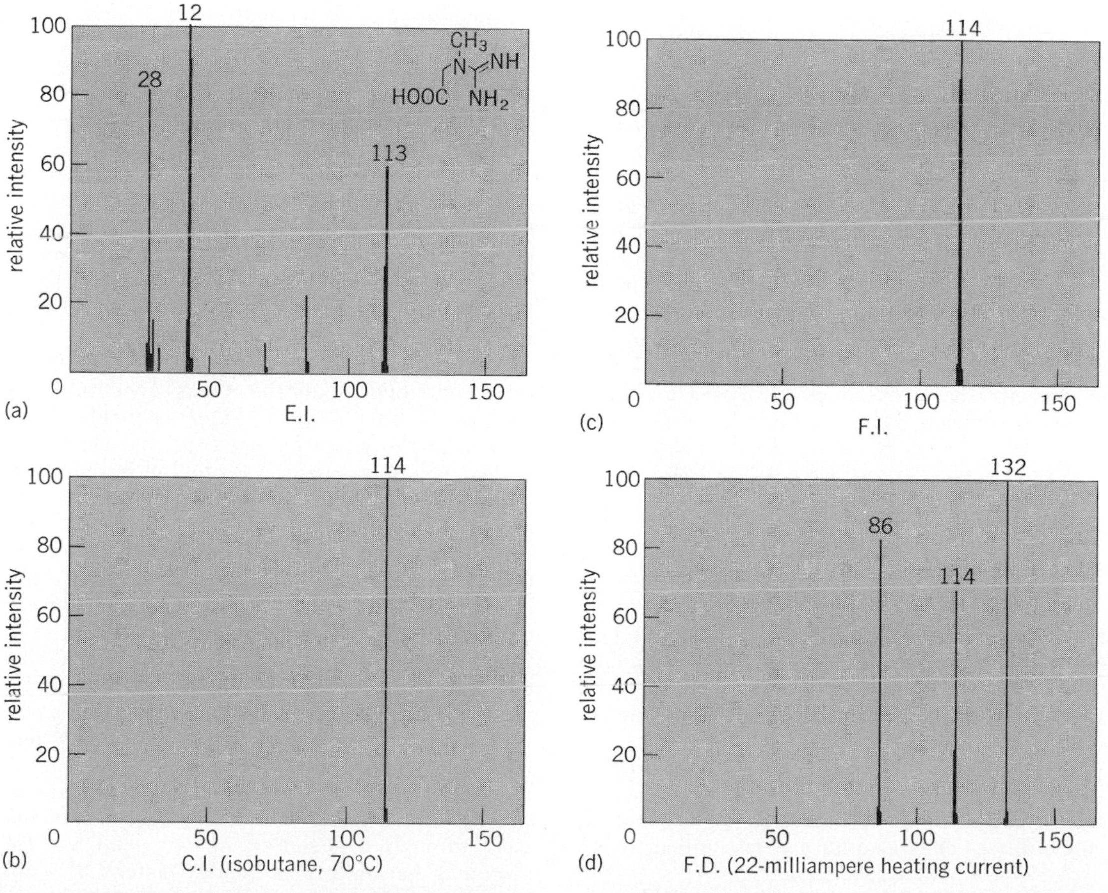

Fig. 5. Mass spectra of creatine (molecular weight 131) obtained by (*a*) electron impact, (*b*) chemical ionization, (*c*) field ionization, and (*d*) field desorption with moderate heating. Both the molecular ion (not observed in *a* and the $M+1$ ion in *b*, *c*, and *d* undergo a loss of water. Field desorption does not always produce the largest ion related to the molecular weight on comparing the four techniques, but this result is typical. (*From H. M. Fales et al., Anal. Chem., 47:207–219, 1975*)

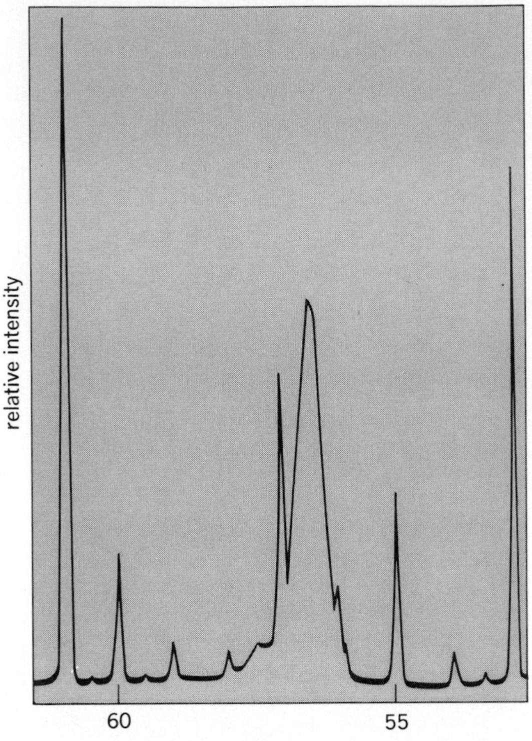

Fig. 6. Scale-expanded portion of acetophenone spectrum showing broad metastable ion at *m/e* 56.5.

Unimolecular MIKE spectra, that is, MIKE spectra of metastable ions, are somewhat dependent on ion internal energy. A less energy-dependent method, and therefore a surer guide to structure, is that of collision-induced dissociation, In this, a collision gas is admitted at a low pressure to the region between the magnetic and electric sector of the reversed instrument. As mass-selected ions travel from the magnetic sector, they collide with the gas molecules, and a fraction of their kinetic energy is transformed into internal energy sufficient to cause rapid decomposition of the ions. The spectra are recorded as in mass-analyzed ion kinetic energy. Charge-exchange processes may also occur on collision, as in reactions (22)–(24). If one

$$\text{A—B—C}^{2+} + \text{N} \rightarrow \text{A—B—C}^{+} + \text{N}^{+} \quad (22)$$

$$\text{A—B—C}^{+} + \text{N} \rightarrow \text{A—B—C}^{2+} + \text{N} + e \quad (23)$$

$$\text{A—B—C}^{+} + \text{N} \rightarrow \text{A—B—C}^{-} + \text{N}^{2+} \quad (24)$$

sets the electric sector at twice, one-half, or minus the voltage which transmits normal ions, the products of reactions (22)–(24) can be detected. One thus generates mass spectra of ions produced by special routes, and the utility of these spectra for interpretive purposes is being explored.

New methods for studying collision-induced decomposition include fragmentation by low-energy collisions in the second stage of a triple quadrupole filter. They also have been studied by ion cyclotron resonance spectrometry.

Applications of computers. The principal applications of computers to analysis have been to data acquisition and structure interpretation. High-resolution spectra contain so much data that the rapid acquisition and presentation of data in a form easily assimilated by the operator has been adapted to the computer. Similarly, in cases where a gas chromatograph effluent passes through the source of the mass spectrometer, the generation of even low-resolution data is so rapid that a dedicated computer is appropriate. A variety of data displays are useful for interpretation. The plot of total ion current versus time produces a reconstructed chromatogram; the plot of ions of a single mass versus time, called a mass fragmentogram, is useful in identifying compound classes among the gas chromatogram peaks if the appropriate mass is chosen, or in identifying compounds directly if some other mass like the molecular weight of a desired component is chosen.

For interpretation two approaches have been used: library searching and training. Library searches of large collections (over 25,000 spectra) by comparison of the spectrum with known spectra yield degrees of closeness of agreement of the unknown and known spectra. Various algorithms for spectral comparison, using the 10 most intense peaks in the spectrum or the two most intense peaks in each 14-mass-unit segment, for example, have been devised, and the minimum amount of information to be supplied for a good chance of identification has been studied. The other approach involves several pattern recognition approaches in which various features of the spectra are correlated with structural characteristics by methods independent of formal theories of mass spectral interpretation; these include learning-machine and factor-analysis approaches. Hybrid techniques, in which the self-trained computer approach is augmented by selected tests derived from the fragmentation theory noted previously, have also been devised.

Analysis of mixtures. For a given instrument, the mass spectrum of a compound serves as a fingerprint of a compound under standard conditions; it varies with temperature, ionizing voltage, and parameters of the construction of the instrument. A mixture of gases may be analyzed because spectra are additive; hydrocarbon mixtures have been so analyzed quantitatively since the 1940s.

A further powerful technique for analysis of mixtures containing unknowns is mass spectral analysis of gas chromatographic effluents either by direct hookup of the gases exiting from the chromatograph, including carrier gas, to the introduction system of the spectrometer or by analysis of trapped effluent fractions.

For electron-impact or charge-exchange ionization of the sample, it is necessary to remove the carrier gas preferentially from the sample; several devices based upon preferential diffusion or effusion have been developed to achieve this. Gas chromatography using methane as carrier can be used directly as a method of introducing reagent gas and sample simultaneously for chemical ionization. Increased sensitivity can be achieved by monitoring the intensities of only a few peaks in the mass spectrum as the chromatograph effluents pass through the source, using rapid voltage switching between these peaks, instead of scanning the whole spectrum repeatedly; this technique is called multiple ion detection. *See* GAS CHROMATOGRAPHY.

Similarly, mass spectrometers may be coupled to liquid chromatographs; the solvent may be removed by evaporation of the solution on a moving belt which next passes into the ion source and is then heated to evaporate the sample, or by freezing solvent on liquid-nitrogen-cooled fingers surrounding the source; or the solvent may be used itself as a reagent gas for chemical ionization. *See* LIQUID CHROMATOGRAPHY.

Ion cyclotron resonance spectroscopy, previously used to study reactions between ions and molecules (and thus the thermodynamics and kinetics of reactions in the absence of solvent), has become of special importance here because of the development of Fourier-transform ion cyclotron resonance (FT/ICR) techniques; extremely high resolution (>500,000) and very rapid acquisition of spectra (millisecond range) suggest FT/ICR applications to wall-coated open tubular column capillary gas chromatography, in which 1-s-wide effluent peaks present problems for other scanning mass spectrometric detection methods.

Collision-induced dissociation forms part of a new technique in which an unseparated mixture, for example, a biological sample, is admitted to the source, perhaps by direct probe. Under chemical ionization conditions, (M + 1) ions of each component are produced. Each of these ions is transmitted in its turn through the magnetic sector, and a collision-induced spectrum of the component is obtained. Mass spectra are thus obtained without prior separation.

Analysis of inorganic solids. The analysis of solid inorganic samples can be made either by vaporization in a Knudsen cell arrangement at very high temperatures or by volatilization of the sample surface so that particles are atomized and ionized with a high-energy spark (for example, 20,000 eV). The wide range of energies given to the particles requires a double-focusing mass spectrometer for analysis. Detection in such instruments is by photgraphic plate; exposures for different lengths of time are recorded sequentially, and the darkening of the lines on the plate is related empirically to quantitative composition by calibration charts. The method is useful for trace analysis (parts per billion) with accuracy ranging from 10% at higher concentration levels to 50% at trace levels. Methods for improving accuracy, including interruption and sampling of the ion beam, are being studied.

Secondary ion mass spectrometry is most commonly used for surface analysis. A primary beam of ions accelerated through a few kilovolts is focused on a surface; ions are among the products sputtered from this surface, and they may be directly analyzed in a quadrupole filter. *See* SECONDARY ION-MASS SPECTROMETRY (SIMS); SPECTROSCOPY; TRACE ANALYSIS. [MAURICE M. BURSEY]

Bibliography: R. G. Cooks (ed.), *Collision Spectroscopy*, 1978; K. Levsen, *Fundamental Aspects of Organic Mass Spectrometry*, 1978; F. W. McLafferty, *Interpretation of Mass Spectra*, 3d ed., 1980; G. R. Waller, *Biochemical Applications of Mass Spectrometry*, 1972, supplement, 1980.

Mass-transfer operation

An operation in chemical engineering that involves transfer of material from one phase to another, or from one place to another within a single phase. Mass transfer may occur for various purposes, for example, to effect a chemical reaction, to obtain a separation of components, or to obtain uniform distribution of material within a phase. The most common types are (1) fluid-fluid transfer, occurring typically in absorption, distillation, and liquid-liquid extraction, and (2) fluid-solid transfer, occurring typically in crystallization and gas adsorption. The subject of mass transfer deals generally with the factors which determine rate of interphase material transfer and with the influence of transfer rate on equipment design and performance. Processes involving mass transfer are often designated as diffusional operations. *See* ADSORPTION; COUNTERCURRENT TRANSFER OPERATIONS; CRYSTALLIZATION; DIALYSIS; DISTILLATION; DRYING; GAS ABSORPTION OPERATIONS; ION EXCHANGE; LEACHING; SOLVENT EXTRACTION; SUBLIMATION.

[CHARLES R. WILKE]

Bibliography: R. E. Treybal, *Mass-transfer Operations*, 3d ed., 1979.

Matrix isolation

A technique for providing a means of maintaining molecules at low temperature for spectroscopic study. This method is particularly well suited for preserving reactive species in a solid, inert environment. Elusive molecular fragments, such as free radicals that may be postulated as important controlling intermediates for chemical transformations used in industrial reactions, high-temperature molecules that are in equilibrium with solids at very high temperatures, and molecular ions that are produced in plasma discharges or by high-energy radiation all can be examined by using absorption (infrared, visible, and ultraviolet), electron-spin resonance, and laser-excitation spectroscopes.

Experimental apparatus. The experimental apparatus for matrix isolation experiments is designed with the method of generating the molecular transient and performing the spectroscopy in mind. Figure 1 shows the cross section of a vacuum vessel used for absorption spectroscopic measurements. The optical windows must be transparent to the examining radiation. The rotatable cold window is cooled to 10–20 K by using closed-cycle refrigeration or 4 K by using liquid helium. The matrix sample is introduced through the spray-on line at rates of 1–5 millimoles per hour; argon is the most widely used matrix gas, although neon,

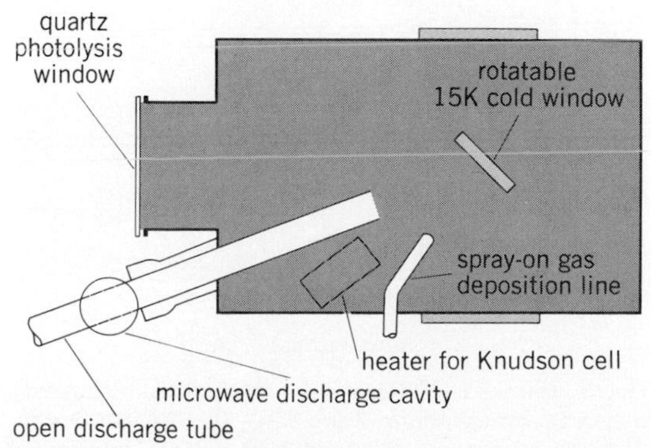

Fig. 1. Cross section of the base of a vacuum vessel that is used for matrix photoionization experiments.

krypton, xenon, and nitrogen are also used. The reactive species can be generated in a number of ways: mercury-arc photolysis of a trapped precursor molecule through the quartz window, evaporation from a Knudsen cell in the heater, chemical reaction of atoms evaporated from the Knudsen cell with molecules deposited through the spray-on line, and vacuum-ultraviolet photolysis of molecules deposited from the spray-on line by radiation from discharge-excited atoms flowing through the tube. For laser excitation studies, the sample is deposited on a tilted copper wedge which is grazed by the laser beam, and light emitted or scattered at approximately 90° is examined by a spectrograph. In electron-spin resonance studies, the sample is condensed on a sapphire rod that can be lowered into the necessary waveguide and magnet. The chemical versatility of the matrix technique can perhaps be best described by considering a number of important examples.

Applications. The first free radical stabilized in sufficient concentration for matrix infrared detection was formyl HCO. Hydrogen iodide was deposited in a CO matrix and photolyzed with a mercury arc; hydrogen atoms produced by dissociation of HI reacted with CO in the cold solid to produce HCO. The infrared spectrum of HCO provided vibrational fundamentals and information about the chemical bonding in this reactive species.

The first molecular ionic species characterized in matrices, $Li^+O_2^-$, was formed by the cocondensation reaction of lithium atoms and oxygen molecules at high dilution in argon. The infrared spectrum exhibited a weak $(O \leftrightarrow O)^-$ stretching vibration and two strong $Li^+ \leftrightarrow O_2^-$ stretching vibrations, as shown at the top of Fig. 2 for the ^7Li and $^{16,18}O_2$ isotopic reaction. The 1-2-1- relative-intensity oxygen isotopic triplets in this experiment showed that the oxygen atomic positions in the molecule are equivalent and indicated an isosceles triangular structure. The ionic model for the bonding in $Li^+O_2^-$ was confirmed by contrasting intensities between the laser-Raman and infrared-absorption spectra shown in Fig. 2.

High-temperature molecules like LiF can be trapped in matrices by evaporating the molecule from the crystalline solid in a Knudsen cell at high temperature or by reacting lithium atoms with fluorine molecules during condensation. The latter method has been used to synthesize the CaO molecule from the calcium atom–ozone reaction for infrared observation of the ground electronic state.

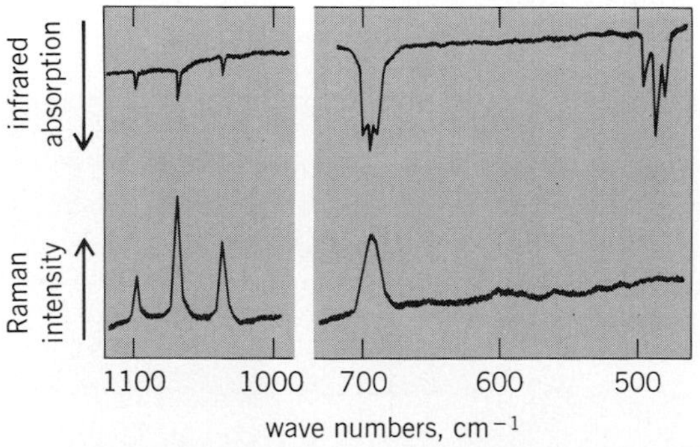

Fig. 2. Infrared and Raman spectra of lithium superoxide, $Li^+O_2^-$, using lithium-7 and 30% $^{18}O_2$, 50% $^{16}O^{18}O$, and 20% $^{16}O_2$ at concentrations of $Ar/O_2 = 100$. The Raman spectrum was recorded using 200 mW of 488-nm excitation and a long-wavelength-pass dielectric filter in the 1000-cm^{-1} region.

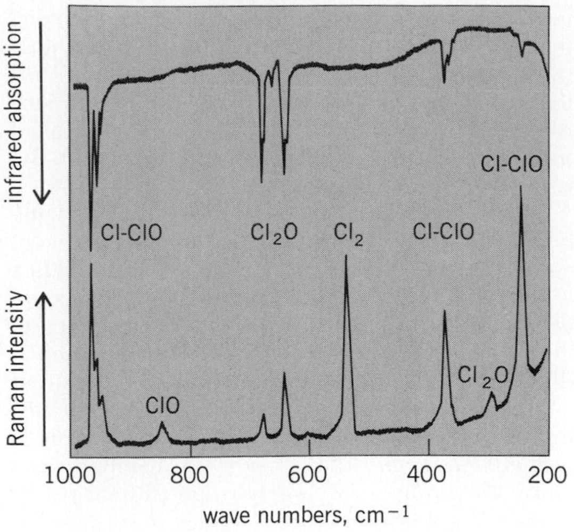

Fig. 3. Infrared and Raman spectra of Cl_2O and its photolysis products. $Ar/Cl_2O = 100$. The infrared spectrum was recorded after 10 min of ultraviolet photolysis. The Raman spectrum was recorded with 300 mW of 488-nm excitation.

A large number of free radicals have been synthesized and trapped using the lithium atom abstraction reaction. For example, the $Li + CCl_4$ reaction produced CCl_3 and LiCl, whereas $Li + CH_2Cl_2$ gave CH_2Cl and LiCl. The C-Cl stretching fundamentals in these chlorocarbon radicals are suggestive of pi bonding between chlorine and the half-filled carbon orbital.

The radical ion Cl_2^- has been formed as the $M^+Cl_2^-$ ion pair and examined by resonance-Raman and optical-absorption techniques. The matrix moderates decomposition of the $M^+Cl_2^-$ species so that laser excitation in the wing of an absorption band can give resonance-enhanced Raman spectra.

Interest in chlorine-oxygen chemistry has been renewed by the possible role of chlorine atom reactions with ozone in the upper atmosphere to give the chlorine oxide free radical. This important free radical was identified in the Raman spectrum of Cl_2O in solid argon following 488-nm laser photolysis, as illustrated in the lower spectrum in Fig. 3. Also shown is an interesting isomerization process initiated by the laser, the asymmetric Cl-Cl-O species, formed upon rearrangement of the symmetric Cl_2O precursor in the solid argon cage. The infrared spectrum of argon/dichlorine monoxide samples after mercury-arc photolysis, shown at the top of Fig. 3, reveals the same band positions for the Cl-Cl-O photoisomer with different relative intensities; the weak infrared–absorbing ClO free radical was not detected in the infrared spectrum.

The important isolated molecular ion C_2^- was produced by vacuum ultraviolet photolysis of C_2H_2 for observation of its visible absorption spectrum at 520.7 and 472.6 nm. The C_2 molecular fragment, produced by photolysis of acetylene, captured an electron from the photoionization of C_2H_2 to give C_2^-. Laser excitations of these samples gives rise to a very intense fluorescence spectrum of this stable molecular anion.

Continuous exposure of a condensing sample to argon-resonance radiation during sample condensation has been used to produce molecular cations for spectroscopic study. In the case of CHF_3, photolysis produced the CF_3 radical which may be photoionized by a second 11.6-eV photon to give CF_3^+. The infrared spectrum of CF_3^+ revealed a very high C-F vibrational fundamental which indicates substantial pi bonding in this planar carbocation. Similar studies with CCl_4 produced a very strong blue absorption band which has been identified as arising from CCl_4^+, the parent molecular ion. The blue absorption was destroyed by visible-light photolysis, showing that the carrier is an extremely unstable species. Matrix photoionization experiments with CF_3Cl yielded the daughter cations CF_2Cl^+ and CF_3^+ and the parent cation CF_3Cl^+. The latter cation photolyzed readily with near-ultraviolet radiation, which helped identify the parent cation.

The matrix isolation technique enables spectroscopic data to be obtained for reactive molecular fragments, many of which cannot be studied in the gas phase. *See* INFRARED SPECTROSCOPY; PHOTOLYSIS; RAMAN EFFECT; REACTIVE INTERMEDIATES.

[LESTER ANDREWS]

Bibliography: L. Andrews, Laser excitation matrix isolation spectroscopy, *Appl. Spectrosc. Rev.*, 11:125–161, 1976; L. Andrews, Spectroscopy of molecular ions in noble gas matrices, *Annu. Rev. Phys. Chem.*, 30:79–101, 1979; H. E. Hallam, *Vibrational Spectroscopy of Trapped Species*, 1973; M. E. Jacox, The stabilization and spectra of free radicals and molecular ions in rare gas matrices, *Rev. Chem. Intermed.*, 2:1–36, 1978.

Melting point

The temperature at which a solid changes to a liquid. For pure substances, the melting or fusion process occurs at a single temperature, the temperature rise with addition of heat being arrested until melting is complete. The direct transition from solid phase to gas phase is not properly called melting but, preferably, sublimation.

Melting points reported in the literature, unless specifically stated otherwise, have been measured under an applied pressure of 1 atm, usually 1 atm of air. (The solubility of air in the liquid is a complicating factor in precision measurements.) The Clapeyron equation for the pressure dependence of the absolute melting temperature T_m is the equation below, where ΔV_f is the change in vol-

$$\frac{dT_m}{dP}=\frac{T_m \Delta V_f}{\Delta H_f}$$

ume, ΔH_f the heat absorbed during the fusion process, and P the applied pressure. Upon melting, all substances absorb heat, and most substances expand; consequently an increase in pressure normally raises the melting point. A few substances, of which water is the most notable example, contract upon melting; thus, the application of pressure to ice at 0°C causes it to melt. Large changes in pressure are required to produce significant shifts in the melting point; a pressure of 10 atm lowers the melting point of ice by only 0.075°C.

A sufficient decrease in temperature at ordinary pressures causes all pure substances except helium to freeze to solids; the lowest normal melting point is that of hydrogen, at 14 K, and one of the highest is that of rhenium, at 3700 K. Liquid helium can be transformed into a solid only by applying a pressure in excess of 25 atm.

For solutions of two or more components, the melting process normally occurs over a range of temperatures, and a distinction is made between the melting point, the temperature at which the first trace of liquid appears, and the freezing point, the higher temperature at which the last trace of solid disappears, or equivalently, if one is cooling rather than heating, the temperature at which the first trace of solid appears. Measurement of the freezing point of a solution and the difference between it and the freezing point of the pure solvent provides a convenient method of determining the molecular weight of a dissolved solute, because the freezing point of a solution is lower than that of a pure solvent. *See* PHASE EQUILIBRIUM; SOLUTION; SUBLIMATION; TRIPLE POINT.

[ROBERT L. SCOTT]

Mendelevium

A chemical element, Md, atomic number 101, the twelfth member of the actinide series of elements. Mendelevium does not occur in nature; it was discovered and is prepared by artificial nuclear transmutation of a lighter element.

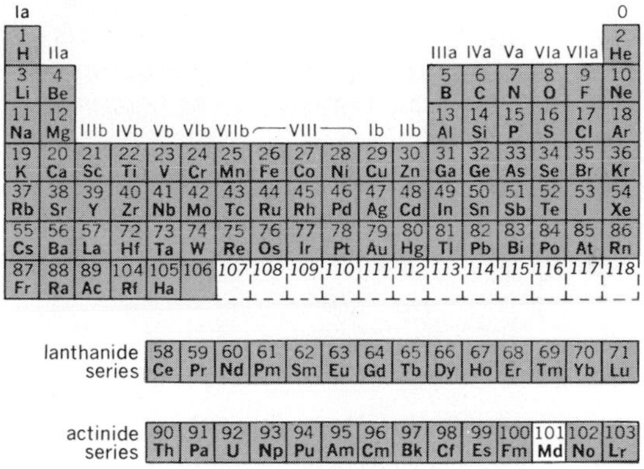

In 1955 A. Ghiorso, B. G. Harvey, G. R. Choppin, S. G. Thompson, and G. T. Seaborg performed a series of classic experiments which constituted the basis for the discovery of all known elements beyond Md. In each experiment only a few atoms of Md were collected by a recoil technique and identified. The discoverers bombarded a small amount of ^{253}Es with helium ions accelerated in a cyclotron. Einsteinium and helium nuclei combined, and a neutron was emitted to give the mass 256 isotope of element 101. These experiments resulted in the discovery of the element which was named in honor of Dmitri Mendeleev, the founder of the periodic table of the elements.

Known isotopes of mendelevium have mass numbers from 248 to 258 and with half-lives from a few seconds to about 55 days. They are all produced by charged-particle bombardments of more abundant isotopes; for example, $^{255}Es + {}^4He \rightarrow {}^{258}Md + n$. The yields in such reactions are very small because the bombardments produce fission products in preference to mendelevium isotopes. Thus the amounts of mendelevium which are produced and used for studies of chemical and nuclear properties are usually less than about a million atoms; this is of the order of a million times less than a weighable amount.

Studies of the chemical properties of mendelevium have been limited to a tracer scale. The behavior of mendelevium in ion-exchange chromatography shows that it exists in aqueous solution primarily in the 3+ oxidation state characteristic of the actinide elements. However, it also has a dipositive (2+) and a monopositive (1+) oxidation state. The oxidation potential $Md^{2+} = Md^{3+} + e^-$ is about 0.2 volt (where the potential $1/2\ H_2 = H^+ + e^-$ is taken as 0.0 volt). In its 3+ state mendelevium exhibits the behavior of the typical tripositive actinide elements, especially the neighboring ones. It is desorbed from cation-exchange resin at a rate different from that of other actinides when a complexing agent is passed through the column of resin. This provides a good basis for the isolation and identification of mendelevium in the presence of other actinide elements. *See* ACTINIDE ELEMENTS; BERKELIUM; CALIFORNIUM; EINSTEINIUM; TRANSURANIUM ELEMENTS. [GLENN T. SEABORG]

Bibliography: *Gmelin Handbuch der anorganischen Chemie: Transurane-Transuranium Elements*, new supplement: vol. 20d, pt. 2, 8th ed., 1975; E. K. Hyde et al., *The Nuclear Properties of the Heavy Elements*, 3 vols., rev. ed., 1971; C. Keller, *The Chemistry of the Transuranium Elements*, 1971; G. T. Seaborg, History of the synthetic actinide elements, *Actinides Rev.*, 1:3–38, 1967; G. T. Seaborg, *Man-Made Transuranium Elements*, 1963; G. T. Seaborg, *The Transuranium Elements*, 1958.

Menthol

A monocyclic, saturated, secondary terpene alcohol with the formula below. Menthol contains three

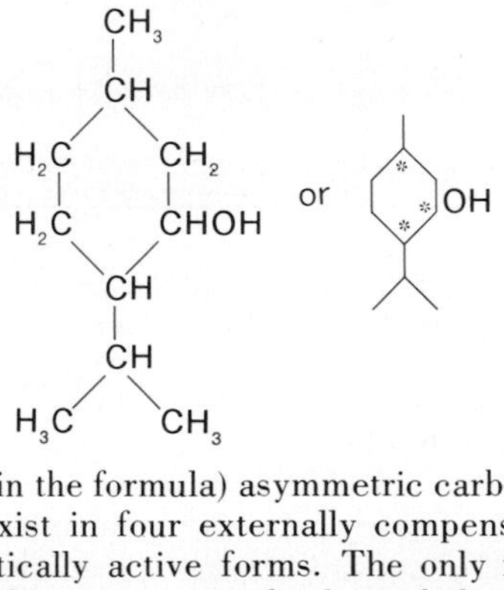

(starred in the formula) asymmetric carbon atoms. It can exist in four externally compensated and eight optically active forms. The only forms encountered in nature are the *l*-menthol and *d*-neomenthol. By far the most important is *l*-menthol, the main constituent (50–65%) in pepperment oil.

Commercial *l*-menthol is isolated principally from the oil of *Mentha arvensis* grown in Japan. The process involves cooling the oil and purifying the crystals formed. It possesses a distinct peppermint flavor and gives the impression of cooling the mouth and skin. Appreciable amounts of the racemic mixture of the optical isomers of menthol are synthesized commercially.

l-Menthol is widely used as a flavoring ingredient in toothpastes, mouthwashes, and ciagarettes, as a rubefacient, and as a cooling agent in chest rubs. *See* TERPENE. [WILLIAM MOSHER]

Bibliography: E. Guenther et al., *The Essential Oils*, 6 vols., 1948–1952, reprint 1972–1976; J. B. Hendrickson, *The Molecules of Nature: A Survey of the Biosynthesis and Chemistry of Natural Products*, 1965.

Mercaptan

One of a group of organosulfur compounds which are also called thiols or thio alcohols and which have the general structure RSH. Aromatic thiols are called thiophenols, and biochemists often refer to thiols as sulfhydryl compounds. The unpleasant odor of volatile thiols causes them to be classed as stenches, but the odors of many solid thiols are not unpleasant.

Mercaptans (1) form salts with bases, (2) are easily oxidized to disulfides and higher oxidation products such as sulfonic acids, (3) react with chlorine (or bromine) to form sulfenyl chlorides (or bromides), and (4) undergo additions to unsaturated compounds, such as olefins, acetylenes, aldehydes, and ketones. The insoluble mercury salts (mercaptides) are used to isolate and identify mercaptans.

The important amino acid cysteine and many pharmaceutical and industrial products contain the mercapto group. Thus, thiosalicylic acid is used in synthesizing the germicide merthiolate; 2-mercaptobenzothiazole is a rubber vulcanization accelerator; 2,3-dimercaptopropanol (British antilewisite) is an antidote for arsenic poisoning; and 6-mercaptopurine is of interest in cancer chemotherapy. The occurrence and removal of mercaptans from petroleum is industrially important. Even the odor of mercaptans finds use; traces of mercaptans added to dangerous gases act as warning agents in case of leaks.

The name thio alcohols suggests that mercaptans are similar to alcohols. Although some properties are analogous, there are decided differences, related to (1) the greater acidities of thiols, (2) the ease with which mercaptans are oxidized, and (3) the ability of mercaptans to enter free-radical reactions. The second difference probably accounts for the absence of thiols, as such, in contrast to alcohols, in nature. *See* ORGANOSULFUR COMPOUND; SULFENYL CHLORIDES.

[NORMAN KHARASCH]

Bibliography: H. Gilman (ed.), *Organic Chemistry*, vols. 1 and 2, 2d ed., 1943, and vols. 3 and 4, 1953; W. A. Pryor, *Mechanisms of Sulfur Reactions*, 1962.

Mercury

A chemical element, Hg, atomic number 80 and atomic weight 200.59. It has been known since ancient times. Mercury is a silver-white liquid at room temperature (melting point −38.89°C); it boils at 357.25°C under atmospheric pressure. It is a noble metal that is soluble only in oxidizing solutions. Solid mercury is as soft as lead. The metal and its compounds are very toxic. With some metals (gold, silver, platinum, uranium, copper, lead, sodium, and potassium, for example) mercury forms solutions called amalgams. With other elements, such as iron, cobalt, nickel, manganese, and silicon, it does not react.

In its compounds, mercury is found in the 2+, 1+, and lower oxidation states, for example, $HgCl_2$, Hg_2Cl_2, or $Hg_3(AsF_6)_2$. Often the mercury atoms

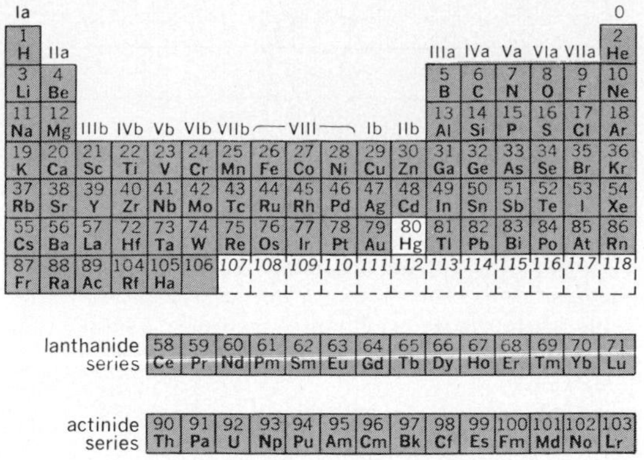

are doubly covalently bonded, for example, Cl—Hg—Cl or Cl—Hg—Hg—Cl. Some mercury (II) salts, for example, $Hg(NO_3)_2$ or $Hg(ClO_4)_2$, are quite soluble in water and dissociate normally. The aqueous solutions of these salts react as strong acids because of hydrolysis. Other mercury (II) salts, for example, $HgCl_2$ of $Hg(CN)_2$, also dissolve in water, but exist in solution as only slightly dissociated molecules. This is related to the tendency of mercury to form covalent compounds. There are compounds in which mercury atoms are bound directly to carbon or nitrogen atoms, for example, H_3C—Hg—CH_3 or H_3C—CO—NH—Hg—NH—CO—CH_3. In complex compounds, for example, $K_2(HgI_4)$, mercury often has three or four bonds.

Uses. Metallic mercury is used as a liquid contact material for electrical switches, in vacuum technology as the working fluid of diffusion pumps, for the manufacture of mercury-vapor rectifiers, thermometers, barometers, tachometers, and thermostats, and for the manufacture of mercury-vapor lamps. Mercury-vapor lamps serve as sources for ultraviolet light because they emit a line spectrum with the following principal lines: 365.0/366.3 mμ, 404.7 mμ, 435.8 mμ, 546.1 mμ, and 577.0/579.0 mμ. For micro gas analyses, mercury is most often used as a sealing liquid for the evolved gases. Very large amounts of mercury are used as the electrode material for the electrolysis of aqueous solutions of alkali halides for the manufacture of chlorine and sodium hydoxide. Also, it finds application for the manufacture of silver amalgams for tooth fillings in dentistry.

Some mercury salts serve as catalysts for organic chemical reactions. Fulminte of mercury, $Hg(CNO)_2$, is used as a primer for explosives. Some complex salts, $Cu_2(HgI_4)$ and $Ag_2(HgI_4)$, find application as temperature indicators, for example, for detecting overheated machine parts, since they change color on heating. A large number of the mercury compounds have been used for hundreds of years as medicines. The compounds of mercury with some organic substances are powerful diuretic substances. $HgCl_2$ or $HgO \cdot Hg(CN)_2$ finds application as a disinfectant in dilutions of 1:1000 in water. The so-called yellow mercury salve contains 5% HgO and is used to treat conjunctivitis. Ammoniated mercury ointment contains 5–10% $HgNH_2Cl$ and is used in the treatment of various skin diseases.

Of importance in electrochemistry are the standard calomel electrode, used as the reference electrode for the measurement of potentials and for potentiometric titrations, and the Weston standard cell, with which the accurate measurement of potential sources is possible. In the calomel electrode and in the Weston standard cell, metallic mercury is used in contact with solutions which contain mercury (I) salts as the solid phase. *See* CALOMEL ELECTRODE.

Natural occurrence. Mercury is commonly found as the sulfide, HgS, frequently as the red cinnabar and less often as the black metacinnabar. Important deposits are found in Spain, Italy, California, Nevada, Oregon, Texas, Mexico, Canada, Brazil, Peru, China, Japan, the Soviet Union, Hungary, Yugoslavia, and Germany. A less common ore is the mercury(I) chloride found in Texas. Occasionally the mercury ore contains small drops of metallic mercury.

The reserve of mercury at Almadén, Spain, has been estimated at 40,000 tons; that in Idria, Italy, at about 20,000 tons. The ore found in Spain has the highest mercury content, with an average of 0.5–1.2% Hg, sometimes as much as 10% Hg.

It has been estimated that the outer 16-km thick layer of the Earth's crust contains 5×10^{-5}% Hg. Mercury is less abundant than platinum, uranium, tantalum, cesium, silver, osmium, palladium, and indium. Nevertheless, it is not regarded as a rare element because it is found in highly concentrated deposits and hence is readily available.

Metallurgical extraction. The separation of mercury metal from ores is accomplished most often by heating in an airstream in a rotary kiln or shaft furnace. In this process the sulfide ore is roasted, reaction (1), or reduced to metallic mercury by the addition of iron, reaction (2), or of quicklime, reaction (3).

$$HgS + O_2 \rightarrow Hg + SO_2 \quad (1)$$

$$HgS + Fe \rightarrow Hg + FeS \quad (2)$$

$$4HgS + 4CaO \rightarrow 4Hg + 3CaS + CaSO_4 \quad (3)$$

The metal vapor is carried over with the combustion gases and condenses in vertical clay pipes having open bottoms and standing in water. The mercury metal collects under the water in a layer of condensation products called mercurial soot, which also contains mercury salts, soot, and tar, and from which the metal is separated by filtration through a cloth. The mercurial soot is pressed with quicklime, whereby additional liquid mercury is obtained. The residue from the compression is roasted with new mercury ore. Final purification of the mercury metal is best done by vacuum distillation, by dropping it into a 1.5-m layer of 5% nitric acid or by pressing it through leather.

The mercury metal is sold in iron flasks 45 cm long and 10 cm thick. Each flask contains 76 lb (34.5 kg) of metal (34.5 kg is the old Spanish hundredweight).

Principal compounds. A number of important compounds of mercury and their properties and major uses are given in the table. *See* TRANSITION ELEMENTS.

Physical and chemical properties. The element mercury has atomic number 80 and atomic weight 200.59. It occurs naturally as a mixture of the following isotopes in the proportions shown: 202 (29.8%), 200 (23.1%), 199 (16.9%), 201 (13.2%), 198 (10.0%), 204 (6.8%), and 196 (0.15%).

The thermal expansion of mercury metal is rela-

Principal compounds of mercury

Compound and formula	Properties and uses
Mercury(II) acetamide, $Hg(NHCOCH_3)_2$	White crystals; soluble in water and alcohol; mp 198°C
Mercury(II) acetate, $Hg(CH_3COO)_2$	White powder; soluble in alcohol and dilute acids; hydrolyzes in boiling water, yielding HgO
Ammonium tetrachloromercurate(II) dihydrate, $(NH_4)_2(HgCl_4)\cdot 2H_2O$	White crystalline powder; soluble in water; slightly soluble in alcohol
Mercury(II) diammine chloride, $Hg(NH_3)_2Cl_2$	Fusible white crystalline precipitate; soluble only in hydrochloric acid with decomposition
Mercury(II) chloride, ammoniated, $HgNH_2Cl$	Infusible precipitate; white powdery lumps or powder; soluble in hydrochloric acid only; earthy metallic taste; consists of chains $(\text{—Hg—NH}_2\text{—})_n^{n+}n\ Cl^-$
Mercury(II) arsenate, $HgHAsO_4$	Yellow powder; insoluble in water; soluble in hydrochloric acid
Barium tetrabromomercurate(II), $Ba(HgBr_4)$	Colorless crystalline powder; very hygroscopic; soluble in water
Mercury(II) benzoate, $Hg(C_6H_5COO)_2\cdot H_2O$	White crystals, slightly soluble in water and alcohol; mp 165°C
Mercury(II) bromide, $HgBr_2$	White rhombic crystals; soluble in alcohol, ether, acetone, and ethyl acetate; sparingly soluble in water; mp 238°C; bp 319°C; specific electrical conductivity at 242°C = 1.45×10^{-4} ohm^{-1} cm^{-1}; vapor pressure at 200°C, 24.1 mm Hg, at 260°C, 200 mm Hg; very high cryoscopic constant (similar to camphor) of 36.7°C for 1 *M* in 1000 g of $HgBr_2$
Mercury(II) chloride, $HgCl_2$	White crystals or powder; soluble in water, alcohol, ether, pyridine, and methylacetate; mp 265°C; bp 303°C; solid substance contains $HgCl_2$ molecules, as do solutions of $HgCl_2$ in water or organic solvents; very strong poison (0.2–0.4 g is deadly; breathing of dust should be avoided)
Copper(I) tetraiodomercurate(II), $Cu_2(HgI_4)$	Dark-red crystalline powder; insoluble in water and alcohol; changes its color from red to black at 71°C, and assumes red color again if cooled to room temperature
Mercury(II) cyanide, $Hg(CN)_2$	Colorless crystalline prisms, darkened by light; soluble in water and alcohol; decomposes on heating into metallic mercury and cyanogen; solution in water contains practically no free ions
Mercury(II) fluoride, HgF_2	Transparent crystals; moderately soluble in water and alcohol; hydrolyzes in water solutions; decomposes at mp 645°C
Mercury(II) iodate, $Hg(IO_3)_2$	Amorphous white powder; soluble in hydrochloric, hydrobromic, and hydroiodic acids, and in water containing sodium chloride or potassium iodide; insoluble in water and alcohol
Mercury(II) iodide, HgI_2	Red crystals, tetragonal, turn to yellow rhombic crystals when heated above 127°C, returning to red modification on cooling; insoluble in water; soluble in ether and benzene; yellow form soluble in alcohol; both modifications soluble in aqueous solutions of sodium thiosulfate or potassium iodide; mp 421°C; bp 349°C
Mercury(II) lactate, $Hg(C_3H_5O_3)_2$	White crystalline powder; decomposes on heating; soluble in water
Mercury(II) nitrate, $Hg(NO_3)_2\cdot H_2O$	Mp 79°C with decomposition; colorless crystals or, if slightly hydrolyzed by water, yellowish powder; soluble in nitric acid; insoluble in alcohol; powerful vesicant
Mercury(II) oxide, HgO	Decomposes above 400°C into mercury and oxygen; soluble in acids; insoluble in alcohol and ether; exists in finely divided form which is more reactive than the crystallized red form; both forms have the same crystal structure with chains $(\text{—Hg—O—})_n$
Mercury(II) oxidcyanide, NC—Hg—O—Hg—CN	Explodes on heating; white crystalline powder; slightly soluble in water; used as antiseptic solution (1 : 5000)
Mercury(II)-chlorate(VII), $Hg(ClO_4)_2\cdot 6H_2O$	Colorless crystals; hygroscopic; soluble in acid solutions
Mercury(II) phosphate, $Hg_3(PO_4)_2$	Heavy white or yellowish powder; soluble only in acids, not soluble in water
Potassium tetracyanomercurate(II), $K_2[Hg(CN)_4]$	Colorless crystals; very poisonous; soluble in water and alcohol
Potassium tetraiodomercurate(II), $K_2(HgI_4)$	Nessler's reagent; soluble in water; crystallizes water-free in yellow crystals; hydrates with 1, 2, or 3 moles of water are known
Mercury(II) salicylate, $Hg(C_6H_4O)COO$	White powder; soluble in warm alkali halide solutions
Silver(I) tetraiodomercurate(II), $Ag_2(HgI_4)$	Yellow powder; becomes red at 40–50°C; insoluble in water and dilute acids; soluble in aqueous solutions of potassium cyanide and potassium iodide
Mercury(II) subsulfate, $2HgO\cdot HgSO_4$	Heavy lemon-yellow powder; insoluble in water; soluble in acids
Mercury(II) sulfate, $HgSO_4$	White crystalline powder; hydrolyzes with water; insoluble in alcohol; soluble in acids
Mercury(II) sulfide, black, HgS	Black powder; insoluble in water, acids, and organic solvents; soluble in sodium sulfide solution; sublimes at 446°C, yielding the red form of HgS, which possesses another crystal structure; the black sulfide is formed by passing hydrogen sulfide gas into mercury salt solutions or by reaction between metallic mercury and sulfur

(*continued*)

Principal compounds of mercury (cont.)

Compound and formula	Properties and uses
Mercury(II) sulfide, red, HgS	Cinnabar; red crystals; insoluble in water, acids, and organic solvents; solvents; sublimes at 446°C; red form of HgS is the main mercury ore
Mercury(II) thiocyanate, $Hg(SCN)_2$	White powder; only slightly soluble in water; soluble in alcohol; decomposes on heating
Mercurochrome, $C_{20}H_7O_5Br_2Na_2HgOH \cdot 3H_2O$	Disodium 2,7-dibromo-4-hydroxymercurifluorescein; green scales; soluble in water; insoluble in alcohol, ether, and chloroform; used in 2% aqueous solution as a general antiseptic
Mercury(I) acetate, $Hg_2(CH_3COO)_2$	Colorless plates; decomposes by light or in boiling water into Hg and $Hg(CH_3COO)_2$; slightly soluble in water; insoluble in alcohol and ether
Mercury(I) bromide, Hg_2Br_2	White powder or tetragonal crystals; becomes yellow on heating and white again on cooling; sublimes at 340–350°C; mp 405°C
Mercury(I) chlorate, $Hg_2(ClO_3)_2$	White crystals; explodes with combustible substances; mp 250°C (decomposition); soluble in water and alcohol
Mercury(I) chloride, Hg_2Cl_2	Calomel; white rhombic crystals; insoluble in water, alcohol, and ether; mp 303°C; bp 384°C; becomes black with ammonia ($Hg_2Cl_2 + 2NH_3 \rightarrow Hg + HgNH_2Cl + NH_4Cl$) by yielding finely divided metallic mercury
Mercury(I) chromate, Hg_2CrO_4	Brick-red powder decomposes on heating to yield Cr_2O_3; insoluble in water and alcohol; soluble in concentrated nitric acid
Mercury(I) iodide, Hg_2I_2	Yellow amorphous powder; becomes greenish in light ($Hg_2I_2 \rightarrow Hg + HgI_2$); orange on heating, but yellow again on cooling; insoluble in water, alcohol, and ether; sublimes at 110–120°C; mp 290°C (partial decomposition); bp 310°C
Mercury(I) nitrate, $Hg_2(NO_3)_2 \cdot 2H_2O$	Short prismatic crystals; soluble in small quantities of warm water, in nitric acid, in boiling carbon disulfide, and in methylamine, and slightly soluble in benzonitrile; hydrolyzes with large quantities of water; mp 70°C with decomposition
Mercury(I) sulfate, Hg_2SO_4	White crystalline powder; almost insoluble in water; soluble in hot sulfuric acid and in dilute nitric acid; decomposes on heating
Mercury dimethyl, $Hg(CH_3)_2$	Colorless volatile liquid; very strong poison; bp 92°C; density at 19°C is 3.084 g/cm³
Mercury diphenyl, $Hg(C_6H_5)_2$	Colorless crystals; mp 122°C; decomposes on heating into mercury and diphenyl
Mercury fulminate, $Hg(CNO)_2$	Dark-brown crystalline powder; explodes when dry under the slightest shock and by heating
Mercury methyl chloride, CH_3HgCl	Colorless plates; mp 170°C; very strong poison
Millon's base, $(Hg_2N)OH \cdot 2H_2O$	Yellow powder; insoluble in water and organic solvents; exchanges OH^- ions with halide ions, when in contact with aqueous alkali halide solutions; consists of NHg_4-tetraeders with positive charge at the N atoms (the NHg_4-tetraeders are linked together like the SiO_4-tetraeders in SiO_2, forming a three-dimensional network) and OH^- ions; the H_2O molecules are in holes of the network
Mercury thiochloride, $Hg_3S_2Cl_2$	Yellow powder; insoluble in water
Mercury oxide chloride, $[O(HgCl)_3]Cl$	Colorless crystals; insoluble in cold water; exchanges 1 Cl^- ion with F^- ions
Mercury(II) oxalate, HgC_2O_4	White powder; insoluble in water; used for Eder's solution (decomposition with light)

tively large, and between 0 and 100°C the thermal expansion approximates that of gases. The specific gravity of metallic mercury is 13.54616 at 20°C. The variation of specific gravity with temperature is shown in Fig. 1.

The vapor pressure of mercury rapidly increases with temperature (Fig. 2). In 1 m³ of air there are, accordingly, 14 mg of Hg at 20°C and 2.42 g at 100°C, if the air is saturated with mercury vapor.

The thermal conductivity of mercury at 0°C is only 2.2% of that of silver; the electrical conductivity at 0°C is only 1.58% of that of silver. By international definition, a column of mercury 1 mm² in cross-sectional area and 106.3 cm long at 0°C has an electrical resistance of 1 ohm.

The surface tension of liquid mercury is 484 dynes/cm, six times greater than that of water in contact with air. Hence, mercury does not wet surfaces with which it is in contact.

Mercury metal and mercury compounds are all diamagnetic because of the electron configuration $5d^{10}$.

In dry air metallic mercury is not oxidized. After long standing in moist air, however, the metal becomes coated with a thin layer of oxide. In air-free

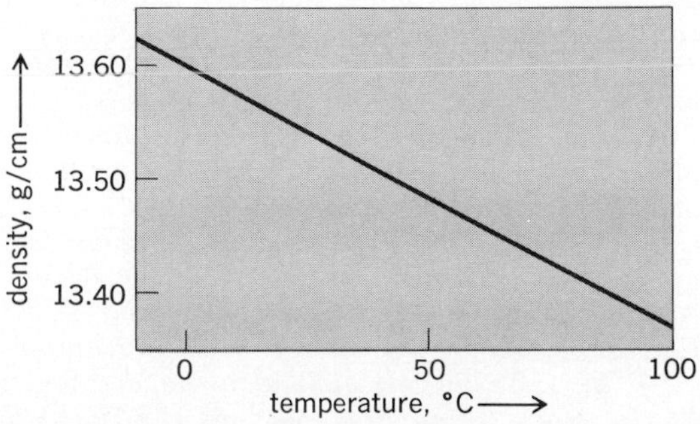

Fig. 1. Variation of mercury density with temperature.

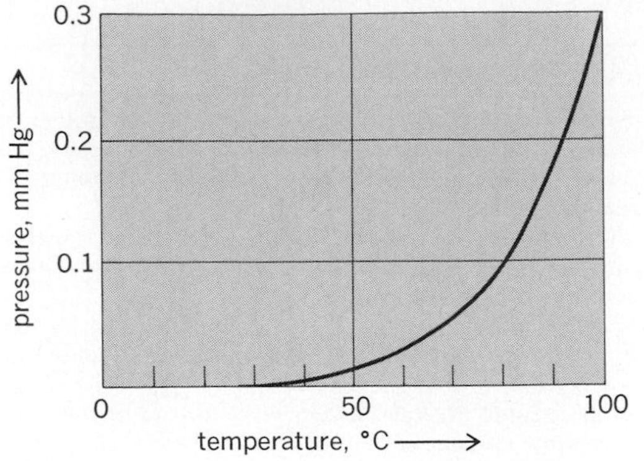

Fig. 2. Graph showing vapor pressure–temperature relationship of mercury.

hydrochloric acid or in dilute sulfuric acid, the metal does not dissolve. Conversely, it is dissolved by oxidizing acids (nitric acid, concentrated sulfuric acid, and aqua regia); the corresponding mercury (II) salt is formed if acid is present in excess, or the corresponding mercury(I) salt if the metal is present in excess. With chlorine and other halogens, mercury reacts quite readily. Also, it combines with sulfur quite easily, when these are ground together, and black mercury (II) sulfide, HgS, is formed. Mercury does not react with phosphorus.

Many mercury salts exhibit catalytic effects. For example, mercury(II) sulfate, $HgSO_4$, catalyzes the oxidation of naphthalene to phthalic acid or accelerates the conversion of acetylene to acetaldehyde.

Colloidal suspensions of mercury metal in water are stable in the presence of a protective colloid. Mercury can be emulsified by grinding it together with fat.

Mercury and almost all of its compounds are quite poisonous to man and animals. In chronic mercury poisoning, reddening and bleeding of the gums, digestive disturbances, deafness, and tremors of the hands occur. For acute mercury poisoning, for example, resulting from drinking a solution of the highly poisonous mercury(II)-chloride (an antiseptic), treatment consists of animal charcoal and milk. In such a case of poisoning, a physician should be called as soon as possible. The compound mercury(I) chloride, which is used as a purgative, is known as calomel.

[KLAUS BRODERSEN]

Bibliography: J. C. Bailar et al., *Comprehensive Inorganic Chemistry*, 1973; D. Breitinger and K. Brodersen, *Angew. Chem.* (internat. ed.), 9:357–367, 1970; F. A. Cotton and G. Wilkinson, *Advanced Inorganic Chemistry: A Comprehensive Text*, 4th ed., 1980; P. A. D'Itri and F. M. D'Itri, *Mercury Contamination: A Human Tragedy*, 1977; Environmental Studies Board, *An Assessment of Mercury in The Environment*, National Academy of Sciences, 1978; R. J. Gillespie et al., *Canad. J. Chem.*, 52:791–795, 1974; *Gmelins Handbuch der anorganischen Chemie*, 8th ed., System no. 34, pts. A and B, 1960–1975; M. W. Miller and T. W. Clarkson, *Mercury, Mecurials and Mercaptans*, 1973; M. J. Taylor, *Metal-to-Metal Bonded States of the Main Group Elements*, pp. 17–44, 1975; A. F. Wells, *Structural Inorganic Chemistry*, 4th ed., 1975.

Metal carbonyl

A compound of a metal combined with carbon monoxide (CO). The structure involves a coordinate link between the metal atom and the CO group, with each CO donating a pair of electrons to the metal. Metal carbonyls are usually low-melting crystalline solids, and several are liquids at room temperature. They are generally soluble in organic solvents but insoluble in water. The primary decomposition products are the metal and carbon monoxide. Metal carbonyls are quite reactive. Halogens, bases, amines, phosphines, cyanides, nitric oxide, acetylenes, olefins, aromatics, cyclopentadienes, alkali metals, and so on react, with either complete replacement of carbon monoxide, or partial replacement to form mixed carbonyls [reaction (1)].

$$Ni + 4CO \rightarrow Ni(CO)_4 \quad (1)$$

Carbonyls are formed by the transition metals of groups V, VI, VII, and VIII of the periodic table (see table). Typically, these compounds are prepared by direct reaction of metal with carbon monoxide or by reduction of metal compound in the presence of carbon monoxide. Representative formation reactions are shown as reactions (2)–(4).

$$2CoO + 10CO \rightarrow Co_2(CO)_8 + 2CO_2 \quad (2)$$

$$CrCl_3 + 3Na + 6CO \rightarrow Cr(CO)_6 + 3NaCl \quad (3)$$

$$WCl_6 + 2Al + 6CO \rightarrow W(CO)_6 + Al_2Cl_6 \quad (4)$$

A number of metal carbonyls are of technological significance. Nickel is simultaneously extracted and purified by reaction of its refined ore with carbon monoxide to form nickel carbonyl, followed by decomposition of the carbonyl. Several other metal carbonyls may be decomposed to yield adherent metal films or powders for powder metallurgy. Manganese, iron, and nickel carbonyls exhibit gasoline antiknock properties in internal combustion engines. In particular, methylcyclopentadienylmanganese tricarbonyl is manufactured in large volume for use as an antiknock agent. Metal carbonyls are, in general, reactive substances, and in the absence of evidence to the contrary, must be regarded as toxic.

Commercially important organic compounds can be made by reactions catalyzed by several metal carbonyls. For example, acetylenes may be either trimerized, inserted into allylic compounds, reacted with alcohol and carbon monoxide to give acrylic esters, or reacted with carbon monoxide to yield cyclopentadienones and quinones. A fuel mixture of hydrocarbons and alcohols may be prepared by synthesis gas (carbon monoxide and hy-

Some simple carbonyl compounds

Compound	Metal
$M_4(CO)_{12}$	Cobalt
$M_7(CO)_{21}$	Osmium
$M(CO)_4$	Nickel
$M_2(CO)_8$	Cobalt, rhodium, iridium
$M_3(CO)_{12}$	Iron, ruthenium, rhodium
$M_2(CO)_9$	Iron, ruthenium, osmium
$M(CO)_5$	Iron, ruthenium, osmium
$M_2(CO)_{10}$	Manganese, rhenium
$M(CO)_6$	Vanadium, chromium, molybdenum, tungsten

drogen) reaction with iron carbonyl (Fischer-Tropsch reaction). The hydroformylation (oxo) reaction of olefins with carbon monoxide and hydrogen to form aldehydes, ketones, and alcohols is catalyzed by cobalt, iron, rhodium, iridium, ruthenium, platinum, and osmium carbonyls. *See* FISCHER-TROPSCH PROCESS; HYDROFORMYLATION. [HYMIN SHAPIRO]

Bibliography: E. W. Abel and F. G. A. Stone, Metal carbonyls, in *Organometallic Chemistry*, vol. 7, Chemical Society, London, 1978; H. Shapiro et al., *Cyclomatic Manganese Compounds*, U.S. Pat. 2,839,552, June 17, 1958, assigned to Ethyl Corp.; I. Wender and P. Pino (eds.), *Organic Synthesis via Metal Carbonyls*, vol. 2, 1977.

Metal cluster compounds

A compound in which two or more metals aggregate so as to be within bonding distance of one another. There is an intense interest in this area of chemistry primarily because of the relationship of metal clusters to catalytic processes. Heterogeneous metal clusters of the transition metals deposited on supports such as alumina, silica, and zeolites have been widely used in catalysis for some time. However, much research involving soluble metal cluster compounds in homogeneous solution has been undertaken. The prognosis is that molecular metal clusters in a homogeneous solution can mimic the heterogeneous counterpart. Moreover, since solution chemistry is more amenable to characterization, the soluble clusters should lead to an intimate understanding of catalytic mechanisms and to the design of very selective catalysts.

Metal clusters in solution fall into two general categories: those encumbered with a coordination sphere of chemically bound ligands (ligated clusters), and those surrounded only by a sphere of solvent molecules (the so-called naked clusters). Clusters of the first class are found primarily in the case of transition metals and are critical to catalysis since the substrate undergoing catalytic transformation must be chemically bound to the cluster some time during the catalytic cycle. *See* HETEROGENEOUS CATALYSIS; HOMOGENEOUS CATALYSIS.

Ligated clusters. Metal carbonyl clusters are a class of ligated clusters of particular interest since they probably are involved in Fischer-Tropsch chemistry, that is, the transformation of carbon monoxide–hydrogen mixtures to hydrocarbons. Some very large clusters have been characterized, such as $[Rh_{13}(CO)_{24}H_3]^{2-}$ (illustration *a*). In most instances it appears that the metal core of ligated molecular metal clusters reflects a closest-packed geometry characteristic of the pure metals. Main-group heteroelements are sometimes found in the interstices, as is hydrogen. In fact, in some clusters hydrogen rapidly moves back and forth from bound surface positions to interstitial positions. Surface ligands, in many instances, also are quite mobile. X-ray crystallographic investigations have shown that a wide variety of ligands and organic substrates undergo bond lengthening or rearrangement; that is, they are activated upon reaction with molecular metal clusters. *See* FISCHER-TROPSCH PROCESS.

Naked-metal clusters. These are known in heterogeneous phases of the main-group metals, transition metals, and mixed metals. For instance,

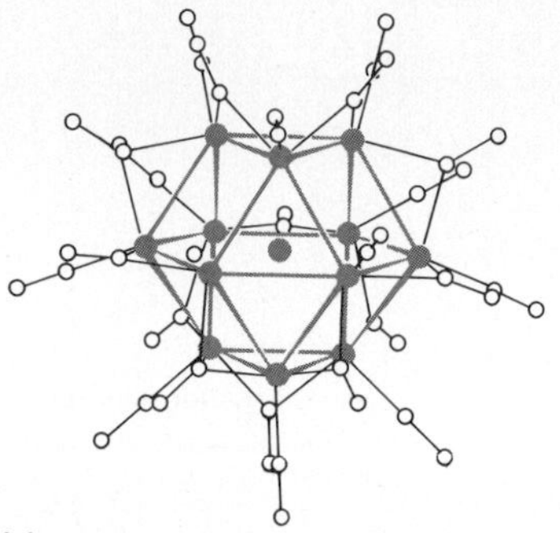

(a)

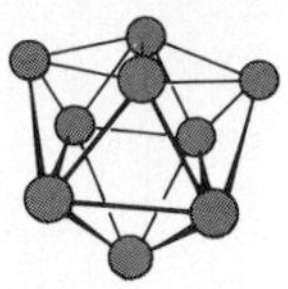

(b)

Metal cluster compound structures. (*a*) $[Rh_{13}(CO)_{24}H_3]^{2-}$, showing the hexagonal close-packed array of Rh atoms (*from S. Martinengo et al., Analogue of metallic lattices in rhodium carbonyl cluster chemistry, J. Amer. Chem. Soc., 100:7097, 1978*). (*b*) The naked cluster Ge_9^{2-}, which is reminiscent of the $B_9H_9^{2-}$ borane.

Pt/Sn, Rh/Sn, and so on are reforming catalysts in the petroleum industry. Aggregates of neutral naked metals have been studied on surfaces of metal oxides and by isolation at very low temperatures in matrices of inert gases or polymers. Cationic naked-metal clusters such as Bi_5^{3+} exist in molten salts. By contrast, anionic naked-metal clusters are obtained when alloys such as Na/Sn, Na/Pb, and Na/Sb are extracted with liquid ammonia or ethylenediamine; examples include Sn_9^{4-}, Pb_5^{2-}, Sb_7^{3-}, and heteroatomic clusters such as $(Sn_{9-x}Pb_x)^{4-}$ where $x = 0$ to 9. Naked clusters of the main-group metals adopt polyhedral structures which are reminiscent of the boranes, as shown in illustration *b*.

Characterization. Although generally it is easier to get a complete characterization of soluble metal clusters by application of a variety of spectroscopic techniques such as ultraviolet–visible, Raman, infrared, electron paramagnetic reasonance, nuclear magnetic resonance, and x-ray crystallography, there have been significant advances in methods of characterization for supported clusters. Advances in instrumentation have made properties such as aggregation, metal-metal distances, atom arrangements, and oxidation states accessible by techniques such as extended x-ray absorption fine structure (EXAFS), x-ray radial distributions, x-ray photoelectron spectroscopy, Mössbauer spectroscopy, ultraviolet reflectance, electron paramagnetic resonance, Raman, and electron microscopy. [RALPH W. RUDOLPH]

Bibliography: E. L. Muetterties et al., Clusters and surfaces, *Chem. Rev.*, 79:91–138, 1979; R. W. Rudolph et al., The nature of naked-metal-cluster polyanions in solution, *J. Amer. Chem. Soc.*, 100: 4629, 1978.

Metal hydrides

A compound in which hydrogen is bonded chemically to a metal or metalloid element. The compounds are classified generally as ionic, transition metal, and covalent hydrides. The periodic table can be arranged according to the type of hydride formed (Fig. 1). Covalent hydrides are of two subtypes, binary and complex. Certain hydrides have achieved a position of modest industrial importance, but most are of theoretical interest only.

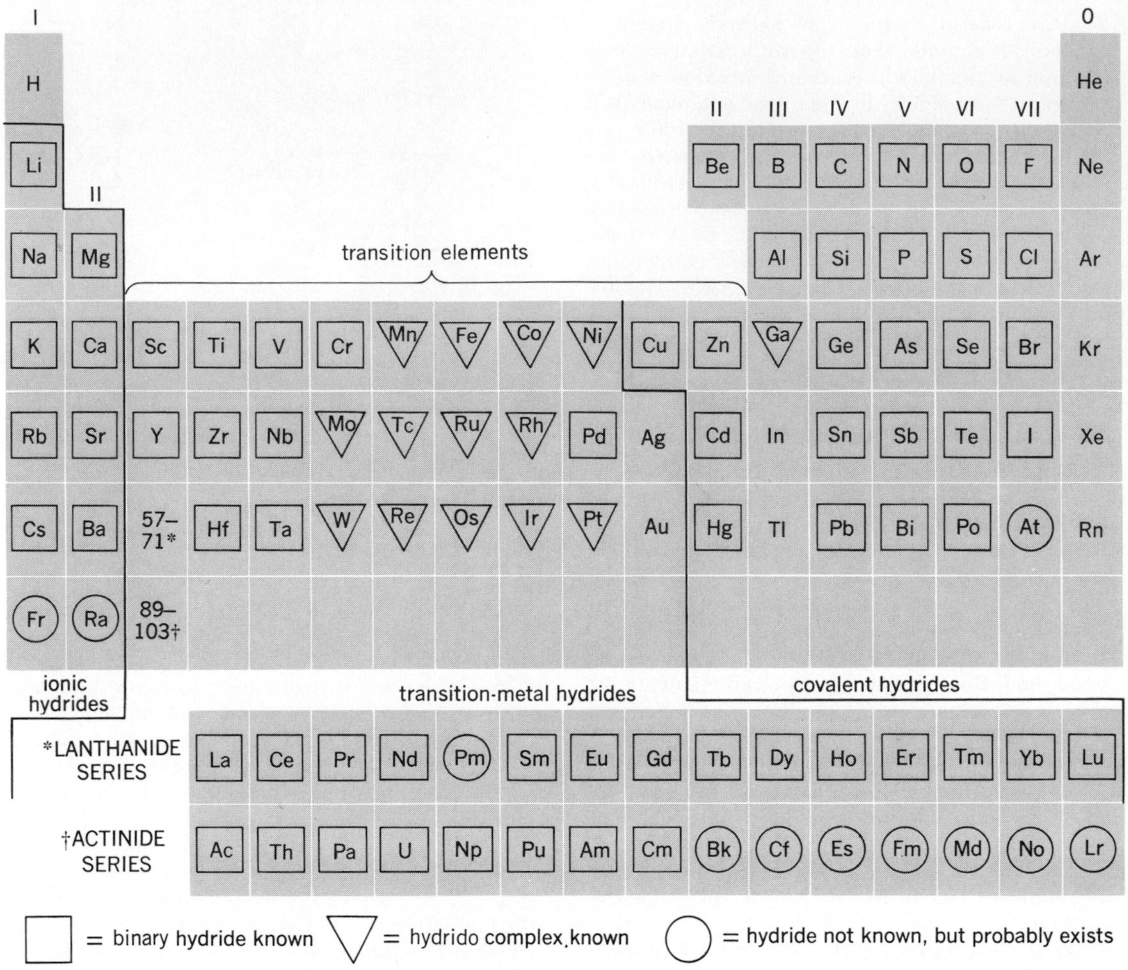

Fig. 1. Periodic table showing hydride types. No hydride or hydrido complex is known for unenclosed elements. Subgroups of elements are not designated by the letters a and b, comforming to modern usage.

Physical properties of representative hydrides

Compound	Formula	Form	Melting point, °C	Density, g/cm³	Boiling point, °C	Temperature at which dissociation pressure is 1 atm,* °C
Lithium hydride	LiH	White crystals	691	0.77	–	~790
Sodium hydride	NaH	White crystals	Decomposes	1.396	–	425
Calcium hydride	CaH_2	White crystals	>1000	1.902	–	960
Titanium hydride	TiH_2	Gray powder	Decomposes	3.78	–	630
Cerium hydride	CeH_2	Dark gray powder or greenish metallic single crystals	1088 for $CeH_{1.73}$	5.45	–	750 for $CeH_{2.2}$
Uranium hydride	UH_3	Black powder	~1050 at 580 atm	10.95	–	440
Lanthanum nickel hydride	$LaNi_5H_6$	Dark gray powder	Decomposes	–	–	~0 (2.5 atm at 25°)
Palladium hydride	$PdH_{0.66}$	Metallic	Decomposes	~10.8	–	25 for $PdH_{0.66}$ −78 for $PdH_{0.83}$
Diborane	B_2H_6	Colorless gas	−165.5	0.438 at bp	−92.5	–
Silane	SiH_4	Colorless gas	−185	0.68 at mp	−111.8	–
Stannane	SnH_4	Colorless gas	−150	–	−52	–
Arsine	AsH_3	Colorless gas	−113.5	–	−55	–
Stibine	SbH_3	Colorless gas	−88.5	2.2 at bp	−17	–
Tellurium hydride	TeH_2	Colorless gas	−51	2.7 at −18°	−4	–
Aluminum hydride	AlH_3	White solid	Decomposes	–	–	–
Copper hydride	CuH	Dark brown solid	Decomposes slowly even at 25°	6.39	–	–

*1 atm = 101.325 kPa.

Examples of hydrides are shown in the table.

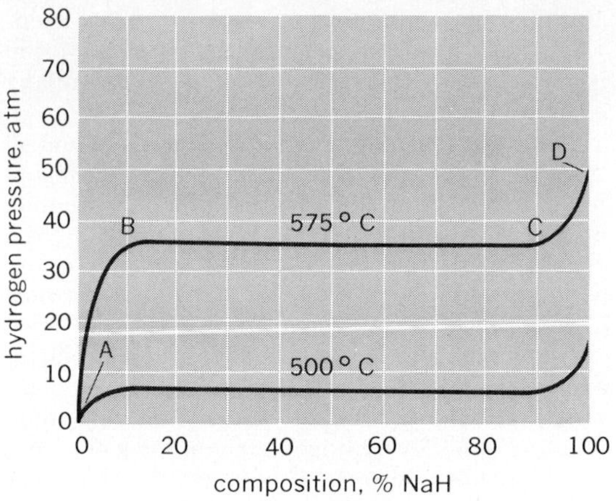

Fig. 2. The sodium-hydrogen system.

Under extreme conditions such as in electric discharges, many metals form volatile, short-lived transient hydrides of the general formula type MH. Although some of these can be prepared experimentally, most are observed only by their spectra. They are important in studying molecular bonding. The action of atomic hydrogen at low temperatures forms surface films of unstable hydrides with many metals. *See* HYDRIDO COMPLEXES; HYDROGEN.

Ionic hydrides. The most reactive metals (alkali and alkaline-earth metals, with the exception of magnesium and beryllium) combine directly on heating with hydrogen gas at a pressure of 1 atm (1 atm = 101, 325 Pa). Magnesium reacts at elevated hydrogen pressure. The reactions of hydrogen with sodium and potassium are slow because of formation of a protective layer of the hydride, but reaction can be accelerated by surface-active agents or catalysts such as anthracene or by stirring the molten metal under paraffin oil while hydrogen is bubbling in. *See* SURFACE-ACTIVE AGENT.

Ionic (or saline) hydrides, when pure, are white, high-melting solids which have simple formulas. They are believed to consist of metal cations and hydride anions H^-. Ionic hydrides result when the difference in electronegativity between hydrogen and the metal is large. *See* ELECTRONEGATIVITY.

The crystal structures of hydrides can be determined by the standard means of x-ray diffraction, which reveals the arrangement of metal ions only. To locate the hydrogen (or its isotope, deuterium), neutron diffraction is used. Such studies confirm that sodium hydride has the same type of structure as sodium chloride. *See* CRYSTAL STRUCTURE.

The thermal stability of the alkali metal hydrides decreases, and their chemical reactivity increases, in the order LiH, NaH, KH, RbH, and CsH. The hydride ion in these compounds has a radius of approximately 1.4 A or 0.14 nm (about the same size as the fluoride ion, F^-). The anion from heavy hydrogen, that is, the deuteride ion D^-, tends to be slightly smaller than H^-, when in an equivalent environment. When molten lithium hydride is electrolyzed, hydrogen is evolved at the anode, since H^- is the migrating species.

A common way to represent a reversible metal-hydrogen system is to plot the composition against hydrogen pressure (called the dissociation pressure) at a fixed temperature. Figure 2 shows the 500 and 575°C isotherms for the sodium-hydrogen system. Pure liquid sodium at 575°C exists at point A, and as hydrogen is admitted, it dissolves in the metal up to the limit at point B. This is a solution of NaH (10%) in molten Na (90%). When more hydrogen is pumped in, a solid appears, whose composition is indicated by point C ($NaH_{0.90}$, that is, sodium hydride deficient in hydrogen). During this stage (points B to C), the hydrogen pressure remains constant at 36.6 atm. The dissociation pressures at the various plateaus may be computed for any temperature by the equation $\log P = (-5960/T) + 8.59$, where P is the pressure in atmospheres and T the absolute temperature. Further introduction of hydrogen finally yields pure NaH at point D. In the plateau region (B to C), each mole of hydrogen which reacts liberates 27 kcal (113 kJ) of heat. More heat is liberated in the cases of lithium hydride (43 kcal or 180 kJ/mole of H_2) and CaH_2(42 kcal or 176 kJ). *See* PHASE EQUILIBRIUM.

The ionic hydrides are all exceedingly reactive reducing agents. Their reaction with water yields the metal hydroxide and hydrogen, which sometimes ignites at the high temperature resulting from the reaction. Sodium hydride dissolved in molten sodium hydroxide is employed in descaling steel and titanium. Calcium hydride is a source of hydrogen in remote areas, for example, to fill meteorological balloons. CaH_2 is also used as a desiccant in transformer oil and in the reduction of certain oxides to metals, such as tantalum.

Transition metal hydrides. This group of compounds is less well understood than the ionic and covalent hydrides. When lanthanum is heated in hydrogen, the gas is readily taken up, forming first a material whose composition is approximately $LaH_{1.86}$ (the "dihydride" phase) and finally LaH_3. The electrical conductivity does not change radically between La and $LaH_{1.86}$, but it falls dramatically from $LaH_{1.86}$ to LaH_3. The hydrogen is generally thought to form hydride ions, with radius about 1.3 A (0.13 nm), but some of the valence electrons remain free or mobile in the dihydride. This accounts for the conductivity of this substance. These mobile electrons are said to be in the conduction band. They are progressively consumed as the dihydride is converted to the trihydride. A single phase exists between $LaH_{1.86}$ and LaH_3, and it is possible to arrest the addition of hydrogen at any intermediate composition, such as $LaH_{2.63}$. Such a substance is sometimes referred to as a solid solution and sometimes as a compound which represents extreme nonstoichiometry. These features (electrical conductivity and nonstoichiometry) characterize the transition metal hydrides. *See* NONSTOICHIOMETRIC COMPOUNDS.

The ionic hydrides, such as NaH, exhibit much more limited nonstoichiometry. When sodium is converted to pure NaH, all electrons in the conduction band are consumed and the product is white and insulating. The hydrides of the rare earth metals are the easiest to describe in terms of an ionic model. The hydrides of the actinide metals are known through curium, the best known being uranium hydride. All are dark-gray powders which ignite spontaneously in the air. Large single crystals of cerium hydride are fairly stable in air.

Copper hydride, CuH, is unique in that it is the only hydride which can be precipitated from aqueous solution. It forms a dark-brown sediment which decomposes when dried. Copper hydride has a hexagonal crystal structure and is essentially a covalent compound.

Titanium, zirconium, vanadium, and niobium all react with hydrogen, forming dihydrides at the limiting compositions. The materials are conducting at all compositions. Except for palladium, the metals to the right of the chromium family in the periodic table show decreased reactivity toward hydrogen. Metals of the vanadium family and also chromium absorb hydrogen almost to the monohydride stage. The transition metals in the manganese, iron, and nickel families dissolve only small amounts of hydrogen, except for palladium, which readily absorbs hydrogen to the limit $PdH_{0.66}$. The palladium-hydrogen system has been much studied, because of its importance in catalytic hydrogenation. Thorium hydride, Th_4H_{15}, becomes superconducting at approximately 8 K. *See* PALLADIUM.

When hydrogen (protium, ^1H) in a metal hydride is replaced with its isotope deuterium, the product (deuteride) is usually less stable. The dissociation pressures of the deuterides are about twice those of the corresponding hydrides at the same temperature in the great majority of cases. The inverse effect is known, however. For example, at 40°C vanadium hydride ($VH_{1.5}$) has a dissociation pressure of 4.8 atm, while $VD_{1.5}$ has a dissociation pressure of 1.8 atm. The uranium compounds of the three isotopes of hydrogen (UH_3, UD_3, and UT_3) are frequently used in the laboratory as sources and reservoirs for these gases and for their purification. Titanium hydride also serves in this capacity. Some of the hydrides serve as a starting point for the synthesis of halides, sulfides, and so forth, of the metal. For example, this is represented by reaction (1). Deuterides and tritides of

$$UH_3 + 3HCl \rightarrow UCl_3 + 3H_2 \qquad (1)$$

actinide metals probably play an important role in thermonuclear (hydrogen) weapons.

Common metals such as steel and copper dissolve small quantities of hydrogen at elevated temperatures. On cooling, the gas comes out of solution and results in severe degradation of the mechanical properties of the metals. This is called hydrogen embrittlement and can be prevented by degassing the metal while it is still molten. Problems from hydrogen embrittlement also arise in high-pressure catalytic units employing hydrogen gas and at electrode surfaces where hydrogen ion is reduced to hydrogen, as in some electroplating processes. The cause of this embrittlement is the introduction of dislocations and stacking faults by the hydrogen atoms. The result is to suppress dislocation glide near cracks, preventing stress relaxation.

A number of alloys or intermetallic compounds are known which react with hydrogen and form ternary hydrides. Examples are $LaNi_5H_6$, Mg_2NiH_4, $AlTh_2H_2$, and $CaAg_2H$. $LaNi_5H_6$ permits an attractive means of portable hydrogen storage. This might be important if hydrogen is generated on a massive scale as a pollution-free energy source for fuel cells and internal combustion engines. Storage of hydrogen as compressed gas or as liquid is not convenient. The energy density of $LaNi_5H_6$ is twice that of liquid hydrogen and 12 times that of the compressed gas. The ternary hydride $LaNi_5H_6$ is readily formed reversibly from lanthanum-nickel alloy and hydrogen. Pure rare-earth metals are not necessary in making the nickel alloy; the commercial material (misch metal, mostly Ce, La, and Nd) is satisfactory. After charging with hydrogen, these materials provide practical gas pressure (a few atmospheres) from −30 to +30°C, and are thus attractive for powering hydrogen-burning engines of automobiles.

Owing to the highly reversible character of these metal-hydrogen systems, they have recently been studied as heat pumps and energy storage devices. To operate an air-conditioning unit, hydrogen is pumped from one metal hydride unit (such as $LaNi_5H_6$), which provides cooling, to an alloy unit ($LaNi_5$), where heat is evolved on formation of hydride. Alternatively, solar or waste thermal energy could be converted into mechanical work.

Another important ternary hydride is the Zr-U-H system, which is employed in pulsed reactors of the Triga type. The role of the hydrogen atom is to moderate fast neutrons. Another type of ternary hydride is illustrated by the thorium carbohydrides, Th_2CH_2 and Th_3CH_4. *See* RARE-EARTH ELEMENTS.

Covalent hydrides. Most evidence indicates that in ionic and metallic hydrides an electron pair is associated primarily with the hydrogen as H^-, while in covalent hydrides the electron pair is shared between the hydrogen atom and an atom of another element. In these compounds hydrogen is considerably smaller (radius 0.3 A or 0.03 nm) than in the ionic hydrides. Covalent hydrides usually consist of small molecules in which case they are gases (SiH_4 and SbH_3), but some form high polymers, in which case they are nonvolatile solids (AlH_3 and ZnH_2). All tend to decompose irreversibly rather easily on heating. Covalent hydrides are generally synthesized indirectly, not from direct combination of the elements. Gaseous hydrides such as SiH_4, PH_3, and AsH_3 can be generated by heating a solid mixture of the corresponding oxide and $LiAlH_4$. *See* CHEMICAL BONDING.

The hydrides of boron deserve special mention, as their bonding posed a difficult problem in chemical theory which has been resolved only recently. In diborane, B_2H_6, four of the hydrogen atoms are bonded to the two boron atoms by ordinary single, covalent bonds. The other two hydrogen atoms are linked to both boron atoms, the four bonds being formed by only two electron pairs. The structure of diborane is illustrated in Fig. 3. The hydrogen atoms which bridge between the two boron atoms do not form an ordinary hydrogen bond but another type usually described as an electron-deficient bridging bond. There are nearly two dozen higher boron hydrides, of which B_4H_{10}, B_5H_9, B_6H_{10}, and $B_{10}H_{14}$ are representative. A new group of compounds, the carboranes, has been discovered in which some of the boron atoms of the higher hydrides are replaced with carbon atoms. An example is $B_{10}C_2H_{12}$. *See* BORANE; CARBORANE.

METAL HYDRIDES

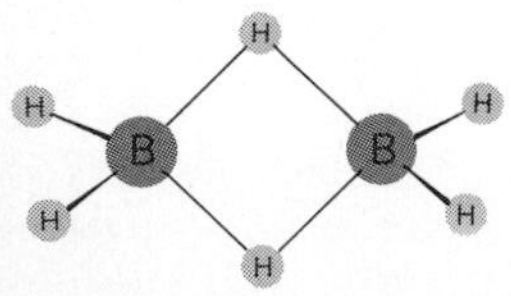

Fig. 3. Structure of diborane, B_2H_6.

Numerous derivatives of the boron hydrides have been prepared, of which sodium tetrahydroborate (or tetrahydridoborate or borohydride), $NaBH_4$ is the most important. It is used as a selective reducing agent in organic chemistry. This hydrido complex is moderately stable in water, especially in alkaline solution. This stability permits its use in recovery of heavy metals from industrial waste streams. *See* HYDROBORATION.

Aluminum hydride, AlH_3, normally exists as an insoluble, nonvolatile polymer. Its derivative, lithium tetrahydroaluminate (or lithium aluminum hydride), $LiAlH_4$, is an important reducing agent in both organic and inorganic chemistry. For example, silane can be prepared as in reaction (2). Lithi-

$$SiCl_4 + LiAlH_4 \xrightarrow{\text{Ether}} SiH_4 + LiCl + AlCl_3 \quad (2)$$

um tetrahydroaluminate reacts vigorously with water. An ether is generally used as the solvent. Sodium cyanotrihydroborate, $NaBH_3CN$, can be used in acidic medium. A family of aluminum-based reducing agents is now available commercially including sodium diethyldihydroaluminate, $NaAlH_2(C_2H_5)_2$; sodium tri-*tert*-butoxohydroaluminate, $NaAlH(O\text{-}t\text{-}C_4H_9)_3$; sodium bis(2-methoxyethoxo)dihydroaluminate, $NaAlH_2(OCH_2CH_2OCH_3)_2$; diethylaluminum hydride, $(C_2H_5)_2AlH$; and di-*iso*-butylaluminum hydride, $(i\text{-}C_4H_9)_2AlH$. All are soluble in aromatic hydrocarbons. They are rather expensive, but their specific reducing powers make them attractive for synthesizing high-value products, such as pharmaceuticals, flavorings, fragrances, dyes, and insecticides.

There are thousands of hydrocarbons, which represent hydrogen compounds of carbon, the first element in group IV. The simplest of these compounds is methane, CH_4. Other IV hydrogen compounds are silane, SiH_4; germane, GeH_4; stannane, SnH_4; and plumbane, PbH_4. This series is progressively less stable, with stannane decomposing at room temperature. Higher hydrides of silicon are known. Si_2H_6 and so on up to $Si_{10}H_{22}$. These compounds, Si_nH_{2n+2}, are analogous to the corresponding alkanes, C_nH_{2n+2}. Various, silane derivatives, such as SiH_3Cl and $(SiH_3)_3N$, have been synthesized. A parallel situation exists in the case of higher germanium hydrides. Unstable distannane, Sn_2H_6. has also been made. Mixed compounds such as SiH_3GeH_3 can be prepared. The group V hydrides show the same stability trends as the preceding group, with stability decreasing in the order NH_3, PH_3, AsH_3, SbH_3, and BiH_3. The VI group, including H_2Se, H_2Te, and H_2Po, also grows less stable down the series, the same trends continuing in the case of hydrogen compounds of the halogens. These compounds in the last two series, such as H_2Se and HI, are ordinarily considered as acids.

[JAMES C. WARF]

Bibliography: R. Bau (ed.), *Transition Metal Hydrides*, 1978; G. G. Libowitz and M. S. Whittingham (eds.), *Materials Science in Energy Technology*, 1979; W. M. Mueller et al. (eds.), *Metal Hydrides*, 1969; E. L. Muetterties (ed.), *Boron Hydride Chemistry*, 1975.

Metalloacid elements

The chemical elements with the following atomic numbers and names: 23, vanadium, V; 41, niobium, Nb; 73, tantalum, Ta; 24, chromium, Cr; 42, molybdenum, Mo; 74, tungsten, W; 25, manganese, Mn; 43, technetium, Tc; and 75, rhenium, Re. These elements are the subgroup members of groups V, VI, and VII, respectively, of the periodic table. In the elemental state all are metals of relatively high density, high melting point, and low volatility. The classification metalloacid elements refers to the fact that their oxides react with water to give somewhat acidic solutions, in contrast to the more typical behavior of the oxides of other metals which yield basic solutions. *See* PERIODIC TABLE.

The acid character of the oxides may be correlated with high positive charges attained by the metal ions in these compounds. (Maximum ionic charges attainable are equal to the respective group numbers 5, 6, and 7.) In aqueous solutions, these ions repel positively charged hydrogen ions and attract negatively charged oxide or hydroxide ions from water molecules in their vicinity. The resulting solutions contain free hydrogen ions and therefore are acidic.

With the exception of technetium, which (because it has no suitably long-lived isotope) probably does not occur naturally in the terrestrial environment, the elements are quite important commercially. Because of its high melting point (3370°C and low volatility, tungsten has long been used to form the filaments of high-temperature incandescent lamps. Vanadium, chromium, manganese, and niobium are essential components of various steels. The resistance of molybdenum and tantalum to chemical corrosion recommends their application in a variety of industrial processes. A thin plate of chromium deposited on corrodible metals gives a bright metallic surface which is remarkably resistant to tarnish. Such chrome plating is used in the manufacture of bathroom fixtures and automobile trim. The protective action of chromium plate is not due to high chemical resistance of the metal itself, but rather to the formation on its surface of an invisibly thin tenacious layer of chromium oxide, Cr_2O_3.

Although technetium is probably not found as a naturally occurring terrestrial element, its spectrum has been observed in certain types of stars. The element is available in small amounts as a product of nuclear fission. It has potential usefulness in the treatment of metal surfaces to inhibit corrosion. *See* CHEMICAL STRUCTURES; TRANSITION ELEMENTS.

[BURRIS B. CUNNINGHAM]

Bibliography: B. H. Mahan, *College Chemistry*, 1966; J. R. Partington, *A Textbook of Inorganic Chemistry*, 6th ed., 1950; P. C. L. Thorne and E. R. Roberts (eds.), *Ephraim's Inorganic Chemistry*, 6th ed., 1955.

Metallocenes

A group of organometallic compounds, bis-cyclopentadienyl metals, which generally possess a sandwich structure (Fig. 1).

Almost all transition metals have been found to

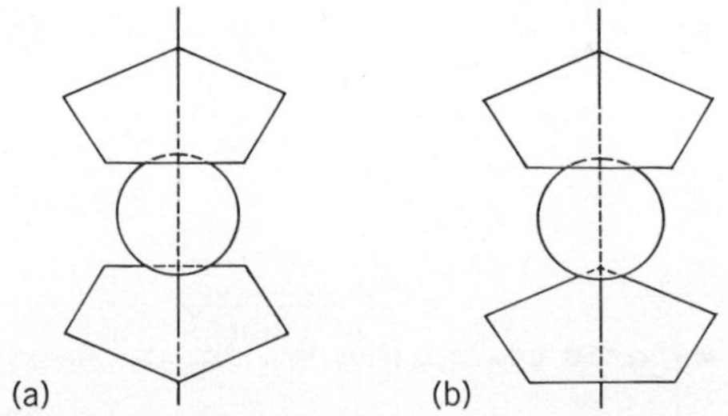

Fig. 1. Metallocene structures. (*a*) Staggered. (*b*) Eclipsed.

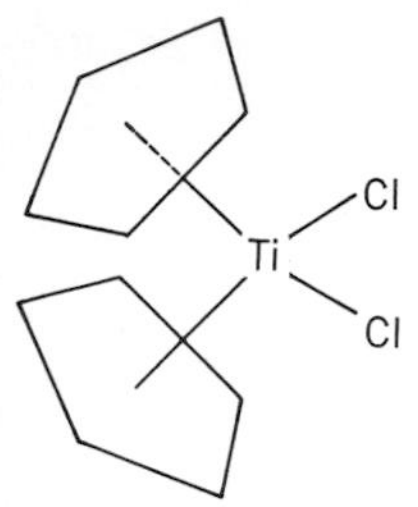

Fig. 2. Titanocene dichloride; two h^5-cyclopentadienyl groups are titled.

form metallocenes. According to the center metal atom, metallocenes are called ferrocenes (Fe), ruthenocenes (Ru), and so on. In a brief definition, both cyclopentadienyl groups bind with the metal atom in π fashion (or pentahapto-h^5).

Preparation. Metallocenes are prepared by several methods. A general laboratory method is the substitution reaction of anhydrous transition metal halides with Grignard reagents (organomagnesium halides) or organolithium in solvents such as diethyl ether or tetrahydrofuran. An interesting development in the preparation is the direct reaction of metal vapor (atomic metals) and cyclopentadiene.

Crystal structure. X-ray diffraction shows that crystalline ferrocene and its substituted derivatives (except bridged ferrocene) have the antiprismatic (staggered) conformation, whereas ruthenocene and osmocene have the prismatic (eclipsed) form which suggests that the inter-ring repulsive forces in these complexes are less than in ferrocene. This is in agreement with their larger inter-ring distances. It is worth noting that ferrocenylruthenocenylketone, h^5-C_5H_5Fe-π-C_5H_4CO-π-C_5H_4Ru-π-C_5H_5, has the same configuration for both metals, namely, about halfway between the eclipsed and staggered structures. The structure of titanocene dichloride is shown in Fig. 2.

Bond strength. The sandwich complexes are thermally stable, and many melt without decomposition at about 170°C. They are stable to hydrolysis and resist catalytic hydrogenation. Ferrocene is even more resistant to hydrogenation than benzene, though it may be reduced, with metal-ring cleavage, by lithium in ethylamine. The hydrogens of the h^5-cyclopentadienyl ligands are normally not labile and do not exchange with D_2O. Deuteroferrocenes may be prepared by reaction between (h^5-$C_5H_5)_2$Fe and $Ca(OD)_2$ at high temperatures, or by proton-deuteron exchange in strongly acidic D_2O solutions.

Thus nickelocene is less stable than ferrocene, which is in agreement with the weaker ring-metal bond in nickelocene, as is suggested by its greater length compared with that of ferrocene. Also, infrared spectra indicate the order of ring-metal bond strength as ruthenocene > ferrocene > nickelocene.

Oxidation. The stability of the bis-h^5-cyclopentadienyls in oxygen varies widely. At room temperature, ferrocene is inert to molecular oxygen, whereas chromocene is pyrophoric in air. The controlled oxidation of h^5-cyclopentadienyl complexes may give h^5-cyclopentadienyl metal oxide derivatives. Frequently these oxidations are reversible.

The oxidation of ruthenocene can occur by a one-electron transfer process, giving the pale yellow ruthenicenium cation. However, chronopotentiometric oxidation gives the cation $[(h^5\text{-}C_5H_5)_2Ru]^{2+}$ by a two-electron transfer process. Osmocene is oxidized by ferric chloride, or chronopotentiometrically to the analogous cation $[(h^5\text{-}C_5H_5)_2Os]^{2+}$. Other species such as the diamagnetic $[h^5\text{-}C_5H_5)_2OsOH]^+$ and $[(h^5\text{-}C_5H_5)_2OsI]^+$ are formed. The order of ease of oxidation of the complexes $(h^5\text{-}C_5H_5)_2M$, M = Fe > Ru ~ Os, has been proposed.

Cobaltocene is readily oxidized, thus losing an electron from an antibonding orbital and attaining the rare-gas configuration. Solutions readily absorb oxygen but, as yet, no oxygen-cobalt complexes of cobaltocene have been isolated. Both cobaltocene and rhodocene may be regarded as pseudoalkali metals in that they readily lose one electron, forming metallicenium cations. The cations are very stable to further oxidation; for example, they are stable in concentrated HNO_3. The cations are, however, susceptible to nucleophilic attack on the h^5-C_5H_5 rings.

18-Electron rule. It has been shown above that the bis-h^5-cyclopentadienyl complexes differ from most organo–transition metal compounds in that they often do not obey the 18-electron rule and many are paramagnetic. Further, for a given metal, two or more oxidation states are sometimes known. Nevertheless, the chemistry of bis-h^5-cyclopentadienyls may be associated to some extent with a tendency to form 18-electron complexes, especially in the case of cobaltocene and nickelocene.

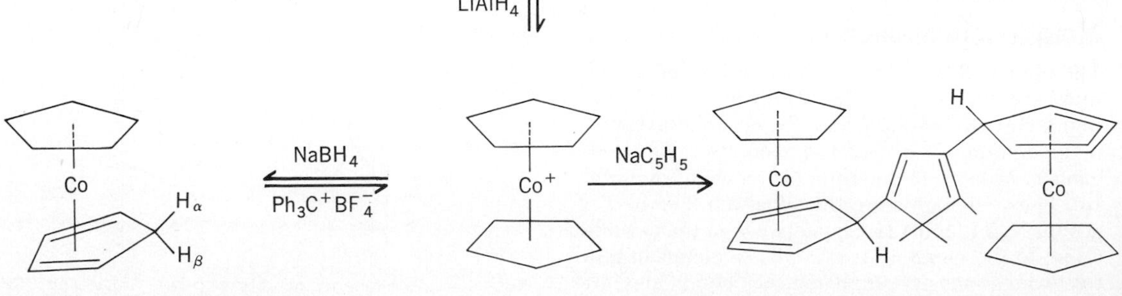

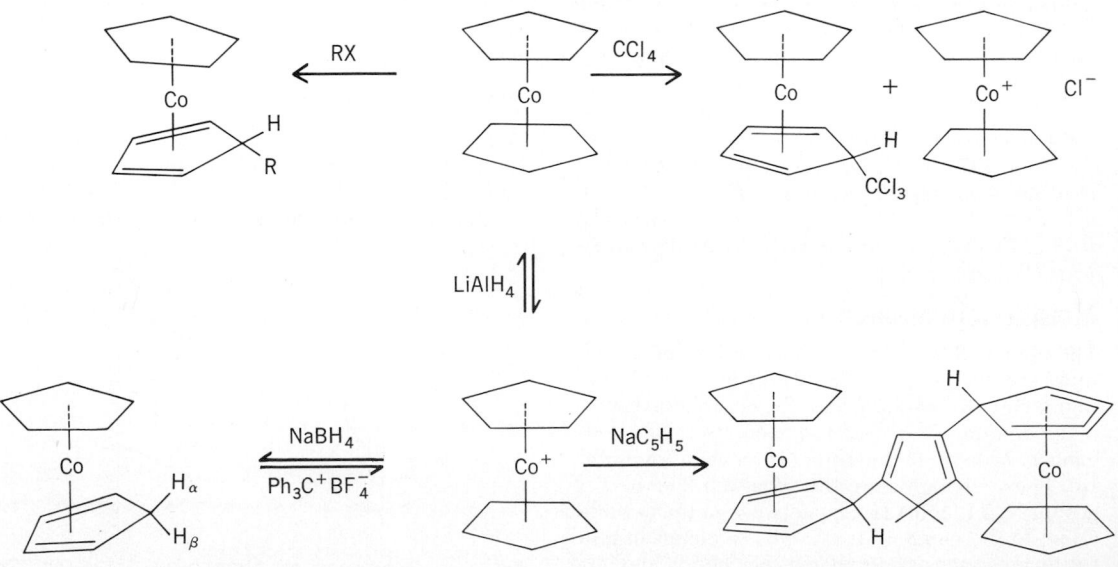

Fig. 3. Reactions of cobaltocene.

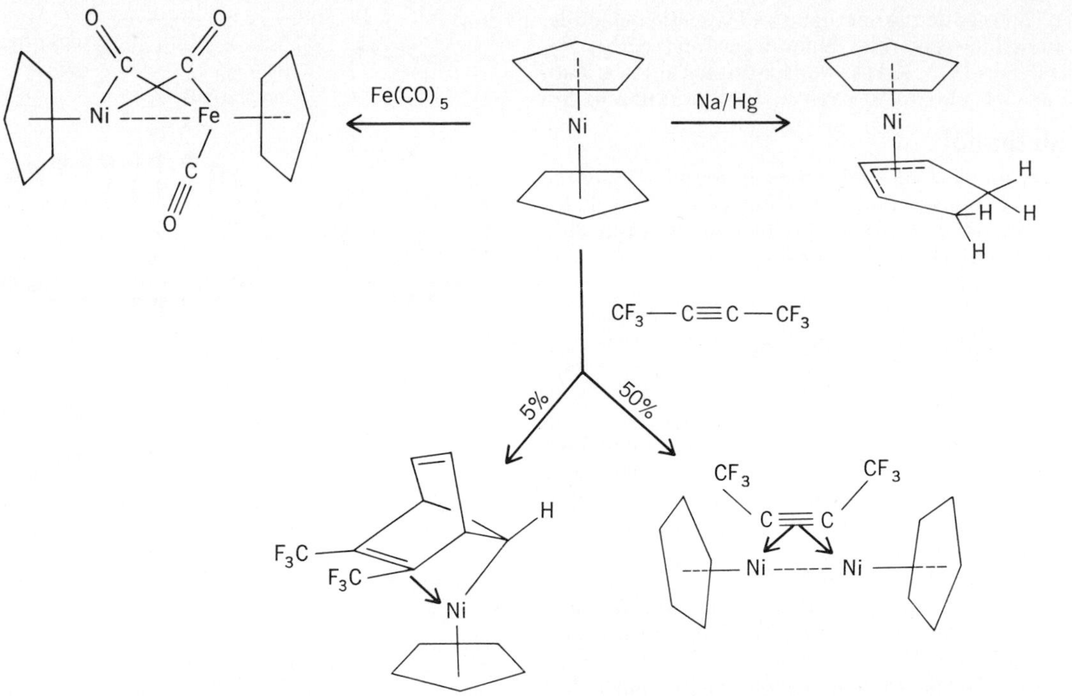

Fig. 4. Reactions of nickelocene.

The bis-h^5-cyclopentadienyls of iron, ruthenium, and osmium have an extensive organic chemistry. They are all 18-electron complexes and are usually stable under the conditions in which ring substitution can be achieved. Some of the reactions of cobaltocene and nickelocene are exemplified in Figs. 3 and 4.

Uses. Titanocene dichloride has been used as olefin (ethylene and propylene) polymerization catalysts. Nickelocene is effective in the dimerization of ethylene to butene-2. Cobaltocene and rhodenocene are also efficient homogeneous hydrogenation catalysts. Metallocenes have been used for coating or plating metal by their vapor-phase decomposition. Applications of metallocenes are versatile and have potential for further developments. The discovery of the catalytic activity of titanocene dichloride in nitrogen fixation under relatively mild conditions is noteworthy. *See* FERROCENE; ORGANOMETALLIC COMPOUND.

[MINORU TSUTSUI]

Bibliography: E. Becker and M. Tsutsui, *Organometallic Reactions*, 2 vols., 1970; E. Becker and M. Tsutsui (eds.), *Organometallic Reactions and Syntheses*, vols. 3–6, 1972–1977; M. L. H. Green, *Organometallic Compounds*, vol. 2, 3d ed., 1968; M. Tsutsui et al., *Introduction to Metal π-Complex Chemistry*, 1970.

Metalloid

An element which exhibits the external characteristics of a metal but behaves chemically both as a metal and as a nonmetal. Arsenic and antimony, for example, are hard crystalline solids that are definitely metallic in appearance. They may, however, undergo reactions that are characteristic of both metals and nonmetals. Certain of their oxides dissolve in either acids or bases, and are said to be amphoteric in character because they behave either as a base or an acid. Many elements form compounds that are amphoteric. However, only when this dualistic chemical behavior is very marked and the external appearance metallic is the element commonly called a metalloid. *See* NONMETAL.

[FRANCIS J. JOHNSTON]

Methane

A member of the alkane or paraffin series of hydrocarbons with the formula shown below.

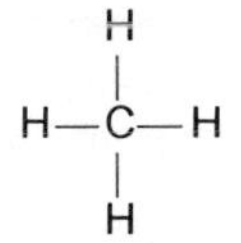

Methane is called marsh gas because it forms by anaerobic bacterial decomposition of vegetable matter in swampy land. Coal miners know it as firedamp because mixtures with air are combustible. It is a major constituent of natural gas (50–90%) and of coal gas. It forms in large amounts in sewage disposal processes, especially in anaerobic digestion. As a liquid it freezes at −182.6°C and boils at −161.6°C.

In addition to its use as a fuel, methane is important as a source of organic chemicals and of hydrogen. Its reaction with steam at high temperatures in the presence of catalysts yields carbon monoxide and hydrogen (synthesis gas), which can be catalytically converted to liquid alkanes (Fischer-Tropsch process) or to methanol and other alcohols. The catalytic reaction of the synthesis gas with olefins yields higher alcohols (the Oxo process), and reaction with steam produces additional hydrogen and carbon dioxide.

The incomplete combustion of methane with air produces finely divided carbon, called carbon black, great quantities of which are used annually as a reinforcing and filling agent in compounding rubber and as a pigment in printing ink.

Chlorination of methane yields methyl chloride, methylene chloride, chloroform, and carbon tetrachloride. *See* ALKANE; FISCHER-TROPSCH PROCESS; HYDROFORMYLATION. [LOUIS SCHMERLING]

Methanol

The simplest alcohol (alkanol). Methanol has the formula shown. Formerly obtained from wood as a coproduct of charcoal production, it was called wood spirit or wood alcohol.

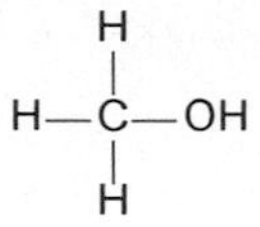

Methanol is a colorless, poisonous liquid with essentially no odor and very little taste. Its lack of odor is an important hazard in handling, and often strong-smelling wood alcohol or kerosine is added so that workers will become aware of leaks or spills. Methanol boils at 64.7°C. It is miscible with water and most organic liquids, including gasoline. It is extremely flammable, burning with a nearly invisible blue flame.

Methanol is produced commercially from hydrogen and carbon monoxide. In the high-pressure process, the synthesis gases are compressed to about 300 atm (30MPa) at 300–400°C over zinc–copper oxide catalyst, sometimes promoted with chromium oxide and alkaline earth salts as stabilizers. The low-pressure process requires lower pressure (50–100 atm or 5–10 MPa) and temperatures (250–300°C) over a copper, zinc, and chromium oxide catalyst. Both processes may be represented by the equation $CO + 2H_2 \rightarrow CH_3OH$.

The catalysts employed in the low-pressure process are sensitive to traces of impurities in the synthesis gases, hence exceptional care must be taken to purify the feed gases. Sulfur, phosphorus, and cyanide are particularly harmful to the low-pressure catalysts. The low-pressure process has the distinct advantage of requiring simpler compressors and less costly reactors.

The chief use for methanol at one time was in the manufacture of formaldehyde. However, methanol has become a major feedstock for the synthesis of acetic acid. Large amounts of methanol are consumed in the production of methyl esters such as methyl methacrylate and terephthalate, methyl chloride, and methyl *tert*-butyl ether, which is used as a gasoline blending agent. Methanol itself is employed as an extractant and solvent for many substances, and is sometimes blended with gasoline in cold weather to reduce condensation problems. [FRANK WAGNER]

Bibliography: J. K. Paul, *Methanol Technology and Application in Motor Fuels*, 1979.

Micelle

A colloidal aggregate of a unique number (50 → 100) of amphipathic molecules, which occurs at a well-defined concentration called the critical micelle concentration. In polar media such as water, the hydrophobic part of the amphiphiles forming the micelle tends to locate away from the polar phase while the polar parts of the molecule (head groups) tend to locate at the polar micelle solvent interface. A micelle may take several forms, depending on the conditions and composition of the system, such as distorted spheres, disks, or rods (see illustration *a*, *b*, and *c*). The dimensions of the particle are derived from those of the amphiphile, for example, a sphere radius of about 2 nm or a rod cross-sectional radius of 2 nm. Frequently the polar head group is a salt which ionizes in polar media. The micelle may have about 30% of the amphiphiles in the ionized state, giving rise to a highly charged particle surrounded by a cloud of counterions, and also counterions bound into the micelle surface. Micelles are formed in nonpolar media such as benzene, where the amphiphiles cluster around small water droplets in the system, forming an assembly known as a reversed micelle (illustration *d*).

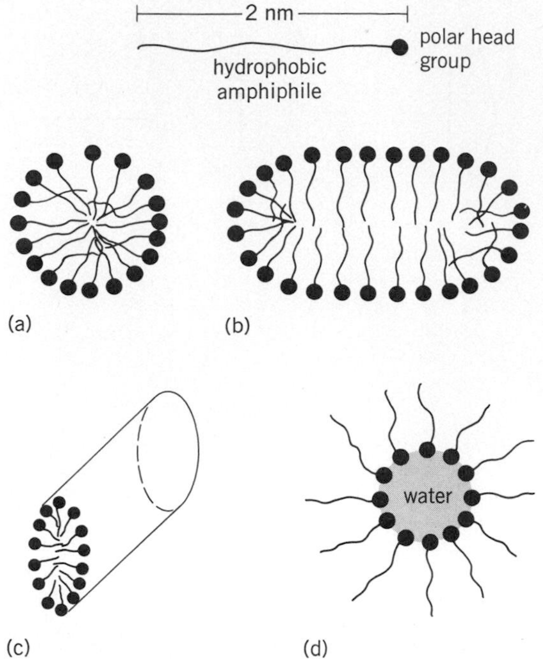

Form of an amphiphile and several forms of micelle: *(a)* spherical, *(b)* disk, *(c)* rod, and *(d)* reversed.

Micellar systems have the unique property of being able to solubilize both hydrophobic and hydrophilic compounds. They are used extensively in industry for detergency and as solubilizing agents. A strong catalytic action is often associated with these systems and is attributed to the clustering of reactants in the micelle, thereby creating high local reactant concentration, and also to the strongly charged surface which influences the transition state of a reaction. These systems play interesting roles in photoinduced reactions, in particular photoinduced electron transfer, where large yields of long-lived ions are observed, a feature of interest in storage of solar energy. Micelles mimic biological systems. *See* CATALYSIS.

[J. KERRY THOMAS]

Bibliography: E. Fendler and J. H. Fendler, *Catalysis in Micellar and Macromolecular Systems*, 1975; K. L. Mittal (ed.), *Micellization, Solubilization and Microemulsions*, 1977; J. K. Thomas, *Acc. Chem. Res.*, 10:133–138, 1977.

Microwave spectroscopy

The study of the interaction of matter and electromagnetic radiation in the microwave region of the spectrum. Microwaves are loosely defined as electromagnetic radiation with wavelengths between about 1 mm and 30 cm or frequencies between 1 and 300 GHz. The wavelengths are comparable to the dimensions of experimental apparatus. Experimental techniques make use of ideas from radio-frequency spectroscopy where wavelengths greatly exceed the dimensions of the apparatus, and also techniques from optics where wavelengths are much smaller than the size of the apparatus. *See* SPECTROSCOPY.

Apparatus. Microwave circuit elements (waveguides, resonant cavities, directional couplers) cannot be characterized by lumped capacitances (electric field regions) or lumped inductances (magnetic field regions), as in the case of radio-frequency circuits. Because of the short wavelengths of microwaves, both electric and magnetic fields are present in most circuit elements. When many wavelengths are present within a microwave circuit element, the momentum of the electromagnetic wave can become well defined, and momentum conservation or phase-matching conditions can have an important bearing on some spectroscopic techniques. An example is the mixing of microwaves and light waves, where the frequency of the mixed wave must equal the sum of the frequencies of the microwave and the light wave, while the momentum of the mixed wave (the inverse wavelength) must equal the vector sum of the momenta of the microwave and the light wave.

The transit times of electrons in ordinary electronic tubes are too long for efficient coupling to the rapidly oscillating fields of the microwave region, and special electronic tubes, klystrons, traveling-wave tubes, magnetrons, and so forth have been designed to overcome transit time limitations. Such tubes are often locked in frequency to a high harmonic of a stable quartz crystal oscillator. Microwave receivers are extraordinarily sensitive, and 10^{-19} W of microwave power can be detected with a good heterodyne system in a 1-Hz bandwidth.

Interaction of microwaves with matter. The interaction of microwaves with matter can be detected by observing the attenuation or phase shift of a microwave field as it passes through matter. These are determined by the imaginary or real parts of the microwave susceptibility (the index of refraction). The absorption of microwaves may also trigger a much more easily observed event like the emission of an optical photon in an optical double-resonance experiment or the deflection of a radioactive atom in an atomic beam.

Microwave energy at a frequency ν is absorbed according to the Bohr frequency condition, Eq. (1),

$$h\nu = E_f - E_i \qquad (1)$$

where h is Planck's constant and E_f and E_i are the energies of the final and initial states of the absorbing system. The initial and final states may be discrete as in the case of rotational states of a molecule, or they may be continuous as in bremsstrahlung in a plasma.

Enhancement of population differences. The characteristic temperature θ of the energy splitting involved in a microwave transition is given by Eq. (2), where k is Boltzmann's constant. At room temperature T, the relative population difference θ/T between the states involved in the transition is a few percent or less. The population difference can be close to 100% at liquid helium temperatures, and microwave spectroscopic experiments are often performed at low temperatures to enhance population differences and to eliminate certain line-broadening mechanisms.

$$\theta = \frac{h\nu}{k} = 0.05\ \mathrm{K} - 15\ \mathrm{K} \qquad (2)$$

The population differences between the states involved in a microwave transition can be enhanced by artificial means. For example, state selection of atoms or molecules in an atomic beam passing through inhomogeneous magnetic or electric fields can lead to very large population imbalances. Such large population inversions can be prepared so that stimulated emission cross sections for microwaves can exceed the absorption cross section. When the molecules or atoms with inverted populations are placed in an appropriate microwave cavity, the cavity will oscillate spontaneously as a maser (microwave amplification by stimulated emission of radiation). Optical pumping can also lead to large artificial population imbalances between the states involved in a microwave transition.

Applications. The magnetic dipole and electric quadrupole interactions between the nuclei and electrons in atoms and molecules can lead to energy splittings in the microwave region of the spectrum. Thus, microwave spectroscopy has been used extensively for precision determinations of spins and moments of nuclei. The 21-cm hyperfine transition of the hydrogen atom is seen from various parts of the Galaxy, and the field-independent component of the 21-cm line is used in the atomic hydrogen maser as a frequency standard. The field-independent component of the 9.192-GHz hyperfine transition in the cesium atom is used as a time standard in many laboratories.

Properties of molecules. The rotational frequencies of molecules often fall within the microwave range, and microwave spectroscopy has contributed a great deal of information about the moments of inertia, the spin-rotation coupling mechanisms, and other physical properties of rotating molecules. The rotational frequencies of water and oxygen molecules are responsible for much of the attenuation of microwaves in the atmosphere. Many microwave transitions in molecules are seen from sources in interstellar space, and microwave astronomy has provided much information about the chemistry and molecular composition of various astronomical objects. Inversion frequencies in molecules, for example, the frequency of periodic motion of the nitrogen atom at the apex of the pyramidal NH_3 ammonia molecule through the three hydrogen atoms at the base, often lie in the microwave region. One of the inversion transitions in ammonia was used in the first maser. *See* MOLECULAR STRUCTURE AND SPECTRA.

Electron-spin resonance. The magnetic resonance frequencies of electrons in fields of a few thousand gauss (a few tenths of a tesla) lie in the microwave region. Thus, microwave spectroscopy is used in the study of electron-spin resonance or paramagnetic resonance. In the simplest cases, the spins may be well isolated in a dilute gas like oxygen (O_2) or cesium vapor. Then the microwave spectrum provides information about the internal spin couplings of a free atom or molecule. For denser gases or condensed phases, the paramagnetic resonance spectrum provides information about interactions between the spins and their environment. Both static interactions, which determine the resonance frequencies of the microwave transitions, and fluctuating random interactions, which determine resonance line widths, are of interest. Typical static interactions are the electrostatic interactions between the spin and the crystal field. Magnetic dipole interactions and spin exchange interactions between spins at neighboring sites can lead to broadening or narrowing of the resonance lines. Particularly strong interactions between neighboring spins occur in ferromagnetic or antiferromagnetic materials, both of which can be investigated by microwave spectroscopy. *See* ELECTRON PARAMAGNETIC RESONANCE (EPR) SPECTROSCOPY.

Cyclotron resonance. The cyclotron resonance frequencies of electrons in solids at magnetic fields of a few thousand gauss (a few tenths of a tesla) lie within the microwave region of the spectrum. The effective mass and the cyclotron frequency of an electron in a solid are greatly modified from those of the free electron by the crystal interactions. Microwave spectroscopy has been used to map out the dependence of the effective mass on the electron momentum.

Plasmas. Gaseous plasmas are strong emitters and absorbers of microwaves. The main coupling mechanisms of charged-particle motion to microwave radiation are bremstrahlung and cyclotron radiation. Thomson scattering is also important in dense plasmas. Microwaves are completely reflected from plasmas where the electron density n is so high that the plasma resonance frequency f given by Eq. (3) exceeds the microwave frequency.

$$f = 8980n^{1/2}\ \text{Hz}\cdot\text{cm}^{-3/2} \qquad (3)$$

Electron densities and temperatures in plasmas can be determined by microwave spectroscopy.

Cosmic microwave radiation. Thermal microwaves at a temperature of about 3 K permeate all space and can be detected by sensitive microwave receivers. These microwaves are believed to be the cooled radiation left over from the big bang at the beginning of the universe.

[WILLIAM HAPPER]

Bibliography: W. Gordy, W. V. Smith, and R. F. Trambarulo, *Microwave Spectroscopy*, 1953; M. A. Heald and C. B. Wharton, *Plasma Diagnostics with Microwaves*, 1965, reprint 1978; D. J. E. Ingram, *Spectroscopy at Radio and Microwave Frequencies*, 1955; C. Kittel, *Introduction to Solid State Physics*, 5th ed., 1976; C. P. Poole, Jr., *Electron Spin Resonance*, 1967; T. L. Squires, *Introduction to Microwave Spectroscopy*, 1963; M. W. P. Strandberg, *Microwave Spectroscopy*, 1954; T. M. Sugden and C. N. Kenney, *Microwave Spectroscopy of Gases*, 1965; C. H. Townes and A. L. Schawlow, *Microwave Spectroscopy*, 1955, reprint 1975.

Mixing

A common operation to effect distribution, intermingling, and homogeneity of matter. Actually the operation is called agitation, with the term "mixing" being applicable when the goal is blending, that is, homogeneity. Other processes, such as reaction, mass transfer (includes solubility and crystallization), heat transfer, and dispersion, are also promoted by agitation. The type, extent, and intensity of agitation determine both the rates and adequacy of a particular process result. The agitation is accomplished by a variety of equipment.

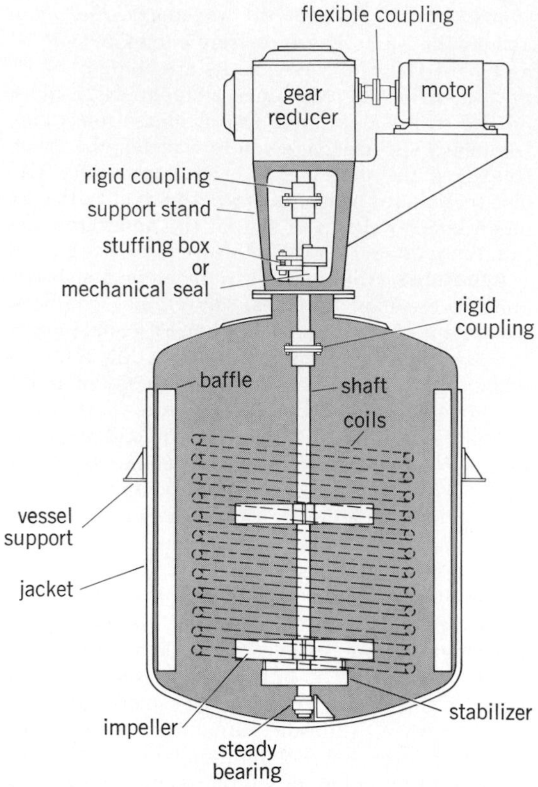

Fig. 1. Typical impeller-type liquid mixer. *(From V. W. Uhl and J. B. Gray, Mixing: Theory and Practice, vol. 2, Academic Press, 1967)*

Most liquid mixing is done by rotating impellers in vertical cylindrical vessels. A typical impeller-type liquid mixer with a variety of features is shown in Fig. 1. The internal features, including the vessel itself, are considered as a whole, that is, as the agitated system. The forces applied by the impeller develop overall circulation or bulk flow. Superimposed on this flow pattern, there is molecular diffusion, and if turbulence is present, also turbulent eddies. These provide micromixing. Solids, granular to powder, are mixed in a variety of contrivances.

Fluid mixing. The range of industrial processes for which mechanical agitation is used is conveyed by the table. They may be batch or continuous. Often several operations are conducted in concert; for example, in the hydrogenation of vegetable oils, the catalyst is suspended, a reaction takes place, and heat is transferred to the jacket wall. The per-

A classification of operations requiring mechanical agitation and mixing

Physical criteria	General kinds of operations	Chemical and mass transfer criteria
Pumping, circulation	Fluid motion	Heat transfer
Blending	Miscible liquids	Reactions
Emulsions	Immiscible liquids	Extraction, reaction
Dispersions	Liquid-gas systems	Absorption, stripping, reaction
Dispersion, suspension	Liquid-solid systems	Dissolving, crystallization

formance criteria for different operations vary, and although several agitation schemes often will accomplish a given process result, certain systems can be shown to be preferable. Here the governing factors generally are the type, size, and speed of the impeller as well as major internals such as baffles.

Fluid motion, both large-scale (bulk circulation) and small-scale (turbulent eddies), is required in turbulent flow. The bulk circulation results when the fluid stream is discharged by the impeller. Turbulence is generated mostly by the velocity discontinuities adjacent to the stream of fluid flowing from the impeller, but also by boundary-and-form separation effects; turbulence spreads throughout the bulk flow and, although attenuated, is carried to all parts of the vessel. It is recognized that some mixing operations should require relatively large bulk or mass flows, whereas others need a high intensity of turbulence (termed shear in this connection). From this it follows that there should be an optimum ratio of flow to shear for a given process result, for example, high flow for blending of miscibles and suspension of solids, and high shear for gas absorption (mass transfer) and dispersion.

Energy must be supplied to produce fluid motion. The power imposed by the mixing impeller is proportional to the flow multiplied by the head developed, which is equated to turbulence (shear).

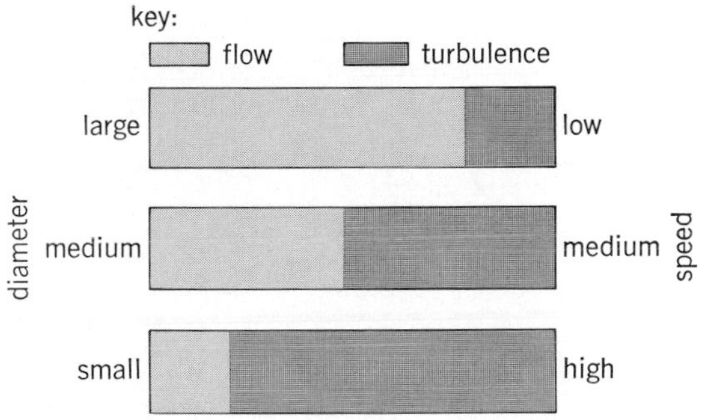

Fig. 2. Effect of impeller size and speed on flow and turbulence (shear) at constant power.

The ratio of the power supplied for flow to that for turbulence can be varied for the same power input and type of impeller; this variation is demonstrated by Fig. 2. Impeller shape also has an effect which can be significant, with a rotating disk representing the extreme case for high turbulence or shear.

In all bench-scale and pilot-plant work where mixing is important, the type of fluid motion in terms of ratio of flow to shear should be found. This becomes a basis for scale-up in design.

Mixing vessels and impellers. Circulation or bulk flow depends on the shape of the container; the fittings it contains; the type, size, and position of the impeller; and, of course, the properties of the fluid. Lateral (radial) and vertical flow currents usually produce the best agitation; these currents must penetrate to all portions of the fluid.

For the laboratory, whenever matters are mixed, cylindrical containers such as beakers should be used. Vertical baffles should be provided for impellers which are centrally located; if a propeller is used in an off-center position, the baffles may be

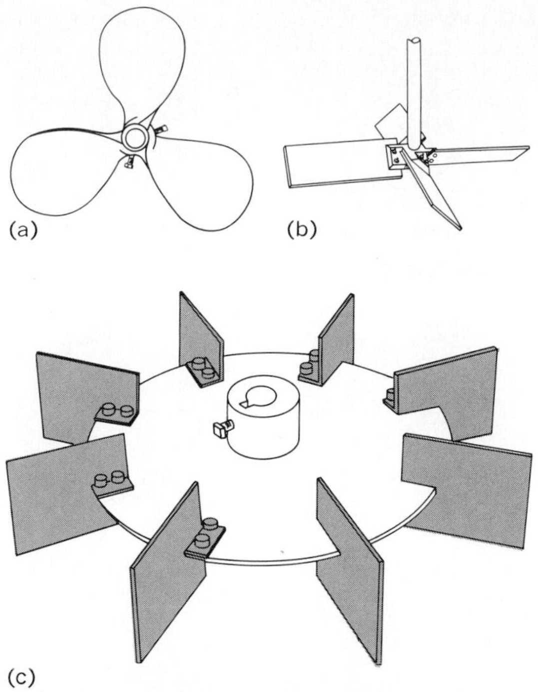

Fig. 3. Common impellers. (*a*) Marine-type mixing propeller. (*b*) Axial-flow (pitched-blade) turbine. (*c*) Flat-blade (Rushton) turbine. (*From R. H. Perry and C. H. Chilton, eds., Chemical Engineers' Handbook, 5th ed., McGraw-Hill, 1973*)

omitted. Round-bottomed flasks are unsatisfactory unless modified by creases blown in the side; this form can be purchased.

The most common and useful types of impellers are the marine-type propeller (Fig. 3*a*), the axial-flow (pitched-blade) turbine (Fig. 3*b*), the flat-blade (Rushton) turbine (Fig. 3*c*), and the flat paddle (generally two vertical blades relatively large compared to the other types). Any of these impellers centrally positioned produces rotating fluid motion (Fig. 4) with a vortex around which liquid swirls. This motion often results in separation or stratification rather than intermingling. Relatively little power can be applied; a dearth of turbulence and of vertical and lateral flow motion results. Inserting projections into the body of the fluid stops the rotary motion, and the vortex disappears. Thus agitation is improved. Such projections at the side of the tank are called baffles (Fig. 5). The propeller and axial-flow turbine with tank-wall baffles (Fig.

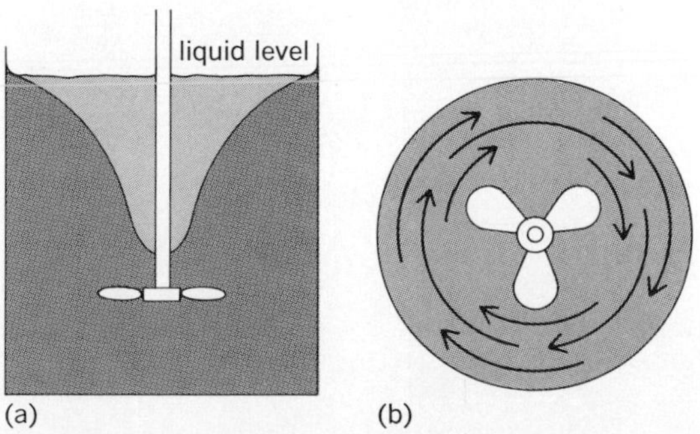

Fig. 4. Swirling flow pattern for impeller of any shape, without baffles. (*a*) Side view. (*b*) Bottom view.

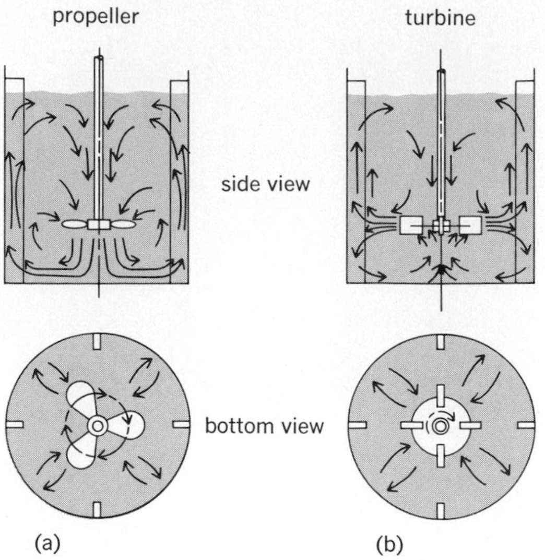

Fig. 5. Flow patterns with baffles. (*a*) Flow pattern for propeller with baffles at tank wall. (*b*) Flow pattern for turbine with baffles at tank wall.

5*a*) will generate an axial-flow pattern; the paddle and Rushton turbine produce radial flow (Fig. 5*b*). These flow patterns are conducive to good agitation.

Swirl can be obviated in laboratory work by using the propeller off-center. Good vertical and lateral flow motion without swirl and surface vortex can be attained by discharging the propeller downward and positioning as in Fig. 6; the exact position is critical but can be found by trial.

Liquids in large tanks, for example, 3,000,000 gal (11,370 m^3), are often blended by a side-entering propeller. Figure 7 shows the correct position for a right-hand, clockwise-rotation propeller; proper angular positioning is critical.

The behavior of low-viscosity liquids has been described above and is shown in Figs. 4–7. High-viscosity liquids, those above 1000 centipoises, will have much less rotary flow; without baffles the motion will approximate that in Fig. 5.

Power and flow. In liquids of low viscosity, power imposed by an impeller in a baffled tank is proportional to the cube of its speed, the fifth power of its diameter, and the density. The fluid regime would be turbulent. For constant power

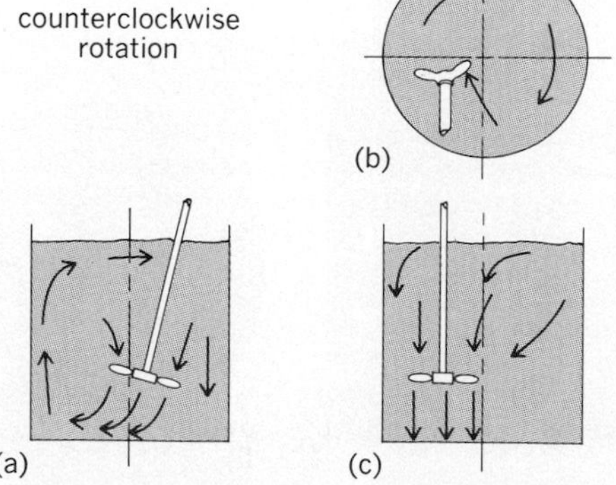

Fig. 6. Flow pattern for top-entering, off-center propeller without baffles: (*a*) front, (*b*) top, and (*c*) side views.

and varying diameter of a given type of impeller, the corresponding speed can be found from the equation $N_r = (1/D_r)^{5/3}$, where N_r is the ratio of speeds and D_r is the ratio of corresponding impeller diameters.

In liquids of high viscosity the power imposed by the impeller is proportional to the viscosity of the liquid, the square of the speed, and the cube of the diameter. Here the fluid regime is laminar. *See* VISCOSITY.

The discharge rate of flow from impellers is proportional to the speed and to the cube of the diameter. Clearly, for the same power there will be greater flow from the large-diameter, low-speed impeller than from a smaller-diameter, higher-speed one (Fig. 2).

High-viscosity liquids. These liquids and pastes often require different techniques from those used for mixing low-viscosity liquids. Special apparatus is necessary to provide for wiping, stretching, and squeezing, because turbulence cannot be generated in such fluids to provide for the small-scale mass transfer necessary to cause interpenetration of substances. There are few quantitative data available which describe the performance of the various types of equipment. An example of typical equipment is shown in Fig. 8.

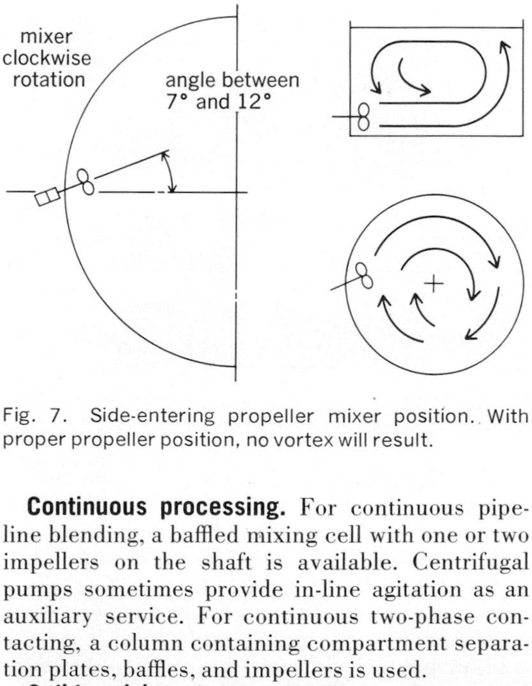

Fig. 7. Side-entering propeller mixer position. With proper propeller position, no vortex will result.

Continuous processing. For continuous pipeline blending, a baffled mixing cell with one or two impellers on the shaft is available. Centrifugal pumps sometimes provide in-line agitation as an auxiliary service. For continuous two-phase contacting, a column containing compartment separation plates, baffles, and impellers is used.

Solids mixing. Solids of different density and size are mixed in tumblers (a double cone turning end on end) or with agitators (a helical ribbon rotating in a horizontal trough). The duration of mixing is an important additional variable because classification and separation often occur after attainment of the desired distribution if the operation is carried on too long.

Equipment. Portable mixers generally range in size from fractional horsepower to 3 hp (2240 W); they are designed for use in open tanks up to 3000-gal (11.4 m^3) capacity and are clamped to the tank shell. Direct-drive units usually run at 3600, 1750, or 1150 rpm or at variable speeds. Gear-drive units run at about 420 rpm or at variable speeds.

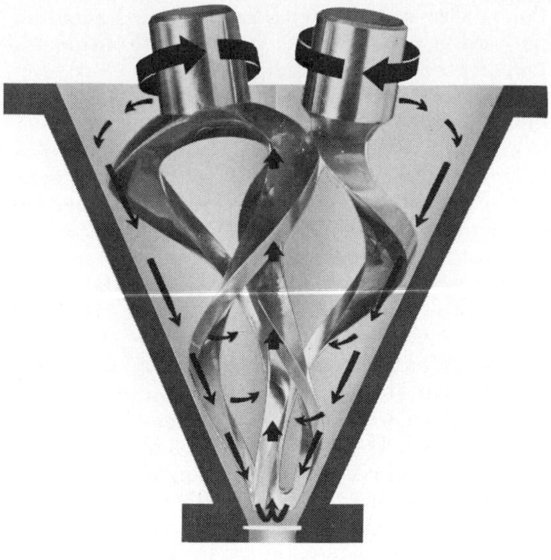

Fig. 8. Ribbon turbine with double spiral, for high-viscosity liquids.

Light-duty, permanently mounted mixers range in size from fractional horsepower to 3 hp, and are used on open or closed tanks up to 3000-gal capacity. Heavy-duty, permanently mounted, top-entering mixers are made up to 500 hp (373,000 W). They are normally designed for speeds of 45–200 rpm, made with independent mounting for the mixer shaft, and connected to the drive by a flexible coupling to protect the speed-reduction gearing. *See* UNIT OPERATIONS.

[VINCENT W. UHL]

Bibliography: F. Holland and F. S. Chapman, *Liquid Mixing and Processing in Stirred Tanks*, 1966; S. Nagata, *Mixing*, 1975; J. H. Rushton and J. Y. Oldshue, *Chem. Eng. Progr. Symp. Ser.*, 55 (25), 1959; Z. Sterback and P. Tausk, *Mixing in the Chemical Industry*, 1965; V. W. Uhl and J. B. Gray, *Mixing: Theory and Practice*, vol. 1, 1966, vol. 2, 1967.

Mixture

An aggregate composed of two or more distinct chemical components which retain their identities regardless of the degree to which they have become mingled. The substituents may be present in any proportion and may form a homogeneous or heterogeneous system. There is no chemical interaction between the components of a mixture, and it is possible, at least theoretically, to separate them by physical means. Attractive or repulsive forces between the different substituents, however, may affect the gross characteristics of the aggregate. For example, the total volume of a mixture of several liquids may not be the sum of the separate volumes. *See* CHEMICAL COMPOUNDS.

[FRANCIS J. JOHNSTON]

Mole

One mole is the mass numerically equal (in grams) to the relative molecular mass of a substance. It is the amount of a substance that contains the same number of molecules as there are atoms in 0.012 kilogram of carbon-12. The mole is an individual unit of mass, that is, it relates only to a given substance. If the relative molecular mass is μ, 1 mole = μ grams, and the M of one mole is μ g/mol.

Moles of different elements or compounds contain the same number of molecules. This number, called the Avogadro number N_0, is 6.023×10^{23}.

Since they contain the same number of molecules, moles of ideal gases occupy the same volume at the same temperature and pressure. At 0°C and 1 atmosphere (101,325 Pa) pressure, this volume, called the molar volume, is 22.4 liters. *See* AVOGADRO NUMBER; GAS; RELATIVE MOLECULAR MASS.

[THOMAS C. WADDINGTON]

Molecular distillation

A process by which substances are distilled in high vacuum at the lowest possible temperature and with the least damage to their composition. In a conventional distillation carried out at atmospheric pressure, molecules that leave a liquid surface as a vapor in their movement toward a condensing surface travel in straight paths until they collide with a molecule of residual gas (usually air), with another vapor molecule, or with the surface of the distilling vessel. These collisions either return the vapor to the liquid or greatly slow its rate of travel from the liquid to the condenser.

If the residual gas is virtually removed from a distillation vessel (for example, to pressures approximating 0.001 mm Hg), and if the condensing surface is located close to the evaporating liquid, then conditions are favorable for molecular distillation (Fig. 1). When the distance between the liquid and the condenser is so spaced that the vapor molecules travel without collision, this path is called the mean free path of the molecule. As an example, the mean free path of di-2-ethylhexyl phthalate, a high-boiling organic ester, is 0.482 cm.

Applicable substances. Materials composed of compounds with molecular weights in the approximate range of 400 to 1200 (there are exceptions) are effectively distilled by the molecular distillation process. Since this method can distill these

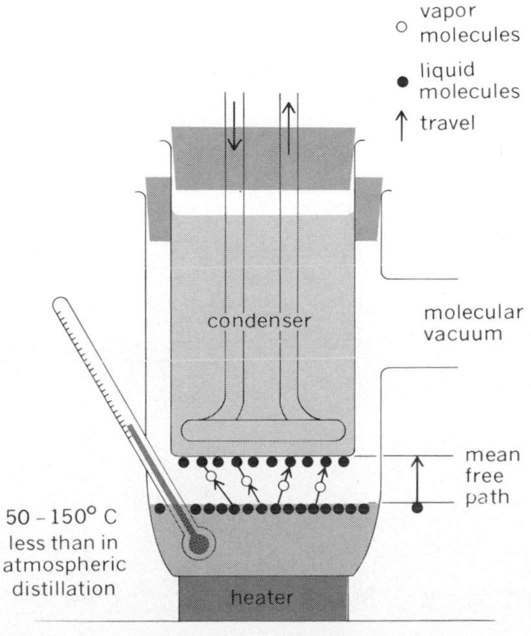

Fig. 1. Schematic diagram illustrating distillation at molecular distillation conditions.

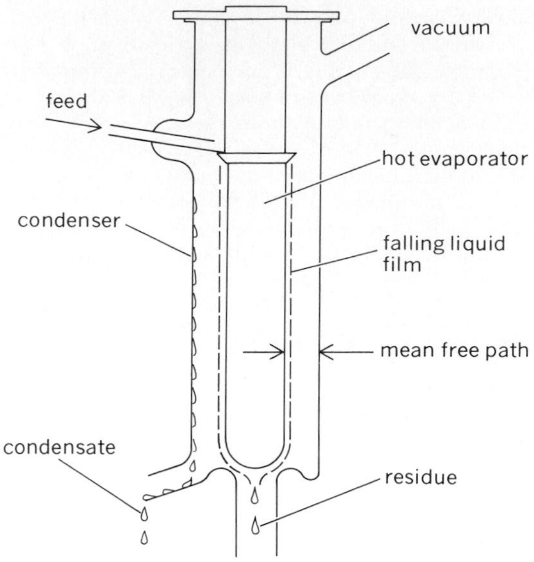

Fig. 2. The falling-film molecular still.

compounds at temperatures 50–150°C lower than by any other means, their thermal decomposition is at a minimum, which is very important when dealing with heat-sensitive materials. Also, a bonus advantage is the elimination of any tendency toward oxidation during distillation, since there is no atmospheric oxygen present.

At molecular distillation pressure, evaporation takes place at all temperatures, for there is no boiling point as there is in atmospheric pressure or in conventional vacuum distillation. Since evaporation is slow at lower temperatures, the liquid temperature is usually raised until the rate of evaporation meets practical requirements.

Much mathematics is involved in the theory of molecular distillation. The bibliography lists several citations covering these mathematical relations. One of the basic formulas for the rate of evaporation under these conditions was developed by I. Langmuir: $w = 0.0583\, P\sqrt{M/T}$, where w = weight evaporated in g/cm^2 of evaporation surface/sec; P = saturated vapor pressure (in mm Hg) of the substance at T, Kelvin temperature; M = molecular weight.

Molecular still types. Among the first to develop a molecular still were J. N. Brönsted and G. von Hevesy. The work of other pioneers in this field (C. R. Burch, H. I. Waterman, E. W. Washburn, and others) culminated in the falling-film molecular still, shown in Fig. 2. K. C. D. Hickman developed the centrifugal molecular still, illustrated in Fig. 3. In this still the material to be distilled, sometimes referred to as the distilland, is fed to the center of a hot, rapidly rotating cone that is housed in a chamber at a molecular distillation vacuum. Centrifugal force, 10 to 100 times that of gravity, spreads the distilland rapidly over the hot surface. Any evaporable material leaves the distilland as vapor, which travels the relatively short distance to liquefy on the condenser. The undistilled portion is caught as residue in a gutter and flows out of the chamber. Since the amount of destruction of heat-sensitive substances is proportional to the duration of exposure to the high temperatures involved in distillations, the almost instantaneous heat exposure followed by cooling, such as occurs in centrifugal molecular distillation, is definitely advantageous, especially when thermally unstable substances are involved. Hickman has calculated that this reduction in temperature exposure time reduces the decomposition by a factor of 10^6 times that which occurs in the best flash vacuum distillation carried out at 1 mm Hg in conventional stills.

Diffusion-ejector and oil ejector pumps, capable of moving relatively large volumes of residual gas at 10^{-3} to 10^{-2} mm Hg, provide the vacuum required in molecular distillation. These pumps use either mechanical or steam ejectors for forevacuum.

Commercial molecular distillation came into extended use in 1937 for distilling vitamin A esters from dogfish liver and shark liver oils. Centrifugal stills were introduced in 1941. This type of still has been made in sizes that process up to 2000 lb of material per hour.

Molecular distillation is widely used in the laboratory for analytical work, as well as in the chemical industry for the improvement of plasticizers, silicone fluids, flavors and fragrances, pharmaceuticals, and edible products. In the latter category, high-purity monoglycerides used in baked goods and confectionery, as well as cosmetics, are purified by molecular distillation.

[EDWARD S. BARNITZ]

Bibliography: G. Burrows, *Molecular Distillation*, 1960; J. Hollo et al. (eds.), *Applications of Molecular Distillation*, 1980; E. S. Perry and A. Weissberger, *Techniques of Chemistry*, vol. 12: *Separation and Purification*, 3d ed., 1978; P. Ridgway-Watt, *Molecular Stills*, 1963.

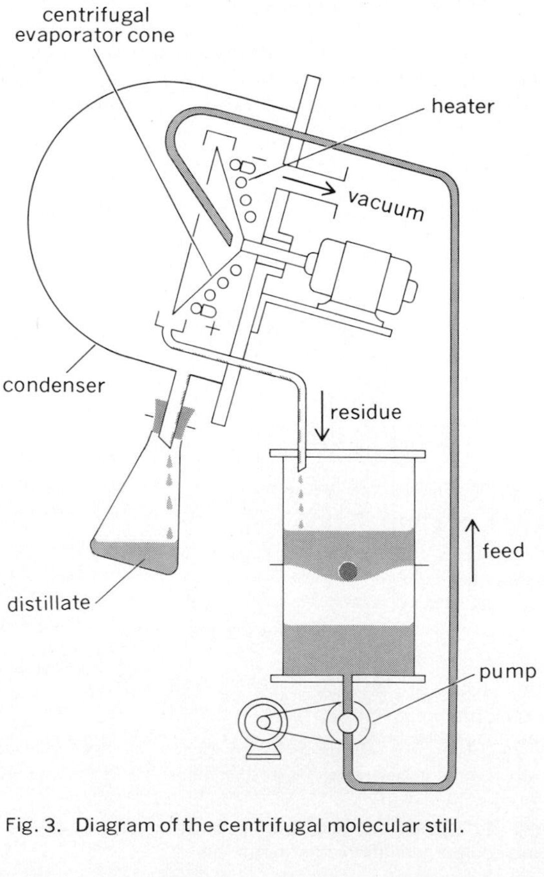

Fig. 3. Diagram of the centrifugal molecular still.

Molecular isomerism

The property of compounds (isomers) which have the same molecular formula but different physical and chemical properties. The difference in properties is caused by a difference in molecular structure (that is, molecular architecture). A typical example is dimethyl ether, CH_3OCH_3, a chemically quite inert gas which condenses at −24°C, and ethyl alcohol, CH_3CH_2OH, a liquid of substantial chemical reactivity which boils at 78°C; both compounds have the molecular formula C_2H_6O.

Classification. Isomers may be classified as constitutional isomers or stereoisomers. Constitutional isomers differ in constitution or connectedness, relating to the question as to which atoms are linked to which others and how. Dimethyl ether and ethanol (Fig. 1) are constitutional isomers. In dimethyl ether each carbon is connected to three hydrogen atoms and the one oxygen atom; the two carbon atoms are thus equivalent. In ethyl alcohol (ethanol) one carbon is linked to three hydrogen atoms and the other carbon; the second carbon is linked to the first carbon, two hydrogens, and the oxygen atom which, in turn, is linked to the sixth hydrogen atom; the two carbon atoms are not equivalent. (Constitutional isomers can often be distinguished, and their constitution recognized, by carbon-13 nuclear magnetic resonance.) Stereoisomers, in contrast, have the same constitution but differ in the three-dimensional array of the atoms in space, called configuration. (In some cases, the difference in three-dimensional arrangement may, however, be due to rotation about single bonds, in which case it is spoken of as a difference in conformation.) *See* NUCLEAR MAGNETIC RESONANCE (NMR).

Constitutional isomers. Constitutional isomers have been subdivided into functional isomers, positional isomers, and chain isomers.

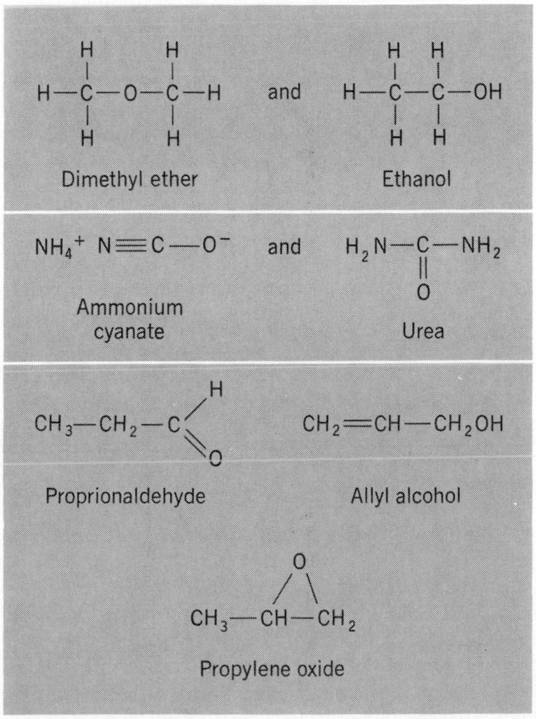

Fig. 1. Functional isomers.

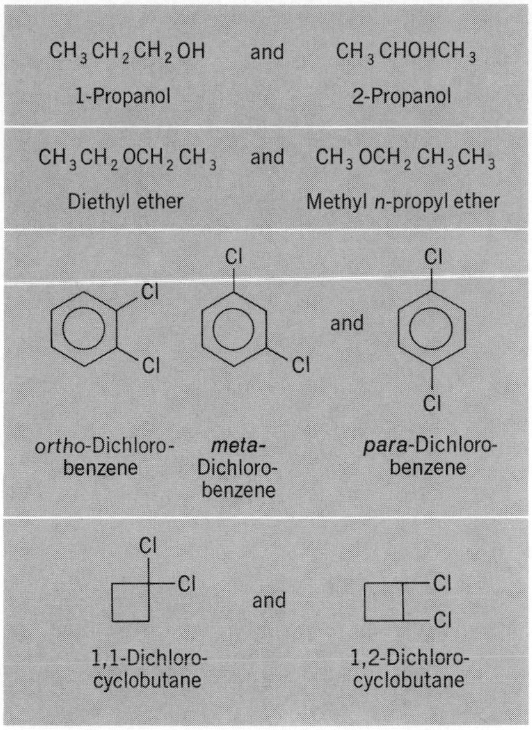

Fig. 2. Positional isomers.

Functional isomers. Functional isomers (Fig. 1) differ in functional group, that is, the group (or groups) most material in determining chemical behavior. The ammonium cyanate–urea pair (CH_4N_2O) shown in Fig. 1 plays an important role in the history of chemistry inasmuch as the conversion of ammonium cyanate to urea by heating, effected by F. Wöhler in 1828, is considered the first example of an organic compound (urea) having been produced in the laboratory from a mineral one (ammonium cyanate). The third example shown in Fig. 1, that of propionaldehyde (propanal), allyl alcohol (2-propen-1-ol), and propylene oxide (methyloxirane), illustrates the fact that functional isomers do not necessarily come in pairs. The three compounds all correspond to the molecular formula C_3H_6O, but the first one has an aldehyde function, the second combines a double bond with an alcohol function, and the third one has an epoxide function. Indeed, the number of possible isomers corresponding to a given molecular formula is generally remarkably large. Thus, even for the relatively simple composition $C_{10}H_{22}$, a saturated hydrocarbon with 10 carbon atoms, there are 75 isomers, and for a hydrocarbon with twice as many carbon atoms—$C_{20}H_{42}$—the number is well over 300,000.

Positional and chain isomers. Positional isomers (Fig. 2) have the same functional group but differ in its position along a chain or in a ring. Closely related are chain isomers which also have the same functional group or groups but differ in the shape of the carbon chain (Fig. 3*a*); quite similar are ring isomers (Fig. 3*b*) which differ in the size of one or more rings. Ring and chain isomers together are sometimes called skeletal isomers.

Properties. It should be emphasized that these

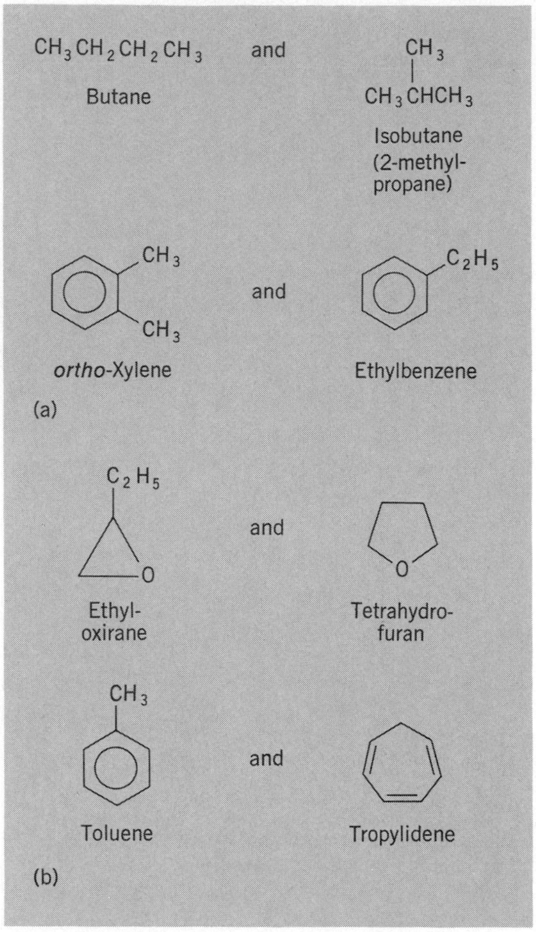

Fig. 3. Skeletal isomers. *(a)* Chain isomers. *(b)* Ring isomers.

subclassifications of constitutional isomers are made for the convenience of the chemist rather than because of any fundamental importance. All

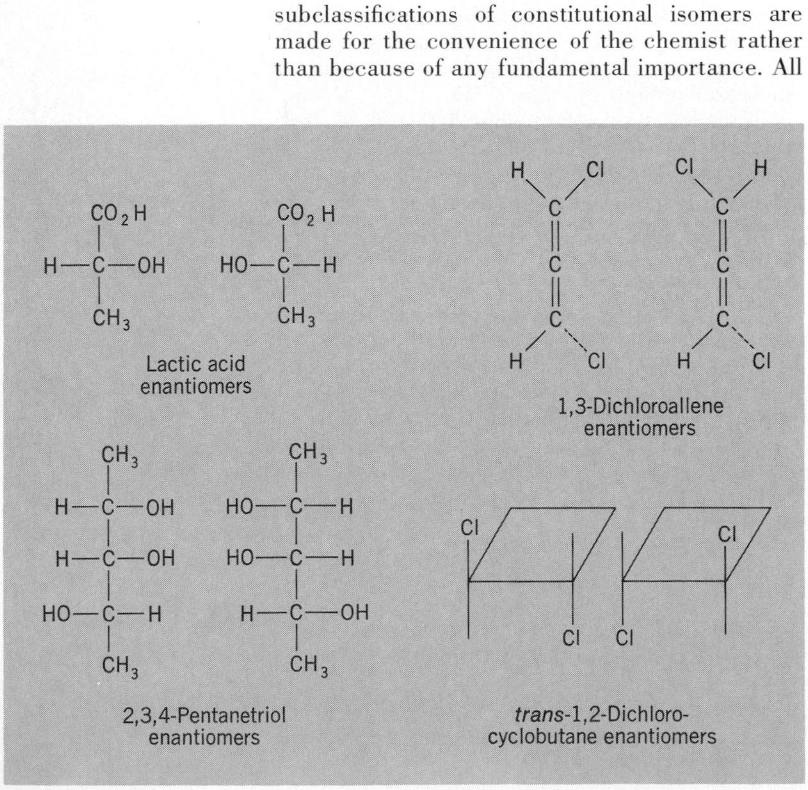

Fig. 4. Enantiomers.

constitutional isomers differ in physical and chemical properties, such as melting and boiling points, density, refractive index, and free energy, as well as in all kinds of spectral properties, such as ultraviolet, infrared and nuclear magnetic resonance spectra, and, to a lesser extent, mass spectra. If crystalline, isomers can generally be assigned their proper structure by x-ray diffraction analysis. The above differences tend to be greatest for functional isomers and more subtle for positional and skeletal isomers. However, the last statement is not universally true; thus, there is a considerable difference between the fairly reactive ethyloxirane and the fairly inert ring isomer tetrahydrofuran shown in Fig. 3*b*. Indeed, it is often not entirely clear when isomers should be called functional and when they should be called positional or chain isomers. Thus acetone (propanone) $CH_3C(=O)CH_3$ and propionaldehyde (propanal) $CH_3CH_2CH{=}O$ may be considered positionally isomeric carbonyl compounds if one chooses not to distinguish between aldehyde and ketone functions. Yet these functions are substantially different (for example, aldehydes are readily oxidized to acids, while ketones are not), so an alternative might be to consider the ketone and the aldehyde as functional isomers. A similar situation occurs with amines: *t*-butylamine, $(CH_3)_3CNH_2$, and *n*-butylamine, $CH_3CH_2CH_2CH_2NH_2$, are clearly chain isomers, whereas diethylamine,

$$C_2H_5N(H)C_2H_5$$

and methylpropylamine,

$$CH_3N(H)CH_2CH_2CH_3$$

should be classified as positional isomers. When one compares *n*-butylamine with diethylamine, however, the situation is less clear-cut. One could argue that these are positional isomers also, but in view of the functional difference (though slight) between primary and secondary amines, they are probably better classified as functional isomers. Such differences in classification are obviously quite tenuous.

Stereoisomers. Compounds which have not only the same molecular formula but also the same constitution (connectivity of the atoms) but which differ in the disposition of the atoms in space are called stereoisomers. Stereoisomers, in turn, are subdivided into two types: those that are mirror images of each other, called enantiomers, and those which are not mirror images, called diastereomers or diastereoisomers.

Enantiomers. These isomers are unique in that they always come in pairs (Fig. 4). Either a molecule is superposable with its mirror image, in which case it does not have an enantiomer, or it is not superposable with its mirror image, in which case it has one and only one enantiomer (since an object can have only one mirror image). Molecules which are not superposable with their mirror images are called chiral; those which are so superposable are called achiral. Enantiomers are much more alike than are other sets of isomers (constitutional isomers or diastereomers); thus they have the same melting point, boiling point, free energy, spectral properties, x-ray diffraction pattern, and so on. This is because their internal relationships are the same; for example, the distances between corresponding atoms are the same, much as the distances between corresponding fingers are the same in the right or left hand. However, enantiomers differ in their behavior toward chiral reagents, much as a right hand and a left hand differ in their relation to a right glove, and they are also different in their behavior toward chiral physical agents, such as circularly polarized light. Thus they differ in circular dichroism and in the direction of rotation which they impart to plane polarized light (optical rotation, optical rotatory dispersion). Such differential physical properties have been called chiroptic properties.

Diastereomers. These isomers have the same constitution but different spatial arrangement and are not mirror images (Fig. 5). They resemble constitutional isomers in that there may be more than two isomers in a set and that their physical, energetic, and spectral properties are generally quite distinct. The example of *cis-* and *trans-*1,2-dibromoethene illustrates cis-trans isomerism in olefins. (It is recommended that the term geometrical isomers which was formerly used for this type of isomers be abandoned, just as optical isomers should no longer be used as a synonym for enantiomers.) 1,3-Dichlorocyclobutane and 1,2-dimethylcyclopropane illustrate diastereoisomerism (also of the cis-trans type) in cyclanes. The 1,3-dichlorocyclobutane exists in two achiral diastereomeric forms, whereas the 1,2-dimethylcyclopropane has a pair of (chiral) enantiomers (trans) which are diastereomeric with the (achiral) cis isomer (called meso because it is an achiral diastereomer in a set also containing chiral species). Pentane-2,3,4-triol illustrates a case with one chiral and two achiral (meso) diastereomers. The enantiomeric pair of the chiral 2,3,4-pentanetriols is shown in Fig. 4.

A set of stereoisomers containing n chiral centers will normally contain 2^n members. Since each member of the set has an enantiomer, there will be 2^{n-1} enantiomer pairs which, in relation to each other, are diastereomeric. However, when there is degeneracy, that is, when two or more of the chiral centers are equivalent (as in the 2,3,4-pentanetriols where the chiral centers at C-2 and C-4 are alike), there will be fewer isomers than the formula predicts. Thus there are only four stereoisomeric 2,3,4-pentanetriols instead of the eight (2^3) predicted by the formula (see Figs. 4 and 5). A general method for counting stereoisomers has been developed.

Limitations. Many isomers are quite stable and if they can be interconverted at all, the barrier between them is quite high (Fig. 6). For example, the barrier between *cis-* and *trans-*2-butene, $CH_3CH{=}CHCH_3$, is of the order of 65 kcal/mol (272 kJ/mol), and such isomers (the 1,2-dibromoethenes in Fig. 5 are another example) are spontaneously interconverted only at very high temperatures. In other cases, however, the barriers are intrinsically quite low or are readily lowered by deliberate or adventitious catalysis, frequently by acids or bases. An example relates to the keto and enol forms of ethyl acetoacetate (Fig. 7*a*). While

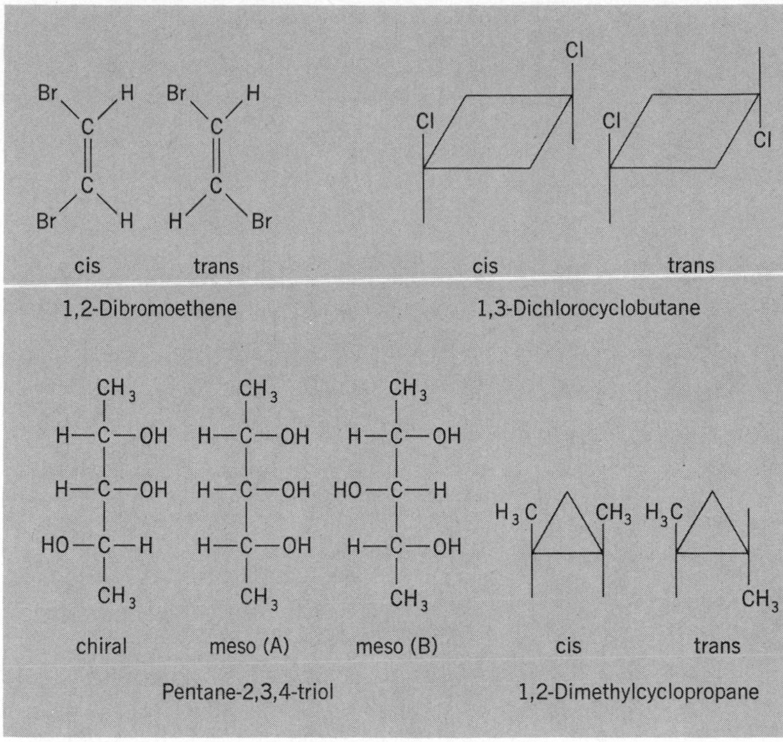

Fig. 5. Diastereomers.

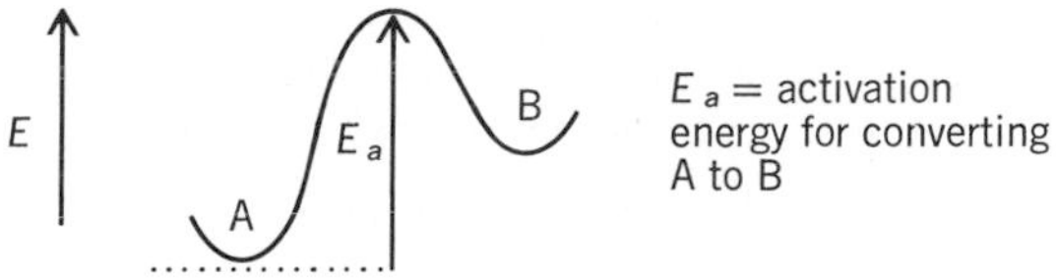

Fig. 6. Energy barriers between isomers A and B.

these isomers can be separated by fractional distillation in clean quartz vessels, they are quickly interconverted in the presence of traces of acids or bases. Such easily interconvertible isomers which differ only in the position of an atom or group (in the case of ethyl acetoacetate, a hydrogen atom which is attached to carbon in the keto form and to oxygen in the enol form) are called tautomers, and the phenomenon is called tautomerism. A closely

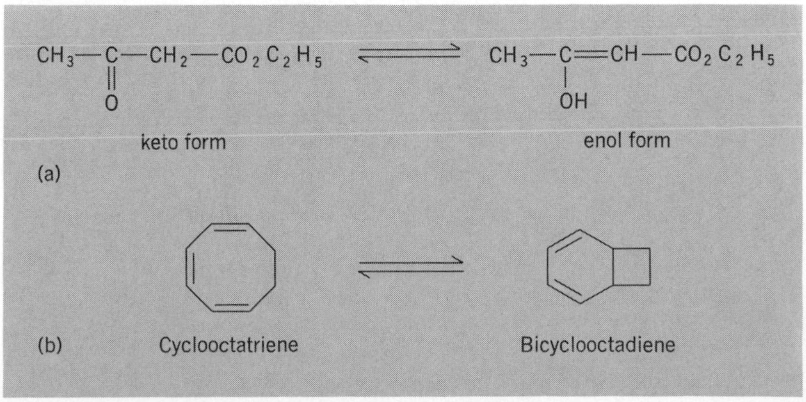

Fig. 7. Easily interconverted isomers. (*a*) Ethyl acetoacetate. (*b*) Cyclooctatriene and bicyclooctadiene.

related type of isomerism in which only bonds shift and atoms remain in place (except for changes in bond distances) is represented by the cyclooctatriene-bicyclooctadiene interconversion (Fig. 7*b*) which takes place readily at room temperature. This kind of rapid isomerization is referred to as valence bond isomerism or valence tautomerism. Two other examples of rapidly interconverting isomers are shown in Fig. 8. The first is chlorocyclohexane which exists in an equatorial and an axial conformation rapidly interconverted by reversal (flipping) of the cyclohexane chair. The barrier to this interconversion is about 10 kcal/mol (42 kJ/mol), which is so low that the two conformational isomers (sometimes called conformers) can be isolated only at temperatures as low as −150°C. However, the two isomers can be seen separately in nuclear magnetic resonance (NMR) spectra below about −65°C (the exact temperature depends on the frequency of the instrument and the nucleus observed). The axial and equatorial conformations of chlorocyclohexane are thus diastereomeric. In 1,2-dichloroethane (Fig. 8), three stereoisomeric conformational minima (two enantiomeric gauche conformations and the achiral anticonformation that is diastereomeric to the two others) can be discerned, but the barrier to rotation is so low (about 3 kcal/mol or 13 kJ/mol) that isolation is out of the question, and even many physical techniques (such as electron diffraction, NMR spectroscopy, and dipole moment measurement) yield only weighted average values of the physical properties of the three conformations. However, the gauche isomers and anti-isomers can be seen distinctly in the infrared spectrum of the substance and are thus clearly different species.

Ultimately there is the question of what will happen if the barrier to interconversion becomes even lower. A point must come where it is no longer possible to speak of two distinct isomeric molecules but where only a single molecule is deemed to exist. This happens when there is no longer an operational way of demonstrating the existence of two separate energy minima. It has been suggested that there are not two isomeric molecules when the barrier (Fig. 6) is lower than the product of the gas constant *R* on the absolute

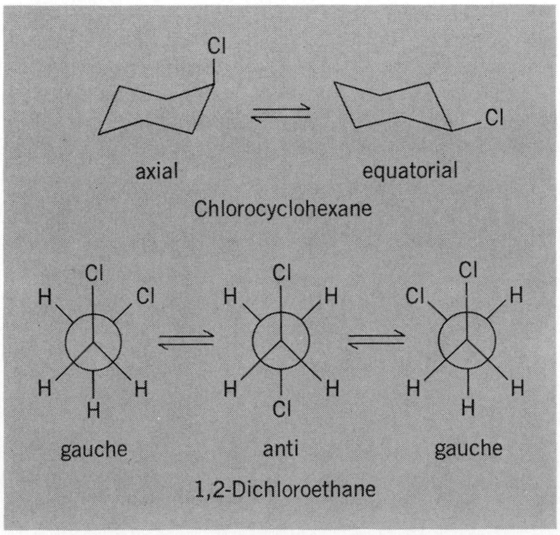

Fig. 8. Conformational isomers.

temperature *T* (about 0.6 kcal/mol or 2.5 kJ/mol at room temperature), since under those circumstances the molecule traverses the barrier in a single molecular vibration, but this limit is clearly somewhat arbitrary. In any case, it is clear that the differentiation between two distinct isomeric molecules and two energy states of a single molecule is not a sharp one in those instances where the energy barrier between isomers is very low. *See* CONFORMATIONAL ANALYSIS; OPTICAL ACTIVITY; STEREOCHEMISTRY; TAUTOMERISM.

[ERNEST L. ELIEL]

Bibliography: E. L. Eliel, On the concept of isomerism, *Israel J. Chem*, 15(1-2):7–11, 1977; J. G. Nourse, Applications of artificial intelligence for chemical inference, *J. Amer. Chem. Soc.*, 101(5): 1210–1216, 1979; J. F. Stoddart in D. Barton and W. D. Ollis (eds.), *Comprehensive Organic Chemistry*, vol. 1, pp. 13–15, 24–26, 1979; P. Uzzell, *Aspects of Isomerism*, 1971.

Molecular orbital theory

A quantum-mechanical (mathematical) model for the electronic structure and chemical bonding of a molecule. Each electron is associated with its particular wave function ψ, which spans the entire molecule. Each molecular orbital (MO), as these wave functions are called, is definable by certain quantum numbers which govern its shape and energy. In addition to the spatial quantum numbers, each electron possesses a spin quantum number which has a value of either $\pm\frac{1}{2}$.

It is reasonable that the form of the molecular orbital in the vicinity of a particular atom should be similar to that of an atomic orbital (AO) located on that atom, since the forces on the electron will be chiefly those due to the nucleus of the atom and other electrons near that nucleus. Hence, a frequently used approximation is to express the molecular orbital as a linear combination of atomic orbitals (LCAO) which include all the atomic centers. Thus, Eq. (1) can be written, where the ϕ_i

$$\psi = \sum_i C_i \phi_i \qquad (1)$$

are normalized atomic orbitals located on the various atoms and the C_i are coefficients which determine the contribution of the atomic orbitals to the molecular orbital. The set of atomic orbitals ϕ_i used to construct the molecular orbital is called the basis set for the molecular orbital.

The energy of the system is given by Eq. (2),

$$E = \int \psi^* \mathscr{H} \psi \, d\tau / \int \psi^* \psi \, d\tau \qquad (2)$$

where $\mathscr{H}$ is the hamiltonian operator which takes into account all the kinetic and potential energy contributions, and $d\tau$ is the volume element for the electron. Since the energy will depend upon the coefficients C_i, application of the variation principle will specify those coefficients for which the energy is a minimum. This approach requires the solution of the secular determinant, Eq. (3), where

$$|H_{ij} - ES_{ij}| = 0 \qquad (3)$$

S_{ij} is the overlap integral, $\int \phi_i^* \phi_j d\tau$, and $H_{ij} = \int \phi_i^* \mathscr{H} \phi_j \, d\tau$. Evaluation of the secular determinant results in *n* values of the energy *E*, where *n* is the number of ϕ functions used to construct the molecular orbital. Each energy value will generate its own set of coefficients C_i, so there will be a set

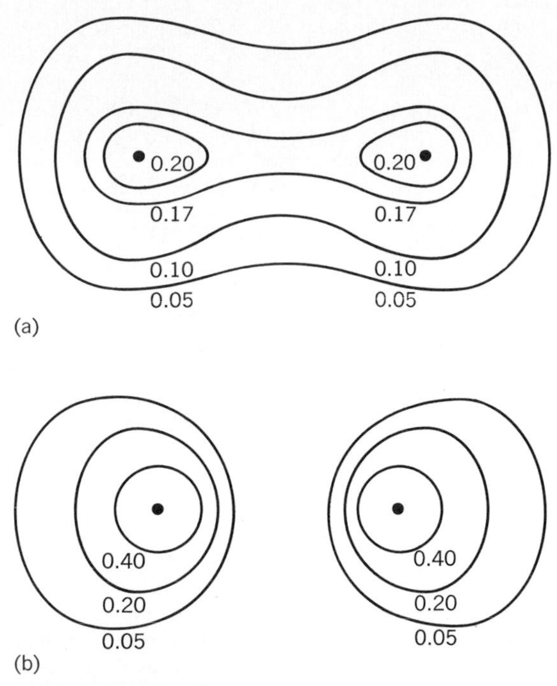

Fig. 1. Contour diagrams for H_2^+. (*a*) Bonding orbital. (*b*) Antibonding orbital.

of n molecular orbitals, ψ_n, one associated with each energy E_n. Therefore, the number of molecular orbitals will be identical with the number of atomic orbitals used. *See* QUANTUM CHEMISTRY.

The H_2^+ molecules. Applied to the simple system H_2^+ in which only 1*s* orbitals on each of the hydrogen atoms *A* and *B* are used to construct the molecular orbitals, the theory results in two molecular orbitals, Eqs. (4) and (5).

$$\psi(b)=\frac{1}{\sqrt{2}}\phi_A(1s)+\frac{1}{\sqrt{2}}\phi_B(1s) \qquad (4)$$

$$\psi(a)=\frac{1}{\sqrt{2}}\phi_A(1s)-\frac{1}{\sqrt{2}}\phi_B(1s) \qquad (5)$$

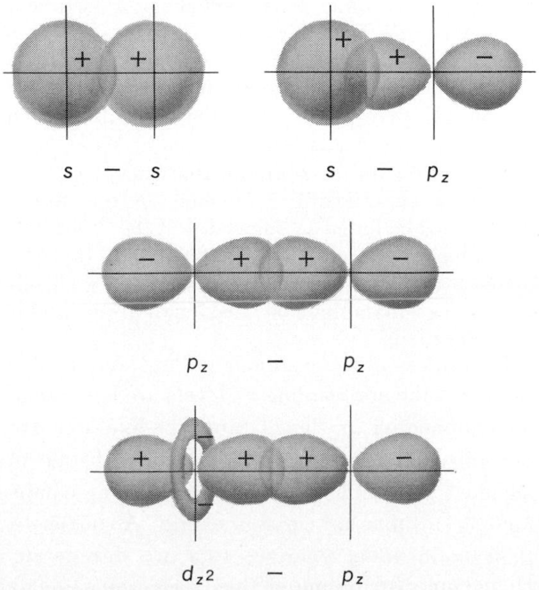

Fig. 2. Atomic orbitals capable of σ-bonding.

Contour diagrams of the electron densities $\psi^*\psi$ for the two functions are given in Fig. 1. The numerical values on the diagrams are for $\psi^*\psi=$ constant evaluated in atomic units. In the case of the bonding orbital ψ_b, the electron density is concentrated between the two nuclei while the antibonding orbital ψ_a possesses a nodal plane in this region and the charge is pushed away from the nuclei. Note that both orbitals have axial symmetry along the line connecting the two nuclei. Such orbitals are called sigma (σ) orbitals and can result from combinations of atomic orbitals whose charge densities prior to bonding lie along the line. Some typical examples of such orbitals are shown in Fig. 2.

Molecular orbitals, bonding and antibonding, can also be constructed by linear combinations of atomic orbitals such that charge densities occur above and below the line connecting the two nuclei, but a nodal plane exists along the line. For example, combinations of p_x orbitals on adjacent atoms can result in bonding and antibonding orbitals as illustrated in Fig. 3. The resultant orbitals are called pi (π) orbitals. The p_y orbitals on each of the atoms in a diatomic molecule are capable of analogous combinations. In addition, π-bonding can occur between p and d or between d and d orbitals, as shown in Fig. 4. Generally, π-interactions are not as strong as σ-interactions, so that the bonding orbital is not lowered nor the antibonding orbital raised in energy as much in π-bonding compared to σ-bonding.

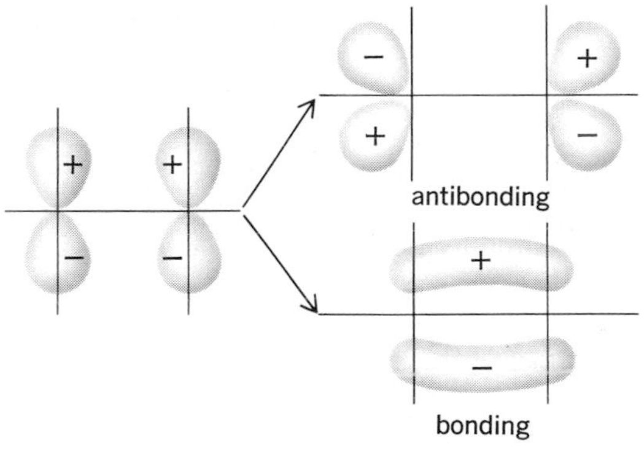

Fig. 3. π-Bonding and antibonding p orbitals.

For the H_2^+ molecule, a qualitative diagram of the two energy levels associated with ψ_b and ψ_a, Eqs. (4) and (5), is given in Fig. 5*a*. The energy of ψ_b is given by Eq. (6), while that of ψ_a is given by

$$E_b=(H_{AA}+H_{AB})/(1+S_{AB}) \qquad (6)$$

Eq. (7). Since H_{AA} and H_{AB} are both negative and

$$E_a=(H_{AA}-H_{AB})/(1-S_{AB}) \qquad (7)$$

S_{AB} is positive and less than one, E_b is lower and E_a is higher than H_{AA}. The single electron of H_2^+ would occupy the E_b energy level.

An analogous figure would be applicable to a description of H_2 except that the numerical values of the various quantities would be different. The

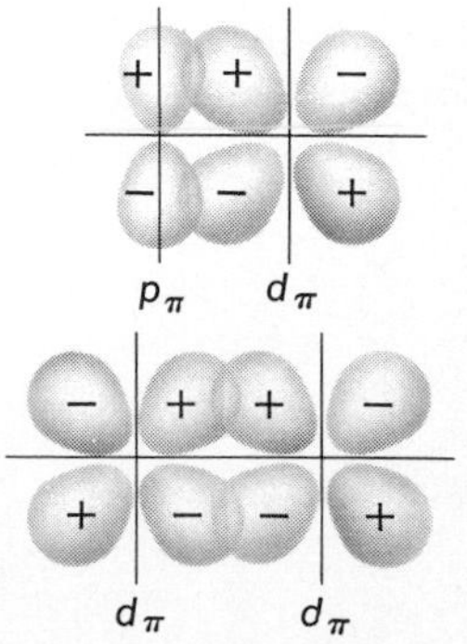

Fig. 4. $p\pi-d_\pi$ and $d\pi-d_\pi$ π-bonding orbitals.

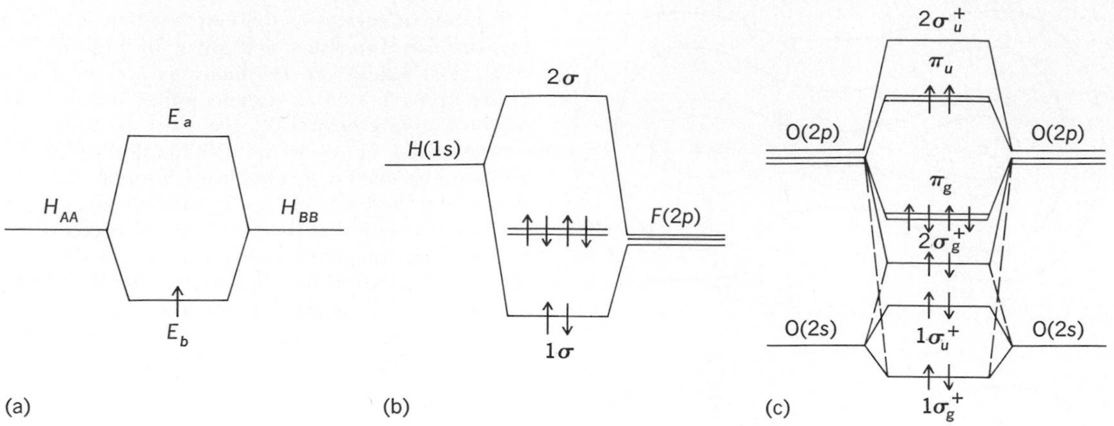

Fig. 5. Energy levels for diatomic molecules. (*a*) Levels for H_2^+. (*b*) Levels for HF. (*c*) Levels for O_2.

two electrons in H_2 can both occupy the E_b level and possess the spatial function ψ_b, but with different spins, $\pm\frac{1}{2}$. The occupation of the bonding molecular orbital by the two electrons constitutes the formation of a single bond between the two atoms. The nonexistence of He_2 can also be rationalized on the basis of Fig. 5*a* since the second set of two electrons would, by the Pauli exclusion principle, be required to occupy the E_a energy levels. Relative to the initial energy levels, E_a is raised to a greater extent than E_b is lowered. Thus the energy of the system is less for the separated atoms than for the combined molecule.

In the previous examples the aufbau, or building up, method was used. That is, the molecular orbital energy level scheme was constructed and the electrons were then placed in the levels in ascending order in keeping with the Pauli exclusion principle. While the general form of the LCAO – molecular orbital method specifies each electron as existing in a molecular orbital, for many qualitative discussions and approximate calculations it is adequate to presume that those electrons that occupy the lower quantum number orbitals in the separate atoms (1*s*, 2*s*, and so on) remain essentially localized on their respective atoms and only the outermost or valence orbitals need be considered in the construction of the molecular orbitals. For example, in the description of SiH_4 one might consider the silicon $1s^22s^22p^6$ electrons as localized on that atom and use only the 3*s* and 3*p* orbitals and electrons in conjunction with the hydrogen 1*s* orbitals and electrons to construct the pertinent bonding molecular orbitals.

Diatomic molecules. The following sections deal with qualitative applications of the theory. In particular the hydrogen fluoride and oxygen molecules are discussed.

Hydrogen fluoride. The pertinent levels for this molecule are given in Fig. 5*b*. The fluorine 1*s* and 2*s* orbitals are presumed to be localized on that atom. The 2*p* valence orbitals are placed below the 1*s* level of hydrogen on the basis of electronegativity considerations. Only one of the 2*p* orbitals is capable of interaction with the hydrogen 1*s* to form a σ-bonding and antibonding set of molecular orbitals, the other two are *nonbonding* since S_{ij} with the hydrogen 1*s* is zero. The six electrons, one from the hydrogen and five from the fluorine, are distributed according to the aufbau principle.

Two electrons in the 1σ level form the single bond, and the other four electrons occupy the nonbonding degenerate (equal energy) fluorine *p* orbitals. Because the energy of the fluorine *p* orbital is closer to the 1σ energy level than the energy of the hydrogen 1*s* orbital, the 1σ molecular orbital has more fluorine than hydrogen character. That is, in the wave function $\psi(1\sigma) = C_1\,\phi\,H(1s) + C_2\,\phi_F\,(2p)$, the coefficient C_2 is larger than C_1. Therefore, the charge density of the two electrons in the molecular orbital is substantial near the fluorine atom, resulting in a partial negative charge on the fluorine and a positive charge on the hydrogen in accord with the polar character of the molecule.

Oxygen. If the 2*s* and 2*p* valence orbitals on each oxygen atom are considered in the basis set for the oxygen molecule, the resultant energy levels are as given in Fig. 5*c*, where the molecular energy levels are labeled $1\sigma_g^+$, $1\sigma_u^+$, πg, πu, and so on, according to group theoretical designations. These symbols characterize certain mathematical properties of the wave functions associated with the energy levels. Traditionally, Greek letters are used for such designations when dealing with small molecules such as O_2 while roman letters are commonly employed for larger molecules. (See the section on group theory in this article.) Both the 2*s* and $2p_z$ orbitals on each atom are capable of σ-bonding interactions so that, strictly speaking, a particular σ-level cannot be designated as pure *s-s* or *p-p* σ-bonding. However, because of the energy separation of the *s* and *p* atomic orbitals, the contribution of, say, the p_z orbitals in $1\sigma_g^+$ are not appreciable and the diagram could have been simplified by the assumption that the pairs of 2*s* electrons were localized on each of the atoms. They were included to suggest that the 2*s* orbitals could have some effect on the position of the $2\sigma_g^+$ level so that its position relative to the π_g levels cannot be unambiguously assigned from simple considerations.

Regardless of the positions of $2\sigma_g^+$ and π_g, it is clear that the antibonding π_u levels are lower than the antibonding $2\sigma_u^+$ level. Since the interaction of the adjacent p_x orbitals is identical with that of adjacent p_y orbitals, both form π-bonding combinations (π_g) that are equal in energy. Analogously, their antibonding energies (π_u) are degenerate. Hence, on distribution of the 12 electrons among the orbitals, the last two can be placed in separate spatial orbitals. Since these two electrons are not required to occupy the same spatial orbital, their spin functions need not be opposed but can be identical, which accounts for the observed paramagnetic triplet (two unpaired electrons) ground state of this molecule. The ability of molecular orbital theory to readily account for the paramagnetism of O_2 was one of its early successes.

Group theory. In dealing with large molecules which possess a substantial degree of symmetry, it is extremely helpful to apply the principles of mathematical group theory in order to simplify the system. By group theory certain linear combinations of atomic functions can be generated such that a very large secular determinant can be factored or simplified to a set of smaller determinants. The functions are said to be symmetry-adapted and form bases for the irreducible representations of the molecular point group.

If a single linear combination acts as a basis function, the representation is generally designated with an A or B, depending on the properties of the function under various symmetry operations. If two functions together form a basis, then the representation is symbolized with an E and the two functions are known to be energetically degenerate. If three functions are basis functions for a threefold irreducible representation, then all three are of equal energy and assigned the symbol T.

Consider the π-electron levels in the benzene molecule (Fig. 6). If the individual p_x orbitals on each of the six carbon atoms were used as basis functions for the secular determinant, a six-by-six determinant would have to be solved. However, by group theory it is possible to construct six new symmetry-adapted basis functions from combinations of the six atomic orbitals. They are Eqs. (8)–(13). In each case N is an appropriate normalization constant.

$$(A_{2u}) = N_{A_{2u}}(\phi_1+\phi_2+\phi_3+\phi_4+\phi_5+\phi_6) \quad (8)$$

$$(B_{2g}) = N_{B_{2g}}(\phi_1-\phi_2+\phi_3-\phi_4+\phi_5-\phi_6) \quad (9)$$

$$(1E_{1g}) = N_{1E_{1g}} \cdot (2\phi_1+\phi_2-\phi_3-2\phi_4-\phi_5+\phi_6) \quad (10)$$

$$(2E_{1g}) = N_{2E_{1g}}(\phi_2+\phi_3-\phi_5-\phi_6) \quad (11)$$

$$(1E_{2u}) = N_{1E_{2u}} \cdot (2\phi_1-\phi_2-\phi_3+2\phi_4-\phi_5-\phi_6) \quad (12)$$

$$(2E_{2u}) = N_{2E_{2u}}(\phi_2-\phi_3+\phi_5-\phi_6) \quad (13)$$

The two functions $(1E_{1g})$ and $(2E_{1g})$ together form basis functions for the irreducible representation E_{1g}. Similarly, $(1E_{2u})$ and $(2E_{2u})$ are basis functions for E_{2u}, while (A_{2u}) and (B_{2g}) are basis functions for the A_{2u} and B_{2g} representations, respectively. Each of the functions is either one of a pair of functions for a doubly degenerate (E) representation or is a basis function for a singly degenerate (A or B) representation, and no two representations are identical. Use of these functions to construct the secular determinant results in zero values for all the elements in the six-by-six determinant except the diagonal elements $H_{ii}-E$. Therefore, the larger determinant has been factored into 6 one-by-one determinants which can be solved directly, $E_i = H_{ii}$. The methods for evaluating each H_{ii} will be covered in the next section.

It should be noted that the π-system of benzene is a simple example since no H_{ij} terms remain after use of the symmetry-adapted functions. Generally this is not the case. Notice also that group theory does not indicate the sequence of the resultant energy levels; for this, one must turn to either rigorous or approximate quantum mechanics.

Approximate methods. Various approximations can be employed to deal with the quantum mechanics of both organic and inorganic systems.

Organic systems. Whether group theory is used or not, one is eventually faced with evaluation of the elements in the secular determinant. One particularly successful theory for qualitative considerations and approximate calculations for organic systems is known as Hückel theory

In the simplest version of this approach, all overlap integrals S_{ij} are considered to be zero except when $i=j$, in which case $S_{ii}=1$. This simplifies all off-diagonal elements in the secular determinant from $H_{ij}-ES_{ij}$ to H_{ij}. The diagonal terms are, of course, $H_{ii}-E$. A further approximation in Huckel theory is that the integral $\int\phi_i\mathscr{H}\phi_j\,d\tau$, where ϕ_i and ϕ_j are atomic orbitals, is zero unless ϕ_i and ϕ_j are on the same or adjacent atoms. In the evaluation of the π-energy levels in benzene, for example, $\int\phi_1\mathscr{H}\phi_2\,d\tau$ is nonzero, while $\int\phi_1\mathscr{H}\phi_3\,d\tau$ would be set equal to zero since carbon atom 1 is not adjacent to carbon atom 3. The nonzero integrals which remain are frequently designated as shown by the Eqs. (14). If all the functions under consideration

$$\int\phi_i{}^*\mathscr{H}\phi_i\,d\tau = \alpha_i$$
$$\int\phi_i{}^*\mathscr{H}\phi_j\,d\tau = \int\phi_j{}^*\mathscr{H}\phi_i\,d\tau = \beta_{ij} \quad (14)$$

are the same, for example, all the π-bonding p orbitals on the carbon atoms in benzene, then the subscripts on α_i and β_{ij} can be discarded. β_{ij} is frequently called the resonance integral.

Application of the Huckel approximations to the symmetry-adapted functions for the benzene π-electron system yields the energy values shown by Eqs. (15)–(18). Since β is negative, the energy levels are $E(A_{2u}) < E(E_{1g}) < E(E_{2u}) < E(B_{2g})$. The six

$$E(A_{2u}) = \alpha + 2\beta \quad (15)$$

$$E(1E_{1g}) = E(2E_{1g}) = \alpha + \beta \quad (16)$$

$$E(1E_{2u}) = E(2E_{2u}) = \alpha - \beta \quad (17)$$

$$E(B_{2g}) = \alpha - 2\beta \quad (18)$$

electrons in the π-orbitals are placed in the A_{1u} and E_{1g} orbitals to give the ground state configuration. $(A_{1u})^2(E_{1g})^4$. Note that the symbols for the irreducible representations become the designations for the spatial quantum numbers. In molecular orbital theory they are the analogs of the angular momentum numbers of atomic theory.

If π-bonding did not occur in benzene, each of the six electrons would occupy a localized p orbital on each atom whose energy would be equal to α. Because of π-bonding, the total energy of the six electrons is given by Eq. (19). Hence, the stabilization of benzene by π-bonding is 8β. If one were to

$$2E(A_{2u}) + 4E(E_{1g}) = 6\alpha + 8\beta \quad (19)$$

incorrectly assume that localized double bonding occurred, that is, that a single Kekule structure of three alternate double bonds properly represented the bonding in benzene, the calculated stabilization would be 6β. The difference between the two values, 2β, is called the resonance or delocalization energy. This energy can be estimated experi-

MOLECULAR ORBITAL THEORY

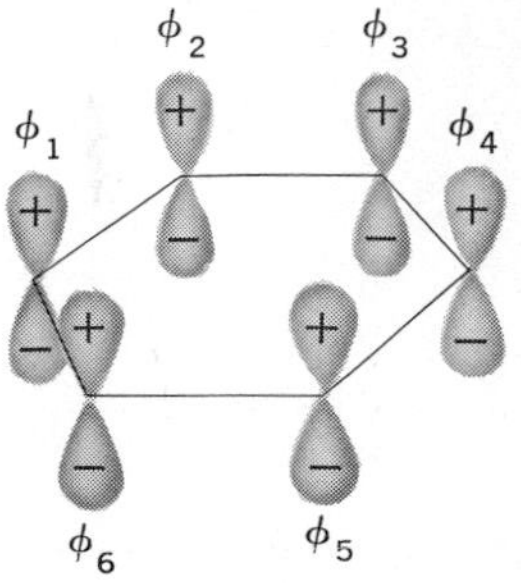

Fig. 6. π-Bonding atomic orbitals on benzene.

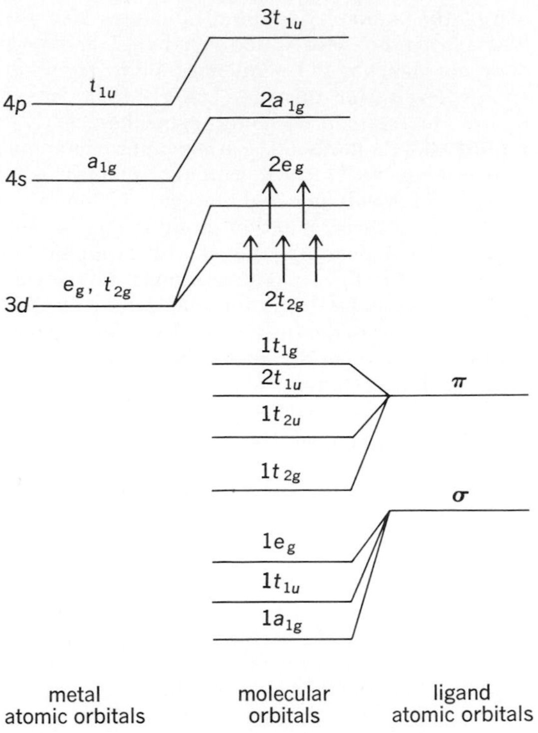

Fig. 7. Diagram of energy levels for π-electron donor complexes, FeF_6^{-3}.

mentally as the difference in the enthalpy of formation of benzene and the energy of formation of the hypothetical Kekule structure using C—C, C=C, and C—H bond energies. The value of β obtained in this way is 18–20 kcal/mole, depending on the choices that are taken for the bond energies. Very similar values of β are obtained by analogous considerations of other aromatic molecules. *See* RESONANCE.

Hundreds of calculations using the Huckel method have been made on a variety of organic molecules including large molecules of biochemical interest. Insight into the chemical and physical properties of the species has been achieved through these calculations. Some difficulty arises in dealing with systems containing several different atoms since the simplifications which could be made in dealing with identical atoms are no longer possible. The α-value for each atom will be different and several different β-terms, depending upon the nature of the adjacent atoms, must be introduced. Many methods have been proposed and are being tested for evaluation of the terms which arise in these more complex systems.

Inorganic systems. Approximate, or semiempirical, methods are also used for the treatment of inorganic systems ranging in complexity from small molecules to transition metal complexes such as FeF_6^{-3} and $Fe(CN)_6^{-3}$. While group theory permits substantial simplifications in dealing with complexes of high symmetry and a qualitative understanding of many systems is possible, their quantitative treatment is still in the developmental stage. Major difficulties arise because of the differences in electronegativities of the various atoms and the high negative charges on the total system, such as the −3 charges on the two complexes cited above. Considerable research is being carried out to elucidate an appropriate method or methods for the quantitative treatment of such species. *See* ELECTRONEGATIVITY.

Even qualitatively, appreciable insight into bonding characteristics and spectral and magnetic properties is possible through molecular orbital considerations. Figures 7 and 8 are the qualitative energy levels for FeF_6^{-3} and $Fe(CN)_6^{-3}$, respectively. The former is typical of π-donor complexes, while the latter is illustrative of π-acceptor complexes. *See* COORDINATION CHEMISTRY.

In π-donor complexes, the valence atomic orbitals of the ligands (the groups or atoms attached to the metal) are fully occupied with electrons prior to bonding and interact to form σ- and π-bonds with the appropriate metal orbitals. The ligand orbitals are presumed to be lower in energy than the metal orbitals so that the bonding combination has greater ligand character and the antibonding level has greater metal character. Of particular importance are the σ- and π-interactions with the transition metal d orbitals.

In octahedral symmetry, the d_{z^2} and $d_{x^2-y^2}$ orbital interact with the σ-orbitals of the ligands to form doubly degenerate bonding ($1e_g$) and antibonding ($2e_g$) combinations. The transition metal d_{xy}, d_{xz}, and d_{yz} orbitals in octahedral symmetry form bonding ($1t_{2g}$) and antibonding ($2t_{2g}$) π-orbitals. Because the σ-interaction is greater than the π-

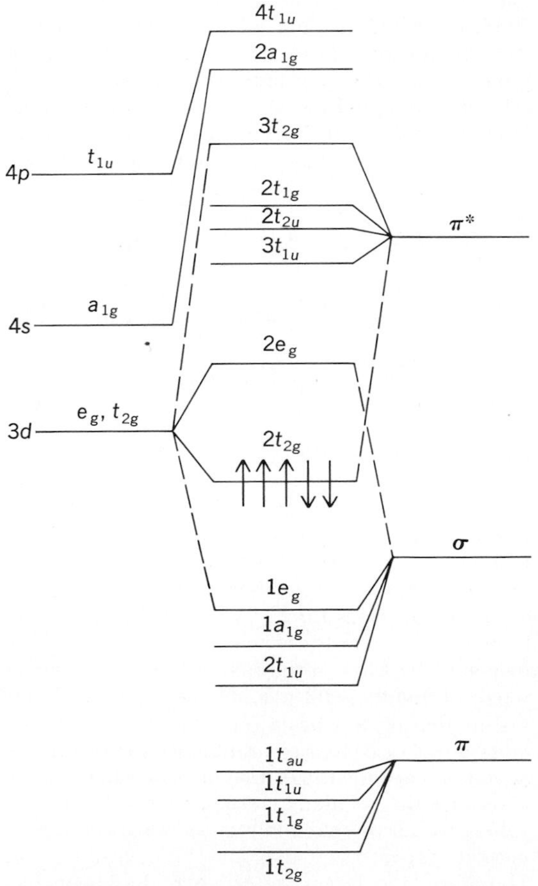

Fig. 8. Diagram of energy levels for π-electron acceptor complexes, $Fe(CN)_6^{-3}$.

interaction, the $2e_g$ level is higher than the $2t_{2g}$ level so that the five d-orbitals which were degenerate in the free metal atom are now split into a set of three ($2t_{2g}$) and a set of two ($2e_g$) levels. A similar rationalization of the splitting of the d orbitals can be made on the basis of crystal field theory. *See* CRYSTAL FIELD THEORY.

In the case of Fig. 7, the fluorine $2p$ orbitals and the metal $3d$, $4s$, and $4p$ orbitals were used in the basis set. To start with, the species as Fe^{+3}, with five $3d$ electrons, surrounded by six fluoride ions with six $2p$ electrons on each fluoride can be envisioned. The 36 fluoride electrons completely fill the energy levels from $1a_{1g}$ through $1t_{1g}$ so that the Fe electrons occupy the $2t_{2g}$ and possibly $2e_g$ orbitals. The $2t_{2g}$ has a threefold spatial degeneracy and hence can hold a maximum of six electrons. In doing so, each set of two electrons would have to possess opposing spins and occupy the same spatial orbital, in which case there is substantial electron repulsion between them.

If such a situation was energetically favorable the five electrons from the iron would have the configuration t_{2g}^5 with three electron spins in one direction and two in the opposite direction. However, in FeF_6^{-3} the energy separation of the $2t_{2g}$ and $2e_g$ orbitals is not very great so that the system has a lower total energy when these five electrons have the configuration $t_{2g}^3e_g^2$. In such a configuration each of the electrons occupies a different spatial orbital and they can have their electron spins in the same direction. Both effects decrease the mutual electron repulsions. Hence, qualitatively, the existence of either high-spin (such as $t_{2g}^3e_g^2$) or low-spin (t_{2g}^5) configurations depends upon the balance between the energy separation of the $2t_{2g}$ and $2e_g$ levels compared to the increased repulsion energy for pairing the electrons in the same spatial orbitals. Notice that in π-donor complexes such as FeF_6^{-3} the π-bonding interaction between the $3d$ orbitals and the occupied p orbitals of the fluorine raises the $2t_{2g}$ level closer to the $2e_g$ level, making high-spin complexes more probable.

On the other hand, Fig. 8 displays the energy levels which result in a typical π-acceptor complex. Prior to bonding with Fe^{+3}, the cyanide ion possesses a triple bond (one σ- and two π-bonds), $:C{\equiv}N:$, in the Lewis model. The energy levels of the ion are given in Fig. 9, which indicates that the last occupied orbitals are in a molecular orbital with substantial carbon character (the two dots in the Lewis model). These electrons are capable of forming σ-bonds to the iron atom. The first set of orbitals above the σ-bonding electrons are empty π-orbitals that are the antibonding (π^*) combination of the two π-orbitals. These antibonding orbitals are of appropriate energy so that they interact strongly with the metal d orbitals (Fig. 8). Since the π^*-orbitals are somewhat higher in energy than the metal $3d$ levels, the interaction lowers in $2t_{2g}$ energy level relative to the metal starting levels. Two effects result in an increased separation between $2t_{2g}$ and $2e_g$ levels favoring low spin configurations (in this case t_{2g}^5). Furthermore, since the π^*-orbitals participate in a molecular orbital of the complex ($2t_{2g}$) which contains electrons, these electrons now partially occupy the

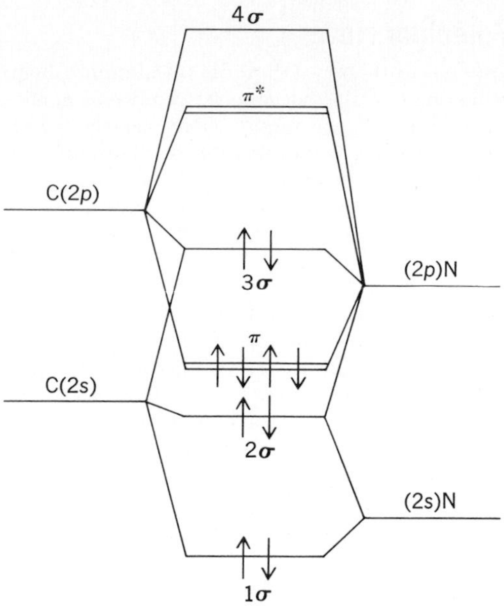

Fig. 9. Energy levels for CN–.

antibonding orbitals and weaken the triple bond between the carbon and the nitrogen. That is, the π^*-orbitals accept the metal electrons (hence the name acceptor complexes). Analogous considerations apply to the bonding of carbon monoxide with transition metals to form metal carbonyl complexes such as $Cr(CO)_6$.

Analyses of the electronic energy levels and the forms of the molecular orbitals for these and related systems have permitted investigators to correlate the results with experimental information ranging from absorption spectra to infrared stretching frequencies to magnetic properties.

Rigorous calculations. The advent of high-speed digital computers has permitted scientists to attempt more mathematically rigorous calculations, based upon the principles of LCAO-MO. Such calculations consider each electron in the potential field of all the electrons and nuclei of the system. Initial assumptions are made concerning the forms of the molecular wave functions. A computation based upon these forms is carried out, and a set of results is obtained. If the final results do not closely conform to the initial assumptions, these assumptions are modified and a new trial calculation is performed. Repeated trials are made until the assumed and calculated wave functions are consistent with one another. This procedure is known as the self-consistent field (SCF) method for LCAO-MO calculations. At present, its applicability is limited by the time and expense required to calculate the hundreds of thousands of integrals which should be evaluated in a rigorous approach. As computer technology improves, exact calculations using SCF-LCAO-MO will lead to an ever-increasing insight into the nature and properties of chemical bonds. [RICHARD F. FENSKE]

Bibliography: W. T. Borden, *Modern Molecular Orbital Theory for Organic Chemists*, 1975; F. A. Cotton, *Chemical Applications of Group Theory*, 2d ed., 1971; C. A. Coulson, *Valence*, 3d ed., 1979; R. L. DeKock and H. B. Gray, *Chemical Structure and Bonding*, 1980; N. H. March (ed.), *Orbital Theories of Molecules and Solids*, 1974.

Molecular sieve

Any one of the crystalline metal aluminosilicates belonging to a class of minerals known as zeolites. These minerals are found widely scattered in nature in relatively small quantities. Synthetic forms of the naturally occurring minerals, as well as many species having no known natural counterpart, have been prepared by a hydrothermal process. An important characteristic of the zeolites is their ability to undergo dehydration with little or no change in crystal structure. The dehydrated crystals are honeycombed with regularly spaced cavities interlaced by channels of molecular dimensions which offer a very high surface area for the adsorption of foreign molecules.

Structure. The basic formula for all crystalline zeolites can be represented as

$$M_{2/n}O:Al_2O_3:xSiO_2:yH_2O$$

where M represents a metal ion and n its valence. In general, a particular crystalline zeolite will have values for x and y that fall into a definite range. For example, in two synthetic varieties of molecular sieve, designated type A and type X, the values of x are typically about 2.0 and 2.5, respectively. When the crystal is fully dehydrated, the value of y for both types is zero. The crystal structure consists basically of a three-dimensional framework of SiO_4 and AlO_4 tetrahedrons (Fig. 1). The tetrahedrons are cross-linked by the sharing of oxygen atoms, so that the ratio of oxygen atoms to the total of silicon and aluminum atoms is equal to 2. The electrovalence of the tetrahedrons containing aluminum is balanced by the inclusion of cations in the crystal. One cation may be exchanged for another by the usual ion-exchange techniques. The size of the cation and its position in the lattice determine the effective diameter of the pore in a given crystal species.

The crystal habit of molecular sieve type X is similar to that of diamond, in which the carbon atoms are replaced by silica-alumina polyhedrons. With alkali metal ions present in the structure, the effective pore diameter is between 9 and 11 A; with the alkaline earth cations present, effective diameter is between 8 and 9 A.

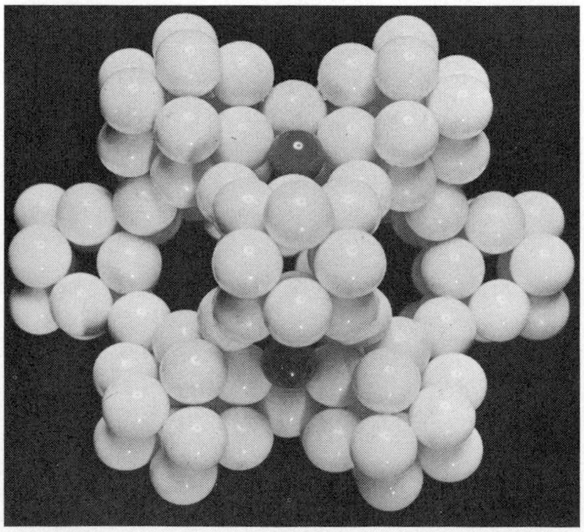

Fig. 1. Molecular sieve type-A crystal model. Dark spheres represent the included cations, and light spheres the SiO_4 or AlO_4 tetrahedrons.

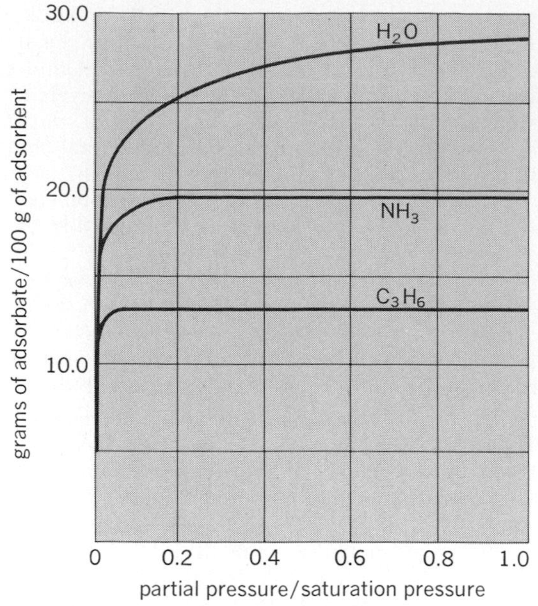

Fig. 2. Isotherms for type-5A molecular sieves at 25°C.

Properties. The properties of molecular sieves as adsorbents which distinguish them from nonzeolitic adsorbents are (1) the relatively strong coulomb fields generated by the adsorption surface and (2) the uniform pore size; the pore size is controlled, in a given crystal species, by the associated cation. The strong surface forces are reflected in the peculiar shape of the adsorption isotherm, the character of the adsorption isobar, and the relatively high heats of adsorption. The isotherm, found by plotting the capacity for a given adsorbate against pressure or concentration at constant temperature, is of the so-called Langmuir type (Fig. 2). The shape of the isotherm is approximately rectangular, rising steeply at low partial pressures or concentrations and leveling off when maximum load is attained. The isobar, found by plotting capacity against temperature at constant pressure, shows that molecular sieves have an unusually high capacity at elevated temperatures. The relatively high exothermic heats characteristic of adsorption on molecular sieves necessitate somewhat higher heat requirements to effect desorption than are necessary with other adsorbents.

The basic characteristics of molecular sieves are utilized commercially in several production and research applications. Their absorption properties make them useful for drying, purification, and separations of gases and liquids. Conversely, molecular sieves can be preloaded with chemical agents, which are thereby isolated from the reactive system in which they are dispersed until released from the adsorbent either thermally or by displacement by a more strongly adsorbed compound. The presence in the crystal lattice of an associated exchangeable metal ion provides the basis for their use as a cation exchange medium. Their chemical composition and crystal structure make them novel catalysts and catalyst supports.

As is predictable from the water-adsorption isotherm (Fig. 2), molecular sieves are capable of drying gases and liquids to extremely low residual water concentrations. The isotherm also shows that, even at low initial water concentrations in a gas or liquid, the relative desiccant capacity is high.

By virtue of the uniform pore size of a given molecular-sieve crystal type, molecules having a minimum projected cross section larger than the effective diameter of the zeolite pore are excluded from the internal surface. Molecules having a minimum projected cross section smaller than the effective pore diameter are adsorbed internally. This phenomenon is utilized in separating molecules of fluid mixtures on the basis of their size or shape. For example, molecular sieve type 5A has an effective pore size such that straight-chain hydrocarbons are adsorbed and thus effectively separated from branched-chain and cyclic hydrocarbons, which are excluded from the pore system in the selective adsorption process.

In general, when two molecules of similar volatility are sufficiently small to enter the pore system, separation is based on the degree of unsaturation or on the polarity of the molecules. The more unsaturated or more polar molecule is more strongly adsorbed. *See* ADSORPTION; GAS CHROMATOGRAPHY; ION EXCHANGE.

[R. L. MAYS]

Bibliography: W. G. Berl, *Physical Methods in Chemical Analysis*, vol. 4, 1961; D. W. Breck, *Molecular Sieves: Structure, Chemistry and Use*, 1974; J. R. Katzer (ed.), *Molecular Sieves II*, 1977; W. M. Meier and J. B. Uytterhoeven (eds.), *Molecular Sieves*, 1973.

Molecular structure and spectra

Until the advent of quantum theory, ideas about the structure of molecules evolved gradually from analysis and interpretation of the facts of chemistry. Chemists developed the concept of molecules as built from atoms in definite proportions, and identified and constructed (synthesized) a great variety of molecules. Later, when the structure of atoms as built from nuclei and electrons began to be understood with the help of quantum theory, a beginning was made in seeing why atoms can combine in definite ways to form molecules; also, infrared spectra began to be used to obtain information about the dimensions and the nuclear motions (vibrations) in molecules. However, a fundamental understanding of chemical binding and molecular structure became possible only by application of the present form of quantum theory, called quantum mechanics. This theory makes it possible to obtain from the spectra of molecules a great deal of information about the nature of molecules in their normal as well as excited states, and about dissociation energies and other characteristics of molecules. For an important aspect of molecular structure which is treated separately *see* CHEMICAL BONDING.

Molecular sizes. The size of a molecule varies approximately in proportion to the numbers and sizes of the atoms in the molecule. Simplest are diatomic molecules. These may be thought of as built of two spherical atoms of radii r and r', flattened where they are joined. The equilibrium value R_e of the distance R between their nuclei is then smaller than the sum of the atomic radii (Fig. 1). However, the nuclei of atoms in two different molecules cannot normally approach more closely than a distance $r+r'$; r and r' are called the van der Waals radii of the atoms. The smallest molecule is hydrogen (H_2), with two electrons whose negative charges equal the positive charges of the two nuclei. Here r is about 1.2 A (1 A$=10^{-8}$ cm) giving $r+r'=2.4$ A, but R_e is only 0.74 A. In HCl $r=1.2$ for H and 1.8 A for Cl, but R_e is only 1.27 A.

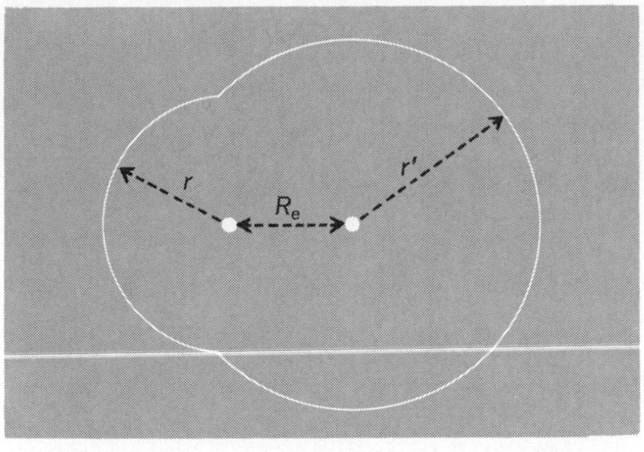

Fig. 1. Diatomic molecule with nuclei at distance R_e apart, built from atoms of radii r and r'.

To describe a polyatomic molecule, one must specify not merely its size but also its shape or configuration. For example, carbon dioxide (CO_2) is a linear symmetrical molecule, the O—C—O angle being 180°. The H—O—H angle in the nonlinear water (H_2O) molecule is 105°. Many molecules which are essential for life contain thousands or even millions of atoms. Proteins are often coiled or twisted and cross-linked in curious ways which are important for their biological functioning.

Dipole moments. Most molecules have an electric dipole moment. In atoms, the electron cloud surrounds the nucleus so symmetrically that its electrical center coincides with the nucleus, giving zero dipole moment; in a molecule, however, these coincidences are disturbed, and a dipole moment usually results.

Thus, when the atoms of HCl come together, there is some shifting of the H-atom electron toward the Cl. A complete shift would give H^+Cl^-, which would constitute an electric dipole of magnitude eR_e, where e is the electronic charge. But in fact the dipole moment is only 0.17 eR_e. This is because the actual electronic shift is only fractional. For further discussion *see* ELECTRONEGATIVITY.

Although in molecules such as H_2, N_2, and CO_2 partial shifts of electronic charge from the original atoms do take place, these necessarily occur so symmetrically that no dipole moment results. Many larger molecules also have zero dipole moments by virtue of high symmetry. Examples are methane (CH_4), uranium hexafluoride (UF_6), and benzene (C_6H_6).

In general, the dipole moment of a neutral molecule is defined as the vector sum of quantities $+Q\mathbf{S}$ for the nuclei and $-e\mathbf{s}$ for the electrons. Here Q is the charge on any nucleus and **S** its vector distance from any fixed point in the molecule; **s** is the average vector distance of any electron from the same point. To calculate a dipole moment with these definitions, quantum mechanics must be used.

However, a study of what is known experimen-

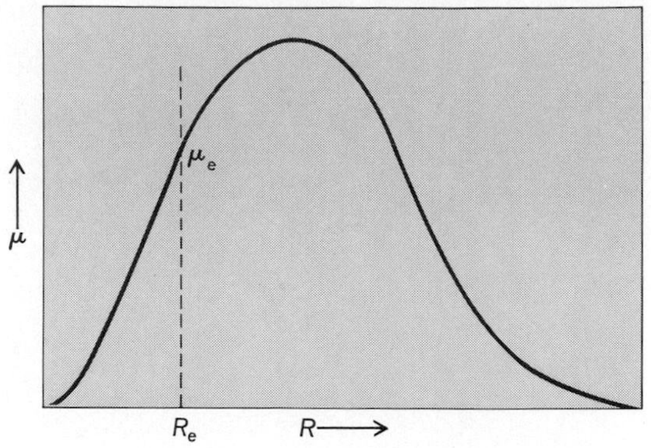

Fig. 2. Electric dipole moment μ of typical diatomic molecule as function of internuclear distance R; μ_e is the dipole moment at the equilibrium distance R_e.

tally about molecular dipole moments has led to useful semiempirical generalizations. Bond moments and group moments have been obtained for various types of chemical bonds and of chemical groups or radicals. By adding these vectorially, the actual dipole moment of a large molecule can often be reproduced fairly accurately. In CH_4 or CO_2, one can assume a moment for each C—H or C═O bond, even though these cancel out vectorially to give a zero resultant. In the linear molecule OCS, the unequal moments of the C═O and C═S bonds give a nonzero resultant. Because of the zero moment of CH_4, the CH bond moment and the CH_3 group moment must be equal and opposite. In CH_3Cl, the total moment can be thought of as the vector sum of the H_3C group moment and the C—Cl bond moment.

When molecules vibrate, their dipole moments usually vary. Figure 2 shows how the dipole moment μ in a diatomic molecule may vary with R; the quantity previously discussed is μ_e, the value of μ at R_e. When μ_e is zero because of symmetry, it remains zero for symmetrical vibrations but, in polyatomic molecules, varies during unsymmetrical vibrations.

Molecules may possess magnetic as well as electric dipole moments; for information on the former *see* MICROWAVE SPECTROSCOPY.

Molecular polarizability. In the preceding consideration of dipole moments, the discussion has been in terms of atoms and molecules free from external forces. An electric field pulls the electrons of an atom or molecule toward it and pushes the nuclei away, or vice versa. This action creates a small induced dipole moment, whose magnitude per unit strength of the field is called the polarizability.

Molecular polarizabilities can be expressed as sums of atomic polarizabilities, plus corrections depending on the types of bonds present. Polarizabilities increase rather rapidly in such series as F, Cl, Br, I, and also from HF to HI, or F_2 to I_2.

Molecular polarizabilities can also be expressed approximately as sums of bond polarizabilities. These polarizabilities are anisotropic, being greater along bonds than perpendicular to bonds.

Molecular energy levels. The states of motion of nuclei and electrons in a molecule, or of electrons in an atom, are restricted by quantum mechanics to special forms with definite energies. The state of lowest energy is called the ground state; all others are excited states. In analogy to water levels, one speaks of energy levels. Excited states exist only momentarily, following an electrical or other stimulus. *See* QUANTUM CHEMISTRY.

Energy levels are either discrete or continuous. The levels of a self-contained atom or molecule are restricted to special, sharply defined values (discrete levels). When an atom or molecule is ionized, that is, when one of its electrons has enough energy to escape completely, the energy can take on any value exceeding the minimum escape energy. One then speaks of continuous levels or of an ionization continuum. Molecules also have dissociation continua, which are discussed later in this article.

Excitation of an atom consists of a change in the state of motion of its electrons. Electronic excitation of molecules can also occur, but alternatively or additionally, molecules can be excited to discrete states of vibration and rotation.

In a diatomic vibration, R varies periodically above and below R_e. The possible vibration energies E_v are given by Eq. (1), where $c\omega_e$ is just

$$E_v = hc\omega_e[(v+1/2) - x_e(v+1/2)^2] + \cdots \quad (1)$$

the small-amplitude vibration frequency, and h is Planck's constant (6.62×10^{-27} erg-sec); x_e is a small quantity which is nearly always positive. The vibrational quantum number v can take whole-number values 0, 1, 2, etc. The $+ \cdots$ in Eq. (1) indicates small correction terms. The zero-point vibration energy $\frac{1}{2}hc\omega_e(1-\frac{1}{2}x_e)$ present in the ground vibration state ($v=0$) is a characteristic manifestation of quantum theory.

The value of $c\omega_e$ depends on the masses m_1 and m_2 of the atoms and the force constant k, as shown in Eq. (2). The frequency $c\omega_e$ (c = speed of light) is

$$c\omega_e = \sqrt{k[(1/m_1) + (1/m_2)]} \quad (2)$$

written in this manner for reasons of convenience in spectroscopic work, where the factor c is usually dropped.

The quantities R_e, k, and the dissociation energy D are the most important properties of a potential curve, which shows how the energy of attraction

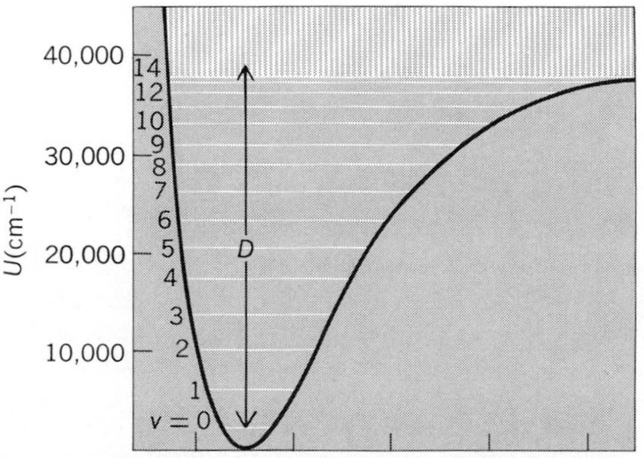

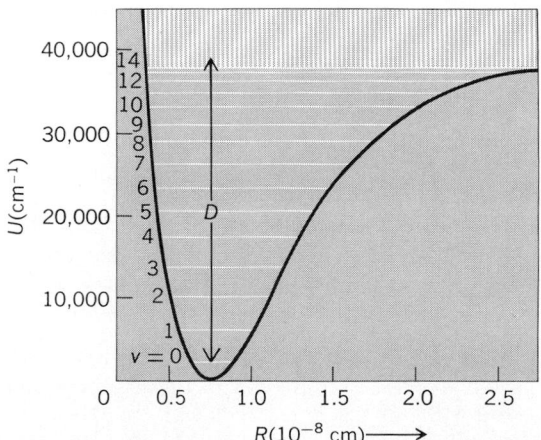

Fig. 3. $U(R)$ curve of ground electronic state of H_2 with vibrational levels and dissociation continuum. D indicates the dissociation energy. Maximum v here is 14. (*After G. Herzberg, Molecular Spectra and Molecular Structure, vol. 1, 2d ed., Van Nostrand, 1950*)

$U(R)$ of the atoms varies with R; k is d^2U/dR^2 taken at R_e. The $U(R)$ curve and vibrational levels for the ground electronic state of H_2 are shown in Fig. 3. Similar curves, but with other R_e, k, and D values, exist for other electronic states and other molecules. Molecules have also repulsive electronic states, whose $U(R)$ curves rise steadily with decreasing R. These are often important for spectroscopy and in atomic collisions. For stable (attractive) $U(R)$ curves, the vibrational levels decrease in spacing as v increases, until finally, as the spacing approaches zero, a maximum v is reached; in Fig. 3 this is 14. After a small gap, a dissociation continuum of energy levels then sets in. Here the atoms have enough mutual kinetic energy to fly apart. For repulsive states, there is only a dissociation continuum, with no vibrational levels. Figure 4 illustrates how strongly vibration level spacings can vary: both k and $1/m$, and therefore $c\omega_e$, decrease from H_2 to O_2 to I_2. Figure 4 likewise illustrates the effect of mass in isotopic molecules.

The total energy of any molecule can be written as Eq. (3). Both the electronic energy E_{el} and vibra-

$$E = E_{el} + E_v + (E_r + E_{fs} + E_{hfs} + E_{ext}) \qquad (3)$$

tion energy E_v can be discrete or continuous. The quantities E_r, E_{fs}, and E_{hfs} denote rotational, fine-structure, and hyperfine-structure energies. The last two appear as small or minute splittings of the rotation levels. The spacings ΔE of adjacent discrete levels of each type are usually in the order given in notation (4).

$$\Delta E_{el} \gg \Delta E_v \gg \Delta E_r \gg \Delta E_{fs} \gg \Delta E_{hfs} \qquad (4)$$

The fine structures of rotational levels differ strongly for different types of electronic states. The simplest diatomic electronic states are called $^1\Sigma$ states, and include $^1\Sigma^+$ and $^1\Sigma^-$ types for heteropolar and $^1\Sigma^+{}_g$, $^1\Sigma^+{}_u$, $^1\Sigma^-{}_g$, and $^1\Sigma^-{}_u$ for homopolar molecules. Most even-electron diatomic and linear polyatomic molecule ground states are $^1\Sigma^+$ states ($^1\Sigma^+{}_g$ if homopolar). The rotational levels of $^1\Sigma$ states have no fine structure; hyperfine structure, because of interaction with nuclear spins, is usually on too small a scale to detect by optical spectroscopy, to which the present article is limited. The E_{ext} term in Eq. (3) refers to additional fine structure which appears on subjecting molecules to external magnetic fields (Zeeman effect) or electric fields (Stark effect). *See* FINE STRUCTURE (SPECTRAL LINES).

The rotational levels of any $^1\Sigma$ state are given by Eq. (5). The quantity B_v is related to the moment of inertia I [$I = m_1m_2R^2/(m_1+m_2)$], and to v, by Eq. (6).

$$E_r = hcB_vJ(J+1) + \cdots \qquad (5)$$

$$B_v = (h/8\pi^2c)\,\overline{(1/I)}_v = B_e - \alpha_e(v + 1/2) + \cdots = B_0 - \alpha_e v + \cdots \qquad (6)$$

The rotational quantum number J can have any whole number value from 0 up, and corresponds to an angular momentum $(h/2\pi\sqrt{J(J+1)})$. The averaging of $1/I$ in Eq. (6) normally results in a slow decrease of B with increasing v (α_e is usually a small positive quantity). The quantity B_e refers to a hypothetical nonvibrating molecule ($R = R_e$).

Figure 5 illustrates how enormously rotational level spacings can vary because of differences in m

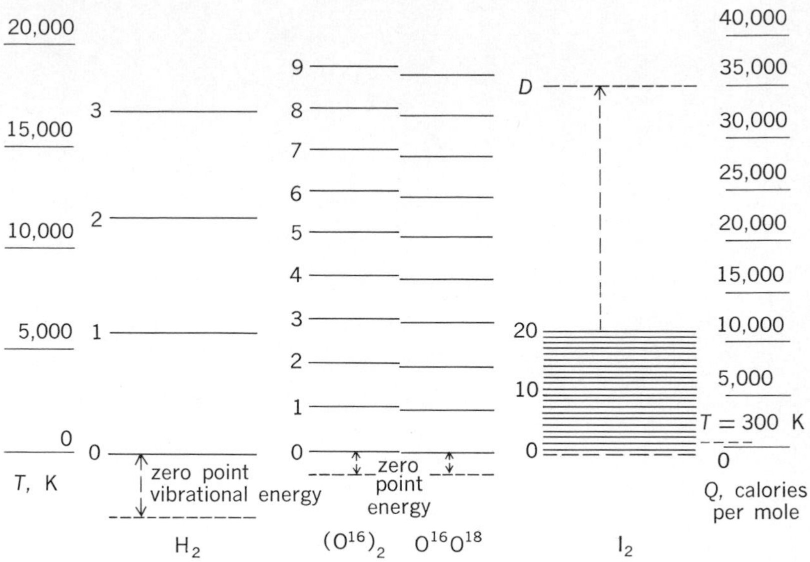

Fig. 4. Lowest vibrational levels of H_2, O_2, and I_2, numbered by vibrational quantum number v. Vibration level spacings decrease with increasing v. Where spacing reaches zero, the molecule dissociates; dissociation level D is indicated for I_2. Energies are given by the scale at right. The scale at left shows the average energy of vibration at various temperatures.

and R_e (both are much greater for I_2 than H_2). The effect of mass for isotopic molecules is illustrated for O_2. Comparison with Fig. 4 illustrates the relation $\Delta E_v \gg \Delta E_r$ mentioned earlier.

Polyatomic molecules have much more complicated patterns of vibrational and (usually) of rotational energy levels than diatomic molecules. The number of normal modes (independent forms) of vibration for a molecule with n atoms is $3n-6$ for nonlinear molecules, and $3n-5$ for linear molecules. Each normal mode is a cooperative vibration of some or all the atoms moving with the same frequency, characteristic of the mode. Sometimes two or even three modes are so related in form that their frequencies are identical. These are called degenerate vibrations.

Figure 6 depicts the normal modes of H_2O and

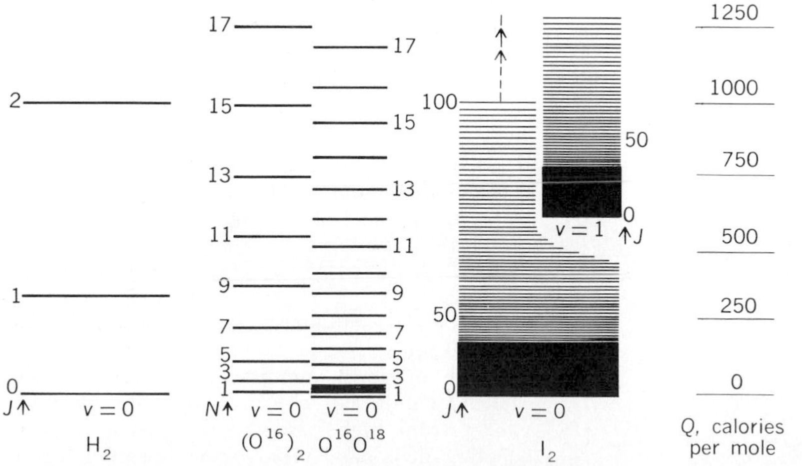

Fig. 5. Lowest rotational levels of H_2, O_2, and I_2. For H_2 and I_2, J is the rotational quantum number, according to Eq. (5) in the text O_2 is in a Hund's case b triplet state, and the rotational levels are designated by N, where the total angular momentum $J = N+1$, N, and $N-1$. This narrow spin tripling is indicated for the $N=1$ level of $O^{16}O^{18}$ only. Energies are given by the scale shown at right.

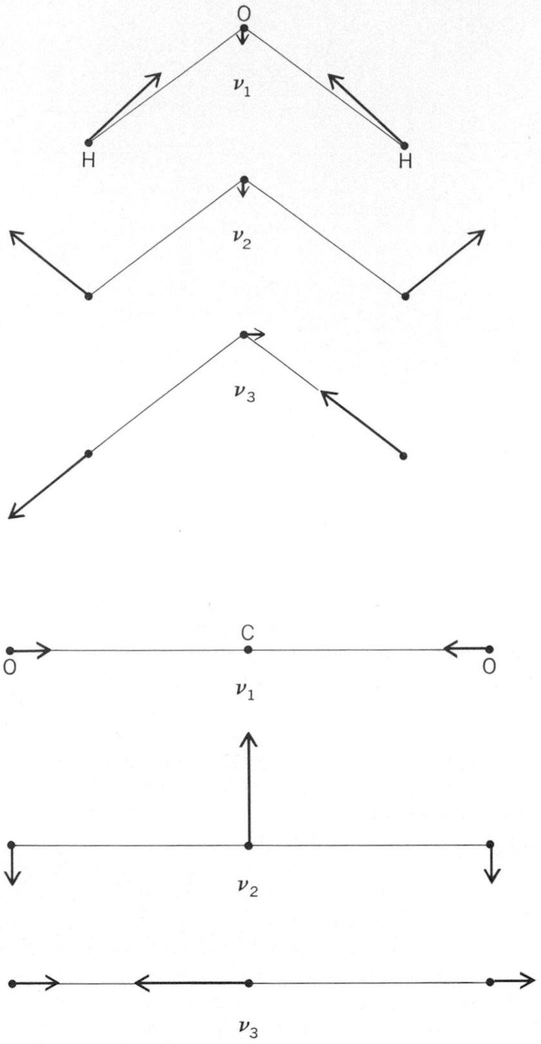

Fig. 6. Normal vibration modes of H_2O and CO_2. Synchronized displacements of atoms occur in proportion to lengths of the arrows. Diagram corresponds to snapshot taken at one phase of vibration.

CO_2. They are labeled by symbols which also denote their frequencies. The arrows indicate the directions of motion of the atoms during one phase of vibration. The CO_2 frequency ν_2 is twofold degenerate: there are two independent modes corresponding to motion in either of two planes at right angles. The other two CO_2 modes, and all three H_2O modes, are nondegenerate.

Molecular spectra. The frequencies $c\nu$ of electromagnetic spectra obey the Einstein-Bohr equation, Eq. (7). The quantities ν, in waves per centi-

$$hc\nu = E' - E'' \qquad (7)$$

meter, or wave numbers (cm^{-1}), will hereafter be called frequencies, as is usual in spectroscopy, although properly only the $c\nu$ are frequencies. Molecular emission spectra accompany jumps in energy from higher to lower levels; absorption spectra accompany jumps from lower to higher levels. Both E' and E'' can be either discrete or continuous levels. If both are discrete, they give a spectrum of discrete frequencies; otherwise, they give a continuous spectrum. Discrete spectra are the main type considered here. Discrete frequencies are usually called spectrum lines because of their appearance when recorded by an optical spectrograph.

Besides its frequency, the intensity and width of a spectrum line are important. Intensities vary over wide ranges. In the extreme case of nearly zero intensity for a spectroscopic transition, the transition is called forbidden. Only a small minority of all pairs of levels yield allowed transitions. These are governed by selection rules derivable from quantum theory.

Under disturbing influences, however, some lines are seen, weakly, which violate these rules. Further, the usual selection rules are electric dipole rules, and additional transitions become very weakly allowed if magnetic dipole, electric quadrupole, and other selection rules are also considered. The following discussion is confined to spectra which obey the electric dipole rules.

Molecular spectra can be classified as fine-structure or low-frequency spectra, rotation spectra, vibration-rotation spectra, and electronic spectra. Low-frequency spectra are discussed elsewhere. *See* ELECTRON PARAMAGNETIC RESONANCE (EPR) SPECTROSCOPY; MICROWAVE SPECTROSCOPY; SPECTROSCOPY.

Pure rotation spectra. Transitions between energy levels differing only in rotational state give

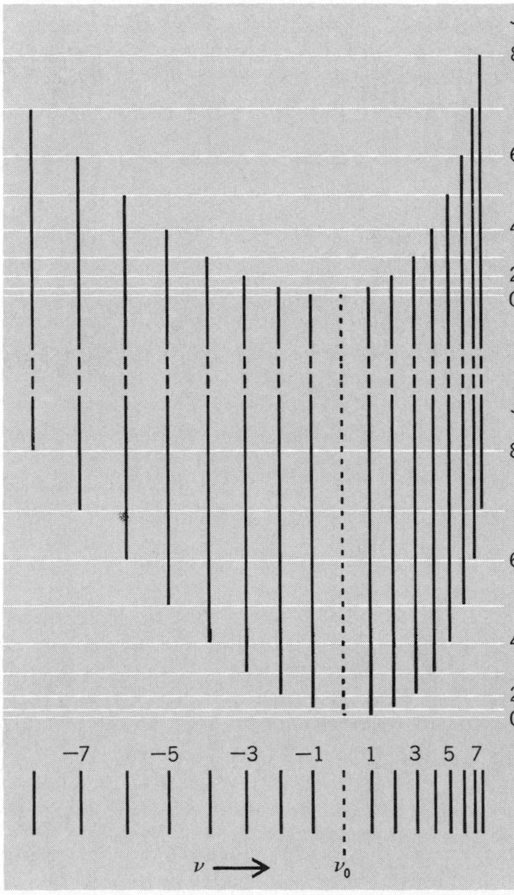

Fig. 7. Relation of band lines (lower part), [see Eqs. (8) and (9)] to rotational levels [see Eq. (6)] for a vibration-rotation band or an electronic band. In the former case, the upper and lower sets of rotational levels belong to two vibrational levels of a $^1\Sigma$ electronic state. In the latter case, they belong to two different $^1\Sigma$ states. Positive *M* values, *R* branch; negative *M* values, *P* branch.

rise to pure rotation spectra. For diatomic molecules in $^1\Sigma$ states, Eq. (5), the relation is given by Eq. (8). The transitions obey the selection rule

$$hc\nu = E'_r - E''_r = hcB_v[J'(J'+1) - J''(J''+1)] + \cdots \quad (8)$$

$\Delta J = 1$ (ΔJ means $J' - J''$). Putting $J' = J'' + 1$, Eq. (9) is obtained. Equation (9) represents a sequence

$$\nu = 2B_v(J''+1) + \cdots \quad (9)$$

of lines spaced almost equidistantly ($2B_v$, $4B_v$, $6B_v$, . . .), and lying in the far infrared or (for small B or low J'') the microwave region. Their intensities are proportional to μ_e^2, where μ_e is the electric moment at R_e (Fig. 2); hence homopolar molecules (H_2, N_2, and so on) show no pure rotation spectra. The intensities are proportional also to the lower-state (v'', J'') level population and to ν (for absorption) or ν^4 (for emission).

Pure rotation spectra of linear polyatomic molecules are like those of diatomic molecules. Polyatomic molecules having $\mu_e = 0$, whatever their shape (examples are CO_2, CH_4, C_6H_6), have no pure rotation spectra. In other cases, the spectra can be obtained using $hc\nu = E'_r - E''_r$ with appropriate E_r expressions and selection rules.

Vibration-rotation bands. Spectra involving only vibrational and rotational state changes lie mainly in the infrared. For a $^1\Sigma$ diatomic state, using Eqs. (1), (5), and (7), Eq. (10) is obtained, with ν_0 defined in Eq. (11). In Eq. (10) B' and B'' mean B_v for v' and

$$\nu = \nu_0(v',v'') + [B'J'(J'+1) - B''J''(J''+1)] + \cdots \quad (10)$$

$$\nu_0 = \omega_e(1-x_e)(v'-v'') - x_e\omega_e(v'^2 - v''^2) \quad (11)$$

v'', respectively. Each band consists of two sets of rotational lines, one on each side of its ν_0. Each line corresponds to a particular rotational transition conforming to $\Delta J = \pm 1$. The two series (branches) have frequencies defined in Eq. (12) for R or positive branch ($J' = J'' + 1$), and in Eq. (13), for P or negative branch ($J' = J'' - 1$). Both can be represented by a single equation, Eq. (14), by letting $M = J'' + 1$ for the R and $M = -J''$ for the P branch.

$$\nu = \nu_0 + 2B''(J''+1) + (B'-B'')(J''+1)(J''+2) \quad (12)$$

$$\nu = \nu_0 - 2B''J + (B'-B'')J''(J''-1) \quad (13)$$

$$\nu = \nu_0 + (B'+B'')M + (B'-B'')M^2 + \cdots \quad (14)$$

Neglecting the term in M^2, Eq. (14) represents a series of equidistant lines with one missing ($M=0$) at ν_0. Figure 7 shows how the line positions are related to the upper (v') and lower (v'') sets of rotational levels.

Since $B' - B''$ is a small negative quantity [see Eq. (6), noting that $v' > v''$], the M^2 term makes the P line spacing increase and the R line spacing decrease slowly as M increases. This is shown, exaggerated, in Fig. 7. At some large M value, the R branch turns back on itself, but usually the lines have beome weak before this value is reached.

The relative intensities of band lines depend primarily on the initial rotational distribution of molecules. More precisely, Eq. (15) holds. Here

$$\text{Intensity} = C(v',v'')\nu^n(J'+J''+1)e^{-B_{in}J_{in}(J_{in}+1)hc/kT} \quad (15)$$

B_{in}, J_{in}, and n in ν^n are B', J', and 4, respectively, for an emission band, and B'', J'', and 1, respectively, for an absorption band. Figure 8 shows diagrammatically how the values of B and T affect the appearance of a typical absorption band ($B' = B''$ has been assumed for simplicity in Fig. 8). Figure 9 shows the appearance of an actual HCl band. The weaker HCl^{37} lines are at slightly lower frequencies than the HCl^{35} lines, mainly because ω_e is smaller [see Eqs. (2) and (11)].

The factor $C(v',v'')$ is largest by far for fundamental bands ($\Delta v = 1$), and falls rapidly with increasing Δv in the overtone bands or harmonics (Δv is $v' - v''$). For fundamental bands, C depends on the slope of the $\mu(R)$ curve (Fig. 2), being approximately proportional to $(d\mu/dR)^2$ taken at R_e. For overtone bands, C depends on the detailed shapes of both the $\mu(R)$ and $U(R)$ curves. Fundamental or overtone bands arising from $v'' > 0$ are called hot bands.

Vibration-rotation absorption bands of liquids and solutions are widely used in chemical analysis. Here the rotational structure is blurred out, and

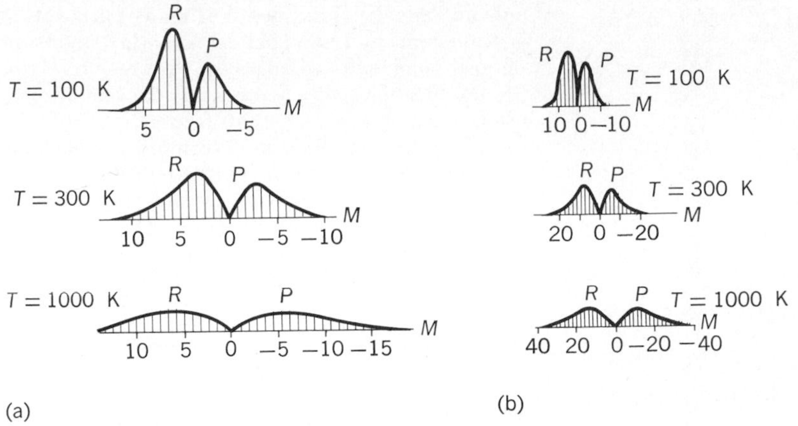

Fig. 8. Intensity distribution at several temperatures for a diatomic absorption band. Line positions are based on Eq. (9) assuming $B' = B''$ for simplicity; frequency increases toward the left (opposite to Fig. 7). (*a*) and (*b*) correspond respectively to B values of HCl ($B = 10.44$ cm^{-1}) and of 2 cm^{-1} (approximately the value for CO, for which $B = 1.93$ cm^{-1}). *(After G. Herzberg, Molecular Spectra and Molecular Structure, vol. 1, 2d ed., Van Nostrand, 1950)*

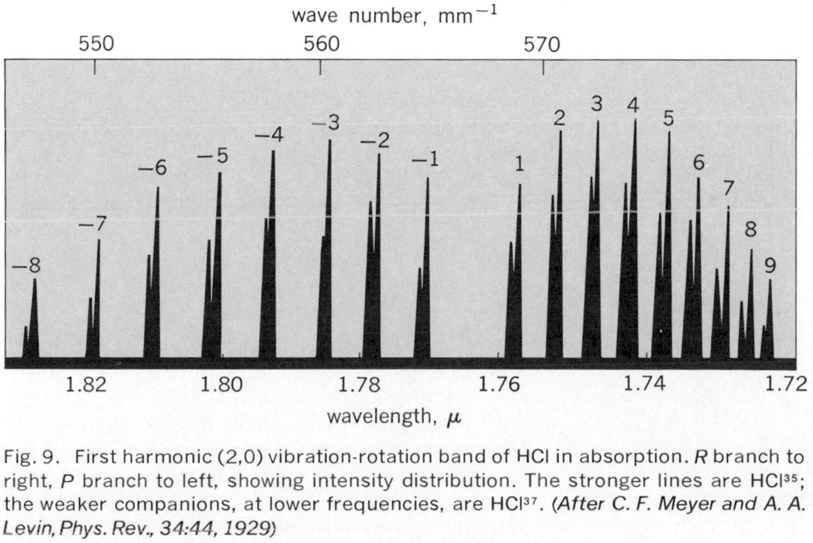

Fig. 9. First harmonic (2,0) vibration-rotation band of HCl in absorption. R branch to right, P branch to left, showing intensity distribution. The stronger lines are HCl^{35}; the weaker companions, at lower frequencies, are HCl^{37}. *(After C. F. Meyer and A. A. Levin, Phys. Rev., 34:44, 1929)*

only an "envelope" is seen. For many purposes, it is sufficient to know empirically the spectrum of each molecule which may be present. Also, groups of atoms which recur in many molecules often have nearly constant frequencies, of use for identification and in determining molecular structures. *See* INFRARED SPECTROSCOPY.

Electronic band spectra. These are the most general type of molecular spectra. The characteristic feature is a change of electronic state. From Eqs. (3) and (7), neglecting fine structure, Eq. (16)

$$\nu = \frac{(E'_{el} - E''_{el}) + (E'_v - E''_v) + (E'_r - E''_r)}{hc} \qquad (16)$$

and (17) are obtained. Diatomic electronic spectra

$$\nu = \nu_{el} + \nu_v + \nu_r = \nu_0 + \nu_r \qquad (17)$$

are often observed in emission, while the electronic spectra of polyatomic molecules are usually absorption spectra. Depending mainly on the magnitude of ν_{el}, electronic spectra occur in the infrared, visible, ultraviolet, or vacuum ultraviolet.

For any one electronic transition, the spectrum consists typically of many bands. These lie in general at frequencies both above and below ν_{el}, since ν_v can be positive or negative. They constitute a band system. Each band consists of numerous rotational lines arranged in two or more branches and lying on both sides of a central position ν_0.

For diatomic molecules, ν_0 depends on a single v' and v'' and, using Eq. (1) for each electronic state, is given by Eq. (18). Since ω_e and $x_e\omega_e$ are

$$\nu_0(v',v'') = \nu_{el} + [\omega'_e(v' + 1/2) - x'_e\omega'_e(v' + 1/2)^2 + \cdots] - [\omega''_e(v'' + 1/2) - x''_e\omega''_e(v'' + 1/2)^2 + \cdots] \qquad (18)$$

now different (often strongly) in the upper and lower states, $\nu_0(v',v'')$ cannot be reduced to as simple an expression as the corresponding Eq. (11) for vibration-rotation bands. Eq. (18) is more convenient when rewritten as Eq. (19), where Eqs. (20)

$$\nu_0(v',v'') = \nu_{00} + (\omega'_0 v' - a'v'^2) - (\omega''_0 v'' - a''v''^2) + \cdots \qquad (19)$$

$$\nu_{00} = \nu_{el} + 1/2(\omega'_e - \omega''_e) - 1/4(x'_e\omega'_e - x''_e\omega''_e) \qquad \omega'_0 = \omega'_e(1 - x'_e) \qquad a' = x'_e\omega'_e, \text{ etc.} \qquad (20)$$

apply. The relative intensities of the bands depend on (1) the initial distribution of molecules among vibrational levels, and (2) the relative transition probabilities from any initial to various final vibrational levels.

The simplest example is the absorption spectrum of a cool gas of low molecular weight, for which all molecules initially have $v''=0$. The spectrum then consists of one "v' progression," a single series of bands with various values of v'; the frequencies are given by $\nu = \nu_{00} + \omega_0 v' - a'v'^2$. For a hot or a heavy gas, additional weaker v' progressions with $v'' > 0$ also appear.

In emission spectra, the initial population usually ranges over a number of v' values, from each of which transitions occur to a number of v'' values, so that the system contains many bands on both sides of ν_{00}. In the special case of fluorescence spectra, the molecule is excited to various v' values by absorbing light; it then emits light belonging to the same (or sometimes another) electronic transition. From each v', it can descend not only to the original v'' but also to various other, mainly larger, values. Hence fluorescence bands lie mainly at lower frequencies than the absorption bands used to excite them.

Relative transition probabilities are governed by the Franck-Condon principle. This takes note of the very great rapidity of electronic motions as compared with those of the far more massive nuclei, and concludes that during the extremely brief time for an electronic transition, the nuclei tend to remain unchanged in their positions and momenta. It is applicable to both polyatomic and diatomic spectra. Consider first a diatomic molecule starting from the $v''= 0$ level of a ground state $U(R)$ curve like the lower curves in Fig. 10. A vertical

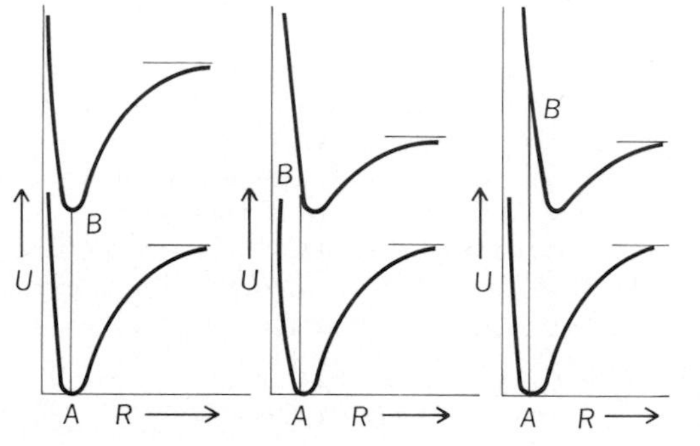

Fig. 10. Diatomic $U(R)$ curves for three cases to explain the vibrational intensity distribution according to the Franck-Condon principle. The asymptote of each curve for large R corresponds to dissociation into atoms, with one or both atoms excited in the case of the upper curves. Starting in each case from the bottom of the lower curve (essentially $v''=0$), the most probable transition in absorption is (*a*) to $v'=0$, (*b*) to $v'=3$ or 4, and (*c*) to the dissociation continuum, as shown by vertical lines. (*After G. Herzberg, Molecular Spectra and Molecular Structure, vol. 1, 2d ed., Van Nostrand, 1950*)

line drawn from the bottom point A ($v''=0$ if zero-point vibration is neglected) to point B on any one of the upper curves corresponds to an electronic absorption transition in which the nuclei have not moved.

In the case of Fig. 10*a*, point B corresponds to $v'=0$, and the conclusion is that this is the most probable v' for $v''=0$. In the case of Fig. 10*b*, point B corresponds to an excited molecule at the inner turning point of a vibration with a v' of possibly 3 or 4, in a typical case. One then concludes (with J. Franck) that the strongest absorption bands for $v''=0$ have $v'=3.0$ and 4.0. To obtain more exact information, a quantum-mechanical calculation (first carried out by E. U. Condon) is necessary.

In the case of Fig. 10*c*, point B corresponds to an energy level in the dissociation continuum above the asymptote of the upper $U(R)$ curve. According to the Franck-Condon principle, the absorption spectrum will have maximum intensity in a continuous range of frequencies, with $hc\nu$ about equal to the energy difference AB. The quantum-mechanical calculation shows that the actual spectrum will extend with appreciable intensity over a range of both higher and lower frequencies than this, including, on the lower-frequency side, a number of high-v' bands. Actual examples of such spectra (a long v' progression followed by a strong continuum) are the far-ultraviolet Schumann-Runge bands of oxygen and the visible bands of iodine. By measuring the frequency at which the continuum begins, one obtains an exact value of the dissociation energy of each of these molecules. In so doing, any excitation energy of the atomic dissociation products to which the upper $U(R)$ curve leads is subtracted.

The Franck-Condon method is useful in understanding intensity distributions and structure in emission as well as absorption band systems. For diatomic spectra, various patterns of intensity as functions of v' and v'' occur, depending largely on the R_e values of the two $U(R)$ curves and, of course, also on the initial distribution among v' levels. Sometimes the upper-state $U(R)$ curve is stable (has a minimum) but the lower state is repulsive. Continuous emission spectra then occur, with the atoms flying apart on reaching the lower state. The H_2 molecule shows such a spectrum, as do rare gas molecules such as He_2 and Kr_2, which are stable only in excited or ionized states.

Molecular electronic states. Before discussing the structures of electronic bands, one must consider the nature of molecular electronic states. Each electronic state has orbital and spin characteristics. The spin quantum number S has a whole-number value if the number of electrons is even, a half-integral value if it is odd. Electronic states with $S=0$ are called singlet states, all others multiplet states. The orbital characteristics differ sharply for linear (including diatomic) and nonlinear molecules.

For linear molecules only, there is a quantum number Λ such that $\pm\Lambda\, h/2\pi$ is the component of angular momentum around the line of nuclear centers. Linear-molecule electronic states can be discussed under three headings: (1) singlet states; (2) multiplet states with strong spin coupling (Hund's case *a*); and (3) multiplet states with weak spin coupling (Hund's case *b*). Strictly speaking, actual multiplet states are intermediate between cases *a* and *b*, or between these and certain other cases called *c* and *d*. The discussion to follow is largely restricted to singlet electronic states.

Singlet states with $\Lambda=0$ include $^1\Sigma^+$ and $^1\Sigma^-$ states: states with $\Lambda=1, 2, \ldots$ are called $^1\Pi$, $^1\Delta$, and so on. In linear molecules with a center of symmetry (H_2, CO_2 and so on), one must further distinguish even and odd (*g* and *u*) states: $^1\Sigma^+_g$, $^1\Sigma^+_u$, $^1\Sigma^-_g$, $^1\Sigma^-_u$, $^1\Pi_u$, $^1\Pi_g$, $^1\Delta_g$, $^1\Delta_u$, etc. The rotational levels of singlet states obey the symmetric rotor formula, Eq. (21). Here J is restricted to integral values equal to or greater than Λ.

$$E_r = hc[BJ(J+1) - \Lambda^2] + \cdots \qquad (21)$$

For $\Lambda > 0$, each rotational level is a narrow doublet (Λ-doubling). Corresponding fine structure [see E_{fs} in Eq. (3)] can usually be detected in electronic bands, but (for ground states only) it can be much more accurately studied in low-frequency spectra. Hyperfine structure [see E_{hfs} in Eq. (3)] is usually too small scale to be detected in electronic band lines, but has been found in a few cases. It is best studied in low-frequency spectra.

Electronic band structures. The simplest electronic bands occur for transitions between singlet electronic states. The possible types of electronic transitions are limited by the selection rule $\Delta\Lambda = 0, \pm 1$. The structures of $^1\Sigma - {}^1\Sigma$ bands are essentially the same as for the $^1\Sigma$-state vibration-rotation bands described earlier. Equations (12) to (15) and Fig. 7, also Fig. 8, for the intensities in absorption are still applicable if Eq. (18) instead of Eq. (11) is used for ν_0, and it is recognized that B' and B'' now belong to two different electronic states.

The quantity $B' - B''$ in Eq. (14), instead of always being a small negative quantity, may now be either positive or negative, and $(B' - B'')/(B' + B'')$ is often fairly large (although it can also be nearly zero). As a result, it is usual in electronic bands to find so-called heads. A head is a position of maximum or minimum frequency; by using Eq. (14) to obtain $d\nu/dM = 0$, one finds $M_{head} = (B' + B'')/2(B' - B'')$. Then, on inserting M_{head} into Eq. (14), one obtains $\nu_{head} = \nu_0 - (B' + B'')^2/4(B' - B'')$. [Since $(B' + B'')/2(B' - B'')$ is not usually a whole number, the actual M_{head} is the nearest whole-number M to that calculated.] According to whether $B' - B''$ is negative or positive, the positive (R) or the negative (P) branch forms the head. Figure 7, if continued to somewhat larger M values, illustrates the formation of an R-branch head at a calculated M of 10.5; the actual head is formed by the two coincident lines $M=10$ and 11.

Although homopolar molecules (H_2, N_2, and so on) have no pure rotation or vibration-rotation spectra, they do have electronic spectra. For homonuclear homopolar molecules, the band lines show alternating intensities. The lines in each branch are alternately stronger and weaker as M increases, this effect being superposed on the otherwise smoothly varying intensity distribution. The alternation ratio depends on the nuclear spin I and has been, in several cases, the means of determining I. When $I=0$, alternate lines are completely missing. Heteronuclear molecules, even if homopolar (for example, HD or $O^{16}O^{18}$) do not show alternating intensities.

Polyatomic electronic spectra. These differ from diatomic electronic spectra because several initial and final vibration quantum numbers are involved, and because the rotational structure (except for linear molecules) is usually much more complicated. However, the detailed structures of the electronic spectra of a number of simple molecules and radicals in the vapor state in emission and in absorption have been studied. Nevertheless, for the most part, the spectra of polyatomic molecules are examined as absorption spectra in solution. The rotational structure is then completely blurred out, but the vibrational structure can be seen.

The Franck-Condon principle is here a useful guide. One of its corollaries, which amounts almost to a selection rule, is that only totally symmetric vibrations (vibrations during which the equilibrium symmetry of the molecule is preserved) undergo quantum number changes. This greatly simplifies the vibrational structure, especially of absorption spectra where most molecules are initially mainly in the $v''=0$ state of all vibrations. One finds then mostly v' progressions of one or a very few totally symmetric vibrations, and combinations of these.

Rather often, polyatomic band systems do not even show obvious vibrational structure. This can happen for any of several reasons: The upper state may involve dissociation; in CH_3I, for example, the first ultraviolet absorption region yields $CH_3 + I$; there may be so many low-frequency, upper-state vibrations that the spectrum looks continuous; or there may be a combination of these and other reasons. Such continuous or pseudocontinuous band systems are often loosely referred to as bands. For complicated molecules, the spectra of several different electronic transitions often overlap strongly so that it is difficult even to separate one system from another system. *See* ELECTRON SPIN; INTERMOLECULAR FORCES; RAMAN EFFECT; RESONANCE.

[ROBERT S. MULLIKEN]

Bibliography: C. N. Barswell, *Fundamentals of Molecular Spectroscopy*, 2d ed., 1973; P. R. Bunker, *Molecular Symmetry and Spectroscopy*, 1979; W. H. Flygare, *Molecular Structure and Dynamics*, 1978; I. N. Levine, *Molecular Spectroscopy*, 1975; J. Steinfeld, *Molecules and Radiation: An Introduction to Modern Molecular Spectroscopy*, 1978.

Molecular weight

The sum of the atomic weights of all the atoms in a molecule. Atomic weights (and therefore molecular weights) are relative weights arbitrarily referred to an assigned atomic weight of 12.0000 exactly for the most abundant isotope of carbon, C^{12}. Prior to Sept. 8, 1960, natural oxygen, a mixture of isotopes, served as the standard of reference with an assigned atomic weight of 16.0000 exactly. Tables of relative atomic weights from which molecular weights are computed apply to elements as they exist in nature, without artificial alteration of their isotopic composition and, further, to natural mixtures that do not include isotopes of radiogenic origin.

Gram-mole. The amount of a substance which has a weight in grams equal to the molecular weight of the substance is called a gram-mole or a gram-molecular weight. The number of molecules in a gram-mole is exceedingly large (6.02×10^{23}) and is called the Avogadro number.

Formula weight. The ultimate analysis of a substance provides precise knowledge of formula weight, namely, the sum of the atomic weights of the atoms in the simplest formula for the substance. The molecular weight is an exact multiple of the formula weight. For example, analysis shows that the simplest formula of benzene is CH and that the formula weight, the sum of the atomic weights of carbon and hydrogen, is 13.02. Each benzene molecule is a hexagonal structure of true formula C_6H_6 and molecular weight 78.11. Some substances, such as sodium chloride, contain no discrete molecules and are best assigned formula weights but no molecular weights. Known molecules range in molecular weight from two (hydrogen) to several billion. The nucleic acid DNA of *Escherichia coli* is a circular molecule with a molecular weight of 2,000,000,000. To indicate the mass of a single, so-called, large molecule such as this, the unit called the dalton is often used by biochemists. A dalton is the mass of a single hydrogen atom; hence the mass in daltons of a single molecule of any substance is numerically equal to the molecular weight of the substance. Though molecules are often called large by this standard, this giant molecule weighing 2,000,000,000 daltons at the same time weighs about 10^{-17} pound (a one-hundred-millionth of a billionth of a pound).

Experimental methods. Molecular weights are determined by a variety of methods, the choice in a particular instance depending on the properties of the substance, the purpose, and the facilities at hand. Historically, molecular-weight determinations were essential in establishing the stoichiometry which forms the basis for the fundamental laws of chemistry; today, in addition to their usefulness in pure science, they are made to interpret chemical reactions, to aid in determining molecular structure and molecular shapes, to give a better understanding of solvent-solute interaction, and to provide data for the design and control of a great variety of useful industrial processes.

When molecular-weight determinations are made of mixtures, the average molecular weight obtained may depend on the method employed. Most polymer samples, for example, contain substances of widely distributed molecular weights and, hence, polymer samples cannot be characterized by a single molecular weight. Instead, various molecular weight averages are defined according to Eq. (1). In this equation N_i is the number of

$$\overline{M}_k = \frac{\Sigma_i N_i M_i^k}{\Sigma_i N_i M_i^{k-1}} \quad (1)$$

molecules of molecular weight M_i. When $k=1$, the equation defines the number average molecular weight. When $k=2$, the equation defines the weight average molecular weight. Except where noted, the methods described here yield number average molecular weights.

Gas-density method. Equal volumes of different gases at the same (low) pressure and temperature contain the same number of molecules (Avogadro's law), and therefore their weights are proportional to their molecular weights. From a precise standpoint the gas-density method of determining molecular weights consists in determining density as a function of pressure at constant temperature. The ratio of density d to pressure p is a linear function of pressure and, at low pressures, every real gas approaches ideal behavior. The ideal or perfect gas equation may be written as Eq. (2). In this equation M is molecular weight, w

$$M = \left(\frac{w}{v}\right)\frac{RT}{p} = \frac{d}{p}RT \quad (2)$$

is weight, v is volume, T is absolute temperature in degrees Kelvin, R is the molar gas constant, d is density, and p is pressure. Values of $(d/p)_{p=0}$ are equal to M/RT. The perfect-gas equation is a useful approximation at atmospheric pressure, but the pressure coefficients of d/p may be quite large and are different for different gases.

The gas-density method may be applied to easily volatile liquids. An approximate value, found in a measurement at atmospheric pressure, often is sufficiently accurate to enable a choice between a formula weight, obtained by precise chemical analysis, and some multiple of it.

Gaseous effusion. Equal volumes of different gases at the same temperature and pressure possess the same kinetic energy of molecular motion; hence, the mean velocities of the molecules of the gases are inversely proportional to the square roots of their densities and thus of their molecular weights. The effusion method of molecular-weight determination involves passing gases through a small hole and measuring the time required for equal volumes of gases, of known and unknown molecular weights, respectively, to pass under the same conditions of pressure and temperature. If u_1 and u_2 are the root-mean-square velocities of the molecules of the two gases which have densities d_1 and d_2 and molecular weights M_1 and M_2, then the ratio of the times t_1 and t_2 taken for each gas is given by Eq. (3). The effusion method is especially

$$\frac{t_1}{t_2}=\frac{u_2}{u_1}=\sqrt{\frac{d_1}{d_2}}=\sqrt{\frac{M_1}{M_2}} \qquad (3)$$

valuable where only small amounts of sample are available.

Vapor-pressure lowering. The vapor pressure of a solvent at constant temperature is decreased by the introduction of a nonvolatile solute in accordance with Raoult's law, which may be written as Eq. (4), where n and N represent the number of

$$p=p_0\frac{N}{n+N} \qquad (4)$$

moles of solute and solvent, respectively, in the solution, and p_0 and p denote the vapor pressures of solvent and solution. For dilute solutions, where Raoult's law is applicable, n is negligible in comparison with N, so that Eq. (5) may be formed.

$$\frac{p_0-p}{p_0}=\frac{n}{N}=\frac{g/m}{G/M} \qquad (5)$$

Here g and G are the weights of solute of molecular weight m and solvent of molecular weight M. Both static methods and dynamic or air-saturation methods may be employed to determine molecular weights by vapor pressure lowering, but there are several sources of serious error. *See* SOLUTION.

Boiling and freezing point methods. Both the elevation of the boiling point and the lowering of the freezing point of a solvent occasioned by introduction of a nonvolatile solute are manifestations of vapor-pressure lowering, and either may be more satisfactorily measured than vapor-pressure lowering itself. In dilute solutions, the magnitude of ΔT_b, the elevation of the boiling point, and ΔT_f, the lowering of the freezing point, are proportional to the molal concentration of the solute. It can be shown that, as a useful approximation, Eqs. (6) and (7) can be formed. Here K_b and K_f are the so-called

$$K_b=\frac{RT_b^2}{L_bN} \qquad (6)$$

$$K_f=\frac{RT_f^2}{L_fN} \qquad (7)$$

ebullioscopic constant and the cryoscopic constant, respectively, of a solvent having a boiling point T_b, a freezing point T_f, a heat vaporization L_b, and a heat of fusion L_f. R is the molar gas constant, N is the number of moles of solute in 1000 g of solvent, and the temperatures are in degrees Kelvin. K_b and K_f are also called the molal elevation of the boiling point and the molal depression of the freezing point. They are the values of ΔT_b and ΔT_f calculated for 1 mole of solute dissolved in 1000 g of solvent but applicable only for much more dilute solutions. Equation (8) follows as a useful approxi-

$$M=K_b\frac{1000}{G}\frac{g}{\Delta T_b}=K_f\frac{1000}{G}\frac{g}{\Delta T_f} \qquad (8)$$

mation for dilute solutions, where M is the molecular weight of the solute and g is the weight in grams of solute dissolved in G grams of solvent. The boiling point T_b varies significantly with pressure changes occurring in the laboratory atmosphere. Hence, ΔT_b, the difference between the boiling point of the solution and the pure solvent, is useful only when the two boiling points are referred to the same pressure.

Osmotic pressure. Though the osmotic pressure π of a dilute solution is given by $\pi V=nRT$, an equation identical in form to the perfect gas equation, it can be shown that the osmotic pressure equation is related to Raoult's law. V is the volume of the solution, n is the number of moles of solute, R is the molar gas constant, and T is the absolute temperature in degrees Kelvin. The molecular weight M is given by Eq. (9), where C is

$$M=\frac{WRT}{\pi V}=\frac{RT}{\pi/C} \qquad (9)$$

the concentration of solute in grams per unit volume. However, for nonideal systems such as those found when dealing with high molecular weight polymers, π/C may be a nonlinear function of C. In osmotic pressure measurement, membranes impermeable to the solute are used. *See* OSMOSIS.

Monomolecular surface films. The equation of state for ideal "gaseous" films is given by Eq. (10).

$$FA=nRT \qquad (10)$$

In this equation F is surface pressure, A is the film area, n is the number of moles of film spread, T is the absolute temperature in degrees Kelvin, and R is the molar gas constant in ergs per mole per degree. Pressure-versus-area determinations are made with a film balance, and a plot of FA versus F, extrapolated to $F=0$, gives the value of nRT, corrected for the area occupied by the molecules themselves and fully appropriate for calculation of molecular weight. *See* MONOMOLECULAR FILM.

X-ray methods. The volume occupied by a molecule in a crystal may be determined by means of x-rays. Multiplication of this volume by the measured density of the crystal and by Avogadro's number gives the molecular weight. The method is especially useful for determining the chemical formulas of complex crystalline substances, for example, ammonium paramolybdate. *See* X-RAY CRYSTALLOGRAPHY; X-RAY DIFFRACTION.

Mass spectrography. Molecular weights may be determined by measurements with the mass spectrograph. For an element such as chlorine, consisting of isotopes, the chemical molecular weight is found by determining the physical atomic weights of the different isotopes and also their relative abundance.

Electron microscopy. When large molecules may be isolated and viewed with the electron microscope, their physical dimensions may be determined and related to molecular weight.

Viscosity measurements. The magnitude of the viscosities of solutions or melts of polymeric substances may be correlated with the molar concentration of the polymeric substances in the solutions or melts. In consequence, viscosity measurements may be used in certain instances to determine molecular weights. The results have been called viscosity-average molecular weights. The method is most common in polymer science and technology because of the ease in performing the measurements and the simplicity of the equipment. There are many sources of errors, however, and no general equation has been developed to cover all situations, even at infinite dilution. *See* POLYMERIZATION; VISCOSITY.

Ultracentrifugation. The molecular weights of large molecules have been determined from sedimentation studies of solutions subjected to intense gravitational fields. In the sedimentation equilibrium method, the solution is centrifuged until an equilibrium is established between sedimentation and diffusion; sometimes this requires several weeks. In the sedimentation rate method, determination is made of the rate of the diffusion of solute molecules in solution into pure solvent in contact with the solution. Concentrations are determined by optical methods (measurement of absorption and refractive index). The equations for calculation of molecular weight are applicable only for dilute solutions. An extrapolated value for the molecular weight M, found at zero concentration by plotting $1/M$ versus concentration, is often reported. Sedimentation studies give so-called weight average, weight-weight average, and Z-average molecular weights, depending on the methods employed.

An ordinary gravity cell has been used successfully to determine the weights of particles greater than 10^8 daltons. The solution is maintained at constant temperature (±0.001°C) until equilibrium between sedimentation produced by gravity is balanced by diffusion. When equilibrium is obtained in an ideal dilute solution in a gravity cell with a constant horizontal cross section and height h, the molecular weight of a monodisperse substance is given by the relation shown in Eq. (11).

$$M=\frac{RT\ln(c_2/c_1)}{(1-\rho v)gh} \tag{11}$$

In this equation c_1 and c_2 are the concentrations at the top and bottom of the cell, respectively, v is the partial specific volume, ρ is the density of the solution, g is the acceleration of gravity, T is the absolute temperature, and R is the gas constant. The weight-average molecular weight M_w is given by Eq. (12), where c_0 is the initial

$$M_w=\frac{RT}{(1-\rho v)gh}\,\frac{c_2-c_1}{c_0} \tag{12}$$

uniform concentration in the cell before sedimentation takes place.

Light scattering. When a beam of light induces electronic transitions in a material, the material serves as a secondary source of light and emits scattered radiation. Molecular weights of polymeric materials in solution have been determined by methods involving determination of light scattering. Weight-average molecular-weight values are obtained.

Velocity of sound. The molecular weight of some pure liquids has been correlated with the velocity of sound in the liquid. The molar sound velocity R in these liquids is given by Eq. (13), where M is the

$$R=\frac{M\gamma^{1/3}}{d} \tag{13}$$

molecular weight, γ is the sound velocity at temperature T, and d is the density at the same temperature. Molar sound velocity R in these liquids is an additive property of atoms and bonds similar to molar refraction N, which is given by Eq. (14), where n is the refractive index.

$$N=\frac{M}{d}\,\frac{(n^2-1)}{(n^2+2)} \tag{14}$$

Chromatographic methods. The characterization of mixtures with respect to both molecular weight and molecular weight distribution can be effected by so-called gel permeation chromatography or gel filtration. Rigid porous gel particles are packed into a glass column, and a solution of the sample is eluted through the column of gel with appropriate solvents. Numerous polymers have responded well to this technique, and it is fast becoming a standard tool. For a homologous series, the relationship between the logarithm of the molecular weight and elution volume may be a straight line. Calibration may be made using narrow fractions with known molecular weights as standards, but calibration of one polymer is not applicable to another. Compared to other methods, the most important advantages of the method are speed, reproducibility for the same kind of polymer on a given gel, and relatively small samples. Thin-layer chromatography has been used where only minute samples are available for study. Molecular weight determinations by chromatographic methods will no doubt become most highly developed in the next decade. *See* CHROMATOGRAPHY.

[JOHN R. ANDERSON]

Bibliography: H. R. Allock and F. W. Lampe, *Contemporary Polymer Chemistry*, 1981; H. Batzer and F. Lohse, *Introduction to Macromolecular Chemistry*, 2d ed., 1979; N. C. Billingham, *Molar Mass Measurements in Polymer Science*, 1977; M. J. Sienko and R. A. Plane, *Chemistry*, 5th ed., 1976.

Molecule

A molecule may be thought of either as a structure built of atoms bound together by chemical forces or as a structure in which two or more nuclei are maintained in some definite geometrical configuration by attractive forces from a surrounding swarm of negative electrons. Besides chemically stable molecules, short-lived molecular fragments called free radicals can be observed under special circumstances, for example, at high temperatures, in electrical discharges, in chemical reactions, and even (but in small quantities) frozen into ordinary solid substances under some conditions. Free radicals are really just highly active molecules. *See* CHEMICAL BONDING; FREE RADICAL; MOLECULAR STRUCTURE AND SPECTRA.

[ROBERT S. MULLIKEN]

Molybdate

A compound containing molybdenum in the 6+ oxidation state and derived from molybdic acid, H_2MoO_4. The normal molybdates, MoO_4^{2-}, are mainly insoluble salts. These salts are soluble in strong acid and form condensed ions or isopolymolybdates, for example, $(NH_4)_6Mo_7O_{24}·4H_2O$. These isopolymolybdates are quite complex, but molybdates also form polymeric compounds with anhydrides of other elements, such as phosphorus, which are called heteropoly compounds. The yellow precipitate which is used in qualitative and quantitative analysis of phosphate is this type of compound and can be written as $(NH_4)_3$-$[P(Mo_3O_{10})_4]$.

The properties of the normal molybdates are similar to those of the sulfates, chromates, and tungstates. The only soluble normal molybdates are the salts of ammonium, sodium, potassium, rubidium, lithium, magnesium, beryllium, and thallium.

If molybdates in an acid solution are carefully reduced, a strong blue color forms. The exact composition of the compounds giving this color is not known. Molybdates are used as pigments, chemical reagents, and corrosion inhibitors. *See* MOLYBDENUM.

[E. EUGENE WEAVER]

Molybdenum

A chemical element, Mo, atomic number 42, and atomic weight 95.95, in subgroup VI of the periodic table. A silver-gray metal, molybdenum derived its name from the Greek "molybdos," meaning "leadlike," but the only physical similarity between the

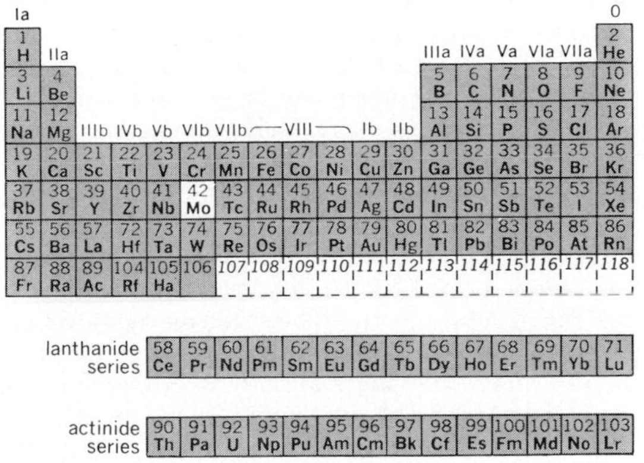

two metals is in terms of density (10.22 g/cm^3 for molybdenum). Molybdenum metal melts at 2610°C. It was discovered by K. W. Scheele in 1778, and metallic molybdenum was first isolated in 1782. It was largely a laboratory species during the 19th century, and no appreciable use was made of the metal until World War I. Reports came to Allied intelligence that the Germans were using "moly steel" in the Big Bertha gun barrels, and the United States moved fast to develop sources and to research molybdenum effects in iron and steel. The reports were subsequently found to be untrue, but the impetus had been provided.

Molybdenum is found in many parts of the world, but relatively few deposits are rich enough to warrant recovery costs. By far the largest and richest deposits occur in the Western Hemisphere, with the United States contributing the major share. Canada and Chile are other important free-world sources. Modest outputs are reported from various locations within the Soviet Union and China. The largest United States mine is at Climax, Colo., an installation situated on the Continental Divide.

Between two-thirds and three-fourths of the free-world supply of molybdenum comes from mines where its recovery is the primary objective of the operation. The remainder is recovered as a by-product of certain copper-mining operations, largely in the United States and Chile.

Extraction and uses. The most abundant mineral is molybdenite (MoS_2), important in both primary and by-product mining operations. Primary operations depend on massive silicified replacement deposits, while by-product mining involves copper deposits containing molybdenite. Less important molybdenum-containing minerals are powellite, $Ca[MoW]O_4$, and wulfenite, $PbMoO_4$. Ore assays vary, but the extensive deposit at Climax yields in the neighborhood of 0.3% molybdenum. Stated in different terms, approximately 6 lb of molybdenum are recovered from each ton of ore that is mined.

Molybdenite is concentrated by first crushing and grinding the ore to particles comparable in size to fine sand, then sending the finely ground material (called pulp) through a series of flotation cells. The cells contain a dispersion of oil in water, and the mineral's affinity for the small oil globules allows it to be floated to the top where it spills over into collecting troughs. Operations recovering molybdenum as a by-product of copper mining produce a concentrate containing both metals. Molybdenite is subsequently separated from the copper minerals by differential flotation.

The steps required to convert molybdenite concentrates into usable products are indicated in the flow chart in the illustration. Principal uses for the various molybdenum-containing products are also noted on the chart. A small fraction of molybdenite concentrate production is purified to a grade of molybdenum disulfide, used in the manufacture of dry lubricants, greases, and oils. Most concentrate production is roasted (heated in an oxidizing atmosphere) to technical molybdic oxide, a product about 95% MoO_3. The roasting process requires precise temperature control to prevent material losses from sublimation and volatization, which occur above about 700°C. Technical-grade oxide may be further purified by heating to about 1000°C, at which temperature molybdic oxide volatilizes readily, thereby effecting a separation from impurities that form more stable oxides. The volatilized oxide (generally 99.97% MoO_3) is collected in bag filters. A second purifying step is sometimes inserted at this stage. It involves dissolving pure oxide in ammonium hydroxide, filtering, and then evaporating to crystallize out the pure compound ammonium molybdate, $(NH_4)_2Mo_2O_7$. Pure oxide or ammonium molybdate may be converted to metal powder by a two-step hydrogen reduction at 680°C (forming MoO_2) and at 1090°C. Molybdenum-metal powder is consolidated into metallic bodies by one of two processes:

1. Arc-casting. Prefabricated or continuously compacted and sintered electrodes made from metal powder are arc-melted in a water-cooled copper mold. The process is usually carried out in a vacuum, although in some instances inert-atmosphere melting is employed. The process is particularly well suited for the preparation of molybdenum-base alloys by virtue of the flexibility

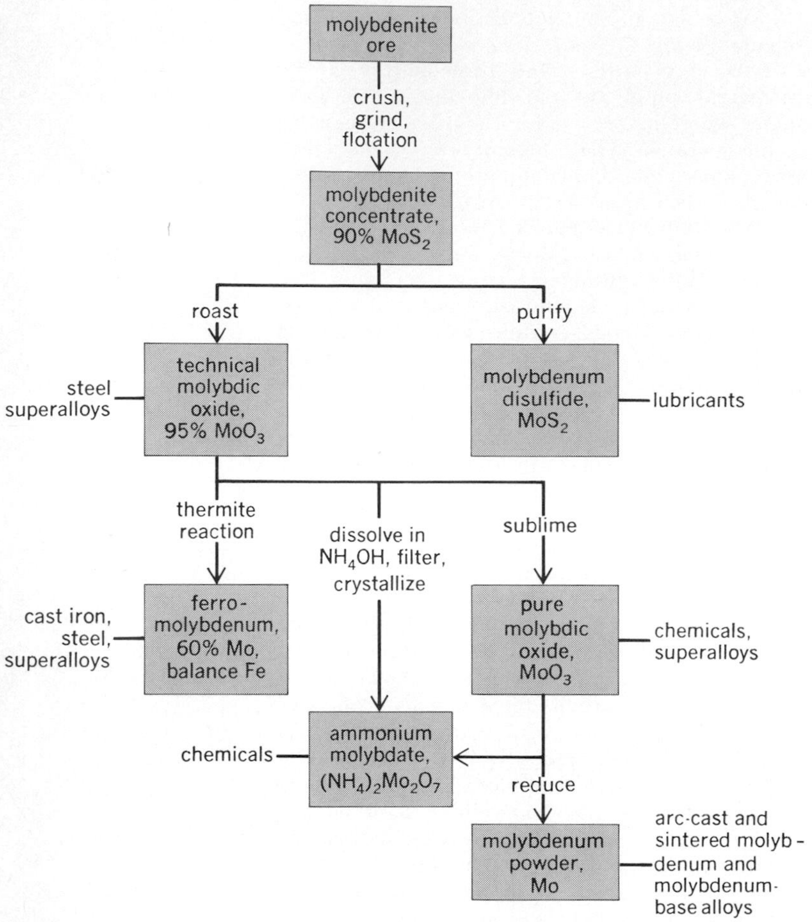

Flow chart for converting molybdenite concentrates into products.

available in altering the composition of the melting electrode. Ingots as large as 1 ton have been made by this process.

2. Powder metallurgy. Molybdenum powder charges are compacted in a mold of the desired size and shape, and the green compacts are sintered (usually in vacuum or hydrogen) to produce bodies generally from 95 to 98% of theoretical density. Further densification is accomplished during subsequent hot-working sequences.

Technical molybdic oxide may also be converted to ferromolybdenum, a product favored by many foundries and steel producers. A thermite process is generally utilized, and a typical charge for the process is generally utilized, and a typical charge for the process contains molybdic oxide, aluminum, ferrosilicon, high-grade iron ore, limestone, and fluorspar. The smelting reaction is exothermic and is complete in a period of approximately 20 min. Some ferromolybdenum is prepared by direct reaction in electric furnaces. This practice is more popular in Europe than in the United States.

Iron-base alloys (the various types of steel, cast iron, and some of the superalloys) account for at least 85% of today's molybdenum consumption. The term alloy steel embraces a wide range of materials, from the low-alloy, constructional steels to the newer ultrastrength compositions, which may have yield strengths as high as 2100 N/mm^2. Molybdenum in amounts from a fraction of 1% to as much as 5% is added to many of the steels included in this broad classification. One of the principal benefits to be derived from molybdenum additions to alloy steels is improved hardenability, that property that determines the hardness pattern throughout a part which is hardened in a given manner.

Hardness and strength are essentially synonymous; thus improved hardenability means that the optimum strength of a given alloy can be more readily achieved. Molybdenum also contributes to strength by a process known as solid-solution strengthening. Here molybdenum atoms dissolve in the lattice of iron atoms, causing an expansion of the lattice and making it necessary to apply more force to cause one layer of atoms to slip past the adjacent layer. In some alloys, molybdenum contributes to strength by bringing about a change in the size and distribution of excess phase constituents, such as carbides. From the practical standpoint, molybdenum additions improve cold formability, toughness, machinability, and weldability of many steel grades.

Molybdenum additions to stainless steel provide three important effects: improved corrosion resistance, improved strength at elevated temperature, and improved weldability. Because of the contributions of molybdenum-containing stainless steels, commercial processes involving the handling of highly corrosive chemicals are feasible today that were known only in the laboratory a few years back. Molybdenum is an important additive to stainless steel used for automotive trim and for architectural members, particularly in coastal areas where salt-laden air is encountered. The more common grades of molybdenum-containing stainless steel contain 2–4% Mo, but special grades may contain as much as 20% Mo.

The principal dividends resulting from molybdenum additions to tool and high-speed steels are improvement in hot strength, in resistance to softening, and in resistance to thermal-cycling effects. Tool steels historically contained tungsten as the major alloying addition. Molybdenum has replaced some or all of the tungsten to the extent that almost 90% of the tool steel produced in the United States today contains molybdenum. The driving force behind the shift to molybdenum was largely economics, but improved toughness and decreased carbide segregation were added benefits. Molybdenum-containing tool steels generally contain 1–10% Mo.

Molybdenum is added to cast iron primarily to improve its strength. A family of rather complex cast irons (and cast steels) has been developed to combat abrasion that is encountered in many forms in today's highly industrialized society. Molybdenum is important in these materials because of its contribution to hardenability and second-phase (carbide) morphology. The economic significance of this class of materials can be appreciated when it is realized that one installation alone, the molybdenum mine and plant at Climax, Colo., budgets some $3,000,000 annually for replacement of parts worn away in its ore-handling and processing equipment. The maximum molybdenum addition to cast iron is in the order of 3%.

Molybdenum is also an important alloying addition to nickel-base, cobalt-base, and iron-nickel-base alloys used for severe heat-resistant and corrosion-resistant applications. Chemical process equipment and aerospace applications are the principal uses in these categories. Molybdenum is found in these alloys at levels up to 30%, and is added mainly because of its contributions to elevated temperature strength and corrosion resistance.

The first uses of molybdenum metal date from the late 19th century and were closely allied to the development of the incandescent light bulb. The increasing use of the metal in recent years has been spurred by improved methods of consolidation and working and by developments in the field of molybdenum-base alloys. The properties of metallic molybdenum that are responsible for increasing applications are high melting point, high strength at elevated temperatures, high modulus of elasticity, high thermal conductivity, good resistance to corrosion, low specific heat, and low coefficient of expansion. The early light-bulb application has been updated to include the use of molybdenum in a variety of electronic devices, comprising tubes, contacts, electrodes, transducers, transistors, and rectifiers, The refractory nature of molybdenum makes it good for certain critical rocket and missile parts. One application receiving widespread acceptance involves the use of molybdenum-base alloys in dies and cores used in the die casting of aluminum and some ferrous alloys. A sprayed coating of molybdenum is applied to automotive piston rings to improve resistance to galling. Table 1 gives selected physical properties for metallic molybdenum.

Molybdenum-containing chemicals such as molybdic oxides and various molybdates are important catalysts in many petroleum and organic chemical reactions, including hydrocracking, hydrodenitrification, hydrodesulfurization, hydrogenation, oxidation, polymerization, and reforming. Sodium and ammonium molybdates used as catalysts exhibit relatively high aqueous solubility which increases their utility and flexibility in this application.

Table 1. Selected physical properties of metallic molybdenum

Property	Value
Melting point	2610°C
Heat of fusion	6.7 kcal/mol
Boiling point	5560°C
Heat of vaporization	117.4 kcal/mol
Heat capacity (298.16 – 1800 K)	$5.48 + 1.30 \times 10^{-3}T$ (in cal/deg mol)
Specific heat (20°C)	0.064 cal/g°C
Thermal conductivity	
200°C	0.298 cal/sec/cm²/cm/°C
1100°C	0.239 cal/sec/cm²/cm/°C
2200°C	0.206 cal/sec/cm²/cm/°C
Mean linear expansion coefficient	
20 – 150°C	5.43×10^{-6}/°C
20 – 1600°C	6.65×10^{-6}/°C
Electrical conductivity (0°C)	34% IACS* (58.62 microhm-cm)
Electrical resistivity	
0°C	5.2 microhm-cm
7250°C	23.9 microhm-cm
1525°C	47.2 microhm-cm
2525°C	78.2 microhm-cm
Magnetic susceptibility	
25°C	0.93×10^{-6} emu/g
1825°C	1.11×10^{-6} emu/g
Lattice	
Type	body-centered cubic
Parameter (25°C)	3.1405 kX† (3.14767A)
Density	10.22 g/cm³
Modulus of elasticity	0.324 N/m²

*International Annealed Copper Standard.
†Kilo X units = 0.100208 nm.

Molybdenum disulfide is an important lubricant that owes its lubricity to a layered lattice structure in which cleavage and shear between adjacent laminae are extremely easy. It is utilized as a dry-film lubricant and is added to greases and oils in suspension to improve lubricating properties. Molybdenum disulfide is stable under a variety of environmental conditions; hence, it is ideally suited for space applications. Certain critical components of the vehicle that landed on the Moon in mid-1969 were lubricated by molybdenum disulfide. This useful compound is also responsible for the increased lubrication cycle in many of today's automobiles, and has been found to reduce gasoline consumption when added as a suspension in crankcase oil. It is interesting that American western pioneers recognized the lubricity of molybdenum disulfide and rubbed the compound, obtained from mineral outcroppings, on the axles of their wagons.

Chemistry. Molybdenum forms compounds in which it displays valence states of 0, 2+, 3+, 4+, 5+, and 6+. Molybdenum as an ionizable cation has not been observed, but cationic species, such as molybdenyl, MoO_2^{2+}, are known to exist. The chemistry of molybdenum is extremely complex and, with the exception of the halides and chalcogenides, very few simple compounds are known.

Molybdenum in life processes. Molybdenum occurs in water, soil, plants, and animals to the extent of a few parts per million. It is one of the seven micronutrients known to be necessary for the growth and development of plants. Most plants cannot survive without molybdenum because it is needed to change nitrogen into forms the plant can use. In general, the amount of molybdenum will determine the amount of nitrogen fixed if other factors are not limiting. From the available evidence, it appears that animals and fish also require molybdenum just as do plants. Molybdenum is an essential constituent of the enzymes xanthine oxidase and aldehyde oxidase, which occur in the livers and intestines of animals, and of hepatic sulfite oxidase.

Oxides and related compounds. A study of the chemistry of molybdenum suggests that oxides should exist having each valence number from 2 to 6. Molybdenum dioxide and trioxide are the most common and most stable; other reported oxides are metastable and are essentially laboratory species prepared by prolonged heating of stoichiometric mixtures of MoO_2 and MoO_3, or MoO_3 and molybdenum powder.

The reported oxides and their selected properties are given in Table 2. Reduction of the trioxide does not appear to proceed stepwise through the intermediate compounds shown in the table. The dioxide is really oxidized to the trioxide at elevated temperatures by air or carbon dioxide.

Chlorine reacts with molybdenum dioxide at 300°C to yield the dioxychloride, MoO_2Cl_2. Molybdenum dioxide is relatively unaffected by nonoxidizing acids, alkalies, and molten salts, but it does react with chlorinated hydrocarbons at elevated temperatures to form molybdenum tetrachloride, $MoCl_4$.

Molybdenum trioxide is perhaps the most important compound of molybdenum. Most of the known chemical compounds containing molybdenum are prepared either directly or indirectly

Table 2. Molybdenum oxides

Empirical formula	Crystal structure	Molecular formula	Name	Color
MoO_3	Orthorhombic	MoO_3	α-Trioxide	White
$MoO_{2.89}$	Triclinic	$Mo_{18}O_{52}$	ζ-Oxide	Blue black
$MoO_{2.89}$	Monoclinic	Mo_9O_{26}	β-Oxide	Violet
$MoO_{2.84}$	Monoclinic	Mo_8O_{23}	β-Oxide	Violet
$MoO_{2.80}$	Tetragonal	Mo_5O_{14}	θ-Oxide	Blue violet
$MoO_{2.78}$	Orthorhombic	$Mo_{17}O_{47}$	κ-Oxide	Red blue
$MoO_{2.75}$	Monoclinic	Mo_4O_{11}	η-Oxide	Wine red
$MoO_{2.75}$	Orthorhombic	Mo_4O_{11}	γ-Oxide	Wine red
MoO_2	Monoclinic	MoO_2	Dioxide	Brown red

from MoO_3. The compound melts at 795°C and boils at 1155°C. It forms transparent, pale green-yellow, orthorhombic crystals. These are flat needles in which the directions of largest and smallest dimensions are (001) and (010), respectively. The space group is *Pbnm*, with four MoO_3 per unit cell. Dimensions of the unit cell are $a=$ 0.3628 nm, $b = 0.13855$ nm, and $c = 0.36964$ nm. Unit-cell volume is 0.2029 nm³, and the calculated crystal density is 4.175 g/cm³.

Molybdenum trioxide reacts with strong acids (notably concentrated sulfuric acid) to form complex cations, such as molybdenyl, MoO_2^{2+}, and molybdyl, MoO^{4+}. These in turn form soluble compounds. Alkali solutions and many metal oxides react readily with MoO_3 to form normal molybdates or polymolybdates, depending on the stoichiometry of the system. The properties of some normal molybdates are given in Table 3.

Mild reduction of an acid solution of a molybdate gives a strong blue color (molybdenum blue) as the initial step in the reduction reaction. Similarly, mild oxidation of acid solutions of molybdenum having a lower valency produces a strong blue color as the last stage in oxidation before the hexavalent stage. Molybdenum blue is important in several colorimetric analytical procedures.

Molybdic acid, H_2MoO_4 (or $MoO_3 \cdot H_2O$) forms a series of stable, normal salts of the types $M_2^{2+}MoO_4$, $M^{2+}MoO_4$, and $M_2^{3+}(MoO_4)_3$. These salts may be prepared by combination of the oxides, neutralization of slurries of MoO_3, or precipitation from molybdate solutions by salts of the desired metal.

Table 3. Properties of normal molybdates

Molybdate	Color	Density	Crystal structure	Melting point, °C
Li_2MoO_4	White	2.66	C_{3i}^2 phenacite $c/a = 1.153$	702
Na_2MoO_4	White	3.28	Hl_1, spinel $a_0 = 9.12$ A	686
K_2MoO_4	White	2.342	Isomorphous with K_2SO_4, K_2CrO_4	919
Rb_2MoO_4	White			958
Cs_2MoO_4	White			936
$CaMoO_4$	White	4.28	C_{4h}^6, scheelite $c/a = 2.19$	965 (dec)
$SrMoO_4$	White	4.662	C_{4h}^6, scheelite $c/a = 2.23$	≤1040 (dec)
$BaMoO_4$	White	4.975	C_{4h}^6, scheelite $c/a = 2.29$	1480
$PbMoO_4$	White	6.811	C_{4h}^6, scheelite $c/a = 2.23$	1065
$MnMoO_4$	Yellow		Monoclinic	
$FeMoO_4$	Dark brown Yellow		Monoclinic	850
$CoMoO_4$	Violet Rose	3.6 (α) 4.5 (β) 4.1 (γ)		1040
$NiMoO_4$	Green	3.5 (α) 4.9 (β) — (γ)	Monoclinic	970
$CuMoO_4$	Light green			820 (dec)
$CdMoO_4$	Light yellow	5.347	C_{4h}^6, scheelite $c/a = 2.174$	(dec)* Approx 900
$ZnMoO_4$	White		Tetragonal pyramids	700
Ag_2MoO_4	White Pale yellow†		O_h^7, spinel,† cubic	483
Tl_2MoO_4	White‡ Yellow§		$a_0 = 9.26$ A	Red heat

*Decomposes to lower oxides. †After fusion. ‡When precipitated. §During fusion.

Polymeric or isopoly molybdates may be formed by acidification of a molybdate solution or, in some cases, by heating the normal molybdates. As a molybdate solution is acidified, polymerization first takes place to form $Mo_7O_{24}^{6-}$ and then octamolybdate, $Mo_8O_{26}^{4-}$. Further acidification results in either the formation of complex molybdenum-oxygen cationic species or the precipitation of molybdic acid.

Hydrogen peroxide reacts with a number of molybdates to form a series of anionic peroxy compounds. Molybdates react with hydrogen peroxide in strongly basic solutions to give the tetraperoxy species $(MoO_8)^{2-}$. In acid solutions, however, species such as $(MoO_7)^{2-}$, $(MoO_6)^{2-}$, and $(MoO_5)^{2-}$ are known to exist.

The heteropoly electrolytes make up a large, fundamental family of salts and free acids, each member containing a complex and high-molecular-weight anion. The best known of these contain 5 to 18 hexavalent molybdenum atoms surrounding one or more central atoms. Examples of these compounds, all of which are highly oxygenated, are $(PMo_{12}O_{40})^{3-}$, $(AsMo_{18}O_{62})^{6-}$, $(TeMo_6O_{24})^{6-}$, and $(PMo_{10}V_2O_{40})^{5-}$, in which P^{5+}, As^{5+}, and Te^{6+} are the central atoms. Approximately 36 elements have been reported to function as central atoms in heteropoly anions. Many of these elements can act as central atoms in more than one series of heteropoly anions.

Hexacarbonyl. Molybdenum hexacarbonyl, $Mo(CO)_6$, in which molybdenum has a valency of zero, may be prepared from metal powder and carbon monoxide at high pressure, or from molybdenum pentachloride, zinc dust, and carbon monoxide in ether at a pressure of 90–120 atm in the absence of oxygen and water. Molybdenum hexacarbonyl decomposes at 150°C without melting, but it is quite stable below this temperature. Toxicity problems have not been encountered during normal handling. The compound reacts with many types of organic chemicals to yield a variety of organomolybdenum compounds.

Halides. Molybdenum forms halides and oxyhalides, representing a wide range in stability. The known halides and oxyhalides of molybdenum are listed in Table 4. The highest halide in each series (MoF_6, $MoCl_5$, $MoBr_4$, and MoI_3) can be made by direct halogenation of molybdenum metal. Few of the compounds are normally monomeric. Lower halides are usually made by reducing the highest member of the series with molybdenum metal, hydrogen, or a hydrocarbon.

Table 4. Halogen compounds of molybdenum

Fluorides	Chlorides	Bromides	Iodides
MoF_6*			
MoF_5*	$MoCl_5$*		
MoF_4	$MoCl_4$	$MoBr_4$	
MoF_3	$MoCl_3$	$MoBr_3$	MoI_3
MoF_2	$MoCl_2$	$MoBr_2$	MoI_2
$MoOF_4$*	$MoOCl_4$*		
	$MoOCl_3$	$MoOBr_3$	
$MoOF \cdot 4H_2O$	$MoOCl$		
	MoO_2Cl		
	$MoOCl_2$		
MoO_2F_2	MoO_2Cl_2*	MoO_2Br_2	
	Mo_2OCl_8		

*Volatile.

Many of the halide compounds are extremely reactive in the presence of water and oxygen and must be handled in an inert atmosphere. Molybdenum pentachloride is converted to $MoOCl_4$ by oxygen at atmospheric pressure and to $MoOCl_3$ by air, the latter being the most stable of the oxychlorides. Molybdenum tetrachloride appears to sublime on heating, but it disproportionates to $MoCl_2$ and $MoCl_5$ on cooling. In acid solution, $MoCl_4$ dissociates to a mixture of trivalent and pentavalent compounds.

Sulfides, selenides, and tellurides. Molybdenum forms a series of homologous compounds with S, Se, and Te that are somewhat similar to the oxides. The sesquisulfide, sesquiselenide, and sesquitelluride (all Mo_2X_3) have been prepared by the direct reaction of the elements at elevated temperatures in sealed, evacuated tubes. Molybdenum sesquisulfide may also be prepared by the direct thermal decomposition of molybdenum disulfide. All of the sesqui- compounds are inert to attack by nonoxidizing acids.

The disulfide, diselenide, and ditelluride (MoX_2) are isomorphous and crystallize in the hexagonal system with a layer lattice structure. As noted earlier, MoS_2 is quite stable, but under strongly oxidizing conditions it is converted to MoO_3. The melting point of MoS_2 has been placed somewhere above 1600°C. It exhibits a hexagonal crystal structure with sixfold symmetry and two molecules per unit cell. A rhombohedral form has been prepared synthetically and even observed in isolated natural occurrences. Molybdenum disulfide has a specific gravity in the range 4.85–5.0. Hardness on the Mohs scale is 1–1.5 (HV 29).

Molybdenum trisulfide is formed when solutions of ammonium tetrathiomolybdate, $(NH_4)_2MoS_4$, are acidified or heated. The latter compound resembles the molybdates and undergoes many similar reactions in which the oxygen and sulfur atoms are interchangeable.

[HUGH MORROW, III]

Bibliography: J. Z. Briggs, *Everyday Uses of Molybdenum,* 1966; Climax Molybdenum Co., *Heteropoly Compounds of Molybdenum and Tungsten,* Bull. no. Cdb–12a, 1969; Climax Molybdenum Co., *Properties of Molybdenum Disulfide,* Bull. no. Cdb–5a, 1962; Climax Molybdenum Co., *Properties of Molybdenum Pentachloride,* Bull. no. Cdb–3a, 1969; Climax Molybdenum Co., *Properties of Molybdic Oxide,* Bull. no. Cdb–1, 1969; Climax Molybdenum Co., *Properties of Simple Molybdates,* Bull. no. Cdb–4, 1962; R. R. Freeman, Everyday uses of molybdenum, *Metal Progr.*, 86:161–168, October 1964; P. Gousseland, Molybdenum, *Eng. Min. J.*, 103–104, March 1975; C. A. Hampel, *Rare Metals Handbook,* 2d ed., 1961; W. E. Latimer and J. H. Hildebrand, *Reference Book of Inorganic Chemistry,* 1951; P. C. H. Mitchell (ed.), *Chemistry and Uses of Molybdenum,* September 1973; U.S. Bureau of Mines, *Mineral Facts and Problems,* Bull. no. 630, 1965.

Monomolecular film

A film one molecule thick; often referred to as a monolayer. Films that form at surfaces or interfaces are of special importance. Such films may reduce friction, wear, and rust, or may stabilize emulsions, foams, and solid dispersions. Thin films

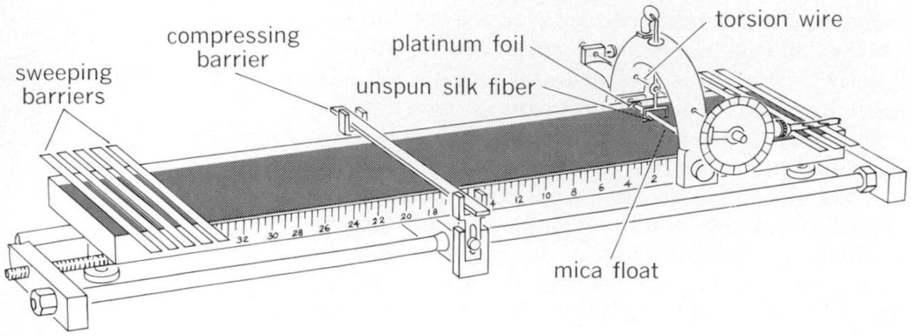

Fig. 1. Schematic drawing of film-balance apparatus.

on water surfaces reduce evaporation losses, which are important in arid regions throughout the world. Nevertheless, the removal of thin films of contaminants is one of many problems in the control of pollution. The broad field of catalysis, which is basic to petroleum refining and many chemical industries, involves chemical reactions that are accelerated in the thin films of reactants at interfaces. Moreover, thin films containing proteins, cholesterol, and related compounds constitute biological membranes, the internal interfaces that control the complex processes of life. *See* CATALYSIS.

In all of these areas, a single monomolecular layer at the interface is the most important. It is held to the adsorbing surface by forces stronger than those that hold any succeeding layer. On solid surfaces, it is the only layer that can be chemisorbed. It may be the site of enhanced chemical reactivity, or the last line of defense.

Monolayers on solids, or at liquid interfaces, may be formed by adsorption from the adjacent bulk phases; the process may show high specificity for particular chemical species. Measurements of the extent of adsorption have historically provided information on the composition and structure of monolayers formed in this way. A variety of surface-sensitive instrumental techniques, such as diffraction and scattering of low-energy electrons, neutrons, and ions, and spectroscopy of adsorbed species, have been brought to bear to obtain information about the structure of the surface layer and chemical perturbations in it. *See* ADSORPTION; SPECTROSCOPY.

In addition, monolayers of a wide variety of substantially insoluble substances can be formed at a liquid-gas interface by allowing them to spread over the surface. The pioneering studies of I. Langmuir and W. D. Harkins in the United States, and N. K. Adam and E. K. Rideal in England, showed how to manipulate, control, and measure the properties of such films at the water-air interface in simple and elegant ways. In their research, and that of many subsequent workers, a variety of specialized experimental techniques have been developed to study these insoluble monolayers.

In order to form spread monolayers which are sufficiently stable to study, a substance must combine low solubility and volatility with some moiety which attracts it to the liquid surface; for films on water, this generally means one or more polar functional groups. Totally nonpolar substances, such as the higher-molecular-weight paraffin hydrocarbons, will not spread on water (although they can spread on liquids of very high surface tension, such as mercury). Typical among the large group of substances which do form insoluble monolayers on water are the long-chain fatty acids and their derivatives such as glycerides, sterols, and many lipid substances of biological origin, including the fat-soluble vitamins and natural pigments such as chlorophyll. Many polar synthetic polymers, including polyvinyl acetate and polymethyl methacrylate, can be made to spread as monolayers on water; so can many proteins, because their tertiary structure unfolds at the air-water interface.

Experimental techniques. The film balance provides basic information on molecular geometry and orientation, location and strength of polar groups, and forces of cohesion and adhesion. With this instrument, the surface pressure (or lowering of surface tension) is measured as a function of the area available to the film-forming molecules, or in other words, their concentration at the surface or proximity to one another.

The apparatus consists essentially of a long, shallow trough filled with high-purity water on which the monolayer is spread, and a float system for measuring the surface pressure (Fig. 1). The

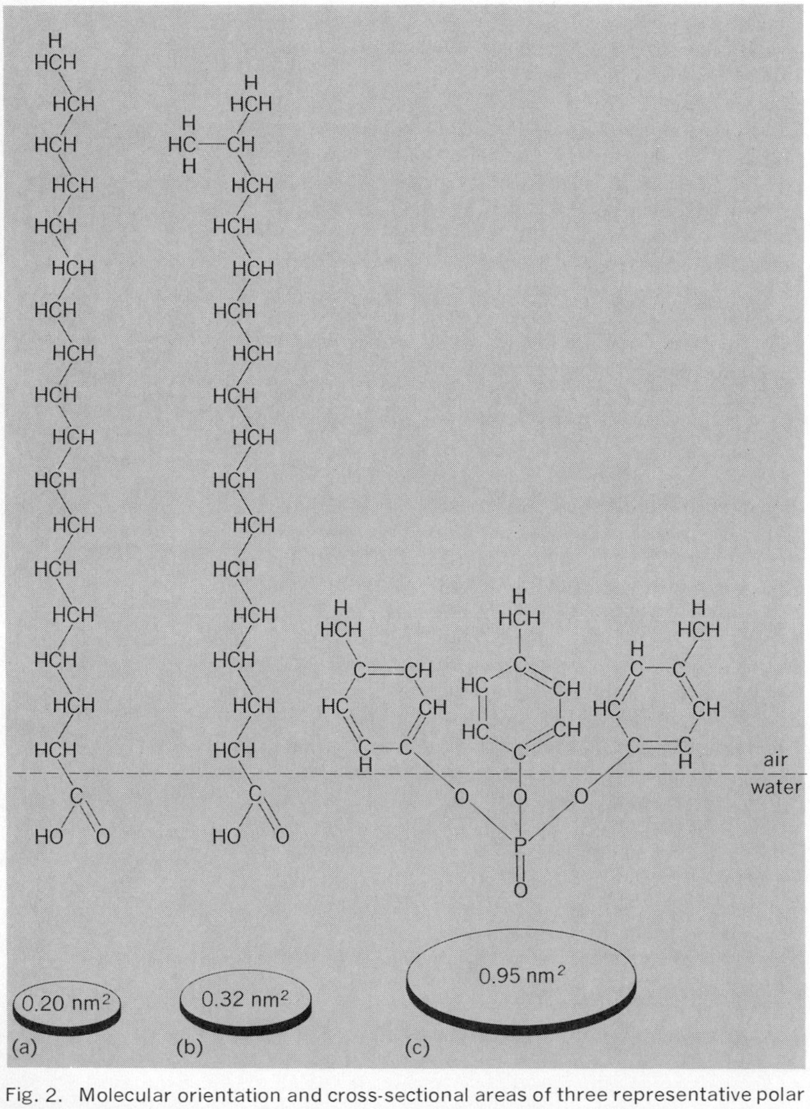

Fig. 2. Molecular orientation and cross-sectional areas of three representative polar organic molecules that are oriented at the water-air interface. (*a*) Stearic acid. (*b*) Isostearic acid. (*c*) Tri-*p*-cresyl phosphate.

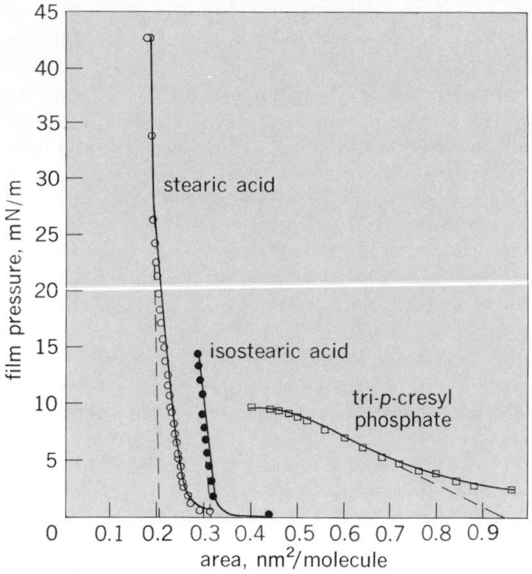

Fig. 3. Pressure-area isotherms.

float is a strip of mica attached to the sides of the trough by thin, flexible platinum foils and to an aluminum stirrup by an unspun silk thread. The stirrup is fixed to a calibrated torsion wire that indicates the surface pressure. Small brass bars or barriers are used for sweeping the water surface free of contamination. A small amount (approximately 0.01 mg) of the film-forming compound in a volatile solvent is spread between the float and the large or main barrier. The barrier is moved gradually toward the float to compress the film. Compression is continued until the pressure remains constant or falls. This procedure at constant temperature provides data for plotting pressure-area isotherms which characterize the films.

Before each experiment the entire apparatus—trough, barriers, float, and platinum foils—are thoroughly cleaned and coated with high-melting paraffin wax (or with Teflon). The trough is filled with water to a height well above the rim. This height is necessary if the sweeping procedure is to be effective and if the monolayer is to be contained and controlled.

Three representative polar organic molecules oriented at the water-air interface are shown in Fig. 2. Stearic acid is the classical compound in monolayer studies; it is the simplest structure representative of thousands of important film-forming compounds. The stearic acid molecule consists of a long, straight hydrocarbon chain and a polar group at one extremity. Also included in Fig. 2 is the structural formula of a very similar molecule, isostearic acid, and a very dissimilar molecule, tri-*p*-cresyl phosphate. The difference between isostearic and stearic acid is very slight—the displacement of the small methyl group (CH_3) at the end of the molecule opposite the polar group. Tri-*p*-cresyl phosphate is greatly different; it has a bulky three-ring hydrocarbon portion attached to a strongly polar phosphate group.

Figure 3 shows the pressure-area isotherms for the three compounds. The surface pressure in millinewtons per meter (mN/m, or dynes/cm) is plotted against the average area per molecule in square nanometers (1 nm² equals 100 square angstrom units). Extrapolation or extension of the steepest part of an isotherm to zero pressure is often used as a measure of molecular area. The point at which the pressure falls or remains constant is called the collapse pressure. Compressibility of the monolayer may be calculated from the slope of the isotherm. Thickness of the monolayer, or the length of the vertically oriented molecules, may be estimated by assuming a density for the monolayer; the volume and area then yield the thickness.

Comparison of the isotherms for stearic and isostearic acids in Fig. 3 demonstrates that the single, small side chain of isostearic acid has increased the cross section from 0.20 nm² for the stearic acid molecule to 0.32 nm² for isostearic acid, an increase of more than 50%. Collapse pressure falls from 42 mN/m to one-third of this value, 14 mN/m. These are indeed striking differences between

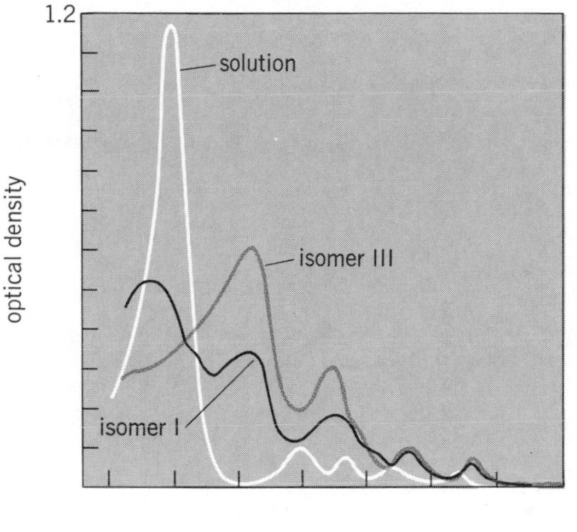

Fig. 4. Absorption spectra of monomolecular films of two isomers of coproporphyrin tetramethyl ester. While the solution spectra are identical, the monolayer spectra are very different, probably because the neighboring polar ($-COOCH_3$) groups on isomer III (shaded) provide a preferred orientation at the water surface.

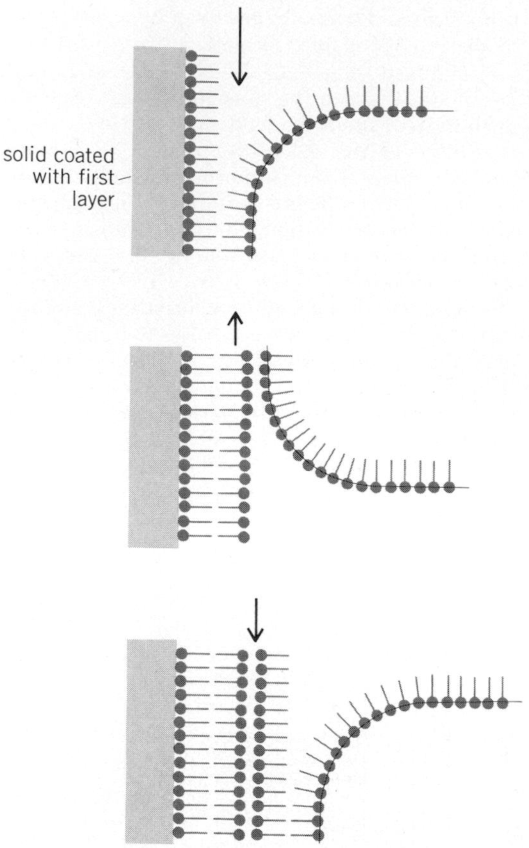

Fig. 5. Successive folding back and forth of a monolayer onto a solid plate, as it is dipped into and out of the liquid in the Langmuir-Blodgett multilayer deposition method. (*From G. L. Gaines, Jr., Insoluble Monolayers at Liquid-Gas Interfaces, copyright © 1966 by John Wiley and Sons, Inc.; used with permission*)

molecules that are extremely difficult to distinguish by most chemical methods.

The curve for tri-*p*-cresyl phosphate reflects a very different molecular structure. The extrapolated area, 0.95 nm^2/molecule, shows the bulkiness of the three-ring group held close to the surface. The low collapse pressure, 9 mN/m, reflects the weakness of such a film. The gradual slope of the curve, or the high compressibility of the film, indicates poor packing of the molecules.

In addition to changing the surface tension, the presence of a spread monolayer alters other properties of the liquid surface. The electric field, flow properties, and optical reflectivity are all more or less altered. Special techniques have been devised to measure the surface potential difference, surface viscosity, and reflectance change due to spread monolayers. Among other properties which have been measured are the ability to damp ripples or to suppress evaporation. While various film-forming substances differ in their effect on all these properties, they are still not understood well enough for the differences to be related to molecular structure in most cases.

Absorption and emission spectra of monolayers of colored and fluorescent substances, such as chlorophyll, have also been measured. Because the molecules in a spread monolayer are held in a preferred orientation at the air-water interface, and are packed close together when the film is compressed, the spectra are sometimes very different from those obtained from solutions. Figure 4 shows an extreme example of this effect. The two isomers of the porphyrin differ only in the position of the substituents at the periphery of the ring system, and their spectra in solution are indistinguishable. When they are spread as monolayers, however, the absorption spectra differ markedly.

Transfer of spread monolayers. Once a monolayer has been formed at the water-air interface, it is possible to transfer it quantitatively to another surface, such as that of a smooth solid. Dipping a clean glass slide, for example, into and out of the film balance trough while the monolayer on it is held at constant surface pressure (by advancing the confining barrier) leads to such a quantitative transfer. The area of monolayer taken up is the same as the surface area of the dipped slide. Properties of the transferred layer such as its wettability and electron diffraction pattern indicate that the molecules on the solid surface retain the preferred orientation which they had on the water surface. Monolayers transferred to solids in this way bear close resemblance in structural details both to the precursor liquid-supported monolayer and to films formed on solids by other processes such as adsorption from solutions, but there are usually some differences. The differences, in structure, tightness of bonding, and so on, depend on the nature of both the monolayer and the solid support.

It is also possible to deposit certain types of monolayers (especially heavy-metal soaps of long-chain fatty acids) sequentially on solid surfaces to form built-up films or multilayers (Fig. 5). Since each monolayer is extremely thin but uniform (for barium stearate, for example, almost exactly 25 A or 2.5 nm), such layer structures are very useful as spacers or thickness gages. Films more than 1000 layers thick have been made. In recent years, much interest has developed in the optical and electrical properties of such structures. It is possible to assemble multicomponent structures in many ways, both by using monolayers containing more than one chemical species and by changing from one kind of monolayer to another at different cycles in the dipping process. Locating different molecules at known small distances and in con-

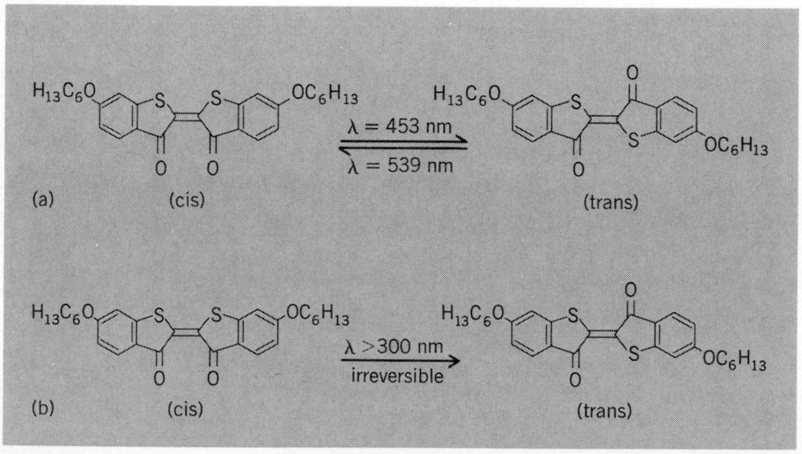

Fig. 6. Thioindigo dye undergoing a photoinduced cis-trans isomerization. (*a*) In solution, blue light (λ = 453 nm) converts it to the trans form, while green light (λ = 539 nm) reverses the reaction. (*b*) In a monolayer, only the cis→trans conversion can occur.

trolled relative orientation has permitted the study of such processes as energy transfer and electron transfer between them.

Chemical reactions. The orientation and close packing of molecules in monolayers may alter patterns of chemical reactivity. Monolayers of fatty acids show highly preferential incorporation of cations (soap formation) with certain salts in the underlying water. For example, a stearic acid film spread on a solution with equimolar concentrations of calcium and magnesium salts may contain 10 times as much calcium stearate as magnesium stearate.

At the air–aqueous solution interface, the electric field, and hence distribution of ions near the liquid surface, is unsymmetrical. This can have a controlling effect on the rates of chemical reactions involving these ions. An interesting example is the rate of hydrolysis of long-chain esters in monolayers, which involves the negatively charged hydroxyl ion from the underlying solution. To begin with, the rate of the reaction can be reduced by increasing the surface pressure, which forces the ester molecules closer together and also changes the surface electrical potential. If charged monolayer-forming molecules, such as long-chain quaternary ammonium compounds (with positive charges), are incorporated in the ester film, the reaction rate can be increased markedly. This results from the increased concentration of OH^- ions induced by the positive charge at the surface. Reactions of this kind are of special interest because they resemble many of the catalytic reactions which occur at membrane surfaces in biological systems.

In extreme cases, even the reaction products which can be formed are altered when molecules are constrained in the special environment of a monolayer or built-up multilayer. For example, certain thioindigo dyes (Fig. 6) undergo reversible light-induced cis-trans isomerization in solution. In monolayer assemblies, however, only the cis to trans transformation seems to be possible; once the trans isomer is formed, illumination cannot reverse the reaction. This effect apparently results from the fact that the trans form occupies less space than the cis form. The isomerization which does occur, therefore, involves a contraction in volume (on a molecular scale), and is irreversible in the built-up film. The surface pressure–area isotherms of the two forms indicate the same thing, since the trans isomer requires less area at the air-water interface, and if a monolayer of the cis isomer is illuminated (at constant surface pressure), it shrinks.

Reactions of the kind just described are fairly easy to study because the reactants and products have intense and very different absorption spectra. In the past, analytical techniques capable of measuring the very small amounts of material present in any convenient area of a single monomolecular layer (typically less than a microgram per square centimeter) were very limited. Modern analytical methods, however, have greatly improved sensitivity, and are being increasingly applied to study monomolecular films. Coupled with improved understanding of physical properties of the films and of liquid interfaces, these newer techniques promise new insights into the behavior of molecules at surfaces and the important biological and technological processes which they control. *See* COLLOID; EMULSION; INTERFACE OF PHASES; SURFACE TENSION.

[GEORGE L. GAINES, JR.]

Bibliography: A. W. Adamson, *Physical Chemistry of Surfaces*, 3d ed., 1976; G. L. Gaines, Jr., *Insoluble Monolayers at Liquid-Gas Interfaces*, 1966; E. D. Goddard (ed.), *Monolayers*, Advances in Chemistry Series, no. 144, 1975; H. Kuhn, D. Möbius, and H. Bucher, Spectroscopy of monolayer assemblies, in A. Weissberger and B. W. Rossiter (eds.), *Physical Methods of Chemistry*, pt. IIIB, 1972.

Mordant

A substance or combination of substances that facilitates the fixing of a dye to a fiber. A mordant enables the production of a more permanent and often deeper color. Metallic salts or hydroxides are most frequently used as mordants.

Certain mordants act directly on the fiber, making it more susceptible to the dye. Fabrics are then pretreated with the mordant before exposure to the dye. For example, cotton is soaked in a mixture of aluminum sulfate, $Al_2(SO_4)_3$; sodium carbonate, Na_2CO_3; and calcium carbonate, $CaCO_3$, before exposure to the dye, alizarin.

Other mordants function through the formation of a complex with the dye. The complex acts as the dyeing agent. Mordant and dye in this case are exposed simultaneously to the fabric. *See* DYE.

[FRANCIS J. JOHNSTON]

Bibliography: C. L. Bird, *Theory and Practice of Wool Dyeing*, 3d ed., Society of Dyers and Colourists, Bradford, England, 1963.

Mössbauer effect

Recoil-free gamma-ray resonance absorption. The Mössbauer effect, also called nuclear gamma resonance fluorescence, has become the basis for a type of spectroscopy which has found wide application in nuclear physics, structural and inorganic chemistry, biological sciences, the study of the solid state, and many related areas of science.

Theory of effect. The fundamental physics of this effect involves the transition (decay) of a nucleus from an excited state of energy E_e to a ground state of energy E_g with the emission of a gamma ray of energy E_γ. If the emitting nucleus is free to recoil, so as to conserve momentum, the emitted gamma ray energy is $E_\gamma = (E_e - E_g) - E_r$, where E_r is the recoil energy of the nucleus. The magnitude of E_r is given classically by the relationship $E_r = E_\gamma^2/2mc^2$, where m is the mass of the recoiling atom. Since E_r is a positive number, the E_γ will always be less than the difference $E_e - E_g$, and if the gamma ray is now absorbed by another nucleus, its energy is insufficient to promote the transition from the nuclear ground state E_g to the excited state E_e.

In 1957 R. L. Mössbauer discovered that if the emitting nucleus is held by strong bonding forces in the lattice of a solid, the whole lattice takes up the recoil energy, and the mass in the recoil energy equation given above becomes the mass of the whole lattice. Since this mass typically corresponds to that of 10^{10} to 10^{20} atoms, the recoil energy is reduced by a factor of 10^{-10} to 10^{-20}, with the important result that $E_r \approx 0$ so that

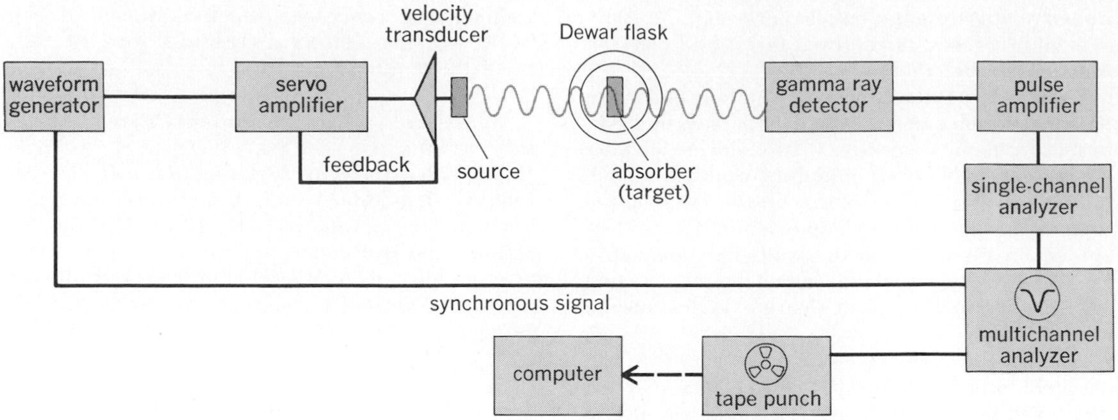

Fig. 1. Experimental arrangement for performing Mössbauer effect spectroscopy. This typical Mössbauer experiment is with ^{57}Fe or ^{119}Sn. (*From R. H. Herber, Mössbauer spectroscopy, Sci. Amer., 225(4):86–95, October 1971*)

$E_\gamma = E_e - E_g$; that is, the emitted gamma-ray energy is exactly equal to the difference between the nuclear ground-state energy and the excited-state energy. Consequently, absorption of this gamma ray by a nucleus which is also firmly bound to a solid lattice can result in the "pumping" of the absorber nucleus from the ground state to the excited state. The newly excited nucleus remains, on the average, in its upper energy state for a time given by its mean lifetime τ (a quantity dependent on energy, spin, and parity of the nuclear states involved in the deexcitation process) and then falls back to the ground state by reëmission of the gamma ray. An important feature of this reemission process is the fact that it is essentially isotropic; that is, it occurs with equal probability in all directions. *See* EXCITED STATE; GROUND STATE.

Energy modulation. Before this phenomenon of resonance fluorescence can be turned into a spectroscopic technique, it is necessary to provide an appropriate energy modulation of the gamma ray emitted in the initial decay process. An estimate of the energy needed to accomplish this can be calculated from a knowledge of the inherent width or sharpness of the excited-state nuclear level. This is given by the Heisenberg uncertainty principle as $\Gamma = h/2\pi\tau$ (h is Planck's constant and τ is the mean lifetime of the excited state). In the case of ^{57}Fe, a nucleus for which resonance fluorescence is especially easy to observe experimentally, $\Gamma = 4.6 \times 10^{-12}$ keV. In order to modulate the emitted gamma-ray energy, which in this case corresponds to 14.4 keV, one can take advantage of the Doppler phenomenon which states that if a radiation source has a velocity relative to an observer of v, its energy will be shifted by an amount equal to $E = (v/c)E_\gamma$. Setting the required Doppler energy equal to the width of the nuclear level and E_γ equal to the nuclear transition energy leads to $v = c(\Gamma/E_\gamma) = 3 \times 10^{10}$ cm/s $\times (4.6 \times 10^{-12}/14.4) =$ 0.0096 cm/s, and relative velocities of this order of magnitude can be used to modulate the gamma ray emitted in a typical Mössbauer transition, that is, to "sweep through" the energy width of the nuclear transition. For the experimental demonstration of this effect and its interpretation in terms of the fundamental physical principles involved, Mössbauer was awarded the Nobel Prize in Physics for 1961. *See* RADIOACTIVITY.

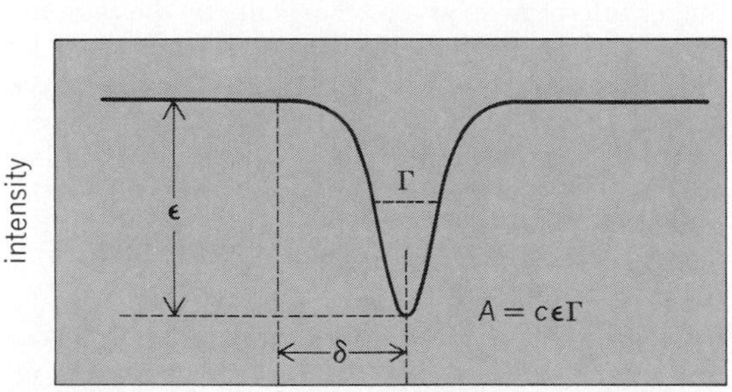

Fig. 2. Mössbauer spectrum of an absorber which gives an unsplit resonance line. The spectrum is characterized by a position δ, a line width Γ, and an area *A* related to the effect magnitude ε.

Experimental realization. The experimental realization of gamma-ray resonance fluorescence can be achieved with the arrangement illustrated schematically in Fig. 1. In a typical Mössbauer experiment the radioactive source is mounted on a velocity transducer which imparts a smoothly varying motion (relative to the absorber, which is held stationary), up to a maximum of several centimeters per second, to the source of the gamma rays. These gamma rays are incident on the material to be examined (the absorber). Some of the gamma rays are absorbed and reemitted in all directions, while the remainder of the gamma rays traverse the absorber and are registered in an appropriate detector which causes one or more pulses to be stored in a multichannel analyzer. The electronics are so arranged that the location (address) in the multichannel analyzer, where the transmitted pulses are stored, is synchronized with the magnitude of the relative motion of source and absorber.

A typical display of a Mössbauer spectrum, which is the result of many repetitive scans through the velocity range of the transducer, is shown in Fig. 2. Such a Mössbauer spectrum is characterized by a position δ of the resonance maximum (corresponding to a minimum in the intensity of the transmitted radiation), a line width Γ, and a resonance effect magnitude ϵ corresponding to the total area A under the resonance curve.

In the case of the Mössbauer active nuclides ^{57}Fe and ^{119}Sn, among others, two additional features which are of great interest to chemists and physicists may be experimentally elucidated. One of these is the quadrupole coupling which is observed if the Mössbauer nuclide is located in an environment where the electric charge distribution does not have cubic (that is, tetrahedral or octahedral) symmetry. Such a spectrum is shown in Fig. 3, in which the magnitude of the quadrupole interaction Δ is equal to $e^2qQ/2$, where e is the electron charge, q is the gradient of the electrostatic field at the nucleus, and Q is the nuclear quadrupole moment. Finally, a Mössbauer spectrum can also give information on the magnitude of the magnetic field H_0 acting on the nucleus through the magnetic hyperfine interaction. This is illustrated in Fig. 4, where only a single resonance line would be observed in the absence of a magnetic interaction.

Moreover, all of these parameters—δ, Δ, Γ, A, and H_0—are temperature-dependent quantities, and their study over a range of temperatures and conditions can shed a great deal of light on the nature of the environment in which the Mössbauer nuclide is located in the sample under investigation. One hundred Mössbauer transitions, involving 43 different elements, have been experimentally observed and reported.

Application. Mössbauer effect experiments have been used to elucidate problems in a very wide range of scientific disciplines, and only a few examples can be cited as representative of the information extracted from such studies.

Nuclear physics and chemistry. One of the narrowest resonance lines which has been observed is that from the 6.8-μs, 6.2-keV gamma transition in ^{181}Ta, and detailed Mössbauer effect measurements using a source of 140-day ^{181}W have shown that the magnetic moment of the spin 9/2 excited state in ^{181}Ta is $+5.35 \pm 0.09$ nm and the nuclear quadrupole moment of this state is $(+4.4 \pm 0.05) \times 10^{-24}$ cm^{-2}. Such data are of considerable use to nuclear physicists in refining models which describe the fundamental interaction forces in the nucleus. Similarly, the 93.26-keV resonance in ^{67}Zn has been used to determine the magnetic moment of the $1/2^-$ state (spin $= 1/2$, negative parity) in this nuclide and leads to the conclusion that $1/2^-$ and $5/2^-$ states can be considered minus-quasiparticle states. Such nuclear information is difficult to obtain by non-Mössbauer effect methods.

Recoilless gamma-ray resonance experiments have been able to provide detailed information concerning excited-state lifetimes involved in the nuclear decay process. The lifetime values for the nuclides ^{119}Sn, ^{197}Au, and ^{73}Ge, among others, are largely based on Mössbauer effect measurements.

Recoilless gamma fluorescence spectroscopy has also been used to study the chemical consequences of nuclear decay, and the lifetimes of the Mössbauer transition (typically about 10^{-8} s) provide a convenient time scale with which one can distinguish rapid electronic relaxation processes (typically 10^{-12} to 10^{-14} s) from atomic translation processes (typically slower than 10^{-6} s), and thus study the chemical fate of an atom which results from the decay of a radioactive parent nuclide.

Solid-state physics. Mössbauer effect spectroscopy has made significant contributions to the study of problems in solid-state physics, especially of the nature of the magnetic interactions in iron-containing alloys and the dependence of the magnetic field on composition, temperature, pressure, and other parameters which are of importance in metallurgical processes, solid-state-device fabrication, the structural use of metals and alloys, and numerous related problems of great practical importance.

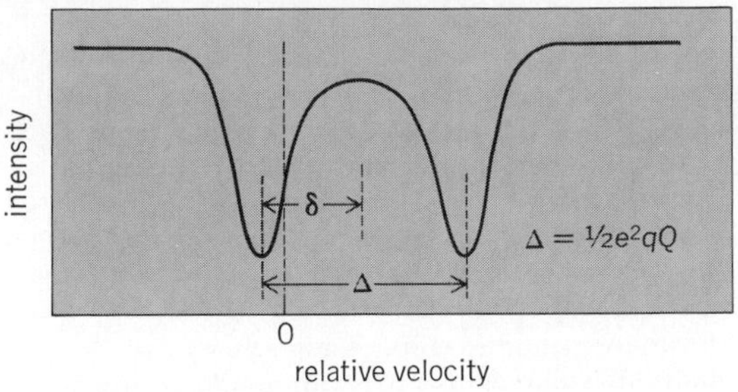

Fig. 3. Mössbauer spectrum of an absorber (containing for example ^{57}Fe or ^{119}Sn) which shows quadrupole splitting Δ.

Combining Mössbauer effect spectroscopy with vibrational spectroscopic studies has led to a clearer understanding of the nature of inter- and intramolecular forces and the relationship of these forces to the properties of polymeric materials.

It has also been possible, using this technique, to study the effect of high pressure and isotropic compressibility on the chemical properties of materials, especially in the case of experiments with ^{57}Fe, ^{181}Ta, and ^{119}Sn. Such studies have led to the design of high-pressure processes in preparative metallurgy and materials science.

At the extremely low end of the temperature scale, Mössbauer effect spectroscopy has been useful in examining the nature of those materials which become superconductive at sufficiently low temperatures and the relationship between chemical composition and structure on the one hand and the superconductive transition on the other, as in the dichalcogen layer compounds $TaS_2 \cdot Sn$ and $TaS_2 \cdot Sn_{1/3}$ (both studied using the nuclide ^{119}Sn). Nb_3Sn, a material widely used in the construction of superconducting magnets, has been subjected to detailed Mössbauer effect investigations.

Structural chemistry. The two Mössbauer nuclides most widely exploited by chemists are ^{57}Fe and ^{119}Sn, although a growing body of data resulting from experiments with ^{129}I, ^{99}Ru, ^{121}Sb, and others has been reported. The position of the reso-

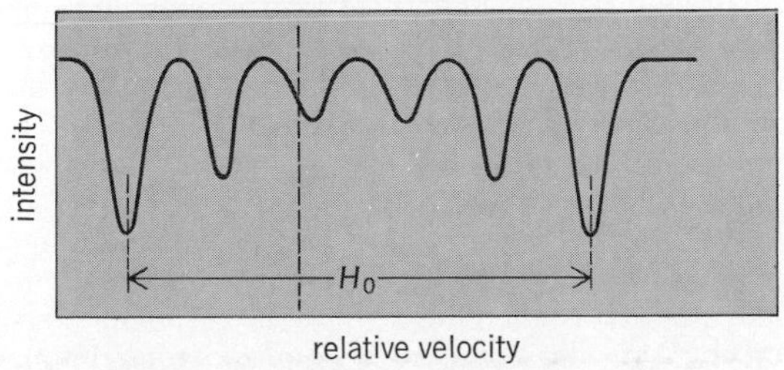

Fig. 4. Mössbauer spectrum of metallic iron showing the splitting of the resonance line by the internal magnetic field ($H_0 = 330$ kG at room temperature).

nance maximum δ, also called the isomer or chemical shift, can be related to the systematics of the electron configuration of the atom, and extensive isomer shift correlations for iron- and tin-containing compounds have been tabulated. In particular, the isomer shift of tin compounds is readily related to the oxidation state, since it has been observed that all stannous (Sn^{+2}) isomer shifts are larger than that observed for metallic tin (β-Sn), while those for stannic compounds (Sn^{+4}) are smaller than this value. This observation allows an assignment of oxidation state to be made on the basis of the isomer shift parameter, as in the two-dimensional layer compound $SnTa_3S_6$ in which the tin atom is clearly identified as a stannous ion, contrary to expectations based on theory. *See* OXIDATION-REDUCTION.

Similarly, the isomer shifts reported for a number of ruthenium compounds, which have been studied using the 89.36-keV resonance in ^{99}Ru, can be correlated systematically with the number of 4*d* electrons involved in the bonding of the metal atom to its nearest-neighbor ligands. Such Mössbauer effect studies have led to a clearer understanding of the nature of "ruthenium red," a trimeric ammonia ruthenium oxide, and a number of other compounds of this relatively rare transition-metal homolog of iron.

In the field of organometallic chemistry, Mössbauer effect spectroscopy has served to clarify the structure of a number of compounds of iron and tin which are of considerable synthetic and industrial importance, including $Fe_3(CO)_{12}$, $[(\pi C_5H_5)_2Fe(CO)_2]_2$, $[(C_4H_9)_3Sn]_2SO_4$, and the organotin thioglycolates which are used as stabilizers in the plastics industry. *See* ORGANOMETALLIC COMPOUND.

Biological science. Many molecules of biological importance, including hemoproteins, iron-sulfur proteins, and iron storage and transport proteins, offer an ideal system in which Mössbauer effect spectroscopy can be used to elucidate the structure and bonding properties of the metal atom in complex systems. The first measurements on such molecules were reported in 1961, and a very large number of iron-containing systems have been studied since then. Paramagnetic iron compounds can be studied at temperatures below the magnetic ordering point (Néel temperature), and it is thus possible, by means of Mössbauer effect spectroscopy, to determine the sign and magnitude of the magnetic field acting on an iron atom in a complex biological material with molecular weights ranging up to 50,000 or more.

It is also possible to study antiferromagnetically coupled iron atoms in biological molecules by carrying out the Mössbauer effect measurements in an external magnetic field over a range of temperatures. Typical of such a study is that of oxidized and reduced putidaredoxin, an iron-sulfur protein (molecular weight = 12,500) which acts as a one-electron transfer enzyme. The Mössbauer experiments on this material clearly showed that in the oxidized material the two irons atoms in the molecule occupy chemically equivalent sites. On one electron reduction, one iron atom remains ferric (Fe^{3+}), while the other becomes ferrous (Fe^{2+}), and the two atoms couple antiferromagnetically to give an electronic ground state of $S = 1/2$. Such detailed knowledge of the chemical behavior of the iron atoms in this molecule can elucidate the action of biological catalysts (enzymes) on the molecular level.

Related fields. Mössbauer effect studies have also played a role in studies in many related fields of science, including archeology, geology, engineering studies, theoretical (relativity) physics, chemical kinetics, and biology. The samples of surface material returned from the Moon by the United States Apollo program have been carefully scrutinized by Mössbauer techniques, as have core samples extracted from deep-drilling experiments on the Earth's outer layer. The geographical distribution of ancient Greek pottery has been traced by making use of characteristic Mössbauer effect data, and the pigments used in painting and decorating have been similarly investigated using this technique.

[ROLFE H. HERBER]

Bibliography: G. M. Bancroft, *Mössbauer Spectroscopy*, 1973; S. G. Cohen and M. Pasternak (eds.), *Perspectives in Mössbauer Spectroscopy*, 1973; H. Frauenfelder, *The Mössbauer Effect*, 1962; V. I. Goldanskii and R. H. Herber (eds.), *Chemical Applications of Mössbauer Spectroscopy*, 1968; N. N. Greenwood and T. C. Gibb, *Mössbauer Spectroscopy*, 1971; I. J. Gruverman (ed.), *Mössbauer Effect Methodology*, vols. 1–9, 1965–1974; International Atomic Energy Agency, *Mössbauer Spectroscopy and its Applications*, 1972; J. G. Stevens and V. Stevens, *Mössbauer Effect Data Index*, 1969–1973; G. K. Wertheim, *Mössbauer Effect; Principles and Applications*, 1964.

Multiple proportions, law of

This law states that, when two elements combine together to form more than one compound, the weights of one element that unite with a given weight of the other are in the ratio of small whole numbers. The law can be illustrated by the composition of the five oxides of nitrogen. One gram (g) of nitrogen is combined with 2.85 g of oxygen in nitrogen pentoxide, N_2O_5; with 2.28 g in nitrogen dioxide, NO_2; with 1.71 g in nitrogen trioxide, N_2O_3; with 1.14 g in nitric oxide, NO; and with 0.57 g in nitrous oxide, N_2O. These numbers are in the simple ratio of 5:4:3:2:1. *See* DEFINITE COMPOSITION, LAW OF.

[THOMAS C. WADDINGTON]

Naphtha

Any one of a wide variety of volatile hydrocarbon mixtures. They are sometimes obtained from coal tar but are more often derived from petroleum. Physical properties vary widely. The initial boiling point may be as low as 80°F (27°C), and end points may reach 500°F (260°C). Boiling ranges are sometimes as narrow as 20°F (11°C) or as wide as 200°F (110°C).

The main process for producing naphthas is fractional distillation. It may be of the extractive type when certain high-quality naphthas are desired. Acid treating, clay treating, and other techniques remove sulfur compounds and improve color, odor, and stability.

Strictly speaking, the refinery streams going into products like gasoline and kerosine are naphthas, and they are so designated within the petroleum industry. The final blended fuels, however, are sold under the more familiar names. The products sold as naphthas find their greatest use as solvents, thinners, or carriers.

Few naphthas are made up entirely of hydrocarbons belonging to one particular family. There is a fairly sharp differentiation, however, between aliphatic and aromatic types.

Aliphatic naphthas are relatively low in odor and toxicity and tend, also, to be low in solvent power, which in some cases is an advantage. In the processing of soybeans, for example, the aim is to extract the oil without extracting the less desirable materials. Naphthas used by dry cleaners likewise require only moderate solvent power. In printing ink, the naphtha is mainly a carrier of the carbon black or other pigments; resins requiring a solvent are present in only minor amounts.

The aromatic naphthas, often described as the high-solvency type, at one time came entirely from coal tar. The development of catalytic cracking and catalytic reforming made petroleum an alternative source. The main components are toluene and xylenes; benzene is less desirable because of the extreme toxicity of its vapors. A major use of these naphthas is as thinners for paints and varnishes, to permit easy brushing. Both varnishes and enamels contain large amounts of gums and resins, and diluents with good solvent action are therefore needed.

The rubber industry also uses naphthas as solvents. The leather industry uses them to degrease skins, the metal industry to degrease metals. Naphthas in insecticides and weedkillers dissolve the toxic agents and often contribute toxic properties of their own. Floor waxes, furniture waxes, shoe polishes, metal polishes, and dry cleaners' soaps are among the many other products in which naphthas are used.

[J. K. ROBERTS]

Naphthalene

A colorless crystalline aromatic hydrocarbon, $C_{10}H_8$, with the familiar odor of mothballs, melting point 80.1°, boiling point 218°C. It is almost insoluble in water but soluble in nearly all organic solvents. Structurally it is best represented as two benzenoid rings fused together (I*a*). In the naming

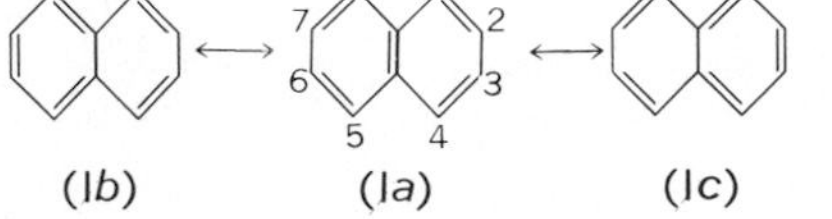

(I*b*) (I*a*) (I*c*)

of naphthalene derivatives, numbers are most frequently used (I*a*) especially for di- and polysubstituted compounds, but the use of α for position 1 and β for position 2 is still encountered.

Occurrence. One gal. of coal tar yields approximately one pound of naphthalene (approximately 120 g per liter). It is present also in certain fractions obtained by the reforming or steam cracking of certain petroleum distillates and is now recovered commercially on a large scale, particularly in the United States. Crude naphthalene may contain a small quantity of sulfur, which can be removed by distilling the crude hydrocarbon from sodium metal.

Reactions. Naphthalene is considered less aromatic than benzene since it is more easily reduced and oxidized, and it shows a greater tendency to react by addition. Substitution reactions occur much more rapidly with naphthalene than with benzene. With sodium in boiling absolute ethanol, it yields 1,4-dihydronaphthalene, while with sodium in boiling amyl alcohol, it affords 1,2,3,4-tetrahydronaphthalene (tetralin). Oxidation of naphthalene with chromic acid yields some 1,4-naphthoquinone (II), while vapor-phase air oxidation over a vanadium pentoxide catalyst yields phthalic anhydride (III), which is important in the manufacture of glyptal resins.

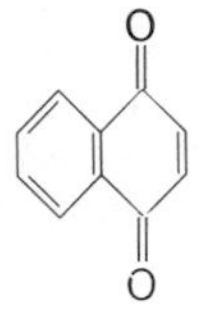

(II)

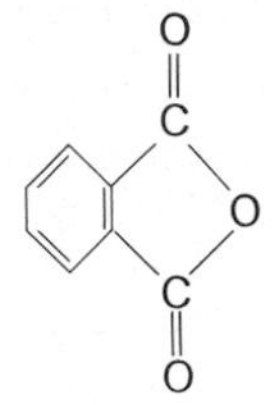

(III)

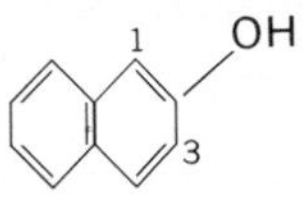

(IV)

The α positions of the naphthalene nucleus are more reactive than the corresponding β positions. Nitration yields 1-nitronaphthalene almost free of the 2-nitro isomer. Halogenation in the presence of a catalyst yields the 1-halonaphthalene. Without a catalyst, chlorination occurs by addition, yielding 1,2,3,4-tetrachloro-1,2,3,4-tetrahydronaphthalene. Sulfonation with concentrated sulfuric acid at low temperatures yields naphthalene-1-sulfonic acid, while at high temperatures naphthalene-2-sulfonic acid, $C_{10}H_7SO_3H$, is obtained. The Friedel-Crafts reaction occurs readily, usually affording mixtures of 1- and 2-substituted naphthalenes.

Bond structure. Naphthalene has been stated to be a resonance hybrid of 42 canonical forms, and of these, the three structures (I*a*), (I*b*), and (I*c*) are the major contributors to the actual structure. The Erlenmeyer formula (I*a*) is commonly used for naphthalene since it clearly indicates which of the peripheral bonds have the greatest double-bond character. As would be predicted from the Erlenmeyer formula, 2-naphthol (IV) is activated at position 1, but not at position 3.

Uses. Of the naphthalene consumed, the majority is converted to phthalic anhydride. Aside from a few percent used in the manufacture of mothballs, the remainder is converted to naphthalene compounds, which are used as dye intermediates, tanning agents, and surface-active agents. *See* AROMATIC HYDROCARBON; POLYNUCLEAR HYDROCARBON. [CHARLES K. BRADSHER]

Bibliography: K. Weissermel and H.-J. Arpe, *Industrial Organic Chemistry*, 1978.

Naphthol

One of the group of phenols in which a hydroxyl group is attached directly to a naphthalene-ring carbon atom. The two simple naphthols are well-known articles of commerce: 1-(or α-)naphthol (I) and 2-(or β-)naphthol (II). The structural formulas

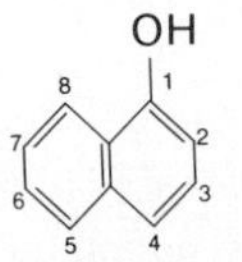

(I)

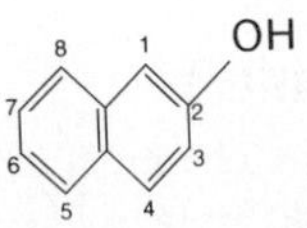

(II)

show also the numbers assigned to the carbon atoms which are part of the ring system. Both isomers are commercially produced from naphthalene by sulfonation and fusion of the resulting sulfonates with caustic soda.

Prior to about 1958 the major use of the naphthols was as intermediates for dyestuffs manufacture, and production of 2-naphthol was very much larger than that of 1-naphthol. Following the commercialization of the carbamate-based insecticides (for example, 1-naphthyl ester of *N*-methyl carbamic acid), this relationship was reversed.

[ROBERT I. STIRTON]

Naphthylamine

One of two organic chemical compounds, nearly insoluble in water, that are used to make various sulfonic acid derivatives to serve as coupling components for azo dye intermediates.

α-Naphthylamine (I; melting point 51°C) is obtained by the reduction of α-nitronaphthalene. β-Naphthylamine (II; m.p. 112°C) is manufactured by heating β-naphthol in an autoclave with a solution of ammonia and ammonium sulfite (Bucherer process). The naphthylamines resemble aniline in their properties.

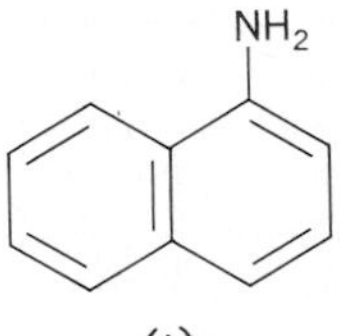

(I)

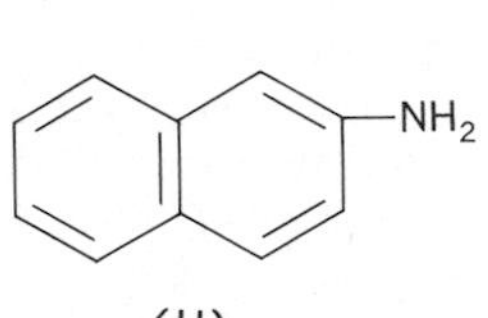

(II)

Heating the sulfate salt of α-naphthylamine yields 1-naphthylamine-4-sulfonic acid (naphthionic acid) in analogy to aniline sulfate. Bucherer's reaction with sodium hydrogen sulfite and sodium hydroxide leads to 1-naphthol-4-sulfonic acid (Neville-Winter acid). By sulfonation of naphthalene, followed by nitration and reduction of the corresponding nitronaphthalene-sulfonic acids, two other acids are obtained: 1-naphthylamine-6-sulfonic acid (Cleve's acid) and 1-naphthylamine-7-sulfonic acid. *See* SULFANILIC ACID.

From β-naphthylamine by various sulfonation and hydrolysis reactions, the following dye intermediates result: 2-naphthylamine-1-sulfonic acid (Tobin's acid), 2-naphthylamine-6,8-disulfonic acid, 2-naphthylamine-1,5,7-trisulfonic acid, 2-naphthylamine-5,7-disulfonic acid, 2-naphthylamine-8-hydroxy-6-sulfonic acid (γ-acid), and 2-naphthylamine-5-hydroxy-7-sulfonic acid (J-acid).

α-Naphthylamine is used to make an effective rat poison, 1-(1-naphthyl)-2-thiourea, sold as ANTU.

β-Naphthylamine is an extremely potent carcinogen. *See* AMINE; ANILINE; DIAZOTIZATION.

[LEALLYN B. CLAPP]

Bibliography: C. R. Noller, *Chemistry of Organic Compounds*, 3d ed., 1965; K. Venkataraman, *The Chemistry of Synthetic Dyes*, 3 vols., 1952–1970.

Neodymium

A metallic chemical element, Nd, atomic number 60, atomic weight 144.24. Neodymium belongs to the rare-earth group of elements. The naturally occurring element includes the isotopes Nd^{142} 27.11%, Nd^{143} 12.17%, Nd^{144} 23.85%, Nd^{145} 8.30%, Nd^{146} 17.22%, and Nd^{148} 5.73%. Nd^{144} is weakly

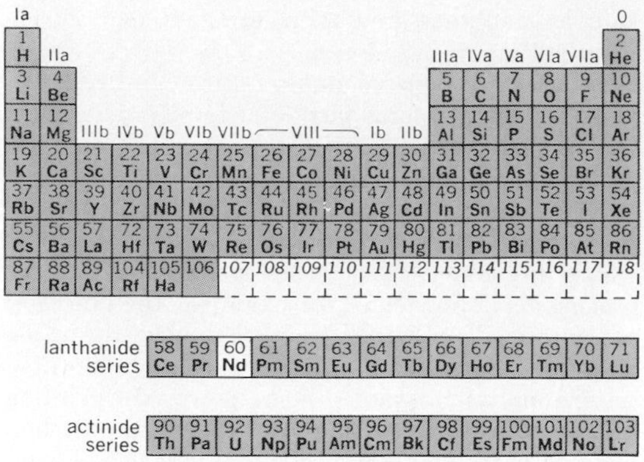

radioactive and emits an α-particle. It has a half-life of 5×10^{15} years. It was discovered by C. F. Auer von Welsbach in 1885, when he separated the so-called element didymium into two fractions, neodymium and praseodymium. The oxide, Nd_2O_3, is a light-blue powder. It dissolves in mineral acids to give reddish-violet solutions. For properties of the metal *see* RARE-EARTH ELEMENTS.

The metal slowly oxidizes in air at room temperatures and is slowly attacked by cold water. The salts have found application in the ceramic industry for coloring glass and for glazes. The glass made with didymium is particularly useful in goggles used by glass blowers, since it absorbs the intense yellow D line of sodium present in the flame. The element has found commercial application in the manufacture of lasers.

[FRANK H. SPEDDING]

Neon

A gaseous chemical element, Ne, with atomic number 10 and atomic weight 20.183. Neon is a member of the family of noble gases in the zero group of the periodic table. The only commercial source of neon is the Earth's atmosphere, although traces of neon are found in natural gas, minerals, and meteorites. *See* INERT GASES.

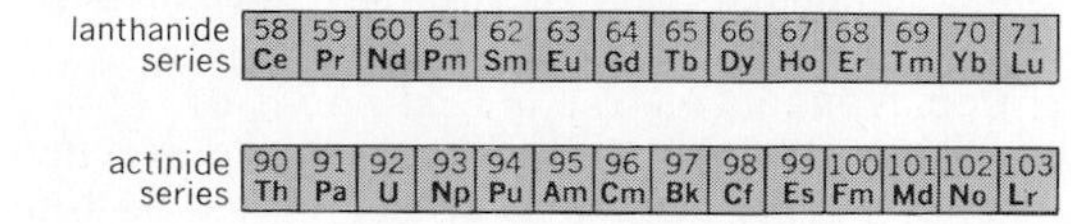

Uses. Considerable quantities of neon are used in high-energy physics research. Neon fills spark chambers used to detect the passage of nuclear particles. Their paths are displayed by a trail of sparks which results from ionization of the neon as the particles pass through the chamber.

Another type of particle detector is the liquid-hydrogen bubble chamber. Since the low density of liquid hydrogen presents a poor target for interaction with the nuclear particles being studied, liquid neon is added to raise the likelihood of such interaction while the liquid hydrogen continues to function effectively as a detector. In this way the bubble chamber can be utilized for a wider variety of experiments than is possible with liquid hydrogen alone.

Experiments have shown that neon is potentially useful in special breathing mixtures for deep-sea diving or space travel. Neon has beneficial properties similar to helium but does not distort voice communications. Neon also has lower thermal conductivity, which reduces the diver's heat loss to the surrounding water.

Liquid neon can be utilized as a refrigerant in the temperature range about 25–40 K. Specific types of infrared detectors and lasers require cooling in this temperature range. Neon is a particularly useful refrigerant because of its high latent heat of vaporization. For some applications where slightly lower temperatures or greater cooling efficiency is required, neon can be used as a solid or a slush.

When neon at low pressure is excited in an electric discharge, it emits a brilliant orange-red color; other inert gases and mercury vapor may be added. Neon is used in lightning arresters; it conducts virtually no current at voltages below its breakdown potential, but when lightning strikes, neon is ionized and allows the current to flow to the ground. Neon discharge tubes are used as overload protection for some electric motors to guard them against damage from surges in current.

Neon is also used in some kinds of electron tubes, in Geiger-Müller counters, in spark-plug test lamps, and in warning indicators on high-voltage electric lines. A very small wattage produces visible light in neon-filled glow lamps; such lamps are used as economical night and safety lights.

Occurrence. Neon constitutes 18.18 parts per million by volume of the Earth's atmosphere, and it is a mixture of three stable isotopes: 90.92% volume neon-20, 0.26% neon-21, and 8.82% neon-22. No naturally occurring radioactive isotopes of neon are known.

It is estimated that about $5 \times 10^{-7}\%$ by weight of the Earth is neon. Neon also occurs outside the Earth; the best estimate is that there are 8.6 times as many atoms of neon as of silicon in the visible universe, silicon being commonly used as a standard for comparison.

Discovery. Neon was discovered in England in 1898 by Sir William Ramsay and M. W. Travers, who found it in the most volatile portion of the mixture of inert elements left after oxygen and nitrogen had been chemically removed from air. The fact that a new element was present was ascertained by the discovery of new lines in the emission spectrum of the residual gas.

Radioactive isotopes. The following radioactive isotopes of neon are known: Ne^{18}, Ne^{19}, Ne^{23}, and Ne^{24}. None of these occurs in nature. They are produced in particle accelerators, such as the cyclotron, or by the neutron bombardment of the appropriate atomic species. All of them have short lifetimes. The isotope Ne^{24} has the longest half-life, 3.38 min.

Properties. Neon is colorless, odorless, and tasteless; it is a gas under ordinary conditions. Some of the other properties of neon are given in the table.

Physical properties of neon

Property	Value
Atomic number	10
Atomic weight (atmospheric neon only)	20.183
Melting point, °C	−248.6
Boiling point at 1 atm pressure, °C	−246.1
Gas density at 0°C and 1 atm pressure, g/liter	0.8999
Liquid density at its boiling point, g/ml	1.207
Solubility in water at 20°C, ml neon (STP)/1000 g water at 1 atm partial pressure neon	10.5

Neon does not form any chemical compounds in the ordinary sense of the word; there is only one atom in each molecule of gaseous neon.

One anomalous property of neon is the gas-to-liquid ratio. Most cryogenic liquefied gases produce between 500 and 800 volumes of gas at ambient temperature. Neon produces over 1400 volumes. This property makes liquid storage and transport particularly convenient and may lead to new uses, especially in space or other environments where compact storage is desirable.

Production. In the production of neon from air, the air is first liquefied. A small amount of it, containing hydrogen, helium, and neon, together with a little nitrogen, remains uncondensed. The nitrogen is removed by low-temperature adsorption. The hydrogen is burned to water, and the residual gas is dried. Helium and neon are then separated by selective adsorption on activated carbon at carefully regulated temperatures and pressures, or by partial condensation of neon at liquid-hydrogen temperature.

Analytical methods. The principal modern methods of detecting and quantitatively determining the neon content in gases are mass spectrometry and gas chromatography.

[ARTHUR W. FRANCIS]

Bibliography: B. L. Smith and J. P. Webb, *The Inert Gases: Model Systems for Science*, 1971.

Neptunium

A chemical element, symbol Np, atomic number 93. Neptunium is a member of the actinide or 5f series of elements. It was synthesized as the first transuranium element by E. M. McMillan and P. Abelson in 1940 by bombardment of uranium with neutrons to produce neptunium-239, according to reaction (1). The lighter isotope ^{237}Np, a long-lived alpha emitter with half-life 2.14×10^6 years, is par-

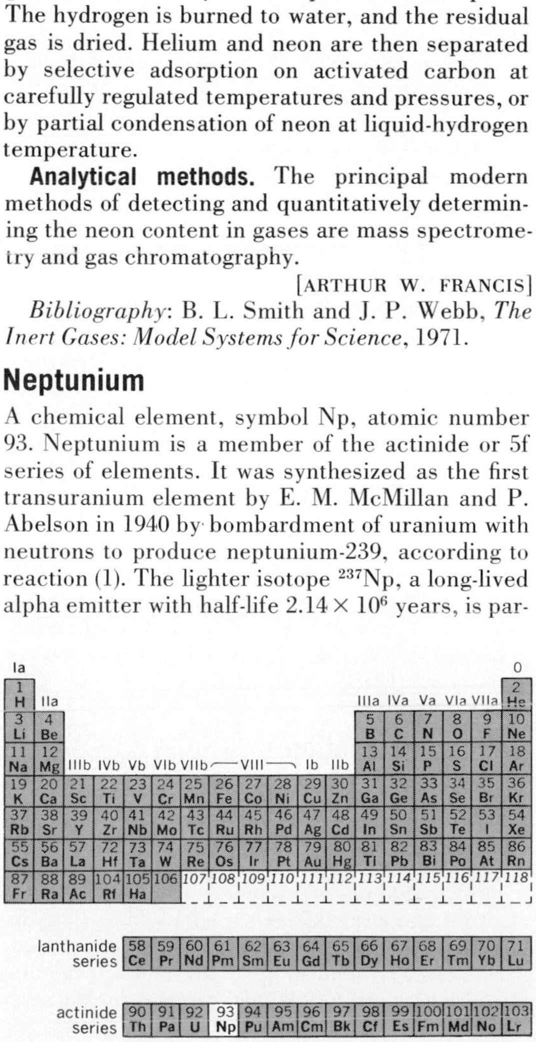

ticularly important chemically. It is also formed from uranium, as in reaction (2).

$$^{238}U(n,\gamma)^{239}U \xrightarrow[23\,min]{\beta^-} {}^{239}Np \xrightarrow[2.3\,d]{\beta^-} \quad (1)$$

$$^{238}U(n,2n)^{237}U \xrightarrow[6.75\,d]{\beta} {}^{237}Np \xrightarrow[2.14\times10^6\,yr]{\alpha} \quad (2)$$

The chemistry of its compounds and oxidation states was first elucidated by microchemical methods. These methods, in conjunction with the x-ray diffraction studies on solid neptunium compounds, provided the means of identification of most of the early neptunium compounds.

The chemistry of neptunium may be said to be intermediate between that of uranium and plutonium. However, in contrast to the behavior of the latter two, the NpO_2^+ ion is of great importance in the aqueous chemistry of neptunium. This is the first stable monocation in the actinides. The electronegative anions O^{2-} and F^- generally form compounds of higher oxidation number with neptunium than do the more polarizable and oxidizable anions Cl^-, Br^-, S^{2-}, and I^-. In aqueous solution, the positive ions of the different oxidation states are stabilized by solution and complex formation.

Soviet chemists extended to alkaline solutions the neptunium chemistry previously largely known in only acid solution. In a major discovery, they found that the hepta-positive state, Np^{7+}, is formed and stabilized by oxygen coordination in media of high OH^- concentrations. Neptunium in the hepta-positive state is a powerful oxidant; oxygen coordination is clearly required since, with pure fluorine, no oxidation beyond NpF_6 is observed. There is a possibility that $NpOF_5$ could be prepared, similar to $OsOF_5$.

Volatile NpF_6 is similar to UF_6, and neptunium forms binary fluorides and chlorides which resemble those of uranium. In all oxidation states, Np ions are just slightly smaller than the corresponding uranium ions, and their compounds are usually isostructural.

Neptunium metal is ductile, low-melting (637°C), and in its alpha form is of high density, 20.45 g/cm³. The metal is known in three modifications similar to uranium. Neptunium metal is reactive and forms many binary compounds, for example, with hydrogen, carbon, nitrogen, phosphorus, oxygen, sulfur, and the halogens. *See* ACTINIDE ELEMENTS; NUCLEAR CHEMISTRY; PLUTONIUM; RADIOACTIVITY; TRANSURANIUM ELEMENTS.

[ROBERT A. PENNEMAN]

Bibliography: K. W. Bagnall, *The Actinide Elements*, 1972; C. Keller, *Chemistry of the Transuranium Elements*, Verlag Chemie GmbH, Weinheim, 1971; W. W. Schulz, *The Chemistry of Americium*, ERDA Critical Review Series, Rep. no. TID-26971, 1976.

Neutron diffraction

The phenomenon associated with the interference processes which occur when neutrons are scattered by the atoms within solids, liquids, and gases. The use of neutron diffraction as an experimental technique is relatively new compared to electron and x-ray diffraction, since successful application requires high thermal-neutron fluxes, which can be obtained only from nuclear reactors. (A thermal neutron is defined as a neutron possessing a kinetic energy of about 0.025 eV.) These diffraction investigations are possible because thermal neutrons have energies with equivalent wavelengths near 1 A and are therefore ideally suited for interatomic interference studies.

The scattering of low-energy neutrons is generally considered a tool for the study of solid-state phenomena, but many significant investigations have also been performed to obtain information necessary for the understanding of nuclear processes. Diffraction techniques have been employed to measure numerous coherent neutron-scattering amplitudes under special conditions, and these determinations have provided details on the interaction between nuclear forces, potential scattering, and resonance effects. Experiments have also helped to establish upper limits for values of a possible small neutron electric charge and a possible small neutron electric dipole moment. The most numerous and important investigations by neutron diffraction have been concentrated on solid-state problems, because these experiments offer unique methods to obtain information on crystallographic properties, magnetic phenomena, and the dynamics of crystal lattices. In its applications to solid-state problems, neutron diffraction is very similar in both theory and experiment to x-ray diffraction, but its importance arises from the significant differences in the scattering of the two types of radiation.

The scattering of x-rays by atoms results from a scattering interaction with the atomic electrons, and the scattering amplitudes are approximately proportional to the atomic number of the scatterer. Since the electrons are distributed within the atom at distances comparable to the x-ray wavelength, interference effects occur which produce an angular distribution of the scattering, usually referred to as a form factor, that is descriptive of the spatial distribution of the electrons. In the scattering of neutrons by atoms, there are two important interactions. One is the short-range, nuclear interaction of the neutron with the atomic nucleus. This interaction produces isotropic scattering because the atomic nucleus is essentially a point scatterer relative to the wavelengths of thermal neutrons. Strong resonances associated with the scattering process prevent any regular variation of the nuclear scattering amplitudes with atomic number. These resonances can cause the scattering amplitudes to have imaginary components of appreciable size, and they can affect the phase changes between the incident and scattered neutron waves so that the scattering amplitudes can be either positive or negative. The other important process for the scattering of neutrons by atoms is the interaction of the magnetic moment of the neutron with the spin and orbital magnetic moments of the atom. The amplitude of the interaction varies with the size and orientation of the atomic magnetic moment, and the intensity of scattering has a form-factor angular dependence that is representative of the magnetic electrons within the atom.

Techniques. Although thermal neutron beams from nuclear reactors have intensities that are lower than those obtained from efficient x-ray tubes, most of the methods developed for x-ray diffraction can be used with neutrons. Furthermore, since the neutron absorption cross section

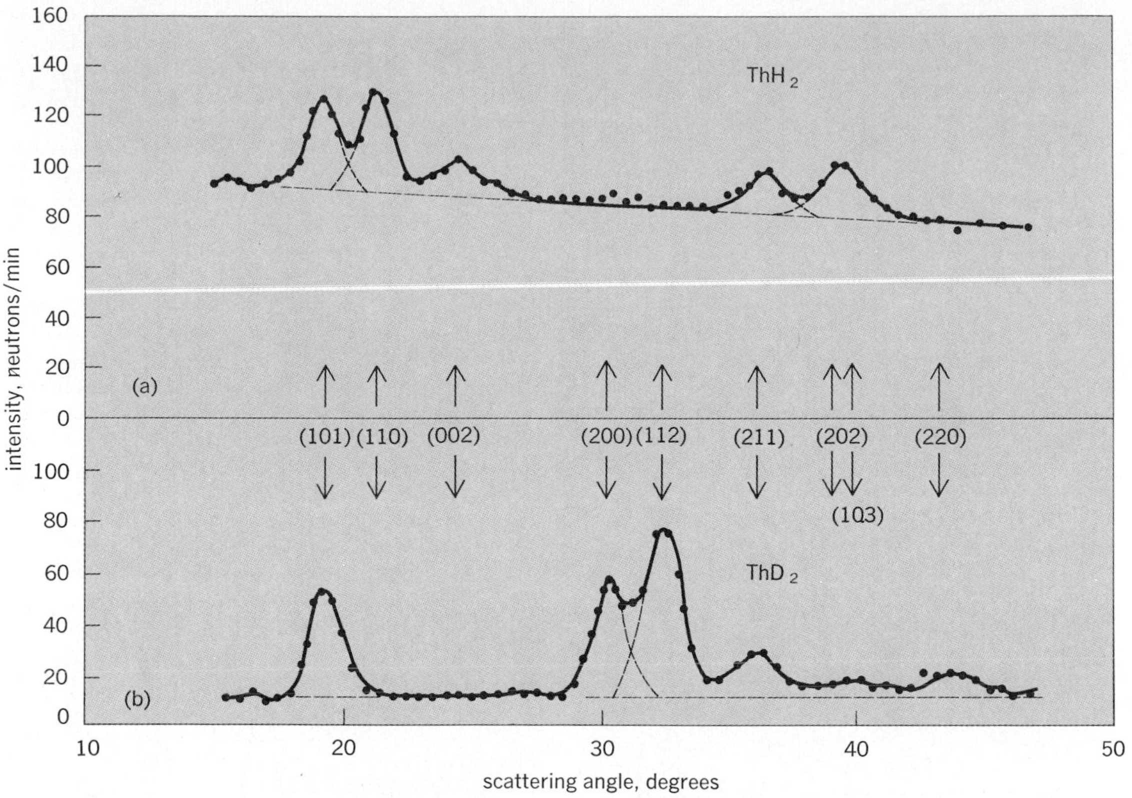

Fig. 1. Neutron diffraction patterns from (*a*) polycrystalline thorium hydride, ThH_2, and (*b*) polycrystalline thorium deuteride, ThD_2. Differences in the patterns are caused primarily by differences in the nuclear scattering from hydrogen and deuterium atoms.

for many materials is very small, diffraction effects can also be investigated by observations of the neutrons transmitted through a sample. In most experiments the sample is irradiated with monochromatic neutrons, and the scattered radiation is measured with a neutron detector, such as a proportional counter filled with boron trifluoride gas, BF_3. In structure determinations and other experiments where there is no energy change in the scattering process, only the angular distribution of the scattered neutrons is required. In inelastic scattering experiments, that is, experiments involving an increase or decrease in neutron energy, the energy distribution of the scattered neutrons must also be measured. The neutron energies can be determined with a crystal spectrometer or by analyzing the time of flight of the scattered neutrons. Both polycrystalline and single-crystal specimens can be examined, and the type of specimen is usually determined by the conditions of the experiment. However, since single-crystal techniques provide much better resolution of the diffraction peaks, this method is required for the study of complicated structures.

For those experiments requiring monochromatic neutron beams, the neutrons must be obtained by isolating a narrow slice of the neutron spectrum from the reactor, because these neutrons have a continuous energy distribution with no pronounced peaks. Monochromatization is usually accomplished by diffraction of the reactor neutrons from large single crystals, but filters and mechanical neutron velocity selectors also can be used. In investigations of certain magnetic properties, it is frequently necessary to use neutron beams that are both monochromatic and polarized. Such beams can be obtained in the monochromating process by using single crystals, because the neutrons scattered under specific conditions from particular ferromagnetic crystals are almost completely polarized. Diffraction experiments with polarized neutrons have been particularly important for precise determinations of magnetic form factors, and techniques using polarization analysis provide a unique method for separating neutrons scattered by magnetic and nuclear interactions.

Another technique for neutron diffraction utilizes pulses containing the entire spectrum of reactor neutrons and employs time-of-flight energy analysis of the neutrons scattered at a fixed angle. This technique is particularly useful with pulsed neutron sources, and it offers definite advantages in certain types of experiments where the range of scattering angles is limited. However, in most investigations with continuous neutron sources, it is not competitive with the conventional diffraction methods. Photographic techniques can also be used in neutron experiments, but they have been restricted almost completely to quick qualitative examinations.

Auxiliary apparatus for controlling the conditions of the samples can be constructed easily because of the relatively low neutron cross sections of most materials. Consequently, many diffractometers and spectrometers have low-temperature cryostats, furnaces, and electromagnets as integral parts of the instruments. Furthermore, the methods of investigation are readily adaptable

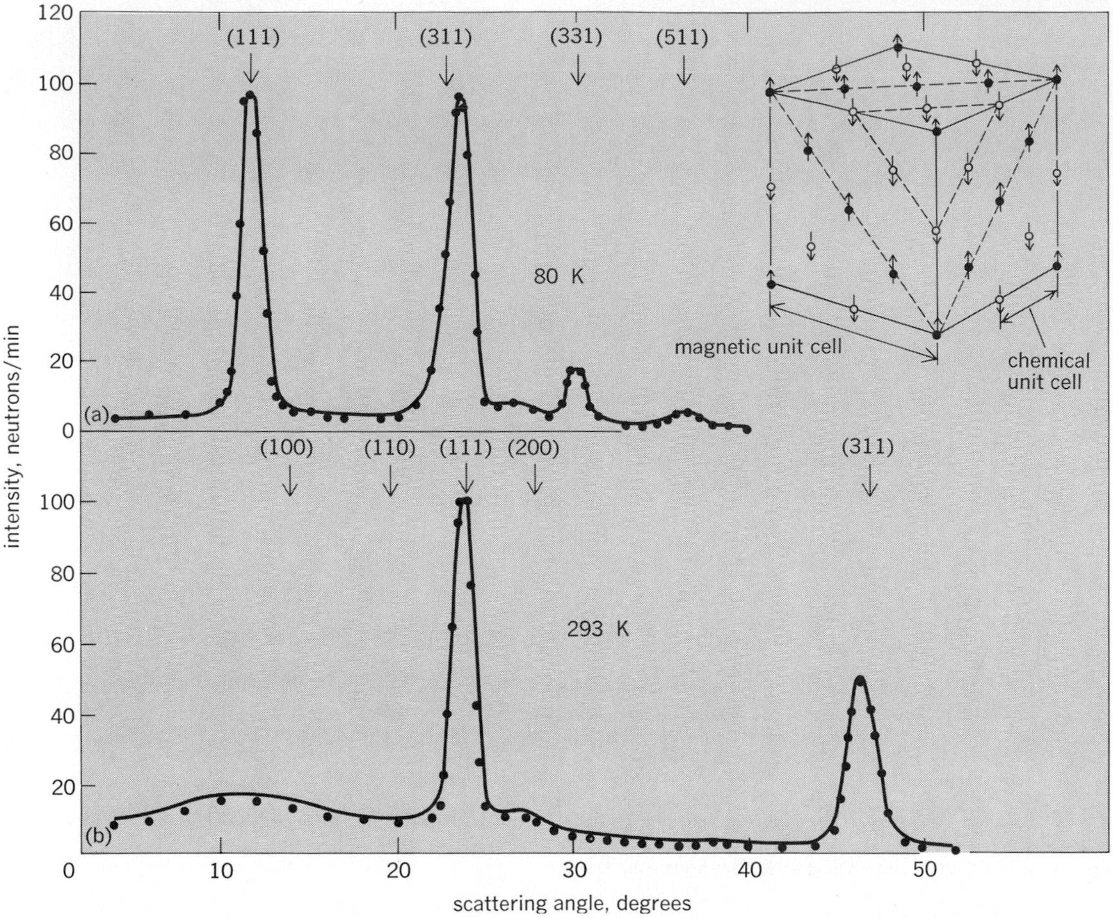

Fig. 2. Neutron diffraction patterns from polycrystalline manganese oxide, MnO, at temperatures (*a*) below and (*b*) above the antiferromagnetic transition at 122 K. At 293 K, only nuclear reflections are observed, while at 80 K, additional reflections are obtained from the indicated antiferromagnetic structure. The atomic magnetic moments in this structure are directed along a magnetic axis within the (111) planes.

to automation, and some of the newer instruments at high-flux research reactors are controlled directly by on-line computers.

Chemical crystallography. Since the nuclear scattering amplitudes for neutrons do not vary uniformly with atomic number, there are certain types of chemical structures which can be investigated more readily by neutron diffraction than by x-ray diffraction. Moreover, since neutron scattering is a nuclear process, when the scattering amplitude of an element is not favorable for a particular investigation, it is frequently possible to substitute an enriched isotope which has scattering characteristics that are markedly different.

The most significant application of neutron diffraction in chemical crystallography is the structure determination of composite crystals which contain both heavy and light atoms, and the most important compounds in this general classification are the hydrogen-containing substances. Since hydrogen and deuterium have neutron scattering amplitudes that are comparable to those of other atoms, their positions in crystals can be determined by this technique, whereas x-ray diffraction usually gives little information about them (Fig. 1). Most of the early investigations concerned relatively simple inorganic compounds, but with the construction of higher-flux reactors and the use of computers in data collection and processing, more complex crystal structures have been examined. The crystal structure of sucrose, $C_{12}H_{22}O_{11}$, which required the measurement and analysis of 2800 independent Bragg reflections, was determined by neutron diffraction, and studies have also been made on some of the simplest biological molecules. In addition to the inherent interest in the materials, all of these investigations have helped to provide a better understanding of hydrogen bonds in crystals. For a discussion of Bragg reflections *see* X-RAY DIFFRACTION.

Neutron diffraction techniques have also been applied to many other compounds with special chemical or physical properties and with unfavorable x-ray scattering amplitudes. These investigations include the ionic displacements associated with ferroelectric transitions, the rotation of molecular groups in compounds, and order-disorder phenomena in alloy systems composed of atoms with almost the same atomic number. Furthermore, since the scattering of neutrons by the nucleus is isotropic, the neutron technique is advantageous in investigations of liquids, gases, amorphous materials, and other structures where the features of the diffraction pattern at large scattering angles are significant.

Magnetic scattering. The interaction of the magnetic moment of the neutron with the orbital and spin moments in magnetic atoms makes neutron scattering a unique tool for the study of a wide variety of magnetic phenomena, because information is obtained on the magnetic properties of the individual atoms in a material. This interaction depends on the size of the atomic magnetic moment and also on the relative orientation of the neutron spin and of the atomic magnetic moment with respect to the scattering vector and with respect to each other. Consequently, detailed information can be obtained on both the magnitude and orientation of magnetic moments in any substance which displays magnetic properties.

Each type of magnetic lattice has a characteristic neutron diffraction pattern. For paramagnetic materials, where the atomic moments are uncoupled and randomly oriented in direction, the magnetic scattering is diffuse. For ordered magnetic lattices the magnetic scattering is found in Bragg reflections. Magnetic reflections from ferromagnetic materials occur superimposed on the nuclear reflections, but for antiferromagnetic materials, in which the atomic moments are oriented with no net magnetization per unit volume, superlattice reflections are observed at other scattering angles, as shown in Fig. 2. Since ferrimagnetic materials have atomic moments with antiparallel components but also possess a net ferromagnetic moment, magnetic reflections are observed at both nuclear and other positions. Thus neutron diffraction experiments can determine the magnetic transition temperature, type of magnetic order, temperature variation of the magnetic order, and detailed magnetic configuration in the ordered lattice. This information is basic to understanding the magnetic exchange interactions that are responsible for producing an ordered magnetic structure.

The investigation of antiferromagnetic and ferrimagnetic substances is one of the most important applications of the neutron diffraction technique, because detailed information on the magnetic configuration in these systems cannot be obtained by other methods. Several hundred antiferromagnetic structures have been investigated, and various types of systems that have been determined are shown schematically in Fig. 3. In most antiferromagnetic substances the magnetic moments are found in truly antiparallel arrays, but more complicated systems have been encountered. Structures have been determined in which the magnetic moments are canted with respect to each other, and a number of systems have been found to have a long-range modulation of the moment distribution. The latter configurations require long-range magnetic interactions for stability, and in the heavy rare-earth metals and alloys which have such configurations, a long-range interaction through the conduction electrons can explain many of their unusual magnetic properties. Similar complex structures have also been observed in certain types of ferrimagnetic materials.

The magnetic moments of a simple ferromagnet can usually be obtained from saturation magnetization experiments, but in substances such as ferromagnetic alloys, although the moments are arranged in parallel alignment, different types of atoms have different moment values. Since magnetic measurements can give only the average moment of the alloy, the determination of the individual magnetic moments of the constituent atoms has been another important aspect of neutron diffraction. Alloys with both ordered and disordered arrangements of the atoms can be studied, and experiments on very dilute concentrations provide information on the effect of magnetic and nonmagnetic impurities.

The form factor for the magnetic scattering of neutrons can be interpreted in terms of the spatial distribution and angular momentum characteristics of the magnetic electrons within the atoms. Determinations of these form factors can be made from either measurements of magnetic intensities in coherent reflections or from measurements of paramagnetic scattering. However, the most precise measurements of this type have been made on reflections from ferromagnetic and ferrimagnetic materials, utilizing polarized neutrons. This technique takes advantage of cross terms in the combined nuclear and magnetic scattering to provide an accuracy not readily obtainable with an unpolarized beam. Measurements on the ferromagnetic iron-group metals have provided detailed maps showing the distribution of magnetic electrons throughout the unit cells.

A variety of changes can be produced in neutron diffraction patterns by application of magnetic fields sufficiently strong to change the orientation of atomic magnetic moments within the sample. The use of magnetic fields in these experiments can therefore provide information on ferromagnetic and antiferromagnetic domains, on the magnetic anisotropy within magnetic structures, and on the

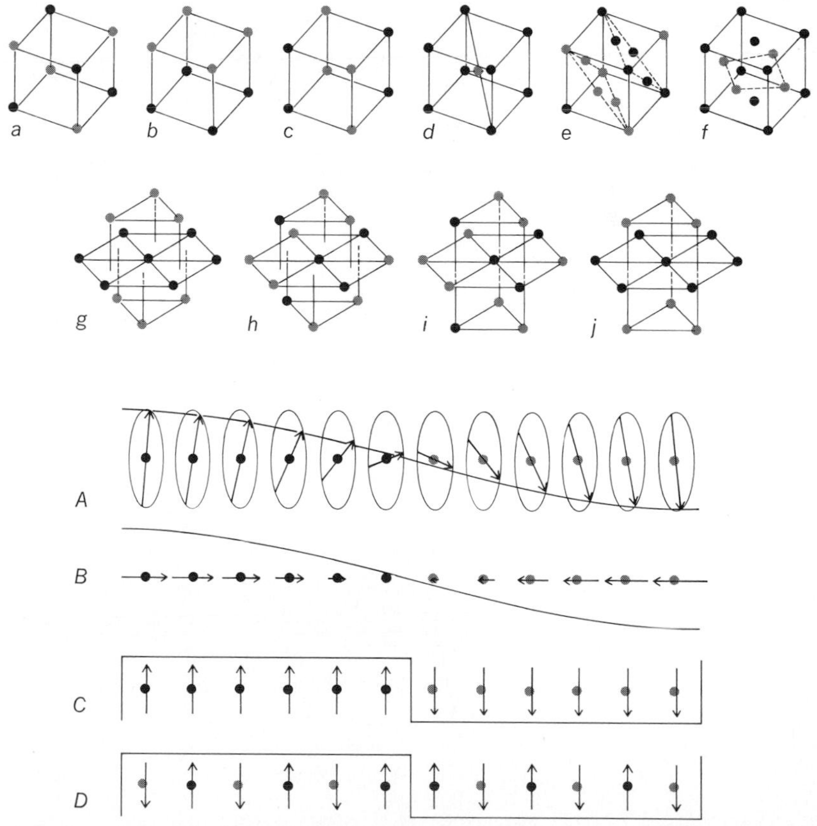

Fig. 3. Schematic representation of various antiferromagnetic systems studied by neutron diffraction. In structures *a* through *j* there is a single magnetic axis and the atomic moments at the darker circles are antiparallel to those at the lighter circles. Figures *A* through *D* indicate types of antiferromagnetism with a long-range modulation of the moment distribution.

nature and strength of magnetic exchange interactions.

Inelastic scattering. Scattering investigations of thermal neutrons, in which the neutrons undergo an energy change, fall into the broad scope of neutron diffraction. However, since both the angular distribution and the energy distribution of the scattered neutrons must be determined, these investigations are different from those usually associated with diffraction experiments. Because of the favorable values of energies and wavelengths associated with thermal neutrons, these measurements provide a method for studying many physical properties of solids and liquids that cannot be studied by any other method. The wavelengths are comparable with atomic separations, and the energies are of the order of the characteristic energies of solids and liquids, so that the energy and momentum changes resulting from many interactions can be measured easily by diffraction techniques. Furthermore, analogous to the case for elastic scattering, these inelastic scattering processes can result from a nuclear interaction or from a magnetic interaction, and the dynamical properties of both atomic systems and magnetic systems can be investigated.

One of the most important uses of inelastic neutron scattering is the study of thermal vibrations of atoms about their equilibrium positions, because lattice vibration quanta, or phonons, can be created or annihilated in their interactions with low-energy neutrons. The measurements provide a direct determination of the dispersion relations for the normal vibrational modes of the crystal and do not require the large corrections necessary in similar x-ray investigations. These measured dispersion relations furnish the best experimental information available on interatomic forces that exist in crystals (Fig. 4). Similar measurements can be made on the quantized motion of magnetic moments about the equilibrium direction in an ordered magnetic lattice, and these magnon dispersion curves can be interpreted in terms of the magnetic forces between atoms. Furthermore, neutron scattering techniques are not restricted to solids but can be used to investigate details of atomic motion in liquids.

With the availability of higher neutron fluxes and more sophisticated techniques, it has been possible to extend these investigations to more difficult problems, such as the effect of impurities on interatomic forces. Localized vibrational modes associated with impurities can be observed, and in certain types of experiments information can be obtained on the degee of spatial localization of the modes in the crystal. Dispersion curves have been measured with sufficient precision to permit observation of additional effects, including the interaction between phonons and magnons and the interaction between phonons and conduction electrons.

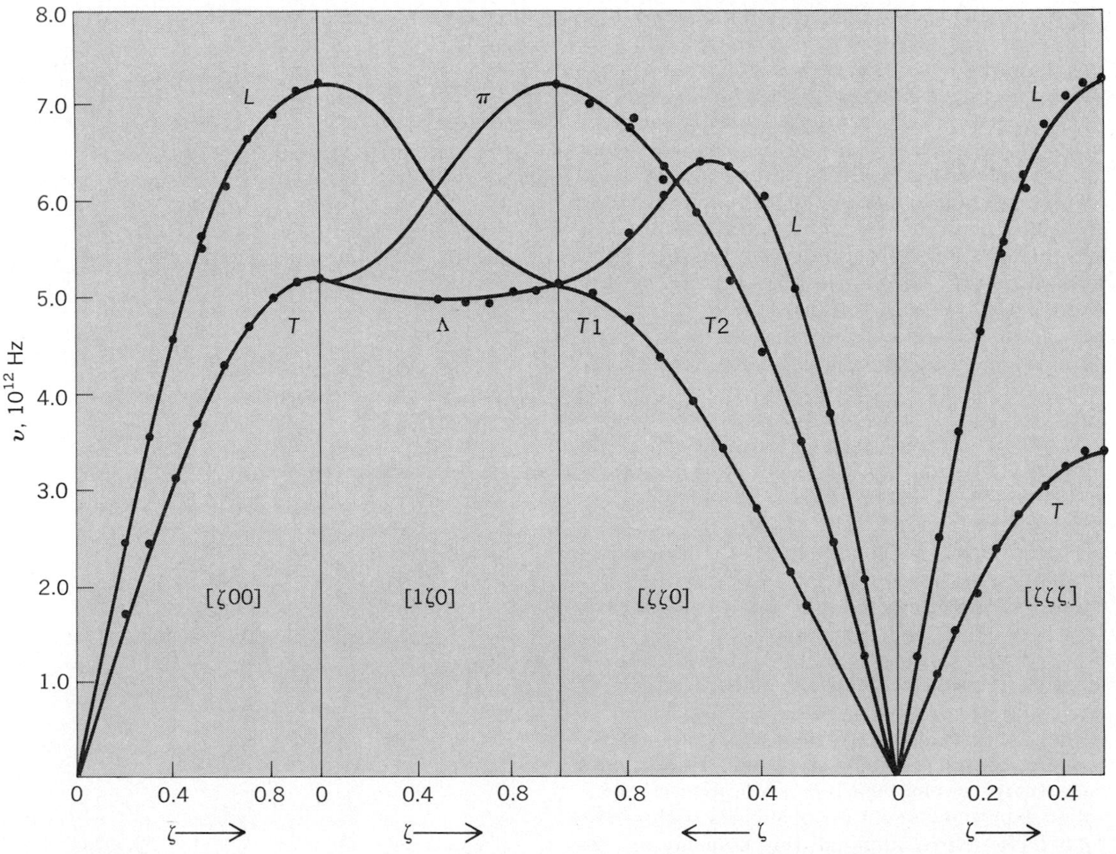

Fig. 4. Phonon dispersion curves for copper at 49°K, which relate phonon frequency ν to phonon wave vector ζ along major symmetry directions indicated in brackets. Solid circles are results from inelastic neutron scattering experiments and smooth curves are calculations based on an axially symmetric interatomic force model extended to six nearest neighbors. (*L* and *T* correspond to longitudinal and transverse modes of vibration, respectively, while π and Λ represent modes of vibration with both longitudinal and transverse components.)

The latter observations may prove particularly useful for determining the Fermi surface in metals as a function of crystallographic direction.

[MICHAEL K. WILKINSON]

Bibliography: G. E. Bacon, *Neutron Diffraction*, 3d ed., 1975; H. Dachs (ed.), *Neutron Diffraction*, 1978; P. A. Egelstaff (ed.), *Thermal Neutron Scattering*, 1965; Y. A. Izyumov and R. P. Ozerov, *Magnetic Neutron Diffraction*, 1970.

Nickel

A chemical element, Ni, atomic number 28, a silver-white, ductile, malleable, tough metal. The atomic mass of naturally occurring nickel is 58.71. Iron, cobalt, and nickel are all members of group VIIIb of the periodic table of elements, and these

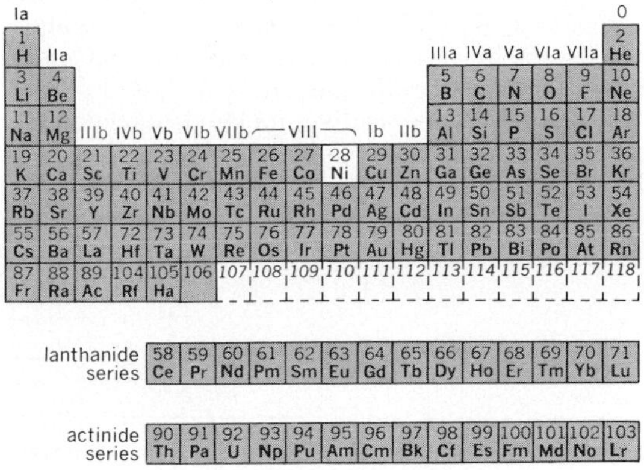
Ia
0
1 H
2 He
IIa
IIIa IVa Va VIa VIIa
3 Li 4 Be
5 B 6 C 7 N 8 O 9 F 10 Ne
11 Na 12 Mg
IIIb IVb Vb VIb VIIb VIII Ib IIb
13 Al 14 Si 15 P 16 S 17 Cl 18 Ar
19 K 20 Ca 21 Sc 22 Ti 23 V 24 Cr 25 Mn 26 Fe 27 Co 28 Ni 29 Cu 30 Zn 31 Ga 32 Ge 33 As 34 Se 35 Br 36 Kr
37 Rb 38 Sr 39 Y 40 Zr 41 Nb 42 Mo 43 Tc 44 Ru 45 Rh 46 Pd 47 Ag 48 Cd 49 In 50 Sn 51 Sb 52 Te 53 I 54 Xe
55 Cs 56 Ba 57 La 72 Hf 73 Ta 74 W 75 Re 76 Os 77 Ir 78 Pt 79 Au 80 Hg 81 Tl 82 Pb 83 Bi 84 Po 85 At 86 Rn
87 Fr 88 Ra 89 Ac 104 Rf 105 Ha 106 107 108 109 110 111 112 113 114 115 116 117 118
lanthanide series
58 Ce 59 Pr 60 Nd 61 Pm 62 Sm 63 Eu 64 Gd 65 Tb 66 Dy 67 Ho 68 Er 69 Tm 70 Yb 71 Lu
actinide series
90 Th 91 Pa 92 U 93 Np 94 Pu 95 Am 96 Cm 97 Bk 98 Cf 99 Es 100 Fm 101 Md 102 No 103 Lr

three metals have many chemical similarities. The six metals of the platinum group (ruthenium, rhodium, palladium, osmium, iridium, and platinum) also belong to group VIIIb, but their physical and chemical properties are mostly distinct from those of iron, cobalt, and nickel.

Some ores and alloys of nickel have been known since antiquity, but the metal was first isolated by A. F. Cronstedt in 1751.

Nickel consists of five natural isotopes having atomic masses of 58 (68% of natural nickel), 60 (26%), 61 (1%), 62 (4%), and 64 (1%). Seven radioactive isotopes have also been identified, having mass numbers of 56, 57, 59, 63, 65, 66, and 67. The half-lives of the radioactive isotopes range from 50 sec for Ni^{67} to 80,000 years for Ni^{59}.

Uses. Most commercial nickel goes into stainless steel and other corrosion-resistant alloys. About 16% of the nickel produced in the United States is used in plating to give hard, tarnish-resistant, polishable surfaces. Conventional plating is still much in favor, but newer techniques such as electroless coating or sintered-slurry coating are used for such applications as turbine blades, helicopter rotors, rolled steel strip, and extrusion dies. Nickel is becoming increasingly important in coins, especially to replace silver. Finely divided nickel is used as a hydrogenation catalyst. Other commercial uses are in ceramics, special chemical vessels, rechargeable nickel-cadmium storage batteries, electronic circuits, a dyeing process for polypropylene, coloring glass green, and the preparation of nickel compounds.

Occurrence. Nickel is a fairly plentiful element, making up about 0.008% of the Earth's crust and 0.01% of the igneous rocks. Appreciable quantities of nickel are present in some kinds of meteorite, and large quantities are thought to exist in the Earth's core. The Sudbury region of Ontario produces about 70% of the free world's nickel. The two most important ores in Ontario are the iron-nickel sulfides, pentlandite and pyrrhotite $(Ni,Fe)_xS_y$. The ore garnierite, $(Ni,Mg)SiO_3 \cdot nH_2O$, is found in New Caledonia and is also commercially important. Other nickel minerals of less importance are niccolite or red nickel ore (NiAs), millerite or yellow nickel ore (NiS), breithauptite (NiSb), chloanthite or white nickel ore ($NiAs_2$), gersdorffite (NiAsS), and ullmannite (NiSbS). In addition to the deposits in Ontario, sulfide ores are fairly widely distributed in Europe and Asia, and some are present in the Northwest Territories of Canada, South Africa, the Soviet Union, and the United States. Some useful nickel ore is found in Cuba as limonite, $(Ni,Fe)O(OH) \cdot nH_2O$. For the most part, the arsenide and antimonide minerals are not important enough to be classed as commercial ores. As with many other metals, the true chemical composition of some of these nickel minerals is not definitely known. Nickel occurs in small quantities (0.1–6 ppm fresh weight) in plants and animals. It is present in trace amounts in sea water, petroleum, and most coal.

Nickel metal. Nickel is of moderate strength and hardness (3.8 on the Mohs scale). When viewed as very small particles, nickel appears black, as do metals in general, but this finely divided state is particularly significant in the catalytic use of nickel. The density of nickel is 8.90 times that of water at 20°C: 1 ft^3 of nickel weighs about 550 lb. Nickel melts at 1455°C and boils at 2840°C. Its electrical conductivity is 24% that of silver; its heat conductivity is 15% that of silver. Nickel has a high enough value of magnetic susceptibility to be classed as ferromagnetic, but it does not equal iron in this respect. Commercial wrought nickel is 99.4% pure. Nickel metal has two crystalline forms: α-nickel crystallizes in the close-packed hexagonal system, and β-nickel is face-centered cubic. Nickel is available commercially as electrolytic nickel, and as ingots, shot, sponge, pellets, powder, or wrought nickel.

Chemical properties. Nickel is only moderately reactive. It resists alkaline corrosion and does not burn in the massive state, although fine nickel wires can be ignited. Specially prepared nickel, consisting of very small, porous particles (pyrophoric nickel), burns spontaneously when exposed to the air. Nickel is above hydrogen in the electrochemical series, and it dissolves slowly in dilute acids, releasing hydrogen and forming the green dipositive nickel ion Ni^{++}. With solutions of oxidizing agents, including strong nitric acid, nickel becomes passive and resists attack. In metallic form nickel is a moderately strong reducing agent. Through physicochemical processes, nickel can take up considerable amounts of hydrogen, as can palladium and platinum, especially when the metals are heated and in the finely divided state. Release of the hydrogen to other substances for chemical reaction is one reason for the catalytic action of these metals. *See* HYDROGENATION.

Some of the atomic and ionic properties of nickel are given in Table 1.

Table 1. Some atomic and ionic properties of nickel

Property	Value
Electronic configuration	$1s^2,2s^2,2p^6,3s^2,3p^6,3d^8,4s^2$
Ionization potential	
First electron loss	7.635 eV
Second electron loss	18.168 eV
Ionic radius	0.072 nm for Ni^{2+}; 0.062 nm for Ni^{3+}
Covalent radius	0.115 nm
Oxidation potentials	$Ni \rightleftarrows Ni^{++} + 2e^-$, $E° = 0.25$ volt $Ni + 2OH^- \rightleftarrows Ni(OH)_2 + 2e^-$, $E° = 0.66$ volt $Ni(OH)_2 + 2OH^- \rightleftarrows NiO_2 + 2H_2O + 2e^-$, $E° = -0.49$ volt
Electronegativity	1.8

Nickel compounds. Nickel is usually dipositive in its compounds, but it can also exist in the oxidation states 0, 1+, 3+, and 4+. Besides the simple nickel compounds, or salts, nickel forms a variety of coordination compounds or complexes. Most compounds of nickel are green or blue because of hydration or other ligand bonding to the metal. The nickel ion present in water solutions of simple nickel compounds is itself a complex, $[Ni(H_2O)_6]^{++}$. The compounds of dipositive nickel closely resemble those of dipositive cobalt, so that chemical separation of the two metals is often difficult.

The complexes of nickel(II) are structurally interesting because their common octahedral arrangement is in some instances in equilibrium with tetrahedral or square-planar structures. *See* COORDINATION CHEMISTRY.

Next to the 2+ oxidation state, the 4+ state is most common for nickel. Nickel(IV) oxide, NiO_2, is probably a definite compound, though it has never been prepared stoichiometrically pure. Complex molybdates containing nickel(IV) are also known, as are barium nickelate(IV), $BaNiO_3$, and the fluoro complex K_2NiF_6. Even though the diarsine, o-$C_6H_4[As(CH_3)_2]_2$, is much less electronegative than fluoride or oxygen, it stabilizes nickel(IV) in the complex $(NiCl_2\{o\text{-}C_6H_4[As(CH_3)_2]_2\}_2)(ClO_4)_2$. Tripositive nickel is not stable as a simple or hydrated ion or in simple salts. Only complexes and possibly the oxide NiO(OH) are known for the 3+ state of nickel. The zerovalent state is exemplified by nickel tetracarbonyl, $Ni(CO)_4$, and other four-coordinate complexes such as $Ni(PCl_3)_4$ and $K_4[Ni(CN)_4]$. Monopositive nickel is uncommon, but it is known for the compound $K_4Ni_2(CN)_6$ and for some phosphine complexes. Some important simple and complex compounds of nickel are listed in Table 2.

Table 2. Some important nickel compounds and their uses

Name	Formula	Uses	Remarks
Nickel(II) sulfate 6-hydrate or 7-hydrate	$NiSO_4 \cdot 6H_2O$ $NiSO_4 \cdot 7H_2O$	In nickel plating; in dip baths for enameling; preparation of nickel compounds and catalytic nickel; a mordant in dyeing; in paints, varnishes, and ceramics	Green or blue soluble compound; commercially most important of nickel compounds
Nickel(II) chloride 6-hydrate	$NiCl_2 \cdot 6H_2O$	Reagent; in electrorefining of catalytic nickel; to absorb ammonia in gas masks	Bright-green soluble compound
Nickel(II) nitrate 6-hydrate	$Ni(NO_3)_2 \cdot 6H_2O$	Reagent; preparation of nickel compounds and catalytic nickel	Emerald green; very soluble; also exists as other hydrates
Nickel(II) oxide	NiO	In production of alloys; in enamel frits and ceramic glazes; in glass manufacture	Green or black; green form insoluble in water, soluble in acids; black form insoluble in water and acids
Nickel(IV) oxide	NiO_2	Oxidizing agent; in Edison storage battery	Black, insoluble
Nickel ammonium sulfate 6-hydrate	$Ni(NH_4)_2(SO_4)_2 \cdot 6H_2O$	Sometimes used in nickel plating	Blue-green soluble compound
Nickel tetracarbonyl	$Ni(CO)_4$	Catalyst; source of carbon monoxide in organic synthesis; source of very pure nickel by decomposition	Colorless, volatile, liquid compound; made by reaction of nickel metal with carbon monoxide; contains nickel in oxidation state 0; more poisonous than carbon monoxide
Bis(dimethylglyoximato) nickel(II)	$Ni(C_4H_7N_2O_2)_2$	Analytical determination of nickel	Red, insoluble complex compound
Ammine nickel(II) cyanide benzene clathrate compound	$Ni(CN)_2 \cdot NH_3 \cdot C_6H_6$	Regenerates benzene when heated or placed in water	Violet-colored compound containing benzene trapped in a molecular-cage structure; other organic compounds form similar clathrates with nickel
Nickel(II) acetate 4-hydrate	$Ni(C_2H_3O_2)_2 \cdot 4H_2O$	Sealer for anodized aluminum; mordant in textile dyeing; reagent in dye preparation; in nickel plating	Blue-green soluble compound
Nickel(II) sulfide	NiS	Analytical determination of nickel	Brown or black insoluble compound
Nickel(II) cyanide 4-hydrate	$Ni(CN)_2 \cdot 4H_2O$	In nickel plating and metallurgy	Green powder or plates; nearly insoluble in water; toxic

Analytical methods. Nickel can be identified qualitatively by precipitation of either the hydroxide or the sulfide. The hydroxide is not soluble in excess of hydroxide ion and is only slowly soluble in dilute hydrochloric acid. If ammonium hydroxide is the reagent used, the precipitate is soluble in excess because of the formation of the blue $[Ni\text{-}(NH_3)_6]^{++}$ complex ion. The sulfide can be precipitated in brown or black form by ammonium sulfide or hydrogen sulfide in neutral or alkaline solution. Cobalt undergoes similar reactions; it may be separated from nickel in acetate buffer solution by precipitating the nickel with dimethylglyoxime. The red nickel dimethylglyoxime may be dried and weighed in the most common quantitative determination of nickel. This red compound is also the basis for a very useful spectrophotometric method.

Precipitation of nickel with α-benzildioxime yields an intense red compound which is even more sensitive evidence of the presence of nickel than the dimethylglyoxime compound. Even a hundredfold excess of cobalt does not interfere with the benzildioxime precipitation. A colorimetric determination of nickel is based on the magenta-colored complex of dithiooxalate ion with nickel.

Nickel may be directly determined in solution by direct titration with ethylenediaminetetraacetate ion (EDTA) in the presence of murexide indicator, or by indirect titration with EDTA of the magnesium displaced by nickel from the magnesium-EDTA complex. Eriochrome Black T is one of several indicators for indirect titration.

X-ray emission analysis has been applied to the determination of nickel in glass and catalysts. Cathode-ray polarography can determine nickel to the range of parts per billion in solution. Atomic absorption spectrometry is useful to about 0.1 ppm. Neutron-activation analysis has determined as little as 0.03 μg of nickel in steel and in ocean sediments. Flame spectrometry is also frequently used to determine nickel in steels.

Other quantitative methods include precipitation of the hydroxide, which is heated and weighed as NiO; precipitation with Grossmann's reagent, dicyandiamidine sulfate; and electrolysis followed by weighing of the nickel as the metal.

Toxicity. Common nickel compounds have produced toxic effects in humans and other animals, but nickel metal is not highly toxic. Furthermore, the traces of nickel in water, soil, and foods are not considered hazardous. However, exposure to larger amounts of nickel can lead to toxicity. Nickel(II) ions bind to nucleic acids and can have significant genetic effects. Certain enzymes are inhibited by nickel ions, whereas others are strongly activated. Alteration of the normal level of nickel in the blood is characteristic of a number of disease states. Of course, ingestion of nickel compounds, other than by prescription, should be avoided. A commonly recognized local reaction to nickel is dermatitis. It has also been reported that inhaled nickel powder or fumes from roasting of nickel ores may be carcinogenic. The volatile compound nickel tetracarbonyl, $Ni(CO)_4$, is outside the relative safety of most nickel compounds; it is extremely poisonous, more so than carbon monoxide, which is one of its constituents. Suggested antidotes are dimercaprol and sodium diethyldithiocarbamate, but in all cases of suspected exposure to nickel tetracarbonyl a physician should be consulted.

[WILLIAM E. COOLEY]

Bibliography: J. A. Dean (ed.), *Lange's Handbook of Chemistry*, 12th ed., 1979; J. R. DiPalma (ed.), *Drill's Pharmacology in Medicine*, 4th ed., 1971; R. H. Dreisbach, *Handbook of Poisoning: Diagnosis and Treatment*, 10th ed., 1980; N. H. Furman, *Standard Methods of Chemical Analysis*, 6th ed., vol. 1: *The Elements*, reprinted 1975; G. G. Hawley, *The Condensed Chemical Dictionary*, 9th ed., 1977; National Research Council Committee on Medical and Biologic Effects of Environmental Pollutants, *Nickel*, 1975; D. Nicholls, in J. C. Bailar, Jr. (ed.), *Comprehensive Inorganic Chemistry*, vol. 3, 1973; M. S. Sienko and R. A. Plane, *Chemistry: Principles and Properties*, 3d ed., 1979; R. C. Weast (ed.), *Handbook of Chemistry and Physics*, 61st ed., 1980.

Nicotine

A liquid alkaloid from the dried leaves of tobacco. Present in *Nicotiana tabacum* and *N. rustica* as citrate or malates in the amount of 2–8%, nicotine is recovered by extraction of cigarette wastes. Chemically it is *(S)*-3-(1-methyl-2-pyrrolidinyl)pyridine, $C_{10}H_{14}N_2$, a colorless to amber oil which takes up moisture from the air very avidly. It has a strong tobacco odor and intensely bitter, burning taste. It decomposes slowly on distillation at atmospheric pressure, but can be smoothly distilled under vacuum. Nicotine has a boiling point of 123–125°C under 15–20 mm Hg, and a density of 1.0097 at 20°C. Quite soluble in water, nicotine forms salts with mineral acids, such as hydrochloride, hydroiodide, and sulfate. The sulfate, which crystallizes as hexagonal tablets that are easily soluble in water or alcohol, is used as an agricultural insecticide.

Nicotine is highly toxic. Symptoms include extreme nausea, vomiting, evacuation of intestines and urinary bladder, mental confusion, and convulsions. The free base can be taken up through the unbroken skin as well as through mucous membranes, so spills of the material on the skin are serious. The oral lethal dose is believed to be about 50–60 mg/kg of body weight.

Nicotinic acid and nicotinamide (one of the B vitamins) are not prepared from nicotine, though they were originally obtained by degradation of the natural alkaloid; they now are synthesized. *See* ALKALOID.

[FRANK WAGNER]

Niobium

A chemical element, Nb, atomic number 41 and atomic weight 92.906, a member of the fifth group of the periodic table and in the 4*d* transitional series. Its valence electron configuration is $4d^45s^1$, which accounts for its maximum oxidation state of V. Oxidation states of IV, III, and II are also known. The similarity in effective ionic radii for Nb^{5+}, 0.70 A, and Ta^{5+}, 0.73 A, accounts for the fact that these elements occur together in the ore, $(Fe,Mn)(Nb,Ta)_2O_6$. Fractional separation during geological processes is evidenced by the occurrence of deposits with varying metal ion ratios. Niobium is also found without tantalum in the mineral pyrochlore, $NaCaNb_2O_6$. Niobium occurs in the Earth's crust to the extent of 2.4×10^{-3}%,

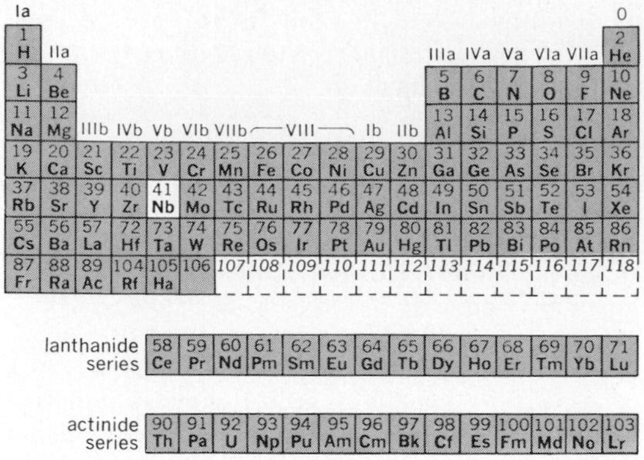

which makes it about 10 times as abundant as tantalum. For a discussion of the metallurgical separation and reduction *see* TANTALUM.

Note should be made of the fact that in the United States this element was originally called columbium. However, the Nomenclature Committee of the International Union of Pure and Applied Chemistry in 1951 adopted a recommendation to name this element niobium. American chemists use this name, but the metallurgists and metals industry still use the name columbium.

Most niobium is used in special stainless steels, high-temperature alloys, and superconducting alloys such as Nb_3Sn. Niobium's low cross-section capture for thermal neutrons of only 1.1 barns makes it suitable for use in nuclear piles.

Niobium metal. Niobium metal crystallizes in the body-centered cubic system, has a density of 8.6 g/cm³ at 20°C, a melting point of 2468°C, and a boiling point of 4927°C. Metallic niobium is prepared by fused electrolysis of K_2NbF_7 or reduction of the oxide by active metals or carbon. Although the standard potential for the reaction shown below is estimated to be +0.62 volt, the

$$2Nb + 5H_2O \rightarrow Nb_2O_5 + 10H^+ + 10e^-$$

metal is quite inert to all acids except hydrofluoric, presumably owing to an oxide film on the surface. Niobium metal is slowly oxidized in alkaline solution. It reacts with oxygen and the halogens upon heating to form the oxidation state V oxide and halides, with nitrogen to form NbN, and with carbon to form NbC, as well as other elements such as arsenic, antimony, tellurium, and selenium.

Niobium compounds. The oxide Nb_2O_5, melting point 1520°C, can be obtained by heating the metal in oxygen. The oxide dissolves in fused alkali to yield a soluble complex niobate, $Nb_6O_{19}^{8-}$, which hydrolyzes to the insoluble oxide below pH 7. The freshly precipitated oxide is a weak acid and dissolves slowly in aqueous alkali. Normal niobates such as NbO_4^{3-} are insoluble. The oxide dissolves in hydrofluoric acid to give ionic species such as $NbOF_5^{2-}$ and $NbOF_6^{3-}$, depending on the fluoride and hydrogen-ion concentration. The highest fluoro complex which can exist in solution is NbF_6^-, although in a very large excess of fluoride ion, salts of NbF_7^{2-} can be crystallized from solution. The NbF_7^{2-} ion is seven-coordinate, with six fluorines at the corners of a trigonal prism and the seventh perpendicular to a rectangular face. The oxide also dissolves in concentrated hydrochloric acid to give species which have been identified as $Nb(OH)_2Cl_4^-$, $Nb(OH)_2Cl_3$, and $Nb(OH)Cl^{3+}$, depending upon the hydrogen-ion and chloride-ion concentrations. The oxide also shows solubility in concentrated sulfuric acid and in solutions of organic dibasic acids as well as hydroxy carboxylic acids. Hydrogen reduction of the pentoxide at 800–1300°C yields NbO_2, and at 1300–1700°C, NbO.

The table lists the melting and boiling points of the niobium pentahalides. Niobium pentahalides may be prepared by direct reaction of the halogen or dry halogen halide with the metal. Niobium pentachloride may also be prepared upon heating the hydrous pentoxide with either thionyl chloride or hexachloropropene at reflux temperatures. Niobium pentafluoride has been shown to be a tetramer in the solid state with the metal atoms in octahedral holes located at the corners of a square linked by linearly bonded fluorine atoms. Viscosity studies suggest that polymeric units are also present in the liquid state. Self-ionization is no more than 1% in the liquid state. Niobium pentachloride and bromide have niobium atoms in adjacent octahedral holes of a close-packed halogen atom lattice yielding dimers in the solid state which persist in solvents such as carbon tetrachloride or nitromethane. In the pentaiodide solid-state structure, the MI_6 octahedra share edges to form an infinite polymer. The anhydrous halides react with oxygen to form oxyspecies $NbOF_3$, NbO_2F, and $NbOX_3$ (X = Cl, Br, I). The latter compounds in the solid state consist of infinite chains of planar Nb_2Cl_6 groups with the trans positions of the CN6 structure occupied by bridging oxygens. In water the pentachloride, bromide, and iodide hydrolyze to the insoluble pentoxide.

Melting and boiling points of the niobium pentahalides, NbX_5

Compound	Melting point, °C	Boiling point, °C
NbF_5	80.0	234.9
$NbCl_5$	212	243
$NbBr_5$	227	272
NbI_5	327(decomposes)	

The reaction of niobium(V) chloride or bromide with the bidentate chelating ligand, *o*-phenylenebisdimethylarsine in nonhydroxylic solvents, yields seven-coordinate NbX_5(diars), while at elevated temperatures in a sealed tube eight-coordinate $NbX_4(diars)_2$ is obtained. Niobium(V) chloride also reacts with tropolone (T) in aqueous solution to yield eight-coordinate species NbT_4^+. With alcohols, the chloride reacts to give dimeric $[Nb(OR)_4]_2$. The metal pentahalides act as Lewis acids. The chloride forms 1:1 adducts with triphenyl derivatives of the group V elements, as well as ethers, alkyl sulfides, cyclic ethers, and thioethers. The alkyl sulfide and ether 1:1 adducts of NbF_5 can be distilled unchanged. With ammonia and amines such as pyridine (py), the pentafluoride forms $NbF_5 \cdot 2NH_3$ and $NbF_5 \cdot 2py$, while with the same ligands the others halides react by ammonolysis and reduction to the IV state, respectively. Dimethylsulfoxide, triphenylphosphine oxide, and triphenylarsine oxide react with niobium pentachloride to abstract oxygen and form $NbOCl_3$ adducts.

The tetrafluoride, chloride, and bromide can be produced by reduction of the pentahalide with niobium metal in a sealed tube with a temperature gradient of 410–350°C. Niobium tetraiodide can be obtained by thermal decomposition of the pentaiodide. The tetrahalides react with moist air to give oxy species and disproportionate at elevated temperatures. The crystal structure of niobium tetrafluoride consists of a cubic closest-packed fluoride lattice with niobium atoms occupying one-fourth of the octahedral holes in such a way that they form a body-centered tetragonal structure. The $NbCl_4$ structure consists of one-dimensional linear chains in which $NbCl_6$ octahedra share trans edges. α-NbI_4 has a similar structure although it is orthorhombic with eight formula units per unit cell and a metal-metal distance of 3.31 A, in contrast to the tetrachloride, which is monoclinic with four formula units per unit cell and a metal-metal distance of 3.06 A. The tetrabromide is isomorphous with the tetrachloride. The metal ions are drawn together in pairs, and the compounds are diamagnetic. The tetrachloride, bromide and iodide form numerous $NbX_4 \cdot 2L$ adducts with monodentate ligands possessing O, S, or N atoms. With bidentate ligands such as diars, and β-keto-enol molecules (AA), 8-coordinate complexes of composition $NbX_4(diars)_2$ and $Nb(AA)_4$ are obtained.

Reduction of niobium pentahalides (Cl, Br, I) with metals at more elevated temperatures yields products exhibiting a range of composition $NbCl_{3.13}-NbCl_{2.67}$, $NbBr_{3.04}-NbBr_{2.67}$, and $NbI_{3.05}-NbI_{2.67}$. Stoichiometric NbI_3 is a metastable phase which decomposes irreversibly at 513°C. Whether the structure of the products of composition ~3X/1Nb possess a chain structure is not known, but products of composition 2.67X/1Nb contain Nb_3 clusters. The gross structures are related to the hexagonal close-packed CdI_2 structure with one-half of the metal ion sites vacated in every alternate row of the metal layer to yield triangular Nb_3 units with metal-metal bonds and inter- and intracluster halogen bridge bonds. The trifluoride, NbF_3, has been reported but in fact may be an oxyfluoride. The dichloride and bromide have been reported but are not well characterized.

Equilibration of Nb_3Cl_8 with niobium metal, or reduction of the pentachloride with sodium amalgam or cadmium metal, results in a polynuclear product of composition Nb_6Cl_{14}. This species contains the $Nb_6Cl_{12}^{2+}$ cation in which the niobium atoms are at the corners of an octahedron and chlorobridges bond along the edges. The equivalent bromo species has been produced by aluminum reduction of the pentabromide. A fluoride of composition Nb_6F_{15} has been reported which contains a similar hexameric cluster. The iodide Nb_6I_{11}, which was obtained as one of the products of the disproportionation of Nb_3I_8, actually contains the $Nb_6I_8^{3+}$ cation in which six niobium atoms are arranged octahedrally with eight iodines located symmetrically above the eight triangular faces. The chlorohexamer is remarkably stable in aqueous solution and can be recovered from solution as $Nb_6Cl_{14} \cdot 7H_2O$ or as $K_4(Nb_6Cl_{12})Cl_6$ and $[(C_2H_5)_4N]_4(Nb_6Cl_{12})Cl_6$. The $Nb_6Cl_{12}^{2+}$ cluster is easily oxidized to the $Nb_6Cl_{12}^{3+,4+}$ in aqueous solution. A metal cluster of composition $M^INb_4X_{11}$ is also known in which two Nb_3 units share a common edge.

The organometallic chemistry of niobium is not well developed. Niobium pentachloride dissolved in diethyleneglycoldimethylether (diglyme), when reduced by sodium in the presence of carbon monoxide, yields a compound of composition $[Na(diglyme)^+{}_2][M(CO)^-{}_6]$. Cyclopentadiene reacts with the niobium carbonyl anion and mercury(II) chloride as an oxidant, to yield π-$C_5H_5Nb(CO)_4$. This product reacts with diphenylacetylene to yield π-$C_5H_5Nb(CO)(Ph_2C_2)_2$ and π-$C_5H_5Nb(CO)_2(Ph_2C_2)$.

Analysis. Niobium of 90% purity or better can be determined by developing the peroxyniobate color with 30% hydrogen peroxide in a solution containing the sample dissolved in concentrated sulfuric acid and phosphoric acid, under which conditions the peroxytantalate color does not appear. The absorption is measured at 325 μm. Atomic absorption spectroscopy may also be used.

[EDWIN M. LARSEN]

Bibliography: F. A. Cotton and G. Wilkinson, *Advanced Inorganic Chemistry*, 4th ed., 1980; D. L. Kepert, *The Early Transition Metals*, 1972.

Nitrate

The negative ion, NO_3^-, derived from nitric acid, HNO_3. Because almost all metallic nitrates are water-soluble, they are not found in nature but are produced from nitric acid. One notable exception to this is the impure sodium nitrate, called Chile saltpeter, which occurs in large amounts because of the aridity and very small rainfall of the Chilean coastal plain.

Because nitrates contain nitrogen in its highest oxidation state (5+), the ion is a useful oxidizing agent. Because of this property, nitrates are often constituents of matches and explosives. Ammonium nitrate will detonate when subjected to shock according to the reaction given below.

$$NH_4NO_3 \rightarrow N_2O + 2H_2O$$

Nitrates are an important source of nitrogen in fertilizers.

The brown-ring test is a common qualitative test for the nitrate ion. A brown ring forms at the juncture of a dilute ferrous sulfate solution layered on top of concentrated sulfuric acid if the upper layer contains nitrate ion. If nitrite ion is present, the brown color appears throughout the solution. The brown color is due to the complex ion $Fe(NO)^{2+}$.

Organic compounds containing the $-NO_2$ group are called nitro compounds and include explosives such as trinitrotoluene (TNT). *See* NITRIC ACID; NITRITE; NITROGEN.

[E. EUGENE WEAVER]

Nitration

An important chemical conversion (unit process) in which a nitro group ($-NO_2$) is introduced into a hydrocarbon compound. Three types of nitration are common, and these are classified according to their chemical structure:

1. C-nitration in which the nitro group attaches itself to a carbon atom, as indicated in reaction (1).

$$\geq C{-}H + HNO_3 \rightarrow \geq C{-}NO_2 + H_2O \qquad (1)$$

2. O-nitration (really esterification) in which an O-N bond is formed to produce a nitrate, as seen in reaction (2).

$$\geq C{-}OH + HNO_3 \rightarrow \geq C{-}O{-}NO_2 + H_2O \qquad (2)$$

3. N-nitration, in which there is formation of a N-N bond, such as shown by reaction (3).

$$>N{-}H + HNO_3 \rightarrow >N{-}NO_2 + H_2O \qquad (3)$$

Nitration of aromatics. Nitration of aromatics is employed commercially for the production of numerous nitroaromatics. The chemical conversion is generally conducted in the liquid phase, and the organic and acid phases are immiscible so that agitation is necessary to provide adequate contact between the reactants. Groups already substituted on the ring affect the electron distribution which, together with steric considerations, affects the ease of nitration, as well as the specific carbon of the ring at which the nitro group attaches itself. Toluene, for example, is nitrated more easily than benzene, but the introduction of one nitro group on the ring hinders further nitration. Nitration reactions are always highly exothermic, and about 15–35 kcal/mole of heat is evolved.

Continuous-flow processes have been adopted to a significant extent since the mid-1960s for many types of aromatic nitrations, whereas batch processes were employed earlier. Such flow processes are widely and in some cases almost exclusively used for the mono- and dinitration of benzene or toluene. As yet, batch processes are probably still used exclusively in the United States for the trinitration of toluene to produce trinitrotoluene (TNT). Flow processes are successfully employed, however, for TNT production in other countries. These flow processes have significant economic advantages in most cases as compared to batch processes.

The Schmid-Meissner and the Biazzi processes are examples of continuous-flow arrangements employing stirred-tank reactors that have been successfully adapted to the nitration of benzene and toluene. Each reactor in these processes is maintained at the optimum temperature, depending on the conversion and acid concentration. Advantages claimed for the processes include smaller and cheaper equipment for the same production; less hazardous operation since the quantities of explosive nitroaromatics, such as dinitrotoluenes or trinitrotoluenes, actually in the system at a given time are small; and better control of operating variables to minimize side reactions. The Biazzi process is also widely used for the production of nitroglycerin.

Tubular reactors are employed to a significant extent in the United States for both the mono- and dinitration of toluene and benzene. These tubular reactors are really nothing more than shell-and-tube heat exchangers, and the reactants flow through the tubes in a turbulent condition. These reactors have the advantage of being relatively cheap to construct and to maintain and yet they are proving to be highly effective.

Nitrobenzene, often employed in the manufacture of aniline and as a solvent, is formed in greater than 99% yield by the nitration of benzene using mixed acids, for example, 39% HNO_3 and 55% H_2SO_4. Good heat transfer must be provided when such strong acids are employed. Sulfuric acid had been considered a dehydrating agent for removal of the water formed during reaction, allowing the reaction to go to completion. This interpretation is incorrect because the reaction is irreversible (the free-energy change is large and negative). Instead, the sulfuric acid acts as a catalyst. When strong sulfuric acid is employed, it reacts with nitric acid to produce nitronium ions, NO_2^+, which are true nitrating agents. Perchloric acid, acetic anhydride, hydrogen fluoride, and boron trifluoride can sometimes be used instead of sulfuric acid in mixed acids to produce nitronium ions. *See* SULFURIC ACID.

Nitrous acid has been reported to be effective in the nitration of easily nitratable aromatics such as phenol and phenolic ethers. In this case the nitrosonium ion, NO^+, is thought to be the ion that initially attacks the aromatic to form a nitrosoaromatic. This latter compound is then oxidized to the nitroaromatic by nitric acid, which is reduced to nitrous acid.

Dinitrobenzene is manufactured for reduction to *m*-nitroaniline or *m*-phenylenediamine. Various chloronitrobenzenes, *p*-nitroacetanilides, nitrotoluenes, and nitronaphthalenes are employed as intermediates. They generally require stronger mixed acids and give lower conversions and yields than does nitrobenzene because of partial oxidation.

TNT is a powerful explosive often produced in a three-step batch nitration to secure economy of the nitrating acids, reduce oxidation, and secure best yields with only two nitrators being necessary. The table gives the acids involved in the steps. The first

TNT acids, composition %

	Mono reaction			Bi reaction		Tri reaction	
Compound	Mixed acid	Spent acid	Cycle acid	Mixed acid	Spent acid	Mixed acid	Spent acid
H_2SO_4	50.4	56.3	54	53.9		40	64.5
HNO_3	14.5	3.6	6.6	13.8	2.5	60	3.9
$NO_2{\cdot}HSO_3$	12.8	14.5	13.3	13			15
Nitrobenzene	2.5	0.6	2.5	11.3	Reduced		14.9
H_2O	19.8	25.0	26.6	8			1.9

nitration is carried out in the so-called mono house at 140°F, introducing toluene under agitation and with cooling into the cycle acid, followed by mono-mixed acid. The use of the cycle acid (last of the spent acid from the previous batch) saves some nitric acid and increases the initial volume to secure better agitation and cooling. The resulting nitration product, so-called mono oil, is conducted to the bi-tri house, where binitration is performed at 90–180°F using bi-mixed acid. The bi-spent acid is settled, drawn off, and fortified to make mono-mixed acid. The binitration product is nitrated in the same kettle to TNT, starting at 180° and ending at 230°F by the use of tri-mixed acid, following which the tri-spent acid is drawn off and sent to the fortifier for further use. The hot tri oil is withdrawn from the nitrator, crystallized in water, and washed in suspension with dilute soda-ash solution followed by acidulated 16% Na_2SO_3 (Sellite) to remove undesirable isomers. After a water wash the TNT is melted, washed, solidified or grained, and flaked or filled into shells. Countercurrent handling of the various mixed acids (after fortification) increases the yield of TNT and greatly decreases cost. Both the HNO_3 and $NO_2 \cdot HSO_3$ of the mono-spent acid are recovered as oxides of nitrogen by passing the spent acid down a tower against a rising column of steam. The oxides are oxidized with air and dissolved in water to produce recycle nitric acid. The sulfuric acid in the bottom of the tower is safely concentrated for further use through countercurrent direct contact with hot combustion gas, thus harmlessly burning the nitrobodies.

Paraffin liquid-phase nitration. Liquid-phase nitration of paraffins is the method used for the production of nitrocyclohexane (previously employed commercially for the manufacture of caprolactam) and of both 2-nitropropane and 2,2-dinitropropane. The reactions involve two immiscible liquid phases and are at the approximate following operating conditions: temperature, 130–200°C; pressure, 800–2000 psi; and nitric acid 60–70%. The pressure must be sufficiently great to maintain the reaction mixture primarily as liquids. Rapid removal of the water from the mixture is reported to be highly beneficial for obtaining high conversions and yields.

Vapor-phase nitration of paraffins. In 1955 the first commercial plant for producing nitroparaffins using a continuous-flow process was completed. Propane is nitrated by spraying cold liquid nitric acid into hot propane. The reactor operates adiabatically since the highly exothermic heat of reaction just balances the latent and sensible heats required to heat the nitric acid to reaction conditions. The four nitroparaffins produced are 1-nitropropane, 2-nitropropane, nitroethane, and nitromethane. These compounds find many uses as solvents, intermediates, and fuels.

All lighter paraffins can be nitrated in the vapor phase, but ethane and especially methane nitrate only with difficulty. Propane and higher paraffins that contain secondary or tertiary hydrogens nitrate readily at 0–100 psi and at approximately 375–440°C in 0.5–2.0 sec when an excess of the alkane is employed. Under these conditions any hydrogen or alkyl group of the alkane can be replaced with a nitro group, and conversions of nitric acids to nitroparaffins are as high as 35–40%. The reaction occurs by a free-radical and chain mechanism, as shown by (4)–(6). Reaction (4) is the chain-initiating step, and (5) and (6) are the chain steps.

$$HNO_3 \xrightarrow{\Delta} \cdot OH + \cdot NO_2 \qquad (4)$$

$$RH + \cdot OH \rightarrow R\cdot + H_2O \qquad (5)$$

$$R\cdot + HNO_3 \rightarrow RNO_2 + \cdot OH \qquad (6)$$

When oxygen or halogens are added to the reaction mixture in appreciable quantities, nitric acid conversions to nitroparaffins as high as 70% are possible. The nitroparaffin product obtained contains a higher concentration of nitromethane when oxygen is used and when the temperature in the reactor is relatively nonisothermal, but more nitropropanes are produced when halogens are employed and with good temperature control. In all cases, but especially when oxygen is employed, some oxygenated hydrocarbons are produced, but no disubstituted nitro compounds are obtained in the vapor-phase process. In the past, difficulty was sometimes experienced in maintaining temperature control in tubular reactors.

Miscellaneous nitrations. Tetryl, ammonium picrate, and cyclonite (RDX) are also widely employed as explosives. Cyclonite is cyclotrimethylenetrinitramine, which is relatively safe and a stronger explosive than TNT. The commercial process used in the United States involves continuous nitration by passing hexamethylenetetramine through a 4-in. pyrex glass pipe approximately 600 ft long and introducing strong nitric acid, ammonium nitrate, and acetic anhydride at appropriate intervals. Cyclonite yields are 70% and higher with strong acetic acid as a by-product. This is shown by reaction (7).

$$C_6H_{12}N_4 + 4HNO_3 + 2NH_4NO_3 + 6(CH_3CO)_2O \rightarrow 2C_3H_6O_6N_6 + 12CH_3COOH \qquad (7)$$

Nitrogen dioxide can be used to nitrate paraffins and olefins, but apparently no commercial processes are employed. Propane reacts with nitrogen dioxide to produce a nitroparaffin product with a high concentration (about 72%) of 2-nitropropane, but conversions are low. At temperatures about 200–250°C the nitrogen dioxide is very effective in attaching secondary C-H bonds. Nitrocyclohexane can be produced in relatively high yields by reaction of cyclohexane and nitrogen dioxide. Nitrogen dioxide also undergoes addition reactions with olefins to produce dinitroalkanes and nitronitrates.

Olefins can be nitrated with nitric acid to produce nitroolefins, nitro alcohols, and the subsequent nitronitrate esters.

The classical Victor Meyer method involving the reaction of silver nitrite with alkyl chlorides is used for the laboratory preparation of aliphatic nitro compounds. A technique employing alkali metal nitrites such as $NaNO_2$, $LiNO_2$, and KNO_2 has been perfected, in which a solvent such as ethylene glycol, water, or other hydrogen-donating solvent is used. Other nitrating agents that find limited laboratory use include nitryl chloride and nitrogen pentoxide. *See* NITRIC ACID; OXIDIZING AGENT.

[LYLE F. ALBRIGHT; R. NORRIS SHREVE]

Bibliography: L. F. Albright and C. Hanson (eds.), *Industrial and Laboratory Nitrations*, 1976; J. G. Hoggett et al., *Nitration and Aromatic Activity*, 1971; A. Marshall, *Explosives: Their History, Manufacture, Properties, and Tests*, 3 vols., 1980.

Nitric acid

A strong mineral acid having the formula HNO_3. Pure nitric acid is a colorless liquid with a density of 1.52 at 25°C; it freezes at −47°C. Nitric acid is used in the manufacture of ammonium nitrate and phosphate fertilizers, nitro explosives, plastics, dyes, and lacquers. The principal commercial process for the manufacture of nitric acid is the Ostwald process, in which ammonia, NH_3, is catalytically oxidized with air to form nitrogen dioxide, NO_2. When the dioxide is dissolved in water, 60% nitric acid is formed. Production of 90–100% nitric acid is based on processes such as the reaction of sulfuric acid with sodium nitrate (an older method of nitric acid manufacture), dehydration of 60% acid, and oxidation of nitrogen dioxide in a solution of dilute nitric acid.

Nitric acid decomposes readily as shown in the reaction below. It is a strong oxidizing agent, oxi-

$$2HNO_3 \rightarrow H_2O + 2NO_2 + \tfrac{1}{2}O_2$$

dizing carbon to carbon dioxide, sulfur to sulfuric acid, and phosphorus to phosphoric acid. It reacts with most metals; products depend on the metal's electromotive series position and nitric acid concentration. *See* AMMONIA; NITROGEN; NITROGEN OXIDES; OXIDIZING AGENT.

[FRANCIS J. JOHNSTON]

Bibliography: F. A. Cotton and G. Wilkinson, *Advanced Inorganic Chemistry*, 4th ed., 1980; F. A. Lowenheim and M. K. F. Moran, *Keye's and Clark's Industrial Chemicals*, 4th ed., 1975.

Nitride

A binary compound of nitrogen with elements less electronegative than nitrogen. Common practice, however, is to exclude azides (such as NaN_3), in which specialized bonding exists among the nitrogen atoms, and the binary compounds with hydrogen, the halogens, and the oxygen group elements. With this limitation, essentially solid nitrides will be encountered.

Direct reaction of group Ia metals with nitrogen yields azides, which can be decomposed by heating (cautiously, to avoid explosion) to form the nitrides Li_3N, Na_3N, K_3N, Rb_3N, and Cs_3N. These nitrides decompose to nitrogen gas and the elements in the vicinity of 400°C; they react with water vapor to liberate ammonia (NH_3) and form the metal hydroxides.

The nitrides of the group IIa elements (Be_3N_2, Mg_3N_2, Ca_3N_2, Sr_3N_2, and Ba_3N_2) may be prepared by direct reaction of the element with nitrogen, as in (1), or with ammonia, as in (2), at elevated temperatures. A convenient preparation is by heating the amides, for example, reaction (3).

$$N_2 + 3Be \rightarrow Be_3N_2 \quad (1)$$

$$2NH_3 + 3Be \rightarrow Be_3N_2 + 3H_2 \quad (2)$$

$$3Ba(NH_2)_2 \rightarrow Ba_3N_2 + 4NH_3 \quad (3)$$

These nitrides begin to melt and decompose to the elements and nitrogen gas in the vicinity of 1000°C, except for Be_3N_2, which melts at about 2200°C. They react with water in the same fashion as noted above for the alkali metal nitrides.

The nitrides of the group IIIa metals and of the lanthanide and actinide elements (ScN, YN, LaN, CeN, PrN, NdN, GdN, ThN, Th_2N_3, PaN_2, UN, U_2N_3, UN_2, NpN, and PuN) are quite stable to temperatures in the vicinity of 1500°C and much higher in some cases. The nitrides of the group IVa metals (TiN, ZrN, and HfN) and of the group Va metals (VN, V_2N, CbN, Cb_2N, TaN, and Ta_2N) are also quite stable; however, those of group VIa (CrN, Cr_2N, Mo_2N, and W_2N) and of group VIIa (Mn_4N, Mn_5N_2, and Mn_3N_2) are less stable. All of these may be prepared by direct reaction of the metal with nitrogen gas. The compositions of a number of these nitrides vary over an appreciable range from the simple definite proportions indicated by the formulas given. When these nitrides do decompose from heating, they yield N_2 gas.

Nitrides of the group VIIIa metals (Fe_2N, Fe_4N, Co_2N, Co_3N, and Ni_3N) are of very low stability and are best prepared by reaction of the metal with ammonia gas. Ni_3N will decompose to the metal and nitrogen gas at 1 atm pressure at about 450°C. Cu_3N, Zn_3N_2, Cd_3N_2 (indirect preparation), and InN are even less stable. GaN, however, is quite stable.

The elements Ag, Au, Hg, Tl, Sn, Pb, Sb, and Bi do not form nitrides. AlN, Si_3N_4, Ge_3N_4, and P_3N_5 are quite stable, and BN is exceptionally so (sublimation point above 3000°C). *See* AZIDE; NITROGEN; OXIDE.

[RUSSELL K. EDWARDS]

Bibliography: K. Jones, *The Chemistry of Nitrogen*, 1975.

Nitrile

One of a group of organic chemical compounds of general formula $RC{\equiv}N$. A nitrile is named from the acid to which it can be hydrolyzed by adding the suffix -onitrile to the acid stem, for example, acetonitrile from acetic acid. An alternative system names the group attached to CN, thus CH_3CN is also named methyl cyanide. In more complex structures the CN group is named as a substituent, cyano.

Nitriles may be identified by a distinctive weak-to-medium infrared absorption band due to triple-bond stretching at 2260–2222 cm^{-1}.

Nitriles are hydrolyzed to acids in either basic or acidic solution, as shown in reactions (1) and (2).

$$RCN + OH^- + H_2O \rightarrow RCOO^- + NH_3 \quad (1)$$

$$RCN + H^+ + 2H_2O \rightarrow RCOOH + NH_4^+ \quad (2)$$

By catalytic hydrogenation in the presence of nickel or cobalt, nitriles are reduced to primary amines, reaction (3). Nitriles are also reduced to

$$RCN + 2H_2 \rightarrow RCH_2NH_2 \quad (3)$$

primary amines by lithium aluminum hydride.

Grignard reagents added to nitriles give ketones (after hydrolysis), as shown in reaction (4).

$$RCN + R'MgX \rightarrow RR'C{=}NMgX \xrightarrow{H_2O} RCOR' \quad (4)$$

The formation of cyanides from alkyl halides is an important chain-lengthening reaction in organic synthesis. The starting compound is most often an alcohol, and the sequence of reactions is shown in (5). The reaction is practical only with primary ali-

$$ROH \xrightarrow{HX} RX \xrightarrow{NaCN} RCN \qquad (5)$$

phatic halides, since the alkali cyanides are fairly strong bases and eliminate HX from secondary or tertiary alkyl halides.

Aromatic nitriles are made by displacement of a diazotized primary amino group with the cyanide group in the presence of copper cyanide or copper powder, reaction (6).

$$ArNH_2 \xrightarrow{HONO} ArN_2^+ \xrightarrow[NaCN]{Cu} ArCN \qquad (6)$$

The dehydration of acid amides or oximes with phosphorus pentoxide or acetic anhydride in either the aliphatic or aromatic series serves as another preparative method, reactions (7) and (8).

$$3RCONH_2 + P_2O_5 \rightarrow 3RCN + 2H_3PO_4 \qquad (7)$$

$$RCH{=}NOH + (CH_3CO)_2O \rightarrow RCN + 2CH_3COOH \qquad (8)$$

Industrially, nitriles are formed by heating carboxylic acids with ammonia and a dehydration catalyst under pressure. The amide is an intermediate but need not be isolated, reaction (9).

$$RCOOH + NH_3 \rightarrow [RCONH_2] \rightarrow RCN \qquad (9)$$

For the preparation of acrylonitrile, which is used on a large scale in the plastics industry, a vapor-phase catalytic ammoxidation of propylene has been developed, reaction (10).

$$CH_3CH{=}CH_2 + NH_3 + 3/2\,O_2 \rightarrow CH_2{=}CHCN + 3H_2O \qquad (10)$$

Acrylonitrile is electrochemically coupled "tail to tail" to give the hydrodimer adiponitrile, which is a key intermediate in the commerical production of nylon-6,6, reaction (11). *See* ACRYLONITRILE;

$$2CH_2{=}CHCN \xrightarrow[2H^+]{2e^-} NC(CH_2)_4CN \qquad (11)$$

AMINE; CARBOXYLIC ACID; OXIME; POLYAMIDE RESINS. [PAUL E. FANTA]

Bibliography: N. L. Allinger, et al., *Organic Chemistry*, 2d ed., 1976.

Nitrite

The negative ion NO_2^- derived from the unstable nitrous acid HNO_2. Because of the intermediate oxidation state of nitrogen in nitrites (3+), the ion can act as either an oxidizing or reducing agent.

Most nitrites are water-soluble. Sodium and potassium nitrites are quite stable and find extensive use in dyestuff manufacture and organic synthesis.

Sodium nitrite is produced by passing nitric oxide and nitrogen dioxide into sodium hydroxide, as in the reaction below.

$$NO + NO_2 + 2NaOH \rightleftharpoons 2NaNO_2 + H_2O$$

When nitrite solutions are acidified by strong acids, this reaction is reversed with the evolution of a mixture of gases, and some nitrate may also be formed in the decomposition.

The presence of the nitrite ion is detected by the formation of the brown $Fe(NO)^{++}$ ion. *See* NITRATE; NITROGEN.

[E. EUGENE WEAVER]

Nitro and nitroso compounds

Nitro compounds are derivatives of organic hydrocarbons having one or more $—NO_2$ groups with nitrogen to carbon bonding. They differ from the oxygen-linked nitrites, which are esters. The group lacks enough electrons to form double bonds with both oxygens. However, both oxygens react alike; hence the bond is regarded as a resonance hybrid of single and double bonds.

Aromatic nitro compounds, known for over 100 years, have been used chiefly as dye intermediates, explosives, and pharmaceuticals. They are formed readily by the reaction of aromatic compounds with nitric acid; H is replaced by the $—NO_2$ group, for example, 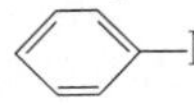$—NO_2$. Nitro groups are electron-attracting; they impede the introduction of further substituents and direct them to meta positions. Nitration is aided by elevated temperature and by the presence of concentrated sulfuric acid, because of the latter's affinity for the water formed as well as its reaction with nitric acid to form nitronium ion (NO_2^+). This ion readily displaces an aromatic H atom. Compounds having electron-donating groups ($CH_3—$ in toluene) are nitrated readily in the ortho and para positions. *See* NITROBENZENE.

Nitro compounds may be reduced by hydrogen to form primary amines. In cool neutral solutions the product may be hydroxylamine, and in alkaline media it may be an azoxy, azo, or hydrazo derivative. Dinitrobenzene may be partially reduced to *m*-nitroaniline. Ortho and para isomers are formed by nitrating chlorobenzene, and subsequently replacing the Cl with NH_2 from ammonia. The use of water instead of ammonia gives 2,4,6-trinitrophenol or picric acid, which is a high explosive. Complete nitration of toluene produces trinitrotoluene (TNT).

Aliphatic nitro compounds are prepared with difficulty and have grown in importance only since the development of vapor-phase nitration of hydrocarbons with nitric acid vapors at 420°C. Other preparative methods include the oxidation of oximes and reaction of alkyl halides with sodium nitrite. The aliphatic nitro compounds, or nitroalkanes, are colorless, high-boiling, and soluble in organic solvents but only slightly soluble in water. The electron-attractive nitro groups decrease C-H bond strength on the adjacent (α-) carbon, making nitroalkanes somewhat acidic and prone to hydrogen displacement. Nitroalkanes may be reduced to amines or hydrolyzed slowly to acids; unlike aromatic nitro compounds, they do not explode readily. *See* NITROPARAFFIN.

Nitroso compounds contain the —NO group attached to carbon or nitrogen. Many are unstable intermediates, for example, nitrosobenzene formed during the reduction of nitrobenzene. Nitrosobenzene can be prepared by oxidizing phenylhydroxylamine with dichromate and sulfuric acid. It is colorless and crystalline, but it forms a green solution. Tertiary amines will react with nitrous acid to form amine salts which lose water to give $R_3N—NO$. Nitroso compounds give identifying red, blue, and white colors with primary, secondary, and tertiary nitro compounds, respectively. *See* NITRATION.

[ALLEN L. HANSON]

Nitrobenzene

A very-pale-yellow liquid with a sweet but sickening odor. The structural formula is shown below.

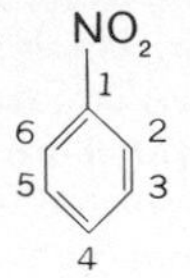

Nitrobenzene boils at 210.9°C and freezes at 5.6 to 5.7°C. It is produced by the nitration of benzene with a mixture of nitric and sulfuric acids.

Nitrobenzene undergoes substitution reactions but requires more vigorous conditions than does benzene. Substitution takes place at the meta (3) position.

Most of the nitrobenzene produced is reduced to aniline, but other dye intermediates including benzidine and metanilic acid are also prepared from it. The great toxicity of nitrobenzene impairs its usefulness as a solvent for organic compounds. It is absorbed through the skin and if the vapor is inhaled, it may produce cyanosis. *See* AROMATIC HYDROCARBON; BENZENE; NITRATION.

[CHARLES K. BRADSHER]

Nitrogen

A chemical element, N, atomic number 7, atomic weight 14.0067. Nitrogen, a gas under normal conditions, is the lightest element of periodic group Va (nitrogen family).

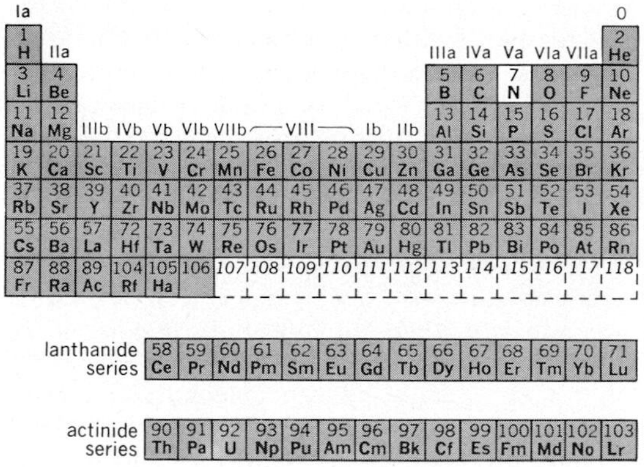

Occurrence. Molecular nitrogen is the principal constituent of the atmosphere (78% by volume of dry air), in which its concentration is a result of the balance between the fixation of atmsopheric nitrogen by bacterial, electrical (lightning), and chemical (industrial) action, and its liberation through the decomposition of organic materials by bacteria or combustion. In the combined state, nitrogen occurs in a variety of forms. It is a constituent of all proteins (both plant and animal) as well as of many other organic materials. Its chief mineral source is sodium nitrate. An important source of this mineral is located in the arid regions of northern Chile.

Preparation. The methods for the preparation of elementary nitrogen may be grouped into two classes, separation from the atmosphere and decomposition of nitrogen compounds. The industrial method for the production of nitrogen is the fractional distillation of liquid air. Nitrogen containing about 1% argon and traces of other inert gases may be obtained by the chemical removal of oxygen, carbon dioxide, and water vapor from the atmosphere by appropriate chemical reagents.

The following chemical reactions have been used to prepare nitrogen.

When a saturated solution of sodium nitrite is mixed with a hot, saturated solution of ammonium chloride, the reaction is shown by (1).

$$NH_4^+ + NO_2^- \rightarrow N_2 + 2H_2O \qquad (1)$$

Ammonia gas is oxidized by passing it through bromine water, and the resulting gaseous mixture separated by passing it through a series of reagents to absorb unreacted bromine, water vapor, and ammonia. The reaction is indicated by (2).

$$2NH_3 + 3Br_2 \rightarrow N_2 + 6H^+ + Br^- \qquad (2)$$

Other oxidants which may be used on ammonium salts include dichromate ion, ozone, fluorine, and manganese dioxide.

The thermal decomposition of very dry barium azide or sodium azide yields spectroscopically pure nitrogen, as in reaction (3).

$$Ba(N_3)_2 \rightarrow Ba + 3N_2 \qquad (3)$$

Ammonia gas will react with hot metal oxides, as shown by (4), to yield nitrogen.

$$3CuO + 2NH_3 \rightarrow 3Cu + 3H_2O + N_2 \qquad (4)$$

Catalytic decomposition of ammonia on hot platinum produces nitrogen and hydrogen.

The reaction of sulfamic acid (or urea) with nitrite ion yields nitrogen, as in (5).

$$NH_2SO_3H + NO_2^- \rightarrow N_2 + HSO_4^- + H_2O \qquad (5)$$

Industrial application. Because of the importance of nitrogen compounds in agriculture and chemical industry, much of the industrial interest in elementary nitrogen has been in processes for converting elemental nitrogen into nitrogen compounds. The principal methods for doing this are the Haber process for the direct synthesis of ammonia from nitrogen and hydrogen, the electric arc process, which involves the direct combination of N_2 and O_2 to nitric oxide, and the cyanamide process. Nitrogen is also used for filling bulbs of incandescent lamps and, in general, wherever a relatively inert atmosphere is required.

Atomic properties. The outer electron shells of the atoms of group Va elements have configurations of the type ns^2np^3. The normal configuration of the nitrogen atom is $1s^22s^22p^3$. Nitrogen is the most electronegative of the elements of this family (3.0 on the Pauling scale) and is a typical nonmetal in its reactions. The ionization energies of the nitrogen atom are sufficiently high (first, 14.54 eV; second, 29.605; third, 47.426; fourth, 77.450; fifth, 97.863; sixth, 551.925) to prevent the nitrogen atom from forming monatomic positive ions under the ordinary conditions of chemical reaction. However, nitrogen atoms may, under some conditions, take up electrons to form N^{3-} ions. The crystal radius of this ion (Pauling scale) is 1.71 A. The covalent radius of tricovalent nitrogen on the Pauling scale is 0.74 A.

Nuclear properties. Nitrogen, as it occurs in nature, consists of two isotopes, N^{14} and N^{15}, in the abundance ratio of 99.635 to 0.365. In addition the radioactive isotopes N^{12}, N^{13}, N^{16}, and N^{17} have been made by a variety of nuclear reactions. The first two are positron emitters and have half-lives of 0.0125 sec and 9.93 min, respectively. N^{16} and N^{17} are electron emitters and have half-lives of 7.35 sec and 4.14 sec, respectively. Unfortunately, none of these has a sufficiently long half-life for convenient use as a tracer. N^{15}, however, has been employed as a tracer by using nitrogen in which N^{15} has been concentrated and by following the reaction by mass-spectrometric techniques.

Molecular nitrogen. At standard temperature and pressure, elemental nitrogen exists as a gas with a density of 1.25046 g/l. This value indicates that the molecular formula is N_2. The N_2 molecule in its ground state has a magnetic susceptibility of -0.430×10^{-6} at 25°C, and therefore has no resultant electronic angular momentum of either the orbital or the spin variety. The electronic formula :N:::N:, indicating a triple covalent bond, is commonly written for the molecule. The electronic configuration of the N_2 molecule in terms of molecular orbital theory is $1s^2$, $1s^2$, $\sigma_{2s}{}^2$, $\sigma_{2s}{}^{*2}$, $\sigma_{2p_x}{}^2$, $\pi_{2p_y}{}^2$, $\pi_{2p_z}{}^2$. The interatomic forces in the N_2 molecule are very high, as indicated by a comparison of the interatomic distance in N_2, 1.10 A, with twice the single-bond covalent radius of nitrogen (2×0.74 A$=$1.48 A). The energy of dissociation of the N_2 molecule into atoms as determined spectroscopically is 225.8 kcal/mole. Spectroscopic studies indicate that at ordinary temperatures molecular nitrogen consists of molecules with symmetrical and antisymmetrical nuclear spins in the ratio 2:1. Because the nitrogen molecule is both very stable and highly symmetrical, intermolecular forces are very small. The phase changes in notation (6) have been determined for molecular nitrogen at standard atmospheric pressure.

$$N_{2(\text{solid }\alpha)\,\text{cubic}} \underset{}{\overset{-237.48°C}{\rightleftharpoons}} N_{2(\text{solid }\beta)\,\text{hexagonal}} \overset{-209.96°C}{\rightleftharpoons} N_{2(\text{liquid})} \overset{-195.78°C}{\rightleftharpoons} N_{2(\text{gas})} \qquad (6)$$

Other physical properties of elemental nitrogen are listed in Table 1.

Table 1. Properties of nitrogen

Property	Value
Heat of transformation (α–β)	54.71 cal/mole
Heat of fusion	172.3 cal/mole
Heat of vaporization	1332.9 cal/mole
Critical temperature	126.26 ± 0.04 K
Critical pressure	33.54 ± 0.02 atm
Density: α-form	1.0265 g/ml at −252.6°C
β-form	0.8792 g/ml at −210.0°C
Liquid	1.1607 − 0.0045T (T= abs temp)

Elemental nitrogen has a low reactivity toward most common substances at ordinary temperatures. At high temperatures, molecular nitrogen, N_2, reacts with chromium, silicon, titanium, aluminum, boron, beryllium, magnesium, barium, strontium, calcium, and lithium (but not the other alkali metals) to form nitrides; with O_2 to form NO; and at moderately high temperatues and pressures in the presence of a catalyst, with hydrogen to form ammonia. Above 1800°C, nitrogen, carbon, and hydrogen combine to form hydrogen cyanide.

When molecular nitrogen is subjected to the actions of a condensed electrode discharge or to a high-frequency electrodeless discharge, it is partially changed to an activated, unstable condition, from which, on standing, it returns to its normal state with the emission of a golden-yellow afterflow. Activated nitrogen is more reactive than ordinary nitrogen, as indicated by reactions (7) to (12).

$$P_4(\text{white}) \xrightarrow{N_2^*} \begin{cases} P_4(\text{red}) \\ \text{Phosphorus nitride} \end{cases} \qquad (7)$$

$$6Na + N_2^* \xrightarrow{150°C} 2Na_3N \qquad (8)$$

$$2NO + N_2^* \rightarrow 2N_2 + O_2 \qquad (9)$$

$$HC{\equiv}CH + N_2^* \rightarrow 2HCN \qquad (10)$$

$$H_2C{=}CH_2 \xrightarrow{N_2^*} \begin{cases} \text{HCN (chief product)} \\ C_2H_6 \\ CH_4 \\ HC{\equiv}CH \\ (CN)_2 \end{cases} \qquad (11)$$

$$\text{Alkyl chlorides} \xrightarrow{N_2^*} \begin{cases} \text{HCN, HCl (chief products)} \\ \text{Olefins, polymers containing C, H, N, and Cl} \end{cases} \qquad (12)$$

Mass spectrometric studies have shown that active nitrogen consists principally of nitrogen atoms in the ground state, and its superior chemical reactivity is believed to result from the presence of these free nitrogen atoms.

Compounds. The elements of the nitrogen family exhibit in their compounds three principal oxidation states, −3, +3, and +5, though other oxidation states are also exhibited. All the elements of the nitrogen family form hydrides of the type shown in formula (13) and +3 oxides, +5 oxides, +3 halides (MX_3), and, except for nitrogen and bismuth, +5 halides (MX_5).

$$\begin{matrix} & H & \\ & \cdot\cdot & \\ H: & M & :H \\ & \cdot\cdot & \end{matrix} \qquad (13)$$

Nitrogen is the most electronegative element of the nitrogen family; its electronegativity on the Pauling scale is 3.0 as compared with 2.1 for phosphorus and 2.0 for arsenic. The chemistry of nitrogen also reflects the fact that the valence shell of the nitrogen atom has a total of only four electron orbitals (one 2*s* plus three 2*p* orbitals), thus limiting the covalence of the nitrogen atom and its coordination number to a maximum of four. The availability of *d* orbitals in the valence shells of atoms of the heavier numbers of the family allow covalencies greater than four. Multiple bond formation by the nitrogen atom requires the use of *p* orbitals in the π-bond and thus further reduces the coordination number below four. Multiple bonding by phosphorus and heavier elements of the family may, in some instances, occur through the use of *d* orbitals without changing the coordination number. The relatively small size of the nitrogen atom limits its coordination number because of steric factors; thus, *o*-nitric acid, H_3NO_4, has not been prepared. Table 2 lists the principal classes of inorganic nitrogen compounds. Thus, in addition to the typical oxidation

Table 2. Compounds of nitrogen

Oxidation state	Examples
+5	N_2O_5, HNO_3, nitrates, NO_2X
+4	$N_2O_4 \rightleftharpoons 2NO_2$
+3	N_2O_3, HNO_2, nitrites, NOX, NX_3
+2	NO, Na_2NO_2, nitrohydroxylamates
+1	N_2O, $H_2N_2O_2$, hyponitrites
0	N_2
−1/3	HN_3, azides
−1	NH_2OH, hydroxylammonium salts
−2	NH_2NH_2, hydrazinium salts, hydrazides
−3	NH_3, ammonium salts, amides, imides, nitrides

states of the family (−3, +3, and +5), nitrogen forms compounds with a variety of additional oxidation states. *See* AMINE; AMMINE; AMMONIA; HYDRAZINE; NITRIC ACID; NITRIDE; NITROGEN OXIDES.

[HARRY H. SISLER]

Bibliography: F. A. Cotton and G. Wilkinson, *Advanced Inorganic Chemistry*, 4th ed., 1980.

Nitrogen complexes

Compounds containing the dinitrogen molecule, N_2, bound to a metal (also called dinitrogen complexes). While it was long known that molecular nitrogen avidly binds to the surface atoms of many metallic phases, no reaction of molecular nitrogen had been observed under the conditions of ordinary solution chemistry until the mid-1960s. Indeed, chemists had been so accustomed to regard nitrogen as a completely inert gas at ordinary temperatures that the formation of coordination compounds with molecular nitrogen aroused general surprise when first reported in 1965–1966. Since then the interaction of molecular nitrogen with coordination compounds has been the subject of intensive research. Today the capability of molecular nitrogen to enter into coordination compounds as a ligand is established beyond any doubt, and it is at least partly understood under which conditions the dinitrogen molecule can lose its customary inertness. *See* COORDINATION CHEMISTRY.

Formation. Outstanding in their ability to form coordination compounds with nitrogen are a number of metals which belong to the group VIII transition metal family (Fig. 1). For each metal of this group, several nitrogen complexes have been identified. Nitrogen complexes of these metals occur in low oxidation states, such as Co(I) or Ni(O). The other ligands present in these complexes besides N_2 are usually of a type known to stabilize low oxidation states; phosphines appear to be particularly prominent bonding partners in this respect. Figure 2 shows the structure of a typical N_2 complex, elucidated by a crystal structure determination. The N-N bond axis in this complex is aimed, within the limits of experimental error, directly toward the position of the metal atom. The Co-N_2 bond length, 1.8 A, is within the normal range of comparable metal-ligand bonds.

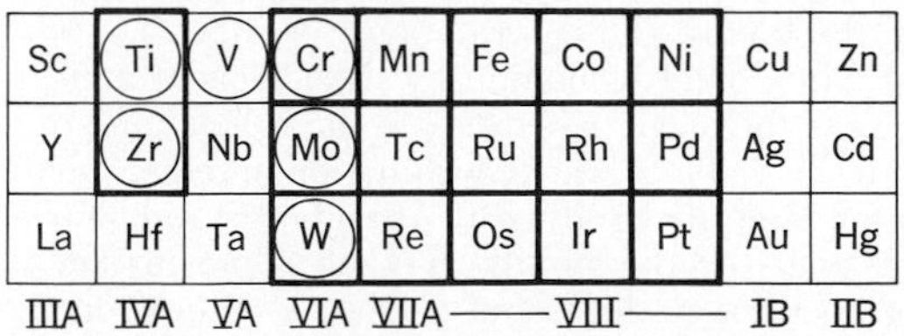

Sc Ti V Cr Mn Fe Co Ni Cu Zn
Y Zr Nb Mo Tc Ru Rh Pd Ag Cd
La Hf Ta W Re Os Ir Pt Au Hg
IIIA IVA VA VIA VIIA ——VIII—— IB IIB

Fig. 1. Transition metals which form coordination compounds with N_2 as ligand are mostly members of groups VI-VIII (boldly outlined squares). Coordination compounds which reduce N_2 to NH_3 in approximately stoichiometric yields are derived from the metals in the left half of the transition series (circles).

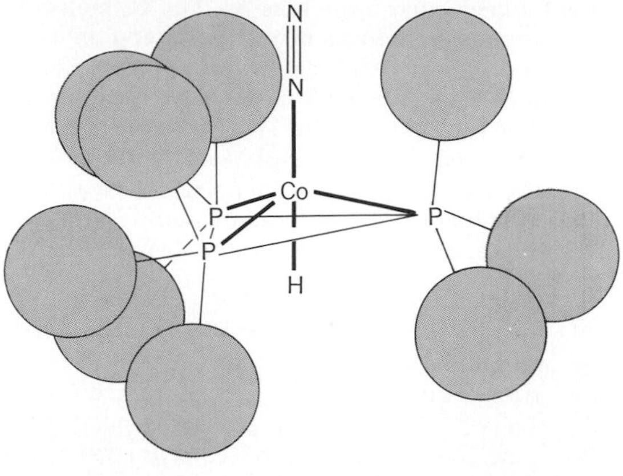

Fig. 2. Structure of a coordination compound with N_2 (circles represent phenyl groups).

The question arises as to the kind of bonding force that is able to attract the neutral, dipole-free, and not easily polarizable N_2 molecule to its position in the coordination sphere of the metal. The usual kind of dative bond between the electron pair on one of the N atoms and an acceptor orbital of the metal (Fig. 3) is likely to be very weak, just as the basicity of N_2 is unmeasureably low. A second and probably more important contribution to the stability of nitrogen complexes is thought to arise from the opposite process—from back-donation of loosely held metal electrons into an acceptor orbital of the N_2 molecule. Such a shift of electron density into the normally unoccupied, antibonding π^* orbitals of the ligand is considered to be an important stabilizing factor in many coordination compounds with unsaturated ligands. While strengthening the metal-ligand bond, the flow of electrons from the metal into these antibonding ligand orbitals would necessarily cause a weakening of the bond between the two N atoms, and this is exactly what is observed. Without exception, the stretching vibration of coordinated nitrogen occurs at a substantially lower frequency than that of free, gaseous nitrogen. Moreover, the known prerequisites for an efficient back-donation explain at least some of the peculiarities of nitrogen coordination. A large number of *d* electrons, usually six or more, has to be available on the metal; in addition, the oxidation state of the metal has to be low so that these electrons are easily released toward the ligand. Accordingly, it is understandable that group VIII metals with their large number of *d* electrons dominate among the known N_2 complexes.

In all the respects discussed above, the N_2 ligand exhibits a strong resemblance to the well-known ligand molecule carbon monoxide, CO, which accommodates the same number of electrons in orbitals of the same type as N_2. Although N_2 by virtue of its symmetry is a much weaker coordinating agent than the lopsided CO molecule, there is no doubt that these two molecules can under certain conditions substitute for each other as ligands. The weaker ligand N_2 appears to be more selective, however. Other ligand partners in the complex, while having to accommodate enough of the metal's excess electrons to stabilize a low oxidation state, should not compete too strongly with N_2 for back-donation of the metal electrons. Phosphines, which are known to draw less heavily on metal *d* electrons than other π acceptors, seem to strike this delicate balance right. That this is not the sole possibility to obtain an appropriate electron distribution, however, is demonstrated by the stability of the complex $[Ru(N_2)(NH_3)_5]^{2+}$, where five NH_3 ligands "condition" the Ru(II) center.

A number of dinitrogen complexes with entirely different structures have been isolated or characterized in solutions and solid matrices. These complexes appear to have their N_2 ligand molecule bound to the metal in an "edge-on" fashion (similar to that in Fig. 4), with both of its N atoms at the same bonding distance from the central metal atom. The general conditions for the occurrence of this type of N_2 complex are not clear at present.

Chemical reactions. Even in the most favorable cases the binding of the dinitrogen molecule to the metal is fairly labile; all the compounds lose their nitrogen on mild heating. Some of the nitrogen complexes are only metastable to loss of dinitrogen even at room temperature; accordingly, they cannot be obtained by direct uptake of gaseous nitrogen. In the synthesis of these metastable complexes, hydrazine or azide compounds serve as a source of nitrogen molecules within the coordination sphere of the metal. Addition of other coordinating agents to the nitrogen complexes usually results in a displacement of N_2 from the metal. The cobalt compound in Fig. 2 exchanges its N_2 ligand quite reversibly for other ligand molecules, such as NH_3 and $H_2C{=}CH_2$. Whereas these ligands are easily displaced again by an excess of N_2, an irreversible exchange occurs with carbon monoxide. The bulky organic groups on the phosphine ligands are likely to interfere with the approach to the metal of all but the slimmest ligands and thereby help the "thin" dinitrogen molecule to maintain or regain its position on the metal in competition with most other ligands.

An interesting question concerns the reactivity of the coordinated nitrogen molecule. Is the weakening of the bond in $N{\equiv}N$, which accompanies coordination and back-donation, sufficient to render the otherwise inert nitrogen molecule susceptible to attack and cleavage? Much systematic research has yet to be done if scientists are to understand the perplexing chemical reaction paths involved in nitrogen fixation.

The conversion of inert, molecular nitrogen to ammonium salts, which can then be utilized by plants in their production of proteins and other indispensable nutrients, is one of the limiting factors of life on Earth. Nature has put the burden of this life-sustaining process on the few strains of microorganisms which possess an enzyme capable of reacting with nitrogen. The enzyme, nitrogenase, actually consists of two protein particles, both of which contain iron in a peculiar form associated with sulfide; one of the enzyme particles also contains molybdenum. There is some support for the hypothesis that this latter particle forms a complex with molecular nitrogen, while the other molybdenum-free enzyme particle serves as a sort of energizer, accepting reducing electron equivalents from ferredoxin and making them available for the N_2- reduction process at a substantially increased negative reduction potential at the expense of adenosinetriphosphate hydrolysis. As for the chemical pathways by which the enzyme entrains its N_2 ligand into its smooth reduction to two NH_3 molecules, there is a still-open debate about the possible occurrence of enzyme-bound intermediates such as N_2H_2 (diimide) and N_2H_4 (hydrazine).

Hardly any better understood is the process by which humans supplement natural nitrogen fixation—the production of fertilizer ammonia from N_2 and H_2 by a high-temperature reaction on metal catalysts. Bonding of nitrogen to the surface of some metals, such as nickel, is spectrally reminiscent of the complexes discussed above and probably involves the same type of end-on coordination of N_2 to surface atoms of the metal. In many other cases, however, there seem to be different types of surface coordination, perhaps with the $N{\equiv}N$ molecule parallel to the surface, which then can probably rearrange directly to two surface-bound nitride particles (Fig. 4). While many metals form this nitride type of surface compounds with nitrogen, the distinction of catalytically active metals is their tendency to detach the nitride

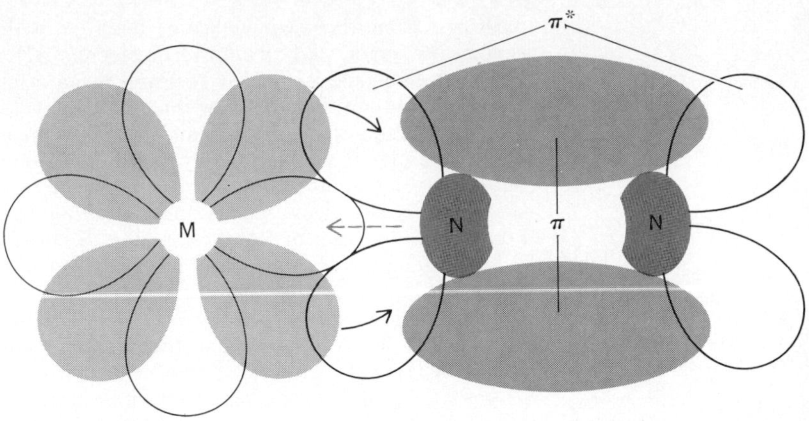

Fig. 3. Factors contributing to the stability of N_2 coordination compounds: dative bond from an electron pair on one of the N atoms to a vacant metal orbital (broken arrow) and the opposite process of back-donation of loosely held metal electrons to the π^* acceptor orbital of N_2 (solid arrows).

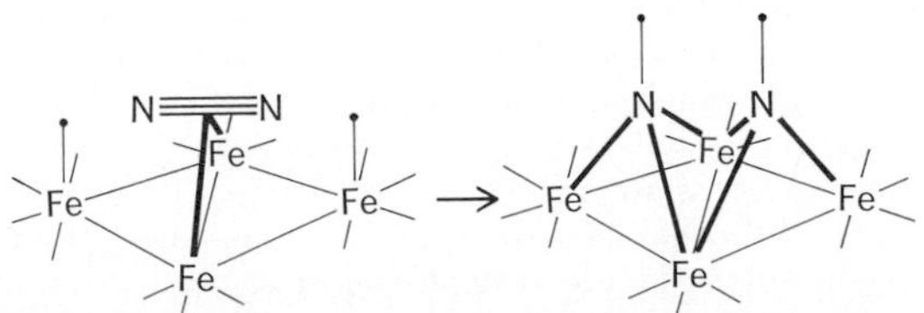

Fig. 4. Conceivable reaction path for the conversion of a surface-bound N_2 molecule to two surface nitride particles. Surface hydrogen may be involved in this reaction (indicated by black dots).

particles from the surface again, together with surface hydrogen particles, to form gaseous NH_3. The metal surface is thereby restored in its original form and is ready to renew the catalytic cycle by coordination of another N_2 molecule. Because of the complexity of these catalyst systems, some important questions have not been completely solved. It is not clear, for instance, whether the breakdown reaction of the coordinated N_2 molecule is just a rearrangement of bonds and electrons between the metal and nitrogen atoms, or if the participation of surface hydrides assists in this breakdown step, which might then involve adsorbed N_2H_2 or N_2H_4 intermediates.

A related problem is encountered with a series of reactions in which an N_2 ligand bound to some metal compound in solution is reduced to NH_3 or N_2H_4. These reactions involve mostly transition metals that are located in the left half of the transition series (Fig. 1). N_2 reductions of this kind have been known since about 1965 to occur in the presence of strong reducing agents. Subsequently some reactions have been found where a mere transfer of protons to the metal-bound (and hence presumably negatively polarized) N_2 molecule of an otherwise stable dinitrogen complex brings about reduction of this molecule to N_2H_4 or NH_3, while leaving the metal behind in a correspondingly increased oxidation state. In view of their obvious relation to enzyme and surface catalysis, it would be highly interesting to establish the relevant intermediates of these reactions—for example, the bonding geometry of the N_2 ligand molecules involved—and the general conditions for the occurrence of N_2 reduction reactions of this kind. The possibility that reactive intermediates such as N_2H_2 or N_2H_4 derived from catalytic reactions of this type might be trapped into useful synthetic reactions is presently under exploration in many laboratories; its realization would be a step forward in the direction of making the vast supply of atmospheric nitrogen more efficiently available for utilization. *See* COMPLEX COMPOUNDS; MOLECULAR ORBITAL THEORY.

[HANS BRINTZINGER]

Bibliography: G. Henrici-Olive and S. Olive, *Coordination and Catalysis*, 1977; M. M. Khan and A. E. Martell, *Homogeneous Catalysis by Metal Complexes*, 2 vols., 1974.

Nitrogen oxides

Chemical compounds of nitrogen and oxygen. Nitrogen and oxygen do not combine when mixed directly (as in air), but they do combine during chemical reactions of compounds containing them. A number of nitrogen oxides can be isolated which differ from one another in the numbers of nitrogen and oxygen atoms present in each molecule.

Table 1 gives data for the five nitrogen oxides which are well established. The structures of these molecules and one laboratory method for the preparation of each oxide are given in Table 2. These structures show only the geometry of the molecules. In most cases the N and O atoms are united by complex (double or triple) bonds.

The existence of three higher oxides has been postulated. They are nitrogen trioxide, NO_3, from reaction of ozone with dinitrogen tetroxide or pentoxide; dinitrogen hexoxide, N_2O_6, from reaction of fluorine with nitric acid; and an oxide NO_4 as an intermediate in the O^{18} isotope exchange between dinitrogen pentoxide and oxygen gases. The identity and properties of these three oxides are not fully established.

Table 1. Oxides of nitrogen and their properties

Name	Stoichiometric formula	Melting point, °C	Boiling point, °C
Nitrous oxide (dinitrogen monoxide)	N_2O	−90.8	−88.5
Nitric oxide (nitrogen monoxide)	NO	−163.6	−151.7
Dinitrogen trioxide	N_2O_3	−103	3.5
Dinitrogen tetroxide (⇌ nitrogen dioxide)	N_2O_4 (⇌ NO_2)	−11.2	21.2
Dinitrogen pentoxide	N_2O_5	41	

Nitrous oxide and nitric oxide. When inhaled, nitrous oxide has anesthetic effects; in small amounts it produces mild hysteria and hence is sometimes called laughing gas. It is colorless, is the least reactive of the oxides, and dissolves in water without chemical reaction. Decomposition into nitrogen and oxygen occurs at an appreciable rate above 560°C.

The equilibrium in reaction (1) lies entirely to

$$N_2 + O_2 \rightleftharpoons 2NO \qquad (1)$$

the left at low temperatures. Some nitric oxide is formed in an electric arc, as in the technical production of nitric acid.

With oxygen or air, nitric oxide is rapidly converted to nitrogen dioxide. Nitric oxide is colorless and is soluble in water without reaction. It is one of the few "odd" molecules which contain an odd number of electrons. Other such molecules (for example, nitrogen dioxide) readily form double

Table 2. Oxides of nitrogen

Formula	Structure	Preparation
N_2O	N—N—O	Heat ammonium nitrate
NO	N—O	Reduce nitric acid with copper
N_2O_3	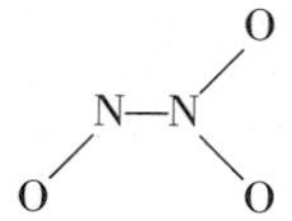	Condense gaseous mixture of NO and NO_2
NO_2	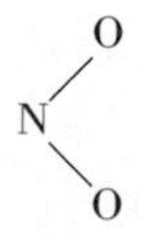	Heat lead nitrate
N_2O_4	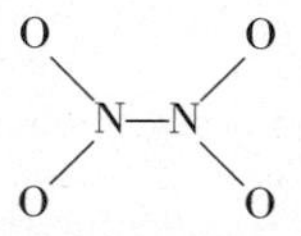	Heat lead nitrate
N_2O_5 Gas	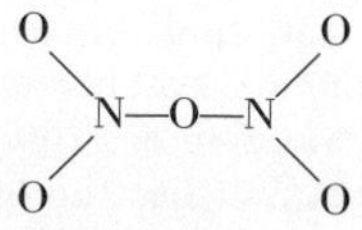	Treat N_2O_4 with ozone
Solid	$NO_2^+ \cdot NO_3^-$	

molecules, but nitric oxide is exceptional. The gas is monomeric, although dimerization occurs in the liquid, and solid nitric oxide (which is blue) is almost entirely in the form of N_2O_2 molecules. As an odd molecule, it has the ability to lose or gain one electron, thus giving the electrically charged ions NO^+ and NO^-. The important nitrosyl compounds contain these ions.

Trioxide. Dinitrogen trioxide exists pure only in the solid state. The dissociation in reaction (2) oc-

$$N_2O_3 \rightleftharpoons NO + NO_2 \qquad (2)$$

curs partially in the blue liquid and almost entirely in the vapor state at room temperature. It is the anhydride of nitrous acid; when the oxide (or an equimolecular mixture of NO and NO_2 gases) is dissolved in an alkaline solution, nitrite ion is produced.

Dioxide and tetroxide. The position of the equilibrium between nitrogen dioxide and dinitrogen tetroxide, reaction (3), depends upon temperature

$$N_2O_4 \rightleftharpoons 2NO_2 \qquad (3)$$

and physical state. The dioxide is a red-brown poisonous gas; the tetroxide is colorless. The colorless solid is entirely in the tetroxide state. In the liquid and gaseous states the tetroxide always contains some dioxide. Thus the liquid tetroxide is brown, although it contains less than 0.1% nitrogen dioxide. The color of the gas becomes more intense with rising temperature; at 100°C the tetroxide is 90% dissociated into dioxide. At temperatures above 600°C further decomposition of nitrogen dioxide into nitric oxide occurs, as shown in reaction (4). Dinitrogen tetroxide reacts readily

$$NO_2 \rightleftharpoons NO + \tfrac{1}{2}O_2 \qquad (4)$$

with water to give an equimolecular mixture of nitrous and nitric acids. As temperature is raised, the nitrous acid decomposes to nitric acid and nitric oxide. These reactions are important in the technical production of nitric acid by catalytic oxidation of ammonia. Dinitrogen tetroxide is an oxidizing agent comparable in strength to bromine, and is employed as such in the lead-chamber process for sulfuric acid. In organic chemistry the tetroxide finds use as a special oxidizing agent (for example, in the production of sulfoxides and phosphine oxides) and as a nitrating agent.

The tetroxide forms molecular addition compounds with many simple organic solvents, for example, esters, ethers, ketones, and nitriles.

Liquid dinitrogen tetroxide, alone or mixed with organic solvents, undergoes self-ionization, as in reaction (5). This is to be compared with the aque-

$$N_2O_4 \rightleftharpoons NO^+ + NO_3^- \qquad (5)$$

ous system shown in reaction (6). For example, liq-

$$H_2O \rightleftharpoons H^+ + OH^- \qquad (6)$$

uid dinitrogen tetroxide attacks some metals (alkali and alkaline-earth metals, zinc, cadmium, and mercury) to produce metal nitrate and evolve nitric oxide. A scheme of reactions has been developed using the liquid tetroxide as a reaction medium, with nitrosyl compounds as acids and nitrates as bases. This medium is therefore valuable for the preparation of anhydrous metal nitrates and nitrato-coordination complexes.

Pentoxide. The ionic nitronium nitrate structure $NO_2^+ \cdot NO_3^-$ found for N_2O_5 in the solid state accounts for its anomalously high melting point. In solution in sulfuric, nitric, or phosphoric acids the oxide has the same ionic structure. Solid dinitrogen pentoxide readily volatilizes, and the molecular type of structure found in the gaseous state is observed also in solutions of the oxide in low dielectric solvents such as carbon tetrachloride and chloroform. Sodium metal reacts with the liquid oxide, liberating nitrogen dioxide and forming sodium nitrate. Gaseous dinitrogen pentoxide decomposes readily, as in reaction (7), and is a strong

$$N_2O_5 \rightleftharpoons 2NO_2 + O_2 \qquad (7)$$

oxidizing agent. With water it is converted to nitric acid. *See* NITROGEN; OXYGEN.

[CYRIL C. ADDISON]

Nitroparaffin

One of a series of aliphatic compounds with relatively high boiling points and high molecular polarity. Such compounds are used as solvents for cellulose derivatives, waxes, and many substances of high molecular weight. Their chief use is in the synthesis of other organic compounds.

Direct nitration of propane at 400°C with nitric acid under 150 lb pressure furnishes the commercially available nitroparaffins: nitromethane, bp, 101°C; nitroethane, bp, 114°C; 1-nitropropane, bp, 131°C; and 2-nitropropane, bp, 120°C.

Primary (RCH_2NO_2) and secondary (RR'-$CHNO_2$) nitroparaffins are stronger acids than water and react with strong bases in aqueous solution to form water-soluble salts. This is shown by reaction (1). The presence of the aci (acid) form can

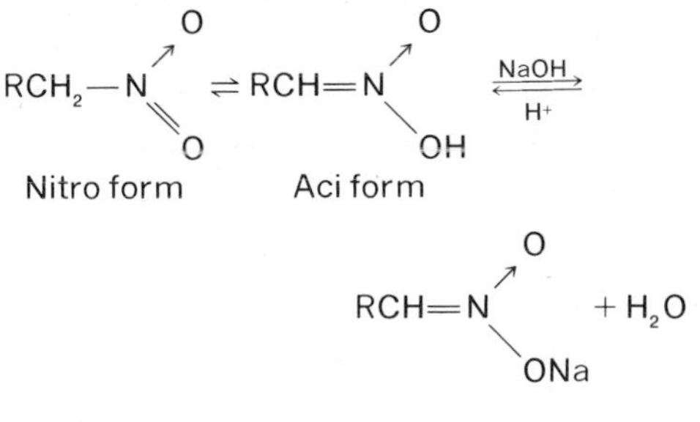

(1)

account for the relatively low pH which is observed in aqueous suspensions of the fairly insoluble nitroparaffins. Tertiary nitroparaffins, R_3CNO_2, which lack an α-hydrogen, do not display this behavior. Nitromethane reacts with strong bases to give salts of methazonic acid as shown in formula (2). *See* TAUTOMERISM.

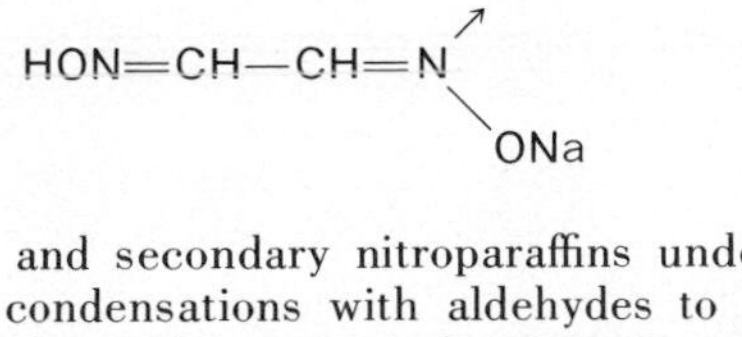

(2)

Primary and secondary nitroparaffins undergo aldol-type condensations with aldehydes to give nitro alcohols, as in reaction (3). With formalde-

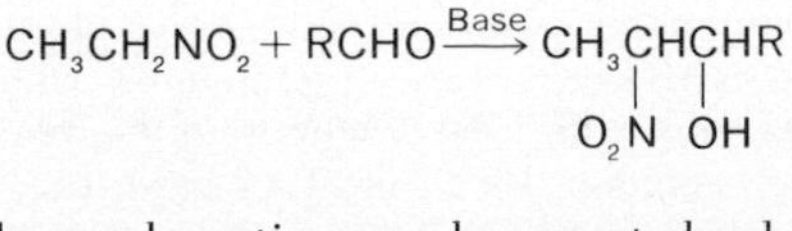

(3)

hyde, the condensation may be repeated as long as

an α-hydrogen remains to yield polymethylol compounds, as in reaction (4). These polymethylol com-

$$CH_3NO_2 + HCHO \rightarrow HOCH_2CH_2NO_2 \xrightarrow{HCHO} (HOCH_2)_2CHNO_2 \xrightarrow{HCHO} (HOCH_2)_3CNO_2 \quad (4)$$

pounds can be reduced to the industrially important amino alcohols. *See* ALKANOLAMINE.

Acid hydrolysis of nitroparaffins yields the corresponding carboxylic acid and salts of hydroxylamine, as in reaction (5). This reaction provides a ready commercial route to hydroxylamine.

$$CH_3CH_2NO_2 + HCl + H_2O \rightarrow CH_3COOH + HONH_3Cl \quad (5)$$

Halogenation of primary and secondary nitroparaffins proceeds readily in alkaline solution. This reaction is analogous to the halogenation of aldehydes and ketones. Only the α-hydrogens are replaced. The 1-chloro- and 1,1-dichloronitroparaffins find use as orchard insecticides.

Primary and secondary nitroparaffins add to α,β-unsaturated systems, for example, base-catalyzed Michael condensations, reaction (6).

$$CH_3CH_2NO_2 + CH_2{=}CHCOCH_3 \rightarrow CH_3\underset{\displaystyle NO_2}{\underset{|}{C}}HCH_2CH_2COCH_3 \quad (6)$$

Polynitroparaffins such as 2,2-dinitropropane increase the octane number in diesel fuels and also reduce the smoke nuisance in diesel motors. *See* NITRATION. [LEALLYN B. CLAPP]

Nobelium

A chemical element, No, atomic number 102. Nobelium, in the actinium series, is a synthetic element produced in the laboratory. It decays by emitting an α-particle, that is, a doubly charged helium ion. Only atomic quantities of the element have been produced to date.

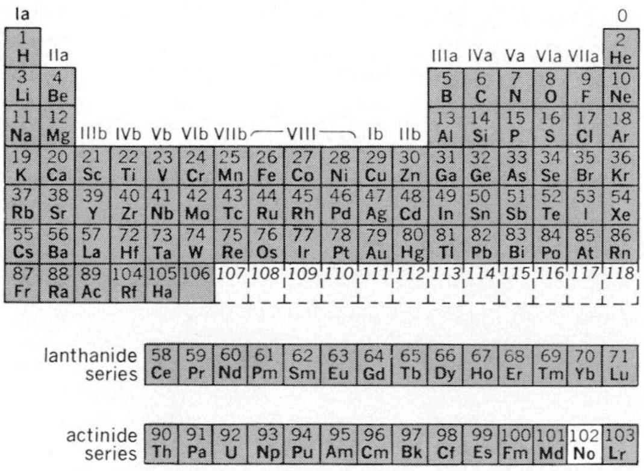

The element is the tenth element heavier than uranium to be produced synthetically. It is the thirteenth member of the actinide series of elements, a rare-earth-like series of elements, which was completed by the 1961 discovery of the element of atomic number 103. *See* LAWRENCIUM; RARE-EARTH ELEMENTS.

The production of element 102 was first reported by an international team of scientists from Great Britain, Sweden, and the United States in 1957. In these experiments the element was made by bombarding isotopes of curium, element 96, with carbon ions accelerated to energies of 60–85 MeV in the Nobel Institute heavy-ion cyclotron. The curium atoms capture some of the high-energy carbon ions to form highly excited atoms of nobelium. These excited atoms lose their excess energy by evaporating neutrons, thus yielding the element of atomic number 102.

Atoms having the same atomic number, and hence identical chemical behavior, but differing in atomic weight are isotopes of a given element. *See* ISOTOPE.

Since the isotopes of this element disintegrate fairly rapidly after their formation, that is, they have a short half-life, special techniques were necessary to characterize the atoms of element 102 shortly after they were formed. A recoil technique developed by research scientists at the Lawrence Radiation Laboratory was employed by all groups working on element 102.

When a high-energy particle, such as a carbon ion, is captured by a curium atom, the resulting compound nucleus is caused to move in a forward direction by the original momentum of the carbon ion with enough velocity to separate it from the stationary curium atoms. Under appropriate conditions these recoil atoms of element 102 can be caught on adjacent catcher foils. By employing such a catcher technique it is possible to separate the atoms of element 102 formed by the nuclear reactions from the curium target and to identify them rapidly.

The international group of scientists reported producing an 8.5 MeV α-emitter with a half-life of about 10 min. Approximately a year later another group of scientists working at the Lawrence Radiation Laboratory, University of California, and using a heavy-ion linear accelerator tried to duplicate the results of the first group, but they were unable to produce the same product. They also irradiated curium with high-energy carbon ions, but this group claimed they had produced an isotope of element 102 with a 3-sec half-life. Initially it was not possible to identify these short-lived atoms directly. However, the process of α-decay gave enough energy to the resultant atoms of element 100 that they in turn recoiled off the catcher foil and were caught on another foil. By varying the collection time of the α-recoil products, it was possible to establish that the parent atoms were decaying with a 3-sec half-life, and chemical identification of the recoil atoms as element 100 indicated the parent atoms were element 102. In later experiments an 8.3 MeV α-activity was reported to be associated with the isotope having the 3-sec half-life. This isotope of element 102 was assigned to mass 254, based on its nuclear properties and the nuclear reaction used to produce it. This group also reported producing another isotope of mass 255 by irradiating californium with boron ions.

A third group of scientists working at the Joint Institute for Nuclear Research, Dubna, Soviet Union, also reported the production of isotopes of element 102. In 1963 they positively identified an isotope of mass 256 made by irradiating uranium atoms with high-energy neon ions using a heavy-ion cyclotron. Subsequently, this group produced the mass-254 isotope of element 102 but found that it emitted an 8.1 MeV α-particle and had an approximately 1-min half-life rather than the 3-sec half-life reported earlier by the California group or the approximately 10-min half-life reported by the first group of scientists. They also reported the preparation of the isotopes of mass 252, 253, and 255.

A series of experiments at the Lawrence Radiation Laboratory confirmed the Soviet results on the mass-254 isotope and, in addition, succeeded in characterizing several new isotopes (mass 251 and 257) of element 102. The California group maintains that their earlier experiment first identified element 102 because they chemically characterized the α-decay recoil atoms as element 100.

Chemical studies, using the mass-255 isotope of element 102, indicate that the divalent state is the most stable species in solution. In contrast to this behavior all the lower actinides (elements 89–101) have the trivalent or a higher valence as the most stable state in solution.

The international team of scientists recommended the name nobelium for element 102. This suggestion has been accepted by the International Union of Pure and Applied Chemistry and by the California group. *See* ACTINIDE ELEMENTS; CALIFORNIUM; CURIUM; FERMIUM; NUCLEAR CHEMISTRY; NUCLEAR FISSION; PERIODIC TABLE; RADIOACTIVITY; TRANSURANIUM ELEMENTS.

[PAUL R. FIELDS]

Nonelectrolyte

A chemical compound which does not conduct electricity either when fused or when dissolved in a solvent such as water. The bonding in nonelectrolytes is covalent. Most organic compounds, with the exception of the acids and amines, are nonelectrolytes, as are many inorganic compounds such as the halides of the nonmetals. However, many inorganic compounds with covalent bonds react chemically with water to produce electrolytic solutions. *See* CHEMICAL BONDING; ELECTROLYTE; SOLUTION.

[THOMAS C. WADDINGTON]

Nonmetal

The elements are conveniently, but arbitrarily, divided into metals and nonmetals. The nonmetals do not conduct electricity readily, are not ductile, do not have a complex refractive index, and in general have high ionization potentials.

The nonmetals vary widely in physical properties. Hydrogen is a colorless permanent gas; bromine is a dark-red, volatile liquid; and carbon, as diamond, is a solid of great hardness and high refractive index. If the periodic table is divided diagonally from upper left to lower right, all the nonmetals are on the right-hand side of the diagonal. Examples of elements which do not fit neatly into this useful but arbitrary classification are tin, which exists in two allotropic modifications, one definitely metallic and the other with many properties of a nonmetal, and tellurium and antimony. Such elements are called metalloids. *See* METALLOID; PERIODIC TABLE.

[THOMAS C. WADDINGTON]

Nonstoichiometric compounds

Chemical compounds in which the relative number of atoms is not expressible as the ratio of small whole numbers, hence compounds for which the subscripts in the chemical formula are not rational (for example, $Cu_{1.987}S$). Sometimes they are called berthollide compounds to distinguish them from daltonides, in which the ratio of atoms is generally simple. Nonstoichiometry is a property of the solid state and arises because a fraction of the atoms of a given kind may be (1) missing from the regular structure (for example, $Fe_{1-\delta}O$), (2) present in excess over the requirements of the structure (for example, $Zn_{1+\delta}O$), or (3) substituted by atoms of another kind (for example, $Bi_2Te_{3\pm\delta}$). The resulting materials are generally of variable composition, intensely colored, metallic or semiconducting, and different in chemical reactivity from the parent stoichiometric compounds from which they are derived. Nonstoichiometry is best known in the binary compounds of the transition elements, particularly the hydrides, oxides, chalcogenides, pnictides, carbides, and borides. It is also well represented in the so-called insertion or intercalation compounds, in which a metallic element or neutral molecule has been inserted in a stoichiometric host. Nonstoichiometric compounds are important in some solid-state devices (such as rectifiers, thermoelectric generators, and photodetectors) and are probably formed as chemical intermediates in many reactions involving solids (for example, heterogeneous catalysis and metal corrosion).

Generality of the phenomenon. Simple stoichiometry, in which a compound A_xB_y is characterized by small integral values for the composition parameters x and y, is strictly speaking a required characteristic only of molecules in the gaseous state. In the condensed state (solid or liquid), unless the simple molecular aggregate of the gas phase clearly retains its identity, x and y no longer need be small integers. Indeed, they may be as large as the number of atoms in a crystal, for example, 10^{22}. Given such large numbers, small but detectable departures from a simple $x:y$ ratio can be achieved, even in cases such as NaCl, without seriously affecting the energetics of the system, provided a mechanism exists for keeping the compound electrically neutral. In the case of sodium chloride, there is no a priori requirement that the number of Na atoms and Cl atoms be exactly identical. As a matter of fact, it is relatively simple to heat a colorless stoichiometric crystal of composition NaCl in sodium vapor and convert it to yellow-brown $Na_{1.001}Cl$. Similarly, heating it in chlorine gas can produce $NaCl_{1+\delta}$. Both of these deviations from stoichiometry result from Schottky and Frenkel defects, such as those which are native in any real crystal. *See* SOLID-STATE CHEMISTRY.

Take-up of excess Na can be achieved, as shown in Fig. 1, by accommodating Na^+ at a cation vacancy and e^- at an anion vacancy. Excess chlorine is similarly accommodated, as shown in Fig. 2, by incorporating the added Cl atom at an anion vacancy. The electron absence, or hole, which distinguishes the Cl^0 from the normal Cl^- ions of the structure, can jump around to any of the other atoms on the anion sublattice. The amount of excess sodium or the amount of excess chlorine that

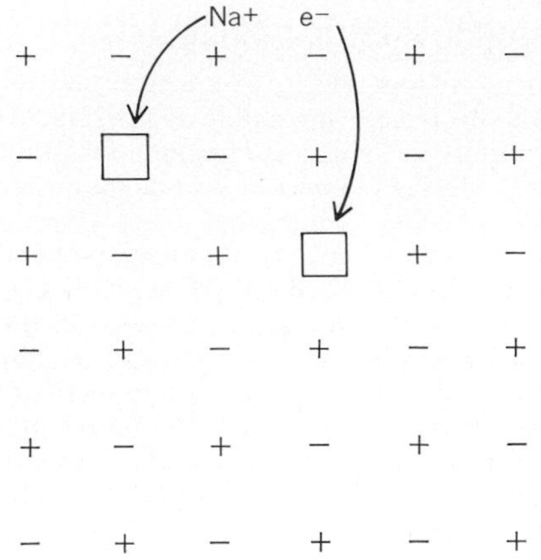

Fig. 1. Formation of $Na_{1+\delta}Cl$ by incorporation of a neutral sodium atom as Na^+ at a vacant cation site and an e^- at a vacant anion site. The +'s represent sodium ions, and the −'s represent chloride ions.

can be accommodated, hence the maximum deviation from stoichiometry, is related to the number of defects in the stoichiometric crystal (intrinsic disorder). Since this intrinsic disorder is an increasing function of temperature, the deviation from stoichiometry that can be achieved is strongly dependent on the preparation conditions, specifically, the temperature at which the preparation is made and the pressure of the constituent elements. Furthermore, since nonstoichiometric compounds are in general prepared by quenching from relatively high temperatures (where atom mobility may be quite high) to room temperature (where diffusion motion is practically negligible), the materials will not usually be in thermodynamic equilibrium with their component elements at room temperature. If the quenching is not efficient, the room-temperature composition will not correspond to an equilibrium state at the preparation temperature, either.

Thermodynamic considerations. A thermodynamic description for formation of a nonstoichiometric compound at a fixed temperature is illustrated in Fig. 3, which shows possible free-energy

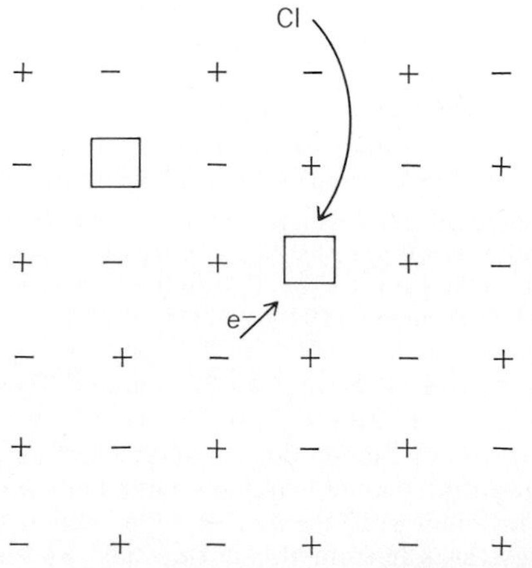

Fig. 2. Formation of $NaCl_{1+\delta}$ by incorporation of a neutral chlorine atom at a vacant anion site and migration of an electron from some chloride ion in the normal structure of the crystal.

relations for the binary system composed of elements A and B forming nonstoichiometric compound $AB_{1+\epsilon}$. The particular case shown corresponds to one in which the compound is completely dissociated in the liquid; that is, there are no specific interactions between A and B in the liquid. At the melting-point equilibrium the chemical potential of each component must be identical in the liquid and solid states. Furthermore, to merit the designation "compound" (as distinct from "solid solution" or "peritectic"), the chemical composition of the solid and liquid phases coexisting at the melting point must be identical. As shown in Fig. 3, this equality is fulfilled not at the stoichiometric composition AB but at $AB_{1+\epsilon}$. This comes about as follows: If in the liquid A and B atoms are completely independent of each other, then starting from AB replacement of some A for B or B for A will have no influence on the enthalpy H and only a slight influence in decreasing the entropy S. The free energy $G=H-TS$ of the liquid will show a rather broad minimum as the composition is varied. In the solid the specific interactions are larger and more important. Replacement of A for B

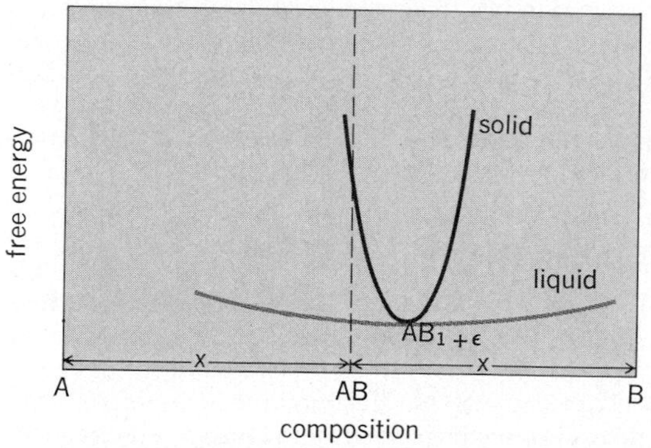

Fig. 3. Free energy versus composition for a two-element system forming the nonstoichiometric compound $AB_{1+\delta}$. Temperature and pressure are constant.

or B for A sharply raises the enthalpy, and the result is a deep minimum in the free-energy curve of the solid. Furthermore, the curve for the solid is generally not symmetric about the ideal composition AB. In the case shown, replacement of A for B raises the enthalpy more than does the replacement of B for A. As a result, the minimum lies in the B-rich region, and the equilibrium compound will be $AB_{1+\epsilon}$. (A similar but parallel argument can be drawn for the case where AB initially has a finite concentration of vacant lattice sites due to Schottky or Frenkel disorder.) *See* CHEMICAL THERMODYNAMICS.

Phase relations. Figure 4 shows possible phase relations involving formation of the compound $AB_{1+\delta}$ from the melt. The solid $AB_{1+\epsilon}$ will crystallize on cooling a melt containing A and B in the ratio $1:1+\epsilon$. On the other hand, if the melt contains, for example, stoichiometric amounts of A and B, the solidification temperature will be less than T_{max}, and the solid first separating will not be $AB_{1+\epsilon}$ but $AB_{1+\epsilon'}(\epsilon' < \epsilon)$. As crystallization proceeds, the liquid composition would change along L_1 and ϵ' would get progressively smaller, eventually even becoming negative. In practice, to keep

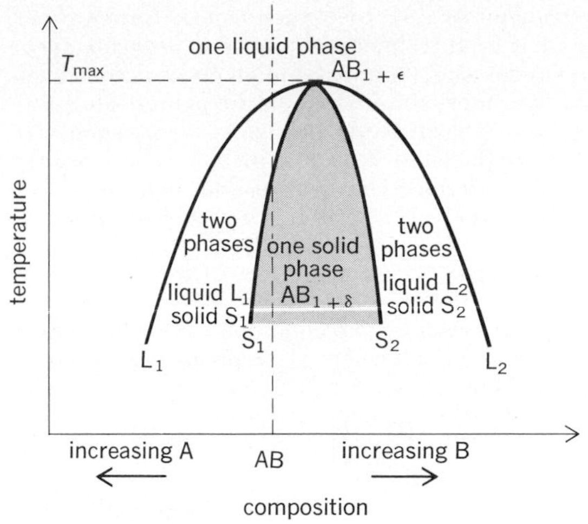

Fig. 4. Temperature-composition plot showing phase relations for formation of the nonstoichiometric compound $AB_{1+\epsilon}$. The tinted region corresponds to a single solid phase of variable composition.

crystallizing a solid of fixed composition other than $AB_{1+\epsilon}$, means would have to be provided to keep the liquid composition constant. *See* PHASE EQUILIBRIUM.

For simple ionic solids, particularly where the ionic polarizabilities are low and a multiplicity of oxidation states is not possible, the range of nonstoichiometry is generally small. Thus alkali or alkaline earth fluorides are notoriously hard to make as nonstoichiometric compounds. Covalent solids, on the other hand, are generally easier to deviate from stoichiometry, especially when the constituent elements are similar to each other. Bismuth telluride, for example, can be made either bismuth-rich or tellurium-rich by simply putting Bi atoms on Te sites or vice versa. (The unexpected *p*-type semiconductivity of $Bi_2Te_{3-\delta}$ is attributable to the fact that the antistructure defect, Bi on a Te site, has placed a five-valent atom on a site calling for six valence electrons, thereby creating a hole.) Transition elements are particularly good at forming nonstoichiometric compounds, partly because of the capability of the ions to exist in several oxidation states and partly because there is an increased possibility of using *d* orbitals either for covalent bonding or for delocalizing electronic charge through *d*-orbital overlap. Still, many of the broad, reportedly homogeneous regions in which transition-metal-compound composition appears to be continuously variable actually contain within themselves a series of closely related compounds. As an example, the system TiO_x $(1.7 < x < 1.9)$ actually includes at least seven discrete oxides, each of which can be described by the common formula Ti_nO_{2n-1}, with *n* ranging from 4 to 10. Similar series have been recognized in W_nO_{3n-1}, W_nO_{3n-2}, and MoO_{3n-1}.

Even single-crystal compositions that do not fit into such homologous series can frequently be explained as arising from the presence of clusters of defects. X-ray, neutron, and electron diffraction techniques have been supplemented by the technique of lattice imaging in transmission electron microscopy to gain information about the structural nature of defect clusters. One such defect cluster, which appears particularly often, is the crystallographic shear (CS) plane, also known as the Wadsley defect. It arises when, along a plane in the crystal, there is a change in the type of polyhedral grouping, as, for example, from corner sharing of octahedra to edge sharing. Each CS plane introduces a departure from the stoichiometry of the parent structure; ordered sets of parallel CS planes at fixed repetitive spacings can lead to block or column structures as found in different homologous series. In some cases, CS planes parallel to two different crystallographic directions, for example, {120} and {130}, may coexist. Given this kind of possible complexity in structure, it is clear that the apparent ranges of homogeneity of phases may well depend upon the wavelength of the radiation used to examine them.

Classification. The simplest way to classify nonstoichiometric compounds is to consider which element is in excess and how this excess is brought about. A classification scheme largely based on this distinction but which also includes some examples of ternary systems is as follows.

Binary compounds:

I. Metal:nonmetal ratio greater than stoichiometric
  (*a*) Metal in excess, for example, $Zn_{1+\delta}O$
  (*b*) Missing nonmetal, for example, $UH_{3-\delta}$, $WO_{3-\delta}$

II. Metal:nonmetal ratio less than stoichiometric
  (*a*) Metal-deficient, for example, $Co_{1-\delta}O$
  (*b*) Nonmetal in excess, for example, $UO_{2+\delta}$

III. Deviations on both sides of stoichiometry, for example, $TiO_{1\pm\delta}$

Ternary compounds (insertion compounds):

IV. Oxide "bronzes," for example, $M_\delta WO_3$, $M_\delta V_2O_5$

V. Intercalation compounds, for example, $K_{1.5+\delta}MoO_3$, $Li_\delta TiS_2$

(Excluded from consideration are the recognized impurity materials, such as $Na_{1-2x}Ca_xCl$, which are best considered as conventional solid solutions wherein ions of one kind and perhaps vacancies have replaced an equivalent number of ions of another kind.) The specific compounds listed here were chosen to be illustrative of structures and properties commonly encountered in nonstoichiometry. They are discussed individually below.

Zinc oxide, $Zn_{1+\delta}O$. Rather small deviations from stoichiometry can be obtained by heating zinc oxide crystals in zinc vapor at temperatures of the order of 600–1200°C. The crystals become red, and their room-temperature conductivity is considerably enhanced over that of stoichiometric ZnO. The red color and the increased conductivity are attributed to interstitial zinc atoms. Since the ZnO structure is rather open, there is ample room for extra Zn atoms in the tunnels of the $P6_3mc$ structure. The activation energy for diffusion of the Zn is only 0.55 eV, supporting the belief the nonstoichiometry arises from interstitial zinc and not from oxygen vacancy. The conductivity of $Zn_{1+\delta}O$ is *n*-type corresponding to action of Zn as a donor. Hall measurements indicate only one free electron from each Zn atom, thus suggesting Zn^+ ions. Similar properties can be produced by heating ZnO in hydrogen gas, but the cause appears to be the addition of H atoms, probably as OH.

Uranium hydride, $UH_{3-\delta}$. Uranium trihydride does not deviate from stoichiometry to any measurable degree at room temperature but does so to a significant degree at high temperatures. For example, at 450°C the existence range is $UH_{2.98\text{-}3.00}$ and at 800°C, $UH_{0.9-3}$. The hydrogen deficiency comes about from hydrogen vacancies. The interaction energy between the vacancies is rather high (4.3 kcal/mole), as there is a great tendency for the vacancies to cluster and cause nucleation of the metal phase as the temperature is lowered.

Tungsten trioxide, $WO_{3-\delta}$. The tungsten-oxygen system shows a variety of materials corresponding to the general formula WO_x, but many of these appear to be discrete compounds, such as $WO_{2.90}$ (which is $W_{20}O_{58}$) and $WO_{2.72}$ (which is $W_{18}O_{49}$), or mixtures of compounds belonging to general series such as W_nO_{3n-2}. Still, it is possible to prepare single crystals of $WO_{3-\delta}$, where δ is continuously variable, provided that δ is very small. One way in which this can be done is to anneal a stoichiometric crystal at, for example, 1100°C for several days in a controlled oxygen pressure. The oxygen pressure can be set by using an appropriate flow ratio of argon and oxygen, or for very low values a mixture of carbon monoxide and carbon dioxide. Although stoichiometric WO_3 is pale yellow, removal of oxygen darkens the crystals first through green, then blue-green, and finally black. The conductivity is *n*-type. Removal of oxygen produces oxygen vacancies, each of which can act as an electron donor. (This can be seen by considering that removal of an oxide ion O^{2-} from WO_3 would disturb electric neutrality, so the two minus charges, or two electrons, would have to be put back into the lattice for each vacancy created.) At low oxygen defect, for example, $WO_{2.999}$, the oxygen vacancies are independent of each other and randomly distributed in the structure. As the defect increases, the oxygen vacancies begin to line up, coalesce, and produce shear planes where the octahedral WO_6 units share edges rather than corners. In $WO_{2.994}$, the average spacing between the CS planes, which are along {120}, would be about 280 A. However, they are observed to be quasi-ordered at spacings of 40–50 A. The oxygen defect in $WO_{3-\delta}$ has been measured with great precision by determining the reducing equivalence of a given sample, as by reaction with $Ag(SCN)_4^{3-}$ to form elemental Ag.

Cobaltous oxide, $Co_{1-\delta}O$. Stoichiometric CoO cannot be made. All preparations whether by dehydration of $Co(OH)_2$, decomposition of $CoCO_3$, or controlled oxidation of cobalt wire lead to a product that is deficient in cobalt. As shown in Fig. 5, the electrical neutrality is maintained by having the doubly positive charge of each missing Co^{2+} ion taken up by the existence of two Co^{3+} ions someplace in the structure. The resulting material is a good *p*-type conductor in which charge transport occurs by transfer of a "hole" from a Co^{3+} to a neighboring Co^{2+}. Since the conductivity depends on the concentration of Co^{3+}, the electric properties of $Co_{1-\delta}O$ are dependent on the oxygen pressure. (In fact, they have been used as a way of measuring oxygen pressures over the enormous range of 10^{-1} to 10^{-13} atm.) How this comes about can be seen from the following considerations: Oxidation of CoO by oxygen can be viewed as resulting from (1) incorporation of an oxygen atom at a surface site, (2) transfer of an electron from each of two interior Co^{2+} ions to the neutral oxygen to form a normal oxide ion, and (3) migration of a Co^{2+} to the surface and creation of a Co^{2+} vacancy in the interior. The net reaction can be written as (1), where $\square$ represents the additional cation

$$\tfrac{1}{2}O_2(g) + 2Co^{2+} \rightarrow O^{2-} + 2Co^{3+} + \square \qquad (1)$$

vacancy created. The condition for equilibrium is shown as Eq. (2) where P_{O_2} represents pressure of the oxygen gas.

$$\frac{[O^{2-}][Co^{3+}]^2[\square]}{P_{O_2}{}^{1/2}[Co^{2+}]^2} = K \qquad (2)$$

Since the concentrations of oxide ions $[O^{2-}]$ and of cobaltous ions $[Co^{2+}]$ are practically constant, they can be absorbed in K. The result is Eq. (3).

$$\frac{[Co^{3+}]^2[\square]}{P_{O_2}{}^{1/2}} = K' \qquad (3)$$

The chemical equation shows that one vacancy is created for every two cobaltic ions formed, so assuming there were negligible vacancies at the start, the concentration of vacancies in the final crystal is just half that of the cobaltic ions, or Eq. (4). Substituting and solving for $[Co^{3+}]$, one gets Eq. (5).

$$[\square] = \tfrac{1}{2}[Co^{3+}] \qquad (4)$$

$$[Co^{3+}] = (2K'P_{O_2}{}^{1/2})^{1/3} \text{ or } \sim P_{O_2}{}^{1/6} \qquad (5)$$

Since the conductivity is directly proportional to the concentration of Co^{3+} ions, it follows that it will show a 1/6th power dependence on oxygen pressure.

If the concentration of vacancies at the start is not negligible but on the contrary intrinsically large, then $[\square]$ is also a constant. In such a case, the conductivity would follow a 1/4th power de-

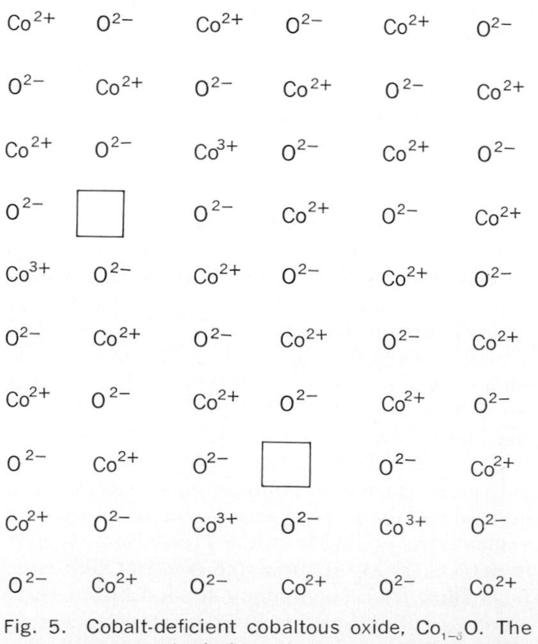

Fig. 5. Cobalt-deficient cobaltous oxide, $Co_{1-\delta}O$. The squares represent missing cobalt atoms. Note the presence of two Co^{3+} to compensate for each Co^{2+} missing.

pendence on oxygen pressure. In practice, plots of log conductivity versus log Po_2 show slopes between 1/4 and 1/6 depending on the biographical defects in the specimen.

Uranium dioxide, $UO_{2+\delta}$. The fluorite structure in which UO_2 crystallizes is capable of great defect concentration; still the nonstoichiometry range accessible $UO_{2.00\text{-}2.25}$ is exceptionally broad. Several independent lines of evidence suggest there are two kinds of defects in this range—oxygen interstitials in the first part, UO_{2+x}, and oxygen vacancies in the second part, U_4O_{9-y}. Precision lattice-constant determination gives two linear segments for the lattice parameter a, as illustrated by relationships (6). Similarly, a plot of the partial molal free

$$\begin{aligned} a &= 5.4705 - 0.094x \text{ for } 0 < x < 0.125 \\ a &= 5.4423 + 0.029y \text{ for } 0 < y < 0.31 \end{aligned} \qquad (6)$$

energy of oxygen in $UO_{2+\delta}$ as a function of δ is best fitted by two straight lines intersecting at $UO_{2.125}$. Finally, a determination of the entropy change ΔS for equation (7) shows $-\Delta S$ tending to very high

$$UO_{2+x} + \tfrac{1}{2}(0.25 - 0.25y - x)O_2 = \tfrac{1}{4}\, U_4O_{9-y} \qquad (7)$$

values (about 200 entropy units (eu) per mole of O_2 reacting) as the composition of the product approaches U_4O_9. Normally, the entropy loss accompanying the fixing of 1 mole of O_2 is only about 35 eu; therefore it is believed that considerable ordering of the interstitial oxygens must be occurring on conversion from UO_{2+x} to U_4O_{9-y}. X-ray and neutron diffraction studies support such a model.

Titanium monoxide, $TiO_{1\pm\delta}$. This material is interesting because of the wide range of composition possible, from $TiO_{0.85}$ to $TiO_{1.18}$, and because of the metallic character observed. The structure is cubic over the whole range and is frequently described as rock salt, though ordering of the cation and anion vacancies makes it more like spinel. Figure 6 shows schematically how the Ti/O ratio can deviate both above and below unity because of vacancy imbalance. Stoichiometric TiO has no imbalance but an unusually large number of vacant lattice sites, amounting to about 15%. This can be determined by noting that the observed density is less than that calculated on the basis of measured x-ray spacings. If TiO is heated in various oxygen pressures, either above or below that corresponding to equilibrium with the stoichiometric material, excess oxygen or excess titanium atoms can be incorporated at the vacant sites. If neutral oxygen is added, the dinegative oxide-ion charge is obtained by converting two Ti^{2+} ions to Ti^{3+} ions; if titanium is added, the Ti^0 moves to a cation site and forms Ti^{2+} by donating two electrons to the metallic orbitals. The metallic behavior of $TiO_{1\pm\delta}$ is attributable to electron delocalization resulting from overlap of the d electrons.

Tungsten bronzes, $M_\delta WO_3$, *and analogs*. The tungsten bronzes are a curious set of compounds, in which alkali metals, alkaline earth metals, copper, silver, thallium, lead, thorium, uranium, the rare-earth elements, hydrogen, or ammonium can be inserted in a WO_3 structure. In the case of sodium the materials are semiconducting for $\delta < 0.25$ but metallic for $\delta > 0.25$. Colors range from blue to violet to coppery to yellow-gold as δ changes from 0.4 to 0.98. The materials are remarkably inert to most reagents at room temperature. They can be prepared by (1) electrolysis of molten Na_2WO_4-WO_3 mixes; (2) heating of Na_2WO_4, WO_3, and W; and (3) vapor-phase reaction of Na with WO_3.

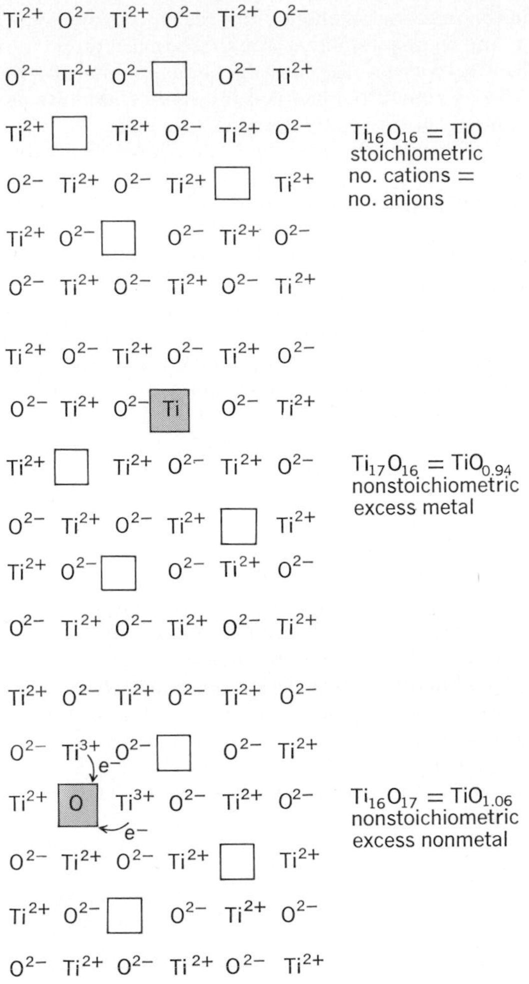

Fig. 6. How nonstoichiometry arises in $TiO_{1\pm\delta}$. The open squares represent vacant lattice sites, and the solid squares indicate added atoms.

The cubic sodium tungsten bronzes, $Na_\delta WO_3$ ($\delta > 0.43$), have a simple structure in which a cube containing tungsten atoms at the corners and oxygen atoms at the edge centers is random-statistically occupied by a sodium atom in the cube center. The sodium atom transfers its electron to the conduction band of the WO_3 host structure, a conduction band which may be formed by overlap of the $5dt_{2g}$ orbitals of the tungsten atoms or by interaction of these orbitals with the oxygen orbitals. Rubidium and potassium form hexagonal tungsten bronzes which become superconducting at temperatures in the range 2–6 K, depending on composition.

Vanadium bronzes, $M_\delta V_2O_5$, are analogous compounds based on insertion of metals in V_2O_5. However, unlike the tungsten bronzes, which are generally metallic and show small temperature-independent paramagnetism, they are semiconductors and normally paramagnetic.

Molybdenum bronzes, $M_\delta MoO_3$, are intermediate in properties. For example, $K_{0.30}MoO_3$ has a low temperature-independent magnetic susceptibility, semiconductive behavior below 180 K, and metallic behavior above 180 K. Somewhat startling in this material is the finding that *n*-type behavior in the semiconducting range changes to *p*-type conductivity in the metallic range, opposite to what is generally observed for semiconductors as temperature rises. Titanium bronzes, $Na_\delta TiO_2$, and platinum bronzes, $Na_\delta Pt_3O_4$, have also been reported.

Intercalation compounds. The intercalation compounds include (*a*) the clathrates, where guest molecules occupy isolated cavities of the host (for example, gas hydrates such as $Cl_2 \cdot xH_2O$, $5.75 < x < 7.66$); (*b*) tunnel compounds, where molecules such as hydrocarbons fit into tunnels of the host structure (for example, urea); (*c*) layer compounds, where molecules such as ammonia or alkali metal atoms are inserted between the layers of a transition-metal dichalcogenide; and (*d*) the zeolites, or molecular sieves, where guest molecules move through a three-dimensional network of tunnels. In all these cases, saturation of site occupancy would lead to stoichiometric products; however, the ratio of guest to host is generally less than saturation and is variable, so the materials can be regarded as nonstoichiometric compounds. The potential number of combinations is enormous. More than a hundred intercalation hosts are already recognized, and just one of these, graphite, is able to accommodate more than 12,000 types of guest.

Typical of the inorganic intercalates is $Li_\delta TiS_2$ ($0 < \delta < 1$). The host, TiS_2, has a layered structure in which a sheet of titanium atoms is sandwiched between two sheets of hexagonally close-packed sulfur atoms. Adjacent sulfur sheets are weakly bonded to each other by van der Waals forces, and guest species can move into this van der Waals gap. Intercalation with lithium can be achieved by exposing TiS_2 to lithium vapor, to lithium dissolved in liquid ammonia, or to *n*-butyl lithium dissolved in a nonpolar solvent. It can also be achieved by electrointercalation, using a cell in which lithium metal is the anode, TiS_2 is the cathode, and the electrolyte is a lithium salt dissolved in an organic solvent such as propylene carbonate. Intercalation proceeds with an expansion of the lattice *c* parameter by about 0.7–0.85 A. The expansion increases with lithium content. Nuclear magnetic resonance studies of $Li_\delta TiS_2$ indicate that the lithium is quite mobile and is almost 100% ionic. The use of $Li_\delta TiS_2$ in lithium batteries is possible.

[MICHELL J. SIENKO]

Bibliography: L. Eyring and M. O'Keeffe (eds.), *The Chemistry of Extended Defects in Non-Metal-Imperfect Crystals*, vols. 1–3, 1974–1975; R. Ward (ed.), *Nonstoichiometric Compounds*, 1963.

Nuclear chemistry

An interdisciplinary field that, in general, encompasses the application of chemical techniques to the solution of problems in nuclear physics. The finding of the naturally occurring radioactive elements and the discovery of nuclear fission are classical examples of the work of nuclear chemists.

Although chemical techniques that are employed in nuclear chemistry are essentially the same as those in radiochemistry, these fields may be distinguished on the basis of the aims of the investigation. Thus, a nuclear chemist utilizes chemical techniques as a tool for the study of nuclear reactions and properties, whereas a radiochemist utilizes the radioactive properties of certain substances as a tool for the study of chemical reactions and properties. There is considerable overlap between the two fields and in some cases (for example, the preparation and study of synthetic elements) a distinction may be somewhat arbitrary. For the application of radioactive tracers to chemical problems *see* RADIOCHEMISTRY. For the chemical effects of radiation on various systems *see* RADIATION CHEMISTRY.

Scope of nuclear chemistry. The chemical identification of radioactive nuclides and the determination of their nuclear properties (half-life, mode and energy of decay, mass number, nucleon binding energies, and so on) has been one of the major activities of nuclear chemists. Such studies have produced an extensive array of radioisotopes, and present studies are concerned mainly with the more difficult identification of nuclides of very short half-life at the boundaries of the "valley of beta-stability" and in the regions of neutron and proton instability, as well as short-lived isomeric states. Nuclear chemical investigations led to the discovery of the synthetic radioactive elements which do not have any stable isotopes and are not formed in the natural radioactive series (technetium, promethium, astatine, and the transuranium elements). The synthesis of element 104 has been announced and attempts to extend the periodic table even further and to prepare larger quantities of the new elements for intensive study are being made by using accelerated heavy-ion bombardments of targets of the heaviest elements available and successive neutron capture reactions in nuclear reactors of very high flux or in proximity to nuclear detonations (atomic and hydrogen bombs). Accelerators capable of accelerating even uranium ions are contemplated, and investigations of the possible existence of superheavy nuclei in a new region of relative stability are planned.

Other major areas of nuclear chemistry include studies of nuclear structure and spectroscopy (involving the determination of the energy spectra of emitted particles and the construction of energy-level diagrams, or decay schemes) and of the probability and mechanisms of various nuclear reactions (involving the determination of excitation functions, that is, the variation of cross section—or probability of reaction—with the energy of the bombarding particle, and the angular and energy distributions and correlations of the reaction products).

Many of these investigations involve the chemical isolation and identification of the products of nuclear reactions or chemical techniques for the preparation of thin, uniform deposits of pure materials for use as bombardment targets and as sources for nuclear spectroscopy. However, the nuclear chemist also carries out many experiments of a purely physical nature, observing the products of nuclear reactions with counters, ionization chambers, nuclear emulsions, and other particle track detectors. Moreover, advances in the technology of nuclear radiation detectors and associated electronic equipment have led to instruments of such sophistication that classical methods of chemical analyses are often not necessary. The borderline between nuclear chemistry and nuclear physics has, in fact, practically disappeared in many areas of investigation, and if a distinction is to be made at all, it may be only that the nuclear physicist tends to concentrate on interactions between the fundamental particles and observes the nucleons and light nuclei (neutrons, protons, deuterons, tritons, and α-particles) emitted in nuclear reactions, while the nuclear chemist examines interactions with complex nuclei and observes the heavier fragments and residual nuclei produced. This distinction is perhaps most meaningful in the area of high-energy nuclear phenomena where the high-energy, or particle, physicist is concerned mainly with the production and properties of and the interactions between the elementary (or strange) particles and their role in the structure of matter. The nature of the interaction of high-energy projectiles with complex nuclei and studies of the products formed is mainly the concern of nuclear chemists.

The contributions of nuclear chemists have not been limited to empirical studies of nuclear reactions and properties. They are also active in theoretical investigations and have made significant contributions to theories of nuclear structure, radioactive decay (particularly α-emission), and nuclear reactions of many types, including fission.

Historical background. Nuclear chemistry began with the work of Pierre and Marie Curie on the isolation and identification of polonium and radium in 1898. The discovery of actinium, thoron, and radon in the next few years, the identification of the α-particle as the helium ion, and the characterization of the relationships in the natural radioactive decay series played an essential role in the development of atomic and nuclear science. The interpretation of the results of these early experiments led to the establishment of the laws of radioactive transformation and the concept of isotopes (E. Rutherford and F. Soddy). The phenomenon of nuclear isomerism was discovered by O. Hahn (1921), who showed that the radioactive species UZ and UX_2 represented different excitation levels of Pa^{234}.

Artificial transmutation reactions were first carried out utilizing the α-particles emitted in the decay of the natural radioactive isotopes as bombarding particles (Rutherford and coworkers, 1919). Such bombardments led to the discovery of artificial radioactivity by Irene Curie and F. Joliot in 1934. The discovery of the neutron (J. Chadwick, 1932) and deuterium (H. Urey and coworkers, 1932) and the development of devices to accelerate the ions of hydrogen (proton), deuterium (deuteron), and helium (α-particle) gave tremendous impetus to the production and study of radioactive nuclides. By 1937 about 200 were known, and the number has grown steadily to nearly 1400 in 1968.

The discovery of nuclear fission by O. Hahn and F. Strassmann in 1939 culminated a dramatic episode in the history of nuclear science and opened the doors to the atomic age. Nuclear chemistry became a mature science during the period 1940–1945, largely as a result of the concentrated effort to produce the atomic bomb. It has continued to grow and expand in scope with the advancing technology in particle accelerators, nuclear reactors, radiation detection instrumentation, mass spectrometers, electromagnetic isotope separators, and computers. *See* NUCLEAR FISSION.

Nuclear spectroscopy. The energy spectra of the radiations emitted by radioactive nuclides are determined by the use of a number of specialized instruments. Alpha-particle spectra may be determined quite well with solid-state detectors coupled with multichannel pulse-height analyzers. The pulse output of the detector is proportional to the energy deposited and the individual events are stored in the analyzer memory. Magnetic spectrometers are used for higher resolution but require sources of higher intensity.

Beta-particle spectra may be examined with proportional or scintillation counters, but precise energy determinations and studies of the shapes of the spectra require magnetic spectrometers.

Gamma-ray and x-ray spectra are very conveniently measured using thallium-activated NaI crystal scintillation counters which give a fluorescent light output proportional to the energy deposited in the crystal. The light is amplified with a photomultiplier tube and converted to a pulse-height that is stored in a multichannel analyzer. Solid-state detectors, particularly lithium-drifted germanium crystals, possess higher resolution and are generally used for studies requiring high accuracy. Bent-crystal spectrometers, which depend on Bragg scattering, are instruments of high resolution but require intense sources of the radioactive nuclide. They are often used in measurements of γ-ray spectra accompanying nuclear reactions—"capture γ-rays" in the (n,γ) reaction, for example.

Studies of nuclear reactions. In a nuclear reaction, a bombarding particle interacts with a target nucleus to produce one or more product nuclei and, perhaps, other particles. (Spontaneous transformations of unstable nuclei, such as α- and β-decay or spontaneous fission, may be considered a special type of nuclear reaction in which no external excitation by bombarding particles is involved.) If the target is thin with respect to the ranges of the product nuclei, or if it is dispersed in a medium such as a nuclear emulsion, the reaction may be studied event by event by recording the emitted nuclei and particles with counters or observing the tracks they leave in nuclear emulsions or other particle track detectors, such as mica or polycarbonate films. Such studies are especially useful for determinations of the angular distributions of the recoil fragments. By suitable in-line arrangements of thin detectors (through which the particles pass, depositing only part of their kinetic energy) and thick detectors (which record the total kinetic energy), the mass and charge of many light nuclei can be uniquely identified. Time-of-flight (velocity) and kinetic energy measurements may also be combined to establish the masses of recoiling nuclei from nuclear reactions, and measurements of the characteristic x-rays may be made to establish the nuclear charge. These measurements require rather sophisticated instrumentation, and they cannot be applied in all cases.

When a thick target is bombarded, the reaction products cannot escape, and an accumulation of many individual events is retained in the target matrix. (Catcher foils are placed adjacent to the target to retain the small fraction of the products escaping from the target surfaces.) The recoil products from thin targets may also be accumulated on catcher foils at various angles to the beam direction for angular distribution studies and recoil range measurements. Chemical separations are employed to identify the elements formed in the reactions and measurements of the radiations made to characterize the isotopes. When a number of isotopes of the same element are produced in the target and cannot be conveniently differentiated on the basis of half-life or radiation characteristics, chemical separations alone may not suffice. Separation of the isotopes can be achieved by the acceleration and deflection of ions in electromagnetic isotope separators. The separated isotopes may be collected and their radiations measured free from any interference. The abundances of stable as well as radioactive species produced in nuclear reactions may also be determined by mass spectrometry. *See* MASS SPECTROMETRY.

Technique of radiochemical analysis. The chemical manipulations and separations of radioactive materials, generally referred to as techniques of radiochemical analysis, differ from the ordinary analytical techniques in a number of respects. Procedures in radiochemical analysis are not required to provide quantitative recovery but are selected for specificity and speed, with reasonably good yields (usually the order of 50% or better) generally sufficing. The criteria of radiochemical purity in a radioactive preparation are somewhat more stringent than those of ordinary chemical purity. Thus, trace quantities of impurities are rarely of importance in ordinary quantitative analyses, but in a radioactive preparation, contamination by trace quantities of radioactive impurities may completely negate the results of an experiment.

The handling of highly radioactive materials presents a health hazard, and special techniques for the manipulation of samples behind shielding walls must be utilized. Some effects of high levels of radioactivity on the solutions, such as heating and the decomposition of solvents which produces bubbling, also may affect normal procedures.

The mass of radioactive material produced in nuclear reactions is usually very small. The concentrations of nuclear reaction products in the solutions of target materials are generally of the order of 10^{-10} *M* or less. Many normal chemical operations, such as precipitation, are not feasible with such small concentrations. Although separations can be carried out with these tracer quantities using such techniques as solvent extraction and ion exchange, it then is difficult to determine the efficiency for the recovery of the product. Moreover, the chemical behavior of such dilute solutions may differ considerably from that normally encountered. For example, radiocolloid formation, adsorption on the walls of vessels and on the surfaces of dust particles and precipitates, and the concentration dependence of some equilibrium constants become prominent at such extremely high dilution. To avoid these difficulties, an isotope dilution technique may be employed in which macroscopic quantities of stable isotopes of the element are added to serve as a carrier for the radioactive species.

The amount of carrier used represents a compromise between considerations of convenience in chemical manipulations and yield determination and the preparation of high specific activity sources in which counting corrections for absorption and scattering of radiations in the sample itself are minimized. Quantities of 10–20 mg are used most often. Chemical procedures are simplified if macroscopic quantities of only a few elements are present. When many elements are produced in a nuclear reaction (in nuclear fission, for example), aliquots of the solution usually are taken for the analysis of each element or small group of elements. It is then necessary to add carriers for only relatively few products of interest. Trace quantities of the other elements present are removed in the chemical procedures by the use of scavenging precipitations of a compound of high surface area, such as iron(III) hydroxide or manganese dioxide, which tend to occlude traces of foreign substances or of a representative precipitate for an insoluble group of elements, such as bismuth sulfide to carry trace quantities of other insoluble sulfides, lanthanum fluoride for insoluble fluorides, or iron(III) hydroxide for insoluble hydroxides. If the element of interest itself forms a precipitate which may occlude traces of other elements, it may be necessary to add holdback carriers for the latter to dilute the effect of radioactive contamination of the product precipitate.

For the isolation of products of high specific activity without regard to yield, the carrier may be an element with chemical properties similar to those of the desired product, but which can be separated from it in the last stages of the procedure, leaving the product essentially carrier-free. Such carriers are referred to as nonisotopic carriers. When it is necessary to determine the yield of a nuclear reaction product, a known quantity of an isotopic carrier must be used. It is also imperative that complete interchange between the valence states of the carrier and the active species be achieved before any chemical separation is begun. In the case of elements which do not have any stable isotopes, or when a carrier-free procedure is desired, a known quantity of an isotopic radioactive tracer may be used. Radiations of the tracer should be easily distinguishable from those of the product. The fractional recovery of the added carrier or tracer then will represent the yield of the product of interest.

Classical precipitation methods of chemical separation are time-consuming, and in order to study short-lived radioactive species, rapid procedures are essential and other techniques are generally employed. Ion-exchange, solvent-extraction, and volatilization techniques have proved most useful, and their development is closely associated with radiochemical investigations. In cases where volatile products (rare gases, halogens, and so on) are formed in the nuclear reaction, a solution of the target may be bombarded in a closed system and the products swept out continuously. Some solid matrices that emanate gases readily may also be used. The products may be chemically separated by passing them through a suitable series of rapid steps in the flowing system before collection in a form suitable for direct measurement of the radiations. Nuclides with half-lives of the order of seconds may be investigated with such on-line techniques. Products recoiling out of thin targets may also be collected in gas streams or on moving tapes for rapid removal from the target and on-line or subsequent analysis. An isotope separator may form part of the on-line system, and in favorable cases the target may serve as the ion source of the separator.

Sample preparation and counting techniques. For studies in nuclear chemistry, the object of the radiochemical separations is the preparation of a pure sample in a form suitable for the radioactive assay of the nuclide of interest or for the determination of its nuclear properties. The detector used will, of course, depend on the type of radiation involved and the kind of information desired.

Alpha particles and fission fragments have short ranges in matter, and to prevent absorption losses samples of less than 100 $\mu g/cm^2$ surface density are generally required. A uniform sample deposit is necessary for accurate α-particle and fission-fragment measurements. This is best accomplished by volatilizing, electroplating, or spraying on metal foils. Samples collected with an isotope separator are well suited for such measurements. Evaporation from solution may also be used if the amount of solid residue is small, but uniformity of the deposit is difficult to achieve. The samples are counted internally in ionization chambers or gas proportional counters or with solid-state detectors.

Beta particles may cover a wide range of energies, and the techniques of sample preparation and counting will vary accordingly. The most commonly used detectors are Geiger, flow-type proportional, and scintillation counters. Samples may be prepared as indicated for α-emitters, in the form of precipitates or filter-paper disks or sample cups, as gases for internal counting, and as liquids. External sample counting usually is employed for convenience whenever feasible.

Gamma radiation is highly penetrating, and the size or form of the sample is generally not very critical. Because of much higher efficiency and resolution, scintillation counters and solid-state detectors have displaced all other types of detectors with γ-radiation.

Whenever possible it is advisable to design experiments so that relative counting of samples will suffice. It is then necessary only to produce the counting conditions for each sample. The determination of absolute disintegration rates is a more difficult task, and special techniques are required. Counters which detect all the radiations emanating from the source (4π-counters) are used, and the samples are either dispersed in the counting medium of a proportional or Geiger counter as a gas, dissolved in the medium of a liquid scintillation counter, or counted as a very thin deposit on a very thin film placed between two identical counters. Beta-gamma coincidence counting may also be used for the determination of absolute disintegration rates when the decay scheme of the nuclide is not too complicated.

Applied nuclear chemistry. The radiochemical and counting techniques outlined above are powerful tools for the study of nuclear reactions and the properties of nuclides. New techniques and instruments are constantly being adapted to the needs of nuclear chemistry and, conversely, investigations in nuclear chemistry have indicated the need and provided stimulation for the development of many new instruments. The techniques of nuclear chemistry have been applied to studies in a number of related fields, and nuclear chemists have contributed to studies in reactor chemistry and physics, isotope production, nuclear engineering, nuclear weapons development, geo- and cosmochemistry, and accelerator beam studies and monitoring as well as to basic studies in chemistry and the life sciences and the industrial and agricultural application of radioisotopes. The field of analytical chemistry has been especially influenced by the technique of activation analysis, which utilizes many of the results and methods of nuclear chemistry.

[ELLIS P. STEINBERG]

Bibliography: G. R. Choppin, *Nuclei and Radioactivity*, 1964; G. R. Choppin and J. Ryberg (eds.), *Nuclear Chemistry: Theory and Applications*, 1980; M. Haissinsky, *Nuclear Chemistry and Its Applications*, 1964; B. G. Harvey, *Introduction to Nuclear Physics and Chemistry*, 2d ed., 1969; G. Hevesy and F. A. Paneth, *A Manual of Radioactivity*, 1938.

Nuclear fission

An extremely complex nuclear reaction representing a cataclysmic division of an atomic nucleus into two nuclei of comparable mass. This rearrangement or division of a heavy nucleus may take place naturally (spontaneous fission) or under bombardment with neutrons, charged particles, gamma rays, or other carriers of energy (induced fission). Although nuclei with mass number A of approximately 100 or greater are energetically unstable against division into two lighter nuclei, the fission process has a small probability of occurring, except with the very heavy elements. Even for these elements, in which the energy release is of the order of 200,000,000 electronvolts (eV), the lifetimes against spontaneous fission are reasonably long.

Liquid-drop model. The stability of a nucleus against fission is most readily interpreted when the nucleus is viewed as being analogous to an incompressible and charged liquid drop with a surface tension. Such a droplet is stable against small deformations when the dimensionless fissility parameter X in Eq. (1) is less than unity, where the

$$X = \frac{(\text{charge})^2}{10 \times \text{volume} \times \text{surface tension}} \qquad (1)$$

charge is in esu, the volume is in cm^3, and the surface tension is in $ergs/cm^2$. The fissility parameter if given approximately, in terms of the charge number Z and mass number A, by the relation $X = Z^2/50\,A$.

Long-range Coulomb forces between the protons act to disrupt the nucleus, whereas short-range nuclear forces, idealized as a surface tension, act to stabilize it. The degree of stability is then the result of a delicate balance between the relatively weak electromagnetic forces and the strong nuclear forces. Although each of these forces results in potentials of several hundred million electron volts, the height of a typical barrier against fission for a heavy nucleus, because they are of opposite sign but do not quite cancel, is only 5,000,000 or 6,000,000 eV. Investigators have used this charged liquid-drop model with great success in describing the general features of nuclear fission and also in reproducing the total nuclear binding energies. *See* SURFACE TENSION.

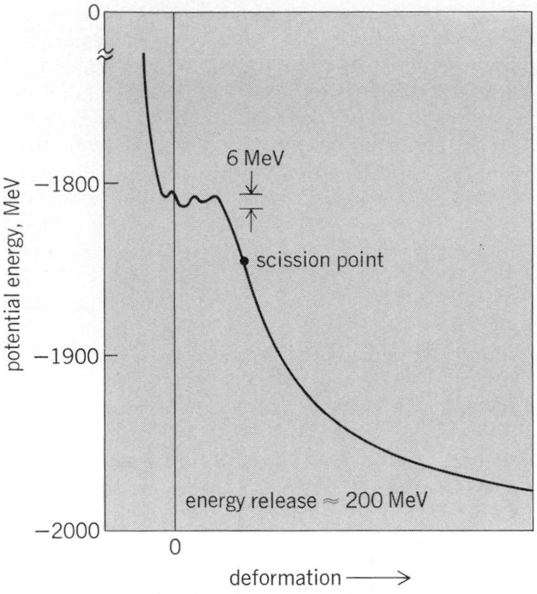

Fig. 1. Plot of the potential energy in MeV as a function of deformation for the nucleus ^{240}Pu. (*From M. Bolsteli et al., New calculations of fission barriers for heavy and superheavy nuclei, Phys. Rev., 5C:1050–1077, 1972*)

Shell corrections. The general dependence of the potential energy on the fission coordinate representing nuclear elongation or deformation for a heavy nucleus such as ^{240}Pu is shown in Fig. 1. The expanded scale used in this figure shows the large decrease in energy of about 200 MeV as the fragments separate to infinity. It is known that ^{240}Pu is deformed in its ground state, which is represented by the lowest minimum of −1813 MeV near zero deformation. This energy represents the total nuclear binding energy when the zero of potential energy is the energy of the individual nucleons at a separation of infinity. The second minimum to the right of zero deformation illustrates structure introduced in the fission barrier by shell corrections, that is, corrections dependent upon microscopic behavior of the individual nucleons, to the liquid-drop mass. Although shell corrections introduce small wiggles in the potential-energy surface as a function of deformation, the gross features of the surface are reproduced by the liquid-drop model. Since the typical fission barrier is only a few million electron volts, the magnitude of the shell correction need only be small for irregularities to be introduced into the barrier. This structure is schematically illustrated for a heavy nucleus by the double-humped fission barrier in Fig. 2, which represents the region to the right of zero deformation in Fig. 1 on an expanded scale. The fission barrier has two maxima and a rather deep minimum in between. For comparison, the single-humped liquid-drop barrier is also schematically illustrated. The transition in the shape of the nucleus as a function of deformation is schematically represented in the upper part of the figure.

Double-humped barrier. The developments which led to the proposal of a double-humped fission barrier were triggered by the experimental discovery of spontaneously fissionable isomers by S. M. Polikanov and colleagues in the Soviet Union and by V. M. Strutinsky's pioneering theoretical work on the binding energy of nuclei as a function of both nucleon number and nuclear shape. The double-humped character of the nuclear potential energy as a function of deformation arises, within the framework of the Strutinsky shell-correction method, from the superposition of a macroscopic smooth liquid-drop energy and a shell-correction energy obtained from a microscopic single-particle model. Oscillations occurring in this shell correction as a function of deformation lead to two minima in the potential energy, shown in Fig. 2, the normal ground-state minimum at a deformation of β_1 and a second minimum at a deformation of β_2. States in these wells are designated class I and class II states, respectively. Spontaneous fission of the ground state and isomeric state arises from the lowest-energy class I and class II states, respectively.

The calculation of the potential-energy curve illustrated in Fig. 1 may be summarized as follows. The smooth potential energy obtained from a macroscopic (liquid-drop) model is added to a fluctuating potential energy representing the shell corrections, and to the energy associated with the pairing of like nucleons (pairing energy), derived from a non-self-consistent microscopic model. The calculation of these corrections requires several steps, namely, (1) specification of the geometrical shape of the nucleus, (2) generation of a single-particle potential related to its shape, (3) solution of the Schrödinger equation, and (4) calculation from these single-particle energies of the shell and pairing energies.

The oscillatory character of the shell corrections as a function of deformation is caused by variations in the single-particle level density in the vicinity of the Fermi energy. For example, the single-particle levels of a pure harmonic oscillator potential arrange themselves in bunches of highly degenerate shells at any deformation for which the ratio of the major and minor axes of the spheroidal equipotential surfaces is equal to the ratio of two small integers. Nuclei with a filled shell, that is,

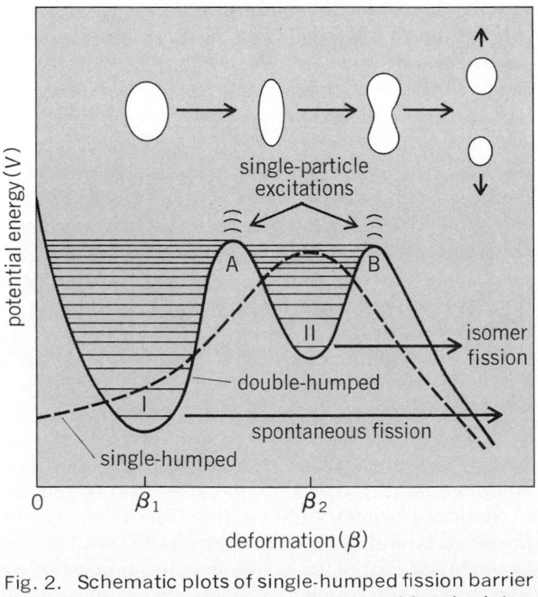

Fig. 2. Schematic plots of single-humped fission barrier of liquid-drop model and double-humped barrier introduced by shell corrections. (*From J. R. Huizenga, Nuclear fission revisited, Science, 168:1405–1413, 1970*)

with a level density at the Fermi energy that is smaller than the average, will then have an increased binding energy compared to the average, because the nucleons occupy deeper and more bound states; conversely, a large level density is associated with a decreased binding energy. It is precisely this oscillatory behavior in the shell correction that is responsible for spherical or deformed ground states and for the secondary minima in fission barriers, as illustrated in Fig. 2.

More detailed theoretical calculations based on this macroscopic-microscopic method have revealed additional features of the fission barrier. In these calculations the potential energy is regarded as a function of several different modes of deformation. The outer barrier B (Fig. 2) is reduced in energy for shapes with pronounced left-right asymmetry (pear shapes), whereas the inner barrier A and deformations in the vicinity of the second minimum are stable against such mass asymmetric degrees of freedom. Similar calculations of potential-energy landscapes reveal the stability of the second minimum against gamma deformations, in which the two small axes of the spheroidal nucleus become unequal, that is, the spheroid becomes an ellipsoid.

Experimental consequences. The observable consequences of the double-humped barrier have been reported in numerous experimental studies. In the actinide region more than 30 spontaneously fissionable isomers have been discovered between uranium and berkelium, with half-lives ranging from 10^{-11} to 10^{-2} s. These decay rates are faster by 20 to 30 orders of magnitude than the fission half-lives of the ground states, because of the increased barrier tunneling probability (see Fig. 2). Several cases in which excited states in the second minimum decay by fission are also known. Normally these states decay within the well by gamma decay; however, if there is a hindrance in gamma decay due to spin, the state (known as a spin isomer) may undergo fission instead.

Qualitatively, the fission isomers are most stable in the vicinity of neutron numbers 146 to 148, a value in good agreement with macroscopic-microscopic theory. For elements above berkelium the half-lives become too short to be observable with available techniques; and for elements below uranium, the prominent decay is through barrier A into the first well, followed by gamma decay. It is difficult to detect this competing gamma decay of the ground state in the second well (called a shape isomeric state), but identification of the gamma branch of the 200-ns ^{238}U shape isomer has been reported. *See* RADIOACTIVITY.

Direct evidence of the second minimum in the potential-energy surface of the even-even nucleus ^{240}Pu has been obtained through observations of the E2 transitions within the rotational band built on the isomeric 0+ level. The rotational constant (which characterizes the spacing of the levels and is expected to be inversely proportional to the effective moment of inertia of the nucleus) found for this band is less than one-half that for the ground state and confirms that the shape isomers have a deformation β_2 much larger than the equilibrium ground-state deformation β_1. From yields and angular distributions of fission fragments from the isomeric ground state and low-lying excited states some information has been derived on the quantum numbers of specific single-particle states of the deformed nucleus (Nilsson single-particle states) in the region of the second minimum.

At excitation energies in the vicinity of the two barrier tops, measurements of the subthreshold neutron fission cross sections of several nuclei have revealed groups of fissioning resonance states with wide energy intervals between each group where no fission occurs. Such a spectrum is illustrated in Fig. 3*a*, where the subthreshold fission cross section of ^{240}Pu is shown for neutron energies between 500 and 3000 eV. As shown in Fig. 3*b*, between the fissioning resonance states there are many other resonance states, known from data on the total neutron cross sections, which have negligible fission cross sections. Such structure is explainable in terms of the double-humped fission barrier and is ascribed to the coupling between the compound states of normal density in the first well to the much less dense states in the second well. This picture requires resonances of only one spin to appear within each intermediate structure group illustrated in Fig. 3*a*. In an experiment using polarized neutrons on a polarized ^{237}Np target, it was found that all nine fine-structure resonances of the 40-eV group have the same spin and parity: $I = 3+$. Evidence has also been obtained for vibrational states in the second well from neutron (n,f) and deuteron stripping (d,pf) reactions at energies below the barrier tops (f indicates fission of the nucleus).

A. Bohr suggested that the angular distributions of the fission fragments are explainable in terms of the transition-state theory, which describes a process in terms of the states present at the barrier deformation. The theory predicts that the cross section will have a steplike behavior for energies near the fission barrier, and that the angular distribution will be determined by the quantum numbers associated with each of the specific fission channels. The theoretical angular distribution of fission fragments is based on two assumptions. First, the two fission fragments are assumed to separate along the direction of the nuclear symmetry axis so that the angle θ between the direction of motion of the fission fragments and the direction of motion of the incident bombarding particle represents the angle between the body-fixed axis (the long axis of the spheroidal nucleus) and the space-fixed axis (some specified direction in the laboratory, in this case the direction of motion of the incident particle). Second, it is assumed that the transition from the saddle point (corresponding to the top of the barrier) to scission (the division of the nucleus into two fragments) is so fast that Coriolis forces do not change the value of K (where K is the projection of the total angular momentum I on the nuclear symmetry axis) established at the saddle point.

In several cases, low-energy photofission and neutron fission experiments have shown evidence of a double-humped barrier. In the case of two barriers, the question arises as to which of the two barriers A or B is responsible for the structure in the angular distributions. For light actinide nuclei like thorium, the indication is that barrier B is the higher one, whereas for the heavier actinide nuclei, the inner barrier A is the higher one. The heights of the two barriers themselves are most reliably determined by investigating the probabili-

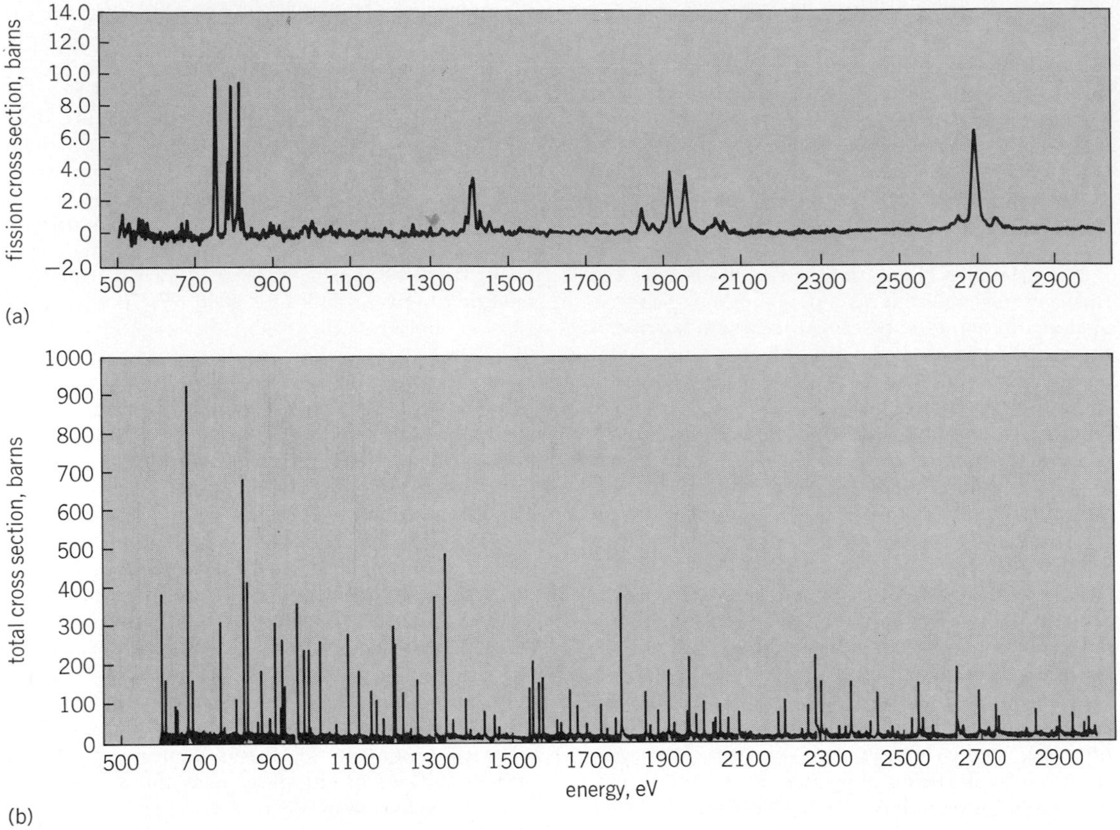

Fig. 3. Grouping of fission resonances demonstrated by (*a*) the neutron fission cross section of ^{240}Pu and (*b*) the total neutron cross section. (*From V. M. Strutinsky and H. C. Pauli, Shell-structure effects in the fissioning nucleus, Proceedings of the 2d IAEA Symposium on Physics and Chemistry of Fission, Vienna, pp. 155–177, 1969*)

ty of induced fission over a range of several megaelectron volts in the threshold region. Many direct reactions have been used for this purpose, for example, (d,pf), (t,pf), and $(^3\text{He}, df)$. There is reasonably good agreement between the experimental and theoretical barriers. The theoretical barriers are calculated with realistic single-particle potentials and include the shell corrections.

Fission probability. The cross section for particle-induced fission $\sigma(y,f)$ represents the cross section for a projectile y to react with a nucleus and produce fission, as shown by Eq. (2). The quanti-

$$\sigma(y,f) = \sigma_R(y)\,(\Gamma_f/\Gamma_t) \tag{2}$$

ties $\sigma_R(y)$, Γ_f, and Γ_t are the total reaction cross sections for the incident particle y, the fission width, and the total level width, respectively where $\Gamma_t = \Gamma_f + \Gamma_n + \Gamma_y + \cdots$ is the sum of all partial-level widths. All the quantities in Eq. (2) are energy-dependent. Each of the partial widths for fission, neutron emission, radiation, and so on, is defined in terms of a mean lifetime τ for that particular process, for example, $\Gamma_f = \hbar/\tau_f$. Here $\hbar$, the action quantum, is Planck's constant divided by 2π and is numerically equal to 1.0546×10^{-34} J s $= 0.66 \times 10^{-15}$ eV s. The fission width can also be defined in terms of the energy separation D of successive levels in the compound nucleus and the number of open channels in the fission transition nucleus (paths whereby the nucleus can cross the barrier on the way to fission), as given by expression (3), where I is the angular momentum and i is

$$\Gamma_f(I) = \frac{D(I)}{2\pi} \sum_i N_{fi} \tag{3}$$

an index labeling the open channels N_{fi}. The contribution of each fission channel to the fission width depends upon the barrier transmission coefficient, which, for a two-humped barrier (see Fig. 2), is strongly energy-dependent. This results in an energy-dependent fission cross section which is very different from the total cross section shown in Fig. 3 for ^{240}Pu.

When the incoming neutron has low energy, the likelihood of reaction is substantial only when the energy of the neutron is such as to form the compound nucleus in one or another of its resonance levels (see Fig. 3*b*). The requisite sharpness of the "tuning" of the energy is specified by the total level width Γ. The nuclei ^{233}U, ^{235}U, and ^{239}Pu have a very large cross section to take up a slow neutron and undergo fission (see table) because both their absorption cross section and their probability for decay by fission are large. The probability for fission decay is high because the binding energy of the incident neutron is sufficient to raise the energy of the compound nucleus above the fission barrier. The very large, slow neutron fission cross sections of these isotopes make them important fissile materials in a chain reactor.

Scission. The scission configuration is defined in terms of the properties of the intermediate nucleus just prior to division into two fragments. In

Cross sections for neutrons of thermal energy to produce fission or undergo capture in the principal nuclear species, and neutron yields from these nuclei*

Nucleus	Cross section for fission, σ_f, 10^{-24} cm²	σ_f plus cross section for radiative capture, σ_r	Ratio, $1+\alpha$	Number of neutrons released per fission, ν	Number of neutrons released per slow neutron captured, $\eta=\nu/(1+\alpha)$
^{233}U	525 ± 2	573 ± 2	1.093 ± 0.003	2.50 ± 0.01	2.29 ± 0.01
^{235}U	577 ± 1	678 ± 2	1.175 ± 0.002	2.43 ± 0.01	2.08 ± 0.01
^{239}Pu	741 ± 4	1015 ± 4	1.370 ± 0.006	2.89 ± 0.01	2.12 ± 0.01
^{238}U	0	2.73 ± 0.04			0
Natural uranium	4.2	7.6	1.83	2.43 ± 0.01	1.33

*Data from *Brookhaven National Laboratory 325*, 2d ed., suppl. no. 2, vol. 3, 1965. The data presented are the recommended or least-squares values published in this reference for 0.0253-eV neutrons. All cross sections are in units of barns (1 barn = 10^{-24} cm² = 10^{-28} m²).

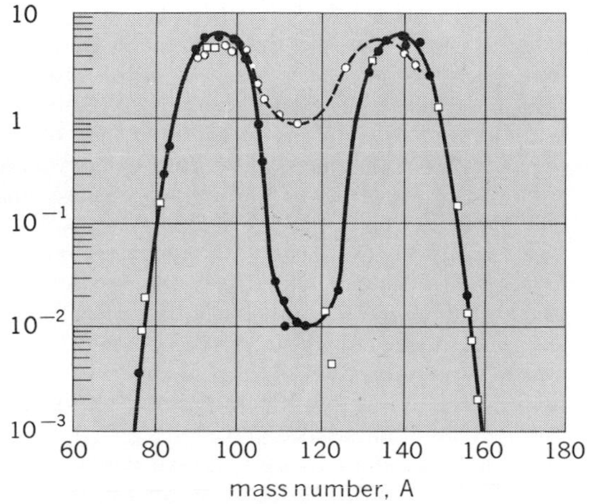

Fig. 4. Mass distribution of fission fragments formed by neutron-induced fission of $^{235}U + n = ^{236}U$ when neutrons have thermal energy, smooth curve (*Plutonium Project Report, Rev. Mod. Phys., 18:539, 1964*), and 14-MeV energy, dashed curve (*based on R. W. Spence, Brookhaven National Laboratory, AEC-BNL (C-9), 1949*). Quantity plotted is 100 × (number of fission decay chains formed with given mass)/(number of fissions).

heavy nuclei the scission deformation is much larger than the saddle deformation at the barrier, and it is important to consider the dynamics of the descent from saddle to scission. One of the important questions in the passage from saddle to scission is the extent to which this process is adiabatic with respect to the particle degrees of freedom. As the nuclear shape changes, it is of interest to investigators to know the probability for the nucleons to remain in the lowest-energy orbitals. If the collective motion toward scission is very slow, the single-particle degrees of freedom continually readjust to each new deformation as the distortion proceeds. In this case, the adiabatic model is a good approximation, and the decrease in potential energy from saddle to scission appears in collective degrees of freedom at scission, primarily as kinetic energy associated with the relative motion of the nascent fragments.

On the other hand, if the collective motion between saddle and scission is so rapid that equilibrium is not attained, there will be a transfer of collective energy into nucleonic excitation energy. Such a nonadiabatic model, in which collective energy is transferred to single-particle degrees of freedom during the descent from saddle to scission, is usually referred to as the statistical theory of fission.

The experimental evidence indicates that the saddle to scission time is somewhat intermediate between these two extreme models. The dynamic descent of a heavy nucleus from saddle to scission depends upon the nuclear viscosity. A viscous nucleus is expected to have a smaller translational kinetic energy at scission and a more elongated scission configuration. Experimentally, the final translational kinetic energy of the fragments at infinity, which is related to the scission shape, is measured. Hence, in principle, it is possible to estimate the nuclear viscosity coefficient by comparing the calculated dependence upon viscosity of fission-fragment kinetic energies with experimen-

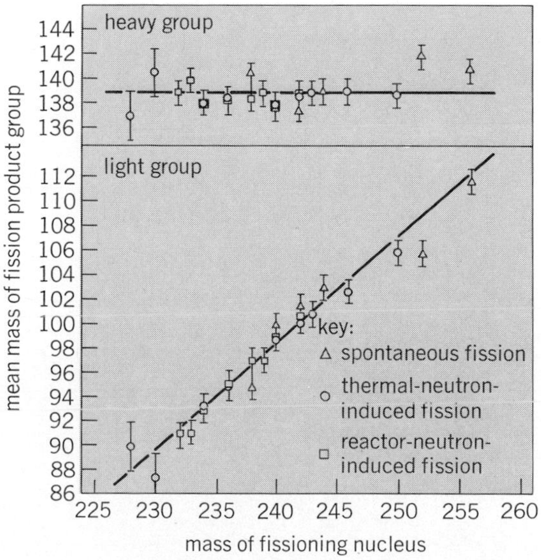

Fig. 5. Average masses of the light- and heavy-fission product groups as a function of the masses of the fissioning nucleus. Energy spectrum of reactor neutrons is that associated with fission. (*From K. F. Flynn et al., Distribution of mass in the spontaneous fission of ^{256}Fm, Phys. Rev., 5C:1725–1729, 1972*)

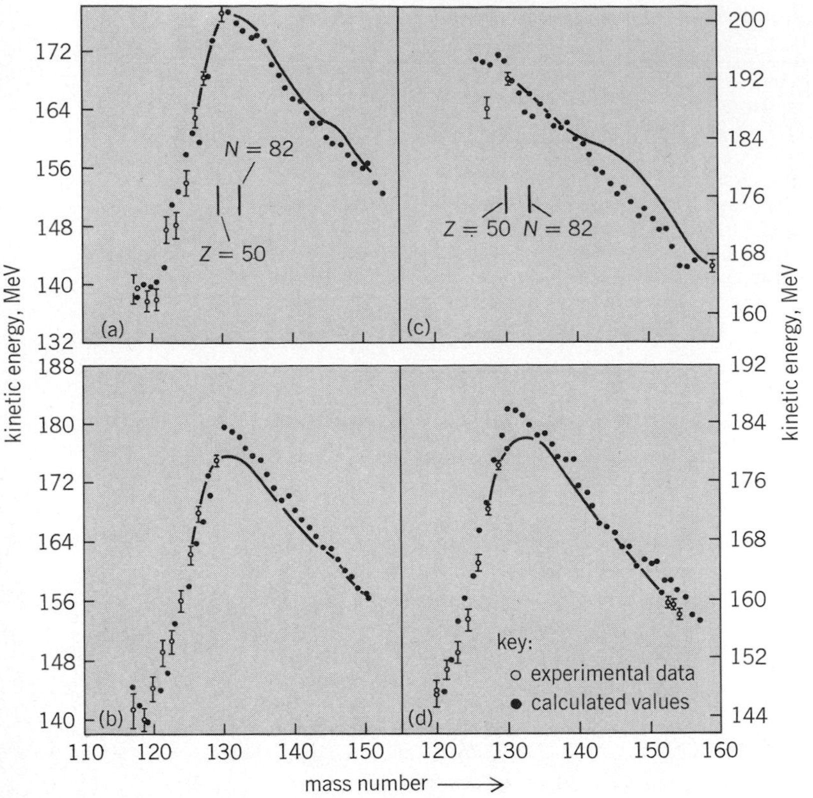

Fig. 6. Average total kinetic energy of fission fragments as a function of heavy fragment mass for fission of (*a*) ^{235}U, (*b*) ^{233}U, (*c*) ^{252}Cf, and (*d*) ^{239}Pu. Curves indicate experimental data. (*From J. C. D. Milton and J. S. Fraser, Time-of-flight fission studies on* ^{233}U, ^{235}U *and* ^{239}Pu, *Can. J. Phys., 40:1626–1663, 1962*)

tal values. The viscosity of nuclei is an important nuclear parameter which also plays an important role in collisions of very heavy ions.

The mass distribution from the fission of heavy nuclei is predominantly asymmetric. For example, division into two fragments of equal mass is about 600 times less probable than division into the most probable choice of fragments when ^{235}U is irradiated with thermal neutrons. When the energy of the neutrons is increased, symmetric fission (Fig. 4) becomes more probable. In general, heavy nuclei fission asymmetrically to give a heavy fragment of approximately constant mean mass number 139 and a corresponding variable-mass light fragment (see Fig. 5). These experimental results have been difficult to explain theoretically. Calculations of potential-energy surfaces show that the second barrier (B in Fig. 2) is reduced in energy by up to 2 or 3 MeV, if octuple deformations (pear shapes) are included. Hence, the theoretical calculations show that mass asymmetry is favored at the outer barrier, although direct experimental evidence supporting the asymmetric shape of the second barrier is very limited. It is not known whether the mass asymmetric energy valley extends from the saddle to scission; and the effect of dynamics on mass asymmetry in the descent from saddle to scission has not been determined. Experimentally, as the mass of the fissioning nucleus approaches $A \approx 260$, the mass distribution approaches symmetry. This result is qualitatively in agreement with theory.

A nucleus at the scission configuration is highly elongated and has considerable deformation energy. The influence of nuclear shells on the scission shape introduces structure into the kinetic energy and neutron-emission yield as a function of fragment mass. The experimental kinetic energies for the neutron-induced fission of ^{233}U, ^{235}U, and ^{239}Pu have a pronounced dip as symmetry is approached, as shown in Fig. 6. (This dip is slightly exaggerated in the figure because the data have not been corrected for fission fragment scattering.) The variation in the neutron yield as a function of fragment mass for these same nuclei (Fig. 7) has a

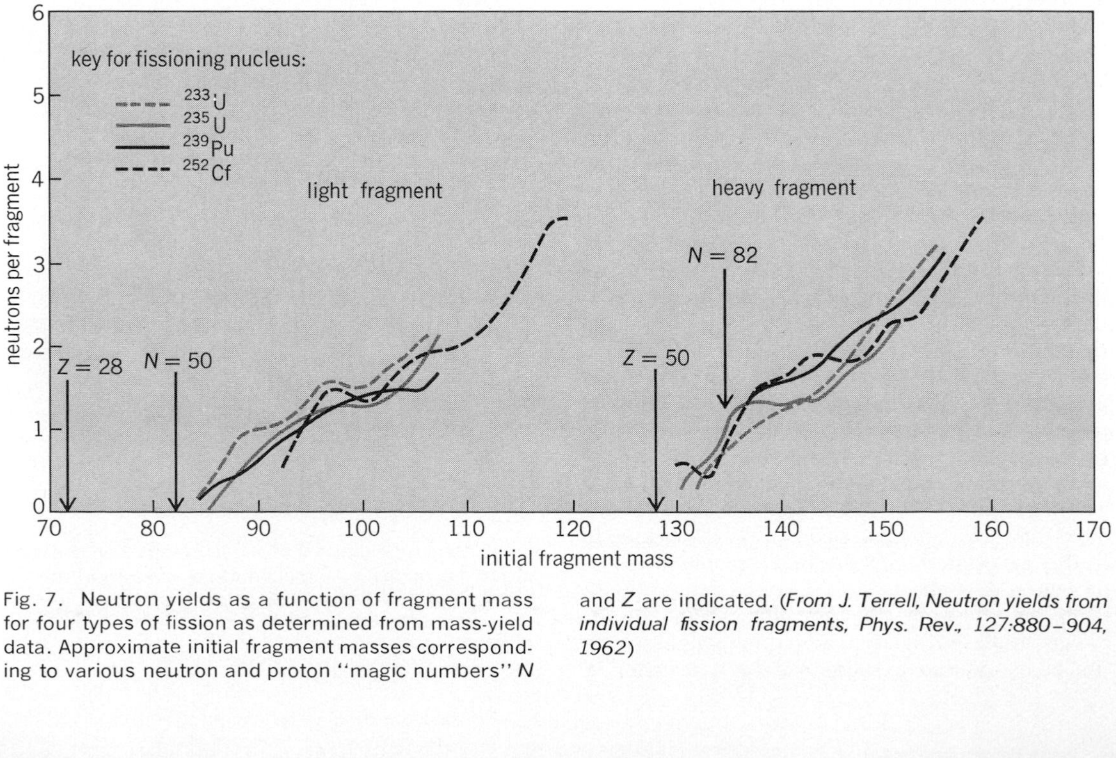

Fig. 7. Neutron yields as a function of fragment mass for four types of fission as determined from mass-yield data. Approximate initial fragment masses corresponding to various neutron and proton "magic numbers" *N* and *Z* are indicated. (*From J. Terrell, Neutron yields from individual fission fragments, Phys. Rev., 127:880–904, 1962*)

"saw-toothed" shape which is asymmetric about the mass of the symmetric fission fragment. Both these phenomena are reasonably well accounted for by the inclusion of closed-shell structure into the scission configuration.

A number of light charged particles (for example, isotopes of hydrogen, helium, and lithium) have been observed to occur, with low probability, in fission. These particles are believed to be emitted very near the time of scission. Available evidence also indicates that neutrons are emitted at or near scission with considerable frequency.

Postscission phenomena. After the fragments are separated at scission, they are further accelerated as the result of the large Coulomb repulsion. The initially deformed fragments collapse to their equilibrium shapes, and the excited primary fragments lose energy by evaporating neutrons. After neutron emission, the fragments lose the remainder of their energy by gamma radiation, with a lifetime of about 10^{-11} s. The kinetic energy and neutron yield as a function of mass are shown in Figs. 6 and 7. The variation of neutron yield with fragment mass is directly related to the fragment excitation energy. Minimum neutron yields are observed for nuclei near closed shells because of the resistance to deformation of nuclei with closed shells. Maximum neutron yields occur for fragments that are "soft" toward nuclear deformation. Hence, at the scission configuration, the fraction of the deformation energy stored in each fragment depends on the shell structure of the individual fragments. After scission, this deformation energy is converted to excitation energy, and, hence, the neutron yield is directly correlated with the fragment shell structure. This conclusion is further supported by the correlation between the neutron yield and the final kinetic energy. Closed shells result in a larger Coulomb energy at scission for fragments that have a smaller deformation energy and a smaller number of evaporated neutrons.

After the emission of the prompt neutrons and gamma rays, the resulting fission products are unstable against β-decay. For example, in the case of thermal neutron fission of ^{235}U, each fragment undergoes on the average about three β-decays before it settles down to a stable nucleus. For selected fission products (for example, ^{87}Br and ^{137}I) β-decay leaves the daughter nucleus with excitation energy exceeding its neutron binding energy. The resulting delayed neutrons amount, for thermal neutron fission of ^{235}U, to about 0.7% of all the neutrons given off in fission. Though small in number, they are quite important in stabilizing nuclear chain reactions against sudden minor fluctuations in reactivity.

[JOHN R. HUIZENGA]

Bibliography: *Proceedings of the 3d IAEA Symposium on Physics and Chemistry of Fission*, Rochester, NY, 1973; R. Vandenbosch and J. R. Huizenga, *Nuclear Fission*, 1973.

Nuclear fusion

One of the primary nuclear reactions, the name usually designating an energy-releasing rearrangement collision which can occur between various isotopes of low atomic number.

Interest in the nuclear fusion reaction arises from the expectation that it may someday be used to produce useful power, from its role in energy generation in stars, and from its use in the fusion bomb. Since a primary fusion fuel, deuterium, occurs naturally and is therefore obtainable in virtually inexhaustible supply (by separation of heavy hydrogen from water, 1 atom of deuterium occurring per 6000 atoms of hydrogen), solution of the fusion power problem would permanently solve the problem of the present rapid depletion of chemically valuable fossil fuels. As a power source, the lack of radioactive waste products from the fusion reaction is another argument in its favor as opposed to the fission of uranium.

In a nuclear fusion reaction the close collision of two energy-rich nuclei results in a mutual rearrangement of their nucleons (protons and neutrons) to produce two or more reaction products, together with a release of energy. The energy usually appears in the form of kinetic energy of the reaction products, although when energetically allowed, part may be taken up as energy of an excited state of a product nucleus. In contrast to neutron-produced nuclear reactions, colliding nuclei, because they are positively charged, require a substantial initial relative kinetic energy to overcome their mutual electrostatic repulsion so that reaction can occur. This required relative energy increases with the nuclear charge Z, so that reactions between low-Z nuclei are the easiest to produce. The best known of these are the reactions between the heavy isotopes of hydrogen, deuterium and tritium. *See* DEUTERIUM; TRITIUM.

Fusion reactions were discovered in the 1920s when low-Z elements were used as targets and bombarded by beams of energetic protons or deuterons. But the nuclear energy released in such bombardments is always microscopic compared with the energy of the impinging beam. This is because most of the energy of the beam particle is dissipated uselessly by ionization and single-particle collisions in the target; only a small fraction of the impinging particles produce reactions.

Nuclear fusion reactions can be self-sustaining, however, if they are carried out at a very high temperature. That is to say, if the fusion fuel exists in the form of a very hot ionized gas of stripped nuclei and free electrons termed a plasma, the agitation energy of the nuclei can overcome their mutual repulsion, causing reactions to occur. This is the mechanism of energy generation in the stars and in the fusion bomb. It is also the method envisaged for the controlled generation of fusion energy.

PROPERTIES OF FUSION REACTIONS

The cross sections (effective collisional areas) for many of the simple nuclear fusion reactions have been measured with high precision. It is found that the cross sections generally show broad maxima as a function of energy and have peak values in the general range of 0.01 barn (1 barn = 10^{-24} cm^2) to a maximum value of 5 barns, for the deuterium-tritium (D-T) reaction. The energy releases of these reactions can be readily calculated from the mass differences between the initial and final nuclei or determined by direct measurement.

Simple reactions. Some of the important simple fusion reactions, their reaction products, and their energy releases in millions of electronvolts (MeV) are given by reactions (1).

$$
\begin{array}{llr}
D + D \rightarrow & He^3 + n + & 3.25\ \text{MeV} \\
D + D \rightarrow & T + p + & 4.0\ \text{MeV} \\
T + D \rightarrow & He^4 + n + & 17.6\ \text{MeV} \\
He^3 + D \rightarrow & He^4 + p + & 18.3\ \text{MeV} \\
Li^6 + D \rightarrow & 2He^4 + & 22.4\ \text{MeV} \\
Li^7 + p \rightarrow & 2He^4 + & 17.3\ \text{MeV}
\end{array} \quad (1)
$$

If it is remembered that the energy release in the chemical reaction in which hydrogen and oxygen combine to produce a water molecule is about 1 eV per reaction, it will be seen that, gram for gram, fusion fuel releases more than 1,000,000 times as much energy as typical chemical fuels.

The two alternative D-D reactions listed occur with about equal probability for the same relative particle energies. Note that the heavy reaction products, tritium and helium-3, may also react, with the release of a large amount of energy. Thus it is possible to visualize a reaction chain in which six deuterons are converted to two helium-4 nuclei, two protons, and two neutrons, with an overall energy release of 43 MeV—about 10^5 kilowatt-hours (kWh) of energy per gram of deuterium. This energy release is several times that released per gram in the fission of uranium, and several million times that released per gram by the combustion of gasoline.

Cross sections. Figure 1 shows the measured values of cross sections as a function of bombarding energy up to 100 keV for the total D-D reaction (both D-D,*n* and D-D,*p*), the D-T reaction, and the D-He^3 reaction. The most striking feature of these curves is their extremely rapid falloff with energy as bombarding energies drop to a few kilovolts. This effect arises from the mutual electrostatic repulsion of the nuclei, which prevents them from approaching closely if their relative energy is small.

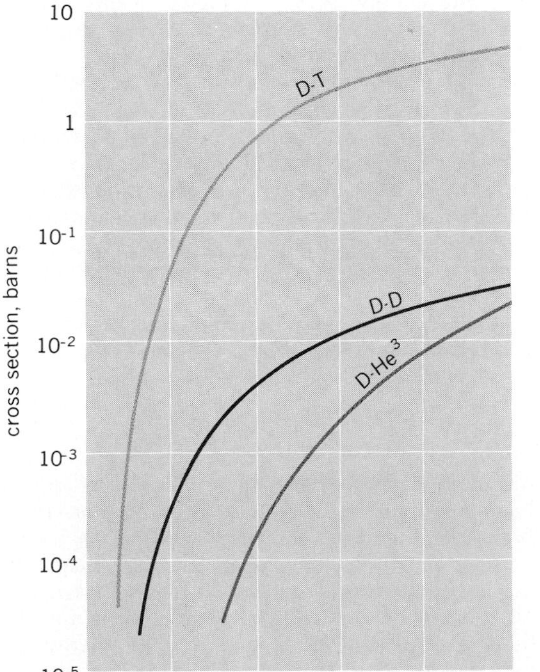

Fig. 1. Cross sections versus bombarding energy for three simple fusion reactions. (*From R. F. Post, Fusion power, Sci. Amer., 197(6):73–84, December 1957, copyright © 1957 by Scientific American, Inc.; all rights reserved*)

The fact that reactions can occur at all at these energies is attributable to the finite range of nuclear interaction forces. In effect, the boundary of the nucleus is not precisely defined by its classical diameter. The role of quantum mechanical effects in nuclear fusion reactions has been treated by G. Gamow and others. It is predicted that the cross sections should obey an exponential law at low energies. This is well borne out in energy regions reasonably far removed from resonances (for example, below about 30 keV for the D-T reaction). Over a wide energy range at low energies, the data for the D-D reaction can be accurately fitted by a Gamow curve, the result for the cross section being given by Eq. (2), where the bombarding energy W is in kiloelectronvolts.

$$\sigma_{\text{D-D}} = \frac{288}{W} e^{-45.8W^{-1/2}} \times 10^{-24}\ \text{cm}^2 \quad (2)$$

The extreme energy dependence of this expression can be appreciated by the fact that, between 1 and 10 keV, the predicted cross section varies by about 13 powers of 10, that is, from 3×10^{-42} to 1.5×10^{-29} cm^2.

Energy division. The kinematics of the fusion reaction stipulates that the reaction can occur only if two or more reaction products result. This is because both mass energy and momentum balance must be preserved. When there are only two reaction products (which is the case in all of the important reactions), the division of energy between the reaction products is uniquely determined, the lion's share always going to the lighter particle. The energy division (disregarding the initial bombarding energy) is as in reaction (3). If reaction (3)

$$A_1 + A_2 \rightarrow A'_1 + A'_2 + Q \quad (3)$$

holds, with the As representing the atomic masses of the particles and Q the total energy released, then Eqs. (4) are valid, where $W(A'_1)$ and $W(A'_2)$ are the kinetic energies of the reaction products.

$$
\begin{aligned}
W(A'_1) + W(A'_2) &= Q \\
W(A'_1) &= Q\left(\frac{A'_2}{A'_1 + A'_2}\right) \\
W(A'_2) &= Q\left(\frac{A'_1}{A'_1 + A'_2}\right)
\end{aligned} \quad (4)
$$

Thus in the D-T reaction, for example, A'_1, the mass of the α-particle, is four times A'_2, the mass of the neutron, so that the neutron carries off four-fifths of the reaction energy, or 14 MeV.

Reaction rates. When nuclear fusion reactions occur in a high-temperature plasma, the reaction rate per unit volume depends on the particle density n of the reacting fuel particles and on an average of their mutual reaction cross sections σ and relative velocity v over the particle velocity distributions.

For dissimilar reacting nuclei (such as D and T), the reaction rate is given by Eq. (5).

$$R_{12} = n_1 n_2 \langle \sigma v \rangle_{12} \quad \text{reactions/(cm}^3\text{)(s)} \quad (5)$$

For similar reacting nuclei (for example, D and D), the reaction rate is given by Eq. (6).

$$R_{11} = \tfrac{1}{2} n^2 \langle \sigma v \rangle \qquad (6)$$

Note that both expressions vary as the square of the total particle density (for a given fuel composition).

If the particle velocity distributions are known, $\langle \sigma v \rangle$ can be determined as a function of energy by numerical integration, using the known reaction cross sections. It is customary to assume a maxwellian particle velocity distribution, toward which all others tend in equilibrium. The values of $\langle \sigma v \rangle$ for the D-D and D-T reactions are shown in Fig. 2. In this plot the kinetic temperature is given in units of kiloelectronvolts; 1 keV kinetic temperature = 1.16×10^7 K. Just as in the case of the cross sections themselves, the most striking feature of these curves is their extremely rapid falloff with temperature at low temperatures. For example, although at 100 keV for all reactions $\langle \sigma v \rangle$ is only weakly dependent on temperature, at 1 keV it varies as $T^{6.3}$ and at 0.1 keV as T^{133}! Also, at the lowest temperatures it can be shown that only the particles in the "tail" of the distribution, which have energies large compared with the average, will make appreciable contributions to the reaction rate, the energy dependence of σ being so extreme.

Critical temperatures. The nuclear fusion reaction can obviously be self-sustaining only if the rate of loss of energy from the reacting fuel is not greater than the rate of energy generation by fusion reactions. The simplest consequence of this fact is that there will exist critical or ideal ignition temperatures below which a reaction could not sustain itself, even under idealized conditions. In a fusion reactor, ideal or minimum critical temperatures are determined by the unavoidable escape of radiation from the plasma. A minimum value for the radiation emitted from any plasma is that emitted by a pure hydrogenic plasma in the form of x-rays or bremsstrahlung. Thus plasmas composed only of isotopes of hydrogen and their one-for-one accompanying electrons might be expected to possess the lowest ideal ignition temperatures. This is indeed the case: It can be shown by comparison of the nuclear energy release rates with the radiation losses that the critical temperature for the D-T reaction is about 4×10^7 K. For the D-D reaction it is about 10 times higher. Since both radiation rate and nuclear power vary with the square of the particle density, these critical temperatures are independent of density over the density ranges of interest. The concept of the critical temperature is a highly idealized one, however, since in any real cases additional losses must be expected to occur which will modify the situation, increasing the required temperature.

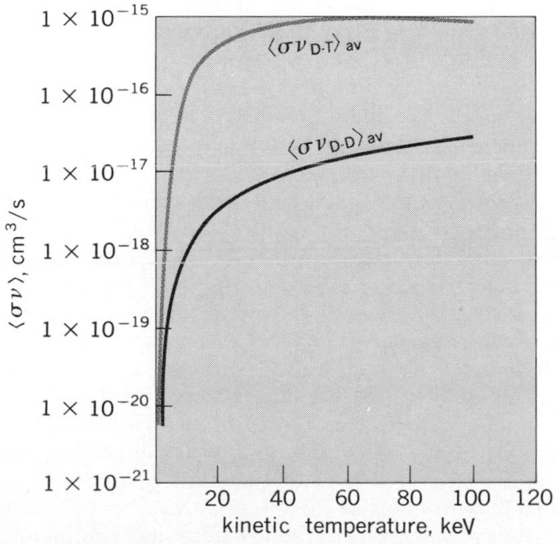

Fig. 2. Plot of the values of $\langle \sigma v \rangle$ versus kinetic temperature for the D-D and D-T reactions.

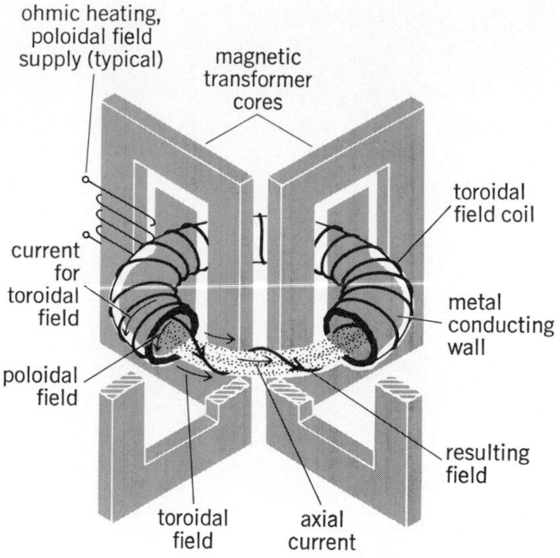

Fig. 3. Schematic illustration of a tokamak device. *(From R. F. Post and F. L. Ribe, Fusion reactors as future energy sources, Science, 186:397–407, 1974)*

FUSION REACTOR

Intense interest in nuclear fusion arises from its promise as a safe and inexhaustible source of energy for the future. Fusion reactors do not yet exist, but studies of the physics and technology that will be needed to construct such reactors have been underway since the 1950s.

The two key problems in achieving net power from a fusion reactor are, first, to heat the fusion fuel charge to its required high temperature, and second, to confine the heated fuel for a long enough time for the fusion energy released to exceed the energy required to heat the fuel to its conbustion temperature, including all relevant losses.

The problem of achieving fusion power is in fact dominated by the quantitative requirements associated with the fusion process. The plasma heating technique employed must be capable of raising the fusion fuel charge to kinetic temperatures of order 100,000,000° or higher. The confinement system must be capable of satisfying stringent requirements on confinement time (which could be as long as seconds in some cases). At the same time it must be capable of sustaining the strong outward gas pressure exerted by the fuel charge. Furthermore, since the rate at which fusion power is generated varies as the square of the fuel density, for any continuously operating fusion reactor engineering limits on heat transfer must be taken into account, thus limiting the fuel density for such systems to a small fraction of the particle density of atmospheric air. At

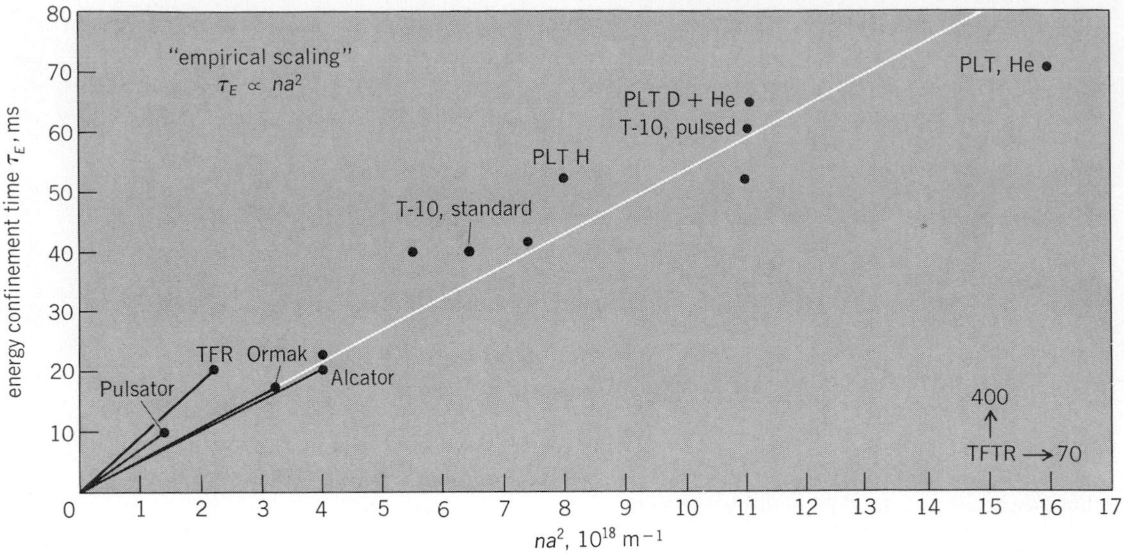

Fig. 4. Plot of "empirical scaling law" results for several tokamak devices, including results projected for the large tokamak fusion test reactor (TFTR).

higher density it is only possible to conceive of pulsed operation—basically microexplosions. The quantitative requirements of fusion therefore strongly limit the possible approaches to fusion.

Two generically different approaches have emerged as constituting the most promising avenues to the eventual achievement of net fusion power, namely, magnetic confinement and pellet fusion.

Magnetic confinement relies on the fact that at fusion temperatures the fusion fuel charge will be completely ionized, that is, it will exist solely in the plasma state. Charged particles can be held trapped by a properly shaped magnetic field, and are thereby isolated from the reactor chamber walls, physical contact with which would instantly cool the plasma.

Pellet fusion aims at the same objective, but by an entirely different route. Here the idea to rapidly heat and compress a tiny fuel pellet, carrying out the entire operation so quickly that fusion can take place before the pellet flies apart—that is, the confinement is properly called "inertial." The major technical effort on pellet fusion is centered on the use of high-powered lasers to accomplish the heating and compression; substantial activity is also being devoted to pellet fusion induced by bombardment of the pellet with very-high-intensity electron beams; and use of heavy ions as the ignition probe is now receiving serious study.

Magnetic confinement. Magnetic confinement of a fusion plasma depends on the nature of the plasma state. The plasma may be viewed as an electrically conducting gas that exerts an outward pressure, or as a collection of free positive and negative charges. The pressure exerted by the plasma can be resisted by the electromagnetic stresses associated with a strong magnetic field; the individual charged particles can at the same time be guided by a properly shaped magnetic field that forces these particles to execute orbits that remain within the vacuum chamber surrounding the plasma without contact with the walls.

Adequate stability of the confined plasma is a prime requirement for effective magnetic confinement; otherwise, particles can escape prematurely, before having a sufficient probability to fuse. Thus, finding means for suppressing the inherent tendency for confined plasma to become unstable has been one of the central goals of nuclear fusion research since its inception.

There have been many types of magnetic confinement systems proposed since the inception of fusion research. Three generic types appear to have the most promise: the tokamak, the mirror and tandem mirror systems, and field-reversed systems.

Tokamak. The tokamak (Fig. 3) is a closed or toroidal (doughnut-shaped) confinement system. It uses confining fields that represent a combination of a strong toroidal field (that is, field lines directed the long way around the toroid), with a weaker poloidal field (field lines circling the short way around the torus). The toroidal field is generated by external coils that encircle the chamber; the poloidal field is generated by a strong toroidal electric current induced to flow in the plasma by transformer action. The field line pattern that results is helical. In the tokamak, the circulating current not only provides the main confining force through the generation of the poloidal magnetic field but also performs the important function of initial heating (ohmic heating) of the plasma.

The strong toroidal field has the main function of stabilizing the plasma against "kinking" magnetohydrodynamic instability modes. The necessity of having a strong toroidal external field has the disadvantage of limiting the beta value of the tokamak to a few percent. Beta, in magnetic fusion, is the ratio of the energy density of the plasma to that of the applied field. High beta is desirable from an economic standpoint to maximize the utilization of the externally generated field, thus minimizing the capital cost of the magnet system relative to the fusion power output.

As a closed device the tokamak has the important advantage that its plasma confinement time increases as the square of the radius, a, of the plas-

NUCLEAR FUSION

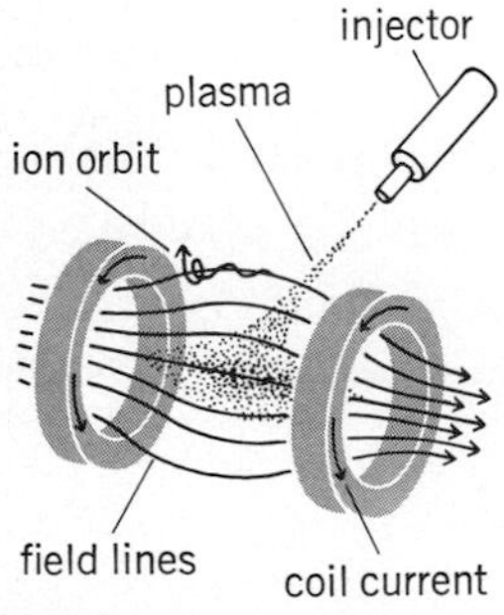

Fig. 5. Schematic illustration of mirror machine using simple mirrors. (*From R. F. Post and F. L. Ribe, Fusion reactors as future energy sources, Science, 186: 397–407, 1974*)

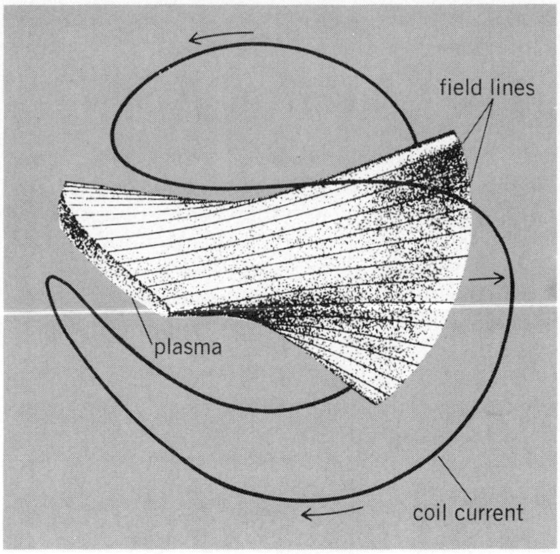

Fig. 6. "Baseball" coil configuration, producing a magnetic well mirror field.

ma column (actually, empirically, as na^2, where n is the plasma density; Fig. 4). This property is a consequence of the fact that as long as gross stability is maintained, plasma particles can be lost only by diffusion across the field, and such losses proceed on a time scale which increases with the square of the characteristic distances involved. Thus, adequate confinement times can always be achieved by scaling up the dimensions. Though large (meters), the plasma diameters for tokamak plasmas as projected from present experiments appear to be acceptable for practical fusion power plants.

Mirror machine and tandem mirror. The mirror machine (Fig. 5) is an open-ended system in which a hot plasma is held trapped by repeatedly reflecting its particles between magnetic mirrors (regions of intensified magnetic field at each end of the confinement chamber).

An important property of magnetic mirror systems is that their fields can be shaped so as to create a magnetic well, that is, a confining field that has a nonzero minimum surrounded by closed contours of increasing magnetic intensity. One type of magnetic well system is the baseball coil (Fig. 6). When confined in a magnetic well, a plasma is restrained from exhibiting any form of gross instability, up to plasma pressures comparable to those of the confining field, that is, up to beta values of order unity.

Mirror systems are in contrast with the tokamak or other closed systems. In the latter there is no need to confine particles as far as their motion along the field lines is concerned. In a mirror machine the required longitudinal confinement is provided by the repelling force exerted on charged particles as they spiral along field lines that are converging, that is, as they move toward regions of increasing field strength. Particles which spiral with sufficiently steep helical pitch angles will be repelled strongly enough to be reflected, that is, trapped, by the mirrors. It follows that this type of mirror system cannot confine an isotropic plasma; only particles whose pitch angles are sufficiently large to lie between the loss cones defined by the strength of the mirror fields relative to the field intensity between the mirrors will be contained.

These systems must therefore rely on the injection of intense beams of energetic neutral atoms to maintain the plasma temperature and density in competition with particle leakage through the mirrors.

An important disadvantage of conventional mirror systems for fusion purposes is that the leakage of particles through the mirrors, arising as it does from collisions that deflect the ions into pitch angles lying within the mirror loss cone, occurs at a rate comparable to the rate of fusion reactions. In this circumstance the energy gain factor Q – the ratio of fusion power to plasma heating power – is at best not much larger than 1, implying an economically unacceptably large fraction of recirculated power needed to maintain the plasma. This deficiency in the mirror concept has stimulated the development of the tandem mirror and the field-reversed mirror concepts, discussed below.

An important aspect of mirror confinement is the positive ambipolar potential of the plasma that arises naturally from its operation. Since the collision frequency for electrons is higher than that for ions (because of the electrons' higher velocities), other factors being equal, they would tend to diffuse into the loss cone and be lost much more rapidly than would the ions. But any such differential loss rate would result in the buildup of a net positive charge in the confined plasma, thus driving it to a positive potential with respect to its surroundings, a potential sufficient to bring the electron loss

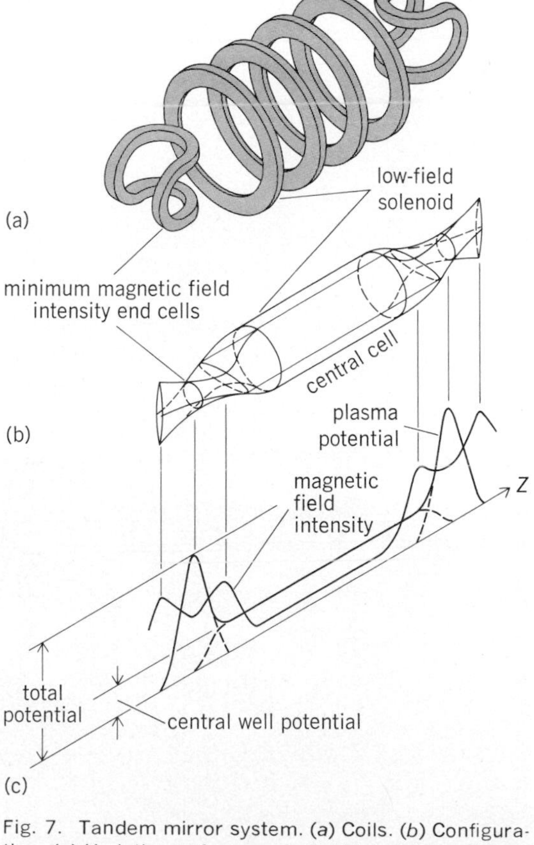

Fig. 7. Tandem mirror system. (a) Coils. (b) Configuration. (c) Variations of magnetic field intensity and plasma potential.

rate to equality with that of the ions.

In the tandem mirror (Figs. 7 and 8) the ambipolar potentials are used to confine a fusion plasma, resulting in greatly improved confinement relative to that in a single mirror cell. A large-volume central chamber, in which the fusion plasma is to be confined, is stoppered at each end by two small-volume mirror cells having a high-ion-temperature plasma. The ambipolar potential of each of these plugs, being relatively more positive than that of the central cell, serves to confine (axially) the ions of the central cell plasma; its electrons are confined by the overall positive potential of the system.

Field-reversal systems. Magnetically confined plasmas exhibit the property of diamagnetism. That is, their confinement necessarily involves the existence of internal electric currents that act to reduce the strength of the confining magnetic field within the plasma relative to the vacuum value it would have in the absence of the plasma. It is in fact these internal diamagnetic currents that, by interacting with the externally generated field, produce a body force balancing the outward-acting pressure of the plasma. If these diamagnetic currents could be employed to provide the major part of the confining force, the confining effect of the external field would be greatly enhanced, with consequent economic and other practical benefits.

The earliest attempted embodiment of this concept was the toroidal pinch, the progenitor of the tokamak, which was unsuccessful because its self-constricted plasma column was subject to rapidly growing kinking instabilities of hydromagnetic origin. However, with increased understanding of magnetic confinement, ways to circumvent such problems in pinch-confined plasmas have been proposed. These ideas, generically describable as field-reversed systems, have taken four distinguishable forms: field-reversing particle rings; the field-reversed mirror; the field-reversed pinch; and self-field tokamaks.

1. Field-reversing particle rings. These forms employ a ring of high-energy charged particles. The particles circle in large orbits in an external mirror field and generate a diamagnetic current that is sufficiently strong to reverse the direction of the field within the ring, thereby producing a poloidal field with closed field lines that traps the fusion plasma.

2. Field-reversed mirror. This is a mirror system where tangentially injected neutral beams maintain ion diamagnetic currents sufficient to create a field-reversed state, thus inhibiting the rate of escape of plasma ions, which must now diffuse across the field-reversed region before they reach the open field lines of the mirror field (Fig. 9).

3. Field-reversed pinch. This is a toroidal pinch in which the pinch (poloidal) field has superposed on it a toroidal field, the direction of which is reversed in the deep interior of the plasma, relative to its direction in the exterior parts.

4. Self-field tokamaks. These are tokamaklike configurations in which the main confining fields, both poloidal and toroidal, are the result of diamagnetic currents flowing within the plasma. An example is the spheromak configuration, in which the plasma and the conducting chamber around it (needed to preserve stability) are roughly spherical.

All of the field-reversed magnetic confinement systems described above have as their objective the more efficient use of magnetic field for fusion confinement purposes. Particularly as exemplified by the field-reversed mirror, the plasma beta val-

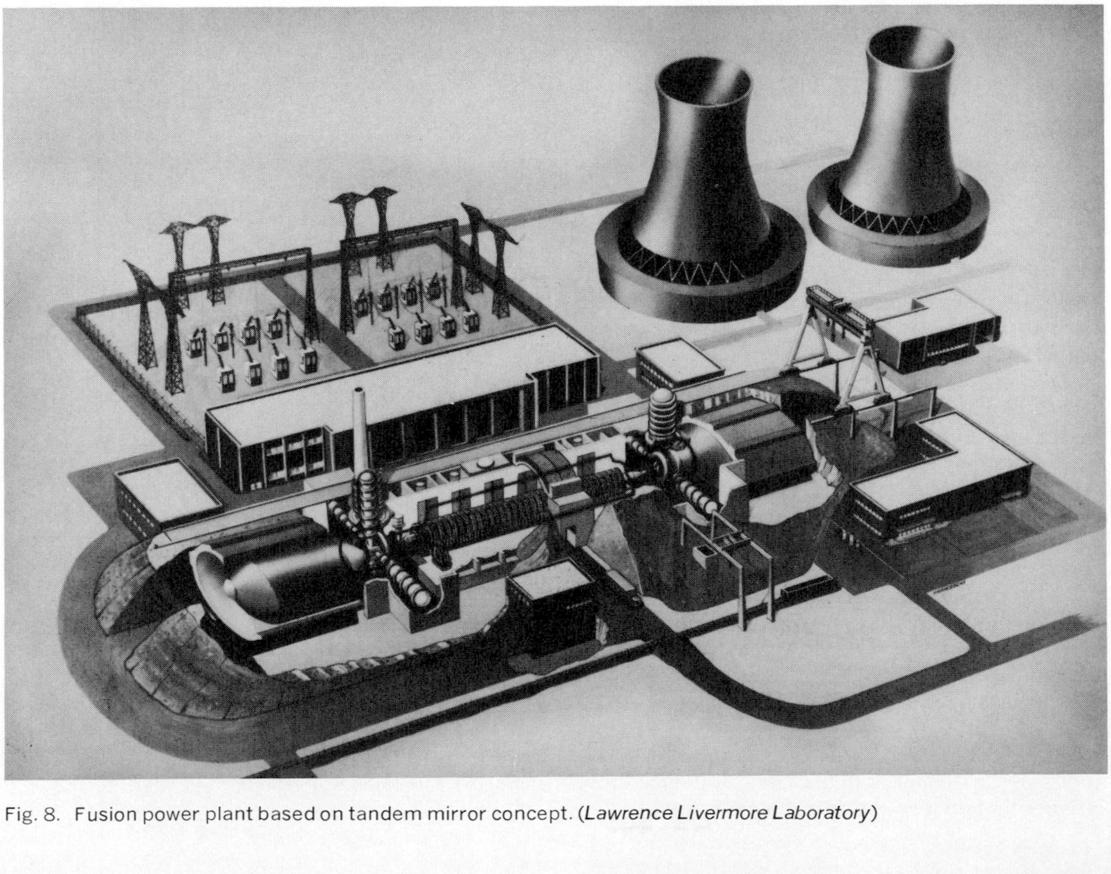

Fig. 8. Fusion power plant based on tandem mirror concept. (*Lawrence Livermore Laboratory*)

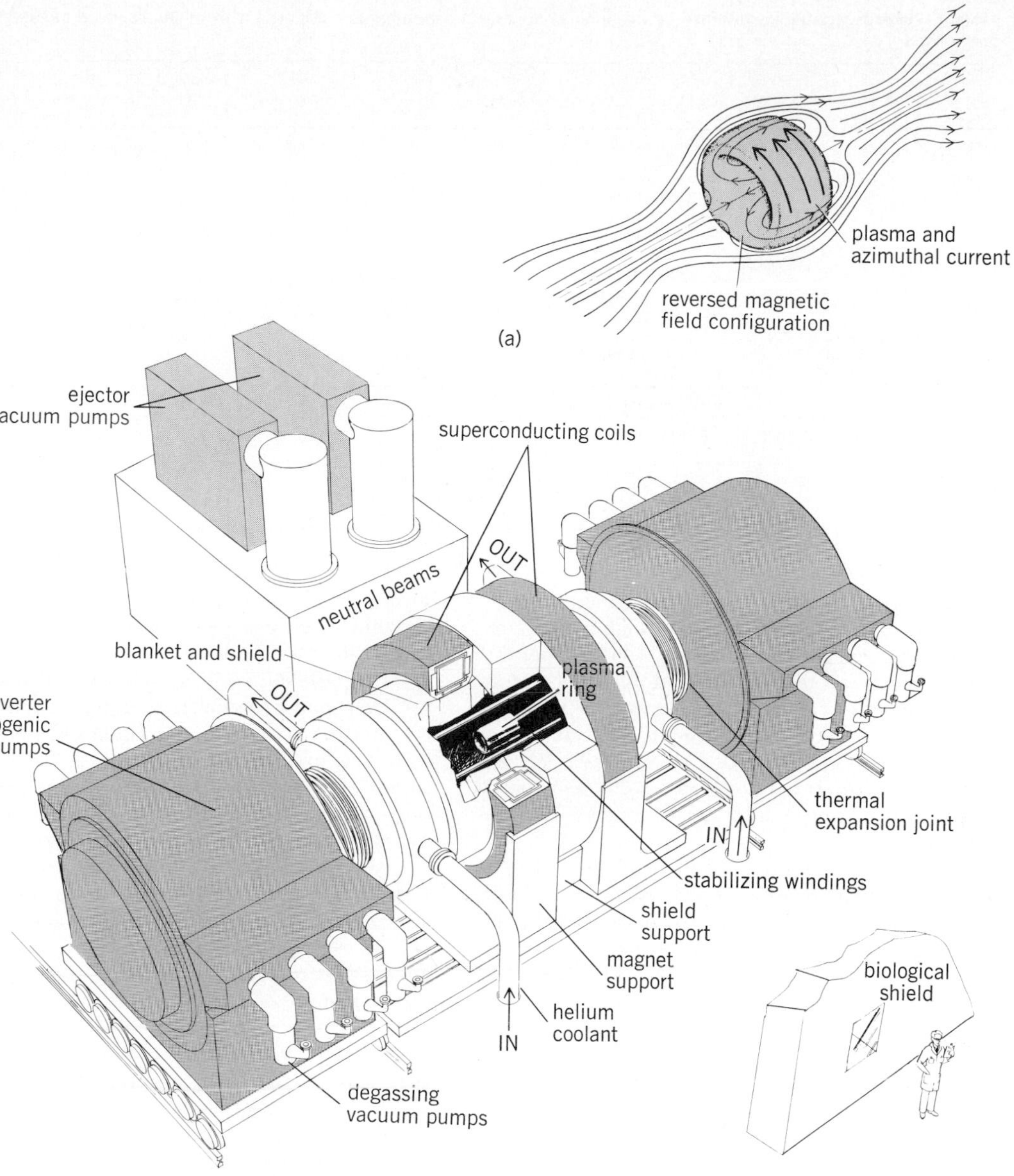

Fig. 9. Field-reversed mirror system. (*a*) Configuration, showing field lines and diamagnetic currents. (*b*) 20-MW demonstration power plant. (*Lawrence Livermore Laboratory*)

ue, as determined relative to the externally generated magnetic field, is very high. This could result in compact fusion power systems with improved economics, that is, lowered capital costs for the magnet and its support structures. For example, calculations for the field-reversed mirror indicate a fusion power density with the D-T reaction of order 500 times that for conventional tokamak. Alternatively, the increased magnetic efficiency could in principle allow the use of so-called advanced fusion fuel cycles, involving the D-^{3}He or other reactions, having special advantages in terms of reduced neutron fluxes or suitability for direct energy conversion. The basic physics understanding is, however, at present too limited to assure the practical success of any of the above systems, although there are indications that all of them can exhibit stable confinement.

Progress. In theory, magnetic confinement has been long perceived as an almost ideal solution to the fusion problem. In practice, over the earlier years of fusion research it was found very difficult to achieve confinement that approached the optimistic theoretical predictions, owing to the ubiquitous problem of plasma instabilities. However, refinement of the theory, coupled with experiments performed with technologically sophisticated, scaled-up apparatus, has led to major advances toward the quantitative goals of fusion power. Plasma temperatures and density–confinement time products (discussed below) have both been brought within a factor of 10 or less of breakeven values, and the next generation of tokamaks will probably come close to, or actually achieve, an energy breakeven situation (fusion energy yield equaling heat energy content of the confined plasma).

Increasingly, the emphasis in fusion research is on problems of practical or economic origin. Scientifically, this means a search for magnetic configurations that will result in smaller, more practical, or more efficient confinement geometries. Tech-

Table 1. Plasma parameters attained by magnetic confinement experiments compared with values believed needed for a fusion power plant

Approach	Device	Location	Plasma diameter (d), cm	Particle density (n), cm^{-3}	Ion temperature (T_i), keV	Confinement time (τ), ms	Lawson product ($n\tau$), s/cm^3	β_{max}
Tokamak	T-4 (1970)	Kurchatov Institute, Moscow	30	3×10^{13}	0.4	10	6×10^{11}	.0006
	TFR (1974)	Fontenay-Aux-rose, Paris	40	4×10^{13}	0.8	15	6×10^{11}	.0016
	Alcator A (1975)	Massachusetts Institute of Technology	18	6×10^{14}	0.8	20	1.2×10^{13}	.006
	PLT (1976)	Princeton University	90	6×10^{13}	1.0	50	3×10^{12}	.005
	PLT (neutral beams, 1978)	Princeton University	90	4×10^{13}	5.0	25	$\sim 10^{12}$	.013
Conventional mirror	PR-6 (1971)	Kurchatov Institute, Moscow	10	2×10^{12}	0.5	0.15	3×10^{8}	.0016
	Baseball II (1972)	Lawrence	20	2×10^{9}	2.0	1000	2×10^{9}	–
	2XIIB (1977)	Livermore	15	$\sim 10^{14}$	13.0	1.	$\sim 10^{11}$	>1.0
Tandem mirror	TMX (1979)	Laboratory	50	2×10^{13}	0.2	~5.	$\sim 10^{11}$	~.2

	Power plant requirements			
	Ion temperature (T_i), keV	Lawson product ($n\tau$), s/cm^3	β	Remarks
Tokamak	15	$>10^{14}$	0.1	~1000 MWe, ignited mode
Conventional mirror	150	$>10^{13}$	0.8	High recirculated power required
Tandem mirror	30	$>10^{14}$	0.5	~500 MWe, ignited mode

nologically, this means that more emphasis is being placed on the problems of high-field superconducting magnet coils, high-power neutral beams, vacuum technique, and materials problems associated with the inner wall of the confinement chamber, which must withstand high fluxes of energetic neutrons as well as relatively high-energy neutral atoms and charged particles (hydrogen and helium ions) escaping from the plasma.

A rough (and sometimes misleading) index of progress toward fusion requirements, mentioned above, is the $n\tau$ confinement factor used in the Lawson criterion, where n is the particle density and τ is the average confinement time of the plasma. This parameter and the plasma ion kinetic temperature T_i at which the confinement is achieved, together provide a useful index of progress toward the goal of fusion power. Table 1 compares values $n\tau$ and T_i achieved in experiments in magnetic confinement with nominal values estimated to be required for a fusion power plant.

Pellet fusion. The basic idea behind pellet fusion (Fig. 10) is the rapid implosion of a high-density fusion fuel pellet to produce a heated core that will fuse before it can fly apart. As usually conceived, the implosion would result from the rapid heating and subsequent ablation of the surface of the pellet, giving rise to an inward-acting reaction force that compresses the core. But for this process to yield a net energy, that is, for it to achieve the required $n\tau$ value, very large compression factors, of order 10,000, are required. That this should be the case can be seen from simple considerations: As matter is compressed spherically, its density (n in the $n\tau$ product) increases as the tube of the radial compression factor. But the confinement time τ, here measured by the time of flight of an average particle out of the compressed core, decreased only as the first power of the radius. Thus $n\tau$ increases with the square of the radial compression factor. However, it is necessary to use tiny pellets to make this approach to fusion technically accessible from the standpoint of engineering limits on the amount of energy deliverable from the pellet drivers (lasers or particle accelerators), and on the amount of fusion energy released from the pellet that can be absorbed in surrounding structures. These limitations, taken together with the Lawson requirement, dictate the need for very large density compression factors, in turn leading to a requirement for very high compression forces [of order 10^{12} atmospheres (10^{17} Pa), ten times greater than those existing at the center of the Sun].

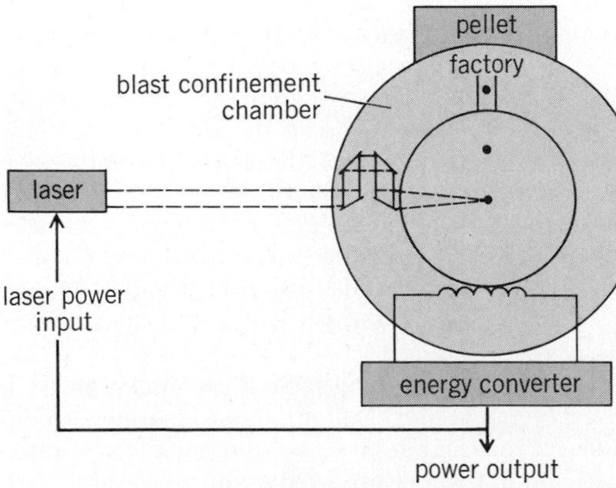

Fig. 10. Conceptual laser pellet fusion power plant.

The problems thereby posed are twofold. First, very high pulsed powers of hundreds of terawatts must be focused down to millimeter dimensions and delivered in times of nanoseconds or less. The implied peak power densities are thus of order 10^{16} W/cm^2, made necessary by the requirement that an almost instantaneous ablation of the outer surface of the pellet should occur before heat can flow into the interior of the pellet; premature heat flow would prevent reaching the required compression factors. This problem can in principle be solved by using a sufficiently large array of high-power lasers, focusing their beams so as to uniformly illuminate the surface of the pellet (Figs. 11 and 12). Alternatively, converging beams of electrons or ions from particle accelerators would be employed.

Second, if the compression process is not carried out with high uniformity, it will go askew; small errors in uniformity can lead to a major reduction in the achievable compression. Sophisticated aiming and timing techniques must be used in the driver, and any instabilities must not be allowed to spoil the symmetry of the compression or lead to undue mixing or preheating.

High temperatures and substantial $n\tau$ values, comparable to those achieved in magnetic fusion, have been attained in pellet fusion experiments. Table 2 lists some of the parameters achieved in laser pellet experiments. Continued progress can be expected, as new and larger facilities come on line, but increase in performance by some orders of magnitude is needed before breakeven can be attained.

Fig. 11. Portion of the Shiva laser pellet fusion facility at Lawrence Livermore Laboratory, showing 6 of the 20 laser amplifier trains, looking at the output end. System is designed to deliver more than 30 TW of optical power in less than 10^{-9} s. (*Lawrence Livermore Laboratory*)

POWER PLANTS

Preliminary studies have been made of the forms that fusion power plants might take, following some of the approaches outlined above. These studies cannot of course be definitive, but they have helped to indicate the sizes, capital costs, and special engineering problems that are likely to characterize fusion power plants, insofar as they can now be visualized.

The types of power plants that have been studied encompass both pulsed and steady-state fusion systems, operating in either a driven mode or one in which plasma ignition would be achieved. A driven fusion power system is one in which the plasma temperature is maintained primarily by a continuous input of energy—for example, by the injection of high-intensity beams of energetic neutral fuel atoms—thereby maintaining the required kinetic temperature and density of the plasma. Electrical power needed to produce the beams would be obtained by recirculating a portion of the electrical output of the plant. A positive power balance, that is, net output power ($Q > 1$), would be possible here only when the recirculated power is less than the electrical output as recovered from the plasma, including not only the energy content of the reaction products but that of the unreacted part (electrons and heated fuel ions); economical power would likely be possible only if the recirculated power fraction were to be much smaller than the power produced (that is, Q much greater than 1). By contrast, a fusion system operated in an ignition mode is one where the energy deposited directly within the plasma by charged reaction products (3.5-MeV alpha particles in the case of the D-T reaction, that is, only 20% of the total fusion energy release) is sufficient to maintain the plasma temperature, including the requirement for heating up new cold fuel particles introduced to maintain the fuel plasma density as fusion "combustion" proceeds. The ignition mode is therefore more demanding with respect to confinement time than is a driven mode, but in principle could be technically simpler, since it does not impose as strict requirements on the efficiencies of the energy recovery and plasma heating systems.

By exploiting the increase in confinement time associated with an increase in plasma radius and extrapolating from the performance of present devices, it appears that large tokamaks could operate in an ignition mode. Conventional and tandem

Fig. 12. Design of Shiva Nova laser pellet fusion experimental facility at Lawrence Livermore Laboratory. System is designed to provide enough power (200–300 TW) and energy (200 kJ) to obtain significant gain from laser fusion targets. (*Lawrence Livermore Laboratory*)

Table 2. Examples of laser fusion facilities

Name	Location	Energy pulse, kJ	Peak power, TW	Compression achieved (× liquid density)
Zeta	University of Rochester	1.2	3–4	7–20
Chrome I	KMS Fusion, Inc., Ann Arbor, MI	1.0	2	7–35
Helios	Los Alamos Scientific Laboratory	5–10	10–20	8–30
Shiva	Lawrence Livermore Laboratory	10	26	30–160

mirror systems, requiring as they do the maintenance of plasma in the mirror cells, would need to be operated in a driven mode. However, in the case of the tandem, it seems that it will be possible to achieve ignition in the central cell plasma, with attendant simplifications and other advantages. Laser pellet systems, although the pellet itself would be expected to ignite, necessarily require recirculated power to initiate the burn. To achieve high net power relative to recirculated power would seem to imply the need for pellet energy gains *(Q)* of 100 or more.

Considering the magnetic confinement approach, systems studies have led to some important conclusions: Fusion power plants based on the tokamak principle in its conventional form will be relatively large both in size and in power output; electrical power outputs of 500 to some thousands of megawatts are likely to be typical. Power plants utilizing the tandem idea might be somewhat smaller in physical size than conventional tokamaks, and possibly also capable of somewhat lower plant electrical outputs than tokamaks, still satisfying economic requirements. Power plants based on field reversal, such as the field-reversed mirror, might be the smallest of all, exhibiting the highest fusion power density while being both compact in size and permitting (at least in the demonstration phase) electrical power outputs as low as tens of megawatts.

Another general result of the design studies is to show the importance of choice of materials and heat transfer characteristics of the inner wall of the containment chamber. The flux of 14-MeV neutrons through this wall coming from D-T reactions in the plasma will cause localized heating, radiation damage, and induced radioactivity. Thus the design of this portion of any D-T fusion power plant can be expected to be of critical importance. The materials chosen need to be picked not only for their resistance to radiation damage but also for minimum activation (that is, minimal yield and short half-life for the neutron-induced radioactivity). It does appear that first-wall materials having the desired characteristics can be developed. Another critical factor is that of the generation of the confining magnetic field. Here the criterion is to achieve the required field (which may be very high for some approaches) at the least capital cost and for the least expenditure of energy. Fortunately, the development of practical high-current-density, high-field superconductors appears to provide an almost ideal solution to this problem.

[RICHARD F. POST]

Bibliography: F. Chen, *Introduction to Plasma Physics*, 1974; N. Krall and A. Trivelpiece, *Principles of Plasma Physics*, 1973; R. F. Post, Controlled fusion research and high temperature plasmas, *Annu. Rev. Nucl. Sci.*, 20:509–558, 1970; G. Schmidt, *Physics of High Temperature Plasmas*, 2d ed., 1979; L. Spitzer, *Physics of Fully Ionized Gases*, 1962.

Nuclear magnetic resonance (NMR)

A phenomenon exhibited by a large number of atomic nuclei which is based upon the existence of nuclear magnetic moments which are associated with quantized nuclear spins. These nuclear moments, when placed in a magnetic field, give rise to distinct nuclear Zeeman energy levels between which spectroscopic transitions can be induced by radio-frequency radiation. Plots of these transition frequencies, called spectra, furnish important information about molecular structure and sample composition.

Spectroscopy. Spinning nuclei with nonzero angular momentum behave like tiny magnets. When placed in an external magnetic field, these tiny nuclear magnets assume certain allowed orientations (Zeeman energy levels) with respect to it. The nuclei can be reoriented by adding energy of the exact transition frequency.

Nuclei excluded from consideration are those with zero angular momentum or spin ($I=0$) and therefore zero magnetic moment (for example, the important C^{12} and O^{16} isotopes). High-resolution methods generally exclude nuclei with I greater than $\frac{1}{2}$, as they possess electrical quadrupole moments which interact with electric field gradients so as to broaden the magnetic resonance signals and prevent resolution of closely spaced resonance lines. High-resolution techniques, therefore, have been limited primarily to the nuclear species of spin $\frac{1}{2}$ (for example, H^1, C^{13}, F^{19}, and P^{31}).

As the separation between the nuclear Zeeman levels is directly proportional to the strength of the perturbing magnetic field, the transition frequency can be varied for a given nucleus by merely changing the applied magnetic field. In this regard, nuclear magnetic resonance (NMR) spectroscopy is unlike other spectroscopic methods, in which the investigator is unable to control the frequency of the spectral transition. Thus an NMR spectrum may be secured by varying the magnetic field to bring the separation of the Zeeman levels into correspondence with a constant irradiating frequency; or the alternative experimental method may be used, in which a constant magnetic field is employed and the irradiating frequency is varied over the range of spectroscopic frequencies.

Chemical shift. It is this unique field-frequency relationship that makes possible the application of NMR spectroscopy in molecular studies. Although identical nuclei have the same frequency dependence upon the magnetic field, a difference in the chemical environment can modify the applied magnetic field, so that all such nuclei in the same sample do not experience the same net magnetic field, and thus they do not resonate at exactly the same frequency. The corresponding spectral shift in the transition frequencies between two such chemically nonequivalent nuclei is referred to as the chemical shift. Being directly proportional to the total applied field, this parameter is recorded in the relative units of parts per million (ppm).

It is convenient to subdivide the chemical shift parameter into a diamagnetic term and a paramagnetic term. Diamagnetism induced by an applied magnetic field is a well-known phenomenon and is attributed to Lamb currents in the molecular electrons. Diamagnetic shielding decreases the field intensity at the nucleus and thereby decreases the separation between the nuclear Zeeman levels. A higher applied field or a lower frequency is now required to attain resonance, and the signal moves to the right (upfield) in the spectrum. Considered simply, this part of the chemical shift is proportional to the electron density in that segment of a molecule in which the magnetic nucleus is found, and therefore reflects in an approximate manner the charge polarization of the molecular electrons. A paramagnetic shift (downfield), attributed to increased magnetic field intensity at the nucleus, is observed in some cases as a result of diamagnetic currents existing in remote anisotropic groups of a molecule. Aromatic systems with their associated ring currents constitute typical examples of such anisotropic groups which enhance the magnetic field in certain regions of space external to the aromatic ring. Finally, in molecules of certain symmetries the magnetic field can remove the quenching of orbital angular momentum associated with electrons involving *p*-orbitals in completed subshells. There is evidence that this paramagnetic interaction may be a significant one in C^{13} and F^{19} magnetic resonance studies. However, theoretical estimates of the magnitude of the several terms in the chemical shift parameter involve considerable difficulty, and the relative importance of the various shielding mechanisms is not completely resolved.

Because nuclei contained in various functional groups have their own characteristic resonance frequencies, NMR has become an important spectroscopic tool for molecular structure determination. Many tables of chemical shifts are available for identification purposes. Figure 1 schematically portrays the distribution of proton chemical shift values for a few selected compounds. Low diamagnetic shielding is observed for the electropositive protons in the two acid compounds. The chemical shifts of less acidic methyl groups are found at higher fields. The shift to lower fields with the addition of an electronegative group is exhibited by the series CH_3Cl, CH_2Cl_2, and $CHCl_3$. Finally, the relatively low field position of the benzene resonance is explained as noted before by a paramagnetic shift resulting from π-electron ring currents. Thus the chemical shifts tell what functional groups may be present in the sample.

Spin coupling. The NMR spectrum of ethyl bromide portrayed in Fig. 2*a* is presented as an example of a moderately high-resolution spectrum in which the resonance peaks of the chemically nonequivalent methylene and methyl protons are separated by a chemical shift of 1.77 ppm. The relative intensities in these two peaks of 2 and 3 reflect the number of hydrogens in the methylene and methyl groups, respectively.

With additional improvement in resolution, each of the ethyl bromide peaks subdivides into the multiplet structure shown in Fig. 2*b*. The methylene resonance is observed to split into a quartet of lines, whereas the methyl peak is replaced by a triplet of lines. Resulting from a nuclear spin-spin interaction between the two sets of protons, the multiplet pattern can be rationalized on the basis

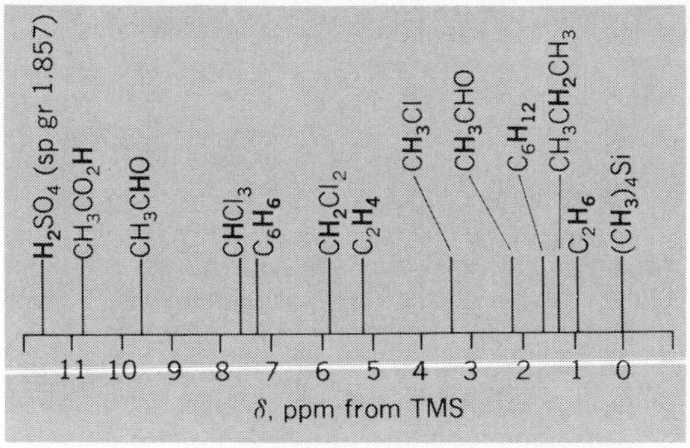

Fig. 1. Chemical shifts portrayed schematically for several representative compounds. Decreasing values of δ correspond to increasing magnetic field in a constant-frequency spectrometer. The scale calibration is obtained from the resonance signal of a small amount of tetramethylsilane (TMS) placed in the sample tube to provide a zero reference point.

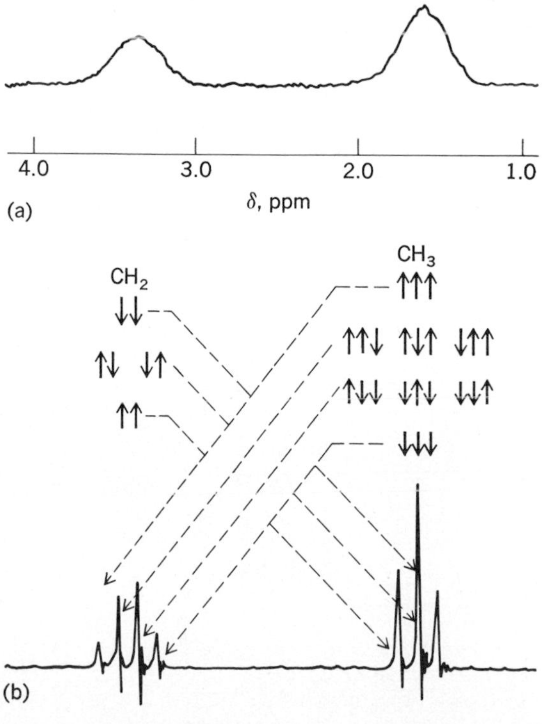

Fig. 2. Nuclear magnetic spectra of ethyl bromide (CH_3CH_2Br), with schematic representation of nuclear spin orientations. (*a*) At a moderate resolution. (*b*) At a high resolution. (*Courtesy of T. Brown, University of Utah*)

of the allowed orientation of the methylene and methyl protons as shown schematically in the figure. Thus, the magnetic field experienced by the methylene protons is perturbed by the four different distinguishable spin orientations exhibited by the methyl protons.

Furthermore, the 1:3:3:1 statistical weights for these orientations are reflected in the intensity pattern of the methylene multiplet. In a like manner the two protons in the methylene group induce a 1:2:1 triplet in the methyl peak. Normally the coupling constant, which measures the multiplet splittings due to spin-spin interactions, attenuates rapidly for protons separated by more than two or three chemical bonds, and only neighboring protons interact significantly. Thus, in addition to the identification of functional groups from the chemical shifts, the multiplicity in the splitting patterns and the magnitude of the coupling constant tells how the groups are arranged in the molecule.

It is not always possible to interpret spectra in the manner indicated by Fig. 2, where the splitting patterns can be explained on the basis of a first-order perturbation of one spin system by a second, neighboring group. Specifically, whenever the spin-spin coupling constant becomes comparable to or larger than the chemical shift parameter, higher-order mixing of the spin states occurs to give spectra of considerably greater complexity. As an example, the spectrum of 1-bromo-2-chloroethane is given in Fig. 3. This spectrum is derived from a molecule differing only slightly from that considered in Fig. 2; yet the spectral features do not resemble the simple pattern shown in Fig. 2*b*. The similarity between the bromine and chlorine atoms results in chemical similarity between the two methylene groups, and the chemical shift between these two sets of protons is reduced to a value comparable with the intramolecular spin-spin coupling constants. Higher-order splitting features are commonly observed in proton NMR spectra, and exact interpretation usually requires detailed numerical analysis with a high-speed digital computer. Further complexity is introduced into spectral features whenever the spin-spin coupling values between the two sets of chemically equivalent nuclei are unequal. This element of complexity, which is referred to as magnetic nonequivalence, is found in Fig. 3, where the inequalities in the coupling constants between protons in the two methylene groups are not eliminated by averaging over the several rotametric conformations existing for this molecule. Were the two methylene groups in a 1,2-disubstituted ethane to have the same chemical shift (either by coincidence or from molecular symmetry in the event that both substituents are identical), then all splittings would vanish and a single resonance line would be observed. Spin-spin interactions between nuclei which are both chemically and magnetically equivalent do not affect the spectral features, and coupling constants for such interactions therefore become unobtainable.

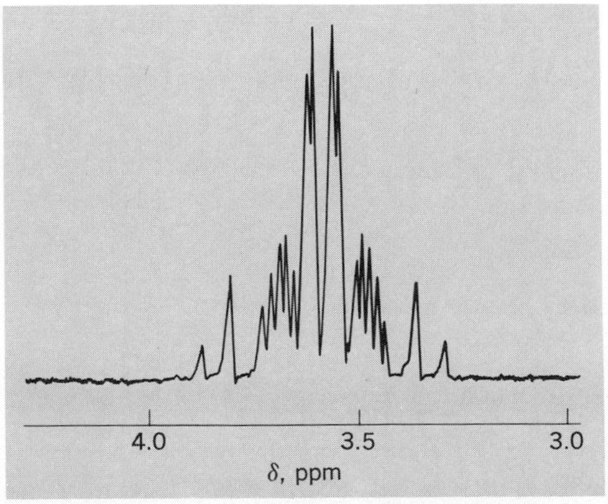

Fig. 3. High-resolution spectrum of 1,2-bromochloroethane ($ClCH_2CH_2Br$), exhibiting higher-order splittings and complexities due to magnetic nonequivalence. *(Courtesy of T. Brown, University of Utah)*

Theoretical interpretation of coupling constants indicates that nuclear spin-spin interactions are transmitted through the molecular electrons. Direct magnetic interactions between nuclei through space are observed in solids to be relatively large, but these coupling terms average to zero in the liquid state under the influence of rapid molecular tumbling. As a result of the quantized orientation of magnetic moments associated with the spin and the orbital angular momentum of electrons, magnetic coupling mechanisms involving the molecular electrons do not average to zero with rapid molecular reorientation. Thus, spin-spin coupling values contribute to a better understanding of the electronic structure of molecules, especially in the areas of electron spin correlation and valence theory.

Spin decoupling. Spectra with higher-order splittings and overlapping multiplets are often simplified by a technique known as spin decoupling. In this procedure the coupling is removed by irradiating the coupled partner with a high-intensity radio-frequency field adjusted to its exact resonance frequency. This changes the polarization of the perturbing nuclear spin system and wipes out the coupling.

The resulting simplifications provide chemical shift data from spectra which are more easily interpreted. Furthermore, information derived with this technique can also be used in obtaining the relative signs of spin-spin coupling constants, which can assume either positive or negative values.

The resonant frequency of the perturbing nucleus can be found by the reverse process, in which the irradiating frequency is swept while the perturbed nucleus is continuously observed.

Nuclear Overhauser effect. The intensities of some resonance signals may be changed by the nuclear Overhauser effect. When one group of nuclei is irradiated, not only are the signals from the coupled nuclei simplified (decoupled) but other nuclei, nearby in space, are also perturbed, causing their signals to change in intensity (Overhauser enhancement). Since the magnitude of this enhancement decreases with the internuclear distance ($1/d^6$), this technique is a powerful test for proposed molecular structures.

Lanthanide shift reagents. Another development aids in spectrum interpretation. Certain organolanthanide reagents can be added to the sample to expand and untangle otherwise overlapped NMR spectra. Figure 4 shows this expansion for the spectra of *cis*-4-*tert*-butylcyclohexanol.

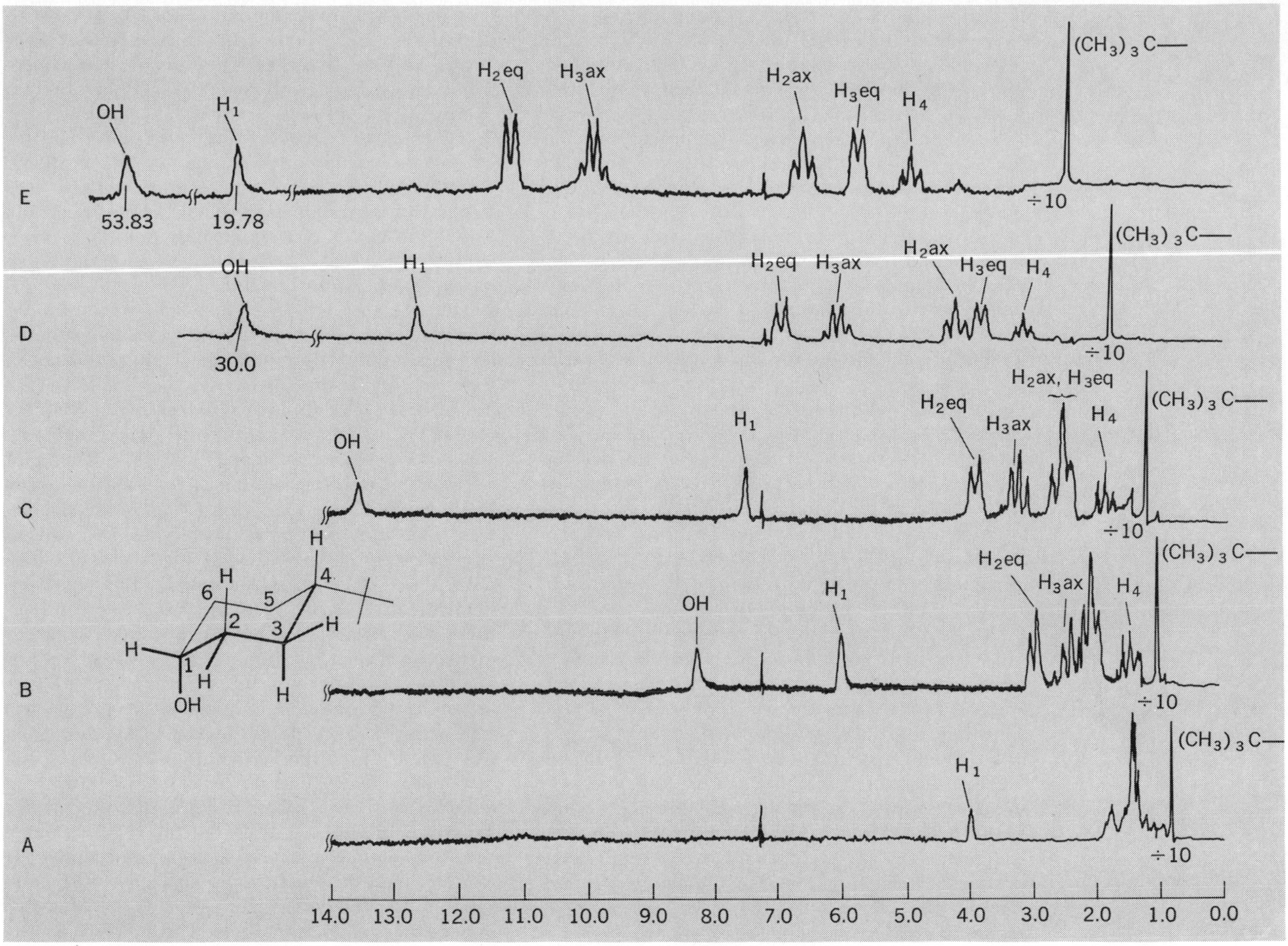

Fig. 4. Resolution of the proton NMR spectrum of *cis*-4-*tert*-butylcyclohexanol in $CDCl_3$ (curve A) by addition of increasing amounts of tris(dipivalomethanato) europium(III) (curves B–E). (*From P. V. Demarco et al., J. Amer. Chem. Soc., 92:5734, 1970*)

The phenomenon is understood as a pseudocontact shift, which means that the lanthanide reagent coordinates with an electron-rich site of the sample molecule, for example, an oxygen atom, and the deshielding effect of the lanthanide ion magnetic dipole, like a beacon, is aimed at this atom. Other nuclei in the sample molecule, for example, protons or carbons, which are observable by the NMR method will therefore be deshielded (shifted downfield in the spectrum) by an amount depending on their position (distance and angle) relative to that beacon. It is possible to quantitate these shifts, and computer programs have been written to fit molecular structures to these measured shifts. N. S. Angerman and coworkers used this new method to determine the three-dimensional structure of the important antimalarial chloroquine in acetone solution, to be compared with the molecular structure in the crystalline state as determined by x-ray crystallography.

Quantitative analysis. The technique of quantitative analysis by NMR is based on the well-established fact that the area under each peak in an NMR spectrum is directly proportional to the number of atomic nuclei causing the absorption peak. The method is applicable to many kinds of atoms. Once the peaks in the spectrum and the atoms in the sample have been related, various types of quantitative analyses are possible, including analysis of numbers of substituent groups in the molecule, analysis of the molecular composition of mixtures, elemental analyses, and molecular weight determinations. If the proper instrument settings have been determined, analyses can be made rapidly and conveniently. The method is nondestructive and is especially useful on dynamic systems. The results have good accuracy and precision.

NMR analysis, based on the intensity of magnetic resonance signals, is applicable to more than 100 gyromagnetic nuclei. It has an exact counterpart in electron paramagnetic resonance (EPR), in which the magnetic species is the unpaired electron. The determination of hydrogen (H^1) is currently most advanced because excellent instruments for proton magnetic resonance are available commercially, and the applications encompass the whole field of organic chemistry. Instruments which are sensitive to fluorine have been improved greatly, but the applications are more specialized. Instruments for C^{13}, B^{11}, P^{31}, H^2, N^{15}, Si^{29}, and O^{17}, where the inherent sensitivity is less, are improving rapidly, and new quantitative applications are expected to follow. Often, nuclei for which no instruments are available can be determined indirectly if they replace or shift the position of a hydrogen signal from the same molecule.

For brevity, the discussion emphasizes application to hydrogen, leaving the reader to develop his own specialized applications involving other nuclei.

The variables determining the intensity of an NMR signal have been thoroughly studied. They can be grouped to include (1) the number of resonating nuclei in the sample, (2) certain design characteristics of the instrument, and (3) appropriate quantum constants for the nuclei. The first variable can be singled out and the others eliminated by performing comparative measurements under carefully selected experimental conditions. The extent of absorption, revealing the number of nuclei present, is then expressed in arbitrary units rather than as the familiar percentage absorption or transmission units used in other forms of spectroscopy. Intensities are measured with respect to one another in the same sample or are referred to those of a reference sample.

Unlike other types of electromagnetic radiation, radio-frequency radiation employed in magnetic resonance experiments can be regarded as constant, for less than about 1% of it is absorbed. Thus the Beer-Lambert law can be disregarded.

To cancel instrumental effects, the spectrometer must have high stability and must be operated in such a way as to avoid errors from saturation. Stability is achieved in the newer internal-locked spectrometers by audiomodulation circuits. Saturation occurs when too many nuclei are excited, significantly decreasing the population of the ground state. The problem arises because nuclei in different molecular environments saturate at different rates. Saturation is therefore minimized by a choice of rapid sweep rate and low radiation power in exchange for some loss in sensitivity. Part of the difference is made up by using more concentrated samples and employing repetitive scans.

Relative intensities are obtained from the areas under the NMR signals in the spectrum. Areas can be measured with a planimeter, or the peaks can be cut out and weighed. If the peaks are very sharp, as they frequently are in NMR, an electronic integrator must be used to avoid errors from exceeding the response time of the graphic recorder. With integrators, the area is read as an accumulated voltage on a capacitor using a digital voltmeter, or the areas may be recorded as a step integral presentation directly above the spectrum, as seen in Fig. 5. Since peaks are not all the same width, heights can only be used if variation in widths is taken into account.

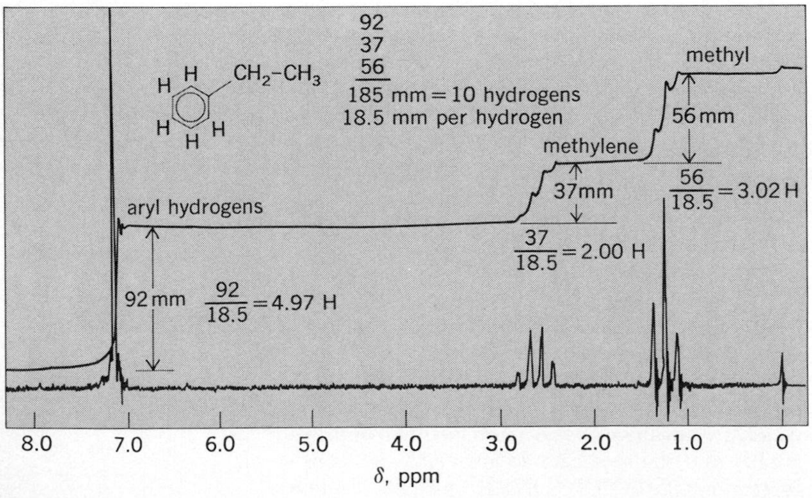

Fig. 5. Step integral representation and NMR spectrum for ethyl benzene.

The areas of strong signals can be measured more accurately than weak ones because there is less relative contribution from background noise. Thus, the reliability of the data depends on the performance of the spectrometer, on the concentration of the sample, and on the shape of the absorption (how the area is distributed).

Proton counting. The method has most often been applied to proton counting, the determination of the relative numbers of hydrogen atoms of each structural type present in the sample. The data are invariably used in molecular structure determination. With a Varian A-60 spectrometer, the overall precision of the count (based on areas measured from the step integral) expressed as standard deviation was 0.6% for concentrated solutions, 2.8% for 0.4 *M* solution, where the absorption occurred as a single peak, and about 10% at the 0.1 *M* level. For example, for toluene, the aryl/methyl-hydrogen ratio was 1.666 ± 0.013 (calculated 1.667); for ethanol the methyl/methylene-hydrogen ratio was 1.498 ± 0.010 (calc 1.500) and the hydroxyl/methylene-hydrogen ratio was 0.504 ± 0.006 (calc 0.500). In another study of 26 samples, the standard error was 0.3% (relative standard deviation from theory).

The method is sometimes used to discover signals which are very broad and consequently too weak to be detected in the absorption spectrum.

A reverse application has been the determination of the location and extent of substitution of other isotopes for hydrogen in isotope-labeled compounds by measuring the decrease in proton signal intensities. The method has frequently been applied to H^2- and C^{13}-enriched molecules, and the precision and accuracy of the results compare favorably with those from mass spectrometry. Active hydrogen is determined similarly, allowing the sample to exchange with deuterium oxide. In a study of seven compounds, the average error in this determination was only 1.6%.

Mixtures. A second type of application is the analysis of the molecular composition of mixtures. Examples include mixtures from chemical reactions, pharmaceutical and natural product mixtures, mixtures of structural isomers, and equilibrium mixtures (where NMR is the method of choice). For the analysis of an *n*-component mixture, resolved peaks are needed for $n-1$ of the components, and the number of contributing nuclei per molecule must be known. Spectra of the pure components are not usually required. If the analysis does not involve closely spaced signals, high resolution will not be required. Better precision is in fact obtained if the resolution is deliberately lowered by not spinning the sample or not degassing the sample to remove the paramagnetic broadening effects of oxygen.

The precision of the results varies with the application. In an analysis of a small amount of by-product in chlorphenesin carbamate, the precision was ±0.5% (standard deviation) and the accuracy was between −0.26 and +0.35% mean error.

Each component need not be identified. In the analysis of petroleum fractions, aromatic, paraffinic, and naphthenic content are determined without identifying the individual components of the mixture. The relative tacticity (the state of being stereochemically tactic) of polymers can be determined without knowing the molecular structure. Moisture content of starches, ground meats, soaps, breakfast cereals, and paper products is regularly determined by low-resolution NMR for quality control of production. Seeds with high oil content have been selected by NMR analysis and later used for propagation—a tribute to the nondestructive quality of the method. One laboratory performs more than 700 moisture analyses during a single day.

The classical example of an equilibrium system is acetylacetone. The enol content, 79.1%, by NMR analysis is accepted as more accurate than the 76% value obtained earlier by the bromine titration method. The method has frequently been applied to inorganic systems. For example, for sodium tripolyphosphate, the pyrophosphate composition is 21 mole % by NMR.

If the intensities are compared with those of an absolute reference standard, absolute assays are possible. In this way, the method has been used to determine total percent of hydrogen or fluorine in a sample. The latter is sometimes difficult to determine by combustion procedures. The standard can be either admixed or observed separately, but must be selected to have a similar response.

The results are as good as most routine combustion analyses on samples of 10% hydrogen or less. In a study of 57 samples, the average error was ±2% relative and the precision was ±0.5%. On eight samples, the average error in fluorine determination was ±1.3% relative.

A related application is the measurement of certain physical constants. Molecular weights and average molecular weights, and the degree of unsaturation expressed as iodine number of unsaturated fats and oils, have been determined by NMR methods using a reference standard for calibration.

Instrument development. The first commercial NMR spectrometer was sold by Varian Associates in 1953. Second-generation instruments were transistorized and had internal-lock stabilization. The introduction of a totally new third generation of instruments with integrated circuits and minicomputers in the early 1970s caused a quantum jump in NMR spectroscopy. In the old continuous-wave (CW) method the narrow radiation frequency was varied slowly to scan the frequency-calibrated spectrum. The new idea is to excite the whole spectrum at once and let the computer untangle the time-based interference patterns of signals coming from the detector by using a mathematical Fourier transformation (FT). The savings in time and increase in sensitivity make for about two orders of magnitude of improvement in sensitivity, permitting the analysis of much smaller samples or nuclei of much less sensitivity such as carbon. *See* SPECTROSCOPY.

Carbon-13 NMR. Chemists have long been interested in the magnetic resonance of the mass-13 isotope of carbon (C^{13}) because carbon is the most fundamental atom in nature and life. Compared to protons, the C^{13} signals are narrower and the range of chemical shifts is about 20 times wider, allowing more detailed analysis of the structural features of fairly large molecules, as shown in Fig. 6. But this attractive applicability is accompanied by certain experimental difficulties that have slowed the development of carbon-13 nuclear magnetic resonance (CMR).

Sensitivity. Carbon-13 behaves much like H^1 in NMR, but two important differences make its signals much more difficult to detect. Compared to H^1, the C^{13} nuclei are much weaker magnets. The magnet strength (magnetogyric ratio γ) is only about 1/4 that of protons, and the NMR experiment depends on the cube of this term; hence the net sensitivity is actually 1/63.

The low natural abundance (1.1%) of the C^{13} isotope in a carbon-containing sample is a mixed blessing. Because the rest of the carbon (C^{12}), having no magnetic moment, is not observable by NMR, the sensitivity is further decreased by a factor of about 100, making the total loss in sensitivity, compared to H^1, about 1/5700. Thus the signals from C^{13} are very weak. The nice feature of the low natural abundance is that chances of finding two adjacent C^{13} atoms in a molecule are so small that C^{13}-C^{13} spin coupling is negligible, and only the coupling with protons need be considered.

Several techniques have been put together to overcome this great sensitivity loss and make CMR practical today. The table compares the effectiveness of these techniques. The most obvious development was to make bigger magnets so that larger samples could be studied. This increase in sample volume gave an 11-fold increase in sensitivity. A. Allerhand described experiments with even larger (20 mm outside diameter) sample cells.

Another development, suggested by D. Grant, is broad-band decoupling of all the protons in the sample. For example, the C^{13} resonance of the six identical carbons of benzene is actually split up into 15 lines because of the coupling of each carbon to one attached hydrogen, two α-hydrogens, two β-hydrogens, and one γ-hydrogen. Thus there is a 15-fold distribution of the net signal intensity from the six C^{13} nuclei, and all the observed lines are that much weaker in intensity. By broad-band irradiation of the whole region where hydrogens normally resonate, all these hydrogens can be decoupled at once. The C^{13} multiplet is collapsed, and all the intensity is concentrated into the single sharp line. This gives a "win back" of 2× for doublets, 3× for triplets, and so on, up to 15× for multiplet signals like those from benzene.

Another development was the observation that the NOE factor for C^{13} is 2.98, considerably more than for H^1. This enhancement comes from interaction of the C^{13} nuclei with nearby hydrogens and varies somewhat with structure. Blanket irradiation of the hydrogens (as described above) gains back as much as a threefold improvement in the intensity of the carbon signals through the NOE enhancement.

There is still a lack in sensitivity by about a factor of 10−100. Another way to improve the sensitivity is to add repetitive spectra and store the re-

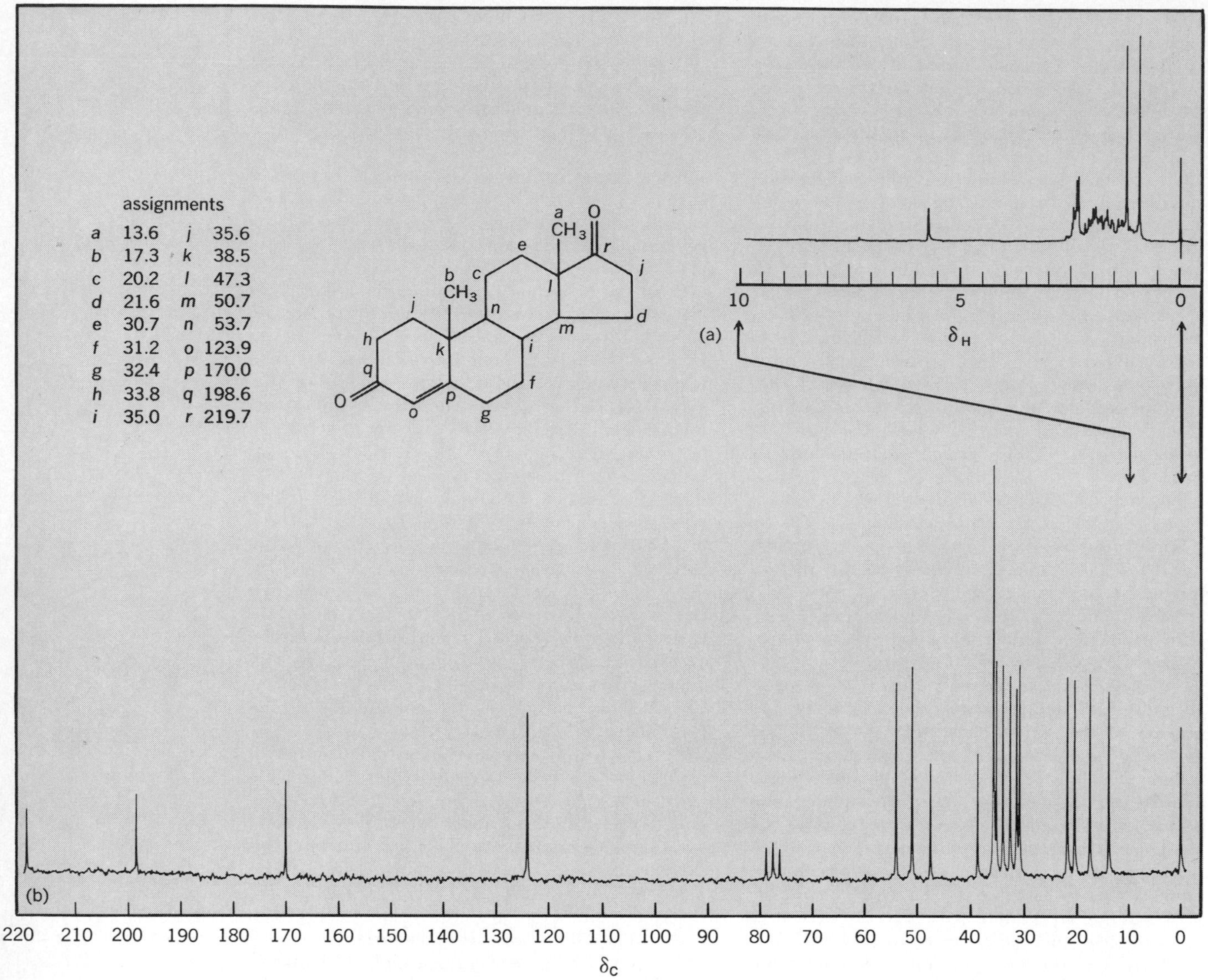

Fig. 6. Comparison of (*a*) H^1 NMR and (*b*) C^{13} NMR spectra of 4-androstene-3, 17-dione dissolved in $CDCl_3$, showing the simplicity and increased resolution of CMR. Assignments are for the CMR spectrum. (*From L. F. Johnson and W. C. Jankowsky, Carbon-13 NMR Spectra, p. 482, copyright © 1972 by John Wiley and Sons, Inc.; used with permission*)

sult in a small digital computer. In this way the signals increase linearly while the noise, being random, increases only as the square root of the number of spectra added. Thus addition of four spectra doubles the sensitivity, nine sums gives a threefold improvement, and so on. With only a little additional gadgetry the system can be made to run unattended. To get adequate sensitivity, several thousand spectra have to be added, requiring several days' time. Often the system fails or the resolution deteriorates during this time and spoils the experiment.

Pulsing FT spectrometers. In 1965, L. Johnson suggested that pulsing was the answer to the elapsed time problem. In this way the spectra could be observed rapidly, and a lot of time would not be wasted sweeping through blank regions of the spectra and collecting nothing but noise. This method requires a digital computer for experimental control, data collection, and calculation of the results. The advent of small dedicated computers made this procedure very attractive for CMR.

In 1969 available results mistakenly indicated that C^{13} relaxation times would be very long compared to those of protons. This would require special intricate methods to avoid the necessity of a long delay between pulses, while waiting for the nuclei to relax so they could be excited again. In 1970 Allerhand showed that only a few special types of carbons have such long spin-lattice relaxation times, and investigations employing rapid summing of C^{13} spectra by pulsing techniques were launched in earnest.

Applications of CMR. Carbon atoms are significantly different from protons, and it might be expected that this would show up in the NMR. Unlike hydrogen, carbon is rarely found on the surface of a molecule and is rarely bothered by intermolecular interactions. Hence, large molecules and viscous or plastic samples do not present a problem in CMR. Therefore, one of the most important uses of CMR is in the determination of molecular structure. The method has been useful in the study of large molecules such as polymers of synthetic or biological origin.

Since the low natural abundance inhibits chances of two such nuclei being adjacent in the same molecule, coupling is not observed in the spectrum, and so the utility of CMR in structure determination comes mostly from the chemical shift information in the NMR spectrum. Furthermore, since carbon has a full set of valence electrons the paramagnetic type of shielding effects become dominant. These shifts are a result of the electronic structure of the carbon atom. Thus doubly bonded carbon atoms resonate in the low-field region of the spectrum, triply bonded atoms in the middle, and singly bonded atoms at high field, with a total spread for most molecules of about 200 parts per million. The shifts also depend to a lesser extent on the electronic contributions from other atoms in the molecule as was observed for H^1 NMR.

Some theoretical treatments have been advanced in which the C^{13} chemical shift in question can be calculated from the additive contributions of substituents in the α, β, γ, and δ positions. One effect which is particularly useful is the steric compression shift. Thus Grant and his students found that a γ-substituent, when oriented so that it is nearby in space, causes the carbon resonance to shift upfield because of compression effect on its electrons. An axial methyl carbon on a cyclohexane ring (Fig. 7*a*) is shielded by the two axial hydrogens at C-3 and C-5 and resonates at a higher field compared to an equatorial methyl carbon (Fig. 7*b*) which does not have this interaction. Numerous examples were found among ortho-substituted aromatics, steroids, and sugars. The effect also appears in alkenes where cis carbons (Fig. 8*a*) are more shielded than trans (Fig. 8*b*). In general an observed upfield shift detects any carbon which can exist, even partly on a time-shared basis, in a gauche or eclipsed orientation with respect to another carbon or heteroatom, and the effect is a powerful tool for making stereochemical assignments by C^{13} NMR.

Considerable progress has been made on the semiempirical correlation of C^{13} shifts with structural features. Since carbon is quite free of intermolecular interactions, substituent effects are reproducible from one kind of molecule to another, and model compounds can be used in the analysis of complex spectra. A structure is considered reasonable when all the chemical shifts are verified by closely related examples.

Very large molecules and solids usually give broad proton resonance lines because the neighboring magnets interact. But with natural-abundance CMR, nearest-neighbor C^{13} atoms are at least 10 A (1 nm) apart, so broadening is only about 10–15 Hz, an amount which is insignificant compared to the width of CMR spectra. Hence the method has shown much success for the determination of molecular structure of synthetic polymers and large biomolecules. With certain restrictions the method can even be applied to solids.

Carbon counting. Since pulsed C^{13} spectra are usually well-resolved singlet signals with the couplings intentionally removed, the technique is often applied to count carbon atoms. Further quantitation, employing integrals of the carbon signals in a spectrum, is usually inaccurate because of saturation problems which are more severe than with protons, and because of variations of the amount of Overhauser enhancement for variously protonated carbons. Where quantitative data are necessary, the Overhauser effect can be removed by use of certain paramagnetic additives, and saturation can be minimized by careful selection of pulse widths and repetition rate so that in favorable cases an accuracy up to ±5% can be obtained.

Tracer studies. Another practical application of CMR is the use of artificially enriched samples for tracer studies of reaction mechanisms and biosynthesis pathways. Since C^{13} is not radioactive, it is easy to handle, and the high resolving power of the C^{13} NMR spectrometer as a detection system is another advantage. *See* TRACE ANALYSIS.

Carbon-13 labeling has also been used for studies of mechanisms of rearrangement reactions. A variation of the tracer technique depends on the isotope shift of C^{13} resonance frequency when attached hydrogens are replaced with deuterium. Thus without elaborate synthesis deuterium can be exchanged at certain positions to identify the attached carbon atoms.

The fast CMR techniques are being applied to studies of the mechanism of rapid free-radical reactions where the appearance of emission- (upside down) or enhanced-absorption lines in the NMR spectrum yields evidence on mechanisms of thermal decomposition of peroxides and azo compounds.

Dynamic effects. Methods have been developed to measure the relaxation times of the different carbon nuclei in the sample by special pulsing sequences. It was shown that the time required for an excited nucleus to dispose of its excess energy to the surroundings (spin-lattice relaxation time T_1) is another very useful parameter for identifying the lines in a CMR spectrum. The T_1 value has also been used to characterize detailed motions of molecules such as ring conversion in substituted cyclohexanes, rotational barriers in substituted benzenes, and segmental motions in large molecules.

The biologists are anxious to characterize the segmental motions in enzymes and have already shown, for example, that the 17 different lysine residues in the protein cytochrome *c* have eight different relaxation times. Thus some lysines are much freer to move about than others, and this information can be correlated with the overall shape of the molecule. [GEORGE SLOMP]

Bibliography: E. D. Becker, *High Resolution NMR: Theory and Chemical Applications*, 2d ed., 1980; N. F. Chamberlain, *The Practice of NMR Spectroscopy with Spectra Structure Correlations for Hydrogen-1*, 1974; P. Diehl, H. Kellerhals, and E. Lustig, *Computer Assistance in the Analysis of High Resolution NMR Spectra*, in P. Diehl, E. Fluck, and R. Kosfield (eds.), *NMR*, vol. 6, 1972; J. W. Emsley et al., *High Resolution Nuclear Magnetic Resonance*, vols. 1 and 2, 1966; L. M. Jackman and S. Sternhell, *Applications of Nuclear*

NUCLEAR MAGNETIC RESONANCE (NMR)

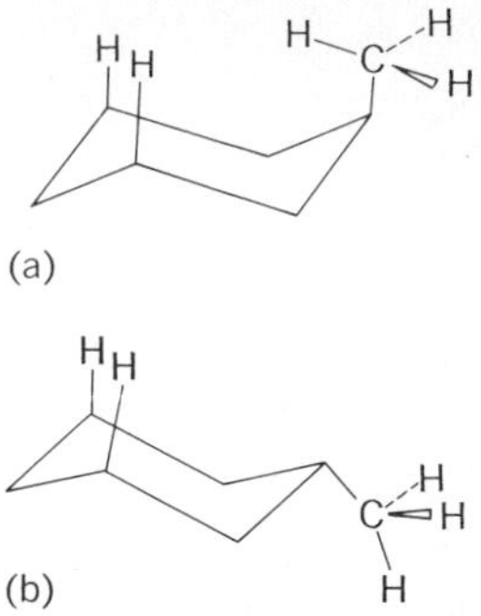

Fig. 7. Two possible configurations of substituents on a cyclohexane molecule, showing proximity of 3- and 5-hydrogens to (*a*) an axial and (*b*) an equatorial methyl group.

NUCLEAR MAGNETIC RESONANCE (NMR)

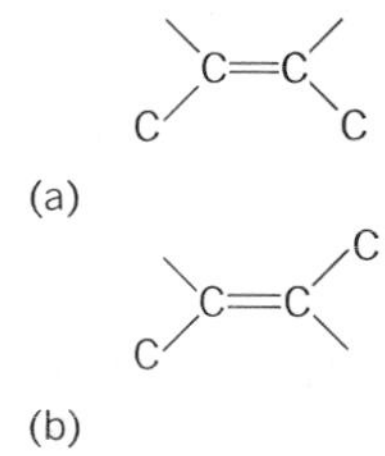

Fig. 8. Alkene configurations showing proximity of (*a*) cis substituents versus (*b*) trans substituents.

Magnetic Resonance Spectroscopy in Organic Chemistry, 2d ed., 1969; F. Kasler, *Quantative Analysis by NMR Spectroscopy*, 1973; G. C. Levy, *Topics in Carbon-13 NMR Spectroscopy*, vol. 1, 1974; C. P. Slichter, *Principles of Magnetic Resonance*, 1980.

Nucleation

The formation within an unstable, supersaturated solution of the first particles of precipitate capable of spontaneous growth into large crystals of a more stable solid phase. These first viable particles, called nuclei, may either be formed from solid particles already present in the system (heterogeneous nucleation), or be generated spontaneously by the supersaturated solution itself (homogeneous nucleation). *See* SUPERSATURATION.

Heterogeneous nucleation involves the adsorption of dissolved molecules onto the surface of solid materials such as dust, glass, and undissolved ionic substances. This adsorbed layer of solute molecules may then grow into a large crystal. Because the crystal lattice of the foreign solid is in general not the same as that of the solid to be precipitated, the first few layers are deposited in a lattice configuration which is strained, that is, less stable than the normal lattice of the precipitating material. The degree of lattice strain determines the effectiveness of a given heterogeneous nucleating agent. Thus, a material whose crystal structure is greatly different from that of the solid to be precipitated will not bring about precipitation unless the solution is fairly highly supersaturated, whereas, if the solution is seeded by adding small crystals of the precipitating substance itself, precipitation can occur at a concentration only slightly higher than that of the saturated solution.

If elaborate precautions are taken to exclude solid particles, it is possible to obtain systems in which the necessary precipitation nuclei are spontaneously generated within the supersaturated solution by the process of homogeneous nucleation. In a solution, ions interact with each other to form clusters of various sizes. These clusters in general do not act as nuclei, but instead, redissociate into ions. However, if the solution is sufficiently supersaturated so that its tendency to deplete itself by deposition of ions onto the clusters overcomes the tendency of the clusters to dissociate, the clusters may act as nuclei and grow into large crystals. The rate at which suitable nuclei are generated within the system is strongly dependent upon the degree of supersaturation. For this reason, solutions which are not too highly supersaturated appear to be stable indefinitely, whereas solutions whose concentration is above some limiting value (the critical supersaturation) precipitate immediately.

Nucleation is significant in analytical chemistry because of its influence on the physical characteristics of precipitates. Processes occurring during the nucleation period establish the rate of precipitation, and the number and size of the final crystalline particles. *See* COLLOID; FLOCCULATION; PRECIPITATION. [DAVID H. KLEIN; LOUIS GORDON]

Nuclide

A species of atom that is characterized by the constitution of its nucleus, in particular by its atomic number Z and its neutron number $A-Z$, where A is the mass number. Whereas the terms isotope, isotone, and isobar refer to families of atomic species possessing common atomic number, neutron number, and mass number, respectively, the term nuclide refers to a particular atomic species. The total number of stable nuclides is approximately 275. About a dozen radioactive nuclides are found in nature, and in addition, hundreds of other have been created artificially.

[HENRY E. DUCKWORTH]

Octane

An alkane having the formula C_8H_{18}. The 18 possible isomers, all of which have been prepared, range in boiling point from 99.3°C (2,2,4-trimethylpentane) to 125.6°C (*n*-octane). One isomer, 2,-2,3,3-tetramethylbutane, is a crystalline solid having a high melting point, 101°C, only 5.5°C lower than its boiling point.

The branched-chain octanes are obtained commercially by the catalytic alkylation of isobutane with the *n*-butylenes or isobutylene. They have high antiknock rating and are therefore valuable constituents of gasoline.

The standard reference fuel (100 octane number) in determining the antiknock rating of gasolines is 2,2,4-trimethylpentane (often incorrectly termed isooctane). It is produced by the hydrogenation of the mixture of 2,4,4-trimethyl-1-pentene and 2,4,4-trimethyl-2-pentene obtained by the polymerization of isobutylene. *See* ALKANE.

[LOUIS SCHMERLING]

Oleate

A salt or ester of oleic acid, whose formula is given below, in which the acid hydrogen of the

$$CH_3(CH_2)_7CH{=}CH(CH_2)_7C({=}O){-}OH$$

carboxyl group is replaced by a metal or an organic radical. Oleates occur in nature chiefly as the glyceryl ester, found in substantial amounts in animal and vegetable fats. A few of the simpler esters have commercial applications in textile, leather, cosmetics, and pharmaceuticals. Alkali-metal oleates are water-soluble and, with the similar stearates and palmitates, are the chief components of toilet and laundry soaps. Other oleates, such as those of aluminum, copper, calcium, mercury, and zinc, are insoluble in water and are used as dry lubricants, paint driers, water repellents, dusting powers, and medicines. *See* CARBOXYLIC ACID; ESTER.

[ELBERT H. HADLEY]

Bibliography: E. S. West and W. R. Todd, *Textbook of Biochemistry*, 4th ed., 1966.

Optical activity

The effect of asymmetric compounds on polarized light. To exhibit this effect, a molecule must be nonsuperimposable on its mirror image, that is, must be related to its mirror image as the right hand is to the left hand. An optically active compound and its mirror image are called enantiomers or optical isomers. Enantiomers differ only in their geometric arrangements; they have identical chemical and physical properties. The right-handed and left-handed forms of a molecule can be distinguished only by their optical activity or by their interactions with other asymmetric molecules. Optical activity can be used to probe other aspects of molecular geometry, as well as to identify which enantiomer is present and its purity.

As an example of optical isomers, consider tartaric acid (Fig. 1), which was one of the first synthetic molecules to be separated into its enantiomers. In this case the asymmetry of each isomer is magnified when trillions of molecules form a crystal; two types of asymmetric crystals are formed.

The physical basis of optical activity is the differential interaction of asymmetric substances with left versus right circularly polarized light. If solids and substances in strong magnetic fields are excluded, optical activity is an intrinsic property of the molecular structure and is one of the best methods of obtaining structural information from a sample in which the molecules are randomly oriented. The relationship between optical activity and molecular structure results from the interaction of polarized light with electrons in the molecule. Thus the molecular groups that contribute most directly to optical activity are those that have mobile electrons which can interact with light. Such groups are called chromophores, since their absorption of light is responsible for the color of objects. For example, the chlorophyll chromophore makes plants green.

Methods of measurement. Optical activity is measured by two methods, optical rotation and circular dichroism.

Optical rotation. This method depends on the different velocities of left and right circularly polarized light beams in the sample. The velocities are not measured directly, but both beams are passed through the sample simultaneously. This is equivalent to using plane-polarized light. The differing velocities of the left and right circularly polarized components yield a rotation of the plane of polarization. A polarimeter for observing optical rotation consists of a light source, a fixed polarizer, a sample compartment, and a rotatable polarizer. A cell containing solvent is placed between the polarizers, and one of them is adjusted to be perpendicular to the other, excluding the passage of light. The solvent in the cell is then replaced by a solution of the sample, and the polarizer is rotated to again exlude passage of light. The optical rotation a is the number of degrees the polarizer was rotated. A positive or negative sign indicates the direction of rotation. Enantiomers have rotations of equal magnitude, but opposite signs. The optical rotation depends on the substance, solvent, concentration, cell path length, wavelength of the light, and temperature. Standardized specific rotations $[\alpha]$ are reported as defined in Eq. (1), where T

$$[\alpha]_\lambda^T = \frac{a}{cl} \tag{1}$$

is the temperature (°C), λ the wavelength (often the orange sodium D line), l the cell path length in decimeters, and c the concentration in grams per milliliter. Alternatively, M_ϕ is defined by normalizing to the rotation for a 1-molar solution, Eq. (2),

$$M_\phi = [\alpha]_\lambda^T \mathrm{MW}/100 \tag{2}$$

where M_ϕ is the molar rotation and MW the molecular weight. For polymers, the mean residue rotation, m_ϕ, may be defined by the right side of Eq. (2) by using the mean residue (monomer unit) weight for MW. The variation of optical rotation with wavelength is known as optical rotatory dispersion (ORD).

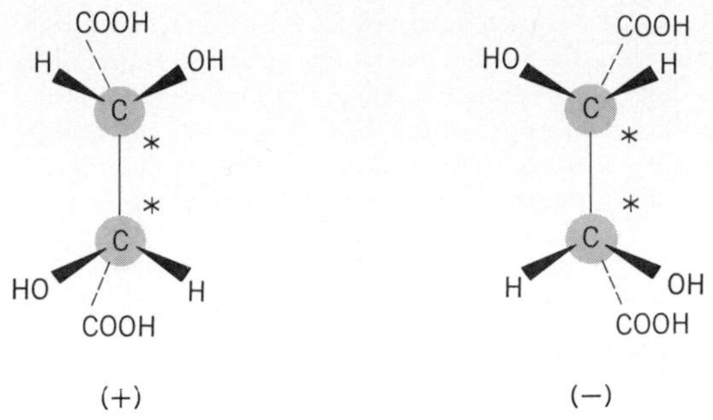

Fig. 1. Enantiomers of tartaric acid.

Circular dichroism. Circular dichroism (CD) is the difference in absorption of left and right circularly polarized light. Since this difference is about a millionth of the absorption of either polarization, special techniques are needed to determine it accurately. Circular-dichroism spectrometers consist of a light source, a monochromator to select a single wavelength, a modulator to produce circularly polarized light, a sample compartment, a phototube to detect transmitted light, and associated electronic components. The modulator rapidly switches (typically 50,000 times per second) between left and right circular polarization of the light beam. The absorption of an optically inactive sample is independent of polarization, so that the light intensity at the phototube is constant; thus a constant direct current is generated. The absorption of an optically active sample depends on the polarization, so that the light intensity at the phototube varies at the frequency of the modulator; thus an alternating current is generated. The circular dichroism is proportional to the amplitude of the alternating current. The proportionality constant is determined through calibration by using a compound of known circular dichroism.

Circular dichroism is reported as a difference in absorption, Eq. (3), or as an ellipticity (a measure

$$\Delta\epsilon = \epsilon_L - \epsilon_R = (A_L - A_R)/(c'l') \tag{3}$$

of the elliptical polarization of the emergent beam), Eq. (4), for a 1-molar solution, where ϵ is the extinc-

$$M_\theta = 3300\Delta\epsilon \tag{4}$$

tion coefficient, A is the absorbance $[\log(I_0/I)]$, subscripts L and R indicate left or right circular polarization, c' is the concentration in moles per liter, l' is the path length in centimeters, I_0 and I are the light intensities in the absence and presence of the sample, respectively, and M_θ is the molar ellipticity. Either $\Delta\epsilon$ or ellipticity, m_θ, may be expressed per residue by making c' the concentration of residues (monomer units). As in the case of optical rotation, enantiomers have circular-dichroism spectra of equal magnitude but opposite signs.

Variation with wavelength. Optical rotation and circular dichroism are two manifestations of the same interactions between polarized light and molecules. They are related by a mathematical transformation. An important difference between the two measurements is the way in which they vary with wavelength. Optical rotation extends to wavelengths far from any absorption of light. Thus colorless substances still have significant optical rotation at the sodium line. However, all groups which absorb light (chromophores) contribute at all wavelengths, and it can be difficult to extract the contribution of a single group. On the other hand, circular dichroism is confined to the narrow absorption band of each chromophore. Thus it is easier to determine the contribution of individual chromophores, information vital to structural analysis. *See* COTTON EFFECT; ROTATORY DISPERSION.

Correlation with molecular structure. In synthesizing enantiomers, chemists focus on an asymmetric center, that is, a locus which imparts asymmetry to the whole molecule. A common asymmetric center is a tetrahedral carbon atom with four different groups attached, such as the carbons marked with asterisks in tartaric acid (Fig. 1). However, in correlating optical activity with molecular structure, the focus is on the three-dimensional arrangement of the chromophores which interact most strongly with light.

As examples, consider the nucleoside adenosine and its dimer (Fig. 2). The most mobile electrons are in the aromatic ring system, the chromophore (Fig. 2*a*). The electrons in the sugar ribose are more tightly bound and interact less strongly with visible and ultraviolet light. However, all the asymmetric centers are in the ribose part of the molecule. For adenosine, light interacting with the aromatic chromophore is only weakly influenced by the asymmetric centers in ribose, so that small circular dichroism bands are observed (Fig. 2*c*).

In the covalently linked dimer of adenosine, the observed circular dichroism bands are about 10 times larger than those of the monomer. In the 240- to 300-nm region of the spectra (Fig. 2*c*), two bands are observed for the dimer, but only one for the monomer. This indicates strong interaction of the two aromatic chromophores, and hence their close proximity in the dimer. Analysis of the circular-dichroism spectra expected for various arrangements of the two chromophores, as well as other types of experimental data, indicates that the aromatic rings are stacked (Fig. 2*b*). The asymmetric centers in ribose cause the formation of the stacked arrangement shown rather than its mirror image. *See* ASYMMETRIC SYNTHESIS.

Stacking of aromatic rings, as exemplified by the adenosine dimer, is a common feature of nucleic acid polymers (DNA and RNA) isolated from biological sources. Slight differences in the stacking geometry gives each of these polymers a characteristic circular-dichroism spectrum. Alterations in the stacking arrangement caused by some pharmacologically active agents can be detected through alterations in the circular-dichroism spectra. These structural changes may in turn be related to the pharmacological action.

A derivative of the amino acid proline (Fig. 3) can be used to illustrate another way in which optical activity depends on molecular structure. In this molecule only the OCN group (amide chromophore) which is in the horizontal plane of the drawing and the hydrogen which is marked $H^{\neq}$ need be considered. By forming the N-H bond, $H^{\neq}$ acquires a charge of about $+1/3$ electron. It has been predicted that such a positive charge will perturb the motion of the electrons in the amide chromophore in a manner which will produce a negative circular dichroism band when the charge is above the plane of the amide group and to the right of the oxygen. Only for the arrangement shown is the magnitude of the circular dichroism band expect-

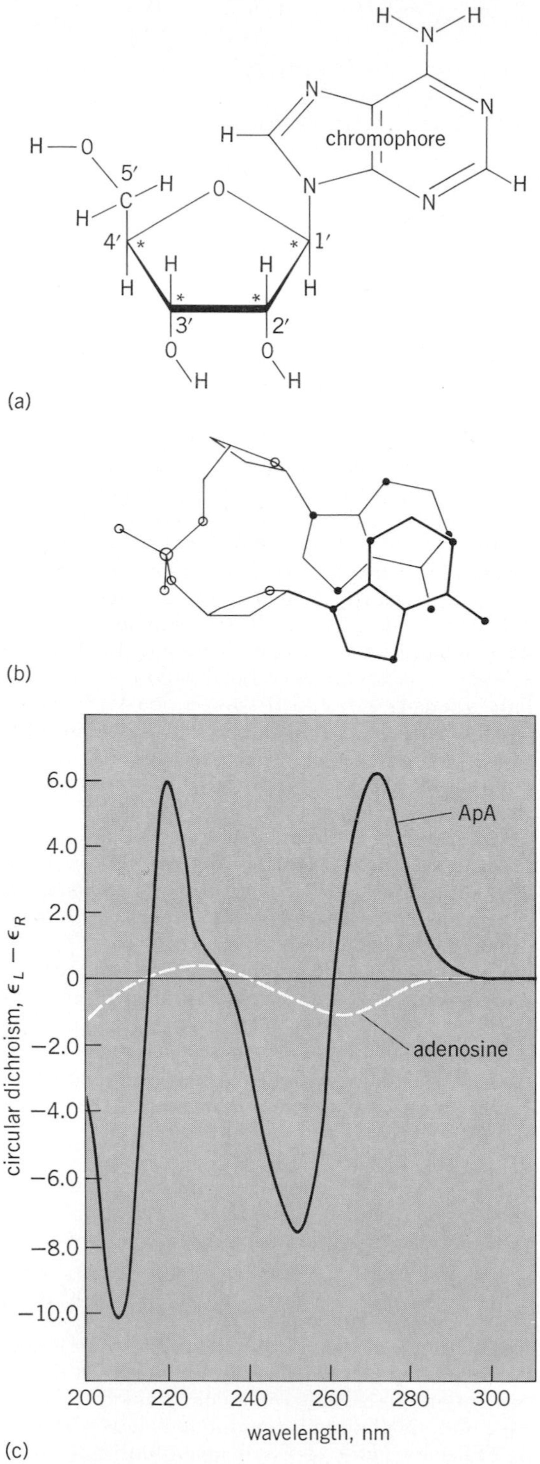

Fig. 2. Adenosine and its dimer. *(a)* Structure of adenosine. Asymmetric centers are marked by an asterisk. *(b)* Stacked arrangement of adenosine dimer (ApA). The 3′ carbon of one adenosine is linked to the 5′ carbon of the other by a phosphate group. (c) Circular dichroism spectra of ApA and adenosine at neutral pH in aqueous solution at room temperature.

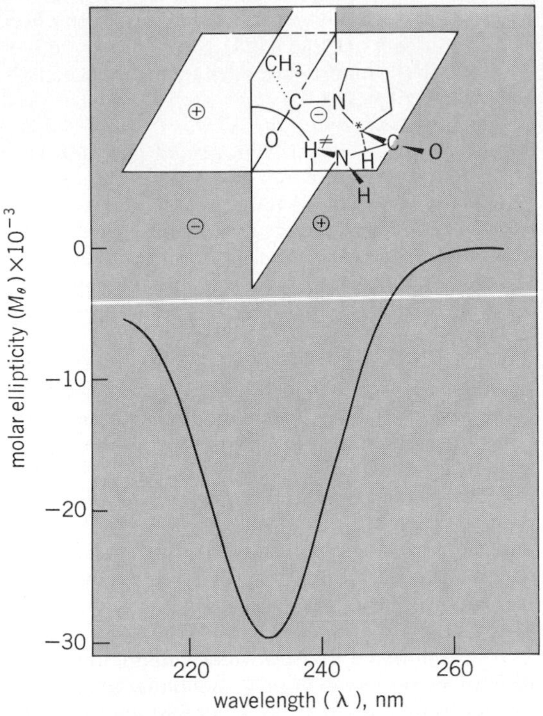

Fig. 3. Folded arrangement of *L*-proline derivative (*N*-acetyl-*L*-proline-amide) and its circular dichroism spectrum in *p*-dioxane solution at room temperature.

ed to be as large as observed. Furthermore, it has been shown that there will be no circular dichroism if $H^{\neq}$ is in either of the two planes shown, and that for $H^{\neq}$ in adjacent quadrants the sign of the circular dichroism band alternates (Fig. 3). For this compound, reflection through the horizontal plane will generate the enantiomer. This would place $H^{\neq}$ in the lower right quadrant and generate a positive circular dichroism band with magnitude equal to that of Fig. 3. *See* STEREOCHEMISTRY.

[VINCENT MADISON]

Bibliography: V. A. Bloomfield, D. M. Crothers, and I. Tinoco, Jr., *Physical Chemistry of Nucleic Acids*, 1974; J. A. Schellman, Symmetry rules for optical rotation, *Accounts Chem. Res.*, 1:144–155, 1968; I. Tinoco, Jr., and C. R. Cantor, Application of optical rotatory dispersion and circular dichroism to the study of biopolymers, *Meth. Biochem. Anal.*, 18:81–203, 1970; R. W. Woody, Optical rotatory properties of biopolymers, *J. Polym. Sci. Macromol. Rev.*, 12:181–321, 1977.

Organic chemical synthesis

The preparation of a desired organic compound by altering the molecular structure of another compound in a rational and predetermined manner. This process may require one or more individual chemical reactions wherein each reaction results in a discrete modification of molecular structure. If multiple reactions are required, the resulting modifications culminate in the formation of the desired compound. Organic synthesis ordinarily denotes the construction of complex organic molecules from more simple ones, although the term is occasionally used to describe a rational process for the preparation of a desired compound by partial degradation of a more complex molecule. The complexity of an organic compound is largely determined by its molecular weight, but the precise arrangement of its constituent atoms is also an important factor. The reactions in organic synthesis may require the use of catalysts or inorganic reagents and frequently involve the combination of two or more molecules of different substances.

Applications. Adaptation of chemical reactions to the stepwise construction of complex compounds of predetermined structure from simple compounds is one of the most important facets of organic chemistry from both a theoretical and a practical point of view. An almost unlimited number of different organic compounds are theoretically capable of existence. Although many organic compounds occur in nature, only a small fraction of the theoretically possible compounds have been isolated from natural sources. Consequently, organic compounds that do not occur naturally or cannot be obtained economically from natural sources must be synthesized. The synthesis of organic compounds forms the basis of the plastics and synthetic fiber industries. Many valuable dyes, flavors, fragrances, vitamins, and drugs cannot be economically produced from natural sources and are also prepared by synthesis. For example, isolation of 200 mg of the important drug cortisone from natural sources requires processing 1000 lb (454 kg) of adrenal glands from about 20,000 cattle. Consequently, the large amounts of cortisone now required for pharmaceutical purposes are prepared by synthesis. Other examples of such compounds include indigo dye, vanillin and menthol flavors, the hormones estrone and testosterone, vitamins A and C, and cinnamaldehyde, which is used by the perfume industry. Many important drugs such as diethylstilbesterol and chlorpromazine have never been isolated from natural sources and must be made by synthesis.

Organic synthesis is also used to confirm the structures of organic compounds that have been postulated on the basis of evidence derived from chemical behavior and physical measurements. Such confirmation of structure by synthesis has been utilized for such diverse compounds as vitamin B_{12}, the hormones aldosterone and insulin, and the alkaloids morphine and lycopodine.

Many synthetically produced organic compounds, which contain combinations of atoms occurring as part of the structures of certain physiologically active natural compounds, have been found to possess the same or higher activity and sometimes lower toxicity than the naturally occurring compounds. For example, procaine has largely replaced cocaine as a local anesthetic; Demerol is frequently used to replace morphine as an analgesic.

Synthetic organic chemistry is the most inclusive branch of this science, for it makes use of the principles developed in all of the others. Especially as it has come to attack more complex problems, has it incorporated more and more of the findings of physical organic chemistry, stereochemistry, and reaction mechanisms.

Classification of methods. The methods of conversion of one organic compound to another are numerous. They have been discovered since 1880 and have been developed and established as general reactions through the continuous studies of many investigators. Most of these synthetic methods fall into five general categories, four of which may be expressed in a very simple form. A sixth category covers methods belonging in the other categories that are used for synthesis of a large special class of substances, polymers.

This classification system is based on a consideration of the overall reaction from the initial to the end product. Although a conversion may appear simple, it is actually complicated in many cases. Several steps, which fall into one or another of the defined categories and which involve unisolated intermediate products, have frequently been demonstrated. In spite of the complex mechanisms often encountered, most organic reactions are so well understood that they may be applied readily and successfully.

Displacement reactions. In these reactions, represented by (1), one functional group in a com-

$$XY + Z \rightarrow XZ + Y \quad (1)$$

pound is replaced by another by a nucleophilic, electrophilic, or free-radical mechanism. To illustrate, *n*-butyl alcohol (n-C_4H_9OH) reacts with phosphorus triiodide (PI_3) to give *n*-butyl iodide (n-C_4H_9I) and phosphorous acid (H_3PO_3). Bromine (Br_2) and benzene (C_6H_6) react to give bromobenzene (C_6H_5Br) and hydrogen bromide (HBr). 2,4-Dinitrochlorobenzene [2,4-$(NO_2)_2C_6H_3Cl$], an inexpensive and important dye intermediate, reacts with sodium fluoride (NaF) to give 2,4-dinitrofluorobenzene [2,4-$(NO_2)_2C_6H_3F$], which is used in structural organic chemistry. The last reaction represents synthesis of a substance of lower molecular weight. *See* SUBSTITUTION REACTION.

Addition reactions. In these reactions, represented by (2), a compound containing an unsat-

$$XY + Z \rightarrow XYZ \quad (2)$$

urated functional group combines with a reagent to give a saturated compound. Propylene (CH_3-$CH{=}CH_2$) will react with hydrogen bromide (HBr) to give either *n*-propyl bromide ($CH_3CH_2CH_2Br$) or isopropyl bromide ($CH_3CHBrCH_3$), depending on the conditions. A ketone such as RCOR′ adds hydrogen cyanide (HCN) to give a cyanohydrin [RR′C(OH)CN]. When two organic molecules combine in this way, the reaction is frequently classified as a condensation reaction. Diethyl malonate [$CH_2(CO_2C_2H_5)_2$] and benzalacetophenone (C_6H_5-$CH{=}CHCOC_6H_5$) combine to give the addition product [$C_6H_5CH(CH(CO_2C_2H_5)_2)CH_2COC_6H_5$].

Elimination reactions. These are the reverse of addition reactions ($XYZ \rightarrow XY + Z$). Some molecule, usually simple in character, is eliminated from a compound with the production of a second compound. Isobutyl chloride [$(CH_3)_2CHCH_2Cl$], upon treatment with ethanolic alkali, loses hydrogen chloride (HCl) with formation of the unsaturated compound isobutylene [$(CH_3)_2C{=}CH_2$]. *tert*-Butyl alcohol [$(CH_3)_3COH$], when heated with an acidic catalyst, loses water with formation of isobutylene [$(CH_3)_2C{=}CH_2$].

Rearrangement reactions. The primary product in many organic reactions ($XYZ \rightarrow YXZ$), either after formation or concomitantly in its formation, may rearrange with or without the loss of some simple molecule. Such reactions furnish additional approaches to certain combinations of atoms which are otherwise difficult to attain. Pinacol [$(CH_3)_2COHCOH(CH_3)_2$], when treated with acidic reagents, loses a molecule of water with formation of pinacolone [$(CH_3)_3CCOCH_3$]. Allyl phenyl ether ($C_3H_5OC_6H_5$), upon heating, rearranges to *o*-allylphenol (o-$C_3H_5C_6H_4OH$).

Condensation reactions. Two or more compounds, usually both organic, react together with or without the elimination of a simple molecule to give a new compound. Reactions of this type are wide in scope and include some that are frequently classified otherwise. Many of the condensation reactions take place by a complex mechanism involving several steps. Representative types are described and illustrated below.

1. Functional groups from each of two compounds may be removed and the two residues combined to form a new compound. Two molecules of bromobenzene (C_6H_5Br) react with copper to give cupric bromide and biphenyl ($C_6H_5C_6H_5$).

2. The same functional groups in two compounds react with each other to give a product of molecular size equal to the sum of the two reacting compounds. Two molecules of aromatic aldehyde (ArCHO) react in presence of cyanide ion to give a benzoin (ArCHOHCOAr).

3. A functional group in one compound reacts with an active position or a functional group in a second compound to bring the two compounds into combination with or without the elimination of some simple molecule. Benzaldehyde (C_6H_5CHO) reacts with acetone (CH_3COCH_3) to give benzalacetone ($C_6H_5CH{=}CHCOCH_3$) and water, through an unstable intermediate hydroxy ketone adduct ($C_6H_5CHOHCH_2COCH_3$). An aliphatic aldehyde (RCHO) reacts with a Grignard reagent (R′MgBr) to form an adduct [RCH(OMgBr)R′] which hydrolyzes to an alcohol (RCHOHR′). *See* CONDENSATION REACTION.

Polymerization reactions. Many molecules of one or more simple compounds react to form giant molecules or polymers, with or without the elimination of some simple substance. Two general types of reactions provide the means of synthesis of the vast majority of polymers. These are illustrated below.

1. A simple compound, usually in the presence of an initiator, may give a homopolymer. Ethylene ($CH_2{=}CH_2$), in the presence of an appropriate initiator, is converted to a plastic polymer, polyethylene ($(\text{—}CH_2CH_2\text{—})_x$). Two different simple compounds with the same functional group may react to give a copolymer with either regularly recurring or randomly distributed units. Vinyl chloride ($CH_2{=}CHCl$) polymerizes with vinylidene chloride ($CH_2{=}CCl_2$) to make a valuable product, Saran [$(\text{—}CH_2\text{—}CHCl)_x\text{—}(CH_2\text{—}CCl_2\text{—})_y$]. Such polymerization reactions involve a succession of reactions of the simple molecule or molecules with each other.

2. Compounds each of which have two or more functional groups may react, with or without the elimination of some simple molecule, to give a condensation polymer. Hexamethylene diamine [$H_2N(CH_2)_6NH_2$] and adipic acid [$HO_2C(CH_2)_4$-CO_2H] react in equimolecular quantities to give water and a polyamide [$\text{—}HN(CH_2)_4NHCO(CH_2)_4$-$CO\text{—}]_x$, commonly known as nylon. *See* ORGANIC CHEMISTRY; ORGANIC REACTION MECHANISM; POLYMERIZATION; STEREOCHEMISTRY; STERIC EFFECT.

[RICHARD A. KRETCHMER]

Bibliography: H. O. House, *Modern Synthetic Reactions*, 1972; R. E. Ireland, *Organic Synthesis*, 1969; *Organic Reactions*, vols. 1–22, 1942–1975; *Organic Synthesis*, vols. 1–52, 1921–1973.

Organic chemistry

The chemistry of the compounds of carbon. Intensive development of this branch of chemistry began about the middle of the 19th century. However, a lapse of several decades occurred between the beginning of the science and the emergence of a clean-cut definition. Despite the comparatively recent development of organic chemistry as a separate branch of the broader field of chemistry, many typical organic compounds have been known and used for centuries. The Old Testament contains numerous references to the physiological effect of ethyl alcohol, a typical organic compound, as a component of fermented grape juice and to the properties of acetic acid present in what is now called wine vinegar. Certain natural dyestuffs, for example, indigo and alizarin, were known to the Egyptians, and the poisonous properties of the hemlock, now known to be ascribable to the alkaloid coniine, were known in the Greek city states. Socrates used an extract of poison hemlock to end his life.

The comparatively late development of organic chemistry is due to the fact that most organic compounds found in nature occur in plant and animal materials as complex mixtures. Methods for separation and isolation of the pure compounds have become available only during the past two or three centuries. By the latter part of the 19th century an impressive number of organic compounds had been isolated from natural sources. Among these are alcohol, urea, uric acid, and many of the organic acids.

Inasmuch as all of these substances were derived from one or another living organism, the idea developed that some vital force was required for the synthesis of compounds by such organisms and the term organic chemistry was coined to denote that branch of chemistry which dealt with the products of living organisms. Despite the fact that a German chemist, Friedrich Wöhler, succeeded in synthesizing urea, a typical organic compound, from ammonium cyanate without the intervention of a vital force in 1828, the concept of the necessity for such a force in the synthesis of organic compounds persisted for several years. Eventually this concept was abandoned and the modern concept of organic chemistry, embracing the chemistry of the carbon compounds, emerged.

From the early 1800s the development of the science was rapid. A firm foundation for the subsequent rapid advances was furnished by the introduction of quantitative analytical methods applicable to the analysis of organic compounds by Justus Liebig and Jean B. A. Dumas and by the development of structural theory by Stanislao Cannizzaro and Friedrich A. Kekule.

Organic chemistry owes its peculiar and important position to the fact that carbon, almost alone among the elements, is capable of uniting with itself indefinitely to form compounds. Other elements, notably boron, display similar tendencies, but carbon forms a far greater number of compounds. Secondly, carbon, almost without exception, displays a constant valence of 4. On these two principles the science of organic chemistry is built. The number of carbon compounds theoretically capable of existence is staggering and this very fact poses problems of major magnitude in connection with nomenclature, molecular structure, and arrangement in space of the atoms of organic molecules.

Organic compounds in general differ from inorganic compounds by the nature of the bonds by which the component atoms of a molecule are united. The valence bonds of most inorganic compounds are of the ionic or electrovalent type in which the outer valence electron shells are filled to a noble gas arrangement by gain or loss of electrons from the constituent atoms with resultant development of charged species (ions).

In contrast, when the outer valence electron shells are filled to a noble gas configuration by sharing rather than by transfer of electrons between two atoms, the bonds so formed are known as covalent or electron pair bonds. These bonds occur in some inorganic compounds, such as ammonia, and in almost all carbon compounds.

Covalent bonds can be altered to give them some ionic character. In general, bonds range from those that are essentially covalent to those that are entirely ionic. There is no sharp boundary between ionic and covalent bonds. However, covalent bonds do share certain characteristics which cause striking differences between the physical and chemical properties of inorganic compounds, such as salts, and those of typical organic compounds. Thus, the inorganic salts in general have high melting points, can be distilled only at extremely high temperatures, are soluble in water but are only sparingly soluble or insoluble in nonaqueous solvents, and conduct electricity when molten or in aqueous solution. In contrast, organic compounds possess relatively low melting points, can frequently be distilled, are sparingly soluble or insoluble in water but soluble in nonaqueous solvents, and with a few exceptions do not conduct electricity.

Over the years certain conventions for expressing molecular structures of organic compounds have been accepted. Actual indication of the shared electrons is cumbersome. Standard practice represents a shared electron bond involving a single pair of electrons (a single bond) by a single line. Double bonds, involving two pairs of shared electrons, are represented by a double line, and triple bonds, involving three pairs of shared electrons, are represented by a triple line. *See* CHEMICAL BOND THEORY; CHEMICAL BONDING; CONJUGATION AND HYPERCONJUGATION; RESONANCE.

Further simplification is often achieved by omission of bond lines when the meaning is obvious, as illustrated below with the hydrocarbon propene.

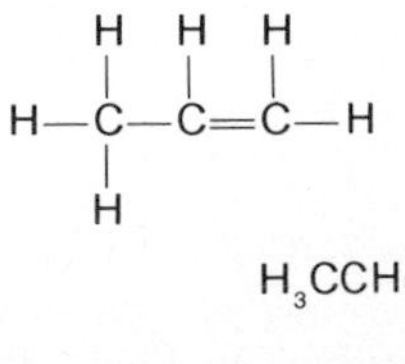

$$H_3C{-}CH{=}CH_2$$

$$H_3CCH{=}CH_2$$

An organic compound containing only single bonds between carbon atoms is said to be saturated; a compound containing one or more multiple bonds between carbon atoms is said to be unsaturated.

The ability of carbon to combine with itself leads to the existence of series of compounds, the formulas of which differ by a constant increment. Such a series, known as a homologous series, is illustrated as follows by the alkanes, each member of which

CH_4 — Methane

CH_3CH_3 or C_2H_6 — Ethane

$CH_3CH_2CH_3$ or C_3H_8 — Propane

differs from the next lower member by the increment CH_2. *See* CHEMICAL STRUCTURES.

CLASSIFICATION OF ORGANIC COMPOUNDS

Because of the great number and variety of organic compounds (well over 1,000,000 are known, and the number identified is constantly increasing), some systematic classification scheme and systematic method of nomenclature for dealing with them becomes mandatory. The problem is complex, and no completely satisfactory system has yet been devised.

In the early days organic compounds were given names which for the most part were derived from the names of the natural sources of the compounds. This practice has carried down to the present, and the use of such trivial names, although convenient, is of little help as far as systematic nomenclature is concerned. Further, at the time of the first isolation of a compound from a natural source, nothing is known about its molecular structure. Adoption of a trivial name is then almost mandatory. However, when the molecular structure of a compound becomes known, assignment of a logical systematic name becomes possible, at least in principle. After many attempts to develop a rational system of nomenclature, the matter was finally placed in the hands of a committee of the International Union of Pure and Applied Chemistry. From the efforts of this committee, an international system, the IUPAC system, is emerging, by which organic compounds can be named in a logical manner. However, a certain amount of nationalism persists and practice does not uniformly conform to the IUPAC rules.

Carbon atoms may combine in the form of long chains, either straight or containing branches, or they may combine to form rings. Further, since carbon is also capable of covalent bonding with other atoms, incorporation of such so-called hetero atoms into carbon rings is possible. This situation furnishes the basis for a broad classification of organic compounds into three main groups depending upon the arrangement of the carbon atoms and the presence or absence of atoms other than carbon in the cyclic compounds. Under these terms of reference, organic compounds may be classified as acyclic compounds, which contain no ring structural arrangements of the constituent atoms; carbocyclic compounds, which contain one or more rings consisting solely of carbon atoms; and heterocyclic compounds, the rings of which contain one or more atoms other than carbon. These principles are illustrated by the following:

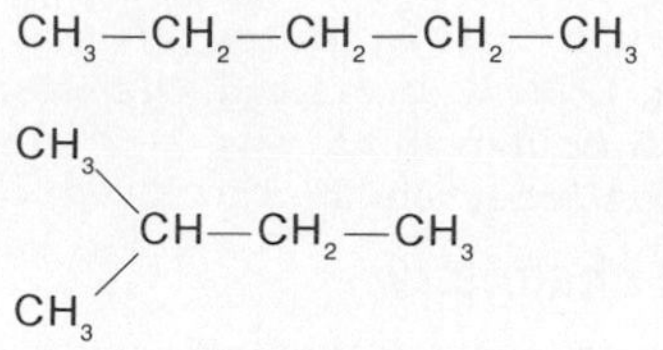

Acyclic

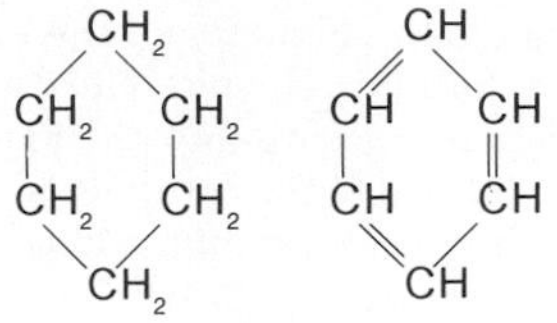

Carbocyclic

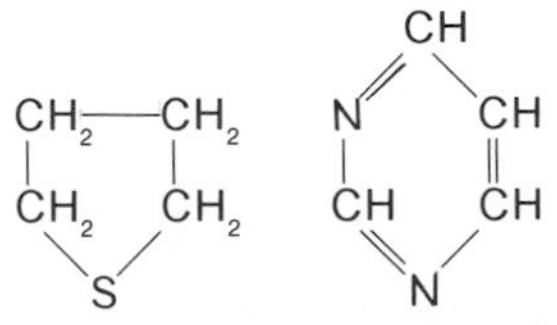

Heterocyclic

Obviously, little progress results by merely subdividing an unwieldy group of compounds into three almost equally unwieldy groups. Further subdivision is mandatory.

Compounds which contain only carbon and hydrogen are known as hydrocarbons. In the hydrocarbons, one or more of the hydrogen atoms, at least in principle, may be replaced by any other atom or group of atoms capable of entering into a covalent bond. Application of this principle furnishes a sound basis for a systematic scheme of further subdivision and nomenclature of the organic compounds. Examples in which such substitution has been performed are the following:

$CH_3—CH_3$ Parent compound

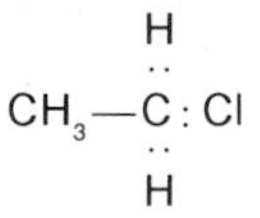

Substitution of hydrogen by chlorine

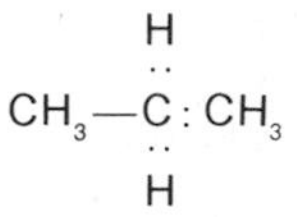

Substitution of hydrogen by $—CH_3$

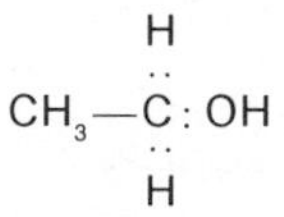

Substitution of hydrogen by —OH

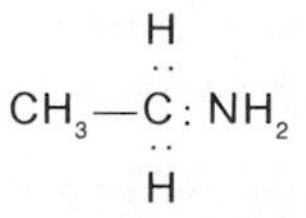

Substitution of hydrogen by $—NH_2$

If the substituent group of atoms is formed from a member of any of the three main classes of organic compounds by loss of one or more atoms of hydrogen, it is known as a radical. An example is the formation of the methyl radical $—CH_3$ from methane, CH_4. If the substituent is a single atom or group of atoms other than a radical, it is known as a functional group, the presence of which confers characteristic chemical properties on the molecule bearing it. Double or triple carbon-to-carbon bonds also come under the heading of functional groups, since their presence changes the chemical properties of the substituted compound from those of the parent substance (usually the saturated hydrocarbon). The common functional groups are:

Name	*Structure*
Halo (chloro, bromo, etc.)	$-Cl$, $-Br$
Hydroxyl	$-OH$
Aldehyde	$-\overset{\displaystyle H}{\overset{\vert}{C}}{=}O$
Carboxyl	$-\overset{\displaystyle O}{\overset{\Vert}{C}}-OH$
Ketone	$\diagdown C{=}O$ $\diagup$
Ether	$-O-$
Amino	$-NH_2$
Cyano	$-C{\equiv}N$
Thiol or mercapto	$-SH$
Sulfonic acid	$-SO_3H$

NOMENCLATURE OF ORGANIC COMPOUNDS

The system of nomenclature for acyclic hydrocarbon derivatives, carbocyclic compounds, and heterocyclic compounds is outlined below.

Acyclic hydrocarbons. The acyclic hydrocarbons are subdivided on the basis of the presence or absence of double or triple bonds. Acyclic compounds containing no multiple bonds are known as alkanes or paraffins; those containing one or more double bonds are known as alkenes or olefins; and those containing one or more triple bonds are known as alkynes or acetylenes.

Alkanes and alkyl radicals. Alkanes are named by reference to the longest straight chain of carbon atoms present. The names always terminate in -ane. Straight-chain alkanes containing 1, 2, 3, or 4 carbon atoms are known by the trivial names methane, ethane, propane, and butane, respectively. Higher members of the series bear a prefix indicating the number of carbon atoms in the alkane. Thus, $CH_3(CH_2)_nCH_3$ is hexane, dodecane, or tetracontane, when *n* is 4, 10, or 38, respectively.

Loss of one hydrogen atom from an alkane results in formation of an alkyl radical which is named by replacement of the suffix -ane of the parent alkane by -yl. Thus $-CH_3$ derived from methane by loss of one hydrogen is a methyl radical; CH_3-CH_2-, similarly derived from ethane, is an ethyl radical. The problem of naming radicals derived from alkanes higher than ethane is complicated, since the hydrogens which may be lost from the alkane are not equivalent. *See* MOLECULAR ISOMERISM.

Branched-chain alkanes are named on the following principle. The longest continuous straight carbon chain represents the parent alkane and the carbon atoms are numbered consecutively, beginning with one end of the chain. The positions of substituent radicals are then indicated by reference to the number of the carbon atom of the parent alkane to which they are linked. Numbering of the parent alkane is done in such fashion that numbers of the carbon atoms bearing the substituents are the lowest (as below). *See* ALKANE.

$$\overset{1}{C}H_3-\underset{\displaystyle CH_3}{\underset{\vert}{\overset{2}{C}H}}-\overset{3}{C}H_2-\overset{4}{C}H_2-\overset{5}{C}H_3 \text{ is 2-methylpentane}$$

$$\overset{6}{C}H_3-\overset{5}{C}H_2-\underset{\displaystyle CH_2CH_3}{\underset{\vert}{\overset{4}{C}H}}-\overset{3}{C}H_2-\underset{\displaystyle CH_3}{\underset{\vert}{\overset{2}{C}H}}-\overset{1}{C}H_3 \text{ is 2-methyl-4-ethylhexane}$$

Alkenes. Alkenes are named by replacing the terminal -ane of the parent alkane by -ene. The position of the double bond is given by a number indicating from which carbon atom of the longest continuous straight chain the double bond proceeds. The lowest number rule obtains, as with branched-chain alkanes. Thus

$$\overset{1}{C}H_8-\overset{2}{C}H{=}\overset{3}{C}H-\overset{4}{C}H_3 \text{ is 2-butene}$$

$$\overset{1}{C}H_2{=}\underset{\displaystyle CH_3}{\underset{\vert}{\overset{2}{C}}}-\overset{3}{C}H_2{=}\overset{4}{C}H_2 \text{ is 2-methyl-1,3-butadiene}$$

See ALKENE.

Alkynes. Alkynes are named by replacing the terminal -ane of the parent alkane by -yne. Otherwise, the rules applying to alkenes hold. *See* ALKYNE.

Acyclic hydrocarbon derivatives. In general, the same principles which govern the naming of branched-chain alkanes, alkenes, and alkynes hold. The parent compound is that which contains the longest continuous straight carbon chain. The position of functional groups is indicated by numbers, with the smallest number principle again controlling, and the nature of the functional substituent is indicated by the suitable prefixes or suffixes.

Halogen compounds. The prefixes fluoro, chloro, bromo, and iodo are used with the name of the parent alkane:

$$CH_3CHClCH_2CH_3 \text{ is 2-chlorobutane}$$

$$CH_3CHBrCH_2CH_2CH_2Br \text{ is 1,4-dibromopentane}$$

Hydroxyl derivatives. These substances are commonly known as alcohols, and many of them are named as such. CH_3OH is methyl alcohol, and CH_3CH_2OH is ethyl alcohol. In the Geneva system they are named by replacing the terminal -e of the parent longest straight-chain alkane by -ol and by indicating the position occupied by the hydroxyl function by a suitable number.

The carbon chain is numbered from the end, which gives the carbon bearing the hydroxyl group the smallest number. Still a third scheme names the higher alcohols as derivatives of the first member of the series, carbinol. Thus

$$CH_3OH \text{ is methyl alcohol (carbinol, methanol)}$$

$$CH_3CH_2\underset{\displaystyle OH}{\underset{\vert}{C}H}CH_2CH_3 \text{ is 3-pentanol (diethylcarbinol)}$$

$$CH_3\underset{\displaystyle CH_3}{\underset{\vert}{C}H}CH_2OH \text{ is 2-methyl-1-propanol (isopropylcarbinol)}$$

Carbonyl derivatives. The functional group $>C{=}O$ is known as the carbonyl group. If one of the unsatisfied valences of carbon is linked to a radical and the other is satisfied by hydrogen, the resulting substance is known as an aldehyde, whereas if the valences of carbon are satisfied by two radicals, the resulting substance is a ketone.

Aldehydes, particularly the lower members of the series, are frequently named by replacing the terminal -ic of the related acid by -aldehyde. Thus CH_3CHO is acetaldehyde by virtue of its relationship to acetic acid, CH_3COOH. By the Geneva system, aldehydes are named by replacing the terminal -e of the parent alkane by -al. The aldehyde carbon atom is considered part of the longest straight carbon chain. Since the aldehyde function of necessity must embrace a terminal carbon atom of the parent alkane, it automatically becomes carbon atom 1 of the chain, so that it is unnecessary to specify its position. Thus

$$\begin{array}{l} CH_3CHCH_2CH_2CHO \text{ is 4-methylpentanal} \\ \qquad\;\; | \\ \qquad CH_3 \end{array}$$

See ALDEHYDE.

Ketones are named by three methods. The lower symmetrical ketones are named by replacing the terminal -ic of the acid which would yield them on pyrolysis by -one. Thus acetic acid, CH_3COOH, on pyrolysis yields acetone, CH_3-$COCH_3$.

A second scheme, convenient for naming mixed ketones, prefixes the word ketone by the names of the radicals joined to the carbonyl carbon atom. Thus $CH_3COCH_2CH_3$ is methyl ethyl ketone. In the Geneva system the terminal -e of the parent alkane is replaced by -one. The carbonyl group position must be shown by a number. Thus

$$\begin{array}{l} CH_3CH_2CCH_2CH_3 \text{ is 3-pentanone} \\ \qquad\qquad \| \\ \qquad\qquad O \end{array}$$

Carboxyl functions. The carboxyl function is characteristic of organic acids. Since many acids were originally isolated from natural products, trivial names associated with their sources are common. For example, HCOOH, formic acid, is named from the Latin *formica* (ant). Branched-chain acids are named frequently as derivatives of acetic acid, CH_3COOH, or as derivatives of the parent straight-chain acid. Thus

$$\begin{array}{l} C_2H_5{-}CHCOOH \text{ is diethylacetic acid} \\ \qquad\quad\;\; | \\ \qquad\quad C_2H_5 \end{array}$$

$$\begin{array}{l} \qquad\;\; \beta \qquad\quad \alpha \\ CH_3CH{-}CHCOOH \text{ is } \alpha,\beta\text{-dimethylbutyric acid} \\ \qquad\;\, | \qquad\;\; | \\ \qquad CH_3 \;\; CH_3 \end{array}$$

This example employs Greek letters to denote positions on a hydrocarbon chain, the principal functional group identified as α, the second as β, and so on. This method is still in use, but the IUPAC system is preferred.

By the IUPAC system, acids are named by replacing the terminal -e of the parent hydrocarbon by -oic acid. The carbon atom of the carboxyl group is always numbered 1, and the longest straight chain bearing the carboxyl group controls. Thus

$$\begin{array}{l} \overset{5}{C}H_3{-}\overset{4}{C}H_2{-}\overset{3}{C}H{-}\overset{2}{C}H_2{-}\overset{1}{C}OOH \text{ is 3-hydroxy-} \\ \qquad\qquad\quad\;\; | \qquad\qquad\qquad\qquad \text{pentanoic acid} \\ \qquad\qquad\quad OH \end{array}$$

See CARBOXYLIC ACID.

Ethers. Ethers are generally named by reference to the alkyl radicals present and addition of the word ether; $CH_3CH_2{-}O{-}CH_3$ is ethyl methyl ether. By the Geneva system they are named as alkoxy derivatives of the parent hydrocarbon. Thus

$$\begin{array}{l} CH_3CH_2CHCH_2CH_2CH_3 \text{ is 3-ethoxyhexane} \\ \qquad\qquad\;\; | \\ \qquad\qquad OCH_2CH_3 \end{array}$$

See ETHER.

Amines. Amines are alkyl derivatives of ammonia. They are classified into primary, secondary, or tertiary amines according to whether 1, 2, or 3 hydrogen atoms of ammonia have been replaced by organic radicals. They are conveniently named by using the names of the radicals attached to nitrogen as prefixes to the word amine. Thus

$$CH_3CH_2{-}NH{-}CH_3 \text{ is ethylmethylamine}$$

The NH_2 group is known as an amino group, and primary amines may therefore be named by treating such a group as a substituent. Thus

$$\begin{array}{l} CH_3CHCH_2CH_2CH_2OH \text{ is 4-amino-1-pentanol} \\ \qquad\;\; | \\ \qquad NH_2 \end{array}$$

See AMINE.

Thiols or mercaptans. When the SH group is at the terminus of a chain, it is convenient to prefix the word mercaptan by the radical with which it is joined. Thus CH_3CH_2SH is ethyl mercaptan. The SH group in the Geneva system is treated as a substituent. Thus ethyl mercaptan is ethanethiol. *See* MERCAPTAN.

Sulfonic acids. These are universally named by an additive system. The term sulfonic acid is added as a suffix to the name of the parent hydrocarbon regardless of the class to which it belongs. CH_3SO_3H then is methanesulfonic acid.

Carbocyclic compounds. The carbocyclic compounds are subdivided into the aromatic compounds and the alicyclic compounds by certain structural features and chemical properties.

Aromatic compounds. The aromatic carbocyclic compounds are characterized structurally by an alternating system of single and double bonds. For the present purpose this somewhat oversimplified picture will suffice. The term aromatic arose because many of these substances were originally isolated from aromatic substances, such as tolu balsam, oil of wintergreen, or oil of cloves. *See* AROMATIC; AROMATIC HYDROCARBON.

The parent aromatic hydrocarbons are almost always known by trivial names, many of which indicate the source from which they were originally obtained. For convenience, it is not customary to write the individual carbon and hydrogen atoms in the structural formulas of aromatic hydrocarbons. Rather, each angular position is assumed to represent a carbon atom, and the presence of sufficient hydrogen atoms to bring the carbon atoms to the quadrivalent state is also assumed. Typical examples are as follows:

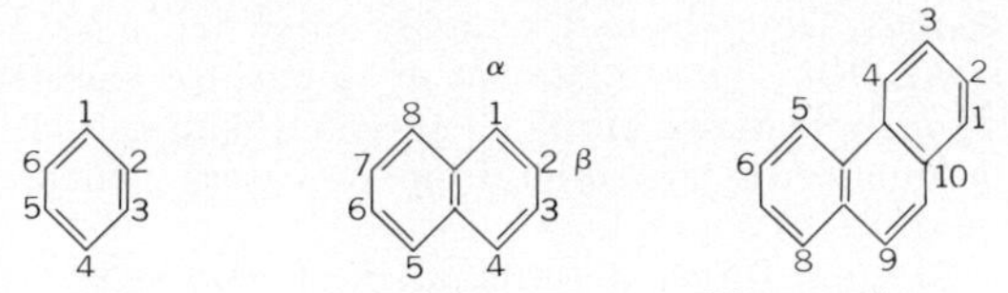

Benzene, C_6H_6 Naphthalene, $C_{10}H_8$ Phenanthrene, $C_{14}H_{10}$

Phenyl, C_6H_5— Naphthyl, $C_{10}H_7$—
Phenanthryl, $C_{14}H_9$—

Radicals

Just as acyclic hydrocarbons can give rise to radicals by the loss of one or more hydrogens, aromatic hydrocarbons can also provide radicals by loss of one or more hydrogens. The radical formed by loss of one hydrogen from benzene is a phenyl radical, and since all hydrogens are equivalent, it is irrelevant which hydrogen is lost. With naphthalene and other polynuclear aromatic hydrocarbons, the positions are not equivalent, and it becomes necessary to specify from which position the hydrogen is missing. This is conventionally done by use of numbers, or in the case of naphthalene, by the use of Greek letters. For example, it is necessary to qualify the names of the two types of naphthyl radicals by use of the terms 1-naphthyl (or α-naphthyl) and 2-naphthyl (or β-naphthyl).

Functional group substituents in aromatic compounds are generally indicated by appropriate adjectives, such as chlorobenzene and nitronaphthalene. The positions occupied by substituents are indicated, when necessary, by numbers. Monohydroxy derivatives of benzene and naphthalene are commonly known as phenol and naphthol. Polyhydroxy derivatives of benzene generally bear trivial names. *See* BENZENE; PHENOL.

Alicyclic compounds. These are cyclic compounds with properties resembling acyclic rather than aromatic substances. They range in complexity all the way from simple mononuclear, or single-ring, hydrocarbons to the complex multiring substances found in the terpenes and steroids. Representative examples are shown.

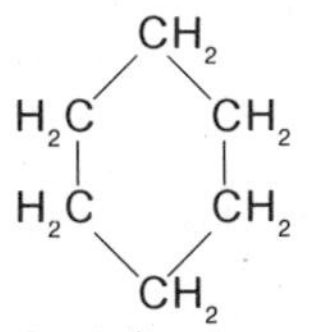

Cyclohexane

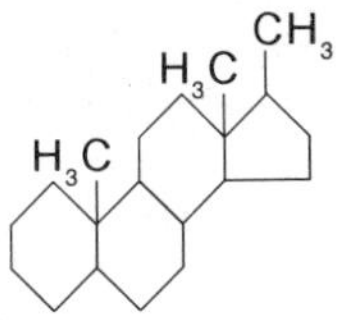

Androstane

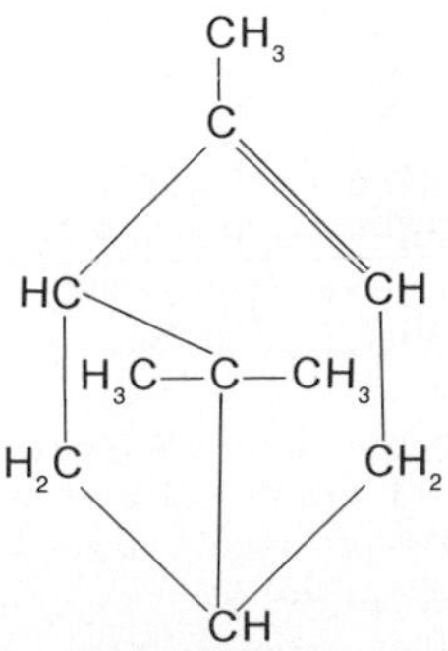

α-Pinene

One or more double bonds may be present in an alicyclic compound, provided they do not occur in an aromatic arrangement.

The problem of nomenclature of the alicyclic compounds is formidable and is complicated by the occurrence of structural, positional, and geometrical isomerism, as well as stereoisomerism. The simple monocyclic hydrocarbons are generally named by use of the prefix cyclo- in conjunction with the names of the alkane of the same number of carbon atoms. Many alicyclic compounds bear trivial names derived from the natural sources of the substances. Definitive rules for naming and numbering many of the polycyclic compounds have been formulated by the International Union of Pure and Applied Chemistry. In general, nomenclature of functional derivatives of alicyclic hydrocarbons utilizes the same terms which are employed in the acyclic series. *See* ALICYCLIC HYDROCARBON.

Heterocyclic compounds. One or more carbon atoms (with or without accompanying hydrogens) in a carbocyclic compound may be replaced by a hetero atom (sometimes in combination with hydrogen), provided that the valence requirements are identical. The resulting substance is a heterocyclic compound. Examples of such replacements are found in replacement of a —CH= in benzene by —N=; of —CH_2 in cyclopentane by —S—, —O—, or —NH—. Although any atom or group of atoms which satisfies the valence requirements may be found in a heterocyclic compound, the most important hetero atoms are O, S, and N. *See* HETEROCYCLIC COMPOUNDS.

A large number of heterocyclic compounds are commonly known by trivial names, many of which reflect their origin. A systematic method which can be applied to the nomenclature of any heterocyclic compound has been developed. In this the related aromatic hydrocarbons are taken as reference standards, the nature of the hetero atom(s) is indicated by a prefix (oxa- for O, thia- for S, and aza- for N), and the position of the hetero atom or atoms is indicated by a number corresponding to the accepted numbering of the aromatic hydrocarbons. Thus

	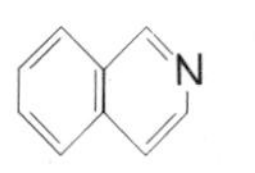	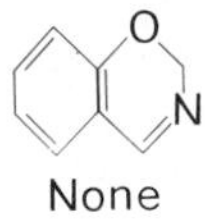
Trivial name:	Isoquinoline	None
Systematic name:	2-Azanaphthalene	1-Oxa-3-aza-2*H*-naphthalene

In compounds carrying more than one hetero atom, numbers are selected so that oxygen bears the lowest number, followed by sulfur and, finally, by nitrogen. The most highly unsaturated compound is taken as the parent. Derivatives in which one or more double bonds are saturated are referred to as dihydro or tetrahydro derivatives of the parent.

No uniformity exists in numbering the positions in heterocyclic compounds. In monocyclic substances, numbers are chosen in such fashion that a single hetero atom bears the number one; if more than one hetero atom is present, oxygen bears the number one, followed by sulfur and then nitrogen. Thus

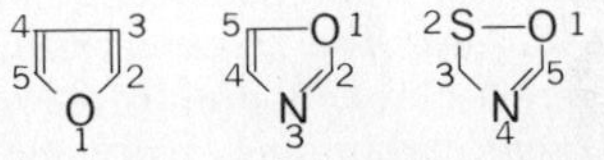

With polyheterocyclic compounds the numbering system is confused. European practice frequently differs from that in the United States, where conventions are undergoing constant revision. The safest practice is to write the complete structural formula and make a visual comparison of the positions of substituents. *See* HALOGENATED HYDROCARBON; QUINOLINE; TERPENE.

[ROBERT C. ELDERFIELD]

Bibliography: International Union of Pure and Applied Chemistry, Definitive rules for nomenclature of organic chemistry, *1957 Report of the Commission on the Nomenclature of Organic Chemistry*, reprinted from *J. Amer. Chem. Soc.*, vol. 82, 1960; J. Rigaudy and S. P. Klesney (eds.), *Nomenclature of Organic Chemistry*, 1978; J. D. Roberts and M. C. Caserio, *Basic Principles of Organic Chemistry*, 2d ed., 1977.

Organic quantitative analysis

The determination of elements, functional groups, or molecules in organic materials. The type of analysis used is determined by the information required. If the total nitrogen content is desired, elemental analysis is used, if only amino nitrogen is desired, then only the amino group is determined. Organic analyses are made on a wide range of materials from pure compounds to mixtures such as blood and fertilizer. The identification of structures in conjunction with purity criteria, analyses of isotopic mixtures of organic molecules, and other examples of structure-related problems are increasingly important analytically. Proton nuclear magnetic resonance and other classes of spectrometric measurement are prominent for such purposes.

Determination of metals. This may be done in two ways. In the first method the sample is moistened with sulfuric acid and heated in oxygen to obtain metal oxides or sulfates, which are weighed. Some metals such as gold and silver are weighed as metals. The second method is based on destruction of the organic portion of the sample by heating with nitric and sulfuric acids, followed by determination of the metals by regular procedures. This method is required for metals which are easily volatilized, such as arsenic.

Carbon and hydrogen. These elements are usually determined simultaneously by burning a sample in a stream of oxygen to form carbon dioxide and water. This is done at 600°C in the presence of platinum as a catalyst. If nitrogen is present in the sample, the gas stream is passed through lead peroxide to remove the oxides of nitrogen. Silver wool adsorbs halogens and oxides of sulfur. The water formed is absorbed on a drying agent, such as magnesium perchlorate and from the increase in weight the amount of hydrogen in the sample can be calculated. Carbon dioxide is absorbed by an alkaline solid such as sodium hydroxide on asbestos fibers, and from the increase in weight the amount of carbon can be calculated.

Carbon alone is determined by wet combustion. The sample is heated in a mixture of sulfuric acid and potassium or silver dichromate. The carbon dioxide formed is determined from the amount of sodium hydroxide with which it combined, or by measuring directly the volume of the carbon dioxide. Reaction chromatography provides a newer basis for rapid carbon and hydrogen analyses on small samples of most classes of organic compounds. The weighed sample is reacted or combusted, and the reaction products are passed through suitable chromatographic columns. The products are measured from the areas beneath their chromatographic peaks.

Oxygen. Direct determination of oxygen in organic materials is accomplished by heating the sample in a stream of nitrogen to form water, oxides of carbon, and hydrocarbons. On passage of this mixture through graphite at 1150°C, all the oxygen is converted to carbon monoxide. The carbon monoxide is oxidized with iodine pentoxide to form carbon dioxide and iodine. Either the iodine or the carbon dioxide may then be measured. This procedure is difficult to perform, so that oxygen in pure compounds is usually obtained by difference rather than by direct measurement.

Nitrogen. Two methods are used to determine nitrogen. In the Dumas method the sample is mixed with copper oxide and heated in a stream of pure carbon dioxide. The elemental nitrogen formed is collected over 50% potassium hydroxide solution, and its volume is measured. This procedure determines total nitrogen in most samples. In the Kjeldahl method the sample is heated in concentrated sulfuric acid containing a catalyst; this procedure converts the nitrogen to ammonia. On the addition of sodium hydroxide followed by boiling, ammonia is distilled into dilute boric acid solution, which is then titrated with acid. The choice of the catalyst is critical because different nitrogen compounds require a variety of catalysts. Esters of nitric and nitrous acids cannot be analyzed by this procedure.

Other elements. Sulfur, halogens, and phosphorus are determined by conversion to sulfuric acid, halogen acids, and phosphoric acids, which are measured by standard inorganic procedures. In the Carius method the sample is heated with fuming nitric acid in a sealed tube. In the combustion method the sample is burned in oxygen, oxides of sulfur are absorbed in neutral hydrogen peroxide, and halogens are absorbed in sodium bisulfite solution. In the bomb method for sulfur the sample is ignited with sodium peroxide and sugar.

Functional groups. Groups such as carboxyl, hydroxyl, nitro, and amide are determined by chemical reactions which are characteristic for each group. These reactions include neutralization, oxidation-reduction, precipitation, condensation, and gas evolution. These methods are based on consumption of a reagent measured by direct titration, or in some cases by determination of excess reagent. For others, a product formed in the reaction is measured. Since most organic compounds are insoluble in water, organic solvents are used. These solvents enhance the acidity or basicity of some groups and also permit some reactions which are not possible in water. Many common organic functionalities can be caused quantitatively to undergo reactions in which water is stoichiometrically consumed or evolved. The titrimetric measurement of such water with Karl Fischer reagent provides a useful basis for determining such functional groups.

Other methods. Many organic compounds absorb energy in the ultraviolet or infrared regions. Since this absorption is characteristic for each type of molecule, a direct analysis based on energy absorption is frequently possible. The variation in mass among hydrocarbons is the basis for the mass-spectrometric analysis of petroleum samples. These and other instrumental methods are based on the properties of molecules.

Physical methods of separation are commonly required to obtain molecules free from interferences. Normal distillation, vapor distillation, azeotropic distillation, ion exchange, chromatography, extraction of solids and liquids, and diffusion have a place in quantitative analysis.

In an analytical perspective, the concurrent separation, characterization, and quantitative measurement of the component molecules from very small samples results from such techniques as mass spectrometry or vapor chromatography. With preconcentration or larger samples, similar techniques detect and measure trace components at ppm (parts per million) to ppb (parts per billion) levels. For complex samples, the preseparated species from a chromatographic column may be introduced at the inlet of a mass spectrometer.

Many enzyme-substrate combinations can be combined with selective-ion electrode responses for potentiometric measurement. For example, the urease-catalyzed hydrolysis of urea produces ammonium ion, which is sensed by an ammonium ion-selective electrode. If an enzyme-gel layer surrounds the electrode, its potential responds to external concentrations of urea. With other techniques, often spectrophotometric, the progress of enzymatic or other organic reactions may be followed with time, and provide a kinetic-based method of analysis. *See* ANALYTICAL CHEMISTRY; QUANTITATIVE CHEMICAL ANALYSIS.

[CHARLES L. RULFS]

Bibliography: T. S. Ma and R. E. Lang, *Quantitative Analysis of Organic Mixtures*, 1979; F. J. Welcher (ed.), *Standard Methods of Chemical Analysis*, vol. 2: *Industrial and Natural Products and Noninstrumental Methods*, 1963, reprint 1975; R. M. Silverstein et al., *Spectrometric Identification of Organic Compounds*, 3d ed., 1974.

Organic reaction mechanism

A detailed description of an overall organic process, R → P, in which the participants and their interconversion rates are characterized. All of the steps from reactants R via possible intermediates I to products P are included, as in reactions (1). To

$$R \underset{-1}{\overset{1}{\rightleftharpoons}} I_1 \qquad I \underset{-2}{\overset{2}{\rightleftharpoons}} I_2 \qquad I_2 \underset{-3}{\overset{3}{\rightleftharpoons}} P \tag{1}$$

characterize the reaction path, one may use a variety of experimental techniques and invoke theoretical and quasitheoretical exclusions.

From an energy viewpoint, a mechanism may be represented schematically in a free energy (G) and reaction coordinate ($Q^\ddagger$) diagram (Fig. 1). Energy minima are associated with observable species (R, P, I), while energy maxima are associated with fleeting transients or activated complexes (AC). Here all of the participants as well as barrier heights, for example, $\Delta G_1^\ddagger = G_1^\ddagger - G(\mathrm{R})$ for step 1 or $\Delta G_{-2}^\ddagger = G_2^\ddagger - G(\mathrm{I}_2)$ of Eqs. (1), are indicated and $Q^\ddagger$ relates in a general way to the progress of a multistep reaction. Unfortunately, this view of a mechanism is somewhat akin to reconstructing a race from snapshots of the horses at the start and finish lines.

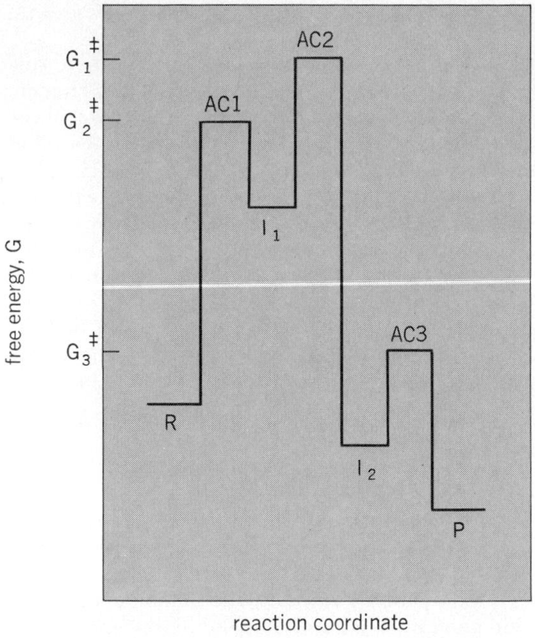

Fig. 1. Progress of a general organic reaction, Eq. (1), in terms of free energy and reaction coordinate.

In a more comprehensive picture, chemical changes among species may be regarded as taking place on a hypersurface representing the changes in potential energy V as a function of atom configurations or a related set of orthogonal coordinates (Q_i). Figure 2 is a portion of this surface which focuses on the change $\mathrm{I}_1 \rightleftharpoons \mathrm{I}_2$. Q_x stands for all of the coordinates except $Q^\ddagger$; the AC is identified with a saddle point or one-dimensional maximum along $Q^\ddagger$. While there is an energy barrier to overcome, the energy is minimized along $Q^\ddagger$.

Individual steps by which changes occur are termed elementary. In such a reaction, one or two—rarely more—molecules become the energized aggregate labeled AC. To illustrate, reactions (2) and (3) show the ACs with partial bonds.

Unimolecular

$$HCCl_3 \rightarrow (HCl_2C \cdots Cl) \rightarrow HCl_2C\cdot + Cl\cdot \tag{2}$$

Bimolecular

$$HO^- + H_3CI \rightarrow (HO \cdots CH_3 \cdots I)^- \rightarrow HOCH_3 + I^- \tag{3}$$

In order to decide which elementary step or steps are involved in a reaction, a variety of techniques may be employed. The exclusion of many conceivable reaction paths makes the task of selecting the probable one easier. At the outset, one must decide what the process is, and determine that it does in fact take place. Then, several of the following tools or devices might be used.

1. An intermediate may be detected: the green radical anion derived from naphthalene, $C_{10}H_8^{\cdot -}$, has a characteristic ESR spectrum.

2. An intermediate may be intercepted or diverted, as indicated in reactions (4) and (5). Though it may be unobserved, reaction (4) as the precursor of reaction (5) is supported by the fact that the starting ketone is labeled (D for H) in D_2O solvent and OD^-.

$$RCOCHR_2 + OH^- \rightleftharpoons RCOCR_2^- + H_2O \tag{4}$$

ORGANIC REACTION MECHANISM

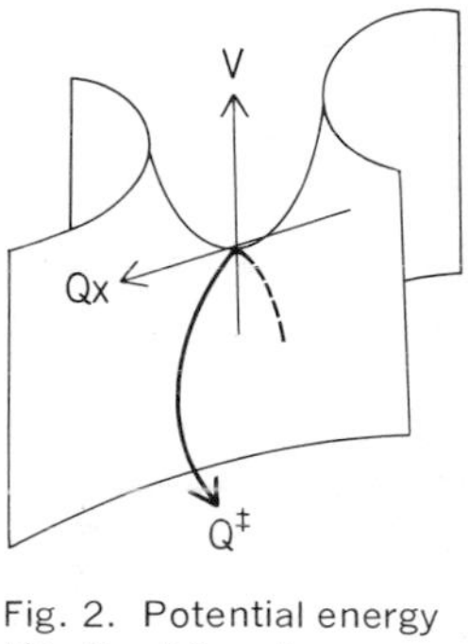

Fig. 2. Potential energy V in the AC region.

$$RCOCR_2^- + Br_2 \rightleftharpoons RCOCR_2Br + Br^- \qquad (5)$$

3. An observed rate law often allows one to rule out all but a few possible mechanisms. For elementary reactions, the mathematical description of the rate process directly reflects the mechanism. For reaction (3), rate $= -d[H_3CI]/dt = k[HO^-][H_3CI]$, in which k is a characteristic rate constant. In nonelementary processes, the mechanism is usually limited by, but not necessarily obvious from, the rate law. Indeed, when a rate expression is not related to overall reaction stoichiometry, several elementary steps must be involved. The fact that the hydrolysis of t-C_4H_9I to yield t-C_4H_9OH has the rate law $-d[C_4H_9I]/dt = k[C_4H_9I]$, and the probable steps, reactions (6) and (7), clearly differentiate this case from that of reaction (3).

$$t\text{-}C_4H_9I \xrightarrow{\text{Slow}} t\text{-}C_4H_9^+ + I^- \qquad (6)$$

$$t\text{-}C_4H_9^+ + H_2O \xrightarrow{\text{Fast}} t\text{-}C_4H_9OH + H^+ \qquad (7)$$

4. Energetics governing the reaction rate provide another constraint. In general, the energy barrier exceeds the free energy for the overall process $\Delta G^{\ddagger} \geqslant \Delta G$ (reaction). Moreover, known or estimated free energies of the elementary steps in a mechanism should not be inconsistent with the observed $\Psi G^{\ddagger}$. *See* CHEMICAL DYNAMICS.

5. Correlations of structure, reactivity, and mechanism in known series are often used to evaluate new mechanistic situations. The change in mechanism from a primary halide, reaction (3), to a tertiary halide, reaction (6), is apropos. Again, one-step cycloaddition, reaction (8), differs from

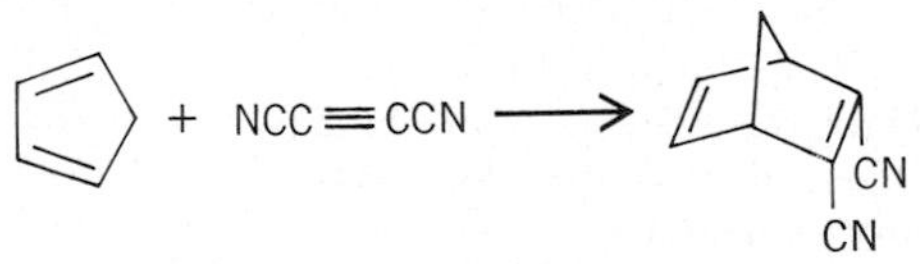

(8)

multistep radical process, reaction (9), for the reason discussed below.

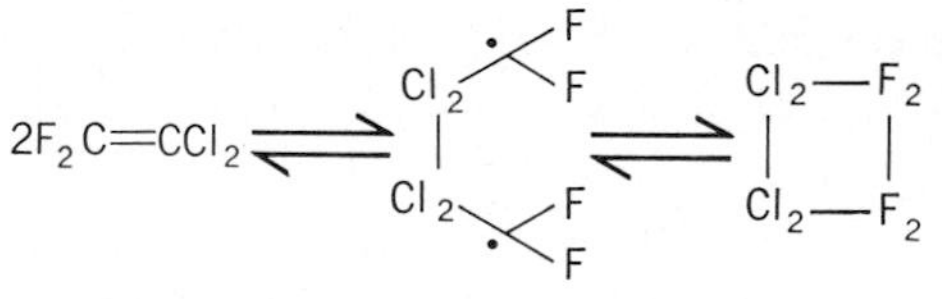

(9)

6. There is a group of useful mechanistic restrictions which might be considered quasitheoretical at most: (*a*) Adapted to the present context, Ockham's razor would caution one to prefer the simplest mechanism consistent with the facts. (*b*) According to the hypothesis of least motion, those paths along which there is the least change are often preferred. (*c*) The notion that simple paths are preferred shows that a conversion involving multiple sites is usually effected in a sequence of small changes rather than a concerted complex change.

7. Increasingly, theoretical calculations, particularly by quantum mechanics, have been applied to mechanistic questions such as: Is this step exclusive or is there competition? Which orientation (reaction site) is preferred? Will the change occur at all? In reaction (9), for example, one of three possible diradicals on the way to the two possible cyclodimers is indicated. Estimates of the relevant potential energy surfaces could lead to a choice of the preferred $Q^{\ddagger}$.

8. In the last category are mechanistic restrictions which are rigorously selective or exclusionary: (*a*) The principle of microscopic reversibility requires that at, and usually near, equilibrium those $Q^{\ddagger}$s traversed in the forward direction are those taken in the reverse direction. Although other paths between reactants and products not at equilibrium may be available, the equilibrium mechanism frequently holds generally for thermalized systems. (*b*) The Murrell-Laidler theorem allows $Q^{\ddagger}$s to intersect at minima (R, P, or I) on the energy surface but prohibits crossings at AC. Branched or forked ACs, as in the following hypothetical substitution, cycloelimination, and

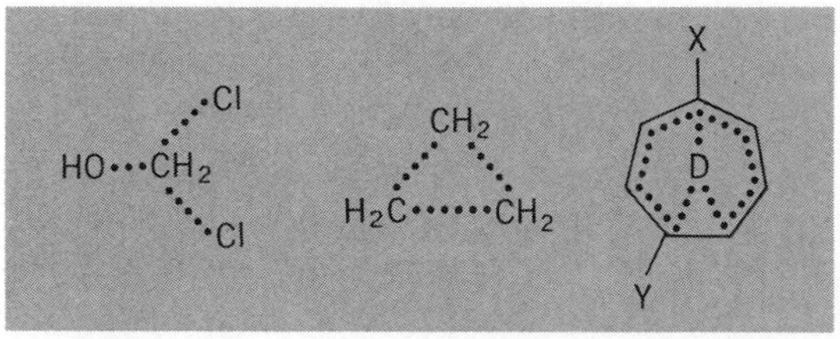

rearrangement, which have been proposed for so-called merged mechanisms, are forbidden. (*c*) The Woodward-Hoffmann rules divide pericyclic reactions into relatively low-energy (allowed) processes in which orbital symmetry is conserved, and high-energy (forbidden) processes in which orbital symmetry is not conserved. A pericyclic process involves a cyclic AC, as in reactions (8) and (10), but

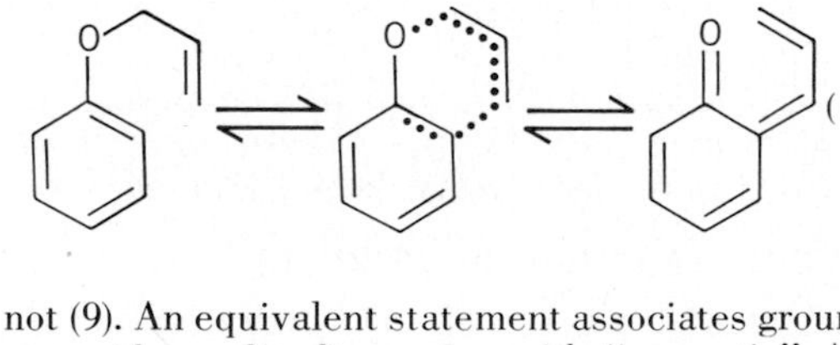

(10)

not (9). An equivalent statement associates ground-state (thermalized) species with "aromatic" AC, and reactions of photoexcited ($h\nu$) species with "antiaromatic" AC. Thus the cycloaddition of reaction (8) and the migration in reaction (10) are allowed reactions involving six electrons (Hückel aromatic). Since a four-electron (Hückel antiaromatic) AC would have to be involved in a one-step cyclodimerization of $F_2C = CCl_2$, it proceeds in steps as in reaction (9). Further, the ring openings (closings) of reaction (11) involve four electrons in

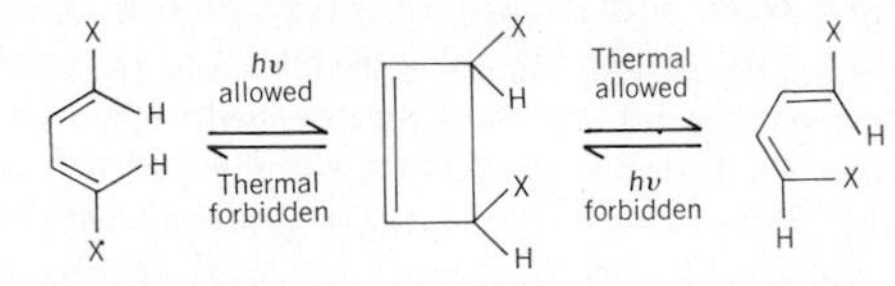

(11)

alternate allowed modes, thermal (Möbius aromatic) or photoexcited (Hückel aromatic). *See* CONDENSATION REACTION; HYDROXYLATION REACTION; STERIC EFFECT; SUBSTITUTION REACTION.

[SIDNEY I. MILLER]

Bibliography: K. A. Connors, *Reaction Mechanisms in Organic Analytical Chemistry*, 1973; J. A. Hirsch, *Concepts in Theoretical Organic Chemistry*, 1974; W. J. le Noble, *Highlights of Organic Chemistry*, 1974; E. S. Lewis (ed.), *Techniques of Chemistry*, 3d ed., vol. 6, pt. 2, 1974; R. E. Stanton and J. W. McIver, Jr., *J. Amer. Chem. Soc.*, 97:532, 1975.

Organoactinides

Organometallic compounds of the actinides—elements 90 and beyond in the periodic table. Both the large sizes of actinide ions and the presence of 5*f* valence orbitals are unique features which differ distinctly from most, if not all, other metal ions.

Organometallic compounds have been prepared for all actinides through curium (element 96), although most investigations have been conducted with readily available and more easily handled natural isotopes of thorium (Th) and uranium (U). Organic groups (ligands) which bind to actinide ions include both π- and σ-bonding functionalities. The importance of this type of compound reflects the ubiquitous character of metal-carbon two-electron sigma bonds in both synthesis and catalysis. *See* CATALYSIS; CHEMICAL STRUCTURES.

Synthesis. The most general preparative route to organoactinides involves the displacement of halide by anionic organic reagents in nonprotic solvents. Some examples of the preparation of cyclopentadienyl, C_5H_5 (structure I), hydrocarbyl (R = CH_3, C_6H_5, and so forth; II), pentamethylcyclopentadienyl, $(CH_3)_5C_5$ (III), allyl, C_3H_5 (IV), and cyclooctatetraenyl, C_8H_8 (V), derivatives are shown in reactions (1)–(5), where M = Th or U.

$$MCl_4 + 3Tl(C_5H_5) \rightarrow \underset{(I)}{M(C_5H_5)_3Cl} + 3TlCl \quad (1)$$

$$M(C_5H_5)_3Cl + RLi \rightarrow \underset{(II)}{M(C_5H_5)_3R} + LiCl \quad (2)$$

$$MCl_4 + 2(CH_3)_5C_5MgCl \rightarrow \underset{(III)}{M[(CH_3)_5C_5]_2Cl_2} + 2MgCl_2 \quad (3)$$

$$MCl_4 + 4C_3H_5MgCl \rightarrow \underset{(IV)}{M(C_3H_5)_4} + 4MgCl_2 \quad (4)$$

$$MCl_4 + 2K_2C_8H_8 \rightarrow \underset{(V)}{M(C_8H_8)_2} + 4KCl \quad (5)$$

These compounds are all exceedingly reactive and must be rigorously protected from air at all times.

Structures. The molecular structures of a number of organoactinides have been determined by single-crystal x-ray and neutron-diffraction techniques. In almost all cases the large size of the metal ion gives rise to unusually high (as compared to a transition-metal compound) coordination numbers. That is, a greater number of ligands or ligands with greater spatial requirements can be accommodated within the actinide coordination sphere. The sandwich complex bis(cyclooctatetraenyl)uranium (uranocene) is an example of this latter type (Fig. 1). In most organoactinides, metal-to-ligand atom bond distances closely approximate the sums of the individual ionic radii. This result argues that bonding forces within organoactinides have a large ionic component. Still, a number of spectral (optical, photoelectron, magnetic-resonance, Mössbauer) and magnetic studies reveal that the covalent character of the bonding cannot be ignored, and that there is considerable overlap of metal and ligand orbitals. *See* COORDINATION NUMBER; METALLOCENES; NEUTRON DIFFRACTION; X-RAY DIFFRACTION.

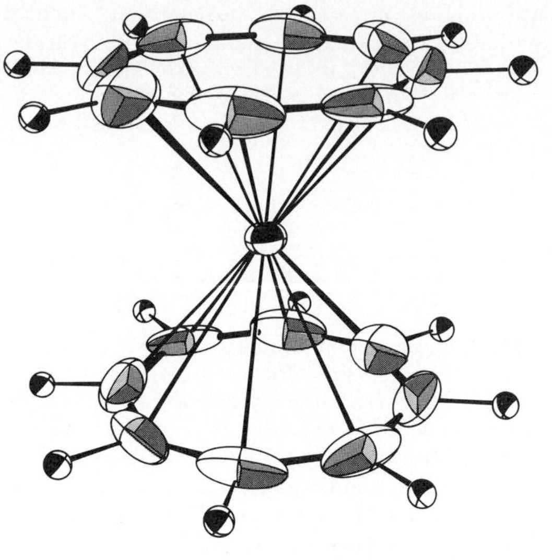

Fig. 1. Molecular structure of $U(C_8H_8)_2$ determined by single-crystal x-ray diffraction. (*From K. O. Hodgson and K. N. Raymond, Inorg. Chem., 12:458, 1973*)

Chemical properties. Studies of organoactinide reactivity have concentrated on understanding the relationship between the identity of the metal, the degree to which the ligands congest the coordination sphere (coordinative saturation), and the types of chemical reactions which the complex undergoes. Employing the methodology of reactions (2) and (3), bis(pentamethylcyclopentadienyl) actinide hydrocarbyl chlorides (VI) and bishydrocarbyls (VII) have been synthesized. These classes of

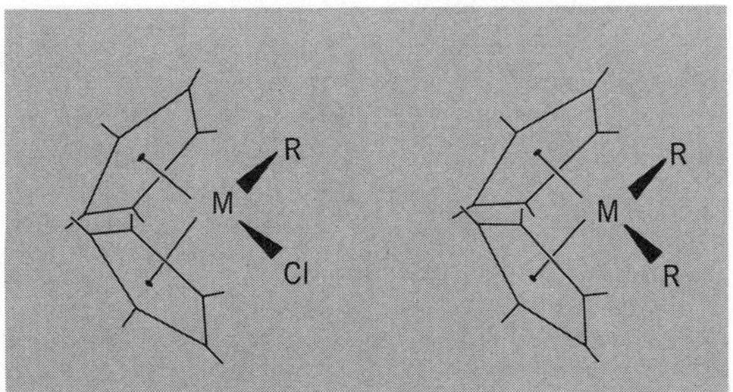

(VI) (VII)

compounds have proved to be some of the most reactive organoactinides, and some of the most informative in terms of revealing new types of reactivity and bonding.

In an atmosphere of hydrogen, thorium-to-carbon and uranium-to-carbon sigma bonds are rapidly cleaved to yield hydrocarbons and the first known organoactinide hydrides, reaction (6). For

$$2M[(CH_3)_5C_5]_2R_2 + 2H_2 \rightarrow \{M[(CH_3)_5C_5]_2H_2\}_2 + 4RH \quad (6)$$

the dimeric thorium hydride (Fig. 2), hydrogen atoms form both terminal (two-center, two-electron) and bridging (three-center, two-electron) bonds to thorium. In solution the organoactinide hydrides are active catalysts for olefin hydrogenation, reaction (7), and for the conversion of C-H bonds in the presence of D_2 to C-D bonds, reaction (8), where M = Th or U.

$$CH_2{=}CHCH_3 + H_2 \xrightarrow{M\text{-}H} CH_3CH_2CH_3 \qquad (7)$$

$$C_6H_6 + n/2D_2 \xrightarrow{M\text{-}H} C_6D_nH_{6-n} \qquad (8)$$

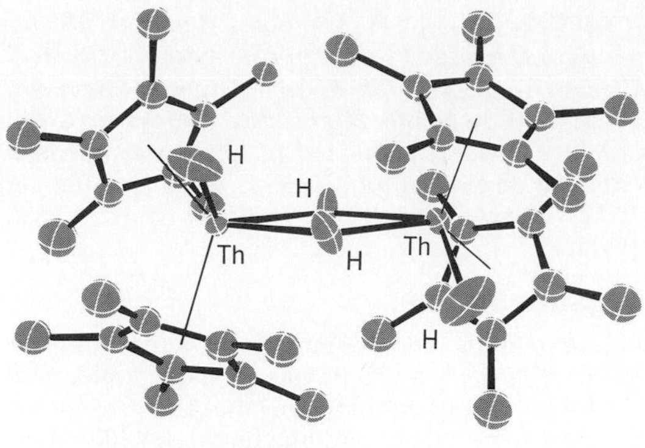

Fig. 2. Molecular structure of dimeric thorium hydride $\{Th[(CH_3)_5C_5]_2H_2\}_2$ as determined by single-crystal neutron diffraction. (*From R. W. Broach et al., Science, 203: 172, copyright © 1979 by the American Association for the Advancement of Science*)

The high oxygen affinity of actinide ions and the unsaturation of the bis(pentamethylcyclopentadienyl) ligation environment give rise to several unusual new types of carbonylation reactions. Thus, the bis(pentamethylcyclopentadienyl) dimethyl derivatives (VIII) react rapidly with CO to produce insertion products in which coupling of four carbon monoxide molecules has occurred to produce two carbon-carbon double bonds and four actinide-oxygen bonds, reaction (9), where M = Th or U. The alkyl chlorides (IX) undergo an insertion reaction to yield acyl [MC(O)R] complexes (X) in which a very strong metal-oxygen interaction takes place, reaction (10). The marked distortion of the bonding in these complexes, away from a classical acyl in which only the carbon atom is bound to the metal, is evident in structural, spectral, and chemical properties. Thus, the metal-oxygen distances are invariably shorter than the metal-carbon distances, the C-O stretching frequencies (about 1460 cm^{-1}) are anomalously low, and the chemistry is decidedly carbenelike. For example, upon heating in solution, the R = $CH_2C(CH_3)_3$ derivative rearranges to a hydrogen atom migration product (XI), and the R = $CH_2Si(CH_3)_3$ derivative to a trimethylsilyl migration product (XII). The pronounced oxygen affinity and coordinative unsaturation of the actinide ions in these environments may serve as models for the active surface sites on heterogeneous CO reduction catalysts. *See* ACTINIDE ELEMENTS; COORDINATION CHEMISTRY; HETEROGENEOUS CATALYSIS; HOMOGENEOUS CATALYSIS; ORGANOMETALLIC COMPOUND.

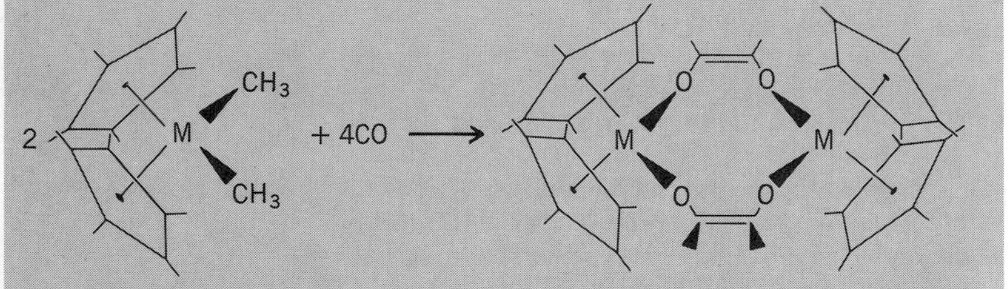

(VIII) M = Th, U (9)

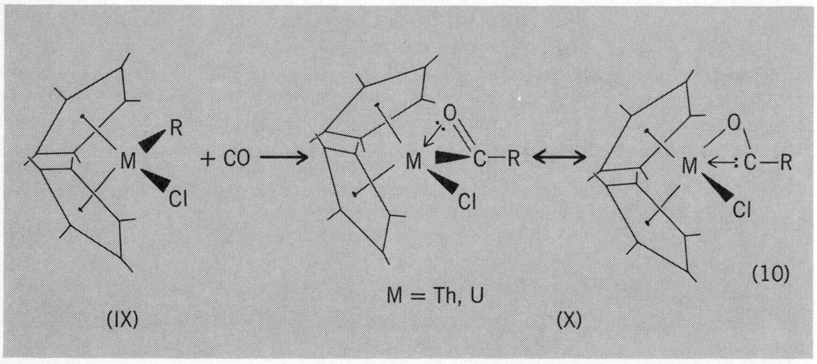

(IX) M = Th, U (X) (10)

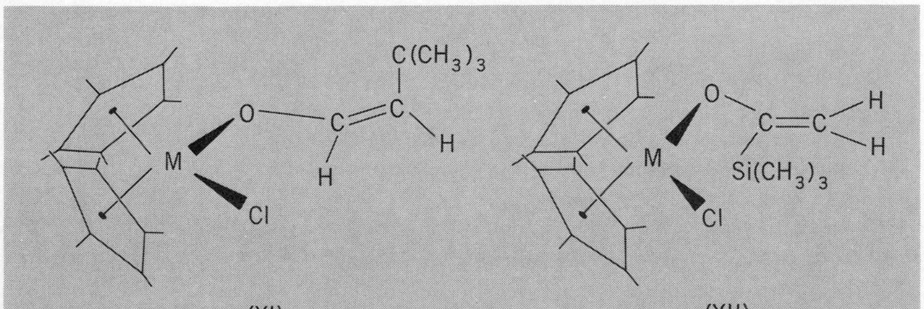

(XI) (XII)

[TOBIN J. MARKS]

Bibliography: E. C. Baker, G. W. Halstead, and K. N. Raymond, *Struct. Bond. (Berlin)*, 25:23, 1976; S. A. Cotton, *J. Organomet. Chem. Libr.*, 3: 189, 1977; T. J. Marks, *Prog. Inorg. Chem.*, 25:224, 1979; T. J. Marks and R. D. Fischer, *Organometallics of the f-Elements*, 1979.

Organoarsenic compound

A derivative of arsenic containing at least one organic radical attached through carbon to the arsenic atom by means of a covalent bond. All organoarsenic derivatives can be derived schematically by formal substitution or oxidation from the primary, secondary, and tertiary arsines. This is shown in notation (1). Depending on the state of

$RAsH_2$ Alkylarsine	R_2AsH Dialkylarsine	R_3As Trialkylarsine
↓	↓	↓
RAsO Alkyl-arsenoxide	$(R_2As)_2O$ Dialkyl-arsinoxide	R_3AsO Trialkyl-arsinoxide
↓	↓	(1)
$RAs(OH)_2$ Alkylarsonous acid	R_2AsOH Dialkylarsinous acid	
↓	↓	
$RAs(O)(OH)_2$ Alkylarsonic acid	$R_2As(O)OH$ Dialkylarsinic acid	

oxidation, organoarsenic compounds are derivatives of tri- or pentavalent arsenic. Although the known types of organoarsenicals are more numerous than those of true organometallic compounds, they do not approach the number of known types of the organo derivatives of phosphorus, which precedes arsenic in group V of the periodic system. A special type of organoarsenic compounds which do not have an equivalent in the organophosphorus series are the arseno compounds, RAs=AsR.

Preparation. The conversion of inorganic arsenic into organoarsenic compounds can be achieved readily in the aliphatic series by the Meyer reaction from trisodium arsenite and an alkylating agent according to reaction (2). The analogous alky-

$$RX + As(ONa)_3 \rightarrow RAs(O)(ONa)_2 + NaX \quad (2)$$

lation of disodium alkylarsonite, $RAs(ONa)_2$, yields sodium dialkylarsinate, $R_2As(O)ONa$; and sodium dialkylarsinite, R_2AsONa, leads similarly to trialkylarsine oxides. The ease with which derivatives of pentavalent arsenic are reduced to trivalent derivatives permits the synthesis of a large variety of organoarsenic compounds from sodium arsenite, and thus ultimately from arsenious oxide, As_2O_3. The alkylation of arsenic trichloride, $AsCl_3$, with metalloorganic compounds, such as Grignard reagents and zinc or mercury dialkyls, presents an alternate method for the formation of a carbon-arsenic bond. Aromatic arsonic acids are conveniently accessible from the diazotized aromatic amines and sodium arsenite by the Bart reaction (3) or by direct substitution of phe-

$$RN_2Cl + As(ONa)_3 \rightarrow RAs(O)(ONa)_2 + N_2 + NaCl \quad (3)$$

nols or amines by arsenic acid according to reaction (4).

$$C_6H_5OH + H_3AsO_4 \rightarrow p\text{-}HOC_6H_4As(O)(OH)_2 + H_2O \quad (4)$$

Use. All organoarsensic compounds exhibit physiological activity. The high mammalian toxicity of the trivalent derivatives makes a number of them potential chemical warfare agents. Lewisite ($ClCH{=}CHAsCl_2$), Adamsite (phenarsazine chloride), phenyldichloroarsine ($C_6H_5AsCl_2$), and ethyldichloroarsine ($C_2H_5AsCl_2$) are some of the chemical agents that are effective as vesicants, irritants, and sternutators. Several of the less toxic derivatives have found use as chemotherapeutics among which Salvarsan (arsphenamine hydrochloride), Atoxyl (sodium arsanilate) and Tryparsamide were of prime importance as the only means for combating certain infectious diseases before the discovery of the antibiotics. *See* ORGANOPHOSPHORUS COMPOUND.

[FRIEDRICH W. HOFFMANN/
THOMAS C. SIMMONS]

Bibliography: G. T. Morgan, *Organic Compounds of Arsenic and Antimony*, 1918; G. W. Raiziss and J. L. Gavron, *Organic Arsenical Compounds*, 1923; H. Zeiss (ed.), *Organometallic Chemistry*, ACS Monogr. no. 147, 1961.

Organometallic compound

One of a group of substances containing carbon-metal bonds. This definition can be broadened to include all compounds of elements that possess metallike properties and that are combined chemically with one or more carbon atoms. In this respect, metal alkoxides, chelates, salts of organic acids, and related compounds are not discussed in this article. *See* CHELATION.

Much of the early research in organometallic chemistry was concerned with the preparation and reactions of organozinc compounds, although the study of organic derivatives of other metals followed in rapid succession. The pharmacological value of many organoarsenic and organomercury compounds stimulated much research on the synthesis and properties of these compounds. The reactions of Grignard reagents and of organolithium and organosodium compounds have found extensive use in synthetic organic chemistry. *See* GRIGNARD REACTION.

Organometallic compounds have found many commercial applications. The use of tetraethyllead as a gasoline antiknock additive is a good example. Organometallic compounds have played an important role in the development of silicone polymers, polyvinylchloride, and polyethylene. *See* TETRAETHYLLEAD.

A systematic classification of all known organometallic compounds is very difficult. For the sake of convenience, however, they can be divided into three main classes. The more electropositive metals of groups I and II of the periodic table form organometallic compounds that are nonvolatile, usually poorly soluble in organic solvents, and essentially ionic in nature. Lithium, beryllium, and magnesium compounds are much less ionic than those of the remaining metals, however. The metals (excepting transition metals) and metalloids of groups III, IV, V, and VI form organometallic compounds which are mainly volatile, soluble in organic solvents, and principally covalently bonded. The transition metals constitute a third main group, possessing bonding to aromatic or aromatic-type groups which is mainly of a special *d*-orbital, or "sandwich," type. *See* CHEMICAL BONDING.

Preparation from free metals. The reaction of metals or metalloids with organic halides is widely used in the synthesis of organometallic compounds, as shown by (1). Sodium, lithium, mag-

$$2M + RX \rightarrow RM + MX \quad (1)$$

nesium, and related metals react with organic halides in this manner to produce organometallic derivatives. Silicon and germanium react readily in the gaseous phase with both alkyl and aryl halides.

Another means of preparing organometallic compounds involves the reaction of highly reactive metals with hydrocarbons that contain active hydrogen atoms, as in (2). The preparations of

$$RH + M \rightarrow RM + H \quad (2)$$

organometallic derivatives of acetylene, triphenylmethane, and cyclopentadiene are examples of this reaction.

The displacement of a metal in an organometallic compound by a more reactive metal to produce a new organometallic compound has been widely used, although reactions of this type are usually reversible, as shown by (3). Organometallic

$$M + RM' \rightleftharpoons RM + M' \quad (3)$$

derivatives of the alkali metals, beryllium, magnesium, and aluminum, have been prepared by this procedure.

Preparation from metal salts. The reaction of organometallic compounds with metal salts has found extensive application in the synthesis of many types of organometallic compounds. Reactions (4) and (5) indicate this synthesis. In general,

$$RM + M'X \rightarrow RM' + MX \quad (4)$$

$$RMX + M'X \rightarrow RM' + MX_2 \quad (5)$$

organic derivatives of more reactive metals react with metal salts of less reactive metals or metalloids by this procedure. Examples include the reactions of alkylsodium compounds, alkyllithium compounds, and Grignard reagents with halides of elements of groups III, IV, V, and VI. Many cyclopentadienyl derivatives of the transition metals can best be prepared in this manner.

Metal salts react with aryldiazonium compounds in the presence of metals to give organometallic halides. Arylmercury and aryltin compounds have frequently been prepared in this way.

Other preparations. These reactions include methods by which an organometallic compound of a given metal is used to prepare other organometallic derivatives of the same metal. The exchange reaction (6) between ethyllithium and diphenylmercury illustrates a form of this reaction.

$$2C_2H_5Li + (C_6H_5)_2Hg \rightarrow (C_2H_5)_2Hg + 2C_6H_5Li \quad (6)$$

The reaction (7) between an organometallic

$$RM + R'H \rightarrow R'M + RH \quad (7)$$

compound and an aromatic hydrocarbon possessing a reactive hydrogen atom is known as metalation. The most commonly used metalating agent, or RM compound, is *n*-butyllithium in ether solution. Many other reactive organometallic compounds have also been used with success.

A large number of organometallic derivatives of aromatic hydrocarbons have been prepared by the reaction (8) of the corresponding aryl halides with a reactive compound such as *n*-butyllithium.

$$n\text{-}C_4H_9Li + RX \rightarrow RLi + n\text{-}C_4H_9X \quad (8)$$

Compounds of group I elements. The alkali metals form organometallic compounds that possess high reactivity. Most alkyl derivatives are colorless solids, except the higher alkyls of lithium, which are liquids. Many aryl derivatives are solids, and have intense colors that depend on the number of aromatic and conjugated groups in the molecule. With the exception of the alkyllithium derivatives, organoalkali compounds are characterized by insolubility in most organic solvents and little tendency to vaporize or melt without decomposition. Such properties indicate a large degree of ionic character in the bonding between the metal and organic group. Because of the small size and high polarizing power of the lithium ion, alkyl derivatives of lithium are largely covalently bonded.

Due to the high reactivity of organoalkali compounds, they must be prepared and used in an inert atmosphere that is free of oxygen, moisture, and carbon dioxide. Organolithium compounds are highly reactive and are extensively used as intermediates in synthetic organic chemistry. Organolithium reagents are in general more reactive than the well-known organomagnesium halides (Grignard reagents). They undergo addition to carbon-carbon multiple bonds, and several important industrial processes for the production of polyolefins, synthetic rubber, and other products are based on this reaction. Reactions of organolithium reagents with carbon dioxide and aldehydes or ketones, followed by hydrolysis, generally produce high yields of corresponding carboxylic acids and alcohols, respectively. Organolithium reagents are frequently employed as precursors for other reactive organic intermediates, such as carbenes, arynes, and ylides, as illustrated in reactions (9)–(11), respec-

$$CH_2Cl_2 + RLi \xrightarrow{-RH} LiCHCl_2 \xrightarrow{-LiCl} :CHCL \longrightarrow \text{products} \quad (9)$$

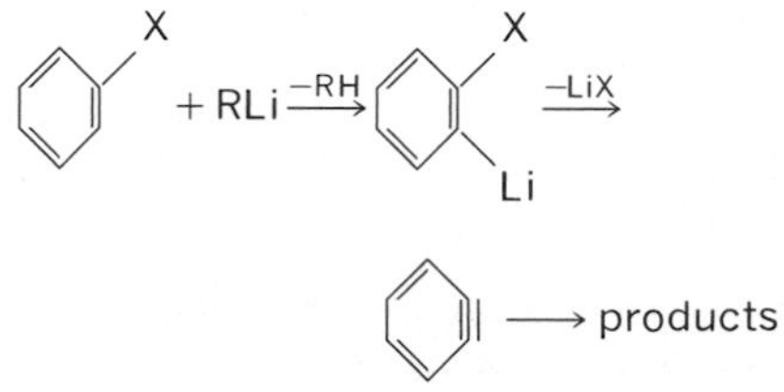

(10)

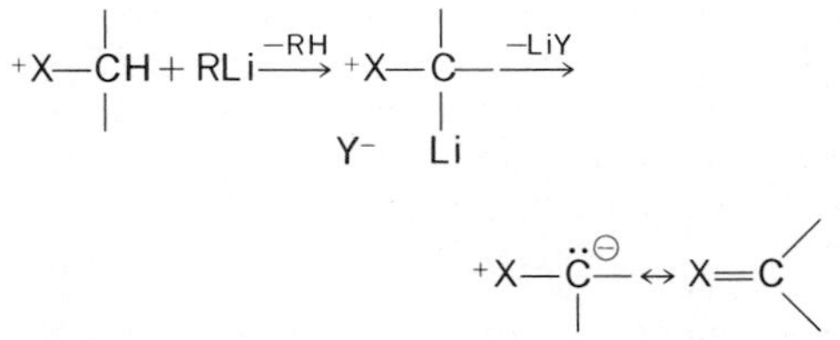

(11)

tively. It is questionable, however, that "free" carbenes and arynes are actually involved in such processes.

The reactivities of organolithium reagents are particularly enhanced by the presence of a chelating ditertiary aliphatic amine such as N,N,N′,N′-tetramethylethylenediamine (TMEDA) in the reaction system (Fig. 1). *See* CHELATION.

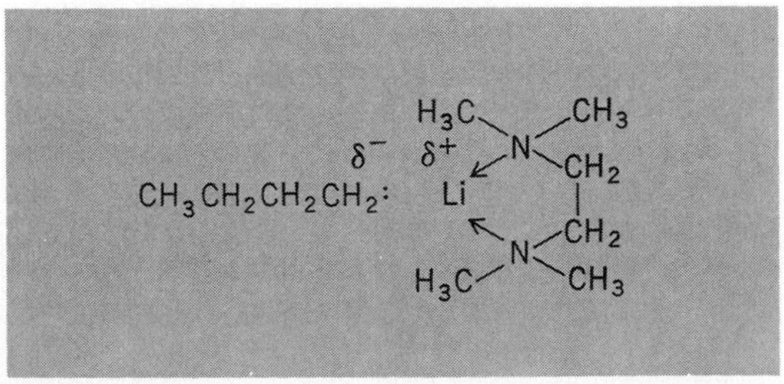

Fig. 1. Structural formula for *n*-butyllithium · TMEDA.

Compounds of group II elements. All elements of group II except radium form organometallic compounds, although compounds of calcium, strontium, or barium have not yet been extensively studied. A remarkable gradation in physical and chemical properties is observed for organometallic compounds of this group. Nearly all these compounds except those of mercury are flammable in the presence of oxygen and water. Alkylberyllium compounds are volatile, although highly associated liquids; an exception is dimethylberyllium. Arylberyllium compounds are solids and are somewhat more stable.

Two main types of organomagnesium compounds are known, alkyl- and arylmagnesium compounds, R_2Mg, and the organomagnesium halides or Grignard reagents, RMgX. The latter compounds are the most widely known organometallic derivatives of group II elements and have found extensive use in organic syntheses. Alkyl- and arylmagnesium compounds are white crystalline solids. They are practically nonvolatile and are insoluble in hydrocarbon solvents. Although these compounds are salt-like, the bonding between the magnesium and the organic group is believed to be largely covalent.

Organozinc and organocadmium compounds were among the first organometallic compounds to be isolated. For many years organozinc compounds were used for synthetic purposes until they were superseded by the more convenient Grignard reagents. Alkylzinc and alkylcadmium compounds are volatile, unassociated liquids; aryl derivatives are white solids with sharp melting points. *See* REFORMATSKY REACTION.

Many organomercury compounds of the type R_2Hg and RHgX have been prepared. The reactivities of these compounds are so low that they are unaffected by water and air. Dialkylmercury compounds are mostly distillable liquids; diarylmercury derivatives are crystalline solids. The bonding in these compounds is essentially covalent. Alkyl- and arylmercury compounds of the type RHgX are stable crystalline solids; their properties depend largely on the nature of the group X.

Compounds of group III elements. Most alkyl and many aryl derivatives of the group III elements are spontaneously flammable in the presence of air. The alkylboron compounds are colorless liquids; the arylboron compounds are crystalline solids. Alkyl- and arylboric acids, $RB(OH)_2$ and R_2BOH, are important substitution derivatives. Compounds containing B-H bonds add to olefins or acetylenes (hydroboration) to form organoboron compounds, which serve as important intermediates to many other types of organic compounds, such as aldehydes, ketones, amines, and alcohols. Organoboron compounds also form many coordination compounds with atoms and molecules that possess electron-donor groups.

Alkylaluminum compounds are colorless liquids of extreme reactivity. Trimethylaluminum has been shown to possess a dimeric structure in contrast to the corresponding boron derivative. The alkylaluminum compounds find important industrial applications as polymerization catalysts. Arylaluminum compounds are crystalline solids. The arylaluminum compounds are also highly reactive. *See* POLYOLEFIN RESINS.

A few alkyl and aryl derivatives of gallium, indium, and thallium are known. In general, these compounds exhibit a decreasing order of reactivity, especially when compared to the corresponding organoaluminum compounds. A recent notable development has been the isolation of a monovalent organoindium compound, cyclopentadienylindium, C_5H_5In.

Compounds of group IV elements. The group IV elements silicon, germanium, tin, and lead form organometallic compounds of the type R_4M. All compounds of this type are covalently bonded, although the carbon-metal bond becomes weaker and more polar as the metal becomes more metallic. A notable feature of elements of this group is their ability to form metal-metal bonds. Alkyl derivatives of group IV metals are liquids; higher alkyl members possess high boiling points and are liquids over wide temperature ranges. Aryl derivatives are high-melting solids. Both alkyl and aryl derivatives are usually unattacked by water and air at room temperature, and are soluble in hydrocarbon solvents.

Compounds of group V and group VI elements. Organic derivatives of phosphorus, arsenic, antimony, and bismuth are well known, although it is debatable whether compounds of the first two elements can be classified as truly organometallic. Group V elements form organometallic compounds of both R_3M and R_5M types and also many mixed organometallic hydrides, halides, and hydroxides. Alkyl derivatives of the type R_3M are liquids of decreasing stability in the case of the more metallic members. A number of pentaphenyl derivatives of the type $(C_6H_5)_5M$ have been reported in which all phenyl groups are covalently bonded, although the fifth phenyl group is not strongly held. *See* ORGANOARSENIC COMPOUND; ORGANOPHOSPHORUS COMPOUND.

Both selenium and tellurium in group VI form alkyl and aryl derivatives of the type R_2M. In some aspects these can be considered to be organometallic compounds. The alkyl derivatives are colorless liquids and the aryl derivatives are low-melting solids; all possess extremely unpleasant odors. A number of diselenides and ditellurides of the type RM—MR are also known.

Compounds of the transition metals. Organometallic compounds of the transition metals have undergone an unprecedented growth in recent years as a result of their novel structures, properties, and reactions. The first compound of this type to be isolated and studied in detail was dicyclopentadienyliron, commonly called ferrocene (Fig. 2*a*). This remarkable compound was found to possess a sandwich structure in which the two cyclopentadienyl rings are coordinated via π-electrons to the iron atom to form a symmetrical, extremely stable structure. It has been suggested that the iron atom is bonded to each cyclopentadienyl ring by a delocalized covalent bond. Dicyclopentadienyliron has been shown to possess highly aromatic properties and has thus opened up an entirely new field of organic chemistry. *See* FERROCENE.

Since the initial discovery of dicyclopentadienyliron in 1951, the synthesis and characterization of cyclopentadienyl derivatives of all the other transition metals as well as many lanthanide and actinide metals followed in rapid succession. Many of these compounds also possess sandwich-type, π-bonded structures, although essentially ionic and covalent types are likewise known.

Organotransition metals containing other π-electron ligands such as olefins, acetylenes, dienes, and arenes are now known. Dibenzenechromium, for example, was characterized several years after the discovery of dicyclopentadienyliron, and has a similar sandwich structure (Fig. 2*b*). Another member of this unique series of organotransition metal compounds is dicyclooctatetraeneuranium,

ORGANOMETALLIC COMPOUND

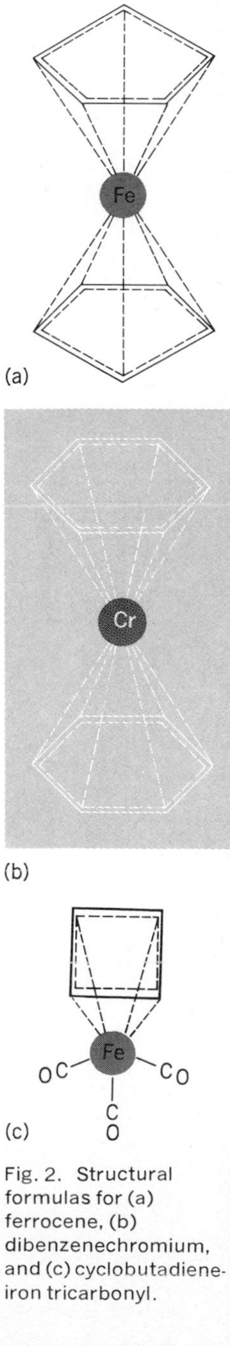

Fig. 2. Structural formulas for (a) ferrocene, (b) dibenzenechromium, and (c) cyclobutadiene-iron tricarbonyl.

or uranocene [$(C_8H_8)_2U$], which was discovered in 1968.

In some cases, transition-metal derivatives of certain unsaturated organic molecules can be isolated in contrast to the parent organic system itself which is highly unstable. An example is cyclobutadiene-iron tricarbonyl (Fig. 2*c*). This compound is not only isolatable and stable, but undergoes aromatic-type substitution reactions and serves as a source for cyclobutadiene when it is treated with oxidizing agents in solution.

A variety of organometallic compounds which contain carbon–transition-metal σ-bonds are also known. Many of these contain other ligands which are also bonded to the metal, such as carbon monoxide, the cyclopentadienyl group, or phosphines. Some specific examples are $CH_3Mn(CO)_5$, $(C_5H_5)_2Ti(C_6H_5)_2$, and $[(C_2H_5)_3P]_2Pd(C_6H_5)_2$. A number of completely alkylated compounds such as $(C_6H_5CH_2)_4Ti$, $[(CH_3)_3SiCH_2]_4Zr$, and $(CH_3)_6W$ have been isolated, suggesting that the carbon–transition-metal σ-bond is more stable than originally believed.

Organometallic compounds of the transition metals are finding increasing application as fuel additives, photostabilizers, antioxidants, burning-rate catalysts, and so on. Organotransition metal compounds also serve as important intermediates in many catalytic processes such as the oxo or hydroformylation reaction, the production of polyethylene, the conversion of ethylene to acetaldehyde, and the reaction of methanol and carbon monoxide to produce acetic acid.

[MARVIN D. RAUSCH]

Bibliography: E. W. Abel and F. G. A. Stone (eds.), *Organometallic Chemistry—A Specialist Periodical Report*, vols. 1–7, 1971–1978; J. P. Collman and L. S. Hegedus, *Principles and Applications of Organotransition Metal Chemistry*, 1980; M. Dub (ed.), *Organometallic Compounds*, 2d ed., 3 vols. plus formula index, 1960–1970.

Organophosphorus compound

One of a series of derivatives of phosphorus that have at least one organic (alkyl or aryl) group attached to the phosphorus atom linked either directly to a carbon atom or indirectly by means of another element (for example, oxygen). The mono-, di-, and trialkylphosphines (and their aryl counterparts) can be regarded formally as the parent compounds of all organophosphorus compounds; see notation (1). Formal substitution of the hydrogen of

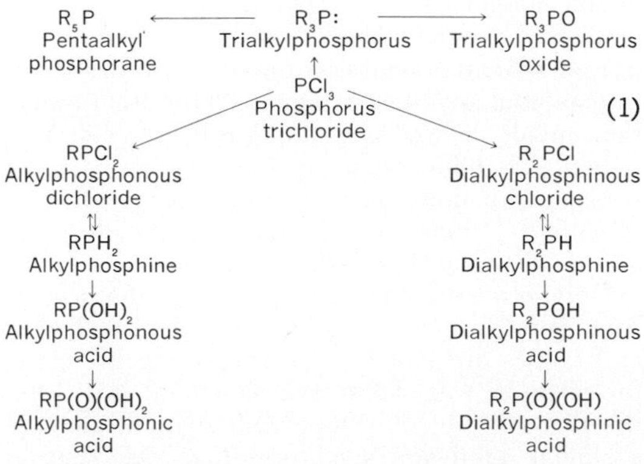

(1)

the phosphines by monovalent groups or atoms leads to a number of basic structures derived from trivalent phosphorus; formal addition of bivalent oxygen leads from the mono- and dialkylphosphines to organophosphorus acids and from the trialkylphosphines to their oxides. Many of these organophosphorus molecules have been synthesized in optically active form, thus serving as stereochemical probes for the study of reaction mechanisms. Formal replacement of the nonbonding electron pair in R_3P: by two substituents gives R_5P. Pentaphenylphosphorane is a stable example in this class.

Considering the large number of organic groups that may be joined to phosphorus as well as the incorporation of other elements in these materials, the number of combinations is practically unlimited. A vast family in itself is composed of the heterocyclic phosphorus molecules, in which phosphorus is one of a group of atoms in a ring system.

Preparation. Organophosphorus compounds may be prepared by a variety of methods. Trialkyl esters of phosphorus acid (trialkylphosphites) react with alkyl halides to form dialkyl alkylphosphonates, in the Michaelis-Arbuzov reaction (2). Since the preparation involves alkyl halides

$$P(OR)_3 + R'X \rightarrow R'P(O)(OR)_2 + RX \qquad (2)$$

and depends on the reactivity of the aliphatic halide, aromatic organophosphorus compounds cannot be obtained in an analogous manner from aryl halides. The preferred method for preparing aromatic compounds is the introduction of a $—PCl_2$ group into aromatic hydrocarbons by means of PCl_3, with anhydrous aluminum chloride as a catalyst. The resulting arylphosphonous dichlorides can react further with the aromatic hydrocarbon to yield the diarylphosphinous chlorides, Ar_2PCl, which are always obtained in addition to the phosphonous dichlorides, but can be made the main product of the reaction by properly choosing reaction conditions.

The reaction of phosphorus halides with organometallic compounds is useful for preparation of trialkyl- or triarylphosphines but is less applicable for making compounds with only one P—C bond. Organophosphorus derivatives with one or more P—H bonds can be added across activated olefin or acetylene bonds to form a P—C linkage. The P—H group can also react with carbonyl compounds to give α-hydroxy derivatives. *See* ORGANOMETALLIC COMPOUND.

Reactions. Organophosphorus compounds frequently serve as valuable synthetic reagents. The Wittig reaction (3) is one of the most useful

$$R_3P{=}CR'_2 + R''_2C{=}O \rightarrow R''_2C{=}CR'_2 + R_3PO \qquad (3)$$

Alkylidene phosphorane   Carbonyl   Alkene

preparative methods known. The alkylidine phosphorane is usually prepared from a phosphonium salt ($R_3P^+CHR'_2Br^-$) and an inorganic or organic base (for example, C_6H_5Li); subsequent treatment with ketones or aldehydes yield olefins. The thermodynamic driving force for this and many reactions involving organophosphorus is formation of the relatively strong P=O bond. Polyacetylenes, carotenoids (vitamin A), and steroid derivatives are more synthetically accessible by employment of the Wittig reaction. Both the structure of the phosphorane and reactive substrate have been varied over a wide range of combinations. *See* ORGANIC CHEMICAL SYNTHESIS.

Uses. Some organophosphorus compounds have been used as polymerization catalysts, lubricant additives, flameproofing agents, plant growth regulators, and insecticides. Hexamethylphosphoramide, $[(CH_3)_2N]_3PO$, is a remarkable polar aprotic solvent used in organic syntheses and capable of forming complexes with organic, inorganic, and organometallic compounds. Cyclophosphamide (CPA) is a phosphorus heterocycle which has been used in the treatment of cancer; it also is an anti-inflammatory agent in diseases. Naturally occurring products with a C—P bond have been found (for example, 2-aminoethylphosphonic acid) in protozoa, in marine life, and in certain terrestrial animals. On the other hand, the high mammalian toxicity shown by some methylphosphonic acid derivatives (inhibitors of the enzyme cholinesterase) limits the usefulness of a number of related though much less toxic materials because of potential health hazards. Organophosphorus compounds were made during World War II for use as chemical warfare agents in the form of nerve gases (Sarin, Trilon 46, Soman, and Tabun). *See* PHOSPHORUS.

[SHELDON E. CREMER]

Bibliography: M. Grayson and E. J. Griffith (eds.), *Topics in Phosphorus Chemistry*, vols. 1–7, 1964–1972; G. M. Kosolapoff and L. Maier (eds.), *Organic Phosphorus Compounds*, vols. 1–7, 1972–1976; S. Trippett (ed.), *Organophosphorus Chemistry*, vols. 1–6, 1970–1975; B. J. Walker, *Organophosphorus Chemistry*, 1972.

Organosulfur compound

One of a group of substances which contain both carbon and sulfur. Oxygen, nitrogen, the halogens, and phosphorus are also often present. Thousands of such compounds are well known.

Many medicinal and natural products are organosulfur compounds, for example, the penicillins, the sulfa drugs, insulin, and amino acids such as cysteine and methionine. The utilizations of organosulfur compounds involve enormous commercial developments in medicinals, detergents, sulfide rubbers and polymers, sulfur dyes, solvents, and agricultural chemicals (herbicides, fungicides, and insecticides). Besides their practical uses, studies of organosulfur compounds helped to establish fundamental theory on the nature of valence bonding, of molecular geometry, and of reaction mechanisms of organic chemistry.

The most common classes of organosulfur compounds are: (1) mercaptans, RSH, also called thiols or sulfhydryl compounds, such as methyl mercaptan, CH_3SH or *tert*-butylmercaptan, $(CH_3)_3CSH$; (2) thiophenols, ArSH, for example, thiophenol, C_6H_5SH, or *p*-thiocresol, *p*-$CH_3C_6H_4SH$; (3) disulfides, RSSR′, where R may be an aliphatic, aromatic, or heterocyclic radical; (4) sulfides, including cyclic sulfides, RSR, for example, methyl ethyl sulfide, $CH_3SC_2H_5$ or ethylene sulfide [formula (I)]; (5) sulfenyl chlorides, RSCl, for example,

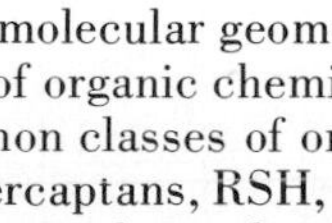
(I)

trichloromethanesulfenyl chloride, Cl_3CSCl; (6) sulfoxides [formula (II)], for example, dimethyl sulfoxide, an interesting solvent: (7) sulfones [formula (III)], for example, bis-(*p*-aminophenyl) sulfone

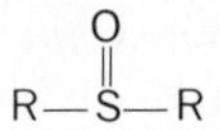
(II)

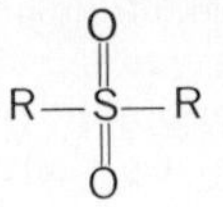
(III)

$NH_2C_6H_4SO_2C_6H_4NH_2$, of interest in the treatment of tuberculosis; (8) sulfonyl chlorides, RSO_2Cl, for example, *p*-toluenesulfonyl chloride, *p*-$CH_3C_6H_4SO_2Cl$; (9) sulfonamides, R—SO_2NH_2, for example, benzenesulfonamide, $C_6H_5SO_2NH_2$; (10) thioaldehydes and thioketones, RC(H)═S and R_2C═S, which generally exist as trimers, for example, trithioacetaldehyde; (11) sulfinic acids, RSO_2H; (12) sulfonic acids, RSO_3H; (13) esters of sulfonic acids, RSO_2OR'; (14) thiol esters [formula (IV)], and thio acids [formula (V)].

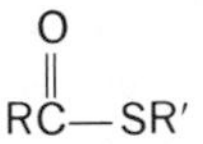
(IV)

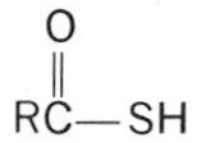
(V)

A large group of organosulfur compounds belongs to the heterocyclic series. In these compounds the sulfur atom is part of a ring system, and many others belong to groups not mentioned above, such as thiocarbonates, dithio acids, sulfimides, sulfamic acids, thioureas, sulfonium salts, mercaptals, mercaptols, organic thiocyanates, and isothiocyanates.

Selenium follows sulfur in the sixth main family of the periodic system and hence has many similarities to sulfur. Organoselenium compounds are also relatively well known; many of them are analogous to the sulfur compounds. *See* DIMETHYL SULFOXIDE; HETEROCYCLIC COMPOUNDS; SULFONAMIDE.

[NORMAN KHARASCH]

Bibliography: T. W. Campbell, H. G. Walker, and G. M. Coppinger, Some aspects of the organic chemistry of selenium, *Chem. Rev.*, 50:279, 1952; F. Challenger, *Aspects of the Organic Chemistry of Sulfur*, 1959; H. Gilman, *Organic Chemistry*, vol. 1, 2d ed., 1943; E. E. Reid, *Organic Chemistry of Bivalent Sulfur*, 6 vols., 1958–1966; C. M. Suter, *The Organic Chemistry of Sulfur: Tetracovalent Sulfur Compounds*, 1944.

Orthoester

A trialkyl derivative of a nonexistent orthoacid, $RC(OH)_3$, with general formula $RC(OR)_3$.

The structure of the orthocarbonates is $C(OR)_4$. Orthoesters are liquids with ethereal odors and are stable to aqueous alkalies but sensitive to acid hydrolysis. By far the most common and most important is ethyl orthoformate, $HC(OC_2H_5)_3$, although ethyl orthoacetate and higher members are known.

Orthoformate esters are manufactured by interaction of chloroform with an excess of the requisite sodium alkoxide; thus ethyl orthoformate results from the reaction of chloroform and sodium ethoxide. Higher orthoesters are usually made by treatment of imidoester hydrochlorides (from nitriles, hydrogen chloride, and alcohol) with excess alcohol. Orthocarbonates cannot be made from carbon tetrachloride; instead, chloropicrin, CCl_3NO_2, is heated with the requisite sodium alkoxide.

Orthoformate esters are frequently used in synthetic organic chemistry, for example, in preparing aldehydes from Grignard reagents. *See* ESTER; GRIGNARD REACTION.

[EVANS B. REID]

Oscillatory reaction

A chemical reaction in which some composition variable of a chemical system exhibits regular periodic variations in time or space. It is a basic tenet of chemistry that a closed system moves inexorably toward an unchanging state called chemical equilibrium. That motion can be described by the monotonic increase of entropy if the system is isolated, and by the monotonic decrease of Gibbs free energy if the system is constrained to constant temperature and pressure. Because of this universal restriction on what is possible in chemistry, it may appear bizarre when electrodes in a solution generate the oscillating potentials shown in Fig. 1.

The species taking part in a chemical reaction can be classified as reactants, products, or intermediates. The concentrations of reactants decrease monotonically, and those of products increase. Intermediates are formed by some steps and destroyed by others. If there is only one intermediate, and if its concentration is always much less than the initial concentrations of reactants, this intermediate attains a stable steady state in which the rates of formation and destruction are virtually equal. The kind of oscillation reflected in Fig. 1 requires at least two intermediates which interact in such a way that the steady state of the total system is unstable to the minor fluctuations present in any collection of molecules. The concentrations of the intermediates may then oscillate regularly, although the oscillations must disappear before the inevitable monotonic approach to equilibrium.

Periodic chemical behavior may be temporal in a uniform system as illustrated in Fig. 1; it may involve simultaneous temporal and spatial variations as in Fig. 2; or it may involve spatial periods in a static system as in Fig. 3.

Well-authenticated examples of periodic chemical behavior have been known for almost a century, but until the 1970s most chemists either did not know about them or deliberately ignored them. Interest has been developing rapidly, but most examples are only poorly understood. The phenomena are classified here according to types of chemical processes involved. Very different classification schemes may become more appropriate in the future. *See* CHEMICAL EQUILIBRIUM; ELECTRODE POTENTIAL; ENTROPY; GIBBS FUNCTION.

Redox oscillators. The systems whose chemistries are best understood all involve an element that can exist in several different oxidation states. Figure 1 illustrates the so-called Belousov-Zhabotinsky reaction, which was discovered in the Soviet Union in 1951. A strong oxidizing agent (bromate) attacks an organic substrate (such as malonic acid), and the reaction is catalyzed by a metal ion (such as cerium) that can exist in two different oxidation states.

As long as bromide ion (Br^-) is present, it is oxidized by bromate (BrO_3^-), as in reaction (1).

$$BrO_3^- + 2Br^- + 3H^+ \rightarrow 3HOBr \quad (1)$$

When bromide ion is almost entirely consumed, the cerous ion (Ce^{3+}) is oxidized, as in reaction (2).

$$BrO_3^- + 4Ce^{3+} + 5H^+ \rightarrow HOBr + 4Ce^{4+} + 2H_2O \quad (2)$$

Reaction (2) is inhibited by Br^-, but when the concentration of bromide has been reduced to a critical level, reaction (2) accelerates autocatalytically until bromate is being reduced by Ce^{3+} many times as rapidly as it was by Br^- when reaction (1) was dominant.

The hypobromous acid (HOBr) brominates the organic substrate to form bromomalonic acid (BrMA), as in reaction (3). Reaction (3) creates the

$$2Ce^{4+} + BrMA \rightarrow 2Ce^{3+} + Br^- + \text{oxidized organic matter} \quad (3)$$

bromide ion necessary to shut off fast reaction (2) and throw the system back to dominance by slow reaction (1).

As other redox oscillators become understood, they fit the same pattern of a slow reaction destroying a species that inhibits a fast reaction that can be switched on autocatalytically; the fast reaction then generates conditions to produce the inhibitor again.

Until 1982 the known examples involved positive oxidation states of Cl, Br, and I. Elements, like N, S, Cr, and Mn, which have many positive oxidation states may be able to participate in other oscillatory reactions.

The temporal traces in Fig. 1 were obtained with a homogeneous stirred solution. If the solution were unstirred but had a gradient in composition, oscillations in different regions would get out of phase and an apparent wave would traverse the medium much the way flashing lights cause a message to move along a theater marquee.

Figure 2 illustrates a still more complex situation. Each light curve is a region of dominance by reaction (2). A combination of reaction and diffusion triggers an advance outward perpendicular to the wavefront into the dark region dominated by reaction (1). Trigger waves annihilate each other when they meet, and Fig. 2 shows two spirals spinning in opposite directions. This kind of behavior has been suggested to explain the fibrillations when a human heart loses its rhythm and degenerates to uncoordinated local contractions

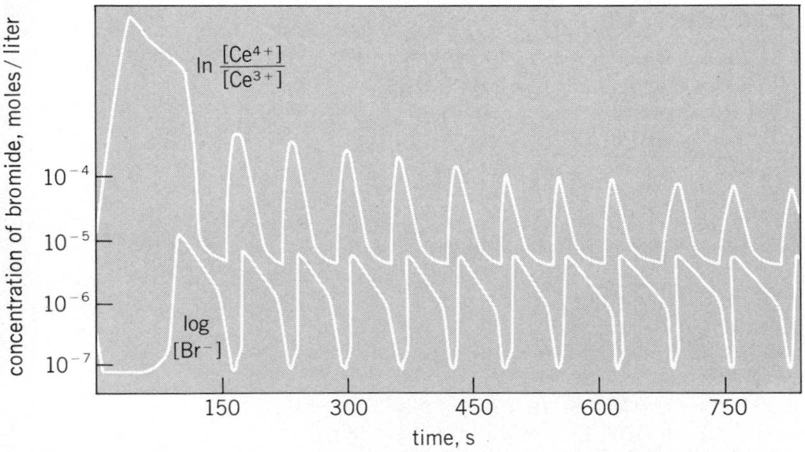

Fig. 1. Oscillatory behavior in a sulfuric acid medium containing potassium bromate, malonic acid, and cerous nitrate. The upper curve is the potential (in unspecified units) of a tungsten electrode and is related to the concentration ratio of cerium(IV)/cerium(III). The lower curve is the potential of an electrode sensitive to bromide ion, and the logarithmic scale at the left relates that potential to the absolute concentration of bromide. (*From R. J. Field, E. Körös, and R. M. Noyes, Oscillations in chemical systems, II. Thorough analysis of temporal oscillations in the bromate-cerium-malonic acid system, J. Amer. Chem. Soc., 94:8649–8664, 1972*)

that result in death if the condition is not rapidly reversed. *See* OXIDATION-REDUCTION.

Nucleation oscillators. If a solution contains a population of identical bubbles or crystals, the resulting steady state is unstable to perturbation. Because surface energy is relatively more important in small bubbles, they tend to dissolve while the larger ones grow. In principle, the equilibrium state consists of a single large bubble or crystal. If a chemical reaction produces a gas or a precipitate, periodicities may be observed. A quantitative treatment is complicated because the solution must become supersaturated before nuclei of the gas or solid phase are formed, and the initial nuclei are so small that they grow very slowly at first.

An example of a gas evolution oscillator is the dehydration of formic acid (HCO_2H) in concentrated sulfuric acid. As carbon monoxide escapes, several times a minute the solution foams up like a shocked glass of beer and then subsides.

Figure 3 illustrates a geologic formation on the east coast of Greenland. A large magma chamber cooled very slowly, and the initially uniform material crystallized in a pattern of regular layers.

Fig. 3. Layers of crystallization from an initially uniform magma in the Skaergaard Intrusion in Greenland. (*Courtesy of A. R. McBirney, University of Oregon, from A. R. McBirney and R. M. Noyes, Crystallization and layering of the Skaergaard Intrusion, J. Petrol., 20:487–554, 1979*).

Thermokinetic oscillators. Many chemical reactions give out heat, and rates are strongly dependent on temperature. If heat is not removed too rapidly from the reactor, reaction rate and temperature may couple to generate oscillations.

The known examples involve highly exothermic reactions like combustion or chlorination of organic compounds. No chemical mechanisms are yet understood in detail. In at least some examples, gradients of temperature and composition in space are important in addition to changes in local rate of reaction.

Reactions on surfaces. Many important industrial reactions take place on the surfaces of catalysts. The occurrence of such a reaction may temporarily alter that surface or its temperature. Such effects sometimes couple to generate oscillations in the rate of reaction. These oscillations may or may not be of value for the reaction being carried out. Specific examples are being studied actively by chemical engineers.

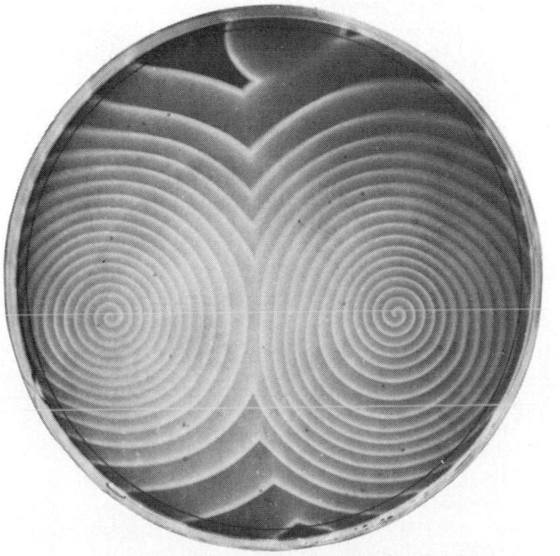

Fig. 2. Rotating spiral bands of oxidation in a thin layer of solution containing the same reagents as in Fig. 1 except that the redox indicator ferrous phenanthroline has been substituted for cerous ion. (*Courtesy of A. T. Winfree, Purdue University, from R. J. Field and R. M. Noyes, Mechanisms of chemical oscillators: Conceptual bases, Acc. Chem. Res., 10:214–221, 1977*).

The surfaces of electrodes may also be influenced by processes taking place on them, and periodic changes in voltage or current are well precedented.

Biological chemistry. Living organisms take in nutrients of high free energy and eliminate waste products of lower free energy. The degradation of those nutrients drives the essential vital processes. The pathways involve intricate couplings so that a decrease in free energy of one species contributes in forming species like adenosinetriphosphate (ATP) having increased free energy. Many important intermediates follow repeated cyclic paths while the nutrients are degraded.

If all of the processes of metabolism took place at the same rate in the same place, the only net effect would be increase in entropy, and life would not exist. Processes involving some intermediates must be separated in space or time from processes involving other intermediates.

Separation in space can be accomplished by membranes permeable to some species but not to others. Such membranes are ubiquitous in biological organisms.

Separation in time can be accomplished by oscillatory reactions that turn component processes on and off much as happens with the Belousov-Zhabotinsky reaction described above. It can hardly be accidental that periodicities are observed in many biological activities.

One of the best chemical examples involves oscillations during oxidation of glucose catalyzed by cell-free yeast extracts. The enzyme phosphofructokinase (PFK) is strongly implicated, and its activity is influenced by the oxidation products.

Aggregation of the slime mold *Dictostelium discoideum* takes place because individual cells move in the direction of a received pulse of cyclic adenylic acid (cyclic AMP) and simultaneously emit a pulse in the opposite direction. Aggregating cells may create spiral patterns resembling those in Fig. 2 except that motion is inward instead of outward.

Circadian rhythms with periods of about 24 h are common in biology. Anybody who has taken a jet halfway around the world becomes well aware of their effects.

Undoubtedly many periodic biological processes of chemical origin are not even recognized. None can yet be described in mechanistic detail. Development of the quantitative explanation characteristic of a mature science is anticipated.

[RICHARD M. NOYES]

Bibliography: R. J. Field and R. M. Noyes, Mechanisms of chemical oscillators, *Acc. Chem. Res.*, 10:214–222, 273–280; 1977; H. Haken, *Synergetics*, 2d ed., 1978; G. Nicolis and I. Prigogine, *Self-Organization in Nonequilibrium Systems*, 1977; R. M. Noyes, Oscillations in homogeneous systems, *Berichte Bunsen Ges. Phys. Chem.*, 84:295–303, 1980.

Osmium

A chemical element, Os, atomic number 76, and atomic weight 190.2. Osmium is a hard white metal of rare natural occurrence.

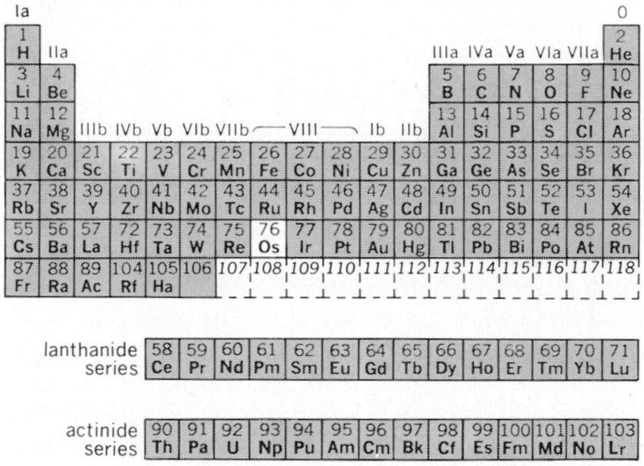

Uses. Osmium, like the other platinum metals, is catalytically active, but only limited use has been made of this property. Osmium tetroxide is used for the hydroxylation of double bonds in the synthesis of certain organic compounds. Particular applications of this reaction occur in the synthesis of cortisone. Osmium tetroxide is also used as a stain for tissue in microscopy. Osmium and its alloys are useful for their hardness and resistance to wear and corrosion. Osmium alloyed with rhodium, ruthenium, iridium, or platinum is used in fountain-pen nibs, phonograph needles, electrical contacts, and instrument pivots. *See* CATALYSIS; PLATINUM.

Chemical and physical properties. Although osmium is very refractory, even at room temperature a blue oxide film, OsO_2, is formed on the metal surface. When heated in air, the oxide dissociates at a relatively low temperature. Even before this dissociation, however, osmium rapidly loses weight by volatilization of the oxides OsO_4 and OsO_3. These volatile oxides are poisonous. This property complicates the analysis and refining of osmium, since losses due to volatilization may be inadvertently encountered. Osmium is readily soluble in hot nitric acid. When fused in alkaline oxidizing fluxes, water-soluble osmates, OsO_4^{--}, are formed. Osmates are also formed by attack of alkaline solutions such as hypochlorites. The chemistry of osmium is very complicated because of the many valences exhibited by the element and the tendency of each of these to form numerous complex ions. Osmium is a very hard metal; it and its alloys are wear-resistant. Pure osmium and alloys in which it predominates are unworkable, so that they must either be used in the cast form or fabricated by powder metallurgy. The table gives values of important physical properties of osmium.

Metallurgical extraction. After osmium has been brought into solution, it may be distilled from nitric acid as the tetroxide. Solids containing osmium often can be roasted in air to volatilize the tetroxide, which is then absorbed in an alcoholic caustic solution. The resulting osmate solution may be precipitated as the sulfide or neutralized and precipitated as the hydroxide. Either precipitate is then reduced in hydrogen to yield the metal.

Principal compounds. Osmium tetrachloride, $OsCl_4$, is a black solid which is insoluble in nonoxidizing acids; it is made by treating osmium with chlorine at 700°C. Osmium tetroxide, OsO_4, is a very-pale-yellow crystalline solid with a melting point of 40°C and a boiling point of 130°C. It is the most important osmium compound, and is made by oxidizing the metal with air, nitric acid, or sulfuric acid. This very poisonous compound is soluble in water and carbon tetrachloride. It is a powerful oxidizing agent. When a potassium hydroxide solution of the tetroxide is treated with alcohol, the osmium is partly reduced, and slightly soluble violet-red crystals of potassium osmate,

Physical properties of osmium

Property	Value
Atomic weight ($C^{12} = 12.00000$)	190.2
Naturally occurring isotopes, with percent abundance in parentheses	184(0.018) 186(1.59) 187(1.64) 188(13.3) 189(16.1) 190(26.4) 192(41.0)
Crystal structure	Close-packed hexagonal
Lattice constant *a* at 25°C, angstroms	2.7341
c/a, at 25°C	1.5799
Thermal neutron capture cross section, barns	15.3
Common chemical valence	4, 6, 8
Density at 25°C, g/cm³	22.59
Melting point, °C	3045
Boiling point, °C	5020
Specific heat at 0°C, cal/g	0.0309
Thermal conductivity, 0–100°C, cal cm/cm²sec °C	0.21
Linear coefficient of thermal expansion, 20–100°C, μin./in./°C	6.1
Electrical resistivity at 0°C, μohm-cm	8.12
Temperature coefficient of electrical resistance, 0–100°C/°C	0.0042
Young's modulus at 20°C, psi, static	81×10^6

$K_2OsO_4 \cdot 2H_2O$, are precipitated.

The dihydrate of osmium dioxide, $OsO_2 \cdot 2H_2O$, is made by neutralizing an alcoholic sodium hydroxide solution of the tetroxide. It is a brown or blue-black insoluble solid. The solid hexachloroosmic acid, H_2OsCl_6, has not been definitely isolated. However, when the tetroxide is refluxed in hydrochloric acid and then treated with ammonium chloride, ammonium hexachloroosmate, $(NH_4)_2OsCl_6$, is precipitated. When heated in hydrogen, this black compound yields osmium.

[HENRY J. ALBERT]

Bibliography: *See* PLATINUM.

Osmosis

The transport of solvent through a semipermeable membrane separating two solutions of different solute concentration. The solvent diffuses from the solution that is dilute in solute to the solution that is concentrated. The phenomenon may be observed by immersing in water a tube partially filled with an aqueous sugar solution and closed at the end with parchment. An increase in the level of the liquid in the solution results from a flow of water through the parchment into the solution. The process occurs as a result of a thermodynamic tendency to equalize the sugar concentrations on both sides of the barrier. The parchment permits the passage of water, but hinders that of the sugar, and is said to be semipermeable. Specially treated collodion and cellophane membranes also exhibit this behavior. These membranes are not perfect, and a gradual diffusion of solute molecules into the more dilute solution will occur. Of all artificial membranes, a deposit of cupric ferrocyanide in the pores of a fine-grained porcelain most nearly approaches complete semipermeability.

The flow of liquid through such a barrier may be stopped by applying pressure to the liquid on the side of higher solute concentration. The applied pressure required to prevent the flow of solvent across a perfectly semipermeable membrane is called the osmotic pressure and is a characteristic of the solution. The walls of cells in living organisms permit the passage of water and certain solutes, while preventing the passage of other solutes, usually of relatively high molecular weight. These walls act as selectively permeable membranes, and allow osmosis to occur between the interior of the cell and the surrounding media. *See* SOLUTION.

[FRANCIS J. JOHNSTON]

Bibliography: R. B. Martin, *Introduction to Biophysical Chemistry*, 1964.

Oxalate

A salt or ester of oxalic acid with the formula

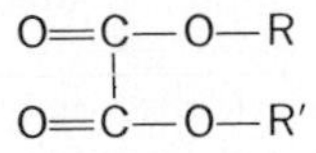

where R and R′ are alkyl groups or metallic ions attached to the acid to produce a series of salts and esters. In general, alkali-metal salts are water-soluble; others are insoluble. Calcium and potassium hydrogen oxalate are found in many plants; the former is also found in urinary calculi. Many salts have practical applications, for example, in pyrotechnics (sodium), blueprinting (potassium-iron), and analytical procedures (ammonium). Esters of the simpler alcohols hydrolyze readily; others are more stable. Diethyl oxalate is used as a solvent, as a dyestuff intermediate, and in plastics. *See* ESTER; OXALIC ACID.

[ELBERT H. HADLEY]

Oxalic acid

A white solid acid melting with decomposition at 189.5°C; it is obtained by careful drying of the dihydrate, which melts at 101.5°C. Oxalic acid (ethanedioic acid), with the structure

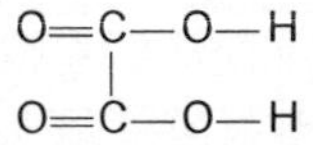

is the first of a series of dicarboxylic acids. Its salts (oxalates) are prevalent in nature, for example, KHC_2O_4 in plants of the oxalis family (wood sorrel) and CaC_2O_4 in eucalyptus bark.

Sodium oxalate is formed when sawdust is fused with sodium hydroxide; sodium formate is formed by a similar vacuum fusion in the presence of hydrogen at 300°C. Oxalic acid is formed directly by the nitric acid oxidation of sucrose or starch.

The acid is liberated from its salts by addition of dilute sulfuric acid. It is used as a bleaching agent for rust and ink stains, in textile and leather production, and as monoglyceryl oxalate in the manufacture of allyl alcohol and formic acid. Easily oxidized, it is determined by titration with $KMnO_4$ in dilute sulfuric acid. When heated with concentrated H_2SO_4, it gives equal volumes of CO and CO_2. *See* CARBOXYLIC ACID; OXALATE.

[EVANS B. REID]

Oxazole

One of a group of organic heterocyclic compounds containing oxygen and nitrogen in the 1 and 3 positions of a five-membered diunsaturated ring. Formula (I) shows the structure and numbering system for a typical member of the group, 1,3-oxazole.

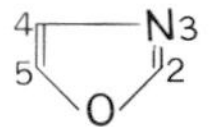
(I)

See AZOLE; HETEROCYCLIC COMPOUNDS.

Properties. The parent compound (I) is a colorless, volatile, weakly basic liquid (bp 69–70°C), with an odor resembling that of pyridine. Oxazole is miscible with water and organic solvents. Mineral acids form salts that tend to dissociate in water. Oxazoles show appreciable resistance to disruption by heat, by acid, and by alkali. The nucleus is susceptible to oxidation, the 4,5 position being the usual point of attack. Hydrogenation over a platinum catalyst or with sodium and alcohol gives tetrahydro derivatives (oxazolidines) or ring-cleavage products.

Preparation. Standard syntheses start with α-acylamidocarbonyl compounds (II), or with α-haloketones (IV). Cyclization of (II) with sulfuric acid or phosphorus pentachloride generates an oxazole (III). The R groups may be aryl or alkyl. The reac-

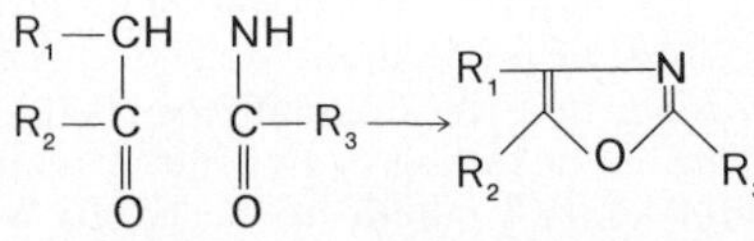
(II) (III)

tion of α-haloketones (IV) with ammonium salts of carboxylic acids gives oxazoles (V). Carboxy ox-

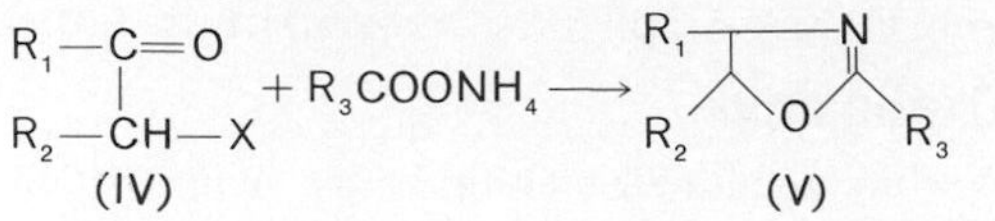

(IV) (V)

azoles obtained from these syntheses can be decarboxylated with relative ease, and therefore serve as useful intermediates in syntheses of carboxyl-free oxazoles.

2-Oxazolines (VII) are formed by cyclization of β-hydroxyalkylamides (VI) or by reaction of β-aminoalcohols with iminoethers (VIII). Hot aqueous acid hydrolyzes 2-oxazolines to *O*-acylethanol-

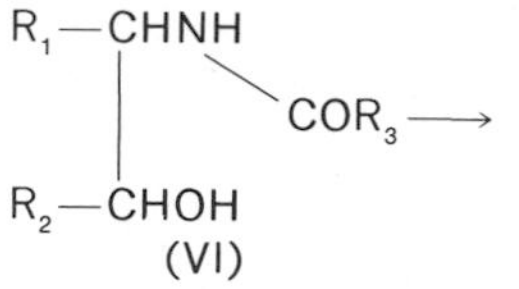

(VI)

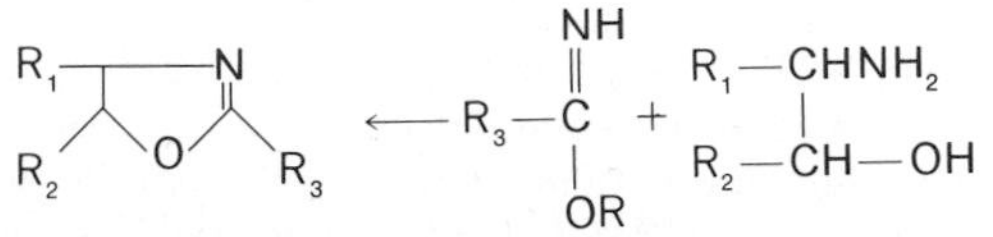

(VII) (VIII)

amines or to ethanolamines. Alkali converts 2-oxazolines to *N*-acylethanolamines or to ethanolamines. 2-Oxazolines are intermediates in the acid- or base-catalyzed interconversion of *O*-acyl- (X) and *N*-acylethanolamines (IX).

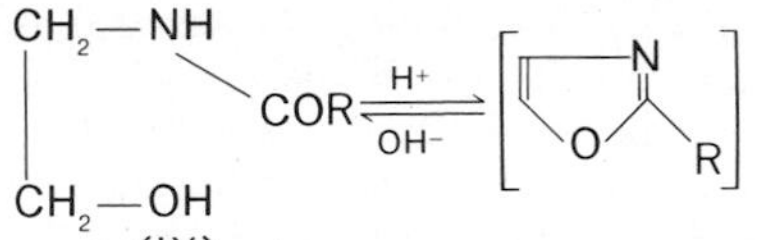

(IX)

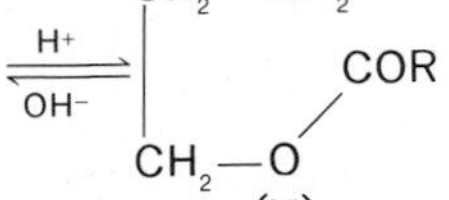

(X)

Tetrahydrooxazoles, or oxazolidines (XI), are formed from ethanolamines and aldehydes. The process can be reversed. Oxazolidines under suitable conditions exist either in equilibrium with, or entirely in the form of, the isomeric imine (XII).

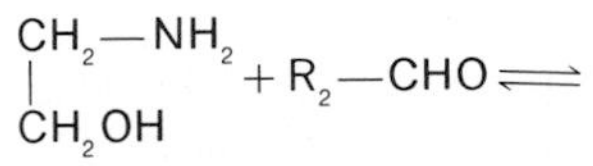

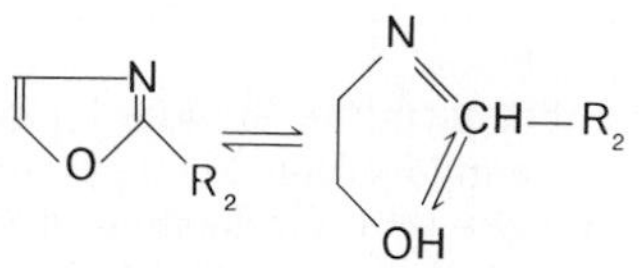

(XI) (XII)

Azlactones, or 5-oxo-2-oxazolines (XIV), are generally formed by cyclization of α-acylamido acids (XIII). Hydrolysis in the presence of acids or bases regenerates the original α-acylamido acid. Alcohols or amines react to give the corresponding α-acylamido ester or amide, respectively. When an aldehyde, generally aromatic, is warmed with acet-

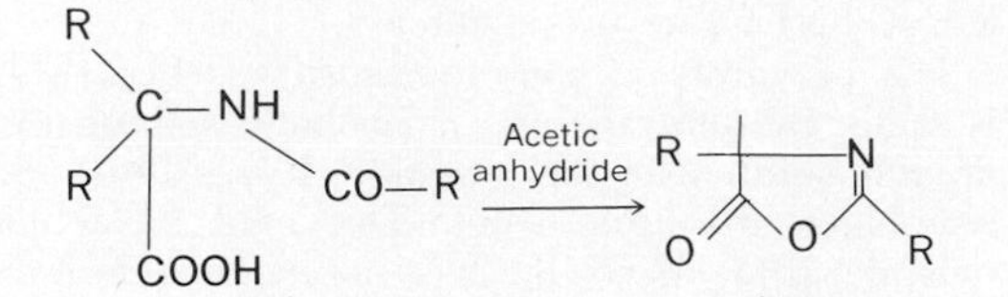

(XIII) (XIV)

ic anhydride and *N*-benzoylglycine, hippuric acid (XV), unsaturated azlactones (XVI) are formed,

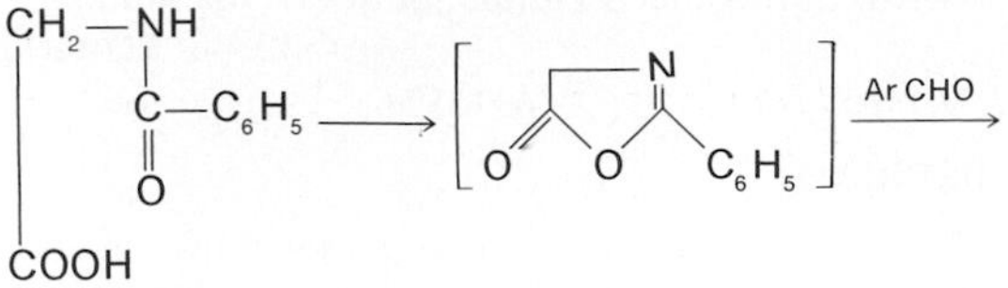

(XV)

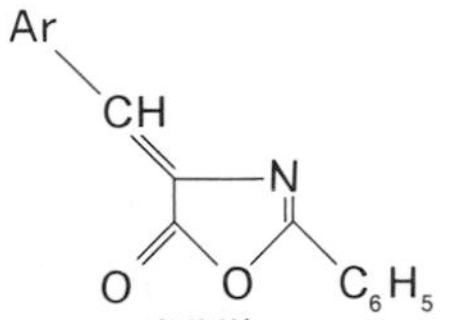

(XVI)

presumably by condensation of the aldehyde with the 2-phenyl-5-oxo-2-oxazoline formed first. Such unsaturated lactones (XVI) by standard conversions furnish several useful products.

[WALTER J. GENSLER]

Bibliography: R. M. Acheson, *An Introduction to the Chemistry of Heterocyclic Compounds*, 3d ed., 1976; H. T. Clarke (ed.), *The Chemistry of Penicillin*, 1949; J. A. Joule and G. F. Smith, *Heterocyclic Chemistry*, 1972.

Oxidation process

Processes in which oxygen is caused to combine with other molecules. The oxygen may be used as elemental oxygen, as in air, or in the form of an oxygen-containing molecule which is capable of giving up all or part of its oxygen. Oxidation in its broadest sense, that is, an increase in positive valence or removal of electrons, is not considered here if oxygen itself is not involved. *See* OXIDATION-REDUCTION.

Most oxidations occur with the liberation of large amounts of energy in the form of either heat, light, or electricity. The stable ultimate products of oxidation are oxides of the elements involved. These oxidations occur in nature as corrosion, decay, and respiration and in the deliberate burning of matter such as wood, petroleum, sulfur, or phosphorus to oxides of the constituent elements. This article deals only with cases where the object of the oxidation process is the manufacture of a chemical product rather than the production of energy.

The principal variables to be considered and controlled in any partial oxidation are temperature, pressure, reaction time (or contact time), nature of catalyst, if any, mole ratio of oxidizing agent, and whether the substance to be oxidized is to be kept in the liquid or vapor phase. Only a narrow range of conditions unique to each substance being oxidized and each product desired will give satisfactory yields. It is also essential to maintain conditions outside the range of spontaneous ignition, to avoid explosive mixtures or the accidental accumulation of unstable peroxides, and to choose materials which not only can resist the environmental conditions but also which do not have adverse catalytic effects or otherwise interfere with the desired reaction. *See* COMBUSTION.

Most oxidations of organic compounds with oxygen appear to proceed through free-radical chain reactions. The specific transient intermediates and sequence of reactions are very complex and are still not completely understood. Many oxidation reactions are autocatalytic. An induction period is therefore often observed during which the concentration of catalyst or intermediate is being built up to the level required for the reaction to proceed at a measurable rate. The reaction rate can continue to increase due to reaction chain branching unless controlled. *See* CHAIN REACTION.

Catalytic effects can be obtained with solid surfaces, generally used in vapor-phase oxidations; with soluble salts, generally used for liquid-phase oxidations; with gases, added in small amounts to the air; or with radiation. Very often catalysts are mixtures in which the action of the major component is modified or maintained by the addition of other agents to the gas, liquid, or solid phase. A limited number of generalizations can be made about oxidation catalysts. The metals, used in the form of their oxides or salts, have a variable valence under the reaction conditions. In vapor-phase reactions the active catalyst is generally deposited on an inert support. Liquid-phase catalysts are usually in the form of a salt, very often containing cobalt or manganese, which is soluble in the organic medium. Inhibitors are also important. Catalyst supports and materials of construction must not have an inhibitory effect and, in liquid-phase oxidations, reactive molecules which terminate free-radical chains, such as phenols, must be avoided. *See* CATALYSIS; INHIBITOR.

In partial oxidations, high selectivity, which may be defined as the equivalents of desired product formed divided by the total moles of feed oxidized, is often made possible by differing degrees of resistance to attack among the various atoms in an organic molecule. Hydrogen atoms attached to aliphatic carbon atoms are more easily oxidized than those attached to carbon atoms in aromatic rings. Among nonaromatics, the ease of oxidation is in the order tertiary > secondary > primary; so that methyl groups are relatively resistant. On the other hand, hydrogen atoms attached to aliphatic carbon atoms are activated by adjacent methyl groups, double bonds, and aromatic rings in roughly that increasing order. These observations are qualitatively illustrated by the examples shown in reactions (1)–(7). Among compounds already containing oxygen, a rough order of decreasing stability under oxidizing conditions is carboxylic acid anhydrides > esters > carboxylic acids > ketones > secondary alcohols > aldehydes.

In reactions (4) and (5) the anhydride group is relatively stable under the reaction conditions, whereas the corresponding carboxyl group is not. In cases where the reaction conditions required to initiate oxidation of the starting material are severe, it is generally necessary to employ short reaction times, for example, 0.001–1 sec, by rapidly quenching the reaction mixture to a temperature at which the desired products can survive. In reaction (6) the hydroperoxide is sufficiently stable under the relatively mild conditions required to oxidize cumene for reaction times of an hour or more to be employed. In reaction (2) the unstable secondary alcohol is prevented from decomposing or oxidizing further by formation of the more stable borate ester. Double bonds themselves appear to be relatively resistant to attack by oxygen. Ethylene can be oxidized to ethylene oxide with silver catalysts at high temperatures, but higher olefins require chemical oxidizing agents, such as peroxides, ozone, or hypochlorite, since oxygen attacks other parts of the molecule first.

Plant and equipment design is of utmost importance in achieving efficient control of the oxidation. Vapor-phase reactors with fixed catalyst beds

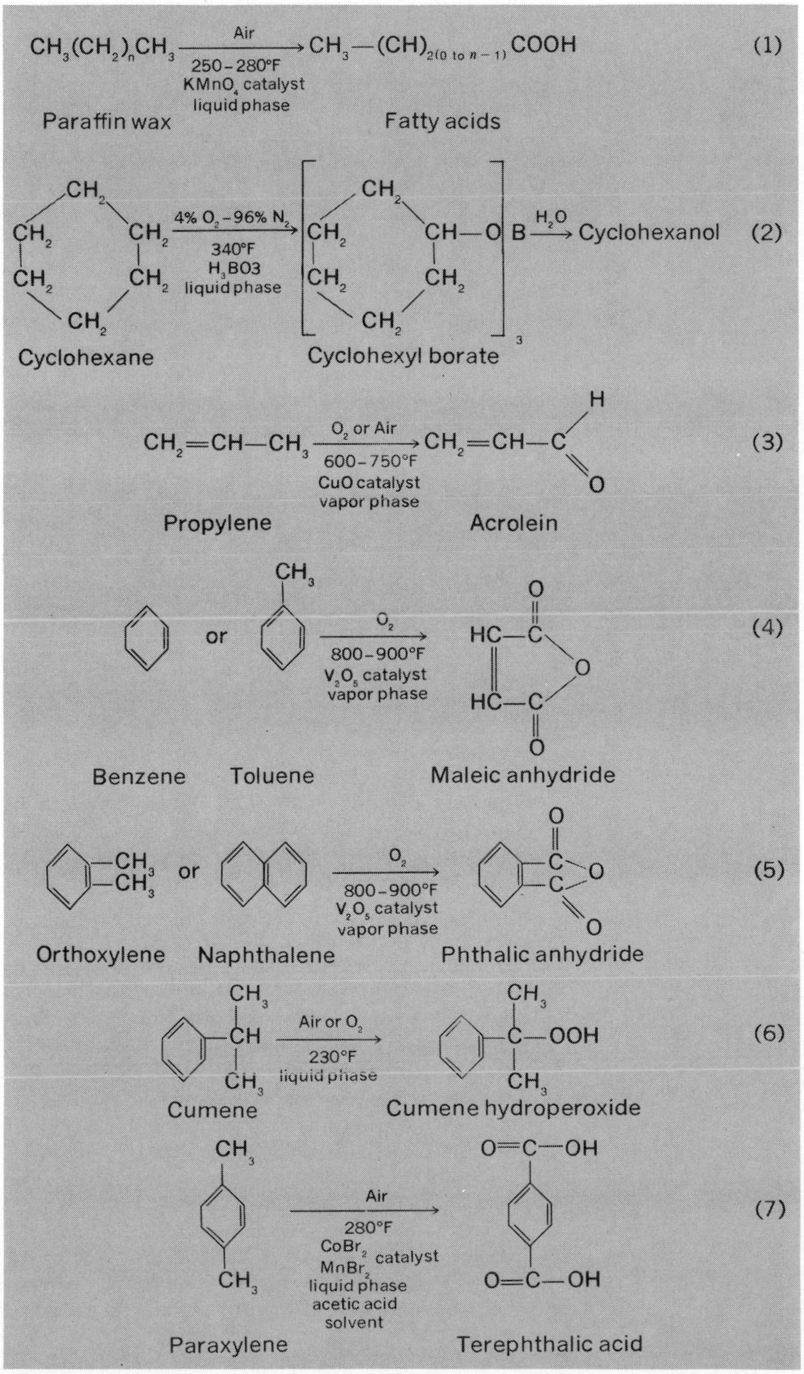

are generally built in the form of tubular heat exchangers if excess air is used. A thermally stable heat exchange medium is circulated outside of the tubes. Molten salts such as sodium nitrate–sodium nitrite mixtures are used for temperatures in the 500–1000°F region. For temperatures between 212 and 600°F, stable organic oils or water under pressure are employed.

If the reaction permits, the catalyst is often used in the form of a fluidized bed; that is, the particle size of the catalyst and support is small enough so that it is suspended in a large vessel by the upward flowing gas. For a given oxidation, reaction time is generally longer than with a fixed bed and the temperature lower. The gas velocity is such that most of the catalyst is not blown out of the reaction zone, and the catalyst which is blown out is removed by cyclone separators or ceramic filters or both and returned. The entire bed is subject to continual mixing, has a high heat capacity, and is easily maintained at a predetermined uniform temperature. Heat is removed by circulating a portion of the hot catalyst through a heat exchanger. Cooling coils can also be put inside the reactor. *See* FLUIDIZATION.

Liquid-phase oxidations are usually carried out in large vessels with internal cooling coils. The inlet air is distributed by spargers or by introducing it at high velocity through a small orifice. Gases leaving the reactor can carry large amounts of the reactants and products. These must be removed by chilling or absorption to minimize losses. In most large oxidation plants gases are incinerated before release to the atmosphere to avoid pollution.

Commercial processes using oxygen or air. Synthesis gas is manufactured by partial oxidation of hydrocarbons with oxygen. The carbon monoxide–hydrogen mixture obtained is used to make primary alcohols from olefins through the Oxo process and for methanol synthesis. Hydrogen is manufactured from synthesis gas by selective oxidation of the carbon monoxide in the mixture to carbon dioxide and by removal of the latter from the hydrogen by combination with an amine. Carbon dioxide is sometimes recovered from the amine as a separate commercial product.

A number of processes for the oxidation of light aliphatic hydrocarbons are in use for the production of mixtures of acetic acid, methanol, formaldehyde, acetaldehyde, and other products in lesser amounts. The processes are reported to use air or oxygen and to operate with and without catalysts over a wide range of temperatures and pressures in both the liquid and gas phases. Processes also have been developed for oxidizing ethylene to acetaldehyde and vinyl acetate. Aqueous systems with palladium–copper salt catalysts are reported. Ethylene is also oxidized to ethylene oxide in the vapor phase with air in tubular reactors with supported silver catalysts. Temperatures are in the range 400–600°F, and pressures are between 100 and 300 psi.

Propylene oxide is manufactured by oxidizing propylene with the hydroperoxides of either cumene or isobutane, as in reaction (6). Isobutene or styrene are formed as by-products. Cumene hydroperoxide is cleaved in acid solution to phenol and acetone. Phenol is also manufactured by oxidation of benzoic acid with cupric salts, which are regenerated in place with air.

In the manufacture of phthalic anhydride, reaction (5), both fixed- and fluid-bed vanadium oxide catalysts are used. Excess air is employed at substantially atmospheric pressure. Processes for oxidizing benzene to maleic anhydride and of durene to pyromellitic anhydride are similar to those for phthalic anhydride.

Large amounts of terephthalic acid are manufactured from paraxylene by liquid-phase oxidation with air in the presence of manganese and cobalt bromide catalysts. Acetic acid is used as a solvent. A similar process co-oxidizes acetaldehyde along with the xylene in the presence of cobalt acetate, forming acetic acid as a by-product. The acetaldehyde acts to assist oxidation of the xylene by way of a peroxide intermediate, which is probably the actual oxidizing agent. Methyl paratoluate is oxidized to methyl hydrogen terephthalate in the liquid phase with soluble cobalt salts. Processes for isophthalic acid are analogous to those for terephthalic.

Many attempts have been made to manufacture fatty acids commercially by oxidation of paraffin wax. Synthetic fats are reported to have been manufactured from them in Germany during both world wars. It is difficult to separate the oxyacid and lactone by-products from the desired fatty acid, however, and the process does not appear capable of competing in a free economy with fatty acids and fats from natural sources. Fatty alcohols are manufactured by air oxidation of aluminum alkyls and hydrolyzing the resulting aluminum alcoholate.

Cyclohexane is oxidized in the liquid phase with air to form a mixture of cyclohexanol and cyclohexanone. This mixture is further oxidized with nitric acid to adipic acid. Another development is the use of boric acid in the liquid oxidation medium to form the borate ester of cyclohexanol. The ester is hydrolyzed to recover cyclohexanol, and the boric acid is recycled. The use of boric acid is claimed to give higher total yields and to minimize the formation of troublesome by-products. This process is also used to make secondary alcohols from paraffins for use as detergent intermediates.

Several oxidation processes to produce acetylene have been developed. In these cases, however, the function of the oxidation itself is to produce temperature of around 3000°F. Acetylene is produced from natural gas or naphtha introduced into the hot gases.

Two commercial oxidation processes involve oxidation of hydrogen chloride to chlorine over a manganese catalyst in the presence of ethylene to form vinyl chloride and the reaction of propylene and ammonia with oxygen or air to form acrylonitrile. Methanol and ethanol are oxidatively dehydrogenated to the corresponding aldehydes in the vapor phase over silver, copper, or zinc catalysts. Acetaldehyde is oxidized to acetic acid in the liquid phase in the presence of manganese acetate and to acetic anhydride if mixtures of cobalt and copper acetates are used. Polyunsaturated oils are oxidatively polymerized at room temperature in the presence of soluble salts of cobalt and other metals of variable valence. The physical properties of certain asphalts are improved by oxidation with air at elevated temperature, and phenol can be converted to a high polymer by direct oxidation. Microbiological oxidation to produce animal feed and sodium glutamate, as well as to treat domestic and industrial wastes, is becoming increasingly important.

Chemical oxidants. Adipic and terephthalic acids are manufactured from cyclohexanol-cyclohexanone mixtures and paraxylene, respectively, by oxidation with nitric acid of about 30% concentration. Organic and inorganic peroxides are used to manufacture higher olefin oxides, glycols, and hydrogen peroxide. Ozone is used to cleave unsaturated fatty acids to form dibasic acids. Other chemical oxidizing agents used for special purposes include the following: sulfur and sulfur compounds, permanganate, perchlorate, hypochlorite, and dichromate. *See* OZONOLYSIS; PEROXIDE.

Inorganic processes. These are generally carried to the highest stable oxidation state and process control is relatively simple.

Vapor-phase oxidations are used to produce major heavy inorganic chemicals, for example, air oxidation of hydrogen sulfide or sulfur dioxide to sulfur trioxide (sulfuric acid); of ammonia to nitric acid; of phosphorus vapor to phosphorus pentoxide (phosphoric acid); of hydrogen chloride to chlorine; and of vaporized zinc to zinc oxide.

Liquid-phase oxidations of inorganic compounds are rare because so few are liquids. Liquid sulfur is burned to sulfur dioxide. At high temperatures mercuric oxide is made from its elements and litharge from molten lead. Air and oxygen, blown through molten iron, make steel by oxidizing such impurities as carbon, sulfur, and phosphorus.

Solid-phase oxidations are applied most commonly to obtain oxides from the elements. High-purity carbon dioxide is made from coke in this way. Mixed lead oxides are purified to the monoxide litharge by roasting. Barium peroxide forms from the oxide. Two of the more powerful and costly inorganic oxidizing agents are obtained by processes involving gas-solid phase reactions. Potassium permanganate is produced by roasting a mixture of manganese dioxide and potassium hydroxide with air in a kiln or muffle furnace. In an analogous way, chromite ore and sodium carbonate yield sodium chromate. [I. E. LEVINE]

Bibliography: N. M. Emanuel et al., *Oxidation of Organic Compounds: Solvent Effects in Radical Reactions*, 1980; N. M. Emanuel, E. T. Denisov, and Z. K. Maizus, *Liquid-Phase Oxidation of Hydrocarbons*, 1967; E. K. Fields (ed.), *Selective Oxidation Processes*, Advances in Chemistry Series, no. 51, 1965; V. Y. Shtern, *The Gas Phase Oxidation of Hydrocarbons*, 1964; M. Sittig, *Organic Oxidation Processes and Products*, 1967; T. Turney, *Oxidation Mechanisms*, 1966.

Oxidation-reduction

An important concept of chemical reactions which is useful in systematizing the chemistry of many substances. Oxidation can be represented as involving a loss of electrons by one molecule and reduction as involving an absorption of electrons by another. Both oxidation and reduction occur simultaneously and in equivalent amounts during any reaction involving either process.

Some important processes which involve oxidation are the rusting of iron or corrosion of metals in general, combustion of hydrocarbons, and the oxidation of carbohydrates (this takes place in a controlled manner in living cells). In each of the foregoing reactions the agent which is reduced is oxygen. Some common important reduction processes are the transformation of carbon dioxide to carbohydrates (this takes place in photosynthesis with water being oxidized), the winning of metals from oxides (carbon is often the reducing agent), electrodeposition of metals (this takes place at the cathode, and an equivalent amount of oxidation takes place at the anode), hydrogenation of fats and of coal, and the introduction of electronegative elements such as oxygen, nitrogen, or halogens into hydrocarbons.

Oxidation number. The oxidation state is a concept which describes some important aspects of the state of combination of the elements. An element in a given substance is characterized by a number, the oxidation number, which specifies whether the element in question is combined with elements which are more electropositive or more electronegative than it is. It further specifies the combining capacity which the element exhibits in a particular combination. A scale of oxidation numbers is defined by assigning to an oxygen atom in a molecule such as SO_4^{2-} the value of 2−. That for sulfur as 6+ then follows from the requirement that the sum of the oxidation numbers of all the atoms add up to the net charge on the species. The value of 2− for oxygen is not chosen arbitrarily. It recognizes that oxygen is more electronegative than sulfur, and that when it reacts with other elements it seeks to acquire two more electrons, by sharing or outright transfer from the electropositive partner, so as to complete a stable valence shell of eight electrons. For compounds of the halogens an analogous rule is followed, but when a halogen atom is in combination with atoms of a more electropositive element, the oxidation number is taken as 1− because only one electron needs to be added to the valence shell to yield a stable octet. The system amounts to a bookkeeping operation on the electrons, so that for this purpose the more electronegative partner is assigned some agreed-upon stable electronic configuration, and after taking into account the total charge on the molecule, the net charge left on the electropositive partner is its particular oxidation number. When the combining capacity of an element toward another one is not completely exhausted in a particular combination, as is the case for oxygen in barium peroxide (BaO_2), the electrons shared between atoms of the same kind are evenly divided between them in carrying out the formal decomposition. Thus in the peroxide unit O_2^{2-}, the oxidation number of oxygen is 1−. This is the net charge left on oxygen in the formal decomposition.

$$[:\underset{..}{\overset{..}{O}}:\underset{..}{\overset{..}{O}}:]^{2-} = 2:\underset{..}{\overset{..}{O}}\cdot^{-}$$

The oxidation number by no means gives a complete description of the state of combination of an atom. Specifically, it is not designed to give the actual charge on an atom in a particular compound. Thus it makes no distinction between fluorine in HF, AlF_3, or NaF, even though the actual charges residing on the fluorine atoms in these three compounds are different.

The utility of the concept is based in part on just this feature because much of the chemistry of these substances can be understood when it is realized that each of them readily yields F^-, as is the case when they dissolve in water. The chemistry of the three substances, in regard to the component fluorine, is concerned with reactions of F^-. Although oxidation number is in some respects similar to valence, the two concepts have distinct meanings. In the substance H_2, the valence of hydrogen is 1 because each H makes a single bond to another H, but the oxidation number is 0, because the hydrogen is not combined with a different element. *See* VALENCE.

The systematization of chemistry based on the concept of oxidation number can be illustrated with reference to the chemistry of iodine. The usual oxidation states exhibited by iodine are 1−, 0, 1+, 5+, and 7+. Examples of substances corresponding to each oxidation state are

7+	IO_4^-, HIO_4, IF_7
5+	I_2O_5, IO_3^-, HIO_3, IF_5
1+	HIO, IO^-, ICl_2^-
0	I_2
1−	I^-, HI, NaI

When the oxidation number of an atom in a species is increased, the process is described as oxidation, no matter what reagent produces it; when a decrease in oxidation number takes place, the process is described as reduction, again without regard to the identity of the reducing agent. The term oxidation has been generalized from its historical meaning, implying combination with oxygen, to combination of an element with an element more electronegative than itself.

When classification by oxidation number is adopted, the reactions fall naturally into two classes. In the first class, no change in oxidation number takes place and, in the second, the class of oxidation-reduction reactions, changes in oxidation number do take place. Some examples of the first class are reactions (1)–(4).

$$I_2O_5 + H_2O \rightarrow 2HIO_3 \quad (1)$$

$$HIO_3 + OH^- \rightarrow IO_3^- + H_2O \quad (2)$$

$$HOI + H^+ + 2Cl^- \rightarrow H_2O + ICl_2^- \quad (3)$$

$$Hg^{2+} + 4I^- \rightarrow HgI_4^{2-} \quad (4)$$

Some samples of the second class are reactions (5)–(8). [In (8) it is implied that electrons are being extracted from I^- by an anode in an electrolytic process.]

$$Cl_2 + 2I^- \rightarrow 2Cl^- + I_2 \quad (5)$$

$$2Fe^{3+} + 2I^- \rightarrow 2Fe^{2+} + I_2 \quad (6)$$

$$16H^+ + 2MnO_4^- + 10I^- \rightarrow 8H_2O + 2Mn^{2+} + 5I_2 \quad (7)$$

$$2I^- \rightarrow I_2 + 2e^- \quad (8)$$

In reactions of the first class, some center regarded as positive undergoes a change in the nature of the groups associated with it, but provided that the group which replaces the electronegative portion is more electronegative than the center, there is no change in oxidation state. Reaction (3) describes the replacement of OH^- on I^+ by Cl^-; both Cl and OH are more electronegative than I. In reactions of the second class, changes in oxidation number occur which may or may not be accompanied by changes in the state of association of the centers in question.

Reactions (5), (6), (7), and (8) illustrate the utility of the concept of oxidation number. A variety of reagents as different in their properties as Cl_2, Fe^{3+}, MnO_4^-, and an anode serve to bring about the change, or oxidation, of I^- to I_2. However, their chemical individuality does not affect the state of the product iodine, and no group characteristic of the oxidizing agent is necessarily transferred in the net change. This situation obtains only for reactions in a strongly solvating medium such as water, which provides the groups that associate with the atom being oxidized or reduced. Thus, when the reactions take place in the solid, it is necessary to specify what iodide is being used, whether sodium iodide, NaI, or silver iodide, AgI, for example, and the properties of the reaction would be dependent on the choice.

In representing an element in a particular environment, it is often convenient to specify only the oxidation state, without attempting to identify all of the groups which are attached to the atom in question. Thus, the iron content of a solution made up by dissolving, say, ferric chloride in water will be composed, among others, of the species Fe^{3+}, $FeCl^{2+}$, $FeOH^{2+}$. Collectively, they are correctly, though of course not fully, described by the notation Fe(III). In this kind of usage, the roman numeral represents the oxidation state.

Oxidation-reduction reactions. In an oxidation-reduction reaction, some element decreases in oxidation state and some element increases in oxidation state. The substances containing these elements are defined as the oxidizing agents and reducing agents, and they are said to be reduced and oxidized, respectively. The processes in question can always be represented formally as involving electron absorption by the oxidizing agent and electron donation by the reducing agent. For example, reaction (6) can be regarded as the sum of the two partial processes, or half-reactions, (9) and (10).

$$2I^- \rightarrow I_2 + 2e^- \quad (9)$$

$$2Fe^{3+} + 2e^- \rightarrow 2Fe^{2+} \quad (10)$$

Similarly, reaction (7) consists of the two half-reactions (11) and (12), with half-reaction (11) being

$$2I^- \rightarrow I_2 + 2e^- \quad (11)$$

$$16H^+ + 2MnO_4^- + 10e^- \rightarrow 2Mn^{2+} + 8H_2O \quad (12)$$

taken five times to balance the electron flow from reducing agent to oxidizing agent.

Each half-reaction consists of an oxidation-reduction couple; thus, in half-reaction (12) the reducing agent and oxidizing agent making up the couple are manganous ion, Mn^{2+}, and permanganate ion, MnO_4^-, respectively; in half-reaction (11) the reducing agent is I^- and the oxidizing agent is I_2. The fact that MnO_4^- reacts with I^- to produce I_2 means that MnO_4^- in acid solution is a stronger oxidizing agent than is I_2. Because of the reciprocal relation between the oxidizing agent and reducing agent comprising a couple, this statement is equivalent to saying that I^- is a stronger reducing agent than Mn^{2+} in acid solution. Reducing agents may be ranked in order of tendency to react, and this ranking immediately implies an opposite order of tendency to react for the oxidizing agents which complete the couples. In the list below some common oxidation-reduction couples are ranked in this fashion:

Strong reducing agent ↑ Increasing reducing power		Weak oxidizing agent ↓ Increasing oxidizing power
	$Mg = Mg^{2+} + 2e^-$	
	$Zn = Zn^{2+} + 2e^-$	
	$H_2 = 2H^+ + 2e^-$	
	$Cu = Cu^{2+} + 2e^-$	
	$I^- = {}^1\!/\!_2 I_2 + e^-$	
	$Fe^{2+} = Fe^{3+} + e^-$	
	$Br^- = {}^1\!/\!_2 Br_2 + e^-$	
	$Cl^- = {}^1\!/\!_2 Cl_2 + e^-$	
	$4H_2O + Mn^{2+} = MnO_4^- + 8H^+ + 5e^-$	
Weak reducing agent		Strong oxidizing agent

A complete list contains the displacement series of the metals. The most powerful reducing agent shown in the list is magnesium, Mg, although this is not the most powerful known. Magnesium is capable of reacting with any oxidizing agent below it in the list to yield Mg^{2+} and to produce the reduced product resulting from the oxidizing agent. Similarly, permanganate ion, MnO_4^-, in acid, the strongest oxidizing agent shown, is capable of reacting with any reducing agent above it in the list. Conversely, an oxidizing agent at the top of the list will not react appreciably with the reducing agent of a couple below it. The list given, containing nine entries, can be used to predict the results of 72 separate experiments (for example, $Mg + Zn^{2+}$ on the one hand and $Mg^{2+} + Zn$ on the other would be counted as separate experiments in arriving at this figure). *See* ELECTROCHEMICAL SERIES; ELECTRONEGATIVITY.

Since the driving force for a reaction depends on concentrations, the concentrations of all reactants and products must be specified, as well as other conditions, in compiling a list such as that given. The order shown obtains for water solutions at 25°C, approximately 1 *M* in all soluble reagents and having gases present at approximately 1 atm pressure. A second limitation on the use of this list lies in the fact that it applies only when the expected reaction products are compatible. Although copper is capable in principle of reducing iodine to form I^- and Cu^{2+} at high concentration, these products are not compatible with each other, but they react to form copper (I) iodide, CuI. Allowance for such features can always be made by incorporating the necessary additional half-reactions into the list. Finally, it must be stressed that the list can be used to predict the results of experiments only for systems which reach equilibrium sufficiently rapidly; it does not serve to predict the rate of reaction. To achieve the reduction of Fe^{3+} by H_2 predicted in the list, it would be necessary to use a catalyst in order to realize the reaction in a reasonable time.

The equilibrium information implied by a table of half-reactions can readily be put into quantitative form. Thus, the standard free-energy change for the reaction of 1 equivalent weight of each reducing agent with some common oxidizing agent can be entered opposite each half-reaction. The numerical values of these quantities will be in the same order as are the half-reactions and can be combined algebraically to yield the standard free energy change, and therefore the equilibrium constant, for any reaction which can be written from the table. *See* CHEMICAL EQUILIBRIUM.

A chemist concerned with reactions of the type under discussion will have a ready vocabulary of facts concerning oxidizing agents and reducing agents, such as their oxidizing or reducing powers, the speed with which they react, and the characteristics which may complicate their application. A typical problem in analytical chemistry is to reduce $Cr_2O_7^{2-}$ to Cr^{3+} in acidic (perchloric acid) solution without introducing elements which are not already present. Metallic reducing agents such as zinc and iron or metal ion reducing agents are immediately eliminated from consideration because the products of oxidation may be difficult to remove from the resulting solution. A solution of hydrogen iodide, HI, would be suitable, except that it would be necessary to take special pains to add it in equivalent amount because excess HI would be difficult to remove (the iodine, I_2, produced by oxidation of I^-, however, is easy to remove by extracting it with a suitable solvent such as carbon tetrachloride). Hydrogen would be ideal (the product of its oxidation, H^+, is already present in solution, and excess reducing agent can easily be removed) except that the rate of reaction would be disappointingly slow. A suitable reducing agent would be hydrogen peroxide, H_2O_2; it reacts rapidly, the product of oxidation is oxygen, which escapes from solution, and excess oxidizing agent is easily destroyed by heating the solution. *See* ELECTRODE POTENTIAL.

Mechanisms. The data needed to predict the outcome at equilibrium of the reaction of most common oxidizing and reducing agents are known. A list of the kind shown above can be extended, and when it is elaborated with entries carrying the quantitative information, accurate calculations of the equilibrium state for all the reactions implied by the table can be made. By contrast, though the rates of reaction are also of great importance, they are much less completely understood and less completely described. To understand the rates of reaction, it is necessary to consider how the reactions take place. To illustrate one of the problems of mechanism, a reaction is selected which, though not nearly as complicated as some, suffices for present purposes. When an aqueous solution containing Fe^{2+} is added to one containing Br_2, reaction (13) takes place. For the final stable products

$$2Fe^{2+} + Br_2 = 2Fe^{3+} + 2Br^- \qquad (13)$$

to be produced in a single step would require that two Fe^{2+} and one Br_2 be brought together in an encounter. This course for the reaction is found to be less probable than one in which a single Fe^{2+} encounters one Br_2. Since in forming stable products the oxidation state of iron increases by one unit while that of a bromine molecule decreases by two (one for each atom), the reaction resulting from the encounter must leave either iron or bromine in an unstable state. Reasonable alternatives for the reactive intermediates produced are represented in reactions (14*a*) and (15*b*), together in each case with a sequel reaction which leads to the correct overall stoichiometry.

$$Fe^{2+} + Br_2 = Fe^{3+} + Br^- + Br \qquad (14a)$$

$$\text{or} \quad Fe^{2+} + Br_2 = Fe(IV) + Br^- \qquad (15a)$$

$$Br + Fe^{2+} = Fe^{3+} + Br^- \qquad (14b)$$

$$\text{or} \quad Fe(IV) + Fe^{2+} = 2Fe^{3+} \qquad (15b)$$

In the present system, the evidence points to the mechanism represented by the first alternative [reactions (14*a*) and (14*b*)]. But even after a reaction such as (13) is resolved into the different steps, and reactive intermediates are identified, there remain questions about how the changes in oxidation state are brought about in the individual steps. Thus, when Fe^{2+} reacts with Br_2, do the two reactants make direct contact—this would involve replacement of a water molecule on $Fe(H_2O)_6^{2+}$, the form which Fe^{2+} adopts in water, by Br_2—or does reaction occur by electron transfer from intact $Fe(H_2O)_6^{2+}$ to Br_2? If the latter occurs, does electron transfer take place over large distances from $Fe(H_2O)_6^{2+}$ to Br_2?

These questions are important not only for inorganic systems of the kind illustrated by the steps making up reaction (13) but also for reactions at electrodes, and for oxidation-reduction reactions catalyzed by enzymes in biological systems. These kinds of questions are under investigation and have been partly answered for certain systems.

Two different kinds of mechanisms are recognized for oxidation-reduction reactions. So-called outer-sphere reactions are easier to understand in a fundamental way and will be described first. Reaction (16) introduces a typical such reaction.

$$IrCl_6^{2-} + Fe(CN)_6^{4-} \rightarrow IrCl_6^{3-} + Fe(CN)_6^{3-} \qquad (16)$$

Here the changes in oxidation state, 4+ to 3+ for Ir and 2+ to 3+ for Fe, take place without bonds to either atom being broken, and in this particular system there is not even much change in bond distances attending the changes in oxidation state. Electron transfer is explicit in such a system, and the electron transfer act is subject to the Franck-Condon restriction. This imposes a barrier to electron transfer in that it requires that the environments (coordination sphere and solvent) readjust prior to electron transfer so that after readjustment the energy of the system is the same whether the electron is on one metal or the other. The rates of reactions such as (16) can be estimated at least approximately by calculating the work to bring the partners together, the work of meeting the Franck-Condon restriction, and assuming that electron delocalization when the partners are in contact is adequate. Greater success has been met in attempts at correlating rates and here the equation developed by R. A. Marcus, relating the rate of reaction such as (16) to the standard free energy change and to the rates of the so-called self-exchange reactions [(17) and (18)], is proving to be

$$Fe^{*}(CN)_6^{4-} + Fe(CN)_6^{3-} \rightarrow {}^{*}Fe(CN)_6^{3-} + Fe(CN)_6^{4-} \qquad (17)$$

$${}^{*}IrCl_6^{3-} + IrCl_6^{2-} \rightarrow {}^{*}IrCl_6^{2-} + IrCl_6^{3-} \qquad (18)$$

useful not only in simple systems but also in understanding electron transfer reactions for large and complex biological molecules.

Much more difficult to systematize and understand is the extensive and very important class of oxidation-reduction reactions in which the changes in oxidation state are linked to bond breaking or bond making. A simple example is provided by reaction (19). Isotopic labeling experiments have

$$ClO_3^{*-} + SO_2 + H_2O \rightarrow ClO_2^{*-} + SO_3O^{*2-} + 2H^+ \qquad (19)$$

shown that in the course of the reaction an oxygen atom originating on ClO_3^- is transferred to the reducing agent. Though the process can formally be represented as involving electron loss by SO_2 and electron gain by ClO_3^-, electron transfer as the means by which the changes in oxidation state are brought about is not at all explicit in a reaction of this kind. These so-called inner-sphere mechanisms operate also for reactions involving metal ions. For example, when $[(NH_3)_5CoCl]^{2+}$ reacts with $Cr(H_2O)_6^{2+}$, the Cr(III) product has the formula $Cr(H_2O)_5Cl^{2+}$, and Cl is transferred from Co(III) to Cr(II) when the former oxidizes the latter. This kind of reaction clearly has much in common with that represented by reaction (19). In the latter system atomic oxygen is formally transferred; in the former, atomic chlorine.

A class of inner-sphere reactions of metal-containing molecules which are now recognized as playing an important role in many catalytic processes involves so-called oxidative addition. Reactions of this kind have been known for a long time, but their significance was not appreciated until interest in homogeneous catalytic processes began to develop. A commonplace example of oxidative addition is provided by reaction (20). It will be not-

$$SnCl_2 + Cl_2 \rightarrow SnCl_4 \qquad (20)$$

ed that in reaction (20) both the oxidation number and the coordination number of Sn increase. Oxidative addition for a strong oxidizing agent such as Cl_2 is not surprising, but the reaction type took on new significance when it was discovered that with a suitable metal complex, oxidative addition can be realized also with hydrogen halides, alkyl halides, and even hydrogen. Among the metal complexes which undergo this kind of reaction are Rh(I), Ir(I), or Pt(0) species with 4 or fewer groups attached. In each case, there is the opportunity for an increase in both oxidation and coordination number. A specific example of a molecule which undergoes oxidative addition with H_2 is $((C_6H_5)_3P)_3RhCl$, which is a useful catalyst for the hydrogenation of alkenes and alkynes. A reaction step in the catalytic sequence is the addition of H_2 to the metal so that the H-H bond is severed; the adduct then reacts with the alkene (or alkyne) transferring two atoms of hydrogen. A substance which will activate H-R bonds (where R is an alkyl radical) in the same way would be desirable. *See* COORDINATION NUMBER; HOMOGENEOUS CATALYSIS.

The fundamental aspects of electron transfer processes in oxidation-reduction reactions have much in common with the electron jump processes in semiconductors. Recognizing this connection is productive both for those interested in chemical effects accompanying electron transfer (that is, in oxidation-reduction processes) and those interested in electron mobility as a subject in its own right.

[HENRY TAUBE]

Bibliography: J. P. Collman and L. S. Hegedus, *Principles and Applications of Organotransition*, 1980; A. Haim, *Account. Chem. Res.*, 8:264, 1975;

W. M. Latimer, *Oxidation States of the Elements and Their Potentials in Aqueous Solution*, 2d ed., 1952; R. G. Linck, *Int. Rev. Sci., Inorg. Ser.* [1], 9:303, 1972; H. Taube and E. S. Gould, *Accounts Chem. Res.*, 2:321, 1969.

Oxide

A binary compound of oxygen with another element. Oxides have been prepared for essentially all the elements except the noble gases. Often, several different oxides of a given element can be prepared; a number exist naturally in the Earth's crust and atmosphere: silicon dioxide (SiO_2) in quartz; aluminum oxide (Al_2O_3) in corundum; iron oxide (Fe_2O_3) in hematite; carbon dioxide (CO_2) gas; and water (H_2O).

Most elements will react with oxygen at appropriate temperature and oxygen pressure conditions, and many oxides may thus be directly prepared. Phosphorus burns spontaneously in oxygen to form phosphorus pentoxide, $(P_2O_5)_2$. Sulfur requires ignition and thereafter burns to sulfur dioxide (SO_2) gas if the supply of oxygen is limited. The relative amounts of oxygen and element available often determine which of several oxides will form; in an excess of oxygen, sulfur burns to form some sulfur trioxide (SO_3) gas. Most metals in massive form react with oxygen only slowly at room temperatures because the first thin oxide coat formed protects the metal; magnesium and aluminum remain metallic in appearance for long periods because their oxide coatings are scarcely visible. However, diffusion of the oxygen and metal atoms through the film becomes rapid at high temperatures, and these metals will burn intensely to their oxides if ignited. The oxides of the alkali and alkaline-earth metals, except for beryllium and magnesium, are porous when formed on the metal surface, and they provide only limited protection to the continuation of oxidation, even at room temperatures. Gold is exceptional in its resistance to oxygen, and its oxide (Au_2O_3) must be prepared by indirect means. The other noble metals, although ordinarily resistant to oxygen, will react at high temperatures to form gaseous oxides.

Indirect preparation of the oxides may be accomplished by heating hydroxides, nitrates, oxalates, or carbonates, as in the production from the latter of quicklime (CaO) by reaction (1), in which

$$CaCO_3 \rightarrow CaO + CO_2 \qquad (1)$$

carbon dioxide is driven off. Gold oxide may be prepared by heating gold hydroxide, as shown by reaction (2). Higher oxides of an element may be re-

$$2Au(OH)_3 \rightarrow Au_2O_3 + 3H_2O \qquad (2)$$

duced to lower oxides, usually at high temperatures, for example, the reduction of tungsten trioxide by tungsten, as shown by reaction (3). Complete

$$2WO_3 + W \rightarrow 3WO_2 \qquad (3)$$

reduction to the element may be performed by other elements whose oxides are more stable, as in the formation of calcium oxide from titanium dioxide by reaction (4).

$$2Ca + TiO_2 \rightarrow Ti + 2CaO \qquad (4)$$

Although the solid oxides of a few metals, such as mercury and the noble metals, can be easily decomposed by heating, for example, reaction (5),

$$2Au_2O_3 \rightarrow 4Au + 3O_2 \qquad (5)$$

most metal oxides are very stable and many constitute important refractory materials. For example, magnesium oxide, calcium oxide, and zirconium dioxide do not melt or vaporize appreciably at temperatures up to 2500°C. A great number of refractories consist of compounds of two or more oxides; silicon dioxide and zirconium dioxide form zirconium silicate by reaction (6).

$$SiO_2 + ZrO_2 \rightarrow ZrSiO_4 \qquad (6)$$

Because so many oxides can be easily formed, studies of them have been most important in establishing relative atomic weights of the elements based on the defined atomic weight for oxygen. Furthermore, these studies were fundamental in forming the basis for the laws of definite proportions and multiple proportions for compounds. It is of special significance that, although any gaseous oxide species must necessarily have a definite oxygen-to-element proportion, a number of solid and liquid oxides can be prepared with proportions which may vary continuously over a considerable range. This is particularly true for oxides prepared under equilibrium conditions at high temperatures. Thus, titanium exposed to oxygen until reaction equilibrium is reached at a number of selected conditions of temperature and oxygen pressure will form the solid oxide TiO. It has the same crystal structure as rock salt; that is, every other site along the three coordinate directions of the crystal will be occupied by titanium atoms and the alternate sites by oxygen atoms (each in its ion form Ti^{++} and O^{--}) to give the simple Ti/O ratio of 1:1. However, with other selected pressure-temperature conditions, oxides of this same structure at every Ti/O ratio from 1:0.7 to 1:1.25 may be prepared. The variable proportions show that variable numbers of oxygen or titanium sites can simply remain vacant in a homogeneous way. The range is referred to as the TiO/O 1:07–1.25 phase or, more loosely, the TiO solid-solution phase.

Most of the nonmetal oxides commonly encountered as gases, such as SO_2 and CO_2, form solids and liquids in which the molecular units of the gas are retained so that the simple definite proportions are clearly maintained. Such oxides melt and boil at low temperatures, because the molecular units are weakly bonded to adjoining molecular units.

Oxides may be classified as acidic or basic according to the character of the solution resulting from their reactions with water. The nonmetal oxides generally form acid solutions, for example, reaction (7), the formation of sulfuric acid. The

$$SO_3 + H_2O \rightarrow H_2SO_4 \qquad (7)$$

metal oxides generally form alkaline solutions, for example, reaction (8), for the formation of calcium

$$CaO + H_2O \rightarrow Ca(OH)_2 \qquad (8)$$

hydroxide or slaked lime. However, given metals of the groups IV and higher of the periodic table will often have basic, intermediate, and acidic oxides. Here the acid character increases with increasing oxygen-metal ratio. *See* ACID AND BASE; EQUIVALENT WEIGHT; OXYGEN.

[RUSSELL K. EDWARDS]

Bibliography: F. A. Cotton and G. Wilkinson,

Advanced Inorganic Chemistry: A Comprehensive Text, 4th ed., 1980; R. P. Elliott, *Constitution of Binary Alloys*, suppl. 1, 1971; M. Hansen, *Constitution of Binary Alloys*, 2d ed., 1958.

Oxidizing agent

A participant in a chemical reaction that absorbs electrons from another reactant. In the process a component atom of this substance undergoes a decrease in oxidation number. In this action as an oxidizing agent, the substance undergoes reduction.

A measure of the effectiveness of a reagent as an oxidizing agent is its reduction potential. This is, in electrochemical terms, the equivalent of the free-energy change for the reduction process. The element with the highest reduction potential (and, therefore, the strongest oxidizing agent) is fluorine, F_2. The half-reaction in which fluorine absorbs an electron from another species in aqueous solution is shown by reaction (1). The reduction potential E,

$$e^- + \tfrac{1}{2} F_{2(gas)} \rightarrow F^-_{(aq)} \quad (1)$$

although strictly applicable only to thermodynamically reversible systems, is a very useful measure of the tendency of a reaction component to undergo oxidation or reduction. For the F_2-F^- pair, it is given at 25°C by Eq. (2), where a_{F^-} is the

$$E = 2.87 - 0.1362 \ln \frac{a_{F^-}}{p_{F_2}^{1/2}}\ V \quad (2)$$

activity of fluoride ion in the solution, and p is the pressure, in atmospheres of fluorine gas. Under standard conditions, $a_{F^-} = 1$ and $p_{F_2} = 1$, the value for E is 2.87 V. This is E^0, the standard reduction potential at 25°C. The F_2-F^- pair constitute a reduction-oxidation couple.

Thermodynamically, a substance is capable of oxidizing another if the corresponding couple has a lower reduction potential. When dichromate ion in acid solution acts as an oxidizing agent, it undergoes the reduction shown in reaction (3). The

$$Cr_2O_7^{2-} + 14H^+ + 6e^- \rightarrow 2Cr^{3+} + 7H_2O \quad (3)$$

standard reduction potential at 25°C for this reaction is 1.33 V. As this reaction does not proceed reversibly, the standard reduction potential represents only a lower limit. For the reduction of acetaldehyde to ethanol, shown in reaction (4), E^0

$$2e^- + CH_3CHO + 2H^+ \rightarrow CH_3CH_2OH \quad (4)$$

is 0.217 V. Dichromate can, therefore, oxidize ethanol according to reaction (5). The correspond-

$$Cr_2O_7^{2-} + 3CH_3CH_2OH + 8H^+ \rightarrow 3CH_3CHO + 2Cr^{3+} + 7H_2O \quad (5)$$

ing electrochemical potential is 1.11 V. The positive value for the resulting potential indicates a spontaneous process. Actually, in this example, the oxidation by dichromate will go beyond the aldehyde stage to produce the carboxylic acid. Dichromate ion in acid solution is a common oxidizing agent used in organic and analytical chemistry.

In oxidation-reduction (redox) systems composed of couples that are not widely separated, the actual reaction that occurs will depend upon the chemical activities in the system. For example, standard potentials for the $Fe^{3+}-Fe^{2+}$ and Ag^+-Ag couples are 0.771 and 0.799 V, respectively. For the redox system shown in reaction (6), the resultant potential at 25°C is given by Eq. (7). When the activity

$$Fe^{2+} + Ag^+ \rightleftharpoons Fe^{3+} + Ag \quad (6)$$

$$E = 0.028 - 0.1362 \ln \frac{a_{Fe^{3+}}}{a_{Fe^{2+}} a_{Ag^+}}\ V \quad (7)$$

quotient $Q = (a_{Fe^{3+}}/a_{Fe^{2+}} a_{Ag^+}) < 1.23$, chemical change will occur from left to right; with $Q > 1.23$, from right to left; and with $Q = 1.23$, the system is at equilibrium and no net chemical change will occur. The above considerations are based only on thermodynamics and allow no prediction concerning the actual rate of the oxidation-reduction process. The practical effectiveness of a given oxidizing (or reducing) agent will depend upon both the thermodynamics and the available kinetic pathway for the reaction process. *See* CHEMICAL THERMODYNAMICS.

Substances that are widely used as oxidizing agents in chemistry include ozone (O_3), permanganate ion (MnO_4^-), nitric acid (HNO_3), as well as oxygen itself. Oxychlorine trifluoride, OF_3Cl, has been suggested and synthesized for possible use as an oxidizer in rocket engines. Organic chemists have empirically developed combinations of reagents to carry out specific oxidation steps in synthetic processes. Many of these utilize transition-metal oxides such as chromium trioxide, CrO_3, vanadium pentoxide, V_2O_5, ruthenium tetroxide, RuO_4, and osmium tetroxide, OsO_4. Oxidation by these species or by specific agents such as sodium perborate, $NaIO_4$, can often be restricted to a single molecular site.

The action of molecular oxygen as an oxidizing agent may be made more specific by photochemical excitation to an excited singlet electronic state. This can be accomplished through energy transfer from a dye, such as fluorescein, which absorbs light and transfers it to oxygen to form a relatively long-lived excited state. *See* PHOTOCHEMISTRY.

In an electrolytic cell, oxidation occurs at the anode, and the designation of a chemical oxidizing agent in the reaction is in general not possible. The electrode acts as an intermediary in the electron transfer. An important example of an anodic oxidation is the Kolbe reaction in which a carboxylic acid anion transfers an electron to the anode, forming a free radical [reaction (8)]. The radicals

$$RCOO^- \xrightarrow{\text{Anode}} RCOO + e^- \quad (8)$$

combine to form a hydrocarbon and CO_2, reaction (9). *See* ELECTROLYSIS.

$$2\,RCOO \rightarrow R_2 + CO_2 \quad (9)$$

Enzymes in living organisms catalyze the exothermic oxidation of carbohydrate, fat, and protein material as a source of energy for life processes. Certain microorganisms are able to oxidize organic compounds at C-H bonds to produce alcohols. These have been suggested as of possible use in cleaning oil spills. *See* OXIDATION-REDUCTION.

[F. J. JOHNSTON]

Bibliography: W. R. Adams, *Photosensitized Oxidations*, 1971; D. Benson, *Mechanisms of Oxidation by Metal Ions*, 1976; L. J. Chinn, *Selection of Oxidants in Synthesis*, 1971; G. S. Fonken and R. A. Johnson, *Chemical Oxidations with Microorganisms*, 1972; S. D. Ross, M. Finkelstein, and E. J. Rudd, *Anodic Oxidation*, 1975.

Oxime

One of a group of chemicals derived from aldehydes (RCH=NOH, aldoximes) or ketones (RR′C=NOH, ketoximes) used for isolation and identification of carbonyl compounds. In general, they are easily purified and have characteristic melting points. The properties of certain oximes have made them industrially important.

Oximes have received considerable attention because of their stereochemistry and their participation in the Beckmann rearrangement.

The discovery of geometrical isomers involving a carbon-nitrogen double bond demonstrated the fact of restricted rotation about such a bond in a manner analogous to that obtaining about a carbon-carbon double bond. However, relatively few pairs of geometric isomers of the oximes, which are conventionally termed syn and anti isomers analogous to the more familiar cis and trans terminology used in carbon-carbon systems, have been isolated. This suggests that interconversion of such isomers involves relatively little energy. Thus, *syn*-benzaldehyde oxime (H and OH in a cis arrangement with respect to the double bond) is converted to the anti (trans) form by ethereal hydrogen chloride solution; reversion to the syn form can be accomplished by irradiation of a benzene solution of the anti form. *See* MOLECULAR ISOMERISM.

In ketoximes the prefixes syn and anti refer to the relative positions of the hydroxyl group and the group adjacent to the prefix [notation (1).]

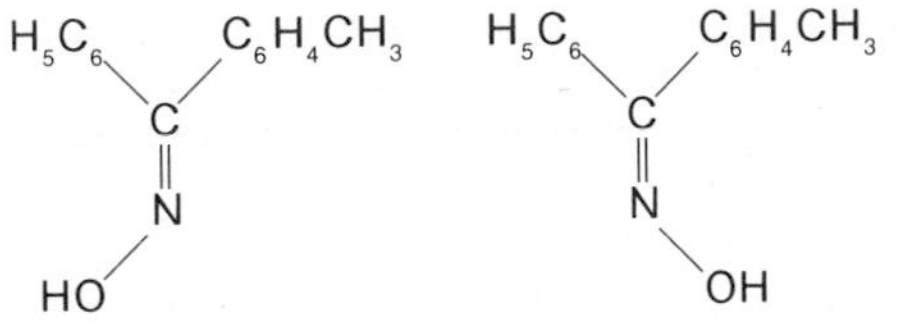

(1)

syn-Phenyl tolyl ketoxime or *anti*-tolyl phenyl ketoxime

syn-Tolyl phenyl ketoxime or *anti*-phenyl tolyl ketoxime

Ketoximes undergo the Beckmann rearrangement under the influence of acidic reagents. In this rearrangement, the substituent anti to the hydroxyl group changes positions with the hydroxyl group with the formation of the lactim form of an amide which immediately tautomerizes to the more stable lactam form. Thus, the oxime of acetophenone (*syn*-methyl phenyl ketoxime) yields the lactim form of the stable acetanilide [reaction (2)].

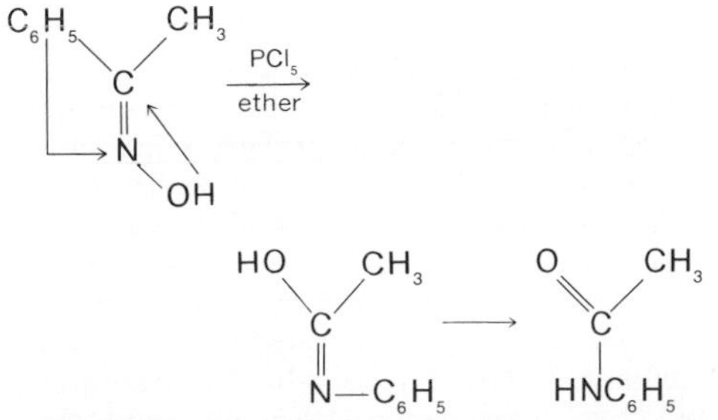

(2)

J. Meisenheimer assigned the presently accepted configurations of the ketoximes largely on the basis of a study of ring-closure reactions involving reactive halogen atoms. For example, as seen in reaction (3), one isomer of the oxime of methyl 2-chloro-5-nitrophenyl ketoxime readily undergoes ring closure with elimination of hydrogen chloride under the influence of sodium hydroxide, whereas the other form gives the same product much more slowly. Therefore, it is concluded that the isomer which undergoes facile ring closure is the anti form and that the resistant isomer has the syn configuration. On rearrangement the anti and syn forms gave (I) and (II), respectively, thus providing a basis for the trans migration of the groups concerned and also providing a basis for the assignment of configuration from the nature of the products of the Beckmann rearrangement.

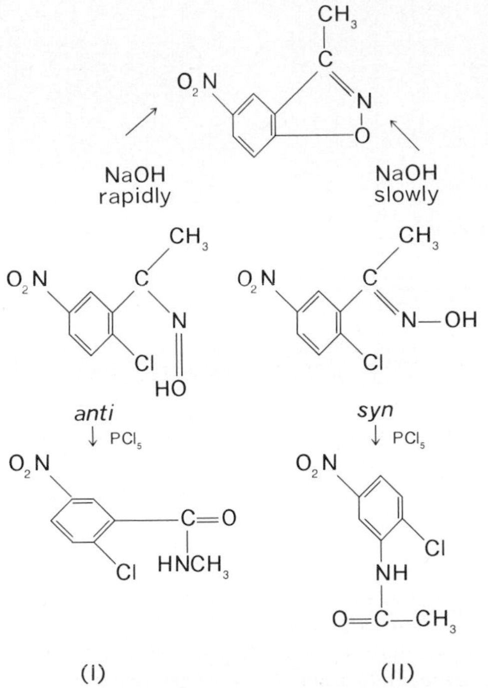

(3)

Cyclohexanone oxime rearranges to the lactam of 6-aminohexanoic acid (caprolactam), a precursor of a polyamide of the nylon type (nylon 6) shown in notation (4).

$$—[NH(CH_2)_5CO]_n— \quad (4)$$

Aldoximes are dehydrated to nitriles by the action of acetic anhydride; oximes may be reduced to primary amines. The lower aliphatic aldoximes find use as antiskinning agents in paints. *See* ALDEHYDE; KETONE.

[LEALLYN B. CLAPP]

Oxygen

A gaseous chemical element, O, atomic number 8, and atomic weight 15.9994. Oxygen is of great interest because it is the essential element both in the respiration process in most living cells and in combustion processes. It is the most abundant element in the Earth's crust. About one-fifth (by volume) of the air is oxygen.

DESCRIPTION AND OCCURRENCE

Oxygen is separated from air by liquefaction and fractional distillation. The chief uses of oxygen in order of their importance are (1) smelting, refining, and fabrication of steel and other metals; (2) manufacture of chemical products by controlled oxidation; (3) rocket propulsion; (4) biological life sup-

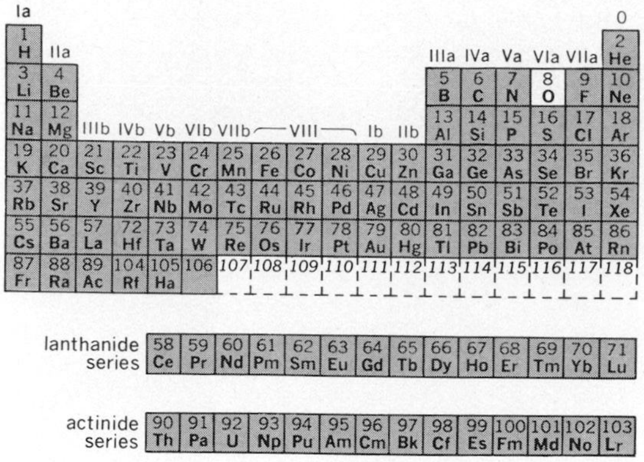
Ia
0
1 H
IIa
IIIa IVa Va VIa VIIa
2 He
3 Li 4 Be
5 B 6 C 7 N 8 O 9 F 10 Ne
11 Na 12 Mg
IIIb IVb Vb VIb VIIb VIII Ib IIb
13 Al 14 Si 15 P 16 S 17 Cl 18 Ar
19 K 20 Ca 21 Sc 22 Ti 23 V 24 Cr 25 Mn 26 Fe 27 Co 28 Ni 29 Cu 30 Zn 31 Ga 32 Ge 33 As 34 Se 35 Br 36 Kr
37 Rb 38 Sr 39 Y 40 Zr 41 Nb 42 Mo 43 Tc 44 Ru 45 Rh 46 Pd 47 Ag 48 Cd 49 In 50 Sn 51 Sb 52 Te 53 I 54 Xe
55 Cs 56 Ba 57 La 72 Hf 73 Ta 74 W 75 Re 76 Os 77 Ir 78 Pt 79 Au 80 Hg 81 Tl 82 Pb 83 Bi 84 Po 85 At 86 Rn
87 Fr 88 Ra 89 Ac 104 Rf 105 Ha 106 107 108 109 110 111 112 113 114 115 116 117 118
lanthanide series
58 Ce 59 Pr 60 Nd 61 Pm 62 Sm 63 Eu 64 Gd 65 Tb 66 Dy 67 Ho 68 Er 69 Tm 70 Yb 71 Lu
actinide series
90 Th 91 Pa 92 U 93 Np 94 Pu 95 Am 96 Cm 97 Bk 98 Cf 99 Es 100 Fm 101 Md 102 No 103 Lr

port and medicine; and (5) mining, production, and fabrication of stone and glass products.

Uncombined gaseous oxygen usually exists in the form of diatomic molecules, O_2, but oxygen also exists in a unique triatomic form, O_3, called ozone. *See* OZONE.

In 1774, Joseph Priestley, an English clergyman who later immigrated to the United States and settled in Northumberland, Pa., observed that mercuric oxide, on heating, yielded a gas that vigorously supported the combustion of a candle. Priestley found that the gas would support respiration and called the gas dephlogisticated air. The name oxygen, meaning acid-former, was given the gas by a group of French chemists in 1787 in recognition of the ability of some oxides, such as the oxides of sulfur, to form acids.

USES

Oxygen is widely used in a variety of applications. While the fraction of oxygen present in the atmosphere is sufficient for many purposes, higher concentrations are necessary to improve some processes.

Metallurgical uses. Oxygen is a component which is used in the metallurgical processes of smelting, refining, welding, cutting, and surface conditioning.

Smelting. Smelting of ore in the blast furnace involves the combustion of about 1 ton of oxygen for each ton of metal produced. When air is used, $3\frac{1}{2}$ tons of nitrogen accompany each ton of oxygen and must be compressed, heated, and blown into the furnace. A large amount of heat is lost with the exhaust gases, which also carry powdered ore and coke away as dust and limit the capacity of the furnace. By removing some or all of the nitrogen, the furnace capacity can be increased, less expensive fuels can be used in place of some of the coke, and fuels can be used more efficiently.

Metal refining. In refining copper and in making steel from pig iron various impurities such as carbon, sulfur, and phosphorus must be removed from the metal by oxidation. If air is blown through the molten metal, as in the Bessemer converter, nitrogen is picked up, limiting the product quality. Nitrogen also carries away a great deal of the heat produced by the oxidation process. Better-quality steel and copper can be produced by injecting pure oxygen into the molten metal until the impurities are completely removed. Oxygen injection can be utilized in the open hearth or electric furnaces. However, new steelmaking equipment has been developed which depends entirely on high-purity oxygen. All the heat for the furnace operation is supplied by oxidation of carbon and other impurities. The technique is called the basic oxygen process. The most common form is known as the L-D process, named after the Austrian cities of Linz and Donawitz, where the procedure was first used in 1951.

Welding, cutting, and surface conditioning. The high-temperature flame of the oxyacetylene torch can be used in welding steel, although most welding is now done by the electric arc process.

In cutting, the point of the steel at which the cutting is to start is first heated by an oxygen-acetylene flame. A powerful jet of oxygen is then turned on. The oxygen burns some of the iron in the steel to iron oxide, and the heat of this combustion melts more iron; the molten iron is blown out of the kerf by the force of the jet. By feeding powdered iron into the oxygen stream this cutting process can be extended to alloys, such as stainless steel, which are not readily cut by oxygen alone and to completely noncombustible materials such as concrete.

Steel ingots normally have oxide inclusions and other defects at the outer surface. After preliminary rolling, the steel in slab or billet form has the surface skin removed to eliminate these defects. This can be most easily accomplished by scarfing. Streams of oxygen from many nozzles are played on all sides of the billet at once. The oxygen burns off the surface defects and some of the steel in a spectacular shower of sparks. The billet is then ready for further rolling. Oxygen scarfing, also known as skinning, became a standard practice in most steel mills.

Chemical syntheses. Several syntheses in the chemical industry involve oxygen. These processes are outlined.

Partial oxidation of hydrocarbons. When natural gas or fuel oil is burned, the heat of combustion first cracks the hydrocarbon molecules into fragments. These fragments usually encounter oxygen molecules within a few hundredths of a second and are oxidized to water and carbon dioxide. However, if the supply of oxygen is carefully controlled and the passage of material through the combustion zone is very rapid, it is possible to freeze the reaction at various stages of completion.

In this manner natural gas (mostly methane, CH_4) can be converted to acetylene (C_2H_2), ethylene (C_2H_4), or propylene (C_3H_6). Ethylene (C_2H_4), in turn, can be partially oxidized to ethylene oxide (CH_2CH_2O).

Syngas production. Reaction of carbon or hydrocarbons with oxygen and steam yields a mixture of carbon monoxide (CO) and hydrogen (H_2), that is, syngas. By use of suitable catalysts, syngas can be recombined to form various organic compounds such as methanol (CH_3OH), octane (C_8H_{16}), and many others. In the presence of other catalysts, carbon monoxide can combine with steam to form more hydrogen and carbon dioxide. After removal of the carbon dioxide, the hydrogen can be used for chemical reactions, such as the manufacture of ammonia (NH_3), hydrogenation of fats, and hydrocracking of petroleum.

Manufacture of pigments. Both titanium dioxide white and carbon black are useful primarily because of the characteristics of their small particles. The size, shape, and surface activity of these particles govern the ability of the material to perform properly as a pigment, bulking agent, or stiffener when blended into other materials. Formation of titanium dioxide or carbon in a flame process produces very fine, useful particles. Carefully controlled addition of oxygen to such burner operations can improve yield and quality of the product. *See* ORGANIC CHEMICAL SYNTHESIS.

Liquid fuel rockets. In rocket engines liquid oxygen is used as an oxidizer either with kerosine or liquid hydrogen fuels. While fluorine could theoretically provide somewhat improved performance in terms of specific impulse, oxygen is very nearly as good, is much cheaper and is easier to handle.

Solid-fueled rockets, based on hydrocarbon polymers that contain sufficient oxidizer to effect self-combustion, dominate the short-range military uses. Liquid-fueled rockets are expected to remain dominant in space work until the full development of nuclear propulsion. The Saturn-Apollo launch vehicle has a fully loaded weight of about 3000 tons of which more than 2000 tons are liquid oxygen. Most of the liquid oxygen consumed by the aerospace industry has been used in the development and proof testing of rocket engines mounted in static test stands. The usage of oxygen in this testing has been in excess of 1000 tons per day.

Biological applications. Oxygen is a fundamental part of many biological processes. A few are described below.

Aerospace and diving. Oxygen is necessary for life support of animals of this planet. Whenever humans desire to live or work in environments low or deficient in oxygen, it is necessary to carry oxygen along to supplement or substitute for the available atmosphere. High-altitude military aircraft normally provide oxygen for the aviators. Commercial transports carry oxygen for emergency use in case of failure of the cabin pressurizing system. Astronauts must of course carry their entire breathing gas requirements with them, which becomes one of the larger load requirements for any extended mission. Divers in shallow water are able to have air transmitted to them from the surface. However, for deeper diving the special breathing gases frequently are carried to the ocean bottom in special diving bells.

Medicine. In medical applications oxygen is provided to patients in amounts up to 15 times normal. This is usually done to reduce the work load of heart and lungs during the course of an infectious disease, during or after major surgery, or during recovery from a heart attack.

Waste treatment. Tests have shown that addition of oxygen to waste treatment plants can assist the biological treatment process. Oxygen is sometimes pumped directly into rivers and streams that would otherwise be overloaded with contamination. With the assistance of the extra oxygen, the stream bacteria are able to decompose the waste rapidly.

Stone, clay, and glass industry. Oxygen has a place in these industries as described below.

Glass manufacture and fabrication. The glass industry uses large quantities of oxygen in the manufacture and shaping of glass. Oxygen additions raise the combustion temperature in the furnace, speeding up and improving control over the melting of glass and its raw materials. Oxygen is used in the burners that heat glass for blowing, shaping, and flame-polishing rough edges.

Mining and quarrying. An oxygen-kerosine burner can be used to heat and shape some types of stone. Granite and similar rocks expand when heated rapidly by such a burner so that the surface cracks loose, or spalls. The hot combustion gases blow the fine chips of rocks away, presenting a fresh surface which is rapidly heated, continuing the process.

In this manner the extremely hard taconite iron ore can be pierced for blast holes more effectively than by conventional drilling methods. Granite for construction and decorative purposes can be quarried by special burners equipped to cut channels through the rock. Slabs of granite can be cut to desired dimension and given an even and pleasing surface using still other burner designs. A rock surface fouled with paint or tarry materials can easily be cleaned by this technique. Artists have used flame shaping to produce statuary.

Cement and kiln operations. In most kiln-type operations, such as manufacture of cement, roasting or sintering ore, and production of refractories, the essential reactions take place at rather high temperatures. When enough heat is provided at the high temperature to carry out the desired reaction, there is more than enough heat to raise the temperature of the fresh feed. Much heat is wasted at lower temperatures where it is not useful to the process. By using oxygen instead of air, the flame temperature is raised and much more heat is available for the high temperature reaction from a given amount of fuel. Extensive tests have shown that large increases in capacity and reductions in fuel consumption are possible. However, certain changes in equipment are needed to achieve all the potential benefits.

Occurrence. About 49.5% by weight of the Earth's crust, including the oceans and atmosphere, is oxygen. Most of this oxygen is combined in the form of silicates, oxides, and water. Water is composed of 88.81% oxygen by weight.

Oxygen also exists outside the atmosphere of the Earth, but since more than 98% of the matter in the visible universe (stars, nebulae, and interstellar space) is composed of hydrogen and helium, the cosmic concentration of oxygen is relatively low.

Dry air contains 20.946% oxygen by volume, and this concentration has been found to be the same at any level between the surface of the Earth and a height of 40 mi. The atoms in atmospheric oxygen consist of three isotopes in the following atomic proportions: 99.759%, oxygen-16; 0.037%, oxygen-17; and 0.204%, oxygen-18. The molecules of oxygen in the air, each of which has two atoms, consist of the statistically expected proportion of the possible combinations of these isotopes, the most abundant molecules being $^{16}O^{16}O$, $^{16}O^{18}O$, and $^{16}O^{17}O$. The isotopic composition of the oxygen in water is slightly different from that in air and varies slightly in samples from different bodies of water (lakes, oceans, and seas).

Even though large quantities of oxygen from the air are continuously being used in respiration, combustion, and other oxidation processes, the concentration of oxygen in the atmosphere remains very nearly constant, chiefly because oxygen is liberated in the process of photosynthesis. In this process carbohydrates are produced by green plants from carbon dioxide and water. The primary source of the free oxygen in the atmosphere is believed by some authorities to have been the decomposition of water vapor by ultraviolet radiation in the upper atmosphere. Almost all the hydrogen formed in this way escaped from the Earth's gravitational field, but the oxygen molecules were too heavy to escape. They remained, therefore, in the atmosphere. This photochemical decomposition of water vapor to produce oxygen gas is still going on.

The following radioactive isotopes of oxygen are known: ^{14}O, ^{15}O, and ^{19}O. These isotopes may be formed in particle accelerators, such as the cyclotron, or by neutron bombardment of the appropriate atomic species; for example, ^{19}O is formed when the nucleus of an atom of stable ^{18}O absorbs a neutron. All three of the radioactive isotopes of oxygen are very short-lived, the one with the longest half-life, that of about 120 sec, being ^{15}O.

Physical properties. Under ordinary conditions oxygen is a colorless, odorless, and tasteless gas. It condenses to a pale blue liquid, in contrast to nitrogen, which is quite colorless in the liquid state. Oxygen is one of a small group of slightly paramagnetic gases, and it is the most paramagnetic of the group. Liquid oxygen is also slightly paramagnetic. Some data on oxygen and some properties of its ordinary form, O_2, are listed in the table.

Properties of oxygen

Property	Value
Atomic number	8
Atomic weight	15.9994
Triple point (solid, liquid, and gas in equilibrium)	−218.80°C (54.35 K)
Boiling point at 1 atm pressure	−182.97°C (90.18 K)
Gas density at 0°C and 1 atm pressure, g/liter	1.4290
Liquid density at the normal boiling point, g/ml	1.142
Solubility in water at 20°C, ml oxygen (STP) per 1000 g water at 1 atm partial pressure of oxygen	30

Before the mass spectrometer was invented and when nothing was known about isotopes, the average weight of the oxygen atoms in oxygen obtained from water was selected by chemists as the standard of weight for the atoms of all elements. This weight was assigned the value 16.0000. It is now known that isotopes exist and that the isotopic composition of many elements is subject to considerable variation. Consequently, there is no longer a good theoretical basis for the system of chemical atomic weights, based on the mixture of oxygen isotopes as they happen to occur in the Earth's atmosphere. Chemists concluded that the single isotope $^{12}C = 12.0000$ should be taken as the standard.

Chemical properties. Practically all chemical elements except the inert gases form compounds with oxygen. Most elements form oxides when heated in an atmosphere containing oxygen gas. Many elements form more than one oxide; for example, sulfur forms sulfur dioxide (SO_2) and sulfur trioxide (SO_3). Among the most abundant binary compounds of oxygen are water, H_2O, and silica, SiO_2, the latter being the chief ingredient of sand. Among compounds containing more than two elements, the most abundant are the silicates, which constitute most of the rocks and soil. Other widely occurring compounds are calcium carbonate (limestone and marble), calcium sulfate (gypsum), aluminum oxide (bauxite), and the various oxides of iron which are mined as a source of iron. Several other metals are also mined in the form of their oxides. Hydrogen peroxide, H_2O_2, is an interesting compound used extensively for bleaching. *See* HYDROGEN PEROXIDE; OXIDE; PEROXIDE.

PRODUCTION AND DISTRIBUTION

Oxygen is produced on a large scale by the liquefaction and fractional distillation of air. A little oxygen is also made by the electrolysis of water, but oxygen produced in this way is more expensive than oxygen distilled from air. Electrolysis of water is not used, therefore, unless there is some special reason, such as a need for the hydrogen that is also produced.

The traditional methods of preparing oxygen in school chemistry courses are (1) heating potassium chlorate with or without addition of a little manganese dioxide or other catalyst; (2) heating mercuric oxide (Priestley's original method); and (3) electrolysis of water to which an electrolyte has been added. When oxygen is needed in the laboratory, however, it is usually obtained from a cylinder of compressed oxygen.

Oxygen is commonly distributed in three ways: (1) Most oxygen is piped directly to users; (2) about 10% is liquefied for transportation and storage in insulated tanks; and (3) about 1% is compressed to high pressure (more than 2000 psi) for transport in steel cylinders or tube bundles.

Oxygen pipelines are usually short since the raw material for air separation is readily available. In industrial areas a single large plant may supply a dozen consumers through a network of pipelines. One such system operating in the heavy industrial area along the south shore of Lake Michigan supplies more than 5000 tons of oxygen daily.

For smaller or intermittent uses or for rocket engines oxygen is produced and distributed as a liquid. In liquid form oxygen is about one-third heavier than water. So long as it is kept at low temperature, the liquid can be stored, transported, pumped, or handled much as any other liquid. To keep heat away from this very cold liquid, the storage and transport tanks use the best possible insulating techniques. Two concentric tanks are constructed. The space between the tanks is filled with a powdered material of low thermal conductivity which is also opaque to radiant heat. The powder-filled space is then evacuated. The combination of vacuum and insulating powder minimizes heat transfer by convection, conduction, and radiation. The result is a container in which liquid oxygen can be transported hundreds of miles with little or no loss. Large liquid oxygen tanks have been mounted on trucks, trailers, and railroad cars. Smaller tanks can be wheeled around by hand. Special lightweight liquid oxygen tanks have been manufactured to permit sick people to carry several hours' supply in a pack about the size of a binocular case.

Generally speaking, liquid transport is preferred to high-pressure gas containers because much more product can be carried per pound of total weight. However, for some applications high gas pressures are desired. For others the use is so intermittent that liquid supply would involve excessive losses during idle periods. For these applications oxygen may be transported at high pressure in steel cylinders. Ordinary cylinders are about 9 in. in diameter, 4 ft in height, and about 150 lb in weight when filled with 240 ft (20 lb) of oxygen. Individual cylinders may be clustered and longer tubes may be mounted on trailers to achieve greater capacity.

Detection and quantitative analysis. The traditional laboratory test for oxygen gas is that it will cause a glowing wooden splinter to burst into flame; this test does not distinguish between oxygen and nitrous oxide.

In laboratory gas-analysis apparatus oxygen is usually determined by absorption in an alkaline solution of pyrogallol or in an ammoniacal solution of copper (I) chloride. The concentration of oxygen in oxygen tents and gas streams is readily determined with oxygen meters that measure the content of the oxygen by its paramagnetism. Oxygen in a mixture of gases may be determined in a gas chromatograph. There are a number of colorimetric tests for traces of oxygen. *See* CALORIMETRY; GAS CHROMATOGRAPHY; OXIDATION-REDUCTION.

[ARTHUR W. FRANCIS]

Bibliography: E. A. Ebsworth et al., *The Chemistry of Oxygen*, 1975; J. W. Giachino et al., *Welding Skills and Practices*, 5th ed., American Technical Society, 1977; O. Hayaishi (ed.), *Molecular Mechanisms of Oxygen Activation*, 1974; H. H. Wasserman and R. W. Murray (eds.), *Singlet Oxygen*, 1979.

Ozone

A powerfully oxidizing allotropic form of the element oxygen. The ozone molecule contains three atoms (O_3), while the more common oxygen molecule has two atoms (O_2).

Ordinary oxygen is a colorless gas and condenses to a very pale blue liquid, whereas ozone gas is decidedly blue, and both liquid and solid ozone are an opaque blue-black color, similar to that of ink. Even at concentrations as low as 4%, the blue color of ozone gas mixed with air or other colorless gas in a tube 1 in. (2.54 cm) or more in diameter and 4 ft (1.22 m) or more long can be seen by looking lengthwise through the tube.

Properties and uses. Some properties of ozone are given in the table. Ozone has a characteristic, pungent odor familiar to most persons because ozone is formed when electrical apparatus produces sparks in air. Ozone is irritating to mucous membranes and toxic to human beings and lower animals. U.S. Occupational Safety and Health Administration standards for industrial workers exposed to ozone on a daily basis limit ozone concentration to 0.1 part per million on the average, with a maximum of 0.3 ppm for short exposures.

High ozone concentrations in liquid- and gas-phase mixtures can decompose explosively when initiated by an electric spark or other high-level energy source. Controlled decomposition to reduce ozone to desirable low concentrations can be accomplished catalytically.

Ozone is a more powerful oxidizing agent than oxygen, and oxidation with ozone takes place with evolution of more heat and usually starts at a lower temperature than when oxygen is used. In the presence of water, ozone is a powerful bleaching agent, acting more rapidly than hydrogen peroxide, chlorine, or sulfur dioxide.

Ozone is utilized in the treatment of drinking-water supplies. Odor- and taste-producing hydrocarbons are effectively eliminated by ozone oxidation. Iron and manganese compounds which discolor water are diminished by ozone treatment. Compared to chlorine, bacterial and viral disinfection with ozone is up to 5000 times more rapid. After treatment, the residual chlorine content leaves a characteristic undesirable taste and odor. In addition, chlorine may yield chloroform and other trihalomethane (THM) compounds which are potentially carcinogenic.

Plants that use oxygen in aerobic digestion of sewage can add ozone treatment at reduced cost. Ozone can be produced more economically from pure oxygen. By proper integration of the facilities, oxygen not transformed into ozone in its generator passes through the ozonization tank into the aerobic digester with very high efficiency.

Ozone undergoes a characteristic reaction with unsaturated organic compounds in which the double or triple bond is attacked, even at temperatures as low as −100°C, with the formation of ozonides; these ozonides can be hydrolyzed, oxidized, reduced, or thermally decomposed to a variety of compounds, chiefly aldehydes, ketones, or carboxylic acids. Double (C=C) bonds are almost always ruptured in this reaction. Commercially ozonolysis (ozonation followed by decomposition of the ozonide) is employed in the production of azelaic acid and certain chemical intermediates used in the drug industry. *See* OZONOLYSIS.

Natural occurrence. Ozone occurs to a variable extent in the Earth's atmosphere. Near the Earth's surface the concentration is usually 0.02–0.03 ppm in country air, and less in cities except when there is smog; under smog conditions in Los Angeles ozone is thought to be formed by the action of sunlight on oxygen of the air in the presence of impurities, and on bad days the ozone concentration may reach 0.5 ppm or more for short periods of time.

At vertical elevations above 20 km, ozone is

Some properties of ozone

Property	Value
Density of the gas at 0°C, 1 atm pressure	2.154 g/liter
Density of the liquid	
−111.9°C	1.354 g/ml
−183°C	1.573 g/ml
Boiling point at 1 atm pressure	−111.9°C
Melting point of the solid	−192.5°C
Wavelength range of maximum absorption in visible spectrum	560–620 nm
Wavelength range of maximum absorption in the ultraviolet spectrum	240–280 nm

formed by photochemical action on atmospheric oxygen. Maximum concentration of 5×10^{12} molecules/cm^3 (more than 1000 times the normal peak concentration at Earth's surface) occurs at an elevation of 30 km.

Intercontinental air transports cruise at altitudes of 12 to 17 km. On flights through northern latitudes, significant concentrations (up to 1.2 ppm) of ozone have been encountered. At these levels ozone can cause coughing and chest pains, especially for cabin attendants who are actively working. Carbon filters and catalytic ozone-decomposing equipment have been installed to eliminate the problem.

Absorption of solar ultraviolet radiation by ozone provides enough energy to raise the temperature of the stratosphere (10–50 km) significantly above that of the upper troposphere. This increase of temperature with increasing height forms a stable layer resistant to vertical mixing. Gases injected into the stratosphere above 20 km may remain 2 years or longer.

By absorbing most of the short-wavelength light, the ozone layer protects human and other life forms. The layer is thinnest at the Equator, where it permits more ultraviolet radiation to reach ground levels in the torrid zone. This is believed to account for the high incidence of skin cancer in equatorial areas.

The dissociation of ozone to oxygen is catalyzed by several chemicals, especially nitrogen oxides and chlorine. Cosmic rays form nitric oxide in the stratosphere. As solar activity causes Earth's magnetic field to increase, cosmic rays are deflected away from Earth. Consequently, there is less nitric acid and more ozone immediately following the maximum phase of the solar activity cycle.

Volcanic eruptions and cosmic rays result in increased levels of chemicals which dissociate ozone. Above-normal levels of these natural events in previous geologic ages are believed to have reduced the ozone layer to 10% of normal. The resulting increase in ultraviolet radiation reaching the Earth's surface may account for the sudden extinction of some species.

Human activities also influence the ozone layer. Nuclear explosions in the atmosphere, and supersonic aircraft cruising at altitudes around 20 km, inject nitric oxide into the stratosphere. A still larger effect may be developing from the release of certain relatively stable fluorocarbons, especially $CFCl_3$ and CF_2Cl_2. This type of compound is believed to remain intact for many years in the atmosphere, permitting the gradual vertical transport from the surface into the stratosphere. Intense photochemical activity decomposes the fluorocarbon molecule, releasing chlorine atoms, each of which may destroy many ozone molecules.

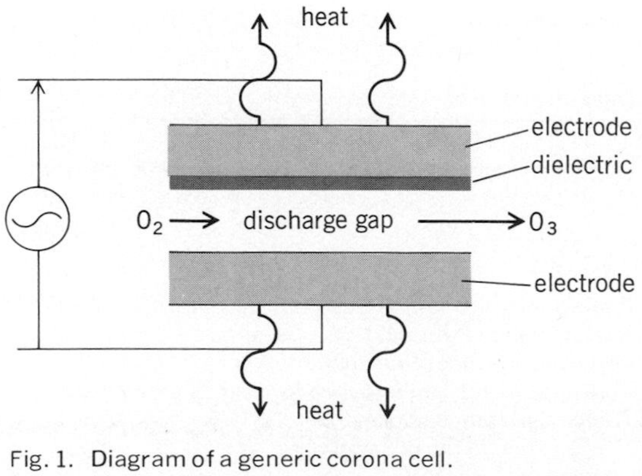

Fig. 1. Diagram of a generic corona cell.

The nature of these events is extremely complex, so that analysis of the observed data is open to varied interpretation. However, the United States government has determined that the potential danger is sufficiently serious to justify some preventive activities. Use of fluorocarbons in aerosol propellants has been prohibited in the United States. Other nations are being encouraged to extend the ban worldwide.

Preparation. The only method used to make ozone commercially is to pass gaseous oxygen or air through a high-voltage, alternating-current electric discharge called a silent electric discharge. First, oxygen atoms are formed as in reaction (1). Some of these oxygen atoms then

$$O_2 \rightarrow 2O \tag{1}$$

attach themselves to oxygen molecules as in reaction (2). The excess energy in the newly formed

$$O + O_2 + M \rightarrow O_3 + M \tag{2}$$

ozone is carried off by any available molecule (M) of gas, thus stabilizing the ozone molecule.

The corona discharge principle employed in all types of commercial ozone generators involves applying a high-voltage alternating current between two electrodes which are separated by a layer of dielectric material and a narrow gap through which the oxygen-bearing gas is passed (Fig. 1). The dielectric is necessary to stabilize the discharge over the entire electrode area, so that it does not localize as an intense arc.

A substantial fraction of the electrical energy is converted to heat. The low volume of gas flowing between the electrodes does not have sufficient capacity to remove this heat. Some external heat sink is necessary, since the decomposition of ozone is accelerated by increasing temperature.

The Lowther cell (Fig. 2*a*) is an example of a modern, plate-type, air-cooled ozone generator. An individual cell is a gastight sandwich consisting of an aluminum heat dissipator, a steel electrode coated with a highly stable ceramic dielectric, a spacer to set the width of the discharge gap, a second ceramic-coated steel electrode with an oxygen inlet, and an ozone outlet passing through a second aluminum heat dissipator. Individual cells are stacked into 30-cell modules (Fig. 2*b*), which are grouped with power supplies and controls into packaged ozonators (Fig. 3). In the concentric-tube type the oxygen or air to be ozonized passes through the annular space (about 2–3 mm across) between two tubes, one of which must be made of a dielectric material, usually glass, and the other may be either glass or a metal which does not catalyze ozone decomposition, such as aluminum or stainless steel. The internal surface of the inner tube and the external surface of the outer tube, when made of glass, are in contact with an electrical conductor such as metal foil, an electrically conducting paint, or electrically conducting water; these conductors act as electrodes. Between 5000 and 50,000 V at a frequency between 50 and 10,000 Hz is then applied across the electrodes. In some commercial ozone generators the inner and outer tubes are both water-cooled; in others only the outer tubes are water-cooled. The latter represents a simpler type of construction, but does not permit as high an input of electrical power as when both tubes are cooled.

The concentration of ozone in the gas stream leaving commercial ozone generators is usually 1–10% by weight. The yield of ozone is better when oxygen is used instead of air. Other factors which increase the yield of ozone in the silent electric discharge are thorough drying of the oxygen or air before it enters the ozonizer, refrigeration, increasing the pressure to a little above atmospheric, and increasing the frequency of the discharge.

A practical method has been developed for distribution of ozone in small quantities convenient for laboratory use. The ozone is dissolved in a liquefied fluorocarbon. The mixture is maintained at low temperature by a jacket filled with dry ice. Under these conditions, ozone decomposition proceeds very slowly, allowing sufficient time for transport to the user and a modest storage time. Rather high concentrations of ozone may be introduced safely into the laboratory in this manner.

Analytical methods. The analytical determination of ozone is usually carried out in the laboratory by bubbling the gas through a neutral solution of potassium iodide, acidifying the solution, and titrating the iodine thus liberated with standard sodium thiosulfate solution. Ozone in a gas stream may be determined automatically and continuously by passing the gas through a cell with transparent windows and measuring the absorption of either visible light or of ultraviolet radiation beamed through the cell. *See* OXIDATION-REDUCTION; OXYGEN.

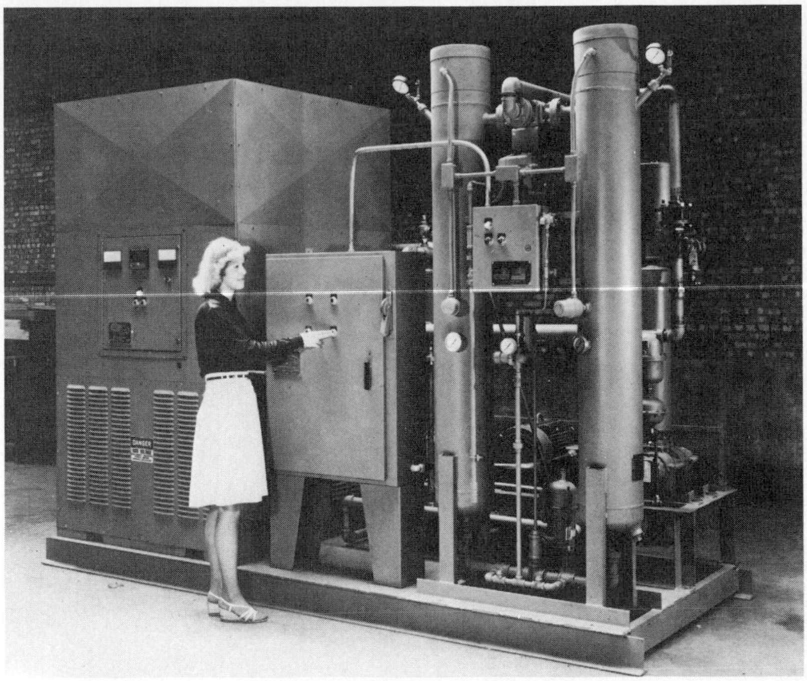

Fig. 3. Packaged ozone generator.

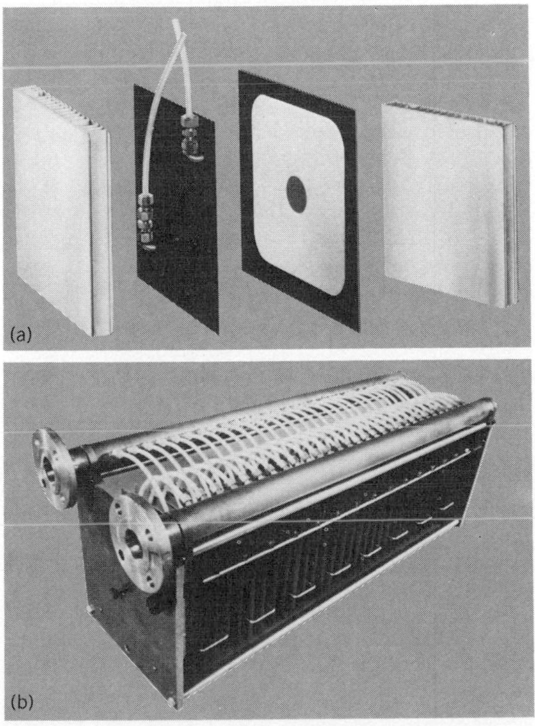

Fig. 2. Lowther cell for ozone generation: (*a*) expanded view of a single cell; (*b*) 30-cell module.

[ARTHUR W. FRANCIS]

Bibliography: S. B. Majumdar and O. J. Sproul, Technical and economic aspects of water and wastewater ozonization, *Water Res.*, 8:253–260, May 1974; J. B. Murphy and J. R. Orr (eds.), *Ozone Chemical Technology*, 1975; National Academy of Science, *Protection Against Depletion of Stratospheric Ozone by Chlorofluorocarbons*, 1979; National Academy of Science, *Stratospheric Ozone Depletion by Halocarbons*, 1979.

Ozonolysis

A process which uses ozone to cleave unsaturated organic bonds. General olefin cleavage was first extensively studied by C. Harries, beginning in 1903, as a technique to determine the structure of unsaturated compounds by identification of the cleavage products.

Generally, ozonolysis is conducted by bubbling ozone-rich oxygen or air into a solution of the reactant. The reaction is fast at moderate temperatures. Intermediates are usually not isolated but are subjected to further oxidizing conditions to produce acids or to reducing conditions to form alcohols or aldehydes. An unsymmetrical olefin is capable of yielding two different products whose structures are related to the groups substituted on the olefin and the position of the double bond.

The presently accepted mechanism of ozonolysis involves the initial formation of an unstable 1,2,3-trioxacyclopentane "primary ozonide" (I) by a 1,3-dipolar cycloaddition of ozone, as shown in one of its resonance structures, with an olefin, in reaction (1). Intermediate (I) readily decomposes to a

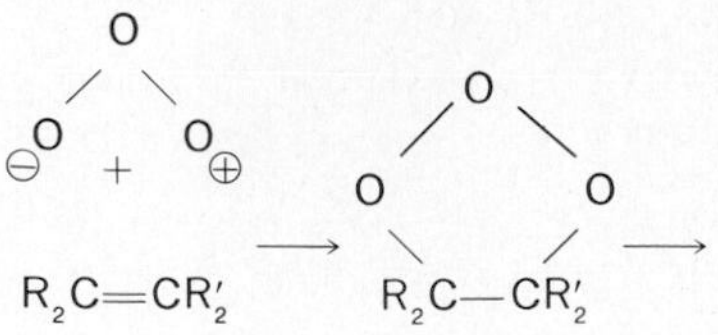

(I)

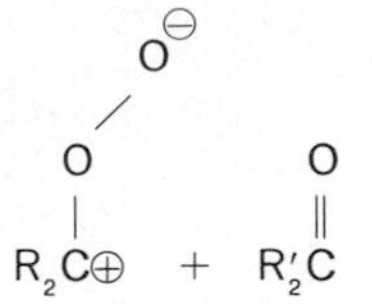

(IIa) (IIb) (1)

zwitterion key intermediate, carbonyl oxide (II*a*), and a carbonyl (II*b*). An important reaction of intermediate (II*a*) is with ketones, for example (II*b*), to form a 1,2,4-trioxacyclopentane (III), called ozonide, as in reaction (2). The intermediate (II*a*)

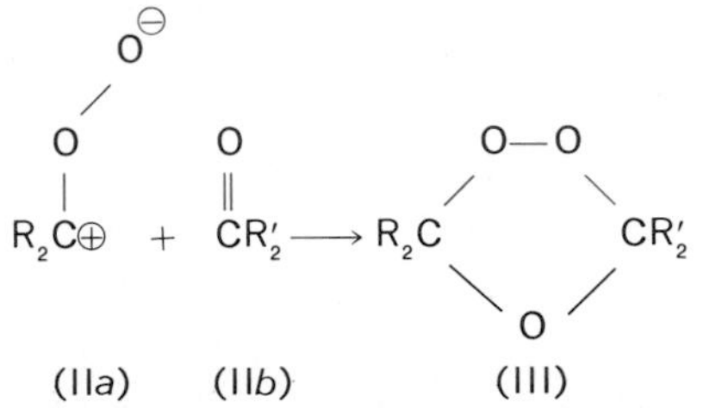

(IIa) (IIb) (III) (2)

can also dimerize to the diperoxide (IV) or polymerize to polymeric ozonides and peroxides. The zwitterion produces oxyperoxides of the general structure (V), where the reaction media contains water ($R'' = H$), ethanol ($R'' = OCH_2CH_3$), or acetic acid ($R''=OOCCH_3$).

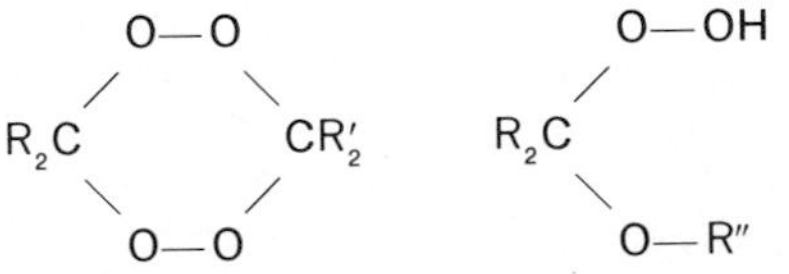

(IV) (V)

Before World War I, ozonolysis was applied commercially to the preparation of vanillin from isoeugenol. Today the only major application of the technique in the United States is in the manufacture of azelaic and pelargonic acids from oleic acid. *See* ALKENE; OZONE.

[ROBERT K. BARNES]

Bibliography: P. S. Bailey, *Ozonation in Organic Chemistry*, vol. 1, 1978.

Palladium

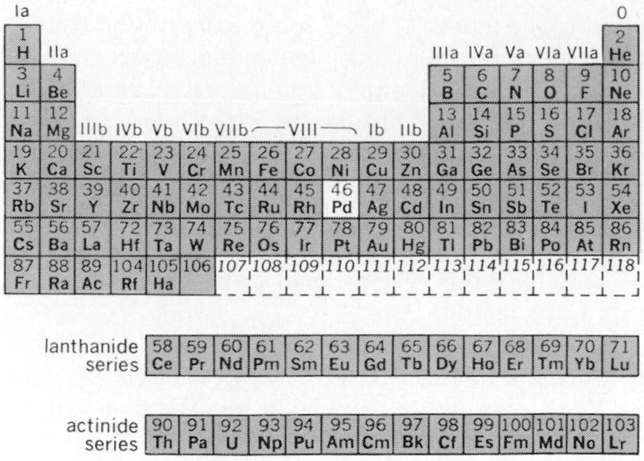

A chemical element, Pd, atomic number 46, and atomic weight 106.4. Palladium is a white and very ductile metal which resembles platinum and follows it in importance and abundance.

Uses. Palladium supported on carbon or alumina is used as a catalyst for hydrogenation and dehydrogenation in liquid and gas-phase reactions. Palladium on alumina is effective for removing traces of oxygen from hydrogen. The palladium catalyzes the reaction of hydrogen with oxygen to form water, which can be readily removed by standard gas-drying techniques. Palladium catalysts may also be used in isomerization and fragmentation reactions. *See* CATALYSIS; CRACKING.

Perhaps the largest single use of pure palladium is for low-current electrical contacts, especially in telephone equipment, where its low contact resistance, among other properties, makes for a highly reliable contact. Other electrical uses include alloys with silver for resistor wire, and with gold and platinum for use in thermocouples. Palladium is unique in the number of metals with which it can be alloyed, and it generally produces ductile solid solutions.

Palladium alloyed with silver, copper, and nickel, with or without manganese, lithium, or boron, is used for brazing. Such alloys show good melting and flow properties without excessive erosion of the materials to be joined. Palladium is used in gold-based dental alloys to improve their strength; in the jewelry field, palladium is used alloyed with 4.5% ruthenium. Palladium-ruthenium alloys have also been used in dental restorations. Addition of about 0.1% palladium to titanium has been found effective in improving the latter's corrosion resistance in HCl. For electrical contacts, printed circuits, and other uses, palladium is readily electroplated in thin or thick deposits. It may also be deposited from organometallic compounds or by suspension in organic vehicles. The latter process may also be applied to palladium alloys; palladium-silver alloy resistors, which include a glassy frit, are made in this way.

Metallurgical extraction. In one refining process, gold, platinum, and palladium are separated from the other platinum metals and silver by solution in aqua regia. Gold and platinum are then removed, resulting in a filtrate containing palladium(II) chloride. This is treated with ammonium hydroxide and heated to form a solution of tetrammine palladium chloride. Dichlorodiammine palladium is then precipitated with hydrochloric acid. This is quite soluble in cold dilute ammonia. Palladium may also be extracted through use of its ability to form insoluble cyanide, iodide, and dimethylglyoxime derivatives.

The separation of palladium, platinum, and rhodium from the other platinum metals may be effected by fusing with lead. Ruthenium, osmium, and iridium are insoluble in lead except at very high temperatures, whereas the other platinum metals, as well as gold and silver, dissolve. *See* PLATINUM.

Physical and chemical properties. Palladium is soft and ductile and may be fabricated into fine wire and thin sheet. On heating at temperatures up to about 800°C, palladium forms a visible oxide tarnish, PdO. This oxide is thin and adherent, not tending to scale or flake off. Above 800°C, the oxide dissociates and the metal is bright if cooled quickly to room temperature. Above the dissociation temperature palladium will gain weight as a result of oxygen solution and will lose weight primarily because of metal vaporization, rather than oxide volatilization, the latter being the case for all the other platinum metals. Certain readily oxidizable elements in palladium, such as chromium or titanium, will precipitate out in the solid state as oxides, reacting with dissolved oxygen (internal oxidation). The values of important physical properties are given in the table.

Hydrogen is readily absorbed by palladium and will diffuse through heated palladium at a relatively rapid rate. This property is used in hydrogen purifiers, which pass hydrogen but no other gas.

Palladium in ordinary atmospheres is resistant to tarnish but will tarnish slightly in outdoor exposure to sulfur-contaminated atmospheres. At room temperature palladium is resistant to hydrofluoric, phosphoric, perchloric, acetic, hydrochloric, and sulfuric acids as gases, but may be attacked by some of them at 100°C or when air is present. The metal is readily attacked by nitric acid, ferric chloride, and hypochlorite solutions and by moist chlorine, bromine, and iodine.

Molten sodium and potassium nitrate do not attack palladium, but molten sodium peroxide, hydroxide, and carbonate show some attack, as does H_2S, above 600°C.

Compounds. Palladium chlorides and related compounds are the most important. These include $PdCl_2$, tetrachloropalladates (Na_2PdCl_4), and tetrammine palladium chloride. Palladium chloride is used in electrodeposition; it and related chlorides are used in refining cycles and as sources of pure palladium sponge in thermal decomposition processes. Palladium nitrate and nitrite compounds are also useful for much the same purposes.

Palladium monoxide, PdO, and dihydroxide, $Pd(OH)_2$, are used as sources of palladium catalysts. Sodium tetranitropalladate, $Na_2Pd(NO_2)_4$, and other complex salts are used as the basis of plating baths. [HENRY J. ALBERT]

Bibliography: *See* PLATINUM.

Physical properties of palladium

Property	Value
Atomic weight $C^{12} = 12.00000$	106.4
Naturally occurring isotopes and % abundance	102 (.96) 104 (10.97) 105 (22.23) 106 (27.33) 108 (26.71) 110 (11.81)
Crystal structure	Face-centered cubic
Lattice constant *a*, at 25°C, angstroms	3.8898
c/a, at 25°C	
Thermal neutron capture cross section, barns	8.0
Common chemical valence	2, 4
Density at 25°C, g/cm³	12.01
Melting point, °C	1554
Boiling point, °C	2900
Specific heat at 0°C, cal/g	0.0584
Thermal conductivity, 0–100°C, cal cm/cm² sec°C	0.18
Linear coefficient of thermal expansion, 20–100°C, μin./in./°C	11.6
Electrical resistivity at 0°C, μohm-cm	9.93
Temperature coefficient of electrical resistance, 0–100°C/°C	0.0038
Tensile strength, 1000 psi	
Soft	21–33
Hard	47–60
Young's modulus at 20°C	
psi, static	16.7×10^6
psi, dynamic	17.6×10^6
Hardness, DPN*	
Soft	37–44
Hard	105–110

*Diamond pyramid number.

Palmitate

A salt (soap) or ester of palmitic acid,

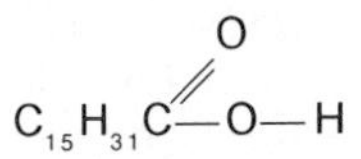

in which the acidic hydrogen has been replaced by a metal or an organic radical. Palmitates occur in nature, chiefly as the glyceryl esters, found in substantial amounts in animal and vegetable fats. The esters of long-chain alcohols are known as waxes. Simple esters have limited uses, chiefly in plastics. Alkali metal palmitates are soluble in water and, with the similar stearates and oleates, are the major components of toilet and laundry soaps. Other metal soaps are used in lubrication greases, pharmaceuticals, and cosmetics, and as waterproofing agents and fungicides. *See* CARBOXYLIC ACID; ESTER.

[ELBERT H. HADLEY]

Pentaerythritol

A white crystalline compound with the formula shown below. Four methylol groups are arranged

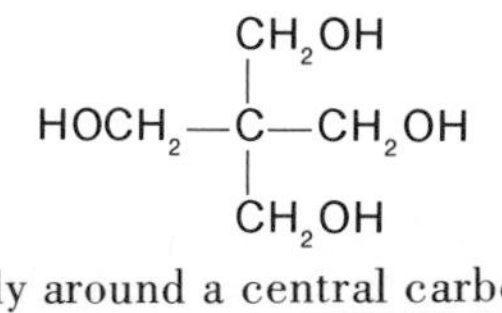

symmetrically around a central carbon atom. It is only slightly soluble in alcohols and other organic solvents but is moderately soluble in cold water and freely soluble in hot water. Pentaerythritol melts at 261–262°C.

Pentaerythritol, also called tetramethylolmethane, is made by reaction of formaldehyde and acetaldehyde in the presence of either calcium hydroxide or sodium hydroxide. During World War II appreciable quantities were used to make the explosive, pentaerythritol tetranitrate (PETN); most of it enters into the manufacture of alkyd resins and other coating compositions. It is also being used to prepare esters with different organic acids for application in specialty lubricants. *See* POLYHYDRIC ALCOHOL. [JOSEPH A. VONA]

Perbromate

The negative ion BrO_4^-. Although perchlorates and periodates, compounds of seven-valent chlorine and iodine, have been known since early in the 19th century, the analogous perbromates have long resisted synthesis, and several theoretical rationalizations were developed to account for their nonexistence. The first successful synthesis of a perbromate, in 1968, resulted from a hot-atom process, the beta decay of radioactive ^{83}Se incorporated into a selenate, as shown in reaction (1). Perbro-

$$^{83}SeO_4^{--} \rightarrow {}^{83}BrO_4^- + \beta^- \quad (1)$$

mates were subsequently prepared by electrolytic oxidation of bromate and by its chemical oxidation with either XeF_2 or molecular fluorine. The oxidation of bromate with molecular fluorine in alkaline solution is the most practical of these methods.

The perbromate ion, BrO_4^-, is tetrahedral, like the perchlorate and meta periodate ions. Perbromic acid is a very strong acid, and its potassium, rubidium, and cesium salts have relatively low solubilities in water. Aqueous solutions of the alkali perbromates are completely stable, while perbromic acid itself is stable up to concentrations of 6*M*. Above this concentration, decomposition to Br_2 and O_2 tends to occur, although an unstable 12-*M* solution of the acid can be prepared. This is an azeotrope corresponding approximately to the composition of the dihydrate $HBrO_4 \cdot 2H_2O$.

The half-reaction (2) has a standard electrode

$$BrO_4^- + 2H^+ + 2e^- \rightleftarrows BrO_3^- + H_2O \quad (2)$$

potential of 1.74 V, making perbromate a slightly stronger oxidant than periodate. However, under ordinary conditions perbromates are very sluggish oxidizing agents, being among the least reactive of all the oxyhalogen compounds. The difficulties encountered in preparing the perbromate ion may be understood in terms of its high electrode potential and its high degree of inertness, which combine to present a substantial activation barrier to its formation.

Perbromyl fluoride, BrO_3F, is the acid fluoride of perbromic acid and is prepared by the reaction of $KBrO_4$ with antimony pentafluoride in anhydrous HF, in accordance with reaction (3). It is a reactive

$$BrO_4^- + 2SbF_5 + 3HF \rightarrow BrO_3F + H_3O^+ + 2SbF_6^- \quad (3)$$

gas at room temperature, but it is the most stable of all the bromine oxides and oxyfluorides.

[EVAN H. APPELMAN]

Bibliography: E. H. Appelman, Non-existent compounds: Two case histories, *Acc. Chem. Res.*, 6:113–117, 1973.

Perchlorate

A compound which contains chlorine in the 7+ oxidation state and which is derived from perchloric acid, $HClO_4$. Perchlorates are more stable than chlorates, chlorites, or hypochlorites but are nevertheless excellent oxidizing agents. On heating, perchlorates decompose into potassium chloride, KCl, and oxygen gas. Because of their oxidizing properties, perchlorates find use in explosives and as oxidizing agents in the laboratory.

Potassium perchlorates can be prepared by heating potassium chlorate, reaction (1). In a side reaction some oxygen is liberated, reaction (2).

$$4KClO_3 \xrightarrow{\Delta} 3KClO_4 + KCl \quad (1)$$

$$2KClO_3 \rightarrow 2KCl + 3O_2 \quad (2)$$

Sodium perchlorate is also prepared commercially by an electrochemical process. Sodium chlorate made from sodium chloride and chlorine is the starting material. The sodium chlorate is then electrolyzed; the net reaction at the anode is shown in reaction (3). *See* CHLORATE.

$$ClO_3^- + H_2O \rightarrow ClO_4^- + 2H^+ + 2e \quad (3)$$

Sodium perchlorate is then converted to perchloric acid or other metallic salts. *See* CHLORINE; HYPOCHLORITE; OXIDIZING AGENT.

[E. EUGENE WEAVER]

Periodate

A salt which contains iodine in the 7+ oxidation state and which is derived from periodic acid. Periodic acid is known in three forms: metaperiodic acid, HIO_4; dimesoperiodic acid, $H_4I_2O_9$; and paraperiodic acid, H_5IO_6. The corresponding salts are also known. Only the periodates of sodium and potassium are important. Sodium metaperiodate, $NaIO_4$, and potassium metaperiodate, KIO_4, are used as oxidizing agents in analytical chemistry. *See* IODINE.

[E. EUGENE WEAVER]

Periodic table

A table of the elements, written in sequence in the order of atomic number or atomic weight and arranged in horizontal rows (periods) and vertical columns (groups) to illustrate the occurrence of similarities in the properties of the elements as a periodic function of the sequence. The concise table used in the element articles in this encyclopedia is shown below.

Ia	IIa	IIIb	IVb	Vb	VIb	VIIb	VIII			Ib	IIb	IIIa	IVa	Va	VIa	VIIa	0
1 H																	2 He
3 Li	4 Be											5 B	6 C	7 N	8 O	9 F	10 Ne
11 Na	12 Mg											13 Al	14 Si	15 P	16 S	17 Cl	18 Ar
19 K	20 Ca	21 Sc	22 Ti	23 V	24 Cr	25 Mn	26 Fe	27 Co	28 Ni	29 Cu	30 Zn	31 Ga	32 Ge	33 As	34 Se	35 Br	36 Kr
37 Rb	38 Sr	39 Y	40 Zr	41 Nb	42 Mo	43 Tc	44 Ru	45 Rh	46 Pd	47 Ag	48 Cd	49 In	50 Sn	51 Sb	52 Te	53 I	54 Xe
55 Cs	56 Ba	57 La	72 Hf	73 Ta	74 W	75 Re	76 Os	77 Ir	78 Pt	79 Au	80 Hg	81 Tl	82 Pb	83 Bi	84 Po	85 At	86 Rn
87 Fr	88 Ra	89 Ac	104 Rf	105 Ha	106	*107*	*108*	*109*	*110*	*111*	*112*	*113*	*114*	*115*	*116*	*117*	*118*

lanthanide series	58 Ce	59 Pr	60 Nd	61 Pm	62 Sm	63 Eu	64 Gd	65 Tb	66 Dy	67 Ho	68 Er	69 Tm	70 Yb	71 Lu
actinide series	90 Th	91 Pa	92 U	93 Np	94 Pu	95 Am	96 Cm	97 Bk	98 Cf	99 Es	100 Fm	101 Md	102 No	103 Lr

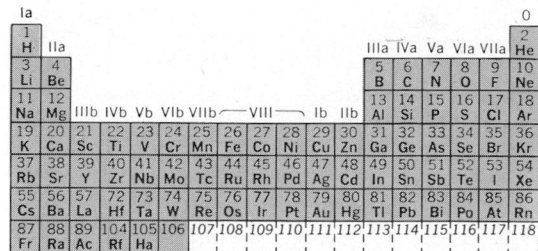

Each element, represented by its symbol and atomic number, occupies a separate square, and the sequential arrangement is in the order of atomic number. The table divides the elements into nine groups, designated by numerical column headings, and seven periods. Seven of the nine groups have frequently been divided into a and b categories to distinguish main-group and subgroup elements. The a and b designations have led to certain ambiguities, however, and other systems are being developed.

Two rows (lanthanum, or rare-earth, series and actinium series), which on the whole are best classified as members of group IIIa, occupy special positions outside the main body of the table, as they cannot be included conveniently in periods six and seven. The elements beyond lawrencium (atomic number 103) presumably should include a third group, the superactinide group, outside the main body of the table. The elements whose atomic numbers are in parentheses have not yet been synthesized and occupy predicted positions in the table. *See* ACTINIDE ELEMENTS; TRANSURANIUM ELEMENTS.

The following sections serve to illustrate some of the correlative features of periodicity.

Valence. In general, elements in the same group display a similar valence, which is numerically equal to the group number. The rule holds most firmly for the main groups I through V, less so for the subgroup elements, and still less for the elements of groups VI, VII, and VIII. Among these latter, the group number valence is more readily achieved by the heavier, as compared with the lighter, elements. (Group 0 elements are distinguished by an almost complete lack of chemical reactivity and hence show zero valence.)

Metals and nonmetals. The periodic table effects a natural division of the elements in their elemental or uncombined state into metals and nonmetals. The nonmetals are confined to the lighter elements of groups IVa through VIIa. Between the area occupied by the typically metallic elements and that occupied by the nonmetals there is a somewhat ill-defined borderland of elements (germanium Ge, arsenic As, antimony Sb, tellurium Te, and polonium Po) whose properties are transitional between the metallic and nonmetallic elements. These borderline elements are sometimes called metalloids.

Acid-base properties of oxides. Associated in a general way with the degree of metallic character of an element is a property called electropositivity, which expresses the tendency of an element to form positive ions by losing electrons. This property is exhibited in highest degree by the heaviest elements of group I and diminishes generally on proceeding to elements which lie above and to the right of this position. Highly electropositive elements form oxides or hydroxides which are strong bases. Potassium hydroxide (caustic potash) is a familiar example. The nonmetals, such as sulfur, attract electrons and hence are termed electronegative. Solutions of the oxides of these elements in water yield acids. Sulfur trioxide, for example, reacts with water to form sulfuric acid.

Atomic volumes. The volume occupied by one gram-atomic weight of an element in the solid state is called its atomic volume. If this property is plotted against atomic number, the resultant curve exhibits periodic maxima and minima, the former being due principally to the metals of group I and the latter to the metals of group VIII.

Other properties. The preceding sections afford only a limited illustration of the immense correlative power of the periodic arrangement. Other properties which have been shown to exhibit analogous trends and periodicities include oxidation potential, heat of formation of type compounds, electrical conductivity, melting point, boiling point, ionic radius, ionization potential, electron affinity, optical spectrum, and magnetic behavior.

The periodic table was developed and largely perfected by Dmitri Mendeleev in the mid-19th century as an empirical correlation between the chemical properties of the elements and their atomic weights. In the light of modern knowledge, no such correlation is to be expected, since the atomic weight is determined almost entirely by the mass of the atomic nucleus which plays only an insignificant role in chemical reactions.

With few exceptions, however, the sequence of elements according to atomic weight is identical with that according to atomic number. This latter quantity is numerically equal to the number of extranuclear electrons in the neutral, or uncharged, atom of an element. It is the electrons, particularly those furthest removed from the nucleus, which determine chemical behavior. The laws of atomic architecture, discovered early in the 20th century, require that in a sequence of atoms of regularly increasing atomic number there be a periodic recurrence in the number and type of electrons in the outermost electronic shell. This forms the true basis of the periodic variation of the properties of the elements.

Few systemizations in the history of science can rival the periodic concept as a broad revelation of the order of the physical world. In the rhythmic pattern of the properties of the elements, the architectural units of the universe, no aspect of their behavior changes in a capricious or wholly novel way. Whatever new elements may be discovered in the future, it is certain they will find a place in the periodic system, conforming to its order and exhibiting the proper familial characteristics. For example, in 1974 scientists at the Lawrence Berkeley Laboratory at the University of California reported the synthesis of a new element, element 106. Their report placed the new substance in the family of elements that includes chromium, molybdenum, and tungsten. The team of scientists did not propose a name for the new element at that time because a group of scientists in the Soviet Union reported the synthesis of another form of element 106 simultaneously. *See* VALENCE.

[GLENN T. SEABORG]

Bibliography: L. Pauling and P. Pauling, *Chemistry*, 1975; R. J. Puddephat, *The Periodic Table of the Elements*, 1977; G. T. Seaborg and E. G. Valens, *Elements of the Universe*, 1958.

Permanganate

The deep purple anion, MnO_4^-, which is derived from permanganic acid, $HMnO_4$. Although the parent acid is stable only in dilute aqueous solution, the salts are well characterized. The permanganates resemble the perchlorates in their oxidizing properties and solubility of both the heavy metal and alkaline-earth metal salts.

Potassium permanganate, the most common permanganate, is produced from a mixture of potassium hydroxide and manganese dioxide which has been oxidized by potassium chlorate, chlorine, or ozone. Permanganates are used as disinfectants, oxidizing agents, wood preservatives, and bleaching agents. *See* MANGANESE; OXIDIZING AGENT.

Peroxide

A chemical compound which contains the peroxy (—O—O—) group, which may be considered to be a derivative of hydrogen peroxide (HOOH). An organic (or inorganic) peroxide is one in which some organic (or inorganic) substituent has replaced one or both hydrogens. Peroxides are used in such diverse reactions as oxidation, synthesis, polymerization, and oxygen generation. Inorganic peroxides include persulfates, hydrogen peroxide (H_2O_2), sodium peroxide, bivalent metal peroxides, and H_2O_2 addition compounds. Organic peroxides include per (oxy) acetic acid, dibenzoyl peroxide, and cumene peroxide.

Inorganic peroxides. Peroxydisulfates, familiarly called persulfates, are produced by electrolytic oxidation of aqueous sulfuric acid or ammonium bisulfate. Persulfuric acid ($H_2S_2O_8$) is not used commercially as such. Ammonium persulfate, a white solid, is used as a polymerization catalyst, dyestuff oxidant, metal etchant, and laboratory oxidant. The potassium salt is used extensively in the manufacture of styrene-butadiene synthetic rubber; smaller quantities are used in hair bleaches.

Peroxymonosulfates, also called monopersulfates, are salts of peroxymonosulfuric acid (Caro's acid). Peroxymonosulfuric acid, made from H_2O_2 and sulfuric acid, is a powerful oxidant and bleach; these properties are shared by the salts. The acid can be used for making wool resistant to shrinkage; the salts are effective bleaching agents for domestic laundering.

Peroxydiphosphates, analogous to peroxydisulfates, have been prepared by electrolytic oxidation of concentrated phosphate solutions; no major technical uses have been reported. Peroxydicarbonates have also been made electrolytically.

Hydrogen peroxide. The most widely used peroxy compound is hydrogen peroxide, a waterlike liquid manufactured as aqueous solutions of 35–90% H_2O_2 by weight; essentially anhydrous H_2O_2 has become commercially available. H_2O_2 is not combustible; water is a safe diluent and coolant. With organic compounds, H_2O_2 can form detonable mixtures; industrial processes guard against their generation. Directions for safe handling and storage of H_2O_2 are available from producers, government agencies, and trade associations.

Hydrogen peroxide is manufactured by electrolytic and organic oxidation processes. The former involves electrolytic production of the peroxydisulfate intermediate, followed by steam hydrolysis to H_2O_2, with regeneration of the original sulfuric acid or ammonium bisulfate raw materials. One organic process uses an anthraquinone dissolved in organic solvents. The quinone is catalytically hydrogenated to the hydroquinone; subsequent aeration of the hydroquinone regenerates the quinone, with simultaneous formation of H_2O_2. The H_2O_2 is water-extracted and concentrated; the quinone is recycled for reconversion to hydroquinone. A second organic process uses liquid isopropyl alcohol, which is oxidized at moderate temperatures and pressures to H_2O_2 and acetone coproducts. After distillation of the acetone and unreacted alcohol, the residual H_2O_2 is concentrated.

Hydrogen peroxide applications include commercial bleaching, dye oxidation, manufacture of organic and peroxy chemicals, and power generation. Bleaching outlets consume more than one-half of the H_2O_2 produced. These outlets include textile mill bleaching of practically all wool and cellulosic fibers, as well as of major quantities of synthetics, and paper and pulp mill bleaching of groundwood and chemical pulps. *See* BLEACHING.

Organic applications include manufacture of epoxides and glycols from unsaturated petroleum hydrocarbons, terpenes, and natural fatty oils. The resultant products are valuable plasticizers, stabilizers, diluents, and solvents for vinyl plastics and protective coating formulations. Production of tonnage organic chemicals may become feasible via low-cost H_2O_2; synthetic glycerol production, using captive H_2O_2, has been planned by one H_2O_2 producer. H_2O_2 outlets for manufacture of peroxides include inorganic compounds such as sodium perborate, and organic compounds such as peracetic acid and dibenzoyl peroxide. Certain peroxides, such as that of sodium, are more economically produced by air oxidation. Power generation applications include use in specialized propulsion units for aircraft, missiles, torpedoes, and submarines. The hot oxygen-steam mixture from catalytically decomposed H_2O_2 powers the feed pumps of many large liquid-propellant rockets.

Metal peroxides. Sodium peroxide (NaOONa), a yellowish powder, is the peroxide produced in second largest amount for direct sale. Annual production in the United States is about 12,000,000 lb (5.4×10^6 kg). Manufacture is via a two-stage reaction of the elements. The sodium monoxide first formed from the reaction of liquid or solid sodium and dried air in rotary steel burners is converted to the peroxide by additional reaction at 250–400°C. From their respective active oxygen contents, Na_2O_2 and H_2O_2 are competitive in price. Aqueous solutions of Na_2O_2 are essentially equivalent to a mixture of caustic soda and H_2O_2. Contact of the powder with skin or combustibles must be avoided to minimize the danger of burns or fire. To smother a fire caused by Na_2O_2, salt or sand instead of water must be used. Major uses of Na_2O_2, as with H_2O_2, are in bleaching processes in the textile and in the pulp and paper industries. In certain areas, the two chemicals are competitive; in others, the choice is dictated by the particular application.

Bivalent metal peroxides that are commercially available include those of barium, calcium, magnesium, strontium, and zinc. Those of barium and strontium can be obtained by roasting the metal in air or oxygen; the others are best made by reaction with H_2O_2 of a solution of a salt or a slurry of oxide or hydroxide. Large-scale commercial uses have not been developed. The barium and strontium compounds are used for coloring flames in pyrotechnics. Calcium peroxide is used as a dough conditioner in the baking industry; magnesium and zinc peroxides are used cosmetically as deodorants and antiperspirants.

Inorganic peroxy anion compounds are readily synthesized from H_2O_2 and solutions of various metal anions (pertitanates, perchromates). Their importance is chiefly in the analytical area; some have importance as catalysts.

Hydroperoxidates are solid addition compounds of H_2O_2 with other materials. Sodium perborate "tetrahydrate" ($NaBO_2 \cdot H_2O_2 \cdot 3H_2O$) and "monohydrate" ($NaBO_2 \cdot H_2O_2$) are convenient sources of H_2O_2 when dissolved in water. Manufacture is by reactions of borax, caustic soda, and H_2O_2. Perborates are used in sizable amounts in household powdered bleaches for the home laundry, and in the textile industry for oxidation of vat and sulfur dyes on cotton and rayon. The hydroperoxidates of sodium carbonate, sodium pyrophosphate, and urea are also available. Fields of use for the hydroperoxidates include cosmetics and photography. *See* HYDROGEN PEROXIDE.

Organic peroxides. As a group, organic peroxides are more hazardous than the inorganic peroxides. Many of the former are flammable or detonable, thus restricting their availability. Some are exceptionally stable (di-*tert*-butyl peroxide). Manufacture is chiefly through reaction of the organic substrate with H_2O_2; air oxidation is also used when feasible.

Peracetic acid [$CH_3(C{=}O)OOH$], prepared from acetic acid and H_2O_2, is the only organic peracid offered commercially (40% by weight in acetic acid). A manufacturing process involving air oxidation of acetaldehyde has been developed. Peracetic acid, as well as performic or perpropionic, may also be generated at the site from H_2O_2, organic acid, and catalyst. Major applications of peracetic acid are in the synthesis of epoxidized and hydroxylated compounds and as a bactericide, fungicide, and sterilizing agent for processing equipment.

Dibenzoyl peroxide, the most important aromatic acyl peroxide, is a white powder, stable at room temperature and explosible with heating. It is manufactured by reaction of benzoyl chloride and alkaline H_2O_2. Major uses include polymer manufacture (0.1–0.2% in the monomer to initiate the polymerization) and flour bleaching (when mixed with phosphates and other ingredients meeting standards for flour-treating formulations).

Cumene hydroperoxide [$C_6H_5C(CH_3)_2OOH$], a colorless to pale-yellow liquid, produced by air oxidation of isopropyl benzene, is no longer used extensively as a polymerization catalyst. However, its ready cleavage in acid solution to phenol and acetone has rendered it a tonnage intermediate in the commercial production of phenol. *See* ELECTROCHEMICAL PROCESS; OXIDATION PROCESS; OXIDATION-REDUCTION; OXIDIZING AGENT; OXYGEN.

[SAMUEL S. NAISTAT]

Bibliography: F. A. Cotton and G. Wilkinson, *Advanced Inorganic Chemistry: A Comprehensive Text*, 4th ed., 1980; D. Swern (ed.), *Organic Peroxides*, vol. 3, 1972, reprint 1980.

Persulfate

A group of compounds more correctly known as peroxysulfates. Persulfates are salts of the two peroxy acids of sulfur, peroxymonosulfuric acid (H_2SO_5) and peroxydisulfuric acid ($H_2S_2O_8$). All these compounds contain sulfur in an oxidation state of 6+, the same as in sulfates. The unusual thing about their structure is the presence of the peroxy group, —O—O—, as shown below.

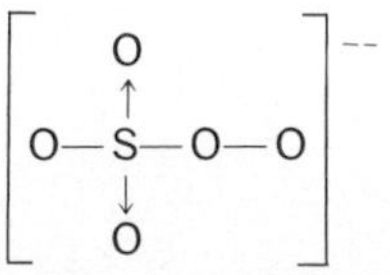
Peroxymonosulfate ion

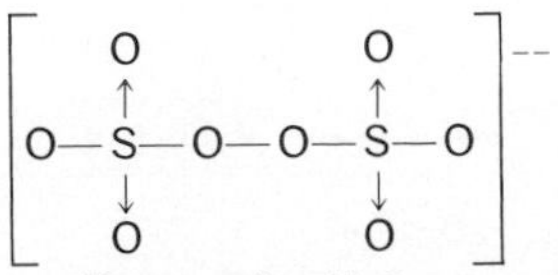
Peroxydisulfate ion

Ammonium peroxydisulfate is produced by the electrolytic oxidation of ammonium bisulfate solution. It is used as an oxidizing agent and a bleach, and is converted to other salts, such as potassium peroxydisulfate which is used as a polymerization promoter. *See* PEROXIDE; SULFUR.

[E. EUGENE WEAVER]

Petrochemical

One of a large number of substantially pure chemical substances produced commercially from petroleum or natural gas. Ordinarily the term does not include hydrocarbon fuels and lubricants nor chemicals produced by other than the processor handling the petroleum raw material. Organic compounds make up the great bulk, as well as number, of petrochemicals, but several inorganic compounds (ammonia, carbon black, sulfur, and hydrogen peroxide) also are produced in large amounts.

Petrochemicals should not be regarded as a particular type or class of chemical, since all of them have been, and many still are, made from other raw materials; for example, benzene, phenol, naphthalene, and acetylene are made from coal, glycerol from fats, ethyl alcohol from agricultural crops, and sulfur from deposits of the element or from metal ores.

A majority of the chemicals once made from other raw materials came to be made entirely, or almost entirely, from petroleum or natural gas. Examples are acetone, originally derived from wood distillation and later by fermentation of agri-

Table 1. Petrochemicals from methane

Basic derivatives and sources	Uses
Ammonia	Agricultural chemicals (as ammonia, salts, urea)
Petroleum sources	Fibers, plastics
Methane hydrogen	Industrial explosives
Refinery hydrogen	Other
Electrolytic, coal	
Carbon black	Rubber compounding
Liquid petroleum	Printing ink, paint, and other
Natural gas	
Methanol	Formaldehyde (mainly for resins)
Petroleum sources	Methyl esters (polyester fibers), amines, and other chemicals
Methane	Solvents
Propane-butane	Other
Coal	
Chloromethanes	Chlorofluorcarbons for refrigerants, aerosols, solvents, cleaners, grain fumigant
Methane chlorination	
Other sources	
Acetylene	Vinyl chloride, vinyl acetate
Petroleum	Chloroprene (neoprene)
Calcium carbide	Chloroethylenes
	Acrylonitrile
	Other chemical uses
	Nonchemical uses
Hydrogen cyanide	Acrylonitrile
	Adiponitrile
	Methyl methacrylate
	Other

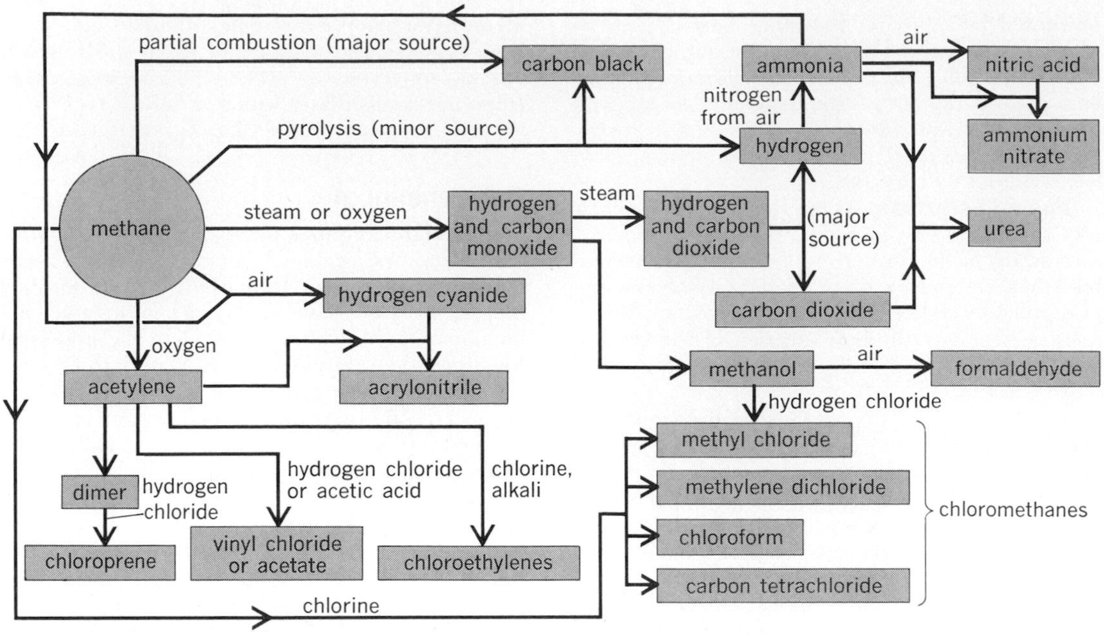

Fig. 1. Petrochemicals from methane.

cultural products; ethyl chloride, originally made from ethyl alcohol produced by fermentation; and butadiene, made solely from ethyl alcohol during World War I and, as a stopgap measure, about half from ethyl alcohol during World War II, but during the period 1946–1980 derived almost completely from petroleum. However, it is anticipated that processes using other raw materials will become increasingly important.

Growth. The overall trend to petroleum and natural gas as the predominant source of organic chemicals in the United States was related to the

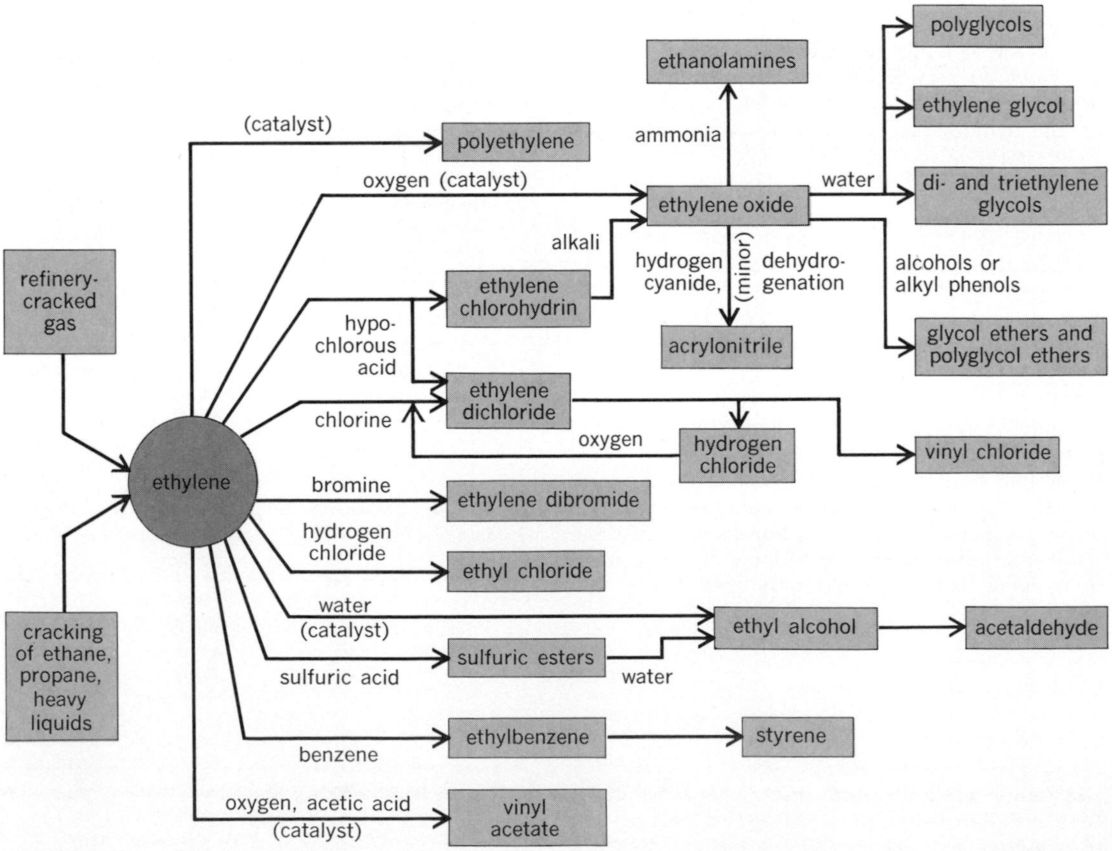

Fig. 2. Petrochemicals from ethylene.

Table 2. Petrochemicals from ethylene

Basic derivatives and sources	Uses
Ethylene oxide Via direct oxidation Via chlorohydrin	Ethylene glycol (polyester fiber and resins, antifreeze) Di- and triethylene glycols Ethanolamines Nonionic detergents Glycol esters Other
Ethyl alcohol (synthetic)	Acetaldehyde Solvents Ethyl acetate Other
Polyethylene Low density High density	 Film, injection molding Blow molding, injection molding
Styrene From ethylene and benzene (or reformate)	Polystyrene and copolymer resins Styrene-butadiene rubber and latex Polyesters Other
Ethylene dichloride By direct chlorination and oxychlorination By-product of ethylene chlorohydrin production	Vinyl chloride Scavenger in antiknock fluids Ethyleneamines and others
Ethyl chloride Ethylene + HCl Chlorination of ethane	Tetraethyllead Minor amounts for ethylations (such as ethyl cellulose)
Ethylene dibromide	Scavenger in antiknock fluid
Acetyls (from ethylene only) Acetaldehyde, vinyl acetate; new technology expected to grow rapidly	Plastics and chemical intermediates
Linear alcohols and α-olefins (ethylene only)	Detergents, plasticizers

abundance of supplies and relatively low cost. Two important inorganic petrochemicals, ammonia and carbon black, are also made from this source.

There were no petrochemicals before 1920 aside from carbon black, which has been made in the United States from natural gas since 1872. The growth of petrochemicals since 1920 was most spectacular in the field of aliphatics, which account for well over half of the total volume of petrochemicals and by far the largest number of the petrochemical-derived compounds. This growth was due mainly to the following five factors:

1. The abundance and relatively low cost of crude oil, natural gas, and natural gas liquids.

2. Advances made in the technology of petroleum refining, spurred especially by the demand for motor gasoline and aviation gasoline. These advances included more efficient fractional distillation and other separation processes, particularly for lower-boiling constituents, and the development of conversion processes to increase gasoline quality and quantity, including thermal and catalytic cracking, hydrocracking, catalytic reforming, hydrogenation, dehydrogenation, isomerization, and alkylation.

3. A demand for chemicals that in many cases could not be supplied in sufficient amounts, or with sufficient assurance of a steady supply and price structure, from other sources—coal, wood, or agricultural products.

4. The rapid growth in demand for synthetic fibers, plastics and resins, and protective coatings.

5. Research, leading to products not before known commercially and to lower-cost processes for established products, often greatly expanding their application. A special characteristic of petrochemicals research has been the development of a series of related derivatives from the primary petrochemicals in order to establish the widest possible market. *See* ALKYLATION; CATALYTIC REFORMING; CRACKING; ORGANIC CHEMICAL SYNTHESIS.

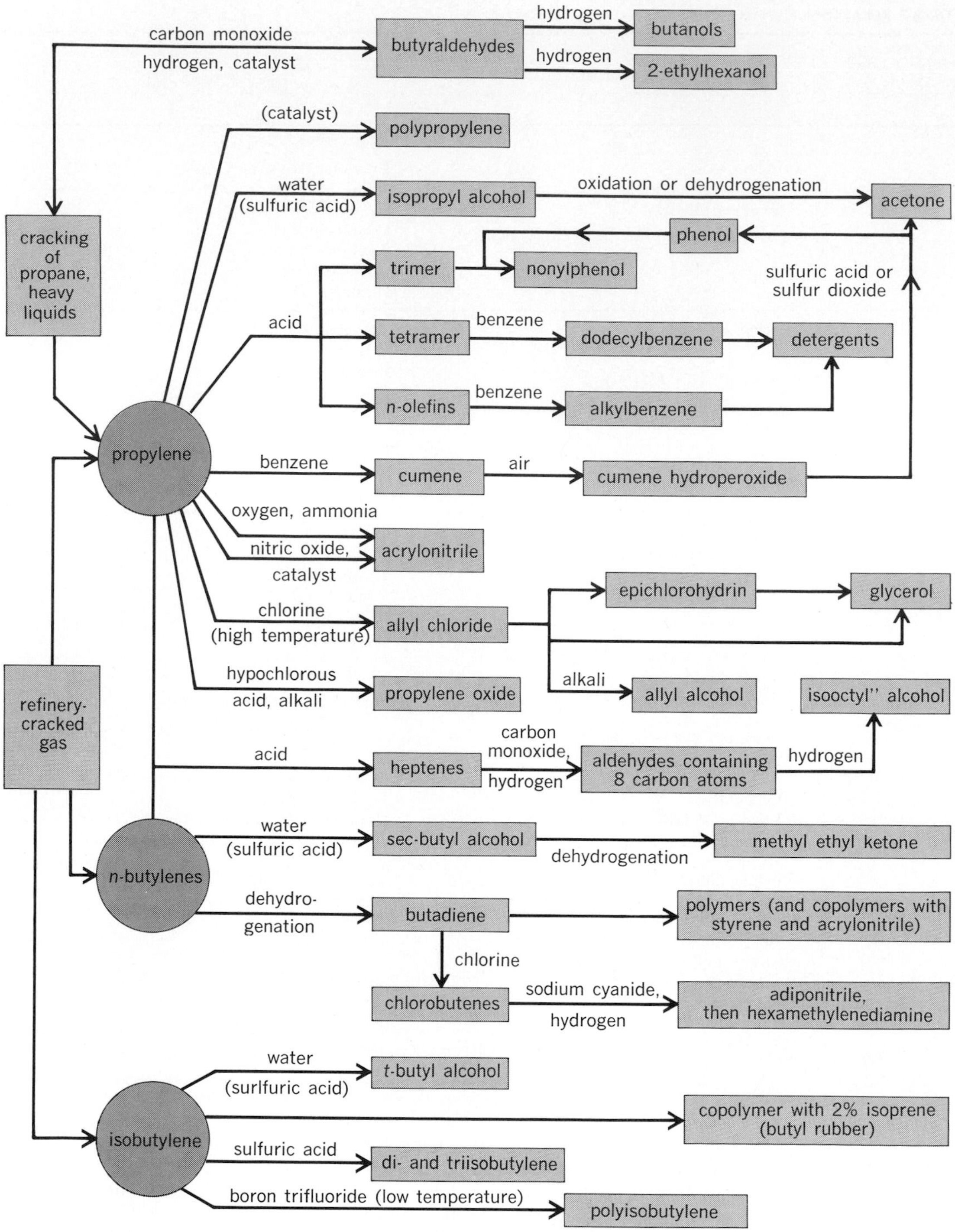

Fig. 3. Petrochemicals from propylene and the butylenes.

Raw material and products. The major operations of the petrochemicals industry are summarized below for each of the principal raw materials. In addition to the products mentioned, a large number of chemicals are made in smaller amounts.

Natural gas, which is predominantly methane but contains significant amounts of ethane, propane, butane, and some hydrogen sulfide and carbon dioxide, is the most important source of raw materials for aliphatic petrochemicals. Essentially all of the ethane separated from natural gas–processing plants is utilized for the manufacture of ethylene by thermal cracking. Relatively small quantities are consumed in the production of acetylene and chlorinated solvents. Over 20% of the total propane made in the United States is used to produce ethylene and propylene. A small percentage is used for oxygenation to produce a great number of oxygenated chemicals, including formaldehyde, acetaldehyde, methyl alcohol, acetone, and acetic acid. Small additional amounts are used to make chlorinated and nitrated solvents.

Butane is an important source of feed stock for the manufacture of butadiene by catalytic dehydrogenation. But the largest volume use for *n*-bu-

Table 3. Petrochemicals from propylene

Basic derivatives and sources	Uses
Isopropyl alcohol	Acetone Solvents, drugs and chemicals
Cumene	Phenol and acetone
Acrylonitrile From propylene	Acrylic and modacrylic fibers Nitrile elastomers and acrylonitrile-butadiene-styrene resins
Polypropylene	Molding, fiber, and film
Propylene oxide	Propylene glycol Dipropylene glycol and polypropylene glycol Polyurethane and other
Oxochemicals Isooctyl alcohol Butyraldehydes (propylene only)	 Phthalate esters Intermediate for butanols 2-ethylhexanol, *n*-butyric acid
Dodecene (tetramer)	Dodecyl benzene Dodecyl phenol
Nonene (trimer)	Decyl alcohol and nonylphenol
Epichlorohydrin (crude and refined)	Glycerol and epoxy resins
Polyisoprene (propylene only) and EP rubber	Elastomers

tane is in gasoline and other fuels.

The heavy product of natural gas processing, natural gasoline, finds use as feed stock in the manufacture of ethylene.

The principal petrochemicals derived from methane are shown in Fig. 1 and Table 1. Methane has been the chief source of methanol, acetylene, chloromethanes, and hydrogen cyanide. *See* METHANE.

Inorganic petrochemicals. Ammonia, produced predominantly by the direct reaction of hydrogen and nitrogen, is by far the largest volume petrochemical. Air is the source of the nitrogen. Hydrogen is produced by two-step steam reforming of natural gas, naphthas, streams from refineries and petrochemicals plants, and partial oxidation of a full range of hydrocarbons, including natural gas, residual oils, coke-oven gas, coal, and lignites. Hydrogen is also produced from the electrolysis of brine. A large proportion of ammonia is converted into ammonium nitrate and other ammonium salts and into urea. The largest use of these, as well as of ammonia, is as fertilizers. *See* AMMONIA.

Carbon black, used primarily in the manufacture of synthetic rubbers, is made almost entirely by partial combustion with insufficient air supply, using natural gas in the old channel process and predominantly highly aromatic petroleum oil and coal tar by-products in the furnace process. The latter process accounts for approximately 76% of carbon black production.

Aliphatic organic compounds. Ethylene is consumed in larger amounts for aliphatic petrochemicals than is any other hydrocarbon. Derivatives of this versatile molecule make up a large segment of the petrochemicals industry, as shown in Fig. 2 and Table 2. About 77% of the ethylene is manufactured from thermal cracking of ethane and propane recovered from natural gas, 10% from refinery thermal and catalytic cracking operations conducted primarily to increase the quantity and quality of gasoline produced, and 13% from thermal cracking of higher-boiling liquids, such as naphtha, gas oil, and natural gasoline. *See* ETHYLENE.

Alternative routes from a single petroleum and natural gas source to certain chemicals are illustrated in Fig. 1 (carbon black, ammonia, methyl chloride) and Fig. 2 (ethylene oxide, ethyl alcohol). Alternative petroleum sources are illustrated by a comparison of Figs. 1–4. For example, vinyl chloride and vinyl acetate are produced commercially from both acetylene (Fig. 1) and ethylene (Fig. 2), and acrylonitrile from acetylene (Fig. 1) and propylene (Fig. 3). Dovetailing operations are illustrated in Fig. 1 by the synthesis of urea using carbon dioxide, coproduced with the hydrogen required for ammonia, and in Table 2 by the production and simultaneous utilization of hydrogen chloride in the manufacture of ethyl chloride and vinyl chloride.

Propylene is produced in large amounts from petroleum refining operations and as a coproduct in the manufacture of ethylene. The major use of propylene is in the manufacture of high-octane gasoline components by alkylation and polymerization. However, large quantities of propylene are consumed in the manufacture of petrochemicals. It is expected that the share of propylene consumed in gasoline will shrink in the future as chemical uses continue to grow rapidly. The principal petrochemicals derived from propylene are shown in Fig. 3 and Table 3. Of these isopropyl al-

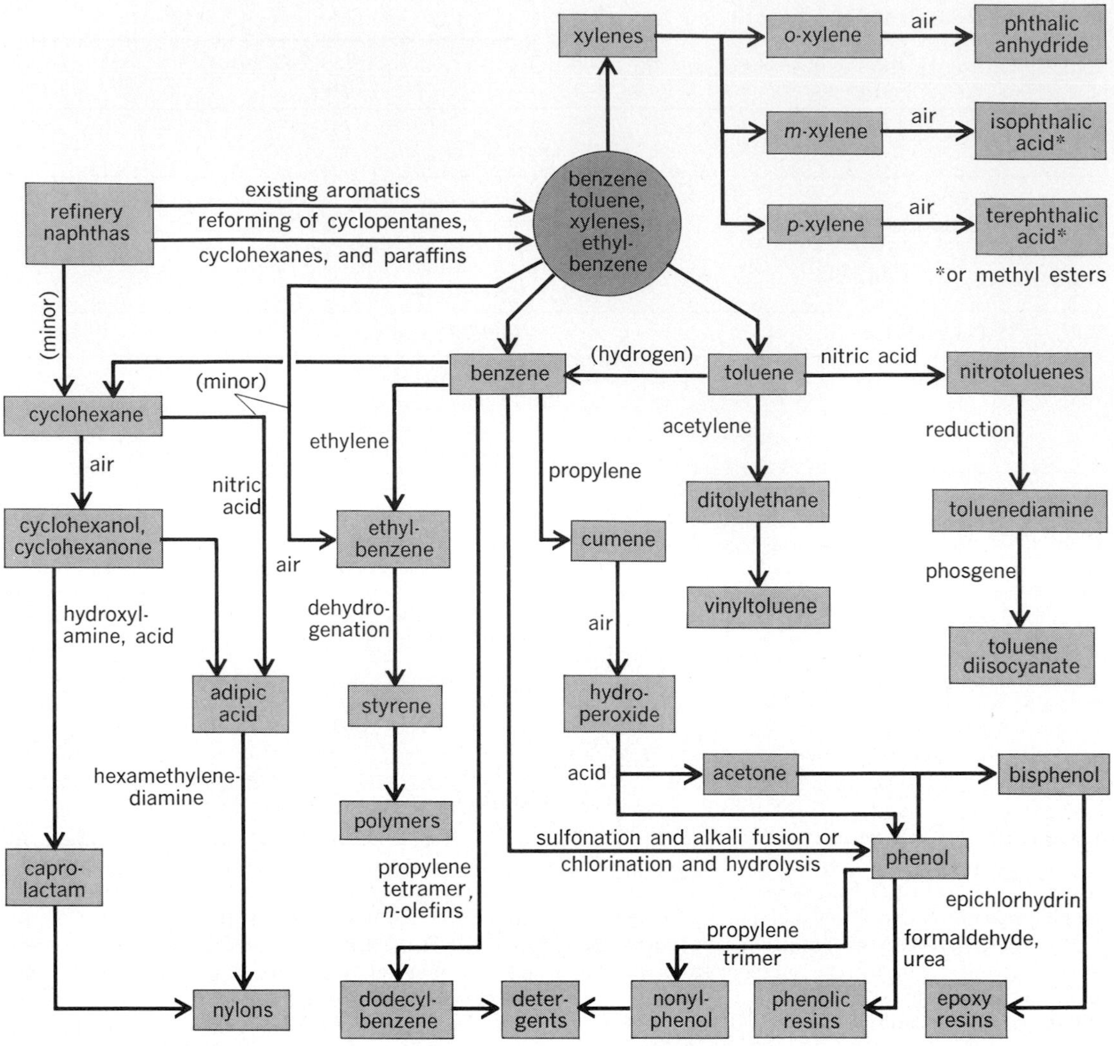

Fig. 4. Cyclic petrochemicals.

cohol is the largest in volume, but in the future it is expected that polypropylene will be the largest, and acrylonitrile, cumene, and propylene oxide will become even more important derivatives. *See* Acrylonitrile; Propylene.

Table 4. Petrochemicals from butanes and butylenes

Basic derivatives and sources	Uses
Butadiene From butylenes From butane By-product	Styrene-butadiene rubber and resins Polybutadiene Adiponitrile Nitrile rubber Acrylonitrile-butadiene-styrene plastics
sec-Butyl alcohol (from *n*-butenes)	Methyl ethyl ketone
Heptenes (from mixture of butenes and propylene)	Isooctyl alcohol
Butyl rubber (from isobutylenes)	Tire and tire products
Polybutenes (from isobutylene)	Lube oil additives Calking and sealing compounds Adhesives Rubber compounding

The C_4 hydrocarbons (derivatives of butanes and butylenes) used in the synthesis of chemicals are important industrially. There is a very large supply of C_4 hydrocarbons available from natural gas liquids and petroleum refining operations. The saturated C_4 hydrocarbons, *n*-butane and isobutane, are primarily derived from natural gas and gasoline. Butanes are also formed during petroleum refining operations, and isobutane is intentionally produced by the isomerization of *n*-butane. The unsaturated C_4 hydrocarbons (butylenes) are formed as by-products during gasoline manufacture, and they are intentionally manufactured by dehydrogenation of the saturated C_4 hydrocarbons. The principal derivatives of *n*-butenes and isobutylene are shown in Table 4 and Fig. 3. Butadiene is the most important C_4 chemical derivative and accounts for about 40% of the total consumption. The *n*-butenes are also used to make methyl ethyl ketone, an important solvent. Isobutylene is used in the manufacture of butyl rubber and polybutene, which is an important lubricating oil additive.

The oxo process (reaction of olefins with carbon monoxide and hydrogen) is an increasingly important process for the manufacture of alcohols. It provides aldehydes from olefins of one less carbon number, and the aldehydes may be converted by hydrogenation to their corresponding alcohols. The principal alcohols obtained are butanols and 2-ethylhexanol from propylene; branched-chain octyl alcohols from the heptene copolymer of propylene and butylene (Fig. 3); branched-chain nonyl alcohols from diisobutylene; and branched-chain decyl and tridecyl alcohols from propylene trimer and tetramer, respectively.

Cyclic organic compounds. Up until 1940 the coal tar industry was the principal supplier of cyclic compounds: benzene, toluene, xylenes, and their derivatives (Fig. 4, Table 4). Petroleum then became the chief source of cyclic organic compounds except for naphthalene, a great deal of which was always derived from coal tar (Table 5). Benzene, toluene, and xylenes are important high-octane components of gasoline. They are manufactured by catalytic reforming and recovered by distillation and extraction processes. Substantial quantities of toluene are hydrodealkylated to benzene. Figure 4 shows the more important of the cyclic derivatives. High polymers (plastics and resins, fibers, and elastomers) constitute the great bulk of the derivatives, with end products including nylon, polyester fiber and film, polystyrene, styrene-butadiene rubber, epoxy resins, phenolic resins, and polyurethanes from isocyanates. *See* POLYMER.

Naphthenic acids are carboxylic acids of substituted cyclopentanes and cyclohexanes; they are present in crude oils. Commercial naphthenic acids lie in the molecular-weight range of 180–350.

Table 5. Cyclic petrochemicals

Chemical	Uses
Benzene	Styrene Cyclohexane Phenol Detergent alkylate Maleic anhydride Aniline DDT
Toluene	Dealkylation to benzene Solvents Toluene diisocyanate Motor and aviation gasoline TNT
Xylenes	*p*-Xylene *o*-Xylene *m*-Xylene Solvents and miscellaneous chemicals Gasoline
Ethyl benzene	Styrene
Cyclohexane	Nylon intermediates Nonnylon uses (cyclohexanone and adipic acid)
Naphthalene	Phthalic anhydride Insecticides β-Naphthol and mothballs Other

Owing to their oil solubility, naphthenates of appropriate metals are used as paint driers, fungicides, and lubricant additives.

Cresylics (alkyl phenols, cresylic acids, and cresols) are made by refining by-product tar acids from coke ovens or gas works, and by recovery from refinery waste streams containing petroleum acids. The lower-boiling fractions of cracked petroleum contain xylenols and smaller amounts of cresols, ethylphenols, trimethylphenols, and methylethylphenols, which are produced at higher cracking temperatures. Cresols are also made synthetically by methylation of phenol to produce *meta-*, *para-*, and *ortho*-cresols and 2,6-xylenol. Their chief uses are in phenolic resins, wire-enamel solvents, phosphate esters, ore-flotation reagents, oil and gasoline additives, antioxidants, and metal cleaners.

Petrochemical sulfur. Sulfur is obtained by the oxidation of hydrogen sulfide, which occurs in some natural gas and most refinery gases, particularly in the off-gas from processes to reduce sulfur content of petroleum liquids (for example, hydrodesulfurization).

Petrochemical sulfur from natural and refinery gas represents about 14% of all sulfur production in the United States. Almost 86% of all sulfur is consumed as sulfuric acid in the manufacture of fertilizer, other chemicals, and steel. The remaining 14% is used in the pulp and paper industry, the manufacture of carbon bisulfide, or as ground or refined sulfur.

[JULIUS D. HELDMAN; W. W. REYNOLDS]

Bibliography: L. F. Hatch and S. Matar, *Hydrocarbons to Petrochemicals*, 1981; R. E. Kirk and D. F. Othmer, *Encyclopedia of Chemical Technology*, vols. 1–5, 3d ed., 1978–1979; J. G. Speight, *Chemistry and Technology of Petroleum: An Introduction*, 1980; R. B. Stobaugh, Jr., *Petrochemical Manufacturing and Marketing Guide*, vol. 1: *Aromatics and Derivatives*, 1967.

pH

A term used to describe the hydrogen-ion activity of a system. It is defined by the expression $pH = -\log a_{H^+}$. Here a_{H^+} is the activity of the hydrogen ion. In dilute solutions, the activity is essentially equal to the concentration, and the pH may be approximately defined as $pH = -\log_{10} [H^+]$, where $[H^+]$ is the concentration of the hydrogen ion in moles per liter. The use of the pH makes negative exponents unnecessary in describing the hydrogen-ion activity. In a system in which the hydrogen-ion activity is 10^{-3} mole/liter, the pH is 3. The term is seldom used to describe solutions in which the hydrogen-ion activity is 1 or greater. The expression defining the pH is closely related to the free energy of the hydrogen ion with respect to a standard reference state. *See* ACID AND BASE; HYDROGEN ELECTRODE; HYDROGEN ION.

[FRANCIS J. JOHNSTON]

Bibliography: R. G. Bates, *Determination of pH: Theory and Practice*, 1964.

Phase equilibrium

A general field of physical chemistry dealing with the various situations in which two or more phases (or states of aggregation) can coexist in thermodynamic equilibrium with each other, with the nature of the transitions between phases, and with the effects of temperature and pressure upon these equilibria. Many superficial aspects of the subject are largely qualitative, for example, the empirical classification of types of phase diagrams; but the basic problems always are susceptible to quantitative thermodynamic treatment, and in many cases, statistical thermodynamic methods can be applied to simple molecular models.

Thermodynamics requires that when two phases, α and β, are free to exchange heat, mechanical work, and matter (chemical species), the temperature T, the pressure P, and the chemical potential (partial molar free energy) μ_i of each particular component i must be equal in both phases at equilibrium. Algebraically, equilibrium exists when $T_\alpha = T_\beta$, $P_\alpha = P_\beta$, $\mu_{i,\alpha} = \mu_{i,\beta}$, and $\mu_{j,\alpha} = \mu_{j,\beta}$.

These conditions of thermal, mechanical, and material equilibrium need not all be present if the equilibrium between phases is subject to inhibiting restrictions. Thus, for a solution of a nonvolatile solute in equilibrium with the solvent vapor, the condition of equality of solute chemical potentials $\mu_{2,\alpha} = \mu_{2,\beta}$ need not apply, since there can be no solute molecules in the vapor phase. Similarly, in osmotic equilibria, in which solvent molecules can pass through a semipermeable membrane, whereas solute molecules cannot, $\mu_{1,\alpha} = \mu_{1,\beta}$ and $T_{1,\alpha} = T_{2,\beta}$, but the solute chemical potentials μ_2 are unequal, as are the pressures on opposite sides of the membrane. *See* OSMOSIS; SOLUTION.

If a system consists of P phases and C distinguishable components, there are $C+2$ thermodynamic variables (C chemical potentials μ_i, plus the temperature and pressure) which are interrelated by an equation for each phase. Since there are P independent equations relating the $C+2$ variables, one needs to fix only $F = C + 2 - P$ variables to define completely the state of the system at equilibrium; the other variables are then beyond control. This relation for the number of degrees of freedom F, or variance, is called the phase rule and was first derived by Willard Gibbs in 1873. It has proved to be a powerful tool in interpreting and classifying types of phase equilibria.

When chemical changes may occur in the system, the number of components C is the number of independent components, that is, the number of components whose amounts can be varied by the experimenter; this is equal to the total number of chemical species present less the number of independent chemical equilibria between them.

An invariant system has no degrees of freedom ($F=0$), for which the number of phases $P = C + 2$. For a one-component system, such an invariant point is a triple point at which three phases coexist at a single temperature and pressure only; for a two-component system, a quadruple point (four phases) would be invariant. *See* TRIPLE POINT.

In a univariant system ($F=1$), $P = C + 1$. With a one-component system, one can fix the temperature at which two phases (liquid and gas, for instance) can coexist in equilibrium; then the pressure (here the vapor pressure) is determined and not subject to external control. A univariant system is described by a line in a phase diagram, for example, a plot of vapor pressure versus temperature. The differential equation, Eq. (1), for such a

$$\frac{dP}{dT} = \frac{\Delta H}{T \Delta V} \tag{1}$$

univariant line in a one-component system was first deduced by B. P. E. Clapeyron in 1834. In the equation ΔH and ΔV are the enthalpy change (heat absorbed) and volume change, respectively, for the transition from one phase to another. *See* VAPOR PRESSURE.

In systems of two or more components, more complicated equations for the univariant line replace the Clapeyron equation, but in special cases they may reduce to the simpler form. For example, in the chemical decomposition of a solid calcium carbonate to solid calcium oxide and gaseous carbon dioxide, reaction (2), the three-phase system

$$CaCO_3(s) \rightleftharpoons CaO(s) + CO_2(g) \tag{2}$$

(two solids and one gas) is univariant (since the number of independent components C is two), the equilibrium pressure is the decomposition pressure of $CaCO_3$, and the ΔH and ΔV of the Clapeyron equation are the enthalpy and volume changes associated with the chemical reaction.

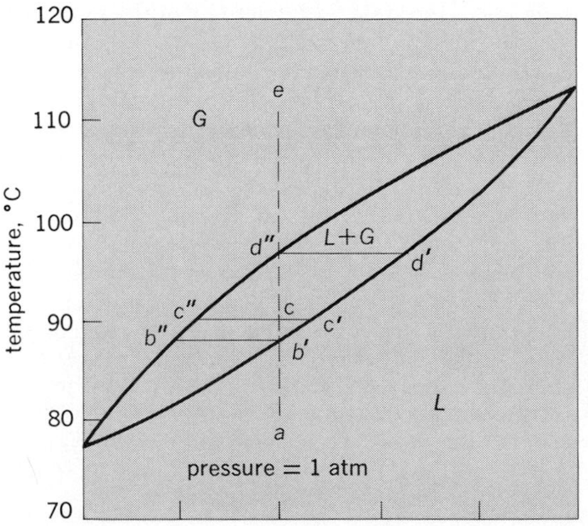

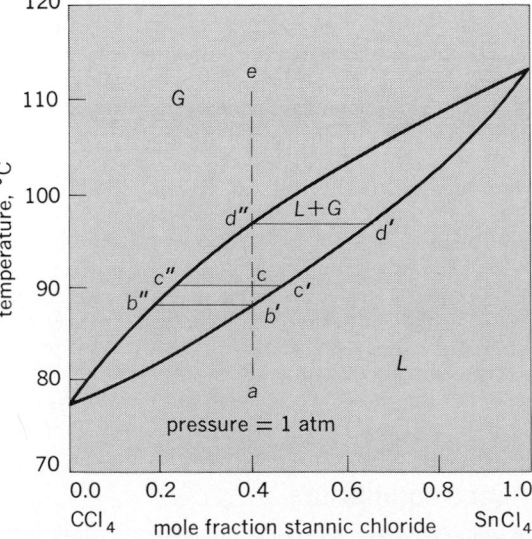

Fig. 1. Binary liquid-vapor temperature-composition diagram for the system carbon tetrachloride + stannic chloride. G = gas (vapor) phase; L = liquid phase; L + G = two coexisting phases.

BINARY SYSTEMS

Phase diagrams of binary systems containing two components are easily classified. Typical examples of the important classes (liquid-gas, liquid-liquid, solid-liquid, solid-solid) have been selected for description.

Liquid-gas equilibrium. In a one-component system, liquid and vapor are in equilibrium at the boiling point. For two-component systems, the two-phase situation is bivariant and more complex. A complete temperature-pressure-composition diagram would be three-dimensional, so most phase diagrams are made for either constant pressure or constant temperature. Figure 1 shows the simplest type of binary liquid-vapor temperature-composition diagram, exemplified by the system carbon tetrachloride – stannic chloride, which forms essentially ideal solutions. The regions labeled G and L are one-phase regions, gas (vapor) and liquid, respectively; the region labeled $L+G$ is a two-phase region in which liquid and vapor coexist. If the temperature of a liquid mixture of 40 mole percent $SnCl_4$ (mole fraction = 0.40) is increased at a constant pressure of 1 atm, the change in the system can be traced along the straight line *ab′cd″e*. At low temperatures, only one phase, the liquid, is present, but at 87.5°C (point *b′*), a vapor phase appears. The composition of this vapor phase is given by point *b″* (mole fraction $SnCl_4$ = 0.18), and the two conjugate phases are connected on the diagram by the tie line *b″b′*. As the temperature is increased further, more vapor is formed; since the vapor is rich in CCl_4, this component becomes relatively depleted in the liquid phase, and the liquid composition moves along the line *b′c′d′*, while the vapor composition moves along the line *b″c″d″*.

At 90°C the overall composition of the two-phase system is represented by point *c*, but the compositions of vapor and liquid separately are given by the two ends of the tie line, points *c″* and *c′*, respectively (mole fractions of 0.22 and 0.47, respectively). The relative amounts of the two phases are given by the lever arm principle of physics. The ratio of the moles of vapor to moles of liquid is given by the ratio of the length *cc′* to the length *c″c*, here 0.07:0.18, or 28%, in the vapor phase. Further increase in temperature produces more and more vapor until, at 97°C, the liquid phase (point *d′*, mole fraction 0.62) has become vanishingly small; at higher temperatures, it disappears, and only the vapor phase (point *d″*, mole fraction = 0.40) remains. Further increase in temperature (along the line *d″e*) is uneventful.

In this simple system, there are no maxima or minima in the liquid and vapor curves; consequently, such liquid mixtures can be separated completely into the two pure components by fractional distillation. Systems with maximum boiling points (acetone + chloroform, Fig. 2) or minimum boiling points (ethanol + benzene, Fig. 3) cannot be so separated into the pure substances. At the maximum or minimum, the composition of the liquid is identical with that of the vapor with which it is in equilibrium; continued boiling will not alter these compositions, so these solutions are called constant-boiling mixtures or azeotropes. It should be noted that the solid and liquid lines are smooth curves tangent to each other at such a point; any phase diagram which shows a sharp corner is thermodynamically incorrent. *See* AZEOTROPIC MIXTURE; DISTILLATION.

Maximum boiling mixtures are associated with negative deviations from ideal behavior; this type of deviation usually arises from strong attractions between molecules of the different species (sometimes called compound formation). Minimum boiling mixtures are associated with positive deviations from ideal behavior; this usually arises when the attraction between two unlike molecules (1-2) is weaker than the average of two like pairs (1-1 and 2-2); extreme examples of this arise when one component may be described as associated. The simpler type of phase diagram (Fig. 1) occurs when the two components mix nearly ideally or when the boiling points are very different.

Liquid-liquid equilibrium. When two liquids are sufficiently different in their intermolecular forces, they may not mix in all proportions but instead be only partially miscible, part of the phase diagram being occupied by a two-phase region of two immiscible liquid phases. Most liquids (but not all) become more miscible as the temperature increases and are completely miscible at the critical solution temperature (also called the consolute temperature). The system aniline + *n*-hexane (Fig. 4) is a typical example of such liquid-liquid immiscibility and critical phenomena. The liquid-liquid phase boundary and the critical solution temperature are only slightly dependent upon pressure. The size of the two-phase region increases with pressure if (as is usual) the two liquids expand on mixing at constant pressure.

Solid-liquid equilibrium. When two substances are completely miscible with each other and form a complete series of solid solutions, the solid-liquid phase diagrams are entirely analogous to the liquid-vapor diagrams illustrated above. Those with no maximum or minimum, usually associated with nearly ideal liquid and solid solutions, such as methane + krypton, are called type I, according to the Bakhuis Roozeboom classification; those with

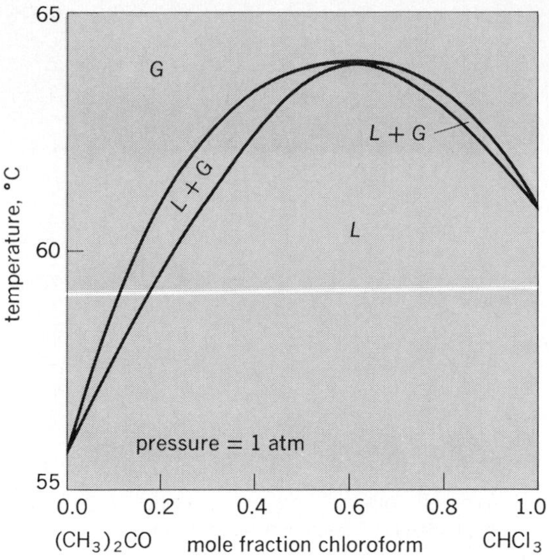

Fig. 2. Temperature-composition diagram for acetone + chloroform, showing maximum boiling point.

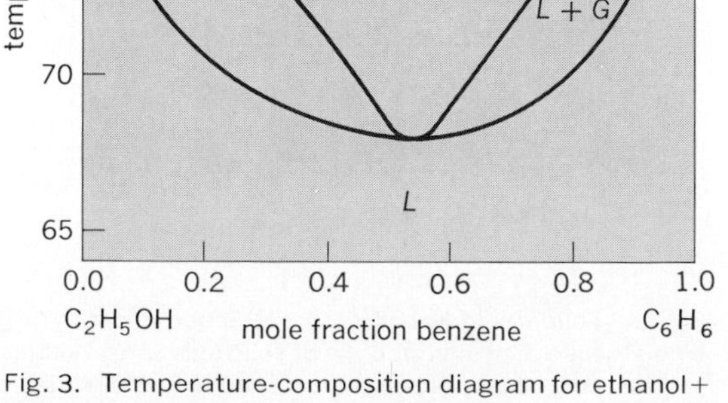

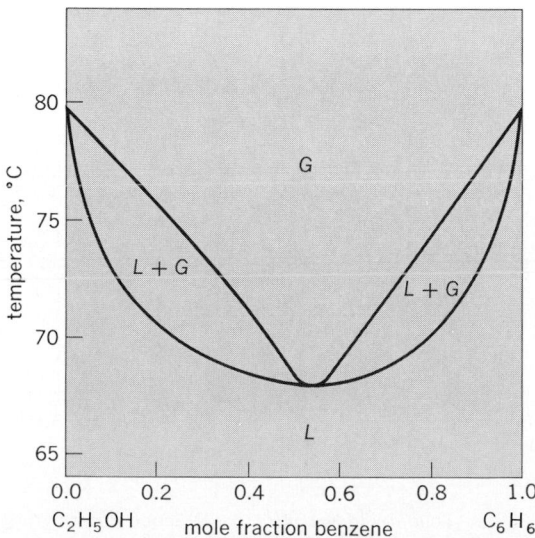

Fig. 3. Temperature-composition diagram for ethanol + benzene, showing minimum boiling point.

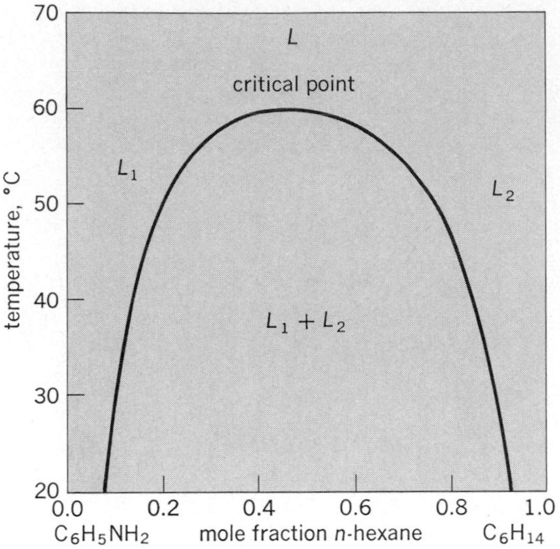

Fig. 4. Liquid-liquid equilibrium for aniline + *n*-hexane.

a maximum melting point, an exceedingly rare type exemplified by *d*-carvoxime + *l*-carvoxime, are type II; and those with a minimum melting point (bromobenzene + iodobenzene), type III. *See* SOLID SOLUTION.

In most binary systems, however, extensive solid solutions are impossible because of the incompatibility of the size, shape, and crystal lattices of the two components. In the absence of solid solution formation, the addition of a solute to a liquid solvent invariably depresses the freezing point, that is, the temperature at which, upon cooling, the first trace of solid solvent appears. This depression of the freezing point of a dilute solution is a convenient method (the so-called cryoscopic method) for determining the molecular weight of a solid solute. *See* MOLECULAR WEIGHT.

As the freezing-point curve of a liquid continues to lower temperatures with higher concentrations of the second component, the conventional roles of solute and solvent become reversed, and one speaks of the solubility of a solute rather than of the depression of the freezing point of the solvent; no point of demarcation exists, and the two situations are in fact two aspects of the same phenomenon.

The second component also has a freezing-point – solubility curve marking the temperature at which the solution is in equilibrium with pure solid 2. The point of intersection of the two phase boundaries is the eutectic point which defines a eutectic temperature and a eutectic composition. Below this temperature, the system consists of two solid phases ($S_1 + S_2$ in Fig. 5*a*).

Solid-liquid phase diagrams are conveniently determined by thermal analysis of cooling curves. An initially homogeneous liquid is cooled gradually, and the temperature plotted against time. Figure 5*a* shows a simple eutectic-type phase diagram, and Fig. 5*b* sketches typical cooling curves obtained for various compositions (those marked *A* to *F* in Fig. 5*a*). Curves *A* and *F* are for the pure components; each shows a single temperature arrest, a horizontal section of the curve for the melting point (between *f* and *m*), at which the temperature remains constant from the time the first bit of solid is formed until the last bit of liquid disappears. *See* THERMOANALYSIS.

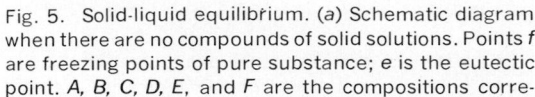

Curves *B* and *E* are for solutions rich in components 1 and 2, respectively. At a temperature below the melting point of the pure substance, the first bit of solvent begins to freeze out; this is indicated by a change in slope (point *f*). The freezing section of the cooling curve (between points *f* and *e*) is not horizontal; as solvent freezes out, the liquid phase becomes richer in solute, and the freezing point is further depressed. When the composition of the liquid phase reaches the eutectic composition, the second component begins to freeze out as well as the first. No further change in liquid composition occurs, so the temperature re-

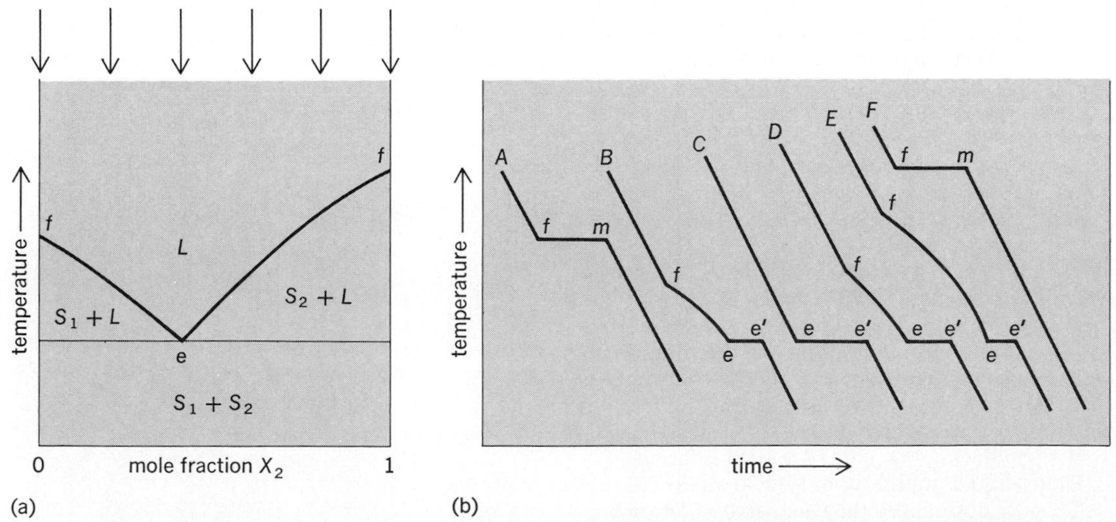

Fig. 5. Solid-liquid equilibrium. (*a*) Schematic diagram when there are no compounds of solid solutions. Points *f* are freezing points of pure substance; *e* is the eutectic point. *A*, *B*, *C*, *D*, *E*, and *F* are the compositions corresponding to the cooling curves of Fig. 5*b*. (*b*) Schematic cooling curves for system shown in Fig. 5*a*; *f* = freezing point, *m* = melting point, *e* = beginning of eutectic freezing, *e'* = end of eutectic freezing.

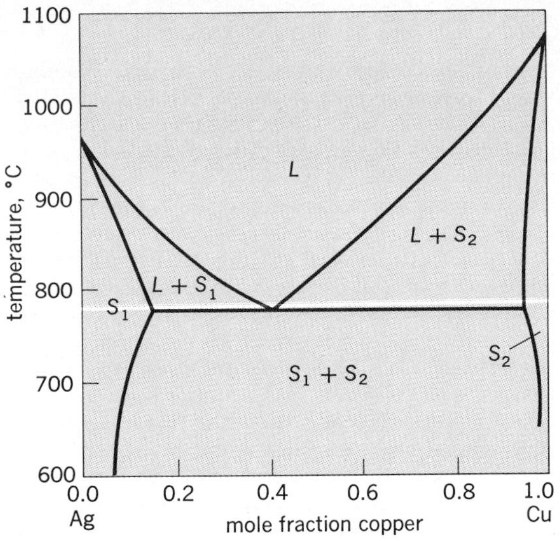

Fig. 6. Solid-liquid equilibrium in the system silver + copper (type-V solid solutions).

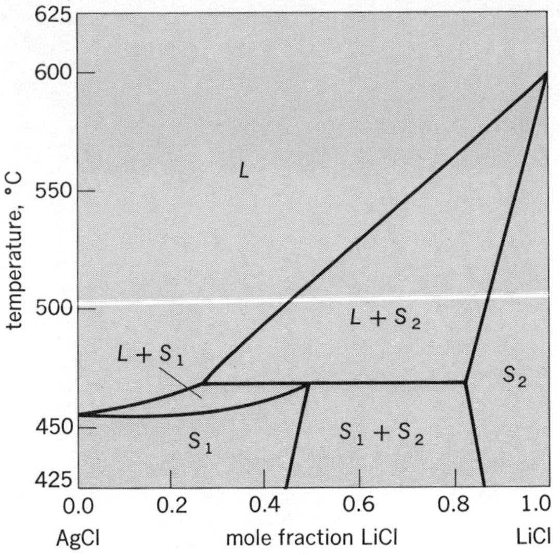

Fig. 7. Solid-liquid equilibrium in the system silver chloride + lithium chloride (type-IV solid solutions).

mains constant until no liquid remains (between points *e* and *e*′). The solid which freezes out at the eutectic (the eutectic mixture) appears superficially very different from either pure solid; it is a mixture of very small crystals of each of the two components which have crystallized together. This microcrystalline two-phase mixture is in no sense a compound.

Curve *D* is a cooling curve for still another composition, and curve *C* is for the eutectic composition. Alone, curve *C* is indistinguishable from that for the freezing of a pure substance; only by combining the information from a series of cooling curves can one construct the whole diagram.

When the components form an incomplete series of solid solutions (partial miscibility), the phase diagram can be of the eutectic type (Bakhuis Roozeboom type V) illustrated by the system silver + copper shown in Fig. 6. Indeed, since the mutual solubilities are never exactly zero, eutectic diagrams, such as that in Fig. 5*a*, are in fact merely extreme examples of this more general case.

When a solid phase upon melting transforms into a liquid phase and a solid phase of different composition, one speaks of incongruent melting and a peritectic-type phase diagram. A simple

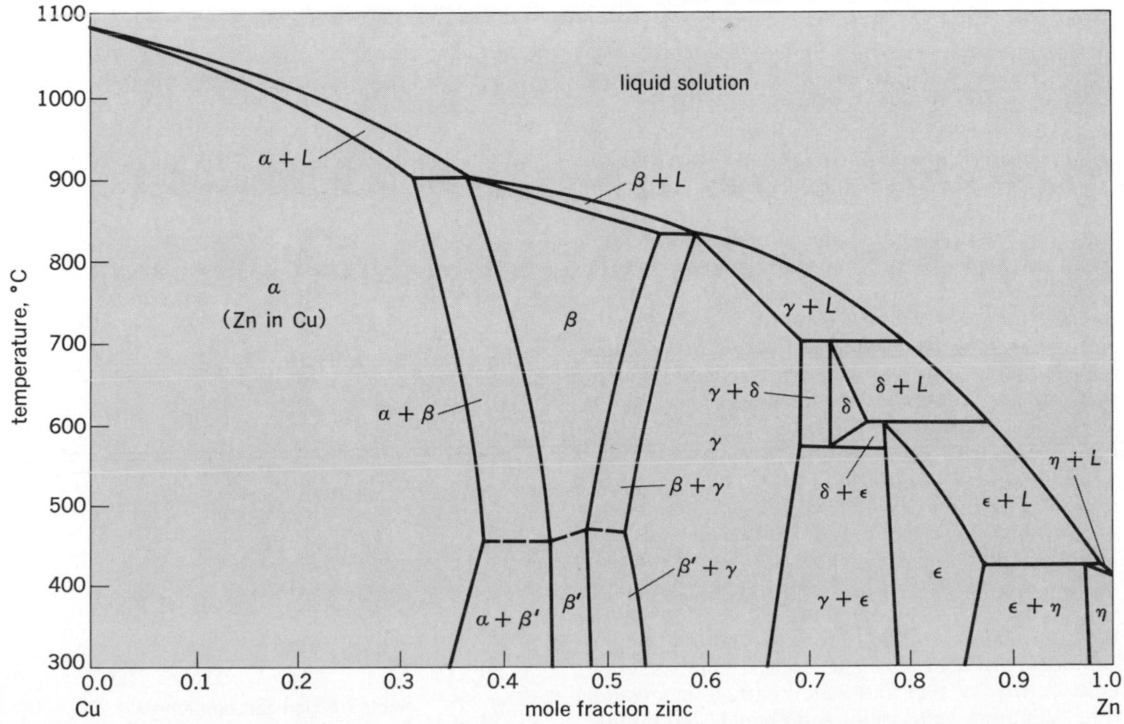

Fig. 8. Temperature-composition diagram for the system copper + zinc (brass). Note the six different solid phases. The dotted line separating β from β' denotes a second-order, order-disorder transition.

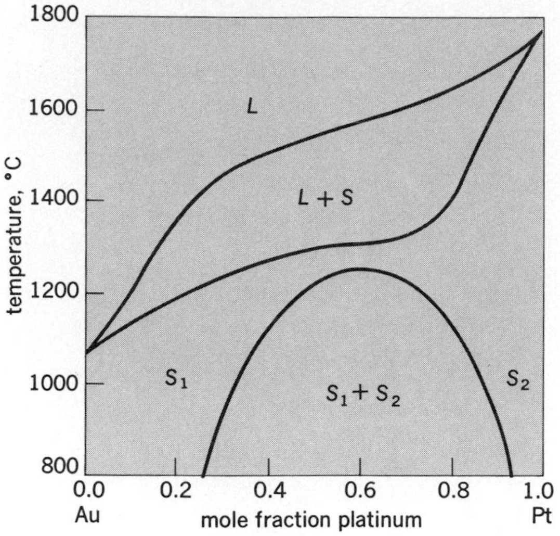

Fig. 9. Solid-liquid and solid-solid equilibria in the system gold + platinum. Critical region of the solid-solid phase boundary nearly touches the solidus curve of the solid-liquid phase boundary. If these touched and coalesced, a type-IV diagram (Fig. 7) would result.

example is the type-IV solid-solution diagram illustrated by the system silver chloride + lithium chloride (Fig. 7); similar phenomena occur in systems without any appreciable solid-solution formation.

Many solid-liquid phase diagrams are complicated by the existence of intermediate crystalline phases of different crystal structure. Usually, these intermediate phases are at compositions close to simple mole ratios of the components and have the two kinds of molecules distributed in a regular arrangement; consequently, it is convenient to call these compounds, even if no specific chemical interactions can be demonstrated unequivocally. There is a rich variety of such complex phase diagrams, especially among binary systems of ionic salts ($CaCl_2$-KCl) and among binary systems of metals (alloys). Figure 8 shows the Cu + Zn system (brass), in which a whole series of crystal structures appears: α-brass has the crystal structure of pure copper (face-centered cubic) with an occasional Zn atom in the lattice; β-brass has a body-centered cubic structure with a Cu-Zn ratio of approximately 1:1; γ-brass has a very complex structure related to the formula Cu_5Zn_9; η-brass has the crystal structure of pure Zn (hexagonal, close-packed); the δ and ϵ phases have still different structures. Note the five peritectic transitions (for example, at 600°C, where the ϵ-solid melts incongruently to the δ-solid and a liquid solution rich in zinc).

Solid-solid equilibrium. These are of several types. One can have solid critical solution temperatures (with solid solutions of the same crystal structure) which are analogous to the more familiar liquid-liquid case (the system gold + platinum shown in Fig. 9). More common are the transitions between one crystalline form and another, which can occur even in pure substances. One such occurs in the system nitrogen + carbon monoxide (Fig. 10 shows solid-solid, solid-liquid, and liquid-vapor transitions for this system.) *See* TRANSITION POINT.

MULTICOMPONENT SYSTEMS

As one proceeds from binary systems to systems with three or more components, the phase diagrams become more complex. Each component adds another dimension to the representation of the phase equilibria. Thus, for three components, two dimensions are required to represent the phase diagrams for a single temperature and pressure; these are conveniently depicted by a triangular diagram in which each vertex represents a pure component. Figure 11 shows a schematic diagram of a ternary system in which three liquid phases can coexist. Even when there are three phases, the system is still bivariant. (An example of such a system is water + succinonitrile + diethyl ether.)

A special case of a three-component system is that in which there are two immiscible solvents and a third component, soluble in both, distributed between the two phases. The ratio of the concentrations of the solute in the two solvents is the distribution coefficient; in dilute solutions, this is independent of the concentration, but at higher concentrations, nonideal behavior of the solute can produce systematic variations of the distribution coefficient which ends in the most concentrated solutions with the ratio of the solubilities in the saturated solutions.

Distribution effects are important in separating similar materials. A small difference in distribution coefficients is amplified by multistage equilibria, such as those used in countercurrent extraction and partition chromatography.

The examples used to illustrate the various types of phase equilibria are not supposed to suggest that these types are restricted to the particular kind of chemical substances shown. In general, examples of each could have been selected from many kinds of substances such as metals, nonmetallic elements, inorganic salts, and organic non-

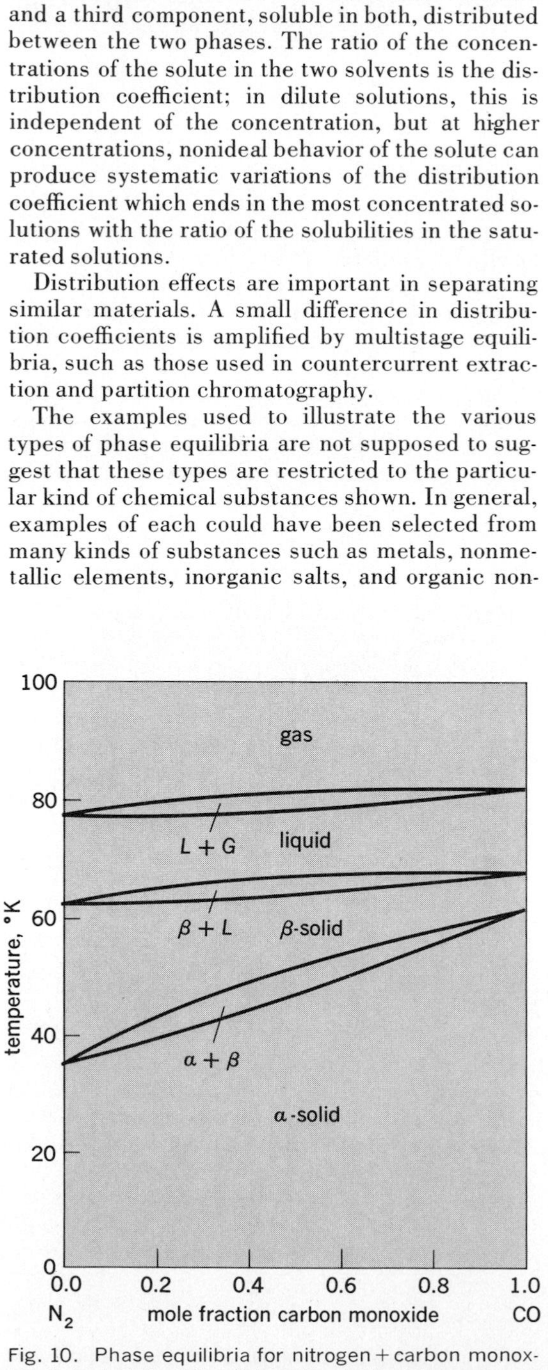

Fig. 10. Phase equilibria for nitrogen + carbon monoxide, being solid-solid, solid-liquid, and liquid-vapor.

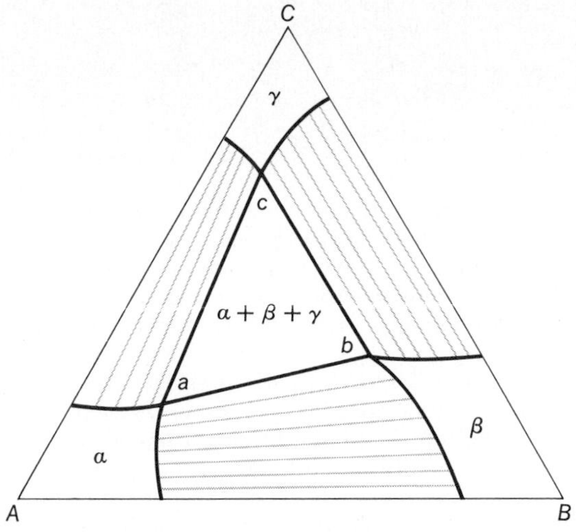

Fig. 11. Schematic diagram of a three-component system at a fixed temperature and pressure. Points *A, B,* and *C* represent the pure liquids. The composition corresponding to a point in the diagram is determined by the positions along a line from each vertex to the opposite side; thus point *b* is 20% *A*, 50% *B*, 30% *C*. Regions α, β, and γ correspond to single phases rich in *A, B,* and *C*, respectively; α+β+γ is a three-phase region; the three saturated solutions have the compositions given by the points *a, b,* and *c*. Three two-phase regions, α+β, α+γ, and β+γ, are indicated by drawing in the tie lines.

electrolytes. *See* CHEMICAL EQUILIBRIUM; CHEMICAL THERMODYNAMICS; COUNTERCURRENT TRANSFER OPERATIONS; CRYSTAL STRUCTURE; EXTRACTION; INTERFACE OF PHASES.

[ROBERT L. SCOTT]

Bibliography: G. M. Barrow, *Physical Chemistry*, 4th ed., 1979; R. Ginell, *Association Theory: The Phases of Matter and their Transformations*, 1979; J. H. Hildebrand and R. L. Scott, *Solubility of Nonelectrolytes*, rev. 3d ed., 1964; W. J. Moore, *Physical Chemistry*, 4th ed., 1972; A. Reisman, *Phase Equilibria*, 1970; J. E. Ricci, *The Phase Rule and Heterogeneous Equilibrium*, 1951; F. E. W. Wetmore and D. J. LeRoy, *Principles of Phase Equilibria*, 1951.

Phenanthrene

A colorless, crystalline hydrocarbon ($C_{12}H_{10}$) which melts at about 100°C and boils at 332°C. Phenanthrene (I) is usually obtained from coal tar,

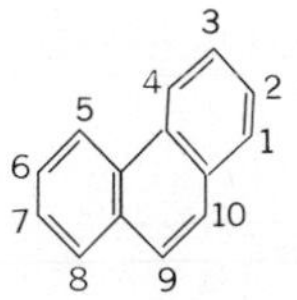

(I)

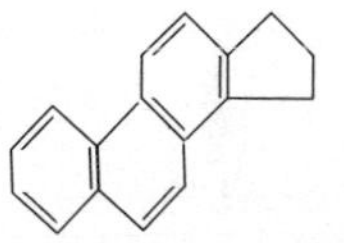

(II)

but it may also be produced by the hydrogenation of coal. Since carbazole and anthracene (usually present in crude phenanthrene) form mixed crystals with it, commercial grades of phenanthrene usually melt at higher temperatures than the pure compound.

Phenanthrene may be hydrogenated in the presence of copper chromite to yield 9,10-dihydrophenanthrene, or it may be oxidized to yield 9,10-phenanthrenequinone. In general, substitution reactions yield a mixture of products which are difficult to separate.

Phenanthrene has little commercial importance, except that it may be hydrogenated to symmetrical octahydrophenanthrene, which in turn may be rearranged and dehydrogenated to anthracene. Phenanthrene is of interest because the nucleus is found in some resin acids and is produced by the degradation of certain alkaloids. Reduction products of 1,2-cyclopentenophenanthrene (II) may be regarded as forming the skeleton of the steroids. *See* AROMATIC HYDROCARBON; POLYNUCLEAR HYDROCARBON.

[CHARLES K. BRADSHER]

Bibliography: L. F. Fieser and M. Fieser, *Natural Products Related to Phenanthrene*, 1949; K. Weissermel and H.-J. Arpe, *Industrial Organic Compounds*, 1978.

Phenol

The simplest member of a class of organic compounds possessing a hydroxyl group attached to a benzene ring or to a more complex aromatic ring system. Phenol itself, C_6H_5OH, may also be called hydroxybenzene or carbolic acid. Pure phenol is a colorless solid melting at 42°C, moderately soluble in water, and weakly acidic (pK 9.9).

Phenol has broad biocidal properties, and dilute aqueous solutions have long been used as an antiseptic. At higher concentrations phenol causes severe skin burns; it is a violent systemic poison.

Structure and nomenclature. Phenol has the structure shown. Simple substituted phenols, such as the three isomeric chlorophenols, are named as indicated, using the ortho, meta, and para prefixes. In more highly substituted phenols the positions of substitution are indicated by numbers. Com-

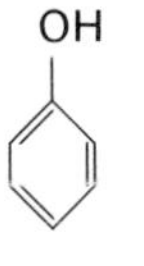

Phenol

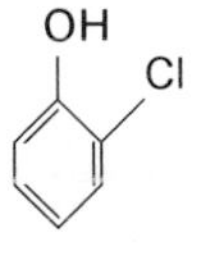

o-Chlorophenol

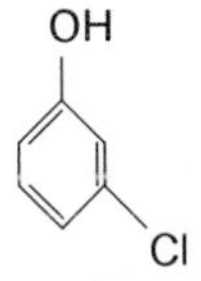

m-Chlorophenol

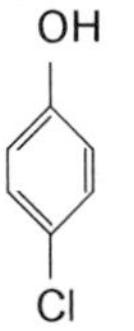

p-Chlorophenol

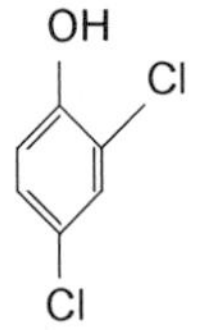

2,4-Dichlorophenol

pounds with more than one hydroxyl group per aromatic ring are known as polyhydric phenols, and include catechol, resorcinol, hydroquinone, phloroglucinol, and pyrogallol.

Reactions. The chemistry of phenol can be discussed under two headings.

Reactions of hydroxyl group. Phenols react readily with alkali hydroxides to form salts, reaction (1), and those phenols containing electronegative substituents (for example, trichlorophenol) are even acidic enough to react with bicarbonate.

$$C_6H_5OH + NaOH \rightarrow \underset{\text{Sodium phenoxide}}{C_6H_5ONa} + H_2O \qquad (1)$$

Like their aliphatic analogs, the alcohols, phenols form esters and ethers. Treatment of phenol with acetic anhydride yields phenyl acetate. The methyl ether of phenol (anisole) can be prepared by treatment of sodium phenoxide with methyl sulfate. Etherification of 2,4-dichlorophenol with sodium chloroacetate, reaction (2), yields 2,4-dichlorophenoxyacetic acid (2,4-D), which is a powerful herbicide.

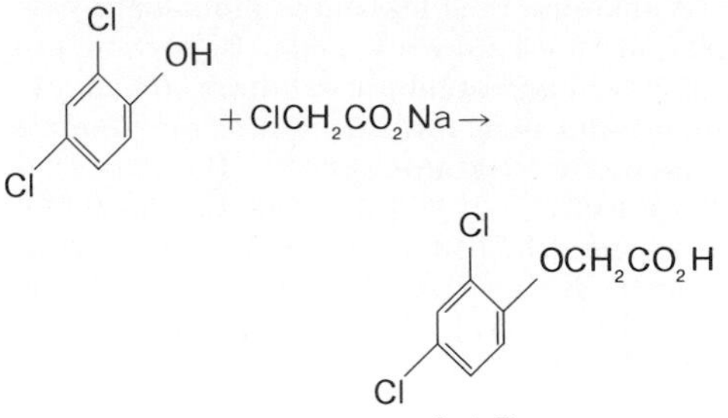

(2)

2,4-D

Reactions of aromatic ring. Under the conditions for electrophilic substitution reactions (for example, nitration, halogenation, sulfonation, alkylation, and diazo coupling), phenol is one of the most reactive aromatic compounds. Such reactions invariably lead to introduction of the substituent into an ortho or para position, as illustrated by the bromination of phenol, reactions (3). At low temperature in carbon disulfide solution, careful addition of one equivalent of bromine leads to the monobromo products of which the para isomer predominates, reaction (3*a*). Treatment of phenol with excess bromine in water yields 2,4,6-tribromophenol, in which both of the ortho and the para hydrogens have been replaced by bromine, reaction (3*b*). Reaction (3*b*) has been used as a fairly sensitive test for phenol, based upon the rapid precipitation of the insoluble tribromo product.

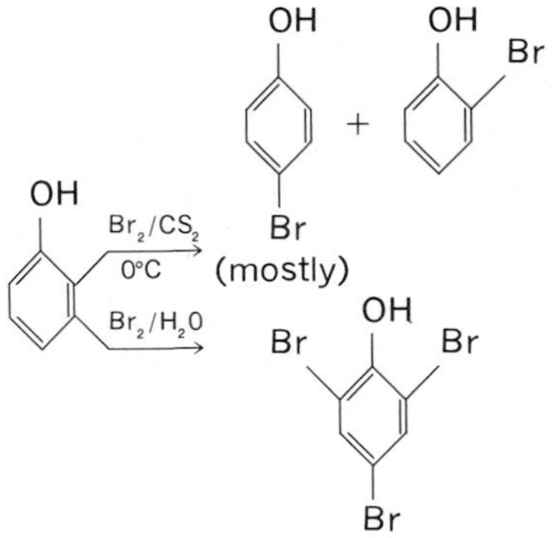

(3*a*)

(3*b*)

Two further examples of electrophilic substitution are the reaction with acetone, reaction (4), to

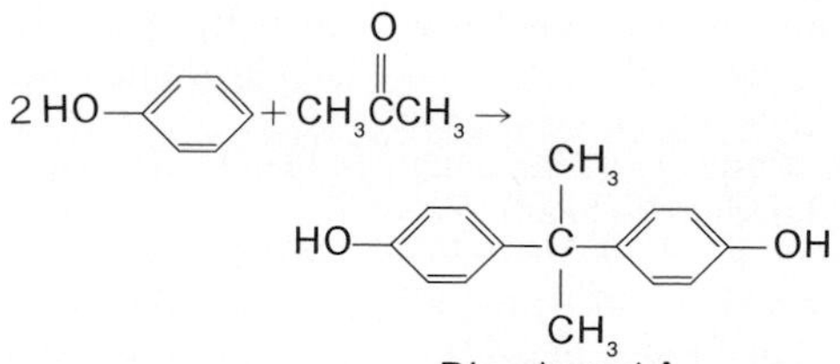

(4)

Bis-phenol A

form bis-phenol A, an important plastics intermediate, and the carboxylation of sodium phenoxide, reaction (5), to yield salicylic acid and thence aspirin.

Phenol is particularly sensitive to oxidation. Quinone is formed by chromic acid oxidation of phenol, but most oxidizing agents convert phenol

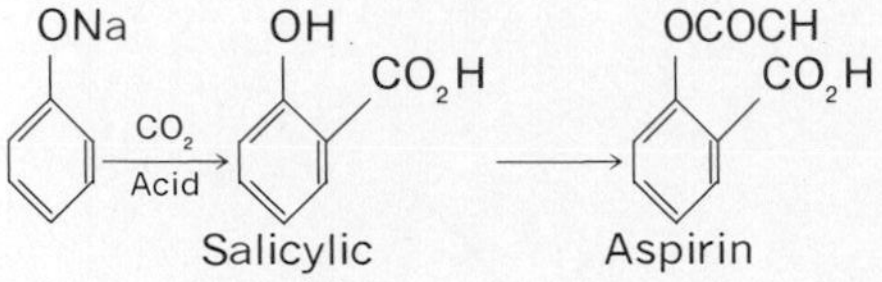

(5)

Salicylic acid   Aspirin

to a complex mixture of products derived from coupling of intermediate phenoxy radicals (oxidative coupling products). Certain 2,4,6-trisubstituted phenols are oxidized to relatively stable free radicals, and this property is used to advantage in the design of phenolic antioxidants. Reduction of phenol to cyclohexanol can be effected by hydrogen in the presence of catalysts such as nickel.

Production. Until World War I phenol was essentially a natural coal tar product. However, synthetic methods have replaced extraction from natural sources. There are many possible syntheses of phenol; six commercially significant ones are listed below.

1. The oldest process, that of sulfonation of benzene followed by neutralization of the sulfonic acid thus produced and fusion with caustic soda, was introduced into the United States about 1915. This is shown by reactions (6) and (7).

$$C_6H_6 + H_2SO_4 \rightarrow C_6H_5SO_2OH \xrightarrow{NaOH} C_6H_5ONa \quad (6)$$

$$C_6H_5ONa \xrightarrow{Acid} C_6H_5OH \quad (7)$$

2. The cumene hydroperoxide process yields the largest production in the United States, reactions (8) and (9). This process has grown very rapidly since its introduction in 1955. The coproduct acetone is of considerable value, which has undoubtedly contributed to the widespread acceptance of the process.

$$C_6H_5-\underset{CH_3}{\overset{CH_3}{|}}{CH} + O_2 \rightarrow C_6H_5-\underset{CH_3}{\overset{CH_3}{|}}{C}-OOH \quad (8)$$

$$C_6H_5-\underset{CH_3}{\overset{CH_3}{|}}{C}-OOH \xrightarrow{Acid} C_6H_5OH + CH_3-\overset{O}{\overset{\|}{C}}-CH_3 \quad (9)$$

3. The Raschig process was introduced in 1940 and involves a first-stage chlorination of benzene using an air–hydrochloric acid mixture, as shown by reactions (10) and (11). The by-product HCl is

$$C_6H_6 + HCl + \tfrac{1}{2}O_2 \rightarrow C_6H_5Cl + H_2O \quad (10)$$

$$C_6H_5Cl + H_2O \rightarrow C_6H_5OH + HCl \quad (11)$$

recycled to the first stage. High temperatures are employed, and high alloy reaction vessels are used to minimize corrosion.

4. The chlorination, or Dow, process was introduced in 1924 and is reasonably straightforward. Starting with chlorobenzene, the process proceeds as in reaction (12). Copper catalysts and high temperatures and pressures are employed.

$$C_6H_5Cl + NaOH \rightarrow NaCl + C_6H_5ONa \xrightarrow{Acid} C_6H_5OH \quad (12)$$

5. A process developed in 1962 starts with toluene and proceeds through oxidation to benzoic

acid, as seen in reaction (13). This is followed by

$$C_6H_5{-}CH_3 + \tfrac{3}{2}O_2 \rightarrow C_6H_5COOH + H_2O \quad (13)$$

further oxidation or decarboxylation in a second step using cupric benzoate as a catalyst, as shown by reaction (14).

$$C_6H_5COOH + \tfrac{1}{2}O_2 \rightarrow C_6H_5OH + CO_2 \quad (14)$$

6. The last process starts with cyclohexane, which is oxidized to a mixture of cyclohexanol and cyclohexanone (mixed oil). The next step is hydrogenation to convert all of the mixture to cyclohexanol. This intermediate is then, by classical methods, ring-dehydrogenated to yield phenol and hydrogen, which is recycled, as shown in reactions (15) and (16). Boron compounds are employed as catalysts or mediators.

$$C_6H_{12} + O_2 \rightarrow \text{(mixed oil)} \xrightarrow{\text{Hydro-genation}} C_6H_{11}OH \quad (15)$$

$$C_6H_{11}OH \xrightarrow{\text{Catalyst}} C_6H_5OH + 3H_2 \quad (16)$$

Uses and derivatives. Phenol is one of the most versatile and important industrial organic chemicals. It is the starting point for many diverse products used in the home and industry. A partial list includes: nylon, epoxy resins, surface active agents, synthetic detergents, plasticizers, antioxidants, lube oil additives, phenolic resins (with formaldehyde, furfural, and so on), cyclohexanol, adipic acid, polyurethanes, aspirin, dyes, wood preservatives, herbicides, drugs, fungicides, gasoline additives, inhibitors, explosives, and pesticides. Because of the rapidly increasing use of many of these derivatives, annual production and sale of phenol in the United States has increased steadily.

Naturally occurring phenols. Phenol and the methyl phenols (cresols and xylenols) are obtained from coal tar. Phenol has been found as a minor constituent of various plant materials, including tobacco leaves and pine needles. Carvacrol, thymol, and vanillin are examples of natural phenols of plant origin. Phenolic rings are also present in such biologically important compounds as the amino acid tyrosine and in estrone, a female sex hormone.

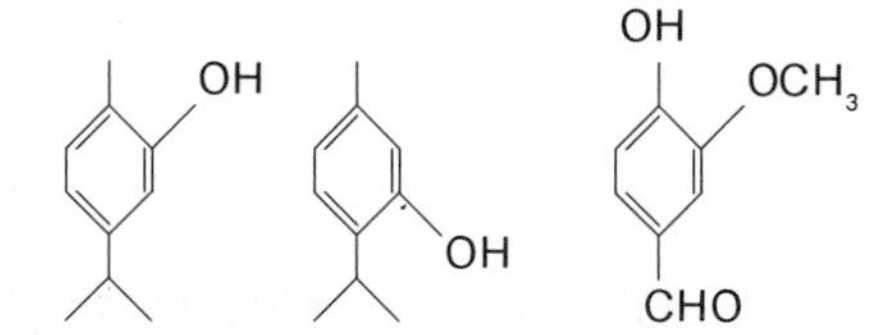

Carvacrol Thymol Vanillin

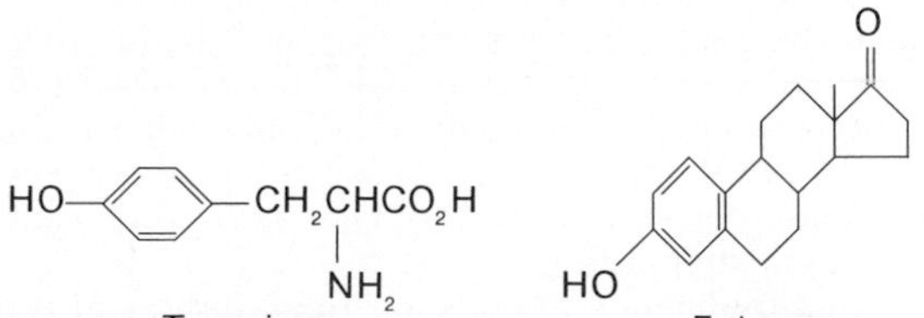

Tyrosine Estrone

See CATECHOL; HYDROQUINONE; NAPHTHOL; PICRIC ACID; RESORCINOL.

[ROBERT I. STIRTON; MARTIN STILES]

Bibliography: D. A. Whiting, Phenols, in D. H. R. Barton and D. Ollis, *Comprehensive Organic Chemistry*, vol. 4, ed. by P. G. Sammes, 1979.

Phenolic resin

One of the condensation products of phenols or phenolic derivatives with aldehydes such as formaldehyde and furfural. The term phenoplasts is sometimes used to refer to the whole group of products. The phenol-formaldehyde resins, developed commercially between 1905 and 1910, were the first truly synthetic polymers and have found wide usage for electrical insulation, molded objects, shell molds for metals, laminates, adhesives, and many other applications. They are characterized by low cost, dimensional stability, high strength, and resistance to aging. The combination of low cost and good properties is reflected in the fact that phenolic resins are produced in greater volume than any other thermosetting resin.

Phenol is prepared by the hydrolysis of chlorobenzene, by the alkali fusion of sodium benzene sulfonate, by the oxidation of toluene, by the dehydrogenation of cyclohexanol/cyclohexane, or by the decomposition of cumene hydroperoxide, as shown in reaction (1).

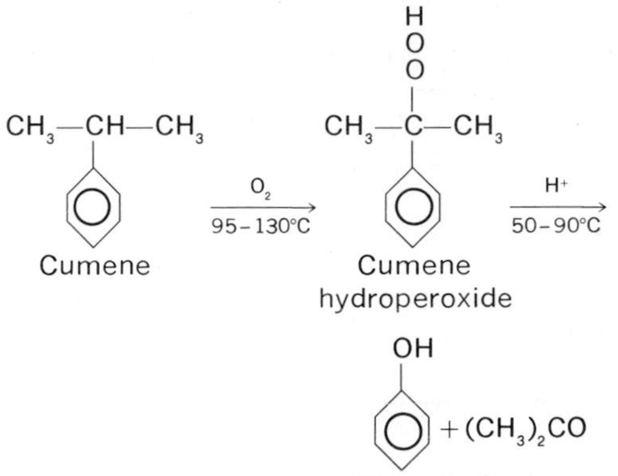

(1)

Resorcinol, obtained by the alkaline fusion of *m*-benzene disulfonic acid, and *m*-cresol from coal tar are also used. The structures are:

m-Dihydroxybenzene or resorcinol *m*-Cresol

Formaldehyde is produced by the oxidation of methane or methyl alcohol, reaction (2), and fur-

$$\underset{\text{Methyl alcohol}}{CH_3OH} + O_2 \xrightarrow[550\text{–}600°C]{\text{Ag, Fe, or Mo catalyst}} \underset{\text{Formal-dehyde}}{HCHO} + H_2O \quad (2)$$

fural is obtained by the hydrolysis of oat hulls. The structural formula of furfural is:

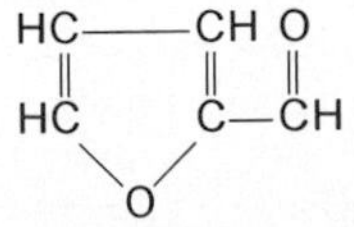

Furfural

See FURFURAL.

Polymerization. In the presence of an acid or base, phenol and aqueous formaldehyde react to form a solution of phenolic alcohols or methylol derivatives with the methylol groups in the ortho and para positions, reaction (3). This reaction takes

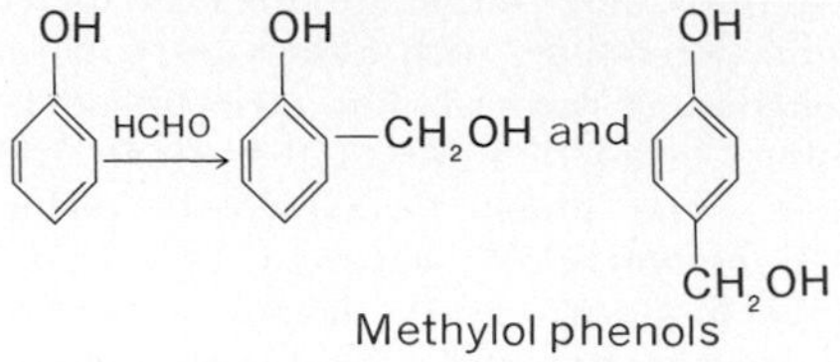

(3)

Methylol phenols

place quickly in a basic medium and slowly in an acidic medium.

The methylol phenols formed initially in a basic medium with formaldehyde in excess condense with each other and with additional formaldehyde to yield an "A-stage" resin or "resole," a brittle resin which is soluble and fusible. The resole resin consists of a mixture of isomers containing free methylol groups, which are available for subsequent cross-linking reactions to form a less-soluble "B-stage" resin. The structural formula for a typical liquid resole is:

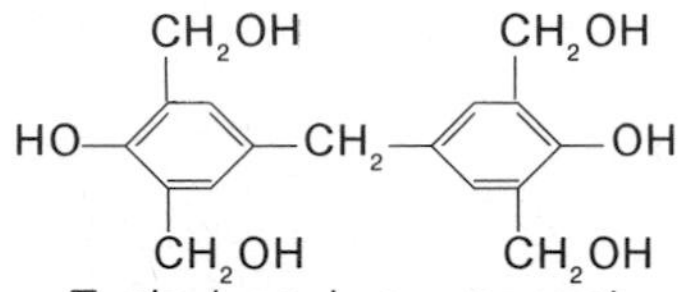

Typical resole component

Many structural variations are possible.

In the presence of acid and less than 0.86 mole of formaldehyde per mole of phenol, the primary alcohols react to yield diphenylmethane polymers called novolacs, which are soluble and fusible and contain an average of 5 or 6 phenol units per molecule. The structural formula for a typical novolac resin is:

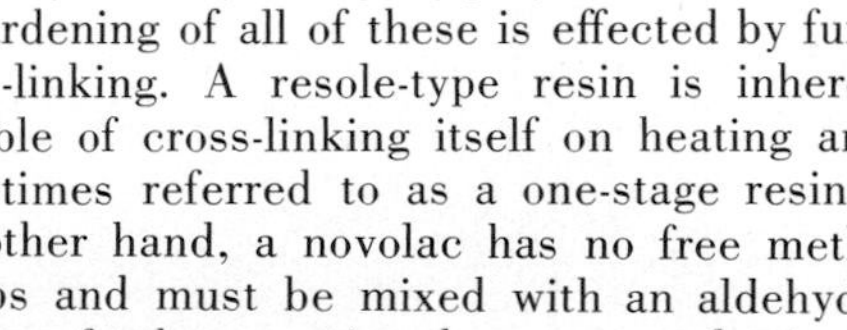

Novolac resin

These resins may also be referred to as A-stage resins. Novolacs may also be reacted with epichlorohydrin to yield epoxy polymers.

Hardening of all of these is effected by further cross-linking. A resole-type resin is inherently capable of cross-linking itself on heating and is sometimes referred to as a one-stage resin. On the other hand, a novolac has no free methylol groups and must be mixed with an aldehyde to undergo further reaction; hence a novolac is sometimes called a two-stage resin.

In the production of phenolic-resin molding compositions, it is common practice to neutralize, concentrate, and dry the B-stage resin; to mix it with fillers and, in the case of a novolac, a curing agent or hardener; and finally to compact it into the form of pellets or briquets. Other ingredients may be present also, such as curing accelerators, pigments, lubricants, and plasticizers. The curing agent for a novolac is hexamethylenetetramine, which at the temperature of molding reacts with water to form formaldehyde and ammonia. In the presence of ammonia and the additional formaldehyde, the B-stage resin cures in the mold to yield a highly cross-linked, insoluble, and infusible C-stage product. This resin is represented as:

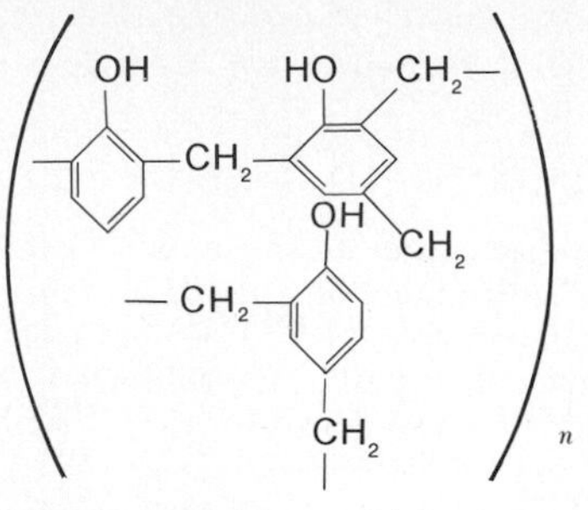

C-stage phenol-formaldehyde r

By use of *m*-phenol derivatives, such as resorcinol or *m*-cresol, resins are obtained which cure rapidly at low temperature because the meta substituents activate the ortho and para positions. An ortho or para alkyl phenol which has only two active sites available for reaction (difunctional) can be used—such as *p-tert*-butylphenol:

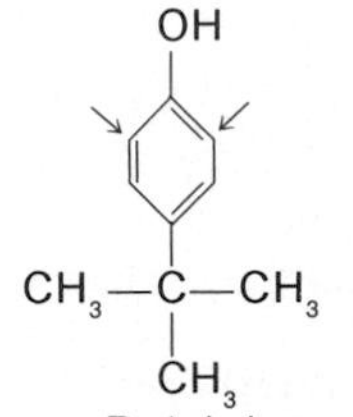

p-tert-Butylphenol

Then oil-soluble, thermoplastic resins are formed instead of the cross-linked materials obtained from the trifunctional phenols just discussed. These products, somewhat more expensive than ordinary phenolic resins, are used in special paint, varnish, and adhesive formulations.

With the use of furfural instead of formaldehyde, the B-stage resin has the unique property of remaining thermoplastic for a relatively long time. The phenol-furfural compositions are useful for molding large complex forms in which extra time is needed for the resin to fill the mold completely.

Fabrication and use. Phenolic resins can be cast from syrupy intermediates or molded from B-stage solid resins. Laminated products can be produced by impregnating fiber, cloth, wood, and other materials with the resin. After heating, laminated sheets can be pressed into any shapes desired. Most of the phenolic plastics can be machined if necessary.

An increasingly important type of phenolic resin product is rigid foam. One type is prepared by incorporating a blowing agent in a curing mixture so that the heat of reaction decomposes the blowing agent. A second type, "syntactic" foam, consists of microscopic hollow spheres of phenolic resin mixed with a curable binder such as an epoxy resin or polyester. Foams of this type have very high strength after curing.

Cured phenolic plastics are rigid, hard, and resistant to chemicals (except strong alkali) and to heat.

Some of the uses for phenolic resins are for making precisely molded articles, such as telephone parts, for manufacturing strong and durable laminated boards, or for impregnating fabrics, wood, or paper. Phenolic resins are also widely used as adhesives, as the binder for grinding wheels, as thermal insulation panels, as ion-exchange resins, and in paints and varnishes. *See* EPOXY RESIN; ION EXCHANGE; PHENOL; POLYMERIZATION.

[JOHN A. MANSON]

Bibliography: J. A. Brydson, *Plastic Materials*, 3d ed., 1975.

Phosphate

A negative ion having the formula PO_4^{3-}. Phosphates are derived from phosphoric acid, H_3PO_4.

The term phosphate is a broad term which encompasses all anions derived from acids containing phosphorus in the 5+ oxidation state, as indicated in the following list (all of those listed are obtained from P_4O_{10} and water):

$(HPO_3)_n$	Metaphosphoric acid
$H_5P_3O_{10}$	Triphosphoric or tripolyphosphoric acid
$H_4P_2O_7$	Pyrophosphoric acid
H_3PO_4	Orthophosphoric acid

The naming of these phosphates is complicated by the fact that the acids contain several hydrogens which can be replaced stepwise by reaction with a base and the fact that phosphates exist as polymers of the simpler acids listed. In the case of the most common acid, orthophosphoric, the salts are named as follows (phosphate means orthophosphate):

NaH_2PO_4	Monosodium phosphate Sodium dihydrogen phosphate Primary sodium phosphate
Na_2HPO_4	Disodium phosphate Sodium monohydrogen phosphate Secondary sodium phosphate
Na_3PO_4	Trisodium phosphate Tertiary sodium phosphate Normal sodium phosphate

The alkali metal phosphates and the primary alkaline-earth metal phosphates are soluble in water, whereas most other metal phosphates are practically insoluble at neutral pH.

A solution of trisodium phosphate is strongly basic and is used as a cleaning compound and water softener. Phosphates are important ingredients in commercial fertilizers. Natural phosphate rock can be converted into a useful fertilizer, superphosphate, by a reaction with sulfuric acid, as shown below.

$$Ca_3(PO_4)_3 + 2H_2SO_4 + 4H_2O \rightleftharpoons Ca(H_2PO_4)_2 + 2(CaSO_4 \cdot 2H_2O)$$

An important use of polymeric phosphates is as an ingredient in synthetic detergents and as sequestering agents.

The phosphate ion gives a yellow ammonium phosphomolybdate precipitate and yellow Ag_3PO_4 precipitate, which serve as analytical tests.

Certain organic phosphates have been used as insecticides and nerve gases. *See* ORGANOPHOSPHORUS COMPOUND; PHOSPHORUS.

[E. EUGENE WEAVER]

Phosphatide

A complex lipid containing phosphorus. The phosphatides, also known as phospholipids, are usually divided into groups on the basis of compounds from which they are derived. For example, glycerophosphatides are derived from glycerophosphoric acid, sphingophosphatides are derived from sphingosine phosphate, and inositol phosphatides are derived from inositol phosphates.

Glycerophosphatides. These phosphatides are derived from glycerophosphoric acid (I), where $R_1 = R_2 = R_3 = H$.

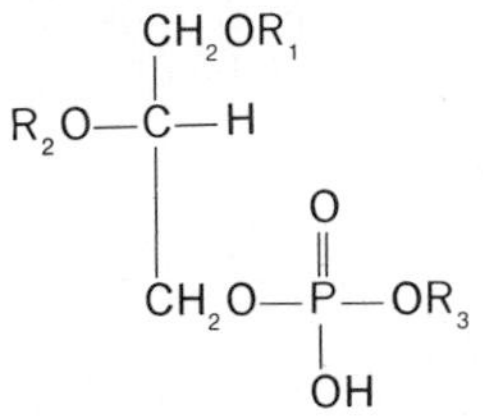

(I)

The following compounds are glycerophosphatides: (1) phosphatidyl ethanolamine, where $R_1 = R_2$ = fatty acid (in ester linkage), R_3 = ethanolamine (in ester linkage); (2) phosphatidyl choline or lecithin, where $R_1 = R_2$ = fatty acid, R_3 = choline; (3) phosphatidyl serine, where $R_1 = R_2$ = fatty acid, R_3 = serine; (4) phosphatidyl inositol, where $R_1 = R_2$ = fatty acid, R_3 = inositol; (5) lysophosphatidyl ethanolamine, where R_1 = fatty acid, R_2 = H, R_3 = ethanolamine; (6) lysophosphatidylcholine, where R_1 = fatty acid, R_2 = H, R_3 = choline; (7) plasmalogens, where R_1 = α,β-unsaturated ether, R_2 = fatty acid, R_3 = ethanolamine, choline, or serine; (8) saturated ether lipids which are analogs of the plasmalogens but contain a saturated ether at R_1; (9) phosphatidic acid, where $R_1 = R_2$ = fatty acid, R_3 = H; (10) phosphatidyl glycerol, where $R_1 = R_2$ = fatty acid, R_3 = glycerol; (11) cardiolipin, consisting of two molecules of (I), $R_1 = R_2$ = fatty acid, joined together through glycerol at R_3.

Phosphonolipids. These compounds contain the phosphonic acid derivative (II), where R is glycerol or sphingosine. The phosphonolipids are present in some natural phosphatide extracts.

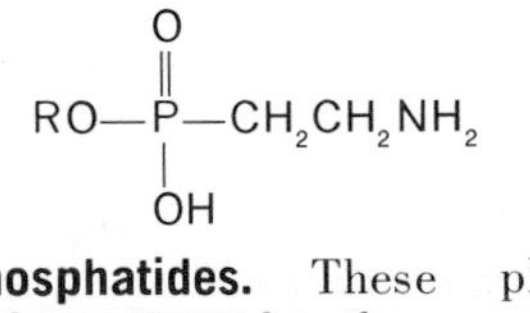

(II)

Sphingophosphatides. These phosphatides contain a sphingosine phosphate residue. They include sphingomyelin (III), the ethanolamine-containing analog of sphingomyelin and phytoglycolipid.

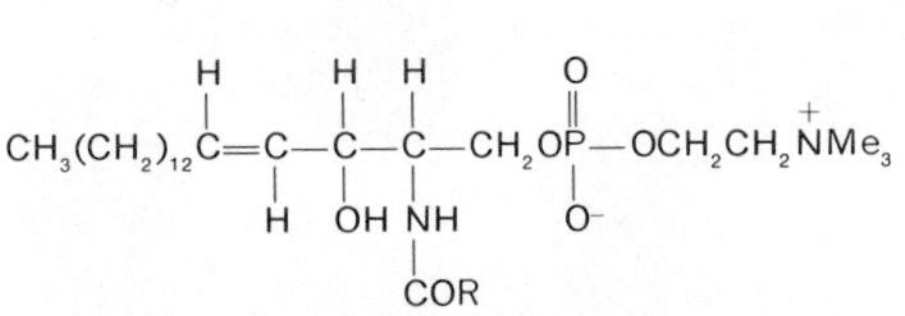

(III)

Occurrence and functions. Phosphatidyl ethanolamine, lecithin, phosphatidyl inositol, and the plasmalogens are present in both plant and animal tissues; phytoglycolipids have been found only in plants; sphingomyelin has been found only in animal tissues. The phosphatides are important components of biological membranes.

Since an individual phosphatide may contain a variety of fatty acid residues, it may be described as pure only with this limitation in mind. Most of the highly unsaturated fatty acids of animal tissue lipids occur in the phosphatides. Phosphatides can act as protective colloids, as wetting and emulsifying agents, and as antioxidants, and are therefore used considerably in the food and petroleum industries. The chief source of commercial phosphatides is soybean.

[H. E. CARTER; R. H. GIGG]

Bibliography: G. B. Ansell et al., *Form and Function of the Phospholipids*, 1973.

Phosphorus

A chemical element, P, atomic number 15, atomic weight 30.9738. Phosphorus forms the basis of a very large number of compounds, the most important class of which are the phosphates. For every form of life, phosphates play an essential role in all

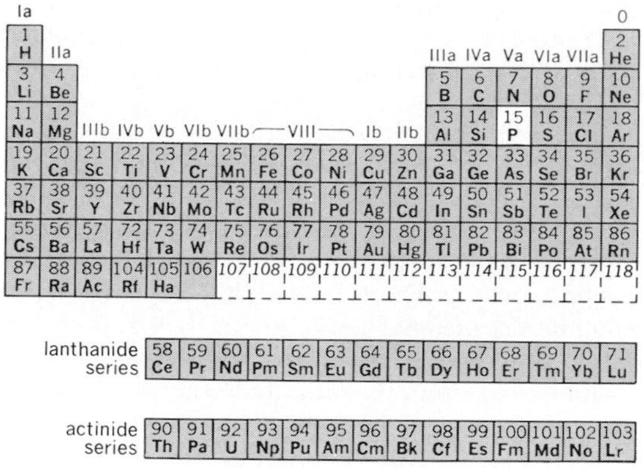

energy-transfer processes such as metabolism, photosynthesis, nerve function, and muscle action. The nucleic acids which among other things make up the hereditary material (the chromosomes) are phosphates, as are a number of coenzymes. Animal skeletons consist of a calcium phosphate.

About three-quarters of the total phosphorus (in all of its chemical forms) used in the United States goes into fertilizers. Other important uses are as builders for detergents, nutrient supplements for animal feeds, water softeners, additives for foods and pharmaceuticals, coating agents for metal-surface treatment, additives in metallurgy, plasticizers, insecticides, and additives for petroleum products. Except for the last four items, these uses involve phosphates. Of the phosphorus compounds utilized in the United States, approximately three-fourths are converted into an impure form of phosphoric acid (wet-process acid) and the remainder into the element (white phosphorus) before further chemical processing into the end products of commerce.

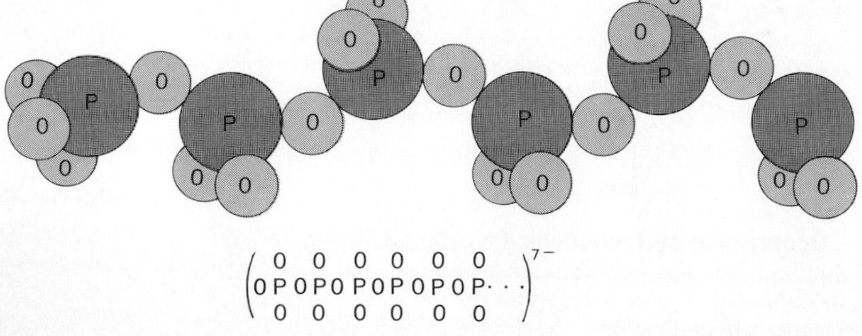

Fig. 1. Long-chain phosphate anion, $(P_nO_{3n+1})^{(n+2)-}$.

Occurrence and manufacture. Of the nearly 200 different phosphate minerals, only one, fluorapatite, is commercially important. Fluorapatite, $Ca_5F(PO_4)_3$, is mined chiefly from large secondary deposits originating from the bones of dead creatures deposited on the bottom of prehistoric seas and from bird droppings in ancient rookeries. In the United States the major phosphate deposits are in Florida, Tennessee, and the Montana-Idaho region. Other important deposits are found in Morocco, Tunisia, and the Soviet Union.

Conversion of phosphate rock (the name given to the common impure form of the mineral apatite) to usable chemicals is accomplished by two major routes: wet acid and elemental phosphorus. In the wet-acid process the phosphate rock is treated with sulfuric acid to obtain a very impure phosphoric acid, plus a precipitate of calcium sulfate. A large body of technology has been developed to achieve easy removal of the calcium sulfate and subsequent concentration and partial purification of the phosphoric acid. Under present-day economic conditions in the United States, the cost of making industrial-grade phosphates via the wet-acid process is about equivalent to the cost of converting the ore to elemental phosphorus and then burning it to give a highly pure phosphoric acid, which is converted into the phosphate. In elemental-phosphorus manufacture phosphate rock, silica, and coke are fed into an electric furnace in which a high-temperature reaction occurs to give the white modification of elemental phosphorus, P_4, a calcium silicate slag, and from iron impurities in the phosphate rock an iron phosphide called ferrophosphorus. Pretreatment of the ore removes most of the fluorine.

Although some elemental phosphorus is used as such in incendiary bombs, in metallurgy, and in the production of organic derivatives and chemicals for matches, most elemental phosphorus is converted to phosphoric acid by reaction with air and water in large burning towers. Elemental phosphorus and phosphoric acid are the starting materials for the synthesis of all other compounds of phosphorus. Phosphoric acid is treated with soda ash, Na_2CO_3, and the resulting orthophosphate composition is then calcined in large rotary converters to make pyro- and tripolyphosphates in very large amounts.

Chemistry. Because of the tremendously large number of compounds based on carbon, descriptive chemistry has been divided into organic, which treats of carbon compounds, and inorganic chemistry, which deals with the compounds of more than 100 other elements. Recent scientific work in phosphorus chemistry indicates that there may be as many compounds based on phosphorus as on carbon; the chemistry of phosphorus therefore may become a major branch of chemical knowledge. In organic chemistry it has been customary to group the varous chemical compounds based on carbon into families which are called homologous series. This can also be done in the chemistry of phosphorus compounds, even though many phosphorus-based families are incomplete. The best known of the families of compounds based on phosphorus is the group of chain phosphates. Phosphate salts consist of cations, such as sodium, along with chain anions which may have 1–1,000,000 phosphorus atoms per anion. A structural representation of the end of a long-chain, stretched-out phosphate anion is given in Fig. 1.

As shown in the figure, the phosphates are based on phosphorus atoms tetrahedrally surrounded by oxygen atoms, with the lowest member of the series being the simple PO_4^{3-} anion (the orthophosphate ion). The family of chain phosphates is based on a row of alternating phosphorus and oxygen atoms in which each phosphorus atom remains in the center of a tetrahedron of four oxygen atoms, as shown in the structural diagram. There is also a closely related family of ring phosphates, a member of which, the trimetaphosphate, is shown in Fig. 2.

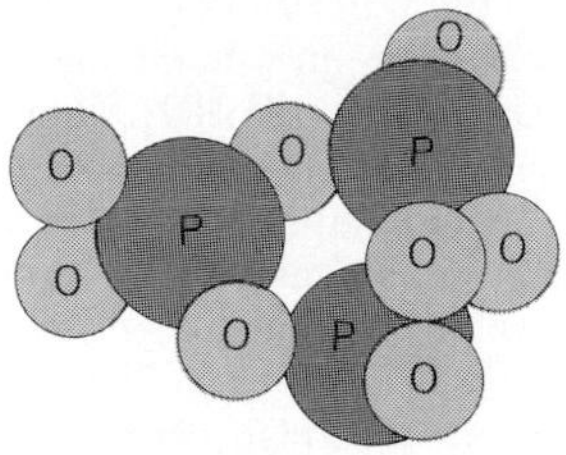

Fig. 2. Ring phosphate anion, $(P_3O_9)^{3-}$.

An interesting structural characteristic of many known phosphorus compounds is the formation of cagelike structures. Such cagelike molecules are exemplified by white phosphorus, P_4, and one of the phosphorus pentoxides, P_4O_{10} (Figs. 3 and 4).

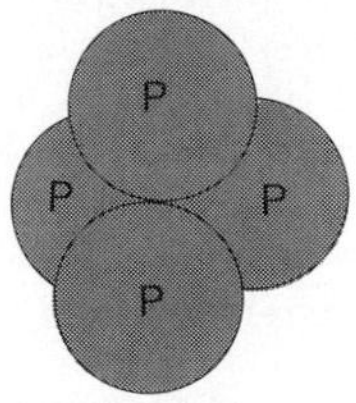

Fig. 3. White phosphorus, P_4.

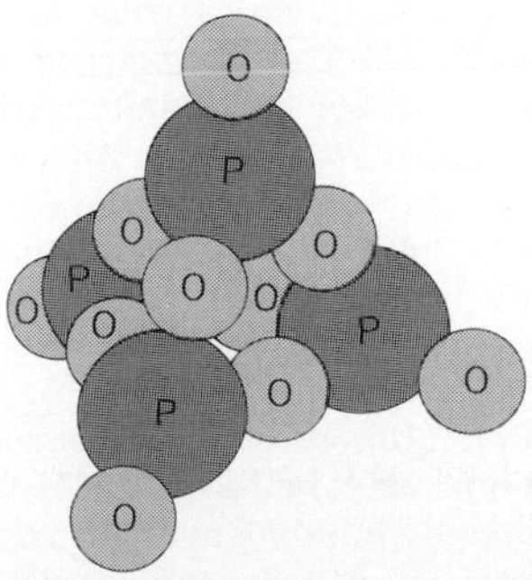

Fig. 4. Phosphorus pentoxide, P_4O_{10}, in vapor state.

Network structures are also common; for example, black phosphorus crystals in which the atoms are bonded together in the form of vast, corrugated planes (Fig. 5).

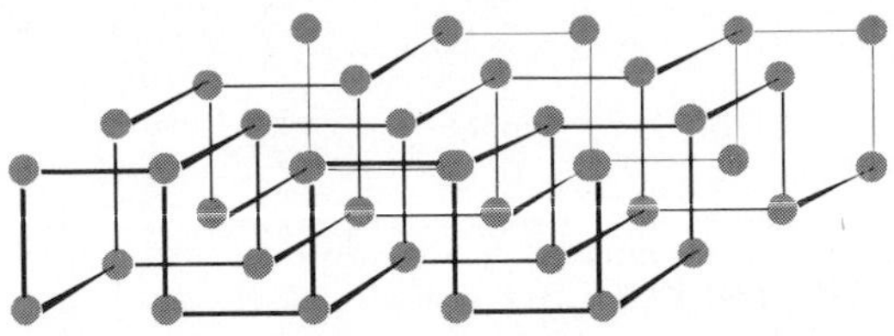

Fig. 5. Black phosphorus, P_n.

In the majority of its compounds, phosphorus is chemically bonded to four neighboring atoms. There is also a large number of compounds in which one of the four neighboring atoms is absent, and in which its place is taken by an unshared pair of electrons. Two typical compounds based on this type of phosphorus are shown in Figs. 6 and 7. In

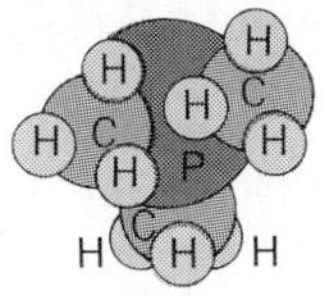

Fig. 6. Trimethyl phosphite, $P(OCH_3)_3$.

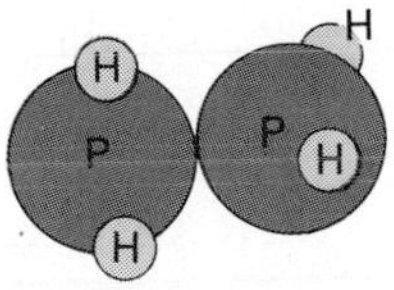

Fig. 7. Biphosphine, P_2H_4

addition to the compounds based on quadruply connected phosphorus and those based on triply connected phosphorus, there are also a few compounds in which there are five or six neighboring atoms bonded to the phosphorus. These compounds are very reactive and tend to be unstable because of the use of *d* orbitals in their σ-bond electronic structure. Examples are given in Figs. 8 and 9.

As in the case for much of inorganic chemistry, structural reorganization plays an important role in the chemistry of phosphorus compounds.

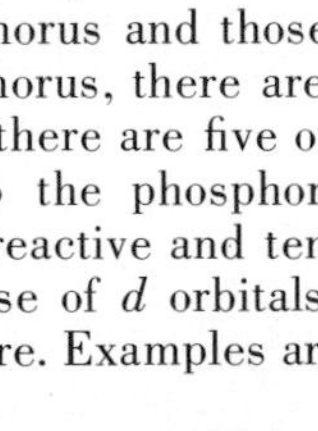

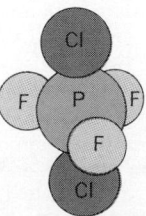

Fig. 8. Phosphorus dichloride trifluoride, PCl_2F_3.

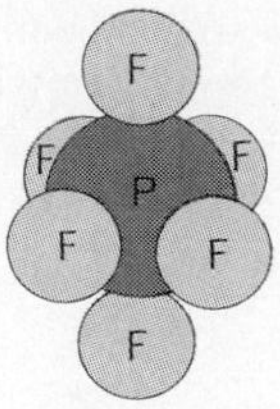

Fig. 9. Hexafluorophosphate anion, PF_6^-.

Thus, for example, when various mixtures of $POBr_3$ and $POCl_3$ are sealed in a glass tube and allowed to come to equilibrium, the intermediate compounds, $POClBr_2$ and $POCl_2Br$, are formed in various amounts depending on the ratio of the starting materials. The $POBr_3$—$POCl_3$ reorganization involves compounds based on a single phosphorus atom to which is bonded one oxygen and three halogen atoms (chlorine and bromine are halogens). Structural reorganization also occurs between various members of a homologous series of compounds. In the polyphosphoryl chloride homologous series, reorganization takes place by exchange of bridging oxygen atoms with chlorine atoms, just as in the $POBr_3$—$POCl_3$ system the exchange is between chlorine and bromine atoms. The various structural units in a polyphosphoryl chloride composition are the monophosphorus compound, $POCl_3$; the end group, $Cl(O)PO_{1/2}$—; the middle group, —$O_{1/2}(Cl)P(O)O_{1/2}$—; and the branching group, $OP(O_{1/2}\text{—})_3$, in which the bridging oxygen atoms are shown as $O_{1/2}$, since they are shared between neighboring phosphorus atoms. A typical structure that can be found in a mixture of polyphosphoryl chlorides is shown in Fig. 10.

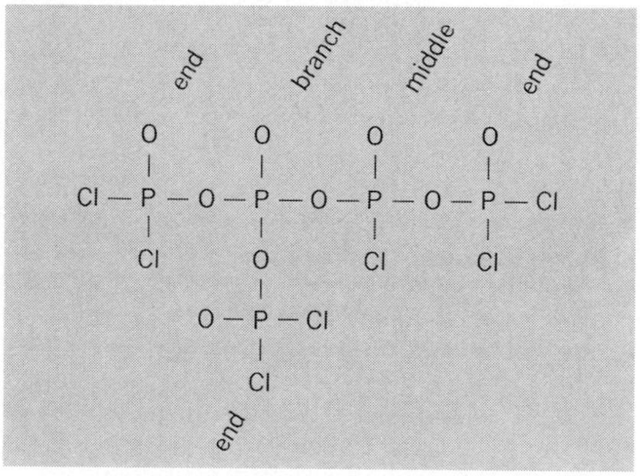

Fig. 10. Isopentaphosphoryl chloride, $P_5O_9Cl_7$.

When various ratios of chlorine to oxygen are employed, the distribution of the structural units changes, as shown in Fig. 10, where *A* stands for the monophosphorus compound, $POCl_3$; *B* for the ends; *C* for the middles; *D* for the branches, and *D'* for the completely branched compound, phosphorus pentoxide. The ends, middles, and branches do not exist by themselves but must be combined together to form chemical compounds. The line labeled *x* in the figure represents the limit beyond which there is a sufficiently large proportion of branching points that infinite wall-to-wall molecular structures become statistically probable. The presence of such wall-to-wall molecular structures in the mixture of various sized and shaped polyphosphoryl chloride molecules leads to high viscosities and noticeable elastic behavior.

In spite of the fact that homologous series and compounds based on a number of phosphorus atoms are emphasized in this article, the extensive chemical literature before 1950 dealing with phosphorus chemistry was restricted almost entirely to compounds thought to be based on a single phosphorus atom (monophosphorus compounds).

During the 1960s and 1970s a large number of organic-phosphorus compounds were prepared. Most of these chemical structures involve three or four neighboring atoms bonded to the phosphorus, but stable structures having two, five, or six neighboring atoms per phosphorus are now also known. Some organic-phosphorus compounds exhibit direct bonding between a phosphorus and a carbon atom, while others have C-O-P, C-S-P, or C-N-P linkages as well as various other atomic arrangements.

Principal compounds and uses. Essentially all of the phosphorus used in commerce is in the form of phosphates. The majority of phosphatic fertilizers consist of highly impure monocalcium or dicalcium orthophosphate, $Ca(H_2PO_4)_2$ and $CaHPO_4$. These phosphates are salts of orthophosphoric acid, which is the monophosphorus compound in the phosphate homologous series. Impure dicalcium orthophosphate for fertilizer use is usually called superphosphate, whereas the impure monocalcium phosphate used in this application is called triple superphosphate.

Two properties of the family of chain phosphates have led to numerous industrial applications for these compounds. These properties are deflocculation of colloidal particles and formation of soluble complexes with cations. The chain phosphates are strongly adsorbed on the surfaces of inorganic solids and, hence, give these surfaces high negative charges. When finely divided particles bear such high charges, they repel each other and are deflocculated, peptized, or dispersed. An interesting example of this phenomenon is found when a plastic clay-water mass is treated with a chain phosphate. By addition of perhaps a few tenths of 1% of sodium tripolyphosphate to a plastic mass of clay suitably rigid for sculpturing, the clay particles are deflocculated so that the mass liquefies to a consistency similar to that of tomato soup.

The formation of soluble complexes with cations has often been described under the term sequestration, because a complexed ion is sequestered or hidden away in the solution so that it no longer exhibits its normal chemical reactions. The calcium and magnesium of hard water are sequestered by the addition of small (stoichiometric) amounts of chain phosphates so that the water is effectively softened. The complexed calcium will then no longer form precipitates with the carbonate or sulfate in the water to give pipe scale, or with soap anions to give, for example, a ring around the bathtub.

The third member of the family of sodium phosphates, sodium tripolyphosphate, is the major compound used in building synthetic detergents to achieve improved cleaning, primarily by dispersing inorganic soil and softening the water. The average household detergent produced in the United States for washing clothes consists of 50% by weight of sodium tripolyphosphate, $Na_5P_3O_{10}$. This compound is used extensively in water softening, as are other members of the homologous series of chain phosphates. The large-volume usage of phosphates in detergent building has led to unwanted growth of algae in inland waters (lakes and rivers) into which the dirty washwaters are discharged. As a result of this fertilizing action, phosphates are considered as water pollutants in those areas where such discharges occur; and steps have been taken to reduce the phosphate content of detergents or, in some sewage treatment plants, to remove it. For reasonably fast-flowing rivers that discharge directly into the ocean, phosphates are not a problem. *See* SURFACE-ACTIVE AGENT; WATER SOFTENING.

An interesting water-softening application is found in "threshold treatment" in which tiny traces of a chain phosphate (much less than would be used in sequestering) are used to prevent the formation of pipe scale from hard waters. This application is related to the dispersing action of the phosphates, because traces of phosphate adsorb on the growing surface of the pipe scale as it begins to form, and this inhibits its further growth.

A major pharmaceutical use of phosphates is in toothpastes, in which dicalcium phosphate is the most popular polishing agent. Monocalcium phosphate and sodium acid pyrophosphate, $Na_2H_2P_2O_7$ (the pyrophosphate is the second member of the phosphate family), are employed as leavening agents in cake mixes, refrigerated biscuits, self-rising flour, and baking powder.

Special mixtures based on orthophosphoric acid, H_3PO_4, are used to phosphatize metal surfaces. In this treatment the surfaces become covered with a thin adhering layer of insoluble orthophosphate salts which protect the metal from corrosion and offer an especially adherent base for painting. Automobile bodies, for example, are generally phosphatized before they are painted to prevent rusting in use. Orthophosphate esters find wide use as plasticizers that have flameproofing properties and as gasoline and oil additives.

The phosphorus compound of major biological importance is adenosinetriphosphate (ATP), which is an ester of the salt, sodium tripolyphosphate, widely employed in detergents and water-softening compounds. Practically every reaction in metabolism and photosynthesis involves the hydrolysis of this tripolyphosphate to its pyrophosphate derivative, called adenosinediphosphate (ADP). The hydrolysis of chain phosphates occurs through splitting of a P—O—P linkage as indicated in the reaction below.

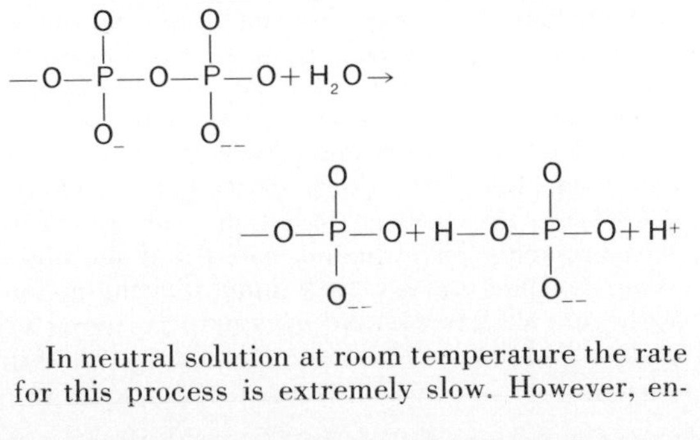

In neutral solution at room temperature the rate for this process is extremely slow. However, enzymes increase the rate many thousandfold. The equilibrium between ATP, water, ADP, and the orthophosphate ion is strongly shifted toward the hydrolysis product, ADP and the orthophosphate ion. Because of these facts, organic reactions in biological systems are naturally controlled so that life can exist.

[JOHN R. VAN WAZER]

Bibliography: J. Emsley and D. Hall, *The Chemistry of Phosphorus*, 1970.

Photoacoustic spectroscopy

A technique for measuring small absorption coefficients in gaseous and condensed media, involving the sensing of optical absorption by detection of sound. It is frequently called optoacoustic spectroscopy. Although the technique dates back to 1880 when A. G. Bell used chopped sunlight as the source of radiation, it remained dormant for many years, primarily because of the lack of suitable powerful sources of tunable radiation. However, the usefulness of optoacoustic detection for spectroscopic applications was recognized early in its development, and pollution monitoring instruments (called spectrophones) dedicated for detection of specific gaseous constituents have been used intermittently since Bell's work.

Methods of measuring absorption. During the transmission of optical radiation through a sample (gas, liquid, or solid), the absorption of radiation by the sample can be measured by at least three techniques. The first one is the straight-forward detection technique which requires a measurement of the optical radiation level with and without the sample in the optical path. The transmitted power P_{out} and the incident power P_{in} are related through Eq. (1), where α is the absorption coefficient and l

$$P_{out} = P_{in}e^{-\alpha l} \qquad (1)$$

is the length of the absorber. With this technique, the minimum measurable αl is of the order of 10^{-4} unless special precautions have been taken to stabilize the source of radiation.

The second of the techniques is the derivative absorption technique where the frequency of the input radiation is modulated at a low radio frequency or audio frequency, ω_m. The transmitted radiation then contains a time-varying component at ω_m, if the optical path contains absorption which has a frequency-dependent structure. (For structureless absorption, modulated absorption spectroscopy does not provide a signal that can characterize the amount of absorption.) For situations where the absorption has well-defined structure, the modulation absorption spectroscopy can be used to measure αl as small as about 10^{-8} for sufficiently high input powers. The ability to measure the small absorption effects is independent of the input and output power levels for the straight-forward measurement technique as long as the noise contributed by the detector is not a factor in determining the signal-to-noise ratio. For the derivative absorption technique, the smallest αl that can be measured varies as $(P_{in})^{-1}$ until the shot noise of the detector begins to be appreciable.

The third technique, optoacoustic detection, is a calorimetric method where no direct detection of optical radiation is carried out but, instead, a measurement is made of the energy, with power P_{abs}, absorbed by the medium from the incident radia-

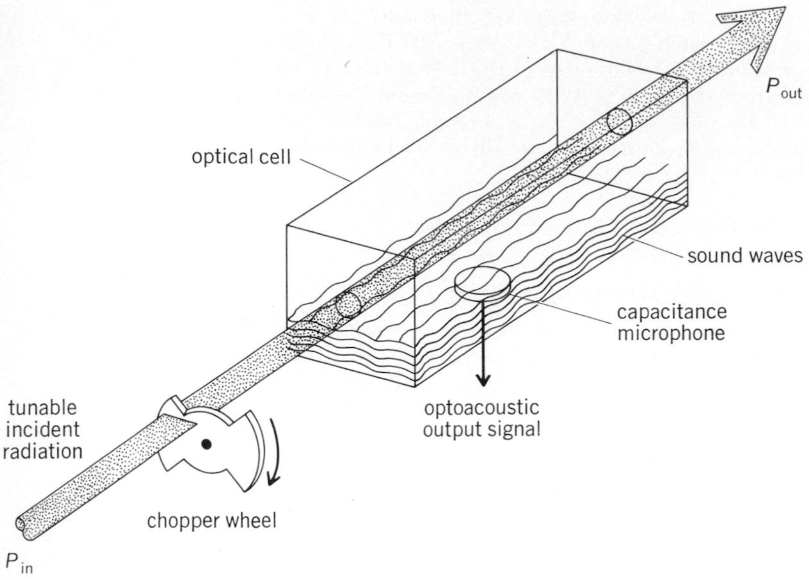

Fig. 1. Optoacoustic cell for gaseous spectroscopy.

tion, Eq. (2). Thus the optoacoustic signal, V_{oa}, is

$$P_{abs} = P_{in}(1 - e^{-\alpha l}) \qquad (2)$$

given by Eq. (3), where K is the a constant describ-

$$V_{oa} = K\{P_{in}(1 - e^{-\alpha l})\} \approx KP_{in}\alpha l \qquad (\text{for } \alpha l \ll 1) \qquad (3)$$

ing the conversion factor for transforming the absorbed energy into an electrical signal using an appropriate transducer. It has been tacitly assumed that the absorbed energy is lost by nonradiative means rather than by reradiation. The optoacoustic detection scheme implies that the absorbed energy will be converted into acoustic energy for eventual detection.

From Eq. (3), the optoacoustic signal is proportional to the incident power and the absorption-length product αl. Thus, for given sources of noise from the detection transducers, the signal-to-noise ratio improves as the incident energy is increased. Put differently, the smallest amount of absorption that can be measured using the optoacoustic technique varies as $(P_{in})^{-1}$ with no limitation on the level to which P_{in} can be increased for detecting small absorptions. Values of αl as small as 10^{-10} can be measured in the gas phase. The techniques generally used for gases and those used for condensed-phase optoacoustic spectroscopy differ in detail somewhat.

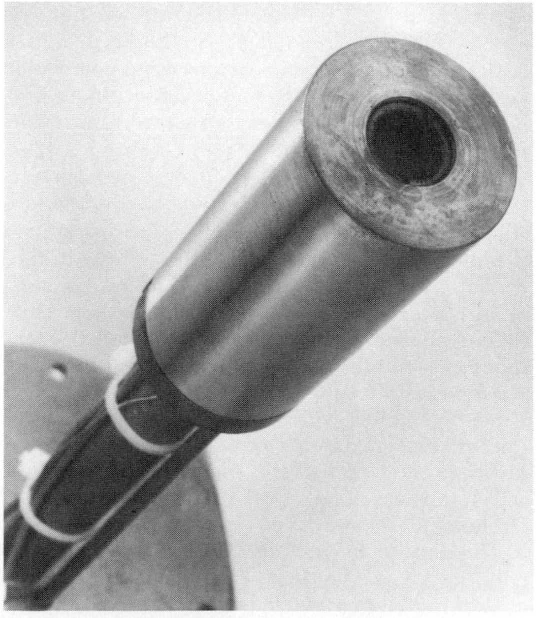

Fig. 2. Sensitive optoacoustic gaseous spectroscopy cell.

Gases. If optical radiation, is amplitude-modulated at an audio frequency, the absorption of such radiation by a gaseous medium that has been confined in a cell with appropriate optical windows for the entrance and exit of the radiation, and nonradiative relaxation of the medium, will cause a periodic variation in the temperature of the column of the irradiated gas (Fig. 1). Such a periodic rise and fall in temperature gives rise to a corresponding periodic variation in the gas pressure at the audio frequency. The audio-frequency pressure fluctuations (that is, sound) are efficiently detected using a sensitive gas-phase microphone. The intrinsic noise limitation to the optoacoustic detection scheme arises from the Brownian motion of gas atoms/molecules, and Kreuzer showed that the minimum detectable absorbed power is $P_{min} \approx 3.6 \times 10^{-11}$ W for a 11.9-cm-long cell. Substituting P_{min} for P_{abs} in Eq. (2), and noting, as in Eq. (3), that $(1 - e^{-\alpha l}) \cong \alpha l$ for $\alpha l \ll 1$, it follows that α_{min} varies as (P_{min}/P_{in}) as indicated above. The usefulness of the optoacoustic detection for measurement of small absorption coefficients became evident with the development of a variety of tunable high-power laser sources which could take advantage of the $(P_{in})^{-1}$ dependence. Using a spin-flip Raman laser tunable in the 5.0- to 5.8-μm range, with a power output of approximately 0.1 W, C. K. N. Patel and R. J. Kerl were able to detect α_{min} of approximately 10^{-10} cm^{-1} for a cell length of 10 cm. These studies used a miniature optoacoustic cell (Fig. 2) with a total gas volume of approximately 3 cm^3. The absorber used was nitric oxide diluted in nitrogen. It is estimated that for a signal-to-noise ratio of approximately 1, and a time constant of 1 s, it is possible to detect a nitric oxide (NO) concentration of approximately 10^7 molecules cm^{-3}, corresponding to a volumetric mixing ratio of approximately 1:10^{12} at atmospheric pressure.

The capability of measuring extremely small absorption coefficients and correspondingly small concentrations of the absorption gases has many applications, including high-resolution spectroscopy of isotopically substituted gases, excited states of molecules and forbidden transitions, and pollution detection. In the last application, both continuously tunable lasers, such as the spin-flip Raman laser and dye lasers, and step tunable infrared lasers, such as the carbon dioxide (CO_2) and carbon monoxide (CO) lasers, have been used as sources of high power radiation. The pollution measurements have demonstrated that the optoacoustic spectroscopy technique in conjunction with tunable lasers can be routinely used for on-line real-time in-place detection of undesirable gaseous constituents at sub-parts-per-billion lev-

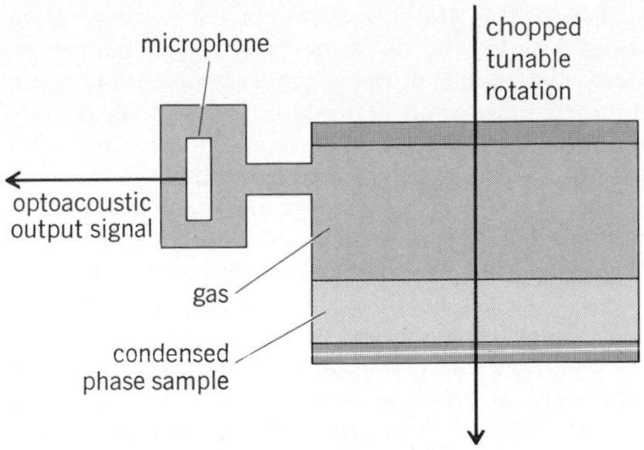

Fig. 3. Arrangement for condensed-phase photoacoustic spectroscopy.

els. Specific examples include the measurement of nitric oxide on the ground and in the stratosphere (where nitric oxide plays an important role as a catalytic agent in the stratospheric ozone balance) and measurements of hydrocyanic acid (HCN) in the catalytic reduction of $CO + N_2 + H_2 + \ldots$ over platinum catalysts. These studies point toward expanding use in the future of optoacoustic spectroscopy in pollution detection. *See* LASER SPECTROSCOPY.

Condensed-phase spectroscopy. A straightforward application of the gas-phase optoacoustic spectroscopy technique to the study of condensed phase (liquid or solid) spectra involves enclosing the condensed phase material within the gas-phase optoacoustic cell (Fig. 3). The "photoacoustic" signals generated in the sample due to the absorption of optical radiation are communicated to the gas-phase microphone via coupling through the gas filling the chamber. The inefficiency of such a system is high because of the very poor acoustical match (coupling efficiency approximately 10^{-5}) between the condensed-phase sample and the gas. In reality, because of the large acoustical mismatch, the detection scheme is really "photothermal" rather than "photoacoustic," and this scheme provides a capability of measuring fractional absorption at a level of approximately 10^{-4} when a continuous-wave laser power of approximately 10 W is used. A more severe drawback of the scheme, however, lies in the difficulty of interpretation of the data because of the intimate dependence of the observed optoacoustic signal from the microphone on the chopping frequency, absorption depth, and heat diffusion depth. However, in spite of its shortcomings, the gas-phase microphone technique for condensed-phase optoacoustic spectroscopy has found application.

A very sensitive calorimetric spectroscopic technique has been developed for the study of weak absorption in liquids and solids. This technique uses a pulsed tunable laser for excitation and a submerged piezoelectric transducer, in the case of a liquid, or a contacted piezoelectric transducer, in the case of a solid, for the detection of the ultrasonic signal generated due to the absorption of the radiation and its subsequent conversion into a transient ultrasonic signal (Fig. 4). The major distinction between the above condensed-phase "photoacoustic" spectroscopy technique and the pulsed-source, submerged or contacted piezoelectric transducer technique, is the high coupling efficiency of approximately 0.2 for the ultrasonic signal in the liquid to the submerged transducer, or an efficiency of approximately 0.9 for coupling the ultrasonic wave in a solid to a bonded transducer. Because of this high efficiency, the pulsed-laser, submerged or bonded optoacoustic spectroscopy technique has been shown to be useful for measuring fractional absorptions as small as 10^{-7} when using a laser source with pulse energy of approximately 1 mJ, pulse duration of approximately 1 μs, and a pulse repetition frequency of 10 Hz. There is room for improvement by increasing the laser pulse energy. There is a possibility of the electrostriction effect giving rise to an unwanted background signal, but this signal is not dependent on the light wavelength, and can be minimized by proper choice of experimental parameters.

The pulsed-laser, submerged or bonded piezoelectric transducer technique has a further advan-

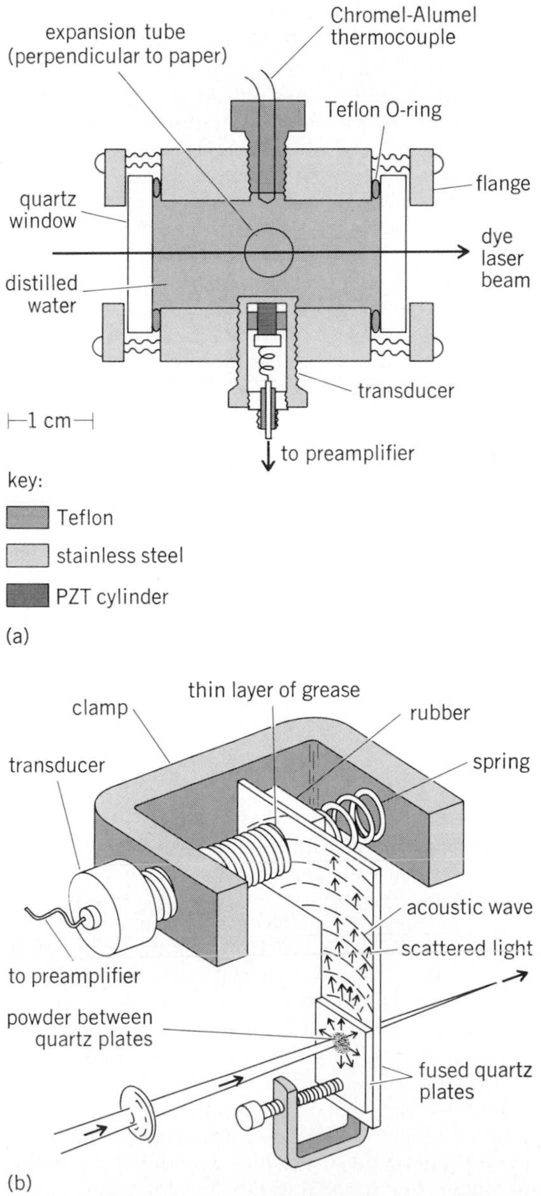

Fig. 4. Arrangement for pulsed-laser (*a*) immersed and (*b*) contacted piezoelectric transducer optoacoustic spectroscopy.

tage that time-gating the ultrasonic signal output can be utilized for the rejection of spurious signals since the sound velocity in condensed media is known and hence the exact arrival time of the real optoacoustic pulse can be calculated. This technique has been used for measurement of very weak overtone spectra of a variety of organic liquids, optical absorption coefficients of water and heavy water in the visible, two-photon absorption spectra of liquids, Raman gain spectra in liquids, absorption of thin liquid films, spectra of solids and powders, and weak overtone spectra of condensed gases at low temperatures. Because of the capability of measuring very small fractional absorptions, the technique is clearly applicable to the area of monitoring water pollution, impurity detection in thin semiconductor wafers, transmission studies of ultra-pure glasses (used in optical fibers for optical communications), and so forth. Further, even though in all of the present studies use is made of only the optical radiation, there is no reason to restrict "optoacoustic" spectroscopy to the optical region. By using pulsed x-ray sources, such as the synchrotron light source or pulsed electron beams, the principle described above for a pulsed-light-source, submerged or bonded piezoelectric transducer, gated-detection technique can be extended to x-ray acoustic spectroscopy and electron-loss acoustic spectroscopy. These extensions are likely to have major impact on materials and semiconductor fabrication technology.

[C. K. N. PATEL]

Bibliography: A. G. Bell, On the production and reproduction of sound by light, *Proc. Amer. Ass. Adv. Sci.*, 29:115–129, 1880; L. B. Kreuzer, The physics of signal generation and detection, in Y.-M. Pao (ed.), *Optoacoustic Spectroscopy and Detection*, 1977; C. K. N. Patel, Spectroscopic measurements of stratosphere using tunable infrared lasers, *Opt. Quantum Electr.*, 8:145–154, 1976; C. K. N. Patel and R. J. Kerl, A new optoacoustic cell with improved performance, *Appl. Phys. Lett.*, 30:578–579, 1977; C. K. N. Patel and A. C. Tam, Pulsed optoacoustic spectroscopy of condensed matter, *Rev. Mod. Phys.*, 53:517–550, 1981; M. B. Robin and W. R. Harshbarger, The opto-acoustic effect: Revival of an old technique for molecular spectroscopy, *Acc. Chem. Res.*, 6:329, 1973; M. Rosencwaig, Photoacoustic spectroscopy of solids, *Optics Comm.*, 1:305, 1973.

Photochemistry

The branch of chemistry concerned with reactions of excited molecules produced by the absorption of light. Chemical reactions involve the breaking (and formation) of chemical bonds which requires energies in the range of 200–600 kJ/mole; this corresponds to the energy of light quanta in the ultraviolet (100–400 nm), visible (400–700 nm) and near-infrared (700–1000 nm) regions of the electromagnetic spectrum. Light of shorter wavelengths (x-rays, γ-rays) has sufficient energy to ionize and to dissociate molecules. The effects of this light constitute the field of radiation chemistry, while light of longer wavelengths (infrared) is not sufficiently energetic to produce electronic excitation in single quanta excitations. *See* INORGANIC PHOTOCHEMISTRY; LASER PHOTOCHEMISTRY; RADIATION CHEMISTRY.

Electronic states of molecules possessing an even number of electrons with spins paired are termed singlet states (superscript 1) or triplet states (superscript 3) if two electrons have parallel spins. Excited singlet or triplet states are often designated $\pi\pi^*$ if they are produced by a transition in which a π-electron in an unsaturated molecule (containing one or more double bonds) is promoted to an antibonding π^*-orbital, or $n\pi^*$ if an electron in a nonbonding (lone-pair) n-orbital is promoted to a π^*-orbital as in the case of a carbonyl compound >C=O. The excited states of an unsaturated molecule may therefore be $^1\pi\pi^*$; $^3\pi\pi^*$; $^1n\pi^*$ or $^3n\pi^*$ in which the double bond is reduced to a single bond. Saturated molecules containing only single (σ) bonds undergo electronic excitation to $\sigma\sigma^*$ states in which a bond is broken to produce atoms or free radicals.

Electronic excitation reduces the ionization potential and increases the electron affinity of a molecule by an increment equal to the excitation energy, and therefore promotes electron transfer reactions. The change in electronic configuration produced by light absorption leads to dramatic changes in chemical properties; thus the $^1\pi\pi^*$ state of 2-naphthol is more acidic than the ground state by a factor of 10^6. Since the Gibbs free energy is also increased by light absorption, electronically excited molecules may undergo spontaneous reactions to products which are thermodynamically inaccessible from the ground state. This is exemplified by the photosynthetic reaction in which the conversion of carbon dioxide (CO_2) and water (H_2O) to carbohydrate $(CH_2O)_n$ and oxygen (O_2) is mediated by excited chlorophyll molecules. The reverse process of respiration leads to a decrease in Gibbs free energy and takes place spontaneously in the absence of light.

The study of photochemical reactions provides a firm molecular basis for the interpretation of those processes arising from the interaction of sunlight with the biosphere which are responsible not only for creating the conditions under which life on the Earth could develop, but also for the provision of food and energy.

Quantum yield. The rate of a photochemical reaction is proportional to the rate at which electronically excited molecules M* are produced; since each absorbed photon (energy $h\nu$) of light produces one excited molecule, this excitation rate is equal to the rate of light absorption or the intensity of light absorbed, I_a. The reaction rate is therefore proportional to I_a, or rate $= \gamma I_a$, where the proportionality constant, $\gamma =$ rate$/I_a$, known as the quantum yield (or efficiency), is characteristic of the reaction under the conditions of examination. The quantum yield of the primary process (that involving the excited molecule) cannot exceed unity and is usually much lower than this since electronically excited molecules revert to the ground state (with or without the emission of luminescence within a period of $\sim 10^{-3}$ s (triplet state) to 10^{-8} s (singlet states). These so-called photophysical processes, summarized in Fig. 1, compete with photochemical reactions, effectively reducing the quantum yield. Additionally, the excited molecule may be quenched by other molecules Q in a process of electron transfer [reaction (1)] or energy transfer [reaction (2)], which leave the molecule M

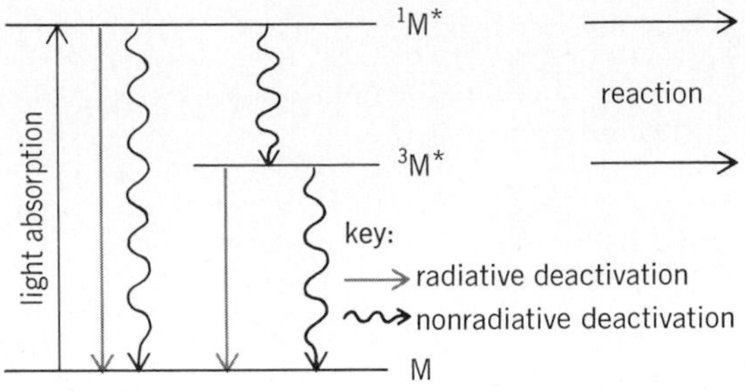

Fig. 1. Radiative and nonradiative deactivation of electronically excited molecule compete with reaction of singlet state $^1M^*$ and of lower-energy triplet state $^3M^*$.

chemically unchanged.

$$M^* + Q \rightarrow (M^+Q^-) \text{ or } (M^-Q^+) \rightarrow M + Q \quad (1)$$

$$M^* + Q \rightarrow M + Q^* \quad (2)$$

On the other hand, if the primary photochemical products are atoms or free radicals, these may undergo secondary nonphotochemical chain reactions which amplify the primary process and lead to overall photochemical quantum yields much greater than unity. An example is reaction (3*a*), in which the primary process [reaction (3*b*)] is followed by the chain propagating steps [reactions (3*c*) and (3*d*)] to produce ~10^6 molecules of HCl

$$H_2 + Cl_2 \xrightarrow{\text{light}} 2HCl \quad (3a)$$

$$Cl_2 + \text{light} \rightarrow 2Cl \quad (3b)$$

$$Cl + H_2 \rightarrow H + HCl \quad (3c)$$

$$H + Cl_2 \rightarrow Cl + HCl \quad (3d)$$

for each light photon absorbed by Cl_2, that is, an overall quantum yield of 10^6. *See* CHAIN REACTION.

Mechanisms. Photochemical reaction mechanisms are deduced from measurements of the dependence of quantum yields on such different reaction variables as reactant concentration and the concentration of added substances. Primarily it is necessary to identify the nature of the electronically excited state responsible for the reaction, which is most often singlet ($^1M^*$) or triplet ($^3M^*$) following which a theoretical examination of the orbital transformations along the reaction coordinate may be attempted.

If the quantum yield is reduced by a selective triplet-state quencher (Q) which introduces the competing process [reaction (4)], then the triplet

$$^3M^* + Q \rightarrow M + {}^3Q^* \quad (4)$$

state $^3M^*$ may be regarded as the reactive state; otherwise the singlet state $^1M^*$ is assigned the role of reactive intermediate. The rate constant of the primary process may then be obtained by monitoring the decay of the intermediate $^{1,3}M^*$ in absorption or emission following flash or laser pulse excitation. *See* EXCITED STATE; FREE RADICAL; QUANTUM CHEMISTRY; TRIPLET STATE.

Photochemical reactions. These are classified as unimolecular, in which the excited molecule itself undergoes chemical change, or bimolecular if the excited molecule reacts with another molecule present.

Unimolecular processes. Such processes which lead to a single product are known as photoisomerization reactions. Examples are the formation of Dewar benzene (I) from benzene (II), shown in reaction (5), which involves a higher singlet state of

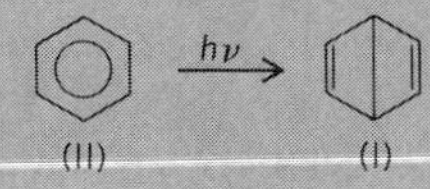

(5)

benzene, and the photochromic reaction of crystalline dinitrobenzylpyridine [reaction (6)]. Ring closure is exemplified by reaction (7), which bears

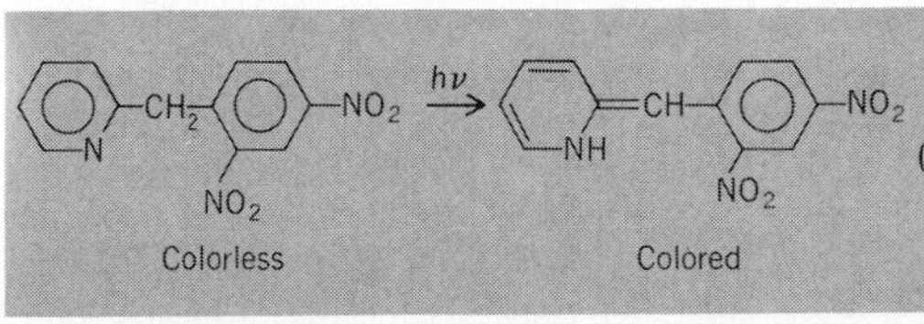

(6)

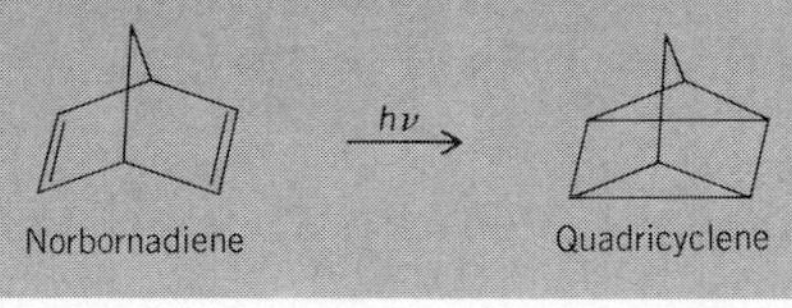

(7)

promise as a solar-energy storage process, whereas ring opening is involved in the photochemical synthesis of vitamin D_2 [reaction (8)]. These reactions proceed via the $^1\pi\pi^*$ state.

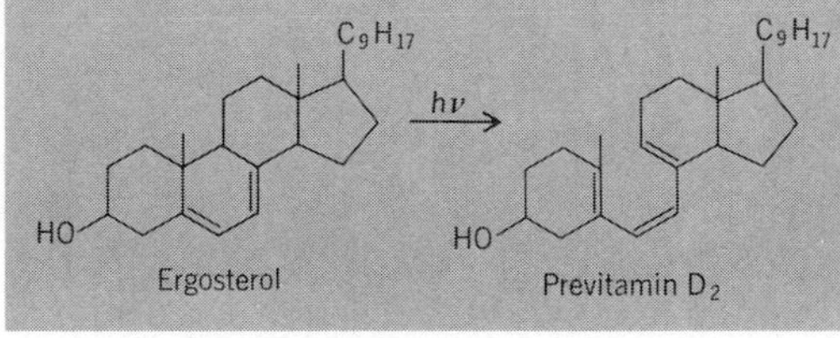

(8)

In a $\pi\pi^*$ state the reduction of a double bond to a single bond permits free rotation of the end groups resulting in cis-trans isomerization as in the case of stilbene [reaction (9)]. The photochemical

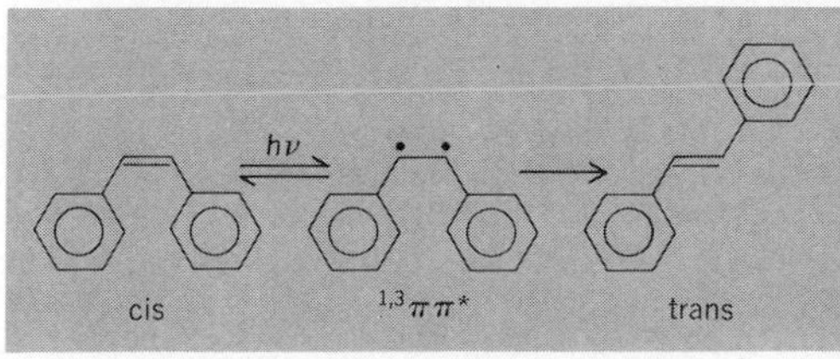

(9)

transformation of 11-*cis*-retinal to all-*trans*-retinal is believed to trigger visual response, and the phototherapy of neonatal jaundice probably involves the cis-trans isomerization of bilirubin, a yellow pigment product of heme degradation, to a water-soluble isomer which can be excreted.

Unimolecular reactions producing two product radicals or atoms as the result of bond rupture are exemplified by the photolysis of ozone O_3 [reaction (10)] and of oxygen O_2 [reaction (11)] in the upper

$$O_3 \xrightarrow{h\nu} O_2 + O \qquad (10)$$

$$O_2 \xrightarrow{h\nu} O + O \qquad (11)$$

atmosphere by absorbing solar radiation of wavelengths shorter than 300 nm. Ozone is reformed by the combination of O_2 molecules and O atoms and exists in photochemical equilibrium in the so-called ozone layer. Concern has been expressed that very stable aerosol spray propellants or Freons such as CF_2Cl_2 which may find their way into the upper atmosphere where the products of photodissociation [reaction (12)] may deplete the ozone concentration by the secondary reaction (13)

$$CF_2Cl_2 \xrightarrow{h\nu} CF_2Cl + Cl \qquad (12)$$

$$Cl + O_3 \rightarrow ClO + O_2 \qquad (13)$$

with the result that part of the solar radiation below 300 nm may reach the Earth's surface. Certain aromatic hydroxy compounds dissolved in water dissociate into ions when excited, for example, 2-naphthol [reaction (14)].

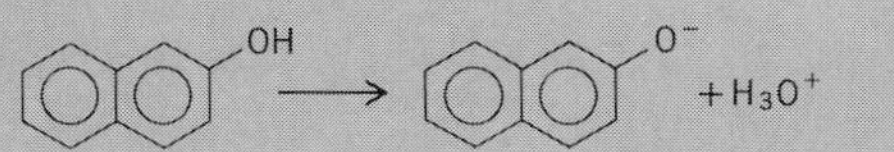

(14)

Bimolecular processes. When these involve electronically excited unsaturated molecules which form a single product the reaction is referred to as photoaddition [reaction (15)]. If the adduct A is the unexcited molecule M itself, the reaction is known as photodimerization [reaction (16)]. In either case the reaction is believed to proceed via the excited adduct MA*, known as an exciplex, or M_2*, an excimer, both of which in certain cases emit a characteristic fluorescence spectrum. The simplest photodimerization reaction is that of ethylene to produce cyclobutane [reaction (17)],

$$M^* + A \rightarrow MA \qquad (15)$$

$$M^* + M \rightarrow M_2 \qquad (16)$$

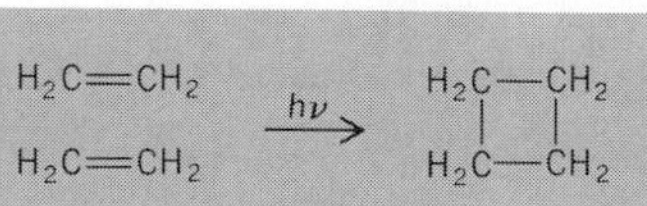

(17)

which is believed to involve the $^1\pi\pi^*$ state of ethylene. This type of process can take place in the solid state if neighboring molecules are suitably oriented in the crystal lattice. For example, the α modification of transcinnamic acid, in which nearest neighbors have a head-to-tail configuration (III*a*), photodimerizes to α-truxillic acid (IV), as shown in reaction (18), whereas the β-crystalline

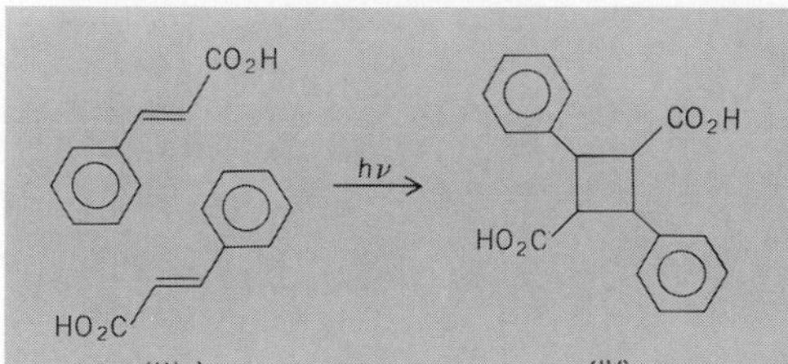

(IIIa) (IV) (18)

modification based on a head-to-head configuration of adjacent molecules (III*b*) produces β-truxinic acid (V), as shown in reaction (19). In the γ modifi-

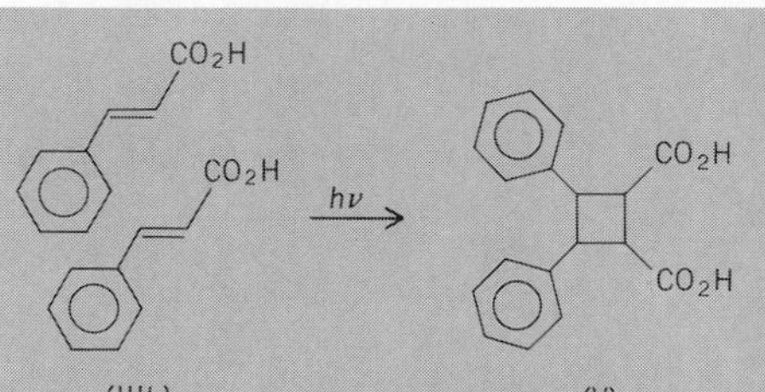

(IIIb) (V) (19)

cation of crystalline transcinnamic acid, the distance between neighboring molecules is too great for dimerization to take place; these examples illustrate topochemical control of product formation by nearest-neighbor orientation. The photodimerization of adjacent thymine residues (VI) on the same strand is responsible for one type of photochemical lesion in DNA [reaction (20)]. The

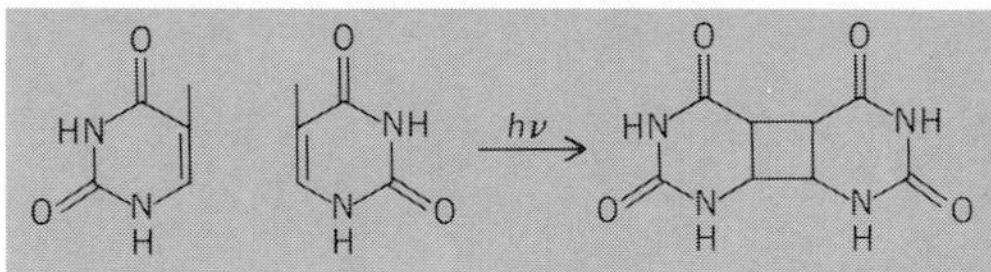

(VI) (VI) (20)

twin strands of DNA may be photochemically cross-linked by psoralen, which can intercalate with two base pairs and form cyclobutane linkages (Fig. 2) which inhibit DNA synthesis and cell division; this is believed to be the molecular basis for the phototherapy of psoriasis following oral ingestion of psoralen derivatives by the patient.

In the presence of light, a light-absorbing sensitizer (S), and molecular oxygen, many unsaturated molecules (A) are converted to peroxides (AO_2), as shown in reaction (21*a*). The primary photochemical process involves the addition of A to singlet

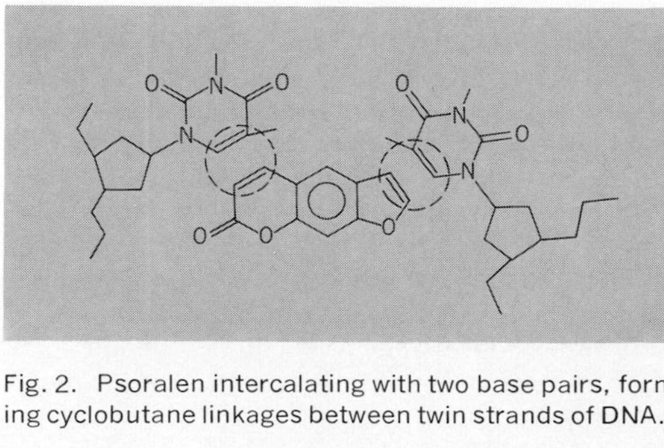

Fig. 2. Psoralen intercalating with two base pairs, forming cyclobutane linkages between twin strands of DNA.

molecular oxygen $^1O_2{}^*$ produced from its triplet ground state $^3O_2$ by energy transfer from the sensitizer triplet state $^3S^*$ [reaction (21*b*)]. The conver-

$$A \xrightarrow[\text{light}]{S/O_2} AO_2 \qquad (21a)$$

$$^3S^* + {}^3O_2 \rightarrow S + {}^1O_2{}^* \qquad (21b)$$

sion of α-terpinene to ascaridole in this way was probably the first commercially exploited photosynthetic reaction (22). Cellular damage under the

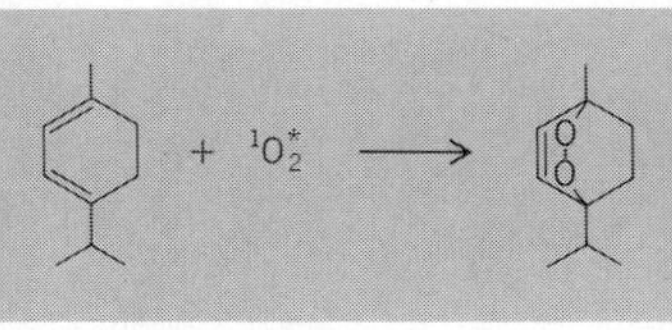

(22)

same conditions is termed photodynamic action, in which the role of $^1O_2{}^*$ is strongly implicated; β-carotene which quenches $^1O_2{}^*$ in the energy transfer process [reaction (23)] is believed to protect

$$^1O_2{}^* + \beta\text{-carotene} \rightarrow {}^3O_2 + {}^3\beta\text{-carotene}^* \qquad (23)$$

plants from self-destruction, and has been successfully used in the treatment of erythropoetic protoporphyria, a hereditary inability to metabolize porphyrins which accumulate in the skin and act as sensitizers in sunlight.

An example of a bimolecular photochemical process which produces two products is afforded by the $^3n\pi^*$ states of certain ketones which abstract H atoms from an adjacent substrate RH [reaction (24) where · denotes free radical]. The

$$\rangle C{=}O(^3n\pi^*) + RH \rightarrow \rangle\dot{C}{-}OH + R\cdot \qquad (24)$$

free radicals thus produced subsequently dimerize to yield the overall reaction products; however, if RH is a polymer chain, secondary reactions of R in the presence of oxygen result in chain breaking, a process which has been exploited in the production of photodegradable packaging and picnic items.

Theoretical aspects. The principle that orbital symmetry is conserved in concerted reactions has afforded an interpretation of many types of photochemical (and thermal) reactions at the molecular level. Ring closure of a π-electron system exemplified by the butadiene derivative may produce either of the two isomers shown [reaction (25)]. The

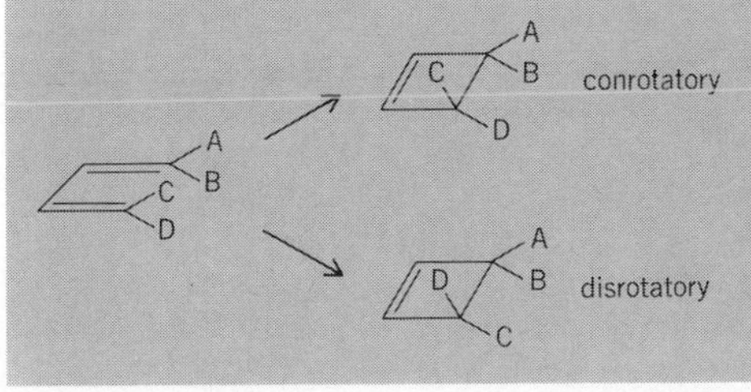

(25)

Woodward-Hoffman rules for this electrocyclic process for a system containing k π-electrons are summarized by

k	Thermal reaction	Photochemical reaction
$4q$	conrotatory	disrotatory
$4q+2$	disrotatory	conrotatory

where $q = 1, 2, 3 \ldots$. The same rules apply to ring opening.

The concerted (bimolecular) cycloaddition reaction of two systems containing m and n π-electrons, an example of which is shown in reaction (26), is photochemically allowed if $m + n = 4q$, and

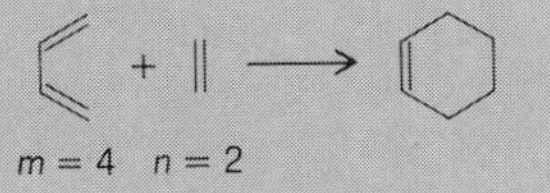

(26)

photochemically forbidden if $m + n = 4q + 2$, where again q is an integer. Rules have also been developed for sigmatropic reactions (atom migration in a π-electron system) and cheletropic (molecular elimination) processes.

The conservation of spin angular momentum limits singlet and triplet energy-transfer processes to those shown in reactions (27) and (28), which can

$$^1M^* + {}^1Q \rightarrow {}^1M + {}^1Q^* \qquad (27)$$

$$^3M^* + {}^1Q \rightarrow {}^1M + {}^3Q^* \qquad (28)$$

be used to identify the photochemically reactive state of M, and often affords an indication of preferred reaction pathways at the molecular level. Since the projected electron spin quantum number has possible values $\pm\frac{1}{2}$, and since the resultant spin angular momentum [$S = \frac{1}{2} - \frac{1}{2} = 0$ for two spin-paired electrons (singlet state) or $S = \frac{1}{2} + \frac{1}{2} = 1$ for two electrons with parallel spin (triplet state)] has the result that the concerted photochemical addition reactions involve either the excited singlet state [reaction (29)] or two excited triplet states [reaction (30)], which is less prob-

$$^1M^*(S=0) + {}^1A(S=0) \rightarrow {}^1MA(S=0) \qquad (29)$$

$$^3M^*(S=1) + {}^3A(S=1) \rightarrow {}^1MA(S=1-1=0) \qquad (30)$$

able. The photochemical addition of triplet and ground states produces the adduct triplet state [reaction (31)] unless it involves an intermediate diradical [reaction (32)] in a nonconcerted process.

$$^3M^*(S=1) + {}^1A(S=0) \rightarrow {}^3MA^*(S=1+0=1) \qquad (31)$$

$$^3M^*(S=1) + {}^1A(S=0) \rightarrow MA(S=\tfrac{1}{2} \pm \tfrac{1}{2}) \rightarrow {}^1MA(S=0) \qquad (32)$$

The formation of organic peroxides by the addition of unsaturated molecules to singlet oxygen $^1O_2{}^*$ (but not the $^3O_2$ ground state) is indicative of a concerted process in which orbital symmetry is conserved. Evidence presented to the effect that thymine dimerization and psoralen binding to DNA base pairs results from triplet excitation, indicates that these are nonconcerted reactions involving a diradical intermediate. See CHEMICAL DYNAMICS; MOLECULAR ORBITAL THEORY. [BRIAN STEVENS]

Bibliography: J. G. Calvert and J. N. Pitts, Jr., *Photochemistry*, 1965; W. A. Noyes et al. (eds.),

Advances in Photochemistry, vols. 1–12, 1963–1979; N. J. Turro, *Modern Molecular Photochemistry*, 1978; R. B. Woodward and R. Hoffmann, *The Conservation of Orbital Symmetry*, 1970.

Photolysis

Chemical decomposition by the action of radiant electromagnetic energy, especially light. Photolysis occurs in certain crystals, notably the silver halides, when they are exposed to radiation. When this occurs, the effect of the radiation is to produce a definite chemical change resulting in the separation of photolytic silver. Photolysis of the silver halides is discussed in this article because of the extensive investigations that have been carried out on these materials and because of their importance in the photographic process. Actually, photolysis occurs in many other materials, such as the lead and thallium halides, zinc oxide, the metallic azides, and in organic compounds such as the oxalates, styphnates, and fulminates.

A photographic emulsion consists of microcrystalline grains of silver bromide, AgBr, or silver chloride, AgCl, embedded in gelatin. Upon prolonged exposure to light, so-called print-out specks of silver form within and on the surface of the grains. Much shorter exposures produce a latent image which can be made visible by the process of development. Experiments have shown that the latent image consists of only a very few atoms of silver in each grain. The high sensitivity of the photographic system comes about because of the enormous gain (10^9) which can be achieved by reduction of each exposed grain that occurs during development.

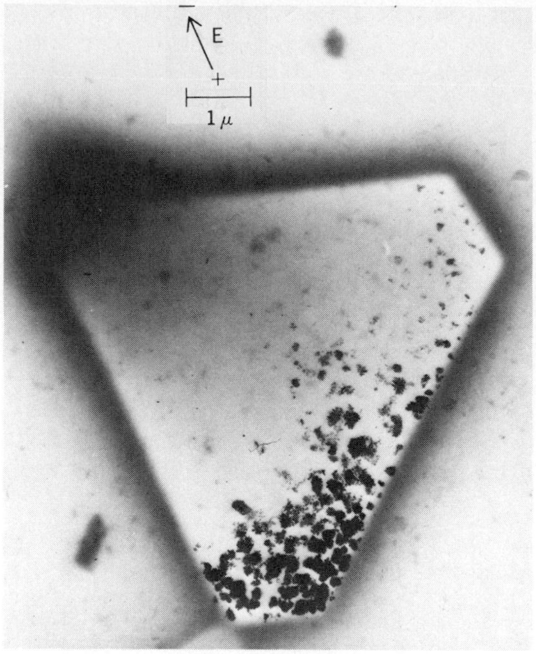

Fig. 2. Electron micrograph of a silver halide emulsion grain exposed to repetitive pulses of light and to an electric field *E*. Cloud in the gelatin is thought to be due to escape of bromine from the grain, as if positive holes were displaced upward and to the left, whereas electrons are displaced downward and to the right by the applied electric field. (*After J. F. Hamilton and L. E. Brady*)

Gurney-Mott theory. This theory of the photographic process proposes a two-stage mechanism as shown in Fig. 1. In the first stage a light quantum is absorbed at a point within the silver halide grain, releasing a mobile electron and a positive hole. These mobile defects diffuse to trapping sites (sensitivity centers) within the volume or on the surface of the grain. In the second stage, the trapped (negatively charged) electron is neutralized by an interstitial (positively charged) silver ion, which combines with the electron to form a silver atom. The silver atom at the sensitivity center is capable of trapping a second electron, after which the process repeats itself, causing the silver speck to grow. The positive holes are assumed to diffuse to the surface without recombining with electrons, where they escape or react with the gelatin.

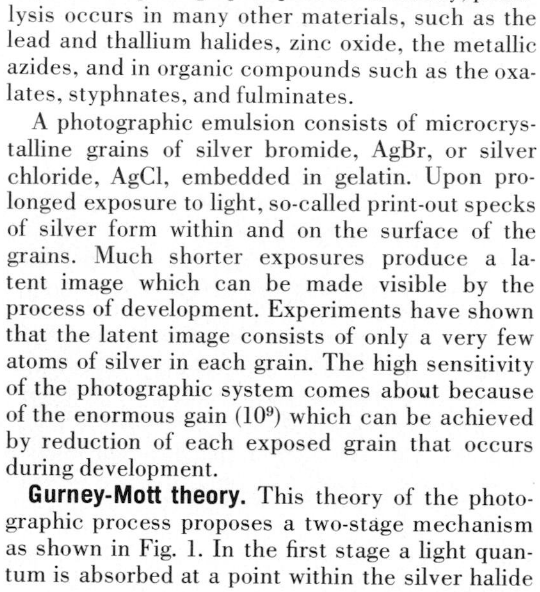

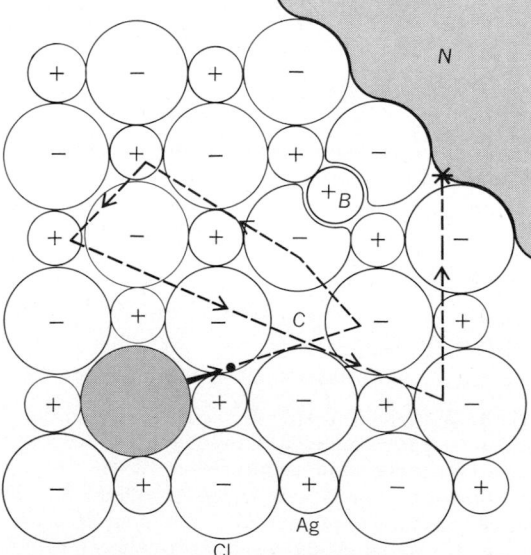

Fig. 1. Schematic representation of Gurney-Mott theory. Broken line shows path of photoelectron which is eventually trapped in vicinity of a silver speck *N*. The interstitial silver ion *B* has come up to neutralize the charge of the electron and thus to add to the silver speck. (*After J. R. Haynes and W. Shockley*)

The early Gurney-Mott theory has been criticized by J. W. Mitchell and others, especially because of the assumed lack of electron-hole recombination. The essential idea of an electronic process linking the initial absorption of light quanta with the formation of the image speck is nevertheless well founded. Figure 2 shows an electron micrograph of an emulsion grain after a print-out level exposure to synchronized light and voltage pulses. The print-out silver specks occur on one side because of the action of the electric field, and there is experimental evidence for positive-hole migration in the opposite direction.

Practical emulsions contain a distribution of silver halide grain sizes and shapes. The slower fine-grain emulsions may have an average grain size of 0.05 μ or smaller, whereas the grain sizes in a high-speed emulsion may be of the order of several microns. The silver halide in a negative emulsion is usually AgBr that contains a small amount of silver iodide. Various substances, for example, gold and silver sulfide, act as sensitizers, and the spectral response can be extended by the addition of dyes. Formation of the latent image in a fast emulsion containing sensitizers is highly complicated and is thought to involve the role of structural imperfections such as dislocations, jogs on dislocations, and twin planes. There is increasing evidence that the surface plays a crucial role.

Large crystals. The photolysis of large crystals of the silver halides and of other compounds such as the lead halides has been studied in considerable detail. When interpreting the results on single crystals, it becomes necessary to distinguish between darkening produced by light in the volume of the crystal and darkening produced near the surface. A pure crystal of AgCl does not darken appreciably upon exposure to light absorbed below the surface. On the other hand, crystals which contain small traces of impurity, particularly copper in the monovalent state, darken with high efficiency up to a saturation level which depends upon the amount of impurity present. Figure 3

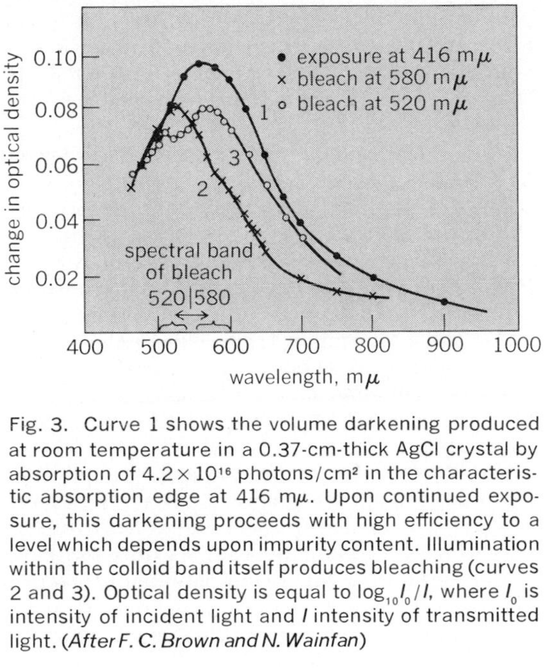

Fig. 3. Curve 1 shows the volume darkening produced at room temperature in a 0.37-cm-thick AgCl crystal by absorption of 4.2×10^{16} photons/cm² in the characteristic absorption edge at 416 mμ. Upon continued exposure, this darkening proceeds with high efficiency to a level which depends upon impurity content. Illumination within the colloid band itself produces bleaching (curves 2 and 3). Optical density is equal to $\log_{10} I_0/I$, where I_0 is intensity of incident light and I intensity of transmitted light. (*After F. C. Brown and N. Wainfan*)

shows the small amount of absorption for darkening of this type, which occurs in the early stages of exposure. Here the extinction of light is caused mainly by absorption. Prolonged exposure, however, produces darkening near the surface, which shows considerable light scattering. The centers responsible for the surface darkening are larger and may be similar to the colloidal metal particles formed in alkali halides by coagulation of *F*-centers during heat treatment.

[FREDERICK C. BROWN]

Bibliography: F. C. Brown, *The Physics of Solids*, 1967; W. E. Garner (ed.), *Chemistry of the Solid State*, 1955; C. E. K. Mees and T. H. James, *The Theory of the Photographic Process*, 1966; J. W. Mitchell, *Rep. Progr. Phys.*, 20:433, 1957.

Phthalic acid

One of the three benzenedicarboxylic acids of formula $C_6H_4(COOH)_2$. Phthalic (or *o*-phthalic) acid, melting point 191°C (sealed tube), is the 1,2 isomer; isophthalic (or *m*-phthalic) acid, melting point 347–348°C, is the 1,3 isomer; terephthalic acid, melting point 425°C (sealed tube), is the 1,4 isomer.

The acids are prepared by permanganate or chromic acid oxidation of appropriate xylenes or by partial chlorination of the xylene, followed by basic hydrolysis and finally oxidation. Phthalic acid, commercially the most important of the three, is manufactured mainly by vapor-phase oxidation of naphthalene over a vanadium pentoxide catalyst at about 480°C, so the product, phthalic anhydride, sublimes from the reaction zone in a state of high purity. *See* OXIDATION PROCESS.

Phthalic acids are used in chemical analysis, in preparation of esters (methyl, ethyl, and butyl phthalates), and in preparation of phthaloyl chlorides, $C_6H_4(COCl)_2$, from the acid and phosphorus pentachloride. Phthalic anhydride can be decarboxylated to benzoic acid and can be used in the synthesis of indigo and derivatives of anthraquinone. Phthalic anhydride reacts at elevated temperature with polyalcohols (ethylene glycol or glycerol) to form polyesters, which are used as plastics. With ammonia, phthalic anhydride gives phthalimide. Terephthalic acid, heated with ethylene glycol, gives polyesters used as synthetic fibers. *See* ACID ANHYDRIDE; CARBOXYLIC ACID; PHTHALIMIDE; POLYESTER RESINS; XYLENE.

[EVANS B. REID]

Phthalimide

The imide of *o*-phthalic acid, also called 1,3-isoindoledione. The melting point of phthalimide is 238°C, and it is only slightly soluble in water. It is a weak acid, $K_a = 5 \times 10^{-9}$. The substance is prepared commercially by the reaction of molten phthalic anhydride and ammonia; in the laboratory phthalic anhydride and either ammonium hydroxide or ammonium carbonate are used.

In the form of its sodium or potassium salt it is widely used in the synthesis of both primary amines and amino acids. It is combined with the malonic ester in synthesis of complex amino acids.

Phthalimide is used as starting material in the synthesis of methyl anthranilate, the active principle in jasmine and orange oils. On gentle treatment with glutamic acid, phthalic anhydride forms phthalimidoglutamic acid. The latter, on vigorous warming with ammonia (about 150°C), undergoes cyclization to give phthalimidoglutarimide. This substance has effective sedative and hypnotic properties and was formerly marketed under the trade name Thalidomide; when it was shown that use during early pregnancy led to fetal malformation, the drug was withdrawn. *See* AMIDE; AMINE; PHTHALIC ACID. [EVANS B. REID]

Bibliography: Merck and Co., *The Merck Index*, 9th ed., 1976.

Physical chemistry

The branch of chemistry that deals with the interpretation of chemical phenomena and properties in terms of the underlying physical processes, and with the development of techniques for their investigation. The term chemical physics is often employed to denote a branch of physical chemistry where the emphasis is on the interpretation and analysis of the physical properties of individual molecules and bulk systems, instead of their reactions. Theoretical chemistry is another major branch, where the emphasis is on the calculation of the properties of molecules and systems, and which uses the techniques of quantum mechanics and statistical thermodynamics. For the present purpose it is convenient to regard physical chemistry as dealing with three aspects of matter: its equilibrium properties, structure, and ability to change.

Chemical thermodynamics. The study of matter in a state of equilibrium constitutes the field of chemical thermodynamics. In particular, chemical thermodynamics provides a technique for discussing the response of a system to a change in the external conditions (such as the shift in the boiling and freezing point of either a pure substance or a mixture when the applied pressure is changed, or when the composition of the mixture is modified), and for rationalizing the energy changes that occur in the course of a chemical reaction. The branch of thermodynamics dealing with the latter is called thermochemistry. Chemical thermodynamics also provides a framework for the determination of the maximum amount of work that may be generated by a system undergoing a specified change, and it therefore provides a way of establishing bounds for the efficiencies of a variety of devices, including engines, refrigerators, and electrochemical cells. Thermodynamics is used in chemistry to assess the position of equilibrium of a chemical reaction (that is, how far it will proceed), and to determine what conditions are necessary in order to optimize the yield of a particular product. The branch of chemical thermodynamics dealing with ionic reactions occurring in the presence of electrodes constitutes the field of equilibrium electrochemistry. *See* THERMOCHEMISTRY.

Chemical thermodynamics is based on the laws of thermodynamics. In chemistry the most important thermodynamic properties are the enthalpy H and the Gibbs function G (which is also called the free energy). The change of enthalpy may be identified as the heat transferred to a system during a specified change under conditions of constant pressure. Tables of enthalpies of materials have been compiled, and from them it is possible to predict the amount of heat available (or required for) a particular reaction. Not the whole of the heat output of a reaction is available to do work (such as mechanical work, or work to drive other reactions toward a particular desired product). The Gibbs function expresses the maximum amount of work, other than work of expansion, that a specified process may generate under conditions of constant pressure, and as such it may be used to assess whether or not one reaction may be used to drive another in an unnatural direction. For example, the assessment of the changes in the Gibbs function that accompany biochemical reactions may be used to discuss the processes that occur in living cells, where the ingestion of food leads ultimately to growth, mechanical work, and nervous activity. *See* ENTHALPY.

The fundamental basis of the Gibbs function lies in the tendency of energy to attain a condition of greatest dispersal. This tendency is expressed most generally in terms of the entropy of a system: as the dispersal of energy increases, so too does the entropy. The concept of entropy may be given a sharp and quantitative definition, and entropies of substances have been determined and tabulated. The crucial feature for the present purpose, though, is that whereas the entropy of the universe increases whenever a spontaneous change occurs, the chemist is normally interested in the changes that occur in the system under investigation. The Gibbs function focuses attention on changes in the properties of the system itself, for G automatically takes into account the changes in entropy of the surroundings. In order to determine whether a specified change has a natural tendency to occur under conditions of constant temperature and pressure, it is only necessary to assess whether that change is accompanied by a decrease in the Gibbs function of the system. If a change is accompanied by an increase of Gibbs function, it may still be achieved by coupling the system to another in which a change is occurring such that the overall change in the Gibbs function of the combined system is negative. Since the Gibbs function depends on the composition of the system, it is possible to use it to predict the composition of the system when it has attained equilibrium (such as when some reaction mixture attains some constant composition).

The crucial equation is $\ln K = -\Delta G_m^\circ/RT$, where K is the equilibrium constant of the reaction (in essence a simple function of the concentration of the components at equilibrium), R is the gas constant, T the absolute temperature, and ΔG_m° the standard molar Gibbs function for the reaction (the change in the Gibbs function under certain specified, standard conditions). This equation lies at the heart of chemical thermodynamics, and is of great practical importance. *See* CHEMICAL EQUILIBRIUM; CHEMICAL THERMODYNAMICS; ENTROPY; FREE ENERGY; GIBBS FUNCTION.

Equilibrium electrochemistry. In equilibrium electrochemistry, attention is focused on systems in which a chemical reaction may release energy by driving electrons through some external circuit. Under certain conditions the potential difference between the two electrodes immersed in the reaction mixture is related to the Gibbs function for the reaction by $E = -\Delta G_m/F$, where F is the Faraday constant, the charge carried by 1 mol of electrons ($F = 96{,}485$ coulombs mol^{-1}). The potential difference under these conditions is denoted E, and is called the electromotive force (emf) of the cell. One object of electrochemistry is to measure the emf of cells, and then to use the tabulated results to discuss the position of equilibria in ionic reactions (via ΔG) and their response to changes of conditions. Practical applications include chemical analysis, the assessment of the power generation and storage capabilities of electrochemical cells and fuel cells, the discussion of tendencies to corrosion, and the analysis of potential differences across biological membranes (such as are responsible for the propagation of nervous impulses). *See* ELECTROCHEMISTRY.

Quantum chemistry. The principal role of quantum mechanics in chemistry is in the discussion of atomic and molecular structure, and in the interpretation of spectroscopic data. In the branch of physical chemistry known as computational quantum chemistry, interest centers on the numerical solution of the Schrödinger equation in order to obtain wave functions and geometries of molecules. In ab initio calculations the computations are done without any appeal to experimental data, and an attempt is made to predict properties from first principles (that is, from the masses and charges of the electrons and nuclei constituting the molecule). In semiempirical calculations (which are identified by initials such as CNDO and MINDO) some of the computational difficulties are circumvented (but with some loss of reliability) by incorporating experimental data. It is now possible to calculate the shape, electron distribution, and spectroscopic properties of large molecules. One application is referred to as quantum pharmacology, where a pharmacologically active molecule is first screened by calculating and analyzing its electron distribution; to some extent this eliminates lengthy and expensive laboratory screening procedures. Computational quantum chemistry is so developed that it is capable of being used to map the changes in the structures of molecules while they are in the course of reaction, when atoms and groups of atoms are being transferred from one molecule to another. *See* QUANTUM CHEMISTRY.

Spectroscopy. Spectroscopic techniques are used not only to identify molecules present in a sample, but also to determine their shape, size, and electron distribution. The techniques fall into four categories as follows.

Absorption spectroscopy. In absorption spectroscopy the basic observation is the amount of radiation absorbed at different frequencies (or wavelengths). When the exciting frequency lies in the microwave region (wavelengths in the vicinity of 1 cm), the absorption arises on account of the rotational excitation of the molecule. Hence the technique may be used to assess the moment of inertia, and therefore the shape and size of the molecule. When the exciting radiation lies in the infrared region (wavelengths in the region of 1000 nm), the vibrational modes of the molecule are excited, and hence the absorption spectrum gives information about the stiffness of bonds. Furthermore, since different groups of atoms absorb in characteristic regions, infrared spectroscopy is also used to identify groups of atoms present in a molecule, and therefore to identify the molecule (this is called fingerprinting). Absorption in the visible and ultraviolet region of the spectrum indicates the occurrence of an electronic transition; that is, a shift of electron density from one region of the molecule to another.

Electronic spectroscopy is used for identification, and provides a basis for the discussion of the processes involved in photochemistry (chemical reactions induced by the absorption of radiation), in fluorescence and phosphorescence, and in laser action. Sometimes the incident radiation is so energetic that its absorption results in the ejection of electrons from the molecule. The analysis of the energies of these electrons leads to information about the energy levels of electrons in the molecules under investigation; these photoelectron spectroscopy techniques are designated uv-PES when ultraviolet radiation expels outer electrons, and x-PES when x-rays are used to eject the more tightly bound electrons (formerly ESCA, standing for electron scattering for chemical analysis).

Emission spectroscopy. In emission spectroscopy the spectrum of frequencies present in light emitted from excited atoms is monitored. Apart from its use to identify the composition of mixtures, emission spectroscopy is also used to determine the state of molecules immediately after they have reacted, and hence gives valuable information about the processes occurring during reaction.

Raman spectroscopy. In Raman spectroscopy light is scattered from molecules. In the process some energy may be transferred from the molecule to the light, or vice versa, with the result that additional frequencies appear in the radiation emerging from the sample. Rotational vibration Raman spectra give information that complements that obtained from the corresponding absorption techniques. *See* RAMAN EFFECT.

Resonance techniques. These techniques constitute the fourth type of spectroscopy. They depend on excitation frequencies of the molecule being brought into resonance with the surrounding radiation by modifying the external conditions. In nuclear magnetic resonance (NMR), the principal example of this technique, the relevant energy levels are those arising from the different possible orientations of the magnetic moments of the nuclei in the molecules, and resonance occurs at radio frequencies. Nuclear magnetic resonance is a powerful technique for structural analysis and identification. Other resonance techniques include electron spin resonance (ESR or EPR) and Mössbauer spectroscopy. *See* ELECTRON PARAMAGNETIC RESONANCE (EPR) SPECTROSCOPY; ELECTRON SPECTROSCOPY; MOLECULAR STRUCTURE AND SPECTRA; MÖSSBAUER EFFECT; NUCLEAR MAGNETIC RESONANCE (NMR); PHOTOCHEMISTRY; SPECTROSCOPY.

Diffraction techniques. Other major techniques for the investigation of molecular structure are based on diffraction. These depend on the observation of the direction through which radiation and particles are scattered when they impinge on a sample. The principal example of these techniques is x-ray diffraction, for through it detailed information about the arrangement of atoms in crystals and even in very complex, large, biologically important molecules may be obtained. Also, there are diffraction techniques based on electrons and neutrons. *See* X-RAY DIFFRACTION.

Other techniques. Other techniques for investigating structure include the electric and magnetic properties of molecules, in particular, the determination of electric polarizabilities and dipole moments (which give information about electron distribution, and are important for the discussion of the dielectric properties), magnetic properties, and the properties based on optical birefringence, such as optical activity and the Faraday effect. In the case of macromolecules and colloids, important structure-analysis techniques include x-ray diffraction, sedimentation rates, and viscosity measurements.

Statistical thermodynamics. Structural properties and thermodynamic properties are brought together by statistical thermodynamics. This major theoretical procedure gives a way of predicting the thermodynamic properties of assemblies of molecules in terms of their individual energy levels. Statistical thermodynamics represents a grand synthesis of the two major aspects of physical chemistry. It provides an understanding of the bulk, macroscopic properties of molecules in terms of the properties of individual molecules, and it also constitutes an important practical technique for calculating otherwise inaccessible properties from readily available spectroscopic data.

Transport processes. The third major branch of physical chemistry is concerned with change: physical change and chemical change. In particular, it is concerned with the rate of change (while thermodynamics is concerned with the possibility of change, and the criteria of spontaneous change). Physical change includes the diffusion of one substance into another, or the migration of ions in an electrolyte solution. The simplest version of the former are the transport properties of gases. These include thermal conductivity and viscosity. They are treated most simply in terms of the kinetic theory of gases, in which a gas is regarded as a swarm of noninteracting points; modern physical chemistry is concerned with the role of intermolecular forces in determining the transport properties. Ion migration gives rise to questions of the mobility of ions in a variety of solvents, and to the description of diffusion processes in general. The application of thermodynamics to change in general constitutes the field of nonequilibrium thermodynamics. *See* GAS; TRANSPORT PROCESSES.

Chemical kinetics. The major aspect of chemistry is change, and change is therefore also a major aspect of physical chemistry. Chemical change may be studied at a variety of levels. Empirical chemical kinetics is the study of reactions in order to determine how their rates depend on the concentrations of the participants in the reaction and on the conditions, mainly the temperature. Investigation of the time dependence of reactions (a time dependence that can be observed from the order of minutes to picoseconds, now that pulsed, mode-locked lasers have entered chemistry) yields a detailed picture of the sequence of molecular transformations involved in a complex chemical reaction. Each step in the sequence depends on the concentration of the participants and an empirical constant referred to as the rate coefficient. The aim of theoretical chemistry is to account for the value of the rate coefficient and its dependence on the conditions. The earliest broadly successful approach led to the Arrhenius rate law, which is still used as a general guide to the way that the temperature affects the rate of a reaction. It asserts that the rate coefficient k varies with temperature as $k = A \exp(-E_a/RT)$, where A and E_a are empirical parameters, the latter being known as the activation energy of the reaction. A wide variety of simple reactions conform to this law, and it can be interpreted on the basis that only molecules possessing sufficient energy are able to undergo reaction when they encounter; the exponential factor indicates the fraction of collisions that satisfy the energy requirement. *See* CHEMICAL DYNAMICS.

Molecular reaction dynamics. Modern approaches to the prediction and explanation of observed reaction rates have been based either on a statistical view of the reaction (the activated complex theory), where the principles of statistical thermodynamics are applied to a system evolving with time, or a fundamental particle dynamics approach, where the trajectories of molecules are calculated as they undergo reaction, techniques then being used to convert these trajectories to values of the rate coefficients. The latter approach constitutes the field of molecular reaction dynamics. Experimental techniques have also been developed for observing individual molecular encounters and reactions; these are based on molecular beams, where a diffuse beam of one reactant is directed into the path of another, and the pattern of molecular scattering, including the electronic and vibrational states of the products, is interpreted in terms of the forces acting between the reactants during the reactive encounter. These techniques bring investigations closer to an atomistic and molecular interpretation of chemistry than anything that preceded them, but the relation of trajectories and rate coefficients and the relation of gas-phase events to those in solution remain open problems.

Surface chemistry. Chemical kinetics is so important that it is rich in applications and extensions. An important extension is to the reactions that occur on surfaces; these are the processes involved in heterogeneous catalysis. The study of surface chemistry breaks down into the analysis of the steps that lead to the affixation of the species to the surface (that is, the study of adsorption processes), the determination of the structure of the adsorbed species, and finally the reactions, and escape, of the adsorbed species. The entire subject, including the related processes occurring at liquid surfaces, constitutes the field of surface chemistry. A special application of surface chemistry is to the stability of colloidal suspensions of species in fluids, and another is to the processes that occur at the interface between an electrode and the solution in which it is immersed. Electrode reactions are governed largely by the rate of electron transfer between the ions in the solution and the metal electrode. The field of dynamical electrochemistry finds important applications in electrochemical power generation and storage, in corrosion, in electrodeposition, and in electrocatalysis. *See* ADSORPTION; COLLOID; HETEROGENEOUS CATALYSIS.

[P. W. ATKINS]

Bibliography: P. W. Atkins, *Physical Chemistry*, 1978; G. M. Barrow, *Physical Chemistry*, 4th ed., 1979; R. S. Berry, S. A. Rice, and J. Ross, *Physical Chemistry*, 1980; H. Eyring, D. Henderson, and W. Jost, *Physical Chemistry: An Advanced Treatise*, 1969 et seq.

Physical law

A physical phenomenon is said to be controlled or governed by a physical law when the phenomenon is one of a broad class of phenomena such that it is possible to formulate some regularity which applies to all members of the class. If the phenomenon is one which can be described in terms of numerical measurement, the law is often formulated in terms of mathematical relations between the numbers obtained by measurement, as in the inverse square law of universal gravitation. However, a mathematical formulation of a natural law is by no means necessary.

It is implicit in the notion of physical law that there are no exceptions, and, unlike law made by humans, it may not be "violated." When a recalcitrant phenomenon is found, a revision of the original statement of the law is needed, for example, by redefining the limits of its application, or by recasting it to encompass the phenomenon.

[PERCY W. BRIDGMAN/GERALD HOLTON]

Picosecond molecular processes

A branch of photophysics/photochemistry/photobiology dealing with various types of molecular energy exchange and chemical reactions occurring on the 10^{-12} to 10^{-9} s time scale. The study of picosecond processes overlaps the subpicosecond regime (less than 10^{-12} s) on one side and the nanosecond regime (about 10^{-9} to 10^{-8} s) on the other. Molecular picosecond experiments are the counterpart of flash photolysis experiments that were carried out mainly in the 1950s and early 1960s. They were about a million times slower than picosecond experiments, reflecting the vast improvements in laser and optoelectronic technology that have taken place since 1965.

A typical experiment is initiated with an ultrashort pulse of energy—radiation (light) or particles (electrons). The study of picosecond molecular processes deals with the rapid changes brought about by the absorption of these ultrashort energy pulses by a system of molecules—gas, liquid, or solid. These changes may be of either a chemical or a physical nature and can be measured by ultrafast detection methods, which are usually based on some type of spectroscopic monitoring of the system. The excited molecules may change their structure internally, they may gain or lose electrons or protons, they may chemically react, they may exchange their energy with other nearby molecules, or they may simply lose memory of the initial dipole direction in which they were excited (dephasing). The important aspect of picosecond processes lies in the fact that elementary motion, such as rotation, vibrational exchange, or charge transfer, takes place on this time scale. These types of studies have therefore provided for the first time a direct look at the relationship between such motion and overall chemical or biological changes. *See* CHEMICAL DYNAMICS.

Ultrashort pulses. Just as a camera with a very fast shutter speed is necessary to obtain a sharp photograph of a fast-moving object, the energy pulses for the study of molecular motions must be very short in order not to blur these motions. Ultrashort pulses of light are produced by mode-locking a laser, while short bursts of electrons can be produced by pulse radiolysis techniques.

Mode-locked lasers. Mode-locking a laser consists of placing in the laser cavity a nonlinear gain device, which amplifies intense light more than weak light. As the light in the cavity passes back and forth through such a device, the intensity at a given instant of time, instead of being spread throughout the entire cavity, tends to pile up in a specific region. Weaker intensity in the other regions is discriminated against because of the nonlinear gain. In a well mode-locked laser, the result is a single intense pulse of light traveling back and forth between the reflectors that define the laser cavity. In practice, one of the cavity reflectors has less than 100% reflectance at the laser wavelength. In this way, some of the energy is allowed to exit the cavity into the laboratory on each reflection so that it can be used in an experiment. The output of a mode-locked laser is therefore a train of very narrow intense pulses, separated in time by the round-trip cavity transient time of the pulse—usually around 10 ns. The time between pulses can be lengthened to any convenient time scale by electrooptical techniques that repetitively deflect pulses from the train.

Lasers used for mode locking may be either continuously operating or flash-lamp-driven. Mode-locked continuous-wave lasers are typically dye lasers pumped by (that is, driven by) a gas laser, such as an argon-ion laser, and provide a continuous train of pulses. Flash-lamp-driven lasers are solid-state or liquid lasers, and the mode-locked pulse trains last about 1 μs. Combinations of pulsed and continuous-wave lasers are possible. For example, frequency-doubled neodymium/yttrium aluminum garnet (YAG) lasers are being used for high amplification of mode-locked continuous-wave dye lasers. Pulse widths from mode-locked lasers typically lie in the range of 1–30 ps: 1 ps (or less) for argon-ion pumped dye lasers, about 4–8 ps for neodymium/glass lasers or flash-lamp-pumped liquid lasers, and 20–30 ps for ruby or neodymium/YAG lasers. Infrared light from a CO_2 laser can also be mode-locked, giving pulse widths of around 30 ps.

The main advantage of continuous-wave lasers for picosecond experiments is that the pulse train can attain high reproducibility for long periods of operation, allowing the results of an experiment to be accurately averaged over many thousands (or millions) of pulses. Pulse widths from mode-locked continuous-wave dye lasers are the shortest that can now be produced, allowing the best time resolution to be obtained. In addition, the output of such lasers, because of the spectral broadness of light emission from dyes, can be tuned over fairly wide wavelength regions. Pulsed lasers have much higher power per pulse than a continuous-wave laser, allowing certain nonlinear optical phenomena, such as two-photon absorption, to be studied. Less averaging is required, or is indeed possible, using this type of laser.

Electrons. Short-duration electron beams can be produced from microwave linear accelerators. Single micropulses of electrons are produced by electronically gating the electron gun. The electrons are bunched by synchronously applied fields and accelerated to around 20 MeV. Pulses of electrons having a time width of about 30 ps can be produced in this way. The total number of electrons in such a pulse is around 10^{10}–10^{11}, while the number of photons in a neodymium/glass laser pulse typically used for molecular studies is 10^{15}–10^{16}. Compensating for this, the energy per electron in the electron pulse is about a million times greater than the energy per photon in the laser pulse. However, the high energy per electron is not usually an advantage when studying molecular events, which typically require no more than a few electronvolts for their initiation. The excess energy not only may lead to complex high energy processes, but can also cause a number of events to take place simultaneously, complicating interpretation.

Detection. Picosecond processes within a molecular system excited by an ultrashort pulse of light are detected by observation of the spectral changes that follow. Either emission or absorption spectroscopy may be used to measure the buildup of concentration of a particular species or the decay of that species. The added parameter of the light polarization can also be used to help detect changes in the orientation of excited molecules or of electric polarization produced in the excited molecular ensemble. Since the events take place on a very fast time scale compared with the time it takes for electrons to travel through an electronic device, special detection techniques are required.

Streak camera. The most direct method of detecting ultrashort events is by means of a streak camera, which operates in some ways like an oscilloscope. Streak cameras capable of resolving two light pulses separated in time by 1 ps have been built. A slit image of the light from the event is focused onto photocathode material deposited on a transparent substrate housed in an evacuated image tube. The resulting electron packet is accelerated and sent through a very fast falling field (a voltage ramp), where it is deflected onto a moderately persistent phosphor screen. The degree of deflection depends on the instantaneous value of the field—electrons arriving early see a larger field and are deflected more than electrons arriving later. Photographing these phosphor streaks, or digitally recording them by means of a multichannel image detector, allows a direct measure of the intensity of the light signal. Thus, the streak camera is capable of directly measuring light intensity from the experimental event as a function of time. Very often the molecular response to a single picosecond pulse of light is a fast rise followed by a slower decay. In that case, the phosphor shows a smear of light with a steep intensity increase on the rising edge and a more gradual intensity decay on the falling edge.

Correlation methods. Two light signals from the same picosecond source (autocorrelation) or from two different sources (cross correlation) when combined in certain media produce a two-photon effect. An example is when two pulses of visible light are combined in a potassium dihydrogen phosphate (KDP) crystal to produce second-harmonic ultraviolet light. The two light signals must be present at the same time in the KDP crystal and must satisfy certain momentum conservation conditions (phase matching). If one pulse is delayed with respect to another, the intensity of the ultraviolet signal will be small, but as the pulses arrive at the detecting crystal more nearly simultaneously, the ultraviolet signal will increase. The amount of ultraviolet light is therefore a measure of how well the two light signals overlap in the crystal. One pulse, typically picked off by a beam splitter from the laser output and with a known width, acts as a standard (sometimes called the probe pulse). The temporal shape of another ultrafast light emission event may be ascertained by measuring the intensity of ultraviolet light as the time relationship between the two pulses is changed. This is accomplished by subjecting the probe pulse to a variable optical path length with a mirror or prism arrangement mounted on a precision translation stage. In this method, ultrafast temporal measurements are unnecessary. The velocity of light itself acts as the yardstick of time.

Interrogation techniques.. The earliest studies of picosecond molecular processes were carried out by using an excitation pulse followed by spectroscopic analysis with a second (probe or interrogation) pulse of different spectral wavelength. This is similar to a conventional flash photolysis experiment using absorption spectroscopy as the detector of short-lived intermediates. The probe pulse is delayed behind the excitation pulse by an optical delay line similar to that used in correlation measurements; or a discrete set of time delays may be obtained on each laser shot by expanding the size of the optical cross section of the pulse and sending it through a step echelon—for example, a stack of displaced microscope slides. The latter technique employs the fact that the velocity of light through glass is n times slower than through air, n being the refractive index of the glass. Crossed spectral and time resolution may be obtained by this technique (and the streak camera technique) by focusing the time-displaced images along the entrance slit of a fast spectrograph.

Biological application. An interesting experimental application of picosecond spectrosopy is in the discovery of the early ultrafast processes in the complex chain of events responsible for animal vision. Furthermore, it has enabled these processes to be put in their important place in this chain. Though various species of higher animals have widely different outward characteristics, the physicochemical processes through which they see are remarkably similar. The overall response of the eye to light may take as long as a few hundredths of a second, yet it has been found that events on the picosecond time scale dominate the early response functions following the absorption of light in the retina. Each link in the chain of events following the absorption of a photon of light is "fingerprinted" by a somewhat different absorption spectrum, making it possible to sort out these events by the interrogation methods of picosecond spectroscopy. The necessity for nature's use of such fast primary processes concerns "winning out" against unwanted competing chemical and energy loss processes—the faster the wanted process, the less the likelihood that unwanted, energy-wasting processes can take place. *See* LASER PHOTOCHEMISTRY; SPECTROSCOPY.

[G. WILSE ROBINSON]

Bibliography: R. Hochstrasser, W. Kaiser, and C. V. Shank (eds.), *Picosecond Phenomena II*, 1980; C. V. Shank, E. P. Ippen, and S. L. Shapiro (eds.), *Picosecond Phenomena*, 1978; S. L. Shapiro (ed.), *Ultrashort Light Pulses*, 1977.

Picrate

One of two types of compound which are derived from picric acid, 2,4,6-trinitrophenol. The hydrogen in the —OH group of the phenol is sufficiently labile to act as an acid and form salts with inorganic bases in the usual manner (for example, sodium picrate).

The second class of picrate is a group of molecular complexes formed when picric acid and aromatic compounds, such as naphthalene, are allowed to react. Although the nature of these complexes is not entirely understood, they are useful in the identification of aromatic compounds, because the complexes are solids with characteristic melting points. These complexes are usually highly colored. *See* PICRIC ACID.

[E. EUGENE WEAVER]

Picric acid

A phenol in which nitro (NO_2) groups are present in the 2, 4, and 6 positions on the ring of carbon atoms. The formula for picric acid is shown below.

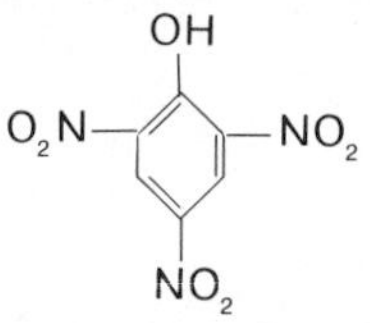

It approaches the mineral acids in acid strength ($pK_a = 0.80$). It is produced by nitration of chlorobenzene to the dinitro derivative, with hydrolysis of the latter, followed by another nitration, reaction (1).

$$C_6H_5Cl \xrightarrow[H_2SO_4]{HNO_3} 2,4\text{-}(O_2N)_2C_6H_3Cl \xrightarrow[Na_2CO_3]{H_2O}$$
$$2,4\text{-}(O_2N)_2C_6H_3OH \xrightarrow[H_2SO_4]{HNO_3} 2,4,6\text{-}(O_2N)_3C_6H_2OH \qquad (1)$$

Another method is by sulfonation of phenol, followed by nitration, reaction (2).

$$C_6H_5OH \xrightarrow{H_2SO_4} 2,4\text{-}(HO_3S)_2C_6H_3OH \xrightarrow{HNO_3} 2,4,6\text{-}(O_2N)_3C_6H_2OH \qquad (2)$$

Picric acid was widely used as a yellow dye for silk before being replaced by modern dyes of greater fastness. It has also been used as an explosive, and can be detonated by both heat and shock. Many salts of picric acid are even more readily detonated than the free acid and must be handled with great care. Storage of picric acid in contact with metal must be avoided because of the danger of salt formation. *See* NITROBENZENE; PHENOL.

[ROBERT I. STIRTON/MARTIN STILES]

Pigment

A finely divided material which contributes to optical and other properties of paint, finishes, and coatings. Pigments are insoluble in the coating material, whereas dyes dissolve in and color the coating. Pigments are mechanically mixed with the coating and are deposited when the coating dries. Their physical properties generally are not changed by incorporation in and deposition from the vehicle. Pigments may be classified according to composition (inorganic or organic) or by source (natural or synthetic). However, the most useful classification is by color (white, transparent, or colored) and by function.

White pigments. These pigments are essentially transparent to visible light. Because of the difference in refractive index between the pigment particles and the vehicles, white pigments refract the light from a multitude of surfaces and return a substantial portion in the direction of illumination without significant change in its spectral composition.

The common white pigments are titanium dioxide, derived from titanium ores; white lead, from corrosion of metallic lead; zinc oxide, from burning of zinc metal; and lithopone, a mixture of zinc sulfide and barium sulfate. Pure zinc sulfide and antimony oxide are less commonly used.

Titanium dioxide may be crystallized in the rutile or anatase form, depending on the method of production. It may be further modified by surface treatment to control the rate of chalking and other properties. Rutile titanium dioxide has a higher refractive index than anatase and therefore higher hiding power, but it has a somewhat yellow color. Anatase titanium dioxide provides a purer white.

White lead pigments are the oldest of white pigments and were used extensively to provide excellent hiding power, flexibility, and durability to interior and exterior paints and enamels. Consumer protection rulings have all but removed white lead paints from the market, because leaded paint particles were ingested by children, with toxic effects.

Zinc oxide and lithopone pigments were extensively used in paint formulation, but have been superseded by titanium dioxide. Pure zinc oxide pigment is rarely used. Antimony oxide pigment is used chiefly in certain fire-retardant paints.

Transparent pigments. The refractive indexes of these pigments are very close to the index of the paint vehicle (about 1.54). They are used to provide bulk, control setting, and contribute to the hardness, durability, and abrasion resistance of the paint film. Because they are commonly used to add bulk to other pigments, they are called extenders. Most transparent pigments are natural minerals reduced to pigment particle size. Among the most commonly used are calcium carbonate (ground limestone, whiting, or chalk), magnesium silicate, bentonite clay, silica, or barites (barium sulfate). Transparent pigments often constitute a substantial portion of a protective coating.

Colored pigments. These pigments are available in a wide variety of colors and properties, depending upon the end use. Several hundred have been used, with the following being the most common.

Red. Iron oxides, often classified by color, include Indian red, Spanish red, Persian Gulf red, and Venetian red, a mixture of iron oxide and calcium sulfate. Other red pigments include cadmium red (cadmium selenide) and organic reds, which are usually coal tar derivatives either precipitated in pigment form (toners) or deposited on a transparent pigment (lakes). Organic reds include toluidines and lithols.

Orange. Chrome orange (basic lead chromate), molybdate orange (lead chromate-molybdate), and various organic tones and lakes are the most common orange pigments.

Brown. Browns are nearly always iron oxides, although certain lakes and toners are used for special purposes.

Yellows. These pigments include natural iron oxides such as ocher or sienna, or synthetic iron oxides, which are stronger and brighter, such as chrome yellow (normal lead chromate) and cadmium yellow (cadmium sulfide), and organic toners and lakes such as Hansa yellow and benzidine yellow.

Green. The most important green pigments are chrome green, a mixture of chrome yellow and Prussian blue; chromium oxide, duller but more permanent; phthalocyanine green, an organic pigment containing copper; and various other organic toners or lakes, often precipitated with phosphotungstic or phosphomolybdic acid.

Blue. The blue pigments include Prussian blue (ferric ferrocyanide, sometimes called milori or Chinese blue, depending upon the shade); ultramarine, an inorganic pigment made by fusing soda, sulfur, and other materials under controlled conditions; phthalocyanine blue, an organic pigment containing copper; and numerous organic toners and lakes.

Purples and violets. These are nearly all organic toners or lakes. Manganese phosphate is a very weak, inorganic purple pigment.

Blacks. The vast majority of black pigments consist of finely divided carbon—carbon black, lampblack, and bone black—usually obtained by allowing a smoky flame to impinge on a cold surface. Black iron oxide and certain organic pigments are used where special properties are required.

Special pigments. Anticorrosive pigments are used to prevent the formation or spread of rust on iron when the metal is exposed by a break in the coating. The most common are red lead, an oxide of lead, and zinc yellow or zinc chromate, a basic chromate of zinc. Other colored chromates are sometimes used. The color of red leads fades rapidly, and the anticorrosive chromates are usually very weak in tinting strength. Metallic lead is sometimes used for anticorrosive paint.

Metallic pigments are small, usually flat particles of metal, prepared for dispersal in coatings. Aluminum is most commonly used because it leafs and forms a smooth, metallic film. The flakes are sometimes colored. Bronze, copper, lead, nickel, stainless steel, and silver appear occasionally. Zinc dust, or powdered zinc, is used more often because of its excellent adhesion to galvanized iron than because of its appearance.

Luminous pigments radiate visible light when exposed to ultraviolet light. Phosphorescent pigments continue to glow for a period after the exciting light has been removed; these are usually sulfides of zinc and other materials, with small amounts of additives which control the phosphorescent properties. Fluorescent pigments lose luminosity as soon as the exciting light is removed; these pigments may be sulfides, although many organic pigments have this property.

Other specialized pigments include pigments which change color at some predetermined temperature, used to indicate hot areas on motors; pigments which give a pearly appearance; and pigments which conduct electricity for printed circuits.

Coarse materials such as pumice are often added when a nonslippery coating is required. Glass beads give a very high degree of refractivity in the direction of illumination and are often used in center-line paints for signs where night visibility is required. Intumescent pigments puff up under heat, giving a fire-resistant coating. *See* DYE.

[C. R. MARTINSON; C. W. SISLER]

Bibliography: B. N. Chapman and J. C. Anderson (eds.), *Science and Technology of Surface Coatings*, 1974; R. Meyers and J. S. Long, *Pigments*, pt. 1, 1975; H. F. Payne, *Organic Coating Technology*, vol. 2, 1961.

Pinene

One of the most important terpenes. α-Pinene (I) is the chief constituent of turpentine and is a component of many essential oils. β-Pinene (II) often

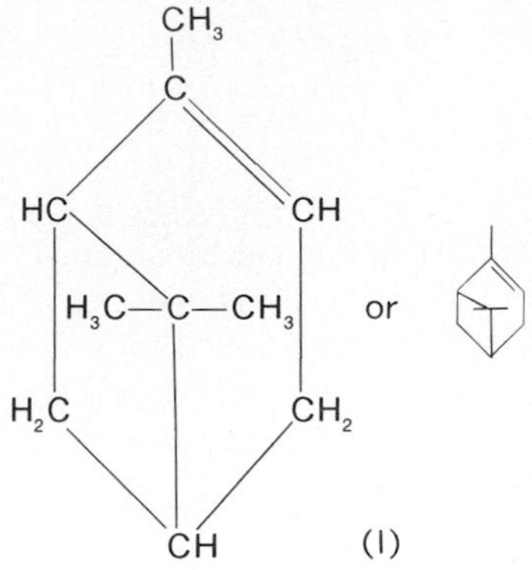

(I)

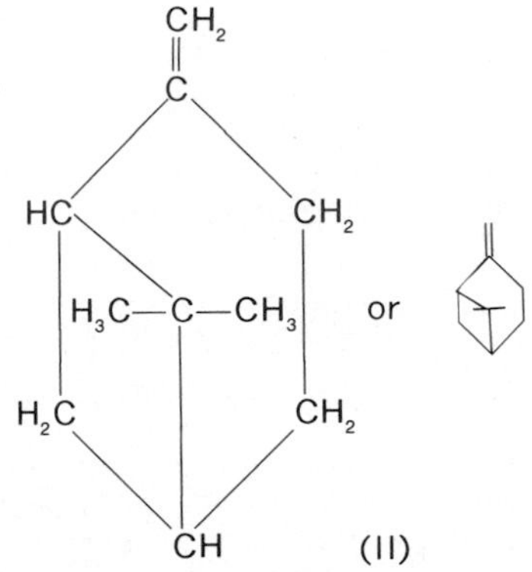

(II)

accompanies it and is usually present in smaller quantities. Both exist as dextro, levo, and racemic forms of optical isomers. Physical properties are listed in the table. *See* TERPENE.

Physical properties of pinene

Property	α-Pinene *d*-, *l*-, and *dl*-	β-Pinene *d*-, *l*-
Boiling point, 760 mm	156°C	164–166°C
Density (20°C)	0.858–0.860	0.87
Refractive index, 20°/D	1.466	1.46–1.48
Specific rotation	$d=+51°$ $l=-51°$	$l=-22°$

The pinenes are usually isolated by fractional distillation. α-Pinene is difficult to separate from β-pinene by distillation, but modern high-efficiency columns permit their ready fractionation.

Both pinenes are colorless oils which resinify on exposure to air. They undergo polymerization in the presence of anhydrous aluminum chloride or boron trifluoride. Because of the heat generated during polymerization, the reaction vessel is usually cooled. The polymerization is quenched by the addition of mineral solvent (naphtha). The polymers are used in the formulation of rubbers, paints, and varnishes. Somewhat different terpene polymers are obtained by treating the pinenes with maleic anhydride.

The chemistry of the pinenes is intricate and complicated by their easy transformation into compounds with different ring structures. These molecular rearrangements permit the pinenes to be used as a convenient starting point for the synthesis of other important chemicals. For example, α-pinene is used in the commercial production of borneol, camphor, terpineol, and terpin hydrate.

[FRANK WAGNER]

Pipet

A tube, usually made of glass, used almost exclusively to deliver accurately known volumes of liquids or solutions. There are two general categories of pipets: volumetric or transfer pipets and the graduated measuring type (see illustration). With volumetric pipets the liquid is sucked up above the mark on the stem above the bulb and the upper end is quickly closed with the index finger; any adhering liquid is wiped from the outside of the lower stem; the level of liquid in the stem is allowed to fall slowly, by regulating the pressure on the finger, until the buttom of the meniscus is tangent to the mark; liquid clinging to the tip is removed; and the pipet is allowed to empty freely into the receiving vessel. After 15 sec or the time specified on the pipet for drainage, the tip is touched and rotated against the inside of the receiver. The liquid remaining in the tip thereafter is not removed. Volumetric pipets, when handled in the described manner, will deliver reproducibly a definite amount of liquid or solution.

A graduated measuring pipet is used in the same general way except that the volume of liquid delivered can be varied by allowing the liquid to drain from one calibration mark to another. These pipets are usually not as accurate in delivering volumes of liquid as are the volumetric type.

For solutions that attack glass, pipets made of various plastics are used. *See* TITRATION; VOLUMETRIC ANALYSIS.

[CLARK E. BRICKER]

PIPET

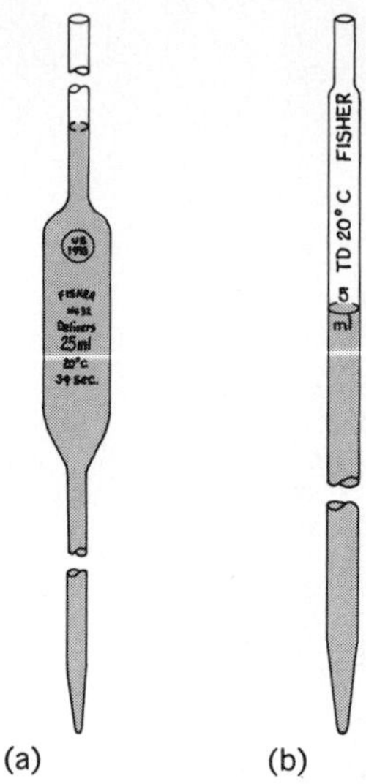

Pipets. (*a*) Volumetric type. (*b*) Graduated type.

pK

The logarithm (to the base 10) of the reciprocal of the equilibrium constant for a specified reaction under specified conditions (for example, solvent and temperature). p*K* values are often more convenient to tabulate and use than the equilibrium constants themselves. The value of *K* for the dissociation of the HSO_4^- ion in aqueous solution at 25°C is 0.0102 mole/liter. The logarithm is $0.008_6-2=-1.991_4$. p*K* is therefore $+1.991_4$. The choice of algebraic sign, although arbitrary, results in positive values for most dissociation constants applicable to aqueous solutions. The concept of p*K* is especially valuable in the study of solutions. *See* CHEMICAL EQUILIBRIUM; IONIC EQUILIBRIUM; PH.

[THOMAS F. YOUNG]

Plasma chromatography

A trace analysis technique in which trace molecules in a gas are detected by subjecting them to an ion-molecule reaction and measuring the ion-molecules produced. It is capable of detecting and identifying a substance present in concentrations far below the parts-per-billion range. The method of plasma chromatography (PC) was first reported early in 1970. Conceived and developed since 1967 by scientists at the laboratories of the Franklin GNO Corp., the instrumentation has now been applied to a wide range of fundamental and analytical studies. Upper atmospheric ion-molecule reactions have been studied as well as many aspects of chemical analysis, such as the identification of specific organic molecules, biomedical analysis of large molecules, and pesticide analysis.

Principle of method. The great sensitivity of the method to trace compounds arises because the ion-molecule reaction occurs at atmospheric pressure, where millions of collisions of reactant ions with trace molecules are possible. This all occurs in the plasma chromatograph, where ion-molecules are created and then separated by drifting them through an electrical field toward a detector. The detector output of ion current versus time presents a plasmagram of the reactant ion and the different ion-molecules formed. Plasmagrams, both positive and negative, resemble chromatograms with a millisecond time scale. This technique uses a shutter grid to inject a discrete ion-molecule pulse into the drift region and a gating grid to produce a recordable scan of the plasmagram (Fig. 1).

In early work the individual ion-molecule peaks of a plasmagram were injected into a mass spectrometer for identification. A correlation was soon found between plasmagram time and molecular weight of the ion-molecule, a discovery which led to simpler instrumentation without a mass spectrometer. Many studies with this simpler instrumentation have produced characteristic "fingerprint" patterns of compounds and have led to detection of specific trace constituents in gaseous mixtures.

Reactant ions are formed from the direct, primary ionization of components in the air used as a carrier for introducing the sample. A drying procedure leaves sufficient water vapor, about 10 parts per million, in the air to provide the major source of reactant ions. The incoming carrier gas passes over a 10-millicurie nickel-63 radioactive beta source which forms primary ions that, by a series of steps, evolve into stable reactant ions of

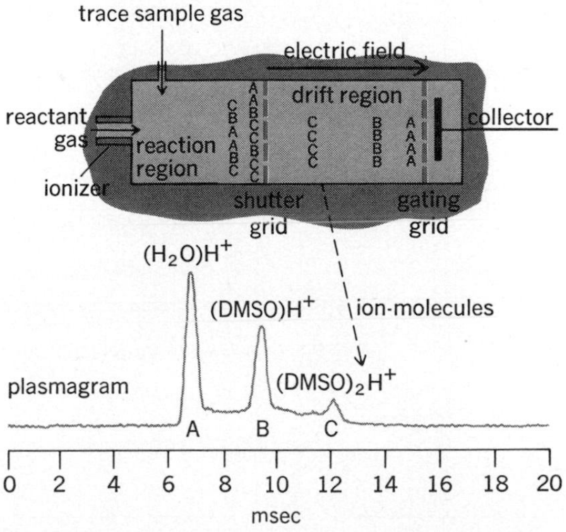

Fig. 1. Simplified diagram of the functions producing a plasmagram of dimethylsulfoxide (DMSO). (*From F. W. Karasek, A drift-mass spectrometer, Res./Develop., 21(12): 25, December 1970*)

the type $(H_2O)_nO_2^-$ and $(H_2O)_nH^+$. When nitrogen carrier gas is used, only electrons appear in the negative mode, while the positive reactant ions are the same as when air is the carrier.

Separation of the ion-molecule complexes is brought about by their different mobilities as they move through an inert gas at 760 torr (101,325 pascals) under the influence of an electrical field. The ion-drift region, where separation occurs, is filled with a flowing inert gas, usually nitrogen, whose flow is countercurrent to that of the carrier gas.

Exploratory analytical studies. An early study was made by F. W. Karasek to determine the qualitative aspects of the plasmagrams obtained for a series of compounds of very similar molecular weights but different molecular structures. The positive plasmagrams of salicylaldehyde, benzoic acid, naphthalene, acetophenone, and phenethyl alcohol show that when concentrations are low, as indicated by the continued presence of a strong reactant-ion peak, a single peak of the ion-molecule complex is formed. This ion-molecule complex is formed by the addition to the organic molecule of the $(H_2O)_nH^+$ reactant. As either the concentration or reactivity of the organic molecule increases, the reactant ion-peak decreases and even disappears, with the resultant formation of higher molecular weight ion-molecule complexes containing multiples of the trace molecules. The appearance of dimer and trimer complexes at higher concentrations is characteristic of the PC technique. Negative plasmagrams occurred only for salicylaldehyde, benzoic acid, and phenethyl alcohol, and mainly by transfer of an electron to negatively charge the trace molecule. These studies show that such compounds display characteristic, identifying positive and negative plasmagram patterns. By continuing such studies one can expect to develop "fingerprint" plasmagrams for all the different classes of organic compounds to provide identification data similar to those given by infrared and mass spectra.

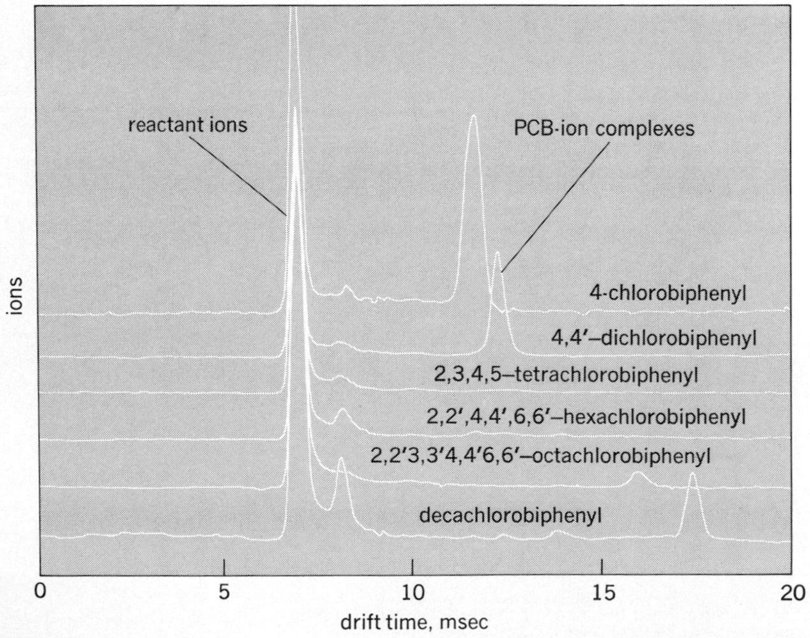

Fig. 2. Composite of the positive plasmagrams for PCB compounds. (*From F. W. Karasek, Plasma chromatography of the polychlorinated biphenyls, Anal. Chem., 43:1982–1986, 1971*)

The polychlorinated biphenyls (PCB) represent a type of compound that has assumed a prominent position in environmental studies because of their high toxicity and mounting evidence of their gross contamination of the global ecosystem. Because of their similarity to the common pesticides such as DDT and their many possible isomers, analysis for traces is usually a difficult and lengthy procedure, involving largely mass spectrometric and gas chromatographic methods. It has been demonstrated that the plasma chromatograph can give a distinct set of positive and negative plasmagrams for these compounds. In the series of PCB compounds shown in Fig. 2, those with the least chlorine substitution give the strongest positive plasmagrams. Those that give weak or no positive plasmagrams give strong negative ones. This information, obtained for trace concentrations below the part-per-billion level, is a valuable aid in detecting and identifying such compounds in the environment.

Fundamental studies. The reactivity of the PCB compounds as evidenced in their negative plasmagrams has a general correspondence to their response in the gas chromatographic electron capture detector. Since its inception the electron capture detector has occupied a unique position in gas chromatographic analysis because of its sensitivity and selectivity. Its importance has led to many studies of its characteristics and the mechanisms by which it functions. The studies of electron attachment phenomena and the mechanisms advanced have done much to further understanding of the capabilities of this detector. The data obtained with plasma chromatography can contribute considerably to knowledge of the fundamental physical processes involved in the electron capture detector. The PC instrument essentially draws out a group of charged particles from a plasma such as that found between the electrodes of the electron capture detector and subjects these charged species to a separation and identification step. Both the positive and negative particles can be examined separately and individually. The plasmagrams indicate the nature and relative quantities of these charged particles. *See* GAS CHROMATOGRAPHY.

By studying the plasmagrams of a series of halogenated aromatic compounds, Karasek verified experimentally the dissociative electron capture mechanism advanced for the electron capture detector by W. E. Wentworth. Using nitrogen gas as the carrier to form reactant ions, only electrons are produced in the negative mode. Upon the addition of trace amounts of halogenated benzene compounds, the plasmagrams revealed the loss of electrons by capture followed by the appearance of the halide ion from the subsequent dissociation. This phenomenon is shown in Fig. 3, where one can see the chloride, bromide, and iodide ions formed. These data also reveal that the dissociative capture reaction does not occur in the case of fluorobenzene.

Plasma chromatography enables one to study almost any ion-molecule reaction that occurs at atmospheric pressure. Typical of these reactions is that of the clustering of water molecules with charged ions. These reactions occur in the upper atmosphere and may have some role in the nucleation formation of water droplets leading to clouds

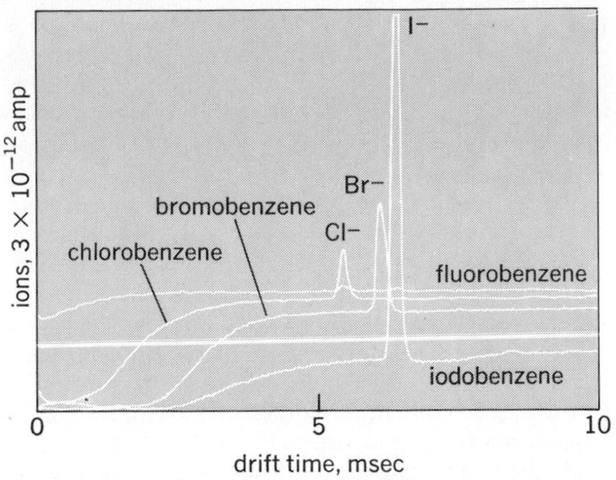

Fig. 3. Composite negative plasmagrams of the monohalogenated benzenes. (*From F. W. Karasek and O. S. Tatone, Plasma chromatography of the mono-halogenated benzenes, Anal. Chem., 44:1758–1763, September 1972*)

and rain. The equilibria of the reactant ions $(H_2O)_nH^+$ and $(H_2O)_nNO^+$ as a function of temperature can be observed with the plasma chromatograph and serve as an example of the ion-molecule reactions being studied.

Interfacing. Much of the exploratory work in plasma chromatography was done with a directly interfaced mass spectrometer so that each ion-molecule peak in a plasmagram could be mass identified. Since the particles are already charged, they can be directed from the drift spectrometer through a single aperture and ion lens into the mass analyzer section of a quadrupole mass spectrometer. From this work an empirical relationship was observed between the plasmagram time of an ion-molecule complex and its mass. This led to the simpler Beta VI and Beta VII PC instruments that have been used for qualitative studies without the use of a mass spectrometer. *See* MASS SPECTROMETRY.

Like a mass spectrometer, the plasma chromatograph functions best with only a single component being observed. An obvious sample source is the peaks from a gas chromatograph. Using a splitting technique which delivers only a fraction of a gas chromatograph peak to the plasma chromatograph, this has been successfully accomplished for a number of compounds such as halogenated aromatics and musk ambrette. Combining the gas chromatographic data with those of the plasmagram provides a unique set of qualitative information.

Future potential. Quite clearly, only the surface of the potential of this method has been explored. As more knowledge about the technique is accumulated and the instrumentation becomes more developed, it can be expected that the analytical capabilities will be extensive. As a tool for fundamental studies of ion-molecule reactions, plasma chromatography is unique and will greatly extend knowledge of these very important reactions. *See* CHROMATOGRAPHY; TRACE ANALYSIS.

[FRANCIS W. KARASEK]

Bibliography: M. J. Cohen and F. W. Karasek, *J. Chromatogr. Sci.*, 8:330–337, 1970; F. W. Karasek, *Anal. Chem.*, 43:1982–1986, 1971; F. W. Karasek, *Res./Develop.*, 21(3):34–37, March 1970; F. W. Karasek, W. D. Kilpatrick, and M. J. Cohen. *Anal. Chem.*, 43:1441–1447, 1971.

Platinum

A chemical element, Pt, atomic number 78, and atomic weight 195.09. Platinum is a soft, ductile, white noble metal.

Occurrence. The platinum-group metals – platinum, palladium, iridium, rhodium, osmium, and ruthenium – are found widely distributed over the Earth. Their extreme dilution, however, precludes their recovery, except in special circumstances.

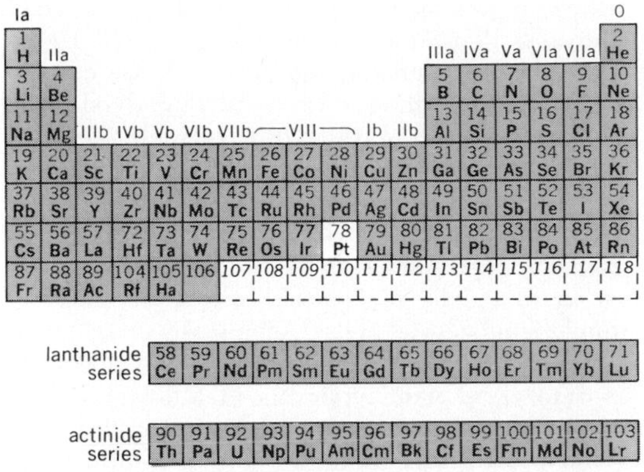

Native platinum found in placer deposits comprises a mixture of the six platinum-group metals. The placer deposits are concentrates resulting from the erosion of ultrabasic rocks and are found in Alaska, Columbia, and the Soviet Union. Historically, placer deposits were the first sources of platinum, found initially in Columbia and then in the Ural Mountains of the Soviet Union. Although these sources are still active, the principal sources today are much less rich lode deposits in Canada, South Africa, and the Soviet Union. These deposits contain small amounts of platinum along with other metals. Such deposits are usually worked for the total metal content, the base metals often making the recovery of the platinum metals economically feasible. In Canada the platinum metals occur in nickel–copper sulfide ores, and the production is dependent entirely upon the demand for nickel. The platinum content of these ores varies somewhat, but is on the order of 1 part per million, practically all of which is recovered.

The platiniferous ores of South Africa contain minor amounts of copper and nickel, but these ores are mined primarily for their platinum content. Deposits similar to those of South Africa are found on the Middle Siberian Plateau in the Soviet Union. Over one-half the world's production of the platinum metals is estimated to come from the Soviet Union. South African gold mines yield small amounts of osmiridium, which contains about 35% osmium and 30% iridium, and iridosmine, which contains about 40% iridium. The principal platinum minerals are platinum arsenide (sperrylite), platinum sulfide (copperite), platinum, palladium, nickel sulfide (braggite), native platinum, osmiridium, and iridosmine.

Small amounts of the platinum metals, palladium in particular, are recovered during the electrolytic refining of copper. It has been predicted that spent nuclear fuels may become a significant source of ruthenium, rhodium, and palladium.

Uses. The platinum-group metals have wide chemical use because of their catalytic activity and chemical inertness. As a catalyst, platinum is used in hydrogenation, dehydrogenation, isomerization, cyclization, dehydration, dehalogenation, and oxidation reactions. Finely divided platinum on aluminum oxide is used to upgrade the octane rating of gasoline by catalytically accelerating a complex series of dehydrogenation, hydrogenation, isomerization, and cyclization reactions. Platinum-rhodium alloy gauzes are used in the catalytic oxidation of ammonia to form nitric acid or oxides of nitrogen. When this process is carried out in the presence of methane, hydrocyanic acid is formed. Platinum is a catalyst in the older contact process for making sulfuric acid, the conversion of carbon monoxide to carbon dioxide by combination with oxygen, the recombination of hydrogen and oxygen to form water, the reduction of nitro groups, and the removal of nitrogen(II) oxide from gas streams by reaction with hydrogen to form nitrogen and water. It is also used as a catalyst in fuel cells to generate electricity directly from chemical reactions. Advantage is taken of platinum's chemical inertness in many forms of laboratory and plant apparatus, including pyrolysis reaction equipment, electrochemical anodes such as those used in perchlorate and persulfate production, spinnerets used for synthetic fiber extrusion, and lining for reaction vessels. *See* CATALYSIS; ELECTROCHEMICAL PROCESS.

In the glass industry, platinum (either pure or alloyed) is widely used at high temperatures to contain, stir, and convey molten glass. Platinum vessels are also used at high temperatures in making crystals, such as rubies grown from corrosive fluxes. Other high-temperature applications include thermocouples and resistance thermometers for temperature measurement, and resistance windings for electrically heated furnaces.

In the electrical industry, platinum is used in contacts and resistance wires because of its low contact resistance and high reliability in contaminated atmospheres.

Platinum is often used in conjunction with other materials to protect them or to be supported by them. To this end, platinum may be metallurgically clad to the base, plated onto the base from aqueous or fused salt baths, or deposited from organometallic compounds or suspensions in organic vehicles, both of which are fired to drive off the organic materials, leaving the metal deposit. Such combinations of platinum and other materials are made where necessary for economic reasons or where the combined properties of the platinum and base are required. For instance, platinum is clad over tungsten for use in electron tube grid wires. Platinum, because of its high thermionic work function, reduces secondary emission from the wire, while the tungsten base provides high-temperature strength.

Alloys. Platinum is soft and ductile; greater strength may be obtained by alloying it with the other platinum metals or base metals. Rhodium, iridium, and ruthenium are effective hardeners of platinum. Rhodium is least effective but is preferred for high temperatures since the oxidation resistance of platinum-rhodium alloys is superior to that of the other alloys. Alloys with up to 40% rhodium are commercially available. Platinum-rhodium alloys are used in glass manufacture 10–20% Rh), including the manufacture of fiber glass, and in the manufacture of thermocouples (10–40% Rh) and ammonia oxidation catalysts and spinnerets (10% Rh). Platinum-iridium alloys with up to 30% iridium are used where hardness and corrosion resistance are required. These uses include electrochemical anodes, electrical contacts, jewelry, surgical tools and implants, and standards for length and weight. Ruthenium is a very potent hardener for platinum, and ruthenium alloys are used in electrical contacts. Base metals are alloyed with platinum for special purposes. A 4% tungsten alloy is used in heavy-duty aircraft spark plugs (lead-corrosion resistance is required), and a 23% cobalt alloy is one of best permanent-magnet materials known.

Physical and chemical properties. Platinum is not affected by atmospheric exposure, even in sulfur-bearing industrial atmospheres. Platinum remains bright and does not visually exhibit an oxide film when heated, although a thin, adherent film forms below 450°C. This oxide film, PtO_2, dissociates at about 450°C, and above this temperature platinum slowly loses weight because of the formation of the volatile oxide, PtO_2. Platinum oxide may also be produced by chemical or electrochemical reactions. Hydrogen or other reducing atmospheres are not directly harmful to platinum at elevated temperatures. However, certain elements in proximity to platinum, such as silicon from refractories or glass, may be reduced by hydrogen, and these elements may then alloy with platinum, thereby degrading it. Platinum may be worked to fine wire and thin sheet, and by special processes, to extremely fine wire. Important physical properties are given in the table.

Platinum is not attacked at room temperature by any single acid or alkali or by aqueous solutions of simple salts and organic materials, but it is readily dissolved by hot aqua regia. Although it is resistant to hydrogen chloride at elevated temperatures, it reacts with chlorine at about 500°C. It may also be attacked by fused alkalies. It is resistant to sulfurous gases, mercury, fused sulfates, chlorides, carbonates, and molten glass. Platinum can be made into a spongy form by thermally decomposing ammonium chloroplatinate or by reducing it from an aqueous solution. In this form it exhibits a high absorptive power for gases, especially oxygen, hydrogen, and carbon monoxide. The high catalytic activity of platinum is related directly to this property. Hydrogen diffuses through heated platinum. Platinum strongly tends to form coordination compounds. *See* CRACKING; HYDROGENATION.

Metallurgical extraction. There is no routine method for the extraction of platinum. The method used is dependent upon the starting material, which may be scrap or used catalyst, slimes resulting from nickel or copper processing, crude platinum, or the very refractory osmiridium or iridosmine. Platinum can be extracted with aqua regia in some cases. In other cases it may be necessary to fuse the ore with a suitable flux and to collect the platinum group metals in a carrier, such as copper or lead. When a hydrochloric acid solution of platinum is oxidized and then made basic, most of the impurities precipitate as hydroxides, and the platinum remains in solution. A similar purification can be obtained by converting the platinum to the hexanitrito complex, which does not precipitate in basic solution. When a hydrochloric acid solution

Physical properties of platinum

Properties	Value
Atomic weight ($C^{12} = 12.00000$)	195.09
Naturally occurring isotopes and % abundance	190, 0.0127% 192, 0.78% 194, 32.9% 195, 33.8% 196, 25.3% 198, 7.21%
Crystal structure	Face-centered cubic
Lattice constant *a* at 25°C, angstrom units	3.9231
Thermal neutron capture cross section, barns	8.8
Common chemical valence	2, 4
Density at 25°C, g/cm³	21.46
Melting point, °C	1772
Boiling point, °C	3800
Specific heat at 0°C, cal/g	0.0314
Thermal conductivity, 0–100°C, cal cm/cm² sec °C	0.17
Linear coefficient of thermal expansion, 20–100°C, μ-in./in./°C	9.1
Electrical resistivity at 0°C, microhm-cm	9.85
Temperature coefficient of electrical resistance, 0–100°C/°C	0.003927
Tensile strength, 1000 psi	
Soft	18–24
Hard	30–35
Young's modulus at 20°C	
psi, static	24.8×10^6
psi, dynamic	24.5×10^6
Hardness, Diamond Pyramid Number (DPN)	
Soft	37–42
Hard	90–95

of platinum(IV) is treated with ammonium chloride, the platinum is almost completely precipitated as ammonium chloroplatinate. The chloroplatinate is readily converted to the metal by thermal decomposition. Platinum can also be reduced to the metal from aqueous solutions of its salts by zinc, magnesium, iron, or aluminum. The platinum in dilute aqueous solutions can also be precipitated as the sulfide and thereby concentrated.

The refining of a specific starting material consists of a combination of these methods. In addition to separating the precious metals from all other impurities, it is necessary to separate them from each other. The refining of precious metals requires a great deal of flexibility on the part of the refiner. It is not unusual for a portion of the material to be recycled in the refinery because of the lack of suitable reactions for quantitative separations.

Principal compounds. Platinum dioxide, PtO_2, is a dark-brown insoluble compound, commonly known as Adams catalyst. It is prepared by fusing chloroplatinic acid with sodium nitrate at 500°C. Solution of this mass in water separates the salts from the insoluble platinum dioxide. Platinum(II) chloride, $PtCl_2$, is an olive-green water-insoluble solid. It is made by heating platinum in chlorine at 500°C or by the thermal decomposition of chloroplatinic acid. Platinum(II) chloride dissolves in hydrochloric acid to form chloroplatinous acid, H_2PtCl_4, which cannot be isolated but which forms soluble salts, such as potassium platinum(II) chloride, K_2PtCl_4. Chloroplatinic acid, H_2PtCl_6, the most important platinum compound, is made by dissolving platinum in aqua regia, destroying the nitric acid by evaporation from a hydrochloric acid solution, and evaporating the solution. The acid is isolated as a hydrate, $H_2PtCl_6 \cdot 6H_2O$. The red-brown crystals are highly soluble in water. Ammonium chloroplatinate, $(NH_4)PtCl_6$, is a lemon-yellow, crystalline, relatively insoluble solid, made by adding ammonium chloride to a solution of chloroplatinic acid. Compounds such as dichlorodiammino platinum, $Pt(NH_3)_2Cl_2$, exhibit cis-trans isomerism because of their planar configuration. *See* IRIDIUM; OSMIUM; PALLADIUM; RHODIUM; RUTHENIUM.

[HENRY J. ALBERT]

Bibliography: American Society for Metals, *The Metals Handbook, Properties and Selection: Nonferrous Alloys and Pure Metals*, 1979; F. E. Beamish and J. C. Van Loon, *Selected Methods of Determining the Seven Noble Metals*, 1977; Bureau of Mines, *Mineral Facts and Problems*, U.S. Department of the Interior, 1976; Bureau of Mines, *Minerals Yearbook*, vols. 1 and 2 (combined): *Metals, Minerals and Fuels*, U.S. Department of the Interior, annual; Engelhard Minerals and Chemicals Corp., *Engelhard Ind. Tech. Bull.*, quarterly; S. I. Ginzburg et al., *Analytical Chemistry of the Platinum Metals*, 1975; Johnson Matthey and Co., Ltd., *Platinum Metals Rev.*, quarterly; P. N. Rylander, *Catalytic Hydrogenation over Platinum Metals*, 1967; E. M. Wise, Platinum group metals and their alloys, in H. H. Uhlig (ed.), *Corrosion Handbook*, 1948.

Plutonium

A chemical element, Pu, atomic number 94; the name is derived from the planet Pluto. Plutonium is a reactive, silvery metal in the actinide series of elements. The first isotope to be identified was produced from the irradiation of uranium by

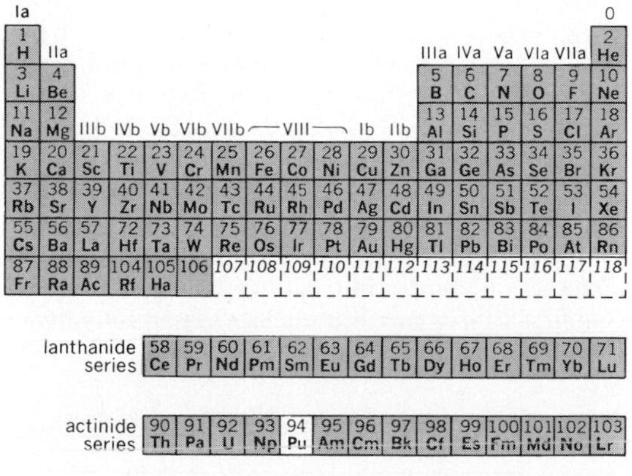

G. T. Seaborg, E. M. McMillan, A. C. Wahl, and J. Kennedy. The principal isotope of chemical interest is ^{239}Pu. It is formed in nuclear reactors by the process shown in (1). Plutonium-239 is fission-

$$^{238}U + n \longrightarrow {}^{239}U \xrightarrow[23.5\ \text{min}]{\beta^-} {}^{239}Np \xrightarrow[2.33\ \text{days}]{\beta^-} {}^{239}Pu \qquad (1)$$

able, but may also capture neutrons to form higher plutonium isotopes ^{240}Pu, ^{241}Pu, ^{242}Pu, ^{243}Pu, and ^{244}Pu. The chemically important isotope ^{239}Pu has a half-life of 24,131 years. The longest-lived species is ^{244}Pu with a half-life of 8.05×10^7 years. Minute quantities of ^{239}Pu are formed in pitchblende and monazite ore by reaction (1), the uranium and plutonium ratio being 10^{11}:1. Primordial ^{244}Pu has been found in the rare-earth mineral bastnasite.

Uses. Plutonium-238, with a half-life of 87.7 years, is utilized in heat sources for space application, and has been used for heart pacemakers. Plutonium-239 is used as a nuclear fuel, in the production of radioactive isotopes for research, and as the fissile agent in nuclear weapons.

Properties. Plutonium exhibits a variety of valence states in solution and in the solid state. Plutonium metal is highly electropositive. The oxidation states III, IV, V, and VI are known in acid solutions. The ions of the IV, V, and VI states are moderately strong oxidizing agents. The III, IV, and VI states can coexist in 1 *M* perchloric acid. In alkaline solutions, certain compounds of Pu(VI), and an unstable Pu(VII) oxidation state, may be formed under strongly oxidizing conditions (ozone). Solid compounds of all of the III, IV, V, VI, and VII valence states have been prepared.

Preparation. Methods for the isolation and purification of plutonium make use of the fact that the element can exist in a multiplicity of oxidation states, each having different chemical properties. The first isolation of plutonium, which was also the first isolation of a synthetic chemical element, was carried out by B. B. Cunningham in 1942. The first sample was weighed on a microbalance and found to weigh 2.77 μg. Special microchemical techniques were used in the early work to characterize the compounds of plutonium.

Carrier precipitation process. For isolation and purification of plutonium on the industrial scale, processes have been developed using carrier precipitation, solvent extraction, and ion exchange. The earliest plant-scale process was a carrier precipitation process developed by S. G. Thompson, which used bismuth phosphate and lanthanum fluoride as the carrier. In more recently developed processes, solvent extraction techniques are being used. These have the advantage that not only plutonium but uranium as well may be readily recovered and decontaminated from fission products. Some of the more important solvents are listed in Table 1. Control of the extraction behavior is obtained by the use of diluents for the solvent, addition of salting agents to the aqueous layer, and control of the solution pH and temperature. In Table 1, diluents and salting agents commonly employed for some of the more important valence states with two of these solvents are listed, and the relative distribution coefficients, defined as the concentration of the metal in the organic phase divided by the concentration in the aqueous phase) are shown in Table 2. The actual values of the distribution coefficients will change with the conditions, but the (approximate) relative values will be maintained.

Table 1. Solvents used in the separation of plutonium and uranium from fission products

Solvent (trivial name)	Diluent	Salting agent
Methyl isobutyl ketone (Hexone)	None	$Al(NO_3)_3$
Tri-*n*-butyl phosphate (TBP)	Kerosine	HNO_3, $Al(NO_3)_3$
Dibutyl ether of ethylene glycol (Carbitol)	None	HNO_3
Dibutyl ether of tetraethylene glycol (Pentether)	None or butyl ether	HNO_3
Triglycol dichloride (Trigly)	None	$Al(NO_3)_3$
Thenoyl trifluoroacetone (TTA)	Benzene or toluene	None

Table 2. Distribution coefficients of uranium, plutonium, and fission products from nitrate solutions of 100-day cooled reactor fuel

Solvent	U(VI)	Pu(VI)	Pu(IV)	Pu(III)	Fission products
Hexone	1.5	7.6	1.6×10^{-2}	4.5×10^{-4}	6×10^{-4}
TBP	8.0	0.6	1.5	2×10^{-2}	2×10^{-3}

The industrial processes are carried out in heavily shielded rows of concrete cells (canyons) in remotely controlled equipment. Charges up to several tons of irradiated fuel may be processed at one time.

Redox process. The redox solvent extraction process employing hexone has superseded the earlier bismuth phosphate separation process. In this procedure the uranium fuel is dissolved in nitric acid. The resulting spent fuel solution is oxidized, and Pu(VI) and U(VI) are separated by extraction from the fission products (codecontamination cycle). Next, the hexone layer is scrubbed to remove impurities, then the solvent is contacted with an aluminum nitrate solution containing a reducing agent (Fe(II) ion, hydroxylamine). The plutonium is back-extracted by this treatment into the aqueous layer as Pu(III), while the uranium remains in the solvent as U(VI) (partition cycle). The aqueous layer may be reoxidized and again extracted as before. By successive cycles of this process the plutonium is purified from fission products to the desired degree. Because the aqueous waste of the process is loaded with aluminum nitrate, and hexone is rather flammable, the redox process was discontinued, and was superseded in turn by a process employing a solution of tributyl phosphate in an aliphatic hydrocarbon solvent.

Purex process. The industrial process employing tri-n-butylphosphate (TBP) as the solvent extractant is the process used in most reprocessing plants. It is called the purex process (purex = *p*lutonium *u*ranium *r*eduction *ex*traction). The general operating procedure is similar to that of the redox process. After dissolution of the fuel, the oxidation state is adjusted so that the plutonium is present as Pu(IV), the uranium as U(VI). The nitric acid concentration is adjusted, and Pu(IV) and U(VI), are extracted into a solution of 30% TBP in an aliphatic hydrocarbon solvent (codecontamination cycle). The TBP phase containing Pu(IV) and U(VI) is scrubbed with nitric acid to remove impurities, and the Pu is removed as Pu(III) by back-extracting and solvent with nitric acid containing a reducing agent [NH_2OH, Fe(II) plus sulfamate, U(IV)], or after electrolytic reduction (partition cycle). Both the uranium and the plutonium are further purified by one or two additional purex extractions (second and third decontamination cycles). The major advantage of the purex system is

Table 3. Properties of plutonium metal

Phase	Symmetry space group	Lattice constants at t°C; a,b,c in angstroms, β in degrees	Atoms per unit cell	Density (20°C), g cm^{-3}	Transition temperature, °C	Linear expansion coefficient* (20°C), α×10^6	Resistivity (20°C) × 10^6, ohm-cm	Temperature coefficient of resistivity†
α	Monoclinic, $P2_1/m$	a = 6.183(1) b = 4.822(1) c = 10.963(1) β = 101.79(1) t = 21°C	16	19.86		54	145	−21
α→β					122 (4)			
β	Monoclinic body-centered, I2/m	a = 9.284(3) b = 10.463(4) c = 7.859(3) β = 92.13(3) t = 190°C	34	17.70		42	110.5	−6
β→γ					207 (5)			
γ	Face-centered orthorhombic, Fddd	a = 3.159(1) b = 5.768(1) c = 10.162(2) t = 235°C	8	17.14		34.6	110	−5
γ→δ					315 (3)			
δ	Face-centered cubic, Fm3m	a = 4.6371(4) t = 320°C	4	15.92		−8.6	103	+7
δ→δ′					457 (2)			
δ′	Body-centered tetragonal, I4/mmm	a = 3.34(1) c = 4.44(4) t = 465°C	2	16.00		−65.6	105	+45
δ′→ε					479 (4)			
ε	Body-centered cubic	a = 3.6361(4) t = 490°C	2	16.51		36.5	114	−7
ε→liq					640°C (mp)			
liquid				16.62		93		
ζ	?	?	?		>330 >0.6 kbar			

*$\alpha = \frac{1}{L}\cdot\frac{L}{T}$ †$\frac{1}{\rho}\cdot\frac{\rho}{t}\times 10^5$

SOURCE: O. J. Wick, *Plutonium Handbook*, vol. 1, Gordon & Breach, 1969; ζ-plutonium data from J. R. Morgan, in W. N. Miner, Plutonium 1970 and Other Actinides, *Nucl. Metallurgy*, vol. 17, Metallurgical Society of AIME.

that the waste generated in the process consists of a nitric acid solution of fission products containing very little solids, from which the nitric acid may be recovered by distillation and recycled.

In both the redox and purex processes, the final plutonium product is a nitric acid solution of Pu(IV). From this solution, plutonium may be isolated in a very pure state by precipitation as the peroxide or as the oxalate, either of which may be ignited to the oxide. Crude plutonium solutions containing other heavy metals may be purified from all such metals (except from thorium) by adjusting the acid concentration to 7–8 *M* HNO_3 and passing the solution through a Dowex 1 anion exchange column. Plutonium is retained as the $Pu(NO_3)_6^{2-}$ anion, which is stripped from the column with 3 *M* hydroxylamine nitrate.

Conversion to metal. To prepare the metal, the nitrate solution of Pu(IV) is precipitated as the peroxide or oxalate, which is ignited to PuO_2. The PuO_2 is hydrofluorinated to PuF_3 or PuF_4 by treating the oxide either with an H_2-HF mixture or with O_2-HF mixture. The fluoride is mixed with calcium metal, and charged into a steel bomb fitted with a ceramic liner. The bomb is closed with a flanged lid and ignited by induction heating. After reduction, the metal is isolated by breaking the liner.

Plutonium metal has unique properties. It exists in six allotropic forms below the melting point (640°C) at ordinary pressure. A seventh allotrope is obtained when the metal is compressed at pressures greater than 0.6 kbar. Particularly puzzling, however, are the contractions which the δ- and δ′-modifications undergo with increasing temperature (see illustration). Noteworthy is the fact that in no phase do both the coefficient of thermal expansion and the temperature coefficient of resistivity have the conventional algebraic signs. If the phase expands on heating, the resistance decreases. The peculiar thermal expansion behavior of the metal prevents the use of unalloyed plutonium metal as a reactor fuel. However, the δ-phase may be stabilized over a wide temperature range by addition of gallium or aluminum (so-called δ-retainers) and may thus be utilized in reactors. A liquid alloy, consisting of plutonium, cobalt, and cerium with a melting point of 415°C, has been used as fuel in the Los Alamos Molten Plutonium Reactor Experiment (LAMPRE). Some of the physical properties of the normal pressure modifications of plutonium are given in Table 3.

Numerous alloys of plutonium have been prepared, and a large number of intermetallic compounds have been characterized.

Principal compounds. Reaction of the metal with hydrogen yields two hydrides: a nonstoichiometric $PuH_{2.0...2.7}$ and stoichiometric PuH_3. The hydrides are formed at temperatures as low as 150°C. Their decomposition above 750°C may be used to prepare reactive plutonium powder.

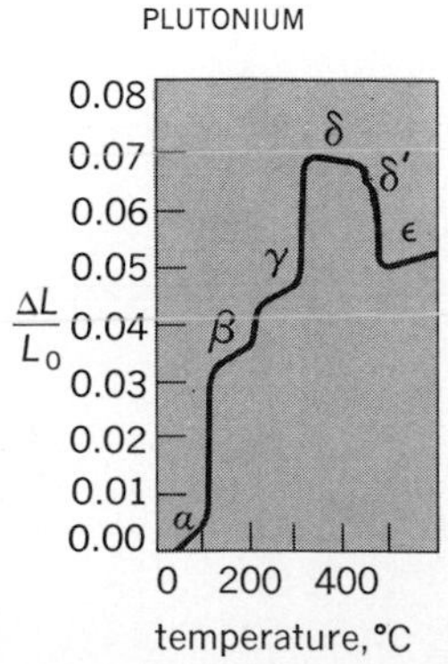

Expansion of high-purity plutonium under conditions of self-heating, L_0 = 0.5 in. (13 mm). (*After E. R. Jette*)

Table 4. Plutonium halides and oxyhalides

Compound	Color	Melting point, °C	Density at 20°C	Crystal structure
PuF_3	Purple	1425	9.32	Hexagonal
PuF_4	Pale brown	1037	7.0	Monoclinic
PuF_6	Reddish-brown	50.75		Orthorhombic
$PuCl_3$	Green	760	5.70	Hexagonal
$PuBr_3$	Green	681	6.69	Orthorhombic
PuI_3	Green	777	6.92	Orthorhombic
PuOF	Metallic	1635	9.76	Tetragonal
PuOCl	Blue-green		8.81	Tetragonal
PuOBr	Green		9.07	Tetragonal
PuOI	Green		8.46	Tetragonal

The most common oxide is PuO_2, which is formed by ignition of hydroxides, oxalates, peroxides, and nitrates of any oxidation state in air of 870–1200°C. PuO_2 crystallizes in a face-centered cubic structure, the exact lattice constant depending on the stoichiometry. Stoichiometric PuO_2 is amber. Deviations from stoichiometry are associated with colors from dark olive green to reddish brown. In the Pu-O system, a number of oxides have been reported which range in composition from $PuO_{1.5}$ to PuO_2, including Pu_2O_3, which is dark blue, hexagonal, and isomorphous with the A-type rare-earth oxides.

A very important class of plutonium compounds are the halides and oxyhalides (Table 4). Plutonium hexafluoride is the most volatile plutonium compound known. It resembles UF_6 and NpF_6 in its properties. PuF_6 is a strong fluorinating agent. Conditions for the preparation of the fluorides are given in reactions (2), (3), and (4). The other hal-

$$PuO_2 + \tfrac{1}{2}H_2 + 3HF \xrightarrow{600°C} PuF_3 + 2H_2O \quad (2)$$

$$PuO_2 + O_2 + 4HF \xrightarrow{550°C} PuF_4 + 2H_2O + O_2 \quad (3)$$

$$PuF_4 + F_2 \xrightarrow{750°C} PuF_6 \quad (4)$$

ides may be prepared by a number of methods. Treatment of PuO_2 with powerful halogenating agents such as CCl_4, PCl_5, SCl_2, and hexachloropropene yields $PuCl_3$. $PuBr_3$ may be obtained by dehydrating $PuBr_3 \cdot 6H_2O$ in a HBr atmosphere, and PuI_3 can be prepared by reacting plutonium metal with either HI, I_2, or HgI_2.

A number of other binary compounds are known (Table 5). Among these are the carbides, silicides, sulfides, and selenides, which are of particular interest because of their refractory nature. These compounds are usually prepared by direct combination of the elements.

Plutonium in any of its valency states forms a large number of fluoroplutonates and oxoplutonates with alkali metal and ammonium fluorides, or alkali metal oxides (Table 6). In addition, numerous oxoplutonates and fluoroplutonates are known of similar composition formed with alkaline earths, lanthanides, and a few transition metals.

A large number of plutonium compounds may be prepared from aqueous solution, either by evaporation or by precipitation. In Table 7 a selection of such compounds is given. Furthermore, numerous compounds of plutonium with organic acids are known, and π-complex compounds such as $Pu(C_5H_5)_3$ have also been prepared.

Properties of the ions in aqueous solution. In aqueous solution, plutonium tends to exist in the IV oxidation state. In strongly oxidizing conditions, Pu(VI) is formed; under strongly reducing conditions, Pu(III) is produced. The redox potentials are so close in their values that pure solutions of intermediate oxidation states undergo self-oxidation and reduction reactions (disproportionation). A typical disproportionation reaction is the one involving Pu(IV), which can be written as (5), for which the equilibrium constant is expressed as (6). K_1, as calculated from the redox potentials, is 0.0089 for Pu(IV) in 1 *M* acid at 25°C.

$$3Pu^{4+} + 2H_2O \rightleftharpoons PuO_2^{2+} + 2Pu^{3+} + 4H^+ \quad (5)$$

$$K_1 = \frac{[PuO_2^{2+}]\,[Pu^{3+}]\,[H^+]^4}{[Pu^{4+}]^3} \quad (6)$$

In solution of acids, such as nitric or hydrochloric acids, whose anions form only weak complexes with plutonium ions, the relative stabilities of the different states are little changed. Qualitatively, it is known that univalent anions, with the exception of fluoride, form relatively weak complexes with plutonium ions in all oxidation states. Higher-valent anions, however, form relatively strong complexes. In general, the relative stabilities of complexes with a given anion decrease in the order

$$Pu^{4+} > PuO_2^{2+} > Pu^{3+} > PuO_2^{+}$$

The ions of the different oxidation states have characteristic colors. Pu^{3+} is bright blue or blue-violet; Pu^{4+} is reddish brown, brownish red, or, as a nitrate complex, brownish green to dark green; PuO^+, pale purple; and PuO_2^{2+}, pink to orange-red. Like the rare earths, they have characteristic absorption spectra with sharp absorption bands. These have been widely used in the analysis of plutonium solutions to determine the relative amounts of each oxidation state present.

Safety precautions. Because of its radiotoxicity, plutonium and its compounds require special handling techniques to prevent ingestion or inhalation. Therefore, all work with plutonium and its compounds must be carried out inside glove boxes. For work with plutonium and its alloys, which are at-

Table 5. Binary compounds of plutonium with elements of groups III, IV, V, and VI

III	IV	V	VI		
PuB	PuC_{1-x}	PuN	$PuS_{0.95}$	PuSe	PuTe
PuB_2	Pu_2C_3	PuP	PuS	Pu_2Se_{3-x}	γ-Pu_2Te_3
PuB_4	PuC_2	PuAs	Pu_3S_4	γ-Pu_2Se_3	η-Pu_2Te_3
PuB_6			Pu_5S_7	η-Pu_2Se_3	$PuTe_{2-x}$
PuB_{12}	Pu_5Si_3		α-Pu_2S_3	$PuSe_{1.8}$	$PuTe_3$
"PuB_{100}"	Pu_3Si_2		$PuS_{1.9}$	$PuSe_{1.9}$	
	PuSi		PuS_2		
	Pu_3Si_5				
	$PuSi_2$				

Table 6. Fluoroplutonates and oxoplutonates of alkali metals and ammonium

Valency	Compound	Type	Color
III	$MPuF_4$	M = Na, K	Blue
	MPu_2F_7	M = K	Blue
IV	MPu_3F_{13}	M = NH_4	Pink to brownish
	MPu_2F_9	M = K	Pink to brownish
	$MPuF_5$	M = Li, NH_4	Pink to brownish
	$M_7Pu_6F_{31}$	M = Na, K, Rb, NH_4	Pink to brownish
	M_2PuF_6	M = Na, K, Rb, Cs, NH_4	Pink to brownish
	M_3PuF_7	M = Na	Pink to brownish
	M_4PuF_8	M = Li, NH_4	Pink to brownish
V	$MPuF_6$	M = Cs	Green
	M_2PuF_7	M = Rb	Green
IV	M_8PuO_6	M = Li	Brown
V	M_3PuO_4	M = Li	Brown
VI	M_2PuO_4	M = Rb, Cs	Brown or black
	M_4PuO_5	M = Li, Na	Brown or black
	M_6PuO_6	M = Li, Na	Dark green
VII	M_5PuO_6	M = Li	Dark green
	M_3PuO_5	M = Rb, Cs	Black

Table 7. Selected plutonium compounds obtained from aqueous solutions

III	IV	V	VI
$PuPO_4 \cdot 0.5H_2O$	$Pu(OH)_4 \cdot xH_2O$	$KPuO_2CO_3$	$PuO_2(NO_3)_2 \cdot 6H_2O$*
$Pu_2(C_2O_4)_3 \cdot 10H_2O$	$Pu(JO_3)_4$	$RbPuO_2CO_3$	$NaPuO_2(CH_3COO)_3$
$PuF_3 \cdot xH_2O$	$PuO_4 \cdot 2H_2O$		$HPuO_2PO_4 \cdot 3H_2O$
$Pu_2(SO_4)_3 \cdot 7H_2O$	$Pu(HPO_4)_2 \cdot xH_2O$		$KPuO_2PO_4 \cdot 3H_2O$
	$Pu(C_2O_4)_2 \cdot 6H_2O$		$NH_4PuO_2PO_4 \cdot 3H_2O$
	$Pu(SO_4)_2$*		$PuO_2C_2O_4$
	$K_4Pu(SO_4)_4 \cdot 2H_2O$		
	$Pu(NO_3)_4 \cdot 5H_2O$*		

*Obtained by evaporation of its aqueous solution.

tacked by moisture and by atmospheric gases, these boxes may be filled with helium or argon. *See* ACTINIDE ELEMENTS; NEPTUNIUM; NUCLEAR CHEMISTRY; TRANSURANIUM ELEMENTS; URANIUM.

[FRITZ WEIGEL]

Bibliography: J. R. Cleveland, *The Chemistry of Plutonium*, 2d ed., 1979; *Gmelin Handbuch der Anorganischen Chemie*, 8th ed., 10 vols., supplement: Transurane, 1973–1979; G. T. Seaborg (ed.), *Transuranium Elements: Products of Modern Alchemy*, Benchmark Papers in Physical Chemistry and Chemical Physics, 1978; O. J. Wick (ed.), *Plutonium Handbook*, 2 vols., 1967.

Polar molecule

A molecule possessing a permanent electric dipole moment. Molecules containing atoms of more than one element are polar except where forbidden by symmetry; molecules formed from atoms of a single element are nonpolar (except ozone). The dipole moments of polar molecules result in stronger intermolecular attraction, increased viscosities, higher melting and boiling points, and greater solubility in polar solvents than in nonpolar molecules.

The electrical response of polar molecules depends in part on their partial alignment in an electric field, the alignment being opposed by thermal agitation forces. This orientation polarization is strongly temperature-dependent, in contrast to the induced polarization of nonpolar molecules. *See* MOLECULAR STRUCTURE AND SPECTRA.

[ROBERT D. WALDRON]

Polarimetric analysis

A method of chemical analysis based on the optical activity of the substance being determined. Optically active materials are asymmetric; that is, their molecules or crystals have no plane or center of symmetry. These asymmetric molecules can occur in either of two forms, *d*- and *l*-, called optical isomers. Often a third, optically inactive form, called meso, also exists. Asymmetric substances possess the power of rotating the plane of polarization of plane-polarized light. Measurement of the extent of this rotation is called polarimetry. Polarimetry is applied to both organic and inorganic materials. *See* OPTICAL ACTIVITY.

The extent of the rotation depends on the character of the substance, the length of the light path, the temperature of the solution, the wavelength of the light which is being used, the solvent (if there is one), and the concentration of the substance. In most work, the yellow light of the D line of the sodium spectrum (5893 A) is used to determine the specific rotation, according to the equation below.

$$\text{Specific rotation} = [\alpha]_D^{20} = \frac{\alpha}{l\rho}$$

Here α is the measured angle of rotation, l the length of the column of liquid in decimeters, and ρ the density of the solution. In other words, the specific rotation is the rotation in degrees which this plane-polarized light of the sodium D line undergoes in passing through a 10-cm-long sample tube containing a solution of 1 g/ml concentration at 20°C.

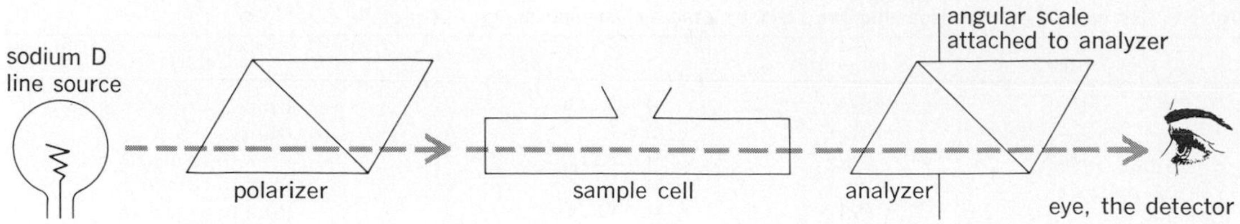

Simplified diagram of a polarimeter.

In the simplified diagram of a polarimeter, light from the sodium lamp is polarized by the polarizer (a fixed nicol prism) before it passes through the cell containing the material being analyzed. After the light passes through the cell, it passes through the analyzer (another nicol prism) and then is detected by the eye or a photocell. A comparison of the angular orientation of the analyzer as measured on the scale with the cell empty and with the cell filled with solution serves to measure the rotation of the polarized light by the sample. This rotation may be either clockwise (+) or counterclockwise (−), depending on the substance in question.

Polarimetry may be used for either qualitative or quantitative analytical work. In qualitative applications, the presence of an optically active material is shown, and then a calculation of specific rotation often leads to the identification of the unknown. In quantitative work, the concentration of a given optically active material is determined either from a calibration curve of percentage of the constituent versus angular rotation or from the specific rotation, assuming the angular rotation to be a linear function of concentration. For this method of analysis to be useful, it is necessary that only one optically active material be present in solution.

Polarimetry is widely used in carbohydrate chemistry, especially in the analysis of sugar solutions. Polarimeters used for this work are specially designed and are called saccharimeters. Other materials often determined by polarimetry are tartaric acid, Rochelle salt (potassium sodium tartrate), various terpenes such as *d*- and *l*-pinene, many steroids, and other compounds of biological and biochemical importance. Since there is great difference between the biological activities of the different optical forms of organic compounds, polarimetry is widely used in biochemical research to identify the molecular configurations.

Optical rotatory dispersion is the measurement of the specific rotation as a function of wavelength. To accomplish this, the sodium lamp is replaced by a monochromator, and a source of continuous radiation. A photocell circuit is substituted for the eye as a detector. In this way, the specific rotation may be determined in the ultraviolet or near-infrared, as well as in the visible region of the spectrum.

The information obtained by optical rotatory dispersion has shown that minor changes in configuration of a molecule have a marked effect on its dispersion properties. By using the properties of compounds of known configuration, it has been possible to determine the absolute configurations of many other molecules and to identify various isomers. To date, most of the applications have been to steroids, sugars, and other natural products, including amino acids, proteins, and polypeptides. *See* COTTON EFFECT; ROTATORY DISPERSION.

[ROBERT F. GODDU; JAMES N. LITTLE]

Bibliography: I. M. Koltoff and P. J. Elving, *Treatise on Analytical Chemistry*, pt. 1, vol. 6, 1965; S. F. Mason (ed.), *Optical Activity and Chiral Discrimination*, 1979; G. Snatzke (ed.), *Optical Rotatory Dispersion and Circular Dichroism in Organic Chemistry*, 1976.

Polarographic analysis

An electrochemical technique used in analytical chemistry. Polarography involves measurements of current-voltage curves, obtained when voltage is applied to electrodes (usually two) immersed in the solution being investigated. One of these electrodes is a reference electrode: its potential remains constant during the measurement. The second electrode is an indicator, and its potential varies in the course of measurement of the current-voltage curve, because of the change of the applied voltage. In the simplest version, so-called dc polarography, the indicator electrode is a dropping mercury electrode, consisting of a mercury drop hanging at the orifice of a fine-bore glass capillary (usually about 0.08 mm inner diameter). The capillary is connected to a mercury reservoir so that mercury flows through the capillary at the rate of a few milligrams per second. The outflowing mercury forms a drop at the orifice, which grows until it falls off. The lifetime of each drop is several seconds (usually 2 to 5). Each drop forms a new electrode; its surface is practically unaffected by processes taking place on the previous drop. Hence each drop represents a well-reproducible electrode with fresh, clean surface.

Apparatus. The dropping-mercury electrode is immersed in the solution to be investigated and placed in a cell containing the reference electrode (Fig. 1). Polarographic current-voltage curves can be recorded with a simple instrument consisting of a potentiometer or another source of voltage and a current-measuring device (such as a sensitive galvanometer). The voltage can be varied by manually changing the applied voltage in finite increments, measuring current at each, and plotting current as a function of the voltage. Alternatively, commercial instruments are available in which voltage is increased linearly with time (a voltage ramp), and current variations are recorded automatically (Fig. 2).

Polarographic curves. Oscillations of current result from growth of the individual mercury drops. Their mean value is usually measured. Three portions can be observed on a typical polarographic curve. At sufficiently positive potentials, only a small current flows. Then, in a region char-

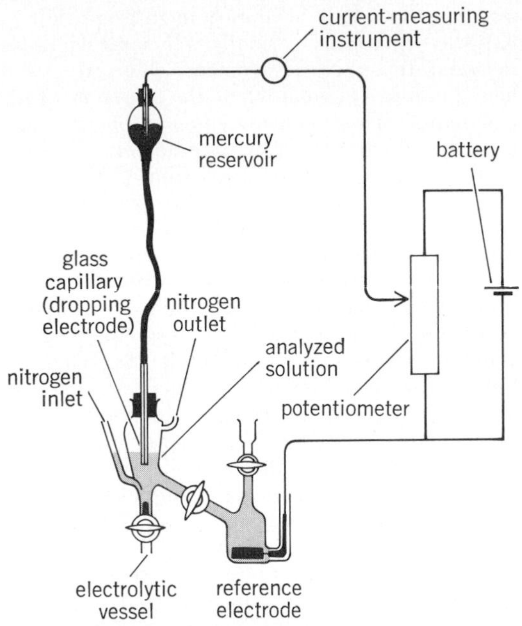

Fig. 1 Polarographic circuit.

acteristic of the particular species and solution, an S-shaped current rise is observed. Finally, at sufficiently negative potentials, the current is again independent of applied voltage over a potential range. This current is called the limiting current and is usually a few microamperes or a fraction of a microampere. The limiting current (also called the height of the polarographic wave) increases usually with concentration of the investigated species in the solution. The wave height can be thus used for determination of concentration (that is, how much of the species is present in the solution). The potential at the point of the polarographic wave, where the current reaches half of the limiting value, is called the half-wave potential (Fig. 2). This potential is characteristic of the species studied (that is, it can be used to confirm which species are present).

Solutions. The solutions investigated polarographically must contain the species to be studied in 10^{-3} to 10^{-6} *M* concentration. In addition, the solution must contain a large excess (50-fold or greater) of a supporting electrolyte which does not react at the electrode in the potential region of interest and which can be a neutral salt, acid, base, or a mixture (such as a buffer). The function of the supporting electrolyte is to reduce the resistance of the solution, to ensure that species are transported to the surface of the electrode by diffusion rather than by migration in the electric field, to keep conditions at the electrode surface unchanged in the course of polarographic electroly sis, and to keep or convert the species into a form most suitable for electrolysis (for example, by complex formation or protonation).

Evaluations. The heights of polarographic waves are measured and compared with heights obtained with standard solutions of the species, the concentrations of which are known. Accuracy varies from 2 to 5%, according to the potential range, shape of polarographic waves, and presence of other components. Under strictly controlled conditions, even 1% accuracy can be achieved.

Polarographically active species. To be studied polarographically, a species must undergo reduction or oxidation at the surface of the mercury drop, must form compounds with mercury, or must catalytically affect electrode processes. From changes in the shape of polarographic curves, it is also possible to determine various surfactants.

To be able to follow a given reducible species, its reduction must occur at more positive potentials than reduction of the component of the supporting electrolyte or solvent. Oxidations can be followed if they occur at potentials more negative than that of oxidation of the solvent, supporting electrolyte, or electrode material. Reduction processes are indicated by cathodic waves (above the zero current line), and oxidation processes by anodic waves (below the zero current line).

Polarography has been used for the determination of most metals (typically lead, copper, zinc, cadmium, iron, uranium, cobalt, nickel, manganese, potassium, sodium, and so forth), some inorganic acids and their anions (for example, iodic and periodic acids, nitrates, and nitrites), and some gases (for example, sulfur dioxide and oxygen). The wave of oxygen forms the basis of a convenient oxygen determination in technical gases, waters, or biological material. Alternatively the reducibility of oxygen means that oxygen must be removed (usually by purging with nitrogen) from solutions to be analyzed for other constituents. Reduction of inorganic species results in the lowering of the oxidation state, which may be accompanied by amalgam formation.

Electrode processes involving organic compounds at the dropping-mercury electrode result in a cleavage or formation of a chemical bond in the investigated organic molecule. Polarographic behavior depends primarily on the nature of the bond involved, even when the molecular environment

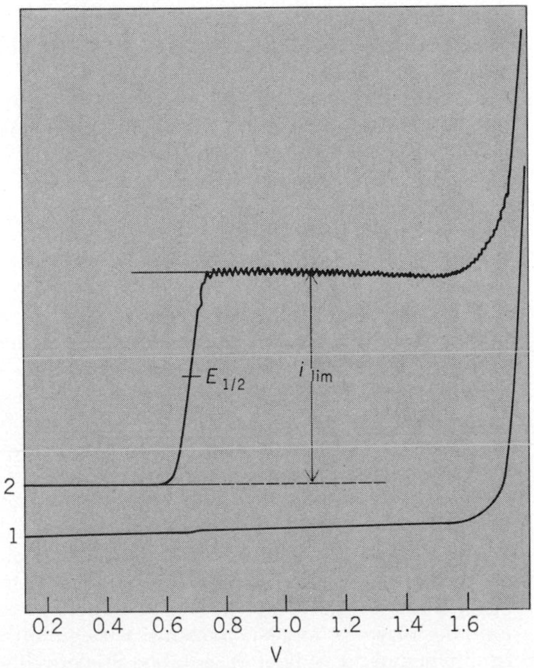

Fig. 2. Polarographic current-voltage curves: 1 = supporting electrolyte; 2 = curve in the presence of an electroactive (reducible) compound; $E_{1/2}$ = half-wave potential; i_{lim} = limiting current (wave height).

(that is, the presence and kind of neighboring groups and other substituents, the type of the molecular frame to which the electroactive group is bound, and the spatial arrangement) also affects the electrode process. The reducibility of multiple bonds (for example, C=O, C=N, C=C, and N=N) or of strongly polarizable groups (such as NO_2, NO, and C-halogen), especially when conjugated, was recognized early. Recently the reducibility of some single bonds such as C—O, C—N, C—S, or C—P when the molecule contains activating groups was proved.

Because mercury ions react with various anions, anodic dissolution current of mercury can be used, for example, for the determination of halides, thiocyanate, cyanide, and sulfide among inorganic species; and of mercaptans, urea and thiourea derivatives, and dithiocarbamates among organic compounds.

Most important among catalytic waves are those which correspond to catalytic hydrogen evolution. Among compounds giving catalytic waves in buffered ammoniacal solutions of cobalt salts are some types of proteins.

The protein molecule must contain a suitably situated thiol or disulfide group and must be able to form a complex with cobalt and ammonia. It is assumed that the proton transferred by this complex undergoes reduction more easily than H_3O^+.

When blood serum is alkali-denatured and the proteins with largest molecular weight are separated by precipitation with sulfosalicyclic acid, the remaining filtrate contains proteins and their fragments, the contents of which differ in the serum of healthy and pathological individuals. Increase of the catalytic wave is observed for cancer and inflammatory diseases, and decrease of the catalytic wave for hepatitis and other liver diseases. In Europe the polarographic test is used in clinical analysis as a general screening test, in connection with nine other tests as a proof of malignancy, and in controlling cure of cancer and measuring the effectiveness of treatments such as surgery or irradiation.

Applications. Polarographic studies can be applied to investigation of electrochemical problems, to elucidation of some fundamental problems of inorganic and organic chemistry, and to solution of practical problems.

In electrochemistry polarography allows measurement of potentials, and yields information about the rate of the electrode process, adsorption-desorption phenomena, and fast chemical reactions accompanying the electron transfer. Since polarographic experiments are simple and not too time-consuming, polarography can be used in a preliminary test designed to find the most suitable model compound for detailed electrochemical investigations.

In fundamental applications, polarography allows one to distinguish the form and charge of the species (for example, inorganic complex or organic ion) in the solution. Polarography also permits the study of equilibria (complex formation, acid-base, tautomeric), rates, and mechanisms. For equilibria established in the bulk of the solution in more than 15 sec, measurement of wave heights of individual components makes possible the evaluation of equilibrium constants. For equilibria that are very rapidly established at the electrode surface, equilibrium constants can be determined from shifts of half-wave potentials. Finally, for some equilibria between these two extremes, which are established in times comparable to the drop time (3 sec), calculation of rate constants is possible (for example, for dehydration of hydrated aldehydes, or for protonation of anions derived from C-acids, such as $C_6H_5COCH_2COCH_3$). In this way, rate constants of very fast reactions of the order 10^5-10^{10} liters $mole^{-1}$ sec^{-1} can be determined.

For slower reactions, rate constants can be found from changes of wave heights with time. Moreover, as some reaction intermediates giving separate waves can be detected, identified, and followed polarographically (if their half-lives are longer than 15 sec), polarography can prove useful in mechanistic studies. Elimination of Mannich bases, hydration of multiple bonds in unsaturated ketones, and aldolization are examples studied.

Finally, polarography can be used for investigation of the relationship between electrochemical data and structure. Reduction of most organic systems, particularly in aqueous media, involves steps with a high activation energy and is therefore irreversible. Half-wave potentials of such systems are a function of the rate constant of the electrode process. This heterogeneous rate constant is frequently influenced by structural effects, as are, in a similar manner, rate constants of homogeneous reactions. Therefore, effects of substituents on half-wave potentials in aromatic systems and in aliphatic systems can be treated by the Hammett relations and the Taft substituent constants, respectively. Among steric effects, for example, steric hindrance of coplanarity or effects of cis-trans isomerisms affect polarographic curves. Polarography also makes it possible to distinguish between some epimers, for example, bearing axial or equatorial halogen.

Practical applications are predominantly analytical procedures. In inorganic analysis, polarography is used predominantly for trace-metal analysis (with increased sensitivity of differential pulse polarography and stripping analysis). In organic analysis, it is possible in principle to use polarography in elemental analysis, but such applications are deservedly infrequent. More frequent are applications in functional group analysis. Either the reaction product (for example, semicarbazone in the determination of carbonyl compounds, or *N*-nitrosoamine in determinations of secondary amines) is measured, or a decrease in concentration of a reagent (for example, chromic acid in determination of alcohols, or mercuric or silver ions in determination of thiols) is followed. Or, alternatively, the polarographic limiting current can be taken as a measure of the amount of a given electroactive group (for example, quinones, aliphatic nitro compounds, phenazines, or thiols).

More frequent is determination of individual compounds. If the species is electroactive, the analysis frequently consists of dissolving the material in a proper supporting electrolyte, recording the waves, and evaluating them. Electroinactive species can be determined indirectly, by converting them into electroactive by a suitable chemical reaction (for example, aromatics by nitration, secondary amines by nitrosation).

The most important fields of application of inorganic determinations are in metallurgy, environmental analysis (air, water, and seawater contaminants), food analysis, toxicology, and clinical analysis. The possibility of being able to determine vitamins, alkaloids, hormones, terpenoid substances, and natural coloring substances has made polarography useful in analysis of biological systems, analysis of drugs and pharmaceutical preparations, determination of pesticide or herbicide residues in foods, and so forth. Polarography also makes possible determination of monomers, catalysts, and even some reactive groupings in polymers.

Other polarographic techniques. To eliminate unwanted charging current and to increase the sensitivity of polarography, the voltage is applied in regular pulses instead of gradually. When the current is measured only during the second half of the pulse, the technique is called pulse or differential pulse polarography. This technique is more sensitive by two orders of magnitude than dc polarography, and in inorganic trace analysis competes with atomic absorption and neutron activation analysis. With smaller dosage of more effective drugs, the sensitivity of differential pulse polarography has found application in drug analysis.

When an alternating voltage of small amplitude (a few millivolts) is superimposed on the dc voltage ramp, the technique is called ac polarography. It is particularly useful for obtaining information on adsorption-desorption processes at the surface of the dropping-mercury electrode. An increase in accuracy of polarographic methods can be achieved when the dropping-mercury electrode (or some other indicator electrode) is used in titrations. At constant applied voltage the current is measured as a function of volume of added titrant in amperometric titrations. The amperometric titration curve frequently has a shape of two linear sections. Their intersection corresponds to titration end point.

Related methods. Methods in which a potential sweep is used (instead of the practically constant potential of the electrode during the life of a single drop in polarography) are frequently called voltammetry. Either an electrode with nonrenewed surface—solid electrode, mercury pool, or mercury drop—is used, or the whole scan is carried out during the life of a single drop when the dropping-mercury electrode is used. The use of solid electrodes (for example, platinum, gold, or various forms of graphite) makes it possible to extend the voltage range to more positive potentials, so that numerous substances which cannot be oxidized when mercury is used can be oxidized. Solid electrodes can be used when they are stationary, rotating, or vibrating, but the change at the electrode surface in the course of electrolysis often decreases the reproducibility of the current-voltage curve compared with polarographic curves.

When extremely small concentrations of amalgam-forming metals (such as lead, cadmium, or zinc) are to be determined, a preconcentration can be carried out. The metal is deposited into a hanging mercury drop or into mercury plated on surfaces of solid electrodes. After a chosen time interval, a voltage sweep is applied, and the dissolution current of the metal ion from the analgam is measured. Such anodic stripping, particularly when combined with the differential pulse technique, allows determination up to 1 part in 10^{12} parts of solution.

When the applied voltage sweep is first increased and then decreased and the plot of dependence of voltage on time resembles a triangle, the technique is called cyclic voltammetry, and offers in particular information about products and intermediates of electrode processes and their reactions.

Finally, when current is applied and potential of an electrode measured, another group of techniques is developed. Most important among them is chronopotentiometry. In this technique, a constant current is applied to the indicator electrode in an unstirred solution, and its potential change is measured as a function of time. The time at which a sudden potential change is observed is called transition time, and its square root is proportional to the concentration of the reacting species. The method has found only limited application but has proven useful for the study of reactions of products of electrolysis. [PETER ZUMAN]

Bibliography: R. N. Adams, *Electrochemistry at Solid Electrodes*, 1969; A. Bard (ed.), *Electroanalytical Chemistry*, vols. 1–11, 1966–1979; J. Heyrovsky and P. Zuman, *Practical Polarography*, 1968; L. Meites, *Polarographic Techniques*, 2d ed., 1965; L. Meites and P. Zuman, *Handbook Series in Organic Electrochemistry*, vol. 4, 1979; L. Meites et al., *Handbook of Inorganic Electrochemistry*, 1980.

Polonium

A chemical element, Po, atomic number 84, a member of group VIa of the periodic table. Marie Curie discovered the radioisotope Po^{210} in pitchblende and named it after her native Poland. This isotope, also known as radium F, is the penulti-

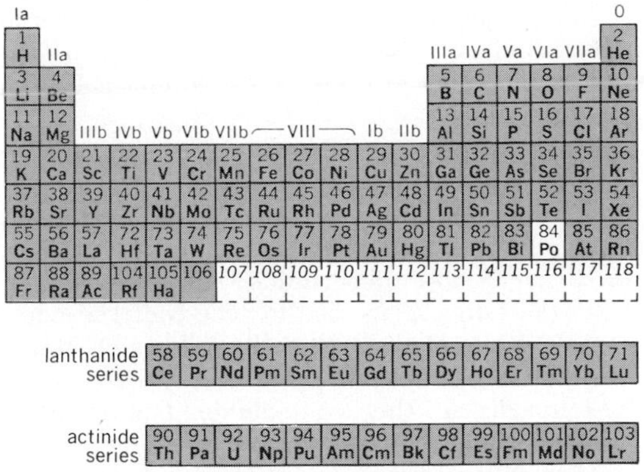

mate member of the radium decay series; pitchblende contains 0.1 mg/ton. All polonium isotopes are radioactive, and all are short-lived except the three α-emitters: Po^{208} (2.9 years), Po^{209} (100 years), both of which are produced by bombarding bismuth with deuterons, and natural Po^{210} (138.4 days), now produced in milligram amounts by the neutron bombardment of bismuth. Polonium is separated from bismuth by spontaneous deposition onto a less noble metal, such as silver, followed by a vacuum sublimation or chemical separation of the deposit.

Uses. Polonium (Po^{210}) is used mainly for the production of neutron sources; for these, the polonium is alloyed with elements such as beryllium which have isotopes of high α,n cross section. It can also be used in static eliminators and, when incorporated in the electrode alloy of spark plugs, is said to improve the cold-starting properties of internal combustion engines.

Properties. Most of the chemistry of polonium has been determined using Po^{210}, 1 curie of which weighs 222.2 μg; work with weighable amounts is hazardous, requiring special techniques.

Polonium is more metallic than its lower homolog, tellurium; two allotropes are known, α-Po (simple cubic) and β-Po (simple rhombohedral) with the phase change $\alpha \rightarrow \beta$ at about 36°C. The metal is chemically similar to tellurium, forming the bright red compounds $SPoO_3$ and $SePoO_3$; a number of polonides are known.

The metal is soft, and its physical properties resemble those of thallium, lead, and bismuth. Valences of 2 and 4 are well established; there is some evidence of hexavalency. Polonium is positioned between silver and tellurium in the electrochemical series.

Compounds. Two forms of the dioxide are known: low-temperature, yellow, face-centered cubic (UO_2 type), and high-temperature, red, tetragonal. It is formed from the elements at 250°C and decomposes at 500°C under vacuum. The black monoxide may be formed in the spontaneous decomposition of $SPoO_3$ and $SePoO_3$. The quadrivalent hydroxide (pale yellow, gelatinous, feebly amphoteric) is precipitated from solutions of polonium salts by alkalies; it is reduced to the metal in alkaline suspension by hydroxylamine, hydrazine, sodium dithionite, and ammonia (liquid or concentrated aqueous solution). The last two reactions may be due to α-radiation effects. The brown, bivalent hydroxide is readily oxidized. Polonium monosulfide (black) is precipitated from acid solutions of polonium salts by hydrogen sulfide (solubility product 5.5×10^{-29}); it decomposes to the elements at 275°C and 10 μ pressure, a property which is used in preparation of pure polonium metal.

The halides are covalent, volatile compounds, resembling their tellurium analogs ($PoCl_4$, yellow; $PoBr_4$, carmine; PoI_4, black; $PoCl_2$, ruby-red; $PoBr_2$, purple-brown; $PoCl_2Br_2$, salmon-pink) with the bivalent state more stable than the quadrivalent; the latter gives rise to face-centered cubic complex salts M_2PoX_6 (X = Cl, yellow; Br, brick-red; I, black), where M is a univalent cation. The cesium salts are the least soluble. The quadrivalent halides in solution are reduced to the bivalent state (pink) by hydrazine, sulfur dioxide, or arsenic(III) oxide (on warming) and to the metal by tin(II) chloride, sodium dithionite, or titanium(III) chloride. Complexes with organic molecules (for example, tributyl phosphate, dithizone, ethylenediaminetetraacetate), nitrosyl chloride, and ammonia are also known.

The increased metallic character of polonium as compared with tellurium is shown in the formation of the salts $Po(SO_4)_2$ (white, hydrated; deep purple, anhydrous) and $Po(NO_3)_4$ (white). Both are readily hydrolyzed, the former to $2PoO_2 \cdot SO_3$ (white when cold and yellow when hot), the latter to a series of basic nitrates which may be polymeric. The basic salts $2PoO_2 \cdot SeO_3$ and $2PoO_2 \cdot CrO_3$ are also known; they are analogous to tellurium sulfate, $2TeO_2 \cdot SO_3$. There is some evidence for a normal chromate [yellow $Po(CrO_4)_2$?] and a hexavalent polonium/chromium complex acid. The acetate, cyanide, iodate, oxalate, and phosphate (all white) have been prepared, and there is evidence for the formation of a vanadate and tartrate.

Determination. Polonium (Po^{210}) is estimated by its α-emission, either by direct counting or calorimetrically; in the latter case, the measurement depends on the heat liberated by the stoppage of the disintegration α-particles within the sample. *See* RADIOACTIVITY; TELLURIUM.

[KENNETH W. BAGNALL]

Bibliography: E. K. Hyde et al., *The Nuclear Properties of the Heavy Elements*, 3 vols., rev. ed., 1971; M. Schmidt et al., *The Chemistry of Sulphur, Selenium, Tellerium and Polonium*, 1975.

Polyacrylate resin

Useful polymers can be obtained from a variety of acrylic monomers, such as acrylic and methacrylic acids, their salts, esters, and amides, and the corresponding nitriles. The most important monomers are shown here.

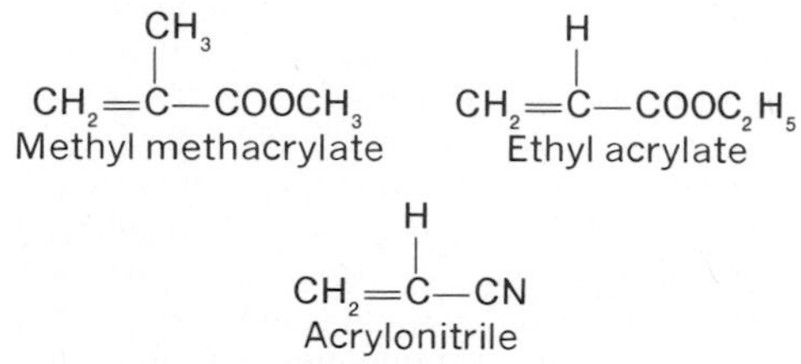

Polymethyl methacrylate, polyethyl acrylate, and a few other derivatives are discussed in this article. *See* POLYACRYLONITRILE RESINS.

Polymethyl methacrylate is a hard, transparent polymer with high optical clarity, high refractive index, and good resistance to the effects of light and aging. It and its copolymers are useful for lenses, signs, indirect lighting fixtures, transparent domes and skylights, dentures, and protective coatings.

The monomer is usually made by the dehydration and methanolysis of acetone cyanohydrin as in reaction (1). Polymerization may be initiated by

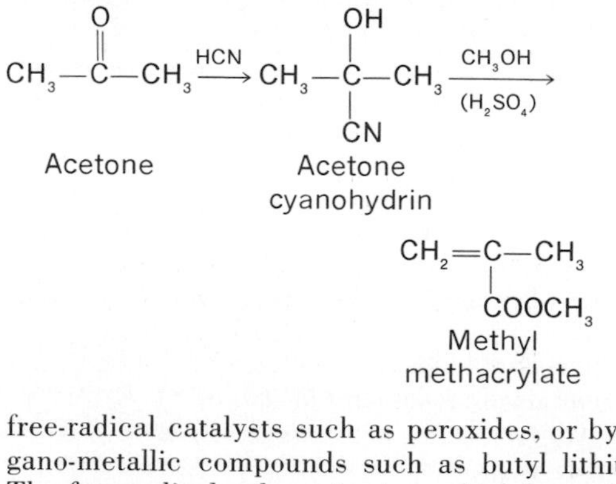

(1)

free-radical catalysts such as peroxides, or by organo-metallic compounds such as butyl lithium. The free-radical polymerization, which is used for commercial production, can be carried out in bulk, in solution, and in aqueous emulsion or suspension. By changing the catalyst, temperature, and solvent in a polymerization it is possible to prepare polymers which are predominantly isotactic, syndiotactic, or atactic, or which have stereoregular blocks. Although solution polymerization is not commonly used, bulk polymerization, usually with comonomers, is frequently employed in various casting operations, as in the formation of sheets, rods, and tubes; in the mounting of biological, textile, and metallurgical test specimens; and in dental applications. A 20% reduction in volume accompanies the conversion of monomer to polymer. This makes it difficult to prepare articles to predetermined dimensions by the cast polymerization technique. The problem is minimized by using a syrup of polymer dissolved in monomer, or a dough of finely divided polymer dispersed in a relatively small amount of monomer.

Molding powders suitable for the injection molding of dials, ornamental fixtures, and lenses may be prepared from the granules produced by aqueous suspension polymerization or from the product of bulk polymerization.

Solutions of polymethyl methacrylate and its copolymers are useful as lacquers. Aqueous latexes formed by the emulsion polymerization of methyl methacrylate with other monomers are useful as water-based paints and in the treating of textiles and leather.

Polyethyl acrylate is a tough, somewhat rubbery product. The monomer is used mainly as a plasticizing or softening component of copolymers. Ethyl acrylate is usually produced by the dehydration and ethanolysis of ethylene cyanohydrine, which can be obtained from ethylene oxide as shown by reaction (2).

$$\underset{\text{Ethylene oxide}}{CH_2\!-\!CH_2\ (\text{—O—})} \xrightarrow{HCN} \underset{\text{Ethylene cyanohydrin}}{\underset{OH}{CH_2}\!-\!\underset{CN}{CH_2}} \xrightarrow[(H_2SO_4)]{C_2H_5OH} \underset{\text{Ethyl acrylate}}{CH_2\!=\!CHCOOC_2H_5} \qquad (2)$$

Polymerization may be effected by catalysts of the free-radical type. Copolymerizations with other monomers are frequently carried out in aqueous emulsion or suspension.

Modified acrylic resins with high impact strengths can be prepared. Blends or "alloys" with polyvinyl chloride are used for thermoforming impact-resistant sheets. Copolymers of ethyl acrylate with 2-chloroethyl vinyl ether are useful rubbers which can be vulcanized in the same way as ordinary polydienes.

Methyl methacrylate is late of interest as a polymerizable binder for sand or other aggregates, and as a polymerizable impregnant for concrete; usually a cross-linking acrylic monomer is also incorporated. The binder systems (polymer concrete) are used as overlays for bridge decks as well as for castings, while impregnation is used to restore concrete structures and protect bridge decks against corrosion by deicing salts.

Other acrylic derivatives are the subject of continuing interest. The butyl and octyl esters of acrylic acid yield rubbery materials, and fluorinated esters have been used as oil-resistant rubbers. Polymers of alkyl-2-cyano-acrylates make excellent adhesives that cure at low temperatures using adventitious moisture as a catalyst. Cross-linkable acrylic copolymers are often used in surface coatings. Polymers of methyl acrylate or acrylamide are water-soluble and useful for sizes and finishes. Addition of polylauryl methacrylate to petroleum lubricating oil improves the flowing properties of the oil at low temperatures and the resistance to thinning at high temperatures. *See* ACRYLONITRILE; POLYMERIZATION.

[JOHN A. MANSON]

Polyacrylonitrile resins

The polyacrylonitrile resins are hard, horny, relatively insoluble, and high-melting materials. Polyacrylonitrile (polyvinyl cyanide) is used almost entirely in copolymers. The copolymers fall into three groups: fibers, plastics, and rubbers. The presence of acrylonitrile in a polymeric composition tends to increase its resistance to temperature, chemicals, impact, and flexing.

Acrylonitrile is generally prepared by one of the following methods.

Catalyzed addition of hydrogen cyanide to acetylene as in reaction (1):

$$CH{\equiv}CH + HCN \xrightarrow{\text{Catalyst}} CH_2{=}\underset{CN}{\underset{|}{CH}} \qquad (1)$$

Reaction between hydrogen cyanide and ethylene oxide as in (2):

$$CH_2\!-\!CH_2\ (\text{—O—}) + HCN \rightarrow CH_2{=}\underset{CN}{\underset{|}{CH}} + HOH \qquad (2)$$

Reaction between ammonia and propylene as in (3):

$$2CH_3\!-\!CH{=}CH_2 + 2NH_3 + 3O_2 \rightarrow 2CH_2{=}\underset{CN}{\underset{|}{CH}} + 6H_2O \qquad (3)$$

The polymerization of acrylonitrile can be readily initiated by means of the conventional free-radical catalysts such as peroxides, by irradiation, or by the use of alkali metal catalysts. Although polymerization in bulk proceeds too rapidly to be commercially feasible, satisfactory control of a polymerization or copolymerization may be achieved in suspension and in emulsion, and in aqueous solutions from which the polymer precipitates. Copolymers containing acrylonitrile may be fabricated in the manner of thermoplastic resins.

The major use of acrylonitrile is in the form of fibers. By definition an acrylic fiber must contain at least 85% acrylonitrile; a modacrylic fiber may contain less, from 35 to 85% acrylonitrile. The high strength, high softening temperature, resistance to aging, chemicals, water, and cleaning solvents, and the soft woollike feel of fabrics have made the product popular for many uses such as sails, cordage, blankets, and various types of clothing. Commercial forms of the fiber probably are copolymers containing minor amounts of other vinyl derivatives, such as vinyl pyrrolidone, vinyl acetate, maleic anhydride, or acrylamide. The comonomers are included to produce specific effects, such as improvement of dyeing qualities.

Copolymers of vinylidene chloride with small proportions of acrylonitrile are useful as tough, impermeable, and heat-sealable packaging films.

Extensive use is made of copolymers of acrylonitrile with butadiene, often called NBR (formerly Buna N) rubbers, which contain 15–40% acrylonitrile. Minor amounts of other unsaturated esters, such as ethyl acrylate, which yield carboxyl groups on hydrolysis may be incorporated to improve the curing properties. The NBR rubbers resist hydrocarbon solvents such as gasoline, abrasion, and in some cases show high flexibility at low temperatures.

In the 1960s development of blends and interpolymers of acrylonitrile-containing resins and rubbers has represented a significant advance in polymer technology. The products, usually called ABS resins, typically are made by blending acrylonitrile-styrene copolymers with a butadiene-acrylonitrile rubber, or by interpolymerizing polybutadiene with styrene and acrylonitrile. Specific properties depend upon the proportions of the comonomer, on the degree of graftings, and on molecular weight. In general, the ABS resins combine the advantages of hardness and strength of the vinyl resin component with toughness and impact resistance of the rubbery component. Certain grades of ABS resin are used for blending with brittle thermoplastic resins such as polyvinyl chloride in order to improve impact strength.

The combination of low cost, good mechanical properties, and ease of fabrication by a variety of methods, including typical metalworking methods such as cold stamping, has led to the rapid development of new uses for ABS resins. Applications include products requiring high impact strength, such as pipe, and sheets for structural uses, such as industrial duct work and components of automobile bodies. ABS resins are also being used to an increasing degree for housewares and appliances and, because of their ability to be electroplated, for decorative items in general. *See* ACRYLONITRILE; STYRENE.

[JOHN A. MANSON]

Polyalkene

One of a class of organic compounds containing two or more ethylenic linkages in the molecule. These compounds are sometimes termed polyenes. They exist in the following three systems:

1. The unsaturated linkages may be directly attached. These compounds are said to possess cumulative unsaturation:

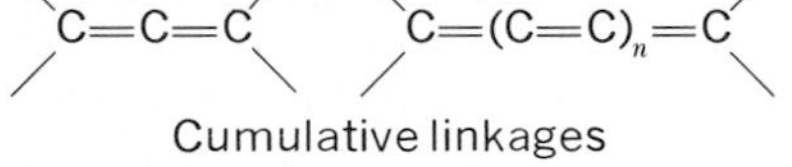

Cumulative linkages

2. The unsaturated linkages may alternate with single linkages, in which case the unsaturation is said to be conjugated:

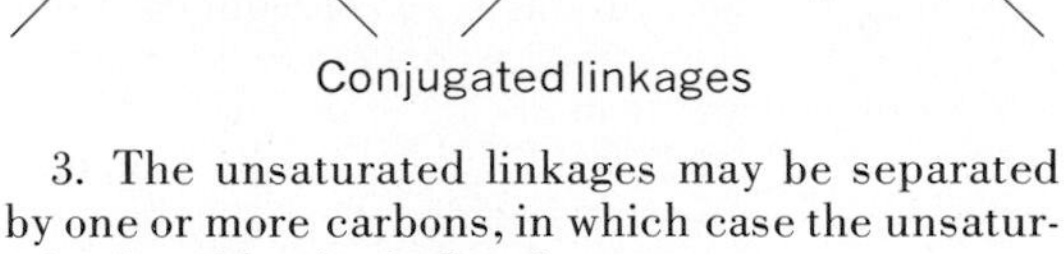

Conjugated linkages

3. The unsaturated linkages may be separated by one or more carbons, in which case the unsaturation is said to be isolated:

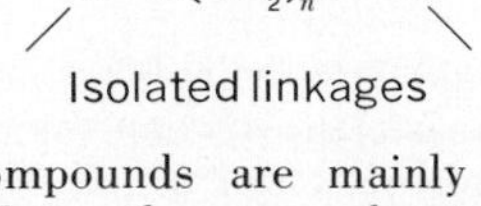

Isolated linkages

Allene compounds are mainly of interest in studies involving the stereochemistry of organic compounds, since it is possible to synthesize optically active compounds which are mirror-images of each other. *See* OPTICAL ACTIVITY.

One double bond of allene reacts with cyclopentadiene in a Diels-Alder synthesis to form 5-methylene-2-norbornene. It is an important third monomer in the synthesis of ethylene-propylene-nonconjugated diene (EPDM) rubbers. *See* DIELS-ALDER REACTION.

Conjugated dienes are the most important group of polyenes because such compounds as butadiene, isoprene, and cyclopentadiene are included in this classification.

Isolated polyenes include the unsaturated hydrocarbon squalene which contains six isoprene units with six isolated double bonds. It is an aliphatic triterpene and is related to the carotenoids such as lycopene, the red coloring matter in tomatoes, and carotene, pro-vitamin A. *See* ALKENE; TERPENE; TRITERPENE.

[CHARLES A. COHEN]

Bibliography: S. Patai (ed.), *The Chemistry of the Alkenes*, 1965.

Polyamide resins

Horny, whitish, translucent, high-melting polymers. Polyamide resins can be essentially transparent and amorphous when their melts are quenched. On cold drawing and annealing, most become quite crystalline and translucent. However, some polyamides based on bulky repeating units are inherently amorphous. The polymers are used for fibers, bristles, bearings, gears, molded objects, coatings, and adhesives. The term nylon formerly referred specifically to synthetic polyamides as a class. Because of many applications in mechanical engineering, nylons are considered engineering plastics.

Brief outlines of the preparations of commercial polyamides by the reaction of dicarboxylic acids with diamines and the condensation of amino acids are given in this article. The most common commercial aliphatic polyamides are nylons-6,6; -6; -6,10; -11; and -12. Other types find specialized application.

Nylon-6,6 and nylon-6,10 are products of the condensation reaction of hexamethylenediamine (6 carbon atoms) with adipic acid (6 carbon atoms), and with sebacic acid (10 carbon atoms), respectively. By heating equimolar proportions of the two reactants, a polymeric salt is formed, as in reaction (1), which, on further heating, yields the polyamide resin shown in the formula below.

$$\underset{\text{Adipic acid}}{n\mathrm{HOOC(CH_2)_4COOH}} + \underset{\text{Hexamethylenediamine}}{n\mathrm{H_2N(CH_2)_6NH_2}} \rightarrow \underset{\text{"Nylon-6,6 salt"}}{[\overset{+}{\mathrm{N}}\mathrm{H_3(CH_2)_6}\overset{+}{\mathrm{N}}\mathrm{H_3}\overset{-}{\mathrm{O}}\mathrm{OC(CH_2)_4CO}\overset{-}{\mathrm{O}}]_n} \quad (1)$$

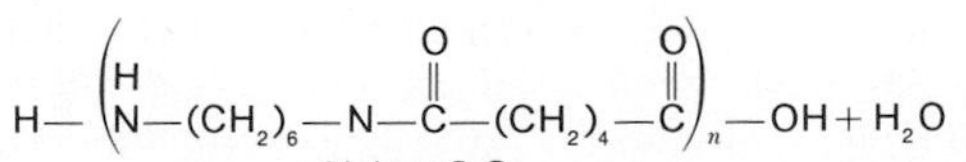

Nylon-6,6

Because the end groups on the polymer can react on further heating, as in melt spinning, it is desirable to add a very small amount of a monoacid or monoamine to the polymerizing mixture in order to prevent the formation of material of very high molecular weight.

Nylon-6,6, nylon-6,10, nylon-6,12, and nylon-6 are the most commonly used polyamides for general applications as molded or extruded parts; nylon-6,6 and nylon-6 find general application as fibers.

Nylon-6 and nylon-11 can be obtained by the self-condensation of ϵ-aminocaproic acid and ω-aminoundecanoic acids, respectively, as shown by reactions (2) and (3).

$$\underset{\epsilon\text{-Aminocaproic acid}}{H_2N(CH_2)_5COOH} \rightarrow \text{Nylon-6} \quad (2)$$

$$\underset{\omega\text{-Aminoundecanoic acid}}{NH_2(CH_2)_{10}COOH} \rightarrow \text{Nylon-11} \quad (3)$$

Each molecule, containing both the amino and carboxylic groups, can condense to yield high polymers by reactions similar to those between the diacids and the diamines.

Nylon-6 is usually made by polymerization of the lactam of ϵ-aminocaproic acid, as in reaction (4).

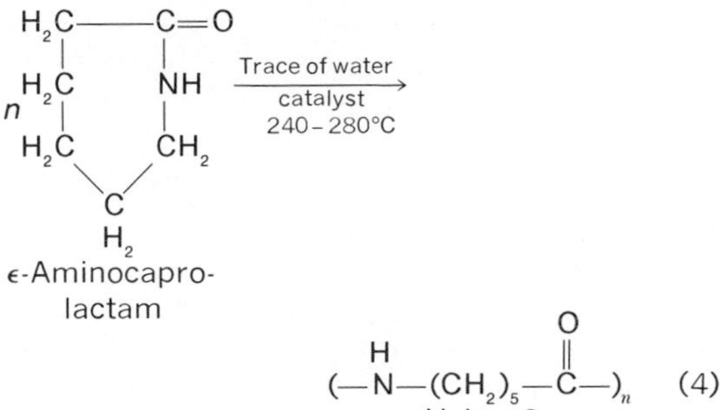

ϵ-Aminocaprolactam → Nylon-6 (4)

Caprolactam (a common term for ϵ-aminocaprolactam) is produced by rearrangement of cyclohexanone oxime, which is prepared by reaction of cyclohexanone with hydroxylamine sulfate.

The lactam process is generally preferred for commercial operation because it is easier to make and purify the lactam than the ϵ-aminocaproic acid. Nylon-12 is prepared by a similar self-condensation of the lactam of aminododecanoic acid, which is prepared by a series of reactions using butadiene as starting material. Lactams may also be polymerized by anionic catalysis. This process makes possible a one-step casting to form large articles, especially from nylon-6.

As a group, nylons are strong and tough. Mechanical properties depend in detail on the degree and distribution of crystallinity, and may be varied by appropriate thermal treatment or by nucleation techniques. Because of their generally good mechanical properties and adaptability to both molding and extrusion, the nylons described above are often used for gears, bearings, and electrical mountings. Nylon bearings and gears perform quietly and need little or no lubrication. Sintering (powder metallurgy) processes are used to make articles such as bearings, and gears which have controlled porosity, thus permitting retention of oils or inks. Nylon resins are also used extensively as filaments, bristles, wire insulation, appliance parts, and film. Properties can also be modified by copolymerization.

Reinforcement of nylons with glass fibers results in increased stiffness, lower creep and improved resistance to elevated temperatures. Such formulations, which can be readily injection-molded, can often replace metals in certain applications. The use of molybdenum sulfide and polytetrafluoroethylene as fillers increases wear resistance considerably. For uses requiring impact resistance, nylons can be blended with a second, toughening phase.

Other types of nylon are useful for specialty applications. Solubility may be increased by interference with the regularity and hence intermolecular packing. This may be accomplished by copolymerization, or by the introduction of branches on the amide nitrogen, for example, by treatment with formaldehyde. The latter type of resin may be subsequently cross-linked. Nylons incorporating aromatic structures, for example, based on isophthalic acid, are becoming more common for applications requiring resistance to very high temperatures.

Aromatic-nylon fibers based on all–para-substituted monomers can be spun from liquid-crystal suspensions to yield fibers that have exceptional moduli and strengths up to ~25,000,000 psi (160 GPa) and 400,000 psi (2.8 GPa), respectively, in the fiber direction. These values may be compared with theoretical moduli and strengths of ~40,000,000 psi (270 MPa) and ~1,000,000–4,000,000 psi (6.9–27 MPa), respectively.

Heterocyclic amid-imide copolymers are also available commercially, and a wide variety of other amide-containing polymers, such as polyesteramides and polysulfonamides, can be synthesized.

The unsaturated fatty acids in vegetable oil, for example, linoleic acid, may dimerize or polymerize to low polymers through their unsaturated groups. The di- or polycarboxylic acids obtained yield polyamides by condensation with di- or polyamines. These products are employed as coatings and adhesives, especially in epoxy and phenolic resin formulations. *See* HETEROCYCLIC POLYMER; POLYETHER RESINS; POLYMERIZATION.

[JOHN A. MANSON]

Bibliography: J. A. Brydson, *Plastics Materials*, 1975; R. W. Lenz, *Organic Chemistry of Synthetic High Polymers*, 1967.

Polyester resins

Polymeric materials in which ester groups

$$-\overset{\displaystyle O}{\overset{\|}{C}}-O-$$

are in the main chains. The aliphatic polyesters tend to be relatively soft, and the aromatic derivatives are usually hard and brittle, or tough. The properties of either group may be modified by cross linking, crystallization, plasticizers, or fillers.

The commercial products are: alkyds, which are used in paints, enamels, and in molding compounds; unsaturated polyesters or unsaturated alkyds which are used extensively with fiber glass for boat hulls and panels; aliphatic saturated polyesters; aromatic polyesters, such as polyethylene terephthalate which is used in the form of fibers and films; and the aromatic polycarbonates.

This article is devoted mainly to these four products. The polydiallyl esters are frequently listed with the polyesters and will be briefly mentioned. However, their polymers are not true polyesters as defined above. For a discussion of polyvinyl esters *see* POLYVINYL RESINS.

Alkyds. The alkyds have been in common use as coatings since World War I. In the beginning, they consisted almost entirely of the reaction products of *o*-phthalic anhydride and glycerol (glyptals), and pigment. Because the functionality of the system is greater than two, a cross-linked insoluble polymer is formed as in reaction (1). The fully cured product is quite hard and brittle. Flexible and tough materials can be produced by incorporation of monobasic acids or monohydroxy alcohols in proportions sufficient to increase flexibility but insufficient to prevent curing. Combinations of conventional vegetable drying oils and alkyd resins represent the basis of most of the oil-soluble paints. For example, by heating a mixture of dehydrated castor oil, the glycerol ester of linoleic acid, with suitable proportions of glycerol and phthalic anhydrid, an oil-soluble polyester is formed. A common oil paint is produced by the addition of thinners, such as aromatic hydrocarbon solvents, a paint drier such as cobalt octoate, and pigments. By exposure to air in the presence of the paint dryer, the unsaturated diene groups of the linoleic ester polymerize to yield a tough, weather-resistant coating. *See* DRYING OIL; POLYMERIZATION.

The drying oil–alkyd described above may be further modified by the inclusion of a vinyl monomer, such as styrene, in the original esterification process. Some of the styrene polymerizes, probably as a graft polymer, and the remainder polymerizes and copolymerizes in the final drying or curing of the paint. Low-molecular-weight liquid condensates are useful as polymeric plasticizers for resins such as polyvinyl chloride. Molding components are also used for such applications as appliance housings and electrical components.

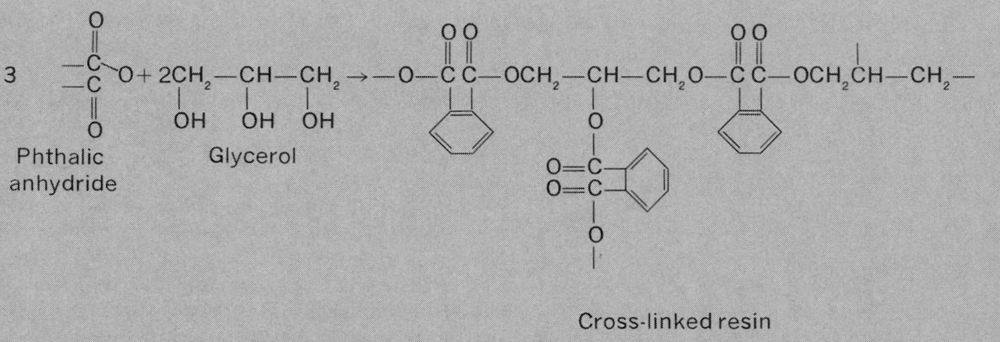

(1)

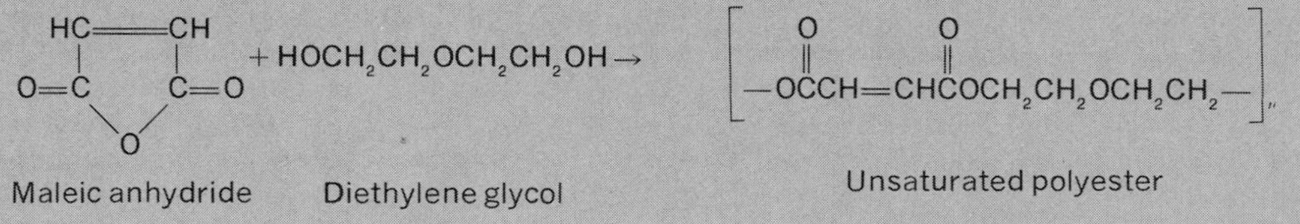

(2)

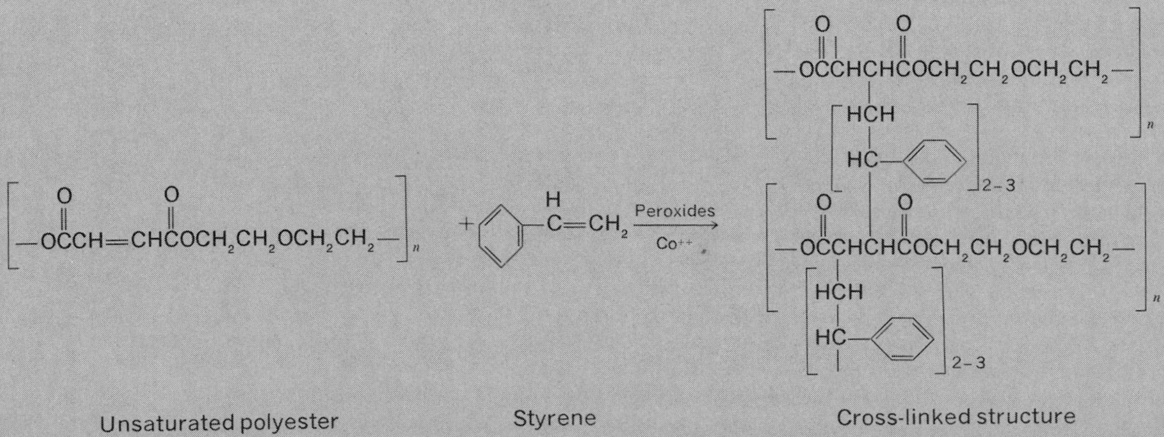

(3)

Unsaturated polyesters. The unsaturated polyesters were developed during and shortly after World War II. In combination with glass fiber, they found immediate applications as panels, roofing, radar domes, boat hulls, and protective armor for soldiers. Structural applications of this type have continued to increase in importance. The compositions are distinguished by ease of fabrication and high impact resistance. *See* POLYMERIC COMPOSITE.

A low-molecular-weight unsaturated polyester intermediate is first produced. The reaction of maleic anhydride with diethylene glycol, reaction (2), is typical. The product is a viscous oil of molecular weight of 2000–4000. Many other acids (or anhydrides), such as fumaric acid, and glycols, such as propylene glycol, may be reacted together in a similar manner.

The low-molecular-weight unsaturated polyester will cross-link in the presence of a peroxide by copolymerization with styrene or other vinyl monomers. The unsaturated maleic group copolymerizes in essentially a 1:2 or 1:3 ratio with styrene. Therefore, several styrene molecules which react effectively join two ester chains together to yield an insoluble cross-linking structure, such as in reaction (3).

The commercially available intermediate unsaturated polyesters usually contain about 30% styrene or other vinyl monomer. On addition of a peroxide or other free-radical catalyst and a paint drier, the copolymerization starts. In this stage, the resin may be handled as a viscous fluid for a few minutes to a few hours, depending upon the activity of the catalyst. The viscous liquid may be applied to glass fiber (with a special surface treatment) in the form of matt, tow, roving, or cloth, with precautions to eliminate air bubbles and to avoid bubbles that may be caused by overheating as a result of too rapid curing. The surface of the glass fiber must have been given a special finishing treatment in advance for the polyester to adhere strongly. Glass fibers treated with a vinyl silane or an organochrome complex are commercially available.

In the absence of the paint drier, oxygen of the air has an inhibiting effect on the curing process with the result that the surface of the product remains soft after the inner portions have hardened. In the presence of a paint drier, such as cobalt naphthenate, this skinning effect is eliminated.

A number of modifications of the composition described above have been made. Other acids, other glycols, and various combinations may be used to vary properties, such as flexibility, of the final product. Blends of thermosetting formulations with thermoplastics like polystyrene show considerable promise as materials for automotive body parts. The chlorinated derivatives have higher resistance to burning. By varying the free-radical initiator, the optimum temperature required for curing may be varied. There are thermosetting molding compositions which have glass fiber as a filler, and a catalyst which is relatively inactive at ordinary temperatures. The mixture is cross-linked in the heated mold by the conventional process for thermosetting molding compounds.

Saturated aliphatic polyesters. Linear polyesters made by the condensation of a diacid such as adipic acid with a diol such as diethylene glycol have long been frequently used as intermediates in the preparation of prepolymers for making segmented polyurethanes. Lactone rings can also be opened to yield linear polyesters; poly-ε-acaprolactone has been used as an intermediate in polyurethane technology, as a polymeric plasticizer for polyvinyl chloride, and in other specialty applications.

Aromatic polyesters. The aromatic polyesters which have achieved general importance are the polyethylene terephthalates,

$$\left[-O-\overset{\overset{\displaystyle O}{\|}}{C}-\underset{}{\bigcirc}-\overset{\overset{\displaystyle O}{\|}}{C}-OCH_2CH_2- \right]_n$$

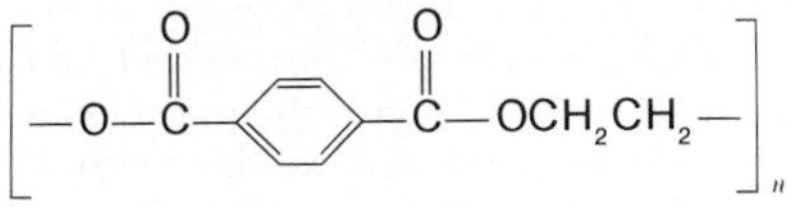

which yield very strong and chemically resistant fibers and films. Polyethylene terephthalate is the principal ingredient of the polyester fibers that are available in the United States and in Europe.

The preparation of the polymer involves several steps. First, the dimethyl or diethylene glycol ester of terephthalic acid is produced and isolated, reaction (4). Dimethyl terephthalate is then converted to

$$HO-\overset{\overset{\displaystyle O}{\|}}{C}-\bigcirc-\overset{\overset{\displaystyle O}{\|}}{C}-OH + 2CH_3OH \rightarrow$$

Terephthalic acid   Methanol

$$CH_3O\overset{\overset{\displaystyle O}{\|}}{C}-\bigcirc-\overset{\overset{\displaystyle O}{\|}}{C}-OCH_3 + 2H_2O \quad (4)$$

Dimethyl terephthalate

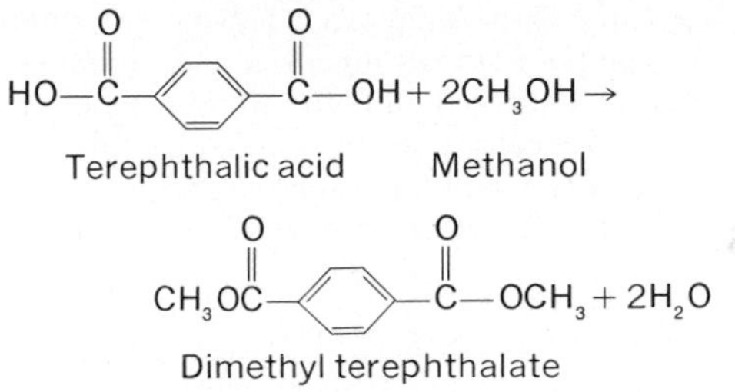

polyethylene terephthalate through ester interchange by heating with ethylene glycol in the presence of a catalyst, reaction (5). Further heating,

$$CH_3O-\overset{\overset{\displaystyle O}{\|}}{C}-\bigcirc-\overset{\overset{\displaystyle O}{\|}}{C}-OCH_3 + 2HOCH_2CH_2OH \rightarrow$$

Dimethyl terephthalate   Ethylene glycol

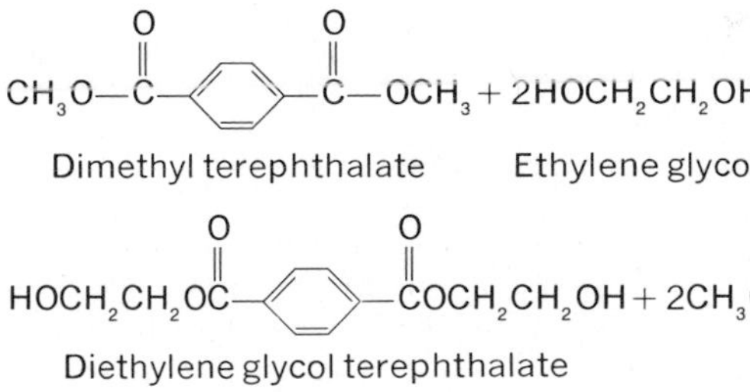

$$(5)$$

$$HOCH_2CH_2O\overset{\overset{\displaystyle O}{\|}}{C}-\bigcirc-\overset{\overset{\displaystyle O}{\|}}{C}OCH_2CH_2OH + 2CH_3OH$$

Diethylene glycol terephthalate

under vacuum, of the condensate eliminates the methyl alcohol and any excess ethylene glycol and low-molecular-weight polymers, and results in the formation of high-molecular-weight, amorphous polyethylene terephthalate. If the diethylene glycol ester is utilized instead of the dimethyl ester, further heating under vacuum yields the polymer with the elimination of the excess ethylene glycol.

Ethylene glycol is obtained by the oxidation of ethylene and terephthalic acid by the oxidation of p-dialkyl benzenes such as p-xylene or p-cymene.

As first produced, the polymer is usually amorphous, but it readily crystallizes on reheating or on extension of the spun filaments or cast or extruded sheets. Polyethylene orthophthalate does not crystallize readily, nor does it yield useful fibers and films. Polyethylene terephthalate does crystallize readily, and has the very high crystalline melting point of 249°C.

The fiber is resistant to mildew and moths. It is used frequently in combination with cotton for women's wear and men's shirts. Its chemical and heat resistance have placed it in demand for sails and cordage. Thermoplastic molding compounds are used extensively to replace glass in nonreturnable beverage bottles.

The film, which is stretched and oriented biaxially, is tough, strong, and insensitive to moisture. It is used for special packaging, as photographic film, in electrical transformers and capacitors, and in high-strength laminates. Useful films and fibers are also prepared using 1,4-cyclohexylene glycol as the glycol.

Newer aromatic polyesters include polybutylene terephthalate, which may be molded or extruded to yield materials that can replace metals or thermoset resins in some automotive, electrical, and specialty applications, especially when reinforced with glass fibers or mineral fillers. Several copolyesters have also been introduced, as well as polyesters based on bisphenol A and isophthalic acid and on *p*-hydroxy benzoic acid. The last-mentioned polymer must be processed by metalforming techniques. Copolymers can be processed conventionally.

Aromatic polycarbonates. These are a strong, tough group of thermoplastic polymers formed most frequently from bisphenol A and phosgene. The products are noted for high softening temperatures, usually greater than 140°C, and high impact resistance, clarity, and resistance to creep.

Reaction (6), between bisphenol A and phos-

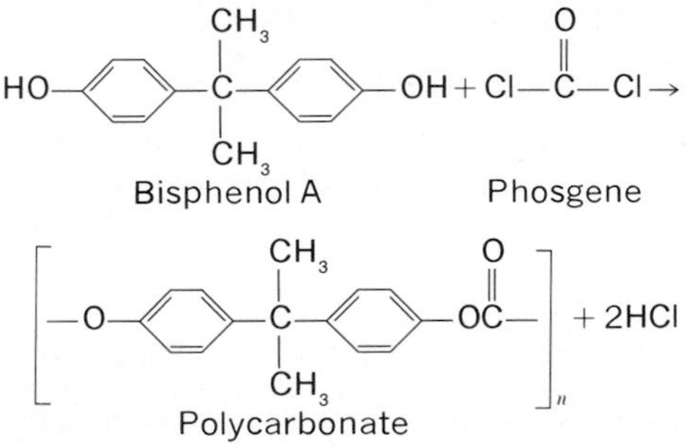

(6)

gene, leads to the polycarbonate and the evolution of hydrogen chloride. Bisphenol A is obtained by the condensation of phenol and acetone, and phosgene is produced by the reaction of carbon monoxide with chlorine. The same type of polycarbonate can also be made by the ester interchange of diphenyl carbonate with bisphenol A.

As formed, the polymer is only slightly crystallized. It is believed that the toughness is due to a balance between high intermolecular attractive forces and local mobility of uncrystallized segments of molecules.

The polymer is usually available as a molding compound. Because of its high strength, toughness, and softening point, the resin, both by itself and as a glass-reinforced material, has found many electrical domestic and engineering applications. It is often used to replace glass and metals. Examples include bottles, unbreakable windows, appliance parts, electrical housings, marine propellers, and shotgun shells. Flame-retardant grades are of interest because of low toxicity and smoke emission on burning. Films have excellent clarity and electrical characteristics, and have been used on capacitors and in solar collectors, as well as in packaging.

Polydiallyl esters. These are polymers of diallyl esters, such as diallyl phthalate, diallyl carbonate, diallyl phenyl phosphonate, and diallyl succinate, in which cross-linked products are produced by polymerization of the allyl groups, as in the case of diallyl phthalate, reaction (7).

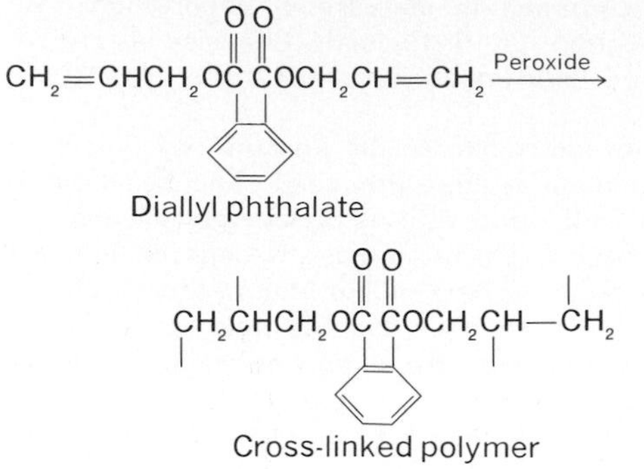

(7)

Thermosetting molding compounds may be produced by careful limitation of the initial polymerization to yield a product which is fusible. Then the polymerization and curing are completed in the final molding operation. Partially polymerized resins and the monomer are useful in making chemically resistant coatings. The monomer is often used as a cross-linking agent for polyester resins.

Major applications are in electronic components, sealants, coatings, and glass-fiber composites.

[JOHN A. MANSON]

Polyether resins

Thermoplastic or thermosetting materials which contain ether-oxygen linkages, —C—O—C—, in the polymer chain. Depending upon the nature of the reactants and reaction conditions, a large number of polyethers with a wide range of properties may be prepared.

The main groups of polyethers in use are epoxy resins, prepared by the polymerization and cross-linking of aromatic diepoxy compounds; phenoxy resins, high-molecular-weight epoxy resins; polyethylene oxide and polypropylene oxide resins; polyoxymethylene, a high polymer of formaldehyde; and polyphenylene oxides, polymers of xylenols.

Epoxy resins. The epoxy resins form an important and versatile class of cross-linked polyethers characterized by excellent chemical resistance, adhesion to glass and metals, electrical insulating properties, and ease and precision of fabrication.

In the preparation of a typical resin, a low-molecular-weight diepoxy compound is first mixed with cross-linking agents, fillers, and plasticizers and then allowed to cure either at room temperature or with the application of heat.

The intermediate diepoxy compounds are condensation products of epichlorohydrin and aliphatic or aromatic diols and are available commercially. An example is the product of the reaction of epichlorohydrin with bisphenol A, reaction (1).

A wide variety of diols may be used in order to obtain a desired balance of properties. The liquid or solid intermediates obtained are quite stable, but do not have useful physical properties until they are polymerized further, by the addition of a curing or cross-linking agent, in a manner similar to addition polymerization. Polymerization may be initiated by catalysts such as boron trifluoride or

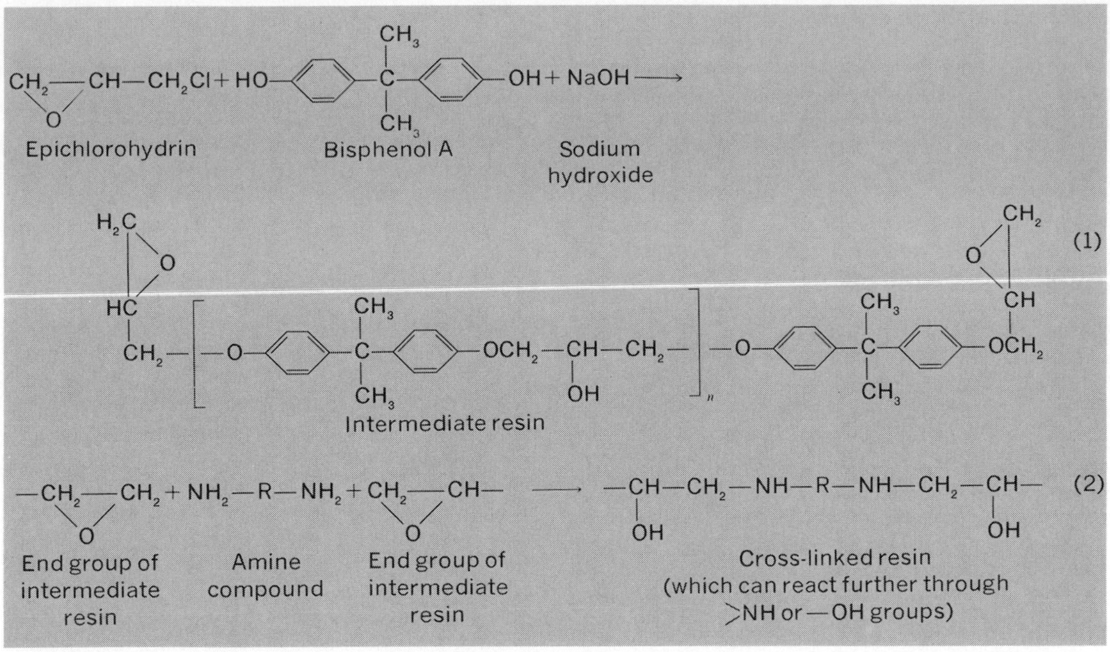

tertiary amines. Alternatively, compounds containing a reactive hydrogen atom such as organic acids, alcohols, mercaptans, primary and secondary amines, and polyamides can also serve as catalysts. Reaction (2) represents this curing process with an amine compound. For special uses, polyols of greater functionalities are often used. These include expoxidized novolacs, epoxidized polyphenols, and glycidyl derivatives of amino-substituted aromatic compounds.

The type of curing agent employed has a marked effect on the optimum temperature of curing and has some influence on the final physical properties of the product. By judicious selection of the curing system, the curing operation can be carried out at almost any temperature from 0 to 200°C.

Various fillers such as calcium carbonate, metal fibers and powders, and glass fibers are commonly used in epoxy formulations in order to improve such properties as the strength and resistance to abrasion and high temperatures. Some reactive plasticizers, such as mercaptans and polyamides, act as curing agents, become permanently bound to the epoxy groups, and are usually called flexibilizers. Rubbery polymers such as carboxy-terminated butadiene-styrene copolymers are added to improve toughness and impact strength.

Because of their good adhesion to substrates and good physical properties, epoxies are commonly used in protective coatings. Because of the small density change on curing and because of their excellent electrical properties, the epoxy resins are used as potting or encapsulating compositions for the protection of delicate electronic assemblies from the thermal and mechanical shock of rocket flight. Because of their dimensional stability and toughness, the epoxies are used extensively as dies for stamping metal forms, such as automobile gasoline tanks, from metal sheeting. Foams are also made.

By combining epoxies, especially the higher-performance types based on polyfunctional monomers, with fibers such as glass or carbon, exceedingly high moduli and strengths may be obtained. Thus a typical carbon-fiber-reinforced epoxy may have a tensile modulus and strength of 75,000,000 psi (500 GPa) and 300,000 psi (2 GPa), respectively. Such composites are of increasing importance in aerospace applications where high ratios of a property to density are desired, as well as in such domestic applications as sports equipment. *See* POLYMERIC COMPOSITE.

The adhesive properties of the resins for metals and other substrates and their relatively high resistance to heat and to chemicals have made the epoxy resins useful for protective coatings and for metal-to-metal bonding.

Polyolefin oxide resins. Polyethylene oxide and polypropylene oxide are thermoplastic products whose properties are greatly influenced by molecular weight. The general method of preparation is represented by reactions (3) and (4). Other cyclic

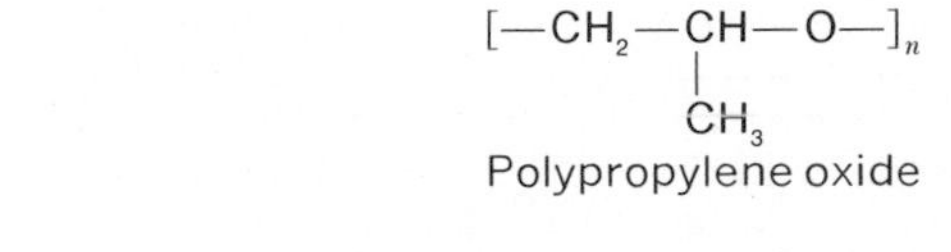

(3)

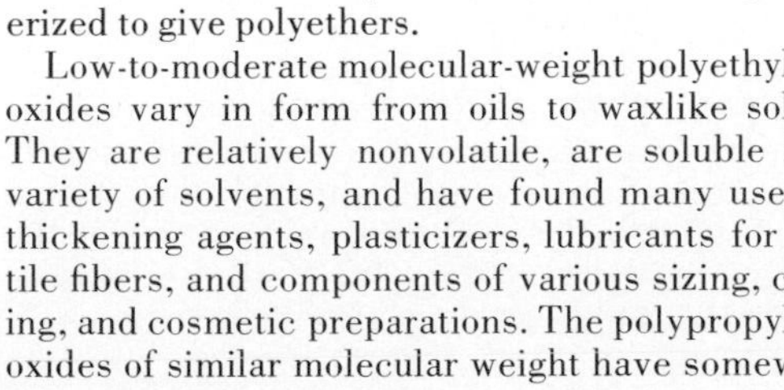

(4)

oxides, such as tetrahydrofuran, can also be polymerized to give polyethers.

Low-to-moderate molecular-weight polyethylene oxides vary in form from oils to waxlike solids. They are relatively nonvolatile, are soluble in a variety of solvents, and have found many uses as thickening agents, plasticizers, lubricants for textile fibers, and components of various sizing, coating, and cosmetic preparations. The polypropylene oxides of similar molecular weight have somewhat similar properties, but tend to be more oil-soluble (hydrophobic) and less water-soluble (hydrophilic).

Nonionic surface-active agents can be prepared from $C_{10}-C_{20}$ fatty alcohols and acids by the condensation of some 5–40 ethylene oxide groups, for example, reaction (5). An interesting commercial

$$C_{11}H_{23}COOH + 20CH_2\text{—}CH_2 \ (\text{—O—}) \xrightarrow{\text{NaOH}}$$

Lauric acid Ethylene oxide

$$C_{11}H_{23}COO(CH_2CHO)_{20}H \quad (5)$$

A nonionic detergent

example of a block copolymer has been produced by polymerization of propylene oxide onto polyethylene oxide to yield a linear chain with sequences of the two compounds, as in reaction (6), in which x

$$[\text{—}CH_2\text{—}CHO\text{—}]_x + yCH_2\text{—}CHCH_3 \ (\text{—O—}) \rightarrow$$

Polyethylene oxide Propylene oxide

$$[\text{—}CH_2\text{—}CHO\text{—}]_x\text{—}[\text{—}CH_2\text{—}\underset{\displaystyle CH_3}{\underset{|}{CH}}\text{—}O\text{—}]_y \quad (6)$$

and y may be in the range 10–100. The hydrophobic-hydrophilic balance necessary for surface activity is achieved because the polyethylene sequence is relatively hydrophilic and the polypropylene sequence is relatively hydrophobic. *See* POLYMERIZATION.

Oil-water emulsions prepared by use of the product have remarkable stability to both hydrophilic and hydrophobic precipitating agents.

While polyalkylene oxides are not of interest as such in structural materials, polypropylene oxides are used extensively in the preparation of polyurethane foams.

Crystalline, high-molecular-weight polypropylene oxides with melting points up to about 74°C have been prepared with three groups of catalysts: (1) solid potassium, (2) complexes of ferric or stannic chloride with propylene oxide, and (3) certain metallic alkyls such as aluminum triethyl. By starting with optically active propylene oxide, an optically active polymer is produced.

Phenoxy resins. Phenoxy resins differ from the structurally similar epoxy resins based on the reaction of epichlorohydrin with bisphenol A mainly by possessing a much higher molecular weight, about 25,000. The polymers are transparent, strong, ductile, and resistant to creep, and, in general, resemble polycarbonates in their behavior. Cross-linking may be effected by the use of curing agents which can react with the —OH groups. Molding and extrusion may be used for fabrication; blow-molded bottles are a potentially important application. The major application is as a component in protective coatings, especially in metal primers.

Polyphenylene oxide resins. Polyphenylene oxide (PPO) is the basis for an engineering plastic characterized by chemical, thermal, and dimensional stability. The polymers are prepared by the oxidative coupling of 2,6-xylenol, using a cuprous salt and an amine as catalyst, as shown in reaction (7). Similar reactions may be used to produce

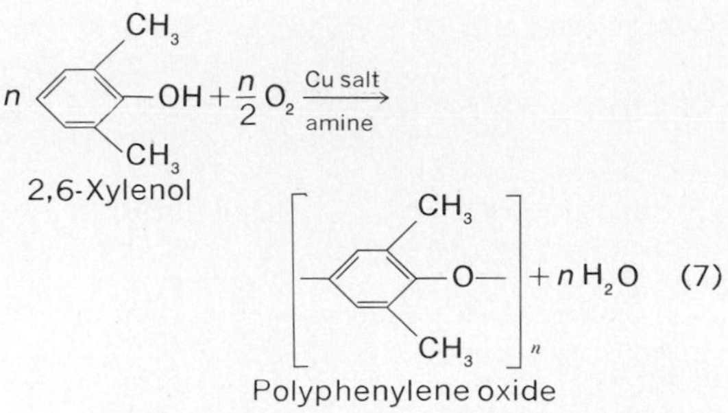

many varied structures. The polyphenylene oxide (PPO) resin is normally blended with another compatible but cheaper resin such as high-impact polystyrene. The blends are cheaper and more processable than PPO and still retain many of the advantages of PPO by itself.

Polyphenylene oxide is outstanding in its resistance to water, and in its maximum useful temperature range (about 170–300°C). In spite of the high softening point, the resins, including glass-reinforced compositions, can be molded and extruded in conventional equipment. Uses include medical instruments, pump parts, and insulation. Structurally modified resins are also of interest and the general family of resins may be expected to replace other materials in many applications.

An analagous polythioether, polyphenylene sulfide, may be prepared by the reaction of p-dichlorobenzene with sulfur or sodium sulfide in a polar solvent yielding a branched polythioether, or by polymerization of metal salts of p-halothiophenols (I), yielding a linear polythioether (II). The combi-

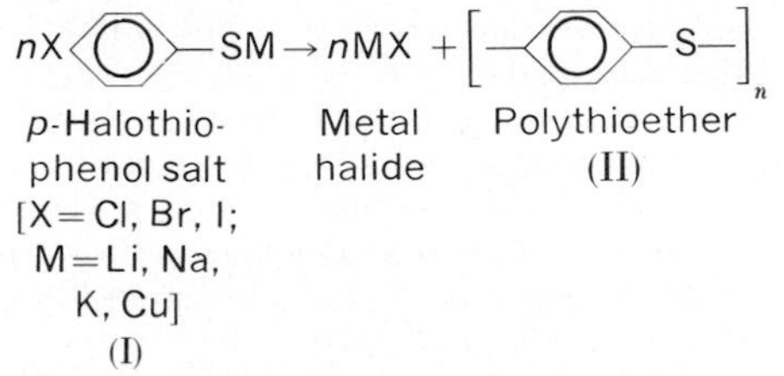

nation of regular structure with the chemically and thermally stable ring-sulfur bond leads to a combination of a high melting point (~290°C), thermal stability, inherent resistance to burning, and resistance to most chemicals. Commercial products can be processed by injection or compression molding, as well as by slurry or electrostatic coating; reinforcement with glass fibers is also commonly practiced. The combination of properties is useful in such applications as electronic pumps, and automotive components and coatings, especially where environmental stability is required.

Polyoxymethylene. Polyoxymethylene, or polyacetal, resins are polymers of formaldehyde. Having high molecular weights and high degrees of crystallinity, they are strong and tough and are established in the general class of engineering thermoplastics. Polymerization is accomplished by the use of a catalyst such as triphenyl phosphine, shown in reaction (8). Since the untreated polymers

$$n\,H\text{—}\overset{\displaystyle O}{\overset{\|}{C}}\text{—}H \xrightarrow[\text{catalyst}]{\text{Triphenyl-phosphine}} [\text{—}CH_2\text{—}O\text{—}]_n \quad (8)$$

Formaldehyde Polyformaldehyde or polyoxymethylene

tend to degrade by an unzipping or depolymerization reaction, the end groups are frequently esterified to block the initiation of degradation. Alternately, small amounts of a comonomer such as ethylene oxide or 1,3-dioxolane may be introduced during polymerization.

Though somewhat similar to polyethylene in general molecular structure, polyacetal molecules pack more closely, and attract each other to a much greater extent, so that the polymer is harder and higher-melting than polyethylene. Polyacetals are typically strong and tough, resistant to fatigue, creep, organic chemicals (but not strong acids or bases), and have low coefficients of friction. Electrical properties are also good. Improved properties for particular application may be attained by reinforcement with fibers of glass or polytetrafluoroethylene.

The combination of properties has led to many uses such as plumbing fittings, pump and valve components, bearings and gears, computer hardware, automobile body parts, and appliance housings. Other aldehydes may be polymerized in a similar way.

A chlorinated polyether based on the cationic polymerization of 3.3-bis(chloromethyl-1-oxacyclobutane) was developed in the 1960s, as an engineering plastic. While it has excellent chemical resistance and chemical stability, other plastics can give as good performance at lower cost. *See* EPOXIDATION.

[JOHN A. MANSON]

Bibliography: J. A. Bryson, *Plastics Materials*, 3d ed., 1975; *Modern Plastics Encyclopedia*, 1979–1980.

Polyethylene glycol

Any of a series of water-soluble polymers with the general formula HO—$(CH_2—CH_2—O)_n$—H. These colorless, odorless compounds range in appearance from viscous liquids for $n = 2, 3, 4$ (with a freezing point of −13 to −4°C) up to waxy solids melting at a 66–67°C maximum for n above about 180. The low-molecular-weight members, diethylene glycol (n = 2) through tetraethylene glycol (n = 4), are produced as pure compounds and find use as humectants, dehydrating solvents for natural gas, textile lubricants, heat-transfer fluids, solvents for aromatic hydrocarbon extractions, and intermediates for polyester resins and plasticizers. These lower polyethylene glycols are distilled by-products of ethylene glycol manufacture.

The intermediate members of the series with average molecular weights of 200 to 20,000 are produced as residue products by the sodium or potassium hydroxide–catalyzed batch polymerization of ethylene oxide onto water or mono- or diethylene glycol. These polymers are formed by stepwise anionic addition polymerization and, therefore, possess a distribution of molecular weights. Examples of commercial uses for products in this range are in ceramic, metal-forming, and rubber-processing operations; as drug suppository bases and in cosmetic creams, lotions, and deodorants; as lubricants; as dispersants for casein, gelatins, and inks; and as antistatic agents. These polyethylene glycols generally have low human toxicity.

The highest members of the series have molecular weights from 100,000 to 10,000,000. They are produced by special anionic polymerization catalysts which incorporate metals such as aluminum, calcium, zinc, and iron, and coordinated ligands such as amides, nitriles, or ethers. These members of the polyethylene glycol series are of interest because of their ability at very low concentrations to reduce friction of flowing water. *See* ETHYLENE OXIDE; GLYCOL; POLYOXYETHYLATION OF ALCOHOL.

[ROBERT K. BARNES]

Bibliography: N. G. Gaylord (ed.), *High Polymers*, vol. 13: *Polyethers*, pt. 1, 1963.

Polyfluoroolefin resins

Resins distinguished by their resistance to heat and chemicals and by the ability to crystallize to a high degree. Several main products are based on tetrafluoroethylene, hexafluoropropylene, and monochlorotrifluoroethylene. Structural formulas of the monomers are shown.

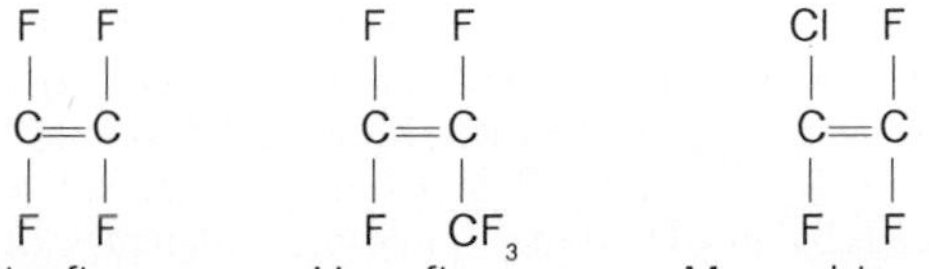

Tetrafluoroethylene (TFE) — Hexafluoropropylene (HFP) — Monochlorotrifluoroethylene (CTFE)

Copolymers of TFE and HFP with each other and of TFE and HEP with ethylene are available commercially. Homopolymers of vinylidene fluoride and copolymers with HFP are also in common use. For a description of polyvinyl fluoride *see* POLYVINYL RESINS.

Polytetrafluoroethylene. Tetrafluoroethylene can be obtained by the pyrolysis of monochlorodifluoromethane, which in turn is obtained from a rather complex reaction between anhydrous hydrogen fluoride and chloroform. This reaction is represented by (1), (2), and (3). Although poly-

$$\underset{\text{Chloroform}}{CHCl_3} + \underset{\text{Anhydrous hydrofluoric acid}}{HF} \xrightarrow{SbCl_5} FCHCl_2 + F_2CHCl + CCl_2F_2 + HCl \quad (1)$$

$$2FCHCl_2 + AlCl_3 \rightarrow F_2CHCl + CHCl_3 \quad (2)$$

$$\underset{\text{Monochlorotrifluoromethane}}{2F_2CHCl} \xrightarrow{700-800°C} F_2C{=}CF_2 + 2HCl \quad (3)$$

merization in bulk can proceed with violence, the monomer can be polymerized readily and conveniently in emulsion under pressure, using free-radical catalysts such as peroxides or persulfates. The polymer is insoluble, resistant to heat (up to 275°C) and chemical attack, and in addition, has the lowest coefficient of friction of any solid. Because of its resistance to heat, the fabrication of polytetrafluoroethylene requires modification of conventional methods. After molding the powdered polymer using a cold press, the moldings are sintered at 360–400°C by procedures similar to those used in powder metallurgy. The sintered product can be machined or punched. Extrusion is possible if the powder is compounded with a lubricating material. Aqueous suspensions of the polymer can also be used for coating various articles. However, special surface treatments are required to ensure adhesion because polytetrafluoroethylene does not adhere well to anything.

Polytetrafluoroethylene (TFE resin) is useful for applications under extreme conditions of heat and chemical activity. Polytetrafluoroethylene bearings, valve seats, packings, gaskets, coatings and tubing can withstand relatively severe conditions. Fillers such as carbon, inorganic fibers, and metal powders may be incorporated to modify the mechanical and thermal properties.

Because of its excellent electrical properties, polytetrafluoroethylene is useful when a dielectric material is required for service at a high temperature. The nonadhesive quality is often turned to advantage in the use of polytetrafluoroethylene to coat articles such as rolls and cookware to which materials might otherwise adhere. A structural derivative containing perfuoroalkoxy side groups (PFA resin) has been introduced. PFA resins behave generally like TFE resins, but can be processed by conventional molding, extrusion, and powder-coating methods.

Polymonochlorotrifluoroethylene. The monomer is prepared by the dechlorination of 1,1,2-trichloro-1,2,2-trifluoroethane as in reaction (4).

$$CCl_2FCClF_2 + Zn\ dust \xrightarrow{C_2H_5OH} CClF{=}CF_2 + ZnCl_2 \quad (4)$$

Polymerization can be carried out in aqueous suspension by the free-radical process in which a combination of a persulfate and bisulfite is used as the initiator.

The properties of polymonochlorotrifluoroethylene (CTFE resin) are generally similar to those of polytetrafluoroethylene; however, the presence of the chlorine atoms in the former causes the polymer to be a little less resistant to heat and to chemicals. The polymonochlorotrifluoroethylene can be shaped by use of conventional molding and extrusion equipment, and it is obtained in a transparent, noncrystalline condition by quenching. Dispersions of the polymer in organic media may be used for coating.

The applications of polychlorotrifluoroethylene are in general similar to those for polytetrafluoroethylene. Because of its stability and inertness, the polymer is useful in the manufacture of gaskets, linings, and valve seats that must withstand hot and corrosive conditions. It is also used as a dielectric material, as a vapor and liquid barrier, and for microporous filters.

Copolymers. Copolymers of TFE and HFP propylene (fluorinated ethylenepropylene, or FEP resins) and copolymers of TFE with ethylene (ETFE) are often used in cases where ease of fabrication is desirable. The copolymers can be processed by conventional thermoplastic techniques, and except for some diminution in the level of some properties, properties generally resemble those of the TFE homopolymer.

Copolymers of ethylene with CTFE can also be processed by conventional methods, and have better mechanical properties than PTFE, FEP, and PFA resins.

Polyvinylidene fluoride. The monomer is prepared by the dehydrohalogenation of 1,1,1-chlorodifluoroethane or by the dechlorination of 1,2-dichloro-1,1-difluoroethane as in reactions (5) and (6). Polymerization is effected using free radical catalysis.

$$CH_3CClF_2 \xrightarrow{-HCl} CH_2{=}CF_2 \quad (5)$$

$$CH_2ClCClF_2 \xrightarrow{-Cl_2} CH_2{=}CF_2 \quad (6)$$

The properties are generally similar to those of the other fluorinated resins: relative inertness, low dielectric constant, and thermal stability (up to about 150°C). The resins (PVF_2 resins) are, however, stronger and less susceptible to creep and abrasion than TFE and CTFE resins.

Applications of polyvinylidene fluoride are mainly as electrical insulation, piping, process equipment, and as a protective coating in the form of a liquid dispersion.

Fluorinated elastomers. Several types of fluorinated, noncrystallizing elastomers have been developed in order to meet needs (usually military) for rubbers which possess good low-temperature behavior with a high degree of resistance to oils and to heat, radiation, and weathering. Fluorinated acrylic rubbers such as polyheptafluorobutyl acrylate and fluorine-containing silicone rubbers have been used to some extent for this purpose. Copolymers of hexafluoropropylene with vinylidene fluoride make up an important class with such applications as gaskets and seals. Copolymers of nitrosomethane with tetrafluoroethylene are showing considerable promise for similar applications. *See* HALOGENATED HYDROCARBON; POLYMERIZATION.

[JOHN A. MANSON]

Bibliography: J. A. Brydson, *Plastics Materials*, 3d ed., 1975.

Polyhydric alcohol

Polyhydric alcohols, or, as they are preferably called, polyols, are compounds containing more than one hydroxyl group (—OH). Each hydroxyl is attached to separate carbon atoms of an aliphatic skeleton. This group includes glycols, glycerol, and pentaerythritol and also such products as trimethylolethane, trimethylolpropane, 1,2,6-hexanetriol, sorbitol, inositol, and polyvinyl alcohol.

Polyols are obtained from many plant and animal sources and are synthesized by a variety of methods. 1,2-Glycols and glycerol are produced by the hydrolysis of epoxides or chlorohydrins. Formaldehyde reacts with acetaldehyde, propionaldehyde, and butyraldehyde to form, respectively, pentaerythritol, trimethylolethane, and trimethylolpropane. Catalytic hydrogenation of sugars produces sorbitol; 1,2,6-hexanetriol is obtained by the reduction of 2-hydroxyadipaldehyde. Saponification of polyvinyl acetate is employed in the industrial manufacture of polyvinyl alcohol, which is represented as:

$$\left(-CH_2\underset{\displaystyle OH}{\underset{|}{CH}}- \right)_n$$

Polyols such as glycerol, pentaerythritol, trimethylolethane, and trimethylolpropane are used in making alkyd resins for decorative and protective coatings. Glycols, glycerol, 1,2,6-hexanetriol, and sorbitol find application as humectants and plasticizers for gelatin, glue, and cork. Explosives are made by the nitration of glycols, glycerol, and pentaerythritol.

The polymeric polyols used in manufacture of the urethane foams represent a series of synthetic polyols which have come into prominence in the 1960s. These polyols, generally polyoxyethylene or polyoxypropylene adducts of di- to octahydric alcohols, cover a molecular weight range of 400 to 6000. *See* GLYCEROL; GLYCOL; PENTAERYTHRITOL. [PHILIP C. JOHNSON]

Polymer

The terms polymer, high polymer, macromolecule, and giant molecule are used to designate high-molecular-weight materials of either synthetic or natural origin. Plastics are relatively stiff at room temperature, rubbers or elastomers are flexible and retract quickly after stretching, and fibers are especially strong filamentary materials. Coatings are generally either plastics or rubbers which have been applied as a thin layer on a substrate. In practice, plastics, rubbers, fibers, and coatings are used as formulations of the polymers with other ingredients such as fillers, pigments, plasticizers, flow improvers, and stabilizers against aging and degradation.

This article discusses polymers as a class and the effects of molecular weight, structure, morphology, and fabrication history upon properties. For a discussion of the general methods of preparation, catalytic processes, and the effects of the conditions of polymerization upon the molecular weight and molecular structure or architecture of the polymeric product *see* POLYMERIZATION. For descriptions of typical synthetic polymers of the condensation type *see* PHENOLIC RESIN; POLYAMIDE RESINS; POLYESTER RESINS; POLYETHER RESINS; POLYSULFIDE RESINS; POLYURETHANE RESINS; SILICONE RESINS; UREA-FORMALDEHYDE-TYPE RESINS. For descriptions of some important members of the addition-type synthetic polymers *see* HYDROCARBON RESIN; POLYACRYLATE RESIN; POLYACRYLONITRILE RESINS; POLYFLUOROOLEFIN RESINS; POLYOLEFIN RESINS; POLYSTYRENE RESIN; POLYVINYL RESINS.

HISTORICAL DEVELOPMENT

The first modified natural polymers, cellulose nitrate and casein-formaldehyde, were commercially produced about 1860, and the first fully synthetic polymer, phenol-formaldehyde, was made about 1910. The major development of present polymer science and technology has taken place since about 1920, while the production of polymeric materials in the United States has grown from a few million pounds to several billion pounds in the same time period.

Interest in the synthesis of products similar to natural products but possessing more useful properties, has been continually stimulated by the successful synthesis of polyamide fibers and rubbers equivalent to natural rubber, and by increasing understanding of the nature of proteins, nucleoproteins, carbohydrates, and enzymes in living tissues.

Several striking advances have been the development of cheap raw materials for plastics and of new polymerization processes, and remarkable advances in understanding the relationships between molecular structure, morphology, and physical and chemical behavior. As properties have been improved, plastics have been developed which can be readily and economically fabricated, and which can be used for hitherto inappropriate engineering purposes such as gears, bearings, and structural members. Such engineering plastics may frequently be used advantageously to replace metals or other materials. *See* POLYMERIC COMPOSITE.

As concern for efficiency in the use of energy and raw materials such as petroleum has increased, polymers continue to maintain a strong position as a class of material. The energy requirements for production of polymers and their composites are often lower than for many traditional materials, especially when the energy required per unit of performance is considered. Also, the proportion of petroleum used for petrochemicals such as monomers, rather than for fuel, is small, about 10%. At the same time, the lightness of polymeric materials serves to reduce the consumption of fuel in vehicles and aircraft, and the corrosion resistance of polymeric coatings on many articles extends service life and conserves material resources.

Interest is increasing in polymers which can withstand even more extreme environments, or possess other specialized properties, for use under the sea, in space, and in biological systems. These interests have emphasized the continuing need for still deeper understanding of bonding (in organic, inorganic, and organic-inorganic systems) and of the implications of bonding and structure to morphology, and, in turn, to the overall response to applied stresses in given environments.

PROPERTIES

The properties of polymeric materials are determined by the molecular properties of the macromolecules, the morphology, and the type of formulation involving plasticizers or fillers. The morphology in turn depends on the conditions of fabrication in which molecular orientation or crystallization may be induced. Properties also depend on the temperature and elapsed time of the measurement.

Molecular properties. These include molecular size and weight, molecular structure or architecture, molecular-weight distribution, polarity, and flexibility of the polymeric chains (or chain segments between cross-links in cured or vulcanized polymers). Molecular properties taken together determine the attractive forces between the molecules, the morphology or arrangement of masses of molecules, and the general behavior of the polymer.

Molecular weight and distribution. The desirable properties of high polymers (strength and resistance to solvents) increase rapidly with increasing molecular weight in the low ranges of molecular weight, and more slowly in the high ranges. On the other hand, the melt viscosity increases with molecular weight in the opposite manner, slowly at first and rapidly in the higher-molecular-weight range. The ease of fabrication (molding, extrusion, and shaping) of polymeric compositions varies inversely with the melt viscosity; that is, the materials become increasingly difficult to mold or extrude at very high values of molecular weight. The optimum molecular weight of a polymer frequently varies for different applications and different methods of fabrication. Commercial products are usually available in several molecular-weight ranges. In all cases, however, the optimum molecular weight must be selected on the basis of the best compromise for desirable properties and ease of fabrication. Techniques of measuring molecular properties and methods of fabrication are briefly discussed at the end of this article.

The molecular-weight distribution represents the particular combination of material of low, medium, and high molecular weight in a given product. The presence of low-molecular-weight portions may lower the softening temperature range of the product, make it more subject to attack by solvents and chemical agents, and lower the melt viscosity or increase the ease of fabrication. The lower-molecular-weight portions behave more or less as plasticizers. The presence of very-high-molecular-weight portions increases the melt viscosity and enhances many properties such as strength.

Frequently, the portions of very high molecular weight are actually insoluble, being highly branched or cross-linked. The insoluble portions show up as "fisheyes" in films and discontinuities in fibers. They also give undesirable nerve (elastic memory) characteristics to compounded, uncured rubber.

Attractive forces. The atoms in the chains are held together by primary valence bonds. If it were possible to apply a force of tension only to the primary valence bonds, then a tensile strength of more than 2,000,000 psi would be observed. In reality, however, under tension the molecules slip past one another and the resistance to that slippage is due to the effective attractive forces (van der Waals forces) between the molecules. Also, flaws reduce strength by initiating cracks.

The effective attractive forces are the result of the polarity of the groups in, or attached to, the chains, and of the degree of fit between chains. Strongly polar groups, such as those containing oxygen, nitrogen, and sulfur, and other polar atoms exert the strongest attractive forces. Bulky side groups attached to the main chains stiffen the chains and also by their bulk may prevent a close fit between chains. Long side chains tend to act as internal plasticizers.

However, if bulky side groups are arranged in a stereoregular way (as in isotactic polymers), then close fit may be possible and crystallization may, in principle, occur. However, prolonged annealing at elevated temperatures may be necessary with very bulky molecules.

An example of the strong attraction of polar groups is found in the following comparison of the properties of polyamides and *N*-methyl polyamides. Nylon-6 is relatively high-melting, hard, strong, and insoluble, whereas the *N*-methyl derivative is lower-melting, less strong, and more readily soluble. Structural formulas of these compounds are shown. The strength of the hydrogen bond between the >N—H group and an oxygen atom is about 3–7 kcal/mole (12 to 29 kJ/mole). By

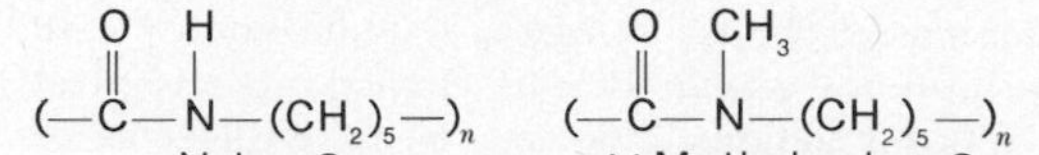

Nylon-6   *N*-Methyl nylon-6

replacing the >N—H by >N—CH_3, the intermolecular attractive forces are reduced to perhaps 2–3 kcal/mole (8 to 12 kJ/mole).

Fit. An example of the effect of the regularity of chains on the fit between them is seen in the comparison of the properties of polyethylene orthophthalate and polyethylene terephthalate, whose structural formulas are shown.

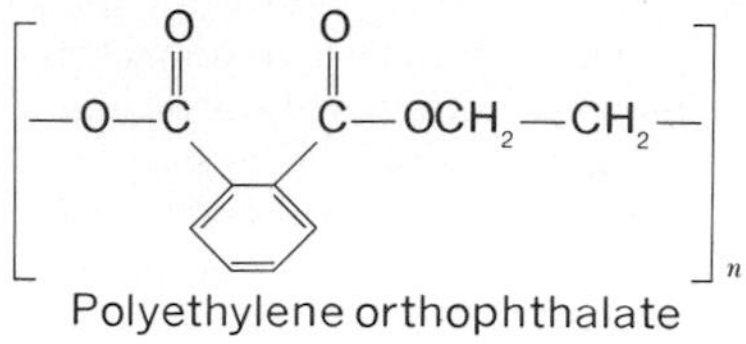

Polyethylene orthophthalate

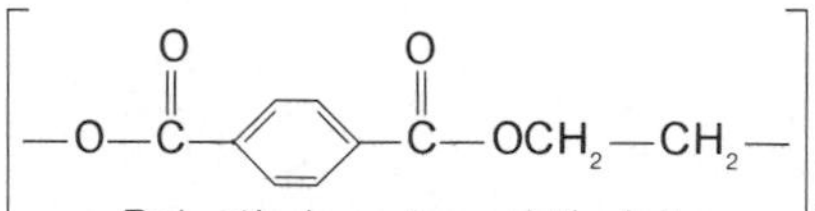

Polyethylene terephthalate

The para derivative, polyethylene terephthalate, has the more regular structure. It melts at higher temperatures, crystallizes more readily, and is stronger and less soluble than polyethylene orthophthalate, even though the chemical nature and polarity of the two polymers are identical.

Other kinds of isomerism also affect fit. Thus trans isomers of 1,4-polydienes have higher melting points than cis isomers, and stereo regular polymers have higher melting points than atactic ones.

Flexibility. The flexibility of linear polymer chains and of the segments between cross-links in cured products is decreased by the presence of polar groups and regularity in the molecular structure. A nonpolar, irregular chain should be the most flexible. Products containing highly flexible chains are rubbery, soft, and nonbrittle, with relatively high resistance to impact and to tear. The mechanical implications are discussed below.

Environmental stability. The stability of a polymer depends on the chemical nature, on morphology, and on molecular properties. Thus polyamides are susceptible to hydrolysis on long exposures to dilute acids at high temperatures. Resistance to high temperatures under oxidizing conditions is enhanced by minimizing the hydrogen content, by increasing the molecular stiffness, and by increasing the content of aromatic structures or heterocyclic rings. *See* HETEROCYCLIC POLYMER.

Morphology. When a close fit between chains is possible, crystallization can take place spontaneously or upon drawing or cooling. The presence of polar attracting groups enhances the tendency to crystallize.

If crystallization is effected from a very dilute solution, it has been found that single, nearly perfect crystals may be obtained from many polymers. This revolutionary discovery, by A. Keller, E. W. Fischer, and P. H. Geil, has stimulated much research on the behavior of crystalline polymers. In single crystals the long molecules fold back on themselves every 100 A (10 nm), or so, so that more or less flat platelets are formed, with the molecules lying perpendicular to the plane of the crystal. Crystallization from the melt leads to much less regular structures, for it is very difficult to disentangle long molecules from each other, and a given molecule may participate in several crystalline units. A melt-crystallized polymer typically contains regions that are rather well ordered (crystallites), tied together by uncrystallized segments of molecules. Such links help diminish the brittleness of a crystalline material. Thus polymers may range from slightly imperfect true crystals to aggregates of slightly crystallized amorphous molecules.

It is also possible to form extended-chain crystallites in which molecules are packed together longitudinally in fibrillar arrays. This may be achieved by the use of special extrusion or spinning techniques. Such highly anisotropic crystallites confer high levels of strength and stiffness in the chain direction.

The physical behavior of a semicrystalline polymer is very dependent on the percentage of crystallinity, on the size of the crystalline units, on the number and nature of intercrystalline links, and on the amount and nature of uncrystallized impurities. The crystalline melting point is also important, and is itself dependent on the degree of perfection in the crystalline units and on the presence of impurities. Good mechanical behavior requires a balance of properties.

Thermomechanical behavior. The degree of Brownian motion of short segments of polymer chains determines the mechanical response, and depends strongly on both temperature and the time scale of the test, as well as on the stress applied. With small stresses at very low temperatures, sequential motion is essentially frozen, and the polymer exhibits Hookean elasticity, the deformation involving primarily the reversible bending of bonds, and the restoring force decreases with increasing temperature. Under these conditions the polymer behaves as a rigid solid. In a nearly perfectly crystalline polymer, the modulus or stiffness remains nearly independent of temperature until the crystalline melting point T_m, where the thermal energy supplied is enough to overcome the binding energy of the crystal lattice. The polymer then melts to a viscous liquid. If the polymer is amorphous, or contains amorphous sequences, another transition is noted as the temperature is raised, the glass transition temperature T_g. This temperature corresponds to the onset of significant segmental motion, and to the change of the amorphous component from a rigid solid to a leathery and finally rubbery material, the rubbery state being unique to high polymers. Rubberlike elasticity is non-Hookean, the restoring force varying directly with temperature; the origin lies in the decreased entropy of the stretched molecules. Still other minor transitions may be noted, for example, transitions corresponding to the onset of motion of small chain segments. These transitions are often related to toughness.

In all the cases mentioned, a constant test time is assumed. If the test time is increased, as it is when the testing rate is decreased, the modulus-temperature curve is shifted toward higher temperatures. In short, testing rate and temperature are reciprocally related in terms of their effect on the modulus curve. Because both viscous and elastic behavior are involved, polymers are said to be viscoelastic bodies. The behavior discussed underlies the classification of polymers. Thus rubbers and plastics are polymers whose transition temperatures are below and above ambient temperature, respectively. By shifting the temperature or testing rate, the behavior can be changed from one type to another.

Molecular structure and morphology also affect the shape of the modulus-temperature curves. Thus, cross-linking inhibits long-range molecular motion so that viscous flow is prevented. Similarly, the crystallites in a semicrystalline polymer give rise to a modulus curve intermediate between that of an amorphous and a fully crystalline polymer; polyblending also gives curves intermediate between those of the two polymers concerned.

Stress-strain behavior. The illustration shows responses of typical rubbers, plastics, and fibers to large stresses. A reinforced or crystallizable rubber exhibits a relatively low value of breaking stress, but high elongation. A ductile plastic such as polyethylene exhibits yielding, drawing, and, at high elongations, some strengthening due to orientation. A brittle plastic such as polystyrene does not yield much, and breaks at a low elongation; a fiber exhibits the highest strength, high stiffness, and low elongation. In all cases, the shape and position of the stress-strain curve are determined by the balance struck between cohesion and segmental mobility at the temperature and testing rate concerned. As with the low-stress behavior discussed above, behavior is sensitive to these last two parameters. Thus a polymer that is brittle at room temperature may become ductile on raising the temperature or lowering the testing rate; similarly, a normally ductile polymer becomes brittle at low temperatures and high testing rates.

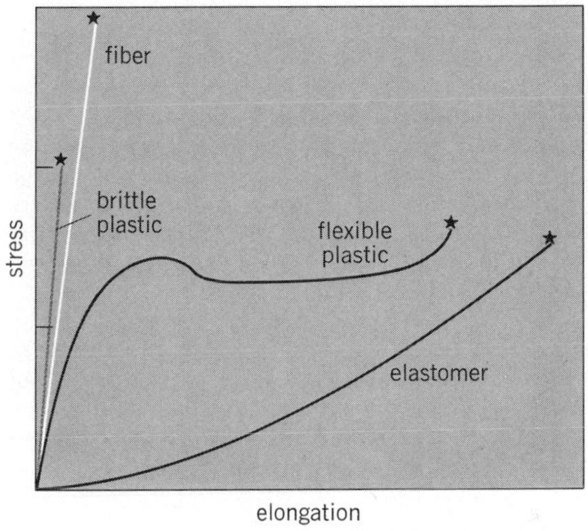

Stress-elongation behavior of typical classes of polymer (not drawn to scale). The star represents the point at which failure occurs.

With rubbers, strength and modulus may be increased up to a point by reinforcement or increased cross-linking, at the expense of elongation, while with plastics, the modulus is increased, but the strength and toughness are often decreased by these factors.

So far only sustained monotonic loading has been discussed. If repetitive or cyclic loads are applied, fatigue develops, and the specimen fails sooner than it would otherwise.

Creep is the term used to describe the slow deformation (increase in strain) of a polymer under constant load (or stress). Stress relaxation refers to the decrease in stress at a constant elongation or strain.

When a polymer is stressed, and then the stress released, some or all of the deformation may be recoverable. The term elastic memory applies to cases in which a polymer is deformed at an elevated temperature, as in the shaping of a sheet into a dome, and then is cooled before the tangled chains have reached an equilibrium condition in the new shape. Strains are said to be frozen in, and at a later time, especially if the product is warmed, these strains cause the product to assume a distorted shape.

Toughness. As plastics are used to an increasing extent in engineering applications, the toughness or the ability to resist fracture assumes considerable importance. Much research is being directed to the understanding of fracture processes, on one hand, and to the improvement of toughness, on the other. In general, toughness appears to require some combination of high modulus with the ability to dissipate applied energy by relaxation of molecular segments or of added materials.

To enable designers to select polymers and design parts for maximum service life, the principles of fracture mechanics are being applied more and more to polymers as well as to metals and ceramics. The central principle is based on Griffith's equation, shown below, where σ_B is the stress at

$$\sigma_B = C\frac{\sqrt{ES}}{a}$$

break, C is a constant, and E is Young's modulus; S is the energy per unit area of crack surface required for catastrophic fracture, and a is the size of a flaw characteristic of the material or introduced by handling. An alternate approach is based on the concept of a critical stress intensity factor for failure, K_c. This factor reflects the magnification of applied stress at the tip of a flow or crack as well as the specimen geometry and flaw size. For an elastic solid, the approach is equivalent to Griffith's, for in such a case $K_c^2 \simeq ES$. A major advantage of using fracture mechanics is that it takes conservative account of the imperfections that are always present in materials. Another advantage is that Griffith's equation and its derivative forms explicitly separate strength into individual parameters that can be characterized as a function of molecular composition and structure.

Other properties. While polymers are electrical insulators, some being used for covering wires and cables, special types such as polyacetylenes doped with certain inorganic species, exhibit semiconductivity.

In general, nonpolar polymers have low dielectric constants, while polar molecules may have high dielectric constants, especially at low frequencies and high temperatures (that is, temperatures $> T_g$). Thermal conductivities tend to be low, making polymers good thermal insulators; foams are even more effective. Permeabilities depend strongly on structure, nonpolar polymers tending to hinder the permeation of water, for example, and polar crystalline polymers tending to hinder the permeation of organic molecules. Lubricity and nonwettability are characteristic of low-surface-energy polymers like fluorinated types.

Measurement of molecular properties. Because of the random, statistical nature of the polymerization process, a distribution of chain lengths, that is, molecular weights, is always formed, and measured molecular weights are necessarily average values. Because of the very high molecular weights of macromolecules, the common methods of measuring properties of solutions (for example, the freezing-point lowering) are frequently not suitable. Among the colligative methods, the osmotic-pressure procedure is most useful and gives number-average molecular weights of reasonable accuracy up to about 500,000. The boiling-point elevation method is suitable for molecular weights up to 5000 – 10,000.

The fact that the amount of light scattered by a solution is a function of the molecular weight of the dissolved particles has provided a means for the determination of weight-average molecular weights up to several million. The light-scattering measurement can also yield valuable information regarding the shape of the molcule in solution.

The determination of molecular weight and also molecular-weight distribution can be accomplished by use of the ultracentrifuge, in which the rates of settling of particles in intense centrifugal fields are measured.

The intrinsic viscosity of a polymer in solution (the viscosity which the unassociated polymer molecules give to the solution) is a function of the molecular weight and is very easily measured. Intrinsic viscosity is commonly used for control purposes, and the values can be converted into molecular weight by calibration with osmotic pressure, light scattering, or sedimentation measurements.

In addition to intrinsic viscosity measurements, the most common technique today is based on gel permeation chromotography, in which a polymer deposited on a column of special beads is eluted with solvent, the higher molecular weights being eluted first. This method yields, with one experiment, a molecular size or distribution curve.

The degree of stereoregularity and its distribution along the polymer molecule may sometimes be inferred indirectly from infrared spectra, but may be determined directly and uniquely by high-resolution nuclear magnetic resonance techniques.

Thermal stability may be characterized by several analytical methods. Thermogravimetric analysis, coupled with differential thermal analysis or differential scanning calorimetry, yields valuable information about degradation. Mass spectrometry and gas chromatography are useful for the study of the products of degradation.

The second-order transition temperature is a time-dependent function. It is conveniently measured by noting the temperature at which the slope of a flexibility or stiffness-temperature curve changes abruptly, or at which a change in the coefficient of expansion occurs. It may also be determined from frequency-dependent measurements of dielectric or mechanical loss, or from thermal changes using differential thermal analysis or differential scanning calorimetry. Measurements of creep or stress relaxation are of particular importance in characterizing viscoelastic behavior as a function of both time and temperature. *See* CHROMATOGRAPHY; MASS SPECTROMETRY; NUCLEAR MAGNETIC RESONANCE (NMR); THERMOANALYSIS.

COMPOUNDING

Plastic masses, rubber formulations, coatings, and other polymeric compositions may contain age inhibitors, strengthening and coloring pigments, flow improvers, and plasticizing or softening agents. Roll-mill, sigma-blade, and dough mixers are generally employed to mix the resin with plasticizers and pigments at temperatures usually between about 50 and 250°C.

Additives. The addition of plasticizers to a polymeric material causes it to be softer and more rubbery in character. Plasticizers are held in association with the polymer chains by secondary valence forces. They separate the molecules, thus reducing the effective intermolecular attractive forces.

Finely divided polar substances may act as strengthening fillers for plastics. The filler surfaces can adsorb links of several polymer chains and produce the effects of cross-linking. Filled polymeric compositions are usually harder, sometimes stronger, more resistant to abrasion, and stiffer than the unfilled products. The effect of strong fiber fillers (glass, carbon, or other materials) on the tensile properties of several polymers is impressive. Products with tensile strengths of up to about 300,000 psi (2 GPa) can readily be obtained.

Age inhibitors are almost always incorporated in polymeric compositions. Oxygen, ozone, light, and electric discharge produce free radicals which cause degradation of the polymer chains. Free-radical inhibitors and light-masking agents are therefore commonly used. *See* INHIBITOR.

Polyblends. These are produced by the addition of small amounts of a rubbery polymer to a polymeric glass. The impact strength of the glass thus is substantially increased. The rubbery polymer is not truly compatible and exists as a finely dispersed separate phase. Interpenetrating polymer networks, in which two different networks are intertwined, constitute an interesting special case of polyblends. The phenomenon has somewhat similar counterparts in inorganic glass technology and in physical metallurgy. The study of polyblends combines the rigor of physics, for example, fracture mechanics, with organic and physical chemistry and the engineering behavior of the systems of interest.

FABRICATION

Polymer formulations can be fabricated into useful forms or articles by a variety of methods.

In the use of molded thermosetting compositions in which heat and pressure are required for the production of sound articles, some form of compression molding is usually employed. The physically compacted composition containing the resin in the fusible stage is forced into a mold cavity of the desired shape and is held under heat and pressure until the curing or vulcanization is complete. When high pressure is not required, as in the preparation of epoxy compositions and polyester-styrene-glass fiber compounds, and in various fast-curing resin laminates, it is still desirable to use moderate pressure to obtain a uniform molding.

Various forms of injection molding are used for the shaping of thermoplastic (permanently fusible) compositions. The composition is first softened temporarily by forcing the compounded resin granules through a heated chamber, after which it is driven into a relatively cold mold. Under proper conditions, the resin remains soft long enough to fill the mold completely and then rapidly hardens as it cools. The injection molding of engineering plastics often makes it possible to replace an assembly of metal parts by one plastic piece. Variations of the injection-molding process are used in the extrusion of films, rods, and pipe and in the spinning of fibers.

Blow molding, a process in which air is blown into a hot tube held in a mold, is used for the manufacture of bottles and other shapes. Certain articles are molded conveniently by rotating a hot mold filled with resin.

In extrusion, a hot melt is forced through a die with an opening shaped to produce the cross section desired. Extrusion and molding techniques may be combined to form foams into sheets and other shapes.

Films are also produced by extrusion of a tube into which air is forced. The process is called bubble extrusion. The pipe expands, because of the air pressure, to a wall thickness equivalent to the film thickness desired. The walls of the expanded bubble are pressed together in nip rolls, and later the large, thin-walled, collapsed pipe is slit to yield flat film. Films and sheets are also produced by calendering, in which the hot resin is forced between tightly fitted rolls.

A technique useful for shaping sheets into various shapes is thermoforming, in which a heated sheet is forced against the contours of a mold by a positive pressure or vacuum.

In the casting process, fluid compositions are poured into molds of the desired shape and then allowed to cool or cure. This process is used for the production of foams and in encapsulation, such as in the protection of electronic components or the mounting of biological specimens. In broad terms, coating, slush molding, and painting may be considered to be casting operations.

Some polymers can be formed by methods similar to those used in metallurgical forming. Forging and cold-forming are adaptable to the mass production of an increasing number of plastics. Sintering techniques and variations are also used for coating metals and forming polymers which cannot be formed by conventional processes.

In the shaping of thermoplastics particularly, the conditions of molding (temperature, time, and pressure) have a marked effect on the properties of the final product. Uneven cooling produces strains, and the flow of the plasticized resin can cause some orientation of the molecules. In bubble molding of film, biaxial orientation is produced. In the drawing of films and fibers, either uniaxial or biaxial orientation may be obtained with equipment similar to the tenters used in the textile industry. Orientation results in a substantial increase in the strength of the product in the direction of stretching; in many polymers, crystallization is induced during cold-drawing or stretching. On occasion, discontinuities or areas of strain are produced by partially orienting the molecules, and the product becomes more subject to stress cracking in the presence of solvents or other agents. Thus the optimum condition of a product results from a judicious combination of mechanical treatment and thermal annealing. *See* CHEMICAL BONDING; COLLOID. [JOHN A. MANSON]

Bibliography: J. J. Aklonis et al., *Introduction to Polymer Viscoelasticity*, 1972; F. W. Billmeyer, *Textbook of Polymer Science*, 1971; A. R. Blythe, *Electrical Properties of Polymers*, 1979; F. A. Bovey (ed.), *Macromolecules: An Introduction to Polymer Science*, 1979; J. A. Brydson, *Plastics Materials*, 1975; C. B. Bucknall, *Toughened Polymers*, 1977; P. J. Flory, *Principles of Polymer Chemistry*, 1953; P. H. Geil, *Polymer Single Crystals*, 1973; H. H. Kausch, *Polymer Fracture*, 1978; H. M. Mark et al. (eds.), *Encyclopedia of Polymer Science and Technology*, 1967– ; H. M. Mark et al. (eds.), *High Polymers*, vols. 1–29, 1949–1977; *Modern Plastics Encyclopedia*, 1975– ; L. E. Nielsen, *Mechanical Properties of Polymers and Composites*, 2 vols., 1974; F. Rodriguez, *Principles of Polymer Systems*, 1970; W. J. Roff and J. R. Scott, *Fibres, Films, Plastics, and Rubber*, 1971; R. J. Samuels, *Structured Polymer Properties*, 1974.

Polymeric composite

Any of the combinations or compositions that comprise two or more materials as separate phases, at least one of which is a polymer. By combining a polymer with another material, such as glass, carbon, or another polymer, it is often possible to obtain unique combinations or levels of properties. Typical examples of synthetic polymeric composites include glass-, carbon-, or polymer-fiber-reinforced thermoplastic or thermosetting resins, carbon-reinforced rubber, polymer blends, silica- or mica-reinforced resins, and polymer-bonded or -impregnated concrete or wood. It is also often useful to consider as composites such materials as coatings (pigment-binder combinations) and crystalline polymers (crystallites in a polymer matrix). Typical naturally occurring composites include wood (cellulosic fibers bonded with lignin) and bone (minerals bonded with collagen). On the other hand, polymeric compositions compounded with a plasticizer or very low proportions of pigments or processing aids are not ordinarily considered as composites.

Typically, the goal is to improve strength, stiffness, or toughness, or dimensional stability by embedding particles or fibers in a matrix or binding phase. A second goal is to use inexpensive, readily available fillers to extend a more expensive or scarce resin; this goal is increasingly important as petroleum supplies become costlier and less reliable. Still other applications include the use of some fillers such as glass spheres to improve processability, the incorporation of dry-lubricant particles such as molybdenum sulfide to make a self-lubricating bearing, and the use of fillers to reduce permeability.

Emphasis on the development of polymeric composites has been stimulated by the need for greatly improved mechanical and environmental behavior, especially on a strength- or stiffness-to-weight basis. Such composites are also often more efficient in their energy requirements for production than traditional materials. The high absolute and specific (per unit of weight) values of properties such as strength and stiffness have made composites ideal candidates for new applications in aircraft and boats, in passenger vehicles and farm equipment, and in machinery, tools, and appliances. Composites based on chemically resistant matrixes are used increasingly in chemical process equipment.

This article discusses mainly the principles of composite behavior with typical examples. For discussion of common thermosetting matrixes *see* POLYESTER RESINS; POLYETHER RESINS. For information on common thermoplastic matrixes, polyblends, polymer concretes, and reinforced rubber *see* POLYACRYLATE RESIN; POLYAMIDE RESINS; POLYMER; POLYSTYRENE RESIN; POLYVINYL RESINS.

Mechanical properties. The behavior of composites depends upon the volume fractions of the phases, their shape, and on the nature of the constituents and their interfaces. With anisotropic phases, the orientation with respect to the direction of stressing or exposure to permeants is also important. In general, given an appropriate preferred direction, the greater the anisotropy the greater the effect on a given property, at least up to some point. Thus, all high-modulus reinforcements will stiffen a lower-modulus matrix, but fibers and platelets are more effective than spheres; a similar role of shape holds for the ability to reduce permeability at right angles to the anisotropic particles. Anisotropic high-modulus inclusions invariably increase strength if adhesion is good, but the effect is more complex with particulate fillers such as spheres. Rubbery inclusions lower the stiffness of a high-modulus matrix, but may enhance toughness by stimulating a combination of localized crazing and shear deformation.

Many fibers, for example, glass and carbon, are very stiff and strong. However, the maximum strength of these brittle materials cannot be realized in practical objects because of a high sensitivity to the inevitable small cracks and flaws which are ordinarily present. Most polymers are much less sensitive to such flaws, even though they are inherently less strong. More energy is needed to fracture a polymer than a ceramic or glass; a crack tends to grow much less readily in a polymer.

If, for example, a mass of a strong but flaw-sensitive ceramic or glass is divided into many parts, typically into a fiber, and embedded in a polymer matrix, a growing crack may break one fiber but its progress may be hindered by the matrix or diverted along the interfaces. Thus, even though the matrix contributes only insignificantly to the total strength, it permits a closer approach to the theoretical maximum strength of the glass. Similar considerations apply to short fibers and, with some qualifications, other forms of reinforcement.

The matrix has several other functions besides the dissipation of energy which would otherwise cause a catastrophic failure. It protects the fiber against damage by mechanical action, such as rubbing, or by environmental agents, such as wa-

Epoxy composites in comparison to metals

Material	Tensile strength, psi (GPa)		Young's modulus, psi (GPa)	
	Actual	Relative to density	Actual	Relative to density
Fiber in epoxy laminate				
High-strength glass	47,000 (0.3)	0.8	5.6×10^6 (38)	80
Boron	83,000 (0.6)	1.2	30×10^6 (210)	430
Carbon	160,000 (1.1)	2.5	33×10^6 (230)	520
Metal				
Steel	320,000 (2.2)	1.2	30×10^6 (210)	109
Titanium	280,000 (1.9)	1.6	17.5×10^6 (120)	109

ter. It must also transfer an applied stress or force to the filaments so that they, being much stronger, can bear most of the load. In order to transfer the stress the matrix must adhere well to the fiber, though the optimum strength of the interfacial bond desired may vary depending on the application.

The strength of such a composite depends on the orientation of the fibers with respect to the applied force and on the nature of the stress (tensile or compressive). In the selection of materials and design for a composite, the ultimate application must be known. For example, strength of a composite based on longitudinally aligned fibers will be greatest in the direction of the fibers. Rupture under tension will require the pulling out of many fibers; this implies a fiber-matrix bond which can yield fairly readily, and in yielding increase the energy required for rupture. In compression, on the other hand, a stronger interfacial bond is needed to prevent buckling. Compromises in design are often necessary. Thus, at the expense of some strength, fibers are often crisscrossed in order to minimize the directionality of strength.

Reinforced thermosetting resins. The most common fiber-reinforced polymer composites are based on glass fibers, cloth, mat, or roving embedded in a matrix of an epoxy or polyester resin. Usually the glass surface is treated with a coupling agent to promote adhesion to the matrix. Fabrication may be effected in several ways. Frequently resin-dipped continuous filaments are wound on a mandrel in one of several patterns. Tapes or bundles of fibers coated with resin can also be positioned by hand or machine in layers, or pulled through a die. After the desired shape is obtained, curing is effected under heat and pressure.

Boron, polyaramid, and especially carbon fibers confer especially high levels of strength and stiffness. Hybrid fiber combinations, for example, glass with carbon, are often used to achieve a desired balance between cost and performance.

Some examples of properties for several experimental composites (unidirectional fibers) in comparison with metals are striking. In particular the table compares tensile strength and modulus for metals and composites. The figures show that relative properties of the polymer composites can exceed those of other materials. Although relative stiffness is less favorable for glass-fiber composites, carbon-fiber composites have a relative stiffness five times that of steel. Because of these excellent properties, many applications are uniquely suited for epoxy and polyester composites, such as components in new jet aircraft, parts for automobiles, boat hulls, rocket motor cases, and chemical reaction vessels.

Reinforced thermoplastic resins. Although the most dramatic properties are found with reinforced thermosetting resins such as epoxy and polyester resins, significant improvements can be obtained with many thermoplastics. Usually short fibers of glass are used—about 1/8 to 2 in. (3 to 50 mm) in length.

Nylon resins are particularly adaptable. Reinforcement of nylon-6,10 with 20% fiber glass raises the tensile strength and heat-deflection temperature from 8500 to 20,000 psi (59 to 138 MPa) and from 150 to 210°C, respectively. Polycarbonates, acetals, polyethylene, and polyesters are among the resins available as glass-reinforced compositions.

The combination of inexpensive, one-step fabrication, by injection molding, with improved properties has made it possible for reinforced thermoplastics to replace metals in many applications in appliances, instruments, automobiles, and tools.

Other composites. The development of other composite systems is receiving much attention. Various matrices are possible; for example, polyimide resins are excellent matrices for glass fibers, and give a high-performance composite. Different fibers are of potential interest, including polymers (such as polyvinyl alcohol), single-crystal ceramic whiskers (such as sapphire), and various metallic fibers. Many new developments may be anticipated in the future.

[JOHN A. MANSON]

Bibliography: L. J. Broutman and R. H. Krock (eds.), *Composite Materials*, 8 vols., 1974; L. J. Broutman and R. H. Krock (eds.), *Modern Composite Materials*, 1967; C. B. Bucknall, *Toughened Polymers*, 1978; H. T. Corten, in E. Baer (ed.), *Engineering Design of Plastics*, 1964; J. E. Gordon, *The New Science of Strong Materials*, 1976; A. Kelly, *Strong Solids*, 1966; J. A. Manson and L. H. Sperling, *Polymer Blends and Composites*, 1976; *Modern Plastics Encyclopedia*, 1975– ; L. E. Nielsen, *Mechanical Properties of Polymers and Composites*, 2 vols., 1974.

Polymerization

The linking of small molecules (monomers) to make larger molecules. Polymerization requires that each small molecule have at least two reaction points or functional groups. There are two distinct major types of polymerization processes, condensation polymerization, in which the chain growth is accompanied by elimination of small molecules such as H_2O or CH_3OH, and addition polymerization, in which the polymer is formed without the loss of other materials. For the many variants and subclasses of polymerization reactions, see the references in the bibliography.

An example of the condensation process is the reaction (1) of ε-aminocaproic acid in the presence

$$\underset{\text{ε-Aminocaproic acid}}{n\mathrm{H_2N{-}(CH_2)_5{-}\overset{O}{\overset{\|}{C}}{-}OH}} \xrightarrow{\text{Catalyst}}$$

$$\underset{\text{Polyamide, nylon-6}}{\mathrm{H}\left[\mathrm{\overset{H}{\overset{|}{N}}{-}(CH_2)_5{-}\overset{O}{\overset{\|}{C}}{-}}\right]_n\mathrm{OH} + (n-1)\mathrm{H_2O}} \qquad (1)$$

of a catalyst to form the polyamide, nylon-6. The repeating structural unit is equivalent to the starting material minus H and OH, the elements of water. A similar product would be obtained by reaction (2) of a diamine and a dicarboxylic acid.

$$\underset{\text{Hexamethylene diamine}}{n\mathrm{H_2N{-}(CH_2)_6{-}NH_2}} + \underset{\text{Adipic acid}}{n\mathrm{HO{-}\overset{O}{\overset{\|}{C}}{-}(CH_2)_4{-}\overset{O}{\overset{\|}{C}}{-}OH}} \rightarrow$$

$$\underset{\text{Polyamide, nylon-6,6}}{\mathrm{H}\left[\mathrm{\overset{H}{\overset{|}{N}}{-}(CH_2)_6{-}\overset{H}{\overset{|}{N}}{-}\overset{O}{\overset{\|}{C}}{-}(CH_2)_4{-}\overset{O}{\overset{\|}{C}}{-}}\right]_n\mathrm{OH} + (2n-1)\mathrm{H_2O}} \qquad (2)$$

In both cases, the molecules formed are linear because the total functionality of the reaction system (functional groups per molecule) is always two. However, if a trifunctional material, such as a tricarboxylic acid, were added to the nylon-6,6 polymerizing mixture, a branched polymeric structure would result, because two of the carboxylic groups would participate in one polymer chain, and the third carboxylic group would start the growth of another. Under appropriate conditions, these chains can become bridges between linear chains and the polymer becomes cross-linked. The arrangements of the chains are shown in Fig. 1.

Some examples of addition polymerization are reactions (3) and (4). The structure of the repeating

$$\underset{\text{Formaldehyde}}{n\mathrm{H_2C{=}O}} \xrightarrow[\text{Initiator (I)}]{\text{Catalyst or}} \underset{\text{Polyoxymethylene}}{[\mathrm{{-}CH_2{-}O{-}}]_n} \qquad (3)$$

$$\underset{\text{Ethylene}}{n\mathrm{CH_2{=}CH_2}} \xrightarrow[\text{Initiator (I)}]{\text{Catalyst or}} \underset{\text{Polyethylene}}{[\mathrm{{-}CH_2{-}CH_2{-}}]_n} \qquad (4)$$

unit is the difunctional monomeric unit, or "mer." In the presence of catalysts or initiators, the monomer yields a polymer by the joining together of n mers. If n is a small number, 2–10, the products are dimers, trimers, tetramers, or oligomers, and the materials are usually gases, liquids, oils, or brittle solids. In most solid polymers, n has values ranging from a few score to several hundred thousand, and the corresponding molecular weights range from a few thousand to several million. The end groups of these two examples of addition polymers are shown to be fragments of the initiator.

If only one monomer is polymerized, the product is called a homopolymer. The polymerization of a mixture of two monomers of suitable reactivity leads to the formation of a copolymer, a polymer in which the two types of mer units have entered the chain in a more or less random fashion. If chains of one homopolymer are chemically joined to chains of another, the product is called a block or graft copolymer:

A—B—A—A—B—B—B
Random copolymer

A—A—A—A—B—B—B—B—B
Block copolymer

A—A—A—A—A (with B—B—B attached to the third A)
Graft polymer

Isotactic and syndiotactic (stereoregular) polymers are formed in the presence of complex catalysts, or by changing polymerization conditions, for example, by lowering the temperature. The groups attached to the chain in a stereoregular polymer are in a spatially ordered arrangement. The configuration of these ordered polymers and the disordered, atactic form is shown in Fig. 2. The regular structures of the isotactic and syndiotactic forms make them often capable of crystallization. The crystalline melting points of isotactic polymers are often substantially higher than the softening points of the atactic product.

In Fig. 2 each carbon atom to which a phenyl group is attached is asymmetrically substituted. For illustration, the heavily marked bonds are assumed to project up from the paper, and the dotted bonds down. Thus in a fully syndiotactic polymer, asymmetric carbons alternate in their left- or right-handedness (alternating d, l configurations), while in an isotactic polymer, successive carbons have the same steric configuration (d or l). Intermediate degrees of stereo-regularity exist, and an atactic polymer is one in which the opposite steric configurations occur in equal frequency and more or less at random. In more complex molecules, more complex stereoregular configurations are possible.

In dienes, cis-trans isomerism may also occur, as well as isomerism with respect to which double bond is opened during polymerization.

Infrared and nuclear magnetic resonance spectroscopy may be used to characterize stereoregularity and other types of isomerism.

The production of *cis*-polyisoprene, "synthetic natural rubber," is also an example of stereoregular polymerization.

CONDENSATION (STEP-GROWTH) POLYMERIZATION

The formation of macromolecules by the condensation process, as in the production of polyamides, polyesters, and polysulfides, requires the elimination of small molecules as noted above; at the same time, strongly polar and strongly attracting groups are produced. Most, though not all, condensations proceed in a stepwise fashion. In the equimolar, stepwise reaction of a dicarboxylic acid

POLYMERIZATION

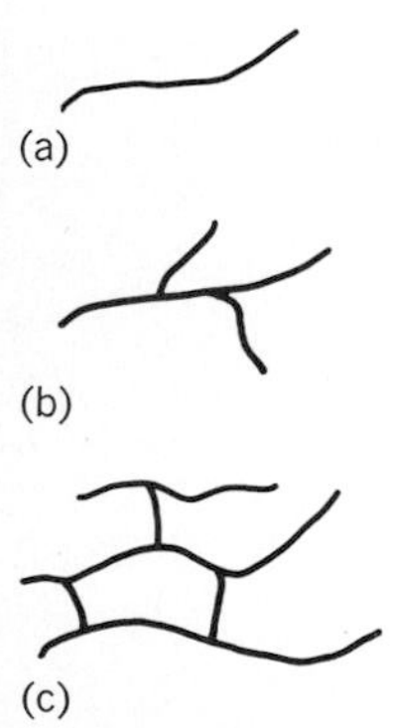

Fig. 1. Polymer chains. (*a*) Linear polymer chain. (*b*) Branched polymer chain. (*c*) Cross-linked polymer chain.

with a diamine, as an example, after 50% of the groups have reacted, the average degree of polymerization (DP) of the polymer formed is 2; after 90% reaction, the DP is 10, and after 99.5% reaction, the DP is 200. If the molecular weight of the repeating unit is 100, then the average molecular weight of the polymer is 20,000. In order to obtain a high-molecular-weight product, the small molecules formed during the condensation must be removed in order for the reaction to approach 100% completion. In practice, the condensations are usually started under moderate conditions of temperature and pressure and completed at high temperature and low pressure to yield linear products in the molecular-weight range of about 5000–30,000. In certain cases, interfacial polymerization is useful. If reactants can be dissolved in immiscible solvents, polymerization may take place at low temperatures and resources. The linear products, thermoplastic condensation resins, are used in fibers, films, coatings, molding compounds, and adhesives.

Useful condensation polymers which are highly cross linked are prepared from low-molecular-weight polyfunctional reaction systems. The condensations of formaldehyde with phenol and with urea, and other condensations, are discussed in the articles on specific products. The reaction of formaldehyde with urea is shown in (5).

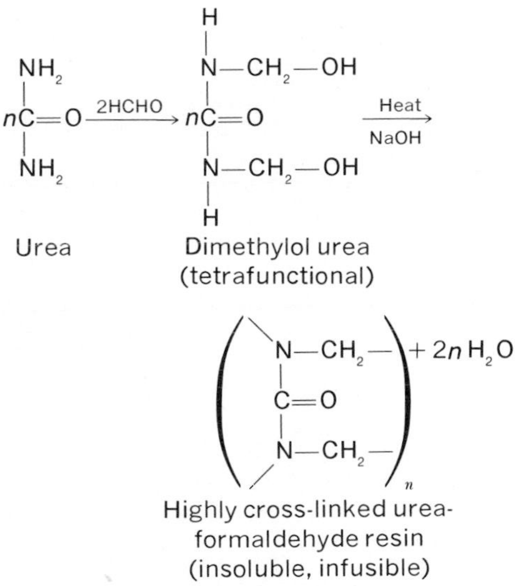

(5)

In the intermediate compound, dimethylol urea is tetrafunctional. The H— and HO— groups on one molecule react with the HO— and H— groups on other molecules to form water. Each urea unit finally becomes bound to other urea units via four methylene (—CH_2—) bridges. Because the final product is cross-linked and infusible, the final shaping operation usually must coincide with the final curing or cross-linking. In practice, the soluble, low-molecular-weight intermediate condensate can be isolated and mixed with strengthening fillers, coloring pigments, and curing catalysts to yield a molding powder. By subjecting the urea-formaldehyde molding powder to heat and pressure in a mold, the curing reaction takes place, some of the water is driven off as steam, and some is adsorbed by the filler. The molded object, such as a piece of dinnerware, is now insoluble and infusible. Because all the molecules are joined

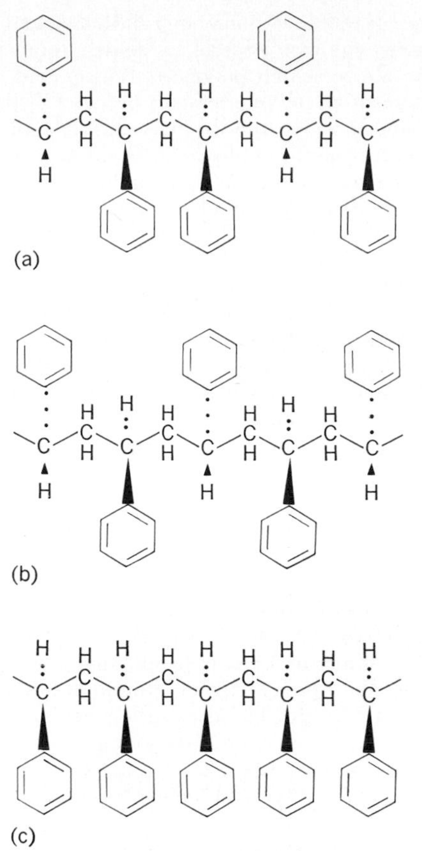

Fig. 2. Spatially oriented polymers. (*a*) Atactic (random; *dlldl* or *lddld*, and so on). (*b*) Syndiotactic (alternating; *dldl*, and so on). (*c*) Isotactic (right- or left-handed; *dddd*, or *llll*, and so on).

together, the molecular weight of a highly cross-linked polymer is a meaningless term since its value is effectively infinite.

The end groups of the polymer molecules are the functional groups that have not reacted at any stage. It is apparent that at exactly 100% conversion in a difunctional condensation, the reaction system would consist of only one molecule. The fact that the end groups of a condensation polymer can always undergo further reaction creates a difficulty in the high-temperature-melt spinning of polyamides and polyesters. To prevent subsequent changes in molecular weight, such as might occur in the melt during spinning at elevated temperatures, a monoacid or monoalcohol (molecular-weight modifier) is added to the original polymerization mixture. The excess of hydroxyl groups, for example, places a limit on the chain growth. That limit is reached when all the acid groups have reacted and all the end groups are hydroxyl. *See* CONDENSATION REACTION.

ADDITION (CHAIN-GROWTH) POLYMERIZATION

Unsaturated compounds such as olefins and dienes polymerize without the elimination of other products. Certain ring structures, for example, lactams or alkylene oxides, may polymerize by an opening of the ring. The molecular weight and structure of the polymer are determined by the reaction conditions, that is, the nature of the catalyst or initiator, the temperature, and the concentration of reactants, monomer, initiator, and modifying agents. In general, a ceiling temperature exists above which polymerization cannot occur. Most addition polymerizations proceed by a chain reaction in which the average chain length of the polymer formed initially is high and may increase further through secondary branching reactions as the polymerization approaches completion.

The molecular-weight range for many useful addition polymers is relatively high, typically, from 20,000 to several million, as compared with the molecular weight range of 5000–30,000 for typical condensation polymers.

The types of catalysis or initiation which are effective for addition (chain-growth) polymerization fall into four groups: (1) free-radical catalysis by peroxides, persulfates, azo compounds, oxygen, and ultraviolet and other radiation; (2) acid (cationic) catalysis by the Lewis acids, such as boron trifluoride, sulfuric acid, aluminum chloride, and other Friedel-Crafts agents; (3) basic (anionic) catalysis, by Lewis bases such as alkali metals and metallic alkyls; and (4) heterogeneous (coordinate) catalysis, by chromic oxide on silica-alumina, nickel, or cobalt on carbon black, molybdenum on alumina, and complexes of aluminum alkyls with titanium chloride. The precise mechanisms of the fourth group are difficult to elucidate. However, evidence suggests that, at least in some cases, the mechanism involves an anionic process modified by coordination of the monomer and polymer with a surface.

It is convenient to discuss the mechanism and experimental methods of free-radical initiation as one subject, and to treat the remaining three types under the heading complex or ionic catalysts. *See* CATALYSIS.

Free-radical catalysis. Among the several kinds of polymerization catalysis, free-radical initiation has been most thoroughly studied and is most widely employed. Atactic polymers are readily formed by free-radical polymerization, at moderate temperatures, of vinyl and diene monomers and some of their derivatives.

At an appropriate temperature, a peroxide decomposes to yield free radicals. In the presence of a monomer, the greater proportion of these radicals adds to the monomer and thereby initiates chain growth. The growing chains may terminate by coupling, by disproportionation, or by transfer with monomer, polymer, or added materials (transfer agents, retarders, and inhibitors). Reactions (6)–(10) illustrate the initiation, propagation, and termination by coupling in vinyl acetate.

If transfer occurs with the unreacted monomer or polymer already formed, higher-molecular-weight branched structures will be produced, and if branching is excessive, insoluble products may be formed. If the radical produced in the transfer process is not sufficiently active to initiate a new

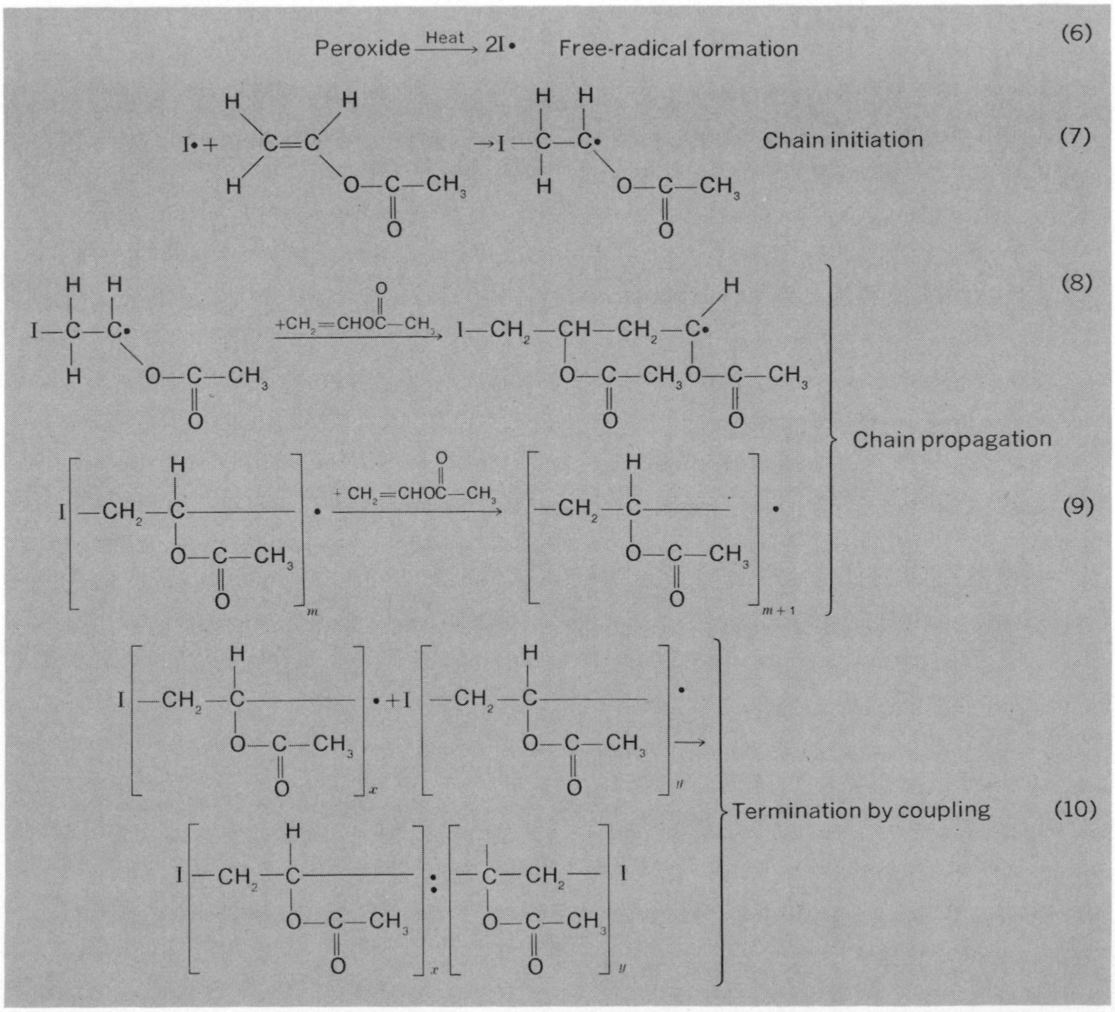

chain, the transfer agent is called an inhibitor or a retarder. Mercaptans (RSH), carbon tetrachloride, and various organic solvents are examples of transfer agents, whereas amines and phenols are frequently used as inhibitors or retarders.

The rate of chemically initiated free-radical polymerization is increased by raising the temperature or increasing the concentration of monomer and initiator, whereas the molecular weight of the polymer is increased by increasing the monomer concentration, by lowering the temperature, and by lowering the concentration of initiator and transfer agents. *See* FREE RADICAL.

Complex or ionic catalysis. Some polymerizations can be initiated by materials, often called ionic catalysts, that contain highly polar reactive sites or complexes. The term heterogeneous catalyst is often applicable to these materials because many of the catalyst systems are insoluble in monomers and other solvents. These polymerizations are usually carried out in solution from which the polymer can be obtained by evaporation of the solvent or by precipitation on the addition of a nonsolvent.

A general mechanism is shown in reactions (11),

Initiative complex:
Complex catalyst + M → *CM**

Initiation:
$$CM^* + M \rightarrow M^*CM \qquad (11)$$

Termination by decomposition of complex:
$$M_x{}^*CM \rightarrow M_x + CM$$

Termination by transfer to monomer:
$$M_x{}^*CM + M \rightarrow M_x + M^*CM$$

in which the growing chain is represented as an activated complex with the complex catalyst, without attempting to specify whether separate ions or free radicals are involved.

A distinguishing feature of complex catalysts is the ability of some representatives of each type to initiate stereoregular polymerization at ordinary temperatures or to cause the formation of polymers which can be crystallized. The polymerization process may often be visualized as the formation of an activated complex of the monomer with the complex catalyst. For stereoregular growth to take place, the entering monomer must collide with the complex and form a new, transient complex in which the new monomer molecule is held in a particular orientation. As reaction takes place, the new monomer assumes an activated condition within the complex catalyst and, at the same time, pushes the old monomer unit out. Chain growth is therefore similar to the growth of a hair from the skin. If conditions favor certain orientations of the old and new monomer units, a stereoregular polymer results. *See* STEREOSPECIFIC CATALYST.

The effect of conditions on rates of polymerization and on molecular size and structure is not yet fully understood. In general, the rate of polymerization is proportional to the concentrations of complex catalyst and monomer. The effect of temperature on the rate depends upon the stability and activity of the complex catalyst at the temperature under consideration. If the complex catalyst decomposes on increasing the temperature, then the rate of polymerization will be reduced. The effect of temperature upon molecular weight also depends upon the stability of the complex catalyst and upon the relative rates of propagation and termination. In some cases, at an optimum temperature of polymerization, the molecular weight depends upon the product of the ratio of the rate of propagation to termination and the monomer concentration, and in other cases, only upon that ratio of rates.

Examples of polymerization with the different types of complex catalysts are briefly described below.

Lewis acids. Carbonium-ion or cationic catalysts such as BF_3, $AlCl_3$, or H_2SO_4 usually require the presence of a promoter such as H_2O or HCl, as in reaction (12). The point of chain growth is a carbonium (positive) ion, which is associated with a negative counter ion. Polymerization in the presence of Lewis acids takes place very rapidly at low temperatures, −100 to 0°C. The order of reactivity of some olefins in Lewis acid catalysis is vinyl ethers > isobutylene > α-methyl styrene > isoprene > styrene > butadiene.

$$\underset{\text{Complex catalyst}}{BF_3\cdot H_2O} + \underset{\text{Isobutylene}}{CH_2{=}C(CH_3)_2} \rightarrow \underset{\text{Initiating complex}}{CH_3C^{+}(CH_3)_2\,{}^{-}BF_3\cdot OH} \qquad (12)$$

Lewis bases. Carbanions of anionic catalysts such as sodium, lithium, and lithium butyl function at moderate to low temperatures, −70 to +150°C. Inert hydrocarbon or ether solvents are generally used as reaction media. The reaction is shown as (13).

$$\underset{\text{Lithium butyl}}{LiC_4H_9} + \underset{\text{Ethylene}}{CH_2{=}CH_2} \rightarrow \underset{\text{Initiating complex}}{(C_4H_9{-}CH_2{-}CH_2^{-})^{+}Li} \qquad (13)$$

It has been suggested that the point of chain growth is a negative or anionic center, although not usually disassociated from its positive counter ion. The order of reactivity of some monomers in carbanion or anionic polymerization is acrylonitrile > methacrylonitrile > methyl methacrylate > styrene > butadiene. *See* HOMOGENEOUS CATALYSIS.

Heterogeneous catalysts. Certain heavy metals or metal oxides on supports and complexes of aluminum alkyls with titanium chloride function as catalysts at moderate to high temperatures, 50–220°C. Inert hydrocarbon or ether solvents are generally used as reaction media. The catalysts may be used in a fixed bed or as a slurry. Two examples are shown as reactions (14) and (15).

$$\underset{\text{Propylene}}{nCH_2{=}CH(CH_3)} \xrightarrow[100-130^{\circ}C]{AlR_3-TiCl_3} \underset{\text{Isotactic polypropylene}}{\left(-CH_2-CH(CH_3)-\right)_n} \qquad (14)$$

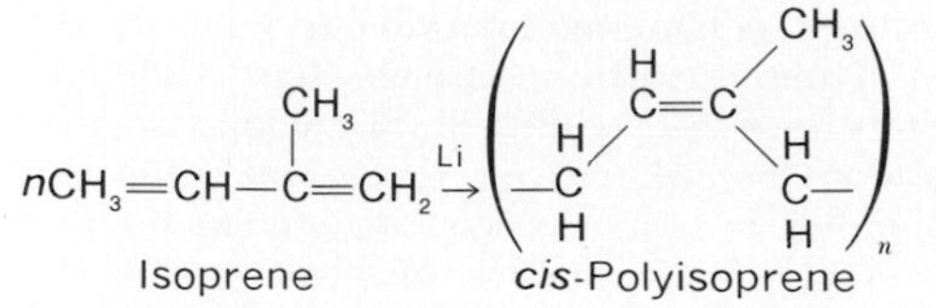

(15)

See HETEROGENEOUS CATALYSIS.

Polymerization processes. There are a number of processes for achieving polymerization, designed for various conditions and end products.

Bulk process. This consists of polymerization of the pure monomer in liquid form. On initiation by heat or light or very small amounts of azobisisobutyronitrile, a very pure polymer can be formed. The monomer and polymer are poor heat conductors; therefore the temperature of bulk polymerization is difficult to control. A further disadvantage is that small quantities of unreacted monomer are difficult to remove from the polymer. Polymerization in solution offers a means of carrying out the polymerization at lower monomer concentrations. Because solvents frequently act as transfer agents, polymerization in solution generally leads to the formation of lower-molecular-weight products.

Aqueous emulsion. This type of polymerization has the advantages of giving a high rate of polymerization, a high molecular weight, and ease of temperature control. A liquid monomer is emulsified in water by use of a surface-active agent, such as soap. The soap micelles provide the polymerization centers. The free radicals (from a water-soluble initiator or growing chains of low molecular weight) diffuse into the soap micelle in which they react to form a relatively linear polymer of high molecular weight. The polymer particles of small diameter, 50–150 nm, are in stable suspension because the soap of the original micelle remains adsorbed in the outer layer of the polymer particle. The rate of emulsion polymerization and the molecular weight of the polymer increase with increasing numbers of micelle particles per unit volume. The product, a stable colloidal suspension of the polymer in water, usually is called a polymer latex or polymer emulsion. Polymer latexes are used directly for water-based paints, for adhesives, and for treating textiles. When polar solvents or electrolytes are added to the colloidal suspension, the polymer coagulates, and it can be separated and dried. In order to produce polymer emulsions which have the desired mechanical and thermal stability, it is frequently necessary to use moderately high concentrations of surface-active agents and protective colloids. Therefore, emulsion polymers are generally less pure than bulk polymers.

Redox initiation. This system was developed for polymerization in aqueous emulsions. In the presence of a water-soluble reducing agent such as sodium bisulfite or ferrous sulfate, the peroxide decomposes more rapidly at a given temperature, and consequently polymerization at useful rates can take place at lower temperatures. By the use of the redox system, the temperature for the commercial emulsion polymerization of styrene and butadiene is lowered from 50 to 5°C to form "cold" rubber, which is higher in molecular weight than the "50°" or "hot" synthetic rubber.

Suspension polymerization. This system offers several advantages. By the use of a very small amount of surface-active materials and mechanical aggregation, the monomer can be dispersed as droplets in the water. The monomer is not colloidally dispersed, but is temporarily broken up into droplets which would coagulate if the stirring were discontinued. In the presence of a peroxide which is soluble in the monomer, bulk polymerization takes place in the droplets. When the polymerization reaches some 15–40%, the droplets containing the dissolved polymer become sticky and may coalesce. Various agents, such as talc and metallic oxides, have been recommended for use in very small amounts to prevent coagulation in the sticky stage. At higher degrees of conversion, the droplets will be transformed to hard balls of polymer containing dissolved monomer. At the completion of the polymerization, the balls or beads settle out. The beads may be dried easily and are ready for use. In the suspension system, the possibility of producing a pure polymer in bulk polymerization is combined with the ease of temperature control in the aqueous emulsion polymerization. There is the additional advantage that the product of suspension polymerization can be easily isolated for use.

Ring opening. In this type, some cyclic monomers can undergo ring opening to yield high polymers; the reaction may proceed by a chain- or step-growth mechanism. Commercially important products are the polymerization of caprolactam, caprolactone, and ethylene oxide.

See ACID AND BASE; ALKENE; CHAIN REACTION; CHEMICAL DYNAMICS; INHIBITOR; INORGANIC POLYMER; ORGANIC REACTION MECHANISM; OXIDATION-REDUCTION; POLYAMIDE RESINS; POLYESTER RESINS; POLYETHER RESINS.

[JOHN A. MANSON]

Bibliography: J. A. Brydson, *Plastics Materials*, 1975; R. W. Lenz, *Organic Chemistry of Synthetic High Polymers*, 1967; H. F. Mark et al. (eds.), *Encyclopedia of Polymer Science and Technology*, 1967; H. Mark et al. (eds.), *High Polymers*, vols. 1–29, 1940–1977; H. Mark and E. H. Immergut (eds.), *Polymer Reviews*, vols. 1–16, 1958–1967 (and successive volumes); *Modern Plastics Encyclopedia*,, 1979– ; G. Odian, *Principles of Polymerization*, 1969; W. J. Roff and J. R. Scott, *Fibres, Films, Plastics and Rubbers*, 1971; C. E. Schildknecht, *Vinyl and Related Polymers*, 1952.

Polynuclear hydrocarbon

One of a class of hydrocarbons possessing more than one ring. The aromatic polynuclear hydrocarbons may be divided into two groups. In the first, the rings are fused, which means that at least two carbon atoms are shared between adjacent rings. Examples are naphthalene (I), which has two six-membered rings, and acenaphthene (II), which has two six-membered rings and one five-membered ring.

In the second group of polynuclear hydrocarbons, the aromatic rings are joined either directly, as in the case of biphenyl (III), or through a chain

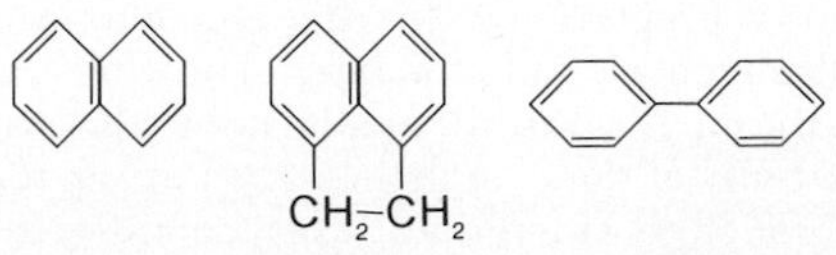

(I) (II) (III)

of one or more carbon atoms, as in 1,2-diphenylethane (IV).

$$C_6H_5CH_2CH_2C_6H_5$$
(IV)

The higher-boiling polynuclear hydrocarbons found in coal tar or in tars produced by the pyrolysis or incomplete combustion of carbon compounds are frequently fused ring hydrocarbons, some of which may be carcinogenic. *See* ANTHRACENE; AROMATIC HYDROCARBON; BIPHENYL; DIPHENYLMETHANE; HEXAPHENYLETHANE; INDENE; NAPHTHALENE; PHENANTHRENE; TRIPHENYLMETHANE. [C. K. BRADSHER]

Polyolefin resins

Polymers derived from unsaturated hydrocarbons containing the ethylene or diene groups. Broadly, polyolefin resins may include virtually all addition polymers; however, the term polyolefin is specifically used for polymers of ethylene, the alkyl derivatives of ethylene (the α-olefins), and the dienes.

The polyvinyl resins, the fluorocarbon polymers, and other addition polymers are covered in other articles. This article includes discussions of the polymers of ethylene, propylene, and isobutylene, and brief mention is made of polymers of other α-olefins and of butadiene, isoprene, and 2-chlorobutadiene. *See* POLYFLUOROOLEFIN RESINS; POLYMERIZATION; POLYVINYL RESINS.

Polyethylene. Polyethylene is a whitish, translucent polymer of moderate strength and high toughness. The available forms range in crystallinity from 50–95%. The physical properties vary markedly with the degree of crystallinity and with the size and distribution of the crystalline regions. The densities d of the products increase with increasing degrees of crystallinity, and it is common to classify the commercial grades as low density ($d \sim 0.92$), medium density ($d \sim 0.93-0.94$), or high density ($d \sim 0.94-0.97$). With increasing crystallinity or density, the products generally become stiffer and stronger, and have higher softening temperatures and higher resistance to penetration by liquids and gases; at the same time, they lose some of their resistance to tear, impact, and stress cracking, and higher temperatures and pressures are needed for molding.

Polyethylene of all types is produced in greater volume than any other plastic. The major uses are as packaging films, containers and bottles, molded articles, electrical insulation, coatings, and pipe. Some foamed products are also made.

Ethylene is produced economically on a large scale by the cracking of aliphatic hydrocarbons found in petroleum. The monomer can be conveniently produced in smaller volumes by the catalytic dehydration of ethanol.

The lower-density polymers are formed by the polymerization of highly purified ethylene at about 150–250°C and 20,000–35,000 psi in the presence of a very small amount of oxygen or organic peroxide. This is represented by reaction (1). At the

$$nCH_2{=}CH_2 \xrightarrow[\text{Oxygen}]{\substack{200°\text{C} \\ 25{,}000\text{ psi}}} \underset{\text{Polyethylene}}{(-CH_2-CH_2-)_n} \quad (1)$$

higher temperatures, the low-density polymer is formed and at the lower reaction temperatures, a medium-density product is produced.

The high-density polymers are formed at relatively low temperatures and pressures (for example, 50–150°C and 100–2000 psi) in the presence of special catalysts, often referred to as stereospecific catalysts. These include oxidized forms of certain heavy metals such as chromium, reduced forms of heavy metals such as cobalt and nickel, and the Ziegler catalyst, usually a complex of an aluminum alkyl and a titanium chloride. Medium-density polymers can also be made by a low-pressure process.

For the low-density material, the softening temperature and the maximum temperature for continuous use are about 105–115°C and 75°C, respectively; the corresponding temperatures for the high-density product are some 25–40°C higher.

Structural studies have shown that the higher-density polymers have highly linear structures and are approximately 85–95% crystalline. The lower-density materials are usually relatively branched (10–30 per 1000 carbon atoms) and are 50–85% crystalline, though newer processes yield more linear polymers. *See* ETHYLENE.

Molecular weight and its distribution are important in determining both processing behavior and applications. For the extrusion of film and blow molding, a high molecular weight is preferred, while a lower molecular weight is suitable for injection molding. A combination of high density with an ultrahigh molecular weight brings polyethylene into the class of other engineering plastics which have superior mechanical properties. Also, by special drawing and orientation processes, exceptional moduli and strengths can be obtained, at least in the laboratory. Very-low-molecular-weight resins, often in emulsion form, are useful in coating formulations and in polishes and finishes.

Polyethylene can be cross-linked by irradiation (electron, gamma, or x-radiation) or by free-radical catalysts such as peroxides. Controlled levels of cross-linking result in improved resistance to stress cracking and heat without sacrifice of good electrical and mechanical properties. Such cross-linked resins are especially useful as wire and cable insulation. Properties can also be modified by the use of fillers and fibrous reinforcements.

Copolymers of ethylene. A wide range of ethylene copolymers can be prepared. Of these, copolymers with 20–30% of vinyl acetate are particularly important. The introduction of such amounts of a bulky monomer results in increased flexibility and clarity and a high degree of compatibility with waxes. Thus, these copolymers are widely used in flexible coating formulations for food and cosmetic packages such as milk cartons.

Several types of copolymers with propylene have been made possible by stereoregular polymerization. Atactic copolymers behave like typical rubbers after vulcanization and reinforcement. The ethylene-propylene rubbers are low in price and have excellent resistance to aging at high temperatures. More stereoregular copolymers, which comprise blocks of ethylene units coupled to blocks of propylene units, are called polyallomers. These polyallomers are useful in applications which require resistance to stress cracking and to flexural fatigue.

An unusual class of copolymer with an ionizable comonomer is the family of "ionomers." As shown by reaction (2), ionomers are prepared by copolym-

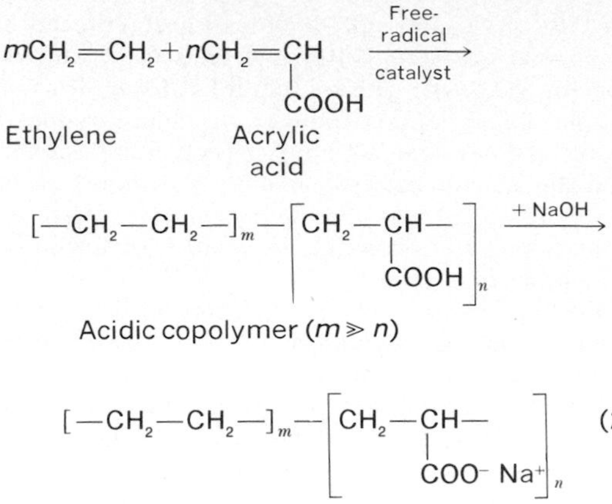

(2)

Copolymer salt, or ionomer

erization of ethylene with small amounts (3–5 mole %) of an unsaturated carboxylic acid, such as acrylic or methacrylic acid, and then ionization of the acid group to yield a metal salt. The ionized groups act as meltable cross-links and confer improved toughness at all temperatures and optical clarity.

Polypropylene. High-molecular-weight, isotactic, highly crystalline polypropylene is generally similar in properties to high-density polyethylene. In comparison with the latter, isotactic polypropylene is harder and stronger, and softens at about 160°C.

Propylene is available at low cost in large quantities from the cracking of petroleum hydrocarbons, and the high-molecular-weight isotactic polymers are formed in the presence of the stereospecific catalyst used in the ethylene polymerization, as in reaction (3).

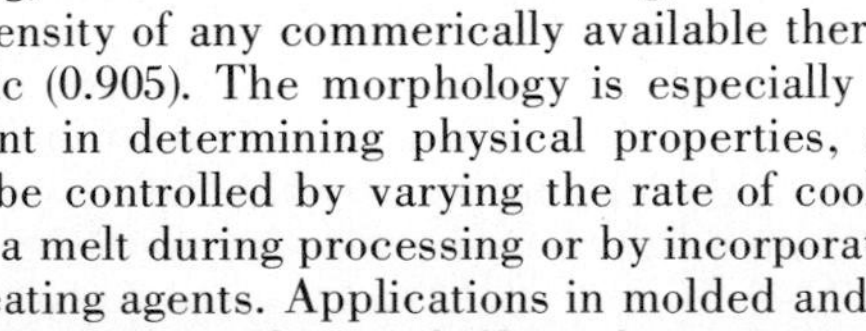

(3)

Polypropylene

The crystalline products, which are about 90–95% isotactic, have an unusual combination of properties: high strength, exceptional resistance to flexing, resistance to stress cracking, and the lowest density of any commerically available thermoplastic (0.905). The morphology is especially important in determining physical properties, and may be controlled by varying the rate of cooling from a melt during processing or by incorporating nucleating agents. Applications in molded and extruded articles, film, and fibers have increased very rapidly.

In order to reduce the tendency toward embrittlement at low temperatures, or to enhance the dyeability of fibers, copolymerization with ethylene or other monomers is frequently advantageous. Block copolymers with small amounts of ethylene (polyallomers) have been discussed above. The wide range of possible copolymer compositions makes it possible to tailor particular compositions and formulations for specific uses. Reinforcements such as asbestos or glass fibers may be used to improve properties such as stiffness.

Fabrication is usually accomplished by extrusion, injection molding, or blow molding. A wide range of applications is possible, including automobile and appliance parts, wire insulation, pipe, and protective liners. The unusual flex resistance makes it possible to mold polypropylene as an integral hinge. Ease of electroplating has led to considerable use for decorative items.

The use of both unoriented and oriented films of homopolymers and copolymers has grown rapidly in such applications as packaging for textiles and food, liners for bags, and heat-shrinkable wraps for records and other articles. Orientation improves strength but decreases tearability and requires coating in order that the film may be heat-sealed.

Many types of fiber products are available, and their high tenacity and good abrasion resistance have led to acceptance for such uses as indoor-outdoor carpeting materials, ropes, and nets.

The low-molecular-weight polypropylene oils formed in the presence of acid catalysts, such as boron trifluoride or phosphoric acid, are useful in the manufacture of gasoline and synthetic detergents, but are not employed in plastics technology. *See* PROPYLENE.

Polyisobutylene. The polyisobutylene polymers vary in properties from low-molecular-weight oils to high-molecular-weight rubbery solids.

The monomer, obtained by cracking petroleum hydrocarbons, readily polymerizes in the presence of acid catalysts, such as boron trifluoride, aluminum chloride, or tin(IV) tetrachloride. This is represented by reaction (4).

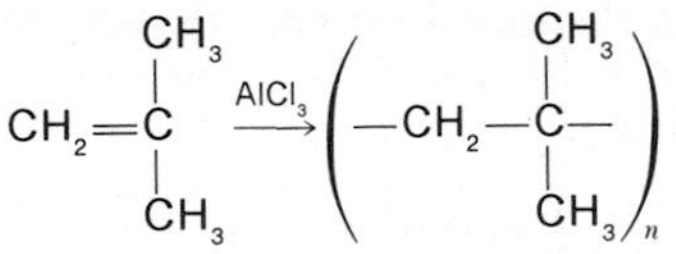

(4)

Polyisobutylene

Polymerization conducted at 0–25°C yields oils which are useful in calking and sealing compositions. At low temperatures, such as −100 to −80°C, rubbery solids are formed. The solids are also useful in calking compositions and adhesive formulations; however, the main use of polyisobutylene is in the form of the copolymer with 2–4% isoprene. The copolymer, known as butyl rubber, can be prepared at −90°C in the presence of methyl chloride as a diluent and aluminum chloride as the catalyst. The product, distinguished by its impermeability to gases and its resistance to aging, is used in automobile tires and tubes.

Polymers of other α-olefins. The discovery of the stereospecific catalysts listed for ethylene polymerization has made possible the formation of high-molecular-weight, isotactic, crystalline polymers of other α-olefins, such as α-butene, α-octene, and α-dodecene. Typical melting temperatures (mp) of the crystalline, isotactic polymers of some of the α-olefins, for example, structures (I), are relatively high. The high melting points of certain of the isotactic crystalline polymers may result from special spatial arrangements necessary to accommodate the presence of bulky branched groups near the polymeric chain. The high-melting

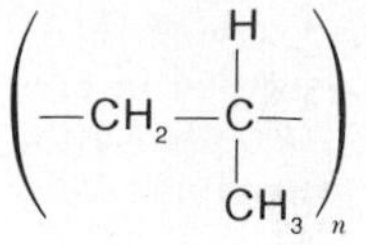
Polypropylene (mp 136°C)

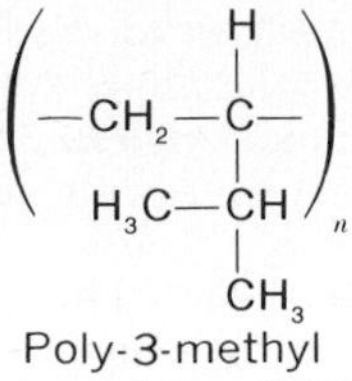
Poly-3-methyl butene-1 (mp 240°C)

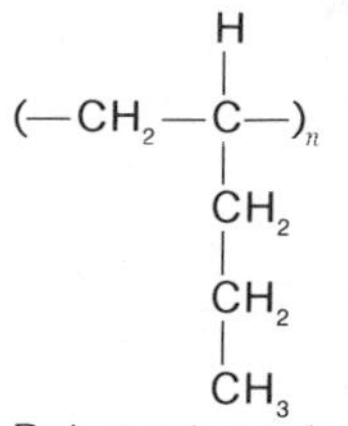
Polypentene-1 (mp 80°C)

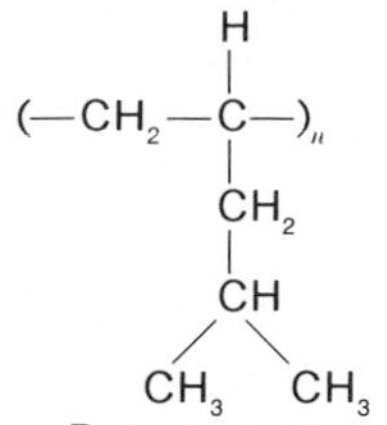
Poly-4-methyl pentene-1 (mp 325°C)

(I)

polymers are of considerable interest because relatively few thermoplastic polymers are available which have high softening temperatures and at the same time can be easily fabricated. Polybutene-1, which is unusually resistant to creep, has shown promise as a material for films and piping. Since poly-4-methylpentene-1 is transparent and resistant to many chemicals as well as to high temperature, it is useful in such applications as chemical and pharmaceutical items and trays for microwave ovens.

Polydienes. Butadiene, isoprene, and 2-chlorobutadiene have gained the widest use of dienes employed in the polymer field. Butadiene and 2-chlorobutadiene have long been used in the production of synthetic rubbers by free-radical catalysis. The isoprene structure has long been recognized to correspond to the repeating unit in natural rubber. Copolymers with styrene and acrylonitrile are discussed separately. *See* POLYACRYLONITRILE RESINS; POLYSTYRENE RESIN.

Special stereospecific catalysts are useful in the polymerization of butadiene and isoprene. Natural rubber consists largely of *cis*-polyisoprene, structure (II), and is characterized by high elasticity and

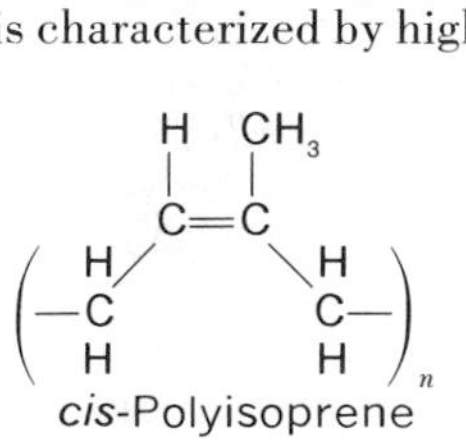
cis-Polyisoprene (II)

low internal friction (on flexing). The polymers of butadiene and isoprene formed by free-radical catalysis contain mixtures of the cis and trans forms, together with some 1,2 or 3,4 structures in the case of polyisoprene, or a mixture of all four forms. The presence of the trans and 1,2 and 3,4 structures shown in (III) causes the rubber to have a lower elasticity and a higher internal friction on flexing.

The application of the stereospecific catalysts to the polymerization of isoprene and butadiene has led to the development of synthetic rubbers which contain high proportions of the cis structure and which are essentially equivalent in properties to

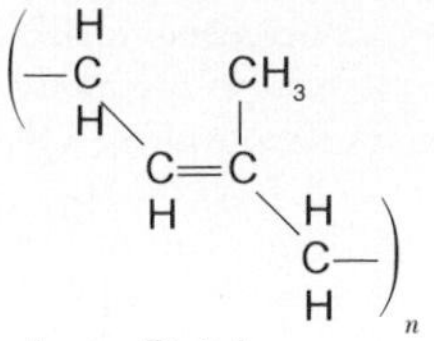
trans-Polyisoprene

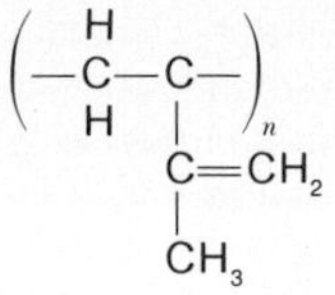
1,2-Polyisoprene (III)

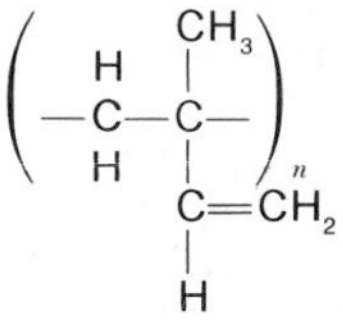
3,4-Polyisoprene

natural rubber. Copolymers of butadiene with styrene can be prepared in such a way that the molecules contain carboxyl end groups. These reactive rubbers are used in specialized applications such as toughening agents for epoxy resins. *See* ALKENE; ISOPRENE.

[JOHN A. MANSON]

Bibliography: J. Brydson, *Plastics Materials*, 1975; *Modern Plastics Encyclopedia*, 1979–1980; R. A. V. Raff and K. W. Doak (eds.), *Crystalline Olefin Polymers*, High Polymers Series, vol. 20, 2 pts., 1964–1965.

Polyoxyethylation of alcohol

The process of reacting an alcohol with ethylene oxide to produce a polyether, as in the reaction below, in which R of the alcohol may be aliphatic

$$ROH + n\,CH_2\overset{O}{—}CH_2 \rightarrow RO—(CH_2CH_2O)_nH$$

or aromatic. The number of moles *(n)* of ethylene oxide reacted may range from 1 to greater than 200. Reaction occurs by a stepwise addition polymerization process usually catalyzed by acids or bases. Average molecular weight of the product alcohols is determined by the moles of ethylene oxide reacted compared to the moles of hydroxyl groups in the starting alcohol. The product often contains significant amounts of unreacted starting alcohol, depending on the relative reactivity of the starting alcohol compared to the reactivity of product alcohols. All unhindered hydroxyl groups of mono- and polyhydric alcohols react, but some may be more reactive than others. Initial reaction rates depend on temperature, catalyst concentration, and alcohol type, and range from fast with primary alcohols to impractically slow with tertiary alcohols.

Oxyethylation processes. Commercial oxyethylation processes may be either batch or continuous.

Continuous. The continuous operation is generally used with lower aliphatic alcohols (such as methanol or butanol) to produce distilled 1-, 2-, and 3-mole oxyethylates. Often a 10-mole excess of alcohol is fed in this process to limit the molecular weight of the products, and the excess alcohol is continuously stripped from the reaction mixture and recycled. Process temperatures range from 100 to 180° C, pressure is required to contain the system, and catalyst concentrations vary from

0.001 to 0.1 mole %. Acid (such as boron trifluoride or sulfuric acid) or base (such as sodium or potassium hydroxide) catalysts are usually employed in commercial operations. The lower oxyethylates, often called glycol ethers, are liquids which find wide use as solvents, jet fuel additives, coupling agents, and brake fluid components, among others.

Batch. Batch oxyethylation is generally carried out on the higher aliphatic alcohols and phenols or when higher polymers, for example the commercially available methoxypolyethylene glycols up to 5000 molecular weight, are desired. The products, very high-boiling, viscous, clear or hazy liquids or low-melting, waxy solids, are usually isolated and used as residue mixtures. Batch operations are run at between 80 to 150° C using about 50–100 psig (345–690 kPa gage) pressure. Catalyst concentrations are similar to continuous operation. Ethylene oxide is fed to maintain reaction pressure until the desired molecular weight is obtained. Three- to nine-mole oxyethylates of the fatty alcohols and alkyl phenols are widely used in nonionic detergents because of their important surfactant properties. The 3-mole adducts of fatty alcohols are also intermediates in the manufacture of anionic surfactants. Esters of methoxypolyethylene glycols find use in detergents and as emulsifying and dispersion agents.

Products. Polyoxyethylation of alcohols can be carried out to produce very-high-molecular-weight products. Depending on the size of the alcohol and the moles of ethylene oxide added, these high-molecular-weight products begin to resemble in properties the simple, high polymers of ethylene oxide, polyethylene glycols. As molecular weights approach 10,000–20,000, oxyethylation side reactions and trace impurities (such as water) lead to more polyethylene glycol contamination and broader than expected molecular-weight distribution. *See* POLYETHYLENE GLYCOL

The molecular weight of alcohol oxyethylates is determined from an analysis of hydroxyl groups by wet chemical methods using phthalic anhydride. A property of polyethers to complex with inorganic salt or iodine has been used to measure low concentrations of polyether in aqueous solutions. *See* ETHYLENE OXIDE; POLYMERIZATION.

[ROBERT K. BARNES]

Bibliography: M. J. Schick (ed.), *Nonionic Surfactants*, 1967.

Poly-p-xylylene resins

Linear, crystallizable resins based on an unusual polymerization of *p*-xylene and derivatives. The polymers are tough and chemically resistant, and may be deposited as adherent coatings by a vacuum process.

Synthesis begins, as in the reaction below, with

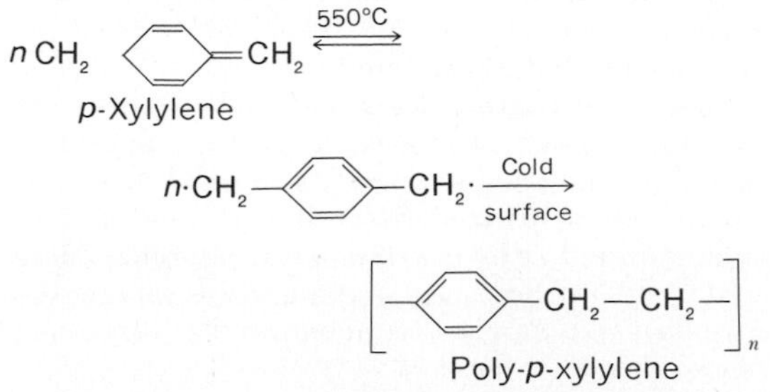

the pyrolysis of xylene to form di-*p*-xylylene. After purification, the di-*p*-xylylene is pyrolyzed to give the monomer *p*-xylylene, a diradical. Condensation of the monomer on cold substrates results in polymerization.

The vapor deposition process makes it possible to coat small microelectronic parts with a thin layer of the polymer. Other potential uses are as coatings for heat-exchange tubes and for encapsulation. Specific properties may be modified by varying the structure of the original xylene, for example, by chlorination. *See* XYLENE.

[JOHN A. MANSON]

Polysaccharide

A class of high-molecular-weight carbohydrates, colloidal complexes, which break down on hydrolysis to monosaccharides containing five or six carbon atoms. The polysaccharides are considered to be polymers in which monosaccharides have been glycosidically joined with the elimination of water. A polysaccharide consisting of hexose monosaccharide units may be represented by the reaction below.

$$nC_6H_{12}O_6 \rightarrow (C_6H_{10}O_5)_n + (n-1)H_2O$$

The term polysaccharide is limited to those polymers which contain 10 or more monosaccharide residues. Polysaccharides such as starch, glycogen, and dextran consist of several thousand D-glucose units. Polymers of relatively low molecular weight, consisting of two to nine monosaccharide residues, are referred to as oligosaccharides. *See* DEXTRAN.

Polysaccharides are either insoluble in water or, when soluble, form colloidal solutions. They are mostly amorphous substances. However, x-ray analysis indicates that a few of them, such as cellulose and chitin, possess a definite crystalline structure. As a class, polysaccharides are nonfermentable and are nonreducing, except for a trace of reducing power due, presumably, to the free reducing group at the end of a chain. They are optically active but do not exhibit mutarotation, and are relatively stable in alkali. *See* OPTICAL ACTIVITY.

The polysaccharides serve either as reserve nutrients (glycogen and inulin) or as skeletal materials (cellulose and chitin) from which relatively rigid mechanical structures are built. Some polysaccharides, such as certain galactans and mannans, however, serve both functions. Through the action of acids or certain enzymes, the polysaccharides may be degraded to their constituent monosaccharide units. Some polysaccharides yield only simple sugars on hydrolysis; others yield not only sugars but also various sugar derivatives, such as D-glucuronic acid or galacturonic acid (known generally as uronic acids), hexosamines, and even nonsugar compounds such as acetic acid and sulfuric acid.

The constituent units of the polysaccharide molecule are arranged in the form of a long chain, either unbranched as in cellulose and amylose, or branched as in amylopectin and glycogen. The linkage between the monosaccharide units is generally the 1,4- or 1,6-glycosidic bond with either the α or β configuration, as the case may be. The branched glycogen and amylopectin contain both the 1,4 and 1,6 linkages. However, other types of linkage are known. In plant gum and mucilage polysaccharides, 1,2, 1,3, 1,5, and 1,6 linkages occur more commonly than the 1,4 type.

In an attempt to systematize the carbohydrate nomenclature, the generic name glycan was introduced as synonymous with the term polysaccharide. This term is evolved from the generic word glycose, meaning a simple sugar, and the ending, "an," signifying a sugar polymer. Examples of established usage of the "an" ending are xylan for polymers of xylose, mannan for polymers of mannose, and galactomannan for galactose-mannose copolymers. Cellulose and starch are both glucans or glucoglycans, since they are composed of glucose units.

Polysaccharides are often classified on the basis of the number of monosaccharide types present in the molecule. Polysaccharides, such as cellulose or starch, that produce only one monosaccharide type (D-glucose) on complete hydrolysis are termed homopolysaccharides. On the other hand, polysaccharides, such as hyaluronic acid, which produce on hydrolysis more than one monosaccharide type (*N*-acetylglucosamine and D-glucuronic acid) are named heteropolysaccharides.

[WILLIAM Z. HASSID]

Bibliography: R. L. Whistler and M. L. Wolfrom (eds.), *Methods in Carbohydrate Chemistry*, vol. 5: *General Polysaccharides*, 1965; A. White et al., *Principles of Biochemistry*, 5th ed., 1978.

Polystyrene resin

A hard, transparent, glasslike thermoplastic resin. Polystyrene is characterized by excellent electrical insulation properties, relatively high resistance to water, high refractive index, clarity, and low softening temperature.

Styrene is produced by the dehydrogenation of ethyl benzene, which in turn is obtained by the alkylation of benzene with ethylene as in reaction (1).

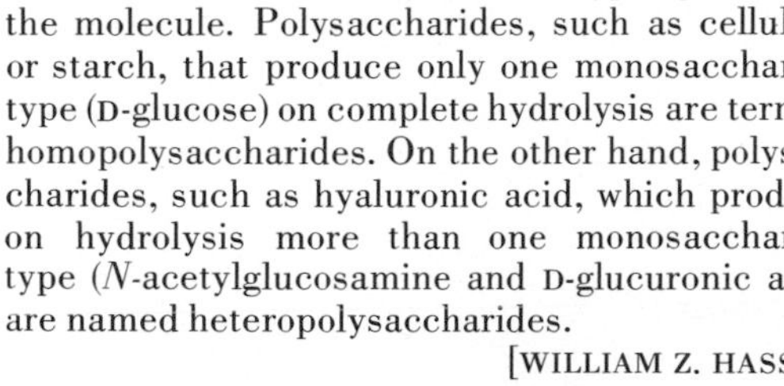

(1)

Free-radical catalysts such as peroxides are often used for polymerization and copolymerization in bulk, solution, and in aqueous emulsion and suspension, as in reaction (2).

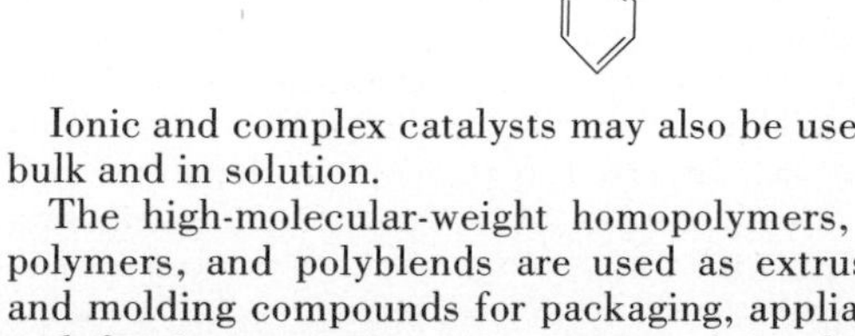

(2)

Ionic and complex catalysts may also be used in bulk and in solution.

The high-molecular-weight homopolymers, copolymers, and polyblends are used as extrusion and molding compounds for packaging, appliance and furniture components, toys, and insulating panels. For a discussion of copolymers of styrene with unsaturated polyesters and with drying oils *see* POLYESTER RESINS.

The copolymer of styrene and butadiene was the major synthetic rubber of World War II. During the 1940s, the redox system of polymerization was developed in which the presence of a reducing agent caused the peroxide to yield free radicals more rapidly at lower temperatures. At the lower temperature of the redox polymerization, a more linear copolymer called cold rubber is obtained in high conversion with improved physical properties. Styrene-butadiene copolymers are still used in large volume for automobile tires and in various rubber articles. *See* POLYMERIZATION.

High-styrene-butadiene copolymers (containing more than 50% styrene) are resinous rather than rubbery. The latexes, as produced by emulsion polymerization, have achieved wide usage in water-based paints.

Copolymers with 20–30% acrylonitrile may be used when a somewhat higher softening point is desired without sacrifice of transparency. Fabrication is, however, more difficult.

By sulfonation of the copolymer of styrene and divinyl benzene, an insoluble polyelectrolyte is produced. This product in the form of its sodium salt is employed as a cationic-exchange resin which is used for water softening.

The effects of blending small amounts of a rubbery polymer, such as butadiene-styrene rubber, with a hard, brittle polymer are most dramatic when the latter is polystyrene. The polyblend may have impact strength greater than ten times that of polystyrene. Even better results may be obtained by graft polymerization of styrene in the presence of the rubber.

Other complex polyblends and interpolymers of styrene, butadiene, and acrylonitrile in various combinations (ABS resins) are important as molding resins which have an excellent balance of properties and low cost. *See* POLYACRYLONITRILE RESINS.

ABS resins and their acrylic counterparts (interpolymers of methyl methacrylate, butadiene, and styrene, or MBS resins) are commonly used as toughening agents for polymers such as polyvinyl chloride.

Block copolymers of styrene with butadiene, the so-called thermoplastic elastomers, are of increasing interest as resins for the injection molding of such articles as toys, footwear components, and pharmaceutical items. These copolymers typically consist of rigid polystyrene end units connected by elastomeric polybutadiene units. They may be fabricated by conventional hot thermoplastic techniques, but on cooling behave like typical elastomers.

The strength of amorphous, atactic polystyrene may be increased by orientation induced by uniaxial or biaxial stretching even though crystallinity is not produced. Oriented products in the form of filaments, sheet, and film are available.

Crystallizable, isotactic polystyrene has been formed in the presence of (1) triphenylmethyl potassium in hexane solution, (2) the Alfin catalyst (sodium allyl–sodium isopropoxide) in hexane and benzene solutions, and (3) a Ziegler catalyst (titanum tetrachloride–aluminum triethyl) in petroleum ether solvent. The softening temperature of the crystalline polymer is substantially greater than that of the amorphous product.

Besides the many applications of styrene in combination with other materials as in rubber and paints, large quantities of the homopolymer and the polyblends are employed in the injection molding of toys, panels, and novelty items, and also in the extrusion of sheets. The sheets are used for panels, or they may be further shaped by vacuum forming for uses such as liners for refrigerator doors.

Polystyrene may also be fabricated in the form of a rigid foam, which is used in packaging, food-service articles, and insulating panels. *See* ACRYLONITRILE; POLYMER.

[JOHN A. MANSON]

Bibliography: J. A. Brydson, *Plastics Materials*, 1975.

Polysulfide resins

Resins that vary in properties from viscous liquids to rubberlike solids. Organic polysulfide resins are prepared by the condensation of organic dihalides with a polysulfide as in the reaction below. By

$$\underset{\text{1,2-Dichloroethane}}{ClCH_2CH_2Cl} + \underset{\text{Sodium tetrasulfide}}{nNa_2S_4} \rightarrow$$

$$\underset{\text{Polysulfide resin}}{(-CH_2-CH_2-\overset{\overset{S}{\|}}{S}-\overset{\overset{S}{\|}}{S}-)_n} + 2nNaCl$$

the use of other dichlorides, such as bis(2-chloroethyl) ether, $ClCH_2CH_2OCH_2CH_2Cl$, by changing the functionality of the sodium sulfide, or by incorporating a trifunctional halide to yield branching, the properties may be varied. Molecular weights can also be adjusted by treatment with chain-breaking disulfides. The condensation is usually conducted in an aqueous medium from which the product may be separated and dried. Many of the polysulfide resins have an odor which is generally characteristic of monomeric sulfur compounds but is usually milder in nature.

The linear, high-molecular-weight polymers can be cross-linked or cured by reaction with zinc oxide. Compounding and fabrication of the rubbery polymers can be handled on conventional rubber machinery. The polysulfide rubbers are distinguished by their resistance to solvents such as gasoline, and to oxygen and ozone. The polymers are relatively impermeable to gases. The products are used to form coatings which are chemically resistant and special rubber articles, such as gasoline bags.

Low-molecular-weight, liquid polymers prepared by cleavage of the —S—S— linkages in the conventional polysulfide are useful as curable ingredients in caulking compounds. In addition, they are used as reactive diluents to reduce the viscosity of epoxy resins and, following reaction with free epoxy groups, to flexibilize and toughen the resin.

The polysulfide rubbers were among the very first commercial synthetic rubbers. Although the products are not as strong as other rubbers, their chemical resistance makes them useful in various applications.

The polysulfide rubbers were among the first polymers to be used in solid-fuel compositions for rockets. *See* ORGANOSULFUR COMPOUND; POLYMERIZATION.

[JOHN A. MANSON]

Bibliography: J. A. Brydson, *Plastics Materials*, 3d ed., 1975.

Polysulfone resins

Engineering plastics whose molecules contain sulfone groups in the main chain. Polysulfones are characterized by high strength, stiffness, toughness, and thermal and oxidation stability. A typical polysulfone is made by reaction of the disodium salt of bisphenol A with *p,p′*-dichlorodiphenyl sulfone in dimethyl sulfoxide and chlorobenzene, as shown below.

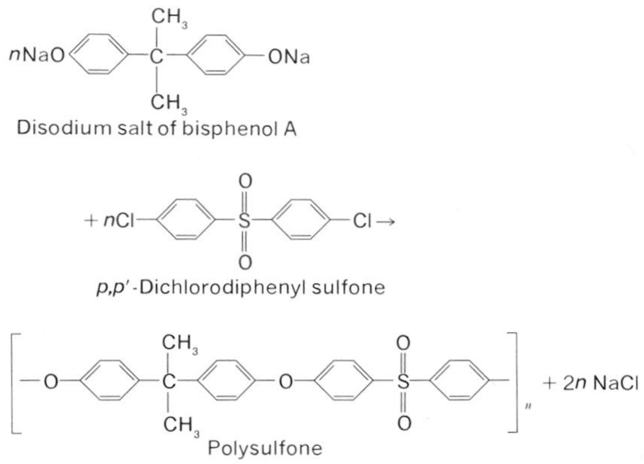

The aromatic structure and the presence of the sulfone groups are responsible for the resistance to heat and oxidation. The polymers can be used over a wide temperature range (−70 to +150°C), and in the presence of acids, alkalies, aqueous media, and nonpolar organic solvents; resistance to creep and dimensional changes due to aging is high. Uses include electronic and automotive parts, hardware and houseware items, and plumbing parts.

Structural variations are also made to raise the service temperature to about 215°C. These include polyethersulfones, in which phenyl groups are alternately linked by ether and sulfone groups, and polyarylsulfones, in which diphenyl and phenyl groups are linked by either and sulfone groups. *See* SULFONE.

[JOHN A. MANSON]

Bibliography: *Modern Plastics Encyclopedia*, 1979–1980.

Polyurethane resins

Resins that can be produced in forms varying from hard, glossy, solvent-resistant coatings, to abrasion- and solvent-resistant rubbers, fibers, and flexible-to-rigid foams. The foams have found the widest use. The more flexible foams are employed as upholstery material for furniture, as rug backing, insulation, and crash pads. The more rigid foams are employed as the core in structural and insulating laminates and as insulation in refrigerated appliances and vehicles.

Polyurethane (or polyisocyanate) resins are produced by the reaction of a diisocyanate with a compound containing at least two active hydrogen atoms, such as a diol or diamine. Toluene diisocyanate (TDI), diphenylmethane diisocyanate (MDI), and hexamethylene diisocyanate (HDI) are frequently employed. They are prepared by the reaction of phosgene with the corresponding diamines, as shown by reaction (1).

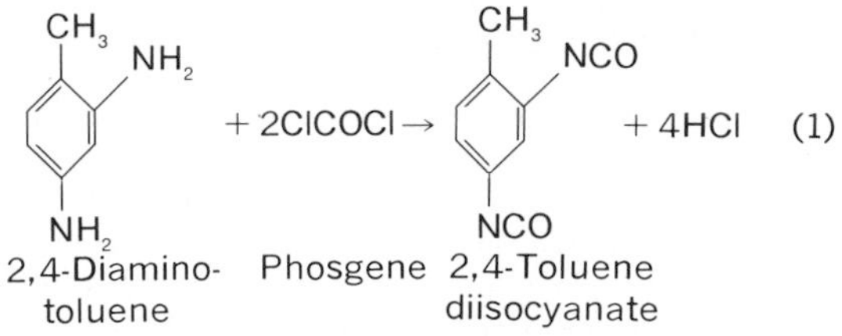

Linear, fiber-forming polymers (I) are formed by the addition of diisocyanates to diols, while cross-

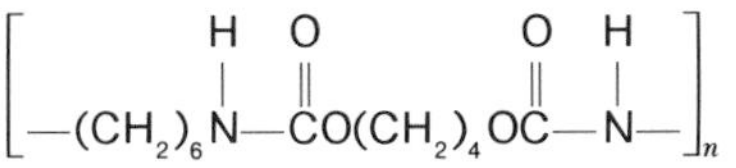

(I)

linking is made possible by the use of polyols or isocyanates having more than two functional groups. Much urethane technology is based on the use of prepolymers in which a low-molecular-weight polyester or polyether is prepared. Some of these have terminal hydroxyl groups, while those reacted with an excess of diisocyanate have terminal isocyanate groups, as in (II), the prepolymer of propylene oxide [with $R(N{=}C{=}O)_2$]. Thus segments of a rubbery block alternate with segments of rigid urethane units. This growth process is called chain extension.

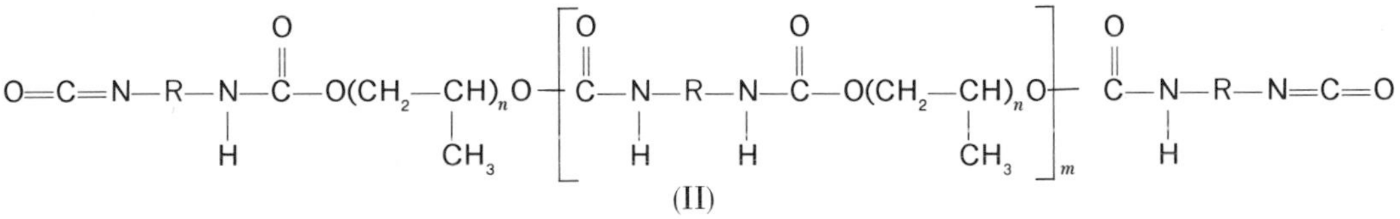

(II)

The addition of hexamethylene diisocyanate to 1,4-butanediol, as in (I), yields a linear, fiber-forming polyurethane (I). The fibers resemble those of polyamides, but are little used due to higher cost. However, unique elastomeric or stretch fibers can be made from polyester prepolymers analogous to the polyether type shown in structure (II), in which rubbery polyester blocks alternate with rigid urethane units; the terminal isocyanate groups provide sites for further chain extension and cross-linking by reaction with suitable reagents such as amines.

Three major types of polyurethane elastomers are available. One type is based on ether- or ester-type prepolymers that are chain-extended and cross-linked using polyhydroxyl compounds or amines; alternately, unsaturated groups may be introduced to permit vulcanization with common curing agents such as peroxides. All of these can be processed by methods commonly used for rubber. A second type is obtained by first casting a mixture of prepolymer with chain-extending and cross-linking agents, and then cross-linking further by heating. The third type is prepared by reacting a dihydroxy ester- or ether-type prepolymer, or a diacid, with a diisocyanate such as diphenylmethane diisocyanate and a diol; these thermoplastic elastomers can be processed on conventional plastics equipment. In general, urethane elastomers are characterized by outstanding mechanical properties and resistance to ozone, though they may be degraded by acids, alkalies, and steam.

A wide variety of tough and abrasion-resistant urethane coatings are available. Many are based on the reaction of castor oil, a triol, with an excess of diisocyanate; the resulting triisocyanate undergoes cross-linking by reaction with atmospheric moisture. Urethane alkyds can also be made by reacting an unsaturated drying oil with glycerol, and then reacting the product with a diisocyanate; curing is effected by atmospheric oxidation of the double bonds. Still other coatings are based on the use of prepolymers. Polyurethanes are also used as adhesives, for example, in the bonding of rubber and of nylon.

For the production of foamed products, advantage is taken by the fact that the isocyanate group can react with water to yield an amine and carbon dioxide, as in reaction (2); low-boiling fluorocar-

$$\underset{}{RNCO} + H_2O \rightarrow \underset{\text{Carbamic acid}}{RNHCOOH} \rightarrow \underset{\text{Amine}}{RNH_2} + CO_2 \qquad (2)$$

bons that will vaporize as the urethane-forming reaction proceeds are often added. Flexible foams are usually made by reacting a polyether polyol with toluene diisocyanate, water, and amine and organotin catalysts in the presence of polymeric fillers; with care, the rates of foaming and cross-linking can be kept in proper balance. Rigid foams are based on low-molecular-weight polyols and diisocyanates such as diphenylmethane diisocyanate in order to increase cross-link density and stiffness; intermediate degrees of stiffness can be obtained by varying the formulation. Rigid foams containing isocyanurate groups are also made, by the trimerization of polymeric isocyanates in the presence of polyols and a blowing agent.

The flexible polyurethanes may be used for coating rubber articles to give them additional resistance to abrasion and solvents. Wire insulated with polyurethane resin can be soldered directly without previously removing the coating because the polymer decomposes at the soldering temperature to yield a clean wire surface. Among these various applications, the uses of the foamed products are developing most rapidly because of the ease of varying the density and flexibility, and the resistance to aging and solvents. *See* POLYESTER RESINS; POLYETHER RESINS; POLYMERIZATION; URETHANE.

[JOHN A. MANSON]

Bibliography: J. A. Brydson, *Plastics Materials*, 3d ed., 1975; *Modern Plastics Encyclopedia*, 1979–1980.

Polyvinyl resins

Polymeric materials generally considered to include polymers derived from monomers having the structure

$$CH_2 = C \big\langle {R_1 \atop R_2}$$

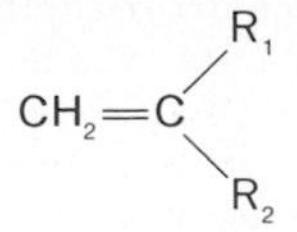

in which R_1 and R_2 represent hydrogen, alkyl, halogen, or other groups. This article refers to polymers whose names include the term vinyl. Of these polymers, several have been used for a number of years, such as polyvinyl chloride, polyvinyl acetate, polyvinylidene chloride, polyvinyl alcohol, polyvinyl acetals, and polyvinyl ethers. Indeed the terms vinyls and vinyl resins are frequently used to refer to the first three polymers of this group. Some polyvinyl resins of more recent origin are polyvinyl fluoride, polyvinylpyrrolidone, and polyvinylcarbazole. For discussions of other vinyl-type polymers see POLYACRYLATE RESIN; POLYACRYLONITRILE RESINS; POLYFLUOROOLEFIN RESINS; POLYOLEFIN RESINS; POLYSTYRENE RESIN.

Many of the monomers can be prepared by addition of the appropriate compound to acetylene. For example, vinyl chloride, vinyl fluoride, vinyl acetate, and vinyl methyl ether may be formed by the reactions of acetylene with HCl, HF, CH_3OOH, and CH_3OH, respectively. Processes based on ethylene as a raw material have also become common for the preparation of vinyl chloride and vinyl acetate.

The polyvinyl resins may be characterized as a group of thermoplastics which, in many cases, are inexpensive and capable of being handled by solution, dispersion, injection molding, and extrusion techniques. The properties vary with chemical structure, crystallinity, and molecular weight.

Polyvinyl acetals. These are relatively soft, water-insoluble thermoplastic products obtained by the reaction (1) of polyvinyl alcohol with aldehydes. Properties depend on the extent to which alcohol groups are reacted. Polyvinyl butyral is rubbery and tough and is used primarily in plasticized form as the inner layer and binder for safety glass. Polyvinyl formal is the hardest of the group;

it is used mainly in adhesive, primer, and wire-coating formulations, especially when blended with a phenolic resin.

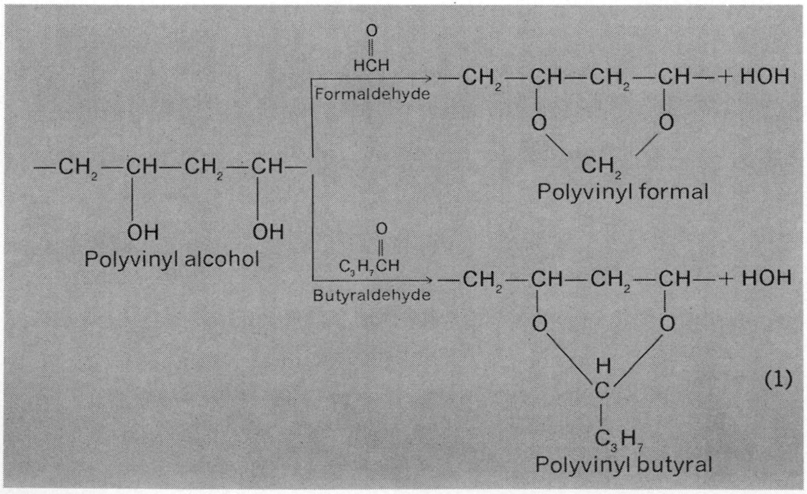

Polyvinyl butyral is usually obtained by the reaction of butyraldehyde with polyvinyl alcohol. The formal can be produced by the same process, but is more conveniently obtained by the reaction of formaldehyde with polyvinyl acetate in acetic acid solution.

Polyvinyl acetate. Polyvinyl acetate is a leathery, colorless thermoplastic material which softens at relatively low temperatures and which is relatively stable to light and oxygen. The polymers are clear and noncrystalline. The chief applications are as adhesives and binders for water-based or emulsion paints.

Vinyl acetate is conveniently prepared by the reaction of acetylene with acetic acid, as in reaction (2). As acetaldehyde has become readily and cheaply available from the oxidation of olefins or hydrocarbons, it has been used to an increasing extent as a starting material. This is shown by reaction (3). Vinyl acetate is also prepared by the

$$CH{\equiv}CH + CH_3COOH \xrightarrow[\text{(catalyst)}]{210-250°C}$$

Acetylene   Acetic acid

$$CH_2{=}CH \atop OCOCH_3 \quad (2)$$
Vinyl acetate

Acetaldehyde   Acetic anhydride

Ethylidene diacetate

$$CH_2{=}CH \atop OCOCH_3 \quad + \text{ by-products} \quad (3)$$
Vinyl acetate

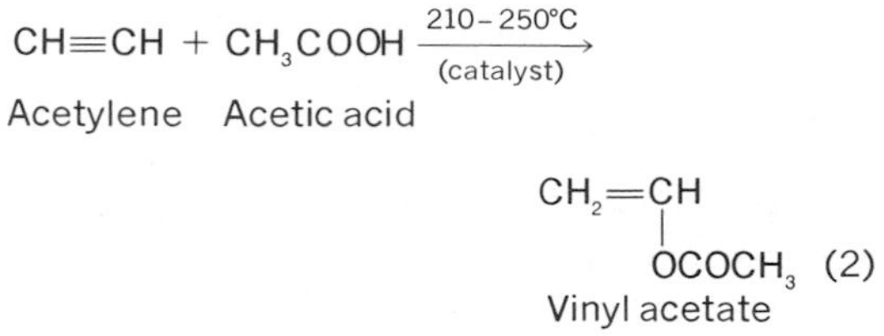

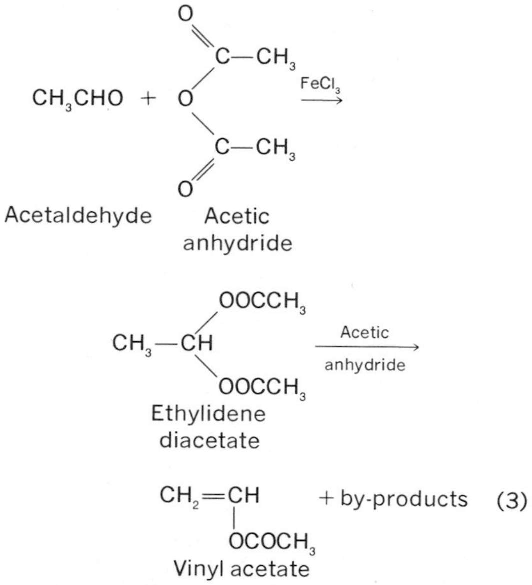

vapor-phase reaction of ethylene, oxygen, and acetic acid using a palladium chloride catalyst.

Polymerization and copolymerization may be conveniently effected by free-radical catalysis in aqueous emulsion and suspension systems. Vinyl acetate copolymerizes readily with various other vinyl monomers; however, it does not copolymerize with styrene by the free-radical process. The polymerization process is represented by (4).

Anhydrous solid polymers and copolymers may be used directly in adhesives.

Aqueous dispersions, produced by the emulsion polymerization process and commonly called polymer emulsions, or latexes, are used for treating

(1)

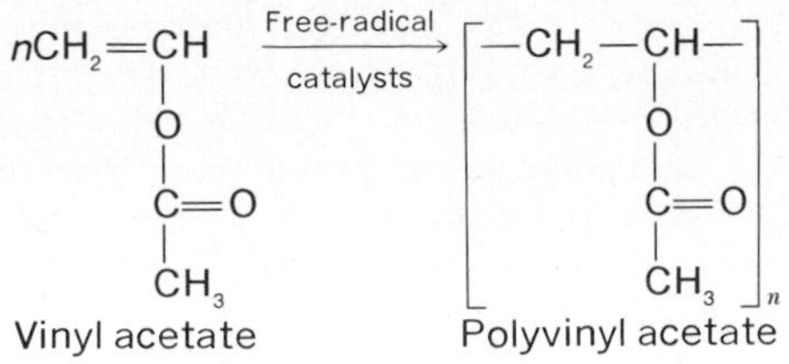 (4)

textiles and paper, as adhesives, and as water-based paints. The water-based paints, prepared by pigmenting vinyl acetate polymer and copolymer emulsion, have achieved wide usage because of low cost of materials, ease of application, and resistance to weathering. As water is removed from the latex by evaporation or absorption, the suspended polymer particles coalesce into a tough film. The character of the film may be modified by the use of comonomers in the original polymerization or by the addition of plasticizers to the final emulsions.

Vinyl acetate is also frequently used as a comonomer with ethylene or vinyl chloride. *See* ETHYLENE; VINYL RESIN.

Polyvinyl alcohol. Polyvinyl alcohol is a tough, whitish polymer which can be formed into strong films, tubes and fibers that are highly resistant to hydrocarbon solvents. Although polyvinyl alcohol is one of the few water-soluble polymers, it can be rendered insoluble in water by drawing or by the use of cross-linking agents.

So far, vinyl alcohol itself, $CH_2{=}CHOH$, has not been isolated; reactions designed to produce the monomer yield the tautomeric acetaldehyde instead. However, the polymer can be produced on a commercial scale by the hydrolysis of polyvinyl acetate, reaction (5).

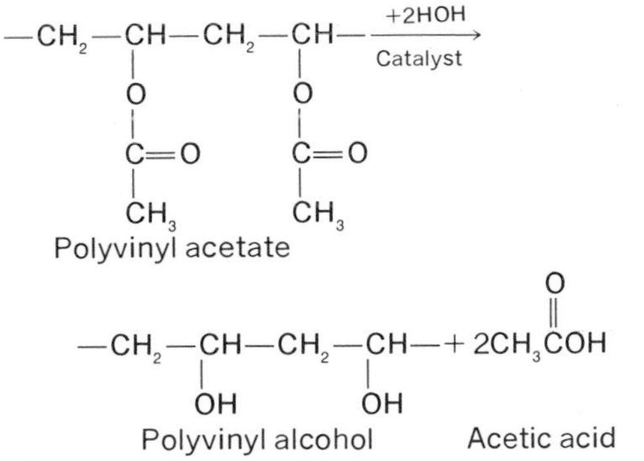 (5)

Two groups of products are available, those formed by the essentially complete hydrolysis (97% or greater) of polyvinyl acetate, and those formed by incomplete hydrolysis (50–90%).

The former, the completely hydrolyzed products, may be plasticized with water or glycols and molded or extruded into films, tubes, and filaments which are resistant to hydrocarbons; these can be rendered insoluble to water by cold-drawing or heat or by the use of chemical cross-linking agents. On cold-drawing, the degree of crystallinity is substantially increased. These products are used for liners in gasoline hoses, for grease-resistant coating and paper adhesives, for treating paper and textiles, and as emulsifiers and thickeners. Insolubilized fibers have found large uses in Japan for clothing, industrial fabrics, and cordage. Films are strong, tough, and relatively impermeable to oxygen but tend to be sensitive to moisture.

The partially hydrolyzed products are generally more water-soluble and less subject to crystallization by drawing. These materials are used as emulsifying agents and thickeners, in steel-quenching solutions, in adhesive formulations, and texile sizes, and in water-soluble films.

Polyvinyl carbazole. Polyvinyl carbazole is a tough, glassy thermoplastic with excellent electrical properties and the relatively high softening temperature of 120–150°C. Polymerization, as represented by reaction (6), can be carried out in

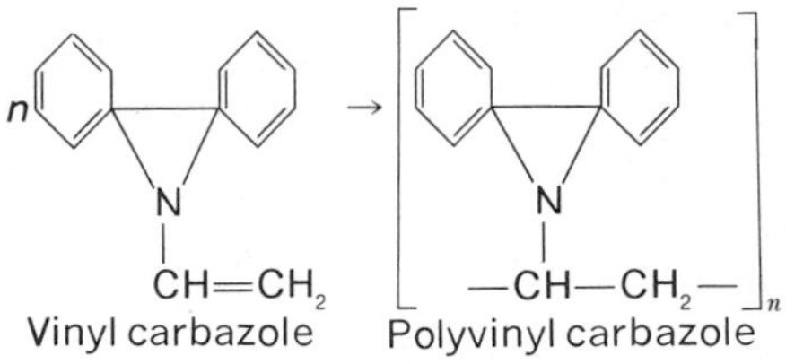 (6)

bulk by free-radical catalysis. Uses of the product have been limited to small-scale electrical applications requiring resistance to high temperatures.

Polyvinyl chloride. Polyvinyl chloride (PVC) is a tough, strong thermoplastic material which has an excellent combination of physical and electrical properties. The products are usually characterized as plasticized or rigid types. Polyvinyl chloride (and copolymers) is the second most commonly used polyvinyl resin and one of the most versatile plastics.

The plasticized types, either soft copolymers or plasticized homopolymers, are somewhat elastic materials which are familiar in the form of shower curtains, floor coverings, raincoats, dishpans, dolls, bottle-top sealers, prosthetic forms, wire insulation, and films, among others.

Rigid polyvinyl chloride products, which may consist of the homopolymer, copolymer, or polyblends, are commonly used in the manufacture of phonograph records, pipe, chemically resistant liners for chemical-reaction vessels, and siding and window sashes.

The monomer is frequently prepared from chlorine, acetylene, and ethylene by a combination of processes which affords complete utilization of the chlorine, as shown in reactions (7) and (8).

$$\underset{\text{Ethylene}}{CH_2{=}CH_2} \xrightarrow{Cl_2} \underset{\text{1,2-Dichloroethane}}{CH_2Cl{-}CH_2Cl} \rightarrow \underset{\text{Vinyl chloride}}{CH_2{=}CHCl} + \underset{\text{Hydrogen chloride}}{HCl} \quad (7)$$

$$\underset{\text{Acetylene}}{CH{\equiv}CH} \xrightarrow{HCl} \underset{\text{Vinyl chloride}}{CH_2{=}CHCl} \quad (8)$$

The polymerization of vinyl chloride, reaction (9),

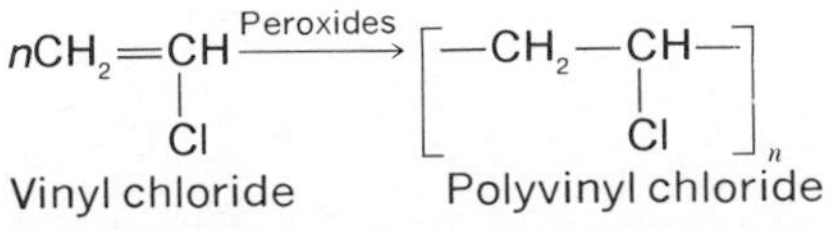 (9)

and its copolymerization with other vinyl monomers may be initiated by peroxide or azo compounds and carried out in bulk or in aqueous emulsion or suspension systems. The structure and properties of the product are quite dependent on polymerization temperature; the lower the temperature, the higher the softening point of the resin. Polymers prepared in bulk or suspension are used for many applications as molded, extruded, or calendered objects. Polymers prepared in emulsion are often molded by dipping or pouring techniques.

Because polyvinyl chloride products have a tendency to lose hydrogen chloride at high temperatures, a stabilizer such as a tin or lead compound is included in the final composition. Fillers are also commonly incorporated.

Copolymers with monomers such as vinyl acetate or propylene can be processed at lower temperatures than the homopolymer. The acetate copolymer is especially useful in floor tiles and phonograph records, though homopolymers are used as well.

Blends or "alloys" of polyvinyl chloride with small amounts of rubbery materials such as the interpolymer of acrylonitrile, butadiene, and styrene (ABS) have been produced for applications such as panels and pipe in which impact resistance, as well as hardness and strength, is desired.

Foams and films are receiving increasing attention, and fibers are used to some extent in Europe.

Chlorination of polyvinyl chloride is sometimes effected to obtain a stiffer resin, at the expense of processability.

Polyvinylidene chloride. Polyvinylidene chloride is a tough, horny thermoplastic with properties generally similar to those of polyvinyl chloride. In comparison with the latter, polyvinylidene chloride is softer and less soluble; it softens and decomposes at lower temperatures, crystallizes more readily, and is more resistant to burning.

Because of its relatively low solubility and decomposition temperature, the material is most widely used in the form of copolymers with other vinyl monomers, such as vinyl chloride. The copolymers are employed as packaging film, rigid pipe, and as filaments for upholstery and window screens.

Vinylidene chloride is normally prepared by the pyrolysis of 1,1,2-trichloroethane. The latter is obtained by the chlorination of 1,2-dichloroethane which, in turn, is formed by the addition of chlorine to ethylene. This preparation is shown by reactions (10) and (11).

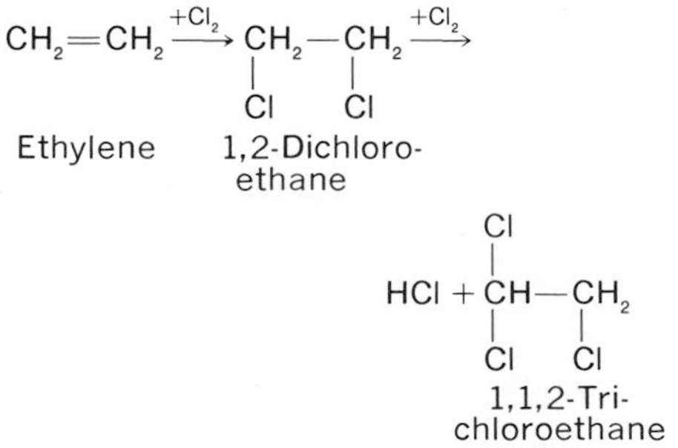

$$CH_2=CH_2 \xrightarrow{+Cl_2} \underset{\substack{| \quad | \\ Cl \quad Cl}}{CH_2-CH_2} \xrightarrow{+Cl_2}$$

Ethylene   1,2-Dichloro-ethane

$$HCl + \underset{\substack{| \quad | \\ Cl \quad Cl}}{\overset{\substack{Cl \\ |}}{CH}-CH_2} \quad (10)$$

1,1,2-Tri-chloroethane

$$\underset{\substack{| \quad | \\ Cl \quad Cl}}{\overset{Cl}{CH}-CH_2} \xrightarrow{Heat} \underset{\substack{| \\ Cl}}{\overset{Cl}{CH_2=C}} + H\dot{C}l \quad (11)$$

1,1,2-Tri-chloroethane   Vinylidene chloride   Hydrogen chloride

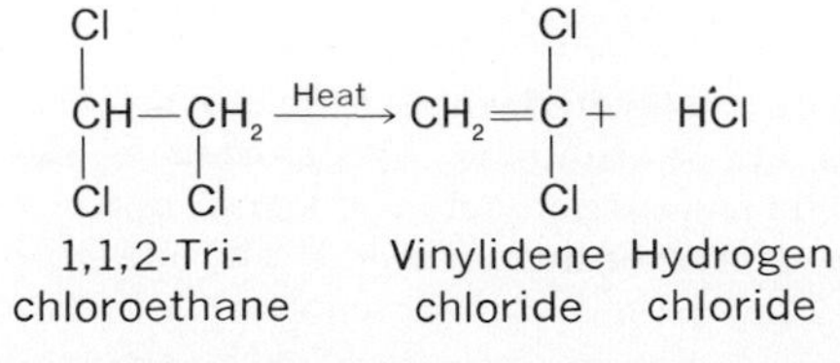

Polymerization as well as copolymerization may be initiated by peroxides and other free-radical catalysts and is most satisfactorily effected by emulsion and suspension techniques. Reaction (12) represents the polymerization.

$$n\underset{\substack{| \\ Cl}}{\overset{\substack{Cl \\ |}}{CH_2=C}} \rightarrow \left[-CH_2-\underset{\substack{| \\ Cl}}{\overset{\substack{Cl \\ |}}{C}}- \right]_n \quad (12)$$

Vinylidene chloride   Polyvinylidene chloride

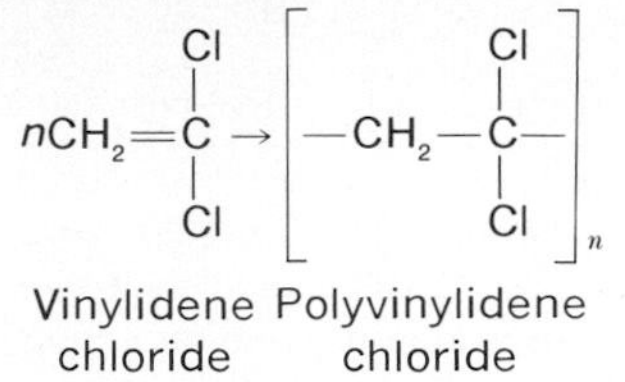

Because of the relatively low decomposition temperature of polyvinylidene chloride, a stabilizer such as an amine is normally included in the composition.

Films of polyvinylidene chloride, and especially the copolymer containing about 15% of vinyl chloride, are resistant to moisture and gases. Also, they can be heat sealed and, when oriented, have the property of shrinking on heating. By warming a food product wrapped loosely with a film of the polymer, a skintight, tough, resistant coating is produced.

By cold-drawing, the degree of crystallinity, strength, and chemical resistance of sheets, filaments, and even piping can be greatly increased.

Polyvinyl ethers. Polyvinyl ethers exist in several forms varying from soft, balsamlike semisolids to tough, rubbery masses, all of which are readily soluble in organic solvents. Polymers of the alkyl vinyl ethers are used in adhesive formulations and as softening or flexibilizing agents for other polymers.

The monomers may be prepared by the reaction of alcohols with acetylene in the presence of alkali. Polymerization may be effected in bulk or solution at temperatures of −100 to +25°C by use of cationic initiators such as boron trifluoride, reaction (13).

$$n\underset{\substack{| \\ O \\ | \\ CH_3}}{CH_2=CH} \xrightarrow[\text{initiators}]{\text{Cationic}} \left[-\underset{\substack{| \\ O \\ | \\ CH_3}}{CH_2-CH}- \right]_n \quad (13)$$

Vinyl methyl ether   Polyvinyl methyl ether

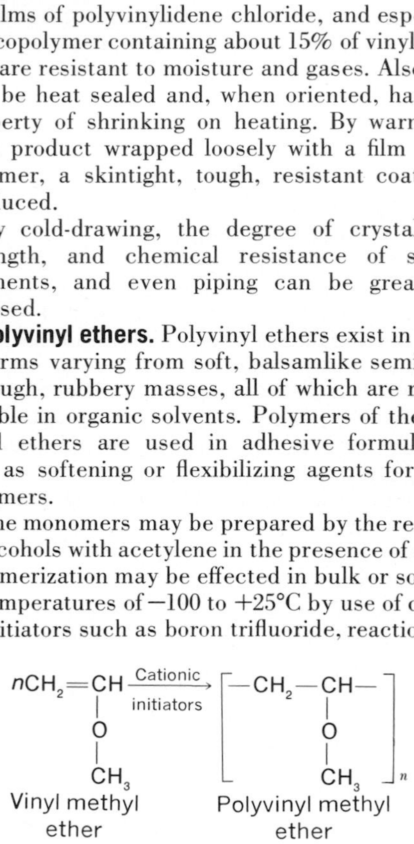

By careful choice of conditions, it is possible to achieve stereoregular polymerizations which yield partially crystalline polymers that are harder and tougher than the amorphous products.

Polyvinyl methyl ether is soluble in cold water but precipitates when the temperature is raised to about 35°C. The other alkyl vinyl ether polymers are insoluble in water. Copolymers of vinyl methyl ether with maleic anhydride are useful in textile applications.

Polyvinyl fluoride. Polyvinyl fluoride is a tough, partially crystalline thermoplastic material which has a higher softening temperature than polyvinyl chloride. Films and sheets are characterized by high resistance to impact and cracking caused by flexing and temperature and by resistance to weathering.

Polymerization can be effected in the presence of oxygen and peroxidic catalysts, reaction (14). Because of the low boiling point (−88°C) and high critical temperature of the monomer, polymerization

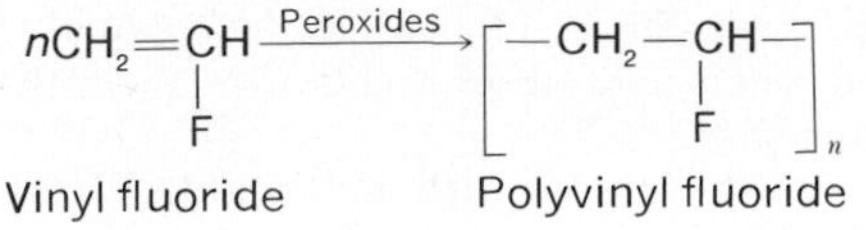
nCH2=CH
F
Peroxides
CH2—CH
F
n
Vinyl fluoride
Polyvinyl fluoride

(14)

is accomplished by use of pressure techniques similar to those employed in the high-pressure process for polymerizing ethylene. Like other polyvinyl halides, polyvinyl fluoride tends to lose the halogen acid at elevated temperatures.

Films are used in industrial and architectural applications. Coatings, for example, on pipe, are resistant to highly corrosive media.

Polyvinyl pyrrolidone. Polyvinyl pyrrolidone is a water-soluble polymer of basic nature which has film-forming properties, strong absorptive or complexing qualities for various reagents, and the ability to form water-soluble salts which are polyelectrolytes. The polymer can be prepared by free-radical polymerization in bulk or aqueous solution, reaction (15). Isotonic solutions were used in Ger-

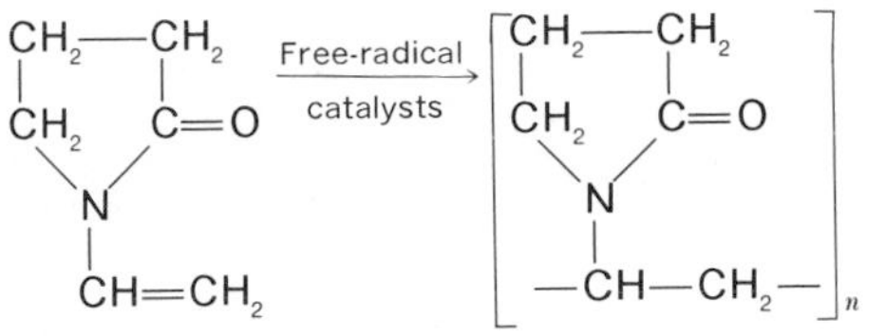
Free-radical
catalysts

(15)

many in World War II as an extender for blood plasma. The main current uses are as a water-solubilizing agent for medicinal agents such as iodine, and as a semipermanent setting agent in hair sprays. Certain synthetic textile fibers containing small amounts of vinylpyrrolidone as a copolymer have improved affinity for dyes.

New resins. New polyvinyl resins are made available, often in the form of copolymers, when effective combinations of monomer preparations, polymerization methods, and property-use relationships are developed. Although the development of new homopolymers is becoming less probable, copolymerization offers many possibilities of obtaining special properties for particular applications. *See* POLYMER; POLYMERIZATION.

[JOHN A. MANSON]

Potassium

A chemical element, K, atomic number 19, and atomic weight 39.102. It stands in the middle of the alkali metal family, below sodium and above rubidium, in group Ia of the periodic table of the elements. It is a lightweight, soft, low-melting, reactive metal. In 1807 Sir Humphry Davy isolated

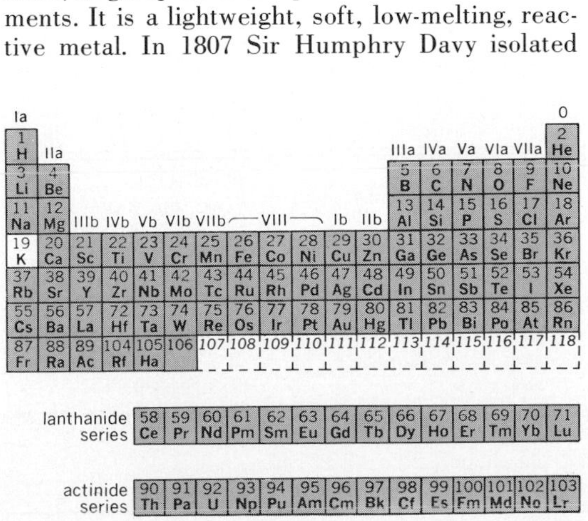
Ia
0
1 H
IIa
IIIa IVa Va VIa VIIa
2 He
3 Li 4 Be
5 B 6 C 7 N 8 O 9 F 10 Ne
11 Na 12 Mg
IIIb IVb Vb VIb VIIb VIII Ib IIb
13 Al 14 Si 15 P 16 S 17 Cl 18 Ar
19 K 20 Ca 21 Sc 22 Ti 23 V 24 Cr 25 Mn 26 Fe 27 Co 28 Ni 29 Cu 30 Zn 31 Ga 32 Ge 33 As 34 Se 35 Br 36 Kr
37 Rb 38 Sr 39 Y 40 Zr 41 Nb 42 Mo 43 Tc 44 Ru 45 Rh 46 Pd 47 Ag 48 Cd 49 In 50 Sn 51 Sb 52 Te 53 I 54 Xe
55 Cs 56 Ba 57 La 72 Hf 73 Ta 74 W 75 Re 76 Os 77 Ir 78 Pt 79 Au 80 Hg 81 Tl 82 Pb 83 Bi 84 Po 85 At 86 Rn
87 Fr 88 Ra 89 Ac 104 Rf 105 Ha 106 107 108 109 110 111 112 113 114 115 116 117 118
lanthanide series
58 Ce 59 Pr 60 Nd 61 Pm 62 Sm 63 Eu 64 Gd 65 Tb 66 Dy 67 Ho 68 Er 69 Tm 70 Yb 71 Lu
actinide series
90 Th 91 Pa 92 U 93 Np 94 Pu 95 Am 96 Cm 97 Bk 98 Cf 99 Es 100 Fm 101 Md 102 No 103 Lr

metallic potassium, by electrolysis, for the first time. It is very similar to sodium in its behavior in metallic form, and its uses are limited by the availability of low-cost sodium in large volume.

Uses. Potassium chloride finds its main use in fertilizer mixtures. It also serves as the raw material for the manufacture of other potassium compounds.

Potassium hydroxide is used in the manufacture of liquid soaps, and potassium carbonate in making soft soaps.

Potassium carbonate is also an important raw material for the glass industry.

Potassium nitrate is used in matches, in pyrotechnics, and in similar items which require an oxidizing agent.

Occurrence. Potassium is a very abundant element, ranking seventh among all the elements in the Earth's crust, 2.59% of which is potassium in combined form. Only oxygen, silicon, aluminum, iron, calcium, and sodium are more abundant. Sea water contains 380 parts per million, making potassium the sixth most plentiful element in solution, exceeded only by chlorine, sodium, magnesium, sulfur, and calcium.

Potassium compounds are found in the historically important deposits at Stassfurt in Germany, which consist of sylvite (KCl) and carnallite ($MgCl_2 \cdot KCl$). In the United States, extensive potassium deposits containing sylvite and polyhalite ($2CaSO_4 \cdot K_2SO_4 \cdot 2H_2O$) are located at Searles Lake, Calif., and Carlsbad, N.Mex. In addition, potash salts are found in France, Spain, Poland, the Soviet Union, and in the Dead Sea.

Metallurgical extraction. Commercial potassium chloride is melted in a gas-fired melt pot and is fed to the exchange column (see illustration) in

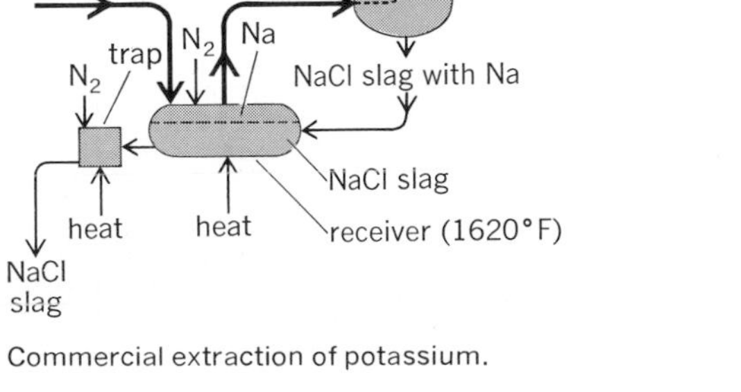

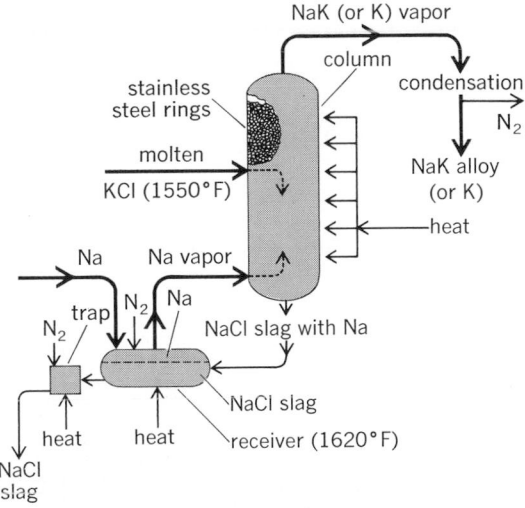

Commercial extraction of potassium.

the commercial manufacture of potassium metal by thermochemical means. The molten potassium chloride flows down over steel Raschig rings in the packed column. It is met by ascending sodium vapors coming from a gas-fired reboiler. An equilibrium is set up between the two, giving sodium chloride and potassium metal as the products. The sodium chloride formed is continuously withdrawn at the base of the apparatus. The column operating conditions may be varied to give practically pure potassium metal as an overhead product or to vaporize sodium along with the potassium to give sodium-potassium (NaK) alloys of varying compositions as products. Potassium metal of over 99.5% purity can be produced continuously.

Unlike lithium and sodium, which are produced by electrolysis, potassium reacts with carbon electrodes, and also can form an explosive carbonyl in electrolysis (or in thermochemical methods using carbon reduction). Therefore, the thermochemical route using the reaction between metallic sodium and potassium chloride has proved most practical and economical.

Physical properties. The physical properties of potassium metal are summarized in the table.

Chemical properties. Potassium is even more reactive than sodium. It reacts vigorously with the oxygen in air to form the monoxide, K_2O, and the peroxide, K_2O_2. In the presence of excess oxygen, it readily forms the superoxide, KO_2 (formerly believed to be K_2O_4).

Potassium does not react with nitrogen to form a nitride, even at elevated temperatures. With hydrogen, potassium reacts slowly at 200°C and rapidly at 350–400°C. It forms the least stable hydride of all the alkali metals.

The reaction between potassium and water or ice is violent, even at temperatures as low as —100°C. The hydrogen evolved is usually ignited in reaction at room temperature. Reactions with aqueous acids are even more violent and verge on being explosive.

Instead of forming the carbide with carbon, potassium forms a rather indefinite solid solution, with the potassium atoms interposed between the layers of the graphite lattice.

Potassium reacts vigorously with the halogens. Lithium and sodium react only superficially with liquid bromine, but potassium detonates in contact with it. Potassium ignites in the reaction with iodine, also.

The reaction of potassium with ammonia gives potassium amide, KNH_2, and hydrogen. Potassium differs from sodium in that an explosive carbonyl is formed when potassium reacts directly with carbon monoxide.

Potassium reacts with many organic compounds, but not with saturated aliphatic hydrocarbons. With some aromatic hydrocarbons, metalation occurs, giving organopotassium compounds. With acetylene, potassium acetylides are formed.

Potassium reacts with alcohols to form alkoxides and hydrogen. Most reactions of potassium with organic carbonyl compounds are very similar to those of sodium. In the form of NaK alloy, potassium is a very effective catalyst for the transesterification reaction involved in the commercial modification of lard.

Availability. Potassium metal is available in one grade of 99+% purity, with sodium as the major impurity. Production is about 50,000 lb/year. The price of the metal varies widely with the quantity ordered. Potassium salts are generally more expensive than the corresponding sodium salts.

Handling. Handling of potassium metal is much the same as that of sodium metal, with two major exceptions. First, the formation of the superoxide, KO_2, causes difficulties because it can react vigorously with hydrocarbons and other organic matter. Second, potassium is generally more reactive than sodium. Potassium forms an explosive carbonyl with carbon monoxide, and the metal detonates in contact with bromine. Usually sodium, potassium, and the sodium-potassium (NaK) alloys are considered to be in the same general class of reactivity, allowing for the chemical differences outlined above and for the liquid (and hence more reactive) nature of the NaK alloys over a wide range of composition. *See* SODIUM.

Principal compounds. Potassium chloride, KCl, is the most important potassium compound. It is not only the form in which potassium is often found in nature, but it is the form in which potash is used as a fertilizer.

Potassium hydroxide, KOH, is also known as caustic potash. It is usually made by the electrolysis of aqueous solutions of potassium chloride.

Potassium carbonate, K_2CO_3, is made from po-

Physical properties of potassium metal

Property	Temperature °C	Temperature °F	Metric (scientific) units	British (engineering) units
Density	100	212	0.819 g/cm³	51.1 lb/ft³
	400	752	0.747 g/cm³	46.7 lb/ft³
	700	1292	0.676 g/cm³	42.2 lb/ft³
Melting point	63.7	147		
Boiling point	760	1400		
Heat of fusion	63.7	147	14.6 cal/g	26.3 Btu/lb
Heat of vaporization	63.7	1400	496 cal/g	893 Btu/lb
Viscosity	70	158	5.15 millipoises	6.5 kinetic units
	400	752	2.58 millipoises	3.5 kinetic units
	800	1472	1.36 millipoises	2 kinetic units
Vapor pressure	342	648	1 mm	0.019 lb/in.²
	696	1285	400 mm	7.75 lb/in.²
Thermal conductivity	200	392	0.017 cal/(sec)(cm²)(cm)(°C)	26.0 Btu/(hr)(ft²)(°F)
	400	752	0.09 cal/(sec)(cm²)(cm)(°C)	21.7 Btu/(hr)(ft²)(°F)
Heat capacity	200	392	0.19 cal/(g)(°C)	0.19 Btu/(lb)(°F)
	800	1472	0.19 cal/(g)(°C)	0.19 Btu/(lb)(°F)
Electrical resistivity	150	302	18.7 microhm-cm	
	300	572	28.2 microhm-cm	
Surface tension	100-150		About 80 dynes/cm	

tassium hydroxide and carbon dioxide. It cannot be made by the Solvay process used for sodium carbonate because potassium bicarbonate is too soluble in ammonium chloride solution.

Potassium nitrate, KNO_3, is made by fractional crystallization of an aqueous solution containing sodium nitrate and potassium chloride.

Analytical methods. As in the case of sodium, the high water solubility of most potassium compounds complicates the analytical determination of potassium. Qualitative detection is usually made by means of the violet potassium flame; the sodium flame, which is usually present as well, is masked by viewing the flame through a cobalt-glass filter.

Gravimetric determination of potassium can be made using sodium triphenylboron, sodium perchlorate, or other reagents. Rubidium and cesium interfere in most of these gravimetric methods when they are present. *See* ALKALI METALS.

[MARSHALL SITTIG]

Bibliography: American Chemical Society, *Handling and Uses of the Alkali Metals*, Advances in Chemistry Series, vol. 19, 1957; I. Fatt and M. Tashima, *Alkali Metal Dispersions*, 1961; C. B. Jackson, *Liquid Metals Handbook: Sodium-NaK Supplement*, 3d ed., 1955; R. N. Lyon (ed.), *Liquid Metals Handbook*, 2d ed., Navexos P-733 (rev.), 1954.

Pour point

The lowest temperature at which an oil will pour when cooled under prescribed test conditions. Because petroleum oils are complex mixtures which become plastic solids when cooled, several solidification temperatures are used, each defined by a definite test procedure. The cloud point is the temperature at which a solid phase separates from solution. For a grease, the dropping point is the temperature at which the plastic solid becomes sufficiently fluid to flow through an orifice. For waxes, the melting point is the temperature at which the material becomes fluid enough to drop from the test thermometer; the congealing point is the temperature at which a sample wetting the thermometer appears to congeal and hence rotate with the thermometer.

[MOTT SOUDERS, JR.]

Praseodymium

A chemical element, Pr, atomic number 59, and atomic weight 140.91. Praseodymium is a metallic element of the rare-earth group. The stable isotope 140.907 makes up 100% of the naturally occurring

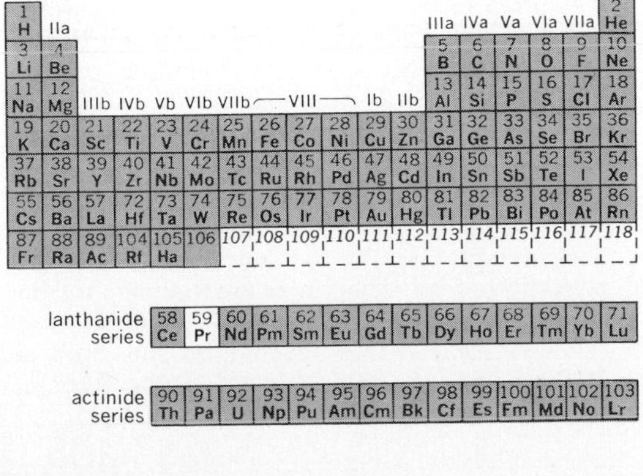

element. It was discovered by C. F. Auer von Welsbach in 1885 when he separated the salts of the so-called element didymium into two fractions, praseodymium and neodymium. The oxide is a black powder, the composition of which varies according to the method of preparation. It is usually considered to be Pr_6O_{11}, although if oxidized under a high pressure of oxygen it can approach the composition PrO_2. It can be reduced in hydrogen to give a pale green Pr_2O_3. The oxide PrO_2 is the only form in which praseodymium has been clearly demonstrated to exist in the quadrivalent form. The black oxide dissolves in acid with the liberation of oxygen to give green solutions or green salts. The salts have found application in the ceramic industry for coloring glass and for glazes. For properties of the metal *see* RARE-EARTH ELEMENTS.

[FRANK H. SPEDDING]

Precipitation

The process of producing a separable solid phase within a liquid medium. In a broad sense, precipitation represents the formation of a new condensed phase, although other terms are often used to describe the process. Thus (1) a vapor or gas condenses to liquid droplets, or more specifically as in meteorology, water vapor in the atmosphere precipitates to form rain, snow, or ice; (2) a substance in the liquid state freezes or solidifies; (3) a dissolved component crystallizes from a supersaturated solution; (4) a new solid phase gradually precipitates within a solid alloy as the result of a slow, inner chemical reaction; or (5) a metal electrodeposits upon the passage of an electrical current through a solution.

In analytical chemistry, precipitation is widely used to effect the separation of a solid phase in an aqueous solution. For example, the addition of a water solution of silver nitrate to a water solution of sodium chloride results in the formation of insoluble silver chloride. Quite often, one of the components in the solution is thus virtually completely separated in a relatively pure form. It can then be isolated from the solution phase by filtration or centrifugation, and the substance determined by weighing. This procedure is known as gravimetric analysis. Precipitation may also be used merely to effect partial or complete separation of a substance for purposes other than that of gravimetric analysis. Such purposes might involve either the isolation of a relatively pure substance or the removal of undesirable components of the solution.

Solubility product constant. The extent to which a component can be separated from solution can be determined from the solubility-product constant obtained by determining the quantity of dissolved substance present in a known amount of saturated solution. This value is known as the solubility. The solubility can be drastically altered merely by adding to the solution any of the ions that make up the precipitate, for example, by adding varying quantities of either silver nitrate or sodium chloride to a saturated solution of silver chloride. Although solubility can be altered over a wide range, the solubility product itself remains practically constant over this same range.

The solubility-product constant can be used to ascertain the quantity of dissolved component remaining unprecipitated in the presence of known concentrations of the ions common to the precipitate. By proper adjustment of the concentration of the added common ion, it is possible to reduce the quantity of dissolved component to a negligible value, although never to zero. It is an extremely important criterion of a method of gravimetric analysis that the quantity of unprecipitated component be negligible, particularly in comparison with the quantity of precipitate formed. The analytical chemist uses the word quantitative to describe such a chemical reaction. *See* SOLUBILITY PRODUCT CONSTANT.

Impurities. The purity of the precipitated solid phase is of major concern, both for the preparation of a desired chemical compound and for a quantitative method of gravimetric analysis. It is not possible for a precipitate to be formed as an absolutely pure compound by chemical reaction within the solution phase. Other soluble substances present in the solution, such as the ions not involved in the structure of the precipitate, tend to accompany the solid phase in varying amounts. This phenomenon is known as coprecipitation. The fraction of the total quantity of such foreign ions coprecipitating may be quite small. Although this fraction depends on experimental variables, it is highly dependent upon the relationship between the solubility characteristics of the desired chemical precipitate and the foreign substance. As a specific example, partial precipitation of iodide (as silver iodide) using silver nitrate would coprecipitate (as silver chloride) only a small fraction of any chloride present, whereas it would coprecipitate (as silver bromide) a larger fraction of any bromide present. The iodide is more insoluble than the bromide, which is more insoluble than the chloride. Knowledge of the relative solubility characteristics of the chemical species present is thus extremely desirable to the chemist, who needs to know whether foreign substances are being collected with the solid or are being left in solution.

A precipitated phase may incorporate foreign ions within its structure in several ways. Best understood of these is isomorphous mixed-crystal formation. Radium and barium sulfates form isomorphous mixed crystals because the two compounds have the same crystal structure and the ionic radii of radium and barium are not greatly different. Thus, radium and barium ions are interchangeable within the crystal lattice to a considerable degree. Silver bromide and silver chloride also form isomorphous mixed crystals. Because of the ease with which interchange can take place within the crystal lattice, isomorphous mixed crystal systems should be avoided if a good separation of the two different ionic species is desired. On the other hand, the property of isomorphism may also be put to good use in concentrating minute traces. For example, barium sulfate precipitated in the presence of minute traces of radium carries with it almost all of the radium. Such procedures are frequently used in the collection of minute traces of radioactive species.

An ion present at high dilution may sometimes be incorporated, apparently by mixed crystal formation, even though such formation would not be predicted on the basis of crystallography and ionic radii. An example of this is the coprecipitation of traces of lead with potassium chloride. This phenomenon is known as anomalous mixed crystal formation, or isodimorphism.

When foreign-ion incorporation cannot be ascribed to isomorphism, coprecipitation may occur by adsorption. Residual charges at the surface of a precipitate attract charged ions in the solution. The adsorption process results in a greater concentration of foreign ions near the precipitate surface than exists in the main body of the solution. Adsorbed foreign ions may remain quite firmly attached to the solid. In fact, they may be covered as succeeding layers of the crystal are deposited and cause imperfections within it. This phenomenon is known as occlusion, although it arises as a result of adsorption. The term occlusion should be distinguished from inclusion, which refers to the mechanical trapping of pockets of solution (and solutes in it) within the precipitate.

Sometimes one substance in a mixture precipitates rapidly, but a second foreign substance then precipitates slowly as a second solid phase. This is not generally considered to be coprecipitation, but is referred to as postprecipitation. The postprecipitation of zinc sulfide with copper sulfide is a typical example.

Methods for reducing contamination. In an effort to reduce contamination by foreign ions, the chemist resorts to various techniques. Precipitation from dilute solution is often effective. Heating the reaction mixture, that is, digesting the solid in contact with the liquid phase, speeds recrystallization processes by which incorporated foreign ions may be returned to the solution phase. Precipitation from homogeneous solution, a technique in which the desired precipitating reagent is formed internally within the solution by chemical synthesis, results in the slow formation of large crystals of small surface area and hence lessens coprecipitation.

If all these methods fail to reduce adequately the quantity of foreign ions incorporated in the solid phase, then reprecipitation is applied. The precipitate is dissolved and reprecipitated by the previous procedure. As in the initial precipitation, most of the foreign ions remain in solution. The process of reprecipitation must be repeated until the quantity of foreign ions present in the precipitate can be disregarded. *See* ADSORPTION; CHEMICAL SEPARATION TECHNIQUES; CRYSTALLIZATION; ELECTRODEPOSITION ANALYSIS; GRAVIMETRIC ANALYSIS; NUCLEATION.

Precipitation from homogeneous solution. In this technique the precipitant is generated in place within the solution phase instead of being added directly, as in the conventional manner.

The substances required to effect precipitation from homogeneous solution can be generated within a solution phase in a variety of ways. For example, if a source of hydroxyl ions is needed to precipitate metallic ions, urea in solution can be hydrolyzed to produce ammonium hydroxide to act as the source. Other anions needed as precipitants can often be formed by the hydrolysis of appropriate esters. Thus, dimethyloxalate can serve as a source of oxalate ions to precipitate thorium, and dimethyl sulfate can be the source of sulfate ions to precipitate lead. There also are methods for the generation of cations. For example, silver ions can be slowly released from a complex ion such as $Ag(NH_3)_2^+$ to precipitate chloride ions. Various

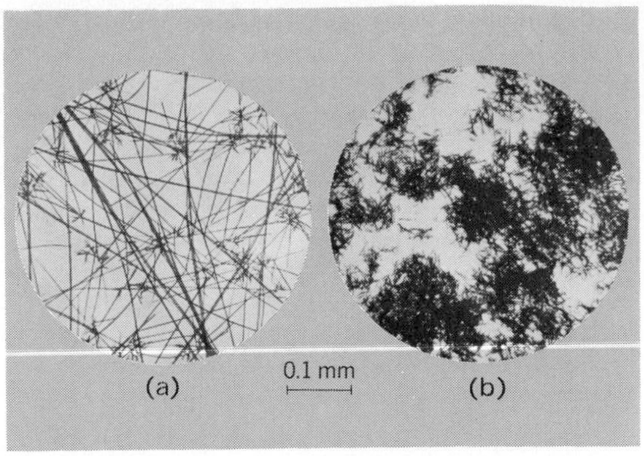

Nickel dimethylglyoximate precipitated by two methods. (a) Precipitation from homogeneous solution. (b) Direct addition to solution. (*From L. Gordon and E. D. Salesin, Precipitation from homogeneous solution, J. Chem. Educ., 38(1):16, 1961*)

investigators have developed ingenious methods for producing the necessary precipitant to effect precipitation from homogeneous solution.

The formation of larger, more perfect, and purer crystals occurs almost without exception when precipitation from homogeneous solution is used in place of direct addition of the precipitant.

The photomicrographs show the difference in appearance of nickel dimethylglyoximate precipitated by two different methods. The precipitate shown in illustration *b* was formed by the direct addition of dimethylglyoxime to a solution containing nickel ions, as in reaction (1). The more perfect

$$CH_3-C=NOH \atop CH_3-C=NOH \; + Ni^{2+} \rightarrow$$

$$CH_3-C=N \underset{Ni}{\overset{OH \cdots O}{\diagdown}} N=C-CH_3 \atop CH_3-C=N \diagup \diagdown N=C-CH_3 \atop O \cdots HO} + 2H^+ \quad (1)$$

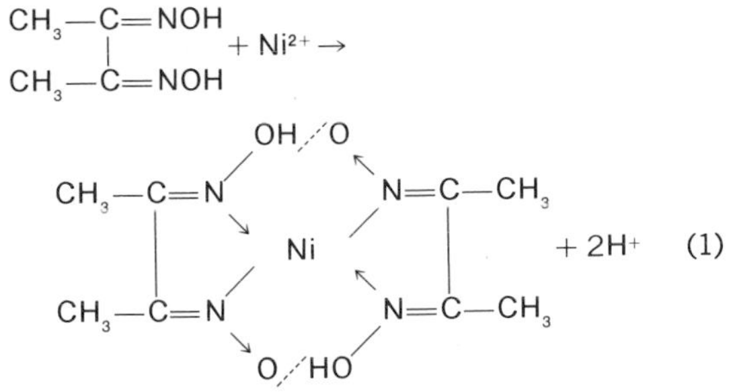

crystals shown in illustration *a* were precipitated from homogeneous solution by synthesizing the necessary dimethylglyoxime from biacetyl and hydroxylamine by reaction (2).

$$CH_3-C=O \atop CH_3-C=O \; + 2H_2NOH \rightarrow$$

$$CH_3-C=NOH \atop CH_3-C=NOH \; + 2H_2O \quad (2)$$

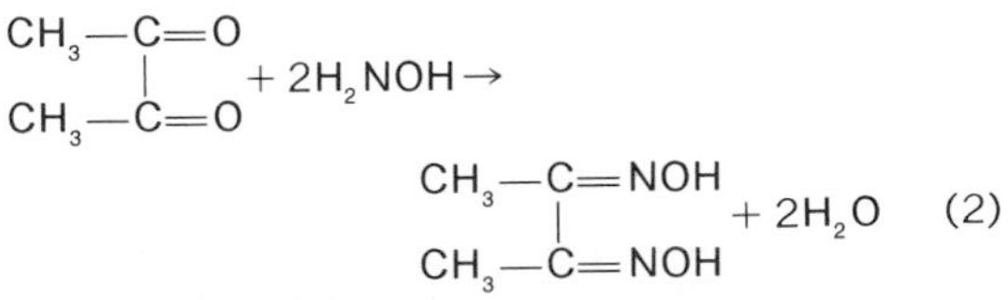

Because the rate at which a precipitant is generated can be controlled very closely, precipitation from homogeneous solution is used as a research technique in studies of the mechanisms of nucleation, precipitation, and coprecipitation. The technique has many applications in gravimetric analysis, in the production of pure chemicals, and in the control of particle size of crystals.

[LOUIS GORDON/ROYCE W. MURRAY]

Bibliography: A. G. Walton, *The Formation and Properties of Precipitates*, 1967, reprint 1979.

Prochirality

The property displayed by a prochiral molecule or a prochiral atom (prostereoisomerism). A molecule or atom is prochiral if it contains, or is bonded to, two constitutionally identical ligands (atoms or groups), replacement of one of which by a different ligand makes the molecule or atom chiral. Examples are shown in Fig. 1.

None of molecules I–IV is chiral, but if one of the underlined pair of hydrogens is replaced, say, by deuterium, chirality results in all four cases. In compound I, ethanol, a prochiral atom or center can be discerned ($C\alpha : CH_2$); upon replacement of H by D, a chiral atom or center is generated, whose configuration depends on which of the two pertinent atoms (H_1 or H_2) is replaced. Molecule V is chiral to begin with, but separate replacement of H_1 and H_2 (say, by bromine) creates a new chiral atom at $C\alpha$ and thus gives rise to a pair of chiral diastereomers. No specific prochiral atom can be discerned in molecules II–IV, which are nevertheless prochiral (III has a prochiral axis).

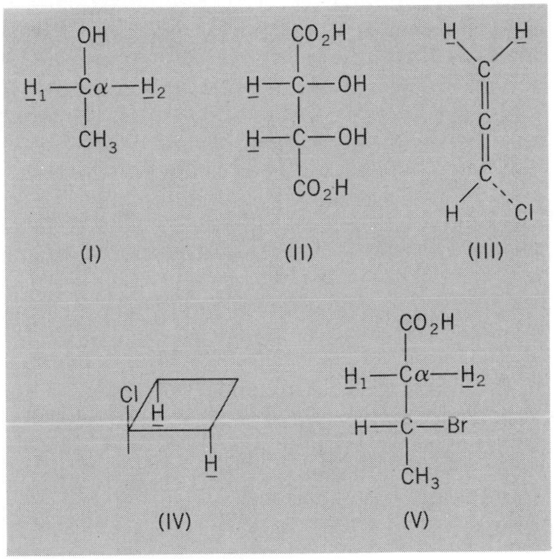

Fig. 1. Prochiral structures.

Faces of double bonds may also be prochiral (and give rise to prochiral molecules), namely, when addition to one or other of the two faces of a double bond gives chiral products. Examples are shown in Fig. 2.

Hydrogenation of VI gives chiral 2-butanol (addition to one face giving rise to one enanti-

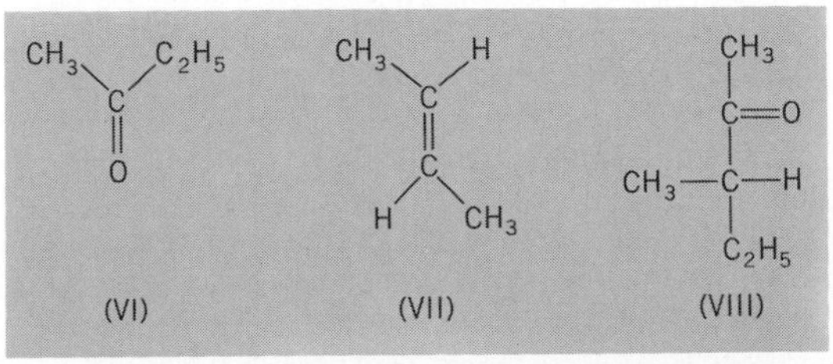

Fig. 2. Structures with prochiral double-bond faces.

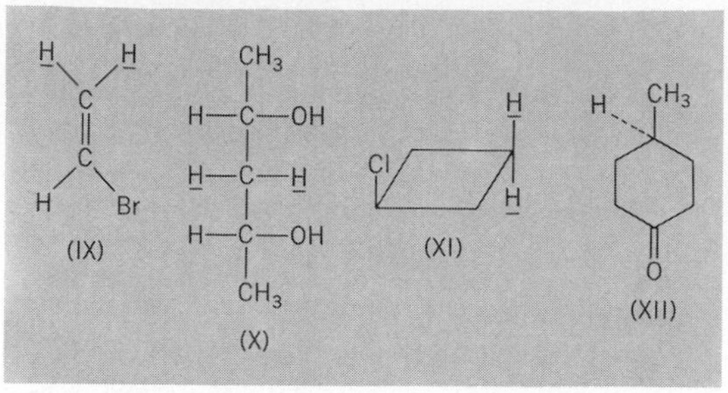

Fig. 3. Structures which can give rise to achiral diastereomers.

omer, addition to the other to the opposite one), and HBr addition to one or the other face of VII gives rise to enantiomeric chiral 2-bromobutanes, $CH_3CHBrCH_2CH_3$. Hydrogenation from one face or the other of VIII also leads to a new chiral center, but in this case diastereomers rather than enantiomers result by addition to the prochiral carbonyl faces.

Prostereoisomerism. Although the term prochirality is widely used, especially by biochemists, a preferred term is prostereoisomerism. This is because replacement of one or other of the two corresponding ligands (called heterotopic ligands; see below) or addition to the two heterotopic faces often gives rise to achiral diastereomers without generation of chirality. Examples are shown in Fig. 3. Replacement of one or another of the underlined pairs of hydrogens in IX by bromine gives rise to *cis*- or *trans*-1,2-dibromoethene; similar replacement in X by OH gives two different *meso*-2,3,4-pentanetriols; analogous replacement in XI by Cl gives *cis*- or *trans*-1,3-dichlorobutane; catalytic addition of hydrogen to the carbonyl double bond in XII gives *cis*- or *trans*-4-methylcyclohexanol. These are stereoisomeric pairs in all cases, but are devoid of chirality. Thus all of the compounds in Figs. 1–3 display prostereoisomerism, but only those in Figs. 1 and 2 display prochirality.

Heterotopic ligands and faces. The pertinent groups (H's) or faces (C=C, C=O) replaced or added to in Figs. 1–3 are called heterotopic (or, more precisely, stereoheterotopic) ligands or faces. If their replacement (see I–IV) or addition to them (see VI–VII) gives rise to enantiomers, they are called enantiotopic. If replacement or addition results in diastereomers (see V, VIII–XII), the ligands or faces are called diastereotopic. The term (stereo)heterotopic comprises both enantiotopic and diastereotopic ligands and faces. Ligands are called homotopic when their separate replacement by different ligands gives rise to identical products, as in $C\underline{H}_3Cl$ or $C\underline{H}_2Cl_2$ (for example, replacing H by Br). Similarly, faces are called homotopic when addition to one or another of them gives rise to identical addition products, as in $(CH_3)_2C{=}O$ or $H_2C{=}CH_2$.

Nomenclature. To name heterotopic ligands, a conceptual replacement is made of one of them by a heavier isotope. If the configuration of the resulting chiral molecule is *R*, the ligand replaced is called *pro-R*; if *S*-configuration results, the ligand is called *pro-S*. Thus replacement of H_1 by D in structure I gives rise to *(S)*-ethanol-1-*d*; hence H_1 is *pro-S* and, by default, H_2 is *pro-R*, a fact which can be established independently in that replacement of H_2 by D yields *(R)*-ethanol-1-*d*. To name heterotopic faces, the Cahn-Ingold-Prelog system (sequence rule) is applied to the three ligands in the plane; if the resulting sequence (highest to lowest precedence) is clockwise, the face is called *Re*; if counterclockwise, *Si*. Thus the front face of the methyl ethyl ketone VI is *Si* ($O \rightarrow C_2H_5 \rightarrow CH_3$) and the front face of the *trans*-2-butene VII is *Re* (${=}C \rightarrow CH_3 \rightarrow H$); the rear faces of these two molecules would have the respective opposite designations (*Re* for compound VI, *Si* for compound VII).

Significance. It is well known that enzymes can usually distinguish between enantiomers; thus liver alcohol dehydrogenase catalyzes the dehydrogenation of *(S)*-lactic acid, *(S)*-$CH_3CHOHCO_2H$, but not of the *R* isomer, to pyruvic acid, CH_3COCO_2H. Similarly enzymes can distinguish between enantiotopic (as well as diastereotopic) ligands or faces, as illustrated by the classical work of F. Westheimer and B. Vennesland (Fig. 4). The fact that yeast alcohol dehydrogenase (YAD) mediates oxidation of *(S)*-ethanol-1-*d* by the coenzyme NAD^+ with total retention of deuterium (that is, clean loss of hydrogen), but of *(R)*-ethanol-2-*d* with integral loss of deuterium (despite the opposing isotope effect), is best explained by saying that the dehydrogenation of ethanol (labeled or not) by YAD-NAD^+ always leads to removal of the *pro-R* hydrogen (H_2 in compound I). Correspondingly, the reverse reaction, reduction of acetaldehyde or acetaldehyde-1-*d* by NAD^2H and NADH, respectively, in the presence of YAD, invariably leads to addition of deuterium or hydrogen to the *Re* face. This type of stereochemical preference is due to the specificity of the fit of the substrate with the active site of the enzyme.

Diastereotopic, but not enantiotopic, groups are in principle distinct (anisochronous) in nuclear magnetic resonance (NMR) spectra. Thus ethanol (I) displays a single (though spin-split) methylene peak, but the underlined protons in V and IX–XI are distinct in their signals. This distinction must be kept in mind in the interpretation of NMR spectra.

In citric acid (XIII) all four marked H's are heterotopic and therefore distinct toward enzymes; only one specific one is implicated in the aconitase-mediated dehydration of citric acid to *cis*-aconitic acid, $HO_2C{-}CH{=}C(CO_2H)CH_2CO_2H$.

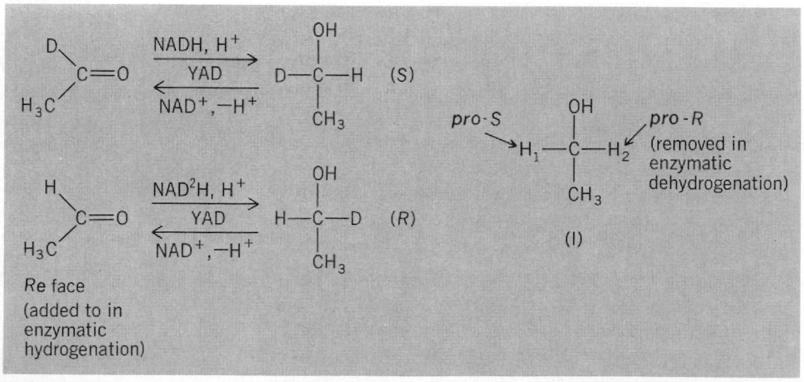

Fig. 4. Reactions illustrating the ability of enzymes to distinguish between enantiotopic ligands or faces.

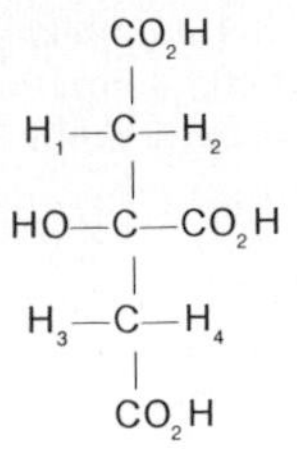

(XIII)

The CH_2CO_2H branch toward which the elimination occurs also can be specifically identified. Moreover, in the biosynthesis of citric acid from oxaloacetate and acetyl-CoA (aldol condensation), the oxaloacetate-derived and acetate-derived branch can be distinguished: the former is *pro-R*, the latter *pro-S*. Since aconitase leads to dehydration toward the *pro-R* branch, it is the *pro-S* branch, that is, the acetyl-CoA derived one, which remains intact. This was demonstrated (by labeling) early in the investigation of the citric acid cycle, and appeared mysterious at the time; the concept of prochirality provides a logical explanation. *See* MOLECULAR ISOMERISM; NUCLEAR MAGNETIC RESONANCE (NMR); STEREOCHEMISTRY.

[ERNEST L. ELIEL]

Bibliography: D. Arigoni and E. L. Eliel, *Top. Stereochem.*, 4:127, 1969; E. L. Eliel, Stereochemical non-equivalence of ligands and faces (heterotopicity), *J. Chem. Educ.*, 57:52–55, 1980; K. R. Hanson, Applications of the sequence rule, I: Naming the paired ligands *g,g* at a tetrahedral atom X_{ggij}, II: Naming the two faces of a trigonal atom Y_{ghi}, *J. Amer. Chem. Soc.*, 88:2731, 1966; W. B. Jennings, Chemical shift nonequivalence in prochiral groups, *Chem. Rev.*, 75:307, 1975; K. Mislow and M. Raban, *Top. Stereochem.*, 1:1, 1967.

Promethium

A chemical element, Pm, atomic number 61. Promethium is the "missing" element of the lanthanide rare-earth series. The atomic weight of the most abundant separated radioisotope is 147. While promethium has a longer half-life, it is much more difficult to accumulate in quantity.

Although a number of scientists have claimed to have discovered this element in nature as a result of observing certain spectral lines, no one has succeeded in isolating element 61 from naturally occurring materials. It is produced artificially in nuclear reactors, since it is one of the products that result from the fission of uranium, thorium, and

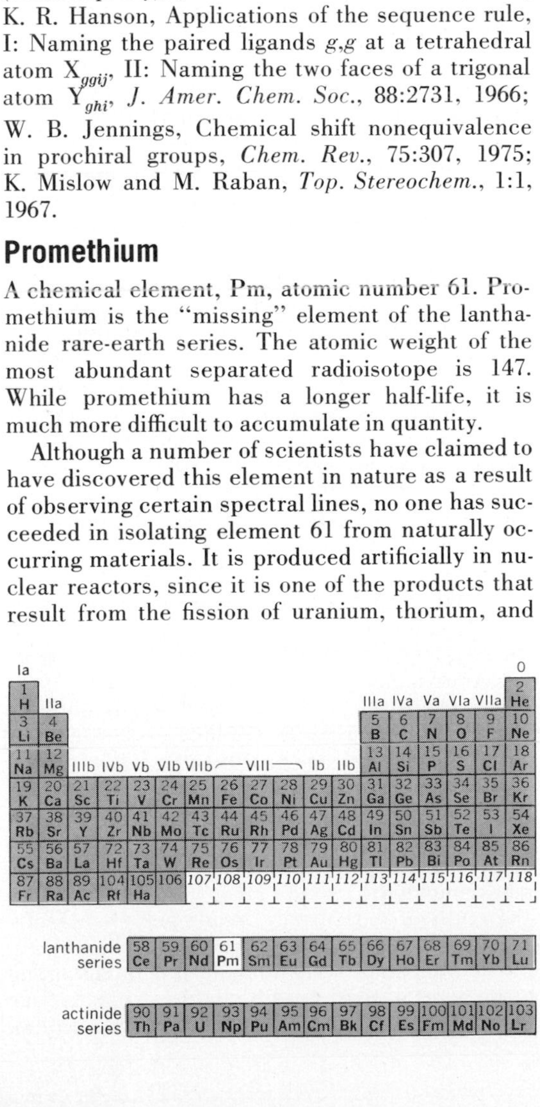

plutonium. In 1945 J. A. Marinsky, L. E. Glendenin, and C. Coryell isolated element 61 from fission product residues. The chemical and metallurgical properties of promethium are very similar to those of neodymium and samarium.

In April, 1958, 2 g of Pm^{147} were recovered at the Oak Ridge National Laboratory. More than 2 kg were subsequently recovered at Hanford.

All the known isotopes are radioactive, and which is most commonly isolated. Its principal Pm^{147}, with a half-life of 2.64 years, is the isotope uses are for research involving tracers. Its main application is in the phosphor industry. It has also been used to manufacture thickness gages and as a nuclear-powered battery in space applications. The oxide, Pm_2O_3, is purple; the nitrate, $Pm(NO_3)_3$, is pink. *See* RARE-EARTH ELEMENTS.

[FRANK H. SPEDDING]

Propane

A member of the alkane or paraffin series of hydrocarbons, formula $CH_3CH_2CH_3$. It makes up 3–18% of natural gas. It is readily liquefied (melting point, −187.7°C; boiling point, −42.1°C), and mixtures with liquefied butane are sold as liquefied petroleum gas (LPG) in cylinders under moderate pressure for domestic fuel.

At temperatures above about 650°C, propane undergoes cracking to ethylene and methane. This reaction is the basis of an important commercial source of ethylene. The reaction is accompanied by some dehydrogenation to propylene. The yield of propylene is increased in the presence of catalyst.

In the petroleum industry, propane is used as a combined solvent and refrigerant for the refining of lubricants and other products. *See* ALKANE; CRACKING.

[LOUIS SCHMERLING]

Propanol

One of the three-carbon saturated, aliphatic alcohols (alkanols). Normal propanol (also known as propyl alcohol, 1-propanol, or ethyl carbinol), $CH_3CH_2CH_2OH$, has chemical properties similar to other primary aliphatic alcohols. It is a colorless, mobile liquid of pungent odor and taste and has the following physical properties: molecular weight, 60.09; boiling point, 97.2°C; melting point, −126.2°C; specific gravity, 0.8036 at 20°C. The compound is miscible with water and is soluble in most organic liquids. Normal propanol is toxic and flammable.

The hydrogenation of propionaldehyde, produced by the hydroformylation of ethylene, is the major commercial route to normal propanol. Its major uses are as a solvent, for production of propyl acetate, and as an intermediate to other chemicals.

Isopropanol, $CH_3CHOHCH_3$ (also known as 2-propanol, isopropyl alcohol, and dimethyl carbinol), is the simplest of the secondary alcohols and is a major industrial organic chemical. It is a colorless, mobile, and toxic liquid (molecular weight, 60.09; boiling point, 82.3°C; melting point, −88.5°C; specific gravity, 0.786 at 20°C) and has a pungent odor and taste. It is miscible with water and soluble in most organic liquids.

It is produced by the catalytic hydration of propylene and is used primarily for the manufacture of acetone by catalytic dehydrogenation. Isopropanol is also used as a solvent, an extractant, an antifreeze, and as a rubbing alcohol.

[P. DWIGHT SHERMAN, JR.]

Bibliography: J. A. Monick, *Alcohols, Their Chemistry, Properties and Manufacture*, 1968.

Propylene

A gas, $CH_3—HC{=}CH_2$, boiling point −48°C, melting point −185°C. All processes for thermal or catalytic cracking of hydrocarbons yield propylene. A typical fraction of a refinery stream containing 2- and 3-carbon molecules is 10–30% propylene. Preparation outside of normal refinery operations is usually by catalytic dehydrogenation of propane. In those refineries having integrated isopropyl alcohol or polymer units for making polymer gasoline, the propylene is utilized in the dilute concentration found in the cracked gas streams. When highly concentrated streams are required, the propylene is recovered in the same manner as ethylene. Major uses for propylene are in the production of isopropyl alcohol, from which acetone is obtained; polypropylene plastics; tripropylene for the alkylation of phenol to produce alkyl phenols, from which are derived nonionic detergents and lubricating oil additives; tetrapropylene for the alkylation of benzene to produce alkyl benzenes, from which are derived alkyl-aryl sulfonate detergents; propylene chlorohydrin for producing propylene oxide from which are obtained detergents, hydraulic fluids, and lubricants; acrolein; allyl alcohol; allyl chloride; epichlorohydrin; synthetic glycerin; butyraldehyde; *n*-butyl alcohol; and isobutyl alcohol. *See* ALKENE; ETHYLENE; POLYOLEFIN RESINS.

[CHARLES A. COHEN]

Bibliography: E. G. Hancock, *Propylene and Its Industrial Derivatives*, 1973.

Protactinium

A chemical element, symbol Pa, atomic number 91. K. Fajans and O. Göhring discovered the isotope of mass number 234 in 1913 and named it brevium because of its short (1.175-min) half-life. However, ^{231}Pa, the parent of actinium, was first

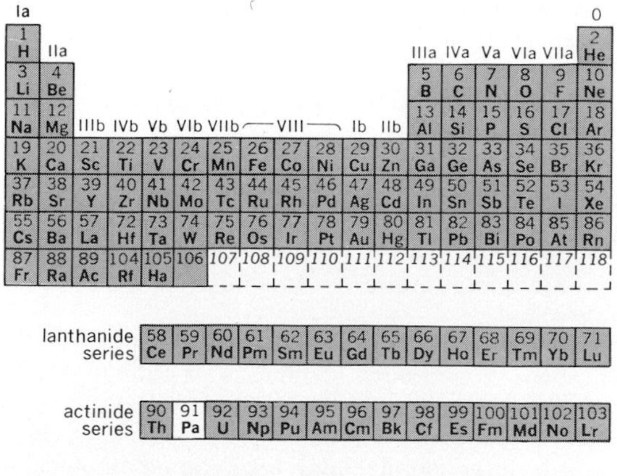

isolated in 1918 by O. Hahn and L. Meitner and, independently, by F. Soddy and J. A. Cranston. *See* ACTINIUM.

Isotopes of mass numbers 216; 217, and 222–238 are now known, all of them radioactive. Only ^{231}Pa, ^{234}Pa (originally termed UZ), and ^{234m}Pa (originally termed UX_2) occur in nature. The most important of these is ^{231}Pa, an α-emitter with a half-life of 32,500 years. Its natural abundance, 0.87×10^{-6} ppm, is strictly limited by that of its primordial ancestor, ^{235}U, which represents only 0.711 wt % of natural uranium. Consequently, only a little more than 100 g of ^{231}Pa has ever been isolated from natural sources. The artificial isotope, ^{233}Pa, is important as an intermediary in the production of fissile ^{233}U. Both ^{231}Pa and ^{233}Pa can be synthesized by neutron irradiation of thorium, according to reactions (1) and (2). *See* RADIOACTIVITY; URANIUM.

$$^{232}\mathrm{Th}(n,2n)^{231}\mathrm{Th} \xrightarrow[25.52\ \mathrm{h}]{\beta^-} {}^{231}\mathrm{Pa} \xrightarrow[32{,}500\ \mathrm{years}]{\alpha} {}^{227}\mathrm{Ac} \qquad (1)$$

$$^{232}\mathrm{Th}(n,\gamma)^{233}\mathrm{Th} \xrightarrow[22.3\ \mathrm{min}]{\beta^-} {}^{233}\mathrm{Pa} \xrightarrow[26.95\ \mathrm{days}]{\beta^-} {}^{233}\mathrm{U} \qquad (2)$$

Protactinium is, formally, the third member of the actinide series of elements and the first in which a 5*f* electron appears, but its chemical behavior in aqueous solution resembles that of tantalum and niobium more closely than that of the other actinides. The predominant oxidation state is +5, but Pa(V) forms no simple cations in aqueous solution. Like Ta(V), it exhibits an extraordinarily high tendency to hydrolyze, polymerize, and adsorb on almost any available surface. This tendency is suppressed in the presence of appropriate concentrations of complexing agents, such as fluoride, sulfate, and oxalate ions. Pa(V) is reduced to Pa(IV) by zinc amalgam or $CrCl_2$ in moderately concentrated solutions of HCl or H_2SO_4. Electrolytic reduction has also been reported. Pa(IV) is rapidly oxidized by air. *See* NIOBIUM; TANTALUM.

Trace amounts of Pa(V) are coprecipitated by most insoluble compounds of Zr(IV) and by the hydroxides of Fe(III), Ti(IV), Ta(V), and Nb(V). Coprecipitation with MnO_2 is often used to concentrate Pa(V). Purification is most commonly performed by extraction from aqueous solution with organic solvents, such as diisopropyl and diisobutyl ketones, diisobutyl carbinol, and tri-*n*-octyl phosphine oxide. Weighable amounts of Pa(V) are precipitated from aqueous solution by hydroxide, iodate, phosphate, peroxide, and phenylarsonate ions. Pa(V) also forms the sparingly soluble double salts K_2PaF_7 and $BaPaF_7$. Pa(IV), unlike Pa(V), is precipitated by fluoride ion, but forms soluble carbonate, citrate, and tartrate complexes.

Metallic protactinium has been prepared by reduction of PaF_4 with Ba or Li vapor at 1300–1400°C and, most recently, by a modified van Arkel technique in which PaC is reacted with 10% of the stoichiometric amount of I_2 required to form PaI_5. The volatile product is then decomposed to yield a deposit of metallic Pa on a tungsten filament heated to more than 1360°C. The metal is silver in color, malleable, and ductile. The crystal structure is body-centered tetragonal, with unit cell dimensions $a_0 = 0.3931$ nm and $c_0 = 0.3236$ nm. Samples exposed to air at room temperature show little or no tarnishing over a period of several months.

The numerous compounds of protactinium that have been prepared and characterized include binary and polynary oxides, halides, oxyhalides, sulfates, oxysulfates, double sulfates, oxynitrates, selenates, carbides, organometallic compounds, and noble metal alloys. *See* ACTINIDE ELEMENTS.

[H. W. KIRBY]

Bibliography: C. Keller (ed.), *Protactinium*. Gmelin Handbuch der Anorganischen Chemie, System-Nummer 51, Ergänzungsband 1, 1977; R. Muxart and R. Guillaumont, *Protactinium*, Compléments au Nouveau Traité de Chimie Minérale, vol. 2, 1974; E. S. Pal'shin, B. F. Myasoedov, and A. V. Davydov, *Analytical Chemistry of Protactinium* (transl.), 1970.

Pyran

One of a group of organic compounds containing a heterocyclic six-membered ring made up of one oxygen and five carbon atoms. Formulas (I) to (V) show pyrans of various kinds. *See* HETEROCYCLIC COMPOUNDS.

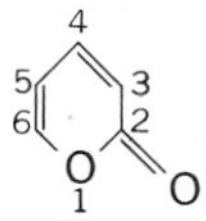

(I)
α-Pyrone

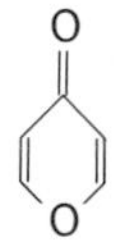

(II)
γ-Pyrone

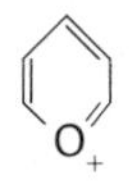

(III)
Pyrylium ion

(IV)
$\Delta^{2,3}$-Dihydropyran

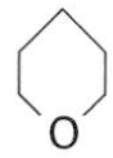

(V)
Tetrahydropyran

All pyranosidic carbohydrates are tetrahydropyran derivatives. Glycals, for example, glucal, are $\Delta^{2,3}$-dihydropyrans. The α-pyrone system is present in cardiac-active steroid derivatives, such as scillaren A, isolated from squill and from a variety of toad. Meconic acid from opium, chelidonic acid from *Chelidonium majus*, and kojic acid from *Aspergillus* molds are γ-pyrones. Maltol is a degradation product from carbohydrate materials.

α-Pyrones as lactones can be saponified, and as dienes can react with Diels-Alder dienophiles. Both α- and γ-pyrones react with ammonia to give the corresponding pyridones. γ-Pyrones can be cleaved with alkali. Other properties suggest that γ-pyrones are resonance hybrids.

Formulas (VI) to (IX) show some benzopyran types. Vitamin E, or α-tocopherol, is a chroman (VI) derivative. Coumarin, or α-chromone (VII), the fragrant body in clover, was formerly used in flavors and perfumes. Dicoumarol, originally en-

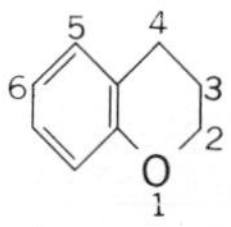

(VI)

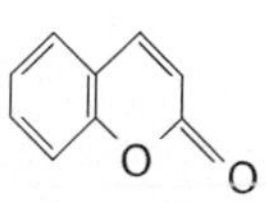

(VII)

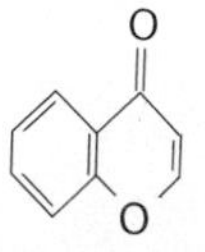

(VIII)

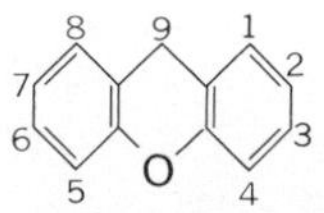

(IX)

countered as the causative agent in hemorrhagic sweet clover disease of cattle, now finds clinical use as a blood anticoagulant. 2-Phenyl-γ-chromone derivatives are leaf pigments. Fluorescein (X), used as a sea marker because of its intense greenish fluorescence; eosin, used as the dye in red ink;

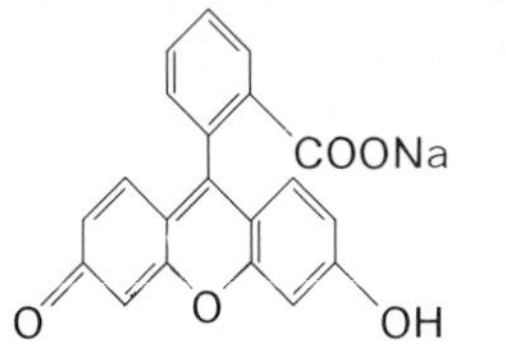

(X)

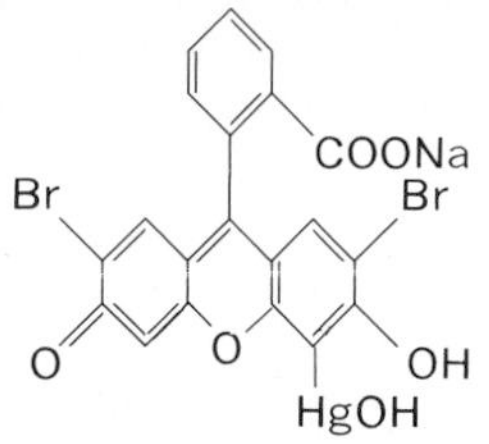

(XI)

and mercurochrome (XI), used as a household antiseptic, are all highly colored derivatives of xanthene (IX). Rotenone (XII), an insecticide from

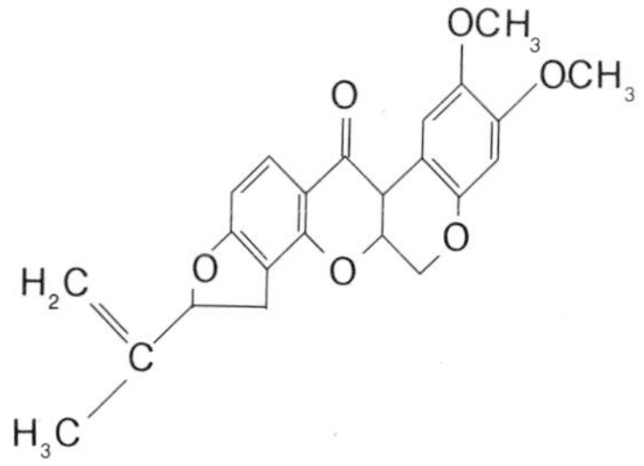

(XII)

Derris elliptica (Malaya) has three oxygen-heterocyclic rings, two of which are pyranoid. The active ingredient in hashish as well as in marijuana—both from hemp (*Cannabis sativa*)—is tetrahydrocannabinol (XIII), a tricyclic pyran derivative.

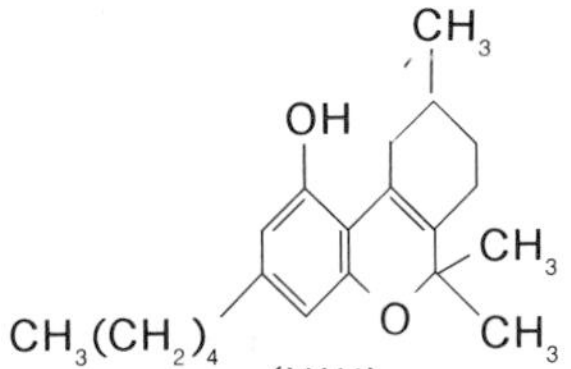

(XIII)

[WALTER J. GENSLER]

Bibliography: R. M. Acheson, *An Introduction to the Chemistry of Heterocyclic Compounds*, 3d ed., 1976; T. A. Geissman (ed.), *The Chemistry of Flavonoid Compounds*, 1962; D. W. Young, *Heterocyclic Chemistry*, 1976.

Pyrazole

One of a group of organic heterocyclic compounds with two nitrogen atoms occupying adjacent positions in a doubly unsaturated five-membered ring. The pyrazole system is resonance-stabilized and aromatic in character. Electrophilic substitution occurs preferentially at the 4 position. The nucleus is resistant to disruption by oxidation. Certain drugs and dyes are pyrazole derivatives. *See* AZOLE; HETEROCYCLIC COMPOUNDS.

The parent compound (see illustration) is a water-soluble, colorless solid, mp 70°C, bp 187°C, with an odor resembling that of pyridine. Pyrazole is both a

PYRAZOLE

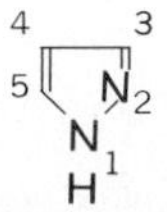

Formula of parent compound for pyrazole.

weak base (pK_a 2.53 at 25°) and a weak acid. The hydrogen at position 1 can be replaced by potassium or by bromomagnesium.

One general synthesis combines 1,3-dicarbonyl compounds with hydrazines; for example, the reaction of acetylacetone with hydrazine, as in (1), gives 3,5-dimethylpyrazole. A more common synthetic method, reaction (2), condenses α,β-unsaturated aldehydes or ketones with hydrazines to give, not fully aromatic compounds, but instead dihydropyrazoles, or pyrazolines. Pyrazolines are also obtained when diazoalkanes react with olefinic compounds, as in the reaction (3) of diazomethane with an ester of maleic acid, to give a pyrazoline. These pyrazolines can be dehydrogenated to pyrazoles. Pyrazoles can be obtained directly when, in the last two reactions, acetylenic unsaturation replaces the olefinic unsaturation. Pyrazolines are prepared by ring syntheses, reactions (2) and (3), or by reduction of pyrazoles. Pyrazolines are susceptible to oxidation. When there is no substituent on either nitrogen, pyrolysis of pyrazolines, as in reaction (4), gives rise to cyclopropanes or substituted olefins.

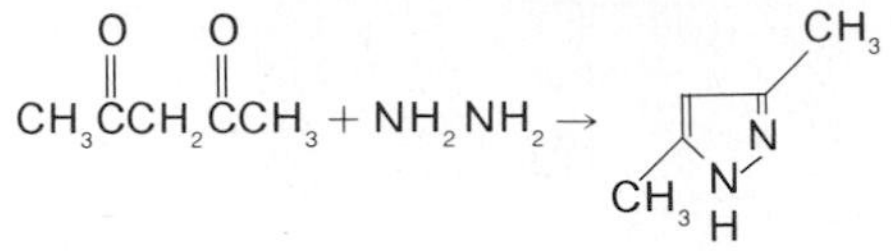

3,5-Dimethyl-pyrazole (1)

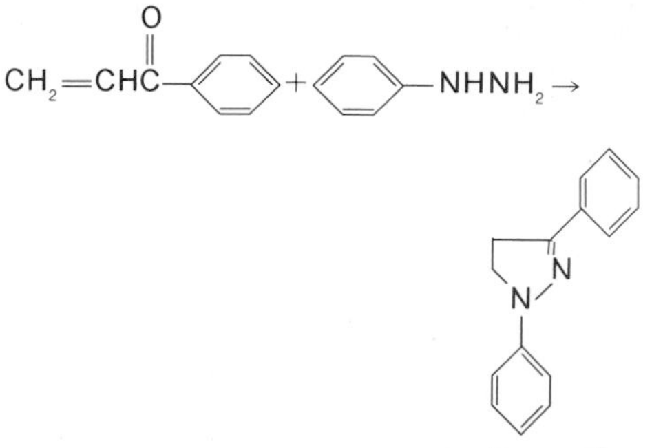

Pyrazoline (2)

$ROOCCH{=}CHCOOR + CH_2N_2 \rightarrow$

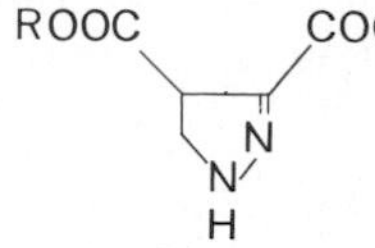

1,3-Diphenyl-pyrazoline (3)

Antipyrine and aminopyrine are pyrazole derivatives of considerable value as analgesics and antipyretics, respectively. The large-scale synthesis of

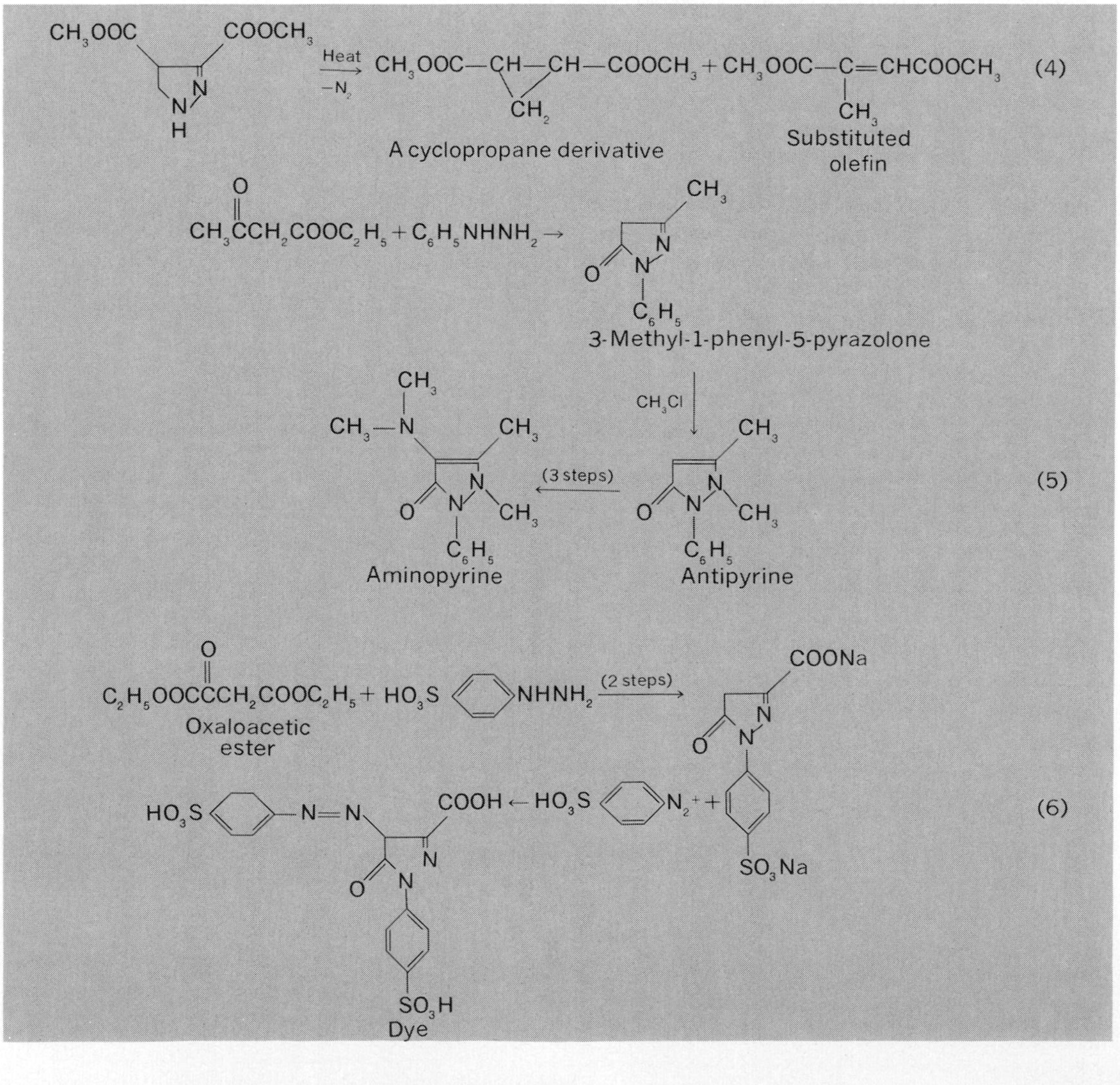

these two materials, reaction (5), starts with the condensation of phenylhydrazine with acetoacetic ester to give 3-methyl-1-phenyl-5-pyrazolone, also known as methylphenylpyrazolone or as Developer Z, followed by methylation to give antipyrine. Subsequent nitrosation, reduction (with zinc), and dimethylation leads to aminopyrine.

Azo coupling of 3-methyl-1-phenyl-5-pyrazolone or of related compounds gives 4-azopyrazolone derivatives, which are of interest as wool, food, and photographic dyes. The synthesis of one such dye starts with oxaloacetic ester and proceeds as indicated in reaction (6).

[WALTER J. GENSLER]

Bibliography: R. M. Acheson, *An Introduction to the Chemistry of Heterocyclic Compounds*, 1976; D. W. Young, *Heterocyclic Chemistry*, 1976.

Pyridine

An organic heterocyclic compound containing a triunsaturated six-membered ring of five carbon atoms and one nitrogen atom. Pyridine (I) and pyri-

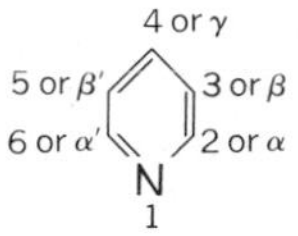

(I)

dine homologs are obtained by extraction of coal tar or by synthesis. The following are available in commercial quantities: pyridine, 2-, 3-, and 4-methylpyridine (also known respectively as α-, β- and γ-picoline), 2,4-dimethyl-, 2,6-dimethyl-, and 3,5-dimethylpyridine (also known respectively as 2,4-, 2.6-, and 3,5-lutidine), 5-ethyl-2-methylpyridine (also called aldehydecollidine), and 2,4,6-trimethylpyridine (also called 2,4,6-collidine). Other pyridine derivatives produced on a large scale include nicotinic acid (pyridine-3-carboxylic acid) for preparation of nicotinamide, nicotine (II) for its insecticidal properties, 2-aminopyridine for synthesis of medicinals, piperidine (hexahydropyridine) as a solvent, and 2-vinylpyridine as a polymerizable monomer. The pyridine system is found in natural products, for example, in nicotine (II) from tobacco, in ricinine (III) from castor bean, in pyridoxine or vitamin B_6 (IV), in nicotinamide or niacinamide or vitamin P (V), and in several groups of alkaloids. *See* HETEROCYCLIC COMPOUNDS.

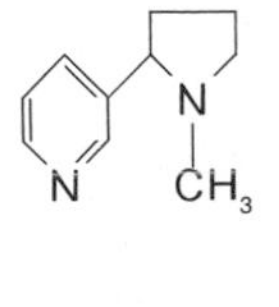

(II)

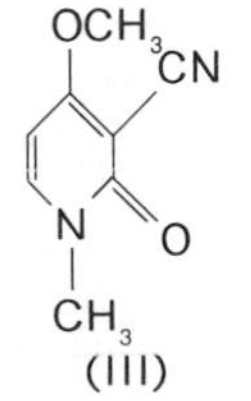

(III)

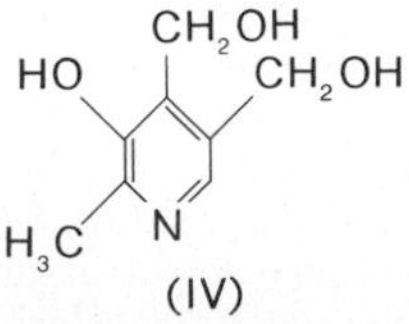

(IV)

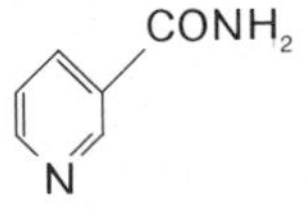

(V)

Properties. Pyridine (I) is a colorless, hygroscopic liquid with a pungent, unpleasant odor. When anhydrous it boils at 115.2 – 115.3°C, its density (20/4) is 0.98272, and n_D^{20} is 1.50920. Pyridine is miscible with organic solvents as well as with water. A constant-boiling mixture, bp 92°C forms with three molecules of water. Dry pyridine is obtained by treatment with barium oxide followed by calcium hydride or phosphorus pentoxide. Pyridine is a tertiary amine (pK_a 5.17 at 25°) that combines readily with Brönsted and Lewis acids. The pyridine system is aromatic. It is stable to heat, to acid, and to alkali. It undergoes electrophilic substitution with difficulty, with the 3-position favored in sulfonation and nitration. On the other hand, it undergoes nucleophilic substitution with ease at the 2-position, and slightly less readily at the 4-position. Its resonance energy is 35 kcal/mole. Pyridine is used as a solvent for organic and inorganic compounds, as an acid binder, as a basic catalyst, and as a reaction intermediate.

Pyridine is an irritant to skin (eczema) and other tissues (conjunctivitis), and chronic exposure has been known to cause liver and kidney damage. Repeated exposure to atmospheric levels greater than 5 parts per million is considered hazardous.

Oxidation of pyridine homologs by nitric acid or by permanganate converts the substituent group in a preparative manner to a carboxylic acid. Reaction at the methyl group of 2- and 4-methylpyridine tends to occur more readily than at the methyl group of 3-methylpyridine. Thus, 2- and 4-methylpyridine condense with benzaldehyde to give styryl derivatives (VI), whereas 3-methylpyridine

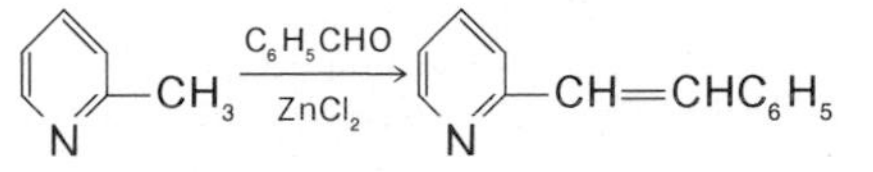

(VI)

does not react. With butyllithium or sodium amide, 2- and 4- methylpyridines metalate to give pyridylmethyl metals (VII), which react normally with carbonyl compounds and with alkyl halides.

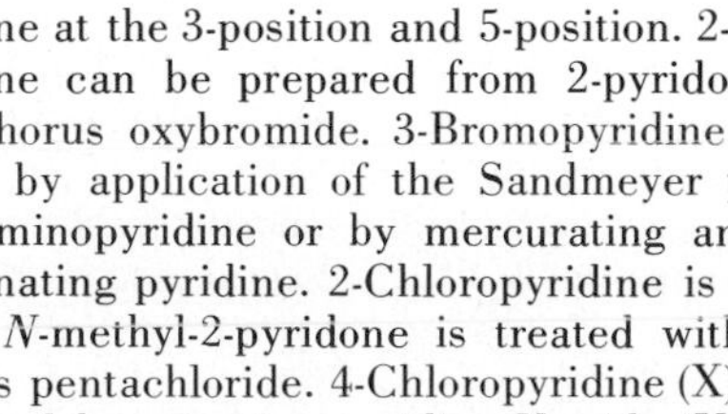

(VII)

Many halogenated derivatives of pyridine are known. Direct bromination at 300 – 400°C places bromine at the 3-position and 5-position. 2-Bromopyridine can be prepared from 2-pyridone and phosphorus oxybromide. 3-Bromopyridine is prepared by application of the Sandmeyer process to 3-aminopyridine or by mercurating and then brominating pyridine. 2-Chloropyridine is formed when *N*-methyl-2-pyridone is treated with phosphorus pentachloride. 4-Chloropyridine (X) can be prepared by nitrating pyridine-*N*-oxide (VIII), exposing the 4-nitropyridine-*N*-oxide (IX) to the action of concentrated hydrochloric acid, and removing the oxide oxygen by iron – acetic acid reduction, but is usually made by heating *N*-(4-pyridyl) pyridinium chloride (from pyridine and thionyl chloride). The 2- and 4-halo substituents are more readily replaced by hydroxy, alkoxy, and amino than the 3-halo substituent. Bromopyridines form Grignard and lithium derivatives that react normally.

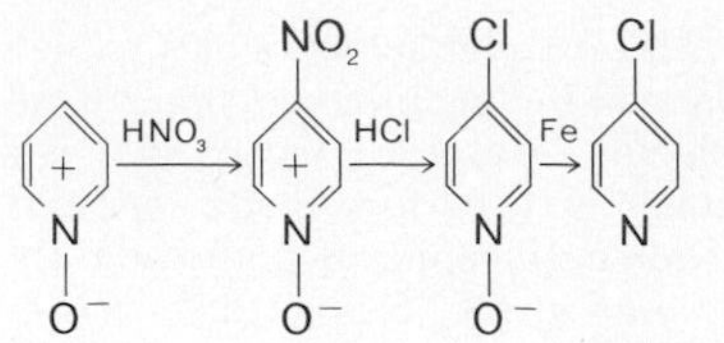

(VIII) (IX) (X)

3-Nitropyridine, from nitration of pyridine, can be converted to 3-aminopyridine. The latter is usually prepared by Hofmann rearrangement of nicotinamide (V). 4-Nitropyridine-*N*-oxide (IX) with iron and acetic acid gives 4-aminopyridine. 2-Aminopyridine, prepared on an industrial scale by direct amination of pyridine with sodamide, is utilized in the manufacture of the antihistaminic, pyribenzamine (XI), and the bacteriostatic agent, sulfapyridine (XII).

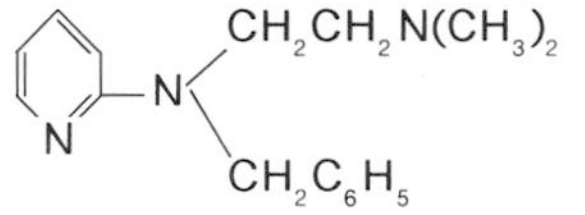

(XI)

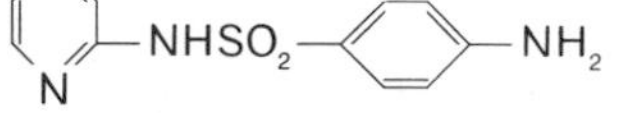

(XII)

In most of its reactions, 3-hydroxypyridine behaves as a normal phenol, just as 3-aminopyridine behaves as a normal aromatic amine. However, hydroxy or amino groups at the pyridine 2- or 4-position show some reactions that are not characteristic of phenols or aromatic amines. Hydroxyl or amino groups on any pyridine position make electrophilic substitution easier and, as ortho-para directing groups, take control of the orientation. 2-Pyridone is prepared from 2-aminopyridine by a diazotization procedure. 3-Hydroxypyridine is produced by sulfonation of pyridine followed by alkali fusion of the pyridine-3-sulfonic acid.

Preparation. Laboratory synthetic methods can lead to pyridines with no oxygen at positions 2 or 6, or to 2-hydroxypyridines, or to 2,6-dihydroxypyridines. The last two pyridine derivatives exist almost entirely in their tautomeric forms, that is, as 2-pyridone (XIII) and 6-hydroxy-2-pyridone, respectively. Glutaconic acids cyclize with ammonia to give 6-hydroxy-2-pyridones. The oxygen at the 2-position and 6-position can be removed by standard conversions with phosphorus oxychloride to the 2,6-dichloro derivative, followed by reductive dechlorination. In this way, for example, β-methylglutaconic acid (XIV) can be converted to 4-methylpyridine (XV). 2-Pyridones (XVII) are formed when 1,3-dicarbonyl compounds (XVI), or their equivalents, react with cyanoacetamide. Subsequent steps remove the cyano groups as well as the oxygen from (XVII) to furnish substituted pyridines (XVIII).

1,5-Dicarbonyl compounds or their equivalents cyclize with ammonia to give pyridines. Thus, the diketone (XIX) from acetoacetic ester and ethyl orthoformate gives the pyridine (XX). In a closely related process, the reaction of ethyl β-aminocrotonate (XXI) with ethoxymethyleneacetoacetic ester (XXII) gives the same product. The Hantzsch synthesis combines four molecules to form an intermediate dihydropyridine derivative (XXIII), which can be readily oxidized to the corresponding

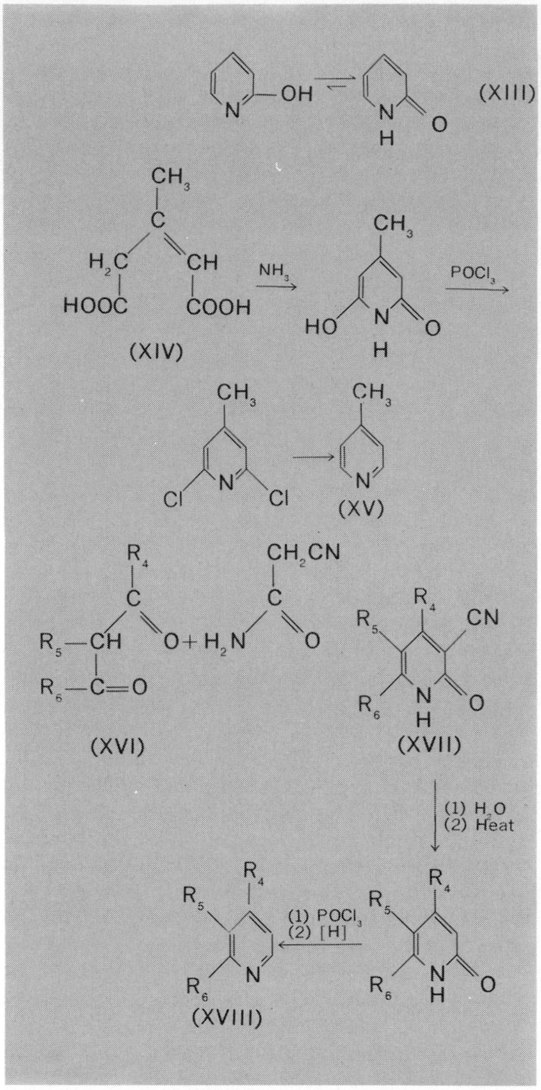

completely aromatic derivative (XXIV). The Chichibabin aldehyde-ammonia synthesis involves the condensation of ammonia with aldehydes and ketones. Generally, mixtures of pyridines are obtained, with the course of the reaction depending on such factors as nature and proportion of reactants, reaction time, temperature, and catalyst. In a commercial process, 5-ethyl-2-methylpyridine (XXV) is obtained in unusually high yields (60–70%) from acetaldehyde and ammonia. Synthetic industrial pyridine and its homologs are prepared from aldehydes and ammonia over a suitable catalyst.

Derivatives. Pyridine compounds containing positively charged nitrogen include simple and quaternary pyridinium salts, acylpyridinium salts, and pyridine-*N*-oxides. The reaction of pyridine with alkylating agents gives quaternary salts, crystalline solids whose aqueous solutions conduct electricity. The quaternary salts are readily oxidized by alkaline ferricyanide to the *N*-substituted-2-pyridones. With acyl halides, pyridine forms *N*-acylpyridinium salts, which are powerful acylating agents for OH, NH, and SH groupings. Pyridine-*N*-oxide (XXVI), prepared by oxidation of pyridine with organic peracids, reacts with phosphorus pentachloride to give 2- and 4-chloropyridine and

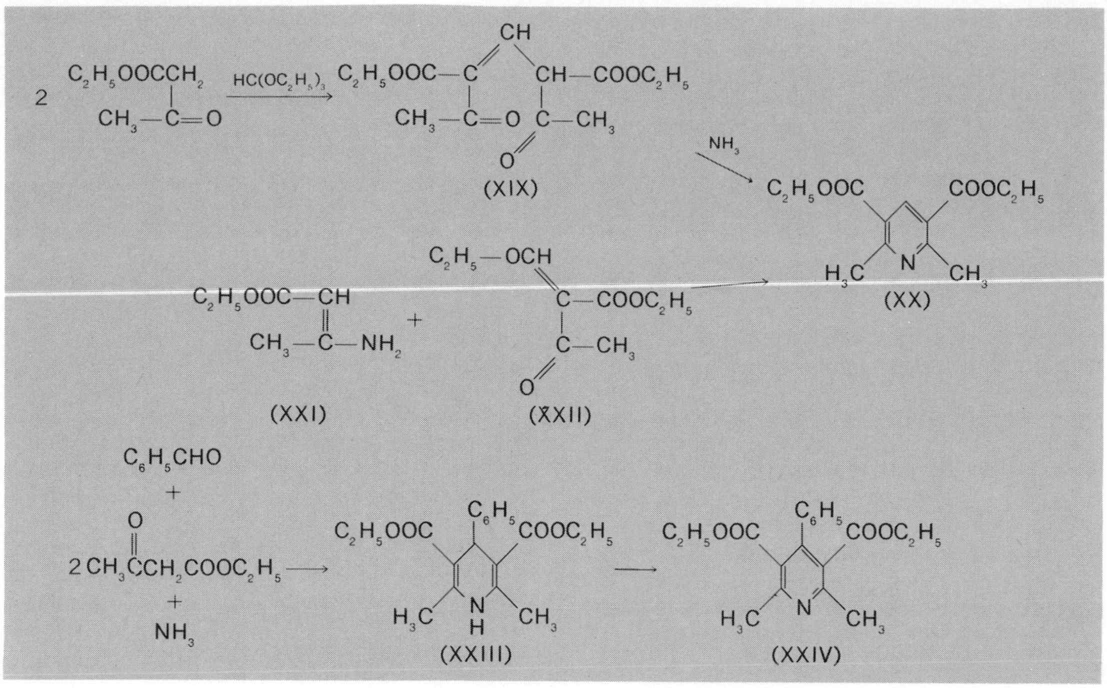

with acetic anhydride to give 2-acetoxypyridine. Pyridine-*N*-oxides nitrate at the 4-position, and by so doing provide a route to 4-substituted derivatives.

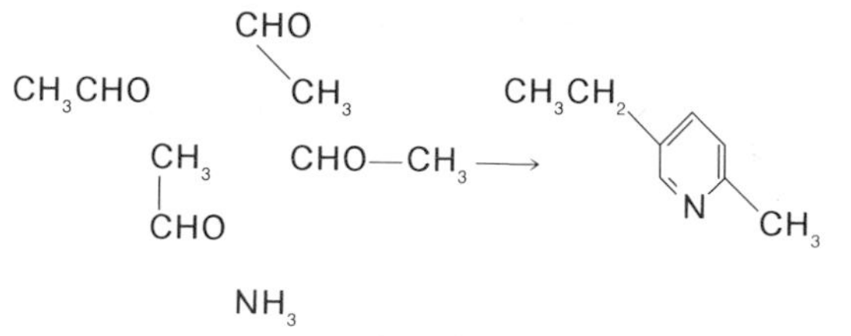

(XXV)

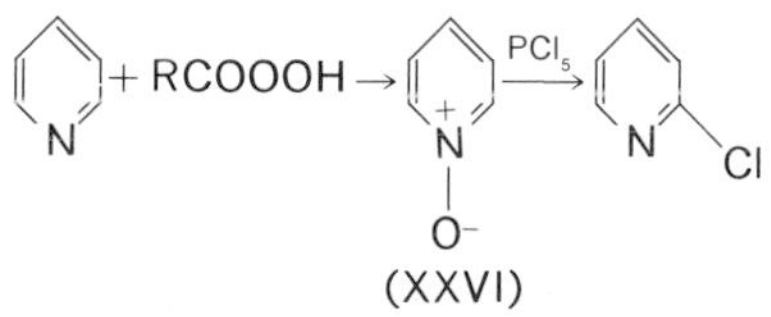

(XXVI)

Pyridine aldehydes are prepared by oxidation of groups already on the ring. Acetylpyridines can be synthesized from pyridine carboxylic esters and ethyl acetate by the Claisen condensation. The reactions of pyridine aldehydes and ketones are normal.

Many pyridine carboxylic acids are known. Their reactions as dipolar materials are not exceptional. Thermal decarboxylation is a standard process, with loss of carboxyl from position 2 easier than from 3 or 4, and loss of carboxyl from position 4 easier than from 3. Nicotinic acid (XXVII) is manufactured by oxidation of 3-methylpyridine (β-picoline) or nicotine; it is also obtained by oxidation of either 2-methyl-5-ethylpyridine or quinoline, followed by decarboxylation of the resulting pyridine dicarboxylic acids. Pyridine-2-carboxylic acid (α-picolinic acid) can be prepared by oxidation of 2-methylpyridine or by carbonation of 2-pyridyllithium. Pyridine-4-carboxylic acid (isonicotinic acid) is obtained by oxidation of 4-methylpyridine or by synthesis from citric acid. See (XXVIII) to (XXIX). The acid hydrazide of isonicotinic acid (Isoniazid) is a tuberculostatic agent.

Although dihydro- and tetrahydropyridines are known, the hexahydropyridines, or piperidines, are the most common reduced forms of pyridine. The piperidines may be prepared by reduction of pyridines or by cyclization of bifunctional compounds, for example, the conversion of 5-bromo-1-aminopentane (XXX) to piperidine.

Piperidine (XXXI), the parent compound, is a colorless, unpleasant-smelling liquid (bp 105.6°C), completely miscible with water. The general properties of piperidine are those of a normal secondary aliphatic amine and, as such, piperidine (pK_a 11.1) is a much stronger base than pyridine (pK_a 5.17). Piperidine carboxylic acids have been

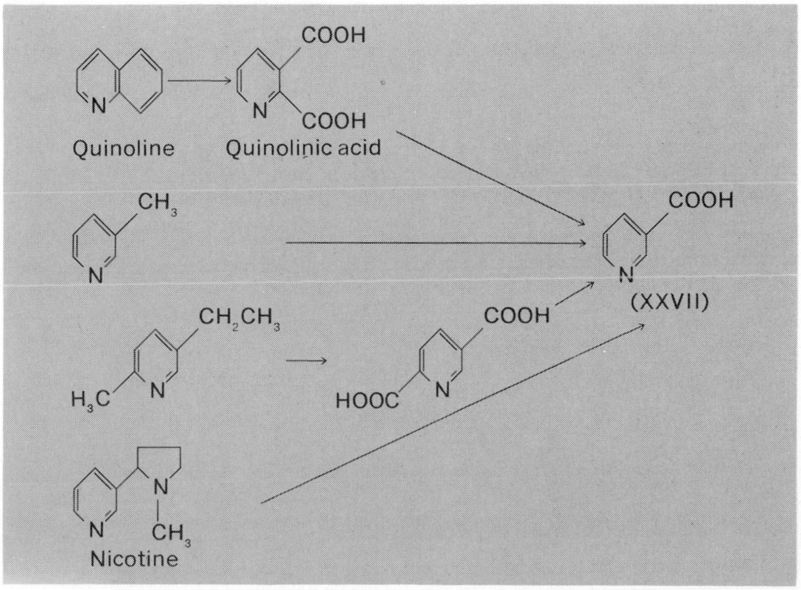

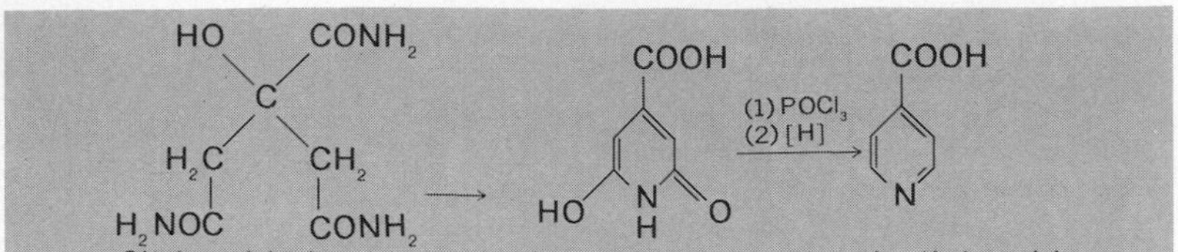

Citric acid triamide (XXVIII) — Citrazinic acid — Isonicotinic acid (XXIX)

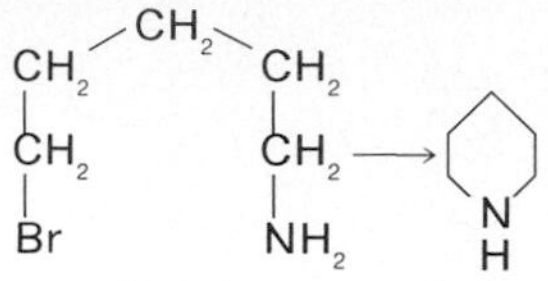

(XXX) (XXXI)

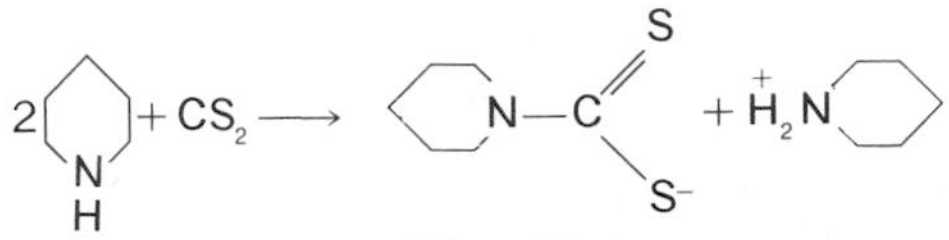

Piperidinium pentamethylene-dithiocarbamate (XXXII)

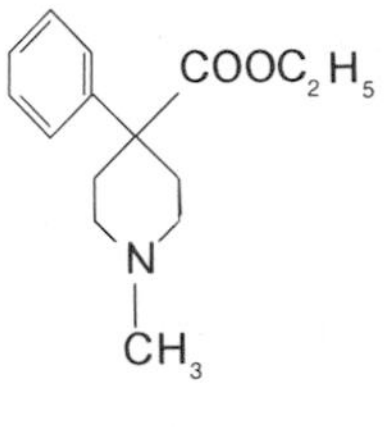

(XXXIII)

investigated in connection with naturally occurring amino acids (for example, pipecolinic acid is piperidine-2-carboxylic acid) as well as with the degradative and synthetic chemistry of quinine. The reaction product (XXXII) from piperidine and carbon bisulfide is a rubber accelerator. 4,4-Disubstituted piperidines such as Demerol (XXXIII) are analgesics. [WALTER J. GENSLER]

Bibliography: R. A. Abramovitch (ed.), *Pyridine and Its Derivatives*, suppl., vols. 1–4, 1974–1975; D. M. Smith, Pyridines, in D. H. R. Barton et al. (eds.), *Comprehensive Organic Chemistry*, 1978.

Pyrolysis

The chemical change of a substance by means of heat alone. Thermal rearrangements into isomers, thermal polymerizations, and thermal decompositions are all included in the term pyrolysis, but it does not include thermal changes that require catalysts or changes that are initiated by other forms of energy. A transformation promoted by ultraviolet radiation, for example, would be photolytic, not pyrolytic. *See* CATALYSIS; PHOTOCHEMISTRY.

Familiar inorganic examples of pyrolysis are shown in reactions (1)–(5).

$$2NaHCO_3 \rightarrow Na_2CO_3 + H_2O + CO_2 \quad (1)$$

Sodium bicarbonate → Sodium carbonate + Water + Carbon dioxide

$$CaCO_3 \rightarrow CaO + CO_2 \quad (2)$$

Calcium carbonate (limestone) → Calcium oxide (quicklime) + Carbon dioxide

$$2HgO \rightarrow O_2 + 2Hg \quad (3)$$

Mercuric oxide → Oxygen + Mercury

$$Ni(CO)_4 \xrightarrow{150°C} Ni + 4CO \quad (4)$$

Nickel carbonyl → Nickel + Carbon monoxide

$$4KClO_3 \rightarrow 3KClO_4 + KCl \quad (5)$$

Potassium chlorate → Potassium perchlorate + Potassium chloride

Prominent industrial examples include cracking of petroleum; destructive distillation of coal or wood; pyrolysis of methane into hydrogen and carbon black; and syntheses of biphenyl from benzene, or styrene from ethylbenzene, or ethylene from ethane. *See* COAL CHEMICALS; COKE; CRACKING.

Radicals as intermediates. Many pyrolytic reactions involve the production of radicals as intermediates, as discussed below.

Paraffinic hydrocarbons (alkanes). Propane and butane are representative paraffins that have been carefully studied for their pyrolytic behavior. Propane changes at 500–600°C bidirectionally into methane and ethylene or into hydrogen and propylene. Butane gives rise to methane plus propylene, ethane plus ethylene, and to a lesser extent hydrogen plus butylene.

To understand these pyrolytic reactions, one should realize that in stable organic molecules each carbon is surrounded by eight electrons, as in the structures for methane and propane shown here, wherein each bond depicts an electron pair.

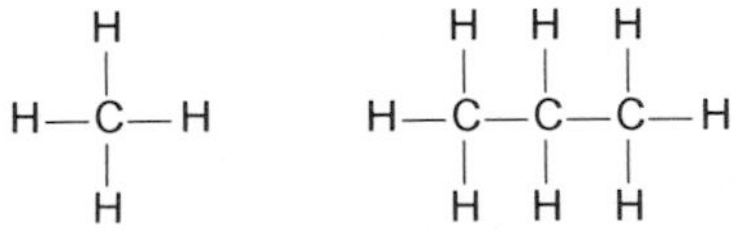

Methane Propane

The energy holding the C—H atoms together in these compounds (bond energy) is about 99 kilocalories per mole (kcal/mole) at room temperature (25°C), whereas the bond energy for each C—C bond is about 83 kcal/mole. From this, methane is more stable to heat than propane (or any other alkane) since it contains only C—H bonds, which are relatively strong. Actually, methane requires a temperature of 1000–1200°C for pyrolytic breakdown at reasonable speed.

The bond energies holding a molecule together become smaller with increasing temperature. If, eventually, one of the bonds comes to zero energy, the molecule falls apart into radicals. One of the carbons in a radical has only seven electrons around it, for example, the methyl radical,

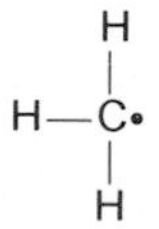

The dot represents one electron. Further on, the word methylene will be used. Methylene represents a carbon with only six electrons around it, as in methylene itself,

$$CH_2 \text{ or } H\!-\!\overset{\displaystyle H}{\overset{|}{C}}:$$

The products from propane or butane mentioned above are the result of initial thermal cleavage of the alkane into radicals, for example, methyl and ethyl from propane, as shown by reaction (6),

$$C_3H_8 \rightarrow CH_3\bullet + C_2H_5\bullet \qquad (6)$$

followed by collision of these radicals (R•) with undecomposed hydrocarbon to form RH and create a new radical, as shown by reaction (7). The $C_3H_7\bullet$

$$C_3H_8 + R\bullet \rightarrow RH + C_3H_7\bullet \qquad (7)$$

radical then breaks into an unsaturated hydrocarbon and a smaller radical or atom, as seen in reaction (8). These new radicals take the place of the

$$C_3H_7\bullet \begin{cases} \nearrow C_2H_4 \text{ (ethylene)} + CH_3\bullet \text{ (methyl radical)} \\ \searrow C_3H_6 \text{ (propylene)} + H\bullet \text{ (hydrogen atom)} \end{cases} \qquad (8)$$

original R• and cause the breakdown to continue. This "chain reaction" is interrupted when the radicals (R• or H•) combine with each other to form R—R, R—H, or H—H.

Kinetic studies have shown that this type of decomposition is unimolecular. Its rate of decomposition, therefore, is independent of the original concentration, and it would not be affected by increased surface in the hot zone.

Tetramethyllead and diphenylmercury. Tetramethyllead, important as an antiknock component in gasoline, was studied by F. Paneth and W. Hofeditz in 1929. They demonstrated that the methyl radical has a real, though brief, existence during high-temperature pyrolysis. On passing its vapor through a quartz tube at low pressure and at 700–800°C, a lead mirror was formed, as in reaction (9).

$$Pb(CH_3)_4 \rightarrow Pb + 4CH_3\bullet \qquad (9)$$

The released methyl radicals, on passing into an unheated part of the tube containing a previously deposited lead mirror, removed the cold mirror by reconverting it into tetramethyllead.

Diphenylmercury is a related organometallic compound that was shown in 1966 to decompose into phenyl radicals at the relatively low temperature of 200°C. If heated in dihydroanthracene, the radicals abstract hydrogen, yielding anthracene ($C_{14}H_{10}$), as shown by reactions (10) and (11).

$$Hg(C_6H_5)_2 \rightarrow Hg + 2C_6H_5\bullet \qquad (10)$$

$$2C_6H_5\bullet + C_{14}H_{12} \rightarrow 2C_6H_6 + C_{14}H_{10} \qquad (11)$$

Aromatic hydrocarbons (arenes). Benzene is more stable thermally than alkanes (except methane). At 700–750°C it changes into biphenyl and some terphenyl as a result of cleavage of C—H bonds, as in reaction (12).

$$2C_6H_6 \rightarrow C_6H_5\!-\!C_6H_5 + H_2 \qquad (12)$$

Toluene decomposes at about 600°C into benzyl radicals which dimerize into 1,2-diphenylethane, via $C_6H_5CH_2\bullet$ and H•, as in reaction (13). At 800–

$$2C_6H_5CH_3 \rightarrow C_6H_5CH_2CH_2C_6H_5 + H_2 \qquad (13)$$

850°C, a much more complex change occurs, yielding benzene, naphthalene, anthracene, phenanthrene, and higher polycyclic hydrocarbons, some of which are carcinogenic.

These same polycyclic hydrocarbons arise at 800–850°C not only from toluene but also from a great variety of organic compounds, including such diverse substances as propylene (a gaseous olefin), methylthiophene (a sulfur-containing heterocyclic), and picoline (a heterocyclic base). The temperature of a burning cigarette also is about 850°C, and tars collected from burning cigarettes also contain these polycyclic hydrocarbons. Obviously, there must be some common intermediates in these several pyrolytic reactions to generate the identical high-temperature products. Further work supplied good evidence that two such intermediates are involved, namely, trimethine (C_3H_3) and tetramethine (C_4H_4). Methine is the name of the hypothetical CH fragment.

If the structural formula of trimethine is drawn, it is seen to be both a radical and a methylene:

$$\bullet CH\!=\!CH\!-\!CH:$$

If tetramethine is expanded in either of two ways, it is either a diradical or a dimethylene:

$$\bullet CH\!=\!CH\!-\!CH\!=\!CH\bullet \quad \text{or} \quad :CH\!-\!CH\!=\!CH\!-\!CH:$$

Dimerization of C_3H_3 yields benzene, C_6H_6. Part of the observed benzene comes by this route; the remainder (from toluene) comes by direct hydrogenolysis of the methyl group. The C_4H_4 would be the precursor of naphthalene, another product, by adding to benzene or toluene and eliminating hydrogen or methane, respectively. Other polycyclics that are formed may be explained in a similar manner.

It is instructive to see how trimethine and tetramethine may arise from toluene. Toluene pyrolyzes initially into benzyl radicals; then the benzene ring of benzyl breaks apart, as shown by reaction (14). Fragment A is tetramethine. Fragment B becomes trimethine by thermal readjustment of one of its hydrogens.

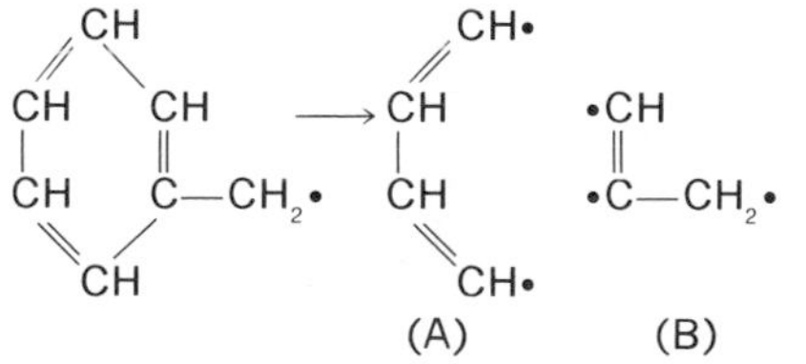
(A) (B)
(14)

Phthalic anhydride. In 1965 it was demonstrated that at 700°C phthalic anhydride decomposes into benzyne (C_6H_4), as in reaction (15). To demonstrate the presence of this short-lived intermediate, the anhydride was mixed with pyridine before pyrolysis. Then, once formed, benzyne reacted with the pyridine to form phenylpyridine, quinoline, and naphthalene, all of which are logically explained from benzyne as an intermediate.

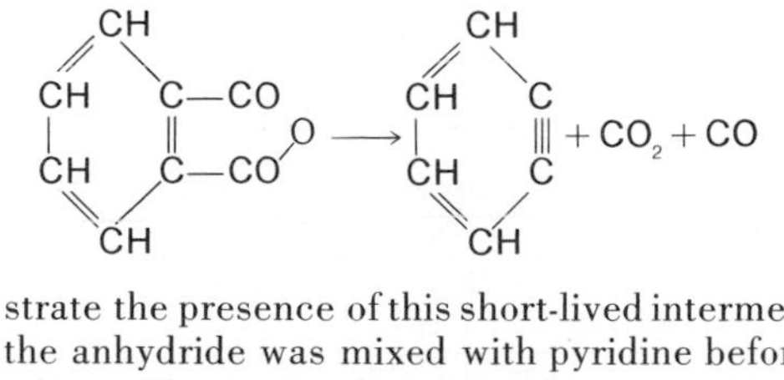
(15)

Acetone. An important synthesis of ketene, $CH_2\!=\!C\!=\!O$, is pyrolysis of acetone at 700°C. No catalyst is known that causes ketene production at significantly lower temperatures. Usually, acetone vapor is passed through a heated tube of glass, quartz, or copper-lined iron, or through an electrically heated Chromel filament (a "ketene lamp"). A radical mechanism applies here also, as represented by reaction (16*a*). Then reaction (16*b*) follows, and this methyl radical becomes the R• in a chain reaction.

$$R\bullet + CH_3COCH_3 \rightarrow RH + CH_3COCH_2\bullet \quad (16a)$$

$$CH_3COCH_2\bullet \rightarrow CH_2{=}C{=}O + CH_3\bullet \quad (16b)$$

Pyrolysis not involving radicals. Many compounds pyrolyze by mechanisms not involving bond rupture into radicals. Instead, pairs of electrons stay together and move as a unit, usually because of some structural feature within the molecule that makes it possible. Lower temperatures are usually observed for such processes than for the ones involving scission into radicals. The following examples illustrate this fact.

Esters. Ethyl acetate decomposes at 500°C into acetic acid and ethylene, isopropyl acetate at 450° into acetic acid and propylene, and tertiary butyl acetate at 350° into acetic acid and isobutylene. It is thus evident that esters of primary alcohols are more stable than those of secondary alcohols which, in turn, are more stable than esters of tertiary alcohols. All three types pyrolyze by a mechanism involving a quasi six-membered ring in the transition state and realignment of electron pairs, as in reaction (17). The term R represents a hydrogen or a methyl group. The cyclic transition state mechanism was first proposed in 1938 to explain pyrolytic processes. Since then, it has been adapted to countless other reactions of organic chemistry.

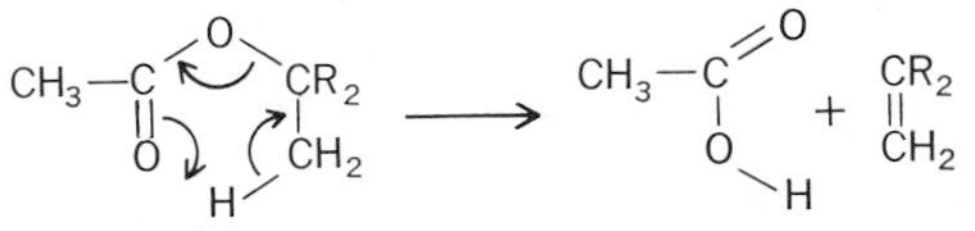
(17)

A methyl ester, which cannot have this kind of transition state because it lacks the second carbon in the alkyl group, must pyrolyze by a free-radical mechanism. This calls for higher temperatures (600–650°C).

Vinyl allyl ether. At 250°C this unsaturated ether rearranges into 4-butenal, a five-carbon aldehyde:

$$CH_2{=}CH{-}O{-}CH_2{-}CH{=}CH_2 \rightarrow CH_2{=}CH{-}CH_2{-}CH_2{-}CH{=}O$$

Here, an original sequence of two or three carbon atoms becomes lengthened to five. A cyclic mechanism is involved here also, as shown by reaction (18).

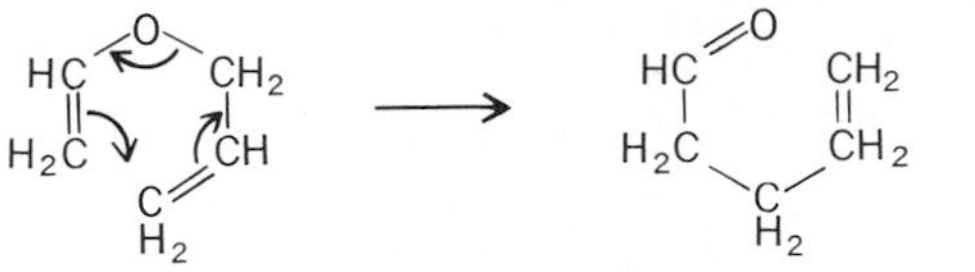
(18)

Malonic acid. The decarboxylation of malonic acid or its homologs into a monocarboxylic acid, as shown by reaction (19), is a valuable reaction for organic syntheses. Temperatures are in the range 150–200°C, and the cyclic mechanism shown by reaction (20) is accepted to explain the result.

$$RCH(COOH)_2 \rightarrow RCH_2COOH + CO_2 \quad (19)$$

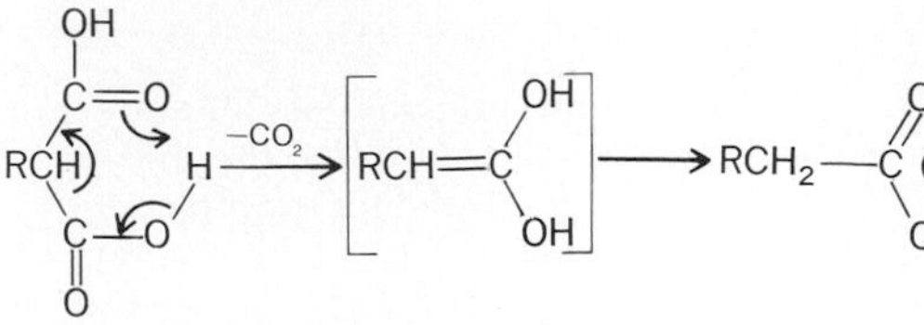
(20)

Diacetone alcohol. This keto alcohol is well adapted to show how catalysts may modify a pyrolytic reaction. If the compound is rigorously purified, diacetone alcohol pyrolyzes exclusively into acetone at 250°C, as shown by reaction (21).

$$CH_3CO{-}CH_2C(CH_3)_2OH \rightarrow 2CH_3COCH_3 \quad (21)$$

This conforms to the cyclic mechanism shown by reaction (22).

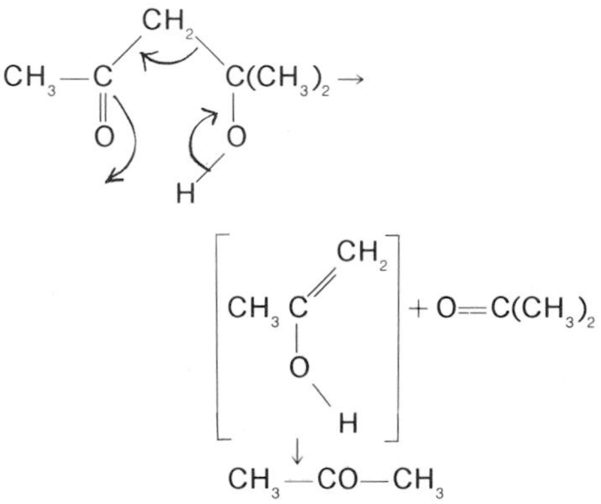
(22)

Addition of a trace of iodine or a drop of sulfuric acid prior to heating changes completely the course of reaction. Mesityl oxide, which is a mixture of $CH_3CO{-}CH{=}C(CH_3)_2$ and $CH_3COCH_2C(CH_3){=}CH_2$, now is the sole product as water is eliminated. The effect of the acid is catalytic, first adding its proton to the alcoholic hydroxyl and forming trace amounts of the conjugate acid of diacetone alcohol, $CH_3COCH_2C(CH_3)_2{-}\overset{+}{O}H_2$. The added proton has so weakened the bond energy of the C—O bond that it breaks with separation of the elements of water. In the ensuing readjustment, mesityl oxide is formed and a proton is lost. With this regenerated acidity, decomposition continues, as shown by reaction (23). This may be considered an example of pyrolysis of the conjugate acid of diacetone alcohol, but not of diacetone alcohol itself.

$$CH_3COCH_2C(CH_3)_2{-}\overset{+}{O}H_2 \xrightarrow{-H_2O} CH_3COCH_2\overset{+}{C}(CH_3)_2 \xrightarrow{-H^+} \text{mesityl oxide} \quad (23)$$

Onium compounds. The conjugate acid of mesityl oxide is an oxonium compound. Its C—O bond was shown to be weaker than the same C—O bond of mesityl oxide itself. The same is true for the conjugate acids of ethers, such as diethyloxonium bromide (made from ethyl ether + HBr), which cleaves into ethanol and ethyl bromide, as in reaction (24).

$$(C_2H_5)_2\overset{+}{O}H\ \overset{-}{Br} \rightarrow C_2H_5OH + C_2H_5Br \quad (24)$$

This weakening is characteristic of all onium compounds. For example, ammonia is synthesized from nitrogen and hydrogen at about 500°C. It is pyrolyzed back to the elements at 700°, but is quite stable at 335°. At 335°, however, ammonium chloride is broken down completely into ammonia and hydrogen chloride. One of the four N—H bonds in the ammonium ion, therefore, must be relatively weak. Tetramethylammonium chloride, $(CH_3)_4\overset{+}{N}\,\overset{-}{Cl}$, pyrolyzes at 360°C into trimethylamine and methyl chloride.

At about 300°C, *N*-isobutylaniline hydrochloride rearranges into *p*-*tert*-butylaniline hydrochloride, as in reaction (25). In this process the bond breaks

$$C_6H_5\overset{+}{N}H_2CH_2CH(CH_3)\ \overset{-}{Cl} \rightarrow (CH_3)_3C{-}C_6H_4{-}\overset{+}{N}H_3\ \overset{-}{Cl} \qquad (25)$$

between isobutyl and nitrogen to form an isobutyl cation which promptly rearranges, as shown by reaction (26), into a *tert*-butyl cation before attacking

$$\begin{aligned} C_6H_5\overset{+}{N}H_2{-}CH_2CH(CH_3)_2 &\rightarrow C_6H_5NH_2 + \overset{+}{C}H_2CH(CH_3)_2 \\ &\qquad\qquad \downarrow \\ &\qquad CH_3\overset{+}{C}(CH_3)_2 \qquad (26) \end{aligned}$$

and becoming attached to the para position of aniline. Reaction (27) describes this last phase.

$$(CH_3)_3\overset{+}{C} + C_6H_5NH_2 \rightarrow (CH_3)_3C{-}C_6H_4\overset{+}{N}H_3 \qquad (27)$$

Ammonium hydroxides are also relatively unstable to heat. The well-known Hofmann degradation shown by reaction (28) involves pyrolysis of a quaternary ammonium hydroxide.

$$R'CH_2CH_2\overset{+}{N}R_3\ \overset{-}{O}H \rightarrow R'CH{=}CH_2 + R_3N + H_2O \qquad (28)$$

Sulfonium salts and hydroxides also show this relative ease of pyrolysis. Distillation of trimethylsulfonium iodide causes it to break into methyl sulfide and methyl iodide, as shown by reaction (29). Triethylsulfonium hydroxide readily pyrolyzes

$$(CH_3)_3\overset{+}{S}\ \overset{-}{I} \rightarrow (CH_3)_2S + CH_3I \qquad (29)$$

into ethyl sulfide, ethylene, and water, as shown by reaction (30), thus paralleling the behavior of quaternary ammonium hydroxides.

$$(C_2H_5)_3\overset{+}{S}\ \overset{-}{O}H \rightarrow (C_2H_5)_2S + C_2H_4 + H_2O \qquad (30)$$

[CHARLES D. HURD]

Bibliography: N. L. Allinger, *Organic Chemistry*, 2d ed., 1976; R. F. Brown, *Pyrolytic Methods in Organic Chemistry: Applications of Flow and Flash Vacuum Pyrolytic Techniques*, 1980; J. Hine, *Physical Organic Chemistry*, 2d ed., 1962.

Pyrometric cone

One of a numbered series of ceramic compositions formed into triangular pyramids of height about four times the base (Fig. 1), and designed so that each will soften and bend under its own weight after a certain heat treatment (Fig. 2).

The cones, approximately 2 in. high, are used to indicate when ceramic ware has been adequately fired and to check on heat uniformity in a kiln. Cones approximately 1 in. high are used in determining the pyrometric cone equivalent of refractories.

Cones are numbered from 022, 021, 020, the most easily fused, through 02, 01, 1, 2, to 42, the most refractory. Where the end points of the cone

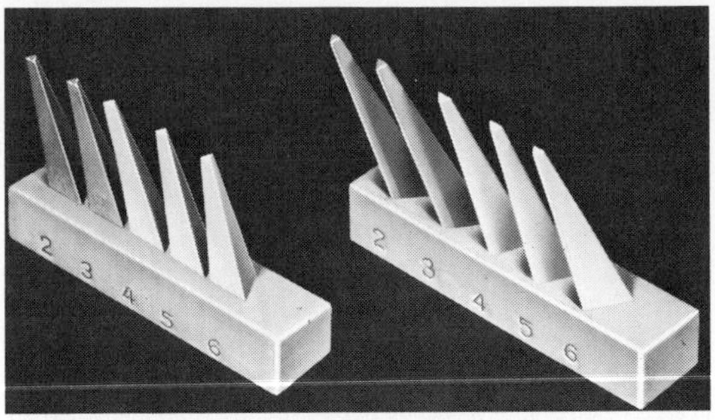

Fig. 1. Pyrometric cones before the firing process, showing two methods of setting in the cone plaque. (*Edward Orton Jr. Ceramic Foundation*)

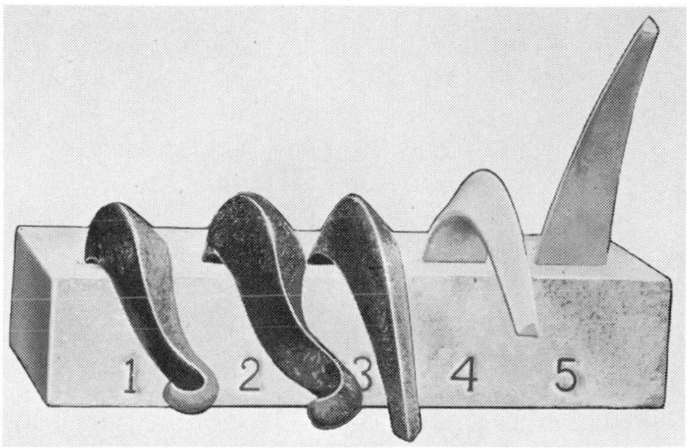

Fig. 2. Pyrometric cones after the heat-treatment process. These cones have been fired to cone 4. (*Edward Orton Jr. Ceramic Foundation*)

series designed by Edward Orton, Jr., were too close together, some have been omitted (for example, 21, 22, 24, 25); where they were too widely spaced, extra cones (31½, 32½) have been added.

A cone is said to have reached its end point when its tip is bent to the level of the base; for a constant rate of temperature rise, this occurs at a definite temperature. The end point indicates a heat treatment depending on both time and temperature; the greater the rate of heating, the higher the equivalent temperature of a given cone. For example, cone 022 is considered equivalent to 585°C (1085°F) when heated at 60°C (108°F) per hour, but it is equivalant to 600°C (1112°F) when heated at 150°C (270°F) per hour.

Cones are most useful for firing ware made of materials similar to those in the cones (clay, feldspar, quartz), in other words, the classical clay products. For radically different materials, such as the magnetic ceramic ferrites, cones are less useful as indicators of maturity.

Cones are designed with end points roughly 20°C (36°F) apart; thus, cones set in different parts of the kiln are a good check on heat uniformity. Even in automatically controlled kilns it is common practice to use cones as a means of checking the firing.

Care should be exercised in using cones, since they are affected by factors other than time and temperature, such as the composition of the furnace atmosphere.

[J. F. MC MAHON]

Bibliography: Edward Orton Jr. Ceramic Foundation, *The Properties and Uses of Pyrometric Cones*, 1951.

Pyrrole

One of a group of organic compounds containing a doubly-unsaturated five-membered ring in which nitrogen occupies one of the ring positions. Pyrrole (I) is a representative compound. The pyrrole system is found in the green leaf pigment, chlorophyll, in the red blood pigment, hemoglobin, and in the blue dye, indigo. Interest in these colored bodies has been largely responsible for the intensive study of pyrroles. Tetrahydropyrrole, or pyrrolidine (II), is part of the structures of two protein amino acids, proline (III) and hydroxyproline, and of hygrine (IV), which is an alkaloid from Peruvian coca.

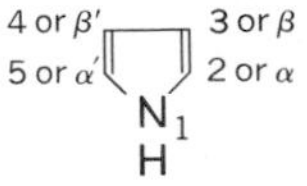

(I)

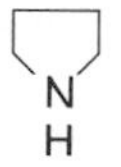

(II)

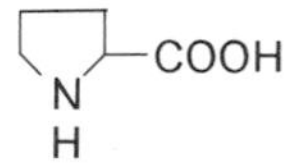

(III)

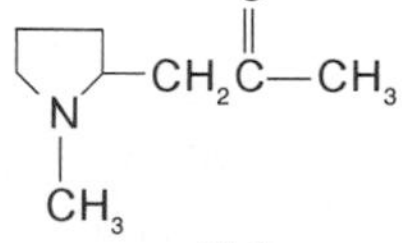

(IV)

See HETEROCYCLIC COMPOUNDS; INDOLE.

Properties. Pyrrole (I) is a liquid, bp 130°C, n_D^{20} 1.5085 (1.5098), and density (20/4) 0.948, (0.969), that darkens and resinifies on standing in air, and that polymerizes quickly when treated with mineral acid. Polyalkyl pyrroles are not so sensitive and negatively substituted pyrroles are even less so. Pyrrole is a planar, aromatic compound, with an experimental resonance energy of 22–27 kcal/mole. Familiar substitution processes, such as halogenation, nitration, sulfonation, and acylation, can be realized. Substitution generally occurs more readily than in the corresponding benzene analog. The entering group favors the 2 or 5 position. Pyrrole, by virtue of its heterocyclic nitrogen, is very weakly basic and is comparable in this respect to urea or to semicarbazide. The hydrogen at the 1 position is removable as a proton, and accordingly, pyrrole is also an acid, although a weak one.

Pyrroles are not as resistant to oxidation as the analogous benzene compounds. Controlled chromic acid oxidation converts pyrroles with or without groups in the α positions to the corresponding maleimides. In this way for example, ethyl-2,5-dimethylpyrrole-3-carboxylate (V) is oxidized to a maleimide derivative (VI). Conditions are available for useful reductions of groups attached to the nucleus. Zinc and hydrochloric acid can reduce the pyrrole ring to dihydropyrrole (pyrroline). Raney nickel under rigorous conditions, or platinum or palladium under milder conditions, serve as catalysts in the hydrogenation of pyrroles to pyrrolidines.

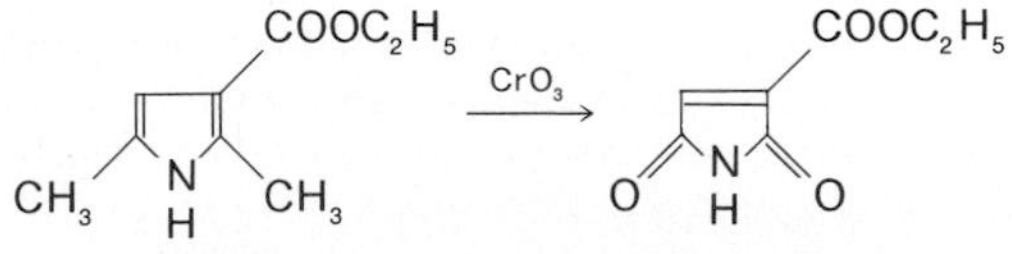

(V) (VI)

Preparation. The Knorr synthesis, probably the most versatile pyrrole synthesis, combines an α aminoketone (VII) with an α-methylene carbonyl compound (VIII). The condensation is carried out

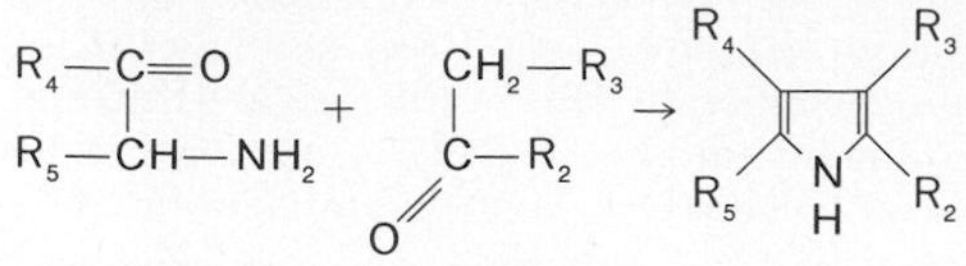

(VII) (VIII)

either in glacial acetic acid or in aqueous alkali. The R groups may be varied widely; however, best results are obtained when R_3 of the carbonyl compounds (VIII) is an activating group, for example, carbethoxy. Another useful general synthesis, the Paal-Knorr synthesis, converts a 1,4-dicarbonyl compounds (IX) by cyclization with ammonia (R_1=H) or with a primary amine to a pyrrole (X).

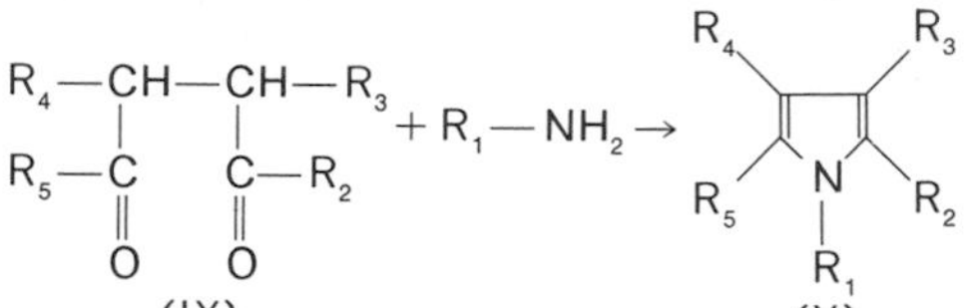

(IX) (X)

Pyrrole itself is obtained by pyrolysis of ammonium mucate (saccharate). Pyrolysis of primary amine salts of mucic acid gives 1-substituted pyrroles.

Derivatives. Chloro-, bromo-, and iodopyrroles have been prepared, with bromopyrroles comprising the largest group. Halopyrroles carrying an electronegative group, such as carbethoxy, are more stable than the same compounds without the group. 2,3,4,5-Tetraiodopyrrole, formed by iodination of pyrrole or of tetrachloromercuripyrrole, is useful as an antiseptic in the same areas as iodoform. Pyrroles with ethylmagnesium bromide react as active hydrogen compounds to give Grignard derivatives. Ethyl formate reacts with such pyrryl Grignard reagents to form pyrrole aldehydes; chloroformic ester reacts to give pyrrolecarboxylic esters. Pyrrole aldehydes can also be formed by formylation of a vacant pyrrole position with hydrocyanic acid and hydrogen chloride. Acylpyrroles, that is, pyrryl ketones, are prepared satisfactorily and under mild conditions by Friedel-Crafts acylation of the pyrrole nucleus.

Pyrrolidine (II), bp 87–88°C, can be prepared by catalytic hydrogenation of pyrrole or by ring-closure reactions. Either 1,4-dibromobutane (XI) or 4-bromobutylamine(XII) can serve as starting material.

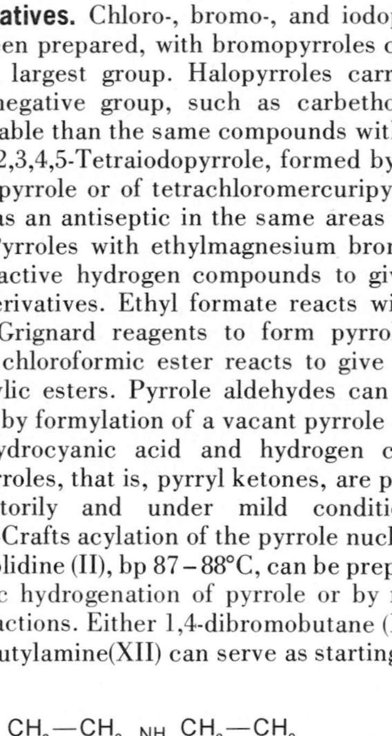

(XI) (XII)

2-Ketopyrrolidine, or pyrrolidone (XIV), is of considerable interest in connection with the preparation of polyvinylpyrrolidone. Pyrrolidone, which can be formed from tetrahydrofuran by autoxidation in the presence of a cobalt catalyst and treatment of the resulting γ-butyrolactone (XIII) with ammonia, is combined with acetylene to form vinylpyrrolidone (XV). Polymerization of this material furnishes polyvinylpyrrolidone, a material of relatively high molecular weight, which is suitable

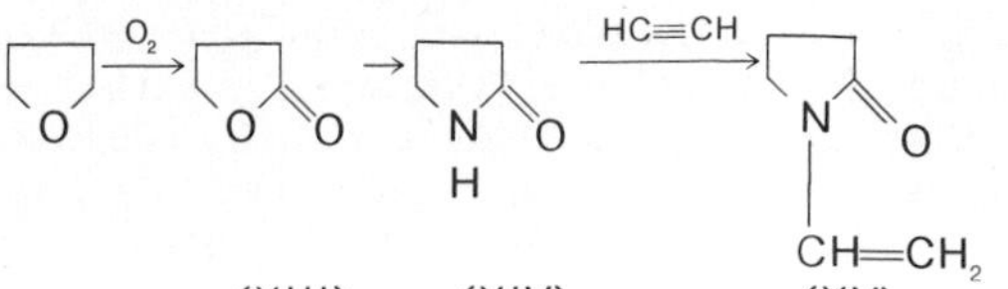

(XIII) (XIV) (XV)

for maintaining osmotic pressure in blood and so acting as an extender for plasma or whole blood. *See* PYRIDINE. [WALTER J. GENSLER]

Bibliography: A. H. Jackson, Pyrroles, in D. H. R. Barton et al. (eds.), *Comprehensive Organic Chemistry*, 1978.

Quantitative chemical analysis

That branch of analytical chemistry which deals with the determination of the relative proportions of constituents in a compound or of components in a mixture. The subject is divided into the broad fields of gravimetric methods, volumetric methods (liquid), gas-volumetric methods, optical methods, electrical methods, and miscellaneous physicochemical methods. The procedures followed in making an analysis may be classed according to (1) the type of constituent determined, (2) the kind of method used, (3) the type of material analyzed, and (4) the amount of constituent present.

Classification of procedures. In an ultimate analysis, the amount of a single element or compound is determined. In a proximate analysis, certain constituents are determined as a group of indefinite relative composition, for example, the determination of ash in a sample of coal.

Methods of analysis may be direct or indirect. In a direct gravimetric method, the desired constituent is converted to a compound of definite, known composition and this compound is weighed. In a direct volumetric method, the desired constituent is determined by measuring the volume of reagent of known concentration required to react completely with the constituent.

In an indirect gravimetric method, a mixture of substances which includes the desired constituent is weighed and then wholly, or in part, converted to some other substance, or mixture of substances, of known composition and weighed. The amount of desired constituent can then be calculated by solving two simultaneous equations that can be set up from the data obtained. Indirect gravimetric methods are usually less precise than direct ones.

In an indirect volumetric method, a measured quantity of reagent is added in excess of the amount required to react with the desired constituent. The excess reactant is then determined by titration, and the amount of it which reacts with the desired constituent is determined by difference.

Methods of quantitative analysis depend greatly upon the nature of the substance being analyzed. For this reason, compilations of tested methods have been prepared, each covering a certain type of material. These compilations are available as reference books and laboratory manuals, and they cover such diverse fields as the analysis of steels, nonferrous alloys, foods, minerals and ores, gases, technical products, and agricultural products.

Methods of quantitative analysis vary with the amount of sample taken and of constituent being determined. A macroanalysis (decigram analysis) uses a sample of about 0.1 – 0.5 gram (g). The analytical balance and volumetric instruments are designed to give a precision of 0.1 milligram (mg) and 0.02 milliliter (ml), respectively.

A semimicroanalysis (centigram analysis) uses a sample of about 0.01 – 0.1 g. In this case, the balance and volumetric instruments give a precision of 0.01 mg and 0.002 ml, respectively.

A microanalysis (milligram analysis) uses a sample of about 1 – 10 mg, and the balance is designed to read to a precision of 0.001 mg (1 microgram, 1 μg, or 1γ). The balance differs from the conventional one in being more delicately constructed of very light material. It is necessary to have it on a firm, vibration-proof foundation, and great precautions are necessary to maintain draft-free air at constant temperature and humidity.

The instruments and apparatus used in microanalytical work are to some extent miniature replicas of macro instruments and apparatus, but in some cases quite different techniques and apparatus are used. Filtration, for example, is unlike that on a macro scale. It is done by inserting into the vessel containing a suspension of the precipitate a filter stick consisting of a glass tube, one end of which is closed by a porous, sintered glass disk. The filter stick and vessel have previously been weighed together and are treated as a unit. Filtration is upward and is carried out by applying gentle suction to the tube. After appropriate washing of the precipitate, the filter stick and vessel are dried and again weighed together. *See* GRAVIMETRIC ANALYSIS.

An ultramicroanalysis (microgram analysis) uses a sample of approximately 0.001 mg. A special quartz-fiber torsion balance is used. It has a capacity of about 20 mg and weighs to about 0.02 μg tolerance.

Calibration of method. Since the accuracy of a quantitative analysis depends in part on the nature of the material and on the nature and quantities of foreign constituents present, calibration of methods of unknown applicability is often necessary.

In the correction factor method, the procedure is applied to a sample containing all the constituents in the given sample except the desired one. Any numerical result obtained is applied as a correction factor to the value obtained on the unknown.

In the synthetic sample method, the procedure of analysis is applied in parallel to the unknown and to a sample containing constituents of the same nature and approximate amounts as those in the unknown, and with the desired constituent added in known amount. Any discrepancy in the value obtained on the synthetic sample is applied to correct the analysis of the unknown. In a variation of this method the proposed procedure is applied to a sample of similar composition obtained from the U.S. National Bureau of Standards. Samples available include various ferrous and nonferrous alloys and many ores and minerals. The values for the certified samples have been obtained by experienced analysts applying independent methods and using every precaution to ensure the highest degree of accuracy. If concordant results are obtained on the standard sample by the proposed method, it can be assumed to be reliable.

Calibration of apparatus. Analytical weights, volumetric glassware, and all other instruments that furnish numerical data should be calibrated.

In an ordinary quantitative analysis, if the same set of weights is used throughout, it is necessary only to determine the relative masses of the weights. That is, the mass of the 10-g weight should be exactly twice that of the 5-g weight. Similarly, the others should be in corresponding proportion to the 10-g weight. The mass of one of the weights is assumed to be correct as marked on the weight, and the correction factor to be applied to each of the other weights to maintain the exact theoretical proportion is determined by experiment. It is more desirable, however, to use as a standard of reference a weight of known mass, such as one certified by the Bureau of Standards.

In calibrating volumetric glassware, the weight of water (or mercury) at a given temperature contained in, or delivered by, the apparatus is found. From the known density of the liquid at the given temperature, the true volume can be calculated and compared to the volume indicated by the marking on the volumetric apparatus. The fundamental standard is the liter which is the volume occupied by 1 kilogram (kg) of water at the temperature of its maximum density (approximately 4°C). The normal temperature for calibrating volumetric glassware in the United States is 20°C, which means that to contain a true liter, a flask must be so marked that at 20°C its capacity will be equal to the volume of water which at 4°C weighs 1 kg in a vacuum.

Several types of calibration are encountered in other branches of analytical chemistry. A common calibration procedure in colorimetry is to bracket the unknown between two standards at known concentration just above and just below that of the unknown. Another is useful in emission spectroscopy, where the density of spectral lines is a quantitative measure of concentration. To eliminate the effects of undesirable variations in the excitation, an internal standard method is used whereby the density of a spectral line of the element of unknown concentration is compared to that of a line of an element of known concentration. The pairs of lines, called homologous pairs, respond in the same way to changes in excitation conditions. A standard addition method is useful in cases where the nature of the sample may preclude the direct use of synthetic samples. Here the slope of a calibration curve (a curve showing corrections to be applied under specified conditions of analysis) is determined from the relative positions on the curve of the value obtained from the diluted sample and that obtained on an equal volume of sample to which has been added a standard solution containing a known additional amount of the constituent being determined.

Graphs are frequently used in analytical chemistry for many purposes. A nomograph is a device by which the numerical result of a given calculation can be read directly from a previously drawn scale or series of scales. It has the advantage over a slide rule in being equally applicable to calculations containing additive and subtractive terms. Since a separate nomograph is needed for each formula to be solved, nomographs are of practical use only when the same type of analytical calculation is made repeatedly.

Other graphs encountered in quantitative analysis are calibration corrections, tabulations of values derived from natural laws, and titration curves in which pH values, conductances, or electrode potential values are plotted against buret readings to establish equivalence points. In such graphs some variables change in logarithmic fashion giving sigmoidal curves; others change in arithmetic proportion and yield straight-line graphs. *See* ANALYTICAL CHEMISTRY; SPECTROCHEMICAL ANALYSIS; SPECTROPHOTOMETRIC ANALYSIS; VOLUMETRIC ANALYSIS. [STEPHEN G. SIMPSON]

Bibliography: J. S. Fritz and G. H. Schenk, Jr., *Quantitative Analytical Chemistry*, 4th ed., 1979; V. Kumar, *Experimental Techniques in Quantitative Analysis*, 1981; H. A. Laitinen and W. E. Harris, *Chemical Analysis*, 2d ed., 1975; S. Siggia and J. G. Hanna, *Quantitative Organic Analysis via Functional Groups*, 4th ed., 1979.

Quantum chemistry

A branch of chemistry concerned with the application of quantum mechanics to chemical problems. More specifically, it is concerned with the electronic structure of molecules. Since 1960 the ease with which the quantum chemist may obtain reliable approximate solutions to the nonrelativistic Schrödinger equation has improved by at least six orders of magnitude. This article presents a brief review of developments in ab initio molecular electronic structure theory since 1960. The term ab initio implies that no approximations have been made in the one- and two-electron integrals, shown in Eqs. (1) and (2), arising from the ordinary nonrelativistic hamiltonian, Eq. (3). In contrast, semiempirical methods resort to various approximate schemes, especially in evaluating the two-electron integrals $(ij|kl)$. The present discussion is restricted to the method which dominates the field of quantum chemistry, namely, the Hartree-Fock or self-consistent-field approximation.

$$I(i|j)=\int \phi_i^*(1)\left\{\frac{-\nabla_1^2}{2}-\sum_A \frac{Z_A}{r_{1A}}\right\}\phi_j(1)\,dv(1) \quad (1)$$

$$(ij|kl)=\int \phi_i^*(1)\phi_j^*(2)\frac{1}{r_{12}}\phi_k(1)\phi_l(2)\,dv(1)\,dv(2) \quad (2)$$

$$H=\sum_i\left\{\frac{-\nabla_i^2}{2}-\sum_A\frac{Z_A}{r_{iA}}\right\}+\sum_i\sum_{j>i}\frac{1}{r_{ij}} \quad (3)$$

Definitions. For closed-shell molecules, the form of the Hartree-Fock wave function is given by Eq. (4), in which $A(n)$, the antisymmetrizer for n electrons, has the effect of making a Slater determinant out of the orbital product on which it operates. The ϕ's are spin orbitals, products of a spatial orbital χ and a one-electron spin function α or β. For any given molecular system, there are an infinite number of wave functions of form (4), but the Hartree-Fock wave function is the one for which the orbitals ϕ have been varied to yield the lowest possible energy [Eq. (5)].

$$\psi_{HF}=A(n)\phi_1(1)\phi_2(2)\ldots\phi_n(n) \quad (4)$$

$$E=\int \psi_{HF}^* H \psi_{HF}\,d\tau \quad (5)$$

The resulting Hartree-Fock equations are relatively tractable due to the simple form of the energy E for single determinant wave functions [Eq. (6)].

$$E_{HF} = \sum_i I(i|i) + \sum_i \sum_{j>i} [(ij|ij) - (ij|ji)] \qquad (6)$$

To make this discussion more concrete, it should be noted that for singlet methylene (the CH_2 molecule), the Hartree-Fock wave function is of the form given in Eq. (7).

$$\psi_{HF} = A(8)\, 1a_1\alpha(1)\, 1a_1\beta(2)\, 2a_1\alpha(3)\, 2a_1\beta(4)\, 1b_2\alpha(5)\, 1b_2\beta(6)\, 3a_1\alpha(7)\, 3a_1\beta(8) \qquad (7)$$

The same energy expression, Eq. (6), is also applicable to any open-shell system for which the open-shell electrons all have parallel spins. This follows from the fact that such Hartree-Fock wave functions can always be expressed as a single Slater determinant. A simple example is triplet methylene, shown in Eq. (8), for which the outer two $3a_1$

$$\psi_{HF} = A(8)\, 1a_1\alpha(1)\, 1a_1\beta(2)\, 2a_1\alpha(3)\, 2a_1\beta(4)\, 1b_2\alpha(5)\, 1b_2\beta(6)\, 3a_1\alpha(7)\, 1b_1\alpha(8) \qquad (8)$$

and $1b_1$ orbitals have parallel spins. For clarity it is often helpful to abbreviate Eq. (8) as Eq. (9). Al-

$$\psi_{HF} = 1a_1^{\,2}\, 2a_1^{\,2}\, 1b_2^{\,2}\, 3a_1\alpha\, 1b_1\alpha \qquad (9)$$

though solution of the Hartree-Fock equations for an open-shell system such as triplet methylene is more difficult than for the analogous closed-shell system, Eq. (7), the procedures are well established.

In fact, methods are available for the solution of the Hartree-Fock equations for any system for which the energy expression involves only coulomb and exchange integrals, Eqs. (10) and (11).

$$J_{ij} = (ij|ij) \qquad (10)$$

$$K_{ij} = (ij|ji) \qquad (11)$$

Open-shell singlets are a class of systems that can be treated in this way, and one such example is the first excited singlet state (of $^1B_1$ symmetry) of methylene, Eq. (12). In addition, these same gener-

$$\psi_{HF} = \frac{1}{\sqrt{2}} 1a_1^{\,2}\, 2a_1^{\,2}\, 1b_2^{\,2}\, 3a_1\alpha\, 1b_1\beta - \frac{1}{\sqrt{2}} 1a_1^{\,2}\, 2a_1^{\,2}\, 1b_2^{\,2}\, 3a_1\beta\, 1b_1\alpha \qquad (12)$$

alized Hartree-Fock procedures can be used for certain classes of multiconfiguration Hartree-Fock wave functions.

Basis sets. To solve the Hartree-Fock equations exactly, either the orbitals ϕ must be expanded in a complete set of analytic basis functions or strictly numerical (that is, tabulated) orbitals must be obtained. The former approach is impossible from a practical point of view for systems with more than two electrons, and the latter has been accomplished only for atoms and for a few diatomic molecules. Therefore the exact solution of the Hartree-Fock equations is abandoned for polyatomic molecules. Instead an incomplete (but reasonable) set of analytic basis functions is adopted and solved for the best variational [that is, lowest energy given by Eq. (5)] wave function of form (4). Such a wave function is referred to as being of self-consistent-field (SCF) quality. For very large basis sets, then, it is reasonable to refer to the resulting SCF wave function as near-Hartree-Fock.

For large chemical systems, only minimum basis sets (MBS) can be used in ab initio theoretical studies. The term "large" includes molecular systems with 100 or more electrons. By 1980, the largest molecule treated by MBS-SCF methods was the carbazole-trinitrofluorenone complex, $C_{25}N_4O_7H_{14}$, with 232 electrons. A minimum basis set includes one function for each orbital occupied in the ground state of each atom included in the moelcule. For the first row atoms B, C, N, O, and F, this means that a minimum basis set includes $1s$, $2s$, $2p_x$, $2p_y$, and $2p_z$ functions.

Traditionally, minimum basis sets have been composed of Slater functions, such as those seen in Table 1 for the carbon atom. However, experi-

Table 1. Minimum basis set of Slater functions for the carbon atom

Label	Analytic form	Exponent ζ*
$1s$	$(\zeta^3/\pi)^{1/2} \exp(-\zeta r)$	5.673
$2s$	$(\zeta^5/3\pi)^{1/2}\, r \exp(-\zeta r)$	1.608
$2p_x$	$(\zeta^5/\pi)^{1/2}\, x \exp(-\zeta r)$	1.568
$2p_y$	$(\zeta^5/\pi)^{1/2}\, y \exp(-\zeta r)$	1.568
$2p_z$	$(\zeta^5/\pi)^{1/2}\, z \exp(-\zeta r)$	1.568

*The orbital exponents ζ are optimum for the ^3P ground state of the carbon atom.

ence has shown that the evaluation of the molecular integrals, Eq. (2), arising when Slater functions are employed is extremely time-consuming. Therefore each Slater function in a minimum basis set is typically replaced by a linear combination of three or four gaussian functions. The resulting chemical predictions obtained with such STO-3G (Slater-type orbital–three gaussian functions) or STO-4G basis sets are usually indistinguishable from the corresponding Slater function results.

Minimum basis sets are inadequate for certain types of chemical predictions. Therefore a basis twice as large, and appropriately designated double zeta (DZ), is often used in theoretical studies. Here, however, it is not as fruitful to expand each Slater function as a linear combination of gaussians. Instead gaussian functions $x^p y^q z^r e^{-\infty r^2}$ are used directly in atomic self-consistent-field calculations and then contracted according to the atomic results. Perhaps the most widely used contracted gaussian double-zeta basis sets are those of T. H. Dunning. His basis has $9s$ and $5p$ original (or primitive) gaussian functions, and is contracted to $4s$ and $2p$. Thus the basis may be designated C($9s5p/4s2p$).

Just as the double-zeta basis logically follows the minimum set, the logical extension of the double-zeta set involves the addition of polarization functions. Polarization functions are of higher orbital angular momentum than the functions occupied in the atomic self-consistent-field wave function. That is, for carbon, $d, f, g, \ldots$ functions will be polarization functions. Fortunately, d functions are far more important than f, f functions are far more important than g, and so on. For most chemical applications a double-zeta plus polarization (DZ + P) basis including a single set of five d functions ($d_{x^2-y^2}$, d_{z^2}, d_{xy}, d_{xz}, d_{yz}) will be quite adequate for first-row atoms.

Structural predictions. Ab initio theoretical methods have had the greatest impact on chemistry in the area of structural predictions. A good illustration of this is the methylene radical for which S. F. Boys reported the first ab initio study in 1960. Boys predicted the structure of triplet methylene to be r_e (C—H) = 0.112 nm, Θ_e(HCH) = 129°. Unfortunately, however, the work of Boys was largely ignored due to the spectroscopic conclusion of G. Herzberg that the lowest triplet state was linear. Herzberg's conclusion was greatly strengthened by a very influential semiempirical study of H. C. Longuet-Higgens, who concurred that the ground state of CH_2 was linear.

It was not until 1970 that a definitive theoretical prediction of the nonlinearity of $^3B_1$ CH_2 appeared. The prediction was swiftly verified by independent electron spin resonance experiments. For many chemists, the structure of triplet methylene was the first genuine example of the usefulness of ab initio theoretical chemistry.

Turning from the specific to the more general, the most encouraging aspect of ab initio geometry predictions is their perhaps surprising reliability. Essentially all molecular structures appear to be reliably predicted at the Hartree-Fock level of theory. Even more encouraging, many structures are accurately reproduced by using only minimum-basis-set self-consistent-field methods. As shown by J. A. Pople and coworkers, this is especially true for hydrocarbons. A fairly typical example is methylenecyclopropane (Fig. 1), with its minimum-basis-set self-consistent-field structure compared with experiment in Table 2. Carbon-carbon bond distances differ typically by 0.002 nm from experiment, and angles are rarely in error by more than a few degrees. Even for severely strained molecules such as bicyclo[1.1.0]-butane, very reasonable agreement with the experimental structure is obtained. It is noteworthy that experimental geometries are available for only half of the C_4 hydrocarbons studied. Thus, for many purposes, theory may be considered complementary to experiment in the area of structure prediction.

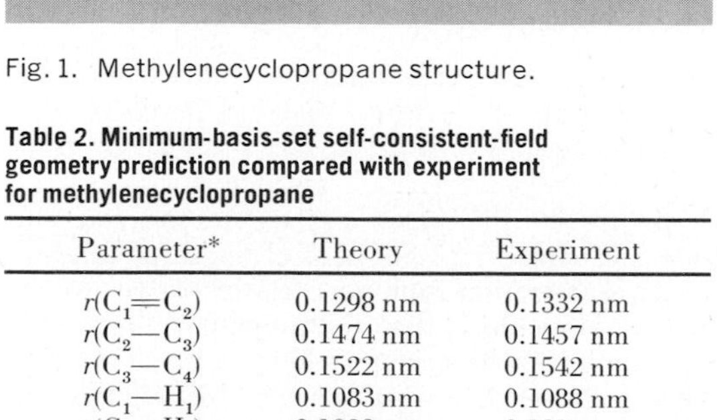

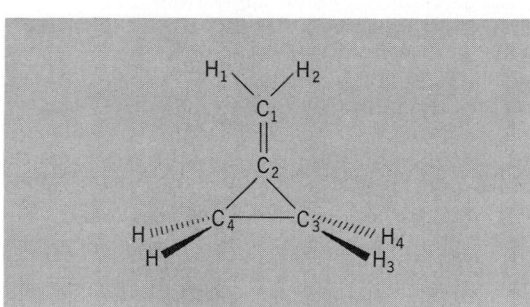

Fig. 1. Methylenecyclopropane structure.

Table 2. Minimum-basis-set self-consistent-field geometry prediction compared with experiment for methylenecyclopropane

Parameter*	Theory	Experiment
$r(C_1{=}C_2)$	0.1298 nm	0.1332 nm
$r(C_2{-}C_3)$	0.1474 nm	0.1457 nm
$r(C_3{-}C_4)$	0.1522 nm	0.1542 nm
$r(C_1{-}H_1)$	0.1083 nm	0.1088 nm
$r(C_3{-}H_3)$	0.1083 nm	0.109 nm
$\theta(H_1C_1H_2)$	116.0°	114.3°
$\theta(H_3C_3H_4)$	113.6°	113.5°
$\theta(H_{34}C_3C_4)$	149.4°	150.8°

*Here r represents the carbon-carbon bond distance; θ represents the bond angle in degrees of H-C-H bonds; the numbers on C and H correspond to the numbered atoms in Fig. 1.

For molecules including atoms in addition to C and H, minimum-basis-set self-consistent-field results are sometimes less reliable. For example, the F_2N_2 molecule has minimum-basis-set self-consistent-field bond distances r_e (N—F) = 0.1384 nm, r_e (N═N) = 0.1214 nm, which are respectively 0.0107 nm longer and 0.0169 nm shorter than the experimental values. Fortunately, vastly improved agreement with experiment results is obtained when a larger basis set is adopted for F_2N_2.

In general, double-zeta self-consistent-field structure predictions are considerably more reliable than those based on minimum basis sets. A noteworthy exception is the water molecule, for which minimum-basis-set self-consistent-field yields a bond angle of 100.0° and double-zeta self-consistent-field predicts 112.6°, compared to the well-known experimental value of 104.5°. More typical are the HF and F_2 molecules, for which the minimum-basis-set, double-zeta, and experimental bond distances are 0.0956, 0.0922, and 0.0917 nm (HF); and 0.1315, 0.1400, and 0.1417 nm (F_2). In fact it can be argued that if the calculations will not be made beyond the Hartree-Fock (single-configuration) approximation, double zeta self-consistent-field is often a reasonable stopping point.

Transition states are typically more sensitive to basis set than equilibrium geometries. This is true because potential energy surfaces are often rather flat in the vicinity of a saddle point (transition state). An example is the carboxime–cyanic acid rearrangement given in reaction (13). Minimum-

$$HONC \rightarrow HOCN \quad (13)$$

basis-set self-consistent-field and double-zeta self-consistent-field transition state geometries are compared in Fig. 2. There it is seen that the minimum-basis-set and double-zeta prediction being presumably the more reliable. It should be noted that for several other transition states, better agreement is found between the two methods. More typical structural variations are ~0.005 nm in internuclear separations and 5° in angles.

As the larger basis sets within the Hartree-Fock formalism are considered, better agreement with experiment is frequently obtained. As implied above, the water molecule bond angle is much improved at the double-zeta plus polarization level,

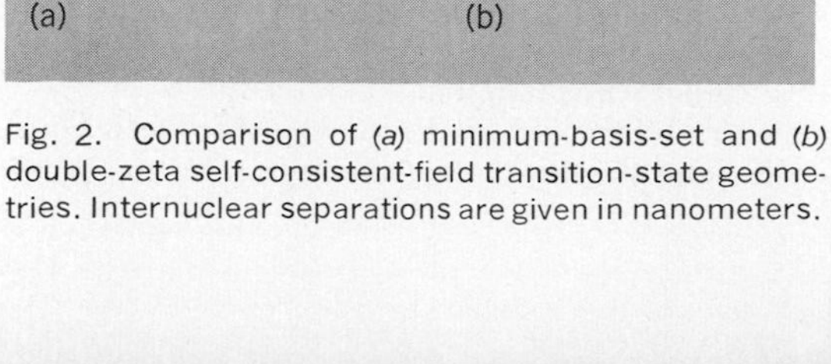

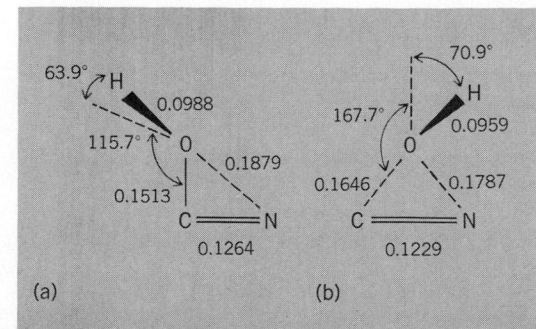

Fig. 2. Comparison of (a) minimum-basis-set and (b) double-zeta self-consistent-field transition-state geometries. Internuclear separations are given in nanometers.

to 106.1°. However, it is often the case that adding polarization functions has only a marginal effect on predicted geometries. A reasonably typical comparison is given by the NH_2F and PH_2F molecules. In this example, the only polarization functions added were sets of *d* functions on the central N or P atom. The only pronounced improvement with respect to experiment is for the P—F separation in PH_2F, and this is improved by 0.0034 nm when *d* functions are added to phosphorus. The good agreement with experiment for NH_2F and PH_2F suggests a high degree of reliability for the comparable NHF_2 and PHF_2 predictions, where no experimental structures have been determined.

An interesting comparison of the three most frequently used basis sets is given in Table 3 for the linear HCNO molecule. The most sensitive geometrical parameter is the N—O bond distance, for which the minimum basis set is 0.0095 nm too long, the double zeta still 0.0056 nm too long, but the double-zeta plus polarization result is in nearly perfect agreement with experiment. For the CN distance, the minimum-basis-set treatment actually gives the best agreement with experiment. The experimental microwave spectrum is difficult to unravel for a quasilinear molecule such as HCNO, and it has been suggested that the double-zeta plus polarization prediction for the C—H distance may be more reliable than experiment.

Table 3. Equilibrium geometry of formonitrile oxide, HCNO, from self-consistent-field theory and experiment*

	Basis set			
Bond distance	MBS	DZ	DZ + P	Experiment
r_e(H—C)	0.1065	0.1049	0.1059	0.1027
r_e(C—N)	0.1155	0.1133	0.1129	0.1168
r_e(N—O)	0.1294	0.1255	0.1201	0.1199

*Values are in nanometers.

Energetic predictions. Among the most chemically important energetic quantities are conformational energy changes, exothermicities or heats of reaction, dissociation energies, and activation energies or barrier heights. In general, only the first of these, and sometimes the second, is reliably predicted at the Hartree-Fock level of theory. In other words, energetic quantities are often sensitive to the effects of electron correlation.

Conformational energy changes are, almost without exception, properly reproduced within the Hartree-Fock formalism. In fact, certain types of barriers, typified by the ethane rotational barrier, are quite satisfactorily predicted at the minimum-basis-set self-consistent-field level of theory. More sensitive problems, such as the ammonia inversion barrier and the rotational barrier of hydrogen peroxide, demand the inclusion of polarization basis functions.

Although Hartree-Fock exothermicities are often unreliable, there is at least one fairly large class of reactions for which consistently good agreement with experiment has been found. Generally speaking, heats of reaction for systems having closed-shell reactants and products are often predicted successfully. More specifically, even better agreement with experiment is found for isodesmic reactions, where the number of bonds of each type is conserved. In fact, reasonable predictions are often made at the minimum-basis-set self-consistent-field level for isodesmic reactions. Further, such information can sometimes be used indirectly (or in conjunction with other thermochemical information) to predict quantities that might be very difficult to evaluate by more straightforward ab initio methods.

The dissociation energies of covalent molecules are generally predicted poorly by single-configuration self-consistent-field methods. Certainly the best-known example is the F_2 molecule, for which the molecular Hartree-Fock energy lies about 1 eV above the Hartree-Fock energy of two fluorine atoms. This problem is often mistakenly attributed to the "perverse" nature of the fluorine atom. In fact, the near-Hartree-Fock dissociation energy of N_2 is 5.27 eV, only about half of the experimental value, 9.91 eV. In addition, the near-Hartree-Fock dissociation energy of O_2 is 1.43 eV, only one-third of the experimental value, 5.21 eV. Thus the Hartree-Fock dissociation energies of covalent molecules are consistently much less than experiment.

Another frequent failing of the Hartree-Fock method is in the prediction of the barrier heights or activation energies of chemical reactions. However, it must be noted that there are many classes of reactions for which Hartree-Fock theory does yield meaningful barrier heights. Two well-studied examples are the isomerizations shown in reactions (14) and (15). For the HNC rearrangement,

$$HNC \rightarrow HCN \tag{14}$$

$$CH_3NC \rightarrow CH_3CN \tag{15}$$

comparison between self-consistent-field (40.0 kcal or 167.4 kJ) and configuration interaction (CI; 36.3 kcal or 151.9 kJ) barriers reveals good qualitative agreement. However, the inclusion of *d* functions in the C and N basis sets appears to be very important. For example, for the methyl isocyanide rearrangement, the self-consistent-field barrier decreases from 60 to 45 kcal (251 to 188 kJ) when polarization functions are added to a double-zeta basis. The remaining discrepancy with the experimental activation energy of 38 kcal (159 kJ) is about equally due to correlation effects and the fact that the zero-point vibrational energy at the transition state is ~ 3 kcal (13 kJ) less than for CH_3NC. Hydrogen isocyanide is one of three interstellar molecules HNC, HCO^+, and HN_2^+) to be identified by ab initio theory prior to their laboratory detection. Generally speaking, unimolecular reactions seem to be treated more reliably by Hartree-Fock methods than bimolecular systems. A second example is the geometrical isomerization of cyclopropane. This system has been studied in considerable detail by L. Salem and coworkers, who totally resolved the structure of the transition state within the full 21-dimensional hypersurface. Although their work involved only a minimum basis set, it goes slightly beyond the self-consistent-field model in that 3 × 3 configuration interaction was included. The predicted barrier height is 53 kcal (222 kJ), in reasonable agreement with the experimental value, 64 kcal (268 kJ).

For many attractive potential energy surfaces, that is, those having no barrier or activation energy at all, Hartree-Fock methods are frequently reliable. An example that has been carefully documented is reaction (16). One of the most important

$$H + Li_2 \rightarrow LiH + Li \quad (16)$$

features of the $H + Li_2$ surface is the fact that the C_{2v} HLi_2 structure is a chemically bound entity. Self-consistent-field theory suggests that the dissociation energy relative to LiH + Li is 20.2 kcal (84.5 kJ), in excellent agreement with the large-scale configuration interaction result of 22.4 kcal (93.7 kJ).

The Hartree-Fock formalism has very powerful predictive capabilities. However, large classes of chemical problems cannot be reasonably described, and a good deal of discretion is required on the part of the theoretician. Although state-of-the-art quantum chemistry has gone well beyond the Hartree-Fock model, the model is certain to remain the cornerstone of electronic structure theory for many years to come. *See* CHEMICAL BONDING; CHEMICAL STRUCTURES; MOLECULAR ORBITAL THEORY; MOLECULAR STRUCTURE AND SPECTRA; RESONANCE.

[HENRY F. SCHAEFER, III]

Bibliography: H. F. Schaefer, *The Electronic Structure of Atoms and Molecules: A Survey of Rigorous Quantum Mechanical Results*, 1972; H. F. Schaefer, *Modern Theoretical Chemistry*, vols. 3 and 4, 1977.

Quasielastic light scattering

Small frequency shifts or broadening from the frequency of the incident radiation in the light scattered from a liquid, gas, or solid. The term quasielastic arises since the frequency changes are usually so small that, without instrumentation specifically designed for their detection, they would not be observed and the scattering process would appear to occur with no frequency changes at all, that is, elastically. The technique is used by chemists, biologists, and physicists to study the dynamics of molecules in fluids, mainly liquids and liquid solutions.

Several distinct experimental techniques are grouped under the heading of quasielastic light scattering (QLS). Intensity fluctuation spectroscopy (IFS) is the technique most often used to study such systems as macromolecules in solution and critical phenomena where the molecular motions to be studied are rather slow. This technique, also called photon correlation spectroscopy and, less frequently, optical mixing spectroscopy, is used to measure the dynamical constants of processes with relaxation time scales slower than about 10^{-6} s. For faster processes, dynamical constants are obtained by utilizing techniques known as filter methods, which obtain direct measurements of the frequency changes of the scattered light by utilizing a monochromator or filter much as in Raman spectroscopy. *See* RAMAN EFFECT.

Static light scattering. If light is scattered by a collection of scatterers, the scattered intensity at a point far from the scattering volume is the result of interference between the wavelets scattered from each of the scatterers and, consequently, will depend on the relative positions and orientations of the scatterers, the scattering angle θ, and the wavelength λ of the light used. The structure of scatterers in solution whose size is comparable to $(4\pi\lambda) \sin \theta/2$ ($\equiv q$) where q is the length of the scattering vector, may be studied by this technique, variously called static light scattering, integrated intensity light scattering, or in the older literature simply light scattering. It was, in fact, developed in the 1940s and 1950s to measure equilibrium properties of polymers both in solution and in bulk. Molecular weights, radii of gyration, solution virial coefficients, molecular optical anisotropies, and sizes and structures of heterogeneities in bulk polymers are routinely obtained from this type of experiment. Static light scattering is a relatively mature field, although continued improvements in instrumentation (mainly the use of lasers and associated techniques) are steadily increasing its reliability and range of application.

Both static and quasielastic light scattering experiments may be performed with the use of polarizers to select the polarizations of both the incident and the scattered beams. The plane containing the incident and scattered beams is called the scattering plane. If an experiment is performed with polarizers selecting both the incident and final polarizations perpendicular to the scattering plane, the scattering is called polarized scattering. If the incident polarization is perpendicular to the scattering plane and the scattered polarization lies in that plane, the scattering is called depolarized scattering. Usually the intensity associated with the polarized scattering is much larger than that associated with the depolarized scattering. The depolarized scattering from relatively small objects is zero unless the scatterer is optically nonspherical.

Intensity fluctuation spectroscopy. The average intensity of light scattered from a system at a given scattering angle depends, as stated above, on the relative positions and orientations of the scatterers. However, molecules are constantly in motion due to thermal forces, and are constantly translating, rotating and, for some molecules, undergoing internal rearrangements. Because of these thermal fluctuations, the scattered light intensity will also fluctuate. The intensity will fluctuate on the same time scale as the molecular motion since they are proportional to each other.

Figure 1 shows a schematic diagram of a typical intensity fluctuation apparatus. Light from a laser source traverses a polarizer to ensure a given polarization. It is then focused on a small volume of

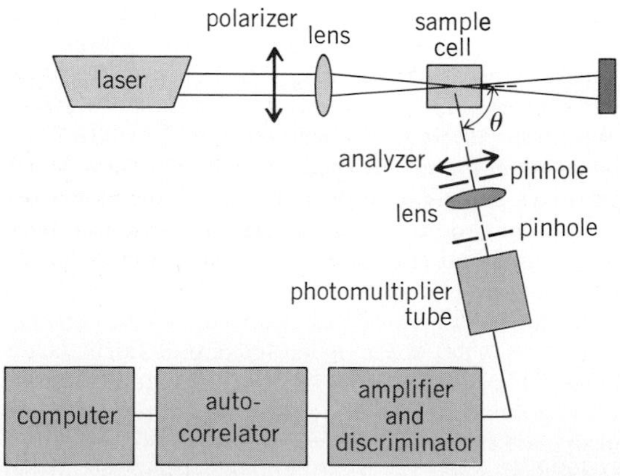

Fig. 1. Schematic diagram of an intensity fluctuation spectroscopy apparatus.

the sample cell. Light from the scattering volume at scattering angle θ is passed through an analyzer to select the polarization of the scattered light, and then through pinholes and lenses to the photomultiplier (PM) tube. The output of the photomultiplier is amplified, discriminated, sent to a photon counter, and then to a hard-wired computer called an autocorrelator, which computes the time autocorrelation function of the photocounts. The autocorrelator output is then sent to a computer for further data analysis.

The scattered light intensity as a function of time will resemble a noise signal. In order to facilitate interpretation of experimental data in terms of molecular motions, the time correlation function of the scattered intensity is usually computed by the autocorrelator. The autocorrelation function obtained in one of these experiments is often a single exponential decay, $C(t) = \exp(-t/\tau_r)$, where τ_r is the relaxation time.

The upper limit on decay times τ_r that can be measured by intensity fluctuation spectroscopy is about a microsecond, although with special variations of the technique somewhat faster decay times may be measured. For times faster than this, filter experiments are usually performed by using a Fabry-Perot interferometer.

Fabry-Perot interferometry. Light scattered from scatterers which are moving exhibits Doppler shifts or broadening due to the motion. Thus, an initially monochromatic beam of light from a laser will be frequency-broadened by scattering from a liquid, gas, or solid, and the broadening will be a measure of the speed of the motion. For a dilute gas the spectrum will usually be a gaussian. For a liquid, however, the most common experiment of this type yields a single lorentzian line with its maximum at the laser frequency $I(\omega) = A/\pi[(1/\tau_r)/(\omega^2 + 1/\tau_r^2)]$. Figure 2 shows a schematic of a typical Fabry-Perot interferometry apparatus. The Fabry-Perot interferometer acts as the monochromator and is placed between the scattering sample and the photomultiplier. Fabry-Perot interferometry measures the (average) scattered intensity as a function of frequency change from the laser frequency. This intensity is the frequency Fourier transform of the time correlation function of the scattered electric field. Intensity fluctuation spectroscopy experiments utilizing an autocorrelator measure the time correlation function of the intensity (which equals the square of the scattered electric field). For scattered fields with gaussian amplitude distributions the results of these two types of experiment are easily related. Sometime intensity fluctuation spectroscopy experiments are performed in what is sometimes called a heterodyne mode. In this case, some unscattered laser light is mixed with the scattered light on the surface of the photodetector. Intensity fluctuation spectroscopy experiments in the heterodyne mode measure the frequency Fourier transform of the time correlation function of the scattered electric field.

Translational diffusion coefficients. The most widespread application of quasielastic light scattering is the measurement of translational diffusion coefficients of macromolecules and particles in solution. For particles in solution whose characteristic dimension R is small compared to $q^{-1} = (4\pi/\lambda \sin \theta/2)^{-1}$, that is, $qR < 1$, it may be shown that the time correlation function measured in a

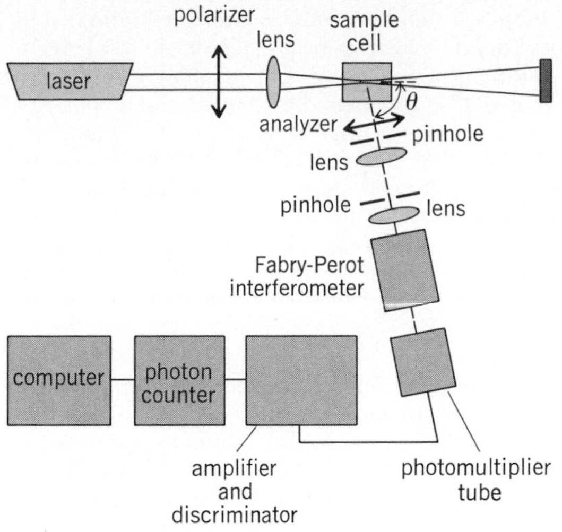

Fig. 2. Typical filter apparatus.

polarized intensity fluctuation spectroscopy experiment is a single exponential with relaxation time $1/\tau_r = 2q^2D$, where D is the particle translational diffusion coefficient. For rigid, spherical particles of any size an intensity fluctuation spectroscopy experiment also provides a measure of the translational diffusion coefficient.

Translational diffusion coefficients of spherical particles in dilute solution may be used to obtain the particle radius R through use of the Stokes-Einstein relation [Eq. (1)], where k_B is Boltzmann's

$$D = \frac{k_B T}{6\pi\eta R} \tag{1}$$

constant, T the absolute temperature, and η the solvent viscosity. If the particles are shaped like ellipsoids of revolution or long rods, relations known, respectively, as the Perrin and Broersma equations may be used to relate the translational diffusion coefficient to particle dimensions. For flexible macromolecules in solution and also for irregularly shaped rigid particles, the Stokes-Einstein relation is often used to define a hydrodynamic radius (R_H).

This technique is routinely used to study such systems as flexible coil macromolecules, proteins, micelles, vesicles, viruses, and latexes. Size changes such as occur, for instance, in protein denaturation may be followed by intensity fluctuation spectroscopy studies of translational diffusion. In addition, the concentration and, in some cases, the ionic strength dependence of D are monitored to yield information on particle interactions and solution structure.

Intensity fluctuation spectroscopy experiments are also used to obtain mutual diffusion coefficients of mixtures of small molecules (for example, benzene–carbon disulfide mixtures) and are also used to measure the behavior of the mutual diffusion coefficient near the critical (consolute) point of a binary liquid mixture. Experiments of this type have proved to be very important in formulating theories of phase transitions.

Rotational diffusion coefficients. Rotational diffusion coefficients are most easily measured by depolarized quasi-elastic light scattering. The instantaneous depolarized intensity for a nonspherical scatterer depends upon the orientation of the scatterer. Rotation of the scatterer will then modulate the depolarized intensity. In a similar way, the frequency distribution of the depolarized scattered light will be broadened by the rotational motion of the molecules. Thus, for example, for dilute solutions of diffusing cylindrically symmetric scatterers, a depolarized intensity fluctuation spectroscopy experiment will give an exponential intensity time correlation function with the decay constant containing a term dependent on the scatterer rotational diffusion coefficient D_R [Eq. (2)]. A depolar-

$$1/\tau_r = 2(q^2D + 6D_R) \qquad (2)$$

ized filter experiment on a similar system will give a single lorentzian with $1/\tau_r$ equal to one-half that given in Eq. (2). For small molecules (for example, benzene) and relatively small macromolecules (for example, proteins with molecular weight less than 30,000) in solution, filter experiments are used to determine rotational diffusion coefficients. In these cases, the contribution of the translational diffusion to τ_r is negligible. For larger, more slowly rotating macromolecules, depolarized intensity fluctuation spectroscopy experiments are used to determine D_R.

Quasielastic light scattering is the major method of studying the rotation of small molecules in solution. Studies of the concentration dependence, viscosity dependence, and anisotropy of the molecular rotational diffusion times have been performed on a wide variety of molecules in liquids as well as liquid crystals.

Rotational diffusion coefficients of very large ($\geqslant$ 100 nm) nonspherical particles may also be measured from polarized intensity fluctuation spectroscopy experiments at high values of q.

Other applications. There are many variations on quasi-elastic light scattering experiments. For instance, polarized filter experiments on liquids also give a doublet symmetrically placed about the laser frequency. Known as the Brillouin doublet, it is separated from the incident laser frequency by $\pm C_s q$, where C_s is the hypersonic sound velocity in the scattering medium. Measurement of the doublet spacing then yields sound velocities. This technique is being extensively utilized in the study of bulk polymer systems as well as of simple liquids.

In a variation of the intensity fluctuation spectroscopy technique, a static electric field is imposed upon the sample. If the sample contains charged particles, the molecule will acquire a drift velocity proportional to the electric field strength $v = \mu E$, where μ is known as the electrophoretic mobility. Light scattered from this system will experience a Doppler shift proportional to v. Thus, in addition to particle diffusion coefficients, quasi-electric light scattering can be used to measure electrophoretic mobilities. *See* ELECTROPHORESIS.

Quasi-electric light scattering may also be used to study fluid flow and motile systems. Intensity fluctuation spectroscopy, for instance, is a widely used technique to study the motility of microorganisms (such as sperm cells); it is also used to study blood flow.

[ROBERT PECORA]

Bibliography: B. Chu, *Laser Light Scattering*, 1974; B. J. Berne and R. Pecora, *Dynamic Light Scattering*, 1976.

Quaternary ammonium salts

Analogs of ammonium salts in which organic radicals have been substituted for all four hydrogens of the original ammonium cation. Substituents may be alkyl, aryl, or aralkyl, or the nitrogen may be part of a ring system. Such compounds are usually prepared by treatment of an amine with an alkylating reagent under suitable conditions. They are typically crystalline solids which are soluble in water and are strong electrolytes. Treatment of the salts with silver oxide, potassium hydroxide, or an ion-exchange resin converts them to quaternary ammonium hydroxides, which are very strong bases, as shown in the reaction below.

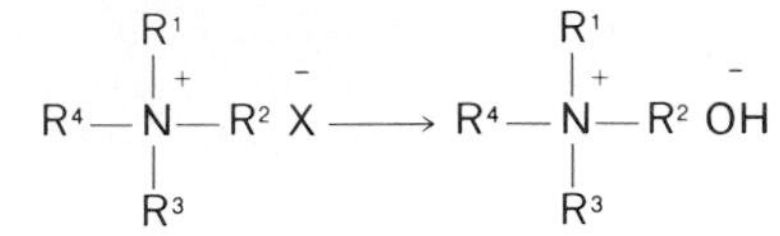

Quaternary ammonium salt   Quaternary ammonium hydroxide

A quaternary salt, methylallylbenzylphenylammonium iodide, containing four different groups on nitrogen, proved to be resolvable (1899), clearly establishing that nitrogen is tetrahedral in these compounds. In general, optically active quaternary ammonium salts racemize readily, suggesting the possibility of an equilibrium between the tertiary amine and the alkyl compound from which it is generated.

Following are some specific examples of useful quaternary ammonium salts and bases: Triton B (I) is used in organic synthesis as a basic catalyst for aldol- and Michael-type reactions. Benzyltriethylammonium chloride is used as a phase-transfer catalyst in organic synthesis to promote the reaction of water-insoluble reagents in an aqueous medium. Quaternary ammonium salts in which one of the alkyl groups is a long carbon chain, such as hexadecyltrimethylammonium chloride (II), are called invert soaps and have useful germicidal properties.

Choline (III) is classified as a vitamin, since the

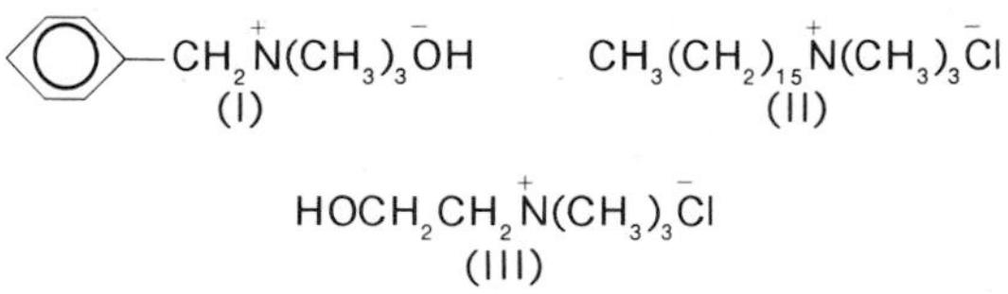

human body makes it too slowly to meet all of its needs and a dietary source is required. Choline is a component of complex lipids such as lecithin, and acetylcholine serves in transmitting nerve signals. *d*-Tubocurarine chloride is a complex quaternary ammonium salt isolated from a tropical plant which is used in medical practice as a skeletal muscle relaxant.

Other quaternary ammonium salts have found use as water repellents, fungicides, emulsifiers, paper softeners, antistatic agents, and corrosion inhibitors. *See* AMINE; AMMONIUM SALT; SURFACE-ACTIVE AGENT.

[PAUL E. FANTA]

Bibliography: N. L. Allinger et al., *Organic Chemistry* 2d ed., 1976.

Quinine

The chief alkaloid of the bark of the cinchona tree, which is indigenous to certain regions of South America. The structure of quinine is shown below.

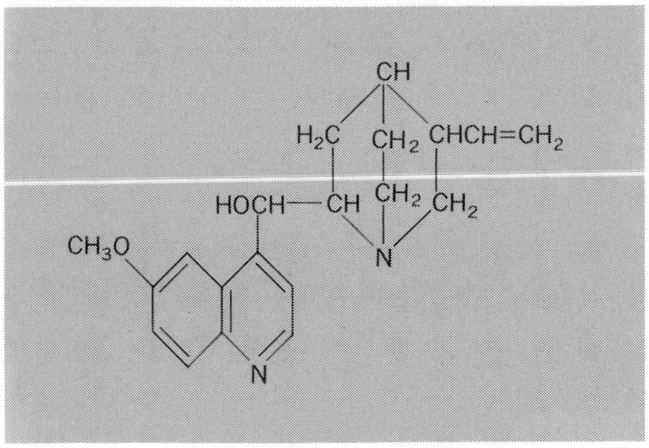

The most important use of quinine has been in the treatment of malaria.

For almost two centuries cinchona bark was employed in medicine as a powder or extract. In 1820 P. Pelletier and J. Caventou isolated quinine and related alkaloids from cinchona bark, and the use of the alkaloids as such rapidly gained favor. Major credit is due to P. Rabe for the postulation of the correct structure of quinine. The difficult laboratory synthesis of quinine by R. Woodward and W. Doering in 1944, although economically unfeasible, corroborated Rabe's structure.

Until the 1920s quinine was the best chemotherapeutic agent for the treatment of malaria. Clinical studies have definitely established the superiority of the newer synthetic antimalarials such as primaquine, chloroquine, and chloroguanide. *See* ALKALOID.

[S. MORRIS KUPCHAN]

Quinoline

One of a group of organic compounds containing a benzene ring fused to the 2,3 positions of pyridine. Positions α, β, and γ are the same, respectively, as positions 2, 3, and 4. Quinoline and some of its homologs are obtained as coal tar extractives. The quinoline ring system appears in synthetic chemotherapeutic agents and in dyes. Quinine and other natural alkaloids also contain the quinoline ring. Quinoline itself is useful as a solvent, a source of nicotinic acid, an acid acceptor and dehydrohalogenating agent, and the starting point in organic syntheses. *See* HETEROCYCLIC COMPOUNDS; PYRIDINE.

Properties. Quinoline (I) is a colorless, steam-volatile liquid, with bp 237.1°C, mp −15°C, specific gravity (15°) 1.09771, and n_D^{15} 1.62928. The solubili-

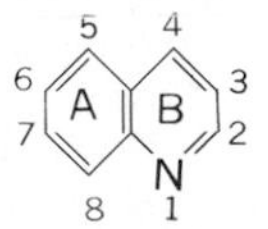

(I)

ty in water at 20°C is 0.65%. Quinoline is a weakly basic (pK_a 4.85 at 20°C) tertiary amine that forms simple salts with acids, and quaternary salts with alkylating agents. The molecule is aromatic in character. It is resistant to disruption by heat, alkali, or acid; it undergoes substitution reactions, and it shows a resonance energy of 69 kcal/mole.

Quinoline is oxidized by nitric acid and other reagents to quinolinic acid (II). In contrast, quinolinium quaternary salts are oxidized by alkaline permanganate to give products of pyridine ring

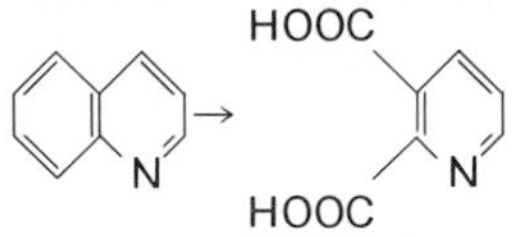

(II)

disruption. Quinoline, by either chemical or catalytic reduction, gives 1,2,3,4-tetrahydroquinoline. For the most part, reactions of such tetrahydroquinolines may be interpreted as reactions of *N*-alkyl orthosubstituted anilines. *N*-Methyl-1,2,3,4-tetrahydroquinoline, or kairolin (IV), is formed by reduction of quinolinium methiodide (III).

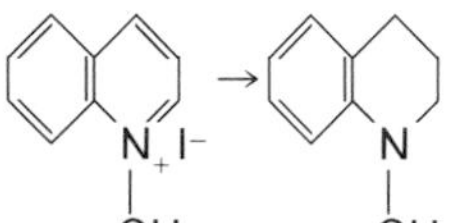

(III) (IV)

The quinoline 2 and 4 positions have properties that distinguish them from the other positions. Methyl groups at these positions are reactive. For example, quinaldine (V) condenses with aldehydes to give 2-vinyl products, and lepidine (VI) reacts with ethyl oxalate to give a pyruvate derivative.

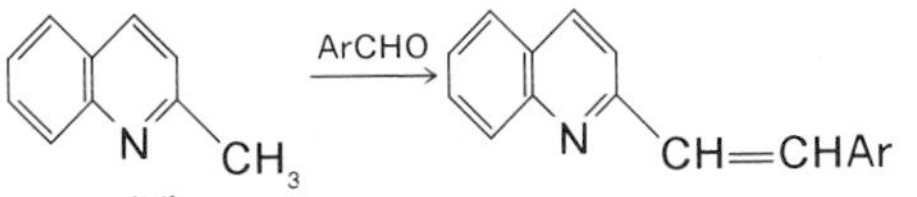

(V)

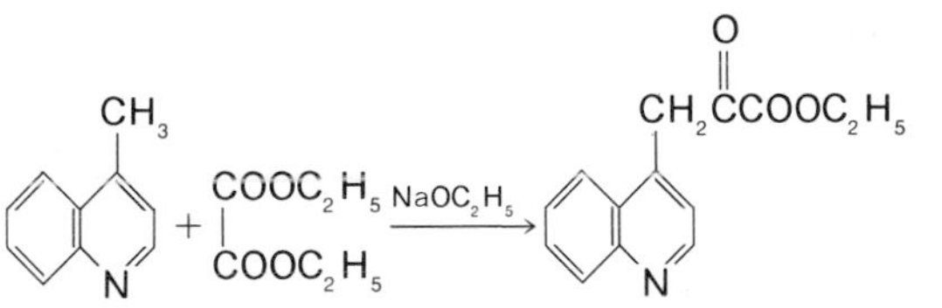

(VI)

Halogens at positions 2 or 4 may be replaced more readily than those at other positions. When the nitrogen carries a plus charge, as in quinolinium quaternary salts, these characteristic features become more pronounced. Quinolinium methiodide easily adds hydroxyl and alkoxyl radicals and carbanions to give 1,2-disubstituted 1,2-dihydroquinolines (VII). Cyanide ion adds readily to the 4 position to give the 1,4-dihydro derivative (VIII),

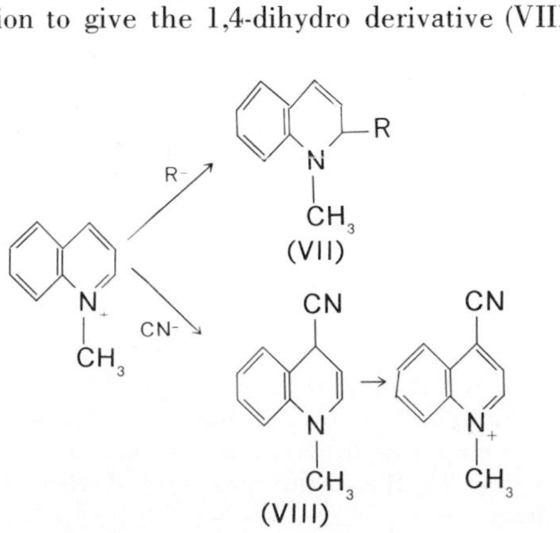

(VIII)

which can be oxidized to a 4-cyanoquinolinium salt. 2-Methylquinolinium ethiodide condenses smoothly with aldehydes; the product with formaldehyde represents one kind of cyanine dye (IX).

N-Benzoylquinolinium cation (X) from the reaction of quinoline and benzoyl chloride reacts easily with cyanide ion to give a useful Reissert compound (XI).

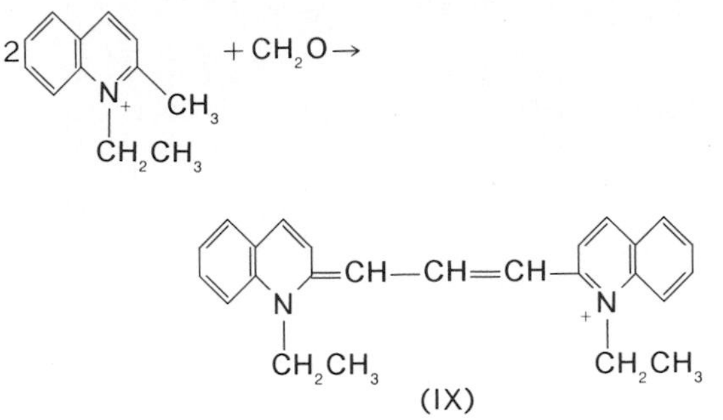

(IX)

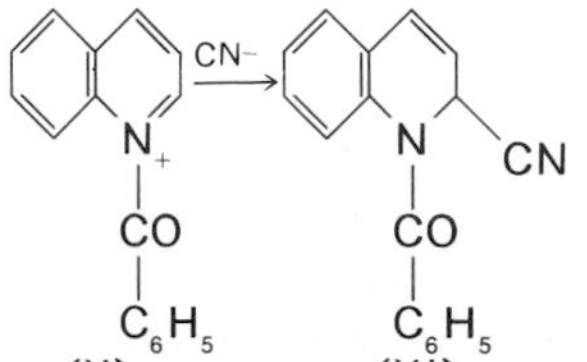

(X) (XI)

Although nitration conditions are known that will place the nitro group at every quinoline position except 2 and 4, 5-nitro- and 8-nitroquinoline predominate when nitric-sulfuric acid is used. Nitroquinolines serve as precursors to aminoquinolines. Sulfonation at 100°C with fuming sulfuric acid gives quinoline-8-sulfonic acid as the main product.

Bromination or chlorination of quinoline in the presence of sulfur occurs at the 3 position. Quinolines halogenated in ring B are formed by ring syntheses starting with halogenated anilines. Diazotization procedures can convert aminoquinolines to the corresponding haloquinolines. 2-Chloroquinoline is prepared by reaction of phosphorus chlorides with 2-quinolone or with 1-methyl-2-quinolone. 4-Chloroquinoline is obtained by the action of phosphorus chlorides with 4-quinolone, or by reaction of sulfuryl chloride with quinoline-*N*-oxide.

Hydroxyquinolines other than 2- and 4-hydroxyquinolines resemble the naphthols in their general behavior, and may be regarded as normal phenols. 2-Hydroxy- and 4-hydroxyquinoline exist almost entirely in the quinolone form and show special behavior. One property of considerable utility is the facile reaction of quinolones with phosphorus halides to give reactive 2- and 4-chloroquinolines. 4-Quinolone, or kynurine, is the decarboxylation product from 4-hydroxyquinoline-2-carboxylic acid (kynurenic acid).

Preparation. Quinoline ring syntheses construct ring B of the quinoline product by operating on anilines or related compounds. A useful classification of quinoline syntheses depends on whether the carbon atom of position 4 in the quinoline product was, as in (XII), or was not, as in (XIII), attached to the ortho position of the starting aniline.

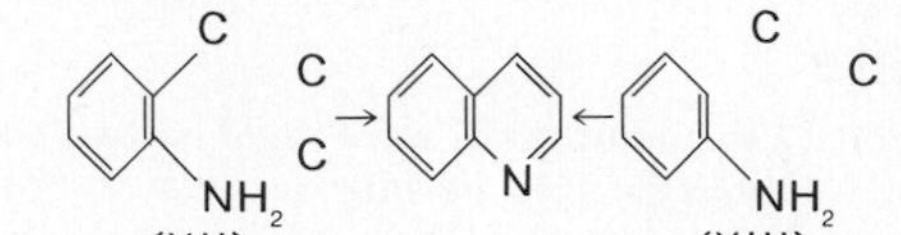

(XII) (XIII)

In the latter category, the Döbner-Miller synthesis combines two molecules of aldehyde with an aromatic amine. In this way, quinaldine or 2-methylquinoline (XIV) is obtained from acetaldehyde and aniline.

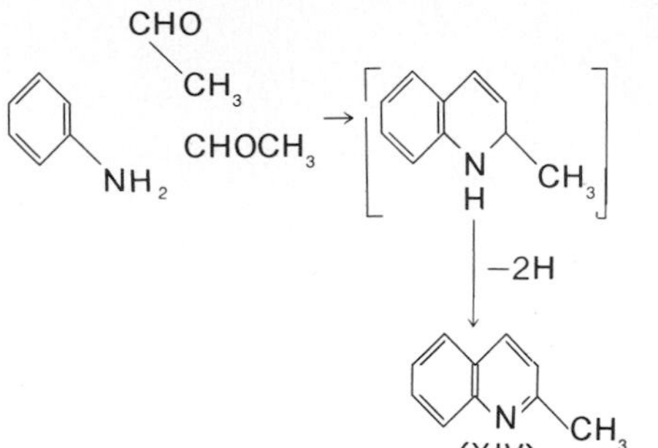

(XIV)

In common with some other quinoline syntheses, the intermediate dihydro stage is not isolated, but is dehydrogenated in place. In the versatile Skraup synthesis aniline and glycerol are heated in the presence of sulfuric acid and an oxidizing agent to form ring B unsubstituted quinolines. With the rationalization that glycerol acts as precursor to the α,β-unsaturated acrolein (XV), the Skraup reaction appears as a modification of the Döbner-Miller process.

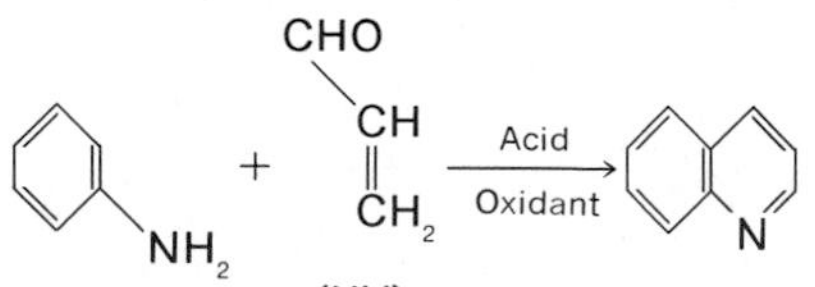

(XV)

Commercial synthetic quinoline is prepared by the Skraup reaction. Syntheses with 1,3-dicarbonyl compounds are also possible. In the Knorr quinoline synthesis, aniline and acetoacetic ester (XVI) react to form acetoacetanilide (XVII), which, with cold concentrated sulfuric acid, cyclizes to give 4-methyl-2-quinolone (XVIII). The reactants can be made to combine in the opposite sense to give intermediate (XIX), which on pyrolysis furnishes 2-methyl-4-quinolone (XX) (Conrad-Limpach method).

When the 1,3-dicarbonyl compound is a diketone, for example, acetylacetone (XXI), the products are 2,4-disubstituted quinolines.

Quinoline syntheses, starting with compounds of type (XII), include the Friedländer method, by which an *o*-acylaniline (XXII) condenses with an α-methylene compound (XXIII), and the Pfitzinger scheme, by which isatin (XXIV) combines with an acetophenone to give a 2-arylcinchoninic acid (XXVI). In the Pfitzinger synthesis, one interpretation has the isatin molecule opening to form the *o*-acylaniline (XXV), which then reacts with the acetophenone according to the Friedländer reaction.

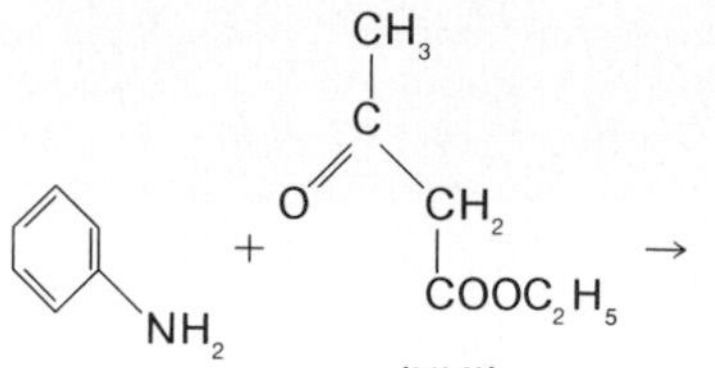

(XVI)

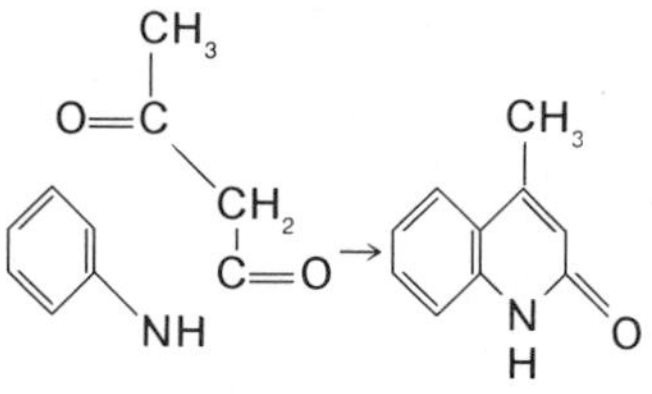

(XVII) (XVIII)

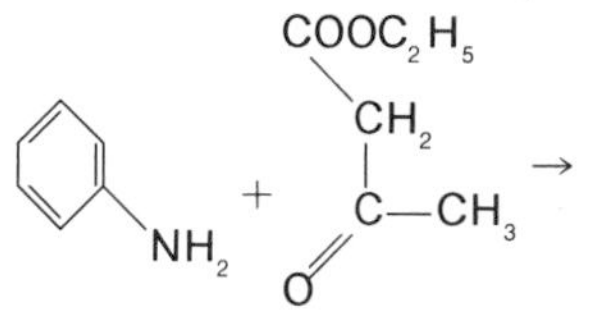

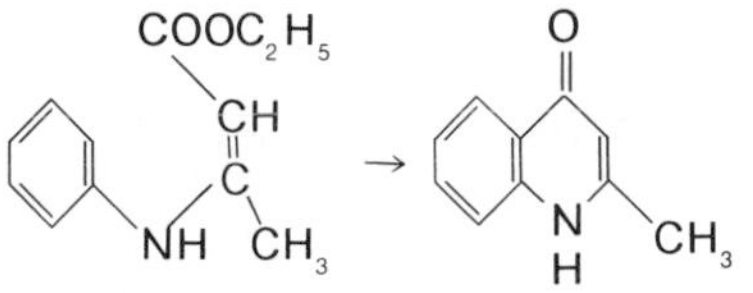

(XIX) (XX)

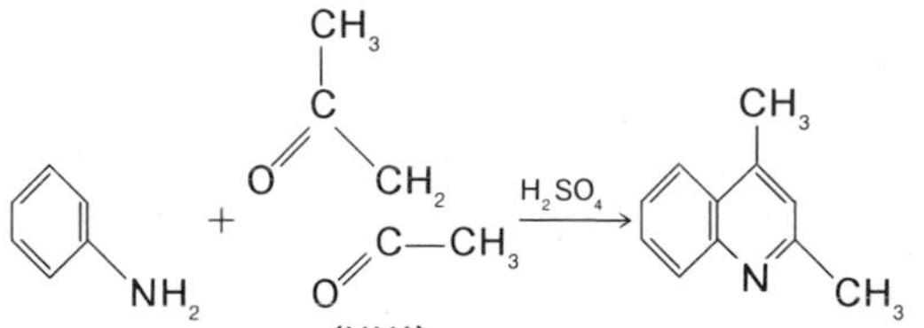

(XXI)

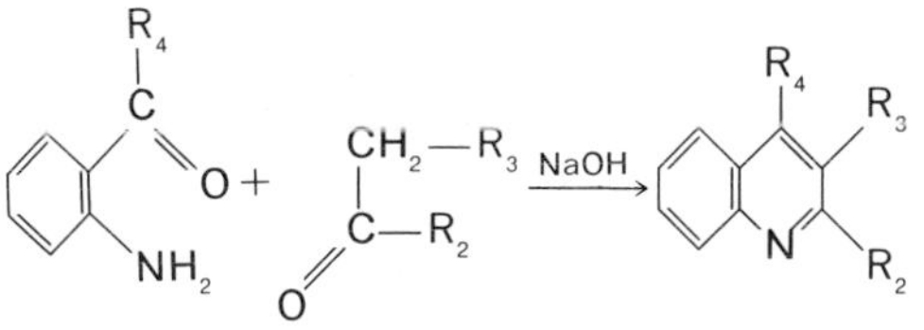

(XXII) (XXIII)

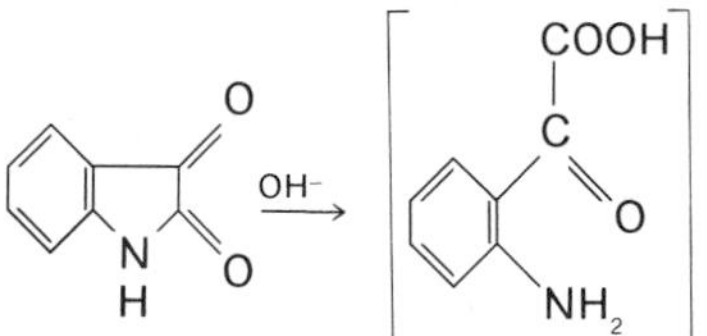

(XXIV) (XXV)

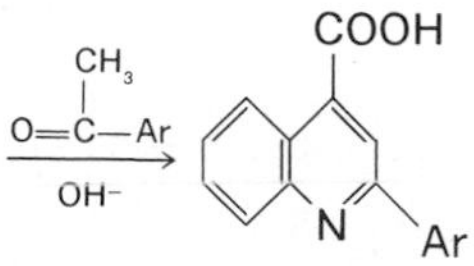

(XXVI)

Important derivatives. 8-Hydroxyquinoline (oxine) is prepared by sulfonation of quinoline, followed by alkali fusion of the resulting quinoline-8-sulfonic acid. 8-Hydroxyquinoline is a reagent of considerable utility in analysis for metals, especially magnesium, zinc, and aluminum. The procedures make use of the chelating properties of oxine with metals; structure (XXXVI) shows magnesium oxinate. 8-Hydroxyquinoline is also a fungicide and antiseptic. 5-Chloro-7-iodo-8-hydroxyquinoline (vioform) and 7-iodo-8-hydroxyquinoline-5-sulfonic acid (chiniofon) are amebicides. Chiniofon is also used in the colorimetric determination of iron and calcium.

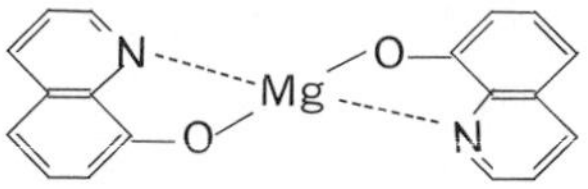

(XXVII)

Quinoline carboxylic acids are prepared by several ring-closure procedures, by oxidation of groups such as methyl already on the ring, or by transformations starting with bromoquinolines. The monocarboxylic acids, with pK_a 4.5–5.0 (in 50% methanol), are somewhat stronger than benzoic acid (pK_a 5.27). Quinoline-8-carboxylic acid (pK_a 7.2) is an exception. The carboxyl groups, especially at the 2 or 4 positions, may be removed by heating.

Quinoline-2-carboxylic acid (quinaldinic acid) is prepared by oxidation of 2-methylquinoline (quinaldine) or 2-styrylquinoline, or by treatment of a Reissert compound (XI) with concentrated hydrochloric acid. 4-Hydroxyquinaldinic acid (kynurenic acid) and 4,8-dihydroxyquinaldinic acid (xanthurenic acid) are two of the several metabolic products of tryptophan. The cinchoninic acids (quinoline-4-carboxylic acids) are prepared by ring closures, by oxidation of 4-substituted quinolines, or by making use of compounds of the type (VIII). A convenient method for preparing cinchoninic acid is given in the conversion of (XXVIII) to (XXX). Nupercaine (XXXI), a potent but somewhat toxic local anesthetic, is prepared from cinchoninic acid derivative (XXIX). 2-Phenylcinchoninic acid (cincophen), prepared by Pfitzinger's isatin method, has been used in treatment of gout, and as an analgesic and antipyretic.

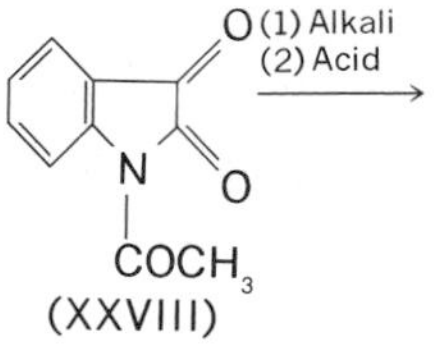

(XXVIII)

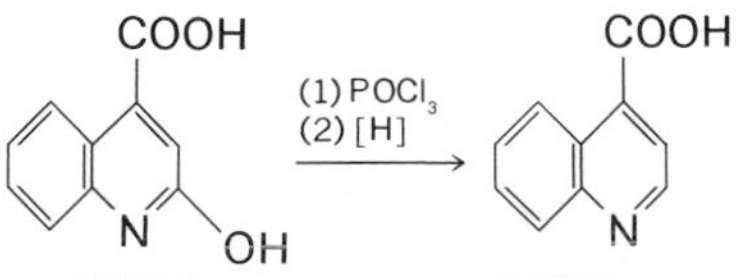

(XXIX) (XXX)

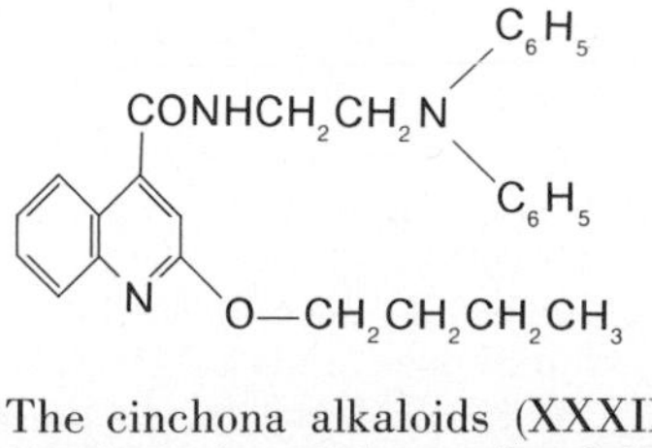

(XXXI)

The cinchona alkaloids (XXXII) are quinoline derivatives; in quinine and quinidine, R = CH_3O; in cinchonine, R = H; and in cupreine, R = OH. Quinine, and to a lesser extent cinchonine, have been used for centuries against malarial fever.

Quinidine, a stereoisomer of quinine, is a useful heart drug.

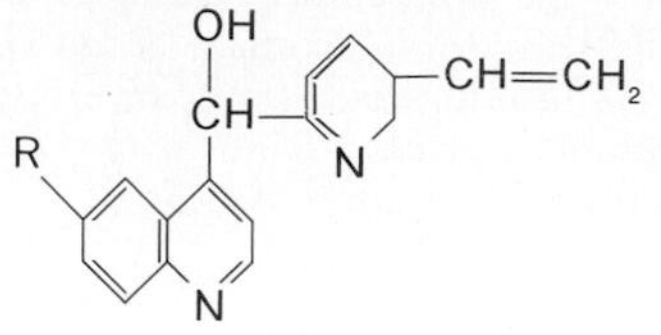

(XXXII)

Derivatives of 4-amino- and 8-aminoquinoline have attracted attention as synthetic antimalarial agents. Chloroquine (XXXIII), one of the more effective 4-aminoquinolines, is formed by combination of 4,7-dichloroquinoline with 4-amino-1-diethylaminopentane. The 8-aminoquinoline derivatives (XXXIV) are prepared by alkylating 8-aminoaminoquinoline with appropriate aminoalkyl halides.

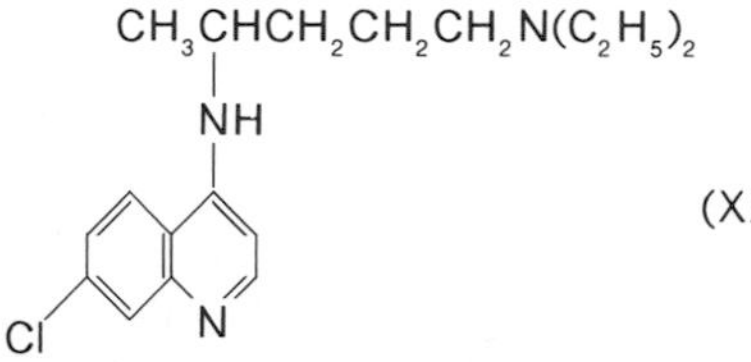

(XXXIII)

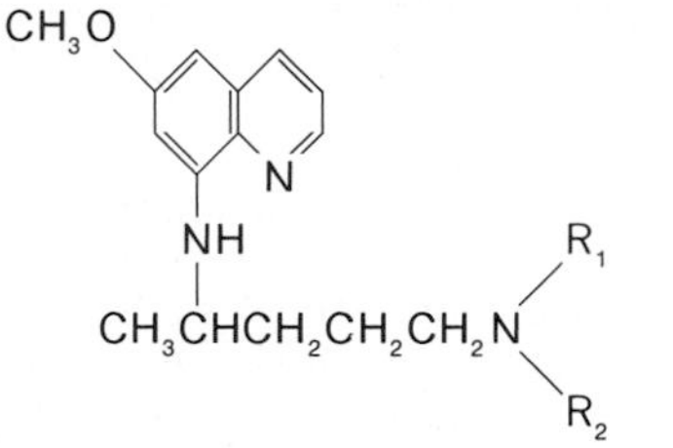

(XXXIV)

Pamaquine (XXXIV), in which $R_1 = R_2 =$ ethyl, isopentaquine in which $R_1 =$ H, and $R_2 =$ isopropyl, and particularly primaquine in which $R_1 = R_2 =$ H, are effective curative antimalarials. The necessary 6-methoxy-8-aminoquinoline for (XXXIV) is obtained from 6-methoxy-8-nitroquinoline, which is formed in a Skraup reaction with 2-nitro-4-methoxyaniline.

The quinoline dyes include those of the cyanine type (IX), azo dyes (XXXV) derived from 2,4-dihydroxyquinoline, and quinophthalones (XXXVI),

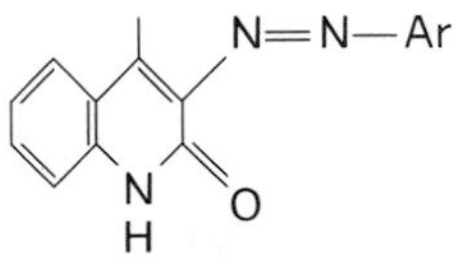

(XXXV)

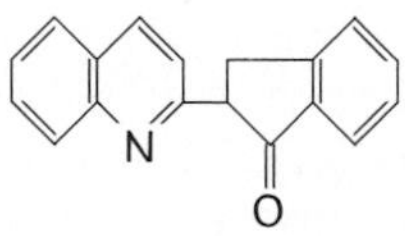

(XXXVI)

from 2-methylquinolines and phthalic anhydride. *See* ISOQUINOLINE; QUININE.

[WALTER J. GENSLER; MAURICE SHAMMA]

Bibliography: J. A. Joule and G. F. Smith, *Heterocyclic Chemistry*, 1972; L. A. Paquette, *Principles of Modern Heterocyclic Chemistry*, 1968; F. D. Popp, Reissert compounds, in A. R. Katritzky and A. J. Boulton (eds.), *Advances in Heterocyclic Chemistry*, vol. 8, 1967; G. Scheibe and E. Daltrozzo, Diquinolylmethane and its analogs, in A. R. Katritzky and A. J. Boulton (eds.), *Advances in Heterocyclic Chemistry*, vol. 7, 1966; D. W. Young, *Heterocyclic Chemistry*, 1976.

Quinone

One of a class of aromatic diketones in which the carbon atoms of the carbonyl groups are part of the ring structure. The name quinone is applied to the whole group, but it is often used specifically to refer to *p*-benzoquinone. *o*-Benzoquinone is also known but the meta isomer does not exist.

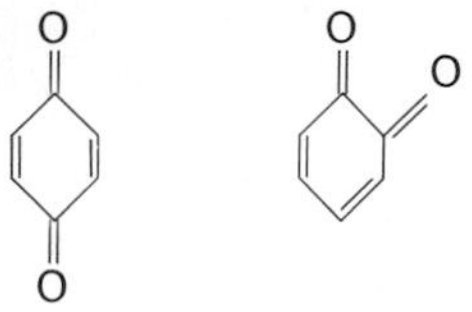

Quinone (*p*-benzoquinone) — *o*-Benzoquinone

Preparation. Quinones are prepared by oxidation of the corresponding aromatic ring systems containing amino (—NH_2) or hydroxyl (—OH) groups on one or both of the carbon atoms being converted to the carbonyl group. *p*-Benzoquinone is prepared by the oxidation of aniline with manganese dioxide, MnO_2, in the presence of sulfuric acid, H_2SO_4. Oxidation of phenol, *p*-aminophenol, hydroquinone, or *p*-phenylenediamine will also produce *p*-benzoquinone. *o*-Benzoquinone is prepared by oxidation of catechol with silver oxide, Ag_2O, in the absence of water. This quinone is much less stable and more reactive than the para isomer.

Three of the several possible quinones derived from naphthalene are known: 1,4-napthoquinone, 1,2-naphthoquinone, and 2,6-naphthoquinone. The

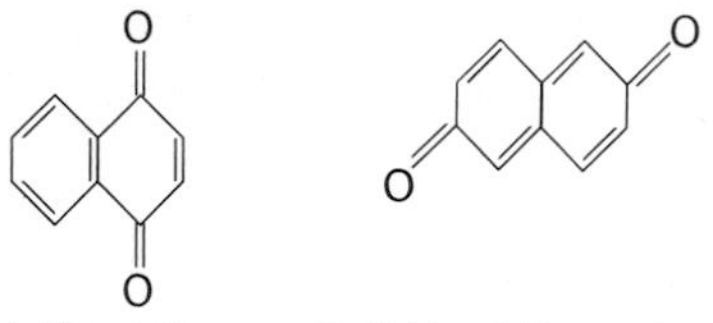

1,4-Naphthoquinone — 2,6-Naphthoquinone

naphthoquinones are prepared by oxidation of the corresponding aminonaphthols. 9,10-Anthraquinone is best prepared, as in reaction (1), by dehy-

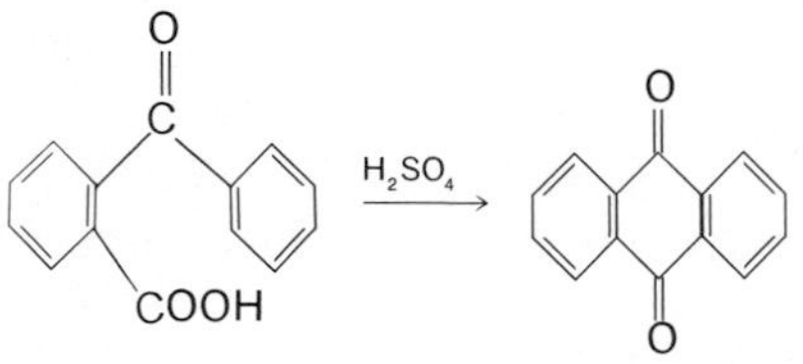

(1)

9,10-Anthraquinone

dration of *o*-benzoylbenzoic acid which is prepared from Friedel-Crafts reaction of benzene and phthalic anhydride. Direct oxidation of phenanthrene with chromic acid yields 9,10-phenanthraquinone, the further oxidation of which gives diphenyl-2,2′-dicarboxylic acid (diphenic acid).

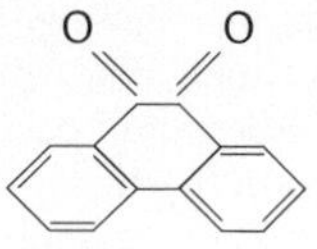

9,10-Phenanthraquinone

Reactions. *p*-Benzoquinone is easily reduced, as in reaction (2), to hydroquinone by a variety of reagents.

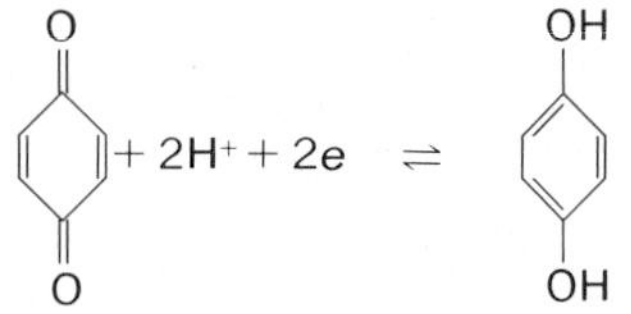

(2)

Hydroquinone

Reaction (2) is reversible, and the position of equilibrium can be made to depend on hydrogen-ion concentration and applied electrical potential.

This system ($E_0 = 0.699$ volt) has been useful for the measurement of hydrogen-ion concentration. The E_0 values for many other quinone-hydroquinone systems have been measured. An intermediate in the reduction of *p*-benzoquinone or in the oxidation of hydroquinone is quinhydrone, a 1:1 molecular complex of these two substances. *See* HYDROGEN ION.

The most characteristic reactions of para quinones are those of the carbon to carbon double bonds and of the conjugated system

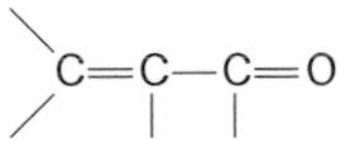

Reaction as a dienophile in the Diels-Alder process, reaction (3), is quite general and occurs under

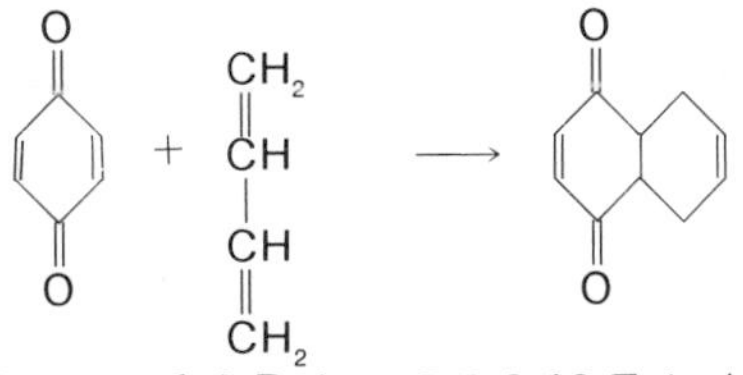

(3)

p-Benzoquinone 1,3-Butadiene 5,8,9,10-Tetrahydro-1,4-naphthoquinone

mild conditions. The remaining >C=C< bond of the quinone ring may also react in the same way. Halogen adds normally to the >C=C< bond as in alkenes. Hydrogen halide, however, adds to the conjugated

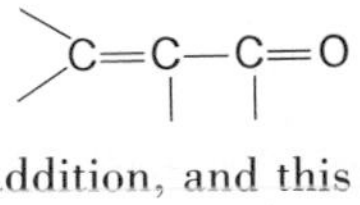

system by 1,4 addition, and this is followed by enolization to a hydroquinone derivative, reaction (4).

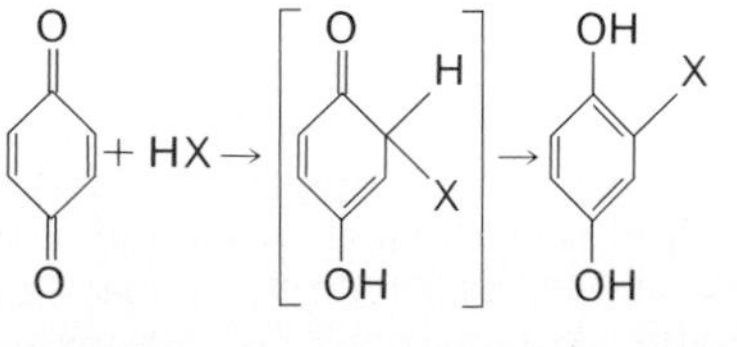

(4)

Malonic ester and acetoacetic ester react similarly with utilization of their active hydrogen atoms. Mercaptans and Grignard reagents give mixtures of normal adducts to the C=O group and 1,4 additions to the system. Simple quinones do not often undergo substitution reactions with the electrophilic reagents commonly used for aromatic systems. Free radicals from decomposition of acyl peroxides or lead tetraacetate substitute as in reaction (5).

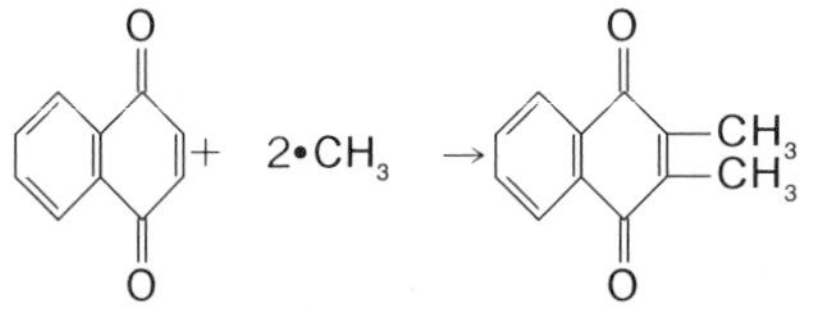

(5)

Methyl radicals

Important naturally occurring naphthoquinones are vitamins K_1 and K_2 which are found in blood

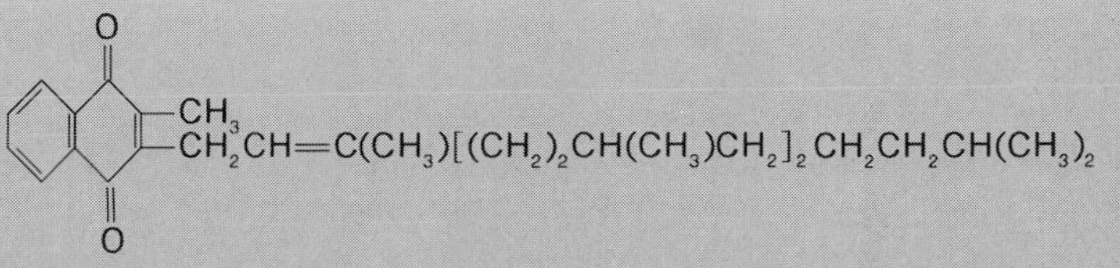

Vitamin K_1

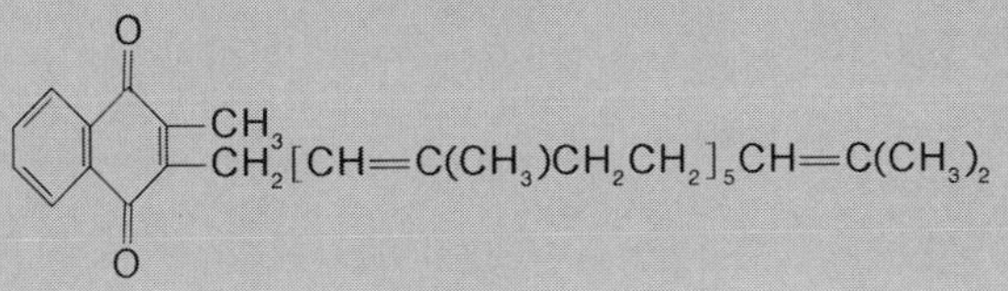

Vitamin K_2

and are responsible for proper blood clotting reaction. The long aliphatic chain has been found unnecessary for the clotting reaction; its replacement by a hydrogen atom gives Menadione or 2-methyl-1,4-naphthoquinone, which is manufactured synthetically for medicinal use.

A number of quinone pigments have been isolated from plants and animals. Illustrative of these are juglone found in unripe walnut shells and spin-

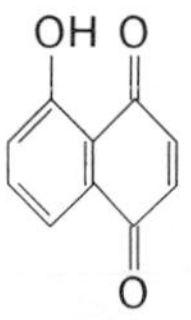

Juglone

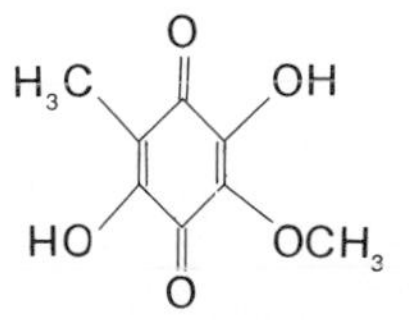

Spinulosin

ulosin from the mold *Penicillium spinulosum*. 9,10-Anthraquinone derivatives form an important class of dyes of which alizarin is the parent type. *p*-Benzoquinone is manufactured for use as a photographic developer. *See* ANTHRAQUINONE PIGMENTS; AROMATIC HYDROCARBON; HYDROQUINONE; KETONE; OXIDATION-REDUCTION.

[DAVID A. SHIRLEY]

Bibliography: R. A. Morton, *Biochemistry of Quinones*, 1965; J. D. Roberts and M. C. Caserio, *Basic Principles of Organic Chemistry*, 2d ed., 1977.

Racemization

The formation of a racemate from a pure enantiomer. Alternatively stated, racemization is the conversion of one enantiomer into a 50:50 mixture of the two enantiomers (+ and −, or R and S) of a substance. Racemization is normally associated with the loss of optical activity over a period of time since 50:50 mixtures of enantiomers are optically inactive. *See* OPTICAL ACTIVITY.

Racemization is an energetically favored process since it reflects a change from a more ordered to a more random state. But the rate at which enantiomers racemize is typically quite slow unless a suitable mechanistic pathway is available, since racemization usually, but not always, requires that a chemical bond at the chiral center of an enantiomer be broken. Racemization of enantiomers possessing more than one chiral center requires that all chiral centers of half of the molecules invert their configurations. *See* ENTROPY.

Mechanisms. Of the several known racemization mechanisms, those involving the temporary formation of intermediates which possess reflection symmetry are the most common. Depending upon the substance and the conditions employed, the intermediate may be a free radical, carbocation, or carbanion. The racemization of an alkyl halide (I) in the presence of a Lewis acid catalyst [reaction (1)] illustrates the intervention of a planar, hence achiral, carbocation intermediate (II). Reversal of the ionization or capture of the cation by a solvent molecule can occur with equal probability to either side of the planar intermediate which then reforms a molecule of tetrahedral geometry (III). When equal numbers of (R)- and (S)-enantiomers are present, racemization is complete.

Chiral compounds with acidic hydrogens at the chiral center are racemized by base with the intervention of carbanions [reaction (2)]. Delocalization of the unshared electrons makes the enolate intermediate planar. For carbanions for which delocalization is not possible, racemization takes place because a nonplanar carbanion usually inverts its configuration readily.

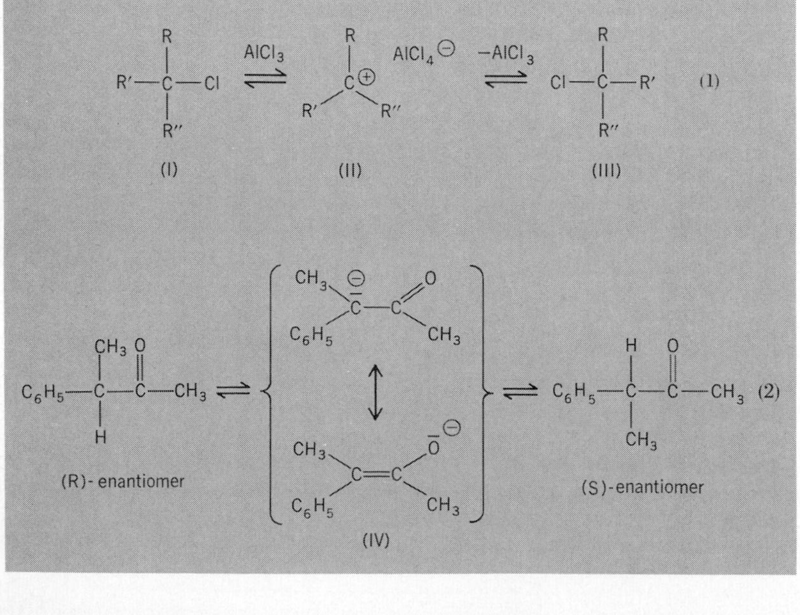

Racemization via free radicals is exemplified by the chlorination of hydrocarbon (V) which involves an intermediate free radical mechanism [reaction (3)]. Product (VI) is racemic. The *t* shown in structures (V) and (VI) stands for tertiary (as intertertiary butyl group) and is a descriptor defining one of the isomers of the butyl group (C_4H_9). Heat can also cause racemization without breaking bonds if it is possible for a molecule to pass through a state which possesses reflection symmetry. For example, the chiral biphenyl (XII) may be racemized thermally by rotation about the central carbon-carbon bond. In this case the benzene rings become coplanar in the transition state (VIII) [reaction (4)]. Nucleophilic displacement on chiral alkyl halides may also lead to racemization via symmetrical transition states if the displacing ion is identical to the leaving ion [reaction (5)]. This type of change in configuration is known as Walden inversion. *See* FREE RADICAL; ORGANIC REACTION MECHANISM; REACTIVE INTERMEDIATES.

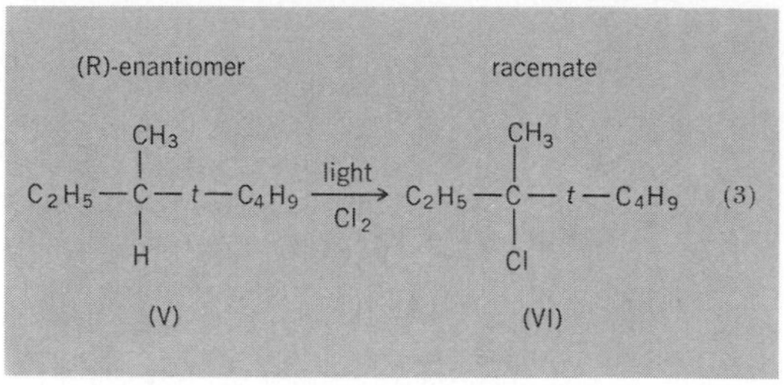

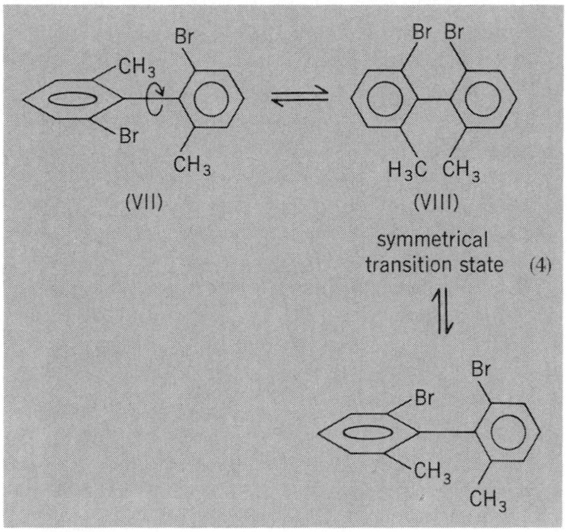

Significance. The observation and study of racemization have important implications for the understanding of the mechanisms of chemical reactions and for the synthesis and analysis of chiral natural products such as peptides. Moreover, racemization is of economic importance since it provides a way of converting an unwanted enantiomer into a useful one. Synthetic medicinal agents are often produced industrially as racemates. After resolution and isolation of the desired enantiomer, half of the product would have to be discarded were it not for the possibility of racemizing the

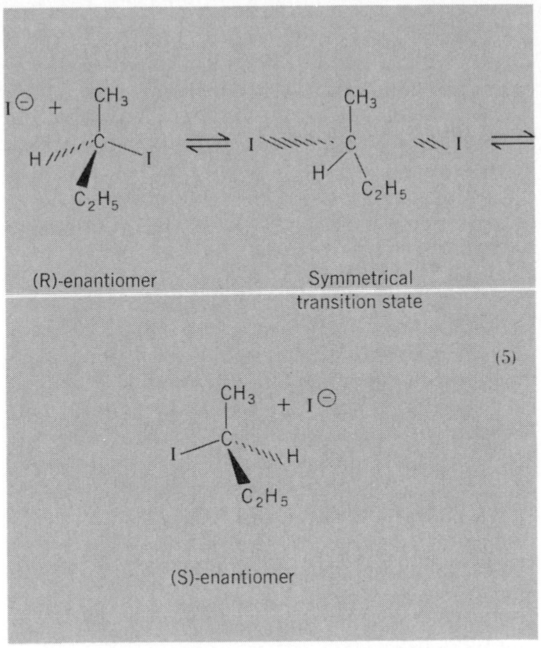

unwanted isomer and of recycling the resultant racemate.

Inversion of configuration at some but not all chiral centers in diastereomeric peptides and sugars does not lead to total loss of configuration. This process, while analogous to racemization, does not lead to complete loss of optical activity. It is called epimerization. Epimers are diastereomers which differ at only one of several chiral centers. Epimerization of amino acids in proteins takes place very slowly on standing, and is used as a means of dating very old protein samples (aminostratigraphy). *See* STEREOCHEMISTRY.

[SAMUEL H. WILEN]

Bibliography: E. L. Eliel, *Stereochemistry of Carbon Compounds*, 1962.

Radiation chemistry

The study of the chemical effects of the absorption of high-energy radiation in matter. High-energy radiation includes the emanation associated with radioactive decay and fission (that is, α-particles, electrons, γ-rays, and neutrons), together with their related atom and fission recoils; and the artificial analogs of such emanations produced by accelerating electrons, protons, deuterons, and helium nuclei, as well as charged nuclei of higher atomic number [currently as high as 18 (argon)] and x-rays.

Sources of high-energy radiations in the laboratory and industry include radioactive nuclides (for example, ^{60}Co, ^{90}Sr, and ^{3}H) and instruments such as x-ray tubes, Van de Graaff generators, the betatron, the cyclotron, and the synchrotron. An electron accelerator known as the linac (linear electron accelerator) has proved particularly valuable for the study of transient species which have lifetimes as short as 16 picoseconds, and another electron accelerator, known as the Febetron, has been used for the study of the effects of single pulses of electrons with widths of several nanoseconds at very high currents.

An experimental apparatus using a single 25-ps pulse of electrons from a linear accelerator is shown in Fig. 1. The short pulse of 13-MeV electrons emerges from a thin window and enters an optical cell containing the chemical system being studied. Before penetrating the cell, part of the electron beam is intercepted by a cell containing xenon gas. A continuous spectrum of light due to the production of Cerenkov radiation emerges from this cell and, after suitable optical delays with respect to the electron pulse, impinges on the sample cell, allowing the absorption spectra and lifetime of the transient species produced by the electron pulse in the cell to be studied. In addition to the optical absorption method, a variety of fast reaction techniques are used to study the reactions of short-lived species produced by such short pulses. These include fluorescence, ionic conductivity, electron spin resonance, nuclear magnetic resonance, resonance Raman scattering, and polarography. *See* ELECTRON PARAMAGNETIC RESONANCE (EPR) SPECTROSCOPY; NUCLEAR MAGNETIC RESONANCE (NMR); POLAROGRAPHIC ANALYSIS; RAMAN EFFECT.

Energy transfer. Energy is transmitted to irradiated material by momentum transfer and by excitation and ionization. The latter two always accompany the former. Momentum transfer is characteristic of processes involving neutrons; it is always involved to some extent in particle effects and is an important contribution to heavy-particle (for example, proton) effects.

In a momentum transfer interaction, the usual effect is the ejection of a nucleus from its molecular or crystalline structure. In the case of solids, this is known as the Wigner effect. When crystalline material is involved, the process is called discomposition. Discomposition results not only from neutron impact, but also secondarily from impacts involving high-energy displaced nuclei. Momentum transfer effects in solids are detected by changes in electronic and thermal conductivity, elastic moduli, and dimensions. In crystalline

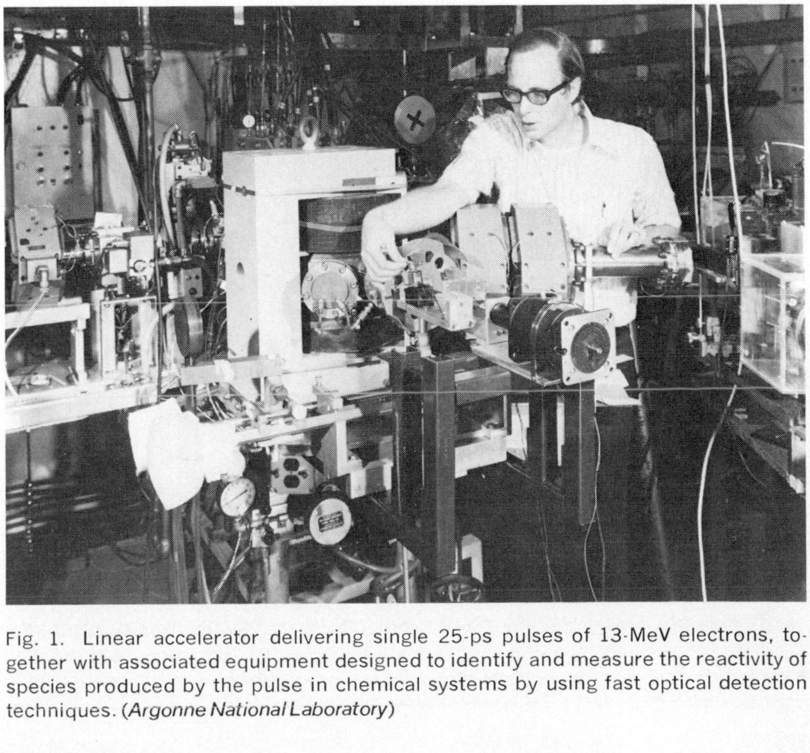

Fig. 1. Linear accelerator delivering single 25-ps pulses of 13-MeV electrons, together with associated equipment designed to identify and measure the reactivity of species produced by the pulse in chemical systems by using fast optical detection techniques. (*Argonne National Laboratory*)

compounds, effects include chemical change, electron trapping, and color production.

Chemical yields are expressed in the older literature as ion-pair yields M/N—a yield being the number of molecules converted or produced per ion pair initially produced by the radiation. The modern literature uses the 100-eV yield G, which is the number of molecules converted or produced per 100 eV of energy absorbed. The term M/N is now used only in cases in which N, the number of ion pairs, is actually determinable from experimental data. A convenient rule of thumb for reading the older literature is $G \simeq 3M/N$. Yields range from values such as $G \simeq 0.01$ (copper phthalocyanin decomposition) to values such as those in notation (1).

$$G(H_2 + Cl_2 \rightarrow 2HCl) \sim 10^5 \qquad (1)$$

Energy input may be determined directly from charge measurements by using Faraday cups when using machine sources or indirectly by chemical dosimetry. The Fricke dosimeter (aerated acidic ferrous sulfate solution) is such a secondary standard for high-energy electrons and for γ and x-rays, where $G(Fe^{2+} \rightarrow Fe^{3+}) = 15.6$. These secondary standards have been calibrated against an absolute measurement of the energy input by using calorimetric methods. The units rad, roentgen, and rep are also used. One roentgen is equivalent to 6.08×10^{13} eV/g of water for x-rays and γ-rays. One gray (Gy) is equal to 1 joule/kg and is equivalent to 100 rads.

Theoretical considerations regarding primary physical processes indicate that for other than momentum transfer, the principal effects are due to fast-moving charged particles. A 1-MeV charged particle, unlike a parent photon which produces only one ionization (as in a Compton process), may produce a total of 10^5 ions (and electrons) and excited molecules. The distribution of such primarily produced entities is inhomogeneous, and is affected greatly by the nature of the radiation and by the state of aggregation of the material irradiated. It is this inhomogeneity of the distribution of activated species which is an important factor differentiating radiation chemistry from photochemistry and thermally induced chemistry. In the latter two, the active species are produced with a homogeneous distribution. In radiation chemistry, it is only when the active species produced by the initial ionizing particle have diffused throughout the reaction volume that homogeneous reaction kinetics can be applied.

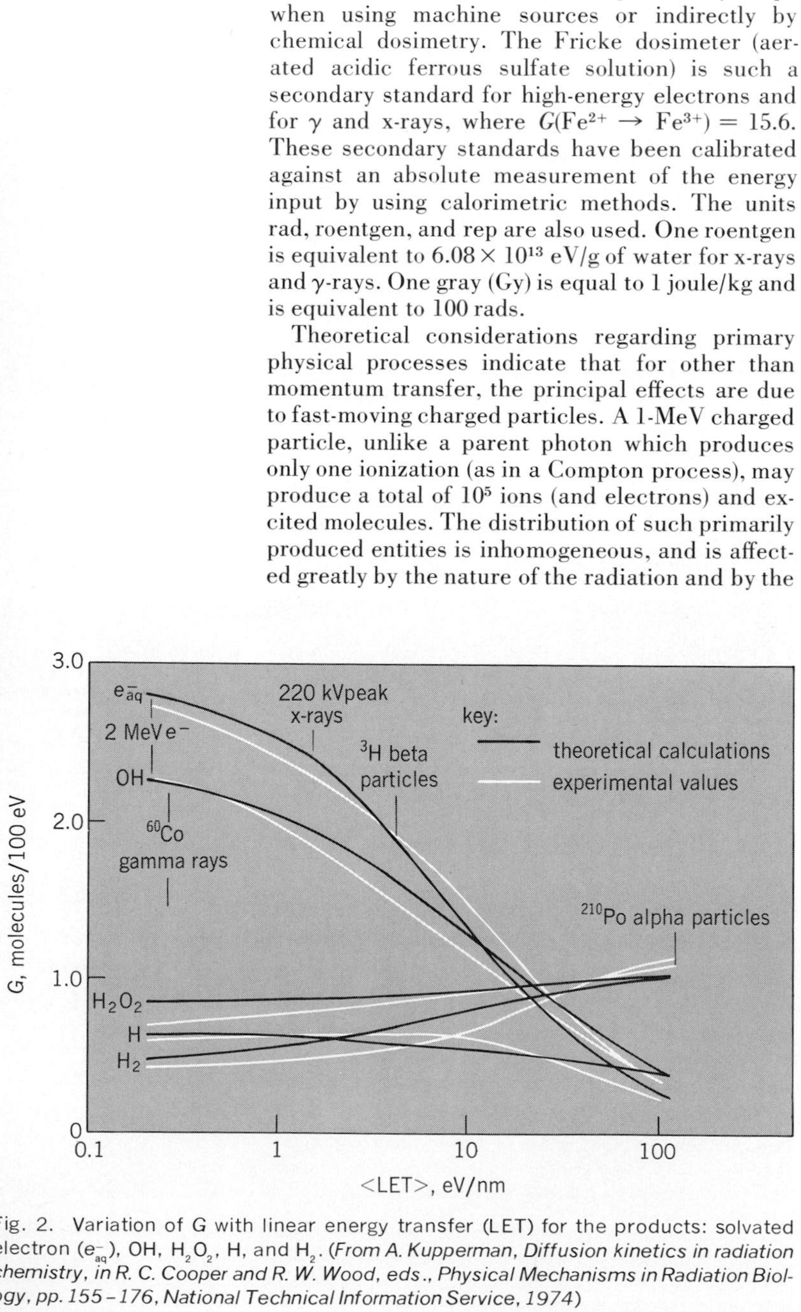

Fig. 2. Variation of G with linear energy transfer (LET) for the products: solvated electron (e^-_{aq}), OH, H_2O_2, H, and H_2. (*From A. Kupperman, Diffusion kinetics in radiation chemistry, in R. C. Cooper and R. W. Wood, eds., Physical Mechanisms in Radiation Biology, pp. 155–176, National Technical Information Service, 1974*)

In condensed systems, ions and excited species tend to be formed in three quantitatively different kinds of groups or clusters: (1) spurs involving energy losses up to about 100 eV; (2) blobs of energy loss of approximately 100–500 eV; and (3) short tracks of relatively high energy of approximately 500 eV to 5 keV. For a 1-MeV electron, the energy is deposited among the entities as follows: 67% spurs, 11% blobs, and 22% short tracks. In liquid water, the average spur contains about three ions and six excited molecules, and is thought to have a diameter of about 2 nm. A spacing of thousands of molecular diameters between spurs is typical for fast electrons such as those produced by ^{60}Co γ-rays. For heavy particles such as polonium α-rays, the spurs overlap to form short tracks representing roughly finite cylindrical regions of high linear energy transfer (LET) to the medium. The existence and distribution of spurs, blobs, and short tracks affect the chemistry in liquid water. The approximate value of G for products at an early stage of the chemical effects in liquid water is shown in Fig. 2. In water vapor, the primary yield of molecules decomposed to radicals is $G \sim 8$.

Liquid water and other polar liquids display particularly interesting phenomena. The existence of solvated electrons with lifetime up to 40 ms in water (depending on the concentration and nature of the solutes present) was first established in the case of aqueous systems; the chemical characteristics of such species have been measured repeatedly by different techniques. The specific rate of the presumed reaction $2e_{aq} \rightarrow 2OH^- + H_2$ in water has been shown to have the surprisingly high value of $4.5 \times 10^9\ M^{-1}\ sec^{-1}$ at pH 10.3. Studies of high-pressure effects in the radiolysis of aqueous systems, as well as theoretical calculations, indicate that the solvated electron in water (contrasted with ammonia) displaces very small volume.

Ionic processes also play a very significant role in the radiation chemistry of low dielectric constants such as cyclohexane. Electrons having a wide range of initial energies (from essentially zero to approximately 1 MeV in most cases) are generated in the liquid by high-energy radiation. Such electrons lose their energy in the medium, and are thermalized at various distances (related to their initial energies) from sibling cations. A certain small fraction corresponding to approximately 3% of the electrons can escape the coulombic field of the sibling cation. Such free ion pairs are produced with $G \sim 0.1$, as shown by electrical conductivity measurements in irradiated alkanes. The remaining 97% (corresponding to $G \sim 4$) of such ion pairs undergo recombination under the influence of their mutual coulombic fields, and are referred to as coupled or geminate ion pairs. For such coupled pairs, thermalization distances may range from about zero to approximately 30 nm and, according to theoretical estimates, have lifetimes for recombination (related to the thermalization distance) ranging from 10^{-7} s to less than 10^{-11} s. Either or both of the ions of such a coupled pair may enter into reactions with solutes present in the system, and thereby determine the nature and yields of certain chemical products.

Representative processes. Processes particularly characteristic of radiation chemistry are represented by reactions (2)–(7) in liquid water (the

$$H_2O \rightsquigarrow H_2O^+ + e^- \quad (2)$$

asterisk denotes a molecule in an excited state). The H_2O undergoes a very rapid proton transfer reaction with a neighboring water molecule [ion-molecule reaction; reaction (3)]. The H_3O^+ be-

$$H_2O^+ + H_2O \rightarrow H_3O^+ + OH \quad (3)$$

comes solvated into $H_3O^+_{aq}$, while the electron ejected in reaction (2) becomes thermalized and solvated into e^-_{aq} and then reacts through (4). The

$$H_3O^+_{aq} + e^-_{aq} \rightarrow H_2O + H \quad (4)$$

net result of reactions (2), (3), and (4) is reaction (5).

$$H_2O \rightarrow H + OH \quad (5)$$

These processes are the major ones in liquid water, but some minor ones occur, such as the dissociation of excited or superexcited molecules, reactions (6) and (7). Some of these reactive species

$$H_2O \rightsquigarrow H_2O^* \quad (6)$$

$$H_2O^* \rightarrow H + OH \quad (7)$$

react with each other before they escape the initial spurs and tracks forming the so-called molecular products (H_2 and H_2O_2), the yields of which are indicated in Fig. 2. When solutes are present, they can react with these H, OH, and e^-_{aq} species.

Other processes characteristic of radiation chemistry are reactions (8)–(12):

Ion molecule reactions:

$$CH_3^+ + CH_4 \rightarrow C_2H_5^+ + H_2 \quad (8)$$

Dissociative capture of an electron:

$$CH_3I + e^- \rightarrow CH_3 + I^- \quad (9)$$

Nondissociative capture of an electron:

$$SF_6 + e^- \rightarrow SF_6^- \quad (10)$$

Charge transfer:

$$C_6H_{12}^+ + C_6H_6 \rightarrow C_6H_{12} + C_6H_6^+ \quad (11)$$

Excitation transfer:

$$C_6H_6^* + p(C_6H_5)_2C_6H_4 \rightarrow C_6H_6 + p(C_6H_5)_2C_6H_4^* \quad (12)$$

A net effect of certain of these reactions may be protection of solvent molecules by energy (or reactivity) transfer to a chemically stable receiver.

Among other processes are some that are observable also in photochemistry, such as radiosensitization, free-radical reactions, and induced internal conversions. Diffusion-controlled reactions of free radicals differ from those of photochemistry, in which radicals are formed initially only in pairs. In radiation chemistry, the existence of spurs can result in primary production of four or more free radicals in close proximity. *See* PHOTOCHEMISTRY.

Radiation effects. Chemical effects of high-energy radiation must be guarded against in nuclear reactors (where effects include radiation corrosion, water decomposition, and the Wigner effect) and in living systems (because of mutations, cancer production, and so on). Such effects may be deliberately employed to induce polymerization of special kinds: to cross-link, to graft, and to thermally stabilize polymers; to sterilize foods, medicinals, and surgical materials; to change the properties of catalysts; and to induce reactions not possible by other means or to induce them under unusual environmental conditions, such as under extremely low temperature, in very thick layers, and in heavy-walled (pressure) vessels. Industries using radiation effects and processing are growing at a rapid rate. Applications include the curing of printing inks, coating of fabrics, wood-plastic combinations for musical instruments, plastic pipes for hot-water lines, and conversion of residual monomers in polymers. [SHEFFIELD GORDON]

Bibliography: M. Burton and J. L. Magee (eds.), *Advances in Radiation Chemistry*, vols. 1–5, 1969–1976; I. G. Draganic and Z. D. Draganic, *The Radiation Chemistry of Water*, 1971; E. J. Hart and M. A. Anbar, *The Hydrated Electron*, 1970; J. Silverman (ed.), Advances in radiation processing, *Rad. Phys. Chem.*, 14(3–6):1–961, 1979; J. W. T. Spinks and R. J. Woods, *An Introduction to Radiation Chemistry*, 1976.

Radio-frequency spectroscopy

The branch of spectroscopy concerned with the measurement of the intervals between atomic or molecular energy levels that are separated by frequencies from about 100 kHz to 1000 MHz (10^5–10^9 sec^{-1}), as compared to the frequencies that separate optical energy levels of about 6×10^{14} sec^{-1}. The importance of radio-frequency spectroscopy lies in the fact that certain specific properties of the nucleus, such as spin, magnetic dipole moment, and electric quadrupole moment, play a relatively major role in determining the intervals between closely lying energy levels; the results of this branch of spectroscopy have been of great importance in determining nuclear properties. *See* MICROWAVE SPECTROSCOPY; SPECTROSCOPY.

[POLYKARP KUSCH]

Radioactivity

A phenomenon resulting from an instability of the atomic nucleus in certain atoms whereby the nucleus experiences a spontaneous but measurably delayed nuclear transition or transformation with the resulting emission of radiation. The discovery of radioactivity by H. Becquerel in 1896 was an indirect consequence of the discovery of x-rays a few months earlier by W. Röntgen, and marked the birth of nuclear physics. Studies of the radioactive decays of new isotopes continue as one of the major frontiers in nuclear research.

In 1934 I. Curie and F. Joliot demonstrated that radioactive nuclei can be made in the laboratory. All chemical elements may be rendered radioactive by adding or by subtracting (except for hydrogen and helium) neutrons from the nucleus of the stable ones. The availability of this wide variety of radioactive isotopes has stimulated their use in science and technology in an enormous number of applications.

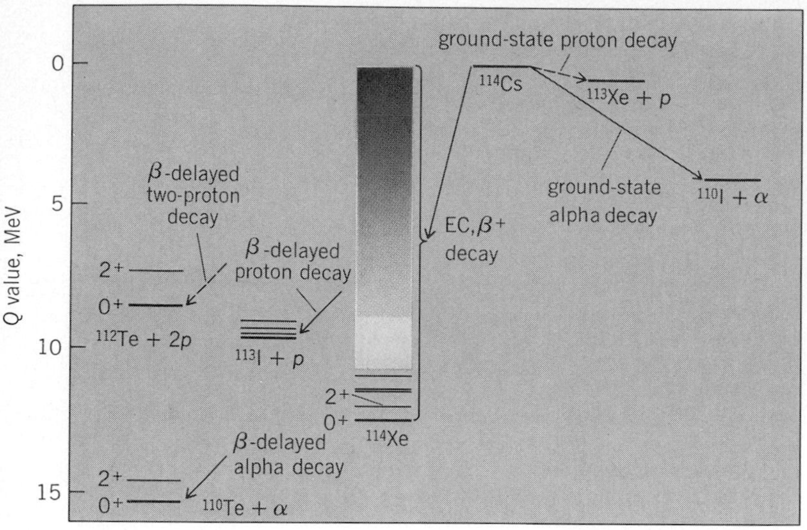

Fig. 1. Decay modes of ^{114}Cs based on Q values from the droplet-model formula for nuclear masses. (*From E. Roeckl, Recent experiments at the GSI on-line separation, in J. H. Hamilton et al., eds., Future Directions in Studies of Nuclei Far from Stability, pp. 397–404, 1980*)

A particular radioactive transition may be delayed by less than a microsecond or by more than a billion years, but the existence of a measurable delay or lifetime distinguishes a radioactive nuclear transition from a so-called prompt nuclear transition, such as is involved in the emission of most gamma rays. The delay is expressed quantitatively by the radioactive decay constant, or by the mean life, or by the half-period for each type of radioactive atom, discussed below.

There are five types of radioactivity commonly found in nearly all elements, each characterized by the particular type of nuclear radiation which is emitted by the transforming parent nucleus (the first five types in Table 1). In addition, there are several other decay modes that are observed more rarely in specific regions of the periodic table, as shown in the lower half of Table 1. Several of these rarer processes are in fact two-step processes, as shown in Figs. 1 and 2. In addition, there are several other processes predicted theoretically that remain to be verified. There is some indirect evidence for double beta decay of ^{130}Te.

In the first type in Table 1, in alpha radioactivity the parent nucleus spontaneously emits an alpha ray; then the atomic number, or nuclear charge, of the decay product is 2 units less than that of the parent, and the nuclear mass of the product is 4 atomic mass units less than that of the parent, because the emitted alpha particle carries away this amount of nuclear charge and mass. This decrease of 2 units of atomic number or nuclear charge between parent and product means that the decay product will be a different chemical element, displaced by 2 units to the left in a periodic table of the elements. For example, radium has atomic number 88 and is found in column II of the periodic table. Its decay product after the emission of an alpha ray is a different chemical element, radon, whose atomic number is 86 and whose position is in column 0 of the periodic table.

TRANSITION RATES AND DECAY LAWS

This section covers radioactive decay constant, dual decay, exponential decay law, mean life, and half-period.

Radioactive decay constant. The rate of radioactive transformation, or the activity, of a source equals the number A of identical radioactive atoms present in the source, multiplied by their characteristic radioactive decay constant λ. Thus Eq. (1) holds, where the decay constant λ has dimensions of second^{-1}. The numerical value of λ

$$\text{Activity} = A\lambda \text{ disintegrations per second} \quad (1)$$

expresses the statistical probability of decay of each radioactive atom in a group of identical atoms, per unit time. For example, if $\lambda = 0.01\ s^{-1}$ for a particular radioactive species, then each atom has a chance of 0.01 (1%) of decaying in 1 s, and a chance of 0.99 (99%) of not decaying in any given 1-s interval. The constant λ is one of the most important characteristics of each radioactive nuclide: λ is essentially independent of all physical and chemical conditions such as temperature, pressure, concentration, chemical combination, or age of the radioactive atoms. There are a few cases where measureable effects are observed for different chemical combinations. One of the largest observed is a 3.2% change in λ for the 24-s isomer in ^{90}Nb. The half-period is inversely proportional to λ.

The identification of some radioactive samples can be made simply by measuring λ, which then serves as an equivalent of qualitative chemical analysis. For the most common radioactive nuclides, the range of λ extends from $3 \times 10^6\ s^{-1}$ (for thorium C′) to $1.6 \times 10^{-18}\ s^{-1}$ (for thorium).

Dual decay. Many radioactive nuclides have two or more independent and alternative modes of

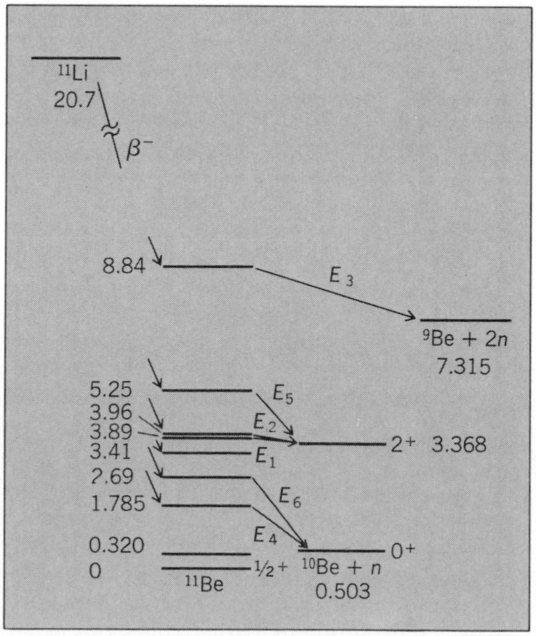

Fig. 2. Observed decay modes of ^{11}Li. Energies in MeV. (*From P. G. Hansen, On-line experiments with high-energy protons: Recent results and possible future directions, in J. H. Hamilton et al., eds., Future Directions in Studies of Nuclei Far from Stability, pp. 405–417, 1980*)

Table 1. Types of radioactivity

Type	Symbol	Particles emitted	Change in atomic number, ΔZ	Change in atomic mass number, ΔZ	Example†
Alpha	α	Helium nucleus	−2	−4	${}^{226}_{86}\mathrm{Ra} \rightarrow {}^{222}_{84}\mathrm{Rn} + \alpha$
Beta negatron	β^-	Negative electron and antineutrino	+1	0	${}^{24}_{11}\mathrm{Na} \rightarrow {}^{24}_{12}\mathrm{Mg} + e^- + \bar{\nu}$
Beta positron	β^+	Positive electron and neutrino	−1	0	${}^{22}_{11}\mathrm{Na} \rightarrow {}^{22}_{10}\mathrm{Ne} + e^+ + \nu$
Electron capture	EC	Neutrino	−1	0	${}^{7}_{4}\mathrm{Be} + e^- \rightarrow {}^{7}_{3}\mathrm{Li} + \nu$
Isomeric transition	IT	Gamma rays or conversion electrons or both (and positive-negative electron pair)‡	0	0	${}^{137m}_{56}\mathrm{Ba} \rightarrow {}^{137}_{56}\mathrm{Ba} + \gamma$ or c.e.
Proton	p	Proton	−1	−1	${}^{53m}_{27}\mathrm{Co} \rightarrow {}^{52}_{26}\mathrm{Fe} + p$ §${}^{114}_{55}\mathrm{Cs} \rightarrow {}^{113}_{54}\mathrm{Xe} + p$
Spontaneous fission	SF	Heavy fragments and neutrons	Various	Various	${}^{238}_{92}\mathrm{U} \rightarrow {}^{133}_{50}\mathrm{Sn} + {}^{103}_{42}\mathrm{Mo} + 2n$
Isomeric spontaneous fission	ISF	Heavy fragments and neutrons	Various	Various	${}^{244f}_{95}\mathrm{Am} \rightarrow {}^{134}_{53}\mathrm{I} + {}^{107}_{42}\mathrm{Mo} + 3n$
Beta-delayed spontaneous fission	(EC + β^+)SF	Positive electron, neutrino, heavy fragments, and neutrons	Various	Various	${}^{246}_{99}\mathrm{Es} \rightarrow \beta^+ + \nu + {}^{246f}_{98}\mathrm{Cf} \rightarrow {}^{138}_{54}\mathrm{Xe} + {}^{107}_{44}\mathrm{Ru} + n$
	β^-SF	Negative electron, antineutrino, heavy fragments, and neutrons	Various	Various	${}^{236}_{91}\mathrm{Pa} \rightarrow \beta^- + \bar{\nu} + {}^{236f}_{92}\mathrm{U} \rightarrow {}^{139}_{53}\mathrm{I} + {}^{94}_{39}\mathrm{Y} + 3n$
Beta-delayed neutron	$\beta^- n$	Negative electron, neutrino, and neutron	+1	−1	${}^{11}_{3}\mathrm{Li} \rightarrow \beta^- + \bar{\nu} + {}^{11}_{4}\mathrm{Be}^* \rightarrow {}^{10}_{4}\mathrm{Be} + n$
Beta-delayed two-neutron (three-neutron)	$\beta^- 2n$ ($3n$)	Negative electron, antineutrino, and two (three) neutrons	+1	−3	${}^{11}_{3}\mathrm{Li} \rightarrow \beta^- + \bar{\nu} + {}^{11}_{4}\mathrm{Be}^* \rightarrow {}^{9(8)}_{4}\mathrm{Be} + 2n$ ($3n$)
Beta-delayed proton	$\beta^+ p$ or (β^+ + EC)p	Positive electron, neutrino, and proton	−2	−1	${}^{114}_{55}\mathrm{Cs} \rightarrow \beta^+ + \nu + {}^{114}_{54}\mathrm{Xe}^* \rightarrow {}^{113}_{53}\mathrm{I} + p$
Beta-delayed alpha	$\beta^+ \alpha$	Positive electron, neutrino, and alpha	−3	−4	${}^{114}_{55}\mathrm{Cs} \rightarrow \beta^+ + \nu + {}^{114}_{54}\mathrm{Xe}^* \rightarrow {}^{110}_{52}\mathrm{Te} + \alpha$
	$\beta^- \alpha$	Negative electron, antineutrino, and alpha	−1	−4	${}^{214}_{83}\mathrm{Bi} \rightarrow \beta^- + \bar{\nu} + {}^{114}_{84}\mathrm{Po}^* \rightarrow {}^{210}_{82}\mathrm{Pb} + \alpha$
Double beta decay§	$\beta^- \beta^-$	Two negative electrons and two antineutrinos	+2	0	¶${}^{130}_{52}\mathrm{Te} \rightarrow {}^{130}_{54}\mathrm{Xe} + 2\beta^- + 2\bar{\nu}$
	$\beta^+ \beta^+$	Two positive electrons and two neutrinos	−2	0	§${}^{130}_{56}\mathrm{Ba} \rightarrow {}^{130}_{54}\mathrm{Xe} + 2\beta^+ + 2\nu$
Double electron capture§	EC EC	Two neutrinos	−2	0	§${}^{130}_{56}\mathrm{Ba} + 2e^- \rightarrow {}^{130}_{54}\mathrm{Xe} + 2\nu$
Two-proton§	$2p$	Two protons	−2	−2	§${}^{114}_{55}\mathrm{Cs} \rightarrow {}^{112}_{53}\mathrm{I} + 2p$
Beta-delayed two-proton§	$\beta^+ 2p$	Positive electron, neutrino, and two protons	−3	−2	§${}^{114}_{55}\mathrm{Cs} \rightarrow \beta^+ + \nu + {}^{114}_{54}\mathrm{Xe}^* \rightarrow {}^{112}_{52}\mathrm{Te} + 2p$
Neutron§	n	Neutron	0	−1	
Two-neutron§	$2n$	Two neutrons	0	−2	

†Excited states with relatively long measured half-lives are called isomeric and are identified by placing the symbol m for metastable after the mass number, as in ${}^{137m}\mathrm{Ba}$. Excited states with essentially prompt decay are identified by asterisks, as in ${}^{11}\mathrm{Be}^*$.

‡Occurs as an additional decay mode when the decay energy exceeds 1.022 MeV.

§Theoretically predicted but not established experimentally.

¶Some indirect evidence for this particular decay has been reported, but one cannot say double beta decay is established.

decay. For example, ${}^{238}\mathrm{U}$ can decay either by alpha-ray emission or by spontaneous fission. A single atom of ${}^{64}\mathrm{Cu}$ can decay in any of three competing independent ways: negatron beta-ray emission, positron beta-ray emission, or electron capture. When two or more independent modes of decay are possible, the nuclide is said to exhibit dual decay.

The competing modes of decay of any nuclide have independent partial decay constants given by the probabilities $\lambda_1, \lambda_2, \lambda_3, \ldots$, per second, and the total probability of decay is represented by the total decay constant λ, defined by Eq. (2). If there

$$\lambda = \lambda_1 + \lambda_2 + \lambda_3 + \cdots \tag{2}$$

are A identical atoms present, the partial activities, as measured by the different modes of decay, are $A\lambda_1, A\lambda_2, A\lambda_3, \ldots$, and the total activity $A\lambda$ is given by Eq. (3). The partial activities, $A\lambda_1, \ldots$,

$$A\lambda = A\lambda_1 + A\lambda_2 + A\lambda_3 + \cdots \tag{3}$$

such as positron beta-rays from ^{64}Cu, are proportional to the total activity, $A\lambda$, at all times.

The branching ratio is the fraction of the decaying atoms which follow a particular mode of decay, and equals $A\lambda_1/A\lambda$ or λ_1/λ. For example, in the case of ^{64}Cu the measured branching ratios are $\lambda_1/\lambda = 0.40$ for negatron beta-decay, $\lambda_2/\lambda = 0.20$ for positron beta-decay, and $\lambda_3/\lambda = 0.40$ for electron capture. The sum of all the branching ratios for a particular nuclide is unity.

Exponential decay law. The total activity, $A\lambda$, equals the rate of decrease $-dA/dt$ in the number of radioactive atoms A present. Because λ is independent of the age t of an atom, integration of the differential equation of radioactive decay, $-dA/dt = A\lambda$, gives Eq. (4), where ln represents the natu-

$$\ln \frac{A}{A_0} = -\lambda (t - t_0) \tag{4}$$

ral logarithm to the base e, and A atoms remain at time t if there were A_0 atoms initially present at time t_0. If $t_0 = 0$, then Eq. (4) can be rewritten as the exponential law of radioactive decay in its most common form, Eq. (5). The initial activity at $t = 0$

$$A = A_0 e^{-\lambda t} \tag{5}$$

was $A_0\lambda$, and the activity at t, when only A atoms remain untransformed, is $A\lambda$. Because λ is a constant, the fractional activity $A\lambda/A_0\lambda$ at time t and the fractional amount of radioactive atoms A/A_0 are given by Eq. (6). In cases of dual decay, the

$$\frac{A\lambda}{A_0\lambda} = \frac{A}{A_0} = e^{-\lambda t} \tag{6}$$

partial activities $A\lambda_1, A\lambda_2, \ldots$, also decrease with time as $e^{-\lambda t}$, not as $e^{-\lambda_1 t}, \ldots$, because $A\lambda_1/A_0\lambda_1 = A/A_0 = e^{-\lambda t}$ where λ is the total decay constant. This is because the decrease of each partial activity with time is due to the depletion of the total stock of atoms A, and this depletion is accomplished by the combined action of all the competing modes of decay.

Mean life. The actual life of any particular atom can have any value between zero and infinity. The average or mean life of a large number of identical radioactive atoms is, however, a definite and important quantity.

If there are A_0 atoms present initially at $t = 0$, then the number remaining undecayed at a later time t is $A = A_0 e^{-\lambda t}$, by Eq. (5). Each of these A atoms has a life longer than t. In an additional infinitesimally short time interval dt, between time t and $t + dt$, the absolute number of atoms which will decay on the average is $A\lambda dt$, and these atoms had a life-span t. The total L of the life-spans of all the A_0 atoms is the sum or integral of $tA\lambda\, dt$ from $t = 0$ to $t = \infty$, which is given by Eq. (7). Then the

$$L = \int_0^\infty tA\lambda\, dt = \int_0^\infty tA_0\lambda e^{-\lambda t}\, dt = \frac{A_0}{\lambda} \tag{7}$$

average lifetime L/A_0, which is called the mean life τ, is given by Eq. (8), where λ is the total radioac-

$$\tau = 1/\lambda \tag{8}$$

tive decay constant of Eq. (2). Substitution of $t = \tau = 1/\lambda$ into Eq. (6) shows that the mean life is the time required for the number of atoms, or their activity, to fall to $e^{-1} = 0.368$ of any initial value.

Half-period. The time interval over which the chance of survival of a particular radioactive atom is exactly one-half is called the half-period T. From Eq. (4), Eq. (9) is obtained. Then the half-period T

$$-\ln (A/A_0) = \ln (A_0/A) = \ln 2 = 0.693 = \lambda T \tag{9}$$

is related to the total radioactive decay constant λ, and to the mean life τ, by Eq. (10). For mnemonic

$$T = 0.693/\lambda = 0.693\tau \tag{10}$$

reasons, the half-period T is much more frequently employed than the total decay constant λ or the mean life τ. For example, it is more common to speak of ^{232}Th as having a half-period of 1.4×10^{10} years than to speak of its mean life of 2.0×10^{10} years or its total decay constant of $1.6 \times 10^{-18}\ s^{-1}$, although all three are equivalent statements of the average longevity of ^{232}Th atoms.

The relationships between T, τ, and λ are summarized graphically in Fig. 3. Any initial activity $A_0\lambda$ is reduced to ½ in 1 half-period T, to $1/e$ in 1 mean life τ, to ¼ in 2 half-periods $2T$, and so on. The slope of the activity curve, or rate of decrease of activity, is $d(A\lambda)/dt = -\lambda dA/dt = -\lambda(A\lambda)$. Thus the initial slope is $-\lambda(A_0\lambda) = -(A_0\lambda)\tau$. The area under the activity curve, if integrated to $t = \infty$, is simply A_0, the total initial number of radioactive atoms. Also, the initial activity $A_0\lambda$, if it could continue at a constant value for one mean life τ, would exactly destroy all the atoms because $(A_0\lambda)\tau = A_0$.

RADIOACTIVE SERIES DECAY

In a number of cases a radioactive nuclide A decays into a nuclide B which is also radioactive; the nuclide B decays into C which is also radioac-

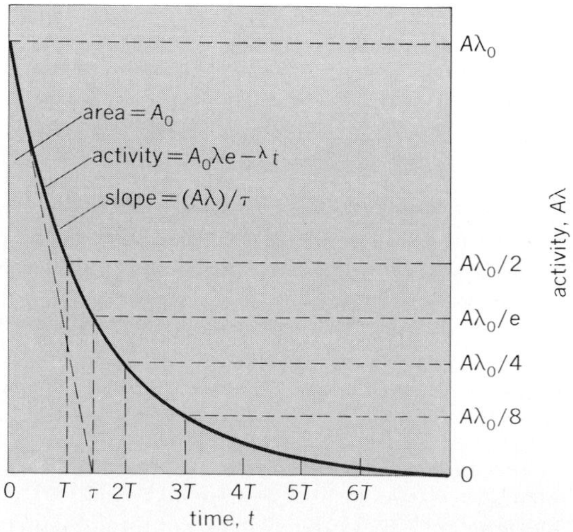

Fig. 3. Graphical representation of relationships in decay of a single radioactive nuclide.

tive, and so on. For example, $^{232}_{90}$Th decays into a series of 10 successive radioactive nuclides. Substantially all the primary products of nuclear fission are negatron beta-ray emitters which decay through a chain or series of two to six successive beta-ray emitters before a stable nuclide is reached as an end product. *See* NUCLEAR FISSION.

Let the initial part of such a series be represented by reaction (11), where radioactive atoms of

$$A \xrightarrow{\lambda_A} B \xrightarrow{\lambda_B} C \xrightarrow{\lambda_C} D \xrightarrow{\lambda_D} \cdots \quad (11)$$

types $A, B, C, D, \ldots$, have radioactive decay constants given by $\lambda_A, \lambda_B, \lambda_C, \lambda_D, \ldots$. Then if there are initially present, at time $t = 0$, A_0 atoms of type A, the numbers $A, B, C, \ldots$, of atoms of types $A, B, C, \ldots$, which will be present at a later time t are given by Eqs. (12)–(14), and the

$$A = A_0 e^{-\lambda_A t} \quad (12)$$

$$B = A_0 \frac{\lambda_A}{\lambda_B - \lambda_A}(e^{-\lambda_A t} - e^{-\lambda_B t}) \quad (13)$$

$$C = A_0 \left(\frac{\lambda_A}{\lambda_C - \lambda_A} \frac{\lambda_B}{\lambda_B - \lambda_A} e^{-\lambda_A t} + \frac{\lambda_A}{\lambda_A - \lambda_B} \frac{\lambda_B}{\lambda_C - \lambda_B} e^{-\lambda_B t} + \frac{\lambda_A}{\lambda_A - \lambda_C} \frac{\lambda_B}{\lambda_B - \lambda_C} e^{-\lambda_C t} \right) \quad (14)$$

activities of $A, B, C, \ldots$, are $A\lambda_A, B\lambda_B, C\lambda_C, \ldots$. General equations describing the amounts and activities of any number of radioactive decay products are more complicated and are given in standard texts.

Figure 4 illustrates the growth and decay of the activity of a short series of radioactive decay products in accord with Eqs. (12)–(14).

Radioactive equilibrium. In Fig. 4 the ratio $B\lambda_B/A\lambda_A$ of the activities of the parent A and the daughter product B change with time. The activity $B\lambda_B$ is zero initially and also after a very long time, when all the atoms have decayed. Thus $B\lambda_B$ passes through a maximum value, and it can be shown that this occurs at a time t_m given by Eq. (15). The

$$t_m = \frac{\ln(\lambda_B/\lambda_A)}{(\lambda_B - \lambda_A)} \quad (15)$$

situation in which the activities $A\lambda_A$ and $B\lambda_B$ are exactly equal to each other is called ideal equilibrium, and exists only at the moment t_m.

If the parent A is longer-lived than the daughter B, as occurs in many cases, then at a time which is long compared with the mean life τ_B of B, the activity ratio approaches a constant value given by Eq. (16), where T_A and T_B are the half-periods of A and

$$\frac{B\lambda_B}{A\lambda_A} = \frac{\lambda_B}{\lambda_B - \lambda_A} = \frac{T_A}{T_A - T_B} \quad (16)$$

of B. When the activity ratio $B\lambda_B/A\lambda_A$ is constant, a particular type of radioactive equilibrium exists. This is spoken of as secular equilibrium if the activity ratio is experimentally indistinguishable from unity, as occurs when T_A is very much greater than T_B.

Equilibrium concepts are applied also between a long-lived parent and any of its decay products in a long series. For example, in a sufficiently old uranium ore, radium ($T = 1620$ years) is in secular equilibrium with its ultimate parent uranium ($T = 4.5 \times$

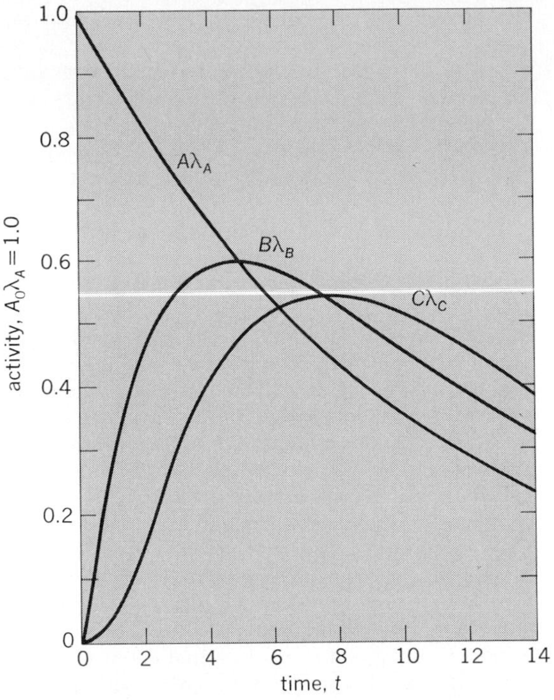

Fig. 4. Growth and decay of the activity $B\lambda_B$ of the daughter product, and $C\lambda_C$ of the granddaughter product, in an initially pure source of a radioactive parent whose activity at $t = 0$ is $A_0\lambda_A$.

10^9 years) although there are four intermediate radioactive substances intervening in the series between uranium and radium. Here, secular equilibrium expresses the fact that the activities of radium and uranium continue to be equal to each other even though the activity of the parent uranium is decreasing with time.

When T_B is comparable with T_A, Eq. (16) shows that the equilibrium ratio will clearly exceed unity; this situation is spoken of as transient equilibrium. For example, in fission-product decay series (17)

$$^{140}\text{Ba} \rightarrow {}^{140}\text{La} \rightarrow {}^{140}\text{Ce} \quad (17)$$

the half-period of ^{140}Ba is 307 hours and that of ^{140}La is 40 h. In an initially pure source of ^{140}Ba the activity of ^{140}La starts at zero, rises to a maximum at $t_m = 135$ h [Eq. (15)], then decreases, and after a few hundred hours is in transient equilibrium with its parent, when the ^{140}La activity [by Eq. (16)] is $307/(307 - 40) = 1.15$ times the activity of its parent ^{140}Ba.

Radioactivity in the Earth. A number of isotopes of elements found in the Earth are radioactive. All known or theoretically predicted isotopes of elements above bismuth are radioactive. Because the Earth is composed of atoms which were believed to have been created more than 3×10^9 years ago, the naturally occurring parent radioactive isotopes are those which have such long half-periods that detectable residual activity is still observable today. As a general rule, one can detect the presence of a radioactive substance for about 10 half-lives. Therefore activities with $T \lesssim 0.3 \times 10^9$ years should not be found in the Earth. For example, present-day uranium is an isotopic mixture containing 99.3% ^{238}U, whose half-period is 4.5×10^9

years, and only 0.7% of the shorter-lived uranium isotope ^{235}U, whose half-period is 0.7×10^9 years, whereas these isotopes were produced in roughly equal amounts in the Earth a few billion years ago. Geophysical evidence indicates that originally some ^{236}U was present also, but none is found in nature now as expected with its half-period of 0.02×10^9 years.

^{238}U decays through a long series of 14 radioactive decay products before ending as a stable isotope of lead, ^{206}Pb. Some of these members of the ^{238}U decay chain have very short half-periods, so their existence in nature is entirely dependent on the presence of their long-lived parent, and thus is a genealogical accident. For example, radium occurs in nature only in the minerals of its parent, uranium. The decay series of ^{235}U supports 14, and the decay series of ^{232}Th supports 10, short-lived radioactive substances found in nature.

A few of the common elements contain long-lived, naturally radioactive isotopes. For example, all terrestrial potassium contains 0.012% of the radioactive isotope ^{40}K, which has a half-period of 1.3×10^9 years, and emits negatron or positron beta-rays and gamma rays in a dual decay to stable ^{40}Ca and ^{40}A. This isotope is the principal source of radioactivity in the normal human being; each human contains about 0.1 microcurie (3.7×10^3 becquerel) of the radioactive potassium isotope ^{40}K.

Table 2 summarizes the radioactive properties of all the well-established cases of radioactivities found in the Earth's surface. Geological age measurements are based on the accumulation of decay products of these long-lived isotopes, especially in the cases of ^{40}K, ^{87}Rb, ^{232}Th, ^{235}U, and ^{238}U.

Laboratory produced radioactive nuclei. With particle accelerators and nuclear reactors, over 1650 radioactive isotopes not found in detectable quantities in the Earth's crust have been produced in the laboratory since 1935, including those of at least 14 new chemical elements up to element 106. Earlier titles of induced or artificial radioactivities for these isotopes are misnomers. Many of these now have been identified in meteorites and in stars, and others are produced in the atmosphere by cosmic rays.

For example, carbon-14 is a negatron beta-ray emitter, with a half-period of about 5600 years, which can be produced in the laboratory as the product of a variety of different nuclear transmutation experiments. Nuclear bombardment of ^{11}B nuclei by alpha rays (helium nuclei) can produce excited compound nuclei of ^{15}N which promptly emit a proton (hydrogen nucleus), leaving ^{14}C as the end product of the transmutation. The same end-product ^{14}C can be produced by bombarding ^{14}N with neutrons, resulting in nuclear reaction (18). This reaction is easily carried out by using

$$^{14}N + \text{neutron} \rightarrow {}^{15}N^* \rightarrow {}^{14}C + \text{proton} \qquad (18)$$

neutrons from nuclear accelerators or a nuclear reactor. This particular transmutation reaction is one which occurs in nature also, because the nitrogen in the Earth's atmosphere is continually bombarded by neutrons which are produced by cosmic rays, thus producing radioactive ^{14}C. Mixing of ^{14}C with stable carbon provides the basis for radiocarbon dating of systems that absorb carbon for times up to about 50,000 years ago (10 half-lives).

Radioactive hydrogen, ^3H, is also formed in the atmosphere from the ^{14}N + neutron $\rightarrow$ ^{12}C + ^3H reaction. Also, ^3H is produced in the Sun, and the Earth's water as well as satellites shows an additional concentration of ^3H from the Sun. Over two dozen radioactive products, ranging in half-life from a few days to millions of years, have been identified in meteorites that have fallen to Earth. The carbon and hydrogen burning cycles that produce energy for stars produce radioactive ^{13}N, ^{15}O, ^3H. At higher temperatures the radioactivities ^7Be and even ^8Be ($T \approx 10^{-16}$ s) help burn hydrogen and helium. In addition to the production of radioactive as well as stable isotopes prior to the formation of the solar system, nucleosynthesis continues to go on in stars with the production of many short-lived radioactive atoms by different processes.

The yield of any radioactivity produced in the laboratory is the initial rate of the activity under the particular conditions of nuclear bombardment. When a target material A is bombarded to produce a radioactive product B whose radioactive decay constant is λ_B, the number of atoms B which are

Table 2. Parent radioactive nuclides found in nature

Nuclide					
Atomic number, Z	Mass number, Z	Percent abundance in nature	Half-period, years	Radioactive transitions observed	Disintegration energy, MeV
19 K	40	0.0117	1.3×10^9	β^-, EC	β^- 1.3–EC 1.5
37 Rb	87	27.83	4.8×10^{10}	β^-	0.3
48 Cd	113	12.2	9×10^{15}	β^-	0.3
49 In	115	95.77	5.1×10^{14}	β^-	0.5
52 Te	130	34.49	2×10^{21}	Growth of $^{130}_{54}Xe$†	1.6
57 La	138	0.089	1.1×10^{11}	β^-, EC	β^- 1.0–EC 1.75
60 Nd	144	23.8	2.1×10^{15}	α	1.9
62 Sm	147	15.07	1.1×10^{11}	α	2.3
62 Sm	148	11.3	8×10^{15}	α	1.99
64 Gd	152	0.20	1.1×10^{14}	α	2.2
71 Lu	176	2.6	3.6×10^{10}	β^-, γ	0.6
72 Hf	174	0.16	2×10^{15}	α	2.5
75 Re	187	62.6	4×10^{10}	β^-	0.003
78 Pt	190	0.013	6×10^{11}	α	3.24
90 Th	232	100.	1.4×10^{10}	α	4.08
92 U	235	0.715	7.0×10^8	α	4.68
92 U	238	99.28	4.5×10^9	α	4.27

†Indirect evidence for $\beta^-\beta^-$ decay.

present after a bombardment of duration t, and their activity $B\lambda_B$, are given by Eq. (19), where the

$$B\lambda_B = \frac{Y}{\lambda_B}(1 - e^{-\lambda_B t}) \qquad (19)$$

yield Y has dimensions equivalent to curies of activity produced per second of bombardment. The yield Y depends on the number of atoms A present in the target, the intensity of beam of bombarding particles, and the cross section, or probability of the reaction per bombarding particle under the conditions of bombardment.

Radioactive transformation series. As noted in Eqs. (12)–(14), many radioactive substances have decay products which are also radioactive. Thus many long chains or series of radioactive transformations are known.

The three naturally occurring transformation series are headed by ^{232}Th, ^{235}U, and ^{238}U. Their genealogical relationships are summarized in Fig. 5 and Table 3.

Each of the naturally occurring radioactive isotopes in these transformation series has two synonymous names. For example, the commercially important radioisotope whose classical name is mesothorium-1 is now known to be an isotope of radium with mass number of 228 and is designated as radium-228 (^{228}Ra). Table 3 summarizes the names, symbols, and some radioactive properties of these three transformation series. However, these chains are not complete, and their uniqueness or importance as chains is an accident of the very long half-lives of ^{232}Th, ^{235}U, and ^{238}U. For example, element 105 of mass 260 has a succession of seven alpha decays and one electron capture and positron decay to ^{232}Th. The special importance of the chains in Table 3 is related to the fact that they were essentially the only early sources of radioactive materials, and more recently to their role in nuclear power.

Transformation series are now known for every element in the periodic table except hydrogen. Chains of neutron-rich isotopes have been produced and studied among the products of nuclear fission. First, heavy-ion-induced reactions and subsequently high-flux reactors have been used to extend knowledge of the elements beyond uranium. Both proton- and heavy-ion-induced reactions have extended knowledge of chains of neutron-deficient isotopes of the stable elements.

ALPHA-RAY DECAY

Alpha-ray decay is that type of radioactivity in which the parent nucleus expels an alpha ray (a helium nucleus). The alpha ray is emitted with a speed of the order of 1 to 2 × 10^7 m/s, that is, about 1/20 of the velocity of light.

In the simplest case of alpha decay, every alpha ray would be emitted with exactly the same velocity and hence the same kinetic energy. However, in most cases there are two or more discrete energy groups called lines, as shown in Fig. 6 in the spectrum of alpha rays from ^{184}Tl and ^{184}Hg decays. For example, in the alpha decay of a large group of ^{238}U atoms, 77% of the alpha decays will be by emission of alpha rays whose kinetic energy is 4.20 MeV, while 23% will be by emission of 4.15-MeV alpha rays. When the 4.20-MeV alpha ray is emitted, the decay product nucleus is formed in its ground (lowest energy) level. When a 4.15-MeV alpha ray is emitted, the decay product is pro-

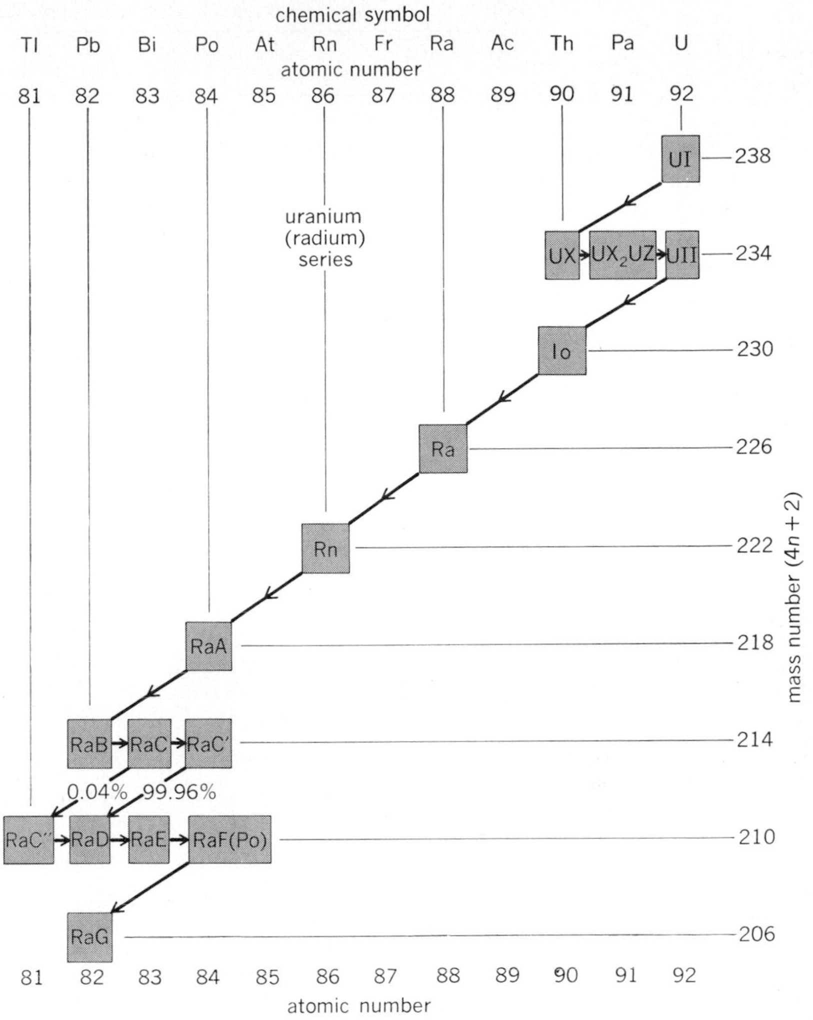

Fig. 5. Main line of decay of uranium series, or $4n+2$ series, of heavy radioactive nuclides, headed in nature by uranium-238. Each member has a mass number given by $4n+2$, where n is an integer.

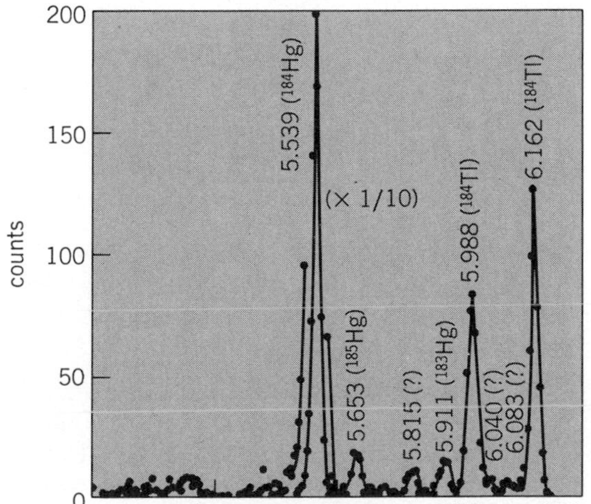

Fig. 6. Alpha groups in the decay of ^{184}Tl (T = 11s) and ^{184}Hg and weak groups from ^{183}Hg and ^{185}Hg, very far off stability (17 neutrons less than the lightest stable thallium isotope). Energies in MeV. (*From K. S. Toth et al., Observation of α-decay in thallium nuclei, including the new isotopes ^{184}Tl and ^{185}Tl, Phys. Lett., 63B:150–153, 1976*)

Table 3. Names, symbols, and radioactive properties of members of the three naturally occurring radioactive transformation series

Conventional name	Conventional symbol	Atomic number	Mass number	Isotopic symbol	Half-period	Type of decay
Uranium (4n + 2) series						
Uranium I	UI	92	238	^{238}U	4.5×10^9 y	α
Uranium X_1	UX_1	90	234	^{234}Th	24 d	β^-
Uranium X_2	UX_2	91	234	^{234m}Pa	1.2 m	It,β^-
Uranium Z	UZ	91	234	^{234}Pa	6.7 h	β^-
Uranium II	UII	92	234	^{234}U	2.5×10^5 y	α
Ionium	Io	90	230	^{230}Th	8×10^4 y	α
Radium	Ra	88	226	^{226}Ra	1600 y	α
Radon	Rn	86	222	^{222}Rn	3.8 d	α
Radium A	RaA	84	218	^{218}Po	3.0 m	α
Radium B	RaB	82	214	^{214}Pb	27 m	β^-
Radium C	RaC	83	214	^{214}Bi	20 m	β^-,α
Radium C′	RaC′	84	214	^{214}Po	1.6×10^{-4} s	α
Radium C″	RaC″	81	210	^{210}Tl	1.3 m	β^-
Radium D	RaD	82	210	^{210}Pb	22 y	β^-
Radium E	RaE	83	210	^{210}Bi	5.0 d	β^-
Radium F	RaF	84	210	^{210}Po	138 d	α
Polonium	Po	84	210	^{210}Po	138 d	α
Radium G	RaG	82	206	^{206}Pb	Stable	Stable
Thorium (4n) series						
Thorium	Th	90	232	^{232}Th	1.4×10^{10} y	α
Mesothorium$_1$	$MsTh_1$	88	228	^{228}Ra	5.8 y	β^-
Mesothorium$_2$	$MsTh_2$	89	228	^{228}Ac	6.1 h	β^-
Radiothorium	RdTh	90	228	^{228}Th	1.9 y	α
Thorium X	ThX	88	224	^{224}Ra	3.7 d	α
Thoron	Tn	86	220	^{220}Rn	56 s	α
Thorium A	ThA	84	216	^{216}Po	0.15 s	α
Thorium B	ThB	82	212	^{212}Pb	10.6 h	β^-
Thorium C	ThC	83	212	^{212}Bi	1.0 h	β^-,α
Thorium C′	ThC′	84	212	^{212}Po	3×10^{-7} s	α
Thorium C″	ThC″	81	208	^{208}Tl	3.1 m	β^-
Thorium D	ThD	82	208	^{208}Pb	Stable	Stable
Actinium (4n + 3) series						
Actinouranium	AcU	92	235	^{235}U	7.0×10^8 y	α
Uranium Y	UY	90	231	^{231}Th	26 h	β^-
Protactinium	Pa	91	231	^{231}Pa	3.3×10^4 y	α
Actinium	Ac	89	227	^{227}Ac	22 y	β^-,α
Radioactinium	RdAc	90	227	^{227}Th	19 d	α
Actinium K	AcK	87	223	^{223}Fr	22 m	β^-,α
Actinium X	AcX	88	223	^{223}Ra	11 d	α
Astatine	At	85	219	^{219}At	0.9 m	α,β^-
Actinon	An	86	219	^{219}Rn	4.0 s	α
Actinium A	AcA	84	215	^{215}Po	1.8×10^{-3} s	α
Actinium B	AcB	82	211	^{211}Pb	36 m	β^-
Actinium C	AcC	83	211	^{211}Bi	2.2 m	α,β^-
Actinium C′	AcC′	84	211	^{211}Po	0.5 s	α
Actinium C″	AcC″	81	207	^{207}Tl	4.8 m	β
Actinium D	AcD	82	207	^{207}Pb	Stable	Stable

duced in an excited level, 0.05 MeV above the ground level. This nucleus promptly transforms to its ground level by the emission of a 0.05-MeV gamma ray or alternatively by the emission of the same amount of energy in the form of a conversion electron and the associated spectrum of characteristic x-rays. Thus in all alpha-ray spectra, the alpha rays are emitted in one or more discrete and homogeneous energy groups, and alpha-ray spectra are accompanied by gamma-ray and conversion electron spectra whenever there are two or more alpha-ray groups in the spectrum.

Geiger-Nuttall rule. Among all the known alpha-ray emitters, most alpha-ray energy spectra lie in the domain of 4–6 MeV, although a few extend as low as 2 MeV ($^{147}_{62}$Sm) and as high as 10 MeV (ThC′). There is a systematic relationship between the kinetic energy of the emitted alpha rays and the half-period of the alpha emitter. The highest-energy alpha rays are emitted by short-lived nuclides, and the lowest-energy alpha rays are emitted by the very-long-lived alpha-ray emitters. H. Geiger and J. M. Nuttall showed that there is a linear relationship between log λ and the energy of the alpha ray.

The Geiger-Nuttall rule is inexplicable by classical physics, but emerges clearly from quantum, or wave, mechanics. In 1928 the hypothesis of transmission through nuclear potential barriers, as introduced by G. Gamow and independently by R. W. Gurney and E. U. Condon, was shown to give a satisfactory account of the alpha-decay data, and it has been altered subsequently only in details. The form of the barrier-penetration equations is such that correlation plots of log λ against $1/\sqrt{E}$ give nearly straight lines.

Nuclear potential barrier. At distances r which are large compared with the nuclear radius, the

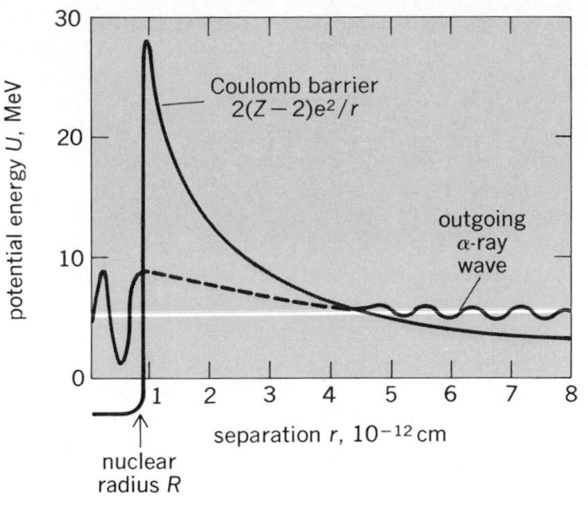

Fig. 7. Schematic of nuclear potential barrier, illustrating emission of an α-ray as a wave which can be transmitted through the barrier.

potential energy of an alpha ray, whose charge is $2e$, in the field of a residual nucleus, whose charge is $(Z-2)e$, is $2(Z-2)e^2/r$. At very close distances this electrostatic repulsion is opposed and overcome by short-range, specifically nuclear, attractive forces. The net potential energy U as a function of the separation r between the alpha ray and its residual nucleus is called the nuclear potential barrier.

One of several operating definitions of the nuclear radius R is the distance $r=R$ at which the attractive nuclear forces just balance the repulsive electrostatic forces. At this distance, called the top of the nuclear barrier, the potential energy is about 25–30 MeV for typical cases of heavy, alpha-emitting nuclei, as indicated in Fig. 7.

Inside the nucleus the alpha particle is represented as a de Broglie matter wave. According to wave mechanics, this wave has a very small but finite probability of being transmitted through the nuclear potential energy barrier and thus of emerging as an alpha ray emitted from the nucleus. The transmission of a particle through such an energy barrier is completely forbidden in classical electrodynamics but is possible according to wave mechanics. This transmission of a matter wave through an energy barrier is analogous to the familiar case of the transmission of ordinary visible light through an opaque metal such as gold: if the gold is thin enough, some light does get through, as in the case of the thin gold leaf which is sometimes used for lettering signs on store windows.

The wave-mechanical probability of the transmission of an alpha particle through the nuclear potential barrier is very strongly dependent upon the energy of the emitted alpha ray. Analytically the probability of transmission T depends exponentially upon a barrier transmission exponent γ according to Eq. (20). To a good approximation, Eq. (21) holds, where $h=6.626\times10^{-34}$ joule-second

$$T=e^{-\gamma} \tag{20}$$

$$\gamma=\left(\frac{4\pi^2}{h}\right)\frac{(Z-2)2e^2}{V}-\left(\frac{8\pi}{h}\right)[2(Z-2)2e^2MR]^{1/2} \tag{21}$$

is Planck's constant, and M is the so-called reduced mass of the alpha particle. For the alpha decay of ^{226}Ra, the numerical value of γ is about 71: hence $T=e^{-71}=10^{-31}$. The first term on the right side of Eq. (21) is about 154 and is therefore the dominant term. When this term is taken alone, $e^{-(4\pi^2/h)(Z-2)2e^2/V}$ is called the Gamow factor for barrier penetration.

Inspection of Eq. (21) shows that the barrier transmission decreases with increasing nuclear charge $(Z-2)e$, increases with increasing velocity V of emission of the alpha ray, and increases with increasing radius R of the nucleus. When the experimentally known values of alpha-decay energy are substituted into Eq. (21), with R about 10^{-12} cm and Z about 90, the transmission coefficient $T=e^{-\gamma}$ is found to extend over a domain of about 10^{-20} to 10^{-40}. This range of about 10^{20} is just what is needed to relate the alpha-disintegration energy to the broad domain of known alpha-decay half-periods. Equation (21) thus explains the Geiger-Nuttall rule very successfully. Figure 8 presents a modern form of the Geiger-Nuttall relationship. The individual points show the measured half-periods and alpha-disintegration energies (alpha-ray energy plus recoil energy) for a number of high-Z emitters of alpha rays. The smooth curves are drawn using the wave-mechanical theory of transmission through nuclear barriers, with a nuclear radius of $R=1.48\times10^{-15}A^{1/3}$ m, where A is the mass number of the alpha-ray decay product. The agreement between experiment and theory is good.

Since 1970, knowledge of alpha-emitting isotopes has been greatly enlarged through the identification of many isotopes far off stability in the region just above tin and in the broad region from neodymium all the way to uranium. For example, fusion reactions between 290-MeV ^{58}Ni ions and ^{58}Ni and ^{63}Cu targets have been used to produce

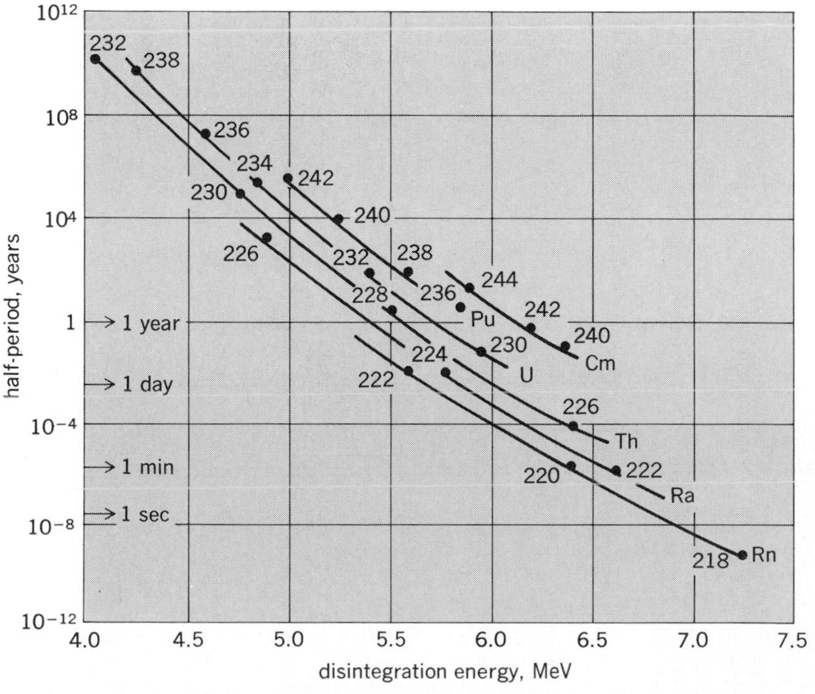

Fig. 8. Systematics of the broad range of half-periods for α-ray decay and their strong dependence on α-decay energy and weaker dependence on nuclear charge. Numbers beside experimental points are mass numbers of parent α-ray emitters. Lines connect parent isotopes and are drawn using wave-mechanical theory of α-ray transmission through nuclear potential barriers.

and study very-neutron-deficient radioactive isotopes, including 12 alpha emitters between tin and cesium. These results provide important data on the atomic masses of nuclei far from the stable ones in nature. These data test understanding of nuclear mass formulas and their validity in new regions of the periodic table.

BETA-RAY DECAY

Beta-ray decay is a type of radioactivity in which the parent nucleus emits a beta ray. There are two types of beta decay established: in negatron beta decay (β^-) the emitted beta ray is a negatively charged electron (negatron); in positron beta decay (β^+) the emitted beta ray is a positively charged electron (positron). In beta decay the atomic number shifts by one unit of charge, while the mass number remains unchanged (Table 1). In contrast to alpha decay, when beta decay takes place between two nuclei which have a definite energy difference, the beta rays from a large number of atoms will have a continuous distribution of energy.

The continuous number-versus-energy distribution of emitted beta rays is illustrated in Fig. 9. For each beta-ray emitter, there is a definite maximum or upper limit to the energy spectrum of beta rays.

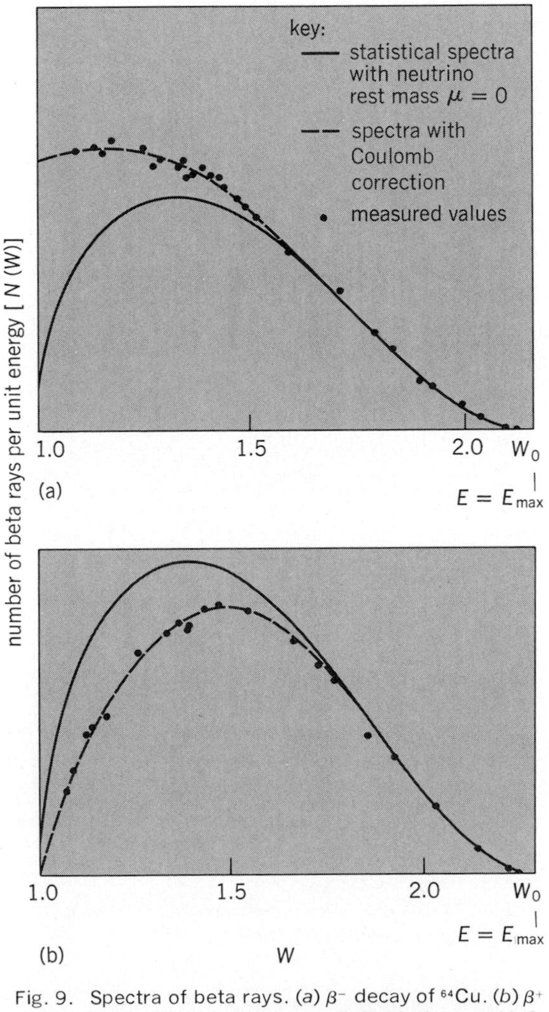

Fig. 9. Spectra of beta rays. (*a*) β^- decay of ^{64}Cu. (*b*) β^+ decay of ^{64}Cu. (*From L. M. Langer, R. D. Moffat, and H. C. Price, The beta-spectrum of Cu*^{64}, *Phys. Rev., 76:1725–1726, 1949*)

This maximum energy, E_{max}, corresponds to the change in nuclear energy in the beta decay. Thus $E_{max} = 0.57$ MeV for β^- decay of ^{64}Cu, and $E_{max} = 0.66$ MeV for β^+ decay of ^{64}Cu. As in the case of alpha decay, most beta-ray spectra are not this simple, but include additional continuous spectra which have less maximum energy and which leave the product nucleus in an excited level from which gamma rays are then emitted.

For nuclei very far from stability, the energies of these excited states populated in beta decay are so large that the excited states may decay by proton, neutron, two-neutron or alpha emission, or spontaneous fission and, it is predicted theoretically, by two-proton emission. In some cases, the energies are so great that the number of excited states to which beta decay can occur is so large that only the gross strength of the beta decays to many states can be studied.

Neutrinos. The continuous spectrum of beta-ray energies shown in Fig. 9 implies the simultaneous emission of a second particle besides the beta ray, in order to conserve energy and angular momentum for each decaying nucleus. This particle is the neutrino. The sum of the kinetic energy of the neutrino and the beta ray equals E_{max} for the particular transition involved except in the rare cases where internal bremsstrahlung or shake-off electrons are emitted. The neutrino has zero charge and presumably zero rest mass, travels at the same speed as light (3×10^{10} cm/s), and is emitted as a companion particle with each beta ray.

Earlier careful measurements of the beta spectra of ^3H established an upper limit for the neutrino rest mass as less than 0.0005 times the rest energy of the electron (Fig. 10). In 1980, however, some indirect evidence and a new ^3H beta spectra measurement yielded evidence for a rest mass much smaller than this limit, but finite. If the neutrino does have a nonzero rest mass, this will have many consequences, although they will not radically change the general features of the beta decay as presented here.

Two forms of neutrinos are distinguished. In positron beta decay, a proton p in the nucleus transforms into a neutron n in the nucleus, thus reducing the nuclear charge by 1 unit. At the time of this transition, two particles, the positron β^+ and the neutrino ν, are created and emitted. The emitted β^+ and ν together carry away the energy E_{max} of the transition and provide for conservation of energy, momentum, angular momentum, charge, and statistics. Thus positron beta decay is represented by reaction (22). Negatron beta decay is a

$$p \rightarrow n + \beta^+ + \nu \tag{22}$$

closely related process, except that a neutron n changes to a proton p in the nucleus, and a negatron beta-ray β^- and its characteristic companion particle, the antineutrino $\bar{\nu}$, are emitted. Thus reaction (23) is written. The antineutrino is the anti-

$$n \rightarrow p + \beta^- + \bar{\nu} \tag{23}$$

particle of the neutrino as the β^+ is the antiparticle of the β^-. They have the same properties of zero charge and zero rest mass, and differ only with respect to the direction of alignment of their intrinsic spin along their direction of motion. In most beta-decay contexts, the term neutrino includes both its forms, neutrino and antineutrino. Because of the fact that the neutron rest mass is greater than the proton rest mass, free neutrons can undergo beta decay (23), but protons must use part of the nuclear energy available to make up the rest mass difference.

The interaction of neutrinos with matter is exceedingly feeble. A neutrino can pass all the way through the Sun with little chance of collision. The thickness of lead required to attenuate neutrinos by the factor ½ is about 10^{18} m, or 100 light-years of lead!

Average beta energy. Charged particles, such as beta rays or alpha rays, are easily absorbed in matter, and their kinetic energy is thereby converted into heat. In beta decay the average energy E_{av} of the beta rays is far less than the maximum energy E_{max} of the particular beta-ray spectrum. The detailed shape of beta-ray spectra and hence the exact value of the ratio E_{av}/E_{max} varies somewhat with Z, E_{max}, the degree of forbiddenness of the transition, and the sign of charge of the emitted beta ray. A rough rule of thumb which covers many practical cases is $E_{av} = (0.40 \pm 0.05)\ E_{max}$, with slightly higher values for positron beta-ray spectra than for negatron beta-ray spectra. The remaining disintegration energy is emitted as kinetic energy of neutrinos and is not recoverable in finite absorbers.

There are other processes that carry off part of the energy of beta decay, including internal bremsstrahlung (gamma rays) and shake-off electrons (atomic electrons). The total probabilities for these additional two processes are the order of 1% or much less per beta decay, and the probability of their emission decreases rapidly with increasing energy so they are mainly low-energy (less than about 50 keV) radiations. In internal bremsstrahlung, through an interaction of the beta ray and the emitting nucleus, part of the decay energy is emitted as a gamma ray. In the shake-off process, part of the beta-decay energy is given to one of the atomic electrons. The gamma rays are not absorbed in matter as easily as the beta rays. In addition, if one tries to absorb the beta rays in matter, the beta rays can interact with the atoms and give off external bremsstrahlung (gamma rays). The number of these gamma rays again is a strongly decreasing function of energy, but their emission extends up to the maximum energies of the beta rays.

Fermi theory. By postulating the simultaneous emission of a beta ray and a neutrino, as in reaction (22), E. Fermi developed in 1934 a quantum-mechanical theory which satisfactorily gives the shape of beta-ray spectra (Fig. 9), and the relative half-periods of beta-ray emitters for allowed beta decays. The energy distribution of beta rays in allowed transitions is then given by Eq. (24).

$$N(W)\,dW = \frac{|P|^2}{\tau_0} F(Z,W)\,(W^2-1)^{1/2}(W_0-W)^2 W\,dW \qquad (24)$$

$N(W)\,dW$ = number of beta rays in energy range W to $W+dW$

$W = 1 + E/(m_0c^2)$ = total energy of beta ray in units of rest energy m_0c^2 = 0.51 MeV for an electron (m_0 = electron mass, c = velocity of light)

$W_0 = 1 + E_{max}/(m_0c^2)$ = maximum energy of the beta-ray spectrum

$|P|^2$ = squared matrix element for the transition, and is of the order of unity for allowed transitions

τ_0 = time constant ≅ 7000 s

$F(Z,W)$ = complex, dimensionless function involving the nuclear radius, nuclear charge, beta-ray energy, and whether the decay is β^- or β^+

Physically this distribution function involves the product of the energy W and momentum $(W^2-1)^{1/2}$ of the beta ray times the energy (W_0-W) and the momentum $(W_0-W)/c$ of the neutrino. The product of these factors gives a "statistical" distribution for the number of beta rays as a function of energy as shown in Fig. 9. The observed spectra show an excess of low-energy β^- and a deficiency of low-energy β^+ particles. This arises because of the Coulomb attraction and repulsion of the nucleus for β^- and β^+. The statistical spectrum is corrected by the Fermi function, $F(Z,W)$, and the new distribution agrees with experiments as shown in Fig. 9.

Equation (24) essentially matches the energy spectra of allowed beta-ray transitions and therefore furnishes one type of experimental verification of the properties of neutrinos. Its counterpart in terms of the beta-ray momentum spectrum is often used for the analysis of spectra, and is given by Eq. (25). The momentum distribution is much more nearly symmetric than its corresponding energy spectrum.

$$N(\eta)\,d\eta = \frac{|P|^2}{\tau_0} F(Z,\eta)\,(W_0-W)^2\eta^2\,d\eta \qquad (25)$$

$N(\eta)\,d\eta$ = number of beta rays in the momentum interval from η to $\eta + d\eta$

$\eta = (W^2-1)^{1/2}$ = momentum of the beta ray in units of m_0c

$F(Z,\eta) = F(Z,W)$ of Eq. (24)

Konopinski-Uhlenbeck theory. After the work of Fermi which explained allowed decay, E. J. Konopinski and G. E. Uhlenbeck in 1941 developed the theory of forbidden beta decay. Allowed decays occur between nuclear states which differ in spin by 0 or 1 unit and which have the same parity. Konopinski and Uhlenbeck developed a theory to describe beta decays where energy is available for decay but the allowed selection rules on spin or parity or both are violated. These beta transitions occur at a slower rate and are called forbidden transitions. In 1949 the theory of forbidden beta decay was confirmed by L. M. Langer and H. C. Price. The orders of forbiddenness, which retard the rate of decay, are: once-forbidden decay when the change in nuclear spin ΔI is again 0 or 1 as in allowed decay, but a parity charge $\Delta\pi$ occurs; once-forbidden unique decay when $\Delta\pi$ changes and $\Delta J = 2$; n-times forbidden decay when $\Delta J = n$, $\Delta\pi = (-)^n$, where $\Delta\pi = -$ indicates a parity change; and n-times forbidden unique decay when $\Delta J = n+1$, $\Delta\pi = (-)^n$. These are illustrated in Table 4. In forbidden decays the first-order allowed matrix elements of the Fermi theory in Eq. (24) vanish because of the selection rules on angular momentum and spin. Then the much smaller

Table 4. Selection rules for beta decay and log fT values

Type	ΔJ	$\Delta\pi$	Log fT	Examples
Allowed (favored)	0 or 1	No	3	n, ^3H
Allowed (normal)	0 or 1	No	4 to 7	^{35}S, ^{30}P
Allowed (l-forbidden)	1	No	6 to 9	^{32}P, ^{65}Ni
Once-forbidden	0 or 1	Yes	6 to 8	^{111}Ag, ^{143}Pr
Once-forbidden (unique)	2	Yes	8 to 9	^{42}K, ^{91}Y
Twice-forbidden	2	No	11 to 14	^{36}Cl, ^{59}Fe
Twice-forbidden (unique)	3	No	12 to 14	^{22}Na, ^{60}Co
Third-forbidden	3	Yes	17 to 19	^{87}Rb, ^{138}La
Third-forbidden (unique)	4	Yes	(~18)	^{40}K
Fourth-forbidden	4	No	~24	^{115}In
Fourth-forbidden (unique)	5	No		

higher-order matrix elements that can be neglected compared to the large allowed matrix elements come into play.

Comparative half-lives, fT. The half-period T of beta decay can be derived from Eq. (24) because the radioactive decay constant $\lambda = 0.693/T$ is simply the total probability of decay, or $N(W)dW$ integrated over all possible values of the beta-ray energy from $W = 1$ to $W = W_0$.

For allowed decays, the matrix elements are not functions of the beta energy and can be factored out of Eq. (24), so Eq. (26) is valid, where f is given

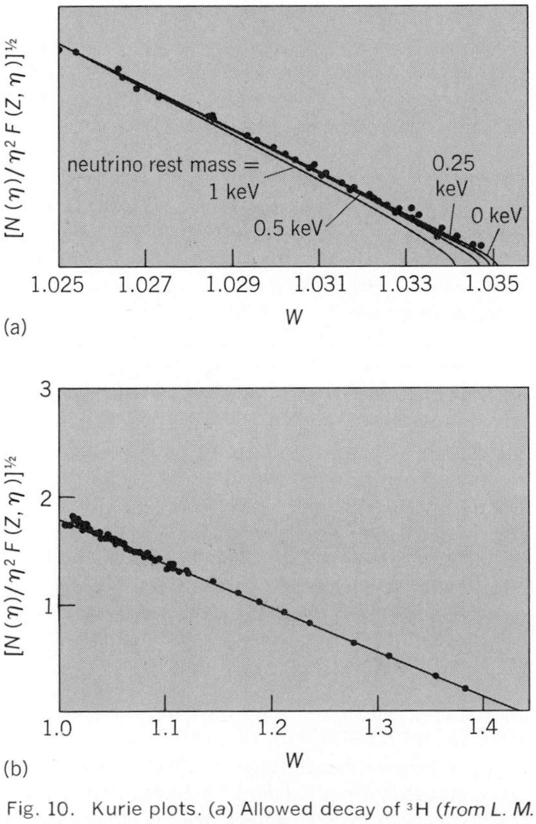

Fig. 10. Kurie plots. (*a*) Allowed decay of ^3H (*from L. M. Langer, and R. J. D. Moffat, The beta-spectrum of tritium and the mass of the neutrino, Phys. Rev., 88:689–694, 1952*). (*b*) Once-forbidden decay of ^{147}Pm (*from J. H. Hamilton, L. M. Langer, and W. G. Smith, The shape of the ^{143}Pr spectrum, Phys. Rev., 112:2010–2019, 1958*).

by Eq. (27), and the constants include $|P|^2$ pf Eq. (24). Equation (26) can be rearranged as Eq. (28).

$$\lambda = 0.693/T = \text{constants} \times f \tag{26}$$

$$f = \int_1^W F(Z,W)(W^2-1)^{1/2}(W_0-W)^2 W\,dW \tag{27}$$

$$fT = \frac{0.693}{\text{constants}} = \text{comparative half-life} \tag{28}$$

For different beta decays, T varies over a range greater than 10^{18} and inversely depends on the beta-decay energy in analogy to the Geiger-Nuttall rule for alpha decay. However, Eq. (28) says that the comparative half-life should be a constant. Indeed it is found experimentally that different classes of beta decay do have very similar fT values. It is generally easier to give the $\log_{10} fT$ for comparison. The groups are illustrated in Table 4 and include, in addition to the forbidden decays, three classes of allowed decays: the favored or superallowed decays of nuclei whose structures are very similar so that the matrix element in the denominator of Eq. (28) is large and log fT is small; normal allowed; and allowed l-forbidden where the total angular momentum selection rule holds, but the individual particle that is undergoing beta decay has a change of 2 units of orbital angular momentum. The matrix elements for each degree of forbiddenness get progressively smaller and so log fT values increase sharply with each degree of forbiddenness. The ranges of these fT values for each degree of forbiddenness are in general so well established that measurements of fT values can be used to establish changes in spins and parities between nuclear states in beta decay.

Kurie plots. For allowed transitions, the transition matrix element $|P|^2$ is independent of the momentum η. Then Eq. (25) can be put in the form of Eq. (29). Therefore a straight line results when

$$\left(\frac{N(\eta)}{\eta^2 F(Z,\eta)}\right)^{1/2} = \text{const}\,(W_0 - W) \tag{29}$$

the quantity $\sqrt{N/\eta^2 F}$ is plotted against beta-ray energy, either as W or as E, on a linear scale. Such graphs are called Kurie plots, Fermi plots, or Fermi-Kurie plots. These are especially useful for revealing deviations from the allowed theory and for obtaining the upper energy limit E_{max} as the extrapolated intercept of $\sqrt{N/\eta^2 F}$ on the energy axis. Practically all results on the shapes of beta-ray spectra are published as Kurie plots, rather than as actual momentum or energy spectra.

Figure 10 shows representative Kurie plots for ^3H and ^{147}Pm. When spectral data give a straight line, such as these, then $N(\eta)$ is in agreement with the Fermi momentum distribution, Eq. (25); and the intercept of this straight line, on the energy axis, gives the disintegration energy E_{max}. In Fig. 10*a*, theoretical curves are given for various values of the neutrino rest mass, and the data points, which are experimental values, lie on the curve corresponding to zero mass.

In addition to allowed decays, all but one known once-forbidden decays have Kurie plots that are essentially linear in energy (Fig. 10*b*). The once-forbidden unique decays have a pronounced characteristic energy dependence for their matrix ele-

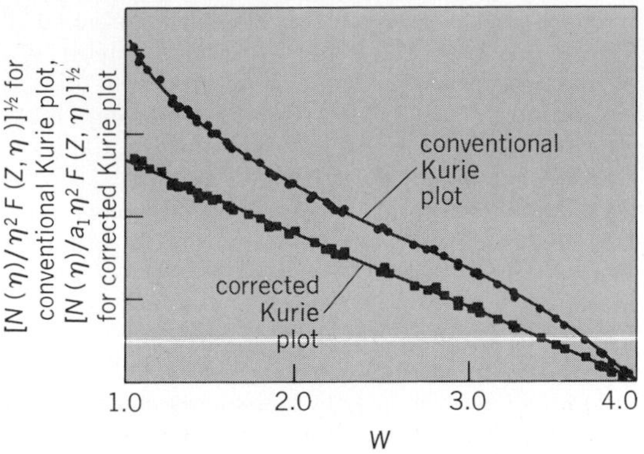

Fig. 11. Once-forbidden spectrum of ^{91}Y: conventional Kurie plot and Kurie plot corrected by the unique shape factor, $a_1 = W^2 - 1 + (W_0 - W)^2$, given by Konopinski and Uhlenbeck, which linearizes the data. (*From L. M. Langer and H. C. Price, Shape of the beta spectrum of the forbidden transition of yttrium 91, Phys. Rev., 75:1109, 76:641, 1949*)

ments, and thus the conventional Kurie plot has a characteristic shape that differs from a straight line (Fig. 11). When the data are corrected by the unique shape factor, given by Konopinski and Uhlenbeck, a linear Kurie plot is again obtained. This unique shape was the key to the discovery of forbidden beta decay by Langer and Price, and Fig. 11 is, in fact, from the data with which they made this discovery. The higher-order forbidden spectra each show different strong energy dependences in their Kurie plots, each characteristic of their degree of forbiddenness.

Double beta decay. When the ground state of a nucleus differing by 2 units of charge from nucleus A has lower energy than A, then it is theoretically possible for A to emit two beta particles, either $\beta^+\beta^+$ or $\beta^-\beta^-$ as the case may be, and two neutrinos or antineutrinos, and go from Z to $Z \pm 2$. Here two protons decay into two neutrons, or vice versa. This is a second-order process and so should go much slower than beta decay. There are a number of cases where such decays should occur, but their half-lives are of the order of 10^{20} years or greater. These are obviously very difficult to detect and have not been seen directly. There is indirect evidence for double beta decay in one case, that of $^{130}_{52}Te$, from the observed buildup of $^{130}_{54}Xe$ in samples. However, it cannot be said with certainty that this process has been shown to occur in nature.

Electron-capture transitions. Whenever it is energetically allowed by the mass difference between neighboring isobars, a nucleus Z may capture one of its own atomic electrons and transform to the isobar of atomic number $Z - 1$ (Table 1). Usually the electron-capture (EC) transition involves an electron from the K shell of atomic electrons, because these innermost electrons have the greatest probability density of being in or near the nucleus.

In EC transitions, a proton p bound in the parent nucleus absorbs an electron e^- and changes to a bound neutron n. The disintegration energy is carried away by an emitted neutrino ν as in Eq. (30).

$$p + e^- \rightarrow n + \nu \qquad (30)$$

The residual nucleus may be left either in its ground level or in an excited level from which gamma-ray emission follows.

EC transitions compete with all cases of positron beta-ray decay. EC has an energetic advantage over β^+ decay equivalent to the mass of two electrons, or 1.02 MeV, because in Eq. (30) one electron mass e^- enters the reaction and is available, whereas in Eq. (24) one electron mass β^+ must be produced as a product of the positron beta-ray decay. For example, in the radioactive decay of $^{64}_{29}Cu$, twice as many transitions go by EC to $^{64}_{28}Ni$ as go by positron beta decay to the same decay product. In the heavy, high-Z elements, EC is greatly favored over the competing β^+ decay, and examples of measurable β^+ decay are practically unknown for Z greater than 80, although there are a large number of examples of electron capture. As the energy for decay increases beyond 1.02 MeV, the probability of β^+ decay increases relative to EC and dominates at several MeV of energy.

Several examples are known of completely pure EC radioactivity in which there is insufficient nuclear energy to allow any positron beta-ray decay. For example, $^{55}_{26}Fe$ emits no positron beta rays, but transforms with a half-period of 2.6 years entirely by EC to the ground level of $^{55}_{25}Mn$. This radioactivity is detectable through the K-series x-rays which are emitted from ^{55}Mn when the atomic electron vacancy, produced by nuclear capture of a K electron, refills from the L shell of atomic electrons. Also, the process of double electron capture, analogous to double beta decay, is theoretically predicted to exist. Here two atomic electrons are captured and two neutrinos emitted.

GAMMA-RAY DECAY

Gamma-ray is a transition between two excited levels of a nucleus, or between an excited level and the ground level. A nucleus in its ground level cannot emit any gamma radiation. Therefore gamma-ray decay occurs only as a sequel of one of the processes in Table 1 or of some other process whereby the product nucleus is left in an excited state. Such additional processes include gamma rays observed following the fusion of two nuclei, as occurs in bombarding ^{58}Ni with ^{16}O to form an excited compound nucleus of ^{74}Kr. This compound nucleus first promptly gives off a few particles like two neutrons to leave $^{72}Kr^*$ or two protons to leave $^{72}Se^*$, both of which will be in excited states which will emit gamma rays. Or one may excite states in a nucleus by the Coulomb force between two nuclei when they pass close to each other but do not touch (their separation is greater than the sum of the radii of the two nuclei). There are also other nuclear reactions such as induced nuclear fission that leave nuclei in excited states to undergo gamma decay.

A gamma ray is high-frequency electromagnetic radiation (a photon) in the same family with radiowaves, visible light, and x-rays. The energy of a gamma ray is given by $h\nu$, where h is Planck's constant and ν is the frequency of oscillation of the wave in hertz. The gamma-ray or photon energy $h\nu$ lies between 0.05 and 3 MeV for the majority of known nuclear transitions. Higher-energy gamma rays are seen in neutron capture and some reactions.

Gamma rays carry away energy, linear momentum, and angular momentum, and account for changes of angular momentum, parity, and energy between excited levels in a given nucleus. This leads to a set of gamma-ray selection rules for nuclear decay and a classification of gamma-ray transitions as "electric" or as "magnetic" multipole radiation of multipole order 2^l, where $l = 1$ is called dipole radiation, $l = 2$ is quadrupole radiation, and $l = 3$ is octupole, l being the vector change in nuclear angular momentum. The most common type of gamma-ray transition in nuclei is the electric quadrupole (E2). There are cases where several hundred gamma rays with different energies are emitted in the decays of atoms of only one isotope.

Mean life for transitions. A reasonably successful approximate theory of the mean life for gamma-ray decay was developed by V. F. Weisskopf in 1951, using the single-particle shell model of nuclei. Figure 12 summarizes the numerical consequences of this theory. An E2 transition of about 1 MeV is expected to take place with a mean life, τ_{el}, or mean delay in the upper level, of about 10^{-11} s. Thus most gamma-ray transitions are prompt transitions, in which the mean life of the excited level is too short to be measured easily. Figure 12 is for electric multipole transitions. The mean life τ_{mag} for magnetic multipoles is of the order of 30 (for $A = 20$) to 150 (for $A = 200$) times longer than τ_{el}.

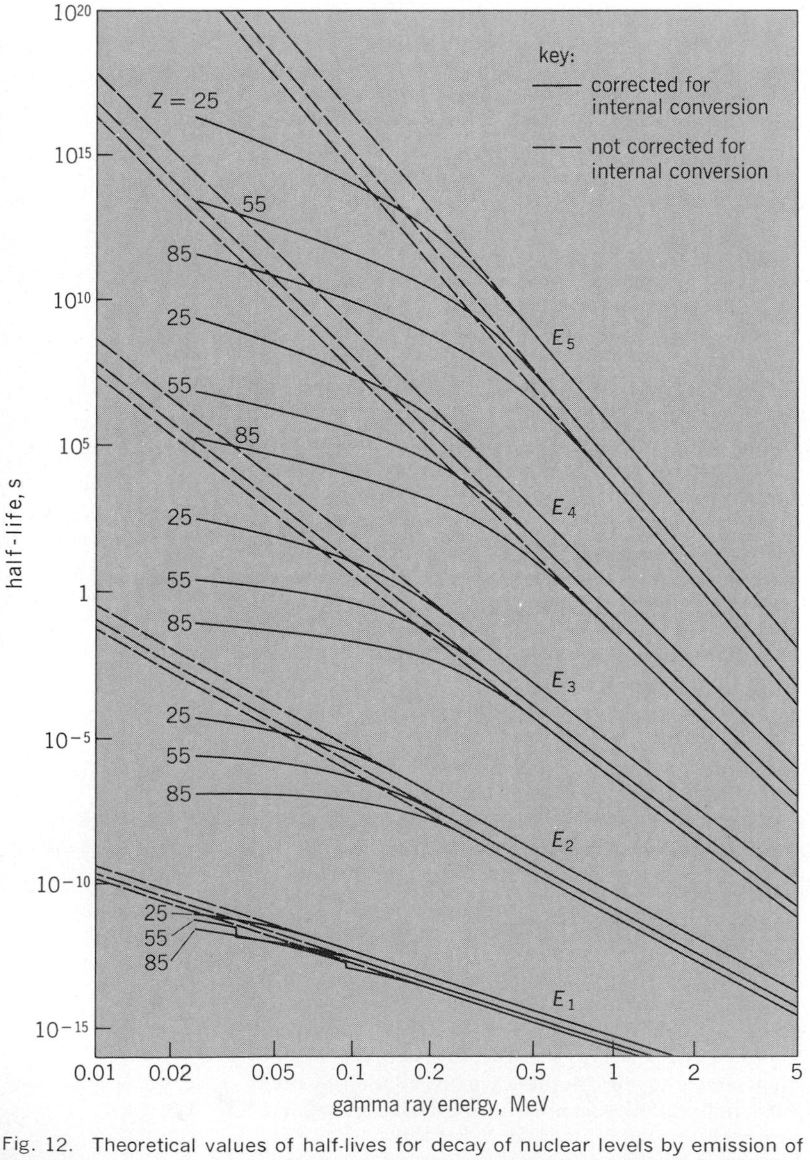

Fig. 12. Theoretical values of half-lives for decay of nuclear levels by emission of gamma rays and conversion electrons, for electric multipoles. *(From A. H. Wapstra, G. J. Nijgh, and R. van Lieshout, Nuclear Spectroscopy Tables, North-Holland, 1959)*

At low energies or high Z, or both, the internal conversion process becomes a very important additional mode of decay that markedly shortens the mean lives of the nuclear levels. In addition, in many cases the structure of the nucleus comes into play and alters the observed mean lives considerably compared to those in Fig. 12. Electric dipole (E1) transitions are generally retarded (longer mean lives) by factors of 10^6 over the Weisskopf estimates of Fig. 12. On the other hand, A. Bohr and B. Mottelson developed a model of collective nuclear motions where E2 transitions are enhanced by factors of 100 or more (shorter τ) over the Weisskopf single-particle estimates, and these predictions are confirmed by experiments. The magnetic dipole (M1) transitions are also often hindered by factors of 100 or more. Measurements of the mean lives for gamma-ray decay provide important tests of nuclear models.

Internal conversion. An alternative type of deexcitation which always competes with gamma-ray emission is known as internal conversion. Instead of the emission of a gamma ray, the nuclear excitation energy can be transferred directly to a bound electron of the same atom. Then the nuclear energy difference is converted to energy of an atomic electron, which is ejected from the atom with a kinetic energy E_i given by Eq. (31). Here B_i

$$E_i = W - B_i \tag{31}$$

is the original atomic binding energy of the particular electron, which is ejected, and W is the nuclear transition energy which would otherwise have been emitted as a gamma-ray photon having energy $h\nu = W$.

The spectrum of internal conversion electrons is then a series of discrete energies, or "lines," each corresponding to an individual value of B_i, for the K, L (L_1, L_2, L_3), M, . . . , electrons in each shell and subshell of the atom. Thus conversion electron spectra are much more complex than gamma spectra. From the spacing of the E_i values in this conversion electron spectrum, it is possible to assign definitely the atomic number Z of the atom in which the nuclear transition W took place. In this way it is known that the conversion electron and the competing gamma-ray emission are sequels and not antecedents of alpha decay, beta decay, and electron-capture transitions. Partial electron spectra, showing K, L, and M shell conversion, and gamma-ray spectra are shown in Fig. 13 for the decay of ^{186}Tl. By comparing the K and L electron intensities of the 402 + 405 and 522 keV transitions with the gamma-ray spectrum, it can be seen that the strong 522-keV electron transition has no gamma ray associated with it. The strong 511-keV gamma ray is from the annihilation of positrons and is not a nuclear transition, and so has no conversion electrons of this energy. One can improve the

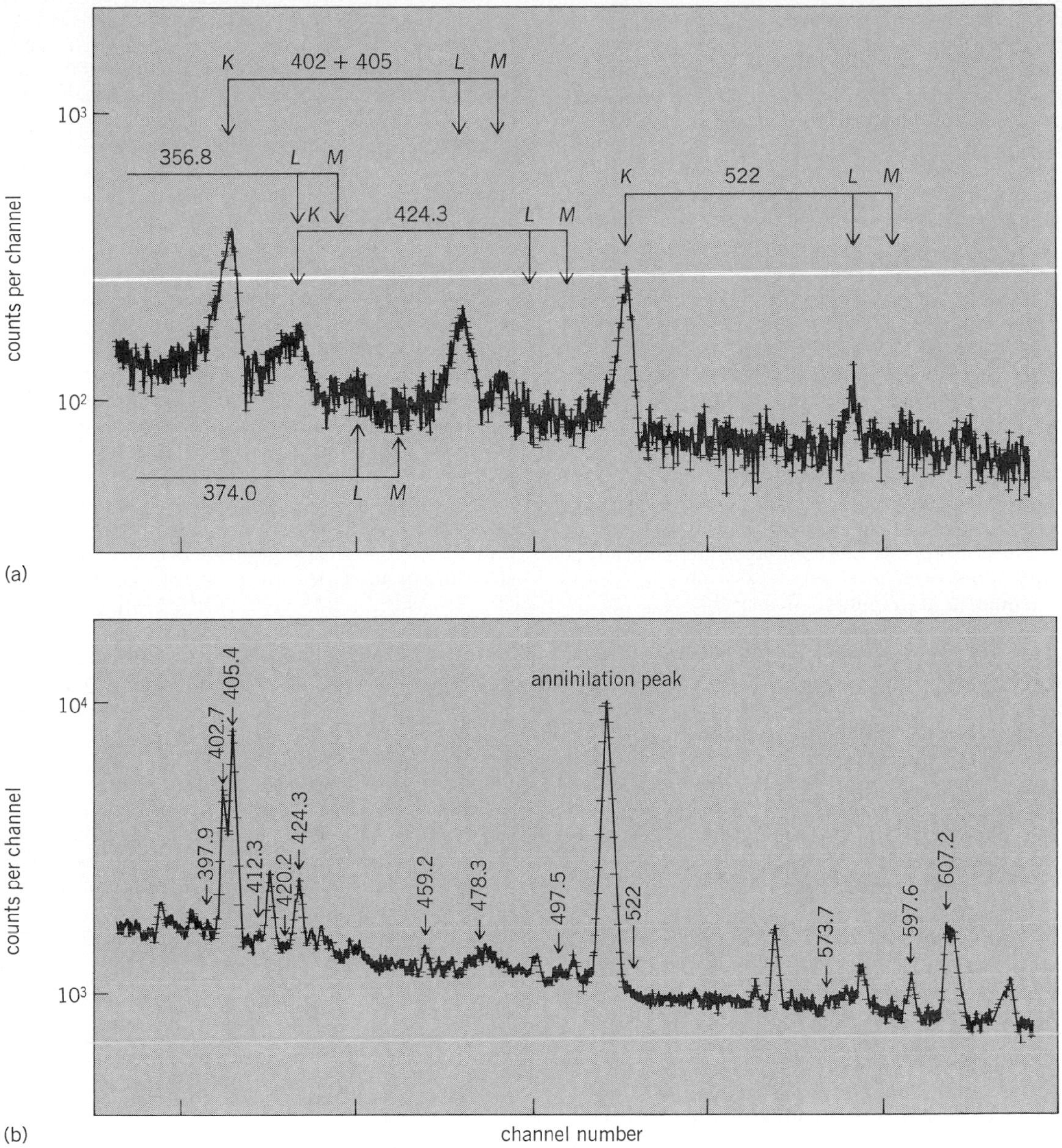

Fig. 13. Spectra from the decay of 30-s half-life ^{186}Tl far off stability (15 neutrons less than the lightest stable thallium isotope). (*a*) Internal conversion electrons. (*b*) Gamma rays. Nuclear transition energies are given in keV. (*From J. H. Hamilton et al., Shape coexistence in ^{186}Hg and the decay of ^{186}Tl, Phys. Rev., C16:2010–2018, 1977*)

energy resolution by factors of 100 – 1000 over that in Fig. 13 with magnetic spectrometers so that, for example, one can separate the lines with different energies from even the five M subshells.

The internal conversion coefficient α is the ratio of the number of transitions proceeding by internal conversion to the number going by gamma-ray emission, for any particular nuclear transformation from an excited level to a lower-lying level. In general, this probability of internal conversion relative to gamma-ray emission increases with increasing atomic number Z, with increasing multipole order 2^l, and with decreasing nuclear deexcitation energy W. In middle-weight elements, for $W = 1$MeV, α is of the order of 10^{-2} to 10^{-4}; while for $W = 0.2$ MeV, α is of the order of 0.1 for electric $l = 2$ transitions, and 10 or larger for electric $l = 5$ transitions.

Radiationless transitions. There are cases where gamma-ray emission is strictly forbidden and conversion electron emission allowed. This occurs when both nuclear states have zero spin and the same parity. The conversion electrons are called electric monopole radiations, E0. These transitions occur because of the penetration of the atomic electrons into the nuclear volume where they interact directly with the nucleus. An example of the E0 decay is shown in Fig. 13. E0 radiation can occur in principle whenever two states have the same spin and same parity, but in practice, E0 decays are found to be very, very small in these cases. There are some exceptions in well-deformed nuclei, as was first shown for ^{154}Gd, where they totally dominate the electron emission for certain transitions that involve large shape changes.

The E0 decays which arise because of the penetration of the atomic electrons into the nuclear volume are thus sensitive measures of changes in shape between two nuclear states, and have played important roles in establishing vibrations of the nuclear shape and the coexistence of states with quite different deformation in the same nucleus. There also are other circumstances where the penetration of the atomic electron into the nuclear volume give rise to additional contributions to the conversion-electron decay. Again these penetration effects probe details of the structure of the nucleus.

Internal pair formation. When the energy of a nuclear gamma-ray transition exceeds 1.022 MeV, twice the rest mass energy of an electron, it is possible for a nucleus to give up its excess energy to an electron-positron pair—a pair creation process. This is a third alternate mode to gamma decay and conversion electron decay. This process becomes more important as the gamma-ray energy increases. It is relatively unimportant below 2–3 MeV of decay energy.

Isomeric transitions. Measurably delayed radioactive transitions from an excited level of a nucleus are known as isomeric transitions. The measurably long-lived excited level is called an isomeric or metastable level or an isomer of the ground level. What constitutes an isomer is not well defined. The terminology grew up when it was difficult to measure mean lives shorter than 10^{-7} s. States with longer mean lives were isomers. Now mean lives down to 10^{-13} s can be measured for many transitions in different nuclei, but these are not generally called isomers. The break point is simply not defined.

Figure 12 shows that if the excitation energy is small (say, 0.5 MeV or less) and the angular momentum difference l is large (say, $l = 3$ or more) then the mean life of an excited level for gamma ray or conversion-electron emission can be of the order of 1 s up to several years.

Most of the long-lived isomers occur in nuclei which have odd mass number A. Then either the number of protons Z in the nucleus is odd, or the number of neutrons N in the nucleus is odd. The frequency distribution of odd-A isomeric pairs, excited level and ground level, displays so-called islands of isomerism in which the odd-proton or odd-neutron number is less than 50, or less than 82. The distribution is one of several lines of evidence for closed shells of identical nucleons at N or $Z = 50$ or 82 in nuclei, and it plays an important role in the so-called shell model of nuclei.

SPONTANEOUS FISSION

This involves the spontaneous breakup of a nucleus into two heavy fragments and neutrons, as shown in Table 1. After the discovery of fission in 1939, it was subsequently discovered that isotopes like ^{238}U had very weak decay branches for spontaneous fission, with branching ratios on the order of 10^{-6}. New isotopes subsequently identified like ^{252}Cf have large (3.1%) spontaneous fission branching. In these cases, the nucleus can go to a lower energy state by spontaneously splitting apart into two heavy fragments of rather similar mass plus a few neutrons. This process liberates a large amount of energy compared to any other decay mode. Thus, ^{252}Cf is becoming important in many applications in medicine and industry as a compact energy source or as a source of nuclear radiation, since the fragments themselves are left in excited states and so emit gamma rays.

An important isomeric decay mode was discovered in the early 1960s in the very heavy elements, spontaneous fission isomers. Here the nucleus in an excited state, rather than emit a gamma ray or conversion electron, spontaneously breaks apart into two heavy fragments plus neutrons exactly as in spontaneous fission. To identify these isomers, the symbol f is often placed after their atomic mass, for example, $^{244f}_{95}Am$. Their half-lives are generally short, 10^{-3} to 10^{-9} s. It is now understood that these fission isomers are states with much larger deformation than the ground states of these isotopes. The Coulomb barrier against fission is in fact a double-hump barrier with the fission isomers in the valley at large deformation (Fig. 14). The study of these fission isomers has provided important tests of understanding of the behavior and structure of nuclei with very large deformation.

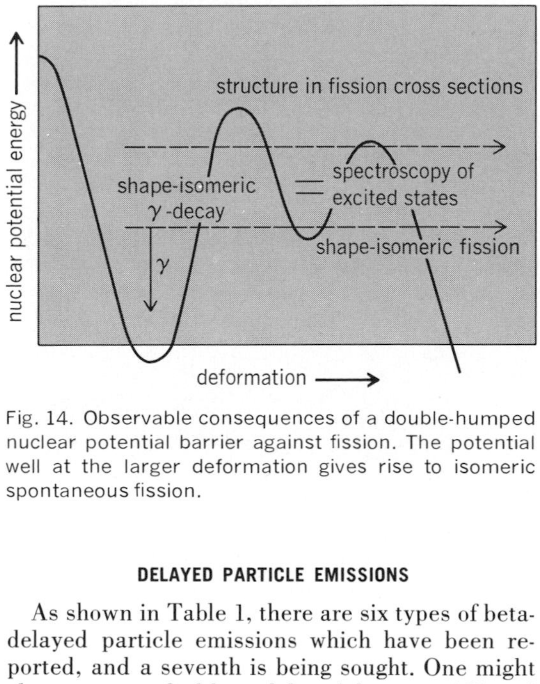

Fig. 14. Observable consequences of a double-humped nuclear potential barrier against fission. The potential well at the larger deformation gives rise to isomeric spontaneous fission.

DELAYED PARTICLE EMISSIONS

As shown in Table 1, there are six types of beta-delayed particle emissions which have been reported, and a seventh is being sought. One might also expect to find beta-delayed deuteron (^2H) and triton (^3H) emissions which are not shown there. There are now over 100 beta-delayed particle radioactivities known. Theoretically, the number of isotopes which can undergo beta-delayed particle emission can exceed 1000. Thus, this mode, which was observed in only a few cases prior to 1965, is among the important ones in nuclei very far from the stable ones in nature. Studies of these decays can provide insights into the nucleus which can be gained in no other way.

Beta-delayed alpha radioactivity. The β^- decay of ^{214}Bi to ^{214}Po leaves the nucleus in such a high-energy excited state that it can emit an alpha particle and go to ^{210}Pb as an alternative to gamma-ray decay to lower levels in ^{214}Po. This is a two-step process with beta decay the first step. After beta decay the nucleus is in such a highly excited state that it can emit either an alpha particle or gamma ray. Several different beta-delayed particle emissions are now known.

β^--delayed alpha emission has been found relatively rarely, but the additional process of β^+-delayed alpha emission has been discovered. In proton-rich nuclei far from stability, the conditions are more favorable for beta-delayed alpha emission because of the excess of nuclear charge, and a number of such β^+-delayed alpha emitters are now known.

Beta-delayed neutron radioactivity. In 1939, shortly after the discovery of nuclear fission, it was proposed that the delayed neutrons observed following fission were in fact beta-delayed neutrons. That is, after the nucleus fissioned, the beta decay of the neutron-rich fission fragments populated high-energy excited states that could promptly undergo dual decay, emitting either a gamma ray or neutron. Beta-delayed neutron emission is illustrated in Fig. 2. The processes of beta-delayed two- and three-neutron emission were discovered in 1979 and 1980 in the decay of ^{11}Li. The former is shown in Fig. 2, and β^-2n decays were subsequently observed in other nuclei.

Beta-delayed proton radioactivities. Proton radioactivity itself is a mode of radioactive decay, generally expected to arise in proton-rich nuclei far from the stable isotopes, in which the parent nucleus changes its chemical identity by emission of a proton in a single-step process. Its physical interpretation parallels almost exactly the quantum-mechanical treatment of alpha-ray decay. It is also theoretically predicted that one can have the simultaneous emission of two protons—two-proton radioactivity (Fig. 1). In addition, one can have β-delayed proton and β-delayed two-proton radioactivities which again ultimately result in emission of protons from the nucleus. These latter processes also occur in quite proton-rich nuclei with very high decay energies; however, they are complex two-step decay modes whose fundamental first step is β-ray decay.

Over 40 nuclei ranging from $^{9}_{6}C$ to $^{183}_{80}Hg$ have been identified to decay by the two-step mode of β^+-delayed proton radioactivity. Figure 15 presents the observed proton energy spectrum arising in the decay of $^{33}_{18}Ar$, with a half-life of 173 ms; it was produced by the $^{32}_{16}S + ^{3}_{2}He \rightarrow ^{33}_{18}Ar + 2n$ reaction. This isotope decays by superallowed and allowed β^+ decay to a number of levels in its daughter nucleus $^{33}_{17}Cl$, which immediately (in less than 10^{-17} s) breaks up into $^{32}_{16}S$ and a proton. More than 30 proton groups arising from the decay of $^{33}_{18}Ar$ are observed, ranging in energy from 1 to approximately 6 MeV and varying in intensity over four orders of magnitude. Although it is normally very difficult to study many β-decay branches in the decay of a particular nuclide—because of the continuous nature of the energy spectrum of the emitted beta particles—it is possible to do so when investigating β^+-delayed proton emitters. The observed proton group energies and intensities can be correlated with the levels fed in the preceding beta decay and their transition rates, thereby permitting sensitive tests via beta decay of nuclear wave functions arising from different models of the nucleus. β^+-delayed two-proton decay is being sought (Fig. 1).

Proton radioactivity. Although proton radioactivity has been of considerable theoretical interest since 1951 and is expected to be a general phenomenon, so far only one example of this decay mode has been observed, because of the experimental difficulties associated with producing extremely proton-rich nuclei. Figure 16 presents the decay scheme of the first nuclide found, in 1970, to decay by proton radioactivity. It is $^{53m}_{27}Co$, where the *m* (metastate) denotes a (relatively) long-lived isomeric state. Because of its very high angular momentum of 19/2 and odd parity, gamma decay is highly forbidden. This mode of decay is essentially the same as that of β-delayed proton emission, except

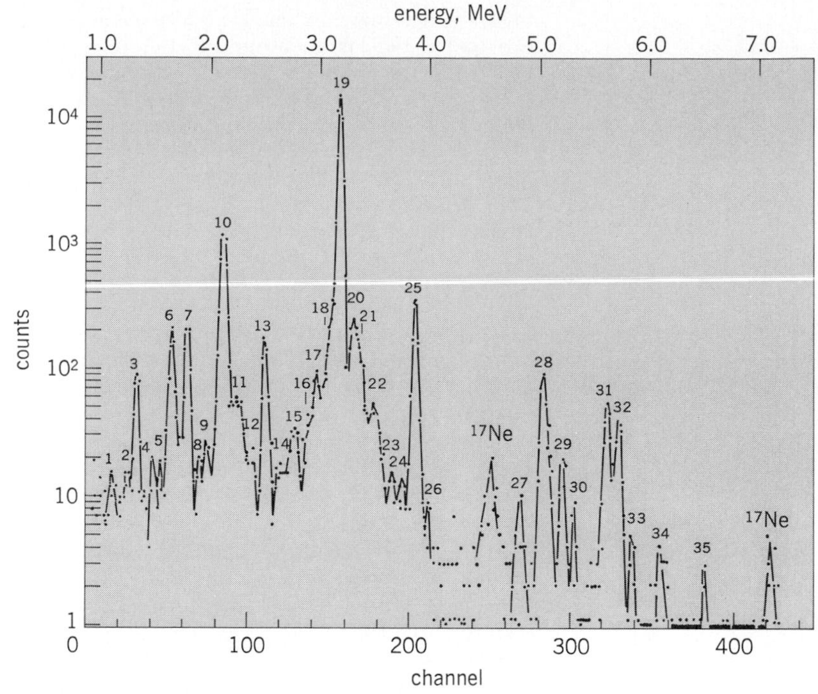

Fig. 15. Spectrum of β^+-delayed protons from the decay of $^{33}_{18}Ar$ as observed in a counter telescope; the proton laboratory energy is indicated at the top. Proton groups are numbered 1 through 35. (*From J. C. Hardy et al., Isospin purity and delayed-proton decay: ^{17}Ne and ^{33}Ar, Phys. Rev., C3:700–718, 1971*)

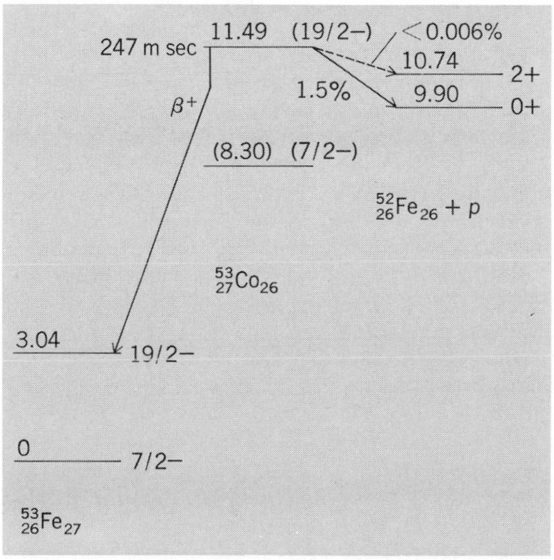

Fig. 16. The decay scheme of $_{27}Co^{53*}$. Numbers to left of levels represent energies in MeV, relative to ground state of $_{26}Fe^{53}_{27}$. Symbols to right of levels are spin and parity. (*From J. Cerny et al., Further results on the proton radioactivity of Co^{53*}, Nucl. Phys., A188:666–672, 1972*)

that now the energy of the excited nuclear level is low, and angular momentum selection rules highly forbid gamma-ray decay so the state lives a relatively long time in comparison to those states populated in beta decay. It was produced in the laboratory by the compound nucleus reactions ${}^{16}_{8}O + {}^{40}_{20}Ca \rightarrow {}^{53m}_{27}Co + p + 2n$ and ${}^{54}_{26}F + p \rightarrow {}^{53m}_{27}Co + 2n$. This 247-ms isomer exhibits two different decay modes: though it predominantly decays by positron (β^+) emission to a similar $19/2^-$ level in ${}^{53}_{26}Fe$, a 1.5% branch in its decay occurs via direct emission of a 1.59-MeV proton to the ${}^{52}_{26}Fe$ ground state. The calculated half-life that ${}^{53m}_{27}Co$ would possess if proton radioactivity were the only decay mode (its partial half-life for this decay branch) is the surprisingly long time of 17 s.

Because of the lower charge on the proton compared to the alpha particle, the Coulomb barrier indicated in Fig. 7 is of less importance in proton radioactivity than in alpha-particle radioactivity; however, the effect of the angular momentum, or centrifugal, barrier is much more significant because of the lower mass of the proton. This can be seen from the mathematical form of the centrifugal barrier, which is $\hbar^2(l+1)/2Mr^2$; here M, the so-called reduced mass of the emitted particle, appears in the denominator; additionally, $\hbar$ is Planck's constant divided by 2π, l is the angular momentum of the emitted proton (or alpha particle), and r is its radial separation. A high centrifugal barrier arises in the decay of ${}^{53m}_{27}Co$ because of the large change of angular momentum of 9 units required for the proton (of intrinsic angular momentum ½) to be emitted from this angular momentum 19/2 isomer (of odd parity), leaving the daughter nucleus ${}^{52}_{26}Fe$ in its angular-momentum zero ground state (of even parity).

Searches for proton and two-proton radioactivities from ground states of nuclei are being carried out. Figure 1 illustrates one of the prime candidates for such decays.

Beta-delayed spontaneous fission. There are also observed beta-decay processes where the excited nucleus following beta decay has a probability of undergoing spontaneous fission rather than gamma-ray decay. This is the same process as in spontaneous or isomeric spontaneous fission. The excitation energy of the nuclear level provides the extra energy to make fission possible. The nucleus splits into two nearly equal fragments plus some neutrons. This process is like isomeric spontaneous fission except that the lifetime of the nuclear level is so short that the level would not normally be called an isomer.

Neutron radioactivity. In very neutron-rich nuclei near the boundary line of nucleus stability, one may find nuclei with ground states which are unstable to the emission of one or two neutrons. Here there is no Coulomb barrier, but one can have a centrifugal barrier that may give rise to one- or even two-neutron radioactivity. These processes for ground states would be very near the limits where nuclei become totally unstable to the addition of a neutron, the neutron drip line, and very difficult to even make much less measure. However, there may be neutron-rich nuclei with high-spin isomeric states where the high spin analogous to the one in ${}^{53m}Co$ gives rise to a large centrifugal barrier. Such isomeric states may undergo one-or two-neutron radioactivity. [JOSEPH H. HAMILTON]

Bibliography: J. M. Eisenberg and W. Greiner, *Nuclear Theory*, 3 vols., 1975; R. D. Evans, *The Atomic Nucleus*, 1955; J. H. Hamilton (ed.), *Internal Conversion Processes*, 1965; J. H. Hamilton et al. (eds.), *Future Directions in Studies of Nuclei Far from Stability*, 1980; J. H. Hamilton and J. C. Manthurathil (eds.) *Radioactivity in Nuclear Spectroscopy*, 1972; W. D. Hamilton (ed.), *Electromagnetic Interaction in Nuclear Spectroscopy*, 1975; J. C. Hardy, Nuclear spectroscopy from delayed particle emission, in J. Cerny (ed.), *Nuclear Spectroscopy and Reactions*, Part C, 417, 1974; I. Kaplan, *Nuclear Physics*, 2d ed., 1963; E. J. Konopinski, *The Theory of Beta Radioactivity*, 1966; C. M. Lederer et al. (eds.), *Table of Isotopes*, 7th ed., 1978; National Bureau of Standards, Tables for the analysis of beta spectra, *Applied Mathematics*, ser. 13, 1952; S. C. Pancholi (ed.), *Gamma-Ray Transition Probabilities*, 1977; F. Rösel et al., *Atomic and Nuclear Data Tables*, 21:91, 1978.

Radiochemistry

A subject which embraces all applications of radioactive isotopes to chemistry. It is not precisely defined and is closely linked to nuclear chemistry. The widespread use of isotopes in chemistry is based on two fundamental properties exhibited by all radioactive substances. The first property is that the disintegration rate of an isotopic sample is directly proportional to the number of radioactive atoms in the sample. Thus, measurement of its disintegration rate (with a Geiger counter, for example) serves to analyze a radioactive compound. With nearly all chemical elements (the most notable exceptions being nitrogen and oxygen, which have no suitable radioactive isotopes), an isotope may be incorporated in a chemical compound, and thereafter, masses of this compound as small as 10^{-6} to 10^{-10} g may be measured with a high precision. Because experimental chemistry depends largely upon analysis, isotopes may be employed in most chemical problems, especially those requiring high analytical sensitivity. The second fundamental property is that the disintegration rate is completely unaffected by the chemical form of the isotope, and conversely, the property of radioactivity does not affect the chemical properties of the isotope. By substituting or labeling a particular atom within a molecule, isotopes can be used to trace the fate of that atom during a chemical reaction. In contrast to physical migration tracer studies, the compounds arising in a reaction must first be isolated in separated pure forms before radioactive assays can be performed.

In general, radiochemical studies can be classified according to whether the use of isotopes represents a convenient or a unique solution to a problem. *See* RADIOISOTOPIC ASSAY.

Convenient applications. These applications usually exploit the high sensitivity of tracer techniques because alternative analytical procedures are slower and often less accurate. The efficiencies of chemical separations, such as those based on selective precipitation, solvent extraction, ion exchange, and electrodeposition reactions, are studied by labeling the desired compound and following the radioactivity during the separations. The rate and extent of adsorption on solid surfaces of either labeled solutions or labeled gases are rapidly determined by assaying periodically the mobile phase, or better, the solid phase. New chemical phenomena, such as the coprecipitation of trace elements and radiocolloid formation, occur at submicro concentrations (10^{-10} g/liter of solution) and may be studied most conveniently with isotopes. The solubility of an "insoluble" precipitate is measured by saturating a solvent medium with a radioactive solid. Similarly, the vapor pressure of a solid is measured by saturating an evacuated volume with vapor or, for pressures below 10^{-4} mm, by effusing vapor into a cooled target, which is later assayed. Qualitative and quantitative analysis for most trace elements present in parts per million or less in a sample is possible by radioactivation analysis. The sample is irradiated in a flux of neutrons or other suitable particles, and the trace element is identified and determined by its induced activity. Depending upon the element, quantitative determination of masses of 10^{-8} to 10^{-12} g is usually possible.

Unique applications. An understanding of diffusion processes is of considerable importance because the rates of many chemical reactions are governed by the rate at which chemical species can diffuse through a medium to the point of reaction. For example, the rate of many electrode processes depends upon the rate of diffusion of electrolyte to the electrode, and the rate of oxidation of copper is determined by the rate of diffusion of copper ions up to the metal surface. If a layer of radioactive copper is sandwiched between two ordinary copper samples, it is found that at elevated temperatures copper ions will diffuse considerable distances within the metal. The rate of diffusion of copper in copper (that is, the self-diffusion rate of copper) can be observed only by the transfer of radioactivity from the labeled region into the unlabeled regions, thin slices being removed from the solid at known distances and then being assayed. The illustration shows some experimental points obtained in such an experiment. The distribution curve is that expected from the integrated form of Fick's law of diffusion, Eq. (1), where c is the concentration

$$\partial c/\partial t = D(\partial^2 c/\partial x^2) \qquad (1)$$

(activity) of the diffusing tracer, t is the diffusion time, x the diffusion distance, and D the self-diffusion coefficient. In addition to solid-state studies with elements, alloys, metallic oxides, and inorganic salts (all of which have important metallurgical implications), self-diffusion experiments are performed with liquids and gases. In all cases, they provide valuable information on the nature of the intermolecular forces which determine the magnitude of D.

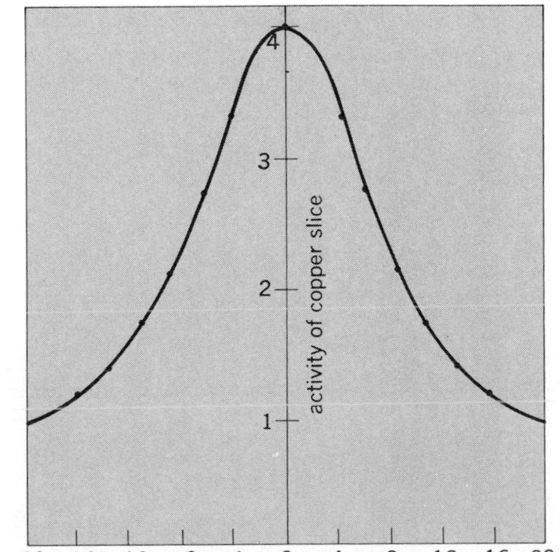

Distribution of radioactive copper after 605 min diffusion at 950°C. The full curve corresponds to $D = 8.92 \times 10^{-10}$ cm²/sec. (*Calculated from the results of M. S. Maier and H. R. Nelson, Self-diffusion of copper, AIME Trans., 147:39–47, 1942*)

Isotopic exchange reactions. When slightly soluble lead chloride crystals are mixed with an aqueous solution of sodium chloride, labeled with chlorine-36, radioactivity rapidly appears in the lead chloride as a result of exchange of chloride ions between the two compounds. Both compounds produce chloride ions on dissociation, and some chloride ions, originating from the sodium chloride, become associated with the lead ions, thereby leading to radioactivity in the lead chloride, as in reaction (2), where s indicates solid and

$$\begin{array}{l} PbCl_2\,(s) \rightarrow PbCl_2\,(soln) \qquad\qquad Na\overset{*}{Cl}\,(soln) \\ \qquad\qquad\qquad \searrow \qquad\qquad\quad \swarrow \\ \qquad\qquad\qquad\quad Pb^{2+} + \{2Cl^-, \overset{*}{Cl}{}^-\} + Na^+ \qquad (2) \\ \qquad\qquad\qquad \swarrow \\ PbCl\overset{*}{Cl}\,(s) \leftarrow PbCl\overset{*}{Cl}\,(soln) \end{array}$$

soln indicates solution. Exchange processes are proceeding continuously, but they can be detected only with isotopes, hence the term isotopic exchange reactions. Exchange reactions may occur between any two species of molecules having a common atom or group. They may be due to a dissociation process (as above) or to a collision between the two species in which chemical bonds are formed and broken (a bimolecular process), as in the exchange of radioactive sulfur (denoted by an asterisk) between sulfite and thiosulfate ions, reaction (3).

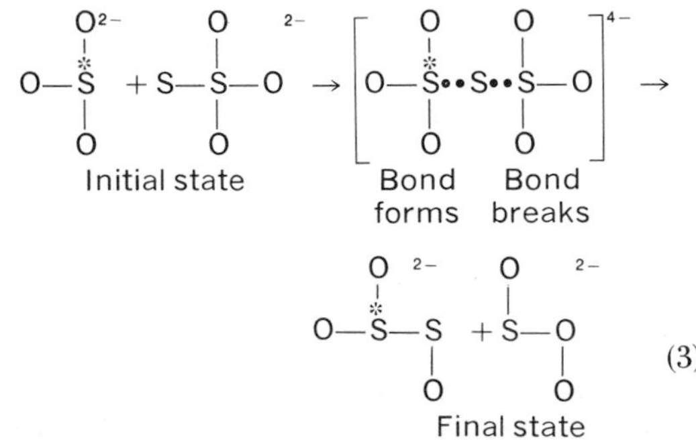

(3)

With two oxidation states of an element, exchange occurs by the transfer of an electron (an electron exchange reaction), for example, reaction (4), although in many cases the electron may actually be transferred by an atom or a group bridging between the reactants. The rates of exchange

$$Fe^{II} + \overset{*}{Fe}{}^{III} \rightarrow [Fe^{II} \overset{e}{\rightarrow} \overset{*}{Fe}{}^{III}] \rightarrow Fe^{III} + \overset{*}{Fe}{}^{II} \qquad (4)$$

reactions are measured by separating the two reactants at various times and determining the fraction F of the total radioactivity in each species. For an initially unlabeled species, F increases exponentially with time to an equilibrium value corresponding to equal concentrations (specific activities) of isotope in each species; for an initially labeled species, F will decrease exponentially with time. As in ordinary chemical kinetics, the exchange mechanism is deduced from the dependence of the exchange rate upon reactant concentrations. Rate studies of exchanges by dissociation provide information concerning ionization (including acid-base equilibria) of both solutes and solvents, the reaction of solvents with molecules (solvolysis), the thermal dissociation of gases, and the dissociation of gases on catalyst surfaces. Bimolecular exchanges are exceptionally important in studies of oxidation-reduction reactions and of substitution reactions of coordination complexes of the transitional elements. The rates of exchange by substitution provide a direct measure of the lability of these coordination complexes.

Isotopic tracer studies. Details of reaction mechanisms are provided by labeling a specific atom within a molecule. Thus, when carboxyl-labeled propionic acid is oxidized in acid dichromate, the product, as anticipated, is only radioactive carbon dioxide, reaction (5). However, the

$$CH_3CH_2\overset{*}{C}OOH \rightarrow HOOC—COOH + \overset{*}{C}O_2 \quad (5)$$

oxidation mechanism is more complex in the case of alkaline permanganate because the isotopic distribution is 75% labeled oxalate and only 25% carbonate. When the masses of the labeling and the normal isotopes differ markedly (notably with hydrogen, carbon, and oxygen), the isotopic molecule will react more slowly than the normal molecule if the reaction mechanism involves significant stretching of the chemical bond to the isotopically substituted atom. Such isotope-effect studies provide mechanistic information.

New elements. The artificially produced transuranium elements and technetium, astatine, and francium are all radioactive, and their chemistry is being elucidated with radiochemical techniques. However, this topic is normally accepted as nuclear chemistry.

Recoil studies. For a discussion of recoil studies *see* NUCLEAR CHEMISTRY. *See also* RADIATION CHEMISTRY; RADIOACTIVITY.

[DONALD R. STRANKS]

Bibliography: J. F. Duncan and G. B. Cook, *Isotopes in Chemistry*, 1968; G. Friedlander and J. W. Kennedy, *Nuclear and Radiochemistry*, 1955; R. I. Overman and H. A. Clark, *Radioisotope Techniques*, 1960.

Radioisotopic assay

An analytical technique including procedures for separating and reproducibly measuring a radioactive tracer. Separations for the assay may be made carrier-free (without stable atoms added) or with a few milligrams of added carrier by techniques such as ion exchange, solvent extraction, distillation, precipitation, electrodeposition, and isotopic exchange. Once separated, the radioisotopes may be measured as gases, liquids, or solids in Geiger-Müller, proportional, or scintillation counters, or in ionization chambers. Each detector has characteristic advantages and disadvantages and is used only after consideration of the amounts and types of radiations emitted by the radioisotope. Weak beta emitters, such as carbon-14, sulfur-35, and particularly hydrogen-3 (tritium), require special techniques to eliminate errors caused by absorption of their rays in the sample.

Geiger and proportional counters are used primarily for measuring beta rays, whereas solid scintillation detectors are used for gamma rays. Liquid scintillation counters eliminate many self-absorption problems with weak beta emitters.

Measurements are usually referred to a standard of known absolute strength. When absolute assay is required, problems of geometry, backscattering, self-absorption, counter efficiency, and the decay scheme of the radioisotope are considered. *See* ACTIVATION ANALYSIS; RADIOCHEMISTRY; RADIOMETRIC ASSAY.

[W. WAYNE MEINKE]

Radiometric assay

An analytical technique that includes procedures for measuring, by tracer methods, elements which are not themselves radioactive. It is particularly useful for small quantities of inorganic ions or complex organic compounds. A typical procedure giving sensitivities comparable to spectroscopy involves quantitative paper chromatography of traces of metals and subsequent exposure to hydrogen sulfide labeled with sulfur-35. Measurement of radiations from the highly insoluble radioactive sulfides is more sensitive than color tests. Complex organic molecules are often treated with iodine-131-labeled pipsyl chloride in this same manner.

In radiometric titrations a radioactive tracer is used to determine the equivalence point in a volumetric determination, where a highly insoluble compound of definite composition is formed. At least one of the solutions used in the titration must be radioactive, because progress of the titration is followed by plotting the radioactivity of the solution versus volume of reagent added. The equivalence point is determined by the intersection of two straight lines on this plot, because a definite change in solution activity appears after equivalent amounts of reagents have reacted. Submicrocurie levels of radioactive tracer suffice for most applications. *See* RADIOCHEMISTRY; RADIOISOTOPIC ASSAY; TITRATION.

[W. WAYNE MEINKE]

Radium

A chemical element, Ra, with atomic number 88. The atomic weight of the most abundant naturally occurring isotope is 226. Radium is a rare radioactive element found in uranium minerals to the ex-

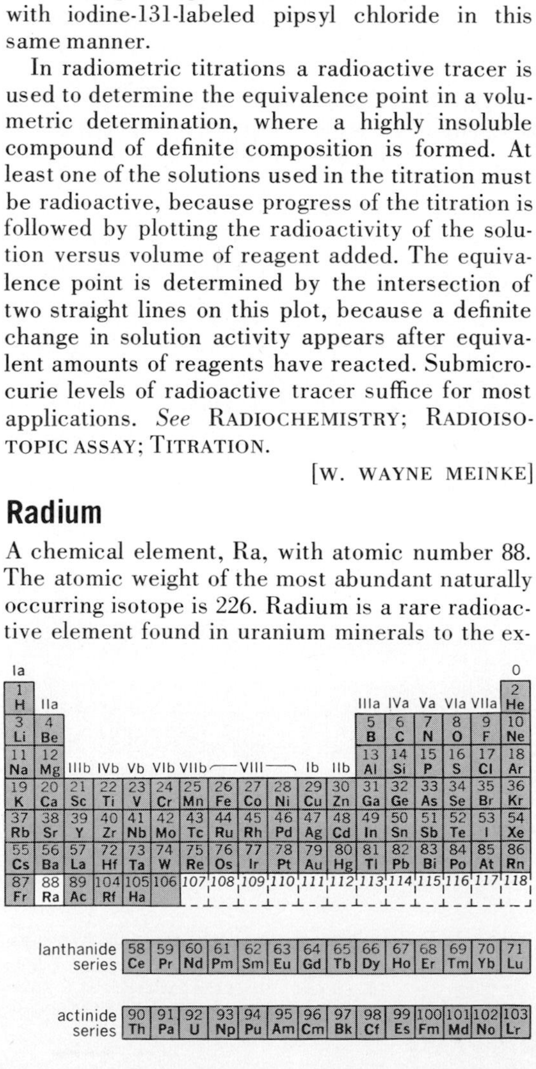

Table 1. Radium isotopes

Mass	Occurrence	Type of decay	Half-life	Energy of radiation, MeV
213	Synthetic	α	2.7m	6.90
219	Synthetic	α	~0.001s	8.0
220	Synthetic	α	0.03s	7.43
221	Synthetic	α	30s	6.71
222	Synthetic	α	38s	6.51
223	Natural, AcX	α	11.2d	5.730 (9%), 5.704 (53%), 5.596 (24%), 5.528 (9%), 5.487 (2%), 5.419 (3%)
		γ		0.026–0.44
224	Natural, ThX	α	3.64d	5.681 (95%), 5.448 (4.6%), 5.194 (0.4%)
		γ		0.241
225	Synthetic	β^-	14.8d	0.31
226	Natural, Ra	α	1622y	4.791
		γ		0.186
227	Synthetic	β^-	41.2m	1.31
		γ		0.291, 0.498
228	Natural, $MsTh_1$	β^-	6.7y	0.012
		γ		0.03
229	Synthetic	β^-(?)	~1m	?
230	Synthetic	β^-	1h	1.2

tent of 1 part for about every 3,000,000 parts of uranium.

Chemically, radium is an alkaline-earth metal having properties quite similar to those of barium. Radium is important because of its radioactive properties and is used primarily in medicine for the treatment of cancer, in the manufacture of self-luminous paints, in atomic energy technology for the preparation of standard sources of radiation, as a source for actinium and protactinium by neutron bombardment, and in certain metallurgical and mining industries for preparing gamma-ray radiographs.

Properties and uses. Biologically, radium behaves as a typical alkaline-earth element, concentrates in bones by replacing calcium and, as a result of prolonged irradiation, causes anemia and cancerous growths. The tolerance dose for the average human being has been estimated at a total of 1 μg of radium fixed within the body.

Because radiations from radium and its decay products preferentially destroy malignant tissue, radium has been used to check the growth of cancer. For this use pure radium compounds are sealed in tubes or needles; or radon, the gaseous decay product of radium, is pumped into small tubes. Radon is safer to administer and its very short half-life decreases the danger of overexposure.

In industrial radiography radium sulfate is generally sealed within concentric spheres or cylinders, the inner one of silver and the outer of aluminum or steel. In each container the quantity of radium ranges from 25 to 1000 mg. Radiographs are used to determine the thickness of the catalyst bed in petroleum cracking units, for the detection of internal flaws in metal castings and other solids, and in the continuous measurement and control of the thickness of metals such as copper, aluminum, brass, and tin in rolling mills.

The use of radium in luminescent paints for watch, clock, and meter dials, and signs visible in the dark depends on its α-radiation striking a scintillator such as zinc sulfide. A standard neutron source for use in research, in analyses of materials by neutron activation, and in the radio-logging of oil wells is prepared from mixtures of radium and beryllium.

Occurrence. Thirteen isotopes of radium are known (Table 1); all are radioactive; four occur naturally; the rest are produced synthetically. Only Ra^{226} is technologically important. It is distributed widely in nature, usually in exceedingly small quantities. The most concentrated source is pitchblende, a uranium mineral occurring in large deposits in the Congo, Canada, and elsewhere, and containing about 0.4 g of radium per ton of uranium. Lower concentrations are found in numerous other minerals such as carnotite from the Colorado Plateau. Detectable traces of radium are found in river, ocean, and other natural waters.

Metallurgy. In the processing of pitchblende for uranium, radium is usually recovered with barium as sulfates in an acid-insoluble residue. A simplified, but typical, method for the extraction of radium from this residue is shown in the illustration. After treatment with a hot concentrated alkali carbonate solution to remove sulfate and the amphoteric metals, the resulting carbonate filter cake is dissolved in dilute hydrochloric acid. The solu-

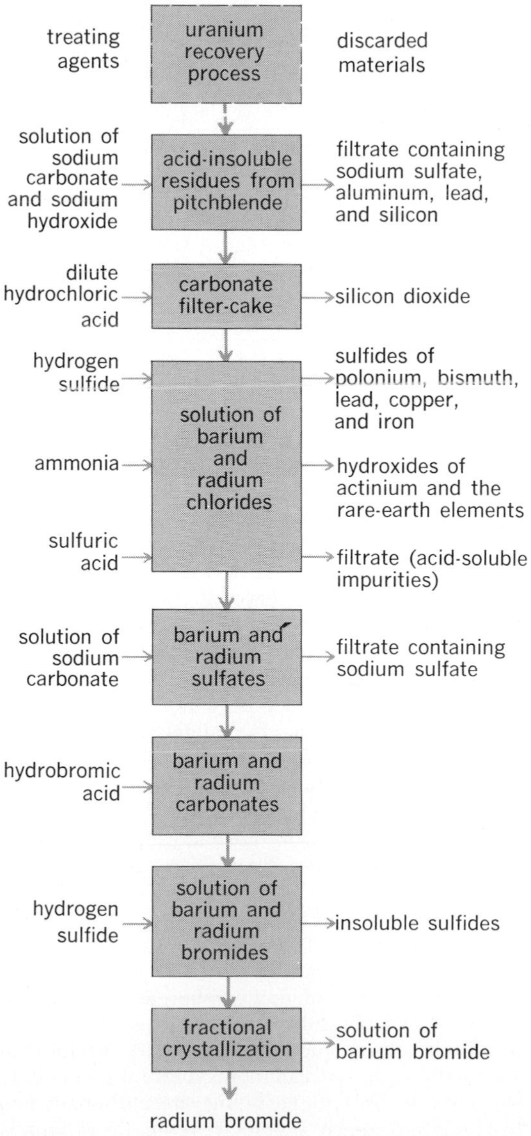

Extraction of radium from pitchblende residues.

tion is treated first with hydrogen sulfide to remove insoluble sulfides, next with ammonia to eliminate metal hydroxides, and finally with sulfuric acid to reprecipitate the sulfates free of acid-soluble impurities. Additional purification is obtained by repetition of the above cycle. Finally, the barium and radium are separated from each other by fractional crystallization until radium bromide of at least 90% purity is obtained.

The winning of the free metal from its compounds has been achieved by electrolysis of a radium chloride solution using a mercury cathode and a platinum-iridium anode. The resulting amalgam is thermally decomposed in a hydrogen atmosphere leaving a residue of pure radium metal.

When freshly prepared, radium metal has a brilliant white metallic luster. Some of its physical properties are shown in Table 2. Chemically, the metal is highly reactive. It blackens rapidly on exposure to air because of the formation of a nitride. Radium reacts readily with water, evolving hydrogen and forming a soluble hydroxide.

Table 2. Physical properties of radium

Property	Value
Atomic number	88
Atomic weight	226.05
Valence states	0, +2
Specific gravity	6.0 at 20°C
Melting point	700°C*
Boiling point	~1140°C
Ionic radius, Ra^{++}	2.45 A (estimated)
Atomic parachor	~140
Decomposition potential†	1.718 volt
Heat of formation of oxide	130 kcal/mole
Magnetic susceptibility	Feebly paramagnetic

*Value of 960°C has also been reported.
†Of normal solutions of radium salts with respect to the normal calomel electrode.

Principal compounds. When first prepared, nearly all radium compounds are white, but they discolor on standing because of intense radiation. Radiation causes a purple or brown coloration in glass on long contact with radium compounds. Eventually the glass crystallizes and becomes crazed. Radium salts ionize the surrounding atmosphere, thereby appearing to emit a blue glow, the spectrum of which consists of the band spectrum of nitrogen. Radium compounds will discharge an electroscope, fog a light-shielded photographic plate, and produce phosphorescence and fluorescence in certain inorganic compounds such as zinc sulfide. The emission spectrum of radium compounds is similar to those of the other alkaline earths; radium halide imparts a carmine-red color to a flame.

All known salts of radium are isomorphous with, and similar to, the corresponding barium salts. Radium forms a sparingly soluble sulfate, chromate, carbonate, and iodate, whereas the chloride, bromide, nitrate, and hydroxide are soluble in water. Most radium compounds are more insoluble in water than the corresponding barium compound, the notable exceptions being the carbonate and hydroxide. A nearly specific reaction for radium is the precipitation of the nitrate from 80% nitric acid. This separates radium quantitatively from all metals except barium, strontium, and lead. Because of its chemical similarity to radium, barium is frequently used as a carrier for minute quantities of radium. Separation of these two elements is extremely difficult, and requires fractional crystallization or precipitation, ion exhange, or solvent extraction with chelating agents. The greatest separation per fractionation step is obtained with bromides or chromates, somewhat less with chlorides and carbonates, and only slight separation with sulfates and nitrates. Complete separation of radium from barium and other metallic impurities has been accomplished by ion exchange on synthetic organic resins, but the method is not easily adapted to separation of macro quantities of radium because the intense radiation causes degradation of the resin which usually results in formation of insoluble radium compounds within the ion-exchange column.

Analysis. Large samples of pure radium are assayed by weighing the chloride or bromide after fusion to remove water, by γ-ray counting and comparison with standard samples, provided that all samples are nearly at radioactive equilibrium, or by calorimetric measurements. Small samples of carrier-free radium may be analyzed by α-ray counting after separating or making correction for the α-emitting activities of daughter elements or by use of an α-ray spectrometer, an instrument which separates α-rays of different energies and permits determination of only those produced by radium. Radium samples in any state of purity may be analyzed by the classical emanation method which depends on the separation and determination of radon. Although the method is indirect, it is extremely sensitive. *See* ALKALINE-EARTH METALS; RADIOACTIVITY; RADON.

[MURRELL L. SALUTSKY]

Bibliography: H. W. Kirby and M. L. Salutsky, *The Radiochemistry of Radium*, National Academy of Sciences–National Research Council, NAS-NS 3057, 1966; J. Sedlet, Radon and radium, in I. M. Koltoff and P. J. Elving (eds.), *Treatise on Analytical Chemistry*, vol. 4, pt. 2, 1966; J. Selman, *Fundamentals of X-Ray and Radium Physics*, 1979; V. M. Vdovenko and Yu. V. Dubasov, *Analytical Chemistry of Radium*, 1975.

Radon

A chemical element, symbol Rn, atomic number 86. Radon is produced as a gaseous emanation from the radioactive decay of radium. The element is highly radioactive and decays by the emis-

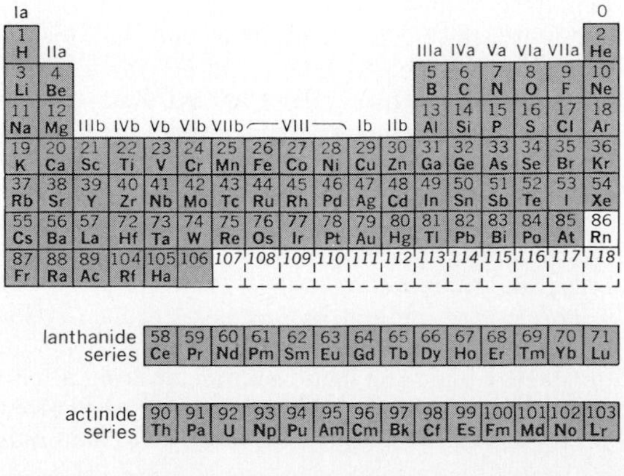

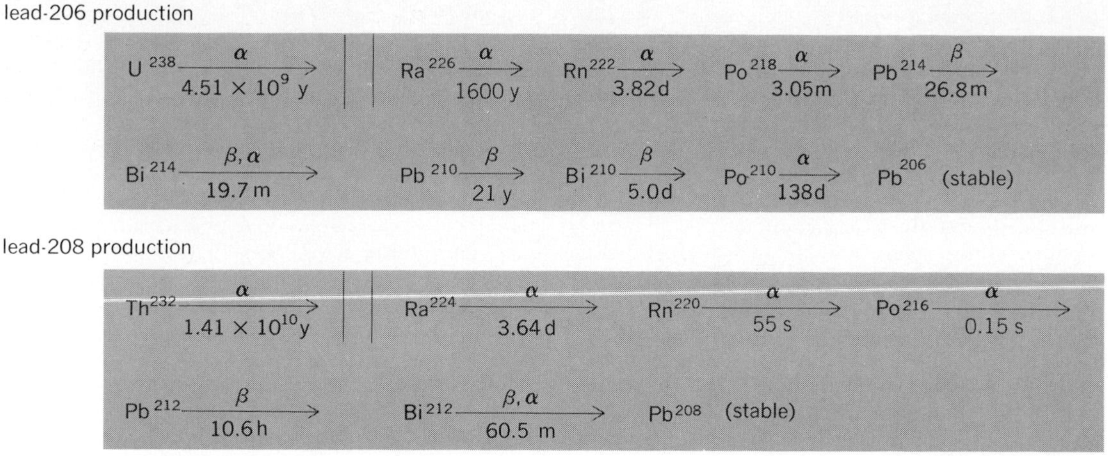

Some of the decay relationships for radon-222 and radon-220.

sion of energetic alpha particles.

Radon is used loosely as the name of element 86 and all its isotopes, although the name emanation (symbol Em) is sometimes preferred for the element. The element radon is the heaviest of the noble, or inert, gas group and thus is characterized by chemical inertness. All isotopes are radioactive with short half-lives.

OCCURRENCE AND SYNTHESIS

Radon is found in natural sources only because of its continuous replenishment from the radioactive decay of longer-lived precursors in minerals containing uranium or thorium. As shown in the illustration of the decay relationships radon (mass number 222, half-life 3.82 days) occurs in the uranium-radium series; thoron (mass number 220, half-life 54.5 sec) is a member of the thorium family. Actinon (mass number 219, half-life 3.92 sec) is found in the actinium series. All three decay by the emission of energetic α-particles. Thoron was discovered in 1899 by R. B. Owens and E. Rutherford, who noted that some of the radioactivity of thorium preparations could be removed in a stream of gas. F. E. Dorn discovered radon in radium preparations in 1900, and F. O. Giesel observed the formation of actinon in actinium compounds in about 1902. The remarkable continuous formation of radioactive gaseous emanations in such samples did much to stimulate thinking on the true nature of radioactivity in the early study of natural radioactivity. *See* RADIOACTIVITY.

The radon formed in minerals is largely trapped within the mineral, but some diffuses out. Radon is detected in surface waters and in streams. The atmosphere near the ground contains radon which has seeped from soil and rocks, all of which contain minute traces of uranium.

In addition to the three natural isotopes, 22 isotopes have been synthesized by nuclear reactions of artificial transmutation in cyclotrons and linear accelerators, but none of these is as long-lived as Rn^{222}. The most stable of them is Rn^{211}, one-half of which disintegrates in 16 hr. It is certain that no isotope of radon will ever be found which is stable or long-lived.

Properties. Any surface exposed to Rn^{222} becomes coated with an active deposit which consists of a group of short-lived daughter products, including radium A, B C, C′, and C″. The radiations of this active deposit include energetic alpha, beta, and gamma rays. Particularly noteworthy are the penetrating gamma radiations of RaC, which are of practical use in radiotherapy, radiography, and other purposes.

The ultimate decay products of radon following the rapid decay of the active deposit include radium D (lead-210), polonium, and finally, stable lead-206. Thoron and actinon also lay down an active deposit on surfaces exposed to them, but thoron, actinon, and their decay products have not had the same general interest and usefulness because of their shorter lifetimes.

Radon possesses a particularly stable electronic configuration ($5s^2 5p^6 5d^{10} 6s^6$; $^1S_0$), which gives it the chemical properties characteristic of noble-gas elements. Its properties are those to be expected by extrapolation from the other noble-gas elements: helium, neon, argon, krypton, and xenon. Some physical properties are the following: boiling point −65°C; melting point variously reported as −71°C and −113°C; ionization potential, 10.6 eV; and ionization potential of Rn^+ ion, 19.9 eV. The spectrum of radon has been extensively studied, and resembles that of the other inert gases. Radon is readily adsorbed on charcoal, silica gel, and other adsorbents, and this property can be used to separate the element from gaseous impurities. The radon is desorbed from charcoal by heating to about 350°C. Radon is appreciably soluble in water and in organic liquids.

Analysis. Radon samples can be analyzed by direct gas counting of α-particles in an ionization chamber; extremely minute amounts can be measured in this way. Radon can also be determined by measuring in suitable counters the gamma radiation of its short-lived daughters. Millicurie amounts can be estimated by comparing the strength of the gamma radiation with that from a calibrated radium standard. *See* INERT GASES; RADIOACTIVITY.

[EARL K. HYDE]

DISTRIBUTION

More than 25 isotopes of radon have been identified. Three of these, ^{222}Rn (radon), ^{220}Rn (thoron), and ^{219}Rn (actinon), occur in nature as members of the uranium, thorium, and actinium series respectively. The rocks and soils of the Earth's crust contain approximately 3 parts per million of ^{238}U, the long-lived head of the uranium series; 12 ppm of ^{232}Th, the head of the thorium series; but only about 0.02 ppm of ^{235}U, the long-lived member of the actinium group. Portions of the uranium and thorium series are shown in the illustration together with their half-lives and modes of decay. The actinium series is of minor importance both because of the relatively low abundance of ^{235}U and because of the short half-life (3.9 sec) of ^{219}Rn (actinon), the noble gas member of the series. The radon isotopes ^{222}Rn and ^{220}Rn are produced in proportion to the amount of the parent present. Some of the newly formed radon atoms which originate in or on the surface of mineral grains escape into the soil gas, where they are free to diffuse within the soil capillaries. Some of the radon atoms eventually find their way to the surface, where they become a part of the atmosphere. Even though thorium (^{232}Th) is generally more abundant than uranium in the Earth's crust, the probability for decay is smaller; hence, the production rate of ^{222}Rn and ^{220}Rn in the soil is roughly the same. Much of the ^{220}Rn decays before reaching the Earth's surface due to its short half-life of 55 sec.

The flux or exhalation of ^{222}Rn from soil to air averages about 7500 atoms per square meter each second. A comparable figure for ^{220}Rn (thoron) is 120 atoms per square meter each second. The longer half-life of ^{222}Rn relative to ^{220}Rn also favors its survival as a trace component of the atmosphere. The ocean surface gives up only about 50 atoms of ^{222}Rn to the atmosphere per square meter each second; hence, the oceans contribute only about 2% of the global burden of ^{222}Rn, estimated at 50 curies per second.

Atmosphere. When radon (^{222}Rn or ^{220}Rn) passes from soil to air, it is mixed throughout the lower atmosphere by eddy diffusion and the prevailing winds. Typical concentrations of ^{222}Rn at 1 m above ground are about 0.2 picocurie per liter (3500 atoms per liter), but this may be exceeded by a factor of 10 or more when radon, like atmospheric pollutants originating at or near the surface, is trapped by temperature inversions. Mean radon levels are found to be higher during those times of year when atmospheric stability is the greatest such as may occur during the fall months. Other approximate ranges of ^{222}Rn concentrations in picocuries per liter are: air over the oceans, 0.0005–0.005; indoor air, 0.5; soil air, caves, 10–300; unventilated uranium mines, 1000–100,000.

The daughter nuclides of ^{222}Rn can be divided into a short-lived group which consists of isotopes of polonium, lead, and bismuth, all having half-lives of one-half hour or less, and a long-lived group made up of ^{210}Pb with a half-life of 21 years and its two immediate daughters, ^{210}Bi (5.0 days) and ^{210}Po (138 days). The decay products of radon, both short- and long-lived, are positively charged heavy metal ions at the instant of formation. As ions or atoms, they soon attach to airborne particulate matter, where they may be removed by decay, by gravitational settling, by action of rain, or by ion migration under the influence of natural atmospheric electric fields. As a result of these removal processes, the short-lived radon daughters are generally present in the air in less than equilibrium amounts—a figure in the range of 10–60% is commonly encountered. ^{210}Pb will in general return to the Earth before it decays, since the residence time of the aerosols to which the daughters attach is of the order of 2 weeks.

While thoron (^{220}Rn) is present typically to the extent of only about 1 atom per liter at 1 meter aboveground, its activity is approximately the same as that of ^{222}Rn because of its much larger decay constant. The isotope that controls the decay character of the ^{220}Rn daughter is the 10.6-hr ^{212}Pb. Because of the rapid decay of ^{220}Rn, it is found mainly near ground level, and its concentration fluctuates rapidly with atmospheric stability. The ^{212}Pb is distributed much more uniformly with height at activity levels only a few percent of that of ^{222}Rn.

Radon and its daughters play an important role in atmospheric electricity. Near the Earth's surface almost half of the ionization of the air is due to ^{220}Rn and ^{222}Rn and their daughter products. The alpha emitters from these chains typically produce about 10,000 ion pairs per second per liter.

Water. Radon is readily soluble in water. Since ground and surface waters are in close contact with soil and rocks containing small quantities of radium, it is not surprising to find radon in public water supplies. The United Nations Scientific Committee on Effects of Atomic Radiation report of 1966 gives typical concentrations of ^{222}Rn in surface water as 10 pCi/liter and in ground water from 100 to 1000 pCi/liter. Water from deep wells and mineral springs may exceed 100,000 pCi/liter. Rivers carry dissolved ^{226}Ra (1600 years) to the oceans, where radioactive equilibrium between ^{226}Ra and ^{222}Rn is approached. Both nuclides may be present in sea water to the extent of about 0.1 pCi/liter. A deficiency of ^{222}Rn exists in water near the air-sea interface, and an excess exists near the ocean floor. ^{222}Rn atoms cannot escape readily from the water surface, with the result that marine air masses contain only about 1% as much ^{222}Rn as air over the continents.

Applications. Radon sealed in small glass or metal tubes has been used extensively in the past for the treatment of cancer. Except for special uses, these radon needles have been replaced by other types of radiation sources in these applications. Radon mixed with beryllium powder is used as a source of neutrons in some laboratories.

The radon isotopes, ^{220}Rn and ^{222}Rn, are used widely in the study of gaseous transport processes both in the underground environment and in the atmosphere. Although ordinary molecular diffusion accounts for most of the transport from soil to air, changes in atmospheric pressure, soil saturation, and pressure gradients across cracks and fissures all furnish mechanisms whereby radon can escape from soil to air. Radon accumulates to high levels of the order of 100 pCi/liter or more in caves unless natural or artificial ventilation occurs. Changes in ^{222}Rn concentrations in spring and well water and in soil and rocks have been suggested as a means of predicting earthquakes.

Turbulent diffusion in the lower atmosphere accounts for most of the measured profiles of radon and daughter products. Long-lived isotopes of radon have been used for determination of aerosol residence times in the lower atmosphere, and as tracers in the motion of air mass systems from the Sahara across the North Atlantic, from continental regions to the trade winds in Hawaii, and in the Southern Hemisphere from the continents to the Antarctic. ^{220}Rn, because of its short half-life, 55 sec, has been used effectively in studies of surface mixing and the atmospheric electrical characteristics within the first meter of the Earth's surface.

Health effects. The tendency of the decay products of radon to attach to aerosols means that these nuclides will be inhaled and deposited in the bronchial epithelium and lungs. The daughter products, therefore, make up the major part of the internal radiation dose from radon. In recognition of this potential hazard, a maximum permissible concentration (MPC) of ^{222}Rn in air breathed by workers in uranium mines and mills and by others similarly employed has been established by the National Council on Radiation Protection and Measurements as 30 pCi/liter. The MPC for the general population is 10 pCi/liter.

The Walsh Healy Act of 1968 established a protection guide for uranium workers at four working level months (WLM) per year, where the working level (WL) is defined as any combination of radon daughters in 1 liter of air that will produce 1.3×10^5 MeV of alpha energy. This would be equivalent to a ^{222}Rn concentration of 100 pCi/liter in equilibrium with its short-lived alpha-emitting progeny. The WLM is a cumulative exposure to a ^{222}Rn concentration of one WL for 170 working hours in one month. Four WLM per year is consistent with a ^{222}Rn level of 30 pCi/liter, assuming equilibrium with the short-lived daughters. Since radon levels up to several thousand picocuries per liter may be encountered in uranium mines, proper ventilation systems and other precautions must be taken to protect the miners. The dose to the lung from ^{212}Pb (10.6 hr) and its daughters in the thorium series is not believed to add appreciably to that received from the ^{222}Rn series.

^{210}Pb, the 22-year nuclide in the ^{222}Rn decay chain, is distributed widely in the biosphere. Tobacco leaves, for example, filter the air circulating over them, trapping ^{210}Pb with the result that cigarettes contain appreciable quantities of ^{210}Pb and its alpha-emitting daughter, ^{210}Po. A similar buildup of these nuclides occurs in the lichens on which reindeer feed. Cigarette smokers and people of certain northland countries who subsist on reindeer meat are two conspicuous groups for whom the long-lived daughters of ^{222}Rn may present a potential health hazard.

In addition to the problem created by radon for miners, uranium mines and tailings piles in some of the western states of the United States are a source of additional radon in the atmosphere and, unless precautions are taken, may add substantially to the radon concentration in drinking water drawn from surface runoff or wells in nearby communities.

There are areas in the world where natural levels of radium and thorium, and hence ^{222}Rn and ^{220}Rn, are from 10 to 100 times above those considered normal. Examples are found in Brazil and the Kerala region of India. Attempts are being made to assess the effects of these anomalies upon the human population in these areas.

[M. WILKENING]

Bibliography: J. A. S. Adams and W. M. Lowder (eds.), *The Natural Radiation Environment II*, vols. 1 and 2, 1972, and *The Natural Radiation Environment III*, Conference Proceedings, 1978; M. Eisenbud, *Environmental Radioactivity*, 2d ed., 1973; H. Israël, *Atmospheric Electricity*, vols. 1 and 2, 1973; C. E. Junge, *Air Chemistry and Radioactivity*, 1963; National Council on Radiation Protection and Measurements, *Maximum Permissible Body Burdens and Maximum Permissible Concentration of Radionuclides in Air and in Water for Occupational Exposure*, NBS Handb. no. 69, 1959; E. R. Reiter, *Atmospheric Transport Processes*, pt. 4: *Radioactive Tracers*, 1978; United Nations Scientific Committee on Effects of Atomic Radiation, *Sources and Effects of Ionizing Radiation*, 1966, 1977; F. Weigel, Radon, *Chem. Zeit.*, 102:287, 1978.

Raman effect

A phenomenon observed in the scattering of light as it passes through a material medium, whereby the light suffers a change in frequency and a random alteration in phase. Raman scattering differs in both these respects from Rayleigh and Tyndall scattering, in which the scattered light has the same frequency as the unscattered and bears a definite phase relation to it. The intensity of normal Raman scattering is roughly one-thousandth that of Rayleigh scattering in liquids and smaller still in gases. *See* TYNDALL EFFECT.

Discovery. Because of its low intensity, the Raman effect was not discovered until 1928, although the scattering of light by transparent solids, liquids, and gases had been investigated for many years before. Prompted by A. H. Compton's observation of frequency changes in x-rays scattered by electrons (Compton effect), the Indian physicists C. V. Raman and K. S. Krishnan examined sunlight scattered by a number of liquids. With the help of complementary filters, they found that there were frequencies in the scattered light that were lower than the frequencies in the filtered sunlight. They then showed, by using light of a single frequency from a mercury arc, that the new frequencies in the scattered radiation were characteristic of the scattering medium. Within a few months of Raman and Krishnan's first announcement of their discovery, the Soviet physicists G. Landsberg and L. Mandelstam communicated their independent discovery of the existence of the effect in crystals. In Soviet literature the phenomenon is referred to as combination scattering, and not Raman effect.

The development of the laser has led to a resurgence of interest in the Raman effect and to the discovery of a number of related phenomena. A beam of laser radiation is intense, polarized, and coherent; it can be made monochromatic, small in diameter, and highly collimated. The laser is therefore nearly ideal for the production of the Raman effect, and other kinds of sources are seldom employed. Many different wavelengths in the visible spectrum and adjacent regions are available. The argon-ion and krypton-ion lasers are most commonly used, since they have high continuous-wave power (1 to 10 W), but tunable dye lasers are also often employed in excitation of resonance Raman scattering.

Raman spectroscopy. Raman scattering is analyzed by spectroscopic means. The collection of new frequencies in the spectrum of monochromatic radiation scattered by a substance is charac-

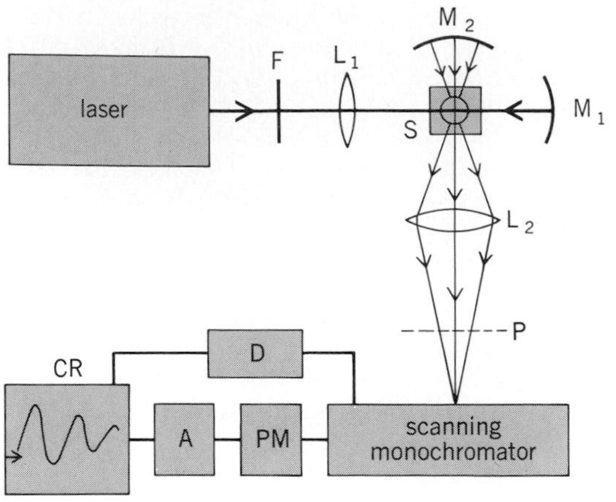

Fig. 1. Laser-Raman spectroscopic system.

teristic of the substance and is called its Raman spectrum. Although the Raman effect can be made to occur in the scattering of radiation by atoms, it is of greatest interest in the spectroscopy of molecules and crystals.

Because of the laser beam's small diameter and high collimation, it can easily be used to excite the Raman effect. A typical optical arrangement is shown in Fig. 1. Monochromatic radiation from the laser impinges on the sample S in an appropriate transparent cell. It may be desirable to condense or expand the laser beam by means of a lens system L_1 and to remove unwanted radiation from the beam by a narrow-band optical filter F. A concave mirror M_1 can return unscattered radiation for a second passage through the sample.

Raman scattering is approximately uniform in all directions and is usually studied at right angles, as shown in Fig. 1. In this way the intense radiation of the laser beam interferes least with the observation of the weak scattered light. This light is collected by a lens system L_2 and focused on the slit of a scanning monochromator, which analyzes it spectroscopically. As the spectrum is scanned, the dispersed radiation from the monochromator is detected by a photomultiplier PM, further amplified and processed electronically at A, and then recorded by a strip-chart recorder CR. The recorder is driven in synchronism with the monochromator by a suitable mechanism D. The concave mirror M_2 may be used to augment the amount of scattered radiation by collecting light scattered at −90° and returning it to the +90° direction. The polarization characteristics of the scattered radiation are frequently of interest, especially since the laser radiation itself is linearly polarized. An analyzing device for evaluating the degree of polarization of the scattered radiation may be inserted at point P.

The appearance of a photoelectrically recorded Raman spectrum of liquid carbon tetrachloride as excited by the red line of the helium-neon laser at 632.8 nm (6328 A, power incident on the sample of about 50 mW) is shown in Fig. 2. Intensity of the scattered light on an arbitrary scale is plotted vertically against the wave number in cm^{-1} measured with respect to the wave number of the exciting line taken as zero. For convenience, it is

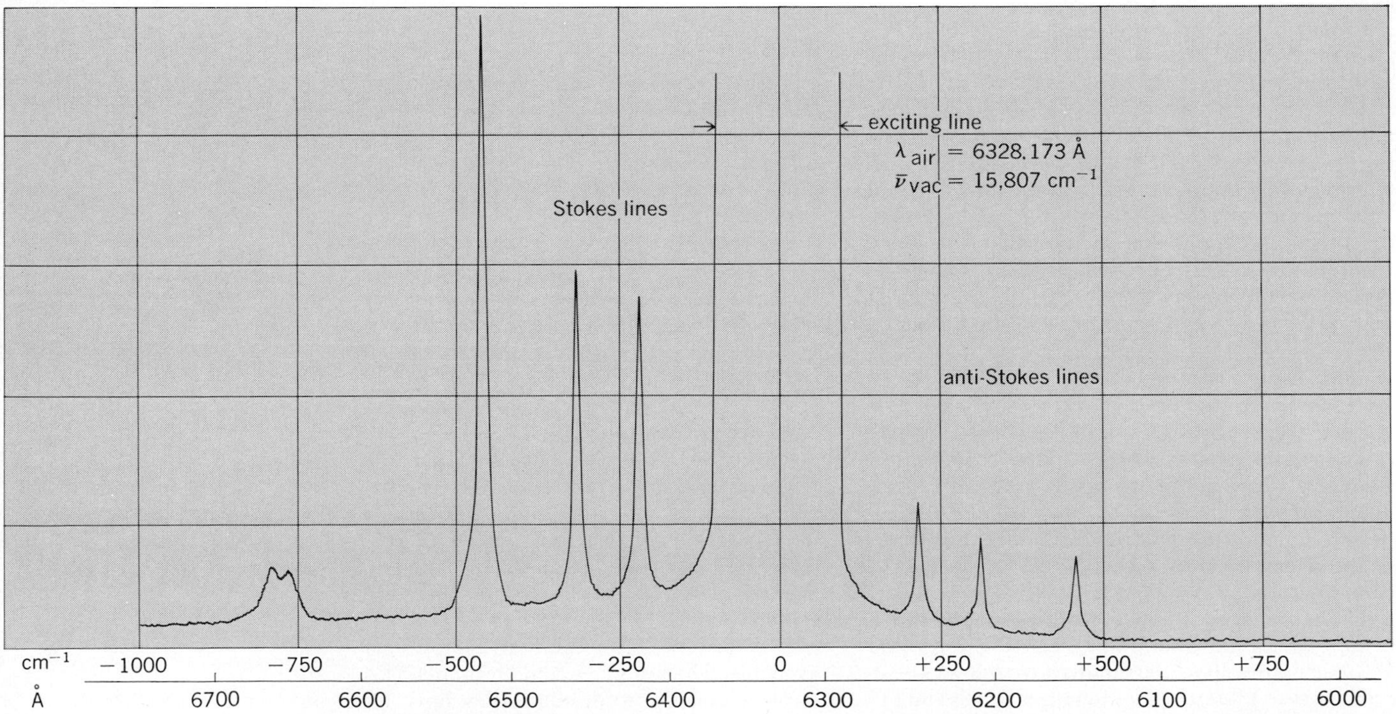

Fig. 2. Photoelectric recording of the Raman spectrum of carbon tetrachloride excited by He-Ne laser line at 6328 A. Intensity of the radiation is recorded vertically against the horizontal wave-number scale (cm^{-1}) measured from the exciting line as zero. The Rayleigh-scattered exciting line is three orders of magnitude more intense than the Raman lines, and its maximum is therefore far off-scale. The Stokes lines appear at lower frequencies, and the less intense anti-Stokes lines at higher frequencies, than those of the exciting line. The lower scale shows wavelengths in angstroms.

usual to express the data of Raman spectroscopy in cm^{-1} rather than frequency units (s^{-1}). (Frequency ν in $s^{-1} = c\bar{\nu}$ in cm^{-1}, where c is the velocity of light in vacuum in cm/s.)

A spectrum of the most intense line in Fig. 2 is shown at higher resolution (smaller spectral slit width $\Delta\bar{\nu}$) in Fig. 3. The line is seen to consist of several closely spaced components. These result from the presence of the two isotopes of chlorine, ^{35}Cl and ^{37}Cl, which produce five isotopic species $C^{35}Cl_n{}^{37}Cl_{4-n}$, $n=0,1,2,3,4$. The line due to the least abundant species, $n=0$, is not visible, but the other four are readily identified. *See* SPECTROSCOPY.

Theory. The mechanism of the Raman effect can be envisaged either by the corpuscular picture of light or from the point of view of the wave theory. Both pictures merge in the basic quantum theory of radiation. The corpuscular model of light scattering envisages light quanta or photons as particles which have linear and angular momenta. On passing through a material medium, these particles collide with atoms or molecules. If the collision is elastic, the photons bounce off the molecules with unchanged energy E and momentum, and hence with unchanged frequency ν. Such a process gives rise to Rayleigh scattering. If the collision is inelastic, the photons may gain energy from, or lose it to, the molecules. A change ΔE in the photon energy by Planck's relationship: $E = h\nu$, must produce a change in the frequency $\Delta\nu = \Delta E/h$. Such inelastic collisions are rare compared to the elastic ones, and the Raman effect is correspondingly much weaker than Rayleigh scattering.

In the wave picture of the effect, the electromagnetic waves which constitute the incoming monochromatic radiation sweep through the material medium. Since the atoms and molecules composing the medium are made up of negatively charged electrons and positively charged nuclei, the electric field of the light waves sets the electrons to oscillating, chiefly with the frequency of the incoming radiation. The oscillating electrons re-create the alternating electric field of the incoming light, thus passing the light wave along through the medium. This process is analogous to the elastic collisions given by the corpuscular picture.

The ability of the electrons and nuclei in a molecule to be displaced by an electric field is called the molecular polarizability α. It is not a simple property of the molecule, but depends in a complicated way on the frequency of the electric field, on the orientation of the molecule, and on the internal motions of the nuclei and electrons. Thus the molecular polarizability α varies periodically with molecular rotation and vibration, and thereby the effect of a light wave on the electrons and nuclei of a molecule can be changed.

When a monochromatic light wave sweeps through a transparent medium containing rotating and vibrating molecules, most of the wave is re-created unchanged by the oscillating electrons, but because of the periodic changes produced in α by rotation and vibration, new frequencies are added to the light wave. The appearance of these new frequencies, whose values are determined by the rotational and vibrational energies of the molecules, is analogous to the result of the inelastic collisions of the corpuscular model. For the wave picture of the Raman effect, the quantity α is the basic quantity. The intensity of the Raman effect depends on the magnitude of the changes produced in α by molecular rotation and vibration, and the number and values of new frequencies (usually expressed as frequency shifts $\Delta\nu$ from the original monochromatic frequency) depend on the variation of α with the frequencies of rotation and vibration.

The temperature of the scattering molecules is an additional factor which affects the intensity of Raman frequencies higher than the exciting frequency (the anti-Stokes lines of Fig. 2). The anti-Stokes lines, having higher frequencies, correspond to photons which have higher energy than that of the exciting light, and this energy must come from the molecules. If the molecules do not have any available vibrational or rotational energy, that is, if they are at the absolute zero of temperature, there is no possibility of inelastic collisions in which energy is transferred from a molecule to a photon. So, anti-Stokes lines vanish at absolute zero. At nonzero temperatures the intensity ratio of an anti-Stokes line to a Stokes line is approximated by the ratio of the number of molecules

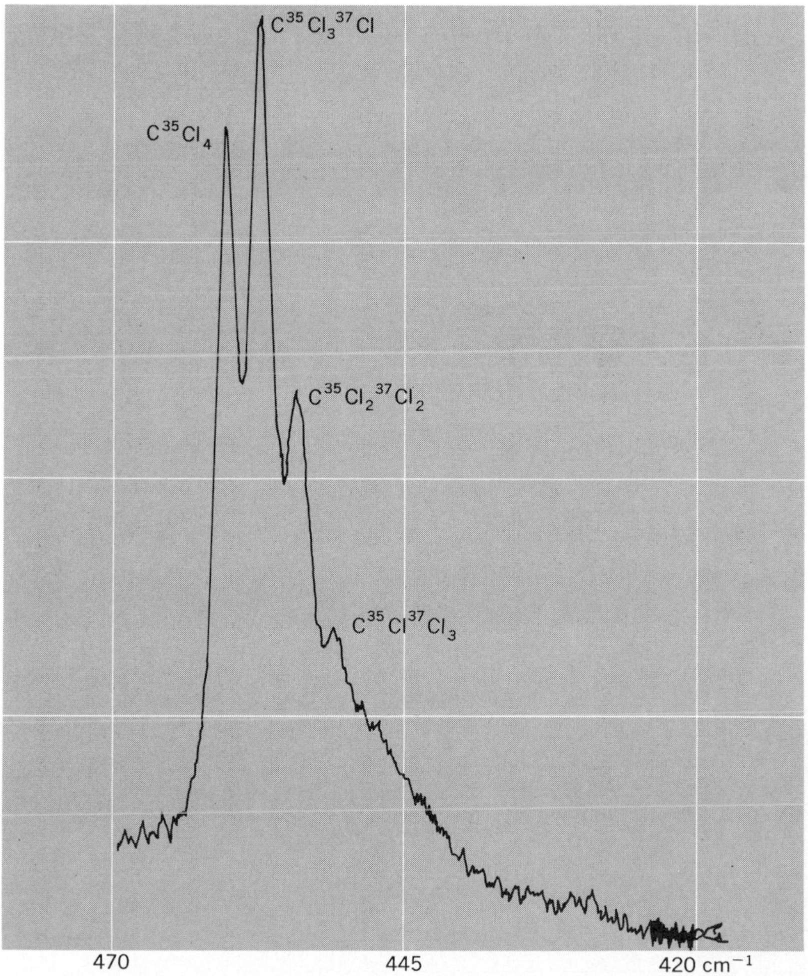

Fig. 3. The Stokes line at −460 cm^{-1} in the Raman spectrum of carbon tetrachloride. This spectrum, taken at about 10 times better resolution than that of Fig. 2, shows lines due to four of five isotopic species resulting from the 3:1 ratio of the chlorine isotopes ^{35}Cl and ^{37}Cl. The instrumental wave-number scale requires a calibration correction to increase the wave numbers by 1.4 cm^{-1}.

which can give up the corresponding energy to the number which can accept it from the light wave.

Special forms. The development of lasers resulted in the discovery of a number of kinds of Raman scattering.

Resonance Raman effect. When the exciting radiation falls within the frequency range of a molecule's absorption band in the visible or ultraviolet spectrum, the radiation may be scattered by two different processes, resonance fluorescence or the resonance Raman effect. Both these processes give much more intense scattering than the normal nonresonant Raman effect. Resonance fluorescence differs from the resonance Raman effect in that the absolute frequencies of the fluorescent spectrum do not shift when the exciting radiation's frequency is changed, so long as the latter does not move outside the absorption band. The absolute frequencies of the resonance Raman effect, on the contrary, shift by exactly the amount of any shift in the exciting frequency, just as do those of the normal Raman effect. Thus the main characteristic of the resonance as compared to the normal Raman effect is its intensity, which may be greater by two or three orders of magnitude.

The resonance Raman effect was anticipated by G. Placzek in 1934 in his pioneering development of the polarizability theory of Raman scattering. It was actually observed before the discovery of lasers, but tunable lasers are the most effective sources for the study of its various aspects. A typical resonance Raman spectrum is shown in Fig. 4, in which oxyhemoglobin is excited by the 568.2-nm (5682 A) wavelength of singly ionized krypton. The top spectrum $I_{\parallel}$ is taken with the polarizer P of Fig. 1 set to pass the components parallel to the direction of laser polarization; the bottom spectrum $I_{\perp}$ is taken with P set to pass perpendicular components. Lines in which $I_{\perp}$ is much greater than $I_{\parallel}$ are said to have inverse polarization and are seen only in the resonance Raman effect; these include the lines at 1305, 1342, and 1589 cm^{-1} and numerous others..

Hyper-Raman effect. The nature of this effect is most easily described in terms of the corpuscular picture of the Raman effect. With an intense laser source, the number of monochromatic photons impinging on the molecules of a medium per unit volume and unit time may be extremely large. If so, the probability that two photons will collide simultaneously with the same molecule is very much larger than in normal scattering, and there is considerable chance that the two photons will unite and be scattered as a single photon of approximately twice the frequency. The rules governing the scattering in such three-photon processes (two incoming and one outgoing photons) are quite different from those for normal (two-photon) Rayleigh and Raman scattering. For example, in molecules that are centrosymmetric, the collision must be inelastic, that is, the molecule must absorb or give up an amount of energy ΔE during the process. The frequency of the scattered photon will therefore not be exactly twice the frequency of the incident photons but will differ from it by $\Delta\bar{\nu} = \Delta E/hc$. Such scattered radiation is called the hyper-Raman effect. Even in molecules that are not centrosymmetric, the likelihood of elastic collisions is much smaller than in the normal case, so that the intensity of hyper-Rayleigh scattering may be substantially weaker than hyper-Raman scattering.

As implied above, the selection rules for the vibrational and rotational transitions in the hyper-Raman effect are different from those of the normal Raman effect. Thus certain transitions are observable in the hyper-Raman effect that are normally forbidden. This is one virtue of the hyper-Raman effect; the other is that it is observed in a spectral region whose frequency is far removed from that of the incoming radiation (and is, in fact, twice that of the latter). The effect is therefore observable without interference from the normal Rayleigh line.

Stimulated Raman effect. The mechanism of the stimulated Raman effect depends on the coherent pumping of the molecules of the sample into an excited vibrational state by the powerful electric field of the laser beam. In view of the large discrepancy of one or two orders of magnitude between the frequency of the vibration and the frequency of the laser, this can be accomplished only if the field of the light wave has a very high value (the threshold power) and if the mismatch in frequency is compensated by the generation of coherent radiation with a frequency equal to that of the laser minus the vibrational frequency. The coherent radiation so produced is called stimulated Raman scattering. It was first observed by R. Woodbury and A. Ng in 1962. They found the effect in liquid nitrobenzene, which they were using as an electrooptical shutter within a laser system.

In addition to its high intensity and its coherence, there are other new features of stimulated Raman scattering. Since the pumping power of the incident laser beam must exceed a certain threshold for the scattering to take place, when the laser power is used up in exciting one vibrational mode, there is insufficient power available to excite other modes. Therefore the stimulated Raman effect usually contains only one frequency, though in rare cases the power may be divided between two vibrational modes of roughly the same thresh-

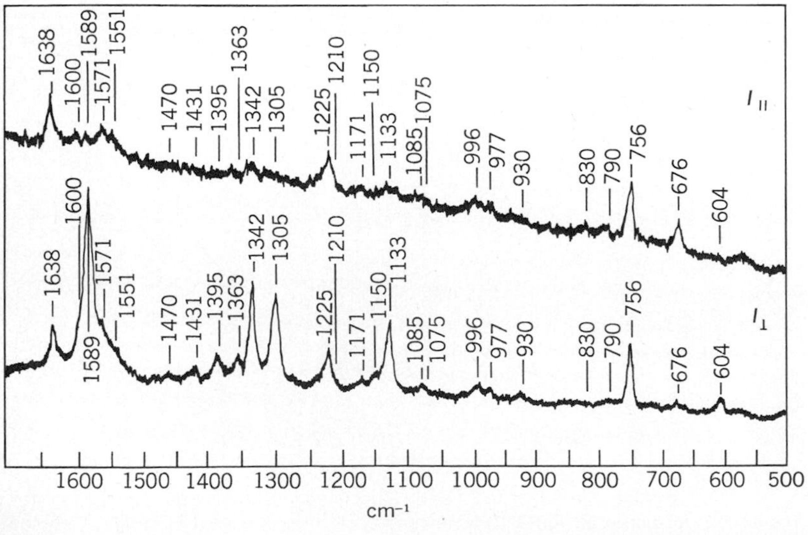

Fig. 4. Resonance Raman spectrum of oxyhemoglobin. (*From T. G. Spiro and T. C. Strekas, Resonance Raman spectra of hemoglobin and cytochrome c. Proc. Nat. Acad. Sci. USA, 69:2622–2626, 1972*)

old. However, the power in the scattered radiation may itself produce further stimulated Raman emission by a repetition of the initial process. This results in a new frequency, which is the laser frequency minus exactly twice the frequency of vibrational mode that is being scattered. This fact shows that the mechanism does not involve a double jump in the vibrational levels; such a double jump would give a frequency shift that is not exactly twice that of the vibrational fundamental because of vibrational anharmonicity.

Another striking and unusual effect in stimulated Raman scattering is the excitation of intense anti-Stokes radiation. This radiation may be even stronger than the Stokes radiation in certain circumstances. Moreover, it can be observed at such low temperatures that the initial populations of the excited vibrational levels needed for normal anti-Stokes Raman scattering are zero. It arises from the above-mentioned pumping of molecules from the ground vibrational state into upper excited states by the initial laser power. These excited molecules can then be pumped by further radiation back into the ground state, with a simultaneous stimulated emission of coherent radiation at a frequency that equals the laser plus the molecular vibrational frequency.

The development of tunable lasers has led to a special technique for stimulated Raman scattering called coherent anti-Stokes Raman spectroscopy (CARS). In this technique, two lasers are used, one of fixed and the other of tunable frequency. The two beams enter the sample at angles differing only by some appropriate small amount (approximately 2°) and simultaneously impinge on the sample molecules. Whenever the frequency difference between the two lasers coincides with the frequency of a Raman-active vibration of the molecules, emission of coherent radiation (both Stokes and anti-Stokes) is stimulated. Thus the total Raman spectrum can be scanned in stimulated emission by varying the frequency of the tunable laser. An advantage of CARS, in addition to the high intensity of the scattering, is that its elevated frequency avoids interference from sample fluorescence, which always has frequencies below that of the exciting radiation.

Applications. Raman spectroscopy is of considerable value in determining molecular structure and in chemical analysis. Molecular rotational and vibrational frequencies can be determined directly, and from these frequencies it is sometimes possible to evaluate the molecular geometry, or at least to find the molecular symmetry. *See* MOLECULAR STRUCTURE AND SPECTRA.

Even when a precise determination of structure is not possible, much can often be said about the arrangement of atoms in a molecule from empirical information about the characteristic Raman frequencies of groups of atoms. This kind of information is closely similar to that provided by infrared spectroscopy; in fact, Raman and infrared spectra often provide complementary data about molecular structure. The complex structures of biologically important molecules, for example, are the subjects of current spectroscopic research. Both normal and resonance Raman spectroscopy are valuable techniques in molecular biology (see Fig. 4). Raman spectra also provide information for solid-state physicists, particularly with respect to lattice dynamics but also concerning the electronic structures of solids. *See* INFRARED SPECTROSCOPY. [RICHARD C. LORD]

Bibliography: N. Bloembergen, *Non-Linear Optics*, 1965; A. J. Clark and R. E. Hester (eds.), *Advances in Infrared and Raman Spectroscopy*, vols. 1–7, 1975–1980; G. Herzberg, *Infrared and Raman Spectra of Polyatomic Molecules*, vol. 2, 2d ed., 1945; D. A. Long, *Raman Spectroscopy*, 1977; M. C. Tobin, *Laser Raman Spectroscopy*, 1971, reprint 1980; E. B. Wilson, J. C. Decius, and P. C. Cross, *Molecular Vibrations*, 1955.

Rare-earth elements

The group of 17 chemical elements with atomic numbers 21, 39, and 57–71; the name lanthanides is reserved for the elements 58–71. The name rare earths is a misnomer, because they are neither rare nor earths. The early Greeks believed that everything in the world was made of four elements: air, earth, fire, and water. The earths were substances which could not be changed with heat by the temperatures then available to the scientist; and in the early part of the 19th century, when the first rare earths were discovered, they resembled the common earths, which were really oxides of magnesium, calcium, and aluminum. Since the rare earths were found in very rare minerals, they were thus called rare earths. They are not rare, however, since cerium is reported to be more abundant in the Earth's crust than tin; yttrium more abundant than lead; and even the scarce rare earths, except promethium, more abundant than the platinum-group elements. All these elements form trivalent bonds, and when their salts are dissolved in water, they ionize to form trivalent ions and the solutions exhibit very similar chemical properties. The elements scandium, yttrium, lanthanum, and actinium in the III column of the extended periodic table show similar properties in aqueous solution. Yttrium and lanthanum are always found associated with the rare earths in nature.

The similarity of properties among the lanthanide rare earths (atomic numbers 58 through 71) originates from the fact that as the atomic number increases in this part of the periodic table, the increased charge on the nucleus is compensated for by electrons which start filling an inner incomplete subshell (4*f*). This subshell can hold up to 14 electrons; there are accordingly 14 elements in the lanthanide series. These extra electrons, however, play almost no role in the valency forces between atoms. The elements with atomic numbers 90 through 103 also occur in the periodic table at a place where a similar inner subshell (5*f*) is being completed. In many ways, these elements resemble the lanthanides. They are frequently referred to as the actinide rare earths. Both groups are usually displayed at the bottom of the periodic table as an appendage of two rows. *See* ACTINIDE ELEMENTS; LANTHANIDE CONTRACTION; PERIODIC TABLE.

Uses. Most of the early uses of the rare earths took advantage of their common properties and were centered principally in the glass, ceramic, lighting, and metallurgical industries. Today these applications use a very substantial amount of the

Table 1. Properties of the rare-earth metals

Symbol	Transitions	Transition temperature, °C	ΔH transition, cal	Lattice parameters, A a	c	c/a	Ionic radius RE^{3+}, A
Sc	hcp (ABAB)	—	—	3.3088	5.2680	1.592	0.732
	bcc	1335	958	—	—	—	—
Y	hcp (ABAB)	—	—	3.6482	5.7318	1.571	0.893
	bcc	1478	1189	4.08	—	—	—
La	dhcp (ABAC)	—		3.7740	12.171	1.6125*	1.061
	fcc (ABC)	310	67	—	—	—	—
	bcc	865	753	4.26	—	—	—
Ce	fcc (ABC)	—	—	4.85	—	—	—
	dhcp (ABAC)	−178	—	3.68	11.92	1.619*	1.034
	fcc (ABC)	~100 (heating)	—	5.160	—	—	Ce^{4+}
		− 10 (cooling)	—	—	—	—	.092
	bcc	726	700	4.12	—	—	
Pr	dhcp (ABAC)		—	3.6721	11.832	1.611*	1.013
	bcc	795	760	4.13	—	—	Pr^{4+}
							.090
Nd	dhcp (ABAC)	—	—	3.6583	11.7966	1.612*	0.995
	bcc	863	713	4.13	—	—	
Pm							0.979
Sm	hcp (nonprimitive)	—	—	3.6290	26.207	1.605	0.964
	bcc	926	744	—	—	—	Sm^{2+}
							1.11
Eu	bcc	—	—	4.5827	—	—	0.950
							Eu^{2+}
							1.09
Gd	hcp (ABAB)	—	—	3.6336	5.7810	1.59	0.938
	bcc	1235	935	4.05	—	—	—
Tb	hcp (ABAB)	—	—	3.6055	5.6966	1.580	—
	bcc	1289	1203	—	—	—	0.923
							Tb^{4+}
		—	—	—	—	—	0.84
Dy	hcp (ABAB)			3.5915	5.6501	1.573	0.908
	bcc	1381	955	—	—	—	—
Ho	hcp (ABAB)	—	—	3.5778	5.6178	1.570	0.894
Er	hcp (ABAB)	—	—	3.5592	5.5850	1.569	0.881
Tm	hcp (ABAB)	—	—	3.5375	5.5540	1.570	0.869
Yb	fcc (ABC)	—	—	5.4848	—	—	0.858
	bcc	795	418	—	—	—	Yb^{2+}
							0.93
Lu	hcp (ABAB)			3.5052	5.5495	1.583	0.848

*The values are the author's opinion of the best average values as of July 1975. The actual values vary slightly in the original literature, depending on the investigator and the purity of the metal at hand. Most of the values were obtained on 99.8–99 9% pure metal, the principal impurities being hydrogen, oxygen, nitrogen, carbon, and tantalum. Some of these values will change slightly as purer metals are produced.

mixed rare earths just as they are obtained from the minerals, although sometimes these mixtures are supplemented by the addition of extra cerium or have some of their lanthanum and cerium fractions removed.

Cerium oxide or the mixed rare-earth oxides make excellent abrasives for glass polishing, and several million pounds of these materials are sold each year for polishing mirrors, television face plates, lenses, and plate glass. The rare earths have long been used in the manufacture of specialty glasses and in the ceramic industry, particularly as glazes.

The elements exhibit very complex spectra, and the mixed oxides, when heated, give off an intense white light which resembles sunlight. Consequently mixtures of the oxides are used in cored carbon arcs, such as those employed in the movie industry.

The rare-earth metals have a great affinity for the nonmetallic elements, as, for example, hydrogen, carbon, nitrogen, oxygen, sulfur, phosphorus, and the halides. Considerable amounts of the mixed rare earths are reduced to metals, such as misch metal, and these alloys are used as "getters" in the metallurgical industry. The rare-earth metals combine with the nonmetals in other metals and alloys, and these impurities are partially removed along with the slag or, as in nodular cast iron, the rare earths convert the carbon inclusions into nodules, and this procedure imparts considerable strength to the iron. Alloys made of cerium and the mixed rare earths are used in the manufacture of lighter flints.

Rare earths are also used in the petroleum industry as catalysts. The rare-earth ions are introduced into zeolite-type oxide catalyst, say in the form of the chloride. The resulting catalyst, which contains about 5% mixed rare-earth oxides, produces petroleum with a higher yield of the desired petroleum fraction.

Since the individual rare earths have been available commercially at reasonable prices, a sizable market has developed for a number of them. This is not surprising, since they represent, as a group, about one-sixth of the naturally occurring elements and about one-fourth of the elemental metals. The fact that these elements form compounds and alloys, many of whose properties change in a predictable manner as one progresses across parts of the series, makes them of particular interest to scientists working in various fields of research.

Table 1. Properties of the rare-earth metals (cont.)

	Melting point, °C	Boiling point, °C	Heat of vaporization ($\Delta H_{v,0}$), kcal/g-atm	Density 25°C, g/cm³	Molal volume 25°C, cc/g-atm	Radius metal atom, A	Electrical resistivity (4.2 K), ohm-cm 10^{-6}	Residual resistivity (4.2 K), ohm-cm 10^{-6}	Compressibility, cm²/kg 10^{-6}
(Sc)	1541	2831	89.9	2.9890	15.041	1.640	52	3	2.26
(Y)	1522	3338	101.3	4.4689	19.894	1.801	59	2	2.68
(La)	921	3457	103.1	6.1453	22.603	1.879	61–80	Superconductor	4.04
(Ce)	799	3426	101.1	6.672	21.001	1.820	70–80	10	4.10
(Pr)	931	3512	85.3	6.773	20.805	1.828	68	1	3.21
(Nd)	1021	3068	78.5	7.007	20.585	1.821	65	7	3.0
(Pm)	1168	2700 (est.)	—	—	—	—	—	—	(2.8)
(Sm)	1077	1791	49.2	7.520	20.001	1.804	91	7	3.34
(Eu)	822	1597	(41.9) $\Delta H° = 29$	5.2434	28.981	1.984	91	1	8.29
(Gd)	1313	3266	95.3	7.9004	19.9041	1.801	127	1	2.56
(Tb)	1356	3123	93.4	8.2294	19.3119	1.783	114	4	2.45
(Dy)	1412	2562	70.0	8.5500	19.0058	1.774	100	5	2.55
(Ho)	1474	2695	72.3	8.7947	18.7533	1.766	88	3	2.47
(Er)	1529	2863	76.1	9.066	18.4499	1.757	71	3	2.39
(Tm)	1545	1947	55.8	9.3208	18.1244	1.746	74	3	2.47
(Yb)	819	1194	36.5	6.9654	24.8428	1.939	28	2	7.39
(Lu)	1663	3395	102.2	9.8404	17.7808	1.735	60	2	2.38

These scientists develop theories about numerous experimental phenomena, and they can use the rare-earth metals, alloys, compounds, and solutions as test materials to see whether the theories still apply when tested on a series of neighboring elements whose properties shift in a known manner.

Hundreds of articles appear each year in the scientific literature on rare earths used in this manner, and this basic research need develops a small but steady market for the individual pure rare-earth metals and compounds.

Since 1964 the television industry has been using highly purified yttrium and europium oxides in ton quantities. A europium-activated yttrium phosphor emits an intense red light when used in a television screen. Use of this type of phosphor greatly improves the color reproduction, and pictures are much more brilliant than when a non-rare-earth phosphor was used. The mercury-activated electric street lights, which are economical to operate, give a purplish-blue color; for about the same energy consumption they can be made to give a brilliant white light by the use of rare-earth phosphors deposited on the glass.

Another very important use of individual rare earths is in the manufacture of solid-state microwave devices widely used in radar and communications systems. Yttrium-iron garnets are especially good, since they transmit short-wave energy with low energy losses. The devices, however, are very small, so the total use of rare earths is not large.

Still another important use of individual rare earths is in the construction of lasers. A high percentage of the patent applications for new lasers involves the use of a rare earth as the active constituent of the laser.

The rare earths show interesting magnetic properties. Alloys of cobalt with the rare earths, such as cobalt-samarium, produce permanent magnets that are far superior to most of the varieties on the market, and many uses for these magnets are developing.

Many phosphors contain rare earths as an ingredient, and barium phosphate–europium phosphor finds applications in x-ray films that form satisfactory images with only half the exposure time of conventional x-rays.

Yttrium aluminum garnets (YAG) are used in the jewelry trade as artificial diamonds. Single crystals of YAG have a very high refractive index, similar to diamond, and when these crystals are cut in the form of diamonds, they sparkle in the same manner as a diamond. Also, they are very hard, so that they scratch glass, and only an expert can tell the difference between a YAG diamond and a real one.

Other small but important uses are mentioned in the articles on the individual rare earths. (For the list of rare-earth elements see Table 1.)

Table 2. Some common rare-earth minerals

Mineral	Crystal form	Formula composition*
Monazite†	Monoclinic	$CePO_4$ with $Th_3(PO_4)_4$
Xenotime	Tetragonal	YPO_4
Gadolinite	Monoclinic	$2BeO \cdot FeO \cdot Y_2O_3 \cdot 2SiO_2$
Bastnasite†	Hexagonal	$CeFCO_3$
Samarskite	Orthorhombic	$3(Fe,Ca,UO_2)_3O \cdot Y_2O_3 \cdot 3(Nb,Ta)_2O_5$
Fergusonite	Tetragonal	$Y_2O_3 \cdot (Nb,Ta)_2O_5$
Euxenite	Orthorhombic	$Y_2(NbO_3)_3 \cdot Y_2(TiO_3)_3 \cdot 1\frac{1}{2}H_2O$
Yttrofluorite	Cubic	$2YF_3 \cdot 3CaF_2$

*Only the most abundant rare earth is listed. Ce minerals are rich in light members of the rare-earth group. Y minerals are rich in heavy members of the rare-earth group.

†One of the two most important minerals used by the rare-earth industry in the United States.

Occurrence. Although the rare earths are widely distributed in nature, they generally occur in low concentrations. They also are found in high concentrations as mixtures in a number of minerals (Table 2). The relative abundance of the different rare earths in various rocks, geological formations, and the stars is of great interest to the geophysicist, astrophysicist, and cosmologist. The development of precision methods for determining the abundance of the various elements present in trace amounts by means of mass spectroscopy, optical and infrared spectroscopy, and radio and nuclear chemistry permits the relative abundance of the rare earths in any sample to be determined with considerable precision. It is found that the deep basic rocks, such as basalt, contain a few parts per million of mixed rare earths, and that the more acid silicate rocks contain a slightly higher concentration. When these two types of rocks come in contact in the molten state, the more acid rocks extract some of the mixed rare earths from the basalt, and the light rare earths are extracted to a larger extent than the heavy ones. Thus, by determining a profile of the relative abundance of the rare earths in given rocks, it can be determined whether they have been molten. Such tests were performed on the various rocks brought back from the Moon. Also, if deposits of rich rare-earth minerals are found in a formation, one can determine the physical condition with regard to temperature and pressure to which the formation must have been subjected in order to form such minerals. The relative abundance of the various rare earths present in stars as determined from their spectra gives clues as to what nuclear processes are taking place in those stars. The numerous scientific papers published in this field make interesting reading, since theories as to the formation of the solar system and the universe are in part based on such data. Monazite, xenotime, and bastnasite are among the more important sources. These minerals are usually concentrated from other rock and minerals by mechanical means, such as flotation or magnetic cross-belt separation methods. The rare earths are leached from the minerals with acid in the case of the phosphate or silicate minerals. Some minerals, such as the columbotantalates, have to be heated with carbon or treated with strong caustic before being leached.

Separation. The mixed rare earths can be separated from the acid solutions by means of oxalate precipitation or by other insoluble precipitates; ignition of the oxalate gives a mixed rare-earth oxide. They are frequently concentrated directly by ion-exchange methods, using the acid leach from the minerals.

The rare earths occur in solution as hydrated trivalent ions whose properties are very much alike; therefore they tend to form mixed crystal precipitates or solid solutions. A single chemical operation only slightly enriches one rare earth over another; thus, to isolate pure compounds of the individual elements by these methods, the processes have to be repeated many times.

Historically, the elements were purified by fractional processes such as fractional crystallization or fractional decomposition. The enormous amount of work involved permitted the separation of only very small amounts; consequently, the pure rare earths were very costly and gained the reputation of being rare. Fractionation methods are still used commercially to separate crude rare earths, particularly for lanthanum and cerium, since cerium can then be separated from lanthanum by taking advantage of the quadrivalent state of cerium. The other members of the rare-earth series, if desired in high purity, are separated by means of ion-exchange processes; if great purity is not desired, liquid-liquid extraction processes can be employed. Some of the rare earths which show anomalous valence are separated from the others by taking advantage of this property.

Yttrium and europium, used in the television industry, are usually separated by means of liquid-liquid extraction. Europium shows the divalent characteristic, and advantage can be taken of this. Yttrium, which occurs in another row in the periodic table, can be shifted either in liquid-liquid extraction or ion exchange along the rare-earth series by means of different complexing agents. Advantage can also be taken of the fact that some minerals contain very little of the heavy rare earths, but this fraction contains the yttrium, and liquid-liquid processes can be quite effective. Furthermore, when ton quantities are desired, the liquid-liquid extraction processes are usually more economical.

Properties. The rare-earth elements are metals possessing distinct individual properties which make them potentially valuable as alloying agents. They are usually reduced thermally by treating the anhydrous halide with calcium, lithium, or other alkali metals and then remelting under vacuum to volatilize the last traces of the reductant. They can also be reduced eletrolytically from fused-salt baths, as is done commercially for cerium and misch metal (a mixed rare-earth metal, mainly cerium, with small amounts of iron present).

Samarium, europium, and ytterbium cannot be reduced by the above methods. These elements form divalent halides when so treated. Fortunately, these metals have very low boiling points so they can be distilled from a mixture of lanthanum or cerium and the oxides of these elements. The pure samarium, europium, and ytterbium are collected on the condenser, and lanthanum or cerium oxides are left behind.

Table 1 gives the properties of the metals. The anhydrous salts show greater differences in properties among the elements than do the hydrated salts.

Many of the properties of the metals and alloys are quite sensitive to temperature and pressure.

They are also different when measured along different crystal axes of the metal; for example, electrical conductivity, elastic constants, and so on. If the properties are plotted against temperature and pressure, they show anomalies whenever a crystal or magnetic transition takes place.

The rare earths form organic salts with certain organic chelate compounds. These chelates, which have replaced some of the water around the ions, enhance the differences in properties among the individual rare earths. Advantage is taken of this technique in the modern ion-exchange methods of separation. *See* CHELATION; CHROMATOGRAPHY; ION EXCHANGE; MAGNETOCHEMISTRY; TRANSITION ELEMENTS.

[FRANK H. SPEDDING]

Bibliography: R. J. Elliott (ed.), *Magnetic Properties of Rare Earth Metals*, 1972; K. Gschneidner and L. Eyring, *Handbook on the Physics and Chemistry of Rare Earths, vol. 1, Metals, vol. 2, Alloys and Intermetallics*, 1978; C. M. Lederer et al., *Table of Isotopes*, 6th ed., 1967; J. E. Powell et al., The separation of rare earths, *J. Chem. Educ.*, 37:629–633, 1960; F. H. Spedding, Rare earths, in *Encyclopedia Britannica*, 15th ed., 1974; F. H. Spedding and A. H. Daane (eds.), *Rare Earths*, 1961.

Reactive intermediates

Unstable compounds which are formed as necessary intermediate stages during a chemical reaction. Thus, if a reaction in which A is converted to B requires that A first be converted to C, then C is an intermediate in the reaction ($A \rightarrow C \rightarrow B$). The term reactive further implies a certain degree of instability of the intermediate; reactive intermediates are typically isolable only under special conditions, and most of the information regarding the structure and properties of reactive intermediates comes from indirect experimental evidence.

In organic reactions the most common types of reactive intermediates are those arising from dissociative reactions, in which carbon has a decreased valence (Table 1). Associative reactions can also give rise to some of the same intermediates, and to others in which carbon has an increased valence (Table 2). *See* VALENCE.

Carbocations. These are compounds in which carbon bears a positive charge. Classical carbocations (also called carbenium ions) are trivalent, and have only six valence electrons. Nonclassical carbocations (also called carbonium ions) are tetra- or pentavalent, and have eight valence electrons. Examples are the methyl cation (classical) CH_3^+, and the methonium ion (nonclassical) CH_5^+.

Classical. The structure of classical carbocations is typically planar. Because of the net electron deficiency, any structural feature which can donate electron density tends to stabilize the carbocation. Resonance electron donation from an adjacent atom having a lone pair of electrons or from an adjacent pi bond or aromatic ring is particularly effective, as in notation (1).

$$^{+}CH_2-\ddot{O}-H \longleftrightarrow CH_2{=}\overset{+}{\underset{..}{O}}-H \quad (1)$$

Reactions of carbocations are dominated by their electron deficiency. Virtually all reactions of carbocations involve bonding to an electron pair, either by bonding to an anion or neutral molecule with a lone pair of electrons, or by bonding to the pi bond of an unsaturated compound, or by eliminating or shifting electrons in an adjacent bond. Thus carbocations are intermediates in a large number of different organic reactions (Table 3). *See* RESONANCE.

Nonclassical. These are carbocations that have properties similar to classical carbocations, but frequently involve modifications in reactivity due to special structural features, as in reaction (2),

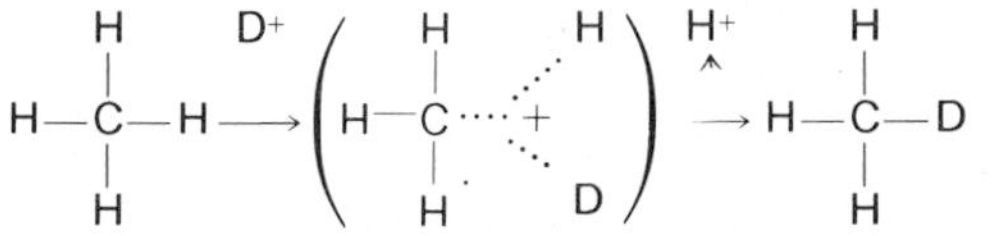
(2)

which shows electrophilic substitution. The key structural aspect of a nonclassical carbocation is a three-center, two-electron bond, a type of bonding commonly encountered in boron chemistry.

Carbanions. These are compounds in which carbon bears a negative charge. A carbanion will always have a positive counterion in association with it; depending upon the particular cation and the stability of the carbanion, the association may be ionic, covalent, or some intermediate combination of ionic and covalent bonding, as in notation (3).

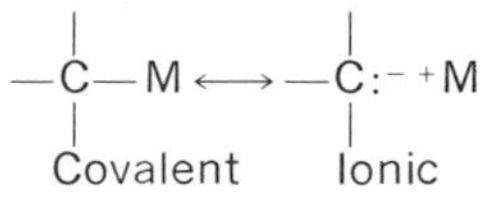
(3)

Compounds involving primarily covalent bonding

Table 1. The most common types of reactive intermediates

Dissociative reaction	Intermediate	Simplest example
$X{-}C{-} \rightarrow X{:}^- + {}^+C{-}$	Carbocation	CH_3^+ (methyl cation)
$X{-}C{-} \rightarrow X\cdot + \cdot C{-}$	Free radical	$CH_3\cdot$ (methyl radical)
$X{-}C{-} \rightarrow X^+ + {}^-{:}C{-}$	Carbanion	$CH_3{:}^-$ (methyl anion)
$X{-}C(Y){-} \rightarrow XY + {:}C\langle$	Carbene	$CH_2{:}$ (methylene)

Table 2. Reactive intermediates produced by associative reactions

Associative reaction	Intermediate
$>C{=}C< + X^+ \rightarrow X{-}C{-}C^+$	Carbocation
$>C{=}C< + X\cdot \rightarrow X{-}C{-}C\cdot$	Free radical
$>C{=}C< + X{:}^- \rightarrow X{-}C{-}C{:}^-$	Carbanion
$>C{=}\ddot{O}{:} + X{:}^- \rightarrow X{-}C{-}\ddot{O}{:}^-$	Tetrahedral intermediate
$-C{-}X + H^+ \rightarrow -C{\cdots}^{+}(H)(X)$	Carbonium ion

Table 3. Some reactions involving carbocation intermediates

Reaction	Reaction type
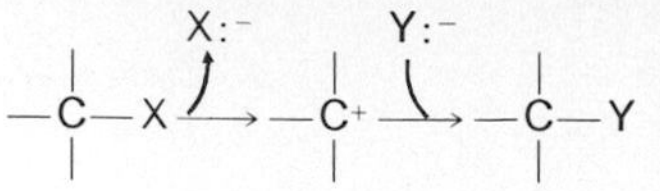	Nucleophilic substitution
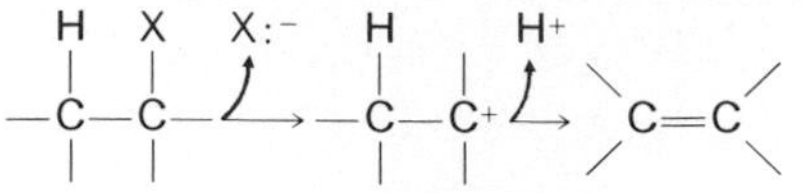	Elimination
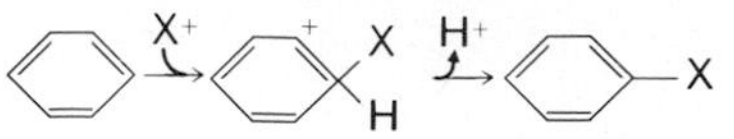	Electrophilic aromatic substitution
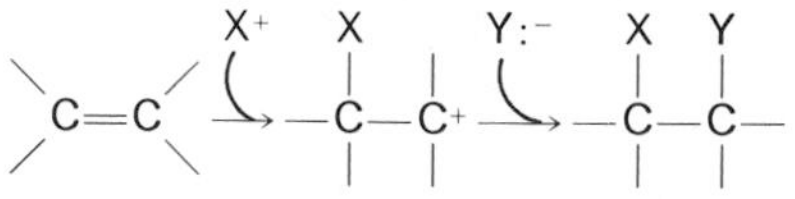	Electrophilic addition
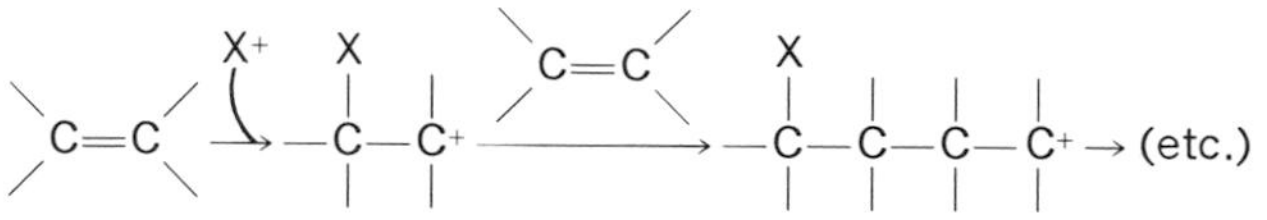	Cationic polymerization

of carbon to a metal are considered organometallic compounds, which frequently have properties similar to carbanions but substantially modified by the presence of the metal. *See* ORGANOMETALLIC COMPOUND.

Carbanions are trivalent, with eight valence electrons. Simple alkyl carbanions assume a tetrahedral structure, and are extremely strong bases. Carbanions are stabilized by structural features which allow electron withdrawal, in particular, inductive effects of halogen substituents or resonance effects of nitro (—NO_2), cyano (—CN), or carbonyl (—C(=O)—) substituents. Such resonance-stabilized carbanions have a planar structure and

Table 4. Some reactions involving carbanion intermediates

Reaction	Reaction types
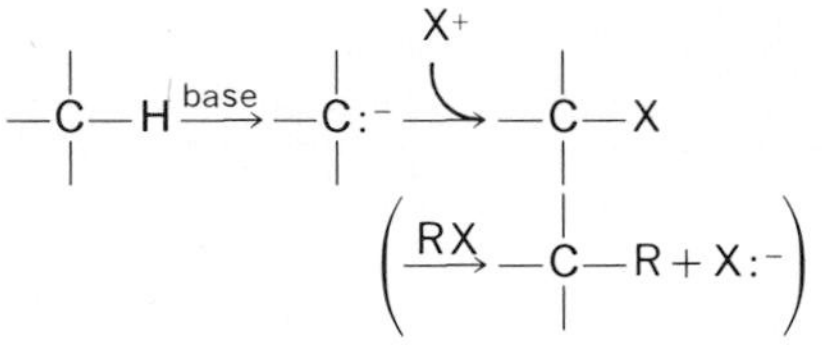	Electrophilic substitution (alkylation)
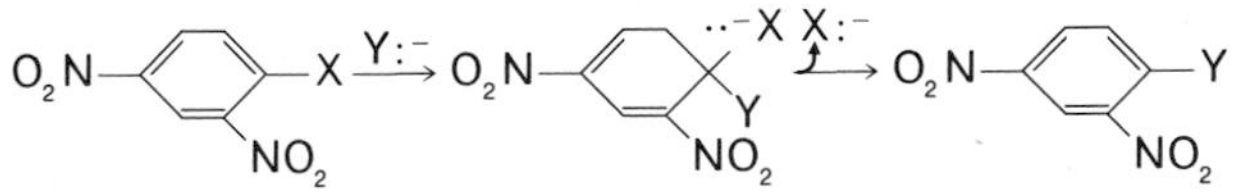	Nucleophilic aromatic substitution
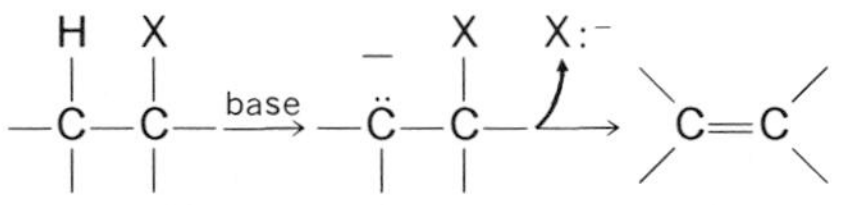	Elimination
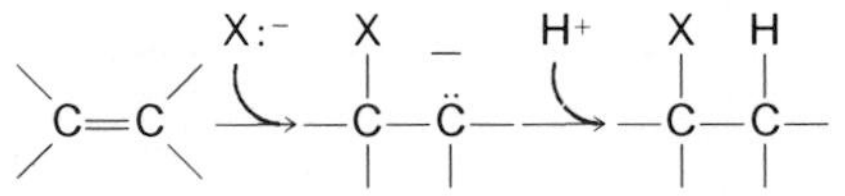	Nucleophilic addition
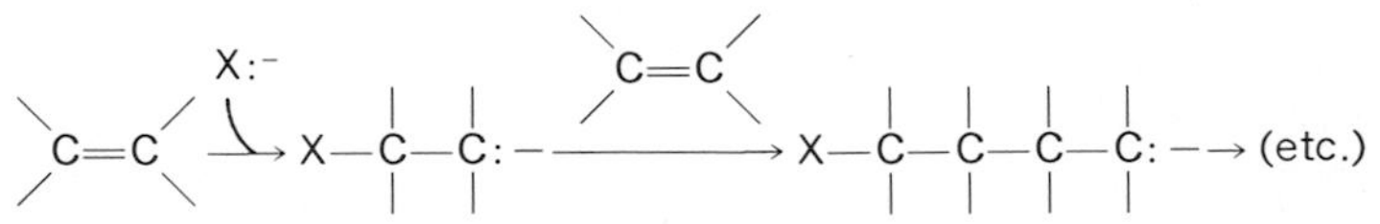	Anionic polymerization
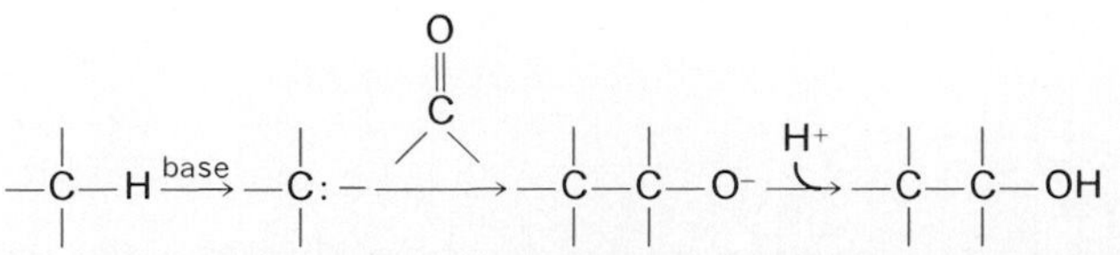	Condensation

are relatively easily formed in acid-base reactions. An example is shown in reaction (5).

$$CH_3-\overset{\displaystyle O}{\underset{\displaystyle H}{C}} + OH^- \rightleftharpoons$$

$$H_2O + {}^-CH_2-\overset{\displaystyle O}{\underset{\displaystyle H}{C}} \longleftrightarrow CH_2{=}\overset{\displaystyle {}^-O}{\underset{\displaystyle H}{C}} \qquad (5)$$

Resonance-stabilized
enolate anion

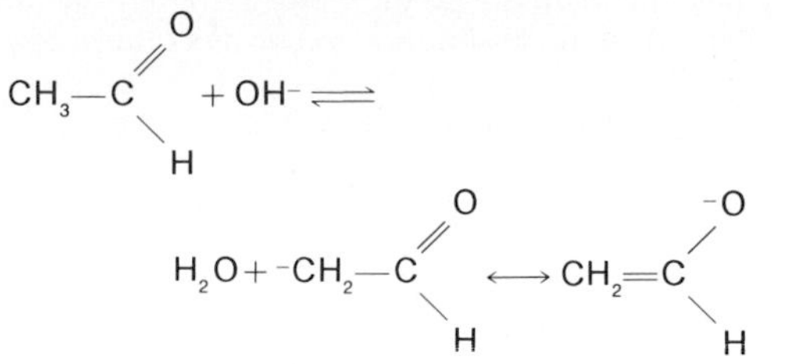

Carbanions are relatively electron-rich; therefore their reactions are predominantly with electron-deficient species (Table 4). The use of carbanions (or organometallic analogs) in alkylation and condensation reactions generates new carbon-carbon bonds and provides some of the most useful reactions for organic synthesis. *See* ALKYLATION; CONDENSATION REACTION.

Free radicals. These are neutral compounds having an odd number of electrons and therefore one unpaired electron. Carbon free radicals are trivalent, with seven valence electrons, and typically assume a planar structure. Free radicals are primarily electron-deficient species and are stabilized by structural features which donate electron density or delocalize the odd electron by resonance. In addition, structural features which prevent the normal reactions of radicals, such as large substituents surrounding the free-radical site, are very effective in increasing radical stability.

An important characteristic of many reactions involving free-radical intermediates is the tendency to follow a chain reaction mechanism. Chain reactions have three fundamental components: initiation, propagation, and termination steps (Table 5). Initiation steps generate the reactive free radical from an appropriate precursor (or radical initiator). Propagation steps utilize the reactive free radical to generate product; propagation steps also regenerate another reactive free radical so that those steps may continue recycling. The overall reaction is the sum of the propagation steps,

which may recycle thousands of times. Termination steps involve the combination of free radicals; each combination effectively terminates two chains.

Other common free-radical reactions include addition reactions, including polymerization. *See* CHAIN REACTION; POLYMERIZATION.

Radical ions. These compounds are charged, have an unpaired electron, and are either radical cations (positively charged) or radical anions (negatively charged). In many cases a radical ion is derived from a stable neutral molecule by addition of one electron (radical anion) or removal of one electron (radical cation); thus, radical ions are frequently encountered as intermediates in electrochemical or other one-electron reduction or oxidation processes, as shown in the pinacol reduction reaction (6). *See* ELECTROCHEMISTRY; OXIDATION-REDUCTION.

$$2(CH_3)_2CO + 2Na \rightarrow 2(CH_3)_2\overset{\cdot}{C}-O^- \rightarrow \qquad (6)$$

Acetone
radical
anion

$$(CH_3)_2\overset{\displaystyle {}^-O}{C}-\overset{\displaystyle O^-}{C}(CH_3)_2$$

$$\downarrow{\scriptstyle 2H^+}$$

$$(CH_3)_2\overset{\displaystyle HO}{C}-\overset{\displaystyle OH}{C}(CH_3)_2$$

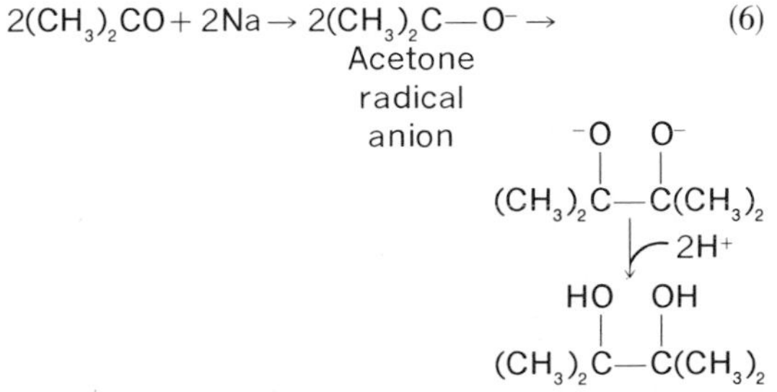

Carbenes. These are compounds which have a divalent carbon. The divalent carbon also has two nonbonded electrons, for a total of six valence electrons. The two nonbonded electrons may have either the same spin quantum number, which is a triplet state, or an opposite spin quantum number, which is a singlet state. The triplet state of carbenes is typically lower in energy than the singlet state. Singlet and triplet carbenes normally undergo the same reactions, but since they must follow different mechanisms, there are differences in reactivity, selectivity, and stereochemical details of the reactions. Typical singlet carbene reactions are concerted, involving a single step, and these reactions usually maintain stereochemistry. Triplet carbene reactions typically proceed through a trip-

Table 5. Free-radical chain reactions

Reaction	Step
$Cl_2 \rightarrow 2\,Cl\cdot$	Initiation
$Cl\cdot + CH_4 \rightarrow CH_3\cdot + HCl$ $CH_3\cdot + Cl_2 \rightarrow CH_3Cl + Cl\cdot$	Propagation
$2\,Cl\cdot \rightarrow Cl_2$ $2\,CH_3\cdot \rightarrow CH_3CH_3$ $CH_3\cdot + Cl\cdot \rightarrow CH_3Cl$	Termination
$CH_4 + Cl_2 \rightarrow CH_3Cl + HCl$	Overall substitution reaction
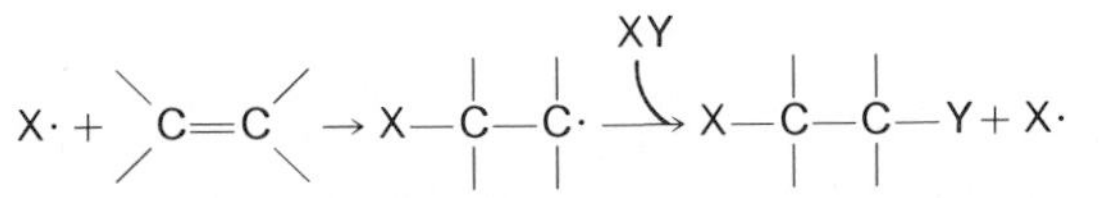 $X\cdot + \,\diagdown{C}{=}{C}\diagup \rightarrow X{-}\overset{\mid}{\underset{\mid}{C}}{-}\overset{\mid}{\underset{\mid}{C}}\cdot \overset{\displaystyle XY}{\curvearrowright} X{-}\overset{\mid}{\underset{\mid}{C}}{-}\overset{\mid}{\underset{\mid}{C}}{-}Y + X\cdot$	Addition by a free radical chain reaction
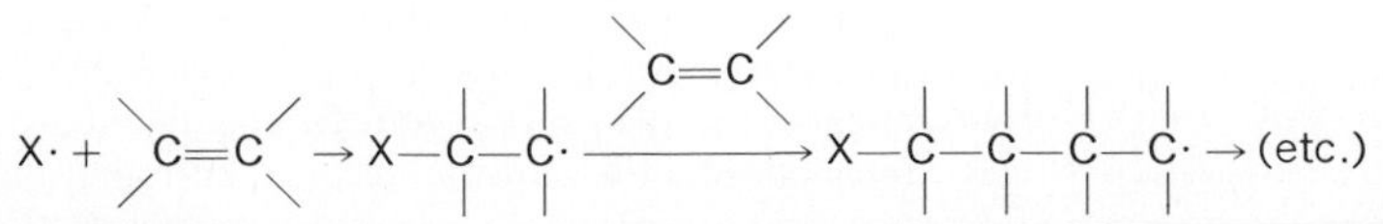 $X\cdot + \,\diagdown{C}{=}{C}\diagup \rightarrow X{-}\overset{\mid}{\underset{\mid}{C}}{-}\overset{\mid}{\underset{\mid}{C}}\cdot \overset{\diagdown C{=}C\diagup}{\longrightarrow} X{-}\overset{\mid}{\underset{\mid}{C}}{-}\overset{\mid}{\underset{\mid}{C}}{-}\overset{\mid}{\underset{\mid}{C}}{-}\overset{\mid}{\underset{\mid}{C}}\cdot \rightarrow (\text{etc.})$	Radical chain polymerization

Table 6. Reactions of carbenes

Reaction	Classification
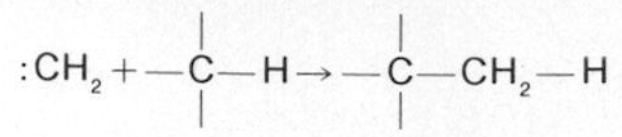	Insertion
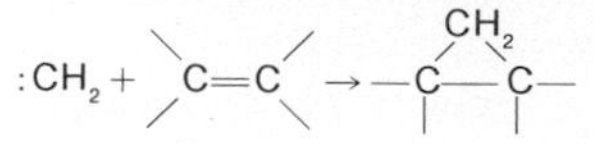	Cycloaddition
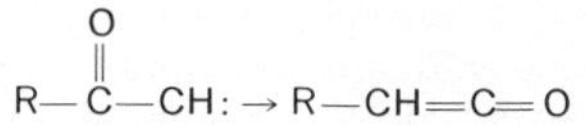	Wolff rearrangement

let radical pair or a triplet biradical intermediate, which frequently randomizes stereochemistry.

Generation of carbenes is most commonly by photolysis or thermolysis of diazo compounds or ketenes, or by alpha-elimination reactions.

Carbenes are extremely reactive because of their electron deficiency. Typical reactions may be simply classified as reactions with single bonds (insertions) and reactions with double bonds (cycloadditions). Cycloaddition reactions are among the best synthetic procedures for making cyclopropane derivatives. Internal rearrangements of carbenes are also common, depending upon the particular structure of the carbene (Table 6).

Carbenoids are carbene complexes which have reactivity similar to carbenes. Frequently the reactivity of carbenoids is moderated in a manner which is very useful synthetically, such as the Simmons-Smith reagent (CH_2I_2-Zn), which undergoes cycloaddition reactions well without any competing insertion reactions. *See* TRIPLET STATE.

Tetrahedral intermediates. These are compounds derived from addition of a nucleophile to a carbonyl group. Depending upon the substituents on the carbonyl group, the initially formed tetrahedral intermediate may give either an addition or a substitution product. Aldehydes and ketones give addition products, whereas carboxyl derivatives (including carboxylic acids, esters, amides, and acid anhydrides) give substitution products. (Table 7).

Other types. There are many kinds of reactive intermediates which do not fit into the previous classifications. Electronically excited states are reactive intermediates in which the electron structure is displaced from the normal ground state, typically as the result of absorption of a photon of light; the chemistry resulting from electronically excited states is called photochemistry. Benzyne is a reactive intermediate derived from an aryl halide by elimination with strong base; benzyne rapidly undergoes nucleophilic addition to give a net result of a substitution, reaction (7*a*), or may undergo cycloaddition, reaction (7*b*). *See* PHOTOCHEMISTRY; SUBSTITUTION REACTION.

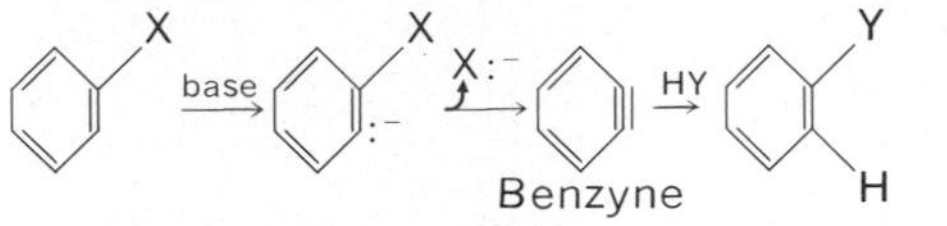

(7*a*)

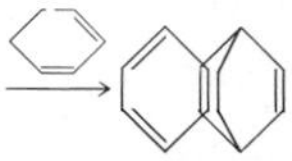
(7*b*)

Other reactive intermediates are simply compounds which are unstable for a variety of possible reasons, such as structural strain or an unusual oxidation state. *See* CHEMICAL DYNAMICS; ELECTROPHILIC AND NUCLEOPHILIC REAGENTS; MOLECULAR ORBITAL THEORY.

[CARL C. WAMSER]

Bibliography: J. March, *Advanced Organic Chemistry*, 2d ed., 1977.

Table 7. Addition and substitution reactions via tetrahedral intermediates

Reaction	Product
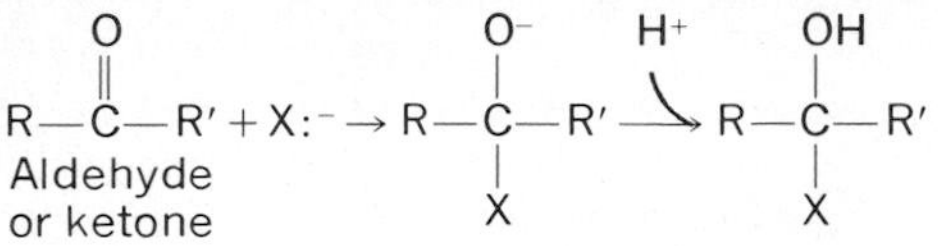 Aldehyde or ketone	Nucleophilic addition
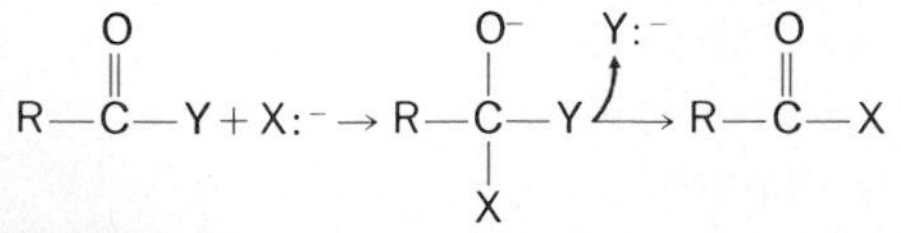 Carboxyl derivative	Nucleophilic substitution

Reagent chemicals

High-purity chemicals used for analytical reactions, for the testing of new reactions where the effects of impurities are unknown, and in general for chemical work where impurities must either be absent or at known concentrations. If the concentration of impurity in any reagent is critical, an analysis should be made.

Methods of purification. Chemicals are purified by a variety of methods, the most common being recrystallization from solution. For many inorganic chemicals a saturated solution is prepared in water at the boiling point. After filtration to remove insoluble matter, the chemical crystallizes out of solution as it cools. The crystals are removed by filtration on a small scale or by centrifugation on a large scale, washed with water to remove impurity-containing solution on the surface, and dried. If the substance to be purified is not appreciably more soluble in hot solution than in cold solution, then one can use isothermal crystallization, the removal of solvent at constant temperature by reducing the pressure. The recrystallization process is not always successful as a purification method. In some cases precipitation of crystals from a saturated water solution by adding a second solvent, for example, ethyl alcohol, is the simplest method. One difficulty is that the impurity crystals may have the same structure as the desired ones, that is, the two crystals may be isomorphous. If water of hydration is present, storage in an atmosphere of known humidity may be necessary to obtain a definite hydrate. Occasionally it is easier to remove the impurity by dissolving it in a solvent in which the desired material is not very soluble.

If the desired chemical is volatile and the impurities are not volatile, sublimation is an effective method of purification. For example, iodine and arsenious oxide are easily purified by sublimation. For liquid chemicals, distillation is an effective procedure. Finally, the simplest procedure may be to synthesize the desired reagent from pure materials; for example, the addition of pure ammonium carbonate solution to pure calcium chloride solution precipitates pure calcium carbonate.

Standards of purity. Commercial chemicals are available at several levels of purity. Chemicals labeled "technical" or "commercial" are usually quite impure. The grade "USP" indicates only that the chemical meets the requirements of the United States Pharmacopeia. The term "CP" means only that the chemical is purer than "technical." Chemicals designated "reagent grade" or "analyzed reagent" are specially purified materials which usually have been analyzed to establish the levels of impurities. The last two classes are the ones usually used in the laboratory. The American Chemical Society has established specifications and tests for purity for some chemicals. Materials which meet these specifications are labeled "Meets ACS Specifications."

A special group of extremely pure chemicals are called "primary standard" reagents. These reagents are usually readily available, easily purified, and unreactive with components of air such as water and carbon dioxide. The total sum of impurities should be less than 0.02%. These reagents are used to determine the concentrations of solutions used in volumetric analysis or for other purposes in which impurities must be quite low in concentration.

Care is necessary to prevent contamination by dust or by other chemicals. Transfers should be made directly from the container by pouring and not with spatulas or other tools. No material should be returned to the container. The maintenance of purity in an opened container is a problem.

Selective and specific reagents. Chemical reagents are often classified on the basis of utility. Many chemicals are general reagents, that is, they may react with many others. For example, any acid will be neutralized to some extent by a base. However, there are some reagents which react with only a limited number of other chemicals. These reagents are called selective reagents. The silver salts of chloride, bromide, iodide, thiocyanate, and a few other ions are insoluble in water; therefore silver nitrate is a selective precipitating reagent for these ions. A limited number of reagents are known which react appreciably with only one ion under specified conditions. These reagents are called specific reagents. The specific property is determined by the product formed. The ion dimensions, the charge densities, and the electron arrangements of the reagent and the ion which react must fall within certain limits or no reaction will occur. Variations in the acidity of the solution and the presence of other ions can change the condition of the ion so that no reaction can occur. Most inorganic reagents are at best selective. Most specific reagents are organic in nature.

These organic reagents have two types of reactive groups. One group forms electrovalent bonds by charge neutralization, and the second group forms covalent bonds by sharing electrons. The products are cyclic because both bonds are to the same ion, and they are called chelate compounds. These chelates are frequently very insoluble in water and intensely colored, making them very useful in analysis. See CHEMICAL SEPARATION TECHNIQUES; CRYSTALLIZATION; DISTILLATION.

[KENNETH G. STONE/CHARLES RULFS]

Bibliography: American Chemical Society, *Reagent Chemicals*: *ACS Specifications*, 4th ed., 1968, 5th ed., 1974; M. Fieser, *Reagents for Organic Synthesis*, 8 vols., 1980.

Reduced-pressure distillation

A technique used to distill high-molecular-weight substances. Materials with relatively low molecular weight, such as water, gasoline, and alcohol, can be readily distilled at atmospheric pressure. However, many materials have much larger molecules and require high temperatures for distillation. There is a limit to this temperature increase, for above 750°F (400°C), thermal decomposition begins. Therefore, to distill a material such as a motor oil whose distillation temperature is well over 750°F (400°C), some other distillation technique must be used.

Theory. There are several operations that will lower the distillation temperature. One of these is the use of steam. Another is the reduction of pressure. If the distillation vessel is put under a vacuum, thus removing atmospheric pressure from the surface of the boiling liquid, the vapor can escape more easily from the the liquid surface and at an appreciably lower temperature (Fig. 1).

The theory involved in vacuum distillation is derived from Dalton's perfect gas law and the Clausius-Clapeyron equation. The form frequently used to express the temperature-pressure relationship is the equation below.

$$\log p = (A/T) + B$$

Here p is the vapor pressure in millimeters of mercury, T is the absolute temperature, and A and B are constants from tables based on experimental data. The vapor pressure data of many substances at a variety of temperatures is based on this equa-

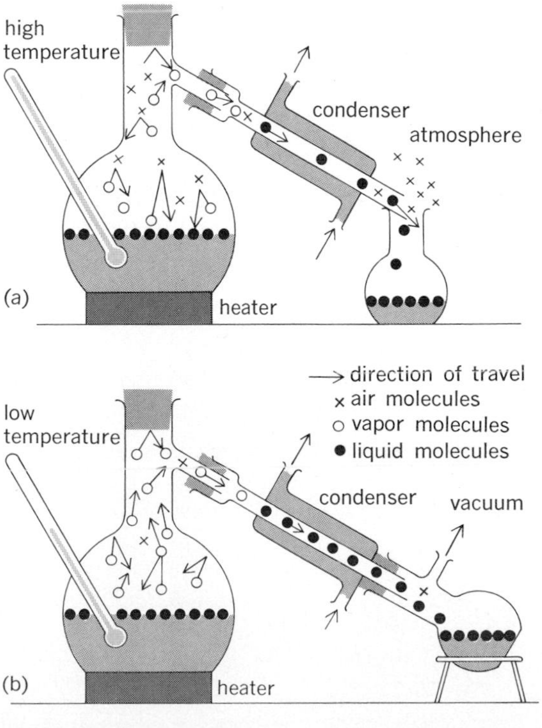

Fig. 1. Two types of distillation. (*a*) At atmospheric pressure. (*b*) At reduced pressure.

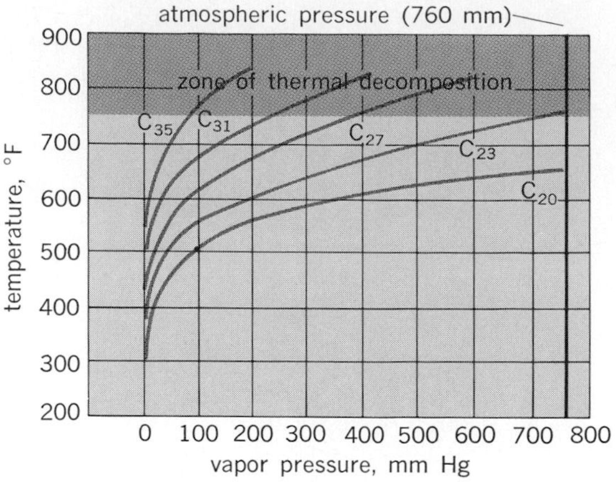

Fig. 2. Effect of pressure on distillation temperature of normal paraffin hydrocarbons. 1 mm Hg = 133.3 Pa. Temperature, °C = 5/9 [(temperature, °F) − 32].

tion. The torr, which is equal to 1 mm Hg absolute, or to 133.3 pascals, is the accepted unit of pressure, especially where vacuum is involved. However, since the millimeter of mercury is the more common unit in vacuum distillation, it is used here.

The effect that a reduction in pressure has on the distillation temperature of a motor oil is illustrated in Fig. 2. This type of oil, a petroleum base hydrocarbon, contains appreciable quantities of molecules ranging from 20 up to 35 carbon atoms (C_{20}–C_{35}) and higher. The temperature–vapor-pressure curves show that, while C_{20} and possibly C_{23} are at about the limit for atmospheric distillation, C_{27} and higher require a vacuum.

Equipment. The three general types of vacuum equipment are classified by the vacuum range they cover. Thus vacuum-distillation equipment covers the range from below atmospheric pressure to 5–20 mm Hg (0.7–3 kPa), wiped-film evaporators from 5–20 mm Hg to 0.2–0.5 mm Hg (30–70 Pa), and molecular distillation apparatus from about 0.025 mm Hg (33 Pa) to 0.001 mm Hg (0.13 Pa). *See* MOLECULAR DISTILLATION.

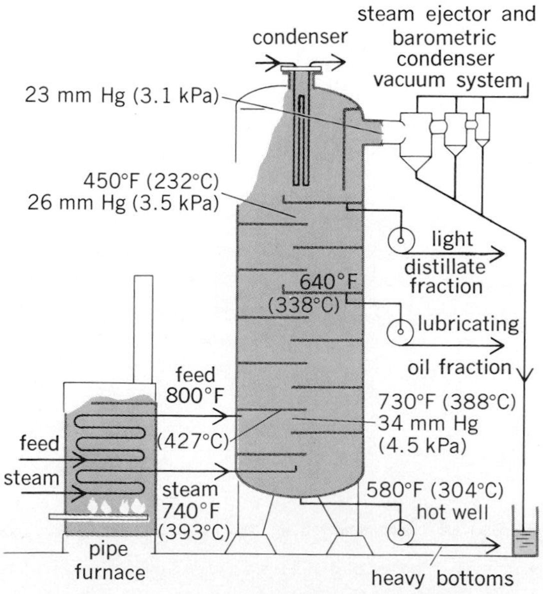

Fig. 3. Fractionating system for petroleum distilling.

Distillation at reduced pressure is widely used as a separation process. Equipment ranges in size from that which will handle micro quantities in the laboratory to plant installations that process thousands of barrels per day. One of the largest uses of this process evolved in the petroleum industry. In the 1920s the petroleum industry developed large vacuum-distillation facilities (usually with steam to assist) to increase the yield of lubricating oil that was otherwise lost as fuel oil. These lube oils amount to 200,000,000–300,000,000 bbl/year (30,000,000–50,000,000 m³/year). This does not take into account the huge quantities of asphalt that are upgraded by this process.

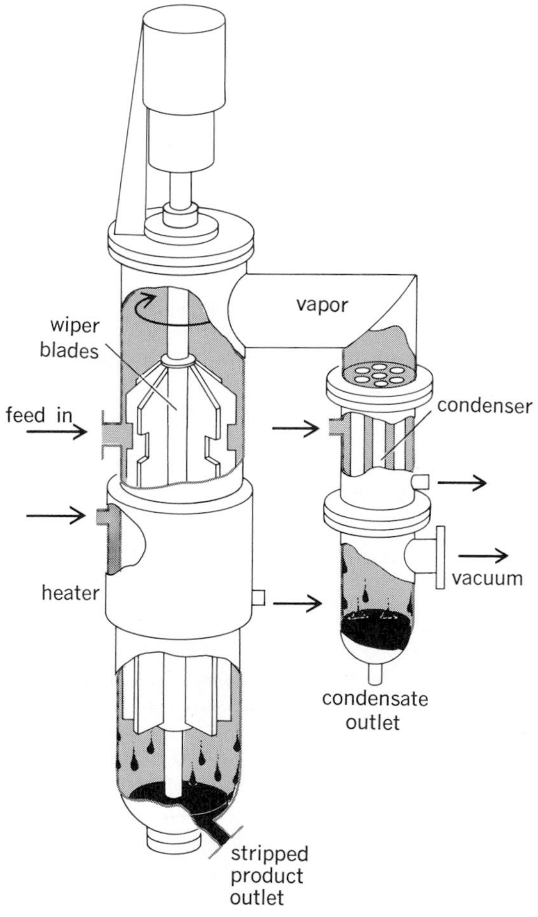

Fig. 4. Wiped-film evaporator.

Figure 3 illustrates the type of continuous vacuum-steam still used in commercial petroleum distillation. A hypothetical set of temperature and pressure conditions are included at pertinent locations. The common method of maintaining vacuum is by the use of steam ejectors combined with barometric condensers.

The same type of distillation equipment, though operated at somewhat lower temperatures, is used to purify and separate fatty acids. Fatty acids are a major ingredient in soap (as distinguished from synthetic detergents) and in lubricating greases. Another large user of vacuum-steam distillation is the edible fat and oil industry. The process is called deodorization, which removes objectionable taste and odor from cottonseed and soybean oils, which are the main constituents in shortenings, and in salad and cooking oils.

The wiped-film evaporator is illustrated in Fig. 4. The material to be distilled enters the top of the vacuum chamber and flows down the wall by gravity. Revolving blades wipe the material over the hot wall. Components that vaporize from the agitated liquid film leave the chamber and are liquefied in an external condenser. The undistilled portion drains out of the bottom of the still. Steam ejectors, with enough stages to produce the required lower pressures, supply the vacuum. While the use and size of the wiped-film evaporator are limited, it finds an important place in certain more specialized fields, especially those where the material is too viscous to flow without mechanical assistance. It is used to concentrate vegetable and fruit juices and pulp, to remove solvents from reaction mixtures, to evaporate liquids containing suspended solids, to remove water from gelatin, and in many similar operations.   [EDWARD S. BARNITZ]

Bibliography: A. E. Bailey, *Industrial Oil and Fat Products*, 3d ed., 1964; A. L. Carter and R. R. Kraybill, Low pressure evaporation, *Chem. Eng. Prog.*, vol. 62, no. 2, February, 1966; A. B. Mutzenburg, N. Parker, and R. Fischer, Agitated thin-film evaporators, *Chem. Eng.*, 72:175–190, Sept. 13, 1965; W. L. Nelson, *Petroleum Refinery Engineering*, 4th ed., 1958; R. H. Perry (ed.), *Chemical Engineers' Handbook*, 4th ed., 1966.

Reformatsky reaction

A reaction which takes place between a carbonyl compound, such as an aldehyde or ketone, and an α-halo ester in the presence of metallic zinc. Hydrolysis of the reaction mixture with dilute acid yields a β-hydroxy ester. The reaction is thought to proceed via an organozinc derivative, analogous to the Grignard reagent, formed by the interaction of the α-halo ester with the zinc. The organozinc compound then adds to the carbonyl group of the aldehyde or ketone (R = alkyl, aryl, or hydrogen; X = iodine or bromine). This sequence is shown in reactions (1)–(3).

The use of zinc in the Reformatsky reaction has the advantage that the organozinc intermediate has little tendency to attack the ester linkage, thus permitting syntheses which would not be possible with the more reactive Grignard reagents. Hence, the Reformatsky reaction is a valuable tool in organic chemistry as a method for preparing β-hydroxy esters and corresponding unsaturated and saturated esters and acids. It also serves as a useful means of lengthening the carbon chain. *See* GRIGNARD REACTION; ORGANOMETALLIC COMPOUND.   [MARVIN D. RAUSCH]

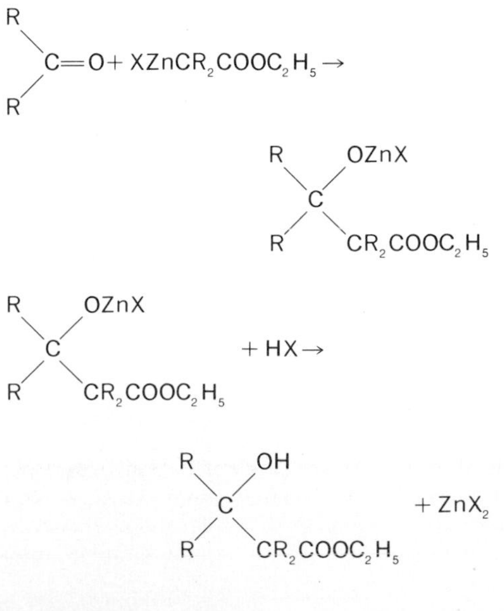

Bibliography: J. D. Roberts and M. C. Caserio, *Basic Principles of Organic Chemistry*, 2d ed., 1977.

Refractometric analysis

A method of chemical analysis based on the measurement of the index of refraction of a substance. As shown in the illustration, when light impinges on the surface of a material at an angle i to the normal to the surface, its direction is changed on passing into the material so that it then travels at an angle r to the normal. The index of refraction is defined in the equation below. It varies as a func-

$$\text{Index of refraction} = \frac{\sin i}{\sin r} = n_{\text{D}}^{20}$$

tion of temperature and wavelength of light, and also of pressure in gases. Refractive indices are usually measured at 20°C, using the yellow D line of the sodium spectrum. The indices of refraction of a few substances are water, 1.333; benzene, 1.5014; chloroform, 1.4464; and acetone, 1.3589.

The most common type of refractometer is the Abbé refractometer. It is simple to use, requiring but a drop or two of sample and allowing a measurement of refractive index to be made in 1–2 min, with a precision of 0.0001. More precise measurements of refractive indices, within 0.00003, may be made by using a dipping or immersion refractometer, the prism of which is completely immersed in the sample. This requires about 15 ml of sample and is widely used to detect trace impurities or to control the quality of a prod-

Zn + XCR$_2$COOC$_2$H$_5$ → XZnCR$_2$COOC$_2$H$_5$ (1)

$$\begin{array}{c} R \\ | \\ C=O + XZnCR_2COOC_2H_5 \rightarrow \\ | \\ R \end{array}$$

$$\begin{array}{cc} R & OZnX \\ & | \\ & C \\ & | \\ R & CR_2COOC_2H_5 \end{array} \quad (2)$$

$$\begin{array}{cc} R & OZnX \\ & | \\ & C \quad + HX \rightarrow \\ & | \\ R & CR_2COOC_2H_5 \end{array}$$

$$\begin{array}{cc} R & OH \\ & | \\ & C \quad + ZnX_2 \quad (3) \\ & | \\ R & CR_2COOC_2H_5 \end{array}$$

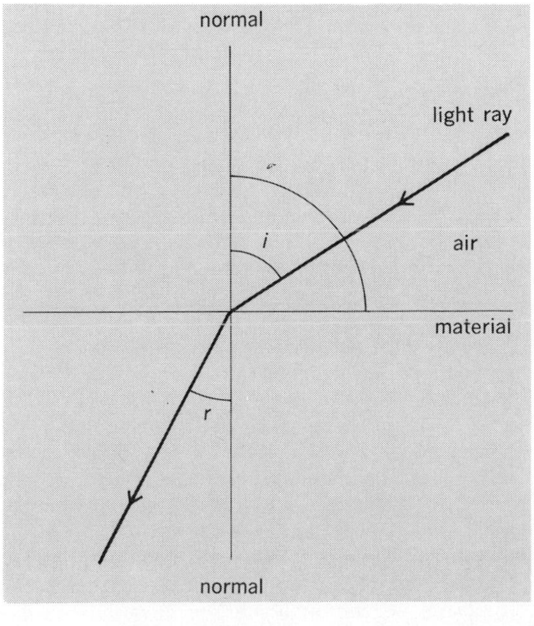

Refraction of light.

uct. The most precise measurements of the refractive indices of gases or solutions containing small traces of impurities are made with an interferometer, based on the interference of light. Its precision is 0.000001. For measurements in flowing systems, differential refractometers, capable of detecting a difference of 10^{-7} refractive index unit, are used.

The measurement of refractive index is used to identify compounds whose other physical constants are quite similar. Because minute amounts of impurities often cause a measurable change in the refractive index of a pure material, refractive index is often used as a criterion for purity. A measurement of refractive index gives information as to the gross amount of impurity; it does not serve to identify the impurity. To give qualitative information, measurements would have to be made at different wavelengths, a rare procedure. In systems containing only two components, such as water and alcohol mixtures or aqueous salt solutions, refractive index is a sensitive and rapid method of determining the composition. Refractometric analysis has also been used for the determination of the water content in milk, the amount of sulfur in rubber, and the isotopic analysis of D_2O–H_2O mixtures.

[ROBERT F. GODDU; JAMES N. LITTLE]

Bibliography: S. S. Batsanov, *Refractometry and Chemical Structure*, 1966; L. Meites (ed.), *Handbook of Analytical Chemistry*, 1963.

Relative atomic mass

The ratio of the average mass per atom of the natural nuclidic composition of an element to 1/12 of the mass of an atom of nuclide ^{12}C. For example, $\mu(Cl) = 35.453$. Relative atomic mass replaces the concept of atomic weight. It is also known as relative nuclidic mass. *See* NUCLIDE.

[THOMAS C. WADDINGTON]

Relative molecular mass

The ratio of the average mass per formula unit of the natural nuclidic composition of a substance to 1/12 of the mass of an atom of nuclide ^{12}C. For example, $\mu(KCl) = 74.555$. Relative molecular mass replaces the concept of molecular weight. *See* NUCLIDE. [THOMAS C. WADDINGTON]

Resonance

A feature of the valence-bond method, which is a mathematical procedure to obtain approximate solutions to the Schrödinger equation for molecules. The term came into use because the procedure is similar to that describing how weakly coupled tuning forks, pendulums, and such resonate, that is, transfer energy back and forth to one another. The valence-bond method is based on the theorem that if two or more solutions to the Schrödinger equation are available, certain linear combinations of them will also be solutions. It has this basis in common with its rival, the molecular orbital method. The valence-bond and molecular orbital approaches are both approximations and, if carried out to their logical and exact extremes, must yield identical results; nevertheless, both are often described as theories. In the valence-bond theory, combinations of solutions represent hypothetical structures of the molecule in question. These structures are said to be resonance (or contributing) structures, and the real molecule is said to be the resonance hybrid (or just simply the hybrid) of these structures. *See* MOLECULAR ORBITAL THEORY.

After it was demonstrated that the Schrödinger equation provided an exact three-dimensional description of the hydrogen atom, and thus that it was the key to the description of all chemical species, the resonance concept was developed to deal with the much more complex problem of finding solutions for multielectron atoms and for the multiatom molecules. The resonance theory was initially more popular than the molecular orbital approach, since it was applicable and understandable in qualitative terms so that the chemist could forego the difficult mathematics altogether. Also, the contributing structures that could be used were the classical ones that had been in use all along, and the resonance theory provided a solution for a molecule which had baffled and preoccupied chemists for a century—benzene. The principal use of resonance still lies in the qualitative description of molecules whose properties would otherwise be difficult to undertand. *See* BENZENE.

Structural formulas. It had long been a tradition in chemistry, especially in organic chemistry, to portray molecules in terms of the symbols of the atoms that are part of the molecules, with lines drawn between them to indicate chemical (covalent) bonds. Thus, methane, which is the principal component of natural gas and which has the formula CH_4, can be pictured to indicate that four hydrogen atoms are bound to a carbon atom (I); alternatively, the slightly more complex structure (II) is often drawn to indicate that the real

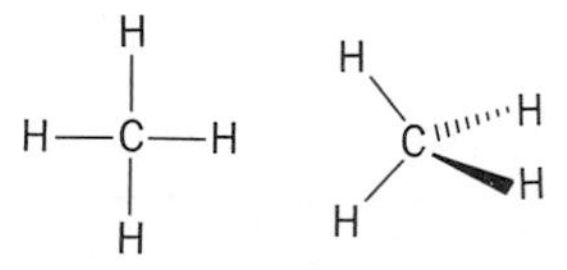

(I) (II)

molecule is not planar, but tetrahedral; that is, the four hydrogen atoms are at the corners of a (hypothetical) tetrahedron which has the carbon atom at the center. The lines drawn between the atoms denote the electron pairs which provide the bonds. Almost all molecules can be represented by such structures, and conversely, correctly devised structures may (with certain exceptions) be expected to correspond to molecules even if as yet they are not known either in nature or in the laboratory.

Formulating a correct structure requires knowledge of the number of electrons each atom has in its outer (valence) shell, and application of the so-called octet rule, which states that most atoms strive in their compounds to reach a total of eight electrons in this shell. Exceptions to this rule are the lightest few atoms—especially hydrogen (H), which can accommodate only two electrons—and the heavier ones which may have more than eight. Thus, the rule is not rigid, but is a generalization that is especially useful in biologically significant molecules, which are primarily made of atoms which obey it, namely, carbon (C), nitrogen (N), and oxygen (O). The isolated atoms, H, C, N, and O, have 1, 4, 5, and 6 electrons, respectively, in their valence shells. Thus methane could also be represented as (III). The carbon atom in this molecule is surrounded by four hydrogen atoms; it shares a pair of electrons with each of these, one

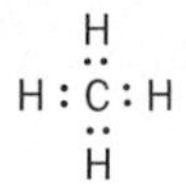

(III)

contributed by the hydrogen atom and one by the carbon. In this way, the needs for all five atoms (two electrons for each H—, and eight for the C atom) are satisfied. In a similar way, the correct structures can be readily devised for such molecules as ammonia, NH_3 (IV), water, H_2O (V), and ethane, C_2H_6 (VI), another constituent of natural gas. The un-

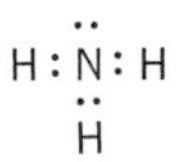

(IV)

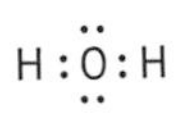

(V)

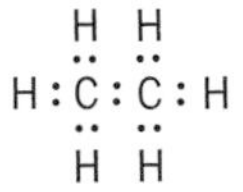

(VI)

shared electron pairs in such representations are often simply omitted, but understood to be there. The structures of such molecules as ethylene, C_2H_4, (VII), acetylene, C_2H_2 (VIII), and formaldehyde, CH_2O (IX), represent slight extensions in

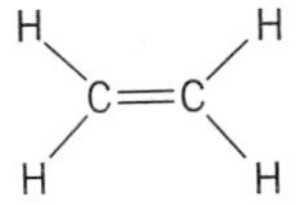

(VII)

(VIII)

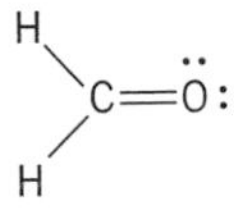

(IX)

that two or even three electron pairs may be used to bind first-row atoms such as C, N, and O together. One of the interesting chemical properties of such molecules containing multiple bonds is their tendency to form single bonds; thus, formaldehyde is readily converted into the new molecule trioxane (X) as shown in the reaction below.

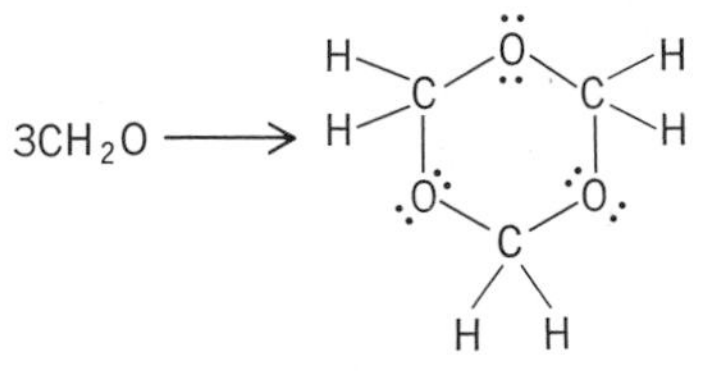

(X)

Benzene structure. Chemists had learned to deduce the atomic composition of molecules and the rudiments of the structural theory in the first half of the 19th century. The extension to three dimensions came in the 1870s, and that of the electron pair followed in the beginning of the 20th century. But through most of this period the substance benzene posed a baffling challenge to organic chemists: in spite of its relatively simple formula, C_6H_6, they were unable to conceive of a suitable structure for it. While a great many structures were proposed, the properties of benzene corresponded to none of them. Thus, it was known that there were three and only three disubstituted benzenes (that is, benzenes in which two of the six hydrogen atoms had been replaced by other monovalent atoms such as chlorine, Cl. Furthermore, although it is difficult to avoid writing several multiple bonds, benzene and its derivatives have little or no tendency to undergo reactions known as additions, which typically convert such multiple bonds into single ones. Thus, structure (XI) could

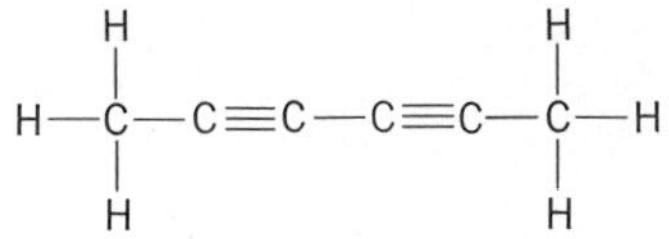

(XI)

be written for benzene, but would not be correct. Such a substance should have only two disubstituted derivatives (with the substituting atoms either at the same or at opposite ends of the six-carbon chain). A compound that does have this structure is known, but it is strongly sensitive to addition reagents.

In the early 1870s F. A. Kekulé proposed a revolutionary idea: benzene must be represented by two structures (XII) and (XIII) rather than one, and

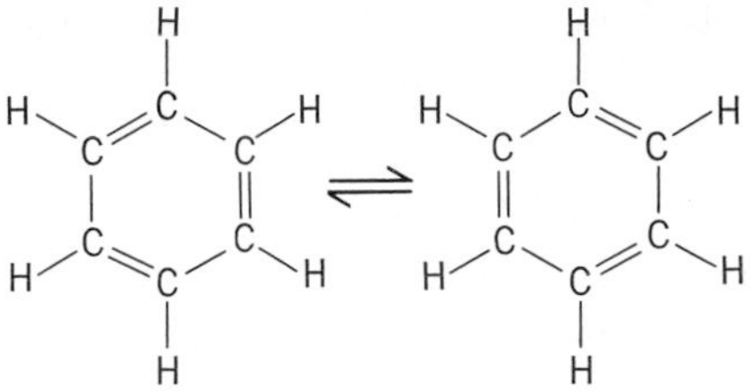

(XII) (XIII)

all compounds containing the benzene skeleton must be subject to a rapid equilibration (oscillation) between the two. Thus, whereas four disubstituted benzenes (XIV–XVII) would be

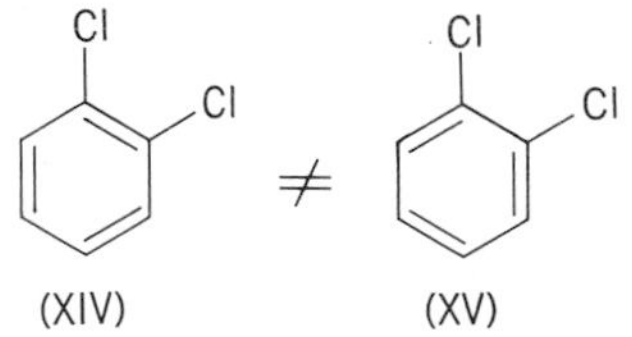

(XIV) (XV)

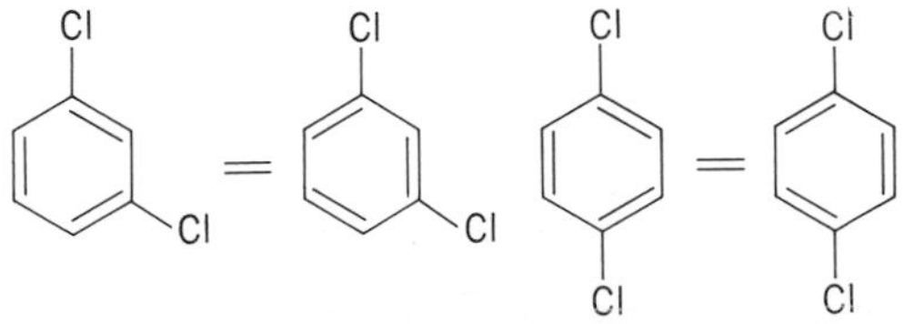

(XVI) (XVII)

expected on the basis of a single structure for benzene, the equilibration between (XIV) and (XV) was proposed to be so facile that only a mixture of the two can be isolated. This was a new development; chemists at that time assumed that every structure that can be drawn must correspond to an isolable compound, and that its chemical conversion to another compound requires some action on the part of a chemist: heating, the addition of chemical reagents, and so on. Many instances of such labile structures are now known; they are described as tautomers. However, benzene is not among them. *See* TAUTOMERISM.

Hybrid structure. Kekulé's description of benzene was not completely satisfactory. While it accounted for the number of isomers, it did not explain why the compound failed to exhibit reactivity indicating the presence of multiple bonds. The problem therefore continued to attract attention, but its final solution had to wait until the advent of quantum mechanics in the early part of this century. In a sense, this solution is an expansion of Kekulé's oscillating pair: the so-called activation energy (the energy which must be imparted to a molecule in order to make it overcome the barrier that keeps it from being converted into another molecule) is negative in the case of benzene with respect to the oscillation, and this molecule therefore exists neither as (XII) nor as (XIII) at any time, but it is in an intermediate form (XVIII) all

(XVIII)

the time. This intermediate structure of benzene is described in terms of Kekulé's structures with the symbol ↔ between them; this is intended to signify that benzene has neither structure, but in fact is a hybrid of the two. The properties of benzene are thereby indicated to be those of neither (XII) nor (XIII), but to be intermediate between the two. Thus, since single bonds are longer than double bonds, it might be expected from the Kekulé structures that the molecule should be irregularly shaped; in fact, benzene is known to have the shape of a regular hexagon. All six of the carbon-carbon bonds are equally long, and their length is intermediate between those of normal single and double bonds. The only property of the hybrid which is not intermediate between those of the hypothetical contributing structures is the energy: the energy of a resonance hybrid is by definition always at a minimum. This fact is responsible for the abnormal reluctance of benzene to undergo addition reactions; such reactions would lead to products that no longer have the resonance energy.

Other structures. Although benzene is the classical example of resonance, the phenomenon is certainly not limited to it; other instances include carboxylate anions such as formate (XIX) and guanidinimum cation (XX). Furthermore, the properties of all compounds are affected by resonance to some degree. Thus, the properties of acrolein (XXI) suggest that the molecule cannot be ade-

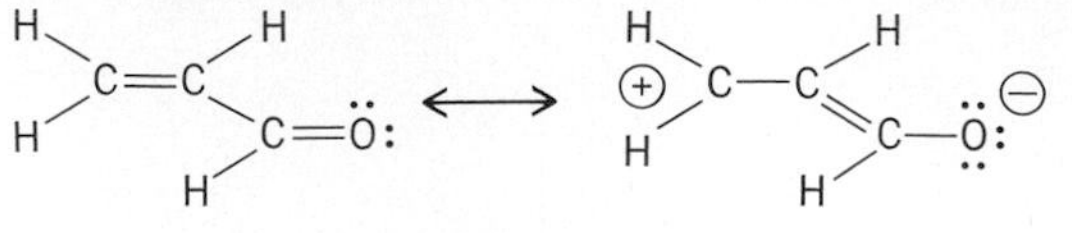

(XXI)

quately described by a single structure: the single carbon-carbon bond length is rather short, and the double bonds are longer than normal. Thus, the dipolar structure is making a contribution. The main difference in the preceding cases is that the contributions are not equal; the weight of the dipolar structure is much less than half. In general, this inequality of weight will occur whenever the two structures drawn are different. *See* CHEMICAL STRUCTURES.

Limitations. The theory of resonance owed its initial success primarily to the fact that molecules could continue to be described in terms of the same structures that chemists had used for a century. However, its use led to new questions that eventually began to undermine its popularity. Thus, its simple language cannot explain why the substance cyclobutadiene (XXII) is extremely un-

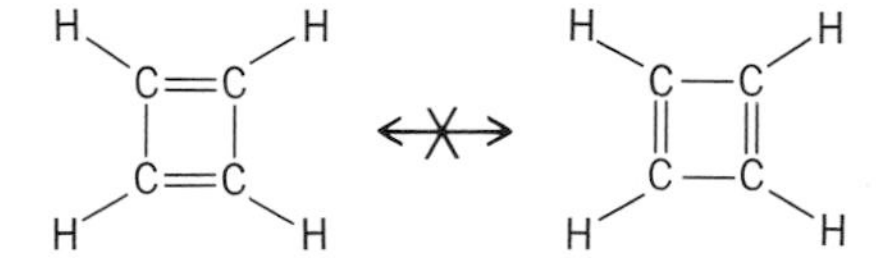

(XXII)

stable: it resisted all attempts to synthesize it for many years, and it readily converts into other substances at temperatures as low as 20 K. It is not obvious why it is apparently not subject to resonance stabilization just as benzene is. The molecular orbital treatment can readily account for this observation; it also provides a much simpler account for the spectral properties of organic compounds. While it represented a more radical departure from the traditional structures and language of classical chemistry, it has been accepted. Although the molecular orbital approach has largely supplanted the valence-bond method, the resonance language remains so convenient that it is still used. *See* CHEMICAL BONDING; CONJUGATION AND HYPERCONJUGATION; MOLECULAR STRUCTURE AND SPECTRA; ORGANIC CHEMISTRY; QUANTUM CHEMISTRY. [WILLIAM J. LE NOBLE]

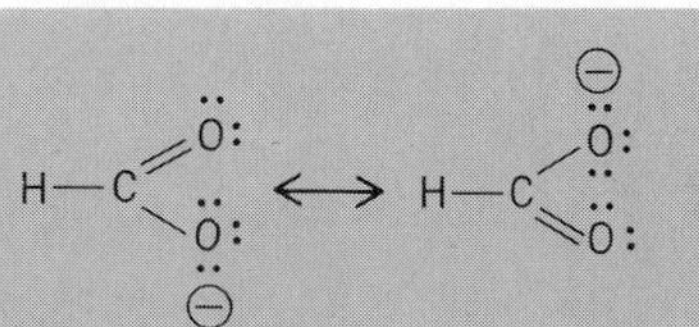

(XIX)

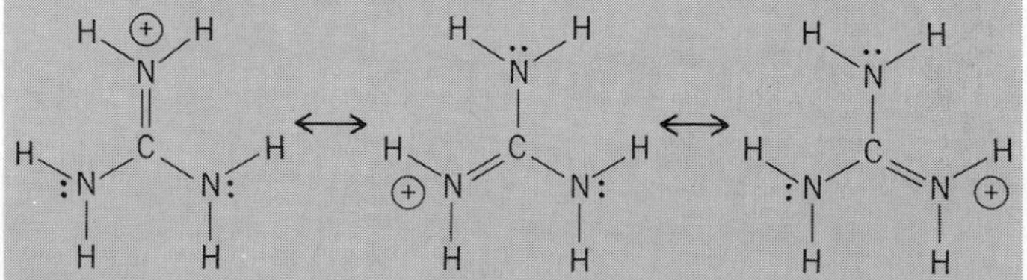

(XX)

Resonance ionization spectroscopy

A form of atomic and molecular spectroscopy in which wavelength tunable light sources are used to remove electrons from (that is, ionize) a given kind of atom or molecule. Laser-based resonance ionization spectroscopy (RIS) techniques have been developed and used with ionization detectors such as the proportional counter to show that single atoms can be detected. Both RIS and one-atom detectors find a wide range of applications in physics, chemistry, and oceanography, and in the environmental sciences.

Theory. When an atom (or molecule) is subjected to a light source that provides photons of an angular frequency ω, these photons can be absorbed by the atom if the photon energy $\hbar\omega$ ($\hbar$ is Planck's constant divided by 2π) is almost exactly the difference in energy between an atom in its normal or ground state and some excited state. Suppose, as in scheme 1 of Fig. 1, that a light source is tuned to a frequency which excites a given kind of atom, A. If the light source is a tunable pulsed laser of very narrow bandwidth, it is highly unlikely that any other kind of atom will be excited. But the atoms which are in an excited state can be further excited to the ionization continuum where electrons are set free, provided that the ionization potential of the atoms is less than $2\hbar\omega_1$. While the final ionization step can occur with photons of any energy above a threshold, the entire process is a resonance process. In sharp contrast with other ionization means—for example, x-rays or radioactive sources—resonance ionization spectroscopy is a selective process in which only those atoms that are in resonance with the light source are ionized. Modern pulsed lasers are excellent tunable sources for resonance ionization spectroscopy; furthermore, they provide enough light in a single pulse to remove one electron from each atom of the selected type. A laser that provides 100 millijoules of photons in a single pulse of 10^{-6} s duration can be tuned to ionize nearly all of the atoms of a given type that may happen to be gas contained in a virtual test tube whose diameter is 1 cm and which can be very long (even meters), consistent with the divergence of the laser beam. *See* LASER SPECTROSCOPY.

Laser schemes. Many laser schemes can be used, as shown in Fig. 1. The notation $A[\omega_1,\omega_1 e^-]A^+$, taken from the standard notation for describing nuclear reactions, is used for the two-step process described above. On the other hand, the frequency of a laser can be doubled to $2\omega_1$, so that scheme 2 requires only one laser, while schemes 3, 4, and 5 involve two lasers to generate photons at frequencies of ω_1 and ω_2. With these five schemes only, it is possible to selectively ionize every known element in the periodic table except two of the noble gases, helium and neon (Fig. 2). *See* NUCLEAR REACTION.

One-atom detection. Both the high selectivity and the extraordinary sensitivity of resonance ionization spectroscopy were demonstrated by pulsing a laser directly through a proportional counter (Fig. 3). It was shown by S. C. Curran, J. Angus, and A. L. Cockcroft in 1949 that an improved version of the 1908 Rutherford-Geiger electrical counter (now known as a proportional counter) can be used to count single electrons at thermal energy. Therefore, if lasers are used to remove one electron from all of the atoms of a selected type, one-atom detection is possible. Proportional counters are normally filled with gases like argon (90%) and methane (10%). A pulsed laser tuned, for example, to 455.5 nm can detect even one atom of cesium without producing background ionization of the counting gas. In the original demonstration of one-atom detection, it was proved that one atom of cesium could be selected out of 10^{19} atoms of the counting gas (argon and methane).

Another important form of one-atom detection involves the time-resolved detection of a single daughter atom in flight following the decay of a parent atom. Thus, it was shown that an atom of cesium could be detected from the fission decay of

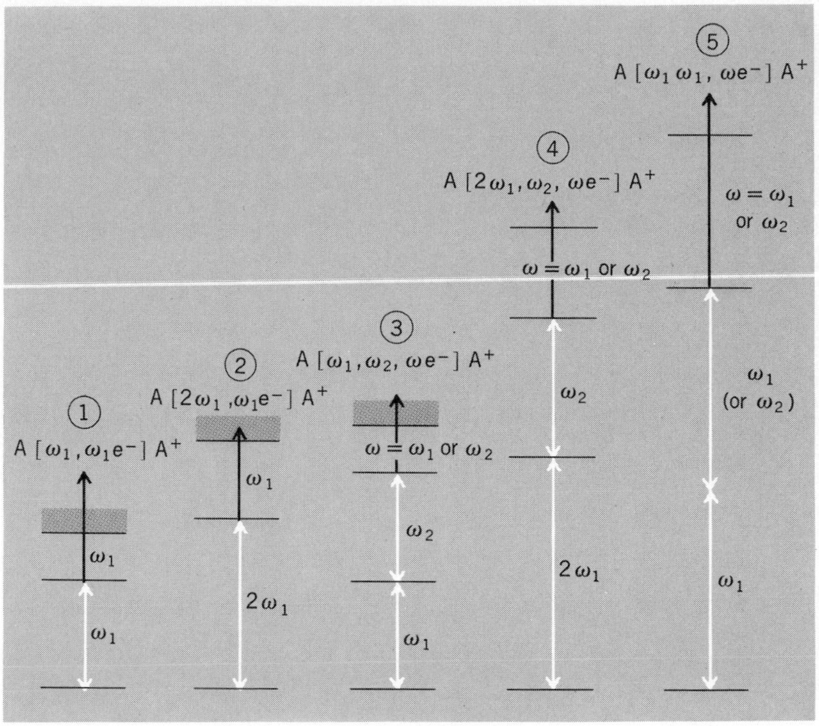

Fig. 1. Various laser schemes used in resonance ionization spectroscopy. *(Office of Health and Environmental Research, U.S. Dept. of Energy under contract W-7405-eng-26 with the Union Carbide Corp.)*

an individual atom of the isotope ^{252}Cf. The energy released in the fission process generated a signal that triggered the laser used to accomplish the resonance ionization spectroscopy process $Cs[\omega_1,\omega_1 e^-]Cs^+$. The success of that experiment proved that daughter atoms can be detected in coincidence with the decay of parent atoms. Such techniques could eventually work for most of the daughter atoms associated with radioactive decay, and could possibly be used to greatly reduce backgrounds in low-level counting facilities. *See* RADIOACTIVITY.

Applications. Resonance ionization spectroscopy and one-atom detection find a variety of interesting applications.

Classical chemical physics. The capability for detecting a population of just a few atoms has made it possible to investigate some problems in classical chemical physics which previously were difficult or impossible. Precision measurements of the diffusion of free atoms among other atoms and molecules have been made in sufficient detail to test the basic diffusion equation in both time and space domains. The determination of rates of reactions of extremely reactive substances (such as alkali atoms) with other atoms or molecules is now possible. Since only a few of these reactive atoms need to be produced, several problems concerning corrosion of the apparatus and the production of complicated chemical by-products are avoided. Study of a population of a few atoms (for example, 10) to observe their statistical behavior has been made. *See* CHEMICAL DYNAMICS.

Modern physics. One-atom detection makes it feasible to detect extremely rare events. Measurements of neutrinos from the Sun are crucial to testing both solar models and neutrino physics. After

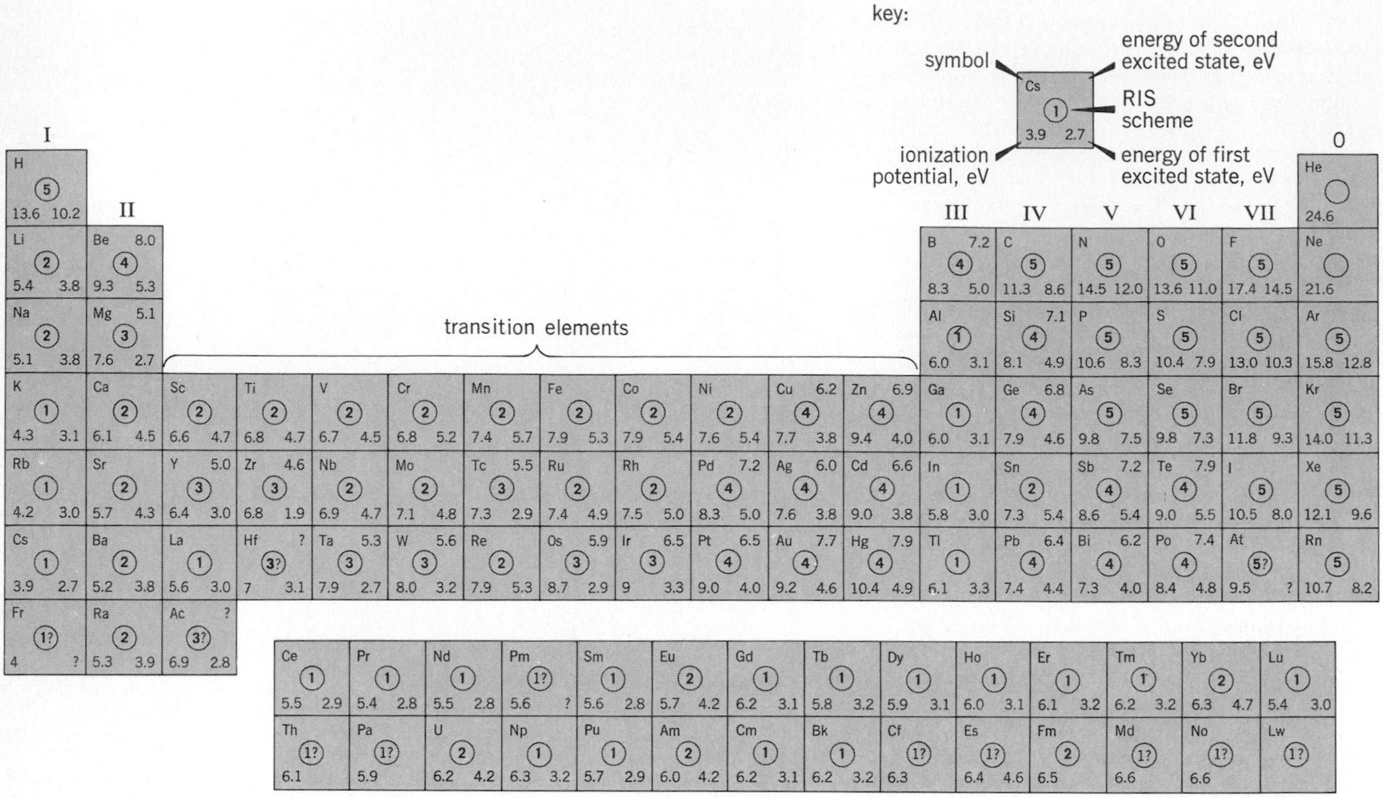

Fig. 2. Resonance ionization spectroscopy schemes for the elements of the periodic table. Numbers identifying schemes are those given in Fig. 1. Subgroups of elements are not designated by the letters a and b, conforming to modern usage. (*Office of Health and Environmental Research, U.S. Dept. of Energy under contract W-7405-eng-26 with the Union Carbide Corp.*)

prolonged exposure to the Sun, the neutrinos may produce on the order of 100 atoms of a particular type in a very large tank. Previously, targets have been rich in ^{37}Cl so that neutrino capture would produce ^{37}Ar, a radioactive atom which can be counted by the standard methods of radioactivity. Resonance ionization spectroscopy and one-atom detection are making possible a much wider variety of neutrino targets. For example, in a lithium-rich target neutrinos produce ^{7}Be, which can be detected by observing daughter (lithium) atoms in time coincidence with the decay of the parent atoms.

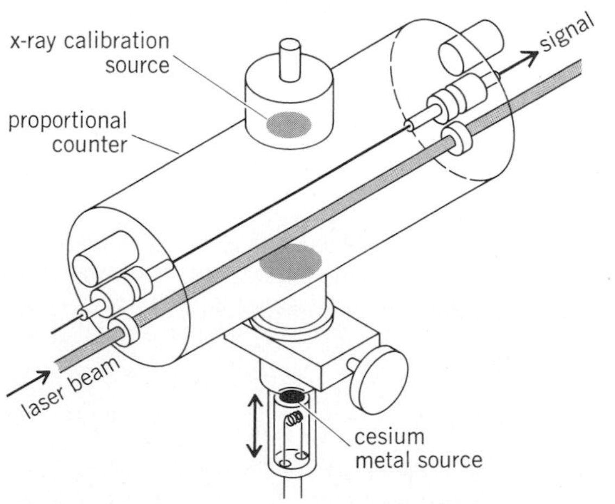

Fig. 3. Apparatus for experiment conducted to prove that resonance ionization spectroscopy can be used to detect a single atom. (*Office of Health and Environmental Research, U.S. Dept. of Energy under contract W-7405-eng-26 with the Union Carbide Corp.*)

Another experiment involves bromine-rich targets, where the neutrino capture produces ^{81}Kr. Radioactive ^{81}Kr can be counted directly (before it decays) by another technique made possible by resonance ionization spectroscopy. Other problems in weak interaction physics are also amenable to the one-atom detection techniques. Some meson interactions with nuclei have extremely low cross sections and thus produce only a few product atoms which, however, can be detected with resonance ionization spectroscopy techniques.

Environmental. Oceanographers have considered the use of ^{39}Ar as a tracer for ocean water circulation. Measurements of ^{81}Kr in the natural environment to obtain the ages of polar ice caps and old groundwater deposits have been suggested. Several techniques made possible by the development of resonance ionization spectroscopy and one-atom detection have been developed for these applications. [G. S. HURST]

Bibliography: S. C. Curran, J. Angus, and A. L. Cockroft, Investigation of soft radiations by proportional counters—I, *Phil. Mag.*, 40:36–52, 1949; R. Davis, Jr., D. S. Harmer, and K. C. Hoffman, Search for neutrinos from the Sun, *Phys. Rev. Lett.*, 20:1205–1209, 1968; G. S. Hurst et al., Resonance ionization spectroscopy and one-atom detection, *Rev. Mod. Phys.*, 51:767–819, 1979.

Resorcinol

A dihydric phenol ($C_6H_6O_2$), also called *m*-dihydroxybenzene. Resorcinol (I) is a water-soluble,

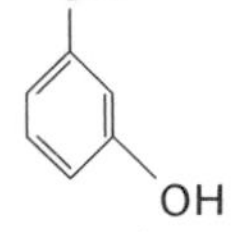

(I)

crystalline compound (m.p. 110°C) that is colorless when pure. Samples may acquire a pink color unless protected from light. Aqueous solutions of resorcinol are mildly acidic (pK 9.2); in this and other respects its properties are similar to those of phenol. Resorcinol is irritating to the skin, particularly that of individuals who have become sensitized, and it is a severe systemic poison.

Resorcinol is highly reactive in electrophilic substitution reactions. Treatment with hexanoic acid and zinc chloride, followed by reduction, yields 4-*n*-hexylresorcinol (II), which is widely used in household antiseptic products. Alkaline solutions of resorcinol react with carbon dioxide to produce, after acidification, β-resorcylic acid, also known as 2,4-dihydroxybenzoic acid (III). Reaction

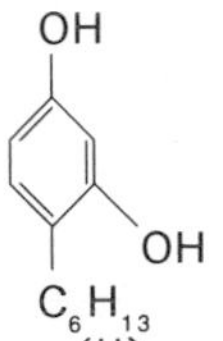

(II)

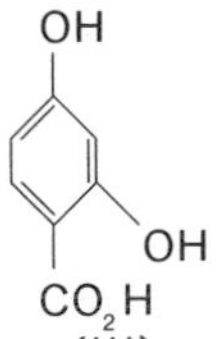

(III)

of resorcinol with phthalic anhydride in the presence of acid catalysts leads to the fluorescent dye fluorescein. The structurally related germicide mercurochrome is also derived from resorcinol.

The largest industrial use for resorcinol is in the manufacture of adhesives and resins. Adhesives based on resorcinol are exceptionally water-resistant, and are used extensively in marine plywood and other products which are constantly exposed to water.

Although resorcinol has been isolated from the degradation of certain plant products, the commercial material is produced synthetically. Industrial production proceeds by methods that resemble those used in the production of phenol, for example, the disulfonation of benzene, followed by fusion of the disulfonic acid with alkali. *See* HYDROQUINONE; PHENOL.

[MARTIN STILES]

L-Rhamnose

A methyl pentose, known also as 6-deoxy-L-mannose and L-mannomethylose. The free sugar, with the structure shown here, occurs in leaves and

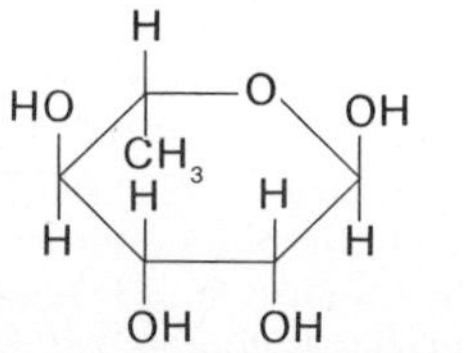

α-L-Rhamnopyranose

flowers of poison ivy (*Rhus toxicodendron* L.). It is a constituent of many of the plant glycosides such as quercitrin and rutin. The type II *Pneumococcus* specific polysaccharides and a wide variety of gums and mucilages contain L-rhamnose, and there is evidence that this sugar is present in a number of bacterial polysaccharides.

L-Rhamnose crystallizes from water or alcohol as α-L-rhamnopyranose monohydrate. Its melting point is 93–94°C, and the specific rotation is $[\alpha]_D$−8.6°, mutarotating to +8.2° (in water). On gentle heating, this sugar forms the anhydrous β-L-anomer that can be crystallized from acetone. *See* OPTICAL ACTIVITY.

Lemon flavin, the main constituent of which is the rhamnoside quercitrin obtained from the bark of an oak species, *Quercus tinctoria* Mich., provides an excellent source of L-rhamnose.

[WILLIAM Z. HASSID]

Rhenium

A chemical element, Re, with atomic number 75 and atomic weight 186.2. Rhenium is a transition element. Its discovery by W. Noddack, I. Tacke, and O. Berg in 1925 represented another outstanding success of the application of Mendeleev's peri-

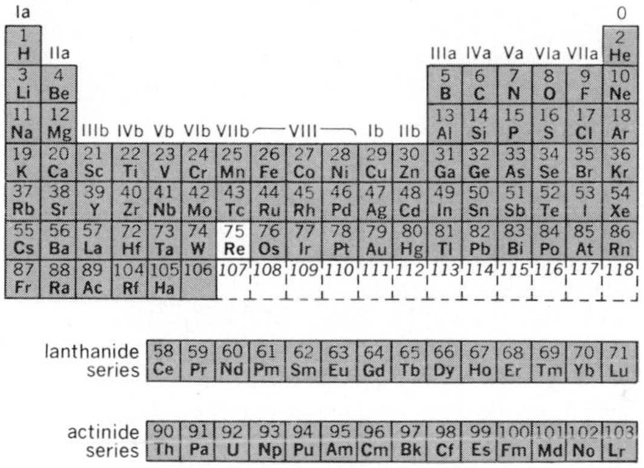

odic law. The blank space in the periodic chart corresponding to dvi-manganese had long occupied the attention of scientists, and many efforts had been made to discover the element whose properties would entitle it to be placed in that position. Noddack, Tacke, and Berg reasoned that it should occur in platinum ores and certain other minerals, notably columbite. *See* PERIODIC TABLE.

Today rhenium is obtained as a by-product of some metallurgical operations and is commercially available in laboratory quantities. Examination of a large number of minerals has shown that rhenium is widely distributed in nature. The most promising source appears to be molybdenum glance, MoS_2. Rhenium metal can be prepared easily by reduction of any of a number of its compounds by hydrogen. It is a dense metal (21.04) with the very high melting point of 3440°C.

Rhenium is similar to its homolog technetium in that it may be oxidized at elevated temperatures by oxygen to form the volatile heptoxide, Re_2O_7; this in turn may be reduced to a lower oxide, ReO_2. The compound ReO_3, as well as several others such as Re_2O_3 and Re_2O, is well known. The heptavalent oxide, Re_2O_7, may be dissolved in water to form the colorless perrhenic acid $HReO_4$, analogous to $HClO_4$, $HMnO_4$, and $HTcO_4$, and a large series of salts corresponding to this acid have been prepared. Perrhenic acid is a strong monobasic acid and is only a very weak oxidizing agent. Complex perrhenates, such as cobalt hexammine perrhenate, $[Co(NH_3)_6(ReO_4)_3]$, are also known.

The halogen compounds of rhenium are very complicated, and a large series of halides and oxyhalides have been reported. The halides and oxyhalides of the higher oxidation states tend to be quite volatile and sometimes are liquid. Tetravalent rhenium also forms a series of double salts corresponding to hexachlororhenic acid, which is analogous to those formed by technetium. Organic complex salts of rhenium, such as nitron perrhenate, can be used in its determination.

Rhenium forms two well-characterized sulfides, Re_2S_7 and ReS_2, as well as two selenides, Re_2Se_7 and $ReSe_2$. The sulfides have their counterparts in the technetium compounds, Tc_2S_7 and TcS_2. *See* TECHNETIUM; TRANSITION ELEMENTS.

[SHERMAN FRIED]

Bibliography: R. Kemmet (ed.), *The Chemistry of Manganese, Technetium, and Rhenium*, 1975.

Rheology

The study of the deformation and flow of matter. The states of matter differ strikingly in their density and in the ease with which they can be deformed. The less dense the state of matter, the more easily deformable it ordinarily is. The viscosity of a gas arises from the crossing-over of molecules from a fast-moving layer into a neighboring more slowly moving layer, and vice versa. Because this crossing-over increases with temperature rise, the viscosity of a gas increases with temperature. Solids and liquids, however, become more fluid with temperature rise. *See* VISCOSITY.

Viscosity and structure. A perfect crystal, according to theoretical calculations, should be orders of magnitude stronger than crystals customarily are found to be. This is because of the presence in actual crystals of various types of imperfections which greatly facilitate their deformation. The thermodynamic properties of crystals are readily calculated by assuming that the atoms form a perfectly ordered lattice. The rheological properties of crystals cannot be calculated from the perfect lattice model, however, but only by considering the number and nature of the lattice imperfections. Besides various types of vacant lattice sites, extra interstitial atoms are frequently present.

The extra fluidity of liquids, as compared with solids, arises from the great increase in the number of imperfections introduced with the 10% expansion in melting shown by normal liquids. Ice is an exception since it contracts by about 10% on melting. This contraction results from a change from the tetrahedral hydrogen-bonded structure displayed by one crystalline form of ice into some denser tetrahedral forms, such as ice III. As a result, water has a viscosity of about 17 millipoises at the melting point, which is close to the normal value for ordinary liquids. (A poise is a unit of viscosity equal to 1 dyne sec/cm^2.) A solid melts at that temperature at which the entropy from imperfections multiplied by the temperature equals the heat of introducing these imperfections.

Metals melt with only 3% expansion instead of the usual 10% expansion of normal un-ionized molecules because the positive ion, which is only about one-third as large as the atom, requires only one-third the space for extra equilibrium positions.

Molten salts, on the other hand, expand about 22% on melting. In this case, new equilibrium positions require about double the usual space. This is probably because a sodium chloride molecule must be accommodated in the added equilibrium position instead of a Na^+ or a Cl^- ion separately.

That the viscosity of a system closely reflects structure is exemplified by the fact that for nearly all simple substances the viscosity of the liquid at the melting point is approximately 2 centipoises. Also, the reciprocal of the viscosity, the fluidity, is a linear function of the molecular volume.

Because holes are necessary to permit viscous flow and since pressure tends to decrease the number of holes in a system, it follows that pressure normally increases viscosity. The pressure coefficient of viscosity indicates that the empty space a molecule requires in order to flow viscously is about one-seventh its normal volume. To provide such vacant sites, an activation energy of about one-third the heat of vaporization is required, as shown by the temperature coefficient of viscosity. The normal effect of pressure rise and temperature drop in increasing viscosity is modified in the rare cases where such changes promote significant modifications in structure. Thus, increasing the pressure on water just above the melting point increases the fluidity of water because the liquid structure is shifted away from the hydrogen-bonded tetrahedral state toward the more fluid ice-III-like structure. At a temperature of about 160°C, sulfur changes from a comparatively fluid, straw-colored, eight-membered ring to a dark, viscous, high polymer, having thousands of atoms in a linear chain. Lubricants likewise are made relatively temperature independent by adding high polymers which are soluble only with difficulty and which unfold with temperature rise, thus counteracting the usual decrease of viscosity of liquids with temperature rise.

When linear high polymers change their state, they introduce holes only in directions normal to the length of the chain. Thus a linear polymer melts in two directions but retains its solidlike properties along the chain length. The result is that high polymers progress by wriggling a segment at a time, as shown by the fact that they exhibit the expected pressure and temperature coefficients of a molecule the size of a segment. The viscosity is much higher for a high polymer because of a negative entropy of activation corresponding to correlation of segment motion. Extensive quantitative correlations of rheological properties are available. *See* POLYMER.

Significant structures in liquids. To obtain quantitative expressions for the viscosity, a model of the liquid state is important. Upon melting, liquids ordinarily expand and become orders of magnitude more fluid. Since x-rays indicate that melting does not materially change nearest-neighbor distances, the expansion must be due to holes in the liquid. These holes should approximate molecular size, since smaller holes would not be as effective in facilitating molecular motion, that is, are wasteful of entropy, and larger holes are unduly wasteful of energy. Melting occurs because a thermodynamically more stable state appears at the melting temperature where the kinetic energy can support the fluidized vacancies, characteristic of the liquid state, against the potential energy of melting, which tends to collapse the holes. Such considerations lead to a model of the liquid in which fluidized vacancies in the liquid closely mirror the vapor molecules in their size, motion, and concentration. The partition function f_l embodies this significant structure theory.

The partition function for argon takes the form of Eq. (1).

With appropriate modifications the partition function can be written for any liquid. Here Eq. (2) is the Einstein partition function for a solid, and

$$f_l=\left[\left(f_s\left\{1+10.7\frac{V-V_s}{V_s}\right\}\exp\left(-\frac{.0052\,E_sV_s}{(V-V_s)\,RT}\right)\right)^{Vs/V} f_g^{(V-Vs)/V}\right]^N \qquad (1)$$

$$f_s=\frac{e^{Es/RT}}{(1-e^{-\theta/T})^3} \qquad (2)$$

Eq. (3) is the partition function of a gas. The significant structure theory yields satisfactory values for all thermodynamic properties of liquids.

$$f_g=\frac{(2\pi mkT)^{3/2}}{h^3}\frac{eV}{N} \qquad (3)$$

In Eqs. (1) and (2), E_s, V_s, and θ are the energy of sublimation, the solid molal volume at melting, and the Einstein characteristic temperature, respectively; V, T, m, h, k, E and N are the liquid molal volume, absolute temperature, molecular mass, Planck's constant, Boltzmann's constant, internal energy, and Avogadro's number, respectively. The two numerical constants, 10.7 and .0052, are dimensionless and are calculated from theoretical considerations.

The viscosity η is defined by Eq. (4). Here f is

$$\eta=\frac{f}{\dot S}=\frac{f\lambda}{u} \qquad (4)$$

the shear stress, that is, the force per unit area along the plane of shear. The equation for rate of shear is $\dot S=u/\lambda$, where u is the relative velocity of contiguous planes of molecules parallel to the shear plane, and λ is the distance between these neighboring planes. If one has a mixture of structures subject to laminar flow in which the fractional area occupied along the shear plane of the ith structure is χ_i, the viscosity is represented by Eq. (5).

$$\eta=\frac{f}{\dot S}=\frac{\Sigma\chi_i f_i}{\dot S}=\sum_i \chi_i\eta_i \qquad (5)$$

In a liquid the gaslike structures occupy a fraction of the surface $\chi_i=(V-V_s)/V$, and the solidlike structures occupy the remaining fraction V_s/V of the planar surface. Thus Eq. (6) for the viscosity of a liquid is formed.

$$\eta=\frac{V_s}{V}\eta_s+\frac{V-V_s}{V}\eta_g \qquad (6)$$

The viscosity of a gas is usually represented by Eq. (7). The significant structure theory of the viscosity of solidlike molecules in the liquid is given by Eq. (8). In the preceding equations N is Avogad-

$$\eta_g=\frac{2}{3d^2}\left(\frac{mkT}{\pi^3}\right)^{1/2} \qquad (7)$$

$$\eta_s=\frac{Nh}{Z\kappa}\frac{V}{V_s}\frac{6}{\sqrt{2}}\frac{\psi}{V-V_s}\exp\left(\frac{aE_sV_s}{(V-V_s)RT}\right)\exp\left(-\frac{P(V-V_s)}{RT}\right) \qquad (8)$$

ro's number; $Z=12$ is the number of nearest neighbor positions, both full and empty; λ is the distance a molecule jumps in viscous flow; $\lambda_2\lambda_3$ is the area parallel to the shear plane occupied by a molecule; d is the molecule diameter and m its mass; and ψ is the vibrational partition function for the degree of freedom which becomes the reaction coordinate in the activated state. The significant structure theory of liquids leads to the rate equation, Eq. (9), for molecules to jump into a neighboring position at a distance λ under zero stress.

$$k'=\kappa\frac{kT}{h\psi}Z\frac{(V-V_s)}{V}\exp\left(-\frac{aE_sV_s}{(V-V_s)RT}\right)\exp\left(\frac{P(V-V_s)}{RT}\right) \qquad (9)$$

This result used in Eq. (4) leads to Eq. (8). For argon the term a has the value .0052 appearing in Eq. (1) and a dimensionless constant calculated from the model. For energy levels of the partition function ψ higher than $E_s/3$, the oscillator is replaced by a translator with a solid volume calculated by using van der Waals's b. The term exp $(P(V-V_s)/RT)$ in Eq. (8) gives only a small change in the fluidity, with an increase in the pressure and temperature at constant volume. Whether it is an increase or a decrease depends on whether P or T increases most rapidly as both increase. The transmission function $\kappa=.375$ is a nonthermodynamic quantity that is chosen to fit the data and is of the expected magnitude. Figure 1 shows agreement with the experimental findings of fluidity, $\theta=\eta^{-1}$,

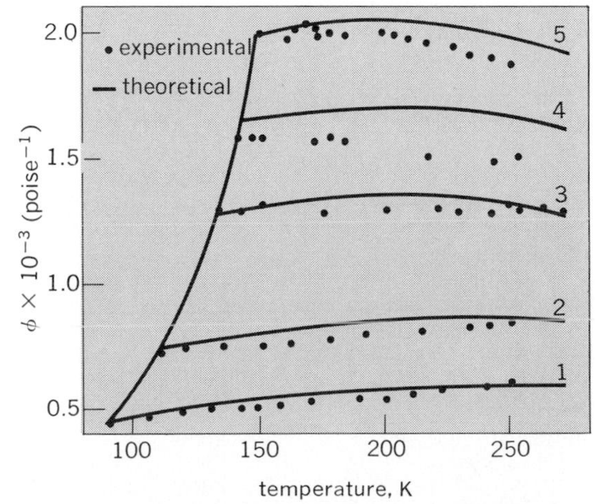

Fig. 1. Fluidities (poise^{-1}) of argon at constant volume versus temperatures (°K). Curve 1 at $V=29.660$ cm^3; curve 2 at $V=33.416$ cm^3; curve 3 at $V=40.59$ cm^3; curve 4 at $V=46.48$ cm^3; curve 5 at $V=52.78$ cm^3. (*Experimental data by N. F. Zhadanova, in T. Ree, T. S. Ree, and H. Eyring, Proc. Nat. Acad. Sci., 48:501, 1962*)

Table 1. Viscosities (millipoise) of argon under its vapor pressure*

T, K	η, calculated	η, observed	Δ, %
94.25	2.91	2.82	3.2
86.90	2.60	2.56	1.6
90	2.32	2.32	0.0
111	1.35	1.37	−2.2
133.5	0.81	0.77	2.6
143	0.65	0.63	3.2
149	0.55	0.50	2.0

*N. F. Zhadanova, Soviet physics, *JETP*, 4:749, 1957.

of N. F. Zhadanova over the entire liquid and gaseous range of argon.

Table 1 shows the agreement between calculated and observed viscosities of argon under its vapor pressure.

High polymer viscosity. Liquid high polymers shear as a result of individual kinetic units jumping in a direction to reduce the stress. If the high polymer is long enough, there will be entanglements and a given molecule will be obligated to "slalom" around a neighbor with which it is entangled. If the small contribution of gaslike molecules to the viscosity in Eq. (6) is neglected, Eq. (10) results.

$$\eta_1 = \frac{\lambda/kT}{2\lambda^2(\lambda_2\lambda_3)k'}\,\frac{V}{(V-V_s)} \qquad (10)$$

Here k' is the rate for a segment to jump into an empty site given in Eq. (9), and should not change with molecular weight after the molecule becomes long enough to flow by segments; $\lambda_2\lambda_3$ is the area of a segment upon which the shear stress acts to make the segment flow and does not change as additional segments are added to the molecule, and λ is the distance between successive layers of molecules normal to the shear plane.

A molecule consists then of $n_1 \equiv M/m_1$ kinetic segments and $n_2 = M/m_2$ tangling segments; that is, there are $(n_2 - 1)$ points about which the molecule has to slalom. Here m_1 is the mass of a kinetic segment, m_2 is the mass of a tangling segment, and M is the molecular weight. The subscript p indicates a property for the polymer and the subscript s the property for the kinetic segment. Accordingly, Eqs. (11), (12), and (13) are expected.

$$(\lambda_1)_p = (\lambda_1)n_s^{1/3} \qquad (11)$$

$$(\lambda)_p = (\lambda)_s/(n_1n_2) \qquad (12)$$

$$(k')_p = (k')_s n_1 \qquad (13)$$

Substituting these results into Eq. (10) leads to the following two cases: For $M < m_2$ Eq. (14) is expected to hold, and for $M > m_2$, Eq. (15) is expected to hold.

$$\eta_p = \eta_s n_1^{4/3} = (\eta_s/m_1^{4/3})M^{4/3} \qquad (14)$$

$$\eta_p = \eta_s\, n_1^{4/3}n_2^{2} = (\eta_s/(m_1^{4/3}m_2^{2}))\, M^{3.33} \qquad (15)$$

Here η_p is the viscosity of the high polymer, and η_s is the viscosity of the kinetic segment. Equation (11) expresses the idea that the effective distance between the centers of gravity of successive layers of molecules normal to the shear plane should increase in proportion to their molecular volume to the one-third power, and therefore to the one-third power of the molecular weight. Equation (12) follows because the motion of a kinetic segment advances the center of gravity of the high polymer containing n_1 segments by the fraction, $1/n_1$, of the amount the kinetic segment itself is advanced. Further, when there is tangling, only one out of n_2 tangling segments is free to choose a course which advances the molecule, since the remaining n_2^{-1} segments must slalom and so do not advance the center of gravity.

F. Beuche, proceeding differently, arrives at a power of the molecular weight equal to unity in Eq. (14) and equal to 3.5 in Eq. (15). The experimental results, as may be seen in Fig. 2 and Table 2, are in reasonable argeement with theory. Equations (11), (12), and (13), and the neglect of any effect of molecular weight on other factors in Eq. (10), thus seem justified, but more important, the model provides a simple method of interpretation of other cases that may arise. For example, simple branching of a high polymer should not greatly affect the dependence of viscosity on molecular weight when the situation is analyzed in terms of kinetic and tangling segments.

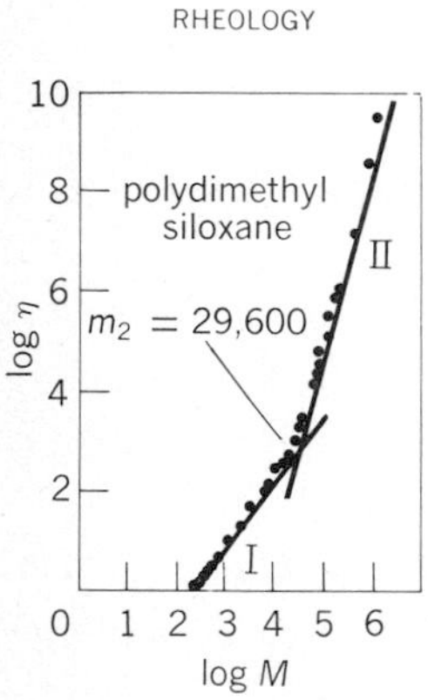

Fig. 2. The curve of log η versus log *M* for polydimethyl siloxane. *(From H. Eyring, T. Ree, and M. Hirai, Proc. Nat. Acad. Sci., 44:1213, 1958)*

Table 2. Effect of molecular weight on viscosity of high polymers*

Polymer	s_1	s_2	$-A$*	$-B$†	Log m_2 Calculated	Log m_2 Observed	Temp., °C
Polyisobutylene	1.75	3.40	13.7	6.30	3.70	4.29	217
Polystyrene	1.58	3.44	13.6	5.03	4.29	4.68	217
Polydimethyl siloxane	1.34	3.70	15.7	5.14	5.28	4.71	25
Polymethyl methacrylate†	1.40	3.40	14.2	5.50	4.35	4.20	60
Polydecamethylene adipate	1.34	3.40	12.1	4.47	3.82	3.68	109
Polydecamethylene sepacate	1.23	3.23	11.4	4.17	3.62	3.67	109
Polydiethylene adipate	1.26	2.92	10.0	4.11	2.95	3.56	109
Poly (ε-capcolactam)							
Linear chain	1.66	3.52	12.8	5.79	3.52	3.78	253
Dichain	1.70	3.50	12.5	5.80	3.35	3.72	253
Tetrachain	1.31	3.24	11.8	4.29	3.76	3.79	253
Octachain	1.31	3.24	12.7	4.75	3.93	4.09	253

*Here s_1 and s_2 are the observed slopes in the regions I and II of Fig. 2 for the substances listed, and are explained theoretically by Eqs. (14) and (15). $A \equiv -\log(\eta_1/m_1^{\frac{4}{3}}\, m_2^2)$. $B \equiv -\log(\eta_1/M_1^{\frac{4}{3}})$.

†25% solution in diethyl phthalate.

Non-newtonian viscosity. Frequently, systems are found which do not obey Newton's equation for viscosity; instead, a small increase in stress is found to disproportionately increase the rate of flow. For such a system one has Eq. (16) for the rate of shear.

$$\dot{S}=\frac{Z\lambda k'}{\lambda_1} < \cos\theta_i > \sinh\left(f\lambda_2\lambda_3\lambda < \cos\theta_1 > /2kT\right) \quad (16)$$

By introducing Eqs. (17) and (18), Eq. (16) becomes Eq. (19) or Eq. (20).

$$\beta^{-1}=Z\lambda k' < \cos\theta_1 > \quad (17)$$

$$\alpha=(\lambda_2\lambda_3\lambda < \cos\theta_i > /2rT) \quad (18)$$

$$\dot{S}=\beta^{-1}\sinh(\alpha f) \quad (19)$$

$$f=\frac{1}{\alpha}\sinh^{-1}(\beta\dot{S}) \quad (20)$$

Here β^{-1} is the mean rate of shear per second in any direction under zero stress, and may be called the intrinsic rate of shear. Thus β, aside from a shear factor near unity, is a relaxation time. The term α^{-1} has the dimensions of stress and may be thought of as an intrinsic stress characterizing the flow unit.

When several different types of flow units occur in a shear plane, with the type i occupying the fraction χ_1 of the area along the shear plane, the viscosity is given by Eq. (5). Combining Eqs. (20) and (5) yields Eq. (21).

$$\eta=\sum_i \frac{\chi_i\beta_i}{\alpha_i}\,\frac{\sinh^{-1}(\beta_i\dot{S})}{(\beta_i\dot{S})} =\sum_i \chi_i\eta_i\,\frac{\sinh^{-1}(\beta_i\dot{S})}{(\beta_i\dot{S})} \quad (21)$$

The factor $\sinh^{-1}(\beta_i\dot{S})/(\beta_i\dot{S})$ approaches unity for low values of the argument $(\beta_i\dot{S})$ and falls off toward zero as the argument increases. For example, two types of flow units, with the following assigned values $\chi_1\beta_1/\alpha_1=3\times10^{-2}$ poise, $\chi_2/\alpha_2=179.3$ (dynes/cm^2), and $\beta_2=7.25$ sec, suffice to fit the experimental data of W. Philippoff for 10% nitrocellulose in 99% butylacetate. Philippoff's data for 11% ethylcellulose in cyclohexanone was also fitted by using $\chi_1/\alpha_1=5\times10^4$ dynes, $\beta_1=1\times10^{-3}$ sec, $\chi_2/\alpha_2=1000$ (dynes/cm^2), and $\beta_2=.50$ sec. The results are plotted in Fig. 3. Clearly, viscosity is a simple concept only for low rates of shear, where a system is newtonian. In the non-newtonian range the intrinsic rates of shear β_i^{-1}, the intrinsic stress α^{-1}, and the fractional area χ_i, covered by units of the kind i, are the properties which characterize a system rather than the viscosity. Equation (21) accordingly applies over the entire newtonian and non-newtonian range.

[HENRY EYRING]

Bibliography: H. J. Cantow et al., *Advances in Polymer Science*, vol. 5, 1968; F. R. Eirich (ed.), *Rheology: Theory and Applications*, 5 vols., 1956–1970; H. Eyring et al., *Statistical Mechanics and Dynamics*, 2d ed., 1979.

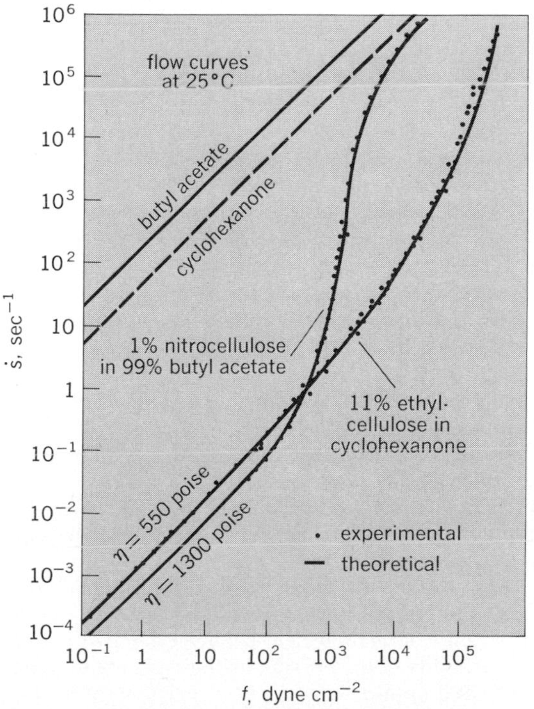

Fig. 3. Flow curves for 1% nitrocellulose in 99% butyl acetate and 11% ethylcellulose in cyclohexanone. Flow curves for solvents are shown by the straight lines of 45° slope. (*From T. Ree, T. S. Ree, and H. Eyring, Ind. Eng. Chem., 50:1036, 1958*)

Rhodium

A chemical element, Rh, atomic number 45, and atomic weight 102.905. Rhodium is a hard, white metal, considerably less ductile than either platinum or palladium, but much more ductile than any of the other platinum metals.

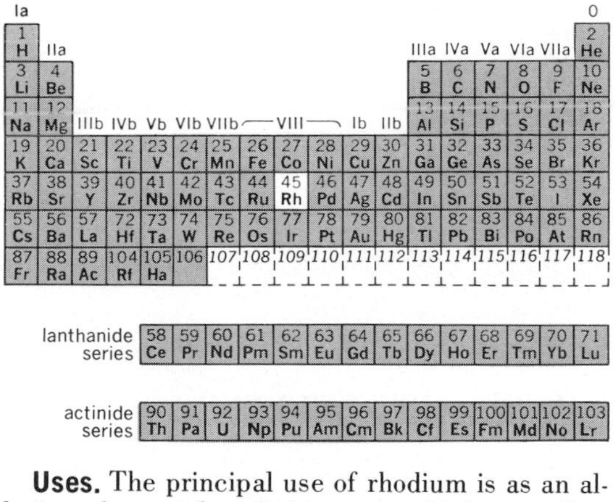

Uses. The principal use of rhodium is as an alloying element for platinum, to which it confers greater mechanical strength and the ability to withstand higher temperatures. Rhodium is an excellent hydrogenation catalyst and is active in catalytic reforming of hydrocarbons. Rhodium may be employed as an electric-furnace winding for use at temperatures too high for platinum-rhodium alloys. It is employed in growing laser crystals at temperatures lower than those requiring iridium.

Rhodium is used in electrical contact applications; its high strength allows use as a combined spring and contact. Rhodium alloys with iridium in all proportions; such alloys are used for thermocouples and high-temperature apparatus which must be used in oxidizing atmospheres above the temperatures at which platinum-rhodium alloys melt. Rhodium is easily electroplated to form a hard, wear-resistant, permanently bright surface. The plated metal is used on both stationary and sliding electrical contacts, for mirrors and reflectors, and as a finish on jewelry. *See* PLATINUM.

Physical and chemical properties. When hot, rhodium is quite ductile; when cold, the metal still has appreciable ductility, although it work-hardens very rapidly. It can be made into fine wire and thin sheet. Rhodium is the whitest of the platinum metals; it remains bright at ordinary temperatures under all atmospheric conditions. Rhodium forms a thin, adherent, protective oxide, RhO_2, on heating; above about 1100°C this oxide dissociates. Rhodium loses weight at higher temperatures through the volatilization of RhO_2. This occurs at about the same slow rate as that of the oxide of platinum. As this rate is much slower than that of iridium oxide and ruthenium oxide loss, platinum and rhodium are preferred precious metals for high-temperature use.

Rhodium is resistant to most common acids, including aqua regia, even at moderate temperatures. It is attacked by hot sulfuric acid, hot hydrobromic acid, sodium hypochlorite, and free halogens at 200–600°C. It is also attacked in varying degrees by fused bisulfates, molten cyanides, molten alkali nitrates, and alkali metal peroxides.

Physical properties of rhodium

Properties	Values
Atomic weight, C^{12} = 12.00000	102.905
Naturally occurring isotopes and % abundance	103 (100)
Crystal structure	Face-centered cubic
Lattice constant *a*, at 25°C, nm	0.38031
Thermal neutron capture cross section, barns (10^{-28} m²)	149
Common chemical valence	3, 4
Density at 25°C, g/cm³	12.43
Melting point, °C	1963
Boiling point, °C	3700
Specific heat at 0°C, cal/g (J/kg)	0.0589 (246)
Thermal conductivity, 0–100°C, cal cm/cm² sec°C (J · m/m² · s · °C)	0.36 (151)
Linear coefficient of thermal expansion, 20–100°C, μ-in./in./°C or μm/(m · °C)	8.3
Electrical resistivity at 0°C, microhm-cm	4.33
Temperature coefficient of electrical resistance, 0–100°C/°C	0.00463
Tensile strength, 1000 psi (6.895 MPa)	
Soft	120–130
Hard	200–230
Young's modulus at 20°C	
psi (GPa), static	46.2 × 10⁶ (319)
psi (GPa), dynamic	54.8 × 10⁶ (378)
Hardness, DPN*	
Soft	120–140
Hard	300

*Diamond pyramid number.

Rhodium is resistant to many molten metals such as gold, silver, mercury, sodium, and potassium, but in contrast to iridium and ruthenium, it is rapidly dissolved by lead and bismuth. The values of important physical properties are given in the table.

Metallurgical extraction. In one method of refining, platinum, palladium, and gold are separated by aqua regia solution, and the remaining rhodium, iridium, ruthenium, and silver are smelted with slag-formers and lead, the latter collecting the precious metals. The lead is enriched by cupeling, and then treated with nitric acid to separate rhodium, iridium, and ruthenium. A sodium bisulfate fusion dissolves rhodium as rhodium sulfate, which can be dissolved in water. Rhodium hydroxide is then precipitated, and redissolved in HCl to form rhodium chloride. This is further processed and finally purified by ion-exchange methods, followed by firing and reduction in hydrogen.

Rhodium compounds. Rhodium trichloride, $RhCl_3$, is a red compound which is insoluble in water. Its low volatility makes it useful in the refining of the element. Rhodium trihydroxide may be formed from the trichloride by boiling with potassium hydroxide. This is soluble in some acids and may be used to produce rhodium salts. Rhodium sulfate, $Rh_2(SO_4)_3 \cdot XH_2O$, is red or yellow and soluble in water. It appears to become complex after being subjected to high temperatures; as such, it forms the basis of rhodium plating baths.

[HENRY J. ALBERT]

Bibliography: *See* PLATINUM.

Rochelle salt

The sodium potassium salt of the *d*-tartaric acid $NaKC_4H_4O_6 \cdot 4H_2O$, also called Seignette salt, the first crystalline solid discovered to possess the properties of ferroelectricity. Such crystals are grown easily and have been widely used, for example, in microphones and phonograph pickup cartridges, because of their large piezoelectric effect.

The crystal structure has orthorhombic symmetry (point group 222) above +24°C and below −18°C. Between these Curie temperatures the crystal is ferroelectric, having a spontaneous polarization along the **a** axis and a spontaneous shear deformation y_z in the (100) plane. This reduces the symmetry to monoclinic. In general, the crystal consists of ferroelectric domains of opposite polarization direction. By applying an electric field along the **a** axis or a shear stress Y_z the spontaneous polarization can be aligned and reversed (hysteresis). Its highest value of 2.5×10^{-3} coulomb/m² is reached at about +3°C. The dielectric constant shows peak values of several thousand at the Curie points. It drops, according to the Curie-Weiss law, on both sides of the ferroelectric temperature range.

For technical applications the anomalously large effects are impaired by the narrow temperature range of ferroelectric behavior and by the limited stability of the crystals with respect to temperature and humidity. [H. GRANICHER]

Bibliography: W. G. Cady, *Piezoelectricity*, 1964; M. E. Lines and A. M. Glass, *Principles and Applications of Ferroelectrics and Related Materials*, 1977; T. Mitsui, *An Introduction to the Physics of Ferroelectrics*, 1976.

Rotatory dispersion

A term used to describe the change in rotation as a function of wavelength experienced by linearly polarized light as it passes through an optically active substance. *See* OPTICAL ACTIVITY.

Optically active materials. Substances that are optically active can be grouped into two classes. In the first the substances are crystalline and the optical activity depends on the arrangement of non-optically active molecular units. When these crystals are dissolved or melted, the resulting liquid is not optically active. In the second class the optical activity is a characteristic of the molecular units themselves. Such materials are optically active as liquids or solids. A typical substance in the first category is quartz. This crystal is optically active and rotates the plane of polarization by an amount which depends on the direction in which the light is propagated with respect to the optic axis. Along the axis the rotation is 29.73°/mm for light of wavelength 5086 A. At other angles the rotation is less and is obscured by the crystal's linear birefringence. Molten quartz, or fused quartz, is isotropic. Turpentine is a typical material of the second class. It gives a rotation of −37° in a 10-cm length for the sodium D lines.

Reasons for variation. In all materials the rotation varies with wavelength. The variation is caused by two quite different phenomena. The first accounts in most cases for the majority of the variation in rotation and should not strictly be termed rotatory dispersion. It depends on the fact that optical activity is actually circular birefringence. In other words, a substance which is optically active transmits right circularly polarized light with a different velocity from left circularly polarized light.

Any type of polarized light can be broken down into right and left components. Let these components be R and L. The lengths of the rotating light vectors will then be $R/\sqrt{2}$ and $L/\sqrt{2}$. At $t=0$, the R vector may be at an angle ψ_r with the x axis and the L vector at an angle ψ_l. Since the vectors are rotating at the same velocity, they will coincide at an angle β which bisects the difference as in Eq. (1). If $R=L$, the sum of these two waves will be

$$\beta=\frac{\psi_r+\psi_l}{2} \qquad (1)$$

linearly polarized light vibrating at an angle α to the axes given by Eq. (2). *See* POLARIZED LIGHT.

$$\alpha=\frac{\psi_r-\psi_l}{2} \qquad (2)$$

If, in passing through a material, one of the circularly polarized beams is propagated at a different velocity, the relative phase between the beams will change in accordance with Eq. (3),

$$\psi'_r-\psi'_l=\frac{2\pi d}{\lambda}(n_r-n_l)+\psi_r-\psi_l \qquad (3)$$

where d is the thickness of the material, λ is the wavelength, and n_r and n_l are the indices of refraction for right and left circularly polarized light. The polarized light incident at an angle α has, according to this equation, been rotated an angle given by Eq. (4).

$$\gamma=\frac{\pi d}{\lambda}(n_r-n_l) \qquad (4)$$

This shows that the rotation would depend on wavelength, even in a material in which n_r and n_l were constant and which thus had no circular dispersion. It is for this reason that the term rotatory dispersion is perhaps ill-defined in much of the literature.

In addition to this pseudodispersion which depends on the material thickness, there is a true rotatory dispersion which depends on the variation with wavelength of n_r and n_l.

From Eq. (4) it is possible to compute the circular birefringence for various materials. This quantity is of the order of magnitude of 10^{-8} for many solutions and 10^{-5} for crystals. It is 10^{-1} for linear birefringent crystals. [BRUCE H. BILLINGS]

Bibliography: P. Crabble, *ORD and CD in Chemistry and Biochemistry*, 1972; T. M. Lowry, *Optical Rotatory Power*, 1935, reprint 1964; G. Snatzke (ed.), *Optical Rotatory Dispersion and Circular Dichroism in Organic Chemistry*, 1976.

Rubidium

A chemical element, Rb, atomic number 37, and atomic weight 85.47. Rubidium is an alkali metal in group Ia of the periodic table. It is a light, low-melting, reactive metal. Little is known about rubidium because it has never been available in quan-

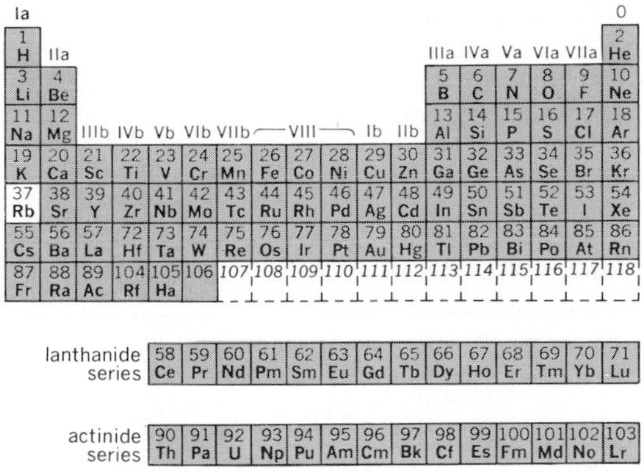

tity at a reasonable price. In 1958 rubidium salts became much more readily available and at lower prices as by-products of lithium chemicals manufacture, and knowledge of the properties and reactions of rubidium metal have developed accordingly. Rubidium was discovered in 1861 by R. Bunsen and G. R. Kirchhoff by interpretation of its spectral lines. Bunsen first prepared free rubidium metal that same year, using an electrolytic method.

Uses. Most uses of rubidium metal and rubidium compounds are the same as those of cesium and its compounds. The metal is used in the manufacture of electron tubes, and the salts in glass and ceramic production. Rubidium compounds are used in treating goiter and syphilis. Rubidium-mercury amalgams have been used as catalytic agents.

Occurrence. Rubidium is a fairly abundant element in the Earth's crust, being present to the extent of 310 parts per million (ppm). This places it just below carbon and chlorine and just above fluorine and strontium in abundance—well ahead

Physical properties of rubidium metal

Property	Temperature °C	Temperature °F	Metric (scientific) units	British (engineering) units
Density	20	68	1.53 g/cm³	95.5 lb/ft³
Melting point	39	102		
Boiling point	688	1270		
Heat of fusion	39	102	6.1 cal/g	10.95 Btu/lb
Heat of vaporization	688	1270	212 cal/g	381 Btu/lb
Viscosity	50	122	6.26 millipoises	4.1 kinetic units
	220	428	3.23 millipoises	
Vapor pressure	294	561	1 mm	0.019 lb/in.²
	628	1162	400 mm	7.75 lb/in.²
Thermal conductivity	39	102	0.07 cal/(sec)(cm²)(cm)(°C)	16.9 Btu/(hr)(ft²)(°F)
	50	122	0.075 cal/(sec)(cm²)(cm)(°C)	18.1 Btu/(hr)(ft²)(°F)
Heat capacity	39–126	102–259	0.0913 cal/(g)(°C)	0.0913 Btu/(lb)(°F)
Electrical resistivity	50	122	23.15 microhm-cm	
	100	212	27.47 microhm-cm	

of chromium, zinc, nickel, copper, and lithium. Sea water contains 0.2 ppm of rubidium, which (although low) is twice the concentration of lithium. Traces of rubidium are found in sea-water plants and animal organisms.

Rubidium is like lithium and cesium in that it is tied up in complex minerals; it is not available in nature as simple halide salts as are sodium and potassium. The major source of rubidium is lepidolite, which may contain up to 3% Rb_2O. Carnallite and pollucite are other minerals in which rubidium is found. These various minerals are found in the United States in California, South Dakota, New Mexico, and Maine. Lepidolite is found in substantial quantities in Rhodesia as well.

Metallurgical extraction. Rubidium metal is not produced on a commercial scale. In the limestone process for the conversion of lepidolite ore to lithium chemicals, however, a mixed alkali carbonate liquor is obtained by carbonation in a submerged combustion evaporator after the separation of the bulk of the lithium values as the hydroxide. Filtration after carbonation removes lithium carbonate and gives a filtrate containing carbonates of potassium, rubidium, and cesium.

The separation of the mixed alkali salts may be effected by a variety of methods, but the most common scheme involves complex formation. The order of ease of complex formation in the alkali metal series is $Cs > Rb > K > Na > Li$. Stannic chloride and zinc ferrocyanide are among the most economical compounds which form insoluble complexes with rubidium. Actually, the bulk of the potassium is first removed by precipitation as the bicarbonate before proceeding to the separation of rubidium and cesium. When stannic chloride is used, the rubidium and cesium chlorostannates are precipitated separately (the cesium salt is more insoluble and comes out first) and are then decomposed to the chlorides with recovery of the stannic chloride. Sodium zinc ferrocyanide may be employed in a similar separation, precipitating first cesium then rubidium. These precipitated ferrocyanides can be decomposed thermally to the carbonates.

Rubidium metal can be made by electrolysis of the chloride, or by thermochemical reduction of the carbonate with a metal such as magnesium or with a reducing compound such as calcium carbide. Magnesium reduction has been employed in many of the small-scale operations which have been practiced to date.

The physical properties of rubidium metal are summarized in the table.

Chemical properties. Rubidium is so reactive with oxygen that it will ignite spontaneously in pure oxygen. The metal tarnishes very rapidly in air to form an oxide coating, and it may ignite. The oxides formed are a mixture of Rb_2O, Rb_2O_2, and RbO_2. The molten metal is spontaneously flammable in air. Rubidium also is reputed to form an ozonide.

Rubidium reacts violently with water or ice at temperatures down to −100°C. It reacts with hydrogen to form a hydride which is one of the least stable of the alkali hydrides. Rubidium does not react with nitrogen. With bromine or chlorine, rubidium reacts vigorously with flame formation. Rubidium dissolves in liquid ammonia in the presence of metallic catalysts or with gaseous ammonia in the absence of catalysts to give rubidium amide, $RbNH_2$, and hydrogen. Rubidium reacts with carbon monoxide to give the carbonyl, RbCO.

Organorubidium compounds can be prepared by techniques similar to those used for sodium and potassium. In general, the behavior of rubidium in organic reactions is probably similar to that of sodium and potassium, but specific data are unavailable because of the cost and unavailability of rubidium metal. This has hindered research on its reactions. For a discussion of handling techniques *see* SODIUM.

Principal compounds. There are really no principal rubidium compounds in the sense of important uses or volume of production.

Analytical methods. Rubidium may be detected qualitatively by the red color obtained when volatile salts are introduced into a gas burner flame. Separation of rubidium compounds from those of the other alkali metals is difficult but can be accomplished by ion exchange. Quantitative determination can be made by flame photometry or by the use of one of various precipitants, such as tetraphenylboron derivatives or the uranyl acetates. *See* ALKALI METALS; CESIUM. [MARSHALL SITTIG]

Bibliography: W. A. Hart et al., *The Chemistry of Lithium, Sodium, Potassium, Rubidium, Cesium, and Francium*, 1975.

Ruthenium

A chemical element, Ru, atomic number 44, and atomic weight 101.07. Ruthenium is a hard, white metal, workable only at high temperatures and then only with difficulty. It is less workable than iridium, more workable than osmium.

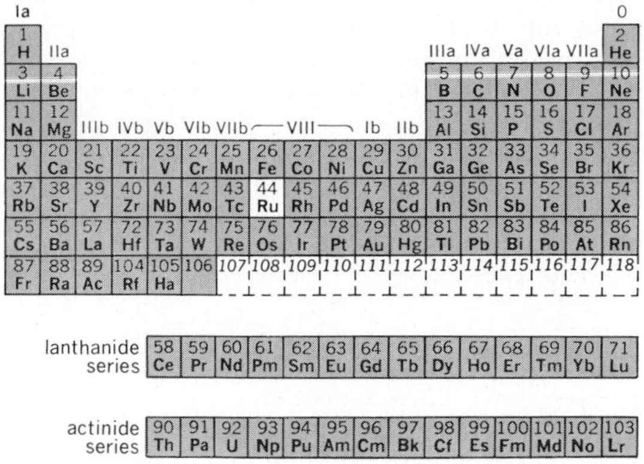

Uses. Ruthenium is an excellent catalyst, useful in reactions involving hydrogenation, isomerization, oxidation, and reforming. Uses of pure ruthenium metal are minor. Ruthenium is an effective hardener for platinum and palladium, used without sacrifice of corrosion resistance. Alloys with large percentages (30–70%) of ruthenium containing other precious or base metals have been used for electrical contacts and applications, where extreme wear resistance and corrosion resistance are required, as in fountain pen nibs and instrument pivots. Ruthenium may be dispersed in a second metallic phase, such as gold, for use in electrical contacts. It may be plated from aqueous or fused-salt baths for use on contacts and other applications, or it may be deposited by thermally decomposing a ruthenium organic compound. Ruthenium dioxide, RuO_2, is conductive. Mixed as a powder with a glassy frit in an organic medium, it may be applied to nonmetallic substrates to form resistor elements. *See* PALLADIUM; PLATINUM.

Physical and chemical properties. Ruthenium is not very ductile when cold, although pure single crystals can be bent easily. The metal can be melted by electric arc or electron beam. Powder metallurgy methods may also be used to consolidate it. Ruthenium must be worked hot, usually at about 1500°C, to relatively thin wire and sheet. Values of important physical properties of ruthenium are given in the table.

On heating up to about 900°C, ruthenium develops a thin oxide film, RuO_2, which persists up to about 1500°C. At about 1000°C the volatile oxides RuO_4 and RuO_3 are simultaneously formed. These oxides are responsible for the loss of weight of ruthenium at about 1000°C, with RuO_3 the more volatile as the temperature is raised. Ruthenium is much more oxidation prone than iridium; this limits the uses of ruthenium, pure or alloyed, at elevated temperatures.

Ruthenium is resistant to the common acids, including aqua regia, at temperatures up to 100°C, and up to 300°C in the case of sulfuric acid. It also resists hydrofluoric and phosphoric acids at 100°C. Chlorine water, bromine water, and iodine in alcohol attack it slightly at room temperature. It is fairly rapidly attacked by sodium hypochlorite and rapidly by fused sodium peroxide; it is also attacked by fused alkaline hydroxides, carbonates, and cyanides. It is resistant to many molten metals, including lead, lithium, potassium, sodium, copper, silver, and gold. It is slightly attacked by liquid bismuth.

Metallurgical extraction. A fusion with sodium peroxide of the residue from previous refining operations, which have removed platinum, palladium, and rhodium, yields water-soluble compounds of ruthenium and osmium. The last two elements may be removed from solution by distillation of their tetroxides; chlorine is streamed through the hot solution, and the resultant gas is passed through hydrochloric acid which reduces the ruthenium to the trichloride. Osmium can then be distilled from the latter solution. Alternatively, hydrated ruthenium dioxide may be precipitated from the original fusion extract by using an alcohol treatment. *See* OSMIUM.

Ruthenium compounds. Potassium ruthenate, $KRuO_2 \cdot H_2O$, is soluble in water and is useful in purifying ruthenium. Ruthenium trichloride, $RuCl_3$, is soluble in water but decomposes in hot water. Potassium hexachlororuthenate, K_2RuCl_6, is produced when ruthenium tetroxide is passed into

Physical properties of ruthenium

Properties	Value
Atomic weight, $C^{12} = 12.00000$	101.07
Naturally occurring isotopes and % abundance	96 (5.51) 98 (1.87) 99 (12.72) 100 (12.62) 101 (17.07) 102 (31.61) 104 (18.58)
Crystal structure	Close-packed hexagonal
Lattice constant a, at 25°C, nm	0.27056
c/a at 25°C	1.5820
Thermal neutron capture cross section, barns (10^{-28} m²)	2.56
Common chemical valence	2, 3, 4, 6, 8
Density at 25°C, g/cm³	12.37
Melting point, °C	2310
Boiling point, °C	4080
Specific heat at 0°C, cal/g (J/kg)	0.0551 (231)
Thermal conductivity, 0–100°C, cal cm/cm² sec °C	0.25
Linear coefficient of thermal expansion, 20–100°C, μ in./in./°C or μm/m/°C	9.1
Electrical resistivity at 0°C, microhm-cm	6.80
Temperature coefficient of electrical resistance, 0–100°C/°C	0.0042
Tensile strength, 1000 psi (6.895 MPa)	Hot swaged bar 72
Young's modulus at 20°C, psi (Pa)	
Static	60×10^6 (4.1×10^{11})
Dynamic	69×10^6 (4.75×10^{11})
Hardness, diamond pyramid number (DPN)	200–350 ($1.4–2.4 \times 10^6$)

a solution of hydrochloric acid and excess potassium chloride. The ammonium salt can be formed in this manner and subsequently reduced to pure metal by heating in hydrogen. Ruthenium tetroxide, RuO_4, may be formed by treatment of the metal with a very strong oxidizing agent such as sodium permanganate, ozone, or chlorine; the oxide is distilled off as described under metallurgical extraction. Ruthenium tetroxide is highly volatile and poisonous. It melts at about 25°C and sublimes at about 100°C. [HENRY J. ALBERT]

Bibliography: *See* PLATINUM.

Saccharin

An organosulfur compound first prepared by Ira Remsen, and also called *o*-sulfobenzoic imide (I). The material used as a sweetening agent (about 500–700 times as sweet as cane sugar) is the sodium salt (II), which passes largely unchanged through the body and is excreted in the urine. The slightly bitter aftertaste of sodium saccharin is mainly that of impurities from the conventional synthesis and can be avoided by syntheses starting

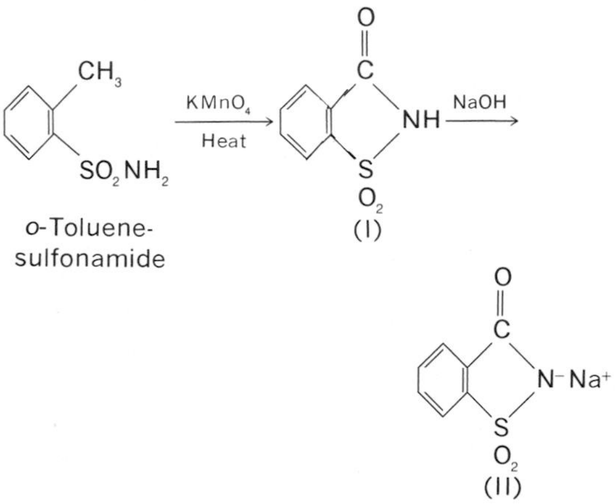

from anthranilic acid or benzothiophene. Saccharin is used in food preparation for low-caloric diets and in diabetes therapy, where normal sugars cannot be tolerated. *See* ORGANOSULFUR COMPOUND; SULFAMATE. [NORMAN KHARASCH]

Salicylate

A salt or ester of salicylic acid having the general formula shown below and formed by replacing the

$$(o)\text{-}C_6H_4(OH)C(=O)\text{—}O\text{—}M \quad \text{or} \quad \text{—}R$$

carboxylic hydrogen of the acid by a metal (M) to give a salt or by an organic radical (R) to give an ester. Alkali-metal salts are water-soluble; the others, insoluble. Sodium salicylate is used in medicines as an antirheumatic and antiseptic, in the manufacture of dyes, and as a preservative (illegal in foods). Salicylic acid is used in the preparation of aspirin. The methyl ester, the chief component of oil of wintergreen, occurs free and as the glycoside in many plants. This ester is used in pharmaceuticals as a component of rubbing liniment for its counterirritant effect on sore muscles. It is also used as a flavoring agent and an odorant. The phenyl ester (salol) and others are used medicinally.

[ELBERT H. HADLEY]

Bibliography: Merck Co., *The Merck Index*, 9th ed., 1976; J. D. Roberts and M. C. Caserio, *Basic Principles of Organic Chemistry*, 2d ed., 1977.

Saline water reclamation

The partial or almost complete demineralization of sea and brackish waters, geothermal brines, wastewaters, and industrial effluents to make fresh water suitable for human or animal consumption, diverse industrial uses, irrigation, recreation, or aquifer recharge—broadly referred to as desalination. The mineral content of these waters (see Table 1) is expressed in units of parts per million (ppm), giving the mineral content in terms of parts by weight of dissolved minerals in a million parts of water or parts per million salinity. The quality requirement for the fresh-water product depends upon its use.

Table 1. Typical mineral content of different waters

Type of feed water	Mineral content, ppm
Brackish water	1000–5000
Sea water	10,000–45,000*
Geothermal brines	3000–20,000
Industrial effluents	500–5000
Municipal waste waters	500–5000

*In some locations up to 70,000.

Standards. The U.S. Department of Health, Education and Welfare Public Health Service Drinking Water Standard of 1962 requires that the total dissolved solids of drinking water used on common carriers engaged in interstate commerce should not exceed 500 ppm when other, more suitable supplies are or can be made available.

Water containing several thousand parts per million of dissolved minerals is consumed by humans in many locations without noticeable ill effects, especially where the rate of perspiration is high. Suitable salinities for irrigation waters depend upon the chemistry of the soil and the mineral requirements of the crop, but generally should not exceed about 1200 ppm, particularly if the sodium content is high. Ruminant animals have developed tolerances for salinities up to 12,000 ppm. Industrial water requirements vary greatly from 10 parts per billion (ppb) for pressurized water reactors to about 1 ppm for boiler waters and semiconductor processing, and up to 35,000 ppm or more for some flushing and cooling operations.

Separation processes. A diversity of approaches are used for the separation of water from saline solutions. Thermal processes effect separation by means of phase changes and include distillation and freezing processes. In the membrane processes, one or more suitably designed organic membranes accomplish the separation process. In a single membrane design, reverse osmosis (RO) pressure forces the fresh water through the membrane. In electrodialysis (ED), a system using multiple membranes, direct current leads to formation of pure water and brine streams and drives the salt ions toward the electrodes through charge-selective membranes. There are also chemical processes. In the ion exchange process, substances are added to exchange the ions in the solution or to precipitate the salts. In the solvent extraction process, chemicals with greater affinity for water can remove the wastes from solutions, a system which has not met with commercial success.

There is no optimum or universal process for demineralization of all water streams. Each so-called contaminated or modified water stream must be evaluated for its peculiarities and an optimal process for purification for a specific end use recommended. Among the criteria to be considered are: salinity, identity of the ions present, presence of other contaminants, nature of the water, availability, energy costs, quantity, and quality.

The absolute minimum theoretical thermodynamic energy required for separating 10^3 gal ($3.8\ m^3$) of fresh water from sea water is 2.8 kilowatt-hours, but the energy requirement of virtually all processes exceeds this value manifold. In fact, desalination is an energy-intensive process. Progress in improving water conversion technology and reducing requisite energy for the purpose of desalination has made major strides.

THERMAL PROCESSES

The separation in thermal processes is accomplished by a phase change caused by distillation or freezing.

Distillation methods. These have developed from comparatively simple systems. Distillation was introduced for shipboard use in 17th century England and finally in 1912 the first land-based unit of the single-effect submerged-tube type was built (Fig. 1*a*). The next generation of equipment consisted of multieffect submerged tube units (Fig. 1*b*). In the single-effect units water is evaporated and condensed once, while in multieffect units several evaporators or effects are utilized so that the latent heat of evaporation can be captured and reused at lower temperature and pressure, making the process more economical. An inseperable problem of submerged-tube distillation plants was the problem of scale. This was the formation of chemical deposits (scale) on the tubes (evaporating and heat transfer surfaces) which materially affected plant performance and had to be removed continuously. Both single-effect and multieffect submerged-tube distillers have become obsolete.

Multistage flash. The goal of avoiding the scaling of submerged-tube desalination plants led to the development of the multistage flash (MSF) process, where water is heated up to 250°C. Polyphosphate, sulfuric acid, or so-called high-temperature additives (Belgard type) are added to the sea water to prevent scaling. Polyphosphate-operated plants have an upper limit of 190°C; acid and high-temperature additive plants, 250°C. Both polyphosphate and acid-dosed plants are gradually being phased out: the first because of low thermal efficiencies, and the second, in order to decrease serious corrosion problems. A process modification, which removes the ions responsible for scaling from the water, has demonstrated operational capabilities approaching 300°C. In all these units, the sea water is heated by counterflow through the system, recovering the heat of evaporation. The preheated water is given an additional amount of heat at the brine heater to bring it to the requisite temperature for the specific version of the process, and is then made to pass through a number of stages at successively lower temperatures and pressures when the sea water flashes (Fig. 2). The condensate from each stage is the product water. This multistage flash process has two versions: the so-called once-through MSF and the MSF with recirculation. The latter is the most widely used distillation process, accounting for over 80% of all land-based desalination plants in the world. Though the process is old, it has proved most reliable. The largest multistage flash complex in the world is in construction in Al Jubail, Saudi Arabia. It will have 2.5×10^6 gal/day capacity. Thus multistage flash plants are expected to be providing distilled water until the year 2000, at least.

Vertical-tube evaporators. In these systems, counterflow preheated sea water, after being brought to boil, is distributed into the top of vertical tubes and allowed to fall as a film down the inside tube wall. Heat transfer through the falling film is excellent. A portion of the sea water evaporates, and this steam vapor is used to heat the outside of the tubes of the following effect, giving

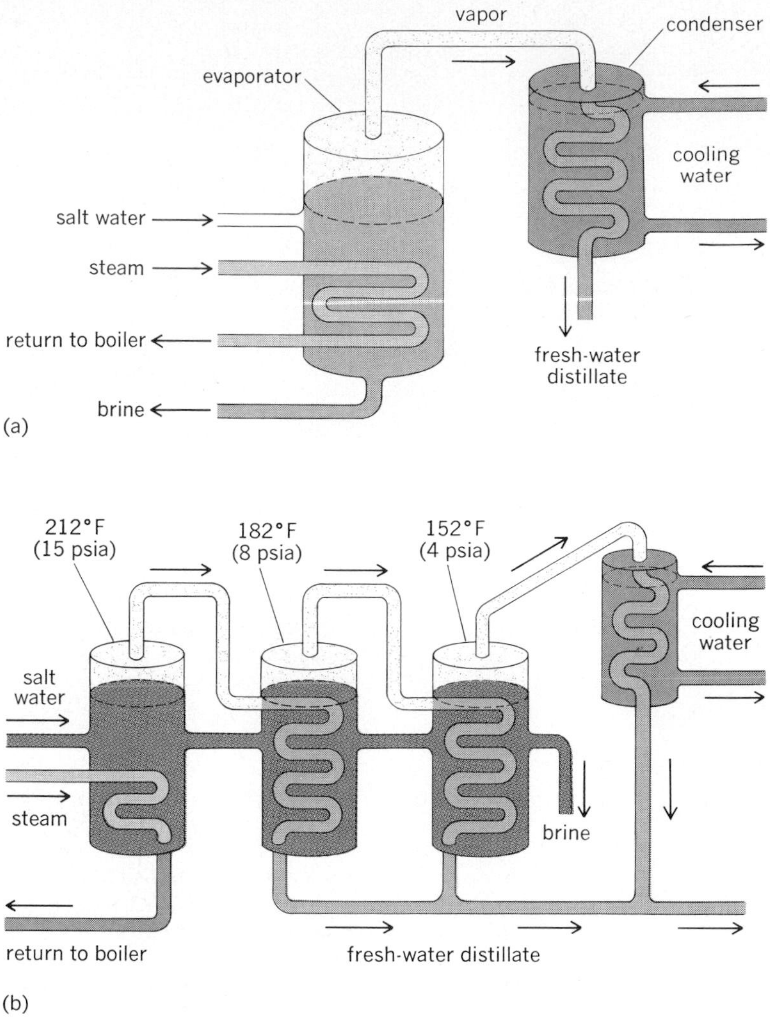

Fig. 1. Schematics of distillation process. (*a*) Single effect. (*b*) Multiple effect.

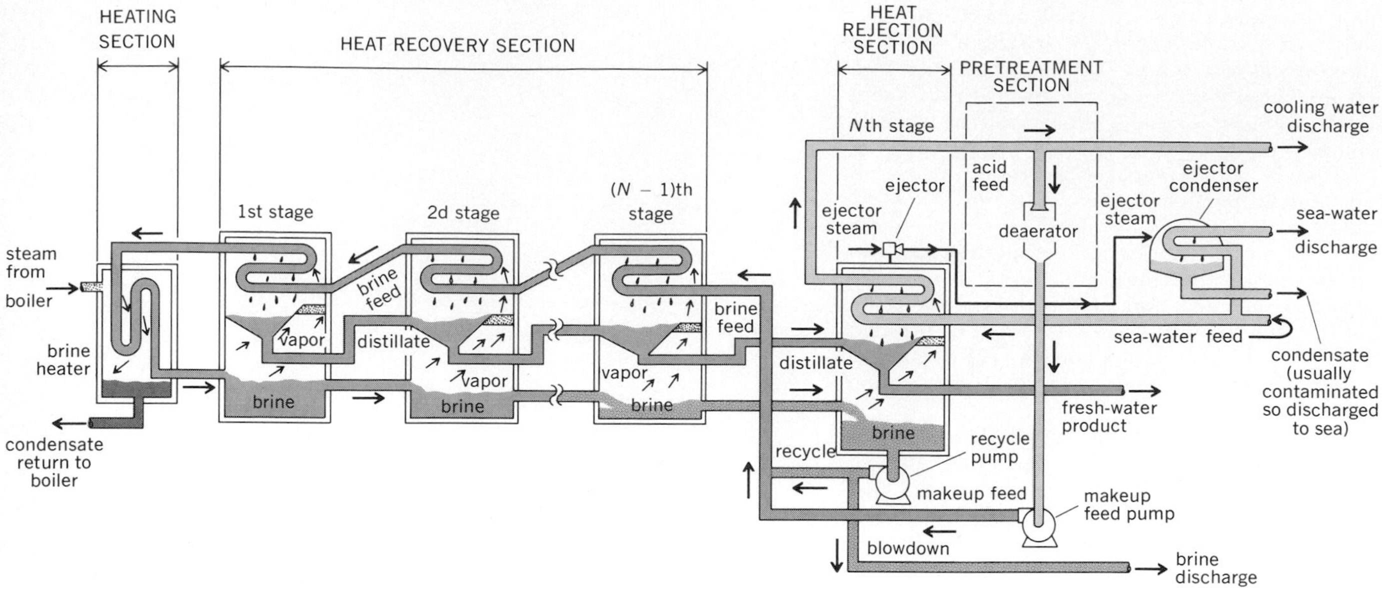

Fig. 2. Simplified flow sheet for multistage flash (MSF) recycle distillation plants. (*From O. K. Buros, The U.S.A.I.D. Desalination Manual, U.S. Agency for International Development and CH2M Hill International Corporation, August 1980, published by IDEA, 1981*)

up its heat of evaporation as it condenses. The condensate is the product water. While vertical-tube evaporation offers potential for high performance, the process has had only marginal field success (Fig. 3). *See* EVAPORATION; EVAPORATOR.

Horizontal-tube multieffect (HTME) design. Another distillation plant system uses the so-called horizontal-tube multieffect (Fig. 4*a*) or the multi-effect stacked design (Fig. 4*b*). These designs offer exciting future possibilities. Several plants of capacity over 10^6 gal/day (3.8×10^3 m³/day) are in service, and the largest facility of this type, a facility producing 10^7 gal/day (3.8×10^4 m³/day), is scheduled for completion in the mid-1980s under joint United States–Israeli partnership.

Vapor compression (VC). Another distillation process is known as vapor compression. There are two versions: the so-called spray-film vapor compression, usually encountered in smaller-type units of 2.5×10^3 to 3×10^4 gal/day (Fig. 5), suitable for facilities such as hotels, industrial plants, and power stations; and vertical-tube vapor compression, which is used for installations in the 10^5 gal/day-and-beyond size (Fig. 6). A variety of vapor compression units are found around the world. The key to all these units is the compressor through which the energy is supplied to the system.

Performance ratio. Distillation plants are rated by performance ratio, the number of pounds of water produced for 10^3 Btu. In operational plants the ratios range from 6 to 16, though 20 and higher is considered possible.

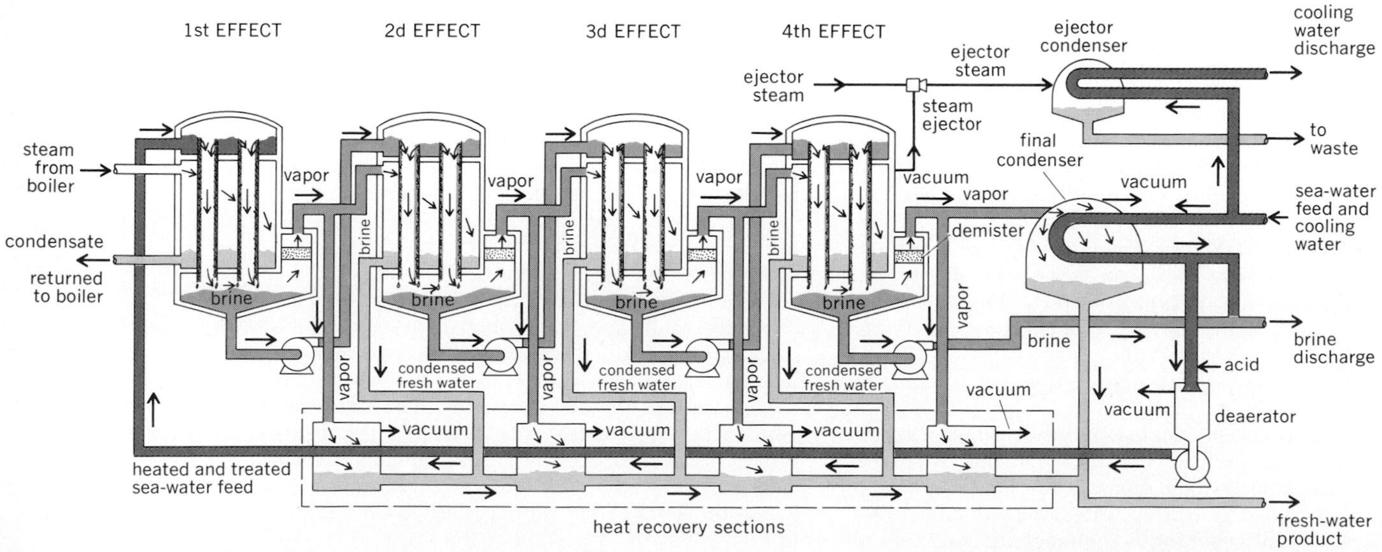

Fig. 3. Simplified flow sheet for a multieffect vertical-tube evaporator (VTE). (*From O. K. Buros, The U.S.A.I.D. Desalination Manual, U.S. Agency for International Development and CH2M Hill International Corporation, August 1980, published by IDEA, 1981*)

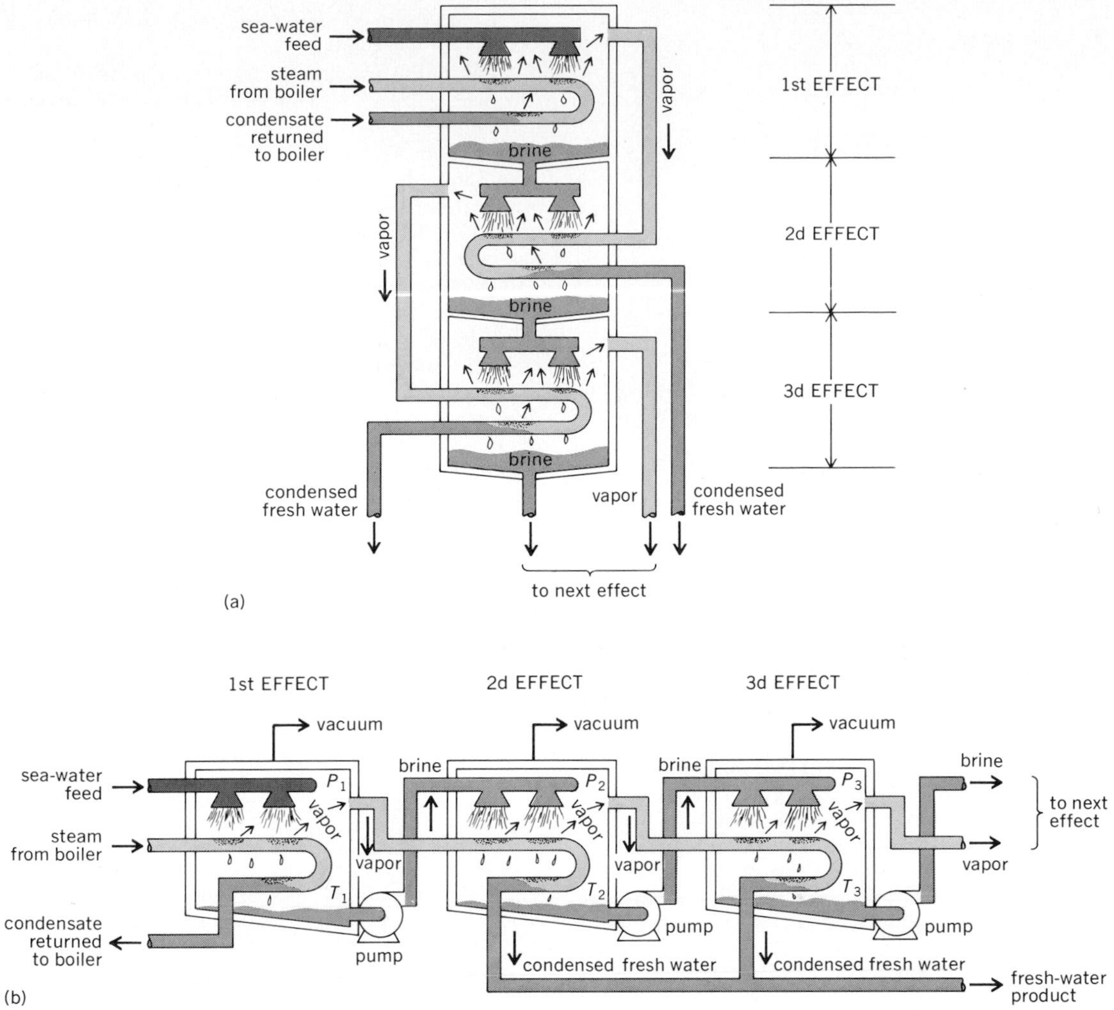

Fig. 4. Multieffect designs. (*a*) Conceptual diagram of a horizontal-tube multieffect (HTME) distillation plant with vertically stacked effects. (*b*) Conceptual diagram of a horizontal-tube multieffect distillation plant. (T = temperature, $T_1 > T_2 > T_3$; P = pressure, $P_1 > P_2 > P_3$). (*From O. K. Buros, The U.S.A.I.D. Desalination Manual, U.S. Agency for International Development and CH2M Hill International Corporation, August 1980, published by IDEA, 1981*)

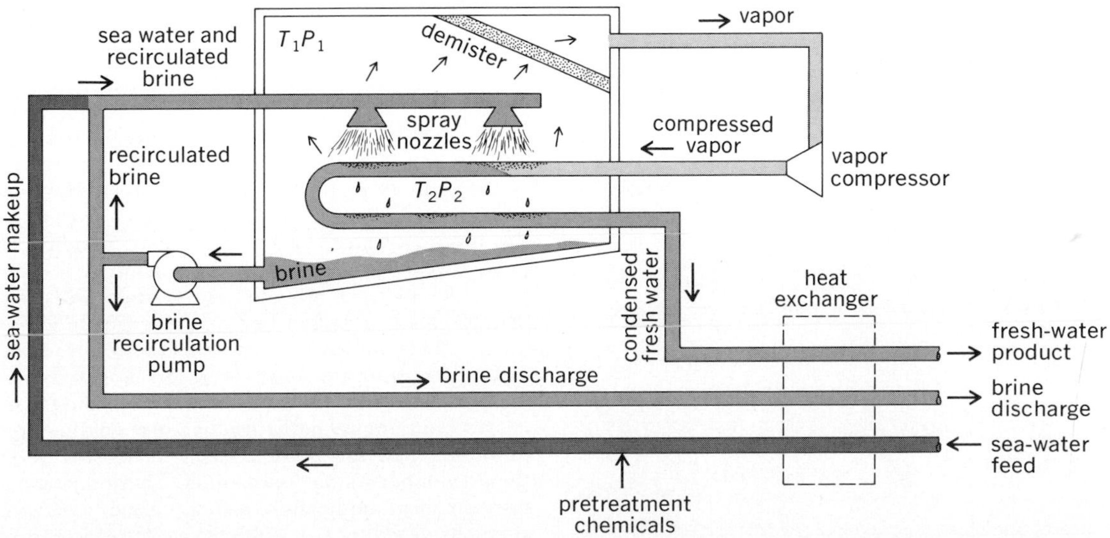

Fig. 5. Simplified flow diagram for an electrically driven spray-film vapor compression process. (T = temperature, $T_2 > T_1$; P = pressure, $P_2 > P_1$). (*From O. K. Buros, The U.S.A.I.D. Desalination Manual, U.S. Agency for International Developmental and CH2M Hill International Corporation, August 1980, published by IDEA, 1981*)

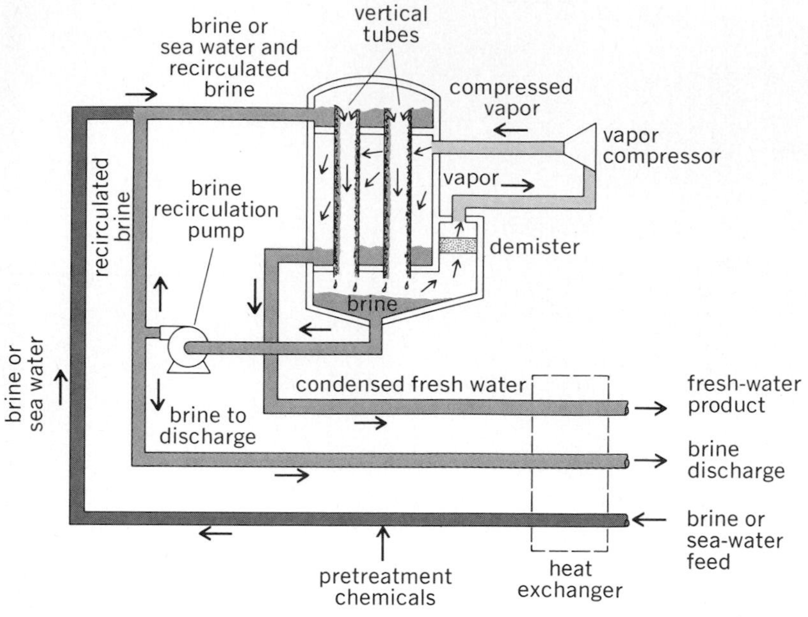

Fig. 6. Simplified flow diagram for a vertical-tube vapor compression process.

Higher performance ratios are obtained at higher capital costs. Vapor compression plants which on the average offer higher performance ratios than multistage flash plants are distinguished by significantly higher operation and maintenance costs.

While the accepted processes predominate, the field itself remains dynamic, and much work directed at developing processes offering better performance and lower cost installations is in progress. Testing and adaptation of new processes are very difficult because water conversion processes with proved performance and reliability are always selected in preference to more efficient but little-known processes.

Dual-purpose plants. The cost of desalination of sea water can be considerably reduced by combining the distillation plant with a power generation plant. This permits the use of waste steam from the power plant in the evaporators. The dual-purpose approach has been successfully applied to fossil-fuel-fired plants and could be adapted to nuclear power plants.

Most of the operational distillation plants around the world are part of fossil-fuel-fired power-water complexes. These provide potable and industrial water and service municipal operations, petrochemical, diverse chemical, pulp and paper, and other processing industries. Nuclear-fueled water plants have been operated on an experimental basis only. *See* DISTILLATION.

Freezing methods. The alternate thermal-based processes are the freezing processes. In principle, they have a great deal to offer, including lower energy consumption and low-temperature operation, simplifying the material corrosion problems which are integral parts of all distillation processes. Pure ice can be frozen from brine and melted to produce fresh water. However, some of the brine clings to the ice crystals' surfaces or is entrapped within them. A freezing process, then, must involve two operations: one to form the pure ice and the second to separate the ice from the brine. One technique (Fig. 7) consists of admitting cold sea water to a chamber under high vacuum in which a portion of the water immediately vaporizes. The evaporation process absorbs heat from the remaining salt water, causing a portion of it to freeze to an ice-brine mixture. This slurry is passed through a separator, where it is washed with a portion of the product water. The water vapor, which has been drawn off from the freezing chamber, is compressed and then brought into contact with the brine-free ice crystals where it condenses, melting the ice. This melted ice is the fresh-water product. Several freezing processes were investigated during the 1970s. In principle at least, freezing processes continue to look attractive, but engineering problems and installation costs have prevented them from becoming commercial. These obstacles may result in applications being limited to solving specialized industrial effluent purification problems. Freeze desalination plants have not been developed on a commercial basis. *See* CRYSTALLIZATION.

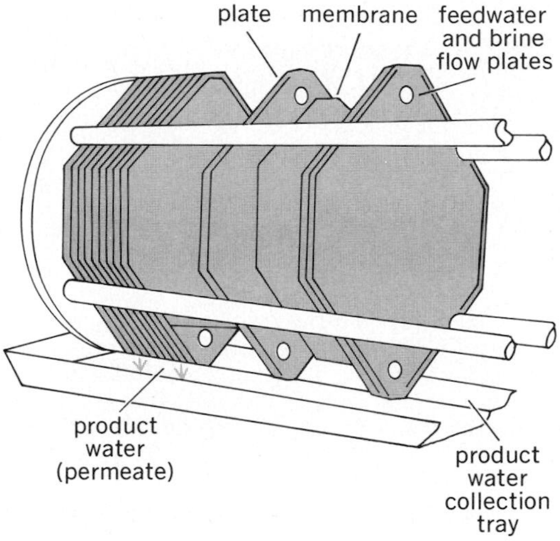

Fig. 8. Construction of a plate and frame membrane.

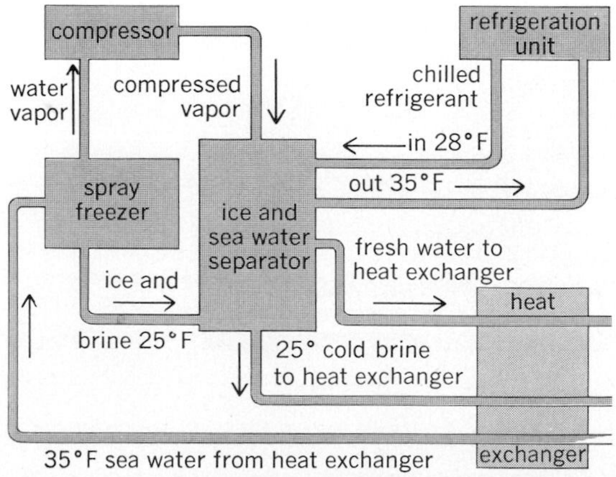

Fig. 7. Freeze-evaporation process. (Temperature, °C) = (5/9) [temperature, °F − 32].

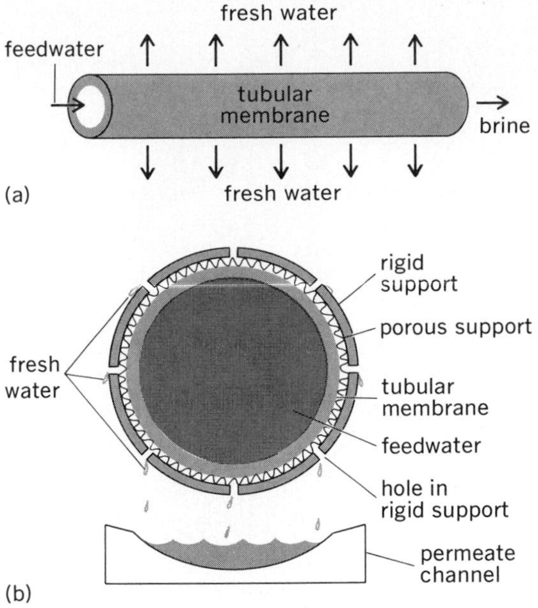

Fig. 9. Construction of a tubular membrane. (*a*) Side view. (*b*) Cross section. (*From O. K. Buros, The U.S.A.I.D. Desalination Manual, U.S. Agency for International Development and CH2M Hill International Corporation, August 1980, published by IDEA, 1981*)

MEMBRANE PROCESSES

Separation of water from saline solutions can be accomplished with the membrane processes of reverse osmosis and electrodialysis.

Reverse osmosis. When pure water and a salt solution are on opposite sides of a semipermeable membrane in a vented container, the pure water diffuses through the membrane and dilutes the salt solution. At equilibrium, the liquid level on the saline-water side of the membrane will be higher than on the fresh-water side. This phenomenon is known as osmosis, and the effective driving force is called osmotic pressure. Its magnitude depends on the characteristics of the membrane, the temperature of the water, and the concentration of the salt solution. By exerting pressure on the salt solution, the osmosis process can be reversed. When the pressure of the salt solution is greater than the osmotic pressure, fresh water diffuses through the membrane in the opposite direction to normal osmotic flow. This is the essence of the reverse osmosis process (RO). *See* OSMOSIS.

A wide variety of polymeric materials, including polyamides and polyimides, have been brought into the membrane field to replace cellulose acetate, which was used in much of the original development work in the 1950s and 1960s. Most of the dense structure film has been replaced with skinned or asymmetric membranes. There are modified configurations in addition to the original plate configuration (Fig. 8). The tubular configuration (Fig. 9) has membranes of tubular shape with a diameter of 0.3–1 in. (0.7–2.5 cm) which are placed inside rigid tubes or pipes. These are separated from the membranes by porous supports. The water is pressurized within the pipes and moves through the membranes, the porous

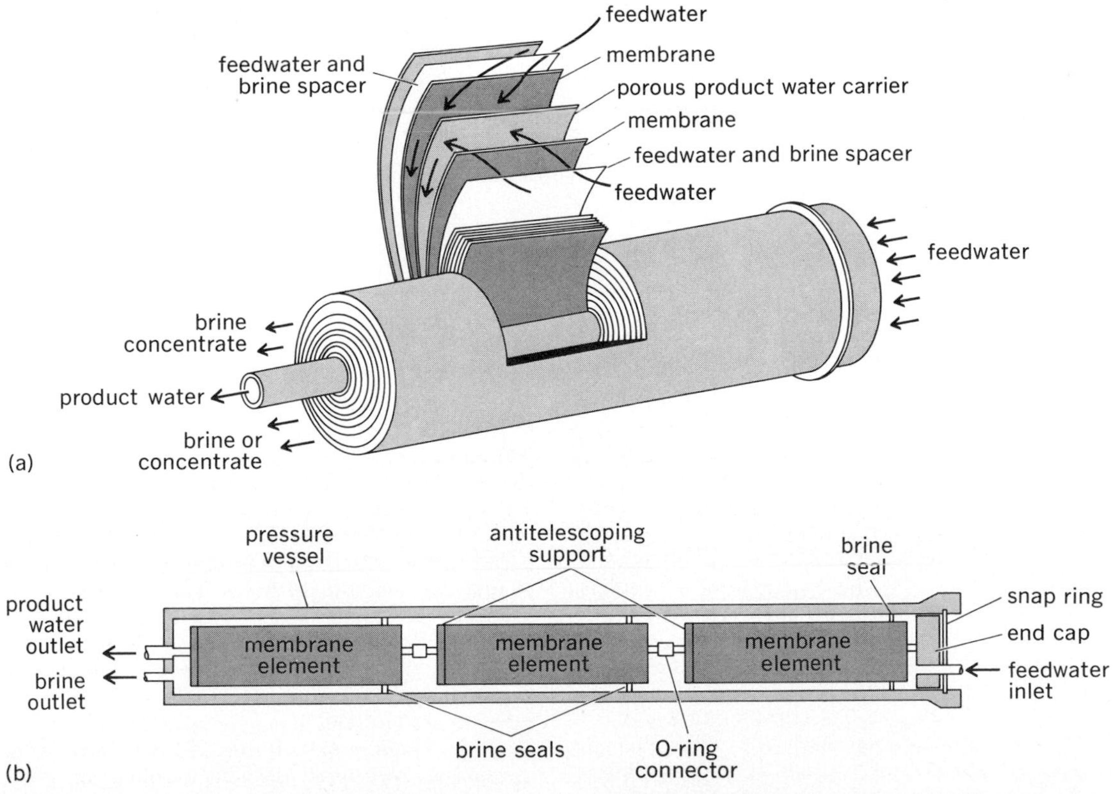

Fig. 10. Spiral membrane elements. (*a*) Cutaway view. (*b*) Cross section of pressure vessel with three membrane elements. (*From O. K. Buros, The U.S.A.I.D. Desalination Manual, U.S. Agency for International Development and CH2M Hill International Corporation, August 1980, published by IDEA, 1981*)

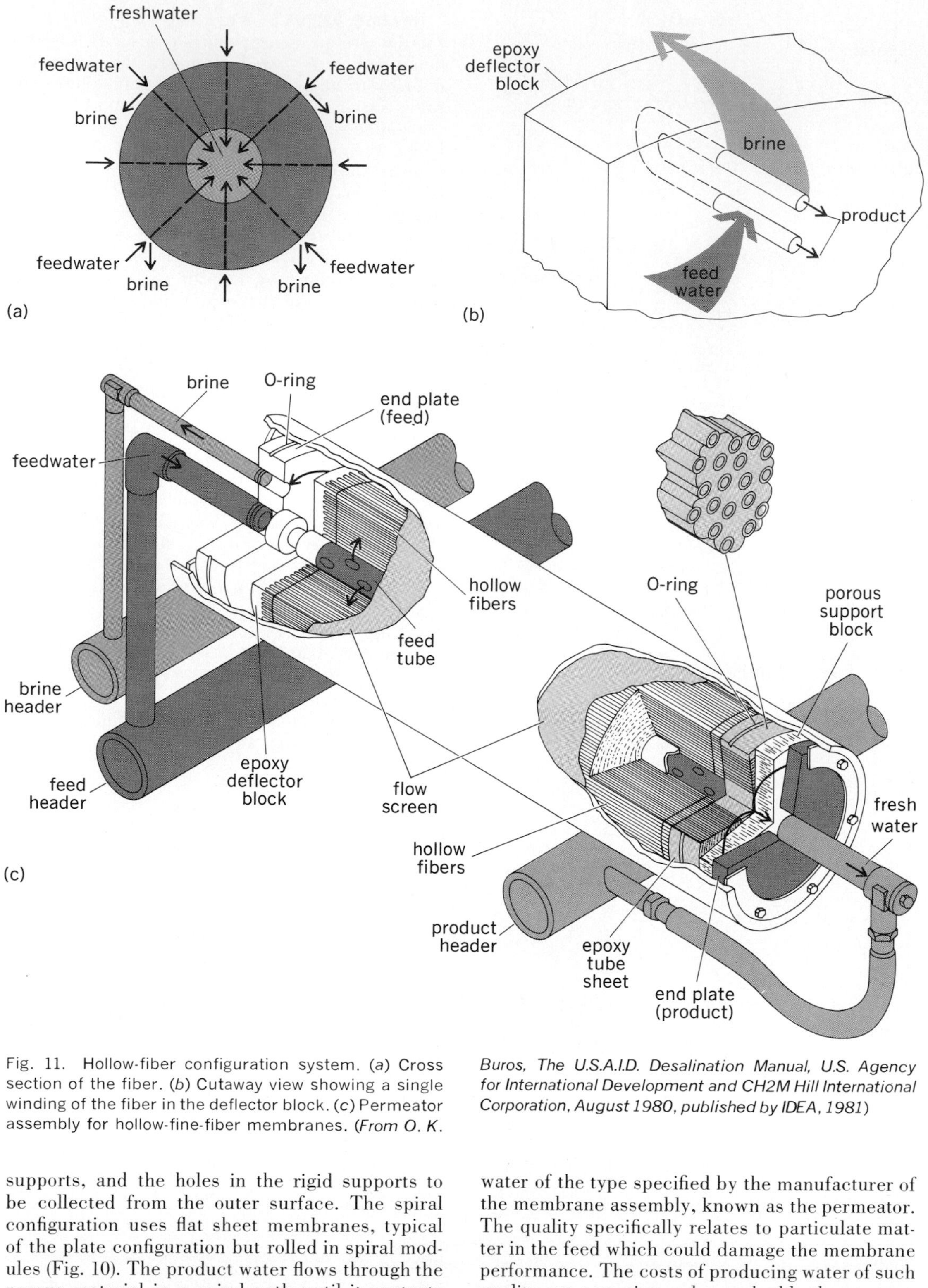

Fig. 11. Hollow-fiber configuration system. (*a*) Cross section of the fiber. (*b*) Cutaway view showing a single winding of the fiber in the deflector block. (*c*) Permeator assembly for hollow-fine-fiber membranes. (*From O. K. Buros, The U.S.A.I.D. Desalination Manual, U.S. Agency for International Development and CH2M Hill International Corporation, August 1980, published by IDEA, 1981*)

supports, and the holes in the rigid supports to be collected from the outer surface. The spiral configuration uses flat sheet membranes, typical of the plate configuration but rolled in spiral modules (Fig. 10). The product water flows through the porous material in a spiral path until it contacts and flows through the holes in the product water tube. The most advanced concept, the so-called hollow-fiber configuration (Fig. 11), employs a design in which the saline water is on the outside and the product flows in the hollow of the fiber.

Reverse osmosis is employed for brackish waters at pressures of 250–400 psi (1.7–2.8 MPa) and sea waters at pressures of 800–1200 psi (5.5–8.3 MPa). Successful and cost-effective processing requires that the feed water be limited to water of the type specified by the manufacturer of the membrane assembly, known as the permeator. The quality specifically relates to particulate matter in the feed which could damage the membrane performance. The costs of producing water of such quality can sometimes almost double the processing cost.

The membranes are judged by several criteria including: flux, the quantity of water which can flow through the membrane; salt rejection, the ratio of product salt concentration to that of the feed-water salt concentration as determined by measuring the total dissolved solids in each stream; recovery, the ratio of the product flow to the feed-water flow. There are differences between brackish- and sea-water conversion. In the case of the first, a 45–55% recovery is possible, which can be increased under special provision to 85–90%, while in sea water the recovery factor is only 20–35%.

Both brackish- and sea-water reverse osmosis are firmly established and rapidly growing water conversion processes. The largest sea-water reverse osmosis facility, producing 3.5×10^6 gal/day (1.32×10^4 m^3/day), is in Jeddah, Saudi Arabia, while the largest brackish-water reverse osmosis plant (capacity of 4×10^7 gal/day or 1.51×10^5 m^3/day) is in Riyadh, Saudi Arabia.

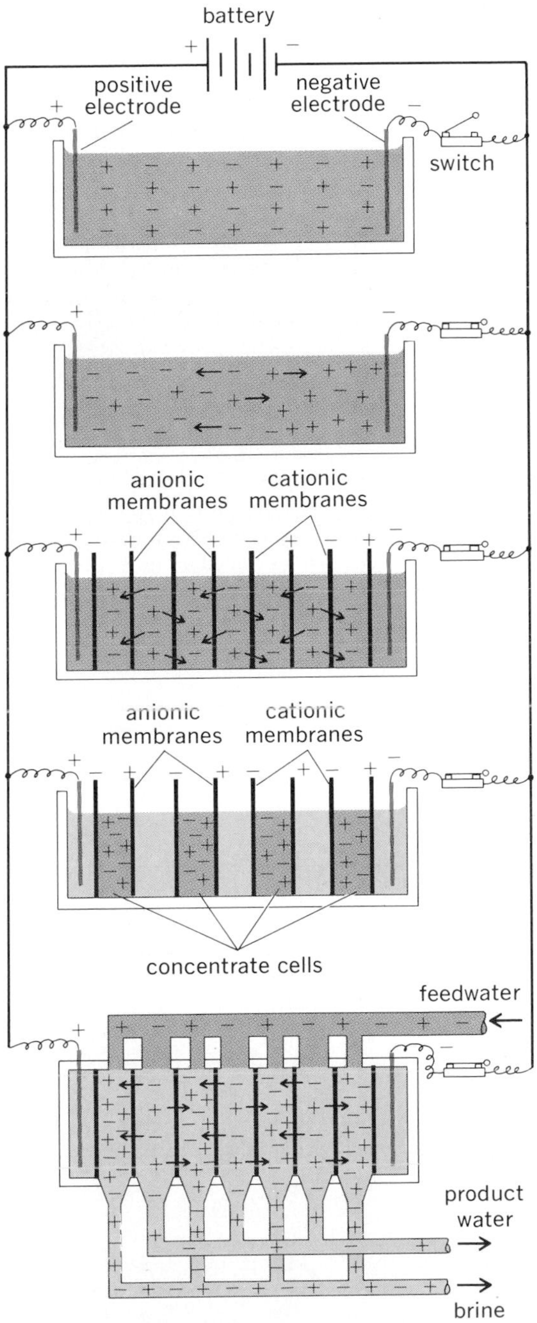

Fig. 12. Movement of ions in the electrodialysis process. (*From O. K. Buros, The U.S.A.I.D. Desalination Manual, U.S. Agency for International Development and CH2M Hill International Corporation, August 1980, published by IDEA, 1981*)

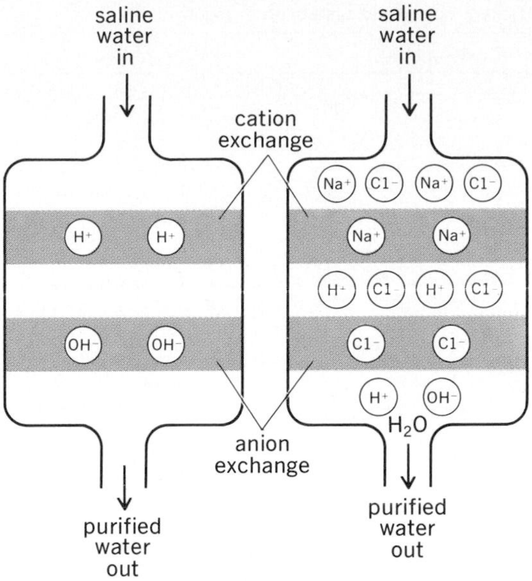

Fig. 13. Ion exchange process.

Electrodialysis. The salts dissolved in saline waters are usually fully ionized, making these waters conductive. When current flows, ions migrate to oppositely charged electrodes. If selectively permeable membranes are suitably located in the path of these ions, they can be redistributed into two streams, one containing an increased ion concentration and the other consisting essentially of pure water, the product. This is the essence of electrodialysis, invented around 1940 and developed commercially in the mid-1950s (Fig. 12). The essence of the process remained virtually unchanged until the so-called polarity-reversal option was introduced into commercial equipment, an advance which materially enhanced performance. Electrodialysis is an accepted and successful brackish-water process used worldwide. In addition, considerable success in adapting the process to sea-water conversion has been reported on a laboratory scale. Some prototype systems are being tested.

CHEMICAL PROCESSES

Of the chemical processes, only the ion exchange has become an integral part of water conversion technology. While it can process higher total-dissolved-solids waters, it is in fact the technique most often recommended for processing of water with total dissolved solids below 500 ppm, as it is economical in that range. It brings such waters down to a very few ppm or even virtually no dissolved solids.

An ion exchanger consists of a porous bed of organic resins that have the ability to exchange ions in the resin through contact with those in the solution to be processed. In the field, both catonic and anionic ion exchangers are used. These can be found placed either in series or in mixed beds. Figure 13 shows schematically how a cation exchanger replaces the sodium ion with a hydrogen ion, in the process converting the dilute salt solution to a dilute hydrochloric acid solution. When this solution passes through the anion exchanger,

Table 2. Component costs of water supply

Component	Description
Source development and collection	Water sheds, intake facilities at rivers or lakes, reservoirs, wells
Transportation	Pipeline, conduit, or canal with pump stations or tank truck, ship, or barge with terminal facilities
Treatment and/or conversion	Treatment: coagulation, settling, filtration, softening, iron and manganese removal, chlorination, and so on; or desalination: distillation, reverse osmosis, electrodialysis, freeze desalination, and so on
Distribution	Distribution mains, laterals, hydrants, and service connections throughout the community as well as distribution systems, storage tanks, and pumping stations
Business functions	Meter reading, billing, collections, accounting, purchasing, engineering, financing, management, and administration

the chloride ions are exchanged with the hydroxide ions. The hydrogen and the hydroxide ions then combine to form pure water molecules. A great variety of resins are available, as are commercial systems designed to use these resins and carry out the ion exchange or deionization process. They are very similar in basic design. As a matter of fact, the water softening device used in many homes is a simple ion exchange device. Ion exchange is used most extensively to solve a wide range of water demineralization problems. The resins must be regenerated, and most of them can be regenerated relatively inexpensively with suitable low-cost chemicals such as NaCl or $Ca(OH)_2$. *See* ION EXCHANGE.

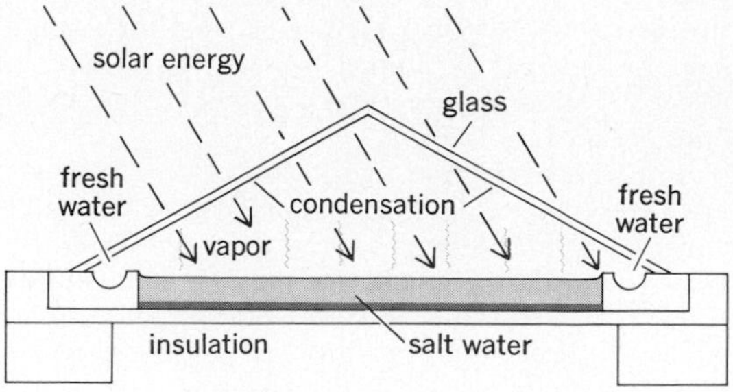

Fig. 14. Simple solar still.

ENERGY SOURCES AND ECONOMICS

Desalination processes could be fueled by energy derived from sources other than fossil fuels or centrally generated electrical power. Solar energy has been the most prominent of these alternate sources of energy for desalination, but geothermal, wind, and even wave power have become potential contenders as energy costs have increased. In fact, solar distillation, the pioneer method, is perhaps the only quasicommercialized process using alternate sources, with a wide variety of designs encountered in operational solar stills around the world. A simple unit is shown in Fig. 14. Only small amounts of desalinated water, up to a few hundred gallons (100 gal = 0.4 m^3) a day, are being produced using solar energy. A major worldwide effort to increase the use of these alternate energy sources for desalination may increase their importance in the production of desalted water. The most promising approach would use a combination of conventional and renewable energy resources. Renewable energy source systems, while having advantages in terms of conservation and replacement of conventional energy sources, will still produce costly water, because these systems are capital-intensive.

There are enormous quantities of fresh water on

Table 3. Energy requirements for desalting processes

Process	Feature	Performance factor, lb water/10^3 Btu*	Btu/lb water*	Pumping energy, kWh/10^3 gal
Distillation	Single-effect waste heat	0.95	1053	15–35
Distillation multistage flash	195°F 215°F 90°C 102°C	8	125	8–10
Distillation multistage flash	250°F 275°F 120°C 135°C	12	83	6–8
Distillation multieffect	250°F 275°F 120°C 135°C Water only	15	67	4–6
Distillation multieffect (low temperature)	160°F 175°F 71°C 79°C	10	100	4–6
Reverse osmosis	45 kWh/10^3 gal	54	18.4	Included
Freezing	50 kWh/10^3 gal	49	20.5	Included
Vapor compression	65 kWh/10^3 gal	38	26.6	Included
Reverse osmosis	5000 ppm feed 12 kWh/10^3 gal	200	5	Included
Electrodialysis	5000 ppm feed 12 kWh/10^3 gal	200	5	Included

*1 lb/10^3 Btu = 4.30×10^{-7} kg/J; 1 Btu/lb = 2326 J/kg; 1 kWh/10^3 gal = 9.51×10^5 J/m^3.

Earth, but they are poorly distributed and leave large areas of the world essentially water-poor. The tremendous growth of population and industry has led to taxing of available water supplies by overdrafts and pollution, even in locations known to have adequate water resources. Such locales must seriously consider the future of their water reserves. Some resolve their water supply problems by importing water from distant locations. Alternately, desalination may become desirable in regions poorly supplied with fresh water but containing surplus saline waters. This is a very expensive choice. In the United States, water costs vary depending on location, with an approximately even division between operating expenses and capital costs.

Table 2 shows the component costs of water supplies. These costs are much lower than desalting capital costs alone, desalination or water conversion being the ultimate water treatment. Desalinated water costs, depending on the quality of the raw feed, the specifics of the process used, and its location, can be as high as 10 times greater than treated fresh water. Table 3 gives the energy requirements for the different desalting processes. In addition, consideration must be given to the costs of heat and pumping. The reported overall costs of any process, including that of water conversion, depend very much upon the specific system of cost accounting. In addition to the specific process, the manner in which items such as equipment, financing, specific depreciation schemes, overhead to be charged, plant location, and other related items can profoundly affect the overall cost of water produced. It must be noted that the desalting costs will affect neither the capital and operating costs of the distribution systems nor the business operating costs, which often account for well over half of the delivered costs for water. In many places, water, specifically desalinated water, is subsidized to a great extent, as the population is often unable to pay the true water costs.

[ROBERT BAKISH]

Bibliography: M. Balaban (ed.), *Desalination* (journal); O. K. Buros, *The U.S.A.I.D. Desalination Manual*, U.S. Agency for International Development and CH2M Hill International Corporation, August 1981; *Proceedings of IDEA Desalination Congresses*, International Desalination and Environmental Association, 1976–1979; *Proceedings of Congress Pure Water from the Sea*, European Federation of Chemical Engineers, 1960– ; K. S. Spiegler and A. D. Laird, *Principles of Desalination*, 2d ed., 1980.

Salt

A compound formed when one or more of the hydrogen atoms of an acid are replaced by one or more cations of the base is called a salt. The common example is sodium chloride in which the hydrogen ions of hydrochloric acid are replaced by the sodium ions (cations) of sodium hydroxide. There is a great variety of salts because of the large number of acids and bases which has become known.

Classification. Salts are classified in several ways. One method—normal, acid, and basic salts—depends upon whether all the hydrogen ions of the acid or all the hydroxide ions of the base have been replaced:

Class	*Examples*
Normal salts	$NaCl$, NH_4Cl, Na_2SO_4, Na_2CO_3, Na_3PO_4, $Ca_3(PO_4)_2$
Acid salts	$NaHCO_3$, NaH_2PO_4, Na_2HPO_4, $NaHSO_4$
Basic salts	$Pb(OH)Cl$, $Sn(OH)Cl$

The other method—simple salts, double salts (including alums), and complex salts—depends upon the character of completeness of the ionization:

Class	*Examples*
Simple salts	$NaCl$, $NaHCO_3$, $Pb(OH)Cl$
Double salts	$KCl{\cdot}MgCl_2$
Alums	$KAl(SO_4)_2$, $NaFe(SO_4)_2$, $NH_4Cr(SO_4)_2$
Complex salts	$K_3Fe(CN)_6$, $Cu(NH_3)_4Cl_2$, $K_2Cr_2O_7$

In general, all salts in solution will give ions of each of the metal ions; an exception is the complex type of salt such as $K_3Fe\ (CN)_6$ and $K_2Cr_2O_7$. In such salts the ionization is entirely as shown by reactions (1) and (2). No detectable quantities of

$$K_3Fe(CN)_6 \rightarrow 3K^+ + Fe(CN)_6^{-3} \quad (1)$$

$$K_2Cr_2O_7 \rightarrow 2K^+ + Cr_2O_7^{-3} \quad (2)$$

Fe^{3+} or Cr^{3+} from these salts exist in solution because of the strong bonding of these ions in the complex ions. However, in those complex salts where the bonding is weak, ions of the metal can be detected; for example, in Na_2CdCl_4, the cadmium complex ion ionizes appreciably as in reaction (3).

$$CdCl_4^{2-} \rightarrow Cd^{2+} + 4Cl^- \quad (3)$$

The elements with unfilled inner electron shells form complex salts readily. The alum type of salt is a sulfate including the univalent cation of a relatively strong base and a trivalent metal ion such as Al^{3+}, Fe^{3+}, or Cr^{3+}.

Double salts include ions of nearly enough the same size to fit into the same crystal lattice.

Hydrolysis. The solutions of some normal salts are neutral, but those of others are acidic or basic. This results from the reaction of the ions of salt with water. This reaction is called hydrolysis. Examples are shown in reactions (4)–(6). The re-

$$CO_3^{2-} + HOH \rightleftharpoons HCO_3^- + OH^- \quad (4)$$

$$CH_3COO^- + HOH \rightleftharpoons CH_3COOH + OH^- \quad (5)$$

$$NH_4^+ + HOH \rightleftharpoons NH_4OH + H^+ \quad (6)$$

sulting solution will be acidic or basic, depending upon whether the hydrolysis produces an excess of hydrogen or hydroxide ion. *See* HYDROLYSIS.

Theories of acids and definition of salts. The development of more general theories of acids and bases in the 20th century has required a broadening of the concept of salts.

The Brönsted theory of acids lays emphasis on the process of the reaction between acids and bases, and not so much on the product except that the products are other acids and bases, as in reactions (7) and (8). Since the Brönsted theory ex-

$$Acid_1 + Base_2 \rightleftharpoons Acid_2 + Base_1 \quad (7)$$

$$HCl + H_2O \rightleftharpoons H_3O^+ + Cl^- \quad (8)$$

tends the proton theory of acids to solvents other

than water, the original definition of a salt must be expanded as follows: A salt is an electrovalent compound that contains some cation other than the solvated proton and some anion other than the anion which is the conjugate base of the solvent. In the water system the salt should not contain the H_3O^+ and OH^- ions alone; in the liquid ammonia system it should not contain the NH_4^+ and NH_2^- ions alone.

In terms of the Lewis theory of acids and bases, compound (I) is an acid, and compound (II) is a base. Hence the salt should be compound (III).

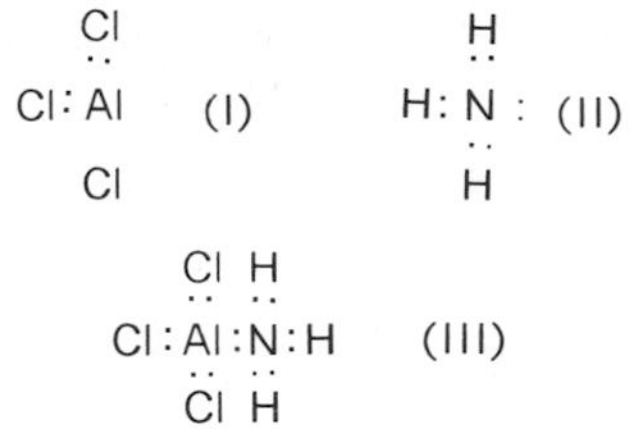

Here the salt is not limited to replacement of the H^+ with a cation of a base. Rather, it is any aggregate of molecules, atoms, or ions joined together with a coordinate covalent bond. Such compounds correctly can be called salts; however, by common parlance, the term salt usually refers to an electrovalent compound, the classical example of which is sodium chloride. *See* ACID AND BASE; CHEMICAL BONDING.

[ALFRED B. GARRETT]

Bibliography: W. T. Lippincott, *Chemistry: A Study of Matter*, 3d ed., 1977.

Salt bridge

A bridge of solution of some salt, usually potassium chloride, which is placed between the two half-cells of a galvanic cell. It is frequently made in tubular form of siphon design. It is used either to reduce to a minimum the potential of the liquid junction between the solutions of the two half-cells or to isolate a solution under study from a reference half-cell and prevent chemical precipitations. When a salt bridge is used to reduce the liquid-junction potential, a saturated solution of potassium chloride is chosen, because the ionic mobilities of the potassium and chloride ions are nearly identical. This reduces the liquid-junction potential to a small value. When a salt bridge is used to isolate a solution from a reference half-cell, various salt solutions may be used for the salt bridge, depending on the circumstances. For example, if the potential of silver in a silver nitrate solution were to be compared with a calomel half-cell, the two half-cells could not be in contact, for then the potassium chloride of the calomel half-cell would precipitate the silver nitrate solution surrounding the silver electrode. In this case, then, a salt bridge of potassium nitrate would be placed between the solutions of the two half-cells. *See* ELECTRODE POTENTIAL; ELECTROMOTIVE FORCE (CELLS).

[WALTER J. HAMER]

Salting-out effect

The decrease in solubility of a species in water resulting from the addition of an electrolyte to the aqueous solution. Gases and organic solids and liquids become less soluble in water in the presence of a salt chemically inert toward them. The solubility of such substances in water depends upon a chemical or physical interaction between solute and water molecules. Ions from a dissolved salt interact through fairly strong electrical forces with water, and become surrounded by a sphere of oriented water molecules. The water held by the salt is then less free to interact with other solute molecules. Some colloidal suspensions may also be precipitated by the addition of a salt to the suspending medium. *See* COLLOID; ELECTROLYTE; ISOELECTRIC POINT; SOLUTION; SOLVENT EXTRACTION.

[FRANCIS J. JOHNSTON]

Samarium

A chemical element, Sm, atomic number 62, belonging to the rare-earth group. Its atomic weight is 150.35, and the naturally occurring isotopes are Sm^{144} 3.09%, Sm^{147} 14.97%, Sm^{148} 11.24%, Sm^{149} 13.83%, Sm^{150} 7.44%, Sm^{152} 26.72%, and Sm^{154} 22.71%. Sm^{147}, Sm^{148}, and Sm^{149} are radioactive and emit α particles. The half-lives are 1.06×10^{11} years, 1.2×10^{13} years, and 4×10^{14} years, respectively.

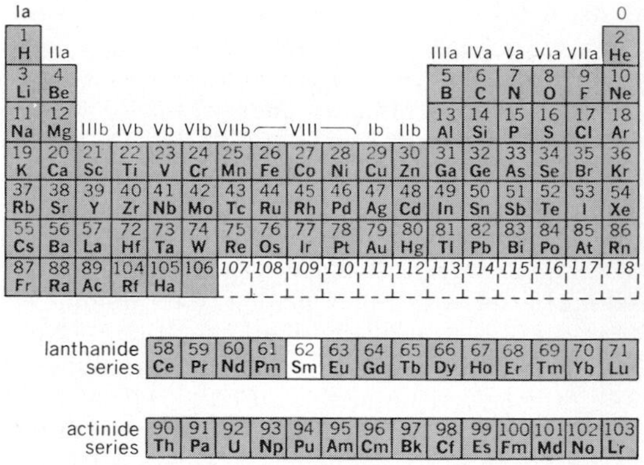

The element was discovered in 1879 by L. de Boisbaudran, and was obtained in the form of very pure compounds by E. Demarçay in 1901. The element can be separated from the other rare earths in its trivalent form by ion-exchange methods. The element also exists in the divalent form, but the divalent ion in solution slowly decomposes water. The metal can be extracted from solution into sodium amalgam but cannot be prepared pure in this manner, because samarium distills with the mercury as an intermetallic compound. Reduction of samarium salts with calcium or alkali metals usually gives the divalent salt. The metal, however, can be prepared in the very pure form by mixing the oxide with lanthanum metal or misch metal which has had the other divalent metals removed, and distilling under a vacuum. Samarium metal has an appreciable vapor pressure at the melting point so that it distills away from the mixture and can be condensed in a very pure form, leaving lanthanum oxide in the residue. *See* RARE-EARTH ELEMENTS.

Samarium oxide is pale yellow, is readily soluble in most acids, and gives topaz-yellow salts in solutions. Samarium has found rather limited use in the ceramic industry, and it is used as a catalyst for certain organic reactions. One of its isotopes has a very high cross section for the capture of neutrons, and therefore there has been some interest in samarium in the atomic industry for use as control rods and nuclear poisons. *See* LANTHANUM. [FRANK H. SPEDDING]

Sampling techniques

In analytical chemistry, the operations required to obtain a laboratory sample from a large quantity of raw material. Most commercial materials are not homogeneous; that is, their compositions vary from portion to portion. If the analysis for a constituent is to be significant, the portion used in the laboratory must have the same composition as that of the original material. The problem of obtaining a representative sample may be more difficult than the problems of analysis.

Sampling of a gas is difficult. Special equipment and procedures are required to obtain a representative, homogeneous portion for analysis. *See* GAS AND ATMOSPHERE ANALYSIS.

Liquids. Homogeneous liquids in tanks or barrels are sampled by dipping several portions from different locations or by drawing off portions from several containers. Liquids flowing in open rivers or in pipes are sampled by dipping from the open stream or by using a bypass system on a pipe. It is important that sufficient liquid be taken to have a representative sample. The sampling of a multiphase liquid must be done so that the portion removed has the same relative amounts of the different phases as the original material. One way to do this is to use a sampling thief, a long tube with holes which can be opened in the liquid and then closed so that portions representative of different levels are obtained. A system with a solid dispersed in a liquid should be agitated before sampling.

Solids. Segregation is a major problem, so usually 0.5–2.0% of the material is taken as the gross sample to be certain of a representative portion.

Metallic materials such as steels, brasses, sheet metals, and wires are sampled by drilling, cutting, milling, or filing if reasonably homogeneous, or by taking very fine drillings from several locations if segregation of components is known to exist. After thorough mixing by rolling the drillings, or particles, on paper, the sample is assumed to be representative. This is not always true, however, as different-size particles may yield different analyses. If the surface is not the same as the interior, for instance, in a case-hardened steel, care is required to distinguish between the two parts of the sample.

Nonmetallic materials, such as coal, ores, rocks, ore veins, and soils, consist of particles of varying size as well as of varying composition. The gross sample is taken by a sampling thief if the material is in bags or barrels. If the material is shoveled, every *n*th shovel is set aside. If the material is handled by conveyor, an automatic arrangement such as a slot or divider removes a definite fraction. Several portions of ore veins or soil must be removed and mixed. The gross sample is crushed with mechanical equipment and piled into a flat cone. Two opposite quarters of the cone are removed, mixed by shoveling if a large quantity or by rolling on cloth if a small amount, and ground to a smaller particle size. This process is repeated until only 8–10 lb (3.6–4.5 kg) is left. This portion is ground in a ball mill to the particle size required for the laboratory sample. Care is necessary to be certain that no impurities are introduced and that no material is lost in the process.

Aliquot portions. Even with care, the final laboratory sample may be somewhat heterogeneous. In this case a portion several times the amount needed for an analysis is dissolved in a suitable solvent, and the sample solution is diluted to a definite volume. A fraction of this solution, usually one-fifth or one-tenth, is taken for analysis. This fraction is called an aliquot.

In general, standard sampling procedures have been recommended for most materials of commerce. *See* ANALYTICAL CHEMISTRY.

[CHARLES RULFS]

Bibliography: W. G. Cochran, *Sampling Techniques*, 3d ed., 1977; D. C. Cornish, *Sample and Sample Process Systems*, 1980; N. H. Furman and F. J. Welcher (eds.), *Standard Methods of Chemical Analysis: Industrial and Natural Products and Noninstrumental Methods*, vol. 1, 1975, vol. 2, reprint of 1963 ed.

Scandium

A chemical element, Sc, atomic number 21, and atomic weight 44.956. In 1871 Dimitri Mendeleev predicted the existence of element number 21 (the first transition element of the first long period) on the basis of his periodic law. The element, a metal,

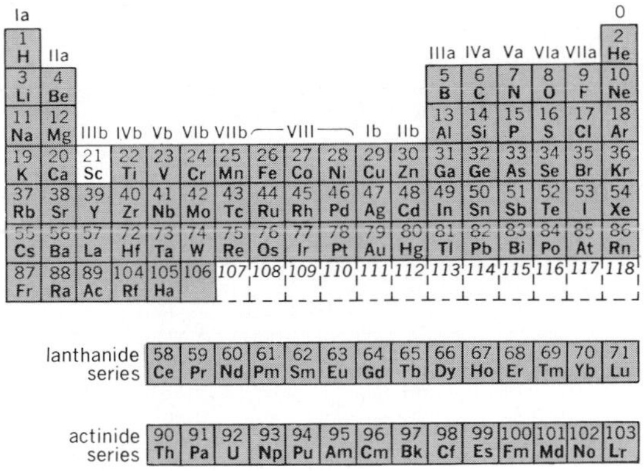

was finally isolated by L. F. Nilson in 1897 from mineral specimens found in Scandinavia. Accordingly, it was named scandium. The physical properties of scandium are listed in Table 1. The thermodynamic properties at various temperatures are given in Table 2.

X-ray properties. The K-emission lines for scandium are as follows:

α_2	α_1	β_1	β_2
Strong	Very Strong	Weak	Not observed
0.303452 nm	0.303114 nm	0.27795 nm	–

The absorption edge is 0.27573 nm, and the excitation potential for characteristic x-ray spectra is K_{Sc} = 4.49 kV. The atomic scattering factor *f* is the ratio of amplitude of wave scattered by an atom to that scattered by an electron. The value of *f* is therefore a function of the number of electrons *Z* in the atom, the angle of diffraction *θ*, and the

Table 1. Physical properties of scandium

Natural atomic weight: 44.956
Outer valence electrons: $3d^1 4s^2$
Atomic radius: 0.162 nm
Ionic radius: $r_{Sc^{3+}} = 0.068$ nm
Atomic volume: 15.028 cm^3/mole
Specific gravity (at 20°C): x-ray 3.02 g/cm^3; arc-melted 3.0 g/cm^3
Melting point: high purity, 1811°C (3260°F)
Boiling point: 2870°C (5198°F)
Sublimation: metallic Sc sublimes at 1600°C and 10^{-4} mm Hg (1.3×10^{-2} Pa)

Vapor pressure: $\log P_{Sc}(\text{mm Hg}) = -\frac{17180}{T(K)} + 8.298$ or, if

$\log P_{Sc}(\text{mm Hg}) = -A/T + B + C \log T(K)$, A = 19,700, B = 13.07, C = 1.0 (1607 K to melting point). (1 mm Hg = 133.3 Pa.)

Specific heat (at 0°C): 6.01 cal/mole/°C [25.1 J/(mole · °C)]
Heat of fusion: 3.37 kcal/mole [14.1 J/(mole · °C)]
Entropy of fusion: 1.86 eu [7.78 J/(mole · °C)]
Heats of vaporization and sublimation: 78.6 to 81.0 kcal/mole [329 to 339 J/mole]
Crystal structure: α-Sc(room temperature to 1335°C) hcp, $a = 0.3308$ nm, $c = 0.52653$ nm
β-Sc(1335°C to mp) bcc, $a = 0.453$ nm

Reports that β-scandium is face-centered cubic, instead of body-centered cubic, are attributed to the formation of fcc ScN on the surfaces of the specimens being examined.
Transformation temperature: nominal impurities may increase the $\alpha \rightarrow \beta$ transformation temperature from 1335°C (1608 K) to 1373 ± 10°C (1646 ± 10 K)
Heat of transformation (99.7% pure Sc): $\Delta H_{tr} = 99.58$ cal/mole (416.6 J/mole)
Entropy of transformation: $\Delta S_{tr} = 0.60$ eu [2.51 J/(mole · °C)]
Debye temperature: $\theta_{Sc} = 304.5$ K
Heat content: $H^\circ_T - H^\circ_{298.15} = A + BT + CT^2 + DT^3$ cal/mole (1 cal/mole = 4.184 J/mole)

Constants for heat-content equation

Phase	A	B	C × 10^4	D × 10^7	α of fit (rms deviation)	Range, K
α-Sc	−1803	5.977	1.177	4.16	23	298–1608
β-Sc	−6195	10.569	–	–	52	1608–1812
Liq. Sc	16,671	–	–	–	–	1812

Linear coefficient of thermal expansion: α-axis, 9.3×10^{-6}/°C; c-axis, 15.2×10^{-6}/°C; average at 400°C, 11.4×10^{-6}/°C
Thermal conductivity (at 47°C): 0.060 cal/(cm · s · °C) or 0.25 J/(cm · s · °C); (at −238°C): 0.033 cal/(cm · sec · °C) or 0.14 J/(cm · s · °C). The Lorenz function for scandium [$k_e = LT/\rho$, where k_e = electronic component of thermal conductivity, L = Lorenz number: 0.58×10^{-8} g-cal-ohm/(s · K^2) or 2.4×10^{-8} g · J · Ω/(s · K^2), T = K, and ρ = electrical resistivity (ohm-cm)], calculated from measured thermal conductivity and electrical conductivity data on the same sample, indicates that there is considerable heat transport by phonons.
Energy of s and p electrons in L shell: $L_1 = 500.4$ eV.
Electrical resistivity (25°C): 60 to 60 μ-ohm-cm depending on purity; (100°C): 78 μ-ohm-cm. The resistivity increases on heating to the transformation temperature but decreases 0.5% during transformation.
Coefficient of electrical resistivity (0 to 25°C): 0.00282/°C
Oxidation potential (for reaction
$Sc(S) \rightleftharpoons Sc^{3+}_{(aq)} + 3e^-$), $E^\circ_{290\,K} = 2.08$ volts

This is lower than for any of the rare-earth elements, but like many of them scandium does not oxidize significantly at room temperature.

wavelength. The atomic scattering factors are as follows:

$\frac{\sin\theta}{\lambda\ (\text{nm})}$	0	1	2	3	4	5	6
$f_{Sc^{3+}}$	18	16.7	14.0	11.4	9.4	8.3	7.0
$\frac{\sin\theta}{\lambda\ (\text{nm})}$	7	8	9	10	11	12	
$f_{Sc^{3+}}$	6.9	6.4	5.8	5.35	4.85	–	

The mass absorption coefficient (μ/ρ) of scandium is as follows: Where μ = reduction in intensity per centimeter, and ρ is the density of a material, the mass absorption coefficient of a material d cm thick depends on the wavelength as follows:

Radiation type	*Wavelength*	*Mass absorption coefficient, μ/ρ*
AgK_α	0.05604	10.5
RbK_α	0.06149	13.8
MoK_α	0.07097	21.1
CuK_α	0.15392	185.
NiK_α	0.16565	222.
FeK_α	0.19344	338.
CrK_α	0.22869	545.

If the intensities of the transmitted and incident beams are I and I_0, respectively, and d = mass/cm^2 of scandium (the absorbing material), the following equation applies:

$$I = I_0 e^{-(\mu/\rho) \cdot \rho d}$$

Isotopes. Isotopes of scandium include ^{40}Sc and ^{51}Sc and one corresponding to every intermediate value. Except for ^{45}Sc, which occurs naturally, the isotopes occur only when generated by nuclear reactions. A typical reaction is $^{40}Ca(p,n)^{40}Sc$, in which each atom of ^{40}Ca (calcium of atomic weight 40) is activated with a proton p which, on being absorbed, stimulates the emission of a neutron n from each product atom. The product of the reaction is ^{40}Sc (scandium of atomic weight 40). The proton beam must have a certain energy, certain energies, or a certain range of energies to activate the reaction, and the emitted beam also has a characteristic energy. Each reaction has a characteristic cross section.

Other reactions that generate scandium isotopes are as follows:

$^{40}Sc(^3He,d)^{41}Sc$	$^{40}Sc(p,\gamma)^{41}Sc$	$^{40}Sc(\alpha,2n)^{42}Sc$
$^{40}Ca(^3He,p)^{42}Sc$	$^{40}Ca(\alpha,d)^{42}Sc$	$^{40}Ca(\alpha,p)^{43}Sc$
$^{45}Sc(p,2n)^{44}Sc$	$^{43}Sc(p,n)^{44}Sc$	$^{45}Sc(n,\gamma)^{46}Sc$
$^{46}Ti(n,p)^{46}Sc$	$^{47}Ti(n,p)^{47}Sc$	$^{46}Ti(p,\gamma)^{47}Sc$
$^{48}Ti(d,^3He)^{47}Sc$	$^{48}Ti(n,p)^{48}Sc$	$^{48}Ti(d,2p)^{48}Sc$
$^{51}V(n,\alpha)^{48}Sc$	$^{48}Ca(^3He,d)^{49}Sc$	$^{48}Ca(^3He,\gamma)^{51}Sc$

This list is far from complete. Some scandium isotopes, notably ^{44}Sc and ^{46}Sc, undergo isomeric energy transitions and orbital K = electron capture by the nuclei. The reaction $^{40}Sc(n,\alpha)^{42}$ K is an example of the production of another element, ^{40}K, by the neutron bombardment of ^{40}Sc. In this case residual scandium can be extracted with tributylphosphate from an 8N HCl solution of the product.

Table 2. Thermodynamic properties at various temperatures*

Temp, K	$H^\circ_T - H^\circ_{298.16}$, cal/mole	C_p, cal/mole·deg	Entropy, S_T, cal/mole·deg	$(H^\circ_T - H^\circ_{298.15})/T$, cal/mole·deg	$-(G^\circ_T - H^\circ_{298.15})/T$, cal/mole·deg
298.15	0.00	6.16	8.20	0.00	8.20
300	1267	6.16	8.24	0.04	8.20
1000	4708	7.46	16.16	4.71	11.46
1200	6258	8.06	17.60	5.22	12.38
1300	7080	8.39	18.24	5.45	12.79
1812(s)	12956	10.57	21.99	7.15	14.00
1812(l)	16325	—	23.85	9.01	14.84

*1 cal = 4.184 J.

Mechanical properties and fabrication. The hardness of scandium, arc-melted, on the Brinell hardness scale is 75–80 (500-kg load, 10-mm penetrator); arc-melted and cold-swaged, 136; from ScF_3, reduced with calcium, 100; from Sc_2O_3, reduced with calcium, 143 (3000-kg load, 10-mm penetrator).

The compressive strength of 99.0% pure Sc is 56,900 psi (392 MPa), and the estimated tensile strength is 45,500–67,000 psi (314–462 MPa), depending on purity. The elastic properties are as follows: Young's modulus, $E = 0.770 \times 10^6$ kg/cm² or 75.5 GPa (at 23°C); shear modulus, $G = 0.294 \times 10^6$ kg/cm² (28.8 GPa); bulk modulus, $K = 0.672 \times 10^6$ kg/cm² (65.9 GPa); Poisson ratio, $\nu = 0.31$. Other values reported are believed to result from some degree of preferred orientation in the samples used for measurement.

Deformation. Being hexagonal, scandium has few slip planes and, if of 99% purity, work-hardens rapidly under cold deformation. This limits the degree of cold-work possible without recrystallization anneals. Hot-working (above about 815°C) can be carried out with large reductions. Scandium that has been vacuum distilled (99.4–99.6% pure) can be rolled to a reduction of 90% without intermediate anneals. Wire 0.4 mm in diameter and foil 0.08 mm thick can be produced.

Machining. The ease with which scandium can be machined has not been reported; but, by analogy with yttrium, scandium should machine as easily as mild steel.

Electroetching. Scandium is successfully electroetched in a solution containing 90 parts lactic acid, 15 parts water, and 6 parts sulfuric acid by volume at 10 volts, 60-cycle alternating current, for 10 sec. The specimen probe and electrode are of stainless steel.

Uses and compounds. Two compounds and the metallic elements are discussed below.

Scandium oxide. The oxide and other compounds of scandium are used as catalysts in the conversion of acetic acid to acetone, in making propanol, and in converting dicarboxylic acids to ketones and cyclic compounds.

Sintered scandia, or scandia-wash coatings on other refractory parts—small crucibles, thermocouple protection tubes, small molds, and stirring rods—should be useful in applications for which the stability of alumina would be marginal or inadequate, for the free energy of formation of Sc_2O_3 is at least 50,000 cal/mole (209 kJ/mole) more negative than that of Al_2O_3 at all temperatures from room temperature to 2000 K. Since yttria is thermodynamically slightly more stable than scandia, chemical or nuclear reasons are usually advanced for preferring scandia for high-temperature uses.

Unstabilized zirconia ceramics normally have a monoclinic crystal structure but transform to cubic on heating. The transformation generally results in cracking. Scandium sesquioxide suitably incorporated in a zirconia ceramic (14 mole atom % Sc_2O_3) effectively prevents transformation cracking, for it maintains the cubic structure (or a tetragonal structure) at room temperature and above. Calcium oxide is used commercially for the same purpose, but the stability of lime is far less than that of scandia.

Scandium sulfate. Treatment with scandium sulfate solution is an economical means of improving the germination of the seeds of many species of plants, including corn, sugarbeets, peas, wheat, and sunflowers. Eight days after corn has been treated with this solution, the roots and shoots have increased the dry weight of the germ by 37 and 78%, respectively. The increases for the other plants are 9–25%. Optional concentrations of $Sc_2(SO_4)_3$ are 10^{-3} *M* to 10^{-8} *M*, depending on the plant.

Metallic scandium. Scandium-47 has a satisfactory half-life for use as a tracer, and it can be prepared carrier-free. The presence of 2.5–25 atom % of scandium in the anode increases the voltage, the voltage stability, and the service life of a nickel alkaline storage battery. Cast iron with graphite as nodules rather than as flakes is stronger and more ductile, and it is often used for crankshafts. Nodular cast iron can be made if 1–5 lb of scandium (in a perforated graphite bell) per ton (or 0.5–2.5 kg per metric ton) of cast iron is plunged to the bottom of a melt. The scandium dissolves in the iron in 0.25–2.5 min. There is no vaporizing, burning, violent reaction, or spattering, and the scandium may be rather impure. It is added as lumps, chips, foil, or pellets. Nodularizing efficiency is retained even if the melt is held for pouring after the scandium treatment. The behavior of scandium in cast iron is therefore analogous to that of cerium.

Occurrence and distribution. The chief scandium ore mineral is thortveitite, (Sc,Y) Si_2O_7, which typically contains Sc_2O_3 34, SiO_2 45, Y_2O_3 10, Al_2O_3 5, Fe_2O_3 3, and La_2O_3 1.5 wt %, plus lesser amounts of oxides of copper, magnesium, manganese, and thorium. Thortveitite is found in granite pegmatites, and it occurs in some ores of tin, of tungsten, and of the rare earths. Scandium is widely distributed in many parts of the world, usually in concentrations much less than 1 wt %, in many different ores, minerals, and other substances. The average scandium content of the Earth is 0.09 wt parts per million (ppm).

Yttrium and lanthanum are often associated with scandium, and a geochemical relationship may exist between scandium and iron in sedimentary rocks.

Scandium replaces Fe^{2+} in the igneous minerals pyroxene, amphibole, and biotite, and up to 1 wt % of scandium occurs in the granite pegmatite minerals beryl, cassiterite, and wolframite. Decreasing contents of scandium in minerals from Madagascar pegmatites are in the order thortveitite, columbite-tantalite (max 4 wt % Sc), and ilmenorutile, muscovite, and biotite (< 1 wt %), and garnet, beryl, and tourmaline. Of the pegmatitic rocks, amphibolites are highest in scandium content. Other igneous rocks typically contain 1–30 wt ppm of scandium.

The detection and estimation of scandium in sea water require neutron activation analyses; the sensitivity of these analyses is measured in parts per billion.

Reduction to metal. Scandium has been reduced from ScF_3 by heating with calcium, zinc, or lithium in an inert atmosphere. When reduced from the oxide with calcium, scandium has a Brinell hardness of 100, which is 20–25 Brinell points higher than that of high-purity scandium. Some tantalum from the melting crucible and nonmetallic impurities cause the excess hardness. Scandium reduced from fluoride with calcium is even harder.

Up to 5 wt % of tantalum can dissolve in molten scandium; but if the scandium is melted rapidly in a tantalum crucible and immediately cast or chilled, the pickup of tantalum will be substantially less than 1 wt %. Tungsten is more nearly inert, but tungsten crucibles are more costly and therefore seldom used. Regardless of the reduction method further purification can be achieved by vacuum-induction melting. First, the more volatile impurities can be distilled out of the scandium; then the scandium can be distilled away from the tantalum and other lower-volatility impurities. The vapor can be condensed at the closed end of a silica tube inverted over the crucible that holds the material to be purified. Scandium can also be purified by zone refining, but the Van Arkel–De-Boer deposition method of purifying many metals is not applicable to scandium.

Table 3. Thermodynamic data for scandium sesquioxide, Sc_2O_3*

Temp, K	Heat content $H_T - H_{298.15}$ cal/mole	Entropy $S_T - S_{298.15}$ cal/(mole·deg)	Free energy F kcal/mole
300			−434
400	2,420	6.95	−418
800	13,500	26.02	−390
1200	25,600	38.29	−361
1600	38,110	47.29	−335
2000	50,750	54.32	−306

*1 cal = 4.184 J. 1 kcal = 4.184 kJ.

Metallic scandium may also be produced by electrolysis of the chloride.

Properties of compounds. Thermal properties of the sesquioxide are as follows. The heat of formation $-\Delta H_{298\,K} = 455.4$ kcal/mole (1905 kJ/mole) of Sc_2O_3; the entropy of formation $S_{298\,K} = 18.4$ cal/°C/mole [77 J/(°C·mole)] of Sc_2O_3; the free energy of formation $-\Delta G_{300\,K} = 434.4$ kcal/mole (1818 kJ/mole) of Sc_2O_3; $-\Delta G_{1000\,K} = 419.6$ kcal/mole of Sc_2O_3; and the specific heat of Sc_2O_3 $C_p = a + bT$, where $a = 23.17$, $b = 5.64$, and $T = 298$ to 2500 K. Other thermodynamic data are given in Table 3. The solid solubility, mole %, of Sc_2O_3 in Fe_2O_3 at temperatures up to 500°C is about 27: that of Fe_2O_3 in Sc_2O_3 at temperatures up to 1000°C is about 45.

Scandium oxide reacts in argon or hydrogen with other oxides as follows:

1. The system Sc_2O_3–CaO contains $CaSc_2O_4$ and forms eutectics at 57.0 mole % $ScO_{1.5}$ (1960°C) and at 73.5 mole % $ScO_{1.5}$ (2015°C).
2. The system Sc_2O_3–La_2O_3 contains $LaScO_{31}$ and forms eutectics at 24 mole % Sc_2O_3 (2000°C) and 73 mole % Sc_2O_3 (2110°C).
3. Sc_2O_3–Y_2O_3 forms metastable $ScYO_3$, but at equilibrium the system is a continuous solid solution.
4. Sc_2O_3–ThO_2 has a eutectic at 83 m/o $ScO_{1.5}$ and 2200°C but no solid solubility.
5. Sc_2O_3–UO_2 has a eutectic at 82 m/o $ScO_{1.5}$ and 2280°C. It also has no solid solubility, but in air 68.5 m/o $ScO_{1.5}$ dissolves in a higher oxide of uranium.

Scandium monoxide. Evidence obtained by flame emission spectroscopy indicates that (ScO) formation is the first step in the oxidation at oxyacetylene flame temperatures of free atoms of scandium. The monoxide can be produced by electrically heating Sc_2O_3 in hydrogen in the presence of suitable amounts of carbon monoxide.

Hydrides. The hydrogen pressure at 700–900°C of hydrides in the composition range $ScH_{0.5}$–$ScH_{2.0}$ is given by the equation below (1 torr = 133.32 Pa).

$$P_{torr} = -[(10.467 \pm 0.552) \times 10^3/T] + 10.527 \pm 0.512$$

In this region the heat of formation of the hydrogen-rich phase, by reaction of 1 mole of hydrogen with the hydrogen-poor phase, is −47.9 kcal/mole (−200 kJ/mole) of H.

Monophosphide. When equimolar amounts of scandium and red phosphorus are sealed in a quartz tube and gradually heated, the two elements react to form scandium monophosphide (ScP), which is produced as a black powder. It has an NaCl structure, an x-ray density of 3.365 g/cm³, and $a_0 = 5.313 \pm 0.005$ A (0.5313 ± 0.005 nm). The compound undergoes no allotropic change when heated from room temperature to 1500°C, nor does it melt on heating to 2000°C. ScP oxidizes in air, increasing in weight at 1200°C by 79%.

Scandium monophosphide is not attacked by water or bases but reacts with acids and can be dissolved in 30% sulfuric acid. If the acidity of the resulting solution is adjusted with ammonia to pH 6 and the scandium ion is held in solution by an excess of complexing agent, the phosphorus can be precipitated as ammonium-magnesium phosphate.

Silicides. Arc-melting scandium with a 1–1.5% excess of silicon in a molybdenum or alumina crucible produces scandium silicide. It can also be made by vacuum arc-melting Sc_2O_3 compacted with silicon. All the scandium silicides, $ScSi_2$, Sc_3Si_5, ScSi, and Sc_5Si_3, are more resistant to oxidation than are the silicides of yttrium or of the rare-earth metals.

Aqueous chemistry. Reactions of scandium derivatives in water are discussed below.

Complexes. Since all three of its outermost electrons function as valence electrons, scandium forms compounds represented by ScX_3 and Sc_2X_3, depending on whether X is a monovalent anion or a bivalent anion. In addition, the trivalent cation, Sc^{3+}, has an even stronger affinity for water than do the rare-earth elements. Accordingly, reactions (1)–(3) occur when Sc^{3+} is in water. The hydrated cation formed in reaction (1) itself ionizes in water,

$$Sc^{3+} + 6H_2O \rightarrow [Sc(H_2O)_6]^{3+} \qquad (1)$$

as shown in reaction (2), to give up a hydrogen ion.

$$[Sc(H_2O)_6]_{aq}^{3+} \rightarrow [Sc(OH)(H_2O)_5]_{aq}^{2+} + H_{aq}^{+} \qquad (2)$$

The water-bonded scandium ion thus behaves as if it were an acid that is only slightly weaker than acetic acid. The resulting divalent cation tends to polymerize, as shown in reaction (3), where $n = 1, 2, 3, \ldots$.

$$n[Sc(OH)(H_2O)_5]_{aq}^{2+} \rightleftharpoons [Sc(OH)(H_2O)_5]_n^{2n+} \qquad (3)$$

The hydroxyl ion and even the bound water in an aqueous complex with a scandium ion can be replaced by many common anions. Such reactions with chloride, sulfate, or acetate ions can produce $Sc(OH)_2Cl{\cdot}XH_2O$, $Sc(OH)SO_4{\cdot}XH_2O$, and $Sc(OH)(CH_3SO_2){\cdot}7.1\,H_2O$. For reaction (4) the value of the first hydrolysis constant is $K_1 = 1.5 \times 10^{-4}$.

$$Sc^{3+} + 2H_2O \rightarrow Sc(OH)^{2+} + 2H^+ \qquad (4)$$

Other complexes formed with Sc^{3+} are: (1) halo, ScF^{2+}; (2) sulfato, $ScSO_4^{+}$; (3) thiosulfato, $Sc(S_2O_3)_3^{3-}$; (4) oxalato, $ScC_2O_4^{+}$ and $Sc(C_2O_4)_2^{-}$; and (5) acetato, $Sc(C_2H_3O_2)^{2+}$, $Sc(C_2H_3O_2)_2^{+}$, $Sc(C_2H_3O_2)_3$, and $Sc(C_2H_3O_2)_4^{-}$.

Reactions. If 10 g/liter of Sc_2O_3 is dissolved in a slight excess of a 1:1 HCl and 20 g of KH_2PO_4 is dissolved in the solution, adjustment of the solution to pH 4 by the addition of sodium acetate solution precipitates $ScPO_4{\cdot}XH_2O$ (where X~3.2). Heating gradually from 20°C to 350°C dehydrates the $ScPO_4$, and heating at 800–830°C makes it amorphous. Analogous procedures with scandia and acid phosphates can precipitate $Sc_2(HPO_4)_3{\cdot}4H_2O$ and $Sc(H_2PO_3)_3$.

The compounds $M_3[Sc(NCS)_6]$, where M = Li, Na, K, Rb, or Cs, result from mixing solutions of $Sc_2(SO_4)_3$, M_3SO_4, and $Ba(SNC)_2$ in the molar ratio $Sc/M/SNC^- = 1:3:6$. Barium sulfate precipitates, forming in the course of 24 hr, and can be filtered off. The compounds can be crystallized in vacuum over H_2SO_4 or $CaCl_2$. All the salts are anhydrous except the lithium salt, which is a tridecahydrate, and the sodium salt, which is a tetrahydrate. The thiocyanate group is bonded to the scandium through the nitrogen alone. The solubilities of the Li, Na, K, Rb, and Cs salts in water at 20°C are, respectively, 78.05%, 75.71%, 73.00%, 71.82%, and 70.48%.

To ScF_3 in solution the addition of NH_4F or KF to $ScCl_3$ forms ScF^{2+} and ScF_6^{3-}. At Sc:F = 1 the addition of one equivalent of KOH precipitates ScF(OH)Cl.

Slow crystallization from scandium solutions containing 70–75, 50–60, and 25–30 wt % N_2O_5 results in single crystals of $Sc(NO_3)_3{\cdot}3H_2O$, $Sc(NO_3)_3{\cdot}4H_2O$, and $Sc(OH)(NO_3)_2{\cdot}3H_2O$, respectively. The first two compounds are monoclinic, the third triclinic.

Halides. Products of reactions of $ScCl_3$ with NaF and with KF at 25°C in aqueous solution are: (for NaF) $Sc(OH)_xF_{3-x}$, $ScF_3{\cdot}0.16H_2O$, $NaScF_4{\cdot}H_2O$, and Na_3ScF_6; (for KF)-$Sc(OH)F_2$, $ScF_3{\cdot}0.16\,H_2O$, $KScF_4{\cdot}0.1H_2O$, and $K_{2.5}ScF_{5.5}{\cdot}0.2H_2O$. The sodium in the formula Na_3ScF_6 may be replaced by lithium or cesium, and the potassium in the formula $KScF_4$ may be replaced by sodium or lithium. Some of the thermal properties of scandium halides are given in Table 4.

Organic complexes. Scandium also forms numerous complexes with organic reagents. The oxinate, $Sc(C_9H_6ON)_2{\cdot}C_6H_7OH$, is an example of organic complexes used in the quantitative determination of scandium.

Colorimetric and spectrophotometric analyses of scandium are made with colored complexes with reagents such as carminic acid, alizarin, sulfonic acid, 1,2,5,8-tetrahydroxy-anthroquinone,

Table 4. Thermal properties of scandium halides

Property	$ScCl_3$	$ScBr_3$	ScI_3	ScF_3
Melting point, °C	960	—	—	1515
Boiling point, °C	967	929	909	2055
Heat of fusion, kcal/mole (kJ/mole)	19.0 (79.5)	—	—	—
Heat of vaporization, kcal/mole (kJ/mole)	65.0 (272)	63.0 (264)	61.0 (255)	—
Heat of sublimation, kcal/mole (kJ/mole) at 1290 K	—	—	—	89 ± 3 (372 ± 12)
Heat of formation, kcal/mole (kJ/mole) of composition, −ΔH(25°C)	215 (900)	—	—	—
Free energy of formation				
kcal/mole of composition, $-\Delta G_f$(27°C)	197.9	163.9	—	—
$-\Delta G_f$(227°C)	187	152.5	—	—
$-\Delta G_f$(727°C)	160	127	—	—
$-\Delta G_f$(1227°C)	140	—	—	—

ammonium purpurate, phenylarsonic acid, and 2,5-dihydroxy-1,4-benzoquinone. The last-mentioned is a sensitive reagent for the qualitative identification of scandium. It has also been photometrically determined after complexing with Arsenazo III and its analogs.

Scandium forms a blue fluorescent complex with salicylaldehyde semicarbazone. The complex can be used for spectrofluorometric determination of scandium, provided interfering ions are removed or masked. Removal of Fe^{3+} and V^{5+} by extraction as cupferrates into chloroform is possible from 10% HCl solution. If the solution is then made concentrated in hydrochloric acid, the scandium can be extracted into butyl phosphate and then back extracted into water. Any appreciable quadrivalent tin or tellurium has to be removed, perhaps by volatilization as a bromide or a chloride, since these elements suppress the fluorescence of the scandium complex. Silver, divalent copper, and nickel cations, if present, can be masked with cyanide and zinc with ortho-phenanthroline. The procedure can determine microgram amounts of scandium.

Scandium also forms complexes with many organic extractants. The stability at 20°C of the complex of scandium formed with ethylenediamine tetraacetic acid (EDTA) is more stable than the complex formed by any of the rare-earth metals with EDTA.

Concentration and analysis. Several techniques are applicable to the analysis of scandium.

Organic extraction. Scandium is amenable to concentration by extraction with dialiphatic acid phosphates, such as $(GO)_2PO(OH)$, symbolized as HDGP where G is a generalized organic group. When used as an extractant HDGP is usually dissolved in an organic solvent—either aliphatic or aromatic. For systems sufficiently undersaturated in scandium the reaction is that given as (5),

$$Sc_{(A)}^{3+} + 3[HDGP]_{2(O)} \rightleftharpoons Sc[H(DGP)_2]_{3(O)} + 3H_{(A)}^{+} \quad (5)$$

where the subscripts A and O stand for aqueous and organic media, respectively. The aqueous phase is usually acidified with HCl. If k is the equilibrium constant for the reaction, the distribution ratio K is Eq. (6), where the amounts in brackets

$$K = k[(HDGP)_2]_{(O)}^{3} / [H^+]_{(A)}^{3} \quad (6)$$

represent concentrations, m/l, in the respective phases.

Other extractants that have proved useful with scandium are Bu_3PO_4 in CCl_4, or in kerosine and a phenyl methyl solution of trilauryl ammonium chloride. The butyl phosphate can extract scandium for thorium, zirconium, yttrium, and other elements. Trivalent actinides extract in the trilauryl solution before scandium, but scandium extracts in it before the lanthanides.

Investigations in the extraction of scandium from hydrochloric acid solution show that dialkyl alkylphosphonates (DAAPh) are 2.2 times as effective as trialkyl phosphates (TAP). Since the relative effectiveness is only 1.5 for iron, repeated extractions with DAAPh should separate scandium from iron.

Carbon tetrachloride is most effective in extracting neutral (un-ionized) species, such as AsI_3, from an acid solution of the corresponding halide. Trivalent scandium ions are therefore not substantially extracted by CCl_4 because they are not neutral.

Precipitation. The precipitation of $Fe(OH)_3$ in one molar ammonium nitrate containing 0.001 g ion/liter of scandium and traces of calcium coprecipitates the scandium at pH 6 but does not remove the calcium. If the scandium is separated by extraction and reextracted into the aqueous phase it can be precipitated with thiosulfate and oxalate. Purification is also possible by 16-hr boiling of 2 M lactic acid containing an excess of the Sc_2O_3 and precipitation of white $MeCH(OH)CO_2Sc \cdot 2.5 \pm 0.5H_2O$ by the addition of 1:2 $Et(OH)$-Et_2O.

Activation analysis. Parts of scandium per 100,000,000 of GeO_2 or Ge can be detected by activation with a thermal-neutron flux of $10^{14}/cm^2$-sec to a fluence of 5.4×10^{18} n/cm^2. After a cooling (decay) period of 30 days the gamma spectra of the irradiated samples and standards are measured with a flat (3 × 3 in. or 75 × 75 mm) cylindrical NaI(Tl) crystal coupled to a 400-channel analyzer. *See* ACTIVATION ANALYSIS.

Leaching. Scandium can be leached at 90–95°C from clay (3 wt ppm Sc) with hydrochloric acid (solid-liquid ratio = 1:5). If sodium fluosilicate is used to separate scandium from the resulting dilute solution, 80% of the scandium can be separated from the solution in the form of a product containing about 2 wt % Sc_2O_3.

Scandium is removed from large amounts of titanium and some impurities (elements of groups IV to VI) by converting the starting material into sulfates, which are then roasted at 600–800°C and water-leached. The solutions are thereupon treated with ammonia to precipitate a scandium concentrate. Similarly, if zirconium concentrates are chlorinated at 850–870°C, aqueous leaching of the residue yields an oxide containing 1.3% Sc_2O_3.

Ion-exchange and chromatography. Sorbing cations from an HCl solution of an oxide mixture on the H^+ form of Dowex 50, eluting first with HEDTA at 90–95°C to eliminate Ti, Zr, and Fe, and finally with EDTA recovers scandium free of rare-earth and alkaline-earth cations. Iron and many other impurities can likewise be eluted from scandium, yttrium, and the rare-earth elements from Varion AP resin with HCl at concentrations in the range of 0.1 to 4 M. After elution of scandium, yttrium, and rare earths with 8 M HCl, triisooctylamine or an inert carrier can be used to chromatographically separate scandium from the rare-earth elements. *See* RARE-EARTH ELEMENTS; TRANSITION ELEMENTS.

[W. D. WILKINSON]

Bibliography: C. T. Horovitz et al., *Scandium: Its Occurrence, Chemistry, Physics, Metallurgy, Biology and Technology*, 1975; F. H. Spedding and A. H. Daane, *The Rare Earths*, 1961, reprint 1971.

Second-order transition

A change of state through which the free energy of a substance and its first derivatives are continuous functions of temperature and pressure, but at which the second derivatives are discontinuous.

For all physical and chemical processes carried out reversibly, the free energy changes continuously. At an ordinary phase transition, such as the boiling of a liquid, the entropy S, enthalpy H, and volume V show sharp discontinuities when plotted as functions of the temperature T or pressure P.

Because all these functions are first derivatives of the free energy, as shown in notation (1), such

$$\begin{aligned} S &= -(\partial G/\partial T)_P \\ H &= [\partial(G/T)/\partial(1/T)]_P \\ V &= (\partial G/\partial P)_T \end{aligned} \qquad (1)$$

phase changes are usually called first-order transitions.

However, for many systems there are points at which the entropy, enthalpy, and volume are continuous, but at which temperature or pressure derivatives, such as the heat capacity $C_p = (\partial H/\partial T)_p$, the coefficient of thermal expansion $\alpha = (\partial \ln V/\partial T)_p$, and the isothermal compressibility $\kappa = (\partial \ln V/\partial P)_T$, show discontinuities. Because these correspond to second derivatives of the free energy, this phenomenon is called a second-order transition.

The illustration shows the typical thermodynamic behavior at first- and second-order transitions. The dotted vertical line for the heat capacity at the first-order transition is a zero-width line of infinite height, representing a finite nonzero area (a Dirac delta function): the heat of transition, absorbed at a single temperature. The dashed lines show metastable phases continued beyond the transition temperature (for example, superheated liquid above the boiling point and supercooled vapor below). Both the low-temperature (α) and high-temperature (β) phases show such extensions beyond a first-order transition, whereas only the β phase shows such an extension at a second-order transition.

Qualitatively all theories of second-order transitions have the following features in common: A system is capable of existing in two forms, one (α) having a lower enthalpy H and a lower entropy S than the other (β). At sufficiently low temperature the enthalpy difference will be the dominant factor and the system will be all α, whereas at sufficiently high temperatures it will be largely or entirely β. If the conditions were such that the α and β forms could not coexist, there would be a first-order transition at the temperature shown by Eq. (2).

$$T = (H_\beta - H_\alpha)/(S_\beta - S_\alpha) \qquad (2)$$

On the other hand, if the change from α to β can take place gradually (that is, if a mixed phase including both forms can exist) and if the energy required to convert an element of the system (a molecule or group of molecules) from α to β decreases as the amount of β increases, a second-order transition will occur. This changeover from α to β in a sense catalyzes itself, so one refers to phenomena of this kind as cooperative. The temperature at which the last trace of α disappears is the λ-point or Curie point; it is, of course, meaningless to extrapolate the $\alpha+\beta$ curve beyond this point. The heat capacity is very large at the λ-point; it is difficult to be sure whether in some systems it may not actually become infinite; in any case the area under the curve (ΔH) is finite. Important examples of second-order transitions are given in the following paragraphs.

Ferromagnetism. In certain metals and alloys (iron and nickel) at low temperatures, the atomic magnets are arranged into ordered groups or domains which can orient in a magnetic field. As the temperature increases, the order within the domains decreases until, at the Curie temperature, all long-range order is gone and only paramagnetic behavior remains.

Order-disorder in crystals. In certain solid solutions (such as β-brass, Cu-Zn), the different atoms are distributed regularly in an alternating arrangement. As the temperature increases, the two kinds of atoms exchange positions until all long-range order is lost at the Curie point, above which the arrangement is essentially random. A similar phenomenon occurs in the solid ammonium halides; each NH_4^+ tetrahedron can have two different orientations. At low temperatures, all have the same orientation; above the Curie temperature, they are distributed randomly between the two.

Liquid helium. Below 2.19 K, helium shows peculiar superfluid properties. At the lowest attainable temperature, all the molecules are in a superfluid state; as the temperature increases, more and more molecules are excited to nonsuperfluid levels until, at the λ-point (2.19 K), the superfluid properties have disappeared and the helium is an ordinary liquid. *See* CHEMICAL THERMODYNAMICS; PHASE EQUILIBRIUM.

[ROBERT L. SCOTT]

Bibliography: J. D. Fast, *Thermodynamics and Phase Relations*, vol. 1, 1965; H. K. Henisch and R. Roy, *Phase Transitions and Their Applications in Materials Science*, 1974; L. D. Landau and E. M. Lifschitz, *Statistical Physics*, 2d ed., 1969; C. N. Rao and K. J. Rao, *Phase Transition in Solids: An Approach to the Study of Physics and Chemistry of Solids*, 1978.

Secondary ion mass spectrometry (SIMS)

A technique for microchemical analysis that permits the visualization of a three-dimensional chemical image of the solid state using primary ions. It is also referred to as ion probing.

The ion probing of a material is accomplished by bombarding the sample with a 5–20-keV beam of primary ions. This process sputters off the surface layers of the sample, producing a variety of secondary species including neutral atoms and molecules, secondary electrons, photons, and positive and negative ions. Mass spectrometric analysis of the positive and negative sputtered secondary ions can provide microchemical characterization with lateral resolution better than that of the electron probe and can produce sampling depths similar to those of the ion scattering and Auger electron spectrometry techniques. In addition, mass spectrometric analysis provides full elemental coverage from hydrogen to uranium, including the capability of isotopic characterization. This combination of ion-bombardment excitation and mass spectrometric detection can provide detectabilities in the range of 10^{-15} to 10^{-19} g. *See* ELECTRON SPECTROSCOPY; MASS SPECTROMETRY.

Ion production. Ion bombardment of a solid can produce positive ions by two different processes—kinetic and chemical. When a primary ion of 5–20-keV energy strikes the surface layers of a material, it transfers kinetic energy to the sample atoms, initiating a collision cascade, which results in sample atoms being ejected from the surface as well as being excited to metastable and ionized states in the outer atomic layers. Any unbound electrons in the sample will have a much higher velocity than

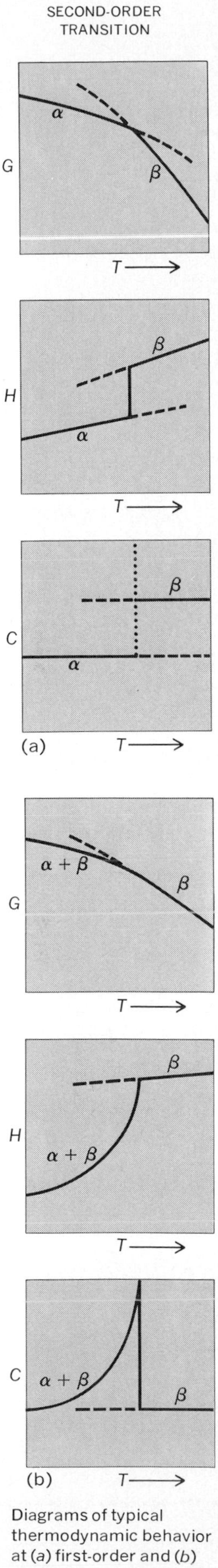

Diagrams of typical thermodynamic behavior at (*a*) first-order and (*b*) second-order transitions.

the solid-state ions have, and all these ions will be neutralized before they can escape into the vacuum. However, an atom can escape from the sample surface as a neutral particle while maintaining its own internal energy in a metastable state. This metastable atom can eject an Auger electron (can quantum-deexcite) in the vacuum above the sample surface and become an ion capable of detection by the mass spectrometer. This is, in essence, the kinetic ionization process, which predominates in the inert gas bombardment of metals and some semiconductors, and in such cases most of the ions which are to be analyzed are produced just outside the sample surface in the vacuum. *See* AUGER EFFECT.

The chemical ionization process depends on the presence of one or more chemically reactive species in the sample to reduce the number of conduction-band electrons available for neutralization of the ions produced in the solid. With the reduction of neutralization events by the presence of this chemical compound, more of the ions produced in the solid state escape the sample surface for mass analysis. Thus, when chemical ionization predominates, most ions are produced in the outer 50 A of the sample (10 A equals 1 nm).

Instrumentation. Ion probe instrumentation ranges from custom-designed, laboratory-constructed devices to general-purpose, commercially available instruments. All these instruments require a primary ion source, primary ion extraction and focusing systems, a sample chamber and sample mounting facilities, a mass spectrometer for mass-charge separation of the secondary ions, and an ion-detection system.

There are three different types of ion probes, or SIMS: (1) The secondary ion mass analyzer provides general surface analysis and depth-profiling capabilities. (2) The ion microprobe mass spectrometer uses primary ions of argon or oxygen from a duoplasmatron ion source that are focused to a 1–2-μm spot before the sample is bombarded. The sputtered secondary ions are extracted into a double-focusing mass spectrometer or spectrograph for mass-charge separation. Ion detection is accomplished by electrical or photographic means or both. The ion microprobe produces a magnified image of elemental or isotopic distributions, in a manner similar to that of the electron probe, using synchronous rastering of the primary ion beam and an oscilloscope. A diagram of an instrument of this type is shown in Fig. 1. (3) The direct-imaging mass analyzer developed by R. Castaing and G. Slodzian is unique among the instruments employed for secondary ion microchemical analysis. In the Castaing-Slodzian design the sample area (10–300 μm in diameter) to be analyzed is bombarded by the primary ions. The secondary ions are extracted and imaged by an electrostatic immersion lens. The image bears a point-to-point relation to the

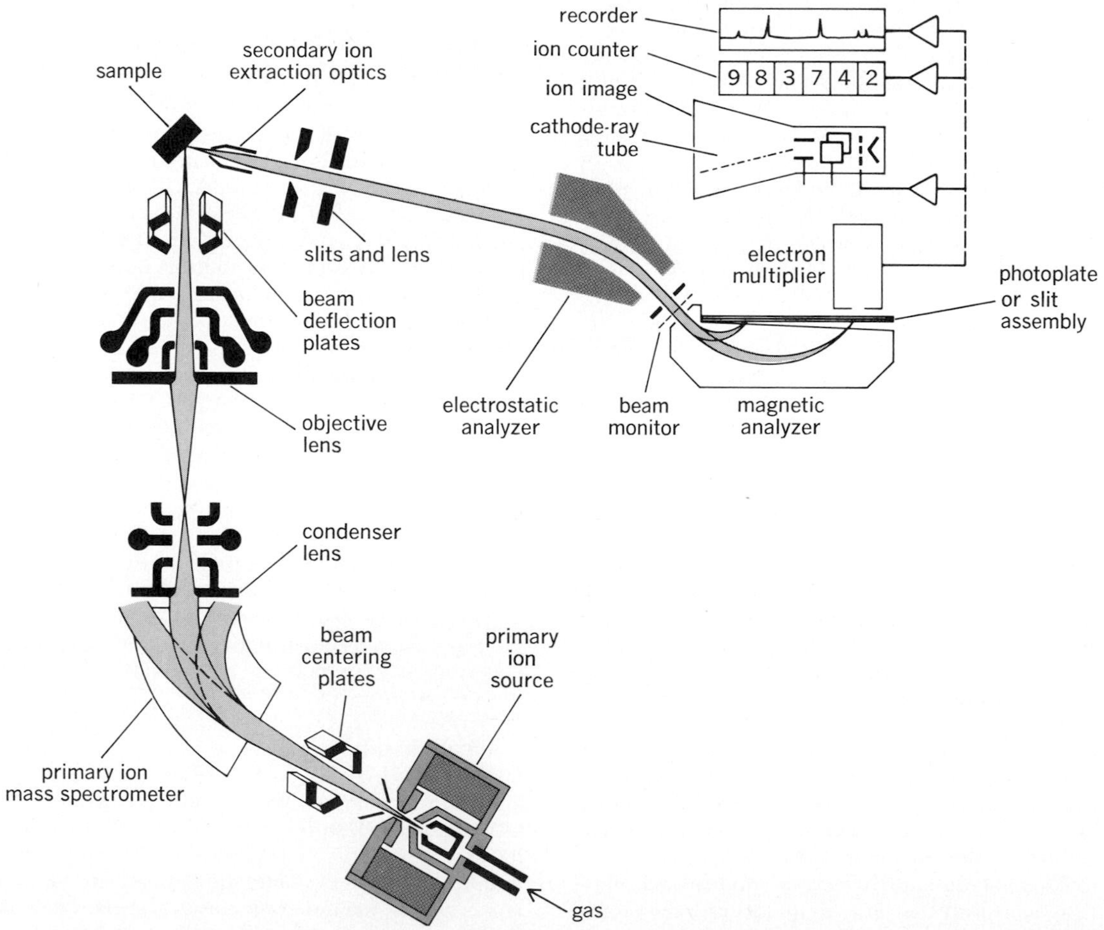

Fig. 1. Ion microprobe. (*From C. A. Evans, Jr., Secondary ion mass analysis, Anal. Chem., 44:67A, November 1972*)

ion's place of origin on the sample surface and is magnified approximately 10 times. The secondary ion image then traverses a magnetic sector, an electrostatic mirror, and a second magnetic sector. This sophisticated system provides mass and energy separation while accomplishing many of the same functions of the double-focusing mass spectrometers used with ion microprobes. The addition of another electrostatic lens and image converter allows the resultant mass-resolved image to be visually observed on a fluorescent screen or recorded on a photographic film for future reference. The same mass-resolved image can be directed onto a scintillator for electrical measurement by a photon-counting system. Secondary ion emission from a localized area, that is, microanalysis, can be quantitatively measured by placing a mechanical aperture in an ion or electron image plane and allowing only those electrons (resulting from the ion-to-electron converter) representing a selected area on the specimen to fall on the scintillator-counting system. The important concept to keep in mind is that the direct-imaging design acquires all data points in the secondary ion image simultaneously, whereas the ion microprobe acquires image data points sequentially.

Analytical applications. The ion probe has been applied to a wide variety of analytical problems. Like the electron microprobe, the ion probe is used to analyze samples on a microscale. This work has included the analysis of small mineral samples or patches in geological and lunar materials for isotopic age dating as well as trace, minor, and major element composition. Other applications include the analysis of airborne particulates and semiconductor devices. The ion probe provides much better elemental sensitivity and broader elemental coverage than does the electron probe in these applications. In addition, the ion probe provides isotopic detection, which is unavailable with the electron probe.

Another feature of the microanalytical capability of the ion probe is the ability to acquire secondary ion images of the elements present in the sample under study. Through the study of these images a great deal of information on the qualitative and semiquantitative elemental distributions can be obtained with 1–2-μm resolution. Figure 2 is the $^{27}Al^+$ image from a Cr-Al microcircuit test pattern. The light areas result from the presence of Al, and the dark regions represent the absence of Al. In this example, Cr is present in the dark regions. The resolution of these images is illustrated by the fact that the lines in the most closely spaced set of lines are 1.5 μm wide and 1.5 μm apart. Secondary ion images are extremely valuable in the study of concentration variations on the micrometer level.

As a result of the layer-by-layer removal of the sample surface by the primary ion beam, subsurface features are exposed and ionized. The continuous monitoring of the secondary ions provides information on the in-depth variation of the sample composition. The in-depth profiling can be used to examine each of the outer monolayers if low primary ion current densities are employed, or to examine variations over the outer 1–5 μm if high primary beam densities are employed.

The realization of optimum depth resolution

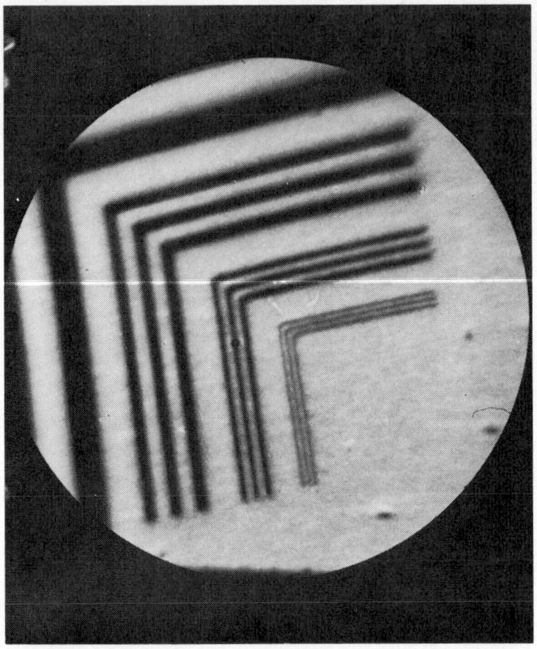

Fig. 2. $^{27}Al^+$ secondary ion image of an Al-Cr microcircuit test pattern. *(From C. A. Evans, Jr., Secondary ion mass analysis, Anal. Chem., 44:67A, November 1972)*

depends on a variety of instrument operating conditions. The most important of these relate to the interaction of the primary ion beam and the sample. During the bombardment process, many of the primary ions interact with several atomic layers and "stir" the upper atomic layers as the ions dissipate their kinetic energy.

It has been shown that at the higher kinetic energies (> 10 keV) the primary ions can actually push some material deeper into the sample and distort the true depth profile. A similar effect has been noted when attempts have been made to profile the outermost 100–200 A of a sample. If a reactive primary ion is being used, it can penetrate below the outermost layer and come to rest some distance into the sample (for example, 100–200 A). Thus the chemical enhancement effect will not be established until that depth is reached by the erosion of the primary beam. As a consequence the ion intensities from the outer 100-A layers cannot be related to those encountered further in. Recently it has been demonstrated that the use of a high residual partial pressure of oxygen can provide a superficial oxygen layer which causes the chemical enhancement effect to occur and allows the continuous comparison of ion intensities.

The other important consideration for high-resolution depth profiling is the shape of the crater produced by the primary ion beam. Since the crater produced by a stationary primary ion beam is gaussian, at any given moment ions are being produced from a variety of depths into the sample rather than from one specific depth. Three methods are generally employed to obtain a flat-bottomed crater, or the effect of a flat-bottomed crater. The first is to appropriately focus the primary beam to obtain a flat-bottomed, vertical-sided crater. This technique is limited since it tends to give very low ion etching rates and is difficult to accomplish. The other two methods depend on the use of primary beam rastering and mechanical aperturing of the secondary ion beam (in the direct-imaging instruments) or electronic aperturing of the secondary ion detector (in the ion microprobe). These techniques produce a flat-bottomed crater with nonvertical sides and allow only those ions from the flat area to be detected.

A variety of materials have been characterized by the above depth-profiling methods. The systems studied include oxides, metals, and a wide number of electronic device—oriented and semiconductor samples.

Future potential. Many application areas of the ion probe technique have not yet been studied. The study of the interstitial elements (C, O, N, and H) in metals will be a very important area. It is also anticipated that many important discoveries will be made as ion probes are brought to bear on biological materials. A major breakthrough for the technique itself will be the development of ion intensity correction methods to obtain quantitative microanalyses. *See* ANALYTICAL CHEMISTRY.

[C. A. EVANS, JR.]

Bibliography: C. A. Andersen and J. R. Hinthorne, *Anal Chem.*, 45:1421, 1973; M. Bernheim, G. Blaise, and G. Slodzian, *Int. J. Mass Spectrom. Ion Phys.*, 10:293, 1972–1973; C. A. Evans, Jr., *Anal. Chem.*, 44:67A, November 1972; J. M. Morabito and R. K. Lewis, *Anal. Chem.*, 45:869, 1973.

Sedimentation

A process based on the settling of solid particles through a liquid. Sedimentation is used in several ways in industrial processes. It may be used to obtain a concentrated slurry from a dilute suspension of a solid in a liquid. This is called thickening. It may also be used to remove solid particles from a liquid to obtain a clear supernatant liquid. This is called clarification. The driving force for the process is the difference in density between the solid and liquid. Ordinarily, sedimentation is accomplished by the force of gravity, and the liquid is water or an aqueous solution. For a given density difference, the rate of settling decreases with particle size. When the particles are fine, or when the density difference is small, gravity settling may be too slow to be practicable; then centrifugal force can be used in place of gravity. When centrifugal force is not adequate, the more positive method of filtration may be used. All these methods of treating solids and liquids fall into the generic group of mechanical separations.

Settling of spheres through fluids. The basic laws of sedimentation are those describing the resistance offered by a fluid to the motion of a solid particle through it. When a constant force is applied to a particle initially at rest in a fluid, the particle immediately accelerates and moves through the fluid. The motion of the particle generates a frictional resistance which in turn reduces the acceleration. When the force moving the particle and the resisting force become equal, the acceleration, by Newton's law of motion, drops to zero, and the particle continues in motion at constant velocity as long as the applied force is active. This steady velocity is called the terminal velocity. Most sedimentation processes are conducted at the terminal velocity.

The law of resistance of a solid particle moving through a fluid is complicated. The resistance depends on all these variables: the diameter and shape of the particle, the viscosity and density of the fluid, the velocity and acceleration of the particle, and the nearness of the particle to other particles and to the wall of the equipment. The basic situation is simplified when the particle is a smooth sphere, when each particle is sufficiently far away from all other particles and from the wall of the equipment so that its motion is unaffected by their presence, and when the particle is moving at constant velocity, that is, without acceleration. Figure 1 shows how, for this case, the terminal velocity u_t is influenced by the acceleration of gravity g, the diameter D_p and density ρ_p of the particle, and the density ρ and viscosity μ of the fluid. Figure 1 is a log-log plot of two dimensionless groups. The abscissa, $D_p u_t \rho/\mu$, is a Reynolds number, and the ordinate in notation (1) is related to a factor

$$(D_p\rho/\mu)\sqrt{gD_p(\rho_p/\rho-1)} \qquad (1)$$

called the drag coefficient. The units used in these expressions may be any self-consistent set, such as the foot-pound-second units suggested in Fig. 1.

To use Fig. 1 to predict a terminal velocity, the magnitude of the ordinate is calculated from known values of g, D_p, μ, ρ, and ρ_p. The corresponding value of the abscissa is read from the solid curve of Fig. 1, and the value of u_t, the terminal velocity, is calculated from the abscissa.

Stokes' law. For Reynolds numbers below about 0.1, the law of settling takes the simple form, called Stokes' law, Eq. (2). This equation is plotted as the straight broken line in Fig. 1.

$$u_t = \frac{gD_p^2(\rho_p-\rho)}{18\mu} \qquad (2)$$

Sedimentation rates in practice. At large Reynolds numbers, which are found with large particles and small viscosities, the terminal velocity is less than that predicted from Stokes' law, and the

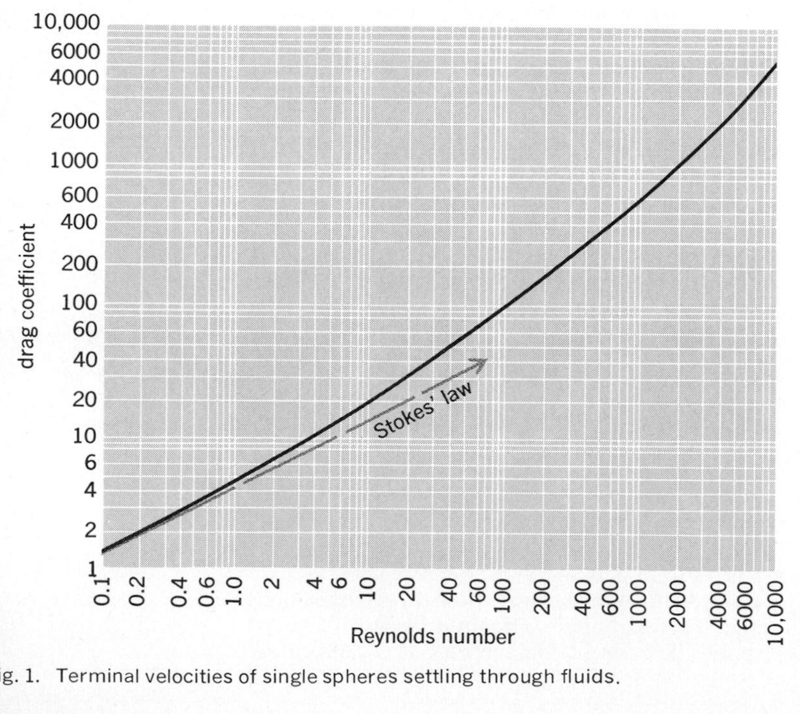

Fig. 1. Terminal velocities of single spheres settling through fluids.

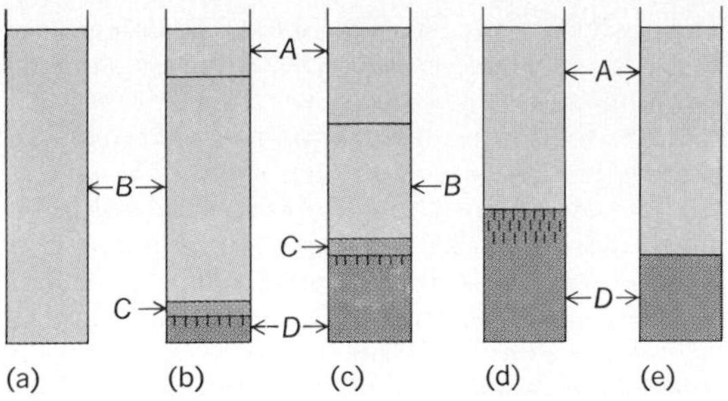

Fig. 2. Stages *a–e* in batch sedimentation of flocculated particles. (*From W. L. McCabe and J. C. Smith, Unit Operations of Chemical Engineering, McGraw-Hill, 2d ed., 1967*)

curved line in Fig. 1 must be used. Also, in practice, the particles are usually very close together, and some are also close to the wall of the equipment. Under these hindered settling conditions, the rate of settling is considerably less than that predicted by the curve of Fig. 1. Actual rates may sometimes be estimated by applying correction factors to the correlation of Fig. 1, but for accuracy, it is best to measure the rates experimentally on the actual sludge to be settled.

Batch sedimentation of flocculated particles. Particles too small to be settled at practicable rates are often amenable to flocculation. This is accomplished by adding agents such as sodium silicate, alum, lime, and alumina. The particles agglomerate into coarse flocs, which act like single large particles, settle at a practicable speed, and leave a clear supernatant liquid behind.

A batch of flocculated pulp passes through several stages during sedimentation. Figure 2 shows the process diagrammatically. Figure 2*a* represents the original homogeneous flocculated pulp ready to settle. In Fig. 2*a–e*, layer *B* is a uniform suspension of the same solid concentration as that of the original pulp, and layer *A* is clear supernatant liquid. Layer *D* consists of flocs resting lightly on one another, with liquid filling the voids between the flocs. Layer *C* is a transition layer, the solid concentration of which varies continuously from that in layer *B* to that in layer *D*. As settling continues, layers *C* and *B* decrease and finally disappear. The end of this stage is shown in Fig. 2*d*. Then, a new effect, called compression, begins. During compression, the weight of the deposit breaks down the structure of the flocs, and some of the liquid in the flocs of layer *D* is expelled as small geysers. The thickness of layer *D* decreases to an equilibrium height called the ultimate height (Fig. 2*e*) and the process stops.

[WARREN L. MC CABE]

Bibliography: Chemical Engineering Magazine, *Separation Techniques II: Gas-Liquid-Solid Systems*, 1980; W. L. McCabe and J. C. Smith, *Unit Operations of Chemical Engineering*, 3d ed., 1975; J. H. Perry (ed.), *Chemical Engineers' Handbook*, 5th ed., 1973.

Sedoheptulose

A seven-carbon ketose sugar originally found in *Sedum spectabile* Bor., a common perennial garden plant. Later it was shown to be widely distributed in the plants of the Crassulaceae family. This sugar, D-sedoheptulose (I), is a significant intermediary compound in the cyclic regeneration of D-ribulose. It also plays an important role as a transitory compound in the cyclic regeneration of D-ribulose for carbon dioxide fixation in plant photosynthesis.

Sedoheptulose can be easily prepared by ex-

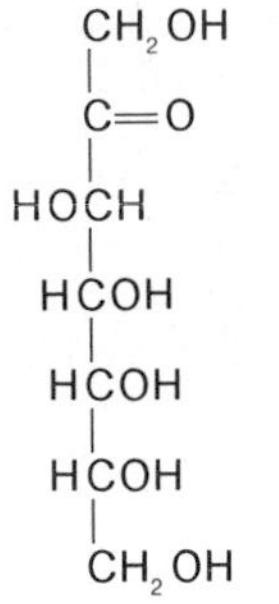

(I)

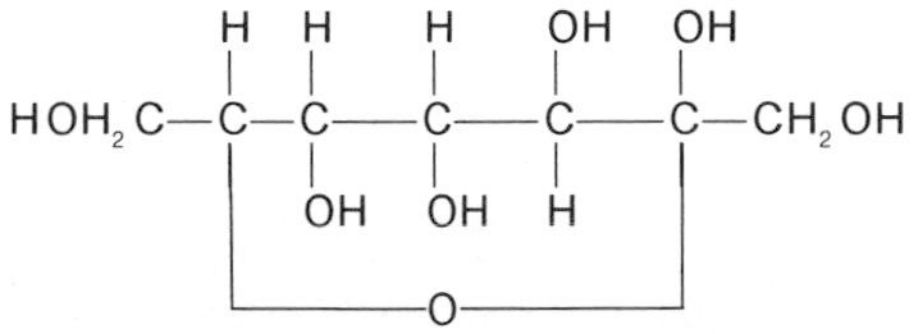

(II)

tracting macerated *Sedum* leaves with water and evaporating to a thick syrup, which is extracted with alcohol. The alcohol containing the sedoheptulose is evaporated to a syrup which is purified by being passed through ion-exchange resin columns. The sedoheptulose itself is not crystallized. However, when the syrup is heated with dilute mineral acid, it forms an equilibrium mixture containing about 20% of the sugar and 80% of a crystalline, nonreducing anhydride, the structure of which is *β*-D-*altro*-heptulopyranose (II).

[WILLIAM Z. HASSID]

Selenium

A chemical element, Se, atomic number 34, atomic weight 78.96, first isolated by J. J. Berzelius in 1817 from sulfur used in the Swedish sulfuric acid industry. The properties of this element are similar to those of tellurium. *See* TELLURIUM.

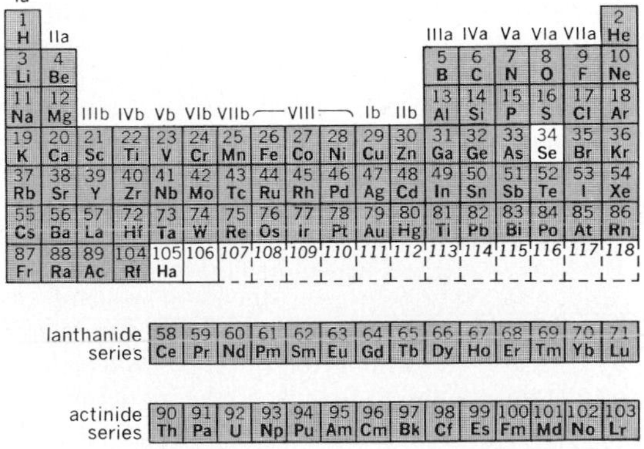

Natural occurrence. The abundance of this widely distributed element in the Earth's crust is estimated to be about 7×10^{-5}% by weight, occurring as the selenides of heavy elements and to a limited extent as the free element in association with elementary sulfur. Examples of the variety of selenide minerals are berzelianite, Cu_2Se, eucairite, AgCuSe, and jermoite, $As(S,Se)_2$. The element's accumulation in certain plants growing in the soils of the United States' Great Plains has given rise in grazing animals to toxic conditions known as blind staggers and alkali disease. Selenium minerals do not occur in sufficient quantity, however, to be useful as commercial sources of the element; instead, selenium-bearing copper sulfide ores found in Canada, the United States, and the Soviet Union represent the primary sources. The anode slimes from electrolytic copper refining account for most of the unrefined selenium-containing material, but significant quantities of the element are recovered from the sludge formed in sulfuric acid production and from the electrostatic precipitator dust accumulated during the processing of certain heavy metals.

Preparation. Selenium-containing materials recovered from the above operations are processed further either by roasting or by the use of hot, concentrated sulfuric acid. In either method selenium dioxide, SeO_2, is formed and subsequently purified by sublimation or conversion to selenious acid, H_2SeO_3, which in turn is reduced by sulfur dioxide in aqueous solution to elementary selenium, as summarized in reaction (1). The United States,

$$H_2SeO_3 + 2SO_2 + H_2O \rightarrow Se + 2H_2SO_4 \quad (1)$$

Canada, and Japan produce the bulk of free-world selenium; in recent years the United States has accounted for 25–45% of the total of about 1000 tons (900 metric tons) per year.

Elemental forms. Of the six known solid allotropes of selenium, the three crystalline forms, α-monoclinic, β-monoclinic, and gray trigonal, are most important. The monoclinic forms are obtained by crystallization from a carbon disulfide solution of the element. The basic structural unit of each of these allotropes is a puckered ring of eight selenium atoms, the α- and β- forms differing only in the way in which these Se_8 molecules are arranged with respect to each other.

The monoclinic forms are less stable than the gray trigonal form; heating monoclinic forms below the melting point of 217°C results in conversion to the gray trigonal form. The gray form, composed of parallel spiral chains of selenium atoms, displays a low electrical conductivity in the dark at room temperature, but with illumination the conduction may rise a thousandfold. An increase in temperature also results in enhanced conductivity.

Selenium also exists in three noncrystalline solid forms: red and black amorphous, and vitreous selenium. The red form, soluble in carbon disulfide or selenium oxychloride, is readily obtained by reduction of a selenious acid solution, as shown in reaction (1). Slow heating at about 30°C transforms this allotrope to black amorphous selenium. Vitreous or glassy selenium is the result of rapid cooling of molten selenium. Details of the structures of these amorphous forms are not known.

Liquid selenium appears black in bulk or reddish-brown in thin films. An increase in temperature results in a decrease in viscosity, and suggests a reduction in the average chain length of the selenium species present. Above the boiling point, 685°C, selenium vapor consists of multiatom molecules, Se_n ($n=2-8$) whose average complexity is reduced with increasing temperature. Thus, at 900°C the vapor's molecular weight suggests Se_2 molecules; at 2000°C it corresponds to monatomic Se.

Uses. Major uses of selenium include the photocopying process of xerography, which depends on the light sensitivity of thin films of amorphous selenium, the decolorization of glasses tinted by the presence of iron compounds, and use as a pigment in plastics, paints, enamels, glass, ceramics, and inks. The use of selenium in rectifiers, important in the 1950s, has declined with increased use of silicon and germanium for this purpose. Selenium is also employed in photographic exposure meters and as a metallurgical additive to improve the machinability of certain steels. Minor uses include application as a nutritional additive for numerous animal species, use in photographic toning, metal-finishing operations, metal plating, high-temperature lubricants, and as catalytic agents, particularly in the isomerization of certain petroleum products.

Principal compounds. Selenium burns in air with a blue flame to give selenium dioxide, SeO_2. The element also reacts directly with a variety of metals and nonmetals, including hydrogen and the halogens. Nonoxidizing acids fail to react with selenium, but nitric acid, concentrated sulfuric acid, and strong alkali hydroxides dissolve the element. Physical properties of some inorganic compounds are given in Table 1.

The only important compound of selenium with hydrogen is hydrogen selenide, H_2Se, a colorless flammable gas possessing a distinctly unpleasant odor, and a toxicity greater and a thermal stability less than that of hydrogen sulfide. The molecule is angular with the H-Se-H bond angle equal to 91°, and the H-Se bond length equal to 1.46 (0.146 nm). The compound is prepared by heating elementary Se in hydrogen; by the hydrolysis of aluminum selenide, reaction (2); or by the action of dilute hydrochloric acid on ferrous selenide, as in reaction (3).

$$Al_2Se_3 + 6H_2O \rightarrow 3H_2Se + 2Al(OH)_3 \quad (2)$$

$$FeSe + 2HCl \rightarrow H_2Se + FeCl_2 \quad (3)$$

Dissolved in water, hydrogen selenide can precipitate many heavy-metal ions as very slightly soluble selenides. Both normal selenides, M_2Se, and acid selenides, MHSe, where M is a metal of oxidation state +1, are known. Aqueous solutions of normal alkali and alkaline earth selenides are capable of dissolving elementary selenium to give polyselenides, M_2Se_x, in a manner comparable to the formation of polysulfides.

Oxides and oxyacids. In addition to direct reaction between selenium and oxygen, the dioxide may be prepared by oxidizing elementary selenium with nitric acid and evaporating the resulting solution, reactions (4) and (5). The dioxide is a volatile

$$Se + 4HNO_3 \rightarrow H_2SeO_3 + 4NO_2 + H_2O \quad (4)$$

$$H_2SeO_3 \xrightarrow{\text{heat}} SeO_2 + H_2O \quad (5)$$

white solid composed of chains of alternating Se and O atoms with the Se-O bond length equal to 1.78 A (0.178 nm). The compound sublimes at 315°C to give a yellow-green vapor composed of discrete SeO_2 molecules. The dioxide dissolves in water to give selenious acid, H_2SeO_3, and is a

Table 1. Physical properties of inorganic selenium compounds

Name and formula	Melting point, °C	Boiling point, °C	Density, g/ml
Hydrogen selenide, H_2Se	−66	−41	2.12 (liquid)
Selenium dioxide, SeO_2	340	315 (sublimes)	3.9
Selenium trioxide, SeO_3	119	sublimes	3.6
Selenic acid, H_2SeO_4	62		2.96
Selenium monochloride, Se_2Cl_2	−85	127 (decomposes)	2.91
Selenium monobromide, Se_2Br_2		225 (decomposes)	3.6
Selenium tetrafluoride, SeF_4	−13	101	2.75
Selenium tetrachloride, $SeCl_4$	305	196 (sublimes)	3.80
Selenium tetrabromide, $SeBr_4$		75 (decomposes)	
Selenium hexafluoride, SeF_6	−39	−47 (sublimes)	3.27 (solid)
Selenium oxyfluoride, $SeOF_2$	15	126	2.80
Selenium oxychloride, $SeOCl_2$	9.5	179	2.44
Selenium oxybromide, $SeOBr_2$	42	217 (decomposes)	3.38
Elementary forms			
Trigonal	217	685	4.82
α-Monoclinic			4.39
β-Monoclinic			4.4
Red amorphous			4.3
Black amorphous			4.3
Vitreous			4.28

strong oxidant reacting with sulfur dioxide, hydrogen, ammonia, phosphorus, or carbon to give elementary selenium.

Selenium trioxide results from the action of dry ozone on selenium dissolved in selenium oxychloride, reaction (6); by the dehydration of selenic acid

$$3SeO_2 + O_3 \xrightarrow{SeOCl_2} 3SeO_3 \quad (6)$$

with phosphorus pentoxide, reaction (7); or by

$$3H_2SeO_4 + P_2O_5 \rightarrow 3SeO_3 + 2H_3PO_4 \quad (7)$$

passing oxygen and selenium vapor through a glow discharge at low pressure and temperature. The trioxide exists in two crystalline forms, a needle-like modification physically similar to asbestos and a cubic form. Reaction of the trioxide with water is vigorously exothermic and yields selenic acid, H_2SeO_4. The latter acid is also prepared in high yields by the fluorination of selenious acid solution, reaction (8). Selenic acid, resembling sulfuric acid

$$H_2SeO_3 + H_2O + F_2 \rightarrow H_2SeO_4 + 2HF \quad (8)$$

in many of its properties, is a strong oxidizing agent reacting, for example, with copper to give copper selenate, $CuSeO_4$, and charring organic matter. Heavy-metal selenate salts such as barium selenate, $BaSeO_4$, are only slightly soluble in water.

Halides. The halides of selenium include monohalides, Se_2X_2(X = Cl or Br), dihalides, SeX_2 (X = Cl or Br), tetrahalides, SeX_4(X = F, Cl, or Br), and hexafluoride, SeF_6. The monohalides are formed by the direct reaction of halogen with selenium. The monochloride, Se_2Cl_2, is a reddish-brown liquid which reacts with water, as in (9). The di-

$$2Se_2Cl_2 + 3H_2O \rightarrow H_2SeO_2 + 3Se + 4HCl \quad (9)$$

halides exist only as gaseous substances; condensation promotes decomposition to the monohalide and free halogen. All the tetrahalides except the iodide have been prepared; direct reaction of selenium and halogen is the preferred synthetic approach, for example, selenium tetrachloride, as in (10). Selenium tetrachloride and tetrabromide

$$Se + 2Cl_2 \rightarrow SeCl_4 \quad (10)$$

are solids at room temperature; the tetrafluoride is a liquid. When volatilized at elevated temperature, the tetrachloride and tetrabromide decompose to the selenium dihalide and free halogen. Gaseous SeF_4 is composed of discrete molecules of the tetrafluoride. The tetrahalides are very easily hydrolyzed, as indicated by reaction (11), for the reaction of $SeBr_4$ with excess water.

$$SeBr_4 + 2H_2O \rightarrow SeO_2 + 4HBr \quad (11)$$

Selenium hexafluoride, SeF_6, results from the reaction of selenium with excess fluorine or other strong fluorinating agents. The molecules of this colorless gaseous compound have six fluorine atoms arranged in an octahedral way about the central selenium atom, with Se-F bond length equal to 1.68 A (0.168 nm). This substance is the least reactive of the various selenium halides; it is stable toward water, air, or sulfur dioxide, but at elevated temperature will react with ammonia, as in (12), or with alkali metals.

$$SeF_6 + 2NH_3 \rightarrow Se + 6HF + N_2 \quad (12)$$

Oxyhalides. Selenium oxyhalide, $SeOCl_2$, is a colorless liquid widely used as a nonaqueous solvent. It is conveniently prepared by reacting selenium tetrachloride and selenium dioxide, as in (13),

$$SeCl_4 + SeO_2 \rightarrow 2SeOCl_2 \quad (13)$$

in a sealed tube or in carbon tetrachloride solution. The oxychloride is readily hydrolyzed, is a strong chlorinating agent, reacts with many metals and nonmetals, and forms addition compounds with many metal chlorides. The corresponding oxybromide, $SeOBr_2$, is an orange solid having chemical properties similar to those of $SeOCl_2$. The oxyfluoride, $SeOF_2$, a colorless liquid with a pungent smell, reacts with water, glass, and silicon, and also forms addition compounds.

Nitride. Tetraselenium tetranitride, Se_4N_4, is an

Table 2. Organic selenium compounds

Type	Example	Properties
Dialkyl selenides, R_2Se	$(CH_3)_2Se$	Bp 58°C, more reactive than ethers
Monoalkyl selenides, RSeH (alkyl selenomercaptans)	CH_3SeH	Bp 12°C
Monoaryl selenides, RSeH (aryl selenomercaptans)	C_6H_5SeH	Bp 183.6°C
Diaryl selenides, R_2Se	$(C_6H_5)_2Se$	Prepared from diaryl sulfones and Se
Cyclic selenoethers, $(CH_2)_nSe$	$(CH_2)_4Se$	5-membered ring; bp 135°C
Selenophene compounds	$(CH)_4Se$	5-membered diene ring; bp 108°C
Selenonium compounds, R_3SeX	$(CH_3)_3SeCl$	Saltlike compounds; R= alkyl or aryl
Polyselenides, R_2Se_2, R_2Se_3	$(CH_3)_2Se_2$	Bp 156°C
	$(C_2H_5)_2Se_3$	Bp 100°C at 26 mm
	$(C_6H_5)_2Se_2$	Mp 63.5°C
	$(C_2H_5)_2Se_8$	Bp 100°C at 26 mm
Organic selenium halides	$(CH_3)SeCl$	
	C_6H_5SeCl	
	$(C_6H_5)CH_3SeBr_2$	Mp 115°C (decomposes)
	$C_6H_5SeBr_3$	Mp 105°C
Selenoxides, R_2SeO	$(C_6H_5)SeO$	Mp 113–114°C
Organic selenium hydroxides	$(CH_3)_3SeOH$	Strong base
	CH_3SeOH	Selenenic acid
	$(CH_3)_2Se(OH)_2$	Decomposes to selenoxide
	$CH_3Se(OH)_3$	Decomposes to RSeOOH (seleninic acid)
Selenones, R_2SeO_2	$(C_6H_5)_2SeO_2$	Mp 155°C
Selenious esters, R_2SO_3	$(CH_3)_2SeO_3$	Bp 60°C at 15 mm
	$(C_2H_5)HSeO_3$	
	$C_2H_5SeO_2Cl$	Bp 175°C
Seleninic acids, $RSeO_2H$	$C_6H_5SeO_2H$	Mp 170°C
Selenic esters, R_2SeO_4	$(CH_3)_2SeO_4$	Bp 100°C at 15 mm
Selenonic acids, $RSeO_3H$	$C_6H_5SeO_3H$	Mp 142°C
Seleno ketones, R_2CSe or dimer	$[(CH_3)_2CSe]_2$	Bp 220–230°C

orange-red solid prepared by the addition of gaseous ammonia to a dilute solution of selenium tetrachloride in carbon disulfide, reaction (14). The

$$12SeCl_4 + 64NH_3 \rightarrow 3Se_4N_4 + 48NH_4Cl + 2N_2 \quad (14)$$

eight-atom molecule has a ring structure similar to that of S_4N_4. It is a very reactive substance, detonating when scratched or heated to 200°C; this property has limited study of its chemical behavior.

Polyatomic cations. For many years it has been known that elementary Se dissolved in sulfuric acid or sulfur trioxide gives green or yellow solutions. Since about 1965, careful study of such solutions has shown the green color to be related to the presence of the Se_8^{2+} ion, an eight-membered ring species, and the yellow color to result from the presence of Se_4^{2+} having four atoms in a ring.

Organic compounds. Compounds in which C-Se bonds appear are numerous and vary from the simple selenols, RSeH, selenenic acids, RSeOH, organyl selenium halides, RSeX, and diorganyl selenides and diselenides, R_2Se and R_2Se_2, to those molecules exhibiting biological activity such as selenoamino acids and selenopeptides. A representative listing of the properties of some of these compounds is given in Table 2. *See* SULFUR.

[JOHN W. GEORGE]

Bibliography: E. Gerlach (ed.), *The Physics of Selenium and Tellurium*, 1980; D. R. Hogg, *Organic Compounds of Sulphur, Selenium, and Tellurium*, vols. 1–5, 1960–1979; M. Schmidt et al., *The Chemistry of Sulfur, Selenium, Tellurium and Polonium*, 1975; R. A. Zingaro and W. C. Cooper, *Selenium*, 1974.

Silicate

A salt or ester of a silicic acid. The alkali-metal silicates are water-soluble, and sodium silicate is a syrupy liquid known as water glass. It is used as an adhesive for corrugated boxes and as a waterproofing and fireproofing agent.

Silicon dioxide is not soluble in water. When solutions of soluble silicates are acidified, a gelatinous precipitate comes down which can best be described as a hydrous oxide and represented by the formula $(SiO_2)_x(H_2O)_y$. This hydrous oxide is weakly acidic and will dissolve in alkaline solutions. Aqueous solutions of monosilicic acid, H_2SiO_3, can be prepared by special techniques.

The basic unit of silicates is the SiO_4^{4-} group which exists in orthosilicates such as Zn_2SiO_4. Other discrete silicate anions contain several of these SiO_4^{4-} groups in groups or rings linked by Si–0–Si bonds. Some of these anions are $Si_20_7^{6-}$, $Si_30_9^{6-}$, $Si_40_{12}^{8-}$, and $Si_60_{18}^{12-}$. Extended chains and sheets of variable sizes are also known. Mica falls in this latter category. In some of these naturally occurring polymers, aluminum atoms replace some of the silicons in the network. *See* SILICON.

[E. EUGENE WEAVER]

Silicon

A chemical element, Si, atomic number 14, and atomic weight 28.086. Silicon is the most abundant electropositive element in the Earth's crust. The element is a metalloid with a decided metallic luster; it is quite brittle. It crystallizes in the diamond lattice, has a specific gravity of 2.42 at 20°C, melts

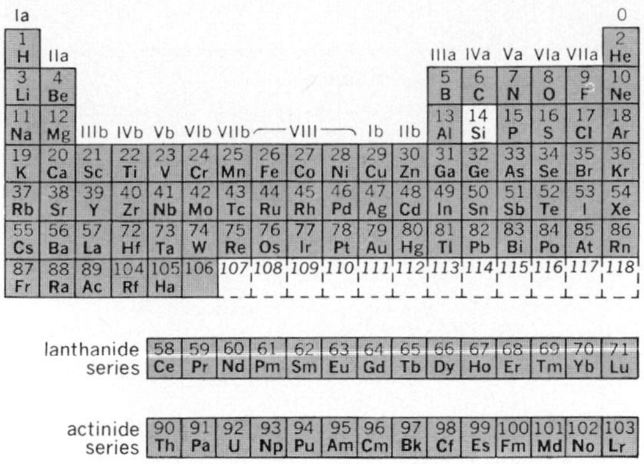

at 1420°C, and boils at 3280°C. The element is usually tetravalent in its compounds, although sometimes divalent, and is decidedly electropositive in its chemical behavior. In addition, pentacoordinate and hexacoordinate compounds of silicon are known. *See* METALLOID.

Uses. Crude elementary silicon and its intermetallic compounds are used in alloying constituents to strengthen aluminum, magnesium, copper, and other metals. Metallurgical silicon of 98–99% purity is used as the starting material for manufacturing organosilicon compounds and silicone resins, elastomers, and oils. Very pure silicon for the electronics industry allows impurity levels of 0.1 part per million, atomic, for carbon and oxygen, and 0.1 part per billion, atomic, for electrically active elements. Due to their extreme purity and high crystal perfection, powdered silicon crystals are used for calibration in x-ray diffraction.

Although miniaturization of semiconductor devices reduced dramatically the size of silicon chips used in integrated circuits, the total commercial use of semiconductor-grade silicon grew rapidly in the 1970s. Improvements in production kept the cost of silicon constant during an inflationary period, while the cost per electronic function on silicon decreased by several orders of magnitude.

Photovoltaic cells for direct conversion of solar energy to electricity use wafers sliced from single crystals of electronic-grade silicon. Fragile wafers and electrical contacts are protected by encapsulating in clear silicone rubber under a clear cover glass in modules designed for a service life of over 20 years.

The semiconductor industry achieved great cost reduction by miniaturizing components, but this path was not open to the makers of solar photovoltaic devices because of the diffuse nature of solar energy. By using optical systems to concentrate sunlight on the cells, there is a reduction in area, and thus the cost of cells required to produce a unit of electrical power. This advantage must be balanced against the cost of concentrators and cooling systems, since silicon cells become less efficient as the temperature rises. Maximum theoretical efficiency of photovoltaic conversion with silicon cells is about 28% at 20°C for sunlight above the Earth's atmosphere at air mass zero (AMO). Practical silicon cells reach about 15% (AMO) efficiency. High costs of production have limited commercial application of photovoltaic cells to spacecraft and high-priority remote installations on Earth.

Efforts aimed at reducing the cost of installed photovoltaic solar energy per peak watt include several approaches to greater economy. Each is accompanied by some loss in efficiency in photovoltaic conversion of sunlight. Applications include use of lower-cost silicon of less than semiconductor-grade purity, use of thin sheets of polycrystalline silicon on a low-cost substrate, and use of amorphous silicon-hydrogen films on a low-cost substrate.

Silicon dioxide, besides its use in tremendous quantities in the ceramic industries, is used as the raw material for making elementary silicon and for silicon carbide. Sizable crystals of it are used for piezoelectric crystals, those thin wafers of quartz which control the frequency of radio oscillators by vibrating at a very exact frequency.

Very accurate digital watches are mass-produced from quartz crystals activated by integrated circuits on tiny silicon chips powered by a small battery—all mounted on fiber-glass-reinforced epoxy circuit boards. Silane coupling agents are used to bond resin to glass in the circuit boards. The contribution of silicon to digital watches may also include silicon solar cells to keep the battery charged.

Fused quartz sand becomes silica glass, used in chemical laboratories and plants as well as an electrical insulator. A colloidal dispersion of silica in water is used as a coating agent and as an ingredient in certain polishes.

There is no widespread use of the simple hydride of silicon. Trichlorosilane ($SiHCl_3$) is made and decomposed in the preparation of very pure transistor-grade silicon. There are some organochlorosilanes (such as CH_3SiHCl_2) which contain silicon-hydrogen bonds, and these are valued because they may be converted into silicone polymers which retain the reactive silane groups. These special silicones are used to impart water-repellent films to textiles and leather.

The organochlorosilanes are used for the manufacture of silicone polymers, which may be resinous (cross-linked structures of high molecular weight), fluid (linear structures with blocking-groups at the ends), or elastomeric (linear structures of very high molecular weight, connected at intervals by the action of curing agents). Such silicone materials are used for electrical insulation, as mold-release agents, for water-repellent coatings, as hydraulic fluids, in polishes and lubricants, in cosmetics, and in a host of other applications. The silicate esters are used as sources of pure silica, in paint formulations, and as heat-transfer fluids.

Properties. Naturally occurring silicon contains 92.2% of the isotope of mass number 28, 4.7% of silicon-29, and 3.1% of silicon-30. In addition to these stable, natural isotopes, artificially radioactive isotopes of masses 25, 26, 27, 31, and 32 are known. The expected stability of an even-even isotope provides only a partial explanation for the prevalence of silicon-28 in the universe, and there must be some further explanation which necessarily becomes an important part of any theory of the genesis of the elements.

Table 1. Heats of formation of silicon, carbon, and germanium compounds*

Carbon compound	kcal/mole	Silicon compound	kcal/mole	Germanium compound	kcal/mole
$CO_2(g)$	94.4	$SiO_2(s)$	201.3	$GeO_2(s)$ (hexagonal)	132.6
		SiC	1.43	(forms no carbide)	
$CH_4(g)$	17.9	$SiH_4(g)$	11.9	$GeH_4(g)$	−21
CI_4	−50. (approx)	$SiI_4(s)$	27.7	GeI_4	+42
$CBr_4(g)$	12.0	$SiBr_4(l)$	91.5	$GeBr_4$	130
$CCl_4(l)$	33.3	$SiCl_4(l)$	149.1	$GeCl_4(l)$	284
$CF_4(g)$	162.5	$SiF_4(g)$	361.3	GeF_4	

*(*g*) gas; (*l*) liquid; (*s*) solid.

Elementary silicon has the physical properties of a metalloid, resembling germanium below it in group IV of the periodic table, and, to a lesser extent, arsenic and boron in the diagonal relationship. It shows an appreciable electrical conductivity, but this is clearly semiconductance in the sense that it increases with rise of temperature. In very pure form silicon is an intrinsic semiconductor, although the extent of its semiconduction is greatly increased by the introduction of minute amounts of impurities. Elements of the third group, such as boron, introduce atoms with a deficiency of electrons into the crystal structure and produce the so-called *p*-type silicon, which conducts electric current by migration of electron vacancies or holes. Similarly, the semiconductance of pure silicon is greatly increased by the introduction of group V elements such as phosphorus or arsenic, but in this case the increased current is carried by migration of extra electrons and the solid solution is called *n*-type silicon.

Silicon resembles the metals in its chemical behavior, and is commonly assigned an electronegativity of 1.8 in the Pauling scale. Close investigation of its relative electronegativity in the sp^3 hybrid state shows it to be about as electropositive as tin, and decidedly more positive than germanium or lead. In keeping with this rather metallic character, silicon forms tetra-positive ions and a variety of covalent compounds in which it is the positive partner of a dipole; it appears as a negative ion in only a few silicides, and, of course, as a positive constituent of oxy acid or complex anions. Its chemical behavior is characterized further by a very high heat of oxidation, so that natural silicon is found in the completely oxidized state as the dioxide or as the silicate minerals. To indicate the extent of its affinity for oxygen and the halogen elements, Table 1 gives the heats of formation of some representative compounds of silicon in comparison with the corresponding compounds of carbon and germanium.

Silicon oxidizes rapidly at room temperature to form a protective layer of silica about 1 nm thick. More complete oxidation begins at 650°C but is not rapid up to about 1200°C. The oxide layer is amorphous to about 1200°C, crystalline (tridymite or cristobalite) above 1200°C, and somewhat volatile above 1600°C. Silicon semiconductor devices are generally protected with a silica layer by oxidizing at 1100–1300°C.

Crystals of the element are insoluble in single dilute or concentrated acids, but a mixture of concentrated nitric and hydrofluoric acids will dissolve the element slowly by converting it to the dioxide and dissolving the dioxide as tetrafluoride. Controlled etching of semiconductor devices is accomplished with hydrofluoric-nitric acid or hydrofluoric-nitric-acetic acid mixtures.

The crystalline element is slowly soluble in concentrated solutions of sodium and potassium hydroxide, in which it liberates hydrogen and forms solutions of the corresponding alkali silicates. Finely divided silicon produced by reduction of its compounds below the melting point of the element is correspondingly more active toward the same reagents.

Several series of hydrides are formed, a variety of halides (some of which contain silicon-to-silicon bonds), and also many series of oxygen-containing compounds which may be either ionic or covalent in their properties.

Natural occurrence. Silicon occurs in many forms of the dioxide and as almost numberless variations of the natural silicates. In abundance, silicon exceeds by far every other element except oxygen. It constitutes 27.72% of the solid crust of the Earth, whereas oxygen constitutes 46.6%, and the next element after silicon, aluminum, accounts for 8.13%. In fact, all the solid material of the Earth's crust, insofar as it has been available for investigation, is known to be siliceous except for the carbonate and phosphate rocks, which of course are not igneous or original. This tremendous abundance of silicon makes it of particular interest as a chemical raw material, and the versatility of its chemical behavior has encouraged more and more uses to be developed through intensive research.

Preparation. In the laboratory, free silicon can be obtained by reducing potassium fluosilicate with metallic potassium at elevated temperatures, followed by washing out the resulting potassium fluoride with water. A somewhat more convenient method consists of the reduction of finely ground or precipitated silicon dioxide with magnesium powder at red heat, followed by dissolution of the resulting magnesium oxide in dilute acid and by washing the resulting brownish powder free of soluble material. Neither of these reactions gives a very pure product, since contamination by the

Table 2. Composition of a commercial silicon

Element	%	Element	%
Silicon	98.53	Manganese	0.04
Iron	0.56	Titanium	0.02
Aluminum	0.31	Other metals	0.08
Calcium	0.12	Oxygen	0.34

original starting materials or by excess reducing agent usually results. Commercial reduction of the dioxide is accomplished electrothermally with carbon at a temperature considerably above the melting point of the silicon, as shown in reaction (1).

$$SiO_2 + 2C \rightarrow Si + 2CO \quad (1)$$

Since quartz, sand, and other natural forms of silica always contain some oxides of iron, aluminum, titanium, and other materials, the silicon so obtained usually is of about 98% purity. A typical analysis of such a commercial silicon is given in Table 2. The impurities may be reduced by grinding the silicon to 50–100 mesh size and leaching exhaustively in a dilute mixture of hydrofluoric and sulfuric acids.

Very pure semiconductor-grade silicon (total impurities less than 1 part per million, or ppm) is obtained by reducing very pure trichlorosilane (preferred) or silicon tetrachloride with hydrogen on a hot silicon rod at 1200°C. The chlorosilanes are prepared from commercial-grade silicon and purified by careful fractional distillation. While further purification of silicon is obtained by pulling a crystal slowly from the melt (Czochralski method) or by moving a molten zone up through the rod vertically, the main purpose of these techniques is to produce single crystals with the desired crystallographic properties.

Principal compounds. Silicon is reported to form compounds with 64 of the 96 stable elements, and it probably forms silicides with 18 other elements. Besides the metal silicides, used in large quantities in metallurgy, silicon forms useful and important compounds with hydrogen, carbon, the halogen elements, nitrogen, oxygen, and sulfur. In addition, useful organosilicon derivatives have been prepared.

Hydrides. The hydrides of silicon are named silanes, the compound SiH_4 being called monosilane, Si_2H_6 disilane, Si_3H_8 trisilane, and so on. Compounds in which oxygen atoms alternate with silicon atoms in the principal part of the structure are called siloxanes, and those with nitrogen between silicon atoms are called silazanes. All other covalent compounds of silicon are considered for the purpose of nomenclature to be derived from these silanes and modified silanes and are named according to substituent groups and their placement along the principal silicon-containing chain or ring. Thus $(C_2H_5)_3SiOH$ is triethylsilanol, $[(CH_3)_2SiO]_4$ is octamethylcyclotetrasiloxane, $C_6H_5SiCl_3$ is phenyltrichlorosilane, CH_3SiHCl_2 is methyldichlorosilane, and so on.

The hydrides of silicon were first investigated thoroughly by Alfred Stock, who prepared them by the reaction of magnesium silicide (from the reaction of silica with excess magnesium at minimum temperature) with dilute aqueous hydrochloric or phosphoric acid. He obtained the saturated series of silanes given in Table 3, all of which could now be derived more readily from the corresponding chlorides by reduction with lithium aluminum hydride in ether solution as shown by reaction (2). The

$$2Si_2Cl_6 + 3LiAlH_4 \rightarrow 2Si_2H_6 + 3LiCl + 3AlCl_3 \quad (2)$$

products must be handled in a well-designed vacuum system, because all the silanes are readily oxidized by air and form spontaneously flammable or

Table 3. Hydrides of silicon and some derivatives

Name	Formula	Melting point, °C	Boiling point, °C
Monosilane	SiH_4	−185.	−112.
Disilane	Si_2H_6	−133.	−14.5
Trisilane	Si_3H_8	−117.	+52.9
Tetrasilane	Si_4H_{10}	−90. (approx)	109. (approx)
Chlorosilane	SiH_3Cl	−118.	−30.4
Dichlorosilane	SiH_2Cl_2	−122.	+8.3
Trichlorosilane	$SiHCl_3$	−127.	31.8
Bromosilane	SiH_3Br	−94.	1.9
Dibromosilane	SiH_2Br_2	−70.1	66. (approx)
Tribromosilane	$SiHBr_3$	−73.5	111.8
Iodosilane	SiH_3I	−57.	45.4
Diiodosilane	SiH_2I_2	−1.0	149.5
Triiodosilane	$SiHI_3$	8.	220.
Disiloxane	$H_3SiOSiH_3$	−144.	−15.2

explosive mixtures with air. All are also attacked by water in the presence of even minute traces of hydroxyl ion to evolve hydrogen and form silicic acid or hydrated silica. The reaction with water is further accelerated by larger amounts of inorganic or organic bases and becomes a dependable method for the quantitative determination of hydrogen bonded directly to silicon, since each silicon-hydrogen bond evolves one molecule of H_2.

The hydrolysis of silanes is believed to take place through the coordination attachment of a hydroxyl ion to silicon, followed by the loss of a hydride ion and its subsequent combination with a proton of water to form molecular hydrogen. A similar process takes place in the hydrolysis of silicon halides, ejecting a halide ion and producing the corresponding hydrohalogen acid along with silicic acid or silica. It is further believed that molecules of water may coordinate in the same way as hydroxyl ions in the initial phase of the reaction and that the demonstrated ability of silicon to expand its valency group to accommodate six covalent neighbors allows this process to go on rapidly. The corresponding mechanism is, of course, not available to carbon, which lacks *d*-orbitals for the expansion of its covalency shell beyond four atoms or groups. *See* INORGANIC POLYMER.

The silicon hydrides decompose thermally at elevated temperatures to liberate hydrogen and deposit spongy, brownish silicon. The most stable hydride, monosilane, decomposes rapidly at 500°C and slowly but appreciably at 250°C, especially in the presence of solid materials such as silica gel or activated alumina. The gas-phase reactions of the silanes are correspondingly limited because of this competing decomposition but all undergo such reactions as addition to the double bond of olefins at slightly elevated temperatures or in the presence of catalysts which generate free radicals. Thus monosilane adds readily to ethylene at room temperature under irradiation with ultraviolet light to form ethylsilane, as shown by reaction (3). A

$$SiH_4 + H_2C{=}CH_2 \rightarrow C_2H_5SiH_3 \quad (3)$$

corresponding reaction occurs in a general way whenever any compound containing a silicon-hydrogen bond is mixed with an alkene or an alkyne under appropriate conditions of concentration, temperature, and catalyst, so that organosilanes and their derivatives become available by this route.

The silanes are oxidized rapidly or explosively by the halogens, but react in controllable fashion with hydrogen halides in the presence of the corresponding aluminum halide to yield halogen-substituted silanes. By control of the concentrations and conditions, progressive substitution with halogen is possible.

Of the substituted hydrides, trichlorosilane (silicochloroform, $SiHCl_3$) is perhaps the best known because it can be prepared readily by the action of hydrogen chloride on elementary, crystalline, or reduced silicon at a temperature of 300–450°C. Since it boils at 31.8°C, it is easily separated from the tetrachloride and other chlorosilanes by distillation. It shows the typical reactions of the silicon-hydrogen bond, as well as those of a silicon chloride. Reaction with olefins or aromatic organic compounds produces organotrichlorosilanes.

Silicon carbide. The reduction of silica with excess carbon under appropriate conditions gives silicon carbide, SiC, which crystallizes in a number of forms but is best known in the cubic-diamond form with spacing a_0 of 0.435 nm (compared with 0.356 nm for diamond). In the pure form, silicon carbide is green (α-hexagonal) or yellow (β-cubic), but the commercial product known as Carborundum is black and has a bluish or greenish iridescence. The carbide is not easily oxidized by air except above 1000°C, and retains its physical strength up to this temperature. For these reasons it is a favorite structural refractory material for the ceramic arts. It also is extremely hard, with a Mohs hardness in excess of 9, and so has found wide application as an abrasive.

Silicon carbide does not melt without decomposition at atmospheric pressure, but does melt at 2830°C at 35 atm (3.5 MPa). High-purity silicon carbide (1 ppm impurities) is made by reducing alkylchlorosilanes with hydrogen at 1100–1400°C. Single filaments of silicon carbide deposited on very fine tungsten wires have been obtained with tensile strengths of about 500,000 psi (3.4 GPa) and modulus of elasticity to about 70×10^6 psi (480 GPa). Chlorine will convert the hot carbide to a mixture of carbon tetrachloride and silicon tetrachloride. The bond energy of the silicon-carbon bond in tetramethylsilane is 72 kcal/mole.

Silicon halides. Silicon tetrachloride, $SiCl_4$, is perhaps the best-known monomeric covalent compound of silicon. It is readily available commercially. It can be prepared by chlorinating elementary silicon, or by the action of chorine on a mixture of silica with finely divided carbon, or by the chlorination of silicon carbide. It is a volatile liquid which fumes in moist air and hydrolyzes rapidly to silica and hydrochloric acid; the mechanism of this reaction, which so sharply distinguishes the tetrachlorides of silicon and carbon, has been given above. Pure silica with very high surface area, produced by this method, is used as a reinforcing filler (white carbon-black) in silicone rubber and as a thickening agent in organic solutions. Silicon tetrachloride reacts readily with alcohols and glycols as shown by (4) to form the

$$SiCl_4 + 4C_2H_5OH \rightarrow 4HCl + Si(OC_2H_5)_4 \quad (4)$$

corresponding ethers, which may also be considered to be esters of silicic acid. It also reacts with Grignard reagents and with the alkyls of zinc, lithium, and mercury to produce organic derivatives of silicon, and has been used as a starting material for the commercial preparation of such compounds.

Some of the principal halides of silicon, together with their physical properties, are listed in Table 4. Although the iodides decompose readily with the liberation of iodine to limit their utility in some reactions, all show the characteristic properties of the silicon-halogen bond such as ready hydrolysis and reaction with alcohols and Grignard reagents. Silicon tetrafluoride complexes readily with hydrogen fluoride to produce fluosilicic acid, H_2SiF_6, which forms a well-known series of fluorosilicate salts. The chlorides, bromides, and iodides do not form complex acids of this type, but pentacoordinate and hexacoordinate silicon is known in some chelate compounds containing bidentate groups. Mixed halides containing more than one kind of halogen are known, and pseudohalides containing cyanate, isocyanate, or thiocyanate groups can readily be prepared by exchange reactions or by reaction of the silicon chlorides with silver cyanate, isocyanate, or thiocyanate.

Silicon nitride. The action of nitrogen on elementary silicon at 1300°C or above produces a refractory silicon nitride of the composition Si_3N_4. The same substance results from the thermal decomposition of ammonia addition compounds of monosilane or silicon tetrachloride, probably by way of the intermediate silicon imide, $Si(NH)_2$, which, like the dioxide, is polymeric. The nitride is inactive chemically, although at high temperatures it reacts with carbon to form silicon carbide. Layers of the nitride deposited (from ammonia and silane) on oxygen-passivated silicon have been used as insulation on electronic components.

Silicon oxides. Silicon dioxide is perhaps best known as one of its crystalline modifications called quartz, colorless crystals of which are also known as rhinestones and Glens Falls diamonds. Purple or lavender-colored quartz is called amethyst, the pink variety is rose quartz, and the yellow type, citrine. Twenty-two different phases of silica have been identified. At 573°C the ordinary or α-quartz changes over reversibly to β-quartz, which has a lower density, and at 867°C, β-quartz changes to a

Table 4. Halides of silicon

Name	Formula	Melting point, °C	Boiling, point, °C
Tetrafluorosilane (silicon tetrafluoride)	SiF_4	−95.7	−65 at 1810 mm
Tetrachlorosilane (silicon tetrachloride)	$SiCl_4$	−70.	+57.6
Tetrabromosilane (silicon tetrabromide)	$SiBr_4$	+5.	153.
Tetraiodosilane (silicon tetraiodide)	SiI_4	121.	290.
Hexachlorodisilane	Si_2Cl_6		147.
Octachlorotrisilane	Si_3Cl_8		216.
Decachlorotetrasilane	Si_4Cl_{10}		150 at 15 mm Hg (2.0 kPa)
Hexabromodisilane	Si_2Br_6	95.	265.
Hexafluorodisiloxane	Si_2OF_6	−47.8	−23.3
Hexachlorodisiloxane	Si_2OCl_6	−28.1	+137.
Hexabromodisiloxane	Si_2OBr_6	+27.9	118 at 15 mm Hg (2.0 kPa)

different crystal modification, β-tridymite. At a still higher temperature, 1470°C, β-tridymite becomes a third modification called β-cristobalite. Both β-tridymite and β-cristobalite have lower temperature α-modifications which have lower optical symmetry, and all the forms can be maintained at room temperature if chilled rapidly from the stable equilibria. Each appears to have its own melting point, but the usual melting point for silica is that of β-cristobalite, about 1723°C. Rapid chilling of the liquid produces silica glass, often incorrectly called quartz glass, a vitreous modification of SiO_2 which has a very low coefficient of expansion and is, of course, isotropic. This silica glass finds many uses because of its resistance to thermal shock and its low electrical conductivity; it is useful in the temperature range below 1000°C, but above this temperature it begins to devitrify and at 1250°C will crystallize quite rapidly.

Because rock crystal has been collected and admired for thousands of years, large and perfectly formed natural crystals of quartz now are very rare. With the growth of radio broadcasting and the electronics industry, piezoelectric crystals cut from perfect specimens of quartz have been used in increasing quantities, to the point of scarcity of natural crystals. As the supply diminished, considerable effort was devoted to the problem of growing crystals of quartz by artificial means. Some success has been achieved by growing the crystals hydrothermally from a solution of silica glass in water containing an alkali or a fluoride, the whole being maintained at a temperature of 200–300°C or higher in a high-pressure bomb. The vitreous silica has a higher solubility in an aqueous medium at the operating temperature than does quartz and so will dissolve in the solution and crystallize out upon a tiny seed crystal of quartz. The solubility of quartz in water and in solutions of hydroxides or fluorides does not decrease sharply as the critical point of water is reached; in fact, the solubility in pure water above the critical point is ample for growing crystals.

A number of natural noncrystalline varieties of silicon dioxide also are known, such as the hydrated silica known as opal and the dense unhydrated variety known as flint. Onyx and agate represent still other semiprecious forms.

Silicon monoxide, SiO, is a brownish powder which can be obtained by heating a mixture of finely divided silica and elementary silicon in the absence of air above 1400°C. It is much more volatile than the dioxide, and is used to form a durable coating on lenses and mirrors and in electron microscopy for shadowing. There is considerable doubt that the monoxide persists as a stable substance at temperatures below 1300°C; it seems to disproportionate into crystallites of silicon and silica if it cools slowly. The monoxide oxidizes readily to dioxide, which of course does not interfere with its use as a lens coating. At a temperature of 300–400°C it also is reactive to halogens and behaves in general as a readily oxidized lower-valent compound of silicon would be expected to do. The only other known compounds of divalent silicon are some nonvolatile liquid lower chlorides, which are also covalently unsaturated and will react (for example) with methyl chloride to form methylchlorosilanes.

Sulfides. Although no monosulfide of silicon has been isolated, silicon disulfide is known in the form of long, flexible acicular crystals which appear to be infinite chains of SiS_4 tetrahedrons. The crystals are mechanically strong in the direction of the long axis but hydrolyze readily to produce silica and hydrogen sulfide.

Esters of silicic acid. As indicated above, the reaction of ethyl alcohol with silicon tetrachloride produces ethyl silicate, a colorless, volatile liquid of pleasant odor which boils at 168°C. Because this can be distilled to a high degree of purity and is made from materials which contain no alkalies or other ionic impurities, it is a favored source of pure silica to be used for the preparation of silicate phosphors and similar materials. It hydrolyzes very slowly in pure water but more rapidly in the presence of a small amount of dilute acid as catalyst, to form hydrated silica and ethanol. The hydrolysis can also be controlled with limited water to give condensed ethyl polysilicates, which have been used as paint vehicles for the protection of masonry and metals.

The reaction of methanol with silicon tetrachloride produces methyl silicate, a liquid of camphor-like odor which boils at 121°C. This might be expected to have the same uses as ethyl silicate, but when made in the manner indicated, it is an extremely dangerous substance to handle. It causes perforating ulceration of the cornea and eventual blindness. The same methyl silicate can be made by the direct action of methanol on elementary silicon at elevated temperatures and in the presence of a copper catalyst, and this product (made in the absence of halogen-containing substances) has been found to have no adverse physiological effect upon experimental animals.

Higher alkyl silicates, as well as some aromatic silicates, are items of commerce and find some use as heat-transfer media because of their long liquid range and their considerable thermal stability. All will hydrolyze in the presence of strong acids and bases, however.

Organic compounds. Important organosilicon derivatives include tetraalkyls and tetraaryls, halides, hydrides, and the organosiloxanes.

Tetraalkyls and tetraaryls. Organic compounds of silicon which contain direct silicon-carbon bonds have been known since about 1860, and some of them show remarkable thermal stability and resistance to oxidation. The tetraalkyl- and tetraarylsilanes can be made by the action of Grignard reagents, zinc alkyls, lithium alkyls, or other alkylating agents on silicon tetrachloride, followed by hydrolysis of the reaction mixture and distillation of the desired compound. A list of some representative compounds, together with their physical properties, is given in Table 5. Of the compounds listed, tetraphenylsilane has the most striking stability; it can be distilled in air at 428°C, and remains quite unchanged. Tetramethylsilane does not decompose until a temperature of about 600°C is reached, but it does oxidize at 350°C or higher.

Organosilicon halides and hydrides. The organic compounds of silicon which contain both directly attached organic groups and silicon-chlorine or silicon-hydrogen bonds are of greater interest than the tetraalkylsilanes because they lend themselves so well to further reactions and hence may serve as chemical intermediates for a wide variety of organosilicon products. The alkyl and aryl chlorosil-

Table 5. Organosilicon compounds of the type R_4Si

Compound	Formula	Melting point, °C	Boiling point, °C
Tetramethylsilane	$(CH_3)_4Si$	α −99, β −102	26.5
Trimethylethylsilane	$(CH_3)_3SiC_2H_5$	–	62.
Dimethyldiethylsilane	$(CH_3)_2Si(C_2H_5)_2$	–	95.8
Methyltriethylsilane	$CH_3Si(C_2H_5)_2$	–	127.
Tetraethylsilane	$(C_2H_5)_4Si$	−69.	153.
Triethylvinylsilane	$(C_2H_5)_3SiC_2H_3$	–	146.
Diethyldiphenylsilane	$(C_2H_5)_2Si(C_6H_5)_2$	–	297.
Tetrapropylsilane	$(n\text{-}C_3H_7)_4Si$	−46.	212.
Tetrabutylsilane	$(n\text{-}C_4H_9)_4Si$	−55.7	157 at 22 mm Hg (2.9 kPa)
Tetraphenylsilane	$(C_6H_5)_4Si$	233.	428
Tetrabenzylsilane	$(C_6H_5CH_2)_4Si$	127.	approx 450
Tetravinylsilane	$(C_2H_3)_4Si$	–	130.6

anes represent one class of such compounds, and these find application as intermediates for the preparation of silicone polymers. In general, organosilicon halides may be made by the action of the classical organometallic reagents, such as lithium alkyls and Grignard reagents, on the halides of silicon according to general reactions (5) and (6).

$$SiX_4 + yRMgX \rightarrow R_ySiX_{4-y} + yMgX_2 \quad (5)$$

$$Si_2X_6 + yRMgX \rightarrow R_ySi_2X_{6-y} + yMgX_2 \quad (6)$$

This substitution method has extreme flexibility and so is widely applicable to laboratory and commercial preparations. For quantity production of a limited number of organochlorosilanes, however, it has given way to the direct synthesis from alkyl or aryl chlorides and elementary silicon. This is shown by reaction (7). For example, methyl chloride

$$2RCl + Si \xrightarrow[\text{Heat}]{\text{Catalyst}} R_2SiCl_2 \text{ and other products} \quad (7)$$

reacts readily with elementary commercial silicon in the presence of copper powder as a catalyst at a temperature in the vicinity of 300°C to produce principally dimethyldichlorosilane, with some methyltrichlorosilane, trimethylchlorosilane, methyldichlorosilane, and a number of other less important products. From this reaction mixture the pure methylchlorosilanes are distilled and used as intermediates for silicone oils, resins, and rubber, since they hydrolyze readily to form the corresponding organosiloxane polymer (silicone) and hydrochloric acid. Because methyl chloride can be made from methanol and hydrochloric acid, the raw materials are seen to be silicon and methanol, which in turn are made from silica, coal, and water as ultimate raw materials. Similarly, phenylchlorosilanes can be made by the action of chlorobenzene on elementary silicon, with silver or copper as catalyst and at temperatures of 350–500°C. The distilled products are cohydrolyzed with methylchlorosilanes to make methyl phenyl silicone polymers which show greater resistance to oxidation than the methyl silicones themselves.

A great variety of organochlorosilanes are known, and many also are available commercially. Table 6 lists a number of the more important ones. By appropriate procedures of synthesis, functional groups may be attached to the organic portions of these organochlorosilanes, or their derivatives, and a growing technology of organofunctional silanes is developing. A major application has been in coupling agents or primers to obtain improved bonding between organic polymers and mineral surfaces. Coupling agents comprise an organofunctional group that is reactive or compatible with a given polymer and groups on silicon that hydrolyze

Table 6. Some organochlorosilanes, R_xSiCl_{4-x}

Name	Formula	Melting point, °C	Boiling point, °C
Methyltrichlorosilane	CH_3SiCl_3	−77.8	65.7
Dimethyldichlorosilane	$(CH_3)_2SiCl_2$	−76.1	70.0
Trimethylchlorosilane	$(CH_3)_3SiCl$	−57.7	57.3
Methylphenyldichlorosilane	$(CH_3)C_6H_5SiCl_2$	–	205.
Ethyltrichlorosilane	$C_2H_5SiCl_3$	−105.6	99.5
Diethyldichlorosilane	$(C_2H_5)_2SiCl_2$	−96.5	129.
Triethylchlorosilane	$(C_2H_5)_3SiCl$	–	143.5
Ethylphenyldichlorosilane	$(C_2H_5)C_6H_5SiCl_2$	–	230.
Vinyltrichlorosilane	$C_2H_3SiCl_3$	–	92.
Divinyldichlorosilane	$(C_2H_3)_2SiCl_2$	–	119.
Propyltrichlorosilane	$n\text{-}C_3H_7SiCl_3$	–	122.7
Dipropyldichlorosilane	$(n\text{-}C_3H_7)_2SiCl_2$	–	175.
Butyltrichlorosilane	$n\text{-}C_4H_9SiCl_3$	–	148.9
Decyltrichlorosilane	$n\text{-}C_{10}H_{21}SiCl_3$	–	183 at 84 mm Hg (11.2 kPa)
Phenyltrichlorosilane	$C_6H_5SiCl_3$	–	201.5
Diphenyldichlorosilane	$(C_6H_5)_2SiCl_2$	–	305.2
Triphenylchlorosilane	$(C_6H_5)_3SiCl$	88.	378.
Benzyltrichlorosilane	$C_6H_5CH_2SiCl_3$	–	216.
Dibenzyldichlorosilane	$(C_6H_5CH_2)_2SiCl_2$	51.	243 at 100 mm Hg (13.3 kPa)
Naphthyltrichlorosilane	$\alpha\text{-}C_{10}H_7SiCl_3$	–	168 at 22 mm Hg (2.9 kPa)

Table 7. Some organosiloxanes

Name	Formula	Melting point, °C	Boiling point, °C
Hexamethylcyclotrisiloxane	$[(CH_3)_2SiO]_3$	64.	134.
Octamethylcyclotetrasiloxane	$[(CH_3)_2SiO]_4$	17.5	175.
Decamethylcyclopentasiloxane	$[(CH_3)_2SiO]_5$	−38.	210.
Dodecamethylcyclohexasiloxane	$[(CH_3)_2SiO]_6$	−3.	245.
Hexamethyldisiloxane	$(CH_3)_3SiOSi(CH_3)_3$	−59.	100.5
Octamethyltrisiloxane	$(CH_3)_8Si_3O_2$	−80.	153.
Decamethyltetrasiloxane	$(CH_3)_{10}Si_4O_3$	−70.	194.
Hexaethylcyclotrisiloxane	$[(C_2H_5)_2SiO]_3$	14.	117. at 10 mm Hg (1.33 kPa)
Octaethylcyclotetrasiloxane	$[(C_2H_5)_2SiO]_4$	−50.	159. at 10 mm Hg (1.33 kPa)
Hexaphenylcyclotrisiloxane	$[(C_6H_5)_2SiO]_3$	190.	295. at 1 mm Hg (133 Pa)
Octaphenylcyclotetrasiloxane	$[(C_6H_5)_2SiO]_4$	201.	335. at 1 mm Hg (133 Pa)

to silanols for bonding to the mineral surfaces. Compounds that find general application as coupling agents include vinyl, methacrylate, epoxy, amine, and mercaptan-functional silanes.

It appears that organosilicon groups may be introduced into a large variety of organic polymers, dyes, drugs, and other products, with results yet to be determined.

The reaction of organochlorosilanes and related substances with alcohols produces the corresponding esters or ethers of the type $R_xSi(OR')_y$. These hydrolyze also to organosiloxanes, but in a controllable way that makes them desirable for some applications.

Silylating agents such as trimethylchlorosilane and pyridine, or hexamethyldisilazane, tetramethyldisilazane, or bis (trimethylsilyl) acetamide are used to prepare volatile silyl derivatives of hydrogen-bonded materials. Active hydrogens are substituted in alcohols, enols, phenols, acids, amides, ureas, and amino acids for preparation of reactive intermediates or to obtain volatile derivatives for separation by gas-liquid chromatography.

Organosiloxanes. The organohalosilanes hydrolyze readily to form organosiloxanes, all of which are polymeric. These cyclic and linear polymers are known as silicones because they were first regarded by F. S. Kipping to be analogs of the organic ketones, but they are decidedly polymeric in composition. The individuality of silicon gives them properties which are very different from those of the ketones.

Table 7 lists a number of pure organosiloxanes and some of their physical properties. In general, the siloxanes are chemically inactive, being unchanged by dilute or even moderately concentrated acids and by most ionic reagents. The siloxane chain is attacked by strong alkalies, however, with the formation of alkali metal salts of the corresponding organosilicates. Hydrofluoric acid also will attack the siloxane chain to produce monomeric organofluorosilanes and water.

Organic groups of the organosiloxanes show a wide variation of chemical inertness or reactivity. The methyl groups in a methyl polysiloxane remain attached at temperatures up to 500°C, where the siloxane chain depolymerizes into volatile, cyclic structures, and phenyl groups are equally firmly attached to silicon. In the presence of oxygen, however, methyl groups begin to be oxidized to formaldehyde at 300°C or higher, whereas phenyl groups are capable of withstanding temperatures of 400–450°C. Higher alkyl groups oxidize at a rate which increases with the chain length, so that the higher alkyl silicones have little applicability at elevated temperatures. In general, negative groups such as halogen, OH, COOH, and NH_2 as substituents in the organic part of the molecule decrease stability of the carbon-silicon bond by their inductive effect upon its polar properties.

Among the interesting physical properties of the methyl and methyl phenyl polysiloxanes is an unusually small dependence of viscosity, dielectric constant, and compressibility upon temperature. The very low temperature coefficient of viscosity of methyl silicone oils, therefore, makes them useful as hydraulic fluids, lubricants, and dielectric fluids, so that they come into service both at very low and very high temperatures in comparison with their organic counterparts. *See* CARBON; GERMANIUM; SILICONE RESINS.

Analysis of compounds. Analytical techniques used with organic compounds may be used with some significant modifications for organosilicon compounds. These modifications relate primarily to the greater reactivity of SiCl, SiH, SiOC, SiOH, and SiN bonds, and the greater stability of the SiC and SiOSi bonds than those of their organic counterparts. Physical methods of analysis, including gas chromatography (GC), nuclear magnetic resonance (NMR) spectrometry, mass spectrometry, and (especially) infrared spectrometry are particularly useful.

Purity of organosilicon materials is often monitored by gas chromatography, although nonvolatile residues will not be detected. The spectroscopic methods produce characteristic "fingerprints" that are useful for identification and analysis of components. Physical properties such as density, refractive index, boiling point, viscosity, and molecular weight often give sufficient information to characterize a material. Chemical analyses for determination of C, H, Si, SiH, and total and hydrolizable halide are used as appropriate, but modifications of conventional analytical procedures are often necessary.

[EDWIN P. PLUEDDEMANN]

Bibliography: F. Boschke (ed.), *Silicon Chemistry One* and *Two*, 1974; R. K. Iler, *The Chemistry of Silica: Solubility, Polymerization, Colloid and Surface Properties and Biochemistry*, 1979; H. A. Liebhafsky, *Silicones under the Monogram: A Story of Industrial Research*, 1978; K. V. Ravi, *Imperfections and Impurities in Semiconductor Silicon*, 1981; E. G. Rochow, *The Chemistry of Silicon*, 1975; A. L. Smith, *Analysis of Silicones*, 1975; M. G. Voronkov et al., *The Siloxane Bond*, 1978.

Silicone resins

Polymers composed of alternating atoms of silicon and oxygen with organic substituents attached to the silicon atoms, as shown in the formula below.

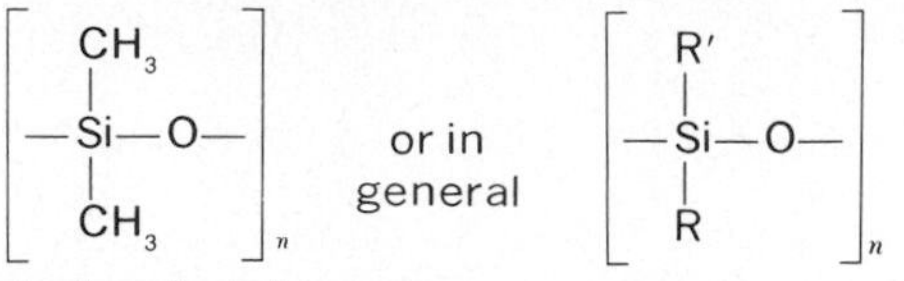

Polydimethylsiloxane

Silicones, also called organopolysiloxanes, may exist as liquids, greases, resins, or rubbers. The distinguishing characteristics of silicone polymers are good resistance to water and oxidation, stability at high and low temperatures, and lubricity.

Preparation. Silicones are obtained by the condensation of hydroxy organosilicon compounds formed by the hydrolysis of organosilicon halides. The required halide can be prepared by a direct reaction between silicon and an alkyl halide or by the reaction between a silicon halide and a Grignard reagent as in (1). The former method is

$$\left.\begin{array}{l} CH_3Cl + Si \xrightarrow[300^\circ C]{Cu, CuO} \\ SiCl_4 + CH_3MgBr \rightarrow \end{array}\right\} \begin{array}{l} CH_3SiCl_3, (CH_3)_2SiCl_2 \\ (CH_3)_3SiCl, (CH_3)_4Si \end{array} \qquad (1)$$

the basis for most commercial processes. Alternate methods may be based on the reaction of a silane with unsaturated compounds, as in (2) and (3). After separation of the reaction products by

$$\underset{\text{Ethylene}}{CH_2{=}CH_2} + \underset{\text{Trichlorosilane}}{SiHCl_3} \rightarrow \underset{\text{Ethyltrichlorosilane}}{CH_3CH_2SiCl_3} \qquad (2)$$

$$\underset{\text{Acetylene}}{CH{\equiv}CH} + SiHCl_3 \rightarrow \underset{\text{Vinyltrichlorosilane}}{CH_2{=}CHSiCl_3} \qquad (3)$$

distillation, organosilicon halides can, in principle, be polymerized by carefully controlled hydrolysis, reactions (4).

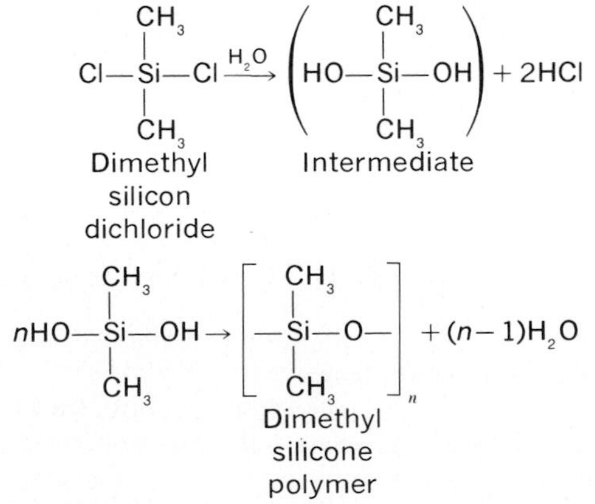

(4)

In practice, the first products are usually low in molecular weight ($n = 2$ to 7), and usually consist of a mixture of linear and cyclic species, especially the tetramer.

Fluids having a wide range of viscosity are prepared by polymerizing further, using a monofunctional trichlorosilane to limit molecular weights to the value desired. Properties such as lubricity and stability can also be varied by partial substitution of phenyl or fluoroalkyl groups for methyl groups. Elastomers are made by polymerization of the purified tetramer using an alkaline catalyst at 100 – 150°C, the molecular weight being controlled by using a monofunctional silane. Curing characteristics and properties may be varied over a wide range by replacing some methyl groups by —H, —OH, fluoroalkyl, alkoxy, or vinyl groups, and by compounding with fillers. To make silicone resins, trichloromethylsilane is copolymerized with the difunctional silane, thus providing sites for subsequent cross-linking. As with the other types of silicone, properties can be varied by partial replacement of some of the methyl groups by other substituents and by the use of reinforcing fillers; the degree of cross-linking depends on the proportion of trifunctional silane.

Other inorganic elements can be introduced into the molecules. For example, boron may be introduced by condensation of dialkylpolysiloxanes containing terminal —OH groups with, for example, boric acid, as in reaction (5).

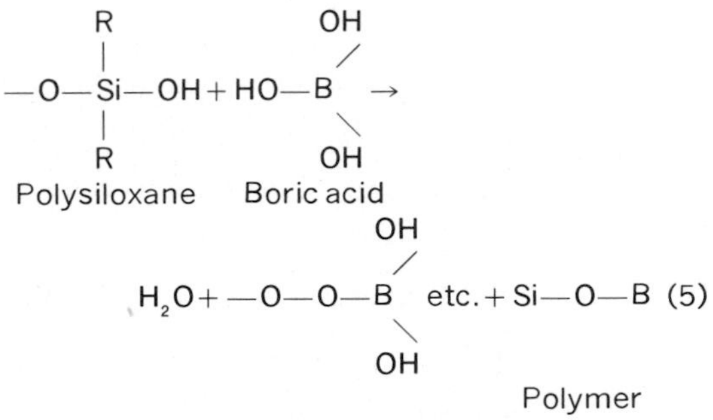

Uses. The wide range of structural variations mentioned makes it possible to tailor compositions for many kinds of applications. Low-molecular-weight silanes containing amino or other functional groups are employed as treating or coupling agents for glass fiber and other reinforcements in order to cause unsaturated polyesters and other resins to adhere more firmly.

The liquids, generally dimethyl silicones of relatively low molecular weight, have low surface tension, great wetting power and lubricity for metals, and very small change in viscosity with temperature. They are used as hydraulic fluids, as antifoaming agents, as treating and waterproofing agents for leather, textiles, and masonry, and in cosmetic preparations. The greases are particularly desired for applications requiring effective lubrication at very high and at very low temperatures.

Silicone resins are frequently selected for coating applications in which thermal stability in the range 300 – 500°C is required. The dielectric properties of the polymers make them suitable for many electrical applications, particularly in electrical insulation which is exposed to high temperatures and as encapsulating materials for electronic devices.

Silicone rubbers are compositions containing high-molecular-weight dimethyl silicone linear polymer, finely divided silicon dioxide as the filler, and a peroxidic curing agent. It has been suggested that cross-linking takes place through reaction of the peroxide with two methyl groups on adjacent chains, either to form a dimethylene bridge or to replace the two methyl groups by a single oxygen bridge. The silicone rubbers have the remarkable ability of remaining flexible at very low temperatures and stable at high temperatures. An application of increasing importance is as implants in the body for cosmetic or prosthetic purposes. Bouncing putty is a silicone rubber containing some Si—O—B groups. *See* INORGANIC POLYMER; SILICON. [JOHN A. MANSON]

Bibliography: J. A. Brydson, *Plastics Materials*, 3d ed., 1975; W. J. Roff and J. R. Scott, *Fibres, Films, Plastics, and Rubber*, 1971.

Silver

A chemical element, Ag, atomic number 47, atomic mass 107.868. It is a gray-white, lustrous metal. Chemically it is one of the heavy metals and one of the noble metals; commercially it is a precious metal. Copper, silver, and gold make up group Ib

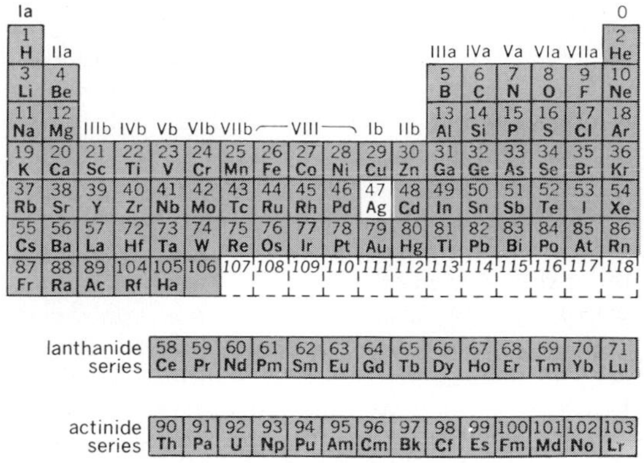

of the periodic table of elements. Silver has been known as a metal since very ancient times; it was mentioned in the books of the Egyptian king Menes, about 3600 B.C., who set its value at two-fifths that of gold. The chemical symbol Ag is derived from the Latin word for silver, *argentum*. Twenty-five isotopes of silver, including nine nuclear isomers, have been reported. Their atomic masses range from 102 to 117. Of the radioactive isotopes, the shortest half-life is that of Ag^{114}, 5 sec, and the longest that of Ag^{108m}, about 5 years. Ordinary silver is made up of the isotopes of masses 107 (52% of natural silver) and 109 (48%).

Uses of the metal. In most of its uses, silver is alloyed with one or more other metals. The traditional use of silver, copper, and gold in coins has caused them to be known as the coinage metals. The coinage of the United States has replaced most of the silver with a high-copper alloy. Silver has the highest thermal and electrical conductivities of all the metals, its electrical resistivity being 1.59 μohm-cm at 20°C. For this reason it is used for electrical and electronic contact points and sometimes for special wiring.

Silver also has well-known uses in jewelry and silverware. It is used in some fuses and medical instruments, in silver solder, and in corrosion-resistant storage batteries. Alloys in which silver is an ingredient include dental amalgam and metals for engine pistons and bearings. A silver surface has some antimicrobial properties, and silver has been used to line sterile containers.

Occurrence. Silver is a rather rare element, ranking sixty-third in order of abundance. It constitutes an estimated 0.000001–0.00001% of the Earth's crust. Sometimes it occurs in nature as the free element (native silver) or alloyed with other metals. Norway has the world's most important deposit of native silver; one piece weighing more than 1500 lb (680 kg) has been found there. For the most part, however, silver is found in ores containing silver compounds. The principal silver ores are argentite, Ag_2S, cerargyrite or horn silver, AgCl, and several minerals in which silver sulfide is combined with sulfides of other metals: stephanite, $5Ag_2S \cdot Sb_2S_5$; polybasite, $9(Cu_2S, Ag_2S) \cdot (Sb_2S_3, As_2S_3)$; proustite, $3Ag_2S \cdot As_2S_3$; and pyrargyrite, $3Ag_2S \cdot Sb_2S_3$.

About three-fourths of the silver produced is a by-product of the extraction of other metals, copper and lead in particular. In addition to this new metal, substantial quantities of silver have been recovered from coins which have been replaced with others containing less silver or none. The recovery of silver from industrial scrap, including photographic residues, is also important.

Silver metal. Pure silver is a white, moderately soft metal (2.5–3 on Mohs hardness scale), somewhat harder than gold. When polished, it has a brilliant luster and reflects 95% of the light falling on it. Silver is second to gold in malleability and ductility. Its density is 10.5 times that of water, so that 1 ft^3 of silver weighs 655 lb (1 m^3 contains 10.5×10^3 kg). Silver melts at 960.8°C (961.93°C on the International Practical Temperature Scale of 1968) and boils at about 2200°C. Gold and silver may be mixed to form true solutions (alloys) in any proportions. The quality of silver, its fineness, is expressed as parts of pure silver per 1000 parts of total metal. Commercial silver is usually 999 fine.

Masses of silver, like those of other precious metals, are measured on the troy scale, which counts 12 oz to the troy pound (1 troy oz = 31.10 g; 1 troy lb = 0.3732 kg).

Silver is available commercially as sterling silver (7.5% copper) and in ingots, plate, moss, sheets, wire, castings, tubes, and powder.

Chemical properties. Although silver is the most active chemically of the noble metals, it is not very active in comparison with most other elements. It does not oxidize at all readily (as iron does when it rusts), but it reacts with sulfur or hydrogen sulfide to form the familiar silver tarnish. Electroplating silver with rhodium will prevent this discoloration. The tarnish may be removed from silver articles by abrasion with a silver cream or polish, which also removes the very thin surface layer of silver that has combined with sulfur. Tarnish may be removed chemically by heating the article in a dilute solution of sodium chloride (table salt) and sodium hydrogen carbonate (baking soda) or placing the tarnished article in contact with a more active metal such as aluminum, which reacts with the sulfur and restores the silver to the metallic state. Silver itself does not react with dilute nonoxidizing acids (hydrochloric or sulfuric acids) or strong bases (sodium hydroxide). However, oxidizing acids (nitric or concentrated sulfuric acids) dissolve it by reaction to form the unipositive silver ion, Ag^+. This ion, which is present in solutions of all simple, soluble compounds of silver, is rather easily reduced to the free metal, as in the deposition of silver mirrors by organic reducing agents. Electroplating of silver involves reduction of com-

Table 1. Atomic and ionic properties of silver

Property	Value
Electronic configuration	$1s^2, 2s^2, 2p^6, 3s^2, 3p^6, 3d^{10}, 4s^2, 4p^6, 4d^{10}, 5s^1$
Ionization potentials	
1st electron loss	7.57 eV
2d electron loss	21.5 eV
3d electron loss	36.1 eV
Ionic radius, Ag^+	0.126 nm
Ag^{++}	0.089 nm
Covalent radius (tetrahedral)	0.153 nm
Oxidation potentials	$Ag \rightleftarrows Ag^+ + e^-$ $E° = -0.799$ volt
	$Ag^+ \rightleftarrows Ag^{++} + e^-$ $E° = -1.98$ volt

plex silver ions. The Ag^+ ion is colorless, but a number of silver compounds are colored because of the influence of their other constituents. Oxygen dissolves in silver to a surprising extent, about 20 parts of oxygen to 1 of silver by volume at the melting point of silver. Even after cooling, the silver retains 0.75 part of oxygen by volume. Some atomic and ionic properties are listed in Table 1.

Compounds. Silver is almost always monovalent in its compounds, but an oxide, fluoride, and sulfide of divalent silver are known. Some coordination compounds of silver, also called silver complexes, contain divalent and trivalent silver. Some of the important compounds of silver are listed in Table 2.

Although silver does not oxidize when heated, it can be oxidized chemically or electrolytically to form silver oxide or peroxide, a strong oxidizing agent. Because of this activity, silver finds considerable use as an oxidation catalyst in the production of certain organic materials. A silver oxide or peroxide anode in conjunction with a zinc cathode in an alkaline electrolyte constitutes an electric battery which will give a large output per unit weight or volume and therefore finds application in special military devices where weight and space are at a premium. This type of battery can be recharged a few times and is therefore a storage battery, but it can withstand only a limited number of cycles of charge and discharge and also has a rather limited shelf life. These features, as well as the cost, restrict its more general use.

Monovalent silver forms a large number of stable coordination compounds. These are often two-coordinate, having two ionic or molecular groups attached to a central Ag^+ ion, as in $[Ag(NH_3)_2]^+$ or $[Ag(CN)_2]^-$. Three-coordinate complexes, such as $[AgCl_3]^{2-}$, are also known, and four-coordinate complexes like $[AgCl_4]^{3-}$ and $[Ag(CN)_4]^{3-}$ probably occur in solution. Silver cyanide, AgCN, is a long-chain coordination compound made up of alternate silver and cyanide ions. Divalent silver can be stabilized against decomposition by complexing the Ag^{2+} ion with the organic compounds *o*-phenanthroline, pyridine, and α, α'-dipyridyl. The trivalent Ag^{3+} ion can be stabilized through complexing with ethylenedibiguanide. All the coinage metals, that is, copper, silver, and gold, complex more readily with substances that can provide nitrogen, sulfur, or halogen atoms for attachment to the metal than they do with oxygen-providing substances. Complexes of silver with hydroxide ion, for example, are not very stable compared with the hydroxide complexes of zinc, which is a good coordinator with oxygen. Accordingly, silver oxide dissolves only slightly in strong solutions of sodium hydroxide, whereas zinc hydroxide dissolves through coordination with hydroxide, displaying the property known as amphoterism.

Analytical methods. Solutions containing silver ion may be readily identified by precipitation of

Table 2. Compounds of silver

Name and formula	Uses	Properties, remarks
Silver nitrate (lunar caustic), $AgNO_3$	Medicinal; preparation of silver compounds, silver mirrors, inks	Colorless, very soluble compound; stains skin; poisonous internally; easily reduced to metallic silver
Diammine silver hydroxide, $[Ag(NH_3)_2]OH$		Soluble coordination compound, formed by adding ammonium hydroxide to silver salt solutions; on standing, forms highly explosive "fulminating silver"
Silver cyanide, AgCN	Electroplating; medicines	Used with excess sodium or potassium cyanide in electroplating to form complex ions $[Ag(CN)_2]^-$ and $[Ag(CN)_3]^{--}$, which are reduced to metallic silver
Silver chloride, AgCl	Photography; ionization detector for cosmic rays	White, insoluble compound; dissolves in ammonium hydroxide to give $[Ag(NH_3)_2]^+$ complex ions; cement for glass; single crystals for infrared absorption cells
Silver bromide, AgBr	Photography	Light yellow, insoluble compound; more resistant to dissolving than is AgCl
Silver iodide, AgI	Cloud seeding; photography	Yellow, insoluble compound; more resistant to dissolving than is AgBr; unit crystals almost identical with those of ice in cloud seeding
Silver sulfide, Ag_2S		Least soluble of all silver salts; black; main component of silver tarnish
Silver oxide, Ag_2O	Medicines; glass polishes and colors; catalyst; battery plates; paints; water purification	Brown powder; almost insoluble in water
Silver orthophosphate, Ag_3PO_4	Photography; catalyst; pharmaceuticals	Yellow photosensitive powder; almost insoluble in water

silver chloride upon addition of hydrochloric acid or a soluble chloride salt. The precipitate may be distinguished from those of lead and monovalent mercury by its ability to dissolve when excess ammonium hydroxide is added, and to reprecipitate when nitric acid is added. Quantitatively, silver chloride, silver bromide, or silver cyanide may be conveniently precipitated, dried, and weighed. Silver ion may also be reduced by electrolysis and weighed as metallic silver. Standard potassium thiocyanate solution may be used to analyze for silver volumetrically.

A method using ethylenediaminetetraacetate ion (EDTA) relies on the displacement by silver of the nickel ion in the $Ni(CN)_4^{--}$ complex, followed by titration with EDTA reagent in the presence of Murexide indicator. Silver may also be determined by polarography in 0.1 *M* KNO_3 electrolyte. Flame photometry is often used to estimate the silver content of alloys. Atomic absorption spectrometry can detect as little as 60 parts of silver ion in 1,000,000,000 parts of solution. Neutron activation analysis has been used to measure 0.00003 μg (30 picograms) of silver in photographic emulsions, steel, and sulfide minerals.

Toxicity. Soluble silver salts, expecially $AgNO_3$, have proved lethal in doses as small as 2 g, though 10 g or more is usually considered a lethal dose for adults. The direct caustic effect on the gastrointestinal tissues is responsible for this acute toxicity. The antidote is NaCl solution to precipitate AgCl. Otherwise silver compounds may be slowly absorbed by the body tissues, with a resulting bluish or blackish pigmentation of the skin (argyria). In the chronic sense silver compounds are not among the more toxic metal salts.

[WILLIAM E. COOLEY]

Bibliography: J. C. Bailar, Jr., et al. (eds.), *Comprehensive Inorganic Chemistry*, 5 vols., 1973; A. F. Cotton and G. Wilkinson, *Advanced Inorganic Chemistry: A Comprehensive Text*, 4th ed., 1980; J. A. Dean (ed.), *Lange's Handbook of Chemistry*, 12th ed., 1978; J. R. DiPalma (ed.), *Drills Pharmacology in Medicine*, 4th ed., 1971; R. H. Dreisbach, *Handbook of Poisoning: Diagnosis and Treatment*, 10th ed., 1980; M. S. Sienko and R. A. Plane, *Chemistry: Principles and Properties*, 5th ed., 1975; M. C. Sneed et al., *Comprehensive Inorganic Chemistry*, vol. 2, 1953; R. C. Weast (ed.), *Handbook of Chemistry and Physics*, 61st ed., 1980; F. J. Welcher (ed.), *Standard Methods of Chemical Analysis*, 6th ed., vols. 1–3, 1962–1966.

Silver chloride electrode

An electrode made of silver, covered or intimately mixed with silver chloride. One method of preparation consists of coating a silver wire or a silver-plated noble metal with silver chloride by electrolysis as an anode in a chloride solution. A second method consists of three steps: (1) pasting silver oxide or oxalate on a platinum helix, (2) reducing the oxide or oxalate to silver by heating to 500°C, and (3) chlorodizing part of the silver to silver chloride. A third method consists of pasting an intimate mixture of silver oxide and silver chlorate on a platinum helix and heating to 500°C. Although these systems are frequently referred to as silver chloride electrodes, they are, in reality, not electrodes until immersed in a chloride solution. The standard potential of the silver chloride electrode is −0.2224 volt relative to the normal hydrogen electrode at 25°C. The potential of the electrode is a logarithmic function of chloride ion; that is, it is reversible to chloride ion. This electrode finds wide use in the study of chloride systems, as a replacement for calomel in half-cells, and as the inner electrode in some glass electrodes. *See* CALOMEL ELECTRODE; ELECTRODE POTENTIAL; HYDROGEN ELECTRODE.

[WALTER J. HAMER]

Single crystal

In crystalline solids the atoms or molecules are stacked in a regular manner, forming a three-dimensional pattern which may be obtained by a three-dimensional repetition of a certain pattern unit called a unit cell. When the periodicity of the pattern extends throughout a certain piece of material, one speaks of a single crystal. A single crystal is formed by the growth of a crystal nucleus without secondary nucleation or impingement on other crystals. *See* CRYSTAL STRUCTURE.

Growth techniques. Among the most common methods of growing single crystals are those of P. Bridgman and J. Czochralski. In the Bridgman method the material is melted in a vertical cylindrical vessel which tapers conically to a point at the bottom. The vessel then is lowered slowly into a cold zone. Crystallization begins in the tip and continues usually by growth from the first formed nucleus. In the Czochralski method a small single crystal (seed) is introduced into the surface of the melt and then drawn slowly upward into a cold zone. Single crystals of ultrahigh purity have been grown by zone melting. Single crystals are also often grown by bathing a seed with a supersaturated solution, the supersaturation being kept lower than is necessary for sensible nucleation.

When grown from a melt, single crystals usually take the form of their container. Crystals grown from solution (gas, liquid, or solid) often have a well-defined form which reflects the symmetry of the unit cell. For example, rock salt or ammonium chloride crystals often grow from solutions in the form of cubes with faces parallel to the (100) planes of the crystal, or in the form of octahedrons with faces parallel to the (111) planes. The growth form of crystals is usually dictated by kinetic factors and does not correspond necessarily to the equilibrium form. *See* CRYSTALLIZATION.

Physical properties. Ideally, single crystals are free from internal boundaries. They give rise to a characteristic x-ray diffraction pattern. For example, the Laue pattern of a single crystal consists of a single characteristic set of sharp intensity maxima. *See* X-RAY DIFFRACTION.

Many types of single crystal exhibit anisotropy, that is, a variation of some of their physical properties according to the direction along which they are measured. For example, the electrical resistivity of a randomly oriented aggregate of graphite crystallites is the same in all directions. The resistivity of a graphite single crystal is different, however, when measured along different crystal axes. This anisotropy exists both for structure-sensitive properties, which are strongly affected by crystal imperfections (such as cleavage and crystal growth rate), and structure-insensitive properties, which are not affected by imperfections (such as elastic coefficients).

Anisotropy of a structure-insensitive property is described by a characteristic set of coefficients which can be combined to give the macroscopic property along any particular direction in the crystal. The number of necessary coefficients can often be reduced substantially by consideration of the crystal symmetry; whether anisotropy, with respect to a given property, exists depends on crystal symmetry.

The structure-sensitive properties of crystals (for example, strength and diffusion coefficients) seem governed by internal defects, often on an atomic scale.

[DAVID TURNBULL]

Bibliography: B. R. Pamplin, *Crystal Growth*, 1975; F. Rosenberger, *Fundamentals of Crystal Growth One*: *Macroscopic Equilibrium and Transport Concepts*, 1979.

Sodium

A chemical element, Na, atomic number 11, and atomic weight 22.9898. Sodium is between lithium and potassium in group Ia of the periodic table. The element is a soft, reactive, low-melting metal with a specific gravity of 0.97 at 20°C. Sodium is

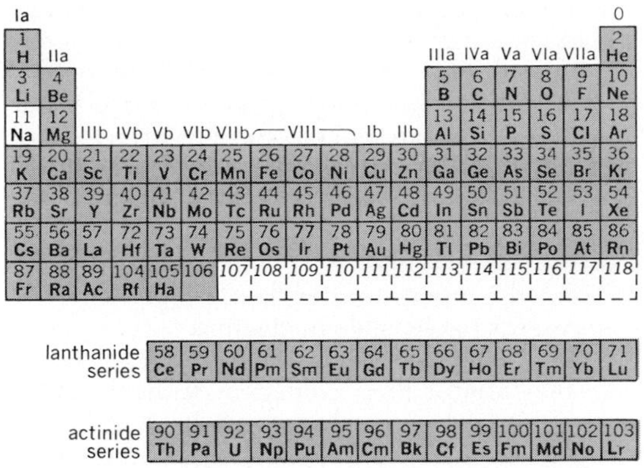

commercially the most important alkali metal. It was named by Sir Humphry Davy, who first isolated it by electrolysis in 1807.

Uses. The largest single use for sodium metal, accounting for about 60% of total production, is in the synthesis of tetraethyllead, an antiknock agent for automotive gasolines. In this process the reaction of a sodium-lead alloy with ethyl chloride gives tetraethyllead, as shown by reaction (1). The unreacted lead is recycled in the process.

$$4PbNa + 4C_2H_5Cl \rightarrow (C_2H_5)_4Pb + 3Pb + 4NaCl \quad (1)$$

A second major use is in the reduction of animal and vegetable oils to long-chain fatty alcohols; these alcohols are raw materials for detergent manufacture. This use has been decreasing in favor of production of such alcohols by high-pressure catalytic hydrogenation.

Another major use is in the reduction of titanium and zirconium halides to the respective metals. Here the use of sodium is increasing at the expense of magnesium as the preferred reducing agent in such operations.

Sodium metal is also used in making sodium hydride, sodium amide, and sodium cyanide, as discussed in the section on inorganic reactions. It is also used in the synthesis of "isosebacic acid," as described under organic reactions. The use of liquid sodium metal as a heat-transfer agent in nuclear reactors is also becoming increasingly important.

Sodium chloride is used in the manufacture of sodium hydroxide, sodium carbonate, sodium sulfate, and sodium metal. In sodium sulfate manufacture, hydrogen chloride is the coproduct. In metallic sodium manufacture, chlorine gas is the coproduct.

Rock salt is used in curing fish, in meat packing, in curing hides, and in making freezing mixtures. Food preparation, including canning and preserving, consumes much salt. Table salt accounts for only a small percentage of sodium chloride consumption, most of it going into the industrial uses outlined above.

Sodium hydroxide is perhaps the most important industrial alkali. Its major use is in the manufacture of chemicals, about 30% going into this category. The next major use is the manufacture of cellulose film and rayon, both of which proceed through soda cellulose (the reaction product of sodium hydroxide and cellulose); this accounts for about 25% of the total caustic soda production. Soap manufacture, petroleum refining, and pulp and paper manufacture each account for a little less than 10% of total sodium hydroxide uses.

Sodium carbonate finds its major use in the glass industry, which takes about one-third of total production. Approximately another third goes into the manufacture of soap, detergents, and various cleansers. The manufacture of paper and textiles, nonferrous metals, and petroleum products accounts for much of the balance.

The major consumer of sodium sulfate (salt cake) is the kraft pulp industry. Increasing quantities of sodium sulfate are being used in the manufacture of flat glass. Other uses of salt cake are in detergents, ceramics, mineral stock feeds, and pharmaceuticals.

Occurrence. Sodium ranks sixth in abundance among all the elements in the Earth's crust, which contains 2.83% sodium in combined form. Only oxygen, silicon, aluminum, iron, and calcium are more abundant. Sodium is, after chlorine, the second most abundant element in solution in sea water.

The important sodium salts found in nature include sodium chloride (rock salt), sodium carbonate (soda and trona), sodium borate (borax), sodium nitrate (Chile saltpeter), and sodium sulfate. Sodium salts are found in sea water, salt lakes, alkaline lakes, and mineral springs. Rock salt deposits occur where salt lakes and ancient seas have existed. In the United States salt domes are the major commercial source of sodium chloride as a raw material for sodium metal manufacture. These deposits are found in Virginia, West Virginia, Michigan, along the south shore of Lake Erie, and along the Gulf Coast of Texas and Louisiana. The salt is removed from the earth both by underground mining and by pumping water down to force up salt brines.

Physical properties. The physical properties of metallic sodium are summarized in the table.

Chemical properties. For convenience, the reactions of sodium are divided into inorganic reactions and organic reactions.

Physical properties of sodium metal

Property	Temperature °C	Temperature °F	Metric (scientific) units	British (engineering) units
Density	0	32	0.972 g/cm³	60.8 lb/ft³
	100	212	0.928 g/cm³	58.0 lb/ft³
	800	1472	0.757 g/cm³	47.3 lb/ft³
Melting point	97.5	207.5		
Boiling point	883	1621		
Heat of fusion	97.5	207.5	27.2 cal/g	48.96 Btu/lb
Heat of vaporization	883	1621	1005 cal/g	1809 Btu/lb
Viscosity	250	482	3.81 millipoises	4.3 kinetic units
	400	752	2.69 millipoises	3.1 kinetic units
Vapor pressure	440	824	1 mm	0.019 lb/in.²
	815	1499	400 mm	7.75 lb/in²
Thermal conductivity	21.2	70.2	0.317 cal/(sec)(cm)(°C)	76 Btu/(hr)(ft)(°F)
	200	392	0.193 cal/(sec)(cm)(°C)	46.7 Btu/(hr)(ft)(°F)
Heat capacity	20	68	0.30 cal/(g)(°C)	0.30 Btu/(lb)(°F)
	200	392	0.32 cal/(g)(°C)	0.32 Btu/(lb)(°F)
Electrical resistivity	100	212	965 microhm-cm	
Surface tension	100	212	206.4 dynes/cm	
	250	482	199.5 dynes/cm	

Inorganic reactions. Sodium reacts rapidly with water, and even with snow and ice, to give sodium hydroxide and hydrogen. The reaction liberates sufficient heat to melt the sodium and ignite the hydrogen.

When exposed to air, freshly cut sodium metal loses its silvery appearance and becomes dull gray because of the formation of a coating of sodium oxide. Sodium probably oxidizes to the peroxide, Na_2O_2, which reacts with excess sodium present to give the monoxide, Na_2O. When sodium reacts with oxygen at elevated temperatures, sodium superoxide, NaO_2, is formed; this reacts with more sodium to form the peroxide. Sodium also forms an ozonide, NaO_3, when ozone is passed into a solution of sodium in liquid ammonia.

Sodium does not react with nitrogen, even at very high temperatures. Sodium and hydrogen react above about 200°C to form sodium hydride. This compound decomposes at about 400°C and cannot be melted. Sodium hydride can be formed by the direct reaction of hydrogen and molten sodium or by hydrogenating dispersions of sodium metal in hydrocarbons. Sodium reacts with carbon with difficulty, if at all, and this reaction may be said to have been adequately studied.

At room temperature fluorine and sodium ignite, dry chlorine and sodium react slightly, bromine and sodium do not react, and iodine and sodium do not react. However, moisture or elevated temperatures will speed the reactions enormously.

Sodium reacts with ammonia, forming sodium amide and liberating hydrogen. The reaction may be carried out between molten sodium and gaseous ammonia. Alternatively, sodium metal reacts with liquid ammonia (−30°C) in the presence of catalysts of finely divided metals. Sodium amide has been used in indigo manufacture (in the condensation of sodium phenylglycinate to sodium indoxyl, which in turn is oxidized by air to indigo). Sodium reacts with ammonia in the presence of coke to form sodium cyanide.

Carbon monoxide reacts with sodium, but the resulting carbonyl, NaCO, is stable only at liquid ammonia temperatures. At high temperatures, sodium carbide and sodium carbonate are formed from carbon monoxide and sodium.

The reactions of sodium with various metal halides to give the metal plus sodium chloride are very important. Thus, titanium tetrachloride is reduced to titanium metal. Similarly, the halides of zirconium, beryllium, and thorium can be reduced to the corresponding metals by sodium. The interaction between sodium and potassium chloride is used in the commercial production of potassium metal.

Organic reactions. Sodium does not react with paraffin hydrocarbons but does form addition compounds with naphthalene and other polycyclic aromatic compounds and with arylated alkenes. It reacts with acetylene, replacing the acetylenic hydrogens to form sodium acetylides. Sodium adds to dienes, the reaction which forms the basis of the buna synthetic rubber process used by both Germany and the Soviet Union in World War II. The addition of sodium to butadiene can also be controlled to give a disodium butadiene dimer, which can be carbonated to give a 10-carbon atom dibasic acid known as isosebacic acid, an important ingredient in plasticizers for vinyl chloride polymers.

The reaction of sodium with alcohols is similar to, but less rapid than, the reaction of sodium with water. Brick sodium, molten sodium, or sodium dispersed in hydrocarbons may be used in the reaction with alcohols, and the alcoholate (alkoxide) products may be handled in solution form, in slurry form in hydrocarbons, or as dry, free-flowing powders.

Sodium reacts with organic halides in two general ways. One of these involves condensation of two organic, halogen-bearing compounds by removal of the halogen, allowing the two organic radicals to join directly. The second type of reaction involves replacement of the halogen by sodium, giving an organosodium compound. Alternatively in this second class of reaction, a metal alloyed with sodium may replace the halogen after the halogen has been removed by sodium; the reaction of sodium-lead alloy with ethyl chloride to give tetraethyllead is an example of this type of reaction.

Sodium can effect the reduction, condensation,

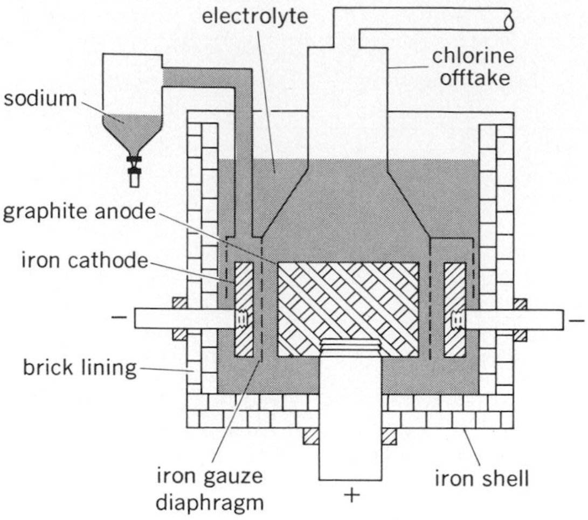

Fig. 1. Downs cell for sodium production. (*C. L. Mantell, Electrochemical Engineering, 4th ed., 1960*)

and alkylation of carbonyl compounds. For these purposes sodium can be used in the form of the metal, as an alloy, as the alkoxide, the amide, the hydride, or as an organosodium compound to effect various specific reactions. Most reactions of sodium with carbonyl compounds proceed through an intermediate formed by reaction of an active hydrogen atom with sodium. This intermediate then reacts with other molecules of the original compounds or with other active compounds present in the reaction mixture.

Metallurgical extraction. Raw sodium chloride either enters the plant as a brine or is fed into a dissolver tank to produce a brine. This brine is treated with sodium hydroxide and ferric chloride and then with barium chloride to remove, among other impurities, any sulfate. The refined brine is then evaporated and the dried salt stored.

Practically all of the metallic sodium is made at the present time in the Downs cell, using a bath consisting of molten sodium chloride plus calcium chloride. The presence of 58–59% of calcium chloride depresses the melting point of pure NaCl from 800°C to 575–585°C. The lower working temperature simplifies cell construction because of less severe operating conditions. *See* CHLORINE.

The Downs cell is a large, refractory-lined, steel vessel (Fig. 1). Graphite anodes project up from the bottom of the cell. These central anodes and the surrounding cylindrical steel cathode define an annular electrolysis zone. The gaseous product of the electrolysis, chlorine, rises into the top of the cell and is removed. Since the product, sodium metal, is lighter (specific gravity 0.88) than the molten bath (specific gravity 2.1), it rises through the bath and is collected under the inverted collector ring and forced by buoyancy up the sodium collector pipe (Fig. 2). Some cooling occurs in this riser pipe, and calcium metal, formed as a coproduct with the sodium, precipitates out and falls back into the cells where it redissolves in the bath mixture. The sodium issuing from the receiver is 99.8% pure.

Availability. Sodium is available in two grades: a commercial grade analyzing 99.8% sodium and a special nuclear reactor grade manufactured, loaded, and shipped under argon gas after being passed through porous metallic filters to remove oxides.

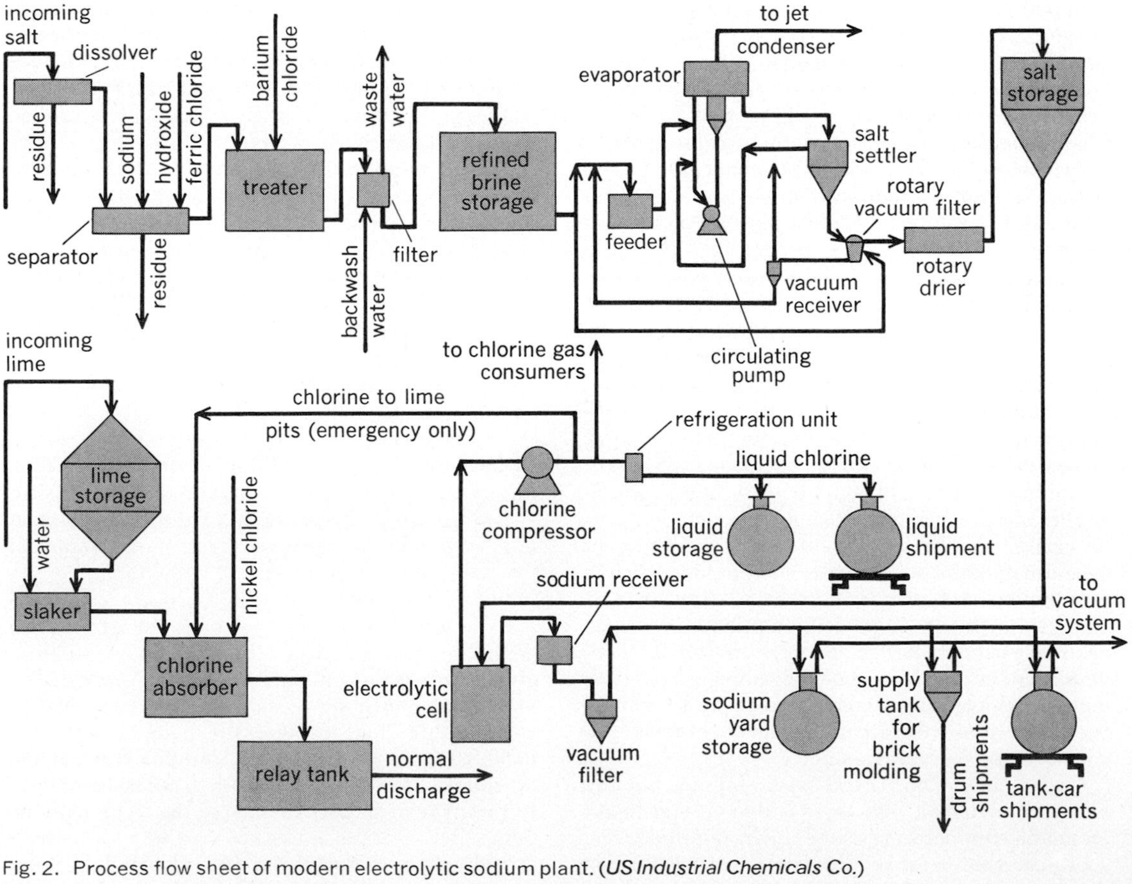

Fig. 2. Process flow sheet of modern electrolytic sodium plant. (*US Industrial Chemicals Co.*)

Sodium is available in containers ranging from 1 to 80,000 lb (0.45 to 36,000 kg) net weight. For laboratory use, 1-lb (0.45 kg) bricks are available in hermetically sealed, key-opening metal cans. Bricks in 1-, 2½-, 5-, 12- or 24-lb (1 lb = 0.45 kg) sizes are shipped in steel pails of about 25 lb (11 kg) net weight and in steel drums of about 300 lb (135 kg) net weight. Sodium is also available cast solid in steel drums. Large consumers purchase sodium cast solid in special railway tank cars of 80,000 lb (36,000 kg) net weight. Hot oil is pumped through channels in the walls of the special car at the destination, and the molten cargo is withdrawn by vacuum.

Handling. Sodium and other alkali metals are no more difficult to handle than many chemicals used every day. There are some hazards involved, however: (1) fire and explosion from hydrogen evolved if the sodium comes in contact with water; (2) caustic soda burns from the residue of a sodium-water reaction; (3) eye injuries from small pieces of sodium or caustic soda; (4) burns about the body from clothing ignited by clinging particles of burning sodium or flammable dispersed sodium; (5) irritation of the eyes and mucous membranes of the nose and throat because of breathing fumes from burning sodium; and (6) flesh burns from contact with metallic sodium.

Safe handling of sodium demands that users observe the following precautions: (1) Keep water away to avoid evolution of hydrogen which can be ignited by the heat of reaction. (2) Use only dry nitrogen (containing less than 1.0 mg/ft^3 water) to blanket sodium. (3) Use dry sodium carbonate (soda ash), dry sodium chloride, or dry graphite as fire-extinguishing agents. Do not use chlorinated hydrocarbons, carbon dioxide, foam, water, or soda-acid. (4) Know the physical and chemical properties of sodium and consider the potentialities of the reaction. Make initial trials of the reaction on a small scale. Control the reaction rate by adding sodium in small increments and by diluting the reactants with inert solvents. (5) Plan the disposal of sodium residues in advance. Dispose of scrap promptly, not allowing it to accumulate. (6) Store sodium in airtight, well-labeled, rustfree metal containers, away from water. (7) Use shallow metal pans on bench tops which can contain spillage in case of accident. (8) Wear clothing suited to the quantity of sodium being handled and to the temperature level at which it is being handled. This means goggles and gloves in most cases. It means full protective clothing for sodium at 1000°F under pressure. It means an impervious apron when dispersions of sodium in hydrocarbons are being handled.

Principal compounds. Sodium chloride, or common salt, NaCl, is not only the form in which sodium is found in nature but (in purified form) is the most important sodium compound in commerce as well. Sodium hydroxide, NaOH, is also commonly known as caustic soda. It readily absorbs water from the atmosphere and must be protected in storage and handling. It is corrosive to the skin and must be handled with extreme care to avoid caustic burns.

Most sodium hydroxide is produced by the electrolysis of sodium chloride solutions in one of several types of electrolytic cells. An older process is the soda-lime process whereby soda ash is converted to caustic soda. This is shown by reaction (2).

$$Na_2CO_3 + Ca(OH)_2 \rightarrow 2NaOH + CaCO_3 \quad (2)$$

Sodium carbonate, Na_2CO_3, is best known under the name soda ash because sodium carbonate occurs in (and once was extracted from) plant ashes. Most sodium carbonate is produced by the Solvay or ammonia-soda process. In an initial reaction, salt is converted to sodium bicarbonate, which precipitates and is then separated. The reaction is represented by (3). Heating of sodium bicarbonate produces sodium carbonate as shown by reaction (4). The carbon dioxide is recycled in the process. Despite its apparent simplicity, the carrying out of this process requires technical skill.

$$NaCl + NH_3 + CO_2 + H_2O \rightarrow NaHCO_3 + NH_4Cl \quad (3)$$

$$2NaHCO_3 \rightarrow NA_2CO_3 + CO_2 + H_2O \quad (4)$$

Sodium sulfate, Na_2SO_4, is also known in the anhydrous form as salt cake. The decahydrate, $Na_2SO_4 \cdot 10H_2O$, is known as glauber salt.

Most sodium sulfate is produced synthetically as a by-product or coproduct in various industries. The Mannheim furnace process for hydrochloric acid manufacture is one important source of salt cake; reaction (5) represents this process. The neutralization of sulfuric acid in rayon and cellophane plants and in other inorganic and organic chemical processes accounts for much production.

$$2NaCl + H_2SO_4 \rightarrow Na_2SO_4 + 2HCl \quad (5)$$

A lesser amount of sodium sulfate is produced from natural sources, such as salt deposits in Wyoming and lake brine in California.

Analytical methods. The determination of the sodium ion in solution is complicated by the high solubility of most sodium salts in water. Thus, quantitative determination of sodium relies heavily on gravimetric techniques using double uranyl salts, such as zinc uranyl acetate, as precipitants in most cases. The yellow color imparted to a flame by sodium ions serves as a sensitive qualitative test for the presence of sodium. *See* ALKALI METALS.

[MARSHALL SITTIG]

Bibliography: F. A. Cotton and G. Wilkinson, *Advanced Inorganic Chemistry*, 4th ed., 1980; O. J. Foust, *Sodium-NaK Engineering Handbook*, vol. 1: *Sodium Chemistry and Physical Properties*, 1972; W. A. Hart et al., *The Chemistry of Lithium, Sodium, Potassium, Rubidium, Cesium, and Francium*, 1975; M. Sittig, *Handling and Uses of the Alkali Metals*, Advances in Chemistry Series, vol. 19, 1957.

Solid solution

The crystalline state, where at least one atomic position may accommodate more than one atomic species. It is easiest to conceptualize the phenomenon by defining structure type, mineral species, and end-member composition. The structure type is a design of distinct-symmetry independent atomic positions possessing an asymmetric unit of atomic positions which must be specified along with the unit cell properties, including axial parameters, crystal system, and space group. The greater effort in crystal structure analysis is establishing unit cell properties and the loci of the atoms in the asymmetric unit, some of whose parameters may have up to three degrees of freedom. Mineral species is defined on the basis of structure type and end-member composition, where each end member in principle is a distinct species. *See* CRYSTAL STRUCTURE.

In the prevalent olivine group of minerals, the two important end-members are fayalite, $Fe_2^{++}(SiO_4)$, and forsterite, $Mg_2^{++}(SiO_4)$. Both belong to the same space group and structure type. The general formula for the silicate olivines can be written $M_2^{++}(SiO_4)$, where M stands for the divalent cations octahedrally coordinated by the oxide anions. When Fe^{++} and Mg^{++} can mix at any ratio over these M sites, a perfect solid solution exists. Another example is the series calcite-magnesite, $CaCO_3-MgCO_3$. Both end members belong to the same structure type, and with increasing temperature, mixing of Ca and Mg increases over the large cation site. At some temperature, mixing of the two cationic species can be complete, as in the olivines. Below this temperature the Ca^{++} and Mg^{++} cations segregate within the crystal and can form a stable ordered species, dolomite, $CaMg(CO_3)_2$, plus an exsolved phase such as calcite.

Ordered arrangements. If an ordered arrangement appears, one of these phenomena may take place. The space group may remain unchanged if the crystal symmetry is not violated through ordering; the space group may be reduced to a subgroup due to the partitioning of two cations into separate positions, which destroys some of the symmetry elements of the space group; the segregation may lead to an unmixing of the structure into two separate phases which usually bear a crystallographic or epitaxial relation with each other. The calcite-magnesite series is an excellent example. Both end members possess the space group $R\bar{3}c$; dolomite possesses the subgroup $R\bar{3}$. Samples can be found with calcite exsolved from a magnesian calcite, a dolomite, or a calcian magnesite. Formerly at crystallization temperature, both Ca^{++} and Mg^{++} were completely mixed at some higher temperature. The composition of the magnesian calcite and the exsolved calcite can be used as a geothermometer to estimate the initial temperature of crystallization in the system $CaO\text{-}MgO\text{-}CO_2$.

Omission solid solution. This type of solid solution can also exist. In the hypothetical composition $\square X_3(TO_4)_2$, $\square$ is a symbol for a vacancy or an empty atomic position in the structure. Imagine the series $\square X_3(PO_4)_2-XX_3(PO_4)_2[=X_4(PO_4)_2]$. The series $\square-X$ is also possible, and frequently structures and compositions are found where the filling of the $\square$ may be only partial; like the join calcite-magnesite, the fraction of occupancy is temperature-dependent. Such models can be also used as geothermometers, providing the thermochemistry of the system is known in detail. One remarkable example is triphylite-sarcopside, $LiFe(PO_4)-Fe_3(PO_4)_2$. Large crystals of triphylite can be found with exsolved sarcopside. Sarcopside possesses a subgroup of the triphylite space group. Triphylite is an ordered olivine structure where Li^+ and Fe^{++} are segregated into two independent atomic positions in the crystal. It is believed that there is exsolution from the series $Li_2Fe_2(PO_4)_2-\square FeFe_2(PO_4)_2$, where at some higher temperature the crystal was completely mixed, that is, $(Li,Fe,\square)_4(PO_4)_2$. Charge balance restricts the admissible ratios for $\square^0:Li^+:Fe^{++}$. *See* THERMOCHEMISTRY.

Occurrence. In mineralogy and metallurgy, solid solution is a frequent and important phenomenon. In metallurgy, examples are the systems Au-Ag-Hg and Fe-Ni. In mineralogy, the most important examples are the systems $CaO-MgO-CO_2$, $Na_2O-CaO-Al_2O_3-SiO_2$, and $MgO-Al_2O_3-SiO_2$. The feldspar series anorthite-albite, $CaAl_2Si_2O_8-NaAlSi_3O_8$, is renowned for a great range of unmixing phenomena at subsolidus temperatures and an astonishing diversity in group-subgroup relations in their crystal structures, even though the framework topology of their structures remains unchanged.

Solid solution is a statistical phenomenon, and implies only that the sites which accommodate distinct ionic species are energetically similar. Since x-ray diffraction experiments (the most important tool) essentially average the atomic populations over their sites according to the space group extinctions, and since about 10^{15} unit cells are involved in the experiment, it is not implied that atoms hop from site to site.

Some general observations have been made about those cations and anions which can form solid solution. Most important are similarity in their ionic radii, and similarity in their electronic structure. Important examples of ions which can form solid solutions are $Fe^{++}-Mg^{++}$, $Fe^{++}-Mn^{++}$, $Mn^{++}-Ca^{++}$, $(OH)^--F^-$, $Mg^{++}-Al^{3+}$, and $Al^{3+}-Si^{4+}$. Coordination numbers must be the same in these cases. *See* X-RAY DIFFRACTION.

Many admired gemstones are results of solid solution. Pure corundum, $\alpha\text{-}Al_2O_3$, is colorless. If doped with small amounts of Cr^{3+}, it is red (ruby), or with $Fe^{++}Ti^{4+}$, it is blue (sapphire). Tourmaline when pure is $NaMg_3Al_6(BO_3)_3(Si_6O_{18})(OH)_4$. Since no chromophores are present, it is colorless. But small amounts of Fe^{3+} color it green (elbaite); $Fe^{++}Fe^{3+}$, blue (indicolite); and Mn^{3+}, pink to red (rubellite). Typical chromophores are ions of the first transition series metals. Another well-known example is the soft, platy mineral vivianite, $Fe_3^{++}(H_2O)_8(PO_4)_2$. When pure it is pale green, but when exposed to air some Fe^{++} is oxidized to Fe^{3+}, leading to a blue vivianite. Extensive oxidation renders it nearly opaque and black, due to homonuclear valence transfer absorption, $Fe^{++}+Fe^{3+} \rightleftharpoons Fe^{3+}+Fe^{++}$. In such a phenomenon, an increase of mixed valences over an equivalent (or structurally adjacent) set of sites increases the absorption of light and the darkness of the color. In these mixed-valence phenomena it is believed that the ions do not jump from one site to another, but that electrons do. In such relations the cations could be written $Fe_x^{3+}Fe_{1-x}^{++}$, and values should be ascribed to x. Calorimetric heats of mixing have been obtained on important geochemical solid solutions such as the feldspars, pyroxenes, and garnets.

[PAUL B. MOORE]

Solid-state chemistry

The science of the elementary, atomic compositions of solids and the transformations that occur in and between solids and between solids and other phases to produce solids. Solid-state chemistry deals primarily with those microscopic features which are uniquely characteristic of solids and which are the causes for the macroscopic chemical properties and the chemical reactions of solids. As with other branches of the physical sciences, solid-state chemistry also includes related areas that furnish concepts and knowledge essential to understandings and explanations of those phenomena which are more characteristic of the subject itself.

The overlap of solid-state chemistry and solid-state physics is extensive. However, the perspectives of the two are sufficiently different that they can easily be identified. In general, solid-state physics treats properties, such as energy and entropy, which are continuously variable in the solid, whereas solid-state chemistry concerns those properties which are discontinuous because of chemical reactions; in a sense, by definition the properties are chemical because they are discontinuous. In another perspective, solid-state chemistry tends to be based on structure in configuration space, whereas solid-state physics tends to be based on momentum space. *See* INORGANIC CHEMISTRY.

STRUCTURES

A study of structures provides a basis for understanding the composition of solids and the reactions in which they participate.

Atomic geometries (configuration). The solid-state chemist is interested in the idealized structures described by space groups as a basis to derive information about bonding. However, the principal interest in structure is in the deviations from these space groups, because they reveal additional information about bonding or because they are the causes (mechanisms) of chemical processes in solids. These idealized structures exist, strictly stated, only at 0 K. At any finite temperatures, these structures are imperfect or distorted in at least two general respects: the atoms or ions vibrate with amplitudes which increase with increasing temperature; and the lattice sites of the space groups become vacant or interstitial positions become occupied, or both occur, with the valencies of the cations or anions becoming altered. *See* CRYSTAL STRUCTURE.

The concentration and spatial distribution of the defects in the solid produce various kinds of distortions of the idealized structures. When the defects interact only weakly and do not aggregate, the distortions are mostly localized near the point defects, which may be either randomly distributed or ordered on a long range. For this distribution, the x-ray or neutron diffraction pattern is essentially that of one of the idealized structures with a changed internuclear distance or lattice parameter and with additional diffraction maxima caused by the long-range order, that is, superstructure. This configuration of point defects occurs especially in several nonstoichiometric phases. In some cases, the defects appear to be aggregated, with the result that a structured distortion occurs. Such a situation exists in cases known as shear structures that occur in WO_3 and MoO_3, where extensive regions appear as an idealized structure connected by a plane in the lattice in which a displacement equivalent to a shear translation exists. *See* NONSTOICHIOMETRIC COMPOUNDS.

Layered structures. Numerous binary compounds or phases have layered structures which in themselves or in combination with other elements can be considered as extensions of the descriptions given above. The layers are displaced relative to each other in such a way that the structures can be classified in the space groups, but when other elements or compounds are incorporated between the layers, the structure is highly distorted. The most widely recognized material to possess the layered structure and the associated solid-state chemistry is graphite (Fig. 1). Carbon atoms in this structure are arranged in hexagons connected along the edges, and thus form a plane throughout the crystal. These planes are stacked in layers, with each layer displaced with respect to its adjacent layers by a distance such that the structure still belongs to one of the space groups. The bonding within a layer is covalent; between the layers it is metallic. Consequently, graphite is highly anisotropic. When it is exposed to bromine, alkali metals, and other chemical substances, these substances, at high temperatures, enter the spaces between the layers to form intercalation compounds; the idealized structure thus becomes highly distorted. A large number of compounds with layered structures are known to form a variety of intercalation compounds. Among them are the disulfides TiS_2 and MoS_2. Two others, which are similar in that additional elements or compounds are located between lattice planes, are the tungsten bronzes and the β-aluminas.

SOLID-STATE CHEMISTRY

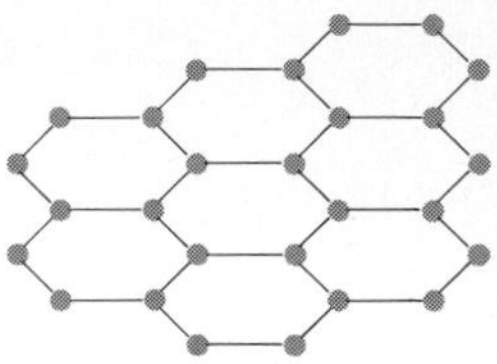

Fig. 1. Structure of graphite. Only a single layer of carbon atoms is shown.

Multicomponent phases. For three or more components, the structural situation is generally complex. The number of possible compounds and phases is extremely large, and so is the variety of structures and distortions. Classification of the distorted space groups is difficult and is still being developed. The actual structures are generally described as distortions of some well-recognized structures, which in themselves are complex. Because many of these complex structures were found in minerals, the distorted structures are frequently classified as being derived from these minerals. Examples of such are spinels, perovskites, bixbyites, feldspars, and garnets.

Glasses. All of the structures described above are based on a lattice with defects but with sufficient long-range order to produce well-defined diffraction patterns. There exists another group or class of solids in which the long-range order is insufficient to yield such patterns. These are the glasses or amorphous solids which generally exist only in metastable conditions. The most readily known among these are the silicate glasses, but there are many others, including some even in metallic systems.

Molecular solids. In all the cases described, there occurs no decisive tendencies toward molecular formation in the solids. However, when the intramolecular forces are great enough, as in organic molecules, the solids are composed of molecular crystals. These solids in general have high volatilities and insufficient defects to cause reactions associated with the transfer of mass in the solids, but they have electronic structures which participate in photochemical reactions in organic solids.

Electronic structures. Aside from combinations near the center of the periodic system where covalency is dominant, the elements exist in solids as ions. For the regular elements, the outer, valence orbitals of the ions are closed, with the inert gas shells having *s*-orbital configurations on the cations and hybridized *p*-orbitals on the anion. In both cases, the ions have filled orbitals and tend to be spherical; they are not highly directional. These ions have no low-lying states, and the crystal fields are insufficient to decouple the spins. Consequently, the electronic structures of solids formed with these ions are simple, and they contribute a basis for solid-state chemistry in terms of ionicity and normal polarization. These solids, which are essentially those formed from groups I, II, VI, and VII in the periodic system, are transparent in the visible and near infrared. In addition, some of the cations of the transition, lanthanide, and actinide groups have closed *p*-orbitals with no low-lying electronic states. These are the scandium and titanium subgroups (thorium is included in the latter). All the other transition, lanthanide, and actinide ions in solids display electronic transitions in their infrared and visible spectra. A large number of the observed transitions are attributable to the fact that the d^n or f^n degenerate states are separated by the electrostatic field of the crystal. In the cases of the lanthanide and actinide ions, some of the observed maxima are also attributable to *f*-to-*d* transitions. The presence of all of these states contributes to anomalous polarization or ionicity and a complexity of the chemistry. *See* CRYSTAL FIELD THEORY.

Photoacoustic spectroscopy with powdered materials can be used to study the spectra of such solids. In this technique, a beam of light from a monochromator is chopped with an audio frequency, and then it is absorbed by the particles in the powder. The energy is absorbed in the electronic transitions and then transferred to vibrational states. The energy stored as heat in these states is transferred to an ambient gas in the form of sound waves of the same frequency as the chopper on the incident light. The sound waves are detected by a microphone. The study of the energies and mechanisms associated with these electron-phonon interactions and with the phonon-electron transfer furnishes microscopic information for the solid-state chemistry associated with charge transfer, polarization, and valence states.

A feature of the electronic structure of solids, which is particularly essential to solid-state chemistry, is the valence state of the ion and the energy required to change the valency in the solid. For the transition, lanthanide, and actinide ions with open *d*- or *f*-orbitals, these energies are sufficiently small that transitions from one valency to another can be promoted by thermal (phonon) activation or the oxidation potentials of anions. The relative concentrations of two valencies in solids can be determined chemically through the stoichiometry or through chemical reactions which presumably do not change the relative concentrations. The known stoichiometries in such solids as Fe_3O_4, Fe_2O_3, and U_3O_8 imply mixed valence states. If these phases are dissolved under nonoxidizing conditions, then the initial ratio of the two valence states can be determined. The question of the existence of two distinct valence states in the solid has been debated for decades. Some have argued that a resonance condition exists so that two well-defined oxidation states cannot be identified. Most of these arguments are based on the results of x-ray diffraction observations, which give no evidence of two different cation sites. Two techniques have been developed which enable direct observation of the valence states in the solid. These are photoelectron spectroscopy and Mössbauer spectroscopy. *See* ELECTRON SPECTROSCOPY; MÖSSBAUER EFFECT.

CHEMICAL BONDING

The structure of a solid is the result of the operation of interatomic or interionic forces and the size and shape of the atoms or ions. Hence, logically, bonding should be described first and structure second. However, the detailed role of the electrons in interionic forces is so complex, and the quantitative aspects of the problem of minimizing the potential energy with respect to all the possible configurations is so difficult, that structures cannot be derived. Rather, it is necessary to derive some information about bonding from structures, cohesive energies, refractive indices, electron binding energies, polarizabilities, and other properties through the use of models. Because of the number of ions and electrons involved, the wave-mechanical formulation of bonding in solids is extremely difficult to solve. Fortunately, some rather simple, classically based models modified by quantum-mechanical concepts have been and continue to be quite useful. These can be classified generally as ionic, covalent, and metallic bonding and combinations of the three. *See* CHEMICAL BONDING; CHEMICAL STRUCTURES.

Ionic bonding. In the ionic model, the solid is composed of positive (cations) and negative (anions) charges located on the respective lattice sites. These imagined point charges create through their collective coulombic interactions a net attractive force. The repulsive forces to balance the attractive forces at equilibrium consist essentially of the overlap of the valence electron orbitals. Additional attractive forces are frequently included in the model. These are occasionally referred to as the dispersive forces because they are the ones related to the dipole and multipole moments in ionic crystals and their indices of refraction. Solids having the most ionic character (ionicity) are those formed from combinations of the most electropositive with the most electronegative elements; thus CsF is highly ionic. In general, the ionicity increases with increasing cationic radius and with decreasing anionic radius.

Covalent bonding. In the covalent model, the electrons in the bond are shared equally between the atomic cores, and the electrons involved are in bound orbitals, frequently referred to as molecular orbitals. In essence, the sources of the attractive and repulsive forces are electrostatic, but they are formulated wave-mechanically. In purely covalent bonding, the atomic cores are identical, as in solids of the inert gases, halogens, oxygen, and sulfur.

Metallic bonding. In the metallic model, the solid is composed of cations on lattice sites surrounded by a uniform negative charge of the conduction or free electrons.

Combination bonding. In all real solids composed of different elements, the bonding consists of admixtures of these models. The bonding in ioniclike crystals composed of the regular metallic elements and the more electronegative elements can be rationalized as admixtures of ionic and covalent bonding. The degree of ionicity or covalency can be evaluated through optical dispersion theories of crystals, wherein the ionicity is evaluated from the index of refraction, the density of valence electrons, and some measure of the separation of the bonding and antibonding orbitals. In these formalisms, the group IV elements carbon, silicon, and so on, are used as the reference for covalency. Thus covalency increases as the pairs of elements in the solid become more nearly the same as group IV elements. In the case of the transition, lanthanide, and actinide elements in valence states having open shells, rationalization of the bonding on the basis of this description is inadequate. The crystal field of the solid removes the degeneracy of the orbitals, and the electrons in the bond can occupy states which increase the strength of the chemical bond. Thus there occurs a crystal field stabilization and the attendant anomalous dispersion accompanying the absorption in the infrared and visible spectra. Low-lying *f*-to-*d* transitions may also contribute to the increased bonding. In a sense, the role of these crystal field splittings and *d* states is to increase the polarizabilities of the cations.

In metals formed by higher-valent cations, starting at least with 3+, the anionic-forming elements can react to produce compounds and phases which still contain free electrons. Thus solids such as subhalides of scandium, the monosulfides and carbides of several elements, and some oxides such as TiO or VO contain cations, anions, and conduction electrons or a bandgap sufficiently small to be *n*-type semiconductors. The roles of the valence states of the cations and the oxidation of the free electrons are illustrated through the variation of the conductivities and lattice parameters of the lanthanide monosulfides and monoselenides. All of these are conductors except those of europium, ytterbium, and samarium. The rationalization of this behavior is that in the metals the cations are in the 3+ valency in all cases except europium, ytterbium, and samarium, in which the valency is 2+. Thus the anionic elements, sulfur and selenium, are reduced by the free electrons to the 2− state so that the three cases with 2+ cations have no free electrons, whereas each of the others has approximately one free electron. The cationic radii and the oxidation-reduction potentials are consistent with this description.

Measurement. One of the more direct ways of measuring the bonding in a solid, and particularly one which measures it at the cationic and anionic sites separately, is by measuring the electron binding energies by photoelectron spectroscopy. Thus, through the ejection of electrons from valence orbitals with x-rays or ultraviolet radiation *hν* and the measurement of the electrons' kinetic energy, their binding energies are determined by reaction (1). In this reaction *cs* represents a cation site; the

$$M^{q+}cs \xrightarrow{h\nu} M^{(q+1)+}cs^{*} + e(g) \qquad (1)$$

asterisk indicates that the cation site is unrelaxed after the ionization; and the electron is in the free gaseous (*g*) state. Solid-state physicists generally refer binding energies to electrons in the Fermi level. In reaction (1) the reference is the free electron; the two differ by the instrumental work function. This technique permits determination of the valence state, *q*+, and the lattice self-potential from which measures of polarization, ionicity, and so on are derived. The lattice self-potential is derived through a comparison of the binding energy with the ionization potential of the gaseous cation.

CHEMICAL COMPOSITION

At finite temperatures, and particularly at high temperatures, the partial vapor pressures of the components in a solid are different. Consequently, in general there occurs a preferential loss of one component so that any solid at equilibrium tends to contain lattice defects and to become nonstoichiometric. However, in a large number of cases the deviations from stoichiometry are not detectable, so that the number of nonstoichiometric phases that can be studied in solid-state chemistry is not extremely large. Among those which have been studied extensively are $Fe_{1-x}O$, CeO_{2-x}, and various metal hydrides, as well as the oxides, carbides, and hydrides of uranium. Because of its use in reactor technology, UO_{2+x} may be the most extensively studied nonstoichiometric phase.

Whenever the structure of a solid is such that an interstitial position of one of the sublattices can be occupied, both interstitials and vacancies occur. However, over the compositional range of nonstoichiometry, one or the other of the defects is usually in the higher concentration. Thus UO_2 has a fluorite structure such that the position at the center of the unit cell can accommodate an oxygen ion with an attendant shift in the oxidation state of the uranium ions. In UO_{2+x} the defects are predominantly oxygen ions on interstitial sites and U^{5+} ions on some of the cation sites. In UO_{2-x}, which exists at temperatures near 2000°C, vacancies on the oxygen sublattice are at higher concentration. In a phase such as $Th_yU_{1-y}O_{2+x}$, thorium ions are substitutional, and in $Pu_yU_{1-x}O_{2\pm x}$ plutonium is substitutional. In the last case, because plutonium has the 3+ and 4+ oxidation states accessible and uranium has the 4+ and 5+ states, the anion composition has an extensive range on both sides of stoichiometry. In metallic carbides, the carbon is often interstitial, and in UC_{1+x} the carbon tends to be incorporated in the UC lattice as a C_2 unit.

The role of the compositional variable in oxidation at the electronic level in the solid is illustrated in the x-ray photoelectron spectra of the valence band region of uranium at four stages of oxidation from metal to UO_3. For uranium metal and dioxide, the intensities of the 5*f*-orbital electrons are the same; no change occurs in the number of 5*f* electrons. Hence, the oxidation of the metal to UO_2 involves only the conversion of the free electrons to bound electrons in the hybridized 2*p*-orbital on the oxygen ion. When UO_2 is oxidized to U_4O_9, however, the intensity of the 5*f* part decreases and that of the 2*p* increases. With further oxidation to U_3O_8, the 5*f* intensities decrease further and the 2*p* intensity increases. Finally, in UO_3, the 5*f* electrons are completely oxidized to 2*p* electrons on the oxygen ion.

One of the measurements, which furnishes information needed to understand the compositional variable in solid-state chemistry, is the observation of partial vapor pressures and their dependence on temperatures are determined. The chemical potentials and the partial molar enthalpies and entropies and their variation with composition are derived from these measurements.

For a binary system, the chemical potential increases monotonically in the single-phase region, and is constant in the diphasic region. The construction and mathematical evaluation of statistical models to describe the chemical potential throughout both regions from the interionic forces, valence states, and defect energies contain conceptual problems associated with discontinuities at the diphasic region. Models can be constructed and evaluated to describe the homogeneous regions. One model is described by Eq. (2), in which

$$\mu = \mu_0 + RT\left(\ln\left(\frac{\theta_i}{1-\theta_i}\right) - E_i - 2\theta_i E_{ii}\right) \quad (2)$$

θ_i is the fraction of interstitial sites occupied, E_i is the energy required to remove an ion from the interstitial site, and E_{ii} is the energy of interaction between two occupied interstitial sites. Above some critical temperature ($T_c = E_{ii}/2R$), this function has a sigmoidal shape with no maximum or minimum. Thus for $T > T_c$, the variation of μ with θ_i (composition) as shown in Fig. 2 represents a monophasic region. At the critical temperature, the isotherm has a zero slope. Below this critical temperature, the function has a maximum, a minimum, and an inflection between them as shown by the isotherm for $T < T_c$ in Fig. 2. Because the slope of μ versus composition in a real system cannot change sign, this behavior is unrealistic. The artifice which is introduced to excuse these van der Waal loops is to construct a horizontal line through the inflection point and use this to represent the diphasic region. A realistic model which describes the evolution to a diphasic region would be one in which the horizontal line is contained in the mathematics. In one technique which has been suggested, the chemical potential is described through the complex (mathematical) variable. In some cases (praesodymium oxides), the compositional isotherm displays hysteresis loops because of defect complexes which form differently in the two directions of μ versus composition.

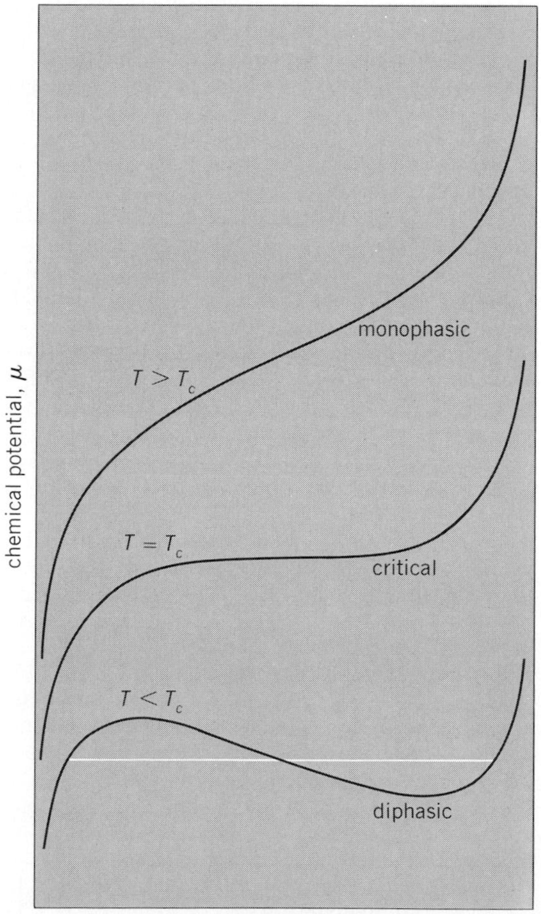

Fig. 2. Three isotherms for the variations of chemical potential with composition of defects.

When a substitutional nonstoichiometric phase is formed between an insulator and a metallic conductor, the electrical conductivity and concentration of free electrons of course change continuously. However, in some cases and perhaps in all cases, the change is so rapid over a small compositional range that the change can be viewed as an insulator-to-metal transition. One system which displays this behavior is $Sm_{1-x}Nd_xSe$. SmSe is an insulator with Sm^{2+} and S^{2-}, and NdSe is a conductor with Nd^{3+}, S^{2-}, and one free electron per unit. As the composition is varied, the conductivity and concentration of free electrons change significantly, but at $x \cong 0.1$ the two change by five orders of magnitude.

CHEMICAL REACTIONS

The mechanisms of chemical reactions within and between solids are through lattice vibrations, lattice defects, and changes in valence states. These are the structural features through which migration of mass, charge, and energy occur. Consequently, diffusion and conductivity are integral, basic parts of solid-state chemistry, even though their quantitative roles in the totality of chemical reactions have not been developed. So long as the solid phase produced during reaction has a density nearly the same as those for the reactants, the microscopic description of the reaction in terms of mass and charge transfer is feasible. However, in a number of reactions the molar volumes differ sufficiently to destroy the integrity of the product, so that much of the reaction then proceeds at interfaces, microcracks, and fissures. Thus the total net mechanism becomes a composite process so complex that a comprehensive theory is lacking. The theories that have been developed are generally discussed under topics such as corrosion. Thus, no formal classification of reactions in solids is universally recognized, or, in fact even cited, in the literature. A categorization which identifies the scope and suggests a basis is the following: condensation, internal processes, interphase reactions, electrochemical reactions, photochemical reactions, and sublimation. Another possible basis of classification is one which recognizes that all reactions occur because of gradients in the chemical potential. These gradients may be caused by gradients in mass, concentration, electrical charge, or energy (heat), or combinations of these.

Condensation. Although it is well recognized that condensation to form solids occurs when the translational energy of the gaseous reactants is dissipated into vibrational and electronic states, the processes involved are sufficiently complex that they are difficult to study and to describe in detail. In general, two kinds of condensation need to be described. In a supersaturated vapor, condensation is a stepwise process in which molecules form in successively larger clusters if the rate of growth of the clusters exceeds the rate of dissociation. At some critical size, the process results in condensation, because the translational energy can be dissipated into the cluster. This process occurs in shock tubes and jets which produce supersaturated vapors and in hydrodynamically flowing gaseous beams in which adiabatic expansion occurs. Another kind of condensation occurs if atoms or molecules in the gas are sufficiently accommodated thermally on a colder substrate. In this case, the growth occurs primarily through the flow of translational energy into the substrate. However, the atoms or molecules bound to the surface can be mobile on the surface so that growth can occur through a stepwise process therein. Usually the process is nonequilibrious, and an amorphous, metastable product is formed. For instance, gaseous silicon monoxide can be condensed on a cold substrate to produce solid, amorphous, metastable silicon monoxide. This is the only known way of producing it. Condensation to produce epitaxial layers on substrates is used to produce various kinds of solid-state devices.

An elementary condensation which can be used to illustrate the processes involved and which represents what could be called the first step in a solid-state chemistry is the formation of a metal. Because the process is one in which the neutral gaseous atom is converted to a collection of cations and free electrons, it involves internal oxidation and reduction, represented by reaction (3).

$$M(---np^6) \rightarrow M^{q+}(---np^{(6-q)})e_q \qquad (3)$$

Thus, in this reaction the gaseous atom with the electronic configuration of the core electrons (———) and the outer six np electron condenses to form something resembling a cation of charge $q+$ associated with q free electrons. In the supersaturated vapor, the critical step probably is the one in which the cluster is large enough to have free or conduction electrons to serve to dissipate the translation energy. When gaseous metal-forming atoms impinge on a metal substrate, the free electrons in the latter serve to transfer the energy, so that the accommodation and condensation coefficients of metal on metal are generally unity. On nonmetalic insulators, however, the coefficients are not unity, and condensation occurs through nucleation on the surface. The significance of recognizing that the condensation to form a metal is a redox process is contained in what is involved upon further oxidation. When the solid metal undergoes further reaction with an anion-forming element, it is the free electrons which are used first, and the cation subsequently oxidized.

Internal processes. One of the simplest reactions that occurs in solids is a phase transition. These transitions may be either first- or second-order. In the first case, a change occurs in the crystal structure, sometimes with the higher-temperature form having an apparently higher symmetry. Second-order transitions are usually order-disorder transitions such as the λ-type transitions which occur on only one of the sublattices.

The primary internal reactions which involve a chemical reaction are those associated with site occupancy, changes in valency, and clustering. Because it has been extensively studied, uranium dioxide serves as a useful illustration. In this phase, the interstitial position (*is*) of the fluorite lattice can accommodate an oxygen ion. Thus, an equilibrium exists between the occupancy of this site and the regular anion sites (*as*), shown in reaction (4). Whenever uranium dioxide is oxidized, the

$$O^{2-}as + is \rightarrow O^{2-}is + as \qquad (4)$$

valence of uranium is increased from 4+ to 5+ so that reaction (5) is the one occurring primarily.

$$2U^{4+}cs + O(g) + is \rightarrow 2U^{5+}cs + O^{2-}is \qquad (5)$$

Reactions similar to these are responsible for the existence of nonstoichiometric phases. The dual valency on the cation also is involved in hole conductivity in p-type semiconductors. Although the precise mechanism involved in hole conduction is not known, some kind of a charge exchange promoted by phonons and involving cations and perhaps anion or anion sites occurs. A plausible one in uranium dioxide is shown in reaction (6).

$$2U^{4+}cs \rightarrow U^{5+}cs + U^{3+}cs \qquad (6)$$

A comparison of the energies associated with these reactions in solids and in gases illustrates the differences in the two cases. In the gas phase, the O^{2-} ion is unstable; in the solids, it is stabilized by the lattice self-potential. The energy associated with reaction (6) in the gas phase equals the difference between the two ionization potentials. Usually, this difference is of the order of tens of electronvolts. However, in the solid, this difference is compensated by the difference in the lattice self-potentials. Consequently, the energy associated with reaction (6), in the solid phase, is only a few electronvolts.

Interphase reactions. Reactions between phases accompanied by product formation with integrity maintained occur via chemical diffusion. This situation occurs when the reactants and the product have the same structure and comparable ionic radii. But even in this case, there generally is a volume change, the concentration of defects changes, and the initial interface between the reactant moves. A classical example is the reaction (that is, diffusion) between copper and copper-zinc alloy. The zinc diffuses more rapidly out of the alloy than the copper diffuses in, and the original interface moves toward the alloy. This phenomenon is known as the Kirkendall effect; it is one of the evidences that diffusion occurs via vacancies and interstitials rather than through positional exchange. *See* PHASE EQUILIBRIUM.

Two well-studied solid-state chemical reactions are sintering and corrosion. In the first reaction, mass is transferred between particles in a solid so that densification and frequently plastic deformation or creep occurs. The details of the transfer between particles is not well understood, but self-diffusion is generally involved. For instance, sintering and creep in uranium dioxide occur at temperatures where self-diffusion of oxygen through the interstitial position becomes significant. At these temperatures, self-diffusion of uranium is insignificant. In corrosion, diffusion through the product layer generally controls the rate of reaction, although electromigration may also be significant. In both cases, the mechanism can involve either the migration of a reacting component from the substrate through the layer or the migration of a corroding component from the external surface. When the layer becomes sufficiently thick and its density is sufficiently different, the layer develops fissures and cracks. Then the mechanism occurs along these, but the primary chemical driving force may still be the same as before.

Electrochemical reactions. There are scientific as well as technological interests in electrochemical reactions in solids. Electrolytic cells with solid electrodes and electrolytes have been used to determine chemical potentials, especially in nonstoichiometric phases. An example is a cell composed of nonstoichiometric oxides, UO_{2+x} and $Fe-Fe_{1-y}O$, as electrodes and stabilized zirconia as the electrolyte: $Fe-Fe_{1-y}O|ZrO_2|UO_{2+x}$. The electrolyte is an oxygen ion conductor, so that the voltage developed is produced by the difference in the chemical potential of oxygen in the two solid electrodes. Rechargeable batteries in which either the electrolyte or the electrodes are solids have been investigated for technological use. Among these are one in which β-alumina is used as an electrolyte and one in which FeS or FeS_2 is used as a cathode and aluminum-lithium alloy is used as an anode. In the first example, sodium and sulfur are used as electrodes; in the second one, molten lithium-potassium chloride is used as an electrolyte. *See* ELECTROCHEMISTRY.

Photochemical reactions. Many of the photoinduced processes which occur in solids can be imagined to be solid-state chemical reactions. The classical ones, of course, are those involved in photographic plates and films. *See* PHOTOCHEMISTRY.

Sublimation. At sufficiently high temperatures, the rate of sublimation of all solids can be measured. A technique which has been especially useful for inorganic solids (and a few organic solids) is the Knudsen effusion technique. Thus the molecular composition of the vapor and the partial vapor pressures are derived through the measurement of the rate at which the saturated vapor effuses through a small orifice in a cell into a vacuum and through the measurement of the mass spectrum of the effusate. The total vapor pressures of several solids have also been determined from measurements of the momentum of the effusate. The partial vapor pressures of most of the inorganic fluorides, chlorides, and oxides and several of the carbides and sulfides have been determined. In a number of cases, the vapor contains complex molecular species, even though the temperature required is high. The presence of these molecules is, in a sense, a consequence of the relatively high vapor pressure and not the high temperature, which of course at constant pressure is a degradative factor. The chemical potentials and the partial molar enthalpies and entropies of sublimation are derived from the partial vapor pressures and their dependencies on temperature. These are direct measures of the bonding in the solid. Although it might be argued that the species observed in the vapor are unrelated to the structure of the solid, it is apparent that there must be a relation because the same basic elements of bonding are involved. Thus ionic solids tend to sublime to ionic molecules; the bonding in gaseous alkali and alkaline earth fluorides and chlorides tends to be ionic. The more covalent oxides of tungsten and molybdenum, WO_3 and MoO_3, sublime to covalently bonded polymers, trimers, tetramers, and pentamers. *See* HIGH-TEMPERATURE CHEMISTRY; SUBLIMATION. [R. J. THORN]

Bibliography: D. M. Adams, *Inorganic Chemistry: An Introduction to Concepts in Solid State Structural Chemistry*, 1974; T. J. Gray et al., *The Defect Solid State*, 1957; N. B. Hannay, *Solid-State Chemistry*, 1967; W. Hayes (ed.), *Crystals with the Fluorite Structure*, 1974; P. Kofstad, *Nonstoichiometry, Diffusion, and Electrical Conductivity in Binary Metal Oxides*, 1972; G. G. Libowitz, *The Solid State Chemistry of Binary Metal Hydrides*, 1965; L. E. J. Roberts, *Solid State Chemistry*, 1972; H. Schmalzried, *Solid State Reactions: Progress in Solid State Chemistry*, 1974.

Solubility product constant

A special type of simplified equilibrium constant (symbol K_{sp}) defined for, and useful for, equilibria between solids (*s*) and their respective ions in solution, for example, reaction (1). For this relatively simple equilibrium, Eqs. (2) and (3) apply.

$$AgCl\,(s) \rightleftharpoons Ag^+ + Cl^- \tag{1}$$

$$[Ag^+][Cl^-] \cong K_{sp} \tag{2}$$

$$(Ag^+)(Cl^-)/(AgCl) = K_{sp} \tag{3}$$

It can be demonstrated experimentally that a small increase in the molar concentration of chloride ion $[Cl^-]$ (produced, for example, by the introduction of NaCl, HCl, or other soluble chloride) causes a reduction in the concentration of silver present as Ag^+. Similarly, an increase in $[Ag^+]$ reduces $[Cl^-]$. The product of the two concentrations is approximately constant as indicated by Eq. (2) and equal to the K_{sp} of Eq. (3). Equation (3) is exact since the variables are activities instead of concentrations. In accordance with the choice of standard state usually made for a solid, the activity of solid AgCl is unity, hence Eq. (4) holds.

$$(Ag^+)(Cl^-) = K_{sp} = 1.8 \times 10^{-10}\ \text{mole}^2\ \text{liter}^{-2} \tag{4}$$

In practice, various complications arise: addition of too much of either ion produces more complicated ions and hence actually increases the apparent concentration of the other ion. Addition of a salt without a common ion (that is, a salt supplying neither Ag^+ nor Cl^-) either may react with Ag^+ or Cl^- or may merely increase the concentration of both ions by a lowering of the mean ionic activity coefficient. (Sodium nitrate at a concentration 0.01 molar increases each concentration by about 10% and the product by 20%.)

It is usually assumed that an aqueous solution saturated with silver chloride contains only Ag^+, Cl^-, and the solvent. Some recent work indicates, however, that about 2.5% of the solute is present as undissociated AgCl. In practice, such effects are usually neglected. Equation (2) is especially useful in the explanation of analytical procedures in which it is desired to add a sufficient quantity of one ion to ensure (virtually) complete precipitation of the other.

An example of a salt of a different charge type is lead iodate $Pb(IO_3)_2$. The solubility equilibrium is represented by reaction (5).

$$Pb(IO_3)_2\,(s) \rightleftharpoons Pb^{2+} + 2IO_3^- \tag{5}$$

At about 25°C, the solubility of this salt is 0.024

g/liter. The mass of 1 gram-formula weight of the salt is 557 g. In the saturated solution, therefore, the concentrations are as in notation (6). Because

$$[Pb^{2+}] = \frac{0.024}{557} = 4.3 \times 10^{-5} \text{ mole/liter}$$
$$[IO_3^-] = \frac{2(0.024)}{557} = 8.6 \times 10^{-5} \text{ mole/liter} \quad (6)$$

the solid is in its standard state, its activity is unity. Hence Eq. (7) applies.

$$[Pb^{2+}][IO_3^-]^2 = K_{sp} = 3.2 \times 10^{-13} \text{ mole}^3/\text{liter}^3 \quad (7)$$

It is important to note that the concentration, 8.6×10^{-5} mole/liter, is the concentration of IO_3^- ion in the solution; it is not twice the concentration of the IO_3^- ion, although it happens to be twice the concentration of Pb^{++}. It is also important to observe that the square of 8.6×10^{-5} mole/liter enters this product because the coefficient of IO_3^- in reaction (5) is 2. *See* GRAVIMETRIC ANALYSIS; IONIC EQUILIBRIUM; PRECIPITATION.

[THOMAS F. YOUNG]

Solubilizing of samples

The process by which difficultly soluble samples are converted into different chemical compounds which are soluble. The sample may be heated in air to evolve volatile components or to oxidize a component to a volatile higher oxidation state with the formation of an acid-soluble form, as in the roasting of a sulfide to form the oxide and sulfur dioxide. Most frequently, the sample is treated with a solvent which reacts with one or more constituents of the sample. The dissolution of organic materials often follows the generalizations suggested by the solubility-based schemes of qualitative organic analysis, such as those described by O. Kamm, R. L. Shriner, and R. C. Fuson.

Solvents. The choice of solvent is determined by the chemical reactions which are required. Reactions used include solvation, neutralization, complex formation, metathesis, displacement, oxidation-reduction, or combinations of these. Most water-soluble salts dissolve by solvation. Basic oxides such as ferric oxide dissolve in aqueous hydrochloric acid by neutralization followed by complex formation. This is shown by reaction (1).

$$Fe_2O_3 + 8H^+ + 8Cl^- \rightarrow 2FeCl_4^- + 2H^+ + 3H_2O \quad (1)$$

Amphoteric oxides dissolve in either acids or bases. Reaction (2) represents the dissolving in base.

$$Al_2O_3 + 2Na^+ + 2OH^- \rightarrow 2AlO_2^- + 2Na^+ + H_2O \quad (2)$$

Aqueous ammonia converts insoluble silver chloride into a soluble complex as shown by reaction (3).

$$AgCl + 2NH_4OH \rightarrow Ag(NH_3)_2^+ + Cl^- + 2H_2O \quad (3)$$

Carbonates are solubilized by treatment with hydrochloric acid to displace carbon dioxide and to form soluble chlorides. Reaction (4) represents this reaction.

$$CaCO_3 + 2H^+ + 2Cl^- \rightarrow Ca^{++} + 2Cl^- + H_2O + CO_2 \quad (4)$$

Many substances are converted to easily dissolved mixtures by metathesis. An example is the boiling of barium sulfate with aqueous sodium carbonate to form barium carbonate and sodium sulfate. The insoluble barium carbonate dissolves easily in hydrochloric acid.

Transpositions of insoluble metal salts to their soluble EDTA complexes, using ethylenediaminetetraacetic acid, are increasingly common. Insoluble phosphates, sulfates, and even silicates have also been transposed by batch digestions with cationic ion-exchange resins as shown by reaction (5).

$$MSO_4 + 2HR \rightarrow MR_2 + H_2SO_4 \quad (5)$$

The metal which is bound to the resin can then be eluted or displaced into solution by treatment with an excess of an indifferent salt solution or with an acid whose anion is compatible with the cation present.

Dissolution of metals and alloys. Metals above hydrogen in the electrochemical series will dissolve in a nonoxidizing acid by reduction of hydrogen ion. Other metals such as copper and lead require an oxidizing acid, usually nitric acid. The treatment of alloys is determined by the constituents present. All of the components of brass and bronze are usually dissolved by nitric acid except tin, arsenic, and antimony, which precipitate as hydrated oxides. Alloy steels are usually dissolved by combinations of hydrochloric, nitric, phosphoric, and hydrofluoric acids, depending on the elements present. Aluminum-base alloys are treated with sodium hydroxide solution and any residues are dissolved in acid. Either total or component-selective anodic dissolution of metals and alloys may be achieved by applying a suitable anodic potential to the massive sample, weighed before and after attack, which is immersed in an electrolyte bath.

Fusions. Many substances do not dissolve at temperatures obtainable in the presence of liquid water. However, many refractory silicates, strongly ignited oxides of beryllium, cerium, plutonium, and so on, are completely attacked by aqueous hydrochloric acid at 300°C. P. Jannasch, E. Wichers, and later workers have developed a sealed quartz-tube procedure which permits this widely useful type of sample attack.

However, fused salt reactions employing temperatures of 400–1100°C are necessary for the attack and decomposition of many types of samples. The material used as the solvent is called a flux, and the process of melting the mixture of dry, solid flux with the sample is called a fusion. Fusion is done in a crucible (usually platinum or nickel) which is not attacked by the flux or the sample constituents. The same types of reactions are used as with aqueous solvents. Sodium carbonate is used to attack acid materials such as silicates and for metathesis reactions with sulfates, as in (6).

$$CaSiO_3 + Na_2CO_3 \rightarrow CaCO_3 + Na_2SiO_3 \quad (6)$$

Potassium bisulfate on fusion yields potassium pyrosulfate, an acid flux, which attacks basic oxides such as alumina and ferric oxide and metals such as chromium and tungsten. This flux must be used in porcelain crucibles. Oxidizing fluxes such as sodium peroxide and mixtures of sodium carbonate with potassium nitrate are used with sulfides, chromium, and tin ores and some silicon samples. These usually require nickel crucibles. Calcium carbonate plus ammonium chloride is used to free the alkali metals from silicates. *See* ANALYTICAL CHEMISTRY.

[CHARLES L. RULFS]

Bibliography: P. J. Elving (ed.), *Treatise on Analytical Chemistry*, vol. 2, pt. 1, 2d ed., 1979; W. F. Hillebrand and E. F. Lundell, *Applied Inorganic Analysis: With Special Reference to the Analysis of Metals, Minerals, and Rocks*, 1980.

Solution

A homogeneous mixture of two or more components whose properties vary continuously with varying proportions of the components. A liquid solution can be distinguished experimentally from a pure liquid by the fact that during transfers into other single phases at equilibrium (freezing and vaporizing at constant pressure) the temperature and other properties vary continuously, whereas those of a pure liquid remain constant. For an apparent exception *see* AZEOTROPIC MIXTURE.

Gases, unless highly compressed, are mutually soluble in all proportions.

A solid solution is, similarly, a single phase whose composition and other properties vary continuously with changing composition of the liquid phase with which it is in equilibrium. *See* SOLID SOLUTION.

Types of intermolecular force. The extent to which substances can form solutions depends upon the kind and strength of the attractive forces between the several molecular species involved. It is necessary to consider the attractive forces exerted by molecules of the following types: (1) nonpolar molecules; (2) polar molecules, that is, those containing electric dipoles; (3) ions; and (4) metallic atoms.

London forces. The theory of attraction between nonpolar molecules, developed by F. London in 1930, is based upon the quantum-mechanical interaction between pairs of electron systems. For two molecules with electrons having frequencies ν_1 and ν_2, polarizabilities α_1 and α_2, and separated by the distance r between centers, the attraction potential is shown as Eq. (1), where h is the Planck

$$\epsilon_{12} = -\frac{3\alpha_1\alpha_2 h}{2r^6} \cdot \frac{\nu_{0,1}\nu_{0,2}}{\nu_{0,1}+\nu_{0,2}} \tag{1}$$

constant. For molecules of the same species, this reduces to Eq. (2). The frequency ν_0 is that corre-

$$\epsilon_{11} = -\frac{3\alpha_1^2}{4r^6} h\nu_0 \tag{2}$$

sponding to $h\nu_0$, the zero-point energy, of the molecule in its unperturbed state. The perturbation by another molecule is related to its perturbation by light of varying frequencies, as seen in the variation of refractive index n with the frequency of light, that is, the dispersion. For this reason London designated these forces as dispersion forces. It is equally appropriate to speak of London forces by analogy with the nearly equivalent term, van der Waals forces. In the case of gases the dispersion n_ν is related to the frequency as in Eq. (3),

$$n_\nu - 1 = C/(\nu_0^2 - \nu^2) \tag{3}$$

where C is a constant. The polarizability α can be determined from the refractive index with the aid of the Lorentz-Lorenz formula. As a substitute for zero-point energy, London proposed the ionization potential.

The model upon which these relations are based is much simpler than the polyatomic molecules in most solutions of interest. In these the potential field is not central and radial; the interaction is between the electrons in the peripheral bonds. A striking example is octamethylcyclotetrasiloxane, whose core of alternating silicon and oxygen atoms is so buried within the eight methyl groups that it behaves toward other molecules essentially as an aliphatic hydrocarbon. The normal paraffins themselves are not symmetrical spherically. Moreover, the electrons in the molecules have many frequencies.

Although attempts have been made to extend London's basic concept to take account of such complexities, only the general implications of the concept as to the characteristics of the London forces are necessary here. These forces are (1) of very short range; (2) additive and nonspecific; (3) temperature independent; (4) operative between molecules of all types, whether nonpolar, polar, or metallic; (5) dependent in magnitude upon the number and "looseness" of the electrons; and (6) ordinarily less than average between molecules of different species. This last property can be seen by comparing ϵ_{12} with ϵ_{11} and ϵ_{22} as given by Eqs. (1) and (2). Eliminating the α terms one obtains Eq. (4).

$$\epsilon_{12} = \frac{(\nu_1\nu_2)^{1/2}}{(\nu_1+\nu_2)/2} \cdot (\epsilon_{11}\epsilon_{22})^{1/2} \tag{4}$$

Most component pairs differ much less in ionization potential than in polarizability, and the factor representing frequencies in this expression is not far from unity. Thus, relation (5) can be written.

$$\epsilon_{12} \approx (\epsilon_{11}\epsilon_{22})^{1/2} \tag{5}$$

This means that the interaction potential between unlike molecules is less than the arithmetic mean of the like potentials.

The pair potentials can be integrated over all the molecules in the pure components as well as the solution to obtain approximate attraction constants, a's, corresponding to the attraction constant of the van der Waals equation. M. P. E. Berthelot proposed the relation $a_{12} = (a_{11}a_{22})^{1/2}$ between the attraction constants of like and unlike species. J. Hildebrand and H. M. Carter found the Berthelot relation to be valid within 1% for seven liquid pairs, for example, $a = 31.21$ for liquid CCl_4 and 64.79 for $SnBr_4$. The calculated geometric mean is 46.46, and a_{12}, observed, is 46.86.

The consequence of this geometric mean relation is that the cohesion in a mixture of two liquids having different cohesion is less than their average. This results in expansion in volume, absorption of heat, and vapor pressures greater than additive upon mixing.

This geometric mean relation is usually adhered to rather well in cases of unlike molecules whose outer electrons are of similar types, such as (1) "N-electrons," nonbonded, as in halogenated paraffins and halogens; (2) π-electrons, as in olefins and aromatics; (3) bonding electrons only, as in H_2, CH_4, and other aliphatics; and (4) fluorochemicals. But deviations are found between molecules whose outer electrons are of different types. Illustrations will be found below in the sections on regular solutions and on solubility of gases.

Dipole interaction. This is the attraction between molecules containing permanent electric dipoles; it includes both the London forces and an electrostatic interaction of the dipoles. The latter depends upon the dipole moments of the molecules; it is temperature dependent because thermal agitation opposes the antiparallel orientation in which the interaction is greatest. Its magnitude depends also upon the geometry of the molecules because it is related to the distance of approach of the dipoles, not the molecular centers; the dipoles of some molecules are buried more deeply than those of others. This is the case with chloroform, which has solvent properties similar to those of carbon tetrachloride, except in a few specific cases. J. G. Kirkwood has expressed the degree of interaction between the dipoles of pure liquids by a g factor. For pyridine, the dipole moment μ is 2.20×10^{-18} cgs units, the dielectric constant ϵ is 12.5, and the dipole interaction g is 0.9. For water, $\mu = 1.84 \times 10^{-18}$, $\epsilon = 78.5$, and $g = 2.7$; for ethyl alcohol, $\mu = 2.80 \times 10^{-18}$, $\epsilon = 24.6$, and $g = 3.0$.

It is the g factor, not the dipole moment or the dielectric constant, that is most significant for understanding solubility relations. Furthermore, in the case of molecules having more than one polar bond, it is the separate polar bonds, not their vector sum of the overall dipole moment, that determine solubility relations. The three isomeric dinitrobenzenes all affect the vapor pressure of benzene virtually to the same extent, even though their dipole moments are quite different.

The substances with the largest g-factors are those that form hydrogen bonds. These have exceptionally high boiling points and are poor solvents for nonpolar substances. These liquids resist penetration by nonpolar molecules. The best-known pairs of incompletely miscible liquids are composed of a nonpolar liquid and water.

Electron donor-acceptor interaction. In the modern theory of generalized acids and bases, initiated by G. N. Lewis, a base is a substance having electrons that may be "accepted" into the vacant orbitals of other molecules, termed acids. This acceptance of electrons takes place reversibly and with little or no activation energy. Typical bases or donors are pyridine, acetone, ether, alkyl bromides, alkyl iodides, alkyl sulfides, iodide ion, thiocyanate ion, and aromatic hydrocarbons. Typical acids are the pure and mixed halogens, sulfur dioxide and trioxide, boron trichloride and trifluoride, aluminum halides, and stannic chloride. R. S. Mulliken and his co-workers have pointed out the close relationship between base strength and ionization potential and elaborated a theory of charge transfer complexes. H. A. Benesi and Hildebrand discovered the strong absorption in the ultraviolet characteristic of such complexes. They found that the basic strength increases in the order benzene < toluene < xylene < mesitylene. R. L. Scott determined that acid strength increases in the order $Cl_2 < Br_2 < I_2 < BrI < BrCl < ICl$.

This type of interaction is specific and saturating, and it reduces the escaping tendencies of the components. It corresponds to $\epsilon_{12} > (\epsilon_{11}\epsilon_{22})^{1/2}$. In cases where it is weak it may reduce but not overcome the opposite effect of unequal London forces.

Ion-ion interaction. The ions in a solid or liquid salt attract and repel electrostatically according to Coulomb's law, but there is also a London force component, and large ions are polarized by smaller ones. This last effect is illustrated by solid silver bromide, which is colored although its ions in aqueous solutions are colorless and whose crystals have the sodium chloride structure. The evidence is that in both the solid and the fused salt the electron cloud of the bromide ion is distorted equally by each of its six neighboring silver ions.

Ion-dipole interaction. In order to dissolve a solid salt, its lattice energy must be supplanted by the ion-ion action of another salt already in the liquid state or by the predominantly electrostatic attraction of a polar solvent or by the specific chemical interaction represented by complex ions.

Ideal solution. It is profitable to deal with actual solutions in terms of their departure from a simple idealized model—a mixture of components having the same attractive fields, which mix without change in volume or heat content. This is analogous to an ideal gas mixture, which is formed with no heat of mixing and in which the total pressure is the sum of the partial pressures. In such a solution the escaping tendency of the individual molecules is the same, whether they are surrounded by similar or by different molecules. Therefore, the combined escaping tendency of all the molecules of species 1, f_1, is given by Eq. (6), where x_1 is the

$$f_1 = f_1^0 x_1 \tag{6}$$

mole fraction of species 1 and f_1^0 is the escaping tendency of the molecules from the pure liquid. For a binary mixture, $x_1 + x_2 = 1$. If the gas imperfections of vapors are disregarded, vapor pressures may be substituted for fugacities to give Raoult's law (1886) in its usual form, $p_1 = p_1^0 x_1$. The total pressure is $P = p_1 + p_2 = p_1^0 x_1 + p_2^0 x_2$.

A more sophisticated derivation than the foregoing requires one to postulate molecules of the same size and shape. The gross structure of the solution containing n_1 plus n_2 molecules of two components is identical with those of the pure liquid components. The number of configurations of the components in the mixture within this structure is $(n_1 + n_2)!/n_1!n_2!$, and the configurational entropy of mixing is, by aid by Stirling's formula, Eq. (7). The substitution of a number of moles of

$$\Delta S = k\left[n_1 \ln \frac{n_1 + n_2}{n_1} + n_2 \ln \frac{n_1 + n_2}{n_2}\right] \tag{7}$$

each, N_1 and N_2, gives Eq. (8). The partial derivative

$$\Delta S = R\left[N_1 \ln \frac{N_1 + N_2}{N_1} + N_2 \ln \frac{N_1 + N_2}{N_2}\right] \tag{8}$$

$(\partial \Delta S / \partial N_1)_{N_2}$ represents the partial molal entropy of transfer of component 1 from pure liquid into an ideal solution of mole fraction x_1, Eq. (9). Because

$$\bar{s}_1 - s_1^0 = -R \ln x_1 \tag{9}$$

the model postulates no change in enthalpy, Eq. (10) applies, and $f_1/f_1^0 = x_1$, which is Raoult's law, where f is fugacity.

$$T(\bar{s}_1 - s_1^0) = -RT \ln (f_4/f_1^0) \tag{10}$$

But to arrive at this conclusion one must assume identical structures in the solution and the pure liquid components. This is very far from the case in solutions of high polymers in ordinary solvents, even though, as with polystyrene in benzene, the heat of mixing is practically zero.

Moderate difference in molal volume between components of high symmetry has little effect, as might be expected from the fact that the radius varies only with the cube root of the volume.

As a foundation for dealing with actual solutions in terms of the deviations of their properties from those of the model, it is necessary to derive other equivalent thermodynamic relationships.

Solubility of a crystalline solid. The fugacity of a crystalline substance f^s at temperature T is less than that of its supercooled liquid f^0 to an extent depending upon its melting point T_m and heat of fusion ΔH^F, as given by Eq. (11). If ΔH^F is assumed

$$\frac{d \ln (f^s/f^0)}{dT} = \frac{\Delta H^F}{RT^2} \quad (11)$$

constant, this gives upon integration Eq. (12). If

$$\ln \frac{f^s}{f^0} = -\frac{\Delta H^F}{R}\left(\frac{1}{T} - \frac{1}{T_m}\right) \quad (12)$$

the heat capacities of the solid and liquid forms are known, the variation of ΔH^F with temperature can be taken into account. This is hardly necessary for the present purpose because the deviations from ideal solubility involve factors that are more uncertain than this.

Molecular weight measurements. If a solid dissolves to form an ideal solution, its heat of solution is the same as its heat of fusion ΔH^F and $f_1^s/f_1^0 = x_1$. Therefore, Eq. (13) is formed. This is the ap-

$$-\ln x_1 = \frac{\Delta H^F}{R}\left(\frac{T_m - T}{T_m T}\right) \quad (13)$$

proximate equation for solubility of a solid that forms an ideal solution. It can be transformed into one much used for determining the molal weight of a solute by the depression of the freezing point of the solvent, here component 1. For $-\ln x_1$, one can write $\ln (1 + N_2/N_1)$. Expanding in powers of N_2/N_1 gives Eq. (14). When $N_2 \ll N_1$, the higher powers

$$\ln\left(1 + \frac{N_2}{N_1}\right) = \frac{N_2}{N_1}\left[1 - \frac{1}{2}\frac{N_2}{N_1} + \frac{1}{3}\left(\frac{N_2}{N_1}\right)^2 - \cdots\right] \quad (14)$$

may be neglected to give Eq. (15), where $\Delta T = T_m - T$.

$$\frac{N_2}{N_1} \approx \frac{\Delta H_1{}^F}{RT_m{}^2}\Delta T \quad (15)$$

By measuring ΔT for a known weight of solute in N_1 moles of solvent, the molal weight of the solute can be calculated. Because of the approximations made and the fact that even good solvents for the solid in question are seldom ideal, the resulting molal weights are not very exact unless extrapolated to $x_2 = 0$ from a series of values.

The lowering of the vapor pressure of a solvent upon the addition of a nonvolatile solute (component 2) may be offset by raising the temperature to restore the pressure of the solvent. These changes are related as in Eq. (16). This relation is

$$x_2 = \frac{\Delta H_1{}^{vap}}{RT_b{}^2}\Delta T \quad (16)$$

far less useful than that for the freezing point depression because the heat of vaporization is much greater than the heat of fusion; therefore, the elevation of the boiling point is much smaller than the depression of the freezing point and also is harder to determine.

Osmotic pressure. One mole of a solvent can be removed from a large quantity of a solution in which its mole fraction is x_1 in two reversible, and hence equivalent, ways. If it is distilled from the solution into pure liquid, the gain in (Gibbs) free energy is $\Delta G_1 = RT \ln (f_1^0/f_1)$. If it is pressed out through a semipermeable membrane against the hydrostatic pressure difference, osmotic pressure ΔP, the gain in free energy is $\Delta P\bar{V}_1$, where $\bar{V}_1$ is the partial molal volume of the solvent. In an ideal solution this is the molal volume. Equating the free energy of these processes gives Eq. (17). Ex-

$$\Delta P\bar{V}_1 = RT \ln \frac{f_1^0}{f_1} = RT \ln \frac{N_1 + N_2}{N_1} \quad (17)$$

panding as before and neglecting the higher powers gives $P\bar{V}_1 \approx (N_2/N_1)RT$ or $PV = RT$, where $V = N_2\bar{V}_1/N_1$, the volume containing 1 mole of solute.

This is the van't Hoff law for osmotic pressure, put forth in 1887. The theoretical basis of Raoult's law, discovered at almost the same time, was not yet appreciated. The formal correspondence between the van't Hoff law and the perfect gas law seemed to lend unique significance to osmotic pressure and elevated the van't Hoff law to the status of an ideal solution law. It is a limiting law only and not valid at high concentrations; it neglects the specific nature of the solvent. The solvent is regarded as furnishing space for a quasi-gas solute. Thus the law cannot cover molecular states in solutions of finite concentration.

The determination of osmotic pressure offers a valuable means for determining molal weights of high polymers in solution, where high weight concentrations correspond to mole fractions so low as to have only minute effects upon the vapor pressure and the freezing point of the solvent. For example, consider a solution of 0.001 mole of solute in 1 mole of benzene; $T_m = 279$ K and $\Delta H_1{}^F = 2370$ cal/mole, ΔT by Eq. (15) would be only 0.065°, but ΔP by Eq. (17) would be 194 mm. The latter is large enough to be measured with some precision.

Nonideal solutions. Unlikeness of the components of a binary mixture leads, as explained earlier, to fugacities in excess of ideal values. The

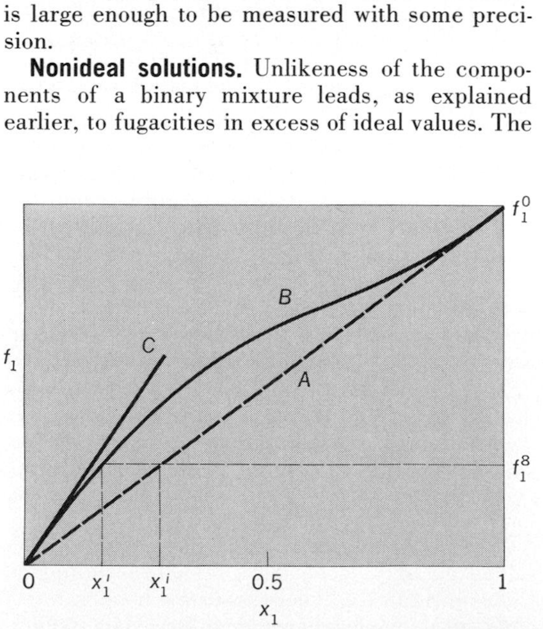

Fugacity and mole fraction. Line *A*, ideal. Line *B*, typical nonideal. Line *C*, Henry's law.

excess is largest when the molecules of one species are surrounded mostly by those of the other, as shown in the illustration. They approach Raoult's law at the upper end and Henry's law, $p_1 = kx_1$ (or $f_1 = k_1x_1$), at the lower end, where k is an experimentally determined constant.

An important relation between the two components is given by the Gibbs-Duhem equation, Eq. (18). If Raoult's law holds in the limit when

$$\left(\frac{\partial \ln f_1}{\partial \ln x_1}\right)_T = \left(\frac{\partial \ln f_2}{\partial \ln x_2}\right)_T \qquad (18)$$

$x_1 = 1$, since $(\partial \ln f_1)/(\partial \ln x_1) = 1$, then also $(\partial \ln f_2)/(\partial \ln x_2) = 1$. Integrating gives $\ln f_2 = \ln x_2 + \ln k_2$ or $f_2 = kx_2$, where k is a constant of integration that cannot be evaluated unless Raoult's law holds for component 1 over the whole range. In that case it also holds for component 2.

The activity in the case of nonelectrolytes is defined as $a_1 = f_1/f_1^0$ and so on. In an ideal solution, $a_1 = x_1$ and $a_2 = x_2$. The activity coefficient is $\gamma_1 = a_1/x_1$ and so on. Alternate, equivalent forms of the Gibbs-Duhem equation include Eq. (19), and $N_1 d\bar{G}_1$

$$(\partial \ln a_1)/(\partial \ln x_1) = (\partial \ln a_2)/(\partial \ln x_2) \qquad (19)$$

$+ N_2 d\bar{G}_2 = 0$, where $\bar{G}$ denotes partial molal Gibbs free energies.

If one component is a crystalline solid, its activity a_s is less than that of the liquid, which is 1, as given by Eq. (12); its maximum solubilities would be x_1^i in the illustration, if in an ideal solvent, and x'_1, if in a real solution.

Regular solutions. There are many mixtures of nonpolar components in which, except in the immediate neighborhood of a critical mixing temperature, thermal agitation suffices to neutralize tendencies to segregate and yields virtually complete randomness of mixing, with a close approach to ideal entropy of mixing, Eqs. (8) and (9).

The enthalpy of mixing can be calculated as the difference between the potential energy of the mixture and the sum of the potential energies of the liquid components. The potential energy of a mole of liquid may be related to the potential between a pair of molecules $\epsilon(r)$. The lattice energy of a crystal is obtained by summation of $\epsilon(r)$ over all of the lattice distances; that of a liquid is obtained by integration over the continuous distribution function $\rho(r)$. The expression for a pure liquid is Eq. (20). Here N_{Av} is the Avogadro number and v is

$$\Delta E_{\text{vap}} = -\frac{2\pi N_{Av}^2}{\text{v}} \int \rho(r)\epsilon(r)r^2\,dr \qquad (20)$$

the molal volume of the liquid. The corresponding expression for the potential of the mixture of N_1 and N_2 moles of the pure components involves their relative sizes. With certain simplifying assumptions, including the geometric mean for $\epsilon_{12}(r)$, Eq. (21*a*) was obtained for the energy of mixing N_1 and N_2 moles of two nonpolar liquids. The corresponding partial molal energy of transferring pure liquid to solution, for component 2, is Eq. (21*b*). Here ϕ_1 denotes volume fraction, neglecting

$$\Delta E_M = \frac{N_1\text{v}_1 N_2\text{v}_2}{N_1\text{v}_1 + N_2\text{v}_2}\left[\left(\frac{\Delta E_1}{\text{v}_1}\right)^{1/2} - \left(\frac{\Delta E_2}{\text{v}_2}\right)^{1/2}\right]^2 \qquad (21a)$$

$$\bar{E} - E_2^0 = \text{v}_2\phi_1^2(\delta_1 - \delta_2)^2 \qquad (21b)$$

Solubility parameters and molal volumes, 25°C

Liquid	Formula	Molal volume, ml	Solubility parameter
Perfluoroheptane	C_7F_{16}	225	5.9
Perfluorotributylamine	$(C_4F_9)_3N$	360	6.0
Perfluoromethylcyclo-hexane	c-$C_6F_{11}CF_3$	196	6.1
n-Heptane	n-C_7H_{16}	147	7.4
Silicon tetrachloride	$SiCl_4$	115	7.6
Cyclohexane	C_4H_{12}	109	8.2
Carbon tetrachloride	CCl_4	97	8.6
Chloroform	$CHCl_3$	81	9.3
Benzene	C_6H_6	89	9.2
Carbon disulfide	CS_2	60	10.0
Bromine	Br_2	52	11.5
Iodine	I_2	59	14.1

expansion on mixing, and the δ's are $(\Delta E_{\text{vap}}/\text{v})^{1/2}$, designated solubility parameters. Because energy and enthalpy are virtually identical for liquids, Eq. (21*b*) may be combined with the entropy of transfer as given by Eq. (9) to give the free energy of transfer, as in Eqs. (22*a*) and (22*b*).

$$\bar{G}_2 - G_2^0 = \bar{H}_2 - H_2^0 - T(\bar{S}_2 - S_2^0)$$

$$RT \ln a_2 = \text{v}_2\phi_1^2(\delta_1 - \delta_2)^2 + RT \ln x_2 \qquad (22a)$$

or

$$RT \ln \gamma_2 = \text{v}_2\phi_1^2(\delta_1 - \delta_2)^2 \qquad (22b)$$

The quantity in square brackets in Eq. (21*a*) is the excess of the arithmetic mean, $\frac{1}{2}[(\Delta E_1/\text{v}_1) + (\Delta E_2/\text{v}_2)]$, over the geometric mean, $[(\Delta E_1/\text{v}_1)(\Delta E_2/\text{v}_2)]^{1/2}$. As mentioned earlier, the geometric mean assumption is amazingly valid for many species of similar electronic structure but fails for others.

Representative values of solubility parameters at 25°C are given in the table. Parameters for substances solid at 25°C have been calculated by Eq. (14*a*) from solubilities of these substances in solvents whose parameters are well determined. With paraffins the solubility data give concordant δ-values a little greater than $(\Delta E_{\text{vap}}/\text{v})^{1/2}$.

[JOEL H. HILDEBRAND]

Bibliography: *Annual Review of Physical Chemistry*, annually; J. H. Hildebrand, J. M. Prausnitz, and R. L. Scott, *Regular and Related Solutions: Solubility of Solids, Liquids and Gases*, 1970.

Solvation

The association or combination of a solute unit (ionic, molecular, or particulate) with solvent molecules. This association may involve chemical or physical forces, or both, and may vary in degree from a loose, indefinite complex to the formation of a distinct chemical compound. Such a compound contains a definite number of solvent molecules per solute molecule.

Solvation occurring in aqueous solutions is referred to as hydration. In aqueous ionic solutions, the highly polar water molecules become oriented about the ions, forming spheres of hydration. As a result, the mobility of an ion under an applied voltage gradient is decreased. The extent of hydration depends upon the size and charge of the ion.

In certain colloidal suspensions, solvation is, to a large extent, responsible for the stability of the sol. Particles in lyophilic sols strongly adsorb on their surfaces one or more layers of solvent molecules. This protective layer prevents the particles from approaching so closely as to adhere. In water, starch and many proteins form suspensions which are highly solvated. Such systems exhibit a viscosity which is markedly higher than that of the pure solvent. *See* COLLOID; ELECTROLYTIC CONDUCTANCE; HYDRATION; SOLUTION.

[FRANCIS J. JOHNSTON]

Solvent

By convention, the component present in the greatest proportion in a homogeneous mixture of pure substances (solutions). Components of mixtures present in minor proportions are called solutes. Thus, technically, homogeneous mixtures are possible with liquids, solids, or gases dissolved in liquids; solids in solids; and gases in gases. In common practice this terminology is applied mostly to liquid mixtures for which the solvent is a liquid and the solute can be a liquid, solid, or gas. *See* SOLUTION.

Three broad classes of solvents are recognized—aqueous, nonaqueous, and organic. Formalistically, the nonaqueous and organic classifications are both not aqueous, but the term organic solvents is generally applied to a large body of carbon-based compounds that find use industrially and as media for chemical synthesis. Organic solvents are generally classified by the functional groups that are present in the molecule, for example, alcohols, halogenated hydrocarbons, or hydrocarbons; such groups give an indication of the types of physical or chemical interactions that can occur between solute and solvent. Nonaqueous solvents are generally taken to be inorganic substances and a few of the lower-molecular-weight, carbon-containing substances such as acetic acid, methanol, and dimethylsulfoxide. Nonaqueous solvents can be solids (for example, fused LiI), liquids (H_2SO_4), or gases (NH_3) at ambient conditions; the solvent properties of the first-named substances are manifested in the molten state, whereas the last-named substances must be liquefied to act as solvents.

NONAQUEOUS SOLVENTS

The classification, often arbitrary, of nonaqueous solvents can be made on the basis of a variety of factors. Excluding utilitarian considerations and classifications based on chemical character, that is, the presence of distinctive groups such as carbonyl in the molecule, useful classification schemes involve the protophilic nature of the solvent and its solvating power. Classification schemes involving nonaqueous solvents inevitably involve considerations of acid-base phenomena. Of the two major theories, the Brönsted-Lowry protonic concept has been the most useful because the early solvents of interest were invariably potential proton donors. However, the Lewis theory of acidity has become very useful for understanding solution phenomena in aprotic systems such as SO_2. *See* ACID AND BASE.

The ability of a solvent to form a so-called onium species is an important factor in defining the nature of the solution phenomena it will support. Onium species can be formed by reaction with a potential proton donor or by self-ionization of the solvent. Four classes of solvents are generally recognized according to their ability to coordinate with the proton—basic, acidic, aprotic, and amphiprotic solvents.

Basic solvents. Such solvents form the onium species most readily and are generally derivatives of ammonia (amines, hydrazines, pyridine, and so on) or water (alcohols and ethers). Hydrogen-containing species in this classification can undergo self-ionization to form the onium species, for example, $2NH_3 \rightleftarrows NH_4^+ + NH_2^-$, but the other substances in this class require the presence of a proton donor to do so, $R_2O + HX \rightleftarrows R_2OH^+ + X^-$.

Acidic solvents. These solvents exhibit a greater tendency to release protons than do basic solvents, and they form onium species with great reluctance. However, it is possible for acidic solvents to undergo self-ionization, for example, $2CH_3CO_2H \rightleftarrows CH_3CO_2H_2^+ + CH_3CO_2^-$, to form the corresponding onium species. Even though a solvent may be classified as acidic, for example, CH_3CO_2H, it is possible to protonate such molecules with a more strongly acidic substance. Thus, the strong mineral acids such as HCl ionize in anhydrous acetic acid to form the onium species, $CH_3CO_2H + HCl \rightleftarrows CH_3CO_2H_2^+ + Cl^-$.

Aprotic solvents. Solvents such as SO_2, also commonly called inert, have very little affinity for protons, and they are incapable of dissociating to give protons. Such aprotic solvents are also called indifferent, nondissociating, or nonionizing.

Amphiprotic solvents. These solvents are capable of either adding or donating protons. Thus, ammonia is amphiprotic because it can lose or accept a proton as shown in the following reactions:

$$(C_6H_5)_3C^- + NH_3 \rightarrow (C_6H_5)_3CH + NH_2^-$$

$$NH_3 + HX \rightarrow NH_4^+ + X^-$$

The classification of a substance as amphiprotic adds little to understanding solvent phenomena, because this designation depends only on the relative strengths of the acids or bases involved. It has always been possible to produce a medium sufficiently acidic to protonate even acidic solvents which, of course, also would be expected to be good proton donors to bases. From this point of view, anhydrous acetic acid is considered amphiprotic. Whether a substance is an acid or base depends upon the character of the solvent in which it is dissolved. For example, urea, which is very weakly basic in water, is a weak acid in the more

Table 1. Autoprotolysis constants of some common solvents at 25°C

Solvent	$\log K_S$	ϵ
H_2O	14.0	78.4
H_2O_2	13	84.2
CH_3OH	16.9	32.6
H_2S	34.5 (−78°C)	8.99 (−78°C)
H_2SO_4	3.6	101
HCO_2H	6.2	58.5 (16°C)
$H_2NC_2H_4OH$	5.1 (20°C)	37.7
$(CH_3)_2SO$	32	46.6
$HC(=O)N(CH_3)_2$	>21	36.7
CH_3CN	28.5	36.0
C_2H_5OH	19.1	24.3
$H_2NC_2H_4NH_2$	15.3	14.2 (20°C)
CH_3CO_2H	14.5	6.1 (20°C)

basic solvent ammonia and a strong base in the acidic solvent CH_3CO_2H.

Amphiprotic solvents exhibit a well-defined self-dissociation, $2HS \rightleftarrows H_2S^+ + S^-$ ($K_S = a_{H_2S^+} \cdot a_{S^-}$), for which an autoprotolysis constant K_S can be measured (see Table 1). It is generally difficult to measure K_S for nonaqueous solvents because apparent self-ionization can arise from the presence of traces of water or other amphiprotic contaminants. *See* IONIC EQUILIBRIUM.

Chemical behavior. Solvents intervene in chemical processes by producing species from solutes that are more reactive than if the solvent were not present. If the solute consists of ions, the energy for the dissolution process is supplied almost entirely from the solvation of ions; this is primarily an electrostatic process. However, two processes may occur when a covalent solute is dissolved: the solvation of molecules or the formation of ions. The solvation of molecular species usually involves dipolar interactions, specific interactions such as hydrogen-bond formation, and formation of covalent bonded species via coordinate covalent bond formation. In many instances it is possible for such intermediate species to undergo ionization. The formation of ions from covalent molecules is a measure of the ionizing power of the solvent, and attempts have been made to correlate this with the dielectric constant of the solvent. Unfortunately, there is a sufficient number of contradictions to make such relationships unreliable.

The extent to which a substance is ionized by neutral donor (solvent) molecules should increase with increasing stability of the cation resulting from nucleophilic attack, $S\!: + X\text{-}Y \rightarrow S\text{-}X^+ + Y^-$, compared with that of the unionized solute. The strength of the coordinate covalent bond found in the species SX^+ is related to the donor ability of S, the acceptor ability of X in the species XY, steric effects, and the magnitude of specific solvent-solute interactions such as hydrogen bonding. The donor strengths, expressed as donicity D_n, of solvent molecules have been defined relative to the reference acceptor $SbCl_5$; the enthalpy of the reaction between $SbCl_5$ and a series of donors in an inert medium is taken as a measure of donicity. The donicity of a solvent has been interpreted as a measure of its donor intensity, nucleophilicity, or Lewis base strength and can be a useful guide in assessing the ionizing power of a solvent. The donicities of some common nonaqueous solvents appear in Table 2. Inspection of the data in Table 2 indicates that there is no necessary correlation between the donicity of the solvent and its dielectric constant ϵ. [J. J. LAGOWSKI]

ORGANIC SOLVENTS

An organic solvent is any organic substance, single or multicomponent, capable of dissolving other substances to form a homogeneous system (solution). Organic solvents are obtained from many sources. The largest volume is obtained from petroleum (plus natural gas). Other important sources are: distillate from the destructive distillation of coal and wood, fermentation, seeds of plants, sap of trees such as the pine, and distillation or extraction from leaves and flowers of plants. About half of the 50 top-ranked chemicals produced in the United States are organic solvents. Some of the physical properties of certain important organic solvents are included in Table 3.

Uses and selection. Organic solvents have many uses in the chemical and allied industries and in scientific research. The largest volume is used in the manufacturing of coatings such as paints, lacquers, varnishes, and printing inks. Substantial amounts are also used in the manufacturing of synthetic fibers and other polymeric materials, which serve as cleaners and degreasers and as the media for chemical reactions. The yield of a chemical reaction may be increased by the selection of the proper solvent; even the course of a reaction may be altered by the selection of the solvent media. Organic solvents are used for the purification of other chemical substances by crystallization. They also are used for solvent extraction and separation by azeotropic distillation.

The choice of a solvent for a specific purpose is based on several of its attributes. There have been several parameters proposed for selecting the solvent ability. These are particularly useful when the solute is a high-molecular-weight substance. Other

Table 2. Donicity (D_n) and dielectric constant (ϵ) of some common solvents

Solvent	D_n	ϵ
1,2-Dichloroethane	0.0	10.1
Sulfurylchloride	0.1	10.5
Tetrachloroethylene carbonate	0.2	9.2
Thionyl chloride	0.4	9.2
Acetyl chloride	0.7	15.8
Benzoyl chloride	2.3	23.0
Nitromethane	2.7	35.9
Dichloroethylene carbonate	3.2	31.8
Nitrobenzene	4.4	34.8
Acetic anhydride	10.5	20.7
Phosphorus oxide chloride	11.7	14.0
Benzonitrile	11.9	25.2
Selenium oxide chloride	12.2	46.0
Monochloroethylene carbonate	12.7	62.0
Acetonitrile	14.1	38.0
Sulfolane	14.8	42.0
Propylene carbonate	15.1	69.0
Benzyl cyanide	15.1	18.4
Ethylene sulfite	15.3	41.0
Isobutyronitrile	15.4	20.4
Propionitrile	16.1	27.7
Ethylene carbonate	16.4	89.1
Phenylphosphorus oxide difluoride	16.4	27.9
Methyl acetate	16.5	6.7
Butyronitrile	16.6	20.3
Acetone	17.0	20.7
Ethyl acetate	17.1	6.0
Water	18.0	81.0
Phenylphosphorus oxide dichloride	18.5	26.0
Diethyl ether	19.2	4.3
Tetrahydrofuran	20.0	7.6
Diphenylphosphorus oxide chloride	22.4	–
Trimethyl phosphate	23.7	6.8
Tributyl phosphate	23.7	6.8
Dimethylformamide	26.6	36.1
Dimethylacetamide	27.8	28.9
Dimethyl sulfoxide	29.8	45.0
Diethylformamide	30.9	–
Diethylacetamide	32.2	–
Pyridine	33.1	12.3
Hexamethylphosphoramide	(38.8)	30.0

Table 3. Physical properties of some organic solvents

Organic solvent	Boiling point		Freezing point		Viscosity, cgs[3], 25°C	Dielectric constant, 25°C
	cgs[1]	SI[2]	cgs	SI		
Benzene	80.100	353.25	5.533	278.683	0.6028	2.275
1,2-Dichloroethane	83.483	356.633	−35.66	237.49	0.730, 30°C	10.36
Methanol	64.70	337.85	−97.68	175.47	0.5445	32.70
1,2-Ethanediol	197.3	470.4	−13.	260[4]	13.55, 30°C	37.7
Acetic acid	117.90	391.05	16.66	289.81	1.040, 20°C	6.15, 20°C
Phenol	181.839	454.989	40.90	314.05	4.076	9.78, 60°C
Acetone	56.29	329.44	−94.7	178.4	0.3040	20.70
2-Propanol	82.26	355.41	−88.0	185.2	1.765, 30°C	19.92
Ethanol	78.29	351.44	−114.1	159.0	1.078	24.55
1,2-Dimethoxybenzene	206.25	479.40	22.5[5]	295.6	3.281	4.09
Fluorobenzene	84.734	357.884	−42.21	230.94	0.517, 30°C	5.42
Pyridine	115.256	388.406	−41.55	231.60	0.884	12.4
2-Ethoxyethanol	135.6	408.8	<−90[6]	—	1.85	29.6
N,N-Dimethylacetamide	166.1	439.2	−20	253	0.838, 30°C	37.78
Dimethyl sulfoxide	189.0	462.2	18.54	291.69	1.996	46.68
2-Nitropropane	120.25	393.40	−91.32	181.83	0.721	25.52

[1]Centimeter-gram-second system of metric units. [2]International System of Units. [3]SI = cgs × 10^3. [4]Freezing point in doubt because of tendency to cool and form a glass. [5]Stable form. [6]Glass.
SOURCE: From J. A. Riddick and W. B. Bunger, *Organic Solvents*, 1970.

than the compatibility of a solvent with the solute and its physical state, several of its physical and thermodynamic properties will be considered in the selection. Some of the more commonly considered properties are given in Tables 3 and 4.

Safety. Both physical and physiological properties of chemicals are used to determine hazards and to set conditions for safe use. The oldest of the special properties is the flash point (Table 4). It is one of the tests used for shipping, storage, equipment design, and building design for fire hazard solvents. It is rather a simple test whose results are a function of vapor pressure and flammability.

The flammable limits (Table 4), often referred to as explosive limits, consist of two parts: a lower explosive limit (LEL) and an upper explosive limit (UEL). The lower explosive limit is the lowest concentration of vapor and air that will explode under standard test conditions. The upper explosive limit is the highest concentration of vapor and air that will explode under standard test conditions. All concentrations of vapor between these limits are explosive. The minimum ignition temperature (MIT) is the lowest temperature at which the vapor of a substance and air will explode under standard test conditions. Thus, a flammable mixture of the vapors of a combustible liquid and air will not ignite unless the ignition source is at, or above, the minimum ignition temperature for that substance. The relatively high minimum ignition temperature of gasoline vapors and the low surface temperature of the glowing end of a cigarette (covered with a coating of ash) is the reason so few explosions have occurred at gas stations.

Physiological properties. Organic solvents may enter the body by oral ingestion, inhalation of the

Table 4. Safety properties of some organic solvents

Solvent	Flash point, K		Flammable limits, % v		Minimum ignition, K	Threshold limit value	
	TCC[1]	TOC[1]	Lower	Upper		Vapor, g/m³	Carcinogenic, mg/m³
Hexane	247	—	1.18	7.43	533	360	—
Benzene	262	—	1.4	7.1	835	30[2]	32
Methanol	285	—	6.72	36.50	1140	260[2]	—
Ethanol	286	—	3.28	18.95	712	1900	—
Phenol	352	—	—	—	988	19[2]	—
Ethyl ether	228	—	1.85	36.50	447	1200	—
p-Dioxane	285	—	1.97	22.25	539	180[2]	1015
Acetone	273	—	2.55	12.80	843	2400	—
Chloroform	nf[3]	—	nf	nf	nf	120	—
Carbon tetrachloride	nf	—	nf	nf	nf	65[2]	—
Trichloroethylene	nf	—	nf	nf	736	535	900
2-Nitropropane	312	—	2.5	—	—	90	—
Nitrobenzene	361	—	—	—	755	5[2]	—
Aniline	349	—	—	—	—	19[2]	—
Pyridine	296	—	1.81	12.40	755	15	—
N,N-Dimethylacetamide	—	350	1.70	18.5[4]	693	35[2]	—
Dimethyl sulfoxide	—	368	—	—	—	—	—
Methyl Cellosolve	315	—	—	—	—	80[2]	—
Epichlorohydrin	—	314	—	—	—	20[2]	5 ppm
Hexamethylphosphoric triamide	—	—	—	—	—	n[2]	>1

[1]Tag closed cup, tag open cup. [2]Can be attributed to cutaneous absorption, including mucous membranes. [3]Nonflammable. [4]At 373 K.

vapors, or absorption of the liquid or vapors through the skin and mucous membranes. Some solvents are metabolized by the body and some are eliminated unchanged. For example, isoamyl acetate (banana oil) is metabolized, but hexane is eliminated unchanged. Other solvents are accumulative, that is, are absorbed and retained by certain tissues, and repeated ingestion may result in organ or systemic damage. Aniline was one of the first solvents recognized as accumulative. Many solvents have had their toxicity determined for animals, and the results are expressed in milligrams of chemical per kilogram of animal required to cause the death of 50% of the test species in a given test under standard test conditions, represented by the symbol LD_{50}.

The threshold limit value (TLV) is the criterion for safe working conditions (Table 4). It has been defined by the American Conference of Governmental and Industrial Hygienists as the conditions under which it is believed that nearly all workers may be exposed repeatedly day after day to airborne concentrations of substances without adverse effect.

Another safety property for chemicals that has become important is the carcinogenicity. Those substances that have been proved to cause cancer in humans or have induced cancer in animals are said to be carcinogenic.

[JOHN A. RIDDICK]

Bibliography: V. Gutmann, *Angew. Chem. Int. Ed.*, 9:843, 1970; J. J. Lagowski (ed.), *The Chemistry of Non-Aqueous Solvents*, vols. 1–5, 1966–1978; J. A. Riddick and W. B. Bunger, *Organic Solvents*, 1970; *Threshold Limit Values*, American Conference of Governmental Industrial Hygienists, Cincinnati, 1977; T. C. Waddington (ed.), *Non-Aqueous Solvent Systems*, 1965.

Solvent extraction

A technique, also called liquid extraction, for separating the components of a liquid solution. This technique depends upon the selective dissolving of one or more constituents of the solution into a suitable immiscible liquid solvent. It is particularly useful industrially for separation of the constituents of a mixture according to chemical type, especially when methods that depend upon different physical properties, such as the separation by distillation of substances of different vapor pressures, either fail entirely or become too expensive.

Industrial plants using solvent extraction require equipment for carrying out the extraction itself (extractor) and for essentially complete recovery of the solvent for reuse, usually by distillation. *See* DISTILLATION; EVAPORATION.

Applications. The petroleum refining industry is the largest user of extraction. In refining virtually all automobile lubricating oil, the undesirable constituents such as aromatic hydrocarbons, which have poor chemical and viscosity-temperature characteristics, are extracted from the more desirable paraffinic and naphthenic hydrocarbons. The principal solvents used are furfural, phenol, and a combination of phenol with propane and cresylic acid; nitrobenzene and 2,2′-dichloroethyl ether are used in minor amounts. Liquid propane is also used preferentially to extract the desirable constituents from unwanted asphaltic compounds.

By suitable catalytic treatment of lower boiling distillates, naphthas rich in aromatic hydrocarbons such as benzene, toluene, and the xylenes may be produced. The latter are separated from paraffinic hydrocarbons with such solvents as liquid sulfur dioxide, furfural, and ethylene glycol to produce high-purity aromatic hydrocarbons and high-octane gasoline.

Gasoline is "sweetened," or freed of its sulfur-containing compounds, by extraction with aqueous caustic solutions containing various naphthenic and aromatic acids, or methanol, to modify the solvent characteristics. Aqueous copper ammonium acetate is used to extract butadiene from other 4-carbon hydrocarbons in synthetic rubber production.

Vegetable oils are separated into relatively saturated and unsaturated glyceride esters with furfural or liquid propane as solvents. The former are edible products; the latter are drying oils used in paints. Fish oils are similarly treated and yield a high-vitamin fraction as well.

In by-product coke-oven plants, phenols and other tar acids are recovered from the ammoniacal liquors with benzene, tricresyl phosphate, butyl acetate, and other solvents in large installations. The pharmaceutical industry uses extraction to separate natural impurities or unwanted chemical by-products from products such as synthetic vitamins, penicillin, Aureomycin, antihistamines, reserpine, and a host of others.

All uranium for atomic energy purposes is freed of its impurities in aqueous solution by extraction into diethyl ether, tributyl phosphate, and other solvents. The reprocessing of atomic energy fuels for the recovery of plutonium, and the separation of many of the other fission products such as the rare-earth metals, utilize solvent extraction extensively. The otherwise hard-to-separate metal pairs, zirconium-hafnium and niobium-tantalum, are separated in quantity with comparative ease by these methods.

Equipment. Extractors bring about direct contact of the feed (solution to be separated) and extracting solvent in order to permit diffusional transfer of the constituents from the feed to the solvent. The rate of transfer depends upon the contact area of the two liquids and the degree of turbulence developed within them. The extractor disperses one of the liquids in the other to produce large surface area, and relative motion to produce turbulence. The extractor must also provide for the subsequent mechanical separation of the dispersion, based upon the different densities of the liquids, to permit withdrawal of the two effluent products, the extract (solvent containing the extracted constituents) and the raffinate (unextracted residue).

Mixer-settlers (Fig. 1) provide for these require-

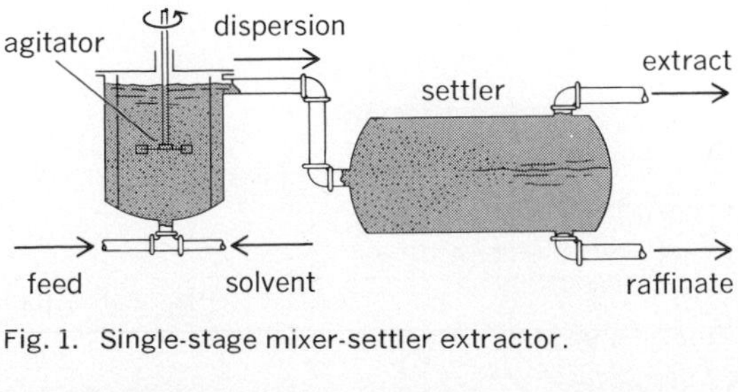

Fig. 1. Single-stage mixer-settler extractor.

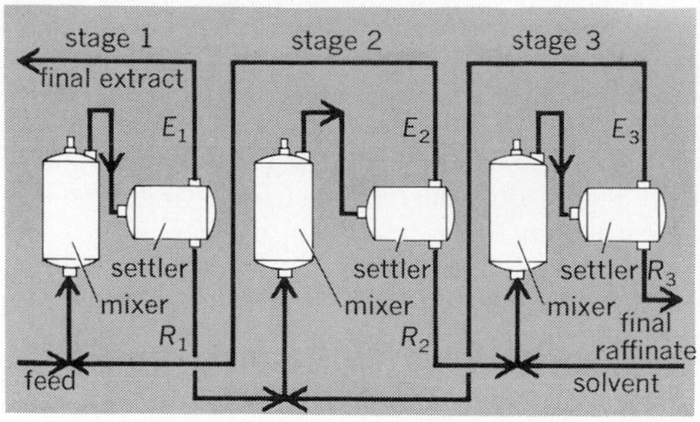

Fig. 2. Diagram of a three-stage countercurrent mixer-settler extractor. (*From R. E. Treybal, Mass Transfer Operations, 2d ed., McGraw-Hill, 1968*)

ments in separate vessels. The feed and solvent flow continuously through the mixer, in which the rotating agitator disperses one of the liquids into small droplets immersed in the other. The size of this vessel must provide sufficient residence time for the liquids that the desired diffusional transfer occurs. The degree of agitation must be intense without, however, producing so fine a dispersion that subsequent settling is difficult. The dispersion flows to the settler, most simply a drum, in which low velocity and lack of agitation promote gravity settling and coalescence of the drops to provide clear effluents.

Since in such single-stage apparatus the extractable substance approaches a concentration equilibrium in the effluents, nearly complete extraction requires a multiplicity of stages. An arrangement for countercurrent interstage flow of the liquids reduces the amount of solvent needed (Fig. 2). The compact modification of Fig. 3 has found particular favor in extraction of radioactive metals from aqueous solutions in processes associated with atomic energy operations.

To reduce the floor space and pump requirements for multistage extractors, a variety of vertical towers is also used. These involve countercurrent vertical flow, under gravity, of one of the liquids in dispersed form through a continuum of the other by virtue of the different liquid densities. A packed tower (Fig. 4*a*) is a cylindrical shell, the bulk of which is filled with manufactured packing, such as rings or saddles, randomly arranged. The more dense liquid, introduced at the top, flows downward as a continuum. The less dense liquid enters at the bottom through small nozzles. The

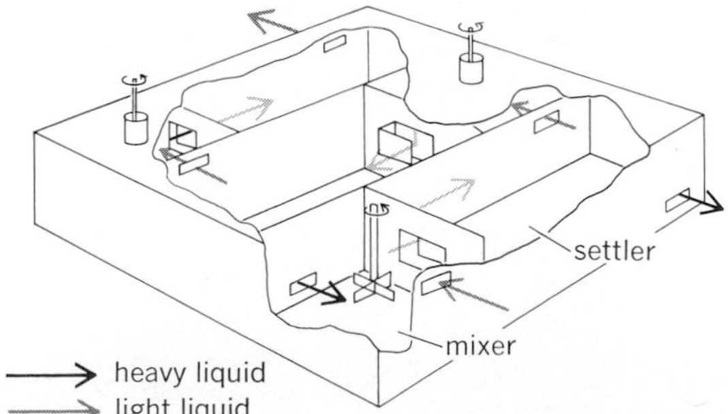

Fig. 3. Three-stage, box-type, mixer-settler extractor.

resulting small droplets rise through the heavy liquid, during which time extraction occurs, and then coalesce into a bulk and leave at the top. The packing serves to maintain the dispersion and provide moderate turbulence. The dispersed liquid may be either feed or solvent, light or heavy. If heavy, the droplets settle downward. Although the liquids are not repeatedly dispersed and settled as in the multistage mixer-settler, nevertheless multistage effects are obtained. Spray towers contain no packing and are not as effective. *See* GAS ABSORPTION OPERATIONS.

In perforated-tray towers (Fig. 4*b*) the light liquid collects in a layer under each tray and is dispersed into droplets by the small perforations. The drops rise through the heavy liquid, which flows across each tray and through the downspouts. The frequent redispersion achieved makes these towers very effective. Alternatively, by turning the

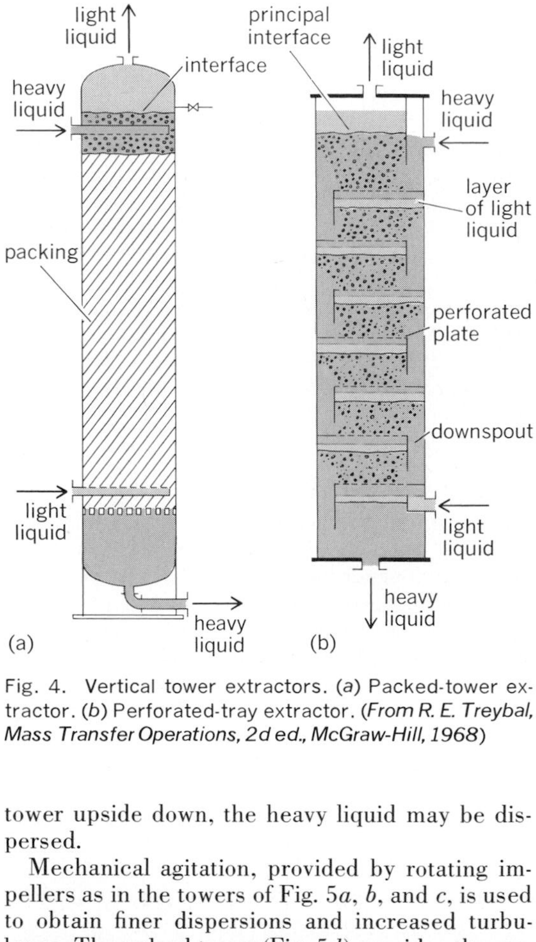

Fig. 4. Vertical tower extractors. (*a*) Packed-tower extractor. (*b*) Perforated-tray extractor. (*From R. E. Treybal, Mass Transfer Operations, 2d ed., McGraw-Hill, 1968*)

tower upside down, the heavy liquid may be dispersed.

Mechanical agitation, provided by rotating impellers as in the towers of Fig. 5*a*, *b*, and *c*, is used to obtain finer dispersions and increased turbulence. The pulsed tower (Fig. 5*d*) provides the mechanical agitation by rapid (20–100 cycles/min), small amplitude (0.25–2 in.), reciprocating motion superimposed upon the natural flow of liquids as they alternately pass through small perforations in the plates. This is particularly useful for handling radioactive liquids, since moving parts may be located in a place of safety.

In all these designs, the tower diameter is governed by the quantity of liquids to be handled, the height by the number of stages of extraction required. Towers up to 15 ft in diameter and 125 ft tall have been built. Auxiliary equipment may include pumps for movement of the liquids, motor-

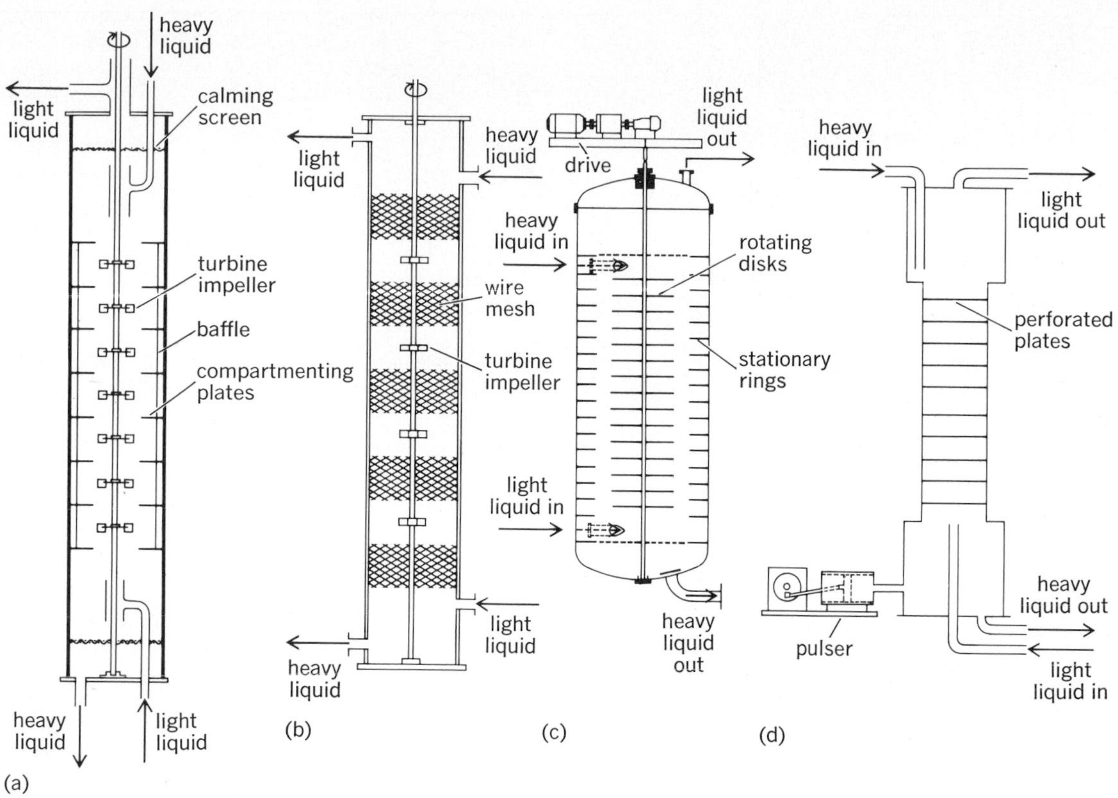

Fig. 5. Types of extractors with mechanical agitation. (*a*) Oldshue-Rushton extractor; (*b*) Scheibel-York extractor (*from R. E. Treybal, Mass Transfer Operations, 2d ed., McGraw-Hill, 1968*). (*c*) Rotating-disk extractor (*from G. H. Reman and R. B. Olney, Chem. Eng. Progr., 51:141–146, 1955*). (*d*) Pulsed extractor (*from T. B. Drew and J. W. Hoopes, eds., Advances in Chemical Engineering, vol. 1, Academic, 1956*).

drives for agitators, valves and flow meters for control of flow rates, and liquid-level control instruments.

The centrifugal extractor (Fig. 6) consists of a series of perforated, concentric rings in a cylindrical drum, the whole rapidly rotated (2000–5000 rpm) on the horizontal shaft. Liquids enter and leave through the shaft; they flow radially and countercurrently in the rotating drum because the effects of density differences are increased by centrifugal force. The particular virtue of this machine is the low residence time of the liquids, which has made it especially useful in the extraction of antibiotic pharmaceuticals from fermentation broths. *See* COUNTERCURRENT TRANSFER OPERATIONS; EXTRACTION; MASS-TRANSFER OPERATION; MIXING.

Laboratory applications. Solvent extraction is carried out regularly in the laboratory by the chemist as a commonplace purification procedure in organic synthesis, and in analytical separations in which the extraordinary ability of certain solvents preferentially to remove one or more constituents from a solution quantitatively is exploited. Batch extractions of this sort, on a small scale, are usually done in separatory funnels, where the mechanical agitation is supplied by handshaking of the funnel.

Natural biological products, such as hormones, serums, vaccines, vitamins, plant extracts, and the like, usually consist of many individual chemical compounds, frequently so similar and complex as to defy ordinary analytical procedures to separate them. Such substances, sometimes initially not even suspected of being mixtures, have been successfully separated into their constituents by the very ingenious laboratory device of L. C. Craig, which is capable of performing tens of thousands of extractions automatically.

Mechanism of extraction. Consider the single-stage extraction of a typical solute from a feed solution with a suitable immiscible solvent, as carried out in the device of Fig. 1. Figure 7, which is characteristic of the chemical nature of the system, shows the equilibrium distribution of solute between the two liquids as the curve *AGHB*. If two

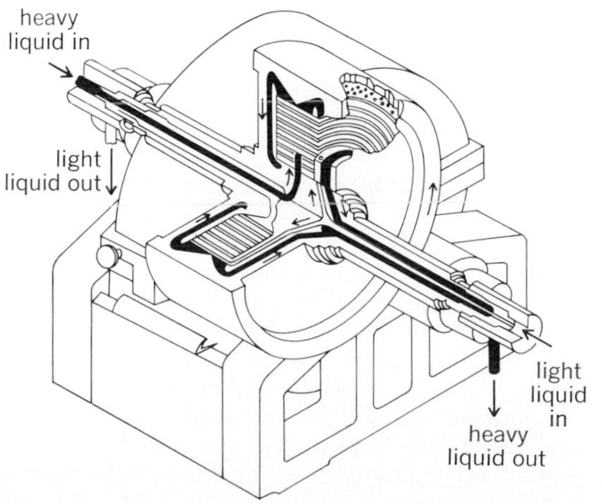

Fig. 6. Podbielniak centrifugal extractor, featuring low residence time of liquid. (*Podbielniak, Inc.*)

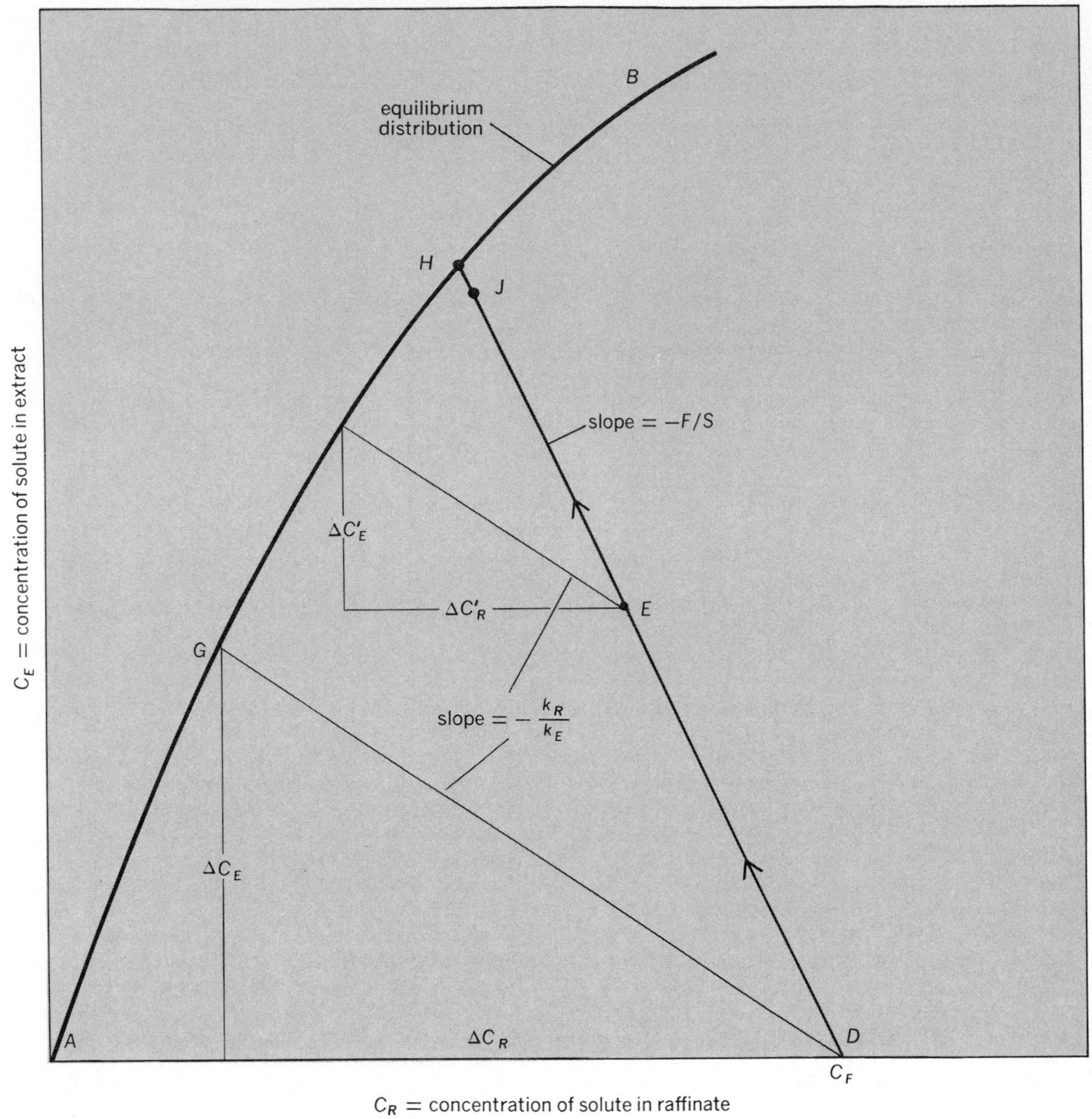

Fig. 7. Single-stage extraction.

solutions whose respective solute concentrations plot as a point on this curve are brought into contact, they will remain at their initial concentrations indefinitely, and no solute transfer will occur. On the other hand, if their concentrations are such that they provide a point removed from the curve, a spontaneous transfer of solute will occur from one liquid to the other, in an effort to bring the system to equilibrium. If the feed solution of Fig. 1, flowing at rate F, is of concentration c_F (point D in Fig. 7), and the solvent, flowing at rate S, contains no solute, transfer of solute will occur from the feed to the solvent. The rate of extraction at any time is proportional to the degree of departure from equilibrium, the proportionality involving the interfacial area A and rate constants (mass-transfer coefficients k_E and k_R) characteristic of the circumstances. As the liquids enter the vessel, the initial rate of extraction is given by the equation below,

$$\text{Rate} = k_E A \Delta c_E = k_R A \Delta c_R$$

where Δc_E and Δc_R are the initial departures from equilibrium at G in the two liquids, as shown in Fig. 7. Because of the necessity of satisfying a material balance (amount of solute removed from the feed equals that taken up by the solvent), the concentrations within the liquids move, not toward G, but instead toward H, whose location is fixed by the relative rates of flow of liquids, as shown. As extraction proceeds, so that the liquids reach point E, the concentration departure from equilibrium is lessened ($\Delta c'_E$ and $\Delta c'_R$) and the rate of extraction is reduced. Theoretically, equilibrium for the effluents at H will never be reached since the rate will then have fallen to zero, but in practice a point J is ultimately reached which can be as close to equilibrium at H as desired, depending upon the time of residence (volume of vessel/volume rate of flow) in the vessel. The degree of approach to equilibrium actually attained is the stage efficiency of the extractor, which may be expressed as the ratio of line lengths DJ/DH.

The interfacial area A is the surface of all the droplets of dispersed liquid in the vessel. Since for reasons of cost it is desired that the rate of extraction be large and the time of residence and vessel volume small, A is made large by creating small droplets through vigorous agitation. In well-agitated vessels, A may be 1000 ft^2 or more for each cubic foot (or 3280 m^2 for each cubic meter) of liquid contents. Too vigorous agitation, which may produce such small droplets that permanent, unsettleable emulsions form, must be avoided, however. The droplets are smallest in the vicinity of the agitator owing to locally high turbulence intensity. They tend to coalesce and become larger in regions remote from the agitator.

The mass-transfer coefficients k_E and k_R are characteristic of the physical properties and the nature of the motion of the extract and raffinate liquids, respectively. If the liquids are entirely quiet, they are related to the molecular diffusion coefficient of the solute through the liquids. The transfer of solute molecules from the bulk of the raffinate, through the interface surface, and into the bulk of the extract, then depends upon the random, thermal motion of the molecules, and their passage is greatly hindered by their frequent collision with the more abundant solvent molecules. The resulting mass-transfer coefficients are then very small.

For liquids in motion, however, the coefficients are very much larger, to an extent depending upon the intensity and scale of turbulence within each liquid. Under these conditions, it is customary to ascribe the transport of solute from the bulk of the raffinate liquid to the interface surface, and ultimately into the extract, to eddies, or relatively large chunks of liquid which move rapidly, carrying with them their contained solute. The influence of the diffusion coefficients is then substantially less.

For a given degree of turbulence as produced by the agitator, a number of additional factors also have influence on the rate of extraction. These are discussed briefly below.

1. The relative motion of liquid droplets and the continuum in which they are immersed result in a transfer of momentum across the interface, and this in turn results in internal circulation of the droplet liquid, leading to relatively large mass-transfer coefficients for the dispersed liquid. The larger this internal circulation the larger the ratio of viscosities of continuous to dispersed liquid. In many systems of industrial importance, however, there are present trace concentrations of substances which tend to adsorb at the interface, producing a rigidity of the interface and a marked reduction in the internal circulation within the droplets, with consequent lowering of the mass-transfer coefficients. In one case, 6×10^{-5} g surface active agent/100 ml liquid reduced the expected rate coefficient by 68%. Other mechanisms have also been proposed to explain such a reduction. Blocking of portions of the interface by the adsorbed molecules and chemical interaction of the adsorbed substance and the extracting solute are two possible explanations.

2. The relative motion of the liquid droplets and the continuum result in unequal mass-transfer rates, and hence unequal solute concentrations, about the drop periphery. Since interfacial tension depends upon concentration, an interfacial tension gradient along the drop surface may result, which brings about a relatively violent surface movement (Marangoni effect). This in turn enhances the mass-transfer coefficient substantially.

3. When two droplets coalesce, the reduction of surface results in a substantial release of energy. This produces an increased relative motion of the droplets and the continuum, in turn leading to increase in the mass-transfer coefficients.

4. Particularly in the extraction of metals from solution, extraction is frequently dependent upon a chemical reaction between the solute metal and a substance in the solvent, to make the metal soluble in the organic solvent. Such reactions may be relatively slow, and the rate of extraction will then be largely governed by the reaction rate.

While the mass-transfer coefficients and the stage efficiency of extraction may be computed with reasonable accuracy for ordinary circumstances, when any of the four conditions described above exist it is presently impossible to predict the rate of extraction with assurance. Resolution of these problems is the object of much current research. [ROBERT E. TREYBAL]

Destraction. These are separation processes based on the application of supercritical gases. In simplified terms, this method is characterized by the following features: (1) Destraction is a high-pressure technique, necessarily requiring special equipment. (2) High-boiling or even nonvolatile material can be dissolved by supercritical gases, whereby a loaded supercritical phase is formed. (3) The ability of a supercritical gas to take up other substances generally increases with increasing density or, at constant density, with increasing temperature. (4) Substances belonging to the same chemical class are taken up into the supercritical phase in the order of their increasing boiling points (for example, olefins in ethylene). (5) The take-up according to boiling points can be overlapped by the selective affinity of the individual substance for the supercritical gas (for example, caffeine in CO_2). (6) The material taken up can be recovered by decreasing the density of the supercritical phase, either by reducing the pressure at constant temperature or by raising the temperature at constant pressure. (7) The phenomena of distillation and extraction are utilized simultaneously; that is, enhancement of vapor pressure and phase separation both play a role. (8) The method can be used in fractionation and is particularly suitable for the isolation of thermally labile substances.

Phenomenological considerations. Distillation and extraction are among the most important separation procedures. Whereas separation by distillation is based upon the different vapor pressures of the components, separation by extraction is based on the properties of the material which determine the intermolecular interaction with the molecules of the extraction agent. Both of these effects are, to a certain extent, united in the general process of separation with supercritical gases. A supercritical gas is one which is above its critical pressure P_c and temperature T_c. Under these conditions the gas cannot be liquefied.

Effective destraction can be obtained in a temperature range of 10–100°C above the critical temperature of the gas being used and in a pressure range of 50–300 atm (5–30 MPa). Gases which are particularly effective are those whose critical temperatures are neither extremely high nor extremely low. Satisfactory results are obtained with the following gases: ethane (T_c 32°C, P_c 48 atm = 4.9 MPa), ethylene (T_c 9°C, P_c 50 atm = 5.1 MPa), propane (T_c 97°C, P_c 42 atm = 4.3 MPa), propene (T_c 92°C, P_c 46 atm = 4.7 MPa), CO_2 (T_c 31°C, P_c 73 atm = 7.4 MPa), and NH_3 (T_c 132°C, P_c 111 atm = 11.2 MPa). NO and N_2O can also be used, but care must be taken since explosions can occur, particularly in the case of N_2O.

The following example illustrates the phenomena: If ethylene is bubbled through paraffin oil at room temperature (23°C, that is, 14°C above T_c) and atmospheric pressure ($\rho \sim 0.0013$ g cm^{-3}), 100 g of ethylene transports about 0.1 g of oil. When the pressure is raised, the amount of oil transported per 100 g of ethylene hardly changes until the critical pressure ($P_c = 50$ atm = 5.1 MPa, $\rho \sim 0.22$ g cm^{-3}) is reached; at this point a dramatic increase in the amount of oil transported is observed, and at 200 atm or 20 MPa ($\rho \sim 0.4$ g cm^{-3}) 100 g of ethylene transports about 25 g of paraffin oil. At 200 atm and 70°C ($\rho \sim 0.32$ g cm^{-3}) the value falls to about 7 g. These effects can be interpreted to a first approximation as follows: The amount transported at atmospheric pressure corresponds roughly to that expected from the partial pressure of the paraffin oil. As the pressure rises, the solubility of ethylene in the oil increases, and is accompanied, at most, by an insignificant increase in the partial pressure as a result of changes in the intermolecular forces. At and just above the critical pressure, the density of the gas increases dramatically as a result of the increased compressibility (the pressure in this case says little about the degree of packing, that is, the density). The high-density gas is now able to take up large quantities of oil, and a loaded supercritical gas phase is formed. Raising the temperature from 23 to 70°C at constant pressure (200 atm) leads to a decrease both in density and in the ability of the gas to transport oil. If, however, the temperature is increased while the density is kept constant by increasing the pressure, then the ability of the gas to take up oil also increases. The large increase in the density above the critical pressure is also accompanied by a dramatic increase in the solubility of the gas in the liquid phase. This facilitates transfer of the oil into the supercritical gas phase.

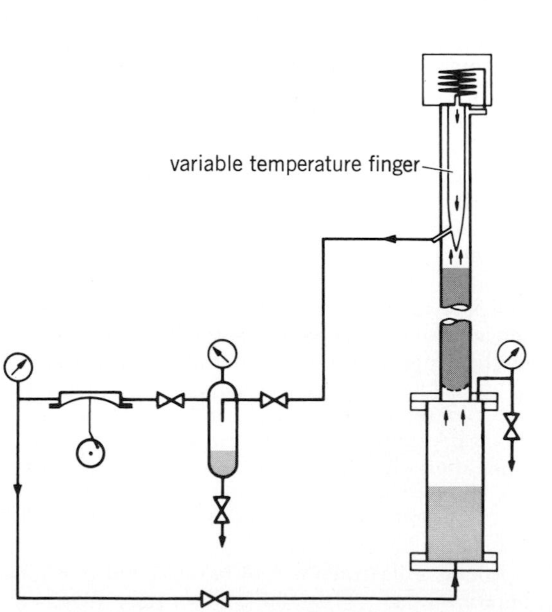

Fig. 8. Apparatus for fractional destraction. (*From Angewandt Chemie, Internat. Ed., 17(10):701–784, Verlag Chemie, Weinheim*)

Fractionation. The observation that an increase in temperature at constant pressure is accompanied by a decrease in density, and hence in the ability to transport material, forms the basis of an important application: mixtures of high-boiling material can be separated. This possibility is illustrated in Fig. 8. The apparatus, which consists of a still (25 liters), a column, and a heated finger, is charged with a mixture of α-olefins (traces C_{14}, 2 liters C_{16}, 2 liters C_{18}, and 7 liters C_{20}), and fractionated with ethane ($T_c = 32$°C, $P_c = 48$ atm = 4.9 MPa) at 45°C and an initial pressure of 60 atm (6.1 MPa) ($\rho \sim 0.2$ g cm^{-3}) with a finger temperature of 85°C ($\rho \sim 0.08$ g cm^{-3}). The supercritical ethane-olefin phase passes through the column, which is filled with copper rings, and reaches the hot finger, where refluxing occurs. The supercritical gas is bled off and partially depressurized to 30 atm or 3.0 MPa ($\rho \sim 0.03-0.04$ g cm^{-3}). The product separates out and is collected (60 ml/hr). The gas is then repressurized. The pressure is slowly increased during the destraction from 60 to 110 atm or 6.1 to 11.1 MPa ($\rho \sim 0.35$ g cm^{-3}). (Increasing the pressure in a destraction is analogous to raising the temperature during a distillation.) A gas chromatographic analysis of the various fractions shows that a successful separation has been accomplished; for example, a fraction containing about 25% of the C_{16} olefin has a purity of 95.5%.

Since the components and the gas are aliphatic hydrocarbons, this separation is based primarily on the differences in vapor pressure of the components and thus resembles a distillation; that is, the components are taken up into the gas phase in the order of their increasing boiling points. This effect can be overlapped by a separation according to the class of compound in those cases in which certain components in the mixture have, for example, little affinity for the supercritical gas and others are taken up preferentially. This separation process can also be applied to materials which are not amenable to fractional distillation.

Practical applications. The fact that destraction can be carried out at relatively low temperatures makes the method particulary appropriate for the separation of thermally labile substances. This is why destraction has received considerable attention from the food industries, for it is possible, for example, to remove fats and oils from vegetable and animal matter under mild conditions without the necessity of a final step in which solvent is removed. A variation of the procedure is to extract the desired product under subcritical conditions, that is, using a liquid gas, and then to separate the extract into its components under supercritical conditions. It has proved possible to remove cocoa butter from cocoa beans, soybean oil from soybeans, the essential oils from spices, as well as the valuable constituents from the hop resins. [GÜNTHER WILKE]

Bibliography: L. C. Craig and D. Craig, in A. Weissberger (ed.), *Techniques of Organic Chemistry*, 2d ed., vol. 3, pt. 1, pp. 149–332, 1957; T.

Ritcey and A. W. Ashbrook, *Solvent Extraction: Principles and Applications to Process Metallurgy*, pt. II, 1979; T. Sekine, *Solvent Extraction Chemistry*, 1977; R. E. Treybal, *Mass-Transfer Operations: Chemical Engineering*, 1980.

Specific gravity

The specific gravity of a material is defined as the ratio of its density to the density of some standard material, such as water at a specified temperature, for example, 60°F, or (for gases) air at standard conditions of temperature and pressure. Specific gravity is a convenient concept because it is usually easier to measure than density, and its value is the same in all systems of units. *See* DENSITY.

[LEO NEDELSKY]

Specific heat

The ratio of the amount of heat required to raise unit mass of a material 1 degree in temperature to the amount of heat required to raise the same mass of a reference substance 1 degree in temperature. Both measurements are made at a reference temperature and in nearly all cases at either constant volume or constant pressure. Water is usually the reference substance. Because the heat capacity of water is nearly unity, the value of specific heat for a material is nearly equal to its heat capacity. Specific heat, as defined here, is a ratio without units, although it is often defined differently. For clarity it is recommended that thermodynamic discussion be carried out in terms of heat capacity instead of specific heat. Also, it is desirable to define heat units in electrical terms. *See* CHEMICAL THERMODYNAMICS; SPECIFIC HEAT OF SOLIDS.

[HAROLD CHRISTIAN WEBER]

Specific heat of solids

When 1 gram (g) of a material absorbs an amount of heat ΔQ and this causes the temperature of the material to increase an amount ΔT, then the ratio $s = \Delta Q/\Delta T$ is often called the specific heat of the material, although other definitions are also used. The heat capacity C of a body of mass M is the product $C = Ms$. The atomic and molecular heats are the heat capacities of a gram-atomic weight and a gram-molecular weight of material, respectively.

The measured heat capacity of solids is usually made at some constant pressure P, such as atmospheric pressure, and is represented by the symbol C_P. The theoretical heat capacity is most often calculated for constant volume V, and is denoted by C_V. The difference $C_P - C_V$ is essentially the heat per degree required to expand the solid against its internal elastic forces. The difference is given by Eq. (1). Here α_V is the temperature

$$C_P - C_V = \alpha_V{}^2 VT/\chi \tag{1}$$

coefficient of volume expansion (at constant pressure), V the volume, T the temperature in K, and χ the isothermal compressibility. The quantities represented by the symbols C_P and C_V are often referred to loosely as specific heats, although they are really heat capacities.

Dulong-Petit law. P. Dulong and A. Petit observed in 1819 that, although the specific heats of the solid elements at room temperature differ widely from one another, the atomic heats are nearly all the same, the values being about 6.3 cal/°C. A theoretical explanation was given by F. Richarz in 1893. It is an extension of the theory of the specific heat of an ideal gas. According to the kinetic theory of gases, the thermal energy of an ideal monatomic gas is the same as its kinetic energy. From this, it was deduced that the atomic heat of such a gas is $3R/2$, where R, the gas constant, is about 2.0 cal/°C. The thermal energy of a solid, however, is the energy of the harmonic motion of the atoms, and this, on the average, is half kinetic and half potential. Richarz then supposed that $3R/2$ is the atomic heat arising from the mean kinetic energy, and $3R/2$ that arising from the mean potential energy, yielding a total atomic heat of $3R$ or 6.0 cal/°C.

The Dulong-Petit law is quite accurate at room temperature. To find s for many solid elements, one need only substitute the atomic weight A from a periodic table into the formula $s = 6/A$. However, it was noticed, even in the 19th century, that there are important exceptions to the law, notably diamond, germanium, and silicon, whose atomic heats at room temperature are considerably smaller than $3R$. Furthermore, many solids showed a decrease in C_V as the temperature was lowered to that of liquid nitrogen, which is 77 K or −196°C.

Einstein theory. The quantum hypothesis which M. Planck introduced into the theory of blackbody radiation in 1900 did not become a general principle until Albert Einstein applied it with success to the photoelectric effect in 1905 and to the theory of specific heats in 1907. In his theory of specific heats, Einstein sought to show that the observed failure of the classical theory, which gives $C_V = 3R$ for the atomic heat, could be explained in terms of the quantum hypothesis.

A so-called Planck oscillator can absorb or emit radiation only in integral amounts $nh\nu$, where n is an integer, h is Planck's constant, and ν is the natural frequency of the oscillator. The temperature is introduced by considering the mean value of the energy ϵ of such an oscillator, using the classical Boltzmann statistics. The result is given by Eq. 2,

$$\bar{\epsilon} = h\nu/[\exp(h\nu/kT) - 1] \tag{2}$$

where k is the Boltzmann constant and T is the absolute temperature.

Einstein's theory assumes that each atom of the solid oscillates with the same frequency ν_E and that this is the frequency observed in infrared absorption studies in crystals. Each atom vibrates in three dimensions and therefore has the mean energy $3\bar{\epsilon}$. The energy E of the solid is $3N\bar{\epsilon}$, if it contains Avogadro's number of atoms N. The quantum hypothesis then leads to Eq. (3). The fre-

$$E = 3Nh\nu_E/[\exp(h\nu_E/kT) - 1] \tag{3}$$

quency ν_E is called the Einstein frequency.

A parameter called the Einstein characteristic temperature Θ_E is defined by equating one quantum of energy $h\nu_E$ to the classical energy kT of an oscillator and denoting the particular value of T obtained in this manner by Θ_E. According to Einstein, the thermal energy Q of the solid is just the energy E of vibration, so that $C_V = dQ/dT = dE/dT$. This yields the Einstein formula of specific (atomic) heats, Eq. (4). Here $y = h\nu_E/kT = \Theta_E/T$ and $Nk = R$, the gas constant.

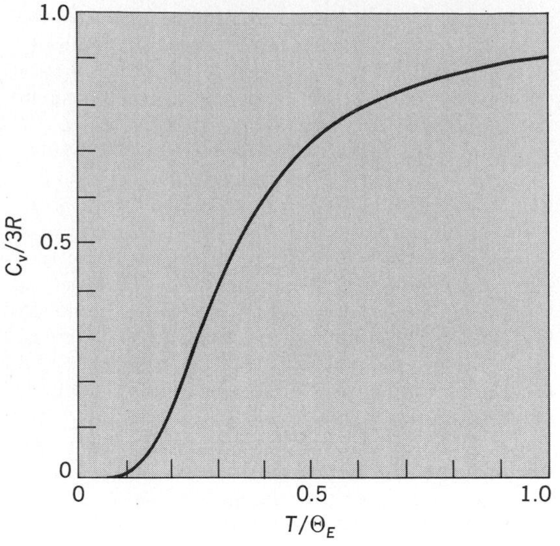

Fig. 1. Einstein specific heat curve.

$$C_V = 3Ry^2e^y/(e^y-1)^2 \quad (4)$$

A plot of $C_V/3R$ versus T/Θ_E is shown in Fig. 1. At $T=\Theta_E$, the value of $C_V/3R$ is 0.92, which means that, at this temperature, C_V has 92% of the Dulong-Petit value. Above this temperature, C_V approaches $3R$ with increasing temperature. Below this temperature, C_V decreases to zero, practically vanishing at $T<0.1\Theta_E$. Einstein's theory thus concludes that C_V is temperature-dependent. Furthermore, the observation that $C_V/3R=0.31$ for diamond at $T=331$ K is explained by stating that diamond has a value of Θ_E equal to about 1800 K, which corresponds to an infrared wavelength of 11 μm.

The prediction contained in the theory that C_V practically vanishes below $T=0.1\Theta_E$ stimulated W. Nernst and his assistants to make experimental investigations of C_V down to 16 K. It was found that C_V is still appreciable at $T<0.1\Theta_E$ for all substances examined; therefore Einstein's theory fails at these low temperatures. However, it appeared from the data that C_V approaches zero at 0 K, in keeping with deductions from Nernst's heat theorem.

Debye theory. The next advance in the theory of specific heats began with the suggestion of E. Madelung and W. Sutherland that the Einstein frequency is equivalent not only to the infrared absorption frequency of the crystal but also to the frequency of the shortest sound wave (or elastic wave) which can propagate through the crystal. This wave travels with the velocity of sound and has a wavelength of about twice the interatomic distance. Since sound waves of longer wavelength can also propagate through the crystal, Madelung made the further suggestion that a whole spectrum of acoustical frequencies should be used in computing C_V rather than just the single frequency ν_E.

In 1912 two theories of the specific heats of solids appeared, incorporating these ideas, one by P. Debye and the other by M. Born and T. von Kármán. Both theories use an acoustical spectrum containing so many frequencies that the spectra can be treated as continuous for purposes of computation. The number of waves (or modes) with frequencies between ν and $\nu+d\nu$ in the solid is thus represented by $g(\nu)\,d\nu$. The energy associated with each of these waves is that of a Planck oscillator, so that one obtains for the total energy E Eq. (5). The two theories differ in the manner of

$$E=\int_0^\infty \frac{g(\nu)h\nu\,d\nu}{e^{h\nu/kT}-1} \quad (5)$$

estimating $g(\nu)$. Only the simpler Debye theory is discussed in this article.

In order to estimate $g(\nu)$, Debye made two assumptions. One is that the solid is a continuous medium. With this idea, $g(\nu)$ is computed in a manner analogous to that employed in the theory of blackbody radiation, resulting in Eq. (6). The

$$g(\nu)=4\pi V\left(\frac{1}{U_l^3}+\frac{2}{U_t^3}\right)\nu^2 \quad (6)$$

symbol U_l represents the velocity of longitudinal sound waves and U_t that of transverse waves. The volume of the solid is V. The second assumption is that the total number of waves is equal to $3N$, where N is the number of atoms in the crystal. This assumption implies that the solid is not really continuous after all and that the shortest permissible wavelengths are those of about two interatomic distances. The restriction is expressed mathematically by Eq. (7), which serves to define a Debye

$$\int_0^{\nu_D} g(\nu)\,d\nu = 3N \quad (7)$$

frequency ν_D. The Debye frequency is the maximum allowable frequency. Thus for $\nu>\nu_D$, $g(\nu)$ is zero and the value of the integral above this limiting frequency is zero. This allows the upper limit in Eq. (5) to be replaced by ν_D.

Debye temperature. It is customary to replace the Debye frequency v_D by the Debye characteristic temperature Θ, defined by the relation $k\Theta = h\nu_D$. From this, and from Eqs. (5), (6), and (7), the energy E is defined by Eq. (8), where $z=h\nu/kT$.

$$E=9R\frac{T^4}{\Theta^3}\int_0^{\Theta/T}\frac{z^3\,dz}{e^z-1} \quad (8)$$

Equation (8) can be integrated and then C_V de-

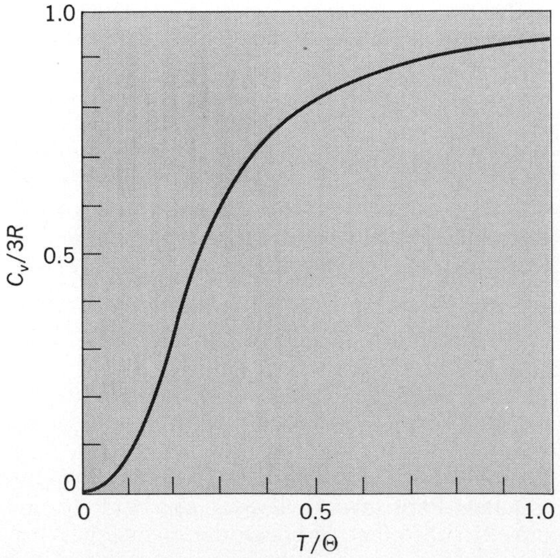

Fig. 2. Debye specific heat curve.

Debye characteristic temperature of solid elements, K

Element	Θ	Element	Θ	Element	Θ
Ar	85	Ga	240	Pd	275
Ag	215	Ge	360	Pr	74
Al	394	Gd	152	Pt	230
As	285	Hg	100	Sb	200
Au	170	In	129	Si	625
B	1250	K	100	Sn, gray	260
Be	1000	Li	400	Sn, white	170
Bi	120	La	132	Ta	225
C, diamond	1860	Mg	318	Th	100
Ca	230	Mn	400	Ti	380
Cd	120	Mo	380	Tl	96
Co	385	Na	150	V	390
Cr	460	Ne	63	W	310
Cu	315	Ni	375	Zn	234
Fe	420	Pb	88	Zr	250

duced from $C_V = dE/dT$. The result is the infinite series given by Eq. (9).

$$C_V/3R = \frac{4\pi^4}{5}\left(\frac{T}{\Theta}\right)^3 - \frac{3\Theta/T}{e^{\Theta/T}-1} + 12\log\left(1-e^{-\Theta/T}\right) - 36\frac{T}{\Theta}\sum_{n=1}^{\infty}\left\{\left[1+\frac{2T}{n\Theta}+\frac{2}{n^2}\left(\frac{T}{\Theta}\right)^2\right]\frac{e^{-n\Theta/T}}{n^2}\right\} \quad (9)$$

As T approaches infinity, the Dulong-Petit value of C_V is obtained. To see this, note that z approaches zero in this limit and that the integrand in Eq. (8) reduces to z^2. Integration then leads to $E = 3RT$, from which $C_V = 3R$. On the other hand, as T approaches 0 K, the Debye T^3 law results, as given by Eq. (10). This law is contained in the first term

$$C_V = \frac{12\pi^4 R}{5}\left(\frac{T}{\Theta}\right)^3 \quad (10)$$

of Eq. (9). The other terms in Eq. (9) contribute less than 1% to C_V at temperatures below $T = \Theta/12$. Figure 2 shows a plot of $C_V/3R$ versus T/Θ as given by the Debye theory. The table lists the values of Θ required to fit the Debye formula for C_V to the experimental data of solid elements in the region near where C_V is about half the Dulong-Petit value. The corresponding values of Θ_E determined in this manner are smaller and are approximately 3Θ/4.

The courses of the two curves shown in Figs. 1 and 2 are quite similar for T above about 0.2Θ. The critical test distinguishing between the two theories must therefore be made at temperatures below about 0.1Θ, where the Debye T^3 law should hold. The T^3 law was first verified by A. Eucken and F. Schwers in 1913 by measuring the heat capacity of a number of insulators. It failed for metals. The reason for this failure is now understood, for A. Sommerfeld's theory of metals (1928) shows that the conduction electrons can make an important contribution to the heat capacity. According to Sommerfeld, there must be a linear term in the temperature included in the expression for C_V in order to account for the electron contribution. Thus Eq. (11) is written. The coefficient γ in

$$C_v = \gamma T + (12\pi^4 R/5)(T/\Theta)^3 \quad (11)$$

the electron term is sometimes called the Sommerfeld gamma. To analyze low-temperature C_V data for metals, C_V/T versus T^2 is plotted. According to Eq. (11), this should give a straight line of slope $12\pi^4 R/5\Theta^3$ and of intercept γ on the C_V/T axis.

Deviations. As experimental measurements became more precise, it was noticed that the data could not be fitted to a Debye curve. One sensitive test is to calculate the Debye Θ for each experimental value of C_V and T, after correcting for the electron contribution. If the data satisfy the Debye formula, then Θ should be independent of the temperature. In most cases, the data do not satisfy this criterion.

A particularly marked deviation from Debye's theory occurs for cadmium. Figure 3 shows a plot of Θ versus T for this element. The data were treated in the following manner. First, C_V was calculated from the measured C_P data, using essentially Eq. (1). Then the 12 items of data below about 3 K were plotted on the basis of Eq. (11) and γ and Θ determined from the straight line graph. Next, all C_V data were corrected for the electron term and then the Θ for each point computed from tables based on Eqs. (8) or (9). The result is plotted in Fig. 3.

Agreement with Debye theory in the case of cadmium exists only below about 3 K, the only part of the curve where Θ is substantially constant. Thus the T^3 law holds (to better than 1%) only below about $T = \Theta/50$, instead of $\Theta/12$ as required by the Debye theory. For most of the solids examined, this limitation of the range of the validity of the T^3 law to the region $T < \Theta/50$ seems to occur. M. Blackman explained this in 1935 on the basis of the lattice dynamics of Born and von Kármán. He gave the estimate $T < \Theta/50$ as the "true" range of the T^3 law for most solids, an estimate which he made when the experimental evidence was still rather meager.

Experimental verification. An important independent check of the theory of specific heats can be made, based on Eq. (6). The velocity of sound is measured for single crystals at temperatures in the "true" T^3 region of specific heats. Since the velocity of sound depends on the direction of propagation through the crystal (an effect known as anisotropy of the velocity of sound), the inverse cube of the velocity must be averaged over all directions. This is done theoretically from the velocity of sound measurements made in several appropriate directions in a single crystal. When this is done, Θ at 0 K can be calculated. The comparison of the values calculated in this manner with those obtained from low-temperature specific heat data has been made for many substances, including copper, silver, and gold. The velocity of sound values of Θ for these last three elements are respectively 345, 226, and 162°, which compare well with the corresponding specific heat values 345, 226, and 165 K.

[JULES DE LAUNAY]

Bibliography: N. W. Ashcroft and N. D. Mermin, *Solid State Physics*, 1976; S. Fluegge (ed.), *Handbuch der Physik*, vol. 7, pt. 1, 1955; C. Kittel, *Introduction to Solid State Physics*, 5th ed., 1976; F. Seitz and D. Turnbull (eds.), *Solid State Physics*, vol. 2. 1956.

SPECIFIC HEAT OF SOLIDS

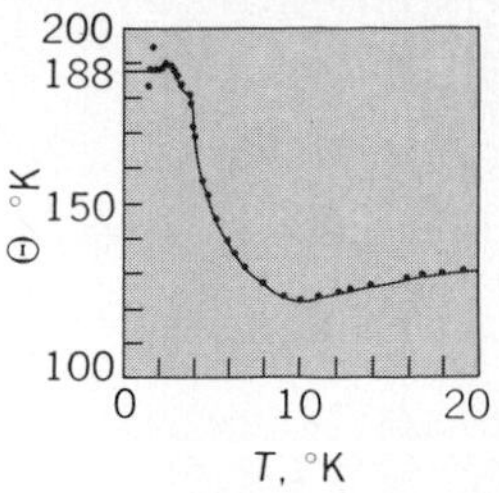

Fig. 3. Plot of Θ versus T for cadmium, illustrating a deviation from Debye's theory. *(After P. L. Smith and N. M. Wolcott, Phil. Mag., ser. 8, 1:854–865, 1956)*

Spectrochemical analysis

Any of a number of techniques for the determination of the presence or the concentration of elemental or molecular constituents in a sample through the use of spectrometric measurements. Spectrometric measurements entail the monitor-

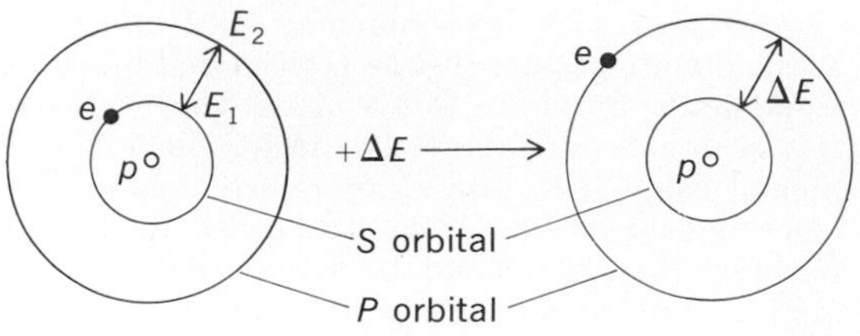

Fig. 1. Change in energy state of the electron in a hydrogen atom when light of wavelength λ_H is absorbed by the atom.

ing of electromagnetic radiation (EMR) as it is caused to emanate from, or interact with, the sample of interest.

Physical basis for measurements. The interaction of an analyte with electromagnetic radiation is based on changes in the level of some characteristic energy state of the analyte, for example, the oscillatory motion of a chemical bond, or the orbital location of a valence electron, or the rotational motion of the magnetic vector of an atomic nucleus. All of these types of characteristic states are quantized in energy by the principles of quantum mechanics. The change in energetic state of an analyte can be caused by the absorption or emission of energy of an amount exactly equal to the difference in energies of the two states.

Electromagnetic radiation is that manifestation of energy which is described by Planck's equation, $E = hc/\lambda = h\nu$, where E is the energy, h is Planck's constant, c is the speed of light, λ is the wavelength, and ν is the frequency of radiation. Energy is thus characterized by discrete wavelengths or frequencies, and the range of all wavelengths is called the electromagnetic spectrum.

Therefore, a spectrometric measurement is the result of the interaction of a particular wavelength and quantity of radiation with some characteristic energy state of an analyte so as to cause the energy state of the analyte to change, as depicted in Fig. 1. Considering hydrogen as an example, the location of the electron (e) of a hydrogen atom in the 1s orbital corresponds to a specific energy, here called E_1. The 1p orbital has a different, higher energy associated with it. If an amount of energy exactly equal to the difference between E_1 and E_2, that is, $\Delta E = E_2 - E_1$, is imparted to the hydrogen atom, then the electron may be raised to the higher orbital. Since $\Delta E = hc/\lambda_H$, the energy transition corresponds to electromagnetic radiation of a specific wavelength, here called λ_H. Therefore, if a beam of light of wavelength λ_H is caused to impinge on a hydrogen atom, that light will be absorbed from the beam as the electron changes orbitals, and the atom will have gained energy. An atom other than hydrogen will have different levels of energy corresponding to the same type of transition; and therefore, a different wavelength of radiation would be used. This is the basis of qualitative analysis. More atoms of any kind will require more light to cause the transitions. The amount of light used is directly proportional to the number of atoms present. This is the basis of quantitative analysis. *See* EXCITED STATE; GROUND STATE.

Spectrometric interactions. There are different types of spectral interaction, or radiative transitions, which can occur, depending upon the condition of the characteristic energy state involved prior to the interaction. The interactions are absorption, emission, and luminescence of radiation, depicted schematically in Fig. 2.

Transitions between the lowest energy level and another level are called resonance transitions. Nonresonance transitions occur between two levels, neither of which is the lowest level. Absorption occurs when a system gains energy through the retention of the energy associated with electromagnetic radiation incident on the system. Emission is the opposite of absorption. A system loses energy, and the energy loss is manifested as electromagnetic radiation. Fluorescence is a process in which energy gained radiatively is immediately lost by a system, also radiatively. Phosphorescence is like fluorescence, but upon gaining energy from incident electromagnetic radiation, the system loses part of it by nonradiative means (for example, through collisions if the system is made up of atoms). The loss of the rest of the energy gained in the absorption process is then emitted as electromagnetic radiation. Since the energy lost radiatively is less than that gained radiatively, the wavelength of the emitted electromagnetic radiation will be longer than that of the incident electromagnetic radiation.

In addition to the monitoring of these transitions, it is possible to monitor the manner in which the polarization of the radiation changes as it is absorbed (circular dichroism) or emitted (optical rotary dispersion). As a consequence, there are many different techniques of spectrometric measurement. They differ depending upon what type of radiation is monitored, what type of transition is involved, and whether some characteristic of the radiation is also observed.

Qualitative and quantitative. The analytical determination of the presence of a constituent (qualitative determination) results from sensing the energetic transition as it occurs. Each particular element or molecule will possess its own energetic characteristics, so that monitoring a spe-

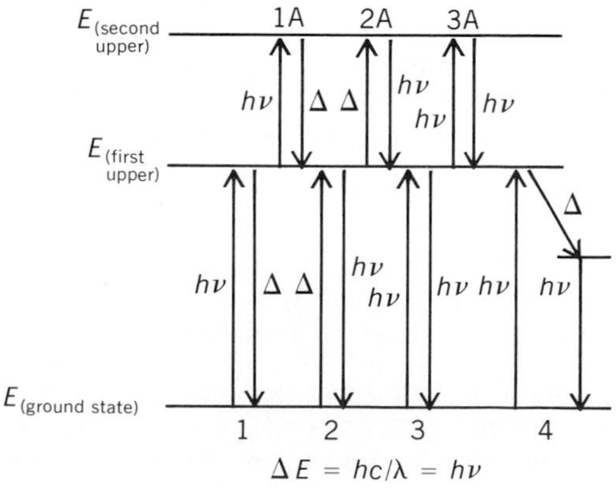

key:

1 = resonance absorption
1A = nonresonance absorption
2 = resonance emission
2A = nonresonance emission
3 = resonance fluorescence
3A = nonresonance fluorescence
4 = phosphorescence
(3, 3A, 4: luminescence)
Δ = any transition mechanism other than radiative

Fig. 2. Types of spectral interaction.

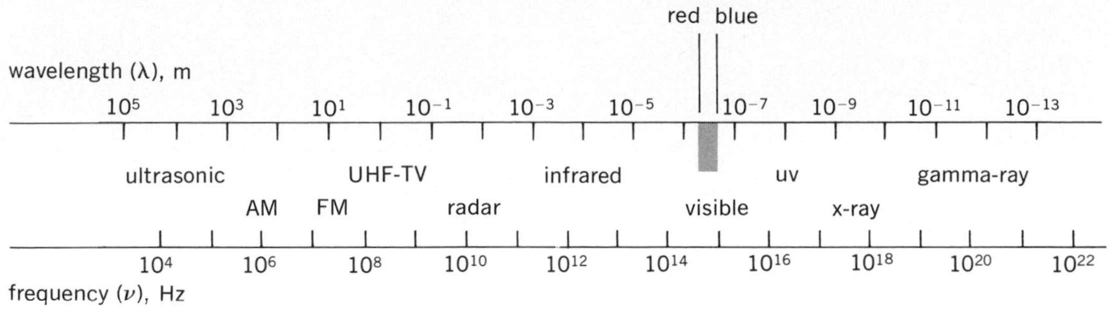

Fig. 3. Electromagnetic spectrum.

cific transition will identify a constituent, exactly analogous to identifying people through the use of fingerprints. The determination of the concentration of a constituent (quantitative determination) is a result of the direct relationship between the amount of radiation which is emitted or absorbed and the amount of element or molecule present.

Electromagnetic spectrum regions utilized. The electromagnetic spectrum extends both higher and lower in energy and wavelength on either side of the visible spectrum, which is only a small portion of it, as shown in Fig. 3. As equipment has been developed for doing experiments in any part of the EMR spectrum, that portion has become available for, and has been used in, spectrometric measurements. As a consequence, there are spectrometric methods which range from gamma-ray spectrometry at the high-energy end of the spectrum, through ultraviolet/visible spectrophotometry, to radio-frequency resonance methods and sonic imaging. These methods, and all others, are separated just by the differences in resolving and detection equipment required, not by any inherent differences in the conduct of a spectrometric measurement. *See* SPECTROPHOTOMETRIC ANALYSIS.

Basic spectrometric measurement. In the basic spectrometric experiment which is performed for spectrochemical analysis, the sample is put into a state in which it will interact in the manner desired with the radiation of choice (Fig. 4) Examples are techniques in which a molecular compound is placed in a magnetic field so that the magnetic vector of the hydrogen nuclei in the molecule will line up with the magnetic field and thus be available to absorb radio-frequency radiation as they precess about the lines of force; or a solution of a sample is sprayed into a flame so that the sample is decomposed, leaving only its atoms, making the atoms available for absorption, emission, or luminescence of light; or the sample may not need preconditioning and only require being held in place. For absorption and luminescence measurements, a source of the radiation of choice, for example, a tungsten filament lamp, a gamma-ray emitter, or a laser, is focused onto the sample so that the radiation may be absorbed. For emission measurements the sample is supplied with external energy which is not radiative, for example, the heat from a flame, or an applied voltage, or momentum from mechanical motion. The sample then emits the gained energy as photons (packets, quanta) of radiation.

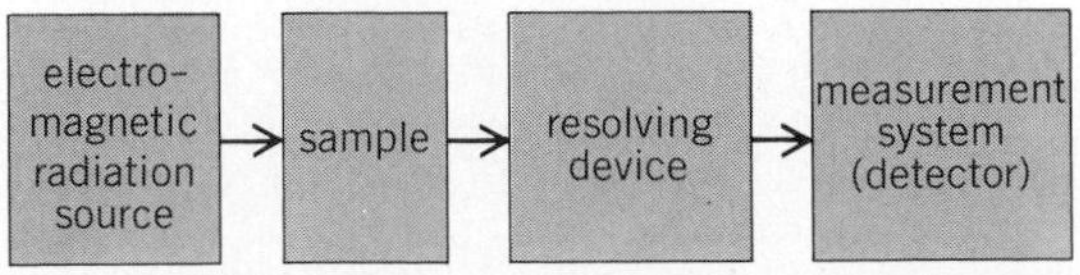

Fig. 4. Schematic diagram of the basic spectrometric experiment.

The next stage of the experiment separates all the wavelengths of radiation present so that they may be measured independently of all others. Finally, there is a device for detecting the amount of radiation present (emission or luminescence) or absent (absorption). The identity of the wavelength or wavelengths gives qualitative information. The amount of radiation gives quantitative information.

Except for rather dramatic differences in the character of the equipment needed to perform the spectrometric measurements (for example, an x-ray source is not at all similar in construction or operation to a radio-frequency transmitter; nor is a gamma-ray detector similar to an ultrasonic receiver), all spectrometric measurements are performed in the sequence described. Spectrochemical analysis follows from relating the spectrometric measurement to the quality or quantity of sample constituent. *See* SPECTROSCOPY.

[ANDREW T. ZANDER]

Bibliography: G. W. Castellan, *Physical Chemistry*, 2d ed., 1971; J. A. Dean and T. C. Rains, *Flame Emission and Atomic Absorption Spectrometry*, vol. 1, 1969; H. H. Willard, L. L. Merritt, and J. A. Dean, *Instrumental Methods of Analysis*, 5th ed., 1974; J. D. Winefordner, *Trace Analysis: Spectroscopic Methods for Elements*, 1976.

Spectrophotometric analysis

A method of chemical analysis based on the absorption or attenuation by matter of electromagnetic radiation of a specified wavelength or frequency. The region of the electromagnetic spectrum most useful for chemical analysis is that between 200 nanometers (nm) and 300 micrometers (μm). Since the sample being analyzed absorbs the radiation, spectrophotometric analysis is sometimes referred to as absorptimetric analysis.

The instruments used in this work are referred to as spectrophotometers. A simple spectrophotometer consists of a source of radiation, such as a light bulb; a monochromator containing a prism or grating which disperses the light so that only a limited wavelength, or frequency, range is allowed to irradiate the sample; the sample itself; and a detector, such as a photocell, which measures the

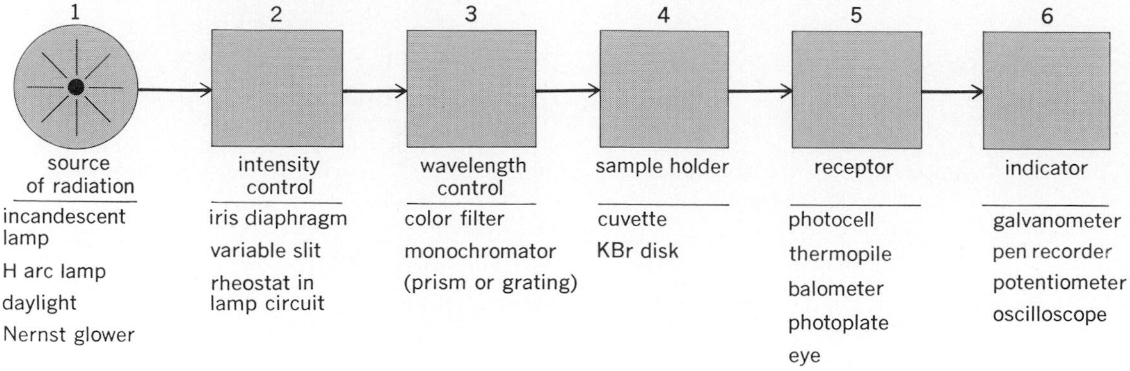

Fig. 1. Block diagram of generalized spectrophotometer.

amount of light transmitted by the sample. See Fig. 1.

Bouguer-Lambert-Beer law. By using a spectrophotometer, the intensity of the light transmitted through an absorbing substance may be compared with the light intensity when no such substance is in the light beam. Two fundamental laws govern the intensity of the light transmitted by an absorbing material. The first law, called the Bouguer-Lambert law, is given as Eq. (1). Here I_0 is the in-

$$\log(I_0/I) = Kb \quad (1)$$

tensity of the light beam with no sample present, I is the intensity of the light beam after passing through the sample, K is a constant depending on the sample and wavelength of the light, and b is the thickness of the absorbing solution. The second law, called Beer's law, is shown by Eq. (2). Here I

$$\log(I_0/I) = K'c \quad (2)$$

and I_0 are as above, K' is a constant depending on the sample and wavelength of the light, and c is the concentration of absorbing material in the sample. Usually these two laws are combined in the form of Eq. (3), where I_0, I, b, and c are as described above,

$$\log(I_0/I) = abc \quad (3)$$

and a is a constant called the absorptivity or extinction coefficient.

Two other terms are commonly used in spectrophotometric analysis. These are transmittance T and absorbance A as defined by Eqs. (4) and (5).

$$T = I/I_0 \quad (4)$$

$$A = \log(1/T) = \log(I_0/I) = abc \quad (5)$$

The absorbance A is directly proportional to the length of the light path through the sample and to the concentration of the absorbing material. It is the term most used in quantitative spectrophotometric work.

The Bouguer-Lambert-Beer law is strictly obeyed only when monochromatic radiation, that is, radiation of a single wavelength or frequency, is used. The monochromators of most commercial spectrophotometers produce radiation which is close enough to monochromatic so that the deviations from the law from this source are minor, except in the infrared region. There are, however, occasional deviations from this law for both chemical and instrumental reasons.

In most quantitative analytical work, a calibration or standard curve is prepared by measuring the absorption of known amounts of the absorbing material at the wavelength at which it strongly absorbs. Such a calibration curve is shown in Fig. 2 for the absorbing material whose absorption spectrum is shown in Fig. 3. The absorbance of the sample is read directly from the measuring circuit of the spectrophotometer. Calibration curves are usually linear, as in Fig. 2. Occasionally, instrumental or chemical factors lead to nonlinear curves.

Absorption spectra. When the transmittance or the absorbance of a sample is measured and plotted as a function of wavelength, an absorption spectrum (Fig. 3) is obtained. This spectrum indicates that the sample transmits the least light at 410 nm and transmits the most around 700 nm. Because of instrumental sensitivity limits, absorption spectra are obtainable only on samples which are relatively transparent to the radiation that is being used.

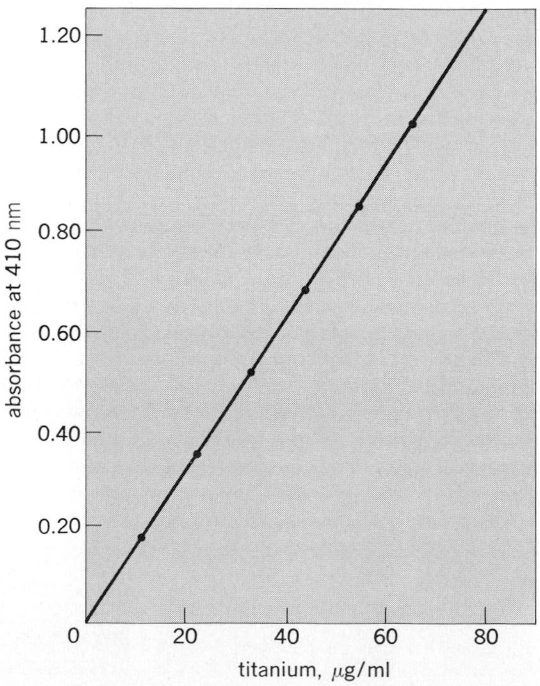

Fig. 2. Calibration curve for determination of titanium by its color formed with hydrogen peroxide.

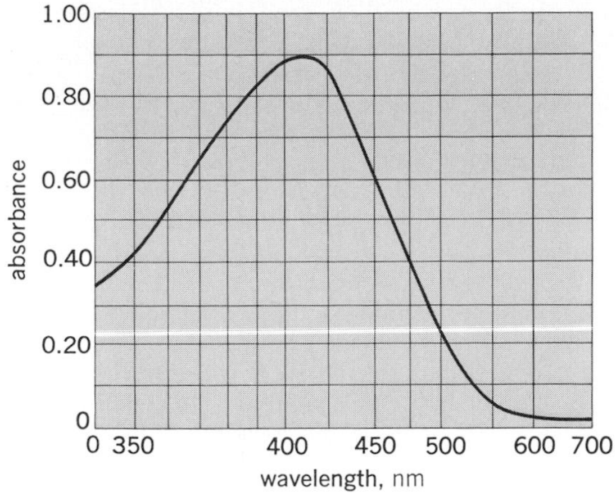

Fig. 3. Absorption spectrum of the peroxytitanate complex in region 340–700 nm.

Reflectance spectra. In opaque samples, such as solids or highly absorbing solutions, the radiation reflected from the surface of the sample may be measured and compared with the radiation reflected from a nonabsorbing or white sample. If this reflectance intensity is plotted as a function of wavelength, it gives a reflectance spectrum. Reflectance spectra are used most often in matching colors of dyed fabrics or painted surfaces; they are used occasionally in qualitative analysis but seldom in quantitative analysis.

Chromogen. A molecule which absorbs radiation in a particular spectral region, usually in the visible or ultraviolet, is called a chromogen.

Chromophore. Groups of atoms within a molecule which are responsible for the absorption of light in the visible or ultraviolet regions are called chromophores. These chromophores are usually resonating structures which absorb at approximately the same wavelength, or frequency, regardless of the molecule to which they are attached. Examples of such groups are the phenyl group, C_6H_5, which absorbs at 270 nm and the azo group N=N, which absorbs at 370 nm.

Auxochrome. Substituent groups which affect the wavelength of the spectral regions of strong absorption of chromophores are called auxochromes. Auxochromes cause two types of wavelength shifts. A shift to longer wavelength, or lower frequency, is called a bathochromic shift. Conversely, a shift to shorter wavelength is called a hypsochromic shift.

Infrared spectrophotometry. The interaction with matter of electromagnetic radiation of wavelength between 1 and 300 μm induces either rotational or vibrational energy level transitions, or both, within the molecules involved. This region from 1 to 300 μm is usually referred to as the infrared. The frequencies of infrared radiation absorbed by a molecule are determined by its rotational energy levels and by the force constants of the bonds in the molecule. Since these energy levels and force constants are usually unique for each molecule, so also the infrared spectrum of each molecule is usually unique. The qualitative analytical use of the infrared region is based on this fact. Because of their individuality, infrared spectra of organic compounds are considered equivalent to, or superior to, the preparation of chemical derivatives for the identification of species in organic chemistry. For this reason the infrared portion of the spectrum is often called the fingerprint region (Figs. 4 and 5).

For radiation sources, infrared instruments usually use a hot filament called a Nernst glower, or a hot carborundum rod called a Globar. Various inorganic prisms are used in the monochromators for different regions, for example, rock salt (sodium chloride) from 2 to 15 μm, potassium bromide from 15 to 27 μm, or cesium bromide from 12 to 40 μm. Gratings are also used in monochromators, either

Fig. 4. A high-resolution infrared spectrophotometer. (*Beckman, Scientific Instruments Division*)

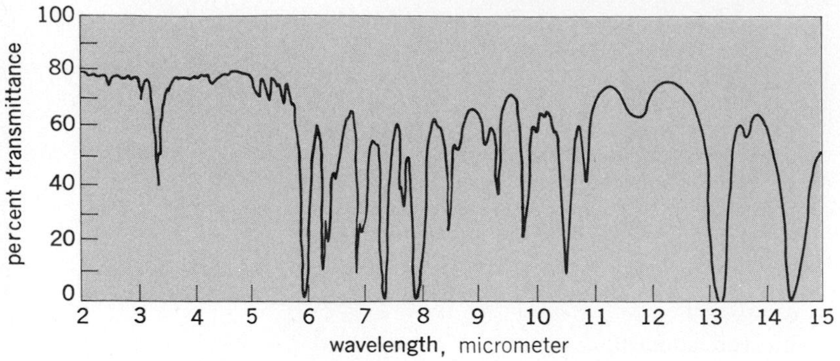

Fig. 5. Infrared spectrum of acetophenone.

alone or in conjunction with prisms. A wide variety of detectors are used; examples are thermocouples and thermistors. These detectors must be very sensitive, as the amount of energy they must detect is quite small. Infrared cells, or sample containers, are prepared from materials transparent in the region of interest, and usually are made of rock salt, potassium bromide, or some other inorganic salt.

Although almost every compound has a unique infrared spectrum, various groupings within a molecule have well-defined regions of absorption. For example, hydroxyl groups in alcohols absorb strongly at 2.8, 7.3, and about 8.5 μm; ester carbonyls absorb from 5.7 to 5.8 μm; and free amino groups absorb at 3.0 and from 6.1 to 6.4 μm. A typical infrared spectrum, that of acetophenone ($C_6H_5COCH_3$), is shown in Fig. 5. The band at 3.3 μm is due to the methyl group (CH_3), that at 5.9 μm to the conjugated carbonyl group (CO), and those at 13.2 and 14.4 μm are due to monosubstituted benzene (C_6H_5).

Quantitative analysis is based on the Bouguer-Lambert-Beer law as applied to a specific absorption band, usually unique to the compound being determined, as shown in Figs. 2 and 3. Occasionally, adequate quantitative results can be obtained using data for a similar compound containing the same functional group as in the material measured. The accuracy and precision of most infrared analyses is $\pm 3-5\%$ of the amount of material present.

Infrared spectrophotometry can be applied to gaseous, liquid, or solid samples. For gaseous samples, cells (sample containers) from 1 cm up to 50 m long are used in order to get enough molecules in the light path to measure. In the long cells, the length is obtained by using mirrors to make the light traverse the cell numerous times before it is measured. Liquid samples are handled in cells whose thicknesses vary from 0.1 mm to 1 cm. The most common solvents for liquids are carbon tetrachloride and carbon disulfide, since these solvents have few absorption bands in the infrared region. Water is a very poor solvent because it absorbs strongly in this spectral region. Solid samples are analyzed by (1) preparing a thin film which may be either a free film or a film cast on a salt plate; (2) preparing a paste or mull by grinding up the solid with a viscous material such as mineral oil (Nujol), which has few infrared bands; or (3) pressing a disk of an intimate mixture of the solid with potassium bromide, KBr. This latter, the so-called KBr disk method, appears to be the best for quantitative infrared work with solids.

The infrared region is used primarily for analyses of organic compounds because they are readily soluble in a desirable solvent and because they have unique and complex spectra. However, work has been done on the infrared spectra of inorganic compounds in the forms of KBr disks or Nujol mulls.

Attenuated total reflectance. A technique that is becoming popular and useful, especially for infrared measurements, is attenuated total reflectance (ATR), also called frustrated internal reflectance (FIR) and internal reflectance spectroscopy (IRS). With this technique, the infrared spectrum of a surface can be obtained without any chemical treatment of the sample; its principle is based on the phenomenon of energy reflection at the interface of two media which have different refractive indexes and are in optical contact with each other.

The sample to be analyzed is positioned on a reflectance prism and placed in the light path of the infrared source. The beam of infrared energy is absorbed at those wavelengths where the sample normally absorbs; at other wavelengths the infrared energy is reflected. The resulting spectrum is similar to that of a transmission spectrum of a material. This technique is used extensively for analyzing coatings, pastes, paints, fibers, and fabrics.

Near-infrared spectrophotometry. This designates work carried out between 0.78 and 3 μm. The instruments in use in this region have quartz prisms in their monochromators and lead sulfide photoconductor cells as detectors. The absorption bands in this region are mainly overtones (harmonics) of bands in the infrared region. These bands are quite sharp and are of great value in quantitative analysis for various functional groups, particularly those containing hydrogen atoms; examples are terminal methylene groups (CH_2), hydroxyl groups, and amines. The cells used in this region are usually made in quartz, and hence are more durable than infrared cells. The near-infrared spectra give some of the same information as those in the infrared region, but occasionally more specifically, more inexpensively, or more rapidly. The accuracy and precision of $\pm 1-3\%$ of the amount present is also somewhat better than in infrared spectrophotometry.

Visible spectrophotometry. The visible region of the spectrum covers the narrow range from about 380 to 780 nm. The spectrophotometers for this region use tungsten lamps as light sources, glass or quartz prisms or gratings in the monochromators, and photomultiplier cells as detectors. Within this narrow portion of the electromagnetic spectrum, a majority of the spectrophotometric analyses are made. The absorption of light in this region is caused by the excitation of the outer electrons of the molecule by the impinging light beam. Figure 3 shows a typical visible absorption spectrum. The substance, peroxytitanate ion, absorbs light in the region below 500 nm; that is, it absorbs violet, blue, and green light and transmits red, orange, and yellow. For analytical work, the wavelength of maximum absorption is usually used, in this case 410 nm. The calibration curve in Fig. 2 was made by plotting the absorption at 410 nm versus the concentration of the absorbing material in the solution. Unknown samples are then analyzed by measuring the absorbance of the solution after appropriate reagents have been added. The amount of material in question, in this case titanium, is obtained from the calibration curve.

Although few materials, in particular inorganic ions, have visible colors, there are spectrophotometric methods utilizing visible colors for most of them. The method used above for titanium is typical. Hydrogen peroxide, when added to a colorless titanium solution, forms the highly colored peroxytitanium complex. Similar reactions are those of thiocyanate ion with ferric iron and of ammonia with copper. In recent years, organic reagents have been prepared which form intense colors with different metal ions. Many of these reagents are so specific that they form colors with but one or two inorganic ions. Examples of these are *o*-phenanthroline which reacts with ferrous ion and 2,2′-biquinoline with copper.

Visible spectrophotometry is used extensively because there are methods available for determining a wide variety of materials, especially inorganic cations, with great sensitivity and selectivity. The equipment is relatively inexpensive ($300–1000).

Visible spectrophotometry is usually carried out with liquid samples and is usually used for quantitative, rather than qualitative purposes. Visible reflectance spectra of solid samples are run where the matching of colors is important or in cases where the samples are opaque. Gaseous samples seldom are run in the visible region since they seldom absorb visible light intensely.

Ultraviolet spectrophotometry. The spectral region from 200 to 400 nm, called the near ultraviolet, is commonly used in chemical analysis. The absorption of ultraviolet radiation by a molecule is usually the result of exciting the outer, or valence, electrons of the molecule in question. The more easily the electrons are excited, the longer the wavelength of the absorption peak.

Ultraviolet spectrophotometers usually have a hydrogen lamp as a radiation source; a quartz prism, or a grating in the monochromator; and a photomultiplier tube as a detector. Quartz or silica cells of 1-mm to 10-cm length are commonly used for the samples.

Simple inorganic ions and their complexes as well as organic molecules can be detected and determined in this region. Useful solvents are water, saturated hydrocarbons, aliphatic alcohols, and ethers. Organic compounds which absorb ultraviolet radiation have at least one unsaturated linkage, such as C=C, C=O, N=N, or S=O, which acts as a chromophore. The wavelength of the absorption peak increases with the degree of unsaturation within the chromophore.

Inorganic groups which absorb in the ultraviolet region owe their activity to numerous valence electrons, as in the complex $FeCl_4^-$, or to electrons in a single atom, possibly hydrated, as in the rare-earth elements.

Quantitative work in the ultraviolet region is common, and most substances obey the Bouguer-Lambert-Beer law over a wide range. In the field of organic analysis, the ultraviolet region is most applicable to aromatic compounds. The spectra of some compounds, such as phenols, may be greatly enhanced by using basic solutions of the samples.

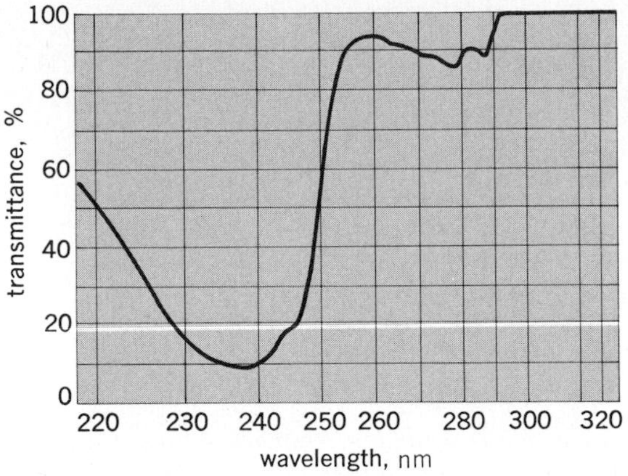

Fig. 6. Ultraviolet absorption spectrum of acetophenone in isooctane.

Since most compounds which absorb in the region have intense bands, it is possible to analyze either dilute solutions or extremely small samples. For example, the spectrum in Fig. 6 is that of 0.01% acetophenone in isooctane in a 1-cm cell. By using longer cells, it is possible to detect 0.00001% acetophenone in isooctane. Many substances absorb much more intensely than acetophenone.

Similarly, ultraviolet spectrophotometry is especially useful for the determination of inorganic ions as simple complexes, such as $FeCl_4^-$, $PtCl_6^{--}$, or I_3^-. Accuracy and precision are usually about ±1–2% of the amount of material being determined.

As indicated above, samples are usually liquids although gases may also be analyzed. Because of relatively greater scattering of short-wavelength radiation, transmission measurements on turbid samples are difficult to interpret, and opaque samples are seldom run by reflectance.

Filter photometry. In filter photometry, the monochromator of the spectrophotometer is replaced by a filter. This filter passes a band of light of a much wider range of wavelengths than those passed by even the poorest monochromator. For the example in Fig. 3, a filter which transmits blue light would be used in order to obtain maximum sensitivity. The filter chosen is usually of the color complementary to that of the solution, that is, the filter is chosen so as to transmit best the light which the sample absorbs most. In the visible region, colored glass or gelatin films containing dyes have been most widely used. Interference filters, based upon selective transmission of radiation through very thin metallic films between two glass plates, usually yield more nearly monochromatic bands and higher transmittance. They are used when their somewhat greater costs can be justified. Absorbing liquids and gases have been used as filters but are generally more cumbersome.

Filter photometers are generally much less expensive than spectrophotometers. Because they do not use monochromatic light, the calibration curves obtained often do not obey the Bouguer-Lambert-Beer law. However, by careful use of calibration curves, filter photometers can give sufficiently accurate and precise results for a wide variety of applications.

Although filter photometers are most often used in the visible region, some filters for the infrared and ultraviolet regions are available. *See* ANALYTICAL CHEMISTRY; MOLECULAR STRUCTURE AND SPECTRA.

[JAMES N. LITTLE]

Bibliography: American Institute of Physics, *The Spectrophotometer*, 1975; Z. Marczenko, *Spectrophotometric Determination of Elements*, 1976; C. N. Rao, *Chemical Applications of Infrared Spectroscopy*, 1964; C. N. Rao, *Ultra-Violet and Visible Spectroscopy: Chemical Applications*, 3d ed., 1975; R. G. White, *Handbook of Ultraviolet Methods*, 1965.

Spectroscopy

An analytic technique concerned with the measurement of the interaction (usually the absorption or the emission) of radiant energy with matter, with the instruments necessary to make such measurements, and with the interpretation of the interaction both at the fundamental level and for practical analysis. Mass spectroscopy is not concerned with the interaction of light with matter, but was so named because the appearance of the data resembles that of the spectroscopic data as defined above.

A display of such data is called a spectrum, that is, a plot of the intensity of emitted or transmitted radiant energy (or some function of the intensity) versus the energy of that light. Spectra due to the emission of radiant energy are produced as energy is emitted from matter, after some form of excitation, then collimated by passage through a slit, then separated into components of different energy by transmission through a prism (refraction) or by reflection from a ruled grating or a crystalline solid (diffraction), and finally detected. Spectra due to the absorption of radiant energy are produced when radiant energy from a stable source, collimated and separated into its components in a monochromator, passes through the sample whose absorption spectrum is to be measured, and is detected. Instruments which produce spectra are variously called spectroscopes, spectrometers, spectrographs, and spectrophotometers.

Interpretation of spectra provides fundamental information on atomic and molecular energy levels, the distribution of species within those levels, the nature of processes involving change from one level to another, molecular geometries, chemical bonding, and interaction of molecules in solution. At the practical level, comparisons of spectra provide a basis for the determination of qualitative chemical composition and chemical structure, and for quantitative chemical analysis.

Early history. In the historical development of spectroscopy, following the fundamental studies of crude spectra of sunlight by Isaac Newton in 1672, certain contributions and achievements are especially noteworthy. The significance of using a narrow slit instead of a pinhole or round aperture so as to produce spectral lines, each one an image of the slit and representing a different color or wave length, was demonstrated independently by W. H. Wollaston in 1802 and by Joseph Fraunhofer in 1814. Fraunhofer made many subsequent contributions to optics and spectroscopy, including first observation of stellar spectra, discovery and construction of transmission diffraction gratings, first accurate measurements of wavelengths of the dark lines in the solar spectrum, and invention of the achromatic telescope. The origin of the dark Fraunhofer lines in the solar spectrum was accounted for by G. R. Kirchhoff in 1859 on the basis of absorption by the elements in the cooler Sun's atmosphere of the continuous spectrum emitted by the hotter interior of the Sun. Further studies by Kirchhoff with R. Bunsen demonstrated the great utility of spectroscopy in chemical analysis. By systematically comparing the Sun's spectrum with flame or spark spectra of salts and metals, they made the first chemical analysis of the Sun's atmosphere. In 1861, while investigating alkali metal spectra, they discovered two new alkali metals, cesium and rubidium. These achievements by Kirchhoff and Bunsen provided tremendous stimulus to spectroscopic researches. The adoption in 1910 of the first international standards of wavelength gave further impetus. These and later standards made possible the measurement of wavelengths of any electromagnetic radiation with unprecedented accuracy. Since World War II, remarkable developments in spectroscopy have occurred in instrumentation, achieved largely through advances in electronics and in manufacturing technology. Direct reading, automatic recording, improved sensitivity with good stability, simplicity of operation, and extended capabilities are features provided by many commercial instruments. Predictably, these developments, by facilitating widespread use of spectroscopic techniques, have had an enormous influence in promoting further developments in both applied and theoretical spectroscopy.

The ultimate standard of wavelength is that of a particular line in the discharge spectrum of krypton-86, defined as 605.61252 nm in dry air at 15°C and 760 mmHg pressure (101.325 kPa). Secondary standards, based upon direct comparison with the krypton-86 line by interferometric methods, cover the range from 244.7 to 703.2 nm.

Spectroscopic units. The change in energy of an ion, atom, or molecule associated with absorption and emission of radiant energy may be measured by the frequency of the radiant energy according to Max Planck, who described an equality $E = h\nu$, where E is energy, ν the frequency of the radiant energy, and h is Planck's constant. The frequency is related to the wavelength λ by the relation $\nu\lambda = c/n$, where c is the velocity of radiant energy in a vacuum, and n is the refractive index of the medium through which it passes; n is a measure of the retardation of radiant energy passing through matter. The units most commonly employed to describe these characteristics of light are:

Wavelength: 1 micrometer (μm) = 10^{-6} m
1 nanometer (nm) = 10^{-9} m (= 10 angstroms)
Frequency: 1 hertz (Hz) = 1 s^{-1}

For convenience the wave number $\bar{\nu}$ (read nu bar), the reciprocal of the wavelength, may be used; for this, the common units are cm^{-1}, read as reciprocal centimeters or occasionally kaysers. This number equals the number of oscillations per centimeter.

Spectral regions. Visible light constitutes only a small part of the spectrum of radiant energy, or electromagnetic spectrum; the human eye responds to electromagnetic waves with wavelengths from about 380 to 780 nm, though there is individual variation in these limits. The eye cannot measure color and intensity quantitatively; even for the visible portion of the electromagnetic spectrum, therefore, instruments are used for measurements and recording.

There are broad regions of the electromagnetic spectrum with associated wavelengths greater or less than the visible region. The ranges are indicated in the table, together with spectrometer components for their analysis and the nature of the process in matter which gives rises to the spectrum of each type.

Origin of spectra. Atoms, ions, and molecules emit or absorb characteristically; only certain energies of these species are possible; the energy of the photon (quantum of radiant energy) emitted or absorbed corresponds to the difference between two permitted values of the energy of the species, or energy levels. (If the flux of photons incident upon the species is great enough, simultaneous absorption of two or more photons may occur.) Thus the energy levels may be studied by observing the differences between them. The absorption of radiant energy is accompanied by the promotion of the species from a lower to a higher energy level; the emission of radiant energy is accompanied by falling from a higher to a lower state; and if both processes occur together, the condition is called resonance.

Transitions. Transitions between energy levels associated with electrons or electronic levels, range from the near infrared, the visible, and ultraviolet for outermost, or highest-energy, electrons, that is, those which can be involved in bonding, to the x-ray region for the electrons nearest the nucleus. At low pressures such transitions in gaseous atoms produce sharply defined lines because the energy levels are sharply defined. Transitions between energy levels of the nucleus are observed in the gamma-ray region. In the absence of an applied electric or magnetic field, these electronic

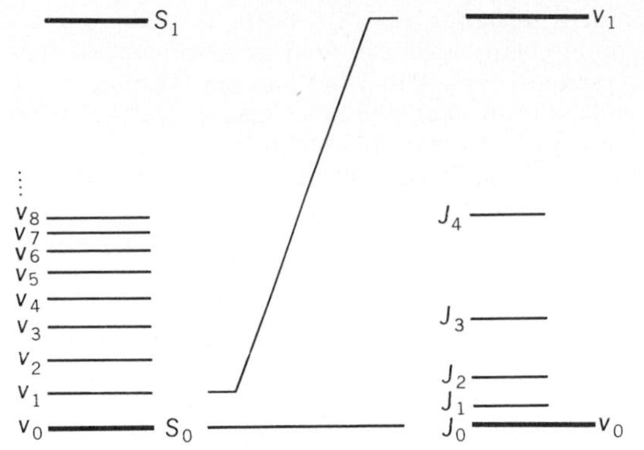

Fig. 1. Spacing of singlet electronic states *S*, vibrational states *V*, and rotational states *J*.

and nuclear transitions are the only ones which atoms can undergo. *See* ATOMIC SPECTROSCOPY.

Electronic transitions in molecules are also observed. In addition, transitions can occur between levels associated with the vibrations and rotations of molecules. Spacings between electronic levels are greater than between vibrational levels, and those between vibrational levels are greater than between rotational levels; each vibrational level has a set of rotational levels associated with it, and each electronic level a set of vibrational levels (Fig. 1). Transitions between vibrational levels of the same electronic level correspond to photons in the infrared region; transitions between rotational levels, to photons in the far infrared and microwave region. Rotational spectra of gaseous molecules consist of sharp lines; vibrational spectra consist of bands, each of which arises from a given transition between vibrational levels altered in energy by increments due to changes in rotation occurring when the vibrational level changes. Likewise, molecular electronic spectra

Principal spectral regions and fields of spectroscopy

Spectral region	Approximate wavelength range	Typical source	Typical detector	Energy transitions studied in matter
Gamma	1–100 pm	Radioactive nuclei	Geiger counter; scintillation counter	Nuclear transitions and disintegrations
X-rays	6 pm–100 nm	X-ray tube (electron bombardment of metals)	Geiger counter	Ionization by inner electron removal
Vacuum ultraviolet	10–200 nm	High-voltage discharge; high-vacuum spark	Photomultiplier	Ionization by outer electron removal
Ultraviolet	200–400 nm	Hydrogen-discharge lamp	Photomultiplier	Excitation of valence electrons
Visible	400–800 nm	Tungsten lamp	Phototubes	Excitation of valence electrons
Near-infrared	0.8–2.5 μm	Tungsten lamp	Photocells	Excitation of valence electrons; molecular vibrational overtones
Infrared	2.5–50 μm	Nernst glower; Globar lamp	Thermocouple; bolometer	Molecular vibrations: stretching, bending, and rocking
Far-infrared	50–1000 μm	Mercury lamp (high-pressure)	Thermocouple; bolometer	Molecular rotations
Microwave	0.1–30 cm	Klystrons; magnetrons	Silicon-tungsten crystal; bolometer	Molecular rotations; electron spin resonance
Radio-frequency	10^{-1}–10^3 m	Radio transmitter	Radio receiver	Molecular rotations; nuclear magnetic resonance

consist of bands due to transitions between electronic energy levels, altered by increments resulting from changes in vibration and rotation of the molecule on changing electronic state. *See* BAND SPECTRUM; INFRARED SPECTROSCOPY; LINE SPECTRUM; MICROWAVE SPECTROSCOPY; MOLECULAR STRUCTURE AND SPECTRA.

External fields. The application of a magnetic or electric field to the sample often separates normally indistinguishable, or degenerate, states in energy from each other. Thus the orientation of the spin of an unpaired electron in an atom, ion, or molecule with respect to an applied magnetic field may have different values, and in the magnetic field the atom or molecule may have different energies. For typical field strengths, the difference in energy between these newly separated energy levels occurs in the microwave region. Similarly, for an atomic nucleus in an ion or molecule, differences in the orientation of the spin of the nucleus with respect to a magnetic field give rise to different energy levels of the nucleus in that field, so that energy differences will be found in the microwave region. The former phenomenon produces electron spin resonance or electron paramagnetic resonance spectra, and the latter produces nuclear magnetic resonance spectra (Fig. 2).

Nuclear magnetic resonance has been advanced by irradiation techniques which permit the identification of solid and other difficult samples, and for the production of three-dimensional plots which permit analysis across the interior of objects without destruction. The implications of this development for medicine in particular are enormous. *See* NUCLEAR MAGNETIC RESONANCE (NMR).

Electronic spectra are also altered by external fields; removal of degeneracy by an externally applied electric field is termed the Stark effect, and removal of degeneracy by an externally applied magnetic field is termed the Zeeman effect.

Spontaneous emission. The spontaneous emission of light from a sample by the decay of an electron from a higher level to the lowest available level is called either fluorescence or phosphorescence. The fundamental difference between these two terms is associated with the orientation of the spin of this electron. If the spin of the electron is such that it falls to the lowest state without a change of spin (the atom or molecule having the same spin before and after the transition), fluorescence occurs, and the process is characterized as allowed and is relatively fast. If the spin is such that the electron can fall to the lower state only with a change dictated by the Pauli exclusion principle (atoms in the initial and final states having different spins), the process is phosphorescence, characterized as a forbidden process, which is relatively slow (Fig. 3). In practice, other factors also govern the time the electron spends in the higher level, and the time intervals associated with each process overlap; that for fluorescence is 10^{-4} to 10^{-8} s, and that for phosphorescence is 10^{-4} to 10 s. The spontaneous emission process thus cannot be completely distinguished merely on the basis of the time interval associated with the emission. Fluorescence is measured at a right angle to the incident beam exciting the sample, in order to avoid background problems.

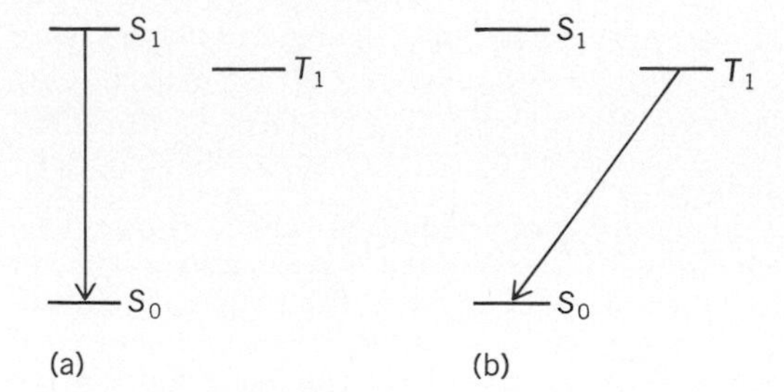

Fig. 3. Spontaneous emission of light by the decay of an electron from a higher to the lowest available level. (*a*) Fluorescence. (*b*) Phosphorescence.

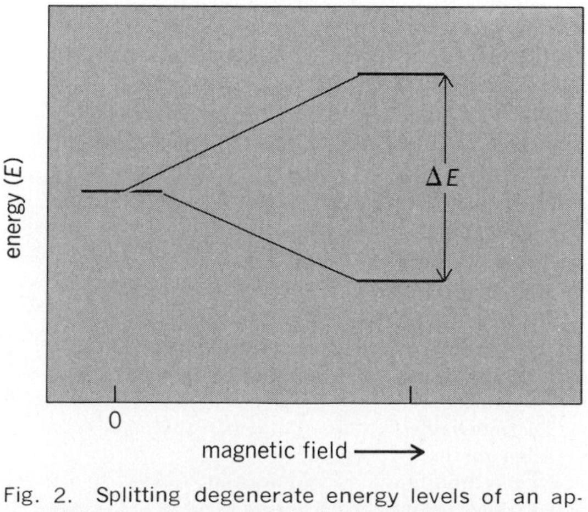

Fig. 2. Splitting degenerate energy levels of an appropriate atom in a magnetic field. Transitions between the separated levels require interaction with a photon of energy ΔE.

Light rotation. Compounds whose molecules have a structure which cannot be superimposed on its reflection in a mirror (asymmetric molecules) rotate the plane of polarized light. If a device can be included in an instrument to plane-polarize the light from the source, that is, to produce light oscillating in only one plane, the amount of rotation of that plane by the sample can be studied as a function of the wavelength of light used in the experiment. A plot of this rotation versus wavelength is called an optical rotatory dispersion curve; such curves are useful in studying fundamentals of light rotation and in establishing the absolute structure of asymmetric molecules empirically. Asymmetric samples whose absorption of the components of plane-polarized light differs are said to exhibit circular dichroism. *See* COTTON EFFECT; OPTICAL ACTIVITY.

Instrumentation. Spectrometers require a source of radiation, a dispersion device, and a detector. For emission spectra, the source may be taken as the sample to be measured itself, although another source of radiation may be needed to excite the sample to an excited state. Absorption spectra require a source to generate appropriate radiation across the range of energies needed for the experiment. These radiant sources produce continuous spectra and those which produce discontinuous spectra.

Continuous spectra. Of the sources listed in the table, most of those from the vacuum ultraviolet region to the far-infrared region produce continuous spectra. Within this type, sources for the near ultraviolet, visible, and infrared regions consist of hot, incandescent solids; the maximum intensity of their emission occurs at a wavelength dependent on their temperature and emissivity. Electrical discharges in gases at higher pressures provide sources of continuous radiation for the vacuum ultraviolet and ultraviolet regions; hydrogen or deuterium gas discharges are especially used for the latter. X-ray continuous emission can be produced by collision of electrons or other particles accelerated through tens of kilovolts with a target. *See* BEAM FOIL SPECTROSCOPY.

Discontinuous spectra. The sources of high-energy radiation which yield discontinuous spectra emit because of discrete processes in individual atoms, ions, or molecules. Such sources may include flames and furnaces to vaporize and atomize samples, and electrical discharges in gases at lower pressures. Other atomizing sources may be produced by high-frequency radio discharges and microwave discharges through a flowing gas. For sources of radiation connected with nuclear processes, solid samples may be used containing the element to be studied, either a radioactive isotope or one which can be made radioactive by particle bombardment. *See* FLAME PHOTOMETRY.

Among these so-called line sources are hollow-cathode lamps, in which a few layers of a cylindrical cathode containing the same element as is to be analyzed are vaporized by collisions with ions generated from rare gases and accelerated electrons. The gaseous atoms, also excited to higher energy states by this process, emit characteristic line spectra useful for analysis of that element. Lasers have become an especially useful source, and have opened new areas of spectroscopy, not only because of the intensity of the radiation supplied to the sample, which permits absorption of several photons in a single process, but also because of the coherent properties of the radiation, which permit the study of events occurring on a picosecond time scale.

The sources for radio-frequency studies are tunable radio-frequency oscillators, designed with special regard for stability. The microwave sources may be spark discharges, magnetrons, or most commonly klystrons. An electron beam traveling through a closed cavity of defined geometry sets up electromagnetic oscillations, of which a certain value will be reinforced as a function of the dimension of the containing unit; the value can be varied somewhat by adjusting the size of the cavity.

Dispersive elements. There are two kinds of dispersive elements, prisms and diffraction elements. The earlier and more commonly available prism has been supplanted in many cases by diffraction elements.

After collimation of the source radiation through a slit, diffraction is achieved in x-ray spectroscopy by the use of a pure crystal. The distances between atomic nuclei in the crystal are on the same order as the wavelength of the radiant energy; under these conditions an efficient dispersion of the radiation over an arc of many degrees is possible. For a distance d between repeating planes of atoms, the wavelength λ will be diffracted through an angle θ related by the Bragg law $n\lambda = 2d \sin \theta$, where n is an integer called the diffraction order. Materials used include gypsum ($CaSO_4 \cdot 2H_2O$), ammonium dihydrogen phosphate ($NH_4H_2PO_4$), and alkali halides (lighter elements). Each of these gives a wide angular dispersion in a portion of the 0.03 to 1.4 nm range of the spectrum commonly used for x-ray studies; the appropriate one must be chosen according to the experiment to be performed.

For spectroscopic techniques using visible light and the regions adjacent to it, ultraviolet and infrared, ruled diffraction gratings are often employed. Because their production has become economical, they have to a large extent replaced prisms, which may require the use of several materials to get adequate angular dispersion and wavelength resolution over the entire range of interest, for example, 2 to 15 μm for an infrared spectrum. The dispersion of a prism increases near the limit of its transparency, and there is always a compromise in efficiency between transmission and dispersion. The resolution, R, defined as the quotient of the wavelength of a line and the wavelength difference between it and another line just separated from it, is given by $T/(dn/d\lambda)$, where T is the thickness of the prism base and $dn/d\lambda$ is the variation of the refractive index with respect to wavelength. Efficient prisms are made of quartz for the ultraviolet, glass for the visible, and various salts for the infrared.

Very-short- or very-long-wavelength spectra are usually produced without a dispersive element; gamma-ray detection uses a wavelength-sensitive detector (a pulse height discriminator, for example), and microwave and radio-frequency detection may use a variable tuning radio receiver; in the latter case the source is tuned to emit a highly resolved frequency as well.

Detectors. Detectors commonly used for the various spectral regions are listed in the table. The devices for the infrared and microwave regions are basically heat sensors. Except for the eye and photographic methods used sometimes in the regions for gamma ray to visible, each detector converts the signal into an electrical response which is amplified before transmittal to a recording device. Direct imaging of the response by vidicon detection has also been used. Otherwise an oscilloscope or oscillograph is used to record very rapidly obtained signals.

Instruments. Spectroscopic methods involve a number of instruments designed for specialized applications.

Spectroscope. An optical instrument consisting of a slit, collimator lens, prism or grating, and a telescope or objective lens which produces a spectrum for visual observation is called a spectroscope. The first complete spectroscope was assembled by Bunsen in 1859 (Fig. 4).

Spectrograph. If a spectroscope is provided

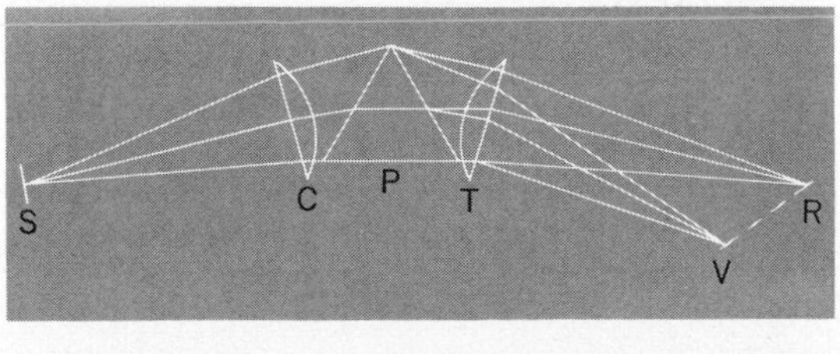

Fig. 4. Prism spectroscope or spectrograph. S, slit; C, collimator lens; P, 60° prism; T, telescope; RV, red to violet spectrum. For producing ultraviolet spectra, the glass items in C, P, and T must be replaced by equal amounts of left- and right-handed crystal quartz.

with a photographic camera or other device for recording the spectrum, the instrument is called a spectrograph. For recording ultraviolet and visible spectra, two types of quartz spectrographs are in common use. One type utilizes a Cornu quartz prism constructed of two 30° prisms, left- and right-handed, as well as left- and right-handed lenses, so that the rotation occurring in one-half the optical path is exactly compensated by the reverse rotation in the other. The other type of quartz spectrograph employs a Littrow 30° quartz prism with a rear reflecting surface that reverses the path of the light through prism and lens, thus compensating for rotation of polarization in one direction by equal rotation in the opposite direction. Thus, in either type the effect of optical activity and birefringence in crystal quartz which produces double images of the slit is eliminated.

Grating spectrographs cover a much broader range of wavelengths (vacuum ultraviolet to far infrared) than do prism instruments. Various-type mountings are employed. The most common mounting for a plane-reflection grating is in a Littrow mount, entirely analogous to that of the Littrow quartz spectrograph. Mountings for concave reflection gratings require that the grating, slit, and camera plate all lie on the Rowland circle (imaginary circle of radius equal to one-half the radius of curvature of the grating) in order to achieve proper focus of spectral lines (slit images) on the plate. Paschen, Rowland, Eagle, and Wadsworth mountings are common in grating spectrographs.

Spectrometers. A spectroscope that is provided with a calibrated scale either for measurement of wavelength or for measurement of refractive indices of transparent prism materials is called a spectrometer. Also, the term frequently is used to refer to spectrographs which incorporate photoelectric photometers instead of photographic means to measure radiant intensities.

Spectrophotometer. A spectrophotometer consists basically of a radiant-energy source, monochromator, sample holder, and detector. It is used for measurement of radiant flux as a function of wavelength and for measurement of absorption spectra. A typical arrangement for a double-beam recording instrument is shown schematically in Fig. 5. The wavelength is scanned by the monochromator at a constant rate (geared to chart speed of recorder) while the detector responds to, and discriminates between, the two alternating beams, one from the sample and the other from the attenuator. The servomechanism drives the attenuator, as well as the recorder pen, so that the intensities of the alternating signals are kept identical. The displacement of the pen on the chart can be calibrated in terms of relative intensity, percent transmittance, or absorbance. Generally a linear attenuator is employed, and the pen displacement is directly proportional to the attenuator displacement and thus the transmittance.

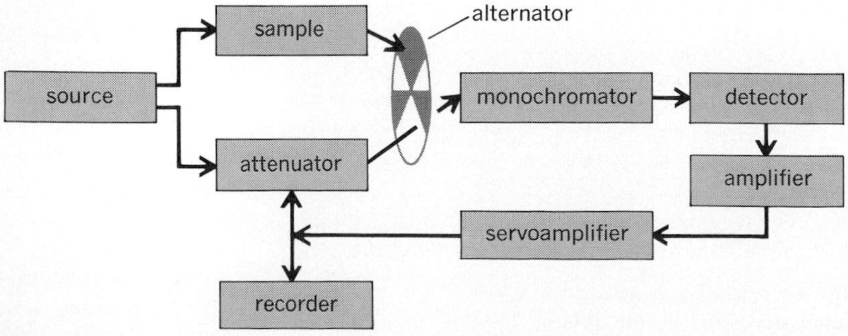

Fig. 5. Schematic diagram of a double beam (in time) spectrophotometer.

Interferometer. This optical device divides a beam of radiant energy into two or more parts which travel different paths and then recombine to form interference fringes. Since an optical path is the product of the geometric path and the refractive index, an interferometer measures differences of geometric path when two beams travel in the same medium, or the difference of refractive index when the geometric paths are equal. Interferometers are employed for high-resolution measurements and for precise determination of relative wavelengths: they are capable of distinguishing between two spectral lines that differ by less than 10^{-6} of a wave.

Quantitative relationships. For practical analysis, the absorption of radiant energy is related to concentration of sample by the relationship $-\log P/P_0 = abc$, where P and P_0 are respectively the power of the light attenuated after passing through the sample and its unattenuated power before reaching the sample, c is the concentration of the sample, b is the path length of the light through the sample, and a is a proportionality constant called the absorptivity of the sample. If c is in moles per liter, the absorptivity is called the molar absorptivity and given the symbol ϵ. This relationship is known as the Beer-Lambert-Bouguer law, or simply Beer's law. For x-ray spectroscopy, the right-hand variables are grouped differently. *See* SPECTROCHEMICAL ANALYSIS; SPECTROPHOTOMETRIC ANALYSIS.

The practical relation between emission of light and concentration is given by the equation $F = k\phi I_0 ebC$, where F is the fluorescence intensity, k is an instrumental parameter, ϕ is the fluorescence efficiency (quanta emitted per quantum absorbed), and the other units are as defined previously. The choice between absorption and fluorescence techniques for determining concentration is sometimes made on the basis of the accuracy of the linear method (fluorescence) versus the versatility of the logarithmic method (absorption). *See* FLUOROMETRIC ANALYSIS.

Other methods and applications. Since the early methods of spectroscopy there has been a proliferation of techniques, often incorporating sophisticated technology.

Astronomical spectroscopy. The radiant energy emitted by celestial objects can be studied by combined spectroscopic and telescopic techniques to obtain information about their chemical composition, temperature, pressure, density, magnetic fields, electric forces, and radial velocity. Radiation of wavelengths much shorter than 300 nm is absorbed by the Earth's atmosphere, and can only be studied by spectrographs transported into space by rockets.

Atomic absorption and fluorescence spectroscopy. This branch of electronic spectroscopy uses line spectra from atomized samples to give quantitative analysis for selected elements at levels down to parts per million, on the average. However, detection limits vary greatly from element to element. *See* TRACE ANALYSIS.

Attenuated total reflectance spectroscopy. Spectra of substances in thin films or on surfaces can be obtained by the technique of attenuated total reflectance or by a closely related technique called frustrated multiple internal reflection. In either method the sample is brought into contact with a total-reflecting trapezoidal prism and the radiant-energy beam is directed in such a manner that it penetrates only a few micrometers of the sample one or more times.

The technique is employed primarily in infrared spectroscopy for qualitative analysis of coatings and of opaque liquids.

Electron paramagnetic spectroscopy. This microwave technique, based on the splitting of electronic energy levels in a magnetic field as shown above, is used to establish structures of species containing unpaired electrons in solution or in a crystalline solid; some qualitative and quantitative analysis is also performed. *See* ELECTRON PARAMAGNETIC RESONANCE (EPR) SPECTROSCOPY.

Electron spectroscopy. This area includes a number of subdivisions, all of which are associated with electronic energy levels. The outermost or valence levels are studied in photoelectron spectroscopy, which uses photons of the far-ultraviolet region to remove electrons from molecules and to infer the energy levels of the remaining ion from the kinetic energy of the expelled electron. This technique is used mostly for fundamental studies of bonding. Electron impact spectroscopy uses low-energy electrons (0–100 eV) to yield information similar to that observed by visible and ultraviolet spectroscopy, but governed by different selection rules. X-ray photoelectron spectroscopy (electron spectroscopy for chemical analysis) uses x-ray photons to remove inner-shell electrons in a similar way. In Auger spectroscopy, an x-ray photon removes an inner electron, and when another electron falls from a higher level to take its place and a third is ejected to conserve the energy gained by the second, the kinetic energy of the third is measured. Both of these techniques are used to study surfaces, since x-rays penetrate many objects for only a few layers. Ion neutralization spectroscopy uses protons or other charged particles instead of photons. *See* AUGER EFFECT; ELECTRON SPECTROSCOPY.

Fourier transform spectroscopy. This family of techniques consists of a fundamentally different method of irradiation of the sample, in which all pertinent wavelengths simultaneously irradiate it for a short period of time and the absorption spectrum is obtained by mathematical manipulation of the cyclical power pattern so obtained. It has been applied particularly to infrared spectrometry and nuclear magnetic resonance spectrometry, and allows the acquisition of spectra from smaller samples in less time, with high resolution and wavelength accuracy. The infrared technique is carried out by using an interferometer of conventional instrumentation.

Gamma-ray spectroscopy. Of special note in this category are the techniques of activation analysis, which are performed on a sample without destroying it by subjecting it to a beam of particles, often neutrons, which react with stable nuclei present to form new unstable nuclei. The emission of gamma rays for each element at different wavelengths with different half-lives as these radioactive nuclei decay is characteristic. *See* ACTIVATION ANALYSIS.

Mössbauer spectroscopy results from resonant absorption of photons of extremely high resolution emitted by radioactive nuclei in a crystalline lattice. The line width is extremely narrow because there is no energy loss for the nucleus recoiling in the lattice, and differences in the chemical environment of the emitting and absorbing nuclei of the same isotope may be detected by the slight differences in wavelength of the energy levels in the two samples. The difference is made up by use of the Doppler effect as the sample to be studied is moved at different velocities relative to the source, and the spectrum consists of a plot of gamma-ray intensity as a function of the relative velocity of the carriage of the sample. It yields information on the chemical environment of the sample nuclei and has found application in inorganic and physical chemistry. Solid-state physics, and metallurgy. *See* MÖSSBAUER EFFECT.

Laser spectroscopy. Laser radiation is nearly monochromatic, of high intensity, and coherent. Some or all of these properties may be used to acquire new information about molecular structure. In multiphoton absorption the absorption of more than one photon whose energy is an exact fraction of the distance between two energy levels allows study of transitions ordinarily forbidden when only a single photon is involved. Raman spectroscopy has also become more common because of the use of lasers as sources, especially with the development of resonance Raman spectroscopy.

Information on processes which occur on a picosecond time scale can be obtained by making use of the coherent properties of laser radiation, as in coherent anti-Stokes Raman spectroscopy. Lasers may also be used to evaporate even refractory samples, permitting analysis of the atoms so obtained, and since the radiation may be focused, such a technique may be applied to one small area or a sample surface at a time, permitting profiling of the analysis across the sample surface. *See* LASER SPECTROSCOPY.

Mass spectrometry. The source of a mass spectrometer produces ions, often from a gas, but also in some instruments from a liquid, a solid, or a material adsorbed on a surface. Ionization of a gas is most commonly performed by a current of electrons accelerated through about 70–80 V, but many other methods are used. The dispersive unit may consist of an electric field, or sequential electric and magnetic fields, or a number of parallel rods upon which a combination of dc and ac voltages are applied; these provide either temporal or spatial dispersion of ions according to their mass-to-charge ratio. *See* MASS SPECTROMETRY; SECONDARY ION MASS SPECTROMETRY (SIMS).

Multiplex or frequency-modulated spectroscopy. The basis of this family of techniques is to encode or modulate each optical wavelength exiting the spectrometer output with an audio frequency that contains the optical wavelength information. Use of a wavelength analyzer then allows recovery of the original optical spectrum. The primary advantage of such a system is that it enables the detec-

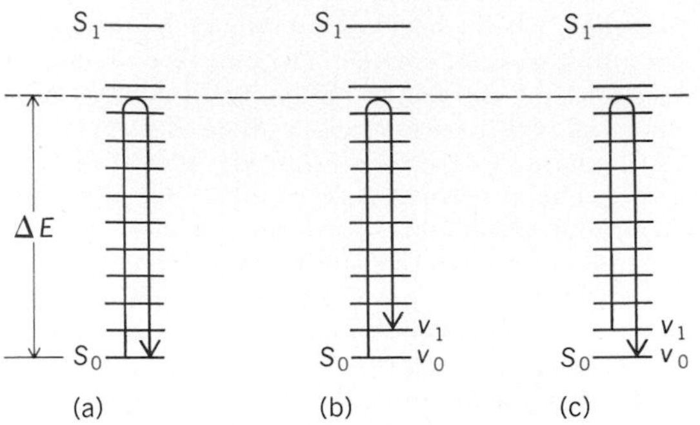

Fig. 6. Interactions of a molecule with a photon of energy ΔE which is not absorbed. (*a*) Rayleigh scattering. (*b*) Raman scattering, producing a Stokes line. (*c*) Raman scattering, producing an anti-Stokes line.

tor to sense all optical wavelengths simultaneously, resulting in greatly decreased scan times and the possibility for signal averaging. Frequency modulation can be achieved, for example, by means of a special encoding mask as in Hadamard transform spectroscopy or by use of a Michelson interferometer as in Fourier transform spectroscopy.

Raman spectroscopy. Consider the passage of a beam of light, of a wavelength not corresponding to an absorption, through a sample. A small fraction of the light is scattered by the molecules, and so exists the sample at a different angle. This is called Rayleigh scattering if the wavelength of the scattered light is the same as the original wavelength; but if the wavelength is different, it is called Raman scattering (Fig. 6). The fraction of photons thus scattered by the Raman effect is even smaller than by Rayleigh scattering. Differences in wavelength correspond to the wavelengths of certain vibrational and rotational processes. If the scattered light has a longer wavelength, the new line in the spectrum is called a Stokes line; if shorter, an anti-Stokes line. The information obtained is of use in structural chemistry, because the rules allowing and forbidding transitions are different from those in absorption spectroscopy. Use of lasers for sources has revived this technique; a related process, resonance Raman spectroscopy, makes use of the fact that Raman probabilities are greatly increased when the exciting radiation has an energy which approaches the energy of an allowed electronic absorption. *See* RAMAN EFFECT.

X-ray spectroscopy. The excitation of inner electrons in atoms is manifested as x-ray absorption; emission of a photon as an electron falls from a higher level into the vacancy thus created is x-ray fluorescence. The techniques are used for chemical analysis. Extended x-ray absorption fine structure, obtained under conditions of high resolution, shows changes in the energy of inner electrons due to the presence of neighboring atoms, and can be used in structural chemistry for the determination of the nature and location of neighbor atoms nearest to that being studied. *See* X-RAY FLUORESCENCE ANALYSIS. [MAURICE M. BURSEY]

Bibliography: O. Howarth, *Theory of Spectroscopy: An Elementary Introduction*, 1973; R. L. Pecsok et al., *Modern Methods of Chemical Analysis*, 2d ed., 1976; H. H. Willard, L. L. Merritt, Jr., and J. A. Dean, *Instrumental Methods of Analysis*, 5th ed., 1974.

Spin label

Molecule which contains an unpaired electron spin which can be detected with electron spin resonance (ESR) spectroscopy. Molecules are labeled when an atom or group of atoms which exhibit some unique physical property is chemically bonded to a molecule of interest. Groups containing unpaired electrons include organic free radicals and a variety of types of transition-metal complexes (such as vanadium, copper, iron, and manganese). Molecules with unpaired electron spins are readily detected with electron spin resonance spectroscopy. Through analysis of ESR spectra, rates of molecular motion and relative orientations of spin-labeled molecules whose motion is restrained by surrounding molecules can be determined. Measurement of rates of molecular motion and molecular orientation has proved to be very important in the study of a variety of types of biologically important problems.

Electron spin resonance. This is a spectroscopic technique which detects transitions between electron spin energy levels. When a molecule possessing an unpaired electron spin is placed in an external magnetic field, the external field interacts with the magnetic moment of the unpaired electron spin, producing two distinct energy states. In one state the unpaired electron is oriented with its magnetic moment along the direction of the external field, while the other state has the moment of the unpaired electron oriented against the external field direction. The separation of these two energy states is about 3 cm in an external field of 3000 gauss (0.3 tesla), and an absorption of radiation can be observed when a sample is irradiated with microwave radiation of about 3 cm wavelength. This absorption is observed as a single line in an electron spin resonance spectrum. The wavelength at which absorption of energy is observed depends on the magnitude of the external field and on a molecular property called the *g* value.

Crystals. If a crystalline sample is used, different *g* values may be observed along the various axes of the crystal. When directional dependence is observed the *g* value is anisotropic, with principal values along the x, y, and z axes of the crystal denoted by g_{xx}, g_{yy}, and g_{zz}. If $g_{xx} = g_{yy}$, the sample has an axial symmetric *g* tensor, and the *g* values are denoted as $g_{\parallel}$ and $g_{\perp}$. The unpaired electron spin is often delocalized over two or more atoms in the spin label. If the nuclei have nuclear spin other than zero, the magnetic moment of the nuclei can interact with the magnetic moment of the electron spin and split the electron spin energy levels. When this is the case, the ESR spectrum shows a series of lines that depend on the number and type of nuclear spins in the molecule. The interaction between the electron and nuclear spins is called a hyperfine interaction, and the separation of the energy levels and ESR lines is called the hyperfine splitting. The magnitude of the interaction between nuclear and electron spins may depend on the orientation of the molecule with respect to the external field. The components of the electron-nuclei anisotropic splitting are denoted by A_{xx}, A_{yy}, and A_{zz}, or $A_{\perp}$ and $A_{\parallel}$ when the molecule has axial symmetry.

Solutions. When a spin-labeled molecule is free

to rotate in solution, the anisotropic g value and anisotropic hyperfine interaction are averaged by rapid reorientation of the label, and one observes the isotropic (angle-independent) g value and hyperfine splitting in the ESR spectrum. As rotation of the molecule is slowed, the anisotropic interactions start to contribute, and the ESR spectrum is changed. A continuous change in the ESR spectrum is observed as the sample goes from the case of free rotation to the case of a rigid sample where motion is very slow. Analysis of ESR spectra permits determination of motional correlation times from about 10^{-12} to 10^{-3} s. In many cases, determination of the direction of motion of the labeled group (that is, rotation in all directions or preferred rotation around a single axis) is also possible.

Biological studies. Analysis of the rate and type of motion of a spin label is important for a wide variety of biological problems. The type of label used in these studies is generally a nitroxide free radical. The unpaired electron spin in a nitroxide radical is delocalized over an N-O group, and an electron-nuclei hyperfine splitting is observed from the spin = 1 nitrogen nucleus. The oxygen atom has a spin of zero. In solution an isotropic nitrogen hyperfine splitting of about 15 gauss (1.5 millitesla) is observed, while in a crystalline solid anisotropic splitting is observed with $A_{zz} = A_{\|} = 32$ gauss (3.2 millitesla) and $A_{xx} = A_{yy} = A_{\perp} = 6$ gauss (0.6 mT). The z direction is defined by the $2p$ orbital of the nitrogen atom.

Enzymes. Nitroxide spin labels have been bonded to a number of enzyme substrates with connecting chains of varying length between the label and the substrate. The labeled substrate is allowed to form a complex with the enzyme, and the ESR spectrum is monitored. The receptor site on the enzyme is often in a type of well within the protein, within which the substrate must fit. If the nitroxide label is connected to the substrate by a short chain, it lies within the well and rotational motion is inhibited. When a sufficiently long connecting chain is used, the nitroxide lies outside the protein and rotational motion is relatively free. Changes in motion of the label can be monitored with ESR, and the transition from hindered to free rotation as a function of the length of the connecting chain can be determined. The distance from the substrate to the label can be calculated from molecular models for labeled molecules with different connecting chain lengths. If the distance at which free rotation starts is determined, the depth of the receptor site well of the enzyme can be calculated. Spin-labeling studies of this type provide a powerful technique for the study of the geometry and dimensions of receptors in enzymes.

Membranes. Spin labels have been used extensively to study the structure of membranes. Biological membranes consist mainly of a phospholipid bilayer with embedded proteins. Other molecules (for example, cholesterol) may also be found in the lipid bilayer, and a variety of types of molecules may be bound to the surface. The phospholipids have polar head groups (hydrophilic) which are on the outside of the membrane, and nonpolar aliphatic carbon chains which are on the inside of the bilayer. The lipids form ordered structures which are readily oriented in an external magnetic field. Spin labels have been bonded to the carbon atoms along the aliphatic chains of the lipids to study internal motion in the membrane bilayer. The ESR spectra of spin labels bound to lipids which have been oriented in an external field show that the lipid can rotate around its long axis, but that head-to-tail rotation is very slow. The relative motional correlation times of labels bound at different positions along the chain can be determined from ESR spectra. These studies show that motion is freer on the inside of the lipid away from the polar head group, indicating greater fluidity on the inside. Membrane structures can be changed by a number of kinds of perturbations, and spin labels have been used to monitor the effect of the perturbation. The structure of the phospholipid bilayer changes with temperature, and spin-labeling studies have been conducted to determine phase transition temperatures. Molecules such as cholesterol or ethanol change the fluidity of the membrane, and spin-labeling studies have been conducted to monitor the effect of chemical perturbations on the fluidity of membranes.

When nitroxide labels are relatively close to one another, the odd electron spins in the nitroxide groups may be interchanged between molecules. This spin exchange reaction dramatically changes the appearance of the ESR spectrum of the labels, since the energy states are averaged by this interaction. This interaction may be used to study the rate of diffusion of labeled molecules from an area of high concentration (strong exchange) to an area of low concentration (weak exchange). This type of investigation has been conducted to determine the rate of lateral diffusion of lipids in membranes. A high concentration of labeled lipid is initially introduced into a membrane, and an ESR spectrum from strongly exchange-coupled spin labels is observed. As the labeled molecules diffuse apart, the ESR spectrum changes to that characteristic of labels with no exchange coupling. By monitoring the change in the ESR spectrum of the label as a function of time, it is possible to determine how fast lipids diffuse through membranes. Spin-labeling studies of these types have provided important information about the structure, organization, and rates of motion in membranes.

Other applications. Spin labels have also been used to study the structure and organization of synthetic polymers and to study phase transitions. Investigations using deuterium substitution for hydrogen with nuclear magnetic resonance detection have been conducted to study problems similar to those investigated with nitroxide labels. Spin-labeling studies allow investigation of a variety of types of problems which cannot readily be studied with other techniques. *See* ELECTRON PARAMAGNETIC RESONANCE (EPR) SPECTROSCOPY; FREE RADICAL; TRIPLET STATE.

[ROBERT KREILIK]

Bibliography: L. J. Berliner (ed.), *Spin Labeling: Theory and Applications*, vols. 1 and 2, 1976, 1979; J. A. Freed, *Annu. Rev. Phys. Chem.*, 23:265, 1972; B. J. Gaffney and C. J. McNamee, *Meth. Enzymol.*, 32B:161, 1974.

Squalene

A triterpene which appears to play a role in the biosynthesis of sterols and polycyclic terpenes. The structure of squalene is shown in notation (I); the dotted lines indicate six isoprene units from which the $C_{30}H_{50}$ hydrocarbon is theoretically condensed. Actually it has been synthesized by the condensation of two molecules of farnesyl bromide (II), and more recently by applying the Wittig reac-

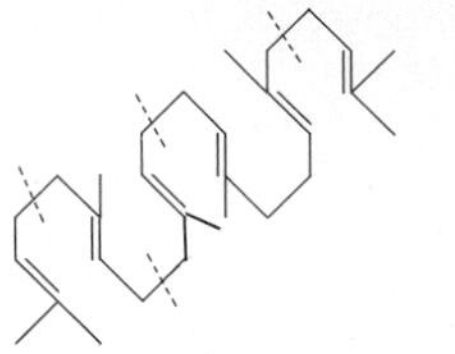

(I)

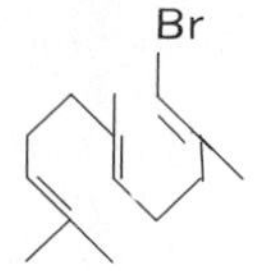

(II)

tion to a pure *trans*-geranylacetone (III). The squalene so formed was probably a mixture of

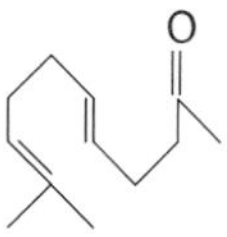

(III)

$$+ (C_6H_5)_3P{=}CH{-}CH_2{-}CH_2{-}CH{=}P(C_6H_5)_3 \rightarrow \text{Squalene}$$

three stereoisomers. Later work showed that purification could be effected by forming a thiourea clathrate. There is an indication that synthetic squalene can be enzymatically converted to lanosterol.

The six olefinic bonds in squalene have been demonstrated to be of the trans configuration. It possesses a faint, pleasant odor and a high boiling point. Decomposition occurs when distilled at ordinary pressures. When exposed to air, squalene absorbs oxygen and resinifies to a viscous mass. It is insoluble in water but soluble in fats and fat solvents. Squalene is found in appreciable quantities in the liver of sharks. Smaller amounts (0.1 – 0.8%) occur in olive oil, wheat germ oil, rice bran oil, and yeast, as well as in human sebum and ear wax. *See* ACETIC ACID; ISOPRENE; TERPENE; TRITERPENE.

[WILLIAM MOSHER]

Stearate

A salt (soap) or ester of stearic acid having the general formula

$$C_{17}H_{35}C({=}O){-}O{-}M \quad \text{or} \quad {-}R$$

and formed by replacing the carboxylic hydrogen by a metal (M) to give a salt, or by an organic radical (R) to give an ester. Stearates occur in nature chiefly as the glyceryl ester, found in substantial amounts in animal and vegetable fats. The esters of long-chain alcohols are known as waxes. Other esters of monohydric and polyhydric alcohols are used in cosmetics, lacquers, nonionic surface-active agents, and plasticizers. Alkali-metal salts are water-soluble and, with the similar oleates and palmitates, are the major components of toilet and laundry soaps. Other metal salts are used in paints, waterproofing, pharmaceuticals, cosmetics, fungicides, and lubricant and grease additives.

[ELBERT H. HADLEY]

Bibliography: M. Windholz (ed.), *Merck Index*, 9th ed., 1976.

Stereochemistry

The study of the three-dimensional arrangement of atoms or groups within molecules and the properties which follow from such arrangement. Molecules that have identical molecular structures (that is, the same kind, number, and sequential arrangement of atoms) but differ in the relative spatial arrangement of component parts are stereoisomers. Inorganic and organic compounds exhibit stereoisomerism. Examples are structures (I)–(VIII).

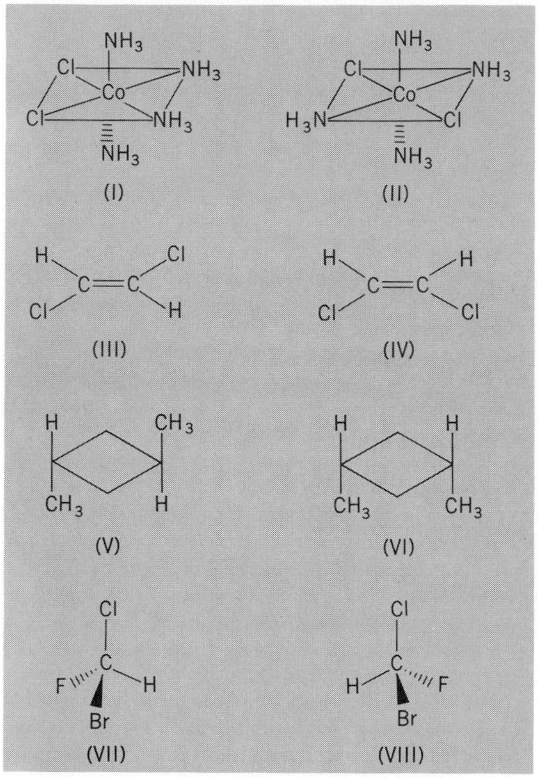

Significance. Stereochemistry has played a significant role in the historical development of theories of molecular structure. The optical activity of substances such as sugar and turpentine was discovered by J. B. Biot in 1813, and L. Pasteur was the first to separate or resolve enantiomers from one another (1848). While Pasteur recognized that enantiomers must differ in symmetry, it was not until 1874 that J. H. van't Hoff and J. A. LeBel were able to relate symmetry, structure, and properties into the hypothesis of the tetrahedral carbon which formed the cornerstone of modern stereochemistry. The foundation of inorganic stereochemistry rests on the coordination theory of A. Werner (1893).

Since the latter part of the 19th century, stereochemical studies have been prominent in the evolution of both inorganic and organic chemistry. These studies have concerned themselves with the determination of configuration, the interconversion of diastereomers and the assessment of their energy differences, and conformational analysis.

Methods for the analysis and separation of stereoisomers have been devised, and stereochemical principles have been applied to the elucidation of reaction mechanisms, to the development of asymmetric syntheses, and to attempts to understand biological processes at the molecular level. *See* ASYMMETRIC SYNTHESIS; COORDINATION CHEMISTRY; RACEMIZATION.

Symmetry. The nature of the stereochemistry of a molecule is determined by its symmetry. The symmetry elements to be considered are: planes of symmetry, axes of symmetry, centers of symmetry, and reflection or mirror symmetry. Two types of stereoisomers are known. Those such as (VII) and (VIII), which are devoid of reflection symmetry—which cannot be superposed on their image in a mirror—are called enantiomers. All other stereoisomers, such as the pairs (I)–(II), (III)–(IV), and (V)–(VI), are called diastereomers. The configuration of a stereoisomer designates the relative position of the atoms associated with a specific structure. The structures of stereoisomers (I) and (II) differ only in configuration. The same is true for (III) and (IV), (V) and (VI), and (VII) and (VIII).

Enantiomers are related to one another as a left hand is to a right hand. Such structures are said to be chiral (Greek *cheir* = hand) or dissymmetric. Unlike diastereomers, which differ in most physical and chemical properties, enantiomers have identical properties other than the sign (+ or −) of their optical activity. This identical behavior exists toward all agents and processes which are themselves achiral. Chiral agents act differently toward enantiomers, however. Biological specificity toward enantiomers is dramatic; for example, the enantiomers of the amino acid leucine have different tastes: one is bitter, while the other is sweet.

Configuration. In order to understand chemical processes involving stereoisomers, it is necessary to know the configurations of reactants and products. The configuration of stereoisomers (I) and (II) are designated cis (chlorines on the same side) and trans (chlorines on opposite sides), respectively; (III) and (V) are trans; (IV) and (VI) are cis. Configurations of diastereomers can often be determined from their physical properties. For example, the isomer of $C_2H_2Cl_2$ with zero dipole moment is (III), while (IV) has a finite dipole moment.

The absolute configuration of chiral molecules such as (VII) and (VIII) is the actual order of atoms about the chiral center (tetrahedral carbon atom) which defines the specific enantiomer. While the three-dimensional picture of a molecule defines its stereochemistry just as the picture of a hand defines its handedness (right or left), it is not convenient or useful to rely on full three-dimensional representations in most cases, and a shorthand notation to designate configuration is widely accepted. Groups and atoms which surround the chiral center are assigned priorities related by unambiguous rules to atomic number, for example, Br > Cl > F > H, and to substitution pattern. The molecule is viewed from the tetrahedral face opposite the carbon substituent with the lowest priority, that is,

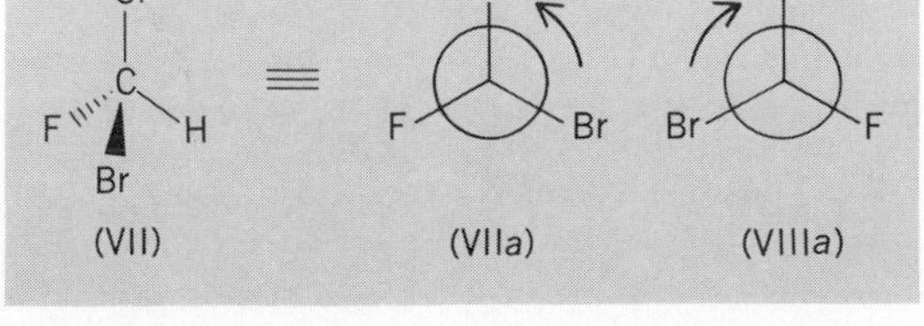

(VII) (VIIa) (VIIIa)

from the face opposite hydrogen in the case of (VII). The counterclockwise descent (VII*a*) from high- to low-priority substituent (Br followed by Cl followed by F) is designated S (Latin *sinister* = left). Enantiomer (VIII) exhibits a similar but clockwise order (VIII*a*) designated R (Latin *rectus* = right). When the order of groups immediately bound to the chiral center is equivocal, the priority is determined by atoms or groups attached to the atoms whose priorities are equivocal.

The same priorities may be used in the designation of the configurations of diastereomers such as (III) and (IV), often called geometric isomers. When the atoms or groups of highest priority lie on opposite sides of the double bond, as in (III), the stereoisomer is designated E (German *entgegen* = opposite). Isomer (IV) in which the reference atoms are on the same side of the double bond are termed Z (German *zusammen* = together). This system is less ambiguous than the cis-trans system which is still applicable in simple cases.

Enantiomers are characterized by the sign of their optical activity at a given wavelength, temperature, and specified solvent. The assignment of configuration to specific enantiomers is carried out by chemical transformations of known stereochemistry, by spectroscopic means such as circular dichroism and optical rotatory dispersion, and by diffraction of single crystals employing anomalous x-ray scattering. From such studies it is known that the lactic acid isomer that is experimentally found to be levorotatory, that is, whose optical activity is (−)- or counterclockwise-rotating, has structure (IX) corresponding to the R configuration (IX*a*). Configuration has no simple connec-

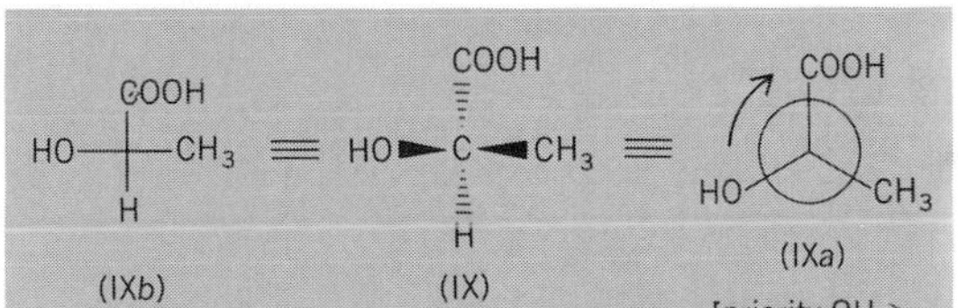

(IXb) (IX) (IXa)

[priority OH > COOH > CH_3]

tion to the sign of the optical rotation. *See* COTTON EFFECT; OPTICAL ACTIVITY; ROTATORY DISPERSION; X-RAY DIFFRACTION.

Two-dimensional (planar) projections of three-dimensional structures are very commonly used to represent configurations. Projection formulas compress the tetrahedral geometry into the plane of the paper. In (IX) the horizontal atoms or groups are forward of the plane (wedges) of the chiral center, and the vertical groups are behind (solid or broken lines). Tipping formula (IX) forward places the COOH group in the plane of the paper and the H atom behind, corresponding to structure (IX*a*). The planar projection of (IX) may be written as (IX*b*). To avoid confusion in their use, these conventions apply to planar projections: they must not be lifted from the plane of the paper; they must not be rotated in this plane through any but integral multiples of 180°; an odd number of exchanges of any pair of substituents is equivalent to transformation into its mirror image, and an even number of exchanges leaves the configuration unchanged.

The relative configurations of stereoisomers applicable to diastereomers are illustrated by the

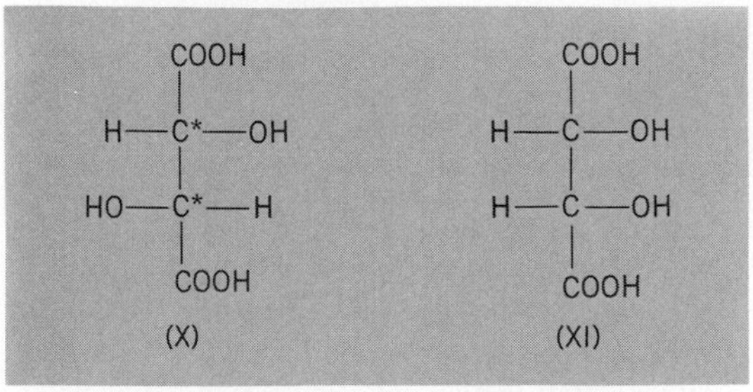

(X) (XI)

tartaric acids (X) and (XI). Structure (X) represents (+)-tartaric acid (both chiral centers * have R configurations), and (XI) represents *meso*-tartaric acid, the diastereomer which is optically inactive (the two chiral centers, one R and one S, are mirror images of one another). Older configurational symbolism (D and L) involving intermolecular comparisons to reference substances such as glyceraldehyde and serine persists, but it is giving way to the R-S designations.

Stereoregular polymers with defined configurations at chiral centers or at carbon-carbon double bonds differ substantially in properties from their diastereomers. Polymeric carbohydrates such as starch and cellulose and hydrocarbons such as rubber and gutta-percha exemplify natural diastereomer pairs. Synthetic polystyrene diastereomers (XII) [isotactic] and (XIII) [syndiotactic] differ only in the configuration at chiral centers. The high-melting, fiber-forming isomer is isotactic polystyrene (XII).

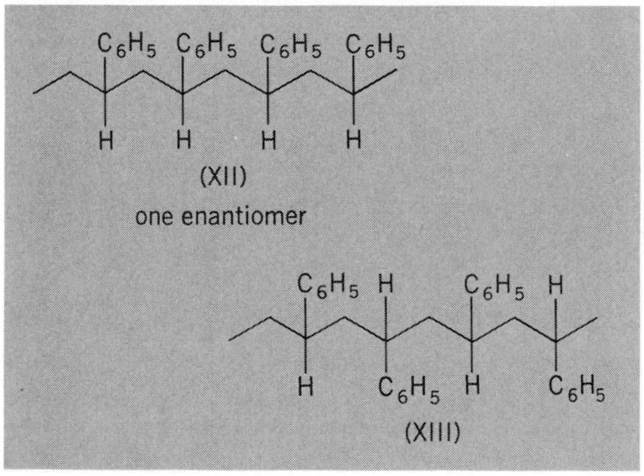

(XII) one enantiomer

(XIII)

Configurational studies are powerful probes in the elucidation of reaction mechanisms. Replacement of substituent atoms or groups in a reaction is often attended by a change in configuration at the chiral center (Walden inversion), or it may result in the loss of optical activity (racemization). In some cases retention of configuration prevails. Such results provide evidence for the geometry of transition states or for the intervention of intermediates (ions or radicals).

Resolution. Most chiral substances occur in nature as only one enantiomer. Few natural products are represented in nature by both enantiomers, and few also exist as mixtures with equal proportions of enantiomers called racemates.

On the other hand, synthesis of the chiral substances in the laboratory or in industry in the absence of chiral agents or conditions results in racemates which exhibit no measurable optical activity, since the activity of one enantiomer cancels that of the other.

The separation of one or both enantiomers from a racemate, called resolution, requires the intervention of an optically active reagent, catalyst, or other chiral influence. The most common resolution procedure involves the reversible conversion of an enantiomer mixture into a pair of diastereomers, with as widely different physical properties as practicable, by reaction with a single pure enantiomer. For example, a (±)-amine reacts with a (+)-acid to give two diastereomeric salts: (+)-R-$NH_3^{\oplus}$(+)-R′$COO^{\ominus}$ and (−)-R-$NH_3^{\oplus}$(+)-R′$COO^{\ominus}$.

These diastereomeric salts may be separated by fractional crystallization and may then be converted to the individual enantiomeric amines by cleavage with a strong base. The (+)-acid resolving agent may be recovered and reused. Covalent diastereomers may also be separated by crystallization or increasingly by gas, liquid, or thin-layer chromatography.

Kinetic resolution takes advantage of differences in rates of reaction of enantiomers with chiral reagents. Enzymes are chiral catalysts which can preferentially catalyze reactions of one enantiomer. Enzymatic resolutions play a major role in the syntheses of substances of biochemical importance. Chromatographic resolution on chiral stationary phases, which bind or dissolve one enantiomer more strongly than the other, constitutes another useful type of resolution.

In resolution by preferential crystallization or entrainment, one enantiomer preferentially crystallizes upon seeding of supersaturated solutions with seed crystals of that enantiomer. Though relatively few racemates are amenable to this type of resolution, it is nevertheless of considerable importance on an industrial scale. *See* CHROMATOGRAPHY; CRYSTALLIZATION.

Conformational analysis. Molecules are not rigid collections of atoms. Torsional stereoisomerism arises as a consequence of the rotation (torsion) of atoms about bonds within molecules. This gives rise to a large number of stereoisomers called conformations which are interconvertible by rotation about carbon-to-carbon single bonds. The existence of two extreme conformations of ethane, (XIV) [staggered] and (XV) [eclipsed], was predicted by K. S. Pitzer in 1936 and later experimentally verified. An infinite number of conformations intermediate between the equilibrium (low-energy) conformation (XIV) and the higher-energy confor-

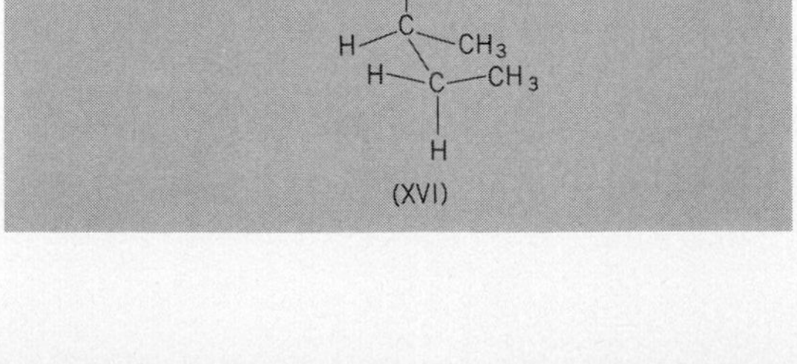

(XIV) (XV)

(XVI)

mation (XV) is possible. At room temperature most ethane molecules resemble (XIV). Though their presence is experimentally readily demonstrated, separation of conformations is normally not possible since the small energy difference between them [3 kcal/mole; 12.6 kJ/mole in the case of (XIV) and (XV) makes their interconversion facile]. The ease of interconversion explains why some enantiomeric structures, such as the *gauche* conformation (large groups, CH_3, 60° apart) of *n*-butane (XVI), are incapable of resolution.

The relationship between the physical and chemical properties of substances and their preferred conformations (conformational analysis) has been the subject of many studies since 1950. Cyclohexane (XVII) has an equilibrium conformation, the chair form (XVIII), which is the structural subunit from which the carbon atom lattice of diamond is constructed. Substituents on six-membered rings are designated as either axial or equatorial, as in (XVIII). Conformational analysis of synthetic hydrocarbon polymers such as polypropylene and of biopolymers such as DNA has led to an understanding of their properties. Much of the biological activity of enzymes, for example, is made possible by the specific conformations adopted by these macromolecules. *See* CONFORMATIONAL ANALYSIS; POLYMER.

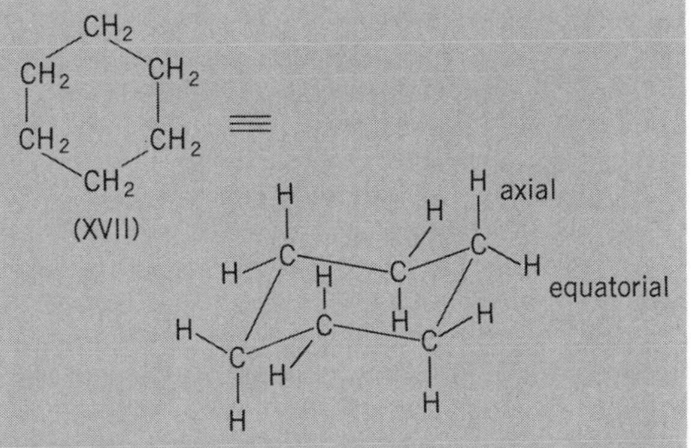

(XVII) (XVIII)

Stereocontrolled synthesis. An understanding of the stereochemical consequences of chemical reactions and the determination of configuration of complex natural products has permitted the total synthesis of many stereochemically complex substances since around 1940. For example, cholesterol and gibberellic acid each contain eight chiral carbon atoms, which makes possible, in principle, $2^8 = 256$ stereoisomers (128 racemates). Disregard for stereochemical consequences during synthesis would lead to extremely complex mixtures requiring repeated separation and to very low yields.

Living organisms unerringly synthesize (biosynthesis) just one of these isomers in fully stereoselective processes. Laboratory and commercial syntheses of substances which have desirable biological properties have increasingly been devised so as to mimic biosyntheses by taking advantage of stereocontrolled reactions. These are of two types: stereospecific reactions and stereoselective reactions. Stereospecific reactions are those in which different stereoisomers are transformed into stereochemically different products. For example, *Z*-2-butene (XIX) [cis] is epoxidized to the *cis*-epoxide (XX) only. The diastereomeric *E*-2-butene [trans] forms only the *trans*-epoxide.

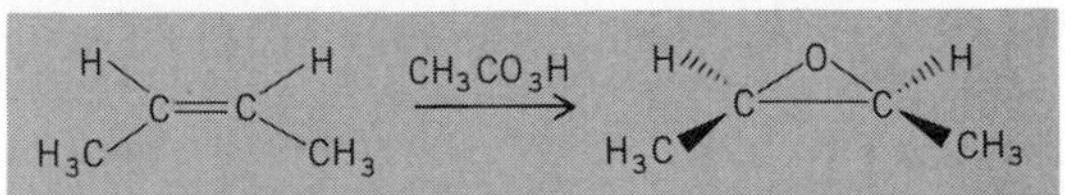

(XIX) (XX)

Stereoselective reactions are those in which a single reactant gives rise to a mixture of diastereomers in which one predominates. Those stereoselective reactions in which new chiral centers are formed in unequal amounts are called asymmetric syntheses. Transformations of optically active compounds into more complex substances which incorporate additional chiral centers are also asymmetric syntheses.

Syntheses of natural products which originate in achiral starting materials require at least one resolution step if nonracemic products are desired. Alternatively, a total synthesis may well originate in a single enantiomer of a compound having but one or two chiral centers. Additional chiral centers are introduced during the synthesis through stereospecific and especially through stereoselective steps. Separation of diastereomers at various stages may be required. *See* STEREOSPECIFIC CATALYST.

[SAMUEL H. WILEN]

Bibliography: R. Bentley, *Molecular Asymmetry in Biology*, 1969–1970; E. L. Eliel, *Elements of Stereochemistry*, 1969; E. L. Eliel, *Stereochemistry of Carbon Compounds*, 1962; K. Mislow, *Introduction to Stereochemistry*, 1965.

Stereospecific catalyst

Stereospecific polymerization catalysts lead to the formation of stereoregular (tactic) polymers, that is, polymers where the centers of steric isomerism in the main chain are arranged in a regular fashion with respect to their configurations. Three factors determine the tacticity of a polymeric chain during its formation: (1) the kind of monomer approach to the growing chain end, (2) the kind of attack of the growing chain end on the double bond (cis or trans opening), and (3) the configuration in the initiation step. In addition to a regular head-to-tail configuration and absence of branching, reactions have to be assumed. The kind of monomer approach is strongly affected by electrical and stereochemical forces, and therefore changes in monomer structure and environment greatly influence polymer tacticity. The intermediate radical or ion can be assumed to have a planar or near-planar structure, and in an uncomplexed form it should be able to rotate freely around its axis with cis and trans addition being equally possible. Upon addition of the next monomer this carbon changes into a tetrahedral structure, thereby creating two isomeric forms, isotactic or syndiotactic. Reaction (1) represents this addition process, with k_i and k_s as the rate constants for the isotactic and syndiotactic polymerization, respectively.

The probability of every placement is strongly dependent on energy considerations. The energy difference of various configurations is relatively small (about 2 kcal/mole). In fact, it is smaller than the energy required for rotation around a carbon-

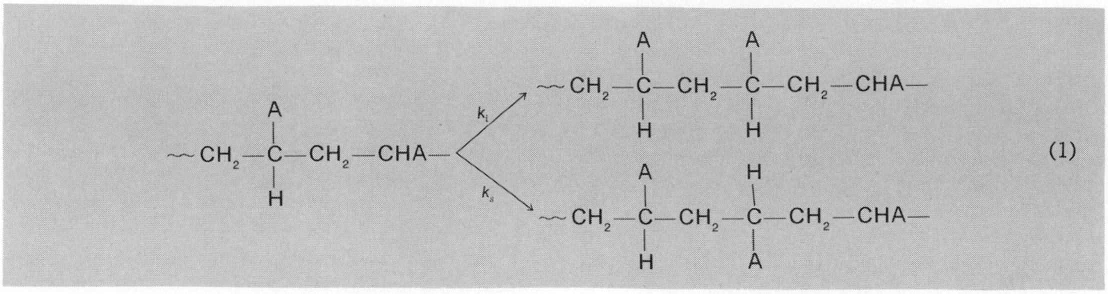

$$\sim CH_2-\underset{\underset{H}{|}}{\overset{\overset{A}{|}}{C}}-CH_2-CHA- \quad \underset{k_s}{\overset{k_i}{\diagdown}} \quad \begin{array}{l} \sim CH_2-\underset{\underset{H}{|}}{\overset{\overset{A}{|}}{C}}-CH_2-\underset{\underset{H}{|}}{\overset{\overset{A}{|}}{C}}-CH_2-CHA- \\[2em] \sim CH_2-\underset{\underset{H}{|}}{\overset{\overset{A}{|}}{C}}-CH_2-\underset{\underset{A}{|}}{\overset{\overset{H}{|}}{C}}-CH_2-CHA- \end{array} \tag{1}$$

carbon single bond in hydrocarbons (3–4 kcal/mole). Consequently, under normal conditions of polymerization both placements are about equally possible and atactic polymers results. Otherwise the smallness of the energy differences allows an effective control over the stereoregularity of the polymer by controlling the polymerization conditions.

Free-radical catalysts. In free-radical polymerization the propagating chain end is the nearest approach to a free propagating species, and therefore steric control will not be very pronounced. The active radical chain end will be essentially planar, and the incoming monomer molecule will have the same probability for attack irrespective of which side it approaches from. Consequently, propagation should lead to atactic polymer unless there is strong steric or electrostatic interaction between the substituents on the chain end and on the monomer. For these reasons, if the monomer is always oriented in the transition state with the substituents trans to each other, the resulting configuration in each growth step will be opposite to that of the previous unit in the chain, and each growth step can be described as a syndiotactic placement. Repetition of a sequence of such steps will give a syndiotactic polymer in which the stereochemical configuration alternates regularly. The situation will become more complex when the mode of addition is influenced by two or more earlier units in the chain, but in general free-radical polymerizations yield syndiotactic polymers. Decreasing reaction temperature usually increases the degree of tacticity because of the lower mobilities of the species.

The steric control in free-radical polymerization can be enhanced if the individual monomer molecules are oriented prior to polymerization, thus forcing the growing polymer chain to proceed in only one way. Monomer orientation can be accomplished by occlusion in urea or thiourea complexes or by adsorption on surfaces such as montmorillonite. The polymerization is then started by a brief exposure to a high-energy electron beam. Polymerizations in canal complexes yielded high-melting crystalline *trans*-1,4-addition polymers from 1,3-butadiene and its substituted derivatives, as well as highly crystalline polymers from vinyl chloride and acrylonitrile.

Ionic catalysts. In ionic polymerization mechanisms the ionic chain end is always associated with the counterion, and the approaching monomer molecule is inserted between both ions. From a mechanistic point of view, a participation of the counterion in the transition state can result in a multicenter reaction with increased stereoregulating power as compared with the one-center reac-

tion generally met in free-radical polymerizations. The influence of the counterion on the mode of monomer placement in the growth step will strongly depend upon the degree of its association with the growing ionic chain end. As proposed by S. Winstein and R. Robinson, ion pairs can be associated to different degrees for which four states, in notation (2), might be selected as being repre-

$$\underset{\text{Covalent}}{R-X} \; \rightleftharpoons \; \underset{\substack{\text{Intimate} \\ \text{ion pair}}}{R-X^+} \; \rightleftharpoons$$

$$\underset{\substack{\text{Solvent-separated} \\ \text{ion pair}}}{R^-//\text{solvent}//X^+} \; \rightleftharpoons \; \underset{\text{Free ions}}{R^-+X^+} \tag{2}$$

sentative. Reaction conditions (for example, choice of solvent or temperature) which favor the formation of solvent-separated ion pairs or of free ions create a transition state similar to the one met in free-radical polymerization. The growing chain end becomes an almost free propagating species, and syndiotactic placements will be favored for the same reasons as pointed out for free-radical polymerizations. For example, predominately syndiotactic poly(methyl methacrylate) is obtained in alkyllithium-initiated polymerization at −60°C if carried out in 1,2-dimethoxyethane wherein the lithium ions are strongly solvated. But when reaction conditions favor the alkyllithium to be in the form of covalent compounds or intimate ion pairs in nonsolvating solvents such as hydrocarbons, isotactic poly(methyl methacrylate) is obtained.

As an example for counterion participation the polymerization of methyl methacrylate to isotactic polymer might be discussed. The driving force for propagation is an initial coordination between the lithium ion, represented in Fig. 1, and M, and the double bond of the monomer molecule. This involves overlap of the olefinic π-electrons with vacant s- or p-orbitals in the lithium and is analogous to the reaction of an olefin with a carbonium ion. The coordination will be followed by an intramolecular rearrangement involving migration of the carbanion R to the most electrophilic carbon atom of the unsaturated molecule. Thereby a new carbon-carbon covalent bond is formed, and the polar metal-carbon linkage is regenerated with the alkyl group increased in length by one monomeric unit. Reaction (3) represents this rearrangement. The growing polymeric alkyllithium will have some enolic character, and the lithium atom will coordinate with the carbonyl oxygen of the penultimate monomer unit.

The intramolecular shielding of one side of the lithium atom in the cyclic intermediate directs the nucleophilic attack by the monomer molecular to

STEREOSPECIFIC
CATALYST

Fig. 1. Overlapping bonds of monomer and lithium ion.

occur from the opposite side, as in reaction (4). This results in a transition state formally analogous to that in an S_N2 reaction. Providing that the incoming monomer molecule always presents the same conformation toward the lithium atom, retention of configuration will be assured by the immediate stabilization of the newly formed enolate system through intramolecular solvation.

In an analogous way the formation of high *cis*-1,4-polyisoprene with alkyllithium initiation might be explained. In this case a pseudo six-membered ring is assumed for the transition state (Fig. 2).

Because of the relatively small ionic radius of lithium and the high proportion of *p*-character in its outermost orbitals, the above mechanisms are especially favored for alkyllithium compounds. This is shown by reaction (5).

The cationic mechanism for the stereospecific polymerization of vinyl ethers in homogeneous media might also involve cyclic intermediates resulting from neighboring group interaction (intramolecular solvation) involving the ether oxygen. Thereby the growing carbonium ion is stabilized and the configuration is retained.

Ziegler-Natta catalysts. Ziegler-Natta catalysts result from the reaction of transition metal halides ($TiCl_4$, $TiCl_3$, VCl_4, $VOCl_3$, and so on) with organometallic compounds (R_3Al, R_2AlCl, R_2Zn, and so on) in hydrocarbon media. The catalyst systems are of heterogeneous nature and consist of a suspension of transition-metal subhalide crystals which are partially alkylated on their surface. For reasons of electroneutrality the transition metal centers situated on the edges of the crystal lattice will not be fully coordinated by halogen atoms but will possess vacant coordination sites. Those vacancies represent a necessary condition for making these transition-metal centers active in polymerization because they provide the possibility for the approaching monomer molecule to become complexed on the transition-metal center.

The stereoregulating mechanism might be depicted as in reaction (6). A titanium center, situated on the edge of a $TiCl_3$ lattice layer, which has been alkylated during the catalyst-forming reaction, possesses one vacant coordination site by which it can accommodate the approaching propylene molecule. The monomer forms a π-complex whereby the double bond approaches the titanium center almost as closely as the halogen atoms. Only small nuclear displacements are necessary for the propagation step during which the alkyl group initially attached to the titanium center becomes attached to the complexed monomer. At the end of the propagation step, alkyl group and chlorine vacancy have mutually exchanged places and the process may now repeat itself. Since the new vacancy is the mirror image of the old one, syndiotactic placements will occur if the chain switches alternatingly

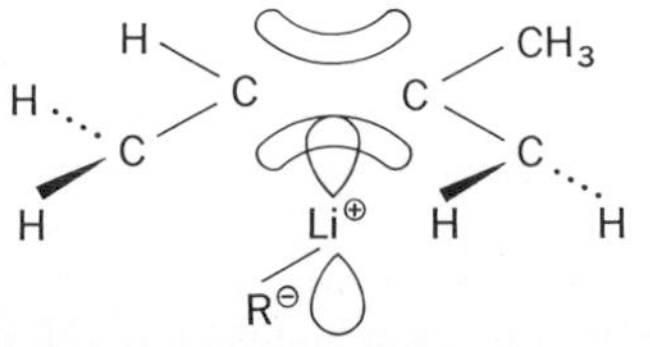

Fig. 2. Bonding in the polyisoprene transition state.

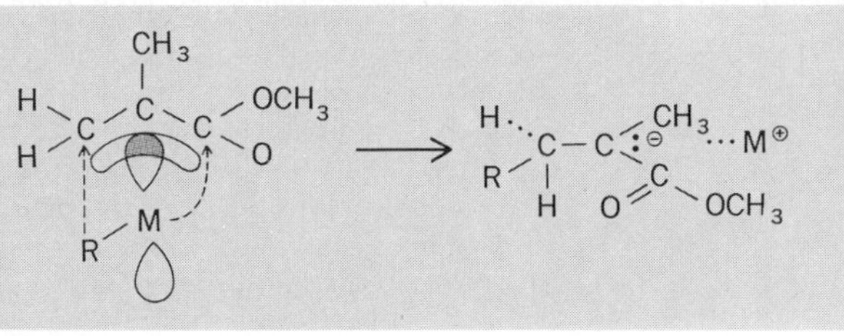

(3)

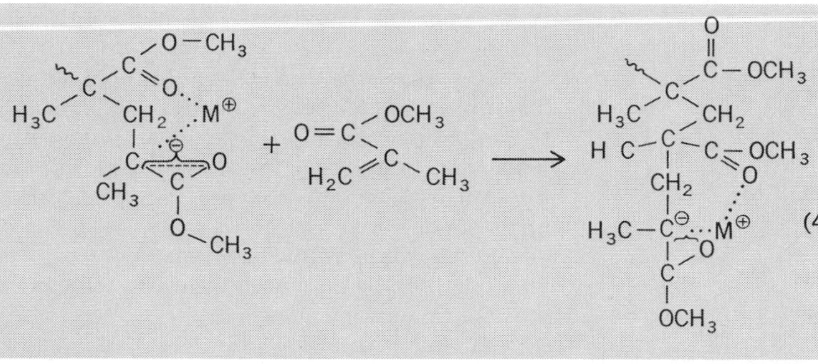

(4)

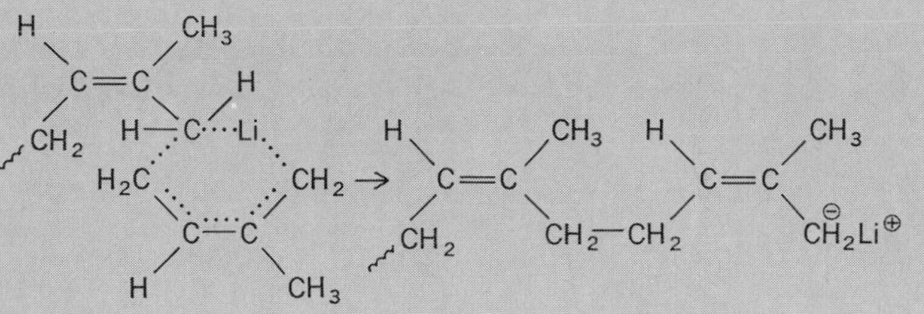

(5)

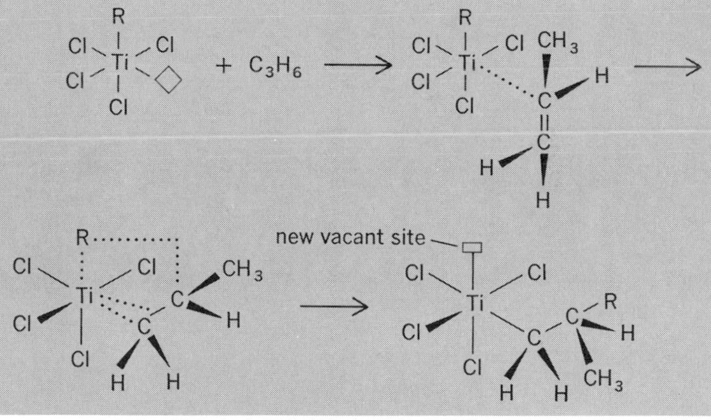

(6)

from one to the other coordination site in each growth step. Steric and energetic considerations revealed that both coordination sites are not equivalent for the growing chain, so one of them will be preferred. In normal polymerizations with Ziegler-Natta catalysts, carried out between 50 and 100°C, the growing alkyl group shifts to the most favorable position before the next monomer molecule is inserted; that is, propagation proceeds in a vacancy of constant configuration which results in a sequence of isotactic placements.

Optically active polymers. Stereoregular polymers do not show optical activity of the main chain, provided conformational phenomena (for example, left- and right-handed helix formation) are disregarded. This fact is obvious for syndiotactic polymers which are built up by regular sequences of *d* and *l* configurations and where only the end groups could contribute to some optical activity. The asymmetric main-chain carbons of an isotactic polymer are optically inactive since every monomeric unit at one side of the macromolecule has its mirror image at the opposite side of the central atom of the chain. Thus a chain of all *d* units, when turned end for end, is superimposable on an all *l*-chain units since both chains are identical except for their end groups. These considerations are not valid if there are optically active side groups present in the polymer. An asymmetric synthesis in the normal sense is not possible from simple α-olefins.

Real asymmetric centers within the main chain might be created in two ways: (1) copolymerization of two monomers whereby real asymmetric carbons are generated because of the chemical differences of neighboring groups; optical activity of the main chain can be obtained by enhancing one configuration in the propagation step; (2) polymerization of disubstituted dienes which can exist in enantiomorphic forms after polymerization; with the aid of optically active catalysts it is possible to enhance one enantiomorphic form with the result that the polymer shows optical activity. *See* ASYMMETRIC SYNTHESIS; CATALYSIS; COMPLEX COMPOUNDS; COORDINATION CHEMISTRY; ORGANIC REACTION MECHANISM; ORGANOMETALLIC COMPOUND; POLYMERIZATION; STEREOCHEMISTRY; SUBSTITUTION REACTION. [ANTON SCHINDLER]

Bibliography: M. P. Doyle and C. T. West (eds.), *Stereoselective Reductions*, 1976; G. Natta and M. Farina, *Stereochemistry*, 1973; R. A. V. Raff and K. W. Doak (eds.), *High Polymers*, vol. 20: *Crystalline Olefin Polymers*, 1965; G. V. Smith (ed.), *Catalysis in Organic Synthesis*, 1978; D. Whittaker, *Stereochemistry and Mechanism*, 1973.

Steric effect

The influence of the spatial configuration of reacting substances upon the rate, nature, and extent of reaction. The sizes and shapes of atoms and molecules, the electrical charge distribution, and the geometry of bond angles influence the courses of chemical reactions.

The steric course of organochemical reactions is greatly dependent on the mode of bond cleavage and formation, the environment of the reaction site, and the nature of the reaction conditions (reagents, reaction time, and temperature). The effect of steric factors is best understood in ionic reactions in solution. The nucleophilic substitution reaction at a saturated carbon atom can serve as an illustration.

Saturated nucleophilic substitution. While the reacting carbon atom is in an electron-deficient state in the transition state (state of highest energy, somewhere between starting material and product) in a nucleophilic substitution process, the reaction can be varied from a two-step, unimolecular ionization to a one-step, bimolecular transformation. The former mode of reaction, a solvolysis or S_N1 process, converts a tetrahedral carbon into a solvated planar carbonium ion intermediate, and hence, leads to racemization (randomization of configuration); for example, reaction (1).

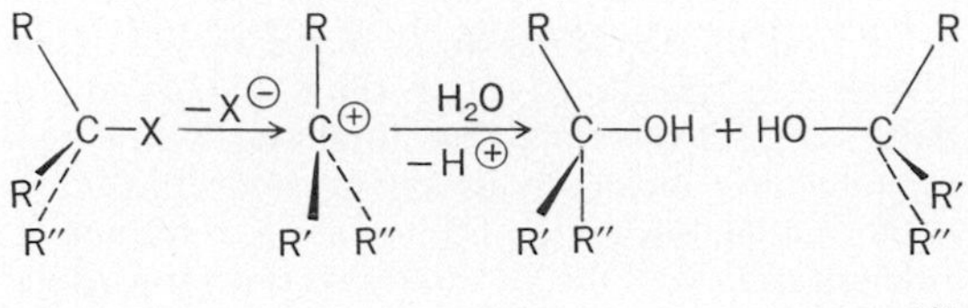

(1)

The second reaction path proceeds by a simultaneous rupture of the old bond and creation of a new one, either by inversion of configuration, a Walden inversion or S_N2 process, for example, reaction (2); or in a few cases by a front-side displacement of part of the substituent already present, an S_Ni process, and hence, retention of configuration, for example, reaction (3).

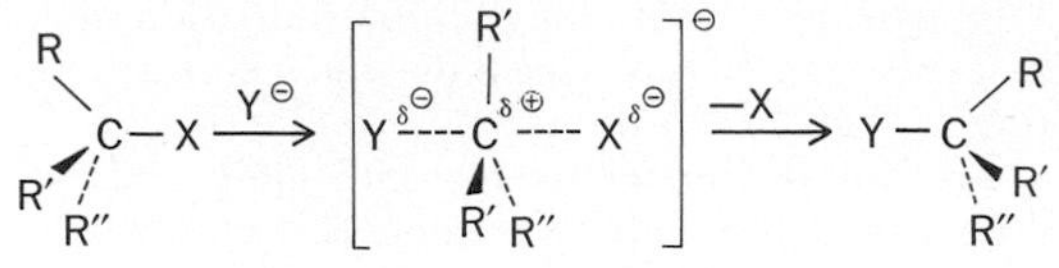

(2)

$$\text{ROH} \xrightarrow{\text{SOCl}_2} \text{ROSOCl} \longrightarrow \underset{\text{(ion pair)}}{\text{R}^{\oplus}\text{OSOCl}^{\ominus}} \xrightarrow{-\text{SO}_2} \underset{\text{(ion pair)}}{\text{R}^{\oplus}\text{Cl}^{\ominus}} \longrightarrow \text{RCl} \quad (3)$$

These substitution reactions are highly solvent-dependent; for example, tert-butyl chloride undergoes solvolysis close to 500,000 times faster in water, a solvent of high ionizing power, than in ethanol, a solvent of low dielectric properties. The nature of R, R′, and R″ is one of the factors determining which of the above pathways a compound prefers for its substitution. The larger the size of these three groups, the greater is the tendency to relieve steric strain by extrusion of X, and thus the more the need for an S_N1 process. The smaller the size of the environment, the greater is the accessibility of the reagent from the back side, and thus the more tendency toward an S_N2 process. As a consequence, tertiary compounds undergo racemization readily, whereas primary systems prefer inversion.

Both the rate and the steric course of an ionic displacement may depend often on the ability of groups adjacent to the reaction site to accommodate a positive charge. Substitution of α-halo ethers and allyl or benzyl halides occurs much faster than a similar reaction of unsubstituted halides because of the intermediacy of the stabilized cations or cationlike transition states in notation (4).

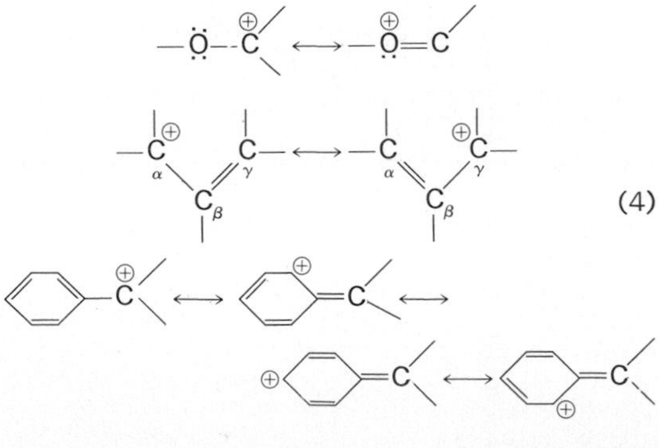

(4)

In view of the charge distribution over more than one atom in these cases, sometimes the incoming reagent forms a bond at a site different from that of the leaving group. As a consequence, an S_N2' process, reaction (5), or an S_Ni' path, reaction (6), may result. Both processes yield retention

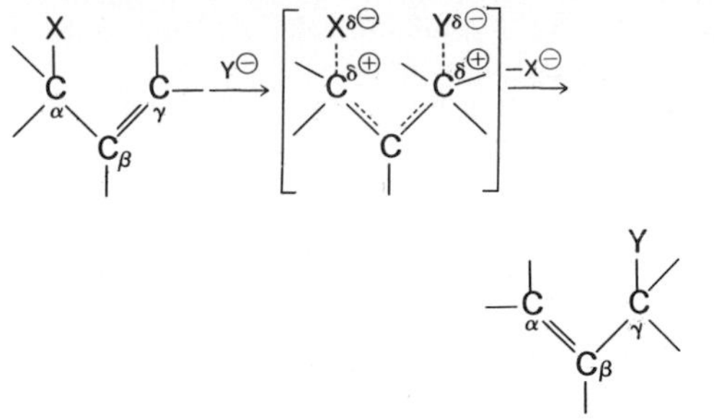
(5)

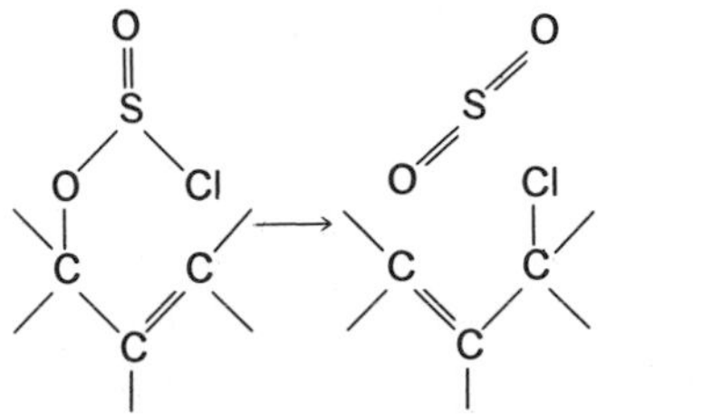
(6)

of configuration; that is, the orientation of the new substituent on its carbon atom is identical with that originally held by the former functional group on a site that is two carbon atoms removed.

In the presence of participating neighboring groups, even solvolyses can lead to retention of configuration. Certain rigidly held homoallyl systems undergo substitution at a site that is three carbons removed from the position of the leaving group, but with retention of configuration, as represented by reaction (7).

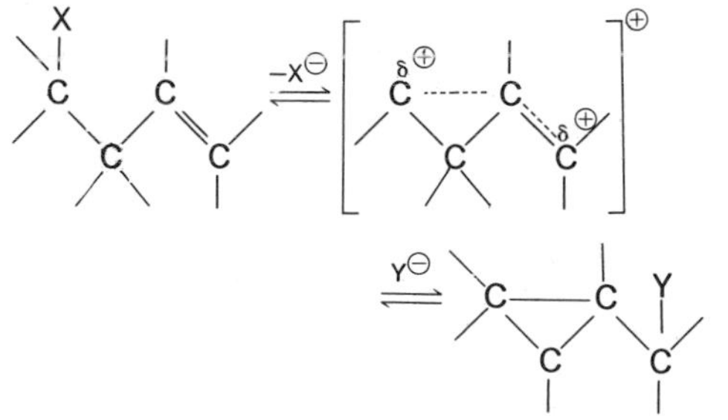
(7)

Double inversion is responsible for the retained configuration of the products of solvolysis of α-halo acid salts, reaction (8).

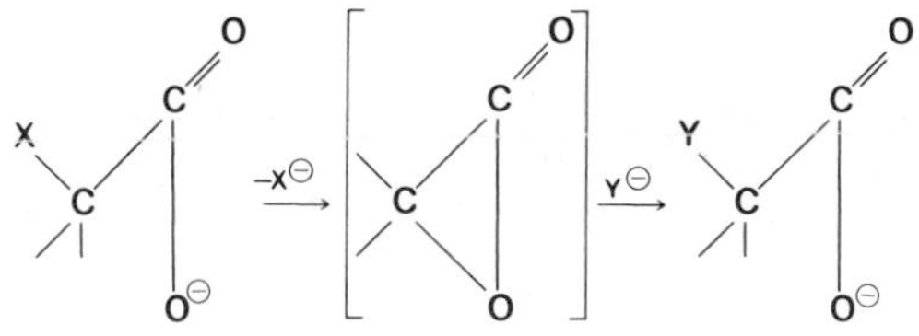
(8)

Solvolysis of *trans*-β-acetoxy systems in nonaqueous media leads to trans products. In the presence of water, cis compounds are obtained, reaction (9).

Base treatment of *trans*-halohydrins leads to *trans*-vicinal glycols. The intermediate epoxide is isolable. Although ring opening of the latter may

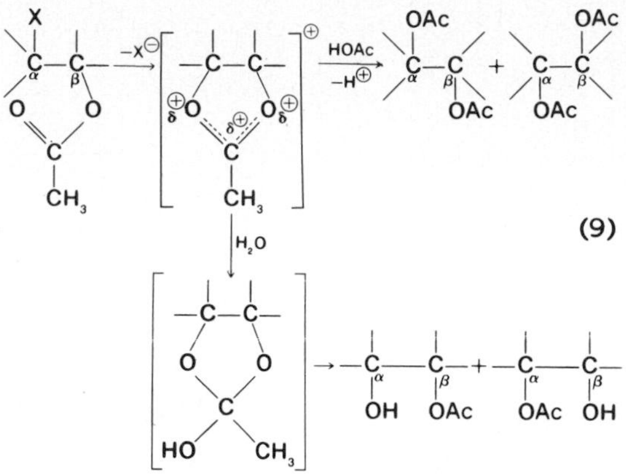
(9)

yield two different trans products, the diaxial one is formed preferentially in cycloalkane cases, reaction (10).

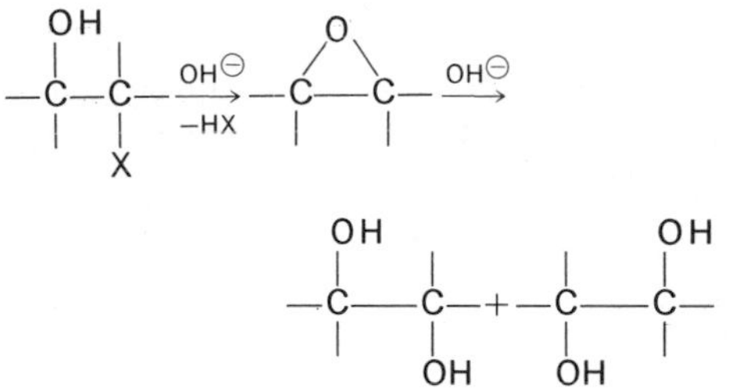
(10)

Rearrangement. Migration of neighboring groups toward the reaction site, resulting in skeletal rearrangements, is a common occurrence. Both the internal displacement of the leaving group by the migrating group and the subsequent external displacement of the migrating group by the solvent or added reagent proceed in a trans sense, that is, by back-side approach. Thus, the overall steric consequence of one migration sequence is retention of configuration.

There are several examples of 1,2-hydride migration, for example, reaction (11).

Transannular hydride shifts are quite similar in nature, as in reaction (12).

The 1,2 migration of phenyl groups proceeds by way of fairly stable phenonium ions, reaction (13).

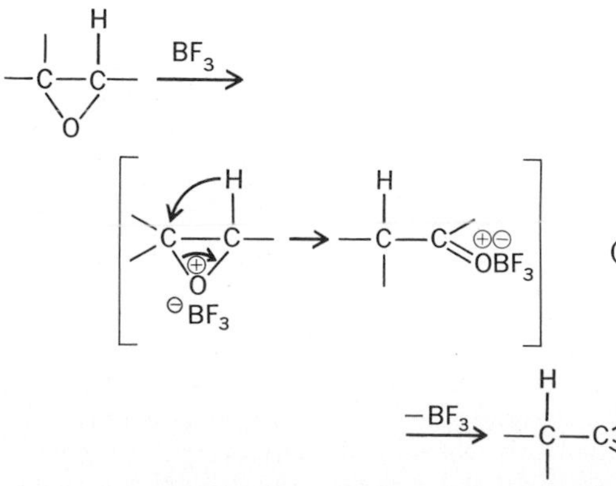
(11)

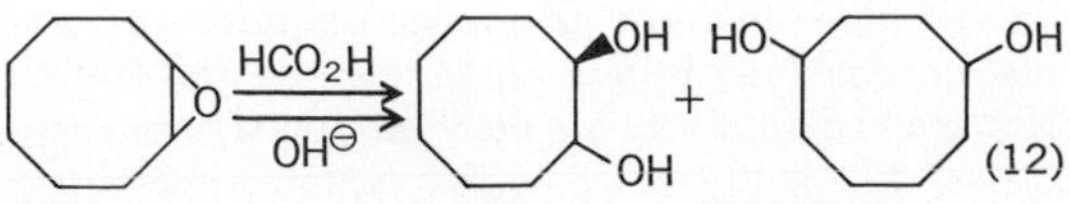
(12)

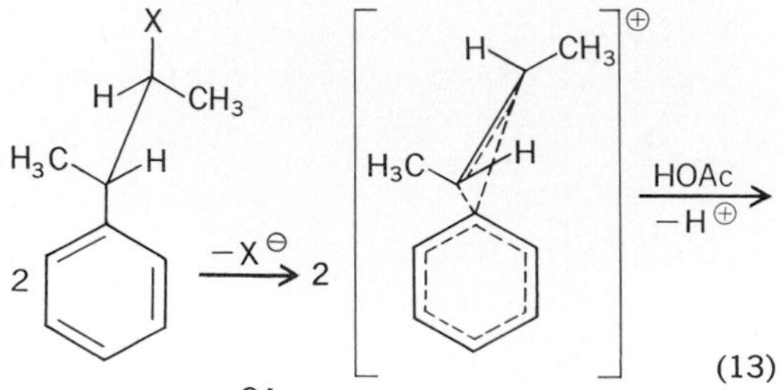

(13)

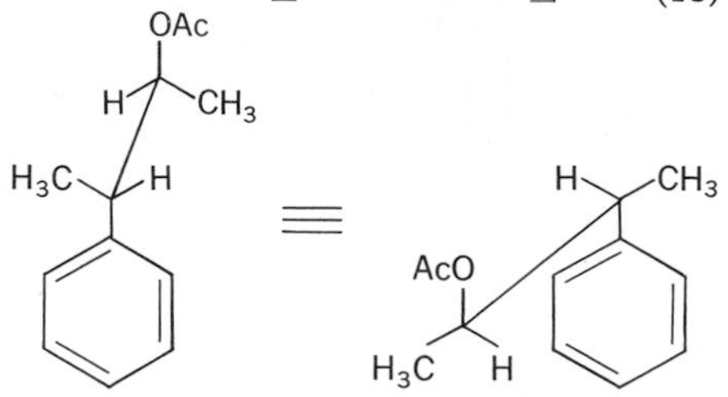

The Wagner-Meerwein rearrangement of saturated neighboring groups shows directional effects similar to those of the above migrations; for example, the conversion of camphene hydrochloride to bornyl chloride is stereospecific, with retention of configuration, reaction (14).

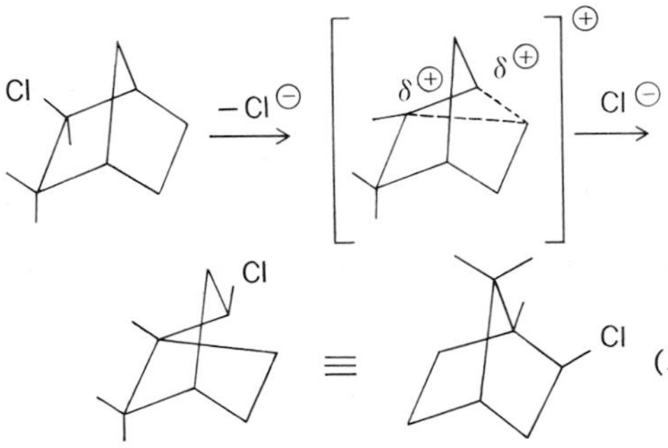

(14)

Neopentyl halides solvolyze to tertiary amyl derivatives, reaction (15).

$$(CH_3)_3CCH_2X \xrightarrow{-X} [(CH_3)_3C\overset{\oplus}{C}H_2 \rightarrow (CH_3)_2\overset{\oplus}{C}CH_2CH_3] \xrightarrow[H^{\oplus}]{ROH} (CH_3)_2\underset{\displaystyle OR}{\underset{|}{C}}CH_2CH_3 \quad (15)$$

Cyclohexane systems with equatorial leaving groups may undergo contraction to five-membered rings, reaction (16).

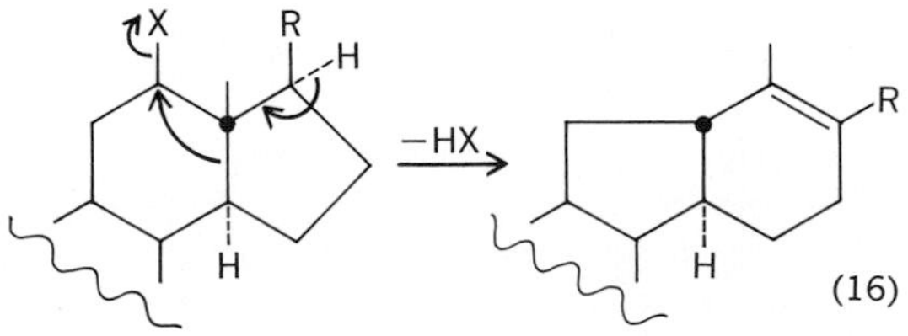

(16)

Organic compounds possessing potential leaving groups at the bridgehead of small bicyclic ring systems undergo substitution processes only sluggishly. In the absence of ready access at the back side of the reaction center, the S_N2 pathway is excluded. The inability of the compounds to form planar carbonium ions precludes a S_N1 route. However, displacements do occur slowly at elevated temperatures, presumably via nonplanar cations.

Unsaturated nucleophilic substitution. Nucleophilic substitution reactions at unsaturated carbon atoms can take place by two possible mechanistic routes, an elimination-addition process and an addition-elimination scheme. The former route is best illustrated by the transformation of aromatic halides into anilines, reaction (17).

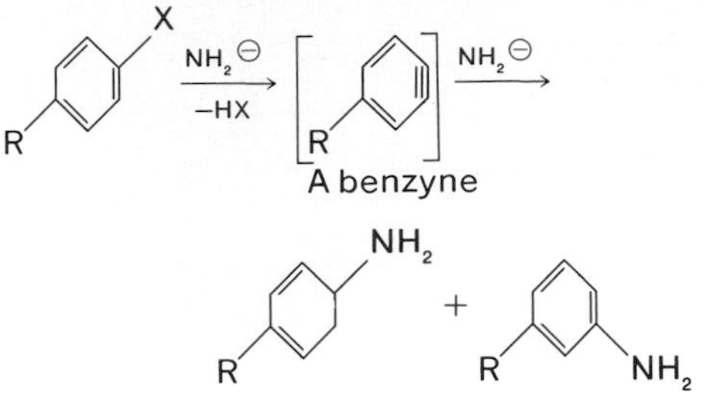

(17)

The latter is encountered in the interconversion of carboxylic acids and their derivatives, for example, reaction (18). Because the central carbon atom

$$RCO_2R' \underset{}{\overset{H_2O}{\rightleftharpoons}} [RC(OH)_2OR'] \overset{-ROH}{\rightleftharpoons} RCO_2H \quad (18)$$

has greater steric requirements in the reaction intermediate than in the starting material, the reaction velocity is strongly dependent on the size and number of neighboring groups; that is, an increase in the bulkiness of R is reflected in a decrease of the rate of the reaction.

The addition-elimination mechanism is portrayed also by the aromatic nucleophilic substitution reaction, for example, (19). In order to be

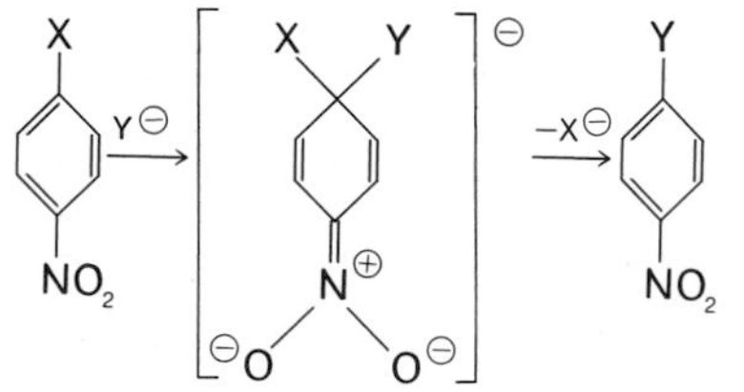

(19)

able to stabilize the reaction intermediate, the all-important nitro group must be coplanar with the benzene ring. As a consequence, ortho substituents, which may block this steric requirement, retard the reaction rate. Unusual aromatic nucleophilic substitutions have been observed in cases where steric hindrance by ortho substituents has prevented addition to aromatic ketones to occur, for example, reaction (20).

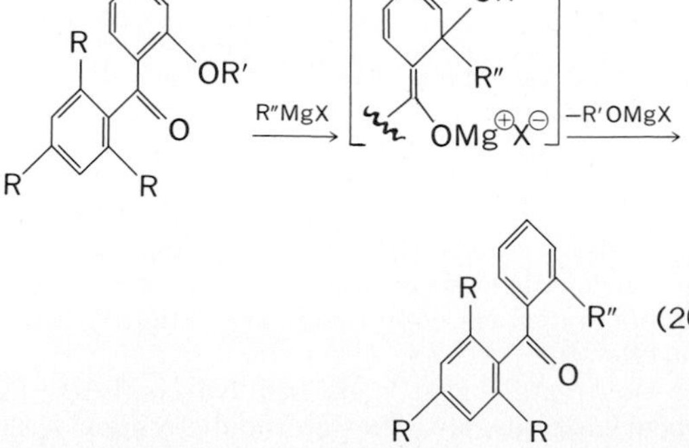

(20)

Addition reactions. The steric course of addition reactions at unsaturated sites depends largely on the reagent. Catalytic hydrogenation, a non-homogeneous process of undetermined mechanism, occurs in a cis manner. In the absence of any steric interference, it leads to thermodynamically stable products. In the presence of steric hindrance, the two new hydrogen atoms are usually introduced on the least hindered side of the unsaturated compounds. However, sometimes some bulky polar groups actually aid, rather than retard, adsorption of the catalyst on their side of the reducing compound, thereby leading to products of opposite configuration.

The oxidation of olefins to vicinal glycols by permanganate salts or osmium tetroxide also proceeds in a cis fashion and also involves the least-hindered side of the reacting substrate. The Diels-Alder reaction behaves similarly, for example, reaction (21). All addition processes, during which

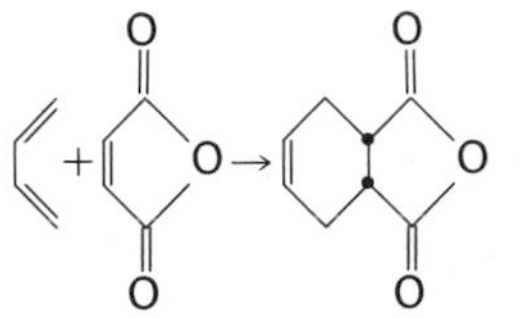

$$\text{(21)}$$

two new bonds are formed more or less simultaneously, yield cis adducts.

Ionic addition reactions of olefins and acetylenes occur in a trans manner. The mode of addition is such as to lead to product via the most stable cations (Markownikoff addition), for example, reaction (22).

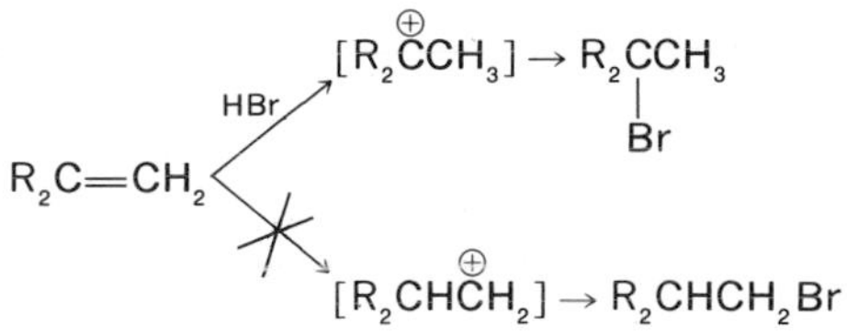

$$\text{(22)}$$

Halogen addition to cyclic olefins leads to trans diaxial dihalides which, on standing, isomerize to the more stable trans diequatorial dihalides, reaction (23).

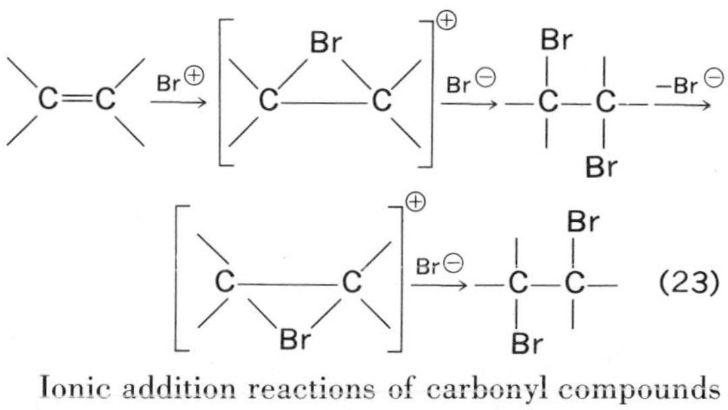

$$\text{(23)}$$

Ionic addition reactions of carbonyl compounds follow a steric course very similar to those of olefins. However, the reagents are mostly nucleophilic, and some reactions are equilibrium processes, for example, reaction (24). The orientation of

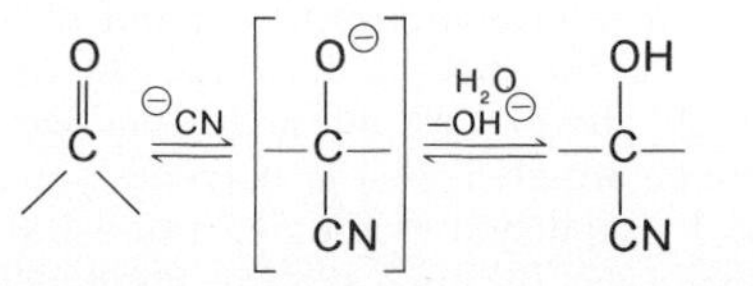

$$\text{(24)}$$

attack and the reaction rate are governed by the

environment of the carbonyl group.

Addition reactions of conjugated carbonyl systems can occur through cation as well as anion intermediates, but they uniformly place the nucleophilic part of the reagents on the β-carbon atoms, for example, reactions (25) and (26).

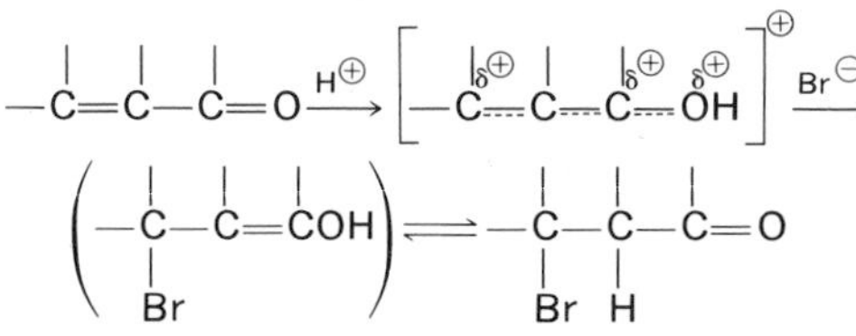

$$\text{(25)}$$

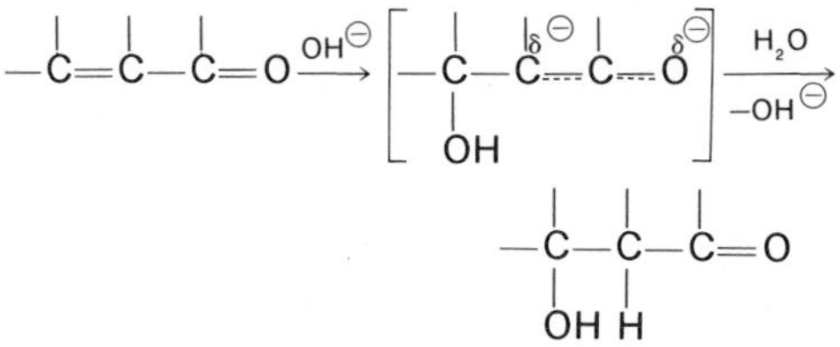

$$\text{(26)}$$

The reaction of carbonyl compounds as enol or enolate anions with electrophilic reagents can take two different courses. If the process is kinetically controlled, the electrophile, a proton, halonium ion, or others, interacts with the substrate on its least-hindered side. However, if the reaction is thermodynamically controlled, then, independent of mechanism, the most stable product is obtained.

Elimination reactions. Elimination reactions can be carried out by pyrolysis of esters, halides, or amine oxides in the liquid or vapor phase. These eliminations always involve a rupture of vicinal cis bonds.

Alternatively, similar cleavage processes can be made to occur ionically in solution, in which case they proceed in a trans fashion. The direction of elimination depends greatly on the molecularity of the process, as well as on the sizes of the leaving group and the attacking base. The two-step, unimolecular cleavage, an $E1$ process, leads predominantly to the more substituted, hence more stable, olefins, for example, reaction (27).

$$CH_3CH_2C(CH_3)_2 \xrightarrow{-Br^{\ominus}} [CH_3CH_2\overset{\oplus}{C}(CH_3)_2] \xrightarrow{-H^{\oplus}}$$

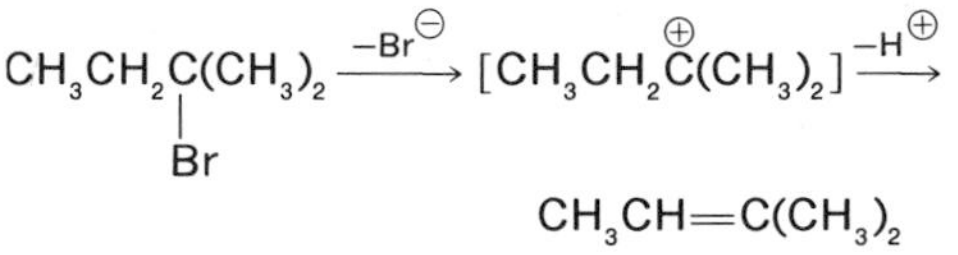

$$CH_3CH{=}C(CH_3)_2 \quad \text{(27)}$$

The one-step, bimolecular elimination (an $E2$ process) of neutral compounds yields similar products, for example, reaction (28). However, an $E2$

$$CH_3CH_2C(CH_3)_2 \xrightarrow{OH^{\ominus}}$$

$$CH_3CH{=}C(CH_3)_2 \quad \text{(28)}$$

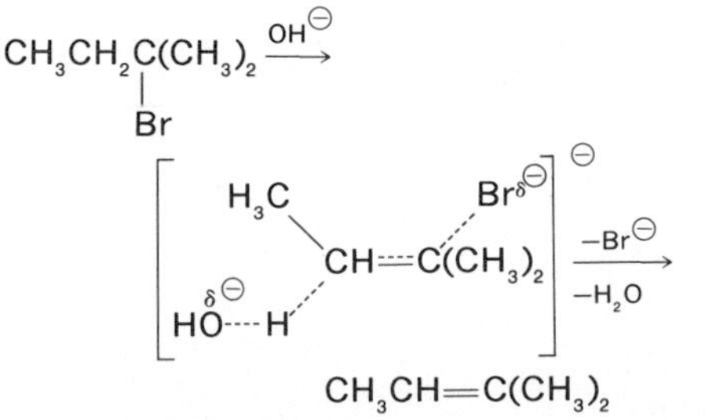

reaction on positive ions, ammonium or sulfonium salts, affords the less substituted olefin in preponderant yield, reaction (29).

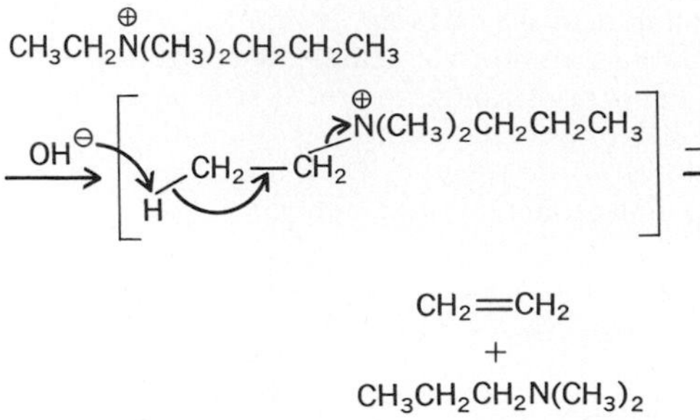

$$CH_3CH_2\overset{\oplus}{N}(CH_3)_2CH_2CH_2CH_3$$

$$\xrightarrow{} OH^{\ominus} \left[\overset{H}{\underset{}{}} CH_2 \overset{}{-} CH_2 \overset{\oplus}{N}(CH_3)_2CH_2CH_2CH_3 \right] \xrightarrow{-H_2O}$$

$$CH_2 {=\!\!=} CH_2$$
$$+ \qquad\qquad (29)$$
$$CH_3CH_2CH_2N(CH_3)_2$$

Reactions leading to the more stable products are said to follow the Saytzeff rule, whereas those yielding less stable olefins obey the Hofmann rule. Because the transition state in the *E*2 reaction is of lowest energy when all atoms involved in the elimination are in a plane, the fastest rates among cyclic compounds are encountered in the cases which permit a diaxial alignment of vicinal substituents.

Many ionic elimination reactions are known which involve the rupture of more than two bonds, for example, reaction (30).

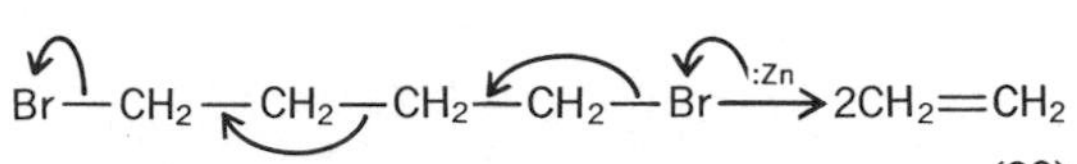

$$Br{-}CH_2{-}CH_2{-}CH_2{-}CH_2{-}Br \xrightarrow{:Zn} 2CH_2{=\!\!=}CH_2$$
$$(30)$$

The *E*2′ processes, eliminations of two groups of carbon atoms separated by an olefinic linkage, appear to be cis in nature, reaction (31).

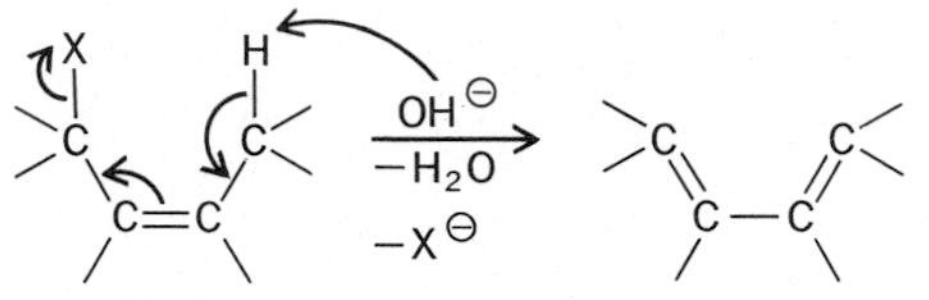

$$(31)$$

Electrophilic substitution. Steric factors have a fair control over the course of the aromatic electrophilic substitution reaction. In reactions of compounds containing ortho-para directing substituents, the *p/o* product ratio is usually greater than 1, and increases with the size of the substituent and that of the reacting species. The rate-accelerating participation of electon-donating groups, located ortho or para to the incoming substituent, in stabilizing the transition state is greatly diminished in the presence of bulky ortho neighbors which would prevent the groups from attaining coplanarity with the benzene, for example, reaction (32).

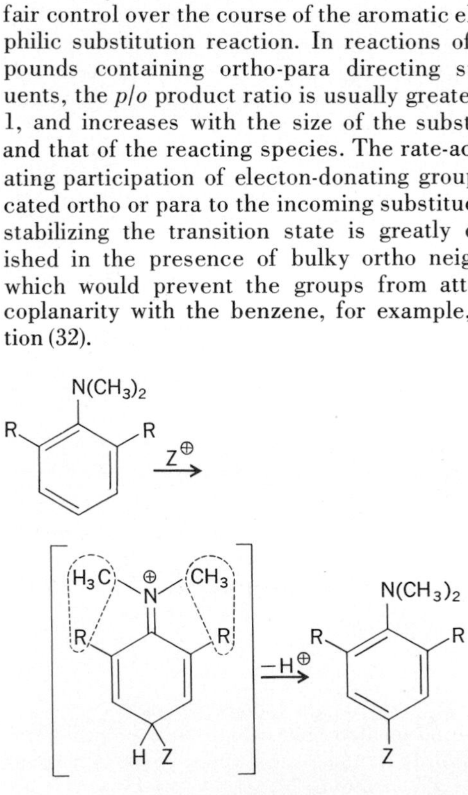

$$(32)$$

See CHELATION; CONFORMATIONAL ANALYSIS;

ORGANIC CHEMICAL SYNTHESIS; ORGANIC REACTION MECHANISM; STEREOCHEMISTRY.

[ERNEST WENKERT]

Bibliography: A. T. Balaban et al. (eds.), *Steric Fit in Quantitative Structure-Activity Relations*, 1980; G. Natta and M. Farina, *Stereochemistry*, 1973; J. D. Roberts and M. C. Caserio, *Basic Principles of Organic Chemistry*, 2d ed., 1977.

Stoichiometry

The interpretation of mass and energy relationships indicated by chemical reaction equations. It is also defined as those calculations used to find the quantities of reactants and products in a chemical reaction.

Principles. These fall into four groups. The first group includes the law of conservation of matter, the law of chemical combining weights, and the law of combining proportions. The second group is based on the law of conservation of energy and includes heat of reaction, heat added, heat lost, and other energy effects associated with the reaction. The third group comprises the equilibrium relationships of the system and includes not only chemical equilibria but also physical equilibria such as gas-liquid systems. The fourth group includes the rate of reaction relationships of both chemical changes and physical changes.

In the chemical laboratory only the first group is usually considered. The second and fourth groups are neglected, or the reactions are forced to completion artificially. The third group is utilized in a limited number of cases. The units used are grams, moles, milliliters, or liters. The types of calculations include weight-weight, weight-volume (either gases or liquids), and thermal relationships. For any calculation a knowledge of the reaction under consideration and a balanced reaction equation are necessary.

Weight-weight problems. In these, a weight of material is given and a weight of product is desired. These calculations are based on the mole ratios shown in the balanced reaction equation.

Example 1: How many grams of calcium carbonate can be prepared from 100 g of sodium carbonate?

$$CaCl_2 + Na_2CO_3 = CaCO_3 + 2NaCl$$

The equation indicates that 1 mole of sodium carbonate will yield 1 mole of calcium carbonate.

$$\frac{100\ g\ Na_2CO_3}{mole\ wt\ Na_2CO_3} = \frac{x\ g\ CaCO_3}{mole\ wt\ CaCO_3}$$

$$x\ g\ CaCO_3 = 100\ g\ Na_2CO_3 \times \frac{mole\ wt\ CaCO_3}{mole\ wt\ Na_2CO_3}$$

$$x\ g = 100 \times \frac{100}{106} = 94.3\ g$$

In this example the desired weight is equal to the given weight multiplied by a definite number, the ratio of molecular weights. This definite number is called a factor or gravimetric factor and is defined as the ratio of the molecular weight of the substance sought divided by the molecular weight of the substance weighed in the correct proportion. The correct proportion is found from a balanced reaction equation by means of an atom which is common to both molecules or by following a series of reactions.

Example 2: How many grams of potassium cyanide are required to complex the ferric ion obtained from 5 g of iron as potassium ferricyanide?

$$Fe \rightarrow Fe^{3+}$$

and $$Fe^{3+} + 6KCN = K_3Fe(CN)_6 + 3K^+$$

The equation says that for each atomic weight (at. wt) of iron, 6 moles of potassium cyanide are required.

$$\text{g KCN} = \text{g Fe} \times \frac{\text{6 mole wt KCN}}{\text{at. wt Fe}}$$

$$= 5 \times \frac{6 \times 65.1}{55.85} = 34.97 \text{ g}$$

Example 3: A method for the isolation and determination of arsenic is based on oxidation to arsenate, precipitation of silver arsenate, and conversion of this to silver chloride which is weighed. If 1.325 g of silver chloride is found, what weight of arsenic was present?

$$As \rightarrow AsO_4^{3-} \xrightarrow{Ag^+} Ag_3AsO_4 \xrightarrow{Cl^-} 3AgCl$$

In this case no common ion is present. However, arsenic which is sought is present in silver arsenate with silver which is weighed. Since the Ag to As ratio is 3:1 in silver arsenate, 3 moles of silver chloride would be weighed for each gram-atom (g-atom) of arsenic present at the beginning.

$$\text{g As} = \text{g AgCl} \times \frac{\text{at. wt As}}{\text{3 mole wt AgCl}}$$

$$= 1.325 \times \frac{74.91}{3 \times 143.4} = 0.2307 \text{ g}$$

Weight-volume calculations. These fall into two types according to whether the volume deals with a gas or a solution. Problems dealing with gases are best solved in terms of molar volumes, the volume occupied by 1 mole of gas at 0°C and 760 mm pressure (standard temperature and pressure, STP). This volume is 22.4 liters. The calculated volumes can be corrected to any stated conditions by use of the gas laws. Gravimetric factors are used as in weight-weight problems except that molar volume will replace molecular weight for one substance.

Example 4: How many liters of dry HCl gas at STP is needed to convert 10 g of ethylamine to ethylamine hydrochloride?

$$C_2H_5NH_2 + HCl = C_2H_5NH_3Cl$$

$$x \text{ liters HCl} = \text{g } C_2H_5NH_2 \times \frac{\text{molar vol HCl}}{\text{mole wt } C_2H_5NH_2}$$

$$= 10 \times \frac{22.4}{45.0} = 5.0 \text{ liters}$$

Calculations for reactions involving solutions are worked in terms of moles also. Molarity, M, is defined as the number of moles of solute per liter of solution. Molarity can also be calculated from the density, grams of solute per milliliter, and the volume in milliliters.

Example 5: How many liters of sulfuric acid, density 1.84 and 96% H_2SO_4 by weight, are needed to convert 75.6 g of ferric chloride to ferric sulfate?

$$2FeCl_3 + 3H_2SO_4 = Fe_2(SO_4)_3 + 6HCl$$

$$x \text{ g } H_2SO_4 = \text{g } FeCl_3 \times \frac{\text{3 mole wt } H_2SO_4}{\text{2 mole wt } FeCl_3}$$

$$V_{\text{liter}}\ H_2SO_4 = x \text{ g } H_2SO_4 \times \frac{1}{\text{density (g/ml)} \times (\text{fraction } H_2SO_4)} \frac{1}{\text{1000 ml/liter}}$$

$$V_{\text{liter}} = 75.6 \times \frac{3 \times 98}{2 \times 163} \times \frac{1}{1.84 \times 0.96} \times \frac{1}{1000}$$

$$= 0.0386 \text{ liter}$$

Normality. In quantitative analysis, volume calculations are cumbersome by the mole method. In order that equal volumes of equal concentration be chemically equivalent, a system based on equivalents is used. Normality, N, is defined as the number of equivalents (equiv) of solute per liter of solution. From this definition the following relationships are derived.

$$N = \frac{\text{no. of equiv}}{\text{liter}} = \frac{\text{no. of milliequivalents (meq)}}{\text{ml}}$$

$$\text{No. of equiv} = N \times V_{\text{liter}}$$

$$= \frac{\text{grams}}{\text{equiv wt}} \frac{V_{\text{ml}} \times \text{density} \times \text{fraction}}{\text{equiv wt}}$$

$$\text{grams} = N \times V_{\text{liter}} \times \text{equiv wt} = N \times V_{\text{ml}} \times \text{meq wt}$$

These relationships are used to solve most volume calculations. The only quantity which varies according to the particular reaction is the equivalent weight. For acids, bases, and salts which are acidic or basic the equivalent weight is the number of grams which contain or react with 1 g-atom of replaceable hydrogen.

Example 6: What are the equivalent weights for example 4?

HCl has one replaceable hydrogen ion, its mole wt is 35.45; $C_2H_5NH_2$ combines with one H^+, its mole wt is 45.07 g.

Example 7: How many grams of NaOH are required to prepare 5 liters of 0.3 N solution? How many milliliters of 0.06 N H_2SO_4 will be neutralized by 75 ml of the NaOH solution?

NaOH reacts with one H^+, and its equiv wt is equal to its mole wt.

5 liters × 0.3equiv/liter × 40g/equiv = 60 g NaOH needed.

$$V_{\text{ml}} \times 0.06\,N = \text{no. of meq } H_2SO_4$$
$$= \text{no. of meq NaOH} = 75 \times 0.3$$

$$V_{\text{ml}} = \frac{75 \times 0.3}{0.06} = 375 \text{ ml } H_2SO_4 \text{ solution}$$

Example 8: If a 2,000-g sample of dilute acetic acid requires 40.00 ml of 0.2250 N base for neutralization, what is the acetic acid percentage?

$$\%\text{ acetic acid} = \frac{\text{g acetic acid}}{\text{g sample}} \times 100$$

$$= \frac{N \times V_{\text{ml}} \times \text{meq wt} \times 100}{\text{g sample}}$$

$$CH_3COOH + OH^- \rightarrow CH_3COO^- + H_2O$$

Then equivalent weight of acetic acid equals molecular weight.

% acetic acid

$$= 40.00 \times 0.2250 \times \frac{60.05}{1000} \times \frac{100}{2} = 27.02\%$$

For oxidation-reduction systems, the equiv wt is found by dividing the mole wt of a substance by its change in oxidation number during a reaction. Oxidation number is the apparent oxidation state or apparent valence of the atom. The rules for calculating oxidation number may be summarized as follows.

Elements have oxidation numbers of zero; single valent elements have an oxidation number of +1; oxygen is −2, except in peroxides in which it is −1; the sum of oxidation numbers for a compound is zero; and two atoms of the same element attached together have a zero contribution. For organic compounds, only the atoms which change need to be considered.

Example 9:

$$\frac{K_2Cr_2O_7}{2(+1)+2(+6)+7(-2)} \rightarrow \frac{Cr_2(SO_4)_3}{2(+3)+3(-2)}$$

$$Cr^{6+} \rightarrow Cr^{3+}$$

The oxidation state of chromium changes by 3 units per atom, $K_2Cr_2O_7$ has 2 Cr atoms; for $K_2Cr_2O_7$ the equiv wt is 1/6 the mole wt.

$$\frac{C_6H_5NHOH}{\text{constant }(0)+0+(+1)+(-1)} \rightarrow \frac{C_6H_5NH_2}{\text{constant }(0)+(-2)+2(+1)}$$

$$N^0 \rightarrow N^{2-}$$

The oxidation state of nitrogen changes by 2 units per atom; equiv wt for C_6H_5NHOH is 1/2 mole wt.

Example 10: How many grams of $KMnO_4$ are present in 1 liter of 0.1 *N* solution to be used in acid medium? How many grams of ferrous ion will be oxidized by this solution?

In acid $KMnO_4 \rightarrow Mn^{++}$, $Mn^{7+} \rightarrow Mn^{2+}$ the oxidation state of manganese changes by 5 units per atom.

$$\text{g } KMnO_4 = N \times V_{\text{liter}} \times \text{equiv wt}$$
$$= 0.1 \times 1 \times 31.61 = 3.161$$

Fe^{2+} is oxidized to Fe^{3+}, change of 1 in oxidation state.

$$\text{g } Fe^{++} = N \times V_{\text{liter}} \times \text{equiv wt}$$
$$= 0.1 \times 1 \times 55.85 = 5.585$$

Example 11. What weight of $Ti_2(SO_4)_3$ is required to reduce 25 g of nitrobenzene to aniline?

$$C_6H_5NO_2 \rightarrow C_6H_5NH_2$$
$$N^{4+} \rightarrow N^{2-}$$

(change of 6 units per atom)

$$Ti_2(SO_4)_3 \rightarrow Ti(SO_4)_2$$
$$Ti^{3+} \rightarrow Ti^{4+}$$

(change of 1 unit per atom)

$$\text{No. of equiv } C_6H_5NO_2 = \text{no. of equiv } Ti_2(SO_4)_3$$

$$\frac{\dfrac{\text{g } C_6H_5NO_2}{\text{mole wt } C_6H_5NO_2}}{6} = \frac{\dfrac{\text{g } Ti_2(SO_4)_3}{\text{mole wt } Ti_2(SO_4)_3}}{2}$$

$$\text{g } Ti_2(SO_4)_3 = \text{g } C_6H_5NO_2 \times \frac{6}{2} \times \frac{\text{mole wt } Ti_2(SO_4)_3}{\text{mole wt } C_6H_5NO_2}$$

$$= 25 \times \frac{6}{2} \times \frac{383.8}{123.1} = 233.8$$

For precipitation and complex-formation titrations the equivalent weight is the number of grams which correspond to a single charge on the metal ion involved. In practice it is often the same as acid-base equivalent weight.

Heat calculations. Occasionally the question of heat must be considered in laboratory work. Heat quantities are usually expressed in units of calories or kilocalories. Heat added, heat of reaction, heat of cooling, and heat uptake by both reactants and products must be considered. In general, if there is no heat loss, the total heat must remain the same.

Industrial stoichiometry. All four groups of principles must be used in industrial stoichiometry. Flowing systems require the use of reaction rates and detailed considerations of heat losses, purity of materials, effects of catalysts, and side reactions. For example, the complete analysis of a plant engaged in the manufacture of sulfuric acid, starting with iron pyrites as the source of sulfur, is a time-consuming task. However, modern chemical industry is based on these detailed considerations. *See* CONCENTRATION SCALES; OXIDATION-REDUCTION; THERMOCHEMISTRY.

[KENNETH G. STONE/CHARLES L. RULFS]

Bibliography: E. Henley (ed.), *Stoichiometry*, 1972; L. K. Nash, *Stoichiometry*, 1966.

Streaming potential

The potential which is produced when a liquid is forced to flow through a capillary or a porous solid. G. H. Quincke (1859) found that the electromotive force produced by the streaming of pure water under a given pressure through a clay plate is independent of the size and thickness of the diaphragm and of the amount of water forced through the diaphragm; the electromotive force is, however, proportional to the pressure.

The streaming potential is one of four related electrokinetic phenomena which depend upon the presence of an electrical double layer at a solid-liquid interface. This electrical double layer is made up of ions of one charge type which are fixed to the surface of the solid and an equal number of mobile ions of the opposite charge which are distributed through the neighboring region of the liquid phase. In such a system the movement of liquid over the surface of the solid produces an electric current, because the flow of liquid causes a displacement of the mobile counterions with respect to the fixed charges on the solid surface. The applied potential necessary to reduce the net flow of electricity to zero is the streaming potential.

The principal objective of streaming potential measurements is the evaluation of zeta potentials at solid-liquid interfaces. Equation (1) may be used

$$\zeta = \frac{4\pi\eta\kappa E}{PD} \qquad (1)$$

for this purpose. Here ζ is the zeta potential, E is the streaming potential, η is the viscosity of the liquid, κ is the conductance of the liquid as it exists in the capillary system, P is the applied pressure, and D is the dielectric constant of the liquid.

An apparatus used by R. A. Gortner for measurement of streaming potentials at cellulose-water and alumina-organic liquid interfaces is shown in Fig. 1. Perforated gold or platinum electrodes e_1

and e_2 are located on either side of a pad of compacted powder or fibers of a selected solid in diaphragm D. Liquid is forced by compressed air to flow from reservoir R_1 through the solid and into reservoir R_2. The potential between the electrodes e_1 and e_2 is measured with an electrometer-potentiometer system. This potential is the streaming potential E.

In systems containing concentrations of electrolyte above 10^{-3} N, streaming potentials are too low to be measured accurately. Then, the current produced by the streaming liquid may be used to evaluate the zeta potential. For capillaries of known dimensions, Eq. (2) applies for the zeta potential.

$$\zeta = \frac{4\pi\eta LI}{DAP} \qquad (2)$$

Here I is the streaming current, L is the length of the capillary, and A is the cross-sectional area of the capillary. For porous solids of unknown capillary dimensions, the ratio L/A in Eq. (2) may be evaluated by measuring the resistance R of the diaphragm impregnated with a liquid of known electrical conductance κ. The ratio is obtained by means of Eq. (3).

$$L/A = \kappa R \qquad (3)$$

The zeta potentials obtained from Eqs. (1) and (2) are valid only when the flow of the liquid through the diaphragm is laminar and when the radius of curvature of the pores is greater than the thickness of the double layer.

Zeta potentials are useful in predicting the stabilities of lyophobic sols such as aqueous colloidal suspensions of oil, clay, gold, metal oxides, and so on. For these colloidal systems, electrostatic repulsion operating between particles of like charge stabilizes the particles against collisions that lead to irreversible coagulation. It is sufficient to say that particle charge is directly proportional to zeta potential and that there is a certain minimum zeta potential (the critical potential) above which a particular type of colloidal sol is stable for an indefinite length of time, and below which coagulation occurs in a relatively short time.

In one versatile streaming current device instead of a capillary, the flow system utilizes a loosely fitted piston, driven by a synchronous motor and reciprocating cam assembly, moving up and down in a plastic block containing a cylindrical, dead-ended bore. This forces the liquid sample to flow back and forth through the annular space between the cylinder wall and the piston. A diagram of the system is shown in Fig. 2.

Since the walls possess an electrical charge, a four-cycle alternating current is generated and collected at the electrodes. The measuring circuit consists of an output transformer (amplifier), a synchronous rectifier, and a microammeter (meter). As given in Eq. (2), the streaming current I is proportional to zeta potential ζ of the cylinder and piston walls.

When a liquid suspension of colloidal particles is placed in the measuring head, the walls of the piston and cylinder take on the charge characteristics of the charged particles by attachment of the particles to the walls. As a consequence, the streaming current developed at the walls reflects the zeta potential of the suspended particles. The attachment of particles to the walls is a rapid and essentially reversible process which makes it possible to follow changes in zeta potential of suspended particles as reagents of various types are added to the suspension.

It is suggested that the device is useful in controlling a wide range of processes involving coagulation and stabilization of colloidal systems in water

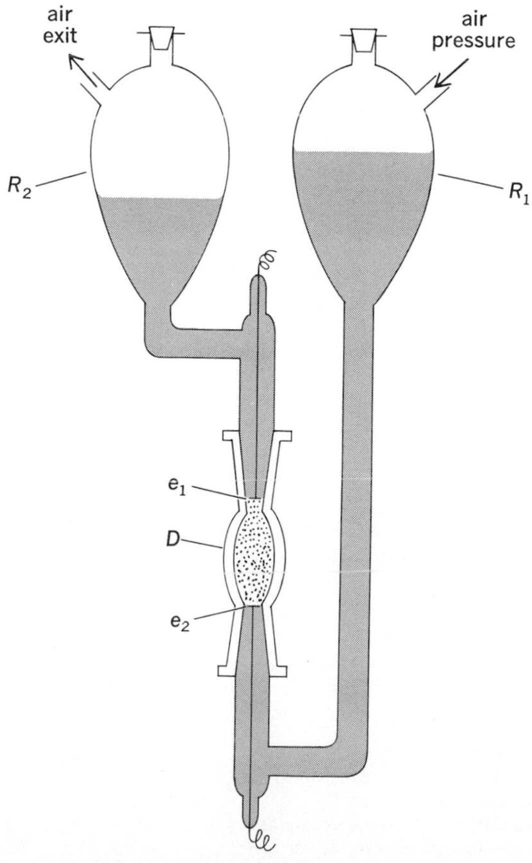

Fig. 1. Streaming potential apparatus.

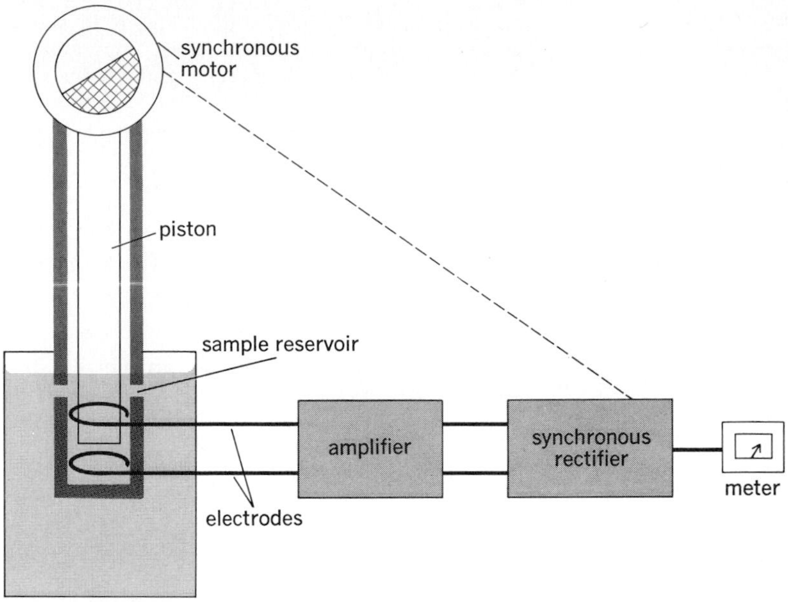

Fig. 2. Streaming current device.

purification, in sewage treatment, and in the manufacture of paper, paint, adhesives, photographic film, pharmaceuticals, cosmetics, and textiles.

[QUENTIN VAN WINKLE]

Bibliography: J. T. Davies, *Turbulence Phenomena: An Introduction to the Eddy Transfer of Momentum, Mass and Heat, Particularly at Interfaces*, 1972; W. F. Gerdes, *A Streaming Current Device*, ISA Meeting, Houston, Tex., May, 1966; K. J. Mysels, *Introduction to Colloid Chemistry*, 1959, reprint 1978.

Strontium

A chemical element, Sr, atomic number 38, and atomic weight 87.62. Strontium is the least abundant of the alkaline-earth metals. The crust of the Earth is 0.042% strontium, making this element as abundant as chlorine and sulfur. The main ores are

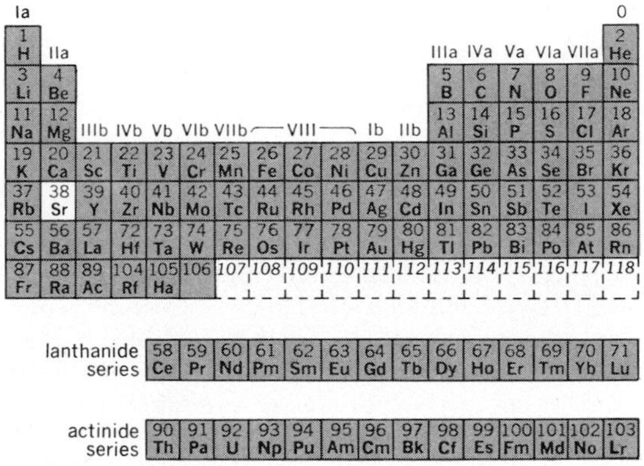

celestite, $SrSO_4$, and strontianite, $SrCO_3$, which are found chiefly in Scotland, Arkansas, and Arizona. Strontianite is colorless to light greenish or reddish, depending upon the impurities. The major contaminants of these ores are iron, aluminum, and calcium. Because calcium and strontium ions have chemical properties which are much alike, they readily replace each other in chemical compounds. This explains the occurrence of calcium in strontium ores and strontium in calcareous parts of plants and animals. Strontium also occurs as a fission product of nuclear reactions in the form of some 16 isotopes, 5 of which are stable.

It was Adair Crawford who, in 1790, first distinguished between naturally occurring strontium and barium carbonates, using a sample which had been unearthed in a lead mine in Strontian, Scotland. This was confirmed in 1792 by Thomas Hope. The name of the carbonate ore and the element itself stem from the place of original discovery.

Uses. Strontium nitrate is used in pyrotechnics, railroad flares, and tracer bullet formulations. Strontium hydroxide forms soaps and greases with a number of organic acids which are structurally stable, resistant to oxidation and breakdown over a wide temperature range, and resistant to disintegration by water and the leaching action of hydrocarbons. Other strontium compounds are used as paint driers and have minor medical uses.

Extraction of the metal. The pure metal was first isolated by Sir Humphry Davy, who electrolyzed a mixture of strontium and mercuric oxides with a mercury pool cathode; the mercury was then distilled away from the amalgam that had formed, leaving silvery globules of the pure metal. The element is produced commercially in the United States on a small scale either by electrolyzing a mixture of potassium and strontium chlorides or by reducing the oxide with aluminum in a vacuum at such a temperature that the strontium distills out. The metal is silvery white and lustrous, and it quickly forms a protective oxide coating in the air. It is softer than calcium and vigorously dissolves in water and acids to yield hydrogen gas. The element combines readily with oxygen to form the oxide when heated but will not form the nitride unless heated above 380°C. The physical properties of the element are given in the table.

Principal compounds. Celestite is used in the modern process for the production of strontium compounds in the United States and is obtained in two grades, 92 and 84–92% strontium sulfate. The ore is first treated with 10% hydrochloric acid and then with water, which leaches out the calcium sulfate and carbonate that are the chief impurities. A small excess of sodium carbonate is then added, and the solution is shaken at 150–160°F for 6 h. after which approximately 85% of the strontium is precipitated as the carbonate. Part of the supernatant liquid is then removed, more sodium carbonate is added, and the liquid is agitated for 10 more hours. The precipitate is then allowed to settle for 4 h. after which the clear supernatant liquid is decanted and the precipitate is washed three times with hot water. The precipitate is taken up again in hydrochloric acid and reprecipitated by the use of sodium carbonate. This precipitate is washed several times with hot water until the supernatant liquid is largely chloride-free, and then it is filtered, pressed dry, and ground to a powder. The carbonate is sold as a fine white powder of various grades for the preparation of other salts.

Strontium is divalent in all its compounds which are, aside from the hydroxide, fluoride, and sulfate, quite soluble. Strontium perchlorate has an appreciable solubility in organic solvents. The nitride is ionic and reacts with water to give ammonia, whereas strontium carbide, SrC_2, releases acetylene upon hydrolysis. The sulfide may be formed by reduction of the sulfate with charcoal, and it undergoes hydrolysis in water to form $Sr(SH)_2$. Strontium is a weaker complex former than calcium, giving a few weak oxy complexes with tartrates, citrates, and so on. The large electrical conductivity of strontium boride, SrB_6, approaches values characteristic of metals.

The reaction of strontium with hydrogen is vigorous and complete at temperatures of a few

Properties of strontium

Property	Value
Atomic number	38
Atomic weight	87.62
Isotopes (stable)	84, 86, 87, 88, 90
Electron configuration	2 8 18 8 2
Ionic radius, nm	0.113
Boiling point, °C	1638 (?)
Melting point, °C	704 (?)
Atomic volume, cm^3/g-atom	34.5
Density, g/cm^3 at 20°C	2.6
Latent heat of vaporization at boiling point, kJ/g-atom	164.0

hundred degrees. The hydride, SrH_2, reacts with alcohols to form the alcoholates and may be used in place of calcium hydride for organic condensation and reduction reactions. The complex alkyl, $SrZn(CH_2CH_3)_4$, has been isolated and has the high chemical reactivity of metal alkyls. $(CH_3CH_2)_2Sr$, which can be prepared only in solution, behaves as a Grignard reagent, as in the reaction

$$C_6H_5-C=CH_2 + (CH_3CH_2)_2Sr \rightarrow$$
$$\underset{\underset{C_6H_5}{|}}{}$$

$$C_6H_5-\underset{\underset{C_6H_5}{|}}{\overset{\overset{Sr-CH_2-CH_3}{|}}{C}}-CH_2-CH_2-CH_3 \xrightarrow{CO_2} C_6H_5-\underset{\underset{C_6H_5}{|}}{\overset{\overset{C-OH}{\|}}{\overset{O}{C}}}-C_3H_7$$

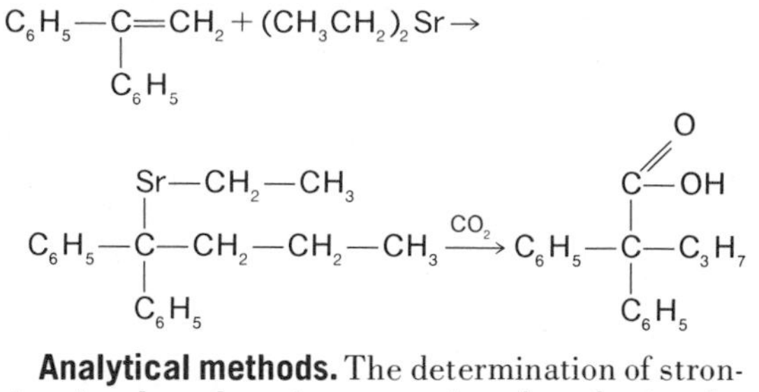

Analytical methods. The determination of strontium involves the prior separation of any barium by precipitation as the insoluble chromate and the subsequent precipitation and weighing of strontium sulfate. Strontium salts can be recognized qualitatively by their vivid crimson flame coloration. *See* ALKALINE-EARTH METALS.

[REED F. RILEY]

Bibliography: F. A. Cotton and G. A. Wilkinson, *Advanced Inorganic Chemistry*, 4th ed., 1980.

Strychnine

The principal alkaloid present in nux vomica, the seeds of a tree native to India, *Strychnos nux-vomica*. It was one of the first alkaloids to be isolated in a pure state in 1818 by P. Pelletier and J. Caventou. The complex structure (see illustration) provided a fascinating problem which was pursued intensively for over a century and was solved only in 1947. The synthesis of strychnine by R. Woodward and his coworkers in 1954 provided a confirmation of the structure.

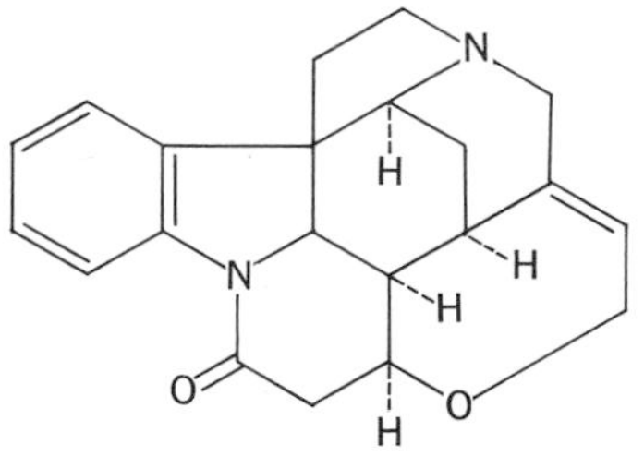

Structural formula of strychnine.

Nux vomica was introduced into Germany in the 16th century as a poison for rats and other animal pests. Strychnine was first employed in medicine in 1540, but it did not gain wide usage until 200 years later and has had an irregular career since then. In the medical practice of an earlier day it had a reputation as a cardiovascular stimulant, respiratory stimulant, and bitter tonic. Present-day opinion, however, holds that the therapeutically desirable effects are obtainable only with doses bordering on the toxic. Pharmacological studies have shown that many of the therapeutic applications of strychnine have little or no rationale. *See* ALKALOID.

[S. MORRIS KUPCHAN]

Sublimation

The process by which solids are transformed directly to the vapor state without passing through the liquid phase. Sublimation is of considerable importance in the purification of certain substances such as iodine, naphthalene, and sulfur.

Vapor pressure. All pure substances, whether solid or liquid, can exist in equilibrium with their vapor states, and the equilibrium pressure of the saturated vapor is called the vapor pressure of the solid or liquid at the temperature in question. In the illustration the area to the right of the line *AOC* comprises the infinite number of pressure-temperature conditions for which a substance exists solely in the vapor state. The area to the left of the line *AOB* represents the field of stability of the solid, in which the liquid or vapor cannot coexist with the solid. Similarly, the area between the lines *OB* and *OC* is the field in which only the liquid phase is stable. The lines define the pressure-temperature conditions for the stable coexistence of pairs of phases. Thus along the line *OA*, the solid and its vapor are in equilibrium, and the vapor pressure of the solid corresponding to any temperature along the line is unique. The single intersection of the three vapor-pressure curves at *O* is called the triple point and represents the only pressure and temperature at which the solid, liquid, and vapor can coexist in equilibrium under the pressure of the vapor alone. It is termed an invariant point, since neither temperature nor pressure may be varied without the disappearance of one of the phases. The lines are univariant, since one variable (either temperature or pressure) may be changed without causing the disappearance of a phase. A system of a pure substance in equilibrium between two phases is therefore said to possess one degree of freedom. The areas (single-phase regions) are bivariant, and possess two degrees of freedom, since both temperature and pressure may be independently varied without the

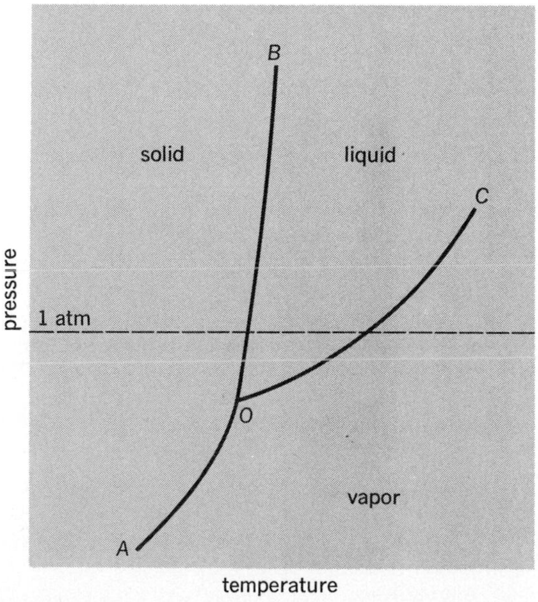

Vapor pressure–temperature diagram prepared for a pure substance.

disappearance of the phase. *See* PHASE EQUILIBRIUM; VAPOR PRESSURE.

Sublimation is a universal phenomenon exhibited by all solids at temperatures below their triple points. For example, it is a common experience to observe the disappearance of snow from the ground even though the temperature is below the freezing point and liquid water is never present. The rate of disappearance is low, of course, because the vapor pressure of ice is low below its triple point. Sublimation is a scientifically and technically useful phenomenon, therefore, only when the vapor pressure of the solid phase is high enough for the rate of vaporization to be rapid. Necessarily, this is a relative consideration.

Triple point. For most substances, the triple point occurs at a comparatively low pressure, and the rate of sublimation is accordingly low. For example, the triple point of water occurs at 0.0075°C and 4.56 mm pressure. For iodine, on the other hand, the triple point occurs at 114.15°C and 90.0 mm pressure. Accordingly, its rate of sublimation at 110°C would be quite high. In fact, the rate of sublimation is fast enough so that the iodine disappears by direct sublimation of the solid before the melting point is reached. The liquid phase is observed only if the vapor is confined in a vessel. For a relatively few substances, the triple point lies above 1 atm pressure, and the vapor pressure of the solid attains atmospheric pressure before the liquid phase appears. Thus, dry ice (solid CO_2) cannot be transformed to liquid CO_2 at atmospheric pressure. Instead, the solid sublimes to gaseous CO_2 without the intervention of the liquid state. The triple point of CO_2 is 5.11 atm at −56.4°C. The vapor pressure of solid CO_2 equals 1 atm at −78°C, on the other hand. Thus the freezing point is higher than the sublimation point, and CO_2 does not possess a normal boiling point.

Energy requirements. Both the vaporization of a liquid and the sublimation of a solid require the absorption of heat to overcome the potential energy of the molecules in the condensed state. The molar latent heat of sublimation is completely analogous to the molar latent heat of vaporization. It is equal to the heat of vaporization of the liquid plus the heat of fusion of the solid. Moreover, the Clausius-Clapeyron equation, Eq. (1), describes

$$\frac{dP}{dT}=\frac{\Delta H_s}{T\,\Delta V} \qquad (1)$$

the variation of the vapor pressure P of the solid with temperature T in similar fashion. Here ΔH_s is the molar latent heat of sublimation, and ΔV is the difference in molar volume of the vapor and solid at the temperature T. If the vapor obeys the ideal gas law ($PV=RT$), Eq. (1) may be put into the form of Eq. (2). Equation (2) may be used to calculate

$$\log_{10}\frac{P_2}{P_1}=\frac{\Delta H_s}{2.3R}\left(\frac{1}{T_1}-\frac{1}{T_2}\right) \qquad (2)$$

the latent heat of sublimation if the vapor pressure is known at two temperatures, or the vapor pressure at a second temperature if the latent heat of sublimation and the vapor pressure are known at one temperature. *See* CHEMICAL SEPARATION TECHNIQUES; EVAPORATION; MASS-TRANSFER OPERATION. [NORMAN H. NACHTRIEB]

Substitution reaction

One of a class of chemical reactions in which one atom or group (of atoms) replaces another atom or group in the structure of a molecule or ion. Usually, the new group takes the same structural position that was occupied by the group replaced.

Substitution reactions involve the attack of a reagent, which is the source of the new atom or group, on the substrate, the molecule or ion in which the replacement occurs. They involve the formation of a new bond and the breaking of an old bond. Substitution reactions are classified according to the nature of the reagent (electrophilic, nucleophilic, or radical) and according to the nature of the site of substitution (saturated carbon atom or aromatic carbon atom). *See* ELECTROPHILIC AND NUCLEOPHILIC REAGENTS.

Systematic names for substitution reactions are composed of the parts: name of group introduced + de + name of group replaced + ation, with suitable elision or change of vowels for euphony. Thus, the replacement of bromine by a methoxy group, as shown in Eq. (1), is called methoxydebromination.

Nucleophilic substitution at saturated carbon. This important class is exemplified by the reactions of alkyl halides with alkoxide ions to form dialkyl ethers, for example reaction (1). Other reac-

$$\underset{\substack{n\text{-Propyl bromide}\\ \text{(substrate)}}}{CH_3{-}CH_2{-}CH_2{-}Br} + \underset{\substack{\text{Methoxide ion}\\ \text{(reagent)}}}{OCH_3^-} \rightarrow \underset{\text{Methyl } n\text{-propyl ether}}{CH_3{-}CH_2{-}CH_2{-}O{-}CH_3} + \underset{\substack{\text{Bromide}\\ \text{ion}}}{Br^-} \qquad (1)$$

tive substrates include alkyl esters of sulfonic acids, quaternary ammonium salts, and tertiary sulfonium salts. Other effective reagents include mercaptide ions, halide ions, carbanions, water, and amines.

Two principal mechanisms have been recognized. The unimolecular or S_N1 mechanism involves two steps: dissociation (usually slow) of the substrate into a carbonium ion and another fragment, and combination (usually rapid) of the carbonium ion with a nucleophilic reagent. An example is the hydrolysis of *tert*-butyl chloride, as shown in reactions (2) and (3). This mechanism is

$$(CH_3)_3C{-}Cl \xrightarrow{\text{Slow}} (CH_3)_3C^+ + Cl^- \qquad (2)$$

$$(CH_3)_3C^+ + 2H_2O \xrightarrow{\text{Fast}} (CH_3)_3C{-}OH + H_3O^+ \qquad (3)$$

favored when the substrate can yield a carbonium ion of comparatively low energy. The bimolecular or S_N2 mechanism involves one step in which formation of the new bond is simultaneous with breaking of the old bond. This mechanism is favored when the reagent has strong nucleophilic character and when the site of substitution is easily accessible to attacking reagents. The reaction of *n*-propyl bromide in (1) is an example of the S_N2 mechanism. The mechanisms of some reactions are intermediate between S_N1 and S_N2.

Electrophilic substitution at saturated carbon. This comparatively minor case is exemplified by the acid cleavage of alkyl mercury compounds, as shown in reaction (4).

$$CH_3HgCl + HCl \rightarrow CH_4 + HgCl_2 \qquad (4)$$

Radical substitution at saturated carbon. Radical substitution usually involves a chain reaction. Such a reaction has three phases: initiaton (in which radicals are generated), propagation (in which most of the actual reaction occurs), and termination (in which radicals are destroyed). These are illustrated by the mechanism of chlorination of methane, which is given in reactions (5).

Initiation: $Cl_2 + \text{photon of light} \rightarrow 2Cl\bullet$

Propagation: $Cl\bullet + CH_4 \rightarrow HCl + CH_3\bullet$
$CH_3\bullet + Cl_2 \rightarrow CH_3Cl + Cl\bullet$

Termination: $CH_3\bullet + CH_3\bullet \rightarrow CH_3—CH_3$
$CH_3\bullet + Cl\bullet \rightarrow CH_3Cl$
$Cl\bullet + Cl\bullet \rightarrow Cl_2$ (5)

The propagation steps occur in a cyclic repetitive fashion: A product of one step is a vital reactant in another step. Hundreds of acts of propagation often occur for each act of initiation or termination.

Nucleophilic substitution at aromatic carbon. An example is the reaction of *p*-fluoronitrobenzene with ammonia, given in (6). In these reactions

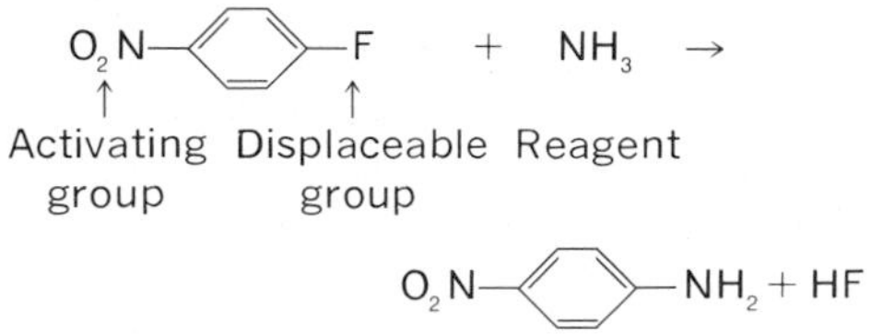

(6)

the usual nucleophilic reagents are effective. Common displaceable groups include the halogens, sulfonyl groups (—SO_2R), ammonio groups (—NR_3^+), the nitro group (—NO_2), and several others which are able to depart with an electron pair as stable anions or molecules. Hydrogen is seldom displaced. Activating groups are usually necessary to obtain reasonable reaction rates. Effective activating groups are those of strong electron-attracting character, such as —N_2^+, —NO_2, —SO_2CH_3, —$COCH_3$, and —CN. The hetero nitrogen atom in pyridine and related heterocycles is also a strong activating structure.

Most nucleophilic substitutions at aromatic carbon occur by a two-step intermediate complex mechanism. In the first step, the reagent becomes covalently attached to the carbon atom at the site of substitution to form a metastable intermediate complex. In the second step, the displaceable group is detached from the same carbon atom. Either formation of the intermediate complex or expulsion of the displaceable group may be the rate-limiting step. Some aromatic nucleophilic substitutions occur via an elimination-addition mechanism involving benzyne intermediates, and a few occur by the S_N1 mechanism.

Electrophilic substitution at aromatic carbon. The mechanism of this type of reaction usually involves two steps and an intermediate reaction complex: The reagent attaches to the carbon atom at the site of substitution, and the displaceable group then detaches. Hydrogen is the group most commonly displaced, but sulfo (—SO_3H), phosphono (—PO_3H_2), carboxyl (—COOH), mercuri (—HgCl, for example), and other groups may also be displaced. Many electrophilic reagents are effective; reactions are classified according to the group introduced, which is derived from the reagent. *See* DIAZOTIZATION; FRIEDEL-CRAFTS REACTION; HALOGENATION; NITRATION; SULFONATION.

Electrophilic aromatic substitutions are activated (accelerated) by electron-releasing substituents such as —OH, —O—CH_3, —NH_2, —NH—CO—CH_3, and —CH_3. Deactivating groups include —NO_2, —SO_3H, —COOH, and —$COCH_3$. Each type, activating or deactivating, acts most strongly on the ortho and para positions. A common practical problem is to predict which of five displaceable hydrogen atoms in a monosubstituted benzene will be displaced by an electrophilic reagent. Activating groups such as —OH are ortho-, para-directing (that is, they direct displacement of an ortho or para hydrogen). Deactivating groups, since they deactivate meta positions least, are meta-directing. Halogen substituents, being mildly deactivating and yet ortho-, para-directing, constitute a special category.

Radical substitution at aromatic carbon. This also occurs by an intermediate complex mechanism, but details are not well understood. An example is the phenylation of chlorobenzene, reaction (7). The phenyl radical may be obtained in various ways, such as by the decomposition of benzoyl peroxide or of *N*-nitrosoacetanilide by heat.

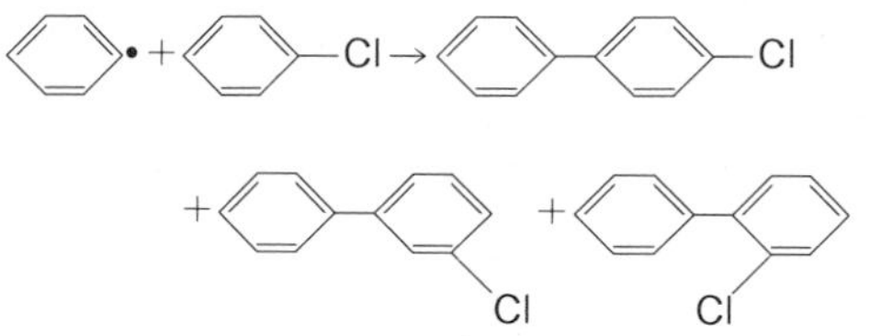

(7)

Nucleophilic substitution at carbonyl carbon. Reactions representative of this type are the hydrolysis of acid chlorides, acid anhydrides, and esters, or the reactions of these substrates with ammonia to form amides. These reactions usually occur by an intermediate complex mechanism, although in some cases the S_N1 mechanism prevails. Thus, common base-catalyzed ester hydrolysis (saponification) occurs as in reaction (8).

$$\underset{\text{Ester}}{R—\overset{\overset{\displaystyle O}{\|}}{C}—O—R'} + OH^- \rightleftharpoons \underset{\text{Intermediate complex}}{R—\overset{\overset{\displaystyle O^-}{|}}{\underset{\underset{\displaystyle OH}{|}}{C}}—O—R'} \rightarrow$$

$$R—\overset{\overset{\displaystyle O}{\|}}{C}—O^- + R'—OH \qquad (8)$$

See ORGANIC CHEMICAL SYSTHESIS; ORGANIC REACTION MECHANISM.

[JOSEPH F. BUNNETT]

Bibliography: F. Badea, *Reaction Mechanisms in Organic Chemistry*, 1976; C. H. Bamford and C. F. Tipper, *Comprehensive Chemical Kinetics*, vol. 12: *Electrophilic Substitution at a Saturated Carbon Atom*, 1973; D. I. Davies and M. J. Parrott, *Free Radicals in Organic Synthesis*, 1978; C. K. Ingold, *Structure and Mechanism in Organic Chemistry*, 2d ed., 1969.

Sulfamate

Organic and inorganic derivatives of sulfamic acid, NH_2SO_2OH.

The inorganic salts which contain the NH_2SO_2-O^- ion are all very soluble with the exception of the basic mercury salt.

The sulfamate ion may be determined quantitatively by measuring the volume of nitrogen liberated during the reaction shown below.

$$NaNO_2 + NH_2SO_2OH \rightarrow NaHSO_4 + N_2 + H_2O$$

Ammonium sulfamate is used to flameproof fabrics and as a weed killer.

Because of their high solubilities, the sulfamates of nickel, copper, and lead have been used in electroplating baths.

Useful organic sulfamates include the cyclohexylsulfamates, which are sweetening agents, and amine salts, which are used as softeners for paper and textiles. *See* SULFAMIC ACIDS.

[E. EUGENE WEAVER]

Sulfamic acids

Organosulfur compounds, $RNHSO_3H$, that are organic derivatives of sulfamic acid, H_2NSO_3H. Many aliphatic examples are known, but free aromatic sulfamic acids, such as phenylsulfamic acid, are not stable, except in the form of salts. A series of stable 2-thiazolylsulfamic acids has been prepared, however, and their stabilities rationalized. Sodium 2-thiazolysulfamate, like sodium cyclohexylsulfamate, is very sweet and has no bitter aftertaste. The cyclohexylsulfamate salts have been widely used as sweetening agents (cyclamates). Their therapeutic safety was, however, challenged and their use was curtailed. *See* ORGANOSULFUR COMPOUND; SACCHARIN.

[NORMAN KHARASCH]

Sulfanilic acid

An aromatic organic compound that contains both an amino group and a sulfonic acid group. It is sometimes called *p*-aminobenzenesulfonic acid. Its formula $H_2N—C_6H_4—SO_3H$ is better written as an inner salt (I).

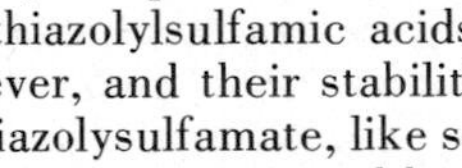

(I)

Synthesis consists of heating aniline sulfate to 190°C until it rearranges. Sulfanilic acid crystallizes with one molecule of water, which it loses at 100°C. Diazotization of sulfanilic acid yields the diazonium salt (II), the starting compound for

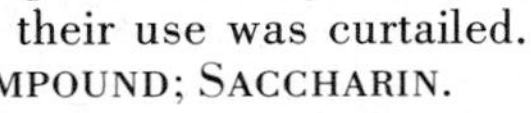

(II)

numerous coupling reactions with phenols or tertiary amines to give azo dyes and acid-base indicators. For example, coupling with dimethylaniline yields methyl orange, *p,p'*-dimethylaminoazobenzenesulfonic acid (III).

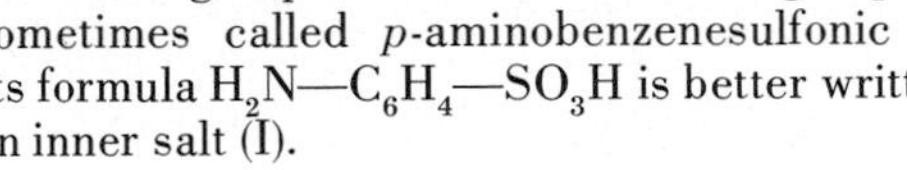

(III)

One important use of sulfonic acid is as a starting material for the preparation of many of the sulfa drugs. *See* ANILINE; ORGANOSULFUR COMPOUND; SULFONIC ACID.

[LEALLYN B. CLAPP]

Sulfate

A negative ion having the formula SO_4^{2-} and derived from sulfuric acid, H_2SO_4. Because sulfuric acid contains two hydrogens, both normal sulfates and bisulfates are known. Of the normal sulfates, most are quite soluble in water, with the exception of those of silver, mercury(I), lead, strontium, barium, and calcium. Sulfate is determined both qualitatively and quantitatively by precipitation as barium sulfate, $BaSO_4$, which is the most insoluble sulfate. Sodium sulfate, obtained as a by-product of the production of hydrochloric acid from salt, is used in making kraft paper, paperboard, and glass. *See* SULFUR; SULFURIC ACID. [E. EUGENE WEAVER]

Sulfenyl chlorides

A group of well-known organosulfur compounds, RSCl. They are highly reactive, but can generally be synthesized and isolated. They undergo numerous reactions in which chlorine is displaced from sulfur by negative ions or bases, for example, cyanide ion, amines, olefins, or acetylenes. The best-known examples are trichloromethanesulfenyl chloride, also called perchloromethyl mercaptan (PMM), Cl_3CSCl, made by controlled chlorination of carbon disulfide, and 2,4-dinitrobenzenesulfenyl chloride, $(NO_2)_2C_6H_3SCl$, a versatile reagent. Reaction of sulfenyl chlorides with amines gives sulfenamides, some of which are useful as rubber vulcanization accelerators. *See* MERCAPTAN; ORGANOSULFUR COMPOUND.

[NORMAN KHARASCH]

Bibliography: N. Kharasch, *Mechanisms of Reactions of Sulfur Compounds*, 5 vols., 1971; S. Oae (ed.), *Organic Chemistry of Sulfur*, 1977.

Sulfide

A negative ion having the formula S^{2-} and derived from hydrogen sulfide or hydrosulfuric acid, H_2S. Because hydrosulfuric is a dibasic acid, sulfides, M_2S, and bisulfides, MHS, are formed. Although most of the metal sulfides are insoluble in water, they dissolve in acids with the evolution of H_2S gas. This escaping gas causes paper moistened with a solution of lead acetate to turn black, a test that can be used to detect the presence of sulfide ion in the original solution.

The insoluble metal sulfides are used to separate and identify metal ions in qualitative analysis. Soluble sulfide solutions are very basic because of the tendency of the sulfide ions to accept a proton from water, as shown in the reaction below.

$$H_2O + S^{2+} \rightleftharpoons HS^- + OH^-$$

See SULFUR. [E. EUGENE WEAVER]

Sulfinic acid

One of a group of organosulfur compounds, RSO_2H, that possess one less oxygen than sulfonic acids and are easily oxidized to the latter. In contrast to sulfonic acids, sulfinic acids undergo self-oxidation-reduction (dismutation), as shown in the reaction below. The compound in brackets is a

$$2RSO_2H \rightarrow RSO_3H + [RSOH]$$

reactive nonisolable intermediate, a sulfenic acid, which decomposes in various ways, including reaction with RSO_2H, to give RSO_2SR (a thiolsulfonate ester). Because of instabilities, sulfinic acids are often prepared and used as salts such as sodium *p*-toluenesulfinate.

Aromatic sulfinic acids are better known and more stable than those of the aliphatic series. Only one free aliphatic sulfinic acid, 1-dodecanesulfinic acid, $C_{12}H_{25}SO_2H$, has been prepared in crystalline form. 1,4-Butanedisulfinic acid has been reported as uniquely stable and readily prepared.

Sulfinic acid salts are used in organic synthesis, as additives to electroplating baths, as redox polymerization catalysts, and as reducing agents. Rongalite, $HOCH_2SO_2Na$, is an important commercial reducing agent. *See* ORGANOSULFUR COMPOUND; POLYMERIZATION. [NORMAN KHARASCH]

Sulfite

A negative ion having the formula SO_3^{2-} and derived from unstable sulfurous acid, H_2SO_3.

Because the sulfur in the sulfite ion has a 4+ oxidation state, it undergoes oxidation to the sulfate ion or reduction to free sulfur or to the sulfide ion. The sulfite ion can be identified by the fact that SO_2 gas is liberated when solutions are acidified, as represented by the reaction below.

$$2H^+ + SO_3^{2-} \rightleftharpoons [H_2SO_3] \rightleftharpoons H_2O + SO_2$$

Because sulfurous acid is dibasic, two series of salts, the normal M_2SO_3 and acid $MHSO_3$, are formed. Most of these salts, with the exception of the alkali metal and ammonium salts, are only slightly soluble. Sodium sulfite is used in dyeing and bleaching and as a preservative; other sulfites are used extensively in the manufacture of certain types of paper. *See* SULFUR. [E. EUGENE WEAVER]

Sulfonamide

One of a group of organosulfur compounds, RSO_2NH_2, that are readily prepared by the reaction of sulfonyl chlorides and ammonia, as in the reaction below. There is great interest in these

$$RSO_2Cl + 2NH_3 \rightarrow RSO_2NH_2 + NH_4Cl$$

substances because of the therapeutic sulfa drugs, but they are also of chemical value. The Hinsberg procedure of organic analysis converts amines to sulfonamides with *p*-toluenesulfonyl chloride as the reagent. A primary amine is distinguishable from a secondary amine because the resultant sulfonamide, $RNHSO_2$-*p*-tolyl, from the primary amine still has an acidic H atom attached to nitrogen and, hence, is soluble in aqueous alkali, whereas R_2N-SO_2-*p*-tolyl is not soluble. *See* AMINE; ORGANOSULFUR COMPOUND; SULFONYL CHLORIDE. [NORMAN KHARASCH]

Sulfonation

A chemical reaction in which a sulfonic acid group, —SO_3H, is introduced into the structure of a molecule or ion in place of a hydrogen atom.

Sulfonation of aromatic compounds is the most important type of sulfonation. This is accomplished by treating the aromatic compound with sulfuric acid, usually containing sulfur trioxide (solutions of SO_3 in sulfuric acid are called oleum, or fuming sulfuric acid). Reaction (1) illustrates

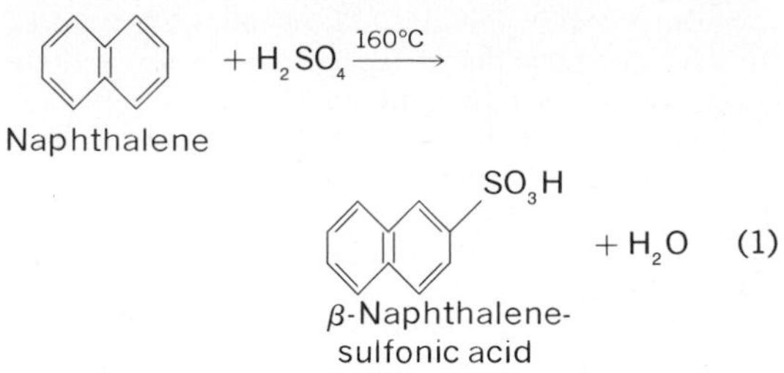

this. This is an electrophilic aromatic substitution reaction; the effective electrophilic reagent is believed to be the SO_3 molecule. The product of sulfonation is a sulfonic acid. [JOSEPH F. BUNNETT]

Sulfonation may also be defined as any chemical process by which the sulfonic acid group, —SO_2OH, or the corresponding salt or sulfonyl halide group, for example, —SO_2Cl, is introduced into an organic compound. These groups may be bonded to either a carbon or a nitrogen atom. The latter compounds are designated *N*-sulfonates or sulfamates.

Sulfation involves the attachment of the —OSO_2OH group to carbon, yielding an acid sulfate, $ROSO_2OH$, or of the —SO_4— group between two carbons, forming the sulfate, $ROSO_2OR$.

Uses of sulfonates and sulfates. Millions of tons of sulfonates are manufactured annually. Most sulfonates are employed as such in acid or salt form for applications where the strongly polar hydrophilic —SO_2OH group confers needed properties on a comparatively hydrophobic nonpolar organic molecule. Some sulfonates, such as methanesulfonic and toluenesulfonic acids, are used as catalysts. A relatively large number of sulfonates are marketed in salt form but used in acid form; such compounds include dyes, mothproofing agents, and synthetic tanning agents. In these cases the salts are applied in acid medium, thereby permitting the free —SO_2OH group of the organic molecule to become attached to the textile fiber or leather. The major quantity of sulfonates and sulfates is both marketed and used in salt form. This category includes detergents; emulsifying, demulsifying, wetting, and solubilizing agents; lubricant additives; and rust inhibitors.

Aromatic sulfonyl chlorides, —$ROSO_2Cl$, are useful for preparing sulfonamides (including sulfa drugs, dyes, tanning agents, plasticizers, and the sweetening agent saccharin).

Sulfonates and sulfates are employed for the preparation of organic compounds devoid of sulfur. Thus phenol, resorcinol, and naphthols are obtained by the caustic fusion of the parent sulfonates, whereas ethanol and isopropanol are obtained by the hydrolysis of sulfated alkenes, for example, ethyl hydrogen sulfate.

Sulfonating and sulfating agents. The principal agents are as follows:

1. Sulfur trioxide and compounds: (*a*) sulfur trioxide, oleum ($H_2SO_4 + SO_3$), and concentrated sulfuric acid (SO_3 + water); (*b*) chlorosulfonic acid (SO_3 + HCl); (*c*) sulfur trioxide adducts with organic compounds (SO_3 + dioxane); (*d*) sulfamic acid.
2. Sulfur dioxide group: (*a*) sulfurous acid and

metal sulfites; (*b*) sulfur dioxide with chlorine or oxygen.

Sulfur trioxide, oleum, and concentrated sulfuric acid are considered together because of their close physical relationship and because they can, in certain cases, be used interchangeably. This group accounts for the preponderant production of aromatic sulfonates.

Sulfur trioxide is theoretically the most efficient sulfonating and sulfating agent, because only direct addition is involved. This direct addition is shown by reactions (2) and (3). However, sulfur trioxide combines with water and aromatic compounds the evolution of so much heat that its activity must be moderated. For this reason the hydrates of SO_3 (oleum and sulfuric acid) are generally used. Here, SO_3 is the true reactive species, the water functioning only as a complexing agent and a solvent. An increase in the water content lowers the activity of the reagent for sulfonation; the reaction rate is inversely proportional to the square of the water concentration. When the SO_3 concentration in the sulfonating agent has been reduced to a critical level, which depends upon the organic compound being treated, sulfonation stops. The critical concentrations (π values) for the monosulfonation of naphthalene, benzene, and nitrobenzene are approximately 52, 64, and 82% SO_3.

$$RH + SO_3 \rightarrow RSO_3H \quad (2)$$

$$ROH + SO_3 \rightarrow ROSO_3H \quad (3)$$

Catalysts. The addition of certain chemicals, usually in small amounts, can have a marked influence on some sulfonations. The addition of mercury changes the orientation in a number of aromatic sulfonations. This is of great importance in the preparation of α-anthraquinone sulfonates. In the absence of mercury compounds, the β-sulfonates are obtained exclusively. About 1% mercury, as metal or salt equivalent, based on anthraquinone used is required. Mercury also affects the orientation in the sulfonation of benzoic acid, phthalic anhydride, and nitrobenzene. In these reactions the quantity of mercury needed is high, and the influence on orientation is only partial.

Aromatic sulfonation, like nitration and halogenation, is a typical electrophilic substitution reaction. Sulfonation, however, differs from these other reactions in two respects: It is reversible, and in certain cases (such as in the sulfonation of naphthalenes) temperature has an important influence on the position of the entering group.

Equipment. Cast iron is resistant to the action of sulfuric acid in the range 75–100% in strength over a fairly wide temperature range and has been a standard material of construction for sulfonation kettles for many years, especially for numerous dye intermediates and for aromatic hydrocarbons. However, it has poor tensile strength and is corroded by oleum or sulfur trioxide. This fault may be controlled in oleum sulfonations by adding acid slowly to the material being sulfonated to keep the acid concentration below the corrosive level.

The use of lined steel vessels combines low cost and high strength with good corrosion resistance. Commonly used linings include glass, enamel, lead, and type 316 stainless steel.

Continuous operations involving special equipment are advantageous only where the reaction rate is fast and where the volume of production (as in benzenesulfonic acid and dodecylbenzene sulfonate) is large and relatively steady. *See* ORGANOSULFUR COMPOUND; SUBSTITUTION REACTION.

[P. H. GROGGINS/ROBERT S. KAPNER]

Bibliography: E. E. Gilbert, *Sulfonation and Related Reactions*, 1965, reprint 1978; E. C. Herrick et al., *Unit Process Guide to Organic Chemical Industries*, 1979.

Sulfone

One of a group of well-known, stable, and generally crystalline organosulfur compounds, RSO_2R'. Sulfones are best prepared by the oxidation of sulfides, and also by the reaction of sulfonyl chlorides, RSO_2Cl, with aromatic compounds in the presence of Friedel-Crafts catalysts, or by reaction of sodium sulfinates and alkyl bromides. Sulfones resist further oxidation, but can be reduced, under special conditions, to sulfides. In the presence of Raney nickel, with absorbed hydrogen, RSO_2R' is reduced to $RH + R'H$ through a carbon-sulfur bond-scission process. Pyrolysis of sulfones frequently involves elimination of sulfur dioxide in a synthetically useful way. For example, alkenyl aryl sulfones smoothly yield alkenes. The sulfone group exerts a strong inductive effect and increases the acidity of α-hydrogen atoms. In $(CH_3SO_2)_3CH$, the acidity becomes very high. The $—SO_2—$ group also deactivates the benzene ring to attack by electrophilic reagents. Numerous practical uses have been claimed for variously substituted sulfones. *See* ORGANIC REACTION MECHANISM; ORGANOSULFUR COMPOUND; SULFIDE; SULFOXIDE. [NORMAN KHARASCH]

This group of chemical compounds is used in the treatment of leprosy. Although 4,4′-diaminodiphenylsulfone, or DDS (Fig. 1), was first synthesized in 1908, it was examined as a possible chemotherapeutic agent only in 1937, when it was found as an impurity in the manufacture of sulfanilamide. Investigators in England and France showed independently that this compound was very effective in curing streptococcal infections in mice. DDS is 100 times more potent than sulfanilamide against *Streptococcus pyogenes* in mice. However, when it was first used in man for the treatment of acute streptococcal infection, it had to be discontinued because of the severe hemolytic anemia which developed. Attempts were made to produce less toxic derivatives, and a number of analogs with substituents on the amino groups were synthesized (Fig. 2).

In 1940 DDS was found to be active against experimental tuberculosis in rabbits, and promin was shown to be effective in guinea pigs infected with human tubercle bacilli. In man, however, promin had disappointing antitubercular activity. It was probably the reports of antitubercular activity in animals that prompted as early as 1941 the trials of

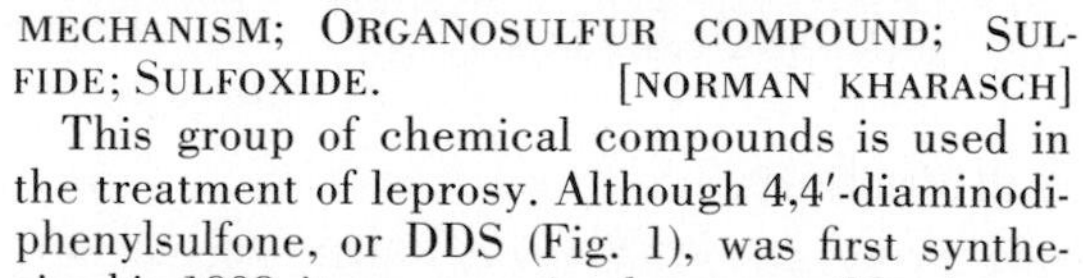

Fig. 1. Formula for 4,4′-diaminodiphenylsulfone (DDS).

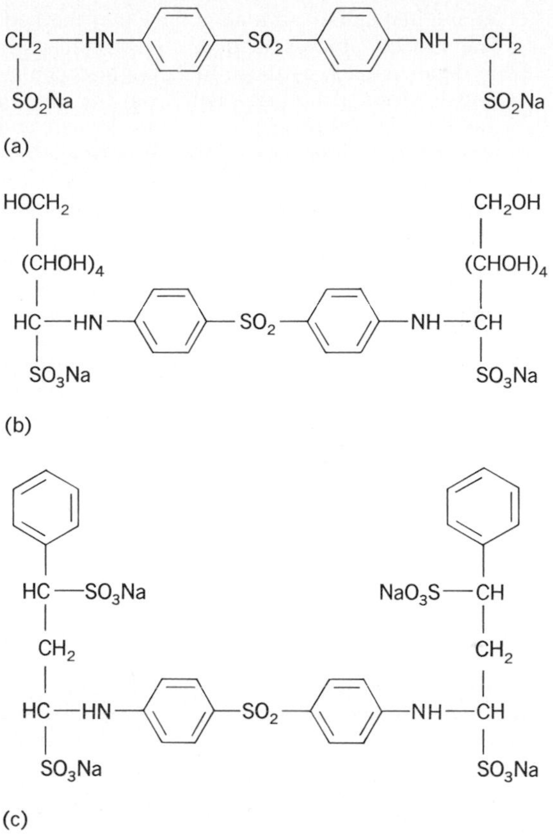

Fig. 2. Formulas for some derivatives of DDS. (*a*) Diasone. (*b*) Promin. (*c*) Sulfetrone.

sulfones in the treatment of leprosy. This marked the beginning of the most significant advance yet made in the treatment of leprosy. DDS is now the most effective treatment for this disease.

[NICHOLAS J. GIARMAN/EMANUEL GRUNDBERG]

Sulfonic acid

One of a group of organosulfur compounds with the formula below. Typical examples are methane-

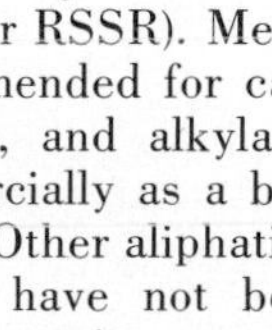

sulfonic acid, CH_3SO_3H, and benzenesulfonic acid, $C_6H_5SO_3H$. They are strongly acidic, water-soluble, nonvolatile, and hygroscopic, and they do not act as oxidizing agents. The aliphatic sulfonic acids are made by oxidizing mercaptans or disulfides (RSH or RSSR). Methane sulfonic acid has been recommended for catalyzing esterifications, hydrolyses, and alkylations. The acid is available commercially as a by-product from petroleum refining. Other aliphatic sulfonic acids are also known but have not been as extensively studied as the aromatic compounds. 10-Camphorsulfonic acid, derived from camphor, is well known, and certain substituted aliphatic sulfonic acids such as ethionic acid, $HOCH_2CH_2SO_3H$, and taurine, $NH_2CH_2CH_2SO_3H$, are of special interest, industrially and biochemically. Hydroxy acids, such as $HOCH_2CH_2CH_2SO_3H$, form sultones analogous to the lactones of hydroxy-substituted carboxylic acids. Fatty-acid esters of ethionic acid and amides of taurine find use as surface-active agents.

Aromatic sulfonic acids are made by sulfonation of aromatic compounds. Sulfuric acid, fuming sulfuric acid ($H_2SO_4 + SO_3$), chlorosulfonic acid ($ClSO_3H$), or sulfur trioxide may be used to introduce the sulfonic acid group. Aromatic sulfonic acids and their derivatives are important industrial chemicals. Of special value is their use as detergents, for example, sodium dodecylbenzenesulfonate. Sulfonated polymers act as cation-exchange resins, and sulfonamide derivatives are valuable pharmaceuticals. The $—SO_3H$ group lends water solubility to many substances, hence increases their usefulness. This application is particularly used in the manufacture of dyes and in some indicators, for example, Congo red or methyl orange. Aromatic sulfonic acids also have applications as emulsifying agents, lubricating-oil additives, and rust inhibitors.

In synthesis, the SO_3H group can be replaced by Br or NO_2 groups by treating with bromine or nitric acid. The sodium salts, $ArSO_3^-Na^+$, yield phenols by fusing with alkali, and acidifying the melt. On fusion with sodium cyanide, they also yield nitriles, as in the reaction below.

$$ArSO_3Na + NaCN \rightarrow ArCN + Na_2SO_3$$

Sulfonation of aromatic hydrocarbons is a reversible process. Hence, treatment of $ArSO_3H$ with superheated steam removes the $—SO_3H$ group and is useful in purifying and separating aromatic hydrocarbons. The greatest utility of sulfonic acids as synthetic intermediates lies in their conversion to sulfonyl chlorides, which have a broad and useful reactivity. *See* ORGANOSULFUR COMPOUND; SACCHARIN; SULFONAMIDE; SULFONATION; SULFONYL CHLORIDE; SULFURIC ACID.

[NORMAN KHARASCH]

Sulfonyl chloride

One of a group of organosulfur compounds, RSO_2Cl, that are useful intermediates for synthesis and analysis. They can be prepared by reactions of sulfonic acids, or their salts, with phosphorus chlorides, or by oxidative chlorination of thiols or disulfides. The synthetic value lies in reductions to sulfinic acids, disulfides, or thiols, and in many displacement reactions of the chloride group: with amines, to sulfonamides; with thiols, to thiolsulfonate esters; with aromatic hydrocarbons, to sulfones (Friedel-Crafts reaction); with fluoride ion, to sulfonyl fluorides; and with alcohols, to sulfonate esters. *p*-Toluenesulfonyl chloride (tosyl chloride) and *p*-nitrophenylazobenzenesulfonyl chloride are often used for making derivatives in qualitative organic analysis. The characteristic melting points of the products identify the unknown amine, alcohol, or thiol. *See* AMINE; ORGANOSULFUR COMPOUND.

[NORMAN KHARASCH]

Sulfoxide

One of a group of organosulfur compounds with the formula below; sulfoxides are produced by con-

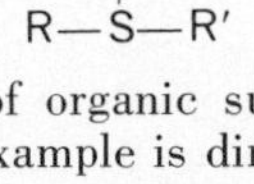

trolled oxidation of organic sulfides (thioethers). The best-known example is dimethyl sulfoxide. If

R and R′ are different, the sulfoxide can exist in two optically active forms. Thus, the sulfur-oxygen bond is not a double bond (as in > C=O), because this would require a planar structure for sulfoxides. A pyramidal structure, with a dipolar $S^+ \rightarrow O^-$ bond, is assumed. Sulfoxides are not as well known as sulfones, but an extensive chemistry of the sulfoxide group is building up. *See* OPTICAL ACTIVITY; ORGANOSULFUR COMPOUND; SULFONE.

[NORMAN KHARASCH]

Sulfur

A chemical element, S, atomic number 16, and atomic weight 32.064. The known stable isotopes and approximate percent abundances in natural sulfur are S^{32} (95.1%); S^{33} (0.74%); S^{34} (4.2%); and S^{36} (0.016%). Sulfur was discovered prior to recorded history. Its elemental character was first recognized by A. L. Lavoisier in 1777.

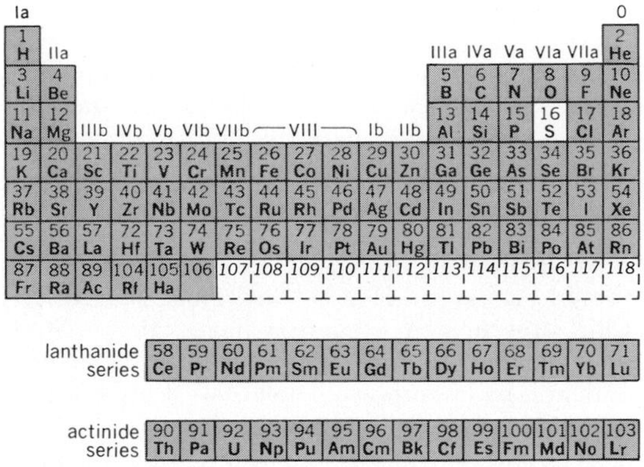

THE ELEMENT

The abundance of sulfur in the Earth's crust is 0.03–0.1%. It is often found as the free element near volcanic regions (impure deposits) in Japan, Sicily, and Mexico. Other deposits are located in New Zealand, Chile, the Soviet Union, Iceland, and Spain. The largest known free sulfur deposits by far are in Texas and Louisiana and are associated with limestone and anhydrite caprock formations over salt domes. Other noteworthy deposits occur in California, Colorado, Wyoming, Nevada, Utah, Mexico, and South America. Combined sulfur exists primarily in sulfates and sulfides such as calcium sulfate dihydrate (gypsum, $CaSO_4 \cdot 2H_2O$), barium sulfate (barite, $BaSO_4$), magnesium sulfate heptahydrate (epsom salt, $MgSO_4 \cdot 7H_2O$), sodium sulfate decahydrate (glauber salt, $Na_2SO_4 \cdot 10H_2O$) (the last two usually occur in mineral springs), strontium sulfate (celestite, $SrSO_4$), lead sulfide (galena, PbS), zinc sulfide (zinc blende, ZnS), copper iron disulfide (chalcopyrite, $CuFeS_2$), iron disulfide (iron pyrites, FeS_2), and mercury sulfide (cinnabar, HgS). It also occurs in mineral springs as hydrogen sulfide (H_2S) and is found in plants and animals as a constituent of such substances as eggs, mustard, garlic, cabbage, horseradish, wool, and hair. It is also found in organic materials such as coal and petroleum, and has even been found in meteorites.

Preparation. The extraction of sulfur is usually carried out by any of three methods. The most important is the Frasch process, developed in 1891 by Herman Frasch. Of less importance are the Sicilian method and a variation of the Claus method.

The Frasch process is used to extract sulfur from deposits such as those in Texas and Louisiana. It consists of boring a hole from the ground surface to the sulfur-bearing calcite deposit and lowering three pipes, concentrically arranged, to the ore bed (Fig. 1). Superheated water (165°C) is forced down the largest (6-in. or 15 cm) pipe into the ore bed where it melts the sulfur (melting point 112.8°C). Compressed hot air is pumped down the smallest (1-in. or 2.5-cm) pipe, and a frothy mixture of molten sulfur, water, and air is forced to the surface through the intermediate (3-in. or 7.5-cm) pipe. As it comes from the well, the sulfur has a purity of 99.5–99.9% and contains virtually no arsenic, selenium, or tellurium.

The Sicilian method consists of piling the sulfur-bearing rock into large mounds called calcaroni, which are ignited at the top. The heat of combustion of the sulfur in the ore causes underlying layers of sulfur to melt; this molten sulfur is poured into molds and is allowed to solidify. It often takes several months for a mound to be depleted, and only about 60% of the sulfur present in the original mound is recovered, because a large part of it is used as fuel during the melting process. This sulfur is impure and is usually refined by distillation. When the sulfur vapor is allowed to solidify directly on the walls of large masonry chambers, it is called flowers of sulfur because of the flowerlike designs in which it deposits. If the chambers are kept above the melting point of sulfur, the vapors condense to liquid sulfur, which is allowed to solidify in wooden molds. This form of the element is called roll sulfur.

Variations of the Claus method are sometimes

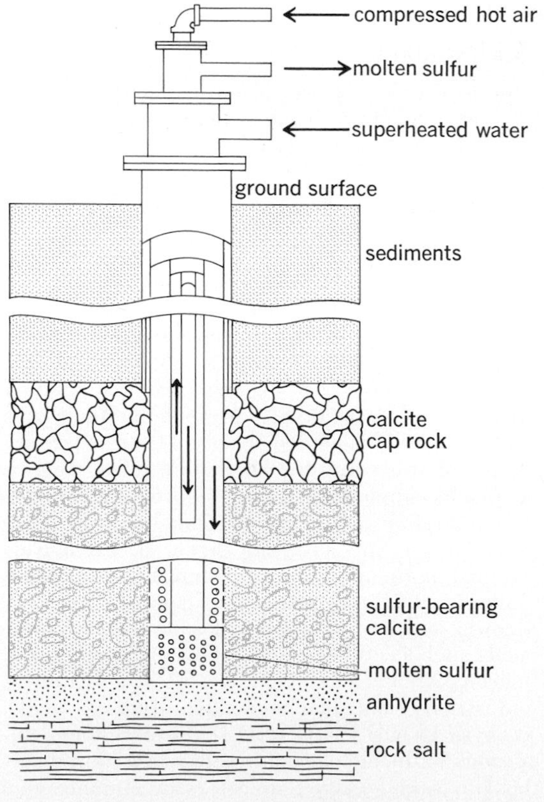

Fig. 1. Frasch process for mining sulfur.

used to obtain sulfur from gaseous hydrogen sulfide, a by-product in the manufacture of many substances. It is based on the partial oxidation of the gas by oxygen in the air to give water and sulfur, according to reaction (1). Sulfur is also obtained

$$2H_2S + O_2 \xrightarrow[\text{catalyst}]{Fe_2O_3} 2H_2O + 2S \quad (1)$$

as a by-produce of many industrial processes by using coke or H_2S to reduce sulfur dioxide in flue gases, as in reactions (2) and (3).

$$SO_2 + C \rightarrow CO_2 + S \quad (2)$$

$$2H_2S + SO_2 \rightarrow 2H_2O + 3S \quad (3)$$

Properties. The allotropes (different crystalline forms) of sulfur have been studied intensively, but as yet the many modifications which exist for every state (gas, liquid, and solid) of elemental sulfur are not fully understood.

Rhombic sulfur. Rhombic sulfur, also called brimstone and alpha sulfur (α-sulfur), is the stable modification of the element below 95.5°C (the transition point), and most of the other forms revert to this modification if allowed to stand below this temperature. The melting point of rhombic sulfur depends on the method of heating the substance and on the nature of the liquid sulfur with which it is in equilibrium. If rhombic sulfur is heated very slowly, it will convert to the monoclinic form, and the melting point obtained will be that for the monoclinic variety. If the heating rate is increased somewhat, rhombic sulfur should ideally come into equilibrium with liquid sulfur only in the lambda form (λ-sulfur), and the melting point is 112.8°C. If the heating is rapid, the rhombic sulfur crystallizes from a melt in which λ-sulfur and mu sulfur (μ-sulfur) are in equilibrium at 110°C. Rhombic sulfur is lemon-yellow, insoluble in water, slightly soluble in ethyl alcohol, diethyl ether, and benzene, and very soluble in carbon disulfide. Its density is 2.06 g/cm^3, and its hardness is 2.5 on the Mohs scale. Its molecular formula is S_8, and its molecular configuration is a ring of eight covalently bonded sulfur atoms in the shape of a puckered crown, with bond distances of 2.12 angstroms (A) and bond angles of 105° (Fig. 2).

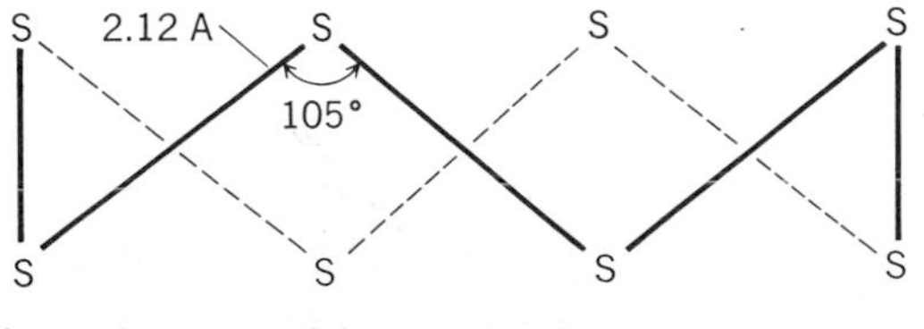

Fig. 2. Structure of the S_8 molecule.

Monoclinic sulfur. Monoclinic sulfur, also called prismatic sulfur and beta sulfur (β-sulfur), is the stable modification of the element above the transition temperature and below the melting point. It crystallizes from molten sulfur in needlelike prisms which are almost colorless. It has a density of 1.96 g/cm^3 and a melting point of 119.3°C. Its molecular configuration is also an eight-atom puckered crown structure, and this form is also soluble in carbon disulfide and insoluble in water.

Plastic sulfur. Plastic sulfur, also called gamma sulfur (γ-sulfur), is formed when molten sulfur at or near its normal boiling point is quenched to the solid state (such as by pouring it into cold water). This form of sulfur is amorphous and is only partially soluble in carbon disulfide. It is thought to be composed of two types of sulfur: λ-sulfur (S_λ, soluble in CS_2) and μ-sulfur (S_μ, insoluble in CS_2). These forms probably exist in liquid sulfur also, because plastic sulfur appears to be only the supercooled liquid. After long standing at room temperature, this form of sulfur reverts to the rhombic form. Plastic sulfur exists as long zigzag chains of sulfur atoms, and if it is strongly stretched, it behaves like rubber because the zigzag chains are oriented in straight lines, and the material becomes fibrous and rigid.

Purple sulfur. Purple sulfur is formed by the sudden cooling of sulfur vapor at an elevated temperature (where the sulfur exists as S_2) to −195°C. This purple modification is believed to be composed of S_2 units. It is unstable and reverts to yellow sulfur upon being warmed to room temperature.

Liquid sulfur. Liquid sulfur exhibits the remarkable property of increasing in viscosity as its temperature is raised. Its color becomes dark reddish-black as its viscosity increases, and both color darkening and viscosity reach a maximum at 200°C. Above this temperature, the color lightens and the viscosity diminishes. It is thought that at the melting point, liquid sulfur is largely S_λ (S_8 rings, yellow), and that as the temperature increases, the percentage of S_μ (polymeric sulfur chains, reddish-black) also increases. At 200°C, there is thought to be a maximum of the highly polymeric sulfur chains which can intertwine to give maximum light absorption and viscosity. Above this temperature, the chains break, are reduced in length, and the viscosity decreases with increasing temperature. There is also believed to be another form of sulfur in solid and molten sulfur called pi sulfur (S_π), but it has not been so extensively studied as the forms already mentioned, and little is known about it. A form of sulfur called rho sulfur (S_ρ), believed to be S_6, has been reported as being obtained by the extraction of acidified aqueous sodium thiosulfate solution with toluene. Little is known of this form of sulfur, however.

Gaseous sulfur. At the normal boiling point of the element (444.60°C) gaseous sulfur is orange-yellow in color. As the temperature is raised, the color becomes deep red and then becomes lighter until, at 650°C, it is straw-yellow. Several molecular species are in equilibrium in gaseous sulfur: S_8, S_6, S_4, and S_2, the proportions varying with the temperature. At the normal boiling point, the vapor is largely S_8; at 750°C, it is largely S_2; above 2000°C, it is largely dissociated into sulfur atoms.

Other forms. Milk of sulfur is a suspension of finely divided, amorphous sulfur in water, obtained by the decomposition of polysulfide solutions with acid. It is soluble in carbon disulfide. Colloidal sulfur, also called delta sulfur (δ-sulfur), is a colloidal dispersion of sulfur in water produced by the action of gaseous hydrogen sulfide on cold, concentrated aqueous solutions of sulfur dioxide, or by decomposing sodium thiosulfate with dilute sulfuric acid. It dissolves quite slowly in carbon disulfide.

Chemical properties and uses. Sulfur is an active element which combines directly with most of the known elements. It can exist in both positive and negative oxidation states, and can form ionic as well as covalent and coordinate covalent compounds. A test for the detection of elemental sulfur is the formation of a red solution when sulfur dissolves in piperidine. All of the important allotropic forms show this behavior.

The uses of sulfur are limited primarily to the manufacture of sulfur compounds. However, large quantities of elemental sulfur are used in the vulcanization of rubber, in lime-sulfur sprays to destroy plant parasites, in the manufacture of artificial fertilizer and certain types of cements and electric insulators, in certain ointments and medicinals, and in the manufacture of gunpowder and matches. Sulfur compounds are used in the manufacture of chemicals, textiles, soaps, fertilizers, leather, plastics, refrigerants, bleaching agents, drugs, dyes, paints, paper, and other products.

PRINCIPAL COMPOUNDS

This section covers first the sulfides and then the oxides and oxy acids of sulfur.

Sulfides. Hydrogen sulfide (H_2S) is the most important compound containing only hydrogen and sulfur. Hydrogen disulfide (persulfide, H_2S_2), hydrogen polysulfide (H_2S_x, $x=3-9$), and their salts have been reported, but are far less well characterized than H_2S. Hydrogen sulfide can be prepared from the elements at elevated temperatures, although it is usually prepared by allowing a sulfide, such as iron sulfide (FeS), to be decomposed at room temperature by an acid such as hydrochloric acid. It is a colorless gas having a foul odor (similar to that of rotten eggs) and is considerably more poisonous than carbon monoxide, but warning of its presence (odor) is usually given before its concentration in the atmosphere is considered dangerous. Its density is 1.5392 g/liter at 0°C; melting point −85.5°C; normal boiling point −60.4°C. It is soluble in water, ethyl alcohol, carbon tetrachloride, and carbon disulfide. It burns in excess oxygen to form water and sulfur dioxide, in deficient oxygen to form water and free sulfur. The gas reacts directly with many metals to form the corresponding sulfides and with many nonmetals to form free sulfur. It is generally regarded as a good reducing agent. In water it behaves as a very weak acid (hydrosulfuric acid). It is used as a precipitating agent for many metal ions having insoluble or slightly soluble sulfides, and as a reducing agent (for example, in the preparation of hydrogen iodide from H_2S and iodine). Hydrogen sulfide can be determined analytically by absorbing the gas in an ammoniacal solution of zinc chloride and titrating with a standard iodine solution.

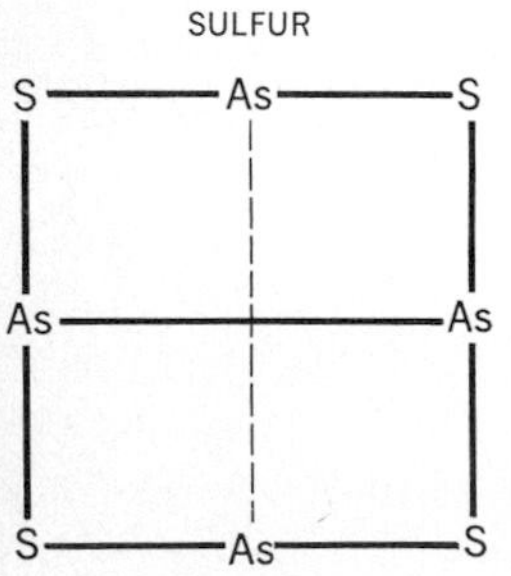

Fig. 3. Cradle structure of realgar, As_4S_4, in vapor state, with two arsenic atoms above and two below plane of four sulfur atoms. Interatomic distances: As—S = 2.23 A, As—As = 2.49 A; bond angles: As—S—As = 101°, S—As—S = 93°.

Hydrogen disulfide (persulfide) is a colorless liquid having a melting point of −89°C and a normal boiling point of 71°C. It is miscible with carbon disulfide, diethyl ether, and benzene, and is decomposed by water, acids, bases, and alcohol. Sulfur dissolves in it to form hydrogen polysulfides having properties similar to those of the disulfide.

Metal sulfides. Metal sulfides can be classified into three categories: acid sulfides (hydrosulfides, MHS, where M = a univalent metal ion), normal sulfides (M_2S), and polysulfides (M_2S_3). The acid sulfides are soluble in water. Those normal sulfides which are soluble undergo hydrolysis in water to give the acid sulfide and usually hydrogen sulfide as well. The ease of hydrolysis of the soluble sulfides increases as the oxidation state of the metal ion increases. Most of the heavy-metal sulfides are only very slightly soluble in water and are precipitated by hydrogen sulfide or ammonium sulfide. Metal acid sulfides and normal sulfides are usually prepared by reaction of the metal salt or hydroxide with hydrogen sulfide or ammonium sulfide, by reduction of the metal sulfates with hot carbon, or by direct combination of the metal with hot sulfur. The alkali and alkaline-earth sulfides are colorless, whereas the heavy-metal sulfides are usually deeply colored. The soluble sulfides are used as reducing agents and in the preparation of sulfur-containing dyes, as depilatory materials and pesticides, and in the tanning of leather and the preparation of phosphors.

Polysulfides are formed by the reaction of free sulfur with solutions of alkali metal sulfides. The intensity of the color of polysulfides generally increases with increasing uptake of sulfur. Metal disulfides and polysulfides (for example, K_2S_x, $x=2-6$) undergo hydrolysis to a much smaller extent than normal sulfides and are decomposed by acids, usually with the deposition of free sulfur as shown by reaction (4). Polysulfides

$$K_2S_2 + 2HCl \rightarrow 2KCl + H_2S + S \qquad (4)$$

are used in analytical determinations of several metal ions. Organic derivatives of polysulfides such as dimethyl trisulfide, $(CH_3)_2S_3$, are known.

Other sulfides. Carbon-sulfur compounds and compounds containing the carbon-sulfur bond are well known. In addition to those previously described, some important compounds are: carbon disulfide, CS_2, a liquid which has a normal boiling point of 46.2°C and a melting point of −111.6°C, and which is an excellent solvent for elemental sulfur and phosphorus; carbon monosulfide, CS, an unstable gas formed by passing an electric discharge through carbon disulfide; and carbon oxysulfide, SCO, formed from carbon monoxide and free sulfur at an elevated temperature, having a normal boiling point of −50.2°C and a freezing point of −138.8°C.

Nitrogen-sulfur compounds which have been characterized are sulfur nitride, N_4S_4 (also called tetranitrogen tetrasulfide), nitrogen disulfide, NS_2, and nitrogen pentasulfide, N_2S_5. They should properly be referred to as nitrides because of the greater electronegativity of nitrogen, although in the literature they are usually called sulfides.

Sulfur nitride is the best characterized of these. It is a yellow-to-red crystalline material which melts at about 178°C and sublimes at reduced pressures and elevated temperatures. It is soluble in carbon disulfide, benzene, ethanol, liquid ammonia, and carbon tetrachloride and reacts with water to form ammonium, sulfite, and pentathionate ions and free sulfur. It also reacts with chlorine to form $N_4S_4Cl_4$. It has the cradlelike structure of arsenic sulfide, As_4S_4 (realgar), with sulfur atoms in the arsenic positions and nitrogen atoms in the sulfur positions (Fig. 3). It can be prepared by the reaction of sulfur with liquid ammonia at or above −11.5°C, as represented by reaction (5). The other nitrogen sulfides are of lesser importance.

$$10S + 16NH_3 \rightarrow N_4S_4 + 6(NH_4)_2S \quad (5)$$

Phosphorus-sulfur compounds which have been characterized are P_4S_3, P_4S_5, P_4S_7, and P_4S_{10}. Their structures are not known with certainty, except for P_4S_{10}, which has a structure analogous to that of P_4O_{10}. All four are yellow crystalline materials that are soluble in carbon disulfide. They are used in converting organic oxygen compounds (for example, alcohols) into the corresponding sulfur analogs. P_4S_{10} is used in the preparation of flotation agents for concentrating sulfide ores. P_4S_3 is used in the manufacture of matches. The compounds can be prepared from the elements. Phosphorus oxysulfide, $P_4S_4O_6$, is a colorless compound which melts at about 102°C and has a normal boiling point of 295°C. It is soluble in carbon disulfide and benzene and is prepared by a reaction between P_4S_{10} and P_4O_{10} at elevated temperatures.

Oxides. The oxides of sulfur which have been characterized have the formulas SO, S_2O_3, SO_2, SO_3, S_2O_7, and SO_4. Sulfur dioxide, SO_2, and sulfur trioxide, SO_3, are of far greater importance than the others. Sulfur monoxide, SO, can be prepared by passing an electric discharge through a mixture of sulfur vapor and sulfur dioxide at low temperatures. It is a gas at ordinary temperatures and produces an orange-red deposit when cooled to the temperature of liquid air. The gas is stable only at reduced pressures, is probably dimeric, and is soluble in thionyl chloride. The solid is probably a long-chain polymer. Sulfur sesquioxide, S_2O_3, is a blue-green solid which is stable only below 15°C. It is prepared by a reaction of free sulfur with excess liquid sulfur trioxide, and it reacts with water to produce free sulfur and sulfurous, sulfuric, and several thionic acids. It appears to be a high polymer, although its structure is not known as yet. Sulfur heptoxide, S_2O_7, is a poorly characterized material which can be prepared by passing sulfur dioxide or sulfur trioxide and oxygen or ozone through an electric discharge. Its structure is not known, although it is believed that a peroxide group, —O—O—, must exist in the compound. Sulfur tetroxide, SO_4, is prepared by passing a mixture of sulfur dioxide and excess oxygen at reduced pressure through a glow discharge. The product is a white solid melting at 3°C (with decomposition) and is a monomer. It is a strong oxidizing agent, and its structure has not yet been proved, although there is little doubt that the molecule contains a peroxide group.

Sulfur dioxide. Sulfur dioxide, SO_2, is a colorless gas with a pungent odor, melting at −75.46°C and boiling at −10.02°C. It has an angular structure, the bond angle (O—S—O) being 119° and the sulfur-oxygen bond distances being 1.43 A. Sulfur dioxide can act as an oxidizing agent (for example, toward hydrogen sulfide, hydrogen, and carbon monoxide) and as a reducing agent (for example, toward permanganate ion). It reacts with water giving an acidic solution (often called sulfurous acid) and bisulfite (HSO_3^-) and sulfite (SO_3^{--}) ions. The equilibrium constant (18°C) for the dissociation to the acid sulfite (HSO_3^-) is 1.6×10^{-2} and for the dissociation of the acid sulfite to sulfite it is 1.0×10^{-7} at the same temperature. At low temperatures, sulfur dioxide forms solvates with metal fluorides, iodides, and thiocyanates ($NaI \cdot 4SO_2$). The dioxide is used as a refrigerant gas because of the ease with which it is liquefied and because of its relatively high heat of vaporization. It is also used as a disinfectant and preservative because of its germicidal properties. It also finds use as a bleaching agent and in the refining of petroleum products. Its major use, however, is in the manufacture of sulfur trioxide and sulfuric acid.

Because it is readily liquefied, sulfur dioxide is also used as a solvent. It is waterlike in many of its properties (dielectric constant = 13.5 at 15°C), and it frequently behaves as weakly dissociated thionyl sulfite, $(SO)SO_3$. Liquid sulfur dioxide is partially miscible with water and also forms crystalline hydrates with it (for example, $SO_2 \cdot H_2O$); it is completely miscible with benzene. Because liquid sulfur dioxide is not a reducing agent, oxidation reactions such as halogenations can be carried out in it. The sulfites of metal ions whose hydroxides are amphoteric in water (for example, Al^{III}) are amphoteric in liquid sulfur dioxide.

Sulfur dioxide is usually prepared by burning sulfur or roasting metallic sulfides in air or oxygen, by the reaction of metals such as copper with concentrated sulfuric acid at elevated temperatures, or by the reaction between a sulfite (Na_2SO_3) or acid sulfite ($NaHSO_3$) and a strong acid (H_2SO_4).

Sulfur trioxide. Sulfur trioxide, SO_3, exists in several forms, and their relationships are still not completely understood. Gramma sulfur trioxide, γ-SO_3, is trimeric and resembles ice in appearance. Its equilibrium melting point is 16.8°C, and it can be prepared by condensing extremely dry sulfur trioxide vapor at −80°C. Beta sulfur trioxide, β-SO_3, is an asbestoslike polymeric substance in which the SO_3 groups are linked in long chains. Its equilibrium melting point is 32.5°C. It is prepared in a manner similar to that for the α form using ordinary sulfur trioxide (which contains some moisture). The condensation yields a mixture of the β and γ forms, and the γ form is removed by distillation. Alpha sulfur trioxide, α-SO_3, is also an asbestos-like solid which is similar to the β form, except that the SO_3 chains are also joined in a layer type of structure. Its equilibrium melting point is 62.3°C, and it can be formed by condensing gaseous sulfur trioxide at liquid air temperatures. The γ and β forms are metastable with respect to the α form, and conversion to the α form is catalyzed by traces of moisture. However, this conversion to a polymer is inhibited by sulfur, tellurium, carbon tetrachloride, and phosphorus oxytrichloride, which are used in commercially available stabilized forms of sulfur trioxide called Sulfans. In the older literature, the γ and α forms are called α and γ, respectively. Liquid sulfur trioxide apparently exists as an equilibrium system between monomeric and trimeric sulfur trioxide. It has a normal boiling point of 44.5°C. Gaseous sulfur trioxide is a monomer, and its structure is a planar equilateral triangle with the sulfur at the center. The O—S—O bond angles are 120°, and the S—O bond lengths are 0.143 nm.

Chemically, sulfur trioxide is extremely reactive. The γ form is the most reactive, and the α form the least. All forms react with water with the liberation of heat to produce sulfuric acid. The reaction with sulfuric acid, H_2SO_4, to form pyrosulfuric acid, $H_2S_2O_7$ (also called fuming sulfuric acid and oleum), is less violent. Sulfur trioxide is a powerful oxidizing agent and will liberate halogens from halides (except fluorides) and will produce carbon or sulfonic acids upon reaction with organic materials. It reacts directly with basic metal oxides to form sulfates and with hydrogen chloride to form chlorosulfonic acid, HSO_3Cl. It is decomposed at high temperatures to sulfur dioxide and oxygen.

Sulfur trioxide is usually prepared by the catalytic oxidation of sulfur dioxide at 400–665°C. Vanadium pentoxide is the catalyst most often used, although platinum metal, the sulfates of nickel and cobalt, and the oxides of iron, tungsten, molybdenum, and chromium can also be used. Small quantities of sulfur trioxide are prepared by the thermal decomposition of certain metal sulfates such as ferric sulfate or of pyrosulfates such as sodium pyrosulfate, $Na_2S_2O_7$. The trioxide is also prepared by the reaction between sulfur dioxide and ozone, O_3, at room temperature and by the reaction between nitric oxide and sulfur dioxide at high pressures and room temperature, as in reaction (6). Sulfur trioxide is used primarily in the preparation of sulfuric and sulfonic acids.

$$2NO + 2SO_2 \rightarrow 2SO_3 + N_2 \quad (6)$$

Oxy acids of sulfur. Although salts (or esters) of all the oxy acids are known, in many cases the free acids themselves have not been isolated because of their instability (see table).

Sulfoxylic acid. Best characterized as the cobalt (II), or cobaltous, salt, sulfoxylic acid precipitates as a brown solid upon treatment of sodium hyposulfite solution, $Na_2S_2O_4$, with cobalt (II) acetate and aqueous ammonia. This is represented by reaction (7). The diethyl ester (C_2H_5—O—S—O—C_2H_5, boiling point 117°C at 733 mm Hg) is a colorless liquid which is not miscible with water. The sulfoxylates are extremely susceptible to oxidation.

$$CoS_2O_4 + 2NH_3 + H_2O \rightarrow CoSO_2 + (NH_4)_2SO_3 \quad (7)$$

Hyposulfurous acid. Hyposulfurous acid may be prepared in aqueous solution by the reduction of sulfurous acid solution with amalgamated zinc, but the solution is unstable. The material behaves as a strong acid (for the first hydrogen), and the solution is a good reducing medium. Hyposulfites (dithionites or hydrosulfites) are more stable as dry solids than as the acid solution. They are also strong reducing agents, and can be prepared by reduction of the corresponding metal bisulfite solution with zinc dust, by electrolytic reduction of the sulfite, or by treatment of an active metal amalgam such as sodium amalgam with dry sulfur dioxide. Metal hyposulfites are used primarily as reducing agents in the dye industry.

Sulfurous acid. Sulfurous acid is not actually known as a pure substance, although the hydrate $SO_2 \cdot 7H_2O$ can be crystallized from concentrated aqueous solutions of sulfur dioxide at low temperatures. Aqueous solutions of sulfur dioxide contain primarily hydrated protons, bisulfite ions (HSO_3^-), and a much smaller concentration of sulfite ions (SO_3^{--}). They are strongly reducing, and can be oxidized to sulfate and dithionate. Such solutions can also behave as oxidizing agents in the presence of strong reductants, such as iodide ion and zinc. Both normal sulfites (Na_2SO_3) and acid (hydrogen) sulfites ($NaHSO_3$) are well known.

The structure of the sulfite ion is pyramidal, with one atom at each corner. Of the normal sulfites, only the alkali metal salts are appreciably

Oxy acids of sulfur

Name and formula	Probable structure	Known forms
Sulfoxylic acid, H_2SO_2	HO—S—OH	Salts, esters
Hyposulfurous acid (dithionous, hydrosulfurous), $H_2S_2O_4$	O HO—S—S—OH O	Acid, salts
Sulfurous acid, H_2SO_3	O HO—S—OH	Acid, salts, esters
Thiosulfurous acid, $H_2S_2O_2$	HO—S—OH S	Esters
Pyrosulfurous acid, $H_2S_2O_5$	O O HO—S—S—OH O	Salts
Sulfuric acid, H_2SO_4	O HO—S—OH O	Acid, salts, esters
Pyrosulfuric acid, $H_2S_2O_7$	O O HO—S—O—S—OH O O	Acid, salts
Thiosulfuric acid, $H_2S_2O_3$	O HO—S—OH S	Salts
Dithionic acid, $H_2S_2O_6$	O O HO—S—S—OH O O	Acid, salts
Polythionic acids (tri-, tetra-, penta-, hexa-), $H_2S_xO_6$ ($x=3-6$)	O O HO—S—$(S)_{x-2}$—S—OH O O (tentative)	Salts
Peroxymonosulfuric acid, H_2SO_5	O H—O—O—S—OH O	Acid, salts
Peroxydisulfuric acid, $H_2S_2O_8$	O O HO—S—O—O—S—OH O O	Acid, salts
Sulfenic acids, RSOH (R = alkyl or aryl group such as CH_3)	HO—S—R	Esters, halides
Sulfinic acids, RSO_2H	O HO—S—R	Acids, esters, halides
Sulfonic acids, RSO_3H	O HO—S—R O	Acids, esters, halides, amides
Thiosulfonic acids, RS_2O_2H	O H—S—S—R O	Salts, esters

soluble, although many metal bisulfites (acid sulfites) are soluble. Both normal and acid sulfites are usually prepared by treatment of the corresponding carbonate or hydroxide with appropriate quantities of sulfur dioxide, and they both liberate sulfur dioxide on treatment with excess acid. Solutions of sulfites dissolve free sulfur to form thiosulfates. Bisulfites can form addition compounds with many organic compounds. The sulfite ion forms many coordination compounds with metal ions, for example, $Na_3[Co(SO_3)_3]$, as well as esters, such as $(CH_3O)_2SO$ (normal boiling point 121.5°C), which are good alkylating agents. Sulfites are used extensively as reducing agents, as addition agents to organic compounds, and in the manufacture of paper from wood.

Thiosulfurous acid. Thiosulfurous acid is known only in the form of its salts, and even these are not very well characterized. They can be formed by the action of a dry sodium alkylate such as $NaOCH_3$ on sulfur monochloride, S_2Cl_2, in ligroin solution. The boiling point of the dimethyl compound is 33°C at 15 mm Hg (2.0 kPa), and of the diethyl compound 67°C at 16 mm Hg (2.1 kPa). A typical structure postulated for these esters is (for the dimethyl ester) $H_3C—O—S—S—O—CH_3$. These substances are stable toward oxidation by atmospheric oxygen and are hydrolyzed by strong bases to give thiosulfates and sulfur.

Pyrosulfurous acid. Pyrosulfurous acid is known only in the form of pyrosulfites, which are usually prepared from aqueous solutions of alkali metal sulfites and sulfur dioxide or by heating alkali metal acid sulfites. The sodium salt is used primarily in the dye, printing, and photographic industries.

Sulfuric acid. This is a colorless, viscous liquid whose melting point is 10.31°C. When this liquid is heated, it gives off sulfur trioxide and begins to boil at 290°C. However, the normal boiling point increases until it reaches 317°C, at which point the acid is 98.54% H_2SO_4. Gaseous H_2SO_4 begins to dissociate into sulfur trioxide and water vapor at about 300°C, the dissociation being 50% complete at 350°C, and essentially complete at 444°C. The acid forms the hydrates $H_2SO_4{\cdot}H_2O$ (melting point 8.47°C), $H_2SO_4{\cdot}2H_2O$ (mp −39.46°C), and $H_2SO_4{\cdot}4H_2O$ (mp −28.25°C). The structure of the H_2SO_4 molecule is a tetrahedron with a sulfur atom at the center and two OH groups and two oxygen atoms at the corners. The sulfur-oxygen bond distances are 0.151 nm. The density of 100% H_2SO_4 is 1.8384 g/ml at 15°C. The acid dissociates essentially completely in water to give a hydrated proton and the acid sulfate (bisulfate) ion. The bisulfate ion dissociates to a large degree in dilute aqueous solution ($pK_2 = 1.7$ at 25°C) to give the normal sulfate anion and another hydrated proton.

The preparation of sulfuric acid is usually carried out by the contact process. A far less important process today is the lead-chamber process because it gives relatively dilute acid (60–78%) which has limited usefulness, whereas the contact process can give acid of any concentration.

In the contact process, sulfur is burned or iron pyrites roasted in air to produce sulfur dioxide, which is then oxidized to sulfur trioxide in the presence of a suitable catalyst (usually vanadium pentoxide or platinum). The sulfur trioxide is absorbed in concentrated sulfuric acid to produce oleum (pyrosulfuric acid, $H_2S_2O_7$), which is then treated with water to produce sulfuric acid of any desired concentration. The schematic diagram of the process shows that the burner gas (about 25% sulfur dioxide, 30% oxygen, and 45% nitrogen) is filtered to remove dust particles, and is cleaned, dried, and treated to remove arsenic (a catalyst poison) before being heated and converted to sulfur trioxide on the catalyst bed (Fig. 4). There are many variations, for example, the Badische, Schröder-Grillo, Mannheim, and Tenteleff processes.

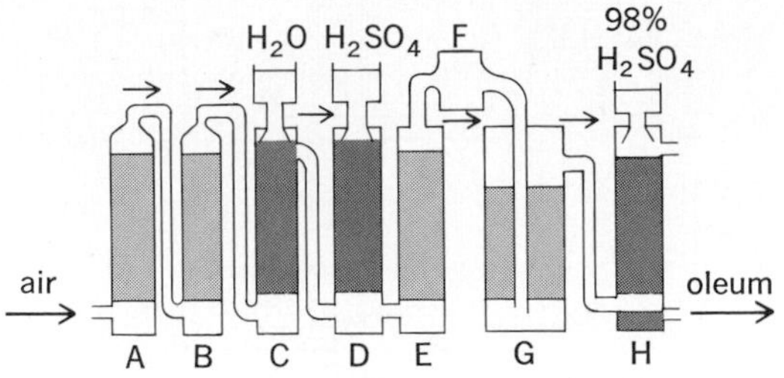

Fig. 4. The contact process. A, sulfur burner; B, dust removal tower; C and D, scrubbing towers to clean sulfur dioxide; E, arsenic removal tower; F, heater; G, catalyst chamber; H, sulfur trioxide absorber tower.

In the lead-chamber process, nitrogen oxides are introduced into the sulfur dioxide–air mixture which is passed upward through a tower about 25 ft (7.5 m) high (called the Glover tower), in which it is sprayed with a sulfuric acid–nitrosyl sulfuric acid ($ONOSO_2OH$) mixture from the Gay-Lussac tower, where reactions (8) and (9) are probable.

$$2ONOSO_2OH + SO_2 + 2H_2O \rightarrow 3H_2SO_4 + 2NO \quad (8)$$

$$4NO + 4SO_2 + 3O_2 + 2H_2O \rightarrow 4ONOSO_2OH \quad (9)$$

The gases (a mixture of SO_2, O_2, NO, and NO_2) then pass through several lead chambers and are sprayed with steam. The reactions which occur here are not fully understood, but it is believed that they take place as in reactions (10)–(12). The gases

$$2NO + O_2 \rightarrow 2NO_2 \quad (10)$$

$$2SO_2 + 3NO_2 + H_2O \rightarrow 2ONOSO_2OH + NO \quad (11)$$

$$2ONOSO_2OH + H_2O \rightarrow 2H_2SO_4 + NO + NO_2 \quad (12)$$

then enter the Gay-Lussac tower where they are sprayed with fairly concentrated sulfuric acid (about 80%). This results in the formation of additional nitrosyl sulfuric acid, reaction (13), which is

$$2H_2SO_4 + NO + NO_2 \rightarrow 2ONOSO_2OH + H_2O \quad (13)$$

then pumped to the Glover tower to complete the cycle. The nitrogen oxides act as catalysts, and because there is some loss, they are replaced periodically by the catalytic oxidation of ammonia. The acid is tapped from the Glover tower and the lead chambers, and is quite impure, containing oxides of arsenic and nitrogen, and many metal salts. Lead-chamber acid is used primarily in the manufacture of fertilizer because removal of these impurities is not essential (Fig. 5).

The chemical properties of sulfuric acid are of considerable importance, and because of them, sulfuric acid has become the largest tonnage manufactured chemical in the world.

The compound is a strong acid in water and reacts with most metals in either the dilute or con-

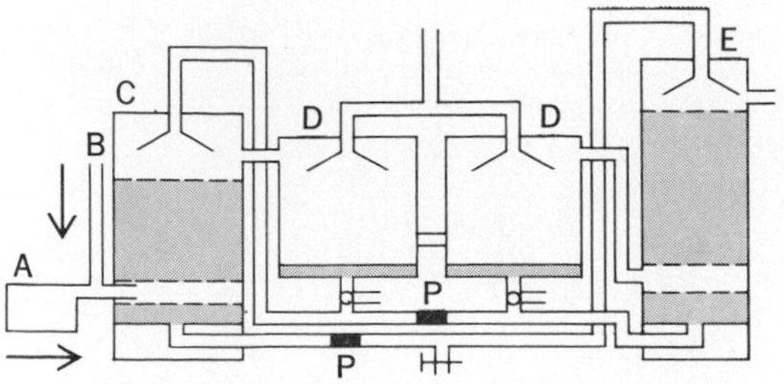

Fig. 5. The lead chamber process. A, sulfur or pyrite burners; B, inlet for nitrogen oxides; C, Glover tower; D, lead chambers; E, Gay-Lussac tower; P, pumps.

centrated form. Iron and steel do not react with the concentrated acid, and so it can be shipped in tank cars. The concentrated acid is a strong oxidizing agent, especially at elevated temperatures, and will react with metals, carbon, sulfur, and other oxidizable materials. Because of its relatively high boiling point (338°C for the 98.3% acid), it can react with salts at elevated temperatures to liberate volatile acids such as HCl. The concentrated acid is a strong dehydrating agent, and reacts vigorously with water with the evolution of much heat (205 cal/g or 858 J/g of H_2SO_4). It will also extract hydrogen and oxygen (to form water) from organic materials such as sugar, wood, and animal tissues, thereby decomposing them and leaving carbon.

The uses of sulfuric acid are varied. It finds primary use in the manufacture of superphosphate fertilizers. Petroleum refining also consumes large quantities of the acid, as does the manufacture of many chemicals, including sulfates, hydrochloric and nitric acids, dyes, drugs, and explosives. The iron and steel, storage battery, paint, plastic, metallurgical, and textile industries also use large quantities of the acid. It is sometimes used as a solvent in chemical research because of its strong hydrogen-ion donating ability. Its use is so widespread that its tonnage production is often used as an indicator of general business conditions in the country.

Both normal sulfates such as Na_2SO_4 and acid sulfates (bisulfates or hydrogen sulfates) such as $NaHSO_4$ are well known. The structure of the sulfate ion is tetrahedral, with the sulfur at the center and an oxygen at each corner. Most sulfates of substances can be prepared by treating the substance with dilute or concentrated sulfuric acid or by oxidizing their sulfites or sulfides. Most normal and acid sulfates are quite soluble in water (notable exceptions are certain alkaline-earth and lead sulfates which are sparingly soluble). Most normal sulfates are thermally stable except at extremely high temperatures. Acid sulfates are converted to pyrosulfates and normal sulfates by intense heat. The sulfate ion can act as a ligand (coordinating agent) in coordination compounds such as $[Co(NH_3)_5SO_4]Cl$. Organic sulfates are known, for example, diethyl sulfate, $(C_2H_5)_2SO_4$, mp −24.5°C, normal bp 208°C, which is used as an alkylating agent. Double sulfates form quite easily, the two most important series being the alums having the general formula $M^IM^{III}(SO_4)_2 \cdot 12H_2O$ (M^I = monovalent positive ion; M^{III} = trivalent positive ion) and the schönites, $M^I_2M^{II}(SO_4)_2 \cdot 6H_2O$. Sulfates are usually determined analytically by treatment with aqueous barium chloride solution (acidified with hydrochloric acid) to give a precipitate of barium sulfate which is washed, filtered, dried, and weighed.

Pyrosulfuric acid. Pyrosulfuric acid is the product of the reaction of equimolar quantities of pure sulfuric acid and sulfur trioxide. Its melting point is 35.15°C. It is an excellent sulfonating agent and loses sulfur trioxide on being heated. It also reacts vigorously with water, liberating a considerable quantity of heat. Alkali metal pyrosulfates such as $Na_2S_2O_7$ can be prepared by heating alkali metal acid sulfates or by a reaction between normal alkali metal sulfates and sulfur trioxide.

Thiosulfuric acid. Thiosulfuric acid is known only in its normal salts. The salts are stable only in the solid state or in neutral or alkaline solution. Thiosulfates are usually prepared by allowing free sulfur to dissolve in a solution of a metal sulfite, by the controlled oxidation of sulfides, or by the action of alkalies on polythionates. Thiosulfates are unstable in acid solution and decompose to free sulfur, pentathionates, and sulfites. The structure of the thiosulfate anion is analogous to that of the sulfate ion, one oxygen atom of the latter being replaced by a sulfur atom. Hydrated sodium thiosulfate (hypo) is used in the photographic industry as a fixing agent to dissolve unchanged silver salts from films and plates. It is also used as an antichlor to remove chlorine from bleached fabrics. Thiosulfate ion acts as a coordinating agent in certain metal coordination compounds such as $Na_3[Ag(S_2O_3)_2]$, and many heavy-metal thiosulfates which are ordinarily insoluble will dissolve in a solution containing excess thiosulfate ion because of the formation of a soluble complex. Thiosulfate is determined analytically by titration with standard iodine solutions (or standard permanganate solutions which are used to liberate iodine from iodides). The reaction, which gives tetrathionate ion, is (14), and the titration is usually carried

$$2S_2O_3^{--} + I_2 \rightarrow S_4O_6^{--} + 2I^- \qquad (14)$$

out until the blue color produced by starch in the presence of free iodine is just destroyed.

Thionic acids. The thionic acids are known as the salts only, except for dithionic acid, of which both the free acid and salts are known. The only structures known with certainty are those for the dithionate anion $(O_3S{-}SO_3)^{--}$ and trithionate anion $(O_3S{-}S{-}SO_3)^{--}$ (the sulfur atoms are nonlinear), but it is assumed that in the other polythionates, additional sulfur atoms are bonded to the central sulfur of the trithionate structure. Dithionic acid and dithionates may be prepared from sulfurous acid and sulfite solutions by oxidation with such oxidants as manganese dioxide, permanganates, and ferric or cobaltic hydroxides. Dithionic acid is stable in dilute solution at room temperature, but undergoes decomposition on being heated. The salts are stable in solution, and all of them appear to be soluble. Polythionates are definitely known as metal salts through the hexathionate, and higher polythionates are believed to exist. They are also water-soluble, and are fair reducing agents, being oxidized to sulfate. They can be prepared by treatment of an aqueous solution of a thiosulfate with sulfur dioxide in the presence of arsenic trioxide. Varying the concentrations of the reactants is a means of forming a higher concentration of one of the polythionates than the others.

Tetrathionates are most easily prepared by oxidation of thiosulfates with iodine. Polythionates are known to be found in Wackenroder's liquid, a solution which contains colloidal sulfur, and which is prepared by passing hydrogen sulfide through sulfurous acid solution.

Persulfuric acids. The persulfuric acids (peroxymonosulfuric acid, called Caro's acid, and peroxydisulfuric acid, called Marshall's acid) are known as the acids and salts, and are usually prepared by the electrolysis of sulfuric acid (or sulfate) solutions or by treatment of chlorosulfonic acid (or salt) solutions with hydrogen peroxide. The monoacid is a hygroscopic crystalline material which melts at 45°C. It is soluble in water, alcohol, ether, and organic acids. One proton is readily lost in water, and the other is strongly held. The diacid is a hygroscopic crystalline substance which melts at 65°C with decomposition. It is hydrolyzed by water to the peroxymonoacid and sulfuric acid, and eventually to hydrogen peroxide and oxygen. It is a powerful oxidant, and both acids (and salt solutions) liberate iodine from iodides readily. Both acids are used to oxidize organic materials, as bleaching agents, and to make hydrogen peroxide.

Sulfenic acids. Sulfenic acids are known as the esters and halides. The ethyl ester, $C_2H_5—S—O—C_2H_5$, can be prepared from $C_2H_5S—CNS$ and sodium ethylate at 0°C. It is a colorless liquid with a foul odor which boils at 108°C (724 mm Hg or 96.5 kPa). It is a weak reducing agent and is oxidized by ethyl hypochlorite to the sulfinic ester, $C_2H_2—S(O)(OC_2H_5)$.

Sulfinic acids. Sulfinic acids are formed by the reduction of the chlorides of sulfonic acids with zinc or by the reaction of Grignard reagents on sulfur dioxide in ether solution. They are unstable in air and are chlorinated by thionyl chloride to give their own acid chlorides, R—S(O)(Cl). Esters of sulfinic acids [for example, $C_2H_5—S(O)(OC_2H_5)$, bp 60°C at 18 mm Hg or 2.4 kPa] are prepared from ester chlorides of sulfurous acid, R—O—S(O)(Cl), and Grignard reagents.

Sulfonic acids. Sulfonic acids (alkyl) are prepared by oxidizing mercaptans (RSH) or alkyl sulfides with concentrated nitric acid, by treatment of sulfites with alkyl halides, or by the oxidation of sulfinic acids. The aromatic derivates (for example, benzenesulfonic acid, $C_6H_5SO_3H$, bp 171°C at 0.1 mm Hg or 13 Pa) are prepared by treatment of aromatic hydrocarbons with oleum. They are stable substances which are usually water-soluble and can be converted into esters, halides, and amides (which have important medicinal properties). Organic materials are frequently sulfonated in order to render them water-soluble.

Thiosulfonic acids. Thiosulfonic acids are known as salts and esters. The salts may be obtained from the chlorides of sulfonic acids and sulfides as in reactions (15) and (16). The salts react with alkyl iodides to form the esters, RSO_2SR.

$$RSO_2Cl + Na_2S \rightarrow RSO_2Na + NaCl + S \quad (15)$$

$$RSO_2Na + S \rightarrow RSO_2SNa \quad (16)$$

Miscellaneous compounds. Other important organic oxygen-sulfur-containing compounds include the sulfoxides, R_2SO (which may be considered as being derived from sulfurous acid), and the sulfones, R_2SO_2 (from sulfuric acid). Aliphatic sulfoxides are usually prepared by the oxidation of sulfides with nitric acid or hydrogen peroxide, and the aromatic derivatives from aromatic hydrocarbons and sulfur dioxide or thionyl chloride in the presence of aluminum chloride. They are usually low-melting solids or oils [for example, $(C_2H_5)_2SO$, mp 5°C]. Aliphatic sulfones are usually prepared by the oxidation of thioethers or sulfoxides with fuming nitric acid or permanganates; and aromatic sulfones, by the action of sulfur trioxide on aromatic hydrocarbons or by the reaction of sulfonic acids with benzene and phosphorus(V) oxide at elevated temperatures. They are stable colorless solids which can be distilled without decomposition [for example, diphenyl sulfone, $(C_6H_5)_2SO_2$, mp 76°C, normal bp 379°C]. Certain disulfones formed by the condensation of ketones and mercaptans followed by oxidation have medicinal value, for example, $(CH_3)_2C(SO_2C_2H_5)_2$, the hypnotic agent sulfonal.

The oxy halides of sulfur may be classified as derivatives of sulfoxylic acid, sulfurous acid (thionyl derivatives), and sulfuric acid (sulfuryl derivatives). Aryl sulfur halides, considered to be sulfoxylic acid derivatives, can be prepared from aryl mercaptans and halogens at low temperatures. An example is phenyl sulfur chloride (C_6H_5SCl), a red oil which boils at 149°C (12 mm Hg or 1.6 kPa). Thionyl halides, such as SOF_2, $SOCl_2$, $SOBr_2$, and SOClF, are all known to have a triangular pyramidal structure with atoms at the corners only. They are low-melting and low-boiling materials (thionyl chloride, $SOCl_2$, mp −99.5°C, normal bp 75.7°C) and are very reactive. Sulfuryl halides (SO_2F_2, SO_2ClF, and SO_2Cl_2) are also low-melting and low-boiling substances; for example, sulfuryl chloride, SO_2Cl_2, has a melting point of −46°C and a normal boiling point of 69.3°C. They are much more stable and less reactive than the corresponding thionyl derivatives, however. A pyrosulfuryl halide ($S_2O_5Cl_2$, pyrosulfuryl chloride) has also been characterized. It melts at −37.5°C and boils at 152.5°C (766 mm Hg or 102.1 kPa).

Other important halogen derivatives of sulfuric acid are the organic sulfonyl halides and the halosulfonic acids. Alkyl and aryl sulfonyl halides, RSO_2X (X = F, Cl, Br), are colorless liquids or solids which usually have high boiling points (phenyl sulfonyl chloride, $C_6H_5SO_2Cl$, normal bp 252°C). The halosulfonic acids, $HOSO_2X$ (X = F, Cl) are known as the acids, salts, and esters. The fluoro compounds are more stable than the chloro compounds. Fluorosulfonic acid, $HOSO_2F$, also called fluosulfonic acid, has a normal boiling point of 162.6°C and can be prepared from KHF_2 and oleum at elevated temperatures. Chlorosulfonic acid, $HOSO_2Cl$ (mp −80°C, normal bp 151°C), can be prepared from hydrogen chloride and sulfur trioxide or oleum. It is decomposed extremely violently by water.

Halogen-sulfur compounds which have been well characterized are S_2F_2 (sulfur monofluoride), SF_2, SF_4, SF_6, S_2F_{10}, S_2Cl_2 (sulfur monochloride), SCl_2, SCl_4, and S_2Br_2 (sulfur monobromide). These compounds have low melting points and low boiling points (S_2F_2, mp −120.5°C, normal bp −38.4°C) and hydrolyze in water, except for SF_6 and S_2F_{10}. Sulfur tetraflucride is a remarkably effective fluorinating agent for organic compounds. Sulfur hexafluoride is quite inert, and this gas is used as a high-voltage insulator. It melts at −50.8°C and sublimes at −63.7°C, and its structure is an octahedron with the sulfur at the center and a fluorine at each corner. The sulfur chlorides are used in the commercial manufacture of rubber; and the monochloride, which is a liquid at room temperature, is also used as a solvent for organic compounds, sulfur, iodine, and certain metal compounds. These halides are usually prepared by direct combination of the elements. *See* ORGANOSULFUR COMPOUND; SELENIUM; TELLURIUM.

[STANLEY KIRSCHNER]

Bibliography: D. J. Bourne (ed.), *New Uses of Sulfur-II*, 1978; F. A. Cotton and G. Wilkinson, *Advanced Inorganic Chemistry*, 4th ed., 1980; S. Oae (ed.), *Organic Chemistry of Sulfur*, 1977; H. Remy, *Treatise on Inorganic Chemistry*, vol. 1, 1956; A. Senning, *Topics in Sulfur Chemistry*, 3 vols., 1977.

Sulfuric acid

A strong mineral acid with the chemical formula H_2SO_4. It is a colorless, oily liquid, sometimes called oil of vitriol or vitriolic acid. The pure acid has a density of 1.834 at 25°C and freezes at 10.5°C. It is an important industrial commodity, used extensively in petroleum refining and in the manufacture of fertilizers, paints, pigments, dyes, and explosives.

Sulfuric acid is produced on a large scale by two commercial processes, the contact process and the lead-chamber process. In the contact process, sulfur dioxide, SO_2, is produced by burning sulfur or a sulfide such as that of iron, FeS_2, in air. The sulfur dioxide is converted to sulfur trioxide, SO_3, by reaction with oxygen in the presence of a catalyst such as platinized asbestos. Sulfuric acid is produced by the reaction of the sulfur trioxide with water. The lead-chamber process depends upon the oxidation of sulfur dioxide by nitric acid in the presence of water, the reaction being carried out in large lead rooms.

The commercial acid may be concentrated to 98.3% by distillation, and pure sulfuric acid obtained by fractional crystallization. Sulfuric acid reacts vigorously with water to form several hydrates, of which the monohydrate, $H_2SO_4 \cdot H_2O$, is relatively stable. The concentrated acid, therefore, acts as an efficient drying agent, taking up moisture from the air and even abstracting the elements of water from such compounds as sugar and starch. Because of the formation of hydrates, the mixing of sulfuric acid and water is accompanied by the evolution of a great amount of heat.

The concentrated acid acts as a strong oxidizing agent because of its tendency to lose an atom of oxygen to form sulfurous acid, H_2SO_3, which readily decomposes to sulfur dioxide, SO_2, and water. Concentrated sulfuric acid reacts with most metals upon heating to produce sulfur dioxide, as shown by reaction (1). Gold reacts least readily.

$$2Ag + H_2SO_4 \rightarrow Ag_2SO_4 + SO_2 + 2H_2O \qquad (1)$$

The concentrated acid decomposes salts of other lower boiling acids, as in reaction (2). It is therefore widely used in the preparation of other acids.

$$H_2SO_4 + NaCl \rightarrow HCl + NaHSO_4 \qquad (2)$$

Sulfuric acid ionizes in water, forming hydrogen (H^+), bisulfate (HSO_4^-), and sulfate (SO_4^{--}) ions. The structural formula of sulfuric acid is usually written as notation (3).

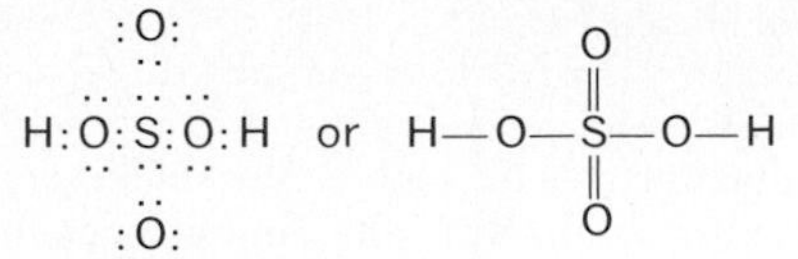

(3)

This structure is but one of several in which the molecule exists. It is therefore a resonance hybrid. *See* SULFATE; SULFUR; SUPERACID.

[FRANCIS J. JOHNSTON]

Bibliography: O. T. Fasullo, *Sulfuric Acid*, 1964.

Superacid

An acid which has an extremely great proton-donating ability. It has proved convenient to define a superacid somewhat arbitrarily as an acid, or more generally, an acidic medium, which has a proton-donating ability equal to or greater than that of anhydrous (100%) sulfuric acid.

Superacids belong to the general class of proton or Brönsted acids. A proton acid is defined as any species which can act as a source of protons and which will therefore protonate a suitable base, as in reaction (1).

$$HA + B \rightleftharpoons BH^+ + A^- \qquad (1)$$

The strengths of acids are often compared by measuring the extent of their ionization in water, that is, the extent to which they can protonate the base water, as in reaction (2).

$$HA + H_2O \rightleftharpoons H_3O^+ + A^- \qquad (2)$$

However, all strong acids are fully ionized in dilute aqueous solution, and they therefore appear to have the same strength. Their strengths are said to be reduced or leveled to that of the hydronium ion (H_3O^+), which is the most highly acidic species that can exist in water. In any case, many of the superacids react with and are destroyed by water. For these reasons, the strengths of superacids cannot be measured by the conventional means of utilizing their aqueous solutions. The acidities of superacids can, however, be conveniently measured in terms of the Hammett acidity function. *See* ACID AND BASE.

Hammett acidity function. This method of measuring acidity is based on the determination of the ionization ratios of suitable weak bases (indicators), usually by means of the change in absorption spectrum that occurs on protonation of the base, although the nuclear magnetic resonance (NMR) spectrum has also been used. The Hammett acidity function (H_0) is defined by Eq. (3),

$$H_0 = \mathrm{p}K_{BH^+} - \log \frac{[BH^+]}{[B]} \qquad (3)$$

where K_{BH^+} is the dissociation constant of the acid form of the indicator and $[BH^+]/[B]$ is the ionization ratio of the indicator. This method was first used by L. P. Hammett in the 1930s to study the acidity of sulfuric acid–water mixtures over the whole range of composition from water to 100% sulfuric acid. Starting with a base of known $\mathrm{p}K_{BH^+}$ in dilute aqueous acid solution, he determined the

pK_{BH^+} values of a series of successively weaker indicators which had overlapping ionization ranges, and hence the H_0 values for the whole range of composition from H_2O to 100% H_2SO_4. In this way 100% sulfuric acid was found to have an H_0 of −11.9. The H_0 scale is joined to the pH scale at low acid concentrations in water, and may therefore be regarded as an extension of the pH scale into regions where the concept of pH is no longer valid. Anhydrous (100%) sulfuric acid ($H_0 = -11.9$) is 10^{13} times as acidic as a 0.1 *N* solution of H_2SO_4 in water (pH = H_0 = 1). Anhydrous sulfuric acid has a much greater acidity (proton-donating power) than any aqueous acid solution because it is composed of H_2SO_4 molecules which are intrinsically much stronger proton donors than H_3O^+.

Sulfuric acid and fluorosulfuric acid. Hammett acidity function (H_0) values for a number of superacids are given in the table. In each case the value refers to the 100% (anhydrous) acid. Each of the superacids in the table is a liquid at room temperature, and each forms the basis of a solvent system. Sulfuric acid undergoes a relatively extensive self-ionization or autoprotolysis according to reaction (4), which is completely analogous to the well-known self-ionization of water [reaction (5)].

Hammett acidity function values for several superacids

Superacid	Formula	$-H_0$
Sulfuric acid	H_2SO_4	11.9
Chlorosulfuric acid	HSO_3Cl	13.8
Trifluoromethane sulfonic acid	HSO_3CF_3	14.0
Disulfuric acid	$H_2S_2O_7$	14.4
Fluorosulfuric acid	HSO_3F	15.1
Hydrogen fluoride	HF	15.1

$$2H_2SO_4 \rightleftharpoons H_3SO_4^+ + HSO_4^- \quad (4)$$

$$2H_2O \rightleftharpoons H_3O^+ + OH^- \quad (5)$$

The acidity of sulfuric acid can therefore be further increased by the addition of a suitable acid, HA, which will protonate sulfuric acid and increase the concentration of $H_3SO_4^+$ ions [reaction (6)], just as an acid such as HCl increases the concentration of H_3O^+ ions in water. The increase in the acidity measured by H_0 for the few known substances that exhibit acid behavior in the solvent sulfuric acid is shown in Fig. 1.

$$HA + H_2SO_4 \rightleftharpoons H_3SO_4^+ + A^- \quad (6)$$

Fluorosulfuric acid undergoes a similar autoprotolysis [reaction (7)]. In principle its acidity could be further increased by the addition of an acid HA which is strong enough to protonate HSO_3F and thus increase the concentration of $H_2SO_3F^+$. No proton acid HA is known which is sufficiently strong to protonate HSO_3F. However, the acidity of HSO_3F can be further increased by the addition of certain pentafluorides and their derivatives. The most effective of the pentafluorides, that is, the strongest acid, is antimony pentafluoride, SbF_5, which ionizes according to reaction (8).

$$2HSO_3F \rightleftharpoons H_2SO_3F^+ + SO_3F^- \quad (7)$$

$$SbF_5 + 2HSO_3F \rightleftharpoons H_2SO_3F^+ + SbF_5(SO_3F)^- \quad (8)$$

Antimony pentafluoride, which is an extremely strong Lewis acid, extracts a fluorosulfate ion from the solvent to form the anion $SbF_5(SO_3F)^-$, leaving an excess proton which is solvated to give $H_2SO_3F^+$. The increase in the acidity of fluorosulfuric acid produced by AsF_5, SbF_5, and the SbF_5 derivative $SbF_2(SO_3F)_3$ is shown in Fig. 2. These media are extremely acidic and truly merit the name superacid. Their acidities reach 10^{20} times that of a 0.1 *N* aqueous sulfuric acid solution. *See* SULFURIC ACID.

Hydrogen fluoride. Although hydrogen fluoride has long been known as a rather highly acidic medium, the difficulties of handling and in particular of obtaining the truly anhydrous material have retarded accurate measurements of its acidity. It has for some time been thought to have an H_0 of −11, but this has been shown to refer to an acid containing traces of water. The truly anhydrous acid has a $-H_0$ of 15. Moreover, this is greatly increased by

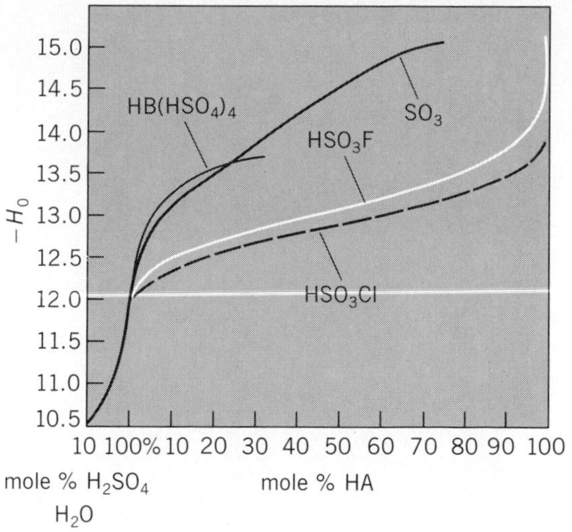

Fig. 1. Hammett acidity function (H_0) values for four systems. (*From R. J. Gillespie, Proton acids, Lewis acids, hard acids, soft acids and super acids, in E. F. Caldin and V. Gold eds., Proton Transfer Reactions, Chapman and Hall, 1975*)

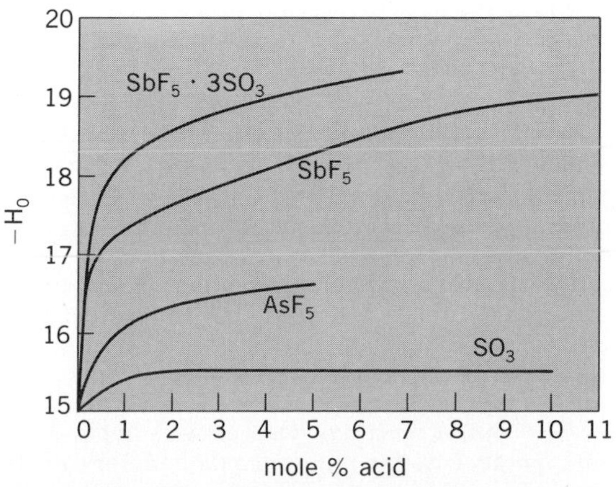

Fig. 2. Hammett acidity function (H_0) values for acids of the fluorosulfuric acid solvent system. (*From R. J. Gillespie, Proton acids, Lewis acids, hard acids, soft acids and superacids, in E. F. Caldin and V. Gold, eds., Proton Transfer Reactions, Chapman and Hall, 1975*)

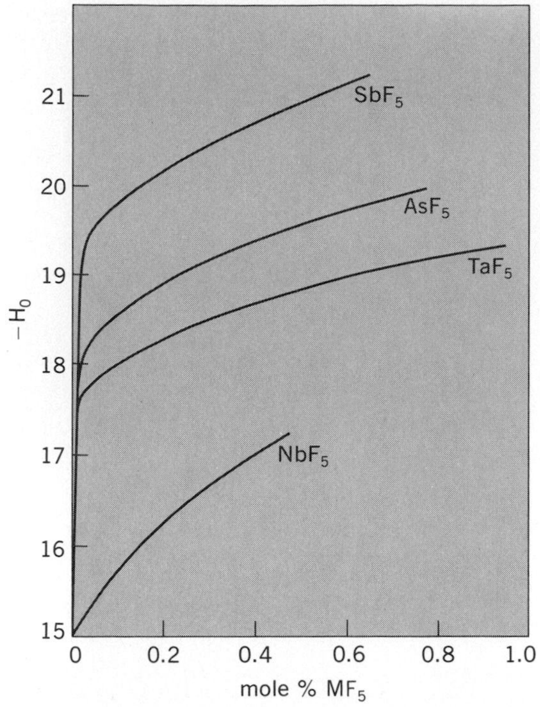

Fig. 3. Hammett acidity function (H_0) values for some $HF-MF_5$ systems.

the addition of pentafluorides and in particular of SbF_5, as may be seen in Fig. 3. Antimony pentafluoride ionizes in HF in a completely analogous manner to its ionization in HSO_3F [reaction (9)]. A

$$2HF + SbF_5 \rightleftharpoons H_2F^+ + SbF_6^- \qquad (9)$$

1% solution of SbF_5 in anydrous HF has a $-H_0$ value of 21, which is the highest value yet measured. There is every reason to suppose that still higher values would be obtained in more concentrated SbF_5 solution, but such studies have been prevented by lack of suitable weakly basic indicators. *See* HYDROGEN FLUORIDE.

Applications. The number of important applications of superacids in both organic and inorganic chemistry has been increasing. Of all the superacid systems described above, fluorosulfuric acid has been the most extensively used for a wide variety of applications. There are several reasons for this. It has a long and convenient liquid range from −89 to 163°C. Its low freezing point, together with a relatively low viscosity, has made it the solvent of choice for nuclear magnetic resonance. Its acidity is very high and can be further increased in an easily controlled manner by the addition of SbF_5. Although it is an extremely strong acid, it can, when free from water, be handled in ordinary glass apparatus. Anhydrous hydrogen fluoride, which has most of the convenient properties of HSO_3F and in fact gives high acidities with SbF_5, cannot, of course, be handled in glass.

Protonation reactions. One obvious application of superacids is for the protonation of very weak bases. A wide variety of organic compounds are protonated in superacid media, including ketones, carboxylic acids, alcohols, ethers, amides, and nitrocompounds. For example, the nuclear magnetic resonance spectrum of a solution of acetamide in fluorosulfuric acid at −80°C shows that under these conditions acetamide is protonated on the oxygen to give the conjugate acid (I).

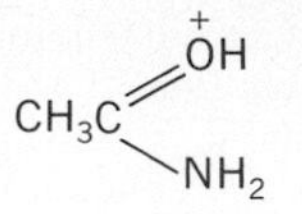

(I)

Acetic acid is protonated in antimony pentafluoride in fluorosulfuric acid (HSO_3F-SbF_5) to give a mixture of two conformers of the conjugate acid, structures (II*a*), 95%, and (II*b*), 5%. At −80°C, ^{13}C

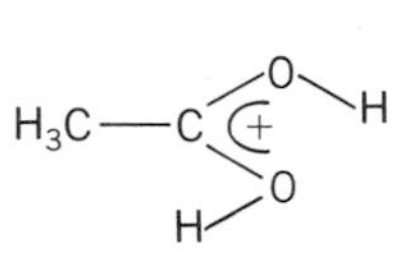

(II*a*)

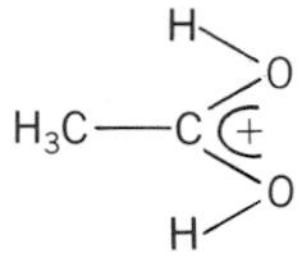

(II*b*)

nuclear magnetic resonance of solutions of inorganic carbonates in HSO_3F-SbF_5 shows a quantitative formation of the conjugate acid of carbonic acid, $C(OH)_3^+$. Only at 0°C does this decompose to CO_2 and H_3O^+.

Many aromatic hydrocarbons are protonated in superacid to produce arenium ions, for example, the conjugate acid of fluorobenzene (III).

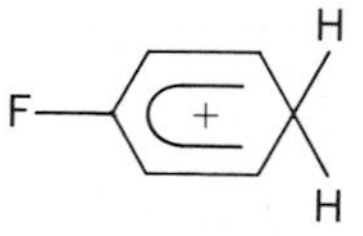

(III)

It is also interesting that many compounds of biological importance—for example, carbohydrates, proteins, and even chlorophyll and vitamin B_{12}—dissolve in anhydrous hydrogen fluoride and appear in many cases to give stable solutions. Presumably these molecules are protonated, probably in many cases more than once. The observation of hydrogen exchange between CH_4 and DSO_3-SbF_5 has been interpreted as indicating the formation of the carbonium ion CH_5^+ at least as an unstable intermediate.

Electrophilic cations. One of the early applications of superacids was the identification of the nitronium ion, the reactive intermediate in aromatic nitronium, which forms in solutions of nitric acid in anhydrous sulfuric acid, as shown in reaction (10). The nitronium ion NO_2^+ is typical of a

$$HNO_3 + 2H_2SO_4 \rightarrow NO_2^+ + H_3O^+ + 2HSO_4^- \qquad (10)$$

large number of electrophilic cations, many of which have previously been postulated as reaction intermediates but which have only a very short lifetime in conventional solvents. Stable solutions of many of these cations can be obtained in superacid media because of the extremely low basicity of such media. More basic solvents such as water immediately destroy these highly electrophilic cations. An example is shown in reaction (11).

$$NO_2^+ + 2H_2O \rightarrow HNO_3 + H_3O^+ \qquad (11)$$

A number of different halogen cations have been prepared in superacid solutions, including I_2^+, I_3^+, Br_2^+, and Br_3^+. Stable crystalline salts of some of these cations have been obtained, for example, $I_2^+Sb_2F_{11}^-$. Such salts are stable only when the anion is derived from a superacid and therefore has an extremely small basicity. The $Sb_2F_{11}^-$ ion is, for example, formed in solutions of SbF_5 in HF [reaction (12)].

$$2HF + 2SbF_5 \rightarrow H_2F^+ + Sb_2F_{11}^- \qquad (12)$$

It is interesting to note that although the simple monatomic ions I^+, Br^+, and Cl^+ have been postulated as intermediates in aromatic halogenation reactions, only polyatomic cations such as I_2^+ and Br_3^+ have been obtained even in the strongest superacids.

A large variety of polyatomic cations of the group VI elements have also been prepared in superacids, including species such as S_4^{2+}, S_8^{2+}, S_{19}^{2+}, Se_{10}^{2+}, Se_8^{2+}, and Te_4^{4+}, many of which have interesting novel structures.

One of the most important applications of superacids had been for the preparation of alkyl carbenium ions. Aryl carbenium ions, for example, $(C_6H_5)_3C^+$, have been known for a long time, but until the 1960s alkyl carbenium ions were known only as unstable reaction intermediates. When dissolved in HSO_3F-SbF_5, *t*-butanol is protonated and dehydrated to give the trimethyl carbenium ion [reaction (13)]. 2-Propanol similarly gives the dimethyl carbenium ion [reaction (14)].

$$(CH_3)_3COH \xrightarrow{H^+} (CH)_3COH_2^+ \xrightarrow{H^+} (CH_3)_3C^+ + H_3O^+ \qquad (13)$$

$$(CH_3)_2CHOH \xrightarrow{H^+} (CH_3)_2CHOH_2^+ \xrightarrow{H^+} (CH_3)_2CH^+ + H_3O^+ \qquad (14)$$

Primary alcohols are protonated in HSO_3F-SbF_5 at low temperatures (−60°C) and then slowly dehydrated, followed by a rapid isomerization of the primary carbenium ion to a more stable tertiary ion [reaction (15)].

$$H_3C{-}CH_2{-}CH_2{-}CH_2OH \xrightarrow[-60^\circ C]{HSO_3F-SbF_5} H_3C{-}CH_2{-}CH_2{-}CH_2OH_2^+$$
$$\downarrow H^+$$
$$(CH_3)_3C^+ \leftarrow [H_3C{-}CH_2{-}CH_2{-}CH_2] + H_3O^+ \qquad (15)$$

Aliphatic ethers are protonated in superacids at low temperatures and subsequently cleaved at higher temperatures to alkyl carbenium ions [reaction (16)].

$$H_3C{-}O{-}CH_2{-}CH_2{-}CH_2{-}CH_3 \xrightarrow[-60^\circ C]{HSO_3F-SbF_5} H_3C{-}\underset{\displaystyle H}{\underset{|}{O}}{-}CH_2{-}CH_2{-}CH_2{-}CH_3 \qquad (16)$$
$$\downarrow H^+$$
$$(CH_3)_3C^+ \leftarrow [CH_3{-}CH_2{-}CH_2{-}CH_2^+] + CH_3OH_2^+$$

Superacids such as HSO_3F-SbF_5 are very effective hydrogen-abstracting agents, allowing the generation of carbenium ions from saturated hydrocarbons [reaction (17)].

$$H_3C{-}\underset{\displaystyle CH_3}{\underset{|}{CH}}{-}CH_3 \xrightarrow{HSO_3F-SbF_5} (CH_3)_3C^+ + H_2 \qquad (17)$$

The possibility of preparing stable solutions of carbenium ions in superacids has led to many detailed studies of the reactions of carbenium ions, including several of considerable industrial importance. For example, the very important alkylation and isomerization reaction of hydrocarbons which proceeds via carbenium ion intermediates can be carried out by using superacids as catalytic reaction media. *See* IONIC EQUILIBRIUM; REACTIVE INTERMEDIATES; SOLUTION.

[RONALD J. GILLESPIE]

Bibliography: R. J. Gillespie, The chemistry of super acid systems, *Endeavour*, 32:3, 1973; R. J. Gillespie, Fluorosulfuric acid and related superacid media, *Acc. Chem. Res.* 1:202, 1968; R. J. Gillespie, Proton acids, Lewis acids, hard acids, soft acids and superacids, in E. F. Caldin and V. Gold (eds.), *Proton Transfer Reactions*, 1975; R. J. Gillespie and M. J. Morton, Halogen and interhalogen cations, *Quart. Rev. Chem. Soc.*, 25:553, 1971; R. J. Gillespie and J. Passmore, Homopolyatomic cations of the elements, in H. J. Emeleus and A. G. Sharpe (eds.), *Advances in Inorganic Chemistry and Radiochemistry*, vol. 17, 1975; R. J. Gillespie and J. Passmore, Polycations of Group VI, *Acc. Chem. Res.*, 4:413, 1971; R. J. Gillespie and T. E. Peel, Superacid systems, in V. Gold (ed.), *Advances in Physical Organic Chemistry*, vol. 9, 1971; G. A. Olah, Carbocations and electrophilic reactions, *Angew. Chem. (Int. Ed.)*, 12:173, 1973.

Supersaturation

A solution is at the saturation point when dissolved solute in it crystallizes from it at the same rate at which it dissolves. Under prescribed experimental conditions of temperature and pressure, a solution can contain at saturation only one fixed amount of dissolved solute. However, it is possible to prepare relatively stable solutions which contain a quantity of a dissolved solute greater than that of the saturation value provided solute phase is absent. Such solutions are said to be supersaturated. They can be prepared by changing the experimental conditions of a system so that greater solubility is obtained, perhaps by heating the solution, and then carefully returning the system to or near its original state. The addition of solute phase will immediately relieve supersaturation. Solutions in which there is no spontaneous formation of solute phase for extended periods of time are said to be metastable. There is no sharp line of demarcation between an unstable and a metastable solution. In fact, the latter is poorly defined and much influenced by many factors such as mechanical shock and the presence of minute quantities of foreign materials. The process whereby initial aggregates within a supersaturated solution develop spontaneously into particles of new stable phase is known as nucleation. The greater the degree of supersaturation, the greater will be the number of nuclei formed. This is a condition to be avoided in gravimetric analysis because of the formation of many small crystals which tend to coprecipitate excessive amounts of foreign ions by virtue of their great surface area. *See* GRAVIMETRIC ANALYSIS; NUCLEATION; PHASE EQUILIBRIUM; PRECIPITATION.

[LOUIS GORDON/ROYCE W. MURRAY]

Supertransuranics

A group of predicted elements beyond the present periodic table of known elements, with atomic numbers around 114, expected to possess half-lives of the order of a year or longer. These are also referred to as the superheavy elements. Although they have not been discovered experimentally, it is generally believed that the difficulty lies in the synthesis of these elements, rather than in their stability once they are made.

Nuclear stability. The underlying physics responsible for the limited extent of the periodic table is the competition between attractive "nuclear forces" among the nucleons (that is, protons and neutrons) and the repulsive electrostatic forces among all the positively charged protons. The limit of the periodic table at an atomic number Z of approximately 106 is then set by the process of nuclear fission, which takes place when the disruptive effect of electrostatic forces overcomes the cohesive effect of the nuclear forces. *See* NUCLEAR FISSION.

The picture began to change when, in 1964, a clearer understanding was reached on the relation between fission half-lives and a well-known property of the nucleus called the magic numbers. If the number of protons (or neutrons) in a nucleus is equal to the proton magic number (or neutron magic number), then such a nucleus displays some special features; one example is that it is spherical in shape. Work of W. D. Myers and W. J. Swiatecki showed that such a nucleus also displays an extra stability against fission. This means that this nucleus would have a longer half-life than would be expected otherwise. From experiments it has been deduced that the proton magic numbers are 8, 14, 28, 50, and 82. Many people have tried to predict the next magic number, and it is believed by many workers in this field that the next proton magic number is 114. The neutron magic numbers are the same as those for protons at lower numbers, but are different at larger numbers. The neutron magic numbers are 8, 14, 28, 50, 82, and 126, with the next predicted number at 184.

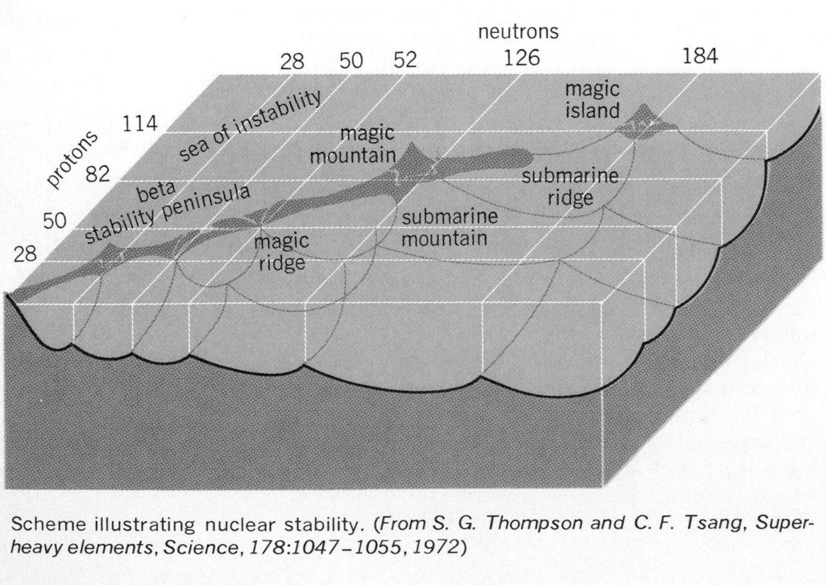

Scheme illustrating nuclear stability. (*From S. G. Thompson and C. F. Tsang, Superheavy elements, Science, 178:1047–1055, 1972*)

The figure illustrates known nuclei by a peninsula surrounded by the sea of instability. Mountains and ridges on the peninsula, representing regions of extra stability against fission, occur at the proton or neutron magic numbers, shown by grid lines. It can be seen from this figure that nuclei exist with a certain ratio of protons and neutrons. If a nucleus has too many protons or too many neutrons, it will decay by emitting a positron or an electron. The superheavy elements occur in this picture as an island of stability a little distance beyond the known peninsula with the center of the island at the proton magic number 114 and the neutron magic number 184. *See* RADIOACTIVITY.

At about the same time as Myers and Swiatecki's work, V. M. Strutinsky developed a method, now named after him, by which a quantitative estimate can be made of the stability of such superheavy nuclei. This method combines two well-known approaches in nuclear physics: the liquid drop model and the shell model of the nucleus. The first systematic calculation of the half-lives of both the known heavy nuclei and the predicted superheavy nuclei was made by S. G. Nilsson, C. F. Tsang, and coworkers in 1969, using the Strutinsky method. Several more refined calculations were made during the period up to 1974, perhaps the most complete version being that by J. R. Nix and coworkers. To a large extent, these different calculations were in agreement.

Possible natural occurrence. One of the surprising results from the calculations of Nilsson, Tsang, Nix, and others is that these superheavy elements may live as long as the age of the solar system. After considering all the major decay mechanisms, they predicted that several of these elements, in particular the element with $Z = 110$, have half-lives of about 10^8 yr. Thus an interesting possibility arises that if these elements were made in nature along with the other known elements during the formation of the Earth, small fractions could have survived the period of time (approximately 4.5×10^9 yr) since the Earth was formed.

It was realized by workers in the field that great uncertainties were involved in these predictions. If the prediction of 10^8 yr is off by two or three orders of magnitude downward, which is quite possible in the present state of the art in theory, no naturally occurring superheavy elements are expected to be found. Also it is an open question whether such elements would have been made at the formation of the solar system. Nevertheless, extensive effort has been expended in looking for these elements in nature.

A search for new elements on the Earth depends on suitable choices of the most promising minerals and ores containing known elements having chemical properties most resembling those of the elements being sought. Using the time-honored method of D. I. Mendelyeev, it is seen from the periodic table that superheavy elements 110, 111, 112, 113, 114, and 115 should have chemical properties similar to those of Pt, Au, Hg, Tl, Pb, and Bi, respectively. Calculations made by using much more sophisticated methods developed in recent years have also been performed for these elements, giving detailed predictions of their chemical properties.

G. N. Flerov's group at Dubna (Soviet Union), S. G. Thompson's group at Berkeley, G. Herrmann's group at Mainz (Germany), and others looked at a variety of ores and minerals, including natural platinum ores, old lead glasses, moon rocks, and manganese nodules collected from the ocean which are found to be particularly rich in metallic minerals. Typically one tries to detect fission events or accompanying neutron emission characteristic of these superheavy elements. However, in some cases, alpha decay detection is performed, since it has been predicted that alpha emission energies of these elements are exceptionally larger than those of known elements. A number of other searches for superheavy elements in nature have also been made, including a search in cosmic rays which may include elements produced in explosions in distant stars or supernovae. None, so far, has given conclusive evidence of their presence. Their absence may be due to one or both of two reasons. First, it is still an open question whether these superheavy elements were synthesized at the formation of the solar system or in the supernovae. Calculations attempting to answer this question involve great uncertainties, since they are very sensitive to the estimated half-lives of nuclei between the known elements and the island of stability. Second, the half-lives of the superheavy nuclei may be a few orders lower than predicted, and they would have disappeared by radioactive decay during the 4.5×10^9 yr since the Earth was formed. These short half-lives are predicted by a study made in 1974 (see table).

A table of predicted half-lives of several superheavy nuclei

Atomic number (Z)	Neutron number (N)	Mass number (A)	Mean life (τ)
108	180	288	10^{-3} sec
110	184	294	10^5 years
112	184	296	100 years
114	184	298	0.1 year
116	188	304	0.5 min
118	192	310	0.1 sec

Assuming that the superheavy elements were produced in nature and have half-lives too short to survive until today, attempts have been made to detect fission products of these elements. Exhaustive experiments have been made to find the tracks caused by these fission fragments in, for instance, lead glasses and moon rocks. No clear evidence for superheavy elements was found. In 1975 Edward Anders and his group at Chicago made a careful radiochemical neutron activation analysis of the Allende meteorite. They concluded that the isotopic distribution of the element Xe in a rare chromium mineral sample of the meteorite indicates the existence of an unknown fissioning element, which may be an extinct superheavy element with $Z=$ 115, 114, or 113.

Artificial synthesis. Perhaps it is too much to expect that these superheavy elements should have such long half-lives, given the uncertainties associated with theories. However, nearly all theories predict these superheavy elements to have half-lives of 1 year, 1 day, or at least 1 minute, which would still be extremely interesting and would allow study with available techniques if these elements could be synthesized. Obviously in this case a big jump must be made over the sea of instability to the island. In other words, a target such as U, Pu, or Cm must be bombarded with a heavy ion such as Ar, Ca, or Kr to make the nuclei fuse together to form a superheavy nucleus. Extensive experiments have been made at both Dubna and Orsay (France), but the results have been negative. Work on further experiments has been undertaken, notably at Berkeley and at Darmstadt (Germany).

Numerous problems may hinder the fusing of two nuclei into a superheavy nucleus. These problems have been given intensive theoretical and experimental study. A better understanding has been achieved in some areas, which will affect the conditions under which one would attempt to produce superheavy elements. [C. F. TSANG]

Bibliography: E. Anders et al., Extinct superheavy elements in the Allende meteorite, *Science*, 190:1262–1271, 1975; T. Johansson, S. G. Milsson, Z. Szymanski, Theoretical predictions concerning superheavy elements, *Ann. Phys.* (Paris), 5:377–416, 1970; J. R. Nix, Predictions for superheavy nuclei, *Phys. Today*, 25(4):30–38, 1972; *Physica Scripta*, vol. 10A, pp. 1–184, 1974; Superheavy elements: Theoretical predictions and experimental generation, *Proceedings of the 27th Nobel Symposium*, Ronneby, Sweden, June 1974; S. G. Thompson and C. F. Tsang, Superheavy elements, *Science*, 178:1047–1055, 1972.

Surface-active agent

A substance that, even though present in small amounts, exerts a marked effect on the surface behavior of a system. These agents are essentially responsible for producing great changes in the surface energy of liquid or solid surfaces, and their ability to cause these changes is associated with their tendency to migrate to the interface between two phases. Consequently, surface-active agents are of potential interest wherever there are solid-solid, solid-liquid, solid-gas, liquid-liquid, or liquid-gas interfaces, and of particular interest at liquid-gas interfaces at which the surface-active agent is a solute whose presence makes the surface properties of the solution greatly different from those of the solvent. *See* INTERFACE OF PHASES.

Mechanism. Soap, for example, when dissolved in small quantities in water, is responsible for greatly decreasing the surface tension of water, and it is this property of soap that accounts for its ability to act as a detergent. In contrast to soap and other related substances that lower the surface energy of a liquid, other solutes, such as inorganic salts, acids, and bases may increase the surface tension of a liquid (see illustration), but their effect in increasing the surface tension is not nearly so great as the effect of those agents that decrease the surface tension. Occasionally the term surface-inactive solutes is applied to these substances whose presence causes an increase in surface tension.

The importance of surface-active agents is indicated by their strategic necessity in such pro-

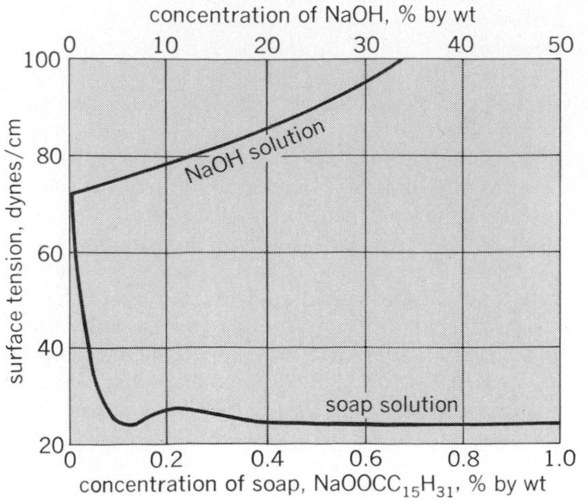

Effect of surface-active agents on water.

cesses as lubrication, wetting, foaming, emulsification, detergency, water repellence, waterproofing, spreading, and dispersion. In lubrication, for example, the oiliness of a hydrocarbon oil can be improved by the addition of a surface-active agent. In order to achieve lubrication between two solid surfaces, a thin film of liquid must be preserved in the space between the two solid surfaces. The viscosity of this liquid film and the ability of the liquid to wet the solid surfaces determine the resistance of the lubricant system to being squeezed mechanically from the region between the two solid surfaces. Addition of fatty acids, fatty oils, metallic soaps, and various derivatives of aromatic and aliphatic hydrocarbons commonly improves the lubricant qualities of mineral hydrocarbons, and these additives are truly surface-active agents.

The mechanism by which surface-active agents alter the surface energy of a solid or liquid is attributed to the dual nature of the molecules or ions of these substances. Within a single molecule or ion of a surface-active agent, there is a group that is lyophilic toward the dispersing medium or solvent, and at a suitable distance within the same molecule or ion, there is another group that is lyophobic toward the dispersing medium. This ability to embody within the same molecular particle two different groups whose properties are diametrically opposed is sometimes termed amphipathy. For example, the surface activity of sodium oleate, $NaOOCC_{15}H_{31}$, is attributed to the combined effect of the hydrophilic ionic carboxyl salt group at one end of the molecule and the hydrophobic hydrocarbon group that constitutes the remainder of the molecule. In a dilute solution of sodium oleate, the solute migrates to the surface where the hydrophobic parts of the molecules can achieve their lowest energy positions as the result of the solvent's striving to exclude the hydrocarbon group from the solution. Even though the external phase is gaseous, the hydrophobic groups find a sufficiently sympathetic environment at the surface of the liquid. If, on the other hand, the external phase were an oil, the hydrocarbon groups would find an even more sympathetic environment, and in either event, the surface energy of the original solvent would be greatly diminished. *See* MONOMOLECULAR FILM.

Classification. Surface-active agents are usually classified in three groups: anionic, cationic, and nonionic types. Anionic types include carboxylate ions such as occur in sodium oleate. The carboxyl group may be attached directly to the hydrophobic group, or there may be an intermediate ester, amide, or sulfonamide linkage. There are a large number of anionic agents derived from sulfuric and sulfonic acids in which the hydrophobic groups attached to them include aliphatic and aromatic groups that often contain substituents of varying polarity, such as halide, hydroxyl, ether, and ester groups.

Cationic surface-active agents are usually derived from the amino group where, through either primary, secondary, or tertiary amine salts, the hydrophilic character may be achieved by aliphatic and aromatic groups that may be altered by substituents of varying polarity. Other nitrogen compounds, such as quaternary ammonium compounds, guanidine, and thiuronium salts, are included in the cationic class.

The third class of surface-active agents, the nonionic type, are organic substances which contain groups of varying polarity and which render parts of the molecule lyophilic, whereas other parts of the molecule are lyophobic. Examples include polyethylene glycol, polyvinyl alcohol, polyethers, polyesters, and polyhalides. In this class are often included certain colloidal substances such as graphite, powdered metals, metallic oxides, clays, macromolecules, and polymers. *See* SURFACE TENSION.

[WENDELL H. SLABAUGH]

Bibliography: M. Ash and I. Ash, *Encyclopedia of Surfactants*, 2 vols., 1980; H. E. Garrett, *Surface Active Chemicals*, 1973; M. J. Rosen, *Surfactants and Interfacial Phenomena*, 1978.

Surface tension

The force acting in the surface of a liquid, tending to minimize the area of the surface. Surface forces, or more generally, interfacial forces, govern such phenomena as the wetting or nonwetting of solids by liquids, the capillary rise of liquids in fine tubes and wicks, and the curvature of free-liquid surfaces. The action of detergents and antifrothing agents and the flotation separation of minerals depend upon the surface tensions of liquids.

Surface energy. In the body of a liquid, the time-averaged force exerted on any given molecule by its neighbors is zero. Even though such a molecule may undergo diffusive displacements because of random collisions with other molecules, there exist no directed forces upon it of long duration. It is equally likely to be momentarily displaced in one direction as in any other. In the surface of a liquid, the situation is quite different; beyond the free surface, there exist no molecules to counteract the forces of attraction exerted by molecules in the interior for molecules in the surface. In consequence, molecules in the surface of a liquid experience a net attraction toward the interior of a drop. These centrally directed forces cause the droplet to assume a spherical shape, thereby minimizing both the free energy and surface area.

From the macroscopic point of view, surface tension may be regarded either as a force exerted normally to a unit length in the surface, or as the work which must be expended upon the liquid to increase its area by unity. Accordingly, surface tension is expressed in centimeter-gram-second (cgs) units of dynes/cm or ergs/cm². From the microscopic point of view, the surface tension (or its equivalent, surface energy) is the reversible isothermal work which must be done in bringing molecules from the interior of the liquid to the surface and creating 1 cm² of new surface thereby.

Most liquids have surface tensions of 20–40 dynes/cm at room temperature, but water has the exceptionally high value of 72.75 dynes/cm at 20°C. Condensed gases such as helium and nitrogen have quite low surface tensions (0.098 dynes/cm at 4.3 K and 6.2 dynes/cm at 90.2 K, respectively). Liquid metals have large surface tensions by comparison: mercury, 470 dynes/cm; and liquid copper at 1131°C has a surface tension of 1103 dynes/cm in hydrogen gas. Small but significant differences in the surface tensions of liquids depend upon the composition of the vapor phase.

In the wetting or nonwetting of solids by liquids, the criterion employed is the contact angle between the solid and the liquid (measured through the liquid) (Fig. 1). A liquid is said to wet a solid if

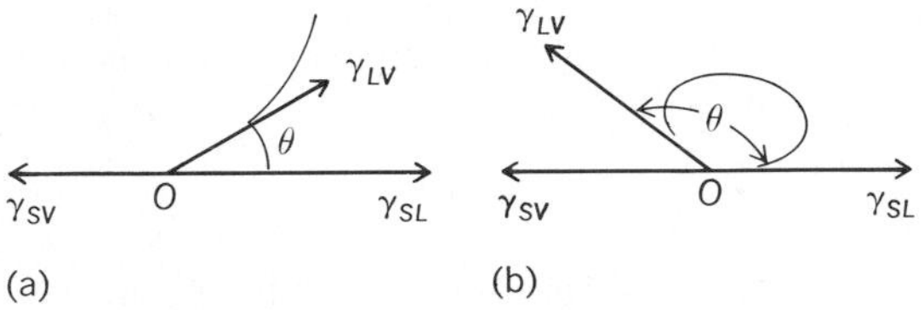

Fig. 1. Contact angle. (*a*) Liquid wets solid. (*b*) Liquid does not wet solid.

the contact angle θ lies between 0 and 90°, and not to wet the solid if the contact angle lies between 90 and 180°. Three interfaces exist when a droplet of liquid contacts a solid, and three corresponding interfacial tensions exist: γ_{SL}, γ_{SV}, and γ_{LV}. The subscripts *S*, *L*, and *V* refer to solid, liquid, and vapor. At equilibrium, a balance of interfacial tensions exists at the line of common contact, which intersects the figures at point *O*. For the case of a liquid which wets the solid ($\theta < 90°$), this equilibrium is expressed by relation (1).

$$\gamma_{SV}=\gamma_{SL}+\gamma_{LV}\cos\theta \qquad (1)$$

Capillarity. Liquids which wet the walls of fine capillary tubes rise to a height which depends upon the tube radius, the surface tension, the liquid density, and the contact angle. In Fig. 2, a liquid of density ρ is shown as having risen to a height *h* in a capillary whose radius is *r*. A balance exists between the force exerted by gravity on the mass of liquid raised in the capillary and the opposing force caused by surface tension. The former is $\pi r^2 h\rho g$, whereas the latter is $2\pi r\gamma$, assuming the contact angle to be zero. It is clear that $h=2\gamma/r\rho g$, and that the capillary rise varies inversely with the tube radius and the liquid density. Liquids which do not wet the capillary walls are depressed in height according to the same equation.

The shape of the free surface of a liquid in a vessel is only an approximation to a plane. In narrow tubes the meniscus of a liquid is concave upward if the liquid wets the tube and, conversely, convex upward if it does not wet the tube. A pressure difference exists between the concave and convex sides of the surface, the excess pressure on the concave side over the convex side being given by relation (2), where r_1 and r_2 are the principal radii

$$p=\gamma\frac{1}{r_1}+\frac{1}{r_2} \qquad (2)$$

of curvature of the surface. The same equation applies for a bubble of gas within a liquid, with the consequence that the vapor pressure p is larger for small bubbles according to relation (3), where p_0 is

$$\ln\frac{p}{p_0}=\frac{2\gamma}{r\rho}\frac{M}{RT} \qquad (3)$$

the vapor pressure over a liquid surface of infinite radius, *R* is the gas constant, and *M* is the molecular weight. *See* VAPOR PRESSURE.

Detergents, soaps, and flotation agents owe their usefulness to their ability to lower the surface tension of water, thereby stabilizing the formation of small bubbles of air. At the same time, the interfacial tension between solid particles and the liquid phase is lowered, so that the particles are more readily wetted and floated after attachment to air bubbles. *See* INTERFACE OF PHASES; SURFACE-ACTIVE AGENT.

[NORMAN H. NACHTRIEB]

Bibliography: American Chemical Society, *Contact Angle, Wettability and Adhesion*, 1964; G. M. Barrow, *Physical Chemistry*, 4th ed., 1979.

SURFACE TENSION

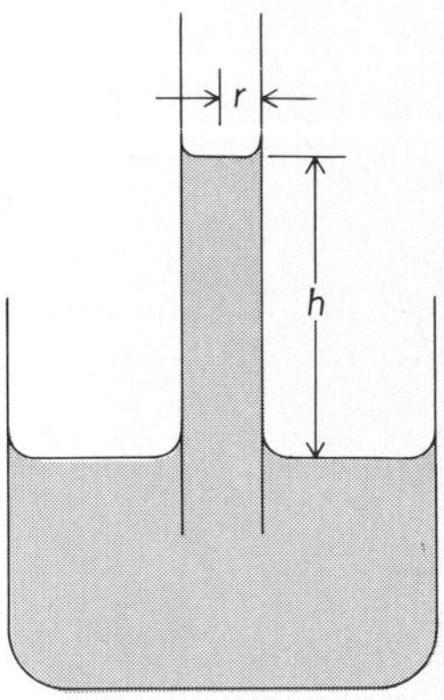

Fig. 2. Rise of liquid in capillary tube.

Tannin

A generic term for a widely occurring group of substances of vegetable origin, capable of rendering raw hides into leather. Common tannin (tannic acid) occurs in oak gallnuts (Turkish nutgall contains 50–60%, Chinese nutgall about 70%); tannins are also present in tea, sumac, oak bark (the word tan itself means oak bark), and mangrove bark. Tannin from the latter source is known as cutch, and is produced on a large scale, especially in Malaya.

The usual method of preparation involves breaking or crushing the bark or gallnuts into small pieces; these are then washed and boiled with water until the tannin has been extracted. After separation of insoluble matter, the thick, reddish-brown, viscous extract is evaporated, leaving the crude tannin as a hard cake. Purification may be effected by extracting the crude material with an alcohol-ether mixture; evaporation deposits the tannic acid as a colorless, noncrystalline mass. Tannic acid may also be prepared by heating gallic acid with phosphorus oxychloride.

Substances capable of tanning, and hence called tannins, are often of greatly different chemical structure; all tannins, however, have the property of converting the gelatin of hides into insoluble nonputrefying material, thus changing the hide into leather. In general, tannins are noncrystalline when solid, but readily soluble in water or alcohol to give colloidal solutions that are strongly astringent and therefore useful in medicine. Tannins have long been used in compounding inks, because they form greenish-black or bluish-black colors with ferric salts.

Tannins may be divided into three main classes: (1) condensed tannins that cannot be hydrolyzed either by acids or enzymes (these include the acacatechin and isoacacatechin tannins and the gambir catechin tannins; all contain highly substituted phloroglucinol nuclei); (2) hydrolyzable tannins, for example, gallotannins, ellagitannins, and caffetannins; and (3) tannins of unclassified nature.

Gallotannin, from which is obtained the tannic acid of commerce and medicine, is present in oak galls. It is a mixture of the gallic acid esters of glucose, one of which is pentadigalloylglucose, the digallic acid moiety having the structure shown. These esters are called depsides.

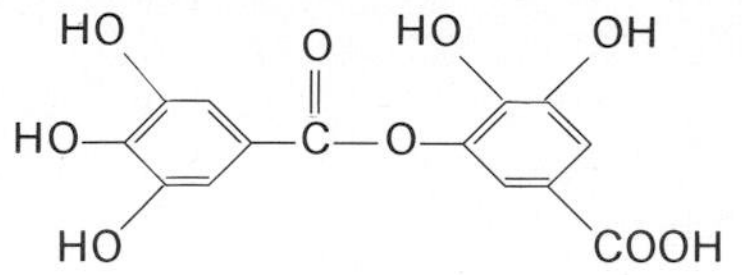

m-Digallic acid

Tannic acid, USP, is a mixture of compounds of gallotannin type. It is a light-yellow powder of very astringent taste, much used in styptic preparations and ointments. The aqueous solution of tannic acid is used in treating burns, because it precipitates the burned protein, forming a nonputrefying, protective layer under which new tissue can grow.

[EVANS B. REID]

Tantalum

A chemical element, symbol Ta, atomic number 73, and atomic weight 180.948. It is a member of the fifth group of the periodic table and is in the 5*d* transitional series. Its valence electron configuration is $5d^3 6s^2$, which accounts for its maximum oxidation state of V. Oxidation states of IV, III, and II are also known. The similarity in effective

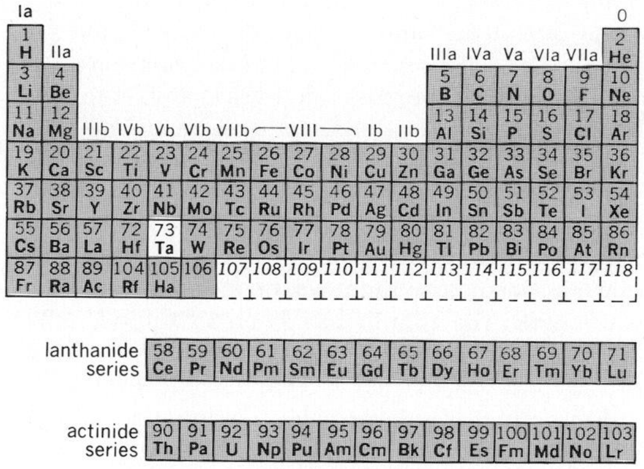

ionic radii for Ta^{5+} and Nb^{5+} (73 and 70 picometers, respectively) accounts for the fact that tantalum is found in nature with its lower-atomic-numbered homolog, niobium, in the following ores: columbite or tantalite, $(Fe,Mn)(Nb,Ta)_2O_6$, depending upon which element is predominant, and euxenite, $(Y,Ca,Ce,U,Th)(Nb,Ta,Ti)_2O_6$. It also occurs by itself in microlite, $(Na,Ca)_2Ta_2O_6(O,OH,F)$. Tantalum occurs in the Earth's crust to the extent of 2.1×10^{-4}%. The principal sources are niobium and tantalum concentrates from Brazil, Canada, Africa, and Spain. The United States supplies are insignificant, although that country is a principal consumer.

Tantalum metal is used in the manufacture of capacitors for electronic equipment, including citizen band radios, smoke detectors, heart pacemakers, and automobiles. An extremely stable film of tantalum oxide acts as an insulator in the capacitor. It is also used for heat-transfer surfaces in chemical production equipment, especially where extraordinarily corrosive conditions exist. Its chemical inertness has led to dental and surgical applications.

Tantalum forms alloys with a large number of metals. Of special importance is ferrotantalum, which is added to austentitic steels to reduce intergranular corrosion. Alloys of composition Ta_3M are formed with tin and cobalt, Ta_2M with cobalt, TaM with nickel, TaM_2 with germanium, chromium, manganese, and iron, and TaM_3 with aluminum, iridium, nickel, and rhodium.

Metallurgical extraction. The coexistence of niobium and tantalum in ores and the similarity in their properties require a fractional method of separating one from the other. The fractional crystallization processes of fluoride salts have been largely replaced by solvent extraction procedures. In one such process, the aqueous solution of metal ions containing 0.5 *N* HCl and 3.3 *N* HF is treated with an organic solvent such as methyl isobutyl ketone. The tantalum species are extracted preferentially into the organic phase, with the niobium species concentrating in the aqueous phase. The organic phase is back-extracted with water to recover the tantalum. Tantalum and niobium are recovered as the oxides from the aqueous phases by complexing the fluoride with boric acid and precipitating the oxide with aqueous ammonia. The precipitate obtained from the initial aqueous layer is 98% niobium oxide; that from the organic layer is 99.5% tantalum oxide. In the absence of hydrofluoric acid and at very high hydrochloric acid concentrations, the niobium concentrates in the organic phase and the tantalum remains largely in the aqueous phase. *See* SOLVENT EXTRACTION.

Tantalum metal. Tantalum metal crystallizes in the body-centered cubic system, has a density of 16.6×10^3 kg/m^3 at 293 K, a melting point of 3223 K, and a boiling point of 5702 K. It has a cross-section capture for thermal neutrons of 21.3 barns. Metallic tantalum is prepared by the fused electrolysis of K_2TaF_7 or by reduction of the oxide with active metals or carbon. A powder is obtained which, after being washed, is compacted into bars and then sintered in a vacuum furnace with the bar acting as the heating element. The bar is then cold-rolled to sheets or wire. Although the standard potential for the reaction shown below is +0.71 V,

$$2Ta + 5H_2O \rightarrow Ta_2O_5 + 10H^+ + 10e$$

the metal in actual practice is quite inert to acid attack except by hydrofluoric acid. It is very slowly oxidized in alkaline solutions. The halogens and oxygen react with it on heating to form the oxidation-state-V halides and oxide. At high temperature it absorbs hydrogen and combines with nitrogen, phosphorus, arsenic, antimony, silicon, carbon, and boron.

Tantalum also forms compounds by direct reaction with sulfur, selenium, and tellurium at elevated temperatures. These compounds possess ranges of composition, $Ta_{1+x}X_2$, which depend on the conditions of synthesis and which exist in numerous polymorphic structures. In these compounds, tantalum sits in trigonal prismatic holes generated by the coincidence of two close-packed layers of sulfurs. It is the variation in the stacking of these resulting slabs which leads to the polymorphism. A wide variety of amines and amides have been inserted between these slabs with a considerable displacement of the interslab distances.

Tantalum compounds. The oxide Ta_2O_5 can be obtained by heating the metal in oxygen or by dehydration of the hydrated oxide. It occurs in a tetragonal form, α, and an orthorhombic form, β. The β form transforms to the α form at 1633 K. The melting points of the α and β form are 2145± 10 K and 2058±30 K, respectively. Tantalum(V) oxide is insoluble in inorganic acids except hydrofluoric acid and concentrated sulfuric acid. Although TaO_x ($x < 2.5$) compositions are in evidence in the oxidation of tantalum metal by oxygen, no discrete TaO or TaO_2 compounds are known. When tantalum(V) oxide is fused with alkali hydroxides or carbonates, insoluble tantalates result which hydrolyze to the oxide upon washing with water. The aqueous potassium hydroxide extract of such a fusion contains a soluble species and, upon treatment with ethanol, a product of composition $K_8Ta_6O_{19} \cdot 16H_2O$ precipitates. The $Ta_6O_{19}{}^{8-}$ anion has been shown to contain six TaO_6 octahedrons with one oxygen common to all six octahedrons, twelve oxygen shared by six octahedrons, and only six oxygen remaining unshared. This ion also exists in aqueous solution and is not subject to further polymerization, depolymerization, or protonation in the pH range 10–13. Upon acidification the hydrous oxide is precipitated. Although it is commonly called tantalic acid, it is insoluble in aqueous bases. The oxide is soluble in hydrofluoric acid, and salts of composition $KTaF_6$, K_2TaF_7, and K_3TaF_8 can be crystallized from aqueous solutions of different fluoride concentrations.

Tantalum is less likely to form oxo species than niobium is. However, oxycompounds of composition M_3TaOF_6 and MTa_2O_5F (where M = univalent cation) have been prepared. Like the corresponding niobium species, the $TaF_7{}^{2-}$ ion is known to be a true seven-coordinate species. The $TaF_8{}^{3-}$ ion is eight-coordinate with the fluorines at the corners of a square antiprism. A Raman spectra study of the Ta(V)—HF—F^- system revealed the presence of $TaF_6{}^-$ and $TaF_7{}^{2-}$ ions in solutions of certain compositions but no evidence of these ions in the octafluoro species.

Anhydrous hexaclorotantalates(V) of composition $MTaCl_6$ (M = alkali metal or NR_4^+, $AsPh_4^+$) are obtained by mixing stoichiometric quantities of the appropriate binary halides in a nonaqueous solvent such as thionyl chloride. Hexabromo and hexaiodo compounds also have been prepared in a nonaqueous solvent.

The pentahalides are all low-melting-point, low-boiling-point compounds which hydrolyze to the oxide when placed in water. They are prepared by direct reaction of the halogen on the metal or by action of the dry hydrogen halides on the metal. Tantalum pentachloride can be prepared by the elevated temperature reaction of liquid carbon tetrachloride on tantalum pentoxide under pressure. Tantalum pentachloride and bromide are dimeric in the solid state and isomorphous with niobium pentachloride. In this structure 10 halides occupy the vertices of two distorted octahedra which share a common edge. The structure of the pentafluoride has the structure of niobium pentafluoride in which tetramers are present. In the gas phase, pentahalides are momeric with a trigonal prismatic structure.

Anhydrous tantalum(V) fluoride, chloride, and bromide react with a wide variety of electron pair donors to form 1:1 adducts, $TaX_5{:}BR_n$. Examples of such donors include, $(C_2H_5)_2O$, $(C_2H_5)_2S$, R_3PO, $(C_6H_5)_3PS$, $(C_6H_5)_3PSe$, RCN, R_3N, and PCl_3O, among others. Tantalum (V) fluoride reacts with XeF_2 in anhydrous hydrofluoric acid to yield the adducts $TaF_5 \cdot 2XeF_2$ and $2TaF_2 \cdot XeF_2$. Adduct formation of TaI_5 has been little studied. The fluoride adducts are the most stable thermally and in general can be sublimed unchanged. Adducts of oxygen-containing Lewis bases and tantalum(V) chloride and bromide decompose to abstract oxygen forming $TaOX_3$ and TaO_2X. In the presence of excess ligand, adducts of these oxy species are formed which contain 2 or 3 moles of base. The oxy species are also obtained by direct reaction with oxygen at elevated temperatures. Alcohols react with tantalum(V) chloride in the presence of ammonia to form $Ta(OR)_5$. These alkoxides are dimers with structures similar to that of Ta_2Cl_{10}. Intermediate compositions, $TaX_{5-x}(OR)_x$ and mixed alkoxycompounds are known. Monomeric compositions, $M[Ta(OR)_6]$ where M = alkali metal cations, are also known. Amines form 1:1 adducts with tantalum pentafluoride, but the pentachloride and pentabromide react with primary and secondary amines to split out HCl to yield products of composition, $TaCl_2(NHCH_3)_3$, $TaCl_3(NHR)_2(NH_2R)$, or $TaCl_3(NR_2)(NHR_2)$. With tetiary amines, $TaCl_5 \cdot 2NR_3$ is formed.

Except for the fluoride, reduction of the pentahalides by pyridine and other cyclic amines provides a room-temperature route for the production of the oxidation-IV halides. The tetrabromide chloride and iodide can be prepared also by reduction of the pentahalides with the metal or aluminum metal at moderate temperature (473–573 K). The tetraiodide is isostructural with α-NbI_4. $TaBr_4$ melts incongruently at 665 K to form the next lower solid, $(Ta_6Br_{12})Br_5$, which in turn melts incongruently within the interval 720–726 K to form the phase $(Ta_6Br_{12})Br_3$. Similarly, $(Ta_6Br_{12})Br_3$ melts incongruently within the interval 944–953 K to yield $(Ta_6Br_{12})Br_2$, which does not melt below 1023 K. The tantalum tetraiodide decomposition to $(Ta_6I_{12})I_2$ occurs in the interval 668–515 K at a very slow rate. These compounds as well as $(Ta_6Cl_{12})Cl_2$ and $(Ta_6Cl_{12})Cl_3$ can also be prepared by reduction of the pentahalide with aluminum foil under appropriate temperature conditions. All these compounds in water yield the cation $Ta_6Cl_{12}{}^{2+}$, which can be oxidized to the charged four ion with Fe(III). A cluster fluorocompound is not known. The structure of the tantalum cluster is similar to that of the corresponding niobium cluster, but the tantalum cluster is subject to distortion, eight halogens bridge the eight faces, and four halogens bridge the edges of a tetragonal pyramid. Three mixed cluster anions, $[(Ta_5MoCl_{12})Cl_6]^{3-}$, $[(Ta_4Mo_2Cl_{12})Cl_6]^{2-}$, and $[(Ta_5MoCl_{12})Cl_6]^{2-}$, have been prepared by reaction of $TaCl_5$-$MoCl_5$ with aluminum in $NaAlCl_4$-$AlCl_3$.

Compounds of composition TaX_3 (X = Cl,Br,I) can be prepared by the reduction of TaX_5 on tantalum in a three-compartment glass reaction tube in which the TaX_5 compartment is heated to 578–593 K, the metal end is heated to 873–893 K, and the intermediate in which TaX_3 is deposited is heated to 638–653 K. The iodide is a stable phase, whereas the chloride and bromide show a range of homogeneity $TaCl_{2.9\text{-}3.1}$. No TaF_3 has been reported.

Tantalum shows only a limited complex ion aqueous chemistry; however, compounds containing typical inorganic anions can be prepared in nonaqueous systems. Thus, the reaction of liquid N_2O_5 on the pentachloride or tetramethylammonium hexachlorotantalate(V) yields $TaO(NO_3)_3$ and $(CH_3)_4NTaO(NO_3)_4$, respectively. The reaction of a phosphoric acid–nitric acid mixture with potassium tantalate(V) at elevated temperature yields $TaOPO_4$, while the normal phosphate, $Ta_3(PO_4)_5$, is obtained by heating the pentachloride with an excess of fused H_3PO_4 at about 1470 K. The normal sulfate, $Ta_2(SO_4)_5$, is obtained by heating sulfur trioxide with the pentachloride in sulfuryl chloride. Some oxysulfate species have also been reported. All of these compounds are unstable with respect to attack by water.

Alkali metal tantalates react with solutions of α-hydroxycarboxylic acids, polycarboxylic acids, and diols to give polymeric species in solution. In many instances, solid products recovered from solution have not been well characterized although $[Ph_4]^+$ $[TaOH(oxalate)_3]^-$ has been isolated and can be recrystallized from alcohol solution. In addition, a large number of tantalum organometallic compounds are known.

Analysis. Tantalum is determined in the presence of niobium by developing the peroxytantalate color in 96% sulfuric acid and measuring the absorbency at 285 nanometers; the niobium can be determined at 365 nm. *See* NIOBIUM; TRANSITION ELEMENTS. [EDWIN M. LARSEN]

Bibliography: J. C. Bailar et al. (eds.), *Comprehensive Inorganic Chemistry*, vol. 3, 1973; *Minerals Yearbook 1976*, vol. 1: *Metals, Minerals and Fuels*, 1978; C. J. Smithells, *Metals Reference Book*, 5th ed., 1976.

Tartaric acid

A compound that possesses two similar asymmetric carbon atoms and occurs in four isometric forms: (1) a dextrorotary form, which rotates plane-polarized light to the right; (2) a levorotary form, which rotates plane-polarized light to the left; (3) a racemate form, which is an optically inactive equimolecular mixture of separate crystals of the dextro and levo forms; and (4) a meso form, which is optically inactive because of internal compensation. Structural formulas of the dextro, levo, and meso forms are shown.

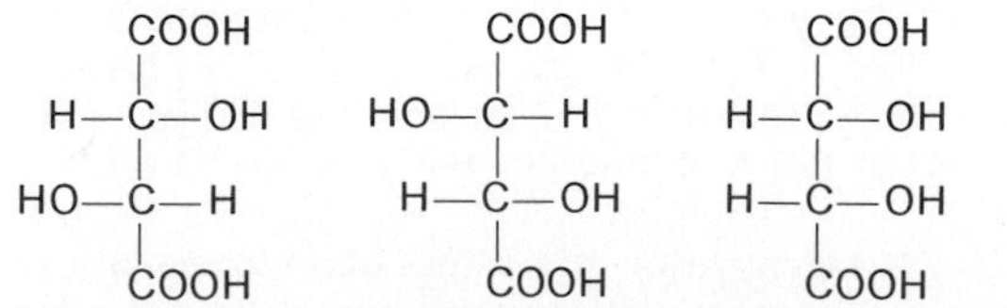

d-Tartaric acid *l*-Tartaric acid *m*-Tartaric acid

The *dl*-tartaric acid is known as the racemic acid. The meso acid is optically inactive by internal compensation; the *dl* acid is optically inactive by external compensation. This asymmetric structural difference among the four acids results in the differences in properties as shown in the table. *See* DIASTEREOISOMER; OPTICAL ACTIVITY.

Properties of tartaric acid

Acid	Melting point, °C	Optical rotation of 20% aqueous solution $[\alpha]^{25}_{D}$	Solubility, g/100 g H_2O at 15°C
Dextro	170	+12°	139
Levo	170	−12°	139
Racemic	206	Inactive	20.6
Meso	140	Inactive	125

The ordinary form, *d*-tartaric acid, is obtained from fermented grape juice as potassium hydrogen tartrate (argol or cream of tartar). Racemization of *d*-tartaric acid with hot sodium hydroxide, or nitric acid oxidation of mannitol or mucic acid, gives racemic tartaric acid. Resolution of racemic tartaric acid via *Penicillium glaucum*, the cinchonine or quinicine salts, or *l*-bornyl hydrogen ester, furnishes *l*-tartaric acid.

The principal uses of tartaric acid are in cream of tartar, Rochelle salt (potassium sodium *d*-tartrate), and tartar emetic (potassium antimonyl *d*-tartrate). *See* CARBOXYLIC ACID; TARTRATE. [EVANS B. REID]

Tartrate

A chemical compound that is a salt or ester of tartaric acid, which is formed by the replacement reaction of the carboxylic hydrogens of tartaric acid by a metal (salt) or organic radical (ester). One or both of the carboxylic hydrogens can be replaced to produce an extensive series of salts, esters, and mixed (double) salts. Tartrates exist in three isomeric forms, two being optically active and one inactive. The two optically active forms can combine to form the racemic compound. Salts are commonly made from crude tartars, obtained as a by-product of the wine-making industry. Many have practical applications, as in cream of tartar (monopotassium salt), Rochelle salts (sodium-potassium salt), tartar emetic (potassium-antimony salt), medicines, and textile dyeing (calcium salt). The common esters are liquids or low-melting solids, but they are not widely used. *See* OPTICAL ACTIVITY; ROCHELLE SALT; TARTARIC ACID. [ELBERT H. HADLEY]

Bibliography: F. A. Lowenheim and M. K. Moran, *Faith, Keyes, and Clark*'s *Industrial Chemicals*, 4th ed., 1975.

Tautomerism

The reversible interconversion of structural isomers of organic chemical compounds. Such interconversions usually involve transfer of a proton (prototropy), but anionotropic (allylic, Wagner-Meerwein) rearrangements may be reversible and so be classed as tautomeric interconversions.

Lactam-lactim tautomerism. A cyclic system containing the grouping —CONH— is called a lactam, and the isomeric form, —COH=N—, a lactim. These terms have been extended to include the same structures in open-chain compounds when considering the shift of the hydrogen from nitrogen to oxygen.

A. von Baeyer first recognized that isatin (I) appears to react in either the lactam (I) or the lactim (II) structure. Thus, a precedent for proving

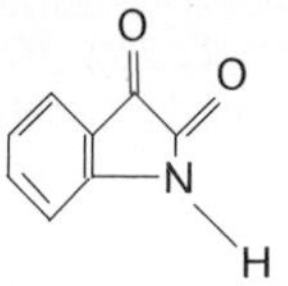
(I)

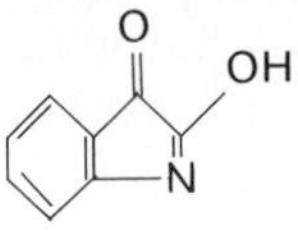
(II)

the existence of lactim-lactam tautomerism—that chemical behavior may be inferred from the structure of a reaction product—was established. Subsequently, spectroscopic techniques were judged more reliable, and it is often possible to determine from the absorption spectrum whether a given substance has either one structure or both.

Keto-enol tautomerism. The molecular grouping —COCH< may in certain substances exist partly or wholly as —COH=C<. The former constitutes the keto form and the latter the enol form. Kurt Meyer first studied the keto, $CH_3CO\text{-}CH_2CO_2C_2H_5$, and enol, $CH_3COH{=}CHCO_2C_2H_5$, forms of ethyl acetoacetate, and recognized them respectively by reactions specific for the carbonyl group and the carbon-carbon double bond. Both forms may be obtained in relatively pure condition, the former by freezing it out of the mixture and the latter by slowly distilling the mixture in quartz apparatus. However, each is slowly converted into the equilibrium mixture of the two. Extensive chemical and spectroscopic studies showed that the enol content of such an equilibrium mixture is a function of the physical state of any given substance. The gas phase or solution in a nonpolar solvent (hexane) favors the enol form, whereas more polar solvents (chloroform, alcohols) repress its formation.

The existence of an enol in an acyclic system requires that a second carbonyl group (or its equivalent, for example, >C=N—) be attached to the same —CN< as an aldehyde or ketone carbonyl. Thus, ethyl acetoacetate tautomerizes demonstrably, but ethyl malonate, $C_2H_5O_2CCH_2CO_2C_2H_5$, does not. Occasionally an enol form exists, these requirements notwithstanding. For example, ethyl pyruvate is partially enolized ($CH_3COCO_2C_2H_5 \rightleftharpoons CH_2{=}COHCO_2C_2H_5$), and α-hydroxy ketones (or aldehydes) exhibit the characteristics of the tautomeric enediols, for example, benzoin, as in reaction (1).

$$C_6H_5COCHOHC_6H_5 \rightleftharpoons C_6H_5COH{=}COHC_6H_5 \quad (1)$$

Where the enol form includes an aromatic ring such as phenol, the existence of the keto form is often not demonstrable, although in some substances such as 4-nitrosophenol (III) and 4-hydroxypyridine (IV) there may be either chemical or

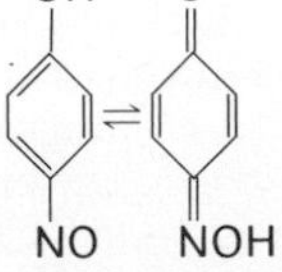
(III)

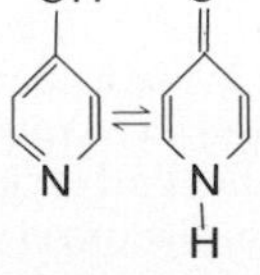
(IV)

spectroscopic evidence for both forms. Closely related to keto-enol tautomerism is the prototropic interconversion of nitro and aci forms of aliphatic nitro compounds such as nitromethane (V).

$$CH_3NO_2 \rightleftharpoons CH_2{=}\overset{\overset{\displaystyle O}{\uparrow}}{N}{-}OH \quad (V)$$

Ring-chain tautomerism. The possibility that an acyclic hydroxyaldehyde may exist in equilibrium with its cyclic hemiacetal was first recognized by Emil Fischer. The failure of glucose to form a normal acetal with an alcohol and the surprising production of two isomeric glucosides instead led to the postulate that carbohydrates exist principally as inner or cyclic hemiacetals in equilibrium with only enough free aldehyde to permit typical aldehyde reactions with reagents which either oxidize the carbonyl group or form derivatives that effectively remove it from the equilibrium, as in reaction (2). The glycosides are formed by the elimination of water between the hydroxyl derived by hemiacetal formation and an alcohol, with two structures being possible, and the hydroxyl lying above or below the hetero ring. *See* GLUCOSE.

$$CH_2OHCHOHCHOHCHOHCHOHCHO \rightleftharpoons$$
$$CH_2OHCHCHOHCHOHCHOHCHOH \quad (2)$$
(ring closed through O between the marked carbons)

In general, tautomeric forms exist in substances possessing functional groups which can interact additively and which are so placed that intramolecular reaction leads to a stable cyclic system. The cyclic form usually predominates (especially if it contains five or six members).

R. P. Linstead showed that certain alkenic acids are tautomeric with their lactones, as in reaction (3), and called this lacto-enoic tautomerism. However, lacto-enoic tautomerism not involving a prototropic shift has also been observed, as in reaction (4).

$$RCH{=}CHCH_2CO_2H \rightleftharpoons RCHCH_2CO \quad (3)$$
(ring closed through O)

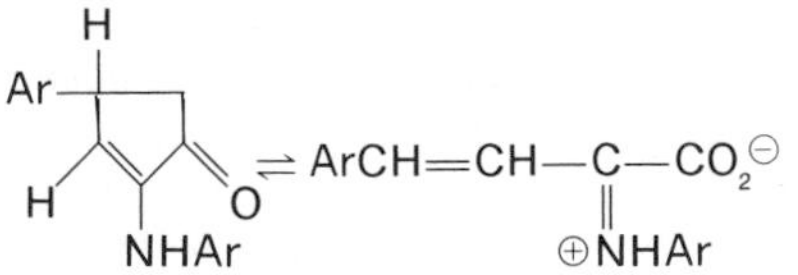
(4)

Still another type of ring-chain tautomerism not involving a prototropic shift is demonstrable by the reactions of phthaloyl chloride, as in reaction (5).

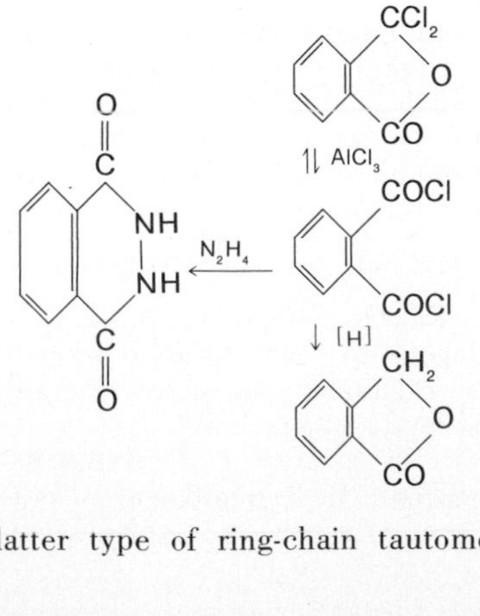
(5)

The latter type of ring-chain tautomerism is

closely related to anionotropic rearrangements such as the allylic, reaction (6), and Wagner-

$$RCHClCH=CH_2 \rightleftharpoons RCH=CHCH_2Cl \qquad (6)$$

Meerwein, reaction (7), which may thus be con-

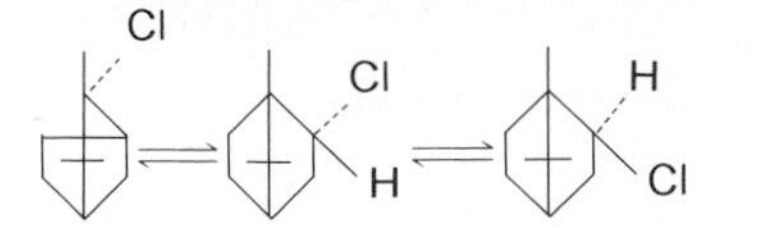

$$(7)$$

sidered examples of anionotropic tautomerism. *See* MOLECULAR ISOMERISM.

[WYMAN R. VAUGHAN]

Technetium

A chemical element, Tc, atomic number 43, discovered by C. Perrier and E. G. Segrè in 1937 as a result of cyclotron bombardment of molybdenum by deuterons. The element was the first made artificially, thus the name technetium, from the Greek for "artificial." It is also produced as a ma-

Ia																	0
1 H	IIa											IIIa	IVa	Va	VIa	VIIa	2 He
3 Li	4 Be											5 B	6 C	7 N	8 O	9 F	10 Ne
11 Na	12 Mg	IIIb	IVb	Vb	VIb	VIIb	——VIII——			Ib	IIb	13 Al	14 Si	15 P	16 S	17 Cl	18 Ar
19 K	20 Ca	21 Sc	22 Ti	23 V	24 Cr	25 Mn	26 Fe	27 Co	28 Ni	29 Cu	30 Zn	31 Ga	32 Ge	33 As	34 Se	35 Br	36 Kr
37 Rb	38 Sr	39 Y	40 Zr	41 Nb	42 Mo	43 Tc	44 Ru	45 Rh	46 Pd	47 Ag	48 Cd	49 In	50 Sn	51 Sb	52 Te	53 I	54 Xe
55 Cs	56 Ba	57 La	72 Hf	73 Ta	74 W	75 Re	76 Os	77 Ir	78 Pt	79 Au	80 Hg	81 Tl	82 Pb	83 Bi	84 Po	85 At	86 Rn
87 Fr	88 Ra	89 Ac	104 Rf	105 Ha	106	107	108	109	110	111	112	113	114	115	116	117	118

lanthanide series	58 Ce	59 Pr	60 Nd	61 Pm	62 Sm	63 Eu	64 Gd	65 Tb	66 Dy	67 Ho	68 Er	69 Tm	70 Yb	71 Lu

actinide series	90 Th	91 Pa	92 U	93 Np	94 Pu	95 Am	96 Cm	97 Bk	98 Cf	99 Es	100 Fm	101 Md	102 No	103 Lr

jor constituent of nuclear reactor fission products or, alternatively, by action of neutrons on Mo^{98}, as in reaction (1). The isotope Tc^{99} is most suitable for

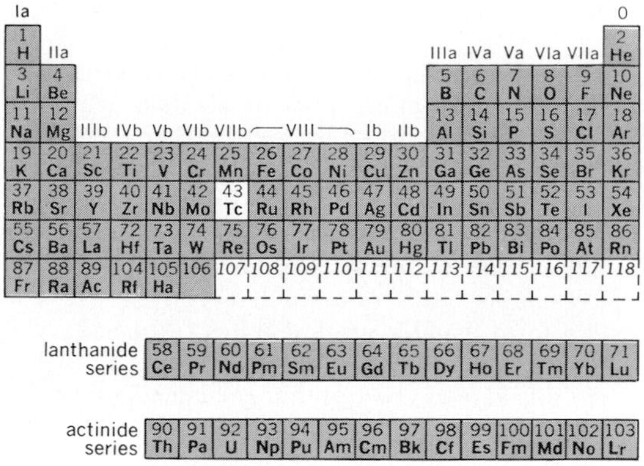

$$Mo^{98}(n,\gamma)Mo^{99} \xrightarrow[67\ hr]{\beta^-} Tc^{99} \qquad (1)$$

chemical investigation because of its long half-life, 2×10^5 years. The chemistry of technetium is very similar to that of rhenium, and corresponding compounds have been prepared in many cases. The metal can be prepared by reduction of the sulfide with hydrogen at temperatures of $1000-1100°C$, and its crystal structure has been found to be isomorphous with that of rhenium, osmium, and ruthenium.

Technetium metal reacts with oxygen at elevated temperatures to form the volatile oxide Tc_2O_7, which is analogous to Re_2O_7. Another oxide, TcO_2, is formed by the decomposition of NH_4TcO_4 at elevated temperatures in vacuum according to reaction (2). Reactions which produce the compounds

$$NH_4TcO_4 \rightarrow TcO_2 + 2H_2O + \tfrac{1}{2}N_2 \qquad (2)$$

$AgTcO_4$, $KTcO_4$, NH_4TcO_4, K_2TcCl_6, and TcS_2 are analogous to those used to form the corresponding rhenium compounds. *See* RHENIUM; TRANSITION ELEMENTS.

[SHERMAN FRIED]

Bibliography: R. Kemmitt (ed.), *The Chemistry of Manganese, Technetium, and Rhenium*, 1975.

Tellurium

A chemical element, Te, atomic number 52, and chemical atomic weight 127.60. The percent abundances of the stable isotopes in natural tellurium are Te^{126}, 18.71%; Te^{128}, 31.79%; Te^{130}, 34.49%; Te^{120}, 0.08%; Te^{122}, 2.46%; Te^{123}, 0.87%;

Ia																	0
1 H	IIa											IIIa	IVa	Va	VIa	VIIa	2 He
3 Li	4 Be											5 B	6 C	7 N	8 O	9 F	10 Ne
11 Na	12 Mg	IIIb	IVb	Vb	VIb	VIIb	——VIII——			Ib	IIb	13 Al	14 Si	15 P	16 S	17 Cl	18 Ar
19 K	20 Ca	21 Sc	22 Ti	23 V	24 Cr	25 Mn	26 Fe	27 Co	28 Ni	29 Cu	30 Zn	31 Ga	32 Ge	33 As	34 Se	35 Br	36 Kr
37 Rb	38 Sr	39 Y	40 Zr	41 Nb	42 Mo	43 Tc	44 Ru	45 Rh	46 Pd	47 Ag	48 Cd	49 In	50 Sn	51 Sb	52 Te	53 I	54 Xe
55 Cs	56 Ba	57 La	72 Hf	73 Ta	74 W	75 Re	76 Os	77 Ir	78 Pt	79 Au	80 Hg	81 Tl	82 Pb	83 Bi	84 Po	85 At	86 Rn
87 Fr	88 Ra	89 Ac	104 Rf	105 Ha	106	107	108	109	110	111	112	113	114	115	116	117	118

lanthanide series	58 Ce	59 Pr	60 Nd	61 Pm	62 Sm	63 Eu	64 Gd	65 Tb	66 Dy	67 Ho	68 Er	69 Tm	70 Yb	71 Lu

actinide series	90 Th	91 Pa	92 U	93 Np	94 Pu	95 Am	96 Cm	97 Bk	98 Cf	99 Es	100 Fm	101 Md	102 No	103 Lr

Te^{124}, 4.61%; and Te^{125}, 6.99%. Tellurium was first isolated by J. F. M. von Reichenstein in 1782.

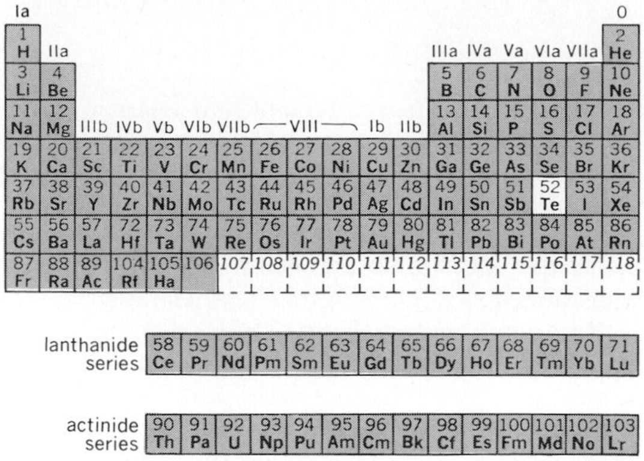

Natural occurrence. Tellurium makes up approximately $10^{-9}\%$ of the Earth's igneous rock. It is found as the free element in central Europe, Colorado, and Bolivia and occurs with selenium in sulfur deposits in Japan. It is more often found as the tellurides sylvanite (graphic tellurium), $(Ag, Au)Te_2$; nagyagite (black tellurium), $(Ag, Pb)_2(Te, S, Sb)_3$; hessite, Ag_2Te; tetradymite, Bi_2Te_3; altaite, $PbTe$; coloradoite, $HgTe$; and other silvergold tellurides, as well as the oxide TeO_2, tellurium ocher. Tellurium is also recovered from the anode slimes obtained during the electrolytic refining of copper.

Preparation. The extraction of tellurium is carried out by the digestion of tellurium-containing materials with hot concentrated sulfuric acid (or hydrochloric acid, for some ores) to convert the material to the tellurite (TeO_3^{--}), followed by treatment with sulfites or sulfur dioxide. This is represented by reaction (1). Tellurium can also be

$$H_2TeO_3 + 2SO_2 + H_2O \rightarrow Te + 2H_2SO_4 \qquad (1)$$

deposited on a lead cathode from a solution of TeO_2 in hydrofluoric and sulfuric acids.

Properties. There are two important allotropic modifications of elemental tellurium, the crystalline and the amorphous forms. The crystalline form has a silver-white color and metallic appearance and is isomorphous with gray selenium B. The electrical conductivity of this form is extremely low and is increased by the presence of small quantities of impurities. Light causes only a small increase in electrical conductivity. This form melts at 449.8°C and boils at 1390°C. It has a specific gravity of 6.25, a hardness of 2.5 on Mohs scale, and a Trouton constant of only 13.2. It is insoluble in all solvents which do not react with it, and its molecular weight at room temperature is not yet known. Between 1400 and 1800°C its formula is Te_2, and the Te-Te distance is 0.26 nm. The amorphous form (brown) has a specific gravity of 6.015. A red colloidal sol of tellurium in water can be obtained by the reduction of telluric acid with hydrazine.

Tellurium burns in air with a blue flame, forming tellurium dioxide, TeO_2. It reacts with halogens, but not sulfur or selenium, and forms, among other products, both the dinegative telluride anion (Te^{--}), which resembles selenide, and the tetrapositive tellurium cation (Te^{4+}), which resembles platinum (IV).

Uses. Tellurium is used primarily as an additive to steel to increase its ductility, as a brightener in electroplating baths, as an additive to catalysts for the cracking of petroleum, as a coloring material for glasses, and as an additive to lead to increase its strength and corrosion resistance.

Principal compounds. Hydrogen telluride, H_2Te, is the only known hydride of tellurium. A colorless gas with an even more offensive odor than hydrogen selenide, it is at least as toxic as this substance. It melts at −51.2°C, boils at −1.8°C, and is less stable thermally than the hydrides of oxygen, sulfur, and selenium, although it is a stronger acid in water than these hydrides. The liquid form has a slightly yellow color, and the solid is colorless. The specific gravity of the liquid at its boiling point is 2.650. The compound is decomposed by light, especially when moist, and it can be prepared by the action of acids on metallic tellurides, especially aluminum telluride, Al_2Te_3. Normal tellurides, for example, Na_2Te, are known and are less stable than the corresponding selenides toward heat. They are strong reducing agents in solution and will reduce tellurite to free tellurium. The alkali tellurides are soluble in water and are attacked by oxygen to give dark red polytellurides, for example, Na_2Te_2. Heavy-metal tellurides are generally insoluble in water.

Halides. Tellurium hexafluoride, TeF_6, is a colorless gas which is formed from the elements; it melts at −37.8°C and sublimes at −38.9°C and is slowly hydrolyzed by water. Hydrolysis is represented by reaction (2). The compound is more reac-

$$TeF_6 + 6H_2O \rightarrow 6HF + H_6TeO_6 \qquad (2)$$

tive than either its selenium or sulfur analog. Reports have been made of lower fluorides, especially tellurium tetrafluoride, TeF_4, and Te_2F_{10}. Tellurium oxyfluoride, $TeOF_2 \cdot \frac{1}{2}H_2O$, is a white, crystalline material formed from anhydrous hydrogen fluoride and tellurium dioxide (other oxy-halides are not well characterized).

Tellurium tetrachloride, $TeCl_4$, is formed by the action of tellurium on excess chlorine, S_2Cl_2 or $AsCl_3$; it is a white, hygroscopic, crystalline substance which boils at 390°C and melts at 225°C; its vapor is monomeric and orange-red up to 500°C, and it has the electrical conductivity of a salt at elevated temperatures; it is soluble in benzene, toluene, and the lower alcohols, but not in ether; its structure is a trigonal bipyramid with one equatorial position occupied by an electron pair. It reacts slowly with water to form TeO_2 and with many organic compounds, and it forms addition products with $2AlCl_3$, $2(C_2H_5)_2O$, $3NH_3$, SO_3, $6NH_3$, and $2SO_3$ per molecule of $TeCl_4$. Tellurium dichloride, $TeCl_2$, is a black solid formed by the action of tellurium on chlorine or $TeCl_4$. It melts at 175°C and boils at 324°C; the liquid has a fairly high electrical conductivity, and the solid and liquid are both reactive; water converts it to Te and H_2TeO_3, and oxygen converts it to $TeCl_4$ and TeO_2.

Tellurium tetrabromide, $TeBr_4$, is an orange-red solid which melts at about 380°C and boils at about 421°C; it is slowly hydrolyzed by excess water to TeO_2 and forms an addition compound with aniline, $TeBr_4 \cdot 2C_6H_5NH_2$. Tellurium dibromide, $TeBr_2$, is a black solid which melts at 210°C and boils at 339°C; it is hydrolyzed by water, as shown by reaction (3).

$$2TeBr_2 + 3H_2O \rightarrow H_2TeO_3 + 4HBr + Te \qquad (3)$$

Tellurium tetraiodide, TeI_4, is a black solid which melts at 259°C, is slightly soluble in acetone and ethyl and amyl alcohols, and is essentially insoluble in carbon tetrachloride, carbon disulfide, ether, and acetic acid; it can be prepared by the reaction of tellurium dioxide and hydrogen iodide.

Complex halides of tellurium have also been prepared, for example, $HTeCl_5 \cdot 5H_2O$, $HTeBr_5 \cdot 5H_2O$, and $HTeI_5 \cdot 8H_2O$; the complex chloride also forms salts; in addition, salts of H_2TeX_6 have been prepared, where X = Cl, Br, and I.

Oxides. The oxides of tellurium are tellurium monoxide, TeO; tellurium dioxide, TeO_2; and tellurium trioxide, TeO_3. The monoxide is reported as a black, amorphous powder which is stable in dry air in the cold but which is oxidized in moist air to the dioxide. On being heated in vacuum, it apparently disproportionates into the dioxide and elemental tellurium. It can be formed by heating the mixed oxide $TeSO_3$, as shown by reaction (4), and

$$TeSO_3 \rightarrow TeO + SO_2 \qquad (4)$$

is also believed to exist at elevated temperatures in equilibrium with its decomposition products. The dioxide is the most stable oxide and is formed when tellurium is burned in air or oxygen or by oxidation of tellurium with cold nitric acid. It has two crystalline forms. One crystallizes from a nitric acid solution as colorless, tetragonal, octahedron-like crystals of specific gravity about 5.8; the other, from molten tellurium dioxide as monoclinic or rhombic needles of about the same density. Tellurium dioxide melts at 452°C, and the white solid becomes a dark-yellow liquid which distills at red heat, apparently without decomposition. It is only very slightly soluble in water and forms a solution which is barely acidic, but which is soluble in concentrated acids (for example, H_2SO_4, HCl, HNO_3) in which it apparently forms salts, and in strong alkalies in which it forms tellurites, such as K_2TeO_3. The dioxide is amphoteric and also forms addition compounds with strong acids, for example, $TeO_2 \cdot 3HCl$ and $(TeO_2)_2 \cdot HNO_3$. The trioxide is an orange-yellow compound formed by heating orthotelluric acid, H_6TeO_6. This oxide decomposes into the dioxide and oxygen at red heat. It is essentially insoluble in cold water but dissolves in hot water after prolonged heating to give orthotelluric acid. It is not attacked by cold acids, but hot, concentrated hydrochloric acid converts it to a mixture of chlorine, the dioxide, and the tetrachloride. Hot, concentrated potassium hydroxide converts it to the tellurate, K_2TeO_4. There are apparently two crystal forms of the trioxide, ordinary α-TeO_3 (sp gr, 5.075) and β-TeO_3 (sp gr, 6.21). The less reactive β form is made by prolonged heating of the α form.

Acids. The important oxy acids of tellurium are tellurous acid and telluric acid. Anhydrous tellurous acid, H_2TeO_3, has never been isolated, al-

Organic tellurium compounds

Type	Example	Properties
Telluromercaptans, RTeH	CH_3TeH	Bp 57°C; prepared from H_2Te and RX in alcoholic NaOR (R = an organic group, for example, CH_3—)
Dialkyl tellurides, R_2Te	$(CH_3)_2Te$	Bp 82°C; prepared from TeX_2 and Grignards; form addition compounds, such as $(CH_3)_2Te \cdot HgBr_2$
Diaryl tellurides, R_2Te	$(C_6H_5)_2Te$	Bp 182°C at 16.5 mm; prepared similarly to dialkyl tellurides
Cyclic tellurides	$(CH_2)_5Te$ $(CH_2)_4Te$	6-Membered ring; bp 82°C at 12 mm; 5-membered ring; bp 166°C
Telluronium compounds, R_3TeX	$(C_2H_5)_3TeCl$ $(C_6H_5)_2CH_3TeOH$	Mp 174°C; prepared from dialkyl tellurides and alkyl halides; moderately strong base
Ditellurides, R_2Te_2	$C_6H_5Te—TeC_6H_5$	Red crystals melting at 53°C
Aryl tellurium monohalides	RTeX	
Dialkyl tellurium dihalides	$(CH_3)_2TeI_2$	
Diaryl tellurium dihalides	$(C_6H_5)_2TeBr_2$	
Alkyl and aryl tellurium trihalides, $RTeX_3$	CH_3TeI_3	Decomposes above 100°C; soluble in acetone and ether to give red solutions
Telluroxides, R_2TeO	$(C_2H_5)_2TeO$	Unstable oil, forming water-soluble salts with HNO_3; formed from dialkyl tellurides and air
Tellurones, R_2TeO_2	$(CH_3)_2TeO_2$	Prepared from dimethyl telluride and H_2O_2; white insoluble solid
Tellurinic acids, RTeOOH	C_6H_5TeOOH	Prepared by oxidation of $(C_6H_5)_2Te_2$ with HNO_3; mp 211°C
Telluric esters, $(RO)_6Te$	$(CH_3O)_6Te$	Mp 86°C; prepared from diazomethane and H_6TeO_6 in absolute alcohol
Telluroketones, R_2CTe	$(CH_3)_2CTe$	Bp 55–58°C at 10–13 mm; prepared from H_2Te and R_2CO in HCl

though treatment of the potassium salt with nitric acid produces white flakes with varying quantities of water of crystallization. Salts of tellurous acids from H_2TeO_3 up to $H_2Te_6O_{13}$ are known. The normal tellurites, for example, K_2TeO_3, are soluble and colorless and tend to be oxidized to tellurates in alkaline solution by air. The acid salts, for example, $KHTeO_3$, are converted by water to the normal tellurites and tellurium dioxide. The higher tellurites are less susceptible to oxidation than the normal tellurites. For example, $K_2Te_4O_9$ is not oxidized by air at 450°C. As a general rule, telluric acid resembles stannic acid. The form H_2TeO_4 has not been isolated, although its salts are known. The ortho acid, H_6TeO_6, can be prepared by the oxidation of tellurium or its dioxide with chromic acid in nitric acid, with aqua regia and chloric acid, or by oxidation of the dioxide in alkaline solution with hydrogen peroxide. There is at least one additional form of the acid, allotelluric acid, or polymetatelluric acid, $(H_2TeO_4)_n$, where n is about 11. The ortho acid is the more stable form and has an octahedral configuration; it crystallizes from water as the 4-hydrate, from which the water can easily be removed. In cold water it is a very weak monomeric acid which polymerizes on being heated until it becomes colloidal. This reaction is reversed on cooling. Telluric acid is more easily reduced than its sulfur and selenium analogs, and it forms normal tellurates, for example, Ag_6TeO_6; acid tellurates, also called alkaline tellurates, for example, $Na_4H_2TeO_6$ and $Na_2H_4TeO_6$; tellurate esters, such as $Te(OCH_3)_6$; and heteropoly acids and salts, such as $H_6[Te(MoO_4)_6]$. The ortho acid has two crystalline forms, a cubic form of specific gravity 3.053 and a monoclinic form of specific gravity 3.071. Normal tellurates can be formed from tellurites by fusing them with potassium nitrate, or by passing chlorine into an alkaline tellurite solution as represented by reaction (5). Tellu-

$$K_2TeO_3 + 2KOH + Cl_2 \rightarrow K_2TeO_4 + 2KCl + H_2O \quad (5)$$

rates are reduced to tellurites by hot hydrochloric acid and to elemental tellurium by sulfur dioxide. The barium salt, $BaTeO_4 \cdot 3H_2O$, is fairly soluble in water.

Mixed compounds. The mixed oxide $TeSO_3$, a red solid, has been obtained from tellurium and sulfur trioxide. It forms the telluride, K_2Te, when fused with potassium cyanide. Carbon sulfotelluride, S=C=Te, has been prepared by passing an arc between carbon and carbon-tellurium electrodes under carbon disulfide in the cold. It is a red liquid with a melting point of −54°C and an extrapolated boiling point of about 110°C. It is quite unstable, being decomposed by light and heat.

Organic compounds. The important organic tellurium compounds are summarized in the table. *See* SELENIUM; SULFUR. [STANLEY KIRSCHNER]

Bibliography: W. C. Cooper and R. A. Zingaro, *Tellurium*, 1971; E. Gerlach (ed.), *The Physics of Selenium and Tellurium*, 1980; K. Irgolic, *Organic Chemistry of Tellurium*, 1975; M. Schmidt et al., *The Chemistry of Sulphur, Selenium, Tellurium, and Polonium*, 1975.

Temperature

A concept related to the flow of heat from one object or region of space to another. The term refers not only to the senses of hot and cold but to numerical scales and thermometers as well. Fundamental to the concept of temperature are the absolute scale and absolute zero and the relation of absolute temperatures to atomic and molecular motions.

Numbers for temperatures, such as 100°C and −15°F, have been used for only about 300 years. By the 17th century, science had developed to the point that, to fully describe the properties of matter, a numerical, quantitative scale of temperature

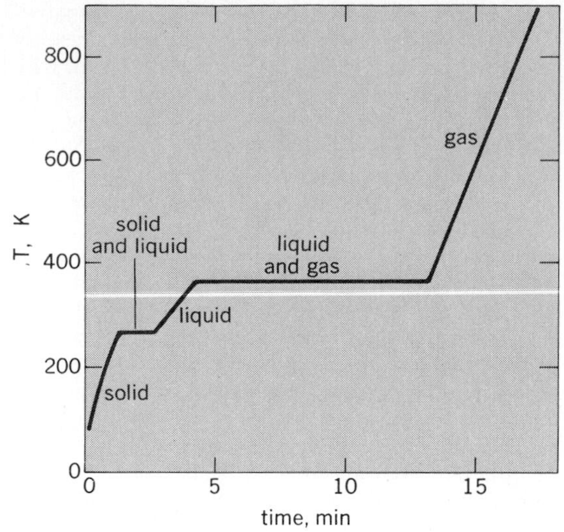

Fig. 1. Temperature of 1 g of H_2O, starting at 0 K, with a constant heat input of 1 cal/s (4.18 J/s).

differences was needed. For example, in 1756 Joseph Black in Scotland discovered that ice does not change temperature when it melts. Almost all substances behave this way; also, the melting temperature depends on the purity of the substance. Thus one reason for devising a thermometer (literally, a meter for temperature) was that with it the composition of matter could be studied.

Temperature measurements are useful for studying molecular motions in material. Figure 1 shows how the temperature of 1 g of H_2O, starting as ultracold ice, changes as heat is added at the constant rate of 1 cal/s (4.18 J/s), assuming that no heat is lost to the surroundings. The plateaus on this graph illustrate Black's discovery that the temperature is constant during a phase change solid-to-liquid or liquid-to-gas. *See* BOILING POINT; MELTING POINT.

Thermometers do not measure a special physical quantity. They measure length (as of a mercury column) or pressure or volume (with the gas thermometer at the National Bureau of Standards) or electrical voltage (with a thermocouple). The basic fact is that, if a mercury column has the same length when touching two different, separated objects, when the objects are placed in contact no heat will flow from one to the other.

Empirical scales. The numbers on the thermometer scales are merely historical choices; they are not scientifically fundamental. The most widely used scales are the Fahrenheit (°F) and the Celsius (°C). The Centigrade scale with 0° assigned to ice water (ice point) and 100° assigned to water boiling under one atmosphere pressure (steam point) was formerly used, but it has been succeeded by the Celsius scale, defined in a different way than the centigrade scale. However, on the Celsius scale the temperatures of the ice and steam points differ by only a few hundredths of a degree from 0° and 100°, respectively. Figure 2 shows how the Celsius and Fahrenheit scales compare and how they fit onto the absolute scales.

These scales have one common value: −40°C = −40°F. This fact can be used to change a temperature from one scale to the other. Given a temperature in °C or in °F, add 40, multiply by 9/5 if converting from °C to °F or by 5/9 if from °F to °C, then subtract 40. Example: Normal human body temperature is 98.6°F. To convert to °C, 98.6 + 40 = 138.6; 138.6 × 5/9 = 77.0; 77.0 − 40 = 37.0°C.

Absolute temperature scale. In 1848 William Thompson (Lord Kelvin), following ideas of Sadi Carnot, stated the concept of an absolute scale of temperature in terms of measuring amounts of heat flowing between objects. Most important, Kelvin conceived of a body which would not give up any heat and which was at an absolute zero of temperature. Experiments have shown that absolute zero corresponds to −273.15°C or −459.7°F. Two absolute scales, shown in Fig. 2, are the Kelvin (K) and the Rankine (°R). *See* ABSOLUTE ZERO.

Interest in temperature and heat flow was stimulated in the early 19th century by efforts to improve the efficiency of steam engines. Out of this came the concept of a Carnot engine. This is not a real machine, but an imagined, ideal, frictionless system. A Carnot engine takes in heat Q_h from a higher temperature source at T_h (kelvins), does work W, and exhausts heat Q_l into a lower temperature source T_l. Two important deductions are: (1) The efficiency of the engine is $W/Q_h = 1 - T_l/T_h$. For example, a Carnot engine operated between the boiling point ($T_h = 373$ K) and the ice point ($T_l = 273$ K) of water has an efficiency of 0.268. A real engine would have an efficiency less than this, but the concept is nevertheless of great importance in engineering. (2) The ratio of temperatures equals the ratio of heats, namely, $T_h/T_l = Q_h/Q_l$. Lord Kelvin suggested that this be the basis for the absolute temperature scale. Define one special system (water at its triple point with ice, liquid, and water vapor present) to have a particular value of absolute temperature (273.16 K). To measure any unknown temperature T_u, operate a Carnot engine between 273.16 K and T_u, measure the heats $Q_{273.16}$ and Q_u absorbed and rejected, and calculate $T_u = 273.16\ \mathrm{K}\ (Q_u/Q_{273.16})$. *See* THERMODYNAMIC PRINCIPLES; TRIPLE POINT.

In practice, absolute temperatures are not measured this way. Instead low-density helium gas and dilute paramagnetic crystals, the most nearly ideal of real materials, allow measurement of temperatures virtually identical with those defined by a

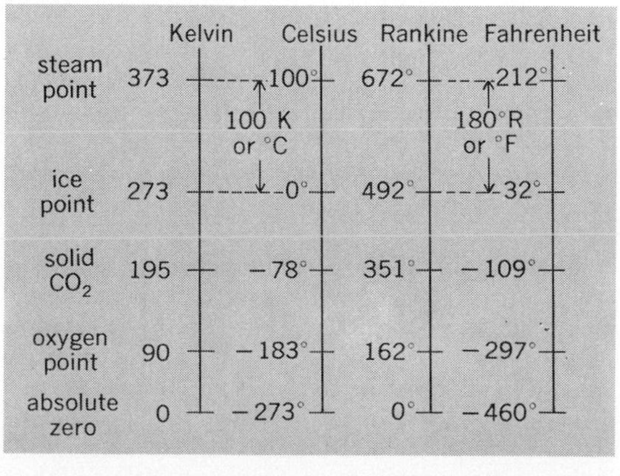

Fig. 2. Comparisons of Kelvin, Celsius, Rankine, and Fahrenheit temperature scales. Temperatures are rounded off to nearest degree. (*M. W. Zemansky, Temperatures Very Low and Very High, Van Nostrand, 1964*)

Carnot process. The advantages are that gas pressures and volumes and magnetic fields and magnetizations can be measured more conveniently and accurately than heat flows.

The measurement of a single temperature with a gas or magnetic thermometer is a major scientific event done at a national standards laboratory. Only a few temperatures have been measured, including the freezing point of gold (1337.91 K), and the boiling points of sulfur (717.85 K), oxygen (90.18 K), and helium (4.22 K). Various other types of thermometers (platinum, carbon, and doped germanium resistors; thermocouples) are calibrated at these temperatures and used to measure intermediate temperatures.

Kinetic temperature. An important aspect of the absolute temperature scale is its relation to the motions of atoms and molecules, whether vibrations as in solids and liquids, or straight path flights with collisions as in gases. There are two important facts here: (1) There is a definite distribution of motions. For example, in a gas, even though the motions are chaotic and a particular molecule changes velocity after each collision with another molecule, at any instant a definite number of molecules have a particular velocity. It cannot be said that the gas is at a definite temperature unless the molecules have this definite distribution of velocities, although different small portions of the gas may have definite, though different, temperatures. The same idea holds for the distribution of vibration frequencies in solids and liquids. (2) A body has a minimum amount of motion energy. It was supposed in the 19th century that this minimum was zero energy, but modern theories and experiments show that the minimum is greater than zero. A body in its lowest energy state cannot give out heat and is at absolute zero.

A system may have several degrees of freedom. The molecules of a gas, besides having straight-line motions, may rotate and vibrate and their electrons may be in different energy levels. When a system is in equilibrium, the energies stored in these different degrees of freedom are related to a common absolute temperature. In several cases, one can measure something related to a particular degree of freedom of a system. Then if the system is in equilibrium, its absolute temperature can be inferred.

Examples of this are measurements of temperatures in the Sun's corona and in the remote regions of the Milky Way Galaxy. The Sun's corona is so hot that atoms lose several electrons, that is, are multiply ionized. This greatly affects the wavelengths of light that these atoms emit. From the measured wavelengths that have been identified, the corona's temperature has been estimated at 2×10^6 K.

Besides light, atoms and molecules can emit radio waves. The straight-line motions of atoms toward or away from Earth-based radio telescopes slightly shifts the received wavelenths. From these Doppler shifts, temperatures from 1 to 100 K have been found in the vast hydrogen atom clouds in the galaxy spiral arms. Rotations of OH molecules influence the populations of energy levels and thus affect the intensities of emitted radio signals. From these also, deep-space absolute temperatures are inferred.

Negative absolute temperatures. Negative Celsius and Fahrenheit temperatures are readily accepted because 0° on these scales is arbitrarily set above absolute zero (Fig. 2.) But the idea that a system could be at, say −50 K, was introduced only in the early 1950s. The concept applies only to systems with a finite number of energy levels; that is, those that can store a finite amount of energy. Thus the translational energy of a gas or the vibration energy of a crystal cannot be at negative absolute temperatures. However, the energies of electron and nuclear magnetic moments (spins) in a magnetic field do have upper limits. Normally there are more spins in lower energy levels, in which case they are at positive absolute temperature. With special techniques one can arrange equal numbers of spins in the energy levels. Then the temperature is infinite K. This does not mean infinite heat can be extracted from such a system. One can go further and cause there to be more spins in the higher energy levels than in the lower, and this situation is described as being at a negative absolute temperature.

The spin system is not in equilibrium with the crystal lattice vibrations which remain at positive temperatures. But at temperatures around a few K, the spins need minutes or hours to exchange heat with the lattice. Negative absolute temperatures are hotter than positive temperatures; this is reasonable since heat flows from the negative temperature spins to the positive temperature lattice.

Extreme temperatures. For reasons involving both pure and applied science, researchers endeavor to achieve extremes of both low and high temperatures. At very low temperatures, phenomena that are well understood become frozen out and new predicted and unpredicted effects are sought. Experimenters have reported holding 2 kg of copper at about 50 microkelvins for a couple of days. Achieving and measuring such temperatures involves using the nuclear magnetic spin system. In the procedure to reach the 50-μK lattice temperature, the copper spins were brought down to 50 nanokelvins. Such research aims to see if nuclear spins will spontaneously align as electron spins do in ferromagnetic materials like iron.

The highest equilibrium temperatures reached on Earth are around 10^7 K in experiments to achieve fusion of hydrogen nuclei. These temperatures have been sustained in very-low-density gases like the Sun's corona for a few seconds. At these temperatures hydrogen nuclei have speeds of about 1.3×10^6 m/s. Such speeds are necessary if the positively charged nuclei are to overcome their electric repulsion force. When the nuclei get close enough together, nuclear forces attract them to fuse and there is a net energy release. The goal of the research is to convert this energy into conventional electrical energy. *See* NUCLEAR FUSION.

[ROLAND A. HULTSCH]

Bibliography: American Society for Testing and Materials, *Evolution of the International Practical Temperature Scale of 1968*, 1974; C. M. Herzfeld (ed.), *Temperature: Its Measurement and Control in Science and Industry*, vol. 3, 1962; K. Mendelssohn, *The Quest for Absolute Zero*, 1966; M. W. Zemansky, *Temperatures Very Low and Very High*, 1964.

Terbium

Element number 65, terbium, Tb, is a very rare metallic element of the rare-earth group. Its atomic weight is 158.924, and the stable isotope Tb^{159} makes up 100% of the naturally occurring element. It was discovered in 1843 by C. G. Mosander, who

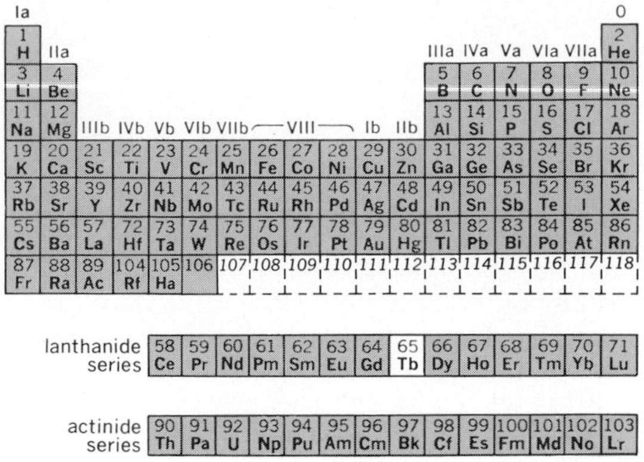

Ia
0
1 H
IIa
IIIa IVa Va VIa VIIa
2 He
3 Li 4 Be
5 B 6 C 7 N 8 O 9 F 10 Ne
11 Na 12 Mg
IIIb IVb Vb VIb VIIb VIII Ib IIb
13 Al 14 Si 15 P 16 S 17 Cl 18 Ar
19 K 20 Ca 21 Sc 22 Ti 23 V 24 Cr 25 Mn 26 Fe 27 Co 28 Ni 29 Cu 30 Zn 31 Ga 32 Ge 33 As 34 Se 35 Br 36 Kr
37 Rb 38 Sr 39 Y 40 Zr 41 Nb 42 Mo 43 Tc 44 Ru 45 Rh 46 Pd 47 Ag 48 Cd 49 In 50 Sn 51 Sb 52 Te 53 I 54 Xe
55 Cs 56 Ba 57 La 72 Hf 73 Ta 74 W 75 Re 76 Os 77 Ir 78 Pt 79 Au 80 Hg 81 Tl 82 Pb 83 Bi 84 Po 85 At 86 Rn
87 Fr 88 Ra 89 Ac 104 Rf 105 Ha 106 107 108 109 110 111 112 113 114 115 116 117 118
lanthanide series 58 Ce 59 Pr 60 Nd 61 Pm 62 Sm 63 Eu 64 Gd 65 Tb 66 Dy 67 Ho 68 Er 69 Tm 70 Yb 71 Lu
actinide series 90 Th 91 Pa 92 U 93 Np 94 Pu 95 Am 96 Cm 97 Bk 98 Cf 99 Es 100 Fm 101 Md 102 No 103 Lr

originally named the oxide erbia; it has been known as terbium since 1877. The element was first isolated in fairly pure form by G. Urbain in 1905. The common oxide, Tb_4O_7, is brown and is obtained when its salts are ignited in air. Its salts are all trivalent and white in color and, when dissolved, give colorless solutions. The only quadrivalent form of terbium known is in the higher oxides, and if Tb_4O_7 is ignited under a high pressure of oxygen, a compound approaching the composition TbO_2 can be prepared. The higher oxides slowly decompose when treated with dilute acid to give the trivalent ions in solution. Although the metal is attacked readily at high temperatures by air, the attack is extremely slow at room temperatures. The metal has a Néel point at about 229 K and a Curie point at about 220 K. For properties of the metal *see* RARE-EARTH ELEMENTS.

Physicists studying the nature of magnetism have a strong interest in the heavy rare-earth metals. Their magnetic behavior is complicated, and terbium has a higher saturation moment in the basal plane than iron. The magnetic behavior is strongly anisotropic, and in the antiferromagnetic region between 220 and 229 K the magnetic moments line up in the *c*-plane of the crystal, but the direction of the moment shows an interesting turn angle between *c*-planes, and the size of this turn angle depends on temperature.

[FRANK H. SPEDDING]

Terpene

A class of natural products having a structural relationship to isoprene (I). Over 5000 structurally determined terpenes are known; many of these have also been synthesized in the laboratory. Historically terpenes have been isolated from green plants, but new compounds structurally related to isoprene continue to be isolated from other sources as well, so the class is also referred to as terpenoids, reflecting the biochemical origin without specification of the natural source. *See* ISOPRENE.

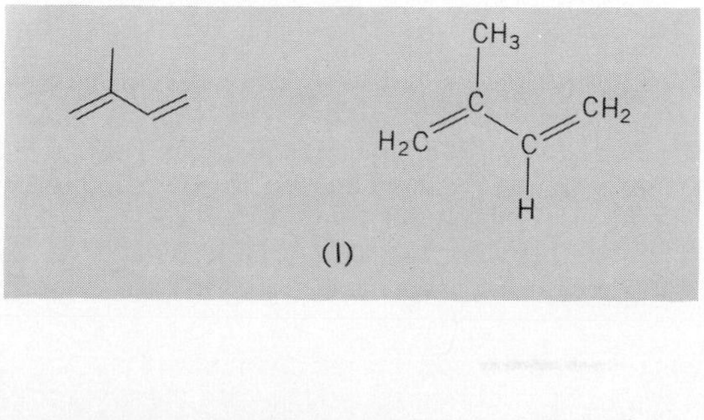

Classification. Terpenes are classified according to the number of isoprene units of which they are composed, as is shown below.

5	hemi-	25	ses-
10	mono-	30	tri-
15	sesqui-	40	tetra-
20	di-	$(5)_n$	poly-

Although they may be named according to the systematic nomenclature and numbering systems set by the International Union of Pure and Applied Chemistry for all organic compounds, it is often easier to refer to terpenes by their common names, which usually reflect the botanical or zoological name of their source.

Biogenesis. The function of terpenes in plants and other organisms is not clear, although they sometimes possess toxic properties linked to the protection of the species. Terpenes are products of secondary metabolism; the key building block, isopentenylpyrophosphate (II), arises from mevalonic acid (III) via hydroxymethyl glutarate (IV). The starting point of this metabolic pathway is believed to be the condensation of two molecules of acetic acid to form acetoacetyl coenzyme A (V). Other biogenetic pathways available for the pro-

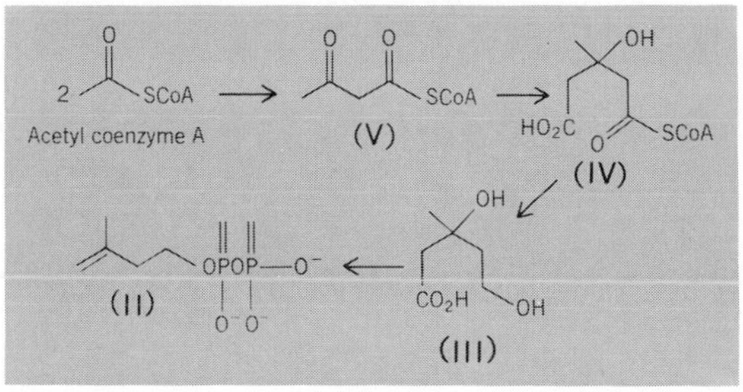

duction of hydroxymethyl glutarate (IV) depend on the particular organism.

Isopentenylpyrophosphate (II) has two reactive sites and can polymerize in a variety of ways to form larger terpenes. Its head-to-tail dimerization is the most commonly encountered process, although head-to-head and tail-to-tail linkages are also found. The multitude of structural types in terpenes arises from intramolecular rearrangements of several basic skeletons formed by the cyclizations of the linear precursors shown in the reactions below, where OPP represents oxygen-alkylated pyrophosphate. *See* POLYMER.

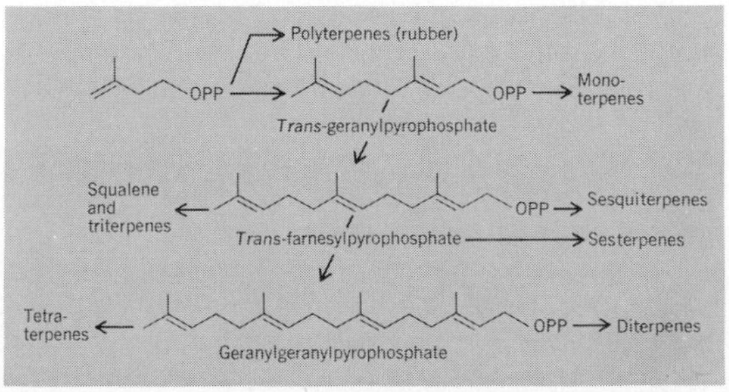

Monoterpenes. Acyclic or cyclic C_{10} hydrocarbons and their oxygenated derivatives are known as monoterpenes. The most common structural types are biogenetic derivatives of geraniol (VI), the main constituent of geranium oil. Menthol (VII) is the chief constituent of peppermint oil; limonene (VIII) composes over 90% of lemon oil; oil of rosemary contains α-pinene (IX); camphor (X) is the main component of sage oil; and other interesting monoterpenes are iridoids, substances which have been isolated from ants, such as iridodial (XI).

Loganin (XII) is the biogenetic precursor of some alkaloids. Its combination with tryptophane, for example, commences the biogenesis of yohimbine alkaloids. Cyclic monoterpenes can be easily oxidized to fully aromatic ring systems such as *p*-cymene (XIII), the constituent of ajowan oil.

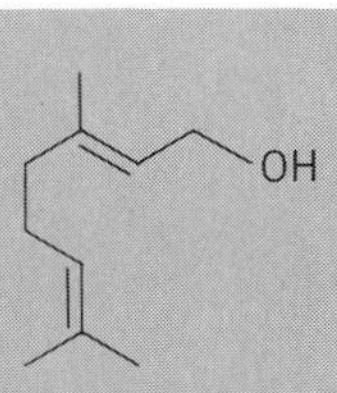

(VI)

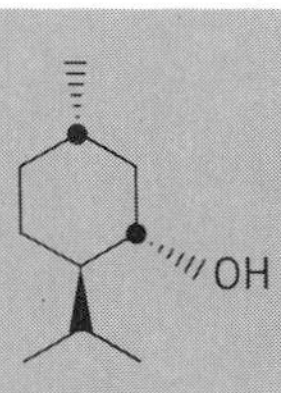

(VII)

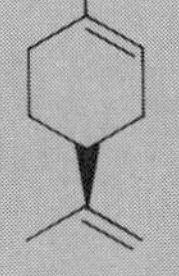
(VIII)

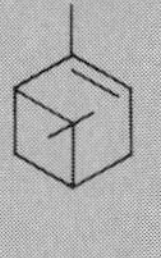
(IX)

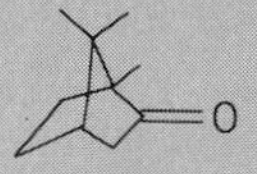

(X)

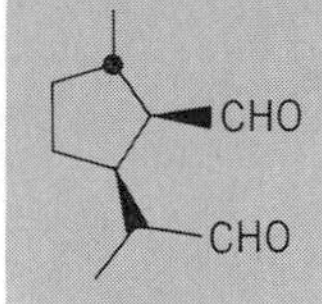

(XI)

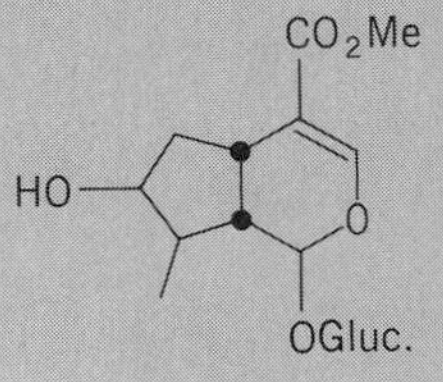

(XII)

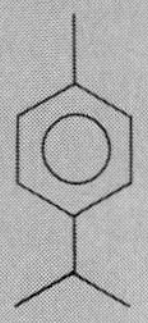
(XIII)

Monoterpenes are widely used in the flavor and perfume industries because of their attractive odors, low molecular weights, and high volatilities. Most are synthesized rather than extracted from plant sources.

Monoterpenes have been well studied, and most of them have been prepared in the laboratory. *See* CAMPHOR; MENTHOL; PINENE.

Sesquiterpenes. These are C_{15} hydrocarbons or their oxygenated analogs. The major categories are mentioned below. They arise from the cyclization of farnesylpyrophosphate and subsequent rearrangements of the resulting carbonium ions (XIV and XV). Almost all known sesquiterpenes can be derived from these two cations.

Many sesquiterpenes are important constituents of the characteristic aromas of plant products, and many others have interesting or useful physiological properties.

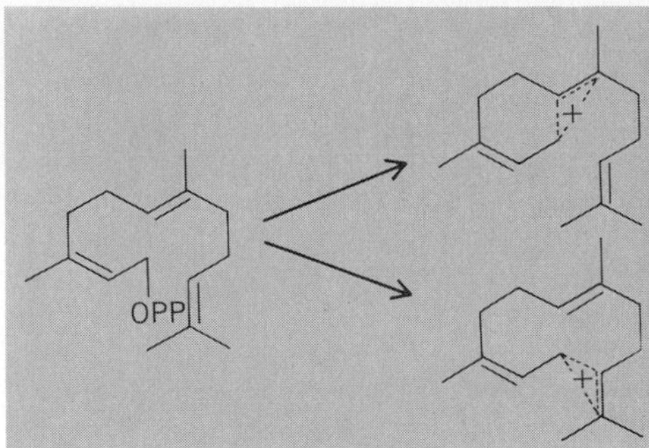

(XIV)
(XV)

Farnesanes and eudesmanes. These are mono- and bicylic terpenes derived from the alkylative cyclization of farnesylprophosphate. They can be exemplified by the structures of abscisic acid (XVI), a plant growth regulator, zingiberine (XVII), the constituent of ginger oil, cadinene (XVIII) from the oil cubebs, and α-santonine (XIX) from plants of the *Artemisia*.

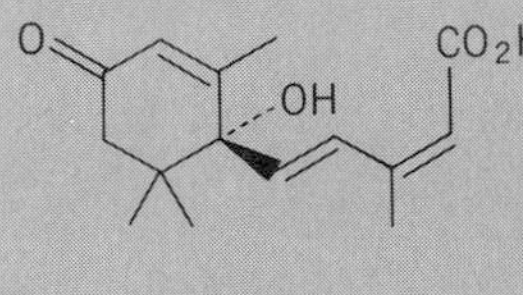

(XVI)

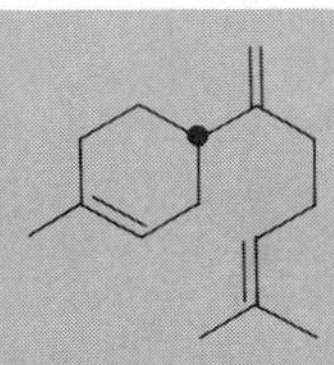
(XVII)

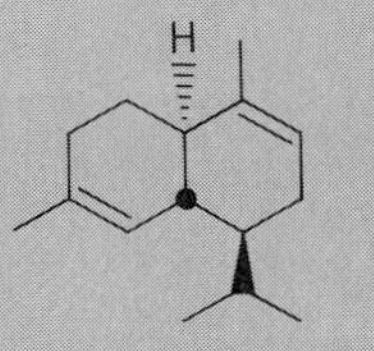

(XVIII)

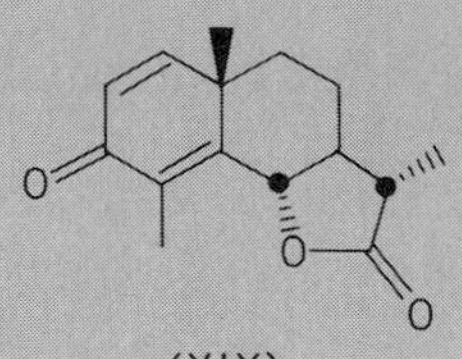

(XIX)

Acoranes, cedranes, chamigrane. These sesquiterpenes are characterized by spirocyclic skeletons. They are obtained from various species of wood, such as the Alaskan cedar. Examples are α-cedrane (XX), β-acorenol (XXI), and β-chamigrene (XXII).

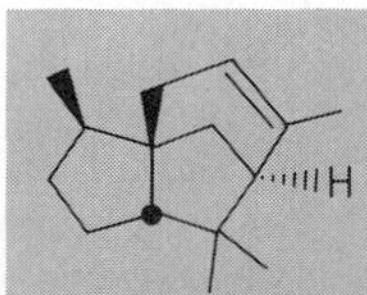

(XX)

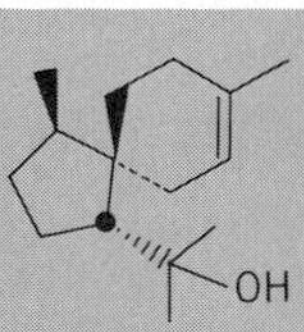

(XXI)

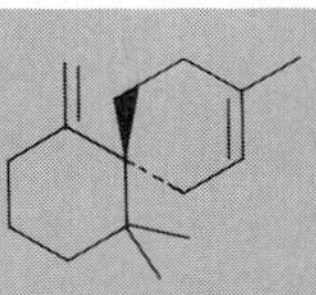
(XXII)

Caryophyllanes, illudanes, humulenes. These are biogenetically related macrocyclic terpenes; they are exemplified by caryophyllene (XXIII) and humulene (XXIV), which occur in hops. Related to illudanes is hirsutic acid (XXV), an important metabolite of certain fungi. Hirsutic acid possesses some antitumor properties.

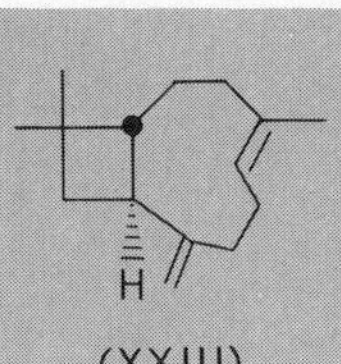

(XXIII)

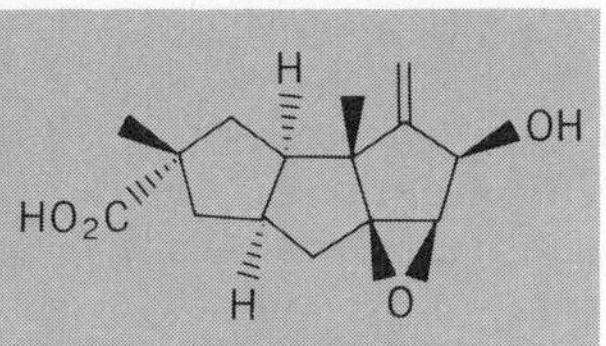

(XXV)

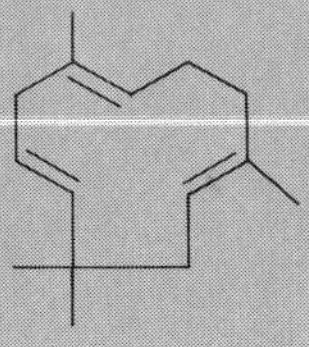

(XXIV)

Himachalenes and longifolanes. These contain fused six- and seven-membered rings, and are derived from decalin skeletons via rearrangements; longifolene (XXVI) and β-himachalene (XXVII) are examples.

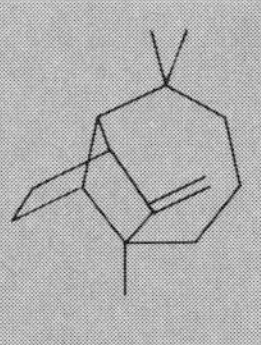

(XXVI)

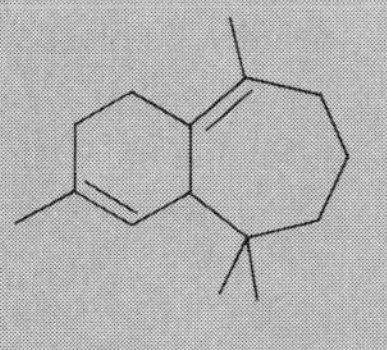

(XXVII)

Perhydroazulenes. The fastest-growing group of identified sesquiterpenenes are the perhydroazulenes. The structures of these compounds were formerly determined by their oxidation to derivatives of the blue hydrocarbon azulene (XXVIII). A representative member of this class is quaiol (XXIX) from quaiacum wood oil.

Quaianes and quaianolides. These constitute the largest single group of sesquiterpenes. New species continue to be isolated from fungi, marine organisms, or plants. The sesquiterpene lactones or quaianolides have remarkable cytoxic properties and are represented by a general structure (XXX).

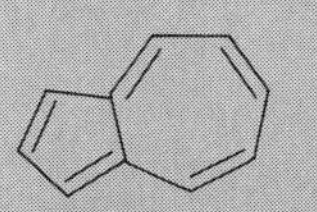

(XXVIII)

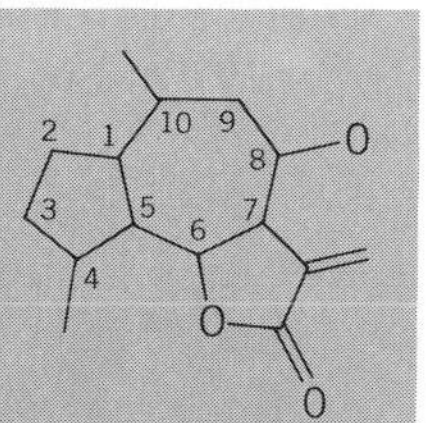

(XXX)

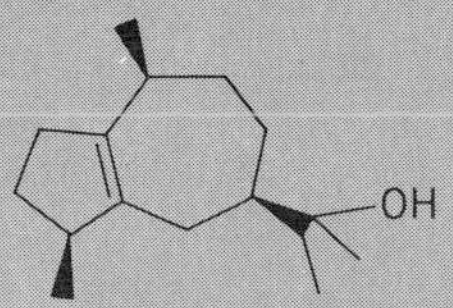

(XXIX)

Germacranes. These are macrocyclic terpene lactones, exemplified by eupassopin (XXXI). The quaianes, quaianolides, and germacranes have been shown to include a number of compounds with promising antitumor activity.

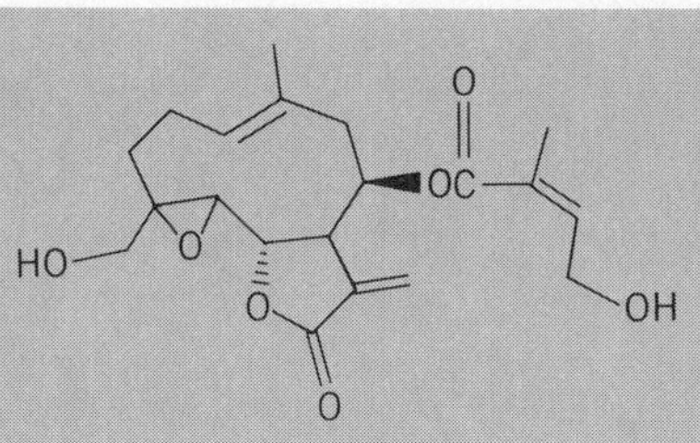

(XXXI)

Diterpenes. These are C_{20} hydrocarbons, and their oxidized derivatives are composed of four isoprene units. They are synthesized by organisms from geranylgeranalpyrophosphate and are utilized in protective coatings of higher plants. Extraction of resins of various coniferous species yields a number of diterpenes. Their skeletons do not exhibit the variety shown by sesquiterpenoids. Abietic acid derivatives are the principal constituent of rosin, a resin obtained mainly from pine with a variety of industrial uses. Diterpene frameworks are found in many complex alkaloids of the delphinium or atisine types. Nitrogen is incorporated into the diterpene nucleus via a carboxylic acid function.

Mild antibiotic activity is associated with some diterpenes. The gibberellins are plant growth promoters. The best-known open-chain diterpene is vitamin A (XXXII). The structure of other diterpenes varies from bicyclic to tetracyclic skeletons.

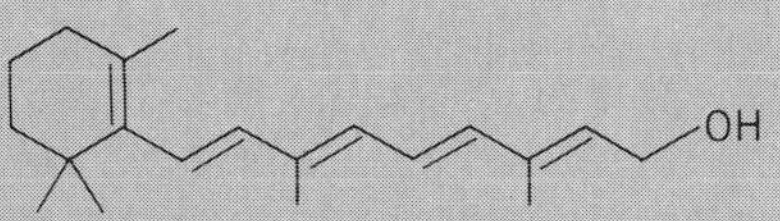

(XXXII)

Labdanes and clerodanes. These are bicyclic resin acids isolated as the "bitter principles" of bark, roots, and stems. Agathic acid (XXXIII), marubiin (XXXIV) from horehound, and clerodane (XXXV) are representatives.

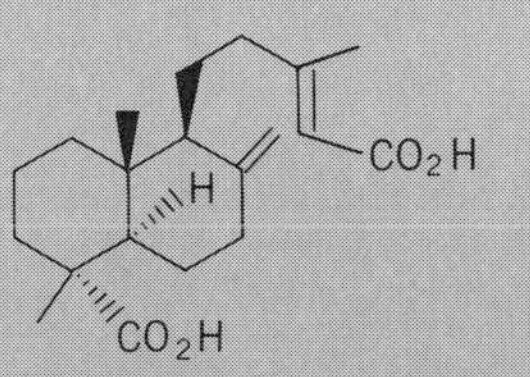

(XXXiii)

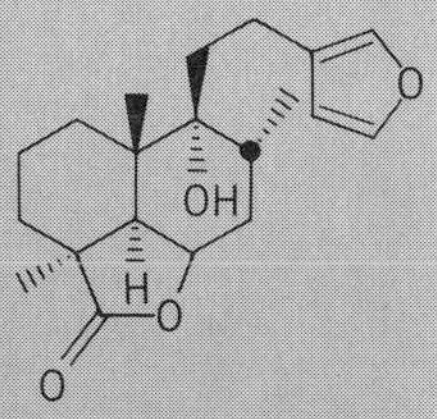

(XXXiV)

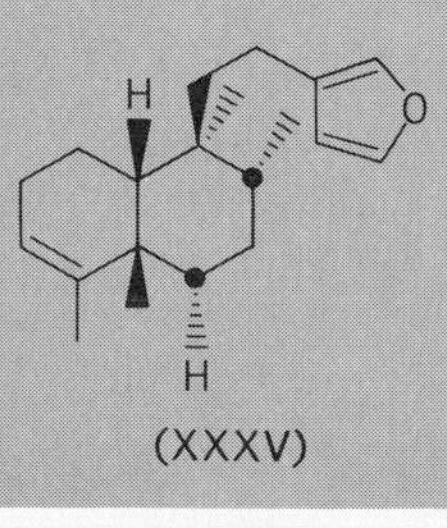

(XXXV)

Tricyclic and tetracyclic diterpenes are typified by the structures of abietic acid (XXXVI; actually an artifact of isolation and not a natural product) from pine resin, and phyllocladene (XXXVII), a typical diterpene constituent of many essential oils.

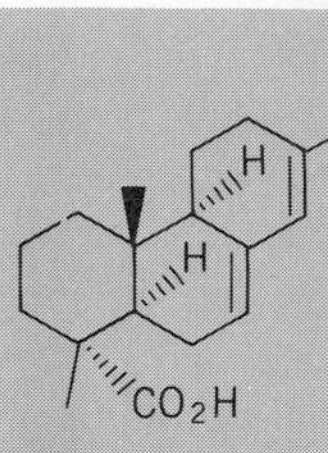

(XXXVI)

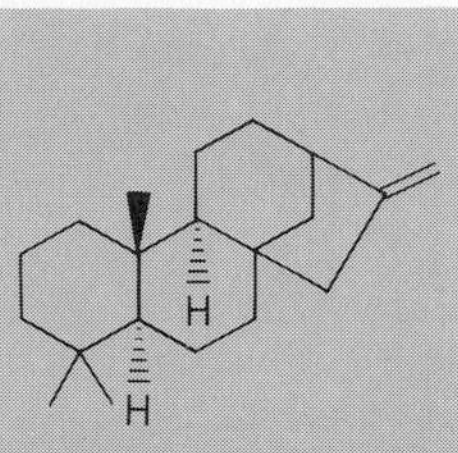

(XXXVII)

Gibberellins. The metabolites of a rice fungus, known as gibberellins, are exemplified by the structure of gibberellic acid (XXXVIII).

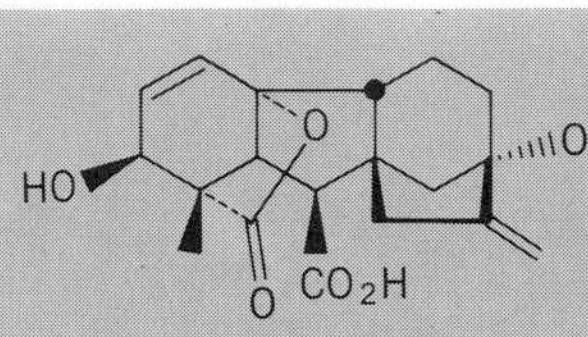

(XXXVIII)

Sesterpenes. These are C_{25} compounds derived from geranylfarnesol pyrophosphate. They can be isolated from protective coatings and waxes of insets or from various fungal sources. Sesterpenes are a relatively new class, the first-known member having been isolated in 1965. Some examples are ophiobolin A (XXXIX), retigeranic acid (XL), and gasgardic acid (XLI).

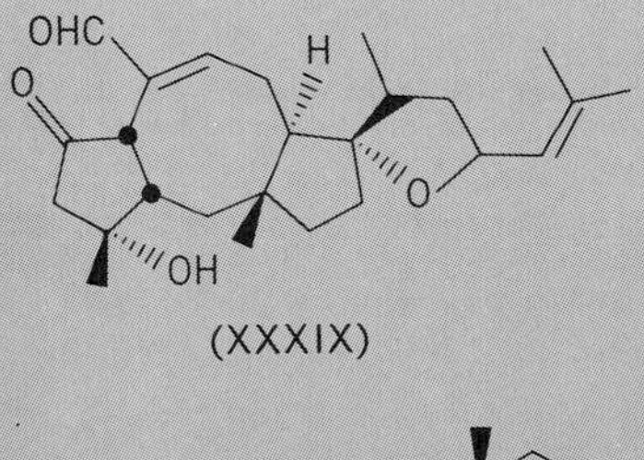

(XXXIX)

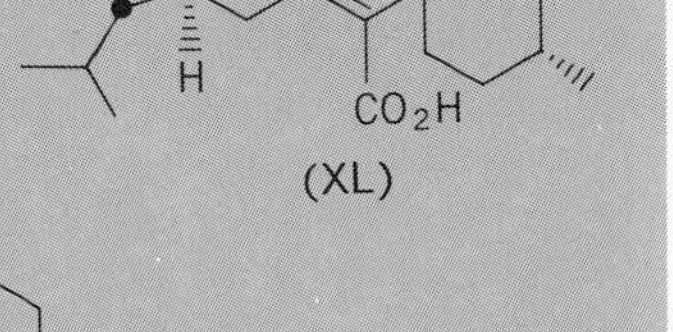

(XL)

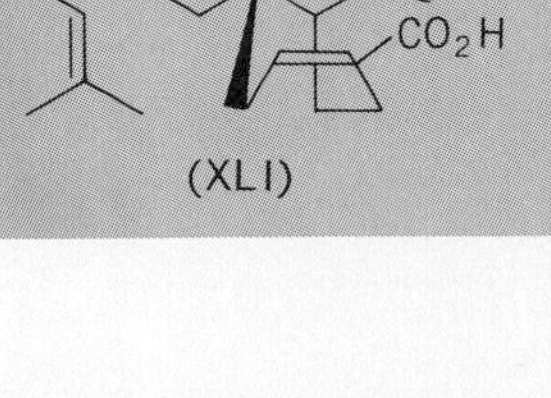

(XLI)

Triterpenes. These have 30 carbon atoms and are composed of six isoprene units. They form the largest group of terpenes. *See* TRITERPENE.

Tetraterpenes. These are C_{40} or higher terpenes. They have a large number of polyene units and therefore do not possess the stereochemical and structural complexity of lower terpenes. Biogenetically, tetraterpenes arise by the dimerization of C_{20} units in a tail-to-tail fashion, with concomitant cyclizations restricted to the termini. Most known tetraterpenes are carotenoid pigments, exemplified by α-carotene (XLII). Carotenes and various oxygenated analogs, found in the chloroplasts of all green plants and in some algae, serve as accessory pigments for photosynthesis. Vitamin A may be regarded as a degraded carotene.

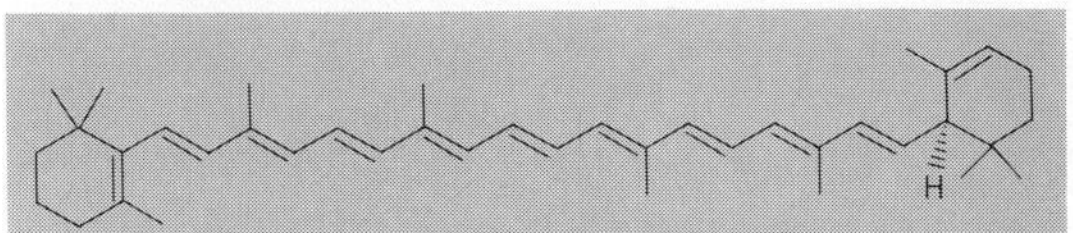

(XLII)

Polyterpenes. These are various natural rubbers from *Hevea*, *Guayule*, and other plants and usually consist of 500–5000 cis-linked isoprene units; gutta-percha contains 100 trans-linked isoprene residues. These polymers arise by a genetic error characteristic of only about 1% of the plant kingdom, whereby species possess the ability to channel all of their mevalonate biogenesis into polymerization.

[TOMAS HUDLICKY]

Bibliography: A. A. Newman, *Chemistry of Terpenes and Terpenoids*, 1972; Specialist Periodical Reports, The Chemical Society, London, *Terpenoids and Steroids*, vols. 1–8, 1978; W. I. Taylor and A. R. Battersby (eds.), *Cyclopentanoid Terpene Derivatives*, 1969; W. Templeton, *Introduction to the Chemistry of Terpenoids and Steroids*, 1969.

Tetraethyllead

An organometallic compound that has found wide commercial application as an additive for motor fuels. Small quantities of tetraethyllead markedly reduce the knocking tendencies of gasoline and thus permit use of higher engine compression ratios. In recent years tetramethyllead has proved equally effective in reducing engine knock.

The combustion of gasoline containing tetraethyllead produces deposits of lead and lead oxide along the cylinder walls. Hence, ethylene dibromide and ethylene dichloride are also added as part of the antiknock fluid to remove the lead as lead halide in the exhaust gases. Automobiles equipped with catalytic converters designed to meet antipollution standards must burn "unleaded" gasoline, since lead deposits tend to poison the platinum and palladium oxidation catalysts used in such systems.

The present commercial process for producing tetraethyllead involves the reaction of ethyl chloride and a lead-sodium alloy at moderate temperatures, followed by steam distillation of the product and recovery of the unused lead. Since tetraethyllead is poisonous and unstable. It must be handled with special care. *See* ORGANOMETALLIC COMPOUND.

[MARVIN D. RAUSCH]

Thallium

A chemical element, Tl, atomic number 81, relative atomic weight of 204.37, a member of group IIIa and the sixth period of the periodic table. The valence electron notation corresponding to its ground state term is $6s^2 6p^1$, which accounts for the maximum oxidation state of III in its compounds. Compounds of oxidation state I and apparent oxidation state II are also known.

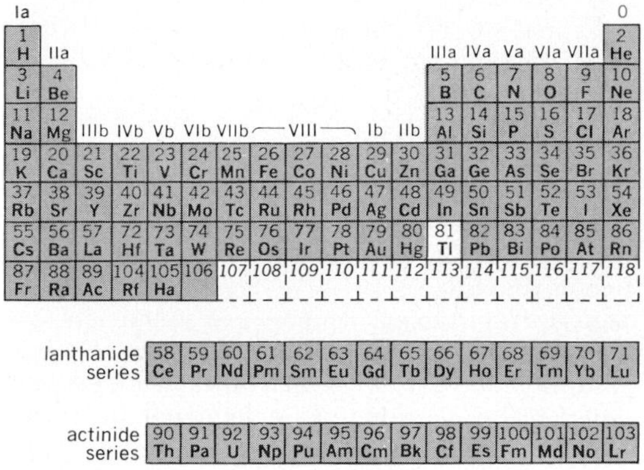

Thallium occurs in the Earth's crust to the extent of 0.00006%, mainly as a minor constituent in iron, copper, sulfide, and selenide ores. Minerals of thallium are considered rare.

Thallium has significant use in electronic components, such as thallium-activated sodium iodide crystals in photomultiplier tubes. It is also used in low-melting-point alloys, optical glass, and in glass seals for enclosing electronic components. Thallium compounds are toxic to humans and other forms of life. When mixed with food they have been used as rodenticides, although a Federal act now prohibits private use of thallium for such purposes.

Thallium metal. Thallium is recovered from the flue dust of roasting operations of sulfide and selenide ores, especially pyrites. It is extracted from these residues and recovered as the metal by electrolytic reduction of sulfate solutions. Thallium is a white, soft metal with a melting point of 302.5°C, a boiling point of 1460°C, and a density of 11.85 g/cm^3 at 20°C. The metal has a hexagonal close-packed structure and a metal radius of 1.71 A.

The metal is capable of being oxidized by hydrogen ion as shown by the standard electrode potential of +0.3363 volts for reaction (1). Usually nitric

$$Tl_{(s)} \rightarrow Tl^{+}_{(aq)} + e^{-} \quad (1)$$

acid is used to dissolve thallium, since thallium(I) chloride and sulfate are not very soluble and their formation interferes with the oxidation reaction. Thallium metal reacts with the halogens and nonmetals to form thallium(I) compounds. The rate of reaction is appreciable even at room temperature.

Thallium compounds. The insolubility of thallium(I) chloride, bromide, and iodide permits their preparation by direct precipitation from aqueous solution; the fluoride, on the other hand, is water-soluble. Thallium(I) chloride resembles silver chloride in its photosensitivity.

Thallium halides are not soluble in aqueous ammonia. The solubility of the halides is enhanced in solutions containing an excess of halide ion owing to the formation of TlX_2^- and TlX_4^{3-} ions. An interesting coordination-number-eight species is obtained when thiourea reacts with thallium(I) compounds. The product has the composition TlX·4Tu and a structure in which each Tl is surrounded by eight sulfur atoms at the corners of an antiprism, with each sulfur bonding to two thallium atoms, thus giving chains of thallium atoms perpendicular to the plane bearing the four sulfur atoms. The physical properties of thallium(I) halides are given in the table.

Thallium(I) oxide is a black powder which reacts with water to give a solution from which yellow TlOH can be crystallized. The hydroxide is a strong base and will take up carbon dioxide from the atmosphere. Thallium(I) sulfide can be precipitated from solution. It reacts with molecular oxygen to give a compound of composition Tl_2SO_2.

Thallium(I) ions can be oxidized to thallium(III) in solution by an oxidizing agent of greater than −1.25 volts, the standard electrode potential for reaction (2). Studies on the rate and mechanism

$$Tl^{+}_{(aq)} \rightarrow Tl^{3+}_{(aq)} + 2e^{-} \quad (2)$$

of reaction lead to the conclusion that this oxidation involves a two-electron transfer process. The magnitude of this potential makes thallium(III) a good oxidizing agent, and in fact it will oxidize water in hot solutions.

When a base is added to a solution of thallium (III), a brown precipitate of Tl_2O_3 is obtained; the hydroxide has been shown to be nonexistent. Thallium(III) oxide decomposes to thallium(I) oxide at 100°C.

The thallium(III) halides are prepared from the oxidation state I halides by reaction with the free halogen. These halides are unstable thermally, decomposing to the oxidation state I halide and the halogen. Thallium(III) fluoride melts at 550°C in an atmosphere of fluorine, but decomposes when heated in air and hydrolyzes in water. The chloride decomposes upon melting at 25°C, and the bromide shows an appreciable partial pressure of bromine even at room temperature. The iodide of composition TlI_3 is apparently $(Tl^+)(I_3^-)$ in the solid state. In solution, the thallium(III) chlorospecies have been shown to be $TlCl^{2+}$, $TlCl_2^+$, $TlCl_3$, $TlCl_4^-$, and $TlCl_6^{3-}$ by Raman spectroscopy. The TlX_4^- (X = Cl, Br, I) is tetrahedral, while (Li, Na)-TlF_4 has a fluoride structure and does not contain TlF_4^- ions. The compound of composition $TlCl_3$ ·

Physical properties of thallium(I) halides, TlX

Halide	Melting point, °C	Boiling point, °C
TlF	327	655
TlCl	430	806
TlBr	456	815
TlI	440	824

2py actually contains a six-coordinate cation $[Tl(py)_4Cl_2^+]$ and the $[TlCl_4^-]$ anion. Thallium(III) is also six-coordinate with bidentate sulfur ligand in the species $Tl(S_2CNEt_2)_3$ and $[Tl(S_2C_2N_2)_3]^{3-}$. Six-coordinate thallium is also encountered in $Tl_2Cl_9^{3-}$, where two octahedra of chlorine atoms share a common face. Products of composition TlX_2, which are obtained from the three-state halides by careful decomposition, in reality are $Tl_3[TlCl_6]$ and $Tl[TlBr_4]$.

Thallium forms organometallic compounds of the following general classes, R_3Tl, R_2TlX, and $RTlX_2$, where R may be an alkyl or aryl group and X a halogen. The R_2TlX compounds are ionic and are very stable, being unreactive toward oxygen of the air and moisture. The R_2Tl^+ ion has been shown to be linear, and is capable of accepting an additional ligand to form a T-shaped species, such as $[Me_2Tlpy]^+$. The cation $(C_6F_5)_2Tl^+$ reacts with dipyridyl to give a compound with apparent coordination number five, $(C_6H_5)_2Tl(dipy)$. The trialkyls on the other hand are quite reactive. Triethylthallium, for instance, which may be prepared from diethyl thallium chloride and ethyl lithium, a yellow liquid with a boiling point of 55°C at 1.5 mm, decomposes at 130°C. Monophenyl dichlorothallium, which is prepared by heating thallium(I) chloride in an aqueous solution of phenylboric acid, tends to decompose to diphenyl compounds and thallium.

A cyclopentadienyl compound of thallium, TlC_5H_5, is prepared with remarkable ease by passing its vapors of cyclopentadiene into an aqueous solution of thallium(I) hydroxide, whereupon cream crystals of the product precipitate. This compound is monomeric in the vapor state but polymeric in the solid state.

Thallium metal dissolves in alcohols to give tetrameric alkoxides, $Tl_4(OR)_4$. The Tl atoms are arranged at the corners of a tetrahedron with the OCH_3 groups perpendicular to the faces and with the oxygen atoms in three coordination to the thallium atoms.

Analysis. Thallium can be determined spectroscopically or in solution by oxidimetry. The oxide, Tl_2O_3, can also be precipitated and carefully dried. *See* GALLIUM; INDIUM. [EDWIN M. LARSEN]

Bibliography: F. A. Cotton and G. Wilkinson, II, *Advanced Inorganic Chemistry*, 4th ed., 1980; R. B. Heslop and K. Jones, *Inorganic Chemistry*, 4th rev. ed., 1976.

Theobromine

An alkaloid prepared from the dried ripe seed of *Theobroma cacao* or made synthetically. It is often extracted from waste products of the cocoa and chocolate industry. Theobromine is a close chemical relative of caffeine (see illustration).

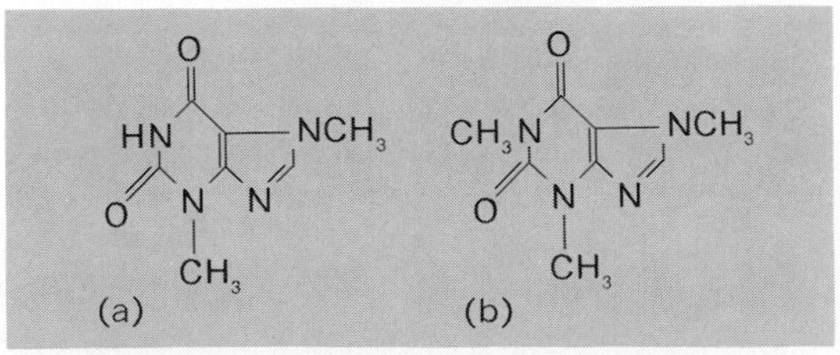

Structural formulas of (*a*) theobromine and (*b*) caffeine.

Theobromine is known to be responsible, in part, for the stimulant action of cocoa and other beverages. In addition, it is used therapeutically as a diuretic in the treatment of various types of edemas. Because of its relatively low solubility, it is used mostly in the form of mixtures which are much more soluble. *See* ALKALOID. [S. MORRIS KUPCHAN]

Thermal expansion

Solids, liquids, and gases all exhibit dimensional changes for changes in temperature while pressure is held constant. The molecular mechanisms at work and the methods of data presentation are quite different for the three cases and are therefore discussed separately in this article.

Expansion of solids. The temperature coefficient of linear expansion α_l is defined by Eq. (1),

$$\alpha_l = \frac{1}{l}\left(\frac{\partial l}{\partial t}\right)_{p=\text{const}} \tag{1}$$

where l is the length of the specimen, t is the temperature, and p is the pressure. For each solid there is a Debye characteristic temperature Θ, below which α_l is strongly dependent upon temperature and above which α_l is practically constant. Many common substances are near or above Θ at room temperature and follow the approximate equation Eq. (2), where l_0 is the length at 0°C and

$$l = l_0(1 + \alpha_l t) \tag{2}$$

t is the temperature in °C. The total change in length from absolute zero to the melting point has a range of approximately 2% for most substances. Typical room temperature values of a_l are given in Table 1.

Table 1. Temperature coefficients of linear expansion for typical substances at room temperature

Substance	Coefficient of linear expansion per °C $\times 10^6$
Aluminum, commercial	24
Copper	17
Diamond	1
Glass, commercial	11
Glass, pyrex	3
Granite	8.3
Ice	50
Iron	12
Invar alloy	0.9
Quartz, crystalline	5
Quartz, fused	0.5
Oak, along fiber	5
Oak, across fiber	54
Rubber, hard	80

Linear, harmonic vibration of the atoms in a solid cannot account for changes in volume, hence this must result from nonlinearity of the thermally excited vibration. The theory of E. Grüneisen takes this into account and shows the coefficient of expansion to be proportional to the constant-volume specific heat of the solid. At low temperatures (small amplitude vibration), the coefficient of expansion approaches zero.

Pure crystals may have different values of α_l

along different axes, but substances such as structural steel have many crystals randomly oriented and are almost free from this effect. At certain temperatures, crystalline substances may change in lattice arrangement, and a sudden change of volume occurs at constant temperature, making α_l momentarily infinity. *See* SPECIFIC HEAT OF SOLIDS.

Expansion of gases. So-called perfect gases follow the relation in Eq. (3), where p is absolute

$$\frac{pv}{T}=\frac{R}{\text{molecular weight}} \tag{3}$$

pressure, v is specific volume, T is absolute temperature, and R is a constant. The magnitude of R, the so-called gas constant, is 1544 ft-lb/(°R)(lb-mole) in the English system, or 8.3144×10^7 ergs/(K)(g-mole) in the metric system. Real gases often follow this equation closely; for example, Table 2 shows values of R at atmospheric pressure and 0°C. *See* GAS CONSTANT.

Table 2. Values for gas constant R at atmospheric pressure and 0°C

Gas	R
Air	1545
Hydrogen	1546
Nitrogen	1543
Oxygen	1544
Methane	1539

The coefficient of cubic expansion α_v is defined by Eq. (4), and for a perfect gas this is found to be $1/T$.

$$\alpha_v=\frac{1}{v}\left(\frac{\partial v}{\partial t}\right)_{p=\text{const}} \tag{4}$$

The behavior of real gases is largely accounted for by van der Waals' equation, Eq. (5), where a and b are constant for a given gas. When the specific volume is large, the effects of these constants

$$p=\frac{RT}{v-b}-\frac{a}{v^2} \tag{5}$$

are unimportant, and the real gas behaves as a perfect gas. In the regions where a and b have a dominant effect it is usually found desirable to use experimentally determined graphs or charts of properties. *See* GAS.

Expansion of liquids. For liquids, α_v is somewhat a function of pressure but is largely determined by temperature. Though α_v may often be taken as constant over a sizable range of temperature (as in the liquid expansion thermometer), generally some variation must be accounted for. For example, water contracts with temperature rise from 0 to 4°C, above which it expands at an increasing rate, as shown by the data in Table 3, which were taken at atmospheric pressure. One approach to this variation is to evaluate the constants α, β, and γ in Eq. (6), where v_0 is the volume

Table 3. Behavior of water at different temperatures

t,°C	Volume expansion, ml/g
−10	1.00186
0	1.00013
4	1.00000
10	1.00027
100	1.007

Table 4. Coefficients of volume expansion of gases

Liquid	$\alpha \times 10^3$	$\beta \times 10^6$	$\gamma \times 10^8$
Ethyl alcohol (99.3% by volume)	1.012	2.20	
500 atm	0.866		
3000 atm	0.524		
Carbon tetrachloride	1.184	0.899	1.351
Mercury	1.182	0.0078	
Petroleum	0.8994	1.396	
Water	−0.06427	8.5053	−6.7900

$$v=v_0(1+\alpha t+\beta t^2+\gamma t^3) \tag{6}$$

at 0°C, and v is the volume at temperature t. Typical values of the coefficients appear in Table 4.

Thermal stresses. When a homogeneous body is subject to constant boundary loads and is raised uniformly in temperature, the stress pattern in it will not change unless its elastic properties change. In general, stresses arise if (1) the body is made up of substances having different coefficients of expansion, (2) changes of boundary dimensions are restrained, or (3) temperature distribution is not uniform. A simple example of the first case is shown in the figure, where the aluminum bar, if heated, would tend to expand faster than the iron bars, thereby putting the iron in tension and the aluminum in compression. Considering the aluminum alone, its change of length would be restrained, and therefore stresses would arise in it. If the aluminum bar were replaced by an iron one and if this one alone were heated, again the other bars would be in tension and the center one in compression. More complex stress patterns may arise in continuous bodies; for example, if the bars were joined along the sides rather than at the ends, shear stresses would arise in the seams. In iron, 360 lb/in.² tensile stress would produce the same elongation as would a temperature rise of 1°C.

THERMAL EXPANSION

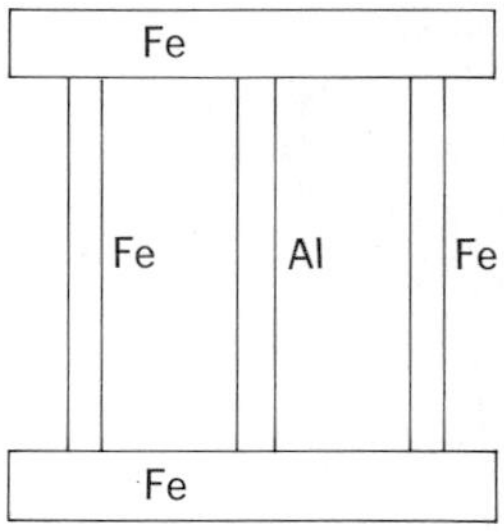

Body composed of substances having different coefficients of expansion. Thermal stresses would arise if it were subjected to heat.

Since one source of temperature variation is the gradient necessary for heat transfer, thermal conductivity and heat capacity may both play a role in determining the stress pattern.

[RALPH A. BURTON]

Thermoanalysis

A group of analytical techniques developed to continuously monitor physical or chemical changes of a sample which occur as the temperature is varied. Thermogravimetry (TG), differential thermal analysis (DTA), and differential scanning calorimetry (DSC) are the three principal thermoanalytical methods. Modern commercial instruments, many incorporating microprocessors, are available for all three types of thermoanalysis.

The theoretical basis of thermal analysis may be approached from either a kinetic or thermodynamic viewpoint. Kinetically the Arrhenius equation (1),

$$\text{Rate}=Ae^{-\Delta E/RT} \tag{1}$$

where A, ΔE, and R represent the preexponential factor, activation energy, and the gas law constant,

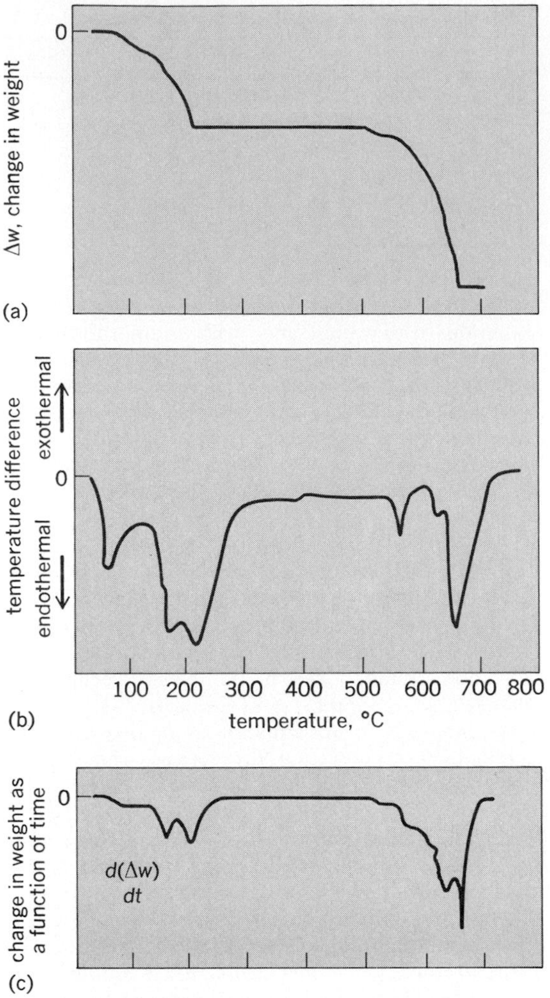

Fig. 1. Thermoanalytical curves for calcium nitrate tetrahydrate, $Ca(NO_3)_2 \cdot 4H_2O$. (*a*) Thermogravimetric analysis. (*b*) Differential thermal analysis. (*c*) Derivative thermogravimetry.

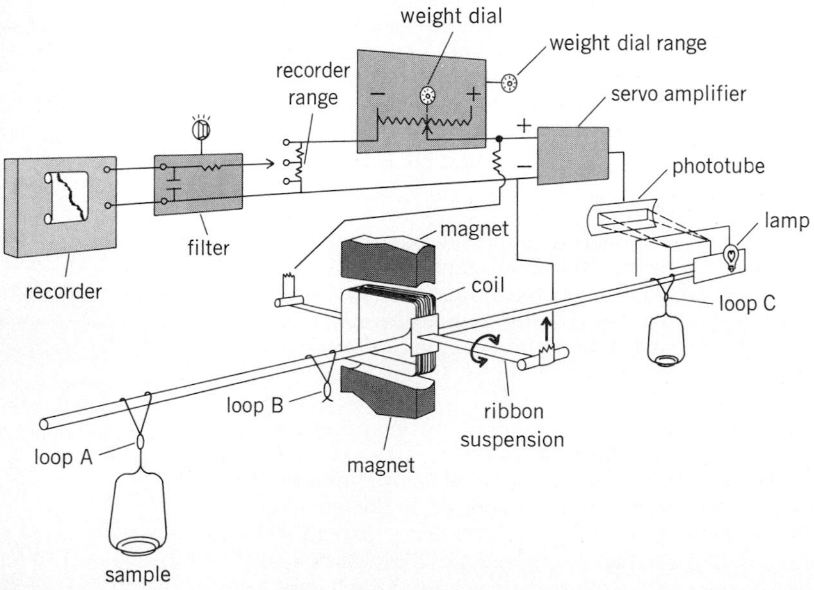

Fig. 2. Thermobalance. (*From W. W. Wendlandt, Thermal Methods of Analysis, 2d ed., p. 68, copyright © 1974 by John Wiley and Sons, Inc.; used with permission*)

respectively, indicates that reaction rates increase with temperature. At some point the rate becomes significant, and a chemical or physical change occurs. Similarly the Gibbs free-energy equation (2), where $\Delta G°$ is the Gibbs free energy, $\Delta H°$ is the

$$\Delta G° = \Delta H° - T\Delta S° \qquad (2)$$

reaction enthalpy, and $\Delta S°$ is the entropy change for the process, shows that the equilibrium constant will change with temperature. If ΔS is positive and temperature is increased, the equilibrium will be shifted to favor formation of products. In many cases a combination of these causes the observed physicochemical process. *See* CHEMICAL THERMODYNAMICS.

Instrumentation for thermoanalysis requires appropriate sample and reference holders enclosed in an oven equipped with a temperature-programming device and the necessary transducers for converting the physicochemical property being measured into electrical signals. Samples may be studied in air or other atmospheres, inert or reactive, at reduced, ambient, or elevated pressures. Most samples are solids, although an increasing number of liquid samples are being analyzed. Exact experimental conditions are chosen to enhance the process under study and ensure reproducibility. Selectable experimental parameters are heating (cooling) rate, temperature range, atmospheric composition and pressure, recorder attenuation and speed, and sample preparation. Each of these may affect the thermogram produced.

Thermograms are plots of the measured property, such as mass or heat evolved, versus the oven temperature (Fig. 1). Under rigorously controlled conditions, thermograms uniquely represent the system under study. Melting, crystallization, decomposition, oxidation, adsorption, absorption, desorption, polymerization reactions, and heat-capacity changes are observable on thermograms.

In addition to traditional methods, where temperature is continuously varied, isothermal analyses may be performed. In this case the sample temperature is rapidly changed from ambient to the desired value, and the appropriate physicochemical property is measured as a function of time. Stability and kinetic studies are done in this fashion to reduce thermodynamic complications.

Thermogravimetry. This method involves measuring the changes in mass of a substance, typically a solid, as it is being heated. Specially designed thermobalances (Fig. 2) are required to continuously monitor sample mass during the heating process. Most balances have a capacity of 1 – 100 mg for the sample and can accurately detect mass changes in the microgram range.

Any type of physicochemical process which involves a change in sample mass may be observed by using thermogravimetry. Mass losses are observed for dehydration, decomposition, desorption, vaporization, sublimation, pyrolysis, and chemical reactions with gaseous products. Mass increases are noted with adsorption, absorption, and chemical reactions of the sample with the atmosphere in the oven.

A typical thermogravimetric curve is shown in Fig. 1*a*. Quantitative gravimetric analyses may be performed due to the precise measure of the mass change obtained. Rates of mass change have been used to evaluate the kinetics of a process and to estimate activation energies. Fine details of these thermograms may also be used to deduce reaction intermediates and reaction mechanisms.

Primary applications of thermogravimetry are to deduce stabilities of compounds and mixtures at elevated temperatures and to determine appropriate drying temperatures for compounds and mixtures. Care must be exercised, since very slow decomposition or dehydration processes are not normally observed and may lead to erroneous conclusions about stabilities.

Differential thermal analysis. This method involves the monitoring of the temperature difference T_D between a sample and inert reference material (usually aluminum oxide) as they are simultaneously heated, or cooled, at a predetermined rate. Thermocouples and thermistors are the most common temperature sensors used for this purpose; they are arranged in an oven (Fig. 3). As enthalpic changes occur, T_D will be positive if the process is exothermic and negative if it is endothermic. The thermogram in Fig. 1*b* shows the endotherms plotted as negative deviations from the baseline.

Differential thermal analysis thermograms are affected by instrumental factors such as furnace design, sample-holder material and geometry, thermocouple size and placement, and instrument response and furnace atmosphere. Sample characteristics such as particle size, thermal conductivity, heat capacity, packing density, and the amount of sample also affect the thermograms. Under rigorously specified conditions, thermograms will uniquely represent the physicochemical characteristics of the sample.

More reactions may be observed with differential thermal analysis than with thermogravimetry. Endothermic physical processes include

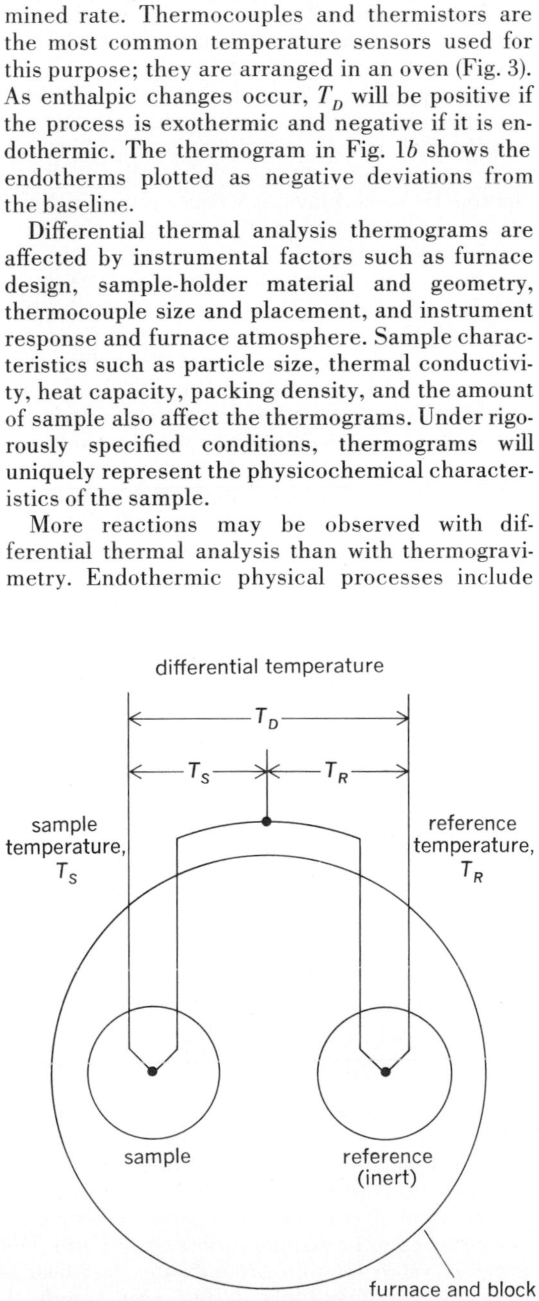

Fig. 3. Schematic diagram illustrating the principles of the apparatus for differential thermal analysis.

crystalline transitions, fusion, vaporization, sublimation, desorption, and adsorption. Endothermic chemical processes include dehydration, decomposition, gaseous reduction, redox reactions, and solid-state metathesis. Exothermic processes include adsorption, chemisorption, decomposition, oxidation, redox reactions, and solid-state metathesis reactions. Both solids and liquids are studied by differential thermal analysis. Hermetically sealed capsules are usually used for liquids and some solids. Other samples are studied in open or crimped pans.

Analytical applications of this technique include the identification, characterization, and quantitation of a wide variety of materials, including metals, clays, minerals, inorganic and organic compounds, polymers, and pharmaceuticals. Characteristic thermograms can be used to determine purity, heats of reaction, thermal stability, phase diagrams, catalytic properties, and radiation damage.

Quantitative analyses may be done by using differential thermal analysis, since the sample mass, enthalpic effect, and peak areas are proportionally related. Instrument calibration at temperatures close to the temperature of the peak of interest is essential for accurate results.

Differential scanning calorimetry. In this method a sample and a reference are individually heated, by separately controlled resistance heaters, at a predetermined rate while they are kept isothermal. Enthalpic processes are detected as differences in electrical energy needed to produce this desired isothermal heating rate. This electrical energy, in millicalories per second, is then plotted versus the sample temperature to obtain the thermogram.

Analytical uses of differential scanning calorimetry are very similar to those of differential thermal analysis. Usually one calibration standard is sufficient to calibrate the entire operating range of the instrument. Differential scanning calorimetry instruments are highly sensitive, and as a result, heat-capacity changes are more easily detected and evaluated. Such evaluation has become important in polymer and biochemical studies. Small (approximately 10 mg) samples are used in most cases, although some instruments have been developed which use up to 1.0 ml of a liquid sample. Fewer instrumental and sample characteristics contribute to distortions of the thermograms as compared to differential thermal analysis. *See* CALORIMETRY.

Other methods. Thermomechanical analysis, thermoluminescence, emanation thermal analysis, thermomagnetic analysis, and evolved-gas analysis are also used in thermoanalysis. Thermomechanical analysis can be used to evaluate the physical stability of structural or electronic components. Thermoluminescence is often used for authenticating ancient objects. Emanation and evolved-gas thermal analysis are used to determine gaseous products, often from destructive oxidation. It is used in the study of fabrics treated with fire retardants. Thermomagnetic analysis can be used to determine Curie points of metals.

In some instances the differential curves are more informative or precise. A differential thermogravimetric curve is shown in Fig. 1*c* Differential curves are produced electronically or mechanically from the original signal. Virtually every method discussed above has been adapted to differential methods. Integration of peak areas is often done by hand, since most peaks are nonsymmetrical and baselines are sloped. *See* CHEMICAL DYNAMICS; PHASE EQUILIBRIUM; THERMAL EXPANSION; THERMOCHEMISTRY; TRANSITION POINT.

[NEIL JESPERSEN]

Bibliography: C. Duval, *Inorganic Thermogravimetric Analysis*, 2d ed., 1963; J. L. McNaughton and C. T. Mortimer, *Differential Scanning Calorimetry*, IRS Phys. Chem. Ser. 2, vol. 10, 1975; W. W. Wendlandt, *Thermal Methods of Analysis*, 2d ed., 1974.

Thermochemistry

A branch of physical chemistry dealing with the heat effects which accompany chemical reactions, the formation of solutions, and changes in the physical state of substances, such as the fusion of a crystalline solid or the vaporization of a liquid. When these processes evolve heat, they are said to be exothermic. Conversely, those absorbing heat are endothermic. A knowledge of the magnitudes of such heat effects is of practical value to the engineer in solving problems of heating and refrigeration and in controlling chemical reactions at suitable temperatures. Such data are also of great theoretical importance to the scientist who wishes to calculate the chemical affinity or free energy for various reactions, hypothetical or real. *See* CHEMICAL THERMODYNAMICS; FREE ENERGY.

These heat effects are usually measured in calories, partly for historical reasons and partly because the figures thereby obtained are of more convenient magnitude than when the joule, the fundamental unit of energy, is employed. The present-day calorie is arbitrarily defined as being equal to 4.184 absolute joules, rather than in terms of the heat capacity of water, which varies with temperature.

Fundamental concepts. The principle of conservation of energy serves as the basis for the fundamental concepts of thermochemistry. Thus, for a chemical system undergoing some change, the increase in its internal energy ΔE is related to the heat q absorbed by the system from its surroundings and the work w done by the system on these surroundings by Eq. (1). In practice, a negative

$$\Delta E = q - w \qquad (1)$$

value for any of these quantities is common and simply denotes a decrease in the energy content of the system, or evolution of heat, or work done upon the system, as the case may be. If the change under consideration takes place at a constant temperature and also at constant volume so that w has zero value, $\Delta E = q$ and may be termed the heat of reaction at constant volume.

The chemist and engineer, however, are often interested in processes which take place at constant pressure, frequently 1 atm (10^5 Pa). In this case, w will be equal to the change in volume ΔV in such a process multiplied by this pressure P, provided no other work is done by, or on, the system. Then the preceding equation may be rearranged to yield $\Delta E + P\,\Delta V = q$. It is now convenient to introduce a new function, enthalpy, which is defined as $H = E + PV$, and accordingly here $\Delta H = q$. Thus, the change in enthalpy measures the heat effect in a process at constant temperature and pressure. Because these are the conditions most frequently encountered in practice, ΔH is a more generally useful value than ΔE in a reaction. Of course, since they differ simply by the $P\,\Delta V$ term, either can be computed from the other.

Experimental results. Many experimental determinations of these heat effects (that is, ΔE and ΔH values) in chemical reactions and other transformations have been made by various calorimetric methods during the past 100 years. Most of the results of the measurements by the pioneers, such as J. Thomsen and M. P. E. Berthelot, involved considerable uncertainties because generally their chemicals were impure and their apparatus often was crude. Since 1930, however, there has been a notable renaissance in thermochemical studies. It has been characterized by the use of extremely pure chemical substances and great improvements in calorimetric equipment and procedures. Consequently, present-day measurements of the heats of combustion of organic compounds, for instance, usually involve uncertainties under 0.05% and in some cases, under 0.01%. The studies of F. D. Rossini, H. M. Huffman, J. Coops, and H. A. Skinner are outstanding examples of such highly accurate work. *See* CALORIMETRY.

In certain cases, it is also possible to evaluate ΔH for a process by indirect, noncalorimetric methods. Thus, some reactions, especially in the field of inorganic chemistry, can be effected in connection with reversible galvanic cells. From a careful measurement of the electromotive force in such cases at several different temperatures, the ΔH value can be calculated by the Gibbs-Helmholtz equation. For a number of reactions, extremely accurate data have been obtained by this method. Likewise, a measurement of the equilibrium constants for a reaction at two or more different temperatures can be used to deduce a moderately accurate value, sometimes at fairly elevated temperatures, by calculation with the van't Hoff equation. Similarly, many heats of vaporization have been derived indirectly from vapor pressure measurements and the Clausius-Clapeyron equation. *See* VAPOR PRESSURE.

These heat effects are frequently recorded in a shorthand fashion by writing a chemical equation for the process involved and the corresponding ΔE or ΔH value, depending on whether the process occurs at constant volume or constant pressure. Thus, the reaction between gaseous hydrogen and oxygen to produce a gram-mole of liquid water at 25°C (298.15 K) is given by reaction (2). Here the

$$\begin{gathered} H_2(g) + \tfrac{1}{2}O_2(g) \rightarrow H_2O(l) \\ \Delta H^\circ_{298} = -68{,}315 \pm 10 \text{ cal} \end{gathered} \qquad (2)$$

abbreviations in parentheses indicate that the hydrogen and oxygen are gases, and the water produced is liquid. The subscript after ΔH represents the absolute temperature for the reaction and the degree symbol indicates that each substance involved is under the standard pressure of 1 atm. The negative value for this ΔH indicates that heat is evolved. The uncertainty in this result is only 10 cal (1 cal = 4.184 J), because this determination was made with extreme accuracy at the U.S. National Bureau of Standards. Because it applies to the process of forming a compound from its elements, this quantity is the standard heat of formation.

Similarly, a few additional thermochemical equations for typical combustion reactions can be shown as (3). The first two reactions involve

$$\begin{gathered} \text{C(graphite)} + O_2(g) \rightarrow CO_2(g) \\ \Delta H^\circ_{298} = -94{,}051 \pm 11 \text{ cal} \\ \text{C(diamond)} + O_2(g) \rightarrow CO_2(g) \\ \Delta H^\circ_{298} = -94{,}504 \pm 23 \text{ cal} \\ CO(g) + {}^1\!/\!{}_2 O_2(g) \rightarrow CO_2(g) \\ \Delta H^\circ_{298} = -67{,}635 \pm 29 \text{ cal} \\ CH_4(g) + 2O_2(g) \rightarrow 2H_2O(l) + CO_2(g) \\ \Delta H^\circ_{298} = -212{,}801 \pm 72 \text{ cal} \\ CH_3OH(l) + {}^3\!/\!{}_2 O_2(g) \rightarrow 2H_2O(l) + CO_2(g) \\ \Delta H^\circ_{298} = -173{,}550 \pm 100 \text{ cal} \end{gathered} \tag{3}$$

the production of carbon dioxide from graphite and diamond, respectively. It is interesting to note that these two crystalline forms of carbon have appreciable differences in their combustion values. Because the end product for each combustion is the same, this means that the enthalpy content of diamond is 453 cal higher per gram-atom than that of graphite. For practical purposes, graphite is usually taken as the standard form of this element. The ΔH values for the combustion of carbon monoxide, methane, and methanol are data of great industrial importance which were quite uncertain for many years.

Law of Hess. It is frequently desirable to calculate the heat effect in a particular reaction for which a direct experimental determination is not available. The thermochemist then has recourse to the law of Hess. This is essentially a corollary of the principle of conservation of energy, although it was discovered a few years earlier, in 1840, by direct experimentation. According to this law, the net heat effect or change in enthalpy in taking a chemical system from a state A to a state B must be the same whether the path between these two states is traversed directly in one step or in a roundabout fashion by two or more steps.

Thus, the law of Hess permits the combination of chemical equations and the corresponding ΔH values to arrive at the ΔH value for the desired reaction. This may now be illustrated by a computation of the heat of formation of methane from its elements. The preceding combustion values can be used in the series of reactions (4). Here, however,

$$\begin{gathered} 2H_2O(l) + CO_2(g) \rightarrow 2O_2(g) + CH_4(g) \\ \Delta H^\circ_{298} = 212{,}800 \pm 72 \text{ cal} \\ \text{C(graphite)} + O_2(g) \rightarrow CO_2(g) \\ \Delta H^\circ_{298} = -94{,}051 \pm 11 \text{ cal} \\ 2H_2(g) + O_2(g) \rightarrow 2H_2O(l) \\ \Delta H^\circ_{298} = -136{,}630 \pm 20 \text{ cal} \end{gathered} \tag{4}$$

the reaction for the combustion of methane must be written backward, also reversing the ΔH sign. Algebraic addition of these thermochemical equations yields reaction (5). The uncertainty in this fi-

$$\begin{gathered} \text{C(graphite)} + 2H_2(g) \rightarrow CH_4(g) \\ \Delta H^\circ_{298} = -17{,}881 \pm 75 \text{ cal} \end{gathered} \tag{5}$$

nal ΔH value has been computed by probability methods from the uncertainties in the contributing data.

Similar calculations may be made for many compounds where it is impractical, or even impossible, to make a direct calorimetric evaluation of the heats of formation. All such calculations deal with the changes in enthalpy accompanying changes in chemical composition or physical state. The absolute values of the enthalpies of the elements or of the resulting compound are not known; such knowledge is not really essential for the practical purposes of the scientist or engineer. It is merely the enthalpy difference that is important. Accordingly, the enthalpy of each element is arbitrarily assigned a zero value in a standard reference state at all temperatures. This reference state is usually the most stable form of the element at room temperature and atmospheric pressure. Thus, carbon is taken as β-graphite, sulfur as the rhombic crystalline form, and the three elements hydrogen, oxygen, and nitrogen in the form of diatomic gas at 1 atm pressure. With this convention, extensive tables of ΔH° values for the formation of various substances, usually at 25°C and 1 atm, have been developed. Noteworthy among such compilations are the tables of *Selected Values* issued by the National Bureau of Standards, the American Petroleum Institute Research Project 44, and the Thermodynamics Research Center Data Project at Texas A & M University.

From these tables of data, the ΔH°_{298} values may now be derived for numerous reactions, some of which are purely hypothetical. One needs only to subtract the enthalpy values for the initial materials from the corresponding value of the reaction product. Thus, for an important reaction for the production of ethyl alcohol from ethylene and water, one finds relations (6) by subtracting the en-

$$\begin{gathered} \begin{matrix} C_2H_4(g) & + & H_2O(l) & \rightarrow & C_2H_5OH(l) \\ (12{,}500) & & (-68{,}315) & & (-66{,}200) \end{matrix} \\ \Delta H^\circ_{298} = -10{,}385 \text{ cal} \end{gathered} \tag{6}$$

thalpies of formation of ethylene and water from that for the alcohol. These contributing values are here placed directly below the respective compounds in the chemical reaction.

Effects of temperature and pressure. Up to this point, only the heat effects in processes at 25°C have been considered. This temperature has become a standard for the recording of thermal data. However, in practice many chemical reactions are carried out at different temperatures, often much higher. The scientist and engineer, therefore, may wish to know the ΔH° value for this second temperature T_2 in a case for which he has the corresponding value to T_1, which is frequently 298 K. For this problem, Kirchhoff's law may be utilized. This law dates from 1858 and can be considered as a corollary of the principle of conservation of energy. It may be stated in the form of Eq. (7), where ΔC_p

$$\Delta H_{T_2} = \Delta H_{T_1} + \int_{T_1}^{T_2} \Delta C_p \, dT \tag{7}$$

represents the difference between the heat capacities at constant pressure of the products and initial substances in the reaction. The use of this equation obviously requires adequate heat capacity data over the temperature range involved and a knowledge of the enthalpy change at one temperature. In practice, the $\Delta H°$ values for many processes vary considerably with temperature. For example, in the case of the formation of methane from its elements, $\Delta H^\circ_{1000} = -21{,}430$ cal, compared with $\Delta H^\circ_{298} = -17{,}881$ cal as previously cited.

Pressure changes also have some influence on the heat effects in chemical reactions. In general, this influence is quite small at ordinary pressures, particularly when only liquids or solids are concerned. However, it can be of considerable importance in certain gas-phase reactions, such as the ammonia synthesis, where pressures of hundreds of atmospheres may be used and the substances then exhibit considerable deviations from ideal gas behavior. Suitable thermodynamic equations can be utilized for estimating such changes in enthalpy with pressure.

Changes in physical state. So far, the discussion of heat effects has been concerned entirely with chemical reactions. However, changes in physical state, such as the vaporization of a liquid or the fusion of a crystalline solid, may be treated in essentially similar fashion. Thus, for the process of converting a gram-mole of methanol from the liquid to the gas as a result of a calorimetric determination, one may write reaction (8). Then, in ac-

$$\begin{gathered} CH_3OH(l) \rightarrow CH_3OH(g) \\ \Delta H^\circ_{298} = 8940 \pm 10 \text{ cal} \end{gathered} \quad (8)$$

cordance with Hess's law, this thermochemical reaction can be combined with that previously given for the combustion of liquid methanol to yield a value for the combustion of the gaseous form as shown in reaction (9). Such a result is valuable for

$$\begin{gathered} CH_3OH(g) + {}^3\!/_2 O_2(g) \rightarrow 2H_2O(l) + CO_2(g) \\ \Delta H^\circ_{298} = -182{,}490 \pm 50 \text{ cal} \end{gathered} \quad (9)$$

developing the thermodynamic treatment of the methanol synthesis.

Solutions. The thermochemistry of solutions can be touched upon only briefly here. The heat effects depend upon the nature of the particular solutions, and sometimes change greatly with the concentration. The solution process for crystalline cetyl alcohol in the closely related liquid *n*-heptyl alcohol is endothermic, and the ΔH approximates the heat of fusion of cetyl alcohol at the temperature involved. It is almost independent of the concentration, because the resulting liquid is nearly an ideal solution.

By contrast, the solution processes for many inorganic substances in water are frequently exothermic as a result of the hydration of ions and other departures from ideality. Two heat quantities commonly occur in such cases, the integral and the differential heats of solution. The integral heat of solution is the ΔH per mole of solute when it is dissolved in a given solvent, such as water, to form a solution of a particular concentration. On the other hand, the differential heat of solution is the ΔH effect when 1 mole of the solute is dissolved in such a large volume of solution of this particular concentration that the concentration is not appreciably changed. These two ΔH values are identical for an infinitely dilute solution. However, they differ by 2000 cal in the case of a 5.0 molal solution of aqueous sulfuric acid. *See* SOLUTION.

[BRUNO J. ZWOLINSKI]

Bibliography: S. W. Benson, *Thermochemical Kinetics: Methods for the Estimation of Thermochemical Data and Rate Parameters*, 2d ed., 1976; D. J. Ives, *Chemical Thermodynamics*, 1971; E. T. Turkdogan, *Physical Chemistry of High Temperature Technology*, 1980.

Thermodynamic principles

Laws governing the conversion of energy from one form to another. Among the many consequences of these laws are relationships between the properties of matter and the effects of changes in pressure, temperature, electric field, magnetic field, and composition. The great practicality of the science arises from the foundations of the subject. Thermodynamics is based upon observations of common experience that have been formulated into the thermodynamic laws. From these few laws all of the remaining laws of the science are deducible by purely logical reasoning. There is a choice as to which few are considered independent laws, from which the remainder may be derived. A modern tendency is to choose basic laws or postulates that are different from those first discovered. Some of these choices are most useful in that the derivation of the remainder may be accomplished very efficiently. However, those laws that arose from the historical development will be discussed here since they are less abstract and lend themselves to a clearer physical interpretation.

One may say that the whole development of thermodynamic principles was completed when three state functions, the absolute temperature T, the internal energy U, and the entropy S, were defined. The zeroth law formalizes the concept of temperature, the first law defines the internal energy, and the second law brings in the concept of entropy as well as the absolute scale of temperature. Finally, the third law describes the behavior of entropy and internal energy as the absolute temperature approaches zero.

For exposition, it is necessary to define a few terms. A *system* is that part of the physical world under consideration. The rest of the world is the *surroundings*. An *open system* may exchange mass, heat, and work with the surroundings. A *closed system* may exchange heat and work but not mass with the surroundings. An *isolated system* has no exchange with the surroundings. A closed or isolated system is sometimes referred to as a body. Those parts of a system spatially uniform and homogeneous are called *phases*. For example, a liquid together with its vapor may be considered a two-phase system. Systems may be made quite elaborate when required, but since focus is on the thermal properties, single-phase isotropic systems not acted upon by electric or magnetic fields are considered so that the only force allowed is that generated by a uniform normal pressure. Such a restriction is not a basic limitation on the generality of thermodynamics but is simply a pedagogical device.

Specification of equilibrium state. The material properties of concern to thermodynamics are the macroscopic properties such as temperature, pressure, volume, concentration, surface tension, and viscosity. Molecular properties such as interatomic distances are not used. The state of a system is specified by all of the macroscopic properties together with their spatial variation. It is a fact of experience, however, that an isolated system approaches a particularly simple terminal state such that the properties are constant and spatially uniform. This simple state is called an equilibrium state. If one confines attention to a given quantity of a single-phase system, the equilibrum state is completely specified by $r+1$ of its properties, where r is the number of components. For a single-component, single-phase system not subject to magnetic or electric fields, one may fix two properties such as pressure and volume; all the remaining properties such as viscosity, surface tension, and so forth then assume fixed values. In other words, any macroscopic property of the system may be expressed as a function of the pressure and volume.

Temperature. It is within the scope of thermodynamics to refine the primitive notion of hotness and coldness into an operational and precise concept of temperature. The equilibrium states of a single-component, single-phase fluid provide a starting point. For such a fluid the equilibrium state is defined by fixing two of its properties. For example, one could construct a mercury-in-glass thermometer that has its pressure held constant; then only one other property could be varied independently. If the volume (height of mercury) is observed at any equilibrium state, there is a 1:1 correspondence between it and any other property excepting the pressure. The degree of hotness is one of these properties.

Also note that thermal equilibrium between different systems exists. For example, if the mercury thermometer is placed in contact with a body of quiet water, the mercury will either expand or contract. The volume change of the mercury will eventually stop, and the properties of the mercury will be constant, indicating an equilibrium state; moreover, the water will also have the constant properties of an equilibrium state. Bodies in thermal equilibrium are said to have the same temperature. Thus, one arrives at a method of measuring the temperature of a body of water.

Now suppose there is a large body of water in thermal contact through a wall with a large body of another fluid such as alcohol, both at equilibrium. It is an experimental fact that the mercury-in-glass thermometer will register the same volume when placed in either of the fluids. This fact is most important if one is to attach meaningful numbers to temperature. This fact of experience is designated as the zeroth law of thermodynamics: If two bodies A and B are separately in thermal equilibrium with C, then A and B are in equilibrium with each other. Thus, a useful empirical temperature measurement based upon the volume of mercury under constant pressure is established.

But the empirical temperature scale is unfortunately unique to the choice of the fluid. The mercury-in-glass thermometer is calibrated by bringing it into equilibrium with an ice-water mixture and then with boiling water, all under a pressure of 1 atm. The two mercury levels are marked 0° and 100°, respectively, and linear interpolation is used to assign numbers between the two fixed points. If one constructs and calibrates another thermometer but uses another fluid instead of mercury, one would find that the numerical values between the two fixed points do not agree. Indeed, water cannot be used as a working fluid at all since it has a minimum value of volume at 4°C and yields anomalous readings in this region. It is necessary to choose some fluid as a standard and calibrate all other thermometers by comparison with the standard. The second law of thermodynamics removes this dependence upon a particular material by defining an absolute temperature scale that is independent of the working fluid. Meanwhile the empirical temperature serves a useful operational purpose.

One could have used a low-pressure gas as the thermometric fluid. The volume could have conveniently been held constant, and the pressure of the fluid would have had a 1:1 correspondence with temperature. Here one would have found that all gases at low pressures yield the same temperature scale. This ideal gas temperature scale proves to be identical with the thermodynamic scale of the second law.

In summary, the relationship between the properties of the equilibrium state, the notion of thermal equilibrium, and the zeroth law have been used to establish the property of temperature. It may be noted in passing that temperature is a state property, and for a given mass of a single-phase, single-component fluid the temperature is a function of pressure and volume as in Eq. (1). This can be inverted as in Eqs. (2).

$$t=t(p,V) \tag{1}$$

$$\begin{aligned} p&=p(t,V)\\ V&=V(p,t) \end{aligned} \tag{2}$$

Internal energy. Thermodynamics does not define the concepts of energy or work but adopts them from the other macroscopic sciences of mechanics and electromagnetism. Also, the conservation of energy is taken as axiomatic. Therefore, if an isolated system is formed from any part of the world, a definite amount of energy will be trapped in the system. The energy resides in the kinetic and potential energy of the trapped molecules. The trapped energy is of a definite quantity because the isolated system cannot gain or lose energy to the surroundings and remains constant because of the conservation principle. This trapped energy is called the internal energy U.

Because of the conservation of energy, the internal energy of a closed system can be altered only by an exchange of energy with the surroundings. There are only three modes by which the exchange can occur: by mass transfer, heat transfer, or work exchange. So for a closed (no mass transfer), adiabatic (no heat transfer) system the change in internal energy ΔU is equal to the work done by the surroundings on the system, as defined by Eq. (3).

$$\Delta U=W_{\mathrm{AD}} \tag{3}$$

Here a convention has been adopted that work done on the system is positive. There is a great mass of experimental information where work has been done on a closed system enclosed within adiabatic walls. Among these are experiments performed by J. Joule more than a century ago. He caused work to be done on an adiabatically enclosed mass of water in several different ways. A measured amount of work was used to drive an agitator in the water, to create an electric current which was then passed through a coil in the water, to compress a gas in a cylinder immersed in the water, and to rub metal blocks together in the water. In Joule's experiments the same temperature increase was always obtained with the same expenditure of work. It may be concluded from Joule's experiments that the expenditure of a given quantity of work always causes the same change of state regardless of how the work is carried out. Both W_{AD} and ΔU are independent of the path. It is concluded that U is a state function. So for a single-phase, single-component fluid Eq. (4) is

$$W_{AD} = U_2 - U_1 = \Delta U \tag{4}$$

written, where U_2 and U_1 depend only on the final and initial state, respectively. Also one may write Eqs. (5).

$$U = U(p,V) \qquad U = (t,p) \tag{5}$$

It is known from experience that the same change in state of a system can be effected by either supplying work to the system in an adiabatic enclosure or by contacting the system through a conducting wall with a higher temperature system. The latter method is a different means of transferring energy than work and is termed heat and given the symbol Q. Measuring the amount of work required to cause the same change in state as an amount of heat enables one to express heat quantities in terms of work quantities. For example, 1 calorie is taken to be the amount of heat necessary to raise 1 g of water 1°C at 15°C and 1 atm. The same change in state can be effected by 4.186×10^7 ergs of work. Therefore, Eq. (6) holds.

$$1 \text{ calorie} = 4.186 \times 10^7 \text{ ergs} \tag{6}$$

The first law of thermodynamics may now be derived by the useful device of a composite system. Imagine a very large system of water that transfers neither heat nor work to the surroundings. Within this large system there is a small system of a cylinder of gas in thermal contact with the water and having a piston connected to the outside. Work can be done on the small system, and it can in turn interchange heat with the large system called a reservoir. Using subscripts s for the small system and r for the reservoir, the process of doing work can be described as in Eq. (7). But $\Delta U_r =$

$$W_s = \Delta U_s + \Delta U_r \tag{7}$$

$Q_r = -Q_s$. Therefore, for the small system that is exchanging both heat and work with its surroundings, one writes Eq. (8), omitting the subscript.

$$\Delta U = Q + W \tag{8}$$

This is the first law of thermodynamics and states that the algebraic sum of heat and work during a process is equal to the change in the state function U. The term $(Q + W)$ is therefore independent of the path taken between the two states. One could, for example, cause 1 g of water to undergo the change in state as in notation (9) by

$$15°\text{C}, 1 \text{ atm} \rightarrow 16°\text{C}, 1 \text{ atm} \tag{9}$$

supplying 1 calorie of heat and no work or by doing 4.186×10^7 ergs of work alone, or one could do a great deal of work and abstract all of this energy in the form of heat excepting 1 calorie. Thus, although $(Q + W)$ is independent of the path, neither Q nor W by itself is independent of the path.

It is important to realize that U is a state function and a property of the system whereas W and Q are not. The work, as well as the heat, simply represents energy in transit. Once the energy is in the system, it is not possible to determine whether it came from heat transfer or work transfer; it is simply internal energy.

The differential form of the first law is given by Eq. (10), where q and w represent small quantities.

$$dU = q + w \tag{10}$$

In general, one may not treat q or w as well-behaved differential coefficients dQ and dW. However, if the change is such that either q or w depends only on the initial and final states and not on the path, Eq. (11) may be correctly written. If two

$$dU = dQ + dW \tag{11}$$

terms are independent of the path, the third must be also. Obviously, if either q or w is zero, one may properly write Eqs. (12). There is a third case

$$\begin{aligned} dU &= dW \\ dU &= dQ \end{aligned} \tag{12}$$

where neither q nor $w = 0$, but nevertheless $dU = dQ + dW$ is still proper. Before treating this interesting case one needs to develop the notion of a reversible process.

Reversible and irreversible processes. Any process that occurs in nature is in agreement with the first law, but many processes permissible by the first law never occur. It has already been noted that systems approach an equilibrium state if left to themselves. There is an overwhelming preference for processes to proceed in one direction. Consider Joule's experiment. A falling weight caused a paddle to do work on an adiabatically enclosed body of water. The total effect of the experiment was to increase the internal energy of the water and to lower the weight. The surroundings remained unchanged. The water temperature increased and the volume increased slightly. There is no way one can reverse this process, that is, restore the water to its original state and raise the weight to its original height without also making some additional change in the surroundings. The process is irreversible.

Consider some other processes occurring within an adiabatic enclosure by examining only the initial and final states. Two blocks of copper are initially at different temperatures and finally at the same temperature which is intermediate between the two initial temperatures. A gas is initially filling just half of a container and finally the whole of the container. Again these processes are irreversible; that is, they cannot be reversed without causing some permanent change in the surroundings. The reverses of these processes do not violate the first law; therefore, there must be some condition other than the conservation of energy which is obeyed by those processes which actually take place.

If one were presented with a description of only the initial and final states of these irreversible processes, as was done in the last two examples, one could unerringly decide from the description which state was initial and which was final. The direction of the process is entirely determined by the nature of the states. It may be expected, therefore, that there is some state function that shows which state precedes the other. The function which tells whether a process is possible or not is the entropy S and will be derived from the information that some adiabatic processes are impossible.

Thermodynamics makes use of an idealization, called a reversible process, that is a limiting case of the natural or irreversible process. The reversible process may be defined as one which can be completely reversed without leaving more than a vanishingly small change in the surroundings. It is a consequence of the definition that a reversible process proceeds through a succession of equilibrium states and may be reversed by an infinitesimal change in the external conditions. Imagine having a cylinder of gas fitted with a frictionless piston. If the piston is moved so slowly that pressure gradients are absent, the gas will be in an equilibrium state at all times. The difference between the gas pressure and the external pressure needs only to be infinitesimal in order to move the frictionless piston. Under the rather restrictive conditions of a reversible process, Eq. (13) may be

$$dW = -p dV \tag{13}$$

quite properly written, where p is the gas pressure and V is the gas volume. The first law now may be written as Eq. (14).

$$dQ = dU + p dV \tag{14}$$

Entropy. The discussion on irreversible processes has led to the second law of thermodynamics, which is just a general statement of the idea that there is a preferred direction for a given process. There are many physical statements of the second law, all being equivalent and leading to the same mathematical statement. The statement of R. Clausius is: "It is *not* possible that, at the end of a cycle of changes, heat has been transferred from a colder to a hotter body without producing some other effect." Lord Kelvin's statement is: "It is *not* possible that, at the end of a cycle of changes, heat has been extracted from a reservoir and an equal amount of work has been produced without producing some other effect."

A specific example of Kelvin's statement may be useful. Work can be converted continuously and completely into heat. For example, work could be expended on rubbing blocks in a large mass of water. The blocks would become infinitesimally hotter than the water and transfer energy to the water by heat flow. The process could be continued indefinitely with the only effect being a complete conversion of work into heat. If, however, heat is converted from the large water reservoir completely into work, some other effect occurs. For example, a gas within a cylinder can be expanded reversibly causing a transfer of heat from the bath to the gas. All of the heat extracted from the bath is converted into work. However, the gas, in this process, has changed its state since its volume is larger. The gas cannot be returned to its original state without undoing the conversion of heat into work already accomplished.

The most efficient way of developing the mathematical consequences of the second law is to proceed from Caratheodory's principle, which can be either taken as another physical expression of the second law or derived from the Clausius or Kelvin statement. Caratheodory's principle is: "In the neighborhood of any equilibrium state of a system there are states which are not accessible by an adiabatic process."

Caratheodory used this principle together with a mathematical theorem that he developed to infer the existence of a state function S and an integrating factor $1/T$, where T is the thermodynamic temperature such that Eq. (15) holds for a reversible

$$dQ_{\mathrm{REV}} = T dS \tag{15}$$

change. The state function S is called the entropy. It can also be shown that the entropy in an adiabatic system increases for an irreversible change and remains constant for a reversible change as in Eq. (16). The implication is that entropy increases for a

$$\Delta S_{\mathrm{AD}} \geq 0 \tag{16}$$

natural change until equilibrium is reached, and then it remains constant at its maximum value.

The first part of the mathematical statement of the second law allows one to write one of the most important thermodynamic equations, Eq. (17). Al-

$$dU = T dS - p dV \tag{17}$$

though this equation was derived for reversible changes, it is valid for all changes. All the quantities are functions of state. Therefore, for a change between two states the integral of the equation will be valid even if the path is not reversible. In other words, for a change from a state characterized by (p_1, T_1) to a state characterized by (p_2, T_2), the values of ΔU, ΔV, and ΔS will all have definite values dependent only upon the two states and independent of how the change came about. From this equation are obtained some of the most fruitful applications of thermodynamics to physical problems.

The second part of the mathematical statement is a concise summary of physical statements on the direction of processes. As a simple example, consider pure heat transfer to a body. The heat transfer causes a definite change of state such that $dQ = dU$. The definite change of entropy is then given by Eq. (18). If heat dQ is transferred from a

$$dS = \frac{dQ}{T} \tag{18}$$

body at temperature T_2 to a body at temperature T_1, the change in entropy is given by Eq. (19).

$$\begin{aligned} dS &= dS_1 + dS_2 \\ &= \frac{dQ}{T_1} - \frac{dQ}{T_2} \\ &= \frac{dQ\,(T_2 - T_1)}{T_1 T_2} \end{aligned} \tag{19}$$

Since dS must be positive or zero, $T_2 > T_1$. Therefore heat flows from the hotter body to the colder body.

Also contained in the second part of the entropy statement is the key idea of equilibrium. The equilibrium state of an adiabatic or isolated system is characterized by entropy being at its maximum value consistent with the physical constraints. Therefore, equilibrium states can be determined by setting $dS = 0$. Also, for a maximum in entropy, $d^2S < 0$. This latter condition leads to the notion of stability that is important in the study of phase equilibrium.

When a system is in thermal contact with its surroundings, the entropy of the system may decrease. For example, a gas being compressed isothermally decreases its entropy, but a greater increase of entropy occurs in the surroundings. The total entropy change is always positive. Clausius stated the first and second laws of thermodynamics as: "The energy of the world is constant. The entropy of the world tends toward a maximum."

By way of completeness the third law of thermodynamics needs comments. In the main body of thermodynamics one is mostly interested in changes of entropy and internal energy between states. However, the third law defines an absolute scale for entropy: The entropy of all perfect crystalline solids is zero at absolute zero temperature. The third law is used primarily in classical thermodynamics for the calculation of absolute entropies which combined with thermochemical data permits the calculation of chemical equilibrium. The foundations of the third law, however, are to be found in molecular theory and require a statistical mechanical treatment.

Summary. By way of summary, for a closed system all of the fundamentals of thermodynamics are contained in notation (20).

$$\begin{aligned} &dU = q + w \\ &dS = \frac{dQ}{T} \qquad \text{reversible change} \\ &dS \geq 0 \qquad \text{for an isolated system} \\ &dU = TdS - pdV \end{aligned} \tag{20}$$

The equations are applicable when work is restricted to volume changes only. But the generalization to include changes of polarization, magnetization, surface area, and so forth is quite straightforward. Also, the equations are not applicable to systems that involve irreversible chemical changes, but here too the extension to include these situations presents no difficulty. In actual application it is convenient to define other state functions in terms of those already introduced, but no additional basic principles are needed. *See* ENTHALPY.

[WILLIAM F. JAEP]

Bibliography: J. Adkins, *Equilibrium Thermodynamics*, 2d ed., 1975; K. Denbigh, *Chemical Equilibrium*, 3d ed., 1971; R. W. Haywood, *Equilibrium Thermodynamics*, 1980; J. S. Hsieh, *Principles of Thermodynamics*, 1974; A. B. Pippard, *Elements of Classical Thermodynamics*, 1966; K. A. Rolle, *Introduction to Thermodynamics*, 2d ed., 1980; M. W. Zemansky and R. Dittman, *Heat and Thermodynamics*, 6th ed., 1981.

Thermodynamic processes

Changes of any property of an aggregation of matter and energy, accompanied by thermal effects. The participants in a process are first identified as a system to be studied; the boundaries of the system are established; the initial state of the system is determined; the path of the changing states is laid out; and, finally, supplementary data are stated to establish the thermodynamic process. These steps will be explained in the following paragraphs. At all times it must be remembered that the only processes which are allowed are those compatible with the first and second laws of thermodynamics: Energy is neither created nor destroyed and the entropy of the system plus its surroundings always increases.

A system and its boundaries. To evaluate the results of a process, it is necessary to know the participants that undergo the process, and their mass and energy. A region, or a system, is selected for study, and its contents determined. This region may have both mass and energy entering or leaving during a particular change of conditions, and these mass and energy transfers may result in changes both within the system and within the surroundings which envelop the system.

As the system undergoes a particular change of condition, such as a balloon collapsing due to the escape of gas or a liquid solution brought to a boil in a nuclear reactor, the transfers of mass and energy which occur can be evaluated at the boundaries of the arbitrarily defined system under analysis.

A question that immediately arises is whether a system such as a tank of compressed air should have boundaries which include or exclude the metal walls of the tank. The answer depends upon the aim of the analysis. If its aim is to establish a relationship among the physical properties of the gas, such as to determine how the pressure of the gas varies with the gas temperature at constant volume, then only the behavior of the gas is involved; the metal walls do not belong within the system. However, if the problem is to determine how much externally applied heat would be required to raise the temperature of the enclosed gas a given amount, then the specific heat of the metal walls, as well as that of the gas, must be considered, and the system boundaries should include the walls through which the heat flows to reach the gaseous contents. In the laboratory, regardless of where the system boundaries are taken, the walls will always play a role and must be reckoned with.

State of a system. To establish the exact path of a process, the initial state of the system must be determined, specifying the values of variables such as temperature, pressure, volume, and quantity of material. If a number of chemicals are present in the system, the number of variables needed is usually equal to the number of independently variable substances present plus two such as temperature and pressure; exceptions to this rule occur in variable electric or magnetic fields and in some other well-defined cases. Thus, the number of properties required to specify the state of a system depends upon the complexity of the system. Whenever a system changes from one state to another, a process occurs.

Whenever an unbalance occurs in an intensive property such as temperature, pressure, or density, either within the system or between the system and its surroundings, the force of the unbalance can initiate a process that causes a change of state. Examples are the unequal molecular concentration of different gases within a single rigid enclosure, a difference of temperature across the system boundary, a difference of pressure normal to a nonrigid system boundary, or a difference of electrical potential across an electrically conducting system boundary. The direction of the change of state caused by the unbalanced force is such as to reduce the unbalanced driving potential. Rates of changes of state tend to decelerate as this driving potential is decreased.

Equilibrium. The decelerating rate of change implies that all states move toward new conditions of equilibrium. When there are no longer any balanced forces acting within the boundaries of a system or between the system and its surroundings, then no mechanical changes can take place, and the system is said to be in mechanical equilibrium. A system in mechanical equilibrium, such as a mixture of hydrogen and oxygen, under certain conditions might undergo a chemical change. However, if there is no net change in the chemical constituents, then the mixture is said to be in chemical as well as in mechanical equilibrium.

If all parts of a system in chemical and mechanical equilibrium attain a uniform temperature and if, in addition, the system and its surroundings either are at the same temperature or are separated by a thermally nonconducting boundary, then the system has also reached a condition of thermal equilibrium.

Whenever a system is in mechanical, chemical, and thermal equilibrium, so that no mechanical, chemical, or thermal changes can occur, the system is in thermodynamic equilibrium. The state of equilibrium is at a point where the tendency of the system to minimize its energy is balanced by the tendency toward a condition of maximum randomness. In thermodynamics, the state of a system can be defined only when it is in equilibrium. The static state on a macroscopic level is nevertheless underlaid by rapid molecular changes; thermodynamic equilibrium is a condition where the forward and reverse rates of the various changes are all equal to one another. In general, those systems considered in thermodynamics can include not only mixtures of material substances but also mixtures of matter and all forms of energy. For example, one could consider the equilibrium between a gas of charged particles and electromagnetic radiation contained in an oven.

Process path. If under the influence of an unbalanced intensive factor the state of a system is altered, then the change of state of the system is described in terms of the end states or difference between the initial and final properties.

The path of a change of state is the locus of the whole series of states through which the system passes when going from an initial to a final state. For example, suppose a gas expands to twice its volume and that its initial and final temperatures are the same. Various paths connect these initial and final states: isothermal expansion, with temperature held constant at all times, or adiabatic expansion which results in cooling followed by heating back to the initial temperature while holding volume fixed.

Each of these paths can be altered by making the gas do varying amounts of work by pushing out a piston during the expansion, so that an extremely large number of paths can be followed even for such a simple example. The detailed path must be specified if the heat or work is to be a known quantity; however, changes in the thermodynamic properties depend only on the initial and final states and not upon the path.

There are several corollaries from the above descriptions of systems, boundaries, states, and processes. First, all thermodynamic properties are identical for identical states. Second, the change in a property between initial and final states is independent of path or processes. The third corollary is that a quantity whose change is fixed by the end states and is independent of the path is a point function or a property. However, it must be remembered that by the second law of thermodynamics not all states are available (possible final states) from a given initial state and not all conceivable paths are possible in going toward an available state.

Pressure-volume-temperature diagram. Whereas the state of a system is a point function, the change of state of a system, or a process, is a path function. Various processes or methods of change of a system from one state to another may be depicted graphically as a path on a plot using thermodynamic properties as coordinates.

The variable properties most frequently and conveniently measured are pressure, volume, and temperature. If any two of these are held fixed (independent variables), the third is determined (dependent variable). To depict the relationship among these physical properties of the particular working substance, these three variables may be used as the coordinates of a three-dimensional space. The resulting surface is a graphic presentation of the equation of state for this working substance, and all possible equilibrium states of the substance lie on this *P-V-T* surface. The *P-V-T* surface may be extensive enough to include all three phases of the working substance: solid, liquid, and vapor.

Because a *P-V-T* surface represents all equilibrium conditions of the working substance, any line on the surface represents a possible reversible process, or a succession of equilibrium states.

The portion of the *P-V-T* surface shown in Fig. 1 typifies most real substances; it is characterized by contraction of the substance on freezing. Going from the liquid surface to the liquid-solid surface onto the solid surface involves a decrease in both temperature and volume. Water is one of the few exceptions to this condition; it expands upon freezing, and its resultant *P-V-T* surface is somewhat modified where the solid and liquid phases abut.

Gibbs' phase rule is defined in Eq. (1). Here f is

$$f = c - p + 2 \tag{1}$$

the degree of freedom; this integer states the number of intensive properties (such as temperature, pressure, and mole fractions or chemical potentials of the components) which can be varied independently of each other and thereby fix the particular equilibrium state of the system (see discussion under Temperature-entropy diagram, below). Also, p indicates the number of phases (gas, liquid, or solid) and c the number of component substances in the system. Consider a one-component system (a pure substance) which is either in the liquid, gaseous, or solid phase. In equilibrium the system has two degrees of freedom; that is, two independent thermodynamic properties must be chosen to specify the state. Among the thermodynamic properties of a substance which can be quantitatively evaluated are the pressure, temperature, specific volume, internal energy, enthalpy, and entropy. From among these properties, any two may be selected. If these two prove to be independent of each other, when the values of these two properties are fixed, the state is determined and the values of all the other properties are also

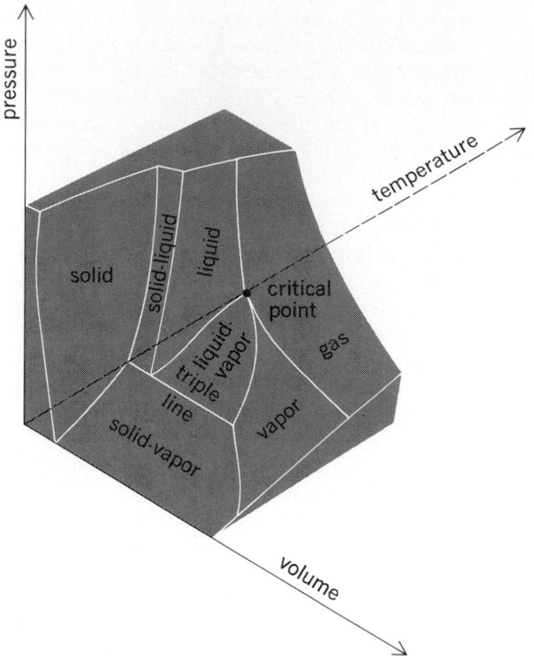

Fig. 1. Portion of pressure-volume-temperature (*P-V-T*) surface for a typical substance.

fixed. A one-component system with two phases in equilibrium (such as liquid in equilibrium with its vapor in a closed vessel) has $f=1$; that is, only one intensive property can be independently specified. Also, a one-component system with three phases in equilibrium has no degree of freedom. Examination of Fig. 1 shows that the three surfaces (solid-liquid, solid-vapor, and liquid-vapor) are generated by lines parallel to the volume axis. Moving the system along such lines (constant pressure and temperature) involves a heat exchange and a change in the relative proportion of the two phases. Note that there is an entropy increment associated with this change.

One can project the three-dimensional surface onto the *P-T* plane as in Fig. 2. The triple point is the point where the three phases are in equilibrium. When the temperature exceeds the critical temperature (at the critical point), only the gaseous phase is possible. The gas is called a vapor when it can coexist with another phase (at temperatures below the critical point). The *P-T* diagram for water would have the solid-liquid curve going upward from the triple point to the left (contrary to the ordinary substance pictured in Fig. 2). Then the property so well known to ice skaters would be evident. As the solid-liquid line is crossed from the low-pressure side to the high-pressure side, the water changes from solid to liquid: Ice melts upon application of pressure.

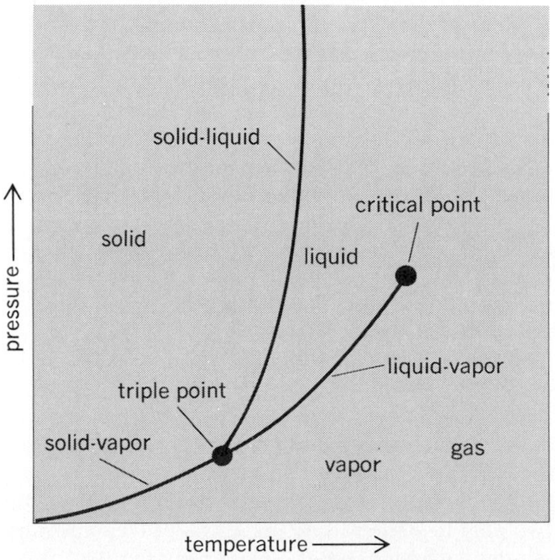

Fig. 2. Portion of equilibrium surface projected on pressure-temperature (*P-T*) plane.

Work of a process. The three-dimensional surface can also be projected onto the *P-V* plane to get Fig. 3. This plot has a special significance: The area under any reversible path on this plane represents the work done during the process. The fact that this *P-V* area represents useful work can be demonstrated by the following example.

Let a gas undergo an infinitesimal expansion in a cylinder equipped with a frictionless piston, and let this expansion perform useful work on the surroundings. The work done during this infinitesimal

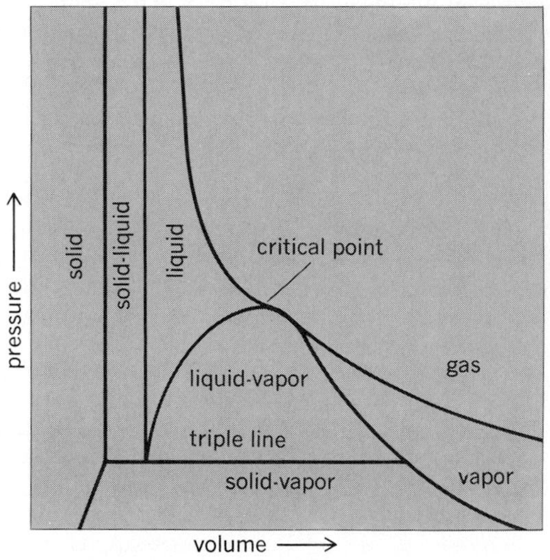

Fig. 3. Portion of equilibrium surface projected on pressure-volume (*P-V*) plane.

expansion is the force multiplied by the distance through which it acts, as in Eq. (2), wherein dW is

$$dW = F\,dl \qquad (2)$$

an infinitesimally small work quantity, F is the force, and dl is the infinitesimal distance through which F acts.

But force F is equal to the pressure P of the fluid times the area A of the piston, or PA. However, the product of the area of the piston times the infinitesimal displacement is really the infinitesimal volume swept by the piston, or $A\,dl = dV$, with dV equal to an infinitesimal volume. Thus Eq. (3) is valid. The work term is found by integration, as in Eq. (4).

$$dW = P\,A\,dl = P\,dV \qquad (3)$$

$$_1W_2 = \int_1^2 P\,dV \qquad (4)$$

Figure 4 shows that the integral represents the

area under the path described by the expansion from state 1 to state 2 on the P-V plane. Thus, the area on the P-V plane represents work done during this expansion process.

Temperature-entropy diagram. Energy quantities may be depicted as the product of two factors: an intensive property and an extensive one. Examples of intensive properties are pressure, temperature, and magnetic field; extensive ones are volume, magnetization, and mass. Thus, in differential form, work has been presented as the product of a pressure exerted against an area which sweeps through an infinitesimal volume, as in Eq. (5). Note that as a gas expands, it is doing work on

$$dW = P\,dV \tag{5}$$

its environment. However, a number of different kinds of work are known. For example, one could have work of polarization of a dielectric, of magnetization, of stretching a wire, or of making new surface area. In all cases, the infinitesimal work is given by Eq. (6), where X is a generalized applied

$$dW = X dx \tag{6}$$

force which is an intensive quantity such as voltage, magnetic field, or surface tension; and dx is a generalized displacement of the system and is thus extensive. Examples of dx include changes in electric polarization, magnetization, length of a stretched wire, or surface area.

By extending this approach, one can depict transferred heat as the product of an intensive property, temperature, and a distributed or extensive property defined as entropy, for which the symbol is S. *See* ENTROPY.

If an infinitesimal quantity of heat dQ is transferred during a reversible process, this process may be expressed mathematically as in Eq. (7),

$$dQ = T\,dS \tag{7}$$

with T being the absolute temperature and dS the infinitesimal entropy quantity.

Furthermore, a plot of the change of state of the system undergoing this reversible heat transfer can be drawn on a plane in which the coordinates are absolute temperature and entropy (Fig. 5). The total heat transferred during this process equals the area between this plotted line and the horizontal axis.

Reversible processes. Not all energy contained in or associated with a mass can be converted into useful work. Under ideal conditions only a fraction of the total energy present can be converted into work. The ideal conversions which retain the maximum available useful energy are reversible processes.

Characteristics of a reversible process are that

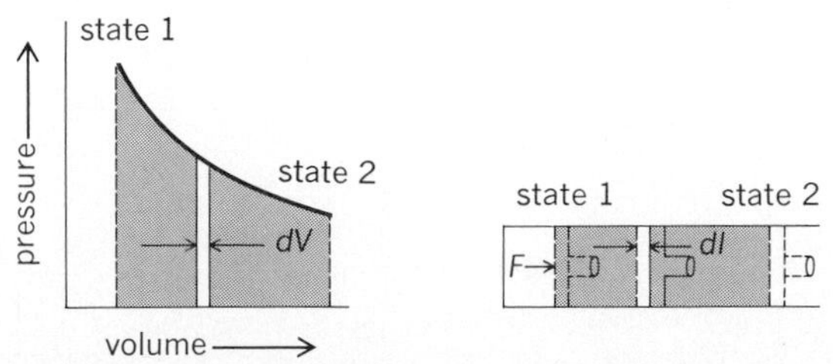

Fig. 4. Area under path in P-V plane is work done by expanding gas against piston.

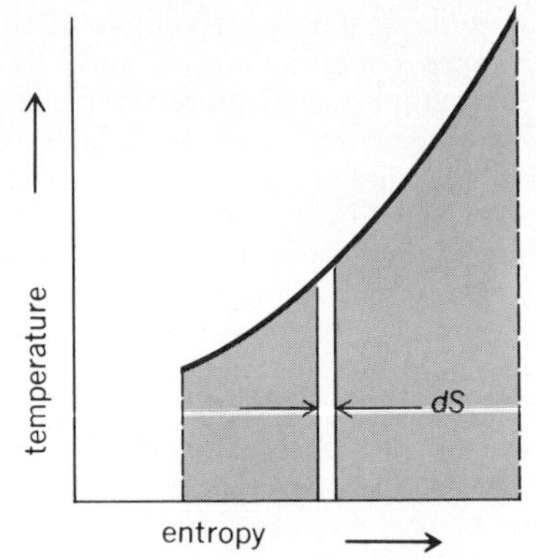

Fig. 5. Heat transferred during a reversible process is area under path in temperature-entropy (T-S) plane.

the working substance is always in thermodynamic equilibrium and the process involves no dissipative effects such as viscosity, friction, inelasticity, electrical resistance, or magnetic hysteresis. Thus, reversible processes proceed quasistatically so that the system passes through a series of states of thermodynamic equilibrium, both internally and with its surroundings. This series of states may be traversed just as well in one direction as in the other.

If there are no dissipative effects, all useful work done by the system during a process in one direction can be returned to the system during the reverse process. When such a process is reversed so that the system returns to its starting state, it must leave an effect on the surroundings since, by the second law of thermodynamics, in energy conversion processes the form of energy is always degraded. Part of the energy of the system (including heat source) is transferred as heat from a higher temperature to a lower temperature. The energy rejected to a lower-temperature heat sink cannot be recovered. To return the system (including heat source and sink) to its original state, then, requires more energy than the useful work done by the system during a process in one direction. Of course, if the process were purely a mechanical one with no thermal effects, then both the surroundings and system could be returned to their initial states.

It is impossible to satisfy the conditions of a quasistatic process with no dissipative effects; a reversible process is an ideal abstraction which is not realizable in practice but is useful for theoretical calculations. An ideal reversible engine operating between hotter and cooler bodies at the temperatures T_1 and T_2, respectively, can put out $(T_1 - T_2)/T_1$ of the transferred heat energy as useful work.

There are four reversible processes wherein one of the common thermodynamic parameters is kept constant. The general reversible process for a closed or nonflow system is described as a polytropic process.

Irreversible processes. Actual changes of a system deviate from the idealized situation of a quasistatic process devoid of dissipative effects. The extent of the deviation from ideality is correspondingly the extent of the irreversibility of the process.

Real expansions take place in finite time, not infinitely slowly, and these expansions occur with friction of rubbing parts, turbulence of the fluid, pressure waves sweeping across and rebounding through the cylinder, and finite temperature gradients driving the transferred heat. These dissipative effects, the kind of effects that make a pendulum or yo-yo slow down and stop, also make the work output of actual irreversible expansions less than the maximum ideal work of a corresponding reversible process. For a reversible process, as stated earlier, the entropy change is given by $dS = dQ/T$. For an irreversible process even more entropy is produced (turbulence and loss of information) and there is the inequality $dS > dQ/T$.

[PHILIP E. BLOOMFIELD; WILLIAM A. STEELE]

Bibliography: H. A. Bent, *The Second Law*, 1965; J. P. Holman, *Thermodynamics*, 3d ed., 1980; M. Mott-Smith, *The Concept of Energy Simply Explained*, 1934; W. C. Reynolds and H. C. Perkins, *Engineering Thermodynamics*, 2d ed., 1977; F. W. Sears and G. L. Salinger, *Thermodynamics, the Kinetic Theory of Gases and Statistical Mechanics*, 3d ed., 1975; K. Wark, *Thermodynamics*, 3d ed., 1977.

Thiazole

One of a class of organic heterocyclic compounds (sometimes specified as 1,3-thiazoles) in which a five-membered diunsaturated ring contains one atom of nitrogen and, in a nonadjacent position, one atom of sulfur. The preferred numbering is shown in formulation (I); however, since other numbering systems have been employed, caution should be exercised in translating names to structures. Dihydrothiazoles are called thiazolines, of which the most familiar are the Δ^2-thiazolines (II). Tetrahydrothiazoles are called thiazolidines (III). Thiazole fused to a benzene ring is benzothiazole (IV). Important thiazole derivatives include vita-

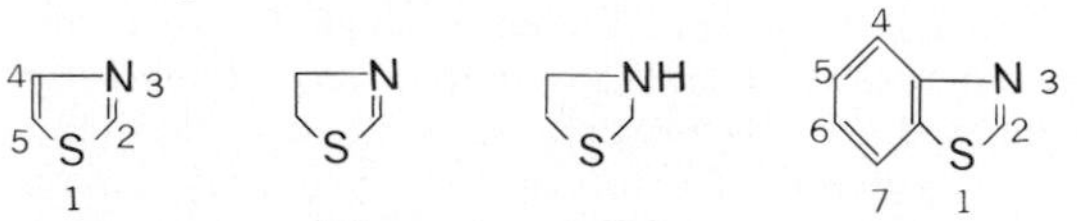

(I) (II) (III) (IV)

min B_1 (thiamine) and sulfathiazole. Several valuable dyes and rubber-vulcanizing accelerators contain the benzothiazole nucleus. Penicillin is a thiazolidine derivative. *See* AZOLE; HETEROCYCLIC COMPOUNDS.

Properties. The thiazole ring is a resonance-stabilized aromatic system. The ring is relatively resistant to hydrolysis and to disruptive oxidation with nitric acid. Bromine-water and permanganate, however, do oxidize the ring. Thiazole can undergo electrophilic substitution such as nitration and sulfonation at the 4 and 5 positions. Further, in line with the aromatic character of thiazole, aminothiazoles form diazonium salts with nitrous acid. *See* AROMATIC HYDROCARBON.

The parent compound, thiazole (I), bp 117°C, is a colorless, water-soluble liquid with an odor resembling that of pyridine. Although thiazole is a weak base (pK_a 2.53), acids form simple salts, and alkylating agents form quaternary salts. Quaternary thiazolium salts, although stable in acid, are decomposed in alkali, and ring rupture occurs.

Amino groups at position 2 can be diazotized. Subsequent coupling to give azo compounds, and replacement of the diazonium group by hydrogen, halogen, hydroxyl, or nitro in standard procedures, are possible. The SH or mercapto group at position 2 can be replaced with hydrogen either by treatment with acidic hydrogen peroxide, or by desulfurization with Raney nickel.

The chemistry of the thiazole 2 position—but in contrast not that of the 4 position—is similar to the chemistry of the pyridine 2 position. Thus, the 2-halothiazoles are convertible with relative ease to 2-hydroxy, 2-alkoxy, 2-amino, 2-mercapto and, by reduction, to 2-hydro derivatives. Thiazole-2-carboxylic acid decarboxylates with relative ease. Further, 2-methylthiazole metalates at the methyl group without difficulty, and also condenses with benzaldehyde.

Preparation. The most versatile general synthesis of thiazoles condenses α-halo carbonyl compounds (V) with thioamides (VI). By suitable

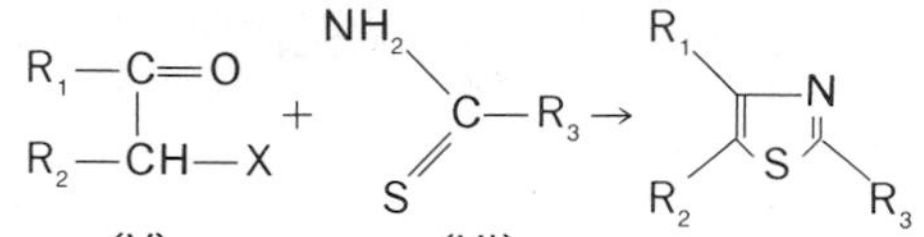

(V) (VI)

choice of the R groups, many kinds of thiazoles can be prepared. The thioamides (VI) used most frequently are thioformamide ($R_3 = H$), thiourea ($R_3 = NH_2$), and dithiocarbamate ion ($R_3 = SH$).

Sulfathiazole (VIII) is one of the useful bacteriostatic sulfa drugs. Reaction of *p*-acetamidobenzenesulfonyl chloride (VII) with 2-aminothiazole

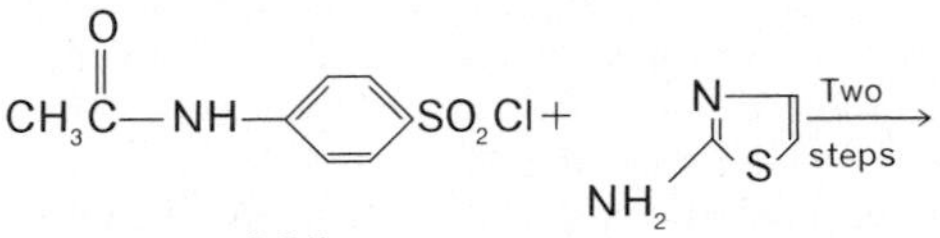

(VII)

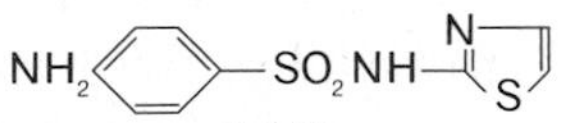

(VIII)

gives the *N*-acetylated sulfathiazole, which on alkaline hydrolysis forms sulfathiazole.

Thiamine includes as part of its structure the thiazole derivative (X). A practical synthesis of this fragment proceeds by condensing the α-chlorinated derivative (IX) of 5-hydroxy-2-pentanone with thioformamide. The thiazole (X), combined with the proper pyrimidine moiety, gives thiamine.

Δ^2-Thiazolines, for example 2-methyl-Δ^2-thiazoline (XIII), are prepared by combining a β-bromoamine (XI) with a thioamide (XII). Thiazolidines (XV) form when aldehydes or ketones react with β-amino mercaptans (XIV). In the laboratory syn-

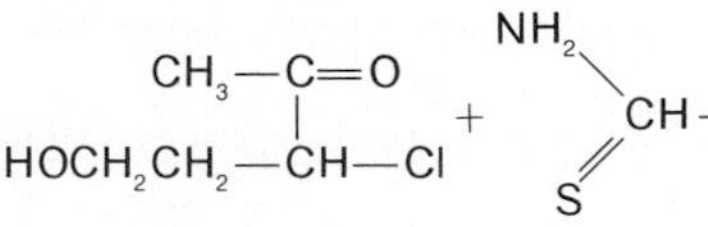

(IX)

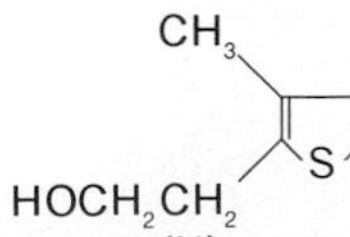

(X)

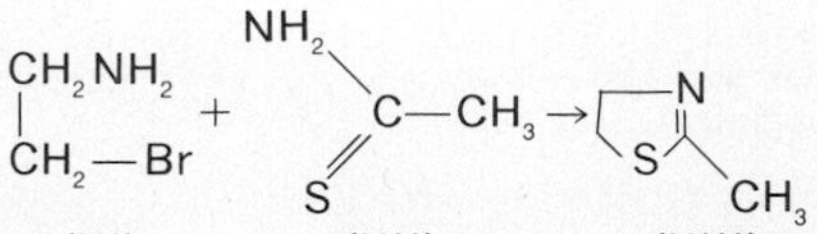

(XI) (XII) (XIII)

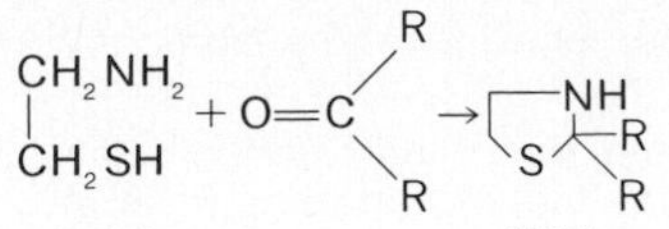

(XIV) (XV)

thesis of penicillin, the thiazolidine ring (XVI) is formed by this process. Rhodanine, or 4-ketothia-

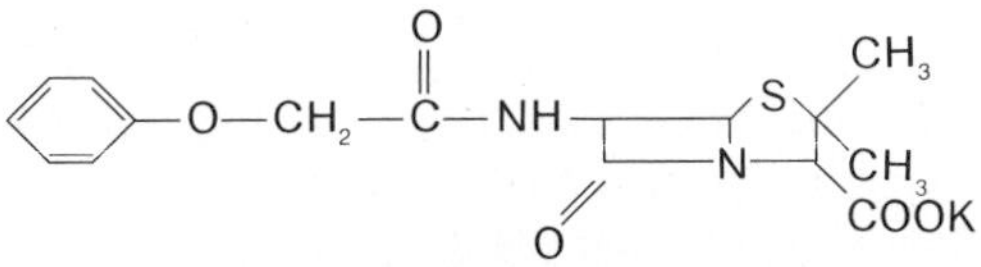

Potassium salt of penicillin V
(XVI)

zolidin-2-thione (XVII), is prepared by the reaction of chloroacetic acid with dithiocarbamate. Condensation of rhodanine with carbonyl compounds gives 5-methylene derivatives (XVIII),

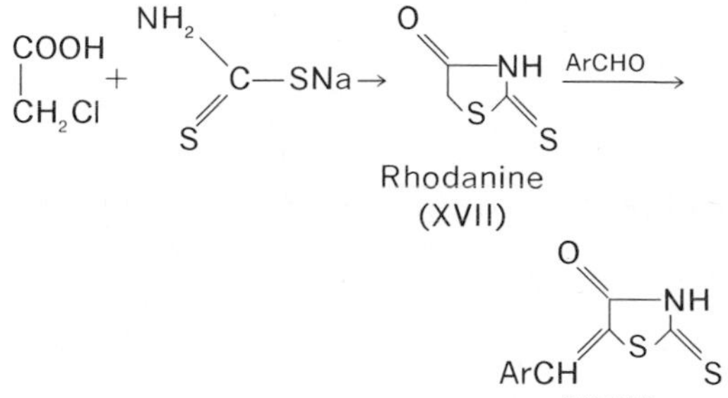

Rhodanine
(XVII)

(XVIII)

which can be converted to a variety of useful products.

Benzothiazole syntheses start with *o*-aminothiophenol (XIX), which, with carboxylic acids, gives 2-substituted benzothiazoles (XX), and with carbon disulfide, gives 2-mercaptobenzothiazole (XXI).

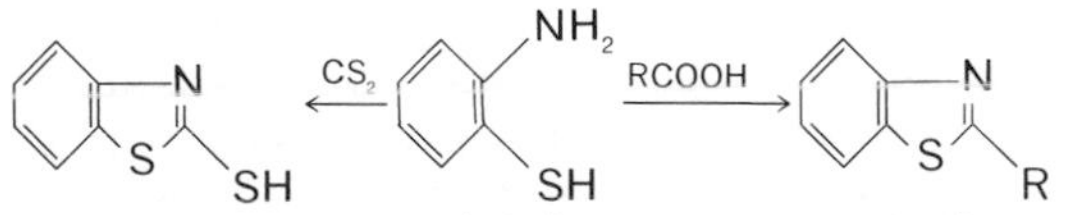

(XXI) (XIX) (XX)

Cyanine dyes, for example (XXII), are formed from the reaction of an aldehyde such as *p*-dimethylaminobenzaldehyde with quaternized 2-methylbenzothiazole. Other benzothiazole dye materials are obtained when *p*-toluidine is fused with sulfur. A mixture of dehydrothiotoluidine (XXIII) and "Primuline dye bases" (XXIV) is formed. Sulfonated

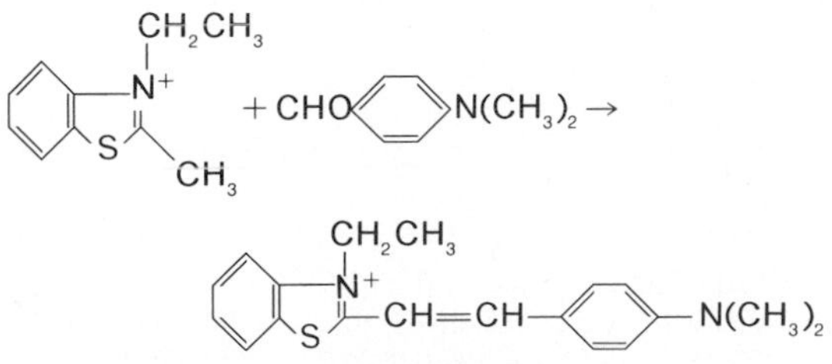

(XXII)

"Primuline dye bases" (XXIV) is a commercial yellow dye which is used after diazotization by application to cotton. Other types of dyes have also been derived from benzothiazoles (XXIII) and (XXIV).

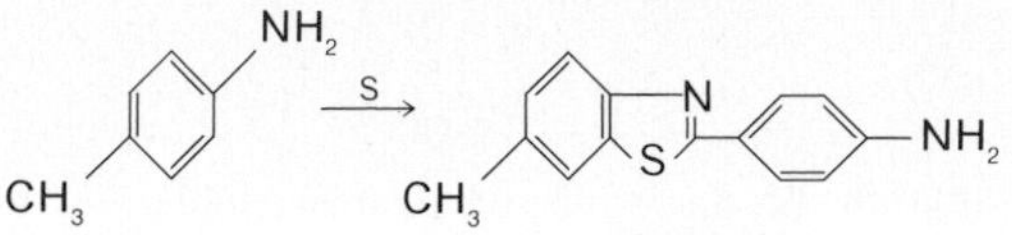

(XXIII)

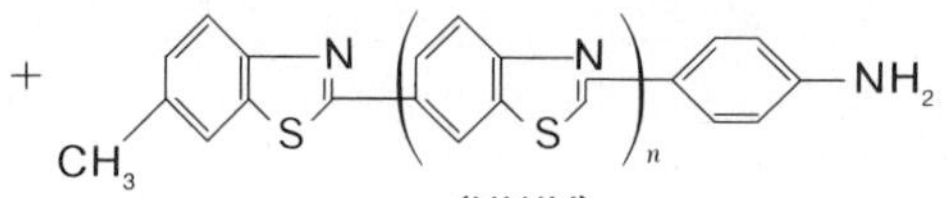

(XXIV)

A commercial preparation of Captax or 2-mercaptobenzothiazole (XXI), a material of considerable value as an accelerator in the vulcanization of rubber, proceeds by heating aniline, carbon disulfide, and sulfur to 200–300°. Benzothiazole-2-sulfenamides (XXV), which are derived from

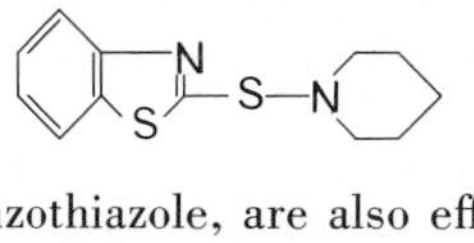

(XXV)

2-mercaptobenzothiazole, are also effective accelerators. *See* DYE; ORGANOSULFUR COMPOUND.

[WALTER J. GENSLER]

Bibliography: R. M. Acheson, *An Introduction to the Chemistry of Heterocyclic Compounds*, 3d ed., 1976; G. M. Badger, *The Chemistry of Heterocyclic Compounds*, 1961; H. T. Clarke, (ed.), *The Chemistry of Penicillin*, 1949; J. Metzger, *Thiazole and Its Derivatives*, 1979.

Thio compounds

Organosulfur compounds in which one or more sulfur atoms replace oxygen. Because of the frequent use of this nomenclature and the large number of structural variations involved, the thio names may appear confusing but are generally understandable in specific cases. Thiols and thiophenols are the sulfur analogs of alcohols and phenols. Dithio and trithio compounds have two or three sulfur atoms, corresponding to the oxygen analogs. Acyl-SH compounds are the thio acids, for example, thioacetic acid, $CH_3C(=O)SH$, of which many are known, and which can be made (1) by acylating H_2S, or its salts, (2) via $RC(=O)OH$ and P_2S_5, or (3) from RMgX and COS (carbon oxysulfide). Dithio acids, $R—C(=S)—SH$, may, for example, be made from RMgX and carbon disulfide, CS_2 (analogous to the reaction of Grignard reagents with CO_2). Trithiocarbonates are derivatives of $S=C(SH)_2$, the sulfur analog of carbonic acid. Dithiocarbamates, related to carbamic acids, are readily obtained in the form of salts (the free acids are unstable) by the reaction, $2R_2NH + CS_2 \rightarrow R_2NC(=S)S^-(H_2N^+R_2)$. Carbon disulfide is important technically for such reactions and also for reactions with alcohols and alkali to give xanthates, as shown in reaction (1). For example, eth-

$$ROH + CS_2 + KOH \rightarrow RO—C(=S)S^-K^+ + H_2O \quad (1)$$

anol yields potassium ethyl xanthate. The synthesis of thiocarbonates often starts from phosgene, $O=CCl_2$, in which the reactive chlorine atoms can be replaced with sulfur groups, by reactions with RSH, RC(O)SH, and HSH.

Many other compounds which carry the thio name are oxygen analogs. For example, thiocyanates, $RSC\equiv N$; isothiocyanates, $R—N=C=S$; thiourea, $(NH_2)_2C=S$; thioamides, $RC(=S)NH_2$; dithioesters, $RC(=S)SR'$; and thio acid chlorides,

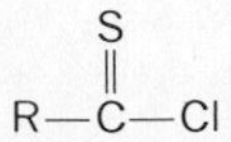

Oxidation of the thiocarbamates yields corresponding disulfides, known as thiuram disulfides,

$$R_2N—C(=S)—S—S—C(=S)NR_2$$

some of which are highly effective vulcanization accelerators.

Thiocyanogen $(SCN)_2$ and chlorothiocyanogen (Cl—SCN) are also of interest. The former has properties similar to those of the halogens and hence is useful as a thiocyanating agent, as in the preparation of sulfenyl thiocyanates, reaction (2).

$$RSH + (SCN)_2 \rightarrow RS—SCN + HSCN \qquad (2)$$

Thiobarbiturates are barbituric acid derivatives in which, generally, an RS group has replaced an alkyl or aryl radical in the 2 position of barbituric acid.

The thio acids, dithio acids, thiocarbamates, dithiocarbamates, thiocarbonates, trithiocarbonates, polythiols, thiocyanates, and many related compounds have a broad and useful set of chemical properties which have been reasonably well developed. Many of these compounds have found extensive industrial applications. *See* MERCAPTAN; ORGANOSULFUR COMPOUND.

[NORMAN KHARASCH]

Thioaldehyde and thioketone

Organosulfur compounds of structure

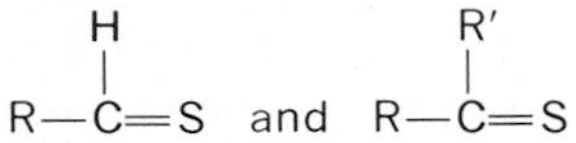

They are isolable only in rare instances because, unlike the oxygen analogs (aldehydes and ketones), thioaldehydes and thioketones tend strongly to polymerize, for example,

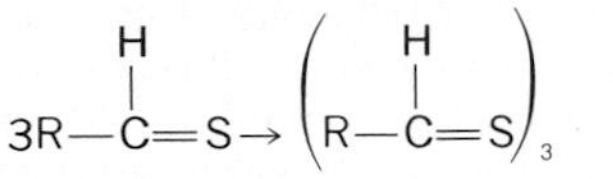

The trimer is a trithiane derivative, with alternate C and S atoms in a six-membered ring, each carbon of which bears R and H groups. The monomeric thiocarbonyl compounds are highly colored. The thiocarbonyl (—C═S) group also is found in thiourea and in heterocyclic thiones, but here tautomeric structures in which —C═S becomes ═C—SH are generally involved. Hence, they are not true thiocarbonyl compounds. *See* ALDEHYDE; KETONE; ORGANOSULFUR COMPOUND.

[NORMAN KHARASCH]

Thiocyanate

One of a group of compounds, both organic and inorganic, which contain the —SCN group and are derived from thiocyanic acid, HSCN. Like cyanic acid, thiocyanic acid may exist in two forms, H—S—C≡N and S═C═N—H. The latter form is called isothiocyanic acid and gives rise to isothiocyanates, which are well characterized as a class of organic compounds.

The inorganic thiocyanates resemble the cyanides and halides because most of the metal salts are water-soluble (except lead, mercury, silver, and copper salts), and many complexes are formed with excess thiocyanate, for example, $[Pt(SCN)_4]^{2-}$ and $[Pt(SCN)_6]^{2-}$.

Potassium thiocyanate can be used to titrate Ag^+ as in the Volhard titration, reaction (1).

$$Ag^+ + SCN^- \rightarrow AgSCN \quad \text{(white precipitate)} \qquad (1)$$

An excess of reagent is detected by the formation of a red complex of iron, reaction (2). The reac-

$$Fe^{3+} + SCN^- \rightarrow Fe(SCN)^{2+} \qquad (2)$$

tion in (2) is used as a very sensitive test for both CNS^- and Fe^{3+} ions.

Sodium thiocyanate is prepared by heating a mixture of sodium cyanide and sulfur. *See* CYANIDE; SULFUR.

[E. EUGENE WEAVER]

Thioether

One of a group of organosulfur compounds that are also called sulfides, RSR′. The simplest, dimethyl sulfide, $CH_3—S—CH_3$, is obtained in large amounts from sulfite waste liquors of wood treatment and is the precursor of dimethyl sulfoxide, a useful solvent and chemical reactant. Some amino acids, such as methionine and lanthionine, are also sulfides. Mustard gas, $ClCH_2CH_2SCH_2CH_2Cl$, formed by the reaction of sulfur dichloride, SCl_2, and ethylene, is a well-known vesicant. The thioethers bear a formal resemblance to the oxygen ethers, and may be synthesized by analogous methods, for example, via alkyl halides and sodium mercaptides, as in the reaction below.

$$RCl + R'SNa \rightarrow RSR' + NaCl$$

Numerous practical uses of sulfides have been claimed, especially as fuel-oil additives, lubricant additives, and agricultural chemicals. *See* ETHER; MERCAPTAN; ORGANOSULFUR COMPOUND.

[NORMAN KHARASCH]

Thiophene

An organic heterocyclic compound containing a diunsaturated ring of four carbon atoms and one sulfur atom. *See* HETEROCYCLIC COMPOUNDS.

Thiophene (I), methylthiophenes, and other alkylthiophenes are found in relatively small amounts in coal tar and petroleum. Thiophene accompanies benzene in the fractional distillation of coal tar. Purification of coal-tar benzene is effected by treatment with concentrated sulfuric acid, which selectively forms water-soluble thiophene-sulfonic acid. Alternately, treatment with aluminum chloride selectively polymerizes the thiophene in the benzene to nonvolatile materials. 2,5-dithienylthiophene (II) has been found in the

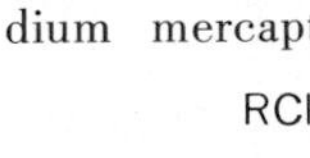

(I)

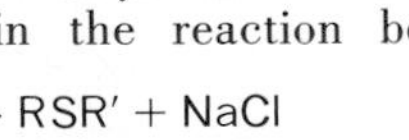

(II)

marigold plant. Biotin, a water-soluble vitamin, is a tetrahydrothiophene derivative.

Properties. The parent compound (I) is nearly insoluble in water (forming 0.02–0.04% solutions at 20°), mp −38.2°C, bp 84.2°C, n_D^{20} 1.5287, and specific gravity (20/4) 1.0644. Thiophene has a resonance energy of 29–31 kcal/mole (121–130 kJ/mole) is stable to heat, and undergoes electrophilic substitutions (nitration, sulfonation, acetylation, halogenation, chloromethylation, and mercuration). Accordingly, thiophene is an aromatic compound. Generally, electrophilic substitutions occur with greater ease than with benzene, but less readily than with furan or pyrrole. The entering group favors the α-position. Thiophenes are stable to alkali and other nucleophilic agents, and are relatively resistant to disruption by acid. *See* AROMATIC HYDROCARBON.

Most oxidative processes (nitric acid, ozone, hydrogen peroxide) involving the nucleus have not proved useful in opening the thiophene ring. Peracetic or perbenzoic acid oxidizes thiophenes such as 3,4-dimethylthiophene (III) to the corresponding sulfones (IV), which behave more as butadiene derivatives than as thiophenes [reaction (1)]. Sodi-

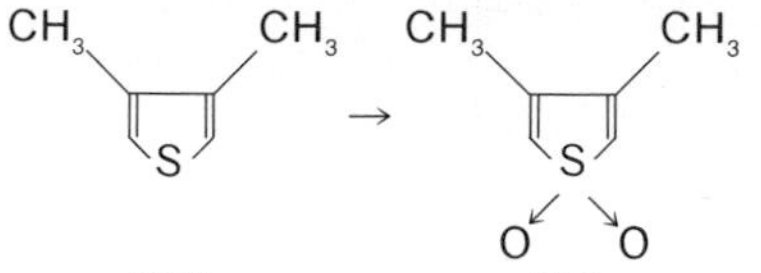

(III) (IV) (1)

um in liquid ammonia and methanol converts thiophene to a mixture of dihydro and acyclic products. Raney nickel strips sulfur from thiophenes in a ring-opening reaction (2), converting (V) to (VI).

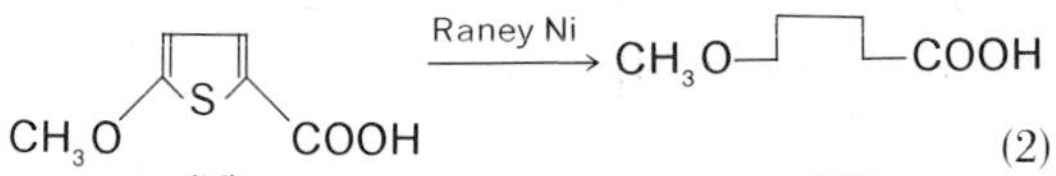

(V) (VI) (2)

Catalytic hydrogenation over molybdenum or cobalt sulfide catalysts at high temperature and pressure, as well as over platinum or palladium catalysts in massive amounts, saturates the ring.

Bromine and chlorine react readily with thiophenes, which undergo both substitution and addition reactions. Control of conditions as well as the possibility of dehydrohalogenation by alkali of the products first formed furnishes halogenated thiophenes in practical preparations. Iodination of thiophene in the presence of mercuric oxide or iodination of mercurated thiophenes gives iodinated derivatives.

Thiophene undergoes the Diels-Alder reaction with the more active dienophiles, such as acetylenedicarboxylic ester, to form benzene derivatives by extrusion of sulfur [reaction (3)].

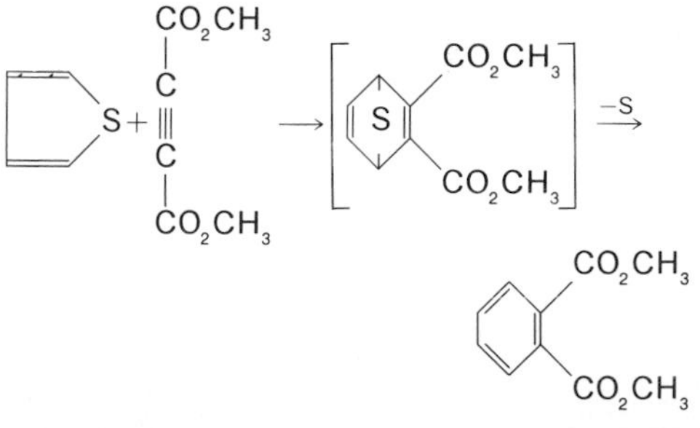

(3)

Preparation. The thiophene ring system is formed by cyclization of 1,4-dicarbonyl compounds in the presence of phosphorus sulfides (for example, 2,5-hexadione gives 2,5-dimethylthiophene; 4-oxo-3-ethylpentanoic acid gives 2-methyl-3-ethylthiophene), or by cyclization of hydrocarbons with sulfur or sulfur compounds at elevated temperatures (for example, the reaction of 2-methylbutadiene with sulfur at 320–420° gives 3-methylthiophene; the reaction of ethylbenzene with sulfur in a bimolecular process gives 2,4-diphenylthiophene). The commercial production of thiophene (I) from readily available butane or butadiene awaits only a large-scale demand. A laboratory synthesis converts sodium succinate to thiophene by heating with phosphorus sulfide.

Alkylthiophenes are prepared by ring synthesis, by alkylation of thienylmagnesium halides with sulfate or sulfonate esters, or by reduction of thiophene ketones. 2-Vinylthiophene, potentially of interest as a polymerizable monomer, can be prepared by reducing 2-acetylthiophene to methyl-2-thienylcarbinol, and dehydrating.

Thiophene aldehydes are prepared by treatment of the thiophene with hexamethylenetetramine (Sommelet process), or with the *N*-methylformanilide–phosphorus oxychloride reagent pair. Friedel-Crafts acylation, often with mild catalysts, gives thiophene ketones in good yields. Thiophene carboxylic acids result from the silver oxide oxidation of thiophene aldehydes, the haloform oxidation of acetylthiophene, and the carbonation of thiophenemetal derivatives. Thiophene aldehydes, ketones, and acids show normal chemical behavior, similar to the corresponding benzene derivatives. *See* ORGANOSULFUR COMPOUND; THIAZOLE.

[WALTER J. GENSLER; MARTIN STILES]

Bibliography: R. Adams et al. (eds.), *Organic Reactions*, vol. 6, 1951; S. Gronowitz, *Recent Advances in the Chemistry of Thiophenes*, in A. R. Katritzky (ed.), *Advances in Heterocyclic Chemistry*, vol. 1, 1963; H. D. Hartough, *Thiophene and Its Derivatives*, 1952; O. Meth-Cohn, Thiophenes, in D. Barton and W. D. Ollis, *Comprehensive Organic Chemistry*, vol. 4, 1979.

Thiosulfate

A negative ion having the formula $S_2O_3^{2-}$, which is derived from thiosulfuric acid, $H_2S_2O_3$, an unstable acid. The usual qualitative test for the thiosulfate ion is to add acid to the substance and watch for the white colloidal sulfur and the evolution of sulfur dioxide when the unstable thiosulfuric acid decomposes. This reaction is shown in (1).

$$2H^+ + S_2O_3^{2-} \rightarrow [H_2S_2O_3] \rightarrow S + H_2O + SO_2 \quad (1)$$

Because sodium thiosulfate is formed by heating sodium sulfite and sulfur, this gives a clue to the fact that the two sulfur atoms in the ion have different roles. The ion is actually related to the sulfate ion, with the substitution of a sulfur atom for an oxygen, as shown in the formula below. The

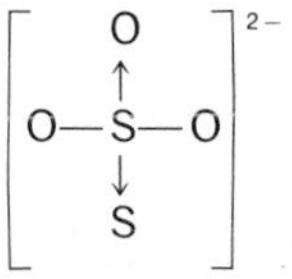

central sulfur atom has an oxidation number of 6+, whereas the second has an oxidation number of 2−.

Sodium thiosulfate, also termed hypo, is used in photography to "fix" films by dissolving the unreacted silver halide. Sodium thiosulfate reacts quantitatively with iodine solutions, as in reaction (2). This reaction can be used in volumetric anal-

$$2S_2O_3^{2-} + I_2 \rightarrow 2I^- + S_4O_6^{2-} \quad (2)$$

ysis. *See* OXIDATION-REDUCTION; SULFUR.

[E. EUGENE WEAVER]

Thiourea

A crystalline, colorless solid (prisms or needles) having formula (I) and melting point 180–182°C.

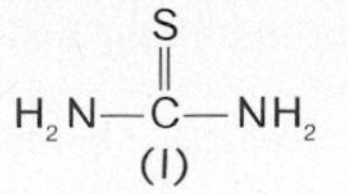

$$\underset{(I)}{H_2N-\overset{\displaystyle S}{\overset{\|}{C}}-NH_2}$$

Thiourea is relatively insoluble in water (1 part in 11) and only slightly soluble in ether. The three most common methods of preparation are (1) heating ammonium thiocyanate, NH_4CNS, to about 180°C, at which point equilibrium with thiourea is established; (2) the action of hydrogen sulfide at about 180°C on calcium cyanamide; and (3) the reaction of dicyandiamide and ammonium sulfide at 60–70°C, reaction (1).

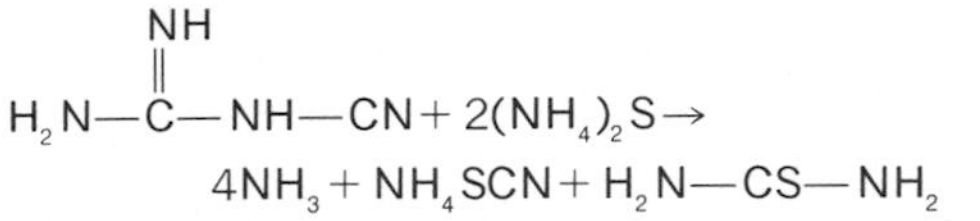

$$H_2N-\overset{\overset{\displaystyle NH}{\|}}{C}-NH-CN+2(NH_4)_2S \rightarrow$$
$$4NH_3 + NH_4SCN + H_2N-CS-NH_2 \quad (1)$$

Thiourea is the sulfur analog of urea, and like urea forms addition compounds with hydrocarbons (branched-chain, alicyclic, and straight-chain of more than 14 atoms). It has been used to protect clothes and furs from insects. *See* UREA.

Chemically, thiourea exists in tautomeric forms, reaction (2), and it often reacts as a tautomeric

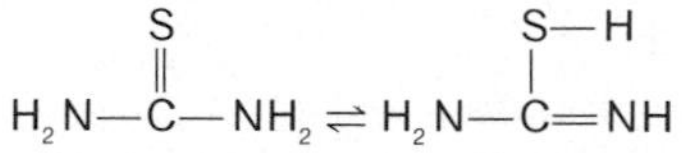

$$H_2N-\overset{\overset{\displaystyle S}{\|}}{C}-NH_2 \rightleftharpoons H_2N-\overset{\overset{\displaystyle S-H}{|}}{C}=NH \quad (2)$$

form, which is known as pseudothiourea. Thus, with an alkyl halide, such as ethyl bromide, ethyl thiouronium bromide (II) is formed. Thiouronium

$$H_2N-\overset{\overset{\displaystyle S-C_2H_5}{|}}{\underset{(II)}{C}}=\overset{\oplus}{N}H_2\overset{\ominus}{Br}$$

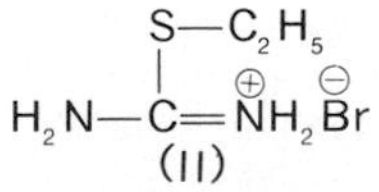

salts are used in the preparation of mercaptans and of alkyl sulfonyl chlorides. Benzyl thiouronium chloride (III) is a valuable reagent for the identifica-

$$H_2N-\overset{\overset{\displaystyle S-CH_2-C_6H_5}{|}}{\underset{(III)}{C}}=\overset{\oplus}{N}H_2\overset{\ominus}{Cl}$$

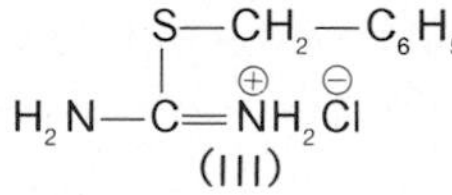

tion of organic acids, because, with the sodium salt of an acid, a crystalline thiouronium salt results.

Condensation of thiourea with substituted malonic esters gives thiobarbituric acid derivatives. Of the thiobarbiturates, with 1-methylbutyl-ethylthiobarbiturate as the sodium salt, sodium pentothal (IV) is much used in anesthetic premed-

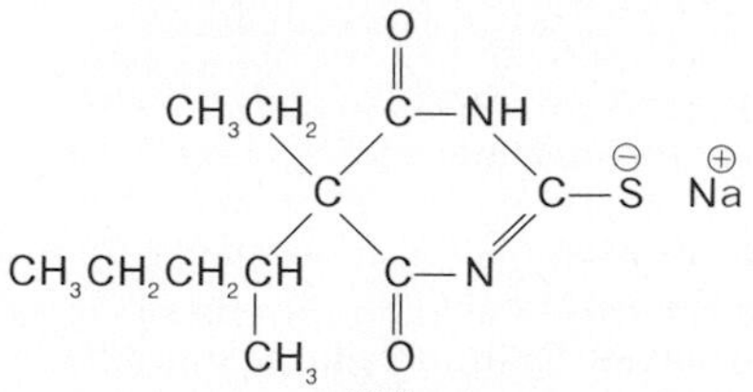

(IV)

ication to reduce the amount of general anesthetic needed. *See* CARBOXYLIC ACID; MERCAPTAN; UREID.

[EVANS B. REID]

Thorium

A chemical element, Th, atomic number 90. It was discovered by J. J. Berzelius in 1828. However, little use was found for thorium before the development of the incandescent gas mantle by C. A. von Welsbach in 1885. Several thousand

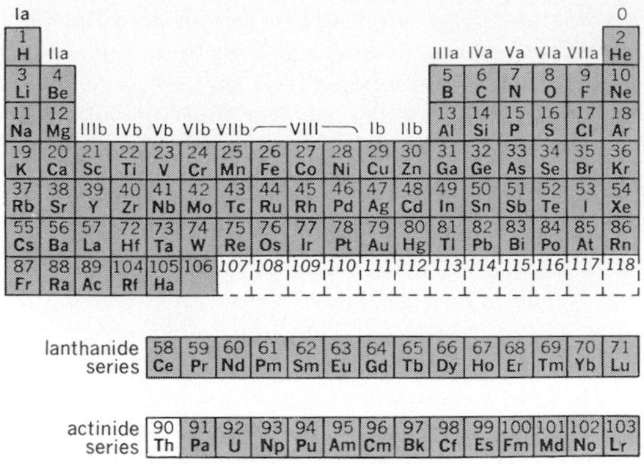

pounds of thorium oxide still go into the annual production of these mantles. For many years thorium oxide has been incorporated in tungsten metal, which is used for electric light filaments; recently, small amounts of the oxide have been found to be useful in other metals and alloys. The oxide is employed in catalysts for the promotion of certain organic chemical reactions. Thorium oxide has special uses as a high-temperature ceramic material. The metal or its oxide is employed in some electronic tubes, photocells, and special welding electrodes. The metal can serve as a getter in vacuum systems and in gas purification, and it is also used as a scavenger in some metals.

Because of its high density, chemical reactivity, mediocre mechanical properties, and relatively high cost, thorium metal has no market value as a structural material. However, many alloys containing thorium metal have been studied in some detail and thorium does have important applications as an alloying agent in some structural metals. Perhaps the major use for thorium metal, outside the nuclear field, is in magnesium technology. Approximately 3% thorium, added as an alloying ingredient, imparts to magnesium metal high-strength properties and creep resistance at elevated temperatures. The magnesium alloys containing thorium, because of their light weight and desirable strength properties, are being used in aircraft engines and in airframe construction.

Thorium can be converted in a nuclear reactor to uranium-233, an atomic fuel. The system of thorium and uranium-233 gives promise of complete utilization of all thorium in the production of atomic power. The energy available from the world's supply of thorium has been estimated as greater than the energy available from all of the world's uranium, coal, and oil combined.

Metallurgical extraction. Processes for thorium recovery generally start by digestion of the monazite sand with either hot concentrated sulfuric acid or hot concentrated caustic. Subsequent chemical treatments, varying greatly even with the same initial treatment, yield a concentrate of impure thorium. This impure concentrate may be further treated by a liquid-liquid extraction process to yield high-purity thorium. For a system consisting of water, tributyl phosphate, nitric acid, thorium, and the associated impurities, an extractor can be set up to remove the thorium with the water-immiscible tributyl phosphate phase, while the impurities are carried away in the aqueous phase. Generally, the purified thorium is back extracted to an aqueous solution and either crystallized from solution as the nitrate or precipitated as the oxalate. From these pure salts the oxide or other compounds of thorium can be prepared.

Because thorium is quite reactive, some difficulty is experienced in preparing thorium metal. Only by electrolysis or by treatment with elements high in the electromotive force series (the alkali and alkaline-earth metals), has good-quality thorium metal been satisfactorily prepared directly from its compounds.

The calcium reduction of ThO_2 has been widely used for many years to prepare thorium metal. In this process, granular calcium metal is mixed with thorium oxide and charged into a lined iron crucible which is then filled with an inert gas and heated to almost 1000°C to form thorium metal powder and calcium oxide. After cooling to room temperature, the thorium powder is recovered by leaching and then drying. Powder metallurgy techniques are employed to obtain massive metal.

The electrodeposition of thorium from a bath, consisting of thorium chlorides or fluorides dissolved in fused alkali halides, yields granular thorium which may be pressed and sintered to give massive pieces of ductile metal.

Large-scale production of thorium metal has been carried out by a bomb process. The charge, consisting of a mixture of thorium tetrafluoride, granular calcium metal, and zinc chloride, is placed in a refractory-lined vessel that is closed by a lid. The charged bomb is placed in a furnace held at about 650°C, where, after several minutes, the charge ignites spontaneously and the resulting reaction yields a slag of calcium fluoride and calcium chloride and an alloy of thorium and zinc. The temperature reached by the reaction in the charge is sufficient to melt the products, and the thorium-rich alloy collects as a molten pool under the liquid slag. The bomb is allowed to cool, and then the solid piece of thorium alloy is removed and cleaned of adhering slag. Next, the zinc is removed by heating the alloy in a vacuum at a temperature of 1100°C, leaving the thorium metal as a sponge. Solid ingots of thorium metal are prepared by vacuum-induction-melting the sponge in a crucible or by shaping the sponge in the form of bars and melting these by consumable electrode arc melting. Good-quality thorium metal can be readily worked to shape by standard methods of fabrication.

Properties. Thorium has an atomic weight of 232. The metal has a density of 11.7 g/cm^3. Good-quality thorium metal is relatively soft and ductile. It can be shaped readily by any of the ordinary metal-forming operations. It must be protected, however, to prevent oxidation in treatments involving high temperatures. The massive metal is silvery in color, but it tarnishes on long exposure to the atmosphere; finely divided thorium has a tendency to be pyrophoric in air.

The atoms of thorium in the metal are arranged in a face-centered cubic system at all temperatures below 1400°C. On heating, the atoms rearrange at this temperature into a body-centered cubic pattern which is stable up to the melting temperature. However, the temperature at which pure thorium melts is not known with certainty; it is thought to be not far from 1750°C.

Thorium is a member of the actinide series of elements, which includes protactinium, uranium, and the synthetic transuranic elements. It is radioactive with a half-life of about 14×10^9 years. It is the first member of the radioactive decay series which in a chain of 10 successive disintegrations (α and β combined) finally terminates as lead-208.

All of the nonmetallic elements, except the rare gases, form binary compounds with thorium. Binary intermetallic compounds have been reported for thorium with beryllium, magnesium, boron, aluminum, and silicon, and with all of the metallic elements in the three long periods of the periodic chart in groups positioned to the right of group VIb. A number of the intermetallic compounds of thorium, especially those with copper, silver, and gold, are quite pyrophoric. A study of the binary alloy systems formed by thorium metal and metals of the IIIb, IVb, Vb, and VIb groups, including the rare earths, shows no evidence of intermetallic compound formation.

Principal compounds. Thorium does not impart any visible spectrum colors to its inorganic compounds or their solutions. With minor exceptions, thorium exhibits a valence of 4+ in all of its salts. Chemically, it has some resemblance to zirconium and hafnium. The most common soluble compound of thorium is the nitrate which, as generally prepared, appears to have the formula $Th(NO_3)_4 \cdot 4H_2O$.

The common oxide of thorium is ThO_2, thoria, which can be obtained by thermal decomposition of the nitrate, hydroxide, oxalate, or other compounds of thorium. A peroxide of thorium, Th_2O_7 with water of hydration, and the hydroxide, $Th(OH)_4$, can be precipitated from solutions of thorium salts.

The halogens form a variety of salts with thorium. Thorium tetrahalides of the general formula ThX_4 (X = halogen), anhydrous and with varying degrees of hydration, are known. $ThOX_2$ and $Th(OH)_2X_2$ with and without water of hydration are known. The halides of thorium also tend to form double salts with other halides, such as those of the alkali metals.

Thorium sulfate can be obtained in the anhydrous form or with 2, 4, 6, 8, or 9 molecules of water of crystallization. A somewhat insoluble basic sulfate forms when a dilute water solution of thorium sulfate is boiled. Double sulfates of thorium with alkali metals or with ammonium are known. The hydrosulfate and the thiosulfate of thorium are water-insoluble compounds.

Thorium carbonates, phosphates, iodates, chlorates, chromates, molybdates, and other inorganic salts of thorium are well known. Thorium also forms salts with many organic acids, of which the water-insoluble oxalate, $Th(C_2O_4)_2 \cdot 6H_2O$, is important in preparing pure compounds of thorium.

Analytical methods. Thorium in small quantities in rocks and other natural sources can be estimated by a study of the radioactivity of the sample. The chemical analysis of materials for thorium, however, generally involves getting the thorium into solution with sulfuric acid. The thorium must then be isolated from other interfering ions, and this may involve a separation by ion exchange or solvent extraction or a precipitation as iodide or pyrophosphate. The final determination of thorium is made by gravimetric, titrimetric, or colorimetric means. The gravimetric method generally depends on subsequent formation of the oxalate, which is calcined to ThO_2 and weighed. A titrimetric determination can be made on a thorium solution by titration with EDTA, an organic chelating agent. The end point of the titration may be detected visually by using an indicator such as xylenol orange. In one colorimetric method, a complex organic compound referred to as Thorin gives, with thorium ion, a color that can be measured to indicate the quantity of thorium present. *See* ACTINIDE ELEMENTS; RADIOACTIVITY.

[HARLEY A. WILHELM]

Bibliography: N. Edelstein (ed.), *Lanthanide and Actinide Chemistry and Spectroscopy*, 1980; International Atomic Energy Agency, *Thorium: Physico-Chemical Properties of Its Compounds and Alloys*, 1975; J. F. Smith et al., *Thorium: Preparation and Properties*, 1975.

Thulium

A chemical element, Tm, atomic number 69, atomic weight 168.934. It is a rare metallic element belonging to the rare-earth group. The stable isotope Tm^{169} makes up 100% of the naturally occurring element. It was discovered by P. T. Cleve in

Ia	IIa	IIIb	IVb	Vb	VIb	VIIb	VIII	VIII	VIII	Ib	IIb	IIIa	IVa	Va	VIa	VIIa	0
1 H																	2 He
3 Li	4 Be											5 B	6 C	7 N	8 O	9 F	10 Ne
11 Na	12 Mg											13 Al	14 Si	15 P	16 S	17 Cl	18 Ar
19 K	20 Ca	21 Sc	22 Ti	23 V	24 Cr	25 Mn	26 Fe	27 Co	28 Ni	29 Cu	30 Zn	31 Ga	32 Ge	33 As	34 Se	35 Br	36 Kr
37 Rb	38 Sr	39 Y	40 Zr	41 Nb	42 Mo	43 Tc	44 Ru	45 Rh	46 Pd	47 Ag	48 Cd	49 In	50 Sn	51 Sb	52 Te	53 I	54 Xe
55 Cs	56 Ba	57 La	72 Hf	73 Ta	74 W	75 Re	76 Os	77 Ir	78 Pt	79 Au	80 Hg	81 Tl	82 Pb	83 Bi	84 Po	85 At	86 Rn
87 Fr	88 Ra	89 Ac	104 Rf	105 Ha	106	*107*	*108*	*109*	*110*	*111*	*112*	*113*	*114*	*115*	*116*	*117*	*118*

lanthanide series	58 Ce	59 Pr	60 Nd	61 Pm	62 Sm	63 Eu	64 Gd	65 Tb	66 Dy	67 Ho	68 Er	69 Tm	70 Yb	71 Lu
actinide series	90 Th	91 Pa	92 U	93 Np	94 Pu	95 Am	96 Cm	97 Bk	98 Cf	99 Es	100 Fm	101 Md	102 No	103 Lr

1878. The salts of thulium possess a pale green color and the solutions have a slight greenish tint. The metal is produced by the reduction of the anhydrous fluoride with calcium or by the vacuum distillation of thulium from a mixture of lanthanum metal and thulium oxide. It has a high vapor pressure at the melting point. When thulium-169 is irradiated in a nuclear reactor, Tm^{170} (half-life 129 days) is formed. The isotope then emits strongly an 84-keV x-ray, and this material is useful in making small portable x-ray units for medical use. No electrical equipment is required and the unit has to be recharged with a reactivated thulium button only every few months. The metal becomes antiferromagnetic at low temperatures. *See* RARE-EARTH ELEMENTS.

The possibility of using the radioactive isotope for a low-energy power source for space applications is under study by the Atomic Energy Commission. [FRANK H. SPEDDING]

Time-of-flight spectrometers

A general class of instruments in which the speed of a particle is determined directly by measuring the time that it takes to travel a measured distance. By knowing the particle's mass, its energy can be calculated. If the particles are uncharged (for example, neutrons), difficulties arise because standard methods of measurement (such as deflection in electric and magnetic fields) are not possible. The time-of-flight method is a powerful alternative, suitable for both uncharged and charged particles, that involves the measurement of the time t that a particle takes to travel a distance l. If the rest mass of the particle is m_0, its kinetic energy E_T can be calculated from its measured speed, $v = l/t$, using the equation below, where c is the speed of light.

$$E_T = m_0c^2\{[1-(v/c)^2]^{-1/2}-1\}$$

$$\approx m_0v^2/2 \qquad \text{if } v \ll c$$

Some idea of the time scales involved in measuring the energies of nuclear particles can be gained by noting that a slow neutron of kinetic energy $E_T = 1$ eV takes 72.3 μs to travel 1 m. Its flight time along a 10-m path (typical of those found in practice) is therefore 723 μs, whereas a 4-MeV neutron takes only 361.5 ns.

The time intervals are best measured by counting the number of oscillations of a stable oscillator that occur between the instants that the particle begins and ends its journey. Oscillators operating at 100 MHz are in common use. If the particles from a pulsed source have different energies, those with the highest energies arrive at the detector first. Digital information from the "gated" oscillator consists of a series of pulses whose number $N(t)$ is proportional to the time-of-fight t. These pulses can be counted and stored in an on-line computer that provides many thousands of sequential "time channels," t_0, $t_0 + \Delta t$, $t_0 + 2\Delta t$, $t_0 + 3\Delta t$, . . . , where t_0 is the time at which the particles are produced and Δt is the period of the oscillator. To store an event in channel $N(t)$, the contents of memory address $N(t)$ are updated by "adding 1."

Time-of-flight spectrometers have been used increasingly for energy measurements of uncharged and charged elementary particles, electrons, atoms, and molecules. Their popularity is due to the broad energy range that can be covered, their high resolution ($\Delta E_T/E_T \approx 2\Delta t/t$, where ΔE_T and Δt are the uncertainties in the energy and time measurements, respectively), their adaptability for studying different kinds of particles, and their relative simplicity.

[FRANK W. K. FIRK]

Bibliography: J. A. Harvey (ed.), *Experimental Neutron Resonance Spectroscopy*, 1970.

Tin

A chemical element, symbol Sn, atomic number 50, atomic weight 118.69. Tin is a member of group IV of the periodic table, and forms tin(II) or stannous (Sn^{2+}), and tin(IV) or stannic (Sn^{4+}) compounds, as well as complex salts of the stannite (M_2SnX_4) and stannate (M_2SnX_6) types.

Evidence of the earliest use of tin dates back

over 4000 years. The ancients found that tin has many properties that other metals do not have. They were quick to realize that it alloys readily with copper to produce bronze. It melts at a low temperature, is highly fluid when molten, and has a high boiling point. It is soft and pliable and is corrosion-resistant to many media. The most important use of tin is for tin-coated steel containers (tin cans) used for preserving foods and beverages. Other important uses are solder alloys, bearing metals, bronzes, pewter, and miscellaneous industrial alloys. Tin chemicals, both inorganic and organic, find extensive use in the electroplating, ceramic, plastic, and agricultural industries.

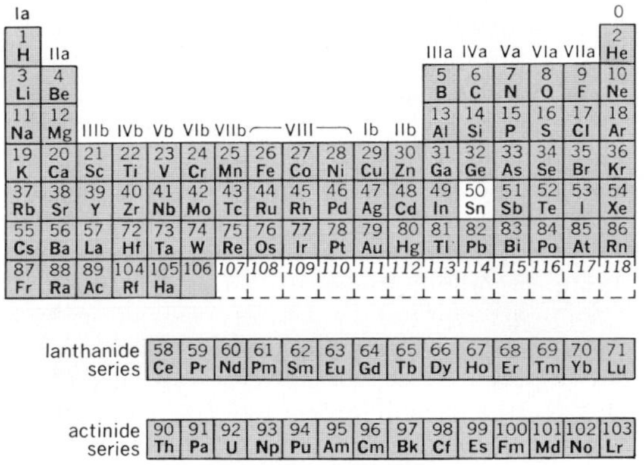

Natural occurrence. The primary tin-producing countries are Malaysia, Indonesia, Thailand, Bolivia, Zaire, Australia, Nigeria, and China. The most important tin-bearing mineral is cassiterite, SnO_2. No high-grade deposits of this mineral are known. The bulk of the world's tin ore is obtained from low-grade alluvial deposits. Because of the efficiency of modern material-handling equipment, tin ore deposits which contain a very low percentage of tin (0.01%) can be recovered economically. Lode deposits which contain higher percentages of tin are located in Bolivia and Cornwall, England, where the cassiterite is associated with granitic intrusives and often with complex sulfide ores.

Mining and concentration. The cassiterite is recovered from alluvial deposits by dredging in a placer, by water jets and gravel pumps on level ground, by hydraulic mining where a head of water permits it, and by open-pit mining. The fine grains of cassiterite have a density 2.5 times that of the gravel, and concentration is a simple matter of screening and gravity separations. The concentrates contain 70–77% tin.

Underground lode deposits in Bolivia are located 12,000 ft (3.7 km) above sea level. Access to these lodes follows the usual pattern of sinking shafts and driving adits. Ore is broken from the working face by drilling and blasting, and waste rock is disposed of belowground. Complex ore dressing methods and high transportation costs make Bolivian tin the most costly to produce.

Properties. Two allotropic forms exist: white (β) and gray (α) tin. Although the transformation temperature is 13.2°C, the change does not take place unless the metal is of high purity, and only when the exposure temperature is well below 0°C. Commercial grades of tin (99.8%) resist transformation because of the inhibiting effect of the small amounts of bismuth, antimony, lead, and silver present as impurities.

Tin reacts with both strong acids and strong bases, but it is relatively resistant to solutions that are nearly neutral. The hydrogen overpotential of tin is 0.85 V for a current density of 0.001 A/cm². Therefore, in a wide variety of corrosive conditions, hydrogen gas is not evolved from tin and the rate of corrosion becomes controlled by the supply of oxygen or other oxidizing agents. In their absence, corrosion is negligible. A thin film of stannic oxide forms on tin upon exposure to air and provides surface protection. Halogen acids attack tin, particularly when they are hot and concentrated. Hot sulfuric acid dissolves the metal, especially in the presence of oxidizing agents. Nitric acid attacks tin slowly when cold and dilute, and more rapidly with rising temperature and concentration. Dilute solutions of ammonium hydroxide and sodium carbonate have little effect on tin, but a strong alkali, such as sodium hydroxide, dissolves tin to form a stannate. Salts that have an acid reaction in solution, such as aluminum chloride and ferric chloride, attack tin in the presence of oxidizers or air. Most nonaqueous liquids, such as oils, alcohols, or chlorinated hydrocarbons, have slight or no obvious effect on tin.

Tin metal and the simple inorganic salts of tin are nontoxic. Tests have shown that concentrations of tin well above allowed limits in canned goods can be consumed without obvious adverse effects on the human system. Some forms of organotin compounds, on the other hand, are toxic. The most important physical constants for tin are shown in the table.

Uses and applications. The United States is the largest consumer of tin. Coatings, alloys, and compounds are the most important outlets. Pure tin and alloys can be applied as coatings to all the common metals by hot-dipping or by electrodeposition techniques. Tin protects metal surfaces that by themselves may oxidize or corrode rapidly in different environments. The coatings also aid in the joining of metals by providing enhanced solderability of

Properties of tin

Property	Value
Melting point, °C	231.9
Boiling point, °C	2270
Specific gravity, α-form (gray tin)	5.77
β-form (white tin)	7.29
Liquid at melting point	6.97
Transformation temperature, °C	13.2
Specific heat, cal/g, white tin at 25°C	0.053
Gray tin at 10°C	0.049
Latent heat of fusion, cal/g*	14.2
Latent heat of vaporization, cal/g	520 ± 20
Heat of transformation, cal/g	4.2
Thermal conductivity, cal/(cm)(cm²)(°C)(s), white tin at 0°C	0.150
Coefficient of linear expansion at 0°C	19.9×10^{-6}
Shrinkage on solidification, %	2.8
Resistivity of white tin, microhms/cm³	
At 0°C	11.0
At 100°C	15.5
Brinell hardness, 10 kg/(5 mm)(180 s)	
At 20°C	3.9
At 220°C	0.7
Tensile strength as cast, psi†, at 15°C	2100
At 200°C	650
At −40°C	2900
At −120°C	12,700

*1 cal = 4.184 J.
†1 psi = 6.895 kPa.

surfaces. Tin provides metallic lubricity to basic metals during forming operations and an attractive appearance to many functional items. In addition, tin coatings offer a clean adherent substrate for paint or lacquers.

Tin-coated steel (tinplate) production in the United States averages about 4,900,000 metric tons per year. About 90% of this material is used for food and beverage containers, while 10% is used for nonfood items such as signs, toys, kitchen wares, and foundry chaplets. A sizable market for tinplate is for the manufacture of beer and soft-drink cans. Tinplate manufacture in the United States is largely a continuous, high-speed electrolytic process, with less than 1% of production from hot-tinning machines. Electrolytic tinplate can be produced from either alkaline or acid electrolytes. Tinplates have either tin on each steel surface or a differential tin-coating thickness. Equally coated tinplates have coating thicknesses of between 3.8×10^{-4} and 15.4×10^{-4} mm on each surface. For differentially coated tinplate, the maximum coating thickness on one surface is 20.7×10^{-4} mm. In can manufacturing, a coating of enamel or lacquer is usually required when thinly coated tinplates are used as construction materials. *See* ELECTROLYSIS.

Hot-dipped tinplate is still used for special corrosive packs, kitchen utensils, hardware items, and automotive parts. For other industrial applications, hot-dip tin coatings are applied to copper wire and sheet, as well as steel and cast iron parts. Examples are tinned copper and copper alloy strip for manufacture of electrical connectors and tinned food processing equipment. Hot-dip tin-lead (terne) coatings find service as coatings for gasoline tanks, roofing materials, electronic applications, radiator water tubes, and component leads.

The plating industry utilizes tin as anodes for the electrodeposition of pure tin and tin alloy coatings. Plated tin, either as a mat or bright finish, provides easily solderable surfaces for steel, copper, or aluminum. Tin alloy coatings (tin-copper, tin-lead, tin-nickel, tin-zinc, tin-cadmium, tin-cobalt) have advantages over single metal plates. They are denser and harder, more corrosion-resistant, brighter or more easily buffed, and more protective to basis metals. Tin-copper (12% tin) has the appearance of 24-karat gold and, when lacquered, serves as an attractive finish for jewelry, trophies, wire goods, and hardwares. Tin-lead electroplates (40–65% tin) have excellent corrosion resistance and solderability, and are well adapted to the plating of printed circuits and electronic parts. Tin-zinc coatings (75% tin) are a good alternative coating for cadmium in particular applications, and they provide galvanic protection to steel in contact with aluminum. Tin-cadmium coatings (25% tin) are especially resistant to salt vapors, and have a number of applications in the aircraft industry and as a coating for fasteners. A tin-nickel coating (65% tin) finds use as an etchant resist in the manufacture of printed circuit boards, as well as an ornamental and highly corrosion-resistant finish for watch parts, scientific instruments, and power connectors. The tin-cobalt alloy (80% tin) has an appearance similar to a chromium deposit, and is used to plate fasteners, ancillary office equipment, hinges, kitchen utensils, hand tools, and tubular furniture.

Principal inorganic compounds. Stannous and stannic tin salts have a number of uses in the field of electroplating, ceramics, and textiles. Inorganic tin compounds produced on a commercial scale include tin(II) (stannous) compounds and tin(II) and (IV) oxides.

Stannous oxide, SnO, is a blue-black crystalline product which is soluble in common acids and strong alkalies. It is prepared by treating stannous chloride with alkali. The precipitated stannous hydroxide is converted to oxide by heating near the boiling point of water at a controlled pH. SnO is thermally stable in air up to 385°C, at which temperature it is converted to stannic oxide. It is used in making stannous salts for plating and glass manufacture.

Stannic oxide, SnO_2, is a white powder, insoluble in acids and alkalies. It is prepared by atomizing tin with high-pressure steam and burning the finely divided metal in oxygen, or by calcination of the hydrated (IV) oxide. It is an excellent glaze opacifier, a component of pink, yellow, and maroon ceramic stains and of dielectric and refractory bodies. It is an important polishing agent for marble and decorative stones.

Stannous chloride, $SnCl_2$, is available in both anhydrous and hydrated forms. The anhydrous salt can be prepared by heating tin in hydrogen chloride gas or by direct chlorination. The hydrated salt is prepared by treating flaked tin with hydrochloric acid, followed by evaporation and crystallization. It is the major ingredient in the acid electrotinning electrolyte and is an intermediate for tin chemicals.

Stannic chloride, $SnCl_4$, a fuming liquid, is prepared by direct chlorination of tin. The pentahydrate is a white solid. It is used in the preparation of organotin compounds and chemicals to weight silk and to stabilize perfume and colors in soap.

Sodium stannate, $Na_2SnO_3 \cdot 3H_2O$, and potassium stannate, $K_2SnO_3 \cdot 3H_2O$, are prepared by dissolving hydrated stannic oxide in alkali. The sodium salt is a by-product of the detinning of tinplate scrap. Stannates are used in alkaline electrotinning baths.

Heavy-metal stannates of lead, barium, calcium, and copper are important in the manufacture of capacitor bodies.

Stannous sulfate, $SnSO_4$, is used in liquor-finishing steel wire and in electroplating. Stannous fluoride, SnF_2, a white water-soluble compound, is a toothpaste additive. Stannous fluoborate, $Sn(BF_4)_2$, is known only in solution. The commercial solution contains 47% stannous fluoborate, and is widely used in tin and tin-lead plating.

Principal organic compounds. Organotin compounds are those compounds in which at least one tin-carbon bond exists, the tin usually being present in the +IV oxidation state. Organotin compounds that find applications in industry are the compounds with the general formula R_4Sn, R_3SnX, R_2SnX_2, and $RSnX_3$. R is an organic group, often methyl, butyl, octyl, or phenyl, while X is an inorganic substituent, commonly chloride, fluoride, oxide, hydroxide, carboxylate, or thiolate.

Most of the production of organotin compounds is for the stabilization of polyvinyl chloride (PVC) plastics. Tin stabilizers are effective in preventing the degradation of the plastic during processing or during prolonged exposure to light or heat. Certain dioctyltin compounds are used for PVC bottles for fruit beverages, vegetable cooking oils, and other household fluids. Clear PVC materials stabilized with organotins are also used for packaging toiletries, confectioneries, pharmaceuticals, and hardware.

A second important area of application of diorganotin (R_2SnX_2) chemicals is in the building industry, where tin-stabilized PVC is used for flooring, siding materials, window frames, and plastic piping. *See* POLYVINYL RESINS.

Most organotin compounds of commercial interest are prepared for the tetraorganotins (R_4Sn) as a starting point. Treatment of R_4Sn compounds with stannic chloride is a widely used procedure for the manufacture of organotin halides. The relative ease and efficiency with which alkyl or aryl groups can be redistributed among the tin atoms is one of the most unusual features of organotin chemistry, and plays an essential role in the production of organotin chemicals. Other derivatives may then be obtained from the halides.

Certain types of trialkyltin and triaryltin compounds possess powerful biocidal properties. Biocidal properties are manifest to a high degree only when the tin atom is combined directly with three carbon atoms, as in trialkyl compounds (R_3SnX); biocidal effects are at a maximum when the total number of carbon atoms attached to Sn is 12. Such compounds are used as fungicides, insecticides, and pest control in agriculture applications. Tributyltin acetate, $(C_4H_9)_3Sn-OOCCH_3$, and bistri-*n*-butyltin oxide, $(CH_4H_9)_3Sn-O-Sn(C_4H_9)_3$, are commercially available for use as antimicrobial agents in the paper, wood preservation, plastics, and textile industries.

Applications of organotin compounds in agriculture require rigid control to avoid crop damage. Marine antifouling paints containing tributyltin oxide or tributyltin fluoride oraganotin-impregnated elastomers for seaway navigational markers, wood preservatives, fungicides for textiles, paper pulp, and rope fibers, as well as bactericides to reduce contamination in hospital environments and water-cooling towers are in wide use.

Catalytic salts. Tin salts of organic acids are important catalysts in the manufacture of urethane foam compositions and various silicone mixtures. Stannous octoate and stannous oleate are most commonly used in these applications.

Analysis. The purity of commercial tin metal is under strict control at the smelters. Standard grade A tin is guaranteed to contain a minimum of 99.8% tin. Except for special purposes, a user rarely needs to know the identity and concentrations of the impurities. However, photometric, chemical, and spectrographic methods are available for their determination in tin.

[JOSEPH B. LONG]

Bibliography: E. S. Hedges, *Tin and Its Alloys*, 1960; W. E. Hoare, *The Technology of Tinplate*, 1965; C. L. Mantel, *Tin*, 1949; Tin Research Institute, Inc. (Columbus, OH), *Tin and Its Uses*, quarterly.

Titanate

A compound obtained when metal oxides or hydroxides are heated with titanium dioxide, TiO_2. Metatitanates of the formulas K_2TiO_3, $ZnTiO_3$, $PbTiO_3$, and $BaTiO_3$ are formed by fusion of TiO_2 with the appropriate metal oxide. Titanates of the Na_4TiO_4 type are often called orthotitanates. The titanates are insoluble in water and are used in the manufacture of porcelain enamels, ceramic dielectrics, temperature-sensitive resistors, and photoconductors. *See* TITANIUM.

[E. EUGENE WEAVER]

Titanium

A chemical element, Ti, atomic number 22 and atomic weight 47.90. It occurs in the fourth group of the periodic table, and its chemistry shows many similarities to that of silicon and zirconium. On the other hand, as a first-row transition element, the aqueous solution chemistry, especially of the lower oxidation states, shows some resemblances to that of vanadium and chromium. *See* TRANSITION ELEMENTS.

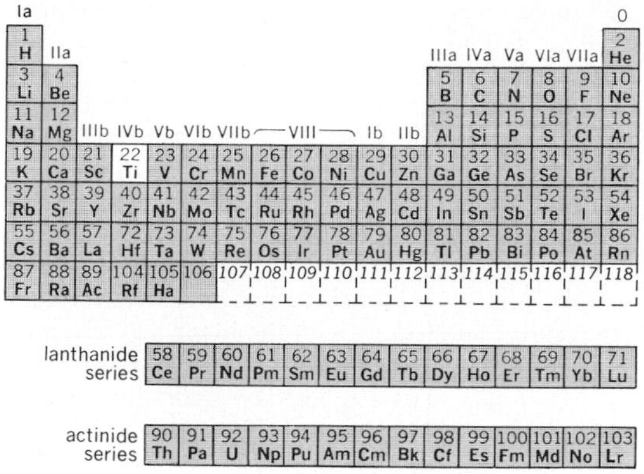

The outer electronic arrangement is $3d^2 4s^2$, and the principal valence state, correspondingly, is 4+; the 3+ and 2+ states are also known, but are less stable. The element burns in air when heated to give the dioxide, TiO_2, and combines with halogens according to reaction (1). It reduces water

$$Ti + 2X_2 \rightarrow TiX_4 \qquad (1)$$

vapor to form the dioxide and hydrogen, and reacts similarly with hot concentrated acids, although it forms the trichloride with hydrochloric acid. The metal absorbs hydrogen to give compositions approaching TiH_2 and forms the nitride, TiN, and the carbide, TiC. The sulfide TiS_2 can be formed by the reaction of $TiCl_4$ with H_2S, and both the lower oxides, Ti_2O_3 and TiO, and sulfides, Ti_2S_3 and TiS, are known.

Salts of all three valence states are known. The yellowish $Ti(SO_4)_2$ is a prominent compound, but except in very highly acid solutions, the titanium hydrolyzes to titanyl ion, TiO^{2+}. There is, for example, a basic phosphate, $KTiOPO_4$. With respect to the lower valence states, the halide salts are the best known and give Ti^{2+} and Ti^{3+} (titanous) ions in solution. The latter is violet, with a broad spectral absorption band at 490 nm; this has been interpreted as being caused by a splitting of the $3d$ orbitals by the electrostatic field of the six water molecules which surround the ion.

Natural occurrence. The dioxide, TiO_2, occurs most commonly in a black or brown tetragonal form known as rutile (a/b ratio = 4.58/2.95 A). Less prominent naturally occurring forms are anatase (also tetragonal, $a/b = 3.73/9.37$ A) and brookite (rhombohedral). Both rutile and anatase are white when pure. The dioxide may be fused with other metal oxides to yield titanates, for example, K_2TiO_3, $ZnTiO_3$, $PbTiO_3$, and $BaTiO_3$. The black basic oxide, $FeTiO_3$, occurs naturally as the mineral ilmenite; this is a principal commercial source of titanium. The dioxide, for example, may be prepared from it by dissolving the ore in sulfuric acid, clarifying, and then partially neutralizing the solution. The titanium hydrogel which forms is then separated from the solution, compacted, dried, and calcined. Depending on the heat treatment, the resulting powdered TiO_2 may be of the anatase (low-temperature) or the rutile (high-temperature) type. The dioxide is refractory and, as rutile, has the unusually high dielectric constant of about 100 (173 parallel to, and 89 perpendicular to, the principal axis) and an index of refraction of 2.7 (as compared to 2.42 for diamond). The anhydrous dioxide, although somewhat acidic as evidenced by the formation of titanates, is quite inert to acids, although the freshly prepared hydrous material is soluble in acid.

Principal compounds. The sesquioxide, Ti_2O_3, may be prepared by the hydrogen reduction of the dioxide; it is acid-soluble, to give solutions containing titanous ion, from which the black $Ti(OH)_3$ may be precipitated by addition of base. In the presence of water, $Ti(OH)_3$ evolves hydrogen to give the dioxide. Titanous ion itself is a good reducing agent. It has found use in volumetric analysis as a quantitative reagent for the determination of ferric and permanganate ions.

The 2+ oxide, TiO, can be obtained by the high-temperature reduction of the dioxide by carbon or various metals. It is basic, but its salts are unstable in water solution because of the strong reducing power of the Ti^{2+} ion.

Among the halogen compounds of titanium, the best known are the tetrahalides, TiX_4. In addition, the complex ions TiF_6^{2-} and $TiCl_6^{2-}$ are well known. Titanium tetrachloride is a light-yellow liquid boiling at 136°C; it may be prepared by the direct reaction of the elements, but one commercial process makes use of the common ore ilmenite as starting material. As in the manufacture of the dioxide, the ore is dissolved in sulfuric acid, but the solid K_2TiCl_6 is precipitated out by saturating the solution with HCl and KCl. The complex salt is then thermally decomposed to give $TiCl_4$. The tetrachloride hydrolyzes with water or moist air to give the dioxide; it reacts with metal trialkyls, such as aluminum triethyl, to form $TiCl_3$ and more complex compounds, and with alcohols to form compounds of the type $Ti(OR)_4$.

The trichloride may also be obtained by reduction of $TiCl_4$ by metals such as silver or zinc, and by electrolysis. The dichloride can be prepared by the thermal decomposition of the trichloride.

Titanium in the 4+ state forms various complex ions, in addition to the TiX_6^{2-} species described above. Chelate compounds with cupferron, 8-hydroxyquinoline, acetylacetone and its derivatives, and quinone are known. Finally, the orange peroxy ion, $TiO_2(SO_4)_2^{2-}$, is formed by the addition of hydrogen peroxide to a solution of the sulfate, and the related acid, H_4TiO_5, is known. The formation of the peroxy complex is the basis for a well-known method for the colorimetric estimation of titanium.

Uses of important compounds. Titanium dioxide is widely used as a white pigment for exterior paints because of its chemical inertness, superior covering power, opacity to damaging ultraviolet light, and self-cleaning ability. Both the rutile and anatase forms are used, but especially the former.

The dioxide has also been used as a whitening or opacifying agent in numerous situations. Examples would be the use as a filler in paper, a coloring agent for rubber and leather products, a pigment in ink, and a component of ceramics. It has found important use as an opacifying agent in porcelain enamels, giving a finish coat of great brilliance, hardness, and acid resistance. Rutile has also been found as brilliant, diamondlike crystals, and some artificial production of it in this form has been achieved. Because of its high dielectric constant, it has found some use in dielectrics.

The alkaline-earth titanates show some remarkable properties. The dielectric constants range from 13 for $MgTiO_3$ to several thousand for solid solutions of $SrTiO_3$ in $BaTiO_3$. Barium titanate itself has a dielectric constant of 10,000 near 120°C, its Curie point; it has a low dielectric hysteresis. These properties are associated with a stable polarized state of the material analogous to the magnetic condition of a permanent magnet, and such substances are known as ferroelectrics. In addition to the ability to retain a charged condition, barium titanate is piezoelectric and may be used as a transducer for the interconversion of sound and electrical energy. Ceramic transducers containing barium titanate compare favorably with Rochelle salt and quartz, with respect to thermal stability in the first case, and with respect to the strength of the effect and the ability to form the ceramic in various shapes, in the second case. The compound has been used both as a generator for ultrasonic vibrations and as a sound detector.

Although somewhat corrosive, liquid titanium tetrachloride has found use in the formation of smokes, especially in World War I, and also in commercial skywriting. On contact with moist air, the compound $TiCl_4 \cdot 5H_2O$ first forms, followed by hydrolysis to the dioxide.

$TiCl_4$ has become an important starting material for the production of titanium metal (by means of magnesium or sodium reduction). Its commercial preparation from ilmenite has already been described. The compound has become very important in the catalytic polymerization of ethylene.

Titanium esters, formed by the reaction of $TiCl_4$ with alcohols, for example, $Ti(OC_nH_{2n+1})_4$, are useful as waterproofing agents for a variety of natural and synthetic fabrics. The tetrabutyl and tetraisopropyl esters hydrolyze in moist air to give the dioxide, and can be used to provide thin, transparent, and adherent coatings. The diacetate, $TiCl_2(O_2C_2H_3)_2$, has been suggested as a flame retardant for cellulose fabrics. The acetylacetonate, $Ti(C_6H_8O_2)_2$, may be used as a cross-linking agent in lacquers so that on drying, the resulting film becomes inert to solvents.

Polymerization catalysis. An important development in the low-pressure polymerization of ethylene has been the use of titanium catalysts. Typical starting materials would be $TiCl_4$ and $Al(C_2H_5)_3$, which then react according to the general scheme shown by reactions (2). At the same time, mixed halide-alkyl complexes of variable composition are formed, for example, $(C_2H_5)_2TiCl_2Al$-

$$TiCl_4 + Al(C_2H_5)_3 \rightarrow Al(C_2H_5)_2Cl + TiCl_3(C_2H_5)$$

$$TiCl_4 + 2\,Al(C_2H_5)_3 \rightarrow TiCl_2(C_2H_5)_2 + 2\,AlCl(C_2H_5)_2 \quad (2)$$

$$TiCl_3(C_2H_5) \rightarrow TiCl_3 + CH_3CH_2\bullet$$

$$TiCl_2(C_2H_5) \rightarrow TiCl_2 + CH_3CH_2\bullet$$

$(C_2H_5)_2$. It is the catalytic activity of such complexes that forms the basis of the well-known Ziegler process for the polymerization of ethylene. This type of polymerization is of great industrial interest since, with its use, high-molecular-weight polymers can be formed. In some cases, desirable special properties can be obtained by forming isotactic polymers, or polymers in which there is a uniform stereochemical relationship along the chain. *See* POLYOLEFIN RESINS; TITANATE.

[ARTHUR W. ADAMSON]

Titration

A quantitative analytical process that is basically volumetric. However, in high-precision titrimetry the titrant solution is sometimes delivered from a weight buret, so that the volumetric aspect is indirect. Generally, a standard solution, that is, one containing a known concentration of substance X (titrant), is progressively added to a measured volume of a solution of a substance Y (titrand) that will react with the titrant. The addition is continued until the end point is reached. Ideally, this is the same as the equivalence point, at which an excess of neither X nor Y remains. If the stoichiometry or exact ratio in which X and Y react is known, it is possibly to calculate the amount of Y in the unknown solution. *See* VOLUMETRIC ANALYSIS.

The normal requirements for the performance of a titration are: a standard titrant solution; calibrated volumetric apparatus, including burets, pipets, and volumetric flasks; and some means of detecting the end point. *See* PIPET; VOLUMETRIC FLASK.

Sometimes the standard titrant solution is prepared by the accurate weighing out of a primary standard, a substance that is stable and readily available in a high state of purity, such as potassium dichromate, anhydrous sodium carbonate, potassium hydrogen phthalate, or silver (which dissolves easily in nitric acid). The weighed material is dissolved, usually in water, and the solution is made up to a fixed volume in a volumetric flask of suitable capacity. The final solution must be homogeneous, a state normally achieved by ample shaking. If better-than-routine accuracy is required, the temperature must be known and constant.

Many valuable titrant solutions cannot be readily and accurately prepared by the direct method. For example, sodium hydroxide solution, much used for titrating acidic substances, is made from a solid that rapidly picks up moisture and so forth from the air. However, a solution of approximately the desired concentration can be prepared. This solution is then standardized by titrating (or being titrated with) another suitable standard solution. Sometimes a known weight of a suitable primary standard is placed directly in the titration vessel, dissolved in a suitable but unmeasured volume of solvent, and then titrated with the solution to be standardized. For example, sodium hydroxide solution is commonly standardized by titrating a known weight of potassium hydrogen phthalate. Analogous procedures are used for other titrant solutions that are made from substances that do not possess primary-standard properties.

Skillful operation may allow the titration to be stopped almost exactly at the end point. However, it is easy to overshoot, or add too much titrant. Provided that the total volume of titrant solution is noted, the titration can often be saved by the process of back titration. In the case mentioned above, a small overshoot of sodium hydroxide could be measured by back titration with standard hydrochloric acid solution. The process of back titration is sometimes used deliberately, especially for slow titration reactions. A known but excessive amount of titrant A is added to the substance to be determined. Then, after a suitable delay to allow for completion of the reaction, the remaining A is back-titrated with a reagent B that reacts rapidly with A. *See* CONCENTRATION SCALES; STOICHIOMETRY.

CLASSIFICATION BY CHEMICAL REACTION

For the purposes of titrimetry, chemical reactions can be placed in three general categories: acid-base or neutralization, combination, and oxidation-reduction.

Acid-base reactions. These titrations involve neutralization of an acid by titration with a base, or vice versa. However, the process is often nonspecific; in the titration of a mixture of nitric and hydrochloric acids, only the total acidity can be found without recourse to additional measurements. This arises because the only real reaction involved in these aqueous systems is the formation of water, reaction (1).

$$H_3O^+ + OH^- \rightarrow 2H_2O \quad (1)$$

A salt derived from a strong base and a very weak acid can often be titrated just as if it were a base. For example, solutions of hydrochloric acid, or of other strong acids, are often standardized by titrating known weights of primary standard sodium carbonate. Sodium hydroxide solutions are often standardized by the use of primary standard potassium hydrogen phthalate, which is a so-called acid salt. *See* ACID AND BASE.

Combination reactions. In titrimetry, attention is usually focused upon the combination of an ion in the titrant with one of the opposite sign in the titrand solution. Sometimes the combination may involve more than two species, some of which may be nonionic. The combination may result in precipitation. A classic example is the determination of chloride ion by titration with silver nitrate solution, when silver chloride is precipitated. Although a reaction may be known to yield a precipitate that has insignificant solubility and constant composition, this does not guarantee that the reaction will be useful titrimetrically. Any practical titration also requires that the reaction involved must not be unduly slow.

The same limitation also applies to complex-formation titrations, where precipitation may be absent or merely incidental. Except in a few special cases, complex-formation titrations were of little importance until the discovery of ethylenediaminetetraacetic acid (EDTA) and related compounds. These titrants not only are powerful complexing agents that combine with very many cations, but also form a single type of complex. This is in marked contrast to complexing agents such as ammonia or cyanide ion, which may yield a mixture of complexes.

With ubiquity comes lack of specificity. This difficulty with the EDTA family of titrants can sometimes be overcome by pH control and by the technique of masking (competitive complexation, whereby some species are prevented from taking part in the titration reaction). Then it may be possible to determine separately some or all of several metal ions in a mixture. *See* COMPLEX COMPOUNDS; PRECIPITATION.

Oxidation-reduction reactions. In so-called redox titrations the titrant is usually an oxidizing agent, and is used to determine a substance that can be oxidized and hence can act as a reducing agent. Because titrants that have usefully strong reducing properties are themselves attacked by oxygen in the air, the reverse procedure, although possible, is less usual. *See* OXIDATION-REDUCTION.

COULOMETRIC TITRATION

Faraday's laws of electrolysis indicate that the extent of an electrochemical reaction is proportional to the total amount of electricity that is passed through the system. The passage of a uniform current for a measured period of time can be used to generate a known amount of a product such as a titrant. This fact is the basis of the technique known as coulometric titration. An obvious requirement is that generation shall proceed with a fixed, preferably 100%, current efficiency. The uniform current is then analogous to the concentration of an ordinary titrant solution, while the total time of passage is analogous to the volume of such a solution that would be needed to reach the end point.

Coulometric titration has several attractive features. Unless the sample to be titrated must be measured by volume, no volumetric glassware is needed. Nor are standard solutions; the titrant is generated during the process. This means that titrants of low stability, such as the strong reducing agent chromium(II) ion, can be employed routinely. The titration is started or stopped by the mere closing or opening of a switch, which can be remote from the titration vessel. This feature is obviously useful when titrating highly radioactive materials. The process can be controlled by an operator who is safely screened from the materials in the titration system. *See* ELECTROLYSIS.

CLASSIFICATION BY END-POINT TECHNIQUES

The precision and accuracy with which the end point can be detected is a vital factor in all titrations. Because of its simplicity and versatility, chemical indication is quite common, especially in acid-base titrimetry. That certain natural pigments such as litmus in an acidic solution change color when the solution is made basic has been known for centuries. Thus an acid solution that contains a small amount of litmus can be titrated with sodium hydroxide solution until the initial red color has just changed to blue. Litmus and other natural pigments, which are usually mixtures of various compounds, have been supplanted by synthetic indicators. These not only have sharper color transitions, but also can be made to suit particular applications.

Indicators. An acid-base indicator is a weak acid or a weak base that changes color when it is transformed from the molecular to the ionized form, or vice versa. The color change is normally intense, so that only a low concentration of indicator is needed. Phenolphthalein is an example of an indicator that acts as a weak monobasic acid according to reaction (2).

$$\underset{\text{(colorless)}}{\underset{\text{Acidic form}}{HIn}} \rightleftharpoons \underset{\text{(red)}}{\underset{\text{Basic form}}{H^+ + In^-}} \qquad (2)$$

Methyl orange acts as a weak base according to reaction (3).

$$\underset{\text{(yellow)}}{\underset{\text{Basic form}}{InNR_2}} + H^+ \rightleftharpoons \underset{\text{(red)}}{\underset{\text{Acidic form}}{InNR_2H^+}} \qquad (3)$$

The working range, or visual color change, of a typical acid-base indicator is spread over about a hundredfold (~ 2 pH units) change in hydrogen ion concentration. Available indicators have individual working ranges that together cover the entire range of hydrogen ion concentration (10^{-1} to $10^{-13}M$, or pH 1 to 13) likely to be encountered in general acid-base titration. For example, the working ranges of phenolphthalein and of methyl orange are pH 8.0–9.8 and pH 3.0–4.4, respectively.

The success of a titration may hinge upon a suitable choice of indicator. Figure 1 shows the titration curves of two monobasic acids (a strong acid, such as hydrochloric acid, and a moderately weak acid, such as acetic acid), each at a concentration of approximately 0.1 *M*, with the strong base sodium hydroxide as titrant. The respective working ranges of methyl orange and of phenolphthalein are also indicated. The curve for the strong acid is

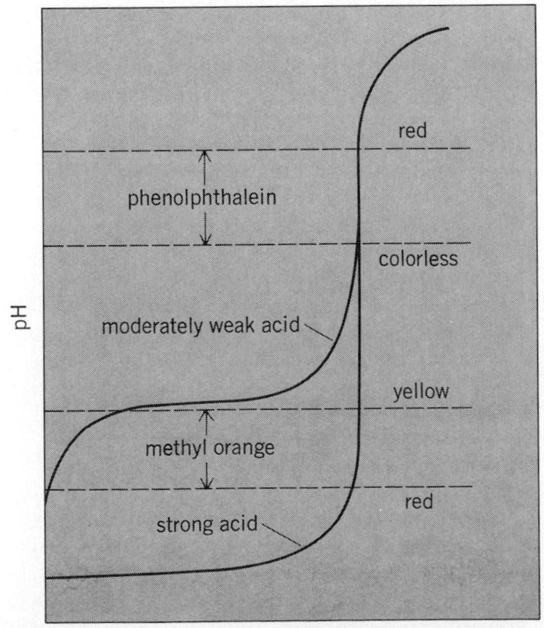

Fig. 1. Titration curves of two monobasic acids (0.1 *M*), with sodium hydroxide as the titrant.

almost vertical in the approximate region pH 4–10. This means that a very small addition of titrant when the pH is approximately 4 causes methyl orange to change from orange to yellow. If phenolphthalein is used in place of methyl orange, the colorless indicator begins its change to the red form when the pH is approximately 8. Theoretically the end point with phenolphthalein requires a little more titrant than that required to reach the methyl orange end point. However, under routine conditions, the difference may be less than calibration-plus-operator error.

In the titration of the weak acid with methyl orange as the indicator, the pH rises to the lower limit of the working range before much titrant has been added. The curve rises slowly within this working range, so that the color change red through orange to yellow is both gradual and complete before all of the acid has been titrated. However, the weak acid curve eventually merges into the strong acid curve. A titration indicated by phenolphthalein therefore yields a sharp end point.

Similar reasoning applies to the titration of a base with an acid solution. If both species are strong, any one of a number of common indicators suffices. The successful titration of a weak base such as ammonia with a strong acid requires an indicator like methyl orange, which has a working range that is low on the pH scale. In the titration of a weak acid with a weak base, or the reverse, no simple indicator is suitable because the titration curve is not steep anywhere near the expected end point. The standardization of a solution of ammonia with standard acetic acid solution would be done indirectly. *See* ACID-BASE INDICATOR; HYDROGEN ION; PH.

In the complexometric titration of a cation, M^{n+}, such as that of magnesium, with EDTA or a similar agent, the concentration of M^{n+} falls most rapidly in the immediate region of the end point. The response resembles that shown by the curve of a strong acid in Fig. 1, but pM^{n+} replaces pH. The indicator is itself a suitable complexing agent that changes color when it is combined with M^{n+}. This is the situation during most of the titration. At or very near the expected end point, the titrant reacts with the small amount of cation in the indicator complex. The color then changes to that of the free agent.

When a solution of an oxidizing agent is used to titrate a substance that is readily oxidized, the electrochemical potential rises most rapidly in the region of the end point. The titration curve is generally similar to that shown for the weak acid in Fig. 1, but potential replaces pH. The chemical indicator is here a substance that undergoes a marked change in color when a suitable potential is reached. Unlike the indicators commonly used in acid-base or complex-formation titrations, oxidation-reduction indicators are often irreversible. Thus, if an excess of oxidizing titrant has been added, the indicator color may not revert to the original if back titration is subsequently attempted.

The important oxidizing titrant potassium permanganate forms an intensely purple solution. Its reduction products are essentially colorless. This titrant is thus self-indicating, provided that the titrand and its oxidation products have little or no color. The end point is taken as the first appearance of a pale but permanent pink color.

Sometimes no suitable chemical indicator can be found for a desired titration. Possibly the concentrations involved may be so low that chemical indication functions poorly. Other situations might be the need for high precision or for the automatic arrest of the titration. Recourse is then made to some physical method of end-point detection.

Potentiometric titration. If a pH meter is used, acid-base titration curves like those shown in Fig. 1 can be plotted or recorded. The meter and its associated electrodes are first standardized by use of a buffer solution of known pH. The electrodes are then immersed in the well-stirred solution to be titrated, and the titration is begun. For routine purposes, interest is in rapid and reasonably close end-point location. Titration is first carried out quite quickly until the meter shows signs of rapid response, and then is slowed so that it can be stopped at the pH jump or fall that marks the end point. Obviously, potentiometric titration can be used for a highly colored titrand solution, in which the response of a color-change indicator could not be seen.

There are two approaches to higher precision. In the first the pH at which the desired end point occurs is determined, and the titration of the actual sample is then arrested exactly at, or very close to, this pH value. The other approach is to stop at the point of steepest slope of the titration curve. This can be done by plotting (or, with suitable instrumentation, sensing) the first derivative ($\Delta pH/\Delta V$) or the second derivative ($\Delta^2 pH/\Delta V^2$) against the volume V of titrant added. In theory this method is applicable to any potentiometric titration, without the need to predetermine the end-point conditions.

By suitable choice of electrodes, these potentiometric methods can also be applied to combination titrations and to oxidation-reduction titrations. The advent of modern ion-selective electrodes has greatly extended the scope of potentiometric titration and of other branches of titrimetry. *See* ELECTRODE POTENTIAL.

Conductometric titration. Several instrumental techniques give rise to titration curves that are essentially linear. Generally, the end point is found graphically from readings taken at a number of points that fall on appropriate linear portions of the curve. One advantage is that poor response in the actual vicinity of the end point does not cause difficulties. Another is good precision, due to the averaging effect of the many individual readings.

The underlying principles of conductometric titration are that the solvent and any molecular species in solution exhibit only negligible conductance; that the conductance of a dilute solution rises as the concentration of ions is increased; and that at a given concentration the hydrogen ion and the hydroxyl ion are much better conductors than any of the other ions.

One example is that of a dilute hydrochloric acid solution titrated with one of sodium hydroxide. The initial conductance, high because the hydrogen ion concentration is high, falls as the titrant is added (Fig. 2). When neutralization is complete, the condutance is that of a solution of sodium chloride. The change in conductance is due to progressive replacement of hydrogen ion by sodium ion, which is a poorer conductor. Further addition

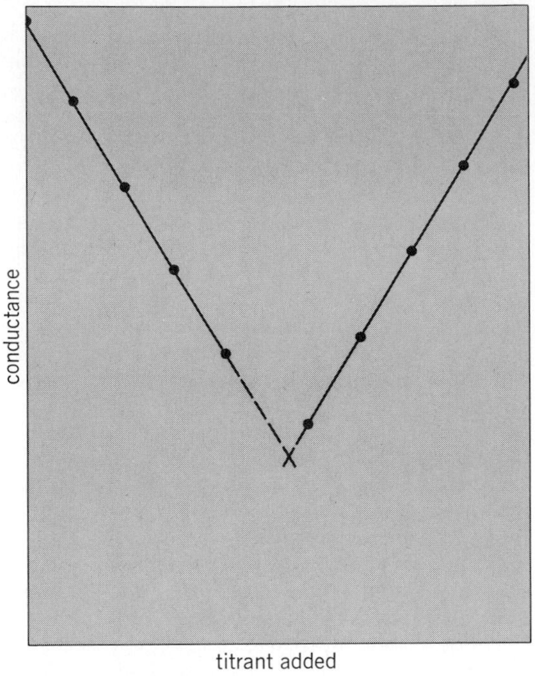

Fig. 2. Conductometric titration curves of a dilute hydrochloric acid solution with sodium hydroxide as the titrant.

of titrant causes the conductance to rise again, principally because the highly conducting hydroxyl ion now remains in the system, instead of undergoing mutual destruction with the hydrogen ion. Addition of titrant naturally dilutes the titrand solution, causing distortion of the linear branches of the titration curve. This diluting effect can be compensated for by a simple arithmetic correction or can be rendered negligible by use of a concentrated titrant solution.

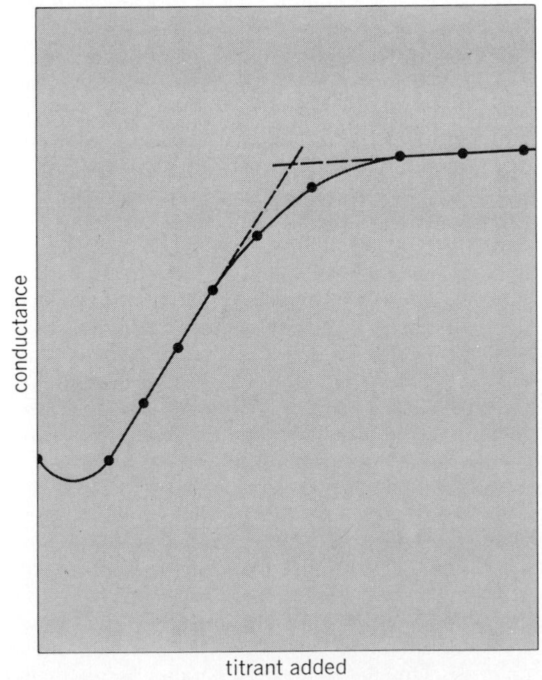

Fig. 3. Conductometric titration curve of acetic acid with ammonia as the titrant.

Conductometric titration is sometimes successful when chemical indication fails. A typical case is the titration of acetic acid with ammonia (Fig. 3). Both reactants are weak electrolytes, but ammonium acetate, the salt produced in the titration, is a strong electrolyte. The rising linear branch is due to the buildup of the ions of this salt. In the region beyond the end point, the addition of more titrant causes very little change in conductance. The concentration of ammonium ion then present forces the added titrant to remain almost entirely in the molecular state. Because of the effects of hydrolysis, the graph is decidedly rounded in the region of the end point. Although an old technique, conductometric titration still finds extensive use in studies of nonaqueous systems.

Certain precipitations can be satisfactorily performed conductometrically. However, conductometry is generally of little use in oxidation-reduction titrations. In fact the technique is difficult to use if the system contains appreciable concentra-

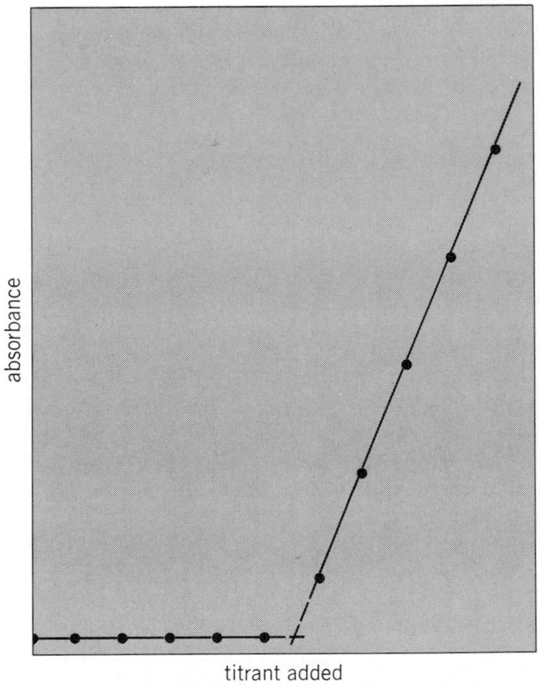

Fig. 4. Spectrophotometric titration curve if iron(II) in dilute sulfuric acid solution with potassium permanganate as the titrant. The curve shows the response when the titrant absorbs beyond the end point.

tions of electrolytes other than those involved in the actual titration. The observed conductance is a function of the total ionic content, while the precision with which the end point can be determined depends upon the relative change in conductance. For example, the angle between the branches in Fig. 2 is less acute if the titrand solution contains sodium nitrate as well as hydrochloric acid. *See* ELECTROLYTIC CONDUCTANCE.

Conductometric titrations are normally carried out with the aid of audio-frequency alternating current. By use of radio frequencies, the titration can be monitored in a cell that has the electrodes on the outside, so that they are not in contact with the titrand solution. However, the titration curves are usually less simple than those encountered in normal conductometric titration.

Spectrophotometric titration. The spectrophotometer is an optical device that responds only to radiation within a selected very narrow band of wavelengths in the visual, ultraviolet, or infrared regions of the spectrum. The response can be made both quantitative and linearly related to the concentration of a species that absorbs radiation within this band. Titrations at wavelengths within the visual region are by far the most common. An example is the titration of iron(II) in dilute sulfuric acid solution with a standard solution of potassium permanganate. The spectrophotometer is adjusted to measure the absorbance at the wavelength of maximum absorbance of the highly colored titrant. Very little absorbance is observed until the end point is reached; iron(II) and the reaction products are essentially colorless and are not "seen" by the spectrophotometer. However, the absorbance rises as titration is continued beyond the end point because the titrant is no longer being destroyed. The titration curve is shown in Fig. 4.

If the titrant does not absorb at the chosen wavelength, the curve is horizontal in the region beyond the end point. If the titrand absorbs but the products do not, the curve has the form of curve A in Fig. 5. Curve B shows the response when neither of the reactants absorb, but the products are strong absorbers.

In some cases, systems that involve no absorbing species can be handled by the addition of an optical indicator. This is a substance that undergoes no change in absorbance until the titrant is present in slight excess over that required for the main reaction. *See* SPECTROPHOTOMETRIC ANALYSIS.

Amperometric titration. By use of a dropping-mercury or other suitable microelectrode, it is possible to find a region of applied electromotive force (emf) in which the current is proportional to the concentration of one or both of the reactants in a titration. Figure 6 shows the separate current-voltage curves, at approximately similar concentrations, of two species X and Y that can react to form nonelectroactive products. For example, X and Y may form an essentially insoluble precipitate. Provided that conditions are suitable, the currents may be anodic or cathodic.

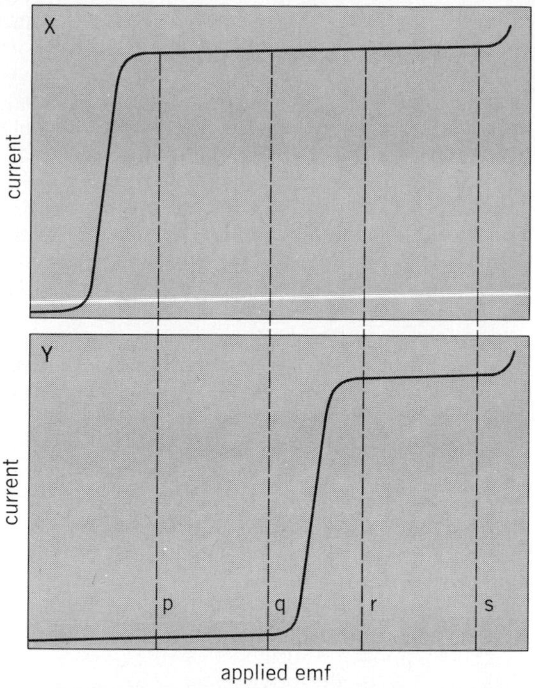

Fig. 6. Current-voltage curves at similar concentrations for two species (X and Y) that can react to form nonelectroactive products.

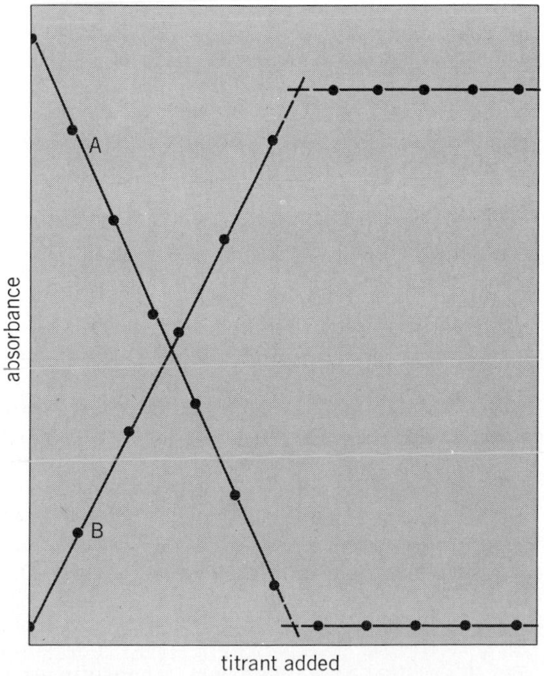

Fig. 5. Spectrophotometric titration curves: A, when the titrand absorbs but the products do not; and B, when neither of the reactants absorbs but the products do.

Suppose that the applied emf is fixed within the range *pq* and that Y is being titrated with X. The current remains at the small residual value until near the end point and rises as the titration is carried past the end point. The general shape of the titration curve is thus similar to that shown in Fig. 4. Titration at a fixed emf within the range *rs* gives a V-shaped curve, because both titrant and titrand are electroactive under these conditions.

Amperometric titration has been applied to all classes of reactions. Foreign ions may be present (and are often deliberately added to suppress migration currents), provided that they neither interfere with the titration reaction nor are electroactive under the chosen conditions.

Biamperometric titration is a closely related technique. An emf that is usually small is applied across two identical microelectrodes, usually of platinum, that dip into the titrand solution. This arrangement, which involves no liquid-liquid junctions, is obviously valuable in nonaqueous titrations, but also finds much use in aqueous titrimetry. The titration curves are nonlinear, so that direct observation, rather than graphing, is normally used to find the end point. *See* POLAROGRAPHIC ANALYSIS.

Thermometric or enthalpimetric titration. Many chemical reactions proceed with the evolution of heat. If one of these is used as the basis of a titration, the temperature first rises progressively and then remains unchanged as the titration is continued past the end point. If the reaction is endothermic, the temperature falls instead of rising. Thermometric titration is obviously applicable to all classes of reactions.

The total temperature change may be small, so that transfer of heat between the titration vessel and its surroundings must be minimized. Temperature changes are usually measured by means of a thermistor, which has both small heat capacity and a high negative coefficient of electrical resistance. The lower specific heats of solvents other than water indicate that thermometric titration is particularly suited to nonaqueous titration. *See* THERMOCHEMISTRY.

Nonaqueous titration. Water, the cheapest solvent, may react with, or may not dissolve, the substances to be used in a titration. Recourse must then be made to nonaqueous titrimetry. However, the main use of this technique is to perform titrations that give poor or no end points in water. Although applicable in principle to all classes of reactions, acid-base applications have greatly exceeded all others.

When an acid AH is dissolved in an amphiprotic solvent SH, and equilibrium is established as shown in reaction (4), if SH is strongly basic

$$AH + SH \rightleftharpoons A^- + SH_2^+ \qquad (4)$$

(protophilic), the equilibrium is forced to the right. All acids, whether weak or strong in water, tend to behave as strong acids. This is termed the leveling effect. Compounds such as phenol are too weakly acidic to be titrated in water. However, the titration can be performed in a solvent such as ethylenediamine. A common titrant is tetrabutylammonium hydroxide in the same solvent.

Analogously, a strongly acidic (protogenic) solvent exerts a leveling effect on basic solutes. Pyridine and other bases that are very weak in water can be nicely titrated in anhydrous acetic acid. The usual titrant is perchloric acid in the same solvent. By suitable choice of solvent, the apparent strengths of individuals in a mixture of acids or bases can sometimes be spread out, so that a succession of separate end points is obtained.

The coefficient of thermal expansion of a typical nonaqueous solvent is several times greater than that of water. Titrations should therefore be carried out at, or very near to, the temperature at which the titrant was standardized. Nonaqueous titrations in which the solvent is a molten salt or salt mixture are also possible.

Automatic titration. Automation is particularly valuable in routine titrations, which are usually performed repeatedly. One approach is to record the titration curve and to interpret it later. This requires a buret of the syringe or pump type that has a constant rate of delivery and is coupled to the recorder to provide the volume-of-titrant axis. Another method is to stop titrant addition or generation automatically at, or very near to, the end point. Although a constant-delivery device is desirable, an ordinary buret with an electromagnetically controlled valve is often used.

The second method is common in automatic potentiometric titration. Overshoot must be prevented if precise and accurate results are needed in minimum time. Titrant addition, rapid during most of the process, is automatically slowed down when the potential is approaching the end-point value. Microcomputer control permits such refinements as the continuous adjustment of the titrant flow rate during the titration. In some cases it is possible to automate an entire analysis, from the measurement of the sample to the final washout of the titration vessel and the printout of the result of the analysis. *See* ANALYTICAL CHEMISTRY.

[JOHN T. STOCK]

Bibliography: L. S. Bark and S. M. Bark, *Thermometric Titrimetry*, 1969; J. G. Dick, *Analytical Chemistry*, 1973; G. W. Ewing, *Instrumental Methods of Chemical Analysis*, 4th ed., 1975; W. Huber, *Titrations in Nonaqueous Solvents*, 1967.

Toluene

A colorless, aromatic hydrocarbon, also called methylbenzene, with the structural formula shown below. It boils at 110.6°C and freezes at −95.0°C.

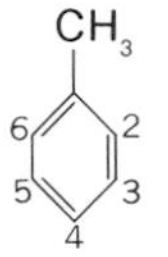

Most of the toluene sold in the United States is produced by the action of catalysts on petroleum hydrocarbons.

The nucleus of toluene, like that of benzene, undergoes substitution reactions. Substitution occurs almost exclusively in the ortho (2) and para (4) positions.

The hydrogen atoms of the —CH_3 group may be replaced by halogen at high temperatures in the presence of ultraviolet light.

Most of the pure toluene isolated is converted to benzene by hydrodealkylation. Other chemical uses for toluene are as an intermediate for the manufacture of tolylene diisocyanate, TNT, benzaldehyde, and benzoic acid. Toluene is also used in the paint industry as a solvent for gums and lacquers.

Prolonged breathing of air containing toluene vapor in concentrations greater than 200 ppm may be injurious to health. *See* AROMATIC HYDROCARBON; BENZENE. [CHARLES K. BRADSHER]

Bibliography: K. Weissermel and H.-J. Arpe, *Industrial Organic Chemistry*, 1978.

Toluidine

Any of three organic isomers, $CH_3C_6H_4NH_2$, homologs of aniline, and having properties very similar to those of aniline. *Ortho*-toluidine boils at 202°C, density 1.008; *meta*-toluidine boils at 204°C, density 0.990; and *para*-toluidine boils at 201°C and melts at 45°C. The three isomeric toluidines are prepared by reduction of the corresponding nitrotoluene with iron and dilute acid. The nitrotoluenes are synthesized by nitration of toluene and fractionally distilled for separation. The toluidines are only slightly soluble in water and are weak bases comparable to aniline. All are used as intermediates in preparing azo dyes. *See* AMINE; ANILINE; NITROBENZENE. [LEALLYN B. CLAPP]

Trace analysis

The determination of the elemental constituents of a sample in which these constituents make up approximately 0.01% of the sample or less. There is no sharp boundary between nontrace and trace constituents. The lower limit is set by the sensitivity of the available analytical methods and, in general, is pushed downward with progress in analytical techniques. A large number of different physical and chemical techniques have been developed for the measurement of the elemental composition at the microgram-per-gram and nanogram-per-gram level, thereby constituting the field of trace analysis.

Goals. Included in the goals of trace analysis are the determination of the bulk or total concentrations of the trace elements; the preconcentration of microconstituents into a small, essentially matrix-free sample to improve detectability or remove matrix interferences; the determination of local concentrations using probe techniques in order to establish the topographical distribution of trace elements in solid samples; and the determination of major and minor species in a minute initial sample. In the past much emphasis has been placed on analyses for bulk concentration, but there has been increased interest in the determination of local concentrations.

Methods. All trace analytical methods can be divided into three component steps: sampling, chemical or physical pretreatment, and instrumental measurement. Depending upon the type of information desired in an analysis and the requirements of sensitivity, precision, and other performance figures of merit, an appropriate instrumental technique is selected. Therefore, it is essential to know the capabilities and limitations of the various methods of instrumental measurement. Once they are known, appropriate steps can be taken in the sampling and pretreatment steps to provide sufficient microconstituent free of interferences and in the appropriate form for the final measurement. In a number of methods the pretreatment step may be omitted, and in others the sampling and measurement occur simultaneously. In spite of possible deviations, these steps are interrelated and require different degrees of emphasis, depending upon the individual analytical situation. In many cases the analyst uses a particular physical or physicochemical method in which manifestations of energy provide the basis of measurement. These methods are indirect in the sense that the emission or absorption of radiation or transformation of energy must be related in some way to the mass or concentration of the species that are being determined. The establishment of these relations almost invariably requires calibration, with the use of standards of known content of the constituent in question.

Techniques. Methods having applicability to trace analysis range from the more classical chemical methods of colorimetric and absorption spectrophotometric analysis to modern instrumental approaches. The wide diversity of methods is apparent from the number of different approaches and the attendant classes and subclasses in the accompanying table. Within any one of these categories there are still more specialized techniques.

Of the various criteria used in the selection of an appropriate trace analytical method, sensitivity, accuracy and precision, and selectivity are of prime importance. Other important considerations, such as scope, sampling and standards requirements, cost of equipment, and time of analyses, are of great practical significance.

Sensitivity and detection limits. The application of any analysis technique to trace problems is governed to a large extent by the analytical sensitivity that can be achieved for the species of interest in a given material. Sensitivity reflects the ability to discern a small change in concentration or amount of species of interest. Detection limit is a closely related, but distinct, term used to indicate the lowest concentration or amount that can be determined with a specific degree of confidence.

The lower limit of detection can be expressed in either of two ways. The absolute limit is the smallest detectable weight of the substance expressed in micrograms, nanograms, and so on. The relative

Trace analysis methods and limits of detection

Method	Limits of detection: Absolute, g	Concentration, ppm
Chromatography		
Thin-layer	10^{-5} to 10^{-3}	
Gas-liquid		10 to 10^{6}
Liquid-liquid		10^{-3} to 10^{0}
Electrochemical		
Coulometry	10^{-9} to 10^{0}	
Ion selective electrode		10^{-2} to 10^{2}
Polarography		
Conventional		10^{0} to 10^{3}
Modern		10^{-3} to 10^{3}
Laser probe microanalysis		10^{2} to 10^{4}
Nuclear		
Neutron activation		10^{-3} to 10^{-1}
Electron probe		10^{2} to 10^{3}
Ion probe		10^{-1} to 10^{1}
Mass spectrometry		
Isotope dilution		10^{-5} to 10^{6}
Spark source		10^{-3} to 10^{1}
Organic microanalysis	$>10^{-5}$	
Optical-absorption		
Atomic		
Flame		10^{-3} to 10^{1}
Nonflame	10^{-15} to 10^{-9}	
Molecular		
UV-visible		10^{-3} to 10^{2}
Infrared		10^{3} to 10^{6}
Microwave		10^{0} to 10^{3}
Optical-emission		
AC spark		10^{1} to 10^{3}
DC arc		10^{-2} to 10^{2}
Electrical plasma		
DC jet		10^{-4} to 10^{2}
RF-induced		10^{-4} to 10^{2}
Microwave-induced	10^{-9} to 10^{-6}	10^{-2} to 10^{1}
Optical-fluorescence		
Atomic		
Flame		10^{-3} to 10^{2}
Nonflame	10^{-15} to 10^{-9}	
Molecular		10^{-3} to 10^{-1}
Optical-phosphorescence		10^{-3} to 10^{2}
Optical-Raman		10^{0} to 10^{5}
Spectrometric-resonance		
Nuclear magnetic		10^{1} to 10^{5}
Electron spin	10^{-9} to 10^{-6}	
Thermal analysis	10^{-5} to 10^{-4}	
Wet chemistry		
Gravimetry	10^{-3} to 10^{-2}	
Titrimetry		10^{-2} to 10^{4}
X-ray spectrometry		
Auger		10^{3} to 10^{5}
ESCA		10^{3} to 10^{5}
Fluorescence		10^{-1} to 10^{2}
Mössbauer		10^{0} to 10^{3}
Photoelectron		10^{0} to 10^{3}

limit is the lowest detectable concentration expressed as a percentage, parts per million, micrograms per gram, micrograms per milliliter, and the like, of the sample. The choice of the absolute versus the relative method is usually made on the basis of convenience or pertinence to the problem.

As a result of widely varying pathways of development of many analytical techniques, there is a lack of consistency in the definition or specification of detection limits in the analytical literature. Consequently, it is difficult to make critical comparisons between methods for a particular element. However, since a particular element may be determined by a number of different techniques, depending upon the matrix in which it is being sought, a summary of the experimental values that have been published is pertinent. These values are given in the table. *See* ACTIVATION ANALYSIS; ANALYTICAL CHEMISTRY; CHROMATOGRAPHY; ELECTROCHEMICAL TECHNIQUES; GAS CHROMATOGRAPHY; SPECTROSCOPY.

[ANDREW T. ZANDER]

Bibliography: W. W. Meinke and B. F. Scribner (eds.), *Trace Characterization, Chemical and Physical*, 1967; G. H. Morrison (ed.), *Trace Analysis: Physical Methods*, 1965; E. B. Sandell and H. Onishi, *Photometric Determination of Traces of Metals: General Aspects*, 1978; J. D. Winefordner (ed.), *Trace Analysis: Spectroscopic Methods for Elements*, 1976.

Transamination

The transfer of an amino group from one molecule to another without the intermediate formation of ammonia. Enzymatic reactions of this type play a prominent role in the formation and ultimate breakdown of amino acids by living organisms, and were first discovered by A. E. Braunstein and M. G. Kritzman in 1937. Most of the reactions studied represent specific instances of reaction (1),

$$\mathrm{R\overset{\displaystyle O}{\overset{\|}{C}}{-}COO^- + R'\overset{\displaystyle NH_2}{\overset{|}{C}}HCOO^- \rightleftharpoons R\overset{\displaystyle NH_2}{\overset{|}{C}}HCOO^- + R'\overset{\displaystyle O}{\overset{\|}{C}}{-}COO^-} \qquad (1)$$

where R and R′ are hydrogen or the organic radicals of the naturally occurring amino acids. Enzymes that catalyze such reactions are widely distributed and are termed transaminases, or aminotransferases. Similar reactions where one of the reactants is not an α-amino acid (or α-keto acid) are also known; a more general formulation of such reactions is shown in reaction (2), where R, R′, R″, and R‴ represent hydrogen or organic radicals.

$$\mathrm{R{-}\overset{\displaystyle O}{\overset{\|}{C}}{-}R' + R''\overset{\displaystyle NH_2}{\overset{|}{C}}HR''' \rightleftharpoons R\overset{\displaystyle NH_2}{\overset{|}{C}}HR' + R''\overset{\displaystyle O}{\overset{\|}{C}}{-}R'''} \qquad (2)$$

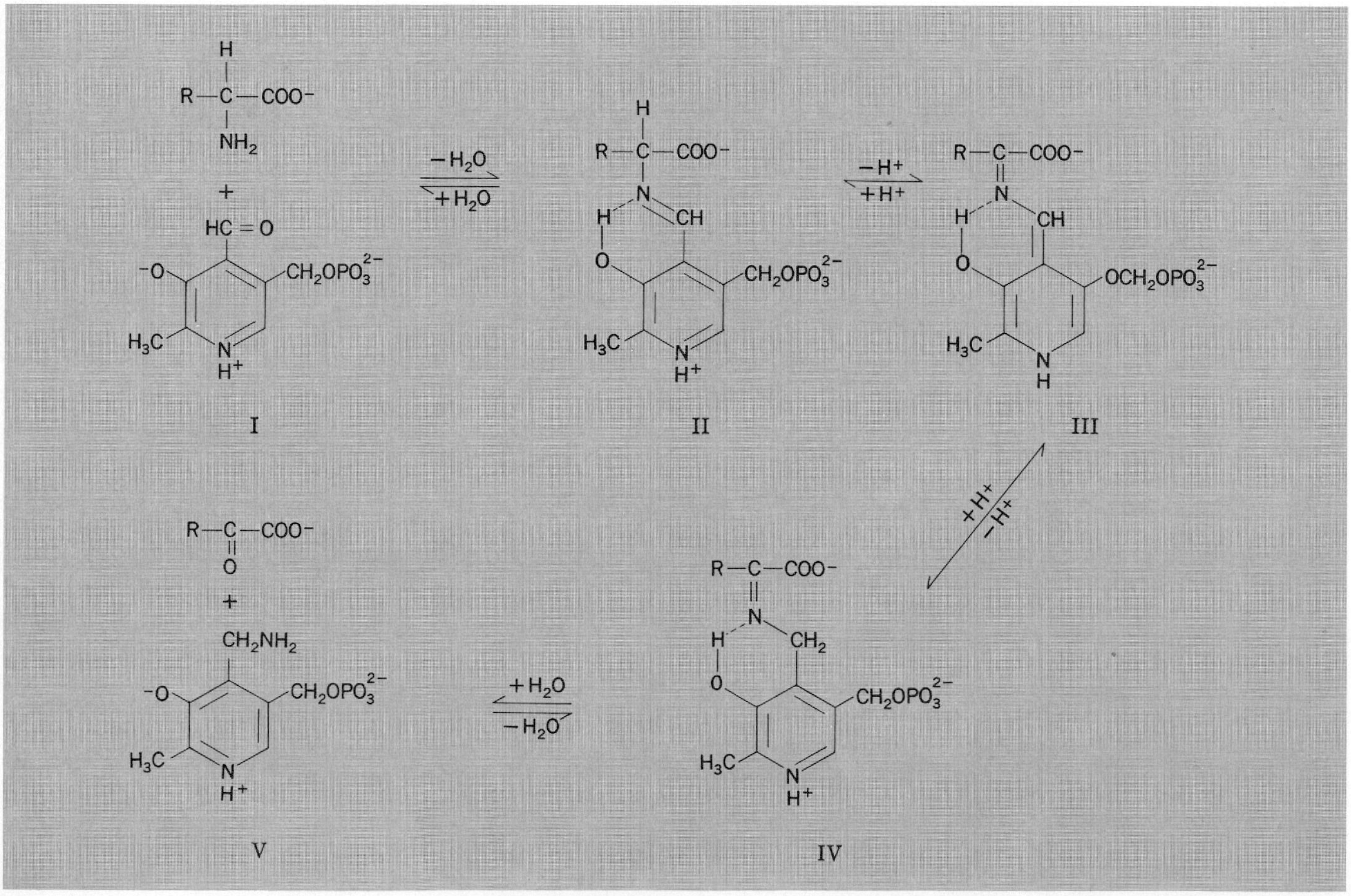

Pathway of reactions for a transamination.

Perhaps the most prominent transamination reactions in higher animals are those in which glutamate is formed from α-ketoglutarate and other amino acids, such as reaction (3).

$$\alpha\text{-Ketoglutarate} + \text{Amino acid}_1 \rightleftharpoons \text{Glutamate} + \text{Keto acid}_1 \quad (3)$$

In conjunction with the action of glutamate dehydrogenase, an enzyme that catalyzes reaction (4), these transamination reactions provide a mechanism (termed transreamination) by which inorganic nitrogen, as ammonia, can be incorporated first into glutamate, by reaction (4), and then into other

$$\text{Glutamate} + \text{DPN}^+ + H_2O \rightleftharpoons \alpha\text{-Ketoglutarate} + NH_3 + \text{DPNH} + H^+ \quad (4)$$

acids, by reaction (3), to form each of the amino acids necessary for protein synthesis. Alternatively, where amino acids are present in excess, reactions (3) and (4) provide a mechanism by which the amino group can be eliminated as ammonia (transdeamination) preliminary to utilization of the nitrogen-free residue as a source of energy. *See* AMINO ACIDS.

All transaminases that catalyze reactions of the type shown in (1) or (3) contain a derivative of vitamin B_6, namely, pyridoxal-5′-phosphate(I) or pyridoxamine-5′-phosphate(V), as an essential coenzyme. Mechanistic studies in both model and enzymatic systems indicate that the coenzyme functions alternately as an amino group acceptor (in the aldehyde form) and an amino group donor (in the amine form), according to reactions (5*a*) and (5*b*). The sum of these two reactions constitutes

$$\text{RCOCOO}^- + \text{Pyridoxamine-P} \rightleftharpoons \text{RCHNH}_2\text{COO}^- + \text{Pyridoxal-P} \quad (5a)$$

$$\text{R'CHNH}_2\text{COO}^- + \text{Pyridoxal-P} \rightleftharpoons \text{R'COCOO}^- + \text{Pyridoxamine-P} \quad (5b)$$

the overall transamination reaction (1), in which pyridoxamine-P and pyridoxal-P (or the enzymes that contain them) serve only as catalysts (see illustration). Catalysis of reactions (5*a*) and (5*b*) by the coenzyme results from formation of intermediary complexes such as II and IV, in which the coenzyme acts as an electron sink, labilizing the α-hydrogen atom of the amino acid or a methylene hydrogen atom of pyridoxamine phosphate, as shown in the illustration, thus forming the common intermediate III, through which all of the reactants and products are in equilibrium. The enzymatic reactions are further catalyzed in ways not yet clear by the protein portion of the enzyme.

[ESMOND E. SNELL]

Bibliography: P. D. Boyer (ed.), *The Enzymes*, vol. 2, 3d ed., 1970; M. Florkin and E. H. Stotz (eds.), *Comprehensive Biochemistry*, vol. 15, 1964; H. R. Mahler and E. H. Cordes, *Biological Chemistry*, 2d ed., 1971; C. J. Suckling and K. E. Suckling, *Biological Chemistry*, 1980.

Transesterification

The group of reactions in which an ester reacts with another compound to form a different ester. The other compound may be an acid, an alcohol, or another ester. These reactions are also known as acidolysis, alcoholysis, and ester interchange, respectively. They are all special cases of esterification and have the characteristics of esterification reactions in general.

A few of the more important applications of transesterification are discussed below. Acrylic and methylacrylic esters of high-molecular-weight saturated and unsaturated alcohols are prepared by alcoholysis of methyl acrylate and methyl methacrylate; polyvinyl alcohol, by alcoholysis of polyvinyl acetate with methyl alcohol; and the prepolymer of polyethylene terephthalate, by alcoholysis of the methyl ester of terephthalic acid with ethylene glycol, because less effort is required to prepare dimethylterephthalate of the required purity than terephthalic acid. Monoglycerides of unsaturated fatty acids are prepared by alcoholysis of the appropriate oil with glycerol. The properties of lard can be improved by combining ester interchange in the natural mixture of glycerides with precipitation of the higher-melting material. Thus, saturated glycerides are preferentially removed.

Transesterification reactions are generally reversible and accompanied by smaller heat effects than are other esterification reactions. The reactions are conducted in the liquid phase, usually in the presence of a catalyst. With catalysts, the common reaction temperature is about 100°C; without a catalyst, it is as a rule about 250°C. One major exception is the directed ester interchange in fats, which is conducted at temperatures around 50°C. The lower temperature permits crystallization of the saturated glycerides. A pressure of 1 atm is normal, and the reaction is pushed to higher conversions by using an excess of the displacing acid or alcohol.

Reaction equilibria. The equilibrium constants, and thus the equilibrium conversion, can be predicted from the data for the formation of the ester reactants and products. Consider the alcoholysis reaction (1).

$$\text{RCOOR'} + \text{R''OH} \rightleftharpoons \text{RCOOR''} + \text{R'OH} \quad (1)$$

The reactions leading to the formation of the two esters from the appropriate acid and alcohol are shown in (2) and (3).

$$\text{RCOOH} + \text{R'OH} \rightleftharpoons \text{RCOOR'} + H_2O \quad (2)$$

$$\text{RCOOH} + \text{R''OH} \rightleftharpoons \text{RCOOR''} + H_2O \quad (3)$$

The respective concentration equilibrium constants are represented as Eqs. (4), so that the equi-

$$K_1 = \frac{[\text{RCOOR''}][\text{R'OH}]}{[\text{RCOOR'}][\text{R''OH}]} \qquad K_2 = \frac{[\text{RCOOR'}][H_2O]}{[\text{RCOOH}][\text{R'OH}]}$$
$$K_3 = \frac{[\text{RCOOR''}][H_2O]}{[\text{RCOOH}][\text{R''OH}]} \quad (4)$$

librium constant for the alcoholysis reaction may be obtained from the ratio of the equilibrium constants (K_3/K_2) for the two esterification reactions. If R″ is an alcohol with a large esterification equilibrium constant relative to R′ for the same acid, RCOOH, a large value will be obtained for K_1. The reaction will be essentially irreversible.

Reaction kinetics and catalysis. The rates of transesterification reactions respond to reaction conditions and molecular structure in a manner similar to that of esterification reactions in general. Catalysts useful for esterification also are good for transesterification. In the presence of an alkaline catalyst, the rates of alcoholysis are very high, being faster even than saponification. Sodium alkoxides are particularly active for alcoholysis and ester interchange and permit practical reaction rates to be obtained at lower temperatures than those of esterification. Because of their basic nature, these catalysts obviously would be unsuitable for other esterifications.

Reaction mechanism. Acid-catalyzed and base-catalyzed transesterification reactions are explained by different mechanisms. Acid-catalyzed reactions follow the same course as other esterification reactions. With base-catalyzed reactions, the basicity of the oxygen of the alcohol appears to be increased so that it attacks the carbonyl carbon directly. Electron-attracting substituent groups in either the acyl or the alkoxy part of the ester increase the reaction rate. Lower activation energies are observed in base-catalyzed reactions. *See* ACIDOLYSIS; ALCOHOLYSIS; ESTER; HYDROLYSIS. [JOHN M. WOODS]

Transition elements

In broad definition, the elements of atomic numbers 21–31, 39–49, and 71–81, inclusive. The symbols of these elements, along with their atomic numbers and valence-shell electronic configurations, are given in the table. The elements are arranged in the order in which they appear in the long, or Bohr, form of the periodic table.

A more restricted classification of the transition elements, preferred by many chemists, is indicated by the heavy border drawn about the central portion of the table. All of the elements in this section of the table have one or more electrons present in an unfilled *d* subshell in at least one well-known oxidation state.

The metals and their uses. All the transition elements are metals and, in general, are characterized by high densities, high melting points, and low vapor pressures. Included, for example, are tungsten and tantalum, melting at 3370 and 3030°C, respectively. At room temperature, the vapor pressure of tungsten is so low that it compares to only one gaseous atom in a volume of space equal to that of the known sidereal universe. In general, those properties related to strong cohesiveness or binding between the atoms in the metallic state, such as high density, extreme hardness, and high melting point, reach a broad maximum in the neighborhood of the fourth member of each series. Within a given subgroup, these same properties tend to increase with increasing atomic weight. Facility in the formation of metallic bonds is demonstrated also by the existence of a wide variety of alloys between different transition metals. The variation in some representative properties of the transition elements as a function of atomic number is illustrated in the graph.

The transition elements include most metals of major economic importance, such as the relatively abundant iron, copper, nickel, and zinc, on one hand, and the rarer coinage metals, copper, silver, and gold, on the other. Also included are the rare and relatively unfamiliar element, rhenium, and technetium, which is not found naturally in the terrestrial environment, but is available in small amounts as a product of nuclear fission.

Chemical properties. In their compounds, the transition elements tend to exhibit multiple valency, the maximum valence increasing from +3 at the beginning of a series (Sc, Y, Lu) to +8 at the fifth member (Mn, Re). For the elements in any vertical column, the highest oxidation state is usually observed in the element at the bottom of the column. Thus the highest oxidation state of iron is +6, whereas osmium attains an oxidation state of +8.

One of the most characteristic features of the transition elements is the ease with which most of them form stable complex ions. Features which contribute to this ability are favorably high charge-to-radius ratios and the availability of unfilled *d* orbitals which may be used in bonding. Examples of such complexes include a very stable cyanide complex of aurous gold $Au(CN)_2^-$, of commercial importance for the recovery of the metal from low-grade ores, a similar complex $Ag(CN)_2^-$, useful in obtaining bright, firmly adherent deposits of silver by electroplating, and numerous ammonia complexes, of which the deep blue $Cu(NH_3)_4^{++}$ is a representative and familiar example, widely used in colorimetric analyses for copper. Vitamin B_{12} is an example of a cobalt (III) complex that is important in nutrition, and hemin, the red pigment of blood, is an important iron(II) complex.

Most of the ions and compounds of the transition metals are colored, and many of them are paramagnetic; that is, when placed in a magnetic field, the magnetic flux within the compound is higher than that of the surrounding field. Both color and paramagnetism are related to the presence of unpaired electrons in the *d* subshell. Excitation of these relatively loosely bound electrons to higher energy states accounts for the absorption of light in the visible region of the spectrum, while the magnetic field associated with the electron spin is responsible for the magnetic behavior of the compounds. Study of the magnetic behavior of compounds and complex ions of the transition elements has contributed much to an understanding

Transition elements, showing atomic number, symbol, and electron configurations in valence shells

21 Sc $3d4s^2$	22 Ti $3d^24s^2$	23 V $3d^34s^2$	24 Cr $3d^54s$	25 Mn $3d^54s^2$	26 Fe $3d^64s^2$	27 Co $3d^74s^2$	28 Ni $3d^84s^2$	29 Cu $3d^{10}4s$	30 Zn $3d^{10}4s^2$	31 Ga $3d^{10}4s^24p$
39 Y $4d5s^2$	40 Zr $4d^25s^2$	41 Nb $4d^45s$	42 Mo $4d^55s$	43 Tc $4d^65s$	44 Ru $4d^75s$	45 Rh $4d^85s$	46 Pd $4d^{10}$	47 Ag $4d^{10}5s$	48 Cd $4d^{10}5s^2$	49 In $4d^{10}5s^25p$
71 Lu $5d6s^2$	72 Hf $5d^26s^2$	73 Ta $5d^36s^2$	74 W $5d^46s^2$	75 Re $5d^56s^2$	76 Os $5d^66s^2$	77 Ir $5d^9$	78 Pt $5d^96s$	79 Au $5d^{10}6s$	80 Hg $5d^{10}6s^2$	81 Tl $5d^{10}6s^26p$

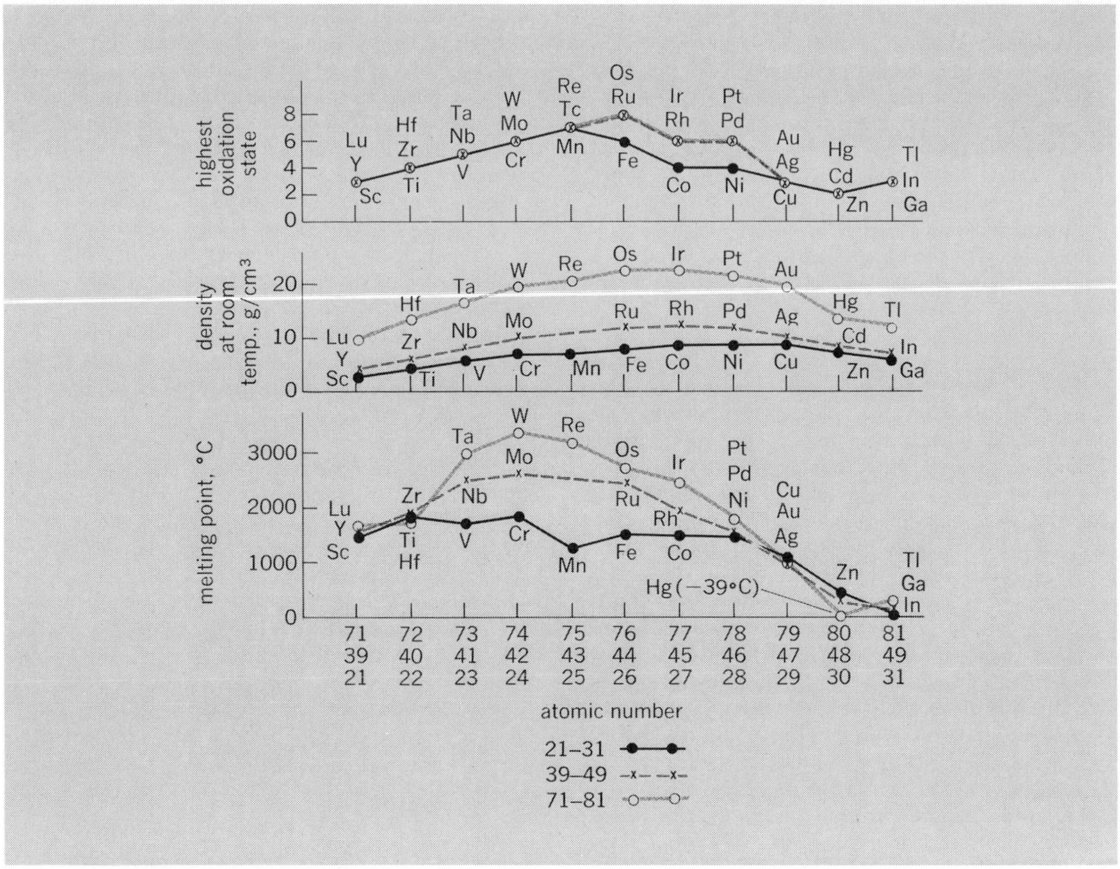

Physical properties of transition elements as function of their atomic number.

of chemical bonding in these elements, since utilization of the *d* electrons in bonding involves electron-pair formation, with consequent cancellation of the magnetic moments and altered magnetic properties. Because of their ability to accept electrons in unoccupied *d* orbitals, transition elements and their compounds frequently exhibit catalytic properties.

Many of the most important catalysts, such as nickel used in hydrogenation, are transition elements.

Broadly speaking, the properties of the transition elements are intermediate between those of the so-called representative elements, in which the subshells are completely occupied by electrons (alkali metals, halogen elements), and those of the inner or *f* transition elements, in which the subshell orbitals play a much less significant role in influencing chemical properties (rare-earth elements, actinide elements). *See* CATALYSIS; COMPLEX COMPOUNDS; COORDINATION CHEMISTRY; MAGNETOCHEMISTRY.

[BURRIS B. CUNNINGHAM]

Bibliography: F. Basolo et al. (eds.), *Transition Metal Chemistry*, vol. 1, 1973, vol. 2, 1977.

Transition point

The point at which a substance changes from one state of aggregation to another. This general definition would include the melting point (transition from solid to liquid), boiling point (liquid to gas), or sublimation point (solid to gas); but in practice the term transition point is usually restricted to the transition from one solid phase to another, that is, to the temperature (for a fixed pressure, usually 1 atm) at which a substance changes from one crystal structure to another.

Some typical examples of transition points are:

$$\underset{\text{(body-centered cubic)}}{\beta\text{-Fe}} \xrightarrow{\text{at 1180 K}} \underset{\text{(face-centered cubic)}}{\gamma\text{FE}}$$

$$\underset{\text{(rhombic)}}{S_8} \xrightarrow{\text{at 369 K}} \underset{\text{(monoclinic)}}{S_8}$$

$$\underset{\text{(monoclinic)}}{CCl_4} \xrightarrow{\text{at 225.5 K}} \underset{\text{(tectragonal)}}{CCl_4}$$

$$\underset{(\beta\text{-rhombic})}{NH_4NO_3} \xrightarrow{\text{at 305.3 K}} \underset{(\alpha\text{-rhombic})}{NH_4NO_3}$$

$$\underset{(\alpha\text{-rhombic})}{NH_4NO_3} \xrightarrow{\text{at 357.4 K}} \underset{\text{(trigonal)}}{NH_4NO_3}$$

Another kind of transition point is the culmination of a gradual change (for example, the loss of ferromagnetism in iron or nickel) at the lambda point, or Curie point. This behavior is typical of second-order transition. *See* BOILING POINT; FIRST-ORDER TRANSITION; MELTING POINT; PHASE EQUILIBRIUM; SECOND-ORDER TRANSITION; SUBLIMATION; TRIPLE POINT.

[ROBERT L. SCOTT]

Transmutation

The nuclear change of one element into another, either naturally, in radioactive elements, or artificially, by bombarding an element with electrons, deuterons, or α-particles in particle accelerators or with neutrons in atomic piles.

Natural transmutation was first explained by

Marie Curie about 1900 as the result of the decay of radioactive elements into others of lower atomic weight. Ernest Rutherford produced the first artificial transmutation (nitrogen into oxygen and hydrogen) in 1919. Artificial transmutation is the method of origin of the heavier, artificial transuranium elements, and also of hundreds of radioactive isotopes of most of the chemical elements in the periodic table. Practically all of these elements also have been artificially transmuted into neighboring elements under experimental conditions. *See* PERIODIC TABLE.

[FRANK H. ROCKETT]

Transport processes

The processes whereby mass, energy, or momentum are transported from one region of a material to another under the influence of composition, temperature, or velocity gradients. If a sample of a material in which the chemical composition, the temperature, or the velocity vary from point to point is isolated from its surroundings, the transport processes act so as to eventually render these quantities uniform throughout the material. The nonuniform state required to generate these transport processes causes them to be known also as nonequilibrium processes. Associated with gradients of composition, temperature, and velocity in a material are the transport processes of diffusion, thermal conduction, and viscosity, respectively. For a large class of materials, the laws which govern the transport processes are quite simple.

Diffusion. Figure 1 shows a sample of a material which is composed of two chemical species. The sample is stationary and has a uniform temperature throughout, but a composition difference is maintained across its two ends, and in this steady state the two species continuously migrate down their concentration gradients. Expressing the composition of the material by means of the molar concentration of one species, c_1 (moles/m^3), it is found that the number of moles of this species which cross unit area of the sample perpendicular to the z direction in unit time known as the flux of mass (J_1), is given by Eq. (1) which is Fick's law of

$$J_1 = -D\frac{dc_1}{dz} \qquad (1)$$

diffusion. The constant of proportionality (D) between the mass flux and the concentration gradient, which depends upon the nature of the material, its temperature, pressure, and composition, is called the diffusion coefficient.

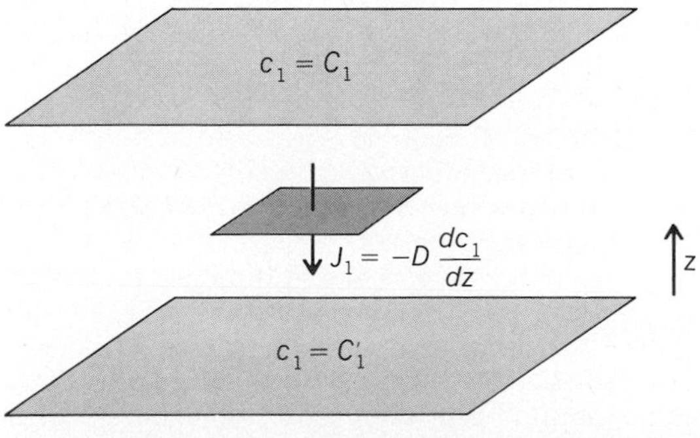

Fig. 1. Diffusion in a sample of material composed of two chemical species. C_1 and C'_1 represent the molar concentration of one of the species in the two planes bounding the material.

The phenomenon of diffusion occurs widely in nature, and it is frequently important in technological applications. For example, the transpiration of the leaves of plants, in which they absorb carbon dioxide from the atmosphere and give off water vapor, is controlled by a diffusion process. The rates of many chemical reactions in fluids which are promoted by catalysts may similarly be controlled by the diffusion of reactants to the active catalyst sites.

Thermal conduction. In Fig. 2, a sample of a material is subjected to a steady temperature difference between two faces perpendicular to the z direction. Under these conditions, energy is continually transported from the hotter face to the colder, and the energy flux, J_q, in the z direction (the energy crossing unit area in unit time) is given by Fourier's law as Eq. (2). The constant of propor-

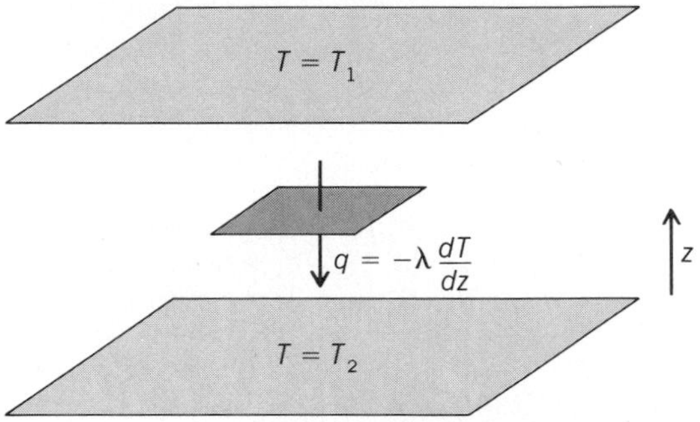

Fig. 2. Thermal conductivity in a sample of material subjected to a steady temperature difference.

$$J_q = -\lambda\frac{dT}{dz} \qquad (2)$$

tionality between the flux and the temperature gradient, dT/dz, is the thermal conductivity coefficient, which again depends upon the material as well as its temperature, pressure, and composition.

Viscosity. The phenomenon of viscosity is associated with the gradient of velocity in a material. Since it is difficult to maintain velocity gradients in solids, the phenomenon is only readily observed in fluids. Because velocity is a vector quantity, Fig. 3 shows a fluid whose upper surface is in contact with a solid boundary which moves with a steady velocity U in the x direction only; the lower surface of the fluid is held stationary. As a result, various layers of the fluid in the z direction move with different x-direction velocities, u. Associated with the motion in the x direction, the fluid possesses a momentum, and the x-direction momentum is transported down the velocity or momentum gradient. The flux of x momentum in the z direction, J_m, is equivalent to a tangential shear stress, τ_{xz}, acting in the negative x direction on each layer of the fluid. This means that a tangential force must be applied to the upper plate to keep it in steady motion. Again the flux is proportional to the imposed gradient and is given by Eq. (3), which is

$$J_m = \tau_{xz} = -\eta\frac{du}{dz} \qquad (3)$$

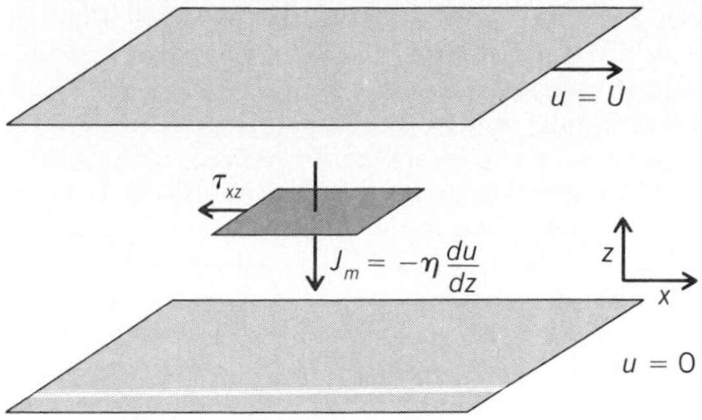

Fig. 3. Viscosity in a fluid whose upper surface is in contact with a solid boundary.

Newton's law of viscosity. The proportionality constant, η, is the viscosity coefficient for the material, and it too depends on the thermodynamic state of the material. The phenomenon of viscosity is revealed whenever a fluid flows near a solid boundary, and it is therefore of significance in almost every aspect of engineering. *See* VISCOSITY.

Thermal processes. Other, more subtle transport processes can occur. For example, in a mixture of two chemical species, the imposition of a temperature gradient leads not only to energy transport, but also to a mass transport which causes a partial separation of the mixture. This phenomenon is called thermal diffusion. Conversely, when diffusion takes place in an initially isothermal mixture as a result of a composition gradient, small temperature gradients can be observed in the material arising from an energy transport. This is the diffusion thermoeffect.

Transport coefficients. The coefficients D, λ, and η are known collectively as transport coefficients. The measurement of these coefficients for materials in solid, liquid, and gaseous phases has been the object of a considerable research effort for many years. The measurements can only rarely be carried out by directly implementing the situations envisaged in Figs. 1 to 3. This is because it is difficult to achieve the one-dimensional gradients of the quantities required when the sample is of a finite size. The exceptions to this are the measurement of thermal conductivity and diffusion in solids, where simple methods have proved effective. In fluids the diffusion coefficient has most often been determined in a time-dependent experiment in which an initial concentration gradient in a mixture is allowed to decay in a closed vessel of known geometry. The approach to equilibrium, which is governed by the diffusion coefficient, is observed with a suitable concentration monitor.

The coefficient of thermal conductivity in fluids is also most accurately determined in a transient experiment. The fluid surrounds a thin, vertical wire which is suddenly heated by an electric current; the rate of the temperature rise of the wire is observed, and the thermal conductivity of the fluid is deduced from it.

The viscosity coefficient of gases and liquids is generally determined by one of two techniques. In the first, the fluid flows through a capillary tube of known geometry, and the pressure difference across the ends of the tube necessary to maintain a given flow rate is determined. This pressure difference is then proportional to the viscosity coefficient of the fluid. In the second method, the damping of the torsional oscillations of a thin, solid, horizontal, circular disk is observed when the disk is suspended in the fluid. The measurement of the logarithmic decrement of the oscillations serves to determine the viscosity coefficient. *See* QUASI-ELASTIC LIGHT SCATTERING.

The results of measurements of the transport coefficients of material are of importance since there are few technological activities which do not involve one or more of the transport processes. However, because the transport coefficients derive their values from the properties and behavior of the atoms and molecules which make up the material, they are also of more fundamental significance. In the particular case of gases at low density (near atmospheric pressure), the kinetic theory of gases has provided an almost complete description of their transport coefficients. In such gases the sole mechanism for the transport of mass, energy, or momentum is by means of free molecular motion. The transport coefficients are therefore a direct measure of the ease of this free molecular motion. Since the molecular motion is hindered only by collisions between pairs of molecules, the transport coefficients are sensitive to the details of such collisions and thereby to the potentials describing the interactions of molecules. Indeed, the transport coefficients of dilute gases have proved a valuable source of information about these interactions. The viscosity and thermal conductivity of coefficients of moderately dense gases are essentially independent of pressure, but the diffusion coefficient is inversely proportional to it. The thermal conductivity and viscosity increase with temperature roughly proportionally, whereas the temperature dependence of the diffusion coefficient varies more nearly as the square of the temperature at constant pressure. These temperature dependencies arise principally from the temperature dependence of the velocity of the free molecular motion.

As the density of the material is increased toward that of a liquid and finally to that of a solid, significant changes in the molecular mechanism of the transport processes occur. The transport by free molecular motion becomes a smaller contribution as the volume available for such motion decreases. In addition, the attractive forces between molecules, which become increasingly significant, tend to inhibit molecular motion. Thus, on the one hand, the diffusion coefficient in condensed phases, which is still determined by molecular motion, is very much smaller (about 10^4 times) than that in low-density gases. On the other hand, the viscosity of liquids is very much greater than that in gases because the attractive forces between molecules make the relative motion of various layers in the fluid much more difficult to achieve. Because increasing the temperature of a liquid increases the average separation of the molecules as well as their energy, the diffusion coefficients for liquids increase rapidly with temperature and, for the same reasons, the viscosity decreases. In the solid the molecules acquire almost fixed positions, and the diffusion coefficient consequently becomes even smaller, whereas the viscosity is practically infinite. [W. A. WAKEHAM]

Bibliography: R. B. Bird, W. E. Stewart, and E. N. Lightfoot, *Transport Phenomena*, 1960; S. R.

de Groot and P. Mazure, *Non-Equilibrium Thermodynamics*, 1969; R. DiPippo, J. Kestin, and J. H. Whitelaw, A high-temperature oscillating disc viscometer, *Physica* 32:2064–2080, 1966; G. C. Maitland et al., *Intermolecular Forces: Their Origin and Determination*, 1980.

Transuranium elements

Those synthetic elements with atomic numbers larger than that of uranium (atomic number 92). They are the members of the actinide series, from neptunium (atomic number 93) through lawrencium (atomic number 103), and the transactinide elements (with higher atomic numbers than 103). Of these elements, plutonium, an explosive ingredient for nuclear weapons and a fuel for nuclear power because it is fissionable, has been prepared on the largest (ton) scale, while some of the others have been produced in kilograms (neptunium, americium, curium) and in much smaller quantities (berkelium, californium, and einsteinium).

The concept of atomic weight in the sense applied to naturally occurring elements is not applicable to the transuranium elements, since the isotopic composition of any given sample depends on its source. In most cases the use of the mass number of the longest-lived isotope in combination with an evaluation of its availability has been adequate. Good choices at present are neptunium, 237; plutonium, 242; americium, 243; curium, 248; berkelium, 249; californium, 249; einsteinium, 254; fermium, 257; mendelevium, 258; nobelium, 259; lawrencium, 260; rutherfordium, 261; hahnium, 262; and element 106, 263.

The actinide elements are chemically similar and have a strong chemical resemblance to the lanthanide, or rare-earth, elements (atomic numbers 57–71). The transactinide elements, with atomic numbers 104 to 118, should be placed in an expanded periodic table under the row of elements beginning with hafnium, number 72, and ending with radon, number 86. This arrangement allows prediction of the chemical properties of these elements and suggests that they will have an element-by-element chemical analogy with the elements which appear immediately above them in the periodic table.

The transuranium elements up to and including fermium (atomic number 100) are produced in largest quantity through the successive capture of neutrons in nuclear reactors. The yield decreases with increasing atomic number, and the heaviest to be produced in weighable quantity is einsteinium (number 99). Many additional isotopes are produced by bombardment of heavy target isotopes with charged atomic projectiles in accelerators; beyond fermium all elements are produced by bombardment with heavy ions. Brief descriptions of transuranium elements follow. They are listed according to increasing atomic number.

Neptunium. Neptunium (Np, atomic number 93, named after the planet Neptune) was the first transuranium element discovered. In 1940 E. M McMillan and P. H. Abelson at the University of California, Berkeley, identified the isotope ^{239}Np (half-life 2.35 days), which was produced by the bombardment of uranium with neutrons according to reaction (1). The element as ^{237}Np was first isolated as a pure compound, the oxide, in 1944 by L. G. Magnusson and T. J. La Chapelle. Neptunium in trace amounts is found in nature, the element being produced in nuclear reactions in uranium ores caused by the neutrons present. Kilogram and larger quantities of ^{237}Np (half-life 2.14×10^6 years), used for chemical and physical investigations, are being produced as a by-product of the production of plutonium in nuclear reactors. Isotopes from mass number 227 to 241 have been synthesized by various nuclear reactions.

$$^{238}U(n,\gamma)^{239}U \rightarrow {}^{239}Np \qquad (1)$$

Neptunium displays five oxidation states in aqueous solution: Np^{3+} (pale purple), Np^{4+} (yellow-green), NpO_2^{+} (green-blue), NpO_2^{2+} (pink), and NpO_5^{3-} (green). The ion NpO_2^{+}, unlike corresponding ions of uranium, plutonium, and americium, can exist in aqueous solution at moderately high concentrations. The element forms tri- and tetrahalides such as NpF_3, NpF_4, $NpCl_3$, $NpCl_4$, $NpBr_3$, NpI_3, as well as NpF_6 and oxides of various compositions such as those found in the uranium-oxygen system, including Np_3O_8 and NpO_2.

Neptunium metal has a silvery appearance, is chemically reactive, and melts at 637°C; the solid metal has at least three crystalline forms between room temperature and its melting point. *See* NEPTUNIUM.

Plutonium. Plutonium (Pu, atomic number 94, named after the planet Pluto) in the form of ^{238}Pu was discovered in late 1940 and early 1941 by G. T. Seaborg, McMillan, J. W. Kennedy, and A. C. Wahl at the University of California, Berkeley. The element was produced in the bombardment of uranium with deuterons according to reaction (2).

$$^{238}U(d,2n)^{238}Np \xrightarrow[2.1\text{ days}]{\beta^-} {}^{238}Pu \qquad (2)$$

The important isotope ^{239}Pu was discovered by Kennedy, Seaborg, E. Segrè, and Wahl in 1941. Plutonium-239 (half-life 24,400 years), because of its property of being fissionable with neutrons, is used as the explosive ingredient in nuclear weapons and is a key material in the development of nuclear energy for industrial purposes, 1 lb (0.45 kg) being equivalent to about 10,000,000 kWhr of heat energy; it is produced in ton quantities in nuclear reactors. The alpha radioactivity and physiological behavior of this isotope make it one of the most dangerous poisons known, but means for handling it safely have been devised. Plutonium as ^{239}Pu was first isolated as a pure compound, the fluoride, in 1942 by B. B. Cunningham and L. B. Werner. Minute amounts of plutonium formed in much the same way as naturally occurring neptunium are present in nature. Much smaller quantities of the longer-lived isotope ^{244}Pu (half-life 83,000,000 years) have been found in nature; in this case it may represent the small fraction remaining from a primordial source or it may be caused by cosmic rays. Isotopes of mass number 232–246 are known. The longer-lived isotopes ^{242}Pu (half-life 390,000 years) and (eventually) ^{244}Pu, produced in nuclear reactors, are more suitable than ^{239}Pu for chemical and physical investigation because of their longer half-lives and lower specific activities.

Plutonium has five oxidation states in aqueous solution: Pu^{3+} (blue to violet), Pu^{4+} (yellow-brown), PuO_2^{+} (pink), PuO_2^{2+} (pink-orange), and PuO_5^{3-} (blue-green). The ions Pu^{4+} and PuO_2^{+} undergo extensive disproportionation to the ions of higher and lower oxidation states. Four oxidation states (III, IV, V, and VI) can exist simultaneously at appreciable concentrations in equilibrium with each other, an unusual situation that leads to complicated solution phenomena.

Plutonium forms binary compounds with oxygen (PuO, PuO_2, and intermediate oxides of variable composition); with the halogens (PuF_3, PuF_4, PuF_6, $PuCl_3$, $PuBr_3$, PuI_3); with carbon, nitrogen, and silicon (including PuC, PuN, $PuSi_2$); in addition, oxyhalides are well known (PuOCl, PuOBr, PuOI).

The metal is silvery in appearance, is chemically reactive, melts at 640°C, and has six crystalline modifications between room temperature and its melting point. *See* PLUTONIUM.

Americium. Americium (Am, atomic number 95, named after the Americas) was the fourth transuranium element discovered. The element as ^{241}Am (half-life 433 years) was produced by the intense neutron bombardment of plutonium and was identified by Seaborg, R. A. James, L. O. Morgan, and A. Ghiorso in late 1944 and early 1945 at the wartime Metallurgical Laboratory at the University of Chicago. By using the isotope ^{241}Am, the element was first isolated as a pure compound, the hydroxide, in 1945 by B. B. Cunningham. Isotopes of mass numbers 237–246 have been prepared. Kilogram quantities of ^{241}Am are being produced in nuclear reactors. The less radioactive isotope ^{243}Am (half-life 7400 years), also produced in nuclear reactors, is more suitable for use in chemical and physical investigation.

Americium exists in four oxidation states in aqueous solution: Am^{3+} (light salmon), AmO_2^{+} (light tan), AmO_2^{2+} (light tan), and a fluoride complex of the IV state (pink). The trivalent state is highly stable and difficult to oxidize. AmO_2^{+}, like plutonium, is unstable with respect to disproportionation into Am^{3+} and AmO_2^{2+}. The ion Am^{4+} may be stabilized in solution only in the presence of very high concentrations of fluoride ion, and tetravalent solid compounds are well known. Divalent americium has been prepared in solid compounds; this is consistent with the presence of seven $5f$ electrons in americium (enhanced stability of half-filled $5f$ electron shell) and is similar to the analogous lanthanide, europium, which can be reduced to the divalent state.

Americium dioxide, AmO_2, is the important oxide; Am_2O_3 and, as with previous actinide elements, oxides of variable composition between $AmO_{1.5}$ and AmO_2 are known. The halides AmF_2 (in CaF_2), AmF_3, AmF_4, $AmCl_2$ (in $SrCl_2$), $AmCl_3$, $AmBr_3$, AmI_2, and AmI_3 have also been prepared.

Metallic americium is silvery-white in appearance, is chemically reactive, and has a melting point of 1176°C. It has two crystalline modifications between room temperature and its melting point. *See* AMERICIUM.

Curium. The third transuranium element to be discovered, curium (Cm, atomic number 96, named after Pierre and Marie Curie), as the isotope ^{242}Cm, was identified by Seaborg, James, and Ghiorso in 1944 at the wartime Metallurgical Laboratory of the University of Chicago. This was produced by the helium-ion bombardment of ^{239}Pu in the University of California 60-in. (152 cm) cyclotron. Curium was first isolated, using the isotope ^{242}Cm, in the form of a pure compound, the hydroxide, in 1947 by L. B. Werner and I. Perlman. Isotopes of mass number 238–250 are known. Chemical investigations with curium have been performed using ^{242}Cm (half-life 163 days) and ^{244}Cm (half-life 18 years), but the higher-mass isotopes ^{247}Cm and ^{248}Cm with much longer half-lives (1.6×10^7 and 3.5×10^5 years, respectively) are more satisfactory for this purpose; these are all produced by neutron irradiation in nuclear reactors.

Curium exists solely as Cm^{3+} (colorless to yellow) in the uncomplexed state in aqueous solution. This behavior is related to its position as the element in the actinide series in which the $5f$ electron shell is half filled; that is, it has the especially stable electronic configuration $5f^7$, analogous to its lanthanide homolog, gadolinium. A curium IV fluoride complex ion exists in aqueous solution. Solid compounds include Cm_2O_3, CmO_2 (and oxides of intermediate composition), CmF_3, CmF_4, $CmCl_3$, $CmBr_3$, and CmI_3.

The metal is silvery and shiny in appearance, is chemically reactive, melts at 1340°C, and resembles americium metal in its two crystal modifications. *See* CURIUM.

Berkelium. Berkelium (Bk, atomic number 97, named after Berkeley, California) was produced and identified by S. G. Thompson, Ghiorso, and Seaborg in late 1949 at the University of California, Berkeley, and was the fifth transuranium element discovered. The isotope ^{243}Bk (half-life 4.6 hr) was synthesized by helium-ion bombardment of ^{241}Am. The first isolation of berkelium in weighable amount, as ^{249}Bk (half-life 314 days), produced by neutron irradiation, was accomplished in 1958 by Thompson and Cunningham; this isotope, produced in nuclear reactors, is used in the chemical and physical investigation of berkelium. Isotopes of mass number 243–251 are known.

Berkelium exhibits two ionic oxidation states in aqueous solution, Bk^{3+} (yellow-green) and somewhat unstable Bk^{4+} (yellow) as might be expected by analogy with its rare-earth homolog, terbium. Solid compounds include Bk_2O_3, BkO_2 (and oxides of intermediate composition), BkF_3, BkF_4, $BkCl_3$, $BkBr_3$, and BkI_3.

Berkelium metal is chemically reactive, exists in two crystal structure modifications, and melts at 986°C. *See* BERKELIUM.

Californium. The sixth transuranium element to be discovered, californium (Cf, atomic number 98, named after the state and University of California), in the form of the isotope ^{245}Cf (half-life 44 min), was first prepared by the helium-ion bombardment of microgram quantities of ^{242}Cm at the University of California, Berkeley. The element was discovered by Thompson, K. Street, Jr., Ghiorso, and Seaborg at the University of California, Berkeley, early in 1950. Cunningham and Thompson, at Berkeley, isolated californium in weighable quantities for the first time in 1958 using a mixture of the isotopes ^{249}Cf, ^{250}Cf, ^{251}Cf, and ^{252}Cf, produced by neutron irradiation. Isotopes of mass number 240–255 are known. The best isotope for the investigation of the chemical and physical properties of californium is ^{249}Cf (half-life 350 years), produced in pure form as the beta-particle decay product of ^{249}Bk.

Californium exists mainly as Cf^{3+} in aqueous solution (emerald green), but it is the first of the actinide elements in the second half of the series to exhibit the II state, which becomes progressively more stable on proceeding through the heavier members of the series. It also exhibits the IV oxidation state in CfF_4 and CfO_2, which can be prepared under somewhat intensive oxidizing conditions. Solid compounds also include Cf_2O_3 (and higher intermediate oxides), CfF_3, $CfCl_3$, $CfBr_2$, $CfBr_3$, CfI_2, and CfI_3.

Californium metal is chemically reactive, is quite volatile, and can be distilled at temperatures of the order of 1100 to 1200°C. It appears to exist in three different crystalline modifications between room temperature and its melting point, 900°C. *See* CALIFORNIUM.

Einsteinium. The seventh transuranium element to be discovered, einsteinium (Es, atomic number 99, named after Albert Einstein), was found by Ghiorso and coworkers in the debris from the "Mike" thermonuclear explosion staged by the Los Alamos Scientific Laboratory in November 1952. Very heavy uranium isotopes were formed by the action of the intense neutron flux on the uranium in the device, and these decayed into isotopes of elements 99, 100, and other transuranium elements of lower atomic number. Chemical investigation of the debris in late 1952 by workers at the University of California Radiation Laboratory, Argonne National Laboratory, and Los Alamos Scientific Laboratory revealed the presence of element 99 as the isotope ^{253}Es. Einsteinium was isolated in a macroscopic (weighable) quantity for the first time in 1961 by Cunningham, J. C. Wallmann, L. Phillips, and R. C. Gatti at Berkeley; they used the isotope ^{253}Es, produced in nuclear reactors, working with only a few hundredths of a microgram. The macroscopic property that they determined in this case was the magnetic susceptibility. Isotopes of mass number 243–256 have been synthesized. Einsteinium is the heaviest transuranium element to be isolated in weighable form. Most of the investigations have used the short-lived ^{253}Es (half-life 20.5 days) because of its greater availability, but the use of ^{254}Es (half-life 276 days) will increase as it becomes more available as the result of production in nuclear reactors.

Einsteinium exists in normal aqueous solution essentially as Es^{3+} (green), although Es^{2+} can be produced under strong reducing conditions. Solid compounds such as Es_2O_3, $EsCl_3$, EsOCl, $EsBr_2$, $EsBr_3$, EsI_2, and EsI_3 have been made.

Einsteinium metal is chemically reactive, is quite volatile, and melts at 860°C; one crystal structure is known. *See* EINSTEINIUM.

Fermium. Fermium (Fm, atomic number 100, named after Enrico Fermi), the eighth transuranium element discovered, was isolated as the isotope ^{255}Fm (half-life 20 hr) from the heavy elements formed in the "Mike" thermonuclear explosion. The element was discovered in early 1953 by Ghiorso and coworkers during the same investigation which resulted in the discovery of element 99. Fermium isotopes of mass number 244–259 have been prepared.

No isotope of fermium has yet been isolated in weighable amounts, and thus all the investigations of this element have been done with tracer quantities. The longest-lived isotope is ^{257}Fm (half-life about 100 days) whose production in high-neutron-flux reactors is extremely limited because of the very long sequence of neutron-capture reactions that is required.

Despite its very limited availability, fermium, in the form of the 3.24-hr ^{254}Fm isotope, has been identified in the "metallic" zero-valent state in an atomic-beam magnetic resonance experiment. This established the electron structure of elemental fermium in the ground state as $5f^{12}7s^2$ (beyond the radon structure).

Fermium exists in normal aqueous solution almost exclusively as Fm^{3+}, but strong reducing conditions can produce Fm^{2+}, which has greater stability than Es^{2+} and less stability than Md^{2+}. *See* FERMIUM.

Mendelevium. Mendelevium (Md, atomic number 101, named after Dmitri Mendeleev), the ninth transuranium element discovered, was identified by Ghiorso, B. G. Harvey, G. R. Choppin, Thompson, and Seaborg at the University of California, Berkeley, in 1955. The element as ^{256}Md (half-life 1.5 hr) was produced by the bombardment of extremely small amounts (approximately 10^9 atoms) of ^{253}Es with helium ions in the 60-in. cyclotron. The first identification of mendelevium was notable in that only one or two atoms per experiment were produced. (This served as the prototype for the discovery of all heavier transuranium elements, which have been first synthesized and identified on a one-atom-at-a-time basis.) Isotopes of mass numbers 248–258 are known. Although the isotope ^{258}Md (half-life 56 days) is sufficiently long-lived, it cannot be produced in nuclear reactors, and hence it will be very difficult and perhaps impossible to isolate it in weighable amount.

The chemical properties have been investigated on the tracer scale, and the element is found to behave in aqueous solution as a typical tripositive actinide ion; it can be reduced to the II and I states with moderately strong reducing agents. *See* MENDELEVIUM.

Nobelium. The discovery of nobelium (No, atomic number 102, named after Alfred Nobel), the tenth transuranium element to be discovered, has a complicated history. For the first time scientists from countries other than the United States embarked on serious efforts to compete with the United States in this field. The reported discovery of element 102 in 1957 by an international group of scientists working at the Nobel Institute for Physics in Stockholm, who suggested the name nobelium, has never been confirmed and must be considered to be erroneous. Working at the Kurchatov Institute of Atomic Energy in Moscow, G. N. Flerov and coworkers in 1958 reported a radioactivity which they thought might be attributed to element 102, but a wide range of half-lives was suggested and no chemistry was performed. As the result of more definitive work performed in 1958, Ghiorso, T. Sikkeland, J. R. Walton, and Seaborg reported an isotope of the element, produced by bombarding a mixture of curium isotopes with ^{12}C ions in the then-new Heavy Ion Linear Accelerator (HILAC) at Berkeley. They described a novel "double recoil" technique which permitted identification by chemical means, one atom at a time, of any daughter isotope of element 102 that might have been formed. The isotope ^{250}Fm was identified conclusively by this means, indicating that its parent should be the isotope of element 102 with mass number 254 produced by the reaction of ^{12}C ions with ^{246}Cm. However, another isotope of element 102, with half-life 3 sec, also observed indirectly in 1958, and whose alpha particles were shown to have an energy of 8.3 MeV by Ghiorso and coworkers in 1959, was shown later by Flerov and coworkers (working at the Dubna Laboratory near Moscow) to be a result of an isotope of element 102 with mass number 252 rather than 254; in other words, two isotopes of element 102 were discovered by the Berkeley group in 1958, but the correct mass number assignments were not made until later. On the basis that they identified the atomic number correctly, the Berkeley scientists probably have the best claim to the discovery of element 102; they suggest the retention of nobelium as the name for this element.

All known isotopes (mass numbers 250–259) of nobelium are short-lived and are produced by the bombardment of lighter elements with charged particles (heavy ions); the longest lived is ^{259}No with a half-life of 58 min. All of the chemical investigations have been, and presumably must continue to be, done on the tracer scale. These have demonstrated the existence of No^{3+} and No^{2+} in aqueous solution, with the latter much more stable than the former. The stability of No^{2+} is consistent with the expected presence of the completed shell of fourteen $5f$ electrons in this ion. *See* NOBELIUM.

Lawrencium. Lawrencium (Lr, atomic number 103, named after Ernest O. Lawrence) was discovered in 1961 by Ghiorso, Sikkeland, A. E. Larsh, and R. M. Latimer using the HILAC at the University of California, Berkeley. A few micrograms of a mixture of ^{249}Cf, ^{250}Cf, ^{251}Cf, and ^{252}Cf (produced in a nuclear reactor) were bombarded with ^{10}B and ^{11}B ions to produce single atoms of an isotope of element 103 with a half-life measured as 8 sec and decaying by the emission of alpha particles of 8.6 MeV energy. Ghiorso and coworkers suggested at that time that this radioactivity might be assigned the mass number 257. G. N. Flerov and coworkers have disputed this discovery on the basis that their later work suggests a greatly different half-life for the isotope with the mass number 257. Subsequent work by Ghiorso and coworkers proves that the correct assignment of mass number to the isotope discovered in 1961 is 258, and this later work gives 4 sec as a better value for the half-life.

All known isotopes of lawrencium (mass numbers 255–260) are short-lived and are produced by bombardment of lighter elements with charged particles (heavy ions); chemical investigations have been, and presumably must be, performed on the tracer scale. Work with ^{260}Lr (half-life 3 min) has demonstrated that the normal oxidation state in aqueous solution is the III state, corresponding to the ion Lr^{3+}, as would be expected for the last member of the actinide series. *See* LAWRENCIUM.

Rutherfordium. Rutherfordium (Rf, atomic number 104, after Lord Rutherford), the first transactinide element to be discovered, was probably first identified in a definitive manner by Ghiorso, M. Nurmia, J. Harris, K. Eskola, and P. Eskola in 1969 at Berkeley. Flerov and coworkers have suggested the name kurchatovium (named after Igor Kurchatov with symbol Ku) on the basis of an earlier claim to the discovery of this element; they bombarded, in 1964, ^{242}Pu with ^{22}Ne ions in their cyclotron at the Joint Institute for Nuclear Research in Dubna and reported the production of an isotope, suggested to be ^{260}Ku, which was held to decay by spontaneous fission with a half-life of 0.3 sec. After finding it impossible to confirm this observation, Ghiorso and coworkers reported definitive proof of the production of alpha-particle-emitting ^{257}Rf and ^{259}Rf (half-lives 4.5 and 3 sec, respectively), demonstrated by the identification of the previously known ^{253}No and ^{255}No as decay products, by means of the bombardment of ^{249}Cf with ^{12}C and ^{13}C ions in the Berkeley HILAC.

All known isotopes of rutherfordium (mass numbers 256–261) are short-lived and are produced by bombardment of lighter elements with charged heavy-ion particles. The isotope ^{261}Rf(half-life 65 sec) has made it possible, by means of rapid chemical experiments, to demonstrate that the normal oxidation state of rutherfordium in aqueous solution is the IV state corresponding to the ion Rf^{4+}. This is consistent with expectations for the first transactinide element which should be a homolog of hafnium, an element that is exclusively tetrapositive in aqueous solution. *See* ELEMENT 104.

Hahnium. Hahnium (Ha, atomic number 105, named after Otto Hahn), the second transactinide element to be discovered, was probably first identified in a definitive manner in 1970 by Ghiorso, Nurmia, K. Eskola, Harris, and P. Eskola at Berkeley. They reported the production of alpha-particle-emitting ^{260}Ha (half-life 1.6 sec), demonstrated through the identification of the previously known ^{256}Lr as the decay product, by bombardment of ^{249}Cf with ^{15}N ions in the Berkeley HILAC. Again the Berkeley claim to discovery is disputed by Flerov and coworkers, who earlier in 1970 reported the discovery of an isotope held to be element 105, decaying by the less definitive process of spontaneous fission, produced by the bombardment of ^{243}Am with ^{22}Ne ions in the Dubna cyclotron; in later work Flerov and coworkers may have also observed the alpha-particle-emitting isotope of element 105 reported by Ghiorso and workers. Flerov has suggested nielsbohrium (named after Niels Bohr, symbol Ns) as the name for element 105.

The known isotopes of element 105 (mass numbers 260–262) are short-lived and are produced by bombardment of lighter elements with charged heavy-ion particles. The isotope ^{262}Ha (half-life 40 sec) makes it possible, with rapid chemical techniques, to study the chemical properties of hahnium. It should exhibit the V oxidation state like its homolog tantalum. *See* ELEMENT 105.

Element 106. The discovery of element 106 took place in 1974 simultaneously as the result of experiments by Ghiorso and coworkers at Berkeley and Flerov, Y. T. Oganessian, and coworkers at Dubna. The Ghiorso group used the SuperHILAC (the rebuilt HILAC) to bombard a target of californium (the isotope ^{249}Cf) with ^{18}O ions. This resulted in the production and positive identification of the alpha-particle-emitting isotope $^{263}106$, which decays with a half-life of 0.9 ± 0.2 sec by the emission of alpha particles of principal energy 9.06 MeV. The definitive identification consisted of the establishment of the genetic link between the element 106 alpha-particle-emitting isotope ($^{263}106$) and previously identified daughter ($^{259}104$) and granddaughter ($^{255}102$) nuclides, that is, the demonstration of the decay sequence: $^{263}106 \xrightarrow{\alpha} {}^{259}\mathrm{Rf} \xrightarrow{\alpha} {}^{255}\mathrm{No} \xrightarrow{\alpha}$. A total of seventy-three $^{263}106$ alpha particles and approximately the expected corresponding number of ^{259}Rf daughter and ^{255}No granddaughter alpha particles were recorded.

The Dubna group chose lead (atomic number 82) as their target because, they believe, its closed shells of protons and neutrons and consequent small relative mass leads to minimum excitation energy for the compound nucleus and therefore an enhancement in the cross section for the production of the desired product nuclide. They bombarded ^{207}Pb and ^{208}Pb with ^{54}Cr ions (atomic number 24) in their cyclotron to find a product that decays by the spontaneous fission mechanism (a total of 51 events), with the very short half-life of 7 msec, which they assign to the isotope $^{259}106$. The identification depends on rather uncertain deductions concerning the nature of the new type of nuclear reactions that the Dubna scientists postulate and new correlations of the dependence of spontaneous fission half-lives on atomic and mass numbers.

On the basis of its projected position in the periodic table, element 106 is expected to have chemical properties similar to those of tungsten (atomic number 74). *See* ELEMENT 106.

Superheavy elements. Although the transactinide elements immediately beyond element 106 are predicted to have very short half-lives, theoretical considerations suggest increased nuclear stability, compared with preceding and succeeding elements, for a range of elements around atomic numbers 110, 115, or 120 because of increased stability predicted to result from closed nucleon shells. The element with atomic number 114 (which should have chemical properties similar to those of lead) seems to show special promise of such relative stability, that is, relatively long half-life. It should be possible to synthesize isotopes of such "superheavy" elements through bombardments of heavy-element targets with intense beams of very heavy ions accelerated in specially constructed heavy-ion accelerators. *See* SUPERTRANSURANICS.

It should be possible to predict the chemical properties of all the transactinide elements with the help of the periodic table. Rutherfordium should chemically be like hafnium, hahnium like tantalum, element 106 like tungsten, 107 like rhenium, and so on across the periodic table to element 118, which should be a noble gas like radon. Beyond element 118, the elements 119, 120, and 121 should fit into the periodic table under the elements francium, radium, and actinium (atomic numbers 87, 88, and 89). At about this point there should start another series, but a special kind of inner transition series, perhaps similar in some respects to the actinide series. However, this series, which may be termed the superactinide series, will be different in that it will contain 32 elements, corresponding to the filling of 18-member and 14-member inner electron shells. After the filling of these shells at element 153, the still higher elements should again be placed in the main body of the periodic table leading to the next noble gas at element 168. The larger atomic numbers are far beyond the predicted region of nuclear stability and hence presumably are not accessible to experimentation. *See* ACTINIDE ELEMENTS; NUCLEAR CHEMISTRY; NUCLEAR FISSION; PERIODIC TABLE; RARE-EARTH ELEMENTS.

[GLENN T. SEABORG]

Bibliography: V. I. Goldanski and S. M. Polikanov, *The Transuranium Elements*, 1973; C. Keller, *The Chemistry of the Transuranium Elements*, 1971; Max Planck Society for the Advancement of Science, *Transurane-Transuranium Elements*, 1975; G. T. Seaborg, The new elements, *Amer. Sci.*, 68:3, 1980; G. T. Seaborg, *Transuranium Elements*, 1958.

Trichloroacetic acid

Trichloroacetic acid, CCl_3COOH, is a colorless, crystalline, deliquescent, highly corrosive acid, with melting point 58°C and boiling point 196–197°C. It is fairly soluble in water and very soluble in alcohol and ether. Trichloroacetic acid is prepared either by nitric acid oxidation of chloral, or by direct chlorination of acetic acid using iodine or phosphorus trichloride as catalyst; it is one of the strongest organic acids known ($K_a = 1.3 \times 10^{-1}$).

Trichloroacetic acid is used as a decalcifier and fixative in microscopy, as a denaturant and precipitant of proteins, and, in the form of dilute solution, in medicine as an astringent and antiseptic. The substance is relatively unstable and decomposes on heating to give chloroform and carbon dioxide. When heated in alkaline solutions, it gives carbonates and chloroform. *See* ACETIC ACID; CARBOXYLIC ACID.

[EVANS B. REID]

Trifluoroacetate

A salt or ester of trifluoroacetic acid, CF_3COOH. Inorganic salts such as sodium trifluoroacetate, CF_3COONa, are obtained by double decomposition reactions. The sodium compound is an intermediate used to prepare fluorinated compounds, herbicides, insecticides, and dyes. Organic salts or esters such as ethyl trifluoroacetate result when the acid is caused to react with ethyl alcohol. Ethyl trifluoroacetate is useful in organic synthesis. The very strong parent acid is used as catalyst for organic esterification and condensation reactions. The acid is, however, expensive. *See* CARBOXYLIC ACID; HALOGENATED HYDROCARBON; ORGANIC CHEMISTRY.

[E. EUGENE WEAVER]

Triglyceride

A simple lipid. Triglycerides are fatty acid triesters of the trihydroxy alcohol glycerol which are present in plant and animal tissues, particularly in the food storage depots, either as simple esters in which all the fatty acids are the same or as mixed esters in which the fatty acids are different. The triglycerides constitute the main component of natural fats and oils.

The generic formula of a triglyceride is shown below, where RCO_2H, $R'CO_2H$, and $R''CO_2H$ rep-

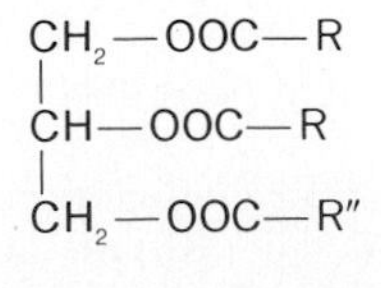

resent molecules of either the same or different

fatty acids, such as butyric or caproic (short chain), palmitic or stearic (long chain), oleic, linoleic, or linolenic (unsaturated). Saponification with alkali releases glycerol and the alkali metal salts of the fatty acids (soaps). The triglycerides in the food-storage depots represent a concentrated energy source, since oxidation provides more energy than an equivalent weight of protein or carbohydrate.

Animal and vegetable triglycerides contain predominantly even-chain-length fatty acids, with palmitic and oleic acids as the main components. Since *n* fatty acids may be esterified in $(n^3 + n^2)/2$ ways into glycerol, and since natural fats contain a variety of fatty acids, the number of component triglycerides of a relatively simple natural fat or oil may be high. Some pure simple and mixed triglycerides have been isolated from natural fats by fractional crystallizations at low temperatures, but in general physical methods are not yet available for the separation of naturally occurring mixtures. Several theories, such as those of even distribution and partial random distribution, have been advanced to account for the distribution of the fatty acids in the triglycerides. Many synthetic triglycerides have been prepared, and the study of the physical properties of these compounds has provided much useful information. Melting-point, x-ray-diffraction, and infrared-spectroscopy investigations have shown that triglycerides may exist in at least three polymorphic modifications. *See* MOLECULAR STRUCTURE AND SPECTRA.

The physical and chemical properties of fats and oils depend on the nature of the fatty acids present. Saturated fatty acids give higher-melting fats and represent the main constituents of solid fats, for example, lard and butter. Unsaturation lowers the melting point of fatty acids and fats. Thus, in the oil of plants, unsaturated fatty acids are present in large amounts, for example, oleic acid in olive oil and linoleic and linolenic acids in linseed soil. Oils are hydrogenated commercially to produce the proper consistence and melting point for use as edible fats. *See* CARBOXYLIC ACID.

[ROY H. GIGG; HERBERT E. CARTER]

Bibliography: C. Hitchcock and B. W. Nichols, *Plant Lipid Biochemistry*, 1971; W. L. Holmes and W. M. Bortz (eds.), *Biochemistry and Pharmacology of Free Fatty Acids*, 1971.

Triphenylmethane

A colorless, crystalline, aromatic hydrocarbon with the structural formula given here, melting at 92.6°C and boiling at 360°C. Triphenylmethane

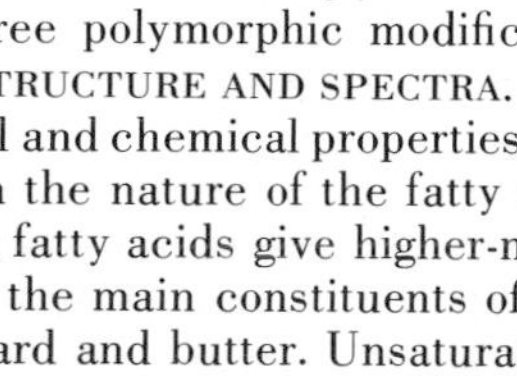

can be prepared from benzene in one or more steps. The one-step preparation involves the action of chloroform, $CHCl_3$, on benzene in the presence of aluminum chloride, $AlCl_3$, an example of the Friedel-Crafts reaction.

Triphenylmethane is easily oxidized to triphenylcarbinol, $(C_6H_5)_3COH$, and reacts with phosphorus pentachloride to form triphenylchloromethane, $(C_6H_5)_3CCl$. The methane hydrogen of triphenylmethane is weakly acidic, and when the hydrocarbon is heated with potassium metal, it evolves hydrogen to yield potassium triphenylmethide, $(C_6H_5)_3CK$.

While triphenylmethane has no industrial importance, some of its amino derivatives, for example, 4,4′-bis(dimethylamino) triphenylmethane, form the colorless leuco bases from which the so-called triphenylmethane dyes are prepared. *See* BENZENE; DYE; FRIEDEL-CRAFTS REACTION; POLYNUCLEAR HYDROCARBON.

[CHARLES K. BRADSHER]

Triple point

A particular temperature and pressure at which three different phases of one substance can coexist in equilibrium. In common usage these three phases are normally solid, liquid, and gas, although triple points can also occur with two solid phases and one liquid phase, with two solid phases and one gas phase, or with three solid phases.

According to the Gibbs phase rule, a three-phase situation in a one-component system has no degrees of freedom (that is, it is invariant). Consequently, a triple point occurs at a unique temperature and pressure, because any change in either variable will result in the disappearance of at least one of the three phases. *See* PHASE EQUILIBRIUM.

Triple points are shown in the illustration of part of the phase diagram for water. Point *A* is the well-known triple point for Ice I (the ordinary low-pressure solid form) + liquid water + water vapor at 0.0099°C (273.16 K) and a pressure of 0.00603 atm (4.58 mm Hg or 611 Pa). In 1954 the thermodynamic temperature scale (the absolute of Kelvin scale) was redefined by setting this triple-point temperature for water equal to exactly 273.16 K. Point *B*, at 251.1 K and 2047 atm (207.4 MPa) pressure, is the triple point for liquid water + Ice I + Ice III; and point *C*, at 238.4 and 2100 atm (212.8 MPa) pressure, is the triple point for Ice I + Ice II + Ice III. At least four other triple points are known at higher pressures, involving other crystalline forms of ice.

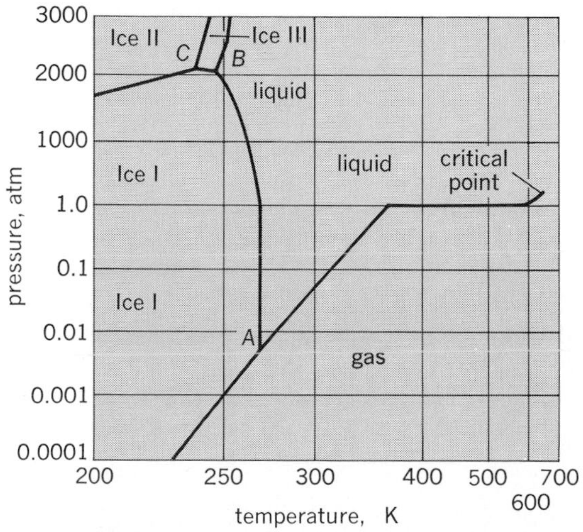

Phase diagram for water, showing gas, liquid, and several solid (ice) phases; triple points at *A*, *B*, and *C*. The pressure scale changes at 1 atm from logarithmic scale at low pressure to linear at high pressure. 1 atm = 6.895 kPa.

For most substances the solid-liquid-vapor triple point has a pressure less than 1 atm (101.325 kPa); such substances then have a liquid-vapor transition at 1 atm (normal boiling point). However, if this triple point has a pressure above 1 atm, the substance passes directly from solid to vapor at 1 atm. *See* SUBLIMATION.

For a two-component system, the invariant point in a phase diagram is a quadruple point at which four phases coexist. The three-phase situation is then represented by a line in the three-dimensional pressure-temperature-composition diagram. *See* BOILING POINT; MELTING POINT; TRANSITION POINT; VAPOR PRESSURE; WATER.

[ROBERT L. SCOTT]

Triplet state

A molecule exists in this electronic state when its total spin angular momentum quantum number S is equal to one. The triplet state is an important intermediate of organic chemistry. In addition to the wide range of triplet molecules available through photochemical excitation techniques, numerous molecules exist in stable triplet ground states, for example, oxygen molecules. Theoretical calculations, furthermore, make predictions concerning the spin multiplicities of the ground states of many prototype organic molecules such as cyclobutadiene, trimethylene methane, and methylene, and indicate that they will be triplets.

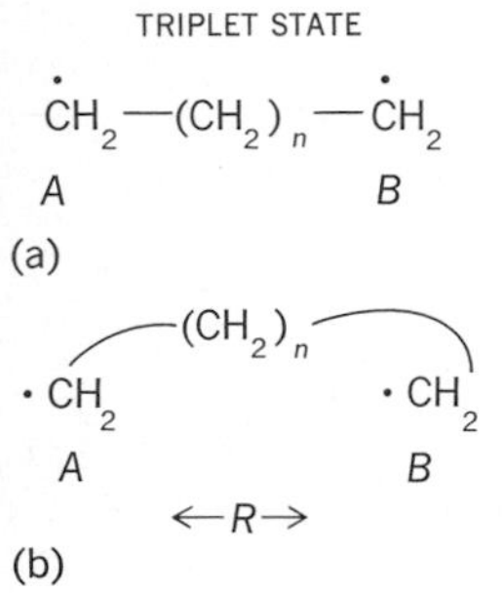

Fig. 1. Two carbon radicals connected by a long methylene chain. (*a*) Biradical state. (*b*) Triplet state.

Practical definition. A good working definition of a triplet state for the chemist is the following: A triplet is a paramagnetic even-electron species which possesses three distinct but energetically similar electronic states as a result of the magnetic interaction of two unpaired electron spins. The several important terms of this definition allow some insight as to the essential features of a triplet. First of all, a triplet is paramagnetic, and should thus display this property in a magnetic field. This paramagnetism serves as the basis for experimental magnetic susceptibility and electron spin resonance studies of the triplet state. However, one can image many paramagnetic odd electron species which are not triplets, for example, nitric oxide. Thus, the criterion that a triplet must also be an even-electron species is apparent.

However, one can imagine paramagnetic, even-electron species which possess (1) only two distinct electronic states or (2) five or more electronic states. The former occurs when the paramagnetism results from two electrons which act as two independent odd electrons. For example, two carbon radicals separated by a long saturated chain will behave as two doublet states if there is sufficient separation to prevent spin interactions. Five or more electronic states result when four or six parallel electronic spins interact (to yield quintet and septet states, respectively). *See* ELECTRON SPIN.

One can now see that conceptual difficulties may arise in differentiating a biradical state (that is, a species possessing two independent odd-electron sites) from a triplet. Suppose two carbon radicals are separated by a long methylene chain as in Fig. 1*a*. If the methylene chain is sufficiently long and the odd-electron centers are so far removed from one another that they do not interact (magnetically and electronically) with one another, then the system is a doublet of doublets, that is, two independent odd electrons or a true biradical. If the methylene chain should be folded (Fig. 1*b*) so that the odd electrons begin to interact (magnetically and electronically) with one another, then at some distance R between the —CH_2 groups the doublet or doublets will become a triplet state. This state will result from the fact that the spin of the electron on carbon A is no longer independent of the spin on carbon B. Since the spins are quantized, selection rule (1) applies, where S is

$$\text{Number of spin states} = 2|S| + 1 \qquad (1)$$

the sum of the spin quantum numbers for the two electrons. This means that either three spin states (if $S=1$ or -1, that is, spins of both electrons on C_A and C_B are the same) or one spin state (if $S=0$, that is, spin of the electron of C_A is paired with that of C_B) will result. The former describes a triplet state and the latter a singlet state.

This leads to a difficulty in terminology: The "triplet state" is not one state but three states even in the absence of an external magnetic field. Indeed, under favorable conditions transitions may be observed between triplet levels at zero external magnetic field. The effect of an external magnetic field is to further split the triplet levels and allow transitions between them to be more easily detected.

Properties. A triplet may result whenever a molecule possesses two electrons which are both orbitally unpaired and spin unpaired. As shown in Fig. 2, orbital unpairing of electrons results when a molecule absorbs a photon of visible or ultraviolet light. Direct formation of a triplet as a result of this photon absorption is a very improbable process since both the orbit and spin of the electron would have to change simultaneously. Thus, a singlet state is generally formed by absorption of light. However, quite often the lifetime of this singlet state is sufficiently long to allow the spin of one of the two electrons to invert, thereby producing a triplet. The following discussion considers the ways in which such a species is unambiguously characteristic. *See* MOLECULAR ORBITAL THEORY.

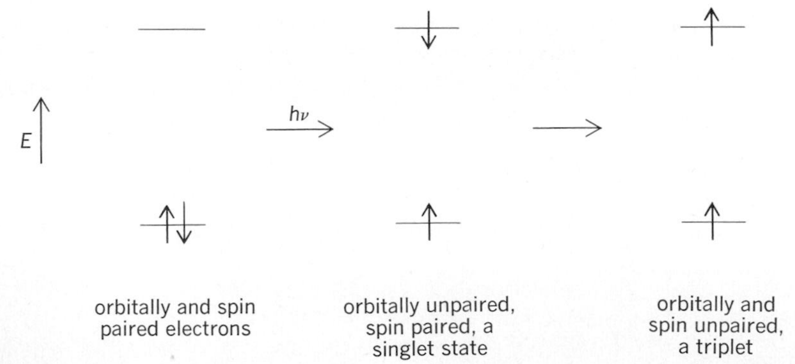

Fig. 2. Simple molecular orbital description of singlets and triplets.

The question to be answered is: What are the general properties to be expected of a molecule in the triplet state? Some of the more important physical properties are: (1) paramagnetism; (2) absorption between triplet sublevels; (3) electronic absorption from the lowest triplet to upper triplets; (4) electronic emission from the lowest triplet to a lower singlet ground state (if the triplet level is not the ground state). The paramagnetism of the triplet results from the interaction of unpaired spins and the fact that an unpaired spin shows a paramagnetic effect (is attracted) in a magnetic field.

Absorption between triplet sublevels may be observed directly by the use of an electron spin resonance spectrometer. *See* ELECTRON PARAMAGNETIC RESONANCE (EPR) SPECTROSCOPY.

The triplet, like any other electronic state, may be excited to upper electronic states of the same spin as the result of light absorption. In favorable cases this may be observed by the method of flash spectroscopy. *See* PHOTOCHEMISTRY.

For most organic molecules the lowest triplet state is an excited electronic state and may emit light and pass to the ground singlet state. Since light absorption to form a triplet from a singlet is improbable, the symmetrically related emission of light from a triplet returning to a ground state is likewise improbable. Indeed, it takes the triplet states of some aromatic molecules an average of about 30 sec to emit light. This phenomenon is known as phosphorescence and is to be contrasted with fluorescence, the emission of light from an excited singlet state returning to a singlet ground state, a process which often occurs in nanoseconds.

Although phosphorescence (long-lived emission) was the first method employed to study triplets, it is not a specific device for establishing whether a long-lived emission occurs from a triplet. For instance, examples are known for which the slow combination of positive and negative sites will generate excited molecules which emit light. In this case the combination reaction may be rate-determining for light emission.

Similarly, absorption from one triplet to another is not a specific method since the precise triplet-triplet absorption characteristics cannot be predicted accurately. It would thus remain to be proven that the absorbing species is indeed a triplet and not some other transient species.

Even paramagnetism is not an infallible probe for a triplet state since free radicals which are also paramagnetic are often produced by the absorption of light.

It appears that electron spin resonance (ESR) is probably the most powerful single method for establishing that a molecule is in its triplet state. The nature of the ESR signals may be predicted and fitted to theoretical relation (2), which describes

$$H = g_0 H \cdot S + DS_z^2 + E(S_x^2 - S_y^2) \qquad (2)$$

the magnetic spin interactions and expected absorptions. Here g_0 is the Landé g factor, D the dielectric constant, and E the electric field strength of the molecule. This particular equation is derived for the special case of molecules with a plane of symmetry and a symmetry axis perpendicular to that plane. However, the important general features of this equation are: (1) the term $g_0 H \cdot S$ which describes the interaction of the external magnetic field H with the unpaired electron spin S; (2) the term $DS_z^2 + E(S_x^2 - S_y^2)$ which describes the spin-spin dipolar interactions along the x, y, and z axes of the molecule. These interactions are indicated in Fig. 3. Thus, from a study of the behavior of a triplet in a magnetic field, information on the electronic distribution in this excited state is obtained. In favorable cases, the nuclear geometry of the triplet may be derived.

[NICHOLAS J. TURRO]

Bibliography: A. Devaquet et al., *Triplet States One*, 1975; F. W. McLafferty, *Interpretation of Mass Spectra*, 1980; N. J. Turro, *Modern Molecular Photochemistry*, 1978; P. J. Wagner et al., *Triplet States: No. 3*, 1976; U. P. Wild et al., *Triplet States Two*, 1975.

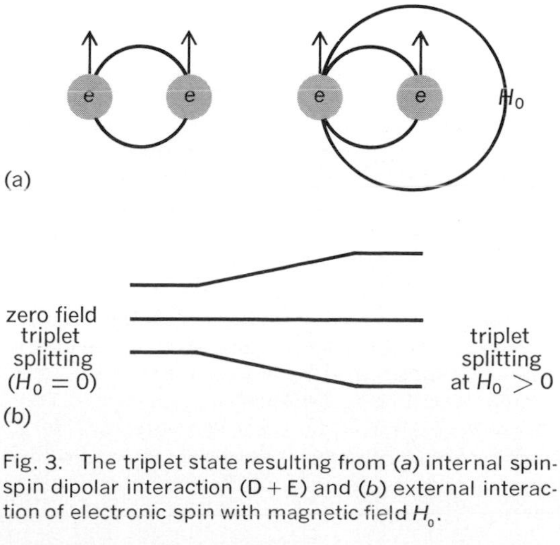

Fig. 3. The triplet state resulting from (*a*) internal spin-spin dipolar interaction (D + E) and (*b*) external interaction of electronic spin with magnetic field H_0.

Triterpene

A hydrocarbon or its oxygenated analog containing 30 carbon atoms and composed of six isoprene units. Triterpenes form the largest group of terpenoids, but are classified into only a few major categories. Resins and saps contain triterpenes in the free state as well as in the form of esters and glycosides.

Biogenetically triterpenes arise by the cyclization of squalene (I) and subsequent skeletal rearrangements.

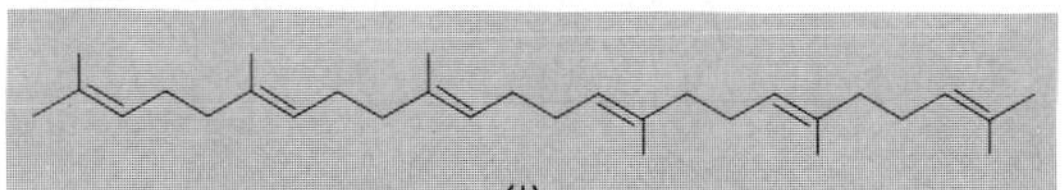

(I)

Squalene can cyclize in five ways, leading to different stereochemical arrangements in the final triterpenoid structure. The conformation of a triterpenoid nucleus is determined in the initial folding of squalene into several chair or boat configurations. Apart from the linear squalene itself and some bicyclic, highly substituted skeletons, most triterpenes are either tetracyclic or pentacyclic compounds. The various structural classes are designated by the names of representative members. The lanosterol and euphol series comprise tetracyclic structures differing only in the stereochemical arrangements around the D ring. Lanosterol (II) occurs in sheep wool, and euphol (III) is obtained from Euphorbium resin. The

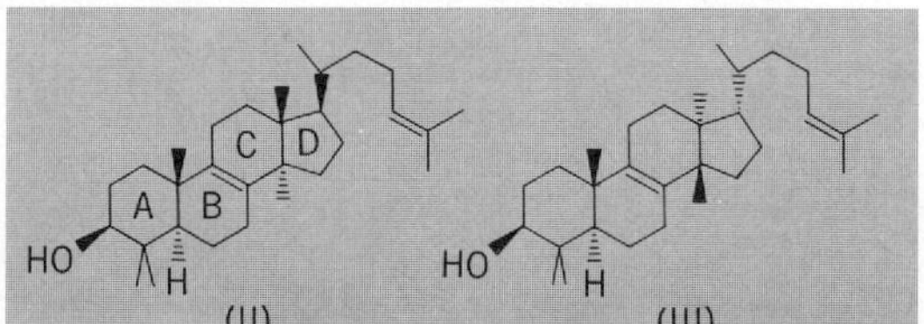

(II) (III)

oleanane (β-amyrin, IV, from grape seeds) and ursane (α-amyrin, V, from Manila elemi resin) se-

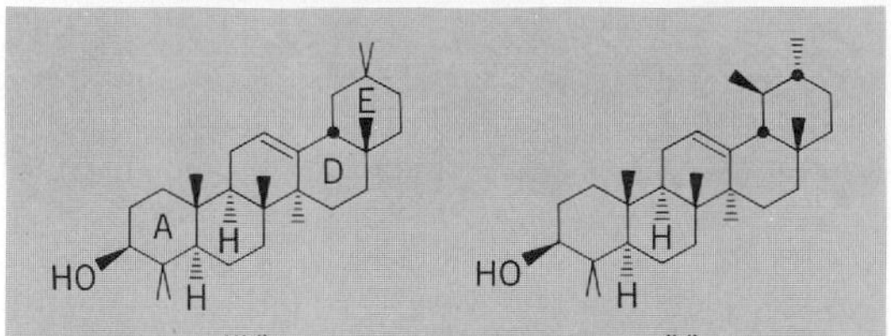

(IV) (V)

ries of triterpenes are all pentacycles differing in the substitution pattern of methyl groups on ring E.

Lupeol (VI) from lupin seeds and hydroxyhopanone (VII) from dammar resin are typical of the

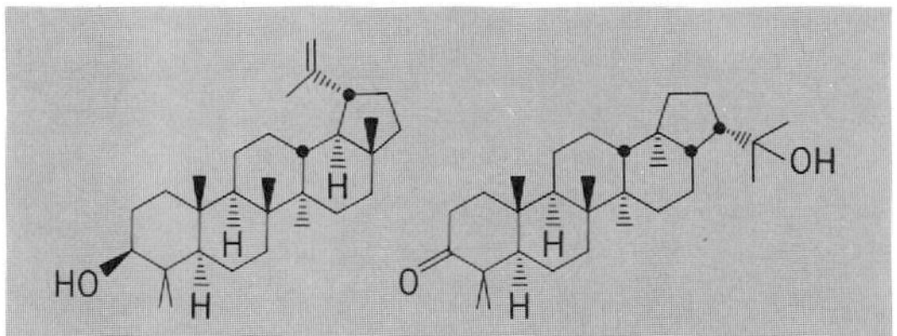

(VI) (VII)

lupane and hopane series. Hydroxyhopanone is one of the few triterpenoids that results from the cyclization of squalene without subsequent rearrangements.

Steroids and sterols are related to triterpenes. These important compounds are classed as nortriterpenes to indicate that they lack some of the 30 carbons of the triterpene skeleton. They have 27 to 29 carbons, and in most cases lack the geminal dimethyl group in ring A and one of the angular methyl groups in ring D of the lanosterol tetracy clic skeleton. Steroids are believed to arise from squalene via lanosterol, followed by the oxidative loss of methyl groups. Because of their significance in mammalian metabolism, steroids are usually treated as a separate class although their terpenoid origin is well understood. Cholesterol (VIII) and cholic acid (IX) are representative ex-

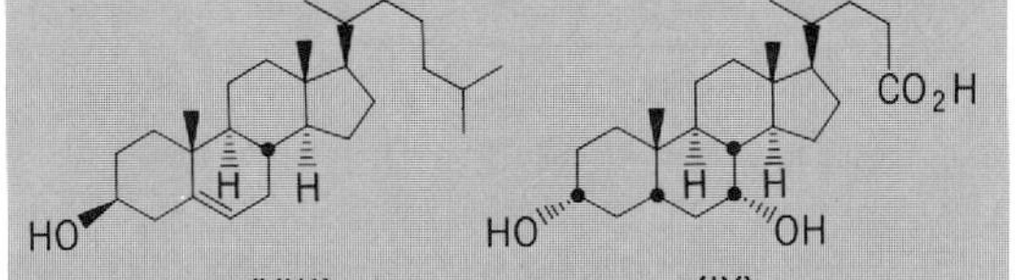

(VIII) (IX)

amples of steroid structures. *See* ISOPRENE; SQUALENE; TERPENE.

[TOMAS HUDLICKY]

Tritium

The heaviest isotope of the element hydrogen and the only one which is radioactive. Tritium occurs in very small amounts in nature but is generally prepared artificially by processes known as nuclear transmutations. It is widely used as a tracer in chemical and biological research and is a component of the so-called thermonuclear or hydrogen bomb. It is commonly represented by the symbol ${}_1H^3$, indicating that it has an atomic number of 1 and an atomic mass of 3, or by the special symbol T. For information about the other hydrogen isotopes *see* DEUTERIUM; HYDROGEN.

Properties. Both molecular tritium, T_2, and its counterpart hydrogen, H_2, are gases under ordinary conditions. Because of the great difference in mass, many of the properties of tritium differ substantially from those of ordinary hydrogen, as indicated in the table.

Properties of tritium and hydrogen

Property	H_2	T_2
Melting point, °C	−259.20	−252.54
Boiling point at 1 atm, °C	−252.77	−248.12
Heat of vaporization, cal/mole	216	333
Heat of sublimation, cal/mole	247	393

Chemically, tritium behaves quite similarly to hydrogen. However, because of its larger mass, many of its reactions take place more slowly than do those of hydrogen. The ratio of reaction rates may be as large as 64:1. These differences in reactivity can give rise to serious errors of interpretation when tritium is used as a tracer for hydrogen.

The nucleus of the tritium atom, often called a triton and symbolized *t*, consists of a proton and two neutrons. It has a mass of 3.01700 atomic mass units (amu), a nuclear spin of ½, and a magnetic moment of 2.9788 nuclear magnetons. It undergoes radioactive decay by emission of a β-particle to leave a helium nucleus of mass 3. No γ-rays are emitted in this process. The half-life for the decay is 12.26 years. The most energetic of the β-particles emitted by tritium have the comparatively low energy of 18.6 kiloelectronvolts (keV); β-particles are completely stopped by 7 mm of air or by 0.01 mm of paper or similar material. The average energy of the β-particles is 5.69 keV.

When tritium is bombarded with deuterons of sufficient energy, a nuclear reaction known as fusion occurs and energy considerably greater than that of the bombarding particle is released. The reaction may be written as (1). This reaction is

$$ {}_1H^3 + {}_1H^2 \rightarrow {}_2He^4 + {}_0n^1 + 18\ \text{MeV} \qquad (1) $$

one of those which supply the energy of the thermonuclear bomb. It is also of major importance in the development of controlled thermonuclear reactors. Enormous quantities of tritium will be required if such reactors are perfected and brought into use as electric power generators.

Natural occurrence. Before the start of thermonuclear weapons testing in 1954, rainwater contained approximately 1–10 atoms of tritium per 10^{18} atoms of hydrogen. Such tritium originates largely from the bombardment of nitrogen in the upper atmosphere with neutrons and protons from cosmic rays, as in reaction (2). Because the half-life

$$ {}_7N^{14} + {}_0n^1 \rightarrow {}_1H^3 + {}_6C^{12} \qquad (2) $$

of tritium is short in comparison with the time required for mixing of the ocean waters, the concentration of tritium in the ocean is much lower than in rainwater. Before 1954 the total amount of tritium on the Earth's surface was estimated at 1800 g, of which about 11 g was in the atmosphere and 13 g in groundwaters. Testing of thermonuclear weapons has resulted in sharp rises in the tritium content of rainwater to values as high as 500 atoms per 10^{18} atoms of hydrogen.

Preparation. Tritium was first produced in the laboratory by bombarding compounds of deuterium with high-energy deuterons, as in reaction (3).

$$_1H^2 + {}_1H^2 \rightarrow {}_1H^3 + {}_1H^1 \qquad (3)$$

A number of other nuclear reactions also give rise to tritium. The most important of these is the absorption of slow neutrons by the lithium isotope of mass 6, according to reaction (4). By irradiating

$$_3Li^6 + {}_0n^1 \rightarrow {}_1H^3 + {}_2He^4 \qquad (4)$$

enriched lithium-6, in the form of an alloy with magnesium or aluminum, with neutrons from a nuclear reactor, tritium may be prepared on a large scale.

Uses. As a result of its production for use in nuclear weapons, tritium has become available in large quantities at very low cost. It is used in admixture with zinc sulfide in the production of luminous paints, which have largely replaced the radium formerly used on watch dials; such mixtures are also used to produce small, permanent light sources. Tritium adsorbed on metals is used in targets for the production of fast neutrons by bombardment with deuterons. Tritium has been much used in hydrological studies, since it is an ideal tracer for water movement. Some studies depend on natural tritium or that introduced by weapons testing; in other cases large amounts of tritium are deliberately added. Investigations include the distribution of groundwater in oil fields; the tracing of springs, rivers, and lakes; water seepage and loss from reservoirs; and the movement of glaciers.

Tritium has also been used as a tracer for hydrogen in the study of chemical reactions. The most widespread use of tritium has probably been in biological research, where it has been used both as a hydrogen tracer and as a molecular label in studies of metabolism, biosynthesis, and cytology. In particular, tritiated thymidine and other nucleotides and nucleosides have been extensively used in studies of the formation of DNA and RNA.

Compounds. Very few compounds of pure tritium have been prepared and studied. Such compounds would undergo decomposition quite rapidly under the action of the tritium β-radiation. Tritium oxide, T_2O, has been prepared by oxidation of tritium gas with hot copper oxide or by passing an electric spark through a mixture of tritium and oxygen. Its melting point is 4.49°C, compared with 0° for ordinary water. Of much greater importance are compounds, especially organic compounds, in which a small fraction of the hydrogen atoms have been replaced by tritium. Such labeled compounds are employed in tracer studies, such as those indicated above. Tritium-labeled compounds may be prepared by ordinary synthetic chemical methods, such as the catalytic addition of tritium-hydrogen mixtures to unsaturated compounds. Tritium may be exchanged for hydrogen in the presence of a catalyst such as platinum or a strong acid. In recoil labeling, a mixture of an organic compound and a lithium salt are irradiated with neutrons in a nuclear reactor; some of the energetic tritons produced are incorporated into the organic compound.

Another important labeling procedure consists of the exposure of an organic compound to tritium gas in a sealed vessel; the tritium β-radiation facilitates the exchange of hydrogen in the compound with tritium in the gas. Some compounds of biological interest have been prepared by growing organisms in tritiated water.

Analysis. Because of its weak β-radiation, tritium is not readily measured by the ordinary Geiger-Müller counter. More efficacious is the introduction of tritium as a gas inside the counting tube. Alternatively, the ionization of a gas caused by the β-radiation may be measured in an ionization chamber, or the tritium compound may be dissolved in a suitable solvent containing a phosphor and the light pulses excited by the β-particles then may be counted with a scintillation counter. Tritium gas containing only small amounts of ordinary hydrogen may be analyzed with a mass spectrometer or by measuring the density of the gas. Because of the very short range of the tritium β-particle, autoradiography, the exposure of radioactive material to a photographic plate, is often used to locate precisely the position of tritium in biological material. *See* HEAVY WATER; NUCLEAR FUSION; RADIOCHEMISTRY.

[LOUIS KAPLAN]

Bibliography: E. Buncel, *Tritium in Organic Chemistry*, vol. 4, 1978; E. A. Evans, *Tritium and Its Compounds*, 1966; International Atomic Energy Agency, *Behavior of Tritium in the Environment*, 1979; M. D. Kamen, *Isotopic Tracers in Biology*, 3d ed., 1957.

Tropolone

A weakly acidic organic compound, $C_7H_6O_2$, having the structure 2-hydroxycycloheptatrienone as shown in (I). Tropolone may be considered as derived from tropone (II) by substitution of a hydroxyl group for the hydrogen atom in position 2.

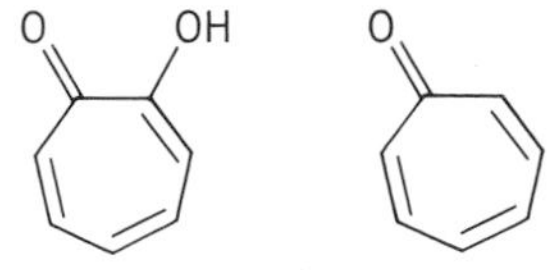

(I) (II)

The acidity of tropolone (pK$_A$ 7.0) is intermediate between that of phenol (pK$_A$ 10) and that of a typical carboxylic acid such as acetic acid (pK$_A$ 4.8). Treatment of tropolone with sodium hydroxide leads to the yellow sodium salt of the delocalized anion (III). Tropolones generally form highly

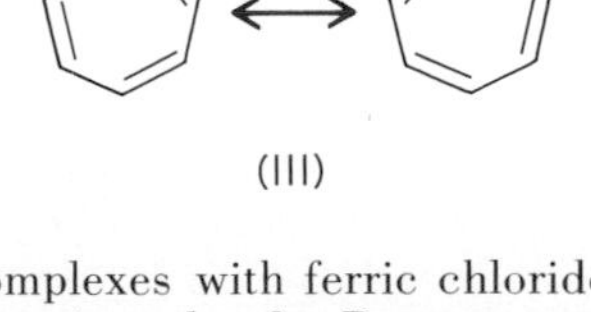

(III)

colored complexes with ferric chloride and other polyvalent cation salts. *See* RESONANCE.

Naturally occurring tropolones. The tropolone ring was first recognized in 1945 as a structural feature of stipitatic acid (IV), a metabolite of cer-

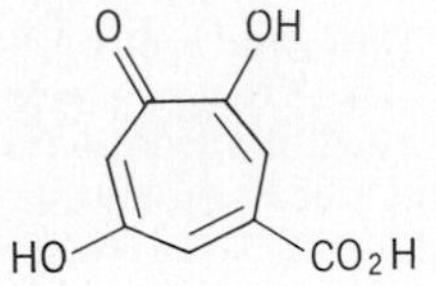

(IV)

tain *Penicillium* mold species. Other natural tropolones were promptly discovered, including the additional mold metabolites puberulic and puberulonic acids and sepedonin. The thujaplicins (isopropyltropolones) occur in several species of *Cupressacea* evergreens and possess fungicidal properties that are believed to protect the wood of such trees against rot. The medicinally important alkaloid colchicine, from meadow saffron (*Colchicum autumnale*), is also a tropolone derivative. *See* ALKALOID.

Aromatic properties. In its reactivity toward substitution reactions, its resistance to hydrogenation, and its formation of complexes with ferric ion, tropolone resembles phenol and other compounds of the aromatic series. The electronic structures of tropone and tropolone are best understood in relation to tropylium ion (V). This highly symmetrical

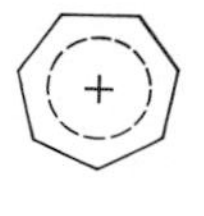

(V)

cation is predicted by theory to form stable salts by reason of its possessing a so-called closed-shell aromatic structure. Dramatic confirmation of the prediction was the isolation of tropylium bromide from treatment of cycloheptatriene (tropilidene) with bromine. Protonation of tropone (II) yields the hydroxy derivative (VI), which can account for the

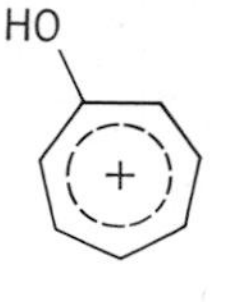

(VI)

remarkable basicity of tropone and for its lack of many properties typical of ketones. *See* AROMATIC HYDROCARBONS; PHENOL. [MARTIN STILES]

Bibliography: D. Ginsberg (ed.), *Non-Benzenoid Aromatic Compounds*, 1959; T. Nozoe, The chemistry of natural tropolones and related troponoids, *Festschrift für Arthur Stoll*, 1957; R. L. Pauson, Tropones and tropolones, *Chem. Rev.*, 55:9–136, 1955.

Tungstate

One of a group of compounds containing tungsten in the 6+ oxidation state and derived from tungstic acid, H_2WO_4. Both the normal tungstates, such as Na_2WO_4, and the molybdates form condensed ions, such as isopolytungstates. These isopolytungstates are usually classified into metatungstates and paratungstates. Examples of these compounds have been formulated as $Na_6W_{12}O_{40}\cdot xH_2O$ and $Na_{10}W_{12}O_{41}\cdot xH_2O$, respectively.

Heteropoly compounds in which another anhydride of an element such as silicon combines with the tungsten to form polymers are also known. Sodium silicotungstate, $Na_4SiW_{12}O_{40}$, is a compound of this type. Most of the polytungstates are soluble, whereas the normal tungstates are sparingly soluble with the exception of the salts of the alkali metals and magnesium.

Tungstates are used to flameproof fabrics and to manufacture phosphorescent screens. *See* TUNGSTEN. [E. EUGENE WEAVER]

Tungsten

A chemical element, W, atomic number 74, and atomic weight 183.85. Tungsten is the heaviest element in the periodic table group VIb. The metal has a body-centered cubic structure and a silver-gray metallic luster. Its melting point of 3410°C is

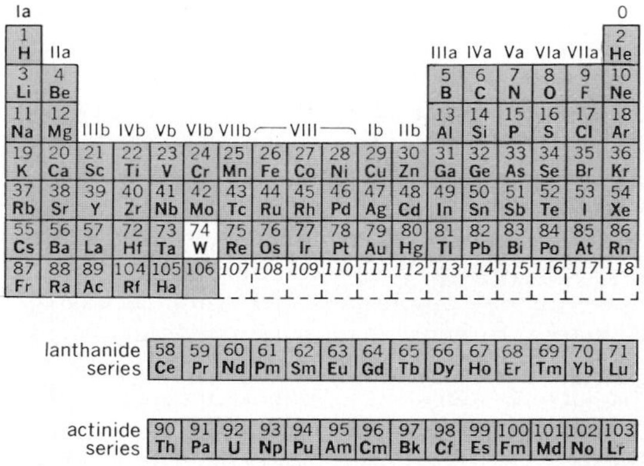

the highest of any metal. The metal exhibits a low vapor pressure, high density (19.3 g/cm³ at 20°C), and superior strength at high temperature in the absence of air, and is extremely hard.

The International Union of Pure and Applied Chemistry (IUPAC) has given this element the alternate name wolfram, but this name has not found acceptance in the United States or Great Britain. *Chemical Abstracts*, of the American Chemical Society, recognizes the term tungsten only.

Chemically, tungsten has two important oxidation states, 2+ and 6+. The following is a tabulation of the natural isotopes and their abundances: 180 (0.2%), 182 (25.8%), 183 (14.2%), 184 (30.6%), and 186 (29.2%).

Uses. Ferrous alloys consume 40% of the tungsten mined. When added to iron or steel, tungsten improves high-temperature strength and hardness. Ferrotungsten, made in electric furnaces by reduction with carbon or with silicon or by the aluminothermic or silicothermic method, is used for high-alloy additions. Tungsten concentrates may be charged directly into the steel melt for alloys of lower tungsten content. Production of nonferrous alloys of the tungsten-chromium-cobalt cutting-tool or hard-facing type consumes 6% of all the tungsten mined. Silver- or copper-tungsten alloys are used for heavy-duty electrical contacts. Alloys for jet-engine and missile applications are under development and will be of future importance. High-density, tungsten-copper-nickel alloys have been utilized for radiation shielding. Tungsten carbide (representing 38% of all W) has replaced diamond for many die and drilling applications.

It is one of the best hard tool materials, retaining its properties at elevated temperatures.

Pure tungsten metal in the form of wire, rod, and sheet (15%) is important to the electric lamp, electronics, and electrical industries. Small shapes of the metal sintered to various porosities or densities have been used as filters and as the backing of probes used in nondestructive testing by measurement of the passage of ultrasound. Other applications are for welding rod, x-ray targets, lead wires, cathodes for power tubes, and automotive and aircraft distributor points.

Tungsten chemicals (1%) have several minor industrial applications. Tungstic oxide and sodium tungstate are important to the metallurgical processing of pure tungsten and are available commercially. Calcium, barium, and magnesium tungstates are used as phosphors. Minor applications of tungsten compounds include: flameproofing materials for fabrics, ink pigments, analytical reagents, and substances used in the glass, ceramic, and tanning industries.

Occurrence. Commercially important tungsten minerals are wolframite, (FeMn) WO_4; ferberite, $FeWO_4$; hübnerite, $MnWO_4$; powellite, Ca(WMo) O_4; and scheelite, $CaWO_4$. Deposits found in China, Burma, Korea, South America (Bolivia), Europe (Portugal), and the United States (California, North Carolina, Nevada, Colorado, Idaho) contain 0.5–2.0% tungstic oxide.

Extraction, reduction, and purification. Concentration of ores to 60–70% oxide is accomplished by crushing, grinding, magnetic or gravity separation, flotation, or special chemical reactions. Wolframite concentrates are converted to soluble sodium tungstate by fusing with sodium carbonate (900–1040°C) or by aqueous caustic digestion. Iron and manganese oxides are removed as insoluble sludge. Acidifying the soluble tungstate solution precipitates tungstic acid. Scheelite concentrates are treated directly with hydrochloric acid to form calcium chloride and insoluble tungstic acid. After washing, the tungstic acid is dissolved in excess ammonium hydroxide, filtered, and evaporated to form pure ammonium paratungstate crystals. Repetition of this operation improves purity.

Heating the paracrystals to 400–500°C produces pure tungstic oxide; heating in the presence of hydrogen reduces the crystals to the blue ($W_{20}O_{58}$) or brown (WO_2) oxide. Further reduction with hydrogen (800–1000°C) produces pure metal powder. Single-step reduction from ammonium paratungstate to metal powder also is possible. The reduction can be controlled to yield a fine powder of 0.5–2.5 μm average particle diameter, or a powder of any coarseness up to 400–500 μm size. Carbon-reduced metal powder of lower purity also is commercially produced.

Vacuum sintering of tungsten powder compacts for 2 hr at 2600°C volatilizes WO_3 impurity and thereby reduces the oxygen content. A typical decrease would be from 310 parts per million by weight (wt ppm) to about 5 wt ppm. If the sintering is done in a hydrogen atmosphere, the WO_3 is reduced with the production of nascent tungsten which sinters relatively easily. The same results obtain at lower temperatures, but more time is required. Carbon can be used for reducing WO_3 to a lower oxide, and hydrogen then used to complete the reduction. Two zone-refining passes can reduce the oxygen content to less than 0.2 wt ppm.

Vapor deposition of tungsten is possible by the pyrolysis of gaseous compounds, such as tungsten carbonyl (WCO_6) which sublimes at 50°C and boils at 175°C, tungsten hexafluoride (WF_6) which boils at 19.5°C, or tungsten hexachloride (WCl_6) which boils at 346.7°C. Vapor deposition of tungsten on a substrate of graphite or copper by heating WF_6 with hydrogen yields a fine-grained deposit. The reaction is:

$$3H_{2\uparrow} + WF_{6\uparrow} \rightleftharpoons \underline{W} + 6HF_{\uparrow} \quad \text{(at 250–2000°C)}$$

The deposits, which are 100% dense, contain less than 50 wt ppm of interstitial impurities and less than 100 wt ppm of metallics. The fluorine content, which is at least 15 wt ppm, increases the ductile-to-brittle transition temperature. If tungsten is deposited on the inner surface of a copper tube and the copper is dissolved in nitric acid, a tube of tungsten will remain. In this way any length of tungsten tubing can be prepared with variations in wall thickness of less than 76 μm. A random grain structure is obtainable by deposition at 800°C, and a columnar structure results at 540°C. The latter undergoes less grain growth on heat treatment.

Tungsten can also be pyrolytically deposited from WF_6 on the inner surface of a copper tube or on a flat plate. A radial grain orientation in the tubular deposit means that the tubing cannot be reduced by drawing without fracture at any temperature up to at least 1000°C. The metal has a ductile-brittle transition temperature of 150°C, and is resistant to grain growth.

Consolidation. Production of consolidated tungsten is generally done by powder processing but is increasingly done by melting. In many instances the powder processing is to prepare electrodes for arc melting or drip rods for electron-beam melting.

For conventional powder metallurgy angular 1–10-μm particles can be compacted without a binder. The fine (1-μm) powder is less dense than the coarse (3–5-μm) powder, and generally contains twice as much nitrogen and six times as much oxygen. Blends of fine and coarse powders weighing up to 5 tons have been compacted. Bar-shaped compacts are commonly hydraulically pressed at 35 kg/mm^2 (25 psi), sometimes presintered at 980–1200°C, machined if necessary, and resistance-sintered in dry hydrogen at 3235°C. The bars are brittle at room temperature and must be hot-swaged for improved ductility. Compacts from coarser powder tend to crack on presintering. Vacuum-induction sintering also is done with the compacts packed in loose tungsten powder in a graphite susceptor.

There is little agreement among fabricators as to purity specifications, which lie in the range: C, 30–200; O, 150–3000; H, 50 maximum; Al, 20–100; Fe, 50–200; Ni, 10–200; Mo, 20–300; and Si, 10–100 wt ppm. Some of the disparities are governed by intended uses and methods. Tungsten powder of the higher purity levels usually is presintered. Only 50 wt ppm of aluminum raises the 30-min recrystallization temperature from 1300 to 1800°C.

During purification by vacuum presintering, ample porosity (bulk and surface) provides exits for escaping gases whose pressure, if they were trapped, would cause swelling.

The rate of densification of tungsten on sintering is accelerated by palladium, infiltrated as palladium chloride, or by the presence of a small addition of nickel. The latter is infiltrated as Zn-Ni alloy from which the zinc is then volatilized.

Nonconsumable electrodes for arc melting are resistance-sintered with the pressed rod suspended from a copper electrode and the lower end immersed in a mercury electrode. The liquid electrode prevents breakage by allowing the ingot to move while shrinking.

Powder compaction and sintering is the only means of consolidating dispersed tungsten alloys containing doping additions, which are added for grain refinement and texture control. Melting would volatilize excessive amounts of doping additions. Agglomerate dispersoids such as ZrO_2 increase grain size and counteract the tendency of a dopant to produce elongated interlocking grains.

Tungsten consolidated by melting is usually of higher purity than tungsten consolidated by powder compaction and sintering. This holds particularly when the reguline (melt consolidated) metals are prepared from sintered compacts by vacuum-melting procedures that selectively volatilize impurities.

Arc-melted tungsten of relatively high purity is obtained with hydrogen-sintered electrodes, straight polarity (electrode negative), relatively large electrodes, a 200-μm flowing-hydrogen atmosphere, and a slow melting rate. The surface quality of small-diameter ingots can be improved by using alternating current or reverse polarity (electrode positive). If the mold-electrode diameter ratio is small, arcing may damage the mold wall. The bottom of the copper mold used for arc-melting tungsten may tend to overheat and burn through unless it is of a hemispherical shape designed to streamline the flow of coolant water. Allowances in the mold diameter should, when necessary, be made for removing a porous surface from the ingot by machining.

In the arc casting of a large-diameter tungsten ingot, excessive chilling results in a thick exterior shell around the molten pool. This collar is apt to have laps and cold shuts, and melting it back into the pool tends to overheat the crucible wall. More of the ingot will be sound if the electrode diameter and corresponding power supply are large enough to keep the skin that separates the melt from the wall reasonably thin.

The rods or ingots to be melted should be 80–100% dense, depending on the intended ingot diameter, the power supply, the desired degree of superheat, and other details. The sintered rods or the ingots are joined end to end by tungsten inert gas (TIG) arc welding to form consumable electrodes. The bulk resistivity should be homogeneous and diametrically opposite welds should be similar in resistance—else uneven heating may cause warpage and breakage.

Consumable-electrode arc melting in vacuum is faster than electron-beam melting with an equivalent amount of power, and should produce fine grains. Duplex electron-beam—vacuum-arc procedures are used to lower the impurity content and yet achieve a suitable grain size, or to achieve a suitable content of alloying elements while reducing the interstitial content.

For best results with tungsten an electron-beam furnace in the megawatt range is used for large-scale melting, and the material to be melted is mechanically fed in the form of a drip rod, rather than as powder. A 2.25-kW unit can, however, produce tungsten ingots up to 10 cm in diameter.

If the concentration of volatiles in the drip rod for electron-beam melting is too high, their evolution will overload the vacuum system and cause arcing. Presintering in vacuum may therefore be necessary. Purification of tungsten by electron-beam melting produces coarse grains, and large-diameter ingots with coarse grains are subject to intergranular cracking. Should the melting time necessary for purification result in an unacceptable degree of coarseness, one or more quick remelts may be necessary. Electron-beam triple melting to form 7.5-cm diameter ingots at 0.05–0.10 μm Hg and 200–260 kW reduces the total interstitial content (C, O, N) from 930 to 120 wt ppm. At this power input the melting rate (2–5 kg/hr) is slow enough for the formation of grains up to 2.5 cm in diameter.

Skull-cast tungsten, which is arc-melted tungsten poured from a "skull" of solid tungsten, is workable and can be upset forged at 1500°C to high reductions. Spin-cast tubes of tungsten of the skull-melted type have been made. The degree of superheating required poses the problem of obtaining mold materials that are sufficiently inert to the molten tungsten. Up to 63% of a 4.5-kg charge has been poured by the skull-casting technique. Because of shrinkage and piping problems with large sections, tungsten and tungsten-alloy shape castings are mostly made with thin walls. If uncoated graphite molds are used, provision should be made for some interaction with them.

Joining and bonding. Manganese containing 16% by weight (wt %) each of nickel and cobalt, and up to 1 wt % of boron (Coast Metal CM62) is an effective tungsten-to-tungsten vacuum-brazing alloy. The bond has substantial strength at 2705°C (4900°F). If tungsten parts are to be soft-soldered, they should be precoated with copper. To minimize loss of ductility in tungsten-to-tungsten spot welds, the spot welding is sometimes done with the parts under water. Solid-state tungsten-to-tungsten diffusion bonds are obtainable in hydrogen at 2350°C with a ram pressure of 1 kg/mm^2 (1400 psi) maintained for 2 hr. Tungsten surfaces to be joined should be mechanically cleaned, chemically etched, or preheated in hydrogen at 650°C (1200°F).

Arc welds or electron-beam welds in powder-metallurgy tungsten have considerable porosity. Those in reguline tungsten at room temperature are perhaps one-half as strong as the base metal, but at the recrystallization temperature the weld and the base metal are equally strong. Welding should be done at a high speed to avoid excessive heat input. For a base metal having a transition temperature of 344°C, the best welds had a transition temperature of 594°C.

Electron-beam welds in tungsten can produce sound joints if heat input is restricted and grain refiners are used. Welds joining tungsten to molybdenum–0.5 wt % titanium alloy are usually ductile.

Tungsten is embrittled more by melting during welding than by high-temperature exposure in a protective atmosphere or under a protective coating. Pressure diffusion bonding can therefore join the metal without damaging it. Two pieces of tungsten can be united in the fraction of a second necessary for resistance welding or for impulse bonding, or joining by autoclave pressure bonding may require 24 hr. The surfaces and protective atmospheres in both cases should be clean.

Deformation and fabrication. Deformation processing of tungsten almost always begins with primary reduction and ingot breakdown by a compressive process, such as extrusion, before secondary fabrication, machining, and surface finishing are practicable. The reduction eliminates porosity of the sintered material; the primary breakdown achieves some grain refinement and is usually the only means of reducing an ingot of tungsten without cracking it. The crude metal is, at the same time, converted into a shape and condition, for example, into tube stock or sheet bar, suitable for secondary working (rolling, swaging, drawing) into final shape.

Tungsten and its alloys usually are hot-cold-worked, since true hot working (at 80% of the melting point) is not practicable. Specimens of tungsten are initially fabricated at 1485°C (2700°F) and, subsequently, at successively lower temperatures. Total reductions greater than 90% are required to obtain suitably low bend-transition temperatures for tungsten sheet 1 mm thick. Stress relief (well below the recrystallization temperature) may be required at different stages of reduction. Tungsten sheet can be produced from arc-melted ingots as follows: (1) extrude as cylinder; (2) forge to sheet bar; (3) recrystallize; (4) roll to sheet with no in-process anneals and at lowest practicable temperatures (92–95% reduction); (5) stress-relieve at 980–1095°C (1800–2000°F). Rolling can reduce the bend transition temperature from 330°C at 40% reduction to 100°C at 95% reduction.

Extrusion. Breakdown extrusion of arc-cast tungsten ingots is done with extrusion ratios less than 5:1. Direct extrusion processes are used with dies having a hardness of up to R_c60. Niobium-clad tungsten can be extruded with the aid of a lubricant at an 8:1 ratio by a conventional low-velocity process at 1705°C (3100°F). Unclad tungsten cannot be extruded with pressures to 105.6 kg/mm^2 (150,000 psi) at a low velocity, even at 1980°C (3600°F) and a reduction ratio of only 6:1. High velocity (635 m/sec) extrusion of unclad tungsten is possible in special equipment at 1650–2095°C (3000–3800°F) with no lubricant and with reduction ratios of 45:1. The high-velocity extrusions are softer, weaker, and more ductile.

Although small-diameter, thin-wall tubing (0.32-cm OD with 0.032-cm wall) cannot be made entirely by extrusion, extrusion of the tube stock is desirable and should be done below the recrystallization temperature. Either sintered or vapor-deposited sleeves can be extruded. High-purity, 475-μm granules of tungsten in a molybdenum jacket and with a molybdenum core have been converted directly to tubing by extrusion at 1250°C.

Extrusion of tungsten tubing cannot readily be done on a fixed steel mandrel, because the latter deforms too readily at the necessary extrusion temperature to give adequate support. Instead, a "filled billet" with a deformable mandrel must be used. The axial core is filled with a material, such as molybdenum, that has the right stiffness to support the tungsten and deform along with it. Eventually the core must be removed by dissolution (for example in hydrofluoric acid), by drilling, or by electric-discharge machining. Numerous other types of filled billets devised for the extrusion of tungsten tubing will not be described. Steel-jacketed billets with tungsten sleeves and molybdenum inner and outer fillers seem promising. Hafnium core centers are also under consideration.

For hydrostatic extrusion, in which the billet is surrounded by a fluid under a high pressure during extrusion, the tungsten billet typically has a hardness of 440 VHN (Vickers hardness number), the extrusion ratio R is 1.2–2.0, and the extrusion pressure in psi is $p = 327 \ln (R + 0.36)$. The billet does not increase greatly in hardness and may have a smooth surface, since there is no need for a rough surface to hold lubricant. A hydrostatic pressure above 10 kilobars (150,000 psi) causes tungsten to behave in a ductile manner even when extruded at a ratio of 14:1 into a receiver at a slightly lower pressure.

Forging. Tungsten is preferably hammer-forged, rather than press-forged, since hammer forging provides less opportunity for premature chilling by conduction. Direct extrusion of tungsten ingots to sheet bar eliminates pickup of contaminants that would result on forging. Despite this, tungsten may be forged to sheet bar after breakdown extrusion if the forging is done in an inert environment. Direct forging of arc-cast ingots involves such small reductions and frequent reheatings as to be impractical. Forging an extrusion can, however, make the microstructure more nearly uniform and is useful in achieving the dimensions desired for finishing.

Ingots of electron-beam-melted tungsten are directly forgeable only if they are fine-grained. Hydrogen-sintered plasma-sprayed tungsten may have to be reduced over 30% by forging if the deposit is to lose its porosity and gain resistance to thermal stress. Careful forging is possible on sintered bars while cooling from 1800 to 1150°C. To avoid radial cracking of the workpiece, the forging must be done at a temperature so high that excessive chilling of the tungsten by the forging die does not occur; yet the tungsten must be forged at a low enough temperature to ensure a substantial introduction of cold work.

Swaging. Sintered tube blanks of tungsten or tungsten–25 wt% rhenium alloy are swaged over a molybdenum mandrel at 1500°C to reduce the outer diameter by 30%, but not the wall thickness. The latter is noticeably reduced, however, by swaging at 1400°C, at which temperature the molybdenum core is stiffer and harder. Short-bladed (chopper-type) dies with 30° entrance angles and 6° exit angles are preferable to long-bladed general-purpose dies. The short-bladed dies are also good for swaging tungsten rod, but dies with 8° entry angles are preferable for pointing tungsten tubing to be drawn.

Drawing. A swage-pointed tungsten tube can be drawn over a stationary, hardened (M-2 steel) mandrel, or preferably over a "Clarite" T-1 class, high-speed tool steel hardened to R_c60. This operation, known as plug drawing, has an ironing action that improves the inner surface of tungsten tubing and prevents the formation of folds or cracks. A threaded end of the plug is attached to the rear end of the drawbench. A brazed-in steel rod extends out from the swage-reduced end for jaws to grip and pull through the die because the tungsten tube is too brittle to grip directly. The tube is lubricated on both sides and is drawn with both tube and tools at 580°C.

In moving-mandrel drawing, a nondeforming mandrel is pushed into the die along with the tube until the tube is squeezed onto the mandrel. After the pass, a reeling operation expands the tube for the mandrel to be withdrawn. Moving-mandrel drawing is useful in reducing the diameter of small-diameter tubing while thinning the wall.

Ductile-core drawing is done with a deformable mandrel that may consist of the core used in filled-billet extrusions. The core must eventually be removed with acid. The process allows higher reductions between anneals than is possible with plug drawing, and it avoids the folding sometimes caused by sinking.

Rolling. Forged extrusions of arc-melted tungsten or of high-purity sintered or swaged tungsten can be rolled to strip or sheet. An increase of 50 wt% of interstitial contaminants occurs on rolling to 75% reduction at 1200–800°C. The following is a typical rolling schedule:

1. Roll sintered billet 1 in. (2.5 cm) thick longitudinally to 0.9 in. (2.25 cm) at 1450–1400°C.
2. Cross-roll to 0.253 in. (0.632 cm) at 1400 down to 1350°C.
3. Stress-relieve 5 min at 1400°C.
4. Roll longitudinally to 0.060 in. (0.15 cm) at 1250 down to 1100°C.
5. Stress-relieve 5 min at 1125°C.

Cross-rolling lowers the bend transition temperature. Material thinner than 1 mm may be rolled at 200–300°C. Tungsten foil 2.5 μm (0.0001 in.) thick can be produced from 12.7-μm stock by a repeated sequence of cold-pack rolling, stress relieving at 950°C in a molybdenum-lined stainless steel jacket, air cooling, placing in a new pack, and cross rolling.

In powder rolling, tungsten powder is rolled to 80–85% dense sheet and then sintered at 1700°C to a density of 90–95%.

Ring rolling of tungsten is a process carried out below 760°C in which one roll bears on the inner surface of a cylindrical workpiece, and an opposing plate or roll bears on the opposite side.

Wire. Tungsten rods are, of course, easier to swage and draw than is tubing. Tungsten carbide and diamond dies, with graphite lubrication, are used for final drawing into fine wire, when that is desired. The reduction imparts enough ductility for coiling the wire into lamp filaments. Heating in service recrystallizes the wire and causes it to become brittle at room temperature. The incorporation of less than 1 wt % of gallium into the original tungsten powder was found to overcome this tendency, but the use of gallium is not yet common practice. At present, traces of potassium, silicon, and aluminum are purposely "doped" into the metal powder, and up to 2% of thoria, ThO_2, is added to inhibit grain growth. This wire is brittle at room temperature and sags at elevated temperatures but has exceptional high-temperature shock strength.

Other fabrication methods. Explosive forming of tungsten sheet is of interest for making asymmetric shapes. The forming is done with an explosive in a hydraulic medium on one side of the specimen on a die, and with a vacuum on the other side (that is, in the die cavity). To provide a high enough temperature, molten aluminum is used as the hydraulic medium.

Tungsten sheet or rods 0.5 mm thick can be readily stamped or sheared at 400°C, or formed by bending at 800°C. Greater thicknesses require higher temperatures.

Spinning tungsten is an operation performed at 815–1095°C in which a tungsten blank sheet is spun to the contour of a rotating mandrel by an external roller device. Intermediate anneals are required.

Machining and grinding. Heating tungsten above the brittle-ductile transition temperature—usually above 260–424°C (500–800°F)—has been looked upon as a prerequisite to the conventional machining operations. Even so, tool wear is high and chipping and cracking on tapping and sawing is common. Although ductile forms of tungsten, produced by calcium reduction, can be hacksawed, electric-discharge machining is a more general answer to problems of chipping and cracking. The operation is slow and usually expensive. The best general method of machining tungsten is to supercool both the probe piece and the cutting tools to temperatures well below zero. Frictional heat cannot then cause a drill or lathe tool to expand enough, relative to the tungsten, to cause chipping or cracking.

Grinding of tungsten ordinarily must be done at wheel speeds slow enough (no more than 2000–3000 surface feet per minute) so that frictional heating will not lead to tensile cooling stresses that result in surface cracking. Electric-discharge grinding methods have been developed that are more rapid and effective than conventional grinding.

Thermal behavior and heat treating. Containers and supports for tungsten workpieces during heat treatment should be compatible with the tungsten. Parts being heated are sometimes seated in tungsten powder or rested upon alumina granules. Zirconium dioxide is suitable for use with tungsten at temperatures up to 3205°C (5700°F). Quenching can be done by dropping a specimen from the heat zone of an induction furnace onto a copper plate that is externally cooled with liquid nitrogen. The specimen will cool below a red heat in about 1 min if pulled out of the susceptor by a wire suspension that passes through a seal, or in 2 or 3 min if the power is simply turned off.

The effect of heating tungsten from room temperature to 3300°C (5975°F) reduces the ultimate strength from 155 kg/mm^2 (220,500 psi) to 3.5 kg/mm^2 (5000 psi). Tungsten heated to 3150°C (5700°F) has a vapor pressure of 0.001 mm Hg. For tungsten at 2200°C (4000°F) in vacuum, the surface recession is only 30 μm/year.

Heat treatment of tungsten is usually to stress-relieve or recrystallize before fabrication. A vacuum or protective atmosphere is ordinarily required during heat treatment. Stress relieving is done to improve fabricability at a temperature above the ductile-brittle transition temperature without increasing it unnecessarily. Recrystallization impairs room-temperature ductility, but eases deformation and eliminates excessive texture.

For 10-min heating at 1000–1400°C (1830-2550°F) the brittle-ductile transition temperature of wrought tungsten sheet remains at 50–65°C

(125–150°F). The same treatment at 1525°C (2775°F) increases the transition temperature to 460°C (1150°F). Increasing the stress-relief time at these temperatures has little effect, but any prolonged heating above 620°C increases the transition temperature. For a 1-hr treatment each degree of increase in stress-relief temperature raises the bend transition temperature by the same amount.

Based on 50% recrystallization in 1 hr, the primary recrystallization temperature of as-rolled tungsten (hardness 464 DPH) is 1340°C (2425°F). Stress relieving 1 hr at 1230°C (2250°F) reduces the hardness by only 30 points, but recrystallizing at 1482°C (2700°F) softens it by a further 90 DPH points. Secondary recrystallization of tungsten is seldom obtained in processing, through occasionally in annealing.

Experience in working with tungsten sometimes indicates the desirability of heating above the recrystallization temperature and working the metal before it has time to recrystallize. If, for example, tungsten that recrystallizes with time at 1500°C is briefly heated to 1600°C, it is plasticized for easy working and the undesirable features of recrystallization are avoided.

Recovery of heavily worked tungsten occurs at 400°C (750°F) by removal of vacancies. This softens the tungsten. Aging 30 min at 649–802°C (1200–1475°F) results in polygonization or subgraining. This strengthens tungsten by blocking the motion of interstitials toward dislocations. Continued heating softens again by causing subgrain growth. Aging at a certain temperature thus activates changes in mechanisms that produce maxima and minima in strength and hardness.

Oxidation and corrosion. Tungsten begins to tarnish in air or oxygen at 400°C (750°F), acquiring a blue-black WO_3 protective layer. At 700°C the oxide changes crystal structure and becomes yellow and nonprotective. The maximum temperature for long-time service of tungsten in the air is thus limited by corrosion considerations to 600–700°C (1110–1290°F). The oxidation behavior and reactions of tungsten with water vapor or humid air require that metal for high-temperature use in such environments be coated or jacketed. The strength and integrity of the enclosures will then be the limiting service factor. Brief exposure of heavy-section tungsten to air during fabrication at 1800°C (3275°F) is not particularly damaging, even though prolonged exposure of tungsten at 1200°C (2190°F) can cause catastrophic oxidation.

Tungsten follows the parabolic rate law for corrosion in oxygen at temperatures up to about 750°C (where the oxide film is protective) and a linear law at temperatures between about 1000 and 1260°C (where the oxide film is protective). At intermediate temperatures the corrosion rates are usually intermediate. At 1260°C (2300°F) the porous, nonprotective film melts, forming a semiprotective layer. For tungsten at 10^{-8} to 10^{-6} atm (10^{-3} to 10^{-1} Pa), the rate of reaction at about 1800–2340°C decreases with increasing temperature and with decreasing pressure.

Alloying tungsten with niobium or tantalum improves its high-temperature corrosion resistance by forming a higher-melting oxide. From 20 to 35 wt % of tantalum lowers the oxidation rate to 10^{-3} times its original value. Niobium is even more effective. Unfortunately any addition of tantalum or niobium in excess of 5 wt % makes tungsten brittle and unworkable.

Compatibility with liquid metals. Tungsten is unaffected by contact with sodium at temperatures up to 600°C (1110°F) and possibly higher. Pressed-and-sintered tungsten is unattacked by zinc, or by immersion in 47.5 m/o magnesium chloride$_2$–47.5 m/o lithium chloride–5 m/o magnesium fluoride (where m/o denotes mole %) at 800°C (1475°F) on 200-hr exposure. Tungsten crucibles have use in the reclamation of uranium-plutonium-fissium, oxidized fuel skulls from EBR-II (experimental breeder reactor) fast-neutron reactor. Sintered crucibles were so thick-walled as to be too expensive, heavy, and unwieldy, and therefore thin-walled crucibles made by shear forming are used. Unalloyed uranium at 1500°C penetrates a 1-mm wall of a tungsten container in 1 min.

Irradiation behavior and nuclear properties. Tungsten has a capture cross-reaction σ_c, for 0.025-eV neutrons of 29.2 barns (2.92×10^{-27} m^2), and (n,γ) cross sections for 30-keV and 65-keV neutrons of 270 and 190 millibarns (2.7 and 1.9 × 10^{-29} m^2), respectively. Tungsten is also a better reflector of fast neutrons than stainless steel or nickel.

Tungsten irradiated to 6×10^{19} n/cm^2 does not change significantly in hardness; however, it increases nearly 40% in electrical resistivity, and changes slightly in linear dimensions and density. Highly wrought tungsten is weakened (that is, annealed). Irradiation to 10^{18}–10^{19} n/cm^2 (E_n − 1 MeV) improves the rupture life of tungsten at 1100°C by a factor of about 2.6 and reduces the creep rate by a factor of about 2.

The principal damage caused by neutron irradiation in tungsten is believed to be the creation of lattice vacancies. Vacancy diffusion is a factor in creep. The activation energy for vacancy diffusion in tungsten is 1.7 eV. Inhibition or retardation of grain growth is, of course, caused by irradiation-induced defects, and the finer the grain the greater the resistance to creep. In addition, fast-neutron irradiated tungsten that has been tested at 1100°C has dislocation loops. These are attributed to the presence of interstitial aggregates consisting mostly of rhenium and osmium produced by transmutation in a thermal-neutron flux.

In high-purity tungsten, metal atoms knocked into interstitial positions by fast neutrons can be annealed out of these positions (for example, to vacancies) at 100–200°C. Relatively small substitutional atoms can trap knocked-on tungsten atoms so firmly that annealing at 1500°C is necessary to restore preirradiation resistance.

Properties of the metal. Chemically, tungsten is relatively inert. It is not readily attacked by the common acids, alkalies, or aqua regia. It reacts with a mixture of concentrated nitric acid and hydrofluoric acid. Molten oxidizing salts such as sodium nitrite attack tungsten rapidly. Gaseous chlorine, bromine, iodine, carbon dioxide, carbon monoxide, and sulfur react with tungsten only at high temperatures. Carbon, boron, silicon, and nitrogen also form compounds with tungsten at elevated temperatures; hydrogen does not.

Table 1 lists several of the important physical properties of tungsten and Table 2, the mechanical

Table 1. Physical properties of tungsten

Property	Value
Melting point	3410°C
Boiling point	5930°C
Debye temperature	305 K
Density, metal powder	2.0–4.8 g/cm³
sintered at 1200°C	10.0–12.0 g/cm³
sintered at 3000°C	17.0–18.5 g/cm³
swaged rod	17.0–19.2 g/cm³
drawn wire	19.3 g/cm³
Specific heat, 0°C	0.03–0.05 cal/g-°C [0.12–0.21 J/(g · °C)]
2600°C	0.06 cal/g-°C [0.25 J/(g · °C)]
Latent heat of fusion	45 cal/g (188 J/g)
Latent heat of vaporization	1180 cal/g (4937 J/g)
Vapor pressure, 1727°C	3×10^{-15} atm (3×10^{-10} Pa)
3410°C (mp)	5.4×10^{-6} atm (5.5×10^{-1} Pa)
4227°C	5.6×10^{-3} atm (5.7×10^{2} Pa)
5727°C	0.9 atm (9×10^{4} Pa)
Thermal expansion, linear	$L = L_0[1 + (4.28t + 0.00058t^2) \times 10^{-6}]$ $t = °C$
Thermal conductivity, 0°C	0.40 cal/(cm)(s)(°C) [1.67 J/(cm · s · °C)]
1227°C	0.28 [1.17]
1827°C	0.40 [1.67] (porous metal)
2227°C	0.50 [2.09] (porous metal)
Electrical resistivity, 20°C	5.6 microhm-cm
927°C	30.2
1827°C	59.0
2727°C	90.4

properties. The data in Table 2 may vary considerably depending upon the history (reduction, purification, and treatment) of the material tested.

Principal compounds. Tungsten compounds embrace oxidation states of tungsten from 2+ to 6+, the higher oxidation states being most stable. Tungsten chemistry resembles that of chromium and molybdenum, which also occupy the same subgroup in group VIb of the periodic table. Tungsten aqueous chemistry is complicated by its tendency to form complex ions.

Table 2. Mechanical properties of tungsten

Property	Value
Tensile strength, sintered ingot	18,000 psi (124 MPa)
swaged rod	500,000–215,000 psi (3447–1482 MPa)
drawn wire	250,000–600,000 psi (1724–4157 MPa)
annealed wire	150,000 psi (1034 MPa)
High-temperature tensile strength, 0.028-in.-diameter (0.7-mm) wire 20°C	430,000 psi (2965 MPa)
800°C	200,000 psi (1379 MPa)
1500°C	50,000 psi (345 MPa)
2800°C	5,000 psi (34 MPa)
Yield strength	Approximately 90% or more of the tensile strength
Rupture strength, 900°C	36,000 psi or 248 MPa (1 hr) 35,000 psi or 241 MPa (10 hr) 32,500 psi or 224 MPa (100 hr)
1200°C	25,000 psi or 172 MPa (1 hr) 22,000 psi or 152 MPa (10 hr) 15,000 psi or 103 MPa (100 hr)
Brittle-ductile transition	100–250°C (wrought) 350°C (recrystallized)
Ductility	0–4% elongation (brittle when recrystallized)
Elongation, 200°C	0% (recrystallized)
400–1000°C	55% (recrystallized)
Reduction in area, 200°C	0.5% (recrystallized)
700°C	75%
Modulus of elasticity	12.8×10^6 psi or 88 GPa (for sintered rod) 53×10^6 psi or 365 GPa (for well-worked wire)
2400°C	32×10^6 psi (220 GPa)
Modulus of rigidity, room temperature	$23.0 \pm 0.2 \times 10^6$ psi (158.6 ± 1.4 GPa) (annealed)
1600°C	18.7×10^6 psi (129 GPa)
Compressibility coefficient	1.67×10^{-3} per ton/in.² or 1.21×10^{4} per MPa (smallest value for all the metals)
Hardness, sintered bar	255 Vickers
swaged bar	400–480 Vickers

Oxides. Tungsten forms four well-defined, stable oxides. Several others have been reported, but their existence is doubtful. Beta-tungsten (reported as W_3O) has not been conclusively identified as an oxide.

Tungsten trioxide (tungsten oxide), WO_3, has yellow triclinic or pseudo-orthorhombic crystals (reported to be monoclinic), and it transforms to the tetragonal form above 740°C. It has specific gravity 7.16 and melting point 1473°C, with tendency to sublime. It is reducible to metal powder by hydrogen at 650°C and above, and by carbon at 1000–1100°C. The oxide is insoluble in water or acids, but forms soluble tungstates in alkali solutions. Its heat of formation is -199 ± 1 kcal/mole (-833 ± 4 kJ/mole).

Other oxides include $W_{20}O_{58}$ ($WO_{2.90}$), dark-blue, thin, monoclinic needle crystals, heat of formation -193 ± 1 kcal/mole (-808 ± 4 kJ/mole); $W_{18}O_{49}$ ($WO_{2.72}$), reddish-violet, monoclinic needle crystals, heat of formation -183 ± 1 kcal/mole (-766 ± 4 kJ/mole); WO_2, tungsten dioxide, brown, monoclinic crystals, specific gravity 12.11. The pure compound is prepared in vacuum or inert atmosphere by the prolonged heating at 950°C of stoichiometric mixtures of W and WO_3. Partial hydrogen reduction of WO_3 yields WO_2 plus a second phase impurity. Its heat of formation is -137 ± 1 kcal/mole (-573 ± 4 kJ/mole).

Acids. Hydration of tungstic oxide produces various acids of which three are well defined.

Tungstic acid, H_2WO_4, is an amorphous yellow powder, specific gravity 5.5. It is insoluble in water or acid solution and soluble in strong alkaline solutions. It precipitates from warm, strongly acid solution and tends to be colloidal if precipitated from weakly acid solution. Boiling with strong acid coagulates the colloid.

Hydrated tungstic acid, $H_2WO_4 \cdot H_2O$, is a white compound precipitated from a cold, acidified solution of H_2WO_4. It is more soluble in water than H_2WO_4; it is converted to yellow tungstic acid on boiling in acid solution; and it tends to be colloidal when washed.

Metatungstic acid, $H_8W_{12}O_{40} \cdot xH_2O$, forms fine yellow crystals that are soluble in water. When heated to 100°C, it is converted to H_2WO_4.

Salts. Tungstates have the general formula, $M_2WO_4 \cdot xH_2O$. Alkali metal and magnesium tungstates are water-soluble; others are insoluble. Insoluble tungstates are prepared by fusing the metal oxide with WO_3 or by precipitation from sodium tungstate solution.

Sodium tungstate, Na_2WO_4, forms water-soluble, white, rhombic crystals, with specific gravity 4.18, melting point 692°C. When crystallized from aqueous solution, it forms a dihydrate. The anhydrous salt is prepared by direct fusion of WO_3 and NaOH.

Ammonium tungstate, $(NH_4)_2WO_4$, cannot be isolated from aqueous solution, and it evolves NH_3 on heating. It is prepared by adding hydrated tungstic acid to liquid ammonia.

Metatungstates, $3M_2O \cdot 12WO_3 \cdot xH_2O$, are water-soluble salts prepared by dissolving tungstic acid in metal tungstate solutions.

Paratungstates, $5M_2O \cdot 12WO_3 \cdot xH_2O$, the structures of which vary with the solution pH, are prepared by precipitation from slightly acid solution. Sodium paratungstate is made by saturating sodium hydroxide solution with WO_3. Ammonium paratungstate, $5(NH_4)_2O \cdot 12WO_3 \cdot 11H_2O$, is a fine white crystal when produced by slow evaporation; rapid evaporation yields platelike crystals. Prolonged boiling converts the salt to a decahydrate. Crystals are water-soluble but decompose in acid or alkali solution. Heating dehydrates the compound and drives off ammonia. Heating in hydrogen reduces the compound to tungsten metal powder.

Heteropoly acids. Many heteropoly compounds have been identified, and two of them are well known.

Phosphotungstic acid, $H_3PO_4 \cdot 12WO_3 \cdot xH_2O$, forms greenish-yellow crystals prepared by evaporation of a solution containing phosphoric and metatungstic acids.

Silicotungstic acid, $Si_2O \cdot 12WO_3 \cdot 26H_2O$, forms pale-yellow, rhombic crystals soluble in water, alcohol, or ether.

Halides and oxyhalides. Numerous halides have been reported. They are generally unstable in air, have low boiling points, and react with water vapor.

Tungsten hexachloride, WCl_6, forms blue to violet, hexagonal crystals, with melting point 275°C, boiling point 346.7°C, specific gravity 3.52. It is soluble in carbon disulfide; it decomposes in water; and it is reducible by hydrogen to lower chlorides or metal. It is prepared by the reaction of dry Cl_2 with metal powder heated to redness; in the presence of moisture, $WOCl_4$ is formed. Other known chlorides are WCl_2, WCl_4, and WCl_5.

Tungsten oxytetrachloride, $WOCl_4$, forms red needles with melting point 211°C, boiling point 227.5°C, specific gravity 11.9. It is soluble in carbon disulfide, and it decomposes in water. It is prepared by the reaction of Cl_2 with W and WO_2 at 400°C, and purified by vacuum distillation at 200°C.

Other tungsten compounds are the bromides, WBr_6, WBr_5, and WBr_2; and the iodides, WI_4 and WI_2. Known fluoride compounds include the highly volatile WF_6 and WOF_2.

Carbides. Two carbides are known; each can be prepared by heating mixtures of the elements to 1400–1600°C. The compounds are gray powders having hardnesses approaching that of diamond. They are insoluble in water, but are attacked by concentrated HNO_3–HF solution.

Tungsten carbide, WC, forms hexagonal crystals, specific gravity 15.6, melting point 2900°C.

Ditungsten carbide, W_2C, forms hexagonal (close-packed) crystals, with specific gravity 17.2, melting point 2850°C.

Other important compounds. These include the carbonyl, nitride, boride, phosphide, silicide, and sulfide.

Tungsten hexacarbonyl, $W(CO)_6$, is a white volatile solid that decomposes into W and CO at 150°C. It is insoluble in water and slightly soluble in organic solvents. Three nitrides have been identified, W_2N_3, WN_2, and W_2N. The known borides are WB_2, WB, and W_2B. The known compounds with phosphorus are WP_2, W_2P, WP, and W_3P_4.

Well-defined tungsten silicides are WSi_2, W_2Si_3, and WSi_3. Tungsten disulfide, WS_2, forms soft gray crystals with specific gravity 7.5. Found in tungstenite ores, it oxidizes in air, is reduced by hydrogen at 900°C to metal powder, decomposes at 1250°C, is insoluble in water, is attacked by fused alkali or HNO_3–HF mixtures, and is prepared by heating H_2S and WCl_6 or by direct reaction of the elements at high temperatures. Ammonium tetrathiotungstate, $(NH_4)_2WS_4$, forms as orange crystals by the reaction of H_2S and ammoniacal tungsten acid solutions.

Alloys. Rhenium is probably the most beneficial single element for alloying with tungsten. When second-nearest-neighbor atoms are considered, the coordination number of a perfect tungsten lattice should be 12 and its electron-atom concentration about 6. In the usual imperfect lattice there are vacancies and grain boundaries where the coordination number is nearer to 10 and the electron-atom ratio is nearer to 5. Such regions are electron-avid and encourage electron donors to migrate to them. If the donors are interstitial atoms, they will form embrittling concentrations at these electron-deficient regions. Rhenium, an electron-contributing substitutional alloying element, also tends to migrate to these lattice imperfections. By preventing the interstitial atoms from doing so, the rhenium increases ductility.

Additional valence electrons reduce stacking-fault energy and thus permit stress relief by extensive twinning without crack formation. But mechanical twinning results in high local stresses and brittle fracture at macrostresses below the theoretical cleavage strength. Although the texture of tungsten sheet is conducive to mechanical twinning, the increased ductility due to the presence of rhenium (28 wt %) increases the ductility of tungsten at room temperature to 300°C. Twin-induced crack formation does not then occur.

An alloying addition of rhenium also improves the oxidation resistance of tungsten at high temperatures and low oxygen pressures. In such circumstances the dissociation pressure of rhenium oxide in a vacuum is greater than the pressure of available oxygen.

Vacuum-melted, as-rolled, polycrystalline tungsten–27 wt % rhenium has some sigma phase, yet the ductile-to-brittle transition temperature for the alloy in either the wrought or fine-grained recrystallized condition is −47° (−52.6°F) and the yield strength is 282 kg/mm² (400,000 psi). Larger-grained, single-phase annealed structures have transition temperatures of 24–46°C (75–150°F). If wrought-recrystallized, and water-quenched before annealing, the room-temperature strength is 50% less, the elongation is 5% and the reduction in area is 3.8%. An alloy of tungsten–5 wt % rhenium–2.2 wt % thorium dioxide is unusually ductile, and has an unusually high recrystallization temperature and an unusually fine grain.

Tantalum containing up to 20 at. % of tungsten is ductile. The other tantalum-tungsten compositions are relatively brittle. Local ordering is responsible for most of the strengthening and embrittling that occur in these alloys.

Small alloying additions of vanadium, zirconium, titanium, and carbon improve the tensile ductility of tungsten–12 wt % niobium at 1927°C (3500°F). If added jointly, they also increase the 1927°C tensile strength from 19.0 kg/mm² (27,000 psi) to 39.9 kg/mm² (57,000 psi). *See* TRANSITION ELEMENTS.

[W. D. WILKINSON]

Bibliography: J. W. Meller, *Comprehensive Treatise on Inorganic and Theoretical Chemistry*, vol. 11, 1931; Mining Journal Books, Ltd., *Tungsten*, 1980; C. L. Robinson, *The Chemistry of Chromium, Molybdenum, and Tungsten*, 1975; C. J. Smithells (ed.), *Metals Reference Book*, 5th ed., 1976; W. D. Wilkinson, *Fabrication of Refractory Metals*, AEC Monograph, 1970; W. D. Wilkinson, *Properties of Refractory Metals*, AEC Monograph, 1969; W. H. Yih and C. T. Wang, *Tungsten: Sources, Metallurgy, Properties, and Applications*, 1978.

Turbidimetric analysis

An analytical method based on the measurement of the attenuation of transmitted light by a solution containing a finely divided precipitate. There is a direct relationship between the amount of light attenuated and the amount of material in suspension. Either visual comparison with known standards or photoelectric detection is used to measure the amount of light transmitted by the turbid solutions. In the photoelectric method, a standard curve of per cent transmittance versus concentration of the material in question is usually prepared and the compositions of unknowns are determined by reference to the curve. The illustration is a schematic diagram of a simple turbidimeter. Turbidity is a function of the concentration and particle-size distribution of turbid material and the path length of the light in the sample.

Turbidimetry is often not very precise since it is difficult to prepare a stable and reproducible suspension of the precipitate. The amount of light scattered and absorbed by the precipitate must also be neither too great nor too small. A precision and accuracy of ±5 – 10% of the amount present is usually obtained, but the method provides a very sensitive means of determining some elements and groups. This is illustrated by the determinations of barium or sulfate as barium sulfate, of phosphate as strychnine phosphomolybdate, and of silver or chloride as silver chloride. Turbidimetric methods are capable of detecting few parts per million of these ions.

In addition to direct measurement or comparison of absorbed light, another common turbidimetric method is the determination of the height of a column of turbid liquid which will obscure a light or an object. The height of liquid is then compared with a solution of known composition or with some standard scale. This type of measurement is used in the Jackson candle turbidimeter to measure the turbidity of water and in the visual Parr-sulfur turbidimeter.

[JAMES N. LITTLE]

Bibliography: I. M. Koltoff and P. J. Elving (ed.), *Treatise on Analytical Chemistry*, pt. 1, vol. 5, 1964; F. D. Snell and C. T. Snell, *Colorimetric Methods of Analysis*, vols. 1 – 4, 3d ed., 1967 – 1971; L. C. Thomas and C. J. Chamberlin, *Colorimetric Chemical Analytical Methods*, 9th ed., 1980.

Tyndall effect

Visible scattering of light along the path of a beam of light as it passes through a system containing discontinuities. The luminous path of the beam of light is called a Tyndall cone. An example is shown in the illustration. In colloidal systems the brilliance of the Tyndall cone is directly dependent on the magnitude of the difference in refractive index between the particle and the medium. In aqueous gold sols, where the difference in refractive index is high, strong Tyndall cones are observed.

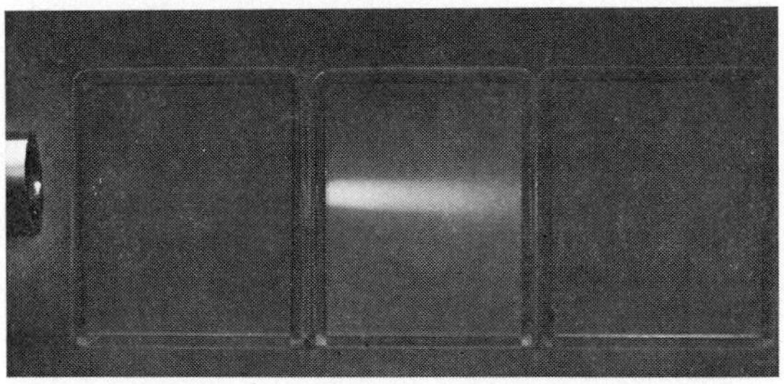

The luminous light path known as the Tyndall cone or Tyndall effect. (*H. Steeves and R. G. Babcock*)

For systems of particles with diameters less than one-twentieth the wavelength of light, the light scattered from a polychromatic beam is predominantly blue in color and is polarized to a degree which depends on the angle between the observer and the incident beam. The blue color of tobacco smoke is an example of Tyndall blue. As

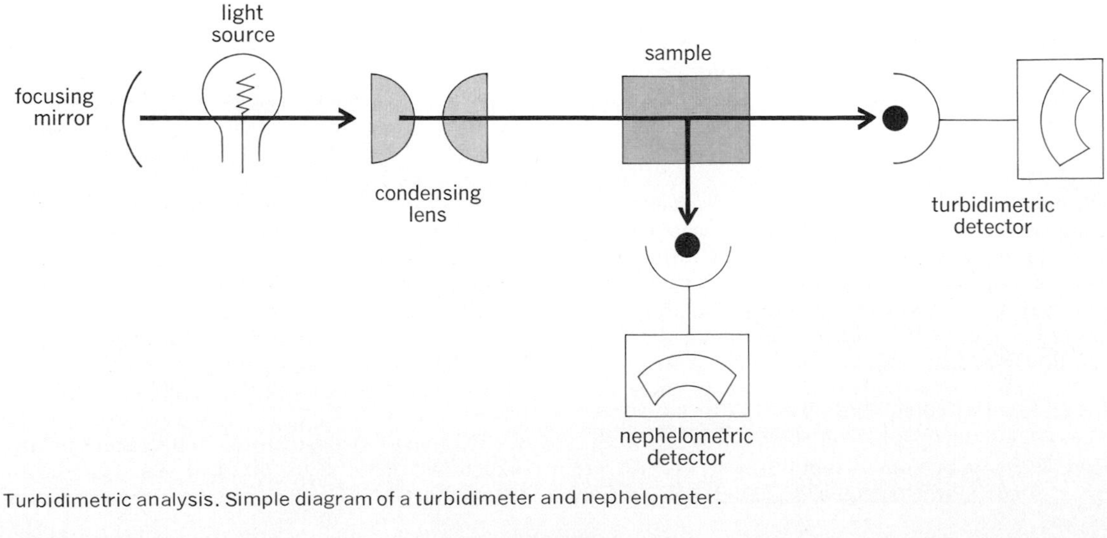

Turbidimetric analysis. Simple diagram of a turbidimeter and nephelometer.

particles are increased in size, the blue color of scattered light disappears and the scattered radiation appears white. If this scattered light is received through a nicol prism which is oriented to extinguish the vertically polarized scattered light, the blue color appears again in increased brilliance. This is called residual blue, and its intensity varies as the inverse eighth power of the wavelength. *See* COLLOID.

[QUENTIN VAN WINKLE]

Ultrafiltration

A filtration process in which particles of colloidal size are retained by a filter medium while solvent plus accompanying low-molecular-weight solutes are allowed to pass through. Ultrafilters are used (1) to separate colloid from suspending medium, (2) to separate particles of one size from particles of another size, and (3) to determine the distribution of particle sizes in colloidal systems by the use of filters of graded pore size.

Ultrafilter membranes have been prepared from various types of gel-forming substances. Unglazed porcelain has been impregnated with gels such as gelatin or silicic acid. Filter paper has been impregnated with glacial acetic acid collodions of varying strengths to produce a series of filters of graded porosity.

A new type ultrafilter membrane is made up of a thin plastic sheet containing millions of tiny pores evenly distributed over its surface. Flow rates of liquids and gases through these membranes are very high because the pore volume is 80% of the total membrane volume and the pores proceed through the filter in a direct path. Nominal pore diameters range from 10 nm to 5.0 μm. *See* COLLOID; FILTRATION.

[QUENTIN VAN WINKLE]

Uranium

A chemical element, symbol U, atomic number 92, atomic weight 238.03; one of the actinide series, in which the 5*f* shell is being filled. The name is derived from the planet Uranus. The valence electron configuration is $5f^36d7s^2$. Uranium was isolated in 1789 by Martin Heinrich Klaproth in a sample of pitchblende from Saxony. In 1841 Eugène-Melchior Péligot showed that the "semimetallic" element obtained by Klaproth was actually the dioxide. Péligot succeeded in preparing the metal by reduction of uranium tetrachloride with potassium. In 1896 Antoine-Henri Becquerel discovered that uranium undergoes radioactive decay. Discovery of the nuclear fission phenomenon by Otto Hahn and Fritz Strassmann in 1939 vaulted uranium from a position of relative obscurity to a role of major importance.

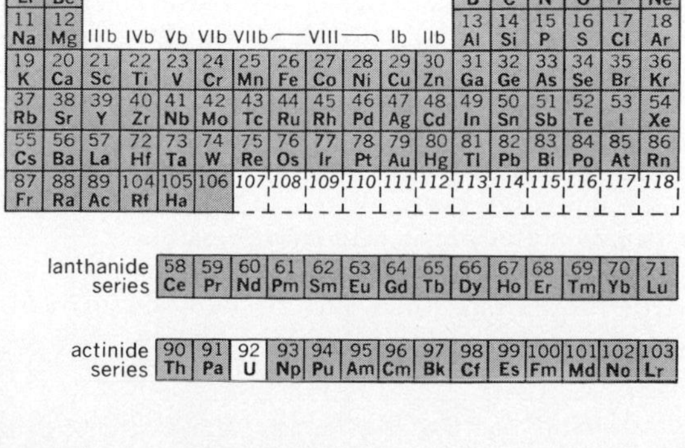

Uranium in nature is a mixture of three isotopes: ^{234}U (0.00054%), ^{235}U (0.72±0.030%), and ^{238}U (99.275%). These abundance values may vary somewhat, depending on the origin or on the degree of depletion of the sample. Half-lives of the three isotopes are $(2.446 \pm 0.007) \times 10^5$ years (^{234}U), $(7.038 \pm 0.005) \times 10^7$ years (^{235}U), and $(4.4683 \pm 0.0024) \times 10^9$ years (^{238}U). Uranium-235, which was discovered by A. J. Dempster in 1936, undergoes fission with slow neutrons to release large amounts of energy. Uranium-238 absorbs slow neutrons to form uranium-239, which in turn decays to fissile plutonium-239 by the emission of two β-particles. Other isotopes of uranium ranging in mass from uranium-226 to uranium-240 have been prepared by radioactive processes. Among these, fissile uranium-233 is obtained by the irradiation of natural thorium with neutrons. Thorium-232, the major component in natural thorium, absorbs slow neutrons to form thorium-233, which decays to fissile uranium-233 by the emission of two β-particles. *See* RADIOACTIVITY.

Natural occurrence. Uranium is believed to be concentrated largely in the Earth's crust, where the average concentration is 4 parts per million (ppm). For comparison, the crust contains 0.1 ppm silver and 0.5 ppm mercury. Basic rocks (basalts) contain less than 1 ppm uranium, whereas acidic rocks (granites) may have 8 ppm or more. Estimates for sedimentary rocks are 2 ppm, and for ocean water 0.0033 ppm. The total uranium content of the Earth's crust to a depth of 25 km is calculated to be 10^{17} kg; the oceans may contain 10^{13} kg of uranium.

Several hundred uranium-containing minerals have been identified, but only a few are of commercial interest. Table 1 summarizes data on some of the more important minerals. All uranium minerals contain lead, which results from radioactive decay of uranium, and thus contain an excess of the isotope ^{206}Pb. Uraninite, as found in pegmatites, usually occurs in rather small amounts which are of little economic significance. The euxenite-polycrase series, brannerite, and davidite are complex pegmatitic minerals. Pitchblende, a variety of uraninite found in hydrothermal veins, is the most important mineral of uranium. It is usually poorly crystalline, contains very little thorium or rare earths, and is frequently found associated with sulfide minerals. Coffinite, first identified in 1951, has become recognized as an important mineral on the Colorado Plateau. The remaining compounds are secondary minerals. Near-surface uranium mineralization invariably consists of oxidized ore.

Prior to 1942, uranium was obtained principally as a by-product of radium mining operations. With the discovery of nuclear fission and the potential of atomic power, the possession of uranium reserves became vitally important. Uranium reserves containing more than 1g U_3O_8/kg of ore, for that part of the world for which statistics are available, are estimated at about 2.2×10^9 kg U_3O_8, and those of the United States are about 10^9 kg U_3O_8. Deposits containing as little as 0.1% uranium are being mined. Some of the largest occurrences are the sandstone-impregnated Colorado Plateau deposits, the Blind River conglomerates (Ontario, Cana-

Table 1. Uranium minerals

Mineral	Chemical composition	Color	Specific gravity, 20°C	Typical occurrence
Uraninite	UO_2 (contains Th, rare earths)	Black	8–10.6	Arendal, Norway
Pitchblende (var.)	UO_{2+x}	Black	6–8	Shinkolobwe, Zaire
Euxenite-polycrase	$(Y,Ca,Ce,U,Th)(Nb,Ta,Ti)_2O_6$	Dark brown	4–6	Nipissing, Ontario
Samarskite	$(Y,Ca,Fe,U,Th)(Nb,Ta)_2O_6$	Black	5–6	Mitchell Co., N.C.
Brannerite	$(Y,Ca,Fe,U,Th)_3(Ti,Si)_5O_{16}$	Black	4–5	Blind River, Ontario
Davidite	$(Fe,Ce,U)(Ti,Fe,V,Cr)_3(O,OH)_7$	Black	4–5	Rum Jungle, Australia
Coffinite	$USiO_4$	Black	5–6	Colorado Plateau
Carnotite	$K_2(UO_2)_2(VO_4)_2 \cdot xH_2O$	Yellow	3–5	Colorado Plateau
Tyuyamunite	$Ca(UO_2)_2(VO_4)_2 \cdot xH_2O$	Yellow	3–4	Ferghana, Turkestan
Autunite	$Ca(UO_2)_2(PO_4)_2 \cdot xH_2O$	Greenish-yellow	3–4	Autun, France
Torbernite	$Cu(UO_2)_2(PO_4)_2 \cdot xH_2O$	Green	3–4	Erzgebirge, Saxony
Uranophane	$Ca(UO_2)_2Si_2O_7 \cdot xH_2O$	Greenish-yellow	3–4	Congo Republic

da), and the reefs of the Witwatersrand (South Africa), from which uranium is produced as a by-product of the gold industry. The vein deposits at Great Bear Lake (Northwest Territories, Canada) and Lake Athabasca (west-central Canada) are also important sources of uranium, but the Shinkolobwe, Zaire, deposits are virtually exhausted. An interesting deposit is the one at Oklo, Gabon, where in primordial times a spontaneous fission chain reaction occurred which caused a shift in the isotopic composition of the uranium in the deposit. In addition to the occurrence mentioned, extensive reserves of low-grade ore (0.005 to 0.02% uranium) exist in phosphate deposits (Florida, Brazil, Soviet Union, and North Africa), in bituminous shales (Soviet Union, Sweden, and Tennessee), and in lignites (the Dakotas).

The Earth's oceans also are a potential source of uranium. It is estimated that in the total volume of the oceans, 1.37×10^9 km^3, a uranium quantity of 4.5×10^9 tons (4.1×10^9 metric tons) is dissolved, probably as carbonate complex. Also, it is estimated that 27,000 tons (24,500 metric tons) of uranium per year are carried to the ocean. There are experiments in progress to isolate uranium from sea water, but these experiments are still in an early development state.

Separation of uranium isotopes. Because of the great importance of the fissile isotope ^{235}U, rather sophisticated industrial methods for its separation from the natural isotope mixture have been devised. Some of these methods date as far back as the Manhattan Project. In the Calutron process, uranium tetrachloride was charged to the ion source of an electromagnetic 180° mass separator (Calutron) and ionized. The beams of the ^{235}U and ^{238}U ions, which were separated in the magnetic field, were collected separately in individual pockets of a collector made of graphite. The pockets were mechanically separated and ignited to burn the graphite, and the separated isotopes were isolated from the ignition residue. Because of its complexity the Calutron process is no longer in use for uranium separation, but it has been utilized to separate stable isotopes and long-lived radioactive isotopes of almost any element in the periodic table. In the gaseous diffusion process, the major industrial process used at the present time, gaseous uranium hexafluoride kept at elevated temperatures is passed through porous barrier tubes contained in so-called converters; $^{235}UF_6$ passes through the barriers slightly faster than $^{238}UF_6$. Thousands of individual successive stages (a cascade) are required to separate pure ^{235}U from the natural mixture. The gaseous diffusion process, which in the United States is operated in three large plants—at Oak Ridge, TN, Paducah, KY, and Portsmouth, OH—has been the established industrial process. Other processes applied to the separation of uranium include the centrifuge proc-
separation of uranium include the centrifuge proc-
ess, in which gaseous uranium hexafluoride is mal diffusion process, the separation nozzle, and laser excitation. However, with the exception of the gas centrifuge method. none of these methods has progressed beyond the pilot plant stage.

Uranium metal. Uranium is a very dense, strongly electropositive, reactive metal; it is ductile and malleable, but a poor conductor of electricity. It is most conveniently prepared by the reduction of a halide (UF_4) with calcium or magnesium in a sealed bomb at 1200–1400°C. The steps involved in preparation of the metal from uranyl nitrate are summarized by reactions (1)–(4).

$$UO_2(NO_3)_2 \cdot 6H_2O \xrightarrow{500°C} UO_3 + 2NO_2 + \tfrac{1}{2}O_2 + 6H_2O \quad (1)$$

$$UO_3 + H_2 \xrightarrow{700°C} UO_2 + H_2O \quad (2)$$

$$UO_2 + 4HF \xrightarrow{550°C} UF_4 + 2H_2O \quad (3)$$

$$UF_4 + \begin{matrix} 2Ca \\ 2Mg \end{matrix} \longrightarrow U + \begin{matrix} 2CaF_2 \\ 2MgF_2 \end{matrix} \quad (4)$$

Selected physical and thermal properties of uranium are listed in Table 2. Uranium metal exists in three crystalline modifications. α-Uranium (25–668°C) is orthorhombic ($a = 0.2854$, $b = 0.5896$, $c = 0.4956$ nm), with four atoms per unit cell, and a density of 19.04 g/cm^3. Its structure is interpreted as a distorted hexagonal lattice containing corrugated sheets of uranium atoms. The β-phase (668–775°C) is a complex tetragonal structure ($a = 1.0754$ nm, $c = 0.5623$ nm), with 30 atoms per cell, and a density of 18.13 at 720°C. γ-Uranium (775–1132°C) is body-centered cubic ($a = 0.3525$ nm), with two atoms per cell, and a density of 18.06 g/cm^3 at 805°C. The β-form can be stabilized at room temperature by addition of small amounts of chromium, the γ-form with molybdenum.

The unique nature of the room-temperature γ-structure curtails solid solution of uranium with many metals. Extensive solid solution without

Table 2. Physical and thermal properties of uranium

Property	Value
Melting point	1132.4 ± 0.8°C
Boiling point	3818°C
Vapor pressure, 1720–2340 K	$\log p\ (\text{atm}) = -\frac{26210 \pm 270}{T} + (5.920 \pm 0.135)$
Heat of fusion	19.7 kJ/g-atom
Heat of vaporization	446.4 kJ/g-atom
Heat of sublimation (0 K)	487.9 kJ/g-atom
Heat of transition $\alpha \rightarrow \beta$	2.791 kJ/g-atom
Heat of transition $\beta \rightarrow \gamma$	4.757 kJ/g-atom
Enthalpy at 25°C	6.3655 kJ/g-atom
Heat capacity at 25°C	27.664 J/K·g-atom
Entropy at 25°C	50.170 ± 0.008 J/K·g-atom
Thermal conductivity (70°C)	0.29 J/cm·s (K)
Electrical conductivity	$2-4 \times 10^4\ (\Omega\cdot\text{cm})^{-4}$

compound formation has been found only with molybdenum and niobium. Aluminum, beryllium, bismuth, cadmium, cobalt, gallium, germanium, gold, indium, iron, lead, manganese, mercury, nickel, tin, titanium, zinc, and zirconium all form one or more intermetallic compounds with uranium. Chromium, magnesium, silver, tantalum, thorium, tungsten, and vanadium, as well as calcium, sodium, and some of the rare-earth metals, form neither compounds nor extensive solid solutions. Many uranium alloys are of great interest in nuclear technology because the pure metal is chemically active and anisotropic and has poor mechanical properties. However, cylindrical rods of pure uranium coated with silicon and canned in aluminum tubes (slugs) are used in production reactors. Uranium alloys can also be useful in diluting enriched uranium for reactors and in providing liquid fuels. Uranium depleted of the fissile isotope ^{235}U has been used in shielded containers for storage and transport of radioactive materials.

Uranium reacts with nearly all nonmetallic elements and their binary compounds. Table 3 lists a number of its reactions. Uranium dissolves in hydrochloric acid to leave a black residue of uranium hydroxy hydride. Addition of fluosilicate prevents formation of this residue. Nitric acid dissolves the metal, but nonoxidizing acids, such as sulfuric, phosphoric, or hydrofluoric acid, react very slowly. Usually a trace of mercuric nitrate tends to catalyze the dissolution. Uranium metal is inert to alkalies, but addition of peroxide causes formation of water-soluble peruranates.

Table 3. Chemical reactions of uranium metal

Reactant (reaction temperature, °C)†	Products	Heat of formation of underlined product, kJ/mole at 25°C
H_2(250)	$\alpha, \underline{\beta\ UH_3}$	−127
O_2(100–350)	$UO_2, \underline{U_3O_8}$	−3572
F_2(250)	$\underline{UF_6}$	−2186
Cl_2(500)	$\underline{UCl_4}, UCl_5, UCl_6$	−1051
Br_2(650)	$\underline{UBr_4}$	−826
I_2(350)	$UI_3, \underline{UI_4}$	−529
B(1650)	$\underline{UB_2}, UB_4, UB_{12}$	−148
C(1500)	$UC, U_2C_3, \underline{UC_2}$	−82
Si(1700)	$U_3Si, U_3Si_2, \underline{USi}, USi_2, USi_3$	−80
N_2(500)	$UN, \underline{U_2N_3}, UN_2$	−709
P(400–1100)*	$\underline{UP}, U_3P_4, UP_2$	−316
S(400)	$\underline{US}, US_2$	−306
H_2O(100)	$\underline{UO_2}$	−1084
H_2S(500)*	$US, \underline{US_2}$	−502
HF(350)*	$\underline{UF_4}$	−1882
HCl(300)*	$\underline{UCl_3}$	−893
NH_3(700)	$\underline{UN}, UN_2$	−303
CH_4(650)*	$\underline{UC}$	−97
CO(750)	UO_2, UC	
CO_2(600)	UO_2, UC	
NO(400)	U_3O_8	
N_2O_4(25)	$UO_2(NO_3) \cdot 2NO_2$	

*Powdered metal. †Values are given for massive metal.

Uranium ions in solution. Four oxidation states of uranium exist in solution, but only two of these, U^{4+} and U^{6+} (or rather UO_2^{2+}), are stable. Aqueous solutions of U^{3+} decompose with hydrogen evolution, even at temperatures as low as 0°C. Solutions of U^{3+} are obtained by electrolytic reduction of U^{4+} or UO_2^{2+} solutions, preferably in sulfuric acid. They have a green color in daylight or fluorescent light, but are deep red in incandescent light.

Green solutions of tetravalent uranium are readily obtained by electrolytic reduction or reduction with strong chemical reductants. They are readily oxidized in air on standing. Although the ion U^{4+} is extensively hydrolyzed, its existence in acid solutions has been proved. The first hydrolysis step is believed to give the monomeric ion $U(OH)^{3+}$. Numerous salts and complexes of uranium(IV) may be prepared from a U(IV) solution.

Pentavalent uranium exists in aqueous solution as the ion UO_2^{+}. It is unstable in solution and tends to disproportionate to U^{4+} and UO_2^{2+}; its lowest rate of decomposition is in the pH range 2–4.

Hexavalent uranium, in the form of the uranyl ion (UO_2^{2+}), is the most stable oxidation state. The lemon-yellow, fluorescent uranyl solutions can be reduced by moderately strong reductants, such as Sn^{2+}, Ti^{3+}, hydrogen in the presence of a catalyst, or sodium dithionite. Hydrolysis of UO_2^{2+} leads to formation of UO_2OH^+; further hydrolysis gives polymeric ions of the type $UO_2[(OH)_2UO_2]_n^{2+}$, involving sheetlike complexes with double hydroxyl bridges.

Extensive studies have been made of complex formation of UO_2^{2+} and, to a lesser extent, of U^{4+}, with many anions (for example, fluoride, chloride, bromide, thiocyanate, nitrate, bisulfate, sulfate, acetate, phosphate, oxalate, citrate, carbonate, and acetyl acetonate). In general, the strength of such a complex is inversely proportional to the strength of the acid from which the complexing anion is derived.

The oxidation potentials of uranium in 1 *M* acid solution are as in notation (5); those in 1 *M* basic solution are as in notation (6).

$$U \xrightarrow{+1.80\ V} U^{3+} \xrightarrow{+0.63\ V} U^{4+} \xrightarrow{-0.58\ V} UO_2^{+} \xrightarrow{-0.06\ V} UO_2^{2+} \qquad (5)$$

$$U^{4+} \xrightarrow{-0.32\ V} UO_2^{2+}$$

$$U \xrightarrow{2.17\ V} U(OH)_3 \xrightarrow{2.14\ V} U(OH)_4 \xrightarrow{0.62\ V} UO_2(OH)_2 \qquad (6)$$

Hydride. Uranium reacts reversibly with hydrogen to form UH_3 at 250°C. Correspondingly, the hydrogen isotopes form uranium deuteride, UD_3, and uranium tritide, UT_3. Because of the low reaction temperatures, and because UH_3, UD_3, and UT_3 are easily decomposed above 430°C to form pyrophoric, finely dispersed uranium powder and the corresponding hydrogen isotope, the reaction is useful both to prepare powdered uranium and to store D_2 or T_2 as the deuteride or tritide in an easily retrievable form.

Oxides. The uranium-oxygen system is characterized by an extremely complicated phase diagram. Table 4 gives a general survey of known uranium oxides. Uranium monoxide, UO, is a gaseous species which is not stable below 1800°C. In the range UO_2 to UO_3, a large number of phases exist. The complexity of the uranium-oxygen system is best understood if one considers that the addition of oxygen to UO_2 produces continually increasing distortions of the original fluorite lattice. The added oxygen (or oxygen deficiency) can be distributed at random to produce a single phase of constant space group and variable stoichiometry, such as $UO_{2\pm x}$, or it can be distributed in an ordered fashion, forming cubic, tetragonal, or monoclinic superlattices. The stoichiometry range of a phase at a particular temperature is a measure of its capacity to add randomly distributed oxygen without change in long-range ordering. When this limit is reached, at least part of the added oxygen atoms become ordered in a superlattice structure, producing a new phase possibly also of variable composition. Such behavior occurs up to $UO_{2.4}$. Above $UO_{2.4}$, an abrupt change to lower-density structures occurs. The phases in the range $UO_{2.4}$ to UO_3 appear to contain uranyl-type bonding for at least part of the uranium atoms present. In this bonding there are two short collinear primary uranium-oxygen bonds with four to six weaker bonds more or less in a plane perpendicular to these bonds.

This general behavior leads to rather strange stoichiometries, and because of the subtle changes from one phase to the next the earlier literature contains many conflicting results. Another peculiarity of the uranium-oxygen system is the fact that UO_3 exists in one amorphous and at least six crystalline modifications, the most stable being γ-UO_3.

The range of each oxide phase is determined by the temperature and by the oxygen partial pressure above the solid. Ignition of any uranium oxide to 750°C in air leads to the formation of U_3O_8.

The dioxide, UO_2, which is obtained by reduction of UO_3 or U_3O_8 with hydrogen or CO at 400–600°C in a fluidized bed, is an important ceramic fuel for nuclear reactors. For this application it may be mixed with plutonium dioxide, and it is compacted into pellets which are then loaded into the fuel tubes.

Uranium trioxide reacts with water to form either $UO_3 \cdot 2H_2O$, α-, β-, or γ-$UO_2(OH)_2$, α-$UO_3 \cdot 0.8H_2O$, or $U_3O_8(OH)_2$, depending on the conditions. All these compounds are yellow or orange. These uranyl hydroxides may also be considered as acids, H_2UO_4 or $H_2U_3O_{10}$. Uranium peroxide, $UO_4 \cdot xH_2O$, where $x = 1$ or 2, precipitates from uranyl salt solutions on addition of H_2O_2. It cannot be dehydrated without decomposition.

Ternary oxide systems of hexa-, penta-, and tetravalent uranium with alkali, alkaline earth, rare earths, and group IV elements have been investigated. Extensive regions of solid solutions are generally encountered. Hexa-, penta-, and tetravalent uranium form ternary oxides (uranates) exhibiting a wide range of composition with other metals. These compounds are insoluble in water but are readily soluble in acids. A survey of typical

Table 4. Survey of defined uranium-oxygen phases

Compound	Symmetry	Color	Density (gcm^{-3})
UO	(Gaseous)	–	–
UO_2	Cubic	Cinnamon brown	10.950
U_4O_{9-y}*	Cubic	Black	(11.299)
$U_{16}O_{37}$	Tetragonal	Black	(11.366)
U_8O_{19}	Monoclinic	Black	11.34
α-U_2O_5	Monoclinic	Black-purple	10.5
β-U_2O_5	Hexagonal	Black-purple	10.76
γ-U_2O_5	Monoclinic	Black-purple	10.36
$U_{13}O_{34}$	Orthorhombic		(8.40)
U_8O_{21}	Orthorhombic		(8.341)
$U_{11}O_{29}$	Orthorhombic		(8.40)
α-U_3O_8	Orthorhombic	Greenish-black	(8.395)
β-U_3O_8	Orthorhombic	Greenish-black	(8.326)
$U_{12}O_{35}$	Orthorhombic		7.72
UO_3 (A)	Amorphous	Orange	6.80
α-UO_3	Hexagonal	Beige	7.3
β-UO_3	Monoclinic	Orange red	8.25
γ-UO_3	Orthorhombic	Yellow	7.80
δ-UO_3	Cubic	Deep red	6.69
ϵ-UO_3	Triclinic	Red	(8.67)
ζ-UO_3	Orthorhombic	Brown	(8.86)

*U_4O_{9-y} is a nonstoichiometric compound, defined by its x-ray structure.

Table 5. Ternary oxides of uranium(V) and uranium(VI) with alkali oxides

Oxidation state	Compound type	Cation
+6	$M_8U_{16}O_{52}$	M = K
+6	$M_2U_7O_{22}$	M = K, Rb
+6	$M_2U_6O_{19}$	M = Li
+6	$M_2U_4O_{13}$	M = Rb
+6	$M_2U_3O_{10}$	M = Li, K, Tl
+6	$M_2U_2O_7$	M = Li, Na, K, Rb, Tl
+6	M_2UO_4	M = Li, Na, K, Rb, Cs
+6	M_4UO_5	M = Li, Na, K, Rb
+6	M_6UO_6	M = Li
+5	MUO_3	M = Li, Na, K, Rb, Tl
+5	M_3UO_4	M = Li, Na
+5	M_7UO_6	M = Li

compounds of this kind is given in Table 5.

For these compounds, solid-state reactions of the type shown in reactions (7) and (8), involving

$$6Na_2CO_3 + 2\,U_3O_8 + O_2 \rightarrow 6Na_2UO_4 + 6CO_2 \quad (7)$$

$$CaO + UO_3 \rightarrow CaUO_4 \quad (8)$$

stoichiometric quantities of the reactants, yield single-phase products, while precipitation from aqueous solutions generally leads to nonstoichiometric mixtures. Addition of hydrogen peroxide and alkali to uranyl solutions leads to the formation of soluble peroxyuranates, for example $Na_4UO_8 \cdot xH_2O$.

Halides. The uranium halides constitute an important group of compounds. Uranium tetrafluoride is an intermediate in the preparation of the metal and the hexafluoride. Uranium hexafluoride, which is the most volatile uranium compound, is used in the isotope separation of ^{235}U and ^{238}U. The halide volatilities increase in the order $UX_3 < UX_4 < UX_5 < UX_6$. Reactions for the preparation of uranium halides are summarized in Table 6. Uranium hexafluoride boils at 56.54°C and melts at 64°C (1140 mmHg or 152 kilopascals pressure). It is a reactive substance and a strong fluorinating agent. Equipment for containing the compound may be constructed of copper, nickel, aluminum, Monel, or fluorine-containing polymers (Teflon, Kel-F). The hexafluoride reacts with water to form UO_2F_2. Uranium tetrachloride, UCl_4, has been used as charge material in the electromagnetic isotope separation process. Like all the other chlorides, bromides, and iodides, it is hygroscopic and soluble in water. The penta- and hexachlorides are also soluble in some nonpolar solvents (CCl_4, CS_2). Besides the binary halides, a number of ternary and quaternary halides of U_{3+} and U^{4+} are known, such as $UBrCl_2$, UCl_2Br_2, and many others. The halides react with oxygen at elevated temperatures to form uranyl compounds and ultimately U_3O_8.

Table 6. Preparation of uranium halides

Compound	Reaction
UF_3	$UF_4 + Al \rightarrow UF_3 + AlF \uparrow$ (900°C)
UF_4	$UO_2 + HF \rightarrow UF_4$ (550°C)
U_4F_{17}, U_2F_9, UF_5	$xUF_4 + yUF_6 \rightarrow U_4F_{17}, U_2F_9, UF_5$ (250°C)
UF_6	$UF_4 + F_2 \rightarrow UF_6$ (350°C)
UCl_3, UBr_3, UI_3	$UH_3 + HX \rightarrow UX_3$ (350°C)
UCl_4	$UO_3 + Cl_2C{=}CClCCl_3 \rightarrow UCl_4$ (210°C)
UCl_5	$UCl_4 + Cl_2 \rightarrow UCl_5$ (500°C)
UCl_6	$UCl_5 \rightarrow UCl_6 + UCl_4$ (125°C in vac.)
UBr_4, UI_4	$U + X_2 \rightarrow UX_4$ (550°C)

Halogeno complexes. Many halogeno complexes of uranium are know. Of particular interest are those based on UF_4 and UF_3, because they may be used as nuclear fuels in molten salt reactors. In particular, $LiF-BeF_2-UF_4-$ and $LiF-ZrF_4-UF_4-$ eutectics are used as fuel materials. Table 7 summarizes the various types of fluoro complexes observed. Many phase diagrams of these systems are known. In the chloro system, complexes of the type UCl_6^- with U^{5+}, and UCl_6^{2-} with U^{4+} are known; in the bromo and iodo system, only a few compounds with the UBr_6^{2-} and UI_6^{2-} ions have been prepared.

Table 7. Ternary fluorides of uranium(VI), (V), (IV), and (III)

Oxidation state	Compound type	Cation
+6	M_3UF_9	M = Na
+6	M_2UF_8	M = Na, K
+6	MUF_7	M = Na, K, NH_4^+, $N_2H_5^+$, NO^+, NO_2^+
+5	M_3UF_8	M = Na, K, Rb, Cs, Tl, Ag
+5	M_2UF_7	M = K, Rb, Cs, NH_4^+
+5	MUF_6	M = Li, Na, K, Rb, Cs, Ag, NH_4^+, H^+
+4	MU_6F_{25}	M = K, Rb
+4	MU_4F_{17}	M = Li
+4	MU_3F_{13}	M = K, Rb
+4	MU_2F_9	M = Na, K
+4	MUF_5	M = Li, Rb, Cs
+4	$M_7U_6F_{31}$	M = Na, K, Rb, Cs, NH_4^+
+4	M_2UF_6	M = Na, K, Rb, Cs, NH_4^+
+4	M_3UF_7	M = Li, Na, K, Rb, Cs
+3	MUF_4	M = Na
+3	M_3UF_6	M = K

Uranyl salts. These salts, which have the cation UO_2^{2+}, are the most common uranium salts. They generally have a yellow to greenish-yellow color and show bright green fluorescence in ultraviolet light. Uranyl nitrate is obtained as the hexahydrate $UO_2(NO_3)_2 \cdot 6H_2O$ (so-called UNH) from dilute nitric acid, as the trihydrate from concentrated nitric acid, and as the dihydrate from fuming nitric acid. The lower hydrates may also be obtained by careful dehydration of the hexahydrate. Anhydrous uranyl nitrate may be prepared by reacting N_2O_5 with UO_3 under strictly anhydrous conditions. Uranyl nitrate is probably the most frequently encountered compound in uranium chemistry.

Uranyl sulfate, also very soluble in water, crystallizes from aqueous solution as the trihydrate, $UO_2SO_4 \cdot 3H_2O$. A monohydrate is formed by careful dehydration or by equilibration in water at 180°C. The anhydrous sulfate is formed by heating the hydrate to 300°C. With alkali sulfates, double sulfates such as $K_2UO_2(SO_4)_2 \cdot 2H_2O$ are formed. Phase diagrams of uranyl nitrate, uranyl sulfate, uranyl fluoride, uranyl carbonate, and uranyl phosphate have been constructed. Acid uranyl phosphate, $HUO_2PO_4 \cdot 4H_2O$, and the corresponding arsenate, $HUO_2AsO_4 \cdot 4H_2O$, form numerous alkali and alkaline-earth salts, many of which occur in nature as minerals. Some of the important uranyl salts of organic acids are the formate $UO_2(HCOO)_2 \cdot H_2O$, the acetate $UO_2(CH_3COO)_2 \cdot 2H_2O$, the sodium double acetate $NaUO_2(CH_3COO)_3$, and the oxalate $UO_2C_2O_4 \cdot 3H_2O$.

Uranium(IV) salts. By electrolytic reduction or the use of suitably strong reducing agents, the yellow UO_2^{2+} solutions may be reduced to green U^{4+} solutions. From these latter solutions, green uranium(IV) salts may be prepared. Of importance is the dark green sulfate $U(SO_4)_2 \cdot 8H_2O$. From the octahydrate, lower hydrates with 6, 5, 4, and 2 H_2O, as well as the anhydrous salt, may be prepared by thermal dehydration in inert atmospheres. Uranium(IV) nitrate is not stable, and it may be obtained only as an *NN*-dimethyl-acetamide complex. $U(NO_3)_4 \cdot 2.5CH_3CON(CH_3)_2$. However, a few unstable salts with the anion $U(NO_3)_6^{2+}$ have been prepared. The oxalate, $U(C_2O_4)_2 \cdot 6H_2O$, is relatively stable.

Uranium(III) salts. From uranium(III) solutions in sulfuric acid obtained by strong electrolytic reduction, uranium(III) sulfate, $U_2(SO_4)_3 \cdot 8H_2O$, precipitates on addition of ethanol as olive-green crystals which are brick red in incandescent light. On heating, the octahydrate may be dehydrated to the orange trihydrate. With alkali sulfates, complexes of the type $MU(SO_4)_2$ (where $M = NH_4$, K, Rb) or $M_5U(SO_4)_4$ (with $M = K$, Tl) may be precipitated. All U(III) salts are extremely sensitive to air.

Behavior in nonaqueous solvents. The solubility of uranium salts, especially uranyl nitrate, in certain organic solvents can be used to separate uranium from other metal ions by solvent extraction. Numerous types of liquid (aqueous)–liquid (organic) separation techniques are employed. Distribution of uranium between the two phases may be controlled by "salting" the aqueous phase with mineral acids or their salts, changing pH in the aqueous phase, or using different types of organic extractants. The three types of organic phase are a neutral extractant, such as diethyl ether, methyl isobutyl ketone (hexone), or tributyl phosphate (TBP); an acidic extractant, such as octanoic acid or thenoyl trifluoro acetone (TTA); and a basic extractant, such as tri-octylamine or trilaurylamine (TLA). In some instances, the organic phase may be diluted with a "carrier" diluent such as kerosine. The most frequently used extraction procedure is the purification of uranium with TBP dissolved in kerosine or hexane. The extraction is based on the formation of a neutral nonionized complex according to reaction (9).

$$UO_2^{2+}(aq) + 2NO_3^-(aq) + 2TBP\,(org) \rightleftharpoons UO_2(NO_3)_2 \cdot 2TBP\,(org) \qquad (9)$$

Quantitative determination. Numerous procedures are available for the quantitative determination of uranium. Macro quantities of uranium may be analyzed by gravimetric or volumetric methods. Gravimetric procedures usually utilize U_3O_8 ignited in air or 8-hydroxy-quinolinate. Volumetric methods are based on reduction of uranium to U(IV) with lead or zinc, followed by titration with an oxidizing agent such as potassium dichromate, ceric sulfate, potassium bromate, or potassium permanganate. Small amounts of uranium may be determined by coulometric, polarographic, colorimetric, fluorescence, or spectroscopic methods. The isotopic composition may be determined by mass spectroscopy or, in the case of ^{235}U, by fission counting. *See* ACTINIDE ELEMENTS; NEPTUNIUM; NUCLEAR FISSION.

[FRITZ WEIGEL]

Bibliography: I. I. Chernyaev, *Complex Compounds of Uranium*, 1966; E. H. P. Cordfunke, *The Chemistry of Uranium*, 1969; *Gmelin Handbuch der Anorganischen Chemie*, System no. 55: *Uran*, including supplementary volumes, 1979; J. J. Katz and E. Rabinowitch, *The Chemistry of Uranium*, NNES VIII 5 (*Survey*), 1951, and TID-5290, vols. 1 and 2 (*Collected Papers*), 1958; National Lead Company of Ohio, *Analytical Chemistry Manual of the Feed Materials Production Center*, TID-7022, vol. 1, books 1–8, and vol. 2, 1964.

Urea

The diamide of carbonic acid. Urea is a white, crystalline, water-soluble compound, with melting point 132.7°C and the formula given below. Important biologically as an end product of normal animal and human protein metabolism, it is also of synthetic value in urea-formaldehyde plastics and is used as a stabilizer for many explosives.

$$H_2N{-}\overset{\overset{\displaystyle O}{\|}}{C}{-}NH_2$$

Urea is made commercially from the partial hydrolysis of cyanamide, reaction (1), and by heating

$$H_2N{-}CN + H_2O \rightarrow H_2N{-}CO{-}NH_2 \qquad (1)$$

carbon dioxide and ammonia under pressure, whereby the equilibrium in reaction (2) is attained.

$$CO_2 + NH_3 \rightleftharpoons HO{-}\overset{\overset{\displaystyle O}{\|}}{C}{-}NH_2 + NH_3 \rightleftharpoons$$

$$\overset{\oplus}{NH_4}\overset{\ominus}{O}{-}\overset{\overset{\displaystyle O}{\|}}{C}{-}NH_2 \rightleftharpoons H_2N{-}CO{-}NH_2 + H_2O \qquad (2)$$

It can also be prepared from phosgene and excess ammonia, reaction (3), and from the rearrangement of ammonium cyanate (F. Wohler, 1828).

$$Cl{-}\overset{\overset{\displaystyle O}{\|}}{C}{-}Cl + 4NH_3 \rightarrow 2NH_4Cl + H_2N{-}CO{-}NH_2 \qquad (3)$$

Urea is used medically as a diuretic, and was used formerly in the treatment of indolent ulcers. Because of its high nitrogen content, it is used as a commercial fertilizer. Straight-chain hydrocarbons, but not branched ones, of six or more carbons combine with urea to form molecular complexes in which the molar ratio of hydrocarbon to urea is proportional to chain length. The complexes are easily decomposed by water; thus, straight-chain hydrocarbons may be separated from branched-chain isomers.

In the commercial production of urea-formaldehyde resins, two molecules of urea are condensed with three of formaldehyde using catalysts such as pyridine, ammonia, or hexamethylenetetramine. The condensation product, a syrup, is mixed with cellulose and coloring matter, and molded to the desired shape under pressure and elevated temperature. In textile treating, or in laminating operations, solutions of the resins are used.

Both methylolurea, $H_2N—CO—NH—CH_2OH$, and dimethylolurea, $HOCH_2—NH—CO—NH—CH_2OH$, are known to be products of base-catalyzed condensations of urea and formaldehyde; it is likely that during the pressure-heating treatment the hard thermosetting resin is formed from progressive three-dimensional dehydrations of these intermediates. *See* AMMONIA; UREA-FORMALDEHYDE-TYPE RESINS.

[EVANS B. REID]

Urea-formaldehyde-type resins

The condensation products obtained by the reaction of urea or melamine with formaldehyde. Resinous condensation products of formaldehyde with other nitrogen-containing compounds, for example, aniline and amides, also belong to this group of resins but have gained only limited utility. Resins derived from the condensation of formaldehyde with urea, melamine, aniline, and *p*-toluene sulfonamide are discussed below.

Urea- and melamine-formaldehyde resins. Because the amino resins possess an excellent combination of physical properties and can be easily fabricated in a variety of colors, they are widely used as adhesives, laminating resins, molding compounds, paper and textile finishes, and surface coatings. Methods of utilizing the resins are generally similar to methods employed for several other condensation resins, such as the phenolic or epoxy resins. First, intermediate condensation resins are prepared; then, after compounding, the combinations are cured to yield a finished product.

The intermediate condensation products are prepared by the reaction of urea or melamine with formaldehyde under neutral or mildly alkaline conditions to form mono-, di-, or polymethylol derivatives, reactions (1)–(3).

If a soluble resin is desired for surface coatings, the dimethylol derivatives are treated with an alcohol in an acidic medium to form ethers, reaction (4).

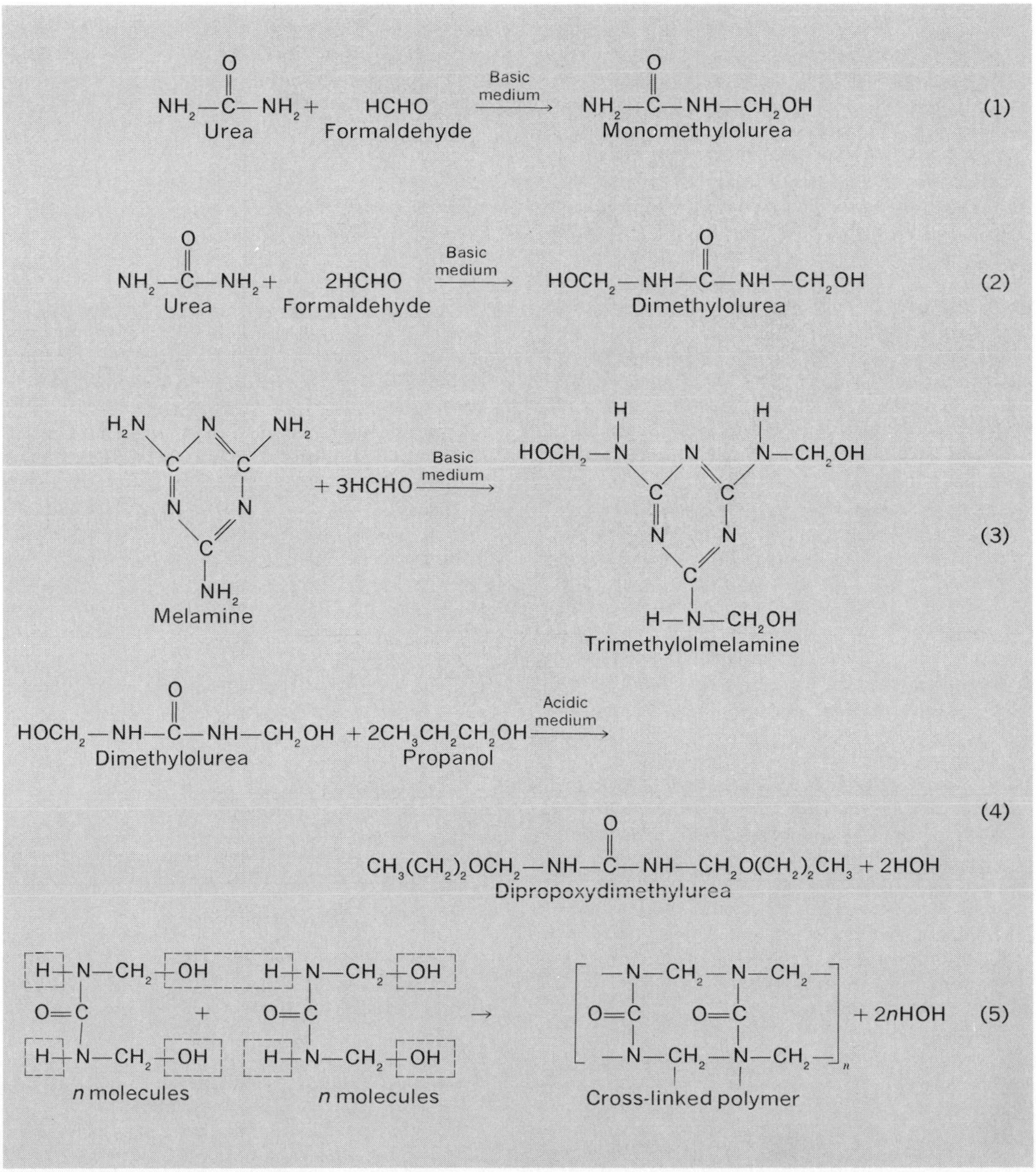

For other purposes, the low-molecular-weight methylol derivatives can then be compounded with fillers, catalysts, plasticizers, and pigments, and cured, usually by the application of heat and pressure. During the curing operation, the hydroxyl portion of a methylol group condenses with a hydrogen atom to yield water. Such a condensation can be represented by reaction (5) for the cross-linking of methylol urea.

Other reactions are also involved in curing. For example, the methylol groups can react with hydroxyl groups of alcohols (added as plasticizers), clays, or cellulose (added as fillers). Methylol groups can also react with each other to form ether linkages.

For immediate curing, a free acid can be used as a cross-linking catalyst. However, for preparation of a stable molding powder that can be stored, a salt or ester that will liberate its corresponding acid at the high temperature used for molding usually serves as the catalyst.

As may be seen from reactions (1)–(5), the completion of the curing reaction depends on the elimination of as much water as possible. Although most of the water is lost as vapor, a hygroscopic filler, such as cellulose, asbestos, or certain clays, is added to absorb the last traces of moisture. In addition, the filler can also react with some of the methylol groups of the resin intermediates.

Although the fabrication and applications of both the urea- and melamine-type resins are generally similar, there are several important differences in properties. The melamine-type resins are more resistant to water, marring, and heat than the urea-type resins, but are, on the other hand, somewhat more expensive.

Curing in a mold results in translucent or opaque products, depending on the nature of the fillers and pigments used. The ease with which color can be introduced into amino resins has made it possible to make very attractive articles, such as dinnerware, buttons, appliance cases, handles, and knobs. In applications where resistance to heat and scratching is important, for example, in dinnerware, the melamine resins are preferred.

Another important application of both resins is in adhesives, especially for lamination of furniture and plywood. Depending on the resin and catalyst used, adhesives that can be cured under a variety of conditions may be formulated. Melamine resins are especially valuable as adhesives for laminates of paper or fabrics, including glass cloth.

Foams have also been developed for use in packaging and as thermal and acoustical insulation.

Soluble urea-formaldehyde and melamine-formaldehyde resins may be used to impregnate cloth or to treat paper. Curing of the resins in the cloth results in improved qualities, such as resistance to creasing or, in the case of melamine resins, in resistance to shrinkage in wools. Curing of the resins in paper results in an improvement in the strength of the paper when wet.

Although unmodified resins are common in the applications cited, modified resins or blends with other resins are often useful in some cases. In combination with alkyd resins, the ether derivatives, especially the ethers of the melamine resins, are components in the formulation of baking enamels that are hard and resistant to water and detergents. Resins modified by the use of an alkyl-substituted urea or melamine are more flexible than the unmodified resin and are thus useful in coating compositions. Blends of urea- or melamine-type resins with resins derived from resorcinol and formaldehyde are sometimes employed in adhesive compositions and as binders for the sawdust or wood chips used in the manufacture of particulate boards.

Another application of blends is the use of urea-type resin blends as impregnants for fabric used in the manufacture of minimum-care or wash-and-wear cotton goods.

Aniline- and sulfonamide-formaldehyde resins. In addition to urea and melamine, other compounds containing $—NH_2$ groups can condense with formaldehyde to form methylol derivatives which are capable of further reaction. Although the corresponding resins have received limited attention, two examples are discussed here, aniline and *p*-toluenesulfonamide.

In a neutral medium, the reaction of aniline with formaldehyde yields a methylol derivative that exists as a cyclic trimer. When the trimer is heated, further condensation results in a resin with the following probable monomer unit:

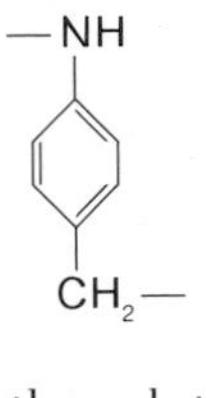

Some cross-linking through the phenyl rings may also occur.

Because of their resistance to the absorption of water, the resins have been used in electrical applications, such as insulation or panels, where their natural brown color is not objectionable.

Similarly, resins can be prepared by curing methylol derivatives resulting from the condensation of *p*-toluenesulfonamide with formaldehyde. The probable monomer unit is as follows:

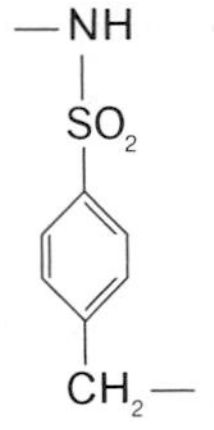

The resulting polymers are less colored than the aniline-type resins and have been employed in surface coatings. *See* CONDENSATION REACTION; FORMALDEHYDE; POLYMERIZATION; UREA.

[JOHN A. MANSON]

Bibliography: *Modern Plastics Encyclopedia*, 1979–1980; W. J. Roff and J. R. Scott, *Fibres, Films, Plastics and Rubbers*, 1971.

Ureid

An acyl derivative of urea, such as acetylurea, $CH_3CO—NHCONH_2$, and diacetylurea, $CH_3CO—NHCONH—COCH_3$. The most important members are cyclic, containing five- or six-atom hetero rings, and are known as parabanic acid and barbituric acid, respectively.

Parabanic acid is formed from urea and oxalyl chloride, reaction (1); barbituric acid or its derivatives (barbiturates) is formed from urea and malonic ester or suitably substituted malonic esters, reaction (2).

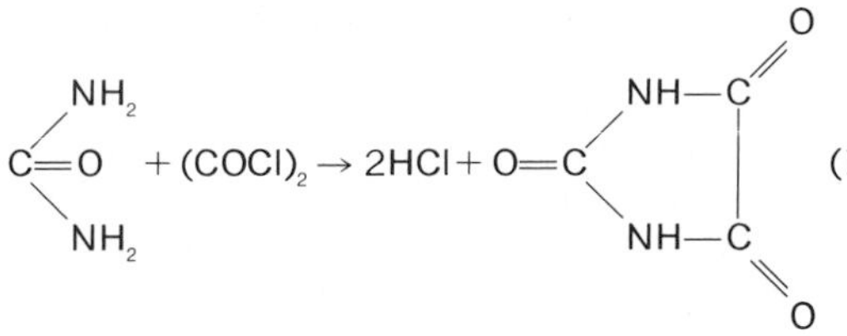

(1)

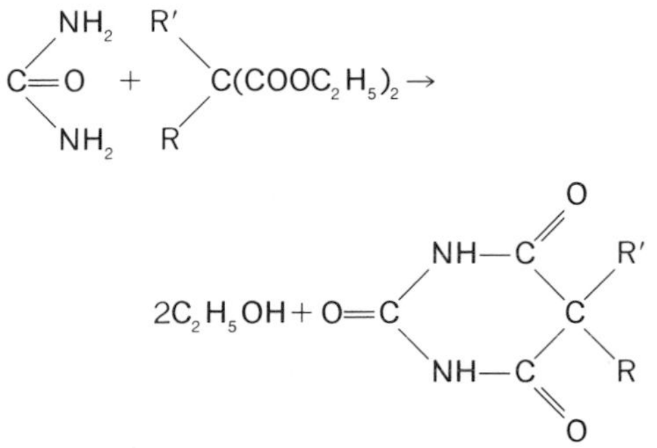

(2)

Barbituric acid (R, R′ = H)
Veronal or barbital (R, R′ = C_2H_5)
Phenobarbital (R = C_2H_5; R′ = C_6H_5)
Amytal (R = C_2H_5; R′ = isoamyl)

The barbiturates are acidic compounds, soluble in base (in which form they are often used). Introduced into medicine about 1900 (Veronal, 1903), the barbiturates are widely used in treating insomnia, epilepsy, hysteria, and as preliminary anesthetics. They should not be used indiscriminately because large doses are fatal, and continued use of small amounts often leads to habituation.

Many naturally occurring substances, for example, the purines (from nucleoproteins of plant and animal cells), and the xanthine alkaloids (from coffee, tea, and cocoa), contain diureid structures, in which two N—C—N units are combined with one C—C—C unit. Some examples follow:

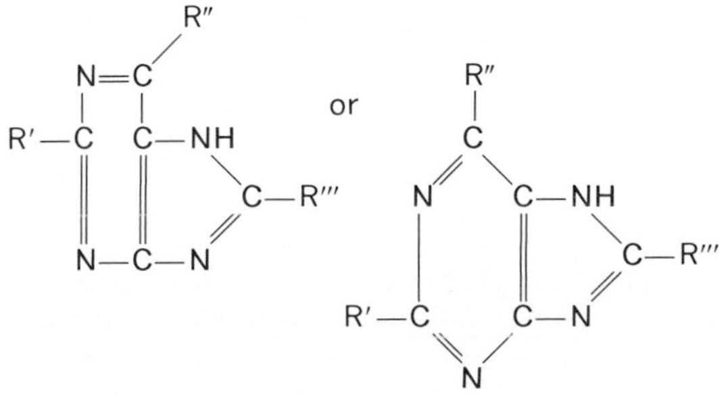

Purine (R′, R″, R‴ = H)
Adenine, 6-aminopurine
(R′, R″ = H; R‴ = NH_2)
Guanine, 2-amino-6-hydroxypurine
(R′ = NH_2; R″ = OH; R‴ = H)
Uric acid, 2,6,8-trihydroxypurine
(R′, R″, R‴ = OH)
Xanthine, 2,6-dihydroxypurine
(R′, R″ = OH; R‴ = H)

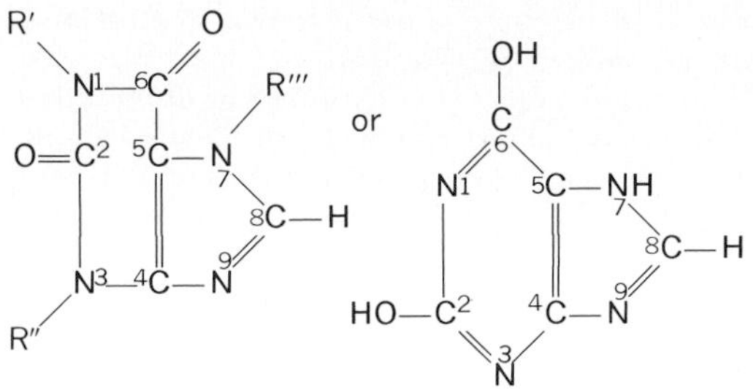

Xanthine (R′, R″, R‴ = H)
Theophylline, 1,3-dimethylxanthine
(R′, R″ = CH_3; R‴ = H)
Theobromine, 3,7-dimethylxanthine
(R″, R‴ = CH_3; R′ = H)
Caffeine, 1,3,7-trimethylxanthine
(R′, R″, R‴ = CH_3)

See UREA.

[EVANS B. REID]

Urethane

Urethane, or ethylurethane or ethyl carbamate, $H_2NCOOC_2H_5$, is the ethyl ester of carbamic acid. It forms white crystals with melting point 48–50°C, boiling point 182–184°C, and it is freely soluble in water, alcohol, and chloroform. Less toxic than other urethanes, ethylurethane is a mild hypnotic and weak diuretic; in water, it forms a neutral solution with a cooling, saline taste. With quinine hydrochloride, it is used as a sclerosing agent for treatment of varicose veins.

Urethane is prepared commercially by two processes. Urea and ethyl alcohol are heated under pressure, reaction (1), and urea nitrate is heated with ethyl alcohol and sodium nitrite, reaction (2).

$$H_2NCONH_2 + C_2H_5OH \rightarrow NH_3 + H_2NCOOC_2H_5 \quad (1)$$

$$H_2NCONH_2 \cdot HNO_3 + C_2H_5OH + NaNO_2 \rightarrow H_2NCOOC_2H_5 + NH_4NO_2 + NaNO_3 \quad (2)$$

See AMIDE.

[EVANS B. REID]

Uric acid

The excretory end product in amino acid metabolism by uricotelic species, including birds, terrestrial invertebrates, and snakes; in purine metabolism by most insects, snakes, lizards, birds, primates (including man), and the Dalmatian dog. Purine bases contain a six-membered pyridine and a five-membered imidazole ring which are derived from formate, CO_2, and portions of the amino acids aspartic acid, glycine, and glutamine. In tissues, purines are bound with pentoses and phosphate to form nucleotides. Nonuricotelic animals oxidize uric acid by the enzyme uricase to a more soluble compound, allantoin. The disease gout is associated with elevated uric acid.

[MORTON K. SCHWARTZ]

Vacuum fusion

A technique of analytical chemistry for determining the oxygen, hydrogen, and sometimes nitrogen content of metals. The method can be applied to a wide variety of metals, the alkali and alkaline earth metals being exceptions. The range of the method extends from 1% down to a few parts per million for oxygen and nitrogen and down to fractional parts per million for hydrogen.

The metal sample is either fused or dissolved in a bath, or flux, of a second metal in a heated graphite crucible supported inside an evacuated glass or quartz vessel (see illustration). Oxygen is released from the metal as carbon monoxide by reaction of oxides or dissolved oxygen with carbon from the graphite crucible at high temperature. Metal nitrides dissociate, although not always quantitatively, to form elemental nitrogen. Hydrogen is evolved as elemental hydrogen.

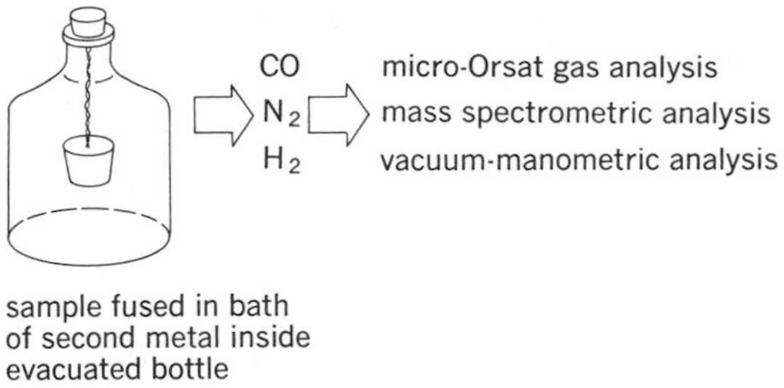

Vacuum-fusion methods of analysis.

The mixture of carbon monoxide (CO), nitrogen (N_2), and hydrogen (H_2) is analyzed to determine individual component concentrations by one of several techniques, including micro-Orsat, mass-spectrometric, and vacuum-manometric procedures. The last is in most general use. Gas quantities are determined by measurement of the pressure of the gas after confinement into a small, calibrated volume. The product of the pressure and volume is proportional to the total number of moles of gas present, regardless of the species of gas molecules. Measurement is first made of the total quantity of CO, N_2, and H_2 collected. These gases are then oxidized to carbon dioxide (CO_2), water (H_2O), and N_2 by passing over hot copper oxide. The CO_2 and H_2O can be removed from the gas mixture successively by selective freezing or by use of chemical absorbents. The decrease in total moles of gas accompanying removal of each species permits determination of the number of moles of CO_2, N_2, and H_2O.

Furnace temperatures and bath conditions must be selected for each metal or alloy to ensure quantitative recovery of oxygen. The required furnace temperatures range from 1650°C for iron, nickel, and low-alloy steels to as high as 1900°C for metals such as titanium, zirconium, tantalum, and thorium. Iron, nickel, and low-alloy steels are usually fused without addition of other metals. High-melting metals require use of a second metal to lower their melting points. The second metal is used as a previously prepared bath or, in some instances, is added simultaneously with the sample metal. Iron, nickel, platinum, and tin are the most generally used fluxing metals.

Nitrogen is not always quantitatively evolved, and values obtained by vacuum-fusion methods for nitrogen are not generally reliable. Hydrogen is always evolved quantitatively upon fusion of the metal in a vacuum. When hydrogen alone is of interest, simpler and more rapid techniques can be used involving vacuum extraction from metal heated to temperatures well below the melting point.

Vacuum-fusion analysis of gases and metals has found its widest industrial application, from a process-control viewpoint, in titanium, tantalum, niobium, zirconium, and molybdenum. Determination of hydrogen in steel is a routine procedure in some applications. As a research tool, the vacuum-fusion techniques have been applied to practically all of these materials except the alkali and alkaline-earth metals.

A new method, inert gas fusion, has been developed for determination of gases in metals. The techniques used are similar to vacuum fusion but substitute inert gases for the vacuum environment. This method is sometimes referred to as argon-fusion. The methods of detection of the gases involved during the fusion are modified somewhat. *See* GAS AND ATMOSPHERE ANALYSIS.

[FRANK C. BENNER]

Bibliography: *Proceedings of the 6th International Vacuum Metallurgy Conference on Special Melting*; O. Winkler and R. Bakish, *Vacuum Metallurgy*, 1971.

Valence

A term commonly used by chemists to characterize the combining power of an element for other elements, as measured by the number of bonds to other atoms which one atom of the given element forms upon chemical combination. The term also has come to signify the theory of all the physical and chemical properties of molecules that specially depend on molecular electronic structure.

Thus, in water, H_2O or

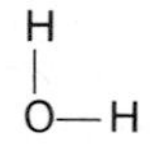

the valence of each hydrogen atom is 1; the valence of oxygen, 2. In methane, CH_4or

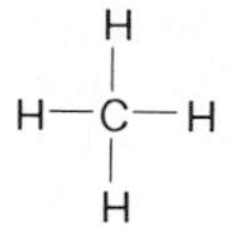

the valence of hydrogen again is 1; of carbon, 4. In NaCl and CCl_4 the valence of chlorine is 1, and in CH_2 the valence of carbon is 2.

Much more is known about a water molecule than that it contains two hydrogen atoms and one oxygen atom. Each OH distance is 0.957×10^{-8} cm and the HOH bond angle is 104°27′. The oxygen and hydrogen ends of the molecule are negatively and positively charged, giving it a dipole moment 1.84×10^{-18} electrostatic units (esu). The molecule absorbs infrared light strongly but is transparent to visible light. Scientists are striving for an understanding of these properties and many more in terms of the fundamental theory of valence. *See* BOND ANGLE AND DISTANCE.

Here the term valency is used as a synonym for valence.

Combining power of an element. By the 1920s the most important facts about atoms had been established experimentally. A neutral atom of atomic number Z comprises a massive nucleus of charge $+Ze$, and Z very light electrons each of charge $-e$, where $e = 4.80 \times 10^{-10}$ esu; most of the space within the atom is empty. Atomic nuclei are immutable through ordinary chemical changes; when one molecule of H_2 combines with one molecule of Cl_2 to give two molecules of HCl, the four nuclei (two hydrogen nuclei, or protons, of charge $+1e$ and two chlorine nuclei of charge $+17e$) are unchanged. It is redistribution of electrons between atoms which constitutes chemical combination. This is what valences of atoms control, and this is what a theory of valence must explain.

Atomic structure. To understand molecule formation, then, one first must understand the electronic structure of atoms. According to Neils Bohr, electrons in an atom move in orbits much like the orbits of planets about a sun, held to the nucleus by electrical attractions for it, prevented from falling into it by centrifugal forces. A special quantum effect is operative at the atomic level, however, which possesses no analogy in the motions of planets; not all orbits are possible for an electron, but only those for which the angular momentum of the electron as it moves about the nucleus is an integral multiple of $h/2\pi$, where $h = 6.63 \times 10^{-27}$ erg-sec is Planck's constant, and for which the energy is similarly quantized. Furthermore, not more than two electrons can move in one orbit at once. *See* QUANTUM CHEMISTRY.

When the consequences of these ideas are worked out, there actually emerges the periodic classification of the elements. To cover just part of the periodic table, occupation of orbits by electrons in the lighter atoms are shown in the table, where the symbol $2p$ stands for three distinct orbits of the same energy and shape but differently oriented in space. The lowest energy orbit is $1s$, forming the K shell. Next in energy are $2s$ and $2p$, making up the L shell. The $3s$ state is still higher, in the M shell. The chemically inert gases helium, He, and neon, Ne, are characterized by closed shells of 2, and $2+8=10$ electrons, respectively. The next inert gas is argon, Ar, with a closed shell of $2+8+8=18$ electrons, followed by krypton, Kr, with $2+8+18+8=36$ electrons, and the others. *See* PERIODIC TABLE.

Rule of eight. Most of the simple facts of valence (though certainly not all) follow from the postulate that atoms combine in such a way as to seek closed-shell or inert-gas structures (rule of eight) by the transfer of electrons between them or the sharing of a pair of electrons between them. Following G. N. Lewis, many molecular structures may be obtained by inspection using these rules. Letting a dot represent an electron,

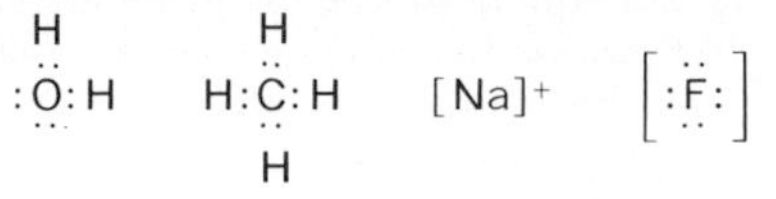

In these electron-dot symbols, the electrons in the K shell are not included for atoms after He, nor are the electrons in the K and L shells for atoms following Ne.

Hydrogen has a valence of 1, because one more electron will give a hydrogen atom an inert gas structure. Carbon can form four bonds because four more electrons give it the neon electronic structure.

Bond types. The bond between two atoms is covalent if one electron in the bonding electron pair comes from each atom, as in H:H or the CH bonds in CH_4. It is coordinate covalent if both electrons come from one atom, as the boron-nitrogen bond in the compound

If there is complete transfer of electrons from one atom to another the bond is electrovalent or ionic, as in sodium fluoride, NaF. Bonds intermediate in type are possible; the bond in hydrogen fluoride, HF, is between covalent and ionic. An ionic bond X^+Y^- will be more stable the less the ionization potential of X and the greater the affinity of Y for electrons, that is, when X is a metallic element from the lower left corner of the periodic table and Y is a nonmetallic element from the upper right corner. Bond type can be inferred from both chemical and physical evidence. *See* CHEMICAL BONDING; ELECTRONEGATIVITY.

Bonds involving one or three electrons are known, but they are rare; H_2^+ and HeH are examples. Multiple bonds between atoms are common and important; examples are the carbon-carbon bond and the carbon-oxygen bonds in ethylene and carbon dioxide:

$$H_2C::CH_2 \quad \text{or} \quad H_2C{=}CH_2$$
$$O::C::O \quad \text{or} \quad O{=}C{=}O$$

For discussion of a bond of special importance in biology *see* HYDROGEN BOND.

Valence electrons are the electrons of an atom that can participate in chemical binding, for example, for H and He the $1s$ electrons, for Li through Ne the $2s$ and $2p$ electrons, and for Na the $3s$ electron.

Oxidation-reduction. As generally used and here defined, the word valence is ambiguous. Before a value can be assigned to the valence of an atom in a molecule, the electronic structure of the molecule must be exactly known, and this structure must be describable simply in terms of simple bonds. In practice neither of these conditions is ever precisely fulfilled. A term not so ambiguous is oxidation number or valence number. Oxidation numbers are useful for the balancing of oxidation-reduction equations, but they are not related simply to ordinary valences. Thus the valence of carbon in CH_4, $CHCl_3$, and CCl_4 is 4; oxidation num-

Electron configurations of some atoms

Atom		Orbit						
		K shell	*L* shell		*M* shell			
	Z	1*s*	2*s*	2*p*	3*s*	3*p*	3*d*
H	1	1	0	0	0	0	0
He	2	2	0	0	0	0	0
Li	3	2	1	0	0	0	0
Be	4	2	2	0	0	0	0
B	5	2	2	1	0	0	0
C	6	2	2	2	0	0	0
N	7	2	2	3	0	0	0
O	8	2	2	4	0	0	0
F	9	2	2	5	0	0	0
Ne	10	2	2	6	0	0	0
Na	11	2	2	6	1	0	0
Mg	12	2	2	6	2	0	0
Al	13	2	2	6	2	1	0

bers of carbon in these three substances are −4, +2, and +4. *See* OXIDATION-REDUCTION.

Quantum theory of valence. The above theory of valence is inadequate in at least three ways. First, it fails to account for many experimental facts, such as why the six C—C bonds in the molecule benzene, C_6H_6, are physically and chemically equivalent, what the electronic structures of the boron hydrides are, why the H—H bond is much stronger than the C—C bond, why CO_2 is a linear molecule but H_2O nonlinear, and what principles govern the rates of chemical combination. Second, the explanations that are offered are not physically satisfying. The stability conferred upon a molecule by the sharing of a pair of electrons by two atoms is established, but what is the real origin of this stability? And third, the theory is not comprehensive or quantitative enough to allow correlation and prediction of the many different properties of molecules. Dozens of properties of molecules can be measured, many to high accuracy. A theory should ultimately, and quantitatively, account for all of these. *See* MOLECULAR STRUCTURE AND SPECTRA.

The quantum theory of valence does not have these faults. It is based on the new precise laws of physics for the atomic domain which were formulated in the 1920s by E. Schrödinger and others, the discipline called quantum mechanics. The quantum ideas of M. Planck and N. Bohr require modification to take care of experimental observations that electrons and other particles at times act like waves. Like waves, they interfere when they are on top of one another in a manner that can be precisely calculated. According to 19th-century physics, an electron moving about a proton would collapse onto it. In the Bohr theory this collapse is prevented by a special quantum hypothesis; in the new mechanics it is prevented by elementary energy considerations. It would be favored by the attractive potential energy of the particle pair, but it turns out to be catastrophic for their kinetic energy. Instead of collapse a compromise is reached; the electron, or wave, is smudged out over a region about the nucleus which defines the atomic size.

The pattern of the periodic table comes out as before. The orbits of Bohr are replaced by new entities, orbitals, which represent not the paths of the electrons but the amplitudes of the electron waves at different points in space. Furthermore, electrons are treated as if they were spinning, but only in two possible ways. The rule that generates the periodic table then is that in an atom no two electrons can occupy the same atomic orbital with the same spin.

In a chemical bond again there is interplay of kinetic and potential energies. An electron pair will tend to be shared by two atoms instead of being located on one of them if that situation is energetically favorable. The region between nuclei is more favorable for the potential energy of electron-nuclear attraction than other regions the same distance from just one nucleus. Moving in this restricted region is not as favorable for the kinetic energy as moving on individual atoms, but the potential energy predominates when a bond is formed. The normal covalent bond may be described as two electrons occupying one molecular orbital, rather than two distinct atomic orbitals, with opposite spins because the exclusion principle is still operative. *See* MOLECULAR ORBITAL THEORY.

When a detailed examination is made of these effects with the new theory, the stabilities of actual molecules and other of their properties can be quantitatively accounted for. In particular, if two atoms approach which have low-energy atomic orbitals which overlap each other in space, and if two electrons are available, the conditions are favorable for forming, with evolution of heat, a chemical bond. It follows that the valence of an atom is given by the number of unpaired electrons it possesses, an old basic rule of valence.

The greater the overlap between two atomic orbitals, the stronger the bond that can be formed with them (criterion of maximum overlapping). This condition may be regarded as determining the shapes of molecules. Two or more orbitals of comparable energy, as 2*s* and 2*p* orbitals, can be combined (hybridized) to give orbitals concentrated along certain directions in space, and these are the orbitals that participate in directed bond formation. In the carbon atom, for instance, the four electrons in the 2*s* and 2*p* subshells are potential valence electrons. The two 2*s* electrons are paired, however, so that to make four bonds possible one of these must be promoted to a vacant 2*p* orbital. Four bonds then are possible, in various directions. Four equivalent bonds can be formed, tetrahedrally directed, as in CH_4. Three bonds in a plane and one other less strong one can be formed, as in $CH_2{=}CH_2$. In this manner Linus Pauling and others have accounted for a multitude of phenomena in stereochemistry.

The peculiar binding in benzene and other aromatic molecules has been explained, together with its consequences for chemical reactivity. The principles governing reaction rates have been formulated and applied.

Research in valence theory through the 1930s and 1940s has led to understanding of a great deal of chemistry, and it has contributed toward acceptance of the language of modern physics as a proper language for chemistry. However, considerable research in this field continues. New substances with new types of bonds are being synthesized constantly (for example, ferrocene, the first "molecular sandwich," in 1951). New physical methods for studying molecules are constantly revealing more intimate details of molecular structure which demand explanation (for example, the new techniques of magnetic resonance). Also, intensive work continues with applications of large digital electronic computing machines to problems of valence. Accurate determination of many properties of molecules containing only light atoms is presently being achieved by such computational methods. *See* CHELATION; CHEMICAL DYNAMICS; CHEMICAL STRUCTURES; CONJUGATION AND HYPERCONJUGATION; COORDINATION CHEMISTRY; CRYSTAL FIELD THEORY; ORGANIC CHEMISTRY; ORGANOMETALLIC COMPOUND; RESONANCE; STEREOCHEMISTRY.

[ROBERT G. PARR]

Bibliography: Chemical Bond Approach Project, *Chemical Systems*, 1965; A. L. Companion, *Chemical Bonding*, 1964; C. A. Coulson, *Valence*, 2d ed., 1962; H. Gray, *Electrons and Chemical Bonding*, 1964; L. Pauling, *General Chemistry*, 2d ed., 1956; L. Pauling, *The Nature of the Chemical Bond and the Structure of Molecules and Crystals*, 3d ed., 1959; F. O. Rice and E. Teller, *The Structure of Matter*, 1949.

Vanadate

A generic term for salts containing vanadium in the 5+ oxidation state and derived from vanadium pentoxide, V_2O_5. Vanadium pentoxide is not very soluble in water, but salts corresponding to the acids, H_3VO_4, $H_4V_2O_7$, and HVO_3, which have been named as the corresponding phosphates (that is, ortho-, pyro-, and metavanadate) are known.

These condensed ions are more ionic in character than the phosphates, which are described as polymers held together by covalent bonds.

Mercury(I) metavanadate, $Hg_2(VO_3)_2$, for example, is a well-characterized salt that is only slightly soluble in water, and hence is used in the quantitative analysis of vanadium. *See* PHOSPHATE; VANADIUM.

[E. EUGENE WEAVER]

Vanadium

A chemical element, V, atomic number 23. It is a metal which has been used primarily as an alloying element for steels and irons since about 1900. Several of the compounds of vanadium are used in the chemical industry, notably in making oxidation catalysts, and in the ceramic industry as coloring agents. It was not until 1950 that this element could be produced in a sufficiently pure form and in large enough quantities to permit its study as an engineering material.

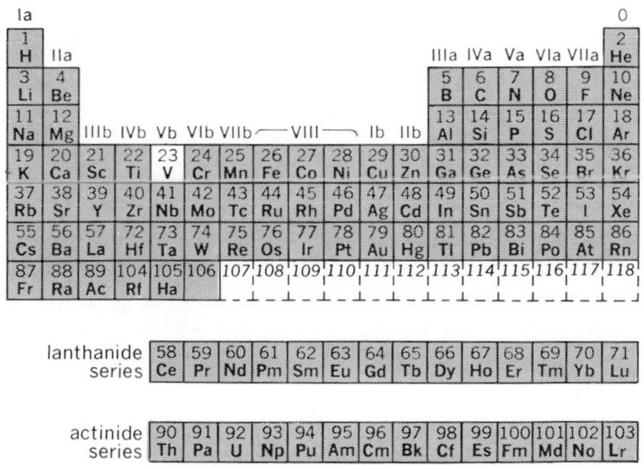

Natural occurrence. Although ores containing small amounts of vanadium are rather widely distributed throughout the world, the most important ones are found in the Western Hemisphere. They include carnotite, roscoelite, vanadinite, and patronite. Carnotite and roscoelite have been of the greatest commercial importance, but other sources are now contributing a larger share. These include the vanadiferous iron ores of South Africa and Finland and the phosphate deposits in the western United States.

Extraction from ores. Vanadium is usually recovered as a coproduct or by-product of other elements with which it is associated. Vanadium and uranium are extracted from carnotite and similar ores by acid leaching or by a roast-quench-leach process. In the latter process the ore is crushed to a size less than 10 mesh and roasted with salt in a multiple-hearth furnace to form the water-soluble sodium metavanadate, $NaVO_3$. The hot ore is then quenched in water or soda ash solution.

Uranium is precipitated first by neutralizing the leach liquor with sulfuric acid. After filtering off the uranium precipitate, vanadium is precipitated by adding more acid. The vanadium precipitates as red cake, a complex sodium polyvanadate corresponding approximately to the hexavanadate, $Na_2H_2V_6O_{17}$. Vanadium and uranium are also extracted from carnotite by leaching the ground raw ore with sulfuric acid. Solvent extraction is then used to separate uranium and vanadium and to enrich the solutions. When solvent extraction is used, vanadium is usually precipitated as a low-soda red cake, approximating the composition $H_4V_6O_{17}$, and in some cases as ammonium metavanadate, NH_4VO_3.

Vanadium in iron ores is recovered in some cases by roasting and leaching the ground ore in a manner similar to that used for carnotite ore. In other cases the ore is smelted to produce pig iron which contains the vanadium. When the pig iron is converted to steel, the vanadium is recovered in a rich slag containing 5–25% V_2O_5. The slag is then processed by a roasting and leaching procedure to extract the vanadium in the form of red cake.

The vanadium in phosphate rock is recovered in ferrophosphorus during the production of elemental phosphorus by electric-furnace smelting. The ferrophosphorus contains 3–7% vanadium and is processed further by roasting and leaching to recover the vanadium.

Red cake contains 85–90% V_2O_5, 2–10% Na_2O, and about 3% H_2O. Solution of red cake with soda ash forms sodium vanadate, $NaVO_3$, which is filtered and precipitated with ammonia to form ammonium metavanadate, NH_4VO_3, a commonly used salt. Heating ammonium metavanadate drives off the ammonia and leaves a purified oxide, V_2O_5. For the production of alloys by smelting, red cake is fused and cast into flake form.

Ferrovanadium from reduction. Ferrovanadium is produced by aluminum or silicon reduction of the oxide in the presence of iron in an electric arc furnace. Aluminum reduction is commonly practiced; the reaction is exothermic and little additional heat from the arc is required. Silicon reduction requires a two-stage reduction to achieve an efficient operation.

Vanadium for metallurgical purposes is also produced by solid-state carbon reduction of vanadium pentoxide in a vacuum furnace. The product contains about 85% vanadium, 12% carbon, and 2% iron.

Vanadium metal from calcium reduction. The best-known process for producing ductile vanadium metal used in the United States is the calcium reduction of the oxide, as developed by R. K. McKechnie and A. U. Seybolt. This process consists of charging pure vanadium oxide, calcium metal, and iodine into a heavy-walled steel cylinder, taking special precautions to exclude moisture. The cylinder is sealed and evacuated, and the reaction between the ingredients is initiated by the application of heat. Within a short time, high enough temperature and pressure are reached to allow the molten droplets of vanadium to collect beneath the calcium oxide–calcium iodide slag to form a single button or regulus. A typical analysis of the product is 99.7% vanadium, 0.10% oxygen, 0.04% nitrogen, 0.008% hydrogen, 0.04% iron, and 0.03% carbon.

Vanadium metal from magnesium reduction. The process for producing pure vanadium by the magnesium reduction of vanadium trichloride is similar to the Kroll process used to reduce titanium tetrachloride to titanium sponge. Vanadium trichloride is produced by the chlorination of ferrovanadium, or, as a later modification, by chlorination of vanadium pentoxide. The trichloride is reduced with magnesium in an atmosphere of argon at about 1550°F (843°C). The resulting sponge is subsequently separated from the magnesium chloride and excess magnesium by melting under argon, then by heating under vacuum, and finally by leaching. A typical analysis of the sponge is 99.7% vanadium, 0.12% oxygen, 0.005% nitrogen, 0.01% hydrogen, 0.01% magnesium, and 0.03% iron.

Properties of ductile vanadium. In its pure form vanadium is soft and ductile. It can be hot- and cold-worked easily, but it must be heated in an inert atmosphere or in a vacuum because it oxidizes readily at temperatures above the melting point of its oxide, about 1225°F (663°C). The strength of vanadium is sensitive to interstitial impurities and varies from 30,000 psi (200 MPa) in the purest form to 80,000 psi (550 MPa) in the commercial form. The metal retains its strength unusually well at elevated temperatures. Vanadium has a relatively low cross section for neutron capture, 5 barns (5×10^{-28} m^2); this property has accounted for considerable interest in connection with the utilization of atomic energy.

The resistance of vanadium to hydrochloric and sulfuric acids is outstanding, and it withstands aerated-salt-water attack better than most stainless steels. Vanadium cannot withstand nitric acid, either dilute or concentrated.

Vanadium has a density of 6.1 g/cm^3 or 0.22 $lb/in.^3$, which is 22% less than that of iron and 28% more than that of titanium. Coefficients of electrical and thermal conductivity are both significantly higher than those of titanium. The elastic modulus of vanadium at room temperature is of the order of 20,000,000 psi (140 GPa).

Alloys. Most of the vanadium produced (86% in 1965) is consumed by the steel industry as an alloying agent. The formation of vanadium carbide when vanadium is added to steel is the basis for many of the unique properties imparted to steel by vanadium. These carbides are extremely hard and wear-resistant; they do not coalesce readily but maintain a state of fine dispersion. A relatively small amount of vanadium (0.06–0.10%) is soluble in the austenitic phase of steels. This small percentage markedly increases the ability of the steel to harden on rapid cooling. Vanadium also forms a stable nitride and it can, in effect, lower the nitrogen content of steel.

Carbon and alloy steels consume more than half of the vanadium produced in the United States. Many plate, structural, bar, and pipe steels contain vanadium to enhance strength and toughness. This family of high-strength steels has been sold in increasingly large quantities because they provide higher strength, without a corresponding decrease in ductility and weldability, for a minimum of additional price over plain carbon steels. The vanadium content of many of these steels ranges from 0.02 to 0.06%, together with higher manganese and often copper. Most of these steels possess higher strength in the as-rolled condition, that is, without heat treatment.

Sheet steels used for deep-drawing auto- and home-appliance parts often contain vanadium to suppress aging. In making such steels ferrovanadium is added to rimming steels, providing a good, nonaging, deep-drawing steel at a lower cost than the aluminum-killed deep-drawing steel.

Vanadium has a promising future as a grain-size control agent in continuously cast steel, replacing aluminum which, to date, cannot be used in this process.

Many large steel forgings contain vanadium in the range 0.05–0.15%; here it acts as a grain refiner, as well as improving the mechanical properties of the forgings. In large steel castings as well as forgings, a small percentage of vanadium is unique in its ability to increase strength and ductility.

Tool steels are the third largest class of vanadium-containing steels. In fact, almost all tool steels contain this element, the amount ranging from 0.10 to 5.00%. Vanadium is required in these steels to ensure the retention of hardness and cutting ability at the elevated temperatures generated by rapid cutting of metals.

Vanadium sometimes is used in cast iron to control the size and distribution of graphite flakes and to improve strength and wear resistance. The most widely used of the titanium-base alloys contains 4% vanadium and 6% aluminum. A high-purity 40:60 vanadium-aluminum alloy is produced for this purpose and now accounts for a significant amount of vanadium.

Principal compounds. Like other elements in the transition group Va, vanadium forms numerous and frequently complicated compounds because of its variable valence. It has at least three oxidation states, 2+, 3+, and 5+. It is amphoteric, mostly basic in the lower oxidation states, acidic in the higher ones. It forms derivatives from more or less well-defined radicals such as VO^{2+} and VO^{3+}. Its oxygen acids readily form condensed acids.

The most important compounds of vanadium are vanadium pentoxide, V_2O_5, and ammonium metavanadate, NH_4VO_3. The pentoxide is commonly sold as the sodium salt of hexavanadic acid, $Na_2H_2V_6O_{17}$. Ammonium metavanadate is formed by adding excess ammonium chloride to an alkaline solution of vanadium pentoxide. Pure vanadium pentoxide, 99.6% V_2O_5, is made by calcining ammonium metavanadate.

Vanadium oxytrichloride is produced commercially for use as a catalyst in making ethylene-propylene rubber. An increase in demand is anticipated.

Sodium metavanadate is also a regularly produced product. This compound is used as an additive for alkaline amine solutions in order to reduce corrosive action in steel piping systems.

Many other compounds have been produced on a smaller scale. Vanadyl sulfate and vanadyl chloride have been in demand rather regularly. Those compounds produced for study include nitrates, acetates, oxalates, nitrides, carbides, bromides, iodides, and fluorides. Metalloorganic compounds made include vanadyl linoleate, oleate, palmitate, phenolate, resinate, and stearate. Ammonium vanadyl tartrate has received attention as a hypocholesteremic agent.

The largest application of vanadium compounds is in the manufacture of oxidation catalysts for the chemical industry. Both ammonium metavanadate and vanadium pentoxide are used for this purpose. Processes which employ such catalysts include the manufacture of polyamides, such as nylon; the manufacture of sulfuric acid by the contact process; the manufacture of phthalic and maleic anhydrides; and various other organic oxidation reactions, such as anthracene to anthraquinone, alcohol to acetaldehyde, dephenylamine to carbazole, and sugar to oxalic acid.

Vanadium compounds have for years been used in the ceramic industry for glazes and enamels. Combinations of vanadium oxide, zirconia, silica, lead, zinc, tin, cadmium, and selenium produce various colors. Vanadium compounds are used in the production of aniline black for the dye industry. *See* METALLOACID ELEMENTS; TRANSITION ELEMENTS.

[TIMOTHY W. MERRILL]

Bibliography: R. Clark and D. Brown, *The Chemistry of Vanadium, Niobium and Tantalum*, 1975.

Van der Waals equation

An empirical equation of state, presented by J. D. van der Waals in 1873, which takes into account the finite size of the molecules and the attractive forces between them. For a homogeneous gas at pressure p, molar volume v_N, and absolute temperature T, the equation is given below.

$$p=\frac{R_uT}{v_N-b}-\frac{a}{(v_N)^2}$$

or

$$\left(p+\frac{a}{(v_N)^2}\right)(v_N-b)=R_uT$$

The terms a and b are constants and evaluated from experimental data. The molar specific volume v_N is equal to the total volume V divided by the number of moles N. R_u is the universal gas constant. If the units for a and b are (atm)(ft^6)/lbmole^2 and ft^3/lbmole, respectively, the value and units for R_u are 0.730 (atm)(ft^3)/lbmole °R. If the units for a and b are (psia)(ft^6)/lbmole^2 and ft^3/lbmole, the value and units for R_u are 10.73 (psia)(ft^3)/lbmole °R. If the units for a and b are (Pa)(m^6)/mole^2 and m^3/mole respectively, the value and units for R_u are 8.31 (Pa)(m^3)/(mole)(K). Values for a and b for a few gases are presented in the table. *See* THERMODYNAMIC PRINCIPLES.

[GEORGE A. HAWKINS]

Bibliography: J. B. Jones and G. A. Hawkins, *Engineering Thermodynamics*, 1960; S. L. Kittsley, *Physical Chemistry*, 3d ed., 1969; W. C. Reynolds and H. C. Perkins, *Engineering Thermodynamics*, 2d ed., 1977; K. Wark, *Thermodynamics*, 3d ed., 1977.

Vapor pressure

The saturation pressures exerted by vapors which are in equilibrium with their liquid or solid forms. One of the most important physical properties of a liquid, the vapor pressure, enters into many thermodynamic calculations and underlies several methods for the determination of the molecular weights of substances dissolved in liquids. For a discussion of the vapor pressure relationships of solids *see* SUBLIMATION. *See also* MOLECULAR WEIGHT; SOLUTION.

If a liquid is introduced into an evacuated vessel at a given temperature, some of the liquid will vaporize, and the pressure of the vapor will attain a maximum value which is termed the vapor pressure of the liquid at that temperature. Although the quantity of liquid remaining does not diminish thereafter, the process of evaporation does not cease. A dynamic equilibrium is established, in which molecules escape from the liquid phase and return from the vapor phase at equal rates. *See* EVAPORATION.

It is important to make a distinction between the vapor pressure of a liquid, as described above, and the pressure of a vapor. The vapor pressure of a pure liquid is a unique and characteristic property of the liquid and depends only upon the temperature. A gas or vapor may, on the other hand, exert any pressure within reason, depending upon the volume to which it is confined, provided it is not in contact with its liquid phase.

Liquid-vapor equilibrium. The relationship between the vapor pressure of a liquid and the temperature is indicated by a phase diagram. The illustration shows the phase diagram for water, the line *OC* being the vapor pressure line for liquid water. Line *AO* is the vapor pressure line (sublimation pressure curve) for ice, and line *BO* is the liquid-solid equilibrium line. Point *O* is called the triple point and is the unique pressure and temperature at which a pure solid, its liquid, and its vapor can coexist in equilibrium under the pres-

Approximate values for the constants a and b

Gas	a (atm)(ft^6)/lbmole^2	a (psia)(ft^6)/lbmole^2	a (Pa)(m^6)/mole^2	b ft^3/lbmole	b m^3/mole
Air	344	5,052	0.136	0.587	3.66×10^{-5}
Carbon dioxide	926	13,600	0.366	0.686	4.28×10^{-5}
Helium	8.57	126	0.00338	0.372	2.32×10^{-5}
Neon	55.4	814	0.0219	0.282	1.76×10^{-5}
Oxygen	350	5,140	0.138	0.510	3.18×10^{-5}

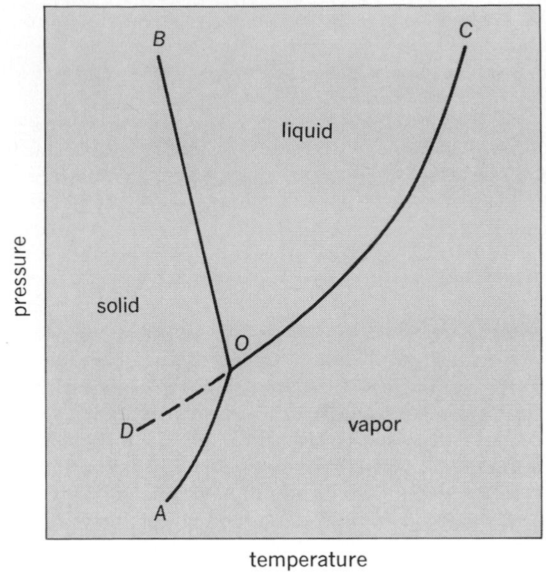

Phase diagram of the water system.

sure of the vapor alone. The triple point of water is not the familiar melting point (0°C), but rather 0.0075°C. The distinction is that the total pressure at the triple point is 4.58 mm (611 Pa), the common vapor pressure of solid and liquid water, whereas the total pressure at the melting point is ordinarily 1 atm (101,325 Pa). Point C is called the critical point and is the point above which there is no distinction between the liquid and gaseous phases. It is not possible to liquefy a gas at temperatures above the critical temperature, regardless of the applied pressure. Necessarily, the surface tension and the latent heat of vaporization become zero at the critical point. For ordinary liquids, the vapor pressure at the critical point is usually about 50 atm. There is no evidence for a critical point on the solid-liquid equilibrium line, and B is meant to indicate a direction, rather than a point; other phases may appear above B, of course, and alter the direction of the line. It is possible to undercool a liquid below its triple point if crystallization nuclei are absent. The vapor pressure of the undercooled liquid is given by the dashed line DO and lies above the equilibrium vapor pressure of the solid. The undercooled liquid is therefore metastable, because the system tends to assume its lowest vapor pressure at equilibrium.

Quantitative relations. For most liquids the relationship between the vapor pressure and temperature can be expressed by an equation having the form of Eq. (1), where a, b, and c are constants.

$$\log P = a + b \log T + \frac{c}{T} \tag{1}$$

The simpler equation Eq. (2), where m and n are

$$\log P = \frac{m}{T} + n \tag{2}$$

constants, is often adequate. The change in the vapor pressure of a liquid with temperature may be expressed by the Clausius-Clapeyron equation, Eq. (3), where L_v is the molar latent heat of vaporization and ΔV is the difference in molar volumes

$$\frac{dP}{dT} = \frac{L_v}{T\Delta V} \tag{3}$$

of the vapor and liquid at the temperature T. Because the molar volume of the liquid is negligible by comparison with that of the vapor except near the critical point, the Clausius-Clapeyron equation may be written as Eq. (4), where V is the molar

$$\frac{dP}{dT} = \frac{L_v}{TV} \tag{4}$$

volume of the gas. If the gas follows the ideal gas law ($PV = RT$), the equation may be further written as Eq. (5), where R is the gas constant. For moder-

$$\frac{1}{P} \cdot \frac{dP}{dT} = \frac{L_v}{RT^2} \tag{5}$$

ate ranges of temperature, the latent heat of vaporization is constant, and the equation may be rearranged and integrated to give the useful form of Eq. (6). From the latent heat of vaporization and the vapor pressure at one temperature, the vapor

$$\log_{10} \frac{P_2}{P_1} = \frac{L_v}{2.3R}\left(\frac{1}{T_1} - \frac{1}{T_2}\right) \tag{6}$$

pressures at other temperatures may thus be calculated.

The vapor pressure of a liquid is lowered when a substance is dissolved in it. At low solute concentrations in nonideal solutions, or at all solute concentrations in ideal solutions, the partial vapor pressure of the liquid in a solution is proportional to the mole fraction of the solvent, Eq. (7), where

$$P_1 = X_1 P_1{}^0 \tag{7}$$

X_1 is the solvent mole fraction and $P_1{}^0$ is the vapor pressure of the pure solvent. Raoult's law, as this important relationship is known, may also be expressed in the form of Eq. (8), where X_2 is the mole

$$\frac{P_1{}^0 - P_1}{P_1{}^0} = X_2 \tag{8}$$

fraction of the solute. The lowering of the vapor pressure of the solvent is proportional to the mole fraction of the solute. This relation provides the basis for the determination of the molecular weights of dissolved substances. *See* CONCENTRATION SCALES.

Vapor pressure measurement. Because the vapor pressures of liquids range widely, a number of methods have been devised for their measurement. The static method consists of introducing an excess of the liquid into an evacuated system at a given temperature, and measuring the vapor pressure with an attached mercury manometer. It is useful for vapor pressures ranging from a few millimeters of mercury up to several atmospheres. The dynamic method consists of introducing the liquid into a system in which the pressure may be varied, and noting the temperature at which the liquid boils at a given pressure. It is useful for moderately high vapor pressures. The transpiration method consists of bubbling a known volume of an inert gas at a definite pressure through the liquid. The quantity of liquid transported into the carrier gas is determined, and the vapor pressure of the liquid is calculated from relation (9), where P_C is

$$P_L = P_C \frac{n_L}{n_L + n_C} \tag{9}$$

the carrier gas pressure and n_L and n_C are the numbers of moles of liquid vaporized and carrier gas, respectively. For the measurement of very low vapor pressures, the Knudsen effusion method may be used. This method is based upon the measurement of the mass of vapor which escapes through a very small hole in a vessel containing a liquid in equilibrium with its vapor. If the diameter of the hole is small compared with the mean free path of the gas molecules, the latter suffer no collisions in their passage through the hole. Equation (10) relates the mass of gas which passes per

$$m = P\sqrt{\frac{M}{2\pi RT}} \tag{10}$$

square centimeter per second through the hole and the vapor pressure, where m is the mass of gas per square centimeter per second and M is the molecular weight. The Knudsen method has been used with radioactive isotopes or with mass spectrometers to determine very low vapor pressures. *See* PHASE EQUILIBRIUM; TRIPLE POINT.

[NORMAN H. NACHTRIEB]

Bibliography: G. M. Barrow, *Physical Chemistry*, 2d ed., 1966; W. J. Moore, *Physical Chemistry*, 3d ed., 1962.

Vinyl resin

An addition polymer made from monomers with the general structure shown, wherein R_1 and R_2

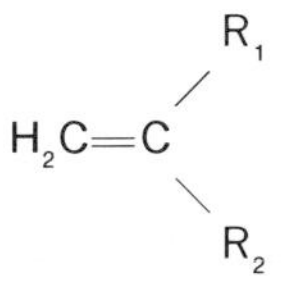

represent the same or different hydrogen, alkyl, halogen, ether, carboxylate, or other substituents. Vinyl resin often refers to any of a large variety of thermoplastics with the $—CH_2CHR_1—$ repeating group. Popularly, vinyl resin refers to a plasticized polyvinyl chloride used as a film or fabric, as for furniture or automobile seat covers. In common chemical usage, vinyl resin implies a polymer of an olefinically unsaturated monomer, whether the double bond is terminal or not. Strictly speaking, vinyl resin applies only to polymers of singly substituted ethylenes, such as vinyl acetate, vinyl chloride, and methyl vinyl ether, and excludes vinylidene and internal olefins. *See* POLYVINYL RESINS.

[FRANK WAGNER]

Vinylogy

A principle used to correlate the properties of organic chemical compounds. Organic compounds that differ from each other by a vinylene linkage (—CH=CH—) are said to be vinylogs of one another. Thus, ethyl crotonate is a vinylog of ethyl acetate and of the next higher vinylog, ethyl sorbate. Their structures are:

$CH_3—CO_2C_2H_5$
Ethyl acetate

$CH_3—CH=CH—CO_2C_2H_5$
Ethyl crotonate

$CH_3—CH=CH—CH=CH—CO_2C_2H_5$
Ethyl sorbate

These esters are the first members of the vinylogous series $CH_3—(CH=CH)_n—CO_2C_2H_5$. The relationship between the members of such a series is known as vinylogy.

The outstanding characteristic common to the members of a vinylogous system is that the influence of the terminal functional groups upon each other persists throughout the series. The readiness with which the methyl group of ethyl acetate loses a proton is ascribed to the enhanced stability of the resulting anion because of the presence of the ester group. Two principal resonance structures of the ion may be written as in reaction (1).

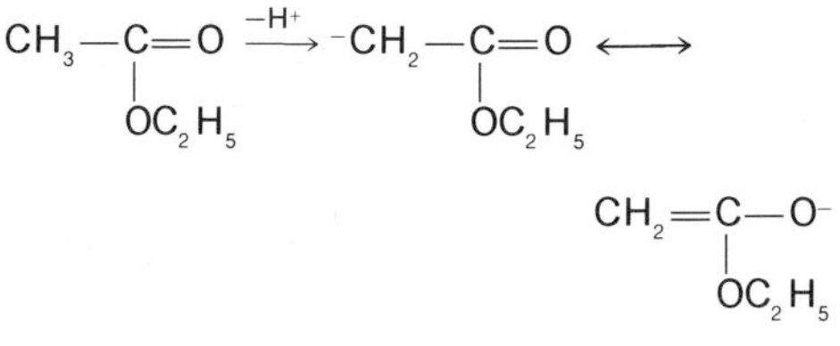

(1)

The anions (enolates) of ethyl crotonate and ethyl sorbate can be formulated in a similar way, reactions (2) and (3). These enolates owe their stability

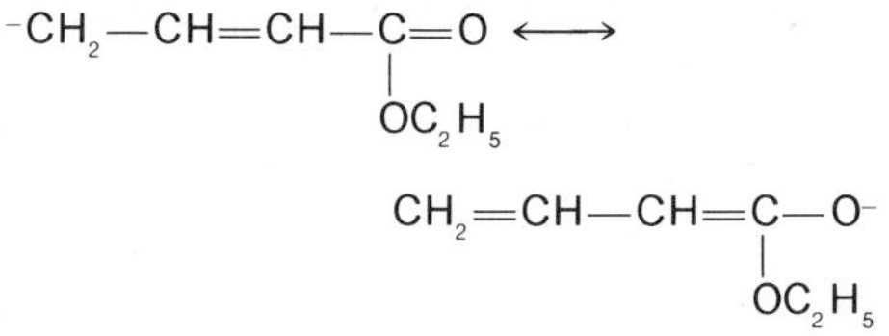

(2)

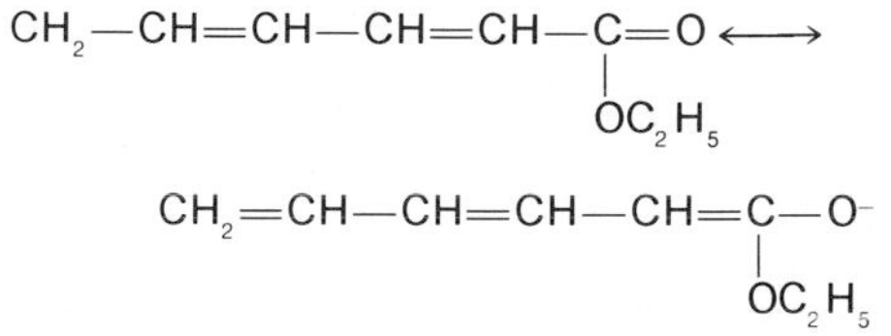

(3)

in part to the resonance made possible by the vinylene linkages.

Similar relationships hold for aromatic systems in which two functions are situated in positions ortho or para to one another. Thus, the reactivities of the methyl groups in *o*- and *p*-nitrotoluene are similar to that of the methyl group in nitromethane, the first member of the vinylogous series; their structures are:

$CH_3—NO_2$ Nitromethane

o-Nitrotoluene

p-Nitrotoluene

The compounds *o*- and *p*-nitrophenol, which may be described as vinylogs of nitric acid, are in fact much stronger acids than phenol itself; these structures are:

$HO—NO_2$ Nitric acid

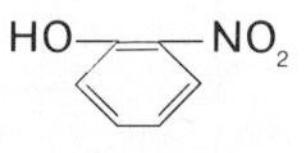

o-Nitrophenol

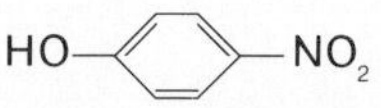

p-Nitrophenol

The principle of vinylogy may be used to explain

why *o*- and *p*-nitrochlorobenzene are readily hydrolyzed to the corresponding nitrophenols, reactions (4) and (5), whereas the meta isomer is not.

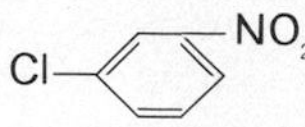

(Not hydrolyzed)

Meta isomer

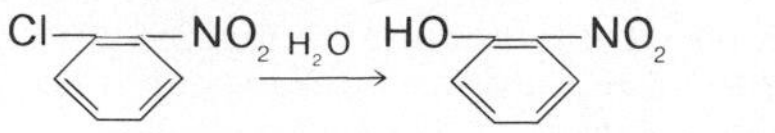

(4)

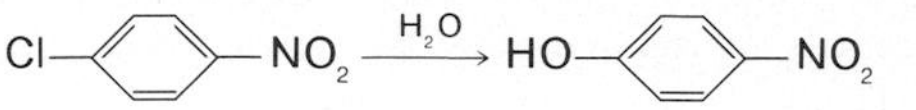

(5)

See CONJUGATION AND HYPERCONJUGATION; RESONANCE.

[REYNOLD C. FUSON]

Viscosity

The resistance that a gaseous or liquid system offers to flow when it is subjected to a shear stress. Viscosity is a measure of the internal friction that arises when there are velocity gradients within the system. For fluids (gases and liquids) its meaning is conceptually and operationally well defined. In the regime of laminar or streamline flow the force required to maintain a velocity gradient, (dv/dx), between planes of fluid of area A is described by Newton's equation (1) and Fig. 1. The

$$f = \eta\, A\left(\frac{dv}{dx}\right) \qquad (1)$$

proportionally constant η is called the viscosity coefficient. Its dimensions are (mass)(length)$^{-1}$ (time)$^{-1}$, and in the cgs system the unit of viscosity is the poise (1 g · cm^{-1} s^{-1}). It is the force per unit area (dynes cm^{-2}) required to sustain a unit velocity gradient (cm s^{-1} cm^{-1}) normal to the flow direction. In the International System (SI) the unit of viscosity is kg · m^{-1} s^{-1}, and is hence larger than the poise by a factor of 10; conversely 1 kg · m^{-1}s^{-1} = 10 poise.

Simple gases typically have viscosities in the range of 100 to 200 micropoise at standard temperature and pressure (273 K, 1 atm or 101,325 Pa), whereas simple liquids under the same conditions have coefficients of viscosity about two orders of magnitude larger. The table lists values for the coefficients of viscosity of selected gases and liquids. The flow characteristics of gases and simple liquids such as water, carbon tetrachloride, and ethyl alcohol are accurately described by Eq. (1), and such fluids are called newtonian fluids. Aqueous suspensions, such as clays, gelatin, and agar, are termed non-newtonian fluids because

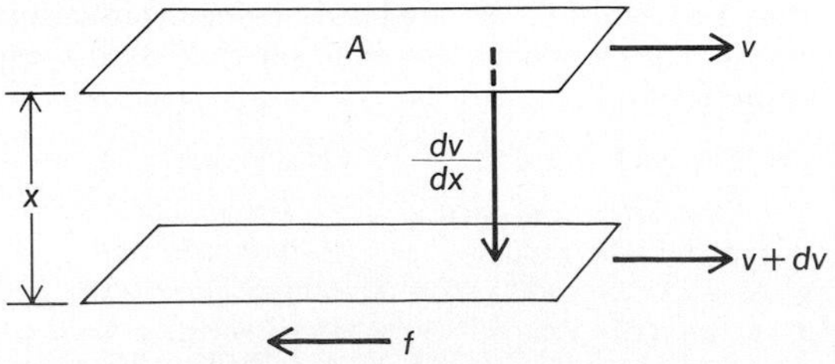

Fig. 1. Viscous shear in fluids.

Coefficients of viscosity of selected gases and liquids

Substance	Temperature, °C	η, poise*
Hydrogen	0	84.2×10^{-6}
Helium	0	186×10^{-6}
Nitrogen	0	167×10^{-6}
Oxygen	0	181×10^{-6}
Water (liquid)	20	10.1×10^{-3}
Ethyl alcohol	20	12.0×10^{-3}
Diethyl ether	20	2.5×10^{-3}
Carbon tetrachloride	20	9.8×10^{-3}
Mercury	20	15.5×10^{-3}
Glycerin	20	10.69
Glass	400	10^{13}
Glass	800	10^{7}

*1 poise = 0.1 kg · m^{-1} s^{-1}.

their viscosities may depend upon the rate of shear and prior treatment. Hydrophilic sols often form extended networks involving water, and their non-newtonian behavior is believed to be due to the breakdown of their structure under shear.

Molecular basis of viscosity in gases. The origin of internal friction (viscosity) at the molecular level is the net transfer of momentum between layers of fluid moving with different velocities in parallel flow by the mechanism of molecular collisions. In this process the directed energy of fluid flow is degraded to random thermal energy (heat).

It was one of the early triumphs of the kinetic theory of gases that established the relationship between the viscosity of a hard-sphere gas and the mean speed of its molecules. If x in Fig. 1 represents the mean free path λ of molecules in hypothetical planes that move with velocities equal to v and v', respectively, an exchange of molecules between the planes will result in the net transfer of momentum per unit time equal to $\frac{1}{3}An\bar{c}(mv - mv')$, where n is the number of molecules per unit volume, m is the mass of a molecule, mv and mv' are the additional momenta of molecules in the planes in consequence of their shear velocities, and $\bar{c}$ is the mean spead of molecules, $(8kT/\pi m)^{1/2}$, where k is Boltzmann's constant and T is the absolute temperature. Since the gas density is $\rho = nm$, the retarding force that resists the shear is given by Eq. (2). Combining Eqs. (1) and (2) gives Eq. (3).

$$f = \tfrac{1}{3}A\rho\bar{c}\left(\frac{dv}{dx}\right)\lambda \qquad (2)$$

$$\eta = \tfrac{1}{3}\bar{c}\rho\lambda \qquad (3)$$

Substitution for the mean free path $[\lambda = (n\sqrt{2}\pi d^2)^{-1}$, where d is the average diameter of a molecule$]$ permits restatement of Eq. (3) as Eq. (4).

$$\eta = \frac{m\bar{c}}{3\sqrt{2}\pi d^2} \qquad (4)$$

More refined calculations for hard-sphere gases replace the factor $\frac{1}{3}$ in Eq. (4) by 0.499, but the functional form is correct. It predicts that the viscosity of gases should be independent of pressure because the mean free path and gas density are affected in opposite ways by pressure, and this prediction is in accord with experiment up to moderately high pressures. Equation (4) also predicts a $T^{1/2}$ dependence of the gas viscosity, which is in fair agreement with experiment. Real gases show a somewhat stronger temperature dependence because their molecules are not ideal hard spheres. Equation (4) has had widespread application to the determination of the diameters of molecules, and the agreement is good but not perfect. Such minor discrepancies as do exist with molecular diameters determined by other methods (such as molar refraction, equation of state, and electron diffraction) are due to the fact that each method probes a somewhat different region of the potential surface of real molecules.

Viscosity of liquids. Momentum transfer between shearing layers also underlies the viscous behavior of simple liquids, but since the mean free path has little meaning for liquids, no simple relation such as Eq. (4) exists for them. In contrast to the behavior of gases, temperature decreases the viscosities of simple liquids and its effect is much larger. The temperature dependence of the viscosity of simple liquids bears no simple relationship to gas kinetic theory, but instead generally follows an exponential law of the form of Eq. (5),

$$\eta = A \exp (B/RT) \tag{5}$$

where A and B are parameters characteristic of the liquid and are reasonably constant over finite ranges of temperature; R is the gas constant, $8.314\ \mathrm{J \cdot mol^{-1} \cdot K^{-1}}$. The form of Eq. (5) is the same as that typically found for transport properties in the defect crystalline state, where the concept of a simple thermally activated process is generally accepted. This similarity in temperature dependence has in the past led to various "hole" theories of the liquid state. These have been based upon analogy with the vacancy model of defect crystals, and the parameter B has been identified with the energy required to create a void of molecular dimensions in the liquid and to move a nearby molecule into it. Such hole theories are now generally thought to be oversimplified, and viscous flow, like diffusion in liquids, is thought to be a highly complex process in which many molecules participate.

Measurements of viscosity as a function of pressure likewise show completely different behavior for gases and liquids. Whereas the former show little dependence of viscosity on pressure in the low-density region, very high hydrostatic pressure generally increases the viscosity of liquids, sometimes quite markedly. In the region of laminar flow, nevertheless, Newton's equation accurately describes the viscous behavior of simple liquids, and the presumption is that the transfer of momentum between shearing layers involves a high degree of correlated molecular motions.

When the shear velocity exceeds a critical value in vessels of a given radius, streamline flow is replaced by turbulent flow. The criterion for the onset of turbulence in a tube of radius r is that the Reynolds number $2r\rho v/\eta$ exceed a certain value (approximately 2000 for normal liquids).

Measurement of viscosity. The laminar flow of both gases and liquids in long narrow tubes is described by Poiseuille's equation (6), where η is the

$$\eta = \frac{\pi (p_1 - p_2) r^4 t}{8Vl} \tag{6}$$

fluid viscosity (poise or $\mathrm{kg \cdot m^{-1}\, s^{-1}}$), where r is the radius of the tube (cm or m), l is its length (cm or m), $(p_1 - p_2)$ is the pressure drop (dynes $\mathrm{cm^{-2}}$ or Pa) across the tube, and V is the volume of fluid ($\mathrm{cm^3}$ or $\mathrm{m^3}$) that flows through the tube in time t (s). Poiseuille's equation is based on the assumption that the layer of fluid in contact with the tube wall is stationary. It provides a basis for the absolute measurement of the viscosity of both gases and liquids.

More commonly for liquids, relative measurements of viscosity are made by use of either the Ostwald viscometer or the falling-sphere viscometer. The former (Fig. 2) consists of two glass bulbs separated by a length of capillary tubing. Liquid is drawn up into the upper bulb, and the time required for its meniscus to fall between calibration marks above and below the upper bulb is accurately measured. A similar measurement is made with a liquid of known viscosity. From Eq. (6), Eq. (7)

$$\frac{\eta_1}{\eta_2} = \frac{\rho_1 t_1}{\rho_2 t_2} \tag{7}$$

follows. Here η_1 and η_2 are the viscosities of the two liquids, ρ_1 and ρ_2 are their densities, and t_1 and t_2 are the corresponding flow times. Equation (7) takes an even simpler form if the viscosity is divided by the density of the liquid. This quantity, called the kinematic viscosity, is measured in units termed stokes ($\mathrm{cm^2\ s^{-1}}$) in cgs units, and in units of $\mathrm{m^2\ s^{-1}}$ in SI.

The falling-sphere viscometer is based upon Stokes' law for the frictional force on a spherical body of radius r falling with constant velocity in a fluid of viscosity η in an unbounded space, Eq. (8).

$$f = 6\pi\eta v r \tag{8}$$

This force is equal and opposite to the net force of gravity acting on the sphere, as in Eq. (9), where ρ

$$f = {}^4\!/_3 \pi r^3 (\rho - \rho') g \tag{9}$$

and ρ' are the densities of a metal sphere and the fluid, and g is the acceleration of gravity.

Equations (8) and (9) lead to the absolute viscosity of the fluid, Eq. (10).

$$\eta = \frac{2gr^2(\rho - \rho')}{9v} \tag{10}$$

As with the Ostwald viscometer, it is simpler to compare the times of fall of the sphere in fluids of known and unknown viscosity, and to use Eq. (11).

$$\frac{\eta_1}{\eta_2} = \frac{t_1(\rho - \rho'_1)}{t_2(\rho - \rho'_2)} \tag{11}$$

Other methods for the absolute or relative measurement of fluid viscosities are based upon the determination of the torque exerted upon a cylinder immersed in a fluid when a coaxial cylinder is rotated with constant velocity, or the damping of the amplitude of an oscillating disk suspended in the fluid by a torsion fiber.

Flow behavior of complex fluids. Many fluids display flow behavior that deviates profoundly from that of simple gases and liquids. This subject is normally treated by the field of rheology. A few examples will serve to indicate the complexity of flow behavior in some fluids. Monoclinic sulfur, whose molecules consist of puckered rings of eight sulfur atoms, melts at 95.5°C to form a simple

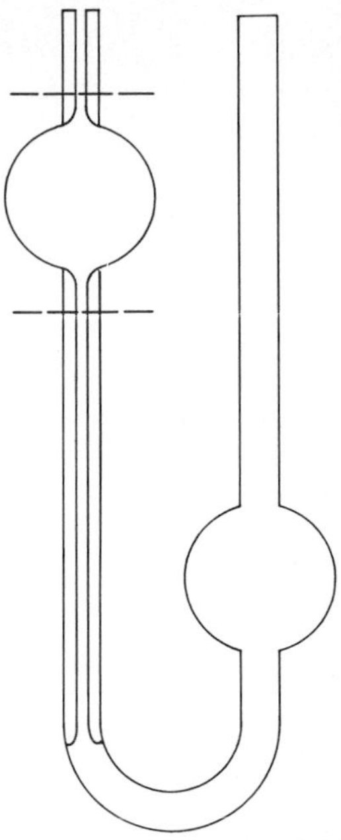

Fig. 2. Ostwald viscometer.

liquid (S_λ) of the same molecularity. Its viscosity is low enough to classify it as a normal liquid, and its viscosity decreases with temperature in the normal manner. Between 160 and 180°C, however, the viscosity increases dramatically by many orders of magnitude, and it appears that ring opening occurs followed by the formation of long-chain polymers. Above this temperature interval, the viscosity again decreases as thermal energy breaks up the long chains into smaller units. The process is highly irreversible, and crystalline sulfur may be recovered only by condensation from sulfur vapor.

Various colloidal dispersions of solids in oil or aqueous media decrease their viscosity when stirred at constant temperature, and revert to their former state of higher viscosity when the shear stresses are reduced. This phenomenon of thixotropy is an essential property of paints that contain solid pigments.

The flow of blood in mammalian vascular systems is non-newtonian, and Poiseuille's law is not obeyed. In part this behavior is attributable to the presence of red corpuscles and other suspended bodies, but the phenomenon is very complex.

Glasses are amorphous solids, structurally much closer to liquids than to crystals. Even at ordinary temperatures they deform under stress over long periods of time, and their viscosity varies over tens of orders of magnitude as the temperature is raised to the softening point. Profound structural changes in the random three-dimensional network and of the dynamical modes of local structural elements take place as the temperature of a glass is increased.

Some adhesives exhibit flow along directions that are not parallel to the direction of stress. Such fluids are anisotropic, and their flow properties are tensors. *See* LIQUID; RHEOLOGY.

[NORMAN H. NACHTRIEB]

Bibliography: P. W. Atkins, *Physical Chemistry*, 1978; J. O. Hirschfelder, C. F. Curtiss, and R. B. Bird, *The Molecular Theory of Gases and Liquids*, 1954; E. A. Moelwyn-Hughes, *Physical Chemistry*, 2d ed., 1964; W. J. Moore, *Physical Chemistry*, 4th ed., 1972.

Volatilization

The process of converting a chemical substance from a liquid or solid state to a gaseous or vapor state. Other terms used to describe the same process are vaporization, distillation, and sublimation. A substance can often be separated from another by volatilization and can then be recovered by condensation of the vapor. The substance can be made to volatilize more rapidly either by heating to increase its vapor pressure or by removal of the vapor using a stream of inert gas or a vacuum pump. Heating procedures include the volatilization of water, of mercury, or of arsenic trichloride to separate these substances from interfering elements. Chemical reactions are sometimes utilized to produce volatile products as in the release of carbon dioxide from carbonates, of ammonia in the Kjeldahl method for the determination of nitrogen, and of sulfur dioxide in the determination of sulfur in steel. Volatilization methods are generally characterized by great simplicity and ease of operation, except when high temperatures or highly corrosion-resistant materials are needed. *See* CHEMICAL SEPARATION TECHNIQUES; DISTILLATION; SUBLIMATION; VAPOR PRESSURE.

[LOUIS GORDON/ROYCE W. MURRAY]

Volumetric analysis

Volumetric analysis, or possibly more correctly titrimetric analysis, is one of the major divisions of quantitative analytical methods. In this method, as the names imply, a titration is made and the amount of the desired component is determined indirectly from the volume and concentration of solution used in the titration.

In most volumetric procedures, the sample to be analyzed is weighed, or an accurately measured volume of a solution of the sample is taken. The sample is dissolved in an appropriate solvent and subjected to some preliminary treatment or separation if necessary, and then the titration is performed on the resulting solution.

During the titration, a chemical reaction occurs between a constituent of the titrant solution and the substance to be determined. In order to be suitable for volumetric titrations, these reactions must be rapid and quantitative. That is, it is necessary to know exactly how many molecules or ions of the substance to be determined react with each molecule or ion of the reagent in the titrant. In volumetric analysis, it is also necessary to determine when the number of reagent ions or molecules equivalent to the substance being determined has been added. This point in the titration is called the equivalence point. The determination of the amount of titrant corresponding to the equivalence point is usually done either by visual means with colored indicators or reagents, or by some physicochemical measurement. When an indicator changes color, the end point of the titration is reached, and at that point, the volume of titrant solution added is measured. Not always, however, are the end point and equivalence point of the titration identical. An indicator, for example, may consume a small amount of titrant before it will change color, or the indicator may not change color until a concentration of reagent considerably higher than that corresponding to the equivalence point is present in solution. In either of these cases, the point where the indicator changes color will not be the same as the true equivalence point. For this reason, it is very essential that the choice of indicator or the method of detecting the end point in a titration be considered carefully in order to minimize the discrepancy between the detected end point and the true equivalence point.

In order to calculate the amount of substance equivalent to the volume of titrant used in the titration, it is necessary to know accurately the concentration of reagent in the titrant. This concentration is usually determined by weighing the amount of reagent added to a definite volume of titrant solution, by titrating a weight of sample of known composition with the titrant, or by titrating a known volume of another solution of known composition. From the volume of titrant used to reach the end point and the concentration of reagent in the titrant, the weight of substance to be determined in the sample is then calculated.

In volumetric analysis, the volume of titrant is

usually measured, although for the most accurate work, the weight of titrant is measured. Both of these procedures fall into the classification of titrimetric analysis. *See* PIPET; QUANTITATIVE CHEMICAL ANALYSIS; SOLUBILIZING OF SAMPLES; STOICHIOMETRY; TITRATION; VOLUMETRIC FLASK.

[CLARK E. BRICKER]

Volumetric flask

A long narrow-necked glass flask which has been calibrated to contain a given volume of liquid when filled so that the bottom of the meniscus is tangent to the ring which has been etched on the neck of the flask (see illustration). Some volumetric flasks have a second calibration mark, higher on the neck of the flask, which is used when the flask is needed to deliver a given volume of liquid. The marks etched on volumetric flasks are accurate only at the temperature at which these calibrations were made, because the volume of a flask increases slightly with an increase of temperature and vice versa. Volumetric flasks are used when it is necessary to dilute a weight or a volume of a substance to some accurately known volume. *See* CONCENTRATION SCALES; TITRATION; VOLUMETRIC ANALYSIS.

[CLARK E. BRICKER]

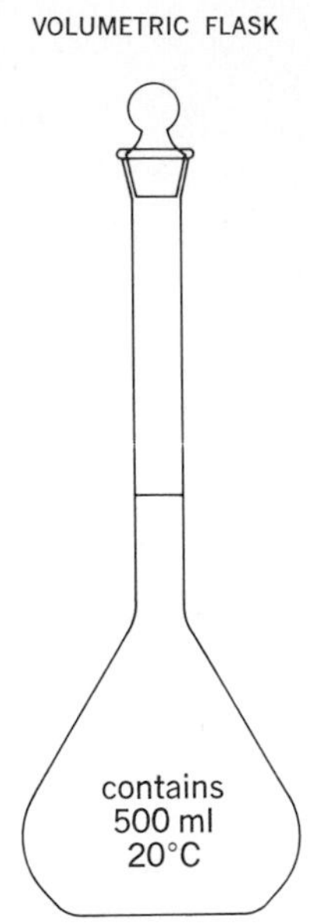

Volumetric flask.

Water

The chemical compound with two atoms of hydrogen and one atom of oxygen in each of its molecules. It is formed by the direct reaction (1) of

$$2H_2 + O_2 \rightarrow 2H_2O \quad (1)$$

hydrogen with oxygen. The other compound of hydrogen and oxygen, hydrogen peroxide, readily decomposes to form water, reaction (2). Water also

$$2H_2O_2 \rightarrow 2H_2O + O_2 \quad (2)$$

is formed in the combustion of hydrogen-containing compounds, in the pyrolysis of hydrates, and in animal metabolism. Some properties of water are given in the table.

Properties of water

Property	Value
Freezing point	0°C
Density of ice, 0°C	0.92 g/cm^3
Density of water, 0°C	1.00 g/cm^3
Heat of fusion	80 cal/g (335 J/g)
Boiling point	100°C
Heat of vaporization	540 cal/g (2260 J/g)
Critical temperature	347°C
Critical pressure	217 atm (22.0 MPa)
Specific electrical conductivity at 25°C	1×10^{-7}/ohm-cm
Dielectric constant, 25°C	78

In the gaseous state. Water vapor consists of water molecules which move nearly independently of each other. The relative positions of the atoms in a water molecule are shown in Fig. 1. The dotted circles show the effective sizes of the isolated atoms. The atoms are held together in the molecule by chemical bonds which are very polar, the hydrogen end of each bond being electrically positive relative to the oxygen. When two molecules near each other are suitably oriented, the positive hydrogen of one molecule attracts the negative oxygen of the other, and while in this orientation, the repulsion of the like charges is comparatively small. The net attraction is strong enough to hold the molecules together in many circumstances and is called a hydrogen bond. *See* CHEMICAL BONDING; ELECTRONEGATIVITY; GAS; VALENCE.

When heated above 1200°C, water vapor dissociates appreciably to form hydrogen atoms and hydroxyl free radicals, reaction (3). These products

$$H_2O \rightarrow H + OH \quad (3)$$

recombine completely to form water when the temperature is lowered. Water vapor also undergoes most of the chemical reactions of liquid water and, at very high concentrations, even shows some of the unusual solvent properties of liquid water. Above 374°C, water vapor may be compressed to any density without liquefying, and at a density as high as 0.4 g/cm^3, it can dissolve appreciable quantities of salt. These conditions of high temperature and pressure are found in efficient steam power plants. *See* CRITICAL CONSTANT; HYDROGEN BOND.

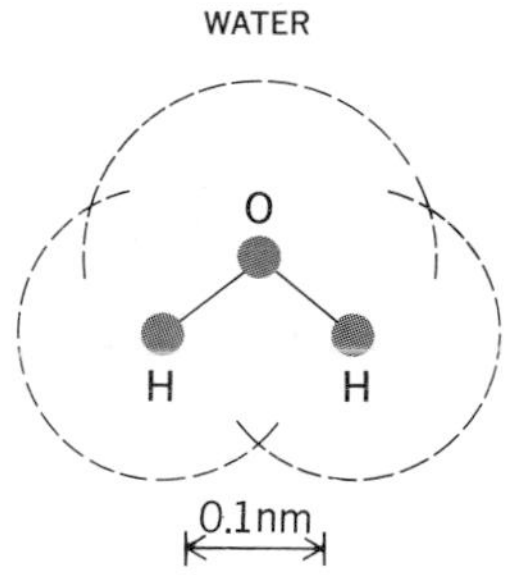

Fig. 1. The water molecule.

In the solid state. Ordinary ice consists of water molecules joined together by hydrogen bonds in a regular arrangement, as shown in Fig. 2. The circles represent only the positions of the atoms, but if the sizes, as indicated in Fig. 1, are superimposed upon the figure, then it appears that there is considerable empty space between the molecules. This unusual feature is a result of the strong and directional hydrogen bonds taking precedence over all other intermolecular forces in determining the structure of the crystal. If the water molecules were rearranged to reduce the amount of empty space, their relative orientations would no longer be so well suited for hydrogen bonds. This rearrangement can be produced by compressing ice to pressures in excess of 2000 atm (14 MPa). Altogether five different crystalline forms of solid water have been produced in this way, the form obtained depending upon the final pressure and tempera-

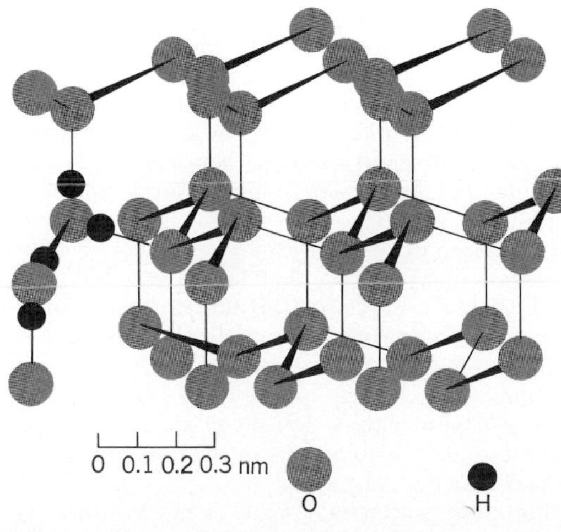

Fig. 2. The structure of ice. The hydrogen atoms are omitted for all but two water molecules.

ture. They are all more dense than water, and all revert to ordinary ice when the pressure is reduced. *See* CRYSTAL STRUCTURE.

In the liquid state. The molecules in liquid water also are held together by hydrogen bonds. When ice melts, many of the hydrogen bonds are broken, and those that remain are not numerous enough to keep the molecules in a regular arrangement. Many of the unusual properties of liquid water may be understood in terms of the hydrogen bonds which remain. As water is heated from 0°C, it contracts until 4° is reached and then begins the expansion which is normally associated with increasing temperature. This phenomenon and the increase in density when ice melts both result from a breaking down of the open, hydrogen-bonded structure as the temperature is raised. The viscosity of water decreases tenfold as the temperature is raised from 0° to 100°C, and this also is associated with the decrease of icelike character in the water as the hydrogen bonds are disrupted by increasing thermal agitation. Even at 100°C, the hydrogen bonds influence the properties of water strongly, for it has a high boiling point and a high heat of vaporization compared with other substances of similar molecular weight. *See* LIQUID.

The electrical conductivity of water is at least 1,000,000 times larger than that of most other nonmetallic liquids at room temperature. The current in this case is carried by ions produced by the dissociation of water according to reaction (4). This

$$H_2O \rightleftharpoons H^+ + OH^- \quad (4)$$

reaction is reversible, and equilibrium is reached rapidly, so there is a definite concentration of H^+ and OH^- ions in pure water. At 25°C, this concentration is 10^{-7} mole/liter of each species or about 10^{14} ions/ml. This concentration of ions is affected by the temperature or by the presence of solutes in the water. *See* ACID AND BASE; HYDROGEN ION.

Pure water, either solid or liquid, is blue if viewed through a thickness of more than 2 m. The other colors often observed are due to impurities.

Solutions in water. Water is an excellent solvent for many substances, but particularly for those which dissociate to form ions. Its principal scientific and industrial use as a solvent is to furnish a medium for purifying such substances and for carrying out reactions between them. *See* SOLUTION; SOLVENT.

Among the substances which dissolve in water with little or no ionization and which are very soluble are ethanol and ammonia. These are examples of molecules which are able to form hydrogen bonds with water molecules, although, except for the hydrogen of the OH group in ethanol, it is the hydrogen of the water that makes the hydrogen bond. On the other hand, substances which cannot interact strongly with water, either by ionization or by hydrogen bonding, are only sparingly soluble in it. Examples of such substances are benzene, mercury, and phosphorus. For discussions of another important class of solutions in water *see* COLLOID; SURFACE-ACTIVE AGENT.

Chemical properties. Water is not a strong oxidizing agent, although it may enhance the oxidizing action of other oxidizing agents, notably oxygen. Examples of the oxidizing action of water itself are its reactions with the alkali and alkaline-earth metals, even in the cold; for instance reaction (5), and its reactions with iron and carbon at

$$Ca + 2H_2O \rightarrow Ca^{++} + 2OH^- + H_2 \quad (5)$$

elevated temperatures, reactions (6) and (7). Reac-

$$3Fe + 4H_2O \rightarrow Fe_3O_4 + 4H_2 \quad (6)$$

$$C + H_2O \rightarrow CO + H_2 \quad (7)$$

tion (7) is used commercially to produce a gaseous fuel from solid coke. The gaseous mixture, $CO + H_2$, called water gas, is formed when steam is passed over coke heated to 600°C.

Water is an even poorer reducing agent than oxidizing agent. One of the few substances that it reduces rapidly is fluorine, but this reaction is complicated. Chlorine is reduced only very slowly in the cold, according to reaction (8).

$$2Cl_2 + 2H_2O \rightarrow O_2 + 4H^+ + 4Cl^- \quad (8)$$

An example of another sort of oxidation-reduction reaction in which water plays an essential role beyond that of the solvent is the disproportionation of chlorine, reaction (9), which is fast and incom-

$$Cl_2 + H_2O \rightarrow HOCl + H^+ + Cl^- \quad (9)$$

plete in neutral solution but goes to completion if base is added. *See* OXIDATION-REDUCTION.

Substances with strong acidic or basic character react with water. For example, calcium oxide, a basic oxide, reacts in a process called the slaking of lime, reaction (10). Sulfur trioxide, an acidic

$$CaO + H_2O \rightarrow Ca(OH)_2 \quad (10)$$

oxide, also reacts, reaction (11). This reaction oc-

$$SO_3 + H_2O \rightarrow H_2SO_4 \quad (11)$$

curs in the contact process for the manufacture of sulfuric acid. Both of these reactions evolve enough heat to produce fires or explosions unless precautions are taken.

Another type of substance with strong acidic character is an acid chloride. Two examples and their reactions with water are boron trichloride, reaction (12), and acetyl chloride, reaction (13).

$$BCl_3 + 3H_2O \rightarrow H_3BO_3 + 3HCl \quad (12)$$

$$CH_3COCl + H_2O \rightarrow CH_3CO_2H + HCl \quad (13)$$

These are often termed hydrolysis reactions, as is the reaction of an ester with water, for instance ethyl acetate, reaction (14). A hydrolysis reaction

$$H_3C\overset{\overset{\displaystyle O}{\|}}{C}OC_2H_5 + H_2O \rightarrow H_3C\overset{\overset{\displaystyle O}{\|}}{C}OH + HOC_2H_5 \quad (14)$$

of a different sort is that of calcium carbide, used in the production of acetylene, reaction (15). *See* HYDROLYSIS.

$$CaC_2 + 2H_2O \rightarrow Ca(OH)_2 + C_2H_2 \quad (15)$$

Water reacts with a variety of substances to form solid compounds in which the water molecule is intact, but in which it becomes a part of the structure of the solid. Such compounds are called hydrates, and are formed frequently with the evolution of considerable amounts of heat. Examples range from the hydrates of simple and double salts, calcium chloride hexahydrate, $CaCl_2 \cdot 6H_2O$, and ammonium aluminum alum, $NH_4Al(SO_4)_2 \cdot 12H_2O$, to the gas hydrates which are stable only at low temperatures, for example, chlorine hydrate, $Cl_2 \cdot 6H_2O$, and xenon hydrate, $Xe \cdot 6H_2O$. *See* CLATHRATE COMPOUNDS; HEAVY WATER; HYDRATE; HYDROGEN; OXYGEN; TRIPLE POINT; VAPOR PRESSURE; WATER SOFTENING.

[HAROLD L. FRIEDMAN]

Water softening

A water-treatment process by which undesirable cations (of calcium and magnesium) are removed from hard waters. The presence of these cations in water is undesirable for household purposes, boiler feed, food processing, and chemical processing, because of reactions that form soap scum, boiler scale, and unwanted by-products.

Hard waters may be softened by precipitation processes, cation-exchange processes, or combinations of these. The choice of a process depends on a number of factors, among which are the composition of the hard water, the end uses of the softened water, the type or types of hardnesses to be removed, the degree of removal required, and the relative processing costs. In the following discussion, hardness is expressed as calcium carbonate equivalents in parts per million (ppm), and lime refers to hydrated lime.

Cold lime-soda processes. The equipment used in these processes (Fig. 1) consists of chemical feeders, a softener unit, and usually filters. Chemicals used are lime (or lime and soda ash) plus a coagulant. Precipitates produced are calcium carbonate and magnesium hydroxide. Lime may be used to reduce the calcium carbonate hardness to about 35 ppm or to reduce both calcium and magnesium carbonate hardness to about 70 ppm. Lime and soda ash will reduce total hardness (carbonate and noncarbonate) to about 70 ppm without using excess chemicals. With excess chemicals, total hardness may be reduced to 16 ppm.

The cold lime-soda process may be combined with the zeolite process in a two-state operation in which the filtered effluent from a cold lime-soda softener is passed through a sodium cation-exchanger (zeolite) water softener. Either lime or lime and soda ash may be used in the first stage. In either case, the residual hardness from most hard waters is reduced to practically "zero" hardness (1 – 2 ppm) by the second stage.

Hot lime-soda process. The equipment used in this process (Fig. 2) consists of one or more chemical feeders, a softener unit, and filters. In addition, an integral of separate deaerator may be employed. Chemicals used are lime and soda ash. Heating is accomplished with steam (usually at 5 – 10 psig), and liberated gases are vented to the atmosphere. Precipitates produced are calcium carbonate and magnesium hydroxide. The latter reduces the silica content, and if insufficient in quantity, dolomitic lime or activated magnesia may be used instead of lime. A slight excess of soda ash (about 30 ppm) is usually employed, and with this, the total hardness is reduced to less than 25 ppm.

Hot lime-soda and phosphate process. In this two-stage process, the settled water from a hot lime-soda softener is treated with sodium phosphate (either in an integral compartment or in a separate settling tank), which precipitates the residual hardness as calcium and magnesium phosphates. The settled water from this second stage is then filtered. Total hardness of final effluent is "zero" (1 – 2 ppm).

Zeolite process. The equipment used in this process (Fig. 3) consists of one or more sodium cation-exchanger (zeolite) water softener units and, in most industrial water softeners, a brine-measuring tank and a wet-salt storage tank or basin. Hard waters are softened by their passage, usually downward, through a columnar bed of a granular or bead-type sodium cation exchanger which removes the calcium and magnesium cations from the water by exchanging for them an equivalent amount of sodium cations. At the end of the softening run, the softener unit is cut out of service and the bed is (1) backwashed; (2) regenerated with a solution of common salt (NaCl), which removes the calcium and magnesium cations by exchanging sodium cations for them, thus restoring the bed to

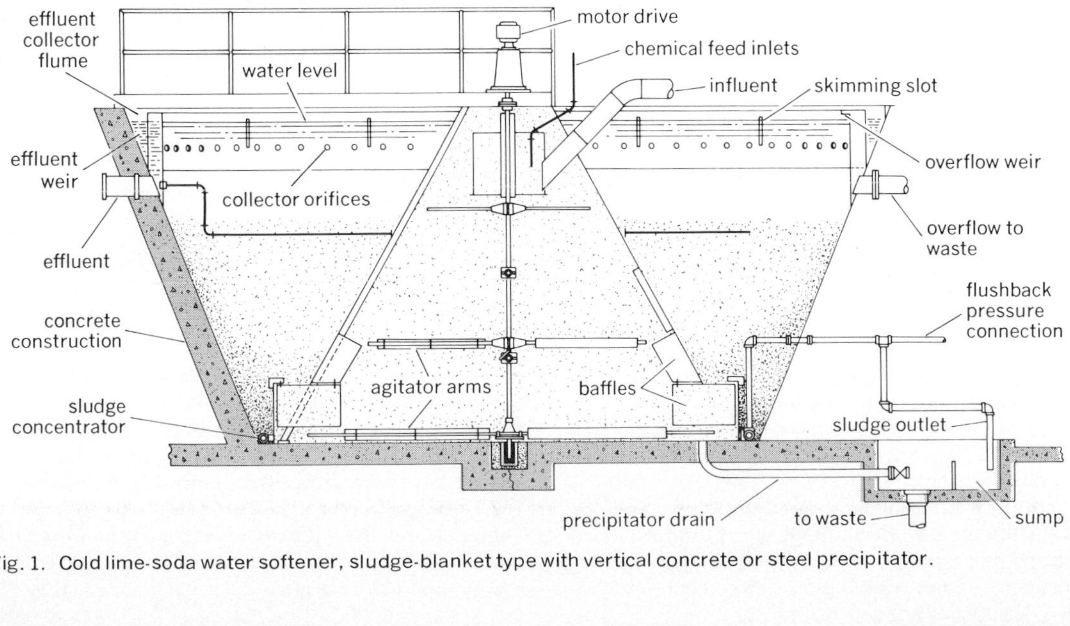

Fig. 1. Cold lime-soda water softener, sludge-blanket type with vertical concrete or steel precipitator.

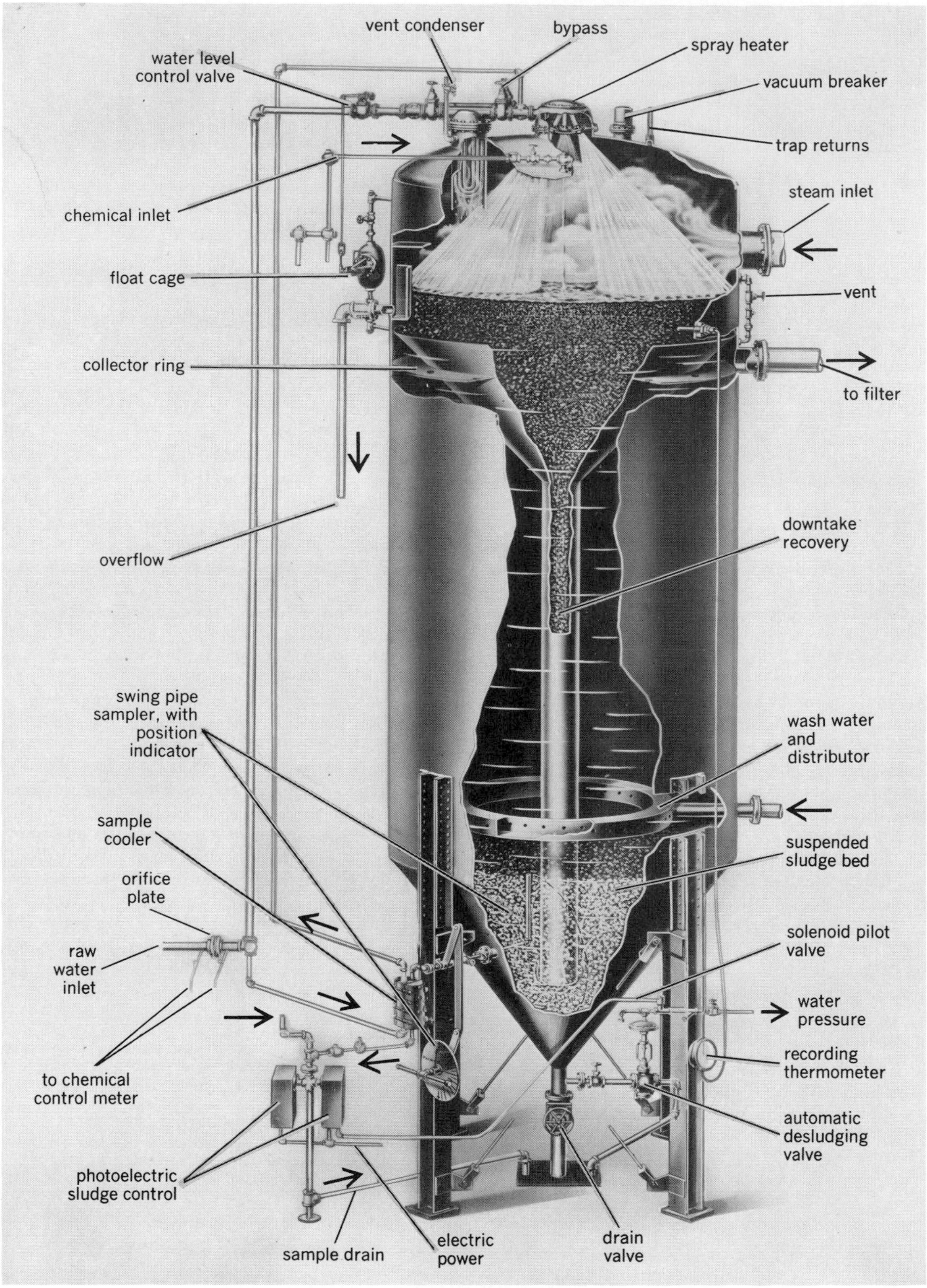

Fig. 2. Hot lime-soda water softener, sludge-blanket type.

its original sodium state; and (3) rinsed free from calcium and magnesium chlorides and excess salt, after which the unit is returned to service. With most hard waters, the hardness is reduced to practically "zero." Waters very high in hardness or sodium salts are not so completely softened.

Hydrogen cation-exchanger process. The equipment for this process consists of one or more hydrogen cation-exchanger units containing a cation-exchange resin in bead form, a dilute acid tank, and a strong acid tank. Operations are similar to those of the above process except that calcium, magnesium, and sodium cations are removed by exchanging hydrogen cations for them; regeneration is effected with a dilute acid, usually sulfuric; and the effluent contains carbon dioxide and a mineral acid content equivalent to that of the sulfate and chloride ions of the hard water. Most of

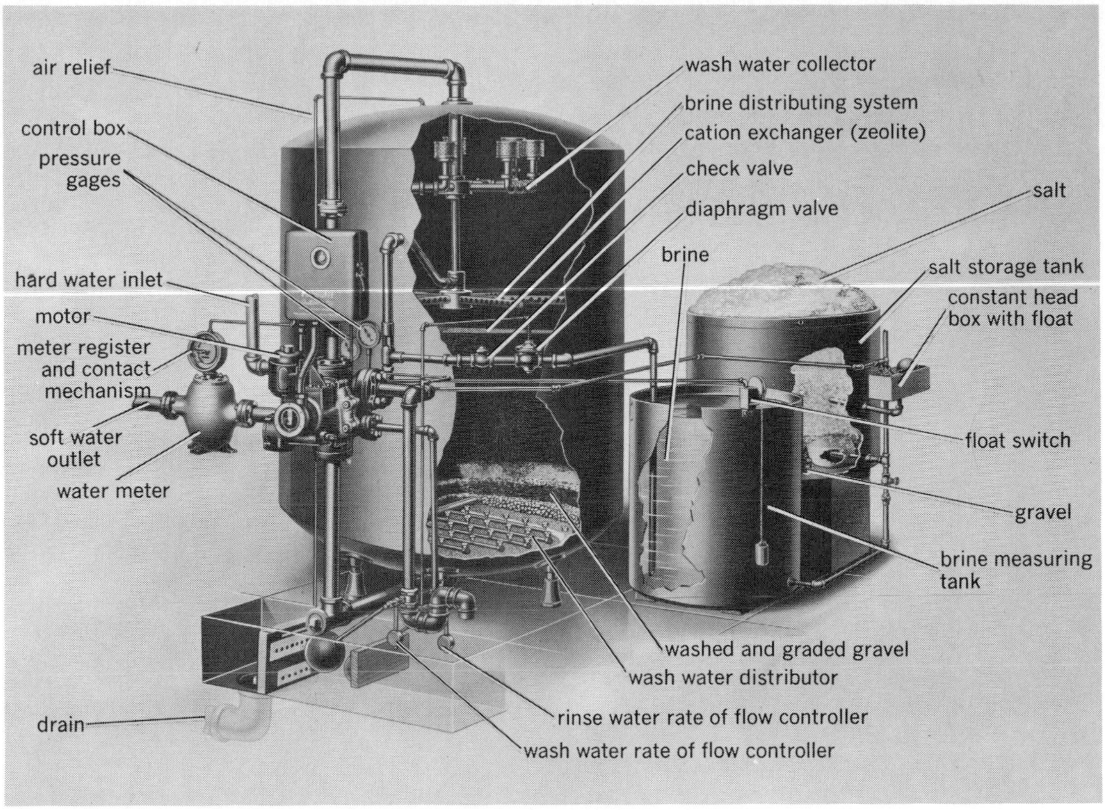

Fig. 3. Sodium cation-exchanger (zeolite) water softener, automatic type.

the carbon dioxide may be removed by aeration or degasification, and the acids may be neutralized with the effluent from a sodium cation exchanger, neutralized with caustic soda, or removed by an anion exchanger. Hardness removal is practically to "zero." *See* ION EXCHANGE; SALINE WATER RECLAMATION; SURFACE-ACTIVE AGENT.

[ESKEL NORDELL]

Weight

The gravitational weight of a body is the force with which the Earth attracts the body. By extension, the term is also used for the attraction of the Sun or a planet on a nearby body. This force is proportional to the body's mass and depends on the location. Because the distance from the surface to the center of the Earth decreases at higher latitudes, and because the centrifugal force of the Earth's rotation is greatest at the Equator, the observed weight of a body is smallest at the Equator and largest at the poles. The difference is sizable, about 1 part in 300. At a given location, the weight of a body is highest at the surface of the Earth; it diminishes with altitude and with the depth below the surface. For example, the weight of a body diminishes by about 0.1% if it is raised 2 mi above the Earth's surface or taken 4 mi below the surface. Weight also depends to a smaller but measurable degree on the density of the Earth's crust below the body. Weight is measured by several procedures. *See* WEIGHT MEASUREMENT.

Since weight is a force, it is expressed in force units. In the United States, the commonest unit of weight is the pound, sometimes written pound force or pound weight, to distinguish it from the mass unit, pound. Pound weight is the weight of a 1-pound mass at a location where the acceleration of gravity is 32.174 ft/sec^2. Where the acceleration of gravity is g, the weight of a 1-pound mass is $g/32.174$ pound weight.

In terms of the international kilogram the pound avoirdupois is now defined as being exactly equivalent to 0.45359237 kilogram. In relation to three smaller avoirdupois weight units, 1 pound equals 16 ounces, 256 drams, and 7000 grains. Since the relation of the pound to the kilogram is divisible by 7, 1 grain equals 0.06479891 gram, exactly.

Besides the avoirdupois pound there is the troy or apothecary pound, in which there are 5760 grains, so that this pound is equal to 576/700 avoirdupois pound. In general, the term pound is taken to mean the avoirdupois pound unless definitely stated otherwise.

The carat is a unit of weight used in evaluating precious stones, and 1 carat equals 200 milligrams.

[HOWARD S. BEAN]

Bibliography: A. V. Astin, *Refinement of Values for the Yard and Pound*, Fed. Regist. Doc. no. 59-5442, July 1, 1959; U.S. Department of Commerce, National Bureau of Standards, *Units of Weight and Measure: Definitions and Tables of Equivalents*, Misc. Publ. no. 286, May, 1967.

Weight measurement

Weight is the resultant force acting on a mass (in a vacuum) due to the Earth's gravitational field corrected for the effect of the Earth's rotation. Units of weight are based upon an acceleration of gravity of 980.665 cm/sec^2 or 32.1740 ft/sec^2. When the weight of an unknown is determined by comparison with a known weight, there is no error in the readings due to gravity variations. The varying

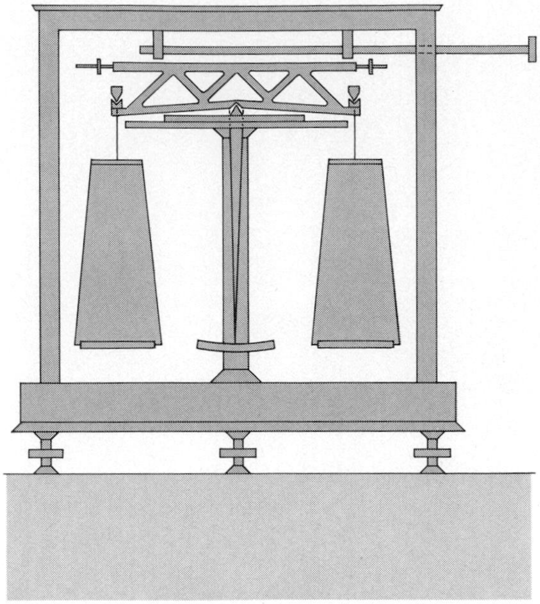

Fig. 1. An equal-arm laboratory balance.

buoyant effect of the atmosphere is negligible when the density of the unknown is approximately the same as that of the standard. In precision weighing, the buoyant effect of air must be considered.

Weight, unlike temperature, pressure, liquid level, and similar variables, is directly related to mass, a basic characteristic of matter. It is widely used as a measure in the transfer of materials, in synthesis and analysis, and in blending and separation. Therefore, it is frequently desirable to measure weight to a greater accuracy than most variables. Weighing devices are designed for a maximum load and a specified accuracy if operated within a given temperature range. They must be rugged enough to withstand service conditions, which may be corrosive and abusive.

Weighing is a dynamic operation, and the time required for a scale or balance to reach an equilibrium condition is important. Mechanical and spring balances are often equipped with fluid dashpots or other devices to critically damp the system, permitting equilibrium to be reached in the minimum time. The more elaborate systems (electronic and pneumatic) have inherently short time con-

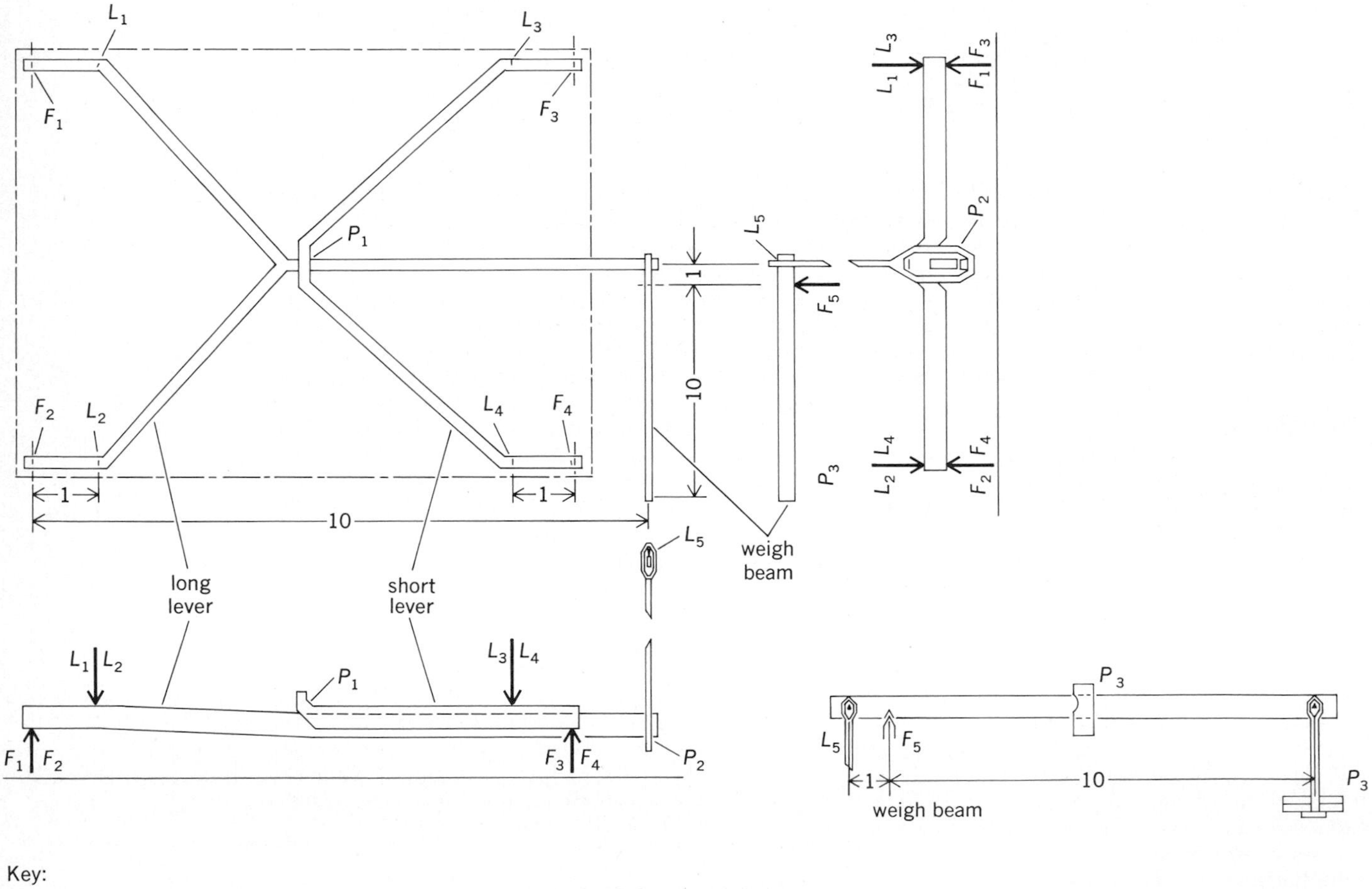

Key:

F_1, F_2 = fulcrums of long lever

F_3, F_4 = fulcrums of short lever

L_1, L_2 = load points of long lever

L_3, L_4 = load points of short lever

P_1 = power point of short lever and secondary load point of long lever

P_2 = power point of long lever

F_5 = fulcrum of weigh beam

L_5 = load point of weigh beam

P_3 = power point of weigh beam and of the combined system

Fig. 2. Lever system of a platform scale. The ratio of ($L_1+L_2+L_3+L_4$) to P_3 is 100:1, as shown by the leverage ratios of the long lever and weigh beam. The multiple-lever system supports the platform so that unequally distributed loads can be accurately weighed.

stants and therefore reach equilibrium quickly. The pneumatic system involves feedback and therefore requires dynamic consideration in the design of the load element to obtain stability. Most weighing devices may be used directly, or with minor modifications, for the measurement of other forces.

Simple balance. The equal-arm balance, shown in Fig. 1, is probably the most common form. These balances are made in many designs and sizes; in some the knife-edge fulcrums are replaced with flexure plates; others have arms of unequal length. Conventionally, the unknown weight is placed on one pan, the known weight on the other. The final securing of a balance is done by adjusting the position of a rider, or small weight, on a bar of the balance arm bridge. The condition of balance is indicated when the pointer swings equal distances from its rest point. Many precision-grade laboratory balances are sensitive to less than 0.01 mg; that is, they detect differences of less than 1 part in 1,000,000. Most general-use balances, whether equal-arm or unequal-arm, provide weighings to between 1 part in 50,000 and 1 part in 100,000. The accuracy, sensitivity, damping characteristics, and load capacity of a balance are determined by its design and physical condition.

Mechanical-type industrial scale. This type incorporates a number of levers with precisely located fulcrums to permit heavy objects to be balanced (weighed) with small, convenient counterweights or counterpoises (Fig. 2). Shaft pivots, knife-edges, cone pivots, and flexures of various types are used at the fulcrum points and are designed to carry maximum load at that point in the lever system with minimum friction. Multiple-lever systems support the platform, or hopper, so that unequally distributed loads can be accurately weighed. Often the counterweights on an industrial scale are separated for the convenience of the operator. The zero-adjusting weight balances the scale with no load on the platform. The tare counterweight, if used, balances the scale carrying an empty container. The load is balanced most often by counterweights of fixed mass at a definite location in the multiple-lever system plus at least one counterweight (poise) which can be moved to increase or reduce its distance from the final pivot. Truck and railway scales frequently use this principle. Although the accuracy of scales of this type varies with the design and physical condition, the commercial accuracy tolerance is 0.1% of the maximum capacity, and the sensitivity of the scale is 0.05%. Somewhat higher accuracy and sensitivity can be achieved by refined designs.

Pendulum-type mechanical scale. This type balances the force of the load by the rotation of a bent lever (Fig. 3). With this construction, the deflection of the load on the scale moves the counterweights through the lever system so that their center of gravity is at a greater distance from the final fulcrum. Thus the increased lever arm of the counterweights automatically balances the load. Normally this movement, through a rack and pinion, moves a pointer on a calibrated dial. Cams are often incorporated in the lever system to linearize the movement of the pointer on the dial. Commercial scales of this type are also accurate to 1 part in 1000.

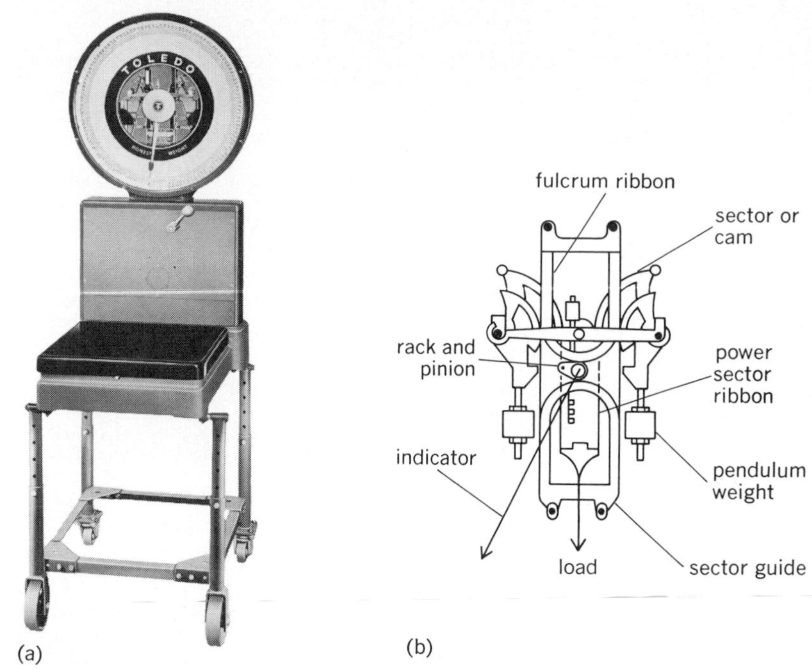

Fig. 3. Pendulum-type mechanical scale. (*a*) Typical commercial model (*Toledo Scale Co.*). (*b*) Schematic detail at pendulum (*from D. M. Considine, ed., Process Instruments and Controls Handbook, McGraw-Hill, 1957*).

Spring scale. This type utilizes the deflection of a spring to measure the load. Two basic designs are illustrated in Fig. 4.

A spring scale, if sensitive enough, can detect and indicate the effect of differences in the weight of a body due to changes in elevation or to variations in distances between the Equator and the poles, as discussed elsewhere. *See* WEIGHT.

Hydraulic systems. In hydraulic systems, such as those in Figs. 5 and 6, the load applied to the load cell piston is converted to hydraulic pressure. The effective area of the piston must be known. The pressure may be measured at a remote point by a pressure-gage, such as a Bourdon tube. Some load cells convert all the force to pressure by means of a slack diaphragm (Fig. 5), which is ac-

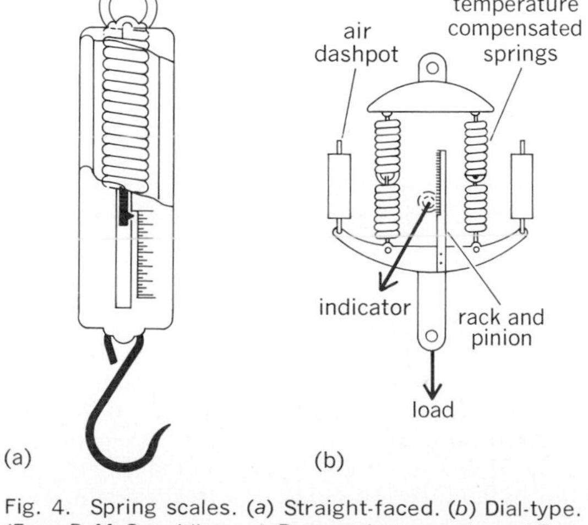

Fig. 4. Spring scales. (*a*) Straight-faced. (*b*) Dial-type. (*From D. M. Considine, ed., Process Instruments and Controls Handbook, McGraw-Hill, 1957*)

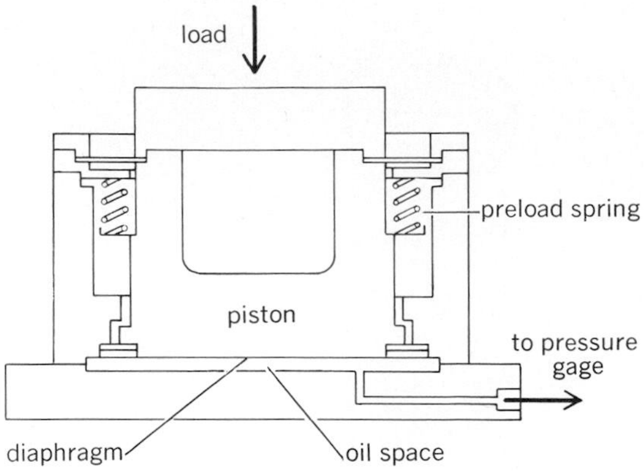

Fig. 5. Hydraulic scale, in which load cell is used. (*From D. M. Considine, ed., Process Instruments and Controls Handbook, McGraw-Hill, 1957*)

curate to 1 part in 400; others carry the bulk of the load directly, using the pressure primarily for transmission purposes (Fig. 6), and are accurate to 1%. Hydraulic weighing systems are temperature-sensitive; while they may be compensated, it is customary to provide a tolerance for temperature variations.

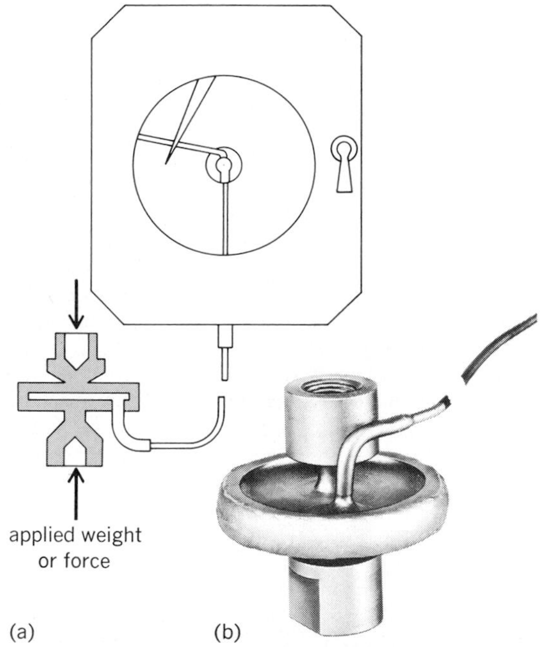

Fig. 6. Volumetric load element. (*a*) Hydraulic system using volumetric load cell. (*b*) Commercial volumetric load element. (*Taylor Instrument Co.*)

Pneumatic systems. Pneumatic systems, such as in Fig. 7, detect the load by a sensitive nozzle and flapper system and balance the load by modulating an air pressure in an opposing capsule. The effective area of the capsule must remain constant; this is accomplished by minimizing the motion of the platform so that the system approaches a null balance. Maximum errors as low as 0.25% are possible at a constant operating temperature.

Electrical weighing systems. These systems, such as in Fig. 8, usually involve the electrical measurement of the elastic deformation of a mechanical element under stress. The strain gage is attached to the weighing element in a manner to produce the maximum resistance change per unit of load. The change in resistance with load is measured and amplified by electronic means, and the load is read on a potentiometer. Usually the elements of the strain gage are the arms of an ac bridge circuit, and the circuit is carefully designed to minimize ambient-temperature and supply-voltage effects, as well as zero drift. The output from two or more strain gage units can be combined

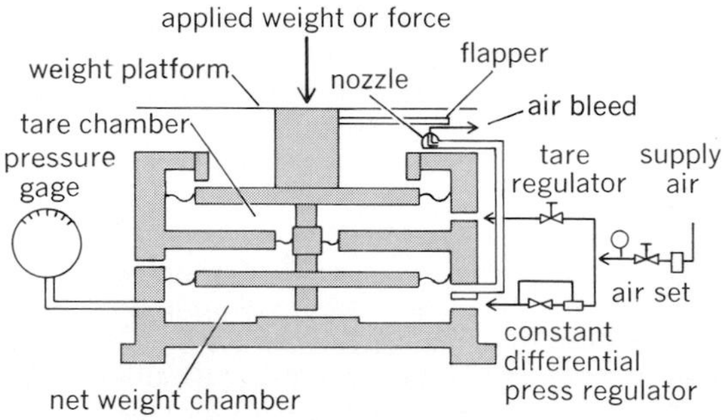

Fig. 7. Pneumatic weighing system. (*From D. M. Considine, ed., Process Instruments and Controls Handbook, McGraw-Hill, 1957*)

electronically to give an aggregate load. This type of unit is adaptable to weighing very large loads. Normal commercial tolerance is about 0.25% at a given temperature, and better accuracy may be achieved through special design.

Other weighers. Many special weighers have been developed to meet the demands of industry. The net weigher is used to weigh products accurately and rapidly for package filling. The check-weigher is used to weigh and divert or accept packages which have been filled. The continuous strip weigher and the continuous product feeder provide continuous measurement of a moving product. Such weighers must have good dynamic characteristics (low mass, high stiffness, short time constant). The maximum error of these weighing devices varies between 1 and 0.25%.

Scales may be of the indicating type (dial or dig-

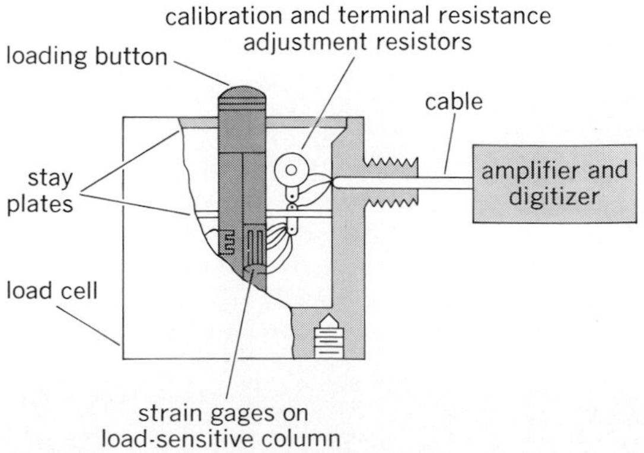

Fig. 8. Electronic scale.

it) or of the recording type (chart or printed tape), or they may be equipped with controlling features such as cutoffs, continuous automatic feeders, and alarms. Scales are also classified by their use, for example, druggists' scales, crane scales, platform scales, or railway scales. Scales are also available for other specialized services, such as counting or inspecting by weight. [HOWARD S. BEAN]

Bibliography: American Society of Mechanical Engineers, *Weighing Scales, Instruments and Apparatus*, pt. 5, 1964; H. Colijn, *Weighing and Proportioning of Bulk Solids*, 1975; *Weighing Machines*, vols. 1 and 2 by T. J. Metcalfe, vol. 3 by E. H. Griffiths, 1970.

Woodward-Hoffmann rule

A concept which can predict or explain the stereochemistry of certain types of reactions in organic chemistry. It is also described as the conservation of orbital symmetry, and is named for its developers, R. B. Woodward and Roald Hoffmann. The rule applies to a limited group of reactions, called pericyclic, which are characterized by being more or less concerted (that is, one-step, without a distinct intermediate between reactants and products) and having a cyclic arrangement of the reacting atoms of the molecule in the transition state. Most pericyclic reactions fall into one of three major classes, examples of which will illustrate the use of the rule.

Electrocyclic reactions. These reactions are defined as the interconversion of a linear π-system, containing n π-electrons, and a cyclic molecule containing $n-2$ π-electrons which is formed by joining the ends of the linear molecule. This is exemplified by the thermal ring opening of a cyclobutene to a butadiene, reaction (1).

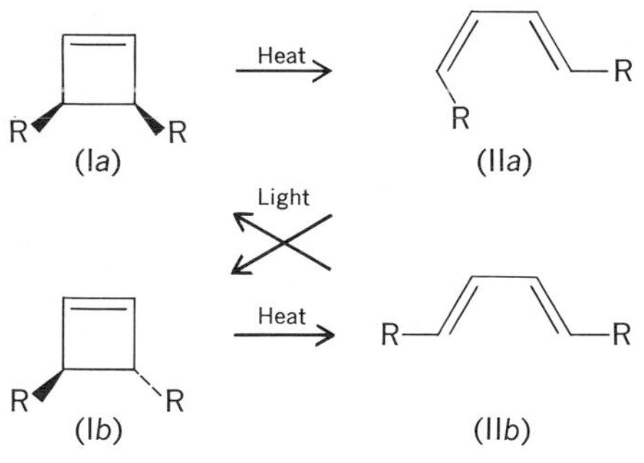

(1)

The reaction is stereospecific in that (I*a*) (R substituents *cis*) gives only (II*a*) and none of (II*b*). Conversely, cyclobutene (I*b*) gives (II*b*) but not (II*a*). This mode of reaction is termed conrotatory, since both R groups rotate in the same direction (clockwise or counterclockwise when viewed edge-on) in going from reactant to product. On irradiation with ultraviolet light the butadienes recyclize to cyclobutenes. This reaction is disrotatory since (II*a*) → (I*b*) and (II*b*) → (I*a*). Similar stereospecificity is observed, in reactions (2) and (3), in the cyclization of substituted hexatrienes (III) and octatetraenes (V) to cyclohexadienes (IV) and cyclo-ocatatrienes (VI), respectively.

(III) (IV) (2)

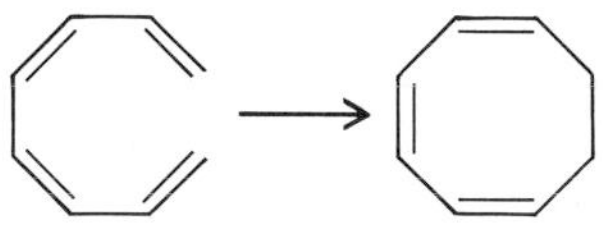

(V) (VI) (3)

To see how these results are explained requires a knowledge of the nature, specifically the symmetry, of the molecular orbitals most involved in the reaction. The relevant occupied orbitals of butadiene and cyclobutene are represented in Figs. 1 and 2. The signs (+ or −) in the orbitals indicate the phase of the wave function in that region of space. *See* MOLECULAR ORBITAL THEORY.

In Fig. 1 the orbitals are labeled S or A according to their symmetry with respect to rotation by 180° about an axis in the plane of the molecules. This element of symmetry is maintained throughout the conrotatory cyclization. Similarly, Fig. 2 shows the same orbitals and their symmetry with respect to reflection through the plane which bisects the molecules as shown. The molecule retains this element of symmetry in its disrotatory cyclization. The fact that in the conrotatory mode of cyclization both butadiene and cyclobutene have the same number of symmetric occupied orbitals (one each) shows that the reaction can take place; that is, it is symmetry-allowed. In the disrotatory mode there are two occupied symmetric orbitals in cyclobutene, but only one in butadiene. Thus in the latter path, orbital symme-

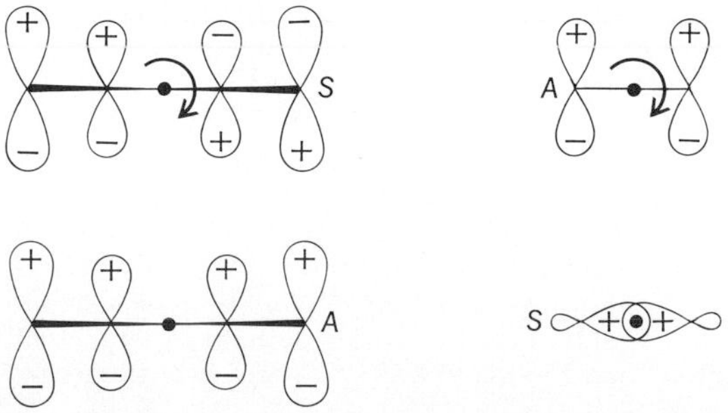

Fig. 1. Conrotatory cyclization: axial symmetry.

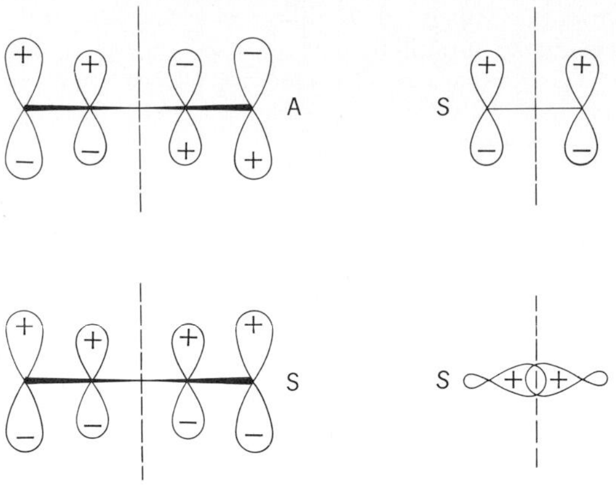

Fig. 2. Disrotatory cyclization: planar symmetry.

try is not conserved, and the reaction is said to be symmetry-forbidden.

The general rules for electrocyclic reactions of polyenes with n π-electrons are summarized in the table, where q is an integer.

Electrocyclic reactions of polyenes

n	Thermal reaction	Photochemical reaction
$4q$ (4,8, . . .)	Conrotatory	Disrotatory
$4q+2$ (2,6, . . .)	Disrotatory	Conrotatory

Cycloaddition reactions. The simplest example of this type of pericyclic reaction is the combination of two unsaturated molecules, ends to ends, to form a cyclic molecule with four fewer π-electrons, for example, reactions (4) and (5). When these reactions are concerted, the Woodward-Hoffmann rule correctly predicts the experimentally observed stereochemistry. A generalization which can be made is that cycloadditions of two or more unsaturated molecules are most facile when there are a total of $4q+2$ π-electrons ($q > 0$). Cycloadditions (or the reverse reactions) involving $4q$ electrons ($q > 0$) are symmetry-allowed, but the stereochemistry of the required transition state is usually sufficiently strained that the reaction proceeds in two steps.

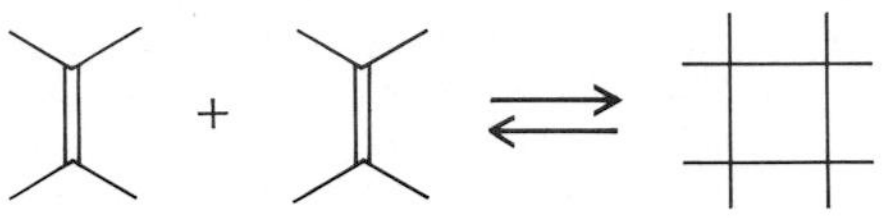
(4)

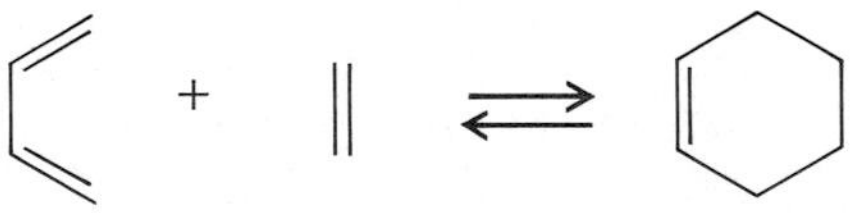
(5)

Sigmatropic reactions. This type of reaction is best defined by reference to the examples shown in reactions (6)–(8). Note that in each reaction there is no overall change in the number of σ- or π-bonds. The name sigmatropic is derived from the fact that there is a change in the location of the σ-bond connecting the upper and lower fragments of the molecule. The individual reactions are classified in terms of the number of atoms from each fragment in the cyclic transition state. Thus the above examples are sigmatropic reactions of order [1,3], [1,5], and [3,3], respectively. All of the above and many other such reactions are known, and in each case the Woodward-Hoffmann rule predicts the experimentally observed stereochemistry.

(6)

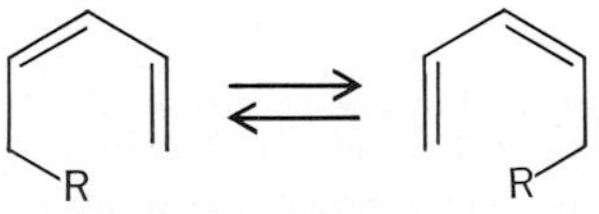

(7)

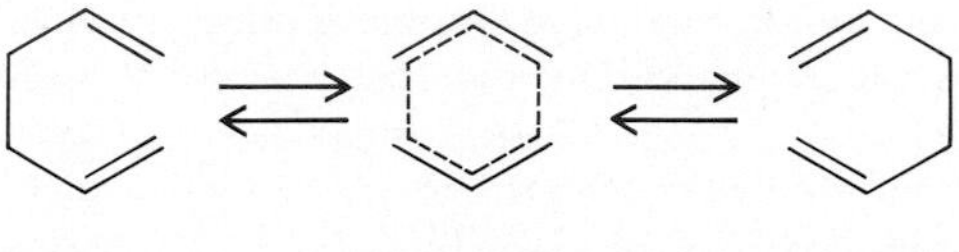
(8)

There are several other miscellaneous pericyclic reactions to which orbital symmetry theory applies. Among these are the cheletropic reactions, for example, (9) and the ene reaction (10). Although the examples have shown simple hydrocarbons as reactants and products, the rules apply also to molecules containing atoms other than carbon and hydrogen. A methylene group may be replaced by an oxygen or nitrogen atom, for example, and the conclusions from orbital symmetry are unaltered.

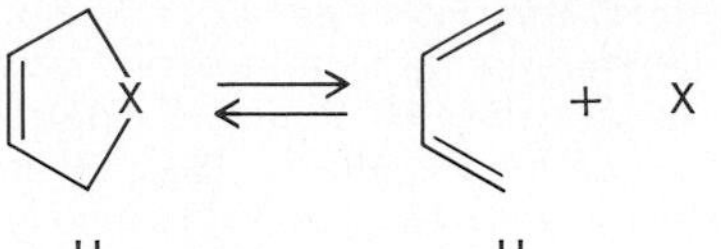

(9)

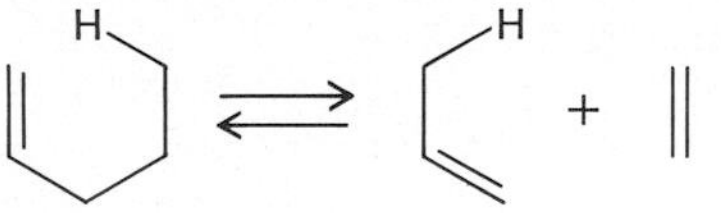

(10)

It should be recognized that the primary explanation for the occurrence of any chemical reaction is found in either the strengths of the reactants' or the products' bonds or both. If the reaction is pericyclic, the conservation of orbital symmetry can affect the rate of the reaction. Thus if two reactions or two mechanisms for the same reaction are possible, the one which is symmetry-allowed will be much faster than the one which is symmetry-forbidden. If, for reasons unrelated to orbital symmetry, no symmetry-allowed reaction is possible for an apparently pericyclic reaction, the reaction may be observed to proceed via a forbidden path. Usually, however, it is found on careful study that the reaction avoids orbital symmetry control by proceeding in two steps via a noncyclic intermediate; that is, it is not pericyclic. *See* ORGANIC REACTION MECHANISM; STEREOCHEMISTRY.

[DAVID L. DALRYMPLE]

Bibliography: T. L. Gilchrist and R. C. Storr, *Organic Reactions and Orbital Symmetry*, 2d ed., 1979; R. Hoffmann and R. B. Woodward, The conservation of orbital symmetry, *Account. Chem. Res.*, 1:17, 1968; R. B. Woodward and R. Hoffmann, *The Conservation of Orbital Symmetry*, 1970.

Work function

A thermodynamic function, also called the work content, Helmholtz free energy, or by the European school, simply the free energy. It is defined as the internal energy E of a system minus the temperature-entropy product, TS, and has a characteristic value for each state of a system. In an isothermal process, the maximum work which can be done by a system is equal to the decrease in its work function. When a process such as a chemical reaction occurs spontaneously at constant temperature and volume, it is characterized by a decrease in the work function. Consequently, the criterion for equilibrium under these conditions is that the work function for the system should be at a minimum. *See* FREE ENERGY; CHEMICAL THERMODYNAMICS.

[PAUL BENDER; WILLIAM A. STEELE]

Wurtz-Fittig reaction

A modified Wurtz reaction in which an aromatic (aryl) halide reacts with an alkyl halide in the presence of sodium and an anhydrous solvent to form alkylated aromatic hydrocarbons. Equimolecular quantities—for example, of bromobenzene and butyl bromide, as in reaction (1), —are reacted at

$$C_6H_5Br + CH_3CH_2CH_2CH_2Br + 2Na \xrightarrow{\text{Dry solvent}} C_6H_5CH_2CH_2CH_2CH_3 + 2NaBr \quad (1)$$

room temperature with a slight excess of two equivalents of sodium metal to yield butylbenzene. A 65–70% yield results. By-products are *n*-octane, $CH_3CH_2CH_2CH_2CH_2CH_2CH_2CH_3$, and biphenyl, $C_6H_5—C_6H_5$.

The more general reaction is shown in (2).

$$C_6H_5X + 2Na + XR \xrightarrow[\text{or benzene}]{\text{Dry ether}} C_6H_5R + 2NaX \quad (2)$$

The reaction is most successful when the halides are bromides. Primary alkyl halides give better results than those obtained from secondary halides. Tertiary alkyl halides do not appear to be suitable.

It is characteristic of the Wurtz-Fittig reaction that the entering alkyl group does not undergo rearrangement. *See* AROMATIC HYDROCARBON; BENZENE; HALOGENATED HYDROCARBON.

[CHARLES K. BRADSHER]

X-ray crystallography

The study of crystal structure by x-ray diffraction techniques. The prediction in 1912 by the German physicist Max von Laue that crystals might be employed as natural diffraction gratings in the study of x-rays was experimentally verified in the same year by W. Friedrich and P. Knipping, who obtained diffraction patterns photographically by the so-called Laue method. Almost immediately after (1913), W. Lawrence Bragg not only successfully analyzed the structures of NaCl and KCl by Laue photographs but also developed a simple treatment of x-ray scattering by a crystal (the Bragg law) which proved much easier to apply than the more complicated but equivalent Laue theory of diffraction. The availability of the first x-ray spectrometer, constructed by his father, William H. Bragg, as well as the substitution of monochromatic (single wavelength) rather than polychromatic x-ray radiation, enabled W. Lawrence Bragg to determine a number of simple crystal structures, including those of diamond; zincblende, ZnS; fluorspar, CaF_2; and pyrites, FeF_2. Since the inauguration of x-ray crystallography as a science, x-ray diffraction has become a powerful tool for the investigation of both the structure of crystals and the nature of x-rays. This article is concerned with the former application. For the theoretical and experimental aspects of x-ray diffraction *see* X-RAY DIFFRACTION.

Structurally, a crystal is a three-dimensional periodic arrangement in space of atoms, groups of atoms, or molecules. If the periodicity of this pattern extends throughout a given piece of material, one speaks of a single crystal. The exact structure of any given crystal is determined if the locations of all atoms making up the three-dimensional periodic pattern called the unit cell are known. The very close and periodic arrangement of the atoms in a crystal permits it to act as a diffraction grating for x-rays. W. Lawrence Bragg treated the phenomenon of the interference of x-rays with crystals as if the x-rays were being reflected by successive parallel equidistant planes of atoms in the crystal. His important equation relating the perpendicular spacing d of lattice planes in a crystal, the glancing angel θ of the reflected beam, and the x-ray wavelength λ is $n\lambda = 2d \sin \theta$. This expression provides the basic condition that the difference in path length for waves reflected from successive planes must be an integral number of wavelengths $n\lambda$ in order for the waves reflected from a given set of lattice planes to be in phase with one another. Instead of referring to the nth order of a reflected beam, modern crystallographers customarily redefine Bragg's equation as $\lambda = 2d_{hkl} \sin \theta$, where d_{hkl} represents the perpendicular interplanar distance between adjacent lattice planes having the Miller indices (hkl). According to this viewpoint, any nth-order diffraction maxima for waves reflected from a set of planes (hkl) with spacing d_{hkl} is equivalent to a first-order reflection due to a parallel set of planes (nh, nk, nl) with the perpendicular distance $d_{nh,nk,nl} = d_{hkl}/n$. For a discussion of Bragg's relation *see* X-RAY POWDER METHODS.

For a crystal with known unit-cell size and shape and with arbitrary orientation in a parallel beam of x-rays of known wavelength, several questions must be answered. First, it must be ascertained what plane, if any, is obeying Bragg's law and is reflecting the rays. Second, the direction of the reflected x-ray must be accurately measured. The Bragg law treatment does not permit these questions to be answered readily, and for this reason the concept of a reciprocal lattice model was introduced. *See* CRYSTAL STRUCTURE.

Reciprocal lattice. The development and application of the reciprocal lattice to x-ray crystallography have been primarily the results of work by P. P. Ewald (1913 and 1921), J. D. Bernal (1926), and M. J. Buerger (1935). In general, a reciprocal lattice consists of a three-dimensional array of points which is related to the crystal lattice (commonly called the direct lattice) in that each set of planes (hkl) in the crystal lattice is represented in reciprocal space by a point denoted by the coordinates hkl (without parentheses). Figure 1 illustrates in two dimensions the reciprocal lattice geometrically produced in the following way from a unit crystal cell. An origin is chosen and given the coordinates 000. To every set of parallel planes of Miller index (hkl) in direct space a reciprocal lattice vector B_{hkl}, which is perpendicular to this set of planes, is constructed from the reciprocal lattice origin at a distance inversely proportional to the interplanar spacing d_{hkl}. Thus, the set of parallel direct-lattice planes (402) with an interplanar spacing d_{402} is represented in reciprocal space in Fig. 1 by the point 402. The normal vector B_{402}, which is directed along the perpendicular of the (402) direct-lattice planes from the origin of the reciprocal lattice to the point with coordinates 402, is of length $|B_{402}| = K/d_{402}$, where K is an arbitrary constant. The value of K which simply scales the reciprocal lattice is normally taken as

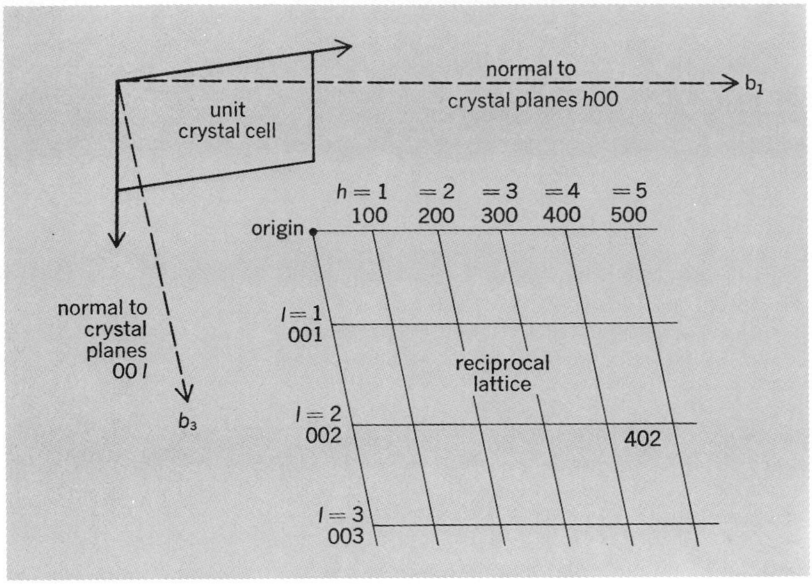

Fig. 1. Reciprocal lattice.

unity or as the wavelength λ; for purposes of this discussion it is taken as 1. All the mathematical definitions of the reciprocal lattice vectors given in the section on reciprocal lattice in the article "X-ray diffraction" in this encyclopedia are based on a K value of unity. It can be shown that this procedure applied to a three-dimensional periodic crystal lattice will result in a three-dimensional reciprocal lattice. Each point of coordinates *hkl* in this reciprocal lattice will be reciprocal to a set of planes (*hkl*) in the direct crystal lattice.

Sphere of reflection. The geometrical interpretation of x-ray diffraction from a crystal can be best interpreted by the use of a "sphere of reflection" in reciprocal space as illustrated in Fig. 2. A sphere of radius $1/\lambda$ (corresponding to $K=1$) is drawn with the direction of the primary x-ray beam (that is, both the incident and transmitted rays) denoted by the unit vector s_0 assumed to travel along a diameter. The crystal C is imagined to be at the center of this sphere of reflection. The origin 000 of the reciprocal lattice is placed at the point 0, where the transmitted beam emerges from the sphere of reflection. A diffracted beam will be formed if the surface of the sphere of reflection intercepts any reciprocal lattice point (other than the origin 0). In Fig. 2 the reciprocal lattice point P lies on the surface of the sphere of reflection. Hence, two important results follow: (1) The set of crystal planes (*hkl*) to which this point is the reciprocal obeys Bragg's law and reflects the incident x-ray beam; (2) the direction of the diffracted rays will be from the point C at the center of the sphere to where the point P lies on the surface of the sphere. The angle of diffraction between the diffracted beam s and the primary beam s_0 is 2θ.

The reciprocal lattice, by virtue of its definition, is tied to the actual lattice insofar as orientation is concerned. As the crystal (and therefore the crystal lattice) is turned about an axis of rotation, the reciprocal lattice, in turning a similar angle about a parallel axis through its origin 0, passes through the sphere of reflection (which is fixed for the primary x-ray beam that is stationary relative to the movement of the crystal). In the rotating crystal method all diffraction maxima that can possibly be recorded by any chosen x-ray wavelength λ must be represented by those points *hkl* of the reciprocal lattice that cut the sphere of reflection. These points of possible diffraction maxima all lie within a sphere of radius $2/\lambda$ called the limiting sphere. This geometrical interpretation of Bragg's law enables one to understand readily both the geometry of reflection of x-rays and the manipulation of all the single-crystal cameras and diffractometers now in use.

Recording techniques. Since x-rays are scattered by the electrons of the atoms, the intensity of each diffracted beam depends on the positions of the atoms in the unit cell. Any alteration in atomic coordinates would result in changes of the intensities relative to one another. Hence, the first important step in a crystal-structure determination is concerned with the collection of the intensity data by a recording technique which effectively measures the intensity at each reciprocal lattice point.

In the early days of x-ray crystallography, the classical Bragg x-ray spectrometer was used with an ionization chamber to collect data on the intensity of x-ray reflections. It was soon found possible and more convenient to use photographic recording for such data collection. Until the early 1960s the great majority of structural determinations were based on photographically recorded intensities which were usually visually estimated. Nevertheless, this process of photographic data collection is time-consuming, and normally several months are needed to record, judge, average, and scale the intensities of three-dimensional film data. The estimated level of accuracy of the relative intensities obtained photographically is 15–20%, which is generally sufficient for the solution of stereochemical and conformational problems but not necessarily for a reliable determination of atomic thermal motion or bonding. Since about 1953 highly stabilized x-ray diffraction units and sensitive, reliable detectors such as Geiger, scintillation, and proportional counters have been developed. This modern instrumentation, coupled with the demand for making the intensity collection from crystals both more rapid and more accurate (especially that from proteins because of their instability and large number of data to be record-

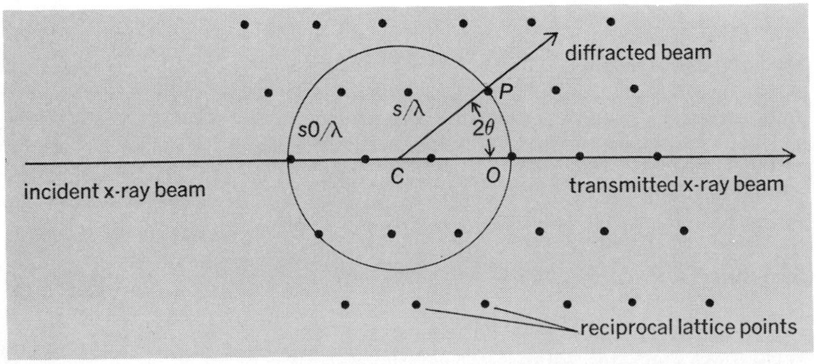

Fig. 2. Sphere of reflection in reciprocal space. Black dots represent reciprocal lattice points. Diffraction occurs when any reciprocal lattice point *hkl* cuts sphere of reflection, as at point P.

ed), has resulted in a gradual return to direct recording by counter techniques. The commercial availability of automatic diffractometers which use computed circuitry to synchronize the movement of the crystal and the detector has effectively revolutionized data collection from crystals such that upward of 200–1000 diffracted beams can be obtained daily with an error factor within 5%.

Structural analysis. Although the size and shape of the unit cell determine the geometry of the diffraction maxima, the intensity of each reflection is determined by the number, character, and distribution of the atoms within the unit cell. The second stage in a structural analysis is the solution of the phase relations among the diffracted beams (that is, the phase problem) from which a correct trial structure can be obtained. The intensity of each diffracted beam is related to the square of its amplitude (that is, $I_{hkl} = k|F_{hkl}|^2$, where $|F_{hkl}|$ is the observed structure factor amplitude). Each diffracted beam has not only a characteristic intensity but also a characteristic phase angle α_{hkl} associated with it which expresses the degree to which the diffracted beam is in plane with the other diffracted beams.

Because the electron density is a real, positive quantity which varies continuously and periodically in a crystal, the electron scattering density $\rho(xyz)$ is derivable from the three-dimensional Fourier series as in Eq. (1), where $\rho(xyz)$ repre-

$$\rho(xyz) = \frac{1}{V}\sum_h \sum_k \sum_l |F_{hkl}| \cos\{2\pi(hx + ky + lz) - \alpha_{hkl}\} \quad (1)$$

sents the electron density at any point with fractional coordinates x, y, z in the unit cell of volume V. Hence, if the characteristic amplitude $|F_{hkl}|$ and its phase α_{hkl} for each diffracted beam are known, the electron density can be calculated at fractional grid points in the unit cell. A properly phased three-dimensional electron-density map effectively provides peaks characteristic of direct images of atoms in the unit cell. The phase problem in x-ray crystallography arises because experimental measurements yield only the magnitudes of the structure factors $|F_{hkl}|$ but not the phases α_{hkl}. Hence, a structural analysis involves a search for the characteristic phases to be utilized together with the observed amplitudes in order to obtain a three-dimensional electron-density map and thereby to determine the crystal structure.

The structure factor itself is related to the scattering by the atoms in the unit cell by Eq. (2),

$$F_{hkl} = \Sigma_n f_n T_n \exp[2\pi i(hx_m + ky_n + lz_n)] \quad (2)$$

where x_n, y_n, z_n are the fractional coordinates of atom n along the three crystallographic axes. The f_n's are the individual atomic scattering factors which are known for an atom at rest. At zero 2θ angle of scattering for which all the electrons of an atom scatter in phase, f_n is equal to the atomic number. The T_n's are the individual modifications of the f_n's as a result of thermal motion. If the positions of the atoms in the unit cell are known, the complex structure factor for each diffracted beam *hkl* can be calculated and both its magnitude and phase obtained by Eq. (3).

$$F_{hkl} = A_{hkl} + iB_{hkl} \quad (3)$$

$$A_{hkl} = \Sigma_n f_n T_n \cos 2\pi(hx_n + ky_n + lz_n)$$
$$B_{hkl} = \Sigma_n f_n T_n \sin 2\pi(hx_n + ky_n + lz_n)$$
$$i = \sqrt{-1}$$

Consequently Eqs. (4) hold.

$$|f_{hkl}| = \sqrt{A^2_{hkl} + B^2_{hkl}} \qquad \tan \alpha_{hkl} = B_{hkl}/A_{hkl} \quad (4)$$

For centrosymmetric crystals which have a center of symmetry located at the corner of each unit cell, the structure factor f_{hkl} simplifies to a real rather than a complex number; that is, since for each atom at x, y, z there must exist a centrosymmetrically related atom at $-x, -y, -z$, it follows that $B_{hkl} = 0$ and $F_{hkl} = \pm A_{hkl}$. The phases of the diffracted beams then are restricted to being either completely in phase (that is, $\alpha_{hkl} = 0°$ corresponding to $F_{hkl} = +A_{hkl}$) or completely out of phase (that is, $\alpha_{hkl} = 180°$ corresponding to $F_{hkl} = -A_{hkl}$) with one another.

Patterson map. Various means of deducing the phase angle of each diffracted beam have been used. One of the most useful general approaches is based on the classic discovery by Patterson in the 1930s that when the experimentally known quantities $|F_{hkl}|^2$ instead of F_{hkl} are used as coefficients in the Fourier series, the maxima in this summation then correspond not to atomic centers but rather to a map of interatomic vectors. This vector map represents a superposition of all interatomic vectors between pairs of atoms translated to the origin of the unit cell. Since the peak height for a given vector between two atoms is approximately proportional to the product of their atomic numbers, the vectors between heavy atoms (that is, those of high atomic number) usually stand out strongly against the background of heavy-light and light-light atom vectors. Consequently, the Patterson map is especially applicable to the structure determination of compounds containing a small number of relatively heavy atoms because approximate coordinates of these heavy atoms can be obtained provided the heavy-atom vectors are correctly recognized in the Patterson map. Computers have enabled the development of a number of powerful techniques in unraveling the Patterson vector map in terms of atomic coordinates; these include multiple superpositions of parts of the Patterson map and "image-seeking" with known vectors.

Fourier electron-density map. Any resulting trial structure consisting of initial coordinates for some of the atoms is often sufficient for location of the other atoms by the application of the method of successive Fourier electron-density maps. The phases calculated from the initial parameters of the presumably known atoms, together with the observed $|F_{hkl}|$, are utilized to compute a density map. New coordinates are obtained not only for the peaks corresponding to these known atoms but also for the other peaks which it is hoped can be interpreted from stereochemical considerations as being due to additional atoms in the structure. The Fourier process is then reiterated, with the new phases calculated from the modified coordinates of the previous set of atoms plus the coordinates of newly located atoms. If a correct distinction between the "true" and the "false" peaks is made, the electron-density function usually converges to give the entire crystal structure.

Statistical method. Another powerful and eminently successful approach to the phase problem, known as the direct or statistical method, makes use of probability theory to generate an adequate set of phases by consideration solely of the structure amplitudes. The mathematical fundamentals of this procedure for direct phase determination are primarily due to the extensive work of J. Karle and H. Hauptman. Although the use of direct methods has been limited almost entirely to centrosymmetric crystals, statistical methods are also meeting with success in the solution of complex noncentrosymmetric crystal structures.

Refinement of parameters. Once the phase problem is solved and the approximately correct trial structure is known, the last step of the structural analysis involves the refinement of the positional and thermal parameters of the atoms. Normally this refinement is carried out analytically by the application of a nonlinear least-squares procedure in which a weighted quantity such as $\Sigma w[|F(hkl)|_{\text{obs}} - |F(hkl)|_{\text{calc}}]^2$ (where the weights w are appropriate to the experiment) is minimized with respect to the parameters. This method of refinement not only gives the best values for the parameters but also provides a means of obtaining an estimation of the statistical errors in the atomic parameters including standard deviations of bond lengths and angles. Although there is no single reliable method for directly assessing the accuracy of a structural determination, a criterion commonly used is the unweighted reliability factor or discrepancy index R_1, defined by Eq. (5)

$$R_1 = \frac{\sum_{hkl} \left| |F(hkl)|_{\text{obs}} - |F(hkl)|_{\text{calc}} \right|}{\sum_{hkl} |F(hkl)|_{\text{obs}}} \qquad (5)$$

as the summation of the absolute difference in the observed and calculated structure amplitudes divided by the summation of the observed amplitudes. The better the structure, including the atomic coordinates and thermal parameters, is known, the more nearly will the calculated amplitudes agree with the observed ones, and hence the lower will be the R_1 value. Discrepancy values of "finished" modern structural analyses found in the literature vary from approximately 0.15 to less than 0.05, depending upon a number of factors, such as the complexity of the structure and the number and quality of the data obtained.

Before the development of large electronic computers the determination of a crystal structure of more than about 20 atoms was not feasible. Nowadays, single-crystal analyses of uncomplicated structures may require the location of as many as 100 atoms, but there is no high correlation between the complexity of the structural determination and the number of atoms involved.

X-ray crystallography provides the quantitative foundation on which much of modern structural chemistry is based. The hundreds of crystal structures analyzed by x-ray diffraction include those of vitamin B_{12} by D. Hodgkin and coworkers (1957), for which she received a Nobel prize in medicine in 1964, and proteins such as myoglobin and hemoglobin, for which J. C. Kendrew and M. Perutz received Nobel prizes in 1962. X-ray crystallography also has been widely used in inorganic and organic chemistry not only to determine the structure of the compound but also in many cases to directly obtain the chemical composition of the compound. The so-called anomalous dispersion technique, first utilized by J. M. Bijvoet in 1951, has enabled x-ray determination of the absolute configuration of a large number of molecules. *See* X-RAY FLUORESCENCE ANALYSIS.

[LAWRENCE F. DAHL]

Bibliography: M. J. Buerger, *Crystal Structure Analysis*, 1960, reprint 1979; J. P. Glusker and K. N. Trueblood, *Crystal Structure: A Primer*, 1972; G. H. Stout and L. H. Jensen, *X-Ray Structure Determination*, 1968; A. J. Wilson, *Elements of Crystallography*, 1970; M. M. Woolfson, *Introduction to X-Ray Crystallography*, 1970.

X-ray diffraction

The scattering of x-rays by matter with accompanying variation in intensity in different directions due to interference effects. X-ray diffraction is one of the most important tools of solid-state chemistry, since it constitutes a powerful and readily available method for determining atomic arrangements in matter. X-ray diffraction methods depend upon the fact that x-ray wavelengths of the order of 1 nanometer are readily available and that this is the order of magnitude of atomic dimensions. When an x-ray beam falls on matter, scattered x-radiation is produced by all the atoms. These scattered waves spread out spherically from all the atoms in the sample, and the interference effects of the scattered radiation from the different atoms cause the intensity of the scattered radiation to exhibit maxima and minima in various directions.

Some of the uses of x-ray diffraction are: (1) differentiation between crystalline and amorphous materials; (2) determination of the structure of crystalline materials (crystal axes, size and shape of the unit cell, positions of the atoms in the unit cell); (3) determination of electron distribution within the atoms, and throughout the unit cell; (4) determination of the orientation of single crystals; (5) determination of the texture of polygrained materials; (6) identification of crystalline phases and measurement of the relative proportions; (7) measurement of limits of solid solubility, and determination of phase diagrams; (8) measurement of strain and small grain size; (9) measurement of various kinds of randomness, disorder, and imperfections in crystals; and (10) determination of radial distribution functions for amorphous solids and liquids.

For the study of crystal structure by x-ray diffraction techniques *see* X-RAY CRYSTALLOGRAPHY.

DIFFRACTION THEORY

When x-rays fall on the atoms of a substance, the scattered radiation is of two kinds: Compton modified scattering of increased wavelength which is incoherent with respect to the primary beam, and unmodified scattering coherent with the primary beam. Because of interference effects from the unmodified scattering by the different atoms of the sample, the intensity of unmodified scattering varies in different directions. A diagram of this variation in direction of intensity of unmodified scattering is called the diffraction pattern of the substance. This pattern is determined by the kinds of atoms and their arrangement in the sample; for simple structures the atomic arrangement is readily deduced from the diffraction pattern.

The atomic scattering factor f is defined as the ratio of the amplitude of unmodified scattering by an atom to the amplitude of scattering by a free electron, which scatters according to classical theory. In general f is a real number which decreases with $(\sin\theta)/\lambda$, where θ is the grazing angle and λ the wavelength, from an initial value $f = Z$, where Z is the number of electrons in the atom. However, if the x-ray wavelength is close to an absorption edge of the atom, f becomes complex. For a definition of absorption edges *see* X-RAY FLUORESCENCE ANALYSIS.

If the electron density in the atom has spherical symmetry and if the x-ray wavelength is small compared to all the absorption-edge wavelengths, Eq. (1) holds. Here $k = 4\pi(\sin\theta)/\lambda$ and $\rho(r)$ is the electron density (electrons per unit volume).

$$f = \int_0^\infty 4\pi r^2 \rho(r) \frac{\sin kr}{kr}\, dr \qquad (1)$$

A crystalline structure is one in which a unit of structure called the unit cell repeats at regular intervals in three dimensions. The repetition in space is determined by three noncoplanar vectors $\mathbf{a}_1\mathbf{a}_2\mathbf{a}_3$, called the crystal axes. The positions of the atoms in the unit cell are expressed by a set of base vectors $\mathbf{r}_n$. The position of atom n in the unit cell $q_1q_2q_3$ is given by Eq. (2). *See* CRYSTAL STRUCTURE.

$$\mathbf{R}_{nq} = q_1\mathbf{a}_1 + q_2\mathbf{a}_2 + q_3\mathbf{a}_3 + \mathbf{r}_n \qquad (2)$$

For a crystal containing $N_1N_2N_3$ repetitions in the $\mathbf{a}_1\mathbf{a}_2\mathbf{a}_3$ directions, the intensity of unmodified scattering is given by Eq. (3). Here I_e is the inten-

$$I = I_e \sum_{nq} f_n \exp\left[\frac{2\pi i}{\lambda}(\mathbf{s}-\mathbf{s}_0)\cdot\mathbf{R}_{nq}\right] \cdot \sum_{n'q'} f_{n'} \exp\left[\frac{-2\pi i}{\lambda}(\mathbf{s}-\mathbf{s}_0)\cdot\mathbf{R}_{n'q'}\right]$$
$$= I_e FF^* \frac{\sin^2[(\pi/\lambda)(\mathbf{s}-\mathbf{s}_0)\cdot N_1\mathbf{a}_1]}{\sin^2[(\pi/\lambda)(\mathbf{s}-\mathbf{s}_0)\cdot\mathbf{a}_1]} \cdot \frac{\sin^2[(\pi/\lambda)(\mathbf{s}-\mathbf{s}_0)\cdot N_2\mathbf{a}_2]}{\sin^2[(\pi/\lambda)(\mathbf{s}-\mathbf{s}_0)\cdot\mathbf{a}_2]} \cdot \frac{\sin^2[(\pi/\lambda)(\mathbf{s}-\mathbf{s}_0)\cdot N_3\mathbf{a}_3]}{\sin^2[(\pi/\lambda)(\mathbf{s}-\mathbf{s}_0)\cdot\mathbf{a}_3]} \qquad (3)$$

sity, at a distance R and angle 2θ, scattered by a free electron according to classical theory, and $FF^* = |F|^2$. For an unpolarized primary beam of intensity I_o, Eq. (4) holds, and $e^4/(m^2c^4) = 7.94$

$$I_e = I_o \frac{e^4}{m^2c^4R^2}\left(\frac{1+\cos^2 2\theta}{2}\right) \qquad (4)$$

$\times 10^{-26}$ cm² if R is expressed in centimeters. Here m is the mass of the electron, c is the velocity of light, and F is the structure factor, a complex quantity given by a summation over all the atoms of the unit cell as in Eq. (5), where $\mathbf{s}_0$ and $\mathbf{s}$ are

$$F = \sum_n f_n \exp\left[\frac{2\pi i}{\lambda}(\mathbf{s}-\mathbf{s}_0)\cdot\mathbf{r}_n\right] \qquad (5)$$

unit vectors in the directions of the primary and diffracted beams.

Laue equations and Bragg's law. The condition for a crystalline reflection is that the three quotients of Eq. (3) exhibit maxima, and this occurs if all three denominators vanish. Expressing the denominators in terms of three integers α, β, and γ, the three Laue equations, Eqs. (6), are obtained. These express the condition for a diffracted beam.

$$\begin{aligned}(\mathbf{s}-\mathbf{s}_0)\cdot\mathbf{a}_1 &= \alpha\lambda\\ (\mathbf{s}-\mathbf{s}_0)\cdot\mathbf{a}_2 &= \beta\lambda\\ (\mathbf{s}-\mathbf{s}_0)\cdot\mathbf{a}_3 &= \gamma\lambda\end{aligned} \qquad (6)$$

It is convenient to introduce the concept of sets of crystallographic planes. As illustrated by Fig. 1, the set of planes with Miller indices hkl is a set of parallel equidistant planes, one of which passes through the origin, and the next nearest makes intercepts $\mathbf{a}_1/h$, $\mathbf{a}_2/k$, $\mathbf{a}_3/l$ on the three crystallographic axes.

X-RAY DIFFRACTION

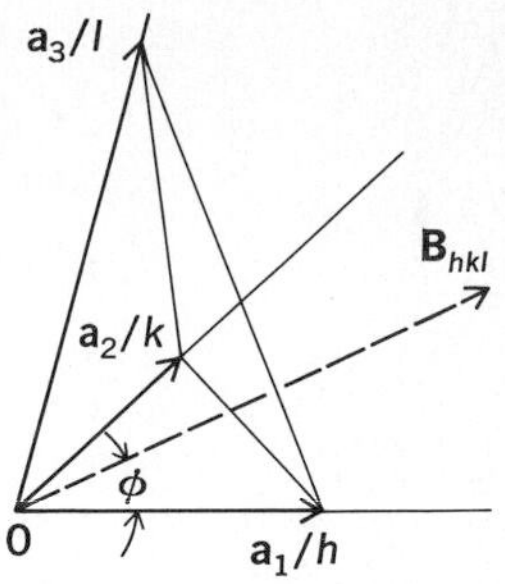

Fig. 1. Crystallographic planes with Miller indices *hkl*.

In terms of sets of planes hkl, the diffraction conditions are expressed by the Bragg law, Eq. (7),

$$\lambda = 2d_{hkl}\sin\theta \qquad (7)$$

where θ is the angle which the primary and diffracted beams make with the planes hkl and d_{hkl} is the spacing of the set. As seen from Fig. 2, the Bragg law is simply the condition that the path

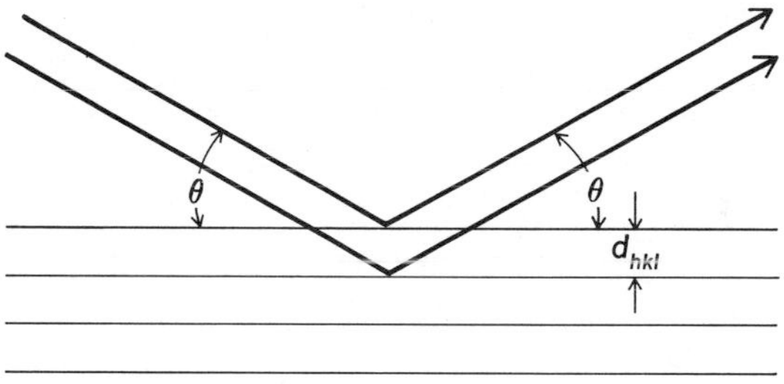

Fig. 2. Interference conditions involved in Bragg law.

difference for rays diffracted from two successive hkl planes be one wavelength. In the early days of x-ray diffraction, the Bragg law was written $n\lambda = 2d\sin\theta$, and $n = 1, 2, 3$ corresponded to first-, second-, and third-order diffraction from the planes of spacing d. That notation has been largely dropped, and instead of being called second-order diffraction from planes hkl, it is called diffraction from the planes $2h, 2k, 2l$. For an extended discussion of the Bragg law *see* X-RAY POWDER METHODS.

Reciprocal lattice. The understanding and interpretation of x-ray diffraction in crystals is greatly facilitated by the concept of a reciprocal lattice. In terms of the crystal axes $\mathbf{a}_1\mathbf{a}_2\mathbf{a}_3$, three reciprocal vectors are defined by Eqs. (8). From these defini-

$$\mathbf{b}_1 = \frac{\mathbf{a}_2\times\mathbf{a}_3}{\mathbf{a}_1\cdot\mathbf{a}_2\times\mathbf{a}_3} \qquad \mathbf{b}_2 = \frac{\mathbf{a}_3\times\mathbf{a}_1}{\mathbf{a}_1\cdot\mathbf{a}_2\times\mathbf{a}_3}$$
$$\mathbf{b}_3 = \frac{\mathbf{a}_1\times\mathbf{a}_2}{\mathbf{a}_1\cdot\mathbf{a}_2\times\mathbf{a}_3} \qquad (8)$$

tions it follows that Eq. (9) is valid. In terms of

$$\mathbf{a}_i \cdot \mathbf{b}_j = \begin{cases} 1 & i=j \\ 0 & i \neq j \end{cases} \qquad (9)$$

integers hkl, the terminal points of the vectors in Eq. (10) generate a lattice of points called the

$$\mathbf{B}_{hkl} = h\mathbf{b}_1 + k\mathbf{b}_2 + l\mathbf{b}_3 \qquad (10)$$

reciprocal lattice. Each point in the lattice is specified by the integers hkl, and the vectors $\mathbf{B}_{hkl}$ represent two important properties of the sets of hkl planes: (1) $\mathbf{B}_{hkl}$ is perpendicular to the hkl planes, and (2) $|\mathbf{B}_{hkl}| = 1/d_{hkl}$. These two relations are readily proved from the geometry of Fig. 1. As seen, $\mathbf{a}_2/k - \mathbf{a}_1/h$ and $\mathbf{a}_3/l - \mathbf{a}_2/k$ are vectors lying in the hkl plane. From Eqs. (9) and (10), Eqs. (11) are

$$\left(\frac{\mathbf{a}_2}{k} - \frac{\mathbf{a}_1}{h}\right) \cdot \mathbf{B}_{hkl} = 0 \qquad \left(\frac{\mathbf{a}_3}{l} - \frac{\mathbf{a}_2}{k}\right) \cdot \mathbf{B}_{hkl} = 0 \qquad (11)$$

obtained, and hence $\mathbf{B}_{hkl}$ is perpendicular to the planes hkl. The spacing of the planes hkl is given by Eq. (12).

$$d_{hkl} = \left|\frac{\mathbf{a}_1}{h}\right| \cos\phi = \frac{\mathbf{a}_1}{h} \cdot \frac{\mathbf{B}_{hkl}}{|\mathbf{B}_{hkl}|} = \frac{1}{|\mathbf{B}_{hkl}|} \qquad (12)$$

Equivalence of the three Laue equations and the Bragg law can be shown as follows: Any vector **r** can be expressed by Eq. (13). Let **r** be the vector

$$\mathbf{r} = (\mathbf{r} \cdot \mathbf{a}_1)\mathbf{b}_1 + (\mathbf{r} \cdot \mathbf{a}_2)\mathbf{b}_2 + (\mathbf{r} \cdot \mathbf{a}_3)\mathbf{b}_3 \qquad (13)$$

$(\mathbf{s} - \mathbf{s}_0)$ and combine it with the three Laue equations and Eq. (13) to obtain Eq. (14). The Bragg law

$$\mathbf{s} - \mathbf{s}_0 = \lambda(\alpha\mathbf{b}_1 + \beta\mathbf{b}_2 + \gamma\mathbf{b}_3) \qquad (14)$$

can be written in vector form as Eq. (15) since the

$$\mathbf{s} - \mathbf{s}_0 = \lambda\mathbf{B}_{hkl} = \lambda(h\mathbf{b}_1 + k\mathbf{b}_2 + l\mathbf{b}_3) \qquad (15)$$

usual form of the Bragg law is simply an equality in the magnitudes of the vectors: $|\mathbf{s} - \mathbf{s}_0| = 2 \sin\theta$ and $|\mathbf{B}_{hkl}| = 1/d_{hkl}$. Comparison of Eqs. (14) and (15) shows that the integers α, β, γ of the three Laue equations are simply the Miller indices hkl of the Bragg law.

The positions of the atoms in the unit cell are represented by a set of atomic coordinates x_n, y_n, z_n such that for atom n Eq. (16) holds. For a Bragg

$$\mathbf{r}_n = \mathbf{a}_1 x_n + \mathbf{a}_2 y_n + \mathbf{a}_3 z_n \qquad (16)$$

law reflection hkl, the structure factor takes the simple form of Eq. (17).

$$F_{hkl} = \sum_n f_n \exp\left[2\pi i(hx_n + ky_n + lz_n)\right] \qquad (17)$$

Integrated intensity. In general, the intensity of a Bragg reflection, as expressed by Eq. (3), is not an experimentally measurable quantity. Other factors, such as the degree of mosaic structure in the crystal and the degree of parallelism of the primary beam, have a profound influence on the measured diffracted intensity for any setting of the crystal. To obtain measurements characteristic of the crystalline structure, it is necessary to adopt a more useful concept, the integrated intensity. For a small single crystal, it is postulated that the crystal is to be turned at constant angular velocity ω through the Bragg law position, and that the total diffracted energy of the reflection is to be measured. The integrated intensity E is then given by Eq. (18), where $d\alpha$ is a change in orientation of the

$$E = \int\int I \frac{d\alpha}{\omega} dA \qquad (18)$$

crystal and dA is an element of area at the point of observation

Most of the equations used in x-ray diffraction studies are derived on the assumption that the intensity of the diffracted beam is so small that any interaction with the primary beam can be neglected. These are classed as the equations for the ideally imperfect crystal. For powder samples, in which the individual crystals are extremely small, and for highly deformed single crystals, if the intensity of the diffracted beam is small, the ideally imperfect crystal is usually a good approximation. For the ideally perfect crystal, it is necessary to use a more elaborate theory which allows for the interaction of diffracted radiation with the primary beam. In general, it is the integrated intensity which is measured, and theory shows that the integrated intensity for an ideally imperfect crystal is larger than that for the ideally perfect crystal. Many of the crystalline samples used for x-ray diffraction studies are not ideally imperfect, and the measured integrated intensity is accordingly less than that predicted by the ideally imperfect crystal formulas which are used in the interpretation. The situation is usually handled by adding a correction factor called the extinction correction to the formulas for the integrated intensity from the ideally imperfect crystal.

Atomic coordinates. To have complete information about a crystalline structure, it is necessary to know all the atomic coordinates $x_n y_n z_n$ of the n atoms making up the unit cell. The atomic coordinates appear in the structure factor as given by Eq. (17), and sometimes the coordinates are obtained directly from structure factor values. Another way is to plot the electron density in the unit cell and infer the atomic positions from peaks in the electron density function. The electron density in the unit cell is given by the triple Fourier series shown in Eq. (19) for which the coefficients are simply the

$$\rho(xyz) = \frac{1}{V}\sum_h \sum_k \sum_l F_{hkl} \exp\left[-2\pi i\left(\frac{hx}{a} + \frac{ky}{b} + \frac{lz}{c}\right)\right] \qquad (19)$$

structure factors F_{hkl}. However, from experimental measurements of either an intensity or an integrated intensity, values for $|F_{hkl}|^2$ can be obtained. These yield the magnitude of F_{hkl} but not the phase. This is the most serious limitation to a straightforward determination of crystalline structures by x-ray diffraction methods. The ambiguity in the phase of F_{hkl} prevents the use of the Fourier plot of Eq. (19) as a general method for determining any crystalline structure.

Simple structures are uniquely determined by combining the x-ray intensity results with space group theory. The space group of a crystal is the repeating spatial arrangement of symmetry elements which the structure displays. Considering all the possible symmetry elements which can exist in a crystalline structure, group theory shows that there are only 230 essentially different possible combinations, and these constitute the 230 space groups. A knowledge of the macroscopic symmetry of the crystal, coupled with the systematically vanishing x-ray reflections, usually determines the space group. The limitations imposed by the space group on the possible atomic positions, coupled with the limitations imposed by the measured $|F_{hkl}|^2$, often allow a complete and unique structure determination for not too complicated structures. For highly complex structures, it is never certain that x-ray diffraction analysis can yield a complete structure determination. Additional techniques such as the isomorphous replacement by heavy atoms, the use of Patterson plots, and the determination of phase relations from inequalities are used with success on some of the complex structures.

Many structures of interest in solid-state chemistry exhibit various kinds of randomness and imperfections. The precise nature of these is sometimes of more interest than the ideal average structure. Randomness and imperfections in a structure show themselves by producing a diffuse intensity in addition to the sharp Bragg reflections. The temperature vibration of the atoms produces a diffuse intensity called temperature diffuse scattering. Quantitative measurements of this scattering lead to values for the velocity of high-frequency elastic waves and to a complete experimental determination of the spectrum of the elastic waves which constitute the thermal vibrations of the crystal. In alloys showing order-disorder changes, the short-range order parameters are obtained from quantitative measurement of the diffuse intensity which results from randomness in the atomic arrangement.

CRYSTALLINE DIFFRACTION

The techniques employed in the study of crystalline substances are discussed below.

Laue method. The Laue pattern uses polychromatic x-rays provided by the continuous spectrum from an x-ray tube operated at 35–50 kV. The transmission Laue pattern is obtained by passing a finely collimated beam through a thin single crystal and recording the diffracted beams on a photographic film placed several centimeters beyond the crystal. For each set of planes *hkl*, θ is fixed, and the Bragg law is satisfied by selecting the proper λ from the primary beam. In a Laue pattern, the different diffracted beams have different wavelengths, and their directions are determined solely by the orientations of the *hkl* planes. Transmission Laue patterns were once used for structure determinations, but their many disadvantages have made them practically obsolete.

On the other hand, the back-reflection Laue pattern is used a great deal in the study of the orientation of crystals. The back-reflection Laue camera is shown schematically in Fig. 3. The polychromatic beam enters through a hole in the x-ray film and falls on a single crystal whose orientation can be set as desired by a system of goniometer circles. Diffracted beams bent through angles 2θ approaching 180° are registered on the photographic film. For cubic crystals it is very easy to read the crystal orientation from a back-reflection Laue pattern, and the patterns find considerable use in the cutting of single-crystal metal ingots.

Rotating crystal method. The original rotating crystal method was employed in the Bragg spectrometer. A sufficiently monochromatic beam, of wavelength of the order of 1 A, is obtained by using

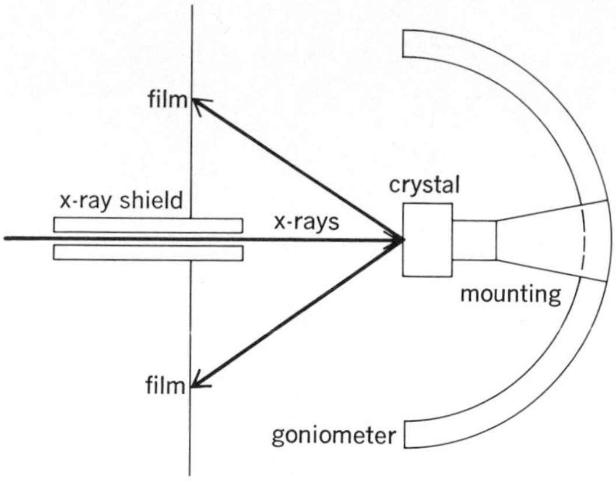

Fig. 3. Schematic of back-reflection Laue camera.

the strong $K\alpha_1\alpha_2$ doublet with a filter which suppresses the $K\beta$ line and much of the continuous spectrum. The beam is collimated by a system of slits and then falls on the large extended face of a single crystal as shown by Fig. 4. Originally the diffracted beam was measured with an ionization chamber, but Geiger counters and proportional counters have largely replaced the ionization chamber. Both the crystal and the chamber turn about the spectrometer axis.

The Bragg spectrometer has been used extensively in obtaining quantitative measurements of the integrated intensity from planes parallel to the face of the crystal. The chamber is set at the correct 2θ-angle with a slit so wide that all of the radiation reflected from the crystal can enter and be measured. The crystal is turned at constant angular speed ω through the Bragg law position, and the total diffracted energy E received by the ionization chamber during this process is measured. Similar readings with the chamber set on either side of the peak give a background correction. For this type of measurement, the integrated intensity E is given by Eq. (20), where P_0 is the power of the primary

$$E=\frac{P_0}{2\mu\omega}\frac{e^4}{m^2c^4}\frac{\lambda^3F^2}{v^2}\left(\frac{1+\cos^2 2\theta}{2\sin 2\theta}\right)\exp\left[-2M\right] \quad (20)$$

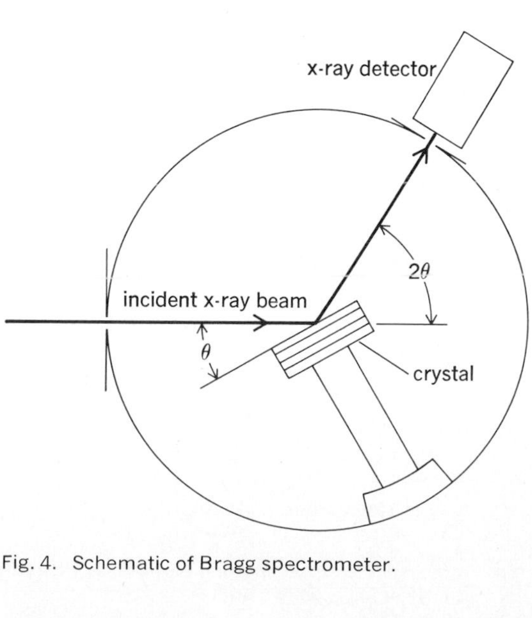

Fig. 4. Schematic of Bragg spectrometer.

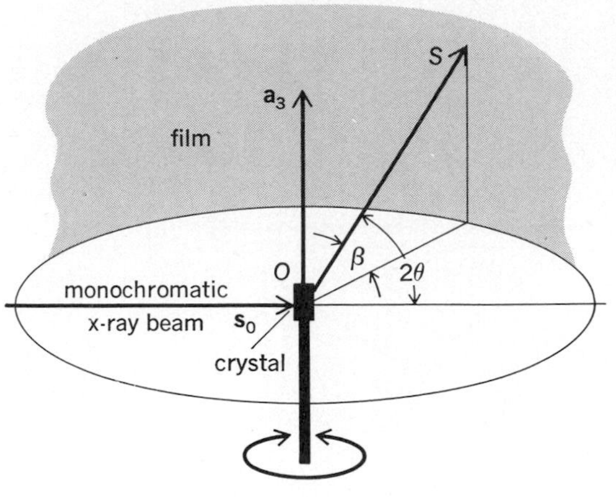

Fig. 5. Schematic of rotation camera.

beam, μ is the linear absorption coefficient in the crystal, ω is the angular velocity of the crystal, λ is the x-ray wavelength, F is the structure factor, v is the volume of the unit cell, and $\exp[-2M]$ is the so-called Debye factor allowing for temperature vibration. When more than one kind of atom is present, this factor must be incorporated in F^2.

Measurements of the integrated intensity E give quantitative values of F^2 directly. When a Geiger counter is used in place of an ionization chamber for this type of measurement, it is necessary to employ a narrow counterslit and traverse the counter through the reflected beam, since the sensitivity of a Geiger counter is not constant over a large window opening.

The rotation camera, which is frequently used for structure determinations, is illustrated in Fig. 5. The monochromatic primary beam $\mathbf{s}_0$ falls on a small single crystal at O. The crystal is mounted with one of its axes (say, $\mathbf{a}_3$) vertical, and it rotates with constant velocity about the vertical axis during the exposure. The various diffracted beams are registered on a cylindrical film concentric with the axis of rotation. For a rotation about $\mathbf{a}_3$ it follows that $\mathbf{s}_0 \cdot \mathbf{a}_3 = 0$, and the third Laue equation gives Eq. (21). The diffracted beams form the elements of a set of cones, and the intersection of these cones with the cylindrical film gives a set of horizontal lines of diffraction spots. This type of pattern is called a rotation pattern, and the horizontal rows of spots are called layer lines. As seen from Eq. (21), the measured values of sin β give directly the length of the axis about which the crystal was rotated, and the layer line in which a spot occurs gives the l index of the reflection. Similar rotations about the other two axes give corresponding information. More elaborate variations of the rotation method, such as those of the Weissenberg and the precession cameras, involve a motion of the film in addition to the rotation of the crystal.

$$\sin\beta = \frac{l\lambda}{|\mathbf{a}_3|} \qquad (21)$$

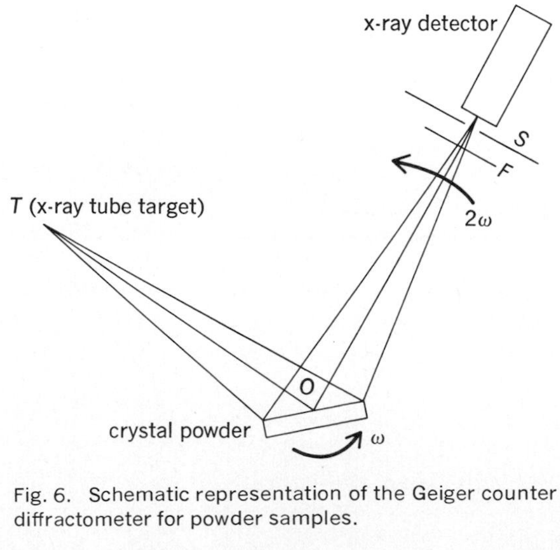

Fig. 6. Schematic representation of the Geiger counter diffractometer for powder samples.

Powder method. The powder method involves the diffraction of a collimated monochromatic beam from a sample containing an enormous number of tiny crystals having random orientation. Since about 1950 an increasing number of powder pattern studies have been made with Geiger counter, or porportional counter, diffractometers. The apparatus is shown schematically in Fig. 6. X-rays diverging from a target at T fall on the sample at O, the sample being a flat-faced briquet of powder. Diffracted radiation from the sample passes through the receiving slit at s and enters the Geiger counter. During the operation the sample turns at angular velocity ω and the counter at 2ω. The distances TO and OS are made equal to satisfy approximate focusing conditions. A filter F before the receiving slit gives the effect of a sufficiently monochromatic beam. A chart recording of the amplified output of the Geiger counter gives directly a plot of intensity versus scattering angle 2θ.

NONCRYSTALLINE DIFFRACTION

For a noncrystalline substance such as a glass or a liquid, a more general expression for the intensity of diffracted radiation is required. If the instantaneous position of each atom in the sample is represented by a vector $\mathbf{r}_n$, the diffracted intensity is given by Eq. (22). A particularly useful variation

$$I = I_e \sum_q \sum_n f_q f_n \exp\left[\frac{2\pi i}{\lambda}(\mathbf{s}-\mathbf{s}_0)\cdot(\mathbf{r}_q-\mathbf{r}_n)\right] \qquad (22)$$

of Eq. (22) given as Eq. (23), is obtained by computing the average intensity $\langle I\rangle$ when the sample as a rigid array is allowed to take with equal probability all orientations in space. In Eq. (23) $k = 4\pi(\sin\theta)/\lambda$ and $r_{qn} = |\mathbf{r}_q - \mathbf{r}_n|$.

$$\langle I\rangle = I_e \sum_q \sum_n f_q f_n \frac{\sin(kr_{qn})}{kr_{qn}} \qquad (23)$$

The fact that there are fairly definite nearest-neighbor and second-neighbor distances in a glass or liquid means that Eq. (23) will show peaks and dips when the intensity is plotted against $(\sin\theta)/\lambda$. Peaks and dips in an x-ray diffraction pattern merely indicate the existence of preferred interatomic distances, not that the material is necessarily crystalline. X-ray patterns of noncrystalline materials are usually analyzed by a Fourier inversion of Eq. (23), which yields a radial distribution function giving the probability of finding neighboring atoms at any distance from an average atom.

[BERTRAM E. WARREN]

Gases. Gases and liquids are found to give rise to x-ray diffraction patterns characterized by one or more halos or interference rings which are usually somewhat diffuse. These diffraction patterns which are similar to those for glasses and amorphous solids, are due to interference effects depending both upon the electronic distribution of each of the individual atoms or molecules and upon their relative positions in the system.

For monatomic gases the only appreciable interference effects giving rise to a distribution of scattered intensities are those produced by the electronic distribution about each nucleus. These interference effects giving rise to so-called coherent intensities are the result of the interference of the individual waves scattered by electrons in different parts of the atom. The electronic distribution of an atom is described in terms of a characteristic atomic scattering factor which is defined as the ratio of the resultant amplitude scattered by an atom to the amplitude that a free electron would scatter under the same conditions. At zero-angle scattering the atomic scattering factor is equal to the atomic number of the atom. The coherent intensity in a given direction is proportional to the square of the atomic scattering factor. If it is assumed that the electronic distribution is spherically symmetrical, the atomic scattering factors can be readily obtained from the observed intensities. For molecular gases the interference effects depend not only on the scattering factors of the atoms but also on their relative positions in the molecule. One can observe only an average intensity scattered over a period of time during which the molecules have taken innumerable positions with respect to the incident beam. Interference effects due to the relative packing of the atoms or molecules can be neglected for dilute gases but not for dense gases.

As in the case of x-ray diffraction by crystals, light atoms such as hydrogen are difficult to detect in the presence of heavy atoms. Because of the shorter exposure times required, electron diffraction rather than x-ray diffraction has been utilized in studies of the structures of gaseous molecules. Both methods appear to be comparable in view of the accuracy of the intensity measurements and the technical difficulties involved.

Liquids. One cannot, as in the cases of dilute gases and crystalline solids, derive unambiguous, detailed descriptions of liquid structures from diffraction data. Nevertheless, diffraction studies of liquids do provide most useful information. Instead of comparing the experimental intensity distributions with theoretical distributions computed for various models, the experimental results are usually provided in the form of a radial distribution function which specifies the density of atoms or electrons as a function of the radial distance from any reference atom or electron in the system without any prior assumptions about the structure. From the radial distribution function one can obtain (1) the average interatomic distances most frequently occurring in the structure corresponding to the positions of the first, second, and possibly third nearest neighbors; (2) the distribution of distances; and (3) the average coordination number for each interatomic distance. The interpretation of these diffraction patterns given by the radial distribution function usually is not straightforward and in general it can be said only that a certain assumed structural model and arrangement is not inconsistent with the observed diffraction data. The models considered represent only a description of the time-average environment about any given atom or molecule within the liquid.

There are great experimental difficulties in obtaining accurate intensity data. The sources of error are many; for a detailed treatment the reader is referred to publications of C. Finback. A brief description of some results obtained by x-ray diffraction of liquids is given below.

Liquid elements. The radial distribution function, first used in a study of liquid mercury, has been applied to a considerable number of liquid elements mainly to compare their physical properties in the liquid and crystalline states. In most cases a lower first mean coordination number is found in the liquid state, exceptions are liquid gallium, bismuth, germanium, and lithium. The radial distribution curves give direct evidence for the existence of molecules in some liquid elements (for example, N_2, O_2, Cl_2, and P_4) and imply the existence of more complicated atomic aggregates in a few cases. Argon and helium have been extensively studied in the liquid and vapor states over wide ranges of temperature and pressure.

Liquid water and solutions. A prime example illustrating the considerable structural information made available from modern x-ray liquid diffractometry investigations is the detailed analysis of liquid water, which revealed the following significant features: (1) There are distinct structural deviations of water molecules from a uniform distribution of distances to about 0.8 nm at room temperature; (2) the first prominent maximum, corresponding to near-neighbor interactions, shifts gradually from 0.282 nm at 4°C to 0.294 nm at 200°C; (3) the average coordination number in liquid water from 4 to 200°C is approximately constant and slightly larger than four; and (4) the radial distribution of oxygen atoms in water at 4°C is not significantly different from that in deuterium oxide at the same temperature. Comparison of calculated radial distribution functions for various proposed liquid water models (which are sufficiently defined at the molecular level) based on those derived from patterns of liquid water have shown that the only realistic model which gives agreement with data from both large- and small-angle x-ray scattering is related to a modification of the ordinary hexagonal ice structure. This solid-state structure is similar to that of the hexagonal form of silicon dioxide, tridymite, with each oxygen atom tetrahedrally surrounded by neighboring oxygen atoms to give layers of puckered six-membered rings with dodecahedral cavities large enough (radius 0.295 nm) to accommodate a water molecule.

In terms of an average configuration, the liquid water phase may be regarded as a "mixture" model comprising network water molecules forming a slightly expanded ordinary ice structure (each oxygen atom forming nearly four hydrogen bonds with neighboring oxygen atoms) and the cavity water molecules interacting with the network by less specific but by no means negligible forces. It must be emphasized that both kinds of water molecules instantaneously exist in environments which are distorted from the average, as implied by sizable root-mean-square variations in interatomic distance. *See* HYDROGEN BOND; WATER.

Radial distribution curves for concentrated $FeCl_3$ solutions indicate a large degree of local ordering of the ions with formation of $Fe^{3+}-Cl^-$ complexes. Studies on metal-metal solutions, colloidal solutions, and molecular solutions have been made. More definite results have been obtained for concentrated solutions of strongly scattering solutes in weakly scattering solvents. Examples are the proof of the existence of a polymeric species in aqueous $Bi(ClO_4)_3$, evidence that in aqueous solution the HgX_4^{2-} anions (X = Cl, Br, and I) are tetrahedral, and definite evidence of ion-pair formation in aqueous BaI_2.

Molten salts. A molten salt is considered to be a loose and expanded imitation of the solid with the same coordination scheme and short-range order. Careful x-ray diffraction studies of a number of molten salts have indicated that melts do not possess such quasi-crystalline structures but instead have quite open structures with a wide variety of individual ion coordinations. Interpretations of radial distribution functions for several other molten salts have been made. Liquid $AlCl_3$ appears to consist mainly of Al_2Cl_6 molecules; liquid SnI_4 is composed of independent tetrahedral molecules. The results for other molten salts are not as conclusive. *See* NEUTRON DIFFRACTION.

[LAWRENCE F. DAHL]

Bibliography: C. S. Barrett and T. B. Massalski, *Structure of Metals*, 3d ed., 1980; A. G. Brown, *X-Rays and Their Applications*, 1975; M. J. Buerger, *Crystal Structure Analysis*, 1960, reprint 1979; N. A. Dyson, *X-Rays in Atomic and Nuclear Physics*, 1973; G. H. Stout and L. H. Jensen, *X-Ray Structure Determination*, 1968.

X-ray fluorescence analysis

A nondestructive physical method used for chemical analyses of solids and liquids. The specimen is irradiated by an intense x-ray beam which causes the elements in the specimen to emit (that is, fluoresce) their characteristic x-ray line spectra. The lines of the spectra are diffracted at various angles by a single-crystal plate which is analogous to the diffraction grating of optical spectroscopy. The elements may be identified by the wavelengths of their spectral lines, which vary in a regular manner with atomic number, and their concentrations may be determined from the intensities of the lines. Counter tubes and associated electronic circuits are used to measure the x-ray intensities.

Unlike x-ray diffraction, the elements (rather than the compounds) are identified, and it is not necessary that the specimen be crystalline. The method supplements optical spectroscopy, being most generally used for concentrations $> 0.1\%$, although with prior wet chemical or physical extraction methods to enhance the peak-to-background ratio, much lower amounts can be detected. *See* X-RAY DIFFRACTION.

Analyses for all elements above about atomic number 12 can be done in a small fraction of the time required by conventional wet chemical methods. Aside from the initial cost of the equipment, x-ray fluorescence spectroscopy, which is also termed x-ray spectrochemical analysis, is inexpensive, and highly trained personnel are not usually required.

Although the method had its origin in the classic work of H. G. J. Moseley in 1913, it was not widely used until about 1950, when the introduction of modern x-ray equipment made it feasible to apply the method to a large variety of analyses. The equipment and methods are quite similar to those employed in x-ray powder diffraction *See* X-RAY POWDER METHODS.

Applications. The x-ray fluorescence method is less sensitive than the optical method, although with prior specimen treatment, analyses can be made in the range of parts per million. Microgram quantities have been determined in milligram samples with standard deviations of 10–15%, using focusing crystal spectrographs. Optical spectrographic methods become difficult to use for concentrations exceeding about 20%, whereas the x-ray method is applicable up to 100%.

The results agree well with those obtained by conventional wet chemical methods while requiring only a small fraction of the time. Thus x-ray fluorescence methods have replaced chemical procedures for routine analyses in many fields.

In metallurgy, for example, analyses of tungsten in a cobalt-base alloy requiring more than 24 hr by wet chemical procedures can be obtained with greater accuracy in 20 min using the x-ray method. The time factor makes possible analyses on a very large scale, such as are required in geochemical problems where the costs for conventional methods would be prohibitive. Many applications to production control processes where speed is essential are now feasible.

The x-ray fluorescence method has proven particularly useful for mixtures of elements of similar chemical properties which are difficult to separate and analyze by conventional chemical methods. The method has been applied to measuring the thickness of thin platings such as tin on steel, and evaporated films. It is used for mineral, ore, and alloy analyses as well as for the control of blending operations. The applications of the method to liquids have been as numerous as to solids.

Basis of the method. The origin of x-ray spectra may be understood from the simple Bohr model of the atom in which the electrons are arranged in orbits within the *K*, *L*, *M*, . . . shells. If enough energy is applied to the atom, an electron may be ejected from one of the inner shells. An electron from one of the outer orbits promptly falls back to the inner shell to take its place, so that the atom returns to its normal state. This action results in the emission of a characteristic x-ray spectral line, that is, a quantum of energy equivalent to the difference of the binding energies of the orbits involved in the electron transition. Because only a limited number of electron orbits may participate, the x-ray spectrum of an element consists of comparatively few lines, in contrast to the much more complex optical spectra.

Moseley's law. Moseley showed that there was a linear relationship between $1/\sqrt{\lambda}$ and Z, where λ is the x-ray wavelength and Z is the atomic number of the element. Figure 1 is a plot of Moseley's law for some *K* and *L* x-ray lines. The *K* lines originate from electron transfers to the *K* shell and the *L* lines from transfers to the *L* shell. The *K* lines have the shorter wavelengths and therefore the higher energies. There are four or five lines in the *K* spectrum ($K\alpha_1$, $K\alpha_2$, $K\beta_1$, and so forth) and a dozen or more lines in the *L* spectrum ($L\alpha_1$, $L\alpha_2$, $L\beta_1$, and so forth). There are also *M* spectra, but

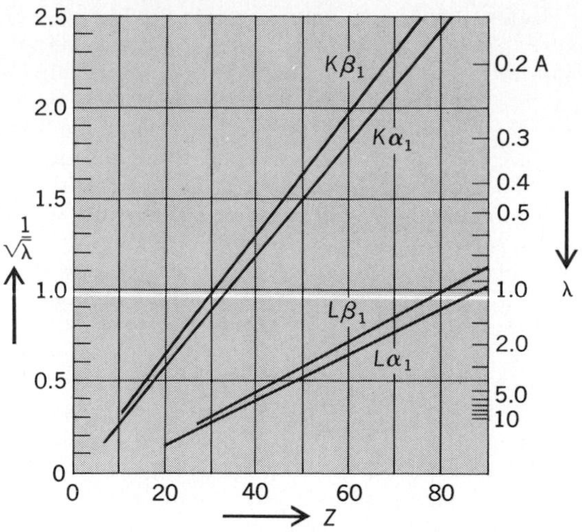

Fig. 1. Plot of Moseley's law, showing dependence of characteristic x-ray line wavelengths on atomic number. (*Philips Tech. Rev., vol. 17, no. 10, 1956*)

these are of very long wavelengths and are rarely used in x-ray analysis. In the filling of vacancies in the electron shells, the ejection of Auger electrons is an alternative process to x-ray emission. *See* AUGER EFFECT.

Production of spectra. A necessary condition for the production of a particular spectrum is that an energy called the critical excitation potential be exceeded. This is the energy required to remove an electron from a particular shell of a given atom. Thus all the L lines of copper appear above 1.1 keV, the K lines of copper above 8.98 keV, and the K lines of silver above 25.5 keV.

There are two general methods for the production of characteristic line spectra. In the first, the direct electron excitation method, primary excitation is used in the conventional x-ray tube, where electrons are focused onto a small portion, called the focal spot, of a pure target element. The generated line spectra are characteristic of the target material, and their intensities increase by nearly the square of the voltage (above the critical potential) and linearly with current. In addition to the line spectrum, there is also generated a continuous spectrum, resulting from the deceleration of the electrons. The intensity and short wavelength limit of the continuous spectrum depend primarily on the applied voltage, as shown in Fig. 2.

The second method for the production of the characteristic line spectrum is the secondary excitation, or fluorescence method, so called from analogy to the optical case. The specimen is placed outside a sealed-off x-ray tube and irradiated by the x-ray beam. All of the line and continuum primary x-rays having energies exceeding the critical excitation potential of the elements in the specimen may cause those elements to emit their characteristic x-ray line spectra. The usable fluorescence spectrum is only about 1/1000 as intense as the direct-electron-excited spectrum (the limitations are mainly due to the geometry of the spectrograph), but the high intensities and great sensitivities of modern equipment and the convenience and simplicity of the method make it much more practical. Although no significant amount of continuous spectrum is generated in x-ray fluorescence, there is always a small amount of scattered x-ray background.

X-ray absorption. Each element has an abrupt change in its x-ray absorption (called the absorption edge) at the wavelength corresponding to the critical excitation potential for a given orbit. An element strongly absorbs x-rays of higher energy, that is, shorter wavelength, than its absorption edge. The amount of absorption is a function of wavelength; absorption decreases gradually below the absorption edge and increases again above the absorption edge. Thus the x-rays of wavelength just shorter than the absorption edge are most efficient in generating x-ray fluorescence, the efficiency decreasing as the exciting wavelength is further removed from the absorption edge. The wavelengths longer than the absorption edge have no effect in exciting fluorescence because they possess insufficient energy. In addition, the fluorescence conversion efficiency falls off with decreasing atomic number, because of, among other things, the Auger effect.

The absorption of x-rays increases rapidly with increasing wavelength, and because the line spectra of the low-atomic number elements are of comparatively long wavelength, these elements can be analyzed only with great difficulty or not at all. The attenuation of x-rays by the x-ray tube window and the air path through which the x-rays pass may be considerable, depending on the wavelengths involved. For example, an air path of 34 cm, which is typical of a modern instrument, absorbs about 22% of Cu$K\alpha$ (wavelength 0.154 nm), 56% of Cr$K\alpha$ (0.228 nm), and 96% of K$K\alpha$ (0.0374 nm). It is therefore necessary to use a vacuum or to bubble helium through an enclosure around the x-ray path to utilize the longer wavelengths. In practice, the analyses of elements above about atomic number 24 (Cr) are usually done in air, while lower atomic numbers down to 12 (Mg) are done in a vacuum or a helium path. Fluorescence analysis of elements below atomic number 12 is not possible with existing equipment.

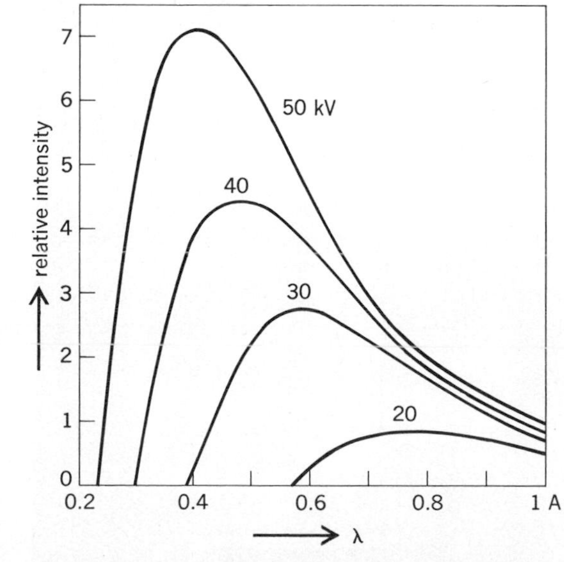

Fig. 2. Continuous spectrum of tungsten target x-ray tube obtained at various peak voltages (full-wave rectification) and same tube current. Measured with silicon 111 crystal analyzer and scintillation counter. (*Philips Tech. Rev., vol. 17, no. 10, 1956*)

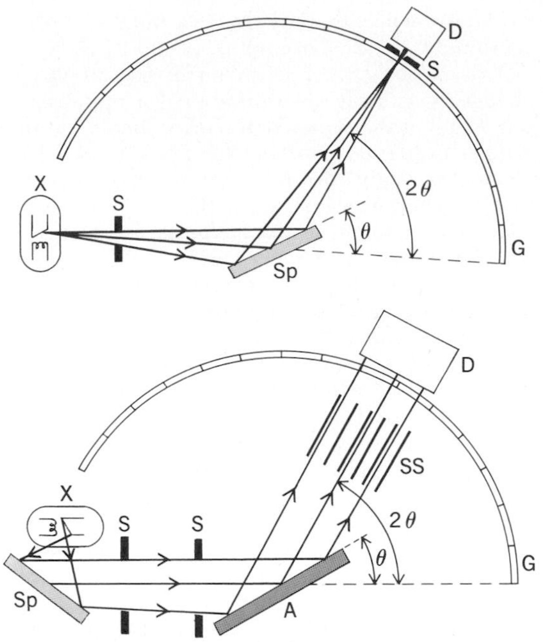

Fig. 3. Schematic drawing of x-ray powder diffractometer (above) and nonfocusing x-ray fluorescence spectrograph (below). X, x-ray tube; S, limiting slits; Sp, specimen; θ, glancing angle of incidence; 2θ, reflection angle; A, crystal analyzer; SS, Soller (parallel) slits; D, counter tube detector; G, goniometer.

Crystal analyzers. A single-crystal plate is used to separate the various wavelengths emitted by the specimen. Diffraction from the crystal occurs according to Bragg's law; $n\lambda = 2d \sin \theta$ where n is a smaller integer giving the order of reflection, λ the wavelength, d the spacing of the particular set of lattice planes of the crystal that are properly oriented to reflect, and θ the angle between those lattice planes and the incident ray. *See* X-RAY CRYSTALLOGRAPHY.

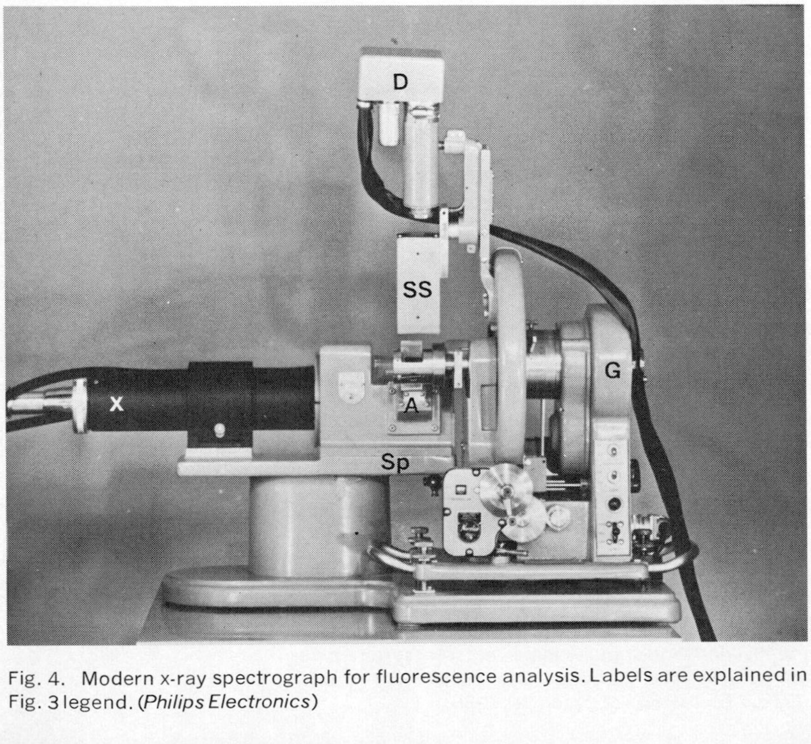

Fig. 4. Modern x-ray spectrograph for fluorescence analysis. Labels are explained in Fig. 3 legend. (*Philips Electronics*)

Reflection occurs only at an angle 2θ with respect to the incident ray, and it is therefore necessary to maintain the correct angular relationship of the crystal planes at one-half the counter tube angle. The goniometer rotates the crystal at one-half the angular speed of the counter tube, and therefore both are always in the correct position to receive the various wavelengths emitted by the specimen, as shown in Fig. 3. For a given d there is only one angle (for each order of reflection) at which each wavelength is reflected, the angle increasing with wavelength.

The angular separation of the lines, or the dispersion $d\theta/d\lambda = n/2d \sin \theta$ increases with decreasing d. It is thus easy to increase the dispersion simply by selecting a crystal with a smaller d. Reducing d also limits the maximum wavelength that can be measured since $\lambda = 2d$ at $2\theta = 180°$; the maximum 2θ angle that can be reached with the goniometer is about 150°.

Although the fluorescent beam from the specimen is not parallel and has a rather large angular aperture, the crystal at any instant reflects only that bundle of parallel rays making the correct θ angle. As the crystal rotates, it reflects other bundles in sequence and thereby collimates the reflected beam. To increase the resolution, that is, decrease the line breadth, it is necessary to limit the angular range over which a wavelength is recorded. Parallel or Soller slits are used for this purpose as shown in Fig. 3. These slits consist of thin (0.001–0.002 in. or 0.002–0.005 cm) equally spaced foils of such materials as nickel and iron, and the angular aperture is determined by the length and spacing. A typical set would have 0.010-in. (0.025-cm) spacing, 4-in. (10-cm) length with angular aperture 0.29°, and cross section about 3/4 in. (1.8 cm) square. The absorption of the foils is sufficiently high to prevent rays that are inclined by more than the angular aperture from entering the counter tube. Two sets of parallel slits may be used, one set between the specimen and crystal and the other between crystal and counter tube. This greatly increases the resolution and peak-to-background ratio, and causes a relatively small loss of peak intensity.

It is essential that the crystal be of good quality to obtain sharp, symmetrical reflections. Unless the crystal is homogeneous, the reflection may be distorted and portions of the reflections may occur at slightly different θ angles. Such effects would decrease the peak intensities of the wavelengths by varying amounts, causing errors in the analysis.

Instrumentation. The x-ray spectrograph (Fig. 4) for fluorescence analysis consists of the primary x-ray tube and its stabilized high-voltage power supply; a ray-proof specimen chamber; the goniometer which carries the crystal analyzer and counter tube; and the electronic circuitry for the counter tube (not shown) consisting of a power supply, linear amplifier, single-channel pulse-height analyzer, scaler, rate meter, and strip-chart recorder.

X-ray tube operation. The primary x-ray tube targets are usually tungsten, molybdenum, or copper. It is usually necessary to avoid the use of a tube whose target is identical with an element in the specimen because the line spectrum from the target is scattered through the system. It is also desirable to select a target whose characteristic lines lie close to the absorption edges of the elements to be analyzed; for example, the *WL* lines and the Cu*K* lines are more efficient in exciting fluorescence in the first transition elements than are the Mo*K* lines.

The focal-spot size of the x-ray tube does not enter directly into the spectrograph geometry. Hence it is usually made two or three times larger than those used in diffraction tubes, and the tube may thus be operated at correspondingly higher power. To obtain the maximum intensity, the specimen should be placed as close to the window of the tube as possible, since the intensity falls off according to the inverse square law. In fluorescence analysis there is an important advantage in operating the x-ray tube at constant dc potential rather than full-wave rectification because x-rays of a certain wavelength are generated only during that fraction of the voltage-time cycle in which the critical excitation potential for that wavelength is exceeded. For example, the critical excitation potentials of the *K* spectra of Ni and Ba are 8.3 and 37.4 keV, respectively. If a primary x-ray tube with tungsten target is operated at 50-kV peak full-wave, the fluorescent intensities of these elements will be only 0.60 and 0.27, respectively, of those obtained at 50 kV in dc operation.

Equipment is normally operated at voltages up to 50–60 kV for fluorescence analysis. These voltages can generate the *K* spectra of all the elements up to the rare earths and the *L* spectra of the higher atomic number elements; equipment operating up to 100 kV or more is required for the *K* spectra of the latter group. Since counter tubes are used, it is essential to have a constant primary intensity and to stabilize the voltage and tube current.

Diffractometry. The fluorescence-analysis instrumentation is similar to that used in x-ray powder diffractometry, as shown in Fig. 4. In diffractometry, x-rays of a single wavelength are used to measure the intensities corresponding to all the *ds* in the polycrystalline specimen. The diffraction pattern is characteristic of the structure of the element or compound. Figure 5*a* and *b* show two x-ray diffractometer diagrams; Fig. 5*c* is an x-ray fluorescence pattern. Figure 5*a* is of a mixture of approximately 2 parts molybdenum and 1 part tungsten. Figure 5*b* shows a solid solution of Mo and W. Each reflection in the two diagrams consists of the Cu$K\alpha_{1,2}$ doublet, a pair of closely spaced lines with a 2:1 intensity relationship. Figure 5*c* is an x-ray fluorescence pattern of Mo and W showing the Mo*K* and W*L* spectra. The Roman numerals indicate second-order reflections.

Counter tubes. Devices such as the Geiger-Müller counter and the scintillation counter are used almost exclusively rather than film as the x-ray detector because they make possible rapid and accurate intensity measurements. In addition to such important characteristics as stability and reliability, counters for fluorescence analysis should have a linear response to avoid intensity corrections at the higher counting rates, a high quantum counting efficiency to obtain the maximum observable counting rate, and good energy resolution to allow discrimination against unwanted wavelengths.

Intensity measurement. X-ray quanta are emitted in a random manner and hence certain statistical rules must be taken into account in using counter tubes for x-ray intensity measurements. There are three basic methods.

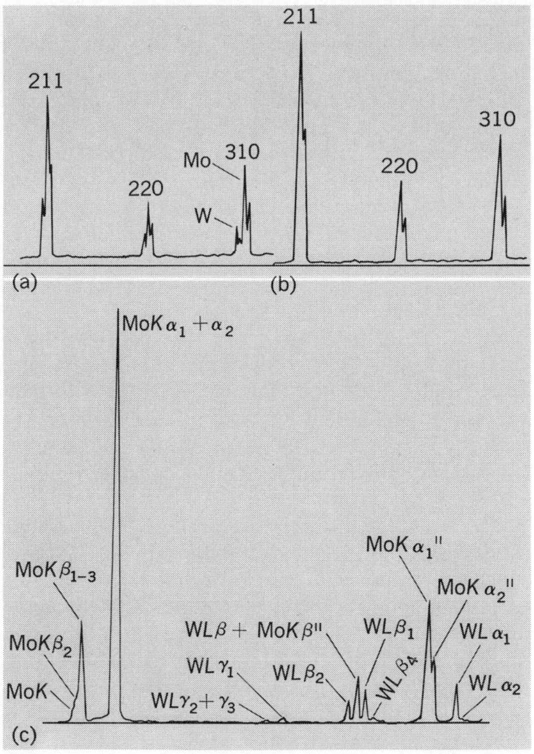

Fig. 5. Powder diffractometer diagrams of (*a*) Mo and W mixture and (*b*) Mo and W solid solution. (*c*) Fluorescence pattern of Mo and W. (*Svensk Kemisk Tidskrift, vol. 68, 1956*)

The fixed-time method uses the scaling circuit to measure the count N during a fixed predetermined time interval t and the counting rate n is determined from $n=N/t$. This method is commonly used when the amount of time available is limited, but has the disadvantage of the error being dependent on the counting rate.

In the fixed-count method, N is predetermined and t is variable. Although the error is then independent of n, long counting times will be required for low counting rates. Both the fixed-time and fixed-count methods are essentially point-by-point manual procedures requiring the setting of the counter tube correctly on the peaks of the lines to be measured.

A commonly used method is to operate the goniometer at a constant angular velocity and to record the output of the rate meter with a strip chart recorder. This introduces a distortion of the line profiles, causing the peaks to be lower than their correct intensity. The decrease of peak intensity is proportional to the product of the time constant and the scanning speed of the goniometer. This method is employed for qualitative and semiquantitative analysis, and once the peaks are located, one of the manual methods may be used for a more precise measurement.

Analytical methods. The specimens may be in the form of powders, briquettes, solids, or liquids. The surface exposed to the primary x-ray beam must be flat, smooth, and representative of the sample as a whole because usually only a thin surface layer contributes to the fluorescent beam. The degree of surface roughness, which is difficult to measure quantitatively, causes losses in intensity and results in errors in the analysis. Consequently, solid samples are generally polished, and then if necessary are lightly etched or specially cleaned to remove contaminants. Powders of large particle size are generally ground and then briquetted, so that the surface will meet the conditions specified.

Normally, if the specimen is of unknown composition, a rate-meter recording is used to make a qualitative or semiquantitative analysis. The reflection angles of each of the lines are read from the chart and the wavelengths computed using Bragg's equation or a table of 2θ versus λ for the specific crystal analyzer used. The elements are identified from the wavelengths, and the concentrations may be estimated from the relative intensities.

For accurate quantitative analysis, the usual procedure is to prepare a number of standard reference specimens of known composition for the various concentrations. The standards are used to prepare working calibration curves in which the intensities are plotted against the concentrations of each element in the required concentration range. The linearity and slope of the calibration curve are dependent upon the x-ray absorption characteristics of the specimen. The greater the differences in absorption of the elements to be analyzed, the closer together in composition should be the standards. In some analyses it is desirable to use an internal standard method, that is, to add a known amount of a reference element to the specimen during preparation. The calibration curve is then prepared by using for the intensity scale the ratio of the intensity of a line of the standard to that of another element of nearly the same absorption.

The intensities may be measured to the required statistical accuracy at the peak of the line using a fixed-count or fixed-time method. If the lines have different shapes, for example, in comparing an unresolved with a partially resolved doublet, the integrated line intensities should be obtained by scanning across the lines and adding together the counts with the scaler.

The precision of the method depends upon many factors: specimen preparation and homogeneity, the accuracy of the calibration curves, the stability of the x-ray source, counting statistics, and the resolution and dispersion of the spectrograph. In some cases, overlapping lines may be difficult to eliminate and may influence the precision. With a good modern, properly adjusted instrument, measurements of elements present in amounts more than 1% can usually be made in a few minutes to a precision of a few percent or better.

[WILLIAM PARRISH]

Bibliography: E. P. Bertin, *Introduction to X-ray Spectrometric Analysis*, 1978; T. G. Dzubay (ed.), *X-ray Fluorescence Analysis of Environmental Samples*, 1977; H. K. Herglotz and L. S. Birks (eds.), *X-ray Spectrometry*, 1978; R. Jenkins, *An Introduction to X-ray Spectrometry*, 1974; H. A. Liebhafsky et al., *X-rays, Electrons and Analytical Chemistry: Spectrochemical Analysis with X-rays*, 1972.

X-ray powder methods

Physical techniques used for the identification of substances and for other types of analyses, principally for crystalline materials in the solid state. In these techniques, a small collimated beam of x-rays is directed onto a small polycrystalline specimen in the form of powder, producing a diffraction pattern that is recorded on film or with a counter tube. This x-ray pattern is a fundamental and uniquely characteristic property resulting from the atomic arrangement of the diffracting substance. Different substances have different atomic arrangements or crystal structures, and hence no two chemically distinct substances give identical diffraction patterns. Identification may be made by comparing the pattern of the unknown substance with patterns of known substances, in a manner somewhat analogous to the identification of persons by their fingerprints. The analytical information is different from that obtained by chemical or spectrographic analysis. X-ray identification of chemical compounds indicates the constituent elements, and shows how they are combined.

The x-ray powder method is widely used in fundamental and applied research; for instance, it is used in the analysis of raw materials and finished products, in phase-diagram investigations, in following the course of solid-state chemical reactions, and in the study of minerals, ores, rocks, metals, chemicals, and many other types of material. The use of powder methods to determine the actual atomic arrangement, which has been so important in the study of chemical bonds, crystal physics, and crystal chemistry, is described in related articles. *See* X-RAY CRYSTALLOGRAPHY; X-RAY DIFFRACTION; X-RAY FLUORESCENCE ANALYSIS.

Instrumentation. Complete automatic equipment for x-ray analysis is available from several companies. A typical counter tube installation is shown in Fig. 1. The high-voltage generator which provides rectified voltage for the x-ray tube is stabilized so that the x-ray intensity varies by less than 1%. The diffractometer goniometer is mounted on the table in front of the x-ray tube window. The electronic circuits mounted in the rack on the right are used to operate the Geiger, proportional, or scintillation counter x-ray detector, and to record the diffraction data. The x-ray tube is sealed off (that is, has a permanent vacuum) and has a water-cooled target and three or four low-energy-absorbing mica or beryllium windows. X-ray cameras may be mounted at other windows.

X-ray spectra. Only about six elements are useful for x-ray tube targets. The spectrum of copper, the most commonly used target, is shown in Fig. 2*a*. It consists of a few lines superimposed on a broad continuum. These lines are the intense Cu$K\alpha$ doublet and the weaker Cu$K\beta$ triplet. In the doublet, the $K\alpha_1$ line at 0.1541 nm is about twice as intense as the $K\alpha_2$ line at 0.1544 nm. The individual components of the triplet are not usually resolved, and appear as a single line of average wavelength 0.1392 nm. The wavelengths of the lines are determined by the target element, and their intensity by the voltage and current applied to the x-ray tube. The other weak lines in this figure are caused by small amounts of tungsten impurity. There are also

L and *M* spectral lines of much longer wavelengths, but these are absorbed by the x-ray tube window. The continuum begins at a wavelength which depends on the applied voltage, rises rapidly to a broad maximum, and then diminishes with increasing wavelength. The intensity of the continuum is mainly dependent on the x-ray tube voltage, but it also increases with atomic number of the target element.

The entire spectrum is scattered and diffracted with varying efficiencies by the polycrystalline specimen. The interpretation of the diffraction pattern is simplified if the $K\beta$ line is eliminated. This is easily accomplished by inserting in the x-ray path a thin nickel foil which almost completely absorbs the $K\beta$ radiation and transmits the $K\alpha$ as shown in Fig. 2*b*. Other filter elements are used for other targets. To decrease the background from the continuum, and thereby increase the peak-to-background ratio of the diffraction pattern, a proportional or scintillation counter may be used in conjunction with a single-channel pulse-height analyzer. When the analyzer window is centered on the Cu$K\alpha$ lines and narrowed to pass about 90% of Cu$K\alpha$ radiation, the recorded spectrum consists almost entirely of Cu$K\alpha$, as shown in Fig. 2*c*.

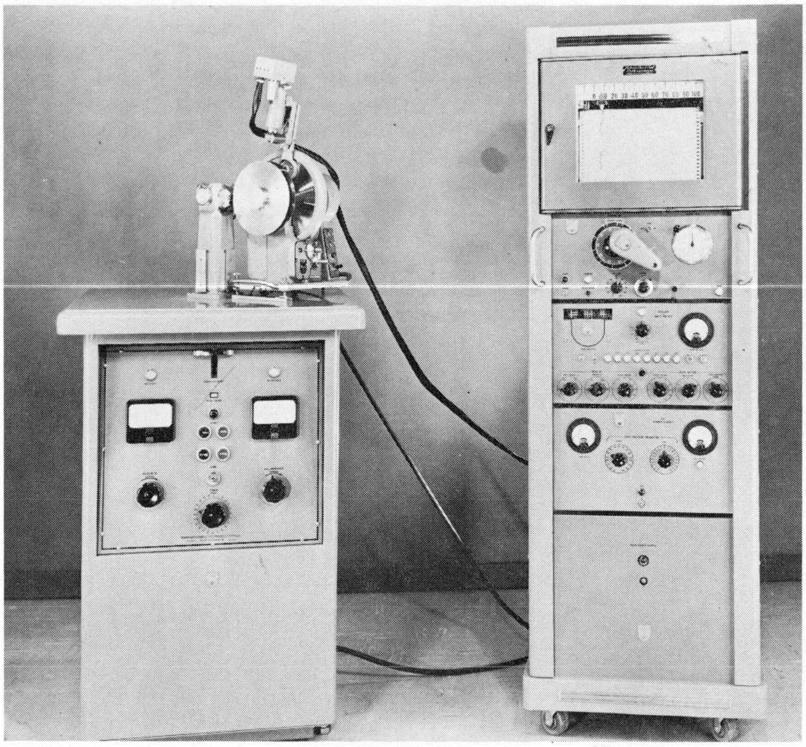

Fig. 1. X-ray diffraction equipment, showing high-voltage control (left) with x-ray tube housing and diffractometer goniometer mounted on table in front of the x-ray tube window, and recording circuits (right). (*Philips Electronics*)

Lattice geometry. The atoms in crystalline substances are arranged in a symmetrical three-dimensional pattern in which some atomic arrangement is repeated by the symmetry of the crystal along straight lines throughout the crystal. The array of points and the light lines in Fig. 3 outline a lattice or framework of a typical crystal in which the third dimension is normal to the plane of the drawing. The smallest group of atoms which has the symmetry of the entire pattern is called the unit cell. There are several ways in which the unit cell is selected. In this case it was drawn parallel to the crystallographic axes *a*, *b*, and *c*. The traces of the various planes (normal to the drawing) are indicated by heavy lines. The method used by crystallographers to identify these planes is as follows: For a given set of planes, count the number of planes crossed from one lattice point to the next along *a*, then repeat the procedure along *b* and *c*. The resulting numbers, called *h*, *k*, *l*, respectively, are known as the Miller indices of that set of planes, and assignment of indices to each line is called indexing the pattern. Miller indices of the (102) planes are called one, zero, two.

The spacings *d* between the planes are related to the Miller indices and the unit cell dimensions. In crystals of the cubic system, the crystallographic axes are normal to each other and have the same length, $a = b = c$, and $d = a/(h^2 + k^2 + l^2)^{1/2}$. Similar relations exist for the other five crystal systems. The relations for the low-symmetry systems, monoclinic and triclinic, are much more complicated. Various types of charts to facilitate indexing cubic, tetragonal, and hexagonal substances are available. *See* CRYSTAL STRUCTURE.

Bragg's law of reflection. The atomic diameters and the wavelengths of x-rays used have approximately the same dimensions, and therefore the crystal acts as a three-dimensional grating for x-rays in a manner analogous to the diffraction of ultraviolet or visible light by a ruled one-dimensional grating.

Under appropriate conditions, the electrons around each atom scatter the incident x-ray beam in a coherent manner and in certain specific directions; the scattering from billions of atoms is in phase at the same time. Shortly after the discovery of x-ray diffraction by M. von Laue in 1912, this complex phenomenon was formulated in a simple relation by W. H. and W. L. Bragg, in which the diffraction is visualized as a reflection from a large number of parallel planes. (This should not be confused with the total reflection of x-rays, which occurs at very small grazing angles from highly polished surfaces.) They showed that when two or more parallel rays of the same wavelength are incident at the same glancing angle θ to a set of atomic planes, the path difference of the reflected rays from adjacent planes is one wavelength. This may be expressed as $\lambda = 2d_{hkl} \sin \theta$, where λ is the wavelength of the incident x-rays and d_{hkl} the interplanar spacing between the (*hkl*) planes of atoms. Thus the conditions for x-ray reflection are very restrictive because there is only one angle θ at which the x-rays of a given wavelength are reflected by a particular set of atomic planes of spacing *d*.

The reflection law may be illustrated by use of Cu$K\alpha$ x-rays and a flat, single crystal section of sodium chloride cleaved parallel to the cube face, $(hkl) = (100)$ with $d = 0.564$ nm. When this crystal surface is correctly oriented on the goniometer at $\theta = 7.8°$, and the counter tube set at $2\theta = 15.6°$, no reflection occurs because the scattered rays from successive (100) planes are out of phase with each other, one plane of atoms exactly canceling the contribution from its neighbor. If the θ angle of the crystal surface is increased to 15.8°, the alternate planes reinforce each other and a very strong reflection is detected by the counter tube at $2\theta = 31.6°$. This is the second-order reflection of (100), or the first-order reflection of the (200) planes, $d = 2.82$ A. The (300) planes also do not reflect, but the (400) planes reflect at $2\theta = 66.3°$, although less strongly than the (200) planes. If other sections were sawed parallel to (111) with $d = 0.326$ nm, and to (220) with $d = 0.199$ nm, reflections of different intensities would be obtained with the counter tube set at 27.4° and 45.5°, respectively.

If longer wavelength x-rays had been used, the reflections would have occurred at correspondingly larger angles and the dispersion would have

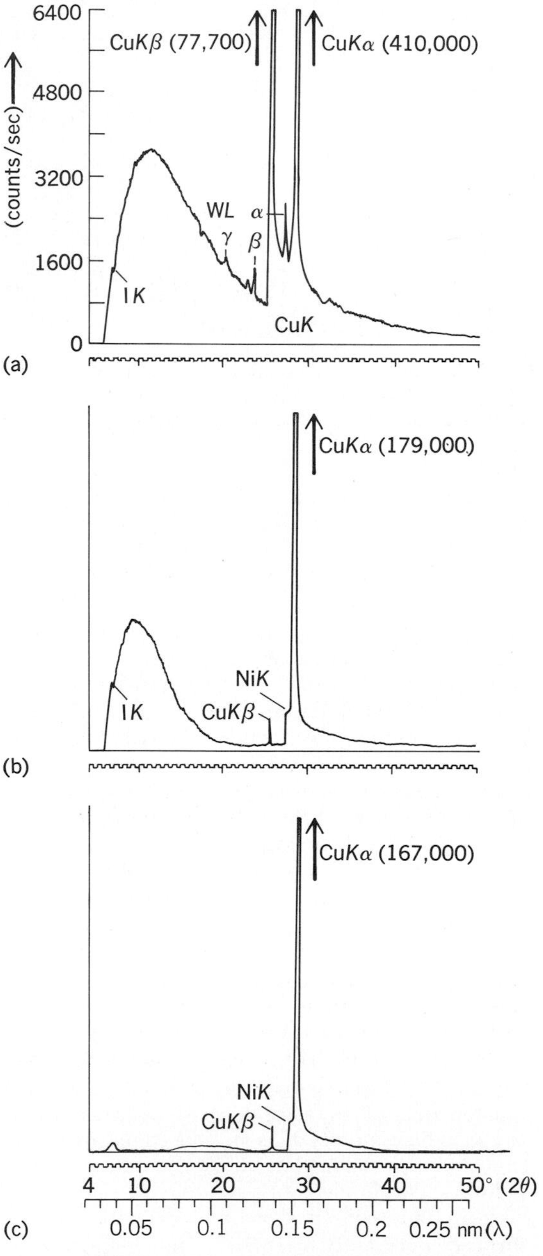

Fig. 2. Spectrum of copper target x-ray tube operated at 40 kilovolts peak (kVp), full-wave rectification. Spectrum recorded using a silicon crystal plate cut parallel to the (111) planes and an NaI(Tl) scintillation counter. (*a*) Spectrum from tube. (*b*) Same as *a* but with nickel filter 15 μm thick. (*c*) Same as *b* but with pulse-height analyzer set to pass about 90% of CuKα. (*Modified from International Tables for X-Ray Crystallography, vol. 3*)

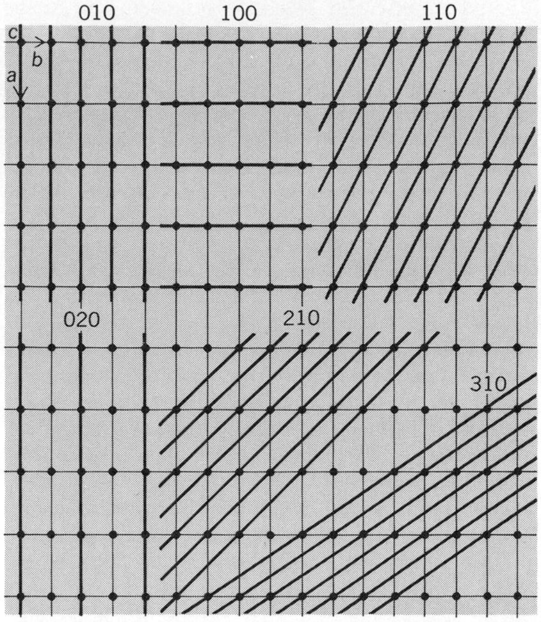

Fig. 3. Schematic drawing of lattice, lattice planes (heavy lines), and their Miller indices. The orthogonal axes *a* and *b* outline the simplest unit cell, and the *c* axis is normal to the paper.

been greater, since the 2θ angle at which the reflection from a given set of planes of spacing d occurs is proportional to $\lambda/\sin\theta$. For a given wavelength, larger d spacings appear at smaller angles. Moreover, if the Bragg equation is differentiated, one obtains $\Delta\theta/\Delta d = -(\tan\theta)/d$, which shows that the shift in line position $\Delta\theta$ due to a change of lattice spacing Δd increases as the tangent of the angle and reaches a maximum at $\theta = 90°$ (reflection angle $2\theta = 180°$), where $\tan\theta$ is infinite. Hence the highest-angle lines in the back-reflection region of the pattern are the most sensitive to changes in the lattice spacings and consequently supply the most accurate data for the measurement of the unit cell dimensions.

Many materials are not available in the form of large single crystals, and moreover it is impractical to obtain all the x-ray reflections from single crystals for identification purposes. If the sample does not already exist in polycrystalline form, it may be pulverized. When a fine-grained powder consisting of thousands of small, randomly oriented crystallites is exposed to the x-ray beam, all the possible reflections from the various sets of atomic planes can occur simultaneously.

Recording of powder patterns. The two principal methods of recording diffraction patterns are by the use of film and with x-ray counter tubes. In the film method, a small amount of the powder is glued to a thin glass fiber and rotated continuously in the center of a powder camera. The latter is essentially a light-tight cylindrical enclosure, usually about 4.5 in. (11 cm) in diameter and provided with collimators to define the x-ray beam. The strip of x-ray film is placed against the inside circumference of the cylinder concentric with the specimen axis. Two holes are punched in the film, one to admit the collimator, and the other to allow the undiffracted beam to pass through. Exposures of 1–4 hours are required. Typical films are shown in Fig. 4.

Each properly oriented crystallite may produce a reflection that appears as a spot on the film. For each set of atomic planes there will be a large number of crystallites of various orientations, each producing its own spot, and all the reflected rays will form the surface of a cone, with its apex at the specimen and subtending the angle 4θ on the film. The various sets of atomic planes thus produce a series of concentric cones and these appear on the film as a series of arcs whose curvature depends on the reflection angle. After the film is developed and dried, it is laid flat and the linear distance measured from the position on the film where the direct beam passed through to each of the arcs. Since λ and the camera diameter are known and since allowance may be made for film shrinkage in development, these linear distances may be converted into angular values, from which d may be calculated for each set of planes by substituting in the Bragg equation given above.

A much more powerful instrument, called the diffractometer (Fig. 1), was developed in 1944; it employs a counter tube instead of film. The specimen is prepared with a flat surface which is exposed to the x-ray beam. The counter tube always points toward the specimen, and the reflection data are recorded line by line. The divergent primary x-ray beam converges, after reflection from the specimen, to a narrow slit in front of the counter tube. By rotating the counter tube at twice the angular speed of the specimen so that the specimen surface always makes an angle θ with the primary beam when the counter tube is at 2θ, sharp well-resolved lines are obtained at all reflection angles. This instrument makes possible the direct and accurate measurement of intensities and reflection angles. The peak-to-background ratios, resolution, and line shapes are far superior to those obtained by the cylindrical film camera just described. The operation is completely automatic.

Identification of crystals. The most important use of the powder method is for the identification of crystalline substances. Because no two different substances give identical powder patterns, an identification may be made by comparison of the unknown pattern with knowns. In the United States, listings of the pertinent information, such as the d values and relative intensities of the lines of many thousands of substances, are published by the American Society for Testing and Materials.

The complexity of the pattern is determined primarily by the symmetry of the substance rather than by its chemical composition. Hence, chemically complex compounds such as $(Cu_{10}Zn_2)Sb_4O_{13}$, $(NH_4)_3AlF_6$, and $2Na_2SO_4 \cdot NaCl \cdot NaF$ have patterns that are nearly as simple as those from Fe or Cu, because all are cubic. Thus the simplest patterns, that is, those with the smallest number of lines, occur in the cubic system and the number of lines increases progressively, with decreasing symmetry becoming greatest in the triclinic system. The number of lines also increases with unit cell size.

One of the most important characteristics of powder patterns is that isostructural substances (that is, substances with the same crystal structure) give similar diffraction patterns. The distribution of spacings and intensities of corresponding lines are nearly the same. The atoms in diamond, silicon, and sphalerite (ZnS) are all tetrahedrally bonded to their neighbors and all have the diamond-type structure. The differences in the spacings and relative intensities are dependent upon the atomic sizes and their x-ray-scattering power. Charts showing the powder patterns of structure types in the cubic, tetragonal, and hexagonal systems have been compiled, and it is often possible to identify the structure type by visual inspection of the patterns.

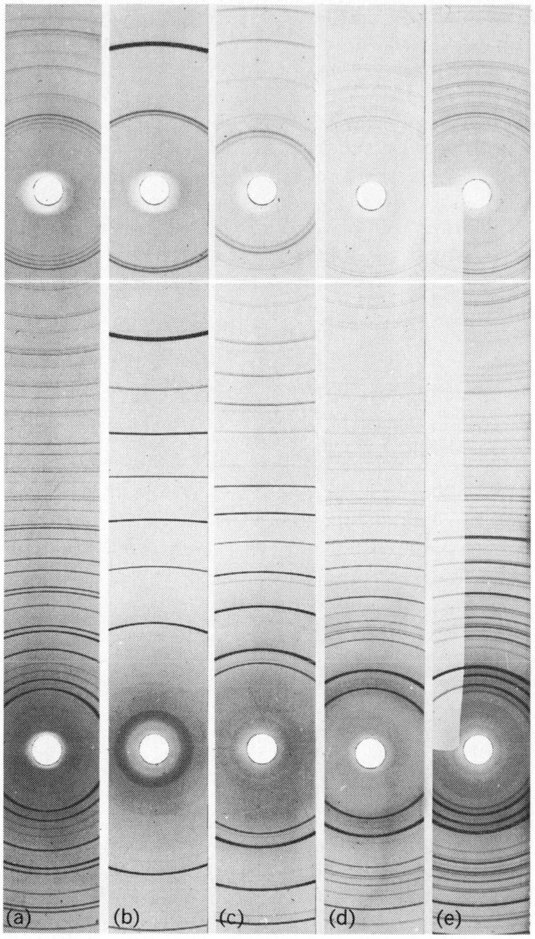

Fig. 4. Films of several polycrystalline substances obtained with a cylindrical powder camera and Cu*Kα* x-rays. (*a*) $Pb(NO_3)_2$. (*b*) W. (c) NaCl. (*d*) SiO_2-α quartz. (e) Same as *d* but with Ni filter covering one side of film. (*Science, 110(2858):369, 1949*)

Many substances may occur in two or more crystal structures; that is, they may have polymorphic forms but the same chemical composition. For example, sphalerite and wurtzite are both ZnS. Such forms may be caused by slight differences in chemical preparation, different heat treatments, and other factors.

When two or more substances are present in the sample, the pattern of each substance appears independent of the others. This makes the identifications more difficult, but mixtures containing as many as six substances have been successfully analyzed. It is also possible to make quantitative analyses of the mixtures by comparing the relative intensities of one or more principal lines of each substance. Usually some reference standards of known chemical compositions are prepared to facilitate the interpretation.

Amorphous substances. Amorphous materials such as glasses and liquids give patterns which consist of only a few broad lines superimposed on a continuous background. Patterns of various amorphous substances closely resemble each other and hence the method is impractical for identification. On the other hand, since the two types of patterns are so different, the method is ideally suited to distinguish between crystalline and amorphous substances and to determine the degree of crystallinity of substances between the two extremes. The study of the progress of devitrification (crystallization) of a glass and similar problems are frequently accomplished with x-rays. Certain chemical and structural properties of liquids might be ascertained from the diffraction patterns when the liquids are frozen to form crystals.

Solid solutions. There are also many smaller changes in the x-ray pattern which may reveal important information. In substitutional solid solutions, for example, atoms of different elements may substitute for one another and occupy the same relative positions as in the pure metals. The substitutions of solute atoms occur on the same lattice sites occupied by the solute atoms, but are randomly distributed. If the atoms are of different size, the average unit cell size will change accordingly. In simple cases, it is possible to determine the chemical composition of intermediate members by measuring the unit cell dimensions because there is often a nearly linear relationship between the two. In interstitial solid solutions, atoms are added to the empty spaces in the structure and there is little, if any, change in the dimensions. *See* SOLID SOLUTION.

Precision measurements of unit cell dimensions, or lattice parameters, are required for many types of solid-state studies. For example, measurements at two known temperatures permit the calculation of the coefficient of thermal expansion. The data have also been used to compute the atomic weight of certain elements with a precision approaching that obtained by chemical methods. When the density of the crystals is known, the data may be used to compute the molecular weight.

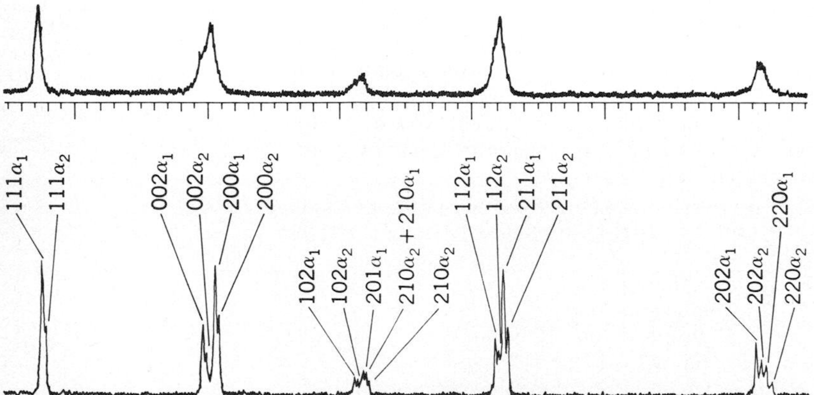

Fig. 5. Diffractometer patterns of strained $BaTiO_3$ (above) and unstrained $BaTiO_3$ (below). Strain has broadened the diffracted lines. (*Modified from Transactions of the Instrumentation and Measurements Conference, Stockholm, 1952*)

In phase-diagram analyses, the x-ray powder method is an extremely valuable tool to determine the limits of solubility and to identify the phases that occur. In solid-state chemical reactions, chemical analysis is of little help because the bulk chemical composition of the starting constituents is the same as the desired end product. X-ray diagrams are therefore used to follow the course of the reaction and to determine if the desired end product has been formed.

Analysis of physical properties. When a substance is strained or plastically deformed, the x-ray lines broaden, and when the strain is removed by annealing, the lines return to their original sharpness. This is illustrated by the diffractometer recordings in Fig. 5. The upper recording shows the pattern of a sample of $BaTiO_3$ which has been strained by crushing, while the lower pattern shows the same substance in an unstrained state. X-ray patterns can thus be used to follow the course of heat treatment or other processes used to remove strains.

If the crystallites do not have a completely random orientation, the line shapes and relative intensities will change accordingly. For example, in rolling thin sheets of metal or in drawing wire, the crystallites align themselves in special ways, depending on the mechanical conditions of the process. The special or preferred orientation of the crystallites gives x-ray diffraction patterns which may be markedly different from those of the random crystallites. Similar conditions may arise in electroplating, where the plating conditions or the substrate may cause the crystallites to have a nonrandom orientation. Comparison of the random and oriented x-ray patterns shows the degree of orientation in the sample.

When the crystallites are very small, for instance, <1 μm, the x-ray lines broaden by an amount which increases with decreasing size. Comparisons with the line breadths of samples of larger crystallite sizes may lead to a measure of the average crystallite size in the specimen.

[WILLIAM PARRISH]

Bibliography: C. S. Barrett and T. B. Massalski, *Structure of Metals*, 3d ed., 1980; B. D. Cullity, *Elements of X-Ray Diffraction*, 2d ed., 1978; H. P. Klug and L. E. Alexander, *X-Ray Diffraction Procedures*, 2d ed., 1974.

X-ray spectrometry

A rapid and economical technique for quantitative analysis of the elemental composition of specimens. It differs from x-ray diffraction, whose purpose is the identification of crystalline compounds. It differs from spectrometry in the visible region of the spectrum in that the x-ray photons have energies of thousands of electronvolts and come from tightly bound inner-shell electrons in the atoms, whereas visible photons come from the outer electrons and have energies of only a few electronvolts. *See* X-RAY DIFFRACTION.

In x-ray spectrometry the irradiation of a sample by high-energy electrons, protons, or photons ionizes some of the atoms, which then emit characteristic x-rays whose wavelength λ depends on the atomic number Z of the element ($\lambda \propto 1/Z^2$), and whose intensity is related to the concentration of that element. Generally speaking, the characteristic x-ray lines are independent of the physical state (solid or liquid) and of the type of compound (valence) in which an element is present, because the x-ray emission comes from inner, well-shielded electrons in the atom. Figure 1 illustrates the removal of one of the innermost, *K*-shell, electrons by a high-energy photon. The photon energy must be greater than the binding energy gas of the electron; the difference in energy appears as the kinetic energy of the ejected electron. The *K*-ionized atom is unstable, and one of the *L*- or *M*-shell electrons drops into the *K*-shell vacancy within 10^{-14} to 10^{-16} s. As this transition occurs, a characteristic x-ray photon is emitted with an energy equal to the difference in energy between the *K* and the *L* (or *M*) shell, or an additional electron, called an Auger electron, is ejected from the atom. Either the x-rays or the Auger electrons may be used for analysis, but in this article the discussion is concerned exclusively with the x-rays: *See* AUGER EFFECT.

Figure 2 shows some of the allowed transitions and the naming of the lines. There is a selection rule in atomic physics which says that only certain ones of the outer electrons are allowed to fill a vacancy in an inner shell. The rule can only be stated in terms of quantum mechanics: the transition must be from one shell to another, and the second (orbital) quantum number must change by ± 1, that is, $p \rightleftarrows s$, $d \rightleftarrows p$, and so on. Figure 3 shows the λ versus Z relationship for the strongest *K*- and *L*-series lines. Figure 2 shows that some of the transitions may come from the valence shell, in which case there will be slight alteration of wavelength or line shape of characteristic lines with valence. This alteration can be measured for multivalent elements such as sulfur by using spectrometers designed for high resolution (Fig. 4).

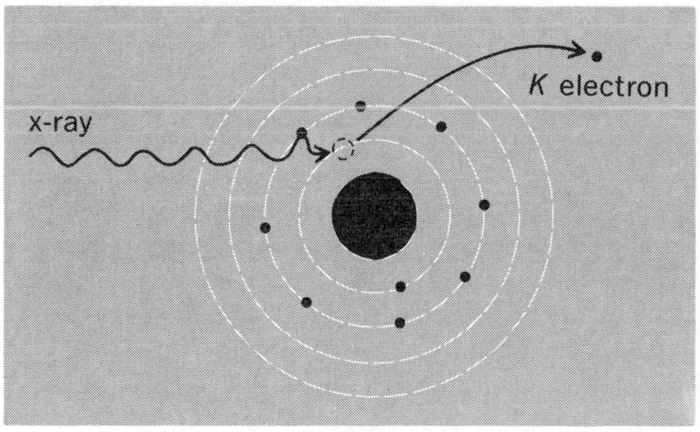

Fig. 1. Removal of a *K* electron from an atom by a primary x-ray photon. (*From L. S. Birks, X-Ray Spectrochemical Analysis, 2d ed., Wiley-Interscience, 1969*)

Spectrum analysis. Photon generation of characteristic spectra is the most common and is called x-ray fluorescence. It is carried out with an x-ray tube as the source of primary radiation. There are two ways of analyzing the spectra: wavelength dispersion and energy dispersion.

Wavelength dispersion. This is shown in Fig. 5*a*. The characteristic emission from the sample is usually excited by a chromium or tungsten target x-ray tube which is operated at 2–3 kW. The emitted radiation is limited to a parallel beam by the blade collimator and is diffracted, one wavelength at a time, by an analyzer crystal. Bragg's law ($n\lambda = 2d \sin \theta$) relates the diffraction angle θ to the wavelength λ for crystal planes with an interatomic-spacing distance d. The term n is the order of the diffraction, 1, 2, 3, etc., as it is in classical optics. As shown in Fig. 3, the characteristic wavelength

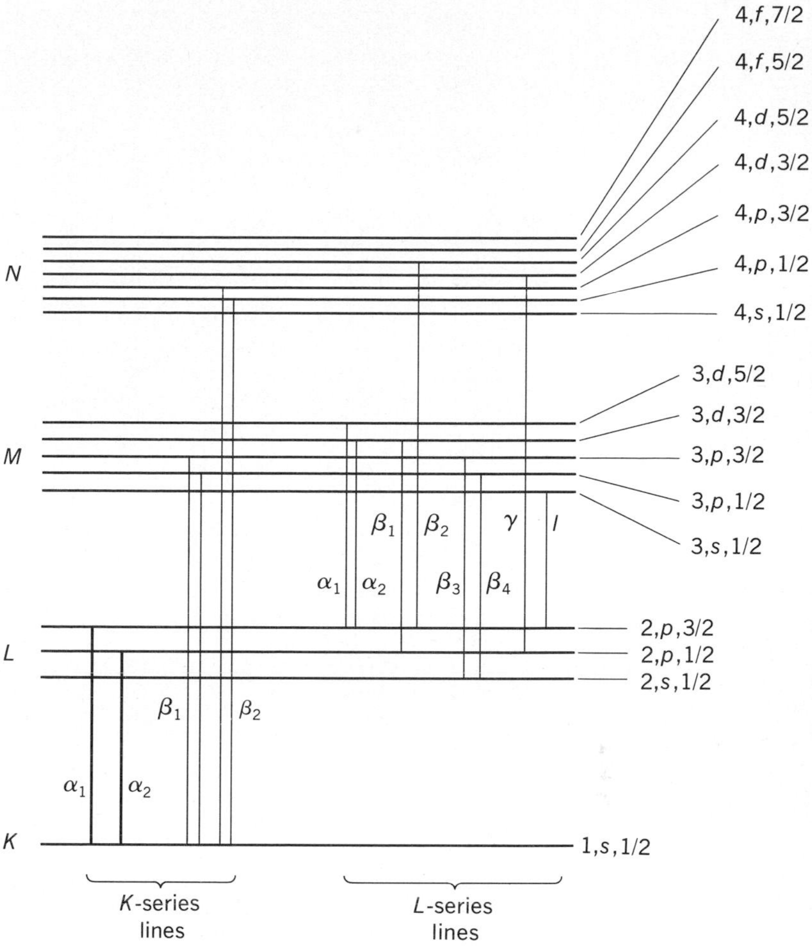

Fig. 2. Partial energy-level diagram showing the transitions leading to the *K* and *L* series lines. (*From L. S. Birks, X-Ray Spectrochemical Analysis, 2d ed., Wiley-Interscience, 1969*)

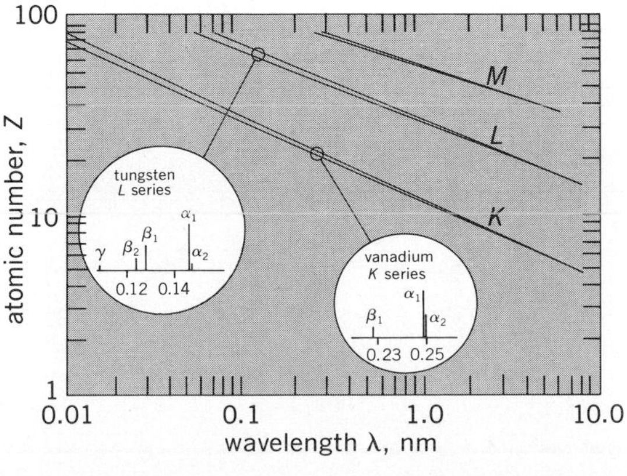

Fig. 3. Atomic number Z versus wavelength λ of the characteristic lines ($\lambda \propto 1/Z^2$). (*From L. S. Birks, X-Ray Spectrochemical Analysis, 2d ed., Wiley-Interscience, 1969*)

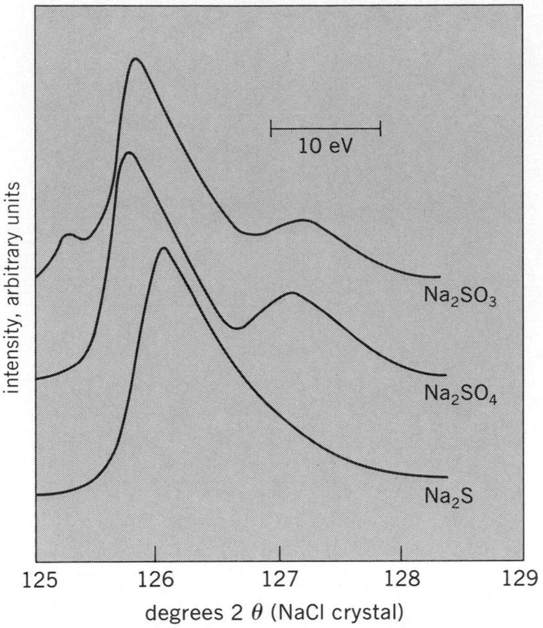

Fig. 4. Variations of the sulfur K_β line with valence state of sulfur. (*From L. S. Birks and J. V. Gilfrich, X-ray fluorescence analysis of the concentration and valence state of sulfur in pollution samples, Spectrochim. Acta, 33B, no. 7: 305, 1978*)

decreases as the atomic number increases. The $2d$ spacing of the analyzing crystal must be greater than the wavelength being diffracted, but if it is too much greater the spectral lines will be crowded toward small θ. It has become the practice to use a crystal of lithium fluoride for elements of atomic number Z greater than 20, a pentaerythritol crystal for Z between 13 and 20, and a potassium acid phthalate crystal for Z between 8 and 13. X-ray fluorescence analysis is not generally suitable below about $Z = 9$, but with special instruments and techniques it can be extended down to about $Z = 5$.

The detectors used for wavelength dispersion are either gas proportional counters or scintillation counters. Both of these count individual photons, but the gas counters are most suitable for wavelengths longer than about 0.2 nm, while the scintillation counters are most suitable for shorter wavelengths. The amplitude of the output pulse for each photon is proportional to the energy of the x-ray photon it represents. However, the statistical variation in the amplitude for each specific photon energy means that characteristic lines from neighboring elements are not resolved by either gas proportional or scintillation counters. Figure 6 shows the resolution for the gas proportional counter alone and with a crystal spectrometer, as well as for the silicon solid-state detector used for energy dispersion analysis, which is discussed below; resolution with a scintillation counter is almost a factor-of-three worse than a gas proportional counter. In wavelength dispersion it is the resolution of the crystal spectrometer, however, which determines the separation between neighboring wavelengths, and the crystal resolution is better than any of the detectors.

Energy dispersion. In this method all of the radiation emitted by the sample enters an energy-sensitive detector, usually a silicon solid-state detector (Fig. 5*b*). Such detectors are operated at liquid-nitrogen temperature to reduce electronic noise and allow an energy resolution of about 150 eV (Fig. 6). This resolution is adequate to distinguish the K_α and K_β lines of a single element, but not adequate to separate the K_β line of one element from the K_α line of the next higher atomic number element (for example, Cr K_β from Mn K_α). Figure 7 compares the wavelength and energy-dispersion spectra from a sample containing a range of atomic number elements. Resolution is better with the crystal spectrometer for most of the spectral range of interest.

In spite of its relatively poor resolution, energy dispersion has become widely accepted because of two advantages it has over wavelength dispersion. First, all the characteristic lines are recorded si-

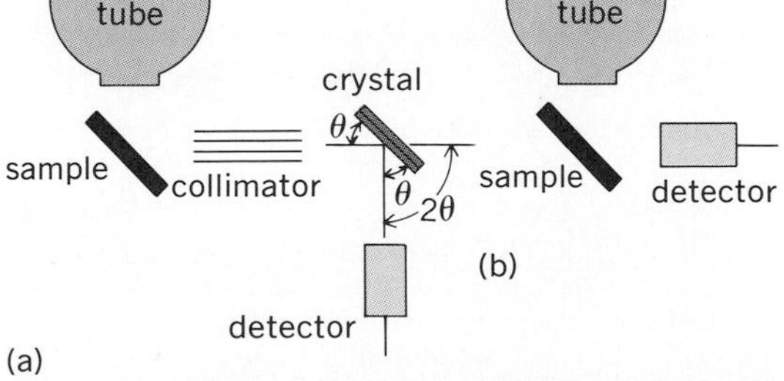

Fig. 5. Wavelength and energy dispersion methods. (*a*) Wavelength dispersion with a crystal spectrometer. (*b*) Energy dispersion with a solid-state detector.

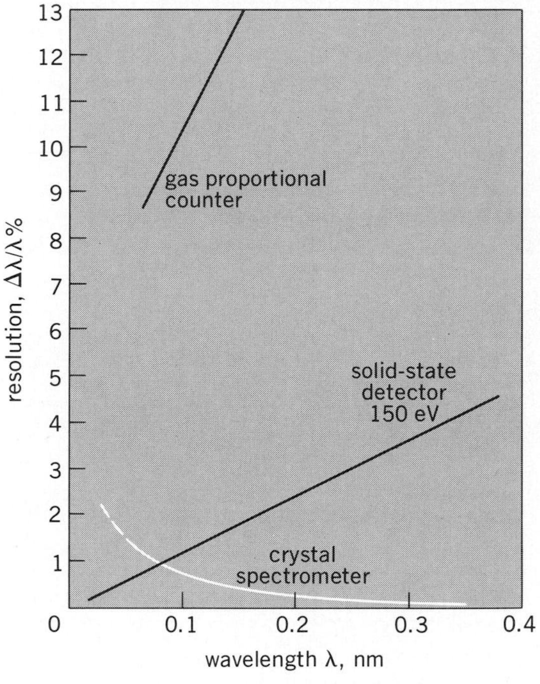

Fig. 6. Resolution of a solid-state detector and a gas proportional counter alone or in conjunction with a crystal spectrometer.

multaneously, which makes the method faster than a scanning crystal spectrometer (but not as fast as the multiple crystal-spectrometer instruments). Second, the solid angle of radiation accepted by the detector is 10–100 times greater than the solid angle accepted by the crystal; this allows the primary-source power to be reduced proportionally. With energy dispersion, it is feasible to use electron excitation at beam currents below 10^{-8} A in the scanning electron microscope, compared to beam currents of about 10^{-6} A required in electron probes used with crystal spectrometers. Likewise, with energy dispersion, it is feasible to use proton or alpha-particle beams from Van de Graaff or cyclotron accelerators. For proton beams, the proton energy must be 1–5 MeV, compared to electron energies of 10–50 keV for the same x-ray yield; for alpha particles, the energy should be even higher, 10–50 MeV. Compared to photon excitation, positive-ion excitation results in lower background intensity and improves the limit of detection by as much as a factor of 10. On the other hand, direct electron excitation results in much higher background intensity because it is generated in the sample rather than merely scattered by the sample; this degrades the limit of detection by a factor of 10–100.

With energy dispersion, it is even feasible to use some of the radioactive isotope sources as the primary radiation to excite the characteristic x-ray spectra in the sample. There are isotopes such as tritium which undergo beta decay and, when used in a metal matrix, produce continuum and characteristic matrix x-rays; other isotopes such as americium-241 undergo alpha decay and produce gamma rays or x-rays. The advantage of isotope sources is their compact size and the elimination of power supplies and so forth. However, advances in low-power air-cooled x-ray tubes (10–50 W) and compact power supplies have largely eliminated the need for isotope sources.

Types of samples and data interpretation. There are two general classes of samples which are analyzed routinely by x-ray spectrometry: thin and bulk samples.

Thin samples. Samples smaller than several mg/cm² such as air-pollution particles collected on a filter, constitute such a thin layer that the primary radiation can penetrate easily and excite emission from each element in proportion to the mass per unit area of that element. Quantitative calibration for each element is accomplished experimentally by determining the sensitivity S in photons/s per μg/cm² for that element. Accuracy (or more properly, precision) depends on the number N of photons counted; the expected standard deviation σ of a measurement is approximated by $\sigma = \sqrt{N}$. Limit of detection C_L depends not only on sensitivity but on the background intensity as well. If the measured intensity of the background is N_B and that at the line peak is N_p, the limit of detection is defined as the amount of material C_L which gives a signal above background of $3\sigma_B$, that is, $N_p - N_B = 3\sqrt{N_B}$ and $C_L = 3\sqrt{N_B}/(S \times t)$, where t is the counting interval in seconds. Thus it is important to minimize the background intensity N_B by careful instrumental design to eliminate scattering from material other than the sample, and also to minimize the mass/cm² of the substrate on which the sample is mounted. With photon excitation and optimized design, the limit of detection varies from about 1 ng/cm² for elements around atomic number 20 to 10–50 ng/cm² for the extremes of high or low Z. The limit is approximately the same for either wavelength or energy dispersion when photon excitation is used. As was stated in the previous portion of this subject area, the limit of detection can be improved by about a factor of 10 with proton excitation, but the analysis is then generally limited to energy dispersion.

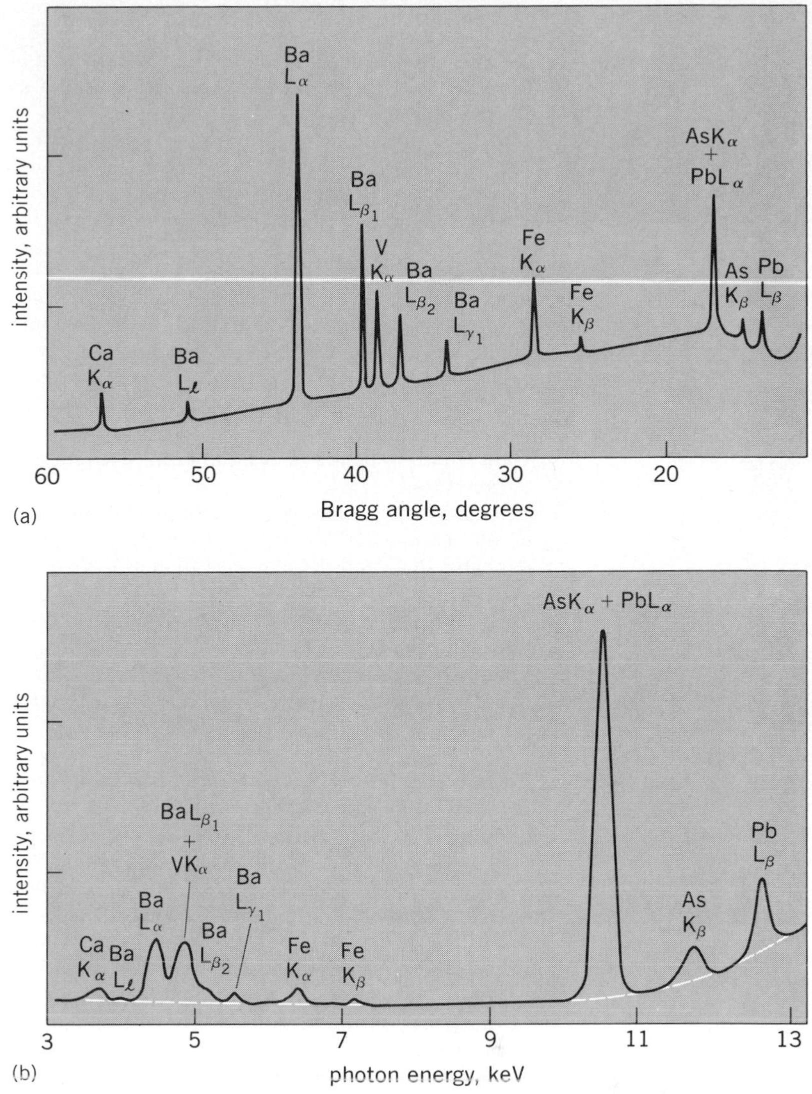

Fig. 7. X-ray spectra of the same sample as measured by (*a*) wavelength dispersion and (*b*) energy dispersion. (*From L. S. Birks, Pinpointing airborne pollutants, Environ. Sci. Tech., 12:150, 1978*)

Bulk samples. These consist of solids, powder, or liquids and present quite a different problem from thin samples. X-ray absorption limits the penetration of the primary radiation through the sample matrix and the depth from which characteristic radiation may emerge. In addition, the characteristic radiation from some of the elements in the specimen may excite the characteristic radiation of other elements by secondary fluorescence. Matrix absorption and secondary fluorescence depend on the sample composition and determine the x-ray intensity versus composition relationship for each element. Thus, instead of a single, linear calibration as was described for thin samples, the calibration is generally nonlinear, and a family of curves is needed for each element as the matrix composition changes.

If I_{ai} is the intensity from a pure sample of element i, and I_{bi} is the intensity from an unknown concentration of element i in a matrix of other elements, then the most useful parameter is the relative x-ray intensity $R_i = I_{bi}/I_{ai}$. To a first approximation, R_i can be expressed in terms of the concentration C_i and the absorption term μ which incorporates both the absorption of the incident primary radiation and the emerging characteristic radiation, as shown in Eq. (1), where A is a con-

$$R_i = AC_i/\mu \quad (1)$$

stant which depends on a number of instrumental parameters.

In x-ray spectrometry the analyst can measure R_i but cannot determine C_i directly, because μ depends on C_i and the concentration of each other element C_j as well. However, it is possible to write an equation containing R_i and C_i in terms of individual absorption and secondary-fluorescence effects of each element on the intensity from each other element. The expression is Eq. (2), where the

$$C_i/R_i = 1 + \Sigma\alpha_{ij}C_j \quad (2)$$

Σ symbol means the sum of a number of terms, one for each other element in the specimen. The coefficient α_{ij} means the effect on element i by the presence of element j; it is often referred to as an influence coefficient and may be determined experimentally by measuring mixtures of elements i and j. The terms may also be calculated from some of the fundamental properties of atoms and radiation.

There are many variations of the mathematical expressions for determining concentration from x-ray intensity; most of them require computers to evaluate. Whatever the method of data interpretation, a few generalizations may be made about the analysis of bulk specimens. The accuracy (precision) is about 1–2% of the amount of an element present for major constituents, but degrades to 5–10% of the amount present for concentrations of 10–100 ppm. The limit of detection is generally about 1 ppm for middle-range atomic number elements, but varies from less than 0.1 ppm for metals in biological tissue to 10 ppm or more for low-Z elements such as carbon in a middle-Z matrix such as steel.

Improvements and limitations. X-ray spectrometry generally does not require any separation of elements before measuring, because the x-ray lines are easily resolved. However, preconcentration methods are sometimes useful as a means for improving the limit of detection. An example is the precipitation or ion-exchange collection of soluble elements in water. Likewise, dilution is sometimes useful to reduce matrix variability or inhomogeneity. An example is the solution of mineral samples in borax.

One limitation of x-ray spectrometry is the progressive difficulty of measurement below atomic number 11. There are several reasons for this, including the strong absorption of such long-wavelength radiation by the x-ray tube and detector windows and also the reduced intensity due to a lower number of x-ray photons emitted per atom ionized (the fluorescent yield factor). Although the elements from boron (5) to fluorine (9) may be measured with specially designed equipment, such measurement cannot be considered routine. In practice, photoelectron spectroscopy and Auger electron spectroscopy are favored for the lower-atomic-number elements, but electron methods may also require special instrumentation and techniques. *See* ELECTRON SPECTROSCOPY; SPECTROSCOPY. [L. S. BIRKS]

Bibliography: E. P. Bertin, *Principles and Practice of X-Ray Spectrometric Analysis*, 2d ed., 1975; L. S. Birks, *X-Ray Spectrochemical Analysis*, 2d ed., 1969; R. Jenkins, *An Introduction to X-Ray Spectrometry*, 1974; H. A. Liebhafsky et al., *X-Rays, Electrons, and Analytical Chemistry*, 1972.

Xenon

A chemical element, Xe, atomic number 54. It is a member of the family of noble gases, group O in the periodic table. *See* INERT GASES.

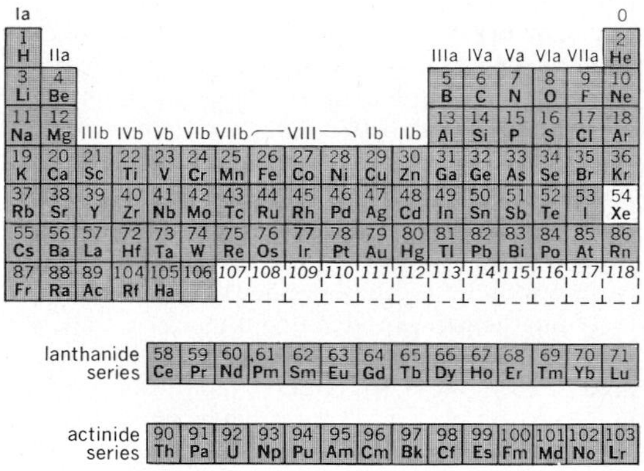

Uses. Xenon is used to fill a type of flashbulb employed in photography and called electronic speed light. These bulbs produce a white light that has a good balance of all the colors in the visible spectrum, and can be used 10,000 times or more before burning out.

A xenon-filled arc lamp gives a light intensity approaching that of the carbon arc; it is particularly valuable in projecting motion pictures.

Xenon readily absorbs radiation, such as x-rays; when xenon is mixed with acetylene, for example, and treated with x-rays, the xenon absorbs the radiation and transmits it to the acetylene; the result is the polymerization of the hydrocarbon.

The relative solubility of xenon in body tissues is high. Experiments on humans have shown that breathing a mixture of 20% oxygen and 80% xenon rapidly produces deep anesthesia useful in surgery; the patients on whom this was tried awoke without unpleasant aftereffects within 2 min of the time administration of the xenon was ended. Because the xenon is chemically inert, it cannot poison the patient, and there is no danger of fire or explosion as there is with ether, ethylene, or other flammable anesthetics. In another medical application, encephalography, xenon injected into the skull gives clearer x-ray pictures than when air is injected; moreover, xenon does not give the patients the long-lasting headache that sometimes results when air is used.

An important development in high-energy physics is the detection of nuclear radiation, such as γ-rays and mesons, by bubble chambers, in which a liquid is kept at a temperature just above its boiling point. Nucleation by the radiation results in bubble formation along the path of the particle. The tracks made by the particles are then photographed. Liquid xenon is one of the liquids used in these bubble chambers.

Xenon is used to fill neutron counters, x-ray counters, gas-filled thyratrons, and ionization chambers for cosmic rays; it is also used in high-pressure arc lamps to produce ultraviolet radiation.

Between 3 and 5% of the fissions in a nuclear reactor using uranium as fuel lead to the formation of xenon-135. An interesting use for this xenon has been suggested. Because it is an effective neutron absorber, it might be purged out of the fuel zone by a stream of helium and passed into a confined space around the reactor to act as a neutron shield.

Occurrence. Traces of xenon are found in minerals and meteorites, but the only commercial source of xenon is the air. Xenon constitutes 0.086 part per million by volume of dry air, and this xenon is a mixture of the following nine isotopes, none of which is radioactive: 129 (26.44%), 131 (21.18%), 132 (26.89%), 134 (10.44%), 136 (8.87%), and 124, 126, 128, and 130 (6.18%). The relative abundances by volume of the isotopes are given in parentheses.

A mixture of stable and radioactive xenon is formed in nuclear reactors in connection with the neutron fission of uranium.

It is estimated that about 3×10^{-9}% of the weight of the Earth is xenon. Xenon also occurs outside the Earth; the best estimate is that there are about 4 atoms of xenon per 1,000,000 atoms of silicon, silicon being the standard of abundance used for the elements in the universe.

Discovery. Xenon was discovered in England by Sir William Ramsay and M. W. Travers in 1898. They separated it by fractional distillation from crude krypton and identified it as a new element from the new lines which they found in the emission spectrum of the residual gas.

Radioactive isotopes. The following radioactive isotopes of xenon are known: Xe^{121}, Xe^{122}, Xe^{123}, Xe^{125}, Xe^{127}, Xe^{129m}, Xe^{131m}, Xe^{133}, Xe^{135}, Xe^{137}, Xe^{138}, Xe^{139}, Xe^{140}, Xe^{141}, Xe^{143}, and Xe^{144} (*m* stands for metastable). As mentioned above, these are produced in the fission of uranium in nuclear reactors; they may also be formed in particle accelerators, such as the cyclotron, or by the neutron bombardment of the appropriate atomic species. The only radioactive isotopes of xenon which are produced in appreciable yield in nuclear reactors are Xe^{133} (half-life 5.3 days) and Xe^{135} (half-life 9.2 hr). The formation of these isotopes is a nuisance in the operation of a reactor, because Xe^{135} and, to a much smaller extent, Xe^{133} readily absorb neutrons; therefore, they gradually "poison" the reactor fuel by slowing down the fission process.

Properties. Xenon is colorless, odorless, and tasteless; it is a gas under ordinary conditions. Other properties are shown in the table.

Xenon is the only one of the nonradioactive noble gases which forms chemical compounds that are stable at room temperature. Xenon also forms weakly bonded clathrates with such substances as water, hydroquinone, and phenol. *See* CLATHRATE COMPOUNDS; XENON COMPOUNDS.

Physical properties of xenon

Property	Value
Atomic number	54
Atomic weight (atmospheric xenon only)	131.30
Melting point (triple point), °C	−111.8
Boiling point at 1 atm* pressure, °C	−108.1
Gas density at 0°C and 1 atm pressure, g/liter	5.8971
Liquid density at its boiling point, g/ml	3.057
Solubility in water at 20°C, ml xenon (STP) per 1000 g water at 1 atm partial pressure of xenon	108.1

*1 atm = 101, 325 Pa.

Production and distribution. Xenon is produced commercially in an air-separation plant. The air is liquefied and distilled. The oxygen is redistilled: the least volatile portion contains small amounts of xenon and krypton, which are adsorbed on silica gel directly from the liquid oxygen. The crude xenon and krypton thus obtained are separated and further purified by distillation and selective absorption, at controlled low temperatures, on activated carbon. Remaining impurities are removed by passing the xenon over hot titanium, which reacts with all but the inert gases.

Xenon could also be obtained from the gases produced in a nuclear reactor. First, the xenon would be separated from the other gases in the reactor and then stored to eliminate radioactive xenon. Because the longest half-life of any of the radioactive xenon isotopes produced in the nuclear reactor is only about 30 days, substantially all radioactivity would cease at the end of about 10 months. The remaining stable xenon could then be purified. At present, the demand for xenon can easily be met by xenon produced from air.

Xenon is sold at atmospheric pressure in sealed glass vessels, and under higher pressures in steel cylinders.

Analytical methods. The principal modern methods of detecting and quantitatively determining the xenon content in gases are mass spectrometry and gas chromatography. Until these methods were developed, it was necessary to separate xenon from other inert gases by selective low-temperature adsorption on activated carbon in order to determine how much xenon was present in a mixture. The older method of detecting xenon is by its characteristic emission spectrum, obtained by passing a gas sample through an electric discharge tube at low pressure and analyzing the light with a spectrometer.

[ARTHUR W. FRANCIS]

Bibliography: I. Asimov, *The Noble Gases*, 1966; B. S. Kirk and A. H. Taylor, Helium-group gases, in R. F. Kirk and D. F. Othmer (eds.), *Encyclopedia of Chemical Technology*, vol. 12, 3d ed., 1980.

Xenon compounds

Compounds in which xenon is one of the combining elements. The chemistry of xenon was established in 1962 when the element was discovered to react both with platinum hexafluoride, PtF_6, and with elemental fluorine. Although xenon exhibits all of the even valence states (II, IV, VI, and VIII), and stable compounds of each of these states have been isolated, xenon chemistry is limited to the stable fluorides and oxyfluorides and their complexes, two unstable oxides, and the aqueous species derived from the hydrolysis of the fluorides. Xenon dichloride has been reported, but its existence is still somewhat questionable. *See* XENON.

Fluorides. The three fluorides, XeF_2, XeF_4, and XeF_6, are thermodynamically stable compounds at room temperature, and they may be prepared simply by heating mixtures of xenon and fluorine at 300–400°C. At these temperatures, F_2 is in equilibrium with a significant concentration of F atoms, and these atoms are probably the reactive species involved in the formation of the xenon fluorides. Usually nickel, Monel, copper, or aluminum reaction vessels are used since these metals are very resistant to fluorine at high temperatures. At 300°C the equilibria among the fluorides may be described by reactions (1)–(3). [One atmosphere =

$$Xe + F_2 \rightleftharpoons XeF_2 \qquad k_1 = 1.02 \times 10^4\ atm^{-1} \quad (1)$$
$$XeF_2 + F_2 \rightleftharpoons XeF_4 \qquad k_2 = 155\ atm^{-1} \quad (2)$$
$$XeF_4 + F_2 \rightleftharpoons XeF_6 \qquad k_3 = 0.211\ atm^{-1} \quad (3)$$

101,325 Pa.] As these equilibrium constants imply, XeF_2 can be prepared by the reaction of fluorine with excess xenon, and XeF_4 can be prepared by the reaction of xenon with a moderate excess of fluorine at a pressure of a few atmospheres, while the preparation of XeF_6 requires a large excess of fluorine and relatively high pressures. The reactions of xenon with fluorine may also be induced at room temperature by any source of energy capable of dissociating the F_2 molecule, such as ultraviolet light, ionizing radiation, or electrical discharges. In fact, pure XeF_2 can be made by the action of sunlight on a glass bulb containing a mixture of xenon and fluorine. *See* FLUORINE.

Properties. Some properties of the simple xenon fluorides and $XeOF_4$ are given in the table. All of

Properties of $XeOF_4$ and the simple fluorides

	XeF_2	XeF_4	XeF_6	$XeOF_4$
Melting point, °C	129	117.1	49.5	−46.2
Vapor pressure at 25°C, mm Hg (Pa)	4.6 (610)	2.5 (330)	28.9 (3850)	28 (3730)
Color of solid	White	White	White	White
Color of vapor	Colorless	Colorless	Yellow	Colorless

*1 mm Hg = 133.

the fluorides are reduced by hydrogen to form xenon and hydrogen fluoride, and their heats of formation have been measured in this manner. From such measurements the average xenon-fluorine bond energy is found to be roughly 30 kcal/mole (1 kcal = 4184 J), which makes the fluorides rather stable compounds.

The xenon fluorides become progressively more vigorous fluorinating agents as one proceeds from XeF_2 to XeF_4 to XeF_6. Thus XeF_6 is so reactive that it slowly fluorinates silica or glass. Xenon diflouride, the least reactive xenon fluoride, has become of interest as a fluorinating agent in organic systems. It substitutes fluorine for hydrogen in aromatic rings and adds fluorine to olefins. Fluorinations with XeF_2 generally appear to proceed by an electrophilic rather than a free radical mechanism.

Oxides. The reaction of XeF_6 with water gives $XeOF_4$; if the reaction is allowed to continue, XeO_3 is formed. XeO_3 is a colorless, odorless, and dangerously explosive white solid of low volatility. Gaseous xenon tetroxide, XeO_4, is formed by the reaction of sodium perxenate, Na_4XeO_6, with concentrated H_2SO_4 according to reaction (4). The

$$Na_4XeO_6 + 2H_2SO_4 \rightarrow XeO_4 + 2Na_2SO_4 + 2H_2O \quad (4)$$

vapor pressure of XeO_4 is about 25 mm Hg at 0°C. It is unstable and has a tendency to explode.

Oxyfluorides. Xenon oxytetrafluoride, $XeOF_4$, is formed by the reaction of XeF_6 with silica or water. It is a rather stable compound, melting at −46.2°C and boiling at 101°C. Xenon dioxide difluoride, XeO_2F_2, forms when $XeOF_4$ reacts with XeO_3. It is also formed as an intermediate in the reaction of $XeOF_4$ with water. It is much less volatile than $XeOF_4$ and is not very stable, tending to decompose to XeF_2 and O_2. Xenon trioxide difluoride, XeO_3F_2, has been made by the action of XeF_6 on perxenates. It is a volatile and explosive compound resembling xenon tetroxide. There is some evidence that $XeOF_2$ is formed as an intermediate in the low-temperature hydrolysis of XeF_4.

Complex addition compounds. The xenon fluorides and $XeOF_4$ form a variety of complex addition compounds. All four react with fluoride-accepting metal pentafluorides, such as SbF_5, TaF_5, and OsF_5, to form fluorine-bridged complexes of 2:1, 1:1, and 1:2 stoichiometry. These complexes appear to contain the cations XeF^+, $Xe_2F_3^+$, XeF_3^+, XeF_5^+, and $XeOF_3^+$. The tetrafluoride shows much less tendency than the other xenon compounds to form such complexes. The reaction between xenon and platinum hexafluoride that inaugurated the modern era of noble-gas chemistry appears to produce complexes of this sort, such as $XeF^+PtF_6^-$. Xenon hexafluoride and $XeOF_4$ also form fluorine-bridged complexes with fluoride ion donors, such as the alkali fluorides and NOF. These complexes may be considered to contain such anions as XeF_7^-, XeF_8^{2-}, and $XeOF_5^-$.

Xenon difluoride reacts with very strong protic acids, such as HSO_3F and $HOTeF_5$, to form such oxygen-bridge species as $FXeOTeF_5$ and $Xe(SO_3F)_2$. It also forms molecular addition complexes with XeF_4, IF_5, and $XeOF_4$. An unusual complex is the species $FXeN(SO_2F)_2$, which results from the reaction of XeF_2 with the acid $HN(SO_2F)_2$, and which is the only compound known to contain a xenon-nitrogen bond. It forms complexes with AsF_5 that have been formulated as $XeN(SO_2F)_2^+AsF_6^-$ and $[XeN(SO_2F_2)]_2F^+AsF_6^-$.

These complexes are all nonvolatile solids of varying stability. The complexes with metal pentafluorides tend to be powerful fluorinating agents. Xenon difluoride dissolved in antimony pentafluoride can react with elemental xenon itself to form the unexpected dixenon cation Xe_2^+.

Aqueous chemistry. Xenon difluoride, XeF_2, dissolves in water to the extent of 25 mg/ml at 0°C. The solutions contain XeF_2 molecules, which decompose gradually according to reaction (5). The

$$XeF_2 + H_2O \rightarrow Xe + \tfrac{1}{2}O_2 + 2HF \quad (5)$$

halftime of decomposition is about 7 h at 0°C and

about 1/2 h at room temperature. The decomposition is almost instantaneous in base or in the presence of a fluoride scavenger, such as Th^{4+}, which suggests that divalent xenon can persist in water only if it is stabilized by attached fluorine atoms.

Xenon difluoride solutions are powerfully oxidizing: they can oxidize iodate to periodate, bromate to perbromate, chlorate to perchlorate, cerous ion to ceric, cobaltous to cobaltic, and Ag^+ to Ag^{2+}. In alkaline solution, XeF_2 oxidizes hexavalent xenon to the octavalent state. Much of the oxidizing behavior of aqueous XeF_2 is probably due to intermediates formed in the course of its reaction with water. These intermediates have yet to be positively identified.

When XeF_4 of XeF_6 reacts with water, solutions of XeO_3 are obtained according to reactions (6) and (7). These solutions are good oxidizing agents,

$$3XeF_4 + 6H_2O \rightarrow XeO_3 + 2Xe + \tfrac{3}{2}O_2 + 12HF \quad (6)$$

$$XeF_6 + 3H_2O \rightarrow XeO_3 + 6HF \quad (7)$$

though not as powerful as XeF_2 solutions. They liberate chlorine from hydrochloric acid, and they slowly oxidize Mn^{2+} to permanganate, and iodine and bromine to iodate and bromate. They also oxidize a variety of organic compounds. In the absence of oxidizable substrates, acidic solutions of XeO_3 are stable.

Xenon trioxide is weakly acidic, and reaction (8)

$$XeO_3 + H_2O = H^+ + HXeO_4^- \quad (8)$$

has an equilibrium constant of about 2×10^{-11}. Alkali metal salts of XeO_3 can be prepared with difficulty, as can alkali fluoroxenates and chloroxenates with the formulas $MXeO_3F$ and $MXeO_3Cl$, respectively.

Sodium perxenate, a salt of octavalent xenon, can be prepared by reacting XeF_6 with strong sodium hydroxide according to reaction (9). It can

$$4Na^+ + 2XeF_6 + 16OH^- \rightarrow Xe + O_2 + Na_4XeO_6 + 12F^- + 8H_2O \quad (9)$$

also be prepared in strong NaOH by oxidizing $HXeO_4^-$ with ozone or by the slow spontaneous disproportionation of $HXeO_4^-$ in reaction (10). The

$$2HXeO_4^- + 4Na^+ + 2OH^- \rightarrow Xe + O_2 + Na_4XeO_6 + 2H_2O \quad (10)$$

salt is virtually insoluble in sodium hydroxide solutions and precipitates out. Sodium perxenate dissolves sparingly in water to give protonated perxenate anions and hydroxyl ions, as shown in reaction (11). Adding acid further protonates the perxenate successively to doubly and singly charged anions.

$$Na_4XeO_6 + H_2O \rightarrow 4Na^+ + OH^- + HXeO_6^{3-} \quad (11)$$

Perxenate solutions are unstable, evolving oxygen and forming hexavalent xenon, as indicated in reaction (12). The reaction is slow in base, but be-

$$HXeO_6^{3-} + H_2O \rightarrow HXeO_4^- + \tfrac{1}{2}O_2 + 2OH^- \quad (12)$$

comes almost instantaneous below pH 7. This reaction provides a method for regenerating XeO_3 from sodium perxenate, which is a convenient compound to store. Perxenates are extremely powerful oxidizing agents, particularly in acid. Perxenate in acid solution very rapidly oxidizes Mn^{2+} to permanganate, Cr^{3+} to dichromate, Ag^+ to Ag^{2+}, and cobaltous ion to cobaltic. Even aromatic hydrocarbons are rapidly oxidized. The oxidations probably proceed via hydroxyl radicals formed by the reaction of perxenate with water.

In addition to the sodium salt, many other perxenates are known, among them salts of lithium, barium, lanthanum, and thorium. Unlike the perxenate solutions, most of the solid perxenates are quite stable. On the other hand, a highly explosive mixed perxenate-xenate, $K_4XeO_6 \cdot 2XeO_3$, has been prepared.

Tentative oxidation potential scheme (13),

Acid solution

$$Xe \text{ —— } 2.10 \text{ —— } XeO_3 \text{ —— } 2.4 \text{ —— } H_4XeO_6$$

$$Xe \text{ —— } 2.64 \text{ —— } XeF_2 \quad (13)$$

Alkaline solution

$$Xe \text{ —— } 1.24 \text{ —— } HXeO_4^- \text{ —— } 1.0 \text{ —— } HXeO_6^{3-}$$

in volts, has been proposed for the aqueous xenon compounds. It has not been possible to synthesize a xenon compound in aqueous solution directly from the element. Elemental xenon can be "fixed" only by reaction with fluorine or reactive fluorides.

Structures. Gaseous xenon difluoride is a linear symmetric molecule. Gaseous xenon tetrafluoride is the only vapor-phase species known to have a square planar configuration.

Sixteen elements are known to form hexafluorides. Of these, only xenon hexafluoride does not have a symmetrical octahedral structure in the vapor phase. The structure of XeF_6 has been described as a slightly distorted octahedron or, more interestingly, as a mixture of distorted and symmetrical molecules. In additions, XeF_6 in the solid and liquid phase is different from other hexafluorides and seems to be made up of rather strongly bonded polymers.

Xenon trioxide is a trigonal pyramid, while the tetroxide is tetrahedral. The $XeOF_4$ molecule is a tetragonal pyramid with the xenon and fluorines in a plane. The XeO_2F_2 and XeO_3F_2 molecules have the oxygen atoms approximately in the plane of the xenon atom, with the fluorine atoms above and below the plane. The perxenate ion, XeO_6^{4-}, is octahedral. [JOHN G. MALM; EVAN H. APPELMAN]

Bibliography: N. Bartlett and F. O. Sladky, in J. C. Bailer et al. (eds.), *Compr. Inorg. Chem.*, vol. 1, pp. 213–330, 1973; H. H. Claassen, *The Noble Gases*, 1966; J. G. Malm and E. H. Appelman, *At. Energ. Rev.*, 7(3):3–48, 1969.

Xylene

One of a group of three isomeric aromatic hydrocarbons. The three isomeric xylenes and ethylbenzene all have in common a molecular weight of 106.2 and the simplified formula, C_8H_{10}. The names, structural formulas, and boiling and melting points of these compounds are as shown in the table.

Originally designated coal tar hydrocarbons, these compounds have, since World War II, been almost exclusively produced from petroleum by hydroforming or catalytic reforming of appropriate naphtha fractions. Well over 90% of total United States supply since 1950 has been from petroleum.

Production. "Pure" *ortho*-xylene and ethylbenzene are commercially separated from refinery

Properties of xylenes and ethylbenzene

Name	Structure	Boiling point, °C	Melting point, °C
o-Xylene	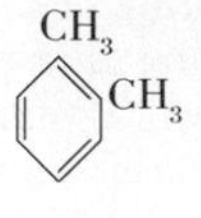	144.2	−25.2
m-Xylene	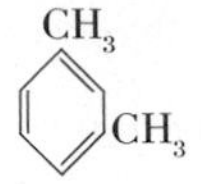	139.1	−47.9
p-Xylene		138.4	13.3
Ethylbenzene		136.2	−95.0

streams by distillation processes. *para*-Xylene is produced in very large quantities by low-temperature fractional distillation. Commercial-scale production of *meta*-xylene has been achieved by sulfonation and hydrolysis, depending on the difference in rates of reaction of the isomers. Other separation schemes have been suggested. Extensive use is also made of catalytic isomerization processes to increase the production of a desired isomer among the three xylenes. Much of total United States production of these materials is marketed as mixed xylenes. This material appears finally either in gasoline or in aromatic-solvents end-use categories at prices well below those commanded by the separated or pure isomers.

Chemical uses. *o*-Xylene is commercially oxidized by air over vanadium oxide catalysts to produce phthalic anhydride as shown in the reaction below. The higher yield in this reaction, compared

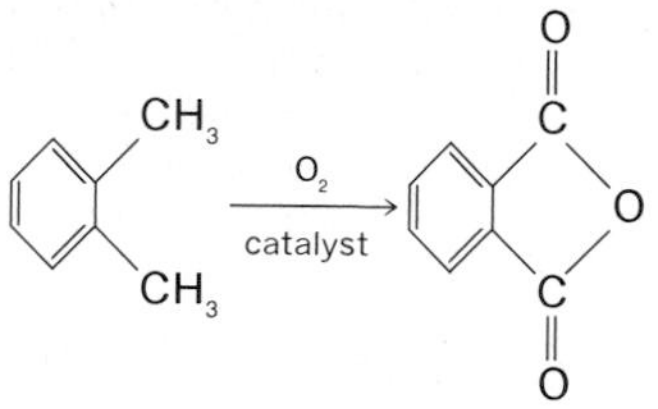

to the traditional route of naphthalene oxidation, resulted in rapid growth of this postwar technology. At least six United States producers of phthalic anhydride are known to use *o*-xylene as raw material.

Phthalic anhydride is one of the most versatile of organic intermediates, finding its way into production of alkyd resins, phthalic ester plasticizers, polyester resins, anthraquinone, phenolphthalein, phthalonitrile, phthalimides, eosin inks and dyes, plus many more third-generation derivatives. *p*-Xylene, like the ortho isomer, has only one significant chemical use. It is oxidized by one of several processes either to terephthalic acid (TA) or to dimethylterephthalate (DMT), both of which are intermediates in the production of, among other things, the high-polymer materials known commonly as polyesters. These have achieved great acceptance under their various trademark names (Dacron, Mylar, Kodel, Terylene, and so on) either as fibers or film formers. *See* POLYESTER RESINS.

m-Xylene, nearly twice as abundant as any of the other isomers, is technologically the least important. Some quantities of isophthalic acid have been produced and marketed at prices approximating those of phthalic anhydride. The chemistry and physics of the oxidation and the purification of the derivatives are of considerable interest. The compounds made from the derived dicarboxylic acid have some properties in common with both the ortho and the para structure isomers. Some pharmaceuticals are derived as second- or third-generation derivatives but are small-volume specialty items. Ethylbenzene, which is not truly an isomer, although it is usually so considered because it has the same simplified formula, is of great commercial importance, being the precursor or raw material from which styrene is manufactured. However, only a small fraction of United States production of ethylbenzene stems from the "virgin" material associated with the xylenes in refinery streams. The overwhelming majority is produced by alkylation of benzene with ethylene. It should be noted that the commercial separation by distillation of the virgin material from *p*-xylene, which boils at only 2.2°C higher, was hailed as a chemical engineering triumph.

A mixture of the xylenes and ethylbenzene is commonly termed mixed xylenes, xylene, or xylol, and has some utility as a solvent both in surface coatings and in chemical manufacture. Such material is usually specified by the boiling range. Thus, the so-called five-degree xylol completely distills over a boiling range of only 5°F (−15°C). A very substantial portion of the total United States production is sold and used in this form. The only other significant use is in gasoline, where excellent antiknock properties are contributed, although the volatility is somewhat below that desired. Safety guidelines recommend against prolonged breathing of air containing more than 100 ppm of xylene. *See* ACID ANHYDRIDE.

[ROBERT I. STIRTON; MARTIN STILES]

Bibliography: *Registry of Toxic Effects of Chemical Substances*, 1976; Stanford Research Institute, *Chemical Origins and Markets*, 1967; R. B. Stobaugh, Jr., *Petrochemical Manufacturing and Marketing Guide*, vol. 1: *Aromatics and Derivatives*, 1967; U.S. Tariff Commission, *Synthetic Organic Chemicals: U.S. Production and Sales*, 1967.

Ytterbium

A chemical element, Yb, atomic number 70, and atomic weight 173.04. Ytterbium is a metal element of the rare-earth group. The naturally occurring stable isotopes are Yb^{168} 0.135%, Yb^{170} 3.03%, Yb^{171} 14.31%, Yb^{172} 21.82%, Yb^{173} 16.13%, Yb^{174} 31.84%, and Yb^{175} 12.73%. Ytterbium compounds were first separated by J. C. G. Marignac in 1878. In 1907 G. Urbain showed that Marignac's ytterbium was composed of two elements, lutetium and what is now known as ytterbium. It has also sometimes been called aldebaranium. The common oxide, Yb_2O_3, is colorless and dissolves readily in

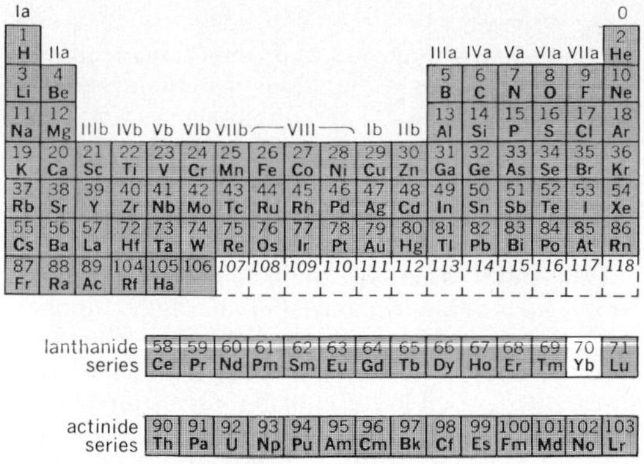

acids to form colorless solutions of trivalent salts which are paramagnetic.

Ytterbium also forms a series of divalent compounds. The divalent salts are soluble in water but react very slowly with water to liberate hydrogen. Ytterbium can be separated from the other rare earths by taking advantage of its divalent state, or it can be removed from other rare earths by extracting it with sodium amalgam. The metal is best prepared by distillation. *See* EUROPIUM; SAMARIUM.

Ytterbium is a silvery soft metal which corrodes slowly in air and resembles the calcium-strontium-barium series more than the rare-earth series, since it probably has two electrons in the conduction bands instead of three. For a discussion of the properties of the metal and its salts *see* RARE-EARTH ELEMENTS.

[FRANK H. SPEDDING]

Yttrium

A chemical element, Y, atomic number 39, and atomic weight 88.905. Yttrium resembles the rare-earth elements closely. The stable isotope Y^{89} constitutes 100% of the natural element, which is always found associated with the rare earths and is frequently classified as one. J. Gadolin discovered the element in 1794, and it was first obtained in high purity by C. G. Mosander in 1843.

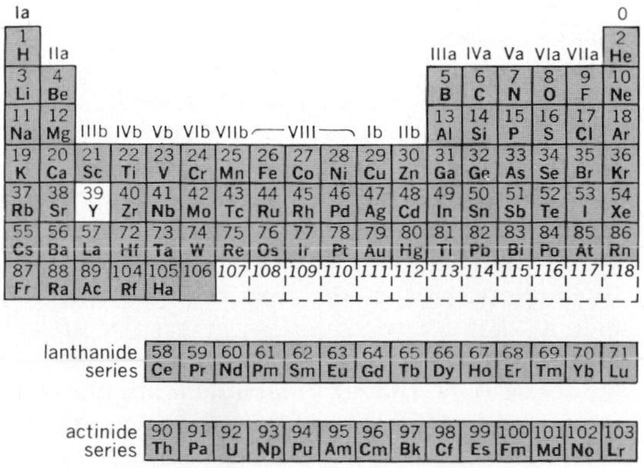

Yttrium forms a white oxide, Y_2O_3, which dissolves in acid to form trivalent yttrium salts. Yttrium has become commercially important since 1964, at which time the demand for highly pure Y_2O_3 exceeded 10 tons (9 metric tons) per year. Yttrium forms the matrix for the europium-activated yttrium phosphors. These phosphors, when excited by electrons, emit a brilliant, clear-red light. The television industry uses these phosphors in manufacturing television screens. It is claimed that this phosphor gives better color reproduction and a much brighter screen than did the older non-rare-earth red phosphor. Also, the yttrium iron garnets, $Y_3Fe_5O_{12}$, and other garnets have found important uses in radar and communication devices. They transmit short-wave energy with very small losses.

Yttrium metal absorbs hydrogen, and in alloys up to a composition of YH_2 they resemble metals very closely. In fact, in certain composition ranges, the alloy is a better conductor of electricity than the pure metal. The density of hydrogen near the YH_2 composition is greater per cubic centimeter than it is in water or liquid hydrogen; therefore, such alloys make excellent potential moderators for nuclear reactors. Also, these alloys can be heated to a white heat (about 2300°F or 1260°C) before the vapor pressure of hydrogen exceeds 1 atm (101 kPa), and therefore the moderator in the reactor can be operated at very high temperatures. Yttrium metal has a low nuclear cross section so it is also a potential structural material for reactors of the future.

Yttrium is used commercially in the metal industry for alloy purposes and as a "getter" to remove oxygen and nonmetallic impurities in other metals. Radioactive yttrium isotopes have been used in attempts at treating cancer. For properties of the metal and its salts *see* RARE-EARTH ELEMENTS. *See also* LANTHANUM; SCANDIUM.

[FRANK H. SPEDDING]

Zinc

A chemical element, Zn, atomic number 30, and atomic weight 65.38. Zinc is a malleable, ductile, gray metal. Because of chemical similarities among zinc, cadmium, and mercury, these three metals are classed together in group IIb of the periodic

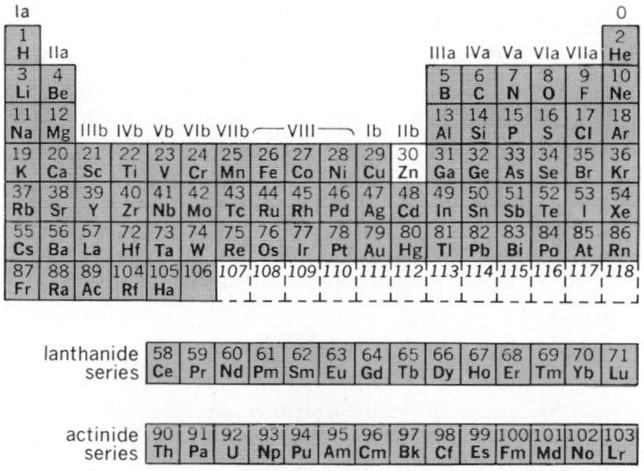

table of elements. Zinc is known to have been present in ancient brass long before it was recognized as a separate element. An alloy found in prehistoric ruins in Transylvania contained 87% zinc. Indian metallurgists produced zinc in the 13th century, and possibly much earlier, by reducing calamine ore with such organic materials as wool. Chinese use of zinc dates from the 15th century. Paracelsus, in the 16th century, is said to have been the first European to recognize zinc as a distinct metallic element and to call it "zinckum." William Champion of England produced zinc in 1743 by reducing zinc carbonate with coke.

Fifteen isotopes of zinc are known, of which five

are stable, having atomic masses of 64, 66, 67, 68, and 70. About half of ordinary zinc occurs as the isotope of atomic mass 64. The half-lives of the radioactive isotopes range from 88 sec for Zn^{61} to 244 days for Zn^{65}.

Uses. Zinc is now in fourth place among all metals in world tonnage produced; only iron, aluminum, and copper exceed it. The most important uses of zinc are in its alloys and as a protective coating on other metals. Coating iron or steel with zinc is called galvanizing, and it may be done by immersing the article in melted zinc (hot-dip process), depositing zinc electrolytically onto the article in a plating bath (electrogalvanizing), exposing the article to powdered zinc near its melting point (sherardizing), or spraying the article with melted zinc (metallizing). A commercial process for continuous coating of a steel strip produces 30 tons of the strip per hour. The mere physical presence of the zinc coat prevents corrosion of iron, and even if breaks in the coat expose portions of the iron, the greater chemical activity of the zinc causes it to be consumed in preference to the iron. Adding small amounts of other metals to galvanizing baths has been found to improve the adhesion and weathering qualities of the coating.

Even such nonstructural materials as cardboard can be zinc-coated by low-temperature flame spraying. Other important uses of zinc are in brass and zinc die-casting alloys, in zinc sheet and strip, in electrical dry cells, in making certain zinc compounds, and as a reducing agent in chemical preparations.

A "tumble-plating" process coats small metal parts by applying zinc powder to them with an adhesive, then tumbling them with glass beads to roll out the powder into a continuous coat of zinc. Rechargeable nickel-zinc batteries offer higher energy densities than conventional dry cells. Foamed zinc metal has been suggested for use in lightweight structures such as aircraft and spacecraft. Some other uses of zinc are in dry cells, roofing, lithographic plates, fuses, organ pipes, and wire coatings. Zinc dust, a flammable material when dry, is used in fireworks and as a chemical catalyst and reducing agent. Radioactive Zn^{65} is used medically in the study of metabolism of zinc, and also in determining rates of wear for zinc-containing alloys.

Zinc is also an essential element in the growth of many kinds of organisms, both plant and animal. A deficiency of zinc in the human diet has been found to retard growth and maturity and to produce anemia. The compound insulin is a zinc-containing protein.

Occurrence. Zinc is one of the less common elements; it has been estimated to make up 0.0005–0.02% of the Earth's crust. It is twenty-fifth in order of abundance among the elements. About one-tenth of the zinc ore mined in the world comes from the United States, and considerably more comes from Canada. The chief ore is zinc blende, marmatite, or sphalerite, ZnS. Other ores include European calamine, $ZnCO_3$; American calamine, $ZnCO_3 + ZnSiO_4 \cdot H_2O$; smithsonite, $ZnCO_3$; willemite, Zn_2SiO_4; zincite, ZnO; and franklinite, $(Zn,Fe,Mn)O \cdot (Fe,Mn)_2O_3$. The last two ores are found almost exclusively in the United States. Distribution of zinc over the Earth's surface, however, is quite wide. Much zinc is also separated as a by-product of processing lead ores. The United States now mines less than a third of the ore needed for its consumption of zinc.

Zinc is present in most foods, especially those high in protein. The zinc content of meats averages 30 ppm. Fresh oysters contain over 1000 ppm. The average human body contains about 2 g of zinc.

Zinc metal. Pure, freshly polished zinc is bluish-white, lustrous, and moderately hard (2.5 on Mohs scale). Moist air brings about a superficial tarnishing to give the metal its usual grayish color. Pure zinc is malleable and ductile enough to be rolled or drawn, but small amounts of other metals present as contaminants may render it brittle. Malleability of even pure zinc is improved by heating zinc to 100–150°C. If heated zinc is mechanically worked, it does not embrittle on cooling. Zinc melts at 420° and boils at 907°C. Its density is 7.13 times that of water, so that 1 ft^3 (0.028 m^3) of zinc weighs 445 lb (200 kg).

As a conductor of heat and of electricity, zinc ranks fairly high. However, its electrical resistivity (5.92 microhm-cm at 20°C), is almost four times that of silver, the best conductor. As a conductor of heat, zinc is likewise only about one-fourth as efficient as silver. At 0.91 K zinc is an electrical superconductor. Pure zinc is not ferromagnetic, but the alloy compound $ZrZn_2$ displays ferromagnetism below 35 K.

Chemical properties. Zinc is a fairly active metal chemically. It can be ignited with some difficulty to give a blue-green flame in air and to discharge clouds of zinc oxide smoke. Zinc ranks above hydrogen in the electrochemical series, so that metallic zinc in an acidic solution will react to liberate hydrogen gas as the zinc passes into solution to form dipositively charged zinc ions, Zn^{++}. This reaction is slow with very pure zinc, but the presence of small amounts of impurities, addition of a trace of copper sulfate, or contact between the zinc surface and such metals as nickel or platinum facilitates formation of gaseous hydrogen and speeds the reaction. The combination of zinc and dilute acid is often used to generate small quantities of hydrogen in the laboratory. Zinc also dissolves in strongly alkaline solutions, such as sodium hydroxide, to liberate hydrogen and form dinegatively charged tetrahydroxozincate ions, $Zn(OH)_4^{--}$, sometimes written as ZnO_2^{--} in the formulas of the zincate compounds. Zinc also dissolves in solutions of ammonia or ammonium salts. The common soluble zinc compounds undergo to some extent the process of hydrolysis, which makes their solutions slightly acidic. The ion Zn^{++} is colorless, so that the relatively few zinc compounds that are not colorless in large crystals, or white as powders, receive their color through the influence of the other constituents. Some of the atomic and ionic properties of zinc are shown in Table 1.

Compounds. Zinc is always divalent in its compounds, except for some of those with other metals, which are classed as zinc alloys. Table 2 lists some of the more important zinc compounds. Most of these are inorganic, since they are much more widely used than the organic zinc compounds.

Zinc also forms many coordination compounds. The zincates are actually coordination compounds,

Table 1. Atomic and ionic properties of zinc

Property	Value
Electronic configuration	$1s^2, 2s^2, 2p^6, 3s^2, 3p^6, 3d^{10}, 4s^2$
Ionization potentials	
1st electron loss	9.39 eV
2d electron loss	17.9 eV
Ionic radius, Zn^{++}	0.072 nm
Covalent radius (tetrahedral)	0.131 nm
Oxidation potentials	$Zn \rightleftarrows Zn^{++} + 2e^-$, $E° = 0.76$ V $Zn + 4OH^- \rightleftarrows ZnO_2^{--} + 2H_2O + 2e^-$, $E° = 1.22$ V

or complexes, in which hydroxide ions, OH^-, are bound to the zinc ions. Ammonia, NH_3, forms complexes with zinc, such as the typical tetrammine zinc ion, $[Zn(NH_3)_4]^{++}$. Zinc cyanide, usually given the simple formula $Zn(CN)_2$, is a coordination compound in which many alternating zinc and cyanide ions are three-dimensionally bound together in a very large molecule. This compound is still widely used in zinc plating, but concern over environmental pollution has led to increasing use of zinc chloride plating baths. In most coordination compounds of zinc, the fundamental structural unit is a central zinc ion surrounded by four coordinated groups arranged spatially at the corners of a regular tetrahedron.

Analytical methods. White zinc sulfide is precipitated by ammonium sulfide or hydrogen sulfide from neutral or alkaline zinc salt solutions. The precipitate is distinguished from those of nickel and cobalt by its lack of color and by its solubility in acid. Sodium, potassium, or ammonium hydroxide precipitates white zinc hydroxide. The solubility of the hydroxide in excess sodium hydroxide distinguishes it from that of iron and manganese, and the solubility in excess ammonium hydroxide from that of aluminum and chromium as well.

Table 2. Compounds of zinc

Name and formula	Uses	Properties, remarks
Zinc oxide, ZnO	Reinforcer and vulcanization activator in rubber tires; pigment; ointment base; in cements and ceramics; in linoleum and other pigments as mold-growth inhibitor; in semiconductors; as photoconductor in copying machines	White, nearly insoluble powder; slightly alkaline in water; turns yellow at 500°C, returns to white on cooling; nontoxic unless inhaled as dust; amphoteric, dissolving in either acid or base
Zinc chloride, $ZnCl_2$ Zinc chloride 4-hydrate, $ZnCl_2 \cdot 4H_2O$	Soldering and welding flux; in dye printing, dye manufacture, and wood treating; as catalyst; in medicines, glues, and cements	Colorless, soluble compound; solutions acidic; when made alkaline, forms a series of zinc oxychlorides; exists in several hydrated forms with less than $4H_2O$ per molecule
Zinc sulfate, $ZnSO_4$	Agriculturally, in arsenical sprays; in rayon manufacture, dyeing, printing; in electrogalvanizing; ingredient in making lithopone pigment	White, soluble compound; commercial form is hydrated
Zinc acetate 2-hydrate, $Zn(CH_3COO)_2 \cdot 2H_2O$	Wood preservative; mordant in dyeing; cross-linking agent in polymerization; in medicines	White, crystalline compound; soluble in water and alcohol
Zinc stearate, $Zn(C_{17}H_{35}COO)_2$	In medicines, cosmetics, lacquers, plastics; as lubricant, mold-release agent, emulsifier, antifoam agent	White, soft powder; insoluble; the most common metal soap
Bis (pyridine 1-oxide, 2-thiolo) zinc, $Zn(C_5H_4NOS)_2$	Antidandruff agent; agricultural fungicide	Also called zinc pyridinethione, zinc pyrithione
Zinc sulfide, ZnS	Component of lithopone pigment; in luminescent form as a fluorescer or phosphor in television screens, oscilloscopes, X-ray apparatus	White, insoluble; only common white sulfide of a metal; will not blacken in sulfide vapors
Zinc chromate, $ZnCrO_4$	In rust-inhibiting coatings; as pigment; in linoleum	Usually a combined zinc hydroxide-chromate such as $ZnCrO_4 \cdot 4Zn(OH)_2$, sometimes contains potassium also; varying shades of yellow
Zinc orthosilicate, Zn_2SiO_4	In television-screen phosphors; as a refractory; in water softeners	Used as both anhydrous form (willemite) and 1-hydrate (calamine)
Sodium zincate, Na_2ZnO_2	Water softener; paper- and cloth-treating agent; flocculating agent in water purification	Formed in strongly alkaline solution
Zinc fluoride, ZnF_2	Catalyst, wood preservative; termite repellent	Rather low solubility in water
Zinc dithionite, ZnS_2O_4	Bleaching agent for wood pulp, rags, clay, sugar, vegetable oils	Also called zinc hydrosulfite
Zinc peroxide, ZnO_2	Prophylactic against wound infections; as vulcanizing agent; in cosmetics	Commercially about half ZnO_2, half ZnO
Zinc orthophosphate 3-hydrate, $ZnHPO_4 \cdot 3H_2O$	Dental cement	Anhydrous form also known, as are many other zinc phosphates
Zinc carbonate, $ZnCO_3$	Nutritional supplement in animal husbandry	White, insoluble powder

Qualitative identification may also be made by ignition of a zinc compound in the presence of cobalt nitrate, which yields a greenish solid.

Quantitatively, zinc may be determined by precipitating the sulfide, igniting it to form the oxide, and weighing. Zinc ammonium orthophosphate, $ZnNH_4PO_4$, may be precipitated, ignited, and weighed as the pyrophosphate, $Zn_2P_2O_7$. Quantitative electrolysis of zinc salts in acetate buffer solution is possible, with the zinc being plated out and weighed as metallic zinc.

Volumetrically zinc may be determined in a direct titration with ethylenediaminetetraacetate ion (EDTA), using Erichrome Black T as an indicator. Zinc in some alloys is separated from nickel, copper, and cadmium by anion exchange as the chloro complex and then titrated with EDTA. Another volumetric method uses standard potassium ferrocyanide solution, with the end point determined potentiometrically or with uranyl acetate indicator.

Zinc in sea water has been determined to parts per billion using the single-sweep cathode-ray polarograph. X-ray emission analysis is useful to determine zinc in glass, polymers, and oils. Atomic absorption spectrometry can detect zinc at 0.02 part per million in soils and fertilizers. Neutron activation analysis has detected as little as 0.003 microgram (3 nanograms) of zinc in blood, semiconductors, sea water, and ores.

Toxicity. Zinc is one of the less hazardous elements, and its compounds are generally of low toxicity. Zinc is, in fact, an essential element for growth and development of the body. Traces of arsenic, lead, cadmium, or antimony in impure zinc are poisonous enough to present a hazard. The ion Zn^{++} is a normal component of some enzymes (carbonic anhydrase) and hormones (insulin). The inhalation of fumes from galvanizing baths and the like sometimes produces "zinc fever," characterized by chills and fever, nausea, and aching. Removal from the source of the fumes brings about complete recovery. Edema of the lungs from fumes over zinc chloride ($ZnCl_2$ smoke) has sometimes proved fatal. Soluble and astringent acidic salts such as $ZnSO_4$ in large doses (for example, 10 g) have caused internal organ damage and death. Zinc containers may be used for storing drinking water but not for food storage.

[WILLIAM E. COOLEY]

Bibliography: B. J. Aylett, in J. C. Bailar, Jr. (ed.), *Comprehensive Inorganic Chemistry*, vol. 3, 1973; F. A. Cotton and G. Wilkenson, *Advanced Inorganic Chemistry*, 4th ed., 1980; J. A. Dean (ed.), *Lange's Handbook of Chemistry*, 12th ed., 1978; J. R. DiPalma (ed.), *Drill's Pharmacology in Medicine*, 4th ed., 1971; R. H. Dreisbach, *Handbook of Poisoning: Diagnosis and Treatment*, 10th ed., 1980; N. H. Furman and F. J. Welcher (eds.), *Standard Methods of Chemical Analysis*, 6th ed., 3 vols., reprinted 1975; J. A. Halsted et al., A conspectus of research on zinc requirements of man, *J. Nutr.*, 104(3), 1974; Z. A. Karcioglu and R. M. Sarper, *Zinc and Copper in Medicine*, 1980; M. S. Sienko and R. A. Plane, *Chemistry: Principles and Properties*, rev. 2d ed., 1974; U.S. Bureau of Mines, *Minerals Yearbook*, annually; R. C. Weast (ed.), *Handbook of Chemistry and Physics*, 61st ed., 1980; F. J. Welcher, *The Analytical Uses of Ethylenediaminetetraacetic Acid*, 1958; *Zinc*, ACS Monogr. no. 142, 1959, reprinted 1971.

Zincate

A negative ion, usually given the formula ZnO_2^{--}, derived from zinc hydroxide. Zinc hydroxide is an amphoteric substance and thus can react with a strong base such as sodium hydroxide to form sodium zincate, as in the reaction below, and can

$$Zn(OH)_2 + 2NaOH \rightarrow Na_2ZnO_2 + 2H_2O$$

also react in the more usual manner with acids to form zinc salts. Zincate solutions are strongly basic, and are used to plate zinc on aluminum prior to copper plating. *See* AMPHOTERISM; ZINC.

[E. EUGENE WEAVER]

Zirconium

A chemical element, Zr, atomic number 40, atomic weight 91.22. Its naturally occurring isotopes are 90 (51.5%), 91 (11.2%), 92 (17.1%), 94 (17.4%), and 96 (2.8%).

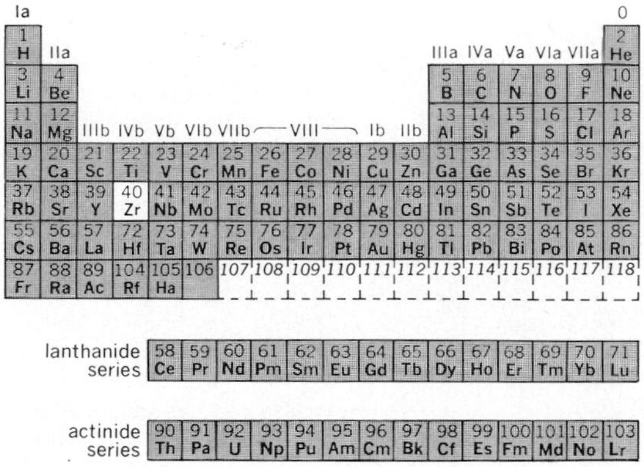

Occurrence. Zirconium is one of the more abundant elements, and is widely distributed in the Earth's crust. Being very reactive chemically, it is found only in the combined state. Under most conditions, it bonds with oxygen in preference to any other element, and it occurs in the Earth's crust only as the oxide, ZrO_2, baddeleyite, or as part of a complex of oxides as in zircon, $ZrO_2 \cdot SiO_2$; elpidite, $Na_2ZrSi_6O_{15} \cdot 3H_2O$; and eudialyte, $Na_{13}(Ca,Fe)_6$-$(Zr,Si)_{20}O_{52}Cl$. Zircon is commercially the most important ore, but baddeleyite also has some importance. There have been numerous reports on the extraction of zirconium oxide from eudialyte in the Soviet Union. Zirconium and hafnium are practically indistinguishable in chemical properties, and occur only together. Most zircon of commerce contains 2 parts by weight of hafnium per 100 of zirconium. Baddeleyite may contain only 0.5 part hafnium per 100 of zirconium. Zircon is recovered from concentrations in certain beach sands, particularly in New South Wales and in Western Australia, but other important sources have been those near Jacksonville, Fla., and in the State of Kerala, India. A flow sheet for recovery of ore from beach deposit is shown.

Uses. Most of the zirconium used has been as compounds for the ceramic industry: refractories, glazes, enamels, foundry mold and core washes, abrasive grits, and components of electrical ceramics. The incorporation of zirconium oxide in glass significantly increases its resistance to alkali. This has been utilized in improving glazes, enam-

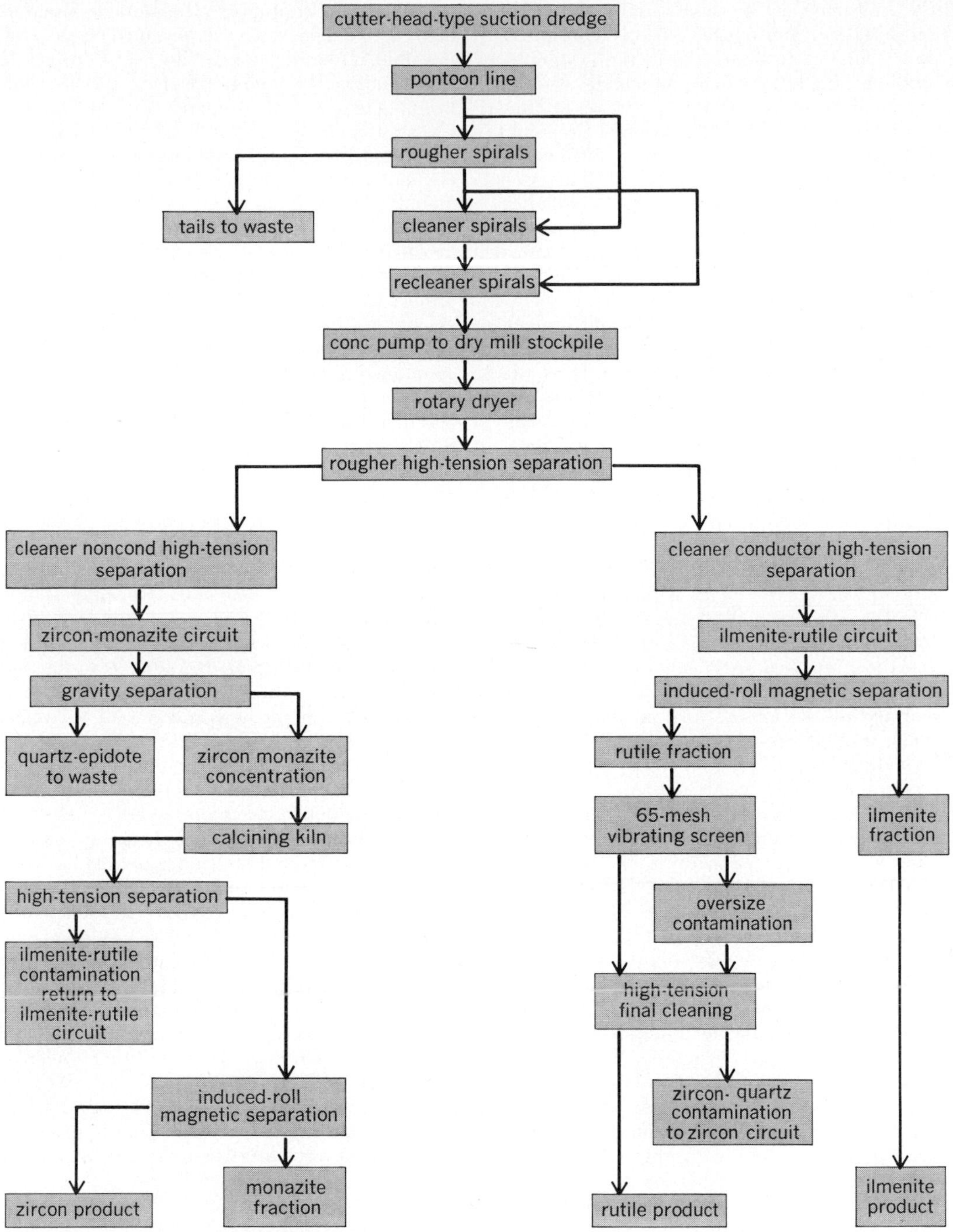

Typical flow sheet for beach deposits of zirconium ore.

els, and glass fibers. The use of zirconium metal is almost entirely for cladding uranium fuel elements for nuclear power plants. Another significant use has been in photo flashbulbs. Some chemical-processing industries use zirconium metal for corrosion-resistant vessels and piping, particularly for withstanding hydrochloric and sulfuric acids. Zirconium chemicals have been used as cross-linking agents for polymers, in treating textiles for water-repellency and withstanding attack by microorganisms, as components of catalysts and paint driers, as bonding agents for such surfaces as oxides and plastics, as precipitants for dyes in the manufacture of pigment dyestuff, as tanning agents, as additives to water pumped into oil-bearing strata to improve yields of wells, as ion exchangers, as protective surfaces on pigments, and in pharmaceuticals such as body deodorants and antidotes for poison ivy.

Production. Impure zirconium metal was produced as early as 1824 by Berzelius, but the metal has little practical value unless it is substantially free of oxygen, nitrogen, and hydrogen, which render it brittle and impossible to fabricate. Ductile metal has been prepared commercially mainly by two methods.

In the Kroll process, zirconium tetrachloride vapor is reduced by liquid magnesium in equipment from which water vapor, oxygen, and nitrogen are rigorously excluded. The steps from mineral to metal ingot are: (1) Zircon is heated with carbon in an arc furnace to form zirconium cyanonitride,

Zr(C,N,O), which is an interstitial solution of carbon, nitrogen, and oxygen (mostly carbon) in the metal. Silicon disappears from the system by evaporation, largely as silicon monoxide. (2) The hot cyanonitride is treated with a stream of chlorine, with formation and volatilization of zirconium tetrachloride. (3) The tetrachloride is purified by resublimation in an inert atmosphere. This step separates it from contaminating oxides, which if carried through to the metal would make it nonductile. (4) The tetrachloride vapor is passed into a chamber containing liquid magnesium. Suitable control of conditions of the reaction leads to recovery of metal sponge, rather than powder, which would be more difficult to handle. (5) The by-product magnesium chloride is separated from the metal by melting and draining it off, and residues are removed by vacuum distillation. (6) The isolated zirconium sponge is crushed, pressed into bars, and arc-melted under an inert atmosphere. The melt is formed into ingots, the quality of which is evaluated by hardness tests.

In the Van Arkel–de Boer process, crude zirconium metal, too high in oxygen content to be suitable for working the metal, is placed in a heated vessel containing iodine and an electrically heated refractory metal wire. The crude metal, at a relatively low temperature, reacts with the iodine to form volatile iodides of zirconium, which are thermolyzed on the heated wire at about 1300°C with deposition of zirconium metal and the liberation of iodine. The metal is recovered as bright, extremely pure metal crystals.

For use as cladding of uranium fuel elements for nuclear reactors, the zirconium must be free of the hafnium which occurred with it in the mineral, since the hafnium has about 600 times the cross section to thermal (slow) neutrons of zirconium. This is usually achieved by subjecting an aqueous solution derived from the zirconium tetrachloride to solvent-solvent extraction. Certain complexes of the metals, particularly thiocyanate, alkylphosphate, and phosphine oxide complexes, can be suitably resolved by this technique. Other methods have been successful in the laboratory but apparently not in industry. These include ion exchange, fractional crystallization of complex fluoride salts, distillation of complexes of zirconium tetrachloride with phosphorus pentachloride or phosphorous oxychloride, and differential reduction of the mixed tetrachlorides. Zirconium tetrachloride is more easily reduced to the nonvolatile trichloride than its hafnium analog.

Properties. Zirconium is a lustrous, silvery metal, with a density of 6.490 at 20°C. It melts at about 1850°C. Estimates of the boiling point from appropriate data have commonly been of the order of 3600°C, but recent observations suggest about 8600°C. Below 862°C its crystal structure is hexagonal close-packed (α form) and above that temperature, body-centered cubic (β-form). Its electrical resistance is 40 μohm-cm at 0°C, about 25 times that of copper. Its cross section to thermal neutrons is 0.18 barn; that of iron is 2.4, nickel 4.5, copper 3.5, aluminum 0.22, magnesium 0.06, and hafnium 105.

The free energies of formation of its compounds indicate that zirconium should react with any nonmetal, other than the inert gases, at ordinary temperatures. In practice, the metal is found to be nonreactive near room temperature because of an invisible, impervious oxide film on its surface. The film renders the metal passive, and it remains bright and shiny in ordinary air indefinitely. At elevated temperatures it is very reactive to the nonmetallic elements and many of the metallic elements, forming either solid solutions or compounds. At 700°C as much as 30 atom % oxygen, 20 atom % nitrogen, and 50 atom % hydrogen will dissolve in the metal. Finely divided zirconium (powder or chips) is extremely pyrophoric and hazardous to handle and store. Spontaneous explosions have been reported for both the wetted sponge and the wet and dry stored scrap. The spontaneous ignition pressure for thin sheets of zirconium metal in oxygen was found to vary from 300 to 750 psi (2.1 to 5.2 MPa), depending on the previous oxygen content of the specimen.

The resistance of the metal to corrosion is generally very good, the corrosion rate being less than a mil per year in acetic, boric, carbolic, chromic, citric, formic, hydrochloric, lactic, monochloracetic, nitric, oxalic, tannic, tartaric, and trichloroacetic acids, in aqueous phenol, in caustic alkali solutions and ammonium hydroxide, and in many solvents and aqueous salt solutions. The metal does not resist hydrofluoric acid. Resistance to hot water and steam is not as satisfactory as required for nuclear power plant use, and the alloy Zircaloy-2 (Zr containing 1.5% Sn, 0.12% Fe, 0.05% Ni, and 0.10% Cr) was developed and used to meet this need. Also a similar alloy, Zircaloy-4, which contains less nickel has been preferred.

Compounds. Zirconium belongs to subgroup IVb of the periodic table, and generally has normal covalency of 4, and commonly exhibits coordinate covalencies of 5, 6, 7, and 8. Zirconium is at oxidation number 4 in nearly all of its compounds. Halogenides in which its oxidation numbers are 3 and 2 have been prepared. While zirconium is often part of cationic or anionic complexes, there is no definite evidence for a monatomic zirconium ion in any of its compounds, but rather for $(4ZrO_2 \cdot 16H_2O \cdot nH^+)^{n+}$ in certain halides, $[ZrO(SO_4)_2]^{--}$ in certain sulfate solutions, and a disproportionation of zirconium tetrachloride into ionic species, reaction (1). This is reflected in melting point and vapor pressure intermediate between those of covalent and ionic chlorides (see table).

$$2ZrCl_4 \rightleftarrows (ZrCl_2)^{++} (ZrCl_6)^{--} \qquad (1)$$

The metal cannot be electroplated from aqueous solutions. Electrowinning from molten salt baths has been achieved, possibly through chemi-

Melting point and vapor pressure of some ionic and covalent chlorides

Chloride	Melting point, °C	Temp., °C, at which vapor pressure is 760 mm (101.325 kPa)
$SiCl_4$	−70	57.6
$TiCl_4$	−30	136
$ZrCl_4$	437*	331
$HfCl_4$	432*	317
$ThCl_4$	770	921

*Under 25 atm (2.53 MPa).

cal reaction with alkali metal liberated by the electric current.

The chemical compounds of zirconium are most conveniently described by assigning them to three categories: those in which the zirconium-containing species is (1) cationic, (2) anionic, or (3) nonionic.

A species of composition $(ZrOCl_2 \cdot 8H_2O)_4$ is obtained when zirconium tetrachloride is dissolved in water, or when precipitated hydrous zirconia is dissolved with hydrochloric acid, and the solution is evaporated to deposit crystals. The zirconium atom has been found to be present in both the crystalline compound and in its aqueous solutions of moderate concentration in a ring consisting of four zirconium atoms alternating with four pairs of bonding oxygen atoms. However, unless meticulously prepared, the product is likely to contain polymers formed by additional bridging of zirconium atoms through oxygen atoms. The polymers have distinctive behaviors, notably toward dyes used in colorimetric analysis. The ring is hydrated and positively charged. The formula of the species is $4ZrO_2 \cdot 16H_2O \cdot nH^+$. Its positive charge is due to the excess of protons. It behaves chemically as a complex cation in (1) precipitating many anions, notably the colored anions of acid dyes; (2) forming complexes in solution with many other anions; (3) reacting with cation-exchange resins; and (4) migrating toward the cathode under the influence of an electric field. Although the empirical formula of the compound suggests it to be a zirconyl salt having the ion ZrO^{++}, structure studies have proved this not to be so. Rather, it is protonically charged zirconium oxide in coulombic balance with chloride ions. The name zirconium oxide hydrochloride has been proposed for the compound, and appears justified by its structure and chemical behavior. Analogous zirconium oxide bromide, iodide, perchlorate, and thiocyanate are known.

When hydrofluoric acid or oxygen-containing anions are added to an aqueous solution of zirconium oxide hydrochloride, they tend to replace the water of hydration, and in some instances the oxo linkages, and to change the electric charge to a negative value. The degree and geometry of polymerization may also be affected. Some empirical reactions and reaction products are illustrated by (2)–(4). The zirconium-containing species of

$$ZrOCl_2 \cdot 8H_2O + 2NaC_2H_3O_2 \rightarrow$$

$$\underset{\substack{\text{Diacetatozirconic acid} \\ \text{(zirconium acetate)}}}{HZrOOH(C_2H_3O_2)_2} + 2NaCl + 7H_2O \qquad (2)$$

$$ZrOCl_2 \cdot 8H_2O + 2H_2SO_4 \rightarrow$$

$$\underset{\substack{\text{Disulfatozirconic acid} \\ \text{(zirconium sulfate)}}}{H_2ZrO(SO_4)_2 \cdot 3H_2O} + 2HCl + 6H_2O \qquad (3)$$

$$ZrOCl_2 \cdot 8H_2O + 2HF + 4KF \rightarrow$$

$$\underset{\substack{\text{Potassium} \\ \text{fluorozirconate}}}{K_2ZrF_6} + 2KCl + 9H_2O \qquad (4)$$

these products generally react with anion-exchange resins and not with cation-exchange resins. They do not precipitate acid dyes. Disulfatozirconic acid and diacetatozirconic acid are good tanning agents for protein substances, whereas zirconium oxide chloride is not. Trisulfatozirconic acid and tetrasulfatozirconic acid are known, as are a variety of hydrolyzates of disulfatozirconic acid, which in the older literature were called basic zirconium sulfates and in the later, polysulfatopolyzirconates. Examples of these are Hauser's salt, $4ZrO_2 \cdot 3SO_3 \cdot 15H_2O$, and the industrially important $5ZrO_2 \cdot 3SO_3 \cdot 15.5H_2O$. Many other anionic complexes of zirconium have been reported. Two have been important in commerce: ammonium carbonatozirconate, $(NH_4)_3ZrOH(CO_3)_2 \cdot 2H_2O$, and sodium trilactozirconate, $Na_2HZrOH(CH_3CHOCO_2)_3$, the first in the cross-linking of polymers such as starch and polyacrylates, and the second as a body deodorant.

Nonionic compounds of zirconium are formed by precipitating or complexing with neutral oxygen-containing compounds or with agents tending to split off ions, as in reaction (5) and (6).

$$ZrOCl_2 \cdot 8H_2O + 2OH^- \rightarrow$$

$$\underset{\substack{\text{Hydrous} \\ \text{zirconia}}}{ZrO_2 \cdot xH_2O} + 2Cl^- + (9-x)H_2O \qquad (5)$$

$$ZrOCl_2 \cdot 8H_2O + 4CH_3COCH_2COCH_3 + Na_2CO_3 \rightarrow$$

$$2NaCl + 10H_2O + \underset{\substack{\text{Zirconium} \\ \text{tetraacetylacetonate}}}{Zr(CH_3COCHCOCH_3)_4} \qquad (6)$$

Hydrous zirconia forms as micelles about 2–5 nm in size. It readily undergoes exchange absorption with many oxygen-containing substances, part or all of the water being replaced by other molecules, such as fatty acids. The so-called zirconium soaps are formed in this manner. Zirconium tetraacetylacetonate is sparingly soluble in water, alcohols, ether, and petroleum ether. It has considerable solubility in aryl hydrocarbons, chlorinated hydrocarbons, and some esters and amines. It is the prototype of a large number of beta-diketonates of zirconium.

Zirconium forms alkoxides by the interaction of zirconium tetrachloride with alcohols in the presence of ammonia or amines. They are generally polymerized or solvated or both, and are capable of forming series of complex hydrolysis products. Analogs of the alkoxides are known in which the alcoholic oxygen is replaced by nitrogen or sulfur, or the carbinol carbon atom is replaced by silicon.

While pi-bonded compounds of zirconium with carbon of organic radicals have been reported for a number of years, there had not been any information on sigma-bonded compounds. The best-known were zircocene dichloride $(\text{pi-}C_6H_5)_2ZrCl_2$ and derivatives, and cyclooctatetraene and allyl compounds. Now sigma-bonded organozirconium compounds have been prepared and characterized, particularly tetrabenzylzirconium. Further, the process called hydrozirconation, has undergone intensive study. It consists of saturating an olefinic bond by its reaction with an organozirconium hydride. The organozirconium group bonds to the less sterically hindered carbon atom. Its ready replacement by other moieties has been shown to be useful in the synthesis of certain isomer-free organic compounds.

Zirconium metal can absorb hydrogen to form a number of phases. The common zirconium hydride of commerce consists of the phase ZrH_2, with about a 10% deficiency of hydrogen. Its chief use is by the military in flares. Other uses are in detonators, brazing metal to ceramic or ceramic to ceramic, and in the manufacture of foam aluminum. The interstitial compounds ZrB, ZrC, and ZrN are high-temperature refractories for use in nonoxidizing environments, the carbide melting at about 3500°C and the others at about 3000°C. They are extremely hard compounds. Zirconium diboride, ZrB_2, is an attrition-resistant, lubricious solid, suitable as a surface for ball bearings. It has shown excellent properties as an electrode material for the electrolysis of molten aluminum compounds. It is exceedingly difficult to convert zirconium carbide, nitride, or oxide, directly to pure metal.

Zirconates are important in certain electrical compositions, particularly in the piezoelectric lead titanate-zirconate and in the high-dielectric-constant compositions consisting of barium titanate and barium zirconate.

Most handling and testing of zirconium compounds have indicated no toxicity. Some have been administered orally to experimental animals in large dosages without observed harm. There has generally been no ill consequence of contact of zirconium compounds with the unabraded skin. However, some individuals appear to have allergic sensitivity to zirconium compounds, characteristically manifested by appearance of nonmalignant granulomas. Inhalation of sprays containing some zirconium compounds and of metallic zirconium dusts have had inflammatory effects, sometimes accompanied by morbidities in organs other than the lungs.

Analysis. Zirconium is the only common element precipitated from hydrochloric acid solution by mandelic acid, and it is one of very few elements precipitated from 10–20% sulfuric acid solution by phosphoric acid. Hydrous zirconia, $ZrO_2 \cdot xH_2O$, is precipitated at a pH of about 2.5, at which few other metal oxides or hydroxides are precipitated. It can be distinguished by characteristic colors imparted to the precipitate by certain dyes, notably alizarin, β-nitroso-α-napthol, and morin. Compounds of zirconium with the dyes Arsenazo III, Catechol Violet, Eriochrome Cyanine R, Neothorin, and Xylenol Orange have proved satisfactory for the detection and spectrometric determination of zirconium. Other dyes useful for analysis continue to be reported in great numbers. Flavones form fluorescent compounds. A standard determination of fluoride ion is performed by running a solution of the unknown into a suspension of the zirconium-alizarin lake until its color is discharged due to formation of fluorozirconate ion.

[WARREN B. BLUMENTHAL]

Bibliography: W. B. Blumenthal, *The Chemical Behavior of Zirconium*, 1958; R. Clark et al., *The Chemistry of Titanium, Zirconium*, and *Hafnium*, 1975; S. V. Elinson, *Analytical Chemistry of Zirconium and Hafnium*, 1972; I. C. Smith and B. L. Carson, *Zirconium*, vol. 3, 1978; P. C. Wailes et al., *Organometallic Chemistry of Titanium, Zirconium, and Hafnium*, 1974.

List of Contributors

List of Contributors

A

Abraham, Dr. Bernard M. *Solid State Science Division, Argonne National Laboratory, Argonne, IL.* HELIUM—in part.

Abrahams, David H. *Dexter Chemical Corporation, Bronx, NY.* DYE—validated.

Abramovitch, Dr. R. A. *Department of Chemistry and Geology, Clemson University.* HETEROCYCLIC COMPOUNDS.

Adams, Dr. Chris, *Unilever Research, Merseyside, England.* IODINE.

Adamson, Prof. Arthur W. *Department of Chemistry, University of Southern California.* TITANIUM.

Addison, Dr. Cyril C. *Professor of Inorganic Chemistry, University of Nottingham, England.* NITROGEN OXIDES.

Albert, Dr. Henry J. *Manager, Physics and Metallurgy Laboratory, Research and Development Department, Engelhard Industries Division, Engelhard Minerals and Chemicals Corporation, Newark, NJ.* IRIDIUM; OSMIUM; PALLADIUM; PLATINUM; RHODIUM; RUTHENIUM.

Albright, Dr. Lyle F. *Department of Chemical Engineering, Purdue University.* HALOGENATION; NITRATION—coauthored.

Allred, Dr. A. Louis. *Department of Chemistry, Northwestern University.* ELECTRONEGATIVITY.

Almond, Dr. Peter R. *Department of Biophysics, University of Texas, Houston.* ISOTOPIC IRRADIATION.

Anderson, Dr. John R. *Vice President, Research and Development, Permutit Research and Development Center, Princeton, NJ.* MOLECULAR WEIGHT.

Andrews, Dr. Lester. *Department of Chemistry, University of Virginia.* MATRIX ISOLATION.

Appelman, Dr. Evan H. *Chemistry Division, Argonne National Laboratory, Argonne, IL.* HYPOFLUOROUS ACID; PERBROMATE. KRYPTON COMPOUNDS; XENON COMPOUNDS—both coauthored.

Atkins, Dr. P. W. *Department of Chemistry, Oxford University, England.* PHYSICAL CHEMISTRY.

Atkinson, Prof. G. *Department of Chemistry, University of Oklahoma.* HYDROGEN ION.

B

Baer, Dr. Donald R. *Deceased; formerly, Jackson Laboratory, Organic Chemical Department, E. I. du Pont de Nemours and Company, Wilmington, DE.* DYE.

Bagley, Dr. Brian G. *Bell Laboratories, Murray Hill, NJ.* AMORPHOUS SOLID.

Bagnall, Dr. Kenneth W. *Research Chemist, Atomic Energy Research Establishment, Harwell, England.* POLONIUM.

Bakish, Dr. Robert. *Department of Engineering Technology, Fairleigh Dickinson University.* SALINE WATER RECLAMATION.

Barnes, Dr. Robert K. *Research and Development Department, Chemicals and Plastics Operations Division, Union Carbide Corporation, South Charleston, WV.* ACETYLENE; ETHYLENE; GLYCOL; OZONOLYSIS; POLYOXYETHYLATION OF ALCOHOL; other articles.

Barnitz, Edward S. *Consultant in Vacuum Engineering, Rochester, NY.* MOLECULAR DISTILLATION; REDUCED-PRESSURE DISTILLATION.

Bashkin, Dr. Stanley. *Department of Physics, University of Arizona.* BEAM-FOIL SPECTROSCOPY.

Basolo, Prof. Fred. *Department of Chemistry, Northwestern University.* COMPLEX COMPOUNDS; COORDINATION CHEMISTRY.

Bayfield, Dr. James E. *Department of Physics, University of Pittsburgh.* ELECTRON CONFIGURATION.

Bean, Howard S. *Deceased; formerly, Consultant on Fluid Metering, Liquids and Gases, Sedona, AZ.* WEIGHT; WEIGHT MEASUREMENT.

Belfit, Robert W., Jr. *Manager, Quality Standards, Quality Assurance Department, Dow Chemical Company, Midland, MI.* CHLORINE—coauthored.

Bender, Prof. Paul. *Professor of Physical Chemistry, University of Wisconsin.* FREE ENERGY; FUGACITY; INTERNAL ENERGY; WORK FUNCTION—coauthored.

Benner, Dr. Frank C. *Assistant Director, Research and Development Department, Norton Company, Worcester, MA.* VACUUM FUSION.

Bigeleisen, Prof. Jacob. *Department of Chemistry, University of Rochester.* DEUTERIUM.

Bikales, Dr. Norbert M. *Chemical Consultant, N. M. Bikales and Company, Livingston, NJ.* CYANOETHYLATION.

Billings, Dr. Bruce H. *Special Assistant to the Ambassador for Science and Technology, Embassy of the United States of America, Taipei.* ROTATORY DISPERSION.

Birks, Dr. L. S. *(Retired) Radiation Technology Division, Naval Research Laboratory, Washington, DC.* X-RAY SPECTROMETRY.

Bloomfield, Prof. Philip E. *Department of Physics, University of Pennsylvania; and City College, City University of New York.* THERMODYNAMIC PROCESSES.

Blumenthal, Warren B. *Blumenthal-Zirconium, North Tonawanda, NY.* HAFNIUM; ZIRCONIUM.

Bobbitt, Dr. James M. *Department of Chemistry, University of Connecticut.* INDOLE.

Bolton, Dr. James R. *Department of Chemistry, University of Western Ontario, Canada.* FREE RADICAL.

Boyd, Dr. Richard H. *Department of Chemical Engineering, University of Utah.* ACID AND BASE—coauthored.

Bradsher, Prof. Charles K. *Department of Chemistry, Duke University.* ANTHRACENE; AROMATIC HYDROCARBON; BENZENE; FRIEDEL-CRAFTS REACTION; INDENE; WURTZ-FITTIG REACTION; other articles.

Brewer, Dr. Charles P. *Shell Development Corporation, Emeryville, CA.* HYDROCRACKING.

Bricker, Dr. Clark E. *Department of Chemistry, University of Kansas.* PIPET; VOLUMETRIC ANALYSIS; VOLUMETRIC FLASK.

Bridgman, Prof. Percy W. *Deceased; formerly, Harvard University.* PHYSICAL LAW.

Brink, Dr. Joseph A., Jr. *Professor and Chairman, Department of Chemical and Nuclear Engineering, Washington State University.* CHEMICAL CONVERSION; CHEMICAL PROCESS INDUSTRY.

Brintzinger, Dr. Hans. *Fachbereich Biologie-Chemie, Universitat Konstanz, West Germany.* NITROGEN COMPLEXES.

Brodersen, Prof. Klaus. *Professor of Inorganic and Analytical Chemistry, University of Erlangen-Nürnberg, West Germany.* MERCURY.

Brodhag, Dr. Alex E., Jr. *Senior Editor, Organic Abstract Editing Department, Chemical Abstracts Service, Columbus, OH.* FURFURAL.

Brooks, Prof. Philip R. *Department of Chemistry, Rice University.* CHEMICAL DYNAMICS—in part.

Brown, Prof. Frederick C. *Department of Physics, University of Illinois*. PHOTOLYSIS.

Brown, Dr. Glenn H. *Professor of Chemistry and Director of the Liquid Crystal Institute, Kent State University*. LIQUID CRYSTALS.

Brown, Prof. Herbert C. *Department of Chemistry, Purdue University*. HYDROBORATION.

Buck, Dr. Richard P. *Department of Chemistry, University of North Carolina*. ION-SELECTIVE MEMBRANES AND ELECTRODES.

Bunnett, Prof. Joseph F. *Department of Chemistry, University of California, Santa Cruz*. ELECTROPHILIC AND NUCLEOPHILIC REAGENTS; SUBSTITUTION REACTION; SULFONATION—in part.

Burg, Dr. William R. *Department of Environmental Health, University of Cincinnati*. GAS AND ATMOSPHERE ANALYSIS.

Bursey, Prof. Maurice M. *Department of Chemistry, University of North Carolina*. MASS SPECTROMETRY; SPECTROSCOPY.

Burton, Prof. Ralph A. *Chairman, Department of Mechanical Engineering and Astronautical Sciences, Northwestern University*. THERMAL EXPANSION.

Burwell, Dr. Robert L., Jr. *Department of Chemistry, Northwestern University*. CATALYSIS; HETEROGENEOUS CATALYSIS.

Busch, Prof. Daryle H. *Department of Chemistry, Ohio State University*. CHEMICAL STRUCTURES—in part.

Butts, Prof. Allison. *Professor Emeritus of Metallurgy and Materials Science, Lehigh University*. COPPER.

C

Cameron, Dr. A. E. *Analytical Chemistry Division, Oak Ridge National Laboratory, Oak Ridge, TN*. ISOTOPE.

Carter, Prof. Herbert E. *Vice Chancellor for Academic Affairs, University of Illinois*. CETYL ALCOHOL; PHOSPHATIDE; TRIGLYCERIDE—all coauthored.

Chinitz, Dr. Wallace. *Department of Mechanical Engineering, Cooper Union*. CHEMICAL FUEL.

Cicerone, Dr. Ralph J. *Director, Atmospheric Chemistry and Aeronomy Division, National Center for Atmospheric Research, Boulder, CO*. HALOGEN ELEMENTS.

Clapp, Prof. Leallyn B. *Department of Chemistry, Brown University*. ALKANOLAMINE; ANILINE; ISOCYANATE; NITROPARAFFIN; OXIME; SULFANILIC ACID; other articles.

Clarke, Ellen. *Columbia Organic Chemicals Company, Inc. Columbia, SC*. FREON; HALOFORM REACTION; HALOGENATED HYDROCARBON—all coauthored.

Cohen, Charles A. *Research Associate (retired), Exxon Research and Engineering Company, Florham Park, NJ*. ALKYNE; ETHYLENE CHLOROHYDRIN; POLYALKENE; PROPYLENE.

Comings, Dr. Edward W. *University of Petroleum and Minerals, Dhahran, Saudi Arabia*. HIGH-PRESSURE PROCESSES.

Cook, Philip H. *Plastics Department, Texas Division, Dow Chemical, U.S.A., Freeport, TX*. GLYCEROL.

Cooley, Dr. William E. *Winton Hill Technical Center, Proctor and Gamble Company, Cincinnati, OH*. GOLD; NICKEL; SILVER; ZINC.

Cooper, Dr. Anthony R. *Dynapol, Palo Alto, CA*. GEL PERMEATION CHROMATOGRAPHY—coauthored.

Cramer, Dr. Friedrich. *Max-Planck Institut für Experimentelle Medizin, Gottingen, West Germany*. CLATHRATE COMPOUNDS—coauthored.

Cremer, Prof. Sheldon E. *Department of Chemistry, Marquette University*. ORGANOPHOSPHORUS COMPOUND.

Cronan, Calvin S. *Consulting Editor, "Chemical Engineering," McGraw-Hill Publications, New York, NY*. CHEMICAL ENGINEERING.

Cunningham, Prof. Burris B. *Deceased; formerly, Radiation Laboratory, University of California, Berkeley*. METALLOACID ELEMENTS; TRANSITION ELEMENTS.

Curtiss, Prof. C. F. *Department of Chemistry, University of Wisconsin*. GAS—coauthored.

D

Dahl, Dr. Lawrence F. *Department of Chemistry, University of Wisconsin*. X-RAY CRYSTALLOGRAPHY; X-RAY DIFFRACTION—in part.

Dalrymple, Dr. David L. *Department of Chemistry, University of Delaware*. WOODWARD-HOFFMANN RULE.

Daniels, Dr. Farrington. *Deceased; formerly, Professor Emeritus, Solar Energy Laboratory, University of Wisconsin*. CHAIN REACTION; HALF-LIFE—in part.

Davis, Dr. J. *Research and Development, Bright Star Industries, Inc., Clifton, NJ*. FUEL CELL—coauthored.

Dekeyser, Prof. Willy C. *Laboratory for Crystallography, Ghent, Belgium*. COORDINATION NUMBER; CRYSTAL STRUCTURE.

Delahay, Prof. Paul. *Department of Chemistry, New York University*. COULOMETER; DECOMPOSITION POTENTIAL; ELECTROCHEMICAL EQUIVALENT; ELECTROLYSIS.

De Vaney, Fred D. *Consulting Metallurgist, Duluth, MN*. MAGNETIC SEPARATION METHODS.

Dexter, Dr. Theodore H. *Research Supervisor, Inorganic Chemistry, Hooker Chemical Corporation Research Center, Niagara Falls, NY*. HYDRAZINE.

Dorfman, Dr. Ralph I. *Senior Vice President and Director, Syntex Research Center, Palo Alto, CA*. DIGITOXIGENIN.

Duckworth, Dr. Henry E. *Department of Physics, University of Manitoba, Canada*. ATOMIC NUMBER; MASS NUMBER; NUCLIDE.

E

Edwards, Prof. John O. *Department of Chemistry, Brown University*. IRON.

Edwards, Dr. Russell K. *Argonne National Laboratory, Argonne, IL*. CARBIDE; NITRIDE; OXIDE.

Elderfield, Prof. Robert C. *Deceased; formerly, Department of Chemistry, University of Michigan*. ORGANIC CHEMISTRY.

Eliel, Prof. Ernest L. *Department of Chemistry, University of North Carolina*. MOLECULAR ISOMERISM; PROCHIRALITY.

Encleson, Stephen C. *Technical Editor, Corporate Communications, Dow Chemical Company, Midland, MI*. MAGNESIUM—coauthored.

Endicott, Prof. John F. *Department of Chemistry, Wayne State University*. INORGANIC PHOTOCHEMISTRY.

Epstein, Dr. W. W. *Department of Chemistry, University of Utah*. DIMETHYL SULFOXIDE—coauthored.

Evans, Dr. C. A., Jr. *Materials Research Laboratory, University of Illinois*. SECONDARY ION MASS SPECTROMETRY (SIMS).

Evans, Prof. Robley D. *(Retired) Department of Physics, Massachusetts Institute of Technology*. HALF-LIFE—coauthored.

Ewing, Prof. George E. *Department of Chemistry, Indiana University*. INTERMOLECULAR FORCES.

Eyring, Dr. Edward M. *Department of Chemistry, University of Utah*. CHEMICAL DYNAMICS—in part.

Eyring, Prof. Henry. *Institute for the Study of Rate Processes, University of Utah*. RHEOLOGY.

F

Faller, Dr. J. W. *Department of Chemistry, Yale University*. FLUXIONAL COMPOUNDS.

Fanta, Prof. Paul E. *Department of Chemistry, Illinois Institute of Technology*. ACID ANHYDRIDE; ACYLATION; ALDEHYDE; DIELS-ALDER REACTION; ESTER; KETONE; other articles.

Feldman, Cyrus. *Analytical Chemistry Division, Oak Ridge National Laboratory, Oak Ridge, TN*. FLAME PHOTOMETRY.

Fenske, Prof. Richard F. *Department of Chemistry, University of Wisconsin*. MOLECULAR ORBITAL THEORY.

Field, Joseph H. *Benfield Corporation, Pittsburgh, PA.* Fischer-Tropsch process.

Fields, Paul R. *Senior Chemist, Argonne National Laboratory, Argonne, IL.* Nobelium.

Firk, Prof. Frank W. K. *Electron Accelerator Laboratory, Yale University.* Time-of-flight spectrometers.

Forster, Dr. Denis. *Monsanto Company, St. Louis, MO.* Homogeneous catalysis.

Fowkes, Prof. Frederick M. *Department of Chemistry, Lehigh University.* Adsorption.

Francis, Dr. Arthur W. *Union Carbide Corporation, Tarrytown, NY.* Argon; Helium—in part; Inert gases; Oxygen; Ozone; other articles.

Fried, Dr. Sherman. *Chemistry Division, Argonne National Laboratory, Argonne, IL.* Actinium; Rhenium; Technetium.

Friedman, Prof. Harold L. *Department of Chemistry, State University of New York, Stony Brook.* Water.

Fulton, Dr. J. W. *Monsanto Company, St. Louis, MO.* Dehydrogenation.

Fuson, Prof. Reynold C. *Department of Chemistry, University of Illinois, Urbana.* Vinylogy.

G

Gaines, Dr. George L., Jr. *Research and Development Center, General Electric Company, Schenectady, NY.* Monomolecular film.

Garrett, Prof. Alfred B. *Department of Chemistry, Ohio State University.* Chemistry; Elements; Salt.

Gauss, Dr. D. *Max-Planck Institut für Experimentelle Medizin, Gottingen, West Germany.* Clathrate compounds—coauthored.

Gensler, Prof. Walter J. *Department of Chemistry, Boston University.* Acridine; Azole; Furan; Isoquinoline; Pyran; Thiazole; other articles.

George, Dr. John W. *Department of Chemistry, University of Massachusetts.* Selenium.

Gergel, Max G. *Columbia Organic Chemicals Company, Columbia, SC.* Freon; Haloform reaction; Halogenated hydrocarbon—all coauthored.

Gerjuoy, Dr. Edward. *Department of Physics, University of Pittsburgh.* Excited state; Ground state.

Ghiorso, Dr. Albert. *Department of Chemistry, Lawrence Berkeley Laboratory, University of California, Berkeley.* Element 104; Element 105; Lawrencium.

Giarman, Prof. Nicholas J. *Deceased; formerly, Department of Pharmacology, School of Medicine, Yale University.* Sulfone—in part.

Gigg, Roy H. *Chemistry Division, National Institute for Medical Research, London, England.* Cetyl alcohol; Phosphatide; Triglyceride—all coauthored.

Giguère, Paul A. *Emeritus Professor, St. Petersburg Beach, FL.* Hydrogen peroxide.

Gillespie, Dr. Ronald J. *Department of Chemistry, McMaster University, Hamilton, Canada.* Superacid.

Gilliland, Prof. Edwin R. *Deceased; formerly, Department of Chemical Engineering, Massachusetts Institute of Technology.* Distillation.

Gleim, Paul S. *Manager, Advanced Products, Texas Instruments, Inc., Dallas, TX.* Germanium.

Glicksman, Dr. Richard. *Solid State Division, RCA, Somerville, NJ.* Electrode potential.

Goddu, Dr. Robert F. *Manager, Fibers and Film Research Division, Hercules, Inc., Wilmington, DE.* Fluorometric analysis; Polarimetric analysis; Refractometric analysis—all coauthored.

Gordon, Prof. Louis *Deceased; formerly, Case Institute of Technology.* Flocculation; Nucleation—coauthored; Precipitation; Supersaturation; Volatilization.

Gordon, Dr. Sheffield. *Chemistry Division, Argonne National Laboratory, Argonne, IL.* Radiation chemistry.

Gränicher, Prof. H. *Laboratory of Solid State Physics, Swiss Federal Institute of Technology, Zurich, Switzerland.* Rochelle salt.

Greensfelder, Dr. Bernard S. *Deceased; formerly, Director of Oil Research, Shell Development Company, Emeryville, CA.* Aromatization; Depolymerization.

Groggins, P. H. *Deceased; formerly, Chemical Division, Food Machinery and Chemical Corporation.* Amination; Sulfonation—in part.

Gross, William H. *Technical Editor, Corporate Communications, Dow Chemical Company, Midland, MI.* Magnesium—coauthored.

Gruen, Dr. Dieter M. *Argonne National Laboratory, Argonne, IL.* Magnetochemistry.

Grunberg, Dr. Emanuel. *Director, Department of Chemotherapy, Hoffmann-LaRoche, Inc., Nutley, NJ.* Sulfone—validated.

Gunkler, Dr. Albert A. *Chief Process Engineer, Midland Division, Dow Chemical Company, Midland, MI.* Halogen elements—coauthored.

H

Hadley, Dr. Elbert H. *Assistant Dean, College of Liberal Arts and Sciences, and Professor of Chemistry, Southern Illinois University.* Acetylation; Acidolysis; Amidine; Chloroform; Lactate; Palmitate; other articles.

Haensel, Dr. Vladimir. *Vice President and Director of Research, Universal Oil Products Company, Des Plaines, IL.* Catalytic reforming—coauthored.

Hague, Wilbur. *Oxy Metal Industries Corporation, Warren, MI.* Cadmium.

Hamer, Dr. Walter J. *Institute for Basic Standards, National Bureau of Standards.* Calomel electrode; Electrode; Electromotive force (cells); Hydrogen electrode; Salt bridge; Silver chloride electrode.

Hamilton Dr. Joseph H. *Department of Physics—Astronomy, Vanderbilt University.* Radioactivity.

Hansch, Dr. Theo W. *Department of Physics, Stanford University.* Laser spectroscopy.

Hanson, Prof. Allen L. *Department of Chemistry, Saint Olaf College.* Acid-base indicator; Concentration scales; Desiccant—in part; Nitro and nitroso compounds.

Happer, Dr. William. *Department of Physics, Columbia University.* Microwave spectroscopy.

Harris, Dr. J. *Postgraduate School of Studies in Chemical Engineering, University of Bradford, England.* Fluids—coauthored.

Harrison, George R. *Deceased; formerly, Dean Emeritus, School of Science, Massachusetts Institute of Technology.* Line spectrum.

Hartley, Dr. A. M. *Department of Chemistry and Chemical Engineering, University of Illinois, Urbana.* Buffers.

Hassid, Prof. William Z. *Deceased; formerly, Department of Biochemistry, University of California, Berkeley.* Arabinose; Dextrin; Polysaccharide; l-Rhamnose; Sedoheptulose.

Hawkins, Prof. George A. *Vice President for Academic Affairs, Purdue University.* Dalton's law; Van der Waals equation.

Heldman, Dr. Julius D. *Shell Development Company, Houston, TX.* Petrochemical—coauthored.

Herber, Prof. Rolfe H. *Department of Chemistry,* Mössbauer effect.

Hildebrand, Dr. Joel H. *Department of Chemistry, University of California, Berkeley.* Solution.

Hirschfelder, Dr. J. O. *Department of Chemistry, University of Wisconsin.* Gas—coauthored.

Hobbs, Dr. Herman H. *Department of Physics, George Washington University.* Crystal.

Hoffmann, Dr. Friedrich W. *Deceased; formerly, Chemical Research and Development Laboratories, Edgewood Arsenal, MD.* Organoarsenic compound.

Holt, Dr. D. L. *Monsanto Company, St. Louis, MO.* Hydroformylation.

Holton, Dr. Gerald. *Department of Physics, Harvard University.* Physical law—validated.

Howells, Dr. T. A. *Director of Continuing Education, Institute of Paper Chemistry, Appleton, WI.* FLAMEPROOFING.

Hudlicky, Dr. Tomas. *Department of Chemistry, Illinois Institute of Technology.* TERPENE; TRITERPENE.

Huizenga, Dr. John R. *Nuclear Structure Research Laboratory, University of Rochester.* NUCLEAR FISSION.

Hull, Dr. McAllister H., Jr. *Department of Physics and Astronomy, State University of New York, Buffalo.* ABSORPTION.

Hultsch, Dr. Roland A. *Department of Physics, University of Missouri.* TEMPERATURE.

Hurd, Dr. Charles D. *Department of Chemistry, Northwestern University.* PYROLYSIS.

Hurst, Dr. G. S. *Department of Physics, Oak Ridge National Laboratory, Oak Ridge, TN.* RESONANCE IONIZATION SPECTROSCOPY.

Hyde, Dr. Earl K. *Lawrence Berkeley Laboratory, University of California, Berkeley.* ASTATINE; FRANCIUM; RADON—in part.

J

Jaep, William F. *Central Research Department, Experimental Station, E. I. du Pont de Nemours and Company, Wilmington, DE.* ENTROPY—in part; GIBBS FUNCTION; THERMODYNAMIC PRINCIPLES.

Jenkins, Prof. F. A. *Deceased; formerly, Department of Physics, University of California, Berkeley.* ATOM; FINE STRUCTURE (SPECTRAL LINES).

Jespersen, Prof. Neil. *Department of Chemistry, St. John's University.* THERMOANALYSIS.

Johnson, Dr. Julian F. *Department of Chemistry and Institute of Materials Science, University of Connecticut.* GEL PERMEATION CHROMATOGRAPHY—coauthored.

Johnson, Philip C. *Research and Development Department, Chemicals and Plastics Operations Division, Union Carbide Corporation, South Charleston, WV.* POLYHYDRIC ALCOHOL.

Johnston, Dr. Francis J. *Department of Chemistry, University of Georgia.* AMMONIUM SALT; BASE; COAGULATION; HYDRATION; MIXTURE; OSMOSIS; SULFURIC ACID; other articles.

Johnson, Dr. James D. *Senior Research Advisor, Pioneering Research, Ethyl Corporation, Baton Rouge, LA.* LEAD—coauthored.

Juvet, Prof. Richard S., Jr. *Department of Chemistry, Arizona State University.* GAS CHROMATOGRAPHY.

K

Kaplan, Dr. Louis. *Senior Chemist, Argonne National Laboratory, Argonne, IL.* HYDROGEN; TRITIUM.

Kapner, Robert S. *Department of Chemical Engineering, Cooper Union.* AMINATION; SULFONATION—both validated.

Karasek, Dr. Francis W. *Department of Chemistry, University of Waterloo, Kitchener, Canada.* PLASMA CHROMATOGRAPHY.

Karger, Dr. Barry L. *Department of Chemistry, Northeastern University.* LIQUID CHROMATOGRAPHY.

Katz, Dr. Joseph J. *Chemistry Division, Argonne National Laboratory, Argonne, IL.* HEAVY WATER; HYDROGEN FLUORIDE; INORGANIC CHEMISTRY.

Kharasch, Prof. Norman. *Department of Biomedicinal Chemistry, University of Southern California.* DISULFIDE; MERCAPTAN; ORGANOSULFUR COMPOUND; SULFINIC ACID; THIO COMPOUNDS; other articles.

Kirby, Dr. H. W. *Mound Facility, Monsanto Research Corporation, Miamisburg, OH.* PROTACTINIUM.

Kirschner, Prof. Stanley. *Department of Chemistry, Wayne State University.* SULFUR; TELLURIUM.

Klein, Dr. David H. *Department of Chemistry, Hope College.* NUCLEATION—coauthored.

Koerker, Frederick W. *(Retired) Technical Expert, Dow Chemical Company, Midland, MI.* CHLORINE—coauthored.

Kreilik, Dr Robert. *Department of Chemistry, University of Rochester.* SPIN LABEL.

Kretchmer, Prof. Richard A. *Department of Chemistry, Illinois Institute of Technology.* ORGANIC CHEMICAL SYNTHESIS.

Kupchan, Prof. S. Morris. *Department of Chemistry, University of Virginia.* ALKALOID; ANTHRAQUINONE PIGMENTS; QUININE; STRYCHNINE; THEOBROMINE.

Kurylo, Dr. Michael J. *National Measurements Laboratory, National Bureau of Standards.* FLUOROMETRIC ANALYSIS—in part.

Kusch, Prof. Polykarp. *Department of Physics, Columbia Radiation Laboratory, Columbia University.* ATOMIC BEAMS; RADIO-FREQUENCY SPECTROSCOPY.

L

Lagowski, Dr. J. J. *Department of Chemistry, University of Texas, Austin.* SOLVENT—in part.

Laitinen, Prof. Herbert A. *Department of Chemistry, University of Florida.* ANALYTICAL CHEMISTRY; ELECTROCHEMISTRY; ELECTROLYTIC CONDUCTANCE.

Larsen, Prof. Edwin M. *Department of Chemistry, University of Wisconsin.* INDIUM; NIOBIUM; TANTALUM; THALLIUM.

Launay, Jules de. *Consultant, Solid State Division, U.S. Naval Research Laboratory.* SPECIFIC HEAT OF SOLIDS.

Laurie, Prof. Victor W. *Department of Chemistry, Princeton University.* BOND ANGLE AND DISTANCE.

Lee, Dr. Roberto. *Monsanto Company, St. Louis, MO.* HYDROGENATION.

Le Noble, Dr. William J. *Department of Chemistry, State University of New York, Stony Brook.* RESONANCE.

Levasheff, V. V. *American Potash and Chemical Corporation, Whittier, CA.* BORON—coauthored.

Levine, Dr. I. E. *Chevron Research Company, Richmond, CA.* OXIDATION PROCESS.

Lewis, Dr. Bernard. *Combustion and Explosives Research, Inc., Pittsburgh, PA.* COMBUSTION—in part.

Liedholm, George E. *(Retired) Department Head, Petroleum Processing, Shell Development Company, Emeryville, CA.* ISOMERIZATION.

Lindsey, J. S. *Liquid Carbonic Corporation, Chicago, IL.* CARBON DIOXIDE.

Lineberger, Dr. W. C. *Department of Chemistry, University of Colorado.* ELECTRON AFFINITY.

Little, Dr. James N. *Research Division, Waters Associates, Milford, MA.* SPECTROPHOTOMETRIC ANALYSIS; TURBIDIMETRIC ANALYSIS. FLUOROMETRIC ANALYSIS; POLARIMETRIC ANALYSIS; REFRACTOMETRIC ANALYSIS—all three coauthored.

Lockhart, Prof. Frank J. *Department of Chemical Engineering, University of Southern California.* COUNTERCURRENT TRANSFER OPERATIONS.

Long, Prof. Franklin A. *Formerly, Department of Chemistry, Cornell University.* ACID AND BASE—coauthored.

Long, Joseph B. *Consultant, Columbus, OH.* TIN.

Lord, Prof. Richard C. *Department of Chemistry, Massachusetts Institute of Technology.* INFRARED SPECTROSCOPY; RAMAN EFFECT.

Luckenbach, Edward. *Exxon Research and Engineering Company, Florham Park, NJ.* CRACKING.

Lynn, Dr. John W. *Director of Technology, Fibers and Fabrics Division, Union Carbide Corporation, New York, NY.* BUTANOL; ETHYL ALCOHOL—in part.

Lyon, Dr. W. S. *Analytical Division, Oak Ridge National Laboratory, Oak Ridge, TN.* ACTIVATION ANALYSIS.

M

McCabe, Dr. Warren L. *Department of Chemical Engineering, North Carolina State University*. SEDIMENTATION.

McMahon, Dr. J. F. *Formerly, New York State Technical Service Program, Alfred, NY*. PYROMETRIC CONE.

Madestau, L. *Chemicals and Plastics Division, Union Carbide Corporation, Bound Brook, NJ*. FORMALDEHYDE.

Madison, Dr. Vincent. *Department of Medicinal Chemistry, School of Pharmacy, University of Illinois, Chicago*. OPTICAL ACTIVITY.

Mahoney, Dr. Lee R. *Chemistry Department, Ford Motor Company, Dearborn, MI*. ANTIOXIDANT; INHIBITOR.

Malm, John G. *Argonne National Laboratory, Argonne, IL*. KRYPTON COMPOUNDS; XENON COMPOUNDS—both coauthored.

Mamantov, Dr. Gleb. *Chemistry Department, University of Tennessee*. FUSED-SALT SOLUTION.

Manson, Dr. John A. *Department of Chemistry, Lehigh University*. HETEROCYCLIC POLYMER; HYDROCARBON RESIN; POLYACRYLATE RESIN; POLYESTER RESINS; POLYMER; POLYURETHANE RESINS; other articles.

Mantell, Dr. Charles L. *President, C. L. Mantell and Associates, Manhasset, NY*. ELECTROCHEMICAL PROCESS.

Margrave, Prof. John L. *Dean of Research, Rice University*. HIGH-TEMPERATURE CHEMISTRY.

Mark, Prof. Harry B., Jr. *Department of Chemistry, University of Cincinnati*. KINETIC METHODS OF ANALYSIS.

Marks, Dr. Tobin J. *Department of Chemistry, Northwestern University*. CHEMICAL STRUCTURES—in part; ORGANOACTINIDES.

Marshall, Prof. William R., Jr. *Associate Dean, College of Engineering, University of Wisconsin*. DESICCANT—in part; DRYING.

Martell, Dr. A. E. *Department of Chemistry, Texas A & M University*. CHELATION—coauthored.

Martinson, C. R. *Corporate Engineering Department, Monsanto Company, St. Louis, MO*. PIGMENT—coauthored.

May, F. H. *Consultant, Kerr-McGee Corporation, Whittier, CA*. BORON—coauthored.

Mays, R. L. *Materials Systems Division, Union Carbide Corporation, Tarrytown, NY*. MOLECULAR SIEVE.

Meggers, Dr. W. F. *Deceased; formerly, National Bureau of Standards*. BAND SPECTRUM.

Meinke, Dr. W. Wayne. *Analytical Chemistry Division, National Bureau of Standards*. RADIOISOTOPIC ASSAY; RADIOMETRIC ASSAY.

Merrill, Timothy W. *Manager, Metallurgical Applications Research, Foote Mineral Company, Exton, PA*. VANADIUM.

Mill, Dr. George S. *Research Chemist, Shell Oil Company, New York, NY*. DONNAN EQUILIBRIUM; EMULSION—both coauthored.

Miller, Dr. Shelby A. *Argonne National Laboratory, Argonne, IL*. FILTRATION; LEACHING.

Miller, Prof. Sidney I. *Department of Chemistry, Illinois Institute of Technology*. ORGANIC REACTION MECHANISM.

Miller, Dr. William H. *Department of Chemistry, University of California, Berkeley*. CHEMICAL DYNAMICS—in part.

Milligan, Dr. W. O. *Robert A. Welch Foundation, Houston, TX*. DONNAN EQUILIBRIUM; EMULSION—both coauthored. GEL; ISOELECTRIC POINT.

Minnis, Dr. Wesley. *Formerly, Assistant to the Director of Research and Development, Allied Chemical Corporation, New York, NY*. DYE—in part.

Moore, Dr. Paul B. *Department of the Geophysical Sciences, University of Chicago*. SOLID SOLUTION.

Morrison, Dr. George H. *Professor of Chemistry and Director, Analytical Facility of Materials Science Center, Cornell University*. CHEMICAL SEPARATION TECHNIQUES.

Morrow, Hugh, III. *Supervisor, Technical Information, Climax Molybdenum Company, Greenwich, CT*. MOLYBDENUM.

Moscowitz, Dr. Albert. *Department of Chemistry, University of Minnesota*. COTTON EFFECT.

Mosher, Dr. William. *Deceased; formerly, Chairman, Department of Chemistry, University of Delaware*. ISOPRENE; MENTHOL; SQUALENE.

Motekaitis, Dr. R. J. *Department of Chemistry, Texas A & M University*. CHELATION—coauthored.

Mottet, Dr. N. Karle. *Professor of Pathology and Director of Hospital Pathology, University Hospital, University of Washington*. COBALT—in part.

Mrak, Dr. Emil M. *Office of Chancellor Emeritus, University of California, Davis*. ETHYL ALCOHOL—coauthored.

Mulliken, Prof. Robert S. *Institute of Molecular Biophysics, Florida State University*. CONJUGATION AND HYPERCONJUGATION; MOLECULAR STRUCTURE AND SPECTRA; MOLECULE.

Murray, Dr. Royce W. *Department of Chemistry, University of North Carolina*. FLOCCULATION. PRECIPITATION; SUPERSATURATION; VOLATILIZATION—all three validated.

N

Nachtrieb, Prof. Norman H. *Chairman, Department of Chemistry, University of Chicago*. EVAPORATION; LIQUID; SUBLIMATION; SURFACE TENSION; VAPOR PRESSURE; VISCOSITY.

Naistat, Dr. Samuel S. *Department of Chemistry, Stephen F. Austin State College*. PEROXIDE.

Nedelsky, Prof. Leo. *Department of Physical Science, University of Chicago*. CONSERVATION OF ENERGY; CONSERVATION OF MASS; DENSITY; SPECIFIC GRAVITY.

Nesbitt, Dr. David J. *Joint Institute for Laboratory Astrophysics and Department of Chemistry, University of Colorado*. LASER PHOTOCHEMISTRY.

Nordell, Eskel. *Director of Technical Information, The Permutit Company*. WATER SOFTENING.

Novotny, Prof. Milos V. *Department of Chemistry, Indiana University*. CHROMATOGRAPHY.

Noyes, Prof. Richard M. *Department of Chemistry, College of Arts and Sciences, University of Oregon*. OSCILLATORY REACTION.

O

O'Dom, Dr. George W. *Research Scientist, TRW, Inc., Redondo Beach, CA*. COULOMETRIC ANALYSIS; ELECTRODEPOSITION ANALYSIS.

Othmer, Prof. Donald F. *Department of Chemical Engineering, Polytechnic Institute of Brooklyn*. FLUIDIZATION.

P

Parr, Prof. Robert G. *Department of Chemistry, Johns Hopkins University*. CHEMICAL BOND THEORY; VALENCE.

Parrish, Dr. William. *Research Staff Member, IBM Research Laboratory, San Jose, CA*. X-RAY FLUORESCENCE ANALYSIS; X-RAY POWDER METHODS.

Patel, Dr. C. K. N. *Bell Laboratories, Murray Hill, NJ*. PHOTOACOUSTIC SPECTROSCOPY.

Pearson, Dr. R. G. *Department of Chemistry, Northwestern University*. CHEMICAL STRUCTURES—in part.

Pecora, Prof. Robert. *Department of Chemistry, Stanford University*. QUASIELASTIC LIGHT SCATTERING.

Penneman, Dr. Robert A. *Los Alamos Scientific Laboratory, Los Alamos, NM*. AMERICIUM; NEPTUNIUM.

Perch, Michael. *Koppers Company, Inc., Monroeville, PA*. COKE.

Perone, Prof. Sam P. *Department of Chemistry, Purdue University*. ELECTROCHEMICAL TECHNIQUES.

Phaff, Dr. Herman J. *College of Agriculture, University of California, Davis*. ETHYL ALCOHOL—coauthored.

Pierotti, Prof. Robert A. *Department of Chemistry, Georgia Institute of Technology*. CHEMICAL THERMODYNAMICS.

Plueddemann, Dr. Edwin P. *Organic Laboratories, Dow Corning Corporation, Midland, MI*. SILICON.

Pollitzer, Dr. Ernest L. *Associate Director of Research, Corporate Research Center, Universal Oil Products Company, Des Plaines, IL*. CATALYTIC REFORMING—coauthored.

Popp, Prof. Frank D. *Department of Chemistry, Clarkson College of Technology*. FERROCENE.

Post, Dr. Richard F. *Lawrence Livermore Laboratory, Livermore, CA*. NUCLEAR FUSION.

R

Rausch, Dr. Marvin D. *Department of Chemistry, University of Massachusetts*. ORGANOMETALLIC COMPOUND; REFORMATSKY REACTION; TETRAETHYLLEAD.

Reid, Prof. Evans B. *Chairman, Department of Chemistry, Colby College*. ANTHRANILIC ACID; CARBOXYLIC ACID; DRYING OIL; ORTHOESTER; PHTHALIC ACID; UREA; other articles.

Renfrow, Dr. William B., Jr. *Department of Chemistry, Oberlin College*. CONDENSATION REACTION.

Reynolds, Dr. W. W. *Manager, Economic Coordination, Chemical Economics, Shell Chemical Company, Houston, TX*. PETROCHEMICAL—coauthored.

Rich, Prof. Arthur. *Department of Physics, University of Michigan*. ELECTRON SPIN.

Richman, Betty. *Staff Editor, "McGraw-Hill Encyclopedia of Science and Technology," McGraw-Hill Book Company, New York, NY*. AROMATIC.

Riddick, Dr. John A. *Baton Rouge, LA*. SOLVENT—in part.

Ries, Dr. Harold C. *Stanford Research Institute, Menlo Park, CA*. ALKYLATION.

Riley, Dr. Reed F. *Quantum Electronics Group, General Telephone and Electronics, Bayside, NY*. ALKALINE-EARTH METALS; BARIUM; CALCIUM; STRONTIUM.

Roberts, J. K. *(Retired) Research and Development Department, Standard Oil Company, Chicago, IL*. NAPHTHA.

Robinson, Dr. G. Wilse. *Department of Chemistry, Texas Tech University*. PICOSECOND MOLECULAR PROCESSES.

Rochow, Prof. Eugene G. *(Retired) Department of Chemistry, Harvard University*. ELECTROCHEMICAL SERIES.

Rockett, Frank H. *Engineering Consultant, Charlottesville, VA*. ALKALI; ALUM; BOYLE'S LAW; CARBON TETRACHLORIDE; LANOLIN; other articles.

Roller, Dr. Duane E. *Deceased; formerly, Harvey Mudd College*. CONSERVATION OF ENERGY.

Rozeanu, Prof. L. *Department of Material Science, Technion, Israel Institute of Technology*. FUEL CELL—coauthored.

Rudolph, Prof. Ralph W. *Department of Chemistry, University of Michigan*. METAL CLUSTER COMPOUNDS.

Rulfs, Dr. Charles L. *Department of Chemistry, University of Michigan*. ORGANIC QUANTITATIVE ANALYSIS; SAMPLING TECHNIQUES; SOLUBILIZING OF SAMPLES. REAGENT CHEMICALS; STOICHIOMETRY—both validated.

Russell, Dr. Allen S. *Vice President, Alcoa Laboratories, Aluminum Company of America, Pittsburgh, PA*. ALUMINUM.

S

Saenger, Dr. Wolfram. *Max-Planck Institut für Experimentelle Medizen, Göttingen, West Germany*. CLATHRATE COMPOUNDS—coauthored.

Salutsky, Dr. Murrell L. *Vice President, Research, Dearborn Chemical Division, W. R. Grace and Company, Lake Zurich, IL*. RADIUM.

Sankey, Dr. Bruce M. *Imperial Oil Enterprises, Ltd., Sarnia, Ontario, Canada*. EXTRACTION.

Schaefer, Dr. Henry F., III. *Department of Chemistry, University of California, Berkeley*. QUANTUM CHEMISTRY.

Schindler, Dr. Anton. *Camille Dreyfus Laboratory, Research Triangle Institute, Research Triangle Park, NC*. STEREOSPECIFIC CATALYST.

Schmerling, Dr. Louis. *Research Associate, Universal Oil Products Company, Des Plaines, IL*. ALIPHATIC HYDROCARBON; ETHANE; HYDROCARBON; KOLBE HYDROCARBON SYNTHESIS; METHANE; other articles.

Schubert, Dr. Jack. *Radiation Health, Graduate School of Public Health, University of Pittsburgh*. BERYLLIUM.

Schwartz, Dr. Morton K. *Department of Biochemistry, Memorial Hospital, New York, NY*. URIC ACID.

Scott, Prof. Robert L. *Department of Chemistry, University of California, Los Angeles*. BOILING POINT; DELIQUESCENCE; FIRST-ORDER TRANSITION; PHASE EQUILIBRIUM; TRIPLE POINT; other articles.

Seaborg, Dr. Glenn T. *Lawrence Berkeley Laboratory, University of California, Berkeley*. ACTINIDE ELEMENTS; BERKELIUM; ELEMENT 106; PERIODIC TABLE; TRANSURANIUM ELEMENTS; other articles.

Sessions, Dr. Richard B. *School of Chemistry, University of Bristol, England*. CROWN ETHERS AND CRYPTANDS.

Shamma, Dr. Maurice. *Department of Chemistry, Pennsylvania State University*. ISOQUINOLINE—in part. QUINOLINE—coauthored.

Shapiro, Hymin. *Research and Development Department, Ethyl Corporation, Baton Rouge, LA*. LEAD—coauthored. METAL CARBONYL.

Shaw, Dr. Robert A. *Department of Chemistry, Birkbeck College, University of London, England*. INORGANIC POLYMER.

Sheft, Dr. Irving. *Chemistry Division, Argonne National Laboratory, Argonne, IL*. FLUORINE.

Sherman, Dr. P. Dwight, Jr. *Research and Development, Chemicals and Plastics, Union Carbide Corporation, South Charleston, WV*. PROPANOL.

Shirley, Prof. David A. *Department of Chemistry, University of Tennessee*. ACETONE; KETENE; QUINONE.

Shreve, Prof. R. Norris. *Professor Emeritus of Chemical Engineering, Purdue University*. NITRATION—coauthored.

Siegbahn, Prof. Kai. *Institute of Physics, University of Uppsala, Sweden*. ELECTRON SPECTROSCOPY.

Sienko, Dr. Michell J. *Department of Chemistry, Cornell University*. NONSTOICHIOMETRIC COMPOUNDS.

Simmons, Thomas C. *Chief, Agents Research Branch, U.S. Department of the Army, Edgewood Arsenal, MD*. ORGANOARSENIC COMPOUND—validated.

Simpson, Prof. Stephen G. *Department of Chemistry, Massachusetts Institute of Technology*. GRAVIMETRIC ANALYSIS; QUANTITATIVE CHEMICAL ANALYSIS.

Sisler, C. W. *Corporate Engineering Department, Monsanto Company, St. Louis, MO*. PIGMENT—coauthored.

Sisler, Dr. Harry H. *Department of Chemistry, University of Florida*. AMMINE; AMMONIA; NITROGEN.

Sittig, Marshall. *Assistant Director, Office of Research Project Administration, Princeton University*. ALKALI METALS; CESIUM; LITHIUM; POTASSIUM; RUBIDIUM; SODIUM.

Skogerboe, Dr. Rodney K. *Department of Chemistry, Colorado State University*. ATOMIC SPECTROSCOPY—in part.

Slabaugh, Dr. Wendell H. *Department of Chemistry, Oregon State University*. INTERFACE OF PHASES; SURFACE-ACTIVE AGENT.

Slomp, Dr. George. *Physical and Analytical Chemistry Division, Upjohn Company, Kalamazoo, MI*. NUCLEAR MAGNETIC RESONANCE (NMR).

Snell, Dr. Esmond E. *Department of Biochemistry, University of California, Berkeley*. TRANSAMINATION.

Souders, Dr. Mott. *Deceased; formerly, Director, Oil Development, Shell Oil Company, Emeryville, CA.* AROMATIZATION; DEPOLYMERIZATION—both validated. FLASH POINT; FRACTIONATING COLUMN; POUR POINT.

Spedding, Dr. Frank H. *Professor Emeritus, Ames Laboratory, Energy Research and Development Administration, Iowa State University.* CERIUM; EUROPIUM; LANTHANIDE CONTRACTION; RARE-EARTH ELEMENTS; YTTRIUM; other articles.

Spindt, Dr. Roderick S. *Gulf Research and Development Company, Pittsburgh, PA.* COMBUSTION.

Standiford, Ferris C. *W. L. Badger Associates, Inc., Consulting Engineers, Ann Arbor, MI.* EVAPORATOR.

Stansbury, Dr. Harry A., Jr. *Group Leader, Research Department, Union Carbide Corporation, South Charleston, WV.* BENZALDEHYDE.

Stauffer, Dr. Randy C. *Halogens Research Laboratory, Dow Chemical Company, Midland, MI.* BROMINE.

Steele, Dr. William A. *Department of Chemistry, Pennsylvania State University.* ENTHALPY; THERMODYNAMIC PROCESSES; WORK FUNCTION—all coauthored.

Steinberg, Dr. Ellis P. *Senior Scientist, Argonne National Laboratory, Argonne, IL.* NUCLEAR CHEMISTRY.

Stevens, Dr. Brian. *Department of Chemistry, University of South Florida.* PHOTOCHEMISTRY.

Stiles, Dr. Martin. *Department of Chemistry, University of Kentucky.* CATECHOL; DIAZOTIZATION; HYDROQUINONE; other articles. PHENOL; THIOPHENE—both coauthored.

Stirton, Dr. Robert I. *Vice President (retired), Roger Williams Technical and Economic Services, Inc., Berkeley, CA.* ISOPROPYLPHENOL; NAPHTHOL; PICRIC ACID. PHENOL; XYLENE—both coauthored.

Stock, Prof. John T. *Department of Chemistry, University of Connecticut.* TITRATION.

Stone, Prof. Kenneth G. *Deceased; formerly, Department of Chemistry, Michigan State University.* REAGENT CHEMICALS; STOICHIOMETRY.

Stranks, Prof. Donald R. *Department of Physical and Inorganic Chemistry, University of Adelaide, Australia.* RADIOCHEMISTRY.

Sully, Dr. Arthur H. *Director, Technical, Special Steels Division, British Steel Corporation, Sheffield, England.* MANGANESE—in part.

Surles, Dr. Terry. *Limnetics, Inc., Milwaukee, WI.* INTERHALOGEN COMPOUNDS.

Sweat, F. W. *Department of Chemistry, University of Utah.* DIMETHYL SULFOXIDE—coauthored.

Swern, Dr. Daniel. *Department of Chemistry, Fels Research Institute, Temple University.* HYDROXYLATION REACTION.

T

Taube, Dr. Henry. *Department of Chemistry, Stanford University.* OXIDATION-REDUCTION.

Taylor, Dr. Barry N. *National Bureau of Standards, Washington, DC.* ATOMIC CONSTANTS.

Teller, Dr. Aaron J. *Teller Environmental Systems, Worcester, MA.* GAS ABSORPTION OPERATIONS.

Thomas, Dr. J. Kerry, *Department of Chemistry, University of Notre Dame.* MICELLE.

Thompson, Dr. Stanley G. *Senior Staff Member, Lawrence Berkeley Laboratory, University of California, Berkeley.* EINSTEINIUM—coauthored.

Thorn, Dr. R. J. *Division of Chemistry, Argonne National Laboratory, Argonne, IL.* SOLID-STATE CHEMISTRY.

Treybal, Robert E. *Deceased; formerly, Department of Chemical Engineering, New York University.* SOLVENT EXTRACTION—in part.

Tsang, Dr. C. F. *Lawrence Berkeley Laboratory, University of California, Berkeley* SUPERTRANSURANICS.

Tsuchiya, Prof. Henry M. *Department of Chemical Engineering, University of Minnesota.* DEXTRAN.

Tsutsui, Prof. Minoru. *Department of Chemistry, Texas A & M University.* METALLOCENES.

Turnbull, Prof. David. *Department of Applied Physics, Harvard University.* SINGLE CRYSTAL.

Turro, Prof. Nicholas J. *Department of Chemistry, Columbia University.* TRIPLET STATE.

U

Uhl, Dr. Vincent W. *Department of Chemical Engineering, University of Virginia.* MIXING.

V

Vanderzee, Dr. Cecil E. *Department of Chemistry, University of Nebraska.* CHEMICAL EQUILIBRIUM.

Van Wazer, Dr. John R. *Department of Chemistry, Vanderbilt University.* PHOSPHORUS.

Van Winkle, Prof. Quentin. *Department of Chemistry, Ohio State University.* COLLOID; DIALYSIS; ELECTROKINETIC PHENOMENA; STREAMING POTENTIAL; TYNDALL EFFECT; ULTRAFILTRATION.

Vaughan, Prof. Wyman R. *Head, Department of Chemistry, University of Connecticut.* ASYMMETRIC SYNTHESIS; CONFORMATIONAL ANALYSIS; DIASTEREOISOMER; ENANTIOMORPH; TAUTOMERISM.

Veillon, Dr. Claude. *Biophysics Research Laboratory, Peter Bent Brigham Hospital, Harvard Medical School.* ATOMIC SPECTROSCOPY—in part.

Vona, Joseph A. *Director, Marketing–Technical Development Laboratory, Celanese Chemical Company, Inc., Summit, NJ.* PENTAERYTHRITOL.

W

Waddington, Prof. Thomas C. *Department of Chemistry, University of Durham.* AVOGADRO NUMBER; CHEMICAL COMPOUNDS; DEFINITE COMPOSITION, LAW OF; ELECTROLYTE; GRAM-EQUIVALENT WEIGHT; ION; other articles.

Wagner, Dr. Frank. *Technical Center, Celanese Chemical Company, Corpus Christi, TX.* ACETALS; ACRYLONITRILE; EPOXIDATION; FLUOROCARBON; METHANOL; other articles.

Wainwright, Howard W. *Coal Research Center, U.S. Bureau of Mines, Morgantown, WV.* COAL CHEMICALS.

Wakeham, Dr. W. A. *Department of Chemical Engineering, Imperial College, London, England.* TRANSPORT PROCESSES.

Waldron, Dr. Robert D. *Director, Research Enterprises, Scottsdale, AZ.* POLAR MOLECULE.

Walton, Prof. H. F. *Department of Chemistry, University of Colorado.* ION EXCHANGE.

Wamser, Prof. Carl C. *Department of Chemistry, California State University.* REACTIVE INTERMEDIATES.

Ware, Dr. B. R. *Department of Chemistry, Syracuse University.* ELECTROPHORESIS.

Warf, Dr. James C. *Department of Chemistry, University of Southern California.* HYDRIDO COMPLEXES; METAL HYDRIDES.

Warren, Prof. Bertram E. *Department of Physics, Massachussetts Institute of Technology.* X-RAY DIFFRACTION.

Wartik, Prof. Thomas. *Department of Chemistry, Pennsylvania State University.* BORANE.

Watson, W. W. *Professor Emeritus of Physics, Yale University.* ATOM; AUGER EFFECT; MASS DEFECT. BAND SPECTRUM; FINE STRUCTURE (SPECTRAL LINES)—both validated.

Waugh, Prof. John L. T. *Deparment of Chemistry, University of Hawaii.* ANTIMONY; ARSENIC; INTERMETALLIC COMPOUNDS.

Weaver, Dr. E. Eugene. *Research Scientist, Product Development Group, Ford Motor Company. Dearborn, MI.* ACETATE; AQUA REGIA; AZIDE; BROMATE; CARBON; FERRIC COMPOUND; SULFATE; other articles.

Weber, Harold C. *Chemical Engineer, Boston, MA*. ENTHALPY; SPECIFIC HEAT.

Webster, Owen W. *Dupont Experimental Station, Wilmington, DE*. CYANOCARBON.

Wegner, Prof. Patrick A. *Department of Chemistry, California State University*. CARBORANE.

Weigel, Dr. Fritz. *Institut für Anorganische Chemie, Universität München, West Germany*. PLUTONIUM; URANIUM.

Weil, Dr. Andrew T. *Consultant, Tucson, AZ*. COCAINE.

Weil, Jonathan F. *Staff Editor, "McGraw-Hill Encyclopedia of Science and Technology," McGraw-Hill Book Company, New York, NY*. ATOMIC MASS UNIT.

Weissman, Prof. S. I. *Department of Chemistry, Washington University* ELECTRON PARAMAGNETIC RESONANCE (EPR) SPECTROSCOPY.

Wendlandt, Prof. Wesley W. *Department of Chemistry, University of Houston*. COPPER CHEMISTRY; MANGANESE.

Wenkert, Prof. Ernest. *Department of Chemistry, Indiana University*. STERIC EFFECT.

Wentorf, Robert H., Jr. *General Electric Research Laboratory, Schenectady, NY*. HIGH-PRESSURE CHEMISTRY.

Wentworth, Prof. R. A. D. *Department of Chemistry, Indiana University*. CRYSTAL FIELD THEORY.

Wilcox, Prof. William R. *Department of Chemical Engineering, Clarkson College of Technology*. CRYSTALLIZATION.

Wilen, Dr. Samuel H. *Department of Chemistry, City University of New York*. RACEMIZATION; STEREOCHEMISTRY.

Wilhelm, Dr. Harley A. *Principal Scientist, EMRR Institute and Ames Laboratory of U. S. Department of Energy, Iowa State University*. THORIUM.

Wilke, Prof. Charles R. *Department of Chemical Engineering, University of California, Berkeley*. MASS-TRANSFER OPERATION.

Wilke, Dr. Günther. *Director, Max-Planck Instituts für Kohlenforschung, Mülheim, West Germany*. SOLVENT EXTRACTION.

Wilkening, Dr. M. *Department of Physics, New Mexico Institute of Mining and Technology*. RADON.

Wilkinson, Dr. Michael K. *Associate Director, Solid State Division, Oak Ridge National Laboratory, Oak Ridge, TN*. NEUTRON DIFFRACTION.

Wilkinson, Dr. W. D. *Consultant in Metallurgy and Nuclear Materials, Santa Barbara, CA*. CHROMIUM; GALLIUM; SCANDIUM; TUNGSTEN.

Wilkinson, Dr. W. L. *Postgraduate School of Studies in Chemical Engineering, University of Bradford, England*. FLUIDS—coauthored.

Williams, Dr. Jack M. *Chemistry Division, Argonne National Laboratory, Argonne, IL*. CHEMICAL STRUCTURES—in part; HYDROGEN BOND.

Wilson, Prof. E. Bright, Jr. *Department of Chemistry, Harvard University*. CHEMICAL BONDING.

Wilson, Dr. Therese. *Biological Laboratories, Harvard University*. CHEMILUMINESCENCE.

Woods, Prof. John M. *School of Chemical Engineering, Purdue University*. TRANSESTERIFICATION.

Y

Yelles, Marvin. *Formerly, Editor, "McGraw-Hill Encyclopedia of Science and Technology," McGraw-Hill Book Company, New York, NY*. EPOXY RESIN.

Yosim, Dr. Samuel J. *Atomics International Division, North American Rockwell, Canoga Park, CA*. BISMUTH.

Young, Dr. Roland S. *Consulting Chemical Engineer, Victoria, British Columbia, Canada*. COBALT—in part.

Young, Prof. Thomas F. *Department of Chemistry, University of Chicago*. AMPHOTERISM; HYDROLYSIS; IONIC EQUILIBRIUM; PK; SOLUBILITY PRODUCT CONSTANT.

Z

Zander, Dr. Andrew T. *Spectrametrics Corporation, Andover, MA*. EMISSION SPECTROCHEMICAL ANALYSIS; SPECTROCHEMICAL ANALYSIS; TRACE ANALYSIS.

Zuman, Dr. Peter. *Department of Chemistry, Clarkson College of Technology*. POLAROGRAPHIC ANALYSIS.

Zwolinski, Dr. Bruno J. *Department of Chemistry, Thermodynamics Research Center, Texas A & M University*. CALORIMETRY; THERMOCHEMISTRY.

Index

Index

Asterisks indicate page references to article titles.

B

C

D

E

H

I

J–K

L

M

N

Q

U

V

W

X

Y

Z